VIRTUAL CAMPUS

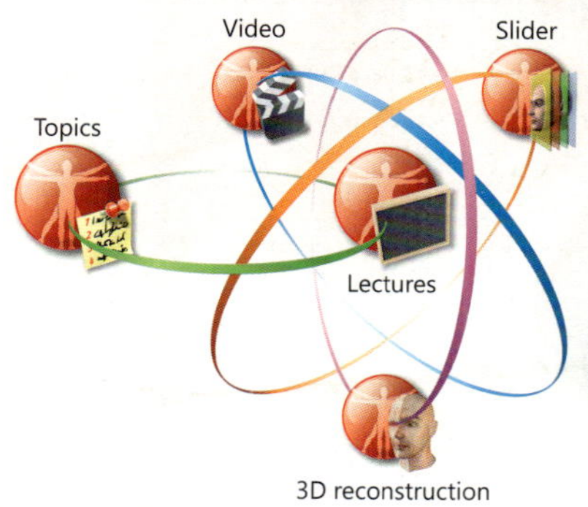

HUMAN ANATOMY *atlas*

FOUR STEPS TO ACCESS DIGITAL CONTENTS

1	CONNECT	Go to the website indicated on the label that includes the access code
2	REGISTER	Register at the Publisher Portal (first time only) to receive a username and password
3	ACCESS	Enter your username and password to access the restricted contents
4	TYPE THE CODE	On the web page dedicated to the work, type in the personal access code hidden under the protection of the label applied to this page

After the first access, digital contents will be available on the website entering your username and password.

The **online access** grants **personal license** limited to a **single user** for each code.
The **access is limited to the individual user** and is not a library or institutional access license.

Sharing of passwords and/or code is not allowed;
any misuse of the personal code will invalidate and lock it.
The access cannot be shared and will expire according as indicated in the license agreement subscribed at the first access.
Further details may be provided upon acceptance of the license agreement. The use of the code is subject to acceptance of the conditions. Any violation or attempted violation of code protection will not be accepted.

To remove the protection use a coin to scrape with light pressure

EE5886004084
www.evirtualcampus.com/en

Hardware and software requirements:
PC with Windows, Macintosh or Linux,
latest generation of internet browser such as Internet Explorer (version 9 onwards), Firefox, Chrome etc.; Internet connection.

Technical Help Desk: by email at support@ediermes.com

Contributors
English Edition

Editors

Giuseppe Anastasi	University of Messina, Italy
Eugenio Gaudio	University of Roma (Sapienza), Italy
Carlo Tacchetti	University Vita-Salute San Raffaele, Milan, Italy

Editor English Edition

Estomih Mtui	Weill Cornell Medicine, New York, USA

Coauthors

Paolo Castano[†]	University Vita-Salute San Raffaele, Milan, Italy
Sergio Castorina	University of Catania, Italy
Ottavio Cremona	University Vita-Salute San Raffaele, Milan, Italy
Raffaele De Caro	University of Padova, Italy
Guido Macchiarelli	University of l'Aquila, Italy
Mario Rende	University of Perugia, Italy
Domenico Ribatti	University of Bari, Italy
Chiarella Sforza	University of Milan, Italy

Other contributors

Demetrio Milardi, University of Messina, Italy, *3D volume rendering images*

Web contents

Estomih Mtui, Weill Cornell Medicine, New York, USA
Angelo Favaloro, University of Messina, Italy
Antonio Esposito, University Vita-Salute San Raffaele, Milan, Italy
Maurizio Vertemati, Department of Biomedical and Clinical Sciences "L. Sacco" University of Milan, Italy

Giuseppe Anastasi
Eugenio Gaudio
Carlo Tacchetti

Estomih Mtui
Editor English Edition

HUMAN ANATOMY
Atlas

Volume 1

GENERAL ANATOMY
MUSCULOSKELETAL SYSTEM

edi·ermes

HUMAN ANATOMY - Atlas - Volume 1 - Giuseppe Anastasi, Eugenio Gaudio, Carlo Tacchetti (eds)
Estomih Mtui (ed English edition)

ISBN 978-88-7051-588-6

A book is the final product of a very complex series of operations that requires numerous tests on texts and images. It is almost impossible to publish a book with no errors. We will be grateful to those who find them and notify us. For enquiries or suggestions about this volume, please use the following address:

External relations - Edi.Ermes srl - Viale Enrico Forlanini, 65 - 20134 Milan (Italy)
Phone +39.02.70.21.121 - Fax +39.02.70.21.12.83

Drawings, Edi.Ermes Archive/Andrea Rossi Raccagni, Marco Fanuli, Raffaella Stilo, Maria Acquaviva, Anna Maria D'Errico, Ettore Orlandi

Printed in November 2017 by Faenza Printing Industries SpA
for Edi.Ermes - viale Enrico Forlanini, 65 - 20134 Milan, Italy
http://www.ediermes.com - Phone +39.02.70.21.121 - Fax +39.02.70.21.12.83

Foreword

The **Human Anatomy – Atlas** is in my opinion the finest Atlas of Human Anatomy that has ever been written. The main factors that make it so unique include the quality of both the anatomical and radiological images and the way they have been organized.

Artists and radiologists working in collaboration with Edi.Ermes in Milan have created a very high quality 3-D reconstructed images from CT scans and MRI's, used in the Atlas. This approach will give the student and the reader a very clear and accurate understanding of the three dimensional presentation of each structure covered in the Atlas.

All the images are accompanied by a very detailed information related to the anatomy and where appropriate clinical correlates have also been included. This approach will assist the student to relate each structure studied with a clinical condition.

In addition, the authorship of this Atlas has organized the images so that they present the flow of information needed by the student to build a clear image of the human anatomy.

An emphasis on cross sectional anatomy and lymphatic drainage has been made in this Atlas by including several illustrations and radiological images.

Each chapter of the Atlas covering both the anatomy and neuroanatomy contains an introduction, dissection movies and interactive slides used in building the anatomical regions with a stratigraphic approach.

It has been a great pleasure to collaborate with Edi.Ermes in preparing the English edition of this Atlas. I wish to thank in particular Raffaele Grandi the CEO of Edi.Ermes for allowing me to be part of the development team for this Atlas of Human Anatomy.

Estomih Mtui, MD
Editor English Edition

Professor of Clinical Anatomy in Radiology
Chief, Division of Anatomy
Department of Radiology
Weill Cornell Medicine
New York, USA

Preface

The study of Human Anatomy requires significant theoretical learning coupled with an indispensable visual and manual contact with the human body. In theoretical learning, graphic representations play an important role in rendering the comprehension more immediate with respect to verbal description, notwithstanding its elaborate detail. Hence, the common need on the part of teachers and students to have clear and explanatory images – teachers for their lessons and students for their personal study. Ever since the beginning of the study of human anatomy – with the drawings of Leonardo da Vinci and Andreas Vesalius – anatomic illustration has been an indispensable tool used alongside dissection to show students the topographic reality of organs and systems.

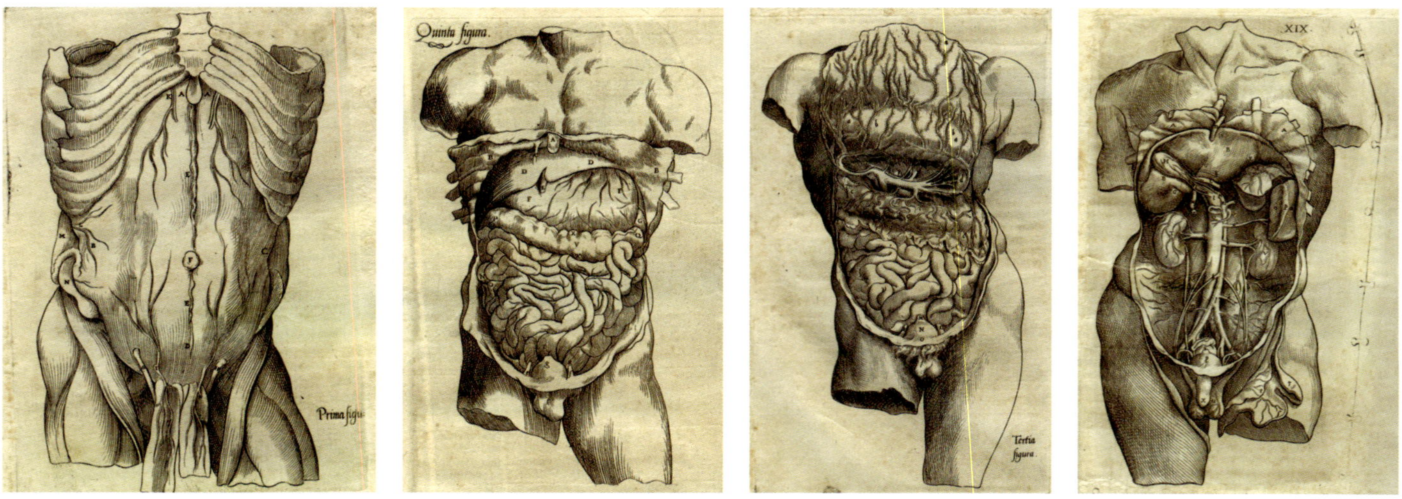

Andreae Vesalii, *Suorum de humani corporis fabrica librorum epitome*, Londini 1545
(courtesy of the Department of Experimental Medicine, Section of Human Anatomy, University of Genoa)

In the last few years, technological progress has enabled computer graphics, image post-processing methods and diagnostic imaging technologies to attain levels of detail and resolution unimaginable even just a decade ago. Combining these approaches, it is now possible to obtain extraordinarily realistic two- and three-dimensional anatomic representations of the living human body. And, with the aid of invasive diagnostic techniques (e.g. endoscopy, laparoscopy or thoracoscopy), one can also accompany these with images and videos of high anatomic quality and comprehensibility.

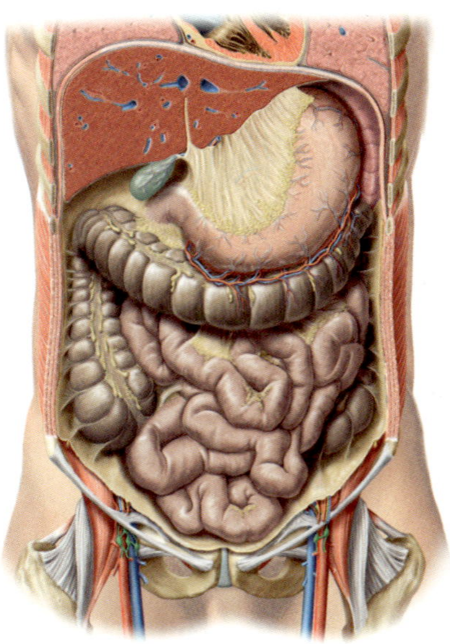

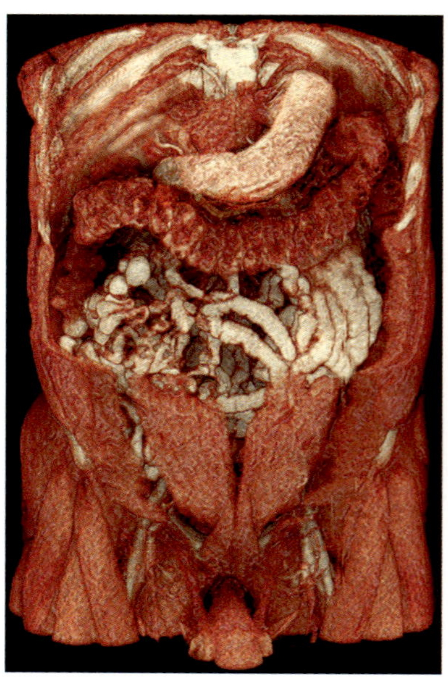

Exemplificative images
from the Atlas of Human Anatomy

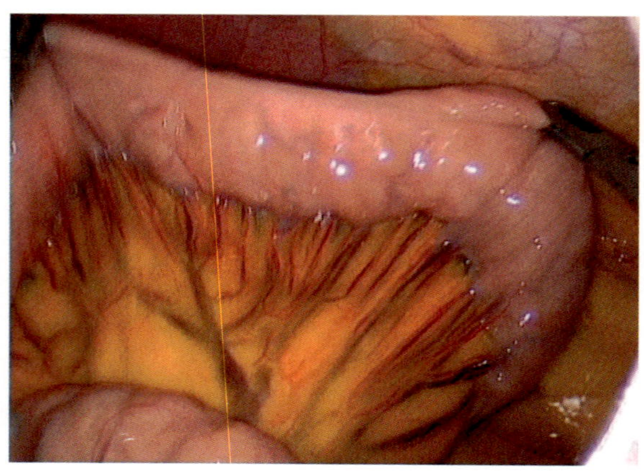

Thus the study of Human Anatomy on the cadaver can profit and gain from that based on the living human body, i.e. what the future doctors and health professionals will eventually be dealing with.

Based on these considerations we have created an innovative Atlas composed of an amply illustrated text and of a web area rich in original content, in which both Anatomy as studied on the cadaver and Anatomy of the living human body find their place within a coordinated, organized framework.

The ***print version of the Atlas*** is organized on a regional and topographic basis. Each chapter is illustrated with the aid of:

- accurately executed and didactically useful drawings
- stratigraphic sequences that allow the reader, page by page, to reconstruct the organization of the various anatomic regions
- realistic CT or MR images of the living human body, elaborated with a program of three-dimensional volume rendering
- photographs of laparoscopic and endoscopic examinations.

The ***"Virtual Campus" web platform*** gathers together a vast catalogue of didactic material that is immediately accessible to the user thanks to its innovative organization with respect to the general world of Anatomy atlases. Each resource proposed is clearly framed within its area of reference; in addition, to shorten research times, the user can take advantage of interactive guided courses that provide full-fledged descriptions of anatomic areas presented in the format of a frontal lesson.

Contents of the web platform include:

- a complete course of Topographic Anatomy, organized as guided video courses that guide the user through interrelated resources by means of specific links that become available during the exposition
- interactive 3D reconstructions of CT or MR examinations of single organs or regions
- commented videos of laparoscopies and endoscopies
- commented videos of anatomic dissections on cadavers
- interactive stratigraphic animations.

Our aim with this work was to create a useful instrument both for students and teachers: as a tool for students in the study of anatomy and an indispensable companion throughout the whole course of their studies leading to graduation; and as an aid for teachers in the preparation of their own lessons.

But not only: it is now statistically proven that the knowledge of Anatomy acquired in the course of studies tends to diminish over time, and so needs to be updated constantly. Hence, this Atlas is also addressed to health professionals as a valid tool of consultation for updating their knowledge.

The particular development of some parts that may appear to be highly specialized (e.g. musculoskeletal biomechanics) makes this Atlas suitable for students and health professionals not only in the area of Medicine and Surgery, but also in the areas of Physiotherapy, Nursing Sciences, Motor Sciences, Dentistry and for all health professional courses.

Hoping that this Atlas will find favor among teachers, students and health professionals, before concluding we would like to express our special sincere thanks to the Publisher, Raffaele Grandi, and to his staff, for having believed in this project and for their great professionalism and enthusiasm demonstrated to us all, and alongside us, in the realization of this work.

We also extend our thanks to all those – colleagues, scientists, teachers and assistants – who have helped us with their advice, materials and critical review, and encouraged us to produce this milestone in the international panorama of publishing.

the Editors, Giuseppe Anastasi, Eugenio Gaudio and Carlo Tacchetti,
and the Authors, Paolo Castano, Sergio Castorina, Ottavio Cremona, Raffaele De Caro,
Guido Macchiarelli, Mario Rende, Domenico Ribatti, Chiarella Sforza

Contents Plan

Volume 1

Volume 2

Volume 3

Constituent Elements of the Atlas

The Atlas deals with the organization of the human body through various means of displaying the complex structures composing it.

Schematic Representations

Simple and immediate summary diagrams present the distribution and drainage pathways of arteries, veins, and lymphatic vessels in a concise and explanatory manner.

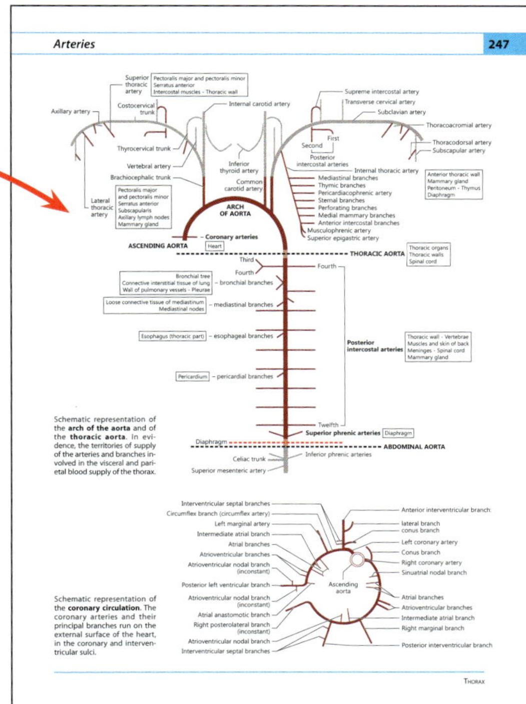

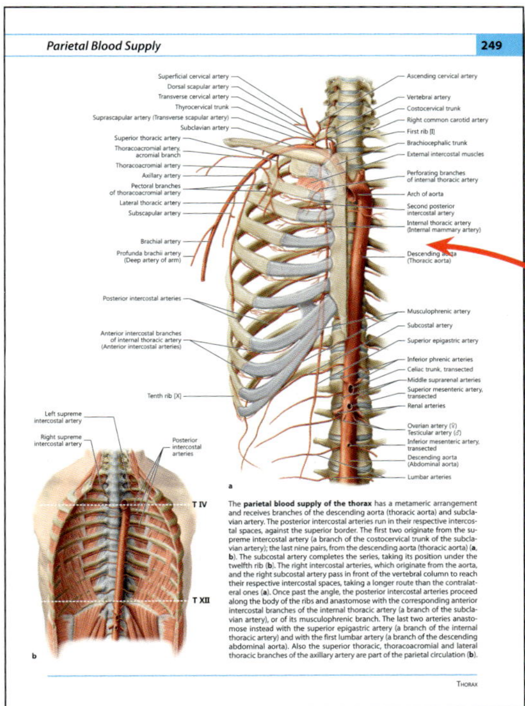

Traditional Pictorial Illustrations

Traditional, top notch iconography with color drawings, masterfully made specifically for this work, basic teaching aids accompanied by extensive explanatory captions.

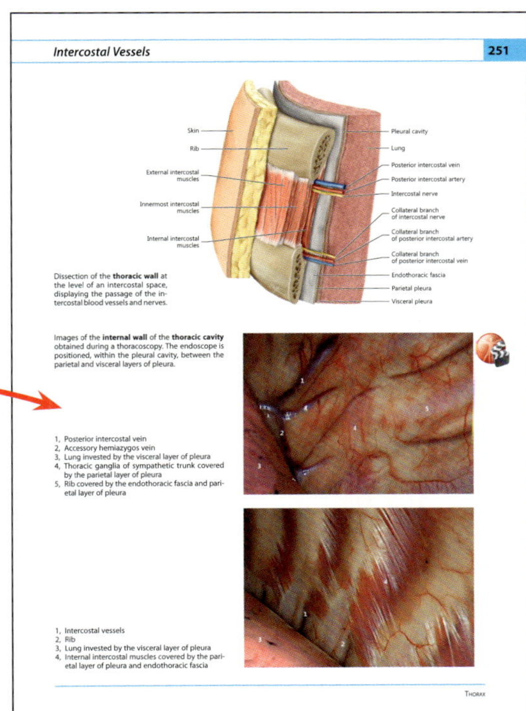

Live Endoscopic Images

Images obtained during diagnostic examinations of thoracoscopy, laparoscopy, and endoscopy provide details of the anatomy of the living.
The images are taken from videos entirely available as digital subsidies on the web platform associated with the Atlas.

Live Three-dimensional Images

The three-dimensional images of the living obtained with 3D volume rendering technique, of CT or MR image sequences help to understand the most refined anatomic details. Stratigraphic sequences and interactive 3D reconstructions are present in the Atlas web platform.

The reader, either a student or a professional, will have access to anatomic images – derived from cadavers, pictorial or obtained from a living human subject, through the use of diagnostic imaging techniques such as NMR or CT, viewed from diverse angles, or projections. In particular, except in the case of particular visualizations which are otherwise indicated, images on a transverse plane obtained with diagnostic imaging techniques are always presented according to a conventional projection from below, with the subject placed in supine position. In contrast, anatomic pictorial images or representations are presented according to a projection from above.

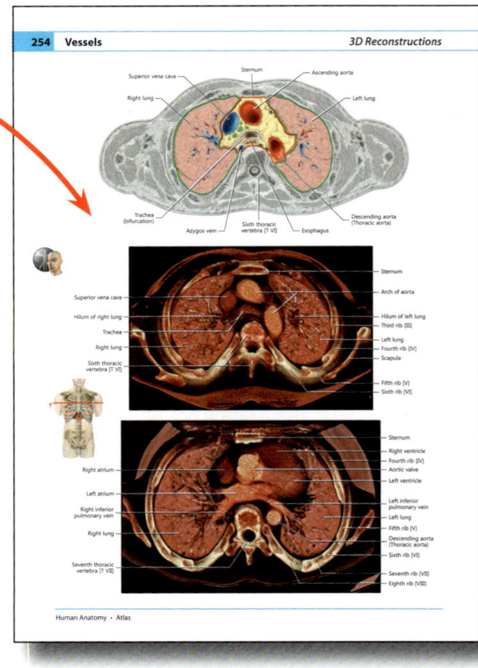

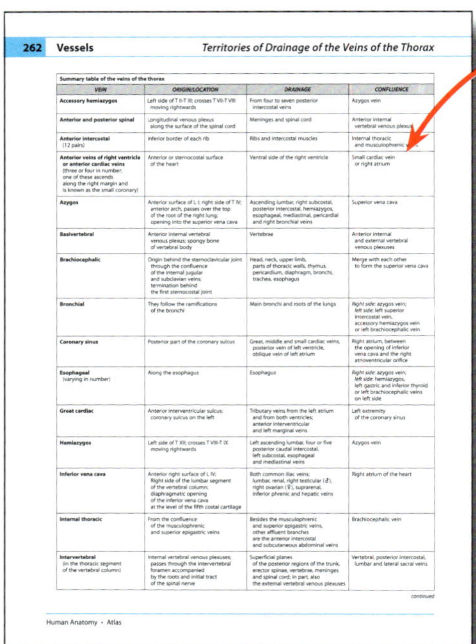

Summary Tables

Numerous tables summarize the boundaries and content of anatomical regions, arteries, veins, nerves, lymph nodes, muscles, joints, and bones, providing and organizing the most relevant information for each item schematically.

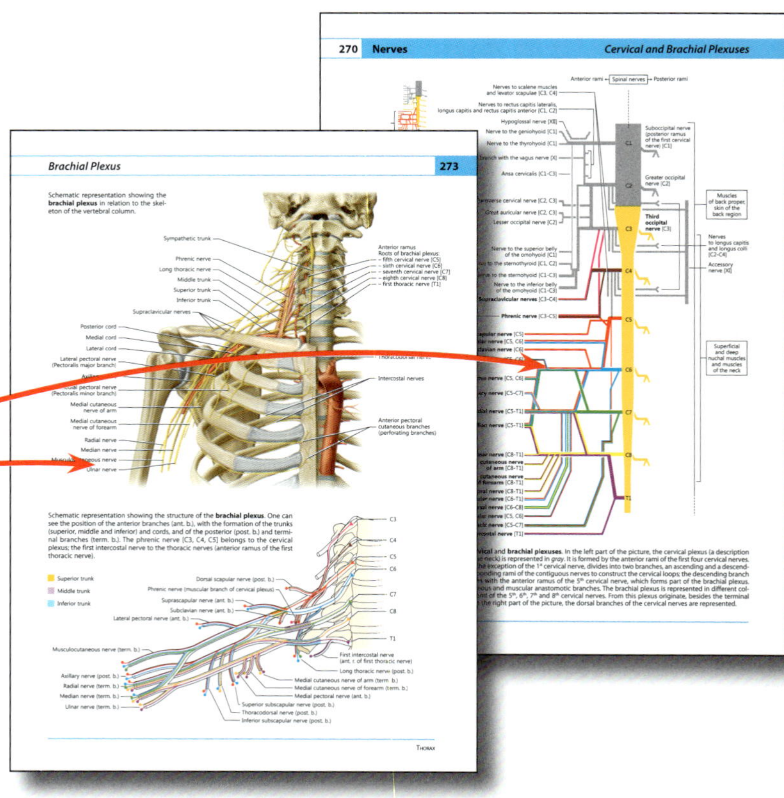

Diagrams Provide Examples of Nerve Plexuses

The organization of the nerve plexuses, sensory, and motor pathways is presented with effective summary patterns that accompany the most complex anatomical drawings.

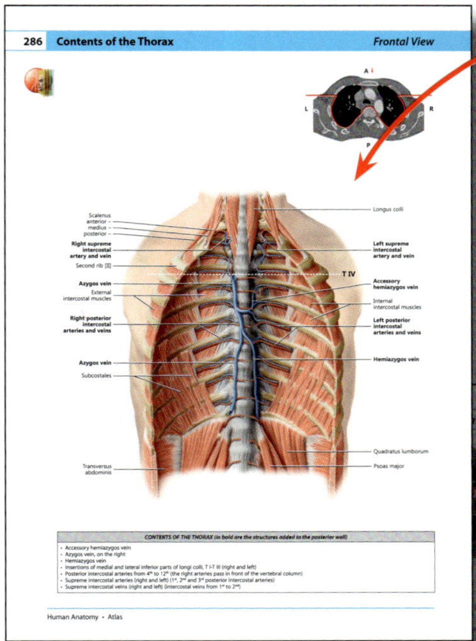

Frontal and Lateral Slider

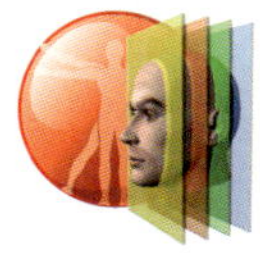

A sequence of drawings allow the student to study the contents of the cavities or the organization of anatomical regions with a stratigraphic multidimensional approach on frontal planes, sagittal planes or successive transverse planes. The different anatomical structures seen in each cavity or region are overlapped and added progressively. Sliders are available in the interactive form in the web platform associated with the Atlas.

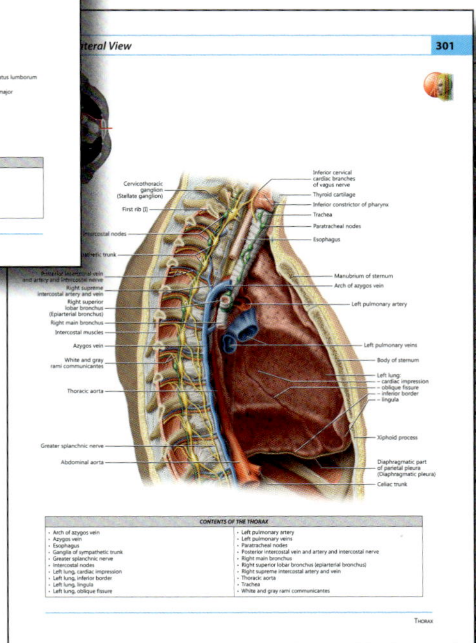

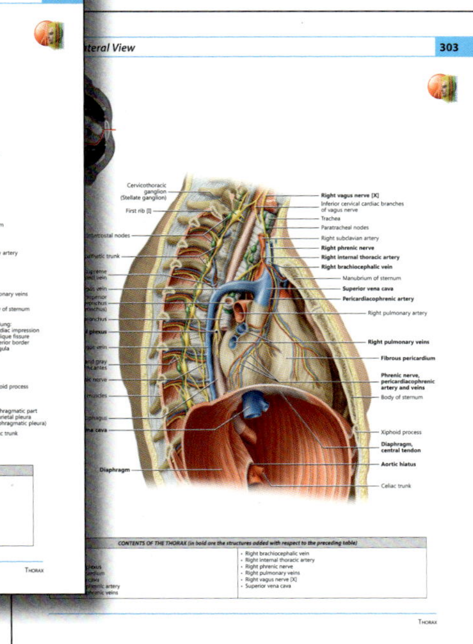

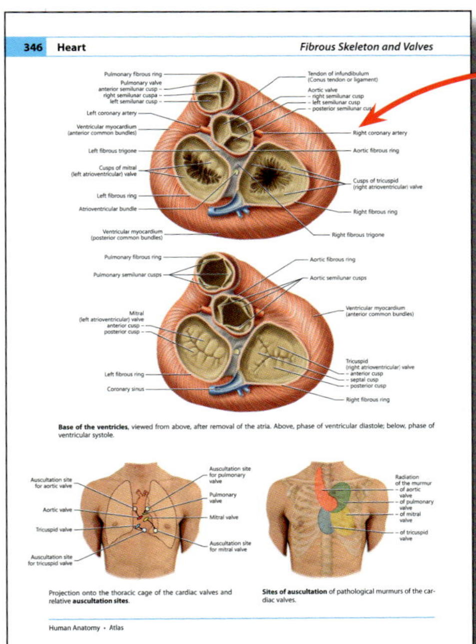

Physiological and Clinical Indications

Physiological and clinical aspects and references were incorporated in order to illustrate the anatomical basics in a logical manner.

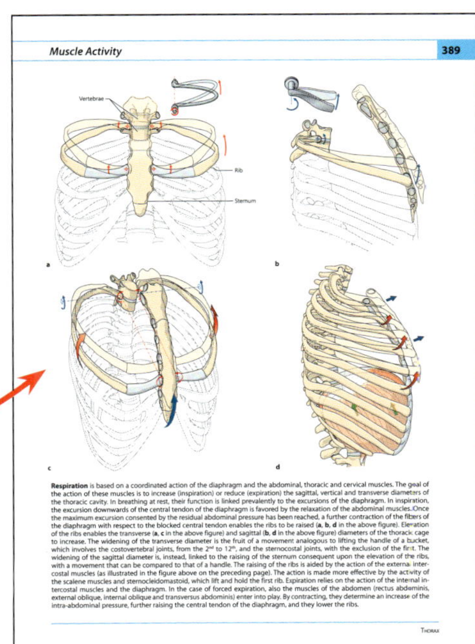

Elements of Biomechanics

Drawings, diagrams, and tables facilitate understanding and memorization of bones, joints, and muscular complexes and joint dynamics.

Table of Contents

GENERAL ANATOMY

MUSCULOSKELETAL SYSTEM

BONES AND JOINTS

MUSCLES

GENERAL ANATOMY

Petros (Athens, 1928 - Milan, 2014), *Sinfonia Dante*, oil on canvas, various materials, cm 202 × 143 and music roll on canvas, cm 28 × 310, 2004

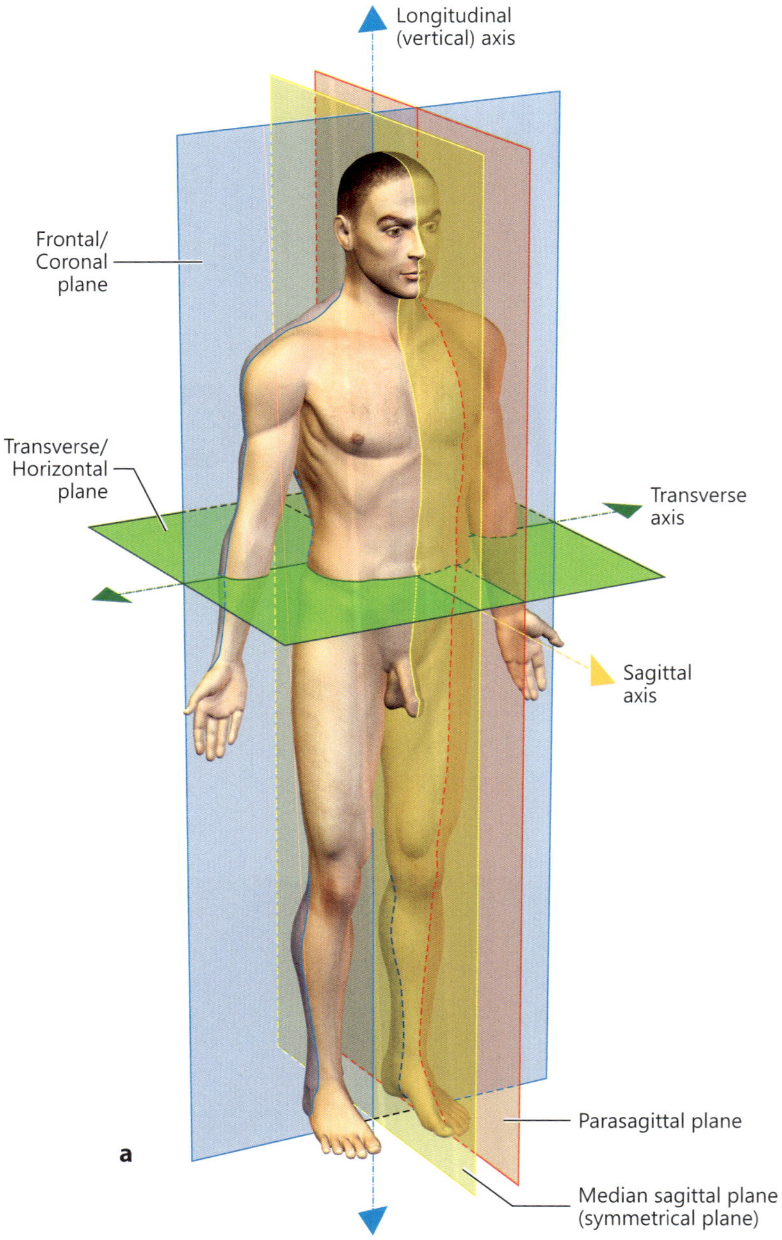

a

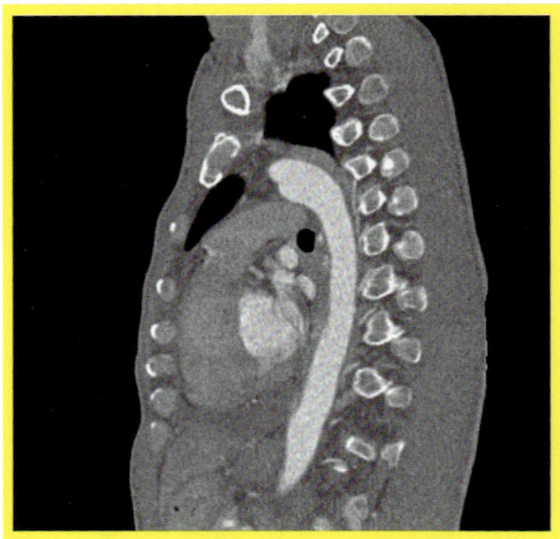

b

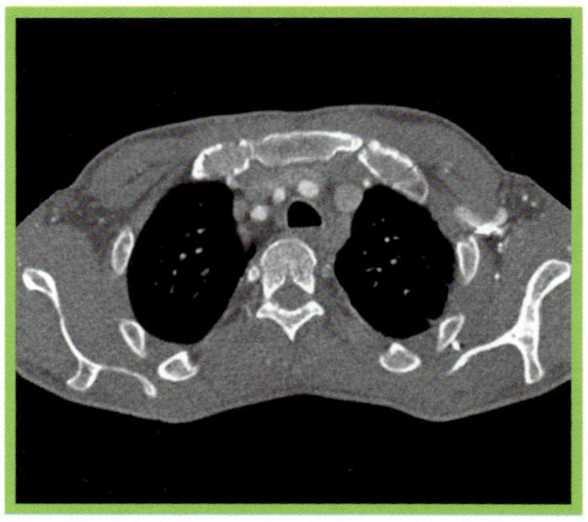

c

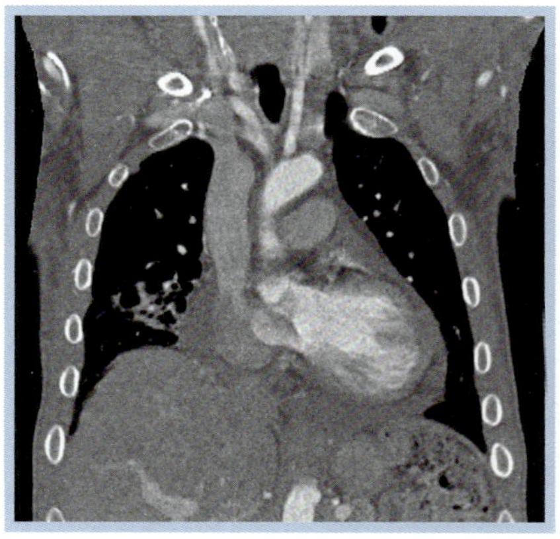

d

ANATOMICAL PLANES AND AXES

The location of every part of the body, viewed in the anatomical position (fig. **a**), may be referenced through three planes set at right angles to each other. The CAT scans (figs. **b**, **c**, **d**) show examples of images obtained in each of these three orthogonal planes.

Sagittal Planes. These are vertical planes with a posteroanterior orientation. The sagittal plane passing through the center of the body (*median sagittal plane* or *symmetrical plane*) defines the bilateral symmetry of the body. An infinite number of *parasagittal planes* can be marked out to the right or left of the median plane. Each sagittal plane provides a point of reference for a right side and a left side. With reference to any given sagittal plane, the relative positions of two parts of the body situated on the same side of that plane can be given using the terms *lateral* (L, fig. **e**) or *medial* (M, fig. **e**). The part farther from the reference plane is described as lateral, and the part nearer to it as medial. The terms *proximal* and *distal* (fig. **f**) are used to indicate the relative position of the two parts of a limb in relation to the point of attachment to the trunk (the glenohumeral joint for the upper limb and the coxofemoral joint for the lower limb). 'Proximal' indicates the point nearer to and 'distal' the point farther from that joint.

Transverse (or Horizontal) Planes. An infinite number of horizontal planes, at right angles to the median sagittal plane, pass through the body at various heights. A part of the body can be defined as *superior* (or *cranial*) or *inferior* (or *caudal*) in reference to any given transverse plane (fig. **g**).

Frontal (or Coronal) Planes. These vertical planes run perpendicular to the transverse planes (fig. **h**). A part of the body can be described as *anterior* (or *ventral*) or *posterior* (or *dorsal*) to any given frontal plane (figs. **i**, **j**). The distal segments of the lower limb are exceptions to this rule, due to the rotation this segment undergoes during fetal development. This is why the term 'dorsum of the foot' refers to the anterosuperior surface of the foot.

The intersection of the reference planes enables us to identify three axes about which the main movements of the body are articulated:

- *Sagittal Axis*: Movements of *lateral inclination* of the spine and of *adduction/abduction* of the limbs
- *Transverse Axis*: Movements of *flexion/extension*
- *Longitudinal Axis*: *Torsional* movements of the spine, *rotation* of the limbs, *pronation/supination* of the distal part of the upper limb (forearm and hand).

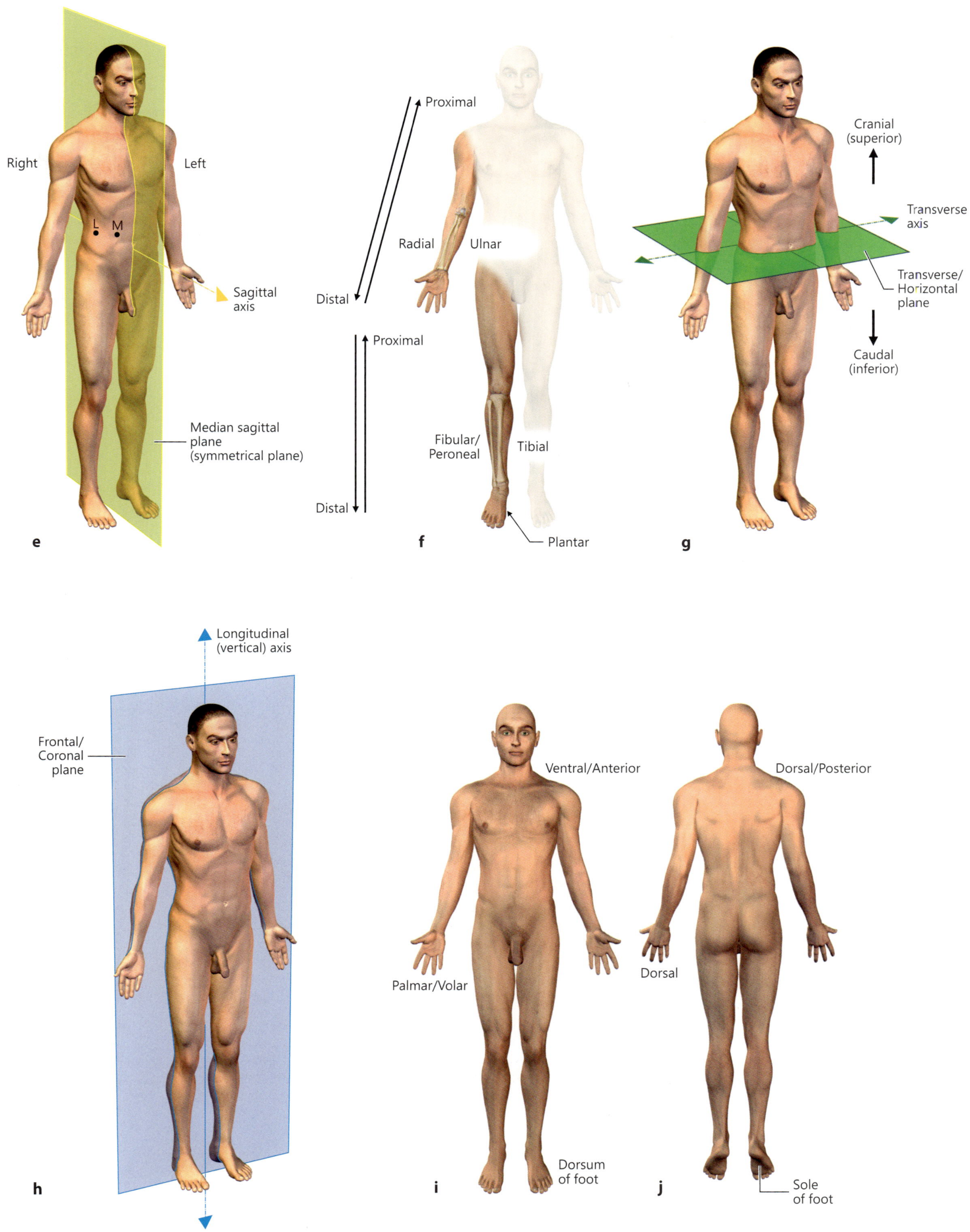
Right
Left
L
M
Sagittal axis
Median sagittal plane (symmetrical plane)
e
Proximal
Distal
Radial
Ulnar
Proximal
Distal
Fibular/ Peroneal
Tibial
Plantar
f
Cranial (superior)
Transverse axis
Transverse/ Horizontal plane
Caudal (inferior)
g
Longitudinal (vertical) axis
Frontal/ Coronal plane
h
Ventral/Anterior
Palmar/Volar
Dorsum of foot
i
Dorsal/Posterior
Dorsal
Sole of foot
j

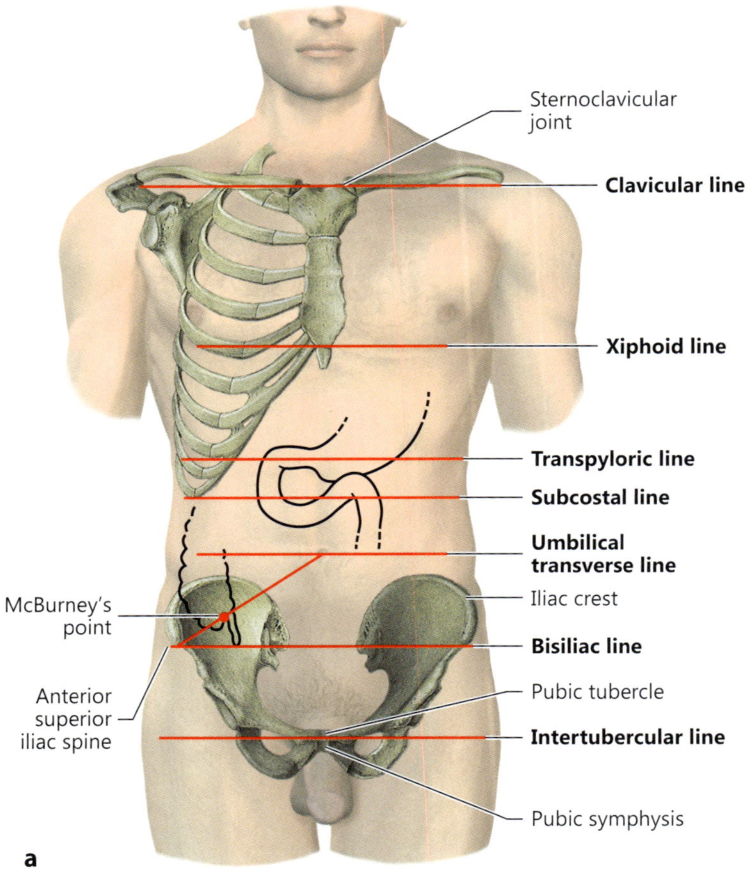

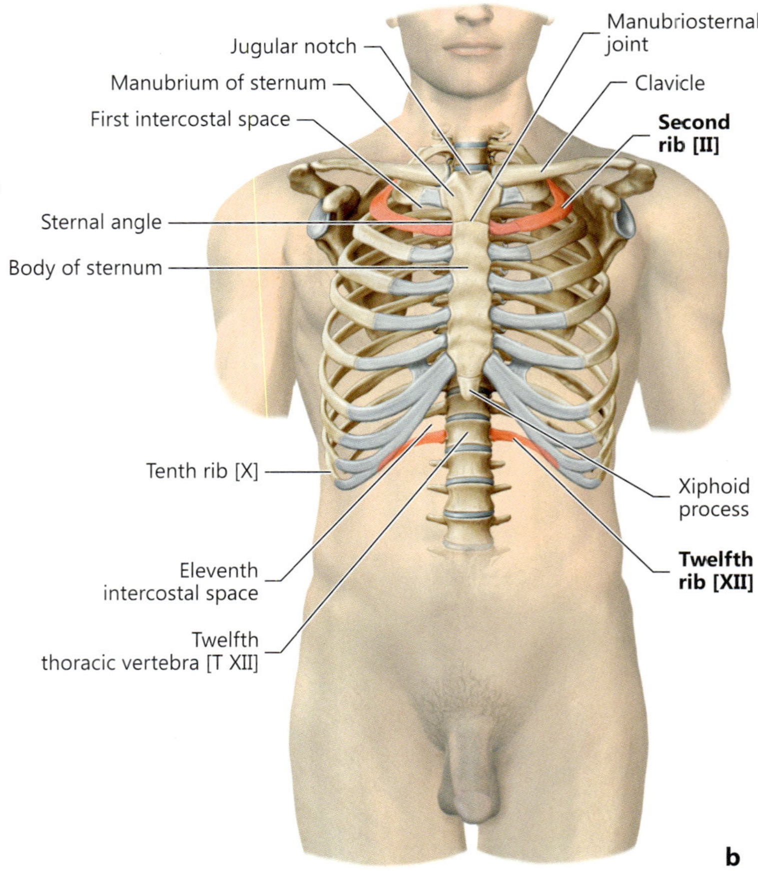

Horizontal surface lines of reference and **topographic landmarks** of the trunk: **a**, and **b**, anterior and **c**, posterior.

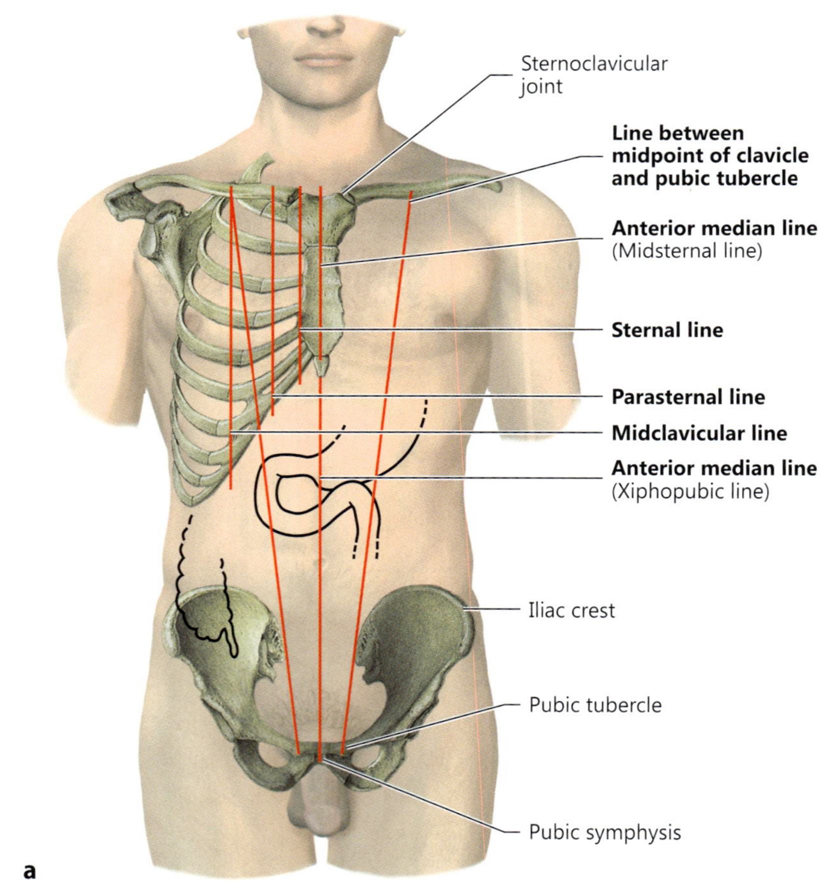

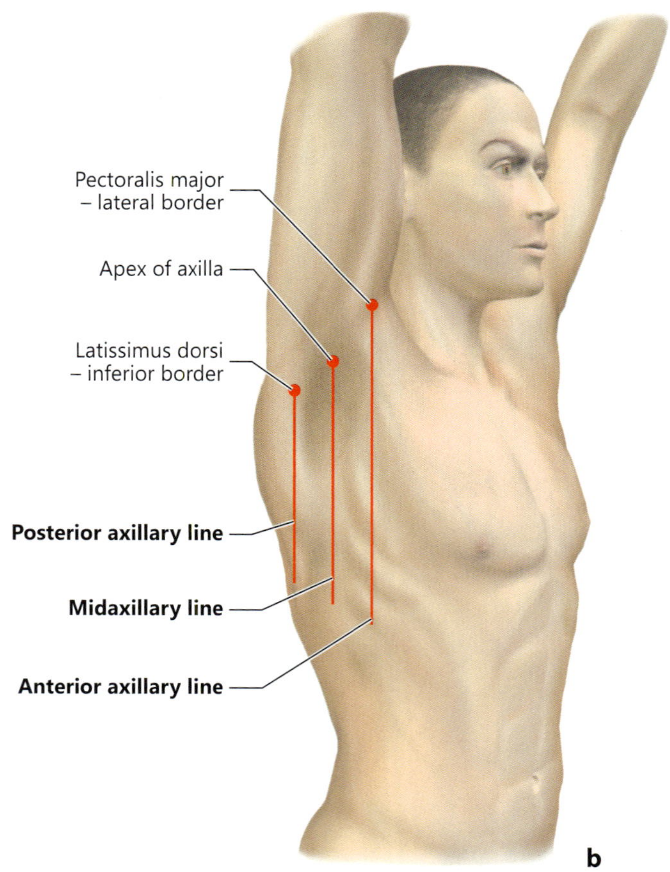

Vertical surface lines of reference of the trunk: **a**, anterior, **b**, lateral and **c**, posterior.

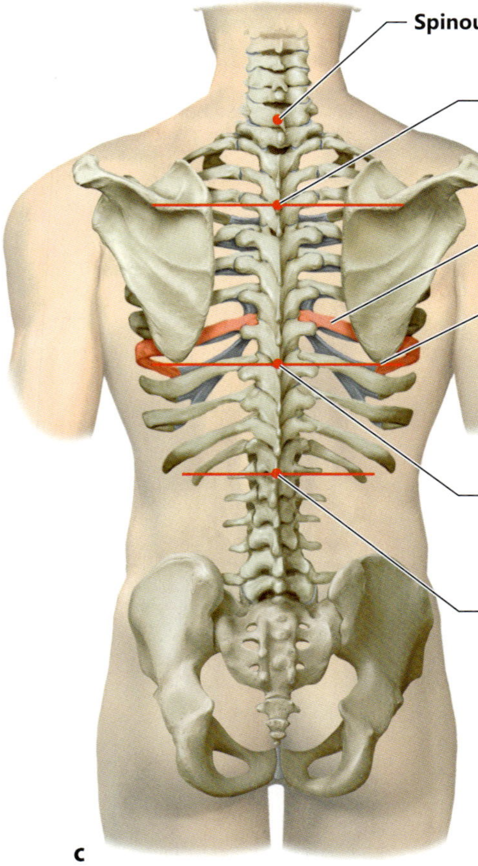

c

Horizontal Lines	
	TOPOGRAPHIC LANDMARKS
Ribs	• 2nd (at the sternal angle) • 8th (at the inferior angle of the scapula)
Intercostal spaces	• from the 1st to the 11th rib, each space positioned between two ribs
Anterior horizontal lines	• Xiphoid line (passing through the point at which the body of the sternum joins the xiphoid process) • Clavicular line (passing through the sternoclavicular joints) • Bisiliac line (passing through the anterior superior iliac spines) • Intertubercular line (passing through the superior pubic tubercles) • Umbilical transverse line (passing through the umbilical region) • Subcostal line (tangential to the costal or subcostal margin – tangent to the 10th rib) • Transpyloric line (midway between the jugular notch of the sternum and the pubic symphysis)
Posterior horizontal lines	• Line passing through the spinous process of T III, connecting the spines of the two scapulae (right and left) • Line passing through the spinous process of T VIII, connecting the inferior angles of the two scapulae (right and left) • Line passing through the spinous process of T XII
Appendix point (McBurney's point)	• Right antimere: oblique line connecting the umbilical region with the anterior superior iliac spine. Union between the lateral third and the two medial thirds

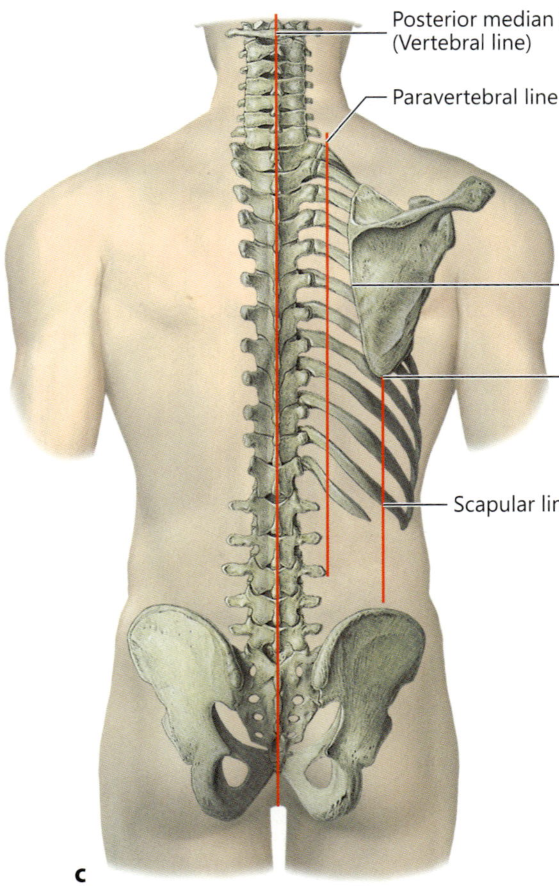

c

Vertical Lines	
	TOPOGRAPHIC LANDMARKS
Anterior vertical lines	• Midclavicular line from the midpoint of the clavicle • Marginosternal line (tangent to the lateral border of the sternum) • Midsternal and xiphopubic lines (anterior median line from the sternum to the pubis) • Parasternal line (3 cm laterally to the midsternal line) • Line from the midpoint of the clavicle to the pubic tubercle
Lateral vertical lines	• Anterior axillary line (from the superior third of the lateral border of the pectoralis major) • Midaxillary line (from the apex of the axilla) • Posterior axillary line (from the inferior border of the latissimus dorsi)
Posterior vertical lines	• Scapular line (passing through the inferior angle of the scapula) • Paravertebral line (2-3 cm laterally to the posterior median line) • Posterior median line (along the spinous processes of the vertebrae)

The grid formed by these **surface reference lines**, which pass vertically and horizontally through easily identifiable topographical landmarks on the surface of the body, is used to define specific points to which a symptom or a part of the anatomy refers. Surface reference lines are also used to delimit specific virtual planes of reference.

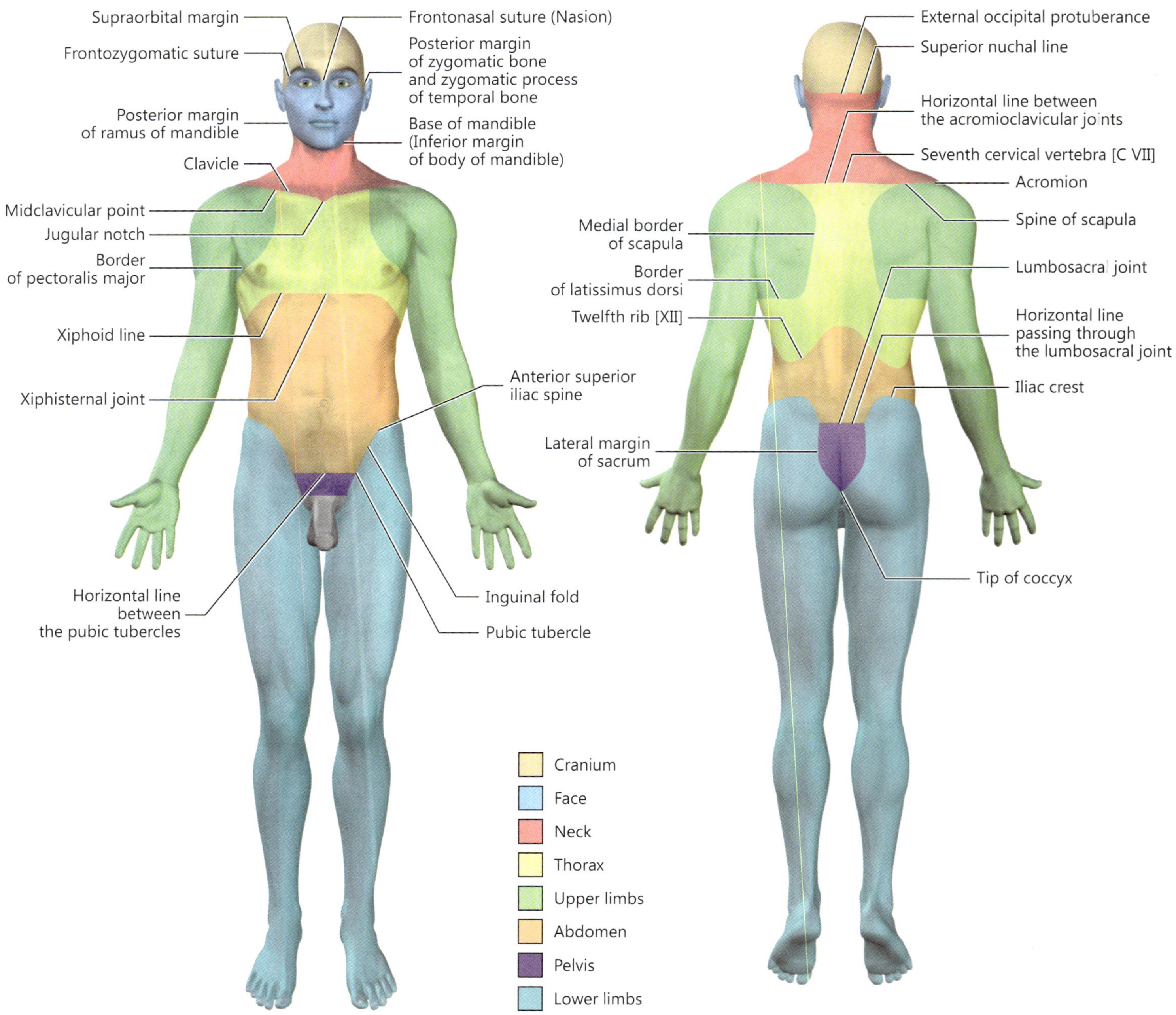

Topographic subdivision of the body surface into regions. Description of the main anatomical **regions of the body**. Anterior (left) and posterior (right) views.

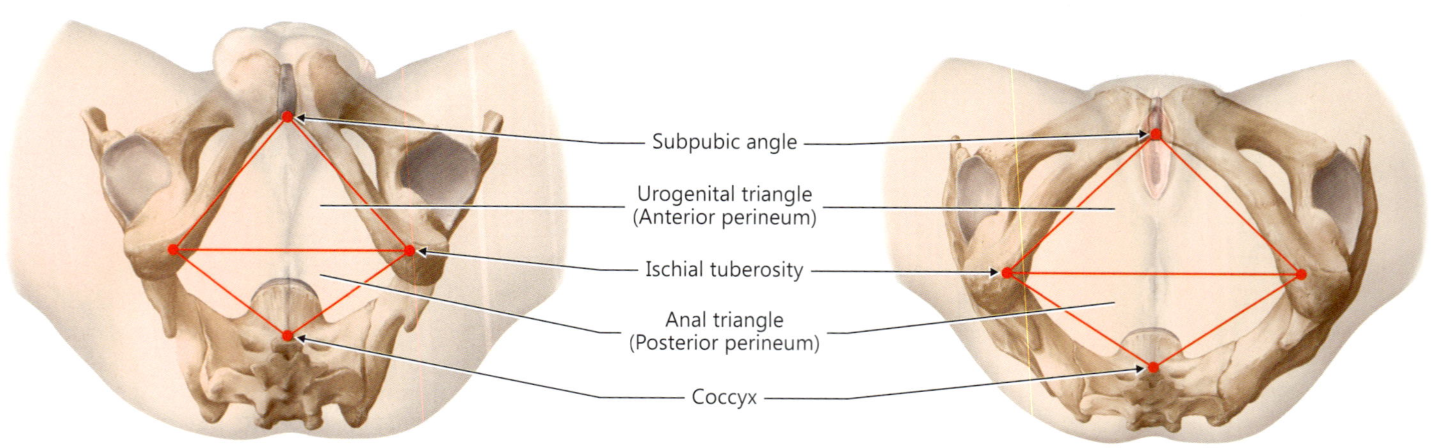

Surface boundaries of the **perineum** with reference to skeletal structures. The tips of the diamond-shaped perineum correspond to the subpubic angle anteriorly, the tip of the coccyx posteriorly and the two ischial tuberosities laterally.

The limits of the surface projection of the thoracic cavity do not correspond to those of the **thoracic cage**. The latter also includes the upper part of the **abdominal cavity**. The diaphragm muscle extends between these two cavities. The dome shape of this muscle causes recesses to be formed between the muscle and the rib cage.
Below, the **anterior trunk wall** as seen from the inside. The recesses mentioned above are indicated with the asterisk.

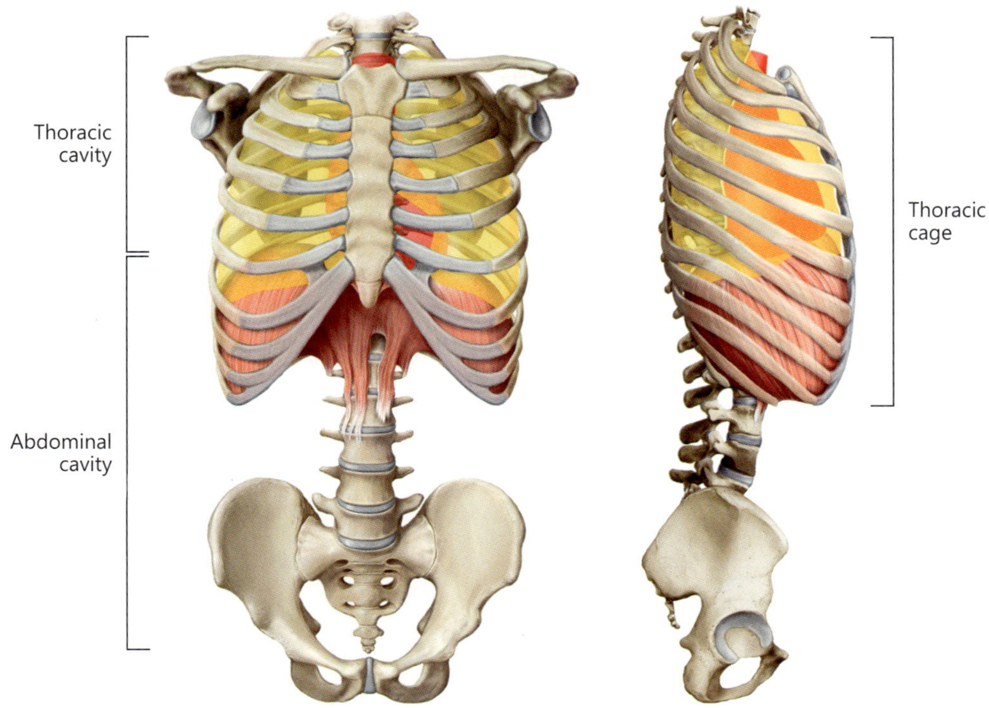

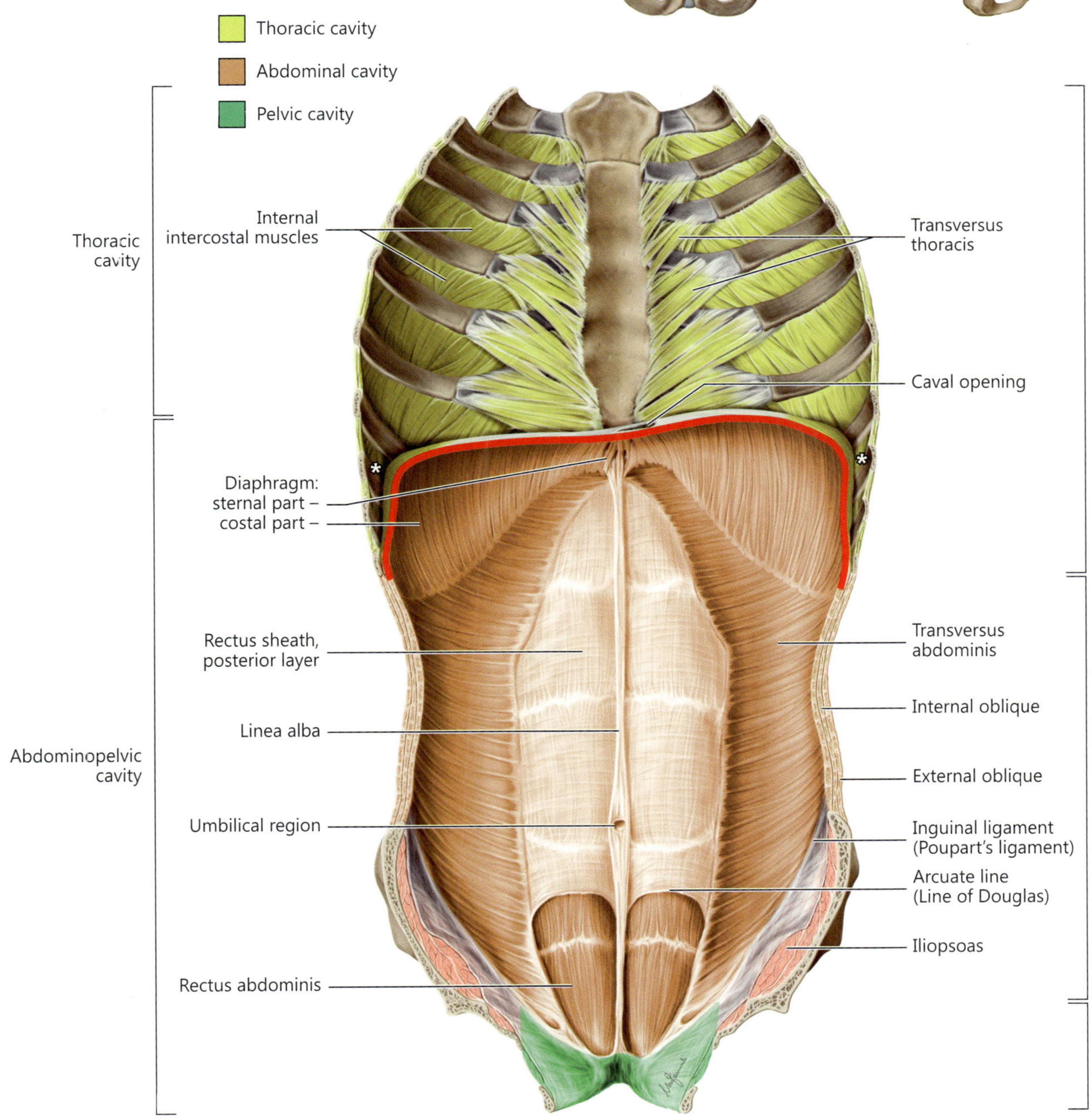

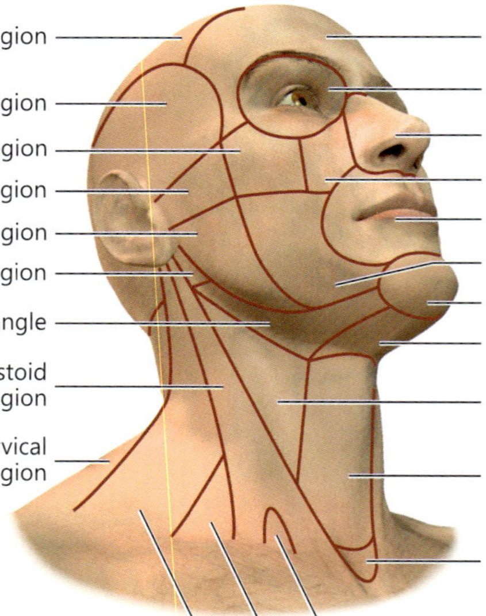

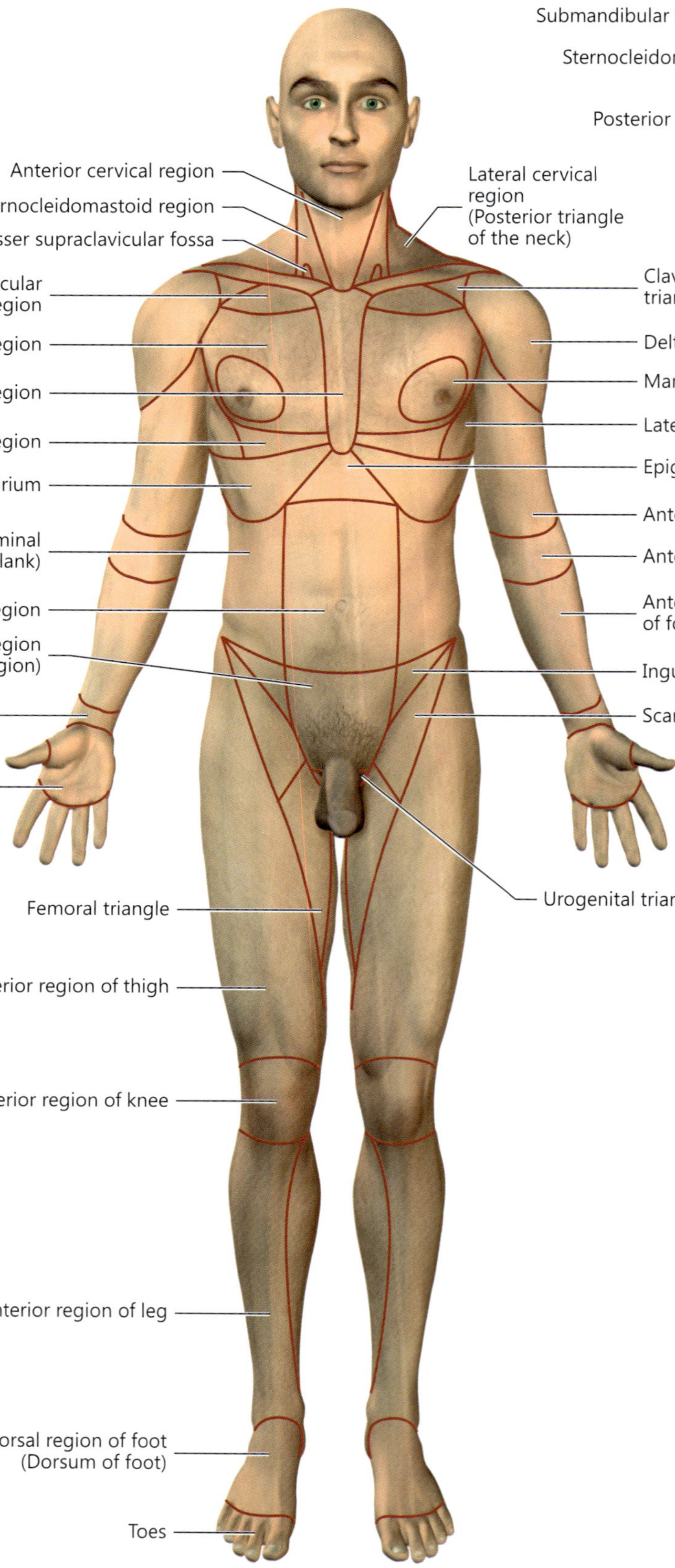

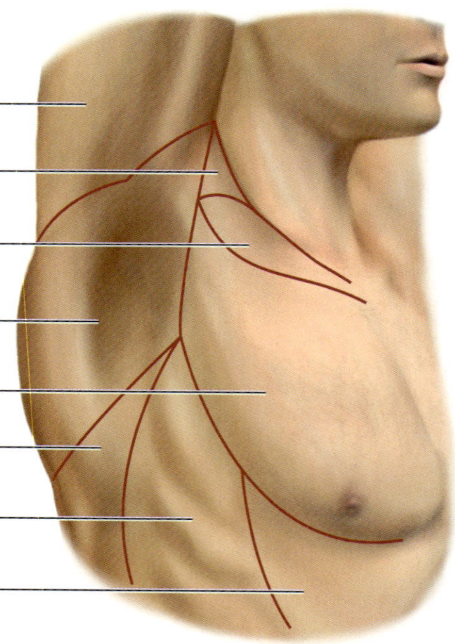

Detailed surface **subdivision of the body into anatomical regions**. Anterior view.

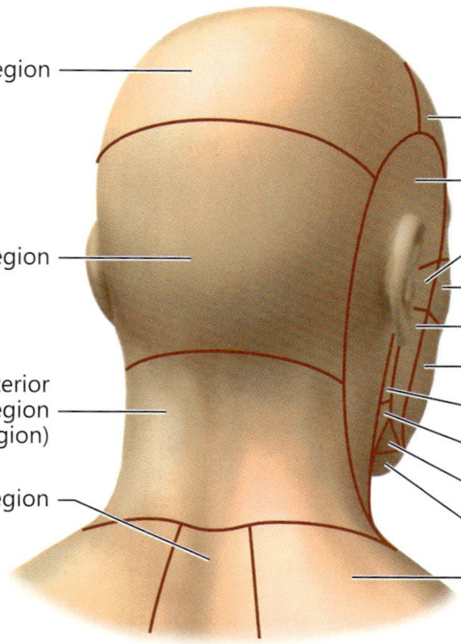

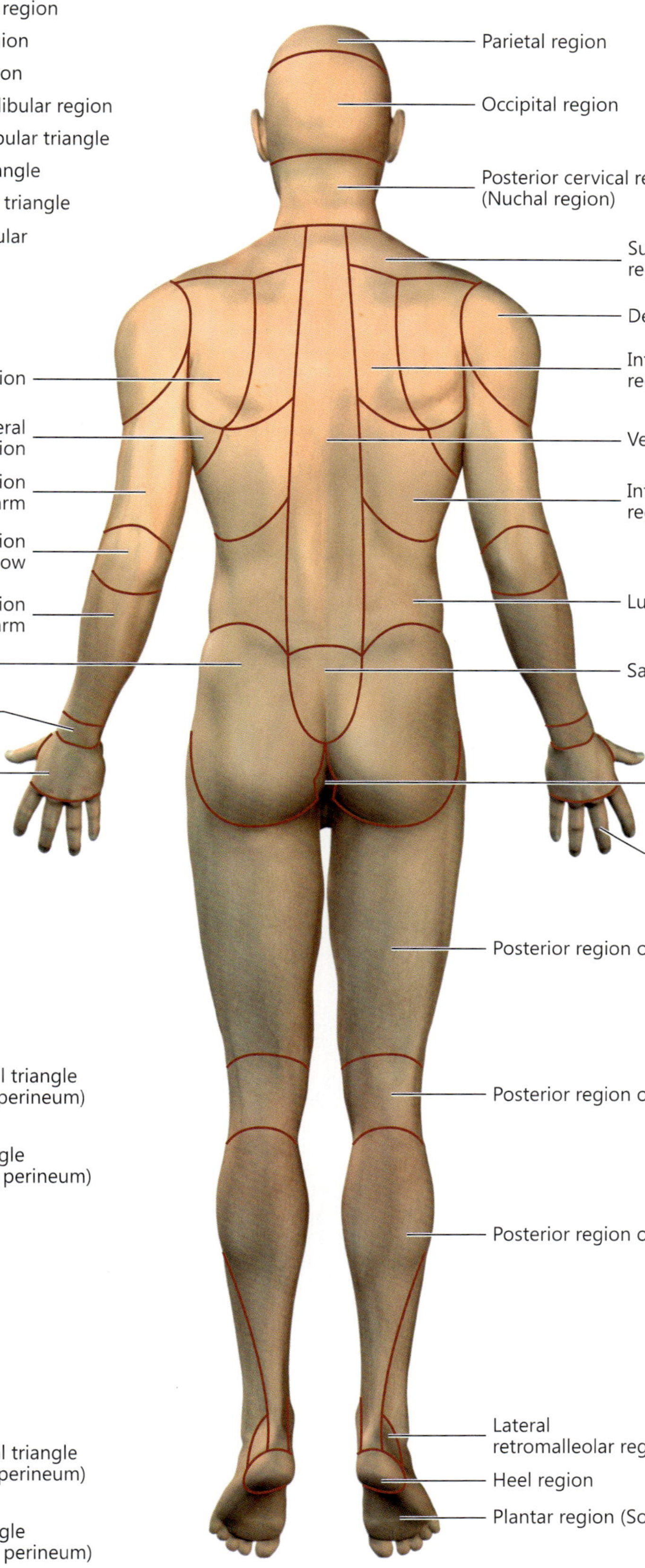

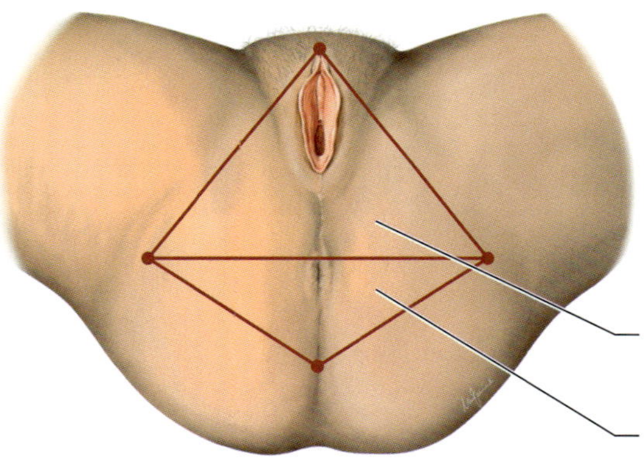

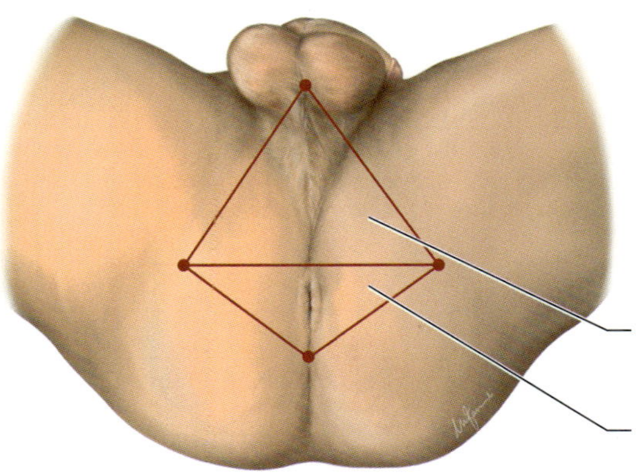

Detailed surface **subdivision of the body into anatomical regions**. Posterior view and view of the perineal region.

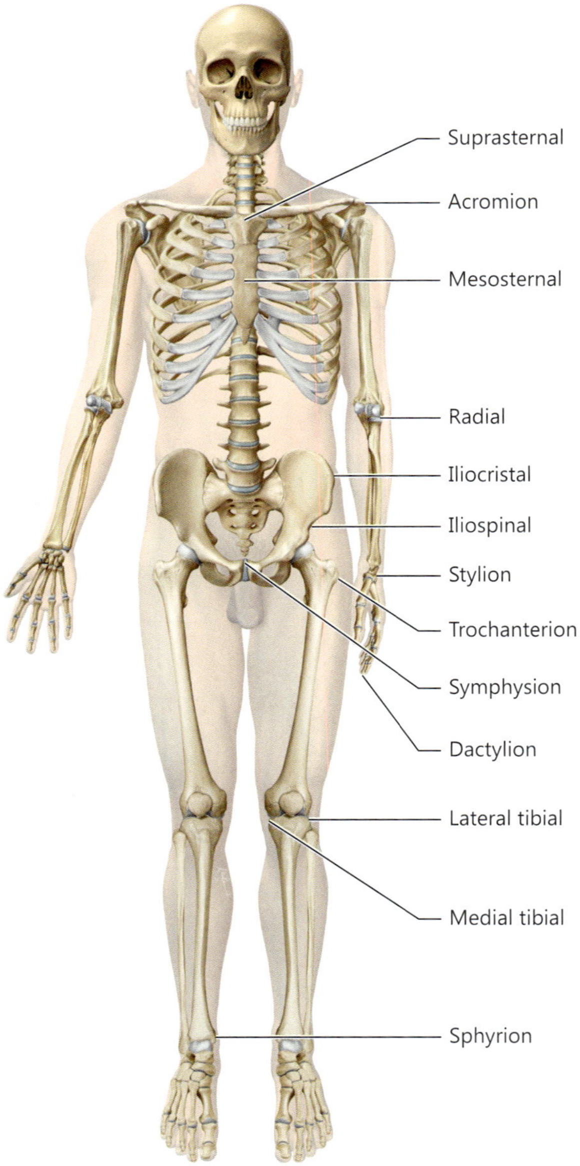

SOMATOMETRIC POINTS ON THE MEDIAN LINE	
Mesosternal (mst)	Median point of the sternum at the junction of the third and fourth sternebra (at the level of the fourth sternocostal joint)
Suprasternal (sst)	Median point on the jugular notch of the manubrium of the sternum
Symphysion (sy)	Median point on the superior border of the pubic symphysis

SOMATOMETRIC POINTS LATERALLY TO THE MEDIAN LINE	
Acromion (a)	Outermost point of the lateral border of the acromion
Dactylion (da)	The most distal point (tip) of the middle finger
Iliocristal (ic)	Outermost point of the iliac crest
Iliospinal (is)	Innermost point of the anterior superior iliac spine
Radial (r)	The most superior point of the head of the radius
Sphyrion (sph)	The most distal point of the medial malleolus
Stylion (sty)	The most distal point of the styloid process of the radius
Medial tibial (mt)	The most superior point on the internal border of the medial condyle of the tibia
Lateral tibial (lt)	The most superior point on the lateral border of the lateral condyle of the tibia
Trochanterion (tro)	Most superior point of the greater trochanter

The **anthropometric points** are subdivided into the cephalometric (craniometric) points, relating to the cranium of a skeleton or the head of a living person (☞ General Anatomy-Head), and the somatometric points, referring to the trunk and limbs (of a skeleton or a living person).
The **craniometric points** are described in detail on pages 68-73.

The following table reports the main superficial topographic areas of the single regions. For a more detailed description of the superficial and deep topography of these regions refer to the specific chapters.

REGIONS OF THE HEAD	*BOUNDARIES OR LOCATION*
Facial	
– buccal (cheek region)	Zygomatic bone (superiorly), base of the mandible (inferiorly), nasolabial sulcus and nasogenian sulcus (medially) and anterior border of the masseter (laterally). Corresponds to outer surface of the buccinator
– infraorbital	Infraorbital margin (superiorly), boundary furrow of the ala of the nose (medially) and the nasolabial sulcus (inferiorly)
– mental (chin region)	Mentolabial sulcus (superiorly) and base of the mandible (inferiorly)
– nasal	Root of nose (superiorly); line from the medial angle of eye to the ala of the nose (laterally)
– oral (mouth region)	Nasolabial sulcus (superiorly) and mentolabial sulcus (inferiorly)
– orbital	Suprapalpebral sulcus (superiorly) and infrapalpebral sulcus (inferiorly)
– parotid (parotid-masseteric)	Zygomatic arch (superiorly), base of the mandible (inferiorly), anterior border of the masseter (anteriorly), anterior border of the sternocleidomastoid and external acoustic meatus (posteriorly)
– zygomatic	Corresponds to the zygomatic bone
Frontal and occipital (occipitofrontal region)	External occipital protuberance and superior nuchal line (posteriorly); continuation of the superior nuchal line through the base of the mastoid process and then the superior temporal line up to the zygomatic process of the temporal bone (laterally); root of the nose and the supraorbital margin of the frontal bone (anteriorly)
Infratemporal	Corresponds to the zygomatic arch
Occipital	☞ Frontal
Parietal	Part of the occipitofrontal region between the coronal suture (anteriorly) and the lambdoid suture (posteriorly)
Temporal	Zygomatic bone, zygomatic process of the frontal bone and temporal crest (anteriorly); superior temporal line up to the asterion (superiorly and posteriorly); horizontal line passing through the zygomatic arch (inferiorly)

REGIONS OF THE NECK	BOUNDARIES OR LOCATION
Anterior cervical (anterior triangle)	Base of the mandible (superiorly); anterior border of the sternocleidomastoid (laterally); anterior median line (medially). The apex of the triangle corresponds to the jugular notch, the base corresponds to the inferior border of the mandible and to a line that joins the angle of the mandible with the mastoid process
Lateral cervical (posterior triangle), and omoclavicular or subclavian triangle and greater supraclavicular fossa	Posterior border of the sternocleidomastoid (anteriorly), anterior border of the trapezius (posteriorly) and clavicle (inferiorly). The greater supraclavicular fossa is contained between the posterior border of the sternocleidomastoid, the medial third of the clavicle and the inferior belly of the omohyoid
Posterior cervical (nuchal region)	Horizontal line that joins the acromions of the scapulae through the seventh cervical vertebra or vertebra prominens (inferiorly); the superior nuchal line and external occipital protuberance (superiorly); anterior borders of the trapezius muscles (laterally)
Suprahyoid (submental and submandibular triangles and retromandibular fossa)	Base of the mandible (superiorly); horizontal line passing through the hyoid bone (inferiorly); anterior border of the sternocleidomastoid (laterally). The retromandibular fossa is included between the ramus of the mandible, the external acoustic meatus and the anterior border of the sternocleidomastoid
Sternocleidomastoid and lesser supraclavicular fossa	Anterior and posterior borders of the sternocleidomastoid. The lesser supraclavicular fossa is circumscribed by the clavicular and sternal heads of the sternocleidomastoid
Carotid triangle	Sternocleidomastoid (posteriorly); superior belly of omohyoid (inferiorly); stylohyoid and posterior belly of the digastric (superiorly)
Muscular triangle (omotracheal triangle or infrahyoid region), and jugular fossa	Horizontal line passing through the hyoid bone (superiorly) and the jugular notch of the sternum (inferiorly); anterior border of the sternocleidomastoid (laterally) The jugular fossa, at the jugular notch, is bordered laterally by the sternal heads of the sternocleidomastoid muscles
Submandibular triangle and retromandibular fossa	☞ Suprahyoid

THORACIC REGIONS	BOUNDARIES OR LOCATION
Infraclavicular fossa (or infraclavicular region), and clavipectoral or deltopectoral triangle	Corresponds to the infraclavicular fossa and extends from the clavicle (superiorly) up to the the second rib (inferiorly), between the sternal line (medially) and the anterior border of the deltoid (laterally). The deltopectoral triangle, also called the infraclavicular fossa, is a triangular depression among the clavicular parts of the pectoralis major (medially) and the deltoid (laterally) and the clavicle (superiorly)
Pectoral (including mammary region) – lateral – inframammary	Second rib (superiorly), sternal line (medially), inferior border of the pectoralis major (inferolaterally). The boundary of the mammary region is given by the circumference of the breast Below the axillary region up to the level of the sixth rib Inferior border of the pectoralis major (superiorly) up to the sixth rib (inferiorly)
Presternal	Jugular notch of the sternum (superiorly), xiphoid process (inferiorly), sternal lines (laterally)
Clavipectoral or deltopectoral triangle	Clavicle (superiorly), clavicular part of the pectoralis major (medially), clavicular part of the deltoid (laterally)
Vertebral (dorsal) [T I - T XII]	Horizontal line passing through the spinous process of the seventh cervical vertebra (superiorly); the line tangential to the base of the sacrum (inferiorly); the vertical line passing through the angle of the rib (laterally)

ABDOMINAL REGIONS	BOUNDARIES OR LOCATION
Epigastric (epigastric fossa)	Xiphoid process (superiorly), right and left costal margins (laterally), horizontal line passing through the medial extremity of the ninth costal cartilage (inferiorly)
Flank (lateral abdominal region or lumbar region)	Costal margin (superiorly); iliac crest and bisiliac line (inferiorly); midclavicular line (medially)
Groin (inguinal region)	Bisiliac line (superiorly), inguinal fold (laterally); midclavicular line (medially)
Hypochondrium	Below the sixth rib up to costal margin
Pubic	Bisiliac line (superiorly), midclavicular lines (laterally), line between the pubic tubercles (inferiorly)
Umbilical	Horizontal line passing through the medial extremity of the ninth costal cartilage (superiorly) midclavicular lines (laterally); bisiliac line (inferiorly)

REGIONS OF THE BACK	BOUNDARIES OR LOCATION
Lumbar (lumbar triangle)	Twelfth rib (superiorly); iliac crest (inferiorly); vertical line passing through the angle of the rib (medially); vertical line from the anterior extremity of the eleventh rib to the iliac crest (laterally)
Sacral region and coccygeal foveola	Line tangential to the base of the sacrum (superiorly); tip of the coccyx (inferiorly); lateral parts of the sacrum (laterally)
Scapular, and suprascapular, interscapular and infrascapular	Superior border of the scapula (superiorly); inferior border of teres major (inferiorly); medial border of the scapula (medially); posterior border of the deltoid (laterally). The suprascapular, interscapular and infrascapular regions are located above, medially and below the scapular region. Their medial boundary is a vertical line passing through the angle of the rib The infrascapular region is bounded laterally by the lateral pectoral region and inferiorly by the lumbar region

PERINEAL REGIONS	*BOUNDARIES OR LOCATION*
Anal triangle (posterior perineum)	Tip of the coccyx (posteriorly); sacrotuberous ligaments (laterally); transverse line that joins the ischial tuberosities (anteriorly)
Urogenital triangle (anterior perineum)	Pubic symphysis (anteriorly); ischiopubic rami (laterally); transverse line that joins the ischial tuberosities (posteriorly)

REGIONS OF THE UPPER LIMB	*BOUNDARIES OR LOCATION*
Anterior – of the arm (or anterior brachial region) – of the wrist (or anterior carpal region, part of the hand region) – of the elbow (or anterior cubital region) and cubital fossa – of the forearm (or anterior antebrachial region)	 Anterior border of the deltoid and inferior border of the pectoralis major (superiorly); a horizontal plane that passes approximately 3 cm above the medial epicondyle and the lateral epicondyle (inferiorly) Horizontal plane that passes above the head of the ulna (superiorly); horizontal plane that passes below the pisiform (inferiorly) Horizontal plane that passes approximately 3 cm above the medial epicondyle and the lateral epicondyle (superiorly); horizontal plane that passes approximately 3 cm below the medial epicondyle and the lateral epicondyle (inferiorly) Horizontal plane that passes approximately 3 cm inferior to the medial epicondyle and the lateral epicondyle (superiorly); horizontal plane that passes above the head of the ulna (inferiorly)
Axillary and axillary fossa	Inferior border of the pectoralis major (anteriorly); inferior border of teres major and latissimus dorsi (posteriorly). The medial and lateral boundaries are delimited by lines that join the anterior and posterior boundaries on the lateral wall of the thorax and on the medial face of the arm
Deltoid	Lateral third of the clavicle, acromion, spine of scapula (superiorly); anterior border of the deltoid (anteriorly); posterior border of the deltoid (posteriorly); deltoid tuberosity of the humerus on which the anterior and posterior borders of the deltoid converge (inferiorly)
Dorsum of the hand (part of the hand region)	Horizontal plane that passes below the pisiform and the tubercle of the scaphoid (superiorly); the line, which is concave proximally, that passes through the digito-palmar folds (inferiorly)
Palm (palmar region, part of the hand region)	☞ Dorsum of the hand
Posterior – of the arm (or posterior brachial region) – of the wrist (or posterior carpal region, part of the hand region) – of the elbow (or posterior cubital region) – of the forearm (or posterior antebrachial region)	 ☞ Anterior of the arm ☞ Anterior of the wrist ☞ Anterior of the elbow ☞ Anterior of the forearm
Scapular	Area of the dorsum included between the superior, medial and lateral borders of the scapula, determined by the muscles inserting on it: trapezius, latissimus dorsi, levator scapulae, rhomboid major and minor, deltoid, supra- and infraspinatus, teres major and minor, subscapularis

REGIONS OF THE LOWER LIMB	*BOUNDARIES OR LOCATION*
Anterior – of the knee (or anterior knee region) – of the thigh (or anterior femoral region) – – femoral triangle – – – Scarpa's triangle – of the leg (or anterior leg region)	 Horizontal line approximately 3 cm above the patella (superiorly); horizontal plane passing through the tibial tuberosity (inferiorly) Iliac crest (superiorly); sartorius (medially); horizontal line passing approximately 3 cm above the patella (inferiorly) Medial border of the sartorius (laterally); inguinal ligament (superiorly); lateral border of the gracilis (medially) Inguinal ligament (superiorly); medial border of the sartorius (laterally); lateral border of the adductor longus (medially) Horizontal plane that passes through the tibial tuberosity (superiorly); line that joins the lateral and medial malleoli passing approximately 2 cm below the intra-articular line of the ankle joint (inferiorly); anterior border of the tibia (medially)
Heel (part of the foot region)	Corresponds to the calcaneal tuberosity
Dorsum of the foot (dorsal region of the foot, part of the foot region)	Line that joins the lateral and medial malleoli passing approximately 2 cm below the intra-articular line of the ankle joint (superoposteriorly); a line, concave posteriorly, that passes through the commissure of the toes (inferolaterally)
Gluteal	Iliac crest (superiorly); gluteal fold (inferiorly); intergluteal cleft and a curved line, concave medially, between the intergluteal cleft and the iliac crest (medially)
Sole (plantar region, part of the foot region)	Calcaneal tuberosity (posteriorly); a curved line, convex anteriorly, that passes through the bases of the toes or the digitoplantar fold (anteriorly)
Posterior – of the knee (or posterior knee region), and popliteal fossa – of the thigh (or posterior femoral region) – of the leg (or posterior leg region)	 Horizontal line approximately 3 cm above the medial and lateral condyles of the femur (superiorly); horizontal plane that passes through the tibial tuberosity (inferiorly). The popliteal fossa appears when the knee is flexed and has a diamond shape formed by two triangles joined at the base: the upper, or femoral triangle, is bounded medially by the semitendinosus and semimembranosus and laterally by the two long and short heads of the biceps femoris; the boundary of the lower or tibial triangle is formed by the two lateral and medial heads of the gastrocnemius Gluteal fold (superiorly); horizontal line running approximately 3 cm above the medial and lateral condyles of the femur (inferiorly) ☞ Anterior of the leg
Lateral retromalleolar (part of the leg region)	Between the lateral malleolus and the calcaneal tendon

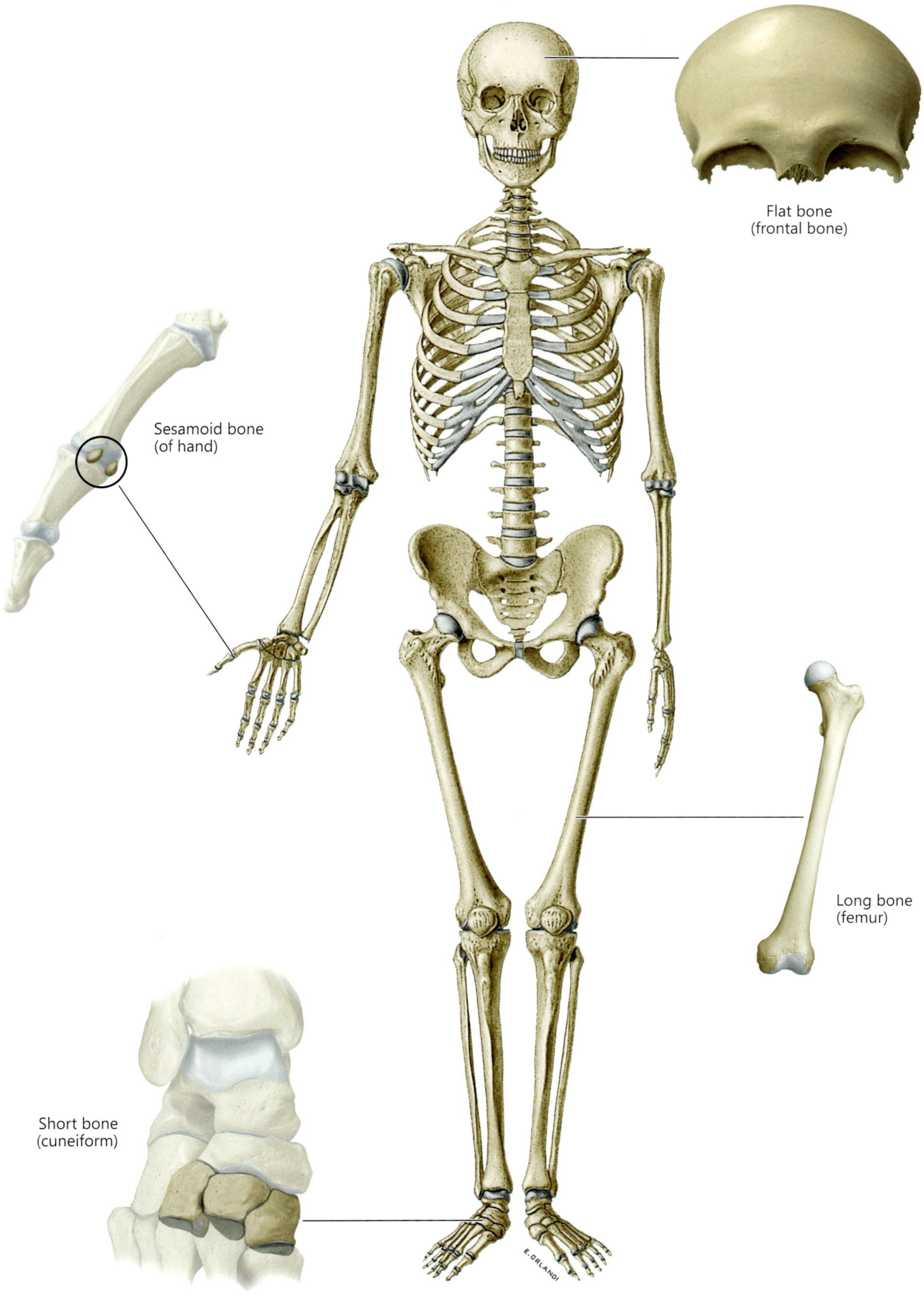

The **adult skeleton** is comprised of approximately 203 bones. This number may vary due to the presence of *supernumerary bones*, or the fusion of bone segments, as may happen later in life. Supernumerary bones may be *sesamoid* (if embedded within a tendon), or *accessory* (resulting from the failed fusion of ossification centers during normal growth processes, for example the flat bones of the cranium). Some bones contain air-filled spaces (for example paranasal sinuses); these are also referred to as *pneumatic bones*. Three bone types can be distinguished morphologically: *long bones, flat bones, short bones*. Some bones of the head, which have both laminar portions typical of flat bones and irregular prismatic portions characteristic of short bones, are defined as *irregular bones*.

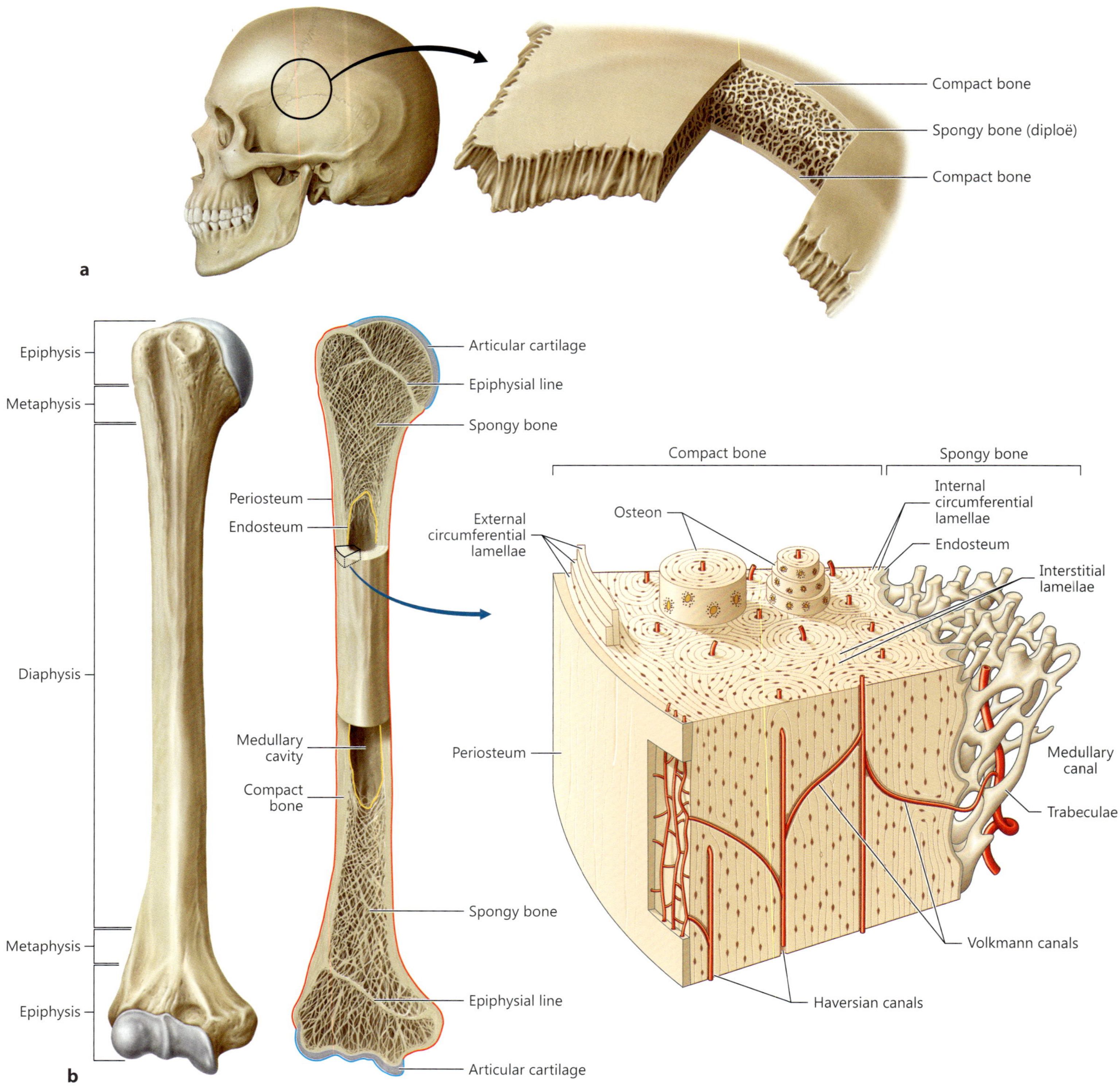

a, The structure of **flat bone** is characterized by greater development in surface width and length compared to thickness. Flat bones comprise two layers of compact tissue (in cranial bones, these are commonly called the inner and outer bony tables of the cranium) enclosing between them an intermediate layer of spongy tissue (called diploë in the bones of the cranium). The two layers of compact tissue fuse at the edges of the bone so that the spongy tissue is completely enclosed. **b**, Schematic representation of a **long bone** (the humerus) and of its structure, displaying the sheath of compact bone that surrounds a medullary cavity along the diaphysis and the spongy bone tissue inside the epiphysis. The medullary cavity and the spongy bone cells contain bone marrow. A framework of trabeculae is arranged in a way that imparts maximum compressive and tensile strength to the spongy tissue, whose structural organization derives from the bone's *lamellae*. Whether the lamellar bone tissue structure is compact or spongy is determined by the complex configurations taken by the combination of several lamellae. Cylindrical lamellae arrange themselves concentrically, one within another, to form the *osteons* of compact bone tissue (in a *concentric* or *Haversian system*). Planar lamellae surfaces conjoin to form *trabeculae*, which knit together into spongy bone tissue through a process of anastomosis. Finally, lamellar layers overlap to form the external and internal enclosing layers that separate compact bone from its external cladding of periosteum and from its internal lining of endosteum. Compact bones receive their nutrition through vessels running along the central canals of the osteon (*Haversian canals*), which are connected to each other by transverse canals (*Volkmann canals*). The lamellae of spongy tissue receive nutrition by means of diffusion.

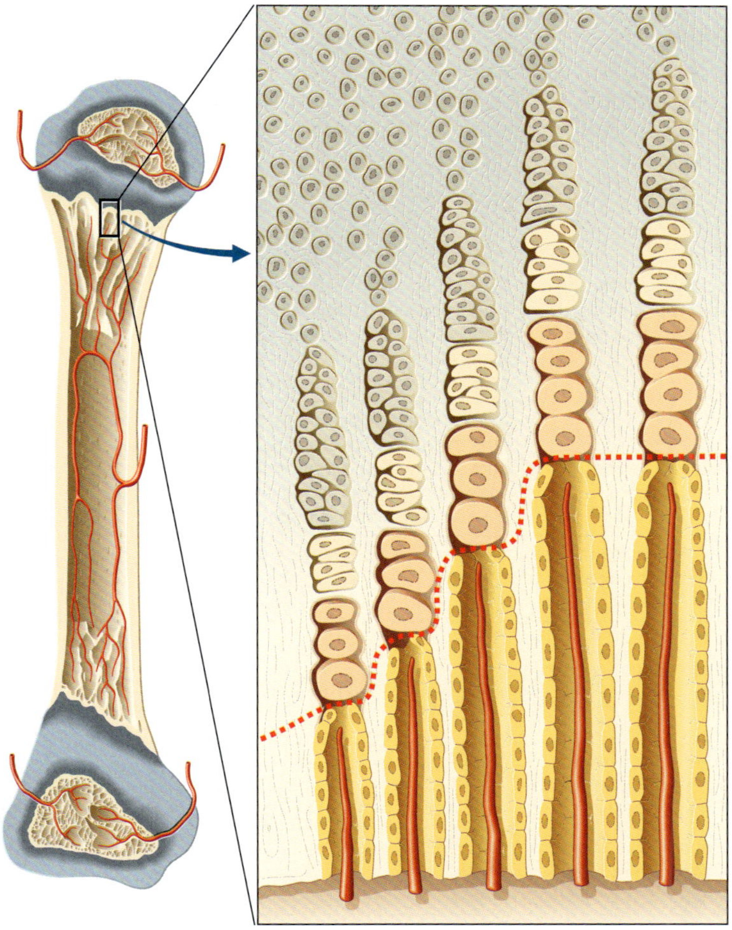

Hypertrophic cartilaginous tissue plays a part in the ossification (**endochondral ossification**) and growth of long bones.

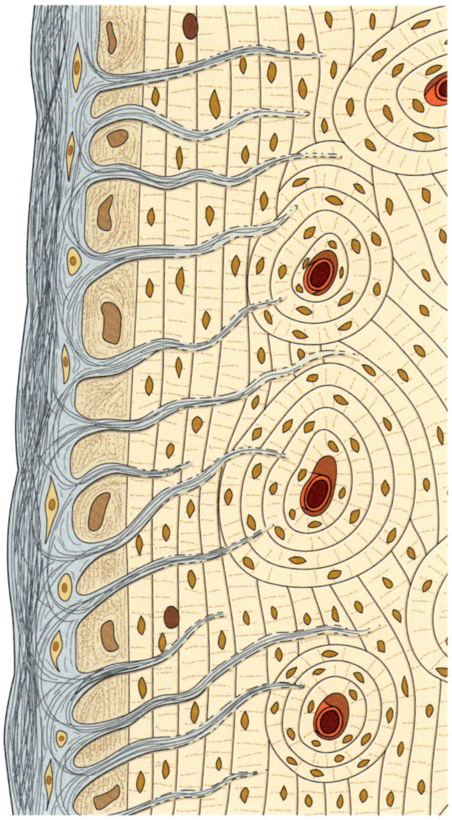

Schematic representation of the fibrous structure of the **periosteum**. Large bundles of fibers (perforating Sharpey fibers) penetrate into the compact bone, passing through the structure of external circumferential lamellae and the peripheral osteons.

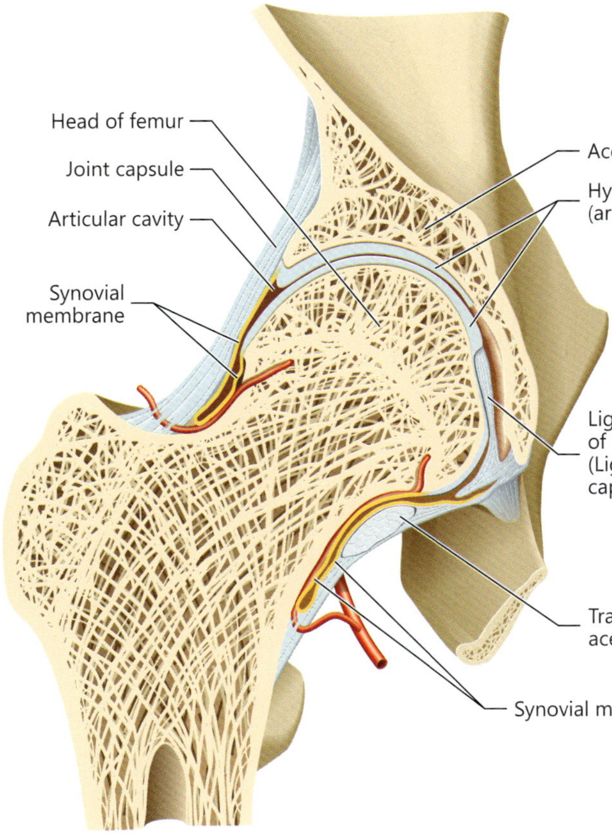

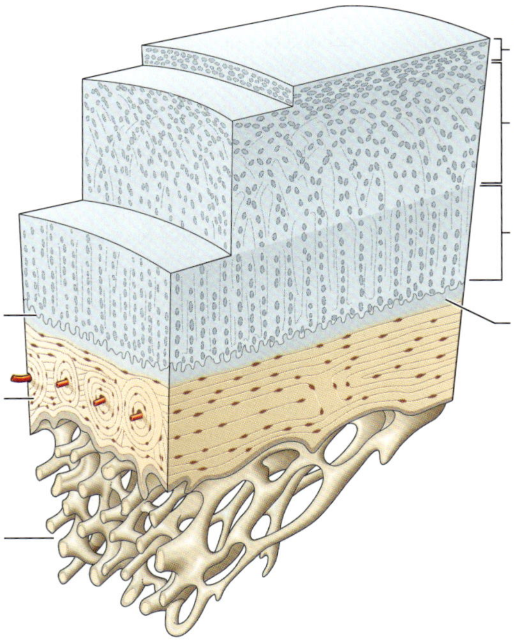

Schematic representation of a **freely movable** (**synovial**) **joint**. The bony surface involved in freely movable joints is lined with a layer of hyaline articular cartilage.

Bones can be classified according to shape, size, structure, and type of ossification process.
The bones making up the skeleton have centers of ossification, whose emergence and processes of fusion follow a chronological pattern that depends on age and on the bone segment concerned, and which also varies from person to person. Similarly, the ossification process of conjugation cartilage takes place at varying rates and depends on the bone segment concerned as well as on gender, the process being completed approximately 1-2 years earlier in women than in men. The entire process of skeletal maturation may be considered complete by around 25 years of age.

TYPES	SHAPE, DIMENSIONS, AND STRUCTURE	OSSIFICATION PROCESS
Long bones	One dimension (length) dominates all others. The epiphyses vary in shape and are formed of spongy bone tissue clad in compact bone tissue. The diaphysis is tubular and formed of compact bone tissue surrounding a central canal (the diaphyseal canal)	Ossification occurs indirectly (bone tissue replaces a template of cartilaginous tissue). The diaphysis displays endochondral and perichondral ossification, while the epiphyses show one or more centers of endochondral ossification
Flat bones	Two dimensions dominate a third, conferring a flattened appearance. They consist of spongy bone tissue internally and compact bone tissue externally. The compact covering tissue may be extremely thin and can be perforated by a biopsy needle (sternum), or very thick and tough (bones of the cranium)	Ossification may be membranous, proceeding directly from connective tissue (as in the calvaria bones of the neurocranium), or indirectly with endochondral ossification centers. The cranial base displays endochondral ossification
Short bones	These have various shapes, without any one dimension being predominant over the others. Often they are made up of spongy bone tissue with a thin covering of compact bone tissue	Ossification is generally endochondral (indirect)

BONE	TYPE AND INITIATION OF OSSIFICATION	COMPLETION OF OSSIFICATION
Long bones		
Femur	Perichondral and endochondral ossification: one diaphysial center in 2nd month. Endochondral ossification: one center for the distal epiphysis shortly before birth and four centers for the proximal epiphysis in the 1st, 3rd, 4th and 7th year	Between 15 and 21 years
Fibula	Perichondral and endochondral ossification: one diaphysial center in 2nd month. Endochondral ossification: one center for the proximal epiphysis in the 5th year and one center for the distal epiphysis in the 2nd year	Between 14 and 21 years
Humerus	Perichondral and endochondral ossification: one diaphysial center in 2nd month. Endochondral ossification: three centers for the proximal epiphysis by the 5th year and four centers for distal epiphysis between the 2nd and 14th year	Between 14 and 22 years
Metacarpals [I-V]	Perichondral and endochondral ossification: centers for the base and the shaft at the 3rd month and center for the head in the 2nd to 3rd year	Between 15 and 16 years
Metatarsals [I-V]	Perichondral and endochondral ossification: centers in the 3rd month for the base and shaft. Endochondral ossification: center for the head between 2nd and 4th year	Between 14 and 16 years
Phalanges (foot)	Perichondral and endochondral ossification: centers for the proximal and middle phalanges between 2nd and 4th year. Centers for the proximal phalanges between 1st and 2nd year with membranous ossification for the distal part	Between 15 and 20 years
Phalanges (hand)	Perichondral and endochondral ossification: proximal and middle phalanges with the respective metacarpals. Membranous ossification occurs in the distal part of the distal phalanges	Between 15 and 20 years
Radius	Perichondral and endochondral ossification: one diaphysial center in 2nd month. Endochondral ossification: 1 center for the distal epiphysis in the 1st year and 1 center for the proximal epiphysis in the 5th year	Between 14 and 22 years
Tibia	Perichondral and endochondral ossification: one diaphysial center in 2nd month. Endochondral ossification: one center for the proximal epiphysis shortly before birth and one center for the distal epiphysis in the 2nd year	Between 14 and 21 years
Ulna	Perichondral and endochondral ossification: one diaphysial center in 2nd month. Endochondral ossification: one center for the distal epiphysis in the 7th year and one center for the proximal epiphysis in the 10th year	Between 14 and 22 years
Flat bones		
Clavicle	Membranous ossification: one center at the end of the 1st month. Endochondral ossification: one center in the 18th year	Between 15 and 28 years
Frontal bone	Membranous ossification: two centers in the 2nd month, others subsequently	After the 15th year
Hip	Endochondral ossification: 3 centers for the three parts. Ilium: one center in the 2nd month; ischium: one center in the 4th month; pubis: one center in the 5th month. Secondary centers in 15th year	Between 13 and 24 years
Lacrimal bone	Membranous ossification: one center in the 3rd month	

continued

(*continued*)

BONE	TYPE AND INITIATION OF OSSIFICATION	COMPLETION OF OSSIFICATION
Mandible	Membranous ossification: two centers in the 2nd month, others subsequently	
Maxilla	Membranous ossification: one center in the 6th week, others subsequently	Between the 10th and 18th year
Nasal bone	Membranous ossification: one center in the 3rd month	
Palatine bone	Membranous ossification: one center in the 2nd month, others subsequently	
Parietal bone	Membranous ossification: two centers in the 2nd month	After the 2nd year
Ribs [I-XII]	Endochondral ossification: one center in the 2nd month starting from the central ribs. Two centers for the head and tubercle in the 16th year	In the 25th year
Scapula	Endochondral ossification: one center in the 2nd month, others subsequently	Between 18 and 21 years
Sternum	Endochondral ossification: one center for the manubrium between the 4th and 6th month. Seven to eight centers for the body at birth	After the 13th year
Vomer	Membranous ossification: two centers in the 2nd month	
Zygomatic bone	Membranous ossification: three centers in the 2nd month	
Short bones		
Calcaneus	Endochondral ossification: nuclei in the 6th month	Between 16 and 19 years
Carpal bones	Endochondral ossification: the ossification centers of the individual bones have the following sequence (more rapid in females): – capitate: 3rd-10th month – hamate: 5th-12th month – triquetrum: 2nd-3rd year – lunate: 3rd-4th year – scaphoid: 4th-5th year – trapezium and trapezoid: 4th-6th year – pisiform: 9th-13th year	
Coccyx [coccygeal vertebrae I-IV]	Endochondral ossification: one center after birth	Between 18 and 25 years
Ethmoid	Endochondral ossification: two centers at birth, others subsequently	After the 15th year
Hyoid bone	Endochondral ossification: at birth	In approximately the 15th year
Incus	Endochondral ossification: in the 4th fetal month	Shortly after birth
Inferior nasal concha (turbinate)	Endochondral ossification: one center in the 6th fetal month	
Malleus	Endochondral ossification: in the 4th fetal month	Shortly after birth
Sacrum [sacral vertebrae I-V]	Endochondral ossification: one center in the 5th month, others subsequently	Between 17 and 23 years
Stapes	Endochondral ossification: in the 4th fetal month	Shortly after birth
Talus	Endochondral ossification: nuclei in the 6th month	Between 14 and 15 years
Tarsal bones	Endochondral ossification: the ossification centers of the individual bones follow the sequence: – cuboid: at birth – navicular: in the 1st year – lateral cuneiform: in the 1st year – medial cuneiform: between the 2nd and 4th year – intermediate cuneiform: between the 2nd and 4th year	Between 14 and 15 years
Irregular bones		
Occipital bone	Membranous ossification: two centers for the squamous part in the 3rd fetal month. Endochondral ossification: from the 2nd month for the basilar and lateral parts	Ossification completed around the 5th year. Between the 17th and 25th year closure of the spheno-occipital syndesmosis
Sphenoid	Membranous ossification: two small centers of the greater wings and of the pterygoid processes. Endochondral ossification: six centers for most of the of bone from the 2nd fetal month	Between the 17th and 25th year closure of the spheno-occipital syndesmosis
Temporal bone	Membranous ossification for the squamous and tympanic parts: the first centers in the 2nd and 3rd month. Endochondral ossification: the other centers from the 5th month onwards	Ossification of the squamous part completed around the 5th year
Vertebrae	Endochondral ossification: three centers in the 3rd month, others subsequently	Between 19 and 25 years
Sesamoid bones		
Patella	Endochondral ossification: numerous small centers after the 3rd year	

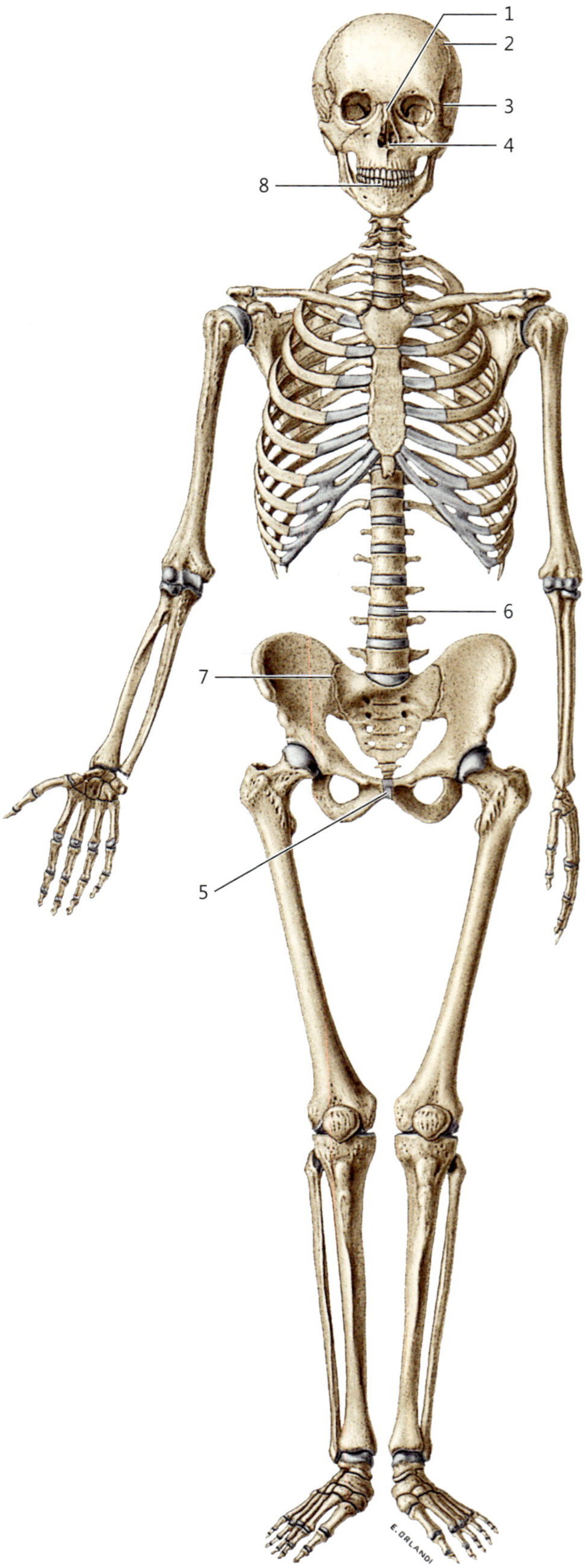

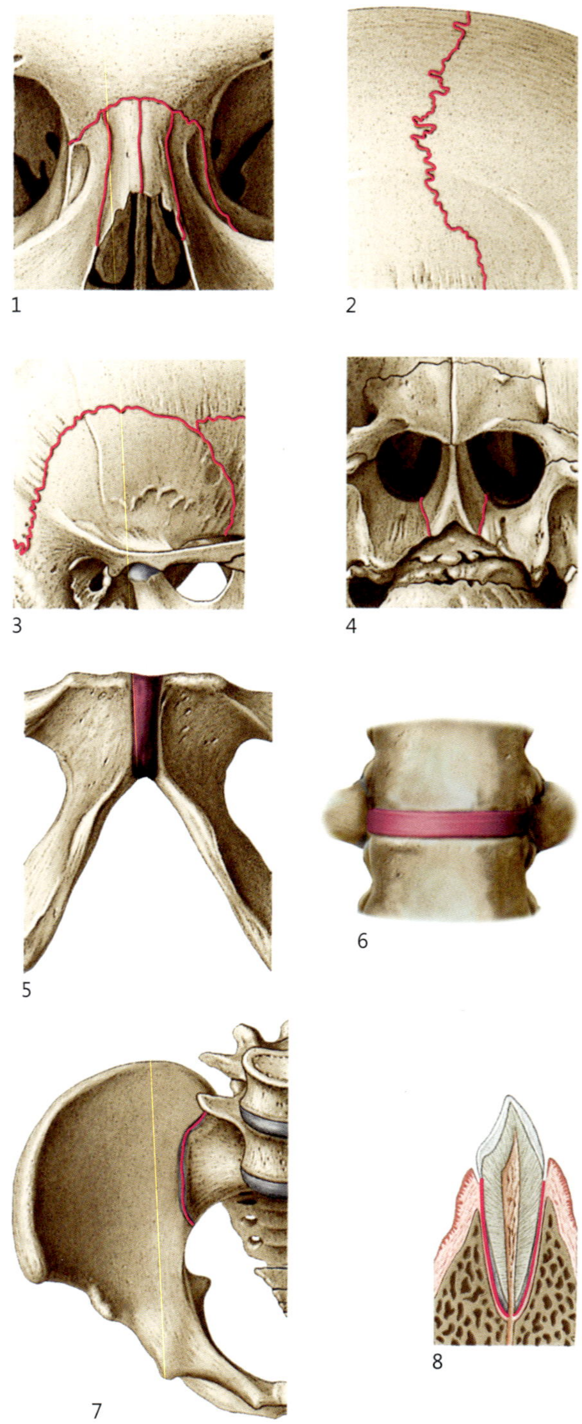

Joints formed between the heads of bones may give rise to absence of movement, to immovable joints or joints with little mobility (synarthroses), or they may permit various degrees of freedom as with (synovial) movable joints (diarthrosis).

Main types of **synarthroses**. 1, Plane suture; 2, serrate suture; 3, squamous suture; 4, wedge-and-groove suture or schindylesis and 5-6, symphyses. Some later types of synarthrosis (for example sacroiliac synchondrosis, 7, and gomphoses between teeth and alveolar cavities, 8) are of an uncertain classification due to their variability, which is linked above all to age. The sacroiliac joint, in particular, can be considered an atypical plane joint, and therefore a diarthrosis, because the heads of the bones form a junction, are free, and are separated by a cavity. This joint performs an involuntary oscillating, rocker-arm movement which is particularly significant in the erect position, and plays a fundamental role in women during childbirth. The movement consists of two stages, one of which, called *nutation*, tilts the base of the sacrum forward and consequently moves the coccyx backward. The other, called *counter-nutation*, does the opposite, retro-positioning the base of the sacrum and moving the coccyx forward.

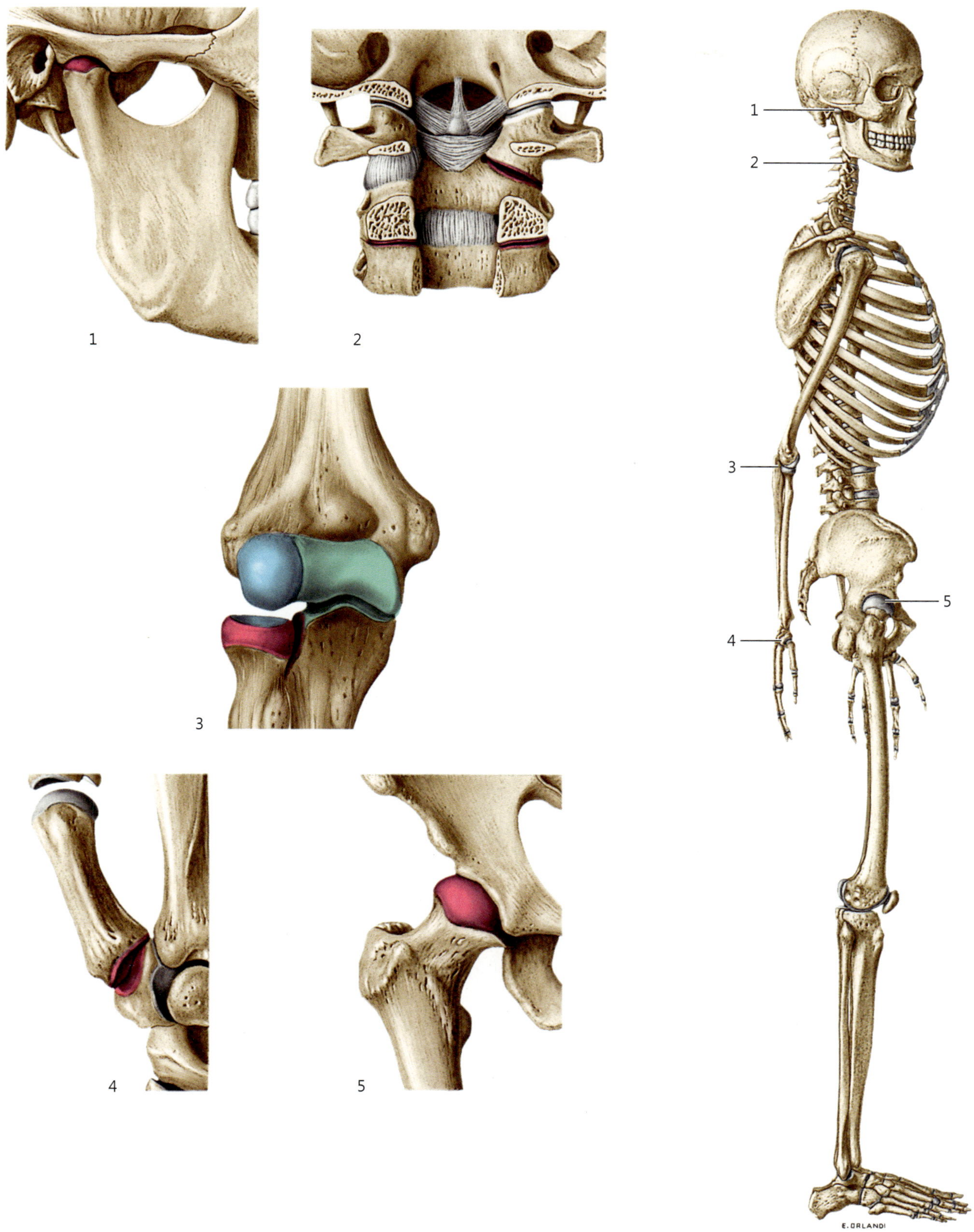

Main types of **diarthroses** and examples. 1, Condylar joint (temporomandibular joint); 2, plane joint (joints between the articular facets of the vertebrae); 3, in *red*, pivot or trochoid joint (proximal radioulnar joint), in *green*, hinge joint or trochlea (humeroulnar joint), in *blue*, condylar joint (humeroradial joint); 4, saddle joint (carpometacarpal joint of the thumb); 5, ball and socket or spheroidal joint (hip joint).

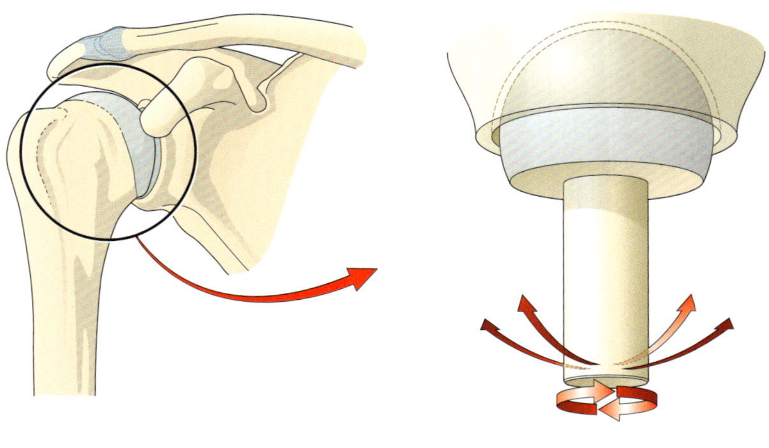

Ball and socket joint (spheroidal joint). The surfaces are shaped like segments of a sphere: concave on one joint head and convex on the other.

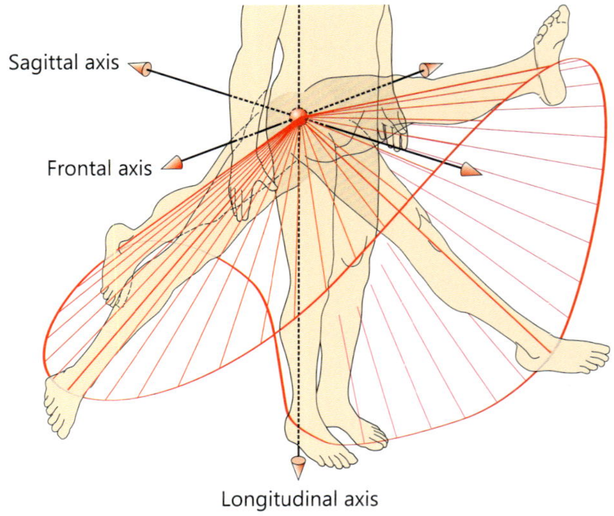

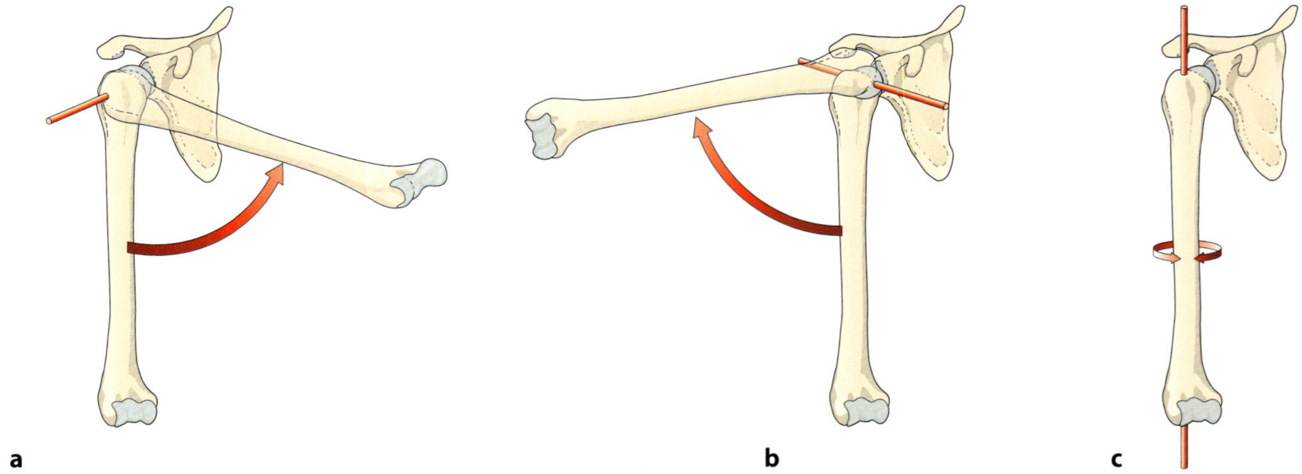

Ball and socket joint (**spheroidal joint**). Multi-axial joint: Possesses three degrees of freedom (e.g., the glenohumeral joint). **a**, **b**, and **c**, show the orthogonal axes along which the following main movements are made: **a**, flexoextension, **b**, abduction-adduction and **c**, internal and external rotation.

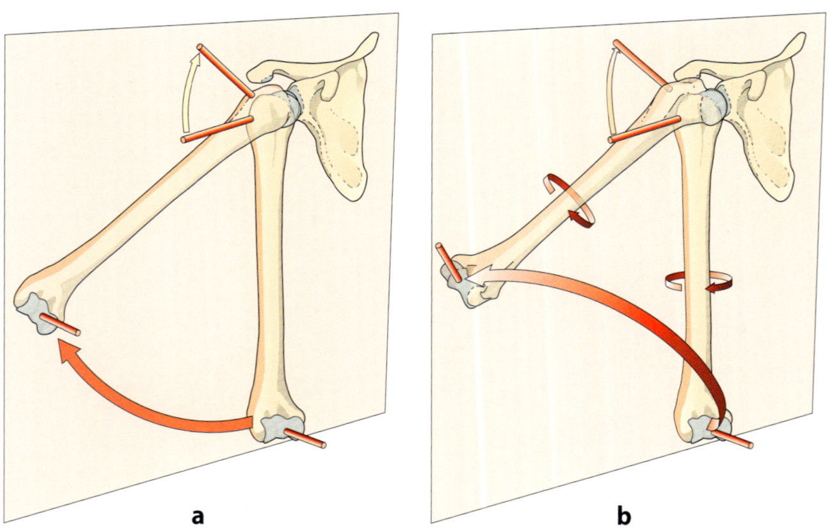

Ball and socket joint (**spheroidal joint**). Simple asymmetrical long bone (glenohumeral joint). Complex movements: **a**, hinge movement and **b**, arc movement. In the first, the mechanical axis moves within one plane; in the second, the axis moves outside the plane shown in the figure and then returns. In addition, the extremities of the mechanical axis trace arc-shaped pathways at the joint and within the ovoid described by the movement.

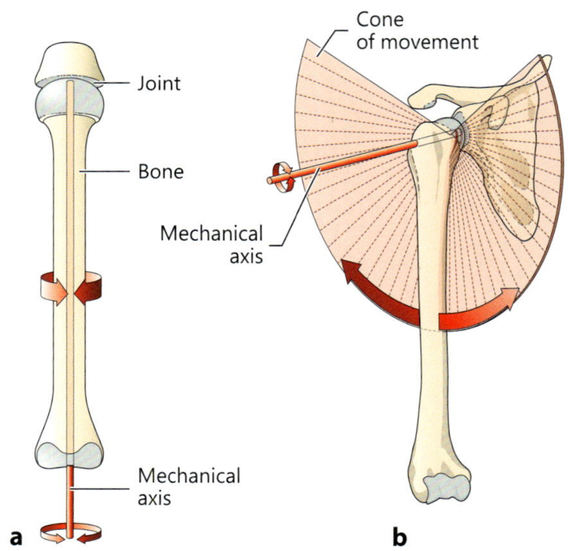

Ball and socket joint (**spheroidal joint**). Mechanical axis: **a**, In a simple long and symmetrical bone with a terminal joint, the mechanical axis corresponds to the longitudinal axis of the long bone. **b**, Bone with eccentric articulation with respect to the longitudinal axis of the diaphysis. The bone's rotation describes part of a conical surface.

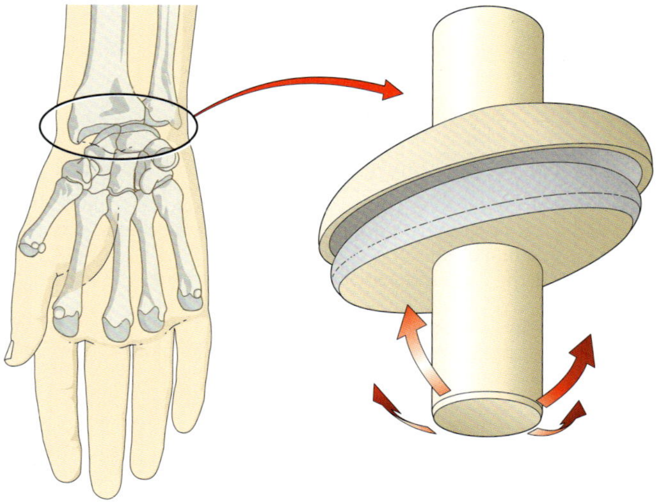

Condylar joint (**ellipsoid joint**). A convex ovoid on one articular head and a concave one on the opposite head, keeping within two orthogonal planes, whose curvatures may diverge.

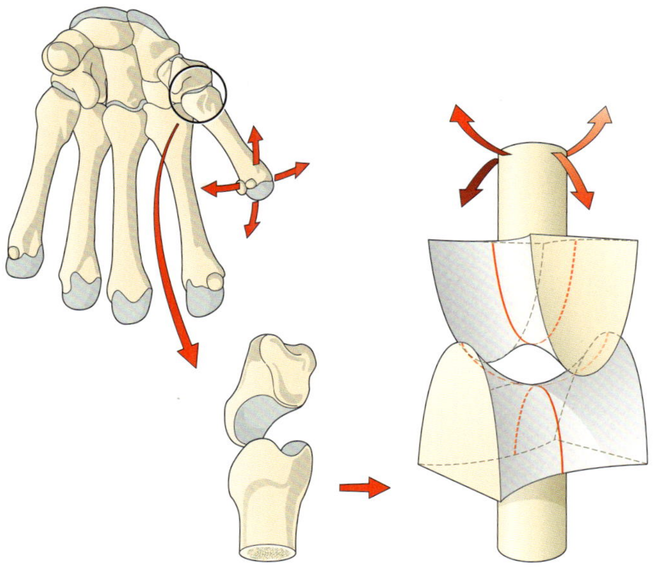

Saddle joint. The two meeting surfaces are concave on one axis and convex on the axis orthogonal to it. The curvature of these two surfaces can vary from "almost planar" to "almost spherical", with wide variations in the radius of curvature from one area to another.

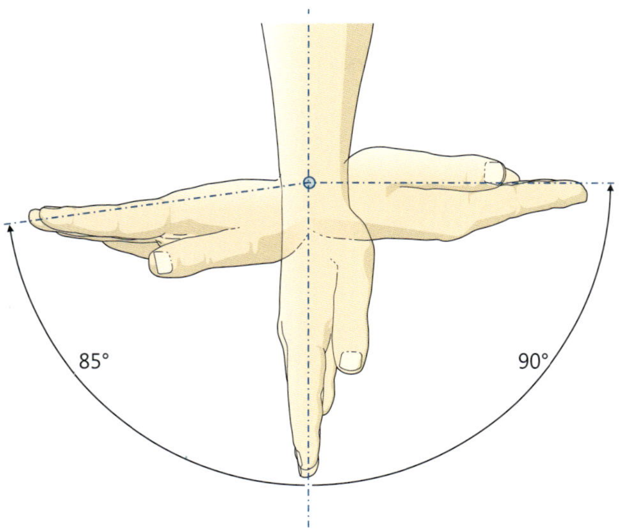

Condylar joint (**ellipsoid joint**). Flexion-extension movement of the wrist.

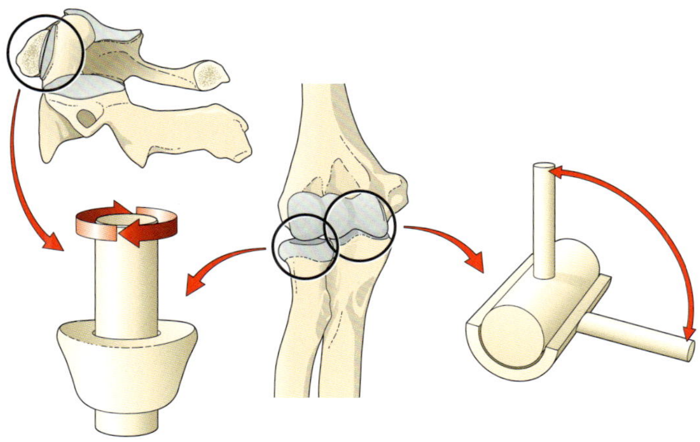

Pivot and **hinge joints**. Elbow joint. Here two cylindrical surfaces meet, one of which is solid and the other hollow. In the pivot joint, the axes of the two surfaces are parallel (left, the proximal radioulnar joint) while they are orthogonal to each other in the hinge joint (right, the humeroulnar joint).

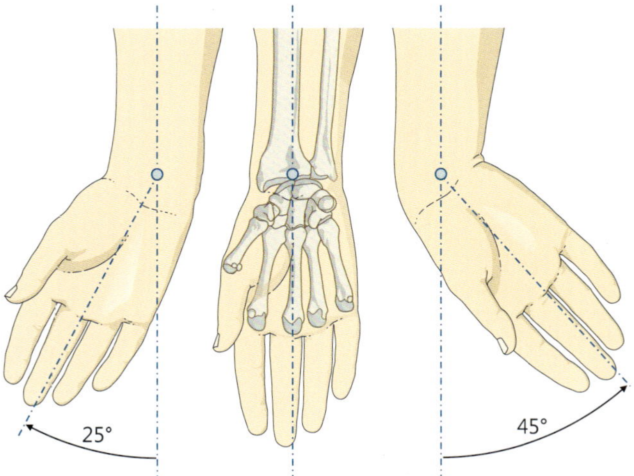

Condylar joint (**ellipsoid joint**). Abduction and adduction movements of the wrist.

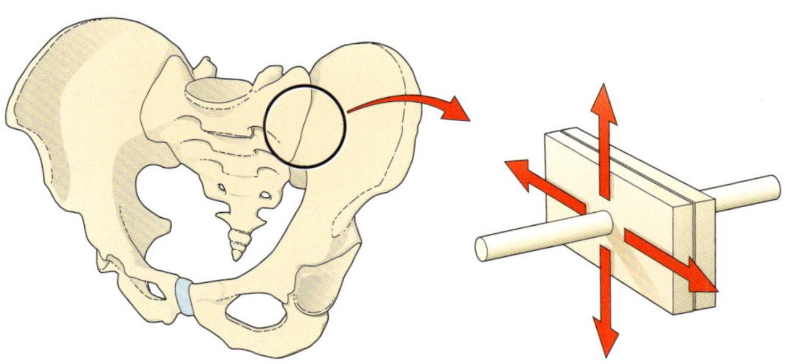

Plane joint. The articular surfaces are planar and slide against each other. The main limit to the extension of the movement is the extension of the joint capsule.

Joints are classified according to the degree of continuity or contiguity between the articular heads: the first type are classified as immovable joints (synarthroses) and the second type as movable joints (synovial joints or diarthroses). There are also joints with little mobility or partly movable joints (amphiarthroses).

SYNARTHROSES	These are classified according to the shape of the joint space and intervening tissue	
NAME	***JOINT SHAPE AND TISSUE***	***JOINTS***
Sutures (syndesmoses)	The bones are joined by fibrous connective tissue. The bones involved undergo direct ossification	*Serrate sutures*: between the frontal bone and the parietal bone (coronal suture); between the two parietal bones (sagittal suture); between the two parietal bones and the occipital bone (lambdoid suture) *Squamous sutures*: between the parietal bones and the squamous part of the temporal bones *Plane sutures*: between the bones of the viscerocranium *Schindyleses* (wedge-and-groove syndesmoses): between the sphenoidal rostrum and the upper border of the vomer
Symphyses	The bones are held together by fibrous cartilage	Manubriosternal symphysis: between the manubrium and body of the sternum Intervertebral discs Pubic symphysis: between the two pubic parts of the hip
Synchondroses	The bones are held together by hyaline cartilage	Spheno-occipital synchondrosis Petro-occipital synchondrosis: between the petrous part of the temporal bones and the body of the occipital bone Xiphisternal joint: between body of the sternum and the xiphoid process Costochondral joints: between the anterior extremity of the ribs and the lateral extremity of the costal cartilages
Gomphoses (dento-alveolar syndesmoses)	The bones are held together by fibrous connective tissue	Between the roots of the teeth and the tooth sockets

SYNOVIAL JOINTS (DIARTHROSES)	These are classified according to the shape of their articular heads. They have articular capsules, articular cartilage, synovial membranes, and ligaments	
NAME	***JOINT SHAPE AND TISSUE***	***JOINTS***
Condylar joints (ellipsoid joints)	The articular surfaces are ellipsoidal. They permit angular movements about the greater axis and limited angular movement about the lesser axis	Atlanto-occipital joint Humeroradial joint Temporomandibular joint Wrist joint: between the distal epiphysis of the radius and the proximal row of carpal bones Midcarpal joint (medial part) Metatarsophalangeal joints
Plane joints	The articular surfaces are relatively level. They allow limited sliding movements	Joint between the superior and inferior articular processes of contiguous vertebrae Costovertebral joints: between head of the ribs and superior and inferior costal facets of the thoracic vertebrae Costotransverse joints: between the tubercle of the rib and the transverse process of the thoracic vertebrae Sternocostal joints: between costal notches and true ribs Acromioclavicular joint Carpal joints (intercarpal joints): between the carpal bones Midcarpal joint (lateral part) Carpometacarpal joints (except the first) Intermetacarpal joints Sacroiliac joint (for some, this is a symphysis) Superior tibiofibular joint Tibiofibular syndesmosis (inferior tibiofibular joint) Cuboideonavicular joint Cuneonavicular joint Intercuneiform joints Cuneocuboid joint Tarsometatarsal joints Intermetatarsal joints: between the bases of the metatarsals
Pivot joints	The articular surfaces are cylindrical with parallel axes. They permit only rotational movements about one of the axes	Median atlantoaxial joint: between the dens and the anterior arch of the atlas [C I] Proximal radioulnar joint Distal radioulnar joint Subtalar joint (talocalcaneal joint): between the talus and the calcaneus
Hinge joints	The articular surfaces are cylindrical with axes perpendicular to each other. They permit only angular movements about the axis perpendicular to the axis of the bone	Humeroulnar joint Interphalangeal joints of hand Knee joint: between the femur, tibia and patella Ankle joint: between the tibia, fibula and talus Interphalangeal joints of foot
Saddle joints	A special kind of condylar joint in which one articular surface is convex and the other has a concave saddle shape. For this reason only angular movements about the greater axis are possible	Sternoclavicular joint Carpometacarpal joint of thumb: between trapezium and first metacarpal Calcaneocuboid joint
Ball and socket joints (spheroidal joints)	The articular surfaces are semi-spherical. They permit angular and rotational movements	Glenohumeral joint (shoulder joint) Hip joint Talocalcaneonavicular joint: between talus, calcaneus and navicular

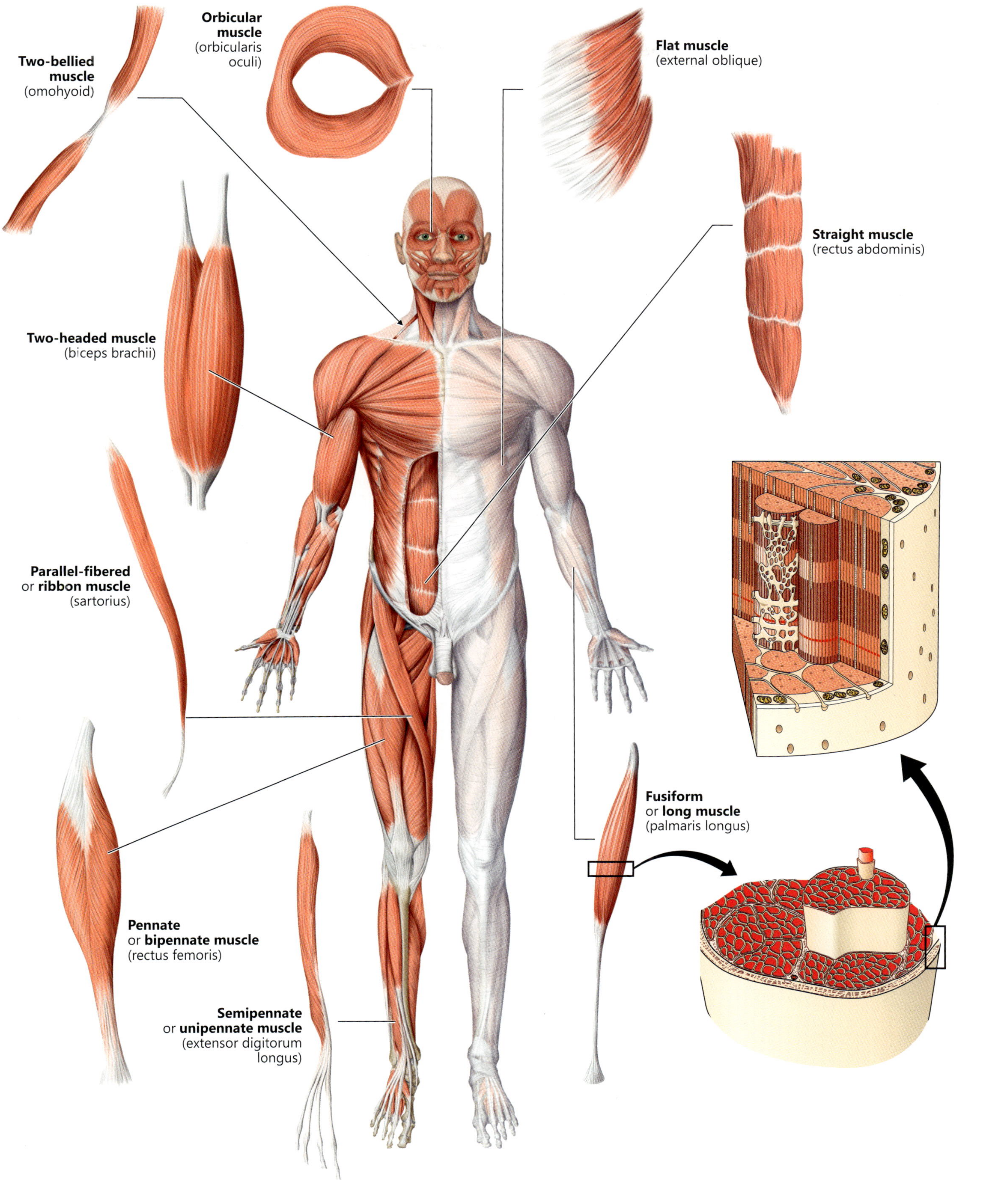

Muscles are classified in two major groups: **superficial muscles** (cutaneous) and **deep muscles** (subfascial). The former are found below the cutaneous layer and adhere to the deep fascia of the dermis. Subfascial muscles are located below the deep fascia and, in most cases, insert onto skeletal bones. A different classification of muscles considers the shape and direction of their fibers (orbicular, ribbon, fusiform, flat), the number of bellies (two-bellied, polygastric), or the number and shape of the insertion heads (two-headed, three-headed, four-headed, pennate or semi-pennate).

Muscle classification and regional subdivision*

MUSCLES OF THE HEAD	STRUCTURE
Antitragicus	Short muscles
Arrector muscle of hair**	Triangular muscles
Auricularis anterior	Short muscles
Auricularis posterior	Short muscles
Auricularis superior	Triangular muscles
Buccinator	Quadrilateral muscles
Caninus ☞ Levator anguli oris	
Ceratocricoid	Short muscles
Chondroglossus	Short muscles
Ciliary muscle	Sphincter muscles
Corrugator supercilii	Short muscles
Depressor anguli oris	Triangular muscles
Depressor labii inferioris	Quadrate muscles
Depressor septi nasi	Short muscles
Depressor supercilii	Short muscles
Dilator narium	Laminar muscles
Dilator pupillae	Radiate muscles
Epicranius	Laminar muscles
Frontal belly ☞ Occipitofrontalis	
Genioglossus	Triangular muscles
Glossopalatinus ☞ Palatoglossus	
Helicis major	Short muscles
Hyoglossus	Quadrilateral muscles
Incisivus labii inferioris	Short muscles
Incisivus labii superioris	Triangular muscles
Inferior longitudinal muscle of tongue	Thin sheat of fibers
Inferior oblique of eyeball	Short muscles
Inferior rectus of eyeball	Ribbon muscles
Inferior tarsal muscle	Flat muscles
Lateral pterygoid	Short muscles
Lateral rectus of eyeball	Ribbon muscles
Levator anguli oris	Quadrilateral muscles
Levator glandulae thyroideae (*inconstant muscle*)	Ribbon muscles
Levator labii superioris	Quadrate muscles
Levator labii superioris alaeque nasi	Short muscles
Levator palati ☞ Levator veli palatini	
Levator palpebrae superioris	Ribbon muscles
Levator veli palatini	Cylindric muscles
Longissimus capitis	Long muscles
Longus capitis	Triangular muscles
Masseter	Quadrilateral muscles

continued

MUSCLES OF THE HEAD (continued)	STRUCTURE
Medial pterygoid	Quadrilateral muscles
Medial rectus of eyeball	Ribbon muscles
Mentalis	Cone-shaped muscles
Musculus uvulae	Short muscles
Nasalis	Ribbon muscles
Oblique muscle of auricle	Short muscles
Obliquus capitis inferior	Rectangular muscles
Obliquus capitis superior	Triangular muscles
Occipital belly ☞ Occipitofrontalis	
Occipitofrontalis	Two-bellied muscles
Orbicularis oculi	Orbicular muscles
Orbicularis oris	Orbicular muscles
Orbitalis (Müller's muscle)	Orbicular muscles
Palatoglossus	Flat muscles
Palatopharyngeus	Flat muscles
Procerus	Laminar muscles
Quadratus labii inferioris ☞ Depressor labii inferioris	
Quadratus labii superioris ☞ Levator labii superioris	
Rectus capitis anterior	Quadrilateral muscles
Rectus capitis lateralis	Cylindric muscles
Rectus capitis posterior major	Triangular muscles
Rectus capitis posterior minor	Short muscles
Risorius	Triangular muscles
Salpingopharyngeus	Long muscles
Semispinalis capitis	Flat muscles
Spinalis capitis	Short muscles
Splenius capitis	Flat muscles
Stapedius	Pennate (bipennate) muscles
Styloglossus	Short muscles
Superior longitudinal muscle of tongue	Thin stratum of fibres
Superior oblique of eyeball	Triangular muscles
Superior rectus of eyeball	Ribbon muscles
Superior tarsal muscle	Flat muscles
Temporalis	Triangular muscles
Temporoparietalis	Laminar muscles
Tensor tympani	Short muscles
Tensor veli palatini	Flat muscles
Trachelomastoid ☞ Longissimus capitis	
Tragicus	Short muscles

continued

* Based on their shape or the their three-dimensional relationship, the number of heads or the insertion of the muscular fibers on the tendon.
** Here, the arrector muscle of hair, associated with the hair follicle, is considered among the muscles of the head, but it is present in almost all the body surface.

MUSCLES OF THE HEAD (continued)	STRUCTURE
Transverse muscle of auricle	Small scattered fibers
Transverse muscle of tongue	Small bundles of fibers
Transversus menti	Triangular muscles
Transversus nuchae	Flat muscles
Triangularis labii inferioris ☞ Depressor anguli oris	
Vertical muscle of tongue	Small bundles of fibers
Zygomaticus major	Ribbon muscles
Zygomaticus minor	Ribbon muscles

MUSCLES OF THE NECK (continued)	STRUCTURE
Thyroarytenoid	Quadrilateral muscles
Thyroepiglottic part of thyroarytenoid	Quadrilateral muscles
Thyrohyoid	Quadrilateral muscles
Trachealis	Flat muscles
Transverse arytenoid (*only unpaired muscle of the larynx*)	Small bundles of fibers
Trapezius	Spiral muscles
Vocalis	Short muscles

MUSCLES OF THE NECK	STRUCTURE
Aryepiglottic part	Small bundle muscle
Cricothyroid	Triangular muscles
Digastric	Two-bellied muscles
Geniohyoid	Cylindric muscles
Inferior constrictor of pharynx	Trapezoidal muscles
Infrahyoid muscles ☞ Omohyoid, Sternohyoid, Sternothyroid, Thyrohyoid	
Lateral cricoarytenoid	Triangular muscles
Longissimus cervicis	Long muscles
Longus colli	Long muscles
Middle constrictor of pharynx	Triangular muscles
Mylohyoid	Quadrilateral muscles
Oblique arytenoid	Two thin bundles of fibers
Omohyoid	Two-bellied muscles
Platysma	Laminar muscles
Posterior cricoarytenoid	Triangular muscles
Scalenus anterior	Long muscles
Scalenus medius	Long muscles
Scalenus minimus	Long muscles
Scalenus posterior	Long muscles
Semispinalis cervicis	Short muscles
Spinalis cervicis	Short muscles
Splenius cervicis	Flat muscles
Sternocleidomastoid	Spiral and cruciate muscles
Sternohyoid	Ribbon muscles
Sternothyroid	Ribbon muscles
Stylohyoid	Long muscles
Stylopharyngeus of pharynx	Long muscles
Superior constrictor of pharynx	Quadrilateral muscles
Suprahyoid muscles ☞ Digastric, Geniohyoid, Mylohyoid, Stylohyoid	

continued

MUSCLES OF THE TRUNK	STRUCTURE
Diaphragm	Flat muscles
Erector spinae	Short muscles
External intercostal muscle	Flat muscles
Iliocostalis	Short muscles
Innermost intercostal muscle	Flat muscles
Internal intercostal muscle	Flat muscles
Interspinales	Short muscles
Intertransversarii	Short muscles
Latissimus dorsi	Spiral muscles
Levatores costarum	Triangular muscles
Longissimus thoracis	Short muscles
Multifidus	Short muscles
Musculi pectinati	Parallel muscles
Pectoralis major	Spiral muscles
Pectoralis minor	Flat muscles
Pleuroesophageus	Flat muscles
Rhomboid major	Flat muscles
Rhomboid minor	Flat muscles
Rotatores	Short muscles
Sacrospinalis ☞ Erector spinae	
Semispinalis thoracis	Short muscles
Serratus anterior	Flat muscles
Serratus posterior inferior	Flat muscles
Serratus posterior superior	Flat muscles
Spinalis thoracis	Short muscles
Sternalis	Flat muscles
Sternocostalis ☞ Transversus thoracis	
Subclavius	Cylindric muscles
Subcostales	Ribbon muscles
Thoracic part of iliocostalis lumborum	Short muscles
Trachealis	Flat muscles
Transversus thoracis	Triangular muscles

MUSCLES OF THE ABDOMEN	***STRUCTURE***
Cremaster	Flat muscles
External oblique	Flat muscles
Internal oblique	Flat muscles
Iliocostalis lumborum	Short muscles
Muscle of Treitz ☞ Suspensory muscle of duodenum	
Psoas minor	Long muscles
Pyloric sphincter	Sphincter muscles
Pyramidalis	Flat muscles
Quadratus lumborum	Quadrilateral muscles
Rectus abdominis	Polygastric muscles
Sphincter of ampulla	Sphincter muscles
Sphincter of bile duct	Sphincter muscles
Sphincter of Oddi ☞ Sphincter of ampulla	
Suspensory muscle of duodenum	Flat muscles
Transversus abdominis	Flat muscles

MUSCLES OF THE PELVIS	***STRUCTURE***
Detrusor of urinary bladder – bladder neck part	Trilaminar muscles
Iliacus	Flat muscles
Iliopsoas	Flat muscles
Internal anal sphincter	Sphincter muscles
Puboprostaticus ♂	Flat muscles
Puborectalis	Flat muscles
Pubovesicalis	Flat muscles
Rectouterinus ♀	Flat muscles
Rectococcygeus	Flat muscles
Rectoperinealis	Flat muscles
Sacrococcygeus dorsalis	Short muscles
Sacrococcygeus ventralis	Short muscles
Sphincter of bladder	Sphincter muscles
Sphincter urethrovaginalis (sphincter of vagina) ♀	Sphincter muscles
Vesical muscle ☞ Detrusor	

PERINEAL MUSCLES	***STRUCTURE***
Bulbospongiosus ♂	Flat muscles
Coccygeus ☞ Ischiococcygeus	
Deep transverse perineal muscle	Flat muscles
Erector penis ☞ Ischiocavernosus	
External anal sphincter	Sphincter muscles
External urethral sphincter	Sphincter muscles

continued

PERINEAL MUSCLES (continued)	***STRUCTURE***
Iliococcygeus	Flat muscles
Ischiocavernosus	Flat muscles
Ischiococcygeus	Flat muscles
Levator ani	Flat muscles
Pelvic diaphragm	Flat muscles
Pubococcygeus	Flat muscles
Pubovaginalis ♀	Flat muscles
Superficial transverse perineal muscle	Flat muscles

MUSCLES OF THE UPPER LIMB	***STRUCTURE***
Abductor digiti minimi	Short muscles
Abductor pollicis brevis	Short muscles
Abductor pollicis longus	Long muscles
Adductor pollicis	Short muscles
Anconeus	Short muscles
Articularis cubiti	Short muscles
Biceps brachii	Two-headed muscles
Brachialis	Flat muscles
Brachioradialis	Long muscles
Coracobrachialis	Long muscles
Deltoid	Multipennate muscles
Dorsal interossei	Pennate (bipennate) muscles
Extensor carpi radialis brevis	Short muscles
Extensor carpi radialis longus	Long muscles
Extensor carpi ulnaris	Long muscles
Extensor digiti minimi	Long muscles
Extensor digitorum	Multi-headed muscles
Extensor indicis	Long muscles
Extensor pollicis brevis	Short muscles
Extensor pollicis longus	Long muscles
Flexor carpi radialis	Long muscles
Flexor carpi ulnaris	Short muscles
Flexor digiti minimi brevis	Short muscles
Flexor digitorum profundus	Multi-headed muscles
Flexor digitorum superficialis	Multi-headed muscles
Flexor pollicis brevis	Short muscles
Flexor pollicis longus	Semipennate (unipennate) muscles
Infraspinatus	Triangular muscles
Levator scapulae	Triangular muscles
Lumbricals (*four in number*)	Cylindric muscles
Opponens digiti minimi	Short muscles
Opponens pollicis	Short muscles

continued

MUSCLES OF THE UPPER LIMB (continued)	STRUCTURE
Palmar interossei (*three in number*)	Semipennate (unipennate) muscles
Palmaris brevis	Short muscles
Palmaris longus	Long muscles
Pronator quadratus	Quadrate muscles
Pronator teres	Flat muscles
Subscapularis	Flat muscles
Supinator	Spiral muscles
Supraspinatus	Triangular muscles
Teres major	Rectangular muscles
Teres minor	Cylindric muscles
Triceps brachii	Three-headed muscles

MUSCLES OF THE LOWER LIMB	STRUCTURE
Abductor digiti minimi	Long muscles
Abductor hallucis	Long muscles
Adductor brevis	Triangular muscles
Adductor hallucis	Pennate (bipennate) muscles
Adductor longus	Triangular muscles
Adductor magnus	Spiral and cruciate muscles
Articularis genus	Short muscles
Biceps femoris	Two-headed muscles
Dorsal interossei	Pennate (bipennate) muscles
Extensor digitorum brevis	Multi-headed muscles
Extensor digitorum longus	Semipennate (unipennate) muscles
Extensor hallucis brevis	Short muscles
Extensor hallucis longus	Long muscles
Fibularis ☞ Fibularis brevis, Fibularis longus and Fibularis tertius	
Fibularis brevis (Peroneus brevis)	Long muscles
Fibularis longus (Peroneus longus)	Pennate (bipennate) muscles
Fibularis tertius (Peroneus tertius)	Semipennate (unipennate) muscles
Flexor accessorius ☞ Quadratus plantae	
Flexor digiti minimi brevis	Short muscles
Flexor digitorum brevis	Multi-headed muscles

continued

MUSCLES OF THE LOWER LIMB (continued)	STRUCTURE
Flexor digitorum longus	Pennate (bipennate) and semipennate (unipennate) muscles
Flexor hallucis brevis	Short muscles
Flexor hallucis longus	Pennate (bipennate) muscles
Gastrocnemius	Two-headed muscles
Gemellus inferior	Ribbon muscles
Gemellus superior	Ribbon muscles
Gluteus maximus	Flat muscles
Gluteus medius	Flat muscles
Gluteus minimus	Triangular muscles
Gracilis	Ribbon muscles
Iliopsoas	Flat muscles
Lumbricals	Long muscles
Obturator externus	Radiate muscles
Obturator internus	Radiate muscles
Pectineus	Flat muscles
Piriformis	Triangular muscles
Plantar interossei (*three in number*)	Semipennate (unipennate) muscles
Plantaris	Triangular muscles
Popliteus	Short muscles
Psoas major	Long muscles
Quadratus femoris	Quadrate muscles
Quadratus plantae	Flat muscles
Quadriceps femoris	Four-headed muscles
Rectus femoris	Pennate (bipennate) muscles
Sartorius	Ribbon muscles
Semimembranosus	Long muscles
Semitendinosus	Long muscles
Soleus	Multipennate muscles
Tensor fasciae latae	Flat muscles
Tibialis anterior	Circumpennate muscles
Tibialis posterior	Semipennate (unipennate) muscles
Triceps surae ☞ Gastrocnemius and Soleus	
Vastus intermedius	Long muscles
Vastus lateralis	Long muscles
Vastus medialis	Long muscles

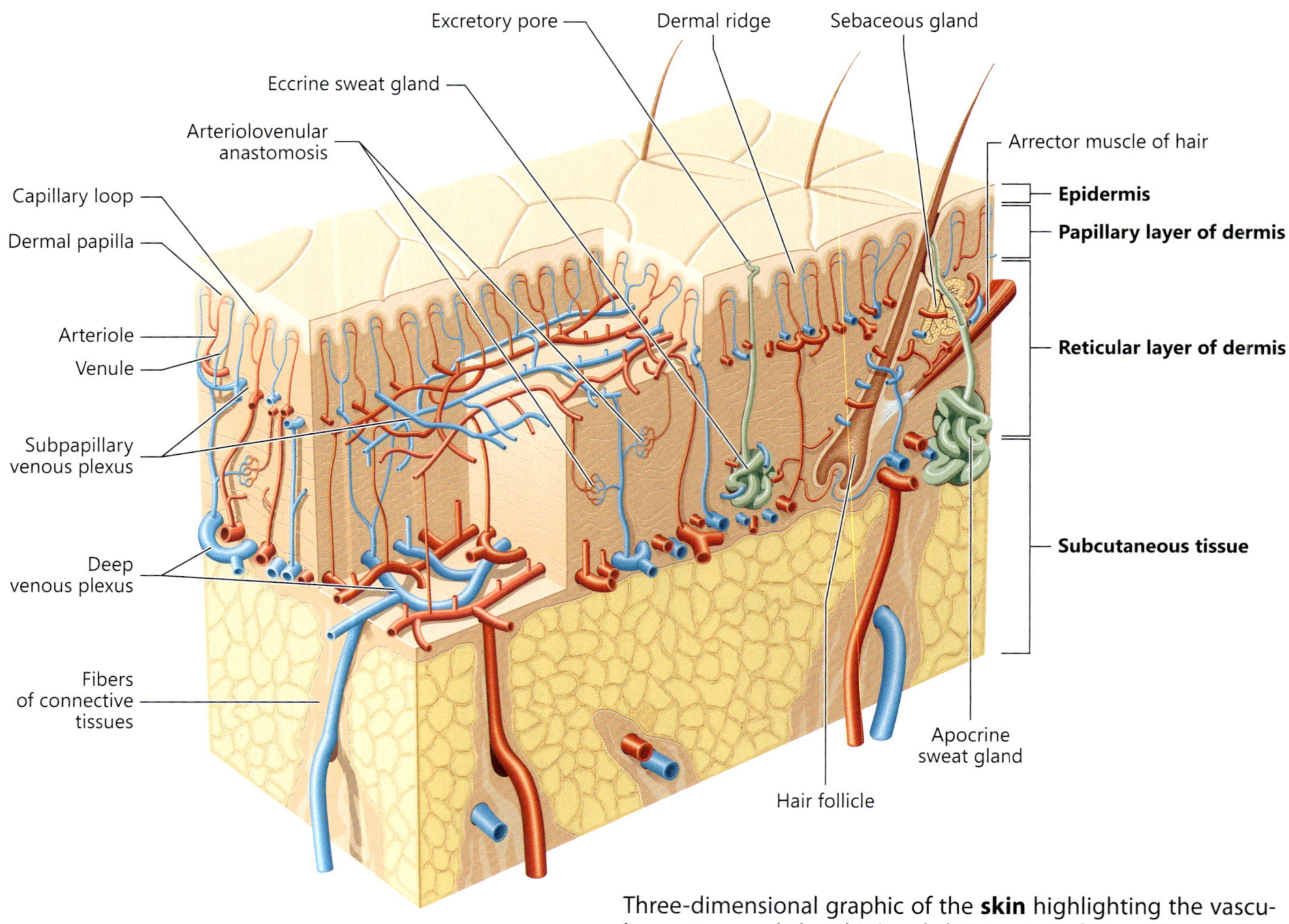

Three-dimensional graphic of the **skin** highlighting the vascular system and the skin's subdivision into layers. Proceeding deeper from the surface, various layers are found in the skin. The epidermis is followed by the dermis, which includes the papillary region, containing capillaries and receptors, and the reticular region, containing blood vessels, lymphatic vessels, hair follicles and sweat glands. The connective tissue of the reticular region is arranged in courses that define the Langer lines. Next comes the subcutaneous layer or *hypodermis*, which is subdivided into the superficial subcutaneous layer, the superficial fascia and cutaneuous muscles (for example the platysma), and the deep subcutaneous layer (containing superficial veins and superficial nerves), followed by the muscle fascia and the osteomuscular layer.

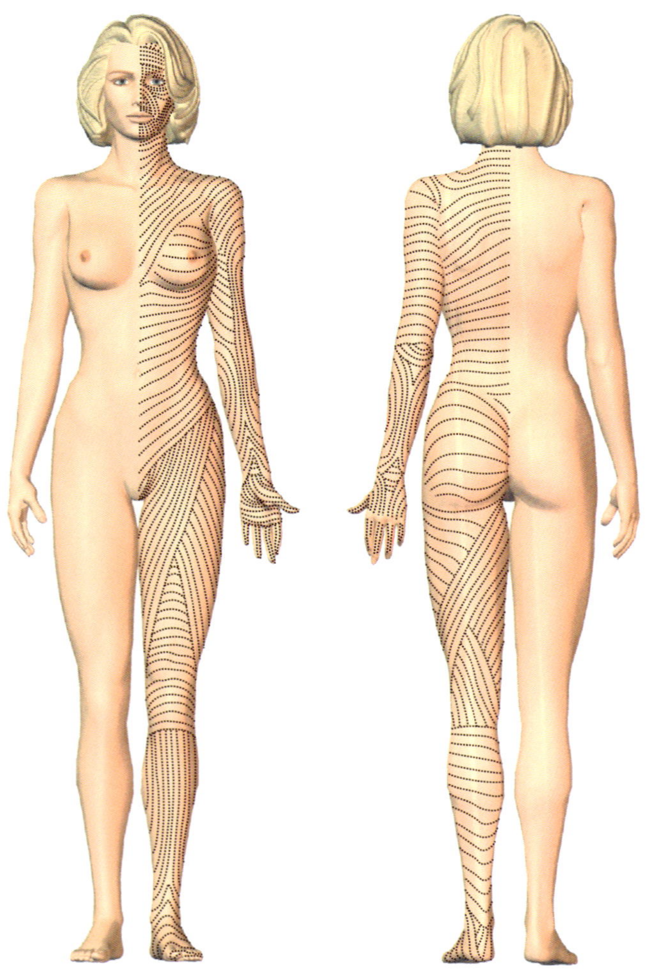

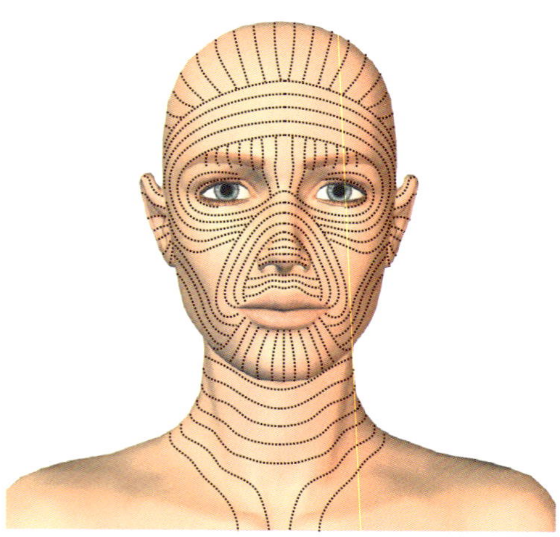

Langer Lines. These are directional lines of cutaneous cleavage corresponding to lines of mechanical tension in the skin (reticular layer).

A series of **appendages** are associated with the skin: nails, pilosebaceous units, sweat glands and mammary glands.

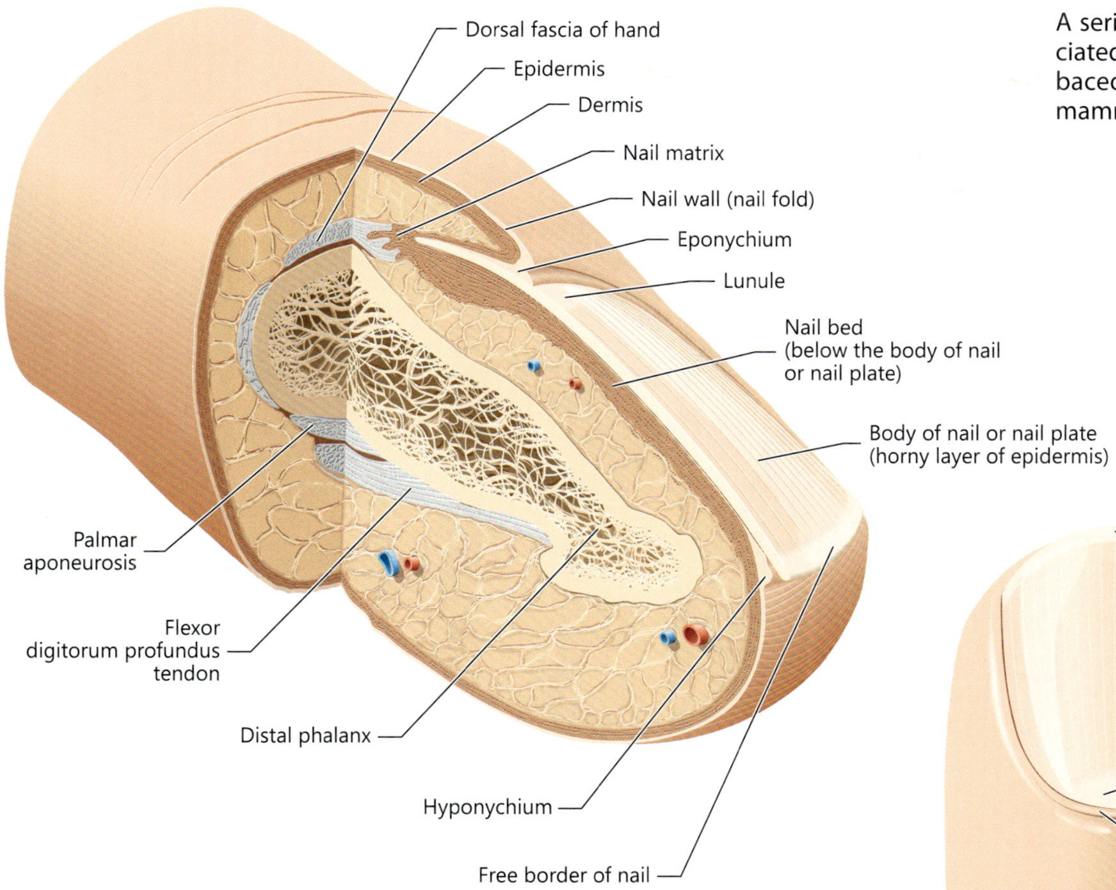

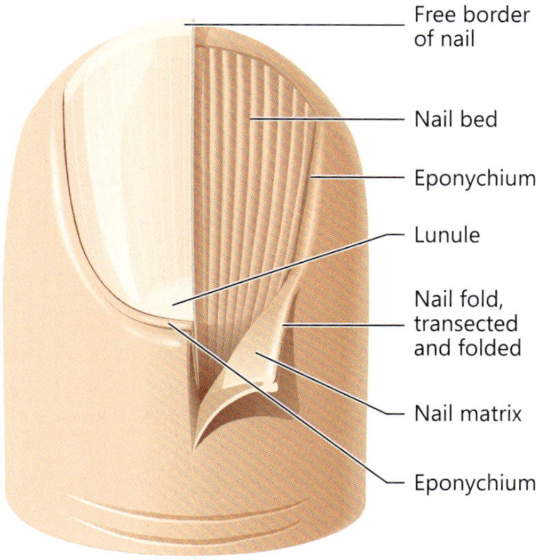

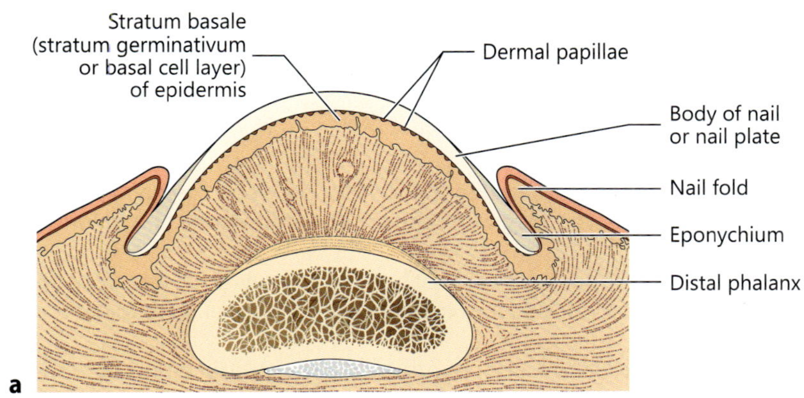

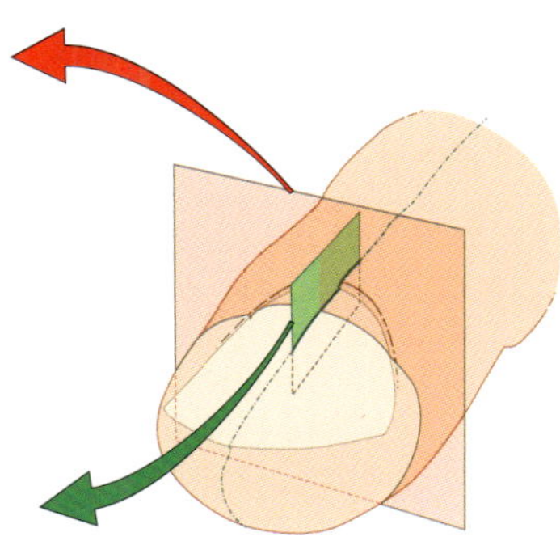

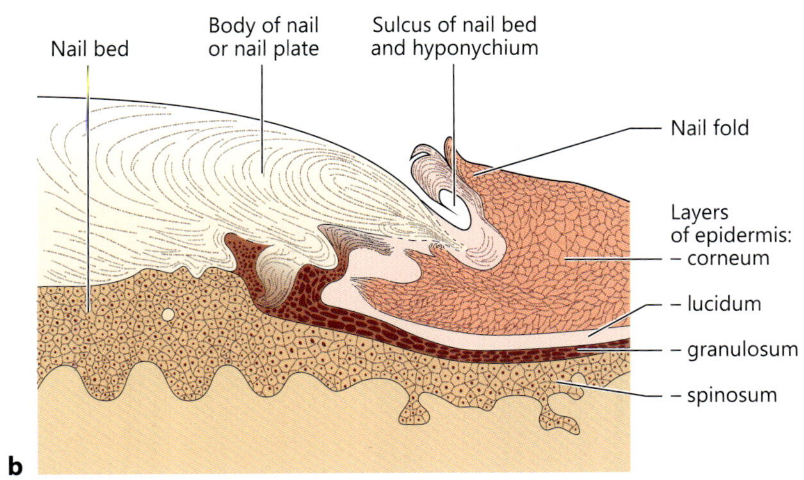

a, Transverse section through the body of the **nail plate**. This shows the structure of the nail and its relations to the nail bed and the nail wall, and the close relationships between the dermis of the nail bed and the periosteum of the distal or ungual phalanx. **b**, Detail of **nail wall** (longitudinal section). Note that the stratum granulosum and stratum lucidum of the epidermis terminate at the nail body and are not found in the nail bed.

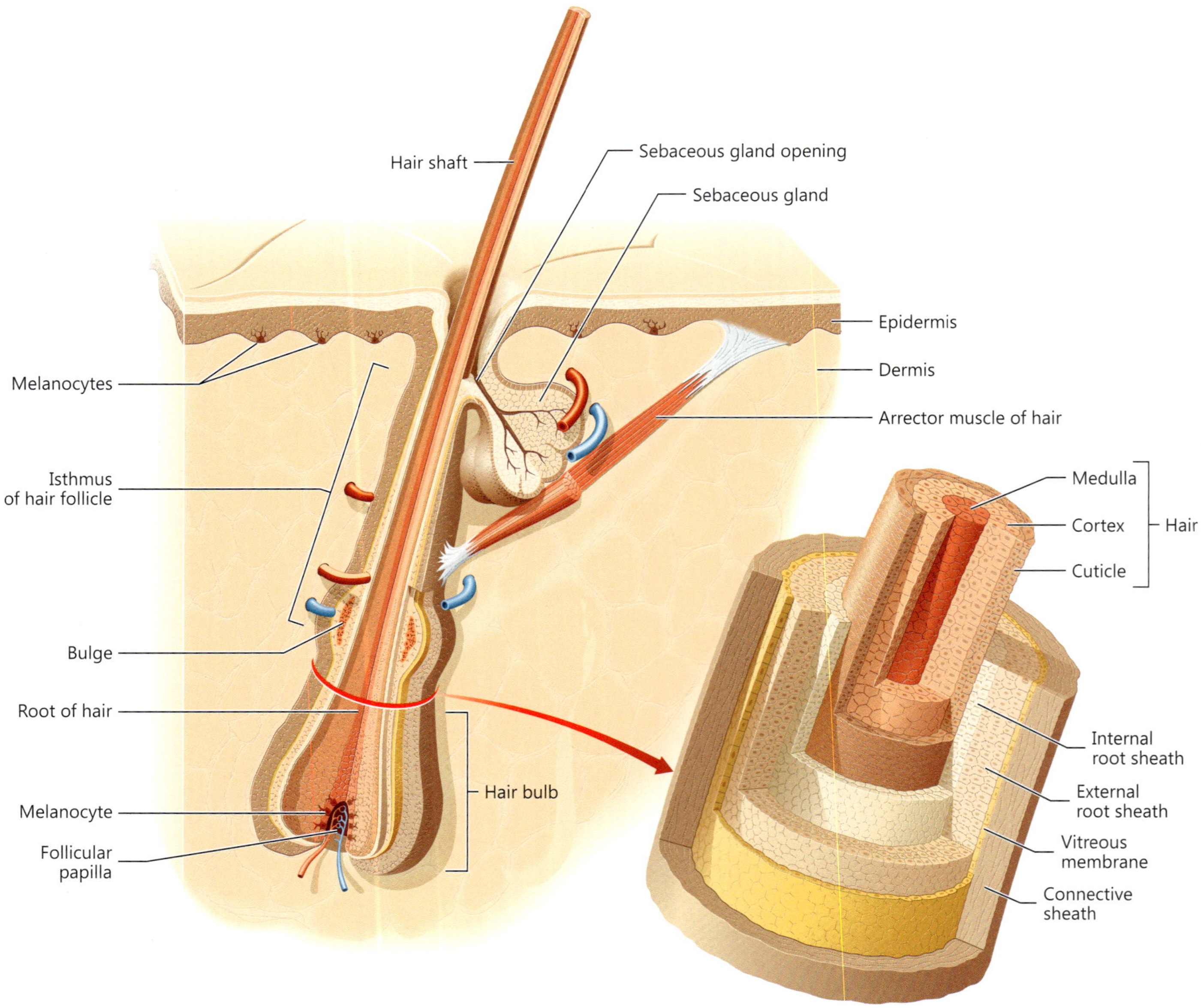

Structural organization of the **pilosebaceous unit**. The sebaceous glands are located in the dermis, between the hair follicles and the arrector muscle of hair, and open into the upper part of the follicle. The diagrammatic representation of a hair in longitudinal section reveals its component parts.

The hair's **bulge** region is located at the same level as the insertion of the arrector muscle of hair. Conditions in this microhabitat favour the presence of stem cells, responsible for regenerating the keratinocytes and melanocytes of the hair, alongside the secretory cells of the sebaceous gland.

In the bulge area, stem cells divide symmetrically to regenerate their population, and asymmetrically to produce undifferentiated progenitor cells, the precursors of the different cell lines. Under normal physiological conditions, the progenitor cells migrate towards the hair papilla region, where they form the hair matrix and, from this, the inner sheaths, root, and shaft.

In the event of skin lesions, the stem cells may migrate towards the surface and give rise to the regeneration of the epidermis and sebaceous glands. It has been known for some time, for example, that second degree burns (those affecting the dermis and epidermis, with the formation of blisters containing exudate) heal much faster in areas of skin covered by hair (for example, the back, scalp, and face) than in more glabrous areas (for example, the volar surface of the forearm or the eyelids).

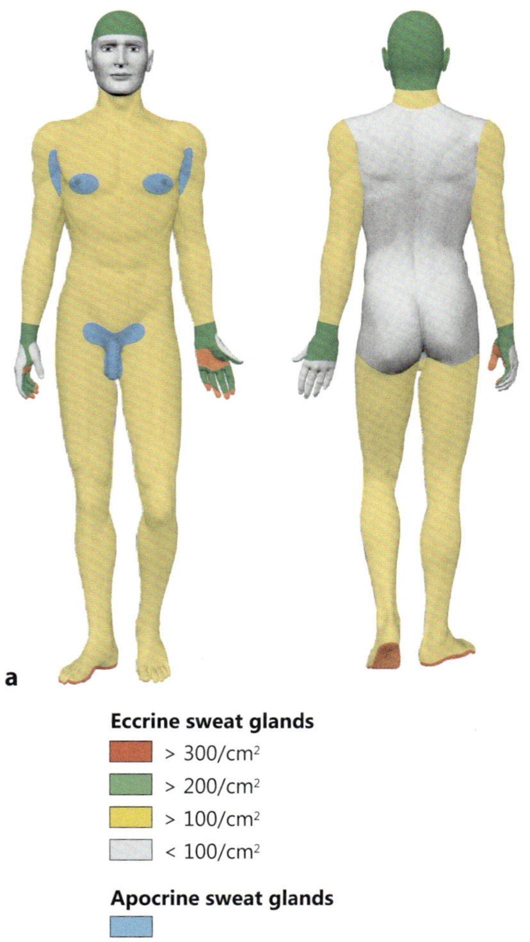

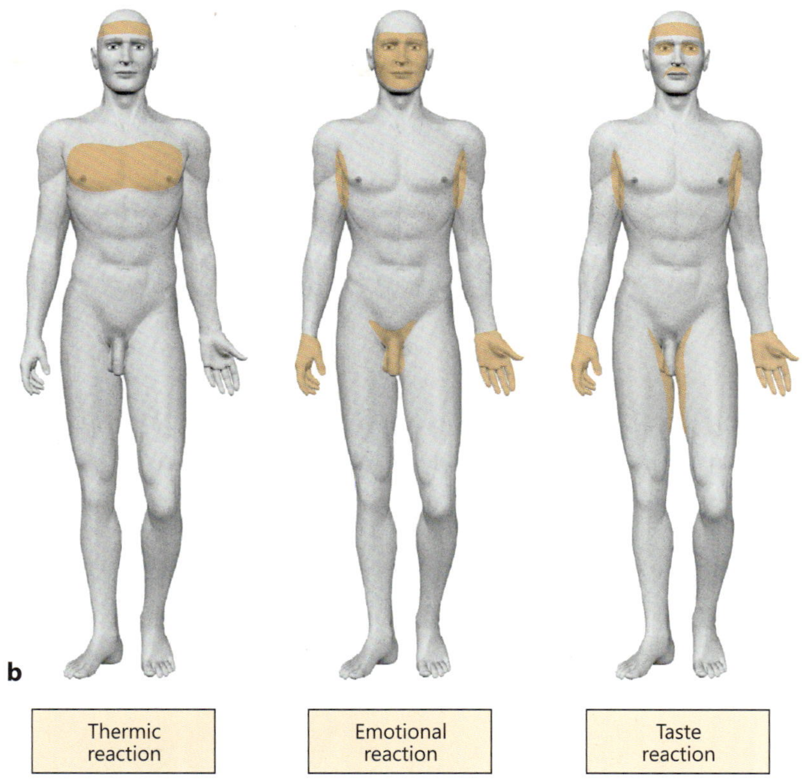

a, The distribution of **sweat glands**. There are two types of sweat gland: eccrine glands, which release secretions directly onto the surface of the skin, and apocrine glands, which open onto the hair follicles. Eccrine sweat glands are distributed across almost the entire surface of the body, varying in number according to area. Apocrine sweat glands are located in the armpits, the mammary areola, the groin and in genital skin. **b**, Activity of the eccrine sweat glands. Distribution areas of sweating subsequent to various stimuli.

SEM photograph of two **sweat glands** (coiled tubular glands), isolated following digestion of dermal connective tissue. The blood capillaries forming a dense network around the glomeruli have been partially preserved.

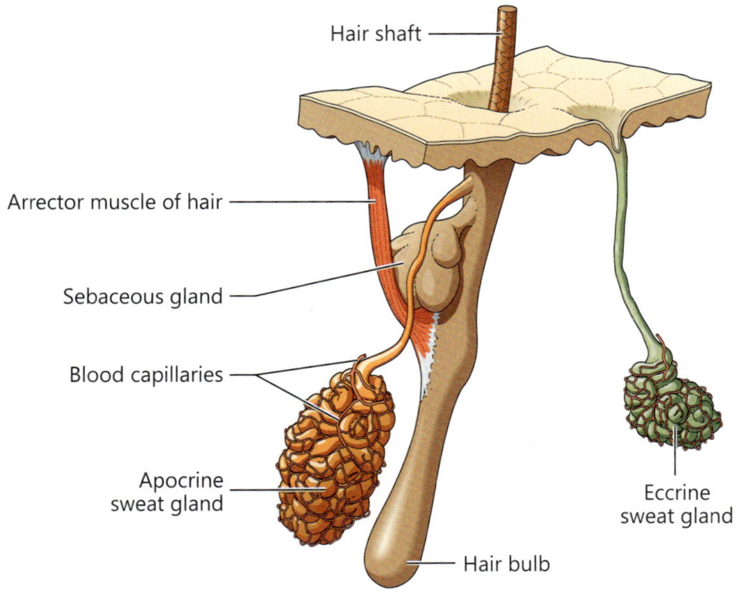

Three-dimensional drawing of a **pilosebaceous follicle** into which both a sebaceous gland and the excretory duct of an apocrine sweat gland flow. The smaller eccrine sweat gland, on the other hand, discharges independently of the follicle.

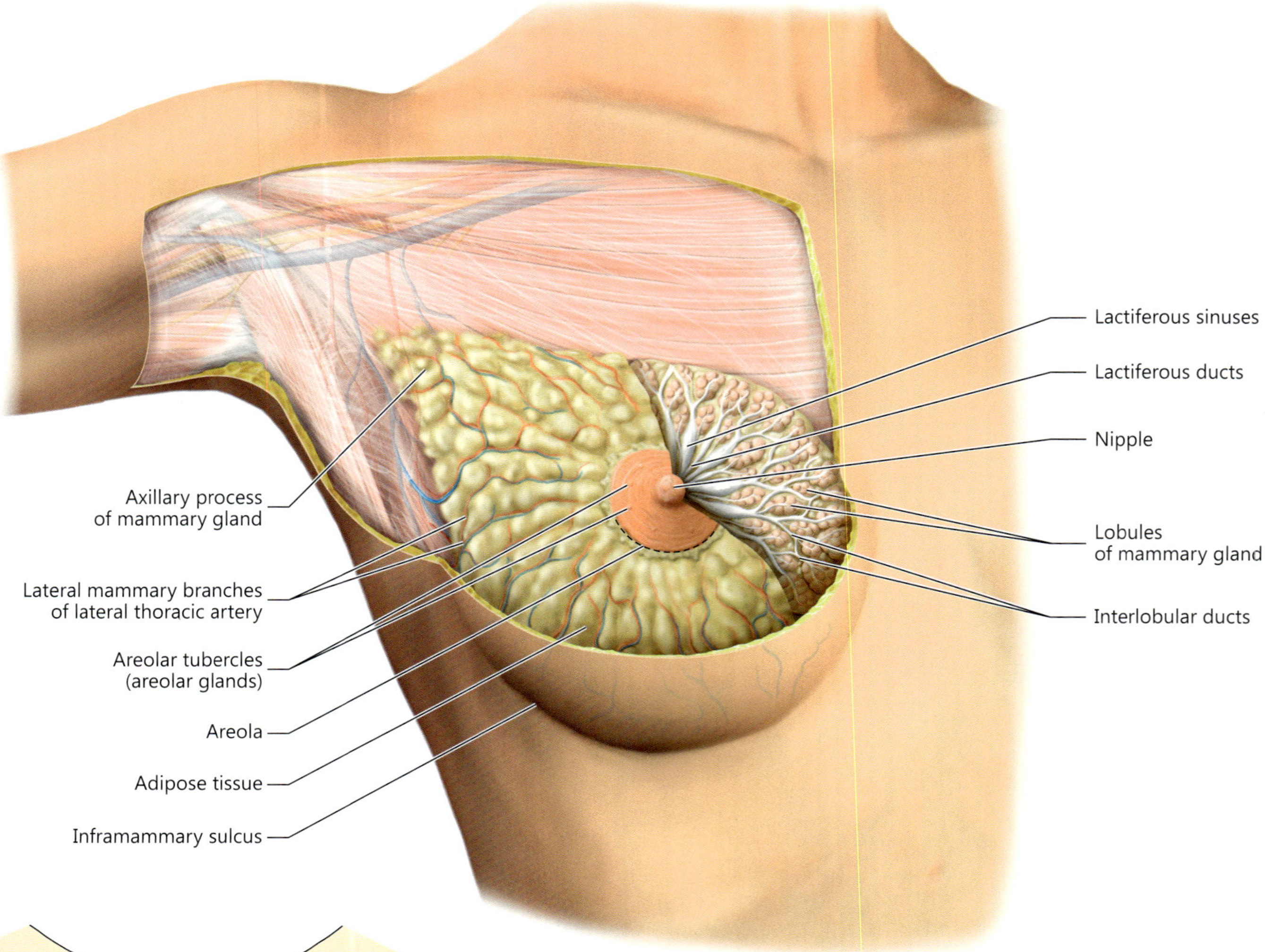

Illustration of the **mammary gland** and the **investing fascia of the pectoralis major**. The medial part of the breast has been sectioned to show the lobular appearance of the gland, the lactiferous sinuses and the lactiferous ducts. The lateral portion shows large amounts of subcutaneous adipose tissue.

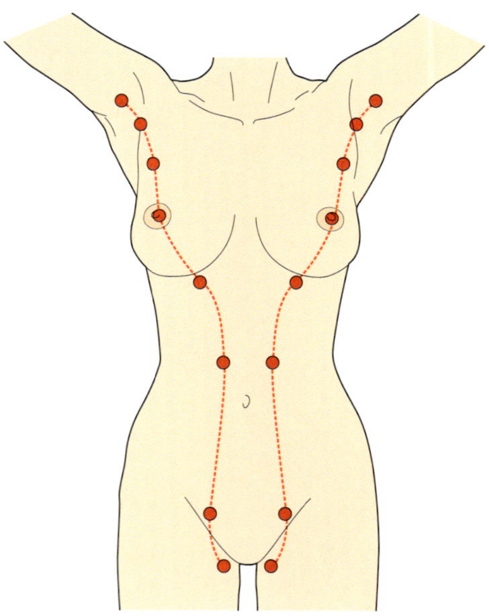

➤

At approximately the fifth week of fetal life, multiple sites for potential mammary gland development are present, distributed along two lines (the **milk lines**) that extend from the axillary space to the groin region. In some mammals, each of these sites can correspond to the development of a breast. In women, these primordia recede during development, with the exception of those destined to become the definitive breasts. Failure to recede may result in the development of supernumerary nipples (polythelia) or supernumerary breasts (polymastia). A frequent remnant of the milk ridge forms the axillary process of the mammary gland.

➤

Three-dimensional illustration of the skin highlighting the **nervous network** and its related **receptors**. Fibers reach the dermal papillae, partly penetrating into the epidermis as free nerve endings. A deep or dermo-hypodermal nerve plexus and a superficial or intradermal one are apparent, which interconnect with each other to varying degrees. Sensory fibers are distributed to the skin and its connected structures making up two groups of formations: free nerve endings and corpusculated nerve endings, or simply *sensory corpuscles*. The latter are constituted of fibers surrounded by sheathings of connective cells: lamellar corpuscles (Pacinian corpuscles) and Golgi-Mazzoni corpuscles in the subcutaneous tissue; Ruffini's corpuscles, Krause's corpuscles and tactile corpuscles (Meissner's corpuscles) in the dermis; tactile menisci (Merkel's disks) between the dermis and epidermis. Efferent sympathetic fibers innervate the glands, the arrector muscles of hair, and vessel walls. The nail bed is innervated by free endings and by receptors of the Golgi-Mazzoni corpuscle type. The hair follicle receives innervation from the base of the bulb up to the epidermis; some nerve fibers extend parallel to the follicle, others surround it, constituting an ordered reticulum.

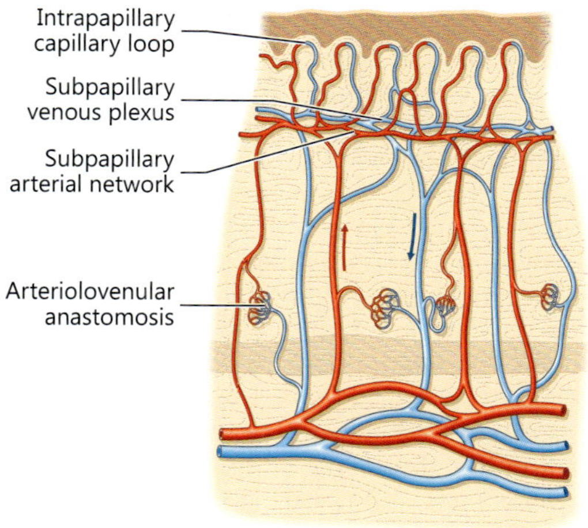

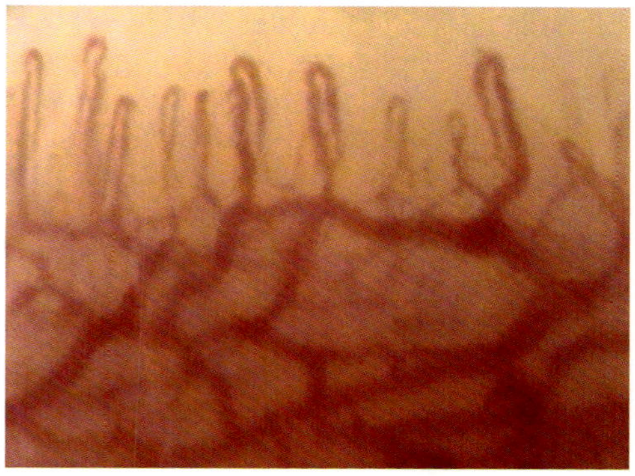

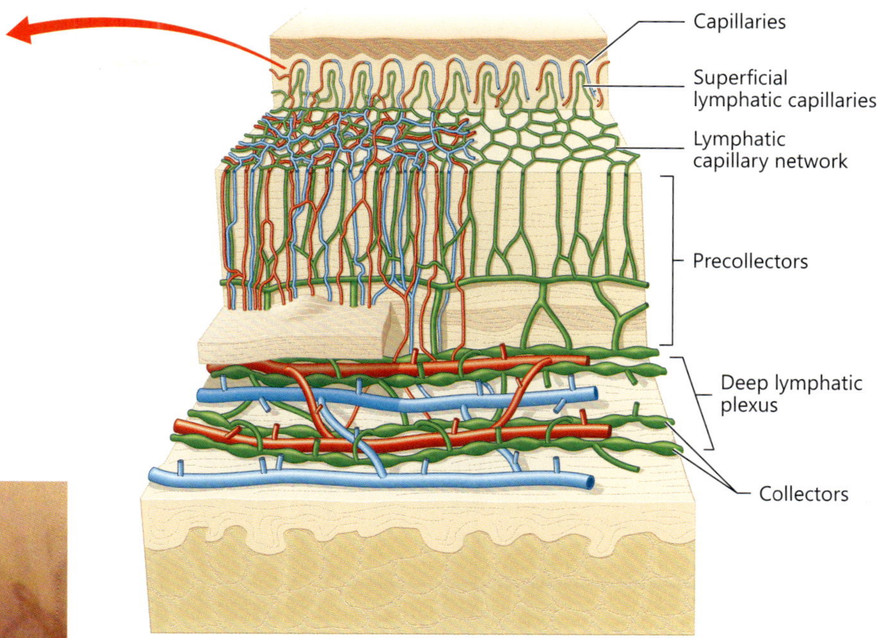

Three-dimensional graphic reconstruction of the **skin** emphasizing blood and lymphatic vascularization. Top left shows only blood circulation to the skin with the arteriovenous shunts. Bottom left shows an image from a capillaroscopy: the capillary loops of the papillary region can be seen.

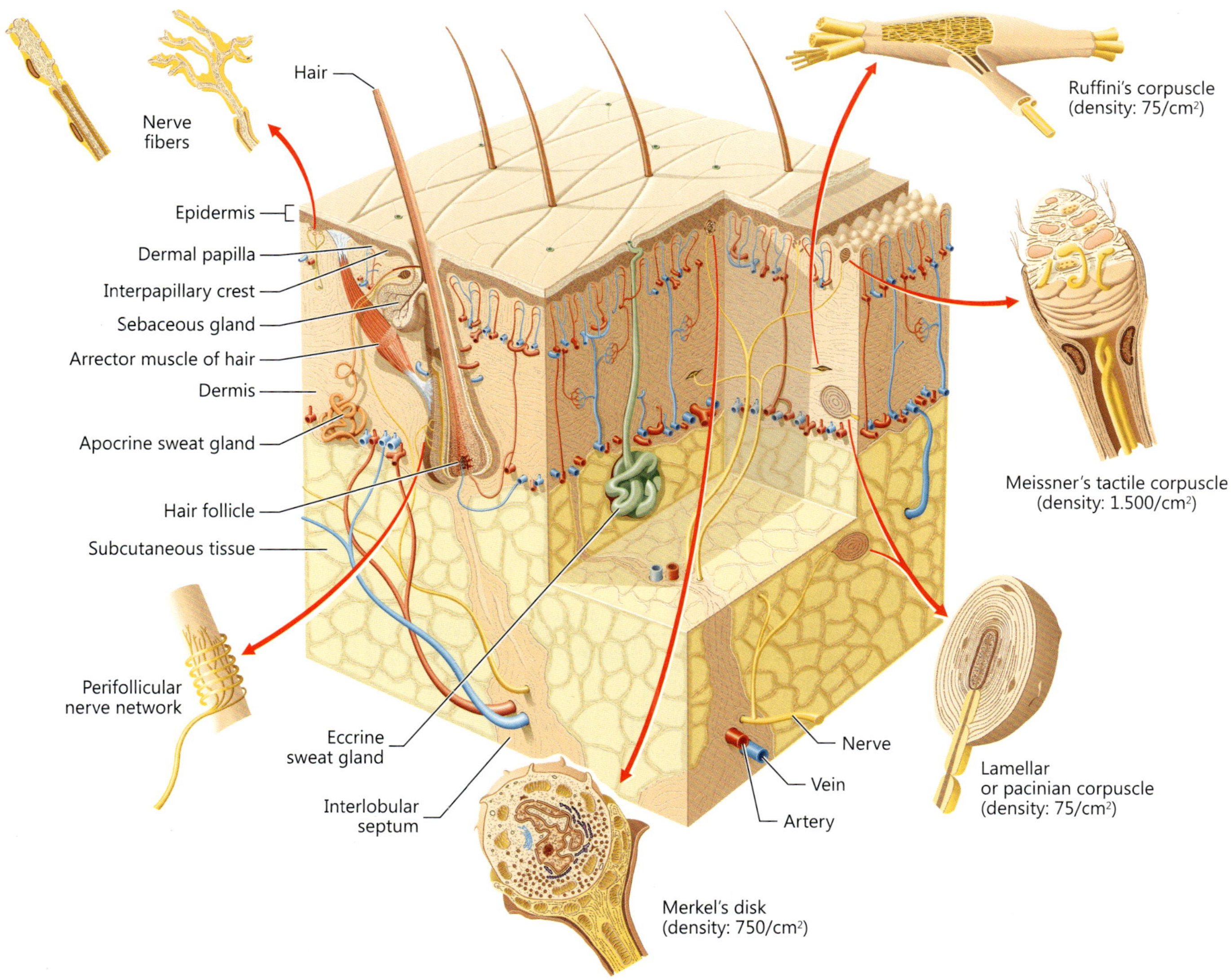

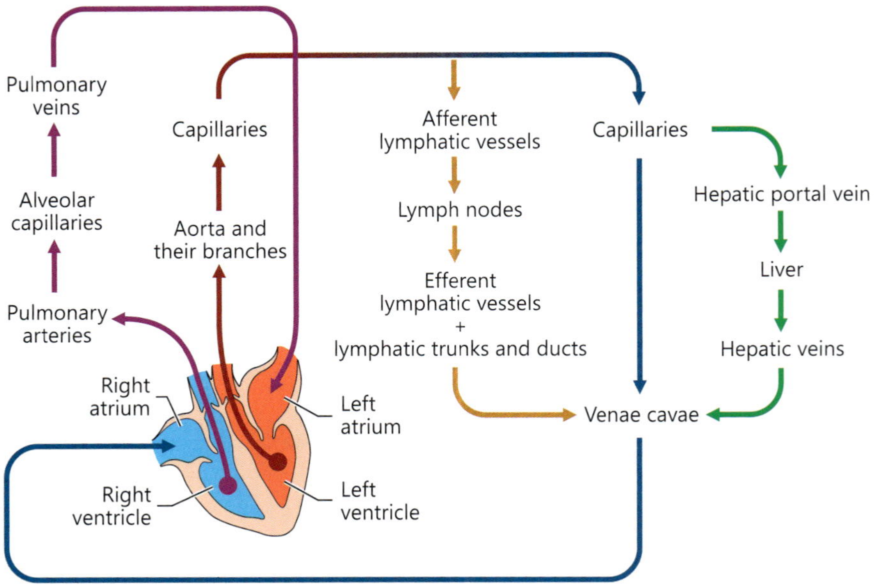

Overview of the **circulatory system**. Blood pumped by the left ventricle of the heart is distributed by the branches of the aorta. Gaseous and metabolic exchanges take place at the level of **microcirculation** between the blood and the extracellular fluid that permeates the tissues. Following microcirculation, blood is drained towards the superior and inferior **venae cavae** and from these to the right atrium of the heart. The superior vena cava drains the body areas above the diaphragm; the inferior vena cava drains the areas below. Before reaching the inferior vena cava, refluent blood from the gastrointestinal tract, collected in the abdominal cavity from the pancreas, biliary tracts, and spleen, is drained through the **hepatic portal system** by the portal vein and its tributaries. Extracellular fluids are drained by lymphatic capillaries, which originate in blind-ended vessels in the body's periphery. This fluid (lymph) is conducted through lymphatic ducts, which gradually increase in size, and passes through lymph node stations before being discharged into the superior vena cava (jugulo-subclavian venous junction).

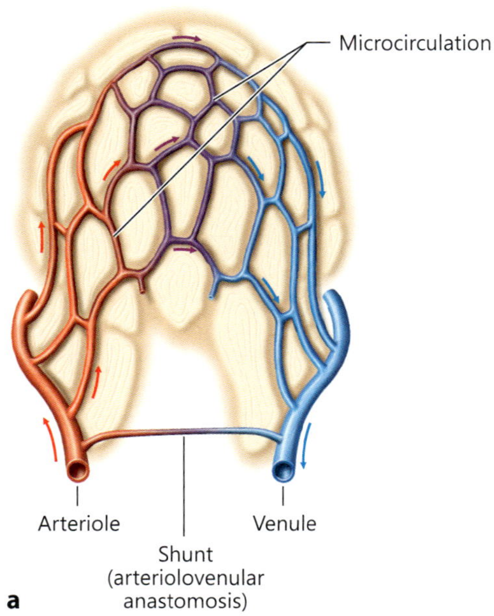

a

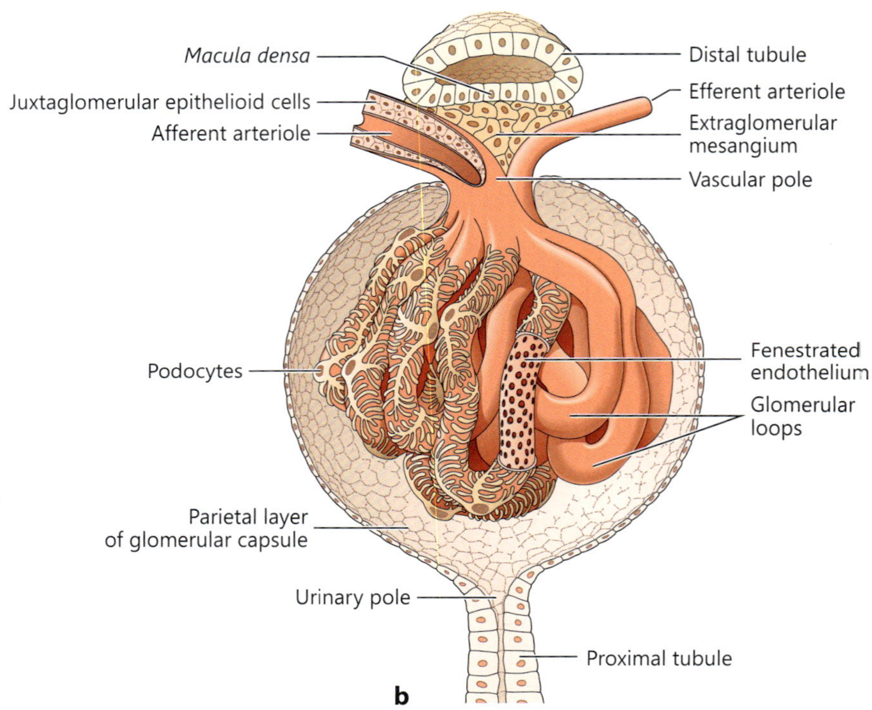

b

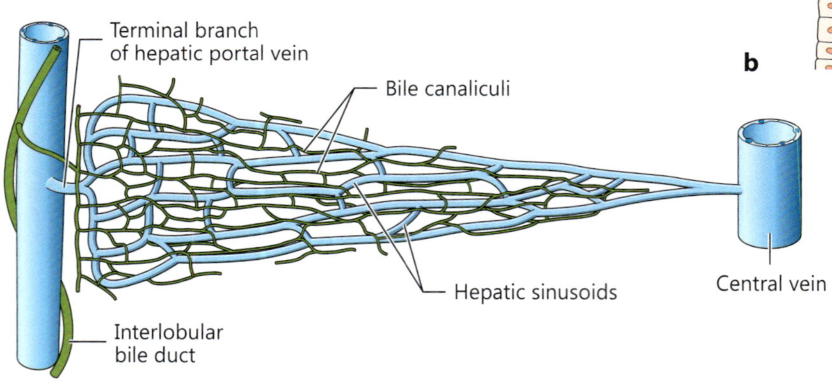

c

A **capillary network** comprises a series of minuscule vessels that originate from an arteriole and merge into a venule where exchanges occur between tissue and the circulation. **a**, It is also possible for blood to pass directly from an arteriole to a venule via a communicating vessel (shunt). **b**, There are also special conditions, known as **retia mirabilia**, where capillary vessels may be found connecting two arterioles, as in the case of an arterial glomerulus, (arterial rete mirabile) or **c**, between two veins, as in the case of the hepatic lobules (venous rete mirabile).

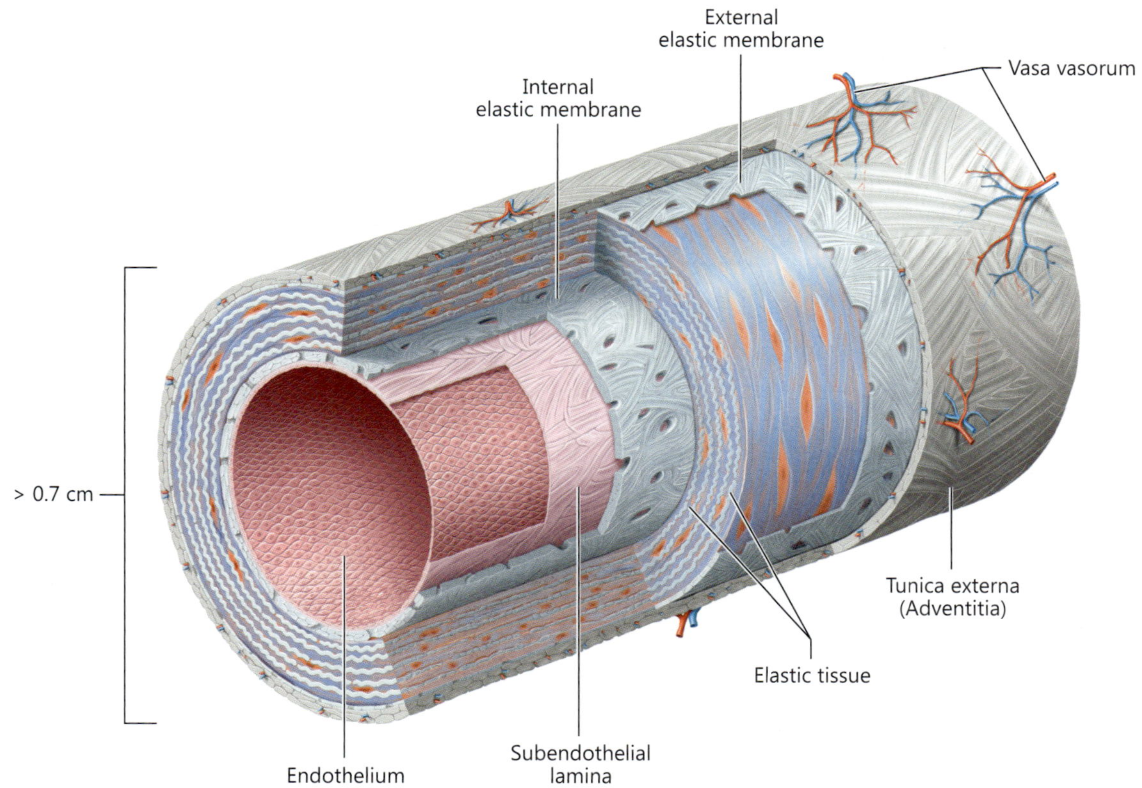

Structure of a **large artery**. Elastic artery: As the arteries branch out, they gradually decrease in size. Larger arteries have a tunica media formed of elastic tissue. This structure enables the artery to stretch with the varying blood pressure during ventricular systole. In the diastolic phase the artery resumes its original shape, promoting blood propulsion in a distal direction.

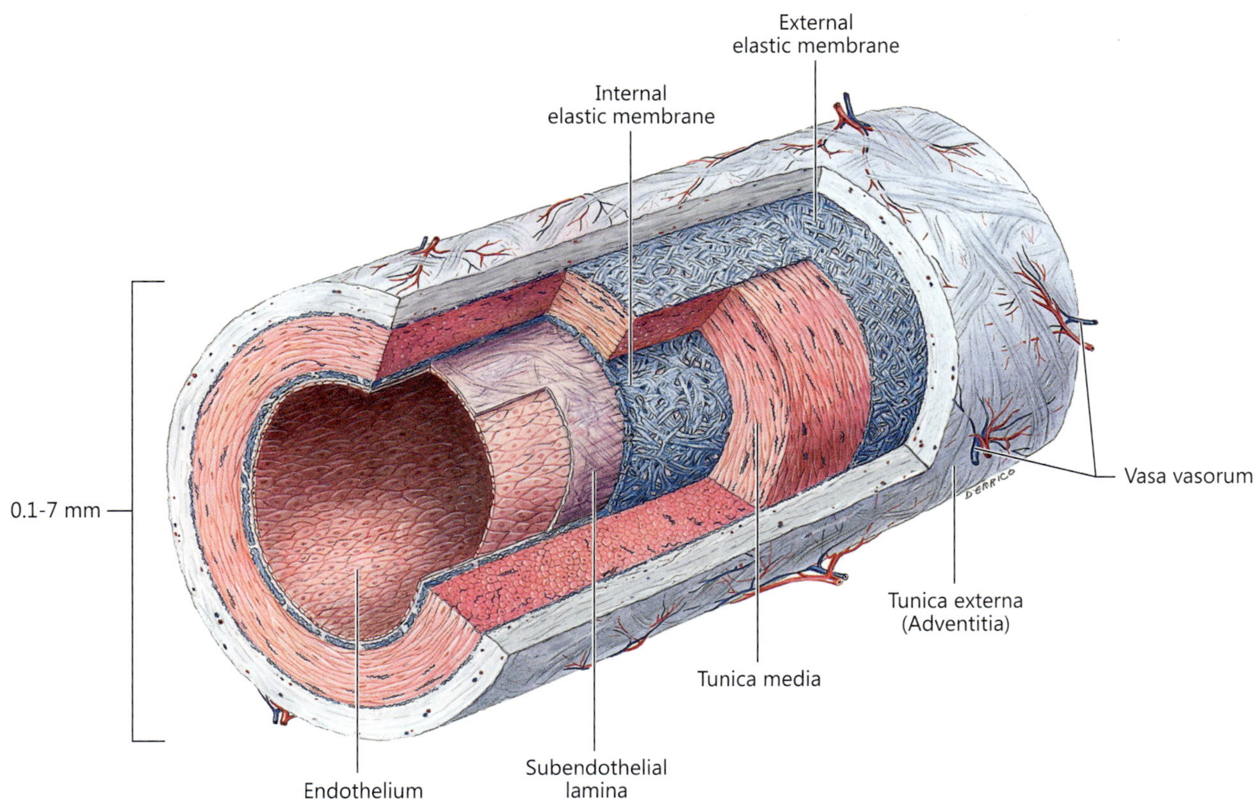

Structure of a **medium-small artery**. Muscular artery: The smaller size arteries have a tunica media formed of smooth muscle tissue. This structure enables the artery to contract and to promote blood propulsion in a distal direction.

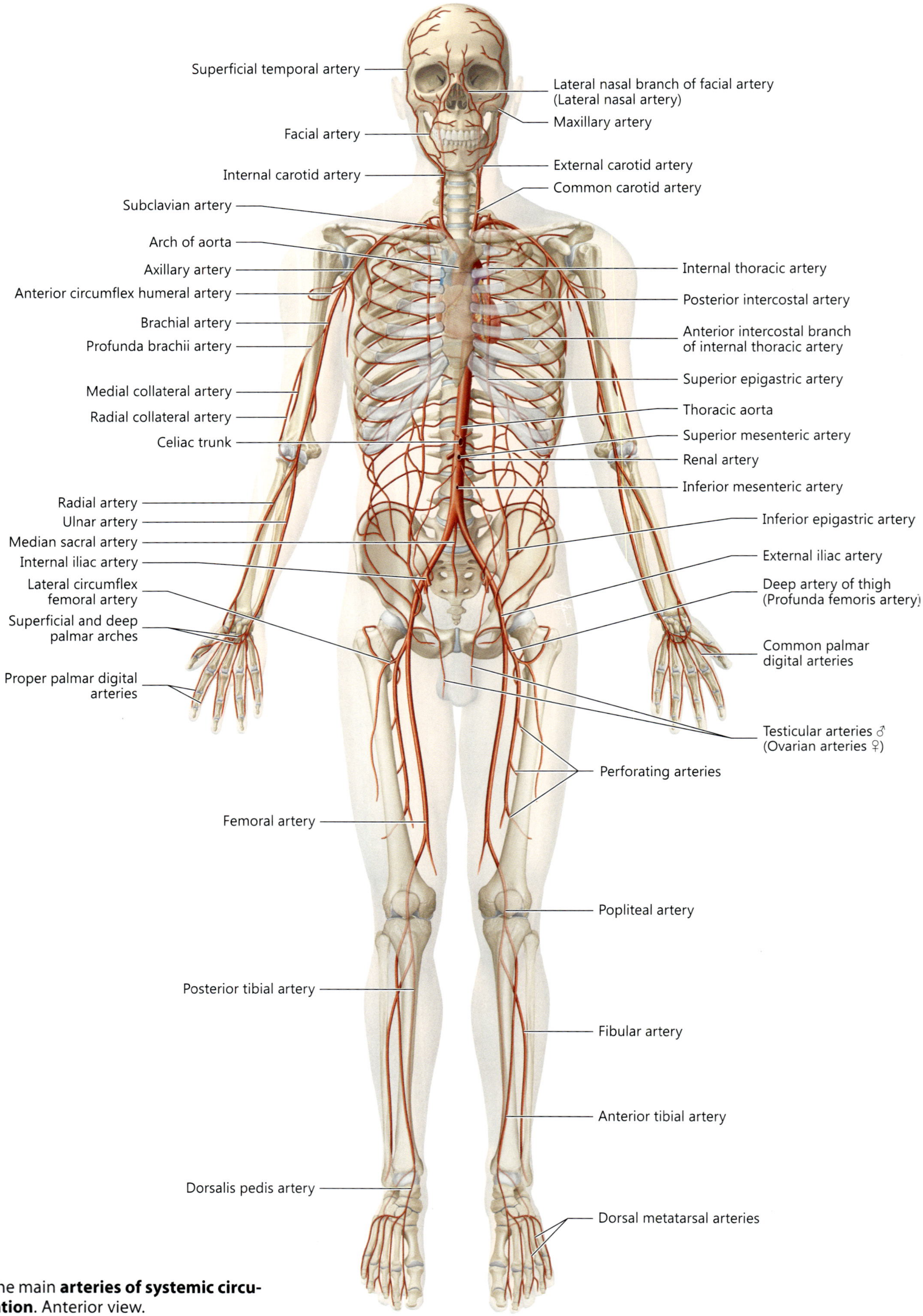

The main **arteries of systemic circulation**. Anterior view.

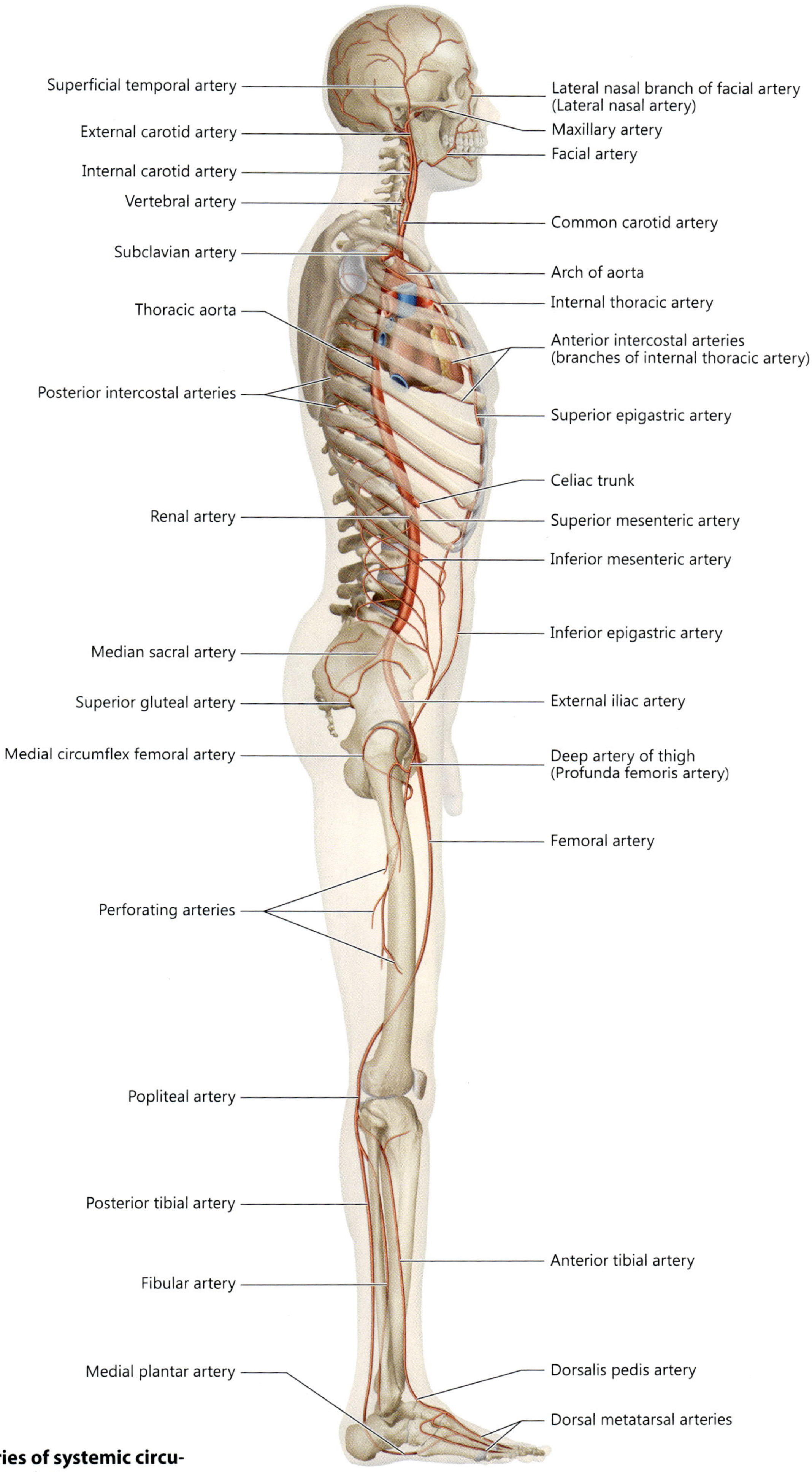

The main **arteries of systemic circulation**. Right lateral view.

Regional subdivision of the arteries (→ branching)*

ARTERIES OF THE HEAD
Arteries of brain (branches of anterior, middle, posterior cerebral arteries, and posterior communicating and anterior choroidal arteries)
Cerebral arterial circle (circle of Willis) (posterior cerebral arteries connected to the internal carotid arteries through the posterior communicating arteries and anterior cerebral arteries interconnected through the anterior communicating artery)

[External carotid artery]**
→ Ascending pharyngeal artery
  → Inferior tympanic artery
  → Posterior meningeal artery
→ Descending palatine artery
  → Greater palatine artery
  → Lesser palatine arteries
→ Facial artery (external maxillary artery)
  → Angular artery
  → Ascending palatine artery
  → Inferior labial branch
  → Lateral nasal branch
  → Submental artery
  → Superior labial branch
→ Lingual artery
  → Deep lingual artery (ranine artery)
  → Dorsal lingual branches (dorsal lingual artery)
  → Sublingual artery
→ Maxillary artery (internal maxillary artery or deep facial artery)
  → Anterior deep temporal artery
  → Anterior tympanic artery (tympanic artery)
  → Artery of pterygoid canal (vidian artery) (branch of maxillary artery)
  → Buccal artery (buccinator artery)
  → Deep auricular artery
  → Inferior alveolar artery (inferior dental artery or mandibular artery)
  → Infraorbital artery
    → Anterior superior alveolar arteries (anterior dental arteries)
  → Masseteric artery
  → Middle meningeal artery (sphenospinal artery)
    → Superior tympanic artery
  → Posterior deep temporal artery
  → Posterior superior alveolar artery (posterior dental artery)
  → Sphenopalatine artery (nasopalatine artery)
    → Posterior lateral nasal arteries
    → Posterior septal branches (posterior septal nasal branches)
→ Occipital artery
→ Posterior auricular artery
  → Posterior tympanic artery
  → Stylomastoid artery
→ Superficial temporal artery
  → Middle temporal artery
  → Transverse facial artery
  → Zygomatico-orbital artery

continued

ARTERIES OF THE HEAD (continued)
Hyaloid artery (*usually disappears in the last month of intrauterine life*)

[Internal carotid artery]**
→ Cavernous part
  → Inferior hypophysial artery
→ Cerebral part
  → Anterior cerebral artery
    → Anterior communicating artery (unpaired artery)
  → Anterior choroidal artery
  → Middle cerebral artery
  → Posterior communicating artery (right and left posterior communicating arteries)
  → Superior hypophysial artery
  → Uncal artery
→ Ophthalmic artery
  → Anterior ciliary arteries
  → Anterior ethmoidal artery
    → Anterior meningeal branch
  → Central retinal artery
  → Dorsal nasal artery (external nasal artery)
  → Lacrimal artery
    → Lateral palpebral arteries
  → Long posterior ciliary arteries (*two in number*)
  → Medial palpebral arteries
    → Posterior conjunctival arteries
  → Muscular arteries
    → Anterior ciliary arteries
    → Anterior conjunctival arteries
    → Episcleral arteries
  → Posterior ethmoidal artery
  → Short posterior ciliary arteries (*from six to twelve in number*)
  → Supraorbital artery
  → Supratrochlear artery (frontal artery)
→ Petrous part
  → Artery of pterygoid canal (vidian artery) (branch of internal carotid artery)
  → Caroticotympanic arteries

[Vertebral artery]**
→ Basilar artery (basilar trunk) (from the confluence of the two vertebral arteries)
  → Anterior inferior cerebellar artery
    → Labyrinthine artery (internal auditory artery)
  → Posterior cerebral artery
    → Postcommunicating part (P2 segment)
      → Peduncular branches
      → Posterior lateral choroidal branches
      → Posterior medial choroidal branches
    → Precommunicating part (P1 segment)
  → Superior cerebellar artery
→ Intracranial part
  → Anterior spinal artery (ventral spinal artery)
  → Posterior inferior cerebellar artery
  → Posterior spinal artery (dorsal spinal artery)

* Certain arteries and branches not mentioned here are shown in the tables attached to the body sections of the Atlas.

[...]** The arteries indicated originate in other anatomical regions.

ARTERIES OF THE NECK

[Brachiocephalic trunk]**
- → Thyroid ima artery (inconstant)

Common carotid artery
- → External carotid artery
  - → Ascending pharyngeal artery
  - → Superior thyroid artery
    - → Sternocleidomastoid branch
    - → Superior laryngeal artery
- → Internal carotid artery
  - → Cervical part

[Inferior alveolar artery]**
- → Mylohyoid branch

[Subclavian artery]**
- → [Costocervical trunk]**
  - → Deep cervical artery
- → Thyrocervical trunk
  - → Ascending cervical artery
    - → Spinal branches (of ascending cervical artery)
  - → Inferior thyroid artery
    - → Inferior laryngeal artery
    - → Esophageal branches (of inferior thyroid artery)
- → Transverse cervical artery#
  - → Superficial cervical artery (superficial branch of transverse cervical artery)
- → Vertebral artery
  - → [Anterior spinal artery (ventral spinal artery)]**
  - → Atlantic part
  - → Cervical part
    - → Spinal branches (of vertebral artery)
      - → Radicular branches (of spinal branches of vertebral artery)
      - → Segmental medullary artery
  - → [Posterior spinal artery (dorsal spinal artery)]**
  - → Prevertebral part

In this table, the transverse cervical artery originates from the subclavian artery; however, it should be considered that it presents a frequent variability among which the origin from the thyrocervical trunk is predominant.

ARTERIES OF THE THORAX

Aorta
- → Arch of aorta (Aortic arch)
  - → Brachiocephalic trunk (innominate artery)
    - → [Right common carotid artery]**
    - → Right subclavian artery
  - → Left common carotid artery
  - → Left subclavian artery
- → Ascending aorta
  - → Left coronary artery
  - → Right coronary artery
- → Descending aorta
  - → Thoracic aorta
    - → Bronchial branches (of thoracic aorta)
    - → Mediastinal branches (of thoracic aorta)
    - → Esophageal branches (of thoracic aorta) (esophageal arteries) (*from four to five in number*)
    - → Pericardial branches
    - → Posterior intercostal arteries (*from third to eleventh artery*)
      - → Dorsal branch (of posterior intercostal arteries)
        - → Spinal branches (of dorsal branch of posterior intercostal arteries)
    - → Subcostal artery (twelfth intercostal artery)
    - → Superior phrenic arteries

[Axillary artery]**
- → Lateral thoracic artery (long thoracic artery or external mammary artery)
- → Superior thoracic artery
- → Thoracoacromial artery (acromiothoracic artery)

Pulmonary trunk
- → Left pulmonary artery
- → Right pulmonary artery

Subclavian artery
- → Costocervical trunk
  - → Supreme intercostal artery (superior intercostal artery)
    - → First posterior intercostal artery
    - → Second posterior intercostal artery
      - → Spinal branches (of second posterior intercostal artery)
- → Internal thoracic artery (internal mammary artery)
  - → Anterior intercostal branches (anterior intercostal arteries)
  - → Bronchial branches (of internal thoracic artery)
  - → Superior epigastric artery
- → Musculophrenic artery
- → Pericardiacophrenic artery
- → [Vertebral artery]**
  - → [Anterior spinal artery (ventral spinal artery)]**
  - → [Posterior spinal artery (dorsal spinal artery)]**

[Subscapular artery]**
- → Thoracodorsal artery

ARTERIES OF THE ABDOMEN
[Anterior spinal artery (ventral spinal artery)]**
Descending aorta → Abdominal aorta → Celiac trunk (celiac artery) → Common hepatic artery (gastrohepatic artery) → Hepatic artery proper → Gastroduodenal artery → Anterior superior pancreaticoduodenal artery → Posterior superior pancreaticoduodenal artery → Retroduodenal arteries → Right gastro-omental artery (right gastroepiploic artery) → Left hepatic artery (left branch of hepatic artery proper) → Interlobular arteries (of liver) → Right gastric artery (pyloric artery) → Right hepatic artery (right branch of hepatic artery proper) → Cystic artery → Interlobular arteries (of liver) → Left gastric artery → Esophageal branches (of left gastric artery) → Splenic artery (lienal artery) → Left gastro-omental artery (left gastroepiploic artery) → Pancreatic branches → Dorsal pancreatic artery → Greater pancreatic artery → Inferior pancreatic artery → Short gastric arteries → Common iliac artery → Inferior mesenteric artery → Left colic artery → Inferior phrenic artery (inferior diaphragmatic artery) → Superior suprarenal arteries → Lumbar arteries → Spinal branch (of lumbar arteries) → Middle suprarenal artery (middle capsular artery) → Ovarian artery (gonadal artery) ♀ → Renal artery → Inferior suprarenal artery → Intrarenal arteries → Arcuate arteries → Interlobular arteries (of kidney) → Interlobar arteries → Superior mesenteric artery → Ileocolic artery → Appendicular artery → Inferior pancreaticoduodenal artery → Intestinal arteries (jejunal and ileal arteries) (*from twelve to fifteen in number*) → Middle colic artery → Right colic artery → Testicular artery (internal spermatic artery or gonadal artery) ♂
[Femoral artery]** → Superficial epigastric artery
[Internal thoracic artery]** → Superior epigastric artery
[Posterior spinal artery (dorsal spinal artery)]**

ARTERIES OF THE PELVIS
[Abdominal aorta]** → [Common iliac artery]** → External iliac artery → Deep circumflex iliac artery → Inferior epigastric artery (deep epigastric artery) → Artery of round ligament of uterus ♀ → Cremasteric artery (external spermatic artery) ♂ → Superficial circumflex iliac artery → Internal iliac artery (hypogastric artery) → Iliolumbar artery → Inferior vesical artery (vesicoprostatic artery ♂ or vesicovaginal artery ♀) → Internal pudendal artery → Inferior rectal artery (inferior hemorrhoidal artery) → Lateral sacral arteries → Spinal branches (of lateral sacral arteries) → Middle rectal artery (middle hemorrhoidal artery) → Obturator artery → Superior gluteal artery → Umbilical artery → Artery to ductus deferens (artery to vas deferens) ♂ → Superior vesical arteries → Uterine artery ♀ → Vaginal artery ♀ → [Inferior mesenteric artery]** → Sigmoid arteries (left inferior colic arteries) → Superior rectal artery (superior hemorrhoidal artery) → [Ovarian artery (gonadal artery) ♀]** → Median sacral artery → Arteria lumbalis ima → [Testicular artery (internal spermatic artery or gonadal artery ♂)]**

ARTERIES OF THE PERINEUM
[Femoral artery]** → Deep external pudendal artery → Superficial external pudendal artery
[Internal iliac artery]** → Inferior gluteal artery → [Internal pudendal artery]** → Artery of bulb of penis ♂ → Artery of bulb of vestibule ♀ → Deep artery of clitoris ♀ → Deep artery of penis (artery of corpus cavernosum) ♂ → Dorsal artery of clitoris ♀ → Dorsal artery of penis ♂ → [Inferior rectal artery (inferior hemorrhoidal artery)]** → Perforating arteries of penis ♂ → Perineal artery → Urethral artery
[Ovarian artery (gonadal artery) ♀]**
[Testicular artery (internal spermatic artery or gonadal artery) ♂]**

ARTERIES OF THE UPPER LIMB
Axillary artery → Anterior circumflex humeral artery → Brachial artery → → Inferior ulnar collateral artery → → Profunda brachii artery (deep artery of arm) → → → Medial collateral artery → → → Radial collateral artery → → → Humeral nutrient arteries → → Radial artery → → → Deep palmar arch → → → → Palmar metacarpal arteries (palmar interosseous arteries) → → → Dorsal carpal branch → → → → Dorsal digital arteries (of dorsal carpal branch of radial artery) → → → → Dorsal carpal arch → → → → Dorsal metacarpal arteries (*three in number*) → → → Nutrient artery of radius → → → Princeps pollicis artery → → → Radial recurrent artery → → → Radialis indicis artery → → Superior ulnar collateral artery → → Ulnar artery → → → Common interosseous artery → → → → Anterior interosseous artery (volar interosseous artery) → → → → → Median artery → → → → Posterior interosseous artery (dorsal interosseous artery) → → → → → Recurrent interosseous artery → → → Common palmar digital arteries (common volar digital arteries) (*three in number*) → → → → Proper palmar digital arteries (collateral digital arteries) → → → Cubital anastomosis → → → Nutrient artery of ulna → → → Superficial palmar arch → → → Ulnar recurrent artery → → → → Anterior branch (of ulnar recurrent artery) → → → → Posterior branch (of ulnar recurrent artery) → Posterior circumflex humeral artery → Subscapular artery → → Circumflex scapular artery → [Thoracoacromial artery]** → → Acromial anastomosis
[Transverse cervical artery]** → Dorsal scapular artery
[Thyrocervical trunk]** → Suprascapular artery (transverse scapular artery)

ARTERIES OF THE LOWER LIMB
[External iliac artery]** → Femoral artery → → Deep artery of thigh (profunda femoris artery) → → → Lateral circumflex femoral artery (lateral circumflex artery of thigh) → → → Medial circumflex femoral artery (medial circumflex artery of thigh) → → → Perforating arteries → → → → Femoral nutrient arteries → → Descending genicular artery (supreme genicular artery) → → Popliteal artery → → → Anterior tibial artery → → → → Anterior lateral malleolar artery → → → → Anterior medial malleolar artery → → → → Anterior tibial recurrent artery → → → → Dorsalis pedis artery (dorsal artery of foot) → → → → → Arcuate artery (branch of dorsalis pedis artery) (inconstant) → → → → → Deep plantar artery → → → → → Dorsal digital arteries → → → → → Dorsal metatarsal arteries (*four in number*) → → → → → Lateral tarsal artery → → → → → Medial tarsal arteries → → → → Lateral malleolar network → → → → Posterior tibial recurrent artery (inconstant) → → → Genicular anastomosis → → → Inferior lateral genicular artery → → → Inferior medial genicular artery → → → Middle genicular artery (azygos articular artery) → → → Patellar anastomosis → → → Posterior tibial artery → → → → Calcaneal branches (of tibial posterior artery) → → → → Fibular artery (peroneal artery) → → → → → Calcaneal anastomosis → → → → → Fibular nutrient artery → → → → → Lateral malleolar branch → → → → → → Calcaneal branches (of lateral malleolar branch of fibular artery) → → → → Lateral plantar artery (external plantar artery) → → → → → Deep plantar arch (plantar arterial arch) → → → → → → Plantar metatarsal arteries (*four in number*) → → → → → → → Common plantar digital arteries → → → → → → → → Plantar digital arteries proper → → → → → Superficial plantar arch (inconstant) → → → → Medial malleolar branches → → → → Medial malleolar network → → → → Medial plantar artery (internal plantar artery) → → → → Tibial nutrient artery (the largest nutrient artery) → → → Superior lateral genicular artery → → → Superior medial genicular artery → → → Sural arteries
[Internal iliac artery]** → [Inferior gluteal artery]** → → Artery to sciatic nerve (ischiatic artery) → [Obturator artery]** → [Superior gluteal artery]**

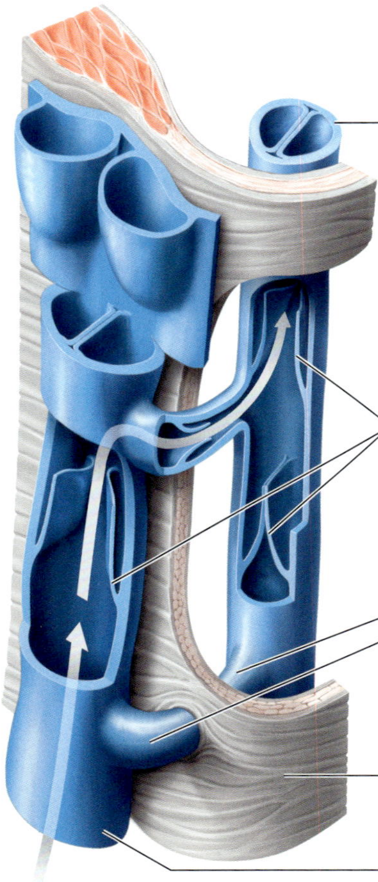

In venous circulation, blood flows from the superficial veins to the deep veins. The major superficial veins drain directly into the deep veins (for example, the great saphenous vein flows into the femoral vein). However, many perforating veins link the two venous levels, allowing **blood flow** to the deep veins along the path of the superficial veins. Direction of blood flow is regulated by valves. Venous return from the subdiaphragmatic regions to the heart is promoted by the compression exercised by muscle bellies on the deep veins that flow together into intermuscular connectival septa (muscular venous pumps) and by the suction exercised on thoracic veins during inhalation.

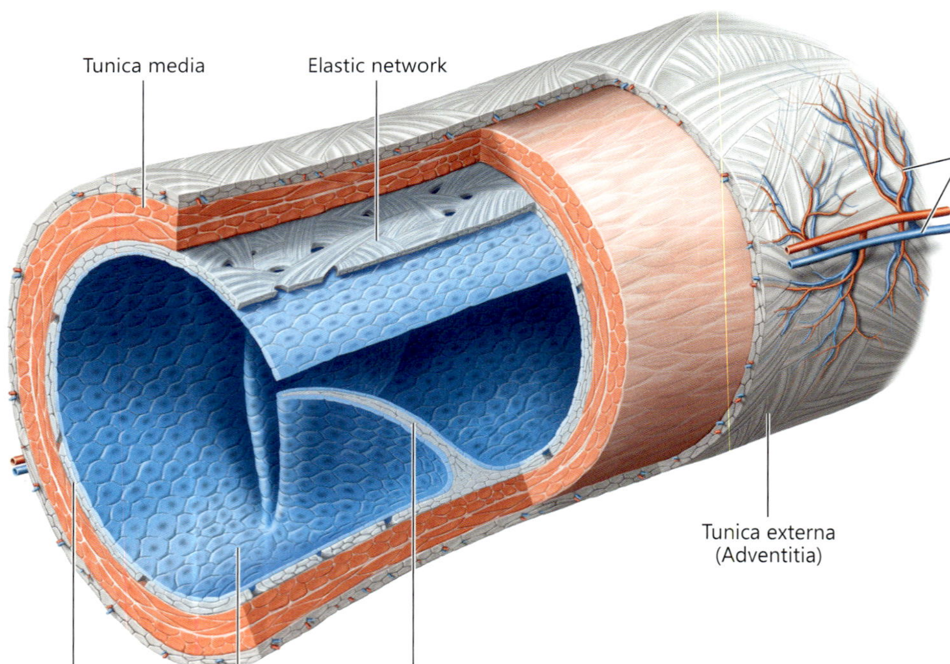

Structure of a **propulsive vein**. Propulsive veins have a tunica media formed of smooth muscle tissue. This structure enables the vein to contract and promote the propulsion of blood towards the heart.

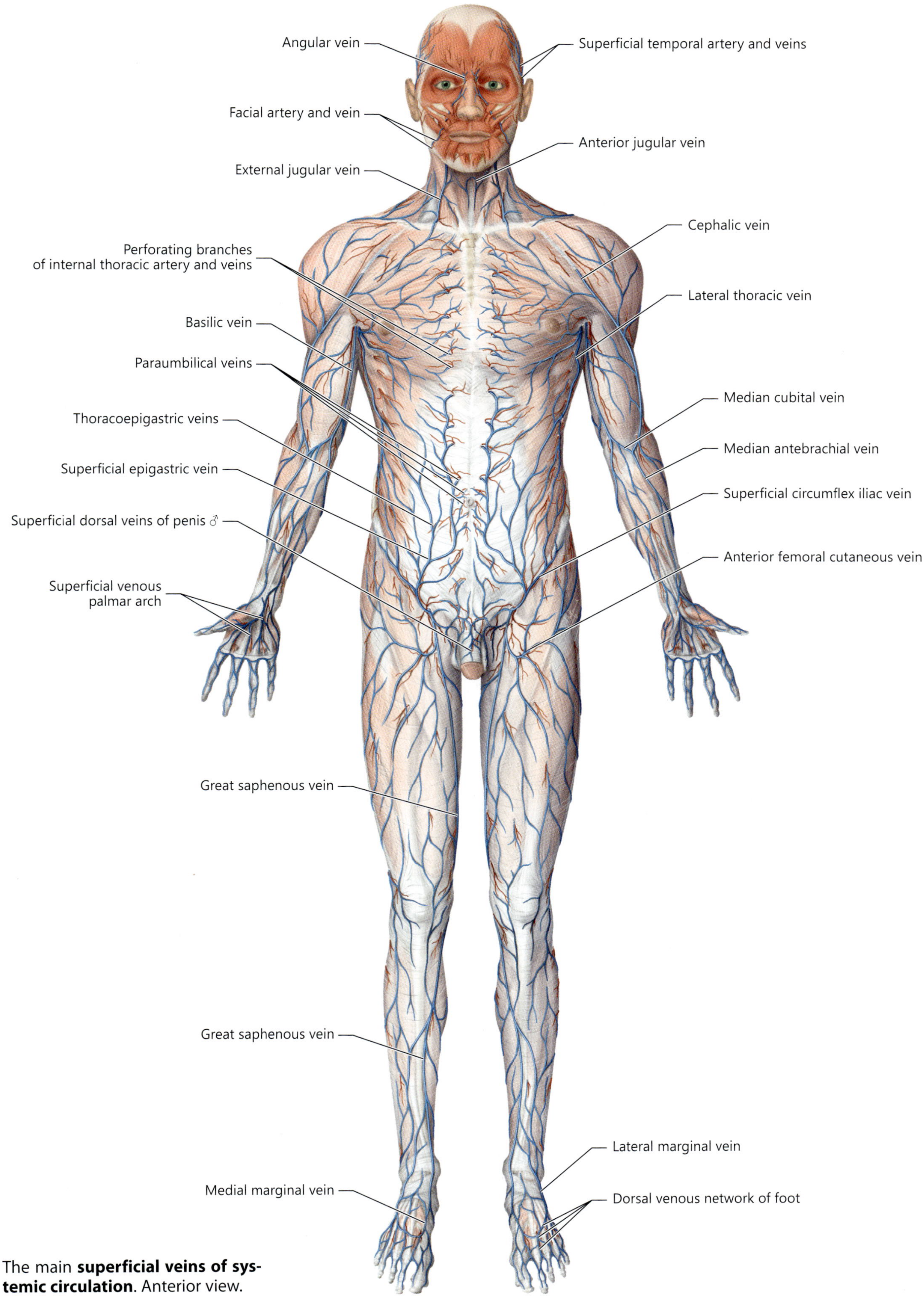

The main **superficial veins of systemic circulation**. Anterior view.

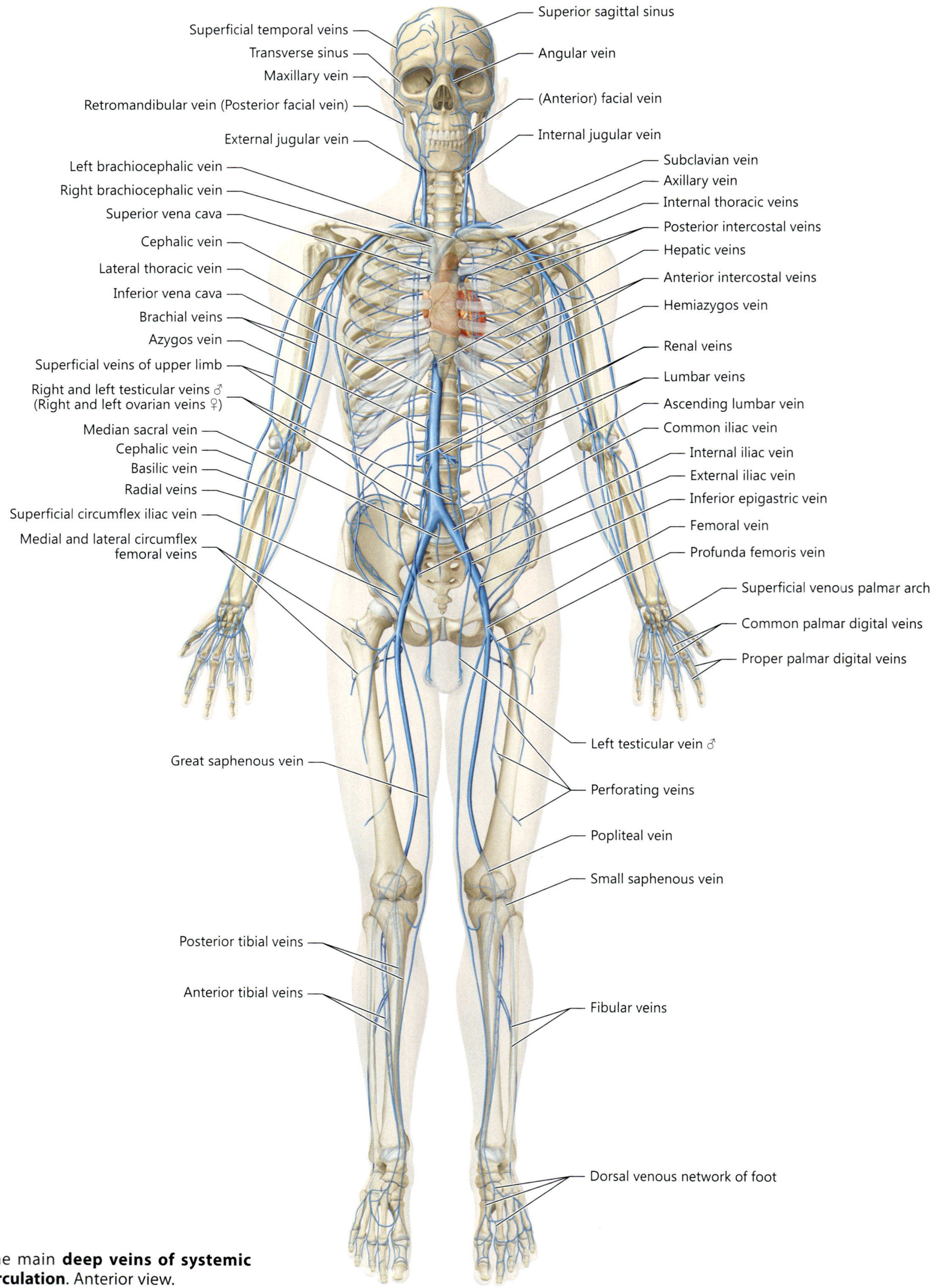

The main **deep veins of systemic circulation**. Anterior view.

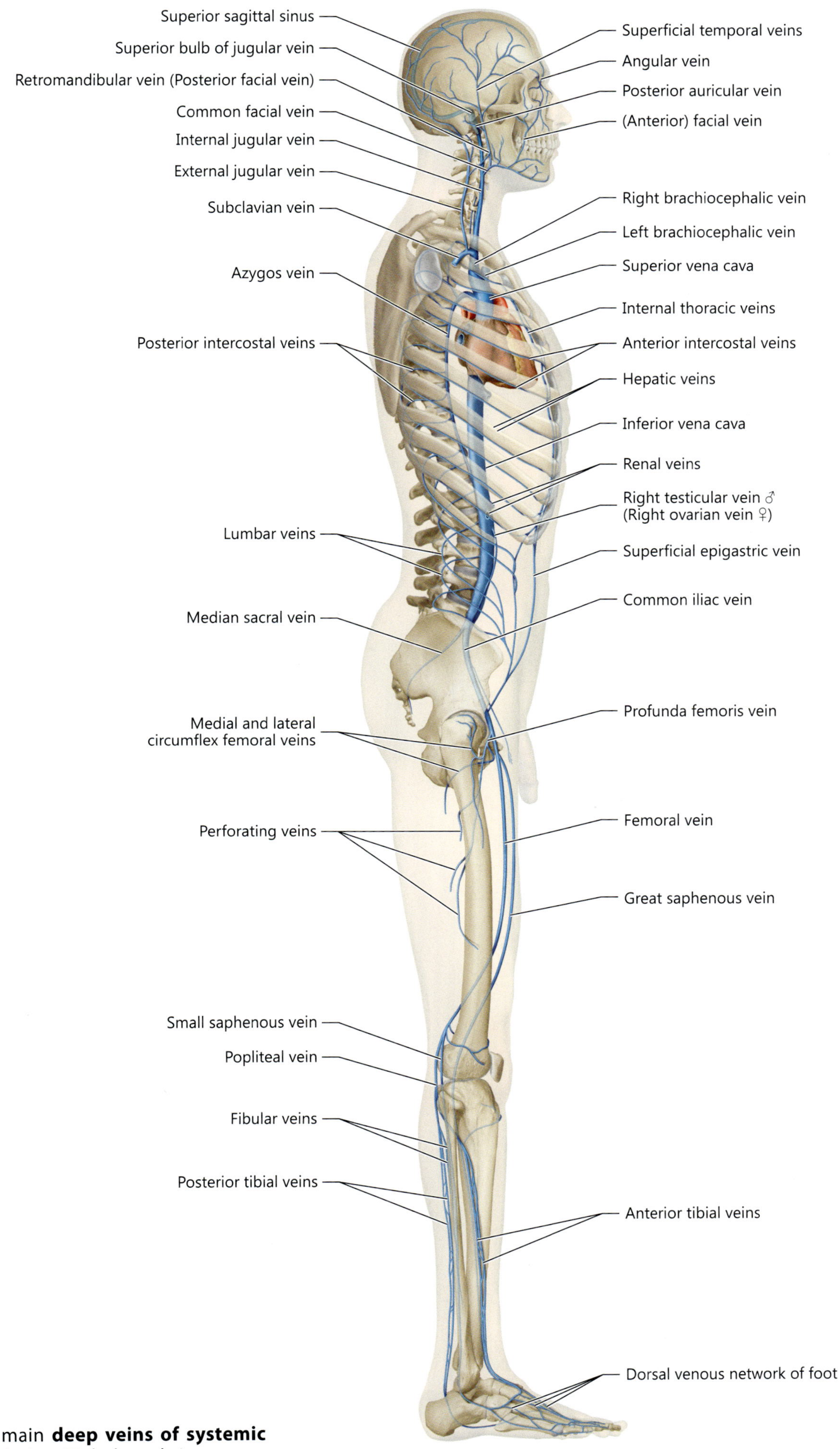

The main **deep veins of systemic circulation**. Right lateral view.

Regional subdivision of the veins (← affluence)*

VEINS OF THE HEAD

Internal jugular vein

- ← Common facial vein [by the union of the facial vein (anterior facial vein) and retromandibular vein (posterior facial vein)]
- ← External jugular vein
  - ← [Anterior jugular vein]**
  - ← Occipital veins
    - ← Mastoid emissary vein
    - ← Occipital diploic vein (*the largest of the diploic veins*)
    - ← Occipital emissary vein
  - ← Posterior auricular vein
  - ← Retromandibular vein (posterior facial vein)
    - ← Maxillary veins (internal maxillary veins)
      - ← Pterygoid plexus
        - ← Anterior auricular veins
        - ← Deep temporal veins
        - ← Emissary veins of foramen lacerum (*two or three in number*)
        - ← Emissary vein of foramen of Vesalius (vein of Vesalius or vein of sphenoidal emissary foramen)
        - ← Inferior alveolar vein (inferior dental vein)
        - ← Middle meningeal veins
        - ← Parotid veins
        - ← Stylomastoid vein
        - ← Tympanic veins
        - ← Vein of pterygoid canal (Vidian vein)
    - ← Middle temporal vein
    - ← Superficial temporal veins
      - ← Parietal emissary vein
    - ← Transverse facial vein
- ← Facial vein (anterior facial vein)
  - ← Angular vein
  - ← Deep facial vein
  - ← External nasal veins (*few in number*)
  - ← External palatine vein
  - ← Inferior labial veins
  - ← Inferior palpebral veins
  - ← Parotid veins (tributaries of facial vein)
  - ← Submental vein
  - ← Superior labial vein
  - ← Superior palpebral veins
  - ← Supraorbital vein
    - ← Frontal diploic vein
  - ← Supratrochlear veins (*generally two in number*)
- ← Inferior petrosal sinus
  - ← Labyrinthine veins (internal auditory veins)
    - ← Common modiolar vein
    - ← Vestibular veins
- ← Lingual vein
  - ← Deep lingual vein (deep vein of tongue or ranine vein)
  - ← Dorsal lingual veins
  - ← Sublingual vein
  - ← Vena comitans of hypoglossal nerve

continued

VEINS OF THE HEAD (continued)

- ← Marginal sinus
- ← Meningeal veins
- ← Transverse sinus
  - ← Confluence of sinuses (torcular herophili)
  - ← Inferior anastomotic vein (small anastomotic vein or vein of Labbé)
  - ← Inferior cerebellar veins
  - ← Occipital sinus
  - ← Petrosquamous sinus
  - ← Posterior temporal diploic vein
  - ← Sigmoid sinus
  - ← Straight sinus
    - ← Deep cerebral veins (internal cerebral veins) (*two in number*)
      - ← Superior choroid vein
    - ← Great cerebral vein (vein of Galen)
      - ← Basal vein (vein of Rosenthal)
        - ← Inferior cerebral veins (superficial cerebral veins)
        - ← Inferior choroid vein
        - ← Inferior thalamostriate veins (inferior striate veins)
    - ← Inferior sagittal sinus
    - ← Superior thalamostriate vein (superior striate vein)
  - ← Superior petrosal sinus
    - ← Cavernous sinus
      - ← Anterior intercavernous sinus
      - ← Basilar plexus
      - ← Inferior ophthalmic vein
      - ← Posterior intercavernous sinus
      - ← Superior ophthalmic vein
        - ← Anterior ciliary veins
          - ← Scleral veins
          - ← Scleral venous sinus
        - ← Central retinal vein
        - ← Ciliary veins
        - ← Episcleral veins
          - ← Conjunctival veins
          - ← Palpebral veins
        - ← Ethmoidal veins
        - ← Lacrimal vein
        - ← Nasofrontal vein
        - ← Vorticose veins (*generally four or five in number*)
    - ← Sphenoparietal sinus (sinus of Brechet)
      - ← Anterior cerebral veins
      - ← Anterior temporal diploic vein
    - ← Superior sagittal sinus
      - ← Superficial middle cerebral vein (superficial Sylvian vein)
      - ← Superior anastomotic vein (great anastomotic vein or vein of Trolard)
      - ← Superior cerebral veins (superficial cerebral veins) (*from eight to twelve in number*)
        - ← Frontal veins
    - ← Vein of vestibular aqueduct
- ← Vein of cochlear aqueduct

* Certain veins, branches, and venous plexuses not mentioned here are shown in the tables attached to the body sections of the Atlas.

[...]** The veins indicated originate in other anatomical regions.

*** External vertebral venous plexuses communicate with the internal vertebral venous plexuses by way of the intervertebral and basivertebral veins and also with the posterior intercostal, lumbar, and lateral sacral veins.

VEINS OF THE NECK

[Brachiocephalic vein]**
← Inferior laryngeal vein
← Inferior thyroid vein
← [Internal jugular vein]**
  ← External jugular vein
    ← Anterior jugular vein (*generally two in number*)
      ← Jugular venous arch
    ← Posterior external jugular vein
    ← Transverse cervical veins (transverse veins of neck)
  ← Middle thyroid veins
  ← Pharyngeal veins (*few in number*)
  ← Sternocleidomastoid vein (sternomastoid vein)
  ← Superior laryngeal vein
  ← Superior thyroid vein
← Posterior jugular vein
← Vertebral vein
  ← Accessory vertebral vein (inconstant)
  ← Anterior external vertebral venous plexus***
  ← Anterior vertebral vein
    ← Ascending cervical vein
    ← Deep cervical vein
  ← Posterior external vertebral venous plexus***
  ← Suboccipital venous plexus

VEINS OF THE THORAX

Brachiocephalic vein
← Bronchial veins (tributaries of brachiocephalic vein)
← [Internal jugular vein]**
← Internal thoracic veins (internal mammary veins)
  ← Anterior intercostal veins (*twelve pairs*)
  ← Musculophrenic veins
  ← Superior epigastric veins
  ← Tracheal veins (*few in number*)
← Left superior intercostal vein
← Mediastinal veins (tributaries of brachiocephalic vein)
← Esophageal veins (tributaries of brachiocephalic vein) (*few in number*)
← Pericardiacophrenic veins
← Pericardial veins (tributaries of brachiocephalic vein) (*few in number*)
← Subclavian vein
  ← [Axillary vein]**
    ← Lateral thoracic vein (long thoracic vein)
      ← Costoaxillary veins
      ← Thoracoepigastric veins
    ← Supreme thoracic vein
  ← Pectoral veins
← Supreme intercostal vein
← Thymic veins

continued

VEINS OF THE THORAX (continued)
[Cephalic vein]** ← Thoracoacromial vein (acromiothoracic vein)

Coronary sinus (large venous channel about 2.25 cm in length)
← Anterior veins of right ventricle (anterior cardiac veins or innominate cardiac veins or veins of Viesseuns) (*from three to four in number*)
← Great cardiac vein
  ← Anterior interventricular vein
  ← Left marginal vein
← Middle cardiac vein (posterior interventricular vein)
← Oblique vein of left atrium (vein of Marshall)
← Posterior vein(s) of left ventricle
← Small cardiac vein (right coronary vein)
  ← Right marginal vein
← Smallest cardiac veins (veins of Thebesius)
  ← Left atrial veins
  ← Left ventricular veins
  ← Right atrial veins
  ← Right ventricular veins

[Inferior vena cava (*the largest vein*)]**
Left inferior pulmonary vein
Left superior pulmonary vein
Right inferior pulmonary vein
Right superior pulmonary vein

Superior vena cava (*the second largest vein*)
← [Azygos vein]**
  ← Accessory hemiazygos vein (superior hemiazygos vein)
    ← First six or seven left posterior intercostal veins
  ← Bronchial veins (tributaries of azygos vein)
  ← [Hemiazygos vein (left azygos vein or inferior hemiazygos vein)]**
    ← Last five or six left posterior intercostal veins
  ← Mediastinal veins (tributaries of azygos vein)
  ← Esophageal veins (tributaries of azygos vein) (*few in number*)
  ← Pericardial veins (tributaries of azygos vein) (*few in number*)
  ← Right and left posterior intercostal veins
    ← Costoaxillary veins
    ← Intervertebral vein
      ← Anterior internal vertebral venous plexus (of Batson)
        ← Anterior spinal veins
        ← Basivertebral veins
        ← Posterior spinal veins
      ← Posterior internal vertebral venous plexus (of Batson)
  ← Right superior intercostal vein
  ← Superior phrenic veins

VEINS OF THE ABDOMEN

Azygos vein
- ← Ascending lumbar vein
  - ← Lumbar veins
- ← Subcostal vein

[Great saphenous vein (long saphenous vein)]**
- ← Superficial circumflex iliac vein
- ← Superficial epigastric vein

[Hemiazygos vein (inferior hemiazygos vein)]**
- ← Left ascending lumbar vein
  - ← Lumbar veins

Hepatic portal vein (about 8 cm in length)
- ← Cystic vein
- ← Inferior mesenteric vein
  - ← Left colic vein
  - ← Sigmoid veins (*few in number*)
- ← Left gastric vein (coronary vein of stomach)
- ← Paraumbilical veins (veins of Sappey)
- ← Posterior superior pancreaticoduodenal vein
- ← Prepyloric vein (vein of Mayo)
- ← Right gastric vein (pyloric vein)
- ← Splenic vein (lienal vein)
  - ← Left gastro-omental vein (left gastroepiploic vein)
  - ← Pancreatic veins (tributaries of splenic vein)
  - ← Short gastric veins (*four or five in number*)
- ← Superior mesenteric vein
  - ← Ileal veins (*generally from ten to fifteen in number,* together with jejunal veins)
  - ← Ileocolic vein
    - ← Appendicular vein
  - ← Jejunal veins (*generally from ten to fifteen in number,* together with ileal veins)
    - ← Right gastro-omental vein (right gastroepiploic vein)
  - ← Middle colic vein
  - ← Pancreatic veins (tributaries of superior mesenteric vein)
  - ← Pancreaticoduodenal veins
  - ← Right colic vein
- ← Umbilical vein

Inferior vena cava (*the largest vein*)
- ← Common iliac vein
  - ← [External iliac vein]**
    - ← Inferior epigastric vein (deep epigastric vein)
- ← Inferior phrenic veins
- ← Lumbar veins (*generally four in number in each side*)
- ← Major hepatic veins (right, intermediate and left hepatic veins)
  - ← Sublobular veins
    - ← Centrolobular veins (central veins of hepatic lobules)
      - ← Interlobular veins
- ← Minor hepatic veins
  - ← Sublobular veins
    - ← Centrolobular veins (central veins of hepatic lobules)
      - ← Interlobular veins
- ← Renal veins
  - ← Interlobar veins
    - ← Arcuate veins
      - ← Deep cortical veins
      - ← Interlobular veins
        - ← Stellate veins (veins of Verheyen)
      - ← Straight veins
  - ← [Left ovarian vein (left genital or gonadal vein) ♀]**
  - ← Left suprarenal vein
  - ← [Left testicular vein (left genital or gonadal or internal spermatic vein) ♂]**
- ← [Right ovarian vein (right genital or gonadal vein) ♀]**
- ← Right suprarenal vein
- ← [Right testicular vein (right genital or gonadal or internal spermatic vein) ♂]**

[Internal thoracic veins]**
- ← Subcutaneous abdominal veins
- ← Superior epigastric veins

VEINS OF THE PELVIS

[Common iliac vein]**
- ← External iliac vein
  - ← Deep circumflex iliac vein
  - ← Inferior epigastric vein
    - ← Pubic vein (accessory obturator vein)
- ← Iliolumbar vein
- ← Internal iliac vein (hypogastric vein)
  - ← [Inferior gluteal veins]**
  - ← Internal pudendal vein
    - ← Inferior rectal veins (inferior hemorrhoidal veins)
  - ← Middle rectal veins (middle hemorrhoidal veins)
    - ← Rectal venous plexus
  - ← Obturator veins
  - ← Prostatic venous plexus ♂
  - ← Sacral venous plexus
    - ← Lateral sacral veins
  - ← [Superior gluteal veins]**
  - ← Uterine veins ♀
  - ← Uterine venous plexus ♀
  - ← Vaginal venous plexus ♀
  - ← Vesical veins
  - ← Vesical venous plexus

[Inferior mesenteric vein]**
- ← Superior rectal vein (superior hemorrhoidal vein)

[Inferior vena cava]**
- ← Ovarian vein (genital or gonadal vein) ♀
- ← Testicular vein (genital or gonadal or internal spermatic vein) ♂

VEINS OF THE PERINEUM

[Common iliac vein]**
- ← [Internal iliac vein (hypogastric vein)]**
  - ← Internal pudendal vein
    - ← Deep veins of clitoris ♀
    - ← Deep veins of penis ♂
    - ← Posterior labial veins ♀
    - ← Posterior scrotal veins ♂
    - ← Pudendal venous plexus
      - ← Deep dorsal vein of clitoris (unpaired vein) ♀
      - ← Deep dorsal vein of penis (unpaired vein) ♂
        - ← Cavernous veins ♂
    - ← Vein of bulb of penis ♂
    - ← Vein of bulb of vestibule ♀
    - ← Urethral veins
- ← Median sacral vein

[Great saphenous vein (long saphenous vein)]**
- ← Anterior labial veins ♀
- ← Anterior scrotal veins ♂
- ← External pudendal veins
- ← Superficial dorsal veins of clitoris ♀
- ← Superficial dorsal veins of penis ♂

VEINS OF THE UPPER LIMB

Axillary vein
← Anterior circumflex humeral vein
← Brachial veins
  ← Basilic vein
    ← Basilic vein of forearm
    ← Median antebrachial vein (median vein of forearm)
      ← Palmar venous network
        ← Proper palmar digital veins
    ← Median cubital vein (median vein of elbow)
    ← Dorsal venous network of hand
      ← Dorsal metacarpal veins
      ← Intercapitular veins
      ← Proper dorsal digital veins
  ← Radial veins
    ← Deep venous palmar arch
      ← Palmar metacarpal veins
    ← Superficial venous palmar arch
      ← Common palmar digital veins
  ← Ulnar veins
    ← Anterior and posterior interosseous veins
    ← Deep venous palmar arch
      ← Palmar metacarpal veins
    ← Superficial venous palmar arch
      ← Common palmar digital veins
← Cephalic vein
  ← Accessory cephalic vein
  ← Cephalic vein of forearm
  ← Dorsal venous network of hand
    ← Dorsal metacarpal veins
    ← Intercapitular veins
    ← Proper dorsal digital veins
  ← Thoracoacromial vein (acromiothoracic vein)
← Circumflex scapular vein
← Posterior circumflex humeral vein
← Subscapular vein

[External jugular vein]** ← Suprascapular vein (transverse vein of scapula)
[Subclavian vein]** ← Dorsal scapular vein

VEINS OF THE LOWER LIMB

[External iliac vein]**
← Femoral vein
  ← Great saphenous vein (long saphenous vein) (*the longest vein*)
    ← Accessory saphenous vein
    ← Medial marginal vein
      ← Dorsal venous arch of foot
        ← Dorsal metatarsal veins (common digital veins)
          ← Proper dorsal digital veins
      ← Dorsal venous network of foot
  ← Perforating veins of the great saphenous vein
  ← Popliteal vein
    ← Anterior tibial veins
      ← Dorsal metatarsal veins
    ← Genicular veins (veins of knee)
    ← Posterior tibial veins (*generally two in number*)
      ← Fibular veins (peroneal veins)
      ← Lateral plantar veins
        ← Deep plantar venous arch
          ← Plantar metatarsal veins
            ← Plantar digital veins
      ← Medial plantar veins
      ← Perforating veins of the great saphenous vein
    ← Small saphenous vein (short saphenous vein)
      ← Lateral marginal vein
        ← Dorsal venous arch of foot
          ← Dorsal metatarsal veins (common digital veins)
            ← Proper dorsal digital veins
        ← Dorsal venous network of foot
        ← Plantar venous arch
          ← Plantar digital veins
          ← Intercapitular veins
        ← Plantar venous network
    ← Sural veins
  ← Profunda femoris vein (deep vein of thigh)
    ← Lateral circumflex femoral veins
    ← Medial circumflex femoral veins
    ← Perforating veins

[Internal iliac vein (hypogastric vein)]** ← Inferior gluteal veins ← [Obturator veins]** ← Superior gluteal veins

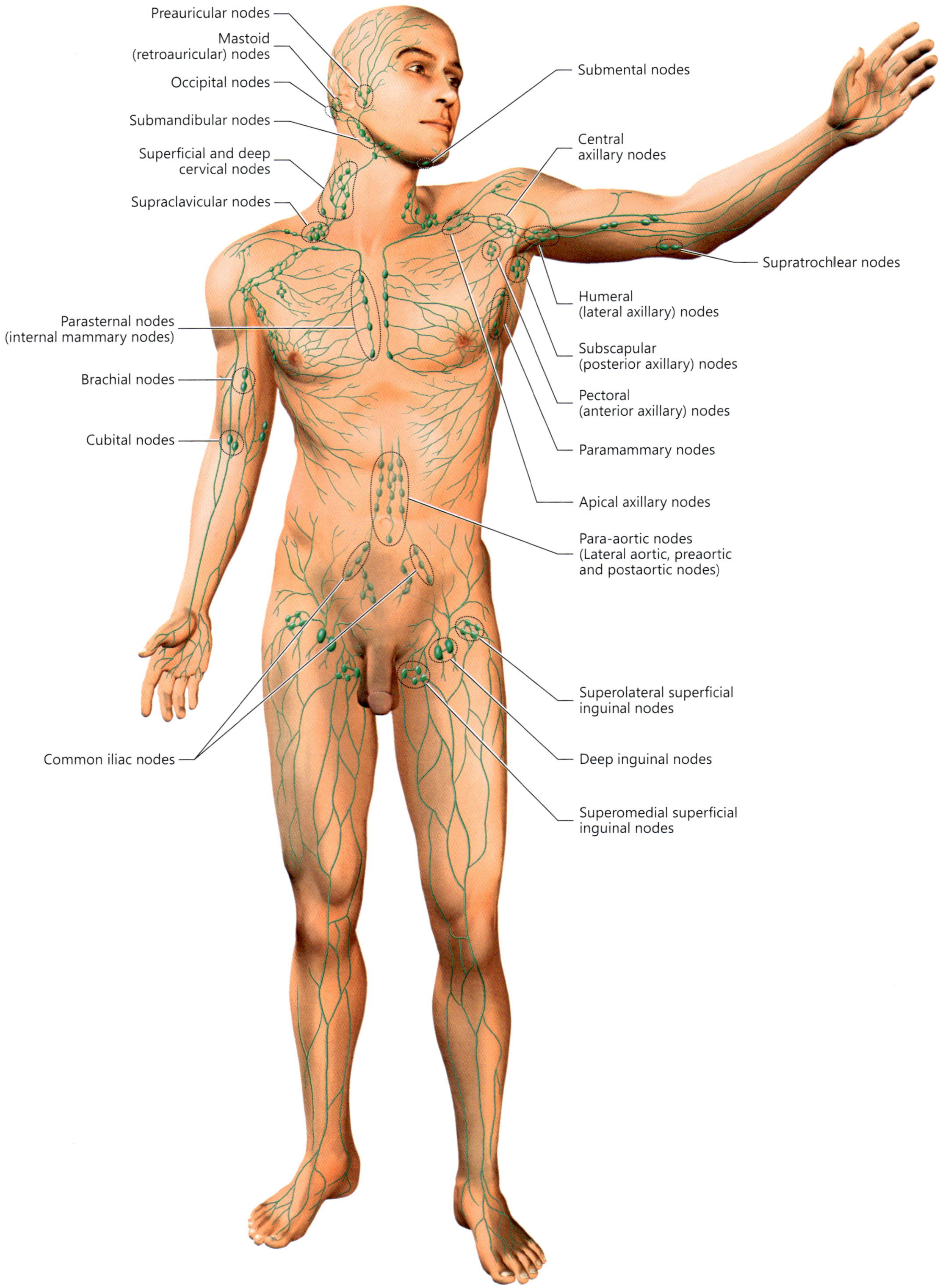

The main superficial and deep **lymph node stations**. Anterior view.

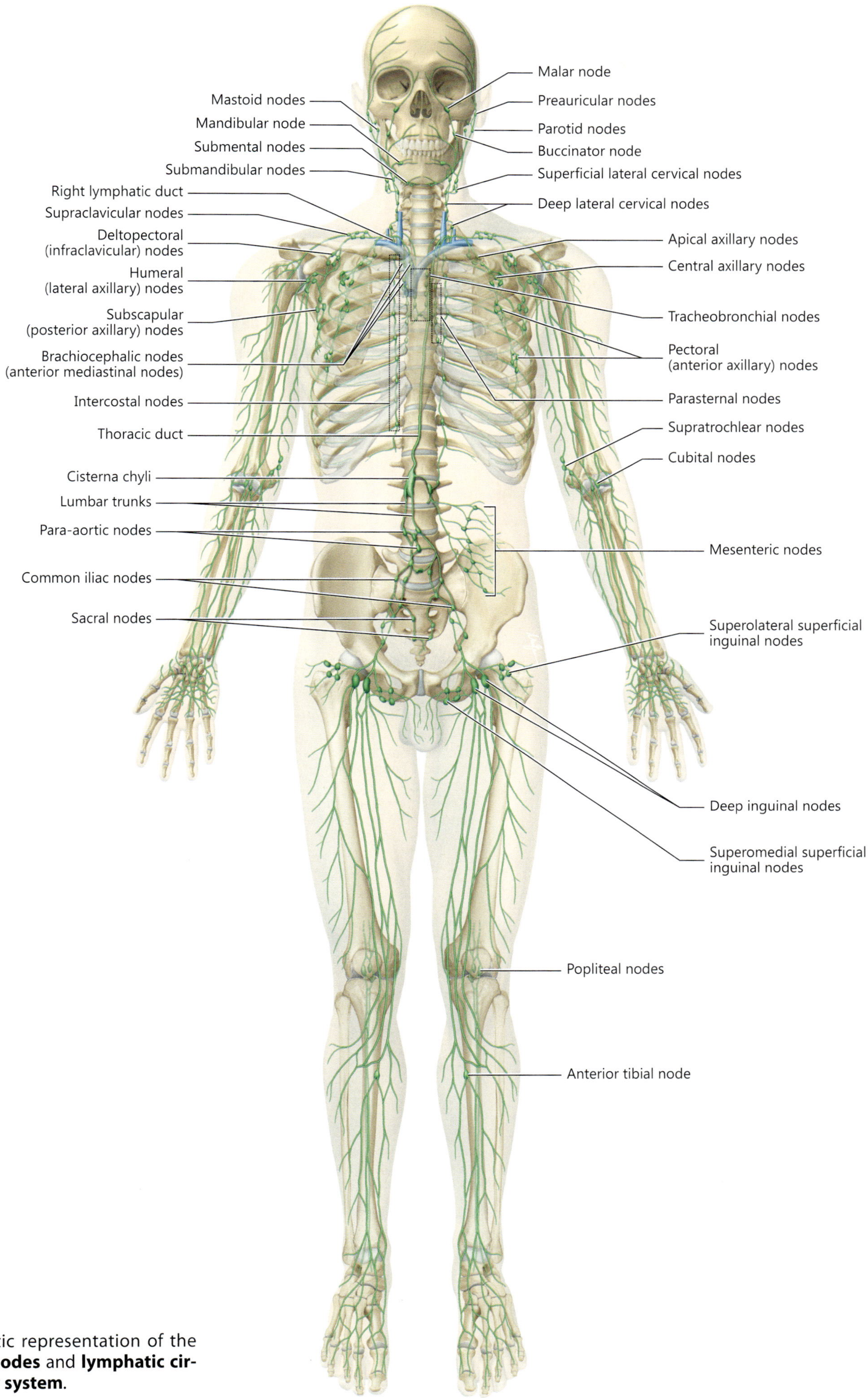

Schematic representation of the **lymph nodes** and **lymphatic circulatory system**.

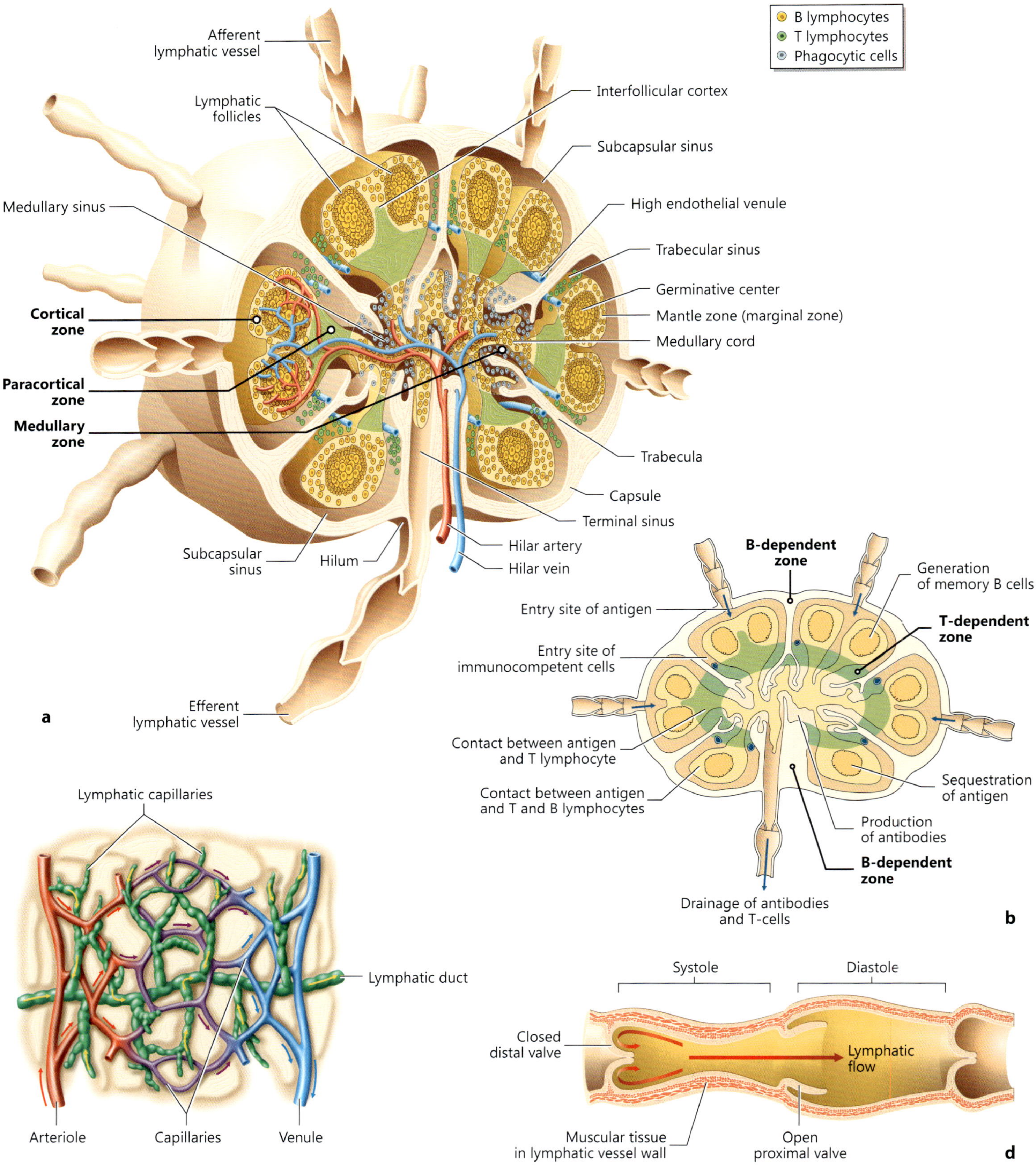

The secondary lymphatic organs comprise the **lymph nodes** along with the white pulp of the spleen and the various other MALT (mucosae associated lymphoid tissue) organs. It is inside these organs that our immune response is mounted. B and T lymphocytes occupy distinct areas (T and B dependent zones). **a**, Structural organization of the lymph node into three main zones: cortical, paracortical and medullary. **b**, Functional organization of the lymph node. **c**, Schematic representation of **microcirculation**. Structural and flow interrelations between microvascular blood and lymphatic flow: lymphatic drainage of tissue takes place through capillary lymph vessels, which begin as blind-ended vessels. **d**, Myogenic activity through contraction and release (systolic/diastolic) of the lymph vessels is responsible for lymphatic motricity and therefore transportation of lymph. The contraction of a **lymphangion** (vascular unit located between two valves) opens the valve below and allows lymph to proceed towards the next lymphangion, which will contract in turn. The valve opens when the pressure difference between the two lymphangions reaches 1-2 centimeters of water.

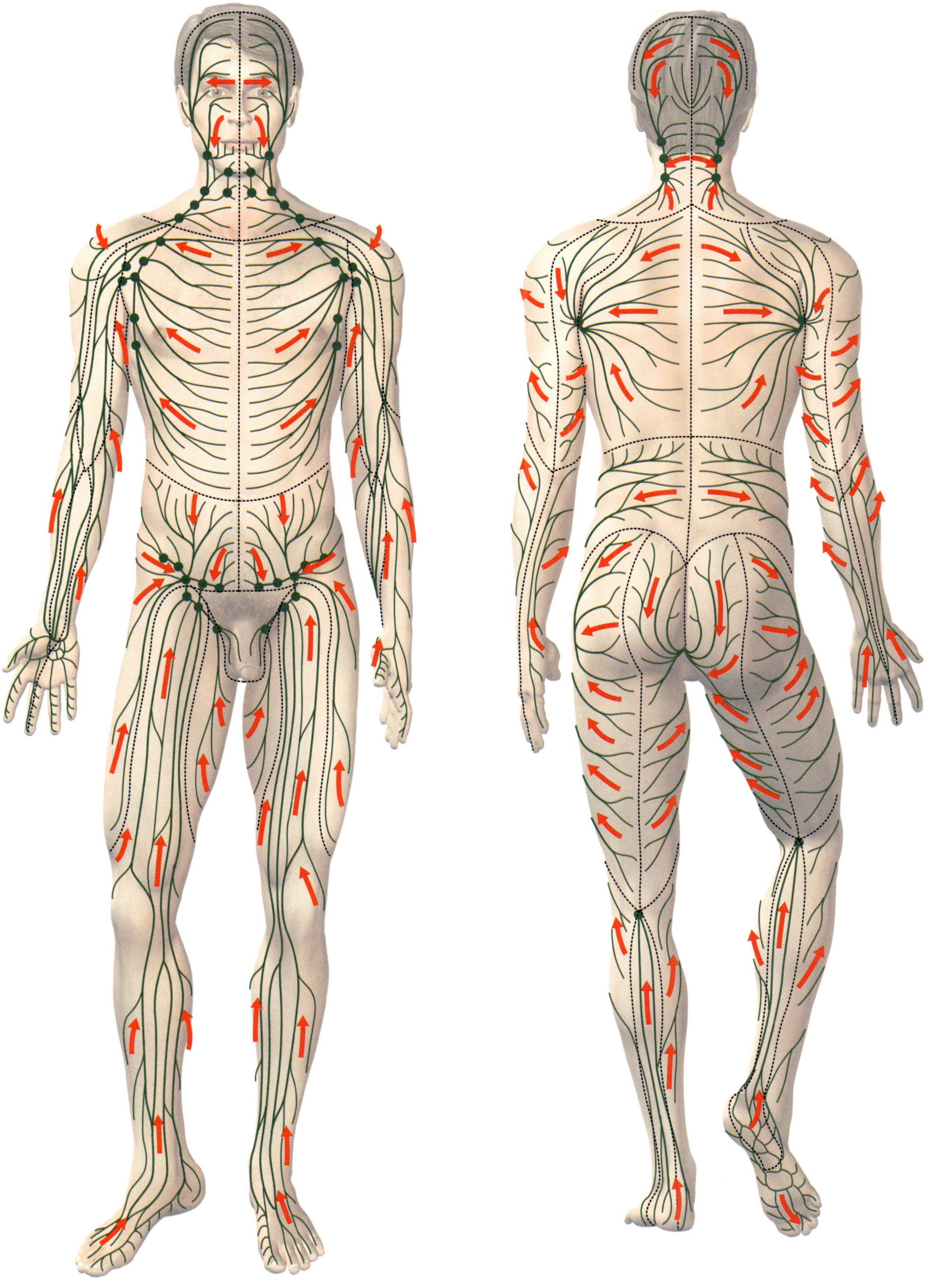

The principal courses of **lymphatic drainage**. As with venous circulation, lymph flows from superficial to deep vessels.

Regional subdivision of the lymph nodes and lymphatic trunks and ducts

LYMPH NODES OF THE HEAD AND NECK	LYMPH NODE GROUP
Anterior cervical nodes – deep nodes – – infrahyoid nodes – – – prelaryngeal nodes – – paratracheal nodes (cervical laterotracheal nodes) – – pretracheal nodes – – retropharyngeal nodes – – – lateral nodes – – – medial nodes – – thyroid nodes – superficial nodes (anterior jugular nodes)	Anterior cervical group Anterior visceral group, Level VI Anterior visceral group, Level VI Anterior visceral group, Level VI Retropharyngeal group Anterior visceral group, Level VI Anterior jugular group, Anterior superficial jugular chain, Vertical chain
Facial nodes (superficial nodes) – buccinator node – malar node (zygomatic node) – mandibular node (inferior maxillary or supramandibular node) – nasolabial (infraorbital node)	Facial group
Lateral cervical nodes – accessory nodes (nodes of accessory nerve) – – retropharyngeal nodes (lateral retropharyngeal nodes) – deep nodes (internal jugular nodes) – – inferior nodes – – – anterior node – – – inferior jugulo-omohyoid node – – – lateral node – – middle nodes – – – jugulo-omohyoid node – – superior nodes – – – anterior node – – – jugulodigastric nodes (subdigastric node) – – – lateral node – supraclavicular nodes (transverse cervical artery nodes)	Lateral cervical group Spinal accessory or posterior triangle group, Vertical chain, Level V, Lateral cervical triangle Lateral internal jugular group, Internal jugular chain, Vertical chain, Lateral cervical triangle Inferior jugular group, Inferior cervical chain, Level IV Anterior internal jugular group Anterior internal jugular group Lateral internal jugular group Middle jugular group, Middle cervical chain, Level III Anterior internal jugular group Superior jugular group, Superior cervical chain, Level II Anterior internal jugular group Anterior internal jugular group Lateral internal jugular group Supraclavicular or transverse cervical group, Supraclavicular inferior horizontal chain, Transverse cervical chain, Level V, Lateral cervical triangle
Lateral pharyngeal node	
Lingual nodes (inconstant) – lateral node – median node	
Mastoid nodes – retroauricular node	Mastoid group, Superior horizontal chain
Occipital nodes – deep node – superficial node	Occipital group, Superior horizontal chain

continued

LYMPH NODES OF THE HEAD AND NECK (continued)	LYMPH NODE GROUP
Parotid nodes – deep nodes – – extraglandular nodes (subfascial nodes) – – – infra-auricular nodes – – – preauricular nodes (anterior to the tragus) – – intraglandular node – superficial nodes – – suprafascial nodes – – – infra-auricular nodes – – – preauricular nodes (anterior to the tragus)	Parotid group, Superior horizontal chain
Retroesophageal node	
Sublingual nodes	Sublingual group
Submandibular nodes – intracapsular node – preglandular node – prevascular nodes – retroglandular node – retrovascular nodes	Submandibular group, Superior horizontal chain, Level I
Submental nodes – anterior node – middle node – posterior node	Submental group, Superior horizontal chain, Level I
Substernocleidomastoid nodes	
Subtrapezoid cervical nodes	
Superficial cervical nodes (external jugular nodes)	Vertical chain

PARIETAL THORACIC LYMPH NODES	LYMPH NODE GROUP
Intercostal nodes	Parietal group
Paramammary nodes	Parietal group
Parasternal nodes (sternal or internal thoracic nodes)	Parietal group
Superior (anterior, middle and posterior) diaphragmatic (phrenic) nodes	Parietal group

VISCERAL THORACIC LYMPH NODES	LYMPH NODE GROUP
Brachiocephalic nodes (anterior superior mediastinal nodes)	Anterior mediastinal group (anterior superior mediastinal group)
Bronchopulmonary nodes (pulmonary nodes) – left bronchopulmonary nodes (hilar nodes or left hilar nodes) – right bronchopulmonary nodes	Tracheobronchial group
Cardiac node	Anterior mediastinal group
Intrapulmonary nodes – interlobar nodes – lobar nodes – segmental nodes – subsegmental nodes	
Juxta-aortic nodes	Posterior mediastinal group
Juxtacardiac nodes	

continued

VISCERAL THORACIC LYMPH NODES (continued)	LYMPH NODE GROUP
Juxtaesophageal nodes (paraesophageal nodes)	Posterior mediastinal group and Station 8
Lateral pericardial nodes	
Lateral vertebral (juxtavertebral) nodes	
Lateropericardiac (juxtaphrenic) nodes	
Node of ligamentum arteriosum	
Node of arch of azygos vein	
Nodes between esophagus and aorta	Posterior mediastinal group and Station 8
Para-aortic nodes	Station 6
Paratracheal nodes (laterotracheal nodes) – superior nodes – right and left inferior nodes	Posterior mediastinal group Station 2 Tracheobronchial group and Station 4
Preaorticocarotid node (preaortic node)	Anterior mediastinal group
Prepericardial nodes – inferior nodes – superior (retrothymic) nodes	
Pretracheal nodes	
Prevascular nodes (retromanubrial nodes)	Anterior mediastinal group and Station 3
Prevertebral nodes	
Pulmonary ligament nodes	Station 9
Retrotracheal nodes	Station 3
Right, intermediate and left mediastinal nodes – anterior nodes (☞ Anterior mediastinal group) – – diaphragmatic nodes – posterior nodes (☞ Posterior mediastinal group) – – diaphragmatic nodes – superior nodes	 Station 1
Subaortic nodes	Station 5
Subcarinal nodes	Station 7
Thymic nodes	
Tracheobronchial nodes – inferior nodes (carinal nodes) – right (hilar nodes or right hilar nodes) and left superior nodes	Tracheobronchial group and posterior mediastinal group

PARIETAL ABDOMINAL LYMPH NODES	LYMPH NODE GROUP
Inferior diaphragmatic nodes	
Inferior epigastric nodes	
Para-aortic nodes (left, intermediate and right lumbar nodes) – lateral aortic nodes (para-aortic nodes, including the right lumbar lateral caval nodes) – preaortic nodes (including the right lumbar precaval nodes) – postaortic nodes (retroaortic nodes, including the right lumbar postcaval or retrocaval nodes)	 Aortic lateral group Preaortic group Retroaortic group

VISCERAL ABDOMINAL LYMPH NODES	LYMPH NODE GROUP
Celiac nodes	Preaortic group
Hepatic nodes – cystic node – node of anterior border of omental foramen	
Mesenteric nodes – inferior nodes – – sigmoid nodes – – superior rectal nodes – superior nodes – – appendicular nodes – – central superior nodes – – ileocolic nodes – – juxtaintestinal nodes – – mesocolic nodes – – – paracolic nodes – – – right/middle/left colic nodes – – prececal nodes – – retrocecal nodes	 Preaortic group Preaortic group
Pancreatic nodes – inferior nodes – superior nodes	
Pancreaticoduodenal nodes – inferior nodes – superior nodes	
Pyloric nodes – retropyloric nodes – subpyloric nodes – suprapyloric node	
Right/left gastric nodes (nodes around cardia)	
Right/left gastro-omental nodes	
Splenic nodes (pancreaticolienal nodes)	

PARIETAL PELVIC LYMPH NODES	LYMPH NODE GROUP
Common iliac nodes – intermediate nodes – lateral nodes – medial nodes – promontorial nodes – subaortic nodes	
External iliac nodes – interiliac nodes – intermediate lacunar node – intermediate nodes – lateral lacunar node – lateral nodes – medial lacunar node – medial nodes – obturator nodes	
Internal iliac nodes – gluteal nodes – – inferior nodes – – superior nodes – sacral nodes	

VISCERAL PELVIC LYMPH NODES	LYMPH NODE GROUP
Pararectal nodes (anorectal nodes)	
Parauterine nodes ♀	
Paravaginal nodes ♀	
Paravesical nodes – lateral vesical nodes – prevesical nodes – postvesical nodes (retrovesical nodes)	

LYMPH NODES OF THE UPPER LIMB	LYMPH NODE GROUP
Anterior and posterior interosseous lymph nodes	
Axillary lymph nodes – apical nodes – central nodes – humeral nodes (lateral nodes) – pectoral nodes (anterior nodes) – subscapular nodes (posterior nodes)	 Apical group (subclavicular group) Central group Lateral group (brachial group) Anterior group (pectoral or thoracic group) Posterior group (subscapular group)
Brachial nodes	
Cubital nodes – supratrochlear nodes (epitrochlear nodes)	
Deep and superficial palmar lymphatic arches	
Deep nodes	
Deltopectoral nodes (infraclavicular nodes)	
Interpectoral nodes (retropectoral nodes)	
Radial lymph nodes	
Superficial nodes	

LYMPH NODES OF THE LOWER LIMB	LYMPH NODE GROUP
Anterior tibial node	
Fibular node (peroneal node)	
Inguinal lymph nodes – deep inguinal nodes – – distal node – – intermediate node – – proximal node – superficial inguinal nodes – – inferior nodes (inferomedial and inferolateral nodes) – – superolateral nodes – – superomedial nodes	 The superficial inguinal nodes are grouped in the shape of a "T" with a vertical chain and medial and lateral oblique transverse arms Vertical chain Lateral oblique transverse arm Medial oblique transverse arm
Popliteal nodes – deep nodes – superficial nodes	
Posterior tibial node	

LYMPHATIC TRUNKS AND DUCTS
Intestinal trunks
Right and left bronchomediastinal trunks
Right and left jugular trunks
Right and left lumbar trunks
Right and left subclavian trunks
Right lymphatic duct (right thoracic duct)
Thoracic duct

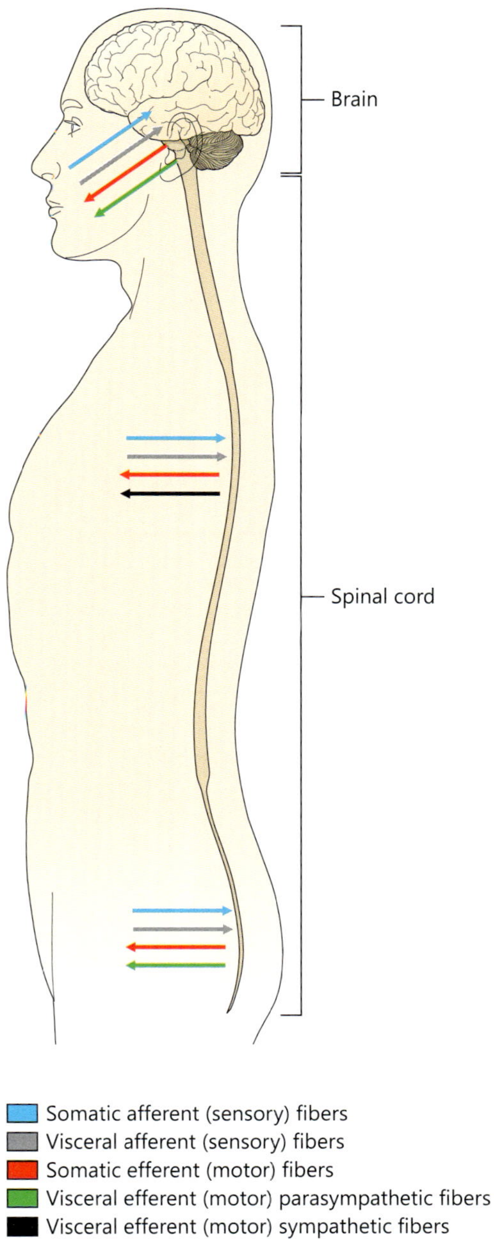

The **neuraxis** consists of the **brain** and the **spinal cord**. The arrows indicate the direction of transmission in the peripheral nervous system from the periphery towards the neuraxis (afferent signals) and from the neuraxis towards the periphery (efferent signals). The different colors indicate different kinds of input or output stimuli from the neuraxis.

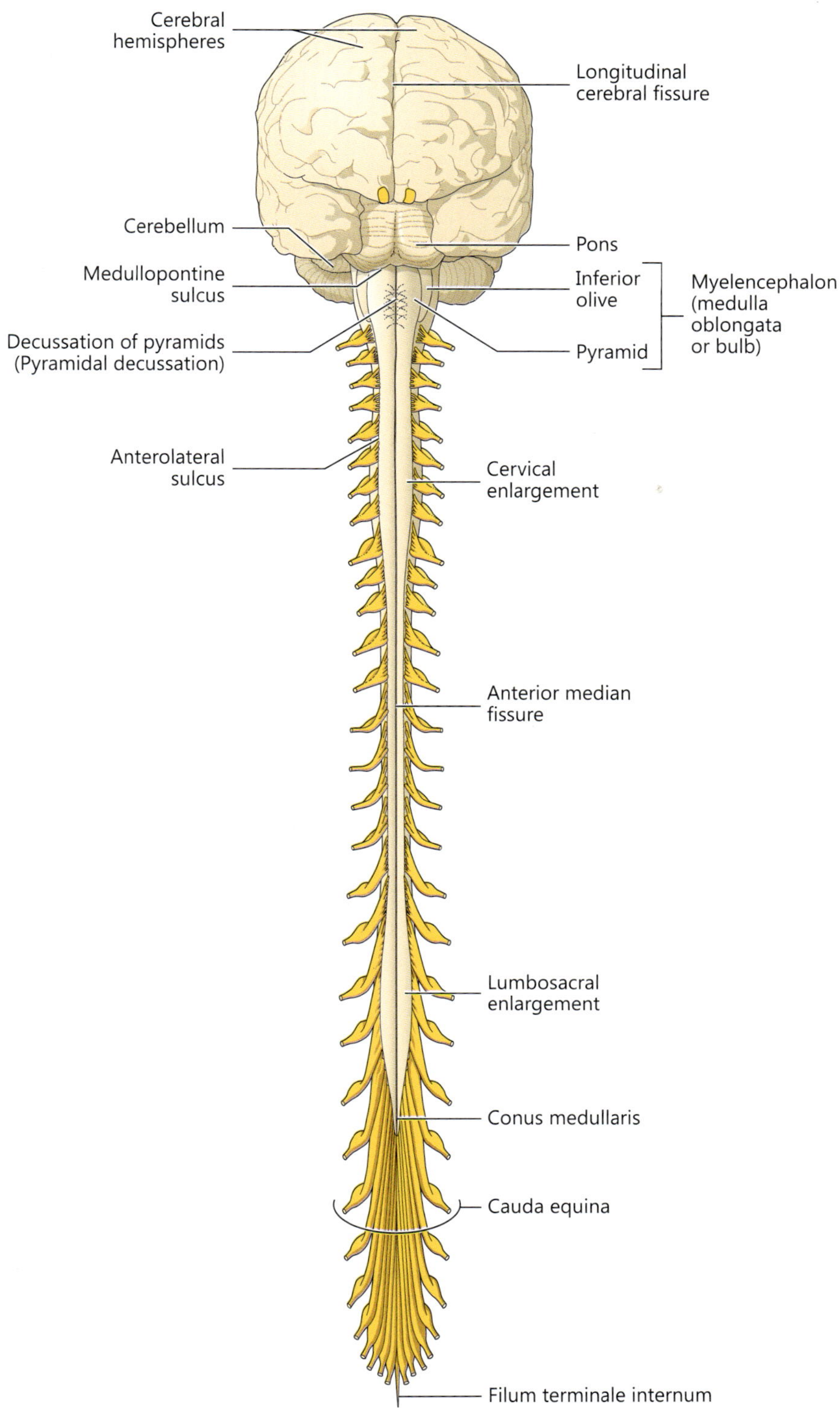

Schematic representation of the **neuraxis**, anterior projection. The spinal cord swells at the cervical and lumbar segments, and terminates in a narrower portion called the conus medullaris. Anteriorly, it is possible to see an anterior longitudinal fissure: the anterior median fissure, and two lateral longitudinal sulci: the anterolateral sulci, from which the anterior roots of the spinal nerves emerge.

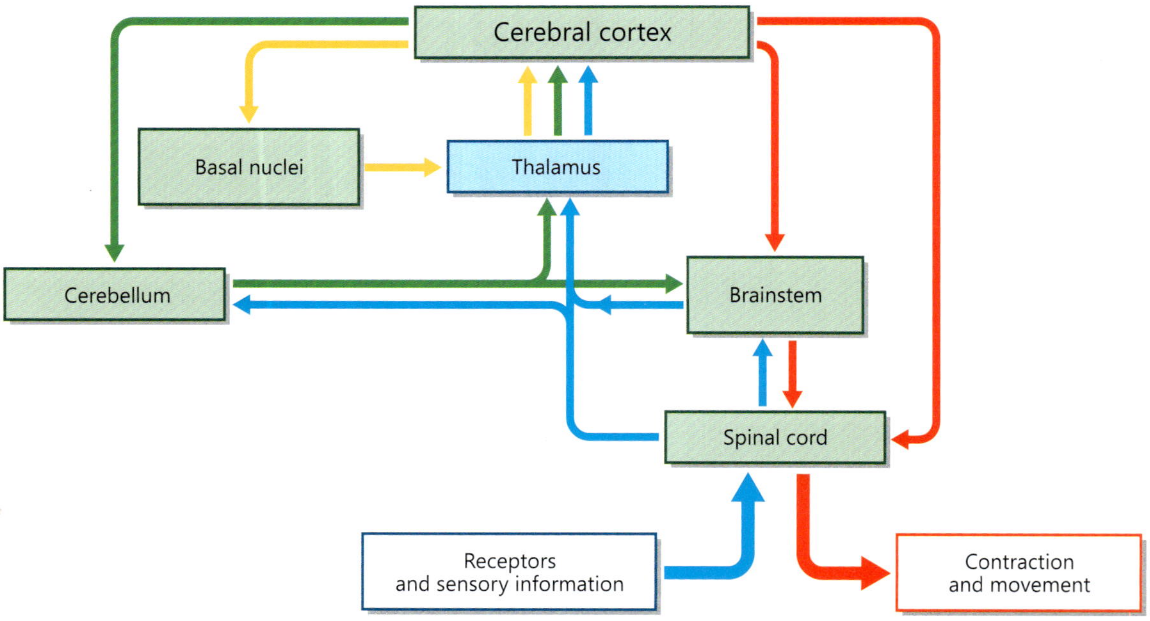

Diagrammatic overview of the main circuit pathways of the **central nervous system**. In detail, this shows the ascending or sensory pathways (in *blue*) which travel from the receptors to the cerebral cortex, and the motor pathways (in *red*) which travel from the cerebral cortex, through the spinal cord, to the effectors. When dealing with sensory information, the cerebral cortex uses basal nuclei circuits (in *yellow*) to select and initiate motor commands, and the circuit paths of the cerebellum (in *green*) to fine-tunes and corrects these even in mid-execution.

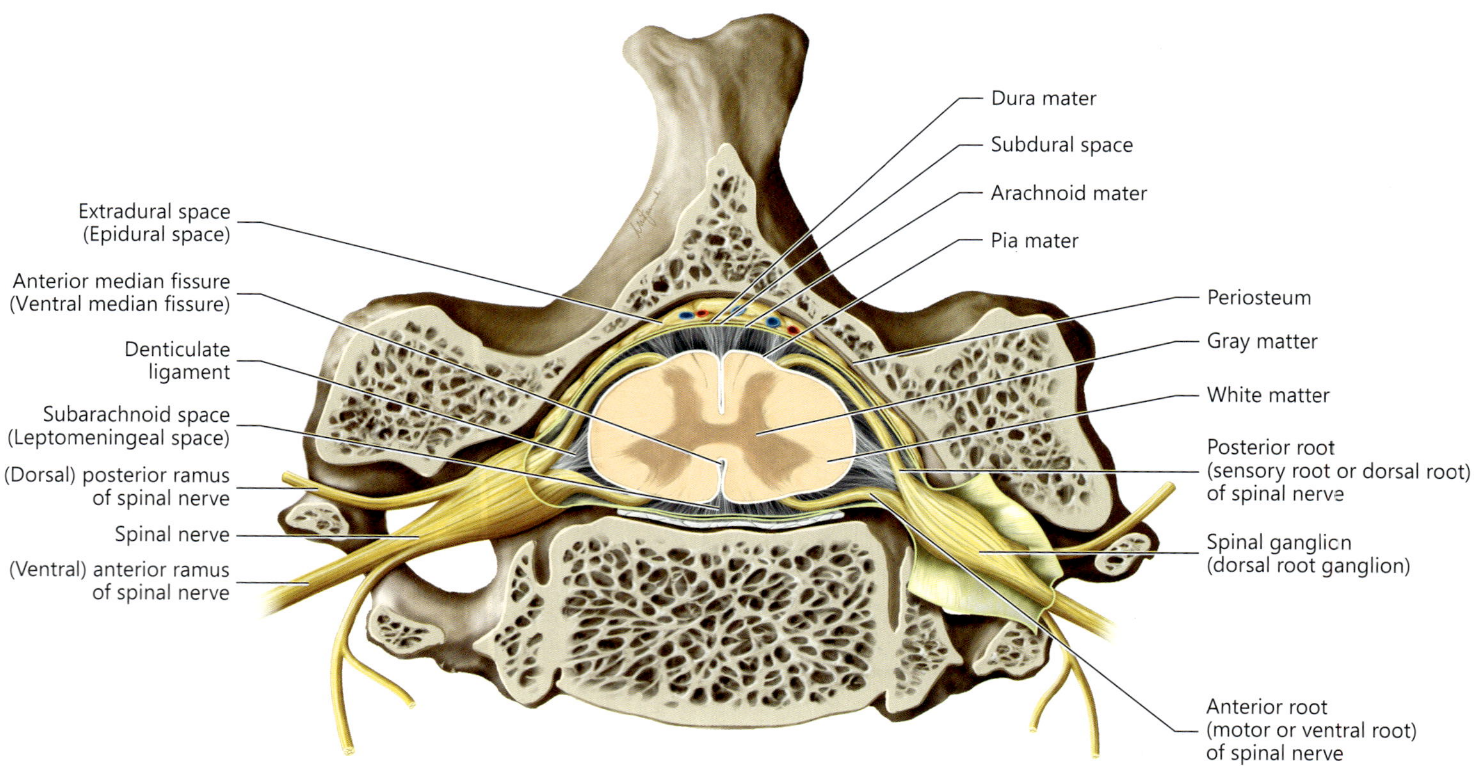

Transverse section of the vertebral column illustrating the **position and relationships of the spinal cord** within the vertebral canal. The meningeal sheath envelops the spinal cord and the root structures of the spinal nerves. The spinal ganglion is located in correspondence of the intervertebral foramen. On exiting the foramen, each spinal nerve divides into an anterior and a posterior ramus.

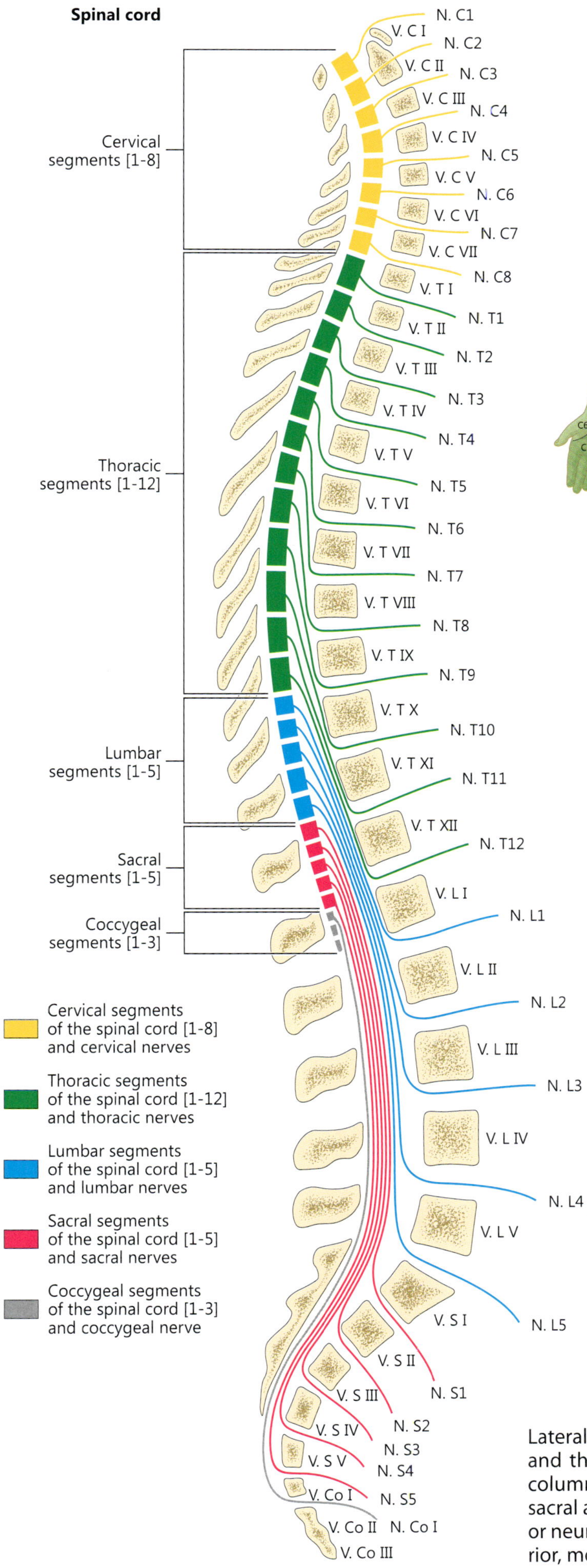

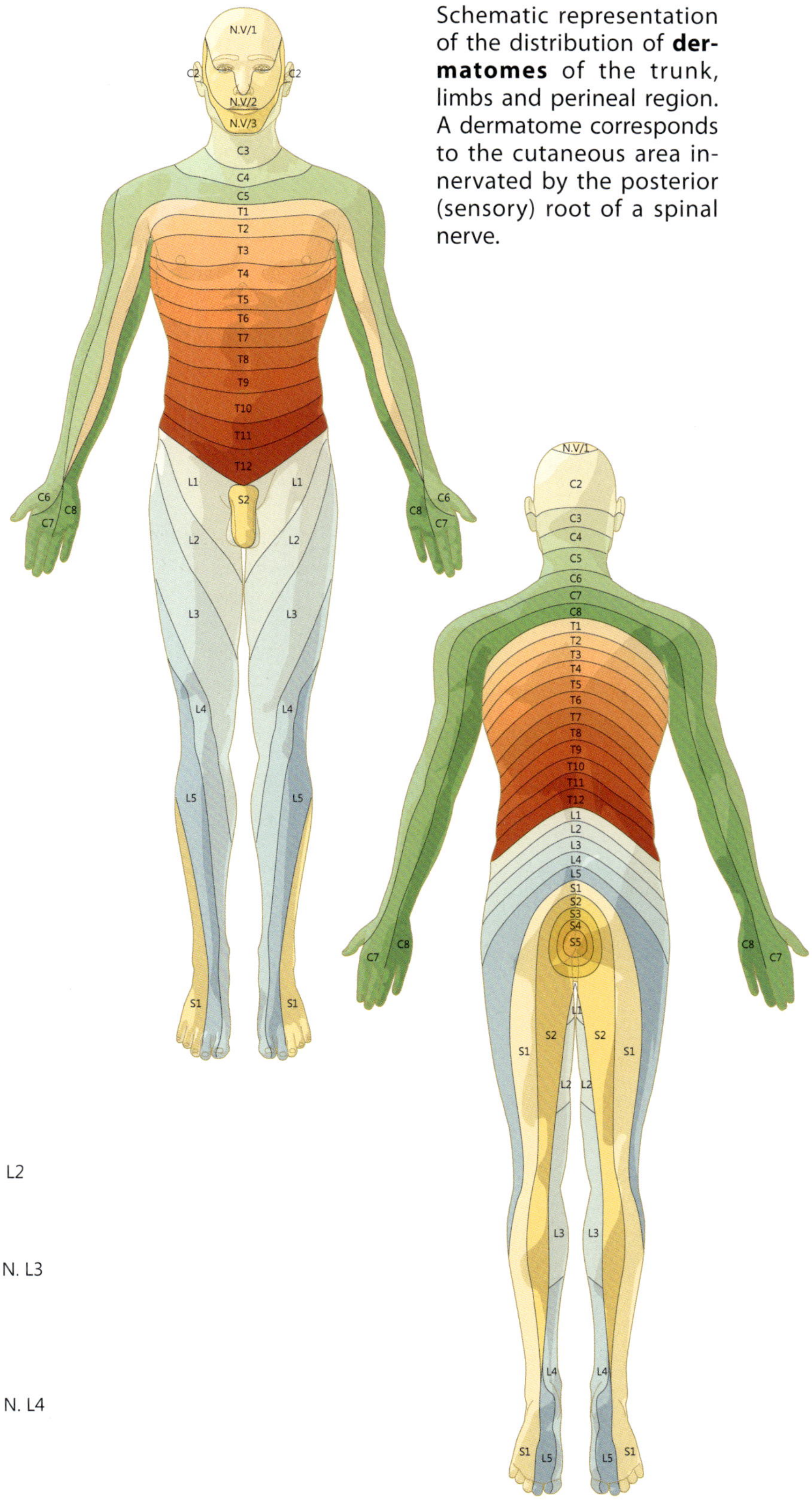

Schematic representation of the distribution of **dermatomes** of the trunk, limbs and perineal region. A dermatome corresponds to the cutaneous area innervated by the posterior (sensory) root of a spinal nerve.

Lateral view illustrating the relationships of the **spinal cord** segments and the **spinal nerves** with the vertebral column. Like the vertebral column, the spinal cord is divided into the cervical, thoracic, lumbar, sacral and coccygeal sections. Each section includes different segments, or neuromeres, which correspond to the bilateral emergence of an anterior, motor root, and the entry of a posterior, sensory root.

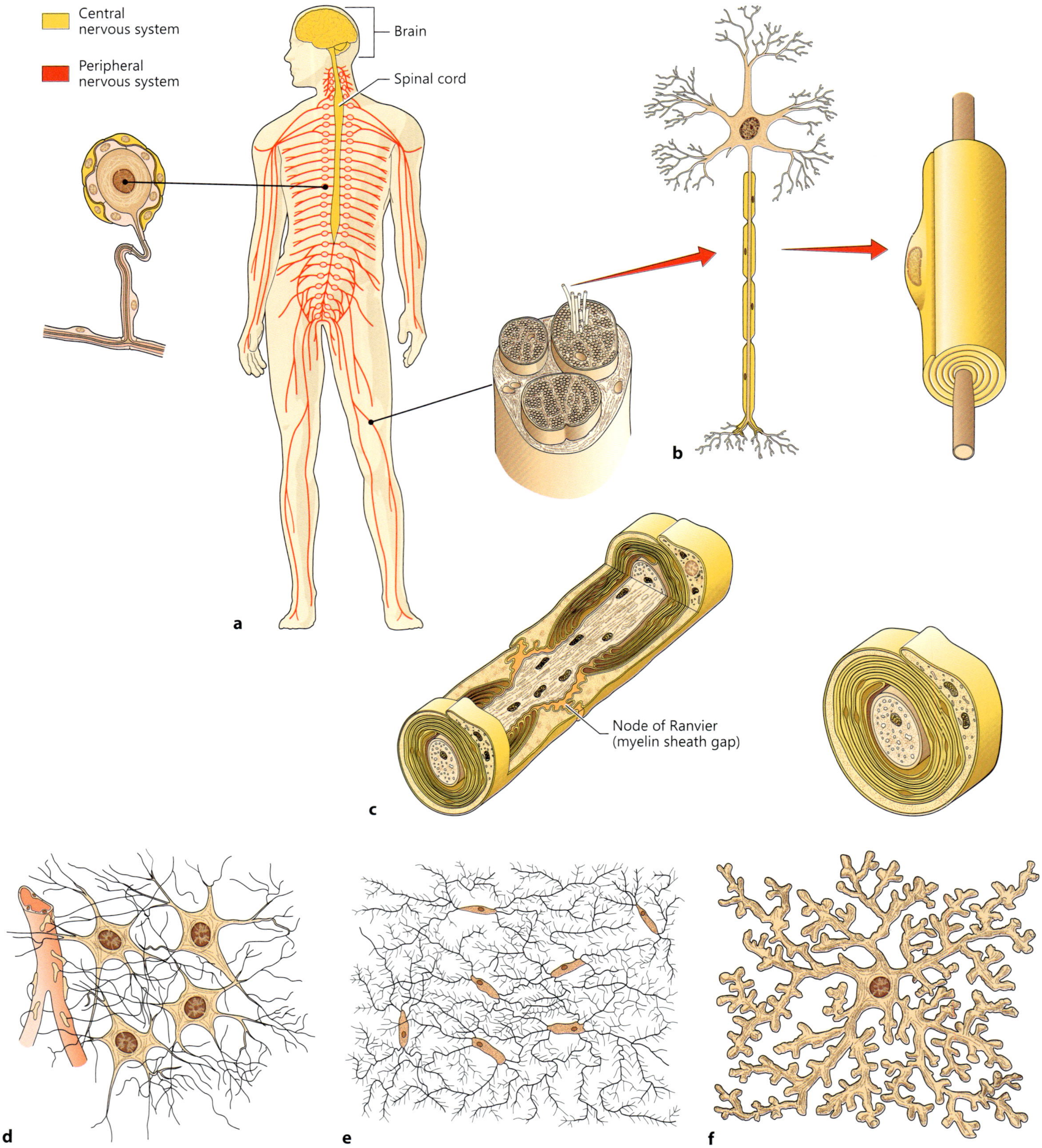

a, Schematic organization of the **central nervous system** (in *yellow*) and of the **peripheral nervous system** (in *red*). This diagram shows a pseudounipolar sensory neuron, where the process from the cell body divides into a T, thus sending out two collateral branches: one longer, towards the periphery or the receptors, and one shorter, projecting to the neuraxis. **b**, Schematic representation of a **multipolar neuron**. Numerous dendrites emerge from the soma while a single neurite or axon emerges from the opposite pole, in a downward direction. The axon is enveloped in a myelin sheath, as shown in detail on the left. **c**, Three-dimensional detailed representation of an **axon** surrounded by its myelin sheath and of a node of Ranvier. **d**, Representation of the relationship between a group of **fibrous astrocytes** and a blood vessel. **e**, Schematic diagram of a group of **microglia** cells, organized into a network. **f**, Diagram of a **protoplasmic astrocyte** whose densely ramified processes give it a rather stubby appearance.

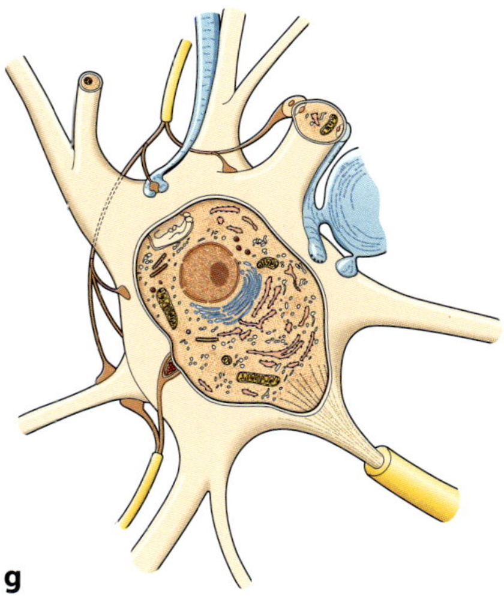

g

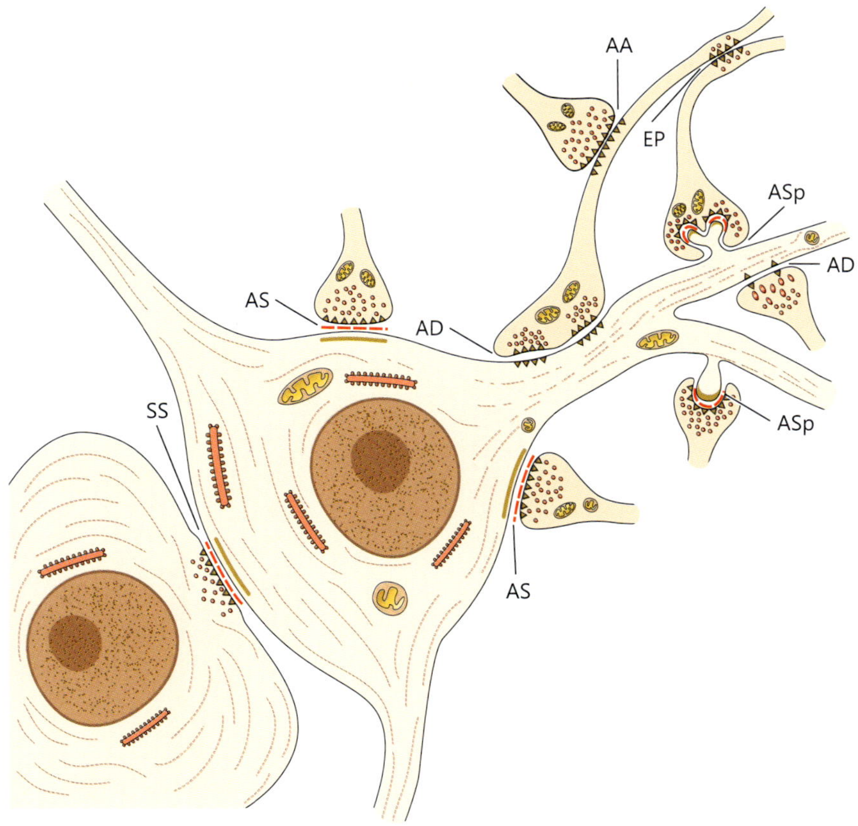

h

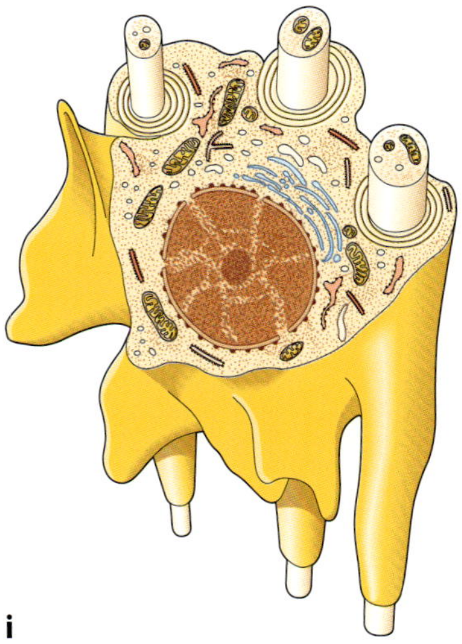

i

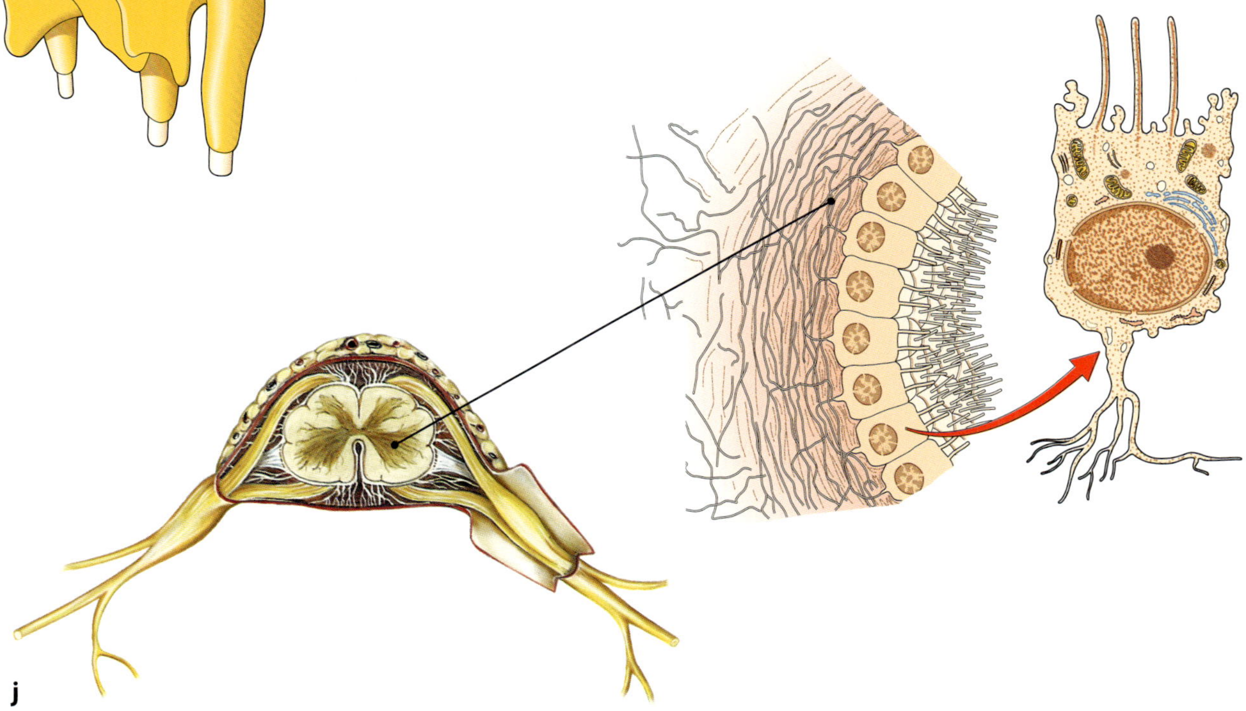

j

g, Drawing of the soma of a **multipolar neuron**, showing different types of synapses. **h**, The most common types of synapse are the axodendritic (AD) synapse, between an axon and a dendrite, the axospinous (ASp) synapse, between an axon and a dendritic spine, the axosomatic (AS) synapse, between an axon and the soma of a neuron, and the axoaxonic (AA) synapse, between an axon and the synaptic bouton of another axon. There are also somato-somatic (SS) synapses between two somas, which are very rare in vertebrates but occur frequently in cell cultures, and *en-passant* (EP) synapses, between two axons along their length rather than at the end of one of them. **i**) Drawing of an **oligodendrocyte** with short processes projecting from its soma to envelop several axons. **j**) Drawing of a transverse section of the **spinal cord**. In the enlargement, the cellular organization of the central canal of the ependyma can be seen. The detail view shows a columnar epithelial cell, with cilia facing into the lumen of the ependymal canal.

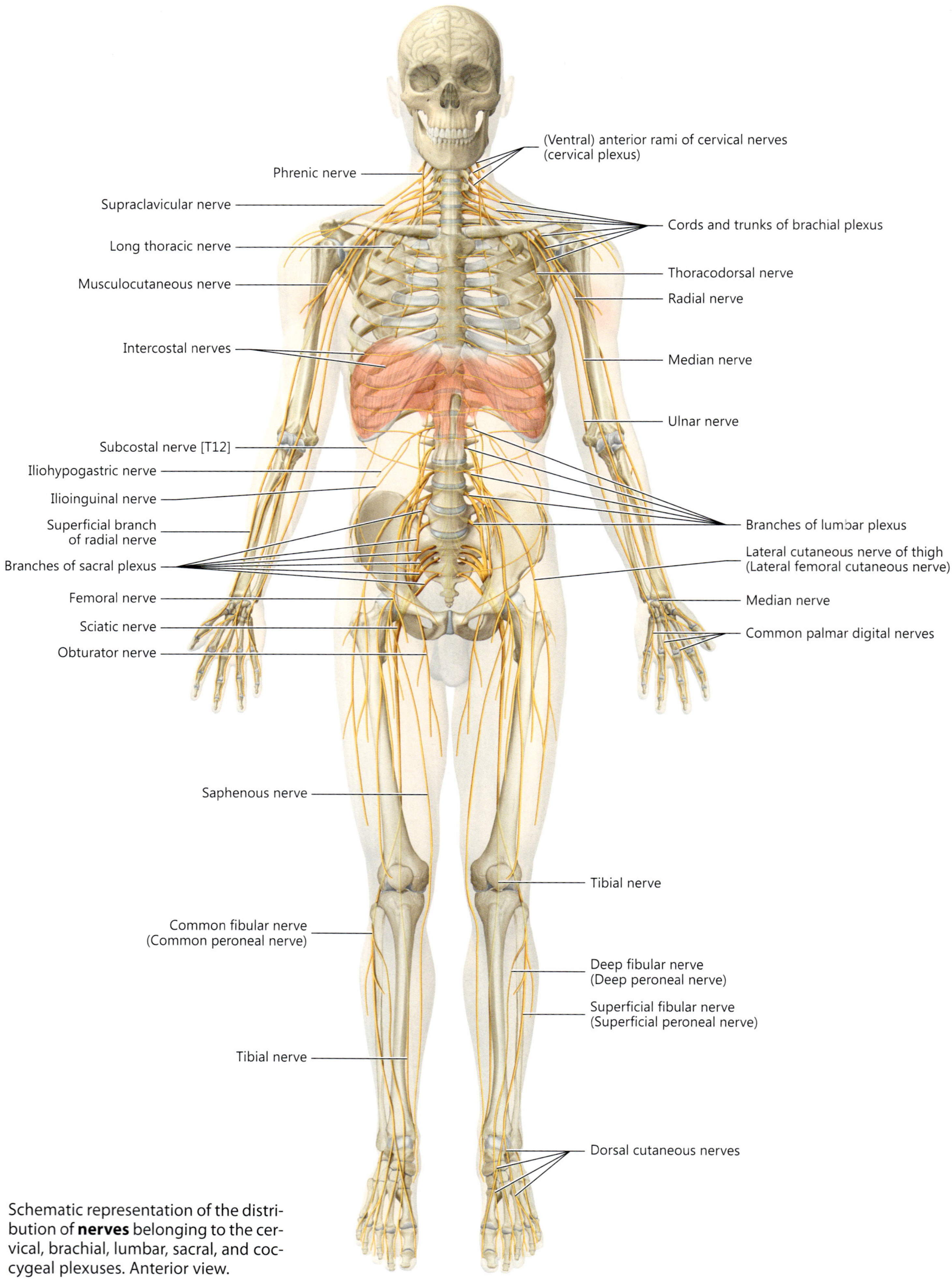

Schematic representation of the distribution of **nerves** belonging to the cervical, brachial, lumbar, sacral, and coccygeal plexuses. Anterior view.

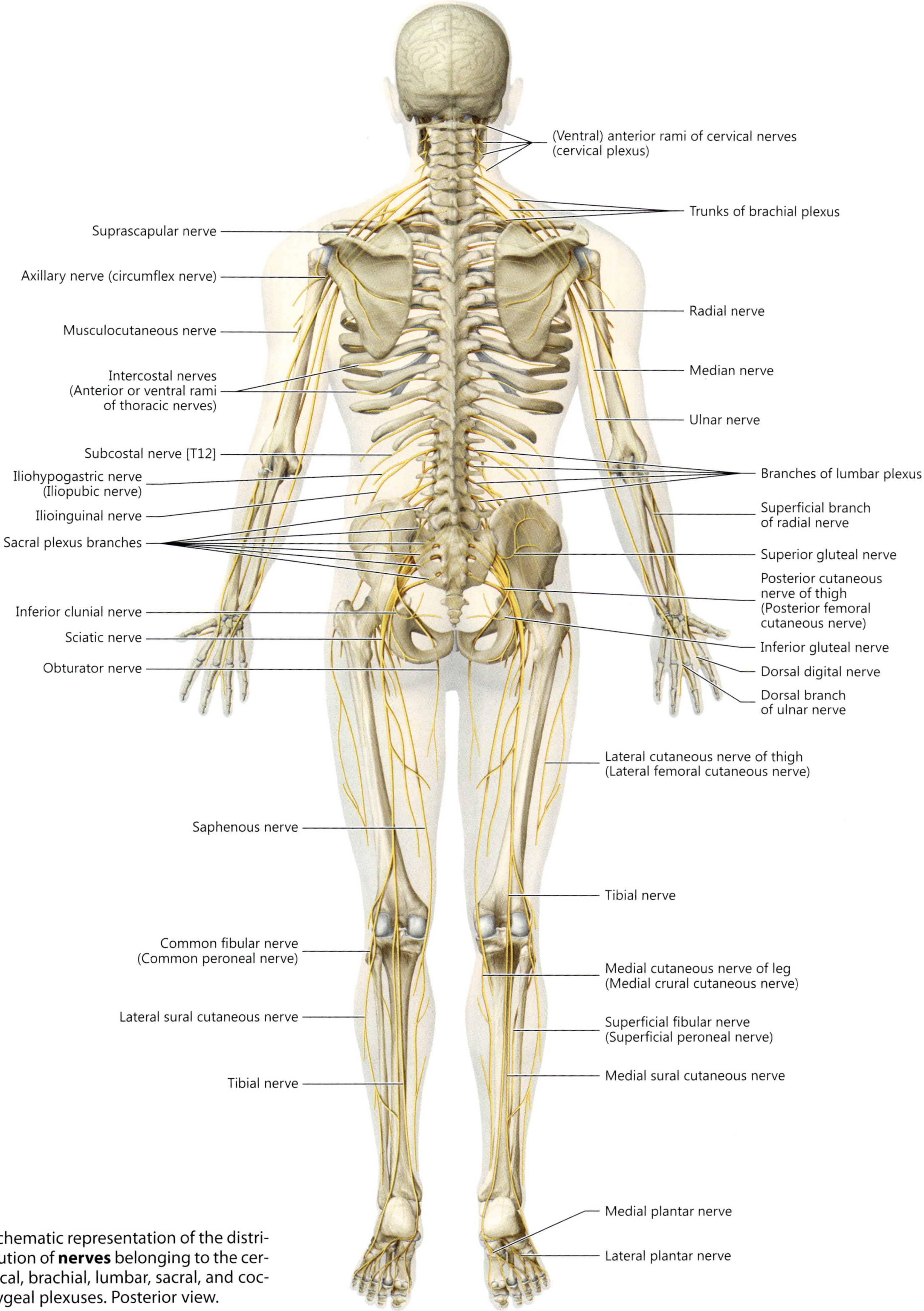

Schematic representation of the distribution of **nerves** belonging to the cervical, brachial, lumbar, sacral, and coccygeal plexuses. Posterior view.

Summary table of regional subdivision of the nerves (→ branching)*

NERVES OF THE HEAD
Abducens nerve [VI] (c.n.)
Caroticotympanic nerves (from tympanic plexus) (symp.nn.)
{Cervical nerves [C1-C8] (*eight pairs of spinal nerves*) (s.nn.)}*** → Great auricular nerve → Greater occipital nerve (posterior ramus of C2) → Lesser occipital nerve → Third occipital nerve (least occipital nerve)
Deep petrosal nerve (symp.n.) → Nerve of pterygoid canal (vidian nerve)**
External carotid nerves (from superior cervical ganglion) (symp.nn.)
Facial nerve [VII] (c.n.) → Cervicofacial trunk → → Cervical branch → → Marginal mandibular branch → Lingual branch (inconstant) → Nerve to stapedius → Posterior auricular nerve → → Auricular branch → → Occipital branch → Temporofacial trunk → → Buccal branches → → Temporal branches → → Zygomatic branches
Glossopharyngeal nerve [IX] (c.n. and p-symp. n.) → Lesser petrosal nerve → Tympanic nerve (Jacobson's nerve)
Hypoglossal nerve [XII] (c.n.) → Superior root of ansa cervicalis → → Ansa cervicalis (ansa of hypoglossal nerve)
Intermediate nerve (Wrisberg's nerve) (c.n. and p-symp.n.) → Chorda tympani → Greater (superficial) petrosal nerve → → Nerve of pterygoid canal (vidian nerve)**
Internal carotid nerve (from superior cervical ganglion) (symp.n.)
Oculomotor nerve [III] (c.n.) → Branch to ciliary ganglion (p-symp.n.) → Inferior branch → Superior branch
Olfactory nerve [I] (c.n.)
Optic nerve [II] (c.n.)
Short ciliary nerves (from ciliary ganglion) (*from six to ten in number*) (symp. and p-symp.nn.)
Terminal nerve [0] (c.n.)
Trigeminal nerve [V] (*the largest cranial nerve*) (c.n.) → Mandibular nerve [Vc; V3] → → Auriculotemporal nerve → → → Anterior auricular nerves (*generally two in number*) → → → Nerve to external acoustic meatus → → Buccal nerve → → Deep temporal nerves (*generally two in number*) → → Lingual nerve → → → Chorda tympani → → → Sublingual nerve → → Inferior alveolar nerve (inferior dental nerve) → → Masseteric nerve → → Mental nerve → → Nerve to lateral pterygoid → → Nerve to medial pterygoid → → Nerve to tensor tympani → → Nerve to tensor veli palatini → → Nervus spinosus (meningeal branch) → Maxillary nerve [Vb; V2] → → Ganglionic branches to pterygopalatine ganglion (sphenopalatine nerves or sensory root of pterygopalatine ganglion)

continued

NERVES OF THE HEAD (continued)
→ → Greater palatine nerve (anterior palatine nerve) → → Infraorbital nerve → → → External nasal branches → → → Inferior palpebral branches → → Lesser palatine nerves → → Meningeal branch → → Nasopalatine nerve (Scarpa's nerve) → → Pharyngeal nerve (Bock's nerve) → → Posterior superior lateral nasal branches → → Posterior superior medial nasal branches → → Superior alveolar nerves → → → Anterior superior alveolar branches (anterior superior dental branches) → → → Middle superior alveolar branch (middle superior dental branch) → → → Posterior superior alveolar branches (posterior superior dental branches) → → Zygomatic nerve → → → Zygomaticofacial branch → → → Zygomaticotemporal branch → Ophthalmic nerve [Va; V1] → → Frontal nerve → → → Supraorbital nerve → → → Supratrochlear nerve → → Lacrimal nerve → → Nasociliary nerve → → → Anterior ethmoidal nerve → → → Communicating branch with ciliary ganglion → → → → Long ciliary nerves (*two or three in number*) (symp.nn.) → → → Infratrochlear nerve → → → → Palpebral branches → → → Posterior ethmoidal nerve → → → → Anterior meningeal branch → → Tentorial nerve (Arnold's nerve)
Trochlear nerve [IV] (*the smallest cranial nerve*) (c.n.)
Vagus nerve [X] (pneumogastric nerve) (c.n. and p-symp.n.) → Auricular branch → Meningeal branch
Vestibulocochlear nerve [VIII] (c.n.) → Cochlear nerve (acoustic or auditory nerve) [VIII] → Vestibular nerve [VIII] → → Inferior part → → → Posterior ampullary nerve (inferior ampullary nerve) → → → Saccular nerve → → Superior part → → → Anterior ampullary nerve (superior ampullary nerve) → → → Lateral ampullary nerve → → → Utricular nerve

LEGEND	
c.n	→ cranial nerve
symp.n.	→ sympathetic nerve
symp.nn.	→ sympathetic nerves
p-symp.n.	→ parasympathetic nerve
p-symp.nn.	→ parasympathetic nerves
s.n.	→ spinal nerve
s.nn.	→ spinal nerves

* Certain nerves, branches or plexus not mentioned here are shown in the tables attached to the body sections of the Atlas.

** Arises from the confluence of the greater petrosal and deep petrosal nerves.

{...}*** The nerves indicated originate in other anatomical regions.

NERVES OF THE NECK
Accessory nerve [XI] (spinal accessory) (c.n. and p-symp.n.)

Cervical nerves [C1-C8] (*eight pairs of spinal nerves*) (s.nn.)
→ Cervical plexus [C1-C5]
→ Ansa cervicalis (ansa of hypoglossal nerve) (C1-C3)
  → Nerves to sternohyoid, sternothyroid and omohyoid
  → Inferior root (descendens cervicalis) (C2, C3)
  → Superior root (C1)
→ Thyrohyoid branch of the first cervical spinal nerve (C1)
→ Nerve to geniohyoid
→ Suboccipital nerve (infraoccipital nerve or posterior ramus of C1)
→ Transverse cervical nerve (transverse cutaneous nerve of neck)

Glossopharyngeal nerve [IX] (c.n. and p-symp.n.) → Carotid branch
Jugular nerve (from superior cervical ganglion) (n.symp.)
Laryngopharyngeal branches (of superior cervical ganglion) (symp.nn.)
Nerve to mylohyoid (branch of inferior alveolar nerve [from mandibular nerve]) (c.n.)
Thyroid branches (mainly from the cervical ganglia of sympathetic trunk) (symp.nn.)

Vagus nerve [X] (pneumogastric nerve) (c.n. and p-symp.n.)
→ Pharyngeal branch
→ Recurrent laryngeal nerve (inferior laryngeal nerve)
  → Pharyngeal branches
→ Superior laryngeal nerve
  → External branch
  → Internal branch

NERVES OF THE THORAX
Ansa subclavia (from inferior cervical ganglion [inconstant] or cervicothoracic or stellate ganglion) (symp.n.)
Cardiac plexus (symp. and p-symp.nn.)

{Cervical nerves [C1-C8] (*eight pairs of spinal nerves*) and first thoracic nerve [T1] (s.nn.)}***
→ Brachial plexus [C5-C8, T1]
  → Lateral pectoral nerve (nerve to pectoralis major and minor)
  → Long thoracic nerve (Bell's nerve)
  → Medial pectoral nerve (nerve to pectoralis minor and major)
  → Subclavian nerve
  → Thoracodorsal nerve
→ Cervical plexus [C1-C5]
  → Medial supraclavicular nerves (anterior supraclavicular nerves)
  → Phrenic nerve
    → Accessory phrenic nerves (inconstant)

Esophageal branches (from thoracic ganglia of sympathetic trunk) (symp.nn.)
Esophageal plexus (symp.nn.)
Inferior cervical cardiac nerve (from inferior cervical ganglion [inconstant] or cervicothoracic or stellate ganglion) (symp.n.)
Middle cervical cardiac nerve (from middle cervical ganglion) (symp.n.)
Pulmonary plexus (symp.nn.)
Superior cervical cardiac nerve (from superior cervical ganglion) (symp.n.)
Thoracic aortic plexus (symp. and p-symp.nn.)
Thoracic cardiac branches (from thoracic ganglia of sympathetic trunk) (symp.nn.)

Thoracic nerves [T1-T12] (*twelve pairs of spinal nerves*) (s.nn.)
→ Intercostal nerves (anterior or ventral rami of thoracic nerves)
  → Anterior pectoral cutaneous branch
  → Intercostobrachial nerves
  → Lateral pectoral cutaneous branch

Thoracic pulmonary branches (from thoracic ganglia of sympathetic trunk) (symp.nn.)

continued

NERVES OF THE THORAX (continued)

Vagus nerve [X] (pneumogastric nerve) (c.n. and p-symp.n.)
→ Bronchial branches
→ Cervical cardiac branches (superior and inferior cervical cardiac branches)
→ Esophageal plexus
→ Pulmonary plexus
→ Recurrent laryngeal nerve (inferior laryngeal nerve)
  → Esophageal branches
  → Tracheal branches
→ Thoracic cardiac branches

Vertebral nerve [from inferior cervical ganglion (inconstant) or cervicothoracic or stellate ganglion] (symp.n.)

NERVES OF THE ABDOMEN
Abdominal aortic plexus (symp.nn.)
Celiac plexus (symp. and p-symp.nn.)
Greater splanchnic nerve (symp.n.)
Hypogastric nerve (symp.n.)
Inferior mesenteric plexus (symp.nn.)
Least splanchnic nerve (lowest splanchnic nerve) (symp.n.)
Lesser splanchnic nerve (symp.n.)
Lumbar nerves [L1-L5] (*five pairs of spinal nerves*) (s.nn.) → Ilioinguinal nerve → Iliohypogastric nerve (iliopubic nerve)
Lumbar plexus [T12, L1-L5] (s.nn.) → Lumbosacral trunk
Lumbar splanchnic nerves (*from two to four in number*) (symp.nn.)
Ovarian plexus ♀ (simp.nn.)
Renal branch (from thoracic ganglia of sympathetic trunk) (symp.n.)
Renal plexus (symp. and p-symp.nn.)
Superior mesenteric plexus (symp. and p-symp.nn.)
Testicular plexus ♂ (symp.nn.)
{Thoracic nerves [T1-T12] (*twelve pairs of spinal nerves*) (s.nn.)}*** → Anterior abdominal cutaneous branch of intercostal nerves → Lateral abdominal cutaneous branch of intercostal nerves → Subcostal nerve [T12]
Ureteric plexus (symp.nn.)

Vagus nerve [X] (pneumogastric nerve) (c.n. and p-symp.n.)
→ Anterior vagal trunk
  → Anterior gastric branches
  → Hepatic branches
→ Posterior vagal trunk
  → Celiac branches
  → Posterior gastric branches
→ Renal branches

NERVES OF THE PELVIS
Genitofemoral nerve [lumbar nerves, L1-L5] (s.nn.) → Genital branch (external spermatic branch ♂)
Inferior hypogastric plexus (pelvic plexus) (symp. and p-symp.nn.) → Deferential plexus (plexus of ductus deferens) ♂ → Prostatic plexus ♂ → Inferior rectal plexus (inferior hemorrhoidal plexus) → Middle rectal plexus (middle hemorrhoidal plexus) → Uterovaginal plexus (Frankenhauser's ganglion) ♀ → Vesical plexus
Pelvic splanchnic nerves (nervi erigentes) (parasympathetic root of pelvic ganglia) (p-symp.nn.)
Sacral nerves [S1-S5] (*five pairs of spinal nerves*) and coccygeal nerve [Co] (s.nn.)
Sacral plexus [L4, L5, S1-S4] (s.nn.) → Pudendal plexus [S2-S4] (s.nn. and p-symp.nn.)   → Pudendal nerve (s.n. and p-symp.n.)

continued

NERVES OF THE PELVIS (continued)
Sacral splanchnic nerves (symp.nn.)
Superior hypogastric plexus (presacral nerve) (symp.nn.)
Superior rectal plexus (superior hemorrhoidal plexus) (s.nn. and p-symp.nn.)

NERVES OF THE PERINEUM
Anterior labial nerves ♀ (branches of ilioinguinal nerve) (s.nn.)
Anterior scrotal nerves ♂ (from ilioinguinal nerve) (s.nn.)
Coccygeal nerve [Co] (s.n.) → Anococcygeal nerve → Coccygeal plexus (s.nn. and p-symp.nn.)
Genital branch (of genitofemoral nerve) (s.n.)
Inferior hypogastric plexus (pelvic plexus) (symp. and p-symp.nn.) → Cavernous nerves of clitoris (major and minor) ♀ → Cavernous nerves of penis (major and minor) ♂
Perineal branches (of posterior cutaneous nerve of thigh or posterior femoral cutaneous nerve) (s.nn.)
{Pudendal nerve [sacral nerves, S1-S5] (s.n. and p-symp.n.)}*** → Perineal nerves → → Dorsal nerve of clitoris ♀ → → Dorsal nerve of penis ♂ → → Inferior anal (rectal) nerves → → Posterior labial nerves ♀ → → Posterior scrotal nerves ♂

NERVES OF THE UPPER LIMB
{Cervical [C1-C8] and thoracic [T1] nerves (s.nn.)}*** → Brachial plexus [C5-C8, T1] → → Infraclavicular part → → → Axillary nerve (circumflex nerve) → → → → Superior lateral cutaneous nerve of arm (superior lateral brachial cutaneous nerve) → → → Medial cutaneous nerve of arm (medial brachial cutaneous nerve) → → → Medial cutaneous nerve of forearm (medial antebrachial cutaneous nerve) → → → Median nerve → → → → Anterior interosseous nerve → → → → Common palmar digital nerves → → → → → Proper palmar digital nerves → → → → Palmar branch of median nerve → → → Musculocutaneous nerve (Casserio's nerve) → → → → Lateral cutaneous nerve of forearm (lateral antebrachial cutaneous nerve) → → → Radial nerve (musculospiral nerve) → → → → Deep branch of radial nerve → → → → → Posterior interosseous nerve → → → → Inferior lateral cutaneous nerve of arm (inferior lateral brachial cutaneous nerve) → → → → Posterior cutaneous nerve of arm (posterior brachial cutaneous nerve) → → → → Posterior cutaneous nerve of forearm (posterior antebrachial cutaneous nerve) → → → → Superficial branch of radial nerve → → → → → Dorsal digital branches (from superficial branch of radial nerve) → → → Ulnar nerve (cubital nerve) → → → → Deep branch of ulnar nerve → → → → Dorsal branch of ulnar nerve

continued

NERVES OF THE UPPER LIMB (continued)
→ → → → → Dorsal digital nerves (from dorsal branch of ulnar nerve) → → → → Palmar branch of ulnar nerve → → → → Superficial branch of ulnar nerve → → → → → Common palmar digital nerves (from superficial branch of ulnar nerve) → → → → → → Proper palmar digital nerves (from common palmar digital nerves of superficial branch of ulnar nerve) → → Supraclavicular part → → → Dorsal scapular nerve → → → Subscapular nerves (inferior/superior) (*generally two in number*) → → → Suprascapular nerve → Cervical plexus [C1-C5] → → Intermediate supraclavicular nerves → → Lateral supraclavicular nerves (posterior supraclavicular nerves)

NERVES OF THE LOWER LIMB
{Thoracic [T12], lumbar [L1-L5] and sacral [S1-S5] nerves (s.nn.}*** → Lumbar plexus [T12, L1-L5] → → Accessory obturator nerve → → Femoral branch of genitofemoral nerve → → Femoral nerve (anterior crural nerve) → → → Anterior cutaneous branches (of femoral nerve) → → → Saphenous nerve → → → → Medial cutaneous nerve of leg (medial crural cutaneous nerve) → → Ilioinguinal nerve → → Lateral cutaneous branch of iliohypogastric nerve → → Lateral cutaneous nerve of thigh (lateral femoral cutaneous nerve) → → Obturator nerve → Medial clunial nerves → Sacral plexus [L4-L5, S1-S4] → → Inferior gluteal nerve → → Nerve to obturator internus → → Nerve to piriformis → → Nerve to quadratus femoris → → Perforating cutaneous nerve → → Posterior cutaneous nerve of thigh (posterior femoral cutaneous nerve or lesser sciatic nerve) → → → Inferior clunial nerves → → Superior gluteal nerve → Sciatic nerve (ischiatic nerve or large sciatic nerve) → → Common fibular nerve (common peroneal nerve) → → → Lateral sural cutaneous nerve → → → Deep fibular nerve (deep peroneal nerve or anterior tibial nerve) → → → → Dorsal digital nerves of foot → → → Superficial fibular nerve (superficial peroneal nerve) → → → → Intermediate dorsal cutaneous nerve → → → → Medial dorsal cutaneous nerve → → Tibial nerve → → → Common plantar digital nerves (from tibial nerve) → → → → Proper plantar digital nerves (from common plantar digital nerves of tibial nerve) → → → Interosseous nerve of leg (crural interosseous nerve) → → → Lateral plantar nerve → → → → Deep branch (of lateral plantar nerve) → → → → Superficial branch (of lateral plantar nerve) → → → → → Common plantar digital nerves (from superficial branch of lateral plantar nerve) → → → → → Proper plantar digital nerves (from superficial branch of lateral plantar nerve) → → → Medial plantar nerve → → → → Common plantar digital nerves (from medial plantar nerve) → → → → Proper plantar digital nerves (from medial plantar nerve) → → → Medial sural cutaneous nerve → → → Sural nerve (short saphenous nerve) → → → → Lateral dorsal cutaneous nerve → Superior clunial nerves

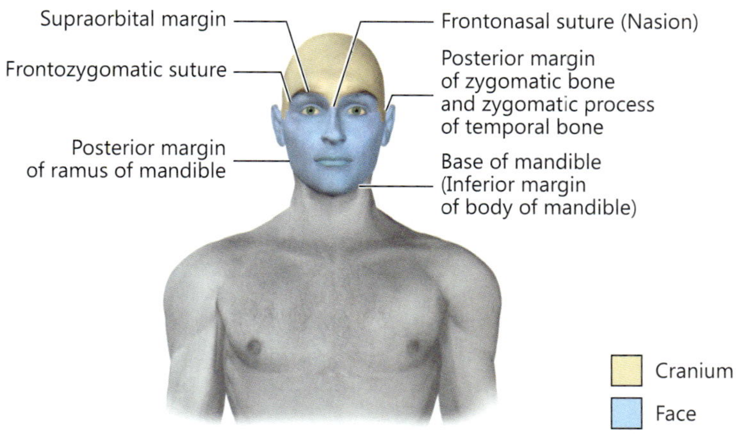

Anterior topographic subdivision of the head.

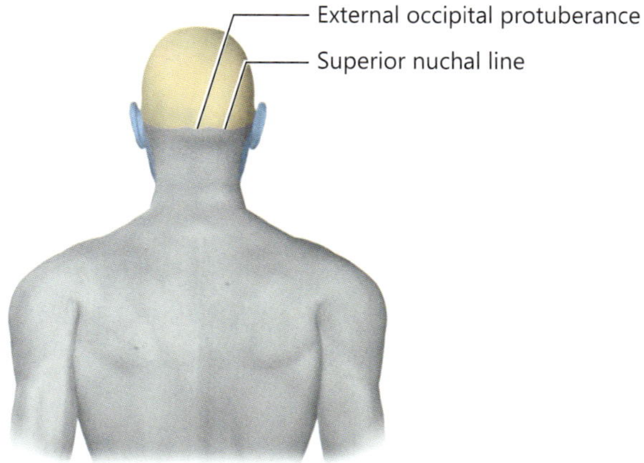

Posterior topographic subdivision of the head.

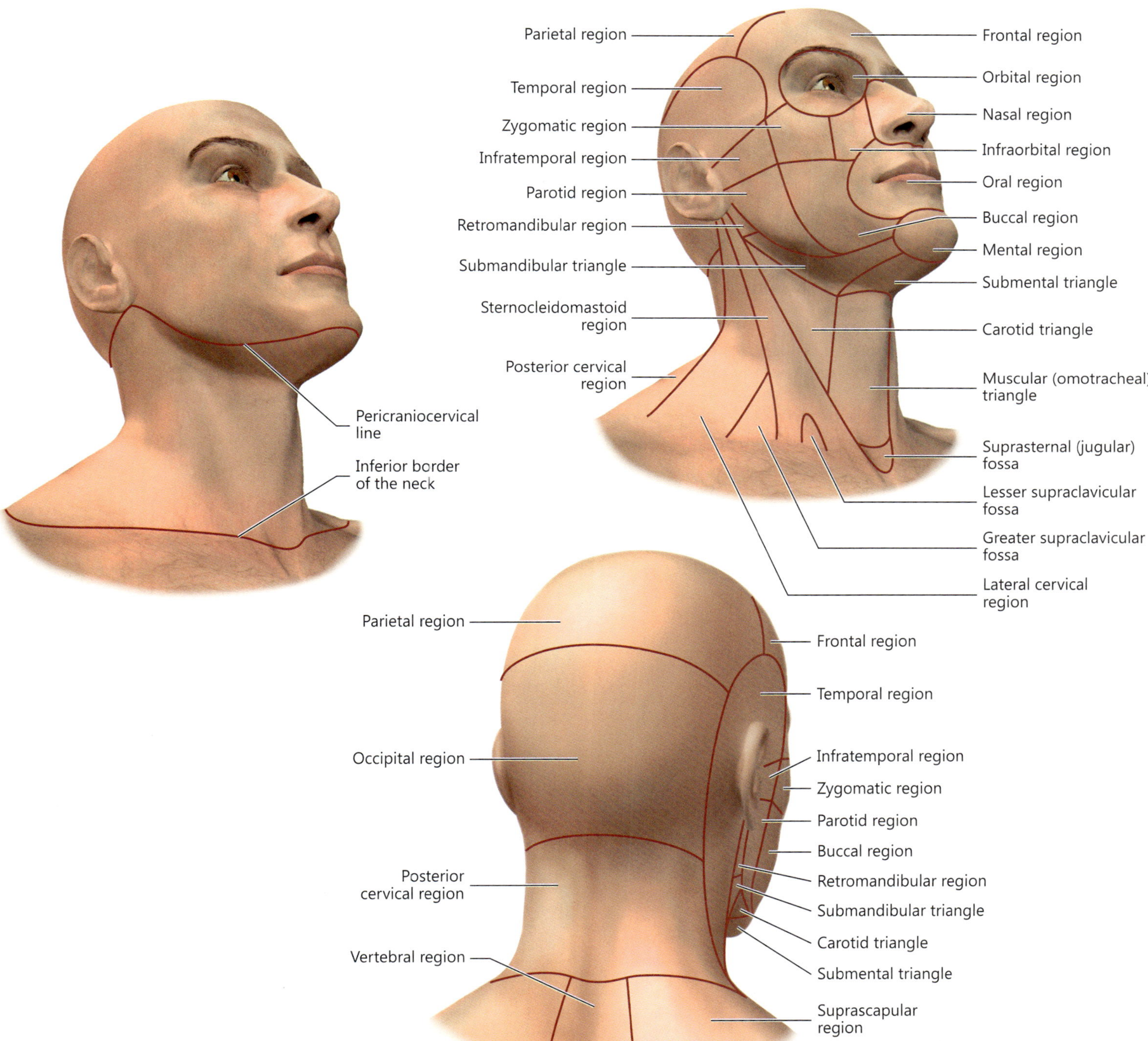

Regional subdivision of the head.

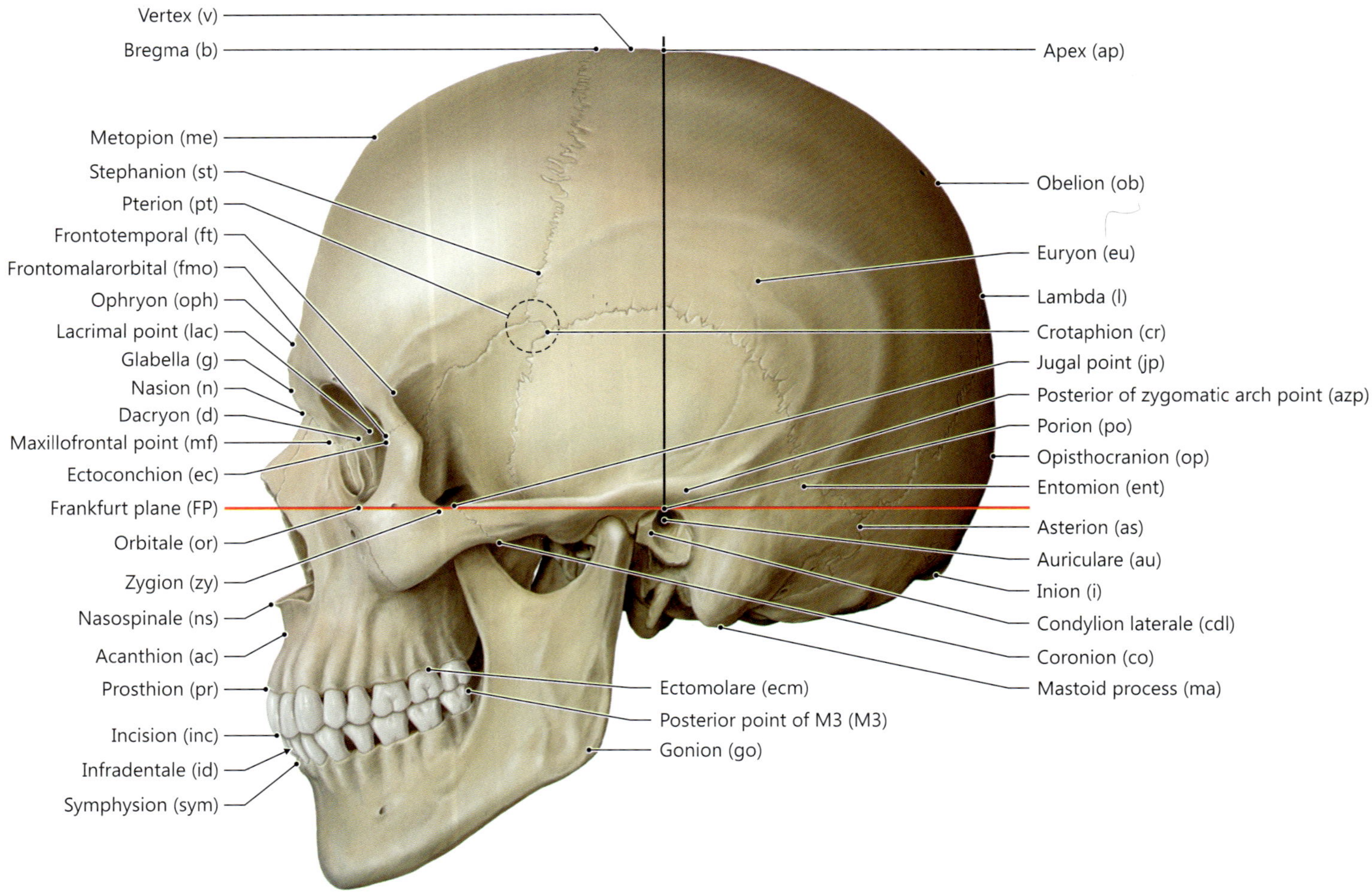

The **anthropometric points** comprise cephalometric (craniometric) points, pertaining to the cranium (in a skeleton) or to the head (of a living person), and somatometric points, relating to the trunk and limbs (of a skeleton or living person).
The **craniometric points** are used to ascertain the diameters of the cranium and, according to position, may be subdivided into median and lateral points. When the head is positioned with the Frankfurt plane horizontal (red line in the lateral view of the cranium, natural head position or NHP), precise and reproducible measurements of the cephalometric parameters may be made. The **Frankfurt plane** (the German plane, Virchow's plane or auriculo-orbital plane) passes through the porion (po) and inferior oribital (or) points.

CRANIOMETRIC LANDMARKS	*REGION*	*POSITION*	*DEFINITION*
Acanthion (ac)	Head	Median	The craniometric point at the anterior extremity of the intermaxillary suture
Apex (ap)	Head	Median	Highest point in a frontal section through the right and left porion with the head in the horizontal Frankfurt plane
Asterion (as)	Head	Lateral	The craniometric point at the union of the parietal, temporal, and occipital bones
Auriculare (au) or auricular point	Head	Lateral	Point at the center of the opening of the external acoustic meatus
Bregma (b)	Head	Median	Point at the union of the sagittal and coronal sutures
Condylion laterale (cdl)	Head	Lateral	The most lateral point of the mandibular condyle (head of the mandible)
Coronion (co)	Head	Lateral	The craniometric point at the tip of the coronoid process of the mandible
Crotaphion (cr)	Head	Lateral	The craniometric point at the tip of the greater wing of the sphenoid
Dacryon (d)	Head	Lateral	Point of union of the lacrimomaxillary and frontonasal sutures
Ectoconchion (ec)	Head	Lateral	Point of the lateral margin of the orbital cavity farthest from the maxillofrontal point
Ectomolare (ecm)	Head	Lateral	Most lateral point of the lateral surface of the alveolar arch of the maxilla

continued

(continued)

CRANIOMETRIC LANDMARKS	REGION	POSITION	DEFINITION
Entomion (ent)	Head	Lateral	The craniometric point at the tip of the angular part of the parietal bone that articulates with the temporal bone
Euryon (eu)	Head	Lateral	Most lateral point of the head; corresponds to the parietal tuber
Frankfurt plane (FP) or auriculo-orbital plane	Head	–	Line between porion and orbitale
Frontomalarorbital (fmo)	Head	Lateral	Point on the lateral margin of the orbit, on the anterior part of the frontozygomatic suture
Frontotemporal (ft)	Head	Lateral	The innermost and most forward point of the superior temporal line, directly above the zygomatic process of the frontal bone
Glabella (g) or nasal eminence	Head	Median	Most prominent point of the frontal bone in the median sagittal plane above the nasal notch and between the superciliary arches
Gonion (go)	Head	Lateral	The most lateral point of the angle of the mandible
Incision (inc)	Head	Median	Point at the incisal margin of the central maxillary incisor teeth
Infradentale (id) or lower alveolar point	Head	Median	The craniometric point between the mandibular incisor teeth where the anterior margins of the alveolar parts are intersected by the median sagittal plane
Inion (i)	Head	Median	Corresponds to the external occipital protuberance
Jugal point (jp)	Head	Lateral	The craniometric point at the union of the frontal and temporal processes of the zygomatic bone
Lacrimal point (lac) or punctum lacrimale	Head	Lateral	Point on the medial wall of the orbital cavity where the posterior lacrimal crest intersects with the frontolacrimal suture
Lambda (l)	Head	Median	Point of union of the sagittal and lambdoid sutures
Mastoid process or mastoidale (ma)	Head	Lateral	Lowest point of the mastoid process of the temporal bone
Maxillofrontal point (mf)	Head	Lateral	Point of union of the anterior lacrimal crest of the maxilla and the maxillofrontal suture
Metopion (me)	Head	Median	The craniometric point midway between the frontal tubers
Nasion (n)	Head	Median	Median point of the nasofrontal suture; corresponds to the root of the nose
Nasospinal point (ns)	Head	Median	Point of the median line on the tangent to the inferior margins of the piriform aperture, at the intermaxillary suture
Natural head position (NHP)	Head	–	The Frankfurt plane is in horizontal position
Obelion (ob)	Head	Median	The craniometric point on the sagittal suture determined by its intersection with the line that joins the parietal foramina
Ophryon (oph)	Head	Median	Above the glabella, the point on the median sagittal plane of a line tangential to the upper limits of the eyebrows
Opisthocranion (op)	Head	Median	Most protrusive posterior point of the cranium
Orbitale or inferior orbital point (or)	Head	Lateral	Most caudal point of the inferior margin of the orbit; the Frankfurt plane passes through this point
Porion (po)	Head	Lateral	Highest point of the superior margin of the external acoustic meatus, through which the Frankfurt plane passes
Posterior point of M3 (M3)	Head	Lateral	Most posterior point of the lower third molar
Posterior zygomatic arch point (Azp)	Head	Lateral	Most posterior point of the zygomatic arch as viewed from behind
Prosthion (pr) or alveolar point	Head	Median	Central point of the alveolar arch of the maxilla
Pterion (pt)	Head	Lateral	Point of union between the frontal bone, the greater wing of the sphenoid, the temporal and the parietal bones; corresponds to the sphenoparietal suture (sphenoidal fontanelle)
Stephanion (st)	Head	Lateral	Point of union between the coronal suture and the superior temporal line
Symphysion (sym)	Head	Lateral	Most anterior point of the alveolar process of the mandible
Vertex (v)	Head	Median	Highest point of the head when the head is positioned with the Frankfurt plane horizontal
Zygion (zy) or malar point	Head	Lateral	Most lateral point of the zygomatic arch

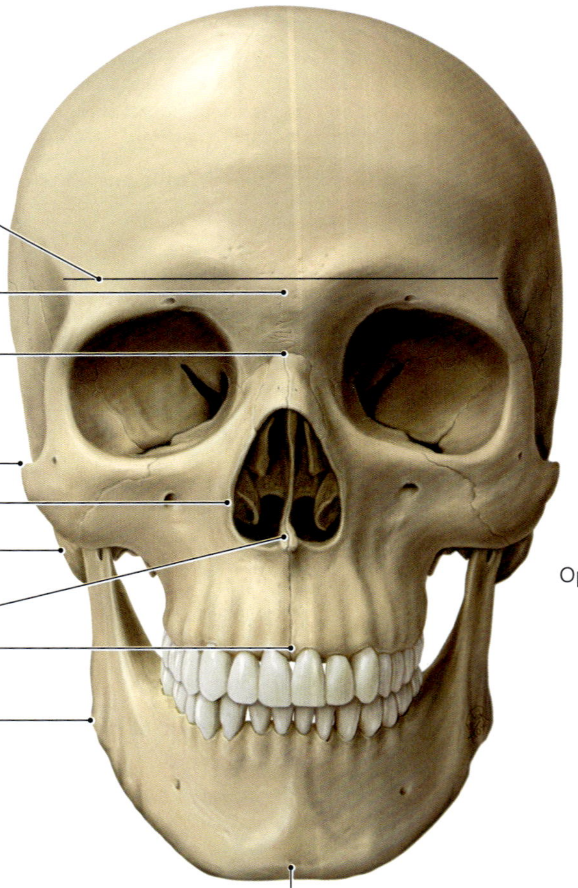

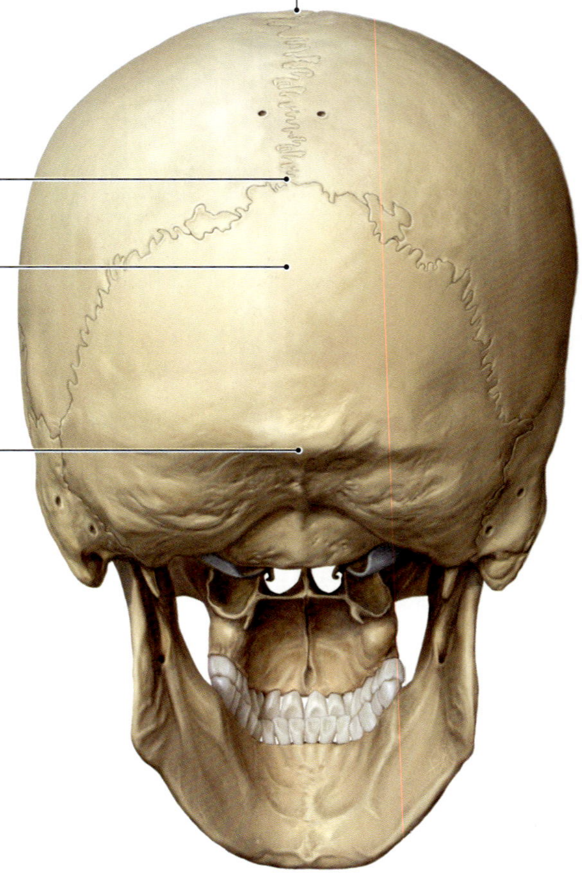

CRANIOMETRIC LANDMARKS	*REGION*	*POSITION*	*DEFINITION*
Bregma (b)	Head	Median	The craniometric point at the union of the sagittal and coronal sutures
Euryon-rhinion (e-r)	Head	Lateral	Most lateral point of the piriform aperture
Glabella (g) or nasal eminence	Head	Median	Most prominent point of the frontal bone in the median sagittal plane above the nasal notch and between the superciliar arches
Gnathion (gn) or menton (me) or mental point	Head	Median	Inferior midpoint of the median line of the mandible; corresponds to the inferior point of the mental symphysis
Gonion (go)	Head	Lateral	Most lateral point of the angle of the mandible
Inion (i)	Head	Median	Corresponds to the external occipital protuberance
Lambda (l)	Head	Median	Point of union of the sagittal and lambdoid sutures
Menton (me) ☞ Gnathion			
Minimum frontal diameter (mfd) or width	Head	Lateral	Minimum distance between the temporal crests of the frontal bone
Nasion (n)	Head	Median	Median point of the nasofrontal suture; corresponds to the root of the nose
Nasospinale (ns)	Head	Median	Point of the median line on the tangent to the inferior margins of the piriform aperture, at the intermaxillary suture
Opisthocranion (op)	Head	Median	Most protrusive posterior point of the cranium
Prosthion (pr) or alveolar point	Head	Median	Central point of the alveolar arch of the maxilla
Zygion (zy) or malar point	Head	Lateral	Most lateral point of the zygomatic arch
Zygomaxillare (zm)	Head	Lateral	Lowest point of the zygomaticomaxillary suture (anterior part of the zygomatic bone)

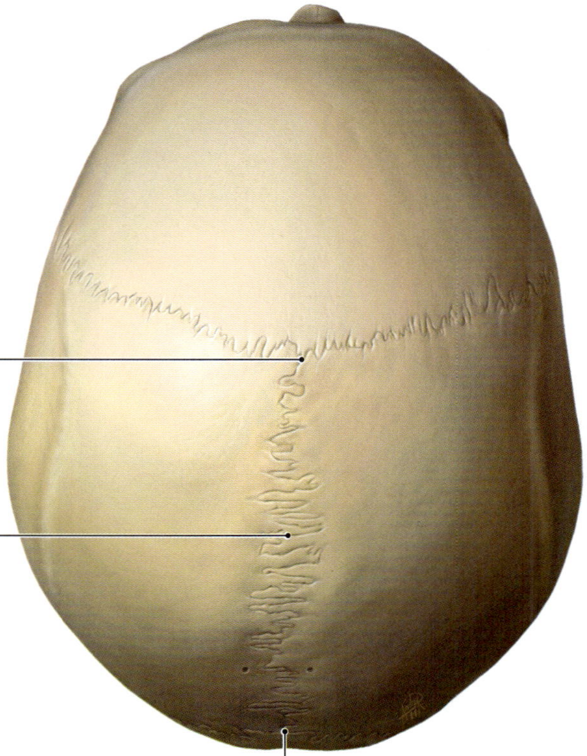

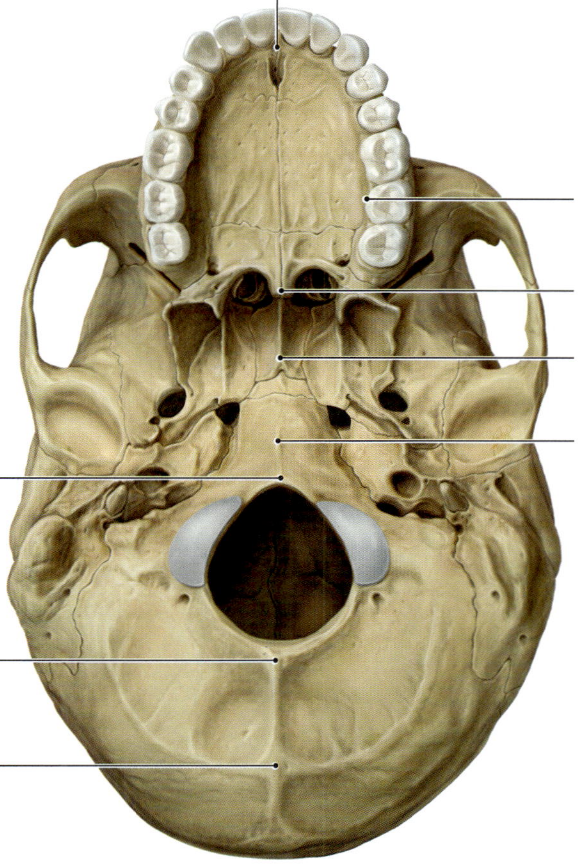

CRANIOMETRIC LANDMARKS	REGION	POSITION	DEFINITION
Basion (ba)	Head	Median	Most anterior point of the anterior margin of the foramen magnum (occipital foramen)
Bregma (b)	Head	Median	The craniometric point at the union of the sagittal and coronal sutures
Endomolare (enm)	Head	Lateral	The craniometric point on the lingual surface of the alveolar process at the second molar tooth
Hormion (ho)	Head	Median	Most posterior point of the vomer
Inion (i)	Head	Median	Corresponds to the external occipital protuberance
Lambda (l)	Head	Median	Point of union of the sagittal and lambdoid sutures
Opisthion (o)	Head	Median	Median point of the posterior margin of the foramen magnum
Orale (ol)	Head	Median	Most anterior point of the hard palate at the central incisor teeth
Sphenobasion (sphba)	Head	Median	Point of spheno-occipital synchondrosis on the external surface of the clivus
Staphylion (sta)	Head	Median	Point of posterior median junction of the palatine bones
Vertex (v)	Head	Median	Highest point of the head when the head is positioned with the Frankfurt plane horizontal

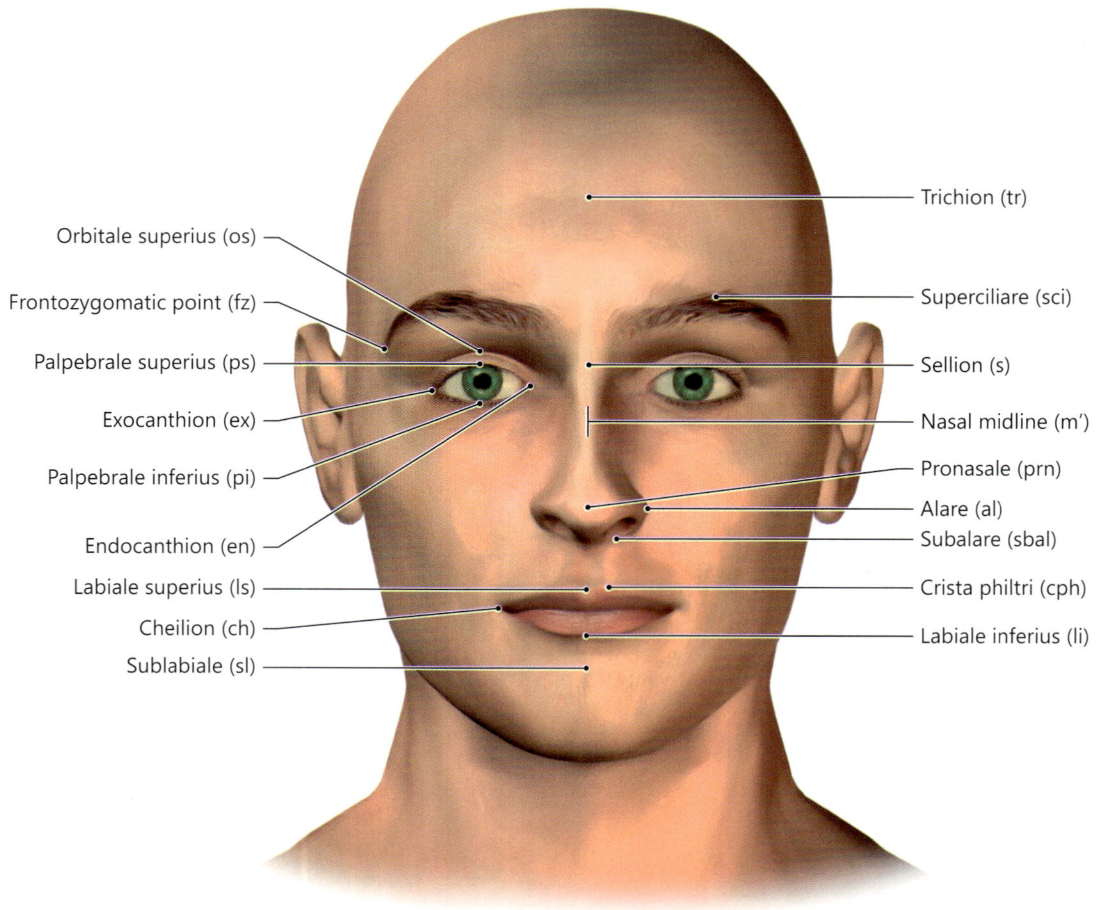
Trichion (tr)
Orbitale superius (os)
Superciliare (sci)
Frontozygomatic point (fz)
Palpebrale superius (ps)
Sellion (s)
Exocanthion (ex)
Nasal midline (m')
Palpebrale inferius (pi)
Pronasale (prn)
Alare (al)
Subalare (sbal)
Endocanthion (en)
Labiale superius (ls)
Crista philtri (cph)
Cheilion (ch)
Labiale inferius (li)
Sublabiale (sl)

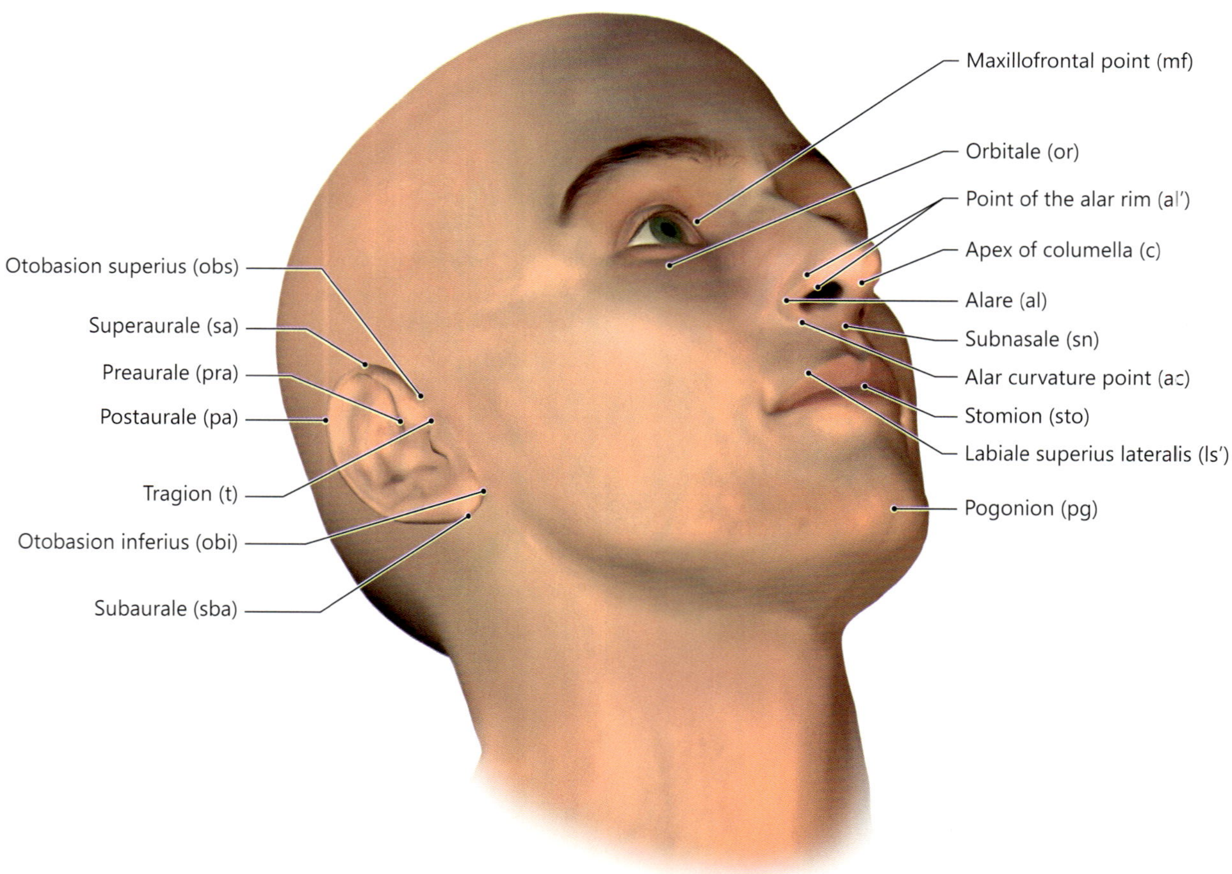
Maxillofrontal point (mf)
Orbitale (or)
Point of the alar rim (al')
Otobasion superius (obs)
Apex of columella (c)
Alare (al)
Superaurale (sa)
Subnasale (sn)
Preaurale (pra)
Alar curvature point (ac)
Postaurale (pa)
Stomion (sto)
Labiale superius lateralis (ls')
Tragion (t)
Pogonion (pg)
Otobasion inferius (obi)
Subaurale (sba)

CRANIOMETRIC LANDMARKS	REGION	POSITION	DEFINITION
Alare (al)	Face	Lateral	Most lateral point of the ala of the nose
Alar curvature point or alar crest (ac)	Face	Lateral	The most posterolateral point of the ala of the nose, located within the fold with the skin of the cheek
Alar rim (al')	Face	Lateral	Anterior and posterior margins of the ala of the nose
Apex of columella (c)	Face	Median	The most anterior point of the columella apex
Cheilion (ch)	Face	Lateral	The most lateral point of the labial commissure
Crista philtri (cph)	Face	Median	Point of philtrum (the vertical depression between the nose and upper lip) immediately above the edge of the upper lip
Endocanthion (en)	Face	Lateral	The internal angle of the palpebral fissure where the eyelids meet; it does not correspond to the caruncle
Exocanthion (ex)	Face	Lateral	Corresponds to the external angle of the palpebral fissure where the eyelids meet
Frontozygomatic point (fz)	Head	Lateral	Most lateral point of the zygomatic suture
Labiale (labrale) inferius (li)	Face	Median	Median point of the inferior vermilion border
Labiale (labrale) superius (ls)	Face	Median	Median point of the superior vermilion border
Labiale (labrale) superius laterale (ls')	Face	Lateral	Point of the superior vermilion border directly below the subalare point
Maxillofrontal point (mf)	Head	Median	Point of union of the anterior lacrimal crest of the maxilla and the maxillofrontal suture
Nasal midline (m')	Face	Median	The median line on the superior part of the dorsum of the nose
Natural head position (NHP)	Head	–	The Frankfurt plane is in horizontal position
Orbitale or inferior orbital point (or)	Head	Lateral	Most caudal point of the inferior margin of the orbit; the Frankfurt plane passes through this point
Orbitale superius (os)	Head	Lateral	Most cranial point of the margin of the orbit
Otobasion inferius (obi)	Head	Lateral	Most caudal point of the attachment of the external ear to the head
Otobasion superius (obs)	Head	Lateral	Most cranial point of the attachment of the external ear to the head
Palpebrale inferius (pi)	Face	Lateral	Central point of the border of the inferior eyelid
Palpebrale superius (ps)	Face	Lateral	Central point of the border of the superior eyelid
Pogonion (pg)	Face	Median	Most anterior point in the center of the soft tissue of the chin
Postaurale (pa)	Head	Lateral	Most posterior point of the free margin of the helix of the external ear
Preaurale (pra)	Head	Lateral	Anterior point of the helix opposite the postaurale point
Pronasale (prn)	Face	Median	Most prominent point of the nasal pyramid
Sellion (s)	Face	Median	Least prominent point of the nasal pyramid (of the nasofrontal angle)
Stomion (sto)	Face	Median	Median point of the labial fissure when the lips are naturally closed
Subalare (sbal)	Face	Lateral	The craniometric point on the inferior margin of the base of the ala of the nose
Subaurale (sba)	Head	Lateral	Most caudal point of the ear lobe
Sublabiale (sl)	Face	Median	Median point of the mentolabial sulcus
Subnasale (sn)	Face	Median	Median point between the base of the nasal septum (between the two nostrils) and the cutaneous part of the upper lip (philtrum)
Superaurale (sa)	Head	Lateral	Most cranial point of the free margin of the helix of the external ear
Superciliare (sci)	Face	Lateral	Central most cranial point of the superior margin of the eyebrow
Tragion (t)	Head	Lateral	The craniometric point corresponding to the tragus apex, the anterior cartilaginous protrusion to the external acoustic meatus
Trichion (tr)	Head	Median	Median point of the anterior hairline

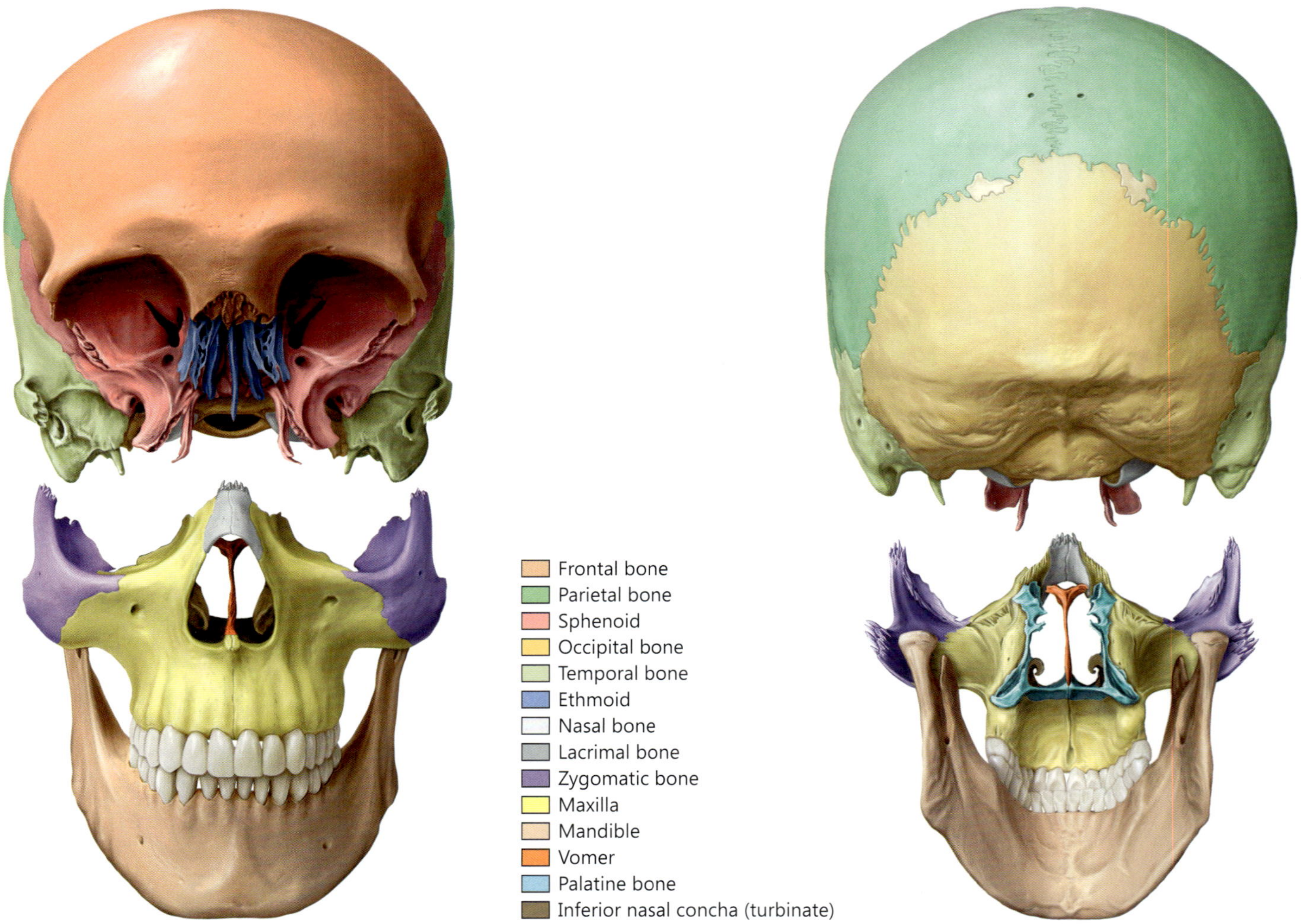

Subdivision of the bones of the cranium into the **neurocranium** (exoskeleton containing the brain with the additional function of protection) and the **viscerocranium** (exo- and endoskeleton with the function of protection and support to internal organs).

Landmarks of the head and neck

	STRUCTURE	*LOCATION*
1	Jugular notch (suprasternal notch)	Rounded depression at the superior border of the sternum, between the medial extremities of the clavicles
2	Thyroid cartilage	Most prominent cartilage of the larynx, located at the level of the 4th and 5th cervical vertebrae; subcutaneous prominence (pomum Adami)
3	Hyoid bone	U-shaped submandibular bone lying above the thyroid cartilage at the level of the 3rd cervical vertebra; the body of the hyoid bone is palpable on the midline below the chin, at the angle formed by the floor of the oral cavity and the anterior part of the neck
4	Angle of the mandible	Angle located medially and distally to the ear lobe
5	Angle of the mouth	Angle formed laterally by the upper and lower lips
6	Nasolabial sulcus	Skin furrow between the ala of the nose and the lip
7	Temporomandibular joint	Palpable anteriorly to the tragus of the external ear when the mouth is open and closed
8	Mastoid process	Bony prominence on the cranium located behind the ear
9	Acromion	Lateral projection of the spine of the scapula at the apex of the shoulder

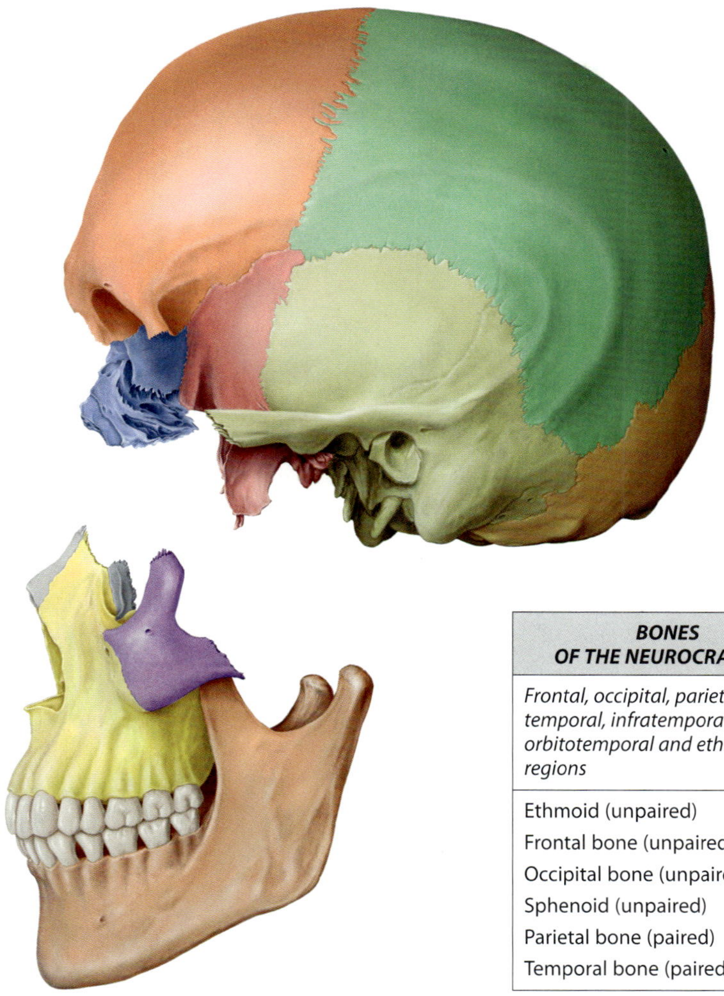

BONES OF THE NEUROCRANIUM	**BONES OF THE VISCEROCRANIUM (FACIAL SKELETON)**	
Frontal, occipital, parietal, temporal, infratemporal, orbitotemporal and ethmoidal regions	*Nasal region*	*Orbital, infraorbital, buccal, parotid, zygomatic, oral and mental regions*
Ethmoid (unpaired) Frontal bone (unpaired) Occipital bone (unpaired) Sphenoid (unpaired) Parietal bone (paired) Temporal bone (paired)	Vomer (unpaired) Inferior nasal concha (paired) Nasal bone (paired) Lacrimal bone (paired)	Maxilla (paired) Palatine bone (paired) Zygomatic bone (paired) Mandible (unpaired)

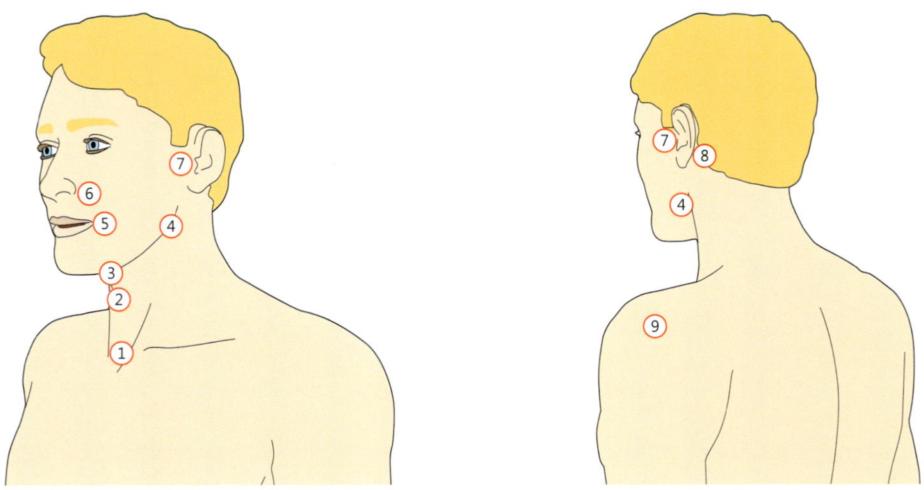

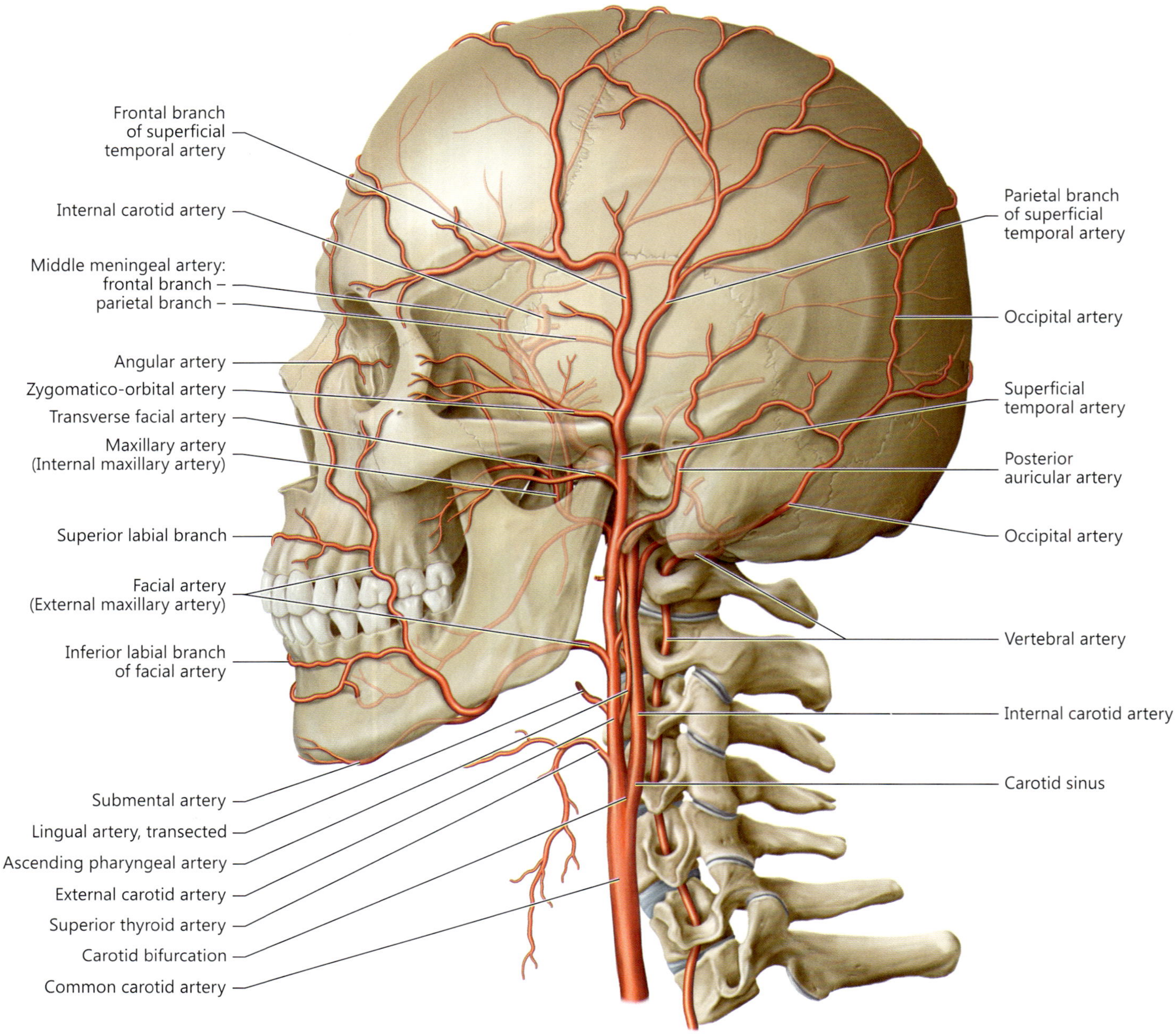

Course of the **external and internal carotid arteries** and the **vertebral artery** with respective collateral and terminal branches.

➤

Three-dimensional images of a living human body, obtained by CT scan using a 3D volume rendering technique, showing the paths of the **arteries supplying the head**.

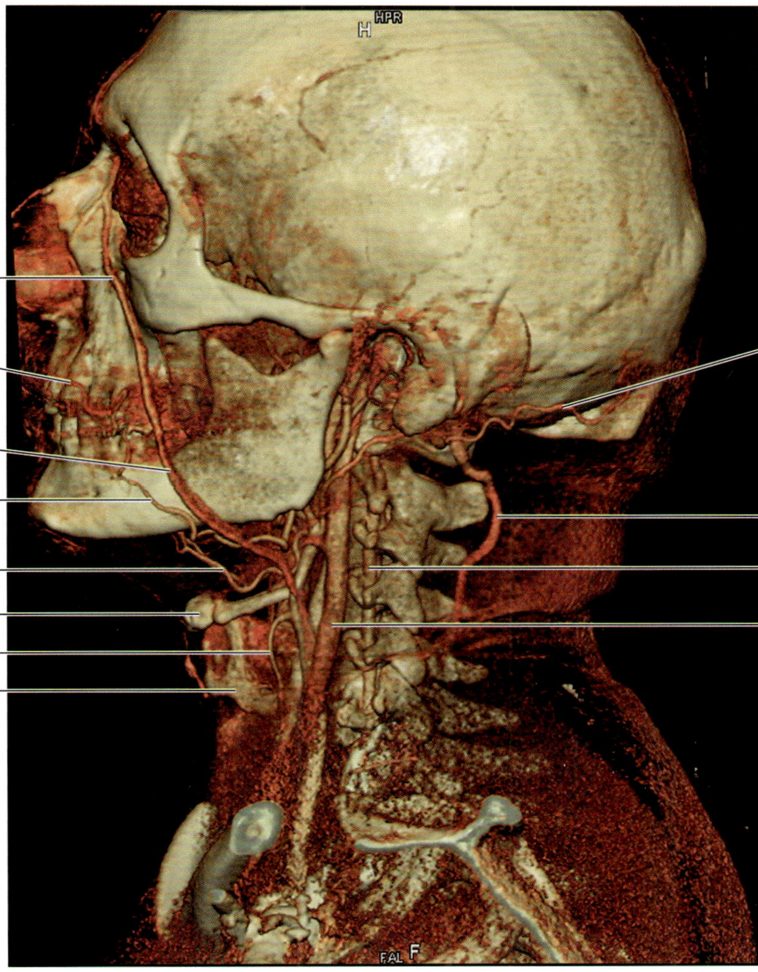
Angular artery
Superior labial artery
Facial artery
(External maxillary artery)
Inferior labial artery
Lingual artery
Hyoid bone
Superior thyroid artery
Thyroid cartilage
Occipital artery
Posterior meningeal artery
Vertebral artery
Internal jugular vein

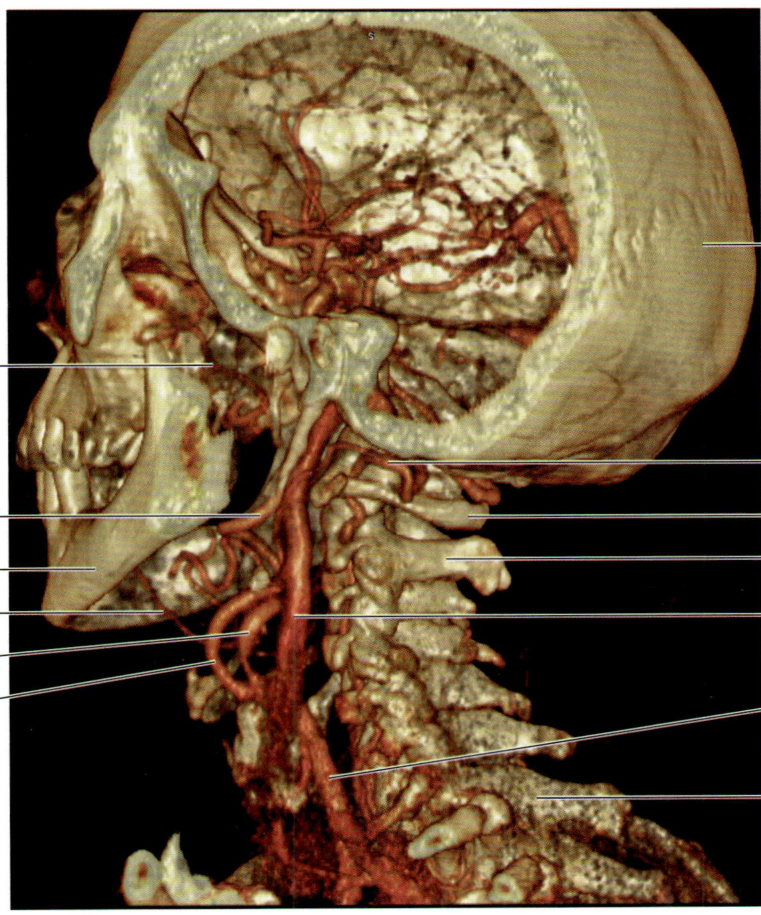
Maxillary artery
Facial artery
Mandible
Lingual artery
Internal carotid artery
External carotid artery
Occipital bone
Vertebral artery
Atlas [C I]
Axis [C II]
Internal jugular vein
Common carotid artery
Seventh cervical
vertebra [C VII]

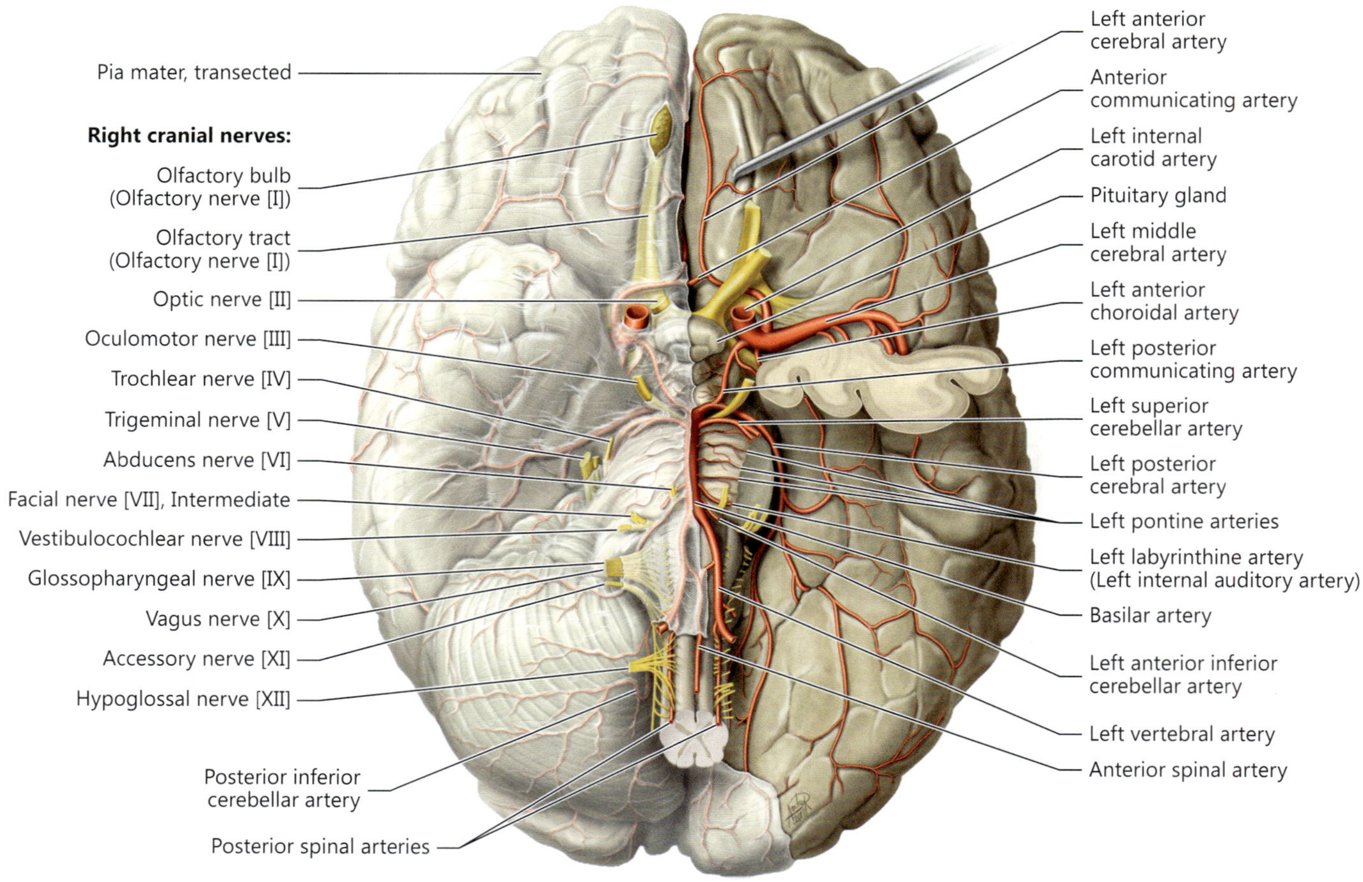

Arteries at the base of the brain. The left cerebellar hemisphere and part of the left temporal lobe have been removed to expose the middle and posterior cerebral arteries. The apparent origins of the cranial nerves are indicated on the left side of the figure.
Below, three-dimensional image of a living human body, obtained by CT scan using a 3D volume rendering technique, showing the **arteries of the cranial base** and **venous sinuses**.

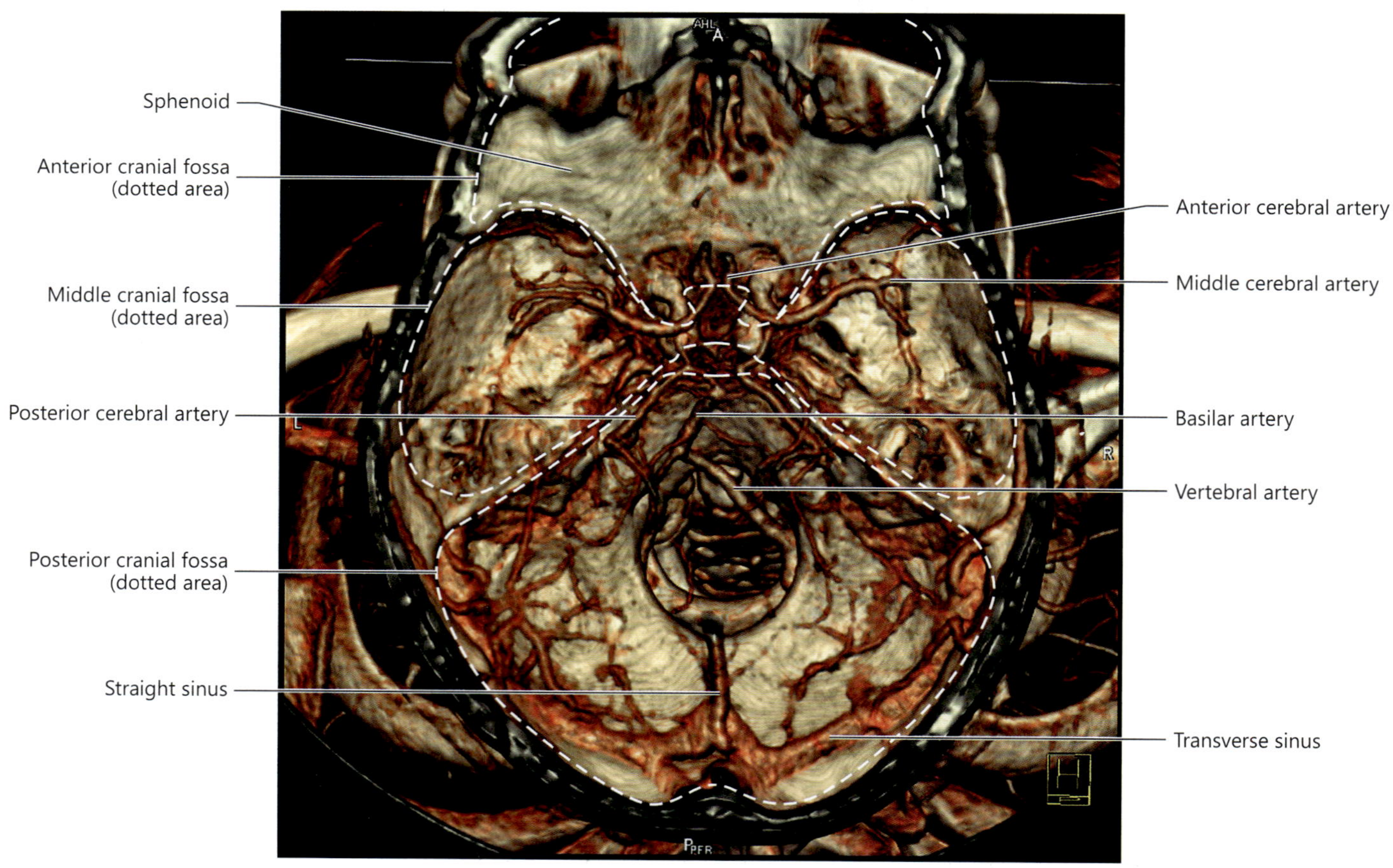

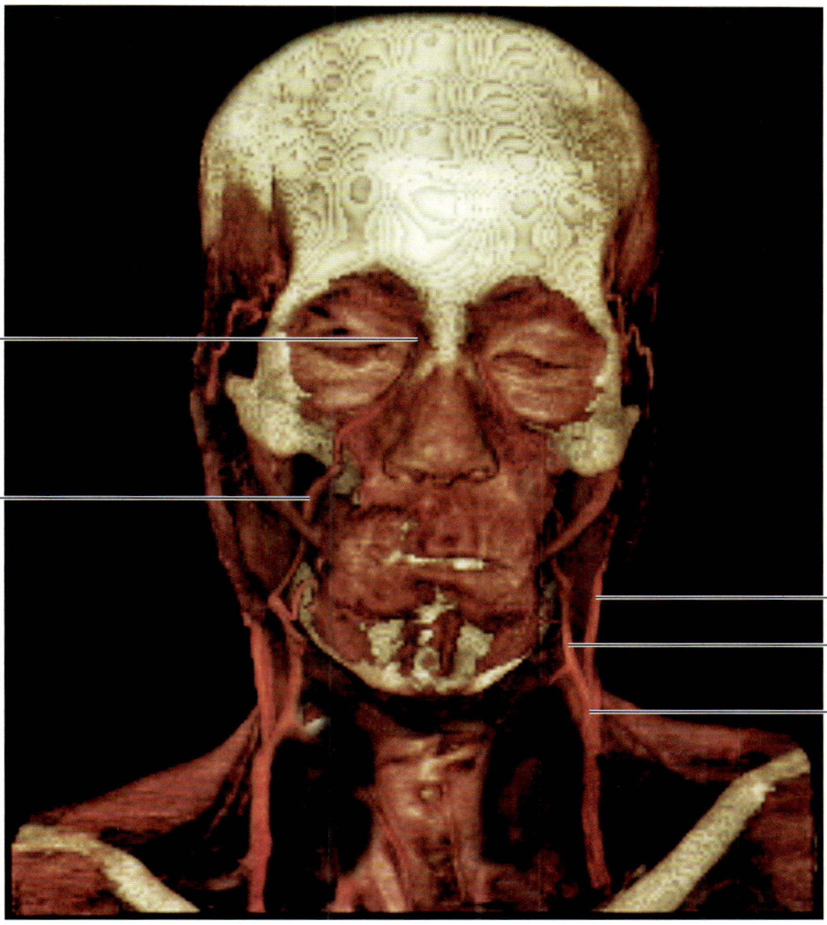

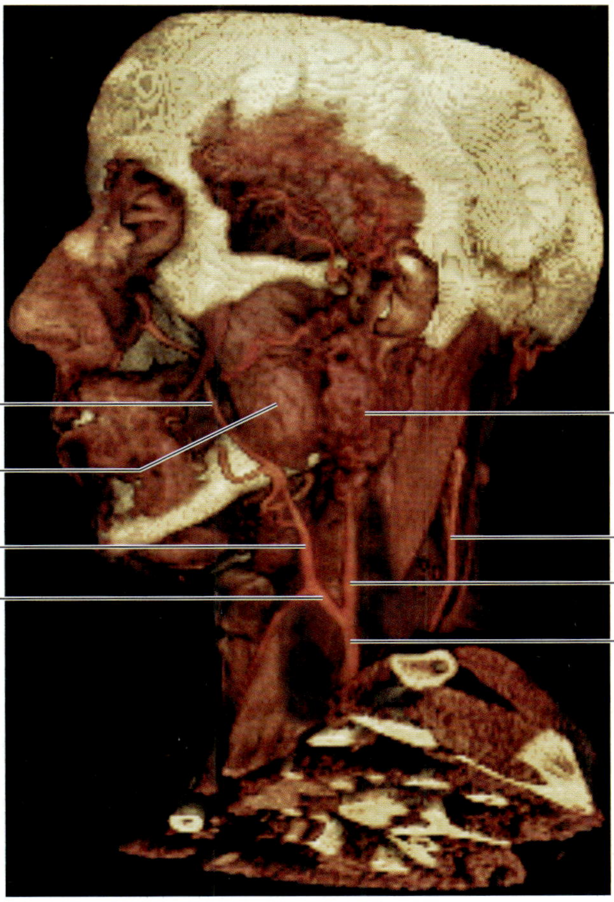

Three-dimensional image of a living human body, obtained by CT scan using a 3D volume rendering technique, showing the **external jugular vein** in the cervical region and the main venous branches tributary to it. Anterior view (**a**) and left lateral view (**b**).

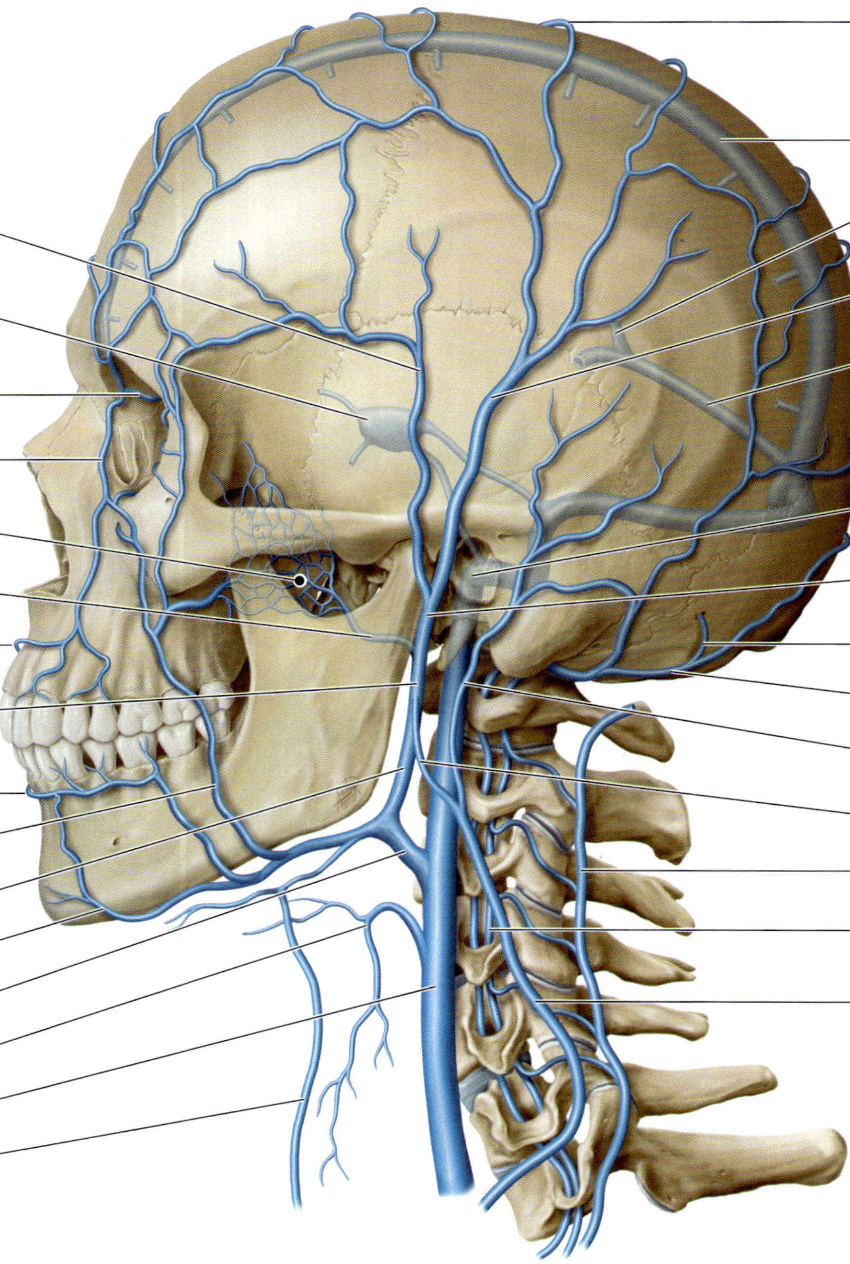

The **dural venous sinuses**, seen in transparency through the cranium, left side. Note their confluence in the bulb of the internal jugular vein. On the surface, some veins of extracranial circulation and their anastomosis with intracranial circulation can be seen.

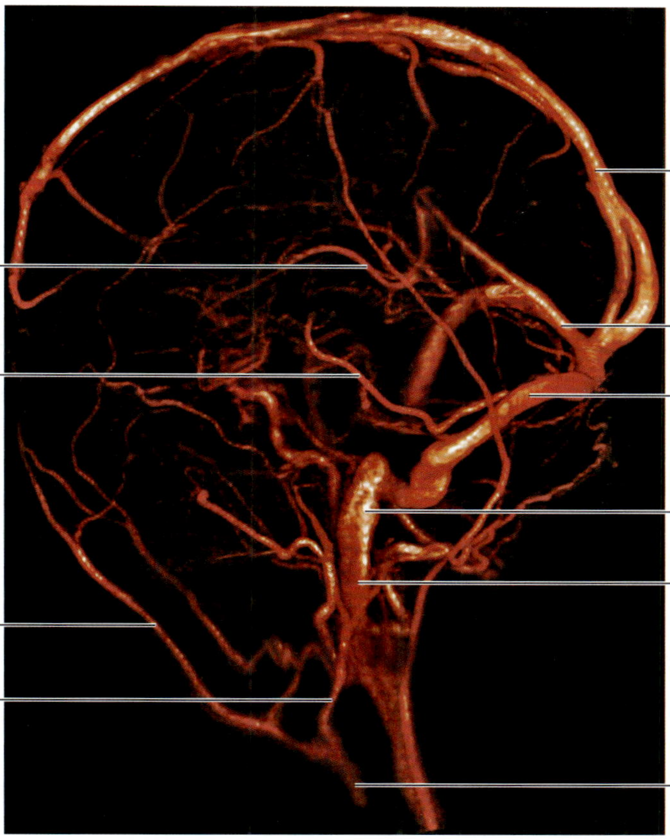

Above, three-dimensional image of a living human body, obtained by CT scan using a 3D volume rendering technique, showing the **dural venous sinuses**, left lateral view. The venous system is conventionally shown in red by the mapping obtained using 3D processing software.
Below, schematic representation of the course of the **dural venous sinuses**.

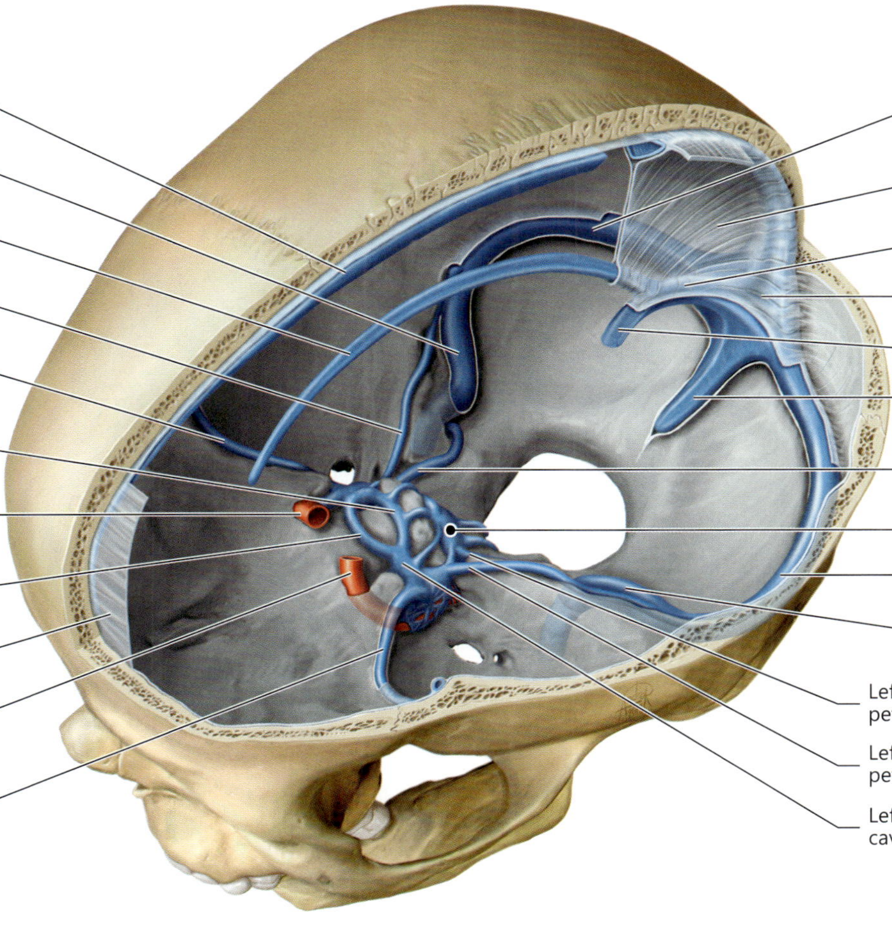

Distribution of **lymph node groups of the head and neck** forming two horizontal chains (superior and inferior) and three vertical chains. The chains are indicated by a dotted black line.

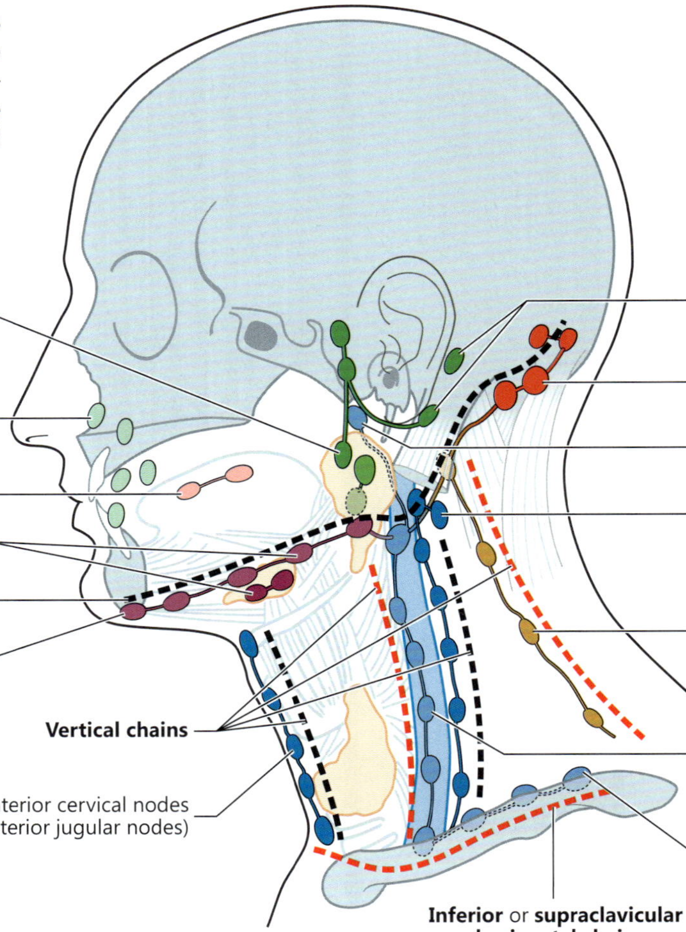

Schematic representation of the **deep lymph node groups of the head and neck**. The different colors indicate the various groupings, the regions they supply and the lymph flow paths. All the lymph nodes in the area lead to the deep cervical nodes, from which the jugular trunk is formed. The arrows indicate the direction of lymph flow.

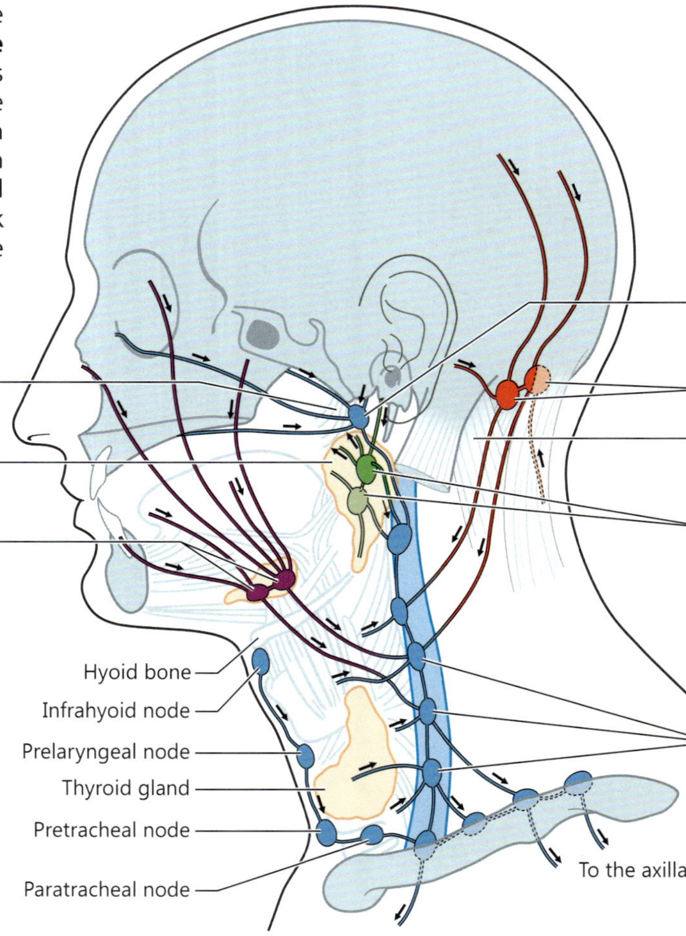

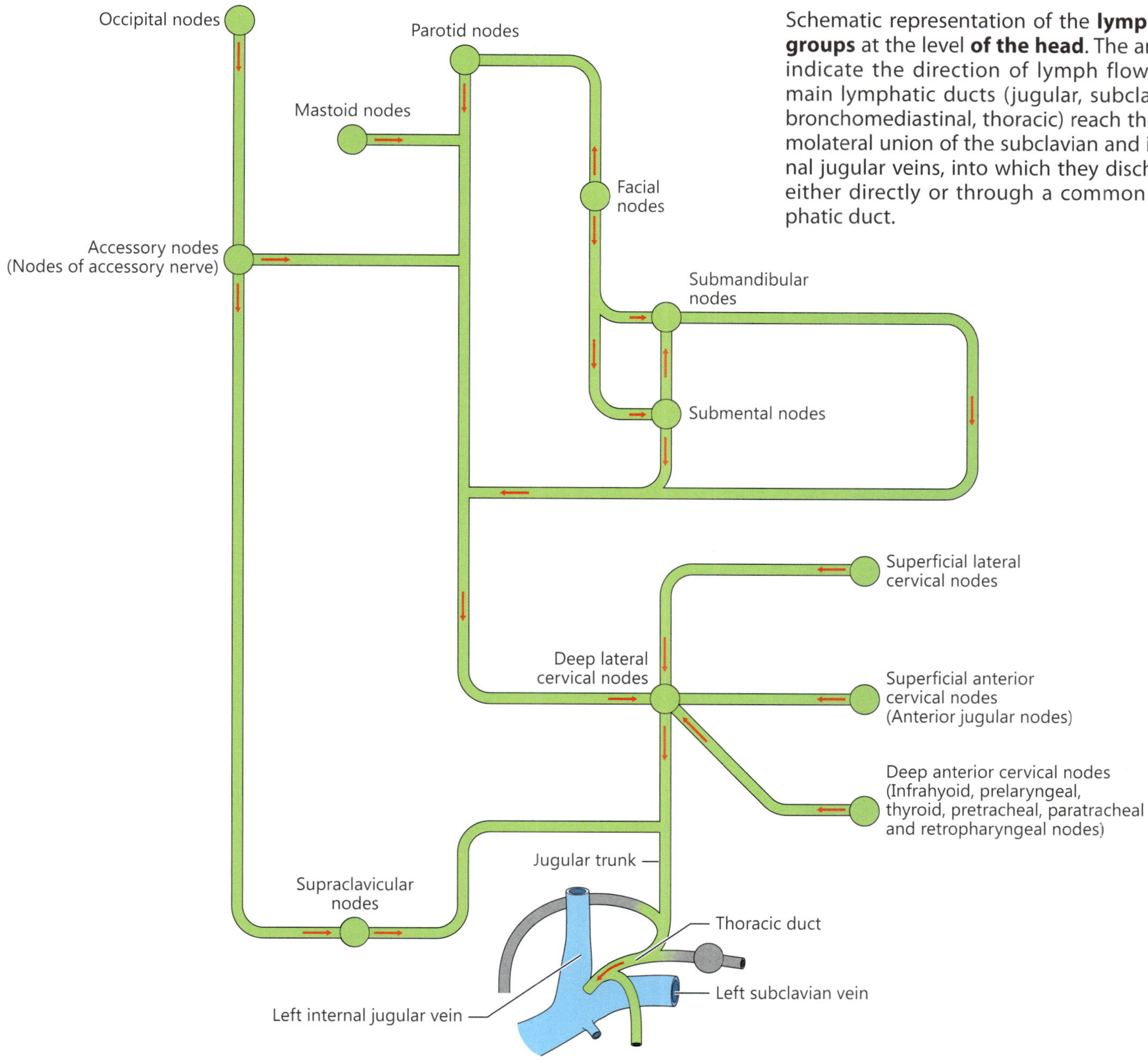

Schematic representation of the **lymphatic groups** at the level **of the head**. The arrows indicate the direction of lymph flow. The main lymphatic ducts (jugular, subclavian, bronchomediastinal, thoracic) reach the homolateral union of the subclavian and internal jugular veins, into which they discharge either directly or through a common lymphatic duct.

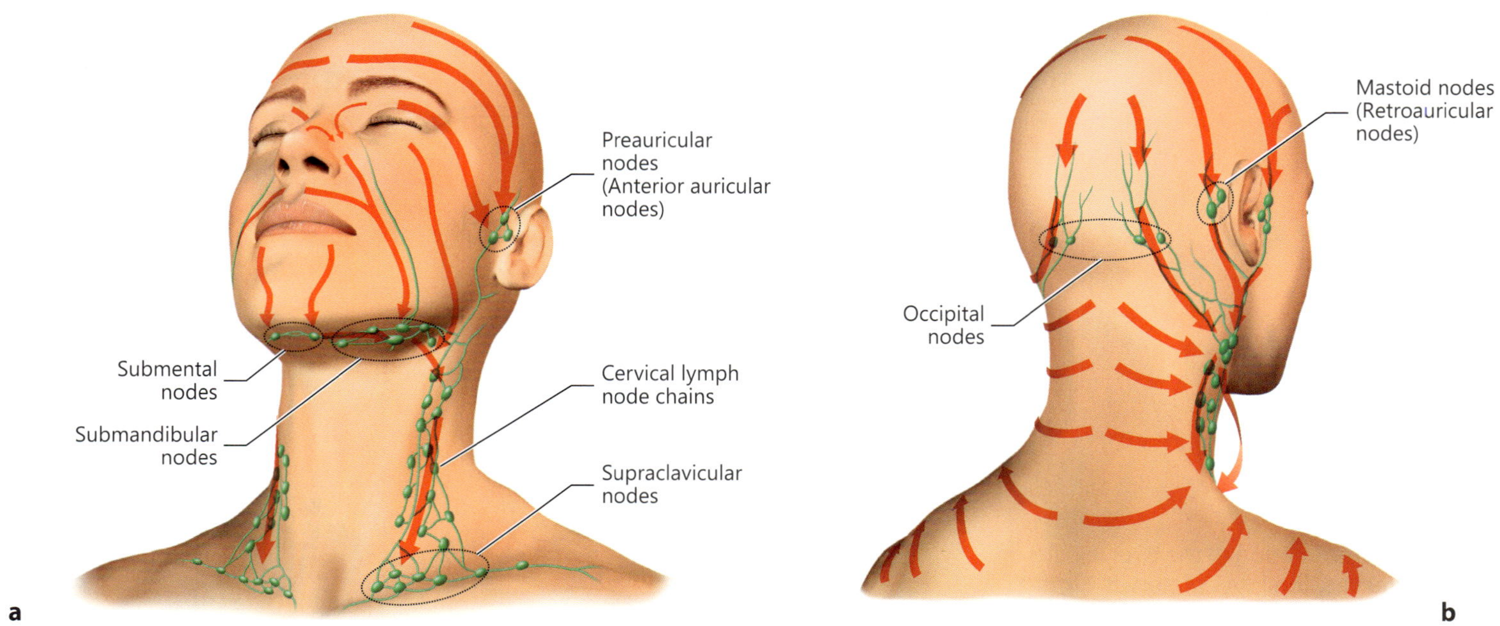

Lymph nodes and **lymphatic drainage from the head** and **neck**. Frontal view (**a**) and posterior view (**b**).

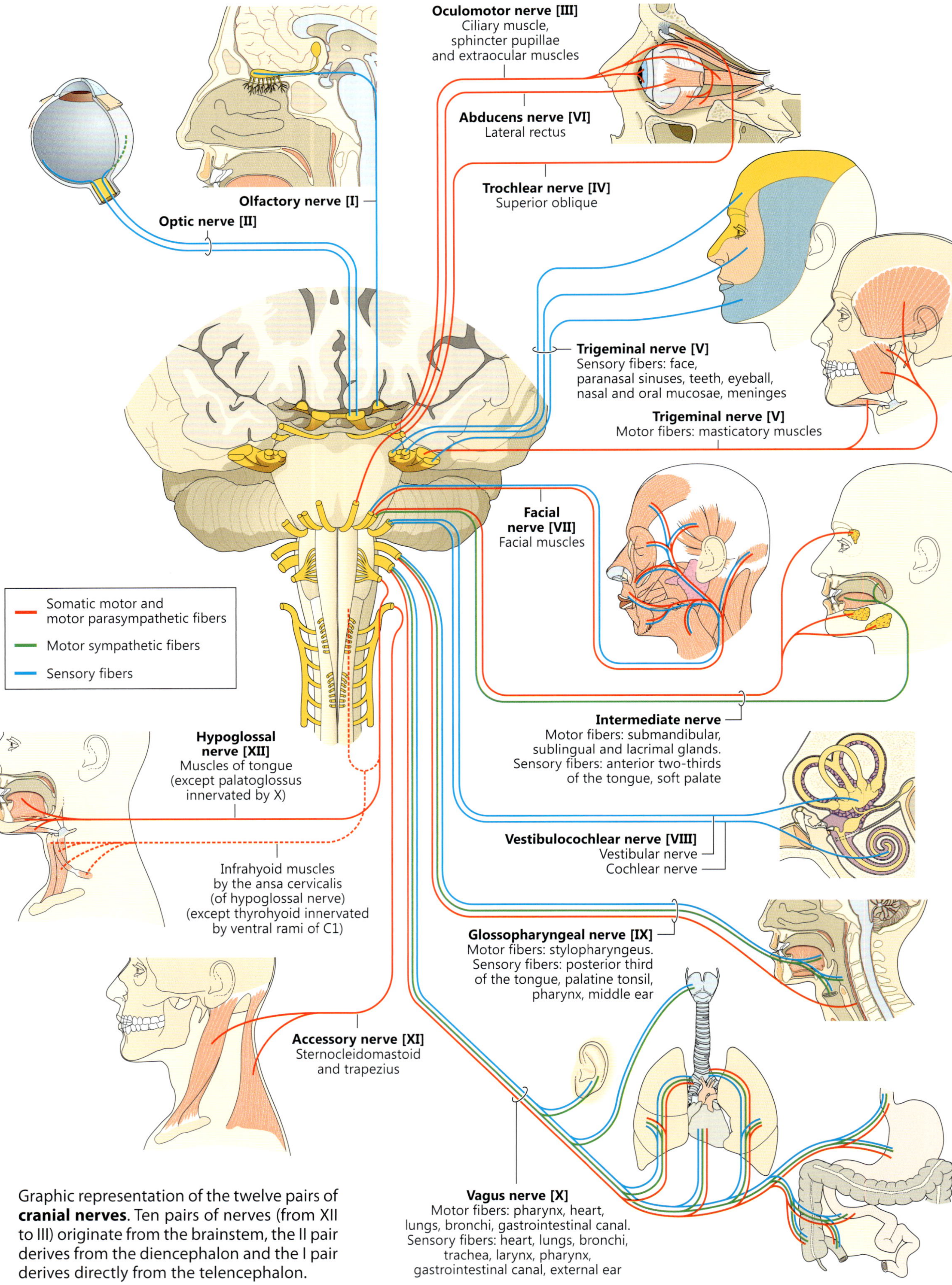

Graphic representation of the twelve pairs of **cranial nerves**. Ten pairs of nerves (from XII to III) originate from the brainstem, the II pair derives from the diencephalon and the I pair derives directly from the telencephalon.

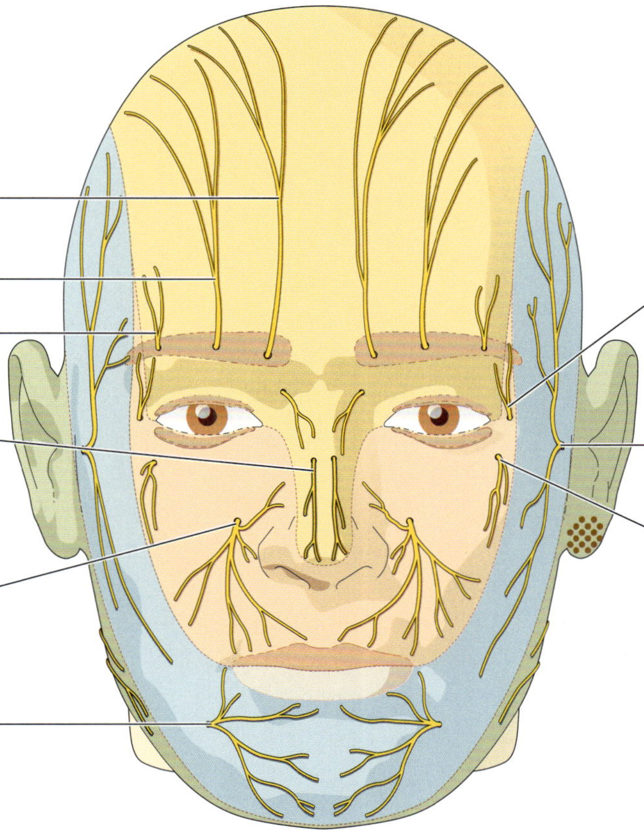

Areas of cutaneous innervation in the head. Anterior view. The areas supplied by the ophthalmic, maxillary, and mandibular nerves are shown in *yellow*, *pink*, and *blue*, respectively. The dotted area of the auricular region is supplied by the facial [VII], glossopharyngeal [IX] and vagus [X] nerves.

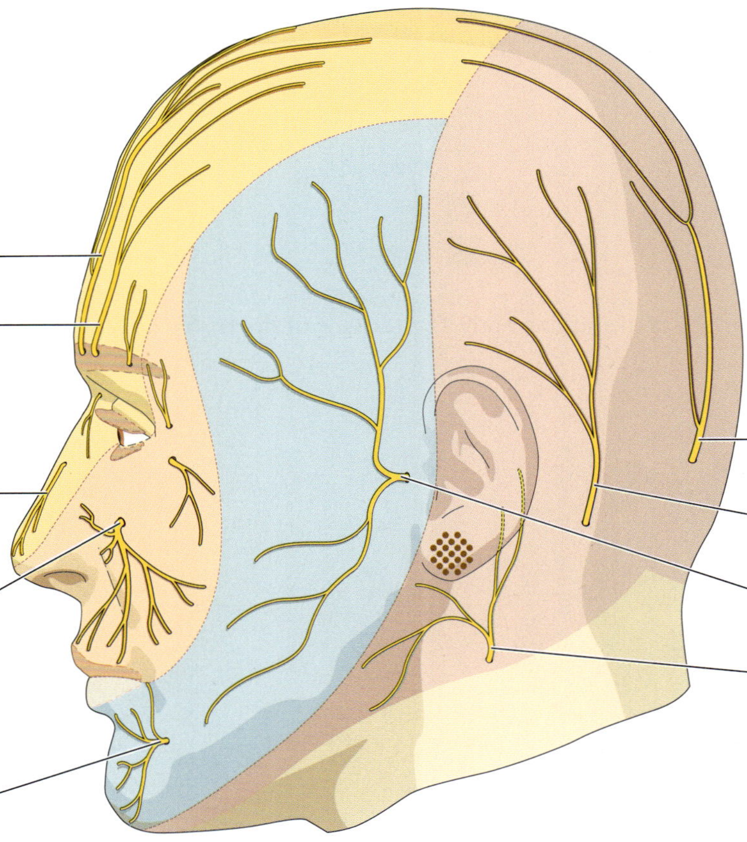

Areas of cutaneous innervation in the head. Lateral view. The three branches of the trigeminal nerve supply the entire face and the anterior half of the cranium. The areas supplied by the ophthalmic, maxillary and mandibular nerves are shown in *yellow*, *pink*, and *blue*. The dotted area of the auricular region is supplied by the facial [VII], glossopharyngeal [IX] and vagus [X] nerves.

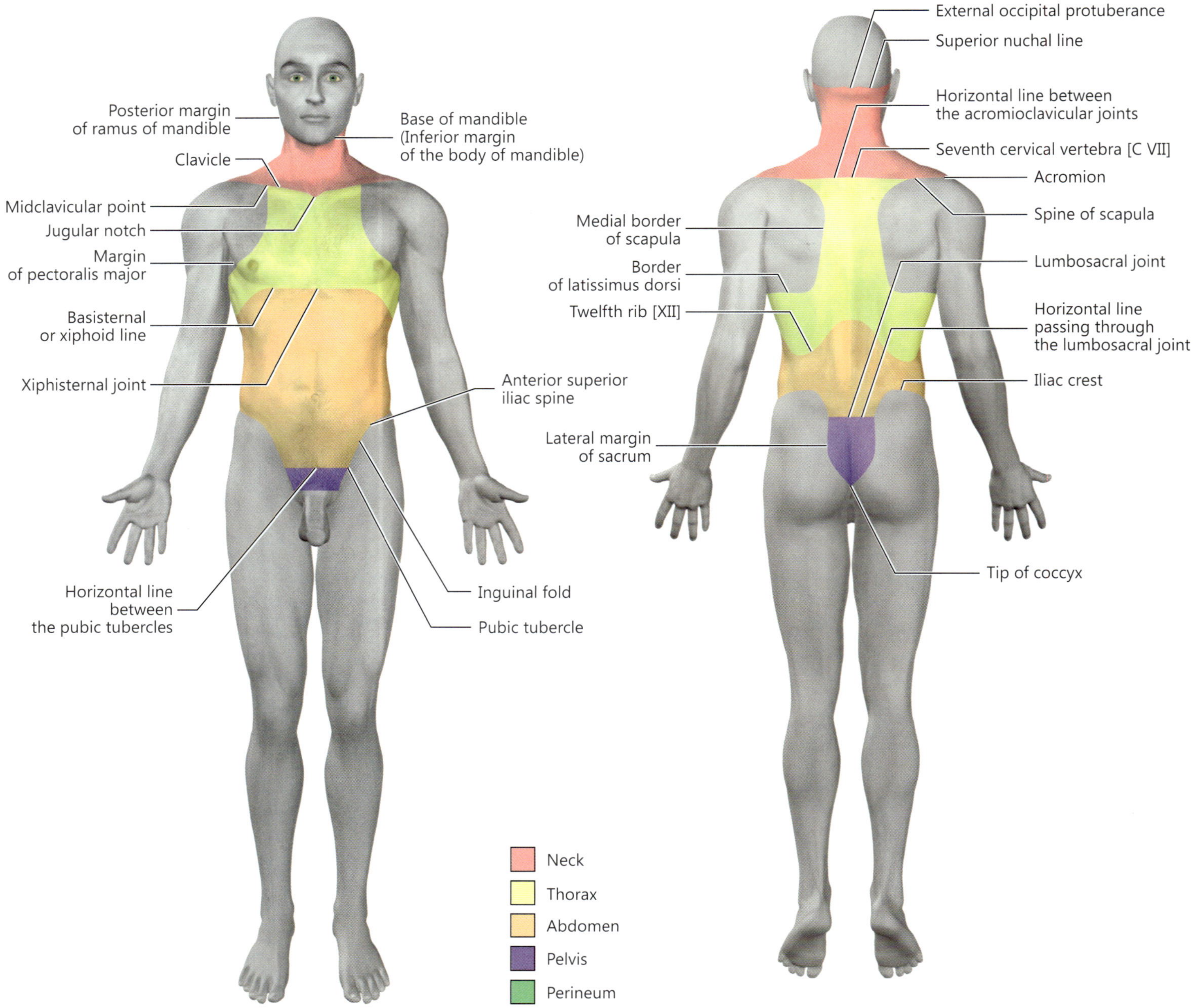

Topographic division of the surface of the human body. The **regions of the trunk** are highlighted. Anterior and posterior views.

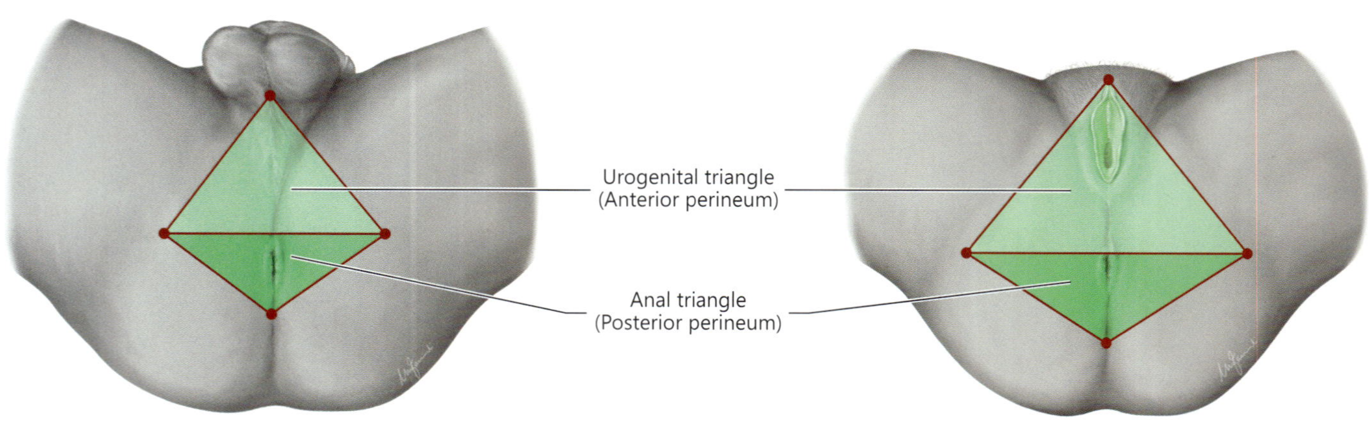

The **perineum** is comprised of the soft tissue enclosing inferiorly the pelvic cavity, and may be subdivided into two triangles, the urogenital triangle (anterior perineum) and the anal triangle (posterior perineum). Appended to the posterior perineum is the anal canal (terminal tract of the rectum); the external genitalia are annexed to the anterior perineum.

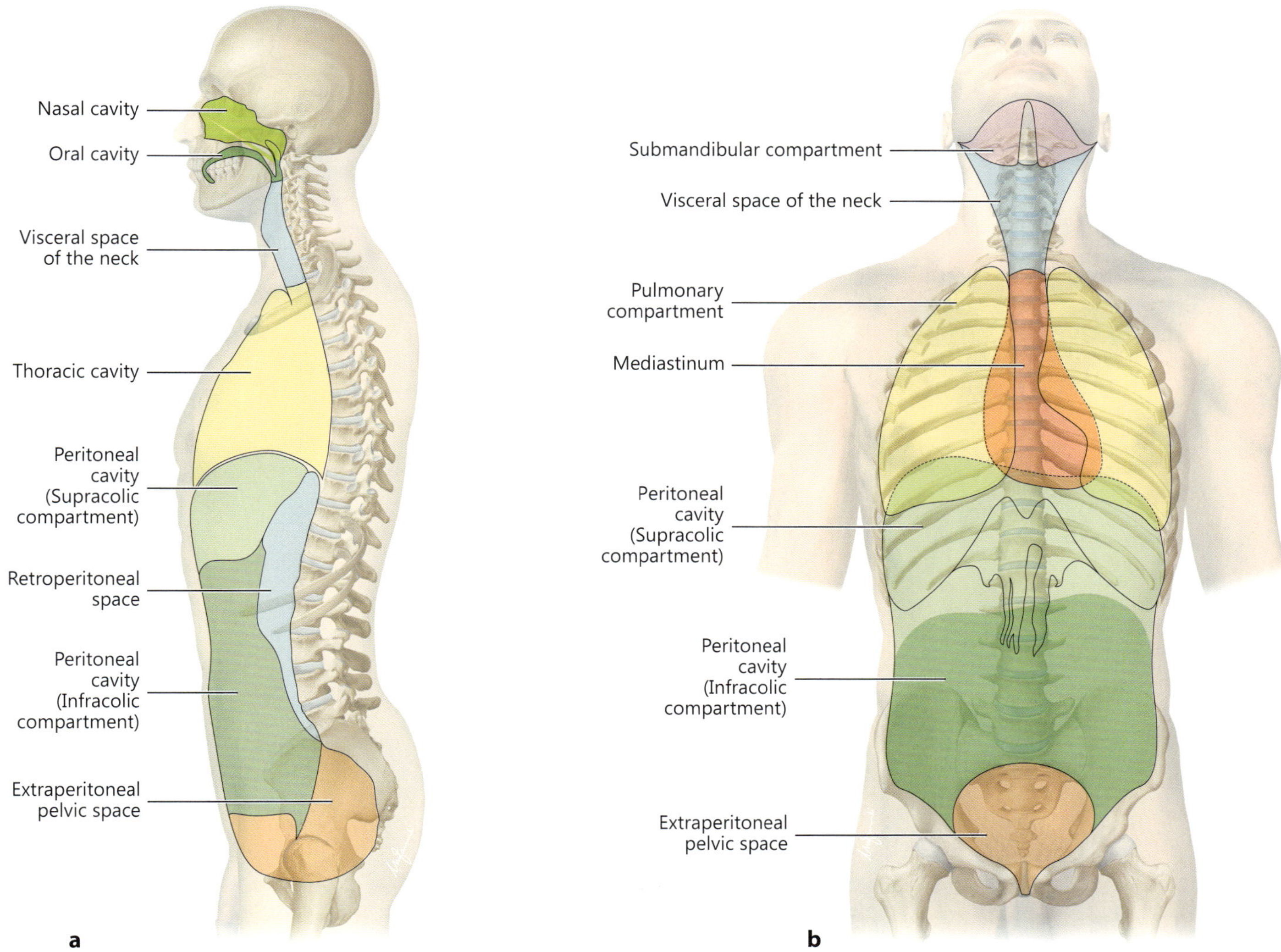

a, Lateral projection of the head and trunk. The **visceral cavities** of the **viscerocranium**, of the **neck**, the **thorax**, the **abdomen** and the **pelvis** are shown. Within the peritoneal cavity, a distinction is made between the supracolic and infracolic compartments. The continuity established between the various splanchnic cavities is also noted. **b**, Frontal projection. **c**, Boundaries and regions of the **anterior abdominal wall** as defined by the intersection of the midclavicular lines with the subcostal and transtubercular planes. The basisternal or xiphoid line represents the approximate level of the central tendon of the diaphragm.

1. Right hypochondrium
2. Epigastric region
3. Left hypochondrium
4. Left flank (left lumbar region)
5. Left groin (left iliac fossa)
6. Pubic region (hypogastric region)
7. Right groin (right iliac fossa)
8. Right flank (right lumbar region)
9. Umbilical region

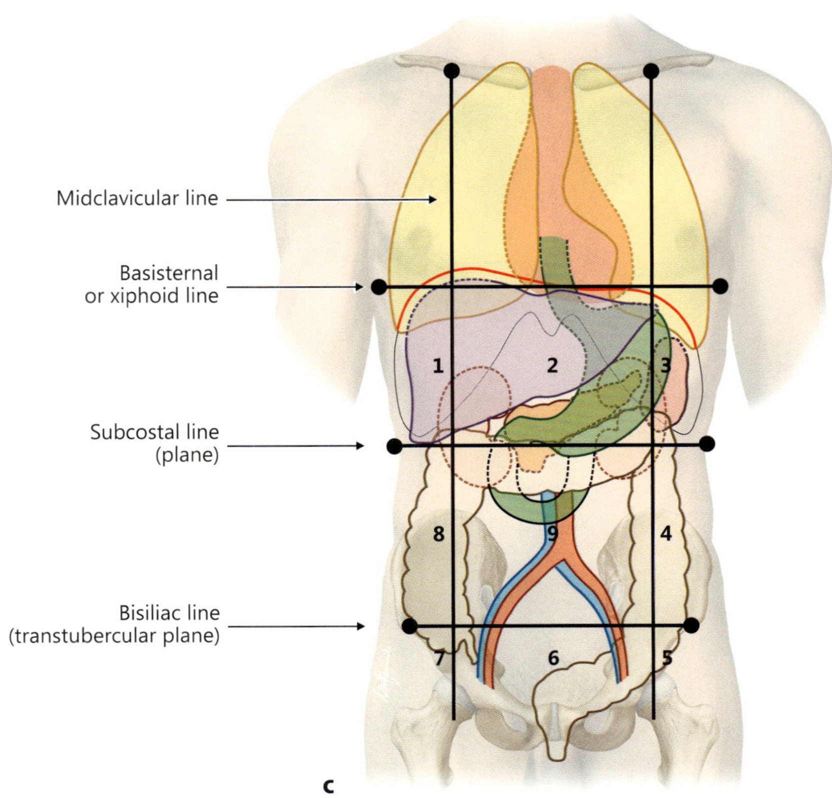

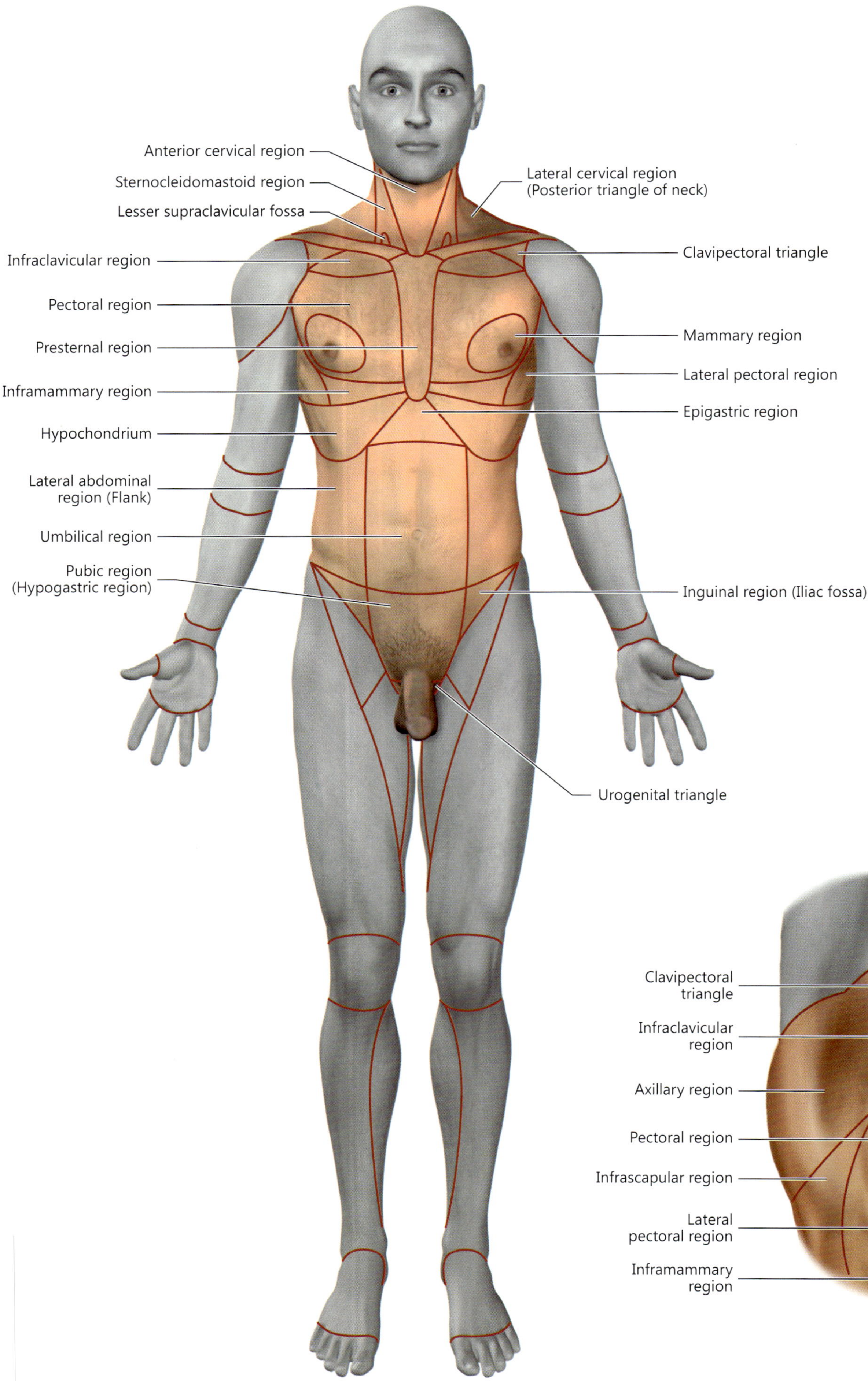

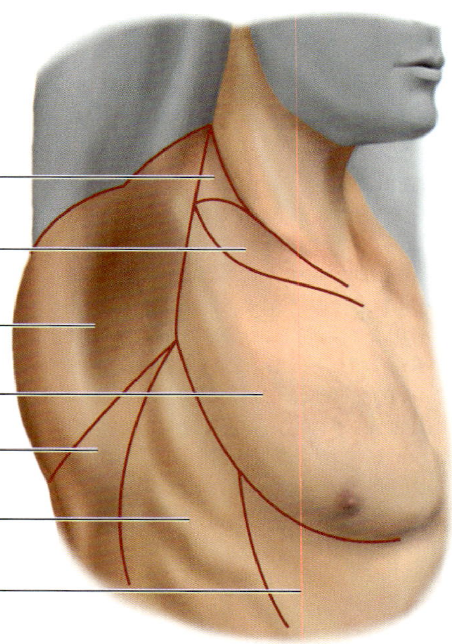

Detailed **subdivision** of the **regions of the trunk**. Anterior view.

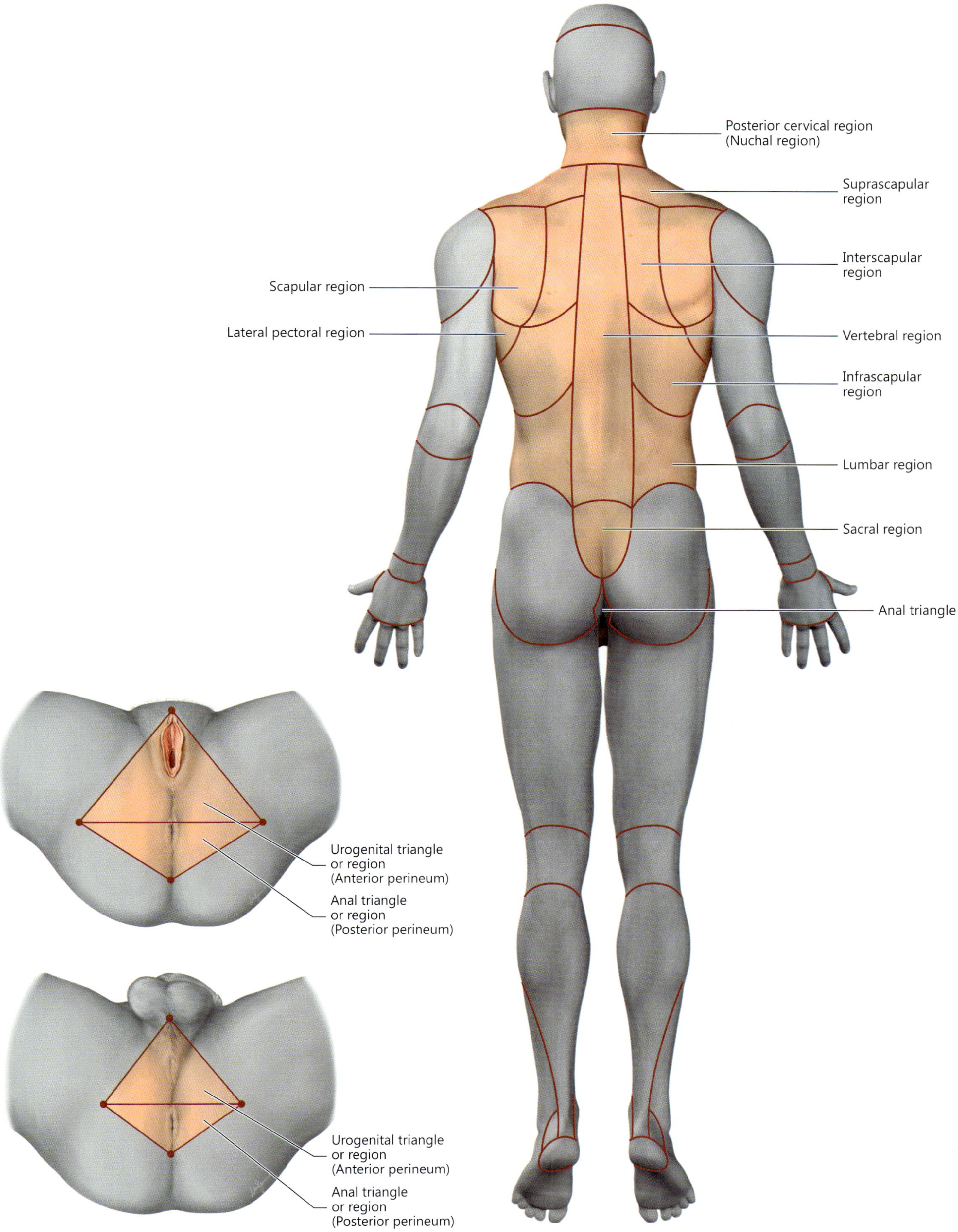

Detailed **subdivision** of the **regions of the trunk**. Posterior view and view of the perineal region.

Anatomical landmarks on the surface of the trunk		
	STRUCTURE	*LOCATION*
1	Jugular notch (suprasternal notch)	Rounded depression at the superior border of the sternum, between the medial extremities of the clavicle
2	Xiphoid process	Inferior part of the sternum
3	Right and left anterior superior iliac spine	Rounded bony prominence at the anterior extremity of the iliac crest
4	Iliac crest	Superior border of the iliac part of the hip bone; convex bony margin whose apex is located at the level of the space between the spinous processes of L IV and L V
5	Right and left posterior superior iliac spine	Rounded bony prominence on the posterior extremity of the iliac crest, palpable beneath the skin in the dimples located on the proximal surface of the buttocks
6	Spinous process of S II	On the central point of a line connecting the two posterior superior iliac spines
7	Inferior angle of the scapula	At the lowest point of the medial border of the scapula
8	Spine of the scapula	Bony margin that obliquely crosses the upper four-fifths of the scapula
9	Spinous process of T VIII	On the median line between the two inferior angles of the scapula
10	Spinous process of T III	In the middle of a line between the origins of the scapular spines
11	Spinous process of vertebra prominens [C VII]	The most prominent spinous process, located at the base of the neck

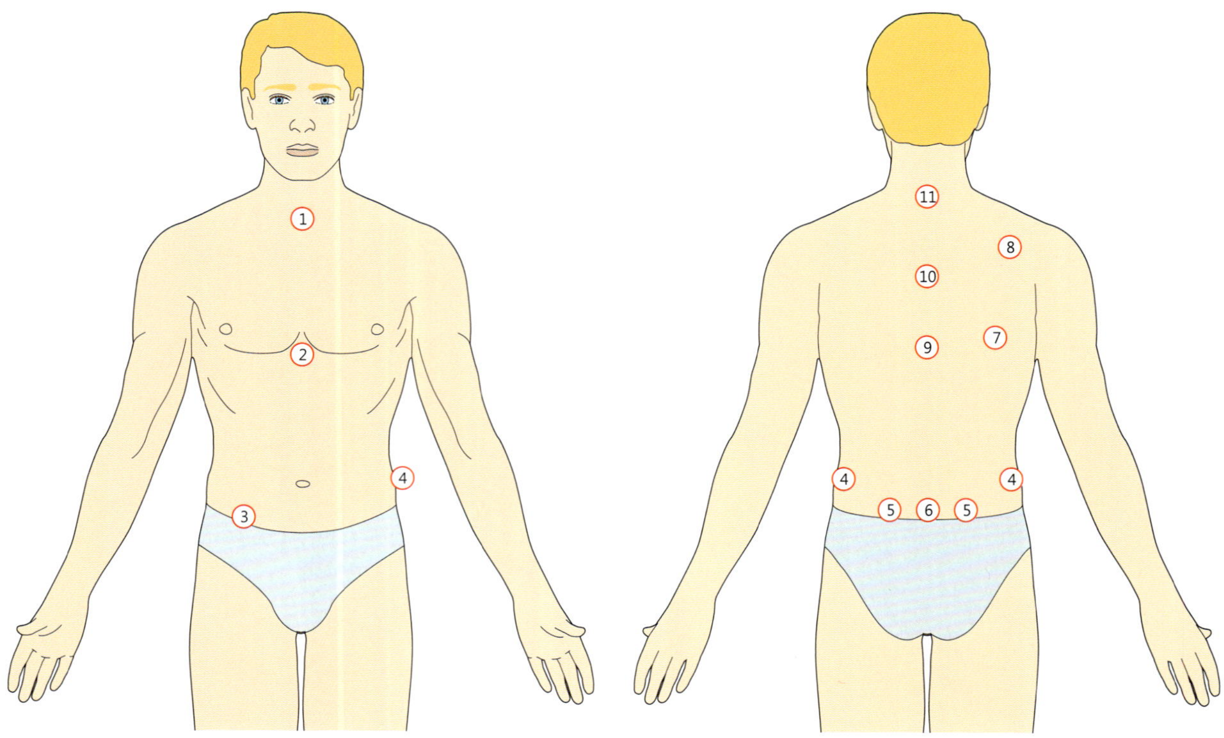

Anatomical landmarks on the surface of the shoulder and pectoral girdle		
	STRUCTURE	**LOCATION**
1	Inion	Process that marks the center of the superior nuchal line
2	Spinous process of vertebra prominens [C VII]	Non-bifid spinous process, particularly long and dorsally prominent, easily palpable at the base of the neck
3	Acromion	Lateral part of the spine of the scapula, at the apex of the shoulder
4	Spine of the scapula	Bony crest passing obliquely over the posterior surface of the scapula between its upper quarter and the lower three-quarters
5	Medial border of the scapula	Laterally to the spinous processes of the thoracic vertebrae, 5-6 cm from them, covering the 2nd to the 7th ribs
6	Inferior angle of the scapula	Inferiorly to the medial border of the scapula
7	Clavicle	Prominent S-shaped bone in the anterosuperior part of the thorax
8	Coracoid process	Prominence directed upwards and outwards distally to the passage between the middle third and lateral third of the clavicle, approximately 2.5 cm from it, in the clavipectoral triangle, deep to the anterior fibers of the deltoid

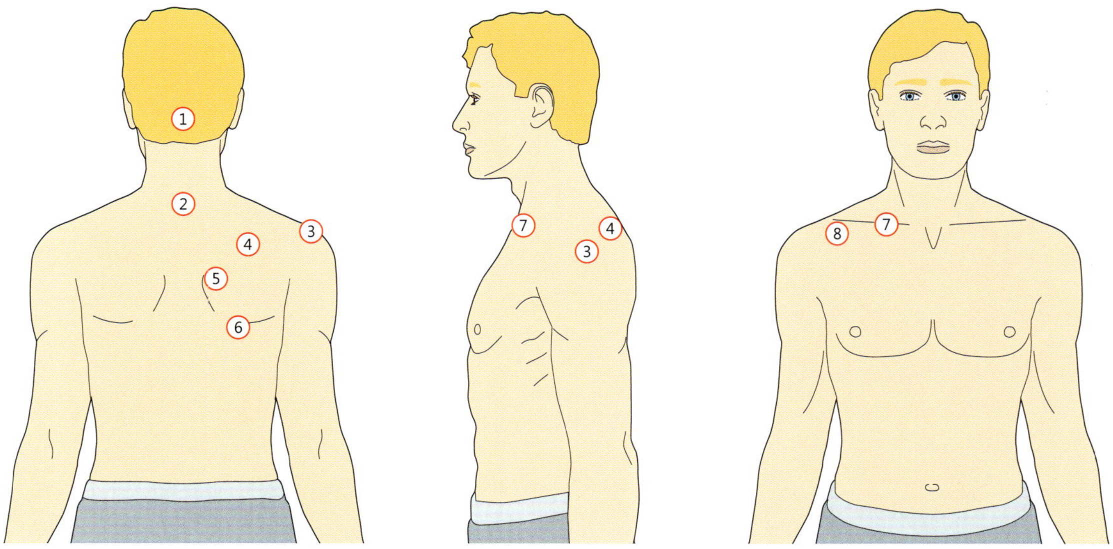

Anatomical landmarks on the surface of the pelvis and hip		
	STRUCTURE	**LOCATION**
1	Iliac crest	Convex bony margin representing the superior border of the ilium; the highest part of the iliac crest is at the level of the space between the L IV and L V spinous processes
2	Anterior superior iliac spine	Rounded bony prominence on the anterior extremity of the iliac crest
3	Iliac tuberosity	Approximately 5 cm above and lateral to the anterior superior iliac spine, along the lateral border of the iliac crest
4	Posterior superior iliac spine	Rounded bony prominence on the posterior extremity of the iliac crest, palpable through the skin at the base of the dimples above the buttocks; the posterior superior iliac spines are at the level of the spinous process of S II

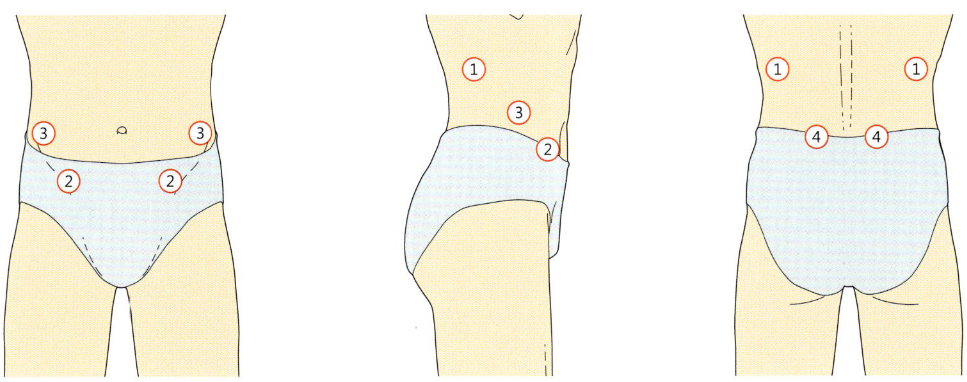

Cavities and spaces of the trunk. Stratigraphic examination starting from the posterior wall in frontal projection. **a**, After removing the organs, it is possible to appreciate the morphology of the posterior walls of the different trunk regions: neck space, thoracic cavity, abdominal cavity and pelvic cavity.

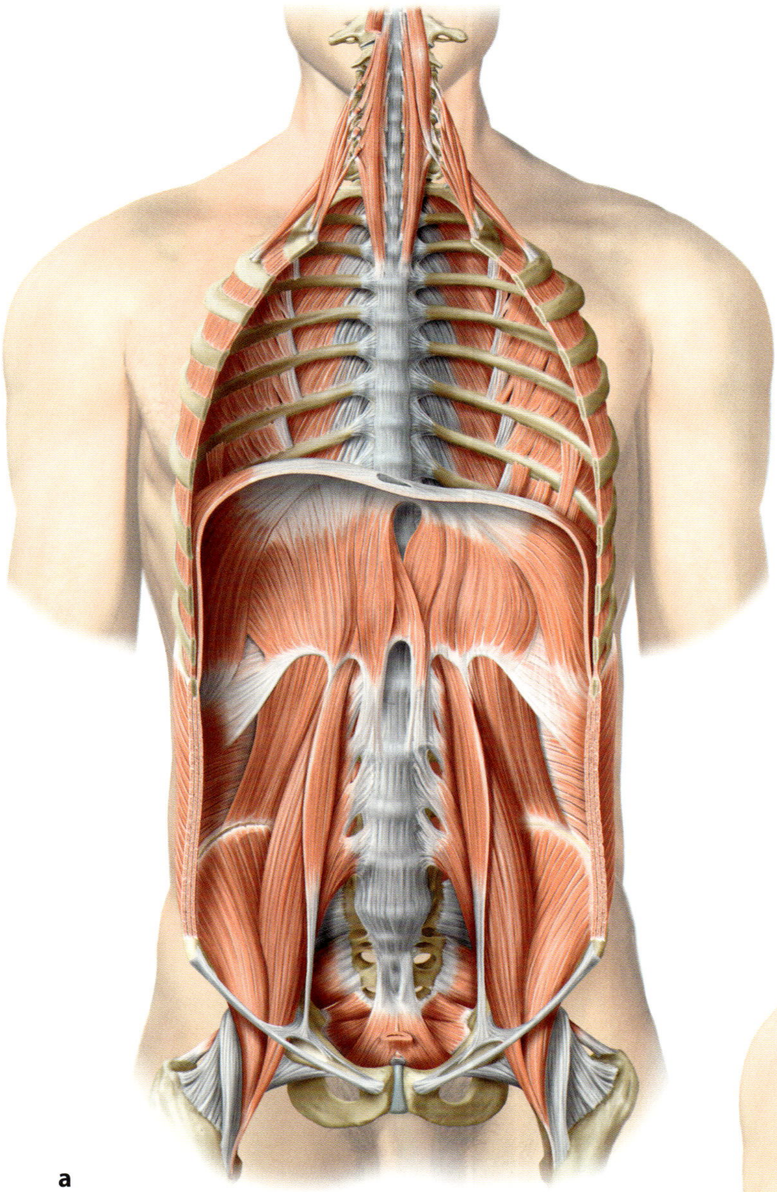

a

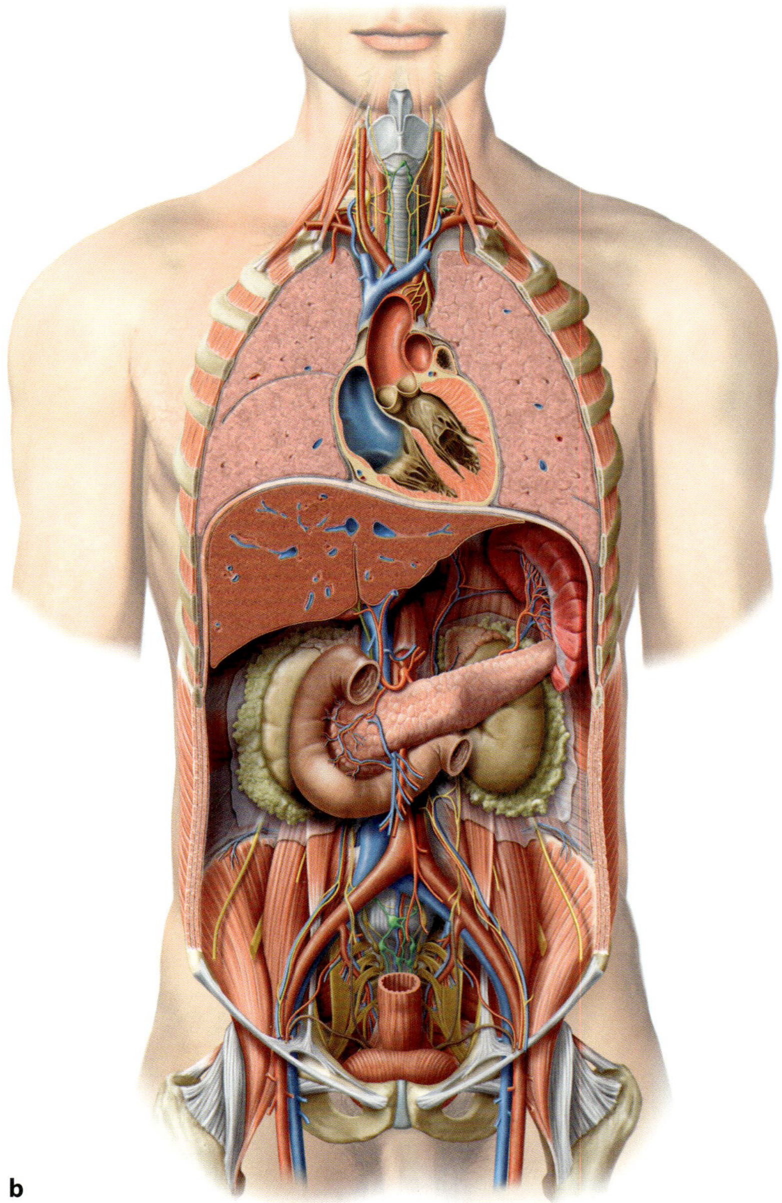

b

Cavities and spaces of the trunk. Stratigraphic examination starting from the posterior wall in frontal projection. **b**, Frontal plane passing through the trachea and the origin of the large arterial vessels of the heart, in the neck and thorax. The pulmonary cavities can be distinguished to the sides of the mediastinum. In the abdominal cavity, the retroperitoneal space contains the primary organs (kidneys, suprarenal glands, ureters, pancreas, and duodenum) and vascular formations.

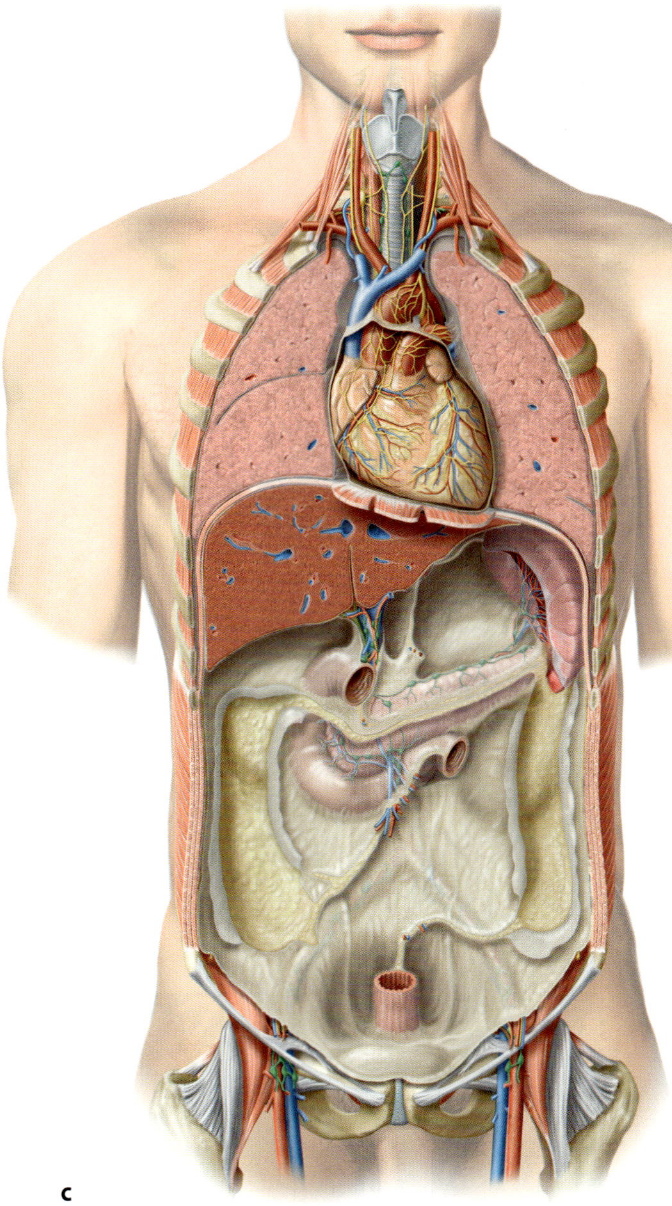

c

Cavities and spaces of the trunk. Stratigraphic examination starting from the posterior wall in frontal projection. **c**, Stratigraphic plane passing through the neck in front of the trachea and through the thorax in front of the heart. In the thoracic region, the organs of the middle mediastinum can be seen: the heart and great vessels of the heart. The trachea is also visible in the posterior mediastinum. The pulmonary cavities contain the lungs and pleurae. At the abdominal level, it can be seen how the posterior parietal peritoneum extends over the posterior abdominal wall and over the organs that lie anterior to it.

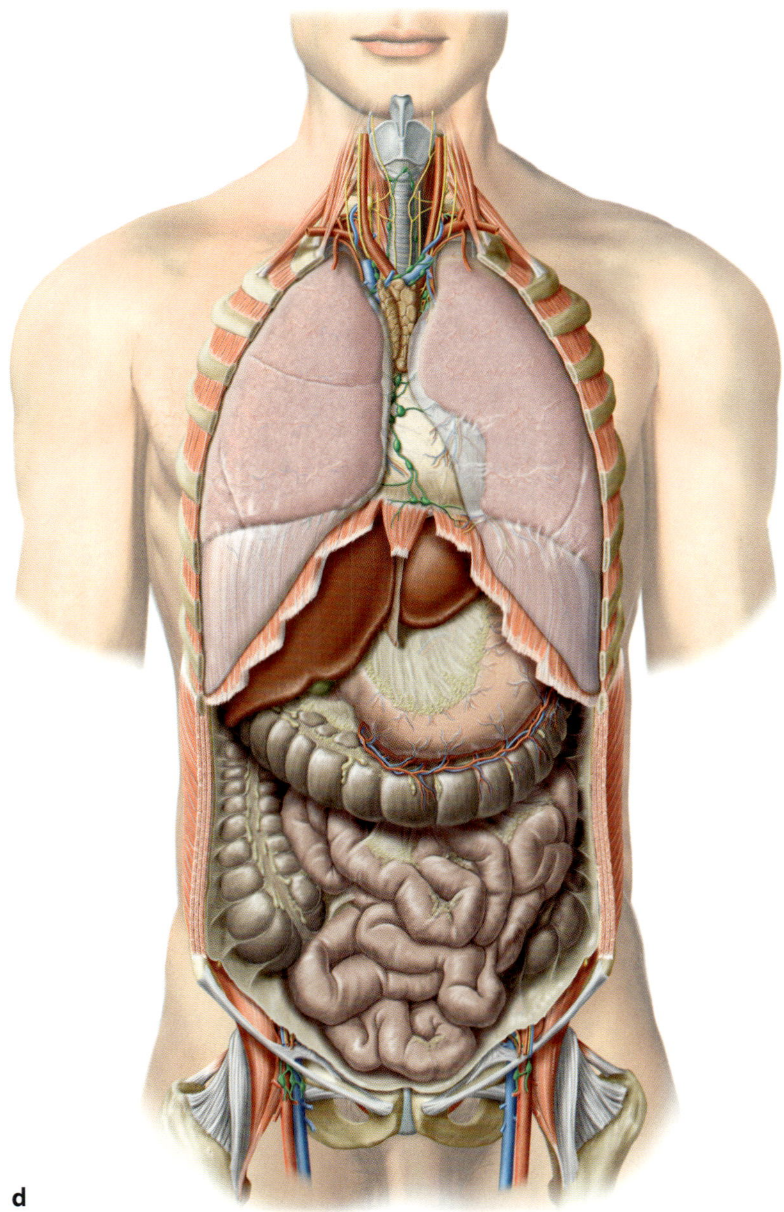

d

Cavities and spaces of the trunk. Stratigraphic examination starting from the posterior wall in frontal projection. **d**, Stratigraphic plane passing through the neck in front of the trachea and through the thorax in front of the thymus and costomediastinal pleural sinuses (costomediastinal recesses). In the abdominal region, the organs contained within the supracolic compartment and infracolic compartment of the peritoneal cavity can be seen.

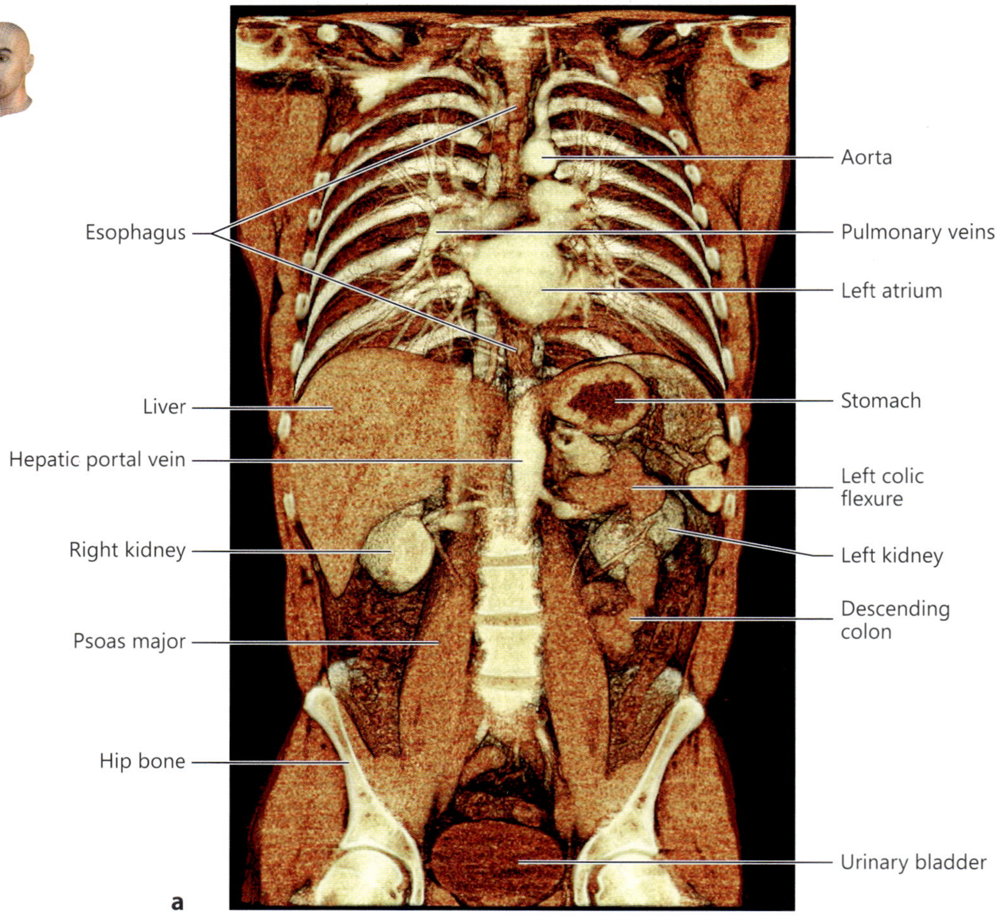

a

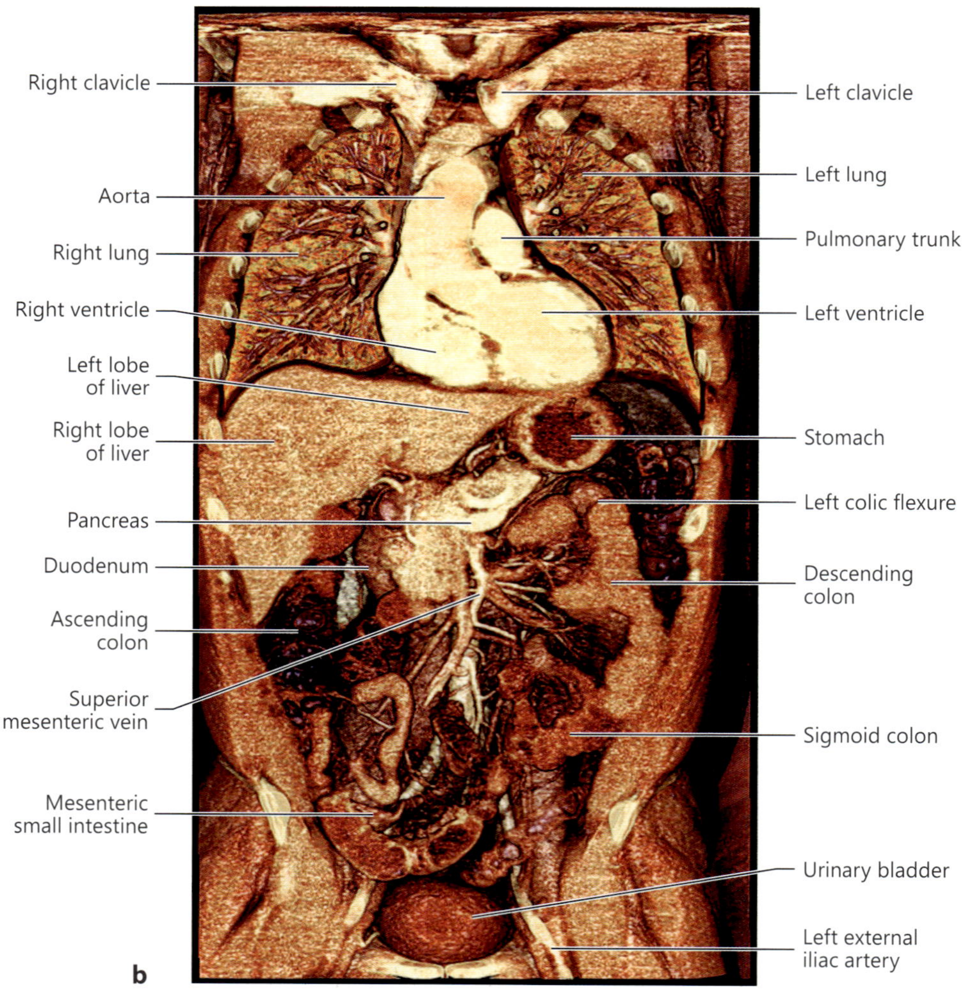

b

Three-dimensional images of a living human body, obtained by CT scan using a 3D volume rendering technique, showing the organs of the **thoracic cavity** and **abdominal cavity**.
a, Frontal projection of the trunk. In the thoracic cavity, it is possible to discern the posterior portion of the left atrium and ventricle, the branching off the bronchial tree and the esophagus. In the abdominal cavity, the relationship of the right kidney to the liver and of the left kidney to the left colic flexure can be observed, as can the arrangement of the psoas muscles. The bladder is visible in the pelvic cavity.

Three-dimensional images of a living human body, obtained by CT scan using a 3D volume rendering technique, showing the organs of the **thoracic cavity** and **abdominal cavity**.
b, Frontal projection of the trunk in a more ventral plane than the previous. In the thoracic cavity, the relation of the heart and larger vessels to the lungs can be seen. In the abdominal cavity, the reciprocal relations between the organs are evident, in particular the course of the sigmoid colon. In the pelvic cavity, the relation of the bladder to the final intestinal loops can also be observed.

Three-dimensional images of a living human body, obtained by CT scan using a 3D volume rendering technique, showing the organs of the **thoracic cavity** and **abdominal cavity**.
c, Frontal projection of the trunk in a more ventral plane than the previous. In the abdomen, the structures of the retroperitoneal space (pancreas) as well as some of the loops of the mesenteric small intestine can be seen. In the thorax, the left ventricle, the right ventricle, the aorta and the pulmonary trunk can be discerned.

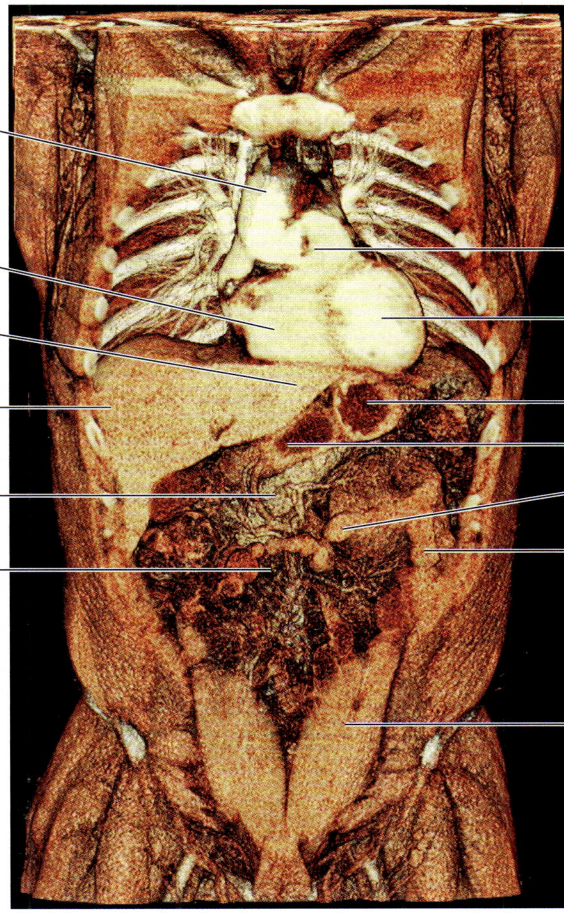

c

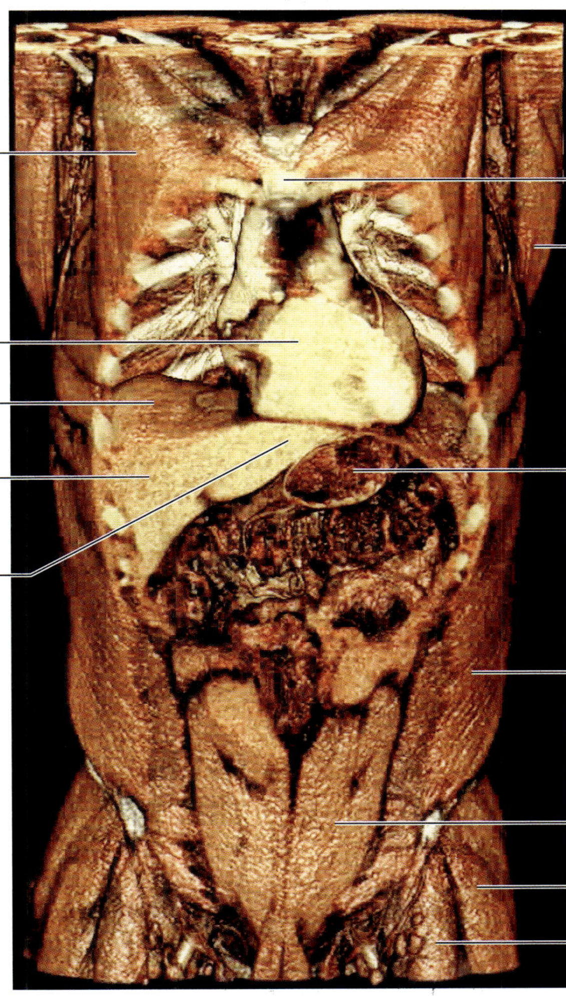

d

Three-dimensional images of a living human body, obtained by CT scan using a 3D volume rendering technique, showing the organs of the **thoracic cavity** and of the **abdominal cavity**.
d, Frontal projection of the trunk in a more ventral plane than the previous. Apart from the reciprocal relations of the thoracic and abdominal organs, the inferior portion of the rectus abdominis, which covers the pelvic cavity and part of the abdominal cavity, can be noted.

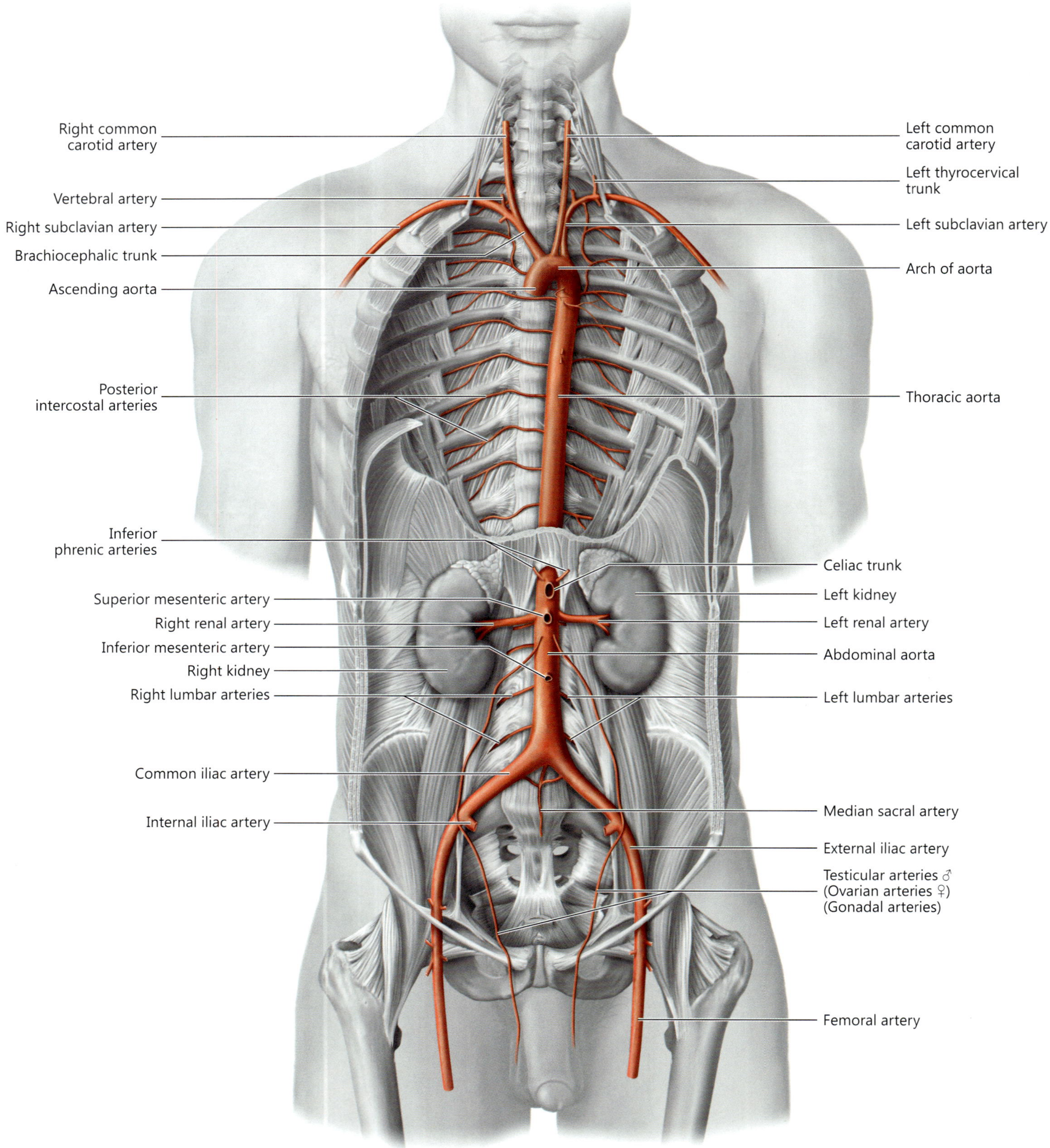

Schematic representation of the **course of the aorta** with its main subdivisions. The different portions of the aorta (ascending aorta, arch of the aorta, thoracic and abdominal descending aorta) send out collateral branches that provide visceral and parietal vascularization to the trunk.

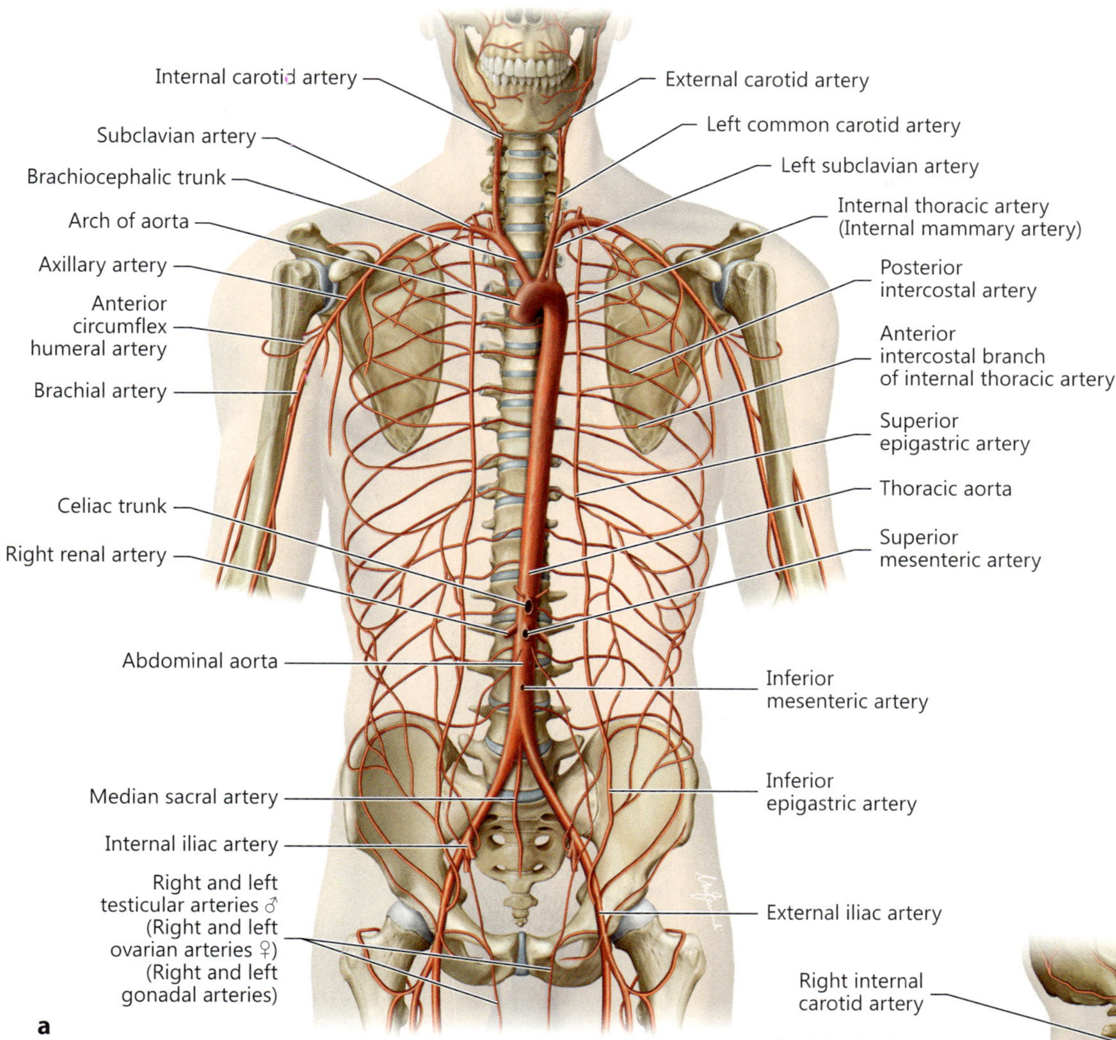

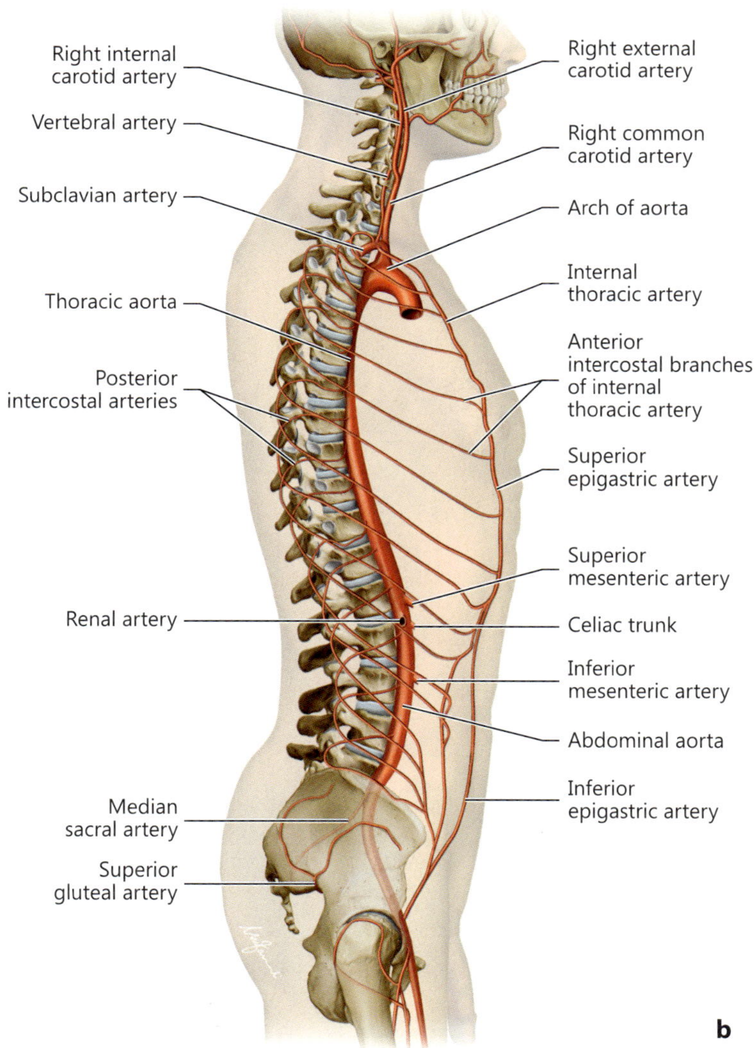

Schematic representation of the **course of the aorta** with its main subdivisions and **vascularization of the trunk wall**. Anterior view (**a**) and right lateral view (**b**).

Arterial Supply - 3D Reconstructions

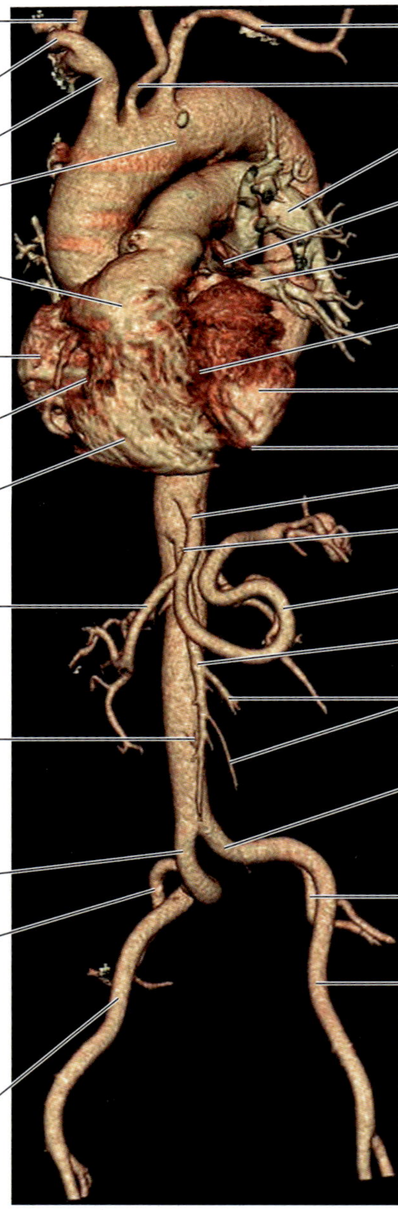

a

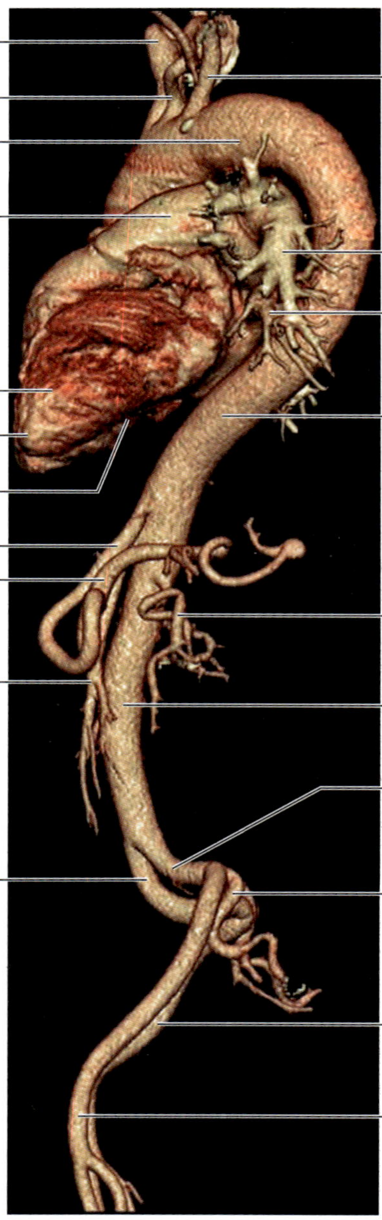

b

Three-dimensional images of a living human body, obtained by CT scan using a 3D volume rendering technique, showing the entire course of the **aorta**, its relations with the **heart** and its **major branches** in anterior (**a**) and left lateral view (**b**).

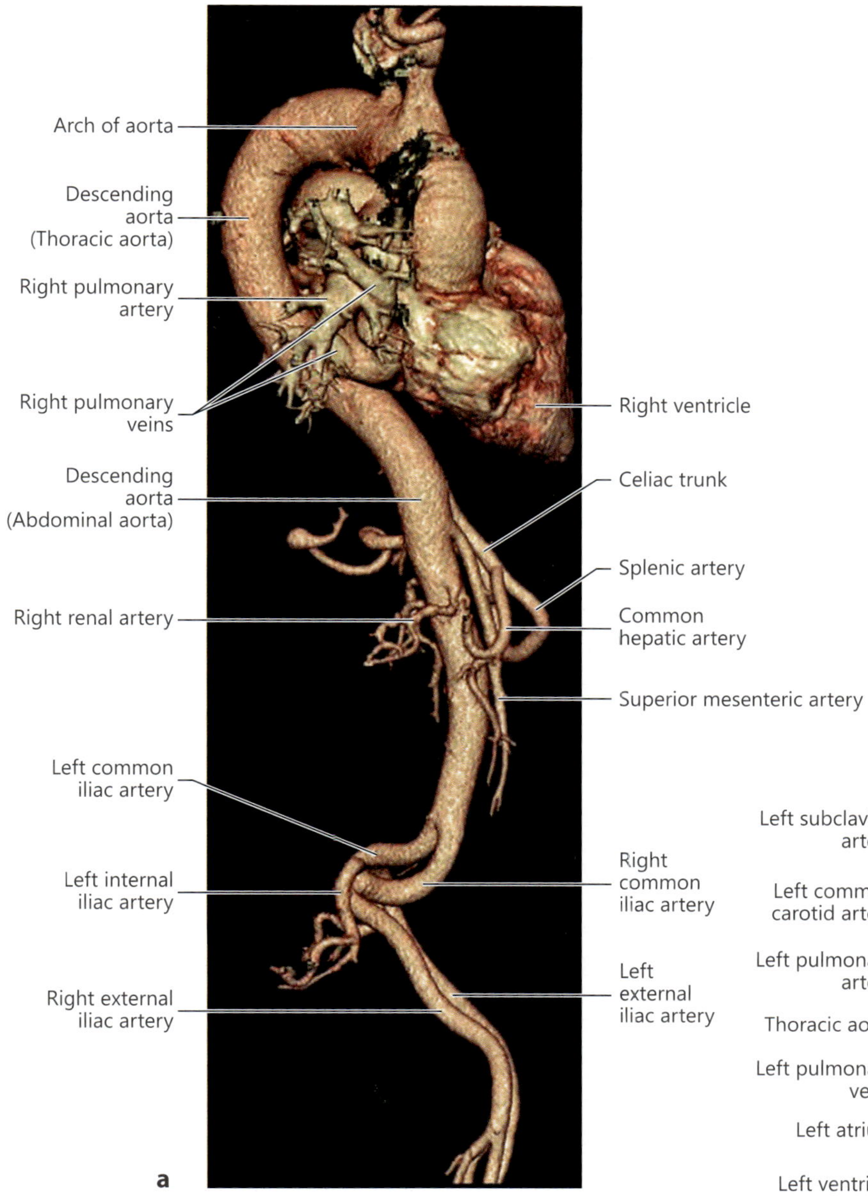

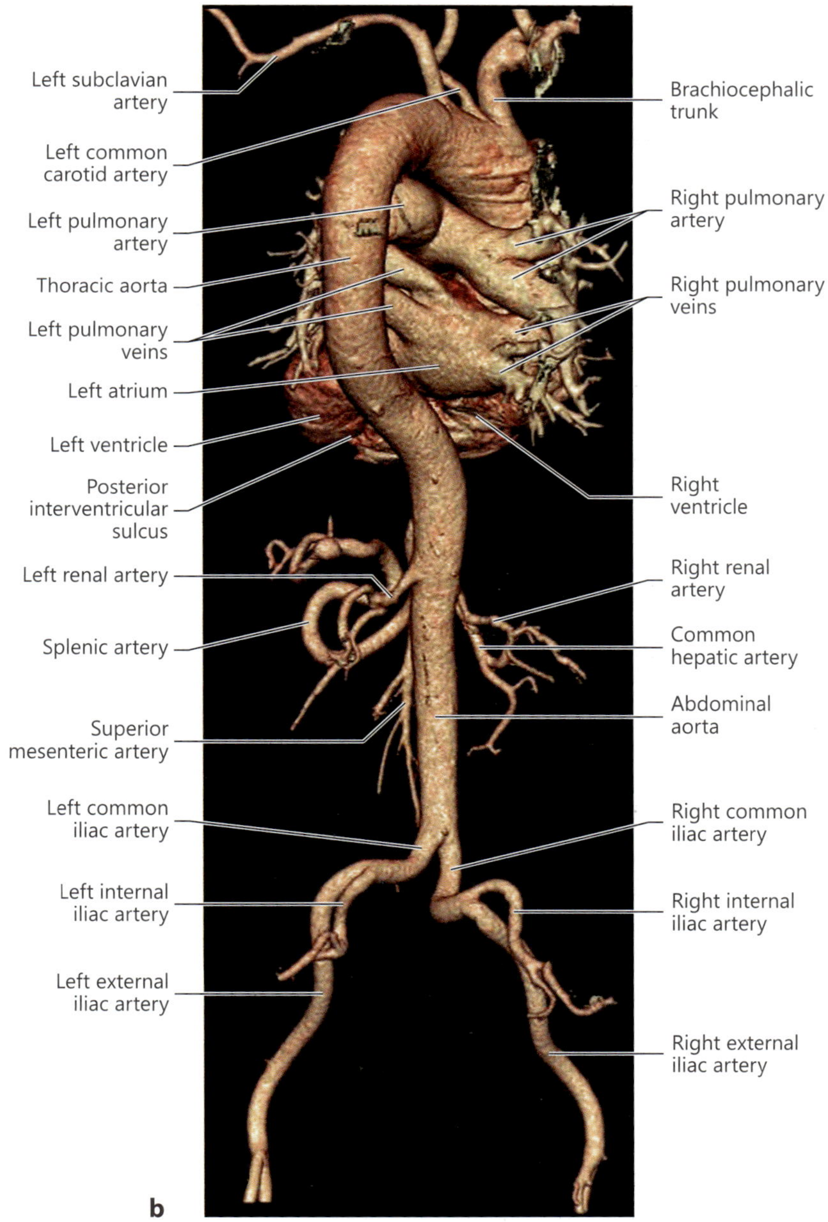

Three-dimensional images of a living human body, obtained by CT scan using a 3D volume rendering technique, showing the entire course of the **aorta**, its relations with the **heart** and its **major branches** in right lateral (**a**) and posterior view (**b**).

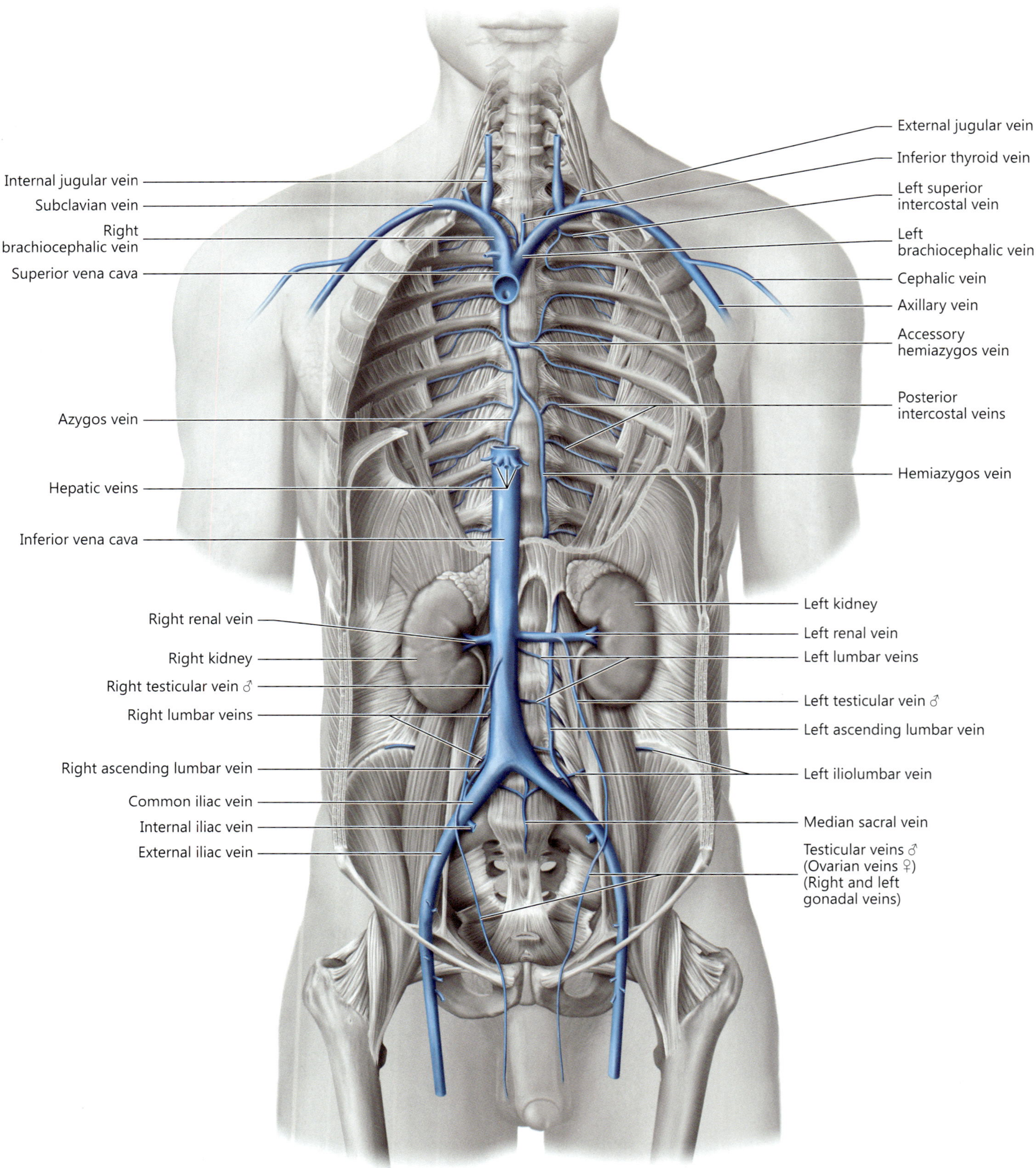

Schematic representation of the system of the **superior** and **inferior venae cavae**, with their respective **collecting veins**. The superior and inferior venae cavae provide visceral and parietal venous drainage to the supradiaphragmatic and subdiaphragmaitc trunk regions respectively. The azygos system of veins (azygos, hemiazygos, and accessory hemiazygos veins) represents important anastomotic circulation between the superior and inferior venae cavae.

Schematic representation of the **superior** and **inferior venae cavae**, their main **tributary veins** and the organization of the **deep parietal venous drainage of the trunk**. Anterior (**a**) and right lateral view (**b**). **c**, Schematic representation of the **portal vein system** and the main **collector branches of the inferior vena cava** with indication of the venous drainage regions. The portal vein drains all the organs of the digestive system contained in the peritoneal cavity as well as the spleen. Having passed through the hepatic circulation, the blood in this circle (splanchnic circulation) flows into the hepatic veins and then into the inferior vena cava.

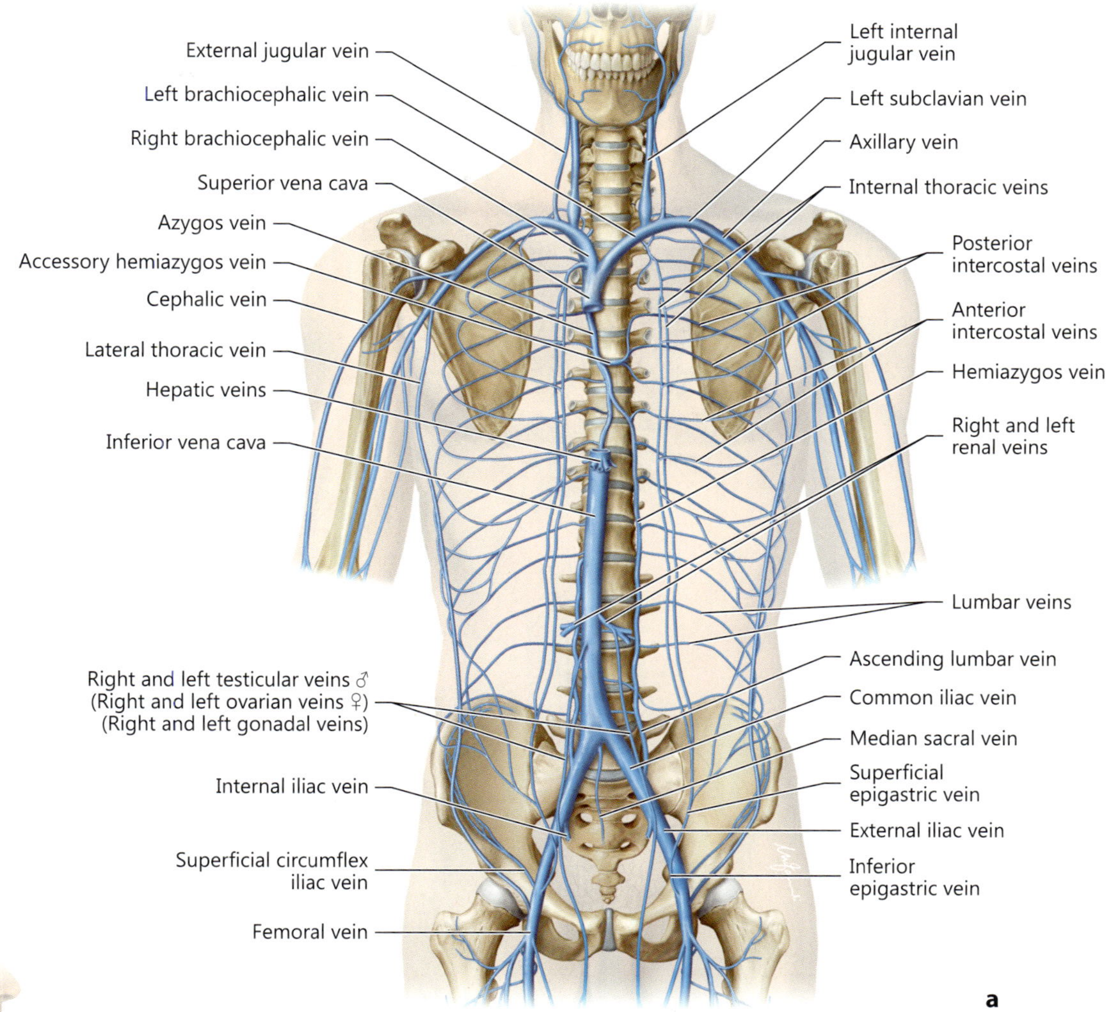

a

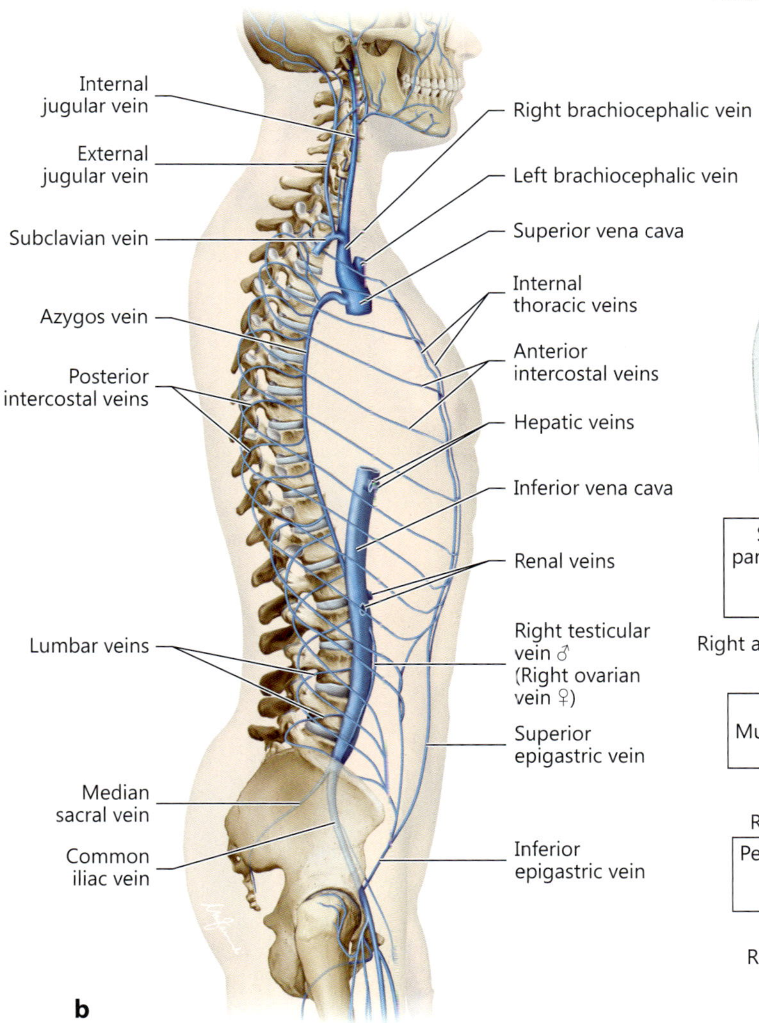

b

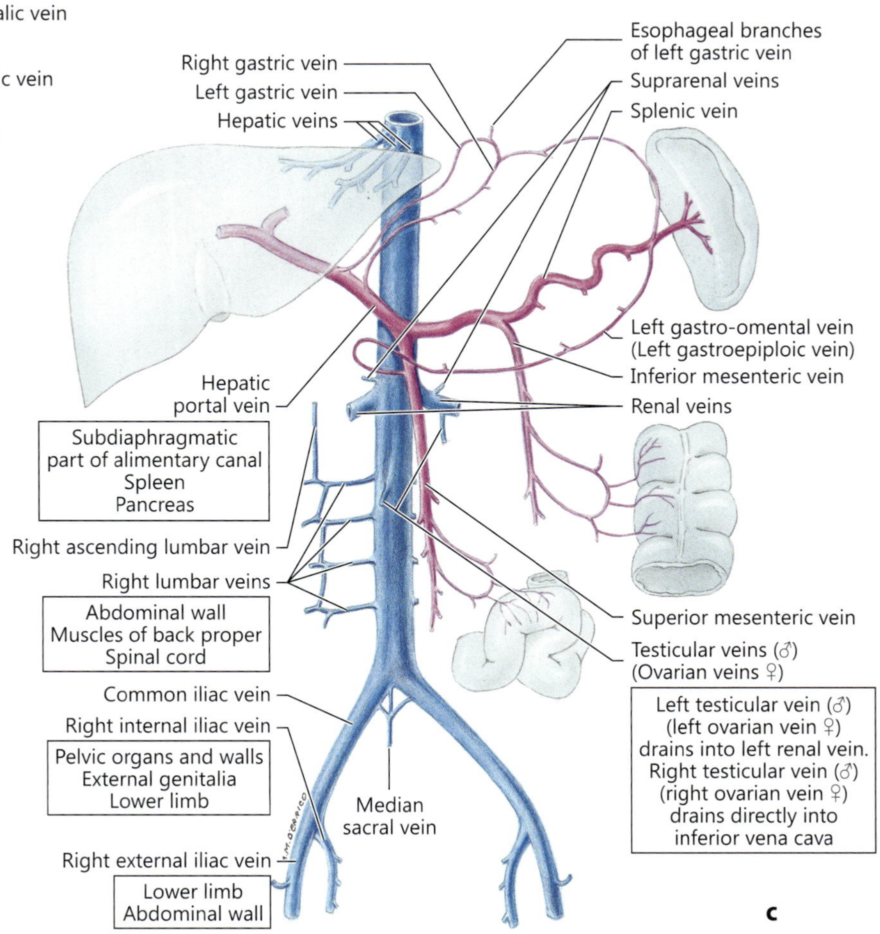

c

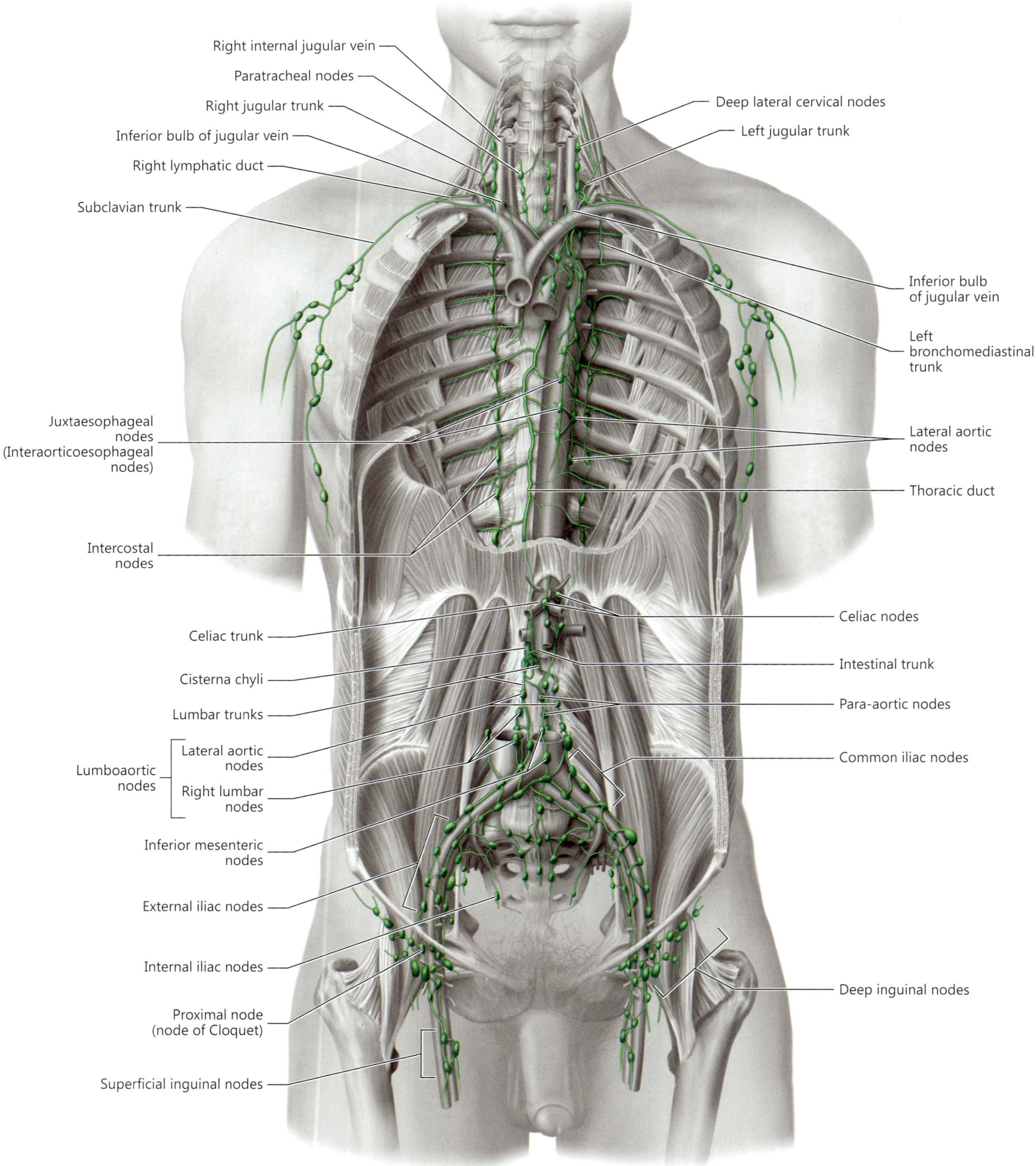

Schematic representation showing the positions of the main **lymphatic vessels of the trunk**, with their respective lymph node groups.

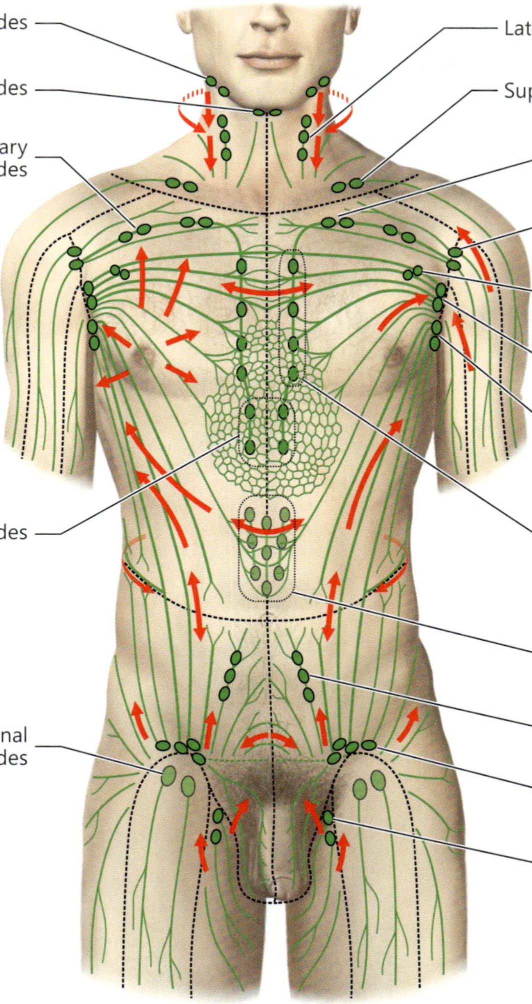

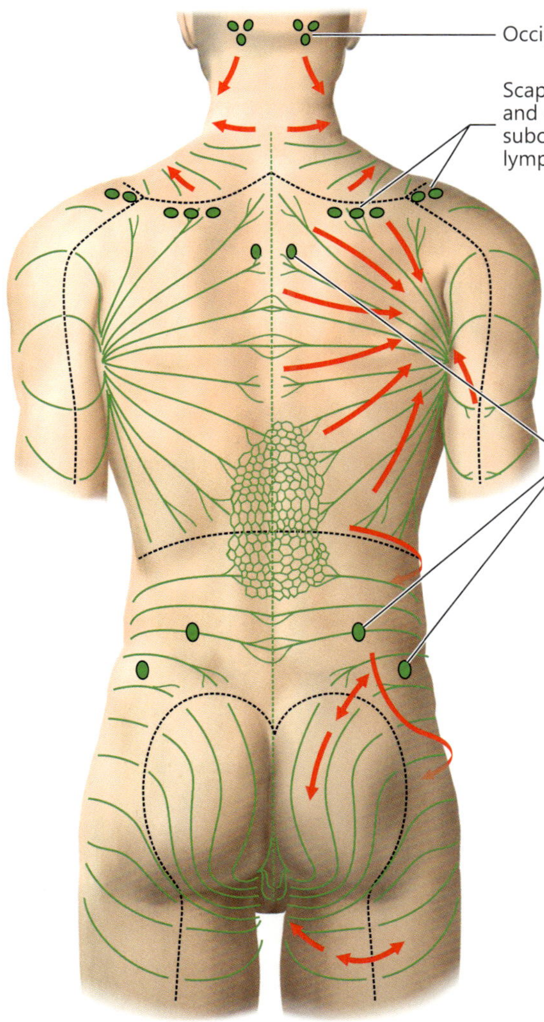

Projection onto the body surface of the main **lymphatic groups** and lymphatic drainage courses. Anterior view (**a**) and posterior view (**b**).

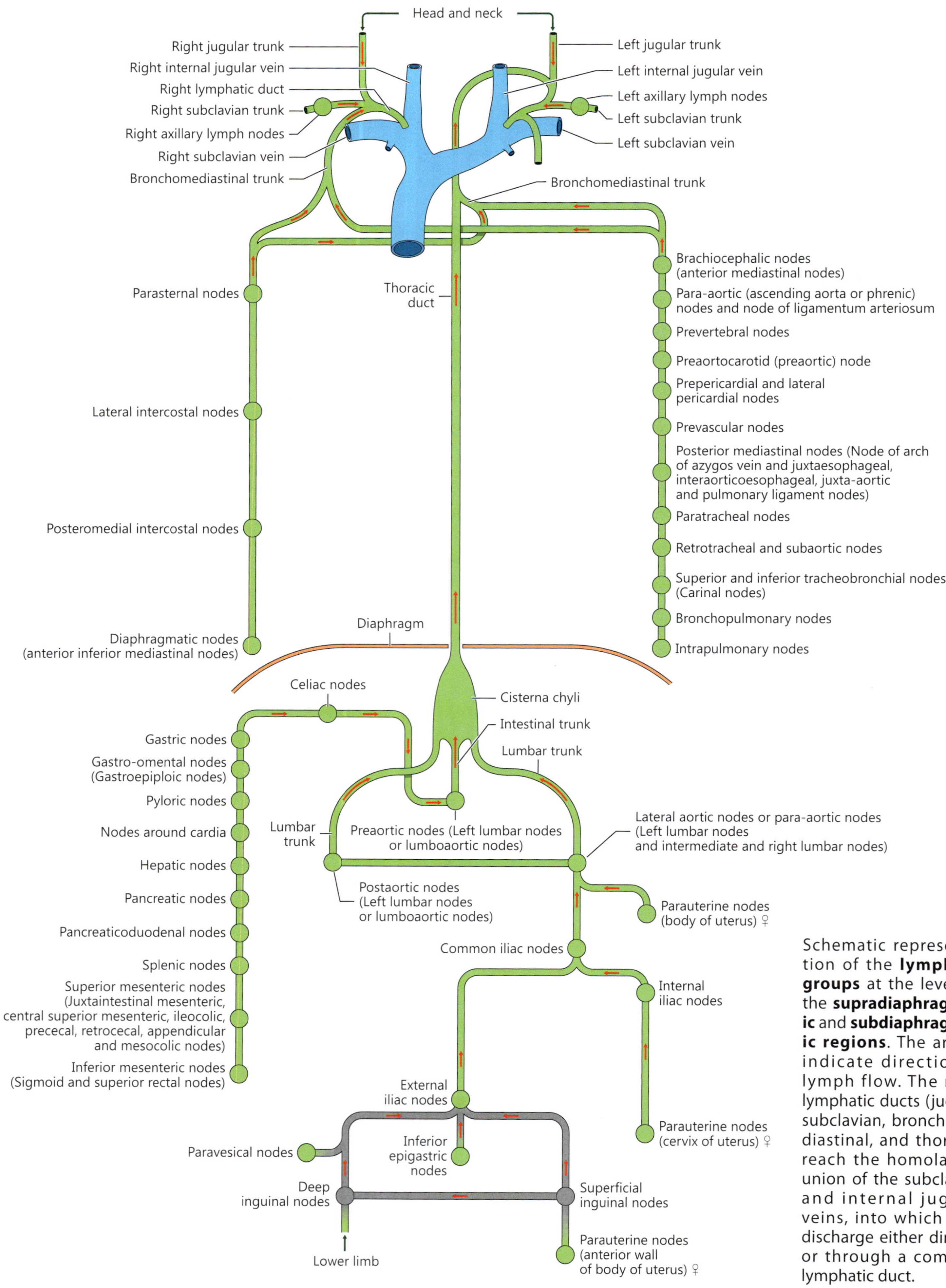

Schematic representation of the **lymphatic groups** at the levels of the **supradiaphragmatic** and **subdiaphragmatic regions**. The arrows indicate direction of lymph flow. The main lymphatic ducts (jugular, subclavian, bronchomediastinal, and thoracic) reach the homolateral union of the subclavian and internal jugular veins, into which they discharge either directly or through a common lymphatic duct.

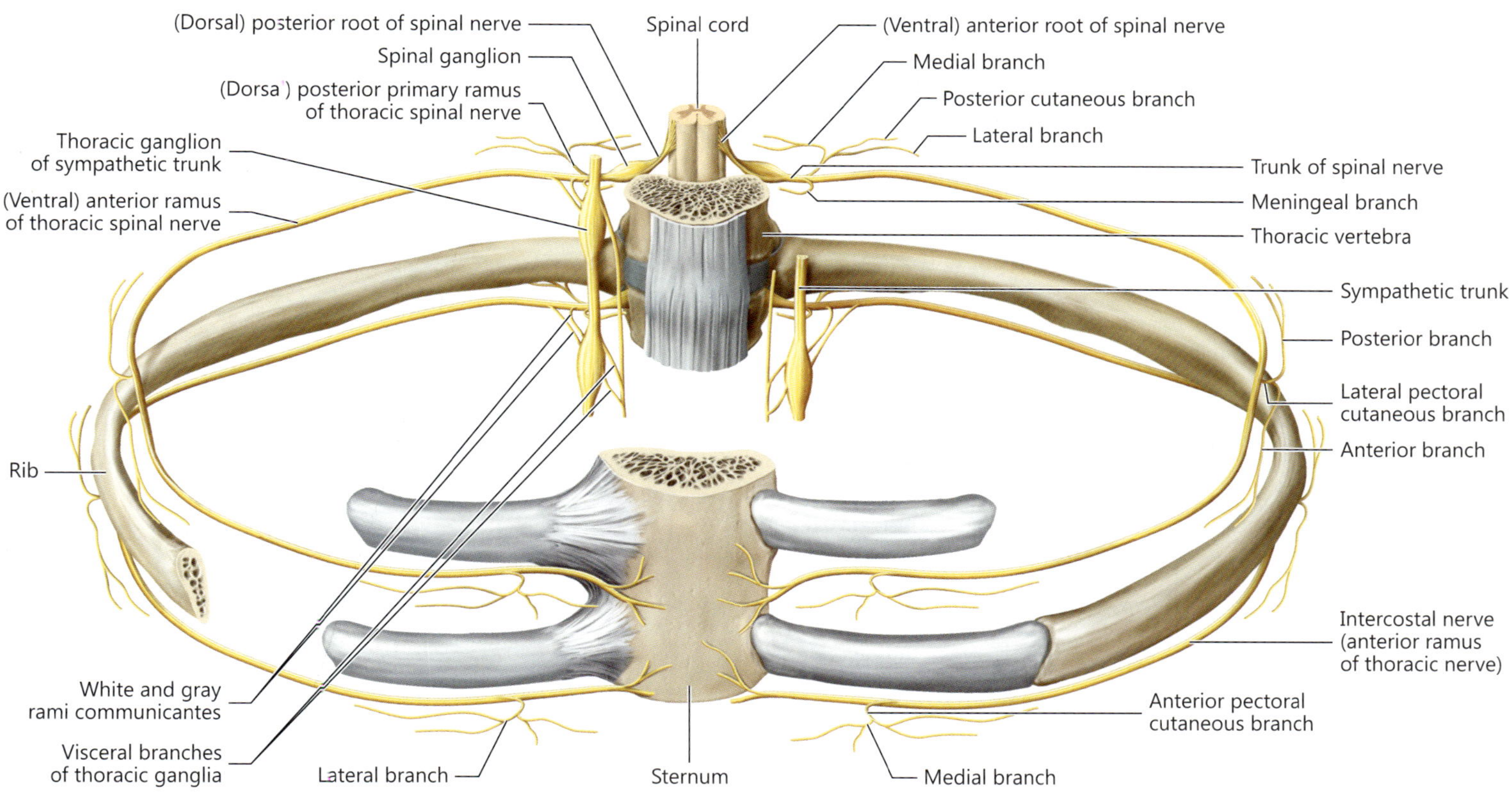

Schematic representation of the course of the **thoracic nerves** in relation to the ribs and sternum. The anterior rami of the thoracic nerves (intercostal nerves) travel forward following the course of the ribs, and send anterior, lateral, and medial cutaneous branches.

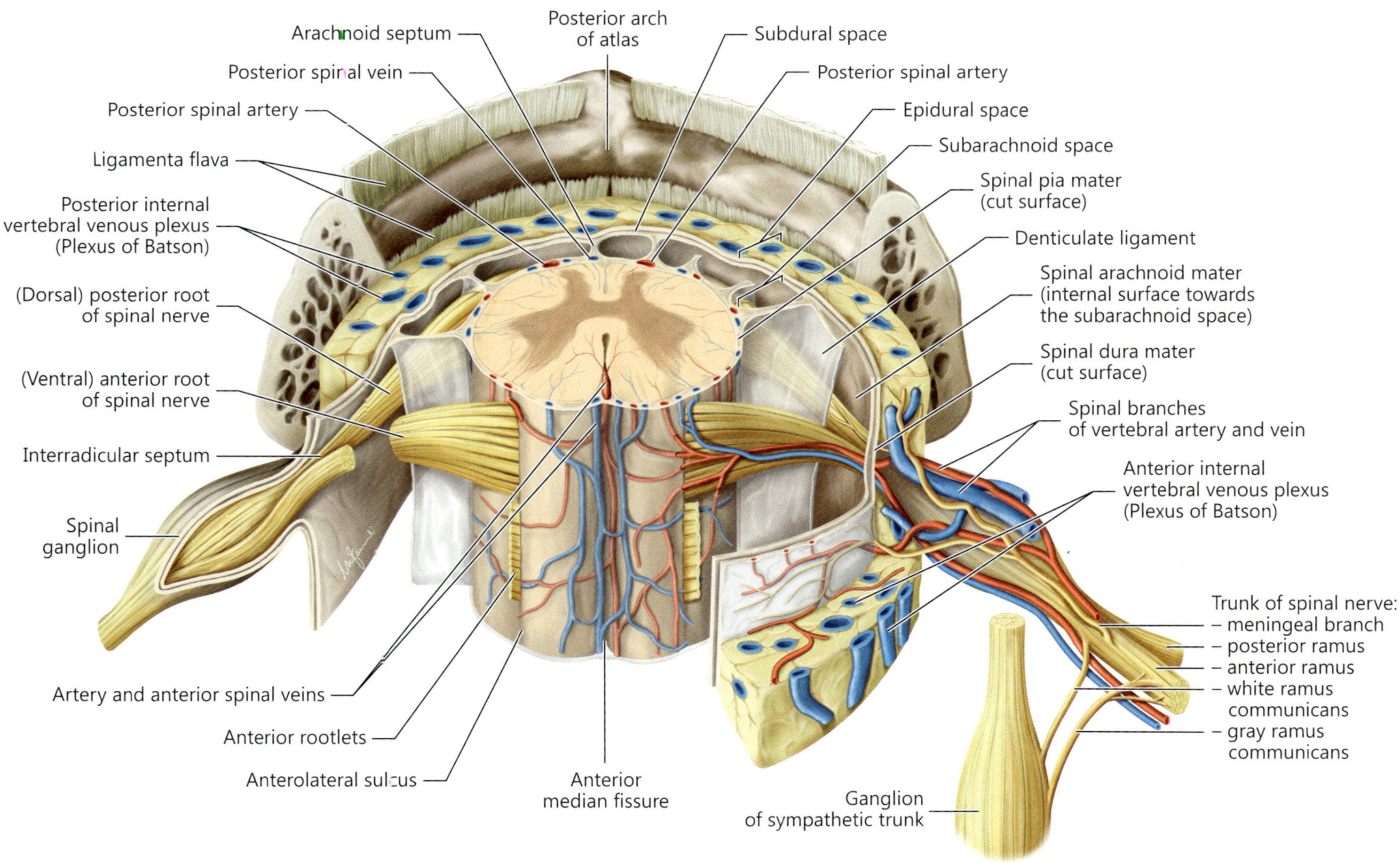

Three-dimensional image of a **transverse section of the spinal cord** passing through the first cervical vertebra. The relations of the spinal cord with the meninges and the spinal vessels can be noted.

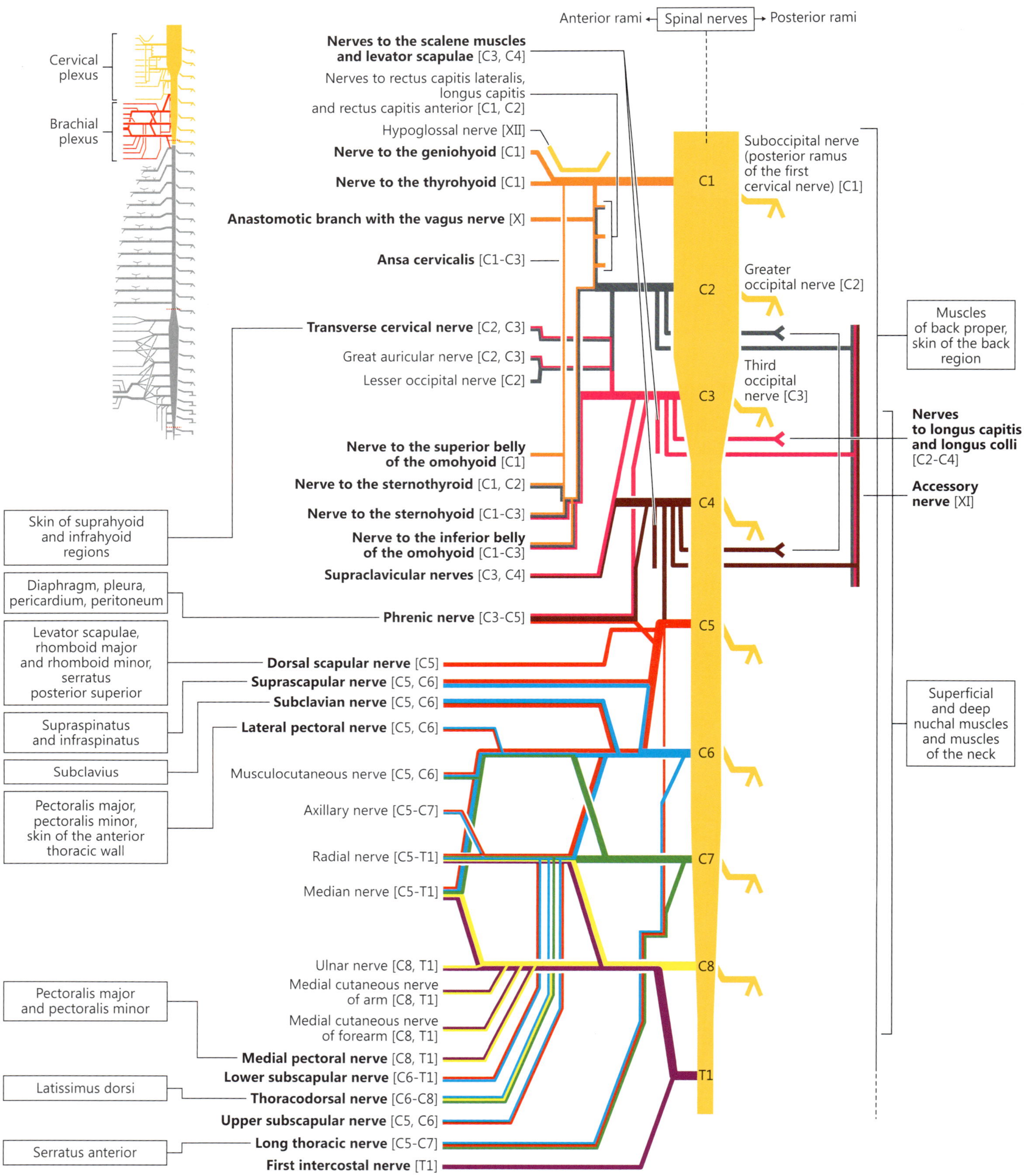

Schematic representation of the **cervical** and **brachial plexuses**. On the left-hand side of the diagram, in *yellow*, the cervical plexus is represented, formed by the anterior rami of the first four cervical nerves. It can be seen how each anterior ramus, excepting that of the 1st cervical nerve, divides into two branches, one ascending and one descending, which join to the corresponding branches of the contiguous nerves to constitute the cervical loops. The descending branch of the 4th cervical nerve anastomoses with the anterior ramus of the 5th cervical nerve, which is part of the brachial plexus. Anastomotic cutaneous and muscular branches originate from the cervical plexus. The brachial plexus is represented in various colors; it comprises the anterior rami of the 5th, 6th, 7th and 8th cervical nerves. The collateral and terminal branches originate from this plexus. The dorsal branches of the cervical nerves are represented on the right-hand side of the diagram.

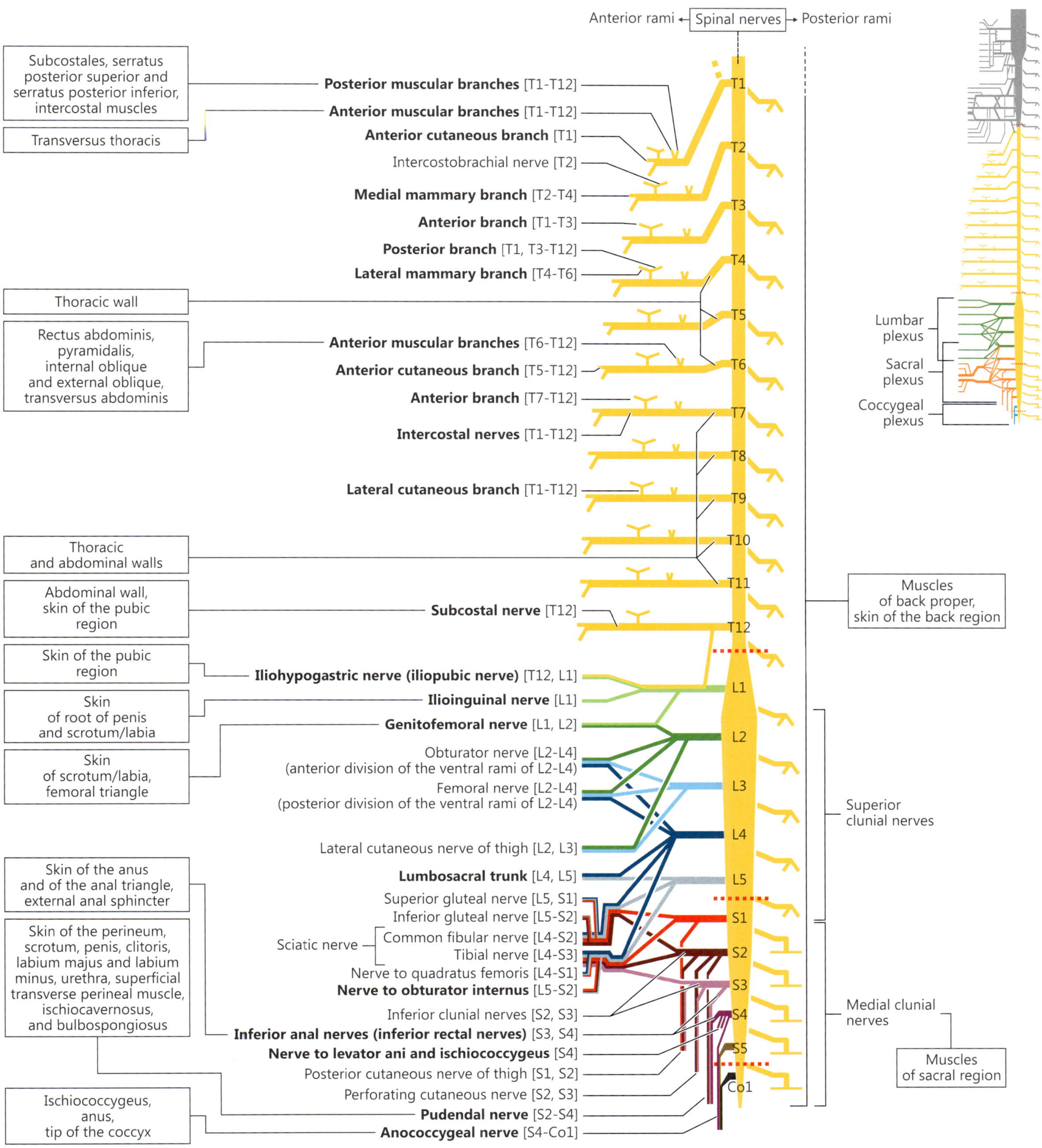

Schematic representation of the **intercostal nerves** and of the **lumbar**, **sacral**, and **coccygeal plexuses**. Proceeding from top to bottom on the left-hand side of the image, we can see primarily the system of intercostal nerves, comprising the anterior rami of the twelve pairs of thoracic nerves. These mixed nerves supply cutaneous innervation to the thoracic-abdominal wall and the thoracic muscles. Below them, the lumbar plexus is represented. It is formed by the anterior rami of L1-L4 with a contribution from T12; the lumbar plexus also gives rise to anastomotic collateral and terminal branches. The lumbar plexus is followed by the sacral plexus, formed by the anterior rami of L5, S1-S4, with a contribution from L4. These roots join in a common trunk from which the sciatic nerve and various collateral branches originate. Below the sacral plexus is the coccygeal plexus, formed by the anterior rami of S5-Co1, with a contribution from S4; it sends out anterior, visceral and posterior branches, the main one being the anococcygeal nerve. On the right side of the diagram, the dorsal branches of the intercostal, coccygeal, lumbar and sacral nerves are represented.

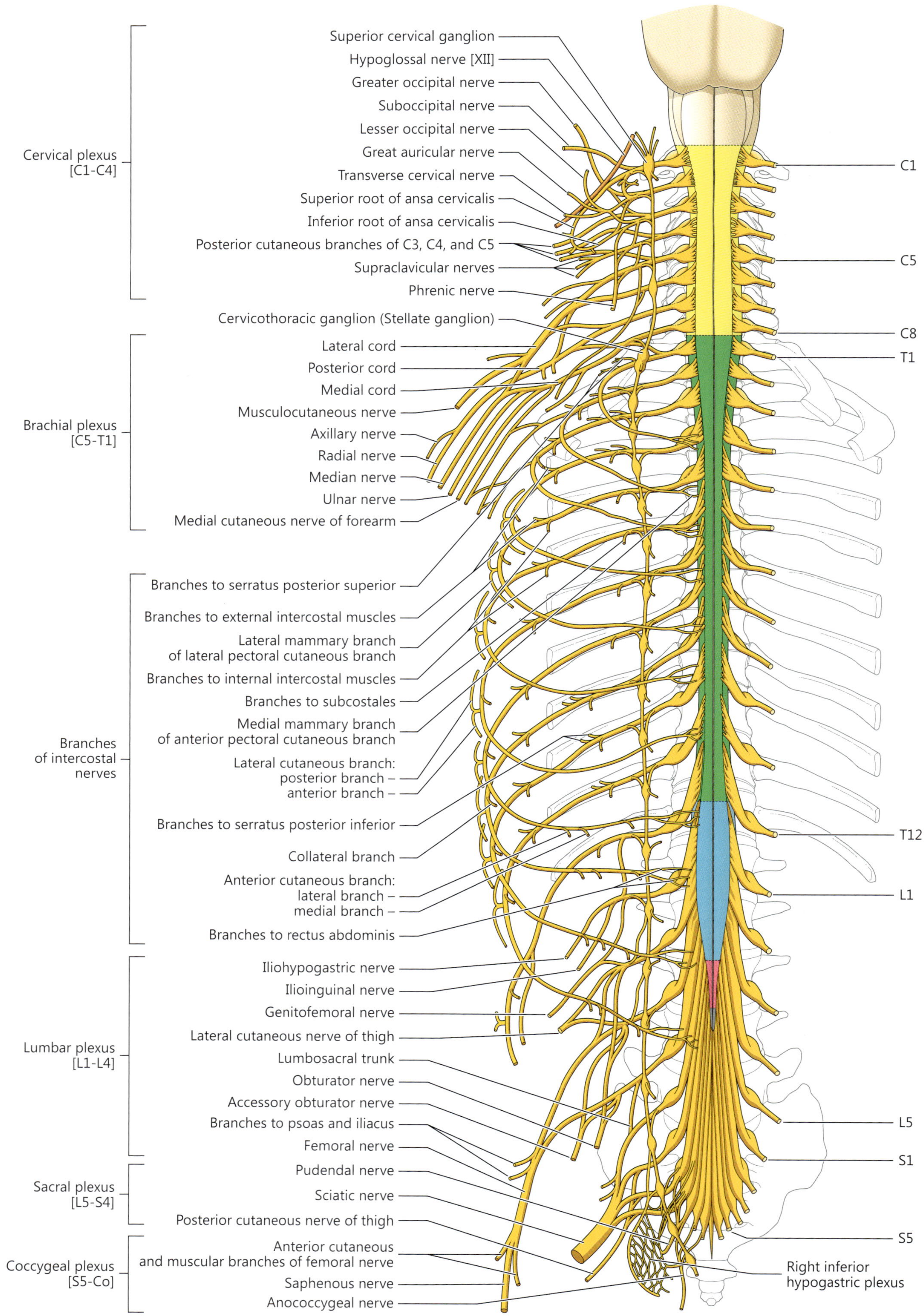
Cervical plexus [C1-C4]
Superior cervical ganglion
Hypoglossal nerve [XII]
Greater occipital nerve
Suboccipital nerve
Lesser occipital nerve
Great auricular nerve
Transverse cervical nerve
Superior root of ansa cervicalis
Inferior root of ansa cervicalis
Posterior cutaneous branches of C3, C4, and C5
Supraclavicular nerves
Phrenic nerve
Cervicothoracic ganglion (Stellate ganglion)
Brachial plexus [C5-T1]
Lateral cord
Posterior cord
Medial cord
Musculocutaneous nerve
Axillary nerve
Radial nerve
Median nerve
Ulnar nerve
Medial cutaneous nerve of forearm
Branches of intercostal nerves
Branches to serratus posterior superior
Branches to external intercostal muscles
Lateral mammary branch of lateral pectoral cutaneous branch
Branches to internal intercostal muscles
Branches to subcostales
Medial mammary branch of anterior pectoral cutaneous branch
Lateral cutaneous branch: posterior branch – anterior branch –
Branches to serratus posterior inferior
Collateral branch
Anterior cutaneous branch: lateral branch – medial branch –
Branches to rectus abdominis
Lumbar plexus [L1-L4]
Iliohypogastric nerve
Ilioinguinal nerve
Genitofemoral nerve
Lateral cutaneous nerve of thigh
Lumbosacral trunk
Obturator nerve
Accessory obturator nerve
Branches to psoas and iliacus
Femoral nerve
Sacral plexus [L5-S4]
Pudendal nerve
Sciatic nerve
Posterior cutaneous nerve of thigh
Coccygeal plexus [S5-Co]
Anterior cutaneous and muscular branches of femoral nerve
Saphenous nerve
Anococcygeal nerve
C1
C5
C8
T1
T12
L1
L5
S1
S5
Right inferior hypogastric plexus

Diagram illustrating the course of the **right** and **left vagus nerves**, their main relations with the aorta and the supra-aortic branches. The main components of the **sympathetic chain** are also shown, with their respective ganglia. In *yellow*, the emergence of the vagus nerve at the level of the posterior sulcus of the bulb; in *blue*, the components of the sympathetic chain.

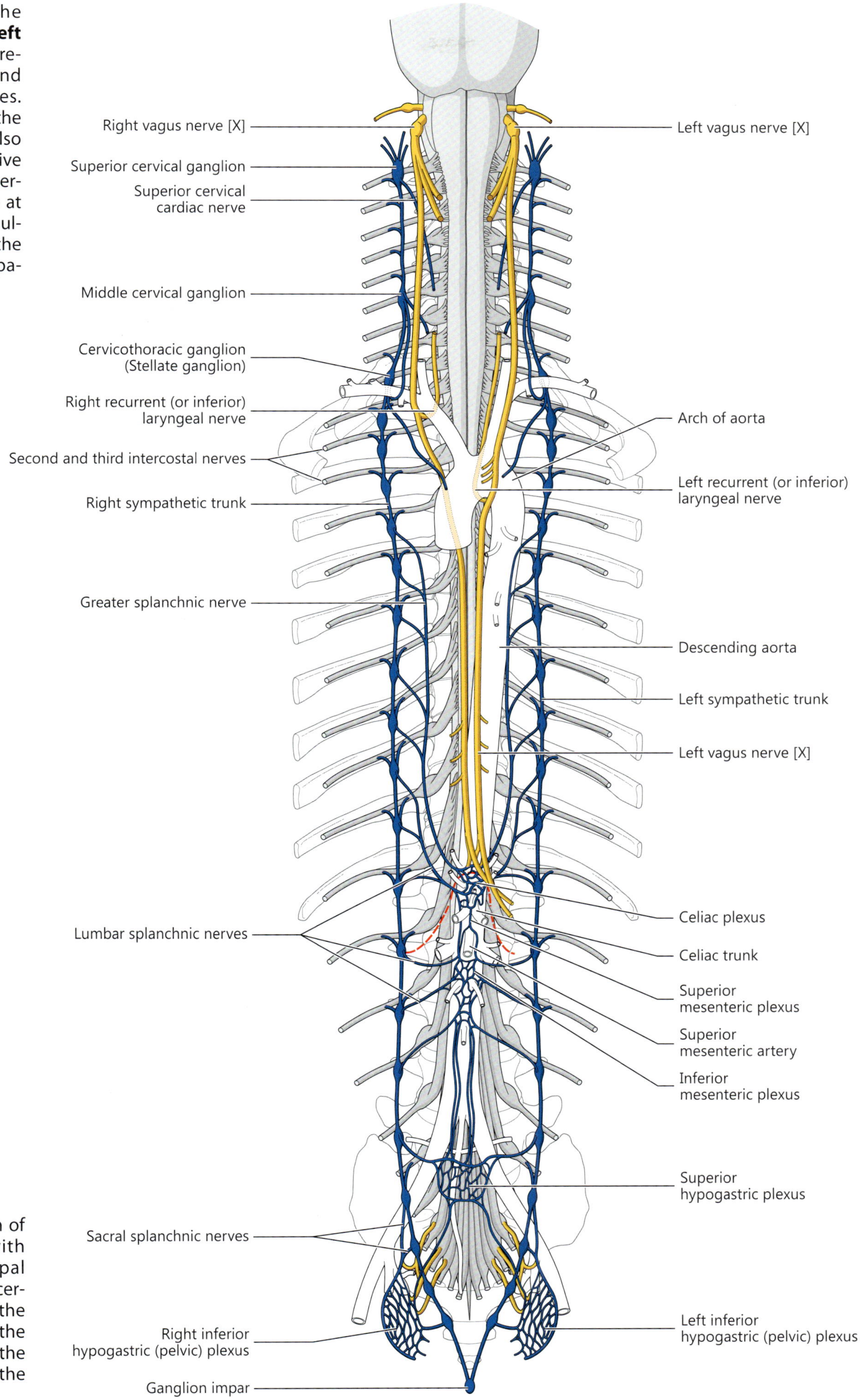

➤ Schematic representation of the **nerve plexuses** with their respective principal branches. In *yellow*, the cervical segment; in *green*, the dorsal segment; in *blue*, the lumbar segment; in *red*, the sacral segment; in *gray*, the coccygeal segment.

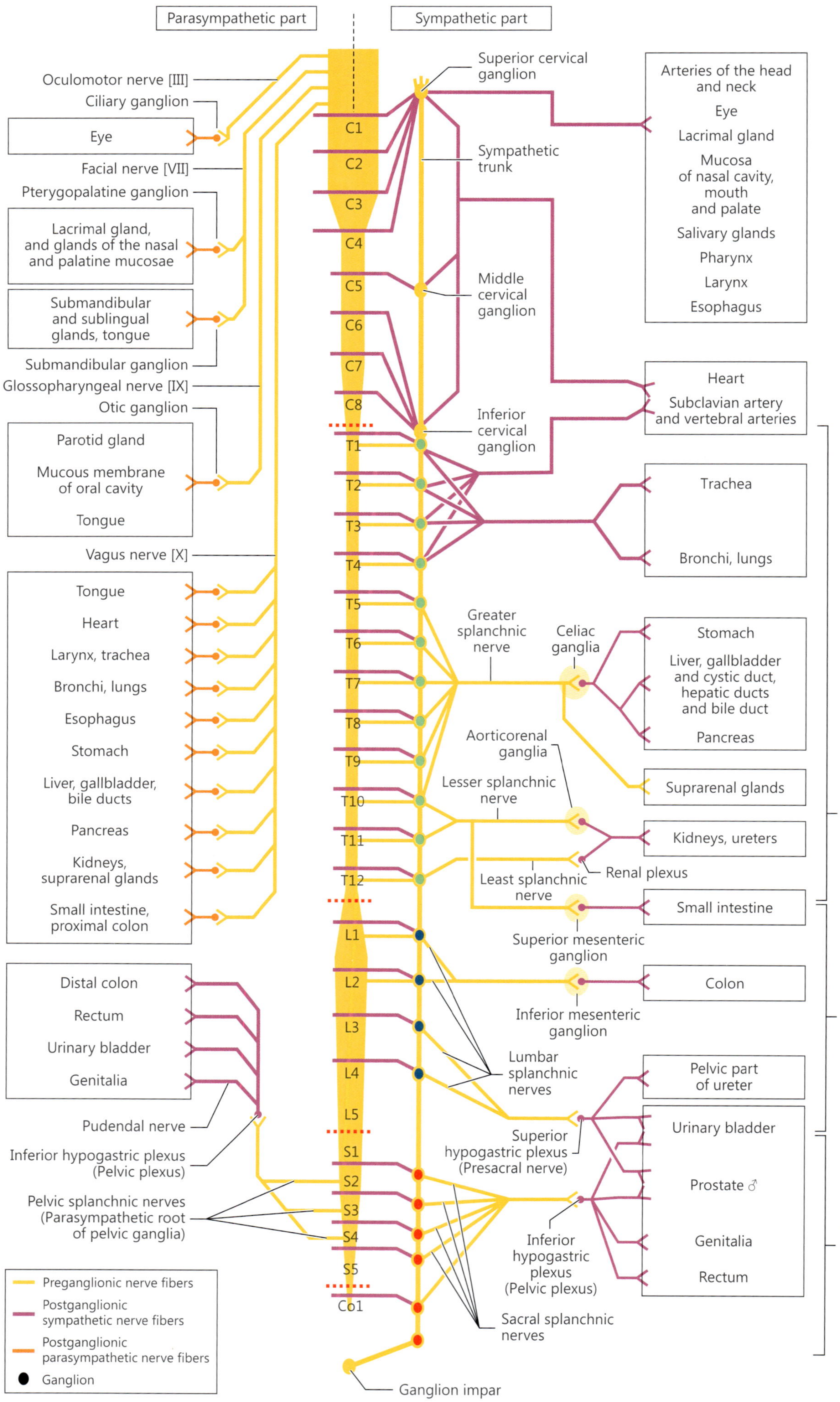

Schematic representation of the **autonomic division** composed of the sympathetic nervous system (on the right) and parasympathetic nervous system (on the left). The autonomic division is the part of the central nervous system that controls the secretory functions of glands and wall motility of internal organs and vessels, thus regulating the vital functions of the organs and all body systems. The organs that are innervated by nerve fibers of the autonomic division are listed, in detail, in the boxes. The preganglionic nerve fibers are represented in *yellow*; in *purple*, the postganglionic fibers of the sympathetic nervous system, and in *orange* those of the parasympathetic nervous system. The parasympathetic (cranial and sacral) nervous system comprises the neuraxial visceral centers situated, respectively, in the brainstem and sacral segment of the spinal cord. Note, in particular, the elevated number of visceral efferent (motor) fibers of the parasympathetic part of the vagus nerve.

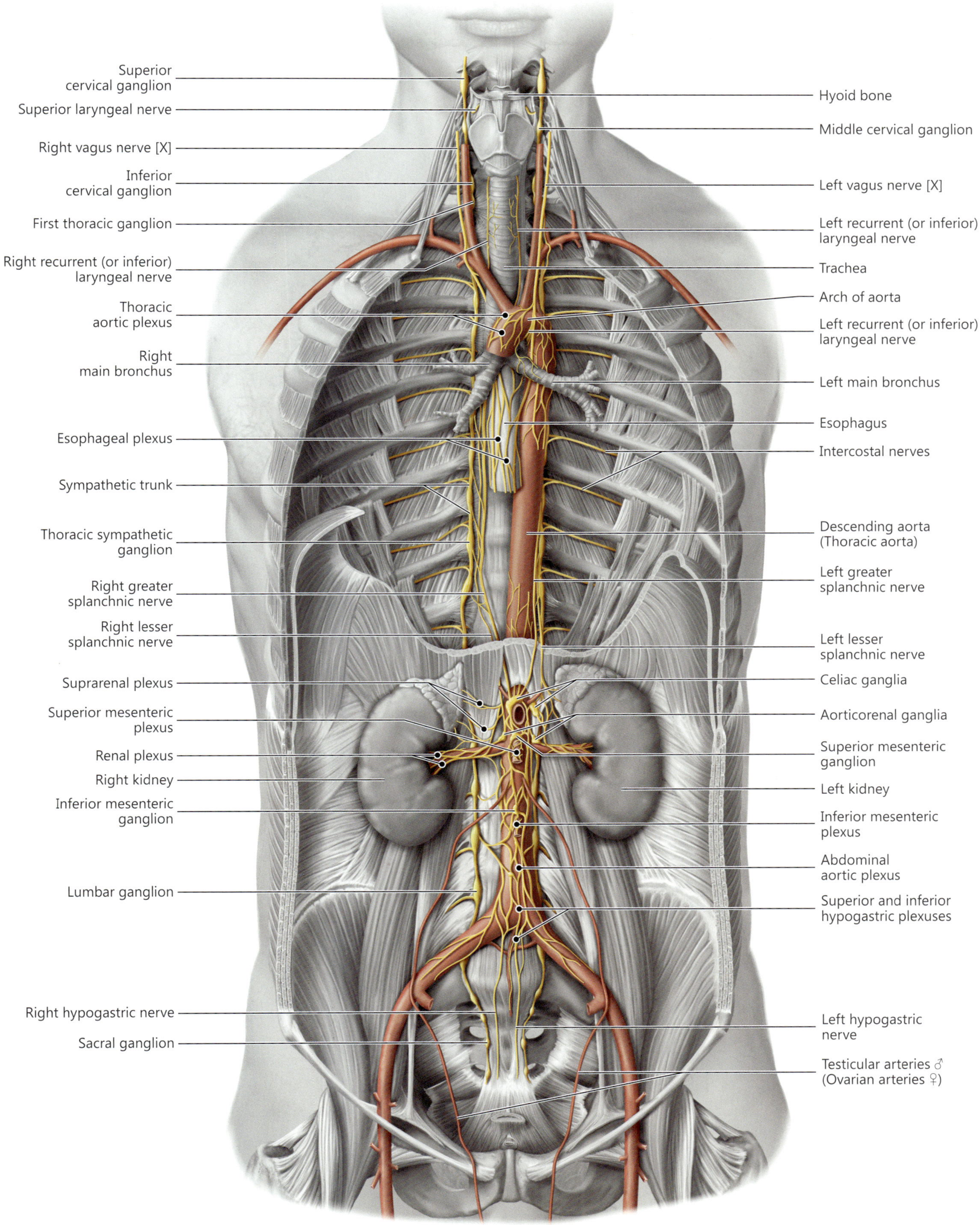

Coronal section of the trunk showing the arrangement of the **autonomic division**.

Superficial innervation of the anterior surface **of the trunk**. In the thorax, the limits of the areas supplied by the lateral and anterior cutaneous branches of the costal nerves can be seen.

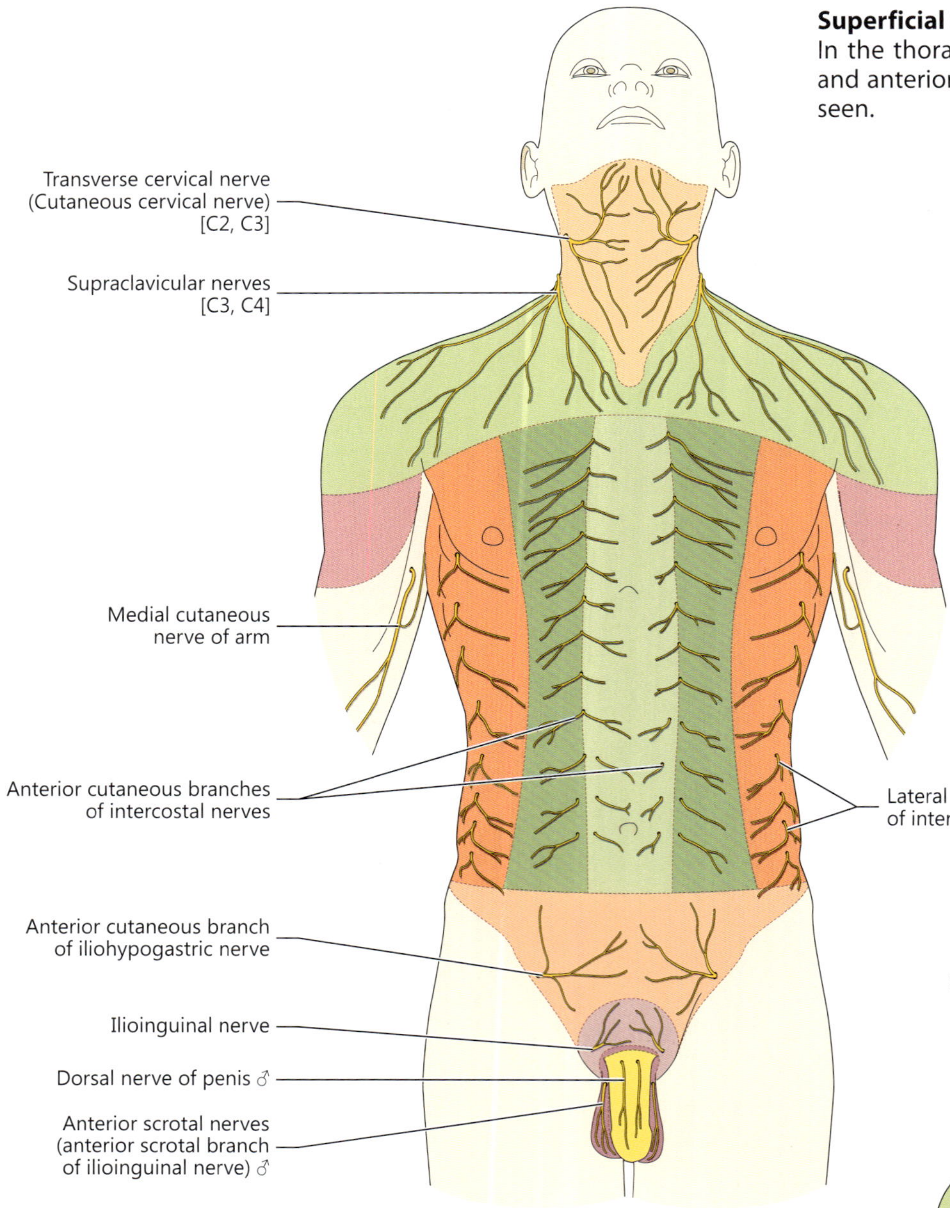

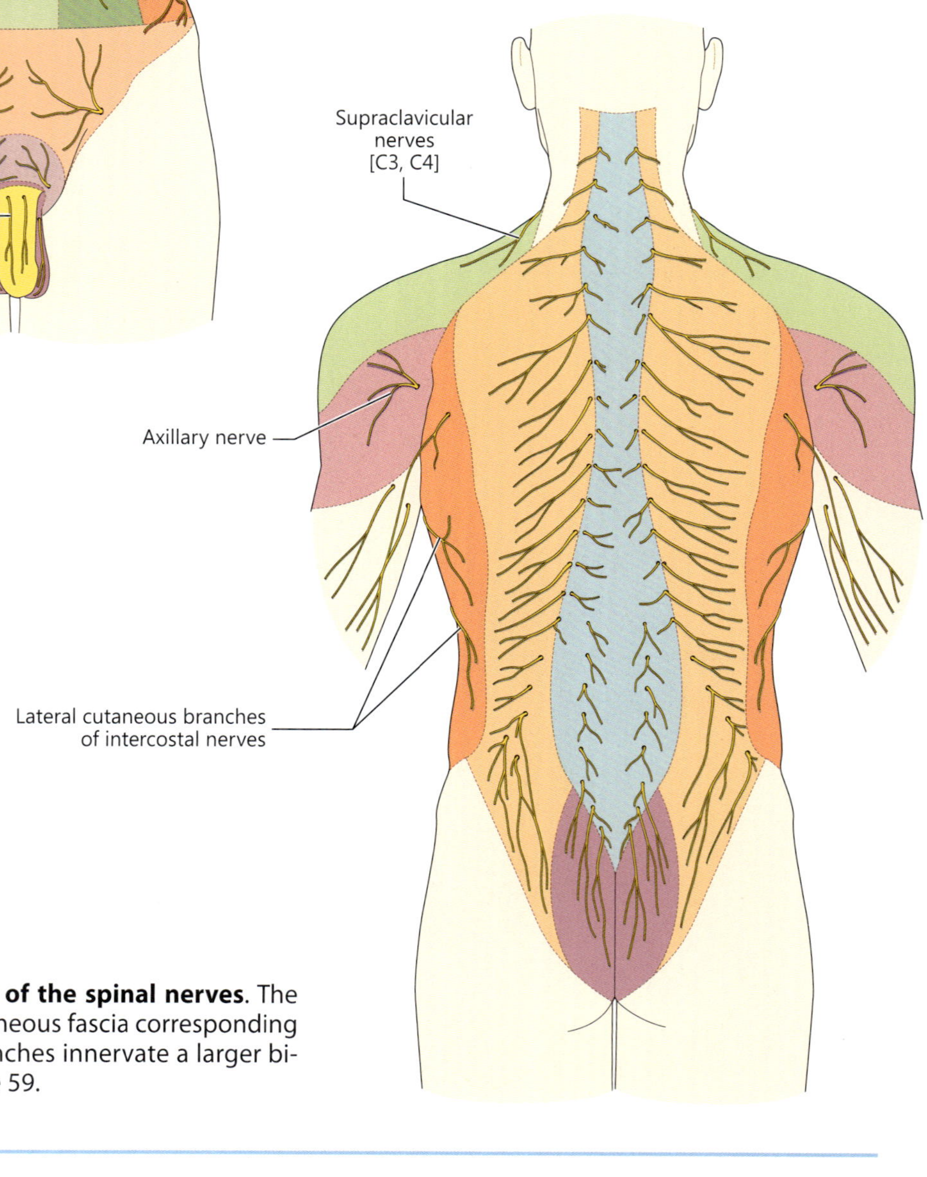

Cutaneous distribution of the **posterior rami of the spinal nerves**. The medial branches supply a narrow median cutaneous fascia corresponding to the vertebral column (*blue*); the lateral branches innervate a larger bilateral area (*brown*). For dermatomes, see page 59.

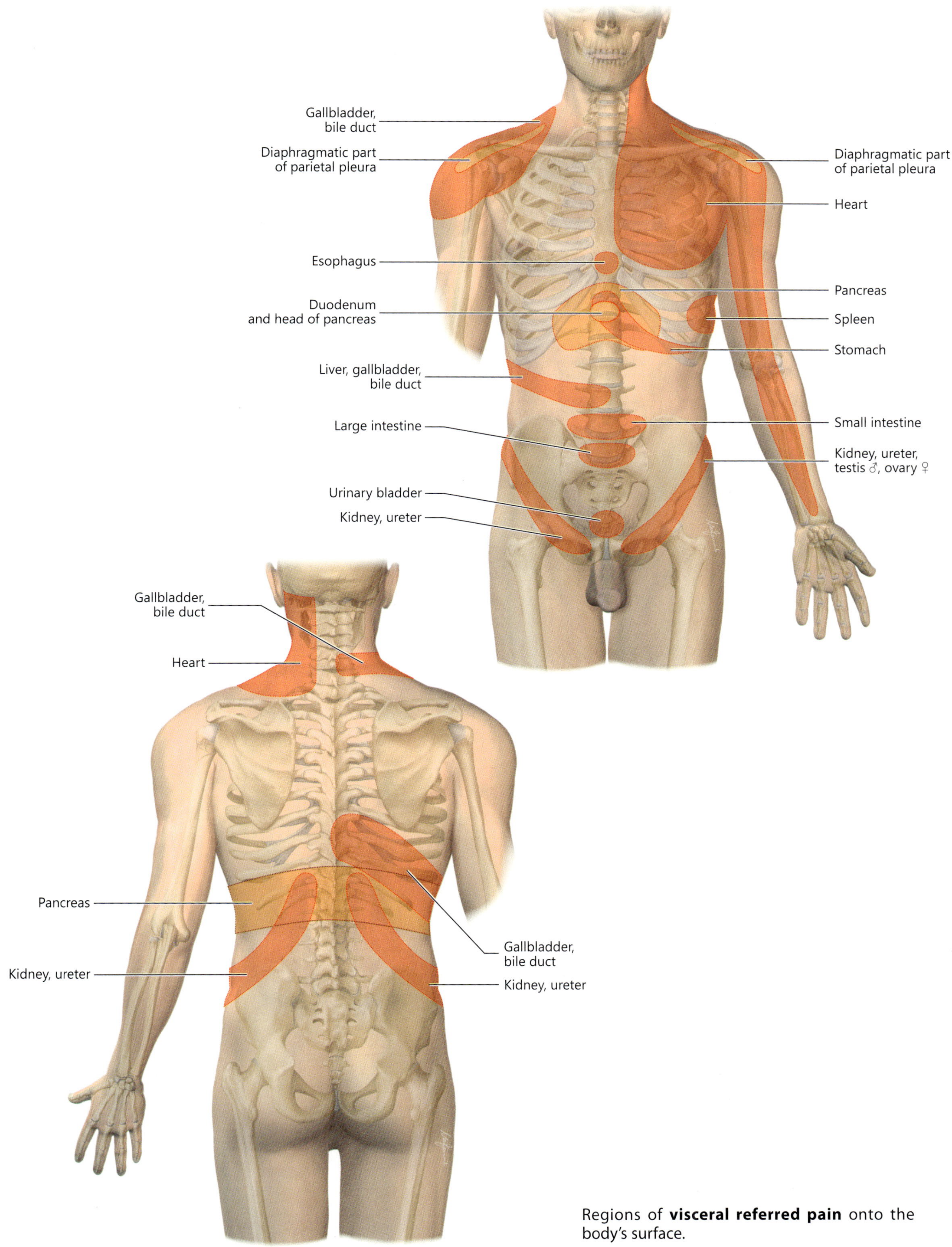

Regions of **visceral referred pain** onto the body's surface.

Visceral Referred Pain

ORGAN	SPINAL SEGMENTS	SITE OF REFERRED PAIN
Bile duct	T6-T9	Epigastric region in the initial stage and subsequently right hypochondrium, or inferior part of the right side of the neck and top right shoulder (in case of irritation of the diaphragm)
Diaphragm	☞ Diaphragmatic part of parietal pleura	
Gallbladder	T7-T9	Epigastric region in the initial stage and subsequently right hypochondrium, or inferior part of the right side of the neck and top right shoulder (in case of irritation of the diaphragm)
Heart	T1-T4	Left hemithorax with radiation to the left shoulder, to the medial surface of the left arm, at times also to the surface of the left forearm and the little finger, along the left side of the neck and head up to the mandible and the occipital bone, and downward to the epigastric region
Kidney	T10-L1	Inferior part of the back, right and left lumbar region is up to the groin
Large intestine – anus (inferior two-thirds) – appendix – cecum – ascending colon – sigmoid colon	 S3-S4 T10-T11 T10-T11 T10-T12 L1-L2	 Perineum Umbilical region initially and right iliac fossa subsequently Umbilical region or right lower quadrant of the abdomen Umbilical region or right lumbar region Left lower quadrant of the abdomen
Liver	T7-T9	Epigastric region initially and right hypochondrium subsequently
Esophagus	T5-T6	Inferior retrosternal and subxiphoidal site
Ovary	T11-L1	From the lumbar region to the groin
Pancreas	T7-T9	Inferior part of the epigastric region or umbilical region initially; paravertebral region adjacent to the inferior part of the thoracic segment of the vertebral column subsequently (peritoneal irritation)
Pericardium	C3-C5	Free or anterior border of the trapezius (irritation of the phrenic nerves)
Pleura – diaphragmatic part – – central – – peripheral – mediastinal part	 C3-C4 T1-T11 C3-C4	 Inferior part of the neck and top of the homolateral shoulder Thoracic wall along the course of the intercostal nerves Inferior part of the neck and top of the ipsilateral shoulder
Small intestine – jejunum – ileal diverticulum – duodenum – ileum	 T6-T10 T10-T11 T5-T8 T7-T10	 Umbilical region or central part of the epigastric region Umbilical region Central part of the epigastric region Umbilical region or central part of the epigastric region
Spleen	T6-T10	Left hypochondrium (peritoneal irritation)
Stomach	T5-T9	Epigastric region and left hypochondrium
Testis	T11-L1	From the lumbar region to the groin
Ureter	T11-L1	From the lumbar region to the groin
Urinary bladder	T11-L2	Pubic region

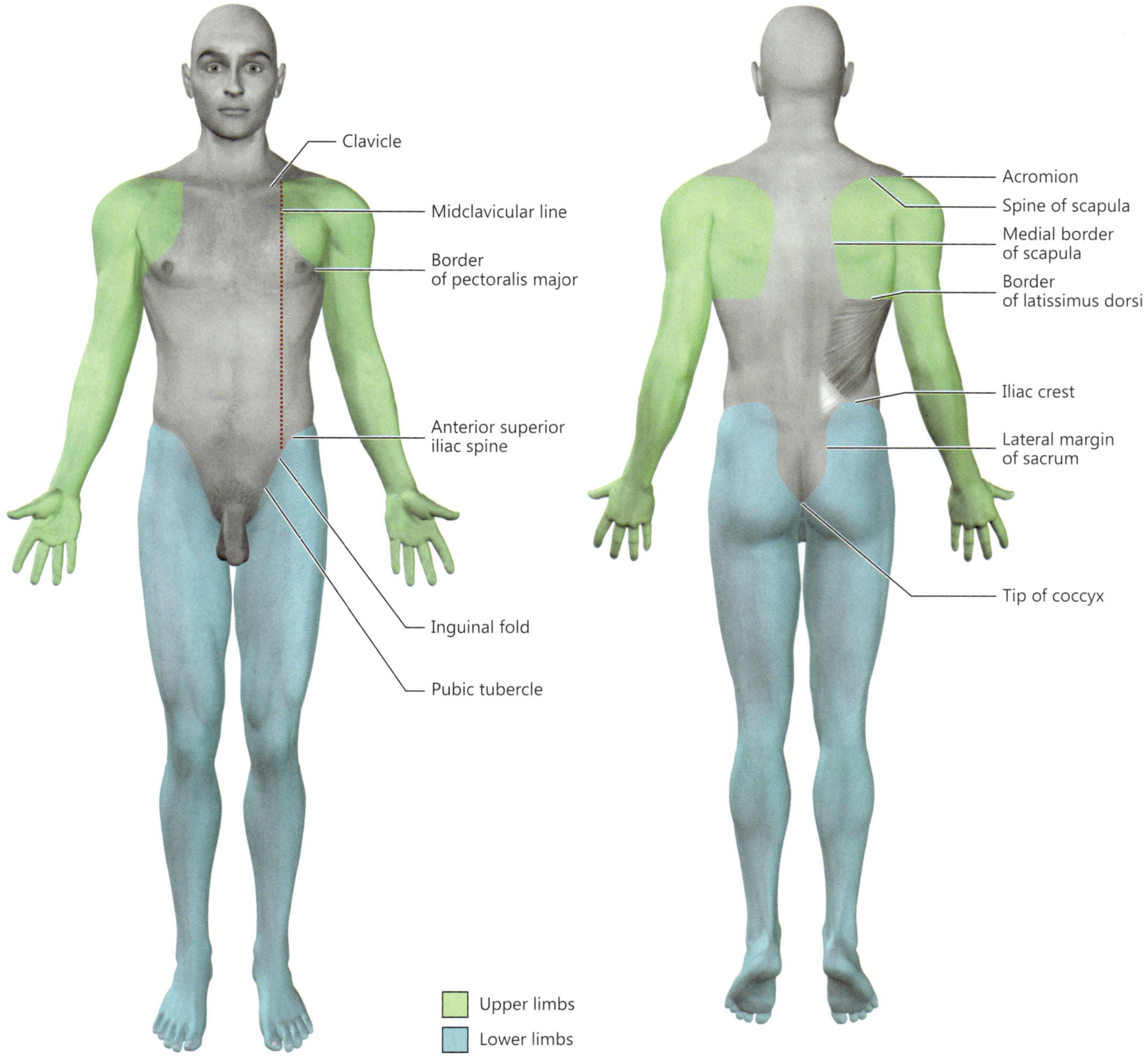

Surface boundaries of the **regions of the upper and lower limbs** with respect to the trunk. For details ☞ Upper Limb, Boundaries, pages 310-311 and Lower Limb, Boundaries, pages 362-363, volume 3.

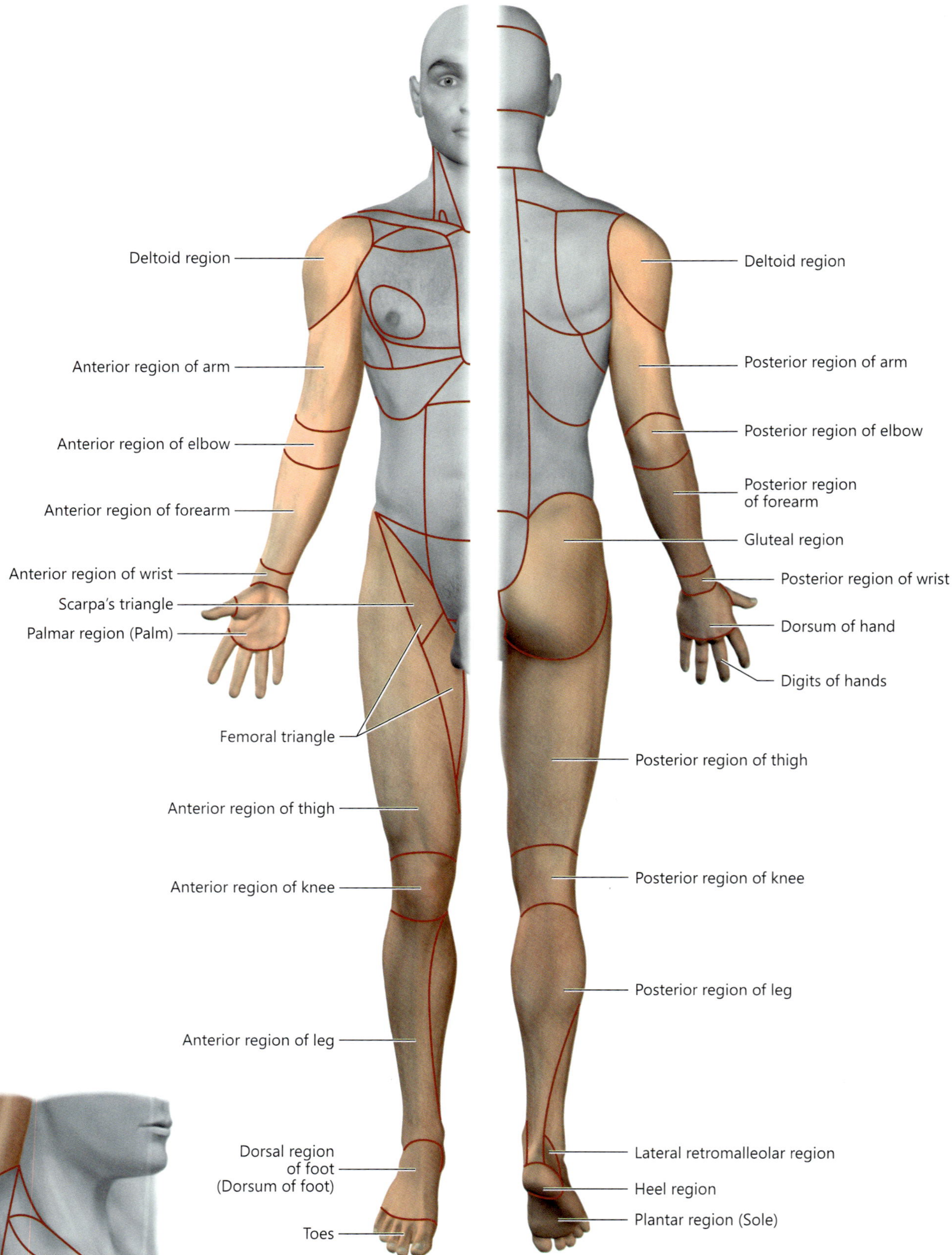

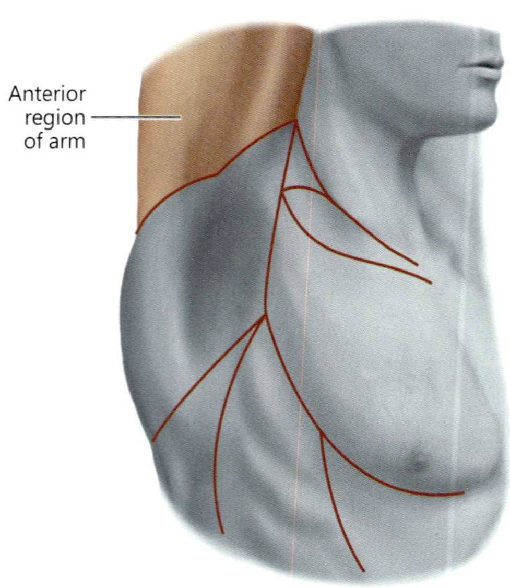

Detailed **subdivision of the limbs** into anatomical regions.

Anatomical landmarks on the surface of the arm, elbow and forearm		
	STRUCTURE	**LOCATION**
1	Acromion	Process by which the spine of the scapula extends laterally to the apex of the shoulder
2	Lateral epicondyle	Lateral prominence of the distal extremity of the humerus
3	Head of the radius	Superior extremity of the radius distally to the epicondyle of the humerus
4	Olecranon	Rough prominence on the proximal extremity of the ulna
5	Radial styloid process	Downward extension of the lateral part of the posterior surface of the inferior extremity of the radius
6	Head of the ulna	Inferior extremity of the ulna
7	Medial epicondyle	Medial prominence of the distal extremity of the humerus
8	Brachial pulse	Palpable on the medial surface of the arm at the proximal level, posterior to the coracobrachialis

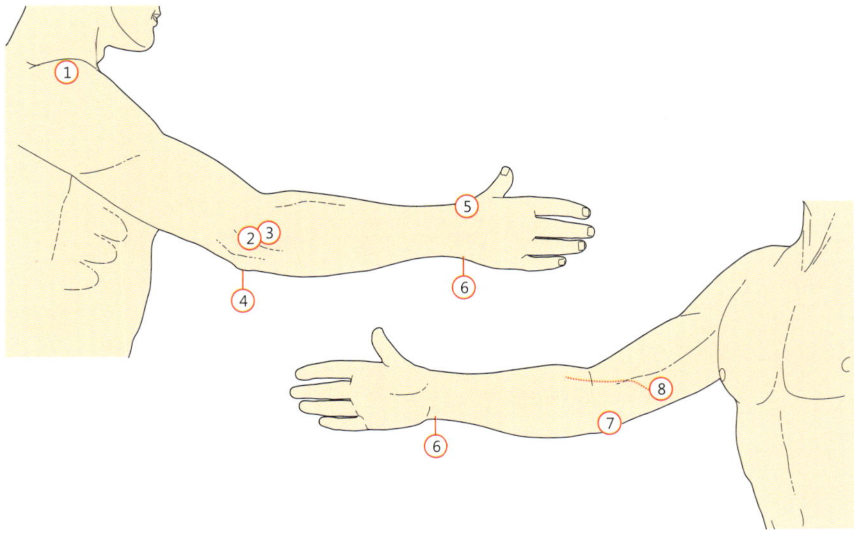

Anatomical landmarks on the surface of the wrist and hand		
	STRUCTURE	**LOCATION**
1	First interdigital space	Cutaneous fold between the 1st finger and the hand
2	Radial styloid process	Bony prominence on the lateral surface of the forearm, at the distal extremity of the radius
3	Metacarpals [I-V]	The bases and shafts are palpable between the tendons of the extensor muscles on the dorsal surfaces of the wrist and hand, while the bony prominences at the base of the fingers are formed by the heads
4	Capitate	In a small proximal concavity at the base of the 3rd metacarpal
5	Ulnar styloid process	Bony prominence on the posterior medial surface of the forearm, at the distal extremity of the ulna
6	Head of the third metacarpal	Distal extremity of the 3rd metacarpal
7	Pisiform	Medial bone of the proximal row of the carpal bones, proximal to the base of the hypothenar eminence
8	Thenar eminence	Muscular prominence on the palm of the hand at the base of the thumb, medially and distally bordered by the longitudinal palmar sulcus
9	Distal palmar crease	Transverse sulcus traveling laterally from the medial margin of the palm of the hand to the interdigital space between the 2nd and 3rd finger
10	Proximal palmar crease	Transverse sulcus traveling medially from the palm of the hand to the hypothenar eminence
11	Hypothenar eminence	Muscular prominence on the medial side of the base of the palm of the hand

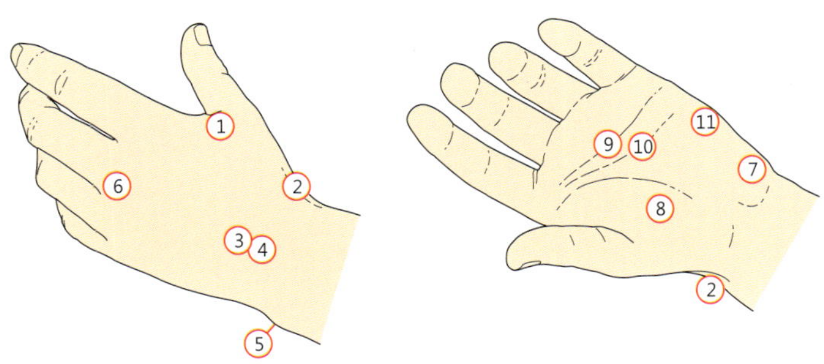

Anatomical landmarks on the surface of the lower limb		
	STRUCTURE	***LOCATION***
1	Ischial tuberosity	When the hip is in passive flexion, this large bony prominence is lateral to the median line of the body just above the gluteal fold – the deep transverse sulcus between the buttocks and the posterior surface of the thigh
2	Greater trochanter of the femur	When the tip of the thumb is placed in the center of the iliac crest, the tip of the middle finger, pointing distally on the lateral surface of the thigh, rests on the superior border of the greater trochanter
3	Patella	Large sesamoid bone, triangular in shape, on the anterior surface of the knee, with proximal base and distal apex
4	Medial and lateral condyles of the tibia	To the two sides of the patellar ligament, the superior margins of the glenoid cavities of the medial and lateral condyles of the tibia are located in depressions formed by soft tissue; following the outlines of the glenoid cavities medially and laterally, the articular fissure of the knee joint can be felt
5	Patellar ligament (patellar tendon)	Extends from the apex of the patella on the tibial tuberosity; its borders are palpable when the knee is extended
6	Head of the fibula	Rounded bony prominence on the lateral surface of the leg at the level of the tibial tuberosity
7	Tibial tuberosity	Bony prominence at the proximal extremity of the anterior border of the tibia, to which the patellar ligament is attached
8	Anterior border of the tibia	Subcutaneous bony crest along the anterior surface of the leg
9	Lateral epicondyle of the femur	Bony prominence on the lateral condyle of the femur
10	Adductor tubercle	Medial prominence on the inferior extremity of the femur behind and above the medial epicondyle
11	Calcaneal tendon (Achilles tendon)	Prominent crest on the posterior surface of the ankle; the margins of the tendon are palpable just above the posterior surface of the calcaneus
12	Lateral malleolus	Distal extremity of the fibula as it protrudes on the lateral surface of the ankle
13	Calcaneus	Posterior part of the tarsus
14	Base of the fifth metatarsal	Small bony prominence in the middle of the lateral border of the foot
15	Medial malleolus	Distal extremity of the tibia as it protrudes on the medial surface of the ankle
16	Tuberosity of the navicular	Approximately 2.5 cm inferior and anterior to the medial malleolus
17	Head of the first metatarsal	Rounded bony prominence on the rounded medial part of the foot, at the base of the great toe

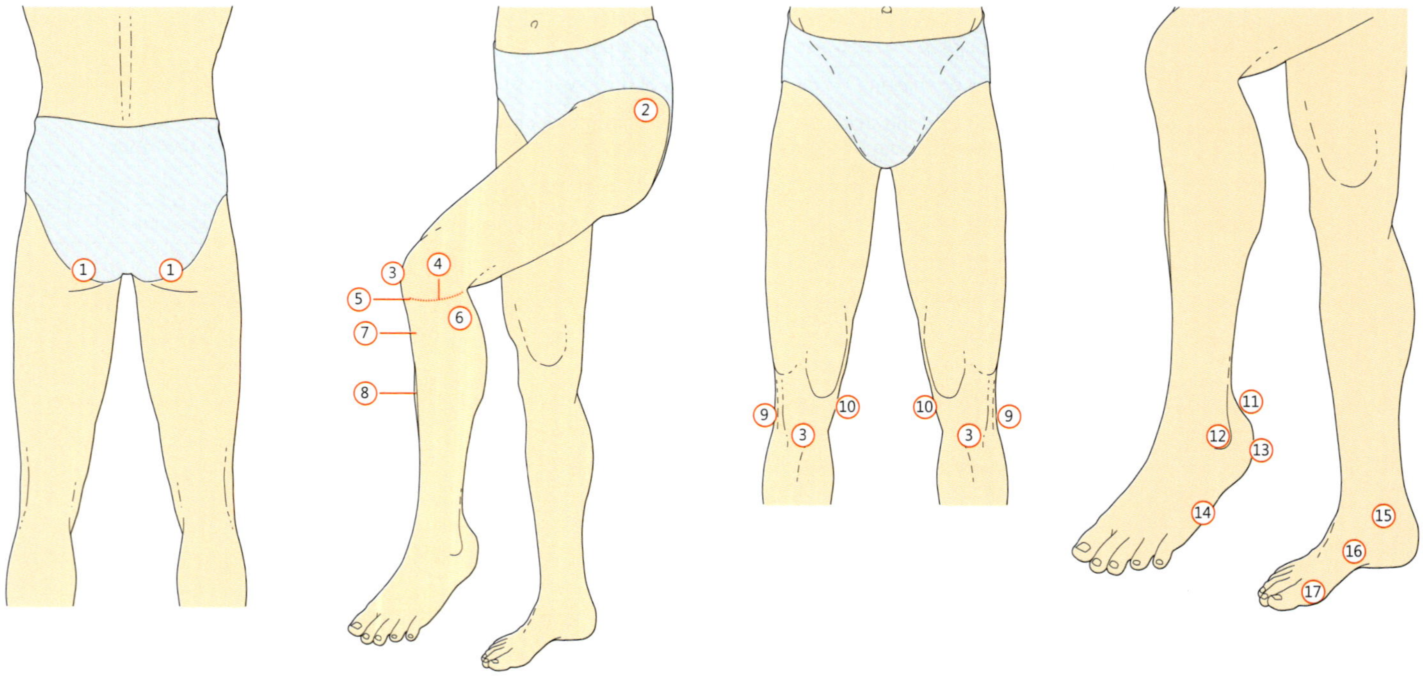

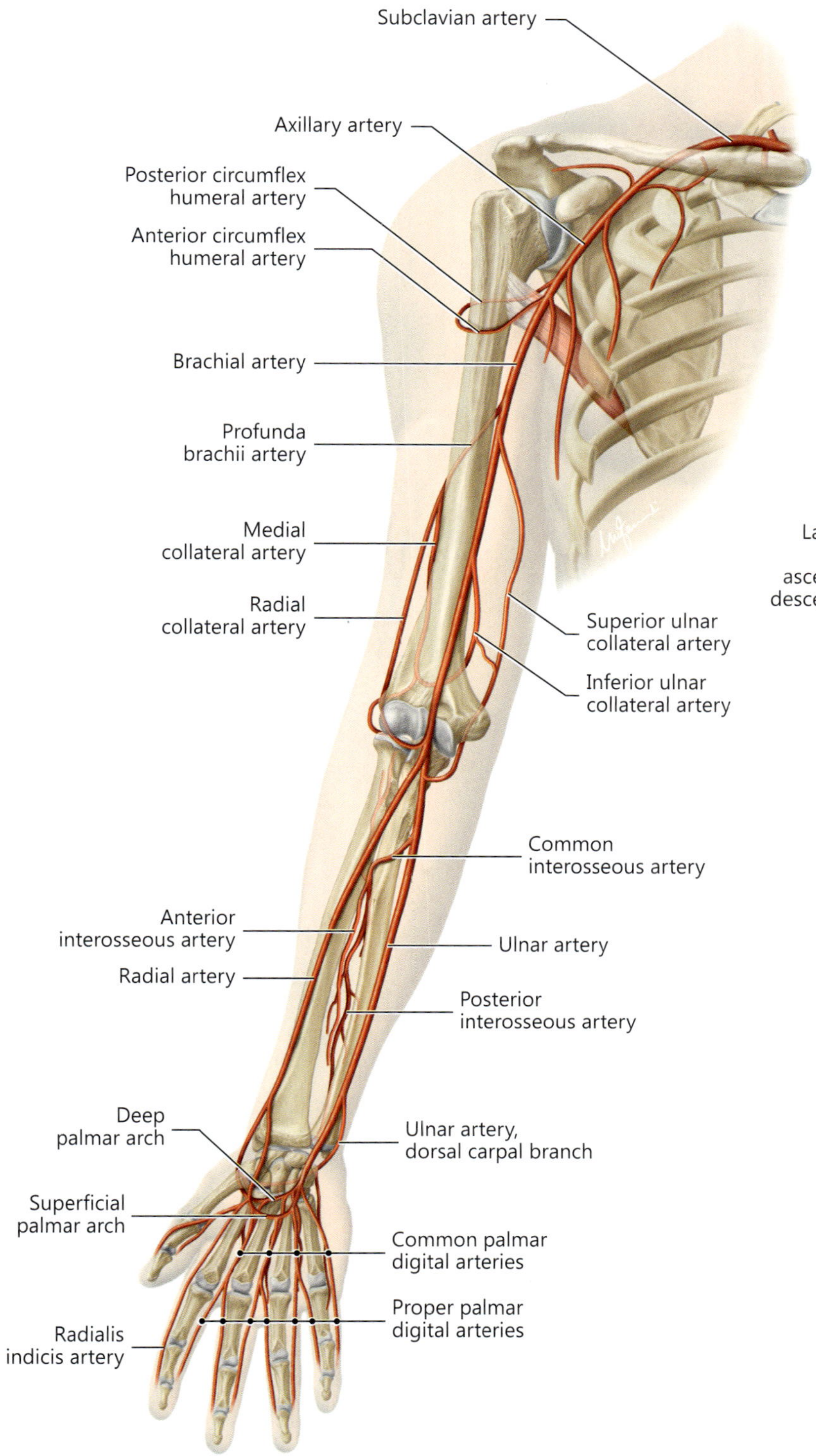

➤ **Arterial supply** of the upper limb.

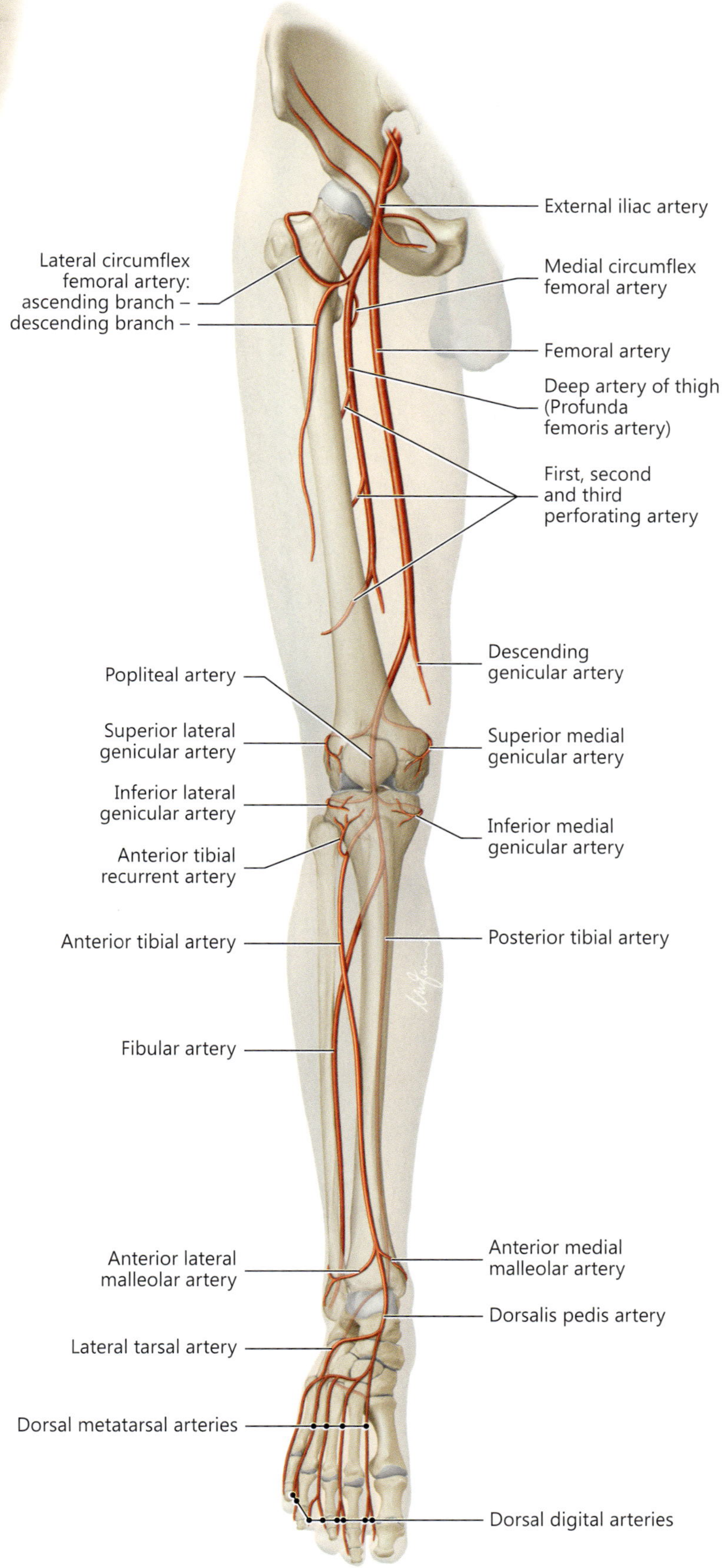

➤ **Arterial supply** of the lower limb.

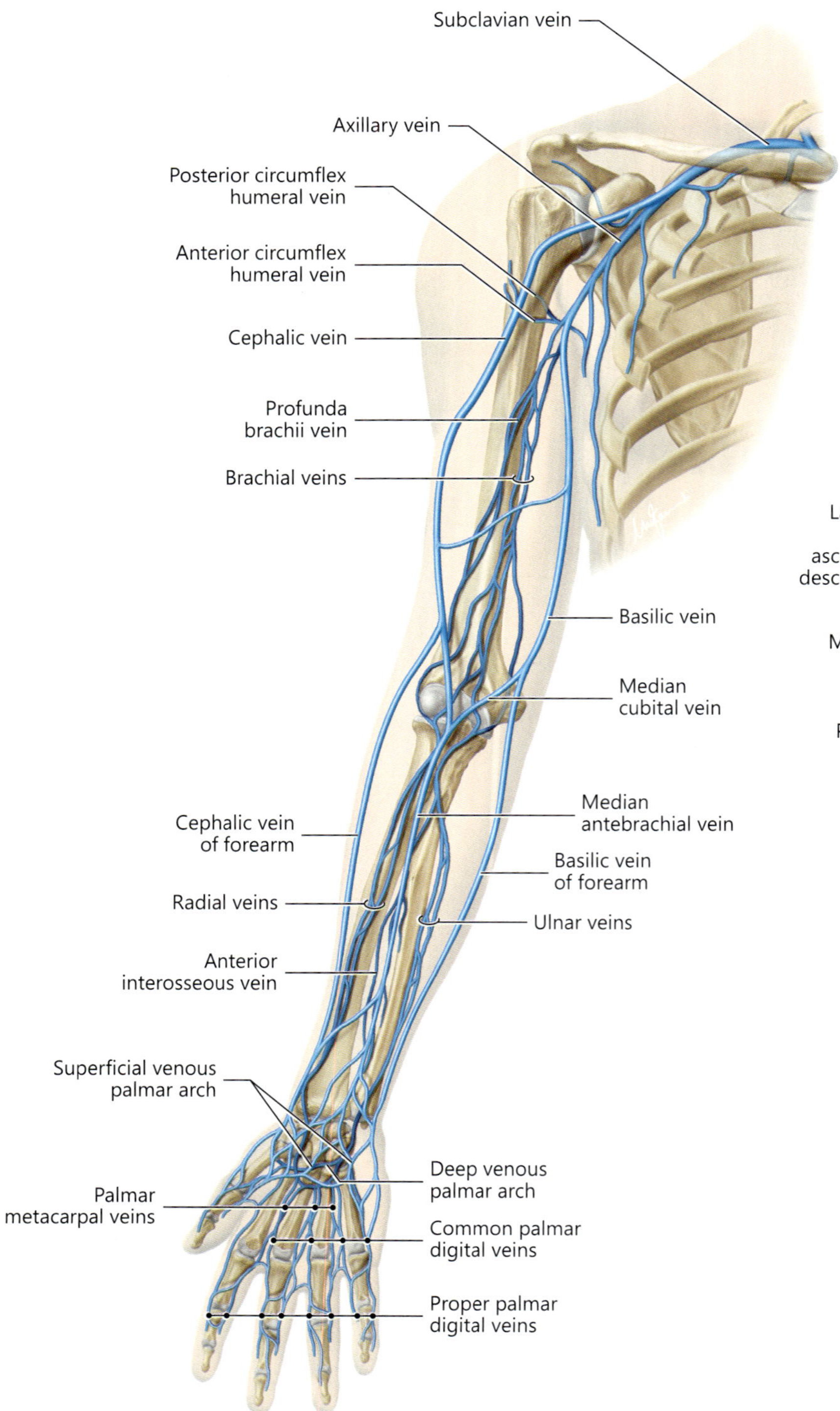

Venous return of the upper limb.

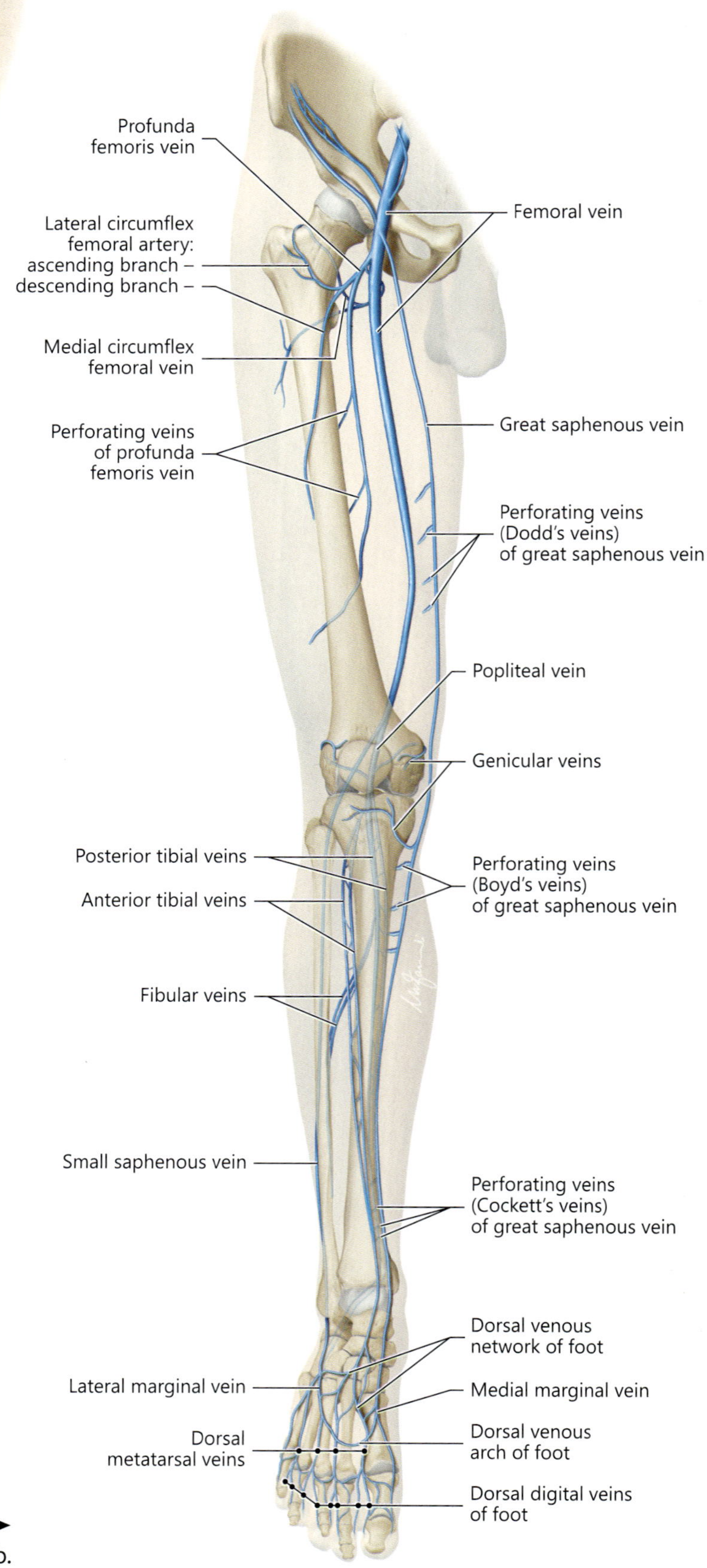

Venous return of the lower limb.

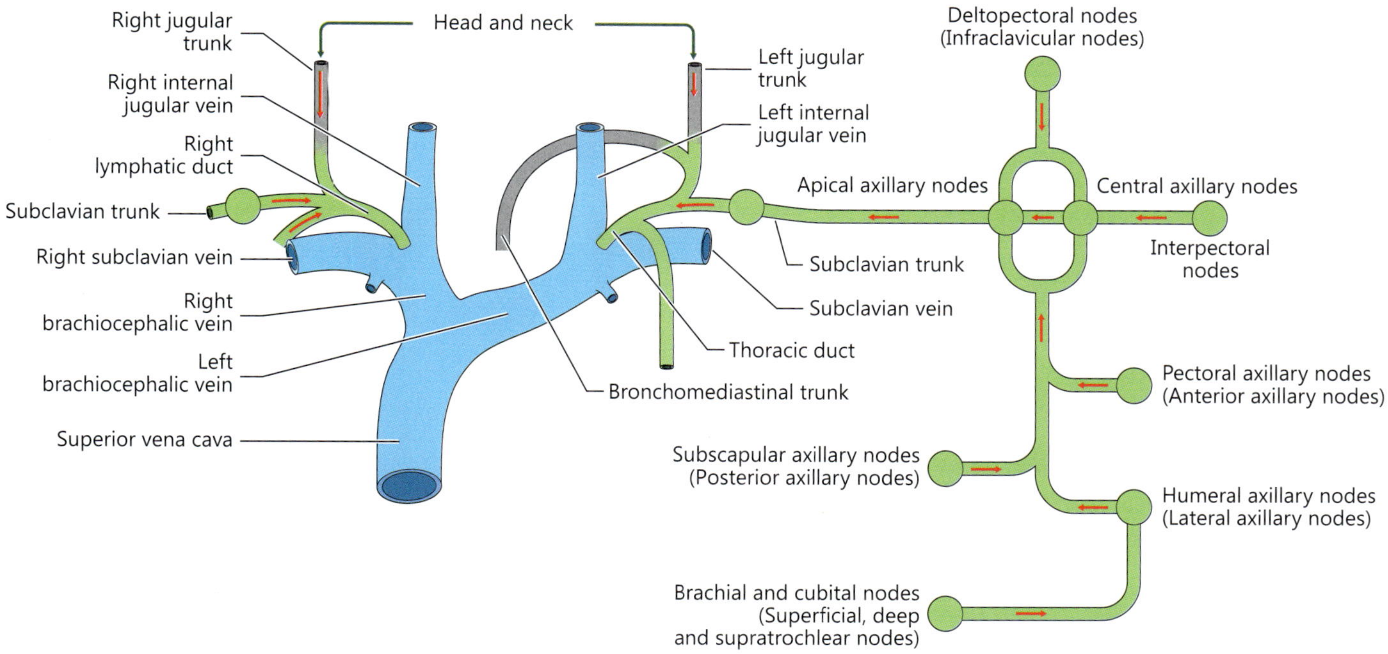

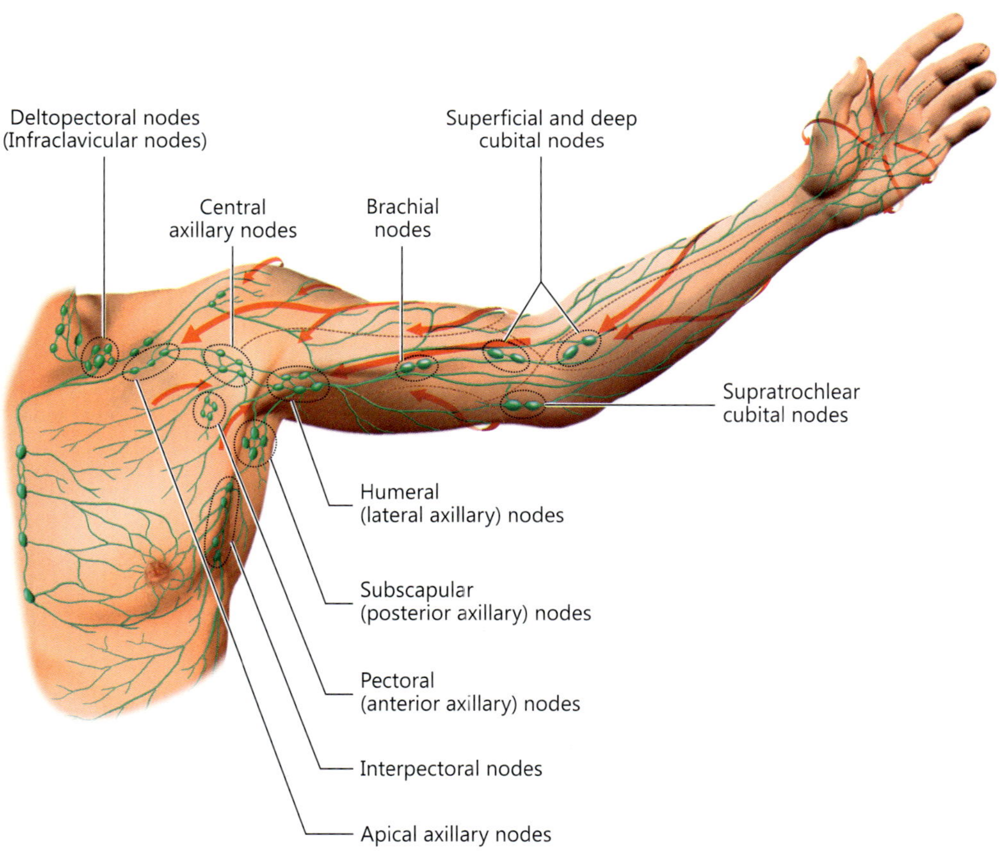

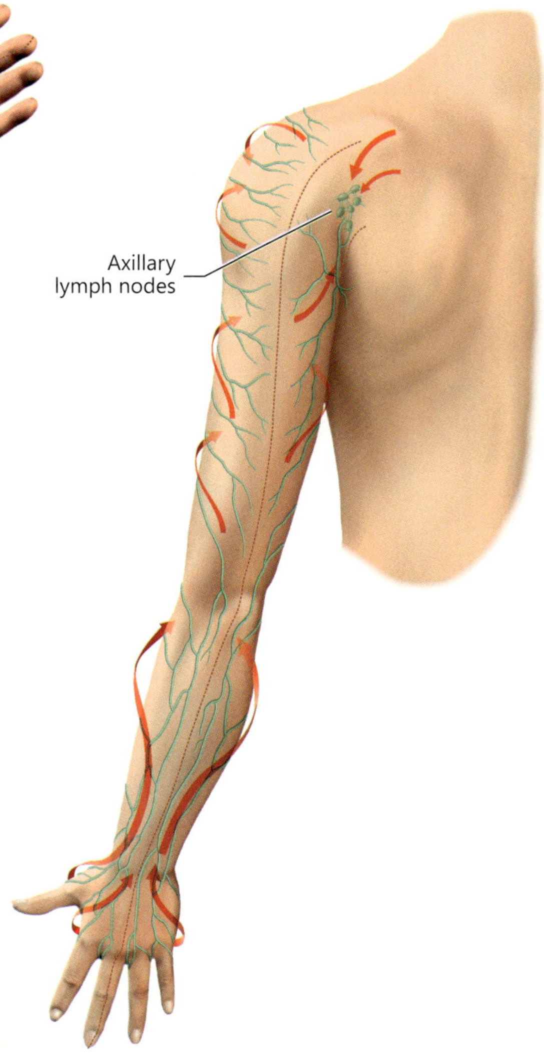

Lymph nodes and lymphatic drainage of the upper limb.

Lymphatic System of the Lower Limb

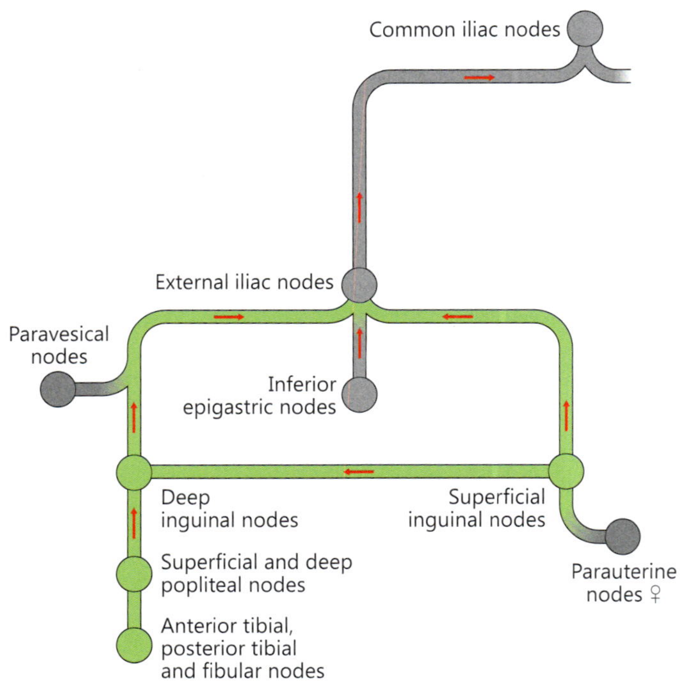

Lymph nodes and lymphatic drainage of the lower limb.

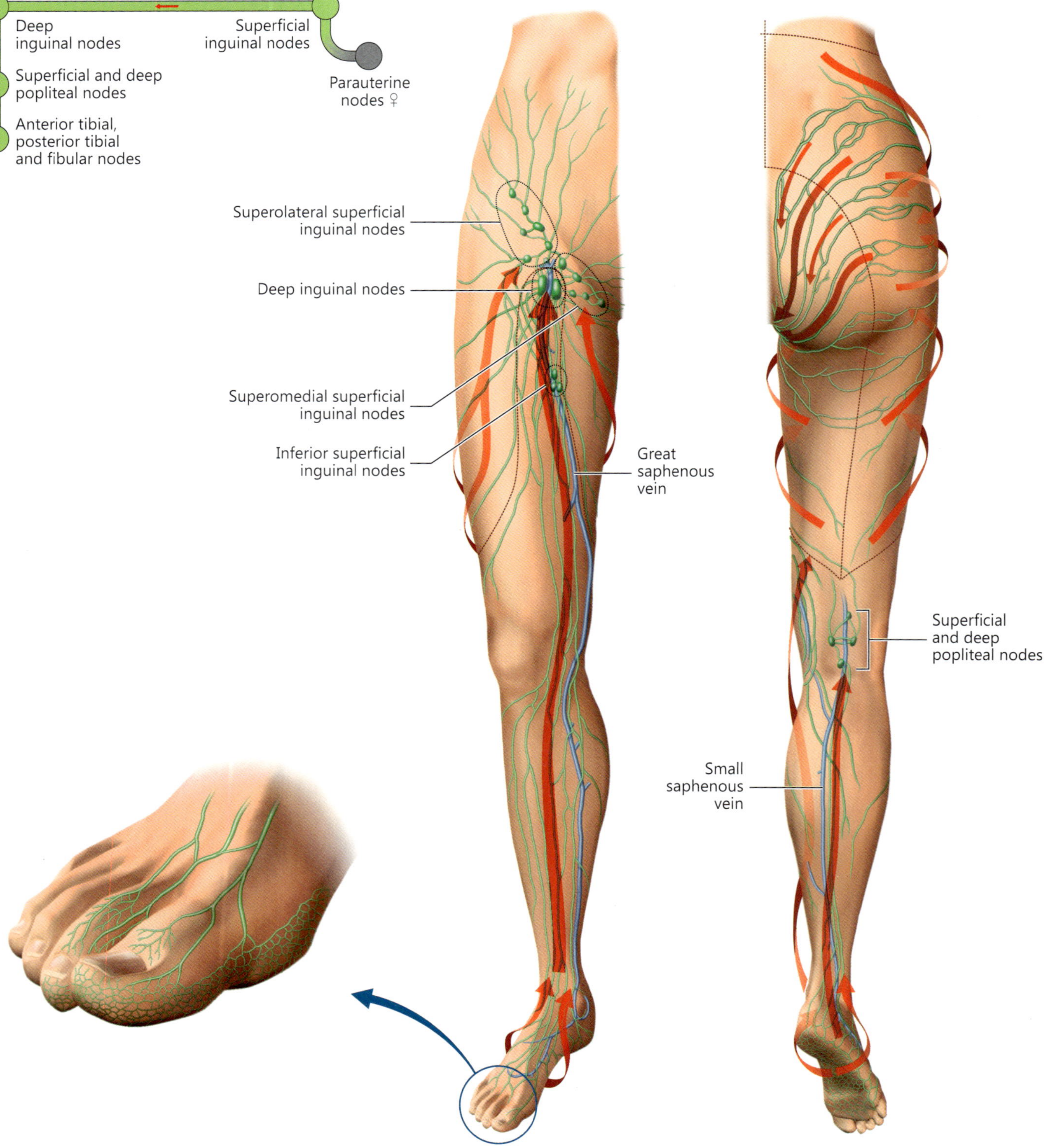

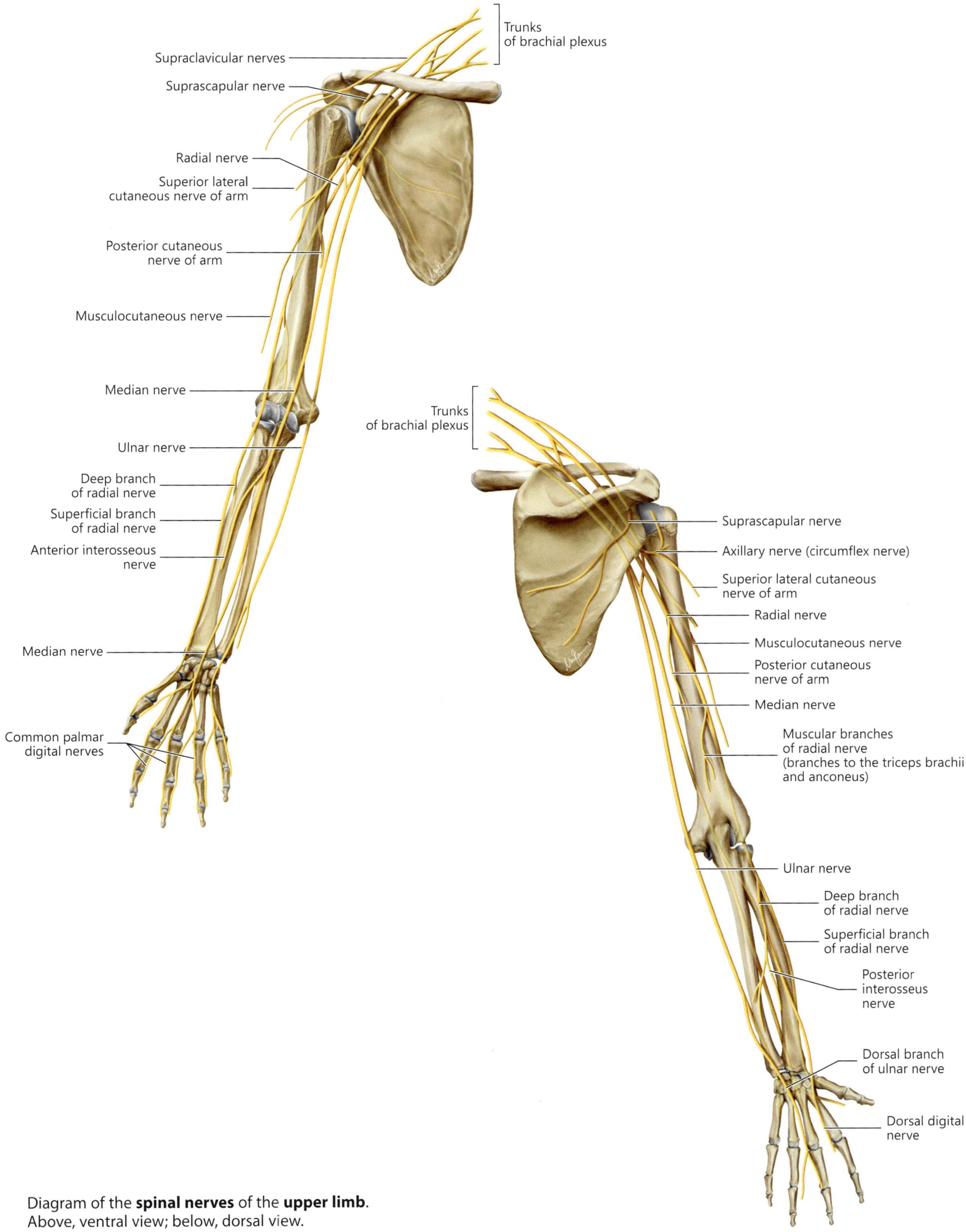

Diagram of the **spinal nerves** of the **upper limb**.
Above, ventral view; below, dorsal view.

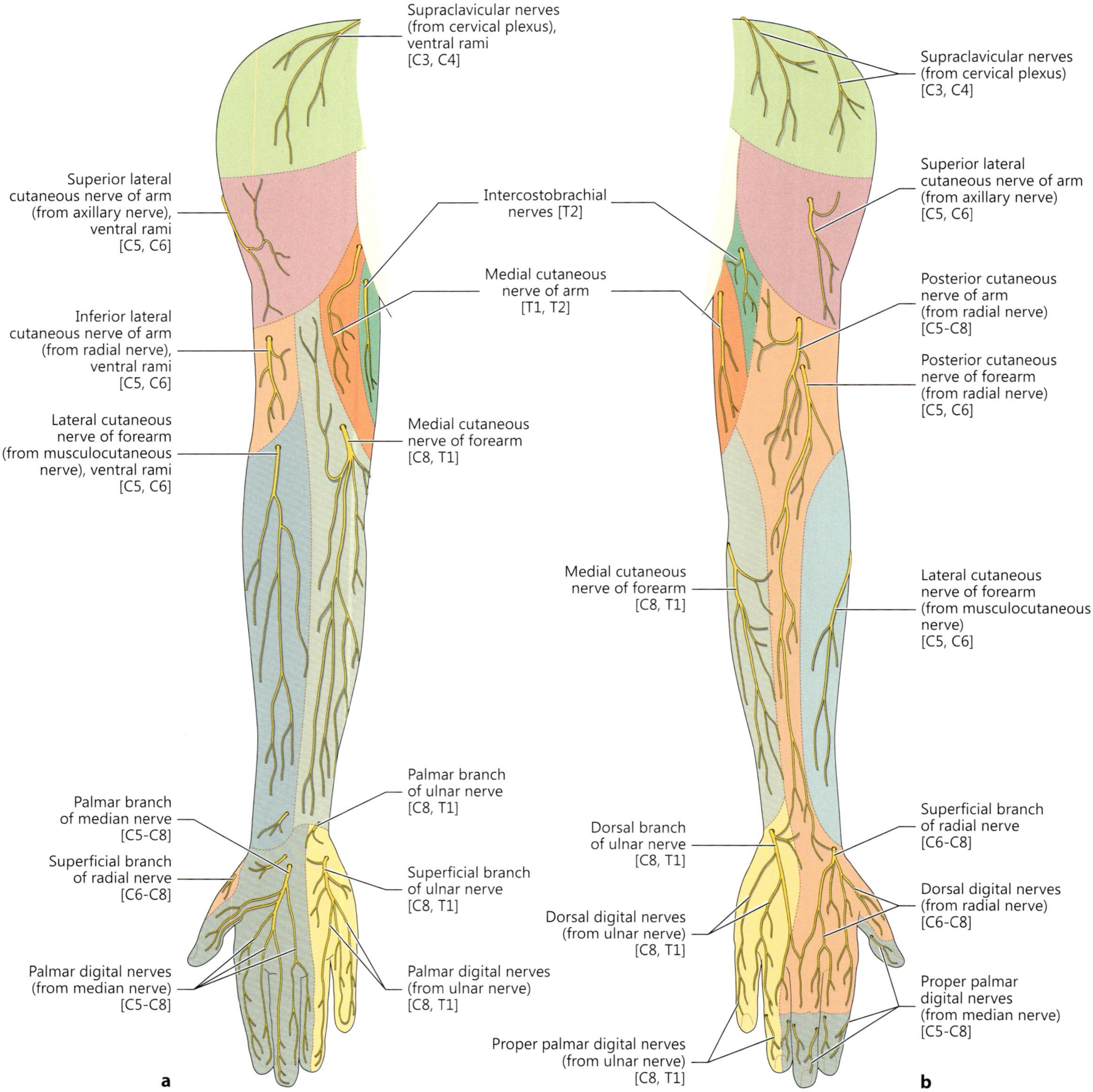

Areas covered by **cutaneous nerves** in the **upper limb**. **a**, Ventral aspect. **b**, Dorsal aspect. The free part of the limb is supplied by medial, lateral and posterior cutaneous nerves of the arm and forearm. Sensory innervation to the hand is dependent on the radial, median, and ulnar nerves. For dermatomes, see page 59.

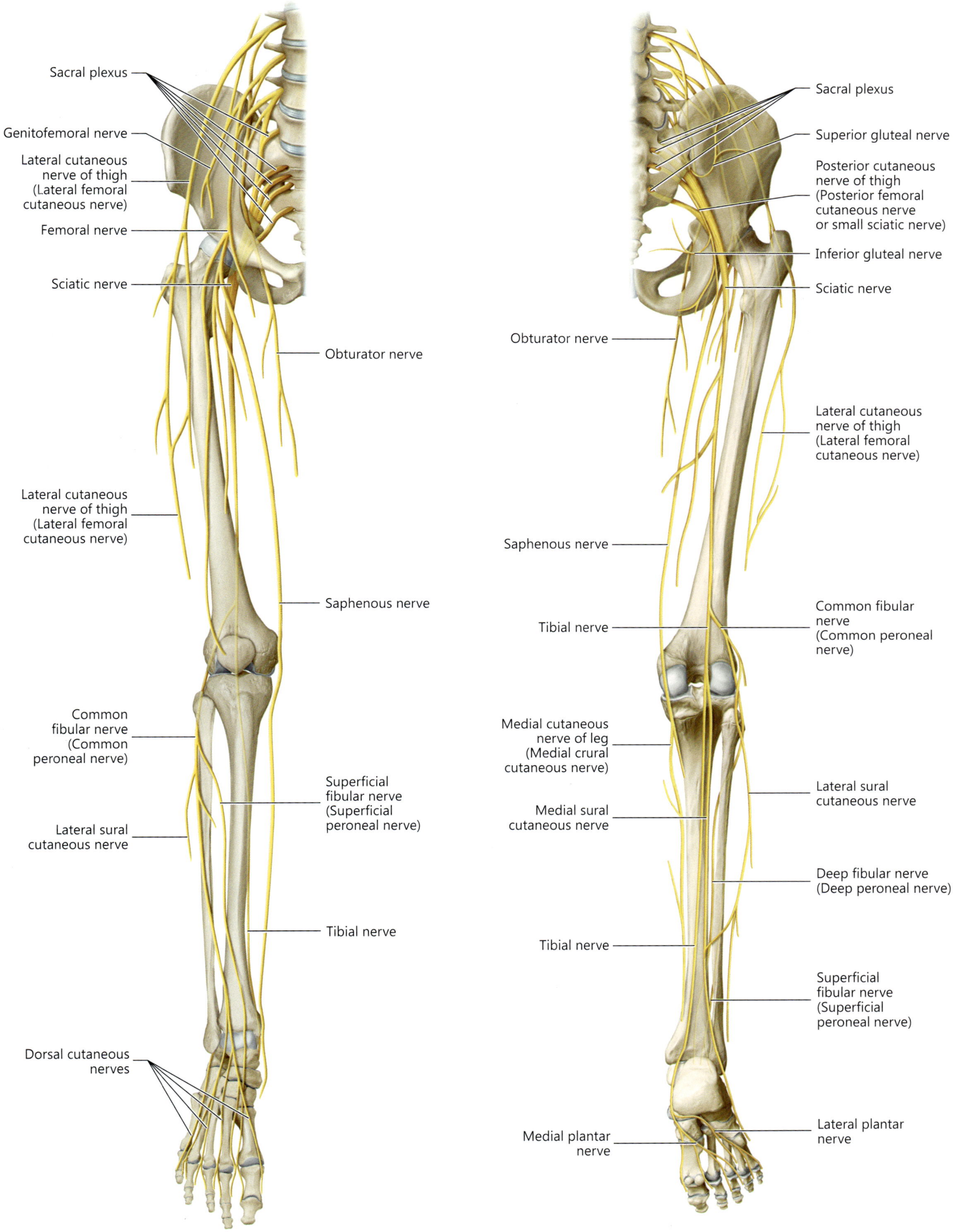

Diagram of the **spinal nerves** of the **lower limb**.

Superficial (Cutaneous) Innervation of the Lower Limb

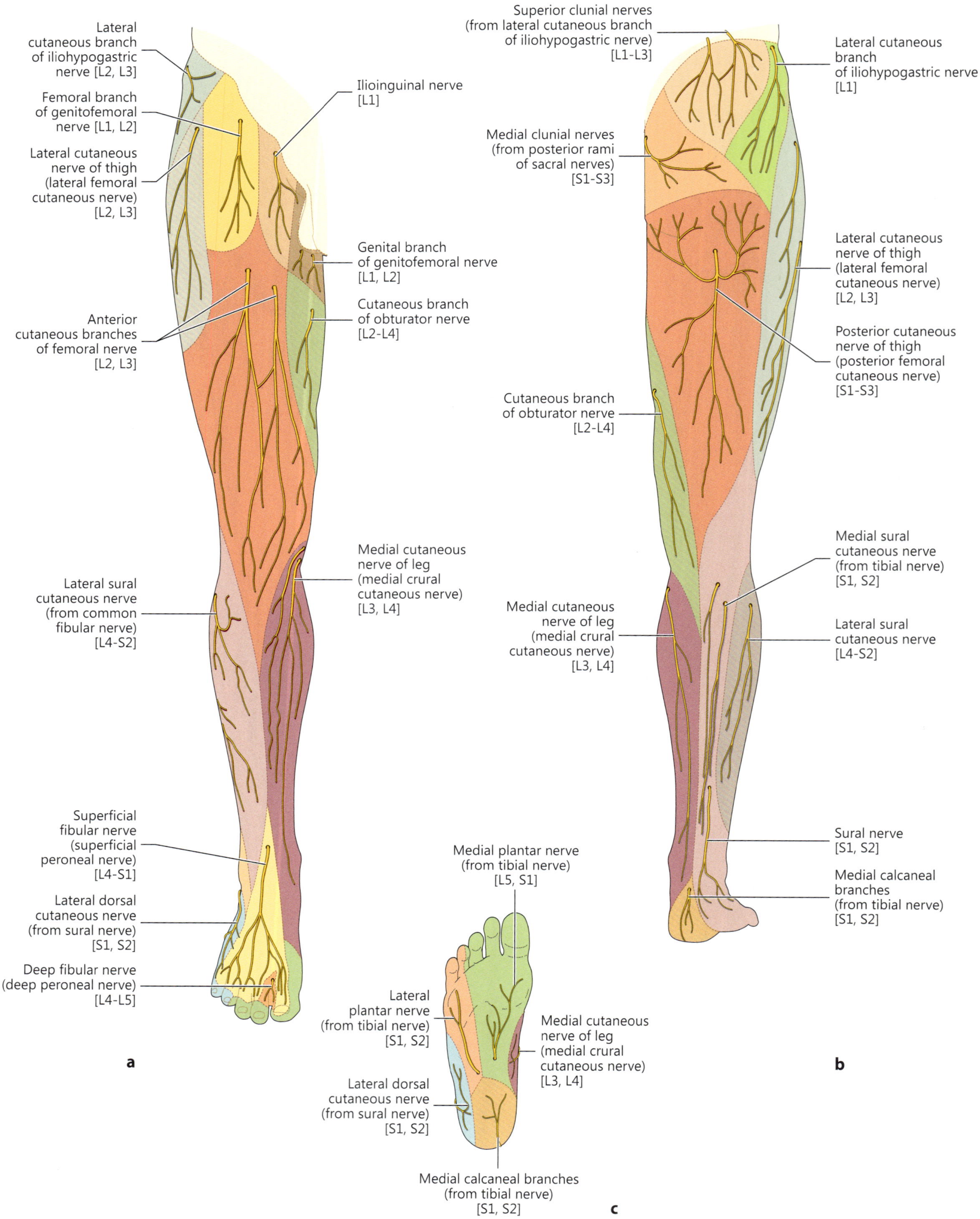

Areas covered by **cutaneous nerves** in the **lower limb**. **a**, Anterior aspect, **b**, posterior aspect and **c**, plantar aspect. For dermatomes, see page 59.

MUSCULOSKELETAL SYSTEM

Petros (Athens, 1928 - Milan, 2014), *Mouse D.M.* (detail), mixed media on canvas-backed paper, cm 65 × 100, 1993

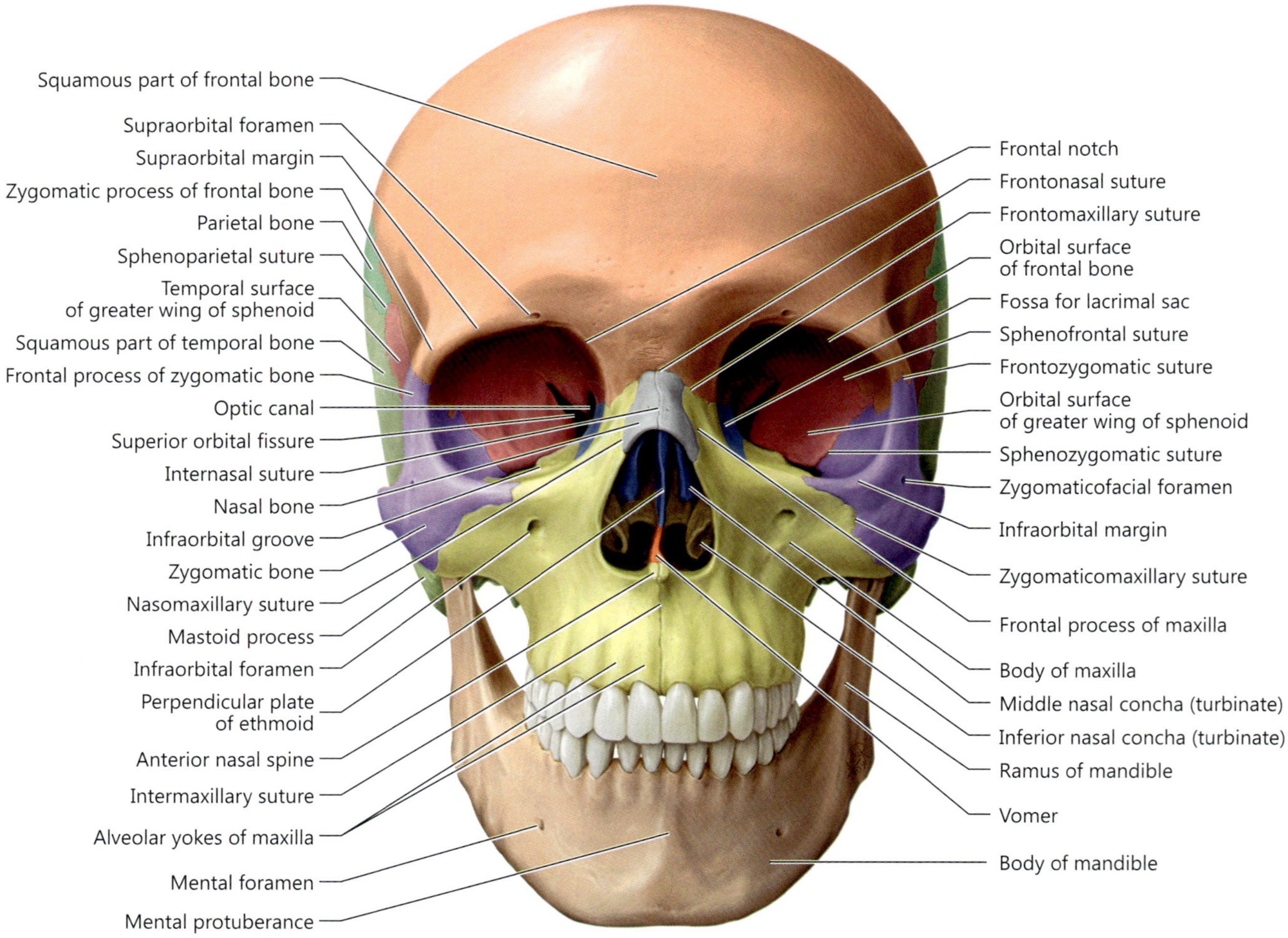

- Frontal bone
- Parietal bone
- Sphenoid
- Occipital bone
- Temporal bone
- Ethmoid
- Nasal bone
- Lacrimal bone
- Zygomatic bone
- Maxilla
- Mandible
- Vomer
- Palatine bone

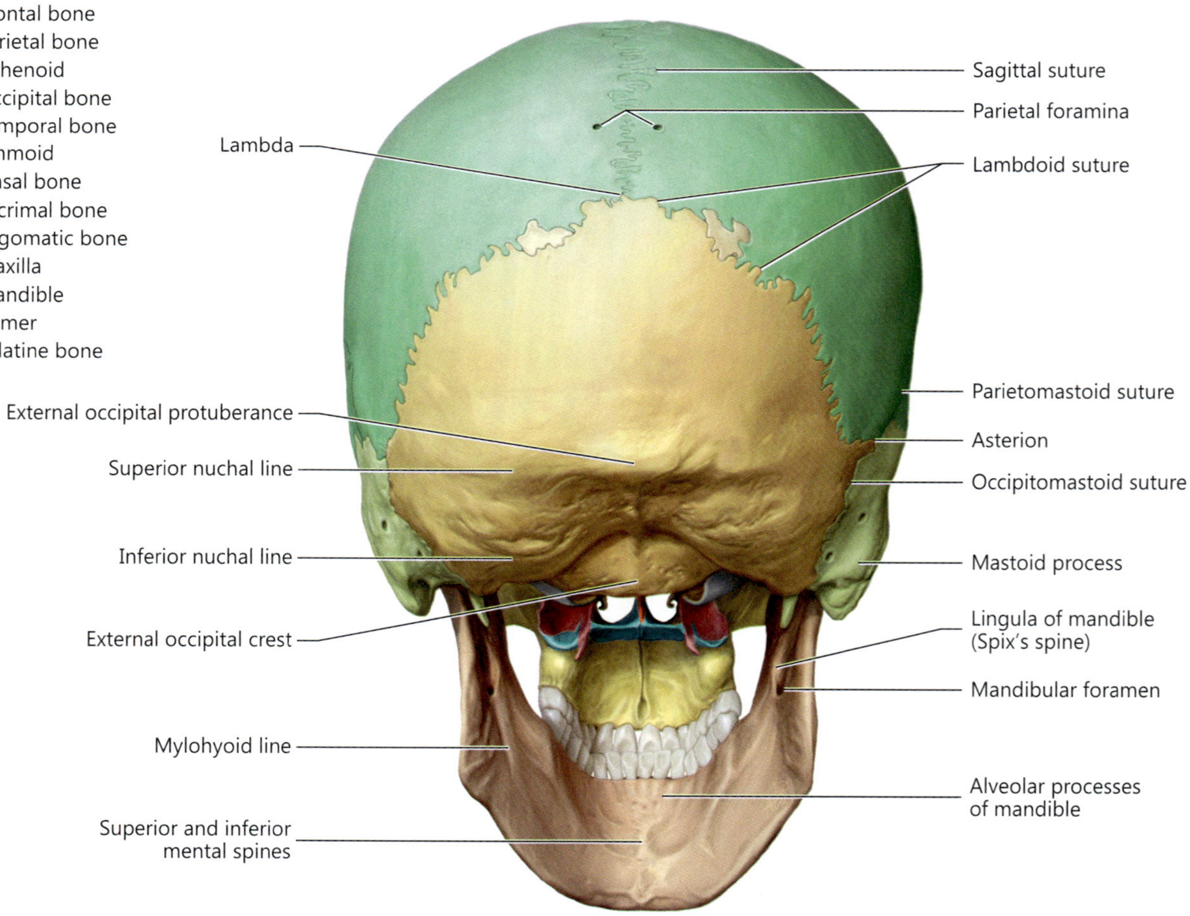

External configuration of the cranium. Above anterior view, below posterior view.

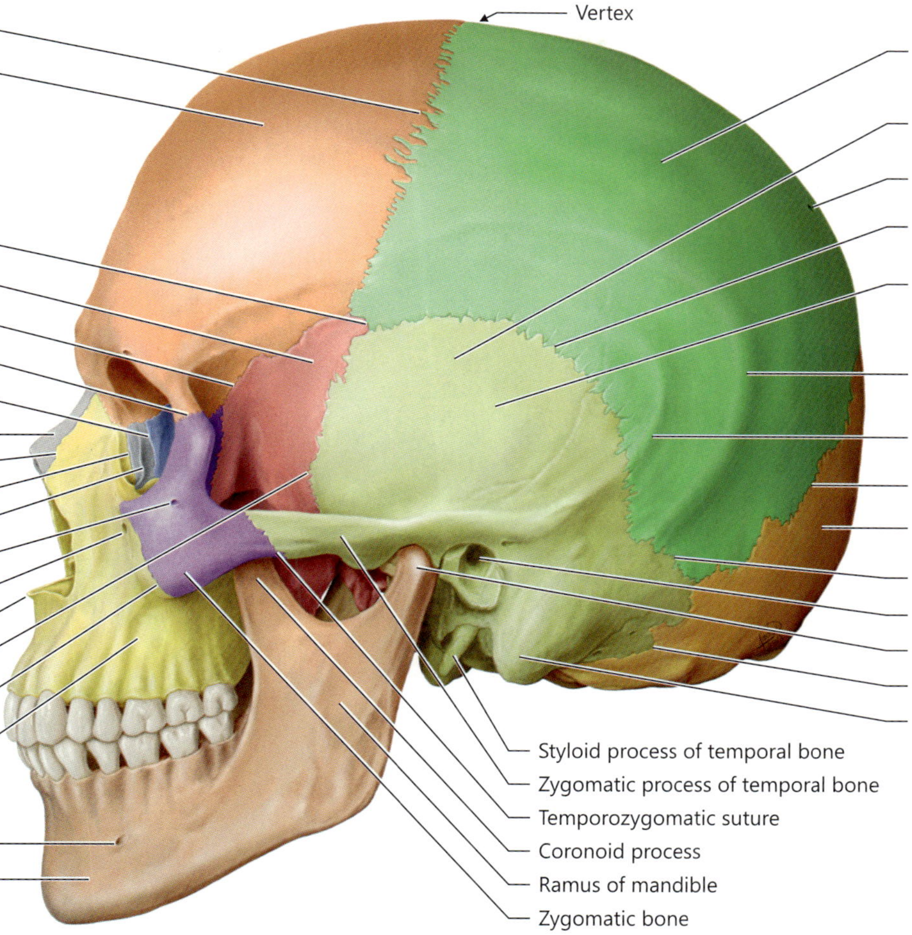

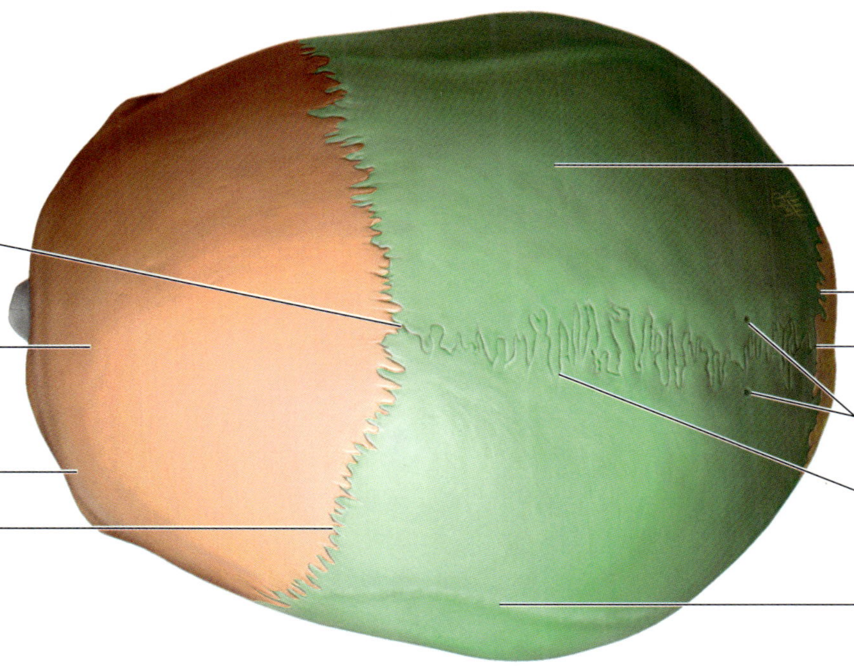

External configuration of the cranium. Above lateral view, below superior view. The outer surface of the calvaria appears smooth. The four bones that constitute it (frontal, paired parietal, and occipital) are mutually joined by the serrate sutures. The convergence points of the three bones anteriorly and posteriorly correspond to the bregma and lambda points in the middle of the coronal and lambdoid sutures, respectively.

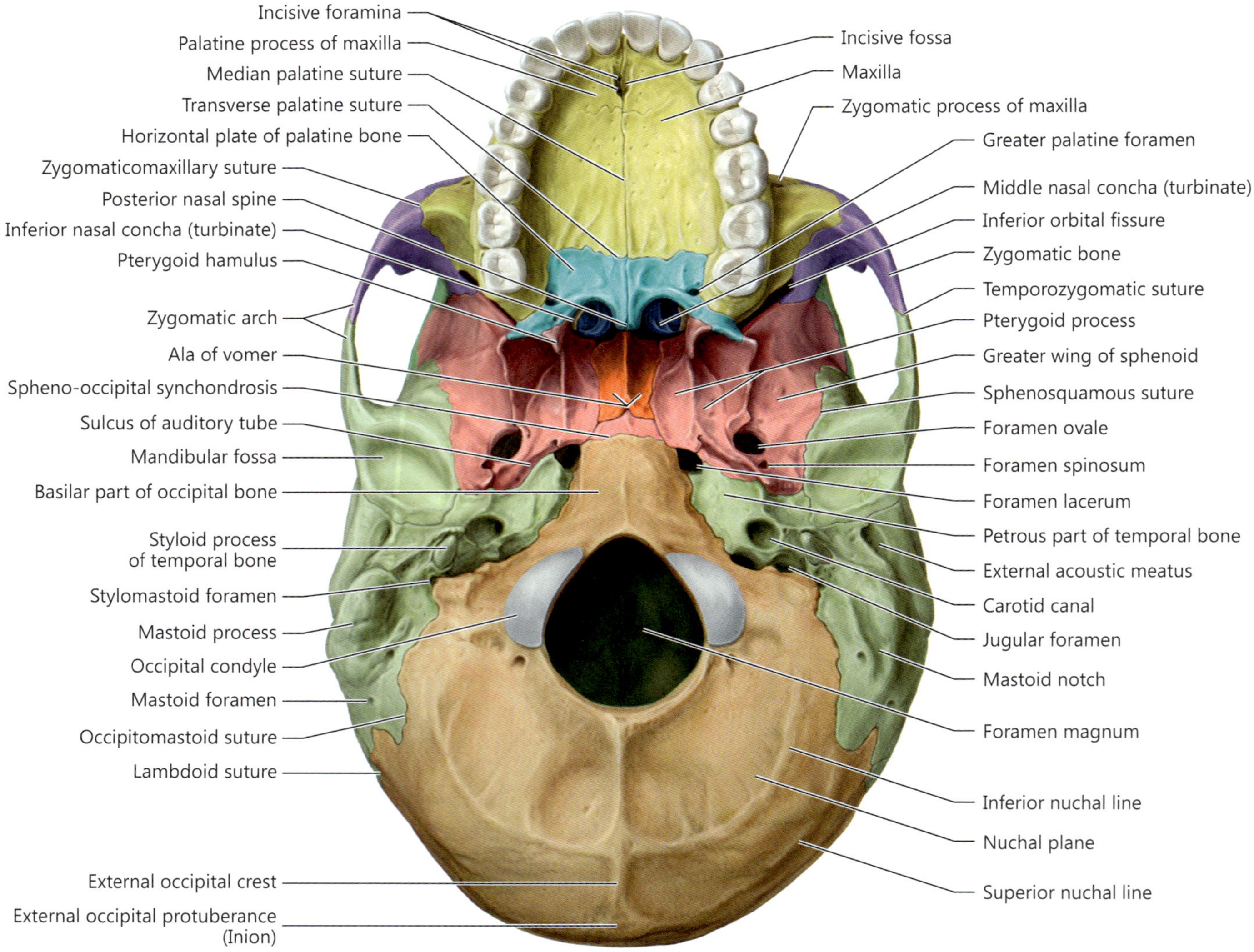

External configuration of the cranium, inferior view. The external surface of the cranial base is rich with foramina and fissures that allow communication between the inside and the outside of the cranium.

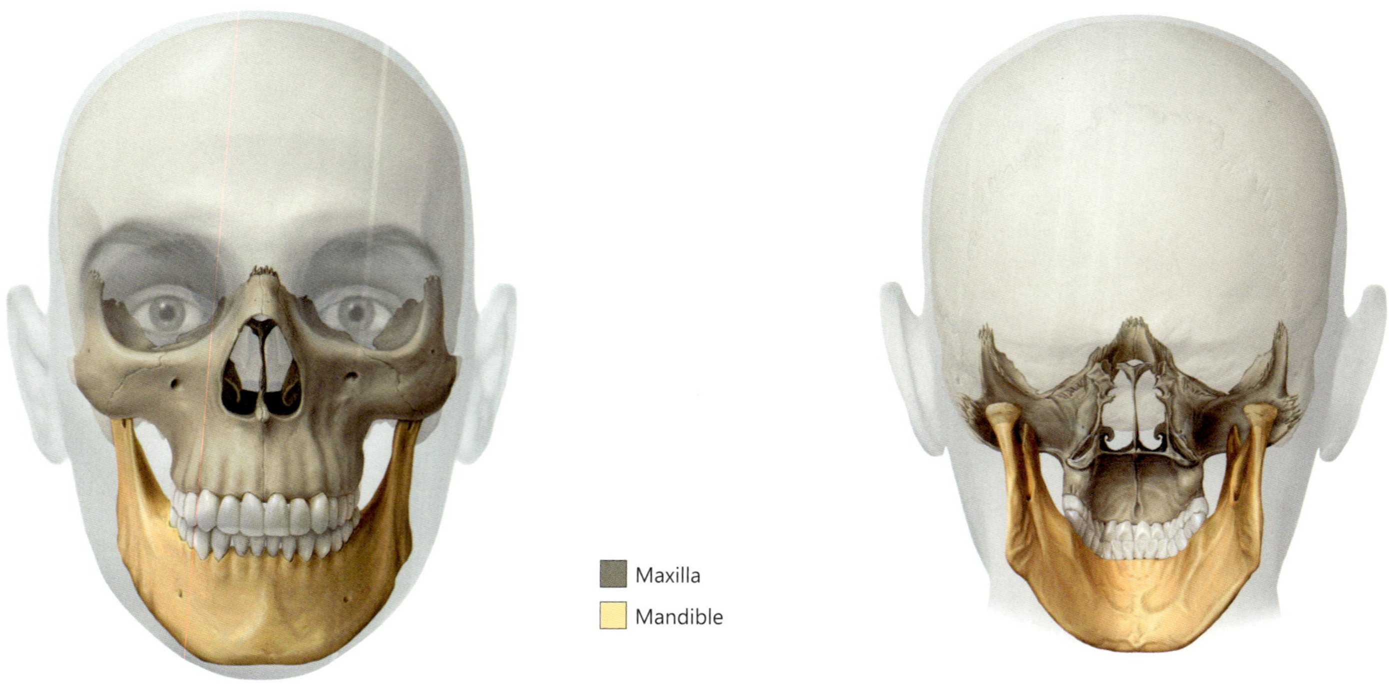

Subdivision of the viscerocranium in **maxilla** and **mandible**.

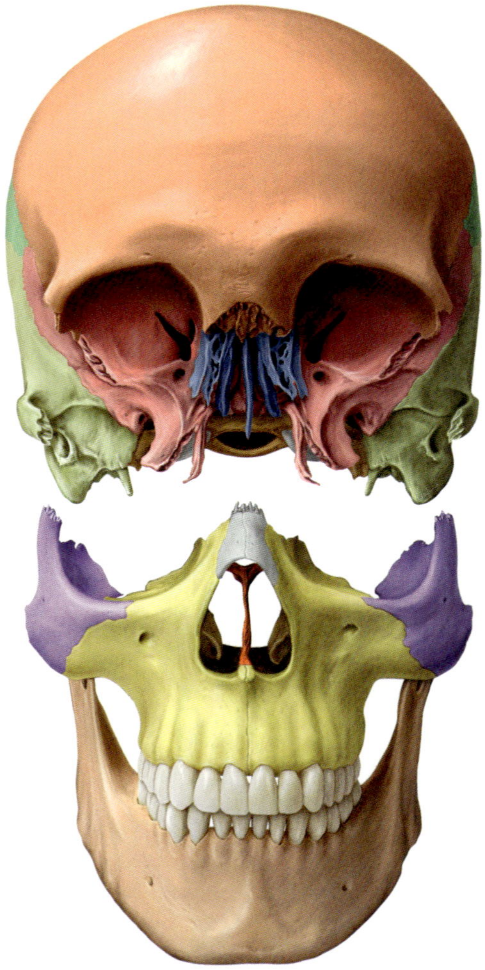

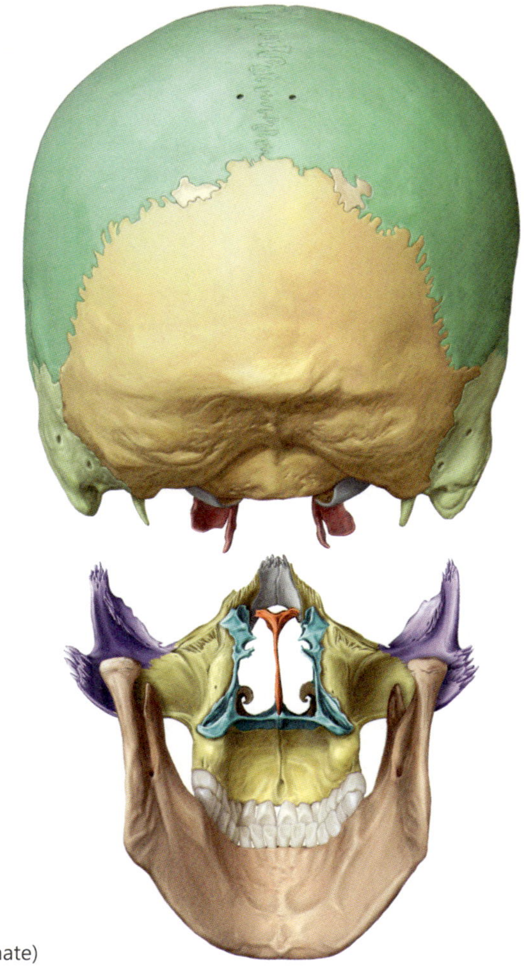

- Frontal bone
- Parietal bone
- Sphenoid
- Occipital bone
- Temporal bone
- Ethmoid
- Nasal bone
- Lacrimal bone
- Zygomatic bone
- Maxilla
- Mandible
- Vomer
- Palatine bone
- Inferior nasal concha (turbinate)

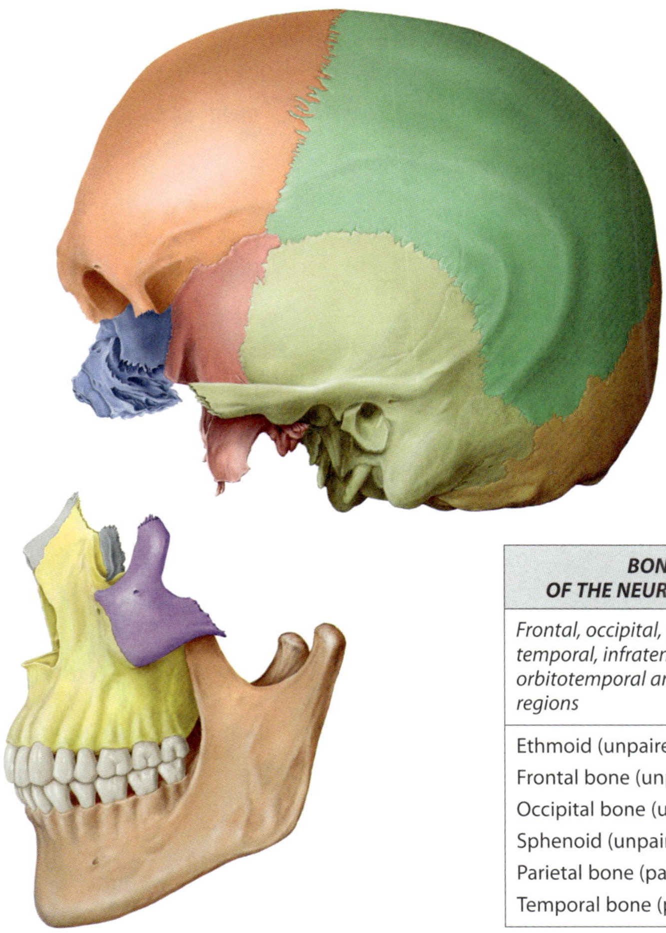

Subdivision of cranial bones into **neurocranium** (braincase, contains and protects the brain) and **viscerocranium** (facial skeleton, provides support and protection of internal organs).

BONES OF THE NEUROCRANIUM	**BONES OF THE VISCEROCRANIUM (FACIAL SKELETON)**	
Frontal, occipital, parietal, temporal, infratemporal, orbitotemporal and ethmoidal regions	*Nasal region*	*Orbital, infraorbital, buccal, parotid, zygomatic, oral and mental regions*
Ethmoid (unpaired) Frontal bone (unpaired) Occipital bone (unpaired) Sphenoid (unpaired) Parietal bone (paired) Temporal bone (paired)	Vomer (unpaired) Inferior nasal concha (paired) Nasal bone (paired) Lacrimal bone (paired)	Maxilla (paired) Palatine bone (paired) Zygomatic bone (paired) Mandible (unpaired)

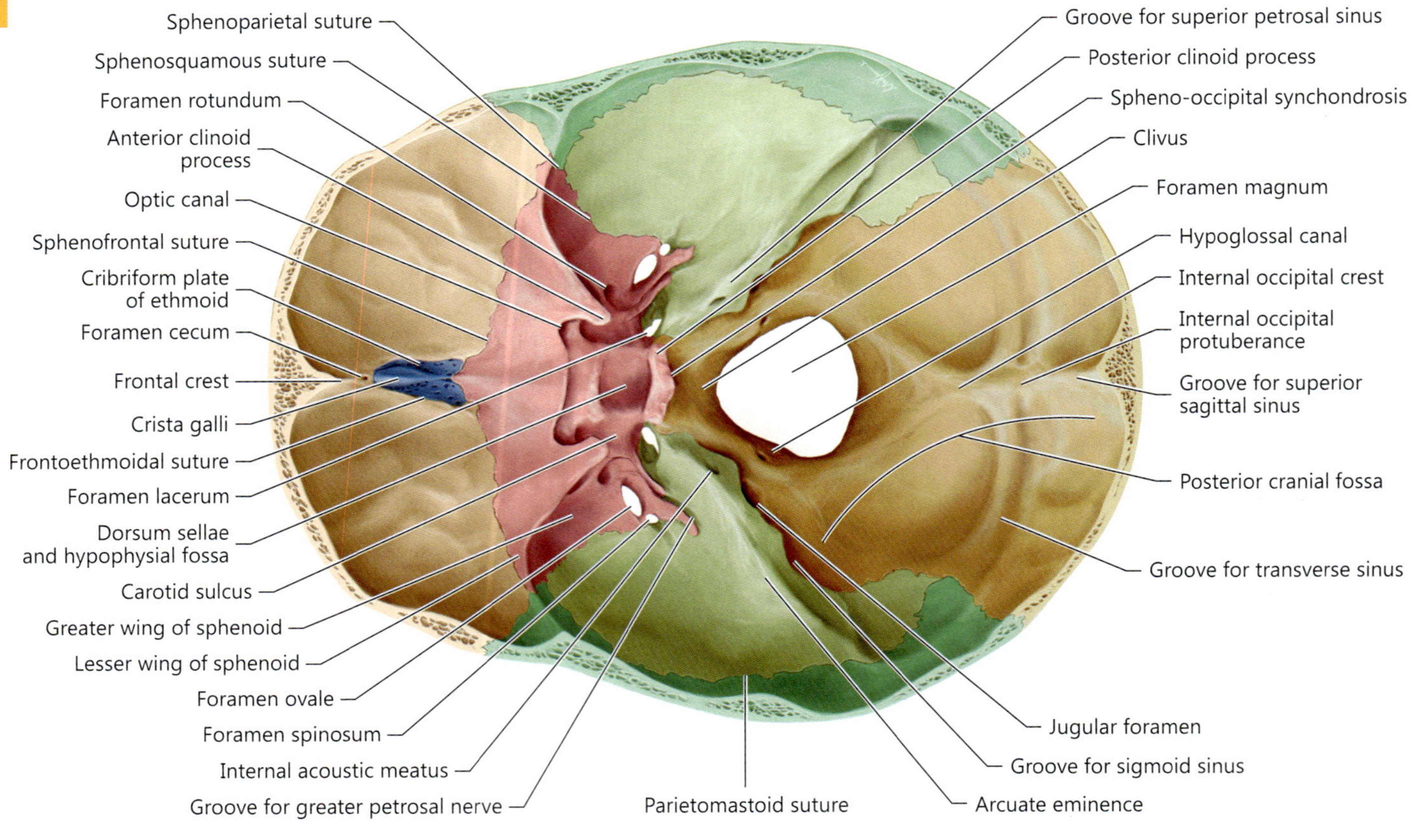

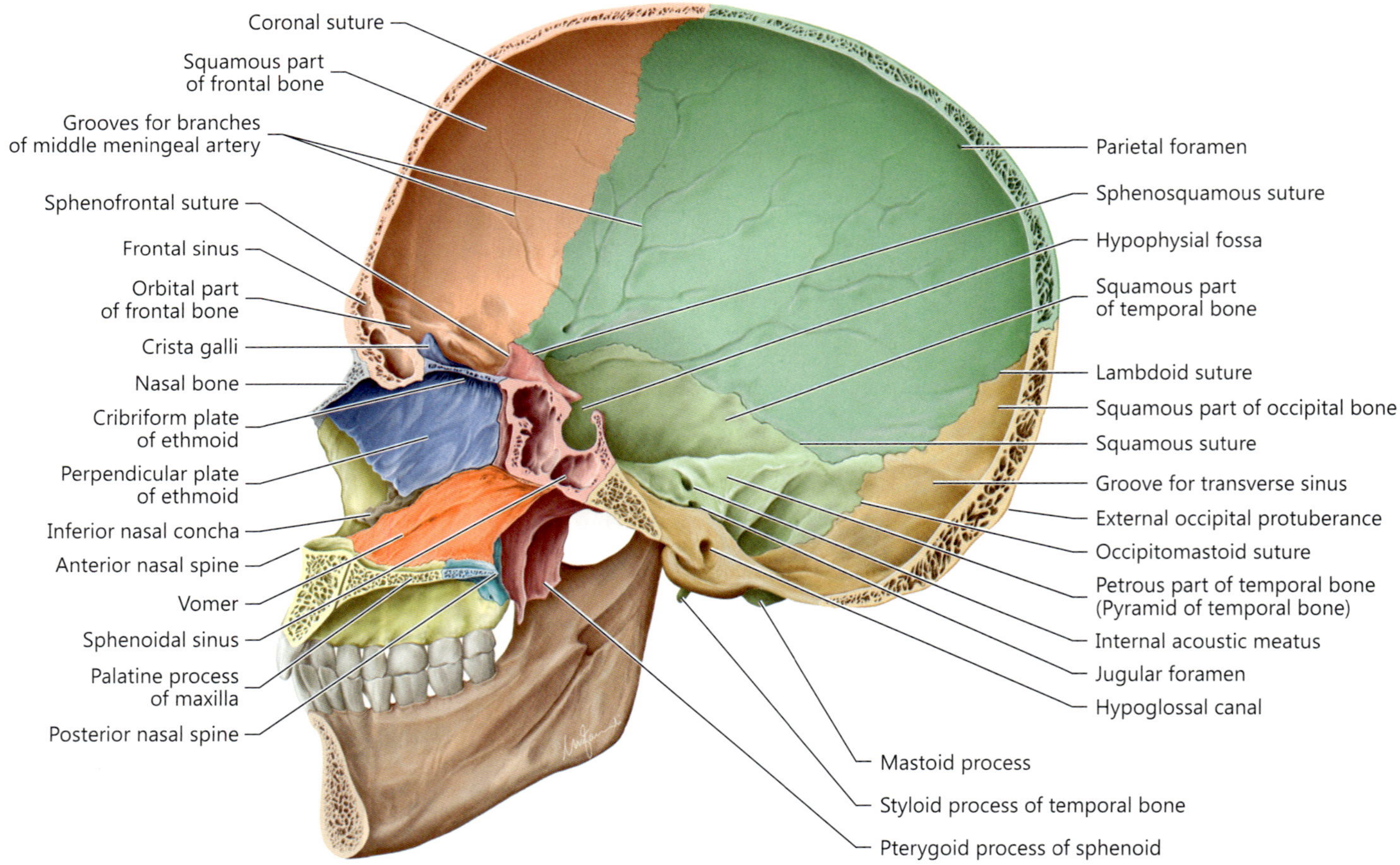

- Frontal bone
- Parietal bone
- Sphenoid
- Occipital bone
- Temporal bone
- Ethmoid
- Nasal bone
- Lacrimal bone
- Zygomatic bone
- Maxilla
- Mandible
- Vomer
- Palatine bone
- Inferior nasal concha (turbinate)

Above, **internal surface** of the **cranial base**; indicating the main foramina and some characteristics of bone morphology. Below, medial **sagittal section of the cranium**.

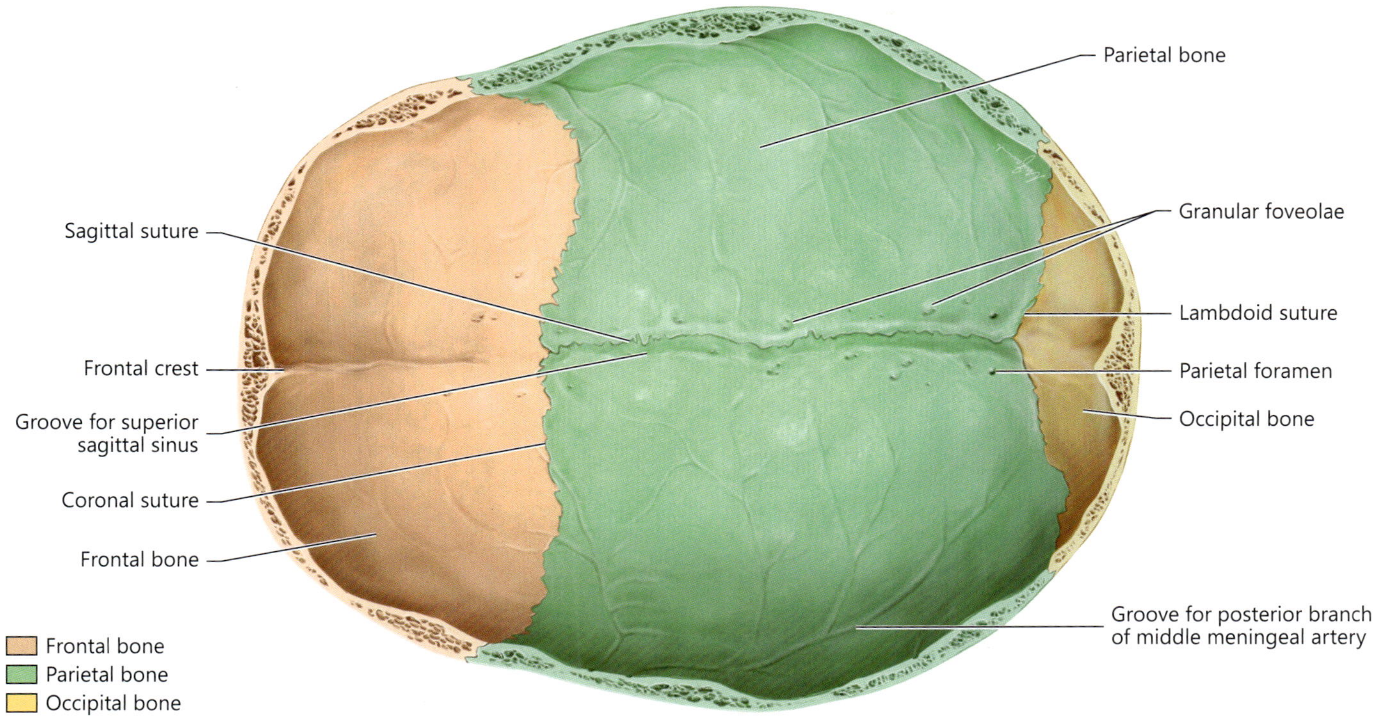

Internal surface of the **calvaria**. In the center there is a slight accentuated depression corresponding to the groove for the superior sagittal sinus; also visible are the vascular impressions and the granular foveolae (of Pacchioni).

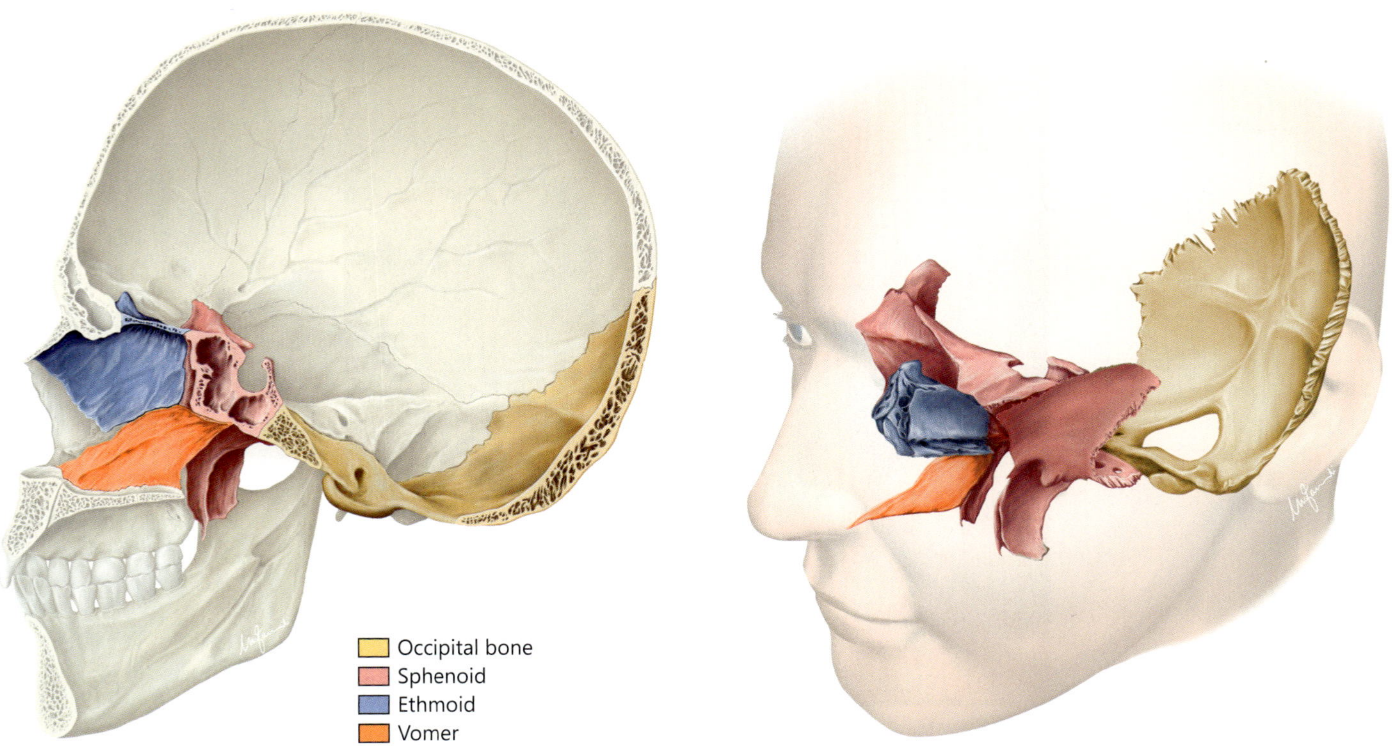

Bones of the **median region of the cranium**.

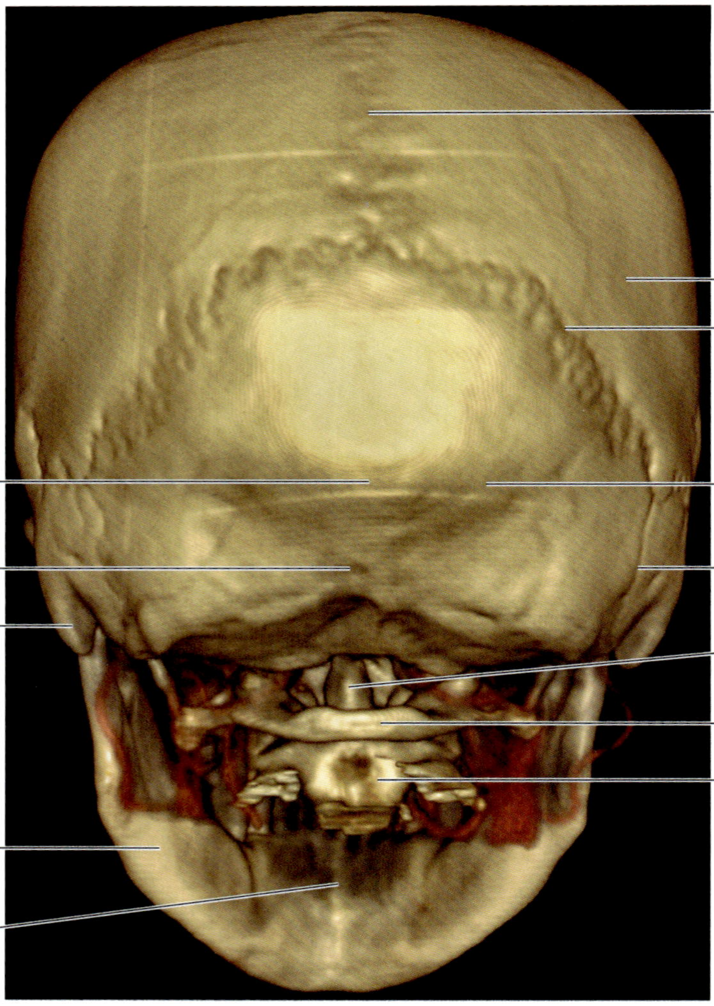

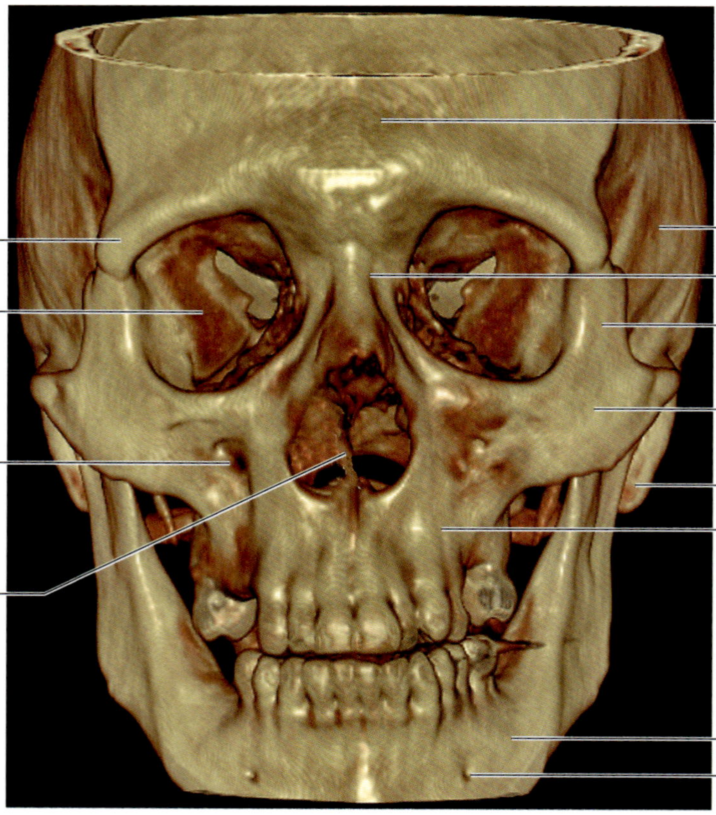

Three-dimensional images of a living human body, obtained using a 3D volume rendering technique, showing the posterior (above) and anterior (below) projections of the **cranial bones**.

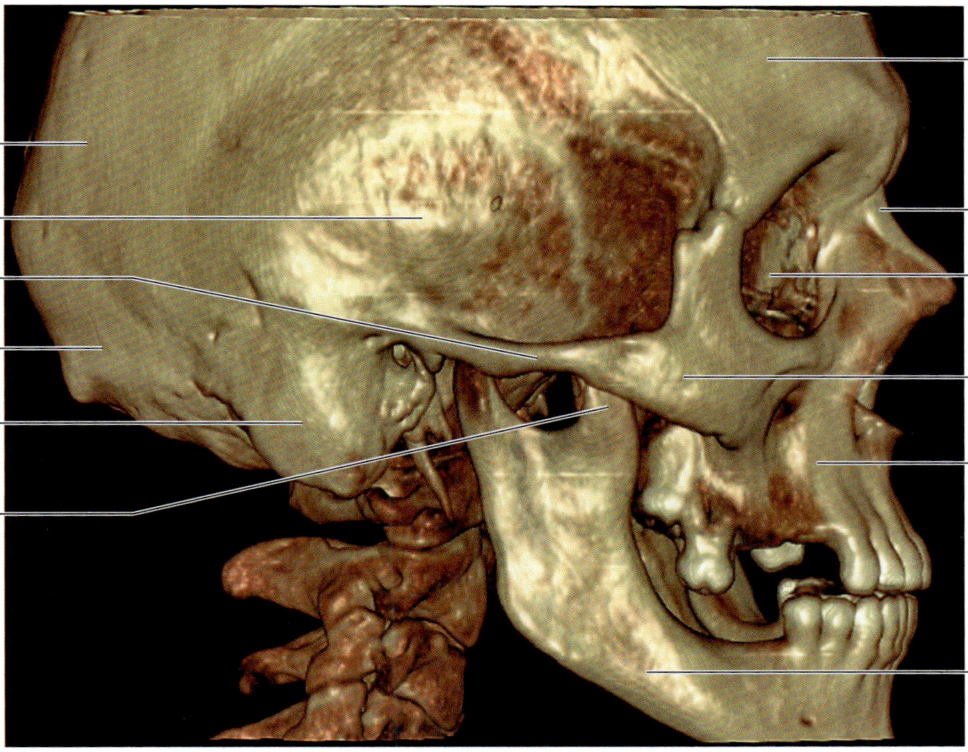

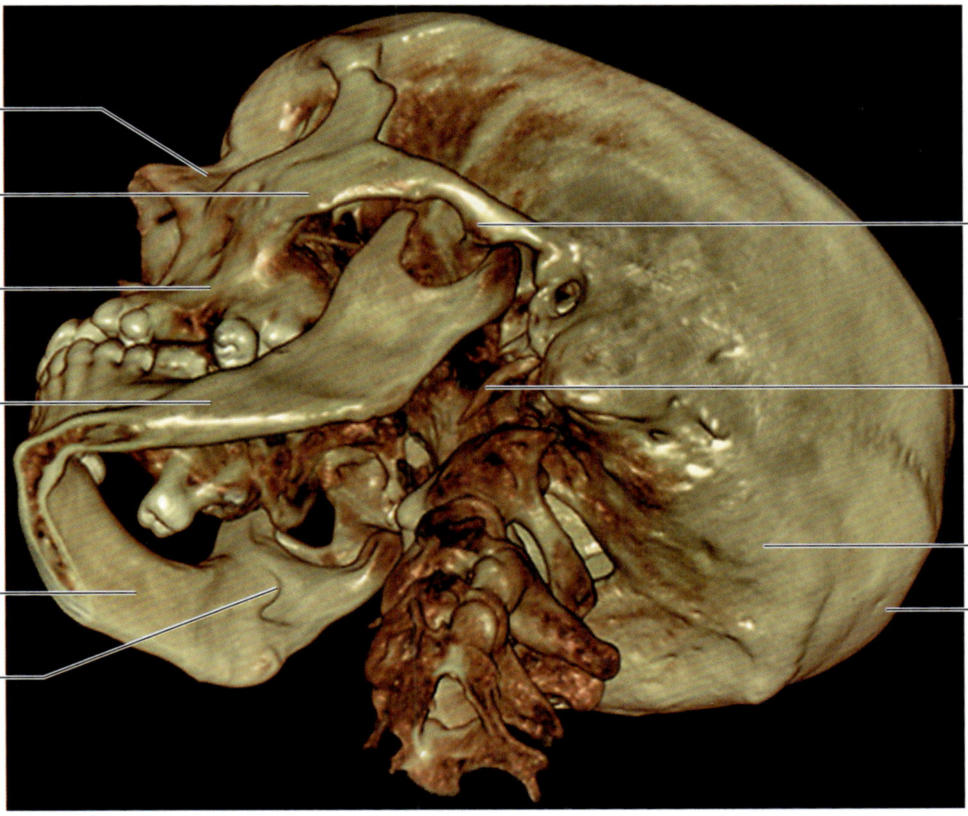

Above, three-dimensional image of a living human body, obtained using a 3D volume rendering technique, showing the right lateral projection of the **cranial bones**.

Below, three-dimensional image of a living human body, obtained using a 3D volume rendering technique, showing the left lateral projection from below of the **cranial bones**.

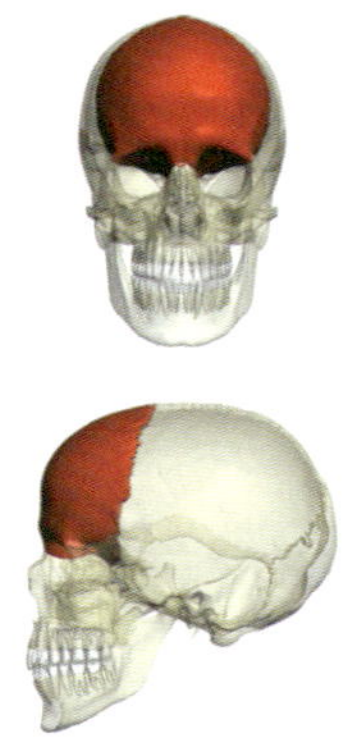

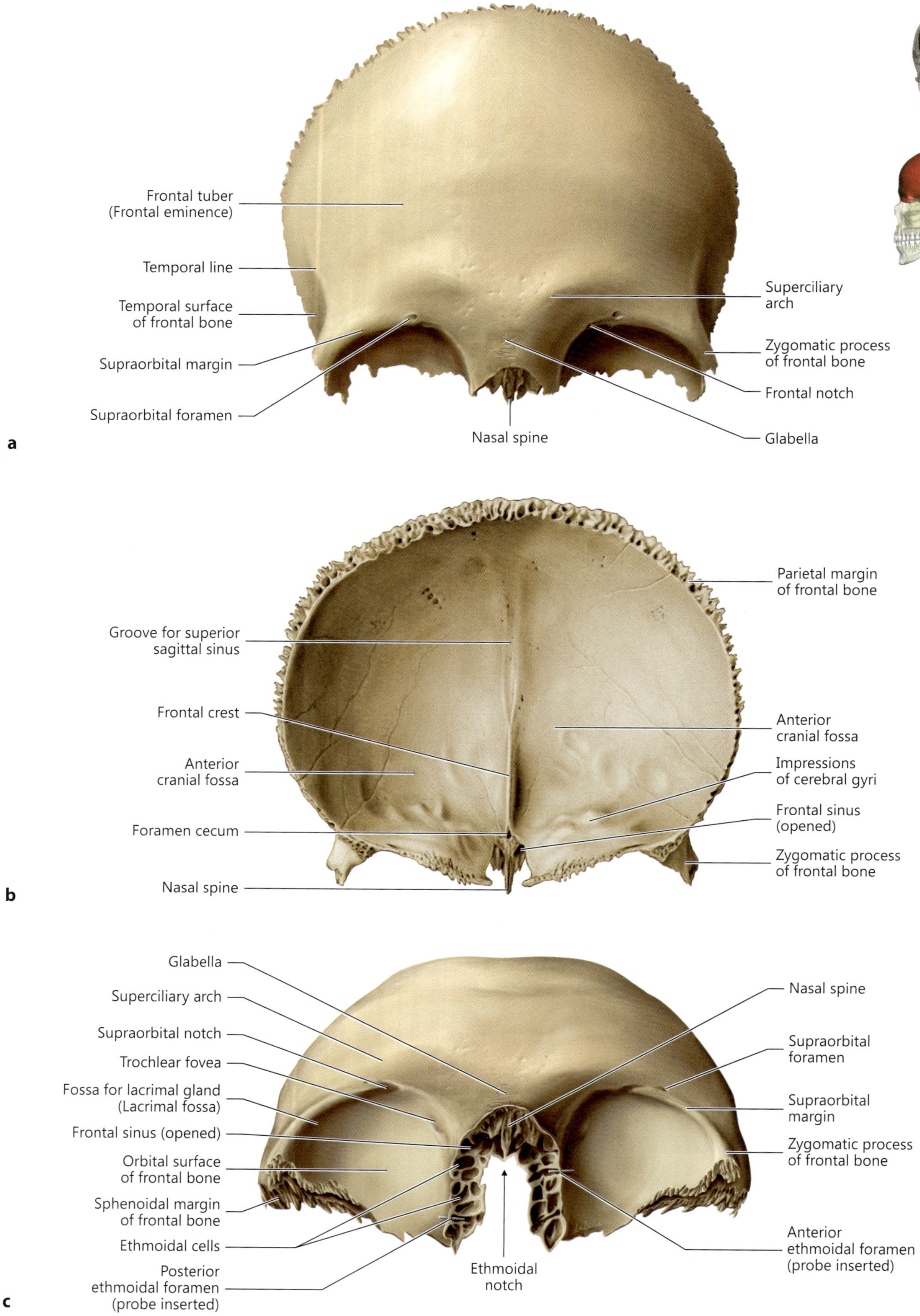

Frontal bone. a, Extracranial surface of the squamous part. **b**, Intracranial surface of the squamous part and of the horizontal part. **c**, Extracranial surface of the horizontal part. On the sides of the ethmoidal notch there are two visible large depressions that form the roof of the orbital cavities.

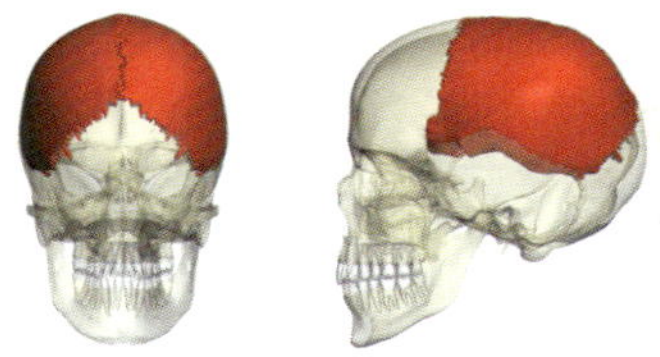

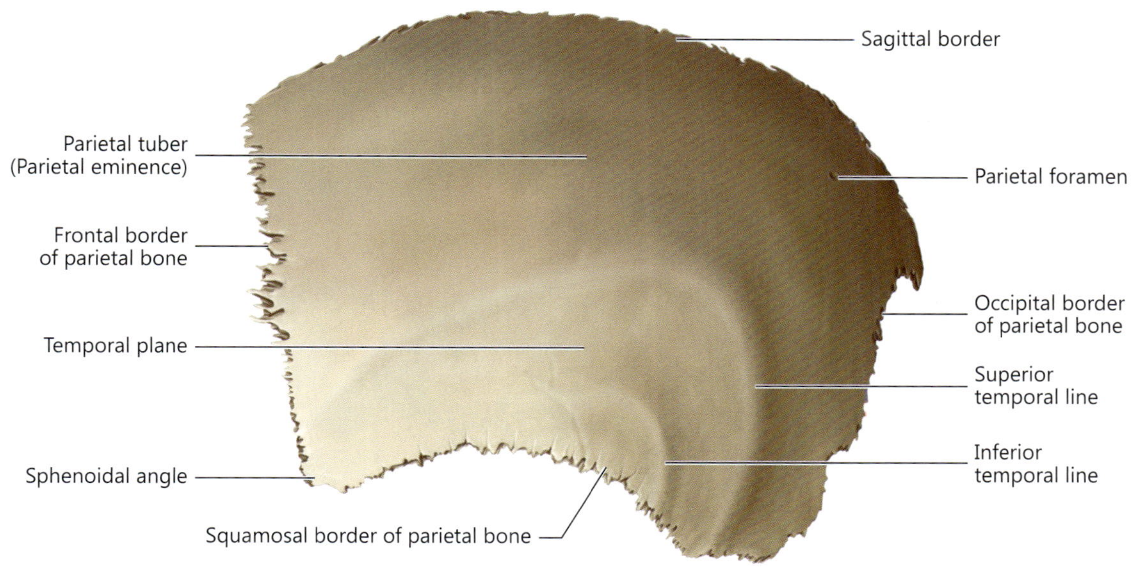

Left **parietal bone**, extracranial surface (external). The bone surface is smooth in the superior part; in its lower part it is crossed by two temporal lines, the superior and inferior temporal lines. Below the inferior temporal line there is a temporal plane.

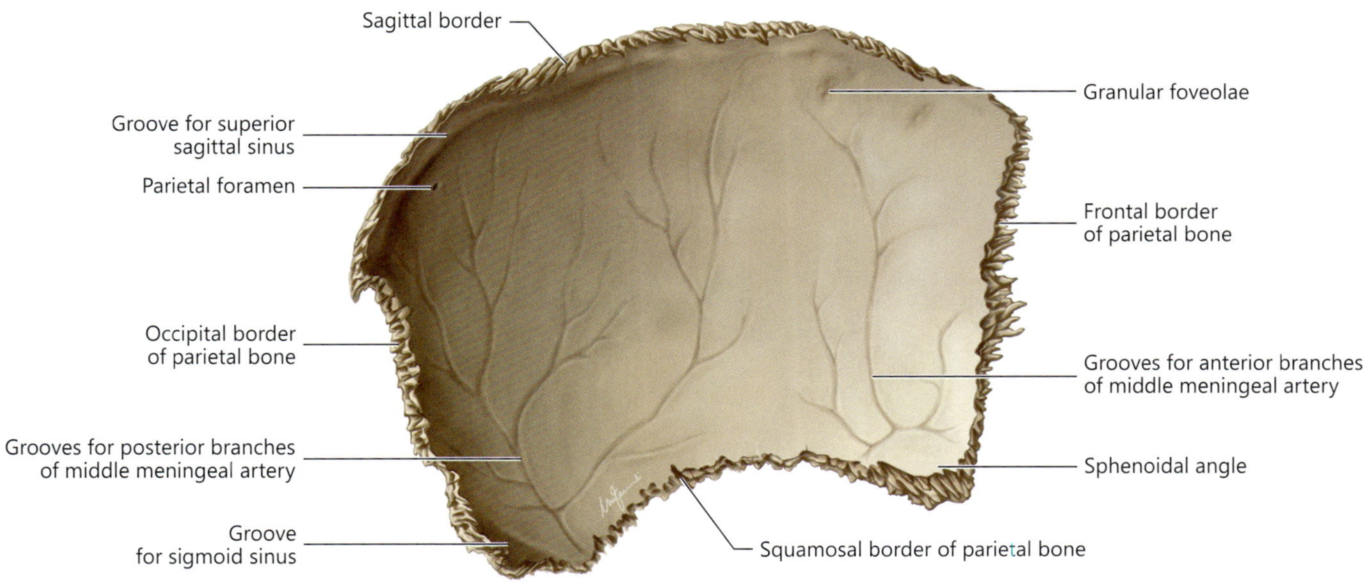

Left **parietal bone**, intracranial surface. Particularly marked is the appearance of numerous vascular grooves (anterior and posterior divisions of the middle meningeal artery), and, above, the granular foveolae (of Pacchioni). The superior border has a semicanal that, by juxtaposing the contralateral bone, marks the sagittal sulcus.

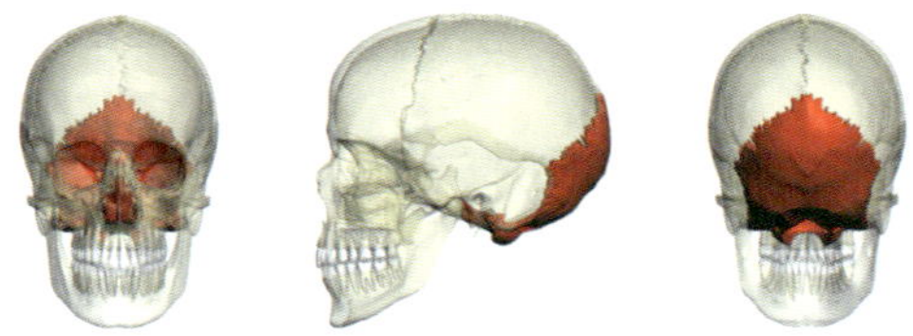

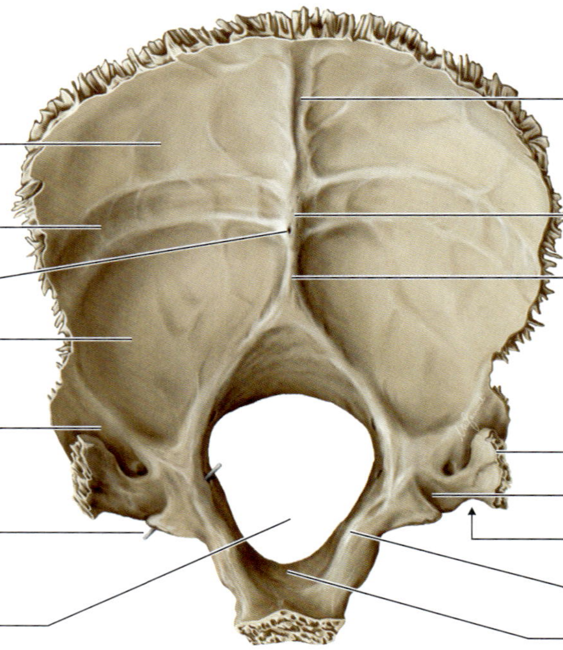

Occipital bone, intracranial surface. A cruciform groove marks the four occipital fossae. Three of the four foramina give passage to the dural venous sinuses (the superior sagittal sinus, transverse sinuses). Near the internal occipital crest runs the occipital venous sinus.

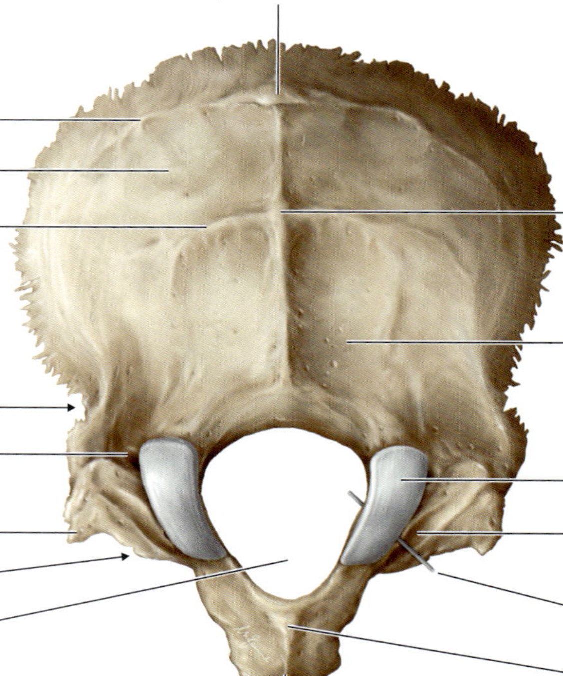

Occipital bone, extracranial surface. The squamous part of the occipital bone appears prominent in the nuchal lines and in the external occipital crest.

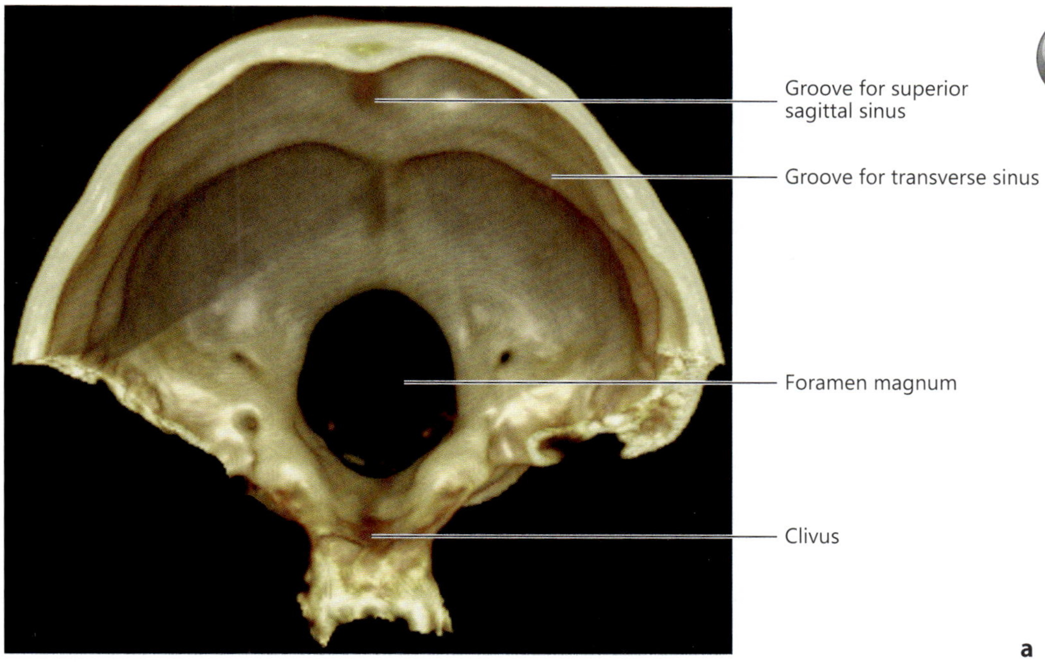

a

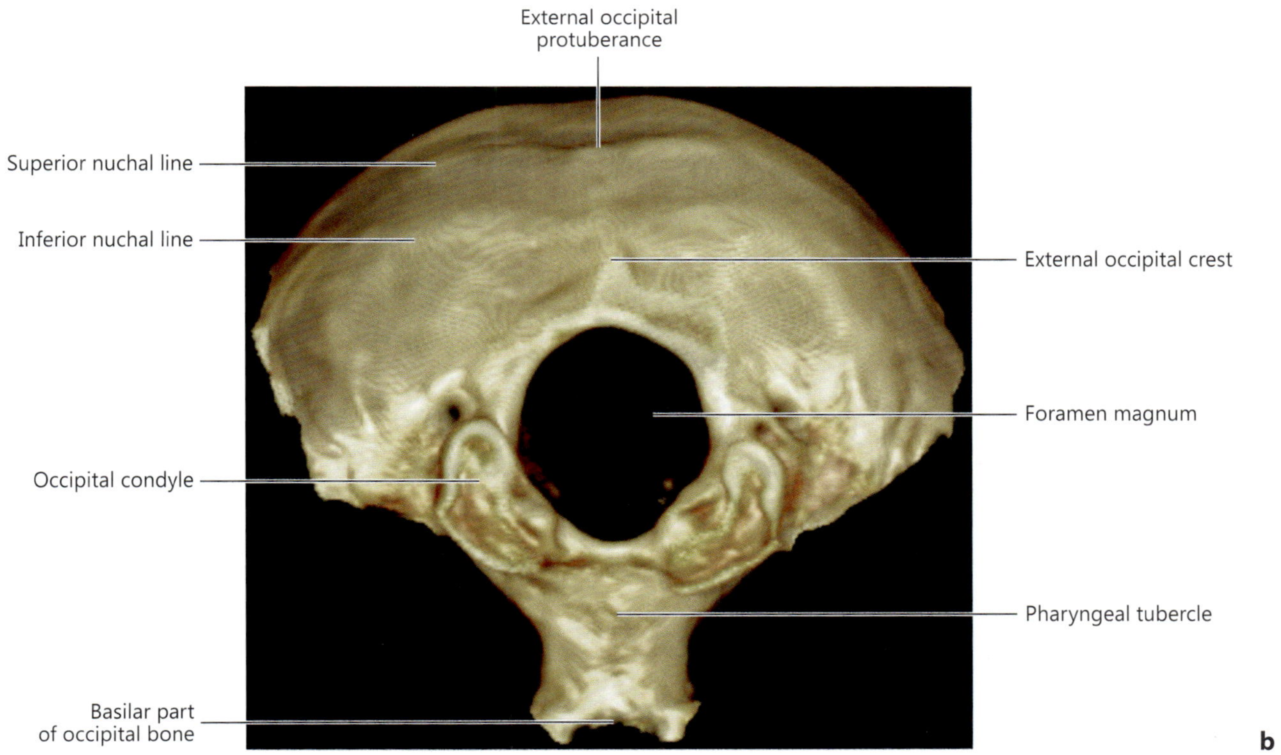

b

Three-dimensional images of a living human body, obtained using a 3D volume rendering technique, showing the intracranial surface (**a**) and the extracranial surface (**b**) of the **occipital bone**.

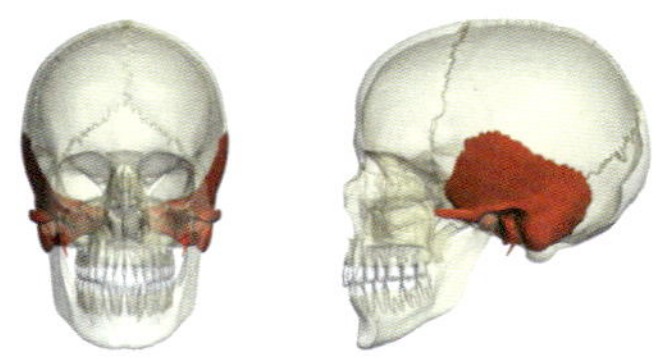

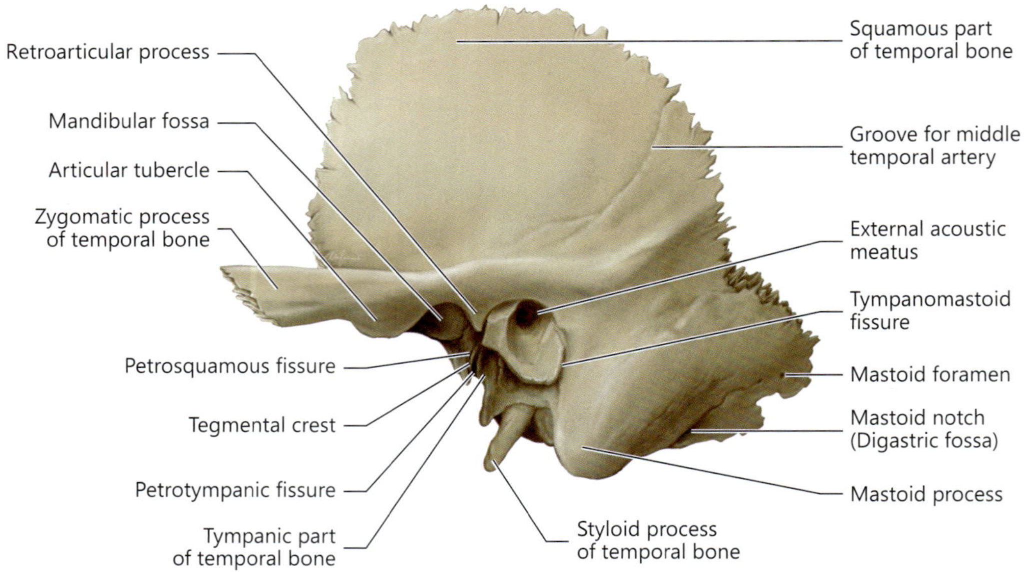

Left **temporal bone**, lateral surface of the extracranial surface.

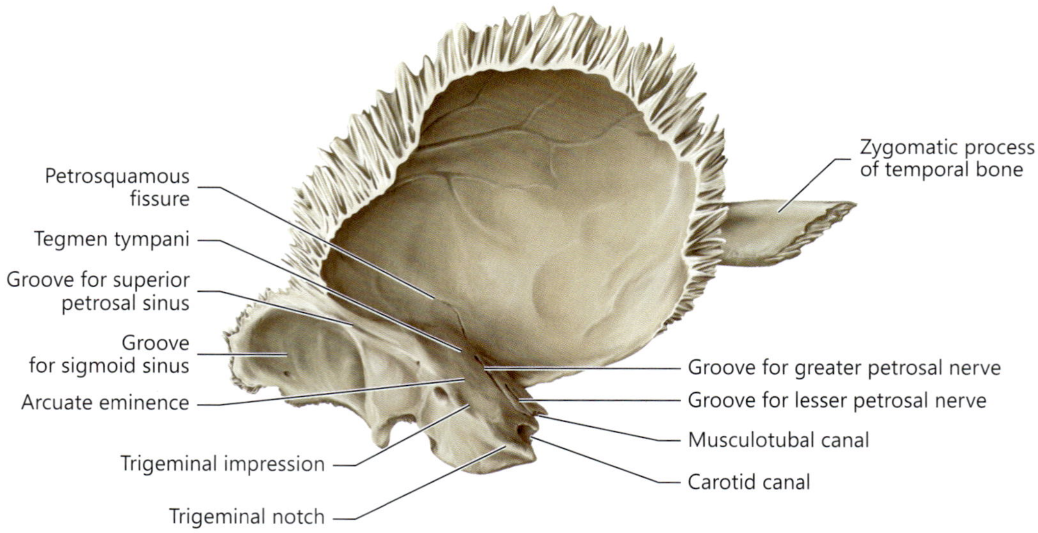

Left **temporal bone**, intracranial surface, top view showing the superior surface of the petrous part (pyramid of the temporal bone).

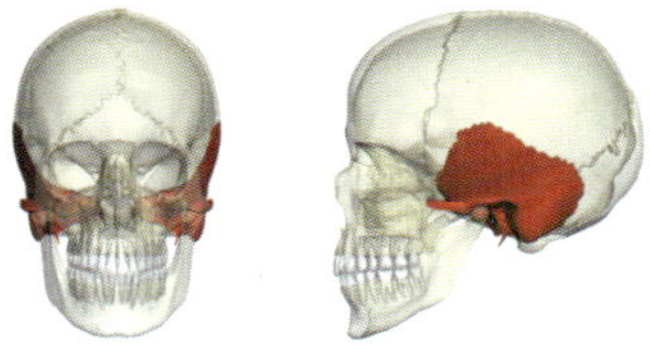

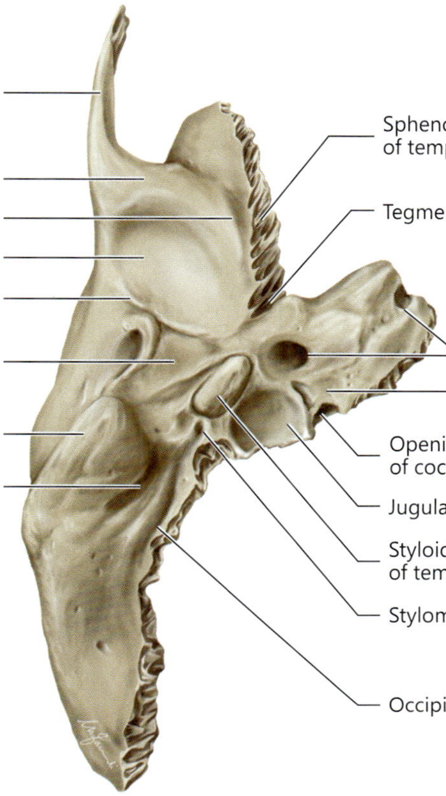

Left **temporal bone**, inferior surface of the extracranial surface.

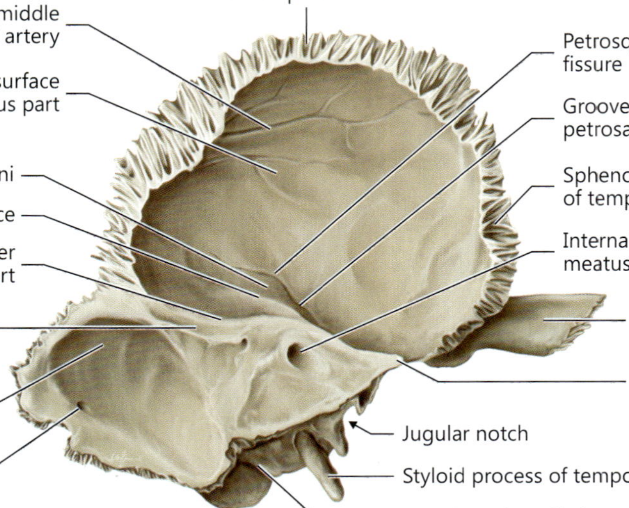

Left **temporal bone**, intracranial surface, view from behind showing the superior and posterior surfaces of the petrous part (pyramid of the temporal bone).

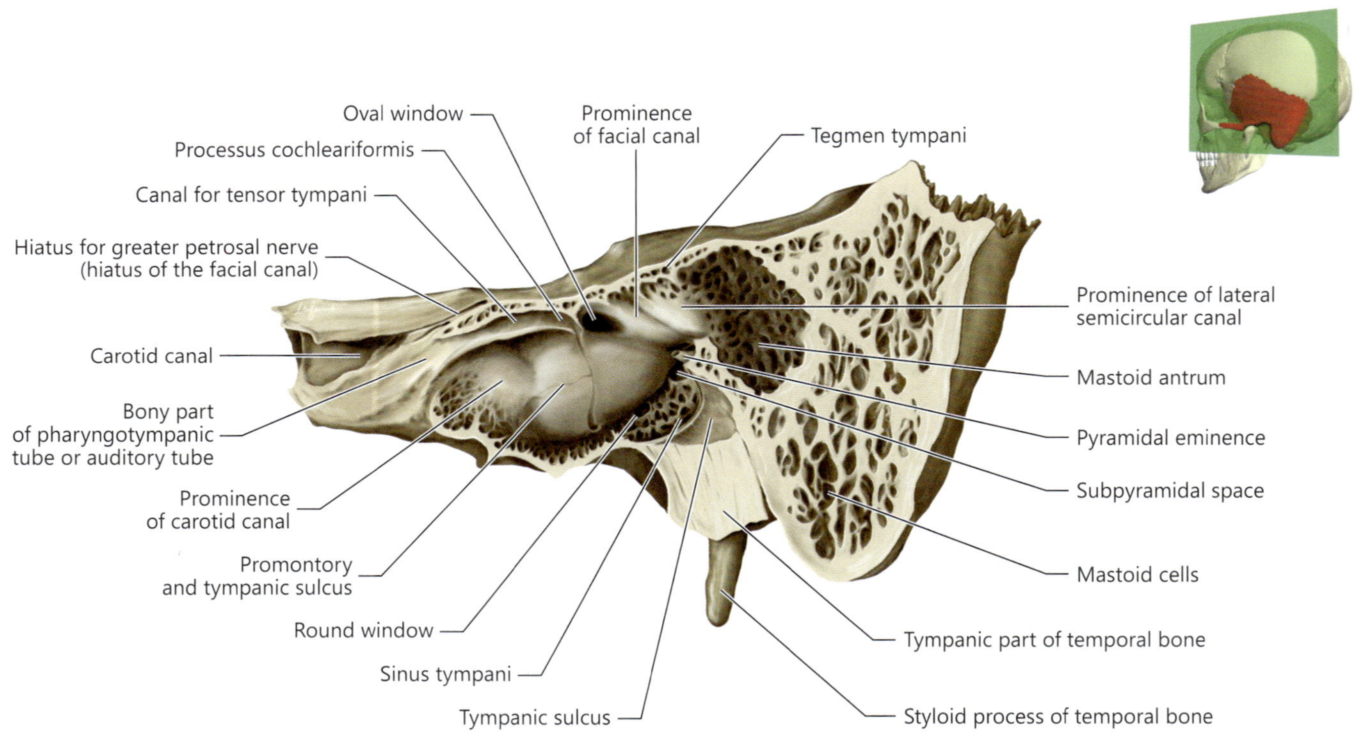

Section of the left **temporal bone** which passes through the **tympanic cavity**; configuration of the medial wall of the cavity.

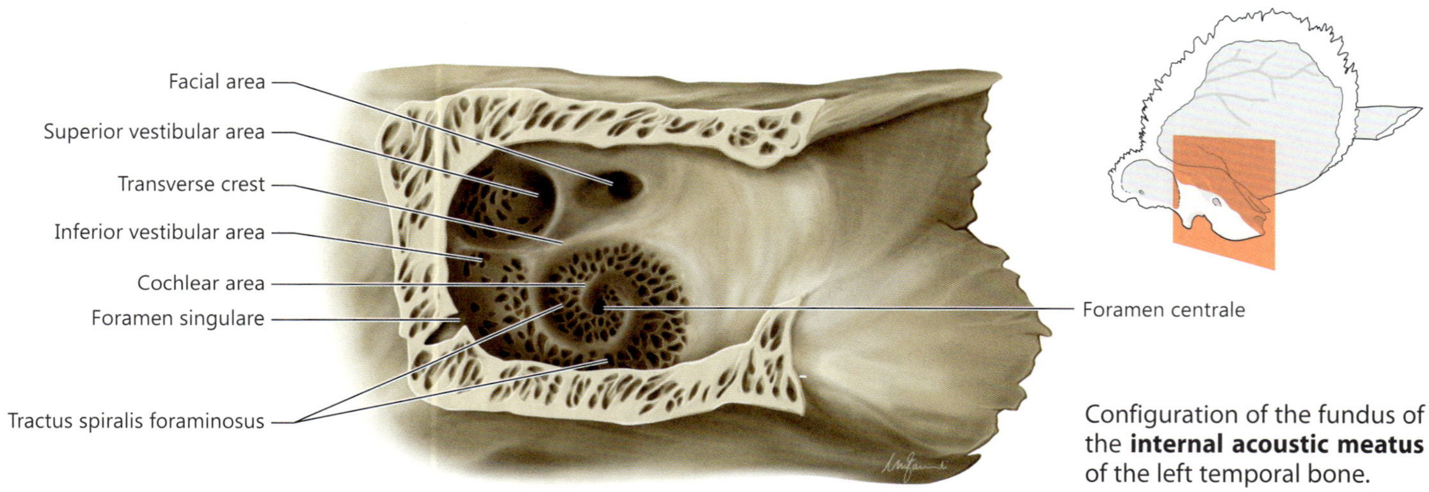

Configuration of the fundus of the **internal acoustic meatus** of the left temporal bone.

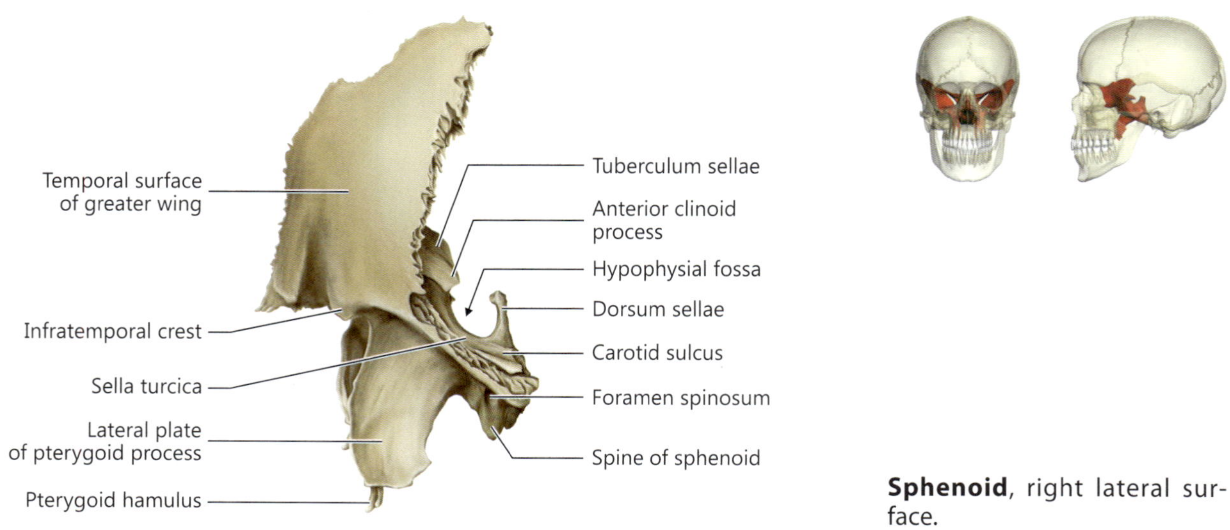

Sphenoid, right lateral surface.

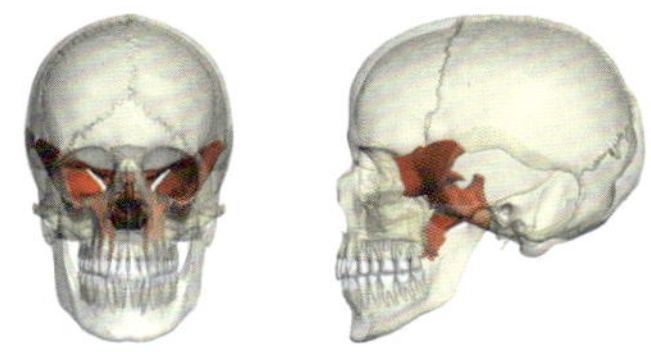

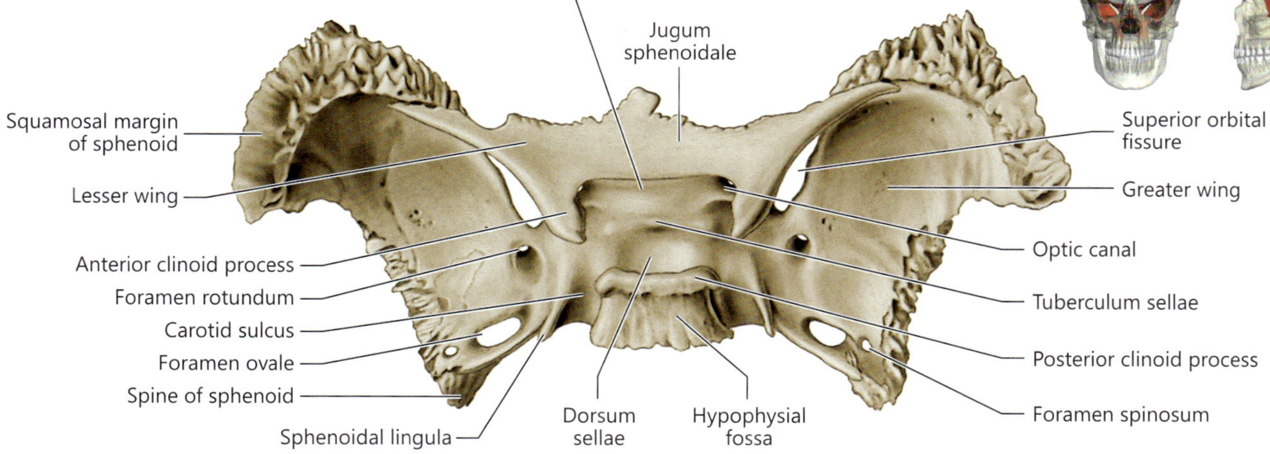

Sphenoid, intracranial surface (top view). The central part corresponds to the body, while the peripheral parts expand into the greater and lesser wings.

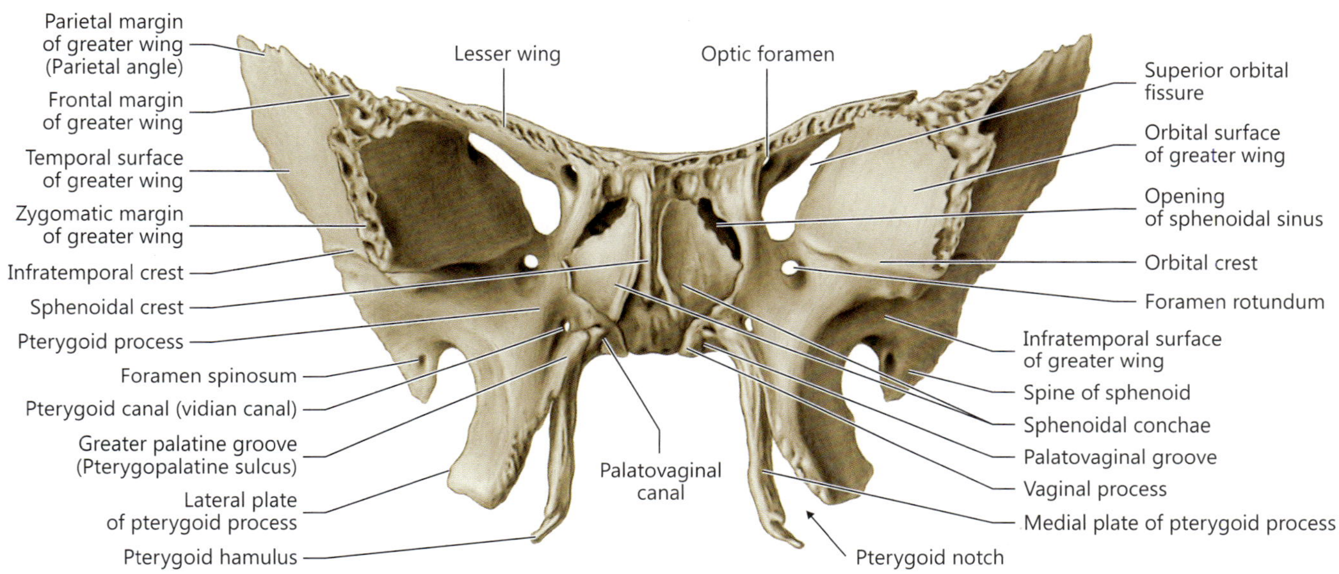

Sphenoid, anterior surface. On the sides of the body, the bone expands into the greater wings, lesser wings, and pterygoid processes.

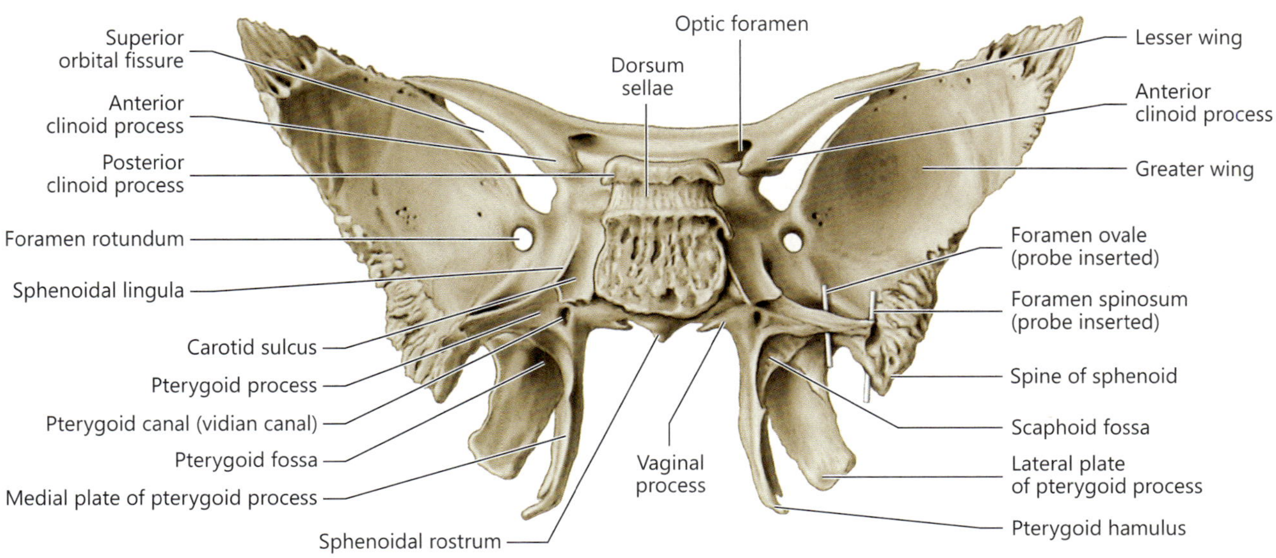

Sphenoid, posterior surface. Two probes have been introduced into the foramen ovale and foramen spinosum.

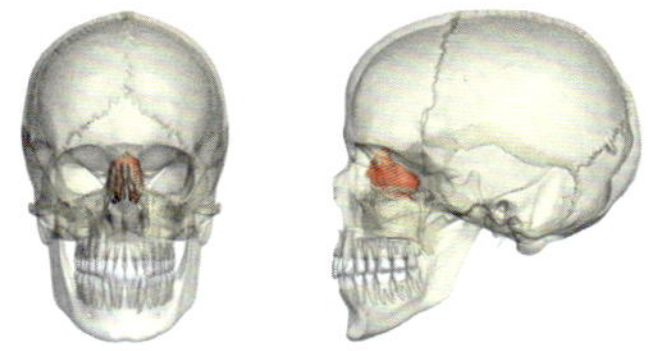

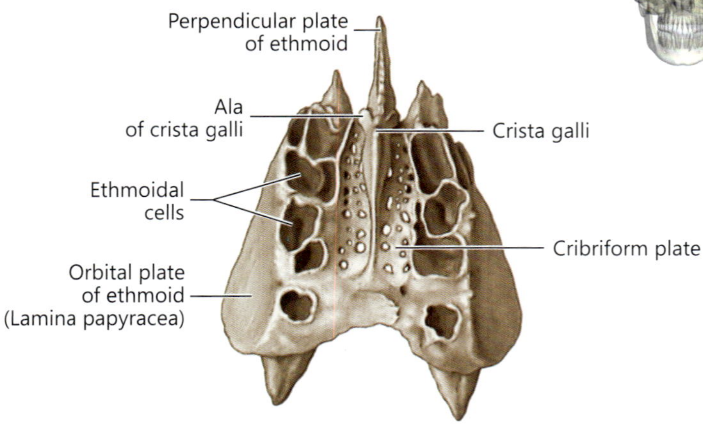

Ethmoid, superior surface. On the side of the cribriform plate, several groups of ethmoidal cells have been opened.

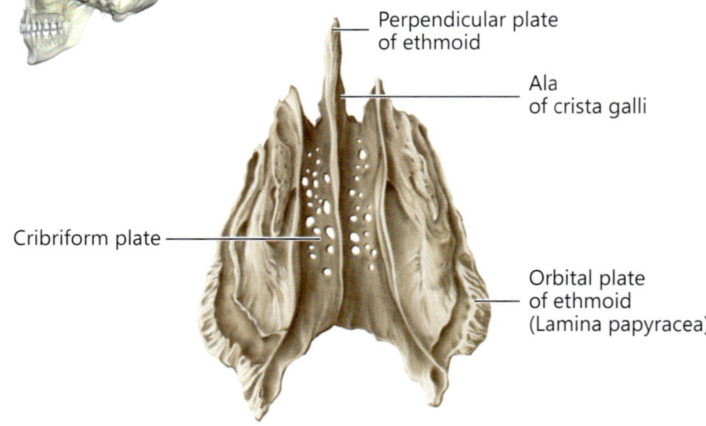

Ethmoid, inferior surface.

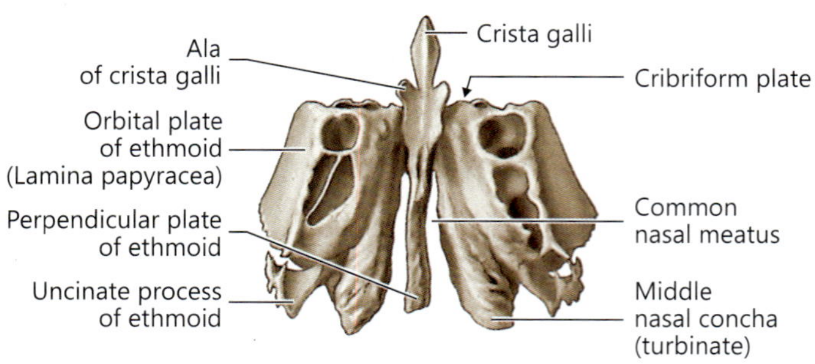

Ethmoid, anterior surface. The perpendicular plate and the two lateral masses are connected to the horizontal cribriform plate.

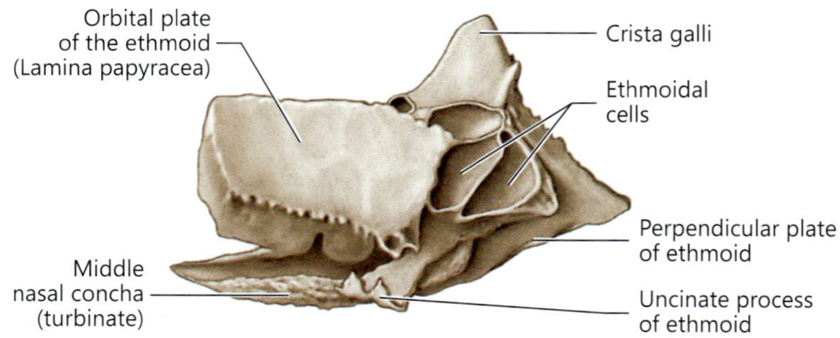

Ethmoid, lateral surface. The ethmoid, with its orbital plate, medially delimits the orbit.

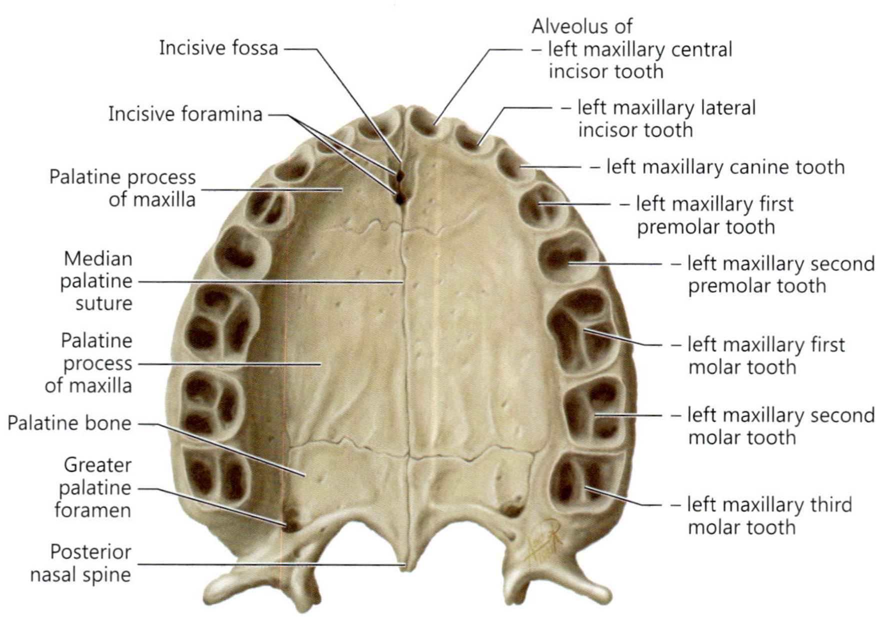

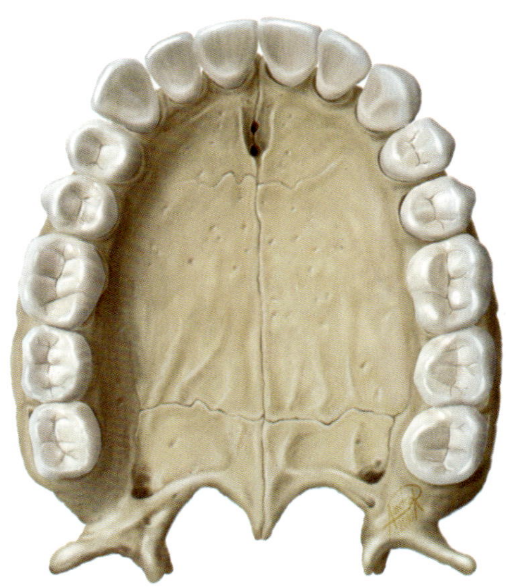

Maxillary arcade. On the left, the superior alveolar arch seen from below. Proceeding from the median line in the lateral direction the alveolar cavities are more and more irregular because of the presence of bone septa that delimit the compartments to accommodate the dental roots of the multiroot teeth.
On the right, maxillary arcade with full dentition.

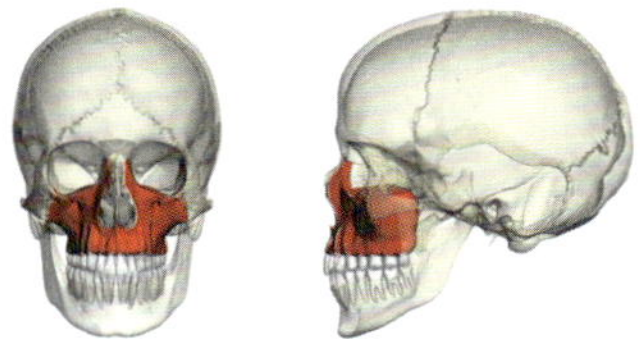

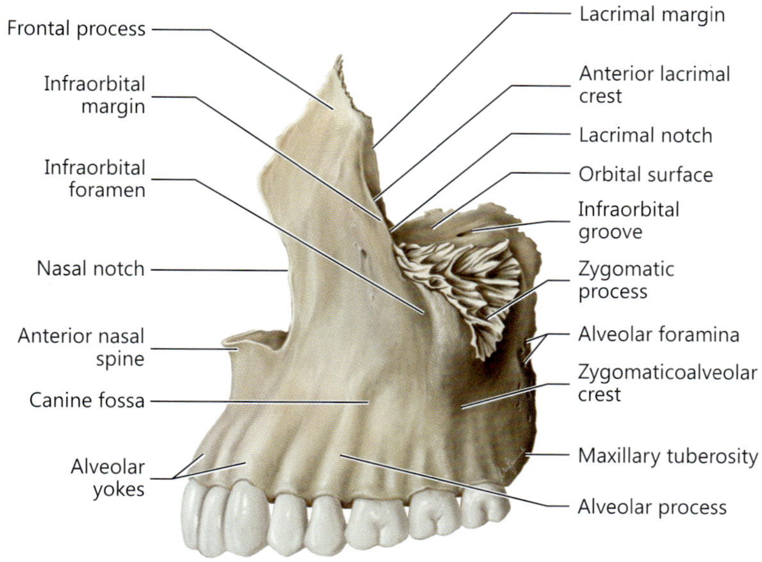

Left **maxilla**, lateral view.

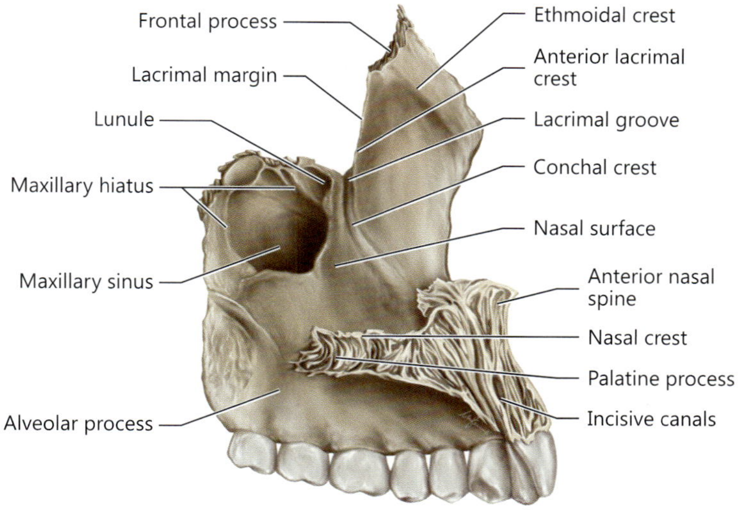

Left **maxilla**, medial view.

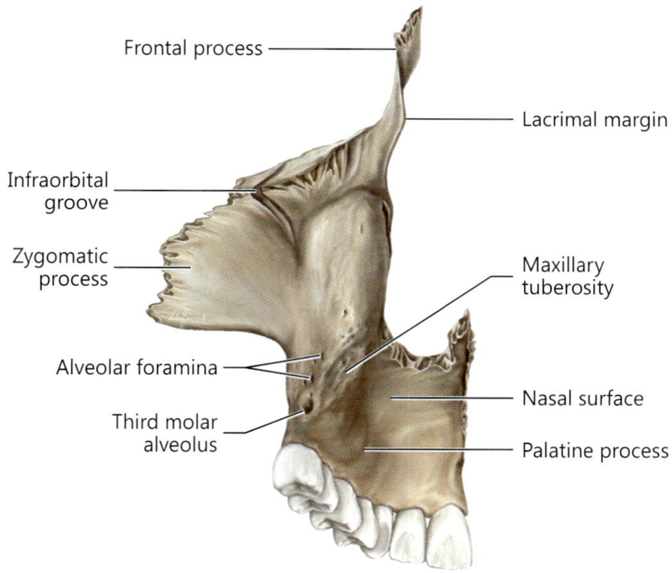

Left **maxilla**, posteroinferior view.

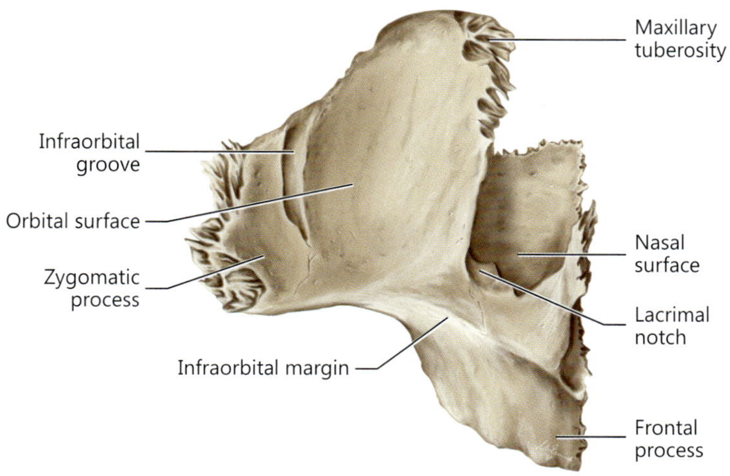

Left **maxilla**, anterosuperior view.

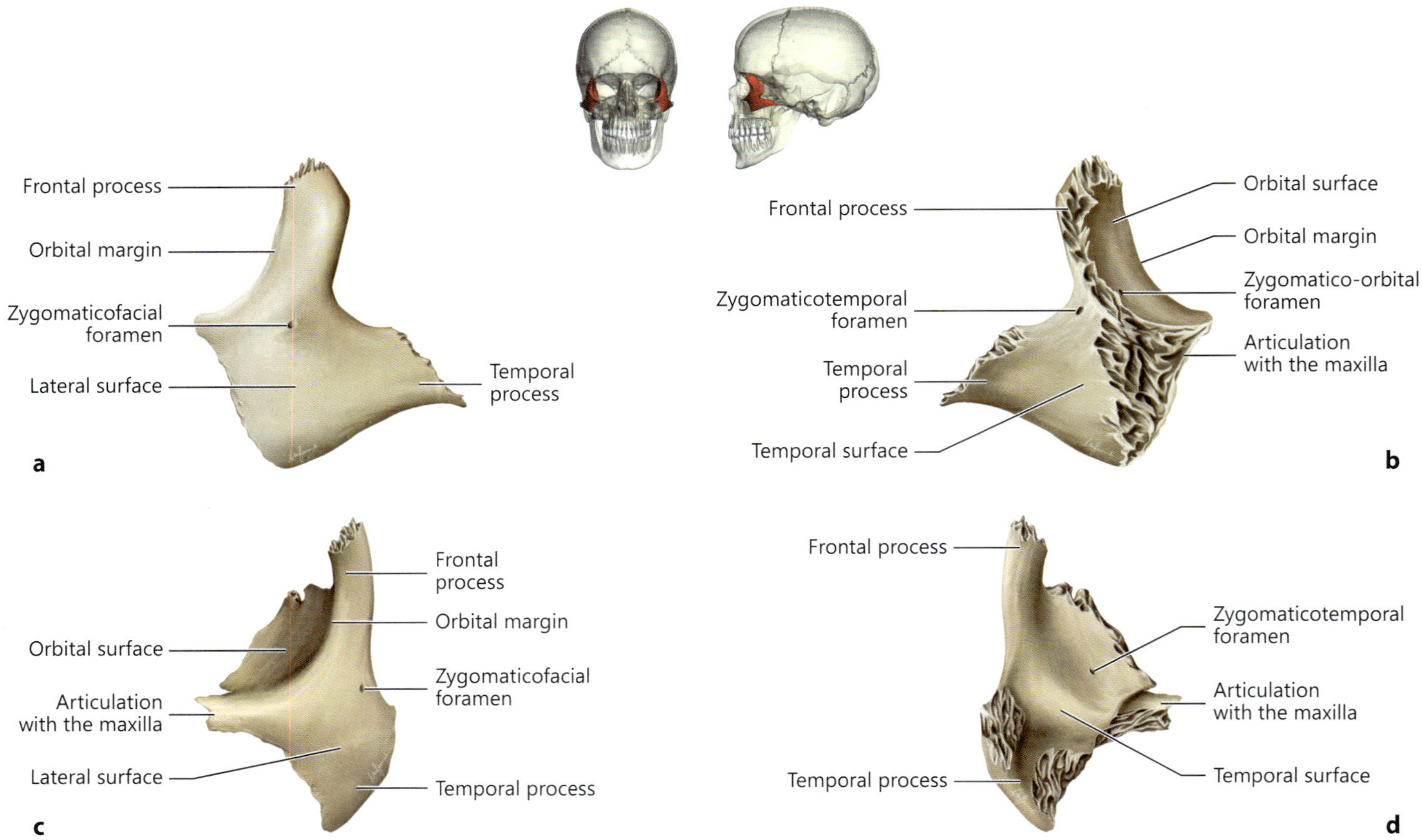

Left **zygomatic bone**. Lateral or malar surface (**a**); medial or temporal surface (**b**); frontal surface (**c**); posterior surface (**d**).

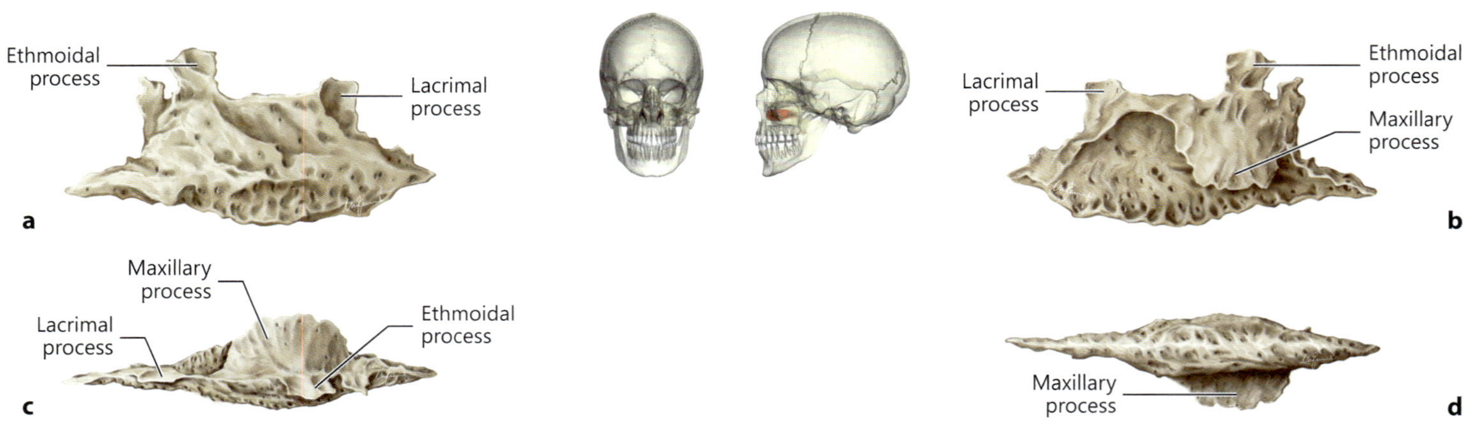

Left **inferior nasal concha** (turbinate). Medial (**a**), lateral (**b**), superior (**c**) and inferior (**d**) surfaces.

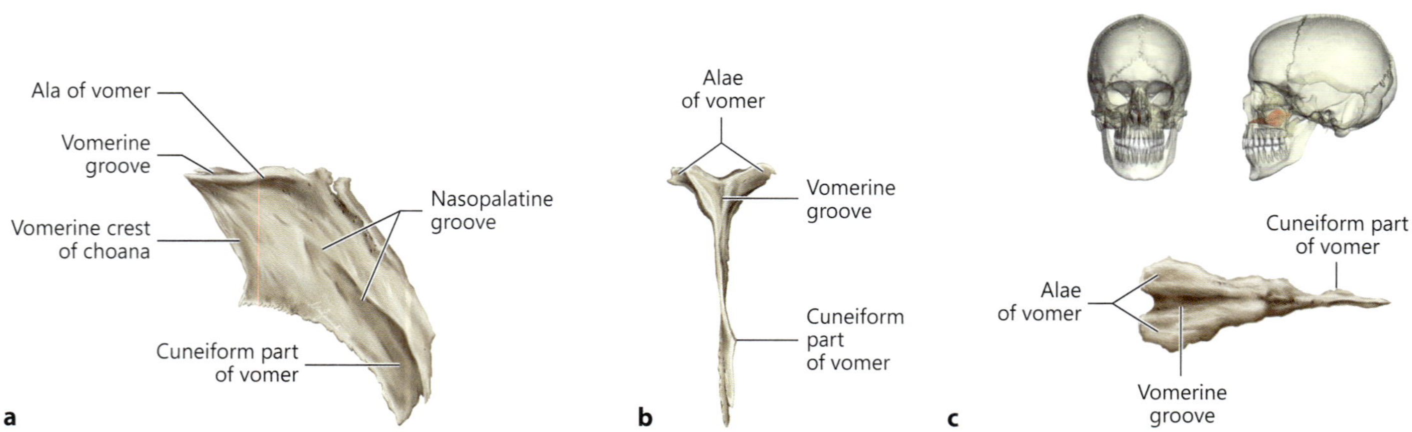

Vomer in lateral (**a**), dorsal (**b**), and superior (**c**) projections. Configuration of the left surface of the bone.

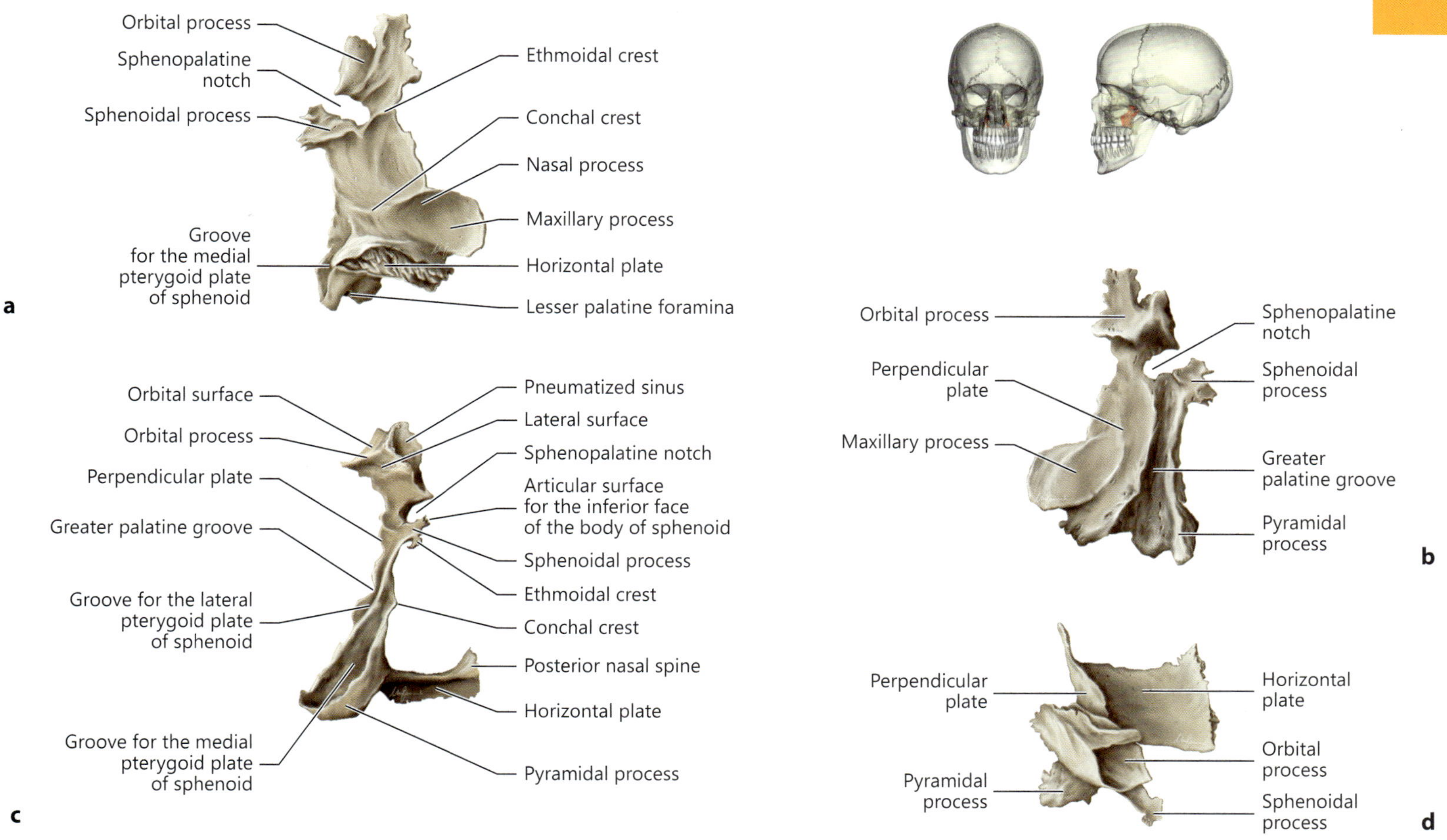

Right **palatine bone**. Nasal surface (**a**), maxillary surface (**b**), posterior margin (**c**), and superior surface (**d**). The horizontal and the vertical parts of the bone meet at a right angle.

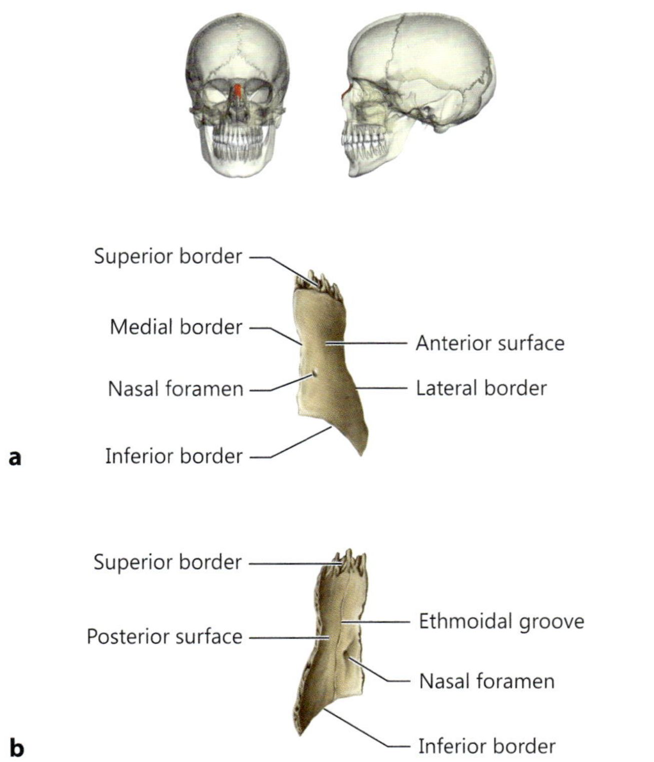

Left **nasal bone**. Anterior (**a**) and posterior (**b**) surfaces.

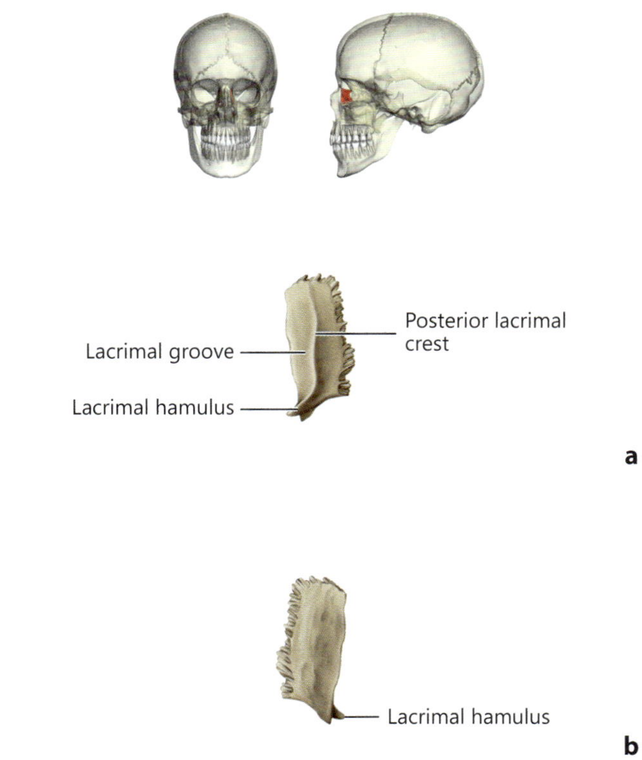

Left **lacrimal bone**. Lateral (**a**) and medial (**b**) surfaces.

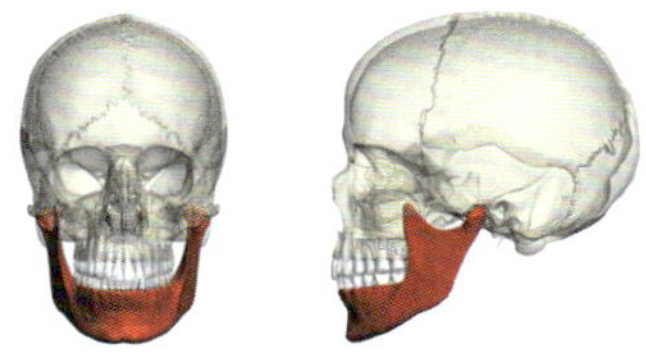

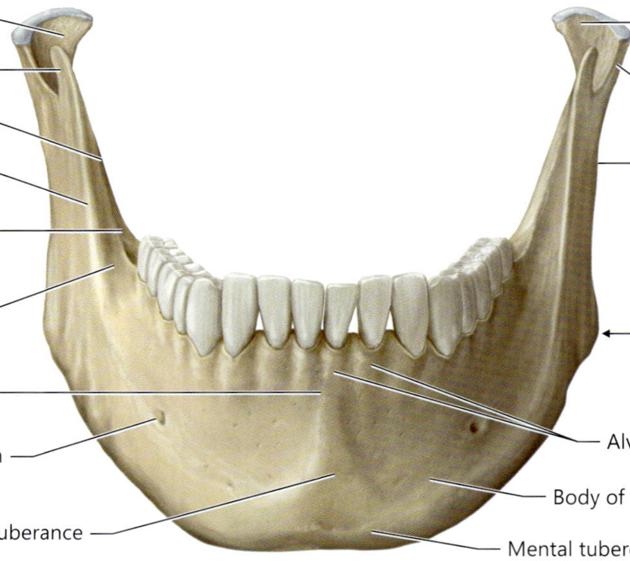

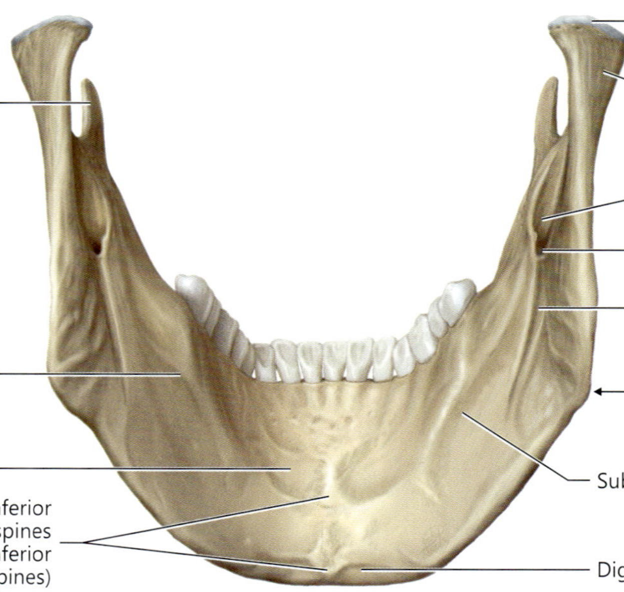

Mandible. **a**, External surface of the mandible in anterior projection; **b**, configuration of the posterior surface of the body of the mandible and the medial surface of the two rami of the mandible.

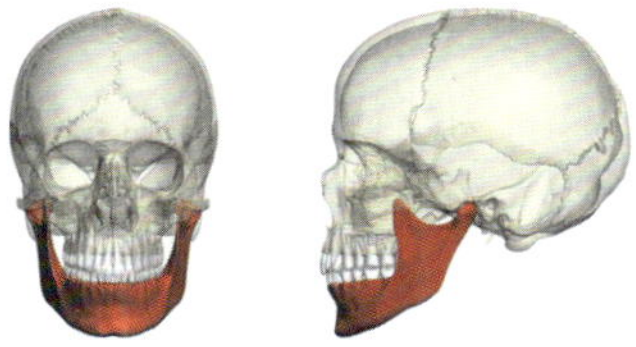

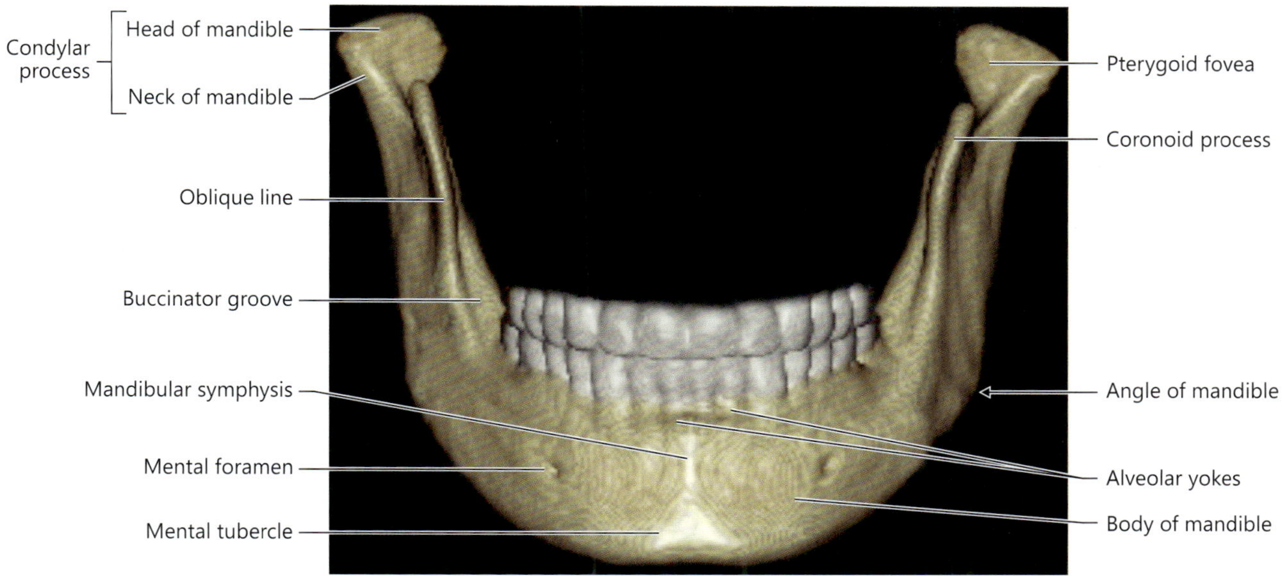

a

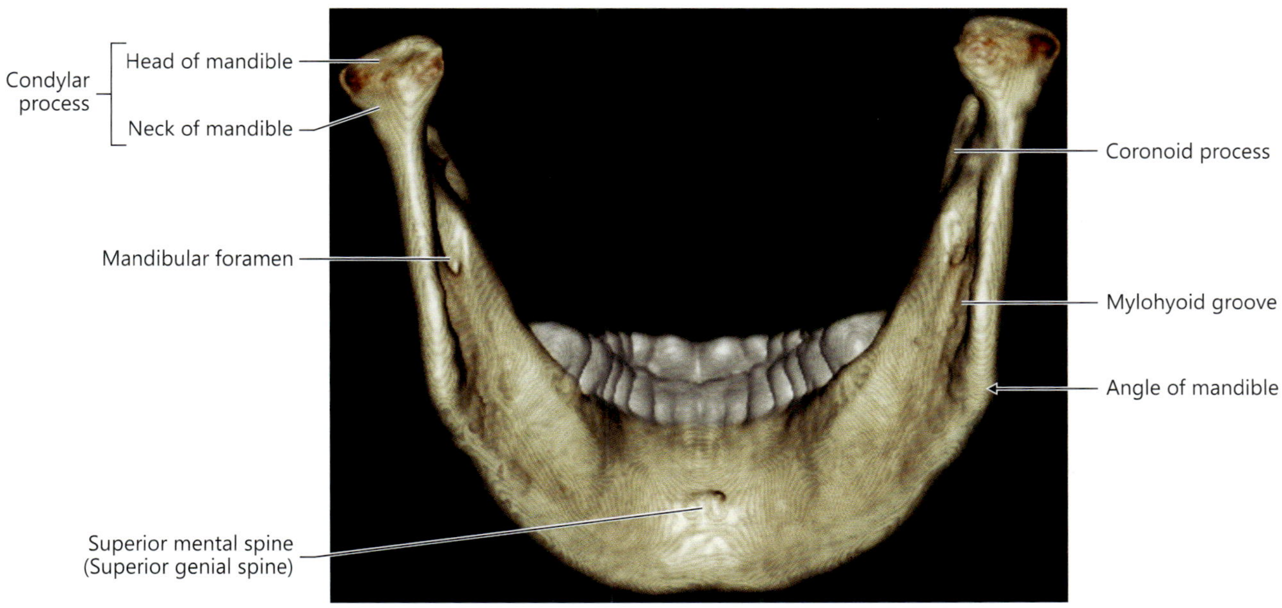

b

Three-dimensional images of a living human body, obtained using a 3D volume rendering technique, showing the external surface of the **mandible** in anterior projection (**a**) and the internal surface in posterior projection (**b**). Part of the teeth of the mandibular arcade is also visible.

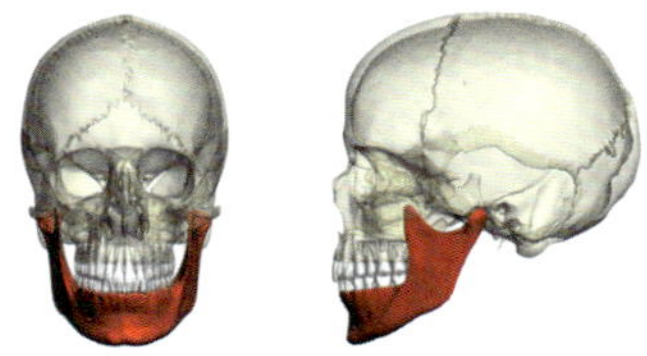

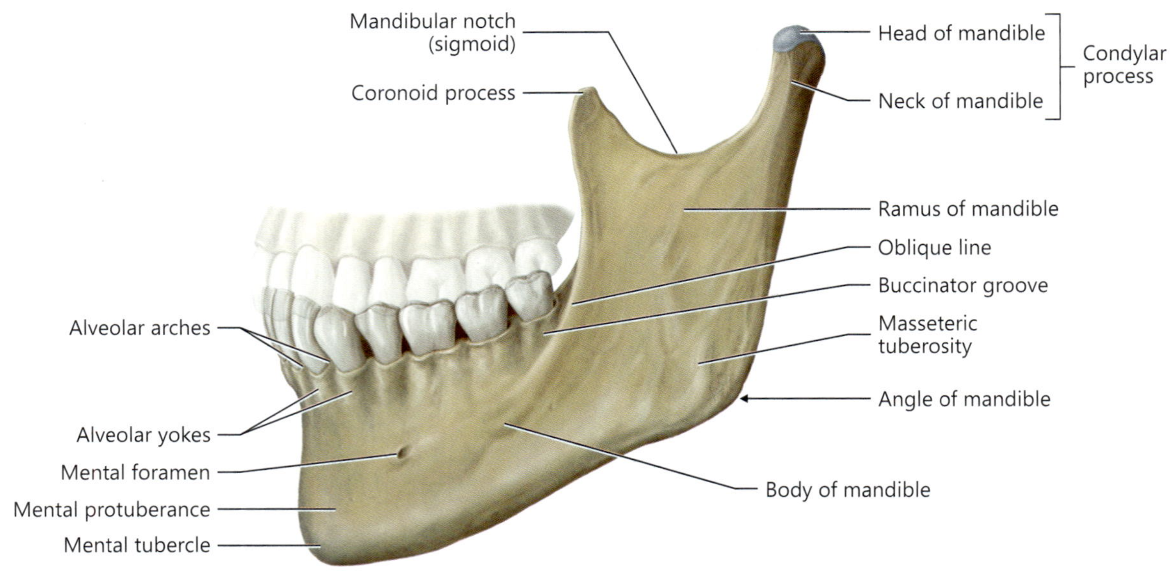

a

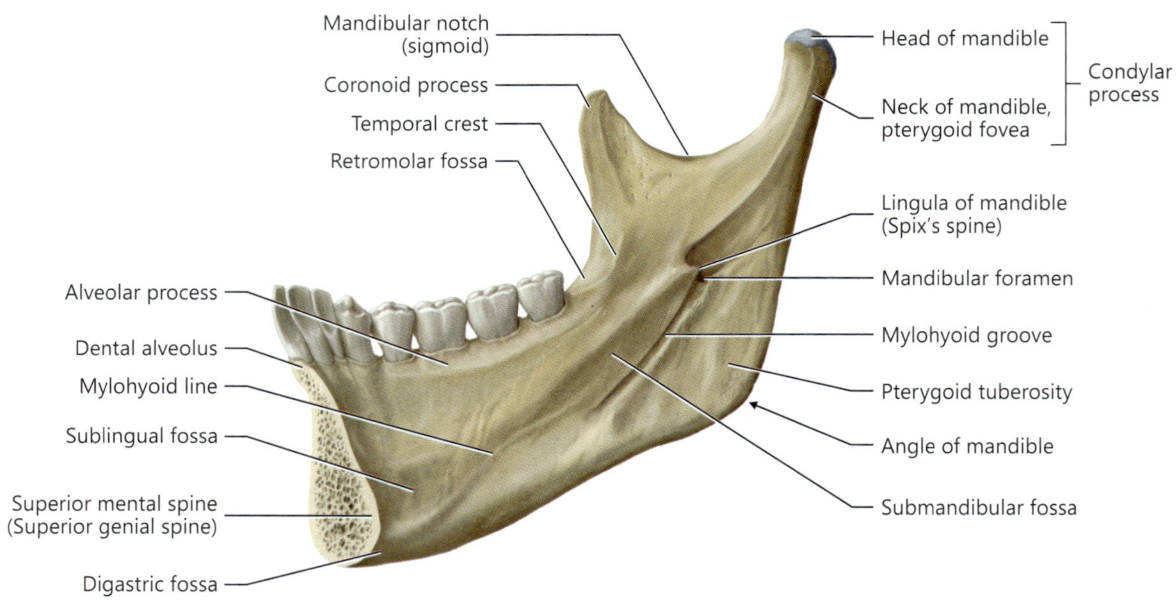

b

Mandible. a, External surface of the mandible in lateral projection. The teeth of the left superior hemiarcade are schematically indicated to illustrate the situation of physiological occlusion; **b**, internal surface of the right mandibular half in lateral projection.

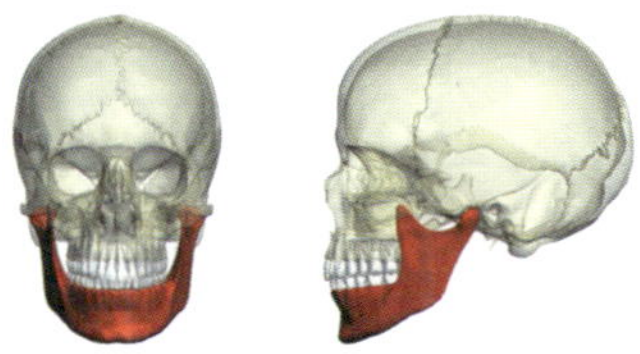

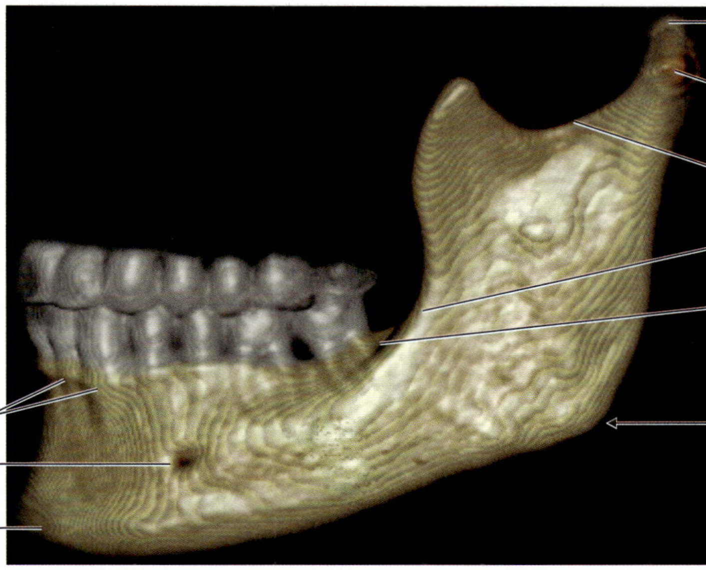

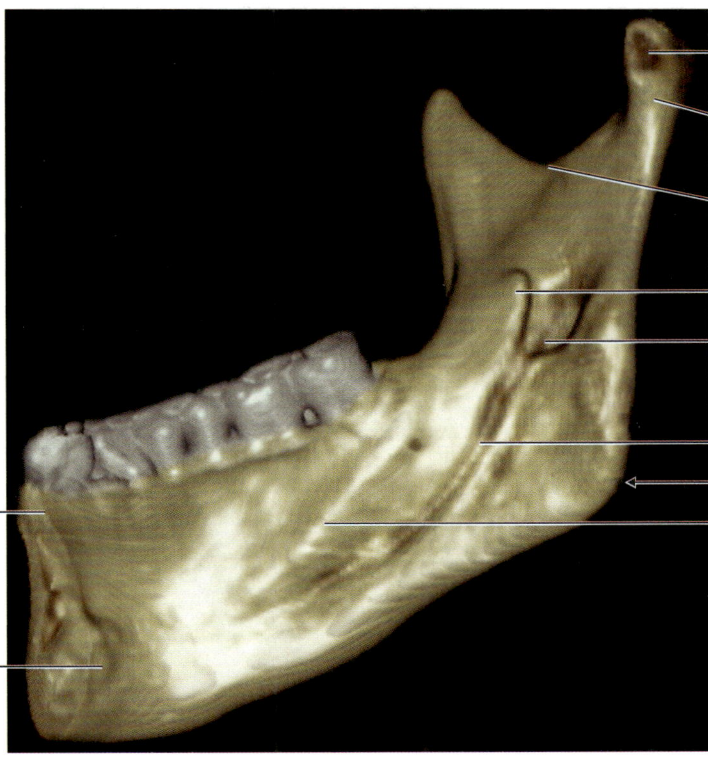

Three-dimensional images of a living human body, obtained using a 3D volume rendering technique, showing the external (**a**) and internal (**b**) surfaces of the **mandible** in lateral projections. Part of the teeth of the mandibular arcade is also visible.

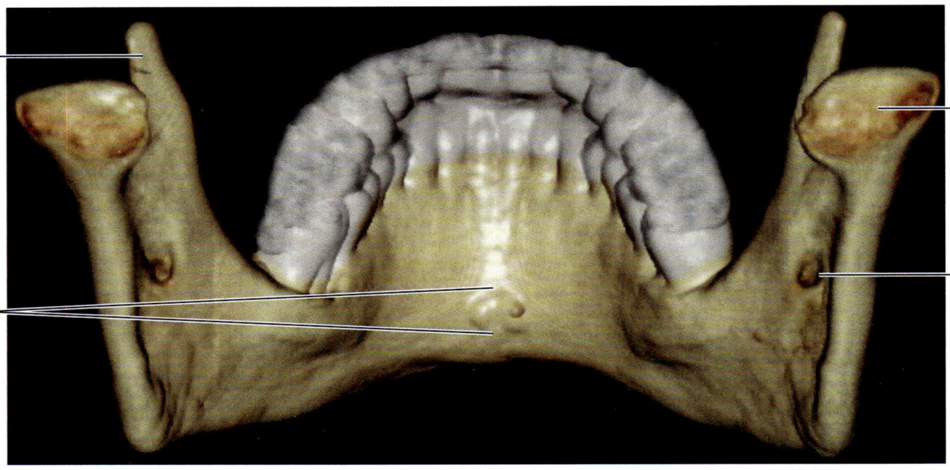

a

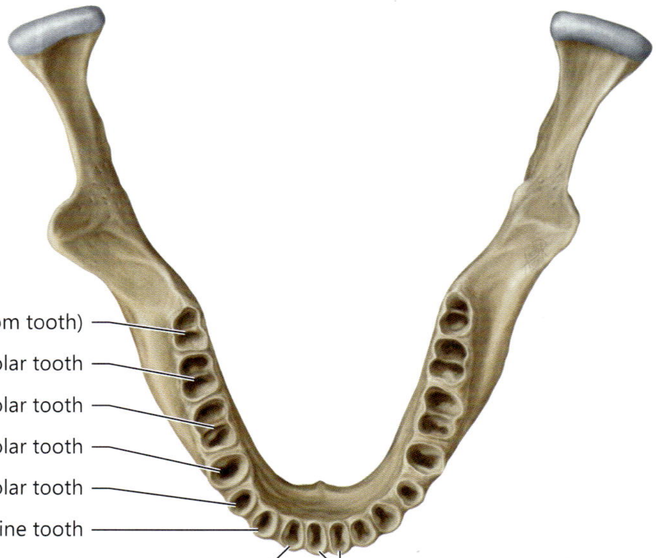

b

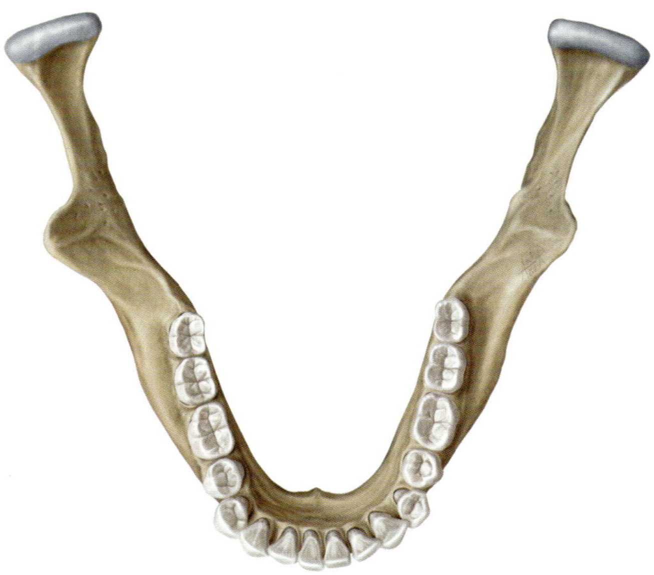

c

Mandible. Three-dimensional image of a living human body, obtained using a 3D volume rendering technique, showing the internal surface of the **mandible** in posterior projection (**a**); **b**, inferior alveolar arch, top view. The alveolar cavities which accommodate the incisors, canines and premolars are not sepimentated, those for the molars seem divided into two concamerations by a septum which runs in the vestibule-lingual direction. The concamerations accommodate the two roots of the inferior molar teeth; **c**, mandibular arcade with complete dentition.

Relation of **Meckel's cartilage** (*gray*) and **body of mandible** during development.

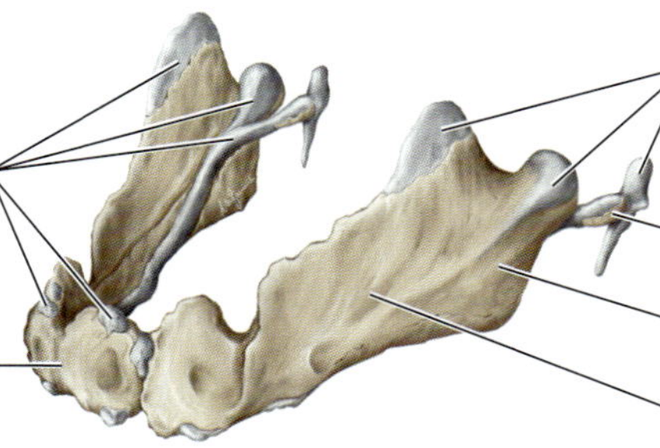

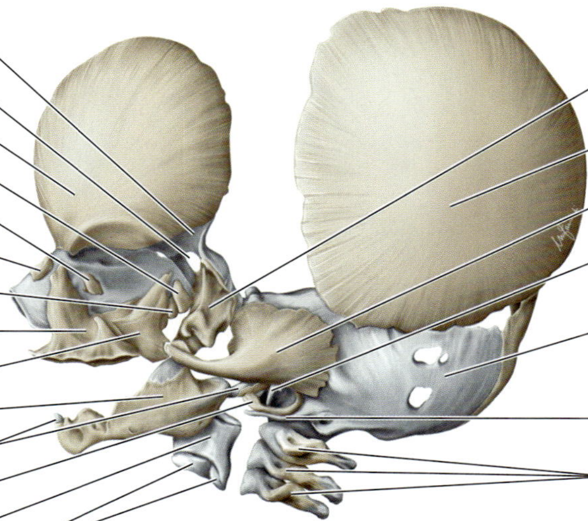

Model of an **early anlage of the cranium** in a human embryo of 8 cm. The portion of the chondral derivation is indicated in *gray*, the portion of the membranous derivation is indicated in color.

Development of the **occipital bone**. In a two-year-old child the various anlages of the bone before their final merger are indicated.

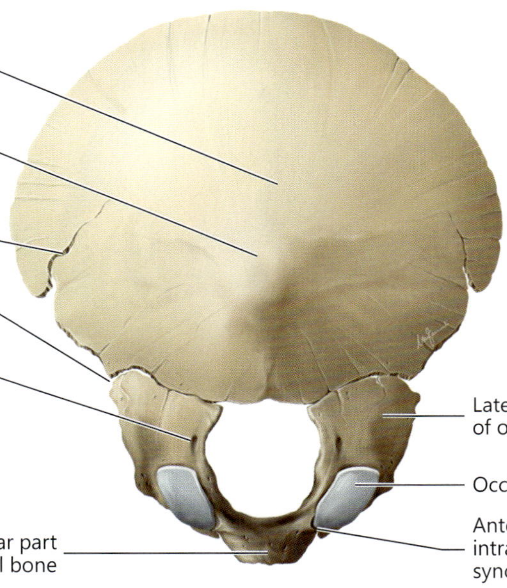

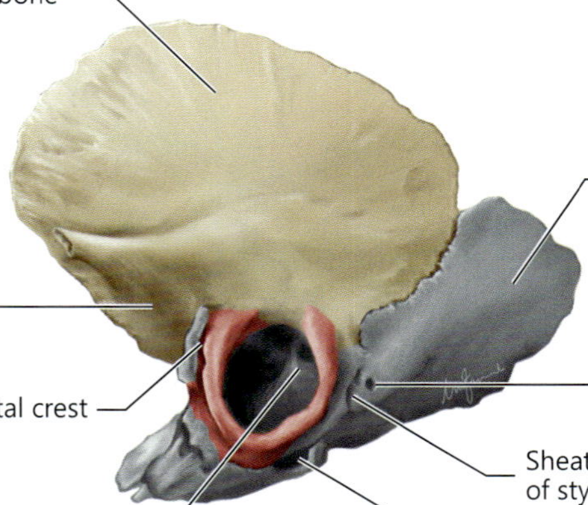

Development of the **temporal bone**. Some anlages of the bone at birth, before their final merger, are indicated. The anlage of the squama is depicted in *yellow*, the petrous part in *gray*, and the tympanic part in *red*.

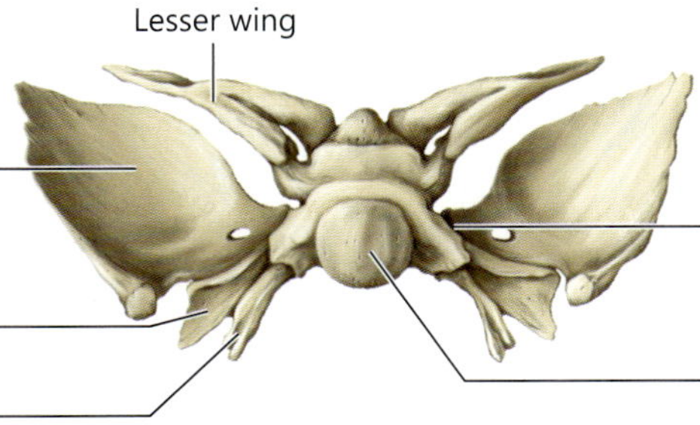

Development of the **sphenoid**. In a newborn the relative immaturity in the development of the different portions is observed. The sella turcica has a slighly pronounced concavity and the pterygoid processes are barely outlined.

Newborn cranium in frontal, inferior, superior, and lateral projections.

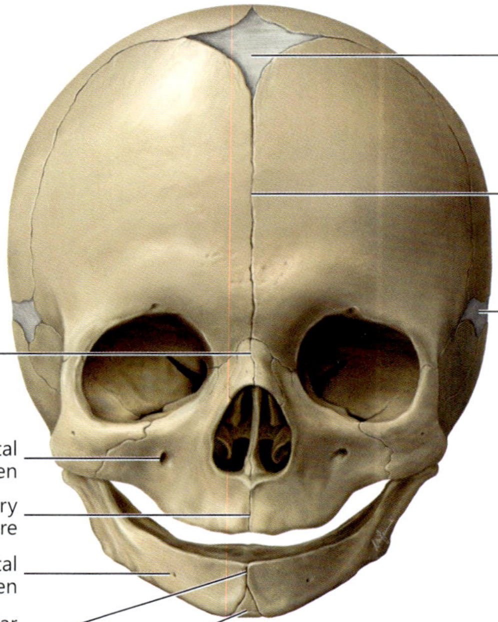

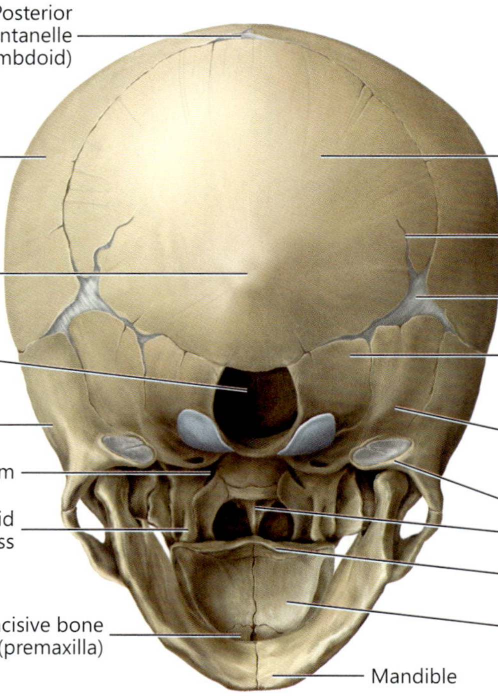

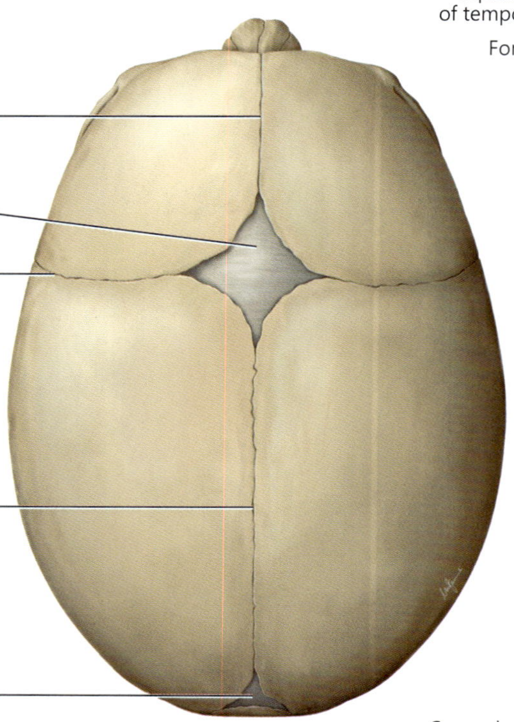

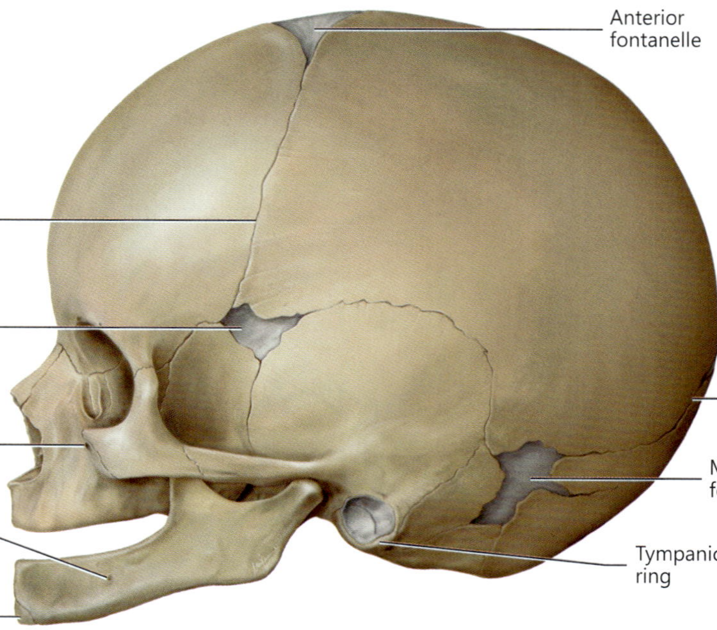

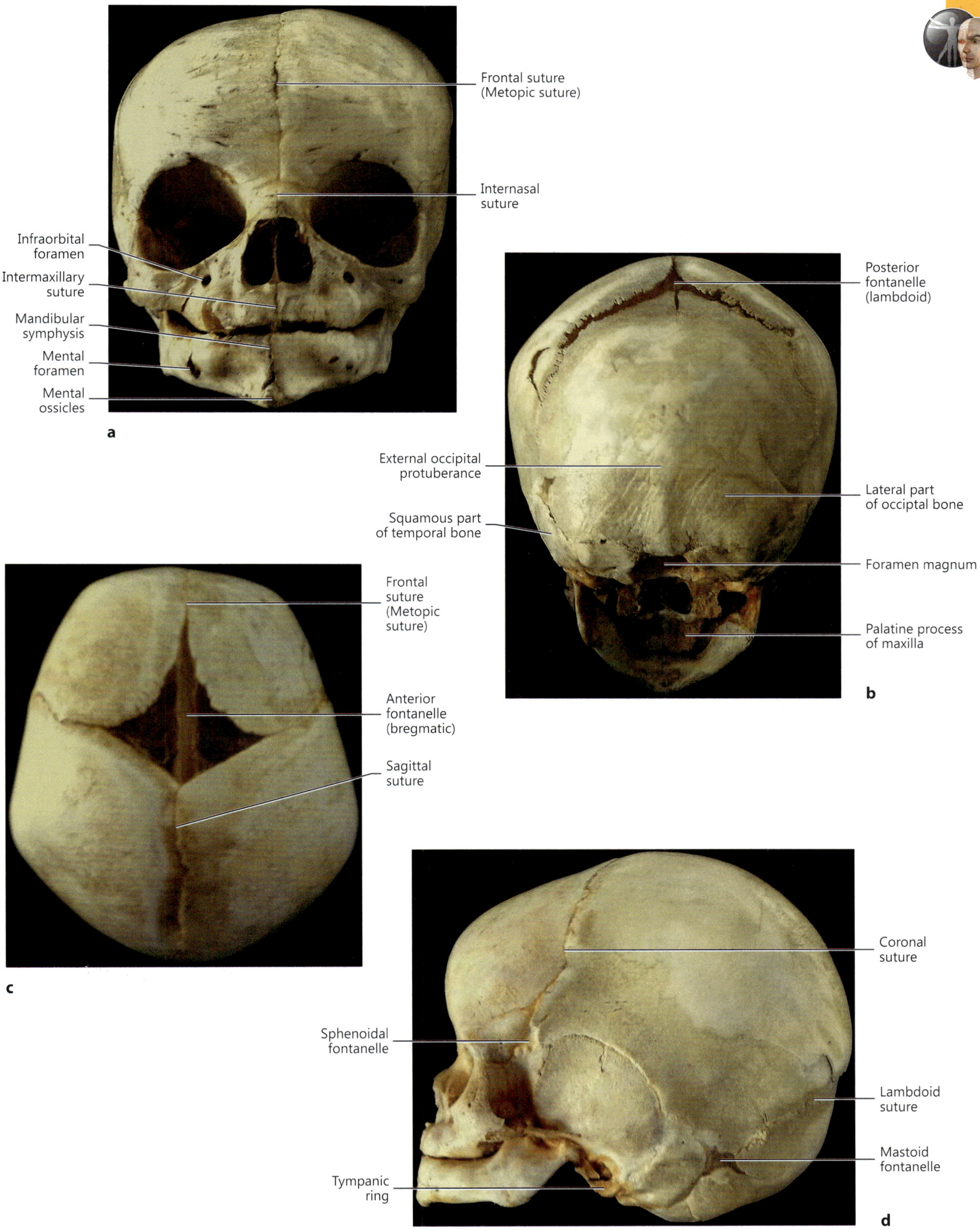

Fetal cranium at the 38th week of gestation. Views in frontal (**a**), posterior (**b**), superior (**c**), and lateral (**d**) projections.

Summary table of the bones and relative joints of the head

BONE	LOCATION	DESCRIPTION	JOINTS
Inferior nasal concha (turbinate)	Cranium	Thin, irregular and curled lamina that extends horizontally along the lateral wall of the nasal cavity	Ethmoid, maxillary, palatine
Ethmoid	Cranium	Unpaired T-shaped bone, constitutes the part of the nasal septum and the roof of the nasal cavities; the curved processes of this bone form the superior and middle nasal conchae	Sphenoid, frontal, both nasal bones, lacrimal, palatine, maxillary, inferior nasal conchae, vomer
Frontal	Cranium	Flat bone that forms the anterior part of the cranium	Ethmoid, sphenoid, maxillary, both nasal bones, parietal, lacrimal and zygomatic
Hyoid	Neck	U-shaped bone situated anteriorly in the neck between the mandible and the larynx	Suspended from the apex of the styloid processes of the temporal bones by ligaments
Incus	Ear	Intermediate bone in the auditory ossicles	Malleus, stapes
Lacrimal	Cranium	The smallest and most fragile bone of the face; similar in shape to a fingernail and is situated in the anteromedial wall of the orbit	Ethmoid, frontal, maxillary, inferior nasal concha
Malleus	Ear	The most lateral of the auditory ossicles, similar in shape to a hammer and consists of a head, a neck, and three processes	Tympanic membrane and incus
Mandible	Face	Horseshoe-shaped bone holding the lower teeth in place	Both temporal bones
Maxilla	Face	The biggest facial bone after the mandible holding the upper teeth in place	Frontal, ethmoid, nasal, zygomatic, lacrimal, vomer, inferior nasal concha, maxillary of the other side
Nasal	Cranium	One of two small oblong bones placed side by side to form the dorsum of the nose	Frontal, ethmoid, nasal contralateral, maxillary
Occipital	Cranium	Single flattened bone that forms the posterior part of the cranial base, presents the foramen magnum	Right and left parietal, sphenoid, right and left temporal, atlas
Palatine	Cranium	Either of the two quadrilateral L-shaped bones that form the posterior part of the hard palate, part of the floor and lateral wall of the nasal cavity and part of the floor of the orbit	Ethmoid, frontal, maxillary, inferior nasal concha
Parietal	Cranium	Pair of bones located between the frontal and occipital bones forming the sides and roof of the cranium	Parietal, contralateral, frontal, occipital, temporal, sphenoid
Sphenoid	Cranial base	Unpaired, irregular bone that constitutes the anterior part of the cranial base and portions of the cranial, orbital, and nasal cavities	Ethmoid, frontal, parietal, temporal, vomer, zygomatic
Stapes	Ear	The most medial of the auditory ossicles that is similar in shape to a stapes	Incus, oval window
Sutural	Cranium	Irregular, isolated bones that are occasionally found along the cranial sutures, most frequently in the lambdoid suture	Generally occipital and parietal
Temporal	Cranium	An irregular bone that consists of three parts: squamous, petrous and tympanic; forms part of the lateral surface and cranial base	Occipital, parietal, zygomatic, sphenoid, mandible
Vomer	Cranium	Thin and flattened, forms the posterior and inferior part of the nasal septum	Ethmoid, sphenoid, both maxillaries, both palatines; also articulates with the septal nasal cartilage
Zygomatic	Cranium	Forms the prominence of the cheekbone and the lateral and inferior walls of the orbit	Frontal, sphenoid, temporal, maxillary

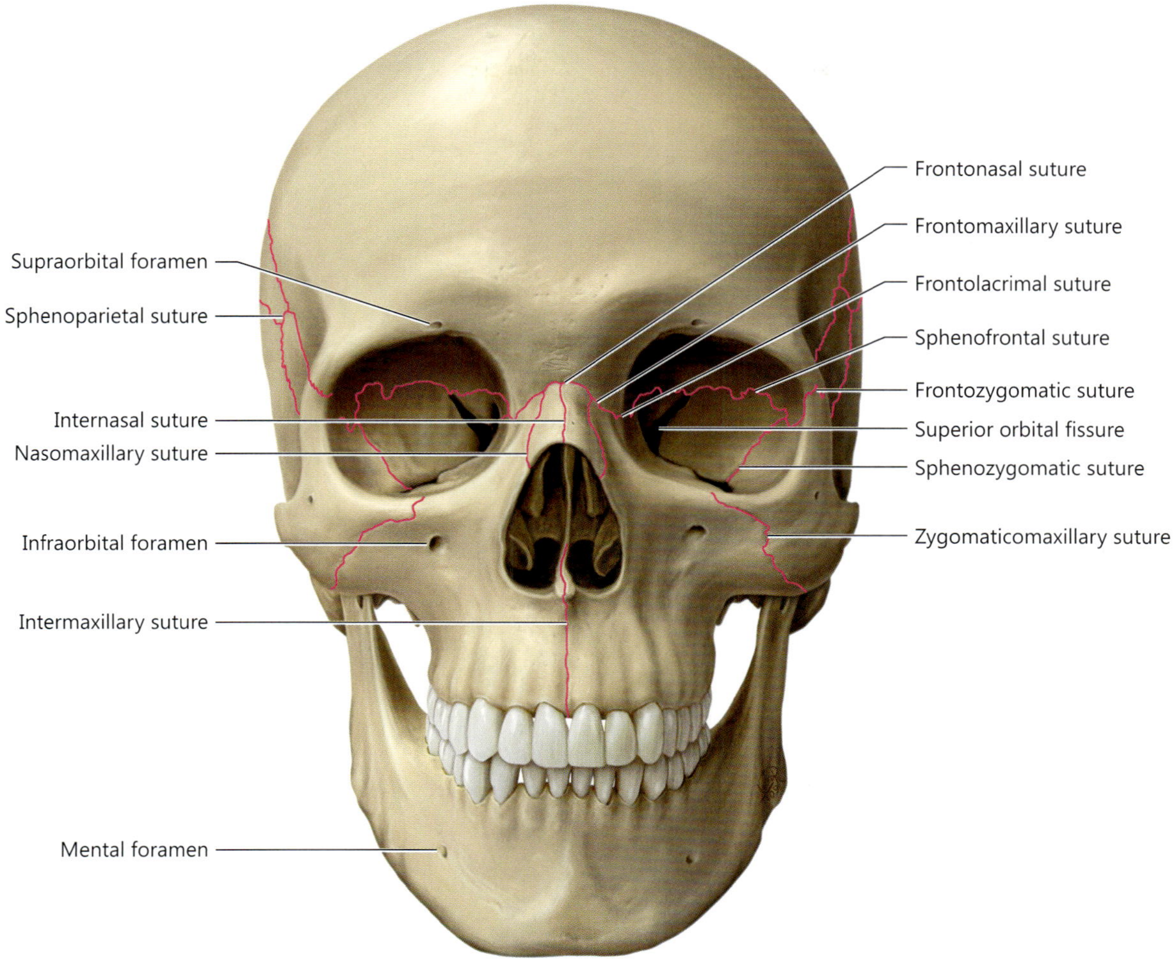

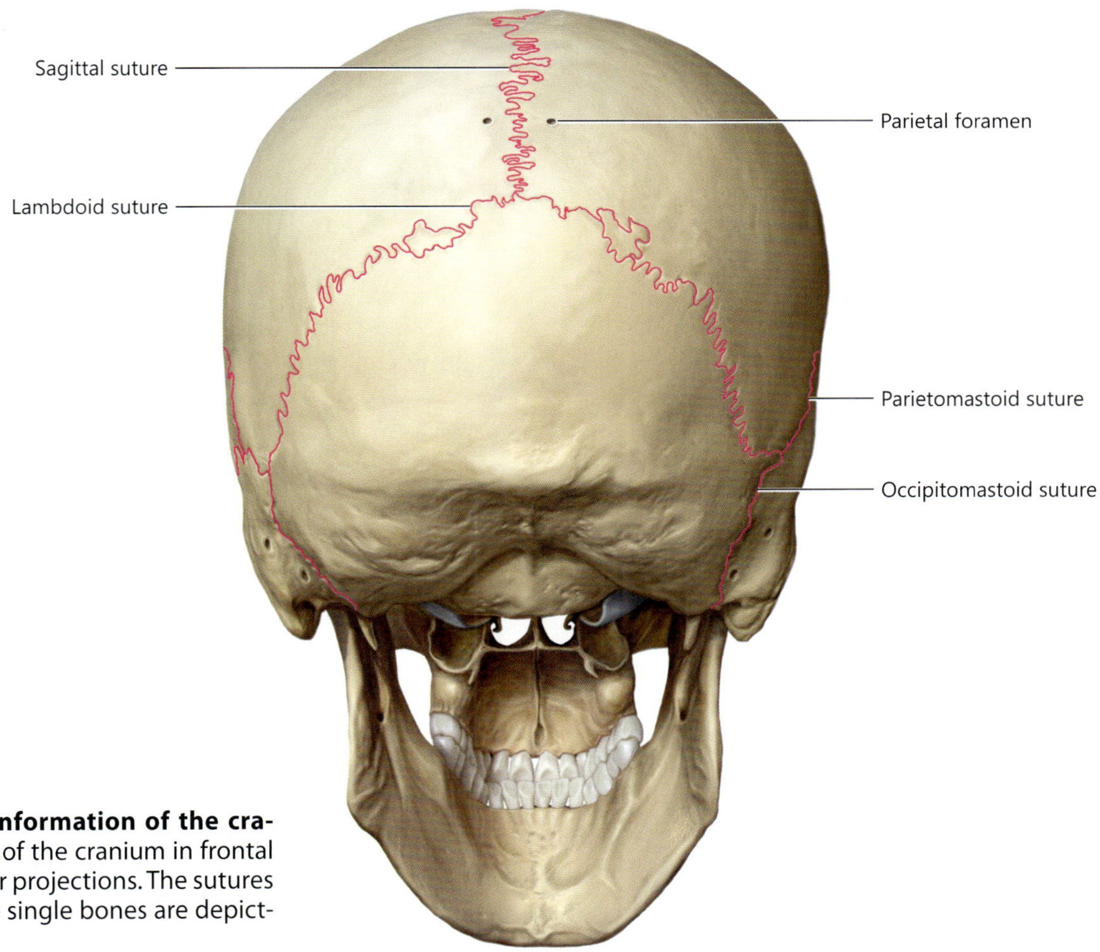

External conformation of the cranium. Views of the cranium in frontal and posterior projections. The sutures between the single bones are depicted in *red*.

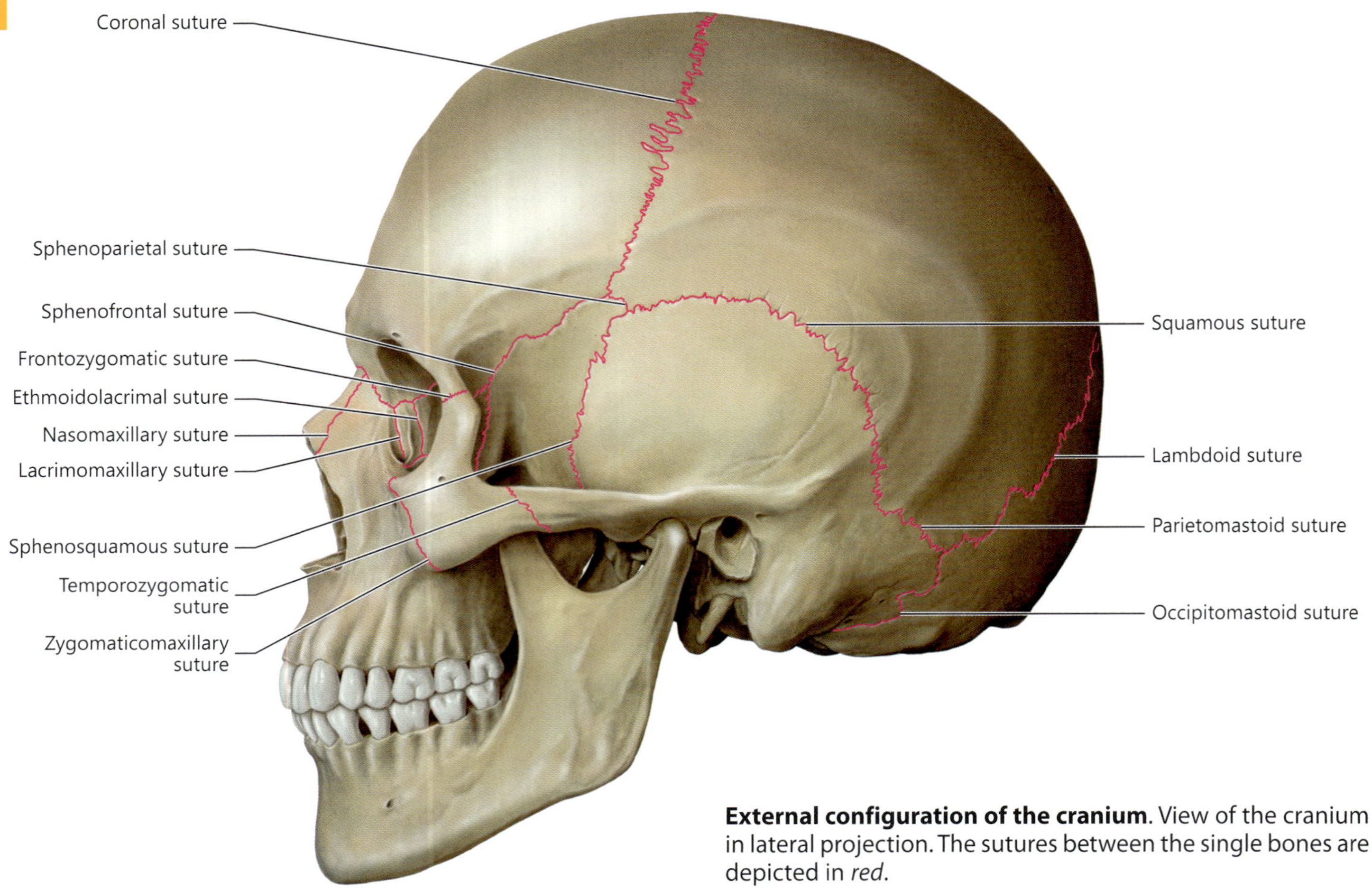

External configuration of the cranium. View of the cranium in lateral projection. The sutures between the single bones are depicted in *red*.

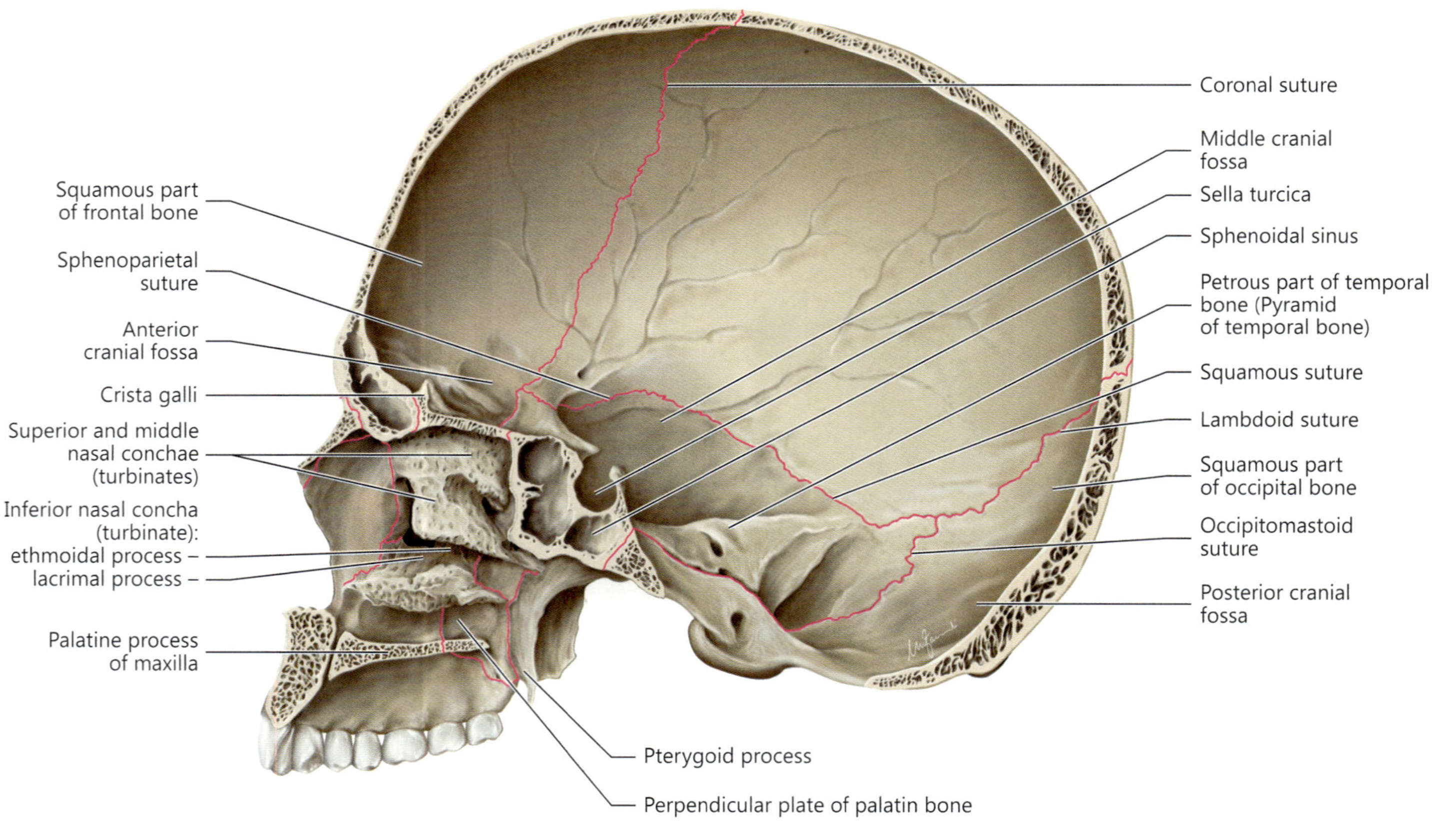

Internal conformation of the cranium in a paramedian sagittal section. The sutures that connect the various bones are highlighted in *red*.

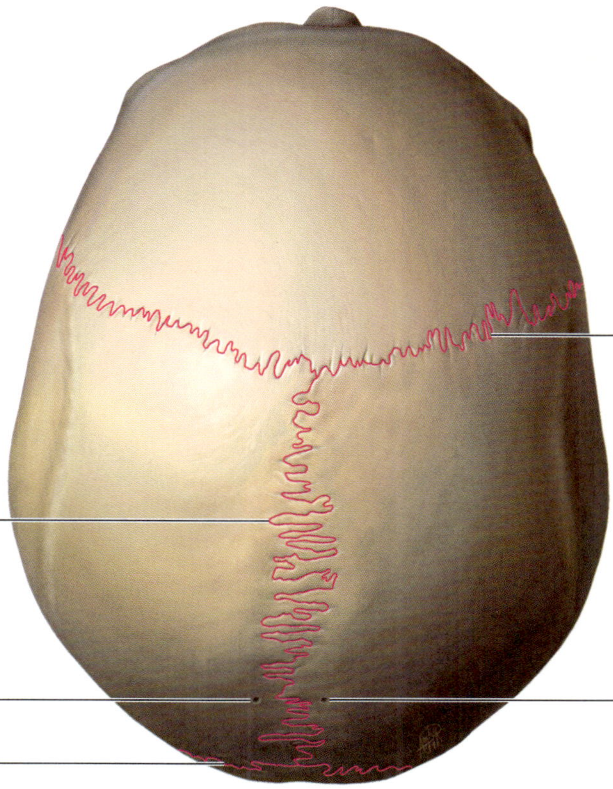

External configuration of the cranium. View of the cranium in superior projection. The sutures between the single bones are depicted in *red*.

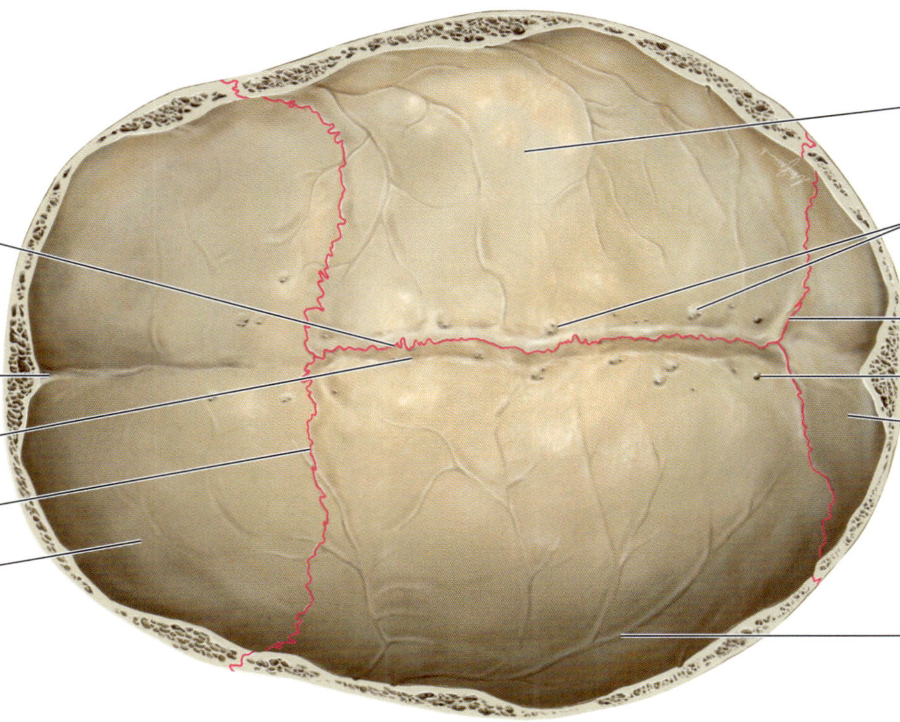

Internal surface of the **calvaria**. In the center there is a slight accentuated depression corresponding to the sulcus for the superior sagittal sinus; also visible are the vascular impressions and the granular foveolae (of Pacchioni). The sutures between the single bones are depicted in *red*.

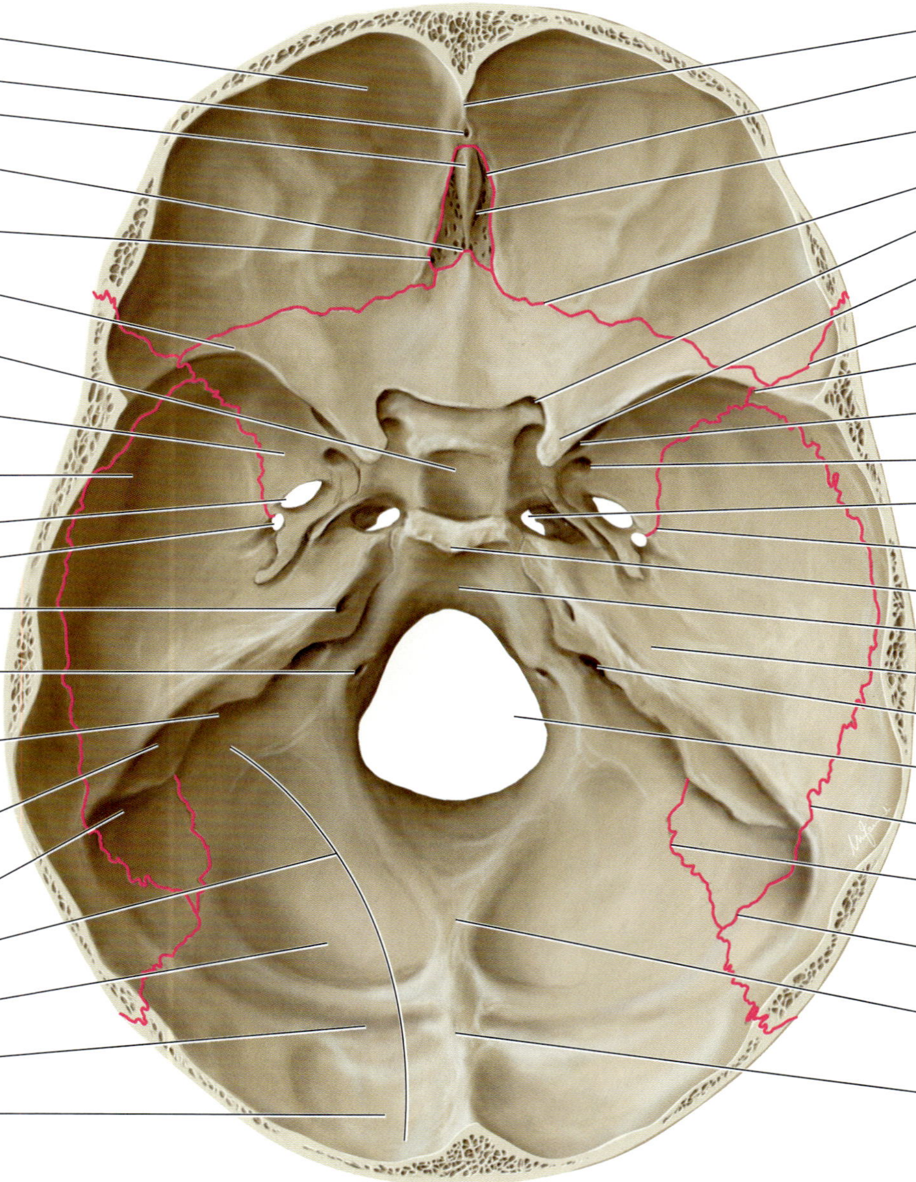

Internal surface of the cranial base; the main foramina and some characteristics of bone morphology are indicated. The sutures between the single bones are depicted in *red*.

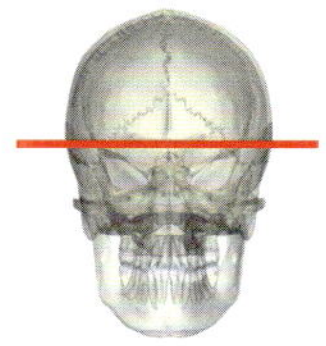

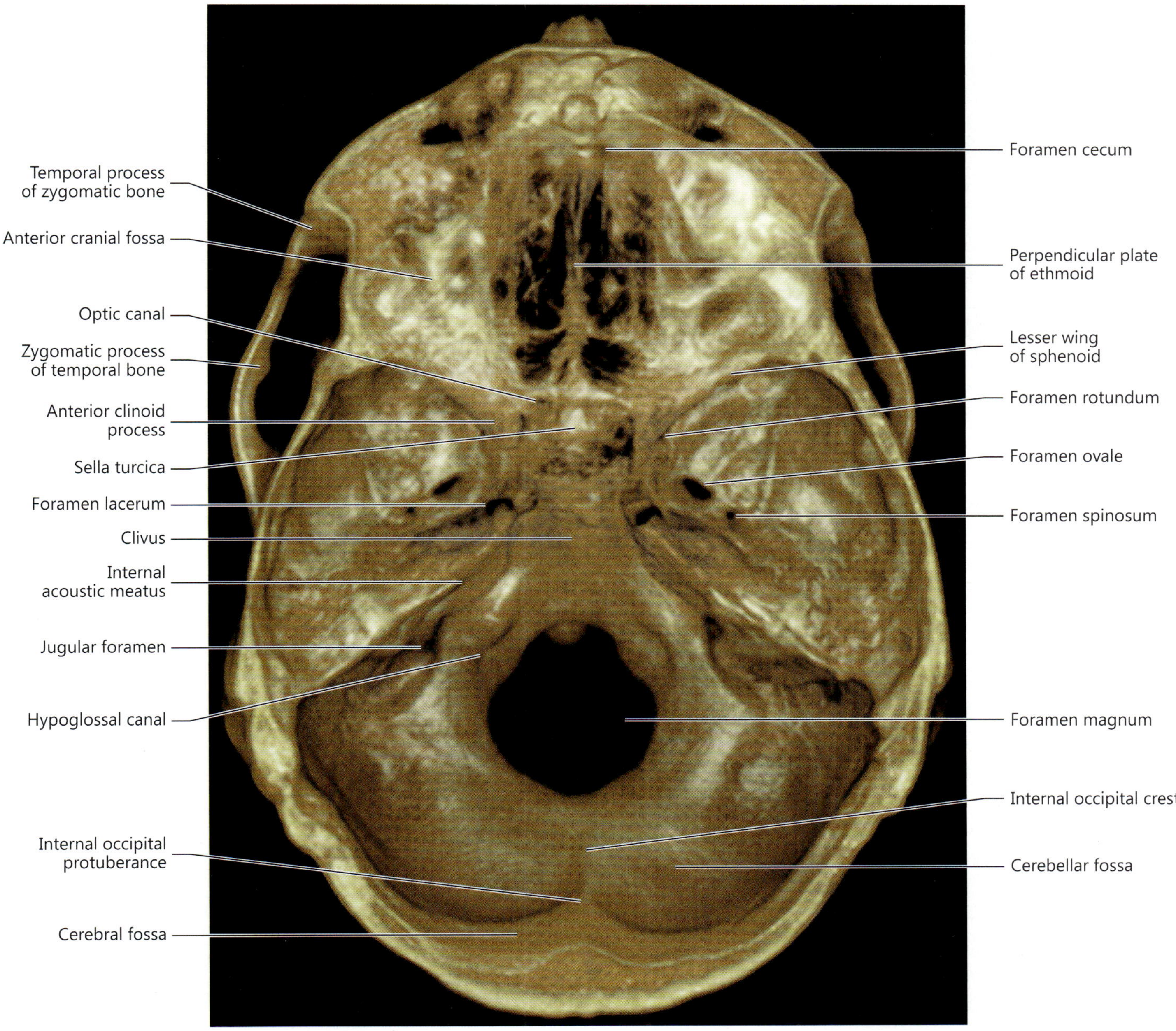

Three-dimensional image of a living human body, obtained using a 3D volume rendering technique, showing the **transverse section of the bones of the cranium**. The cranium has been transected just above the orbital surface of the frontal bone.

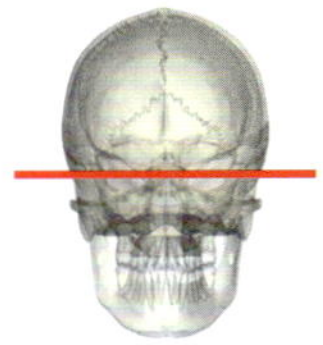

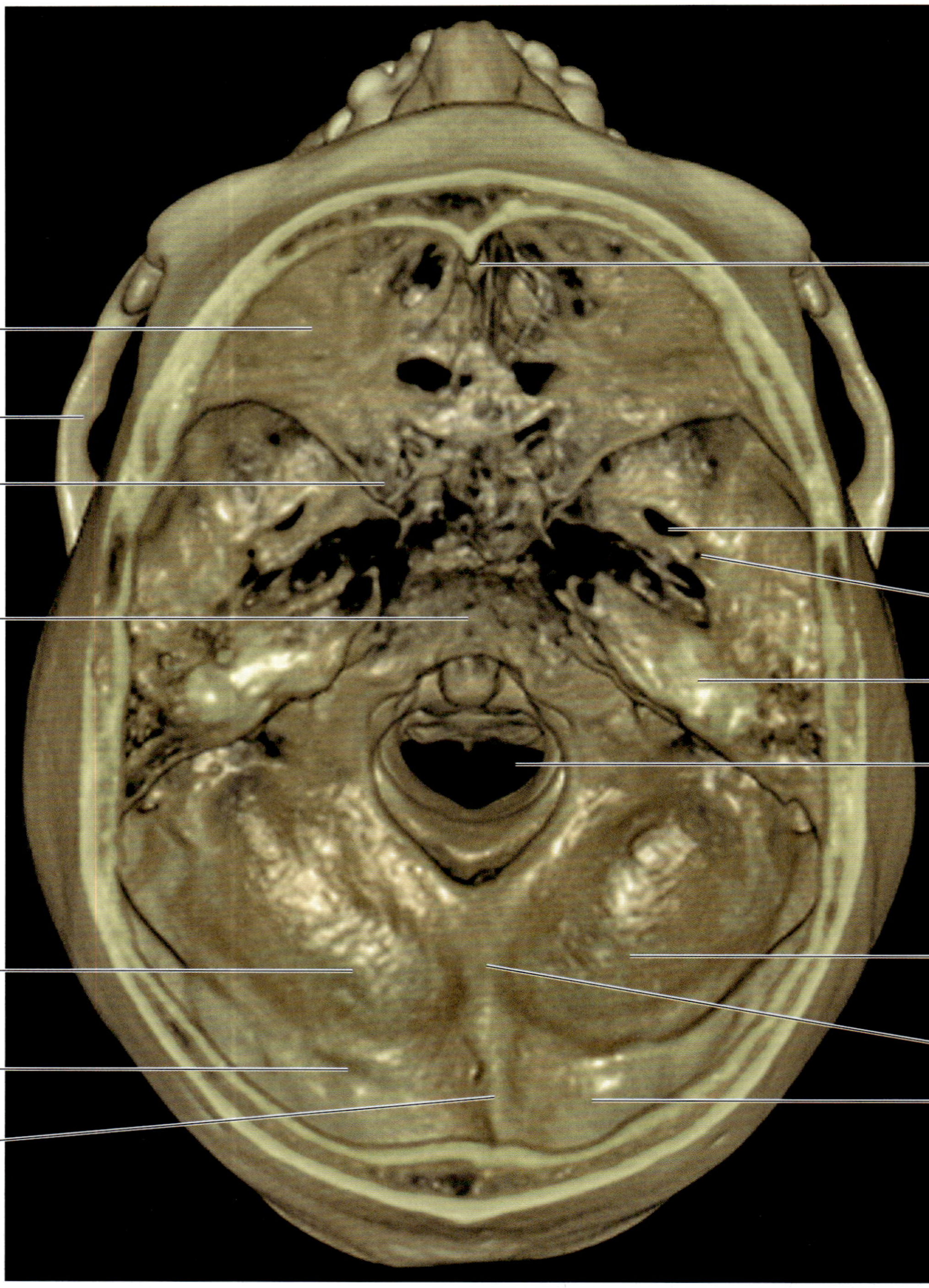

Three-dimensional image of a living human body, obtained using a 3D volume rendering technique, showing the **transverse section of the bones of the cranium**. The cranium has been transected just above the crista galli of the ethmoid.

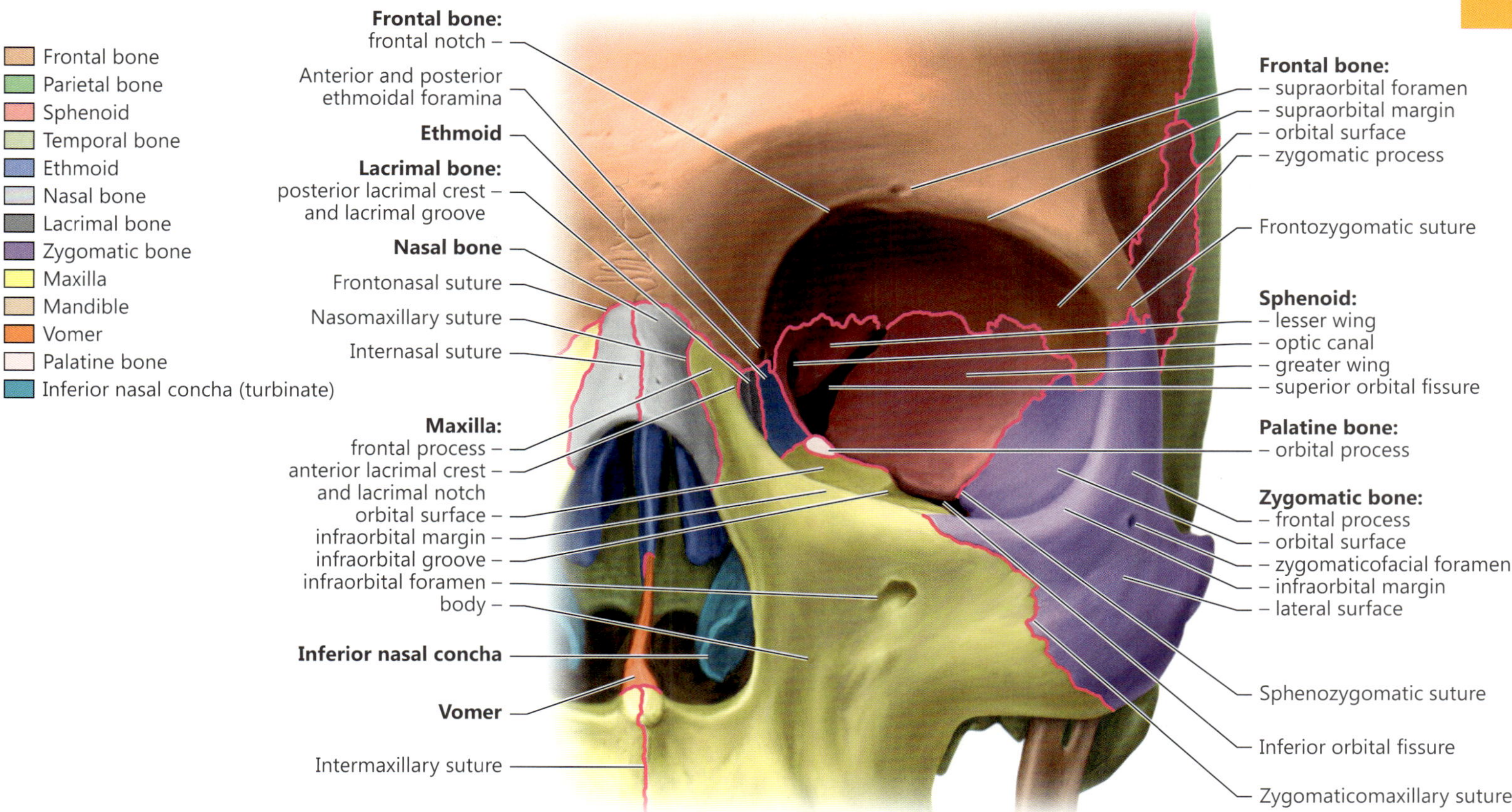

Left orbital cavity, anterior projection. The bones are highlighted in different colors.

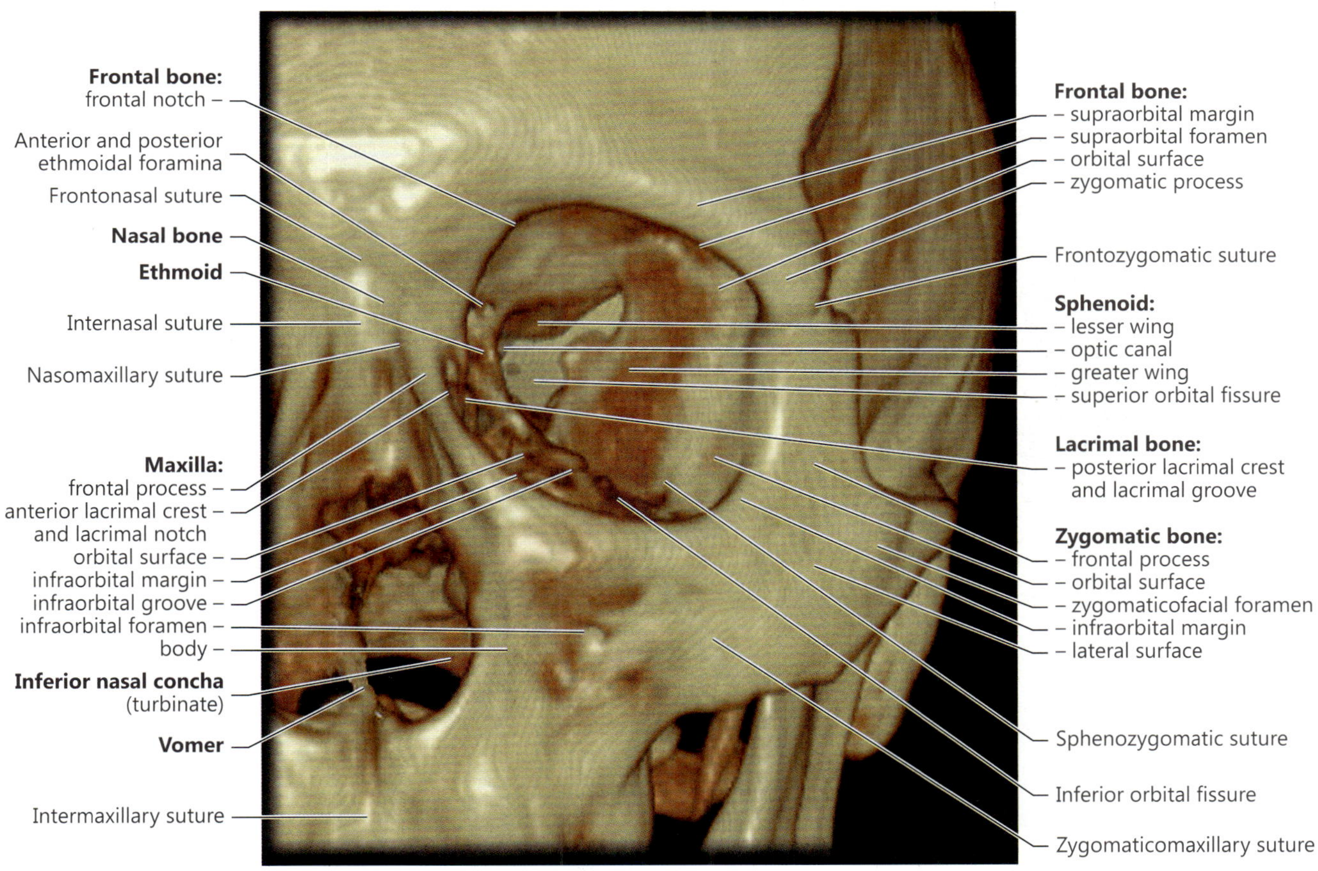

Three-dimensional image of a living human body, obtained using a 3D volume rendering technique, showing the bones that form the **left orbital cavity** in frontal projection.

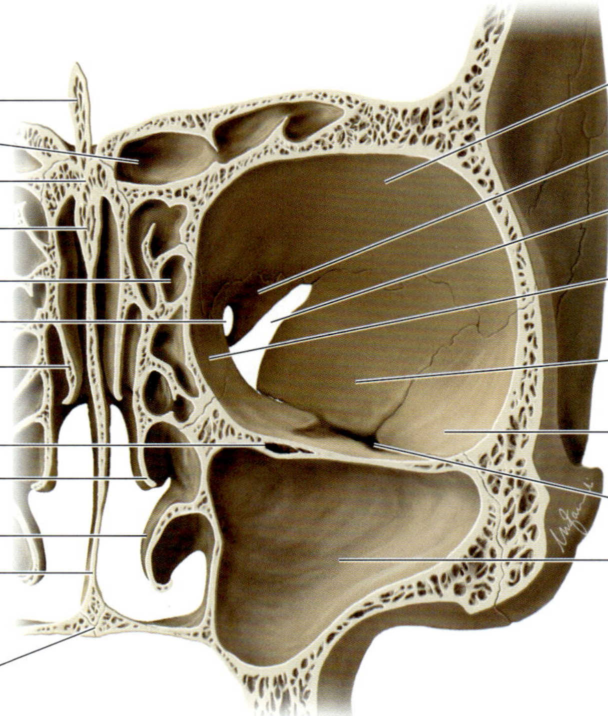

Frontal section of the **nasal cavities** and left **orbital cavity**. Anterior view.

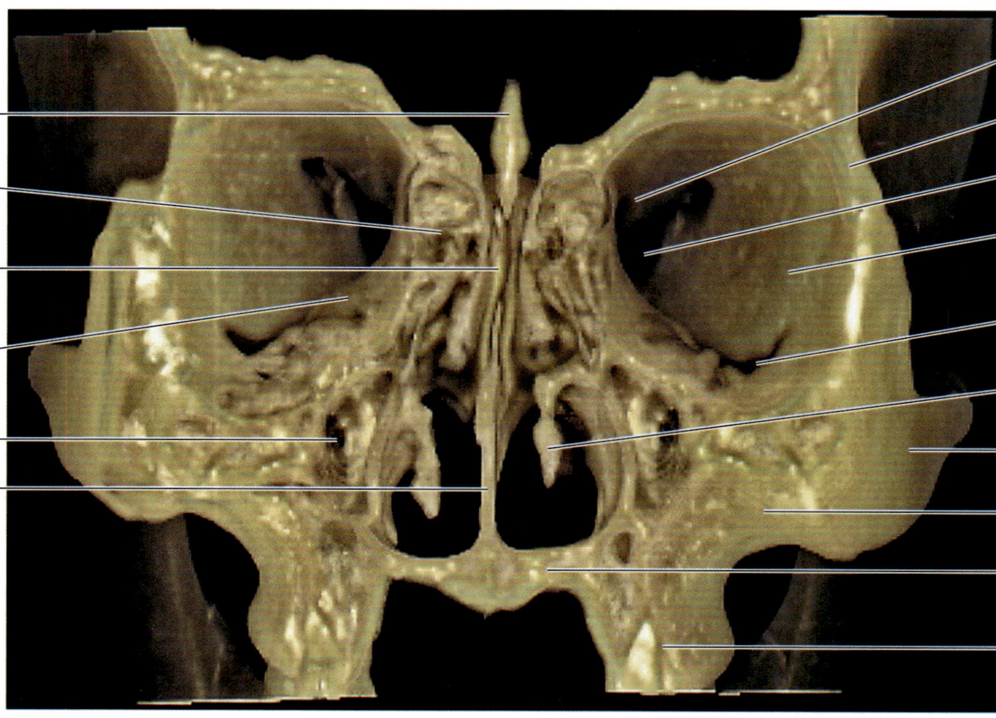

Three-dimensional image of a living human body, obtained using a 3D volume rendering technique, according to a coronal section plane passing in front of the crista galli. The **nasal cavities region** and the **orbital cavities region** are particularly observed.

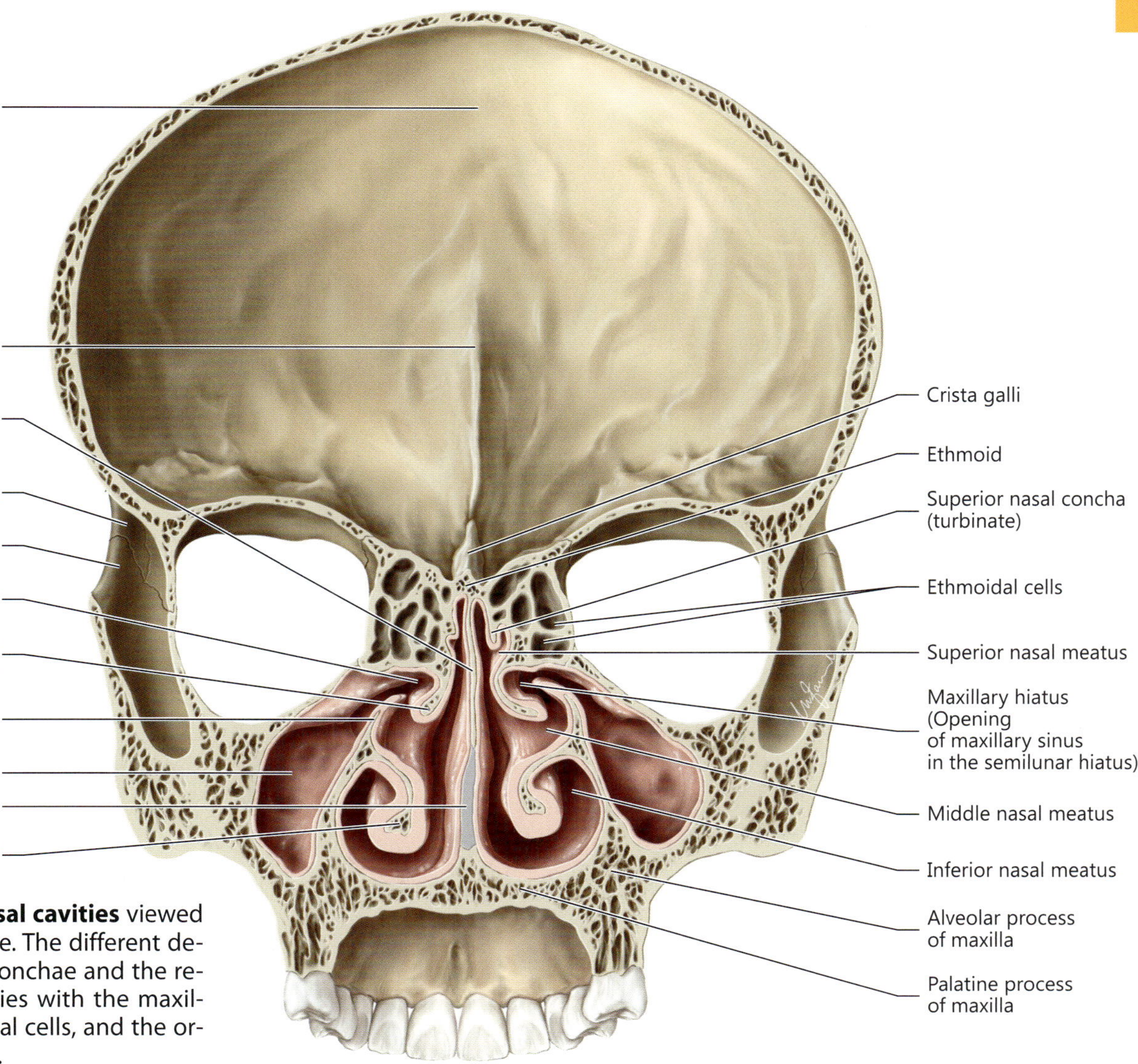

Frontal section of the **nasal cavities** viewed from the posterior surface. The different development of the nasal conchae and the relations of the nasal cavities with the maxillary sinuses, the ethmoidal cells, and the orbital cavities can be seen.

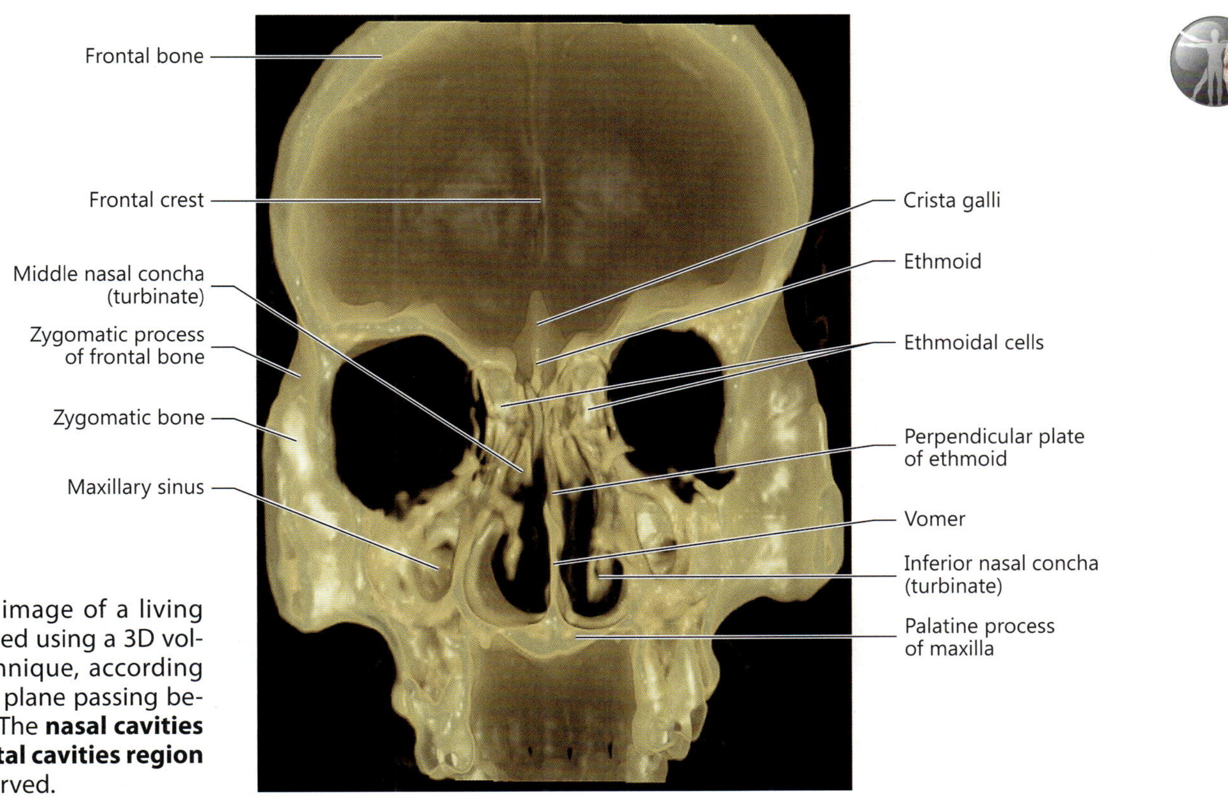

Three-dimensional image of a living human body, obtained using a 3D volume rendering technique, according to a coronal section plane passing behind the crista galli. The **nasal cavities region** and the **orbital cavities region** are particularly observed.

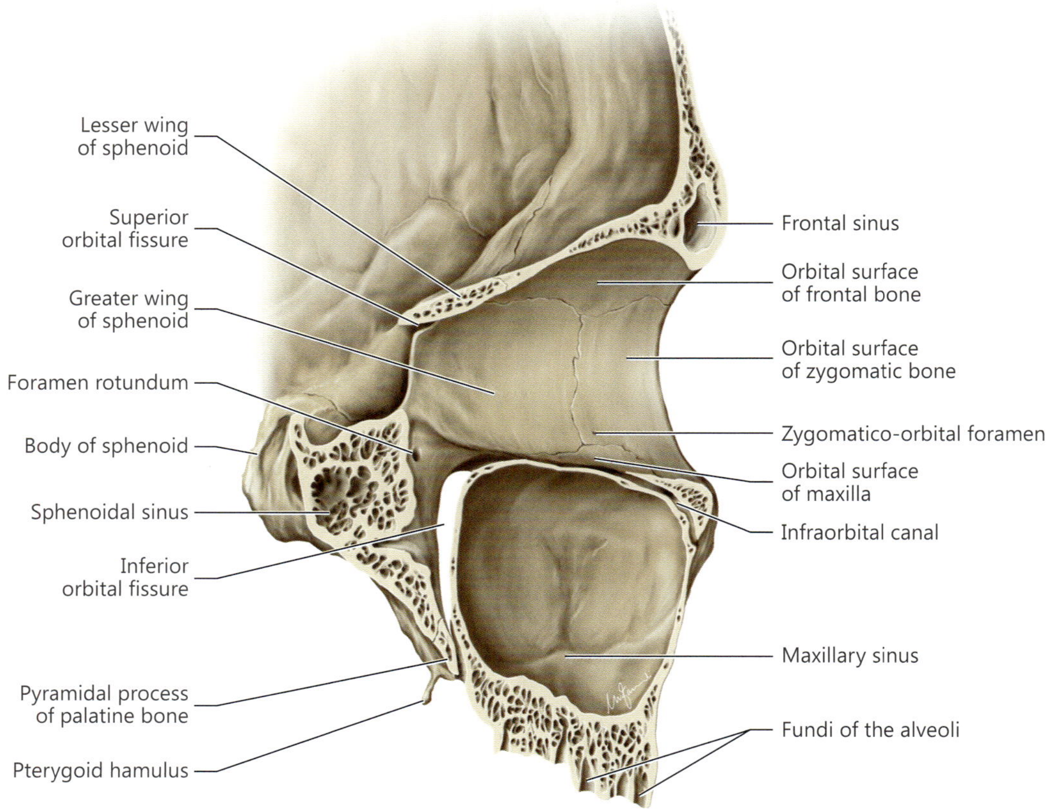

Sagittal section of left **orbital cavity**. The lateral wall and the underlying maxillary sinus cavity can be seen.

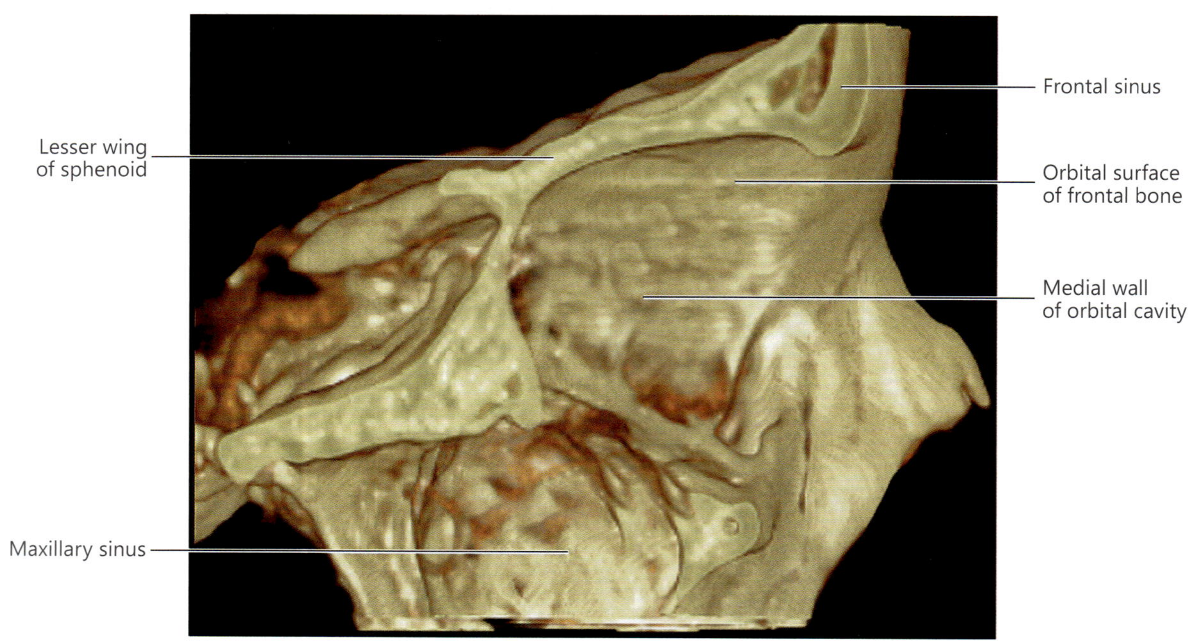

Three-dimensional image of a living human body, obtained using a 3D volume rendering technique, according to a paramedian sagittal section plane showing the right **orbital cavity** and the **maxillary sinus**.

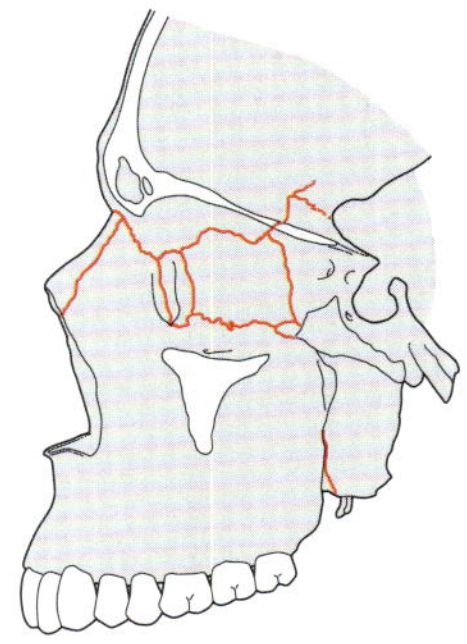

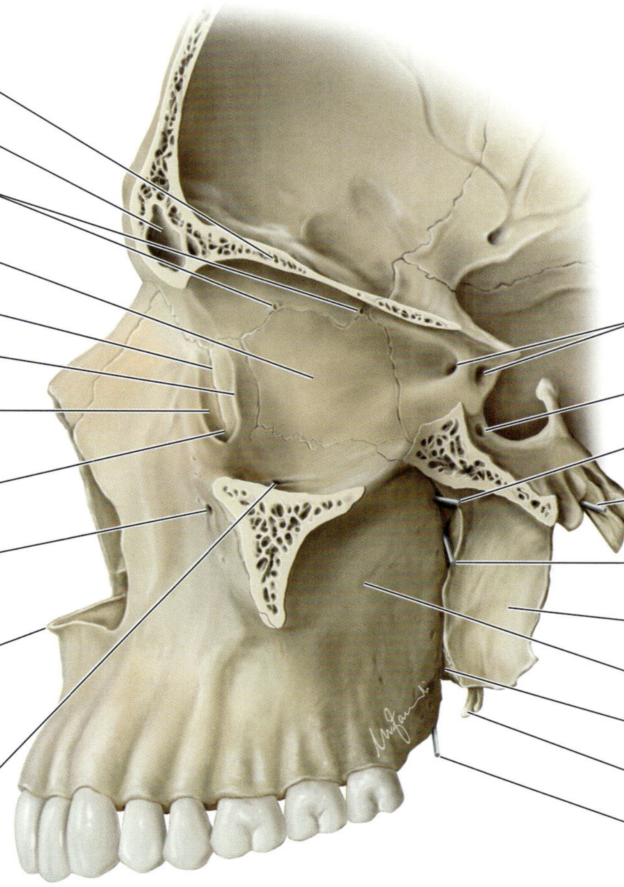

External surface of the **viscerocranium** seen after demolition of a large part of the orbit in the paramedian sagittal section plane; the medial wall of the orbital cavity and of the infratemporal fossa along with the pterygopalatine fossa are visible. The sutures between the various bones are depicted in *red* in the top image.

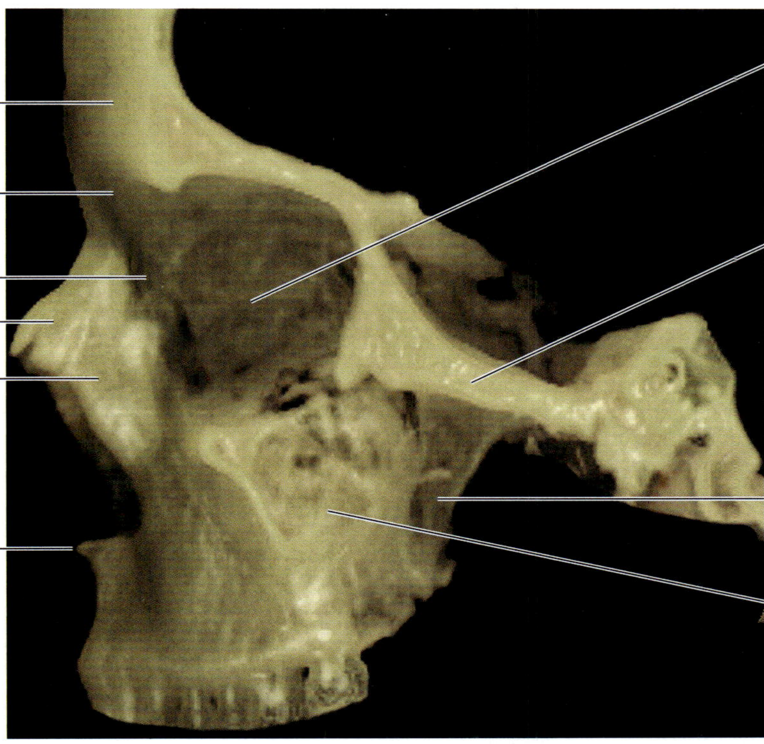

Three-dimensional image of a living human body, obtained using a 3D volume rendering technique, according to a paramedian sagittal section plane tangent to the zygomatic arch. The medial surface of the left orbital cavity and the **viscerocranium** are observed.

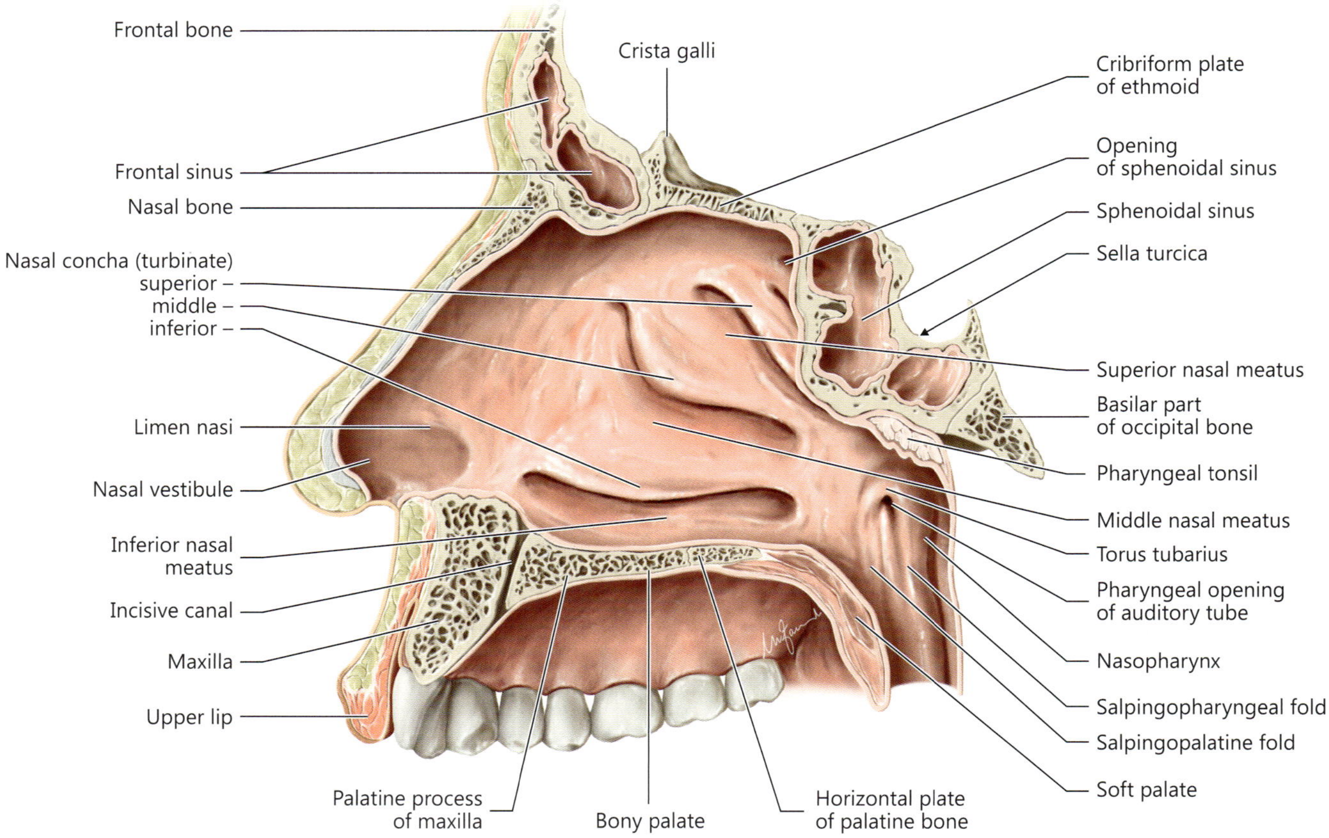

Internal configuration of the lateral wall of the right **nasal cavity**. The nasal mucosa, overlapping the osteocartilaginous skeleton, reduces many of the roughness and irregularities that the skeletal components present.

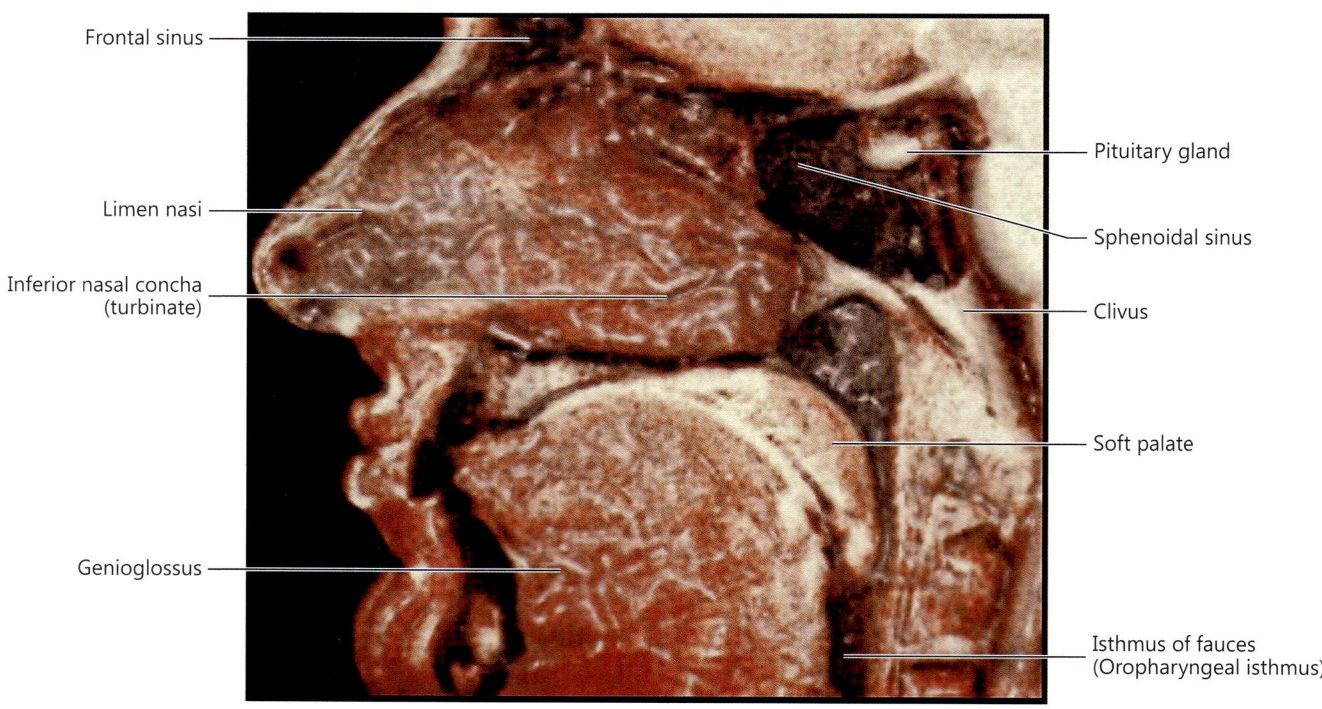

Three-dimensional image of a living human body, obtained using a 3D volume rendering technique, according to a sagittal plane showing the internal configuration of the right **nasal cavity** and the **oral cavity**.

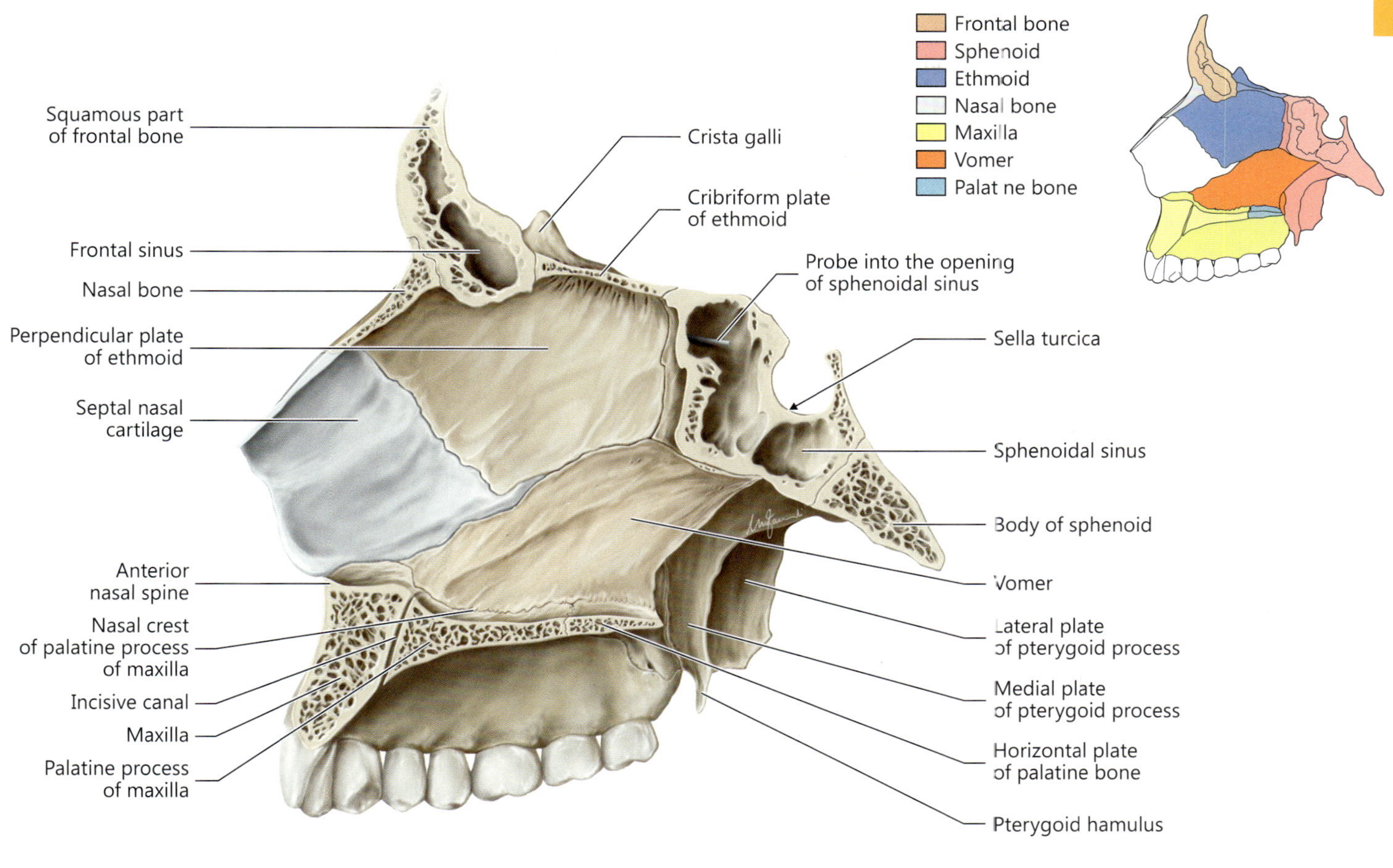

Nasal septum. Skeleton of the medial wall of the right nasal cavity.

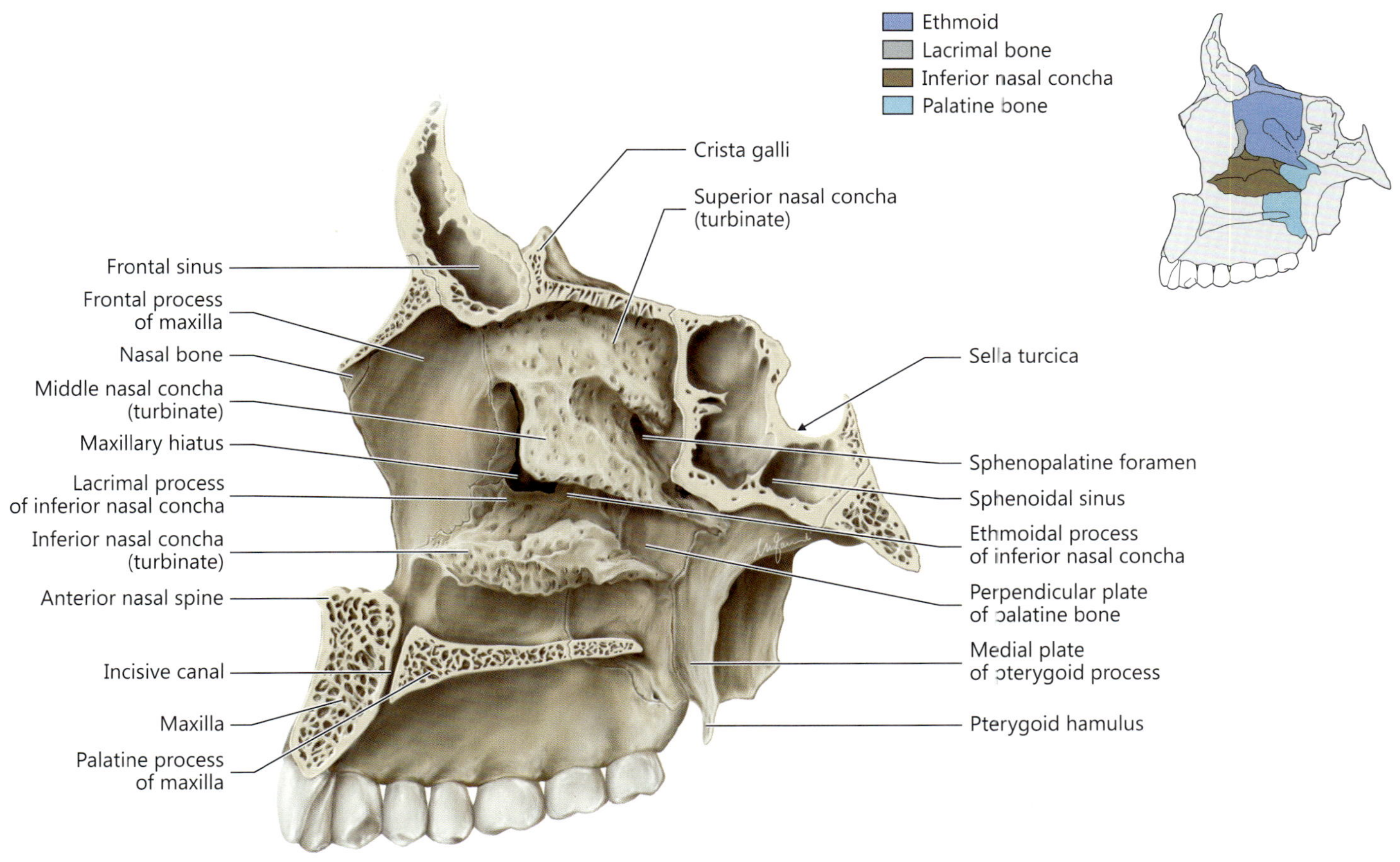

Skeleton of the lateral wall of the right **nasal cavity**.

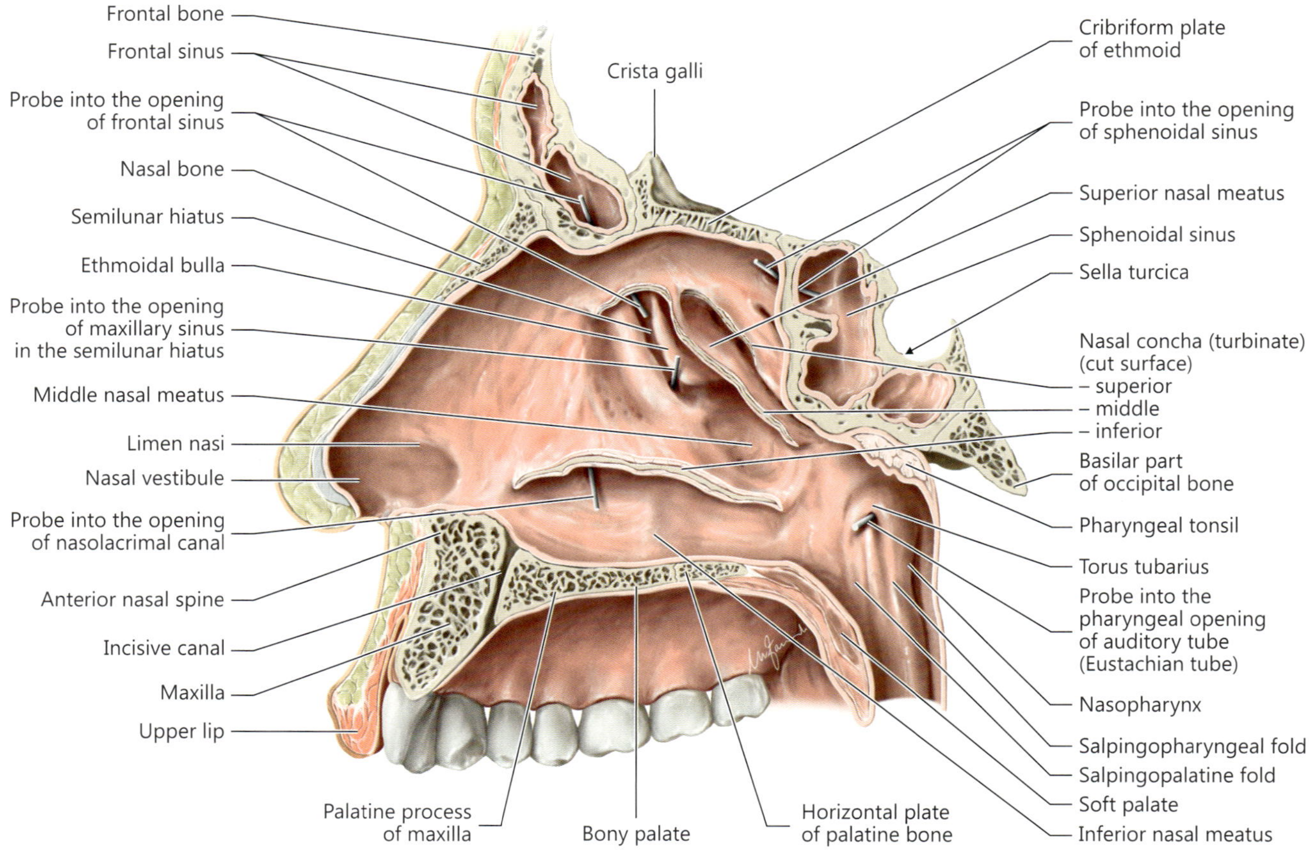

Lateral wall of the right **nasal cavity** seen after partial demolition of the middle and inferior nasal conchae. The three probes of the superior plane indicate, from left to right, the frontal, maxillary, and sphenoidal sinus outlets. The two inferior probes mark the opening of the nasolacrimal canal into the inferior nasal meatus and the opening of the auditory tube into the nasopharynx.

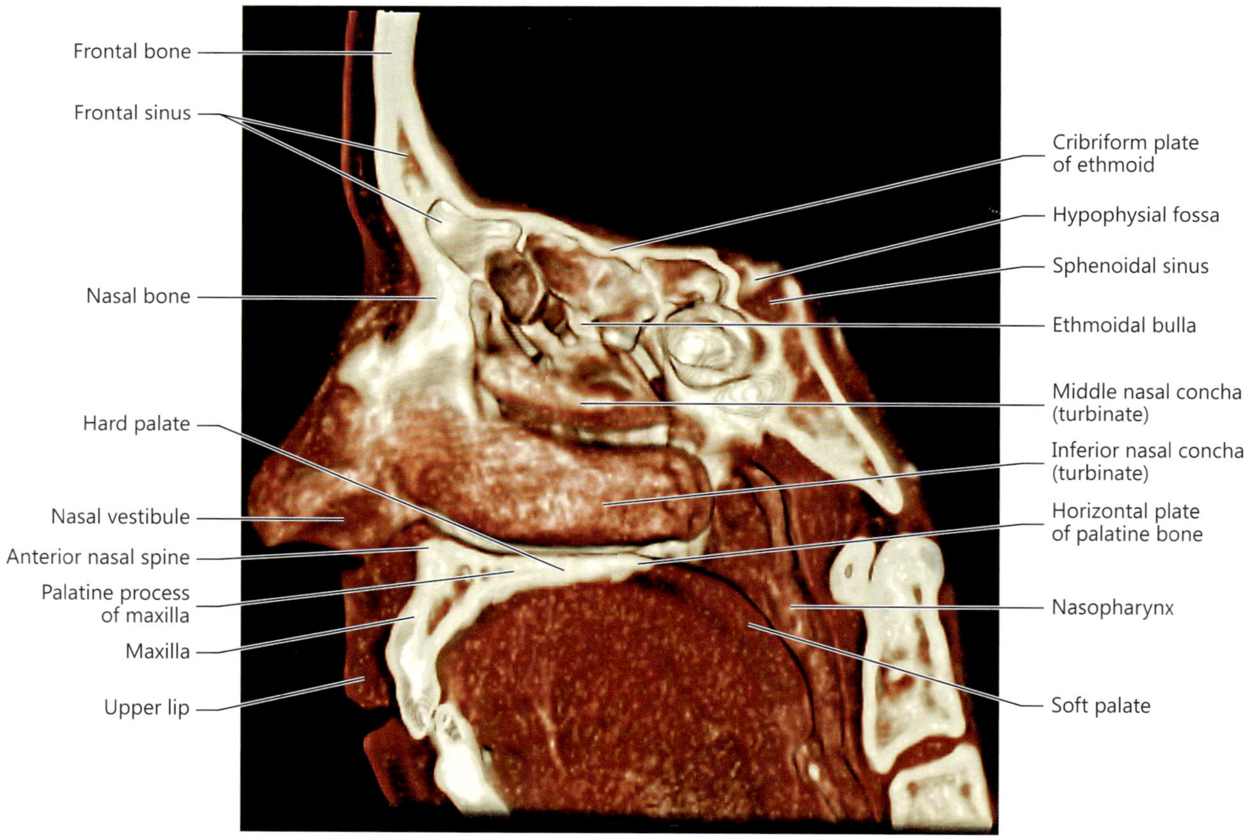

Three-dimensional image of a living human body, obtained using a 3D volume rendering technique, according to a sagittal plane showing in detail the **nasal cavity** and the **oral cavity**.

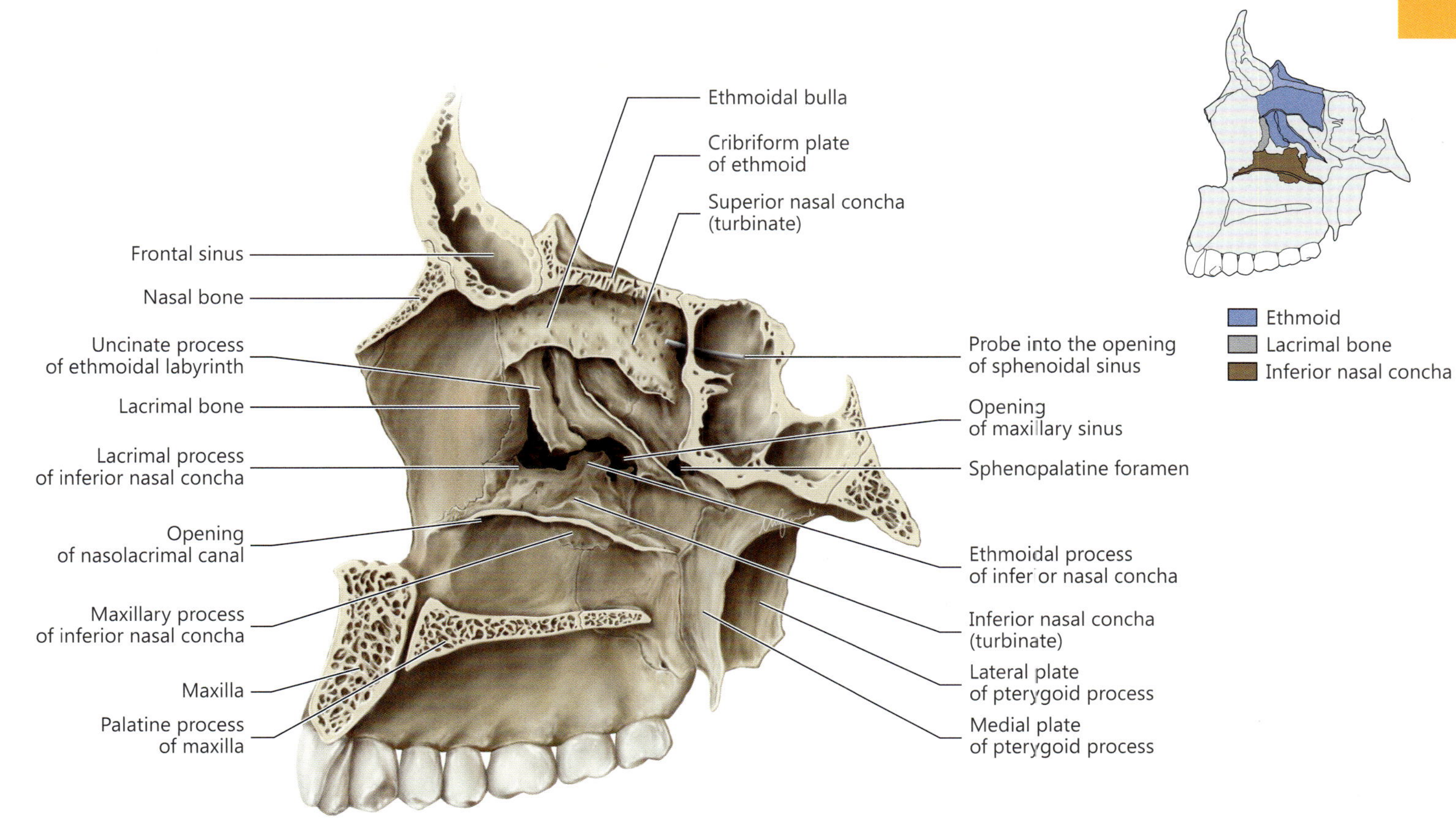

Lateral wall of the right **nasal cavity**, after demolition of the middle nasal concha and part of the inferior nasal concha. The skeletal element that limits the opening of the maxillary sinus in the middle nasal meatus is highlighted. The probe penetrates into the sphenoidal sinus which opens into the nasal cavity at the level of the superior nasal concha.

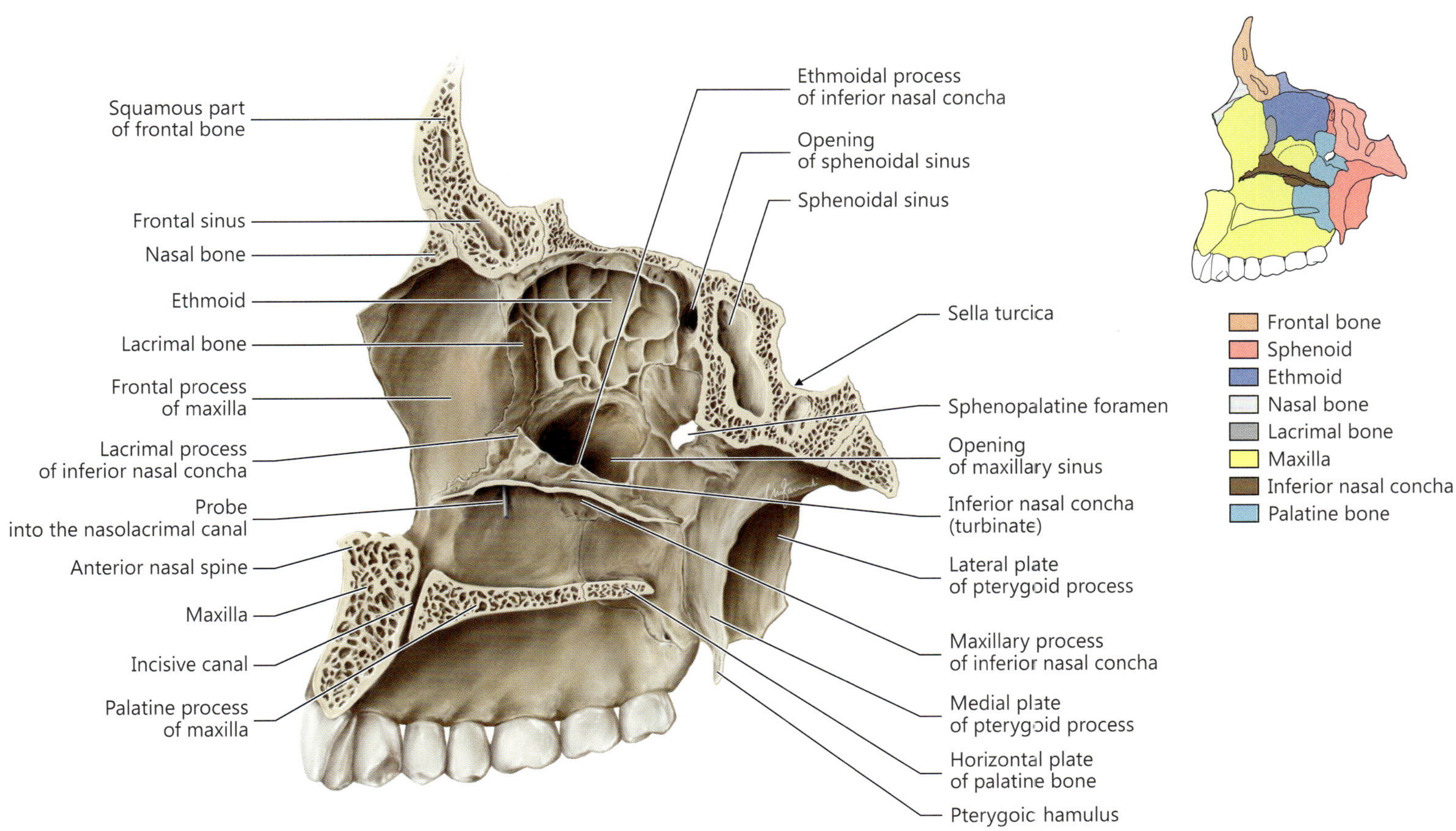

Lateral wall of the right **nasal cavity**, after demolition of the conchae and the ethmoidal labyrinth. The probe is into the naso-lacrimal canal which opens into the inferior nasal meatus.

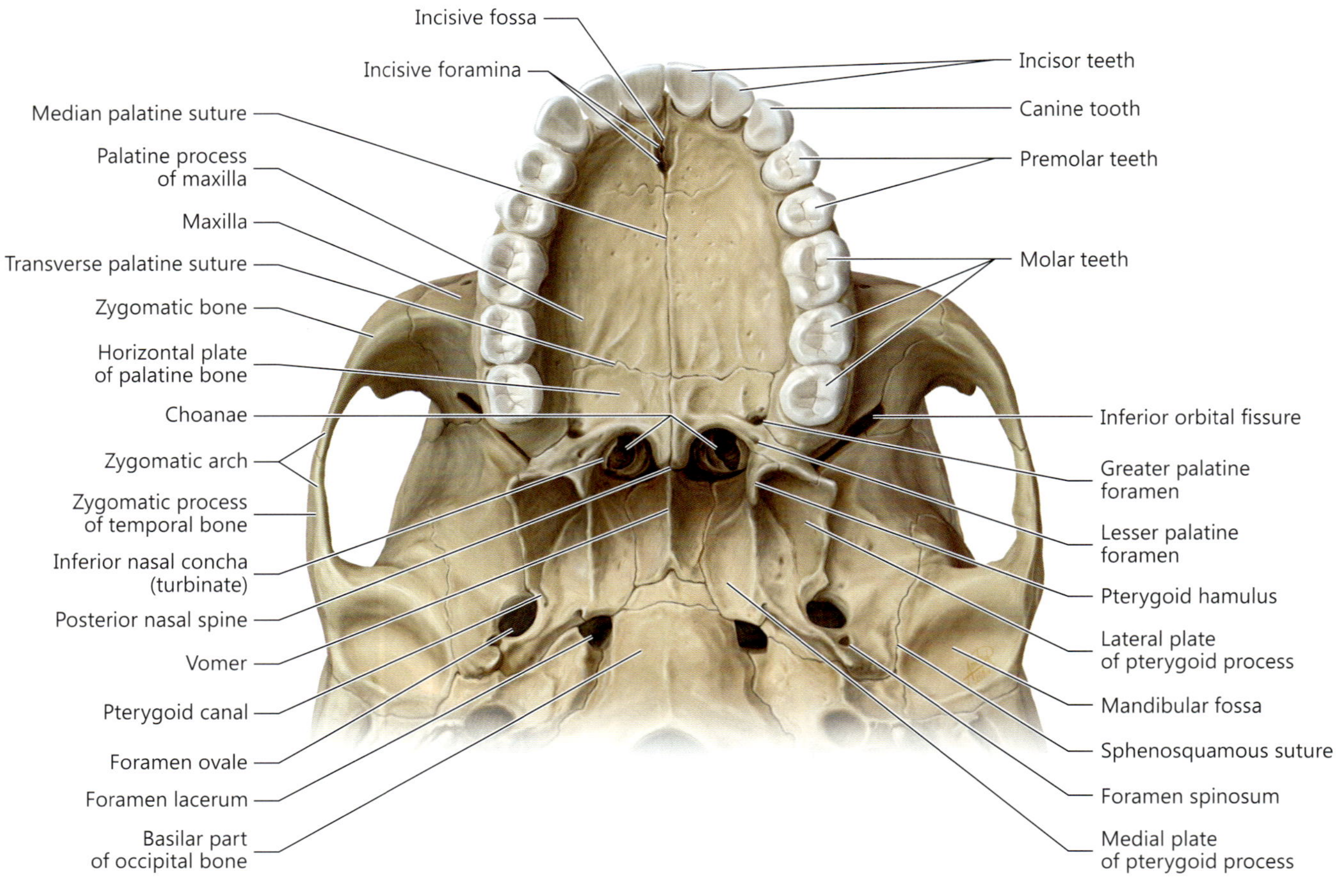

External configuration of **anterior cranial base**, inferior view.

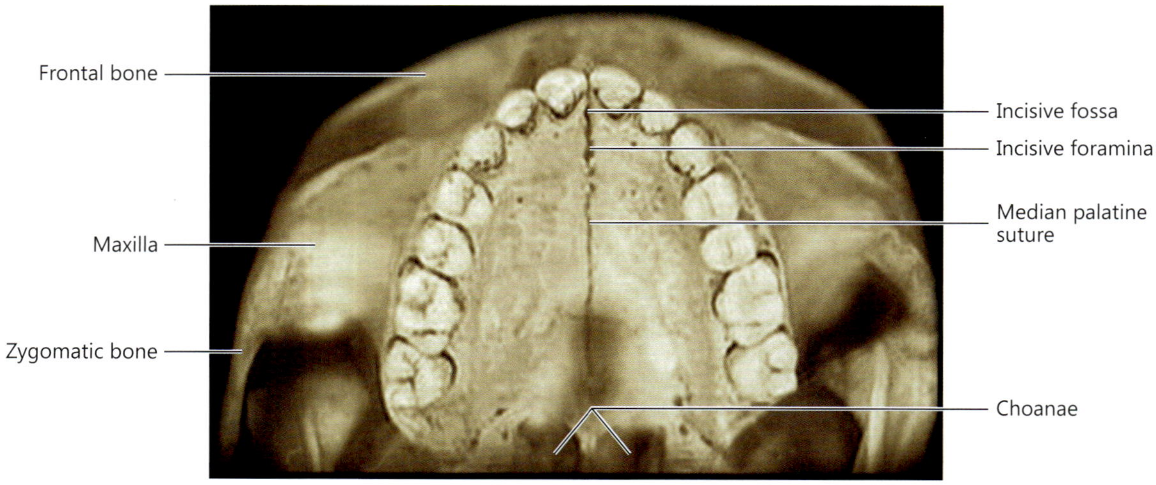

Three-dimensional image of a living human body, obtained using a 3D volume rendering technique, showing the **anterior cranial base** in inferior view.

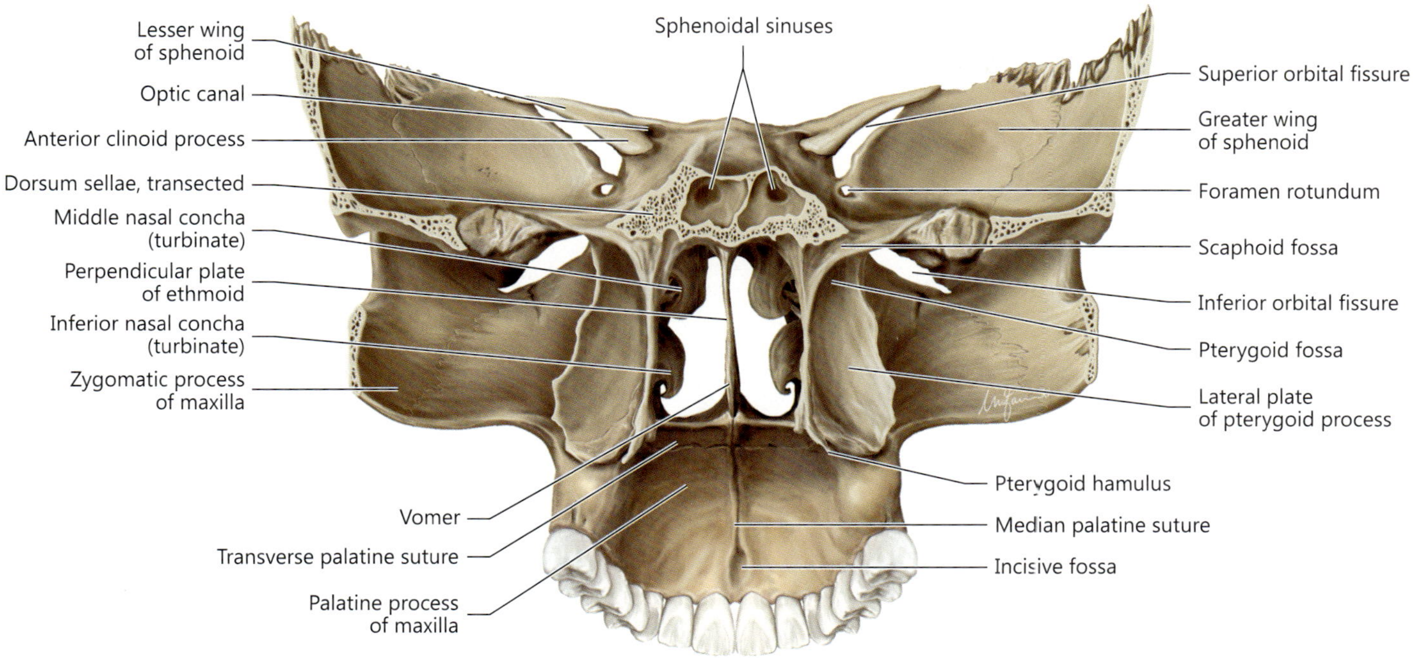

Configuration of the **hard palate** and **nasal cavity**, posterior view.

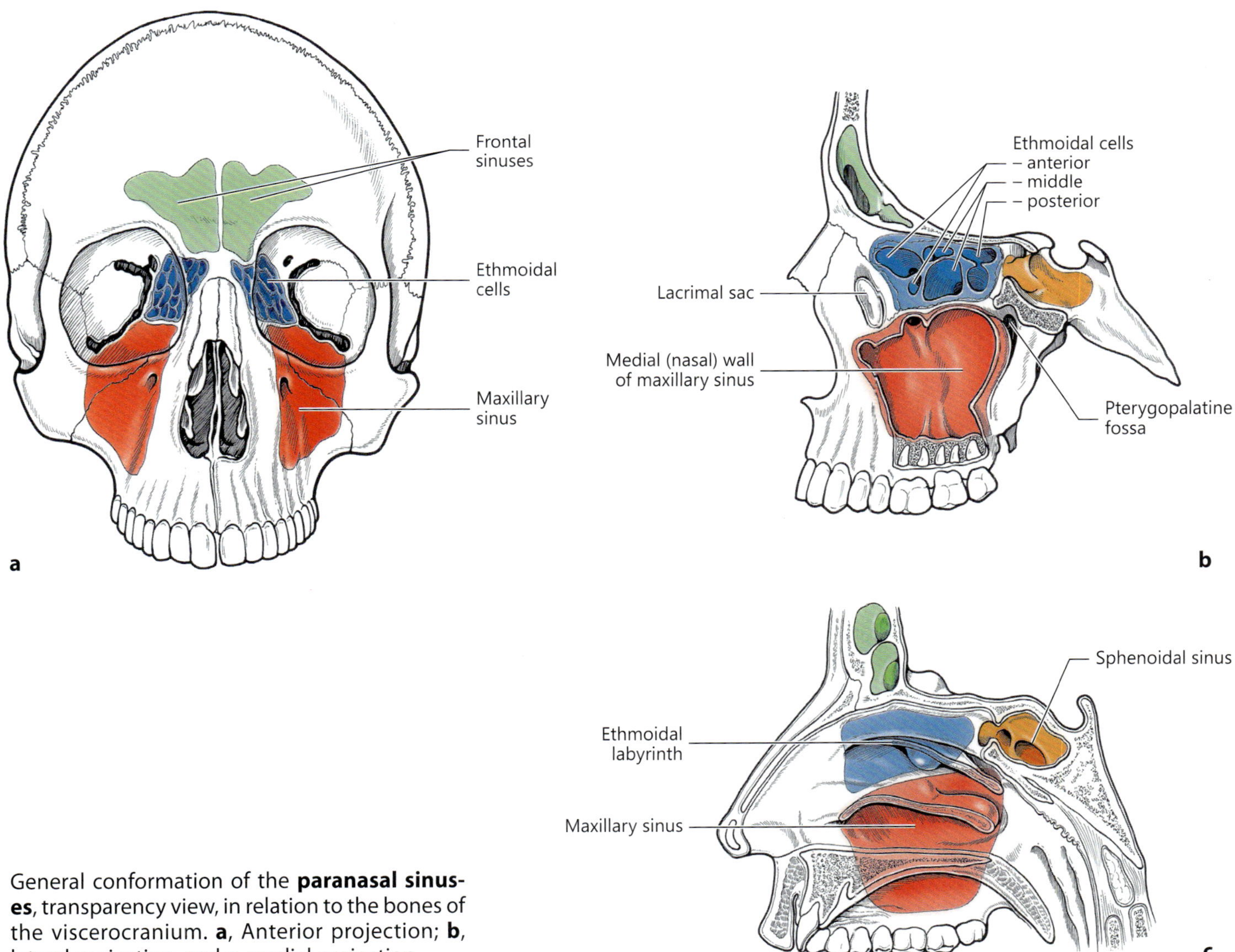

General conformation of the **paranasal sinuses**, transparency view, in relation to the bones of the viscerocranium. **a**, Anterior projection; **b**, lateral projection, and **c**, medial projection.

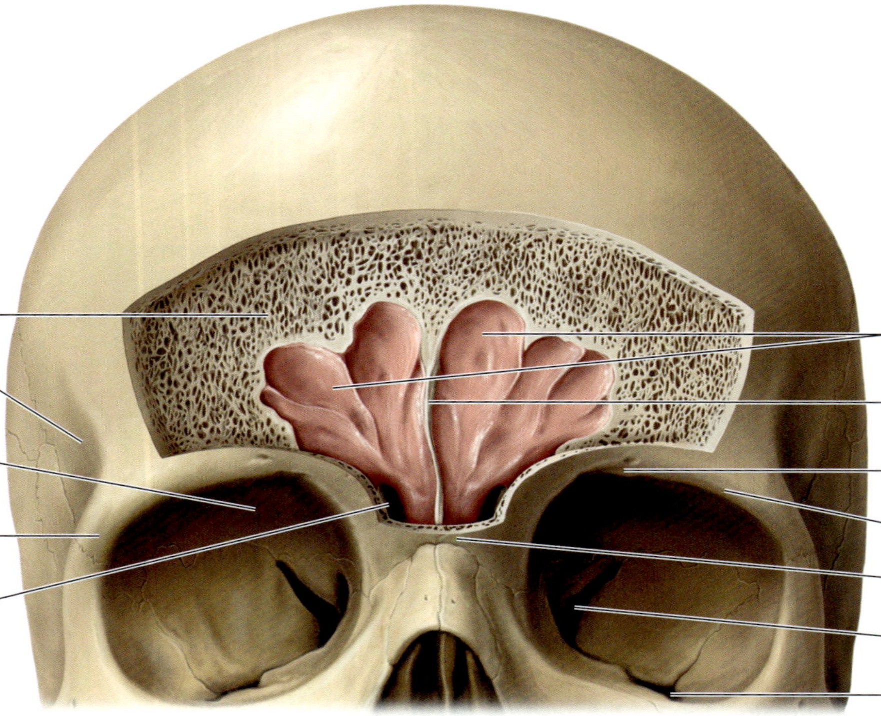

Internal configuration of the **frontal sinuses**, after demolition of their anterior bony wall. Note the median septum which separates the right paranasal sinus from the left one.

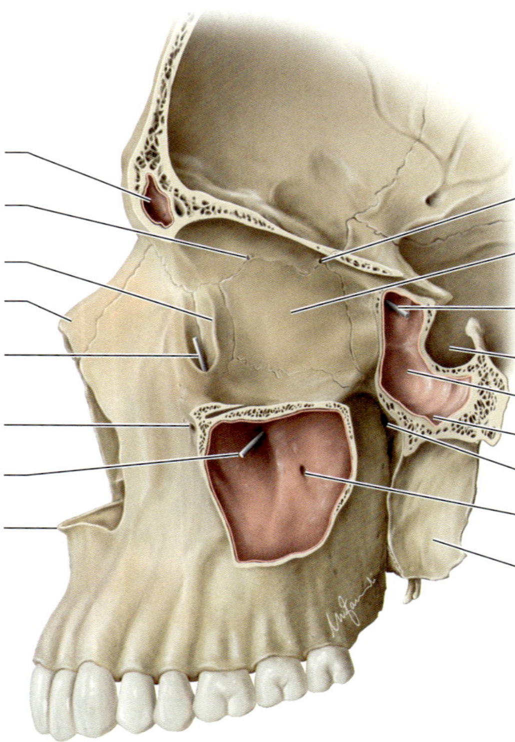

Maxillary sinus and left **sphenoidal sinus** shown through a paramedian sagittal section of the cranium. The probes indicate the opening of the sphenoidal sinus (left) and the main opening of the maxillary sinus (right). It is also note the accessory maxillary ostium.

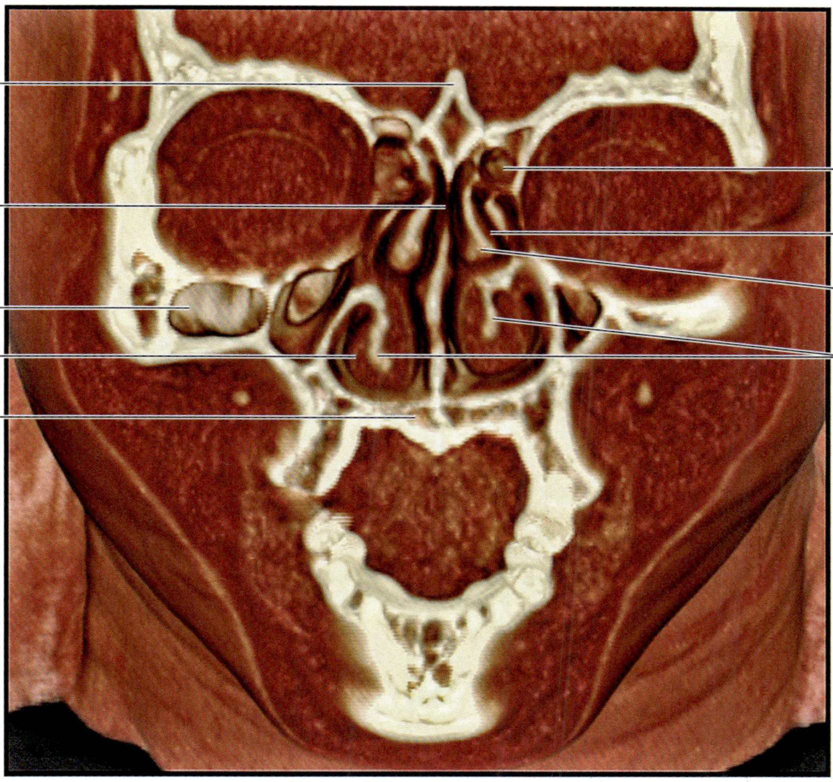

Three-dimensional image of a living human body, obtained using a 3D volume rendering technique, according to a coronal plane tangent to the anterior margin of the frontal process of the zygomatic bone. Note the **nasal cavities**, the adjacent **orbital cavities** and the configuration of the **maxillary sinuses**.

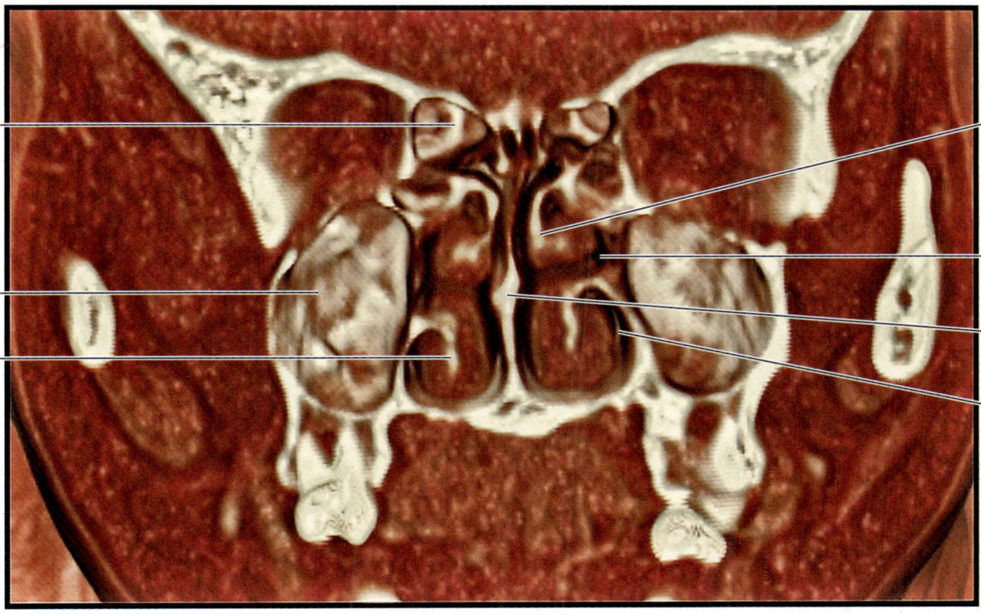

Three-dimensional image of a living human body, obtained using a 3D volume rendering technique, according to a coronal plane passing a few millimeters in front of the posterior surface of the maxillary sinus. In addition to the **maxillary sinus**, note the **ethmoidal cells**.

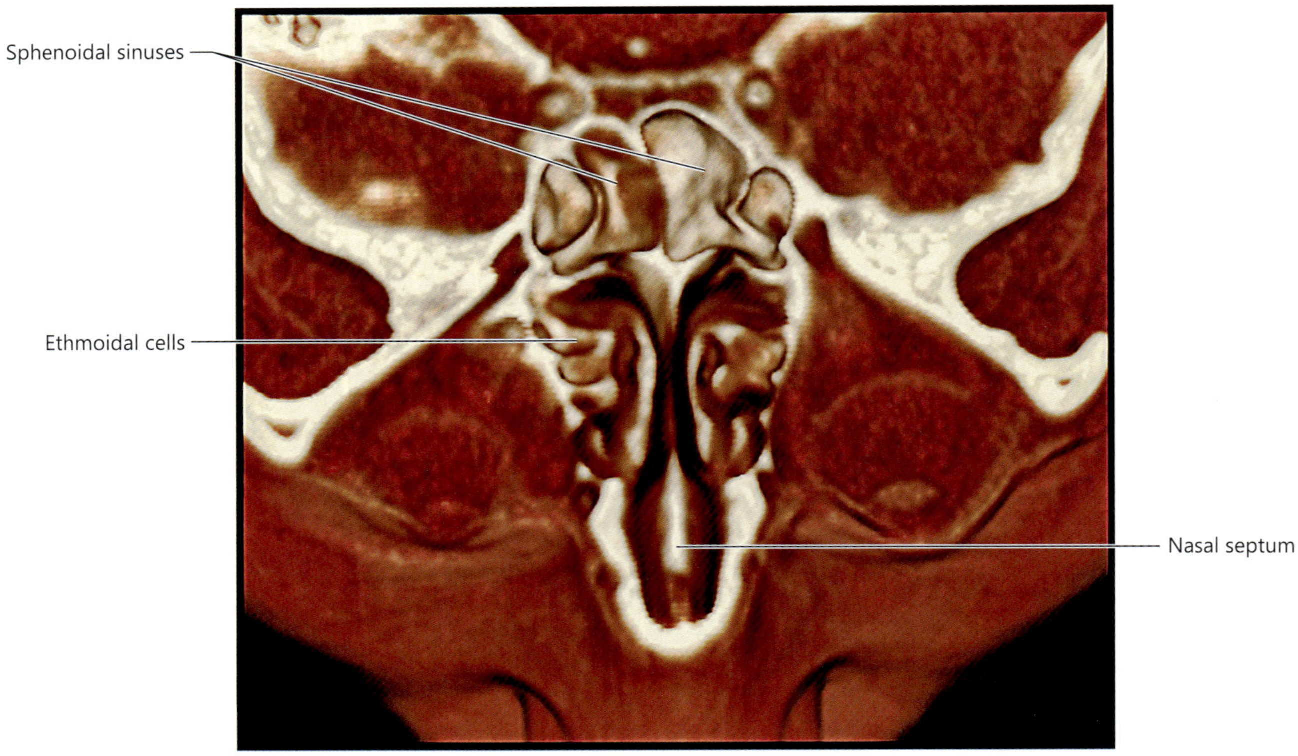

Three-dimensional image of a living human body, obtained using a 3D volume rendering technique, according to a transverse plane passing a few millimeters above the inferior surface of the **sphenoidal sinuses**. Note the ethmoidal cells located in front of the sphenoidal sinuses.

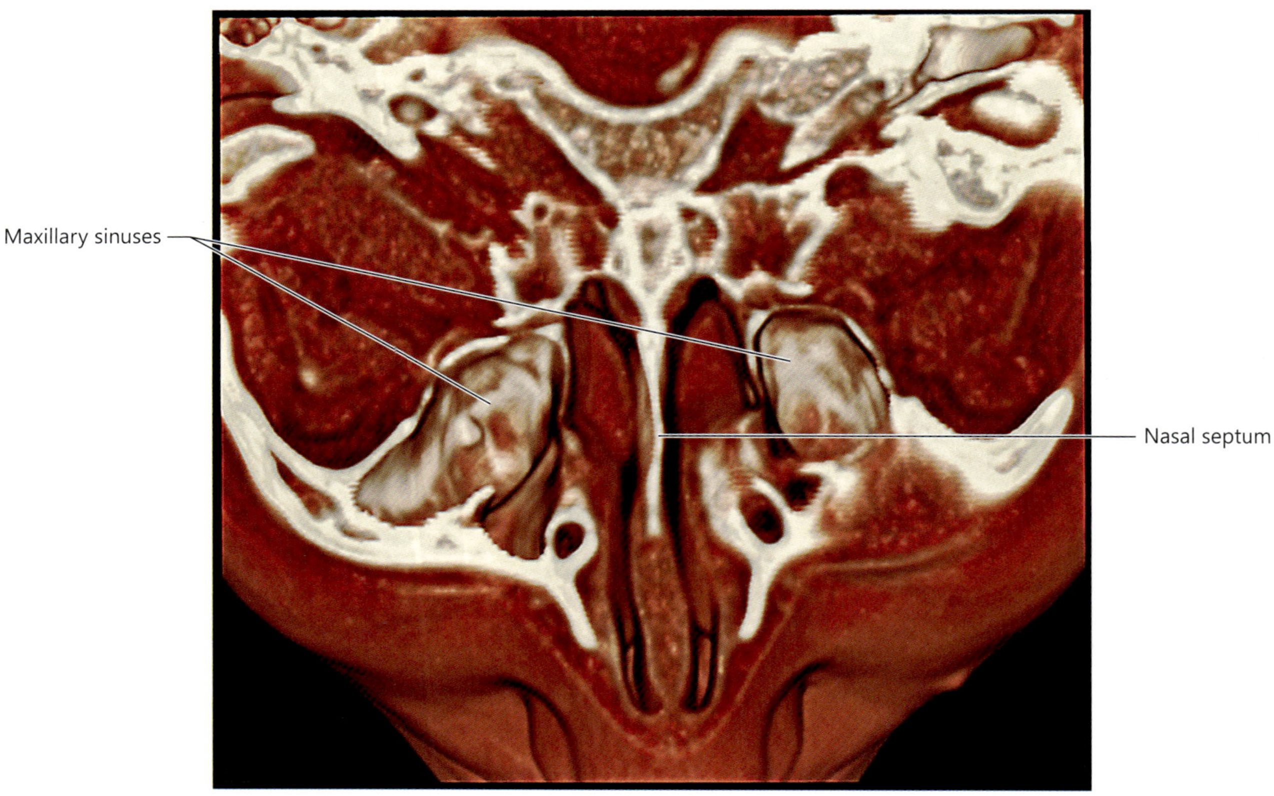

Three-dimensional image of a living human body, obtained using a 3D volume rendering technique, according to a transverse plane passing a few millimeters below the superior wall of the **maxillary sinuses**. In addition to the **maxillary sinuses**, note the conformation of the **nasal septum**.

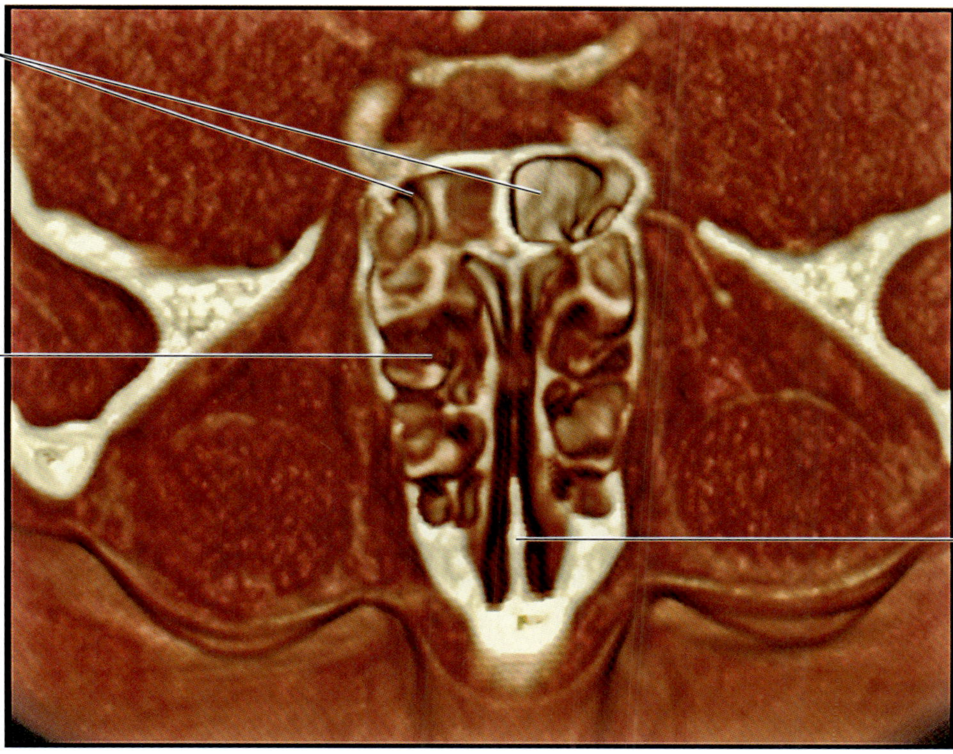

Three-dimensional image of a living human body, obtained using a 3D volume rendering technique, according to a transverse plane passing a few millimeters below the cribriform plate of the ethmoid. In addition to the **sphenoidal sinuses**, note the numerous small cavities (**ethmoidal cells**) in the ethmoidal labyrinth.

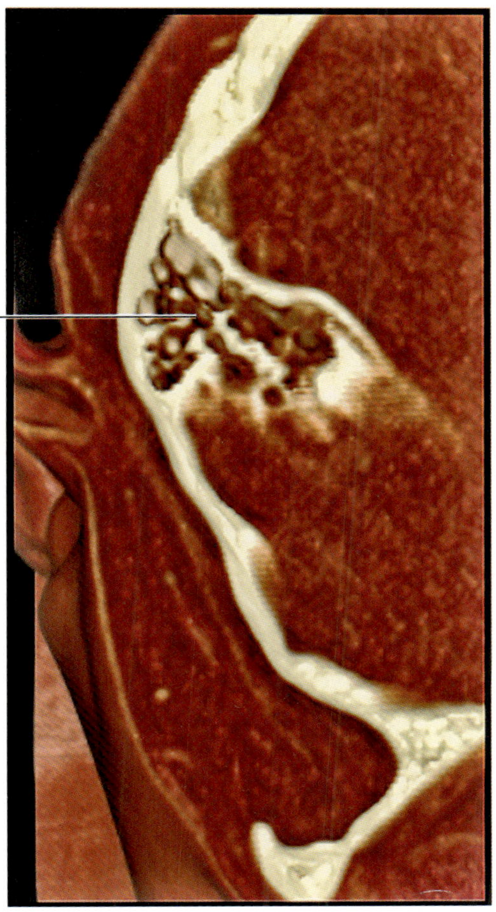

Three-dimensional image of a living human body, obtained using a 3D volume rendering technique, according to a transverse plane passing a few millimeters below the cribriform plate of the ethmoid. Note the internal conformation of the **mastoid process**.

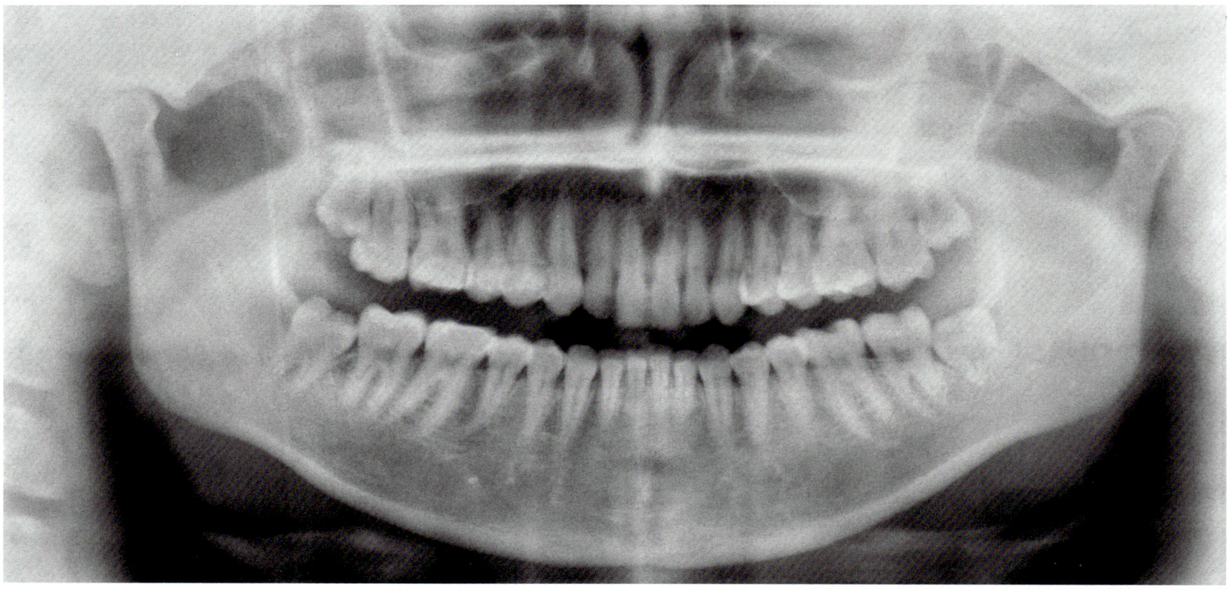

a

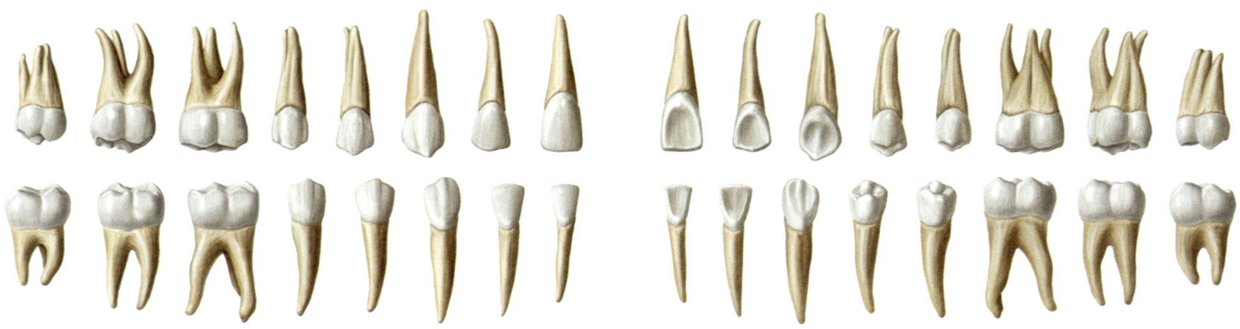

b

c

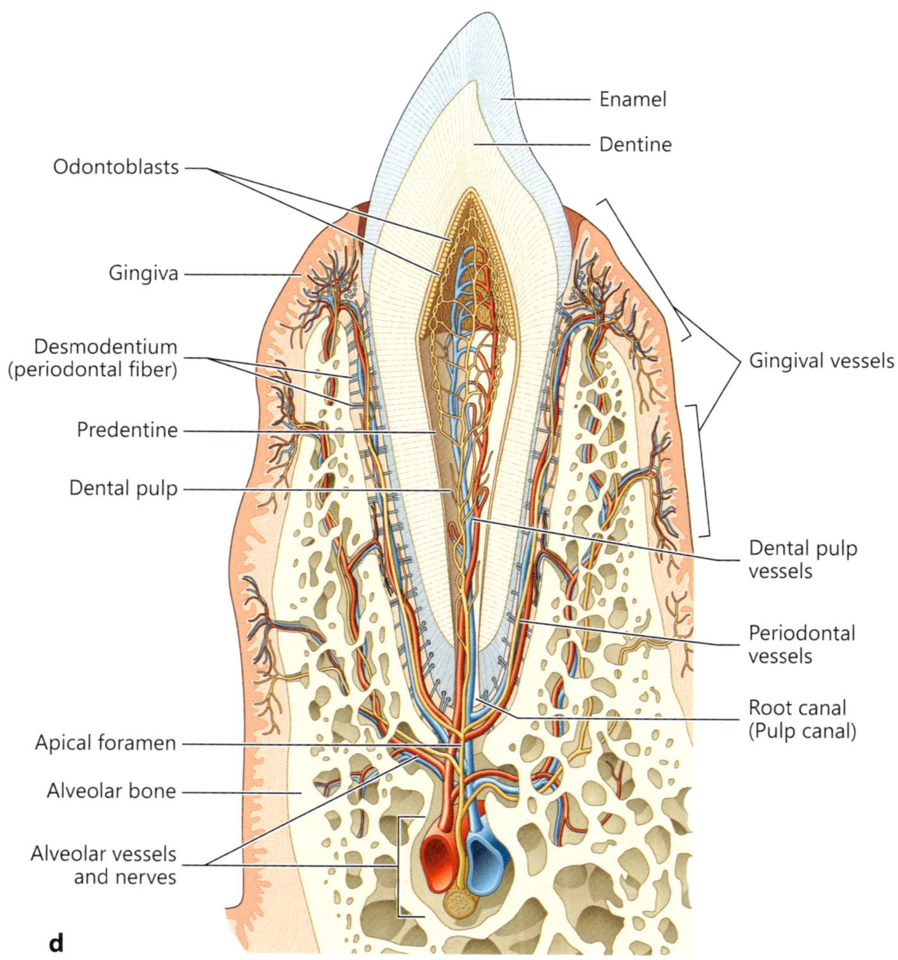

d

Permanent dentition. a, Maxillary and mandibular alveodental arches in a panoramic radiograph; **b**, permanent teeth of the superior hemiarch seen from the vestibular and lingual surfaces. The premolars present a reference of bifid root; the molars are three-rooted; **c**, permanent teeth of the inferior hemiarch seen from the vestibular and lingual surfaces. The premolars have one root while molars have two roots; **d**, oral tissue irrigation. Note the blood vascular organization and the innervation of the desmodentium and dental pulp.

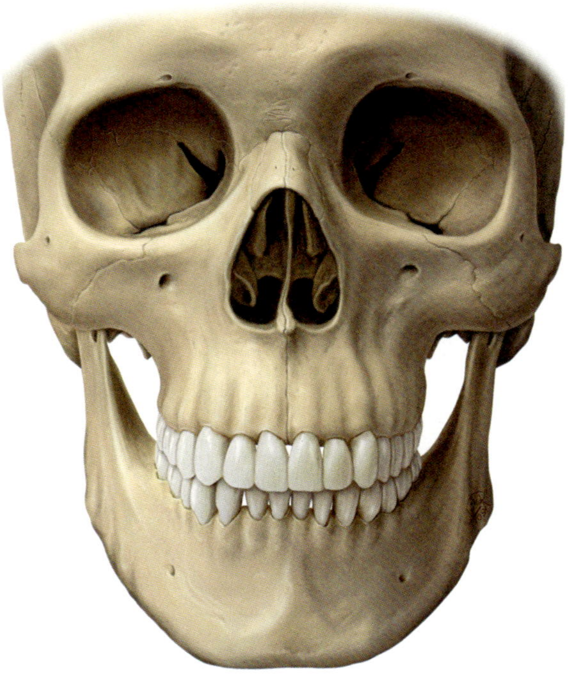

a

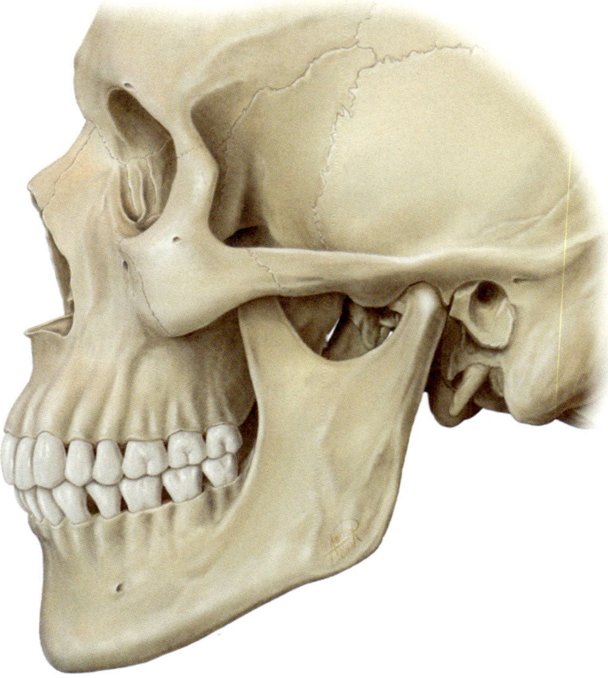

b

Permanent dentition. Permanent teeth physiological occlusion seen in frontal (**a**) and left lateral (**b**) projections. Note that the relation between the crown of the upper and lower teeth is set for cusps and grooves or fossae. **c**, Permanent teeth physiological occlusion seen in posterior projection. Note that the front teeth of the upper dental arch are situated anteriorly compared to the corresponding teeth of the lower dental arch.

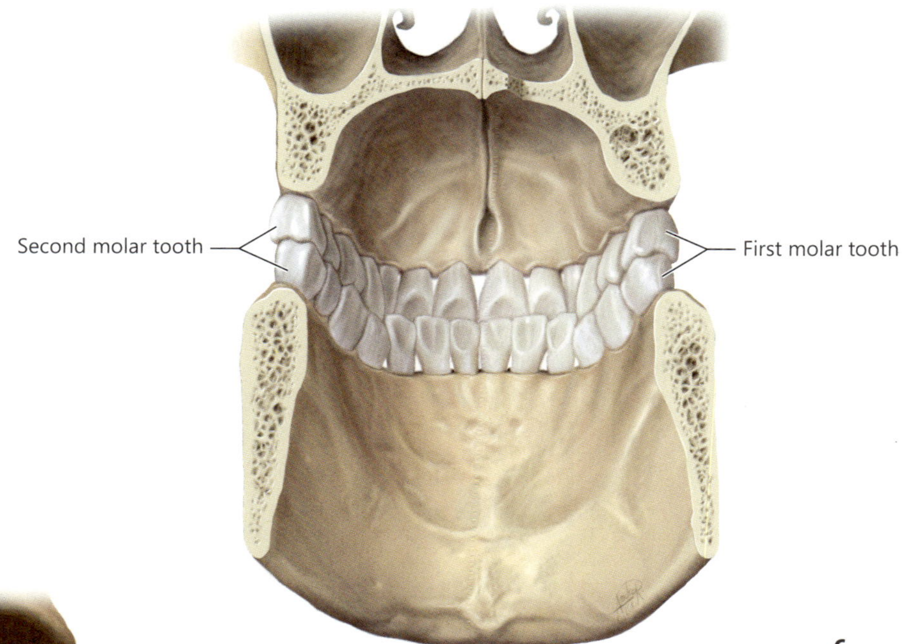

c

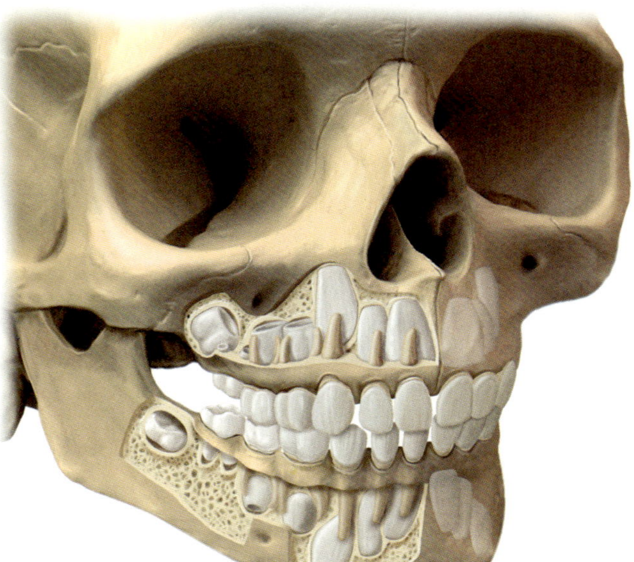

Deciduous dentition. The permanent teeth are preceded by the deciduous teeth. These 20 teeth followed one another in each of the four quadrants, starting from the median line in the following order: 2 incisors, 1 canine, and 2 molars, whose position corresponds to that of the definitive premolars.

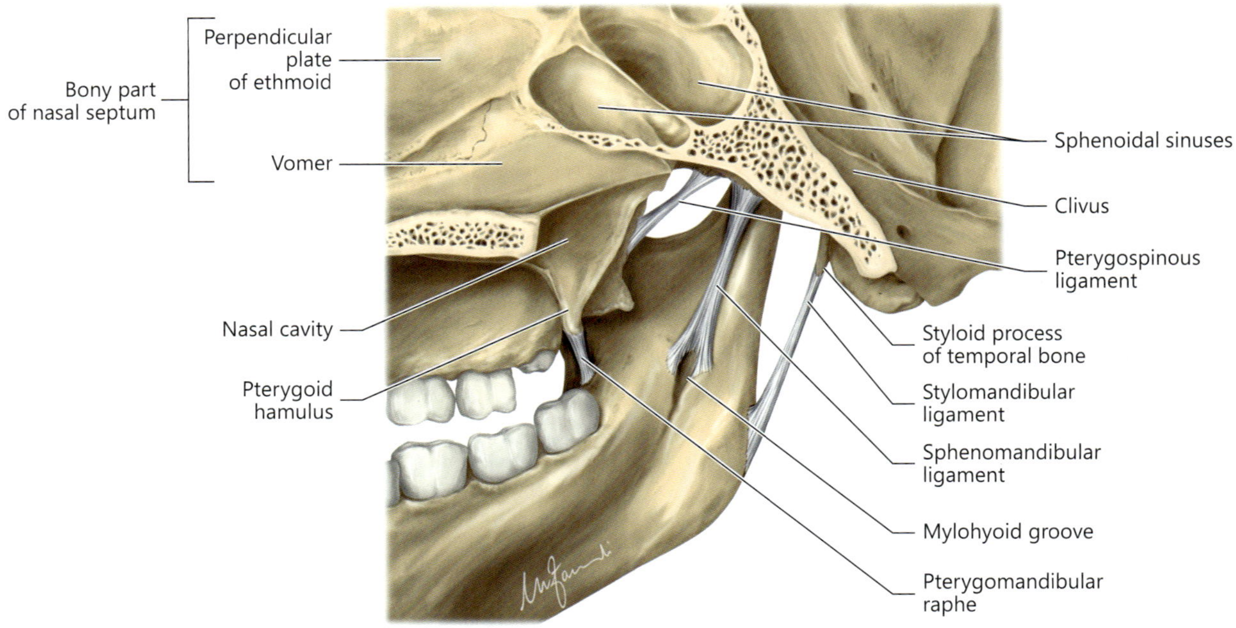

Temporomandibular joint, internal view.

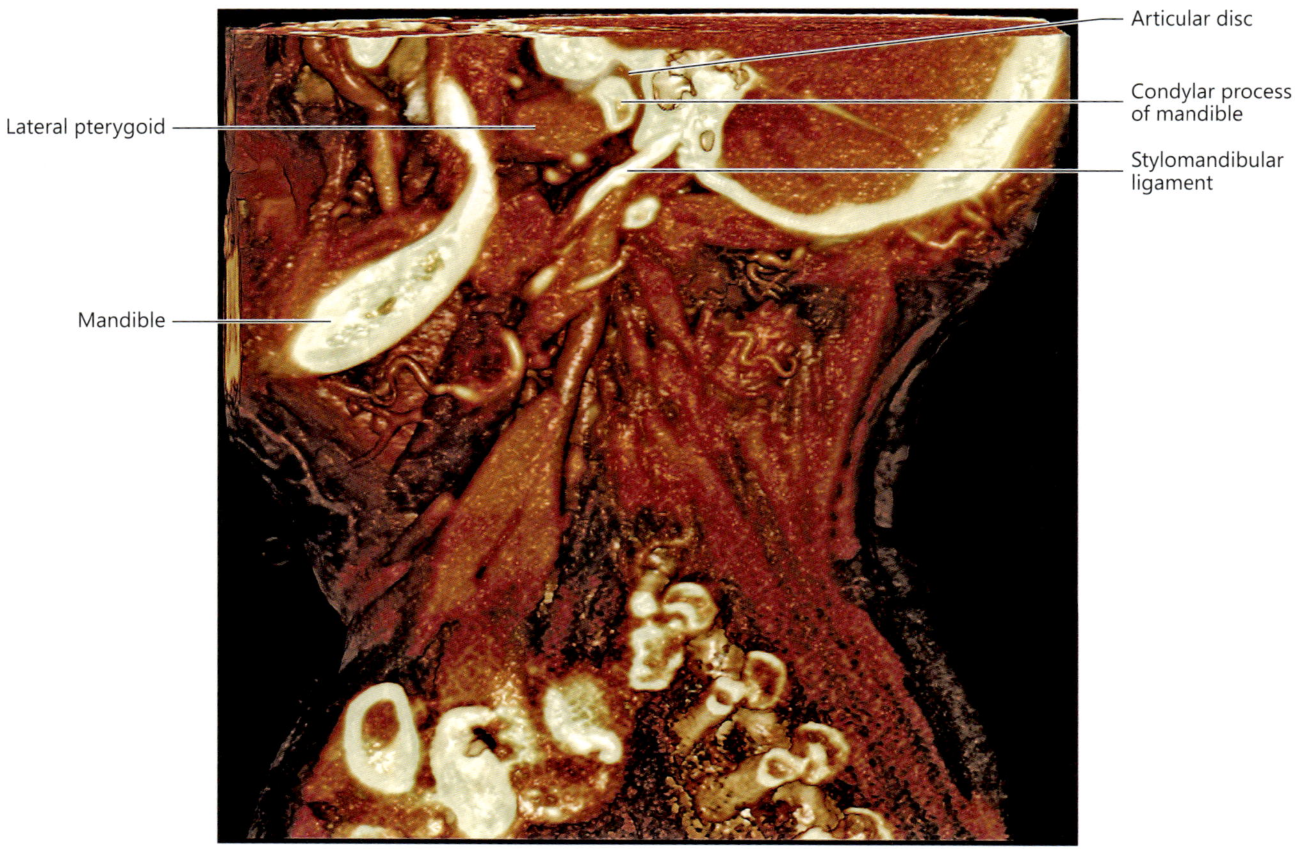

Three-dimensional image of a living human body, obtained using a 3D volume rendering technique, according to a sagittal plane showing the **stylomandibular ligament**, the **lateral pterygoid** and the **articular disc**.

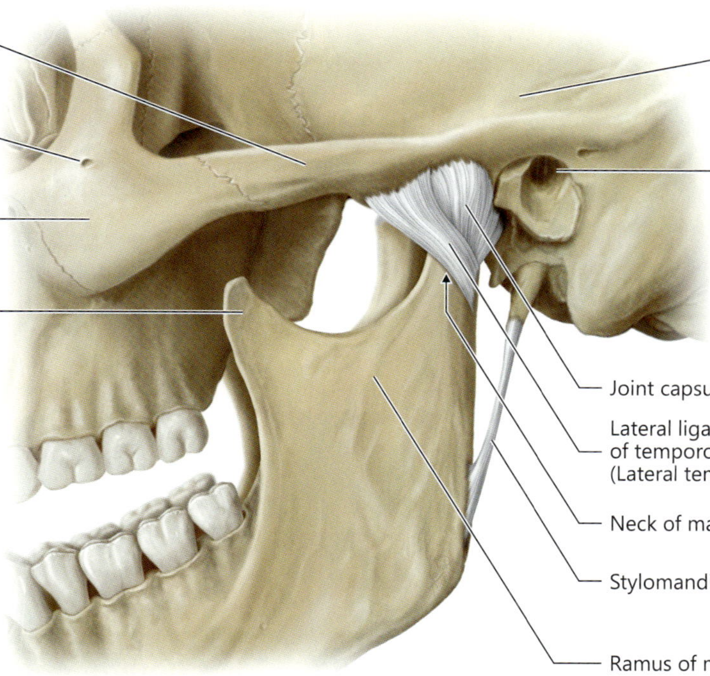

Temporomandibular joint. Left lateral view: capsule and ligaments.

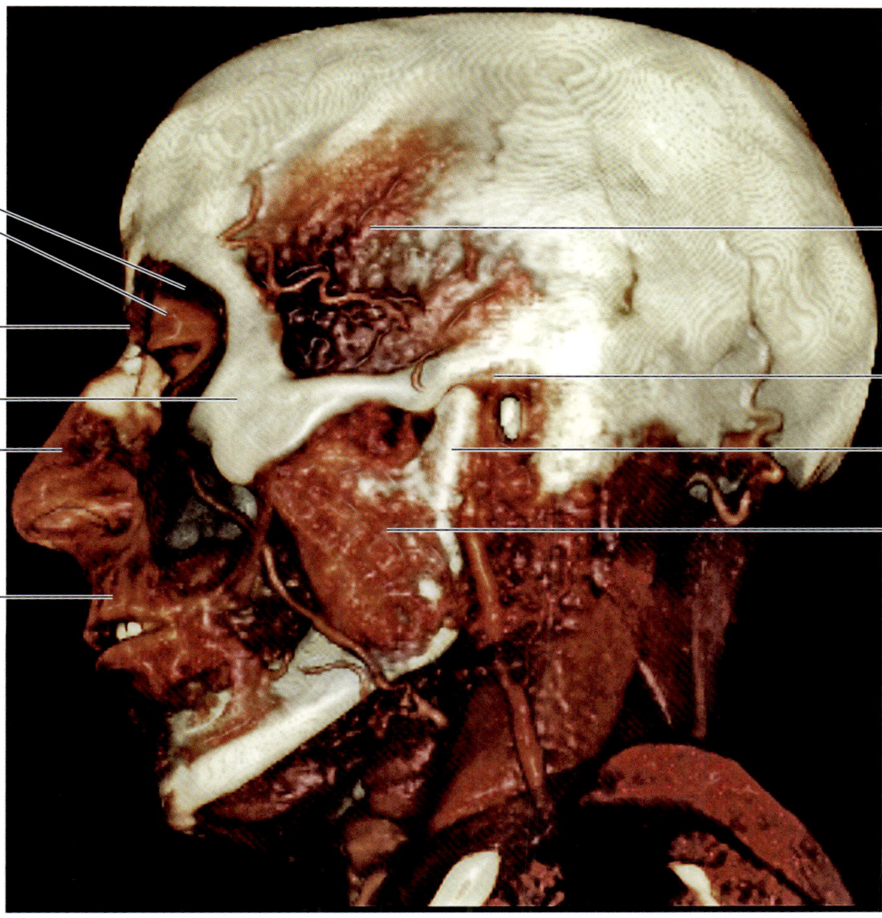

Three-dimensional image of a living human body, obtained using a 3D volume rendering technique, in lateral projection showing the left **temporomandibular joint** and certain **facial** and **masticatory muscles**.

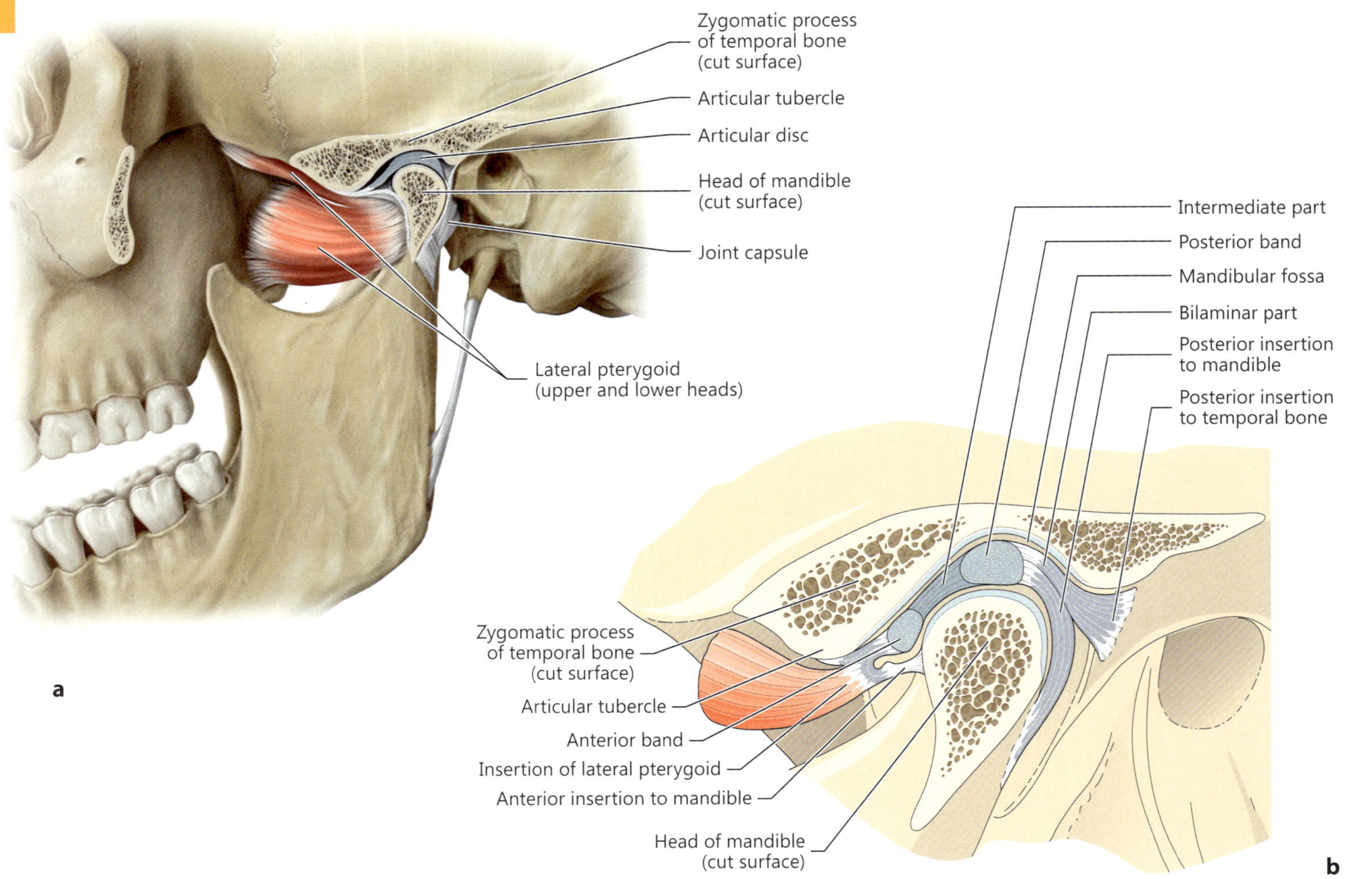

Temporomandibular joint. Left lateral view: **a**, sagittal section (the lateral pterygoid has been left in situ); **b**, articular disc detailed scheme.

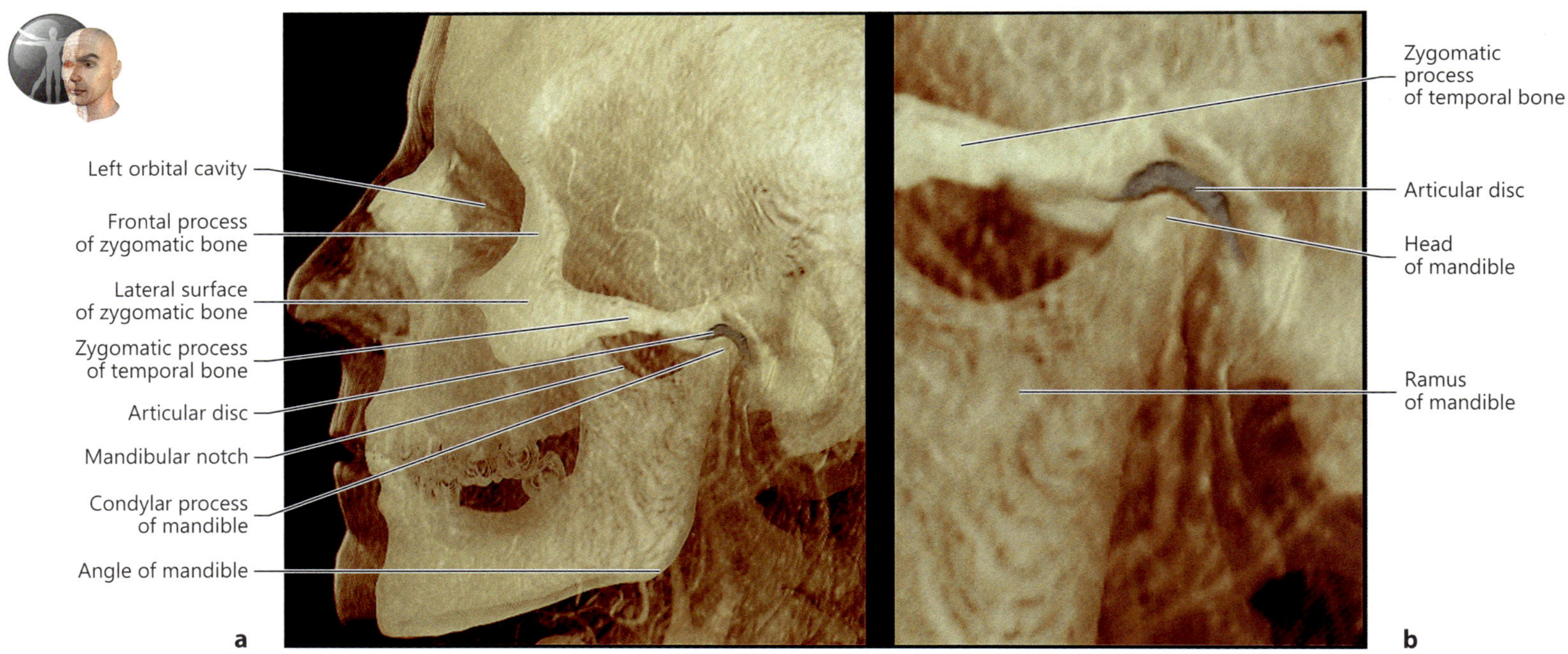

Three-dimensional image of a living human body, obtained using a 3D volume rendering technique, in lateral projection showing the left **temporomandibular joint** (**a**) and the detail of the **articular disc** (**b**).

Three-dimensional image of a living human body, obtained using a 3D volume rendering technique, in lateral projection showing the **disc of the left temporomandibular joint**.

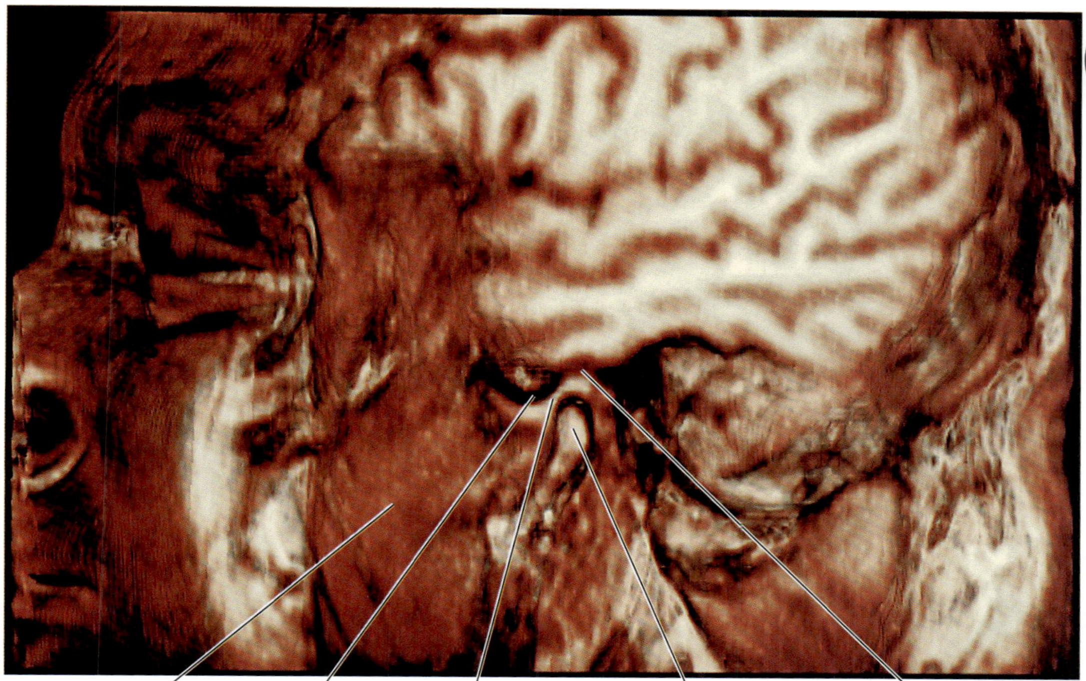

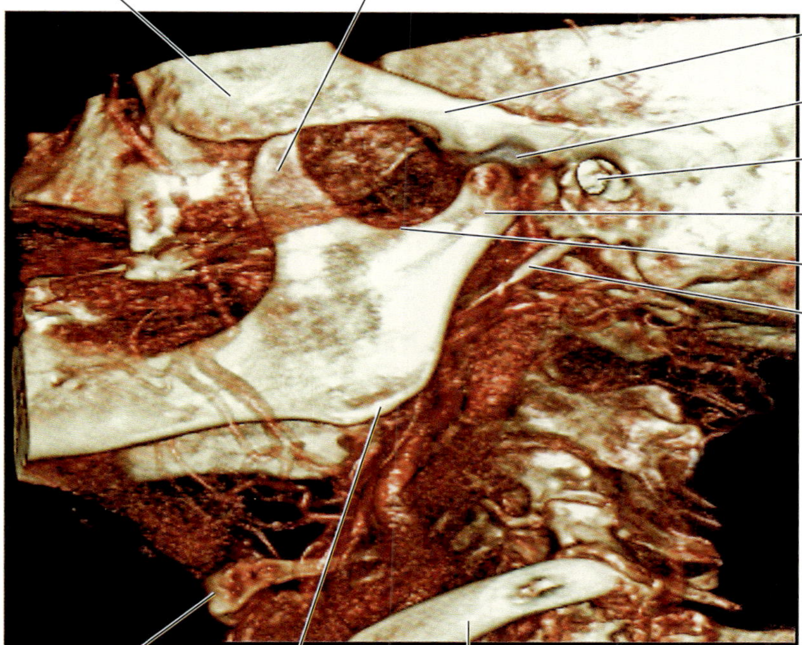

a

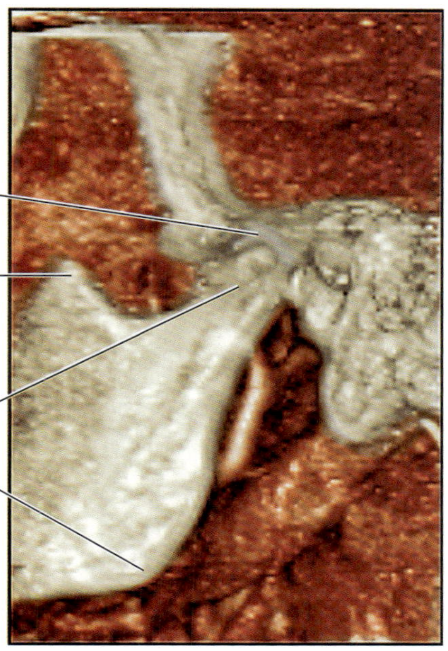

b

Three-dimensional image of a living human body, obtained using a 3D volume rendering technique, in lateral projection showing the **articular disc** interposed between the two articular surfaces (**a**); detail of the disc (**b**).

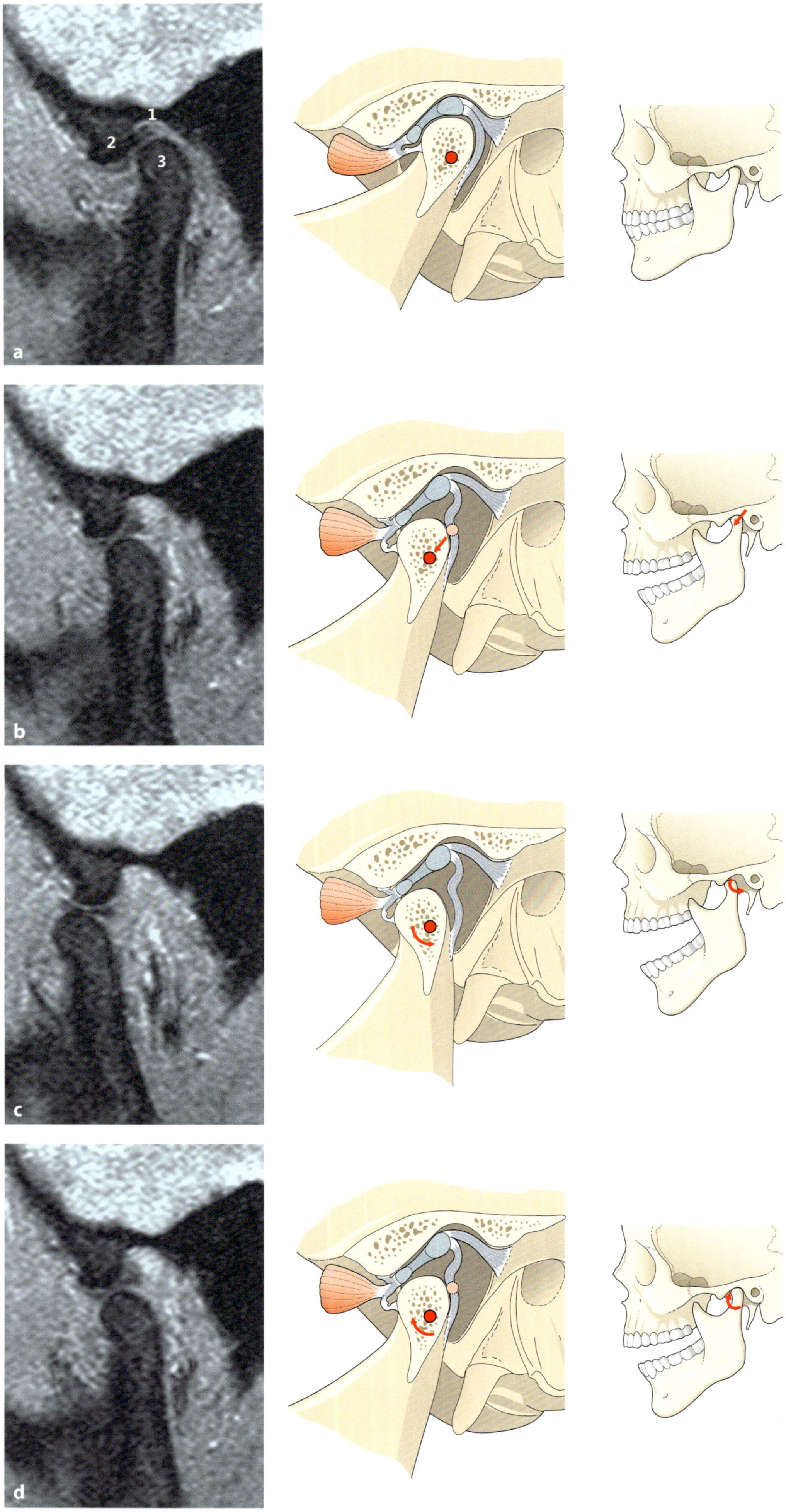

Temporomandibular joint. Magnetic resonance imaging (T_1-weighted image on the sagittal plane) of temporomandibular joint with closed (**a**) and gradually opened mouth (**b-d**). **1**, Mandibular fossa; **2**, articular tubercle; **3**, head of the mandible.

Summary table of the joints of the head	
JOINTS	***BONES INVOLVED***
	SYNARTHROSIS
	Fibrous joints
Syndesmosis or gomphosis	
Dentoalveolar	Between the roots of the teeth and the dental alveoli of maxilla and mandible
Sutures	
Incisive or anterior palatine (not always present)	Between incisive bone (premaxilla) and palatine processes of maxillae
Coronal	Between frontal bone and parietal bones
Ethmoidoconchal	Between ethmoid and inferior nasal concha
Ethmoidopalatine	Between palatine bone and ethmoid
Frontal or metopic or mediofrontal (not always present)	Between the two halves of the frontal bone
Frontoethmoidal	Between frontal bone and ethmoid
Frontolacrimal	Between frontal bone and lacrimal bone
Frontomaxillary	Between frontal bone and maxilla
Frontonasal	Between frontal bone and nasal bone
Frontozygomatic or zygomaticofrontal	Between frontal bone and zygomatic bone
Intermaxillary	Between the maxillae
Internasal	Between the nasal bones
Interpalatine	Between the palatine bones
Lacrimoconchal	Between lacrimal bone and inferior nasal concha
Lacrimoethmoidal	Between ethmoid and lacrimal bone
Lambdoid	Between parietal bones and occipital bone
Maxilloconchal	Between maxilla and inferior nasal concha
Maxilloethmoidal	Between ethmoid and maxilla
Maxillolacrimal	Between lacrimal bone and maxilla
Maxillonasal	Between nasal bone and maxilla
Maxillopalatine	Between palatine bone and maxilla
Median palatine (intermaxillary and interpalatine)	Between the maxillae and between the palatine bones
Occipitomastoid	Between occipital bone and mastoid process of temporal bone
Palatoconchal	Between palatine bone and inferior nasal concha
Parietomastoid	Between parietal bone and mastoid process of temporal bone
Sagittal	Between the parietal bones
Sphenoethmoidal	Between sphenoid and ethmoid
Sphenofrontal or frontosphenoidal	Between sphenoid and frontal bone
Sphenomaxillary	Between sphenoid and maxilla
Sphenoparietal	Between sphenoid and parietal bone
Sphenosquamous	Between sphenoid and squamous part of temporal bone
Sphenozygomatic	Between sphenoid and zygomatic bone
Squamous	Between the squamous part of temporal bone and parietal bone
Squamomastoid or petrosquamous (not always present)	Between the squamous and petrous parts (mastoid process) of temporal bone
Squamotympanic or tympanosquamous	Between the squamous and tympanic parts of temporal bone
Trasverse occipital (variable)	At the level of the superior nuchal line between the occipital and nuchal planes of squamous part of occipital bone
Transverse palatine	Between palatine process of maxilla and horizontal plate of palatine bone
Vomeroethmoidal	Between ethmoid and vomer
Vomeropalatine	Between palatine bone and vomer
Zygomaticomaxillary or infraorbital	Between zygomatic bone and maxilla
Zygomaticotemporal	Between temporal bone and zygomatic bone
	Cartilaginous joints
Synchondrosis	
Anterior intraoccipital or basioccipital (not always present)	Between the basilar and lateral parts of occipital bone
Petro-occipital	Between the petrous part of temporal bone and occipital bone
Posterior intraoccipital (not always present)	Between lateral part and squamous part of occipital bone
Sphenoethmoidal	Between sphenoid and ethmoid
Spheno-occipital	Between sphenoid and occipital bone
Sphenopetrosal	Between sphenoid and petrous part of temporal bone
	DIARTHROSIS (SYNOVIAL JOINTS)
Atlanto-occipital joint	Between atlas and occipital bone
Temporomandibular joint	Between temporal bone and mandible

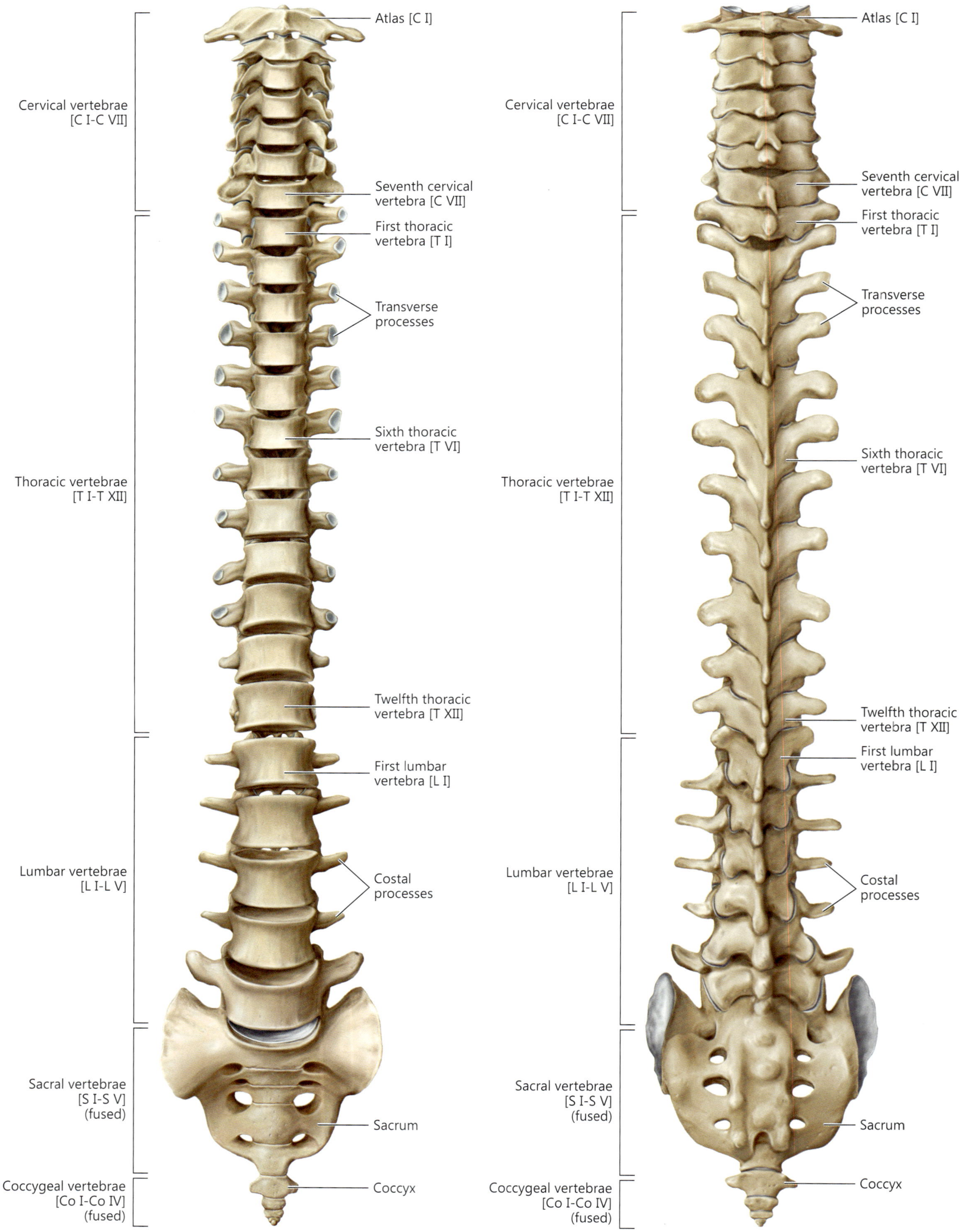

Preparation indicating the **vertebral column** in anterior and posterior projections.

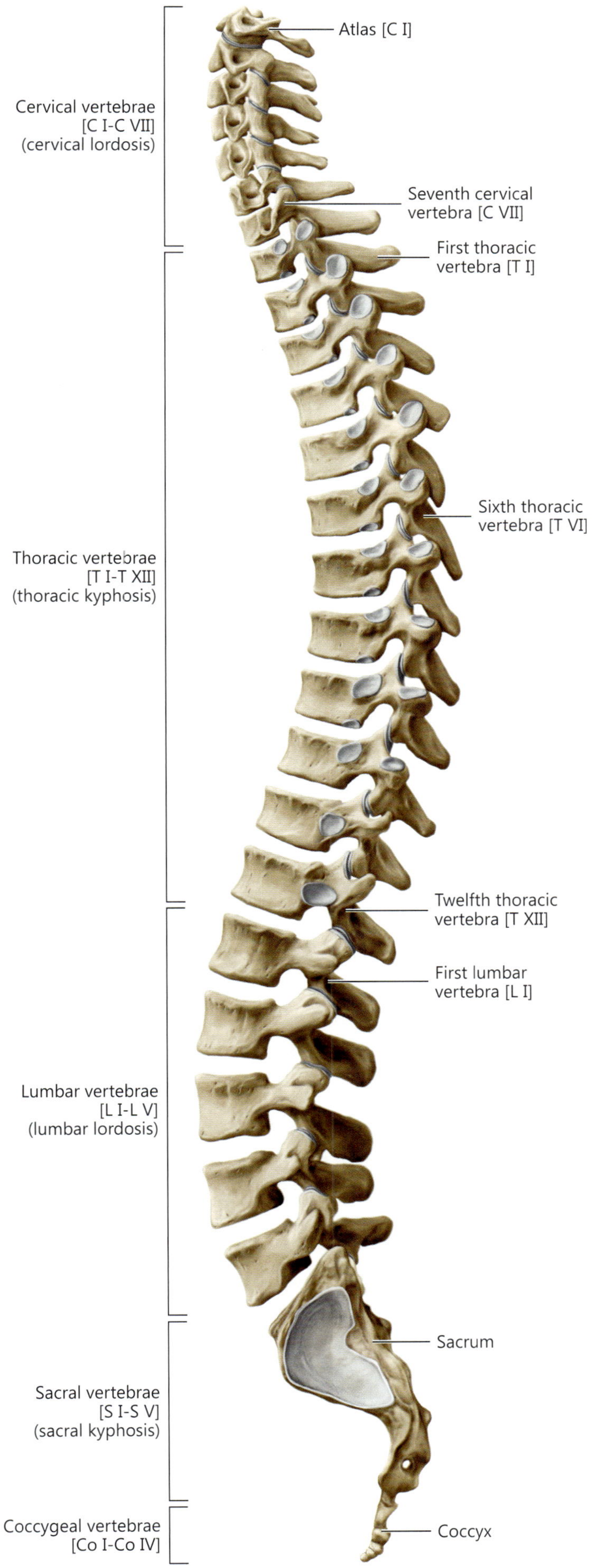

Preparation indicating the **vertebral column** in left lateral projection.

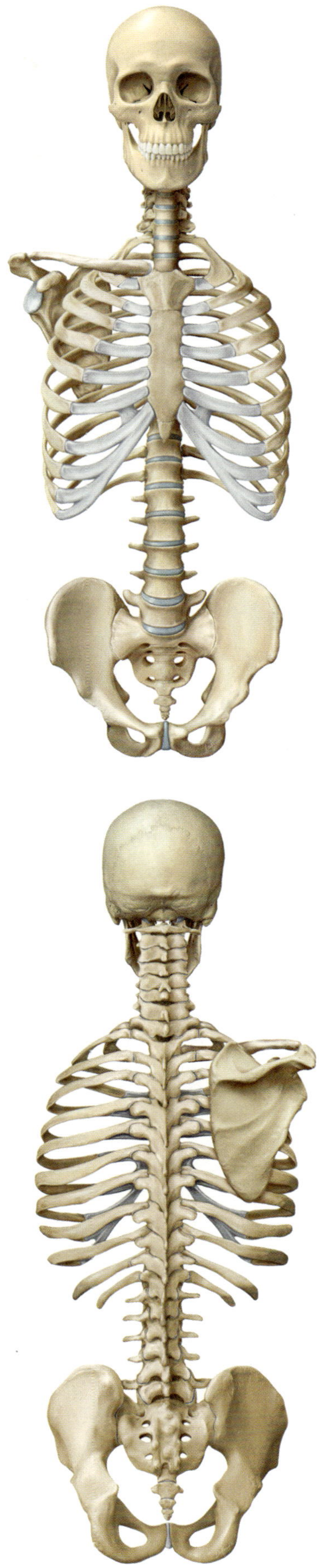

Skeleton of the head and trunk. Anterior and posterior projections. The bones of the right scapular girdle and the pelvic girdle are also present.

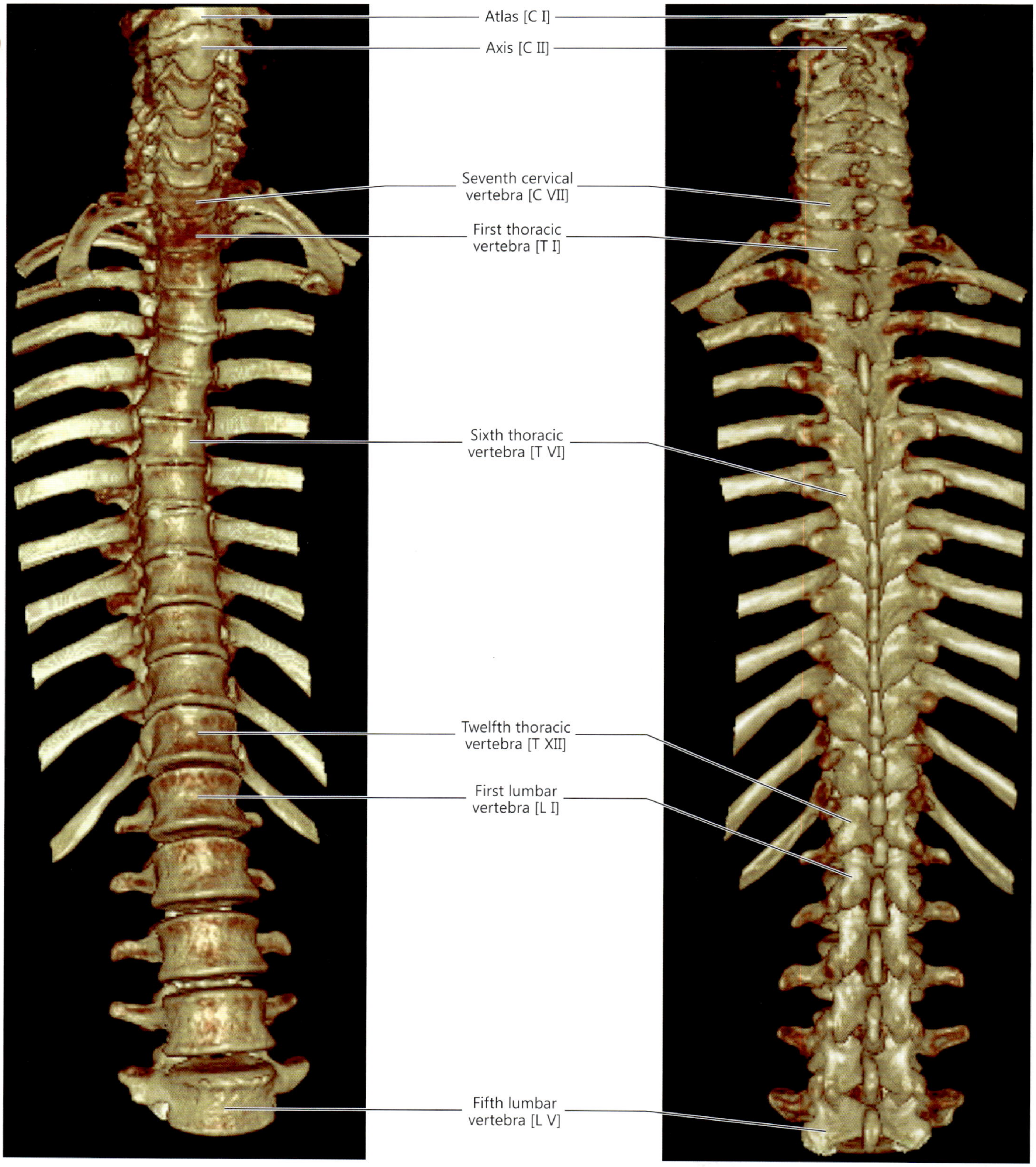

Three-dimensional images of a living human body, obtained using a 3D volume rendering technique, showing the cervicothoracolumbar portion of the **vertebral column** and part of the ribs in anterior and posterior projections.

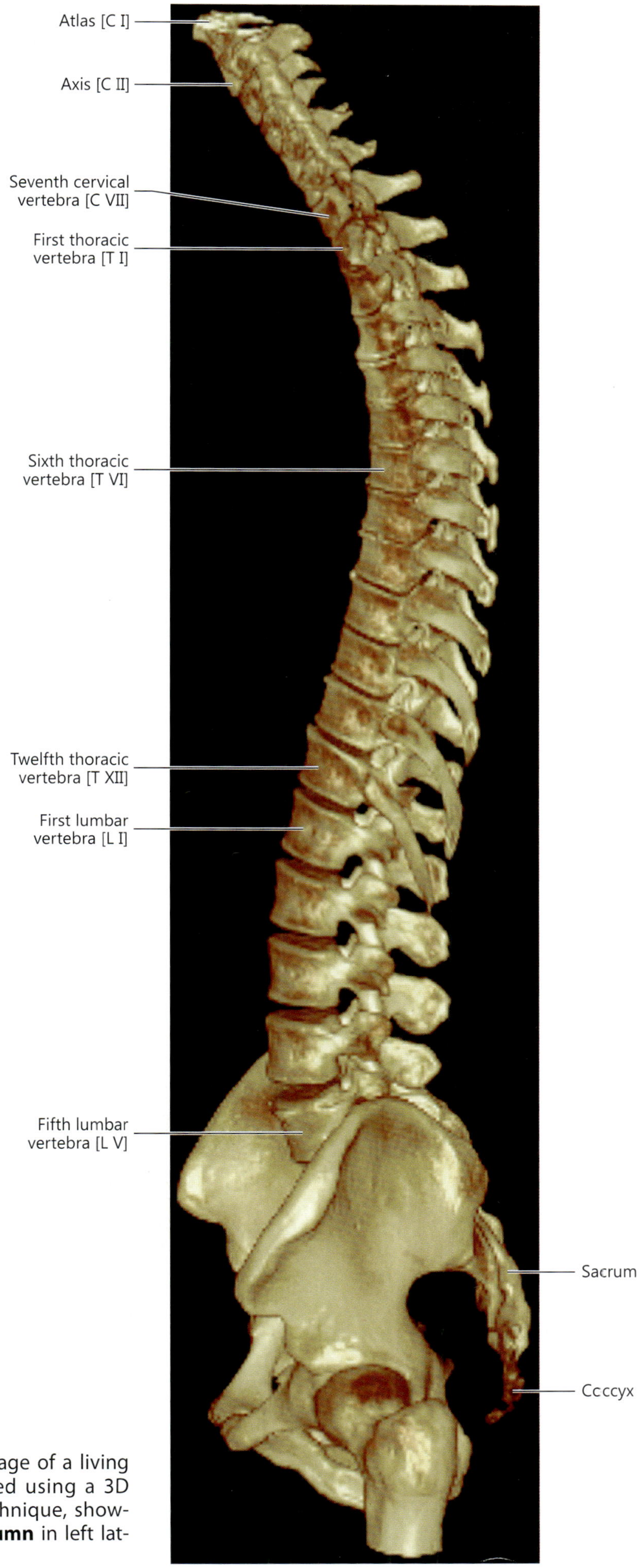

Three-dimensional image of a living human body, obtained using a 3D volume rendering technique, showing the **vertebral column** in left lateral projection.

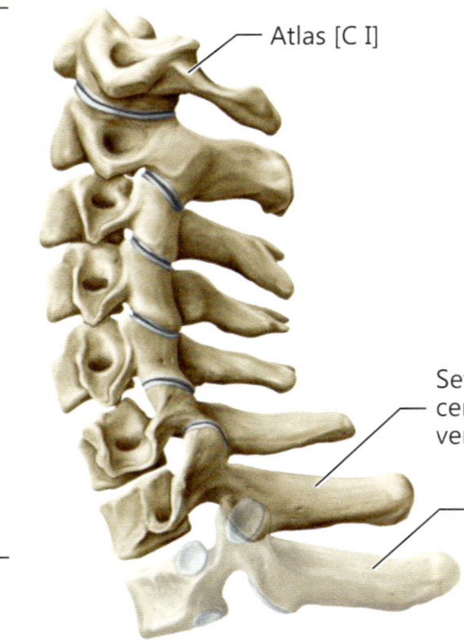

a

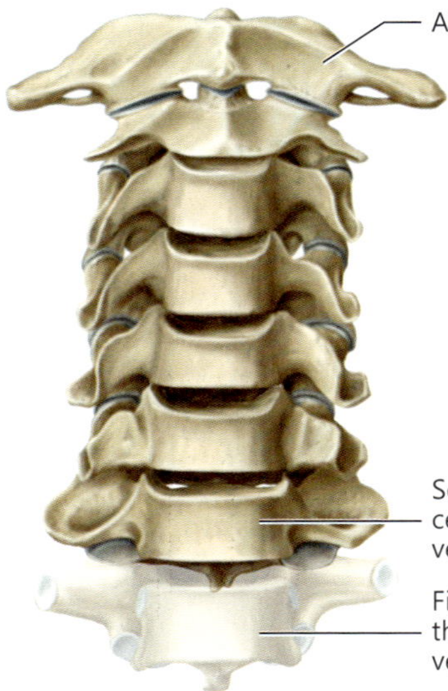

b

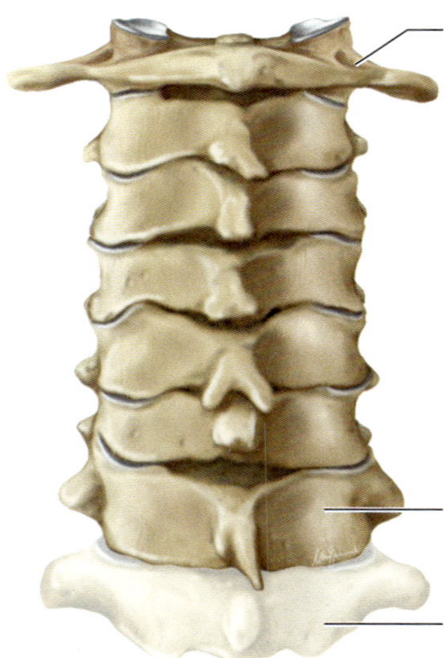

c

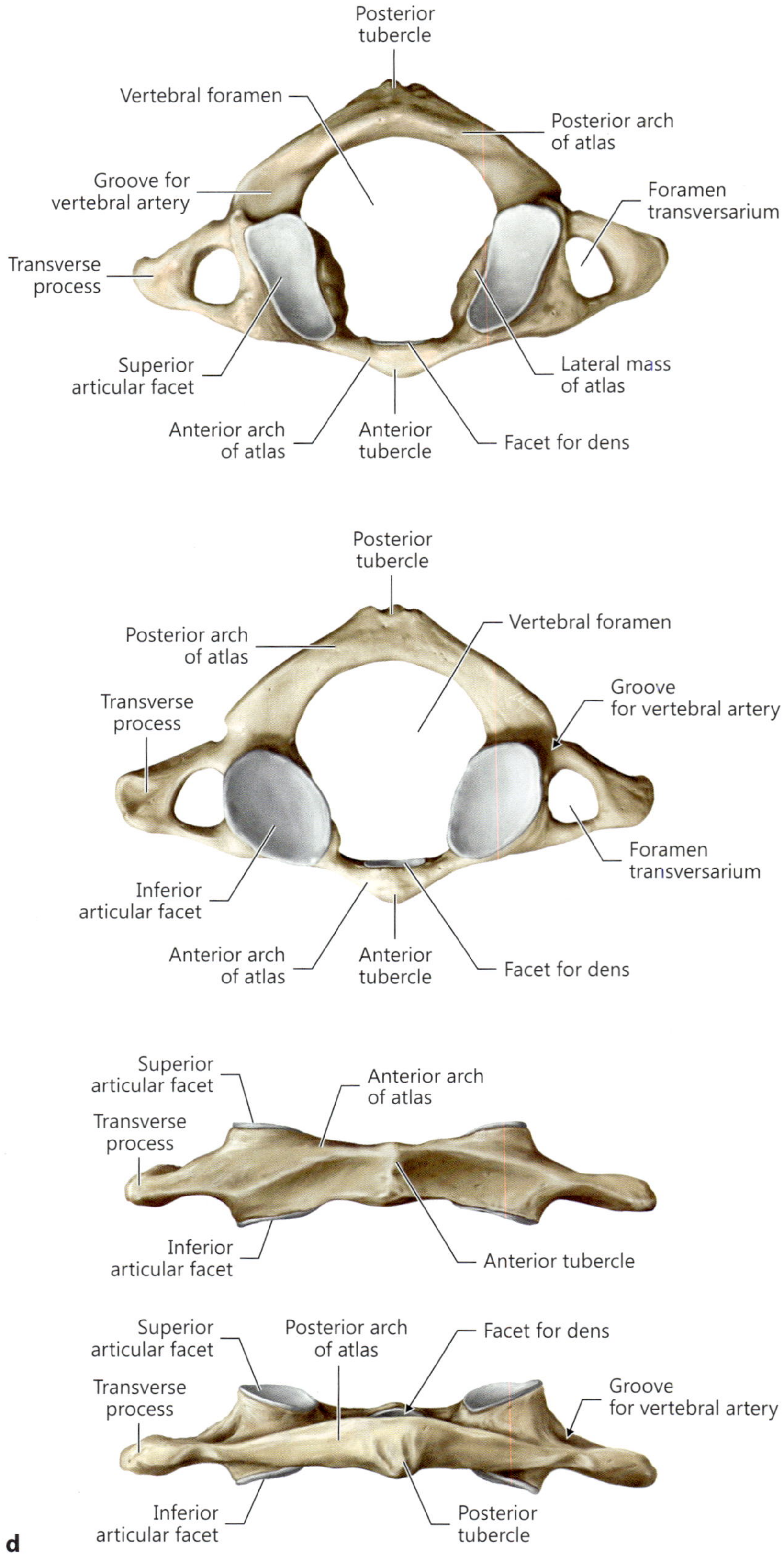

d

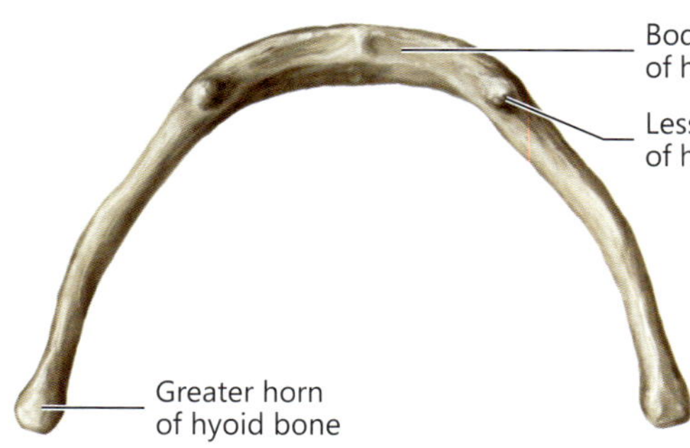

g

Cervical column: lateral (**a**), anterior (**b**), and posterior (**c**) views; **d**, atlas, superior, inferior, anterior and posterior surfaces; **e**, axis, left lateral surface, superior and anterior surfaces; **f**, third cervical vertebra, left lateral surface, superior and anterior surfaces; **g**, external surface of the **hyoid bone** in superior, lateral, posterior and anterior projections.

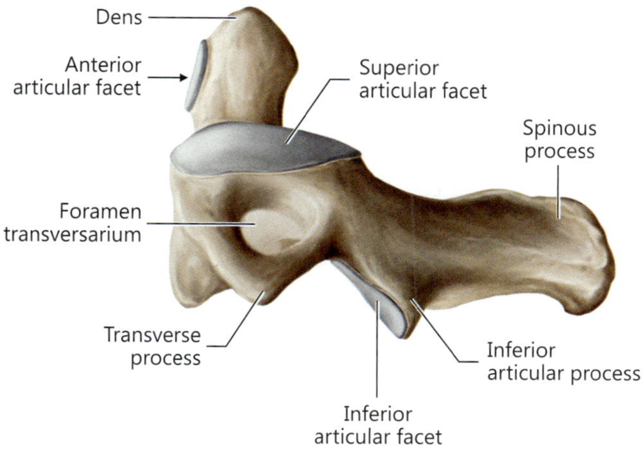
Dens
Anterior articular facet
Superior articular facet
Spinous process
Foramen transversarium
Transverse process
Inferior articular process
Inferior articular facet

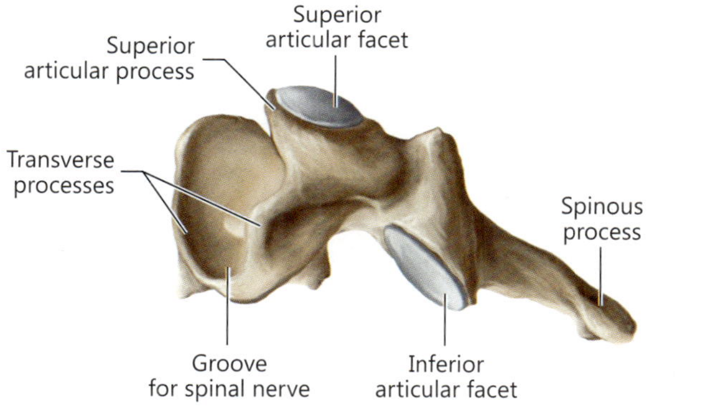
Superior articular process
Superior articular facet
Transverse processes
Spinous process
Groove for spinal nerve
Inferior articular facet

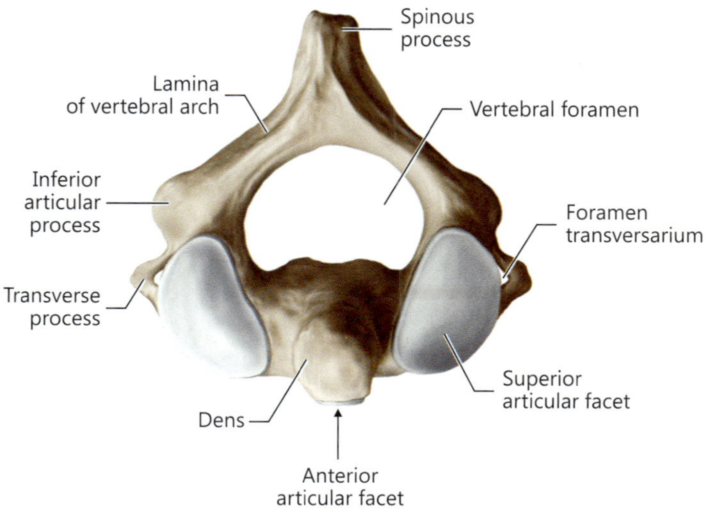
Spinous process
Lamina of vertebral arch
Vertebral foramen
Inferior articular process
Foramen transversarium
Transverse process
Superior articular facet
Dens
Anterior articular facet

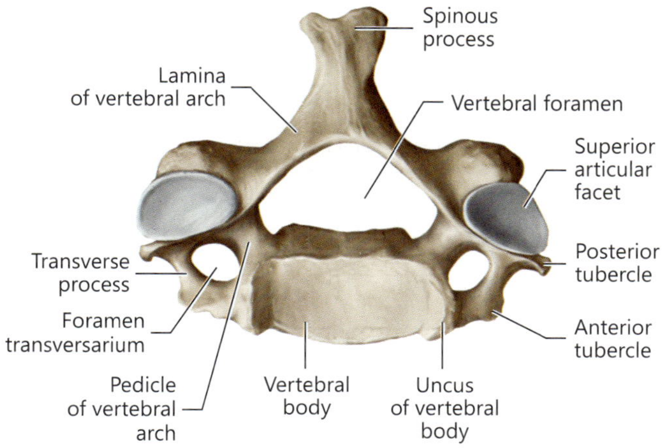
Spinous process
Lamina of vertebral arch
Vertebral foramen
Superior articular facet
Posterior tubercle
Transverse process
Anterior tubercle
Foramen transversarium
Pedicle of vertebral arch
Vertebral body
Uncus of vertebral body

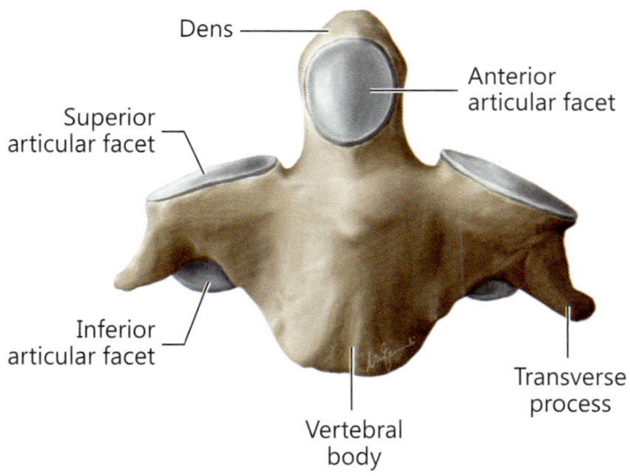
Dens
Anterior articular facet
Superior articular facet
Inferior articular facet
Transverse process
Vertebral body

e

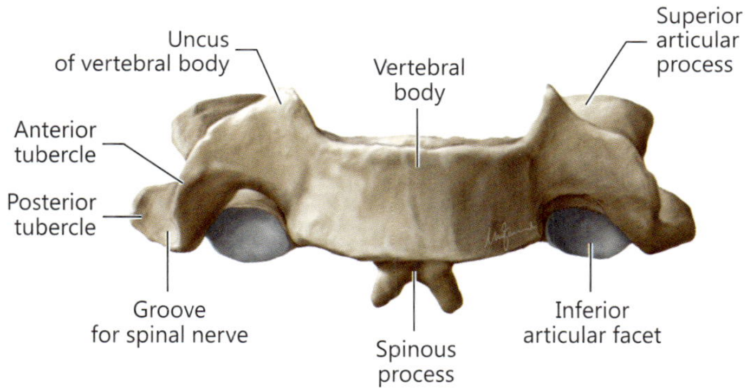
Uncus of vertebral body
Vertebral body
Superior articular process
Anterior tubercle
Posterior tubercle
Groove for spinal nerve
Spinous process
Inferior articular facet

f

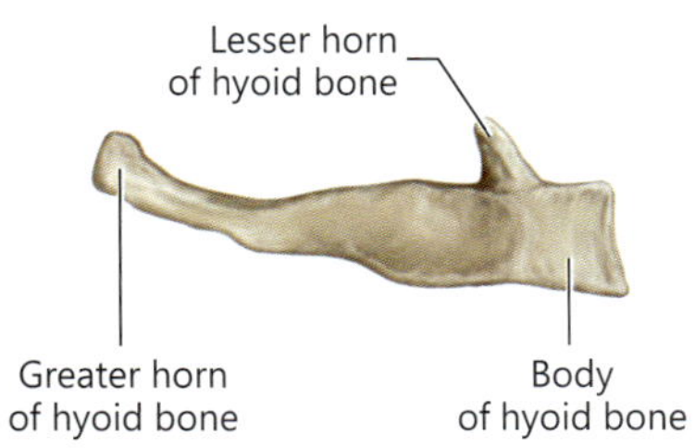
Lesser horn of hyoid bone
Greater horn of hyoid bone
Body of hyoid bone

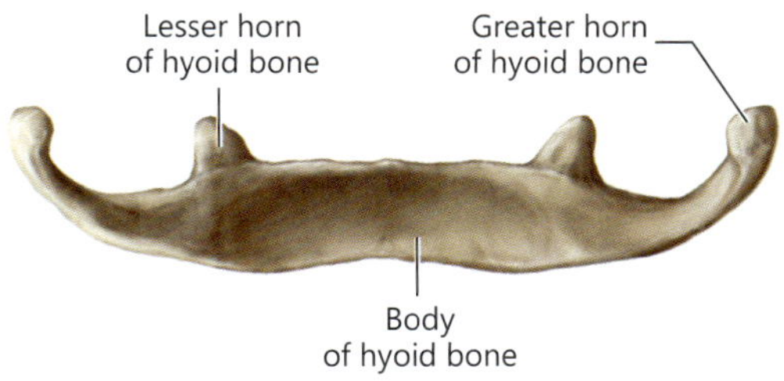
Lesser horn of hyoid bone
Greater horn of hyoid bone
Body of hyoid bone

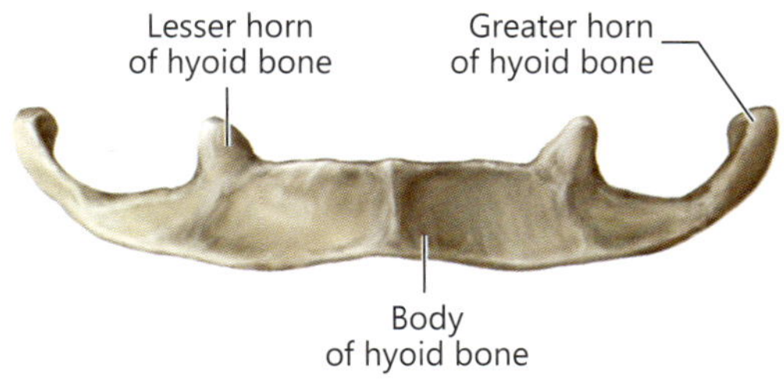
Lesser horn of hyoid bone
Greater horn of hyoid bone
Body of hyoid bone

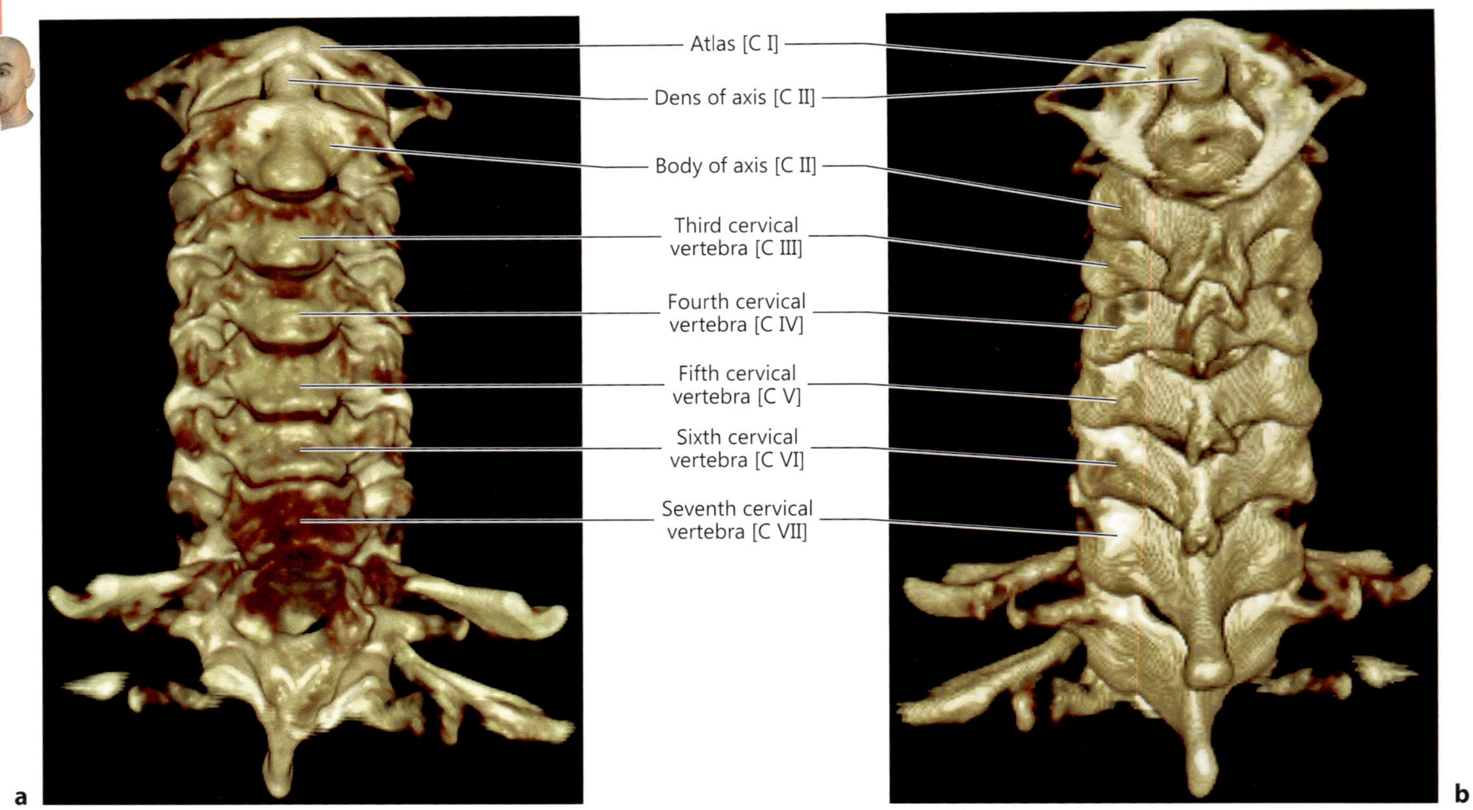

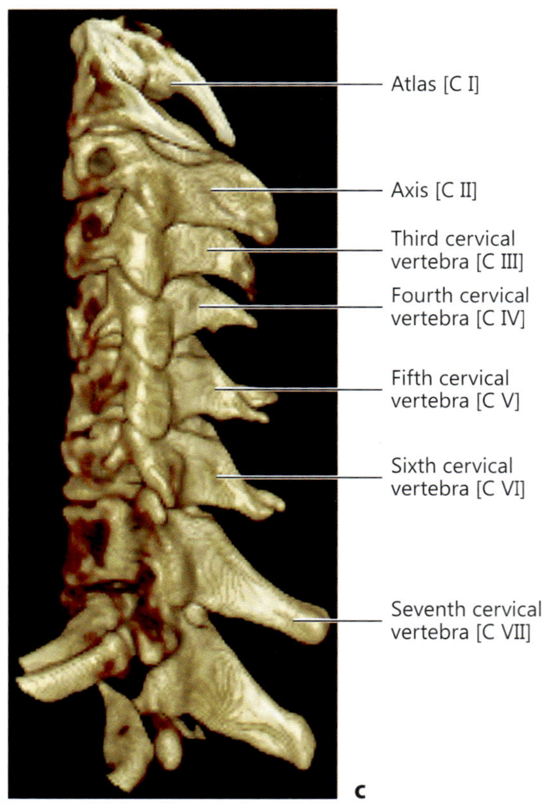

Three-dimensional images of a living human body, obtained using a 3D volume rendering technique, showing the **cervical column** in anterior (**a**), posterior (**b**), and left lateral (**c**) projection.

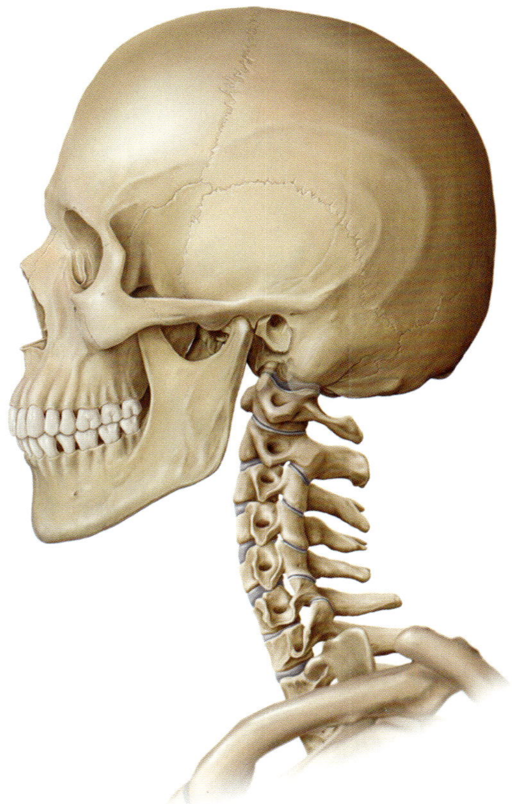

Head and **cervical column** in left lateral projection. In all the articular areas the color of the cranial bones has been graphically attenuated.

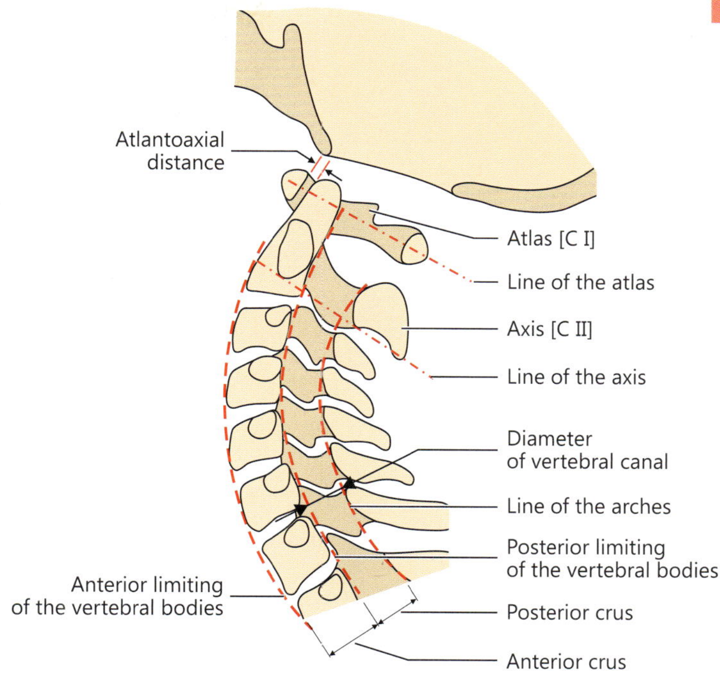

Cervical column in left lateral projection.

- The anterior limiting of the vertebral bodies, the posterior limiting of the vertebral bodies and the line passing through the vertebral arches form parallel arches
- The distance between the last two lines is 15-20 mm
- The horizontal line passing through the atlas and the one between the inferior margin of the vertebral arch and the spinous process of axis are parallel.

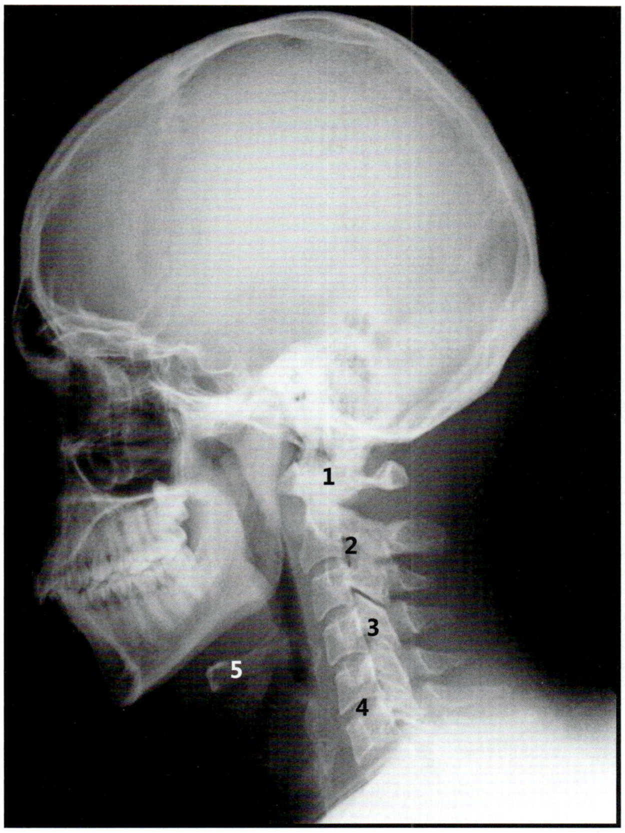

Radiography of the **head** and **cervical column** in left lateral projection. 1, atlas; 2, axis, 3, intervertebral foramen; 4, intersomatic space; 5, hyoid bone.

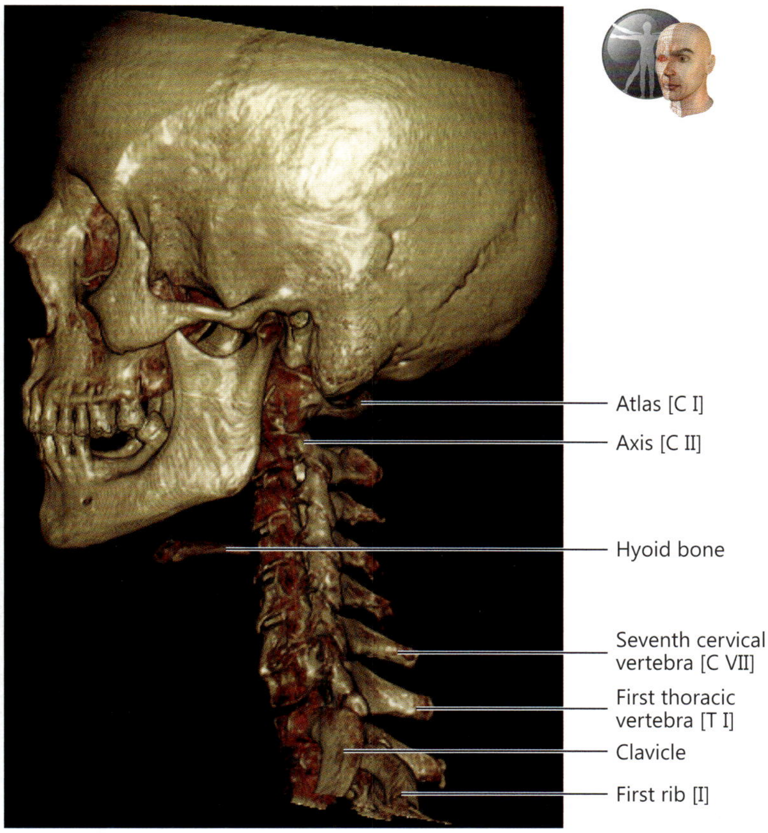

Three-dimensional image of a living human body, obtained using a 3D volume rendering technique, showing the **head** and **cervical column** in left lateral projection.

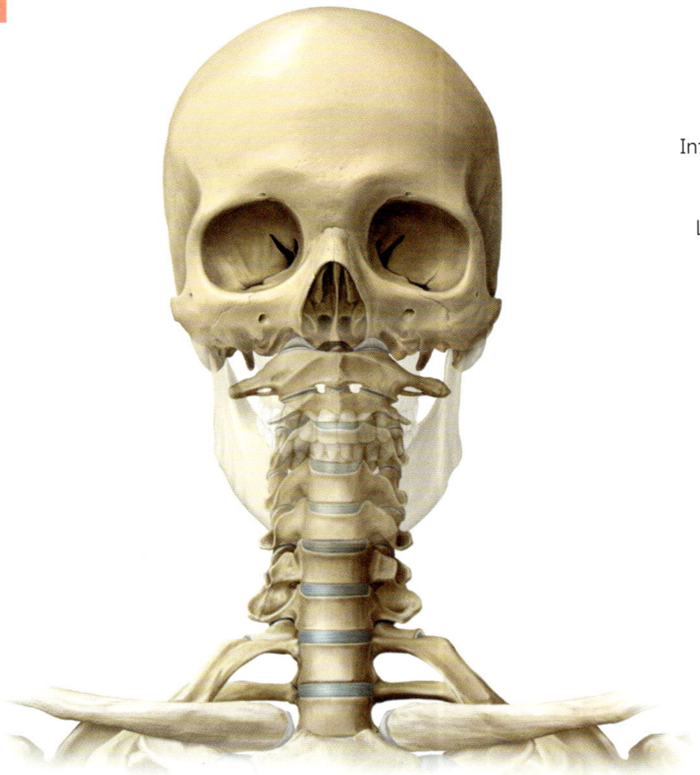

Head and **cervical column** in anterior projection. Mandible and maxilla are colorless to show the bones of the vertebral column.

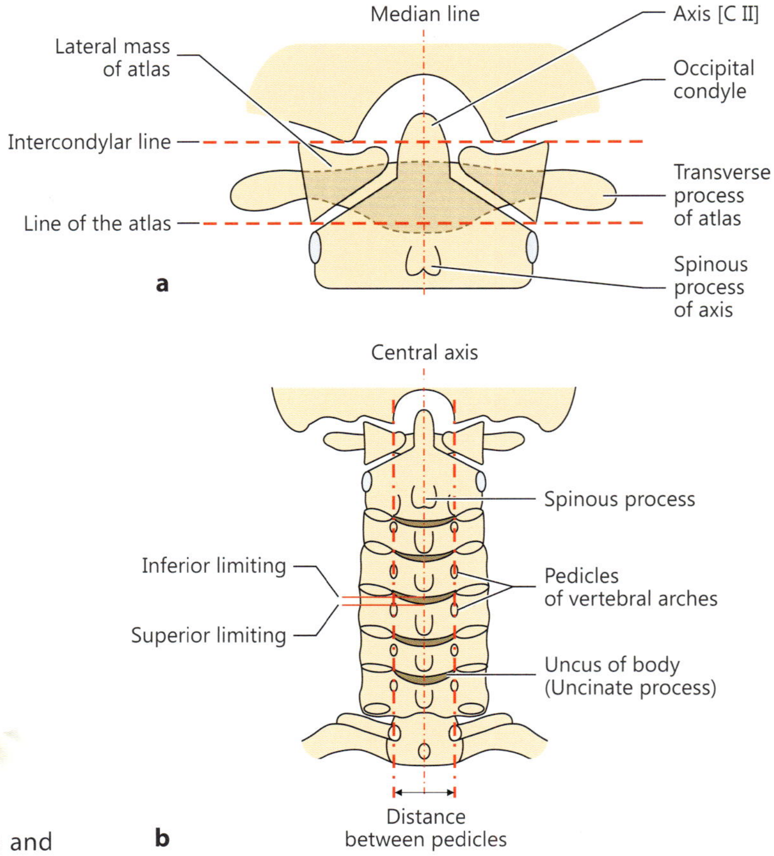

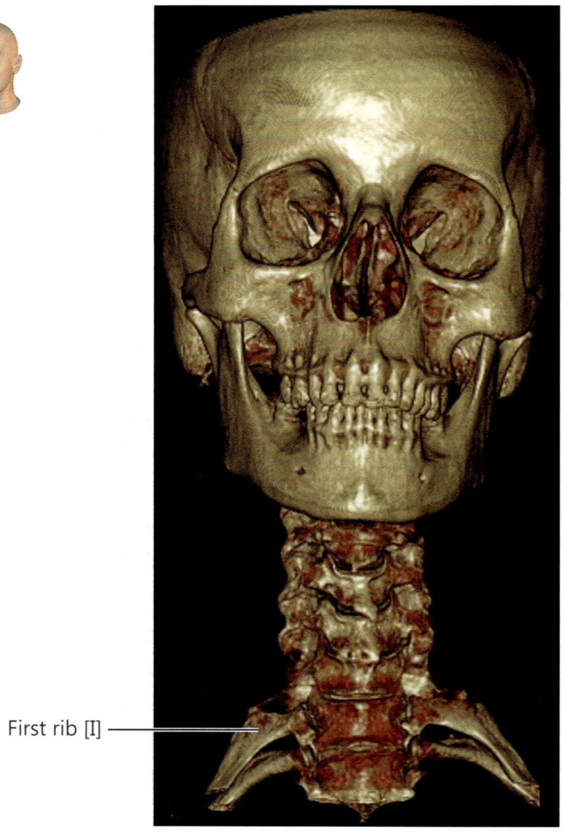

Three-dimensional image of a living human body, obtained using a 3D volume rendering technique, showing the **head** and **cervical column** in frontal projection.

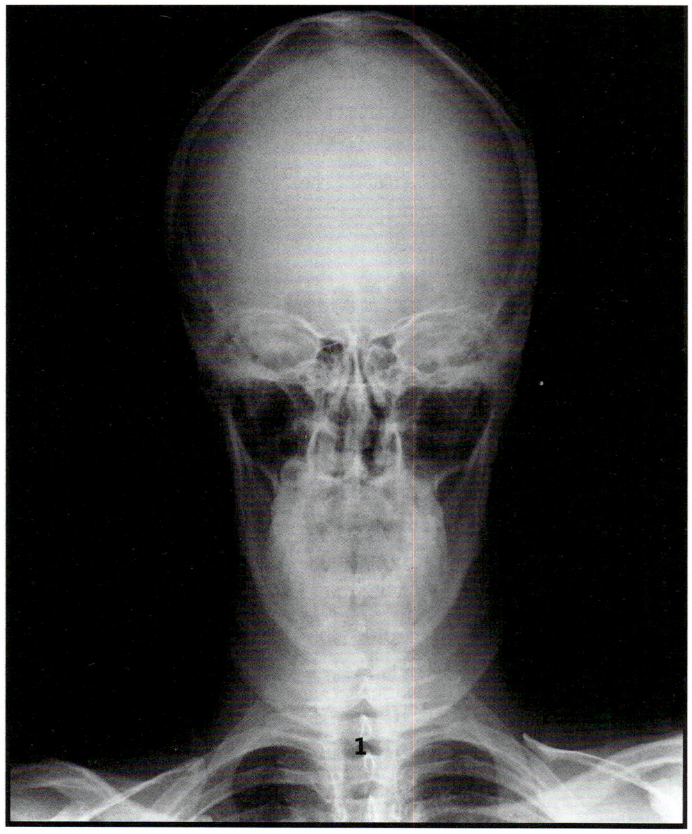

Radiography of the **head** and **cervical column** in anterior projection. 1, hyperlucency of intersomatic spaces.

➤
a, Anteroposterior projection of the **upper cervical column**. **b**, Anteroposterior projection of the **lower cervical column**.
Radiology of the cervical column
Anteroposterior projection

- The dens and the spinous process of the axis are placed on the median line
- The occipital condyles and the lateral masses of the atlas are parallel and symmetrical
- The transverse processes of the atlas have the same length and are located at an equal distance from the occipital bone
- The line passing through the lower limits of the occipital condyles (intercondylar line) is parallel with the line passing through the lower margins of the lateral masses of the atlas (line of the atlas)
- The distance between the atlas and axis is 3 mm
- The inferior and superior limiting of the vertebral bodies are parallel
- The spinous processes are aligned along the central axis
- The pedicles of vertebral arches are aligned and arranged symmetrically at equal distances from the median line
- The distance between the pedicles is 24-33 mm.

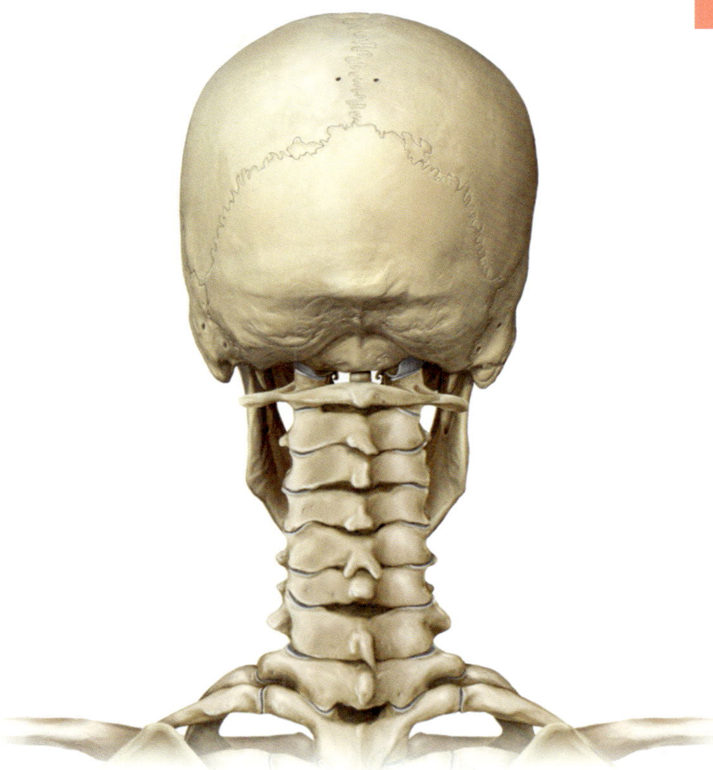

Head and **cervical column** in posterior projection.

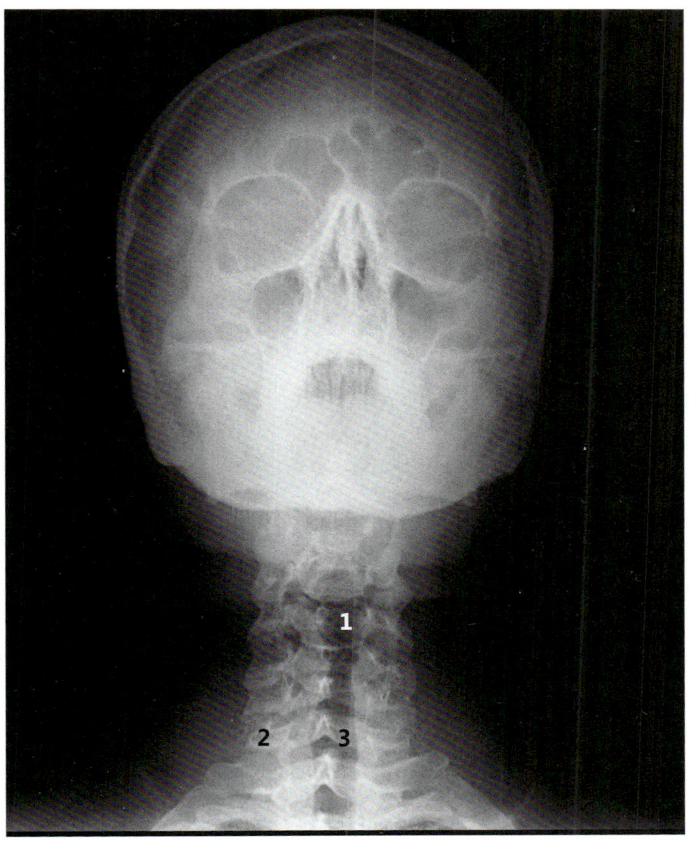

Radiography of the **head** and **cervical column** in posterior projection. 1, laryngotracheal air column; 2, vertebral body; 3, spinous process.

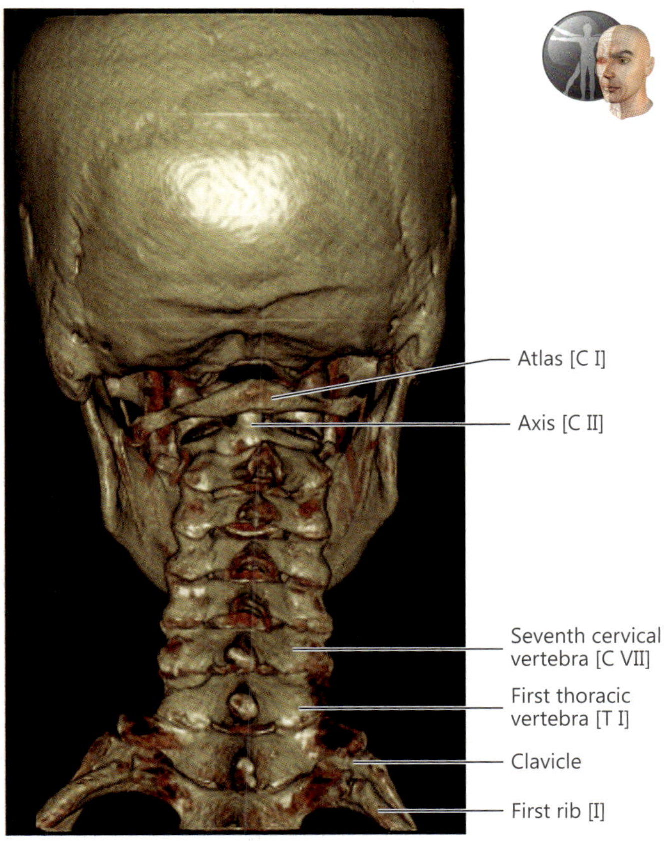

Three-dimensional image of a living human body, obtained using a 3D volume rendering technique, showing the **head** and **cervical column** in posterior projection.

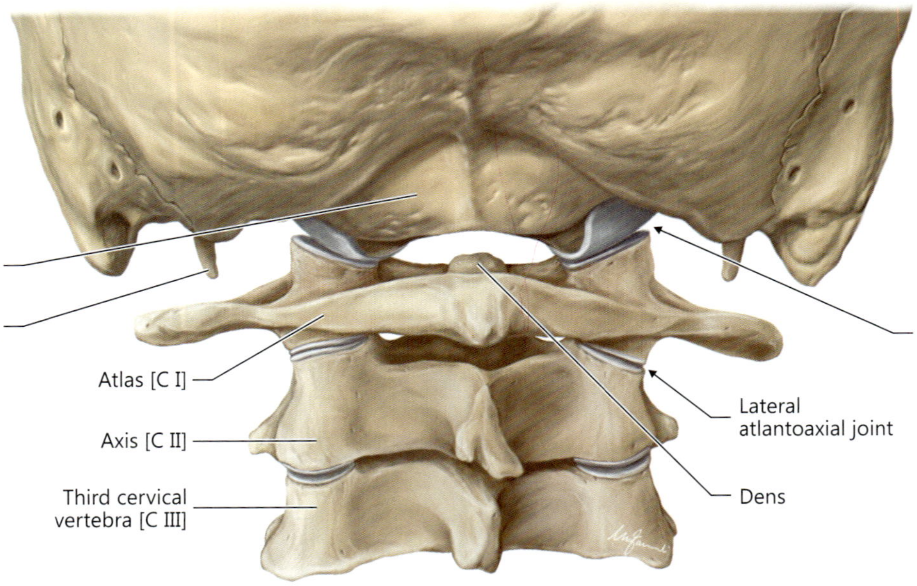

Atlanto-occipital and **atlantoaxial joints**, posterior view.

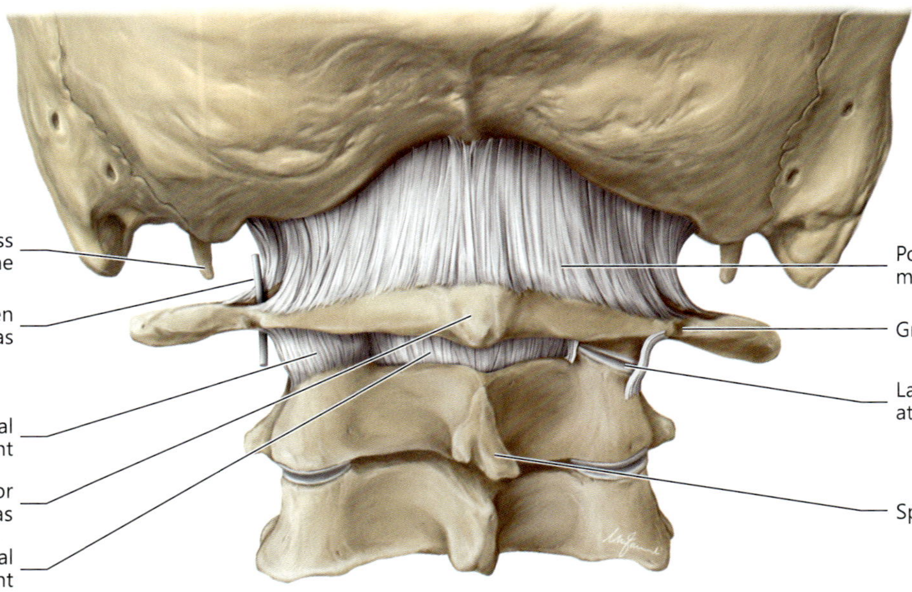

Atlanto-occipital and **atlantoaxial joints**, posterior projection.

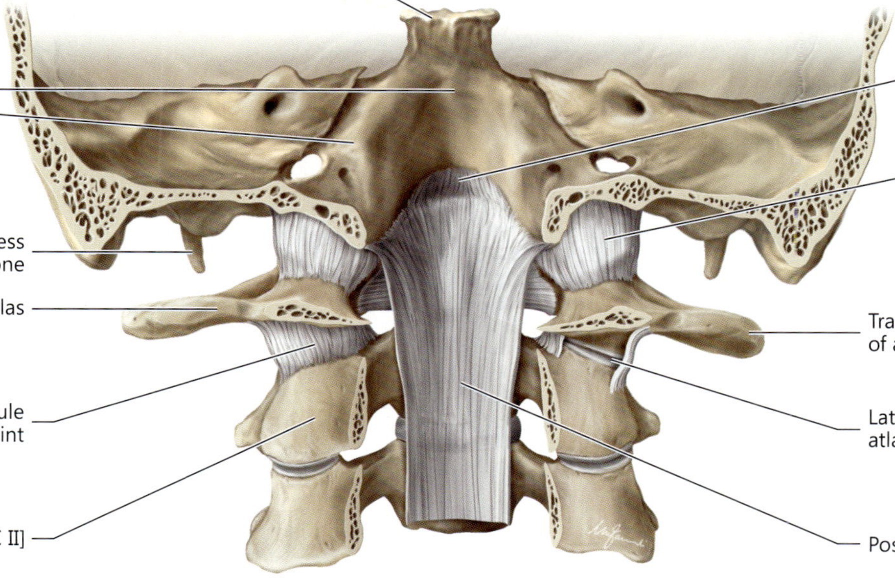

Atlanto-occipital and **atlantoaxial joints** in posterior projection after demolition of the posterior part of the occipital bone and superior cervical arches; the vertebral canal has been opened.

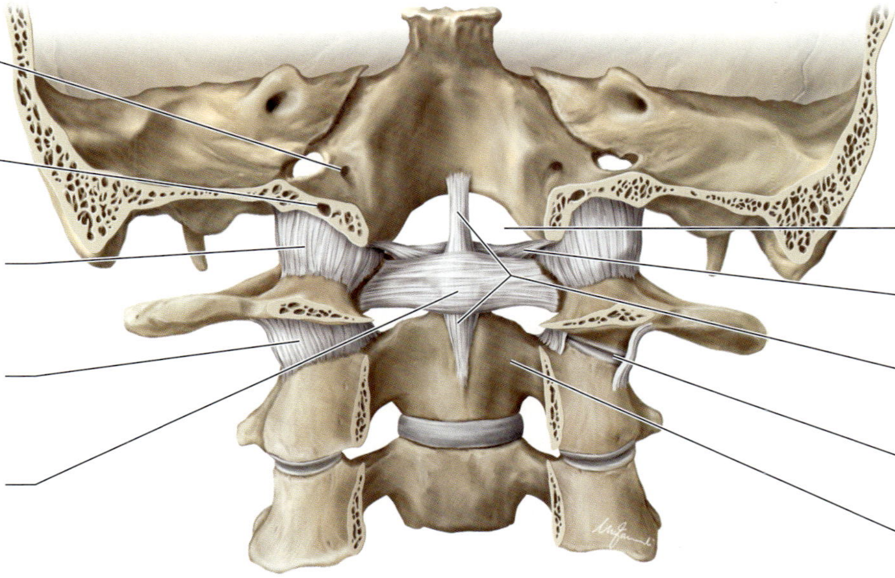

Atlanto-occipital and **atlantoaxial joints** in posterior projection, after posterior longitudinal ligament removal.

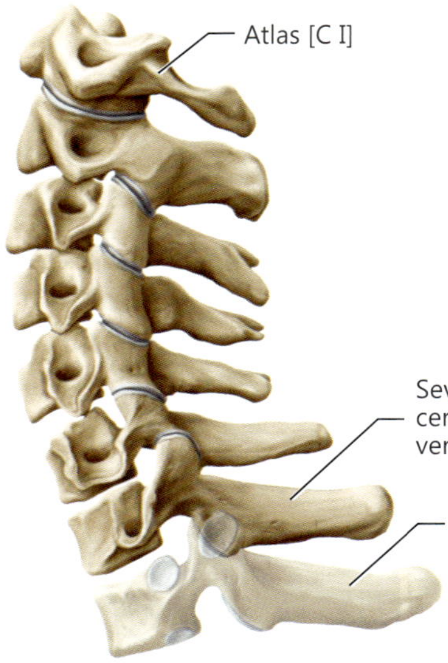

Cervical vertebrae [C I - C VII], lateral view.

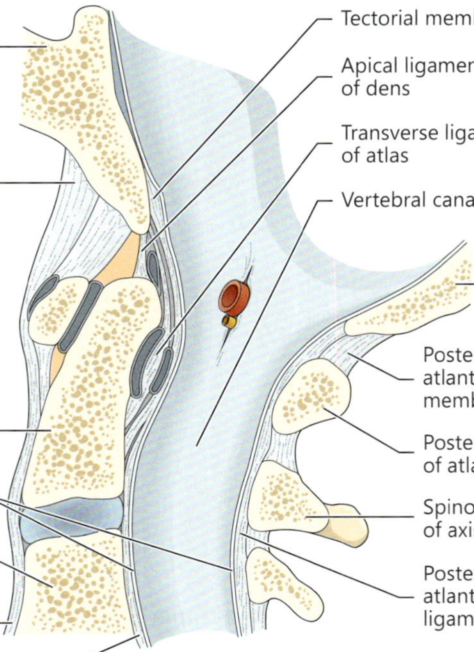

Cervical column ligaments, sagittal view.

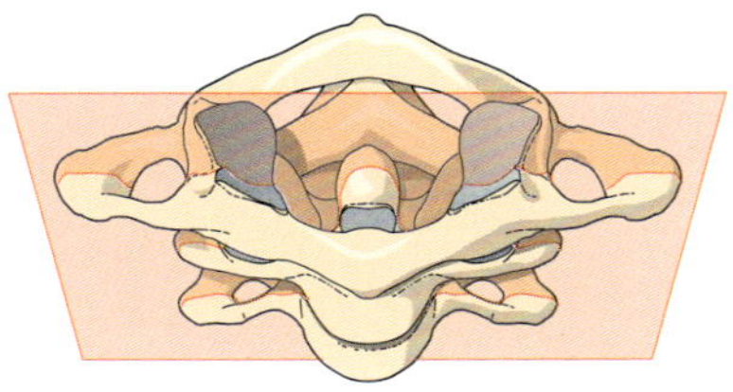

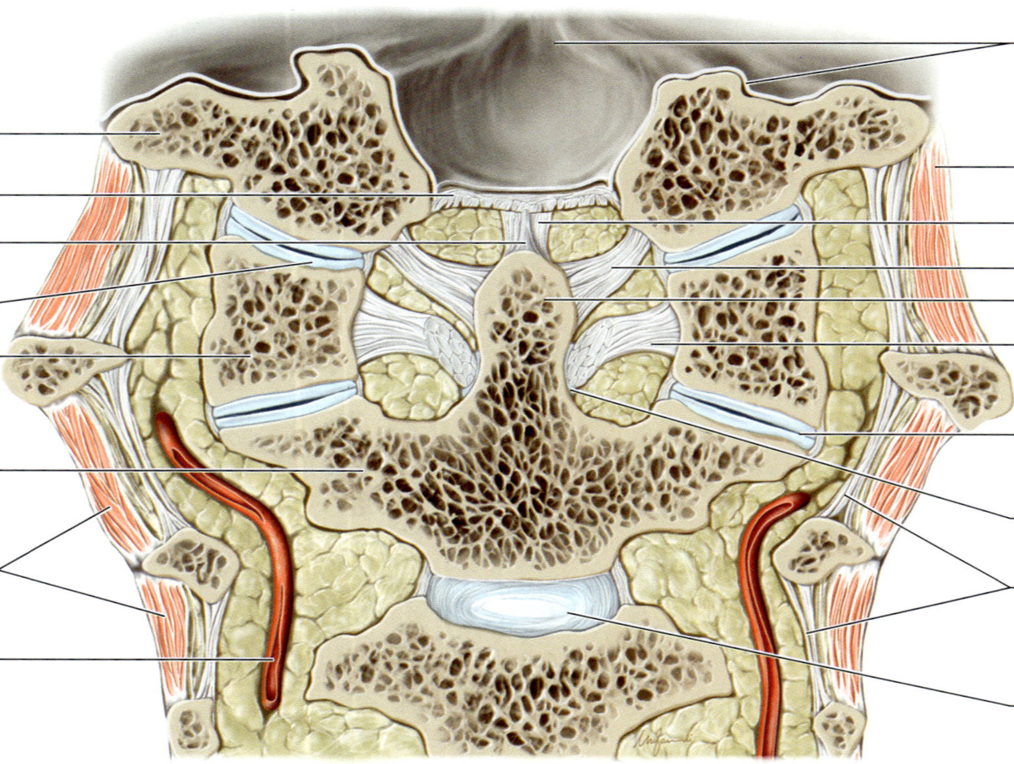

Joints and means of union between **occipital bone**, **atlas**, and **axis**, in a frontal section passing through the foramen magnum and the axis.

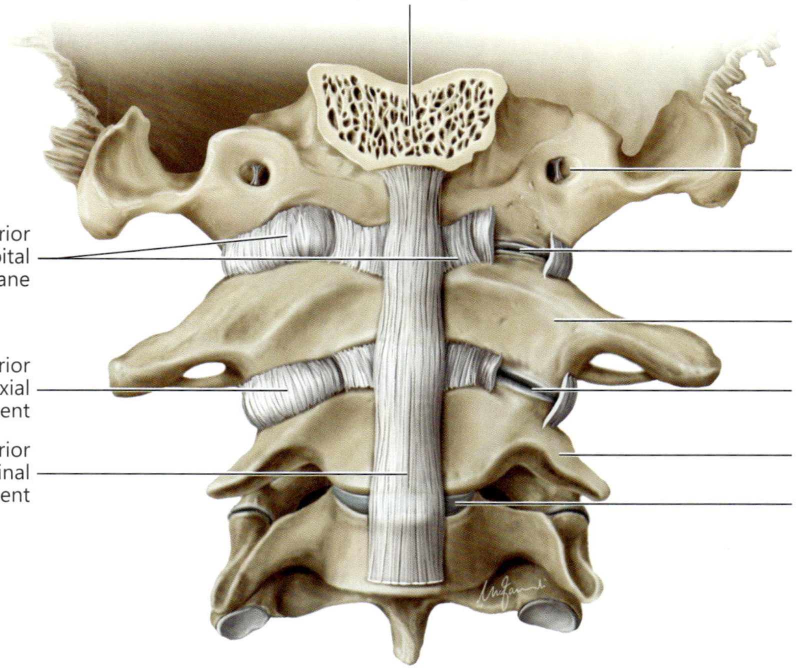

Atlanto-occipital and **atlantoaxial joints** in anterior projection.

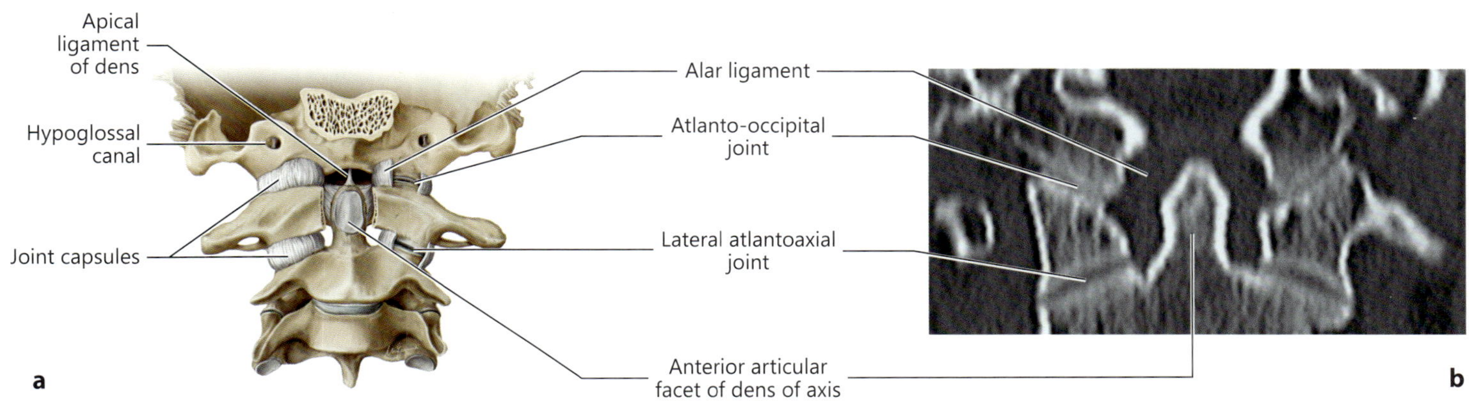

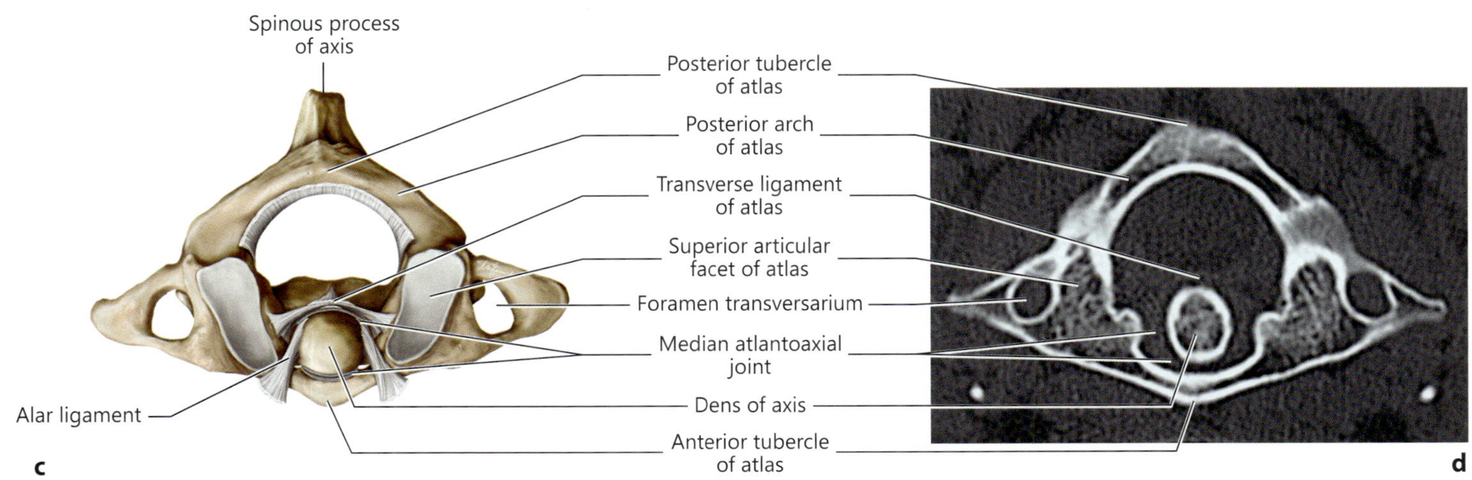

Atlanto-occipital and **atlantoaxial joints**. **a**, Anterior projection after demolition of anterior tubercle of atlas; **b**, computed tomography: coronal plane electronic reformatting in correspondence with the dens of axis; **c**, superior projection of the median atlantoaxial joint; **d**, computed tomography: axial plane electronic reformatting at the level of the atlanto-axial joint.

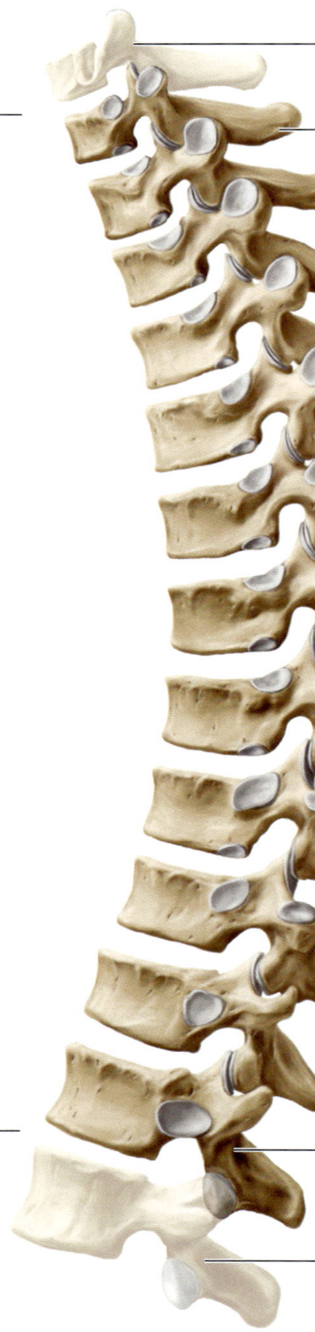

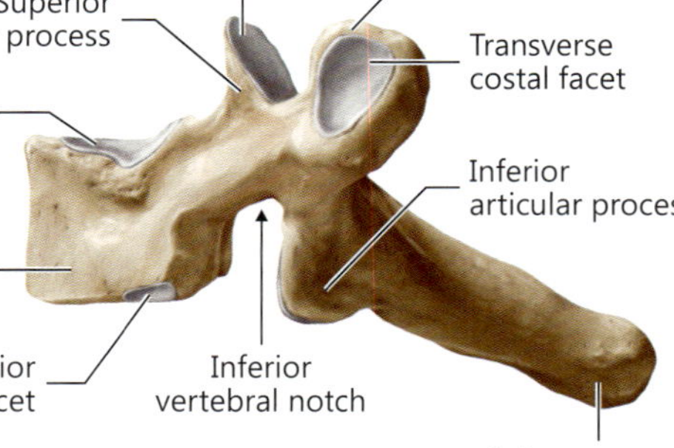

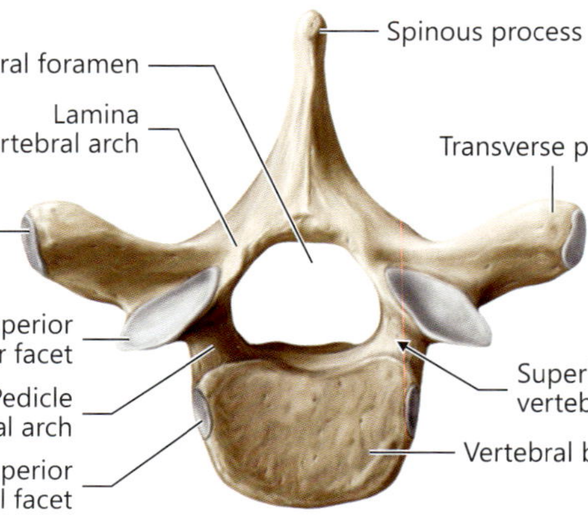

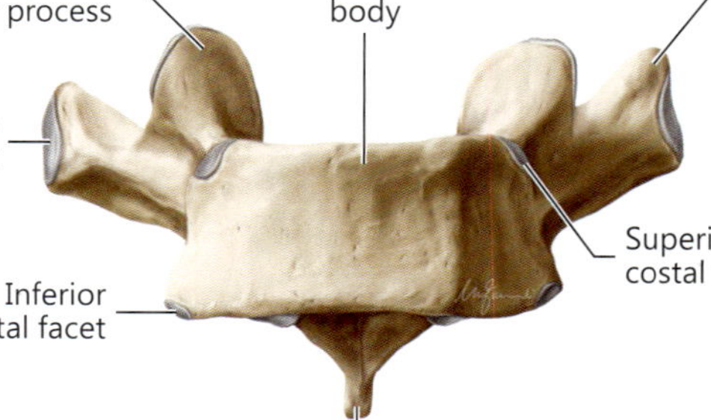

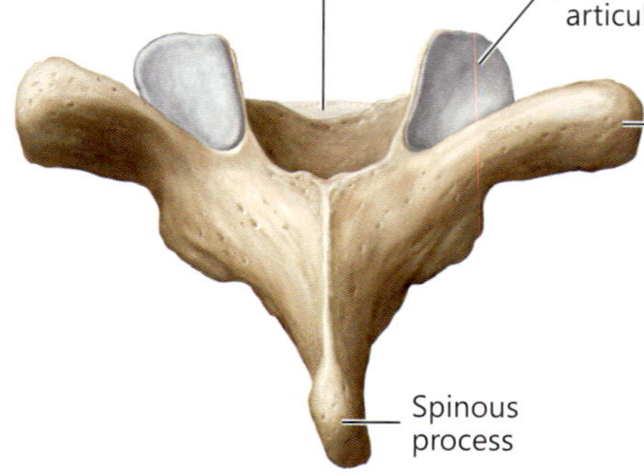

Thoracic column. Left lateral, superior, anterior and posterior views of the following vertebrae: **a**, second thoracic vertebra; **b**, sixth thoracic vertebra; **c**, twelfth thoracic vertebra.

a

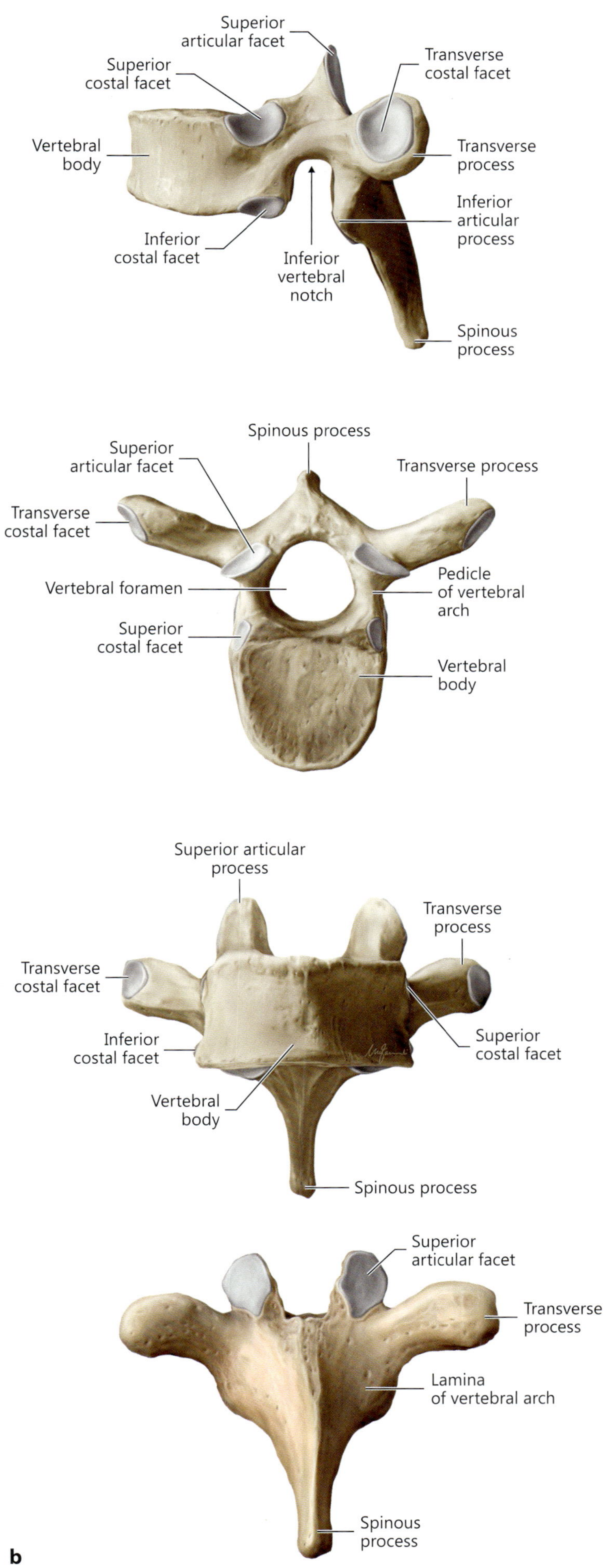

Superior articular facet
Superior costal facet
Transverse costal facet
Vertebral body
Transverse process
Inferior articular process
Inferior costal facet
Inferior vertebral notch
Spinous process
Spinous process
Superior articular facet
Transverse process
Transverse costal facet
Vertebral foramen
Pedicle of vertebral arch
Superior costal facet
Vertebral body
Superior articular process
Transverse process
Transverse costal facet
Inferior costal facet
Superior costal facet
Vertebral body
Spinous process
Superior articular facet
Transverse process
Lamina of vertebral arch
Spinous process

b

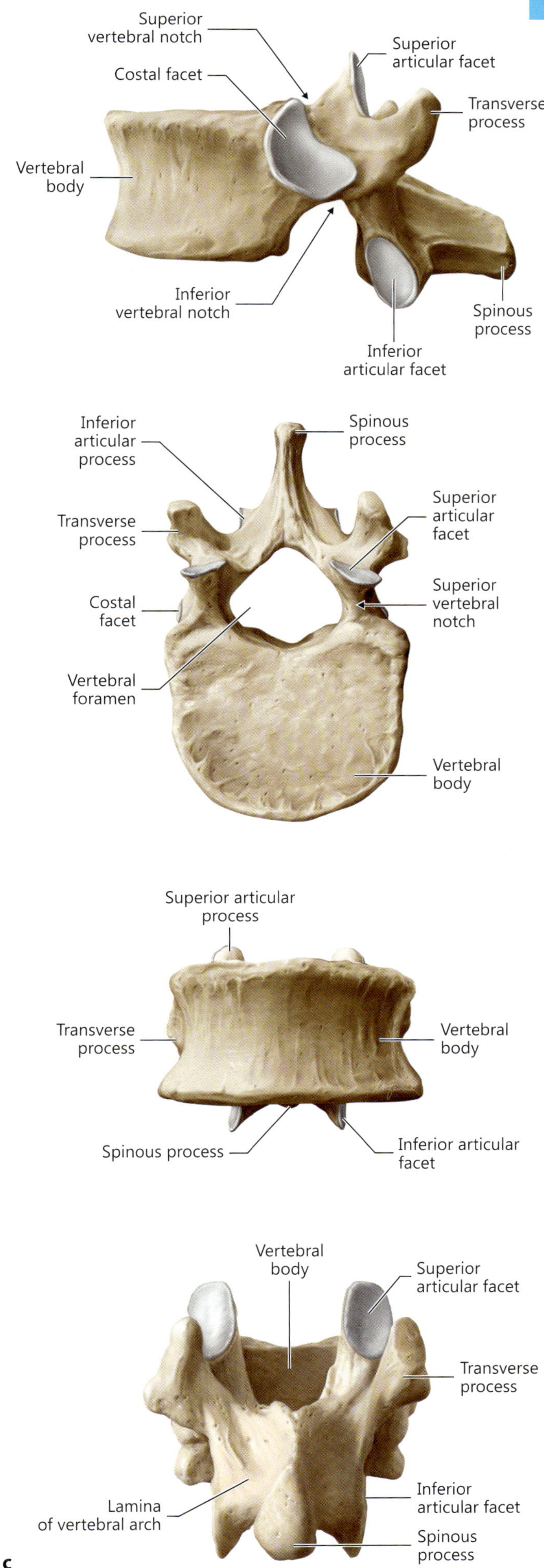

Superior vertebral notch
Superior articular facet
Costal facet
Transverse process
Vertebral body
Inferior vertebral notch
Spinous process
Inferior articular facet
Inferior articular process
Spinous process
Superior articular facet
Transverse process
Superior vertebral notch
Costal facet
Vertebral foramen
Vertebral body
Superior articular process
Transverse process
Vertebral body
Spinous process
Inferior articular facet
Vertebral body
Superior articular facet
Transverse process
Inferior articular facet
Lamina of vertebral arch
Spinous process

c

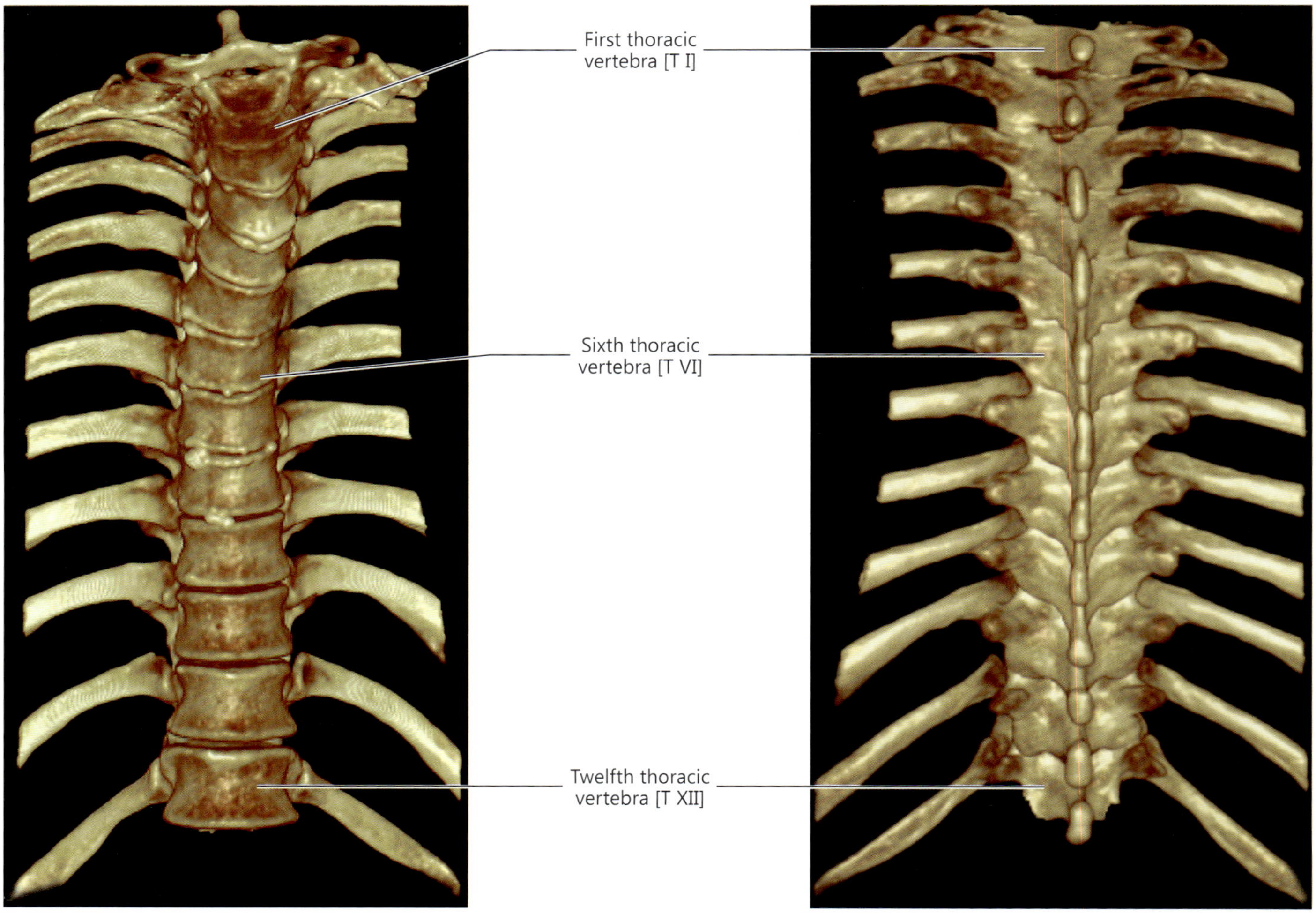

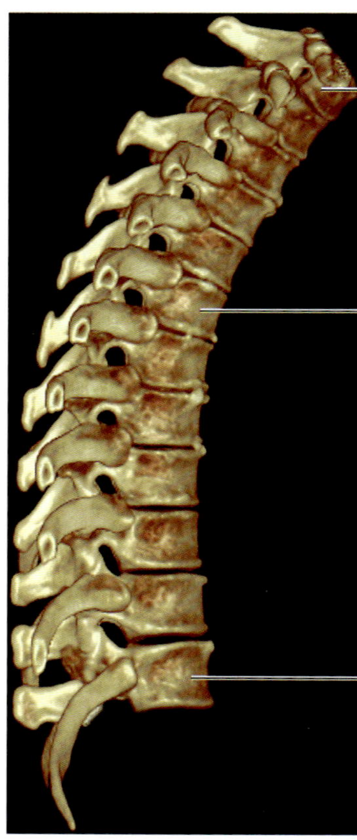

Three-dimensional images of a living human body, obtained using a 3D volume rendering technique, showing the **thoracic column** and part of the ribs in anterior (**a**), posterior (**b**), and right lateral (**c**) projection.

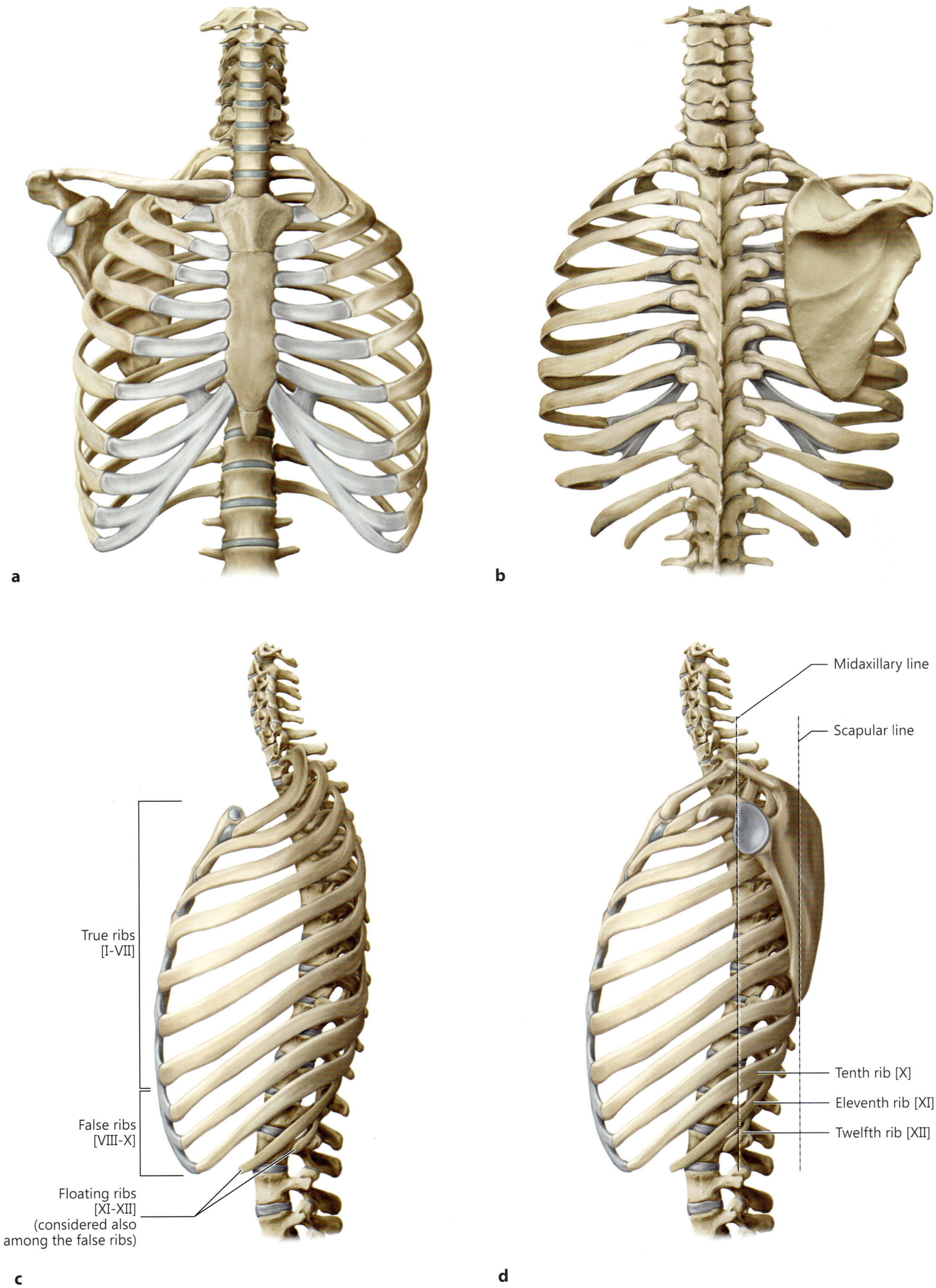

Thoracic cage in **a**, anterior; **b**, posterior; **c** and **d**, left lateral projections.

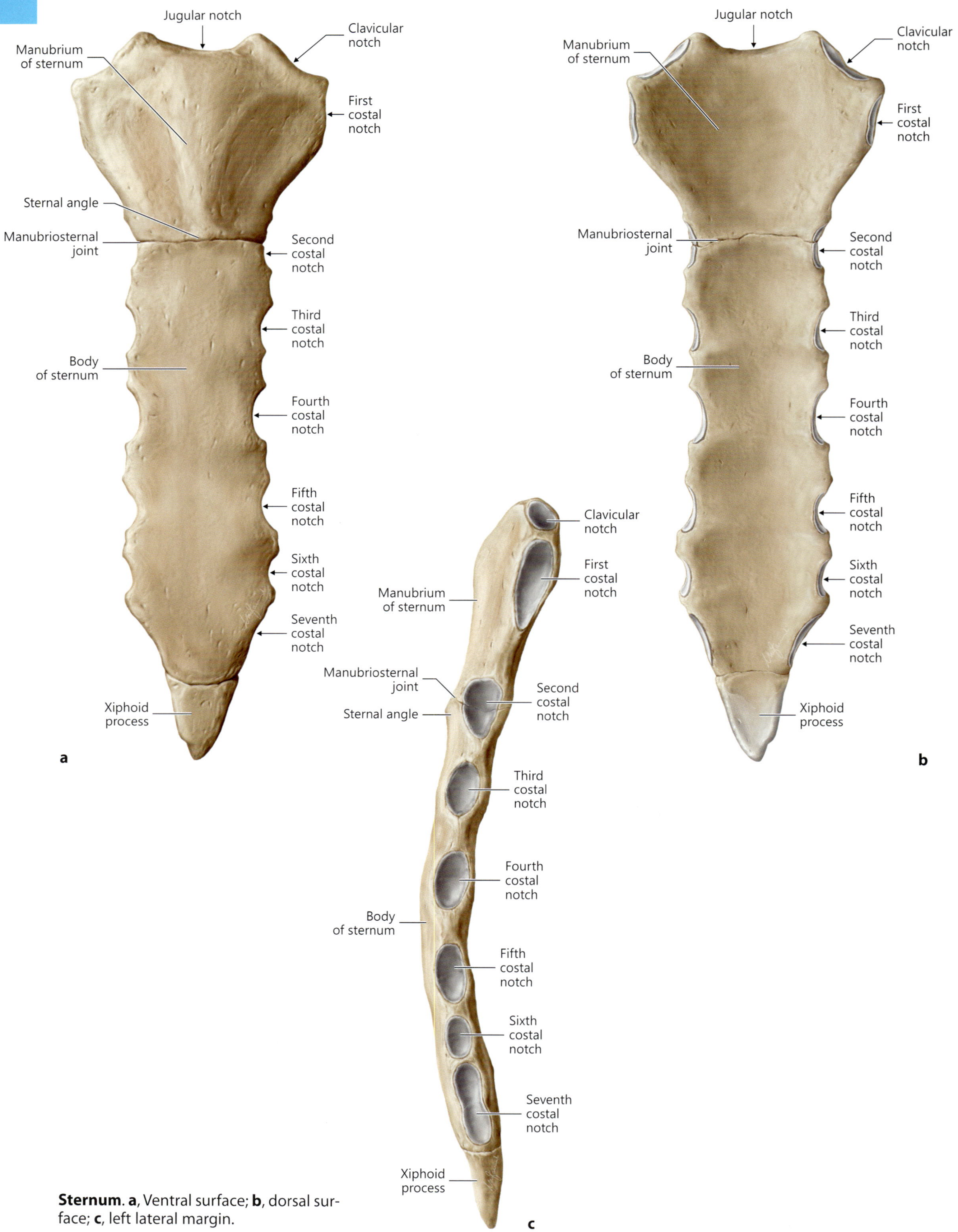

Sternum. a, Ventral surface; **b**, dorsal surface; **c**, left lateral margin.

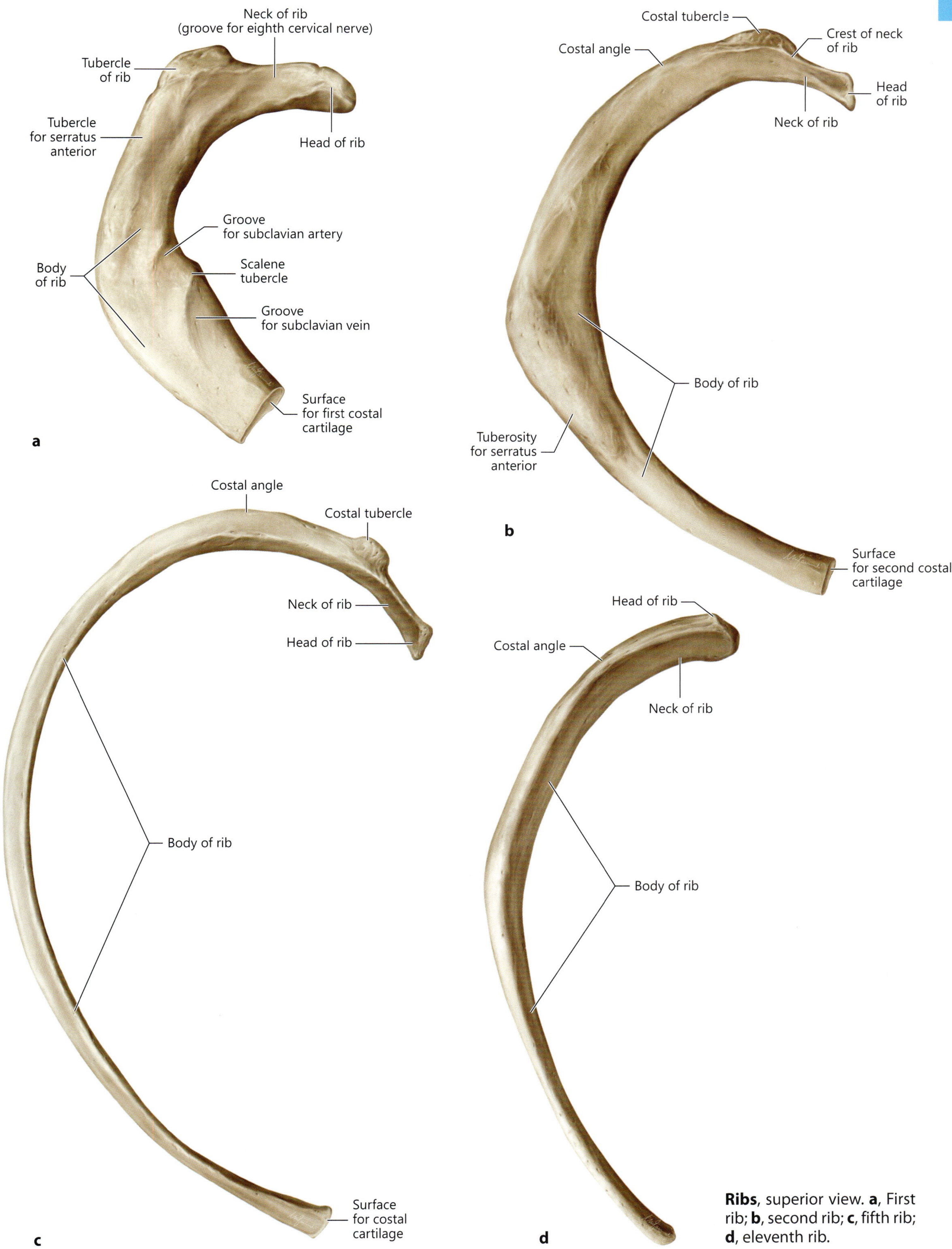

Ribs, superior view. **a**, First rib; **b**, second rib; **c**, fifth rib; **d**, eleventh rib.

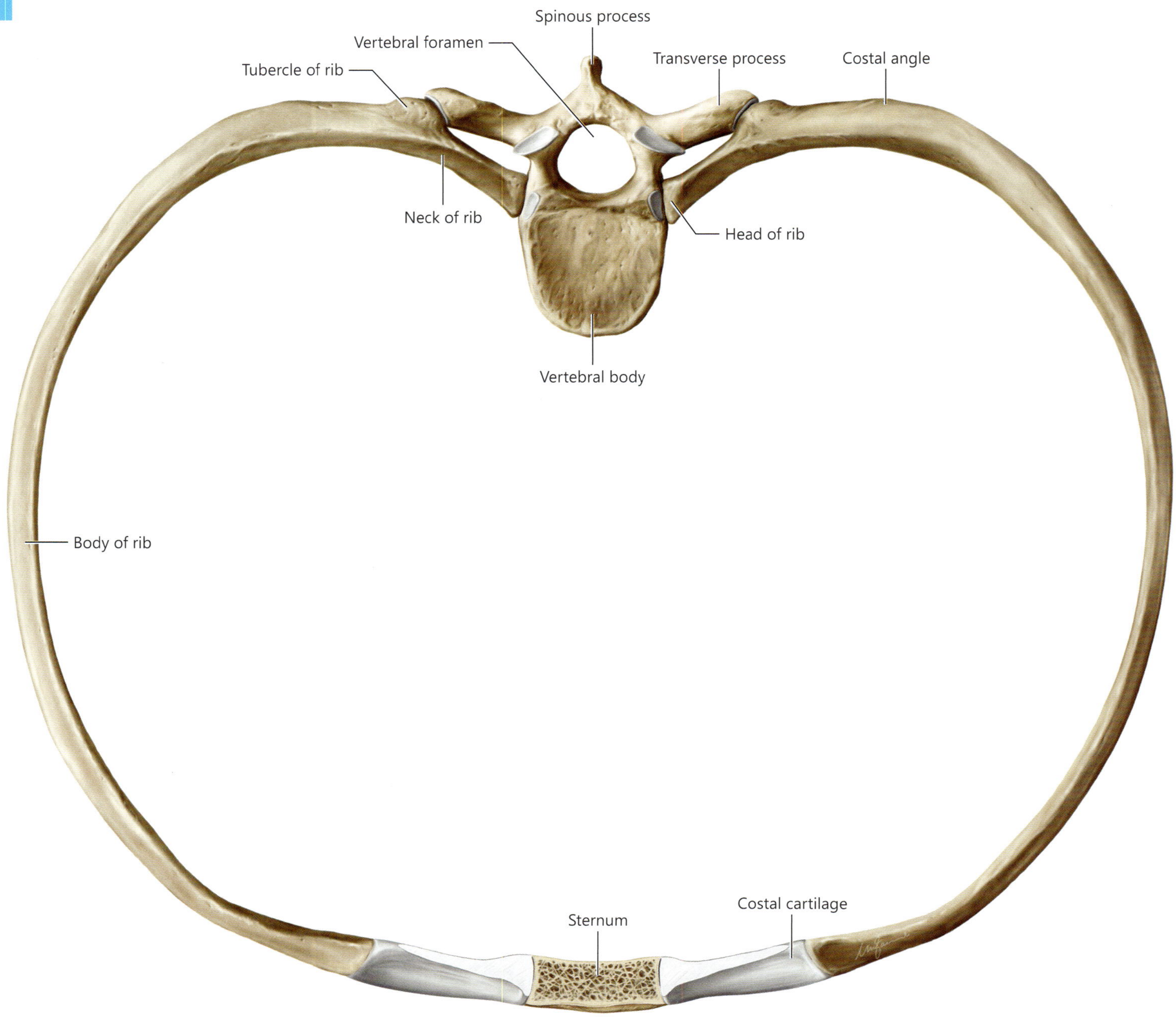

Costovertebral joints at the level of the fifth thoracic vertebra.

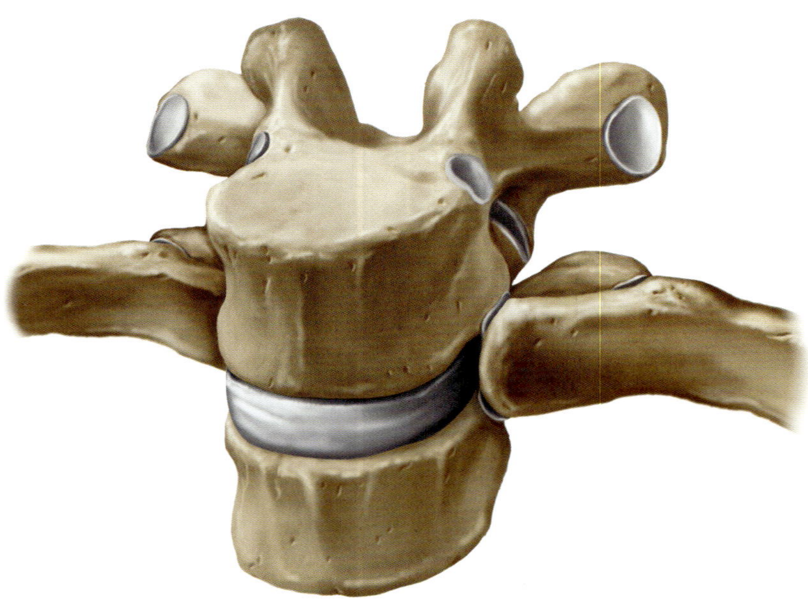

Costovertebral joints. Except for the first, the eleventh, and the twelfth rib, the ribs articulate through the head of the rib with two contiguous vertebral bodies.

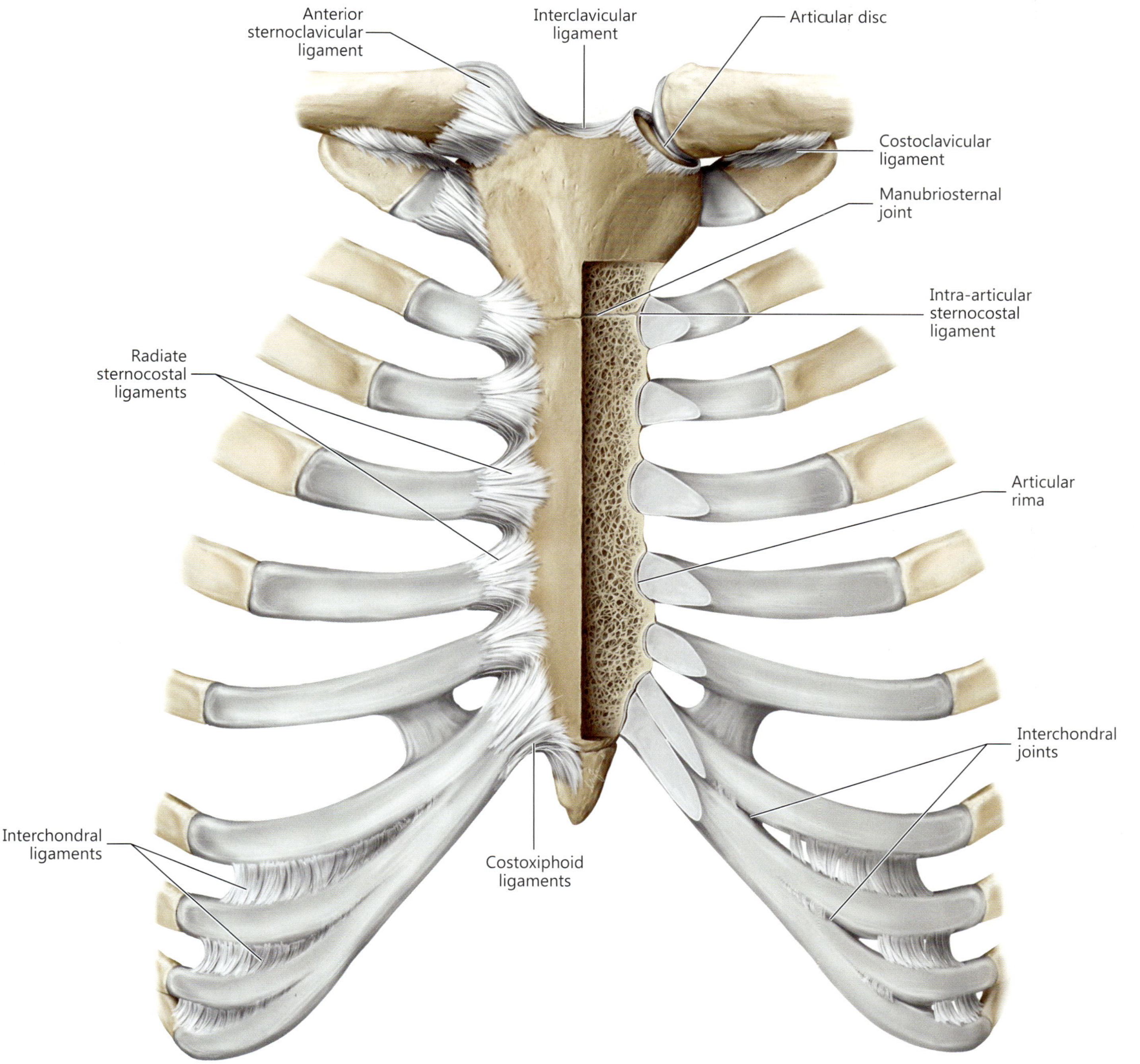

Sternocostal joints.

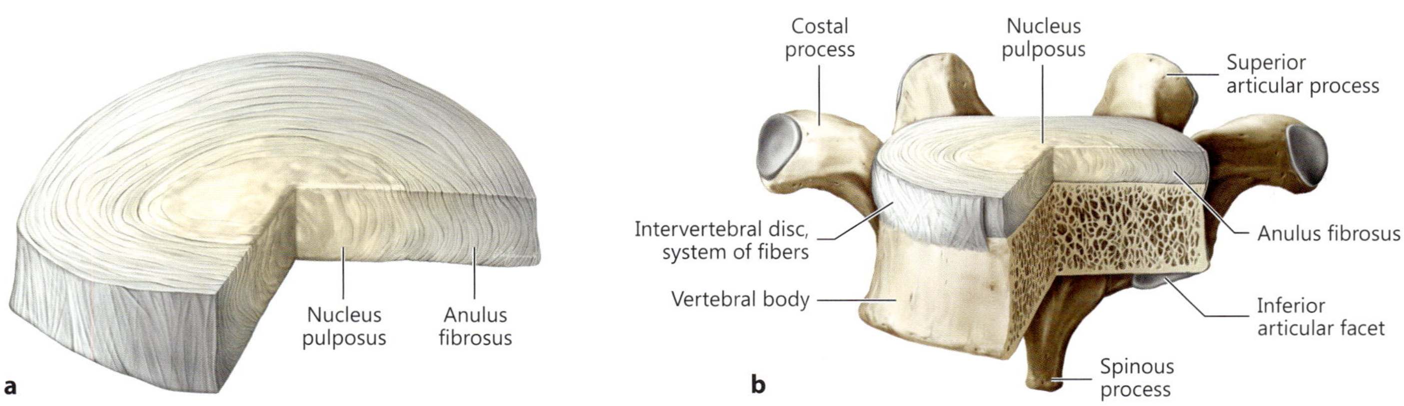

Intervertebral disc. **a**, Sagittal section; **b**, relation with the vertebra.

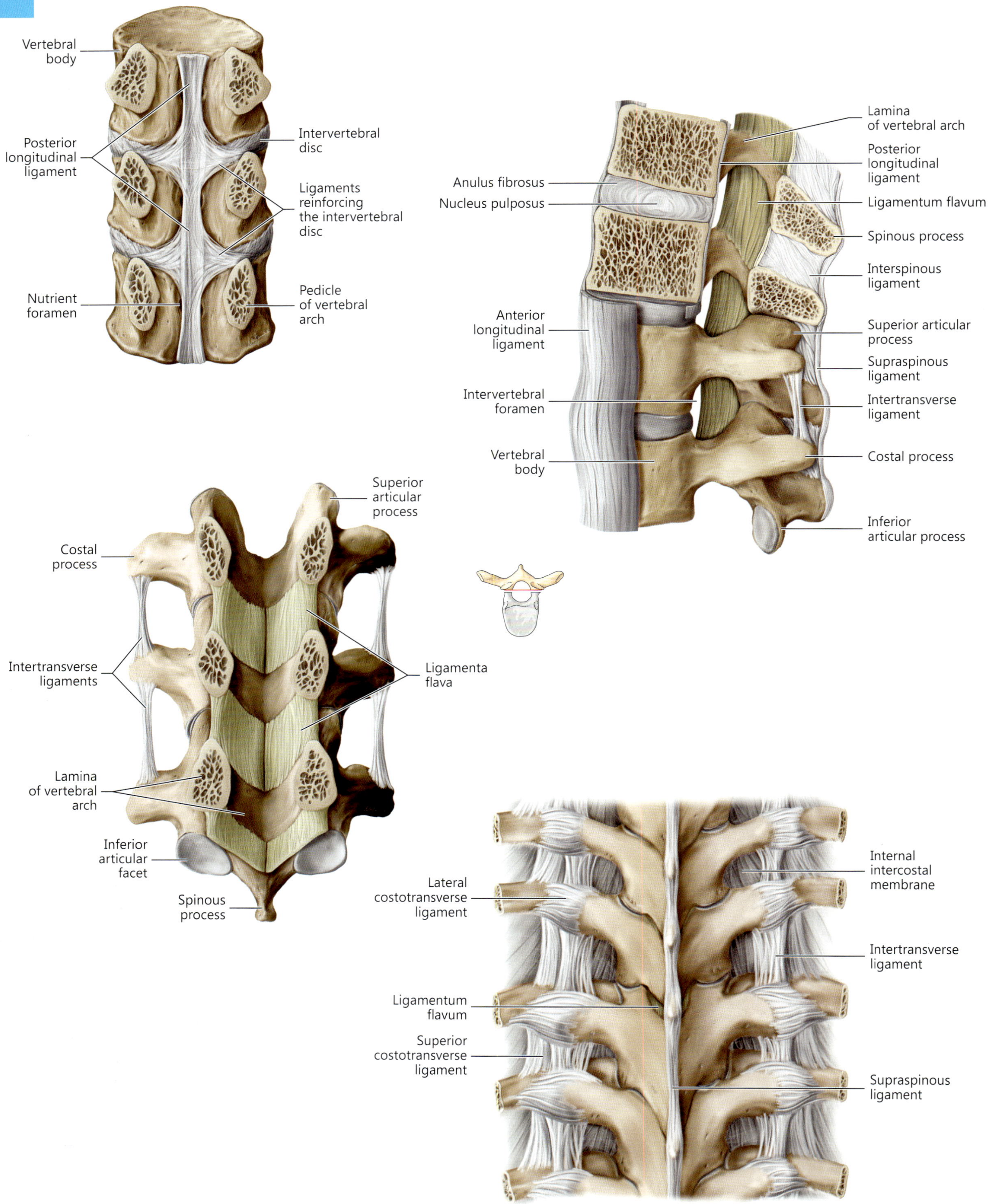

Ligaments of the vertebral column.

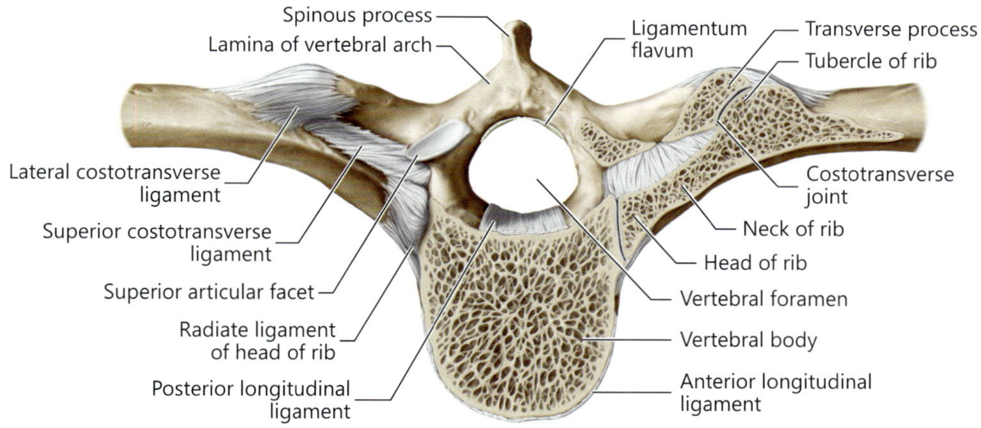

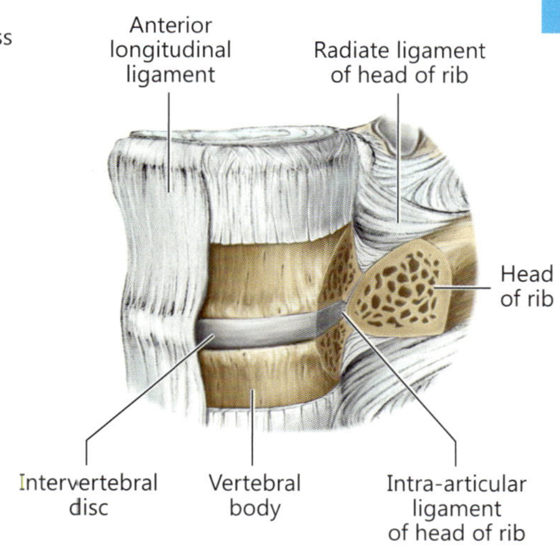

Ligamentous apparatus of the **costovertebral joints**.

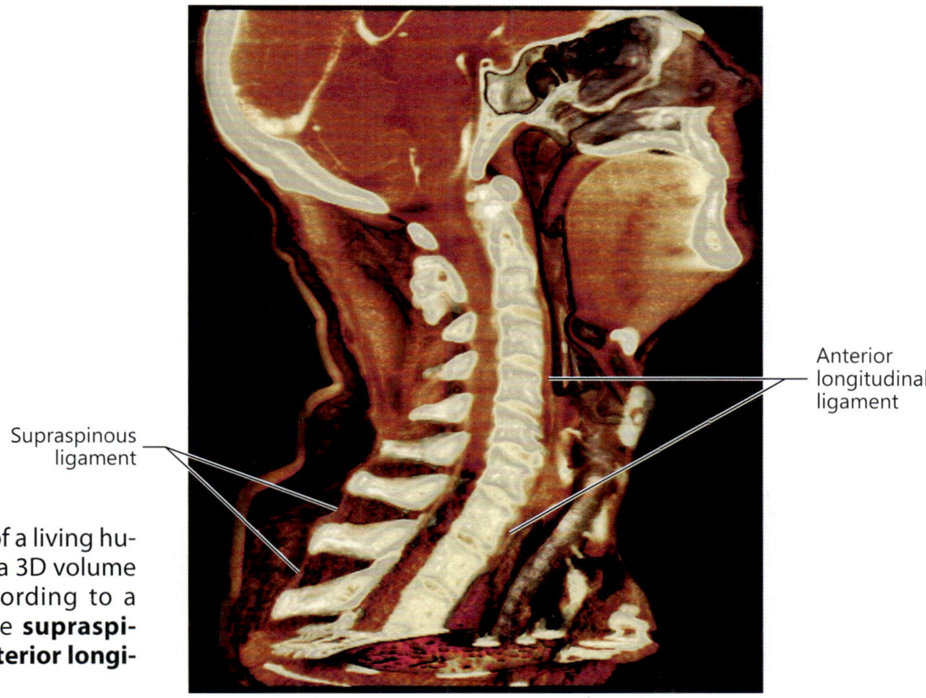

Three-dimensional image of a living human body, obtained using a 3D volume rendering technique, according to a sagittal plane showing the **supraspinous ligament** and the **anterior longitudinal ligament**.

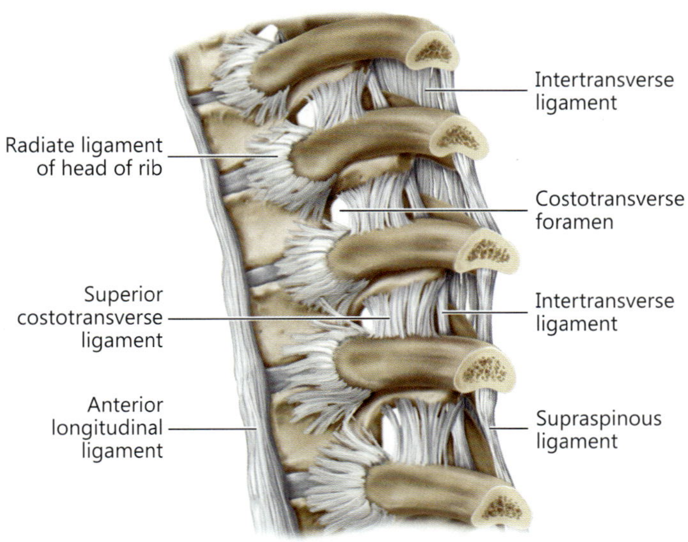

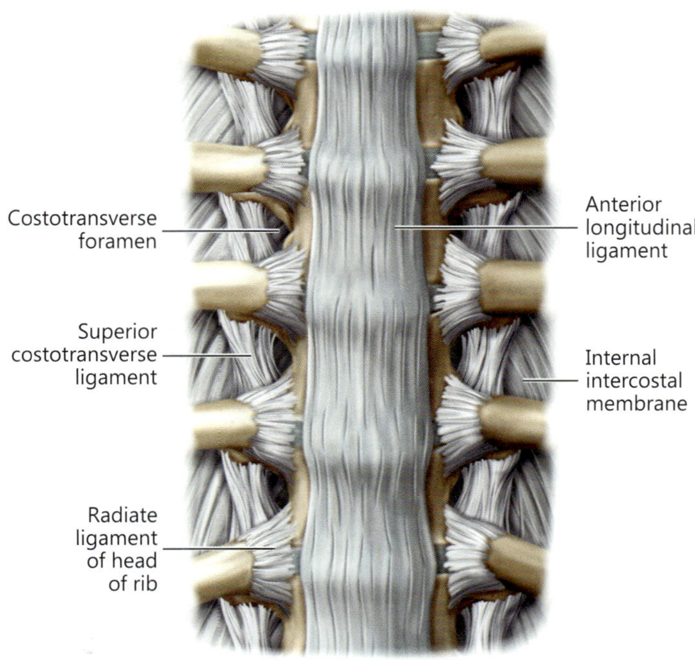

Costovertebral joints.

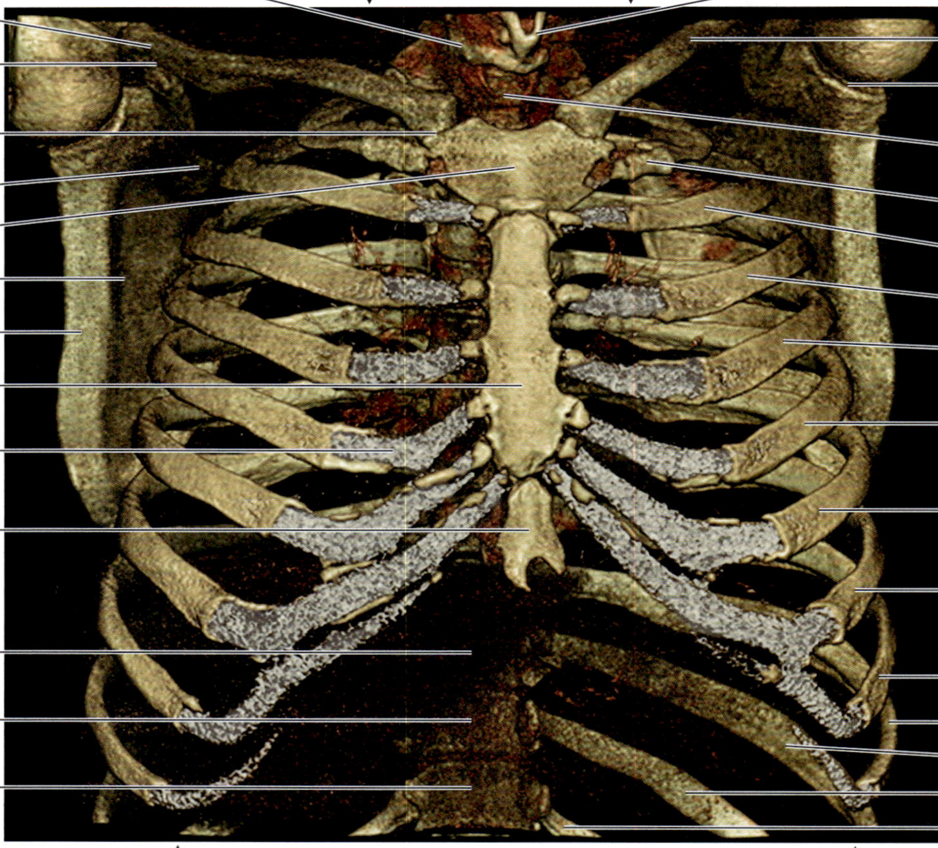
Superior thoracic aperture
Cricoid cartilage
Thyroid cartilage
Acromioclavicular joint
Coracoid process
Sternoclavicular joint
Suprascapular notch
Manubrium of sternum
Subscapular fossa
Scapula, lateral border
Body of sternum
Costal cartilage
Xiphoid process
Tenth thoracic vertebra [T X]
Eleventh thoracic vertebra [T XI]
Twelfth thoracic vertebra [T XII]
Clavicle
Glenohumeral joint
First thoracic vertebra [T I]
First rib
Second rib
Third rib
Fourth rib
Fifth rib
Sixth rib
Seventh rib
Eighth rib
Ninth rib
Tenth rib
Eleventh rib
Twelfth rib
Inferior thoracic aperture

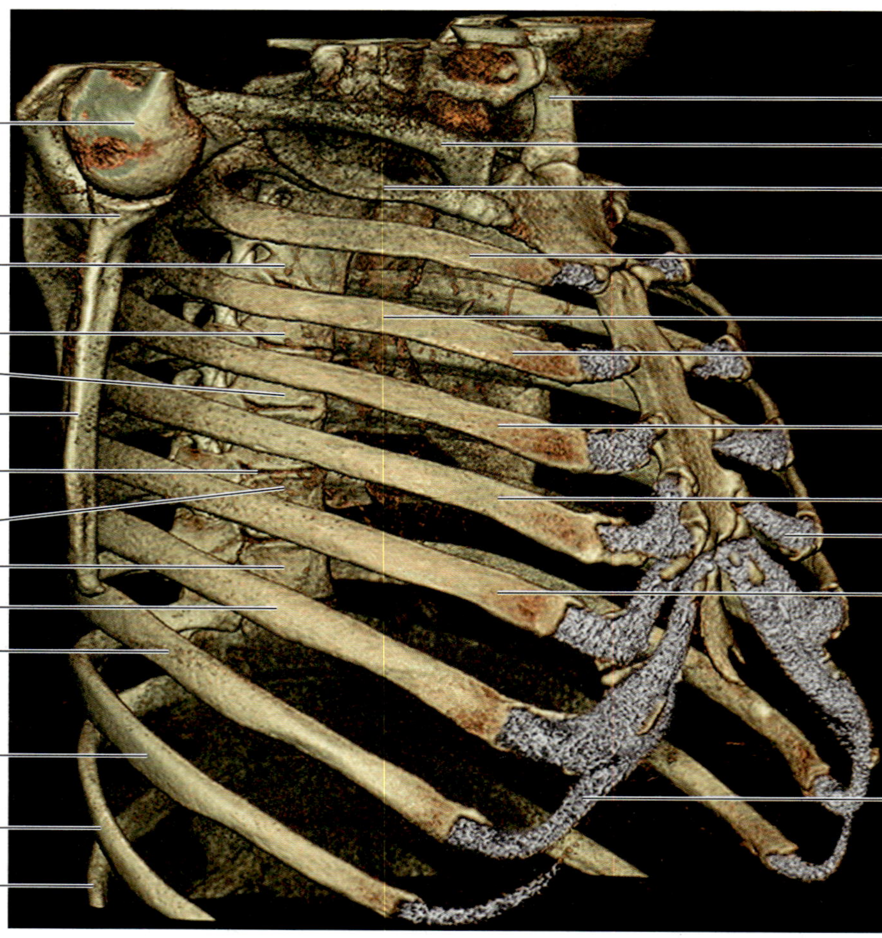
Head of humerus
Infraglenoid tubercle
Fourth thoracic vertebra [T IV]
Fifth thoracic vertebra [T V]
Sixth thoracic vertebra [T VI]
Scapula, lateral border
Intervertebral disc
Eighth thoracic vertebra [T VIII]
Ninth thoracic vertebra [T IX]
Seventh rib
Eighth rib
Ninth rib
Tenth rib
Eleventh rib
Left clavicle
Right clavicle
First rib
Second rib
Costal margin
Third rib
Fourth rib
Fifth rib
Costal cartilage
Sixth rib
Costal margin

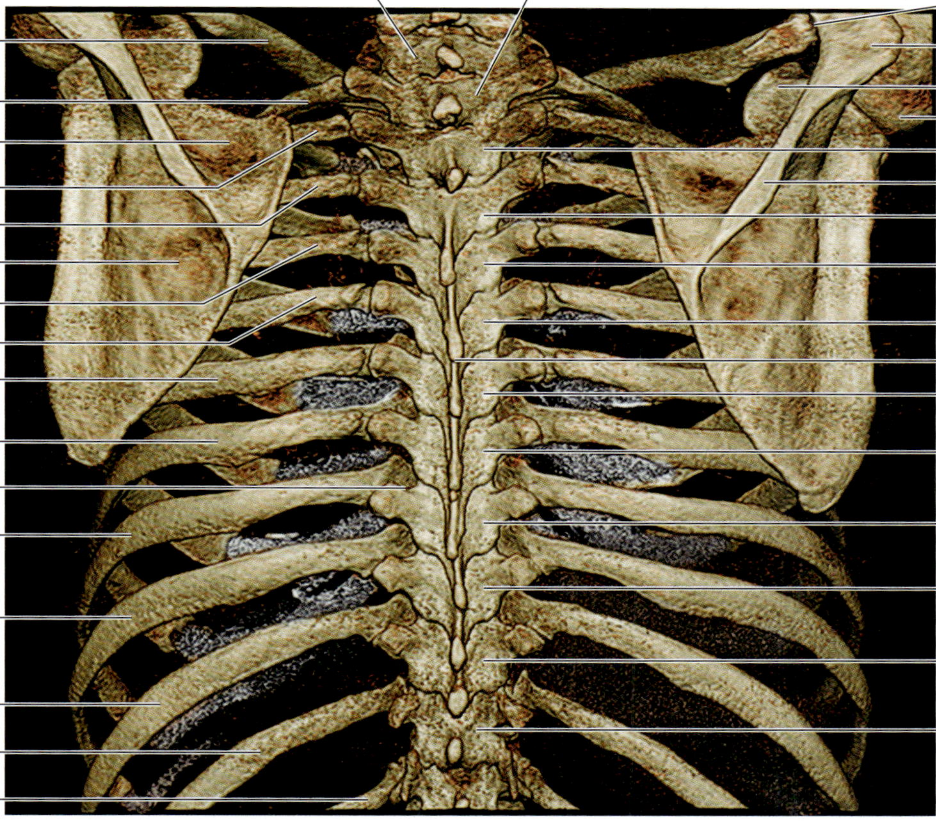

Three-dimensional images of a living human body, obtained by CT scan using a 3D volume rendering technique, of the **thoracic cage**. Above posterior projection, below bottom view.

➤
Previous page, above anterior projection, below right anterior oblique projection.

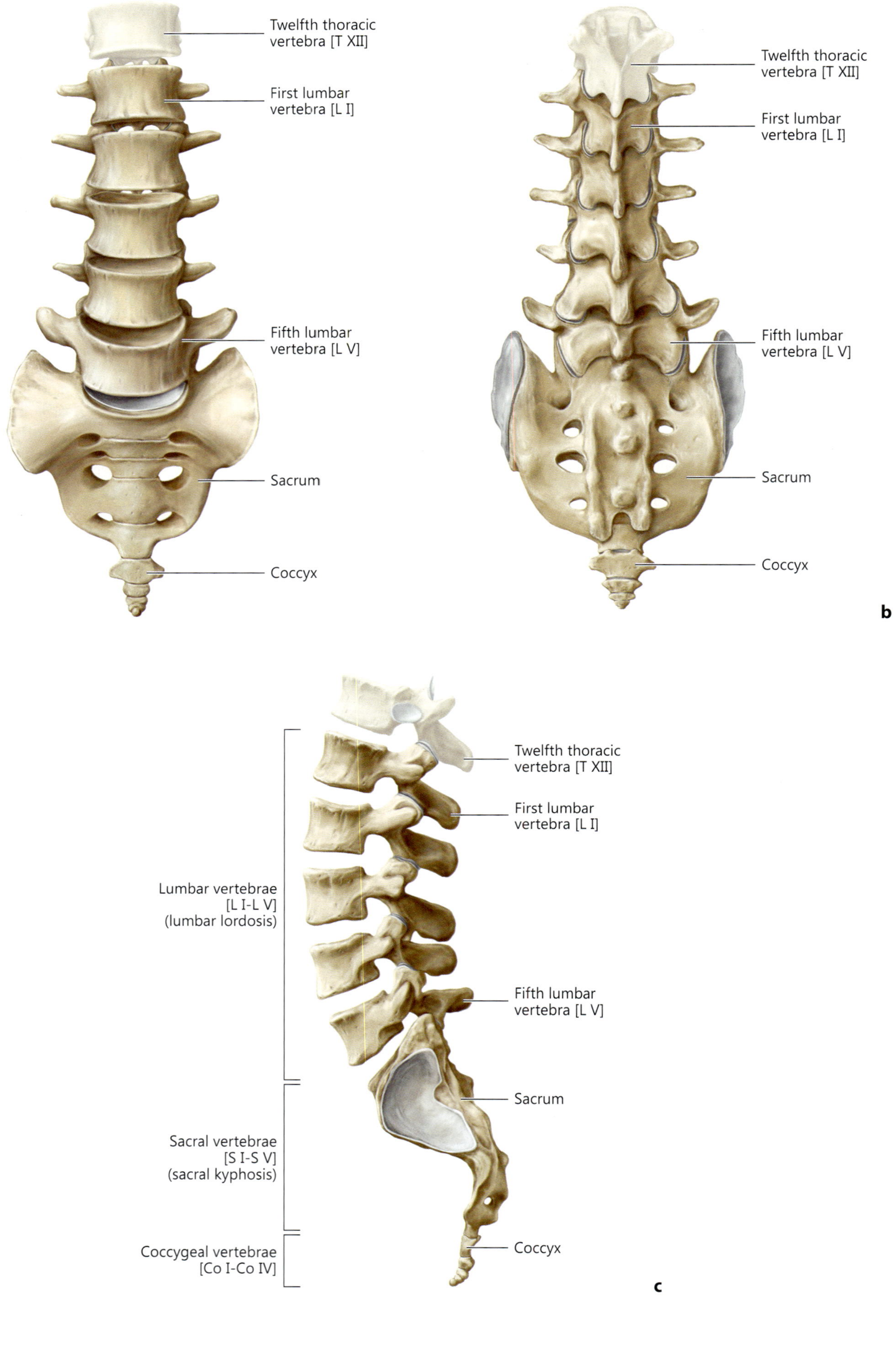

Lumbar column. Anterior (**a**), posterior (**b**), and left lateral (**c**) views.

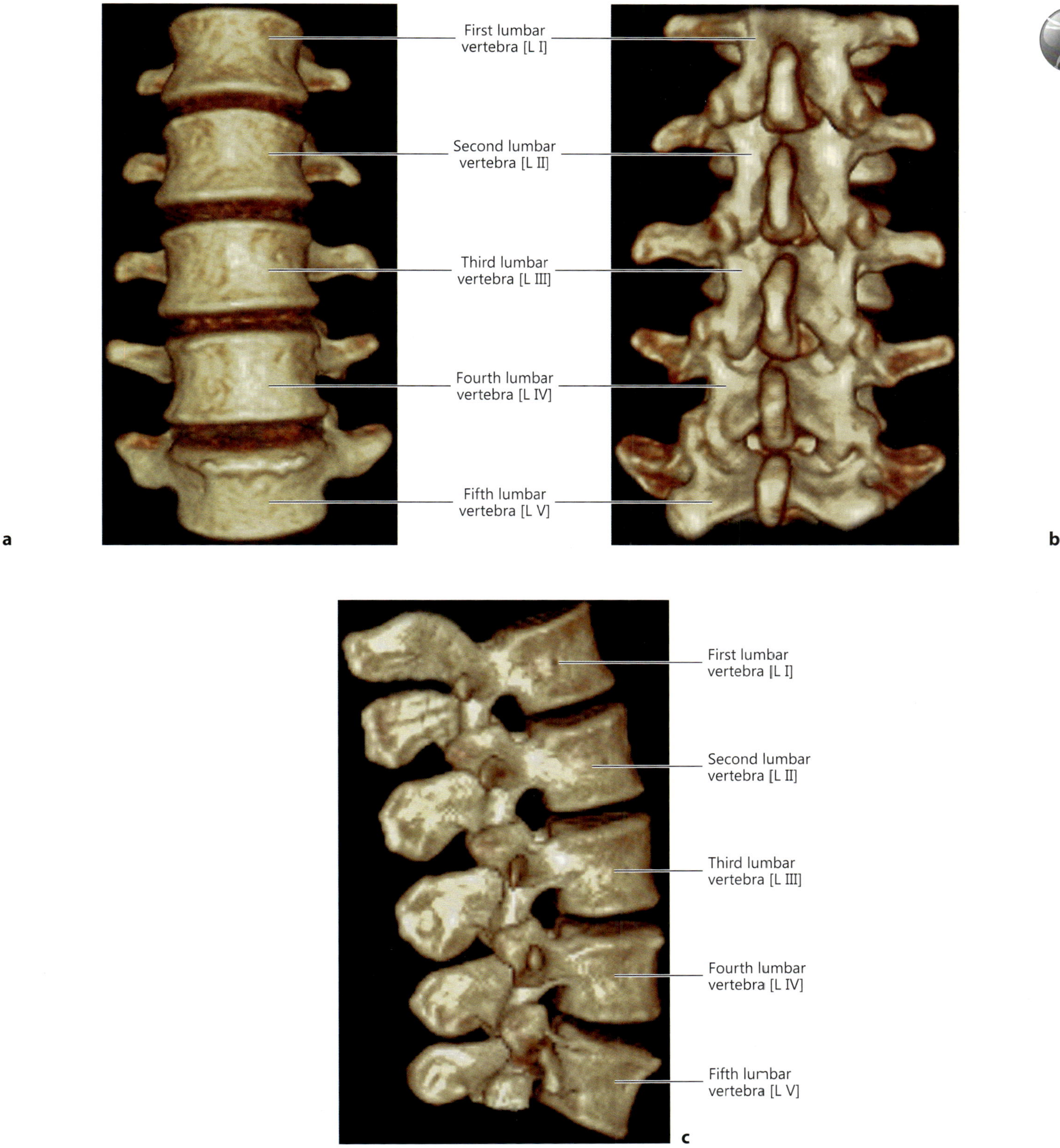

Three-dimensional images of a living human body, obtained using a 3D volume rendering technique, showing the **lumbar column** in anterior (**a**), posterior (**b**), and right lateral (**c**) projection.

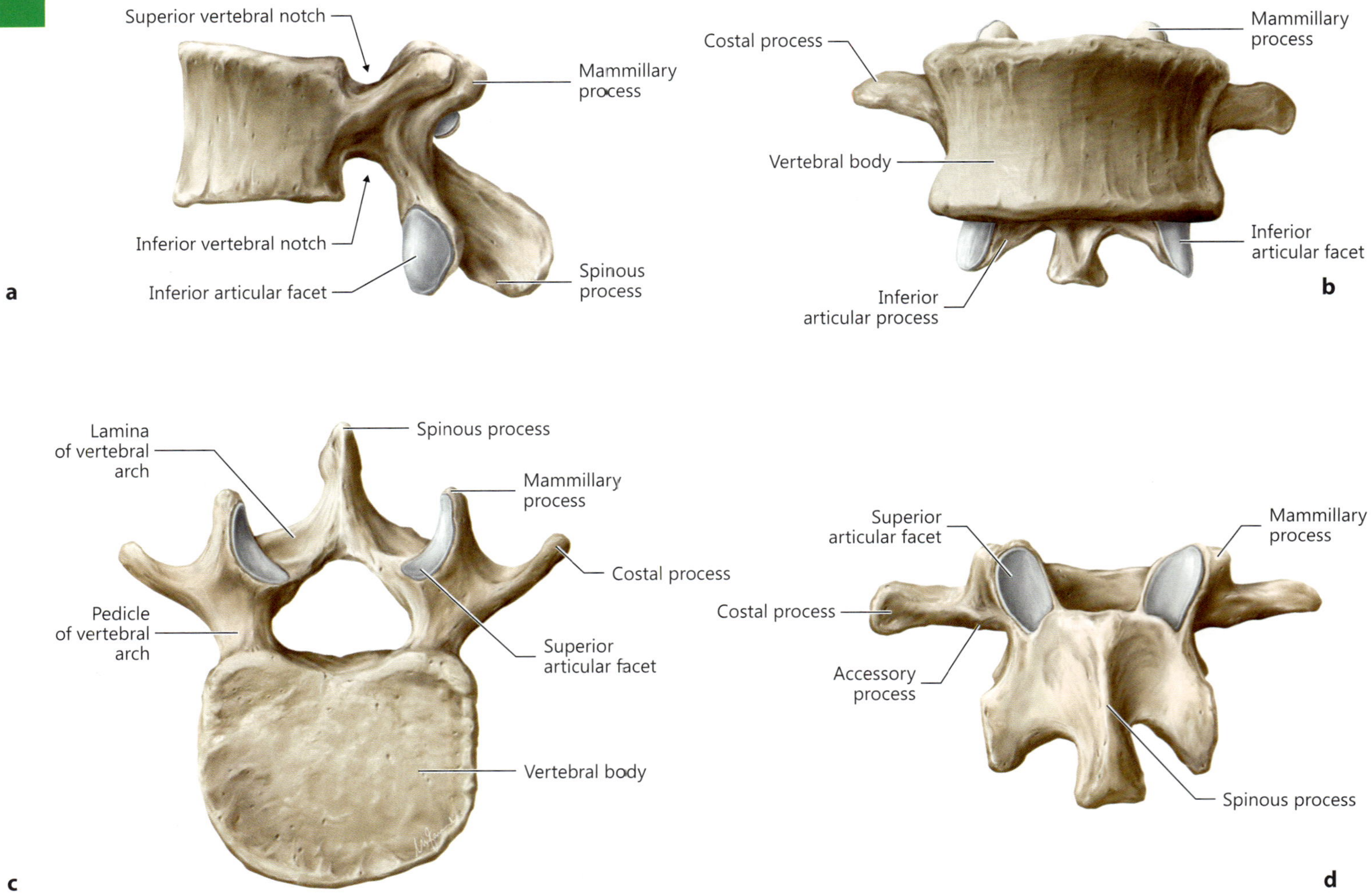

Third lumbar vertebra. Left lateral (**a**), anterior (**b**), superior (**c**), and posterior (**d**) views.

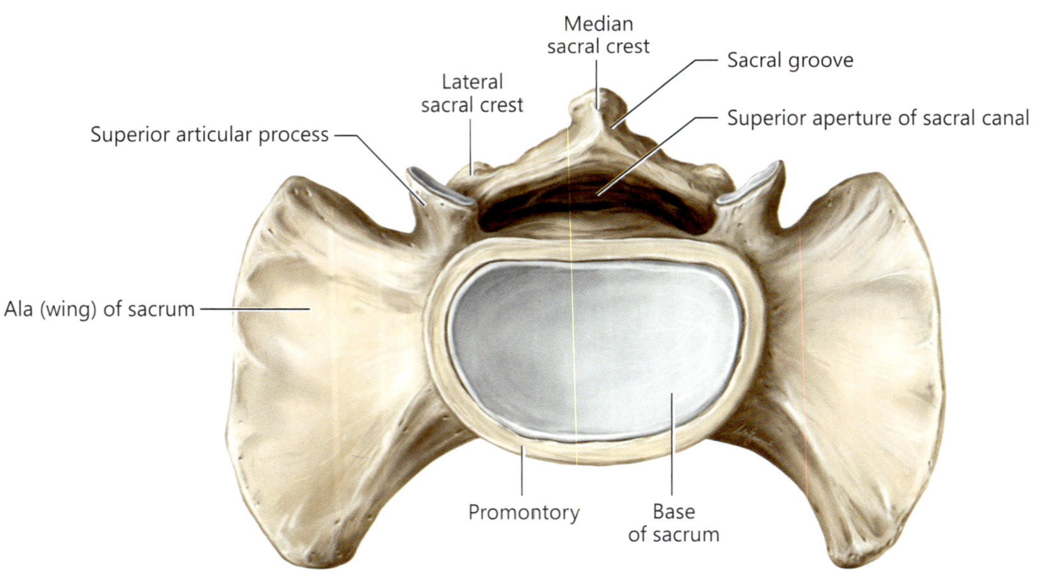

Sacrum. Superior surface view.

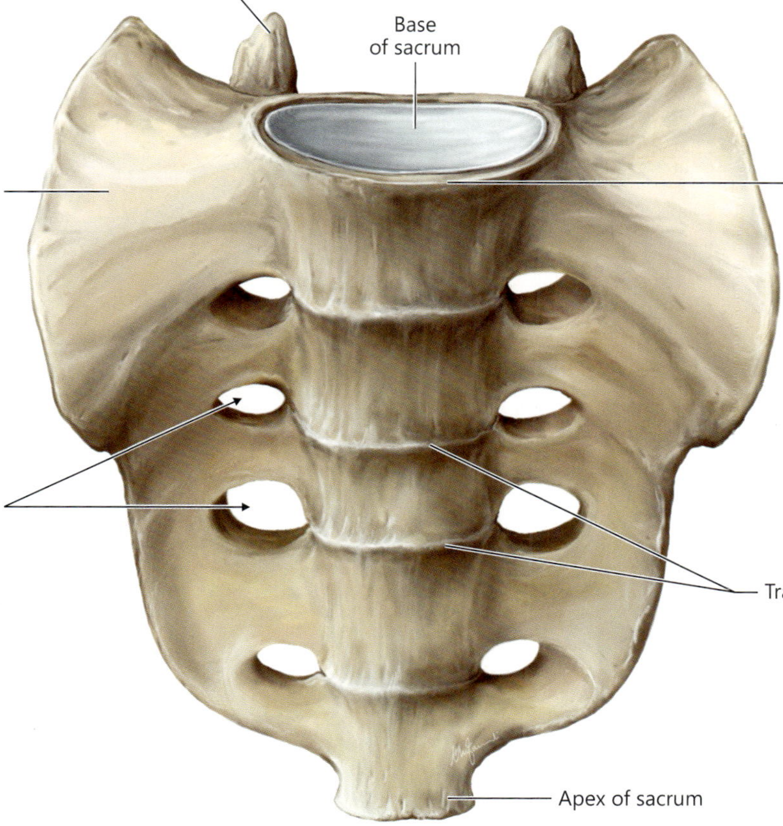

Sacrum. Anterior surface view. The upper part of the bone is broadened into the wings; a number of transverse ridges in the middle mark the welding between the primary osseous metamers.

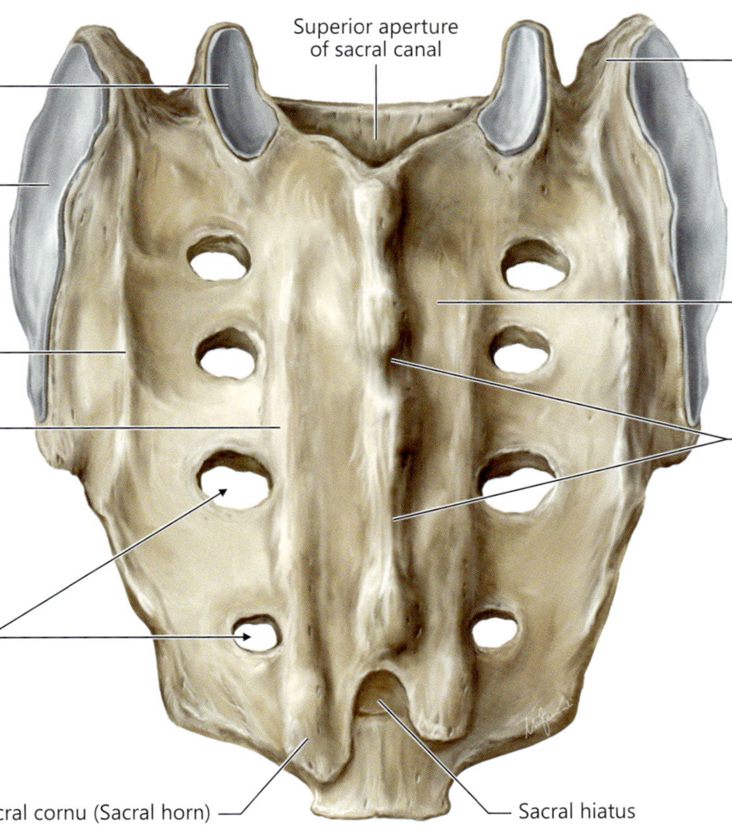

Sacrum. Posterior surface view. The bone surface appears prominent in the center of the median sacral crest and rough laterally where it provides insertion to numerous muscles.

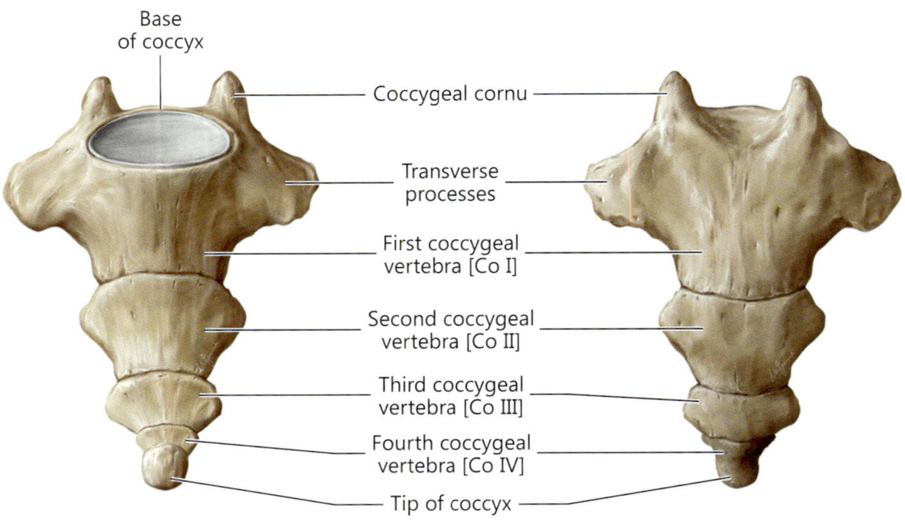

Posterior and anterior surface of the **coccyx**. Note the vestiges of primitive bone segments.

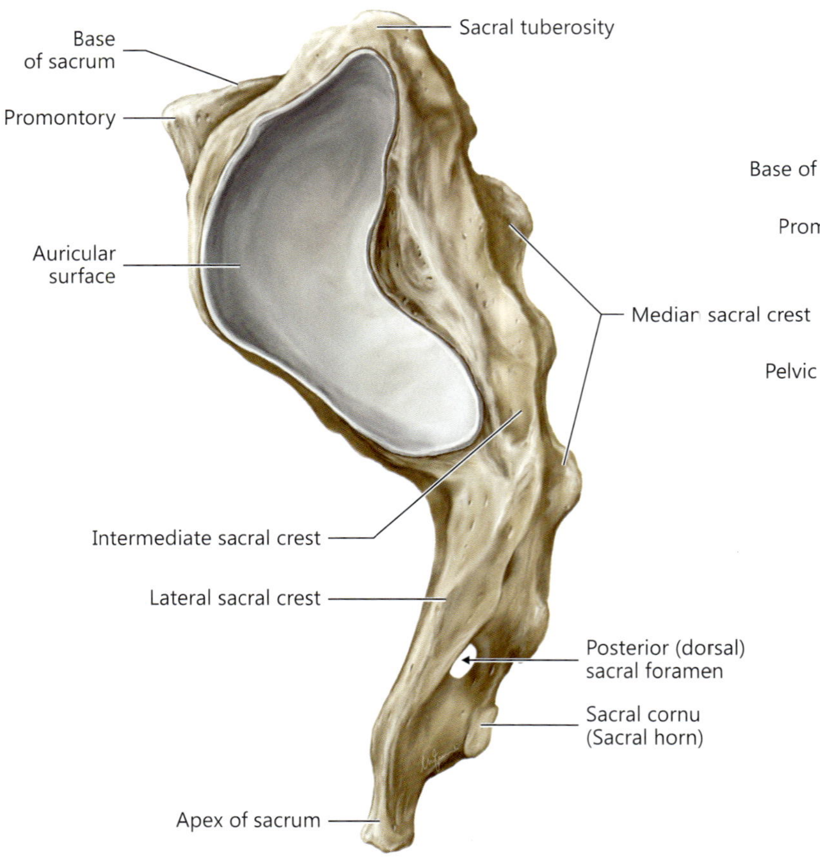

Left lateral surface of **sacrum**. This surface broadens towards the top into the auricular surface; behind this articular surface lies the prominence of the sacral tuberosity.

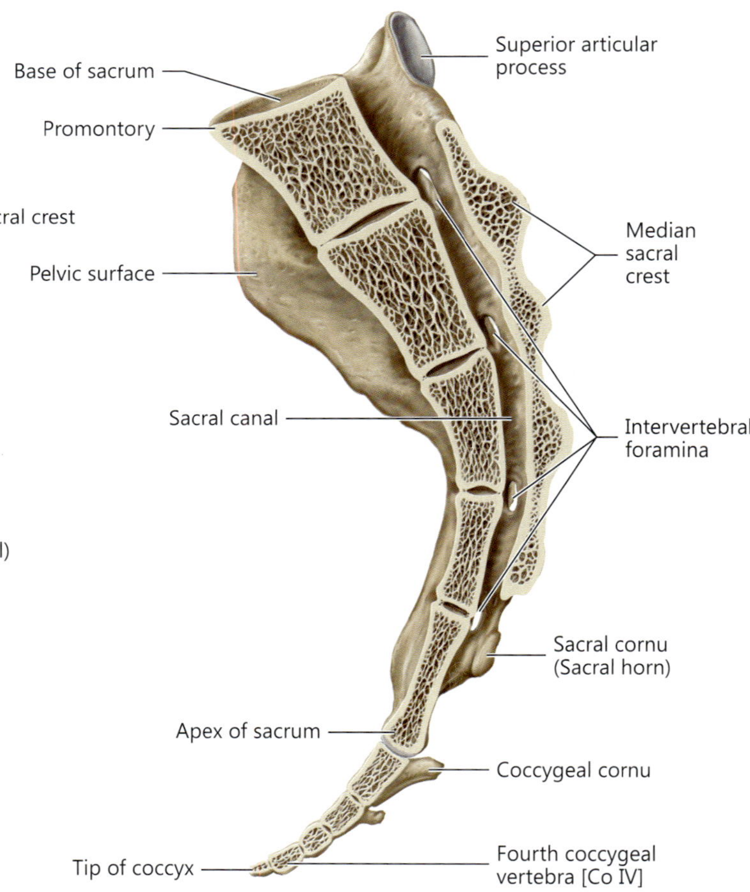

Sacrum and **coccyx**. Sagittal section passing through the sacral foramina.

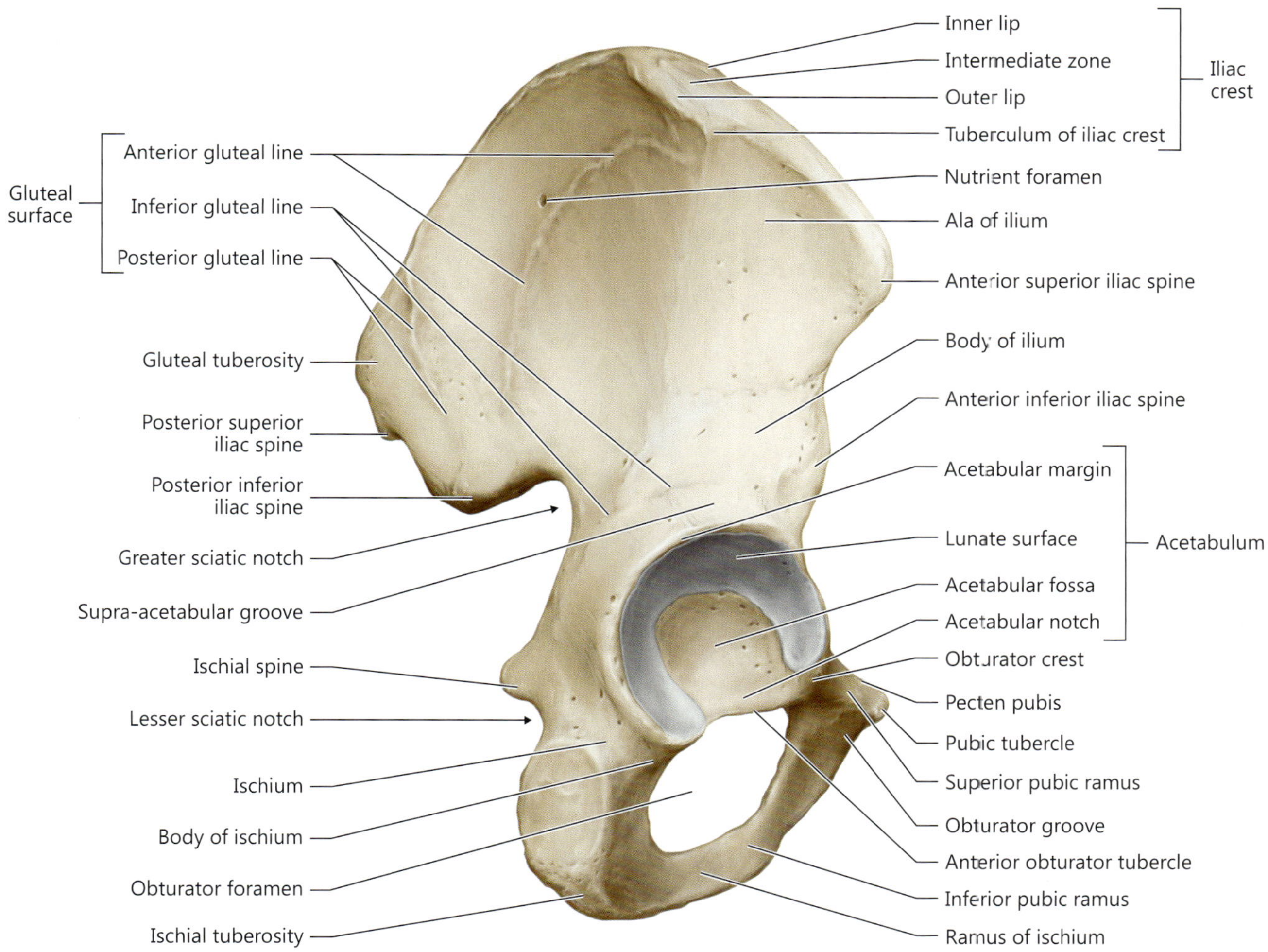

Hip. External view of the right hip bone. The hip bone is a flat, even bone, the result of a merger of three anlages which are distinct from each other until puberty: ilium, ischium and pubis. On the surface of the iliac portion, the reliefs formed by the gluteal lines are observed. In the acetabular fossa there is a meeting point of the ilium (*pink*), ischium (*blue*) and pubis (*green*).

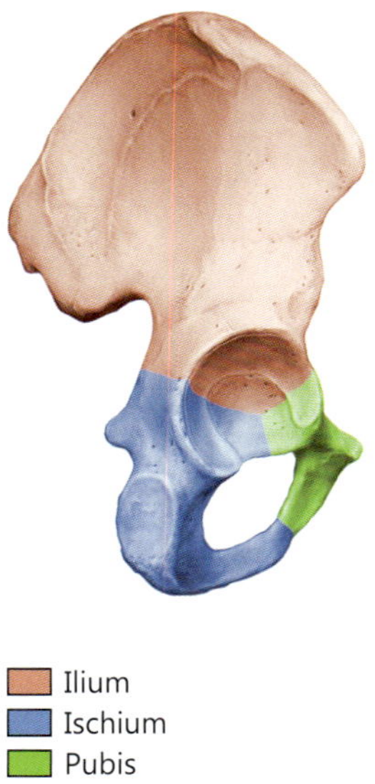

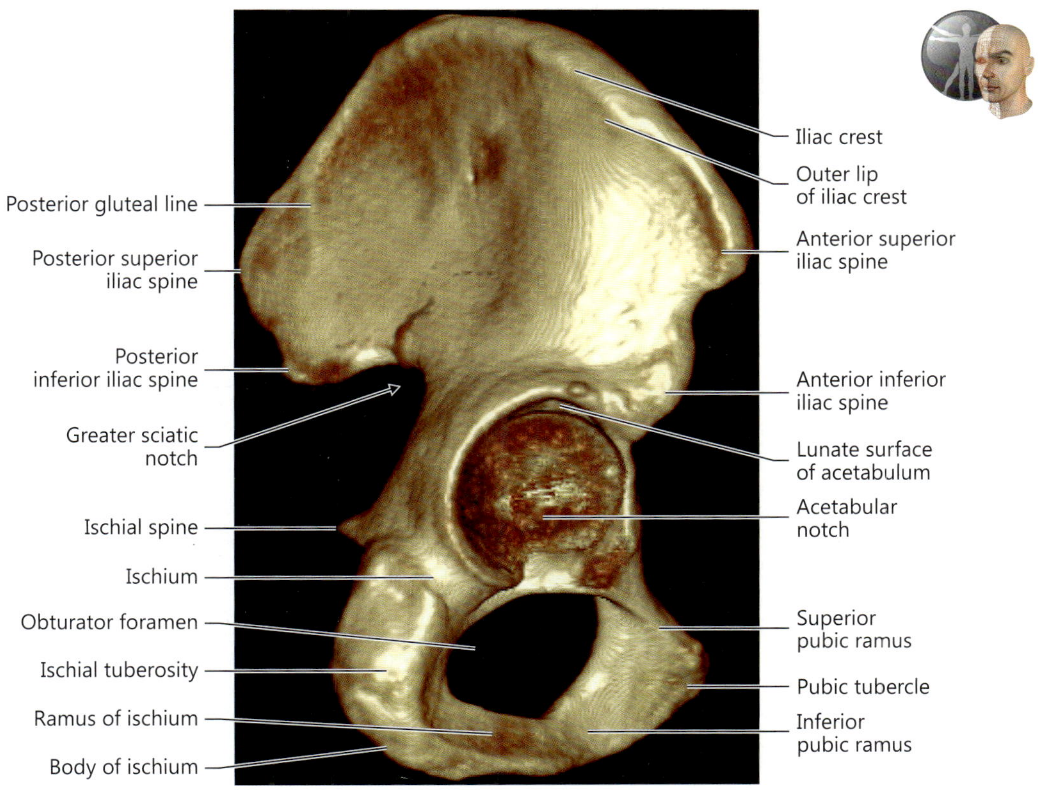

Three-dimensional image of a living human body, obtained using a 3D volume rendering technique, showing the right **hip bone** in lateral projection.

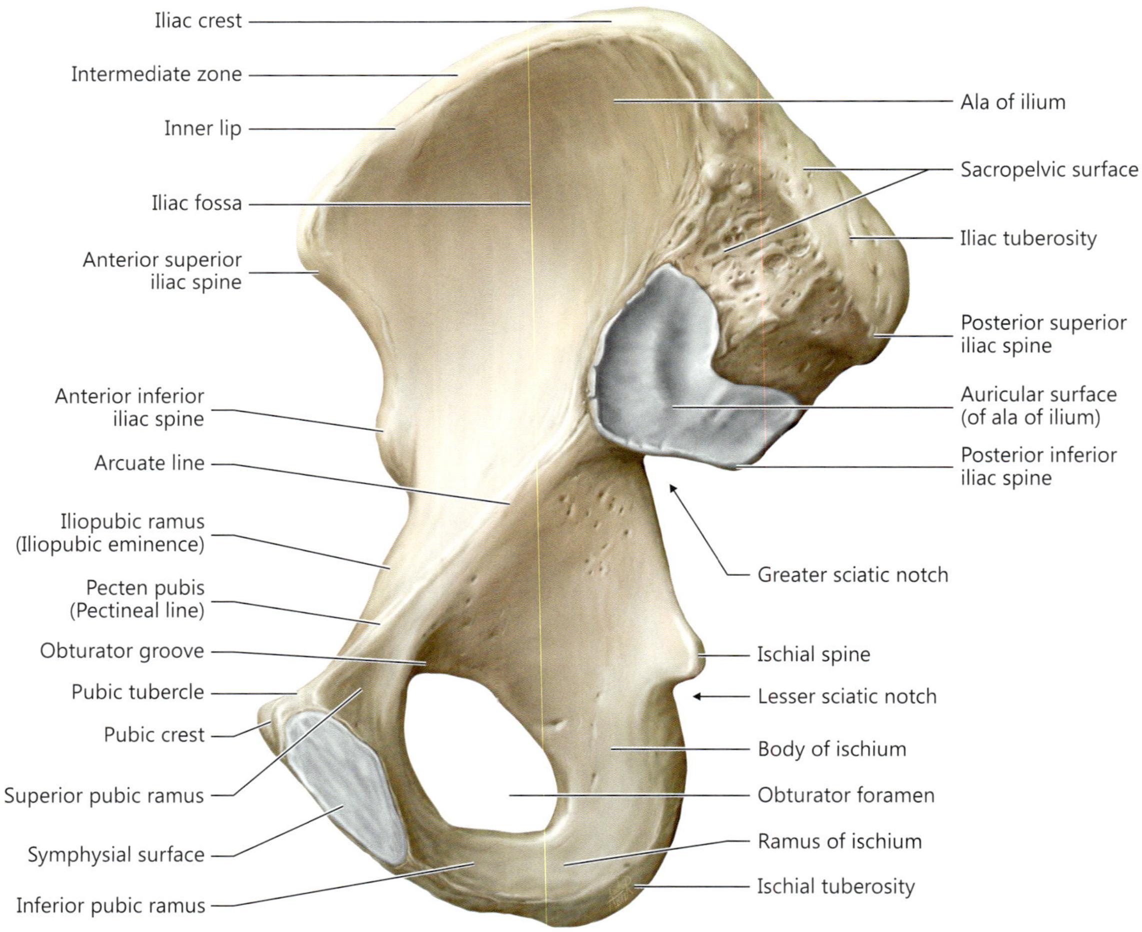

Hip. Internal view of the right hip bone. The arcuate line marks the limit between the iliac portion (*pink*) of the hip bone that is part of the abdomen and the ischial (*blue*) and pubic (*green*) portions that are part of the walls of the small pelvis.

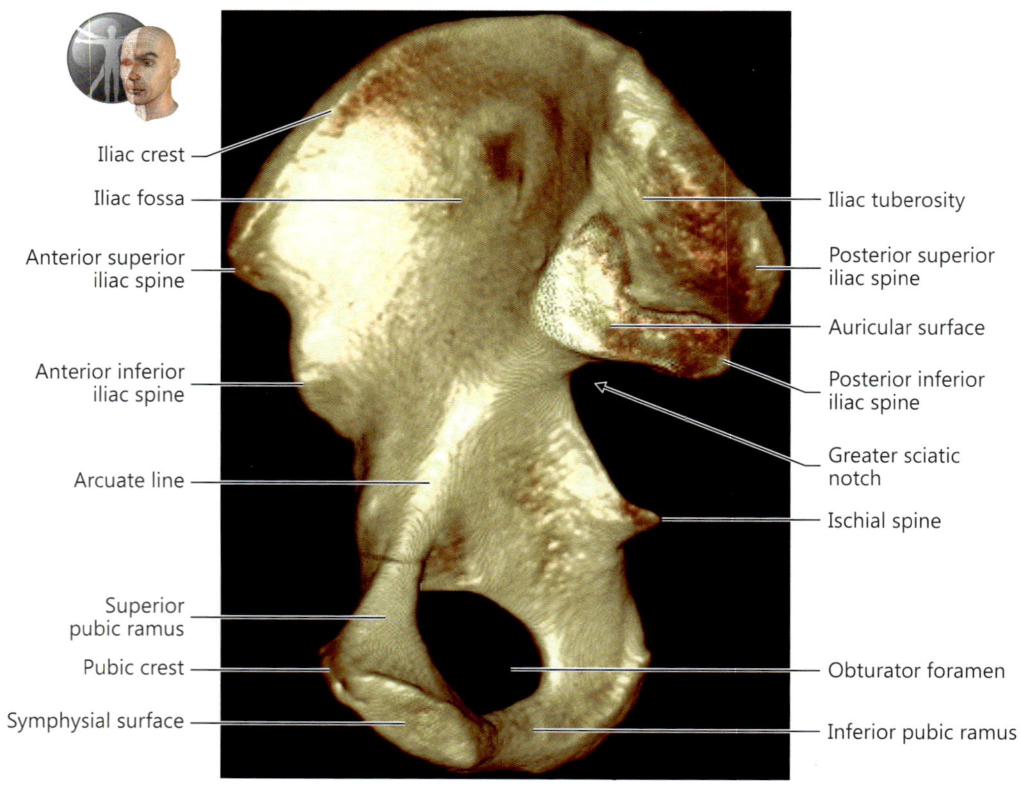

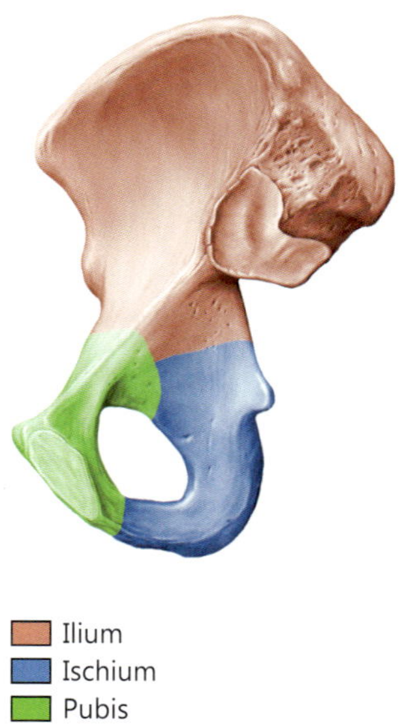

Three-dimensional image of a living human body, obtained using a 3D volume rendering technique, showing the right **hip bone** in medial projection.

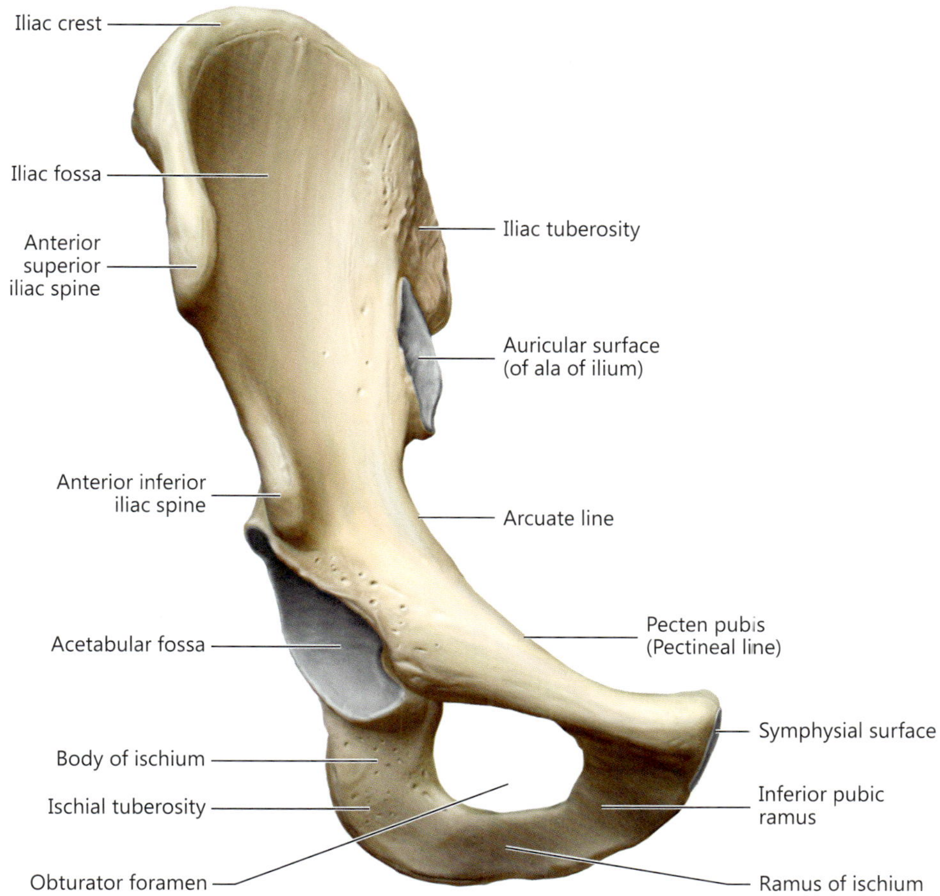

Hip. Frontal view of the right hip bone.

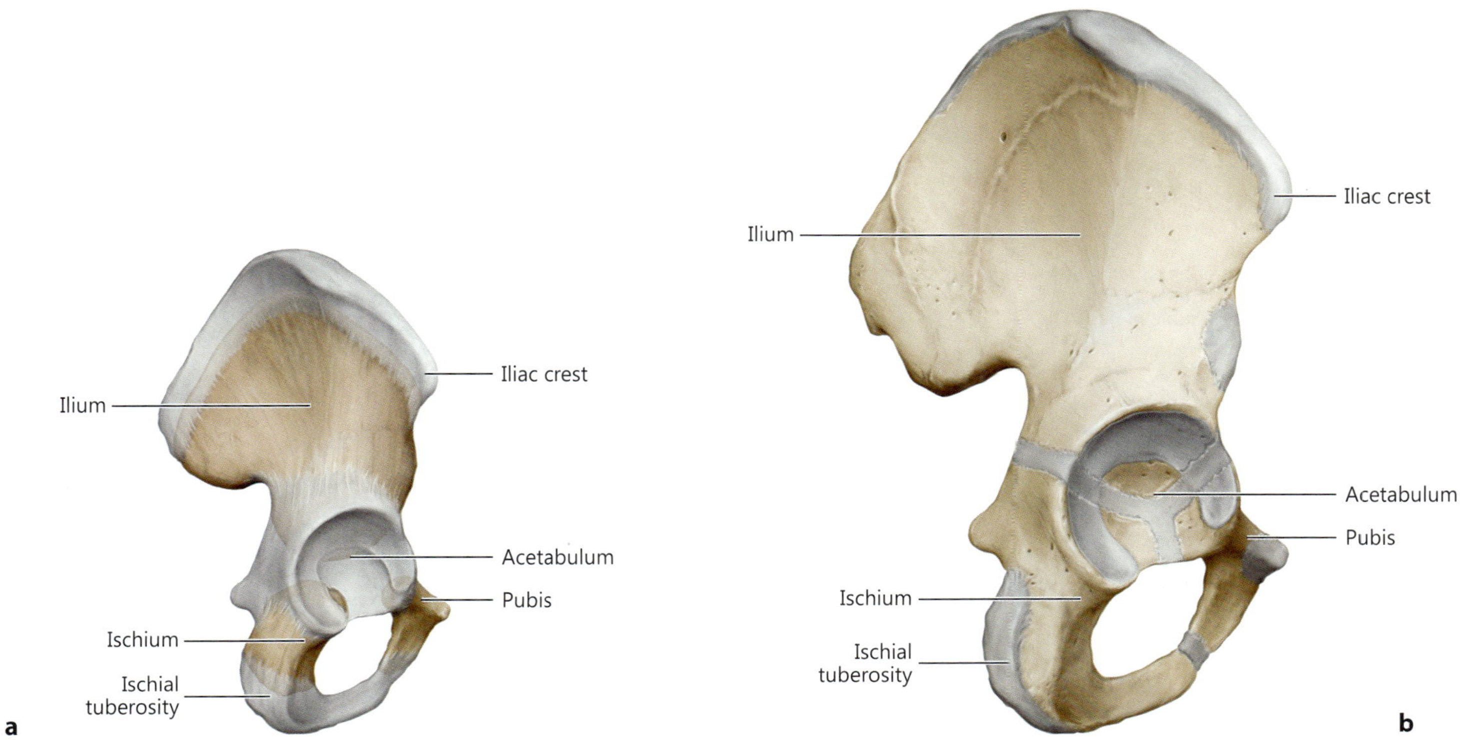

Hip bone at birth (**a**) and at the age of ten (**b**). The areas that have not yet been ossified (cartilaginous) are highlighted in *gray*.

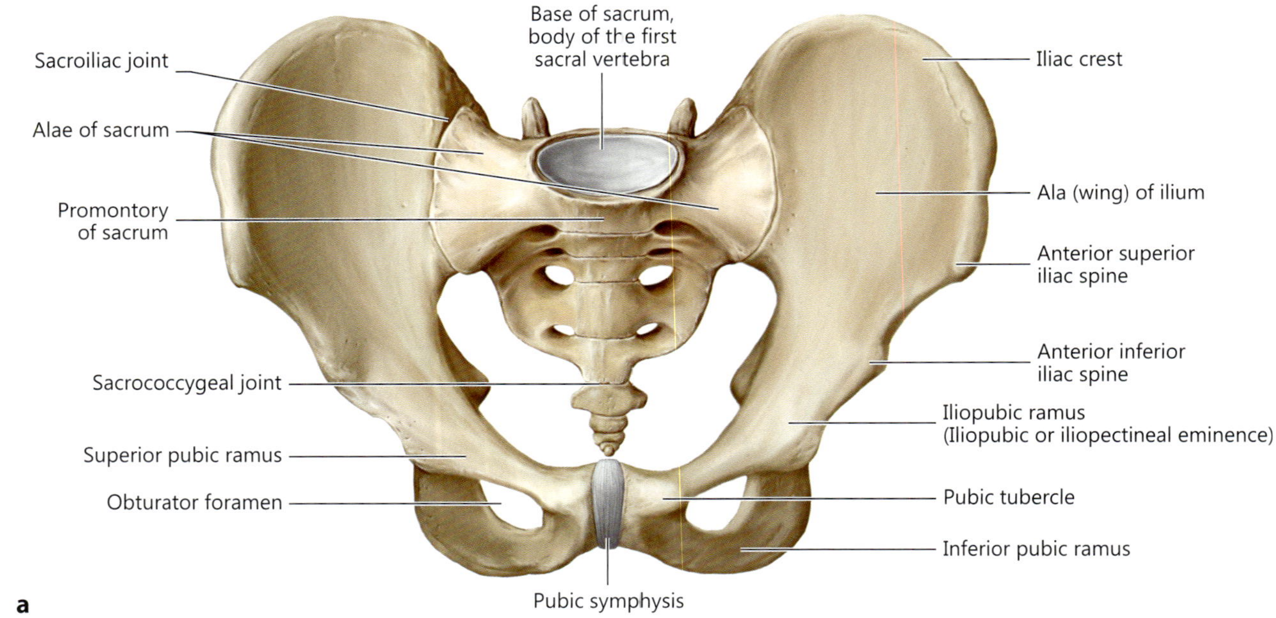

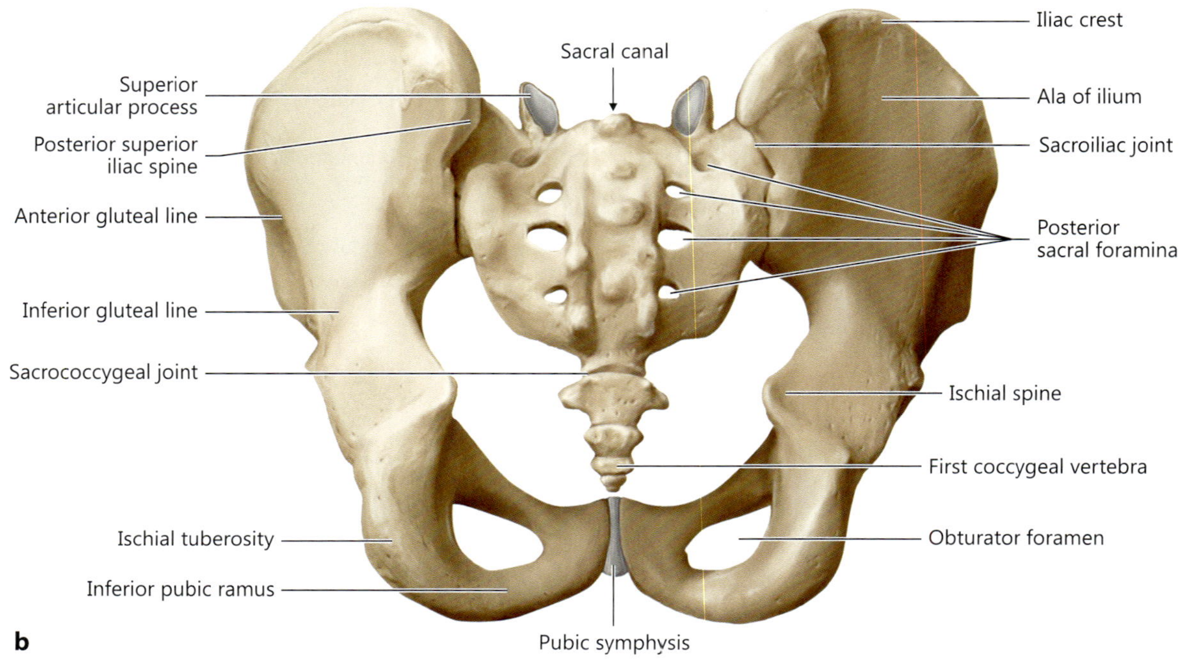

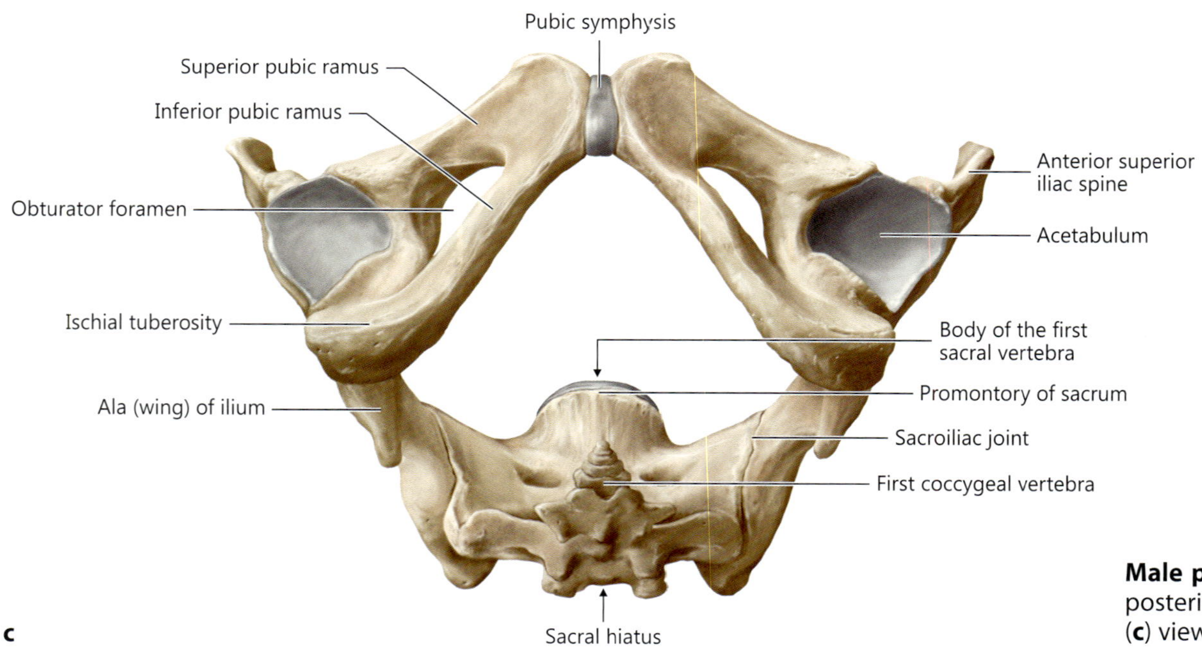

Male pelvis. Anterior (**a**), posterior (**b**), and inferior (**c**) views.

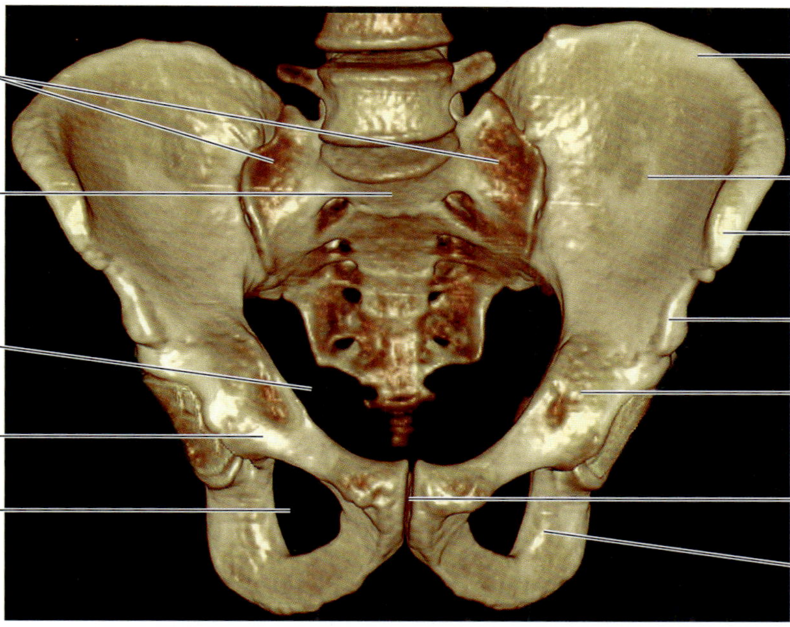

a

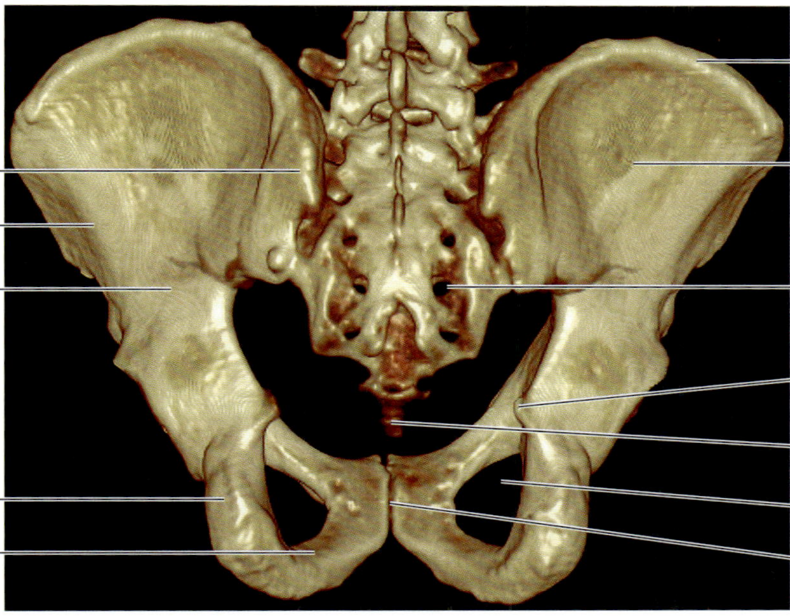

b

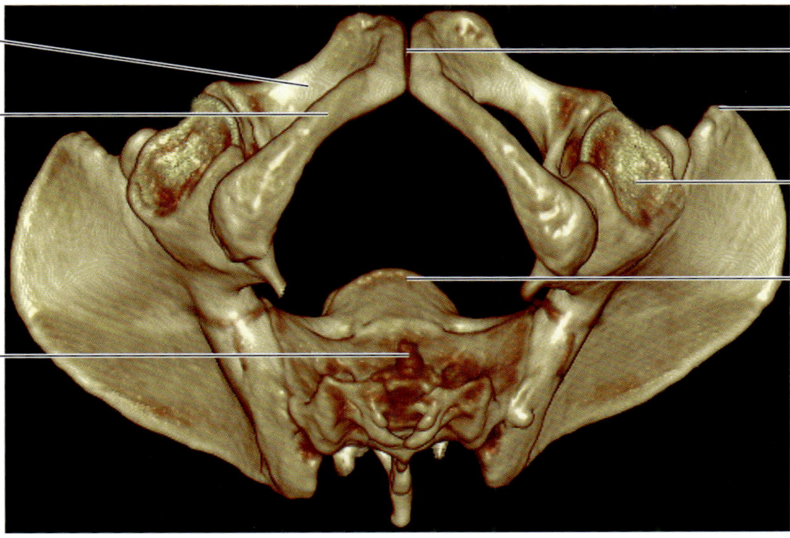

c

Three-dimensional images of a living human body, obtained using a 3D volume rendering technique, showing the **pelvis** in anterior (**a**), posterior (**b**), and inferior (**c**) projections.

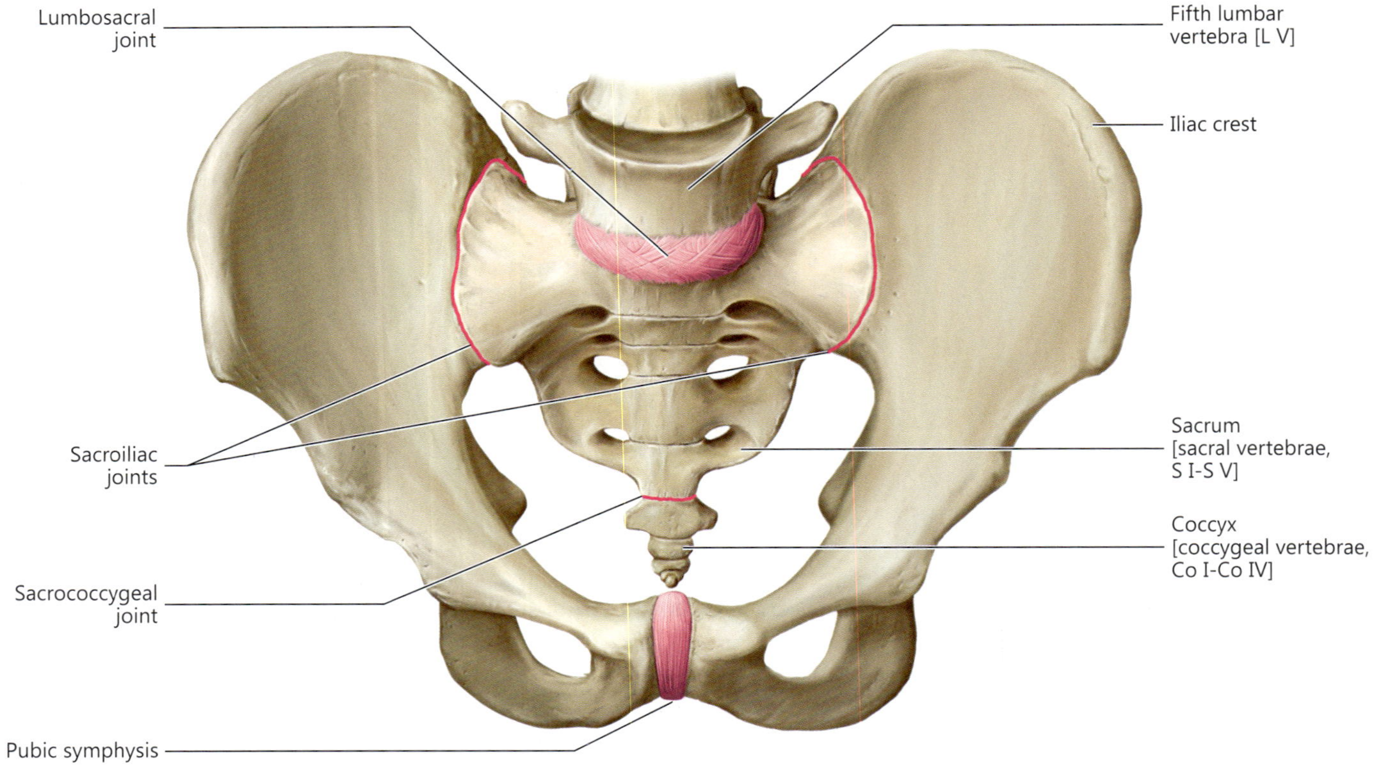

Male pelvis. The joints of the **pelvic girdle** in anterior view are highlighted in *red*.

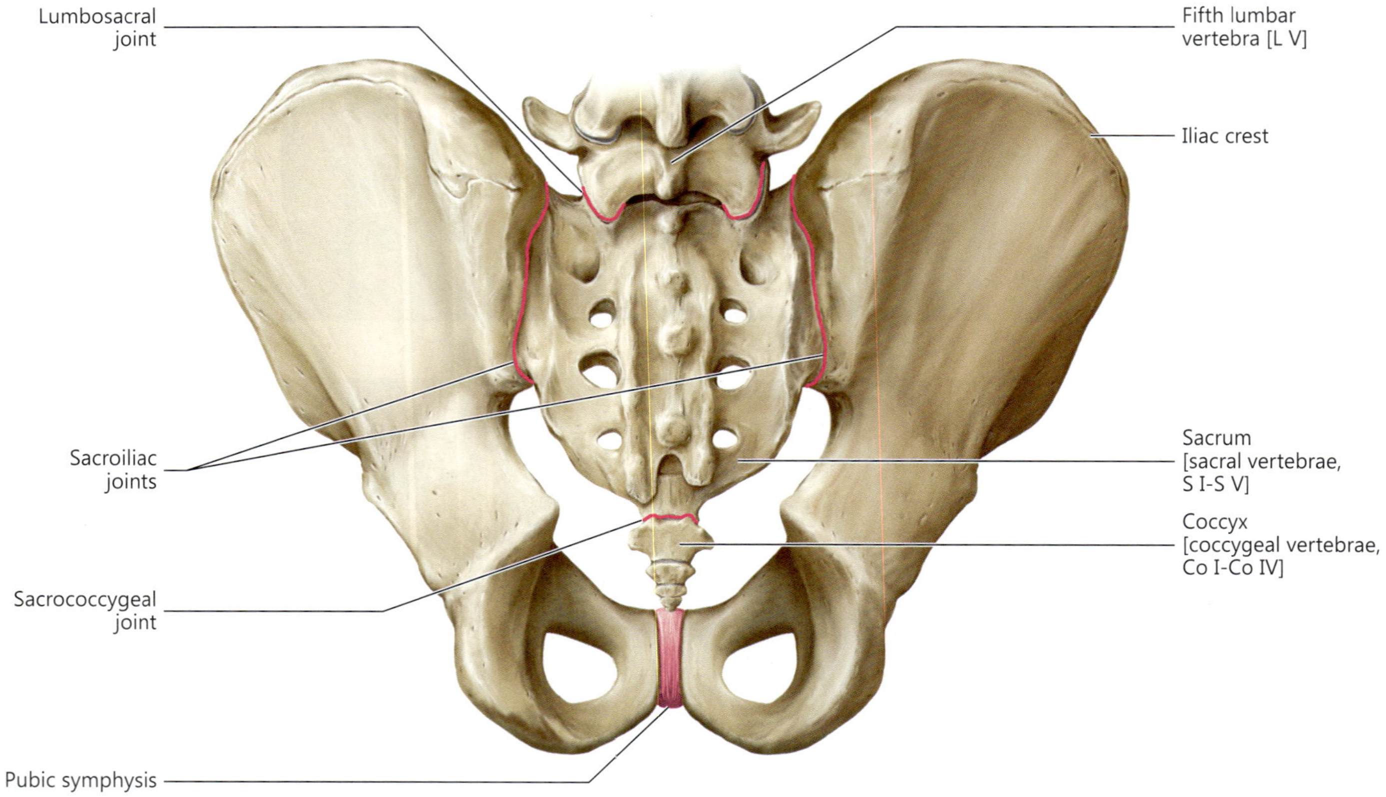

Male pelvis. The joints of the **pelvic girdle** in posterior view are highlighted in *red*.

Male pelvis X-ray image in anteroposterior view. The diameters of the pelvic inlet can be also delineated.

1, Sacroiliac joint
2, Sacrum [sacral vertebrae, S I-S V]
3, Transverse diameter of pelvic inlet
4, Superior pubic ramus
5, Ramus of ischium
6, Pubic tubercle
7, Pubic symphysis
8, True conjugate (anteroposterior diameter of pelvic inlet)
9, Greater trochanter
10, Anatomical neck of femur
11, Head of femur
12, Ala of ilium (wing of ilium)
13, Iliac crest
14, Oblique diameter of pelvic inlet

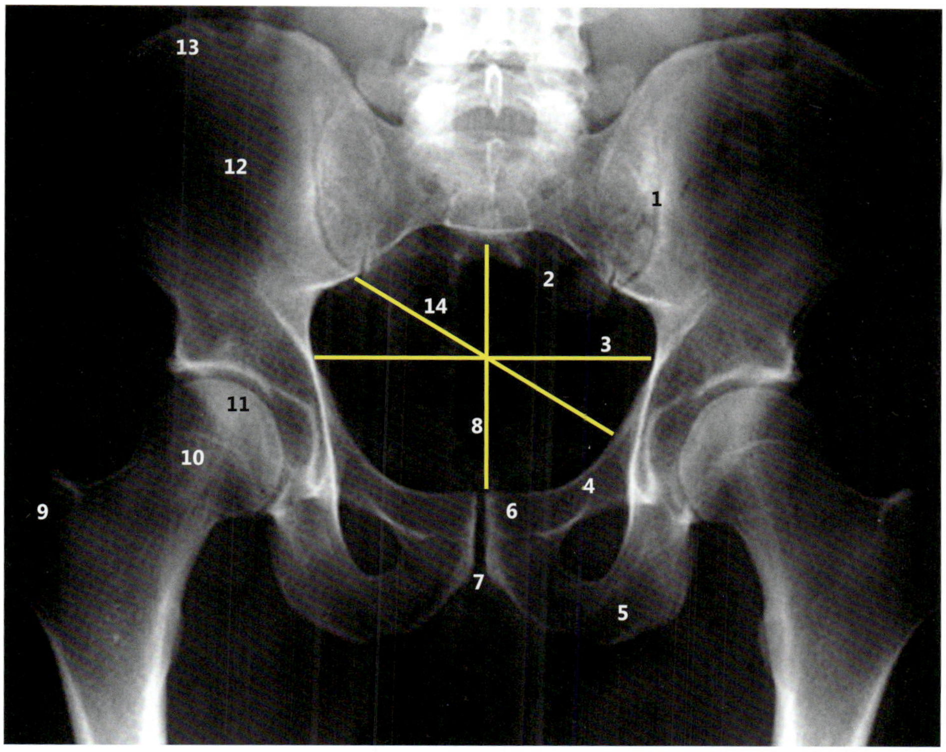

Female pelvis X-ray image in anteroposterior view. The diameters of the pelvic inlet can be also delineated.

1, Sacroiliac joint
2, Sacrum [sacral vertebrae, S I-S V]
3, Transverse diameter of pelvic inlet
4, Superior pubic ramus
5, Ramus of ischium
6, Pubic tubercle
7, Pubic symphysis
8, True conjugate (anteroposterior diameter of pelvic inlet)
9, Greater trochanter
10, Anatomical neck of femur
11, Head of femur
12, Ala of ilium (wing of ilium)
13, Iliac crest
14, Oblique diameter of pelvic inlet

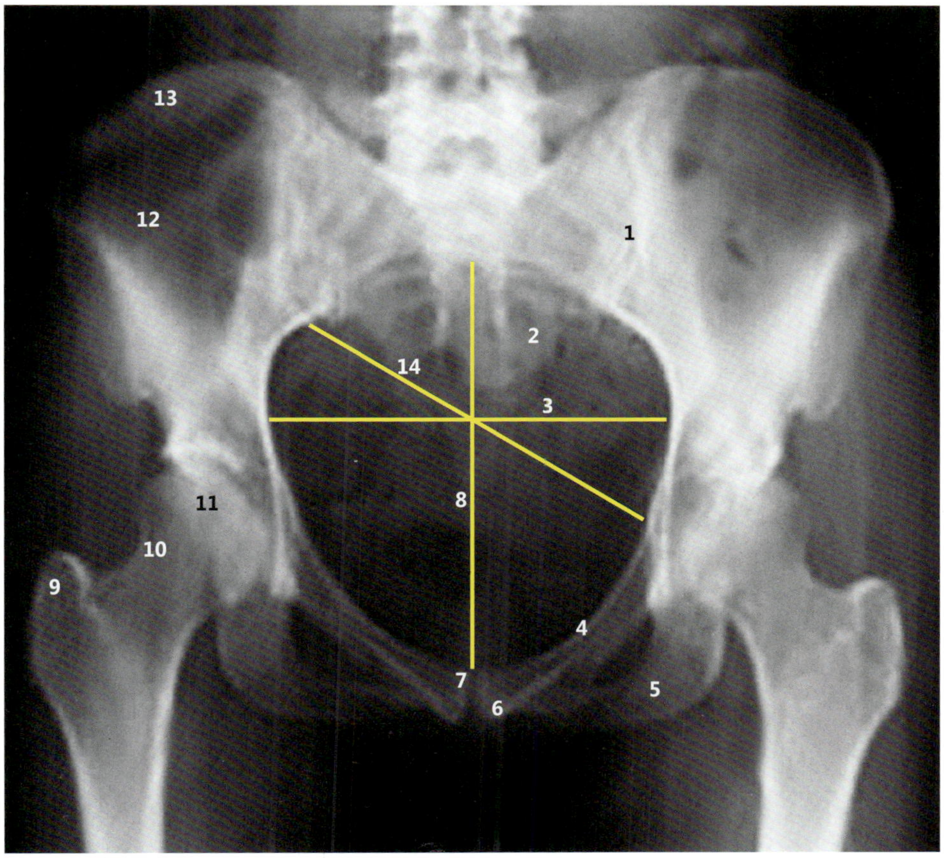

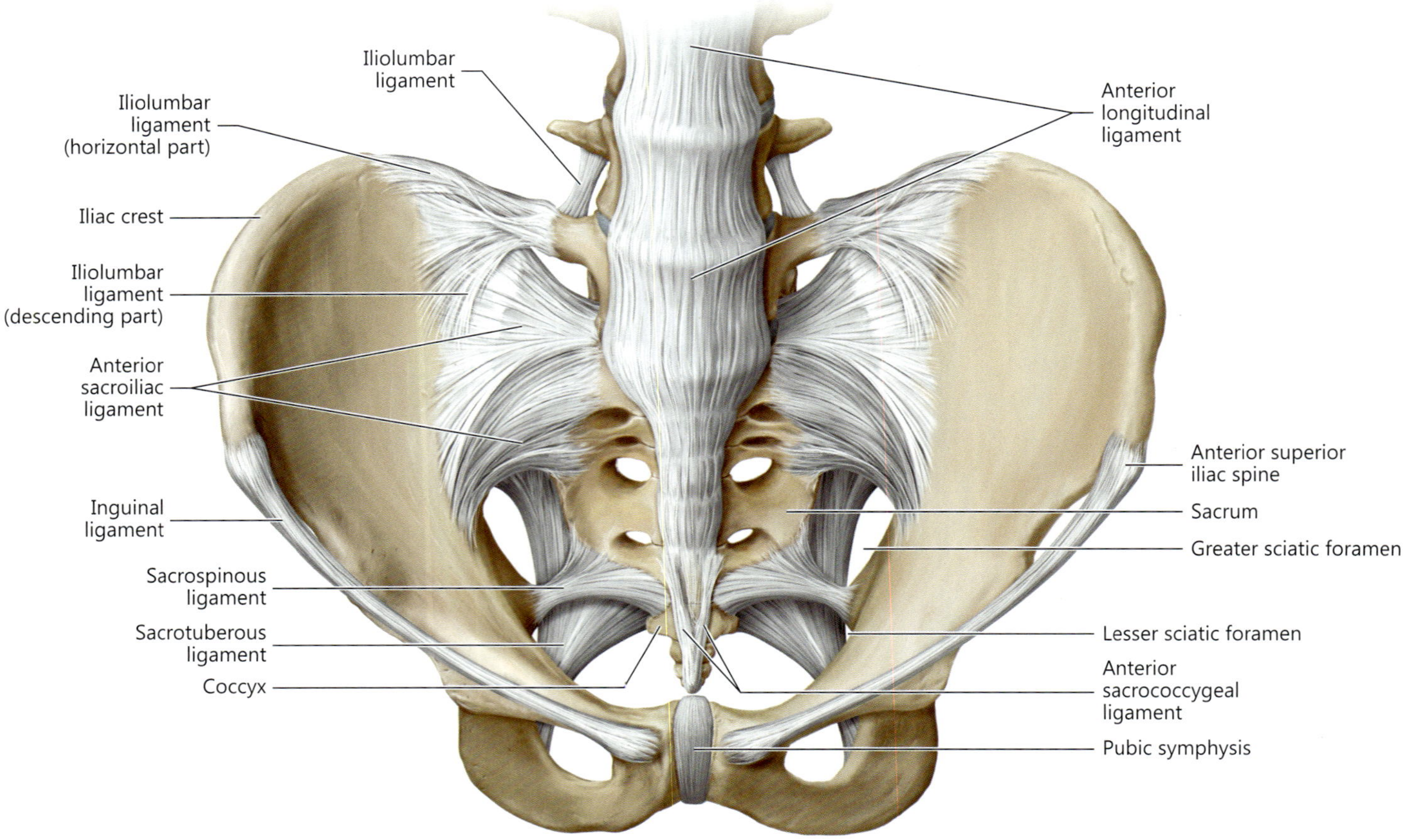

Ligaments of the male pelvis, anterior view.

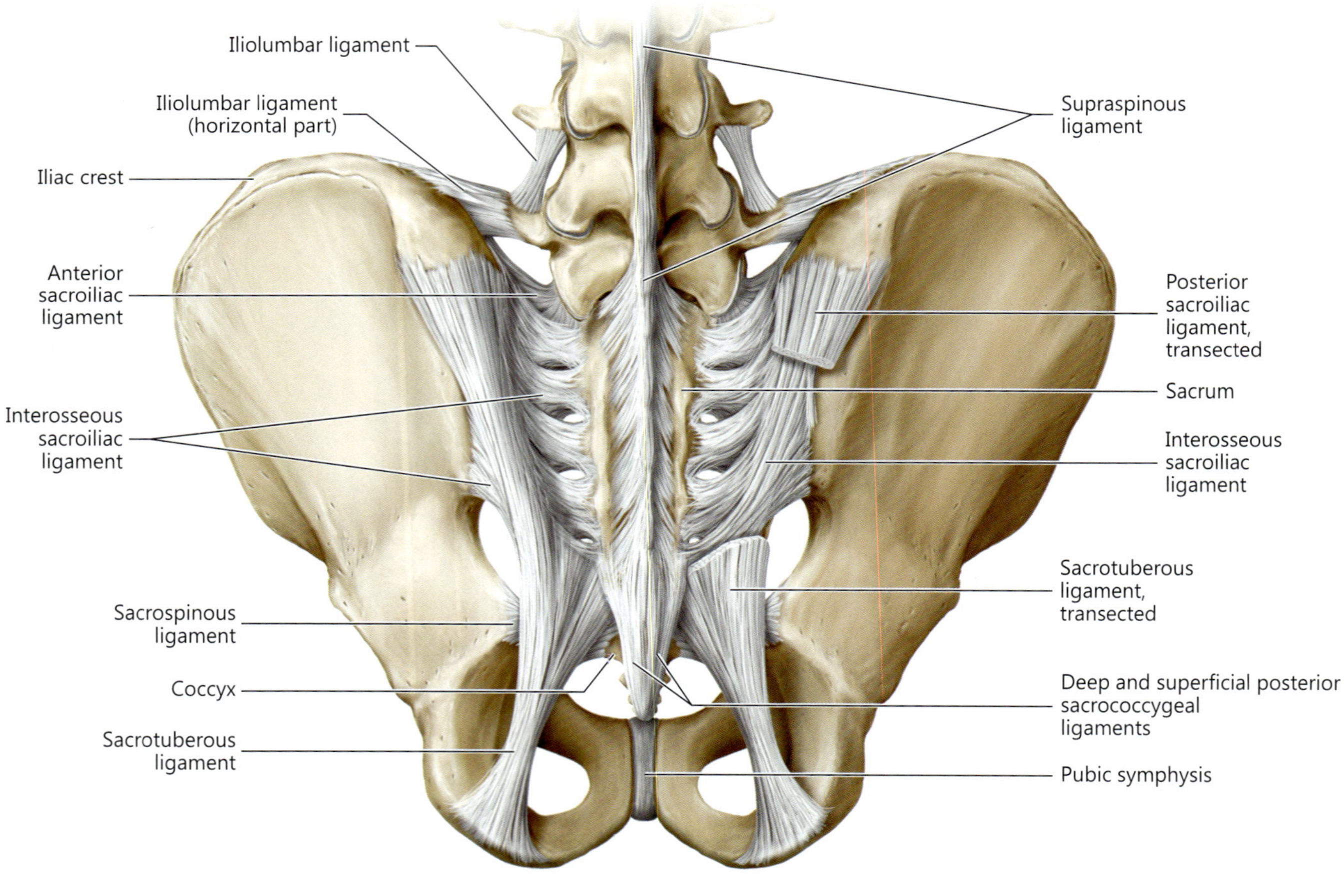

Ligaments of the male pelvis, posterior view. The sacrotuberous and posterior sacroiliac ligaments on the right side have been transected to show a deeper plane.

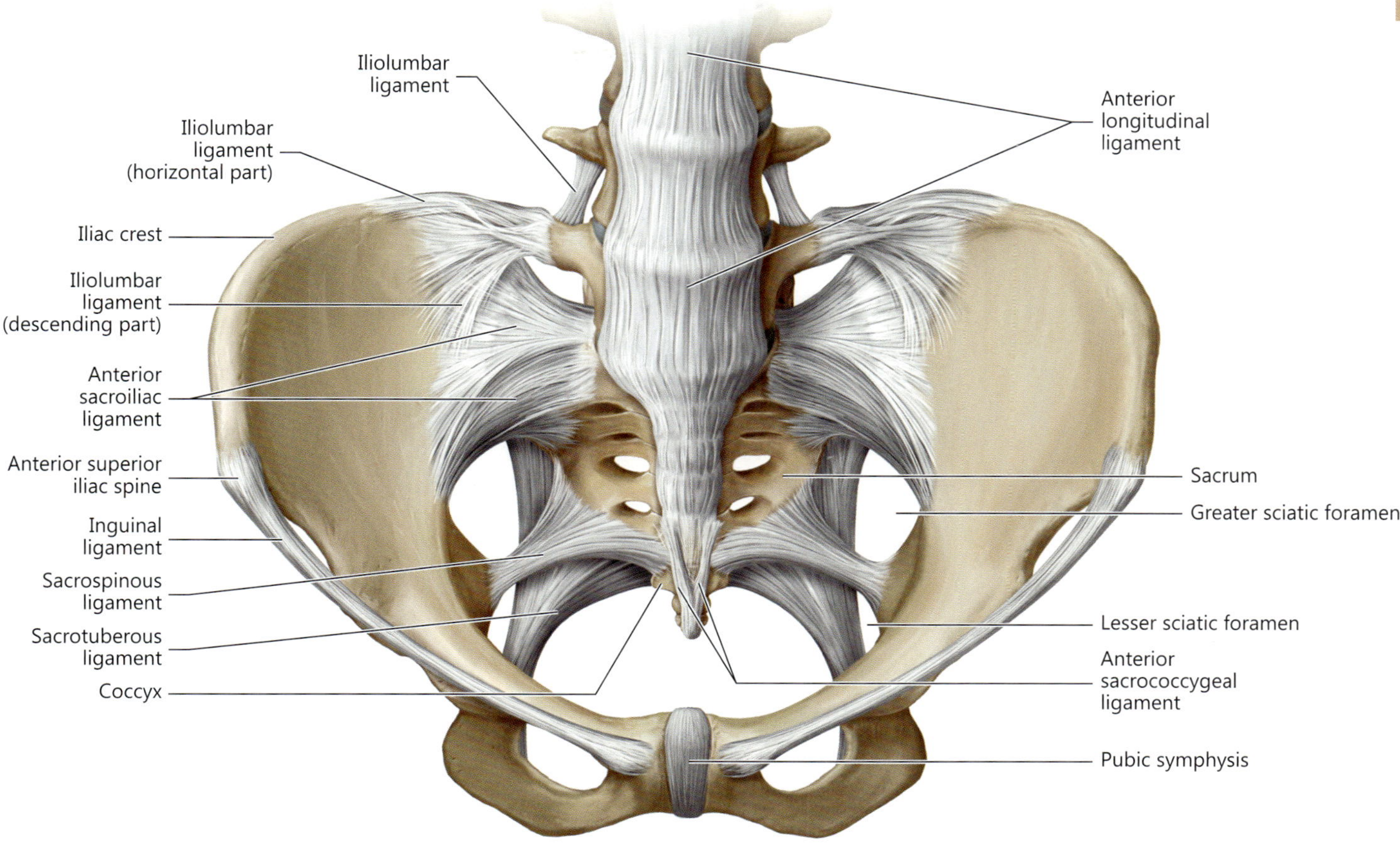

Ligaments of the female pelvis, anterior view.

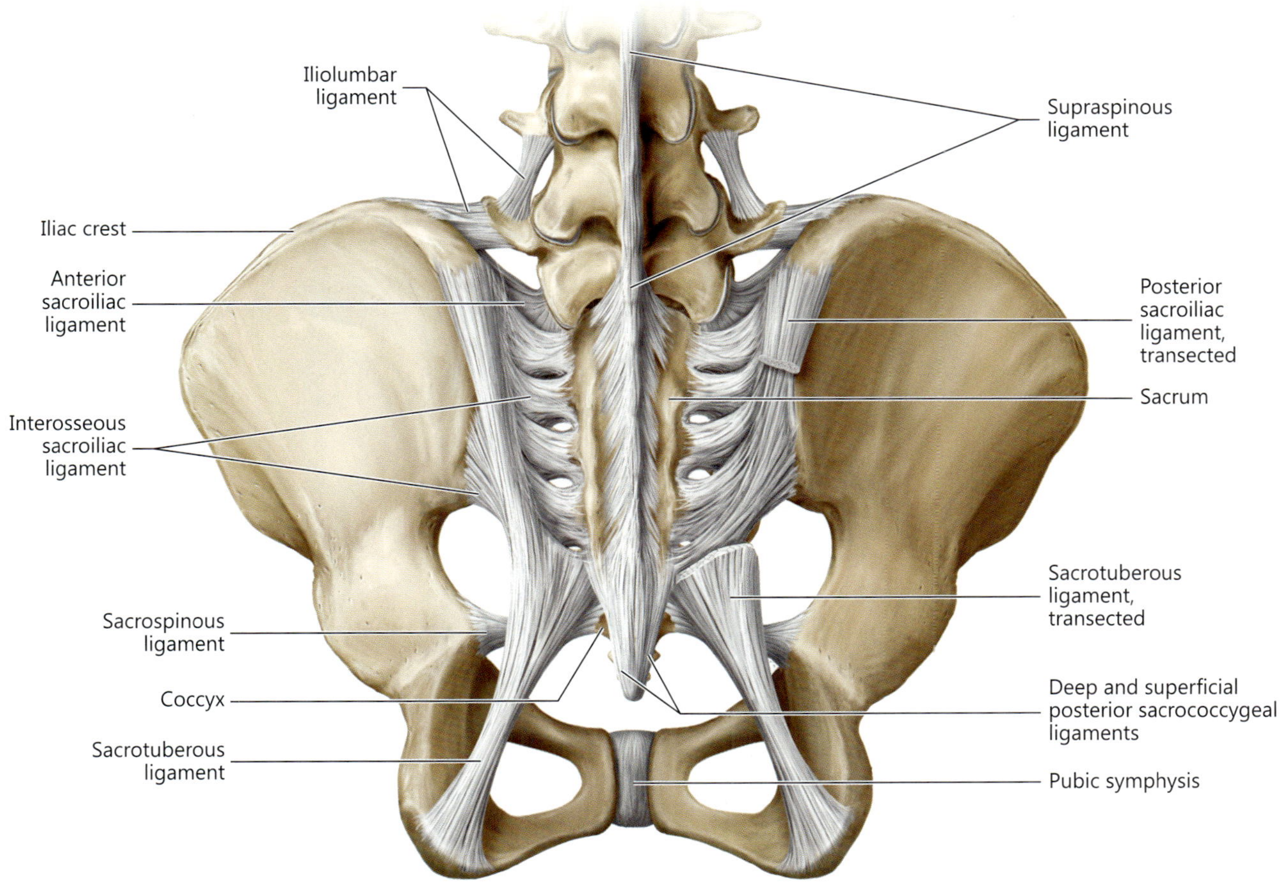

Ligaments of the female pelvis, posterior view. The sacrotuberous and posterior sacroiliac ligaments on the right side have been transected to show a deeper plane.

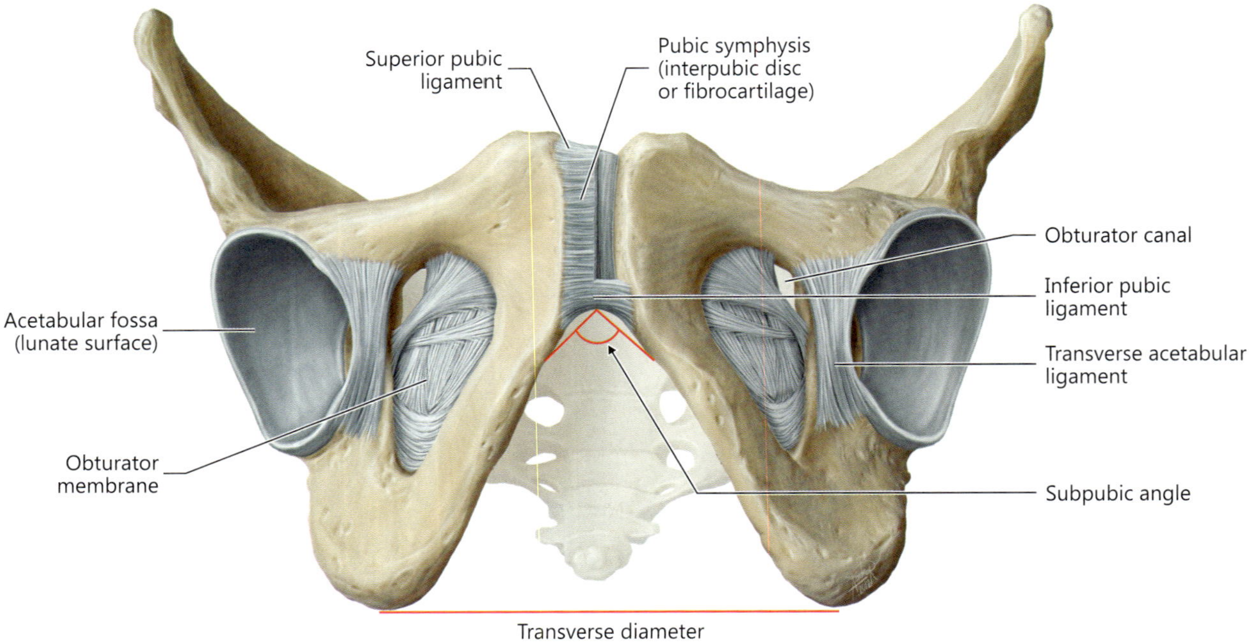

Male pelvis, anteroinferior view. In this projection certain sex-related characteristics of pelvic conformation such as, the amplitude of the subpubic angle, the obturator foramen shape, the orientation of the acetabular fossae and the inclination of the iliac wings are observed.

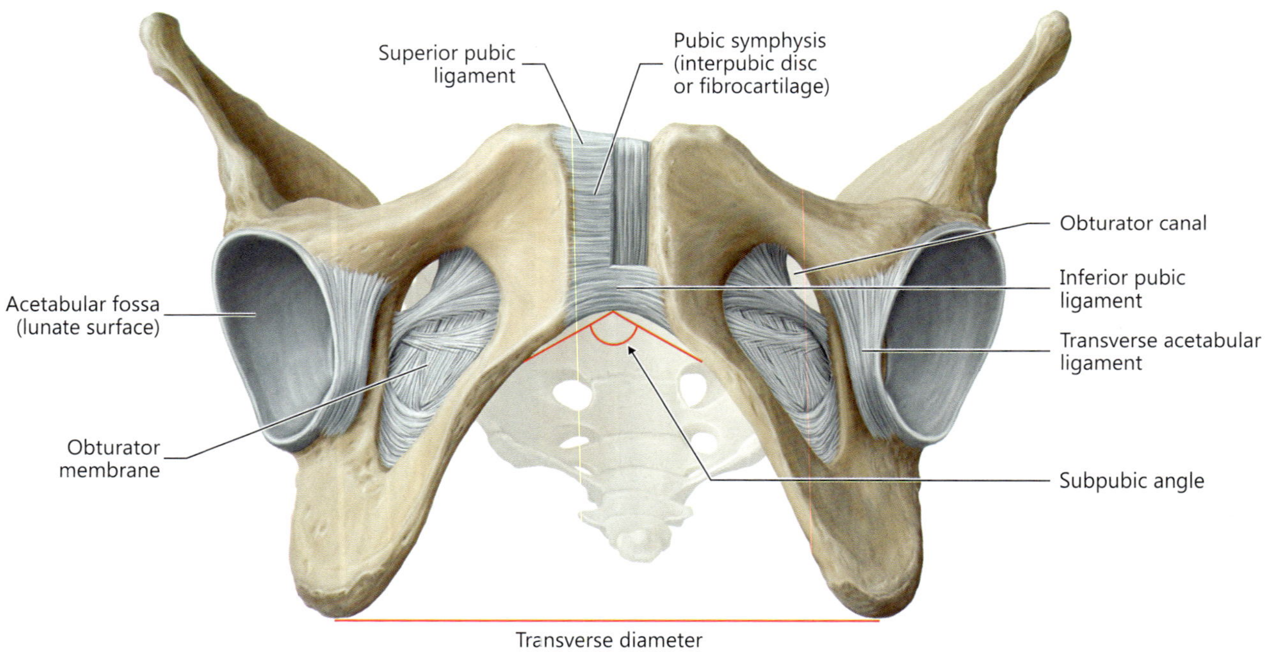

Female pelvis, anteroinferior view. In comparison to the previous illustration, note the increased amplitude of the subpubic angle and the distance between the two ischial tuberosities (transverse diameter of the pelvic outlet); that is the distance indicated by the *red* segment.

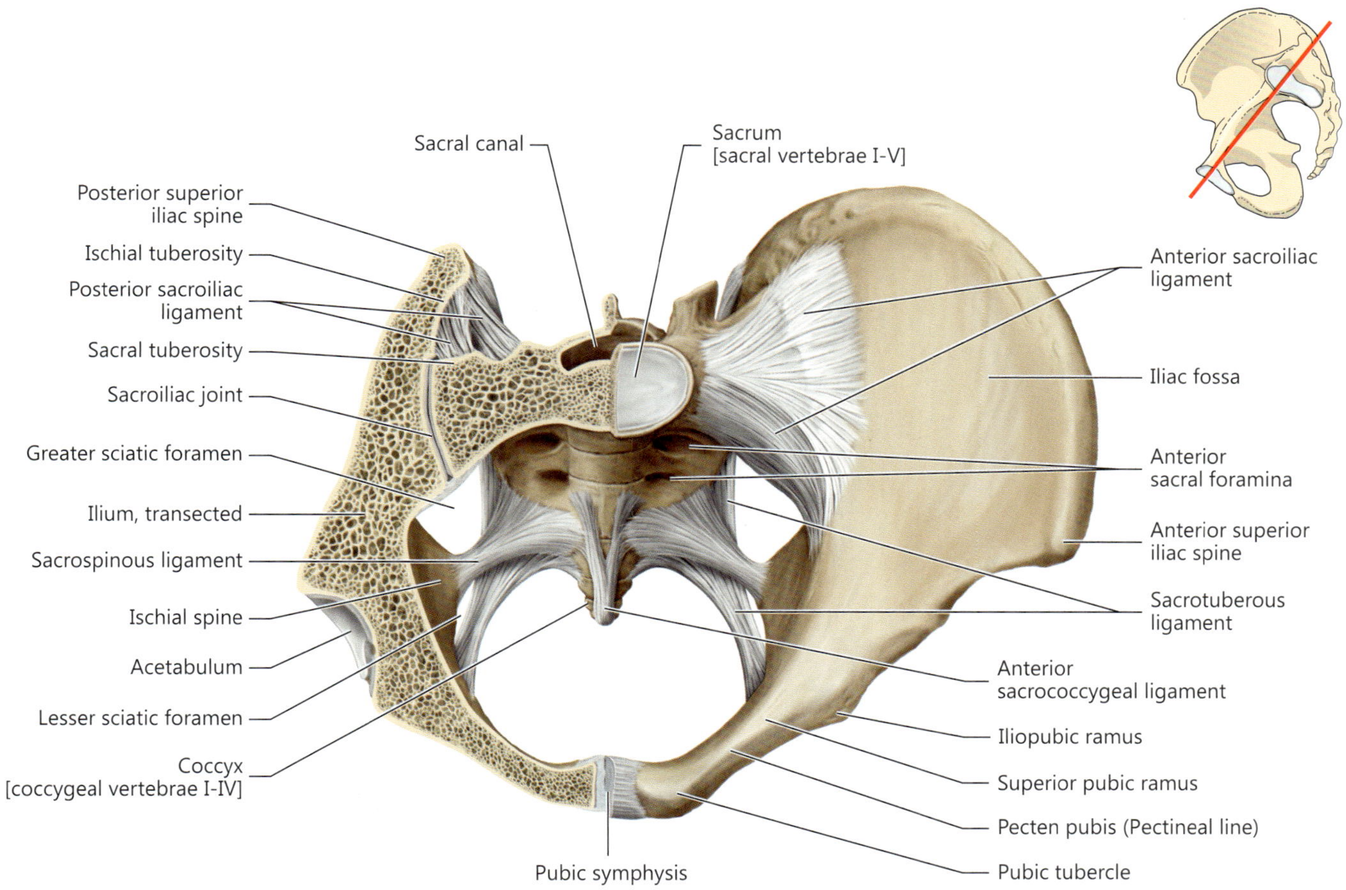

Joints and ligaments of the **pelvis** (anterosuperior view). On the left, the sacrum, the ilium and pubis have been transected in a transverse oblique plane toward the front, to show the disk of the sacroiliac joint (cut surface) and the posterior sacroiliac ligament.

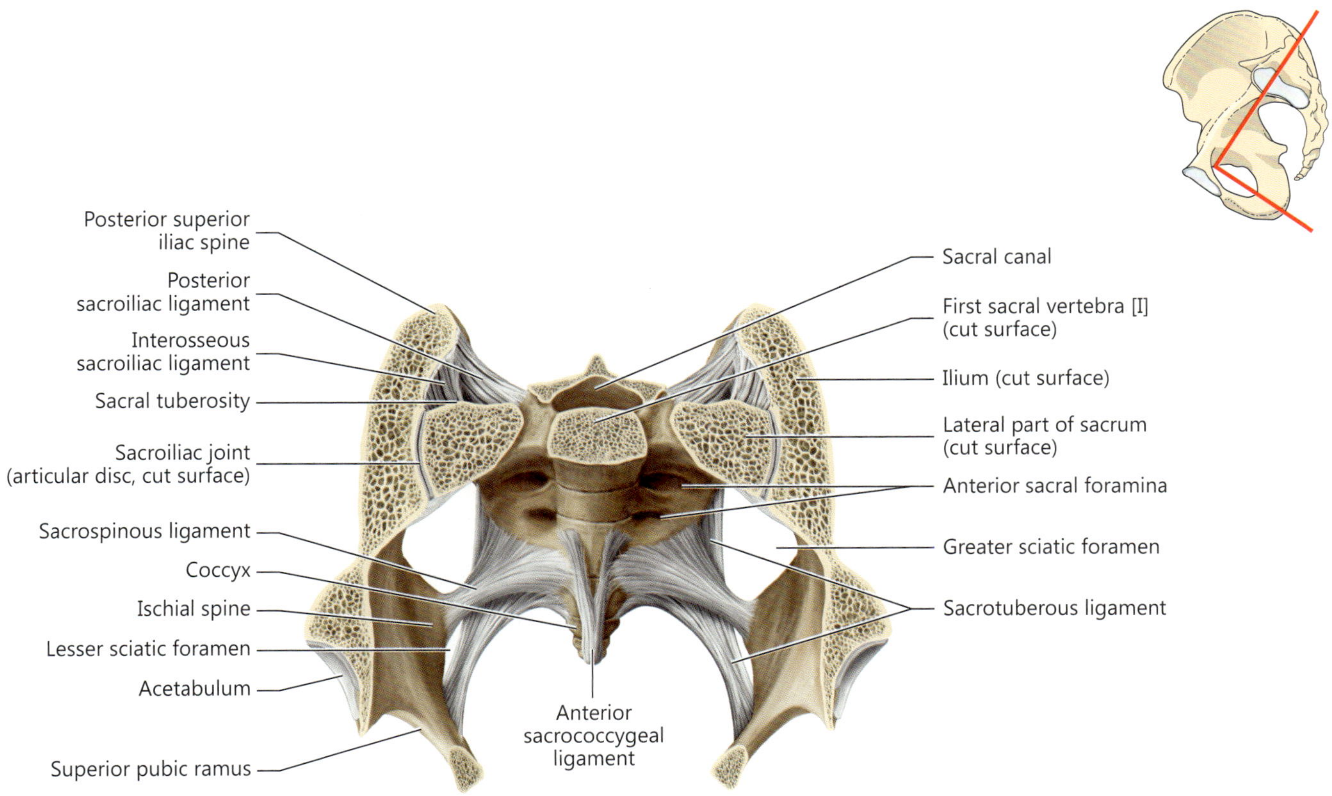

Transverse section passing through the first sacral segment showing the relations between **sacrum** and **hip bones** at the level of the sacroiliac joints.

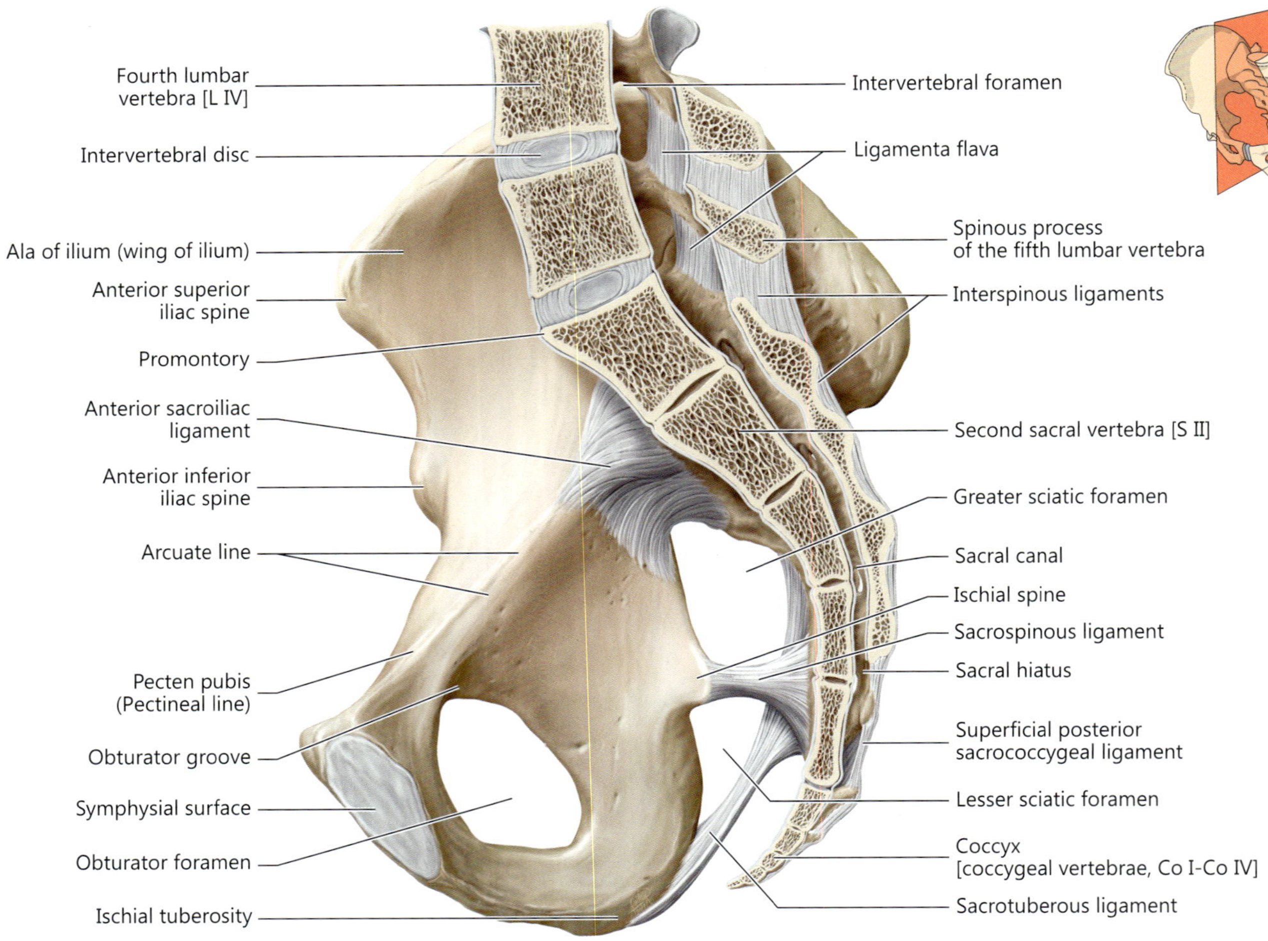

Median sagittal section of the **pelvis** passing through the pubic symphysis. The sacrospinous and sacrotuberous ligaments help to define the great and lesser sciatic foramen.

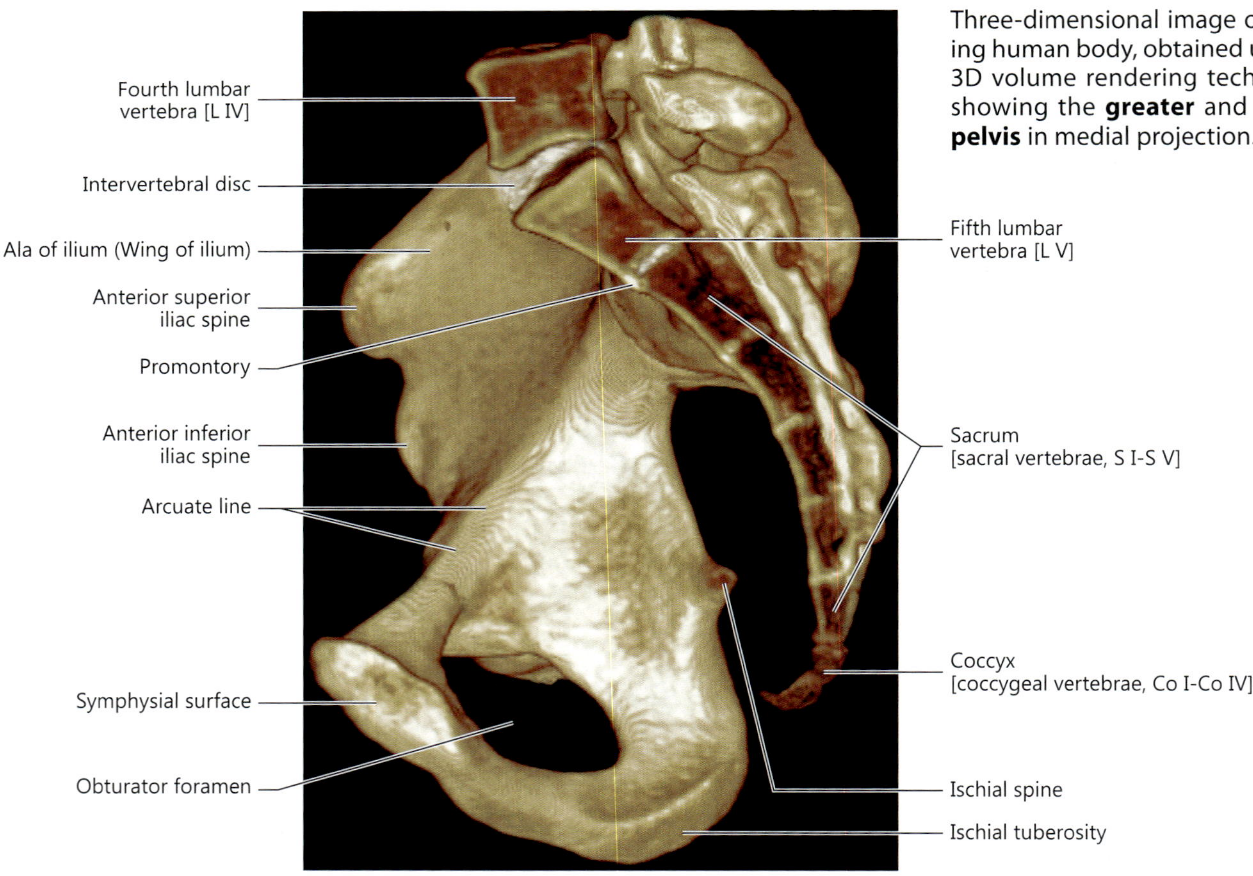

Three-dimensional image of a living human body, obtained using a 3D volume rendering technique, showing the **greater** and **lesser pelvis** in medial projection.

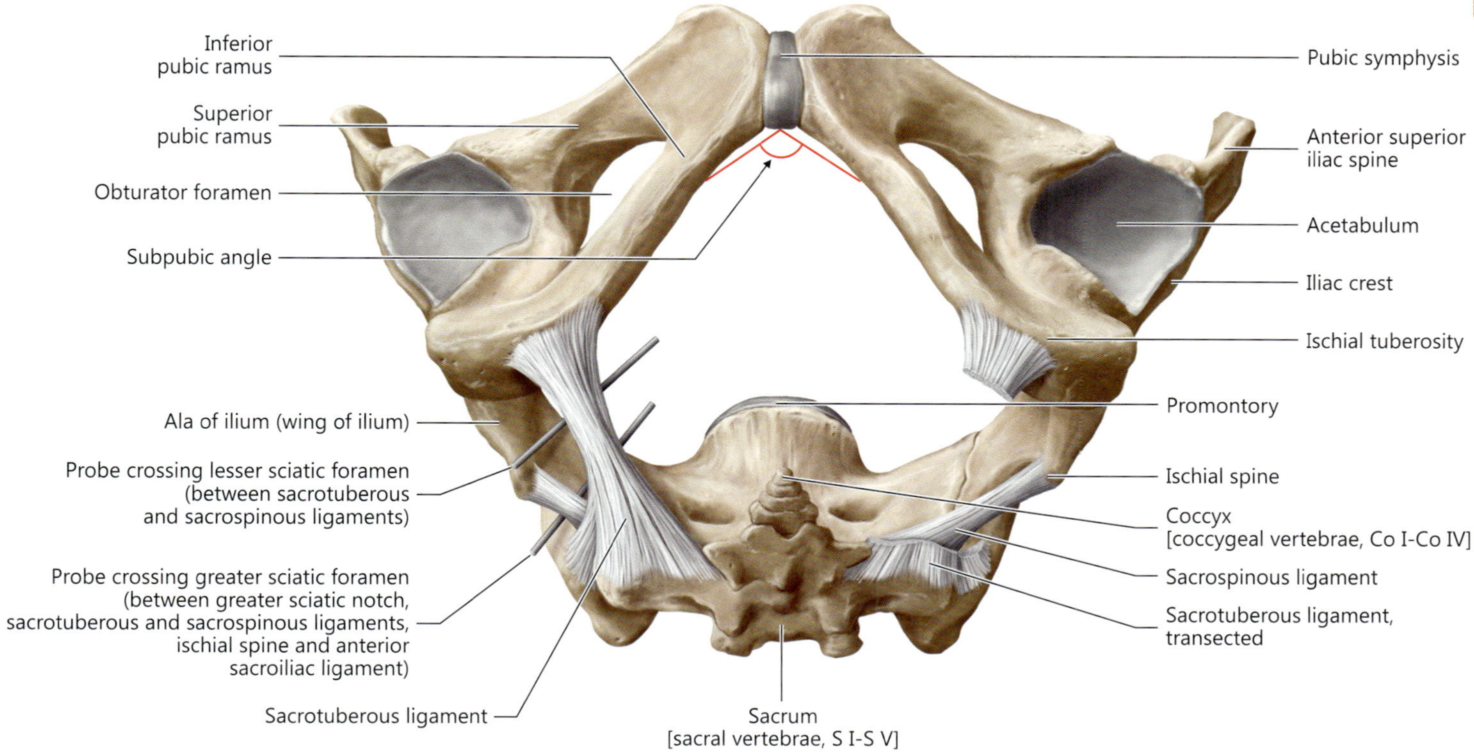

Male pelvis, view from below. Note the contour and the width of the pelvic inlet. In the left half the sacrotuberous ligament has been transected to note the location of the sacrospinous ligament.

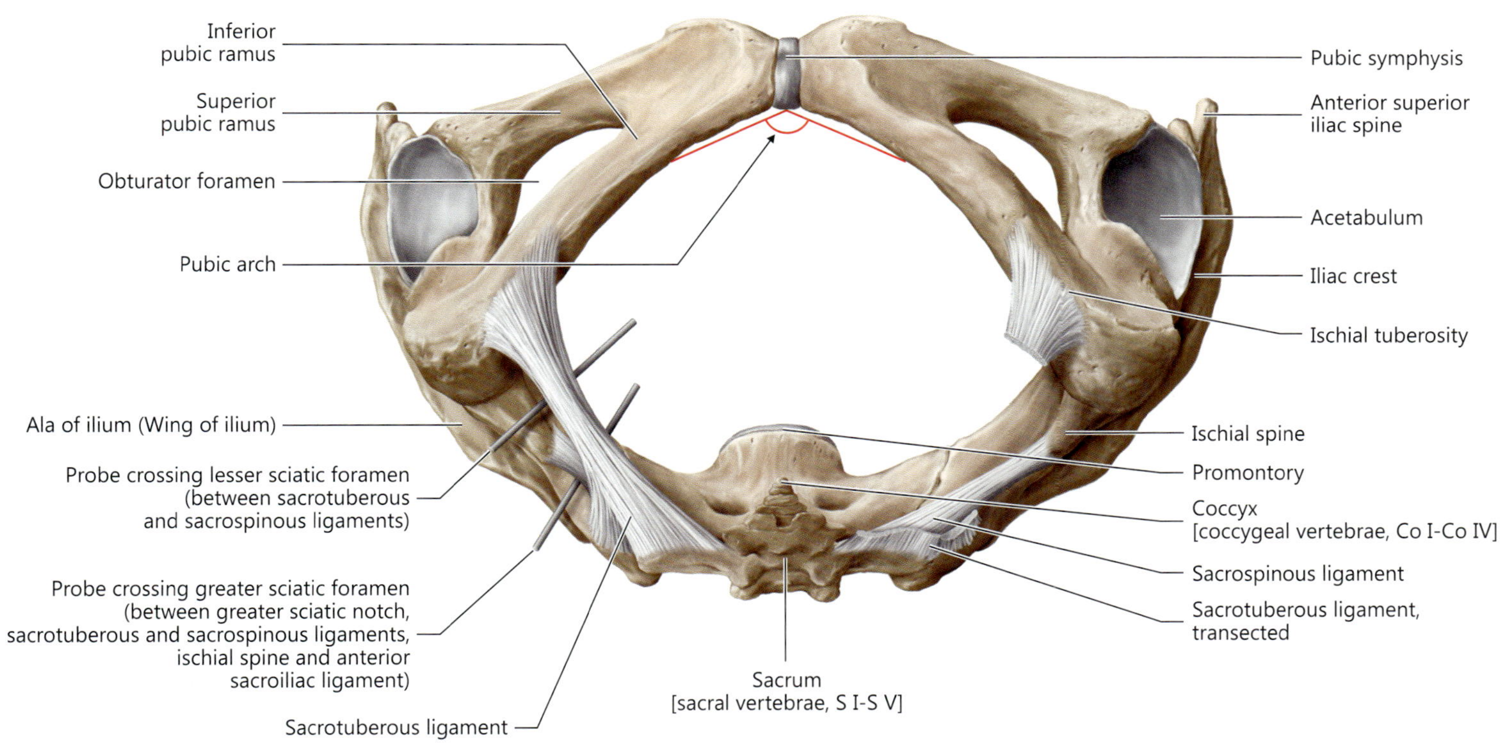

Female pelvis, view from below. In comparison to the male pelvis, the pelvic inlet is wider. In the left half the sacrotuberous ligament has been transected to note the location of the sacrospinous ligament.

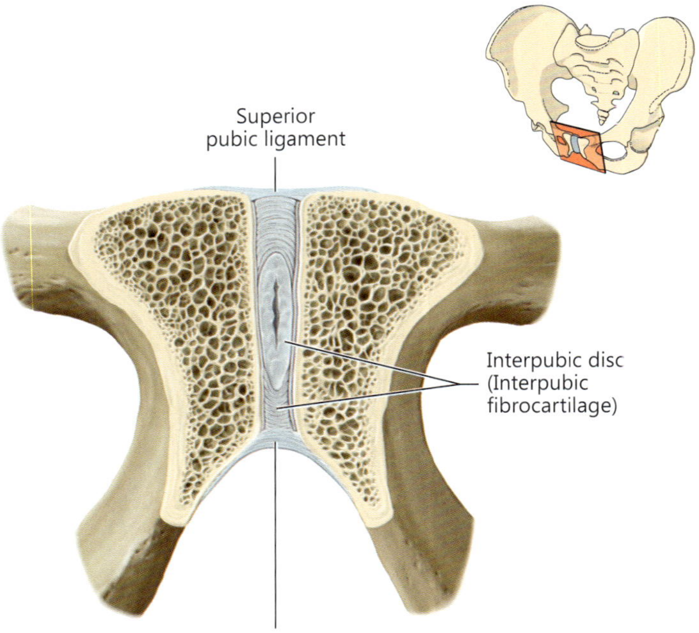

Frontal section of the pelvis passing through the **pubic symphysis**. Note the intra-articular means of union and the superior and inferior pubic ligaments.

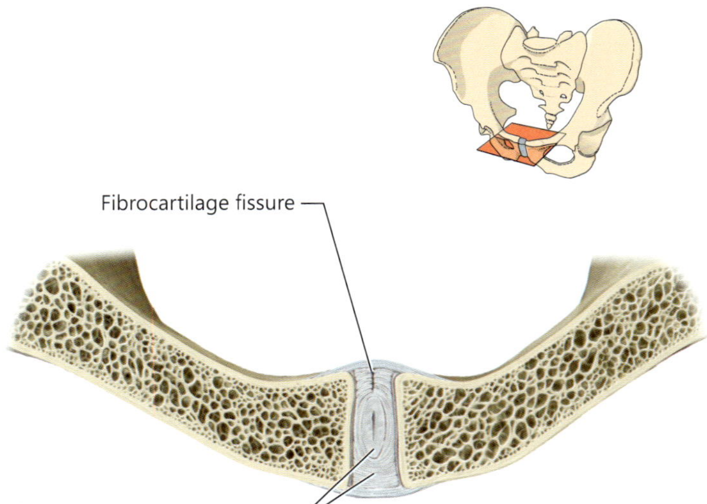

Pubic symphysis. The area between the symphysial surfaces of the pubic bones, covered with hyaline cartilage, is occupied by the interpubic disc (interpubic fibrocartilage), which in its posterosuperior portion may be subject to reabsorption with the formation of a fissure or cavity.

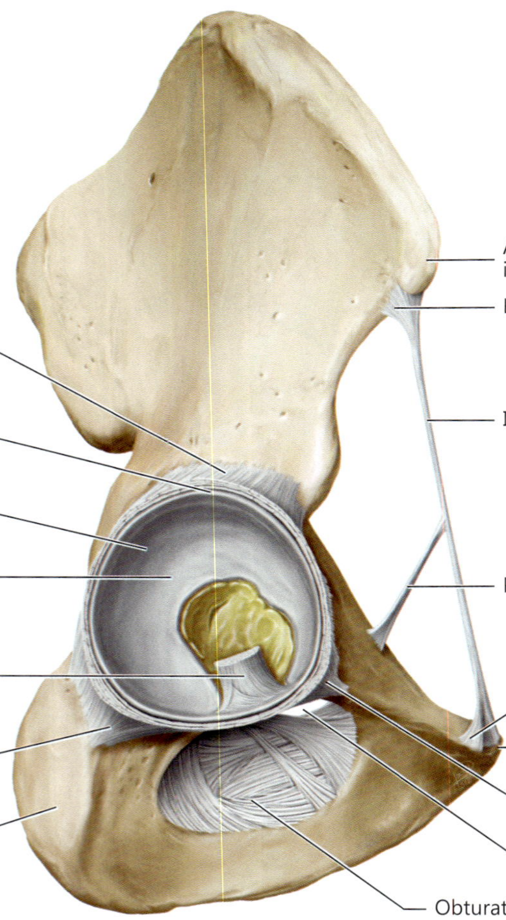

Right **coxofemoral joint**. Note the articular cavity of the hip bone, the acetabulum. This cavity is enlarged by the acetabular labrum which passes through the acetabular notch by turning it into a cup.

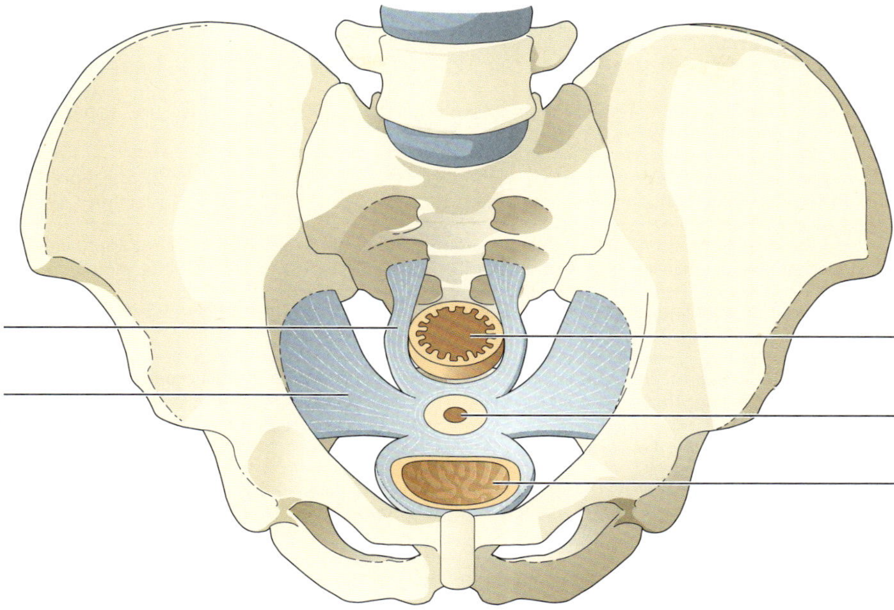

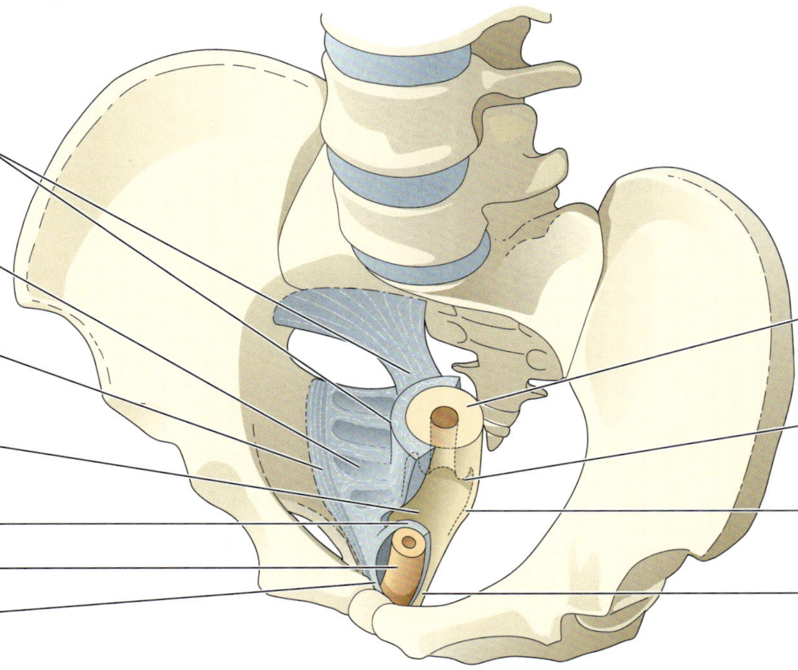

The **suspensory system of the pelvis**, consisting of a lateroposterior coupling system (cardinal and uterosacral ligaments) (**a**) and an anterior coupling system (pubourethral, vesicopelvic and urethropelvic ligaments) (**b**).

Internal pelvimetry	
	PELVIC INLET (THE FETAL HEAD PASSES THROUGH THE PELVIC INLET AT THE BEGINNING OF LABOR)
Anatomical conjugate or anteroposterior diameter	Unites the superior margin of the pubic symphysis to the promontory; measures approximately 11 cm
True conjugate	Drawn between the posterior surface of the pubic symphysis and the promontory; represents the useful space for the passage of the fetus head; measures 10.5 cm; can be deduced by subtracting 1.5 cm from the diagonal conjugate
Diagonal conjugate	Unites the promontory to the inferior margin of the pubic symphysis; measures approximately 12 cm; can be directly measured by introducing the index and middle fingers into the vagina up to touch the promontory
Transverse diameter (of the pelvic inlet)	Unites the most lateral points of the pelvic inlet, measures approximately 13.5 cm
Oblique diameter	Drawn between the iliopubic ramus of one side to the opposite sacroiliac joint; it is the diameter that normally engages the head of the fetus and measures approximately 12 cm
	PELVIC OUTLET
Median conjugate	Unites the inferior margin of the pubic symphysis to the apex of the sacrum; measures approximately 11.5 cm; it is the most important anteroposterior diameter because the position of the coccyx, due to its flexibility, is subject to variations
Straight conjugate or anteroposterior diameter	Unites the inferior margin of the pubic symphysis to the tip of the coccyx; measures approximately 9.5 cm
Bisciatic diameter	Drawn between the two ischial spines; measures approximately 10.5 cm
Transverse diameter of the pelvic outlet (bituberous diameter)	Drawn between the inner aspects of the two ischial tuberosities; measures approximately 11 cm
	EXTERNAL PELVIMETRY (EXTERNAL MEASURES USEFUL FOR INFORMATION ON INTERNAL MEASURES AND PROPORTION BETWEEN DIFFERENT PARTS OF THE PELVIS)
External conjugate	Unites the spinous processes of the fifth lumbar vertebra to the superior margin of the pubic symphysis; measures approximately 20 cm
Intercristal distance	Unites the most lateral points of the iliac crest; measures 26-28 cm
Interspinous distance (interspinous diameter)	Unites the two anterior superior iliac spines; measures approximately 26 cm
Intertrochanteric distance	Unites the most protruding points of the great trochanter of the femur; measures approximately 32 cm

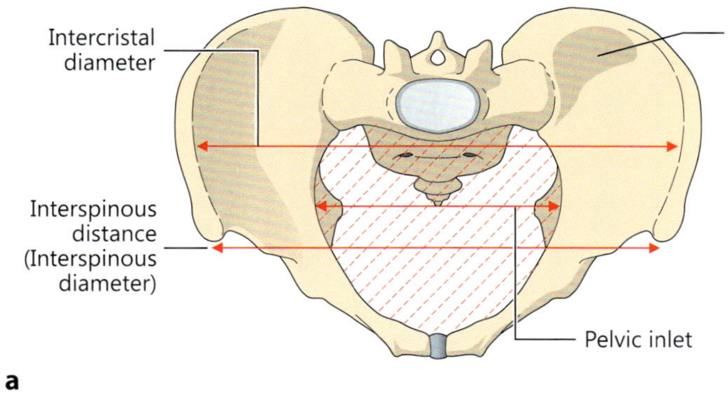

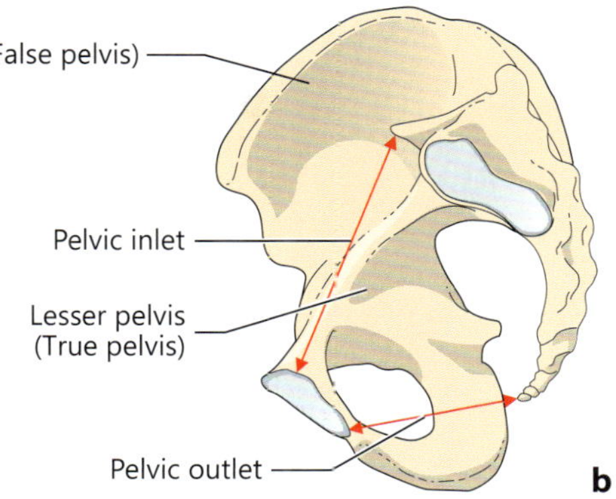

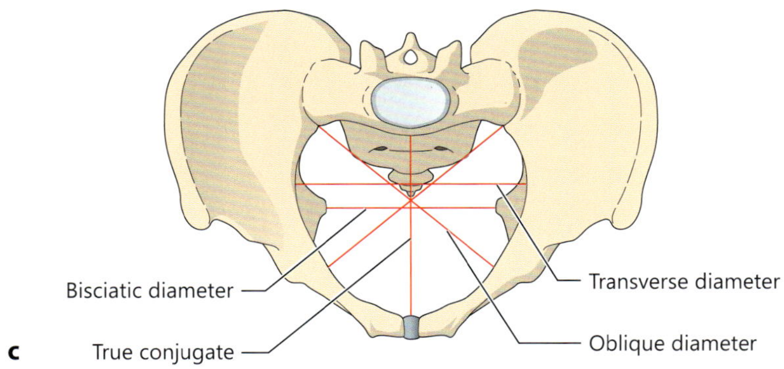

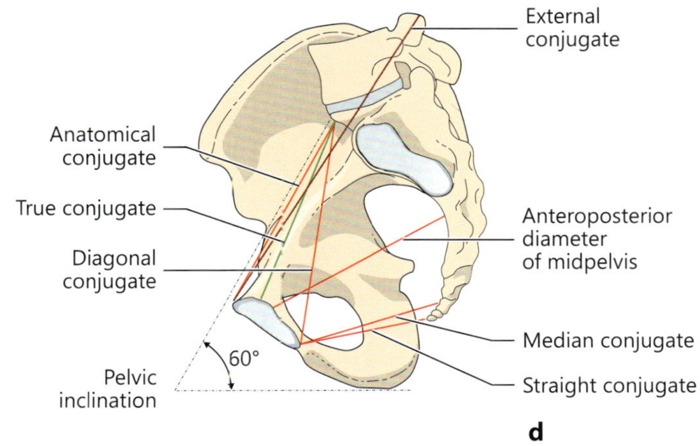

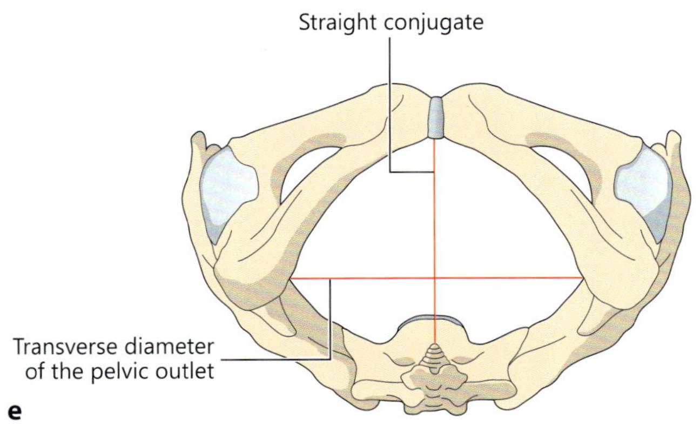

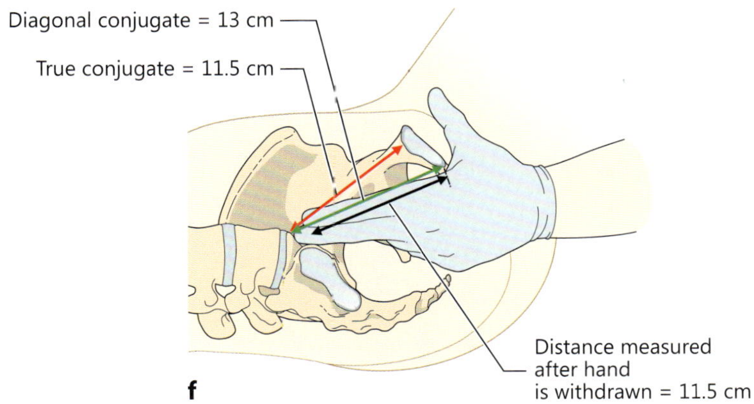

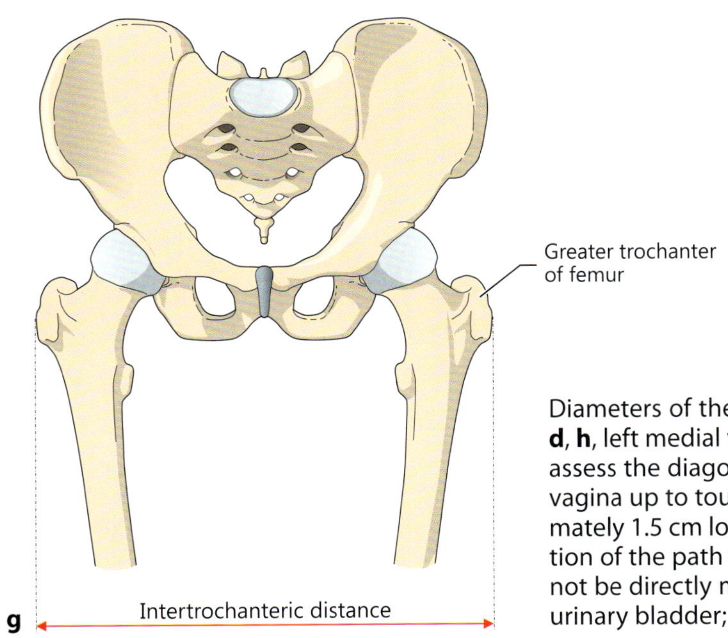

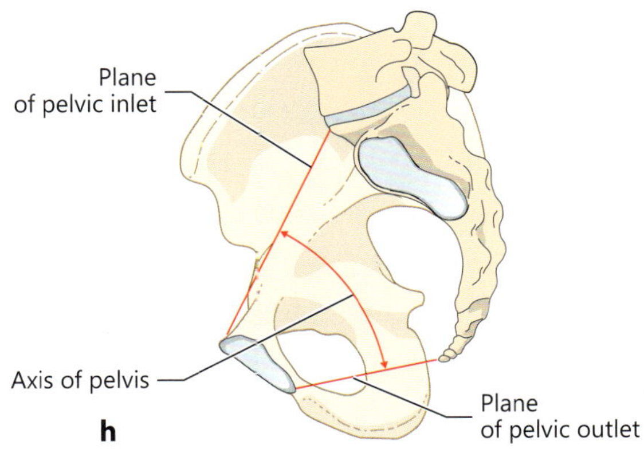

Diameters of the **female pelvis** and the sizes of the **pelvis**. **a**, **c**, superior view; **b**, **d**, **h**, left medial view; **e**, inferior view; **f**, right medial view, with examining hand to assess the diagonal conjugate. The index and middle fingers are inserted into the vagina up to touch the promontory of sacrum. The diagonal conjugate is approximately 1.5 cm longer than the obstetric conjugate (the narrowest deformable portion of the path through which the fetal head must pass during birth), which cannot be directly measured, during vaginal examination, due to the presence of the urinary bladder; **g**, anterior view.

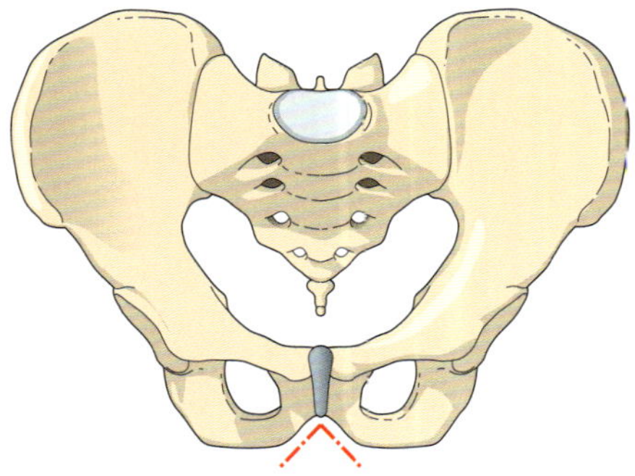

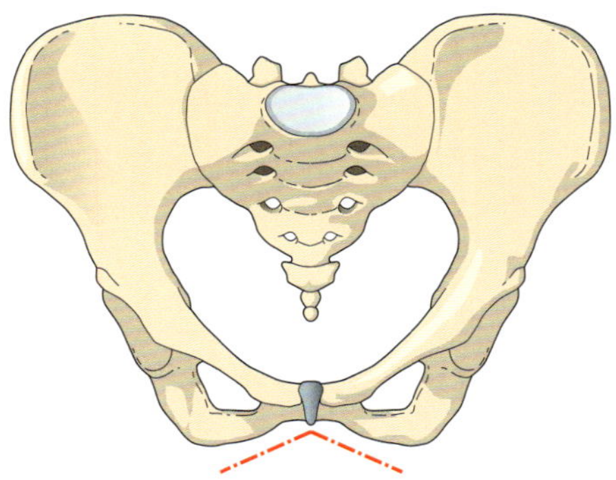

CHARACTERISTICS OF THE PELVIS	MALE (♂)	FEMALE (♀)
General structure	Thick and heavy	Thin and light
Greater pelvis	Elongated and deep	Shorter and shallower
Lesser pelvis	Narrow and deep, tapered	Wide and shallow, cylindrical
Pelvic inlet	Heart-shaped, narrow	Oval and rounded, wide
Pelvic outlet	Relatively small (compared to the female)	Relatively wide (compared to the male)
Pubic arch and subpubic angle	Narrow (<70°)	Wide (>80°, ~110°)
Obturator foramen	Round	Oval
Acetabulum	Wide	Small
Greater sciatic notch	Narrow (~70°); inverted V-shaped	Almost 90°

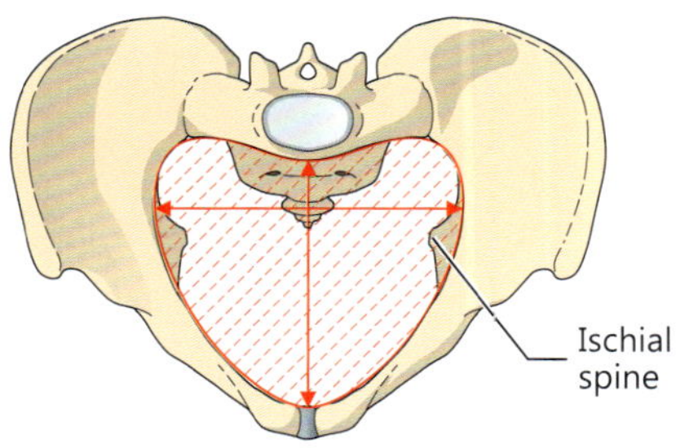

a Android

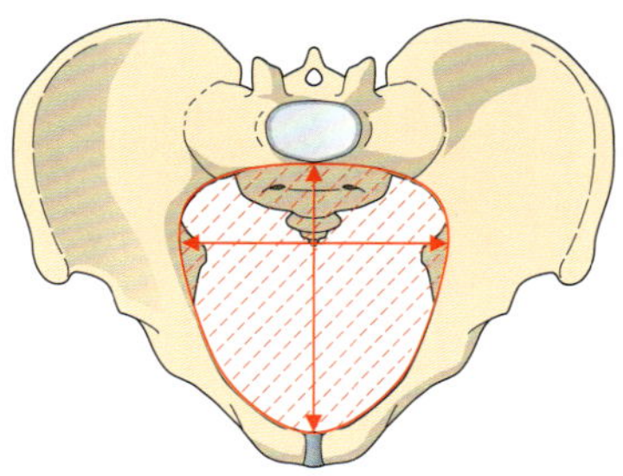

b Anthropoid

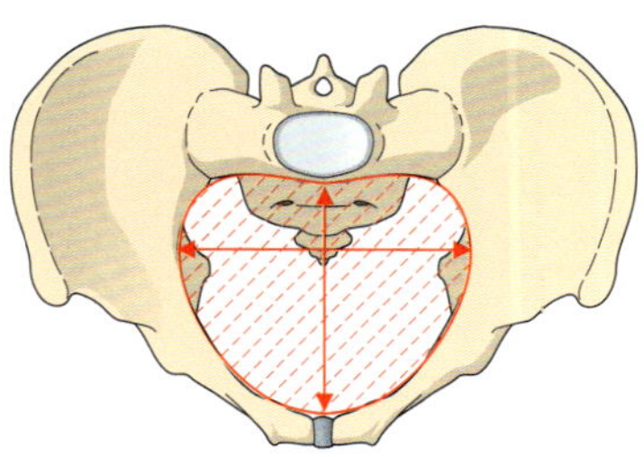

c Gynoid

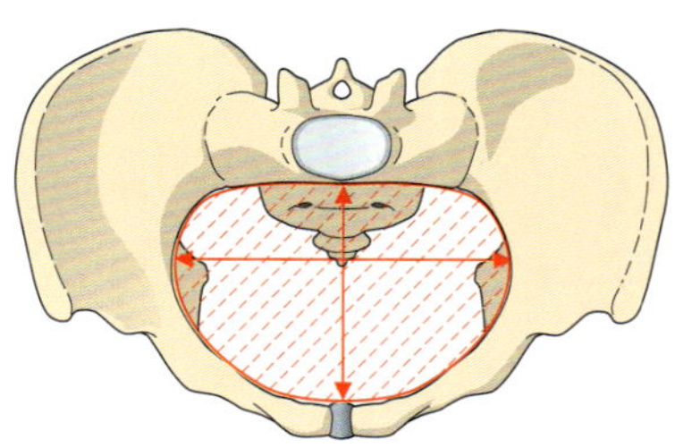

d Platypelloid

Pelvis. The android (**a**) and anthropoid (**b**) types are more common in males, while the gynoid (**c**) and android (**a**) types prevail in the Caucasian female, the gynoid (**c**) and anthropoid (**b**) types in African females; finally, the platypelloid type (**d**) is rare in both sexes. The gynoid pelvis is a normal female pelvis. Sex determination can be made through the identification of human skeletal remains. One of the first structures to be considered is the pelvis because sexual differences are usually clearly visible. Such findings may also be completed on portions or fragments of the pelvis.

Summary table of the bones and relative joints of the trunk			
BONE	*LOCATION*	*DESCRIPTION*	*JOINTS*
Atlas	Neck	First cervical vertebra	Occipital (superiorly), axis (inferiorly)
Axis	Neck	Second cervical vertebra	Atlas (superiorly), third cervical vertebra (inferiorly)
Cervical vertebrae	Skeleton of the neck	Seven segments of the vertebral column; the smallest of the vertebrae; presence of a foramen in each transverse process	With the adjacent vertebrae; the first with the occipital bone; the seventh with the first thoracic vertebra
Clavicle	Shoulder	Flat and curved bone horizontally placed above the first rib	Sternum, scapula, cartilage of the first rib
Coccyx	Lower part of the trunk	Consists of three to four rudimentary vertebrae of which only the first one is not merged	Sacrum
Ilium	Pelvis	The upper part is expanded of the hip bone, divisible into a body and a wing	Sacrum, femur, ischium and pubis
Ischium	Pelvis	Inferior and dorsal part of the hip bone, divisible into a body and a ramus	Femur, ilium, pubis
Lumbar vertebrae	Back, inferior part	Five segments of the vertebral column; the largest vertebrae of the mobile part of the vertebral column	With the adjacent vertebrae; the fifth vertebra with the sacrum
Pelvis	Bone formation, which resembles a basin, consists of two hip bones, the sacrum and the coccyx		
Pubis	Pelvis	Anterior and inferior portion of the hip bone	Ilium, ischium, femur
Ribs	Thorax	Twelve pairs of thin bones, elongated and curved that form the posterior and lateral walls of the thorax	All articulate posteriorly with the vertebral column; the seven superior pairs articulate anteriorly with the sternum through the costal cartilages; the five inferior pairs articulate anteriorly with the costal cartilages; the last two pairs are free to the ventral extremity (floating ribs)
Sacrum	Lower part of the trunk	Large, triangular bone at the dorsal part of the pelvis. It consists of five fused vertebrae	Superiorly attached to the last lumbar vertebra, laterally with the ilium, inferiorly with the coccyx
Sternum	Thorax	Long flat bone shaped like a dagger that constitutes the ventral wall of the of the thorax; it consists of three parts: manubrium, body and xiphoid process	Both clavicles and the seven pairs of costal cartilages
Thoracic vertebrae	Back, posterior part	Twelve segments of the vertebral column; articular facets on the side of each vertebral body and the first ten have articular facets on the transverse processes as well	With the adjacent vertebrae, with the heads and the tubercles of the ribs (except the 11th and 12th)

Unit of motion

The unit of motion corresponds to the FSU (functional spinal unit, or motion segment).

In the space between two vertebrae there are:

- the joints of vertebral arches (in addition to joint capsules there is also the ligamentum flavum)
- the spinal canal and intervertebral foramen (with the spinal nerves, the meningeal branches and vessels)
- the intervertebral disc (in addition to the cartilage surfaces of the vertebral bodies, with the margins of the vertebral endplates, the nucleus pulposus, and the fibrous ring, there are also the anterior and posterior longitudinal ligaments).

The structures of the unit of motion are coordinated both anatomically and functionally. A lesion at the expense of one of its components also affects other structures.

The unit of motion can be divided into two parts:

- anterior crus (vertebral body and intervertebral disc) with support function (absorbs axial compressive forces, such as the body loading which is discharged through the column of the vertebral bodies and then transmitted)
- posterior crus (joints of the vertebral arch) with function of direction of movement (allowing or blocking movements permitted by the shape of the vertebral arch and limited by the ligamentous structures, the position of the joints and fibrous ring).

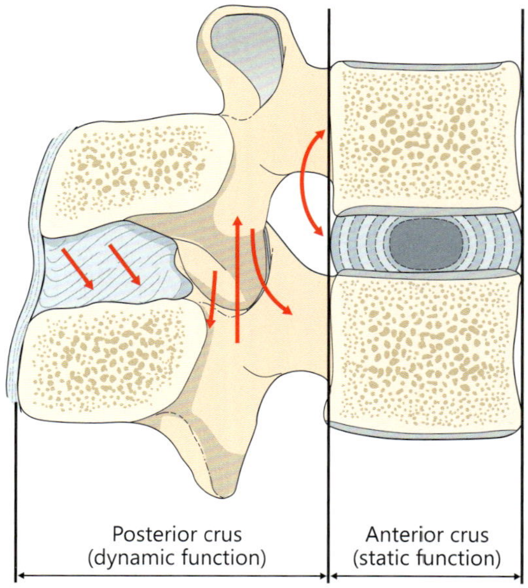

Intervertebral discs

There are 23 intervertebral discs, which are missing between the occipital bone and atlas and between the atlas and axis. Together these constitute about one quarter of the column height.

The intervertebral disc consists of:

- anulus fibrosus
- nucleus pulposus.

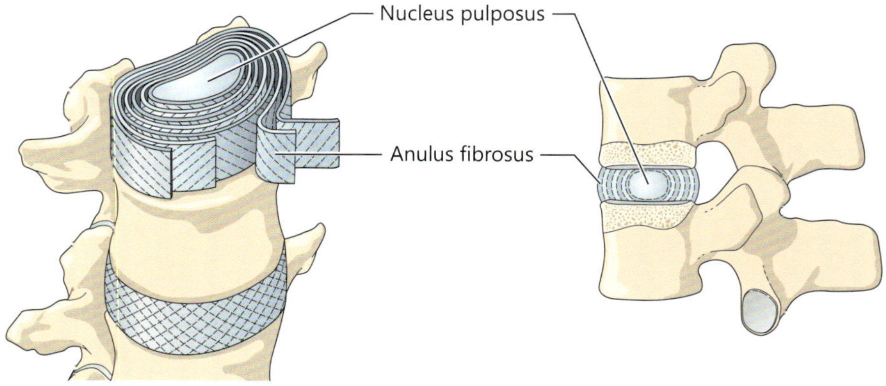

Shape of intervertebral discs

Since the vertebral bodies have similar dimensions, except those thoracic which present a slight wedge shape with increased posterior height, the physiological curves of the vertebral column largely depend on the shape of the discs and not that of the vertebral bodies.

The height of intervertebral discs is lower at cervical level, medium at thoracic level and higher at lumbar level.

Since the higher the ratio between the height of the disc and height of the vertebral body, the higher the mobility, the most mobile portion of the vertebral column is the cervical, followed by the lumbar and thoracic.

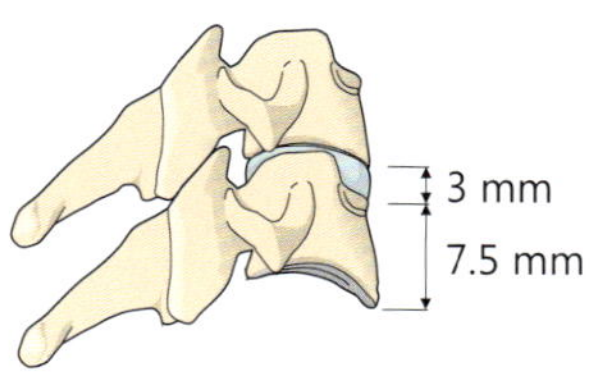

Cervical vertebrae [C I-C VII]

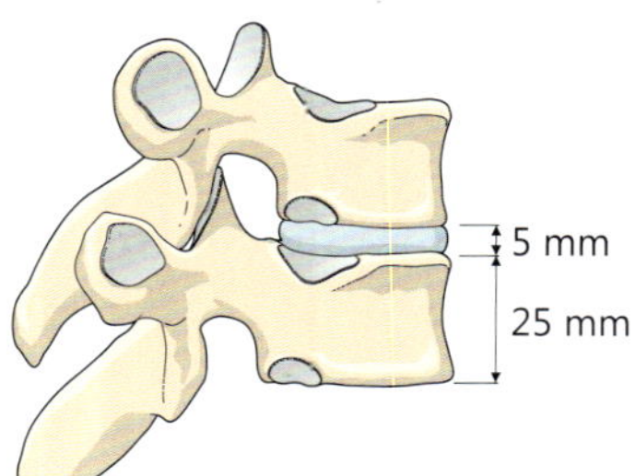

Thoracic vertebrae [T I-T XII]

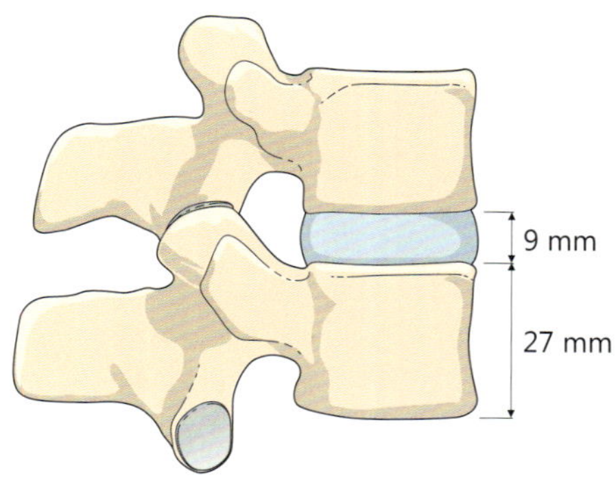

Lumbar vertebrae [L I-L V]

CERVICAL AND LUMBAR PORTIONS OF THE VERTEBRAL COLUMN	THORACIC PORTION OF THE VERTEBRAL COLUMN
Increased anterior height of the intervertebral disc	Increased posterior height of the intervertebral disc

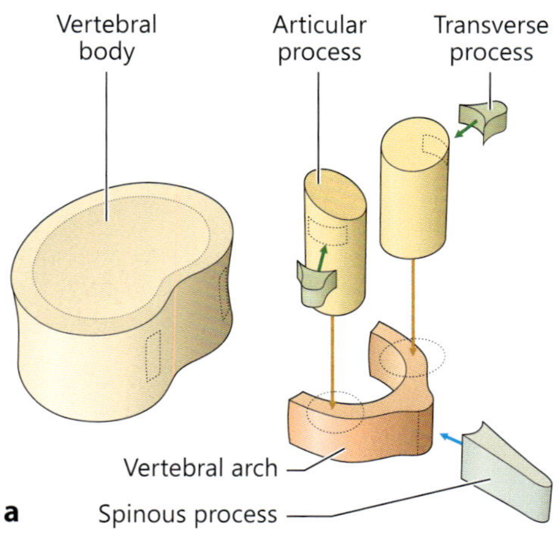

Facets of articular processes
Geometrization of parts of a vertebra. **a**, Single parts: vertebral body, vertebral arch, articular, transverse and spinous processes; **b**, vertebral body and arch (with the articular, transverse and spinous processes); **c**, vertebra in toto.

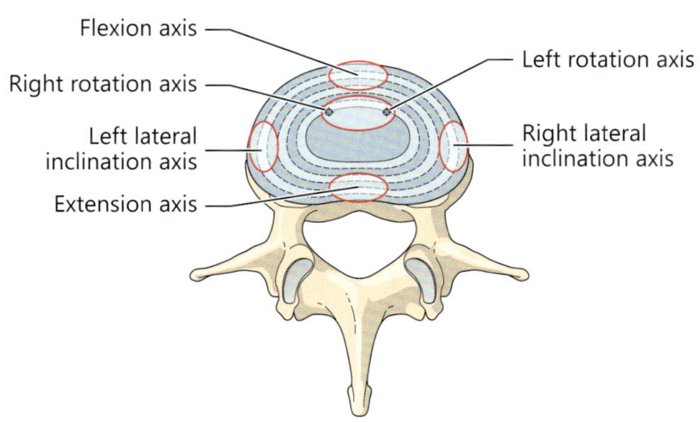

Axes of movement of the intervertebral disc
The axes of movement depend on the capacity of the nucleus pulposus to adjust. Cannot be accurately predetermined.

The orientation of the articular facets of the vertebral processes vary in cervical, thoracic and lumbar regions of the vertebral column, affecting the mobility of different segments of the vertebral column.

ARTICULAR FACETS	INCLINATION WITH RESPECT TO THE PLANE	
	SAGITTAL	HORIZONTAL
Cervical	45°	Backward open arch (almost coinciding with the frontal plane)
Thoracic	60°	Forward open arch (20° per side)
Lumbar	90°	Backward open arch (45° per side)

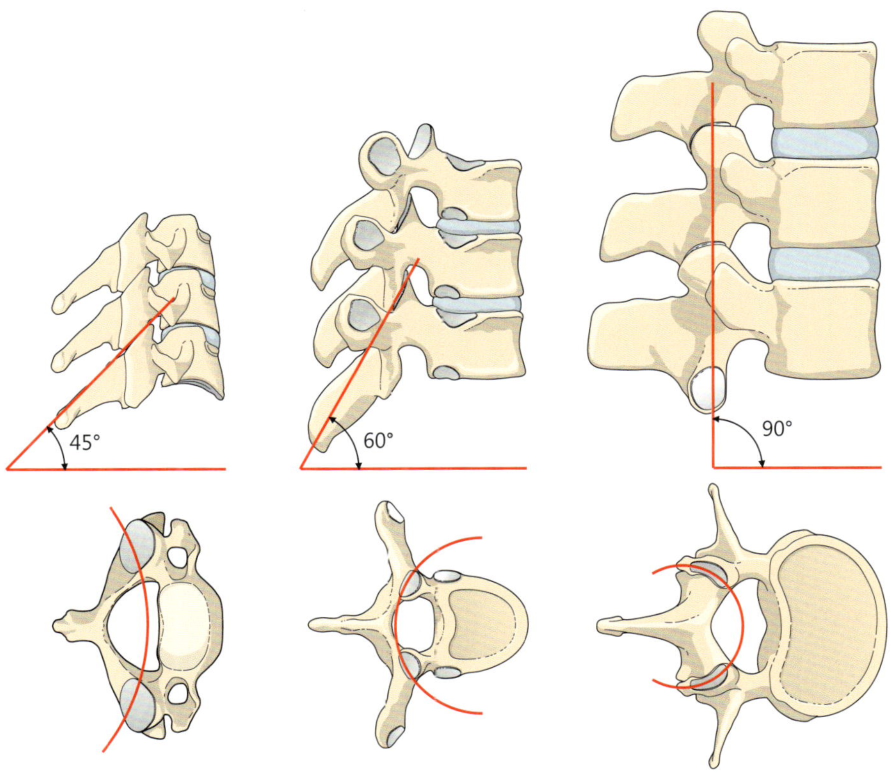

Angle difference between the epiphyseal joints of the cervical, dorsal and lumbar regions of the spine.

Movements of the upper cervical column

The joints between occipital bone, atlas, and axis form a functional unit.

Flexion

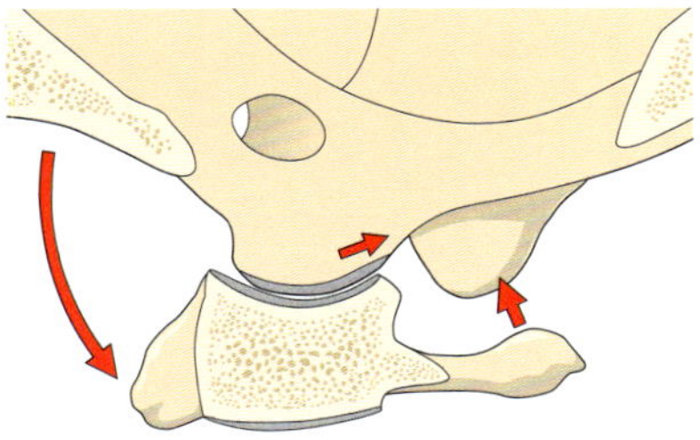

Flexion of the atlanto-occipital joint.

Atlanto-occipital joint

The occipital condyles slide on the superior articular facets of the atlas moving in posterior direction, increasing the distance between the occipital bone and posterior arch of the atlas.

Altogether the entire cervical column flexes 40°.

Rotation

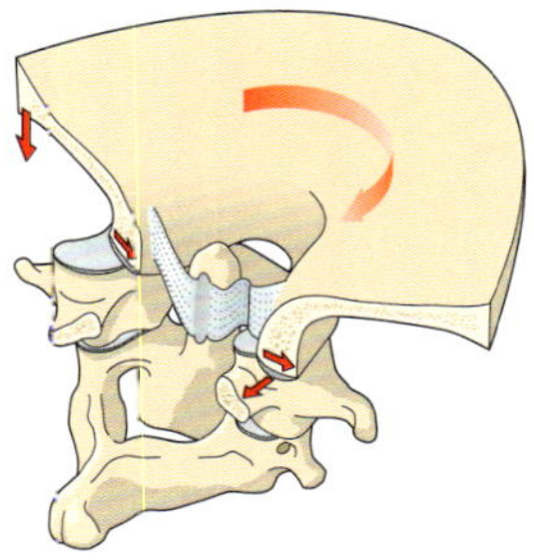

Rotation combined with lateral inclination of the atlanto-occipital joint.

Atlanto-occipital joint

- The occipital bone rotates around the atlas because the contralateral condyle slides forward and the ipsilateral condyle slides backward
- The contralateral alar ligament is put into tension and exerts traction on the corresponding condyle by bringing it toward the dens of axis causing a slight contralateral inclination
- The amplitude of movement is 5°.

Extension

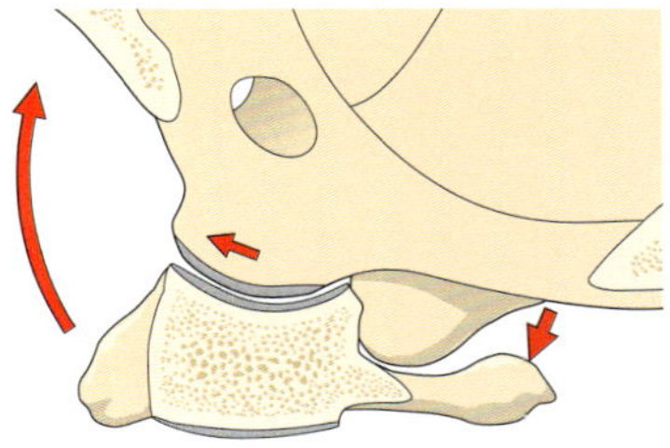

Extension of the atlanto-occipital joint.

Atlanto-occipital joint

- The occipital condyles move in anterior direction
- Approach between the occipital bone and the posterior arch of the atlas.

Altogether the entire cervical column extends 45°.

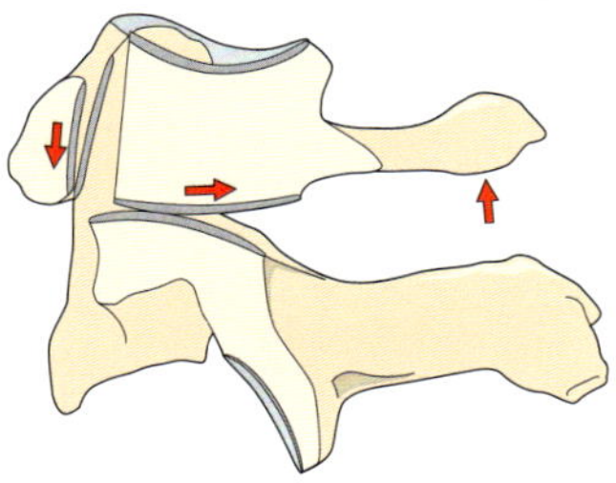

Flexion of the atlantoaxial joints.

Atlantoaxial joints

- The inferior articular facets of the atlas slide backward and upward
- The fossa of the dens of axis moves downward and away from the dens
- The distance between the posterior arch of the atlas and the spinous process of C II increases.

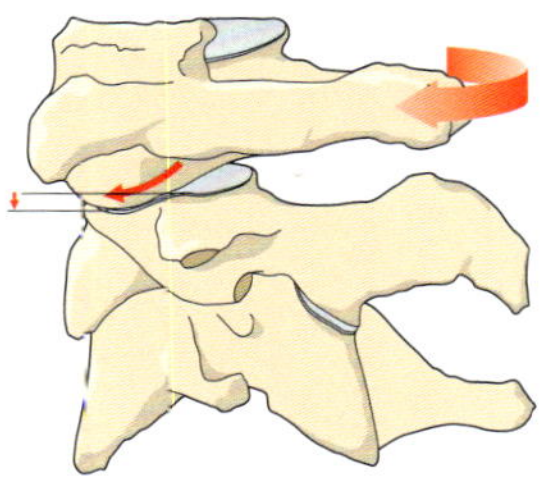

Rotation of the atlantoaxial joints.

Atlantoaxial joints

- The osteofibrous ring which encloses the dens of axis, formed by the transverse ligament of the atlas, rotates around the dens
- The lateral mass of the ipsilateral rotation of the atlas moves posteriorly and in the opposite direction (anteriorly) with respect to the contralateral
- Starting from 20° of rotation there is a lowering of the atlas towards the axis caused by the convexity of the articular surfaces
- The amplitude of movement reaches 40°.

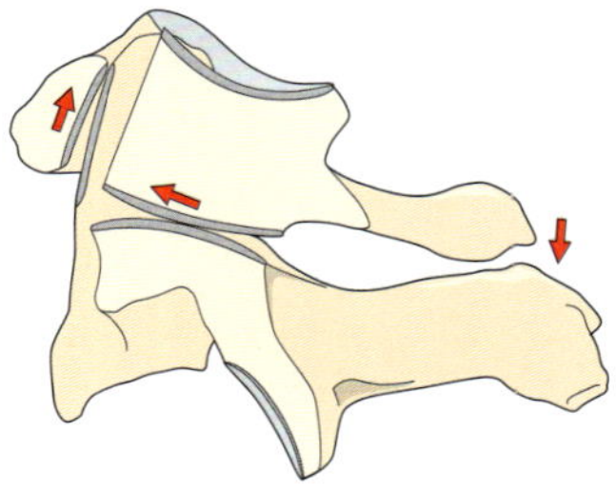

Extension of the atlantoaxial joints.

Atlantoaxial joints

- The inferior articular facets of the atlas slide forward and upward
- The fossa of the dens of axis moves upward, anteriorly approaching the dens and posteriorly moving away from it.

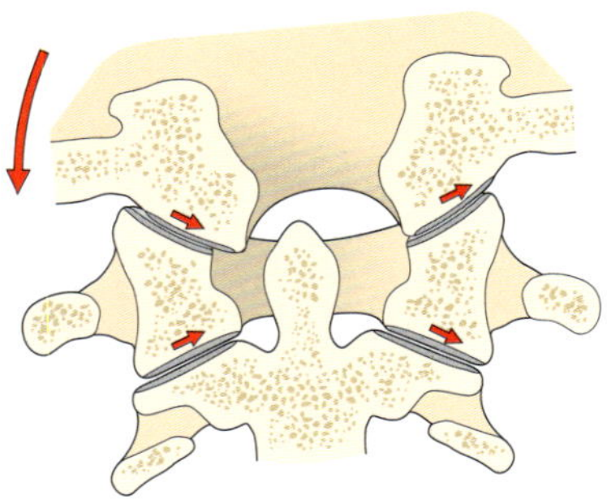

Lateral inclination of the atlanto-occipital and atlantoaxial joint.

Lateral inclination

Atlanto-occipital joint

Modest movements of the occipital condyles in the lateral and medial direction (do not exceed 3-5° in one or the other direction).

Atlantoaxial joints

The movements of the dens of axis in lateral direction are so modest that cannot be measured and are accompanied by rotation.

Movements of the lower cervical column

Flexion

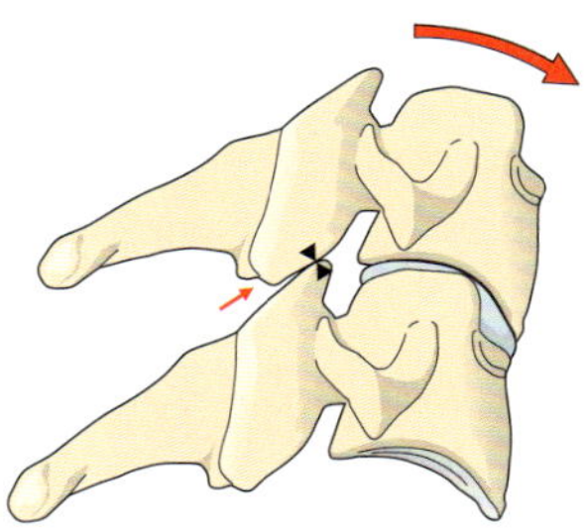

Flexion of the lower cervical column.

- The articular facets move away from each other and the articular space widens with the formation of an open angle posteriorly
- When the range of motion is maximum, there is a distance of only two fingers between the chin (mouth closed) and the sternum.

Extension

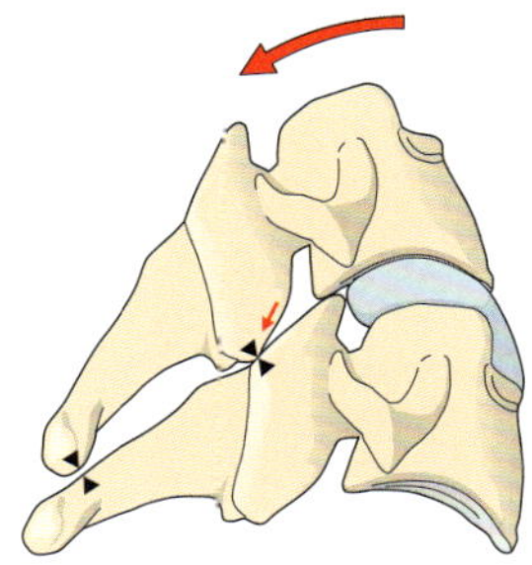

Extension of the lower cervical column.

- The articular facets slide over one another by moving upward and forward
- The anterior capsuloligamentous structures inhibit the movement, which is also impeded by the contact between the spinous processes and the posterior parts of the articular facets
- When the range of motion is maximum, the nasofrontal line is localized in a plane forming an angle of 30° with respect to the transverse axis.

Rotation

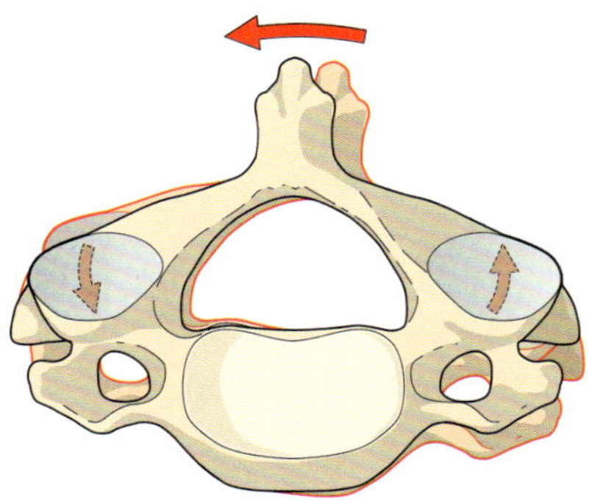

Rotation of the lower cervical column.

- The movement is accompanied by ipsilateral inclination of the superior vertebra
- The orientation of the articular facets of the more cranial vertebrae form an angle of 40-50° between the vertical axis and the axis of rotation, enabling a greater lateral inclination with respect to the more caudal vertebrae (the lateral inclination between C VII and T I is only 10°)
- The amplitude of rotation is greater in flexion than in extension
- The maximum amplitude of movement is 40°.

Lateral inclination

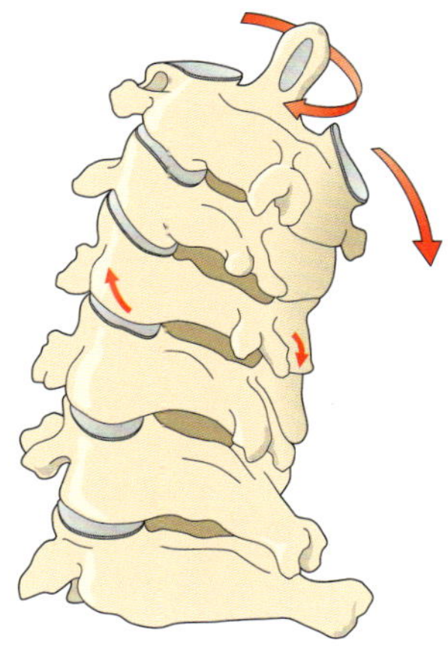

Combination of movements in lateral inclination and consensual rotation of the lower cervical column.

- The movement is hindered by the orientation of the articular processes and by the presence of the semilunar processes, accompanied by a ipsilateral rotation whose amplitude varies from segment to segment (5° between C II and C III and even less between C VII and T I)
- The maximum amplitude of movement is 50°.

Delmas index

$$\frac{h \times 100}{l}$$

Where h = body height (measured on a x-ray in the upright posture from the first cervical vertebra to the base of the sacrum) and l = length of the column (fully extended or straightened, i.e. the sum of the measurements of the various segments of the vertebral column carried out on a x-ray of the column).
a, 94-96 corresponds to an index of 'normality' with medium curves (normal type);
b, <94 corresponds to a lower index with increased spinal curves (dynamic type);
c, >96 corresponds to a higher index with decreased spinal curves (static type).

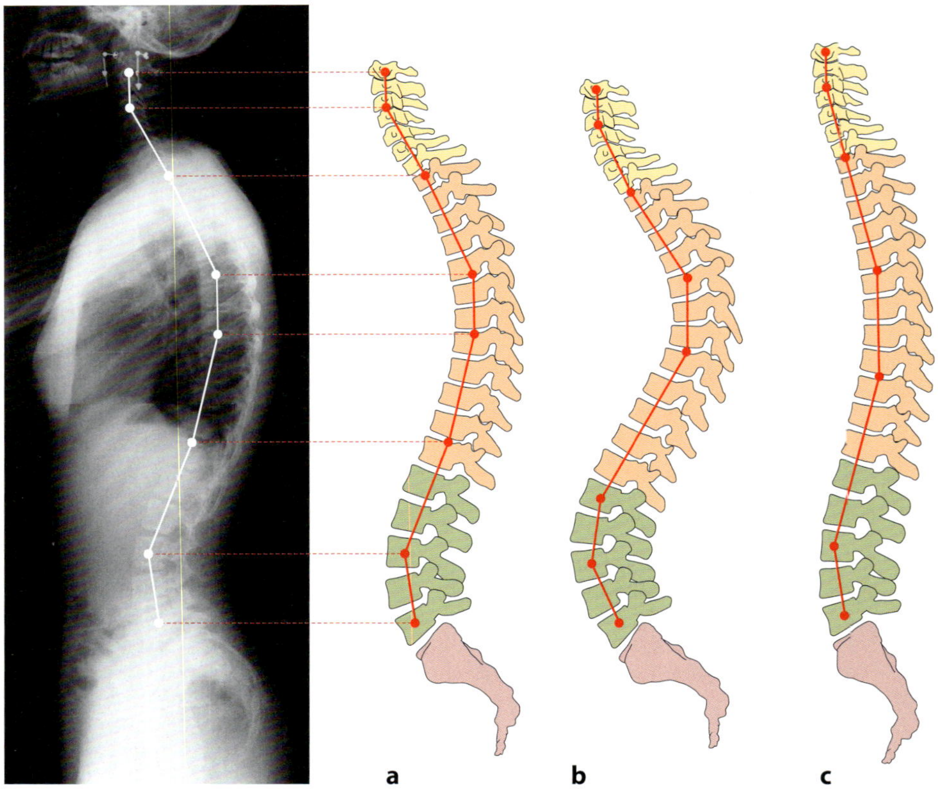

Radiology

Anteroposterior projection:

- the pedicles of the vertebral arches appear as oval shadows on the sides of the column
- the vertebral canal increases in width in the craniocaudal direction.

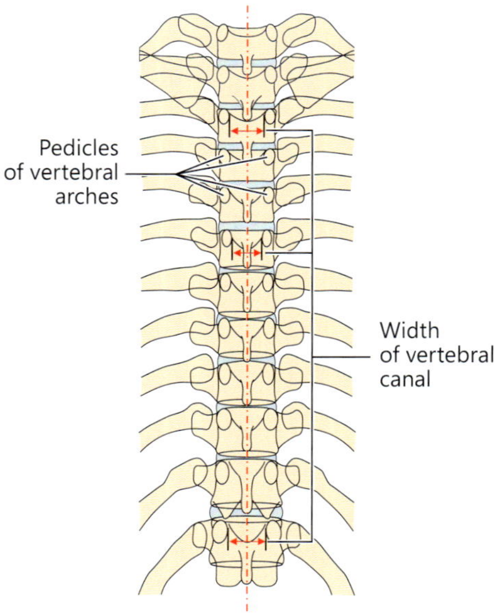

Lateral projection:

- if one traces perpendicular lines to the superior and inferior margins, respectively, of the vertebral bodies of T III and T XI, the acute angle formed by the orthogonal to these perpendicular lines has a width of 25° (angle of kyphosis)
- the height of intervertebral discs increases in craniocaudal direction.

Note: the following shall be considered normal: a cervical lordosis of 35°, a dorsal kyphosis of 35° and a lumbar lordosis of 40°.

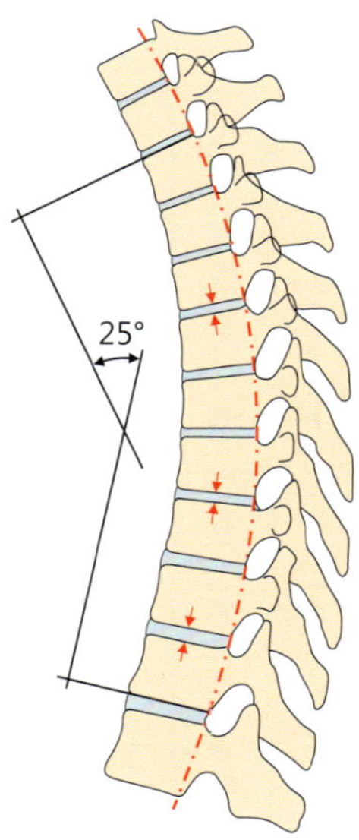

Vertebral resistance to pressure

Resistance elements:

- compact bone of the vertebral body, located peripherally
- direction of the trabeculae (horizontal, oblique, vertical) of spongy bone of the vertebral body, located internally.

The image shows an area of least resistance, triangular in shape, with the base corresponding to the anterior surface of the vertebral body. At this stage there is a lack of oblique trabeculae which are directed backwards towards the arch from the top and bottom of the vertebral body.
The anterior part of the vertebral body is therefore more subjected to crushing or fracture due to compressive forces.

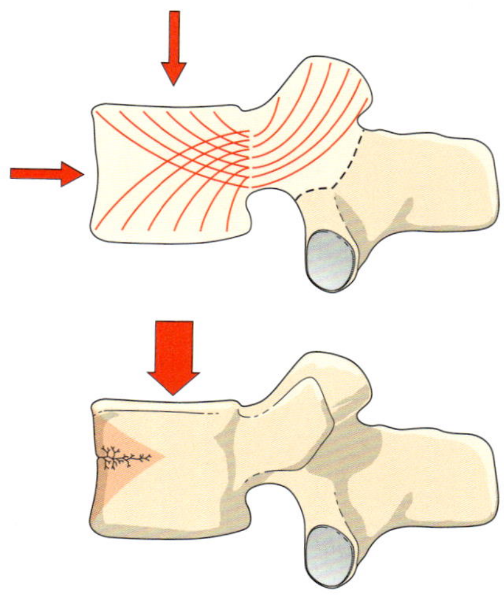

Modifications of the intervertebral disc during movement
For example, forward flexion of the trunk:
- the vertebral body leans forward
- the intersomatic space assumes a wedge shape
- the anterior collagen fibers of the fibrous ring are compressed
- the posterior collagen fibers of the fibrous ring are stretched
- the nucleus pulposus slides backwards, dragging the elastic fibers of the fibrous ring
- the posterior space between the vertebral bodies increases.

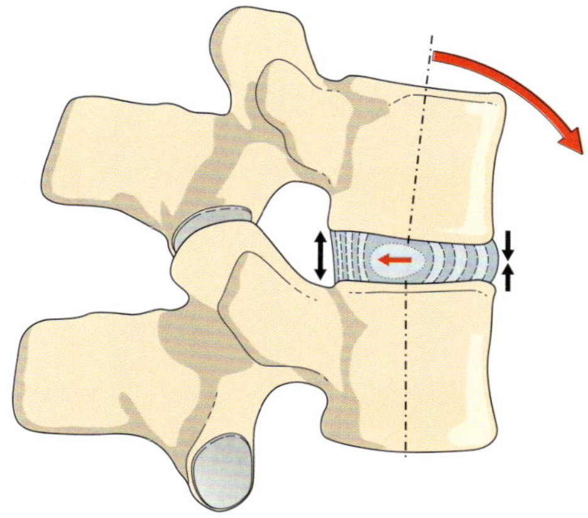

Behaviou- of intervertebral disc during flexion.

Alterations in the intervertebral disc

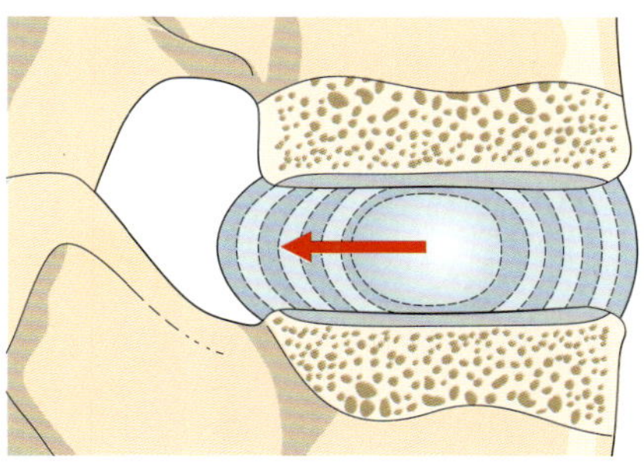

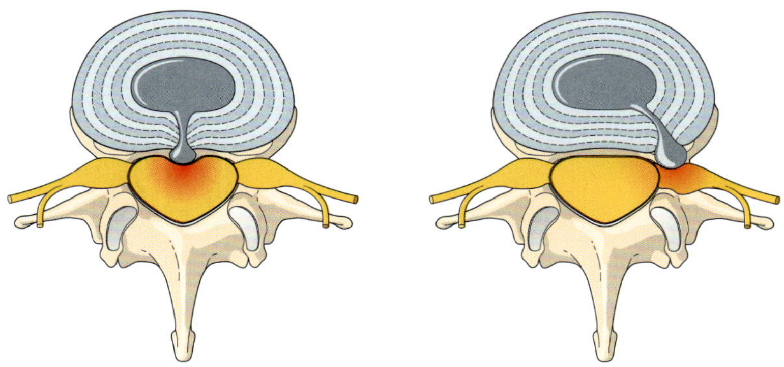

Protrusion

Definition	Intervertebral disc progress over the posterior margin of the intervertebral bodies.
Etiopathogenesis	Degeneration of the collagen fiber bundles of the fibrous ring (due to overload). Formation of fissures in the layers of collagen fibers of the fibrous ring. Displacement of the nucleus pulposus that seeps into the fissures formed by moving the fiber layers (as a result of an unbalanced load).

Herniation

Definition	Displacement of the nucleus pulposus in the direction of the spinal canal or spinal nerve.
Etiopathogenesis	Degenerative formation of fissures in all the layers of collagen fibers of the fibrous ring.

According to the severity of the displacement of the disc, the herniations can be also classified into:
- contained subligamentous hernia
- protruding subligamentous hernia
- extruded hernias (in situ)
- extruded hernias (migrated).

Degeneration of the intervertebral disc
- Alterations of the lines of diffusion of fluids in the disc with reduction of hydration
- Loss of uniform load distribution and capacity to adapt to different stimuli
- Reduction of space between the vertebral bodies
- Increased stress of the joints of the vertebral arch
- Origin of osteophytes from the margins of the vertebral bodies.

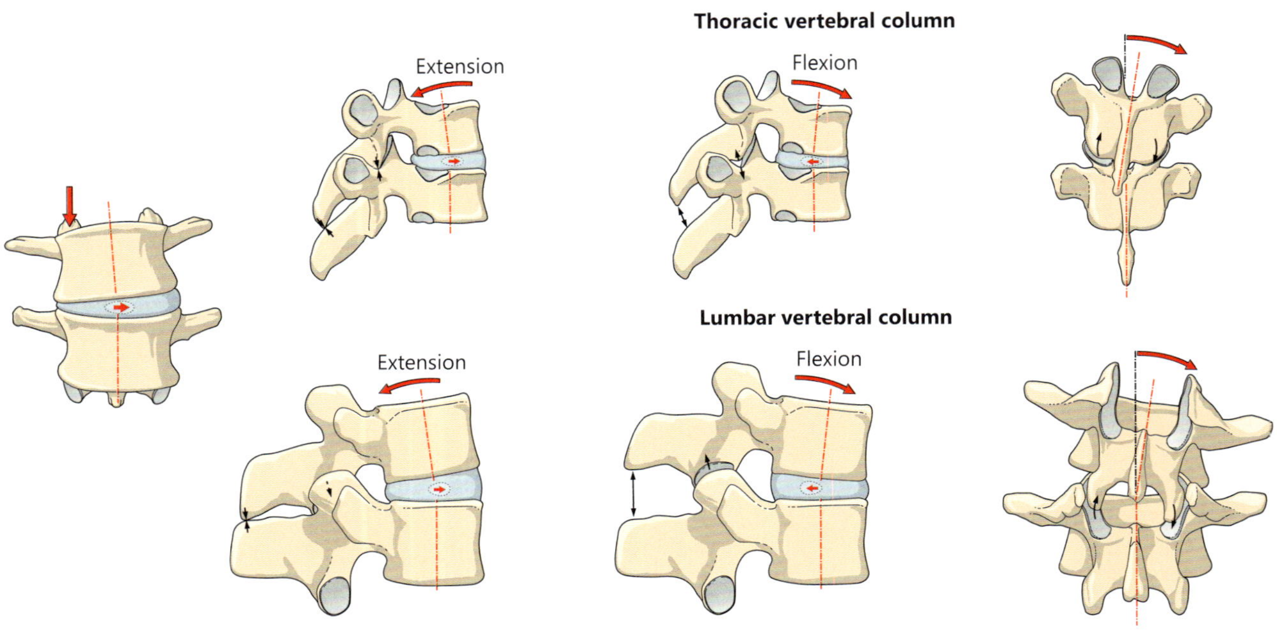

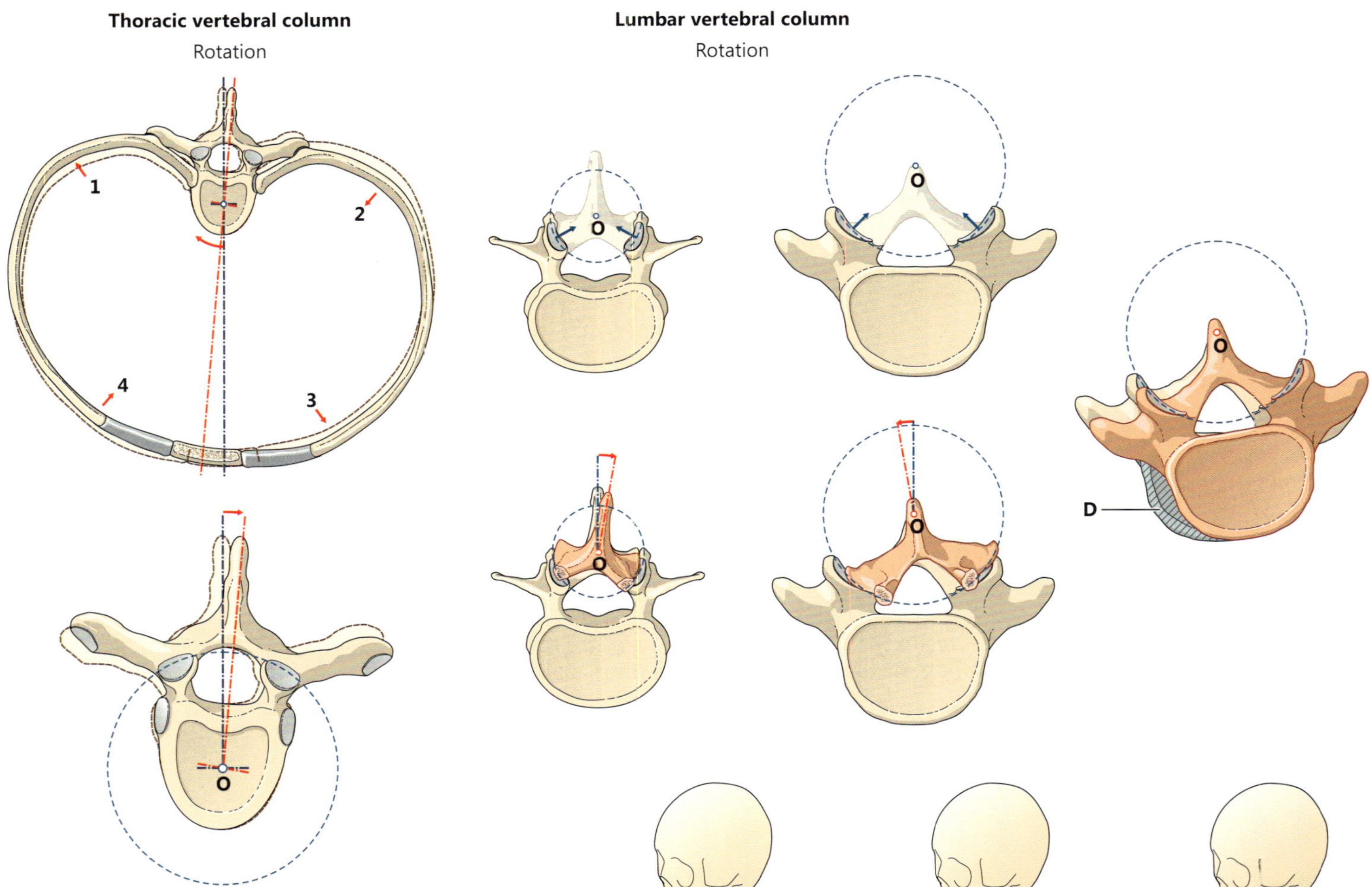

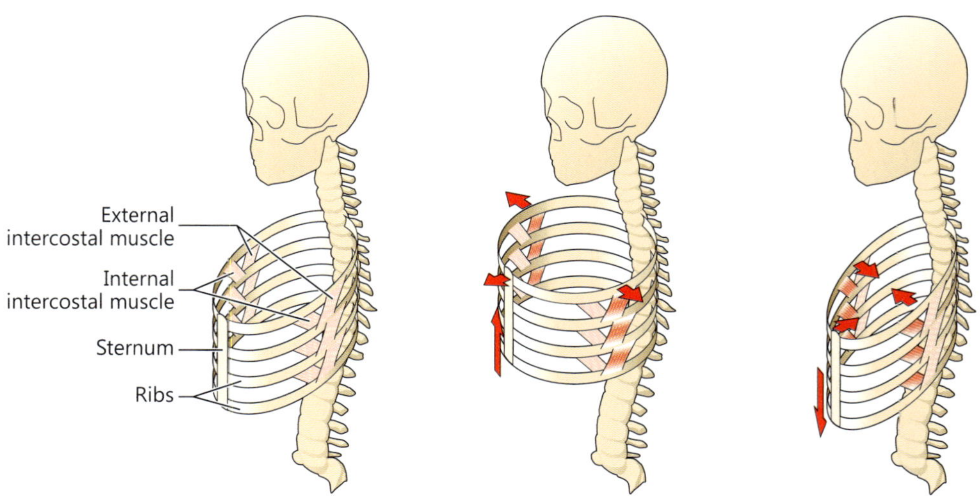

Rib movement during the dynamics of respiration.

Flexion and extension of the vertebral column

Plane of movement	Sagittal
Axis of rotation	Transverse
Flexion	• cervical 40° (F_C) • thoracic 40° (F_T) • lumbar 60° (F_L)

Lumbo-pelvic rhythm

The lumbar flexion is accompanied by the movement of the pelvis around the bicondylar axis (or coxofemoral), which extends the range of motion; otherwise, it would not exceed half of the indicated values.

Lumbar flexion	• 70% at the level of L V - S I • 20% at the level of L IV - L V • 10% at the level of L I - L II/L II - L III/L III - L IV
Extension	• cervical 45° (E_C) • thoracic 15° (E_T) • lumbar 30°-35° (E_L)
Trunk flexion (thoracolumbar)	100°-105° (F_{TL})
Trunk extension (thoracolumbar)	45° (E_{TL})

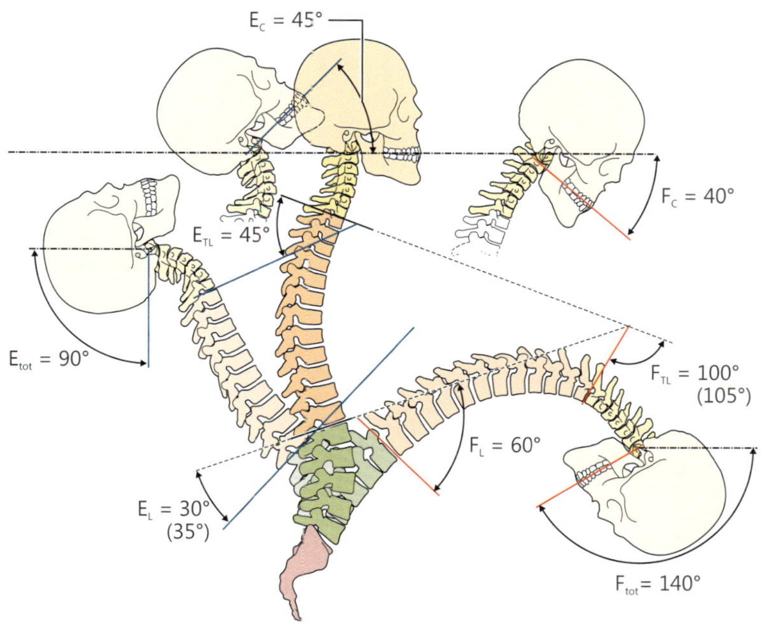

In a normal subject, note a total flexion and extension movement of the vertebral column equal to 230° (total flexion of 140° + total extension of 90°); in a flexible subject such value can reach 250°.

FORMATIONS THAT LIMIT THE MOVEMENT OF	
FLEXION OF THE VERTEBRAL COLUMN	***EXTENSION OF THE VERTEBRAL COLUMN***
• posterior longitudinal ligament • ligamenta flava • interspinous ligaments • supraspinous ligaments • intertransverse ligaments • ligamentum nuchae • superior fascicle of the lumbosacral ligament • articular capsules of the posterior joints of the column • anterior obstruction of the disc • posterior traction of the fibrous ring • trunk extensor muscles	• anterior longitudinal ligament • compaction of posterior articular facets and spinous processes • anterior traction of the fibrous ring • posterior obstruction of the disc • trunk flexor muscles

Lateral inclination of the vertebral column

Plane of movement	Frontal
Axis of rotation	Sagittal
Lateral inclination	• cervical 40° • thoracic 25° • lumbar 15°
Lateral trunk inclination (thoracolumbar)	40°

The lateral inclination of the vertebral column shows the same amplitude in both sides. The existence of scoliosis may limit considerably the inclination in one or both sides.

Physiological rotation of the vertebrae

During the lateral inclination of the vertebral column there can be determined a physiological rotation of the vertebrae that, by expanding the radius of the arc of a circle produced by the lateral inclination of the vertebral column, limits the movement.

Formations that limit the movement of lateral inclination of the vertebral column:

- physiological rotation of the vertebral bodies
- tension of the intertransverse and flava contralateral ligaments and interspinous ligaments
- articular capsules of the posterior joints of the contralateral vertebral column
- tension of contralateral lumbosacral ligament
- contralateral tension of the fibrous ring
- obstruction of the disc ipsilaterally.

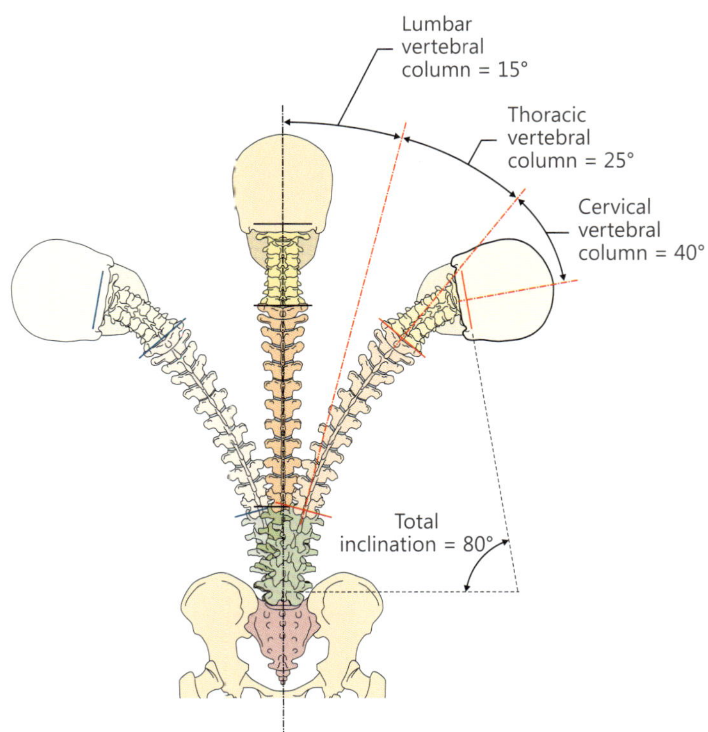

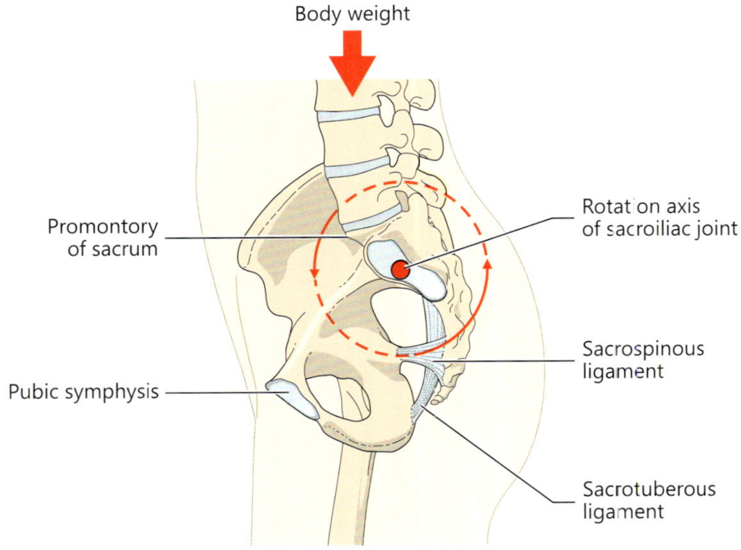

The picture shows the **axis of rotation for the sacroiliac joint**. The body weight, transmitted to the sacrum anteriorly to the rotation axis, tends to push the superior part of the sacrum inferiorly, thus causing the upward rotation of the inferior part of the sacrum. Such tendency is opposed by strong ligaments, the sacrotuberous and the sacrospinous. During the upright posture, the body weight is transferred from the axial skeleton to the ileum and further to the femur, while in a sitting position is transferred to the ischial tuberosity.

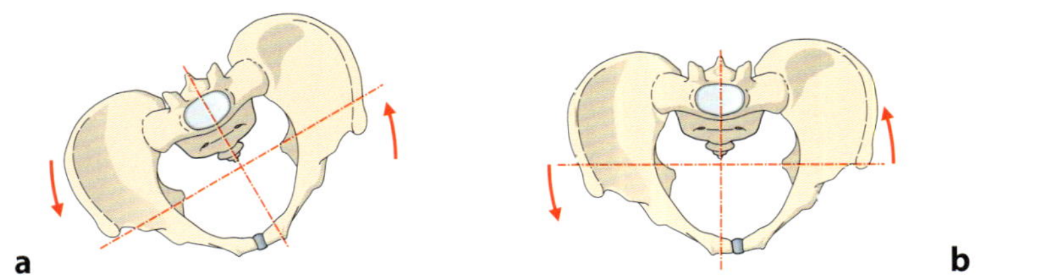

Left **rotation of the pelvis** (**a**); internal rotation of the right hemipelvis and external rotation of the left hemipelvis (**b**).

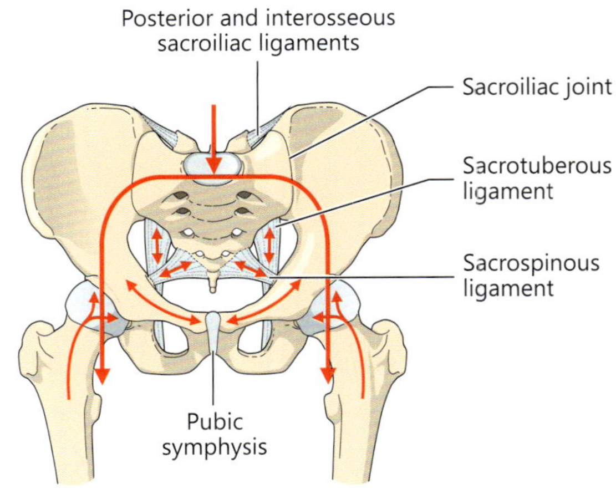

Front view of the pelvis. **Load distribution** within the **pelvis** and lower limbs (arrows towards the bottom). Response to the stress of the ground (arrows towards the top). The arrows in opposite directions indicate the stresses caused by traction forces or pushing on the ligaments and bones.

Other pelvic movements

Lateral inclination of the pelvis (around a sagittal axis) – gradient to the right or left.
Rotation (around a longitudinal axis)

- Of the entire pelvis:
  - left
  - right
- Internal of the right hemipelvis and external of the left hemipelvis
- External of the right hemipelvis and internal of the left hemipelvis

Translational movement of the pelvis – anterior, posterior or lateral.

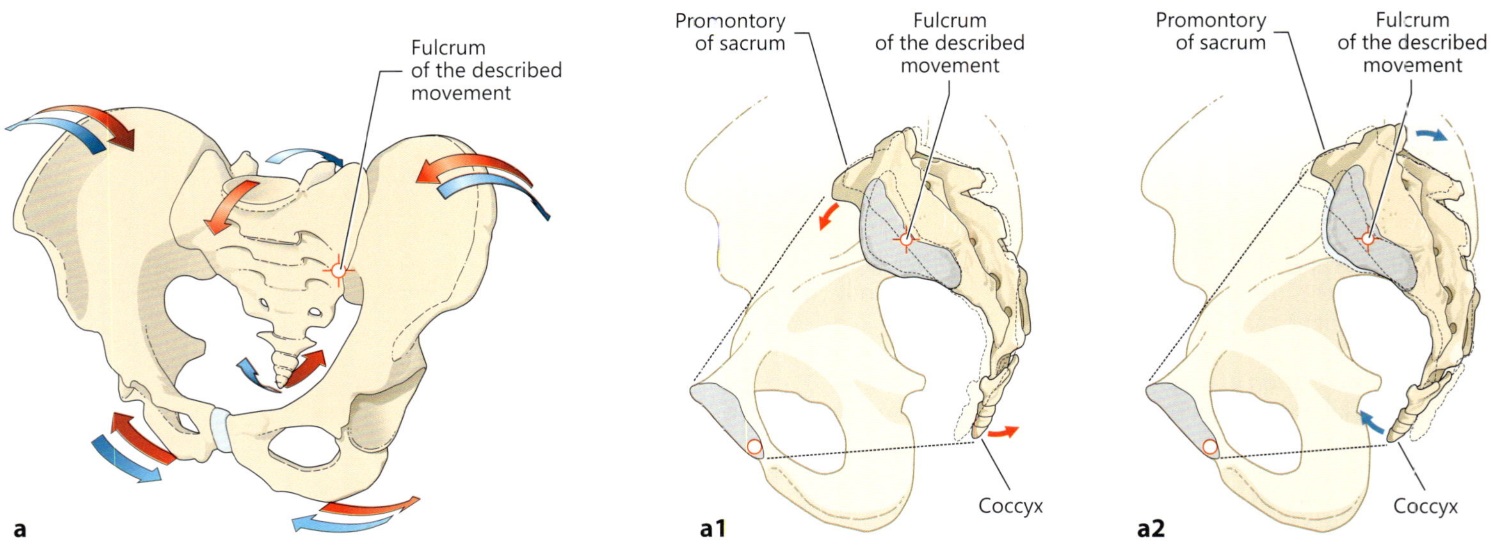

Nutation (**a1**) and **counternutation** (**a2**) movements of the sacroiliac joints (**a**). Note the variations in distance with the pubis. In nutation movements the base of the sacrum rotates forward and downward, the apex rotates upward and backward, the iliac wings approach the median line, the ischium moves away from the median line and the distance between the two ischial tuberosities increases, the distance between the pubis and the promontory decreases and the distance between the pubis and coccyx increases: in nutation the two inferior diameters increase and the superior one decreases, therefore the fibers of the pubococcygeus extend. In counternutation movements the base of the sacrum rotates backward and upward, the apex rotates downward and forward, the iliac wings move away from the median line, the ischial tuberosities approach: the two inferior diameters decrease, and the superior one increases, therefore the fibers of the pubococcygeus relax.

Summary table of the joints and the ligaments of the trunk (neck, thorax, abdomen, pelvis and perineum)		
JOINT, LIGAMENT OR MEMBRANE	***TYPE OF JOINT***	***BONES INVOLVED***
	JOINTS OF THE VERTEBRAL COLUMN	
Syndesmoses		
Anterior longitudinal ligament Interspinous ligaments Intertransverse ligaments Ligamenta flava (yellow ligaments) Ligamentum nuchae Posterior longitudinal ligament Supraspinous ligament Transverse ligaments of the vertebral column		
Synchondroses		
Intervertebral joints – intervertebral disc – – fibrous ring – – nucleus pulposus	Symphyses	Between the vertebral bodies
Lumbosacral joint – iliolumbar ligament	Symphyses	The fifth lumbar vertebra and first sacral vertebra
Sacrococcygeal joint – anterior sacrococcygeal ligament – deep posterior sacrococcygeal ligament – lateral sacrococcygeal ligament – superficial posterior sacrococcygeal ligament	Symphyses (in general) or synovial joint (occasionaly)	Apex of the sacrum and the base of the coccyx
Synovial joints (diarthroses)		
Lateral atlantoaxial joints	Plane joints	Atlas and axis
Median atlantoaxial joint – alar ligaments – cruciate ligaments of the atlas – – longitudinal fascicles – – transverse ligament of the atlas – apical ligament of dens – tectorial membrane	Pivot joint	Anterior arch of the atlas and dens axis
Zygapophysial joints (between the articular processes)	Arthrodial joints	Between the articular processes of the vertebrae
	JOINTS OF THE THORAX	
Syndesmoses		
External intercostal membrane Internal intercostal membrane		
Synchondroses		
Costochondral joints	Synchondroses	Anterior extremities of the ribs and lateral extremities of the costal cartilages
Costosternal synchondroses	Synchondroses or symphyses	Seventh costal cartilage and sternum
Synchondroses of the first rib	Varieties of synarthrosis (improperly synchondroses)	First costal cartilage and sternum
Sternal joints		
– xiphisternal joint	Synchondroses	Body and xiphoid process of the sternum
– manubriosternal joint	Symphyses or synchondroses	Manubrium and body of the sternum
Synovial joints (diarthroses)		
Costovertebral joints		
– costotransverse joint – – costotransverse foramen – – costotransverse ligament (interosseous) – – – lateral – – – superior – – lumbocostal ligament	Plane joint	Costal tubercle and transverse process of the vertebra
– joint of the head of the rib – – intra-articular ligament of the head of the rib – – radiate ligament of the head of the rib	Double plane joint	Head of the rib and vertebral body
Interchondral joints	Synovial joints	Between the costal cartilages of the sixth rib to the ninth rib
	Particular types of junctions (fibrous bundles)	Between the costal cartilages of the eighth rib to the tenth rib
Sternocostal joints – costoxiphoid ligaments – intra-articular sternocostal ligament – radiate sternocostal ligaments – sternal membrane	Double plane joints	Costal cartilages from the first to the seventh rib and costal incisures of the sternum

continued

Summary table of the joints and the ligaments of the trunk (neck, thorax, abdomen, pelvis and perineum)		*(continued)*
JOINT, LIGAMENT OR MEMBRANE	***TYPE OF JOINT***	***BONES INVOLVED***
	JOINTS OF THE PELVIS	
Syndesmoses		
Obturator canal Obturator membrane		
Pubic Symphysis – inferior pubic ligament – interpubic disc – superior pubic ligament	Symphysis	Between the pubic bones
Sacroiliac joints – anterior sacroiliac ligament – greater sciatic foramen – interosseous sacroiliac ligament – lesser sciatic foramen – posterior sacroiliac ligament – sacrospinous ligament – sacrotuberous ligament – – falciform process	Symphysis or amphiarthrosis or plane joints	Sacrum and iliac bones

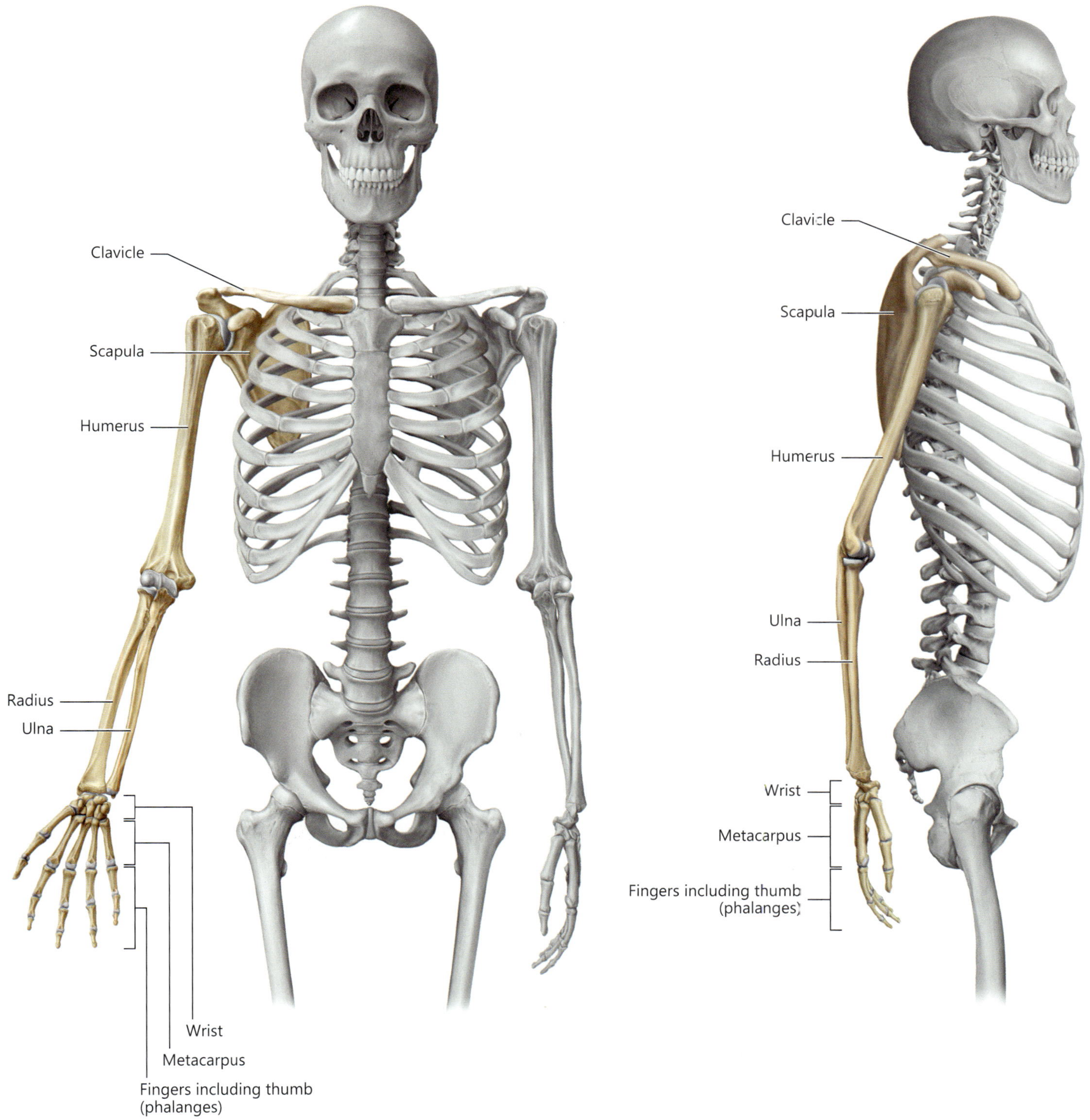

Osseous configuration of the **upper limb**. Anterior view (on the left) and lateral view (on the right).

View of the right **upper limb** from the scapula to the hand.

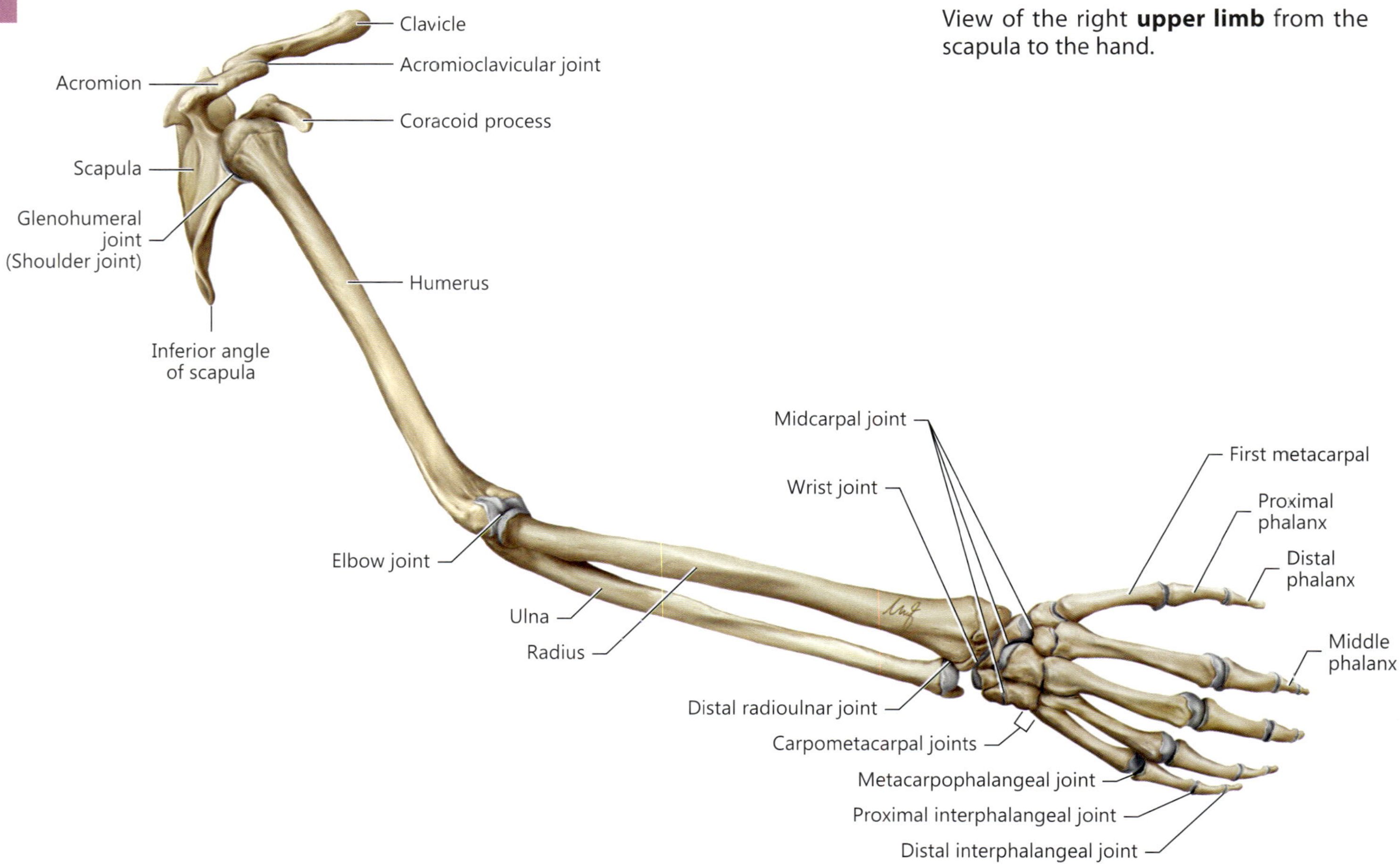

Three-dimensional image of a living human body, obtained by CT scan using a 3D volume rendering technique, showing the right **upper limb**.

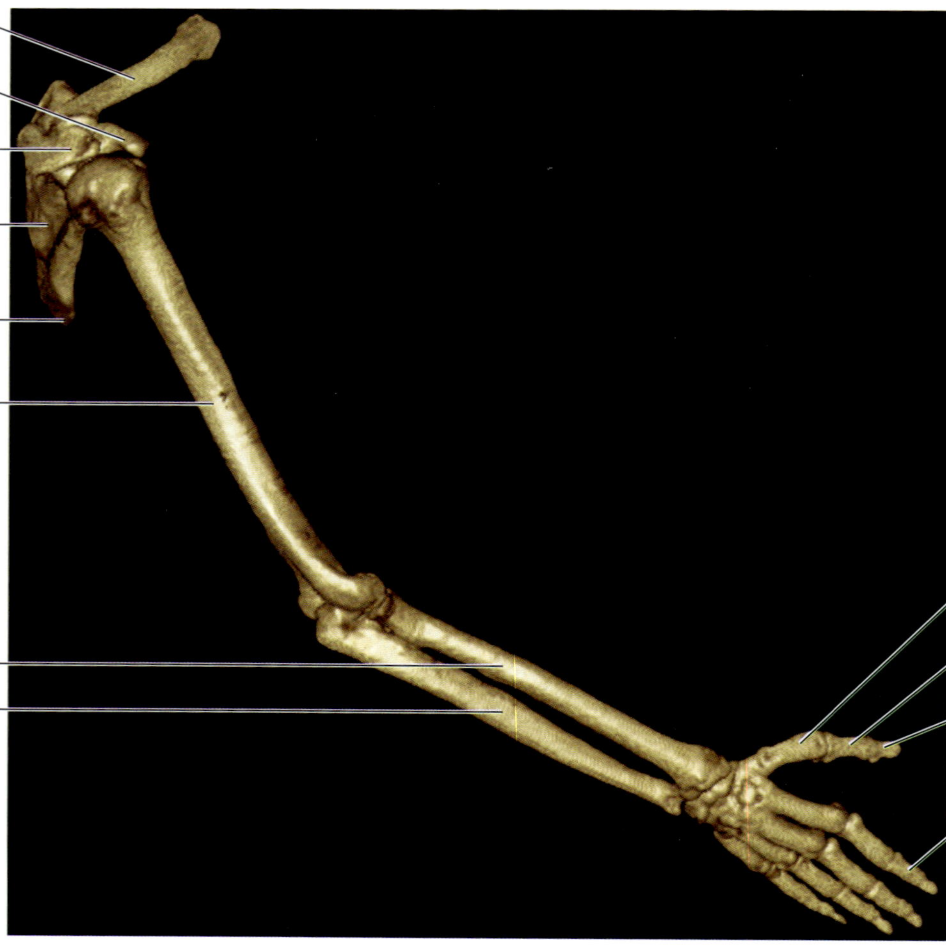

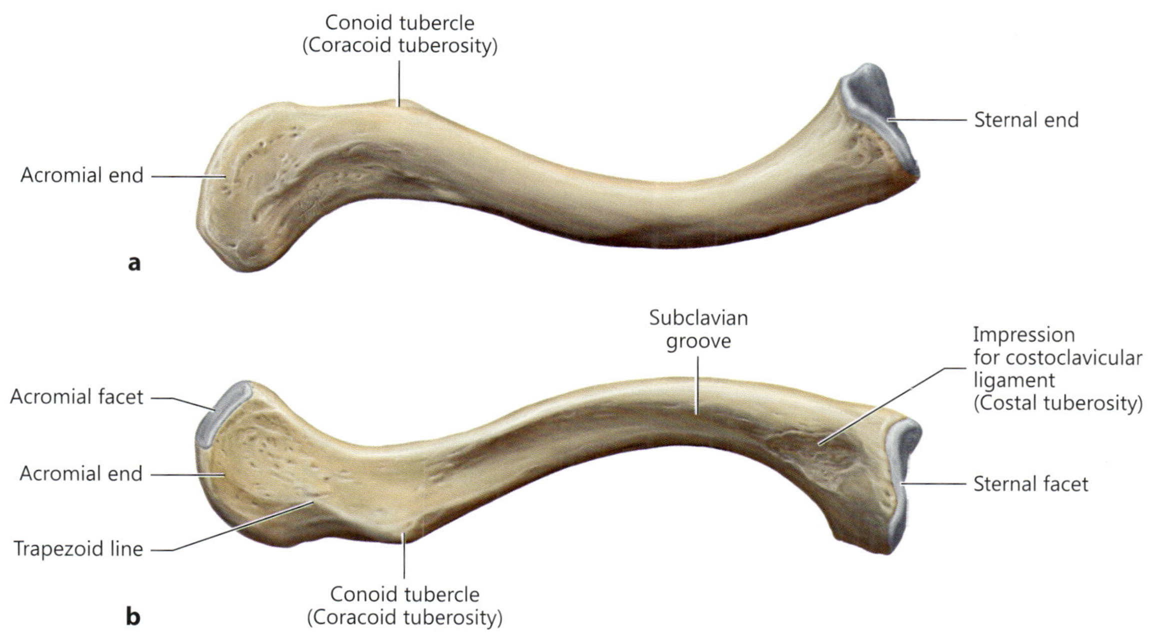

Clavicle. **a**, View from above; **b**, view from below.

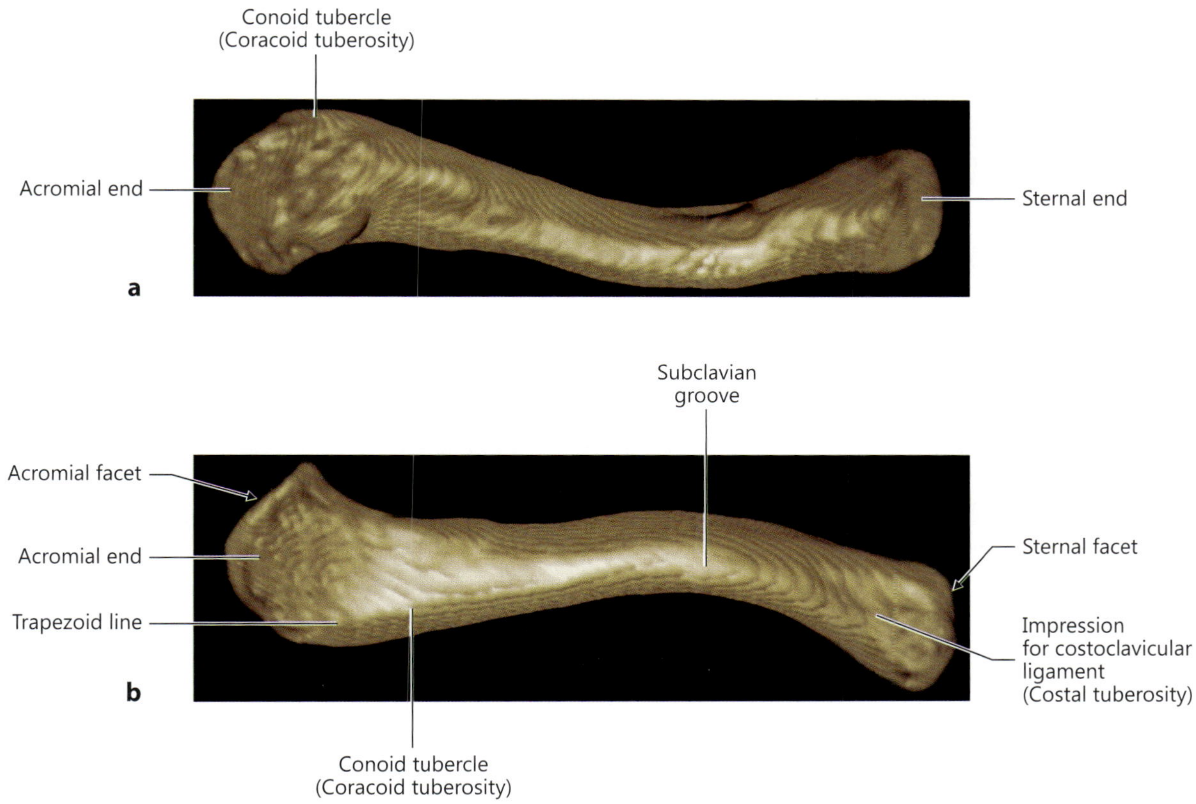

Three-dimensional images of a living human body, obtained by CT scan using a 3D volume rendering technique. **Clavicle**, superior and inferior view.

Anterior surface of the right **scapula**. The majority of this surface is occupied by the **subscapular fossa**.

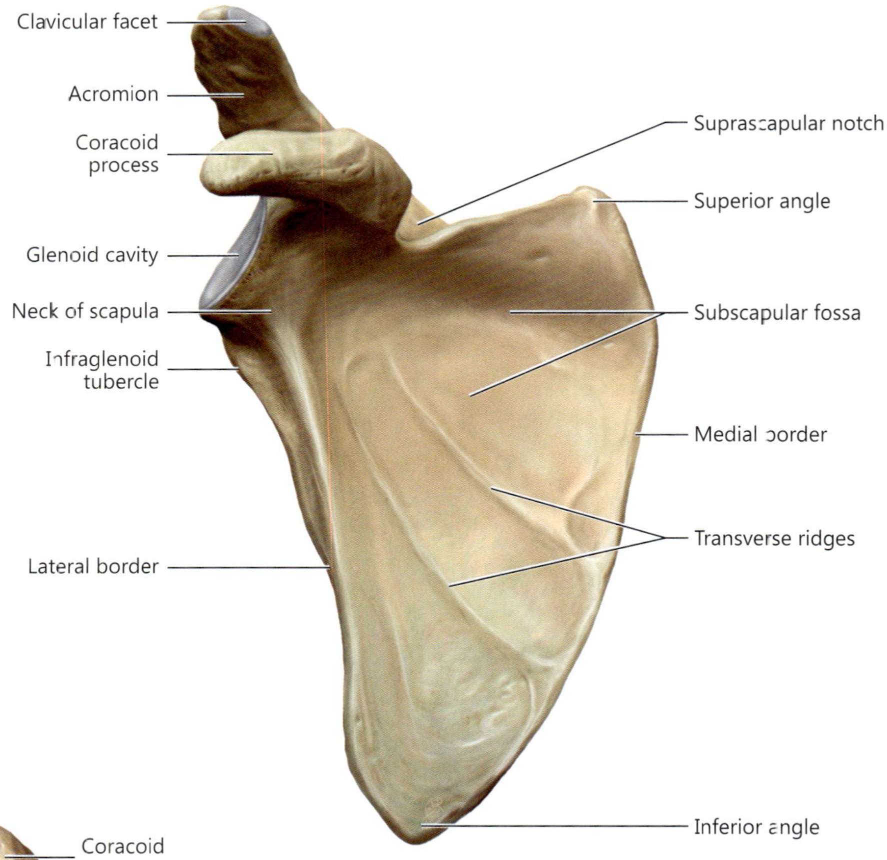

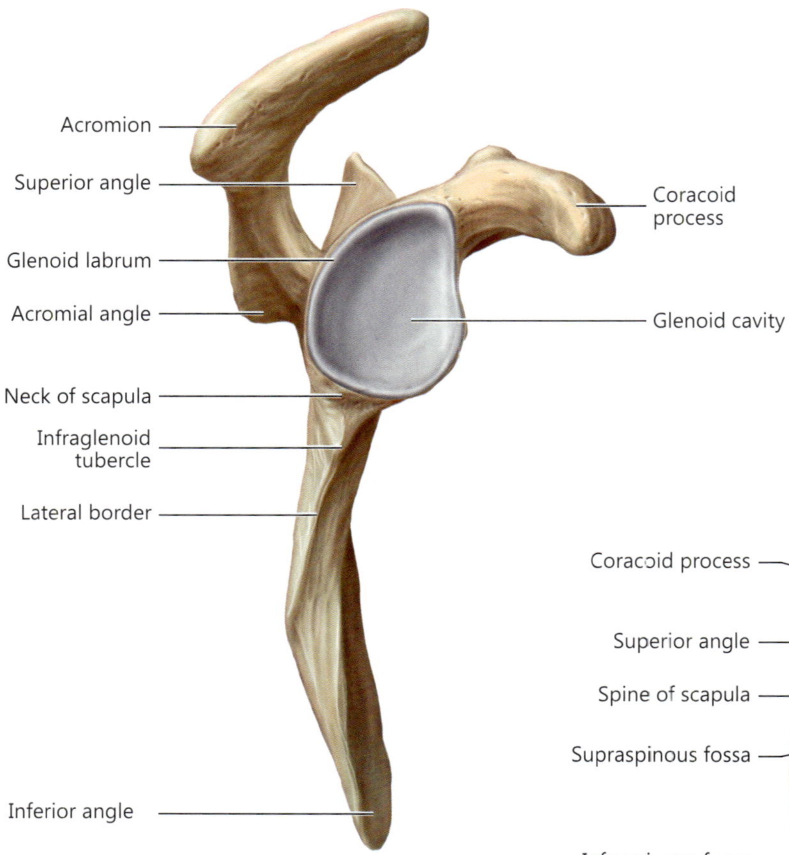

Medial view of the **scapula**. Here you can see the **glenoid cavity**, which receives the head of the humerus to form the glenohumeral joint.

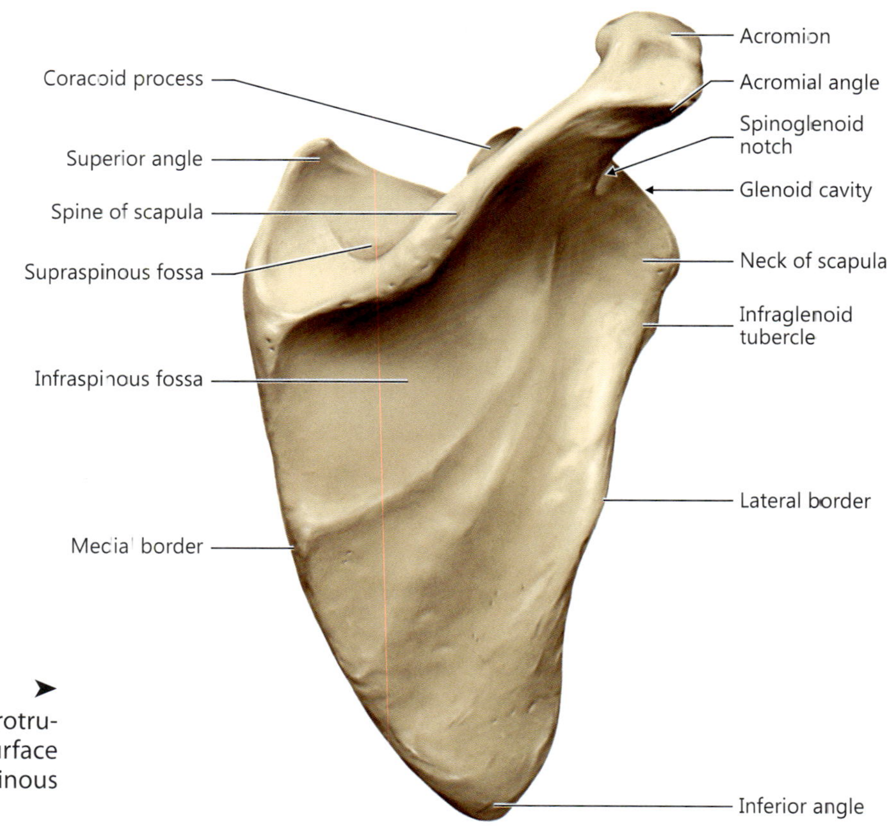

Posterior surface of the right **scapula**. The protrusion of the **spine of the scapula** divides the surface of the bone into the supraspinous and infraspinous fossae.

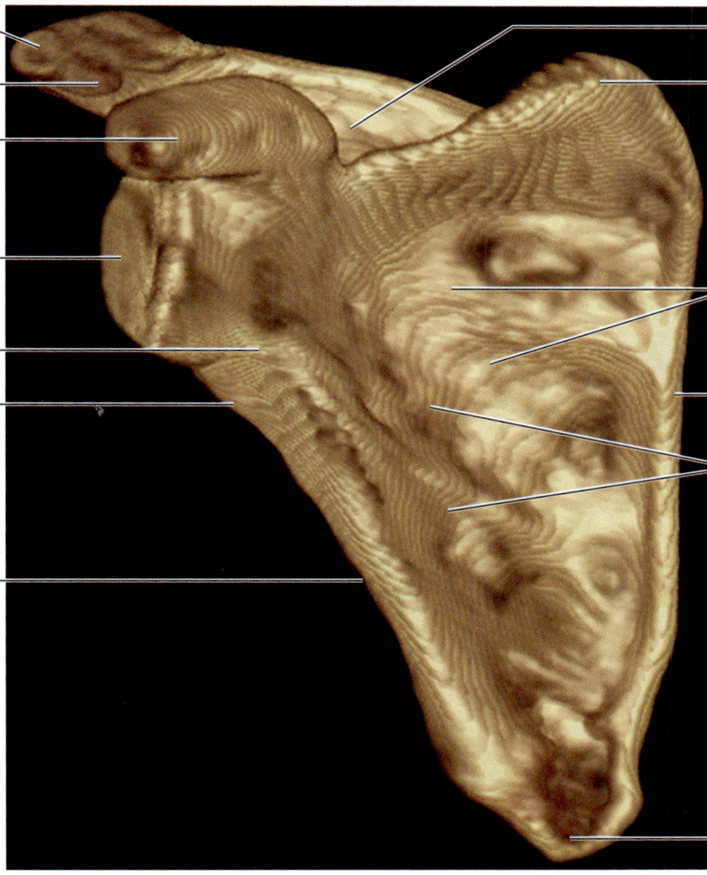

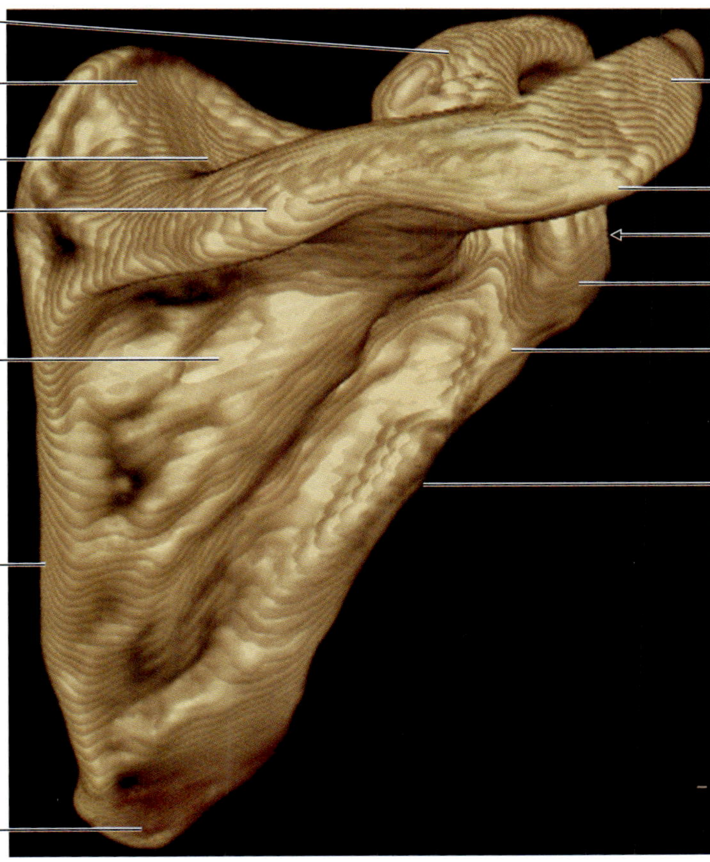

Three-dimensional images of a living human body, obtained by CT scan using a 3D volume rendering technique. Right **scapula**, anterior view (above) and posterior view (below).

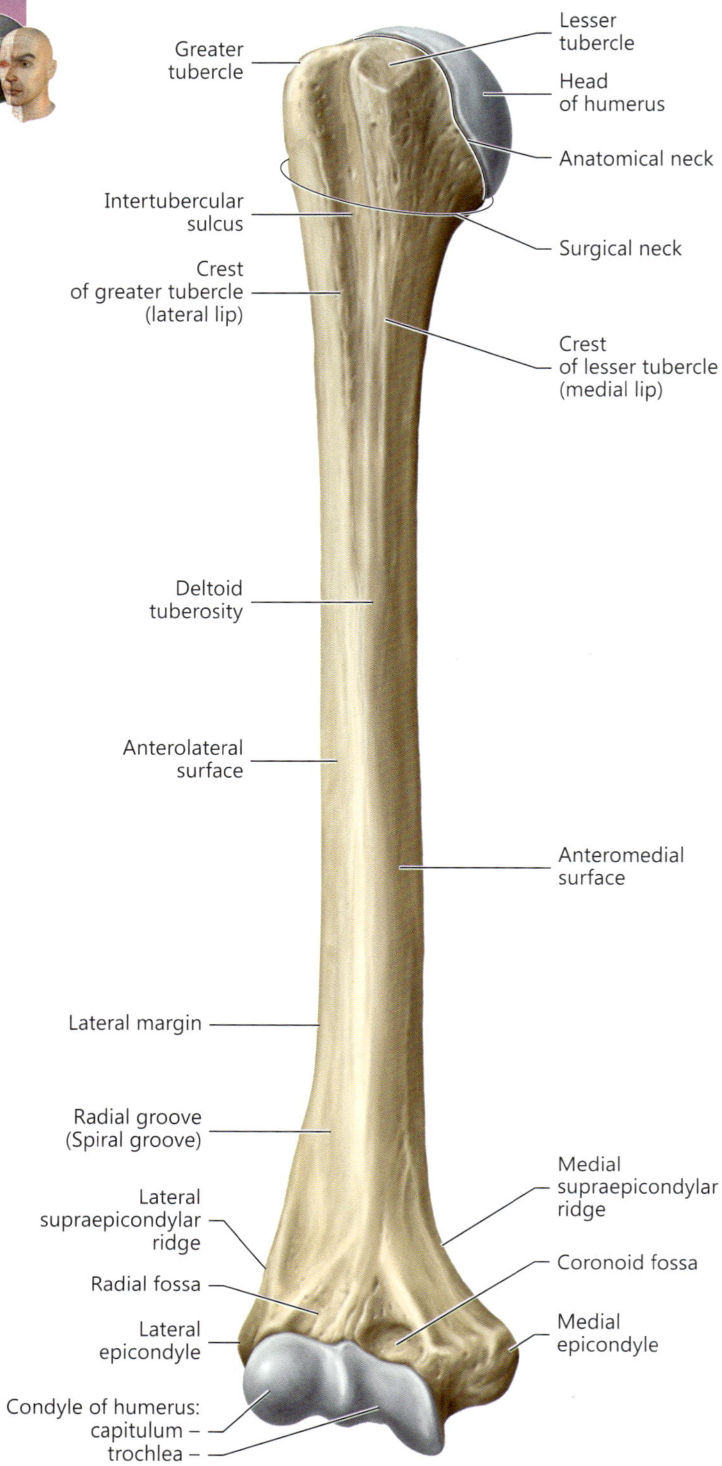

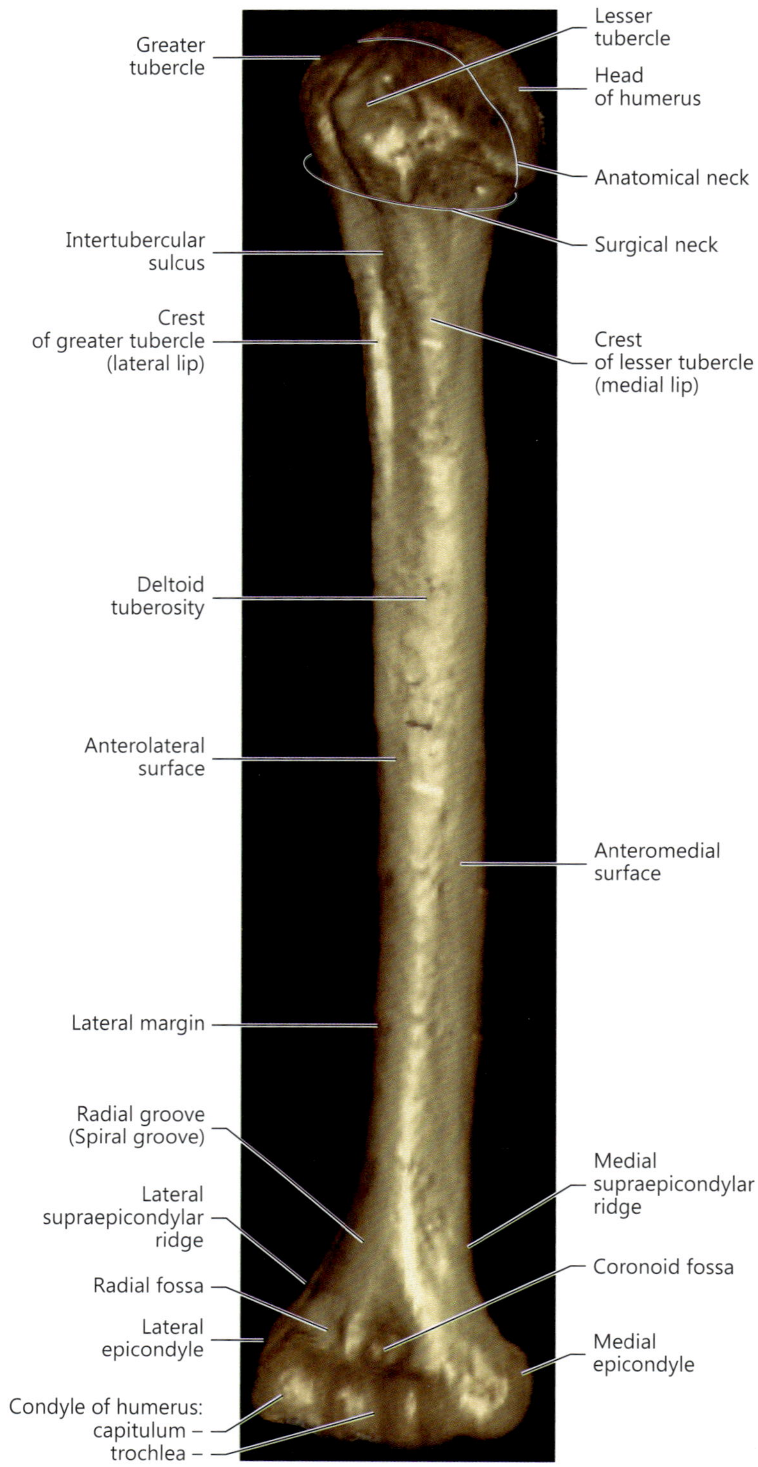

Right **humerus**. On the left, view of the anteromedial surface. On the right, three-dimensional image of a living human body, obtained by CT scan using a 3D volume rendering technique, of the same projection.

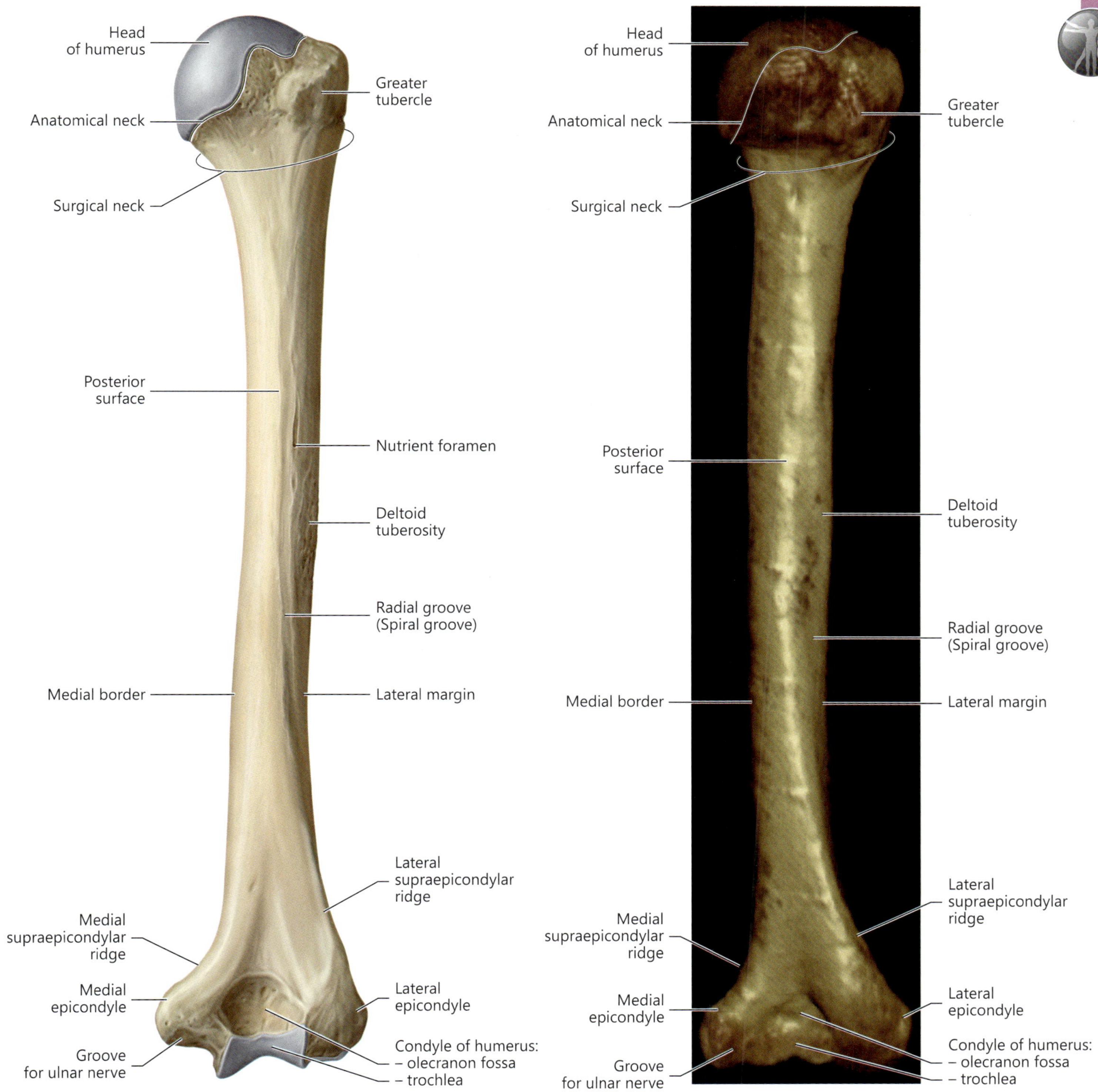

Right **humerus**. On the left, view of the posterior surface. On the right, three-dimensional image of a living human body, obtained by CT scan using a 3D volume rendering technique, of the same projection.

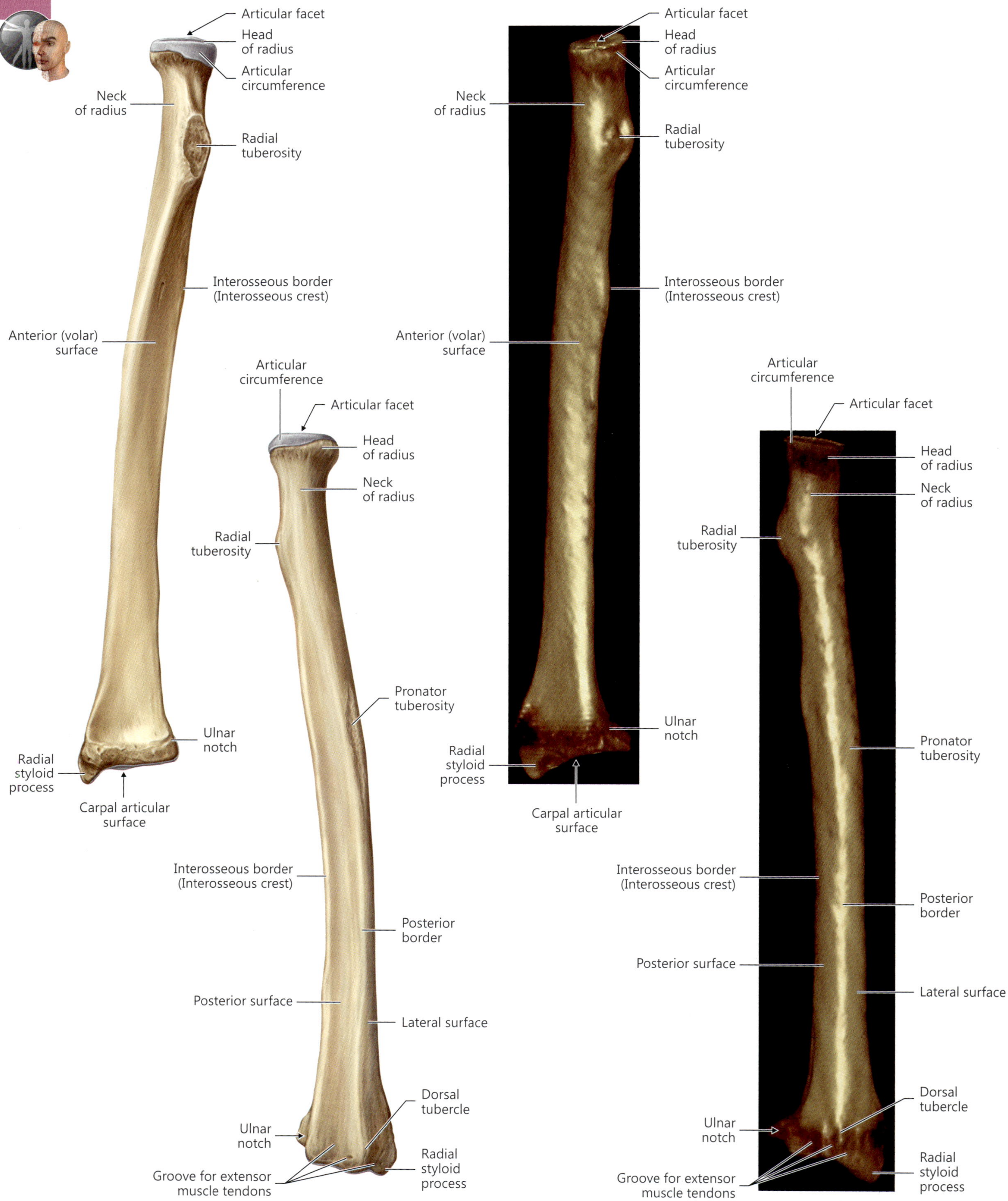

Right **radius**. On the left, view of the anterior, or volar, and posterior, or dorsal, surfaces. On the right, three-dimensional images of a living human body, obtained by CT scan using a 3D volume rendering technique, of the same projections.

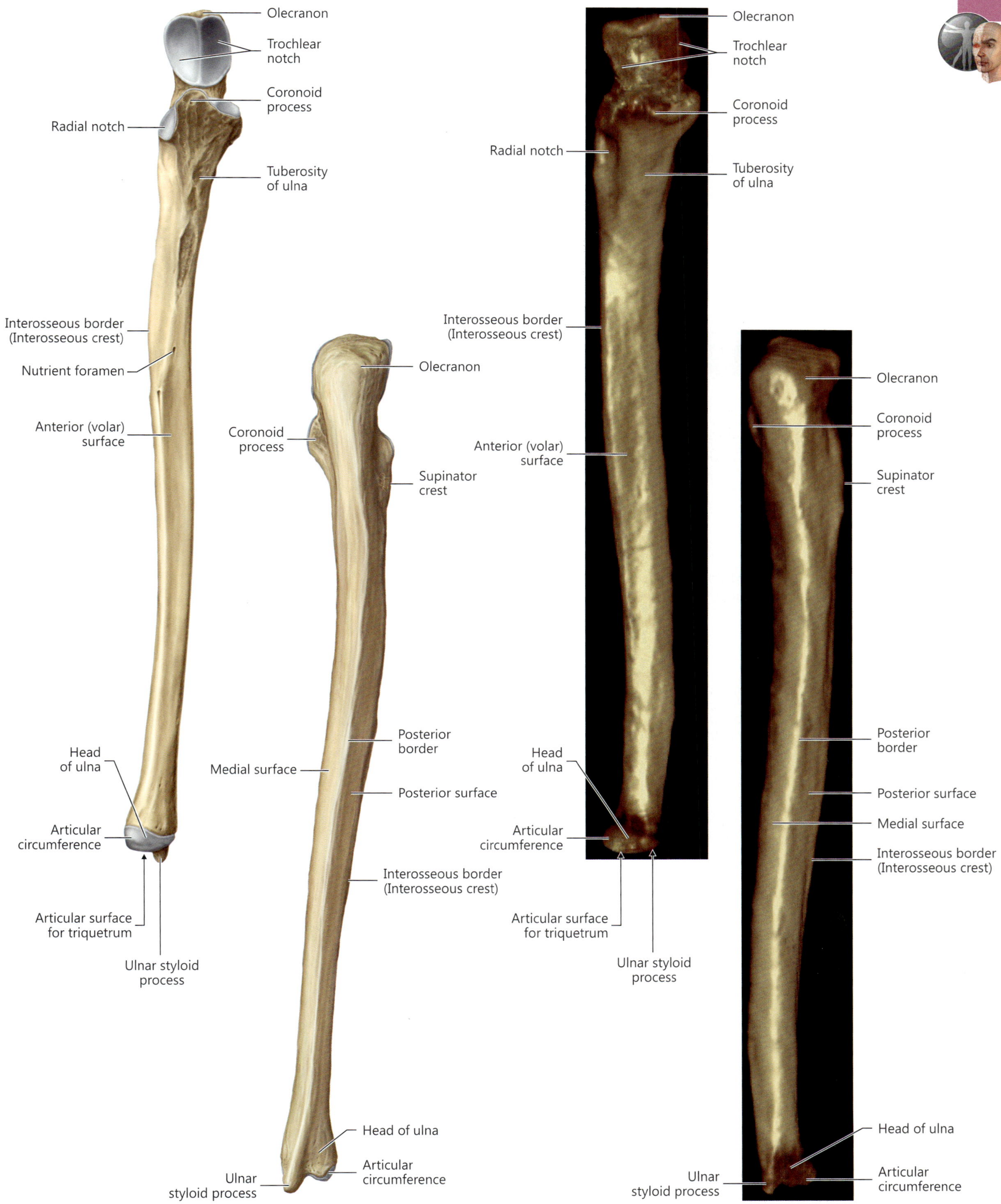

Right **ulna**. On the left, view of the anterior, or volar, and posterior, or dorsal, surfaces. On the right, three-dimensional images of a living human body, obtained by CT scan using a 3D volume rendering technique, of the same projections.

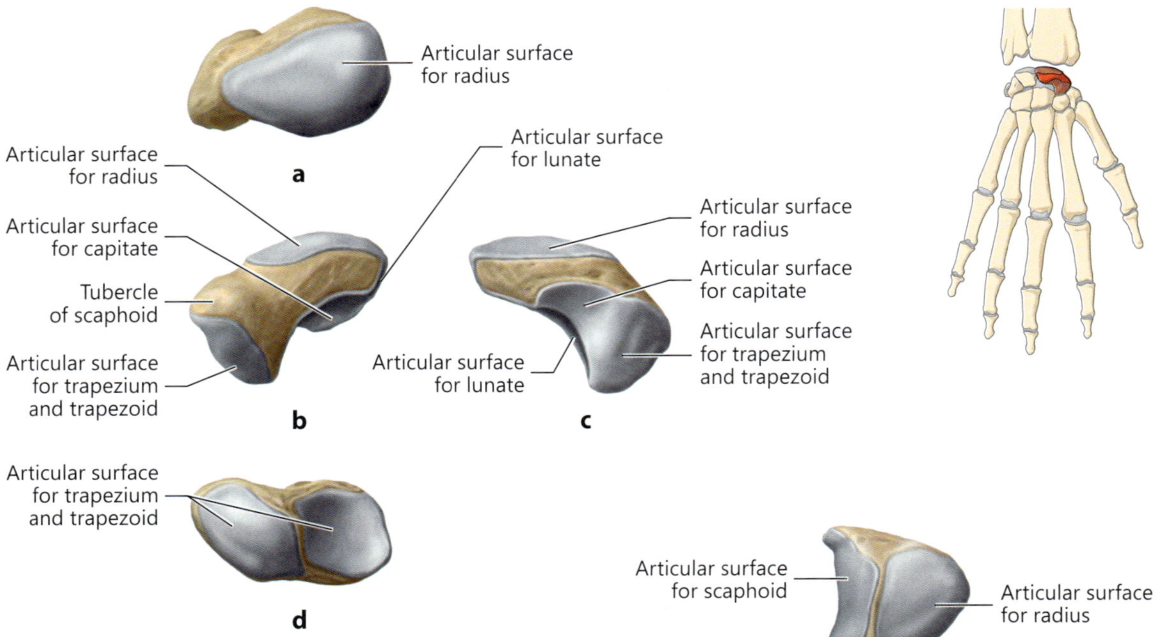

Scaphoid (**navicular**), right hand. **a**, Proximal surface; **b**, palmar surface; **c**, dorsal surface; **d**, distal surface.

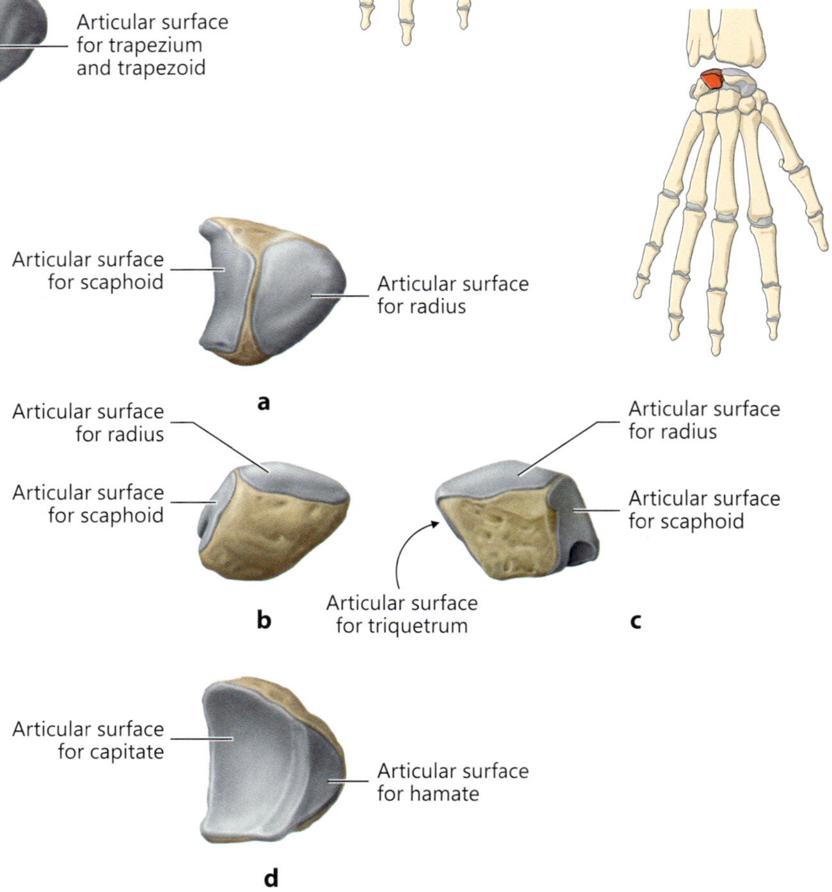

Lunate. **a**, Proximal surface; **b**, palmar surface; **c**, dorsal surface; **d**, distal surface.

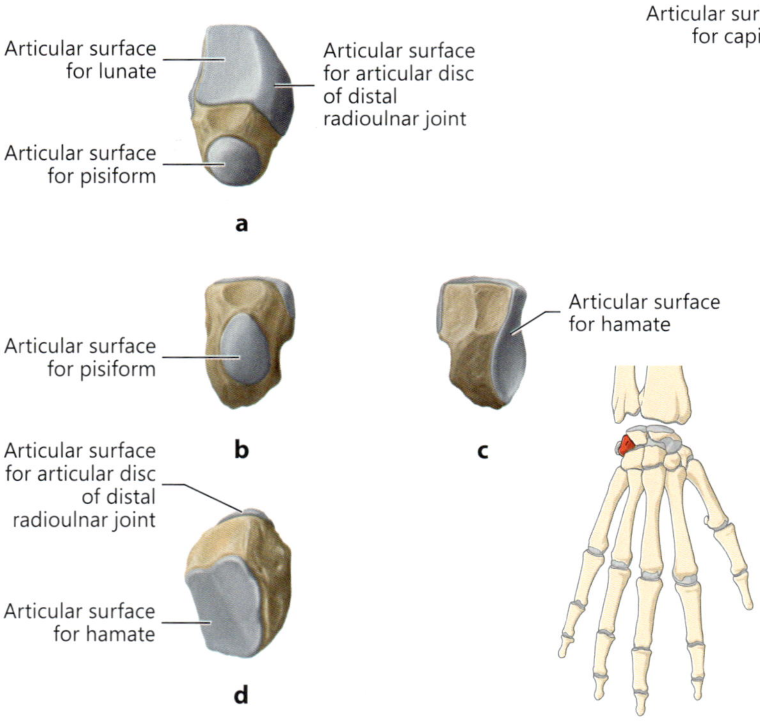

Triquetrum. **a**, Proximal surface; **b**, palmar surface; **c**, dorsal surface; **d**, distal surface.

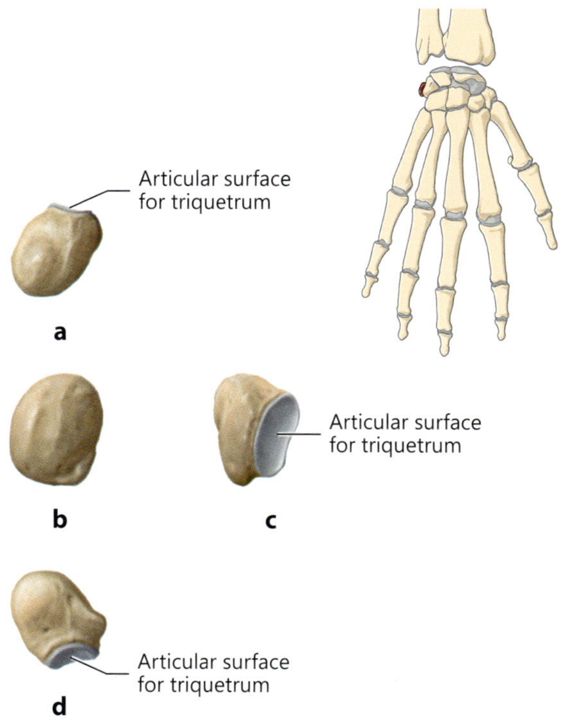

Pisiform. **a**, Proximal surface; **b**, anterior surface; **c**, posterior surface; **d**, distal surface.

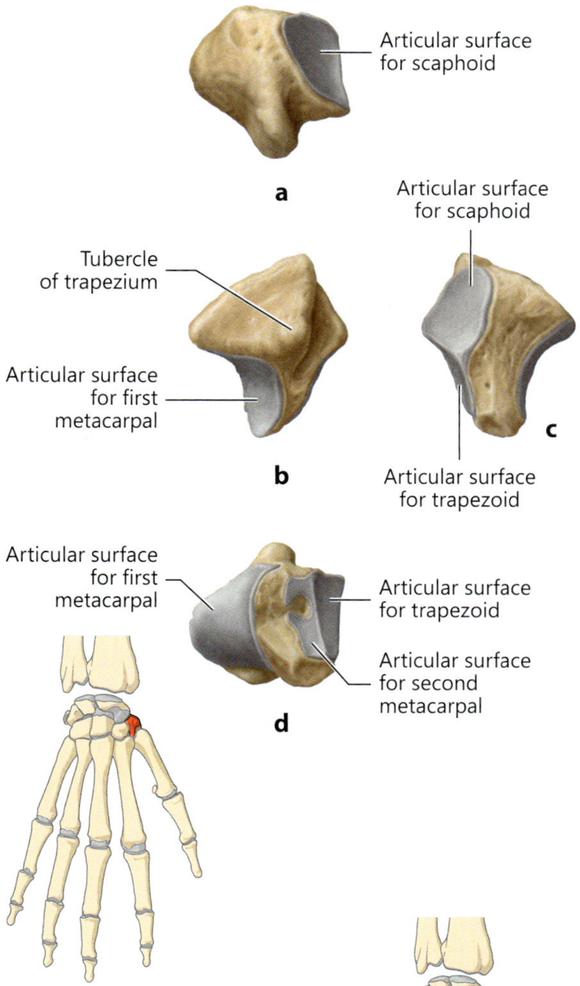

Trapezium. **a**, Proximal surface; **b**, palmar surface; **c**, dorsal surface; **d**, distal surface.

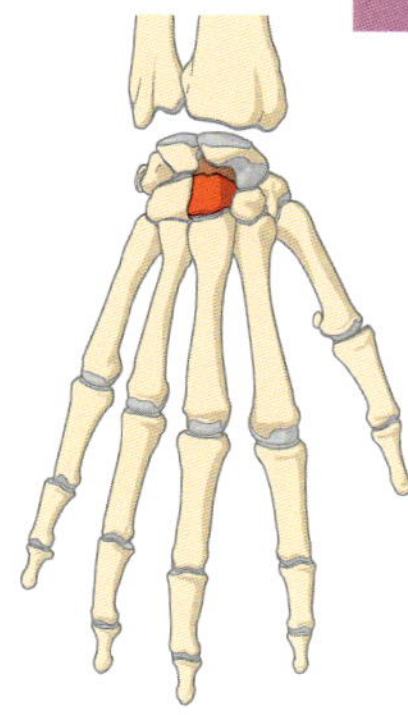

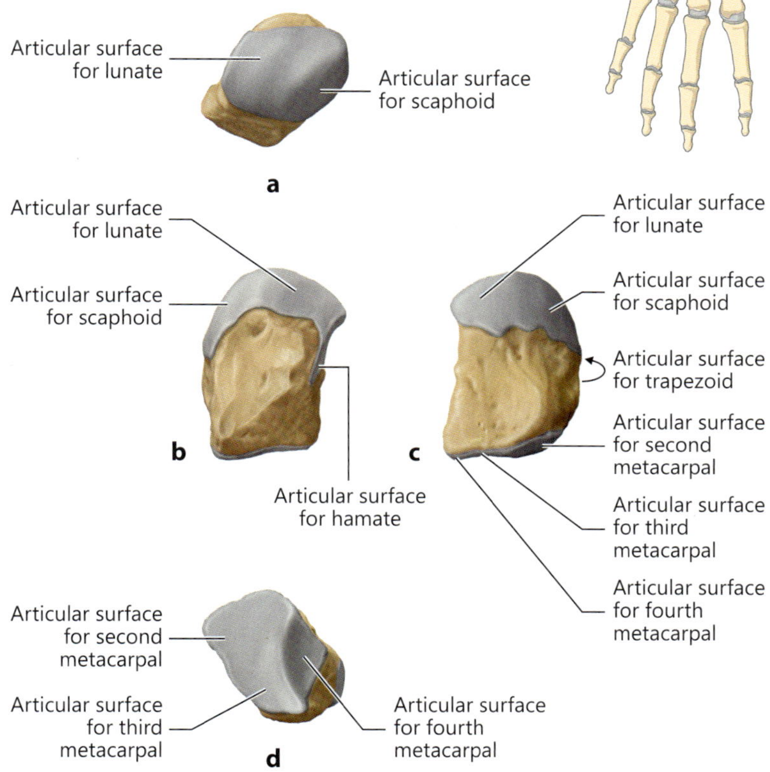

Capitate. **a**, Proximal surface; **b**, palmar surface; **c**, dorsal surface; **d**, distal surface.

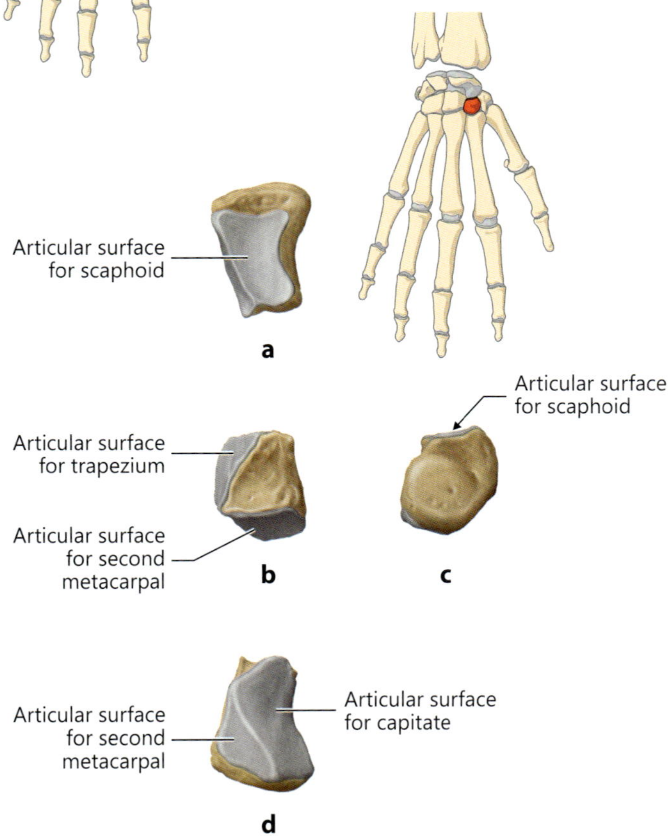

Trapezoid. **a**, Proximal surface; **b**, palmar surface; **c**, dorsal surface; **d**, distal surface.

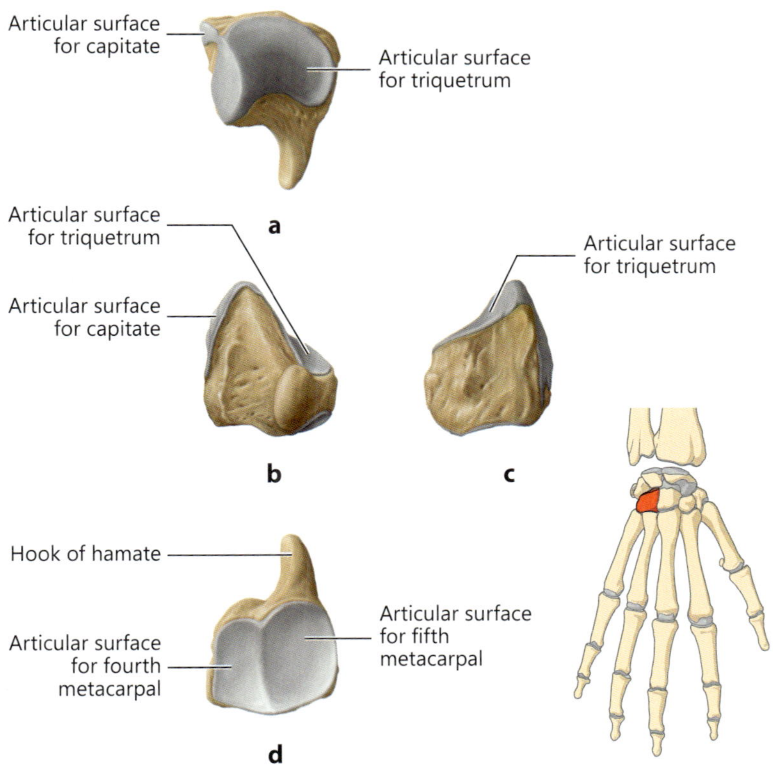

Hamate. **a**, Proximal surface; **b**, palmar surface; **c**, dorsal surface; **d**, distal surface.

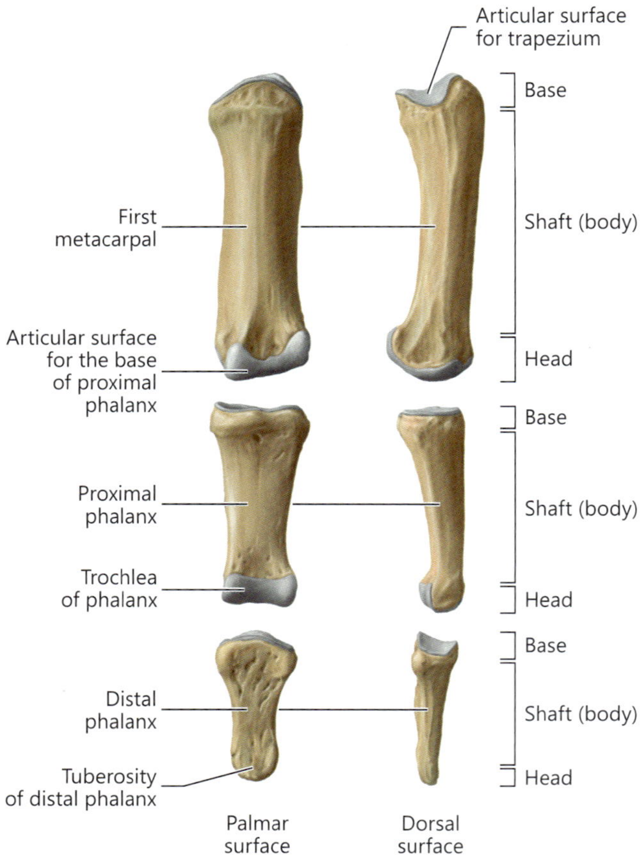

Bones of the thumb.

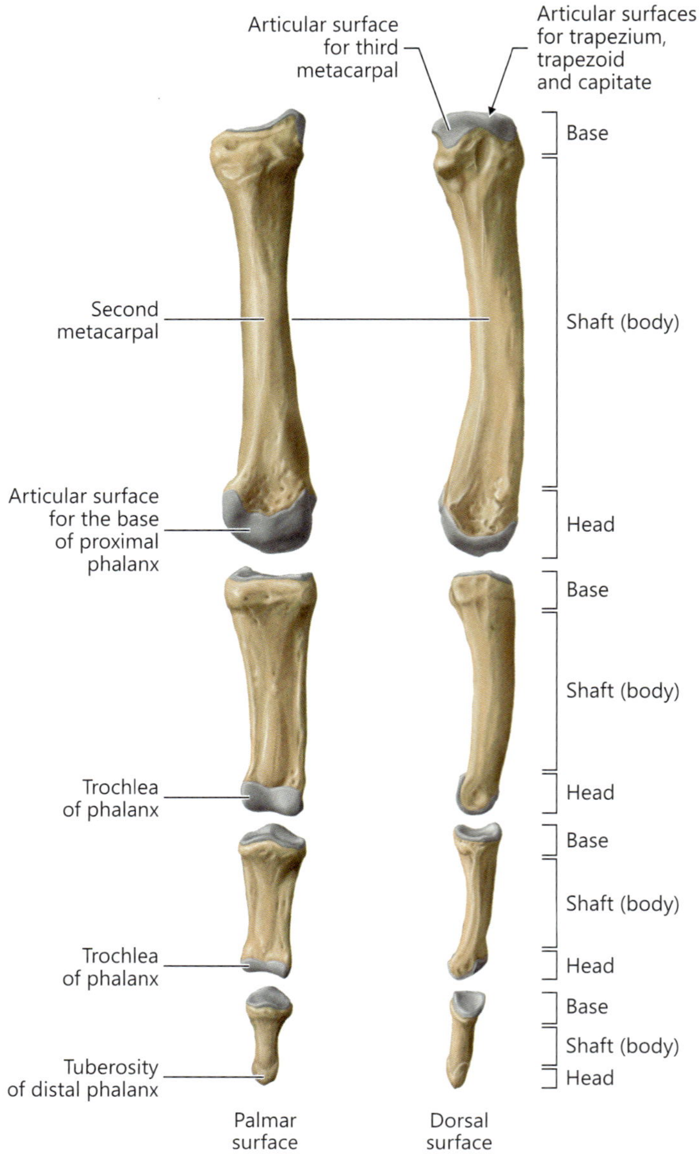

Bones of the index finger.

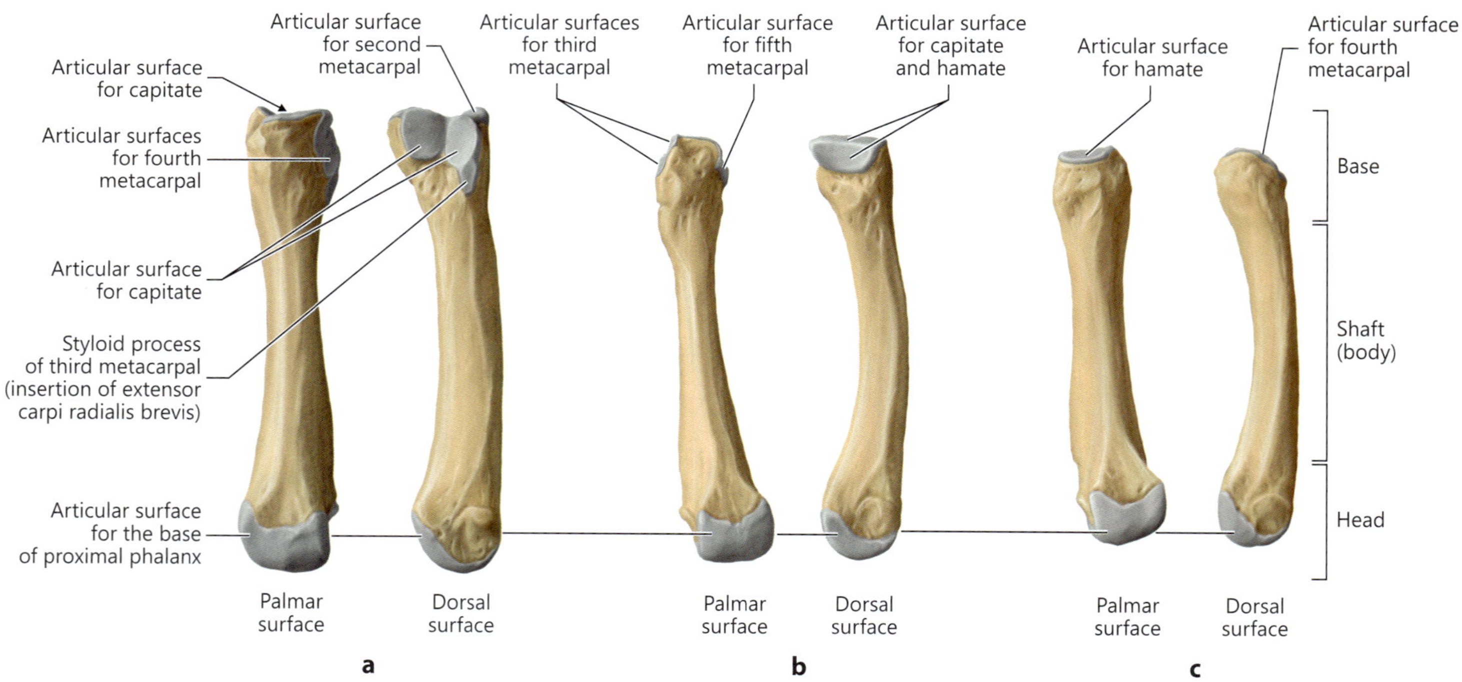

Third (**a**), fourth (**b**) and fifth (**c**) **metacarpals**.

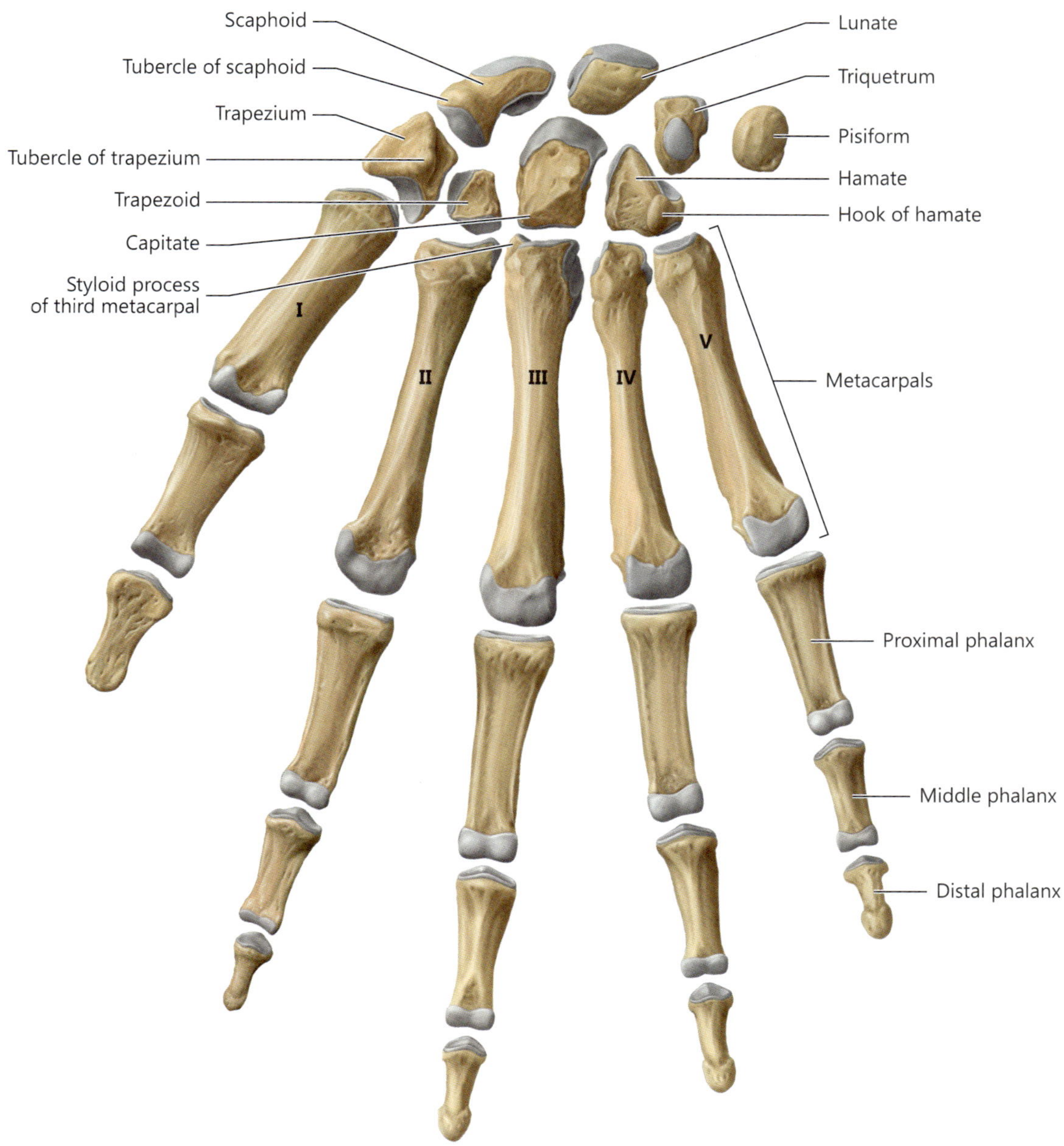

Hand, palmar (or volar) view. The bones have been separated to reveal the articular surfaces.

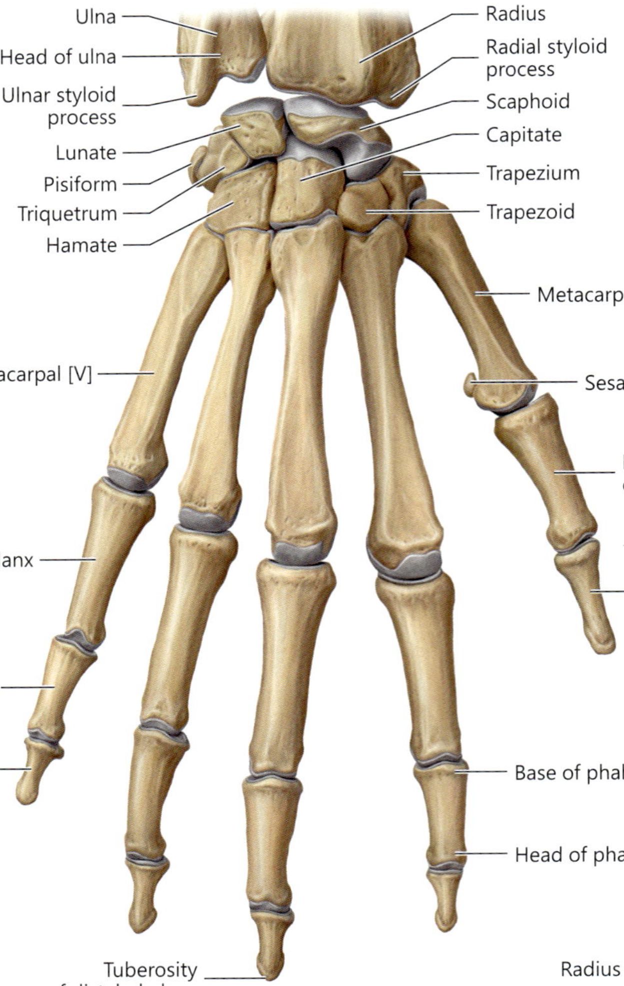

Right **hand**. Dorsal view of the skeletal structure. There is a sesamoid bone located at the metacarpophalangeal joint of the thumb.

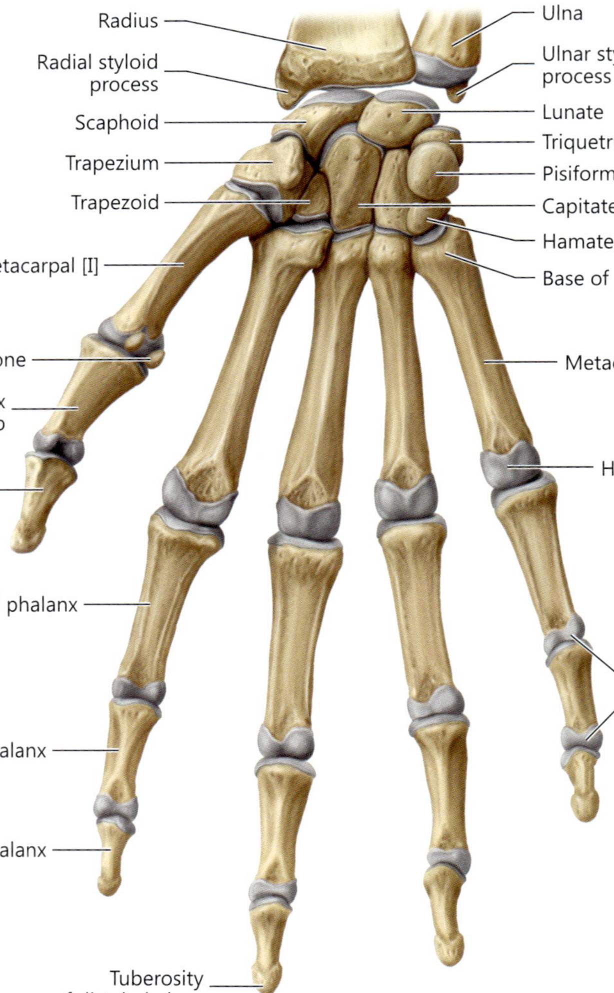

Right **hand**. Palmar view of the skeletal structure. There is a sesamoid bone located at the metacarpophalangeal joint of the thumb.

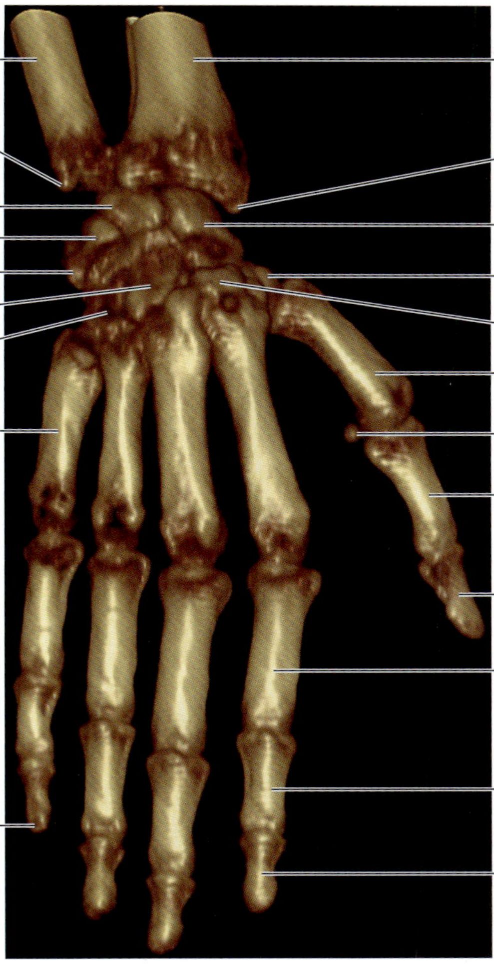

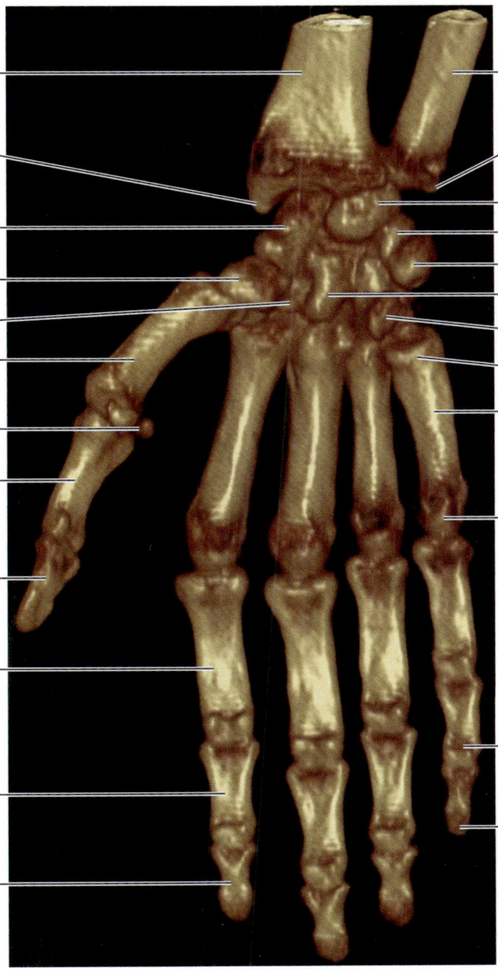

Three-dimensional images of a living human body, obtained by CT scan using a 3D volume rendering technique, of the right **hand**. Dorsal view (above), palmar view (below).

➤
Right **hand**, lateral view.

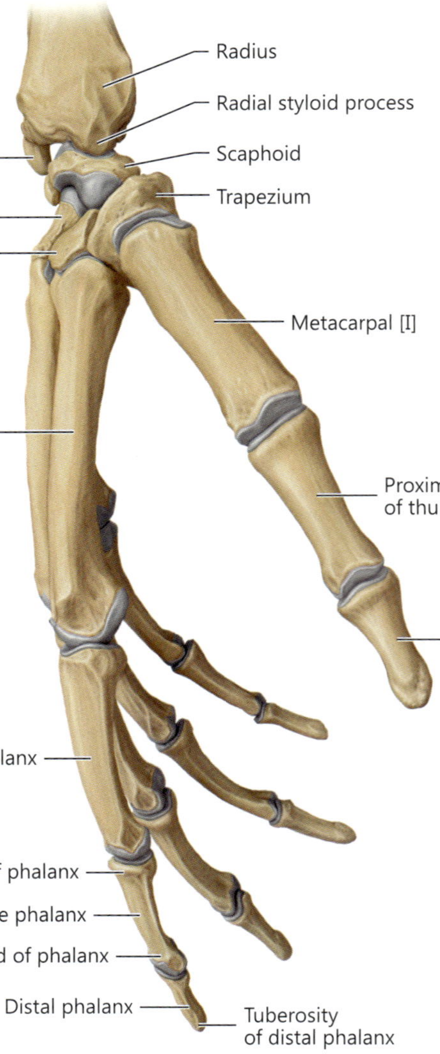

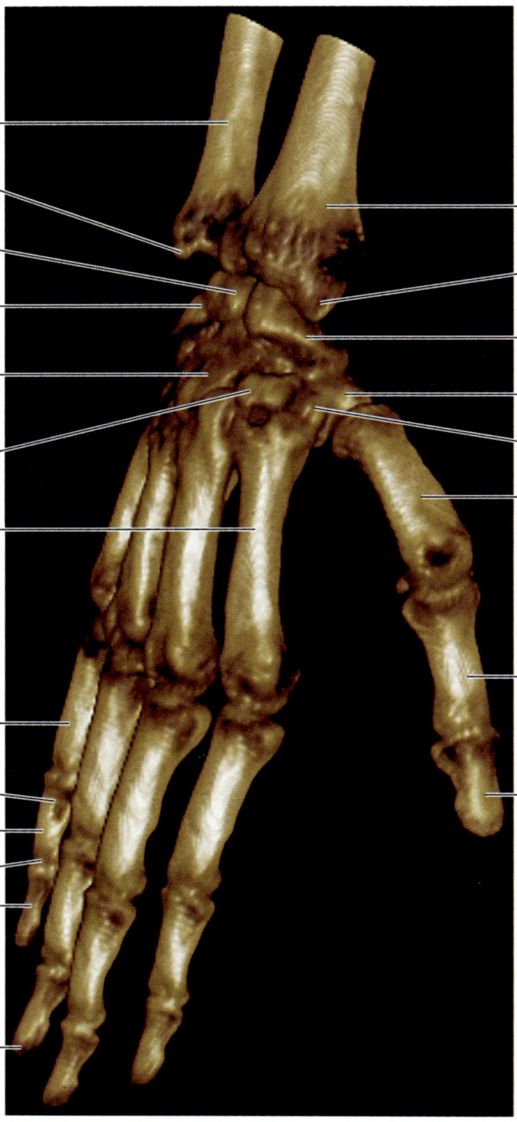

➤
Three-dimensional image of a living human body, obtained by CT scan using a 3D volume rendering technique, showing the **hand**, lateral view.

Summary table of the bones and relative joints of the upper limb

BONE	LOCATION	DESCRIPTION	JOINTS
Capitate	Wrist	The largest of the carpal bones at the center of the distal row	Lunate, second, third and fourth metacarpals, scaphoid, trapezoid, and hamate
Carpal bones	Wrist	There are eight in total, arranged in two rows: scaphoid, lunate, triquetrum and pisiform; trapezium, trapezoid, capitate, and hamate	
Clavicle	Pectoral girdle	Curved, elongated bone located horizontally above the first rib	Sternum, scapula, cartilage of the first rib
Hamate	Wrist	The most medial in the row of distal carpal bones; different from an hamate process (hamulus) which projects from its palmar surface	Lunate, triquetrum, capitate, fourth and fifth metacarpals
Humerus	Arm	The longest and largest bone of the upper limb located between the shoulder and the elbow	Scapula, radius, and ulna
Lunate	Wrist	At the center of the proximal carpal row between the scaphoid and the triquetrum	Radius, capitate, hamate, triquetrum, scaphoid
Metacarpus metacarpals	Hand	Five thin bones of the hand, each consisting of a body and two extremities (head and base); they are numbered starting with the thumb	Trapezium, trapezoid, hamate, capitate; distally with the respective proximal phalanges
Phalanges of the hand	Hand	Small, long bones, two in the thumb and three in the other fingers	Proximal row of phalanges with the corresponding metacarpals and with the intermediate phalanges; intermediate phalanges with the proximal and distal phalanges; distal phalanges with the intermediate phalanges
Pisiform	Wrist	The most medial of the bones of the proximal carpal row; the smallest of the carpal bones	Triquetrum
Radius	Forearm	Lateral bone of the forearm; the proximal extremity is small and is only minimally involved with the elbow joint; the distal extremity is flared and constitutes the largest part of the wrist joint	Humerus, ulna, lunate, scaphoid
Scaphoid	Wrist	Large bone of the proximal carpal row located on the side of the thumb	Radial, trapezium, trapezoid, capitate, lunate
Scapula	Shoulder	Large flat triangular bone that forms the dorsal part of the pectoral girdle	Clavicle, humerus
Trapezium	Wrist	The most lateral of the four bones of the distal carpal row	Scaphoid, first metacarpal, trapezoid, second metacarpal
Trapezoid	Wrist	The smallest bone of the distal carpal row	Scaphoid, second metacarpal, trapezium, capitate
Triquetrum	Wrist	Pyramid-shaped; it is the second in the proximal carpal row starting from the side of the little finger	Lunate, pisiform, hamate, articular disc of the distal radioulnar joint (triangular fibrocartilage disc)
Ulna	Forearm	Medial bone of the forearm, parallel to the radius	Humerus, radius

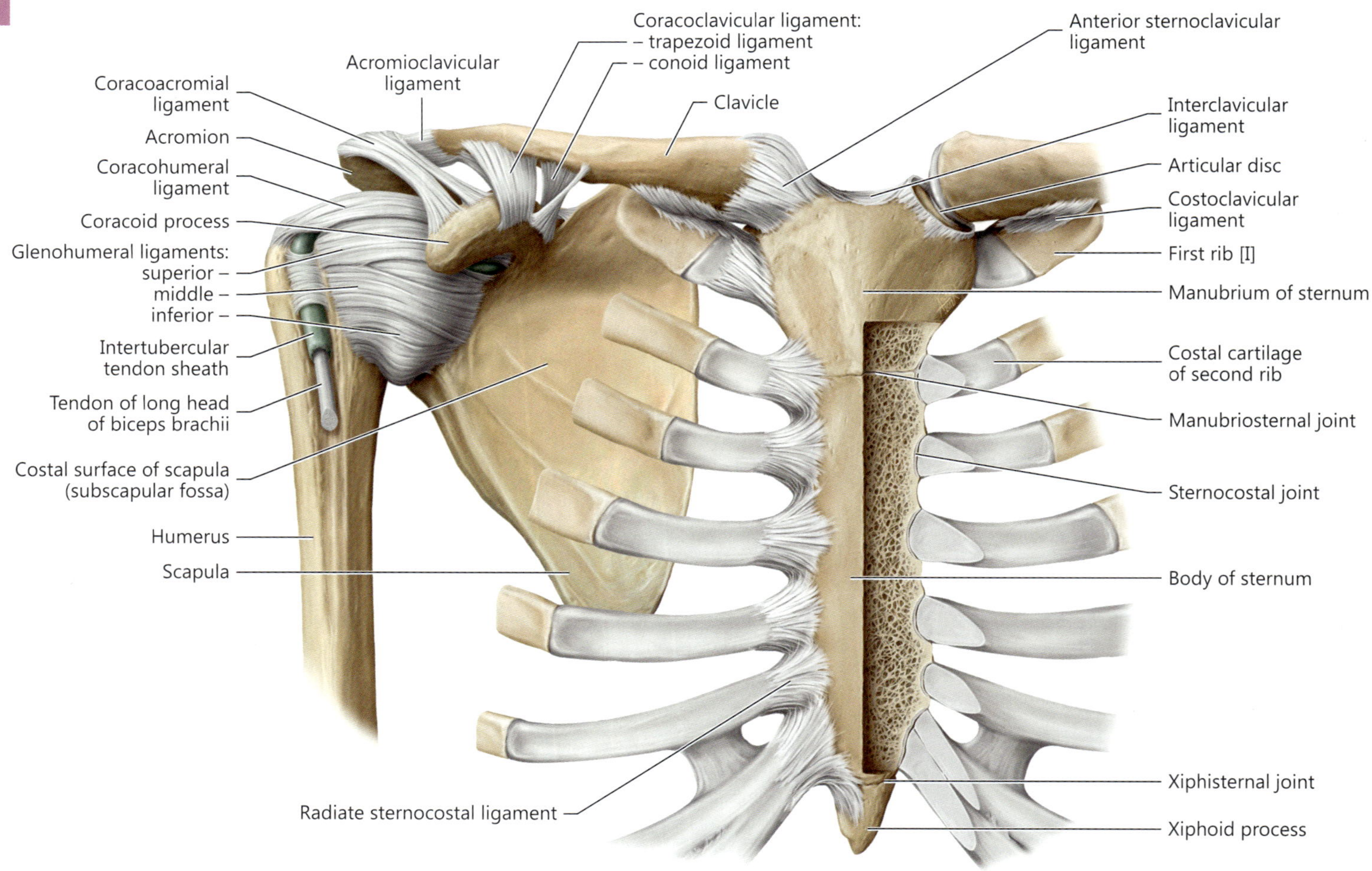

Right **glenohumeral (shoulder) joint**. Anterior view in which the relationships with the bones of the thorax are shown.

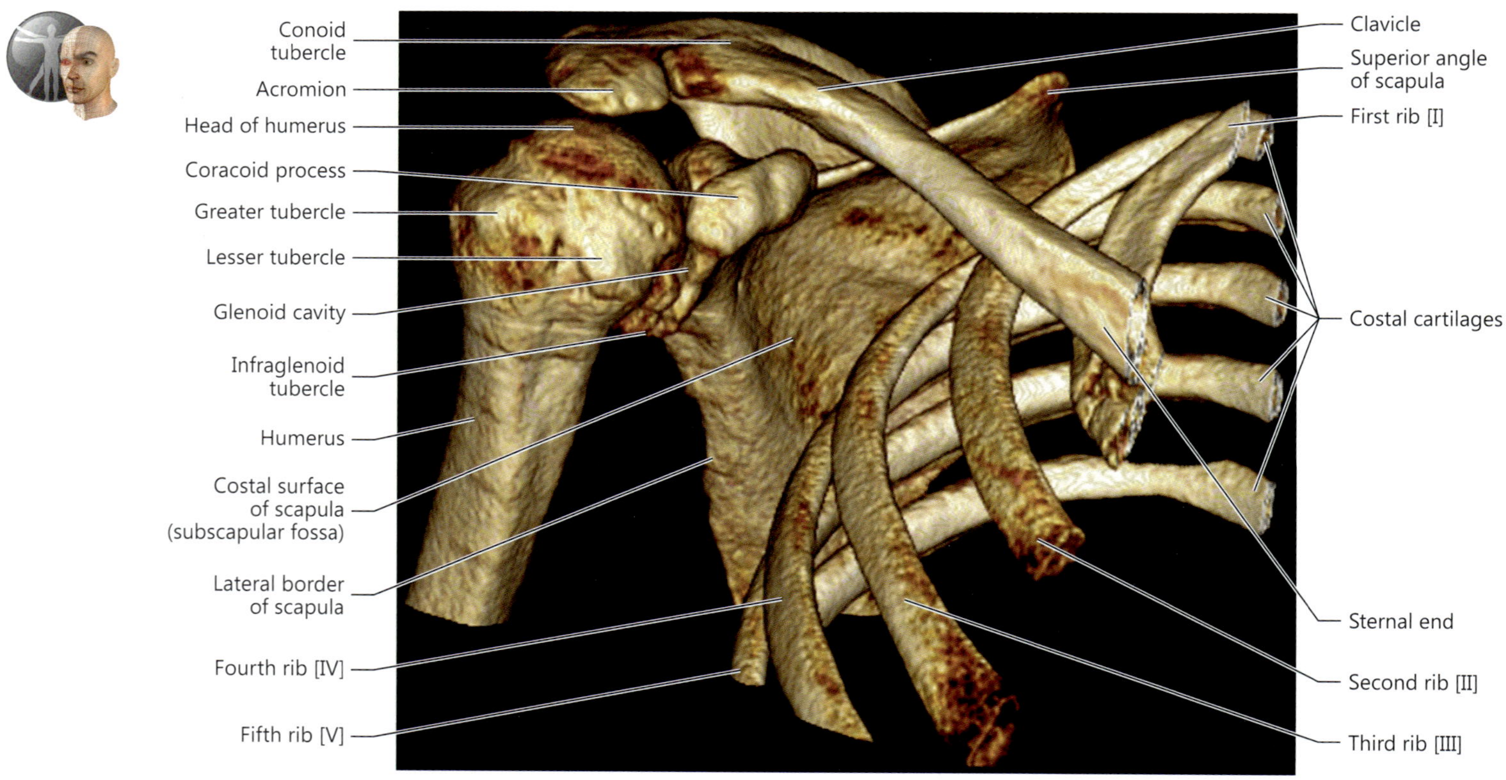

Three-dimensional image of a living human body, obtained by CT scan using a 3D volume rendering technique, showing the right **glenohumeral (shoulder) joint**, superoanterior view.

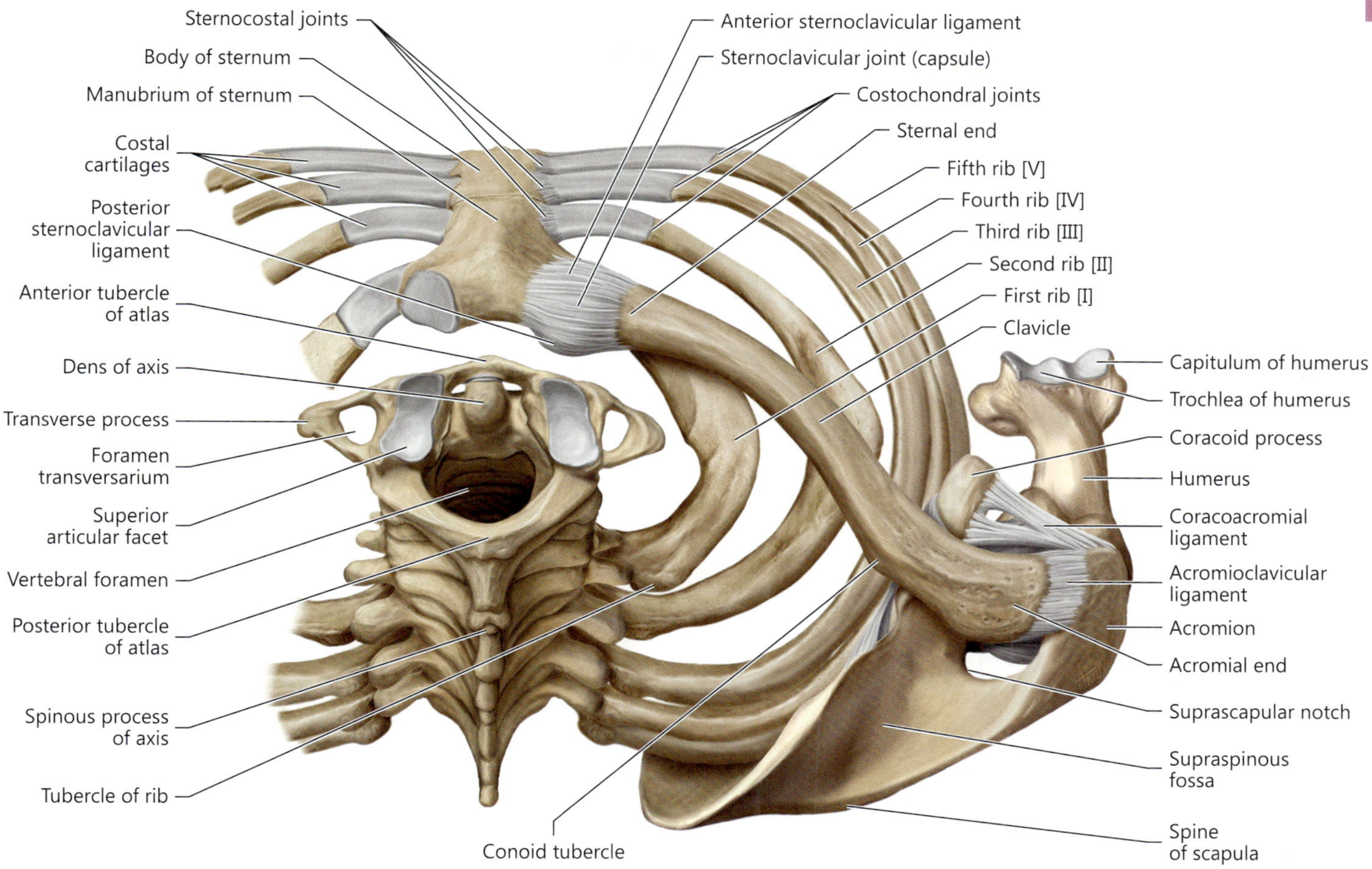

Right **glenohumeral (shoulder) joint**. Superior view in which the relationships with the rib cage are shown.

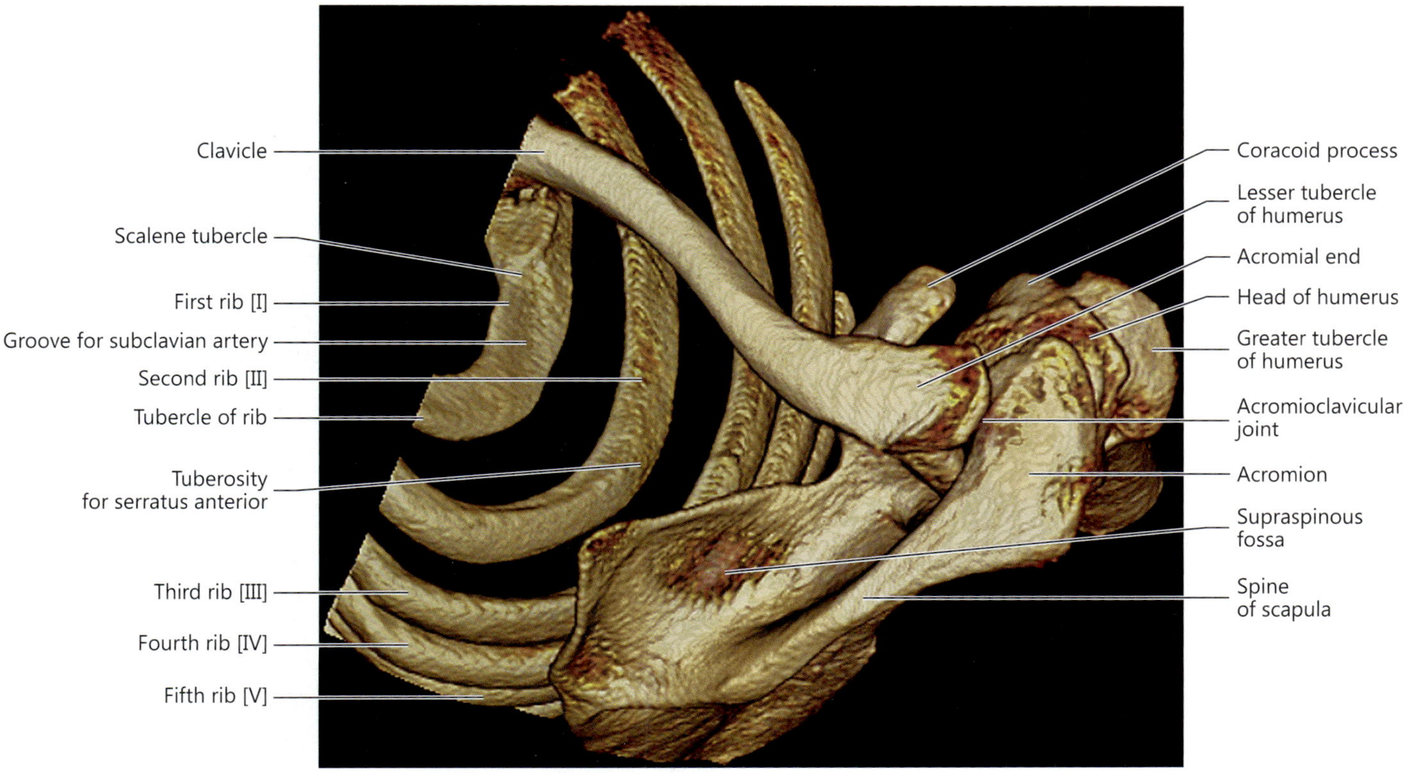

Three-dimensional image of a living human body, obtained by CT scan using a 3D volume rendering technique, showing the right **glenohumeral (shoulder) joint**, superior view.

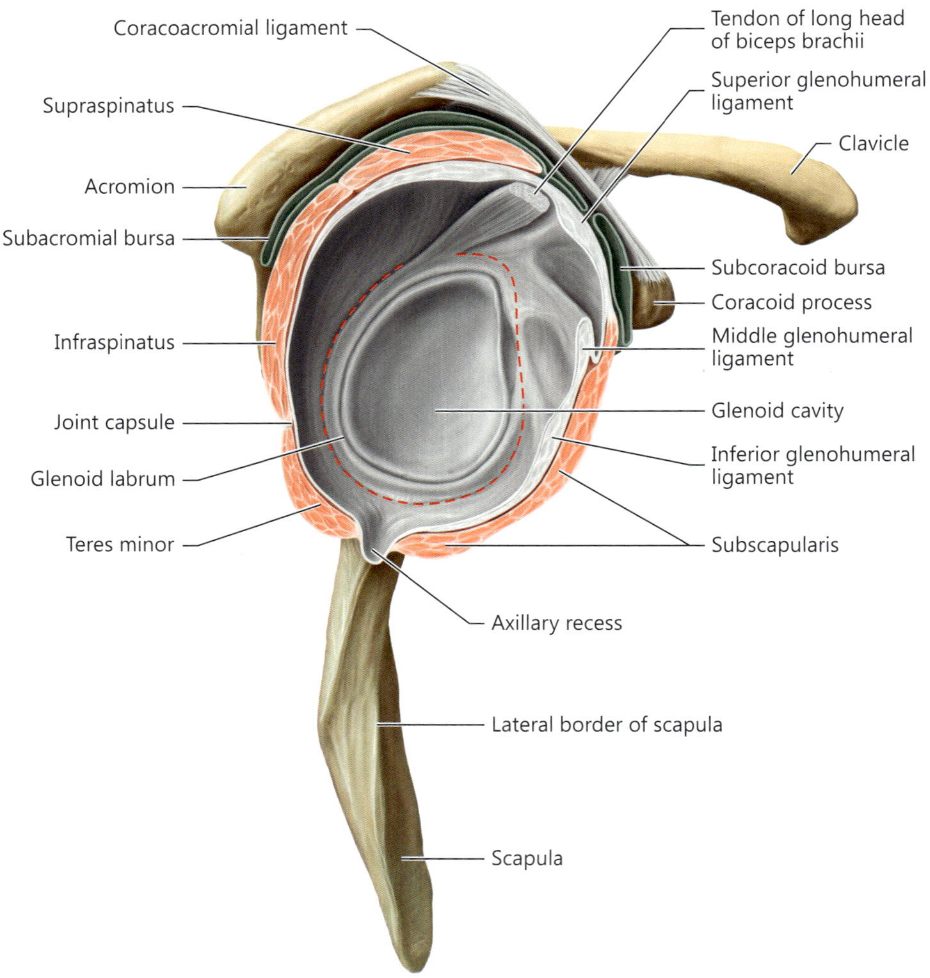

Acromioclavicular joint and **ligaments of the scapula**. On the articular surface which is involved in the glenohumeral joint, the glenoid cavity, a red line marks the limit reached by the articular cartilage.

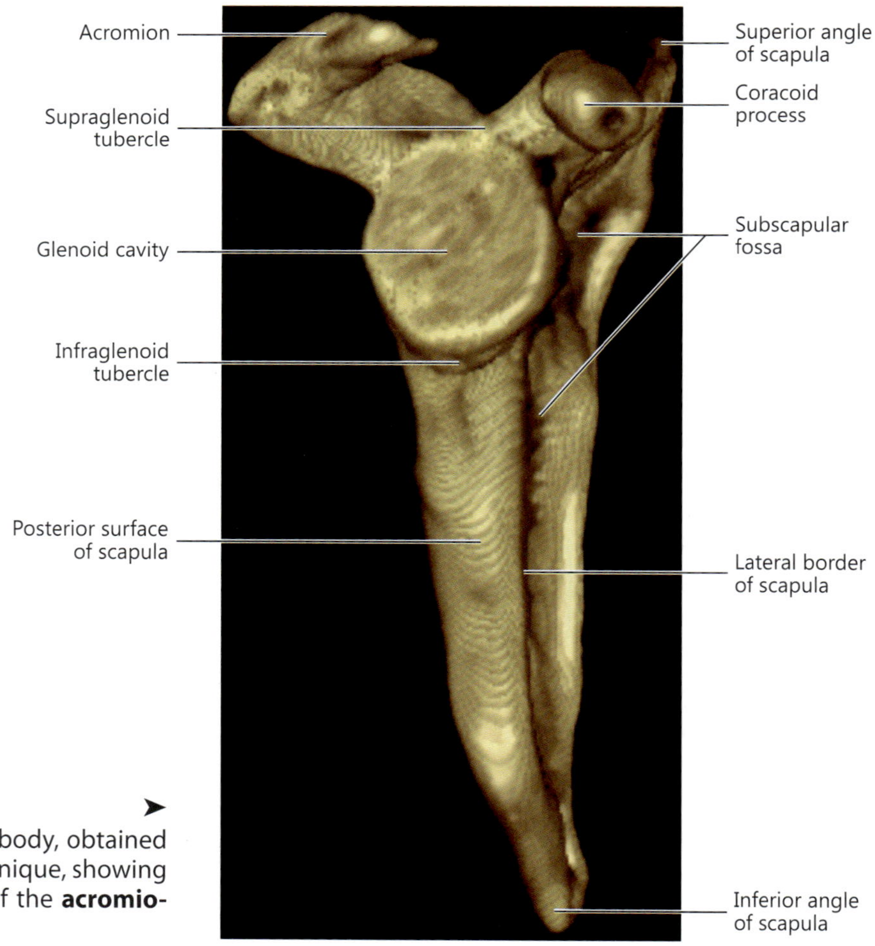

Three-dimensional image of a living human body, obtained by CT scan using a 3D volume rendering technique, showing the medial border of the articular surfaces of the **acromioclavicular** and **glenohumeral joints**.

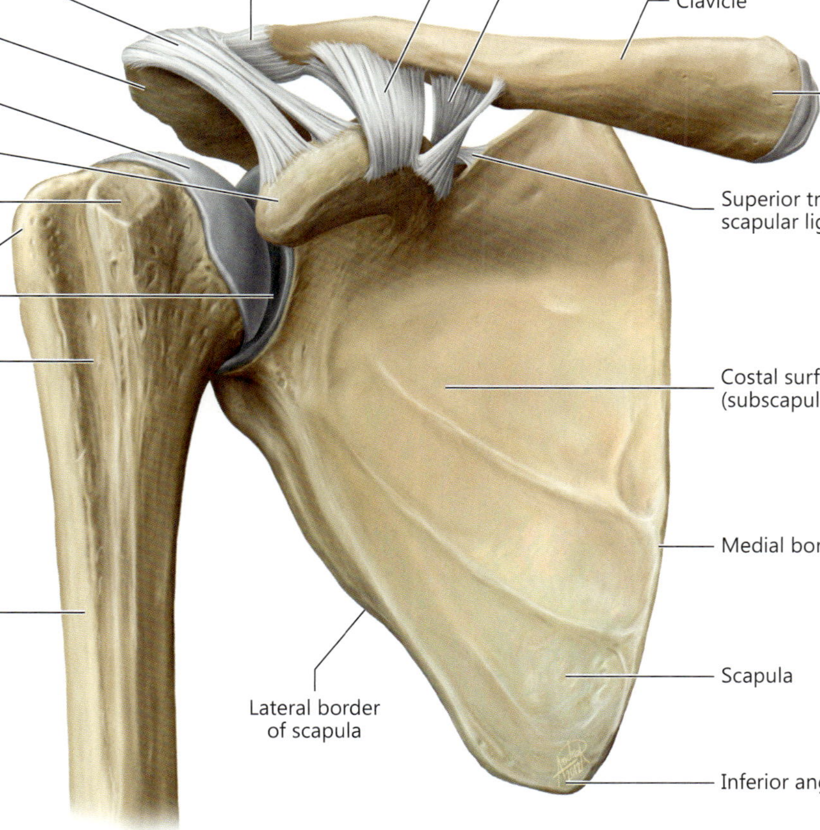

Right **glenohumeral joint** (shoulder joint), seen from the front. The head of the humerus is positioned in the glenoid cavity without the capsular ligaments.

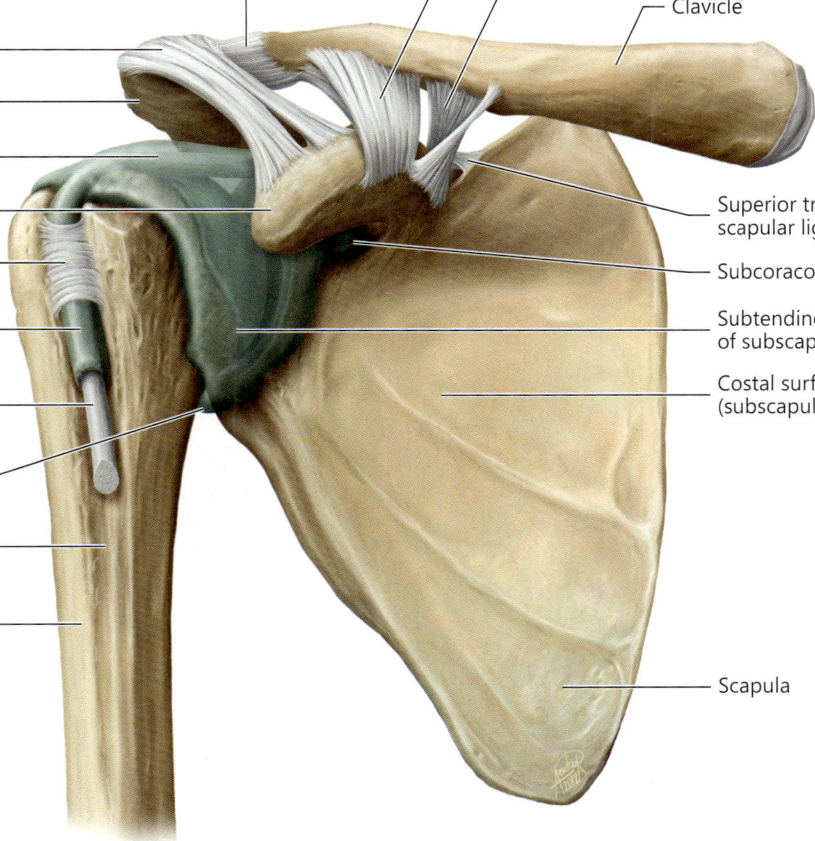

Synovial bursae of the **glenohumeral joint**. Anterior view.

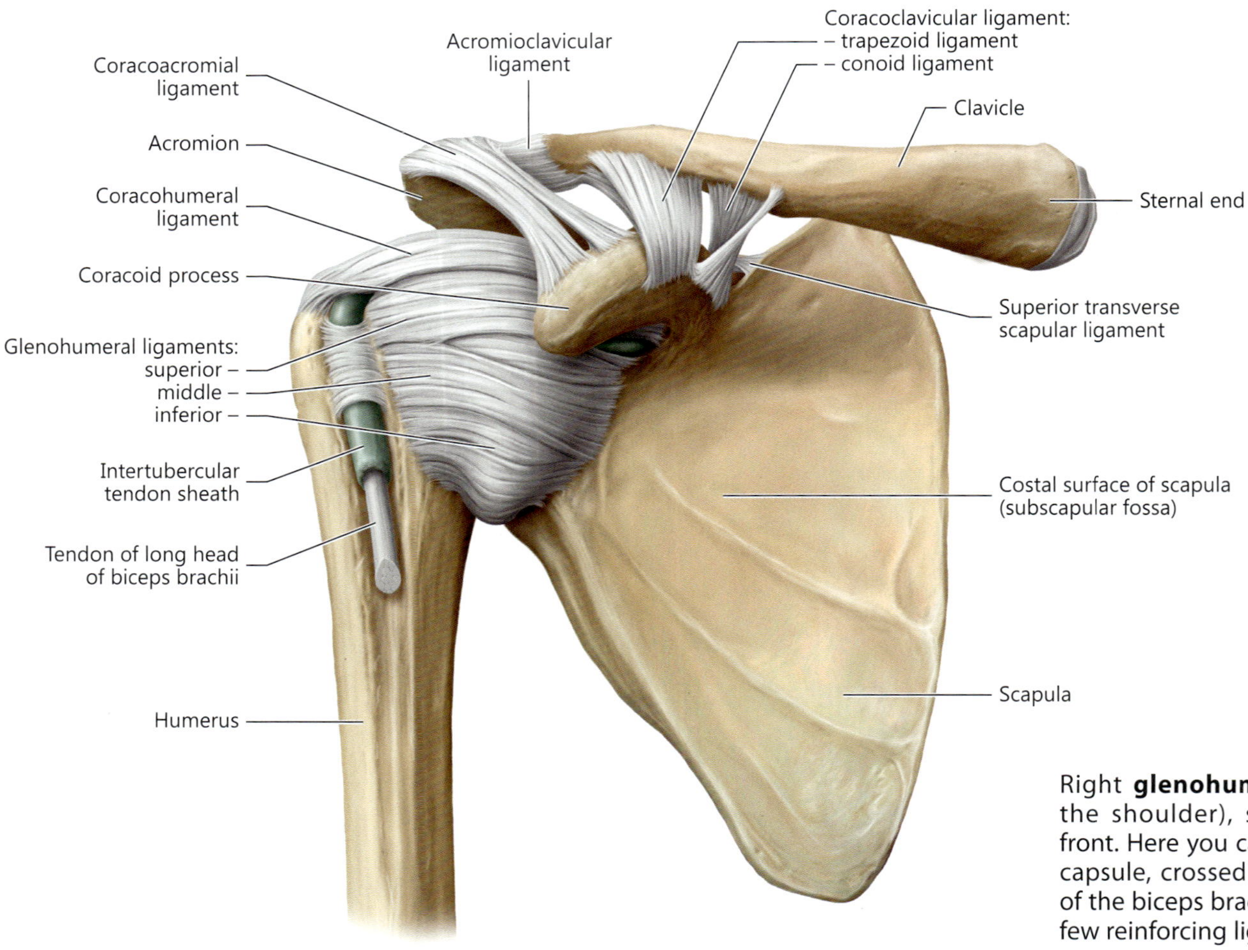

Right **glenohumeral joint** (of the shoulder), seen from the front. Here you can see the joint capsule, crossed by the tendon of the biceps brachii, as well as a few reinforcing ligaments.

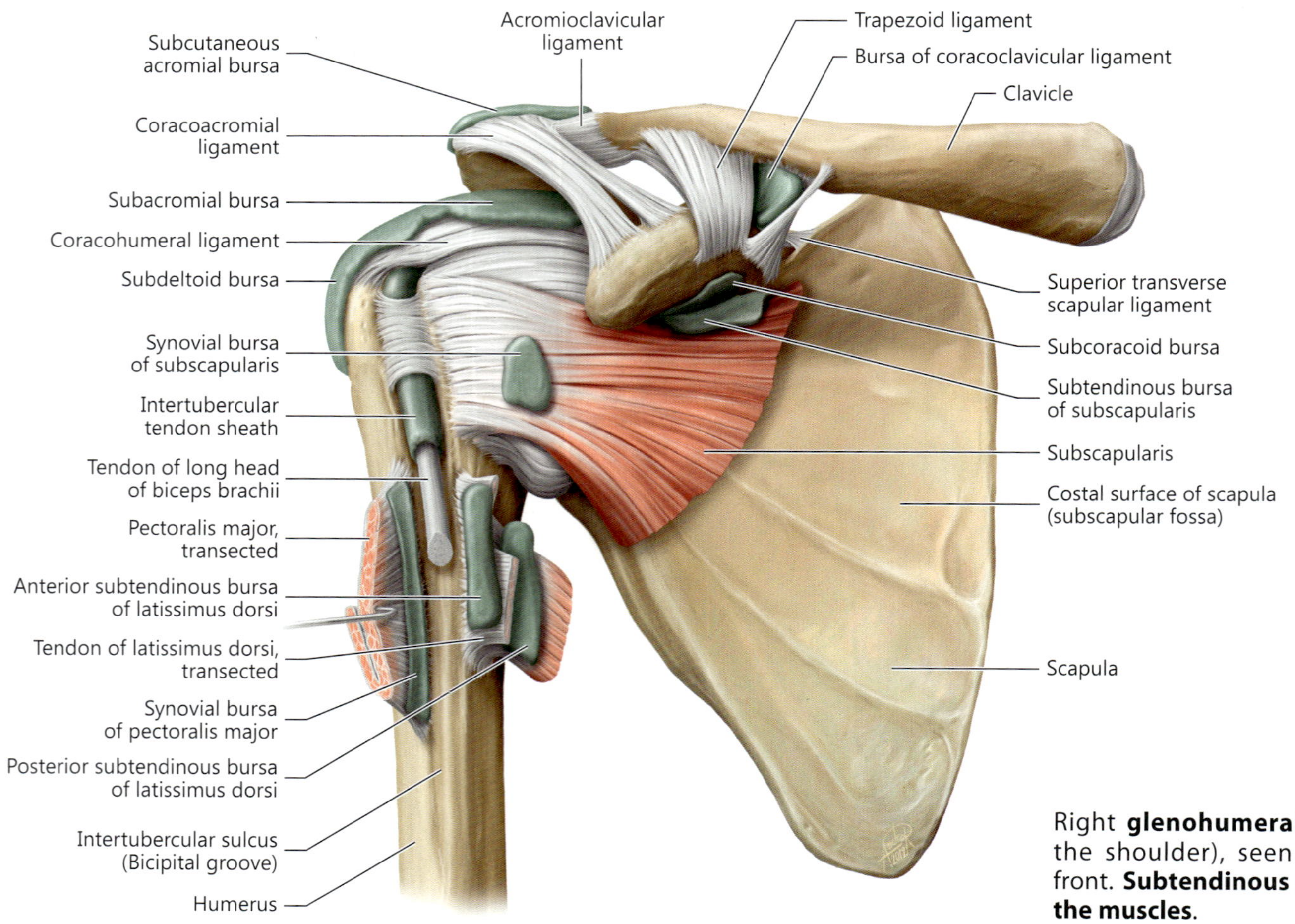

Right **glenohumeral joint** (of the shoulder), seen from the front. **Subtendinous bursae of the muscles.**

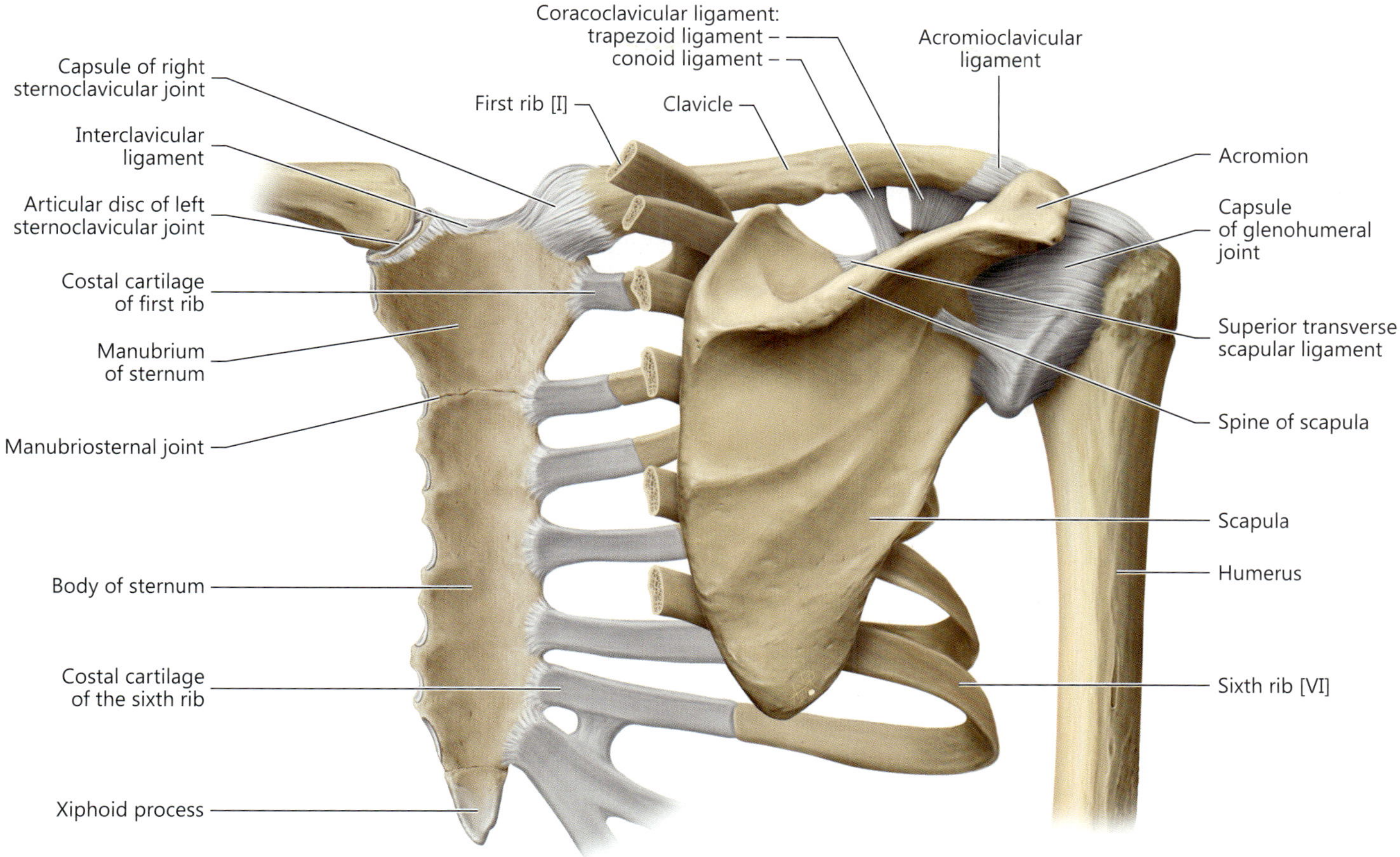

Right **glenohumeral joint** (of the shoulder). Posterior view in which the relationships with the bones of the thorax are shown.

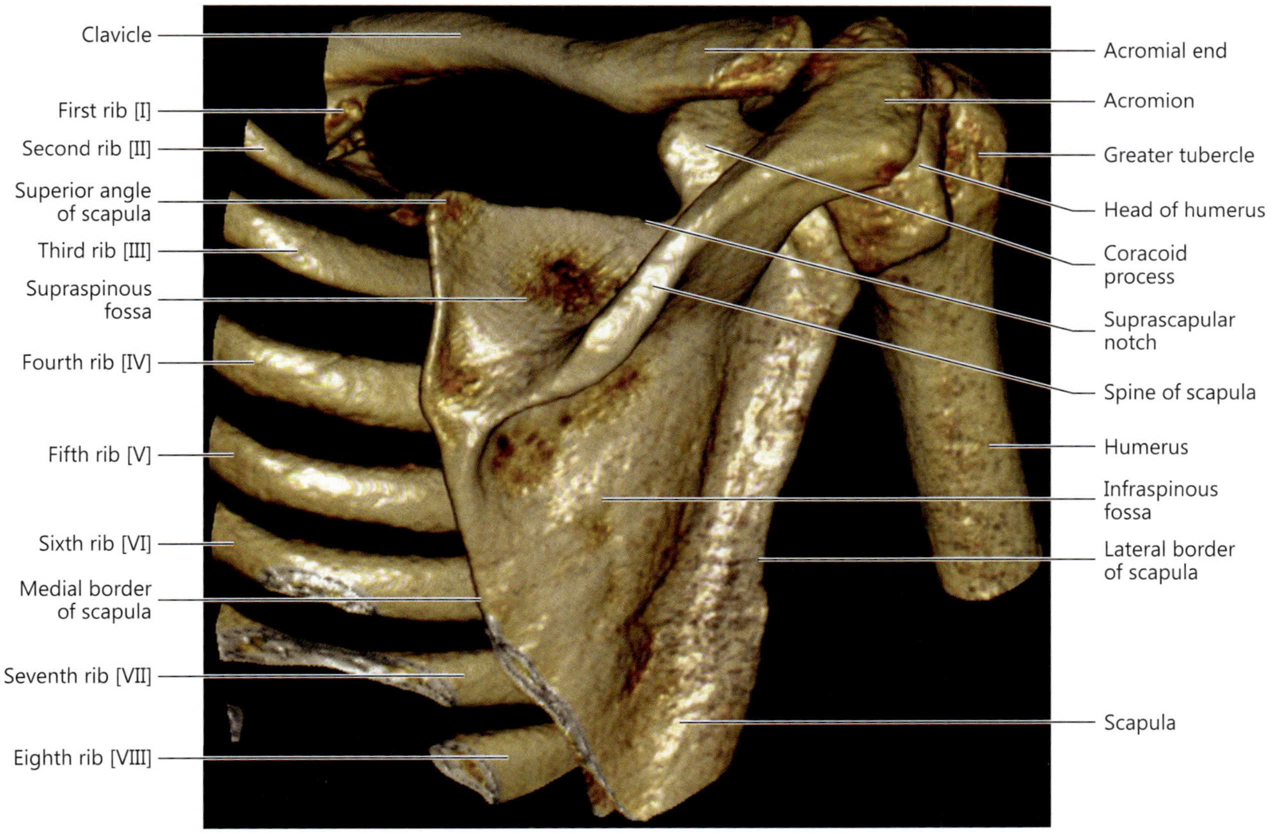

Three-dimensional image of a living human body, obtained by CT scan using a 3D volume rendering technique, showing the right **glenohumeral joint** (of the shoulder), posterior view.

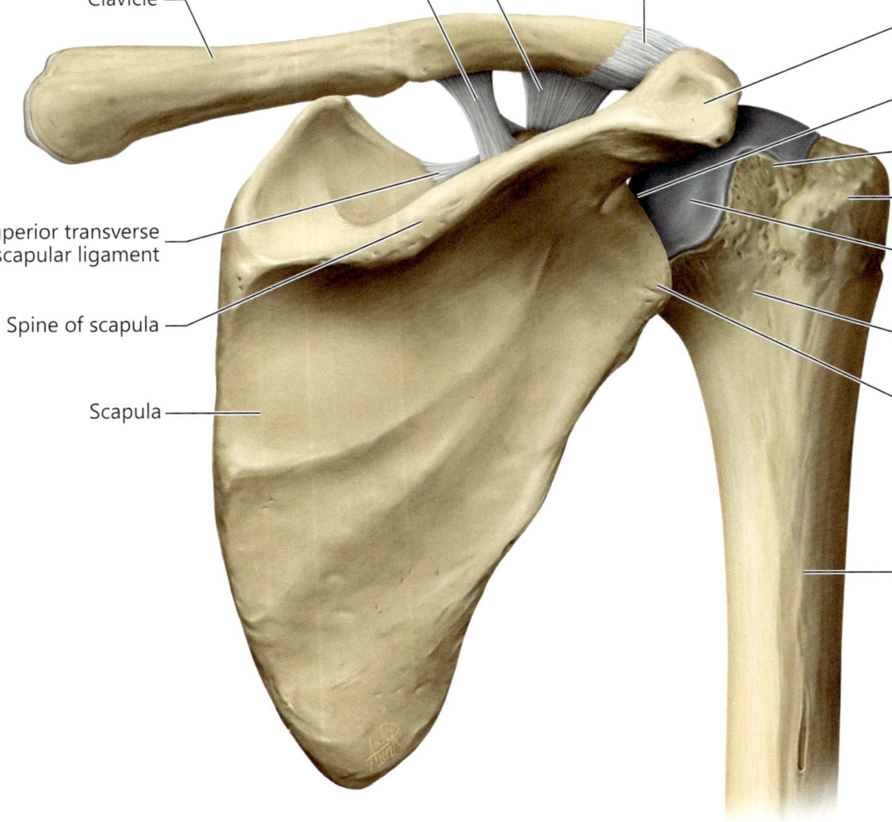

Right **glenohumeral joint** (of the shoulder), posterior view. The head of the humerus is positioned in the articular cavity without the capsular ligaments.

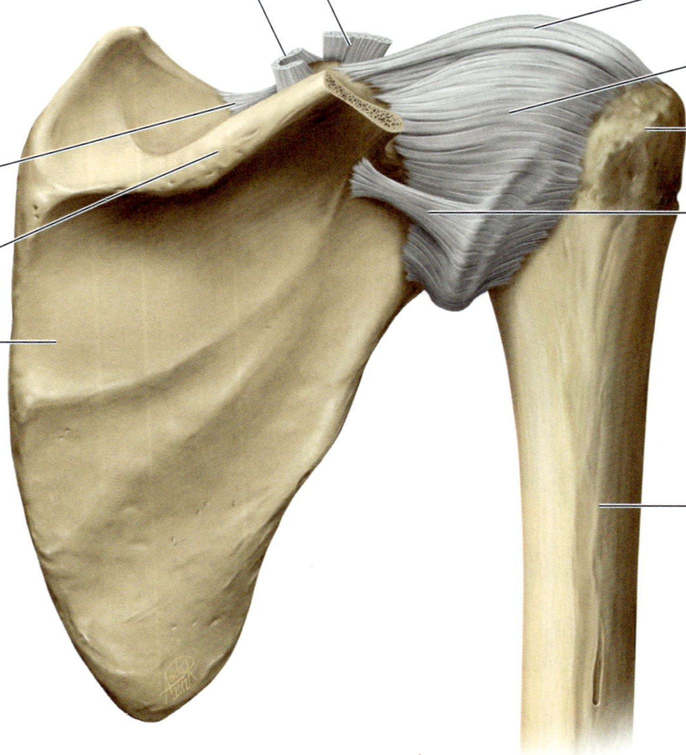

Right **glenohumeral joint** (of the shoulder), seen from behind. The laxity of the joint capsule is indicated by the folds formed therein when the limb is adducted.

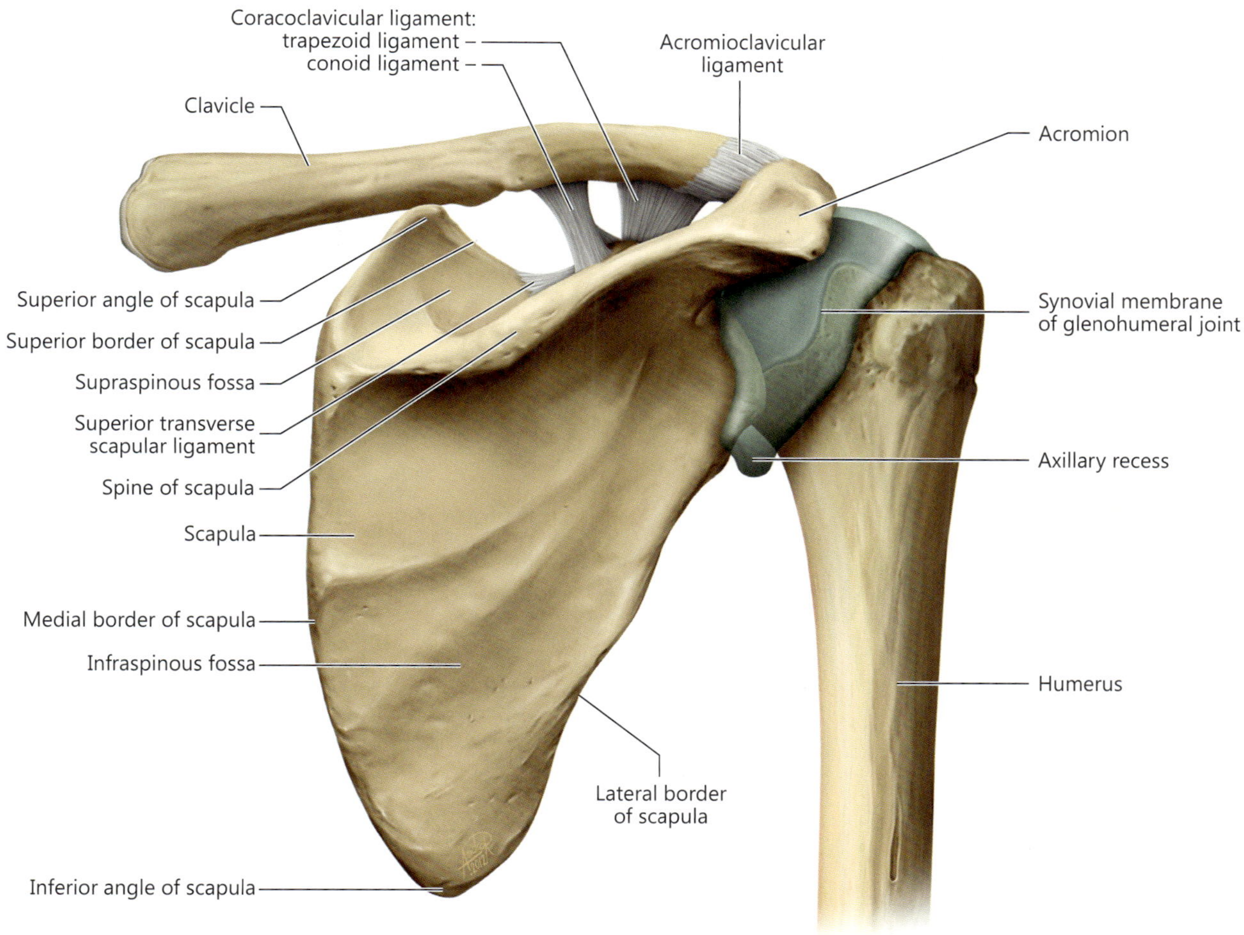

Synovial bursae of the **glenohumeral joint**, posterior view.

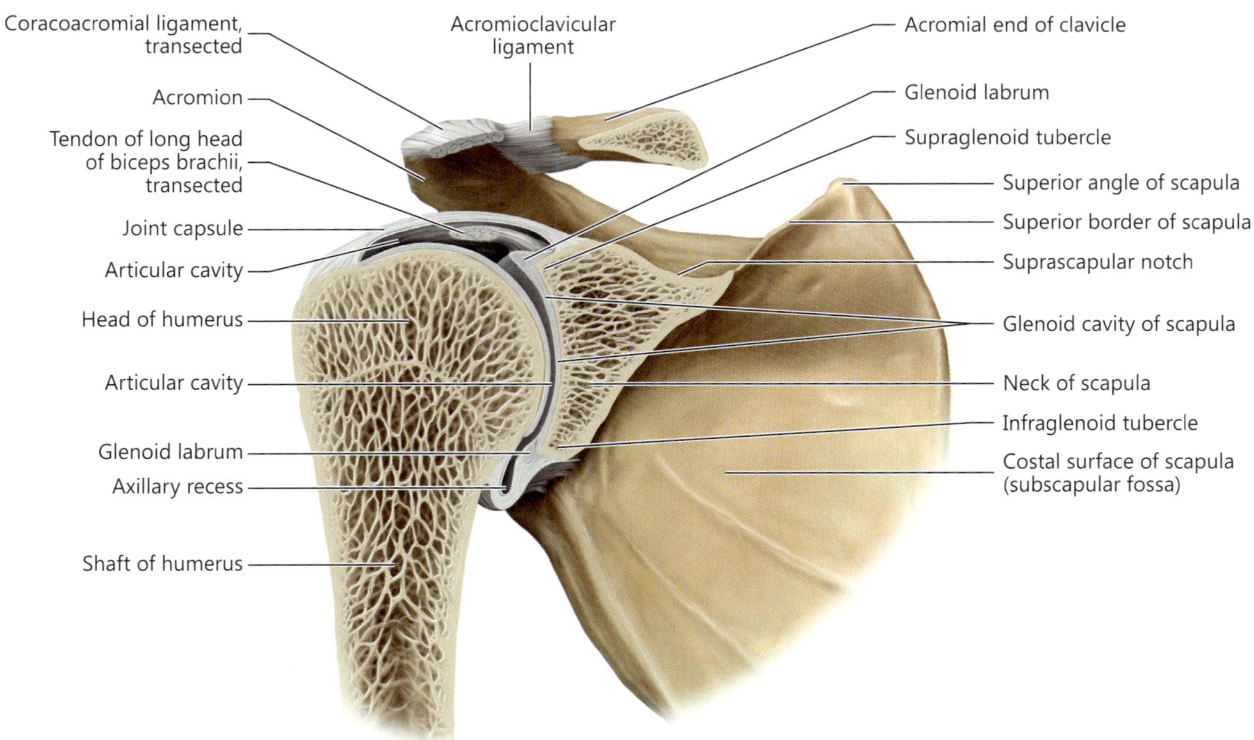

Right **glenohumeral joint** (of the shoulder) transected along the frontal plane passing through the humerus and the spine of the scapula and seen from above. Here you can see the intra-articular portion of the tendon of the long head of biceps brachii.

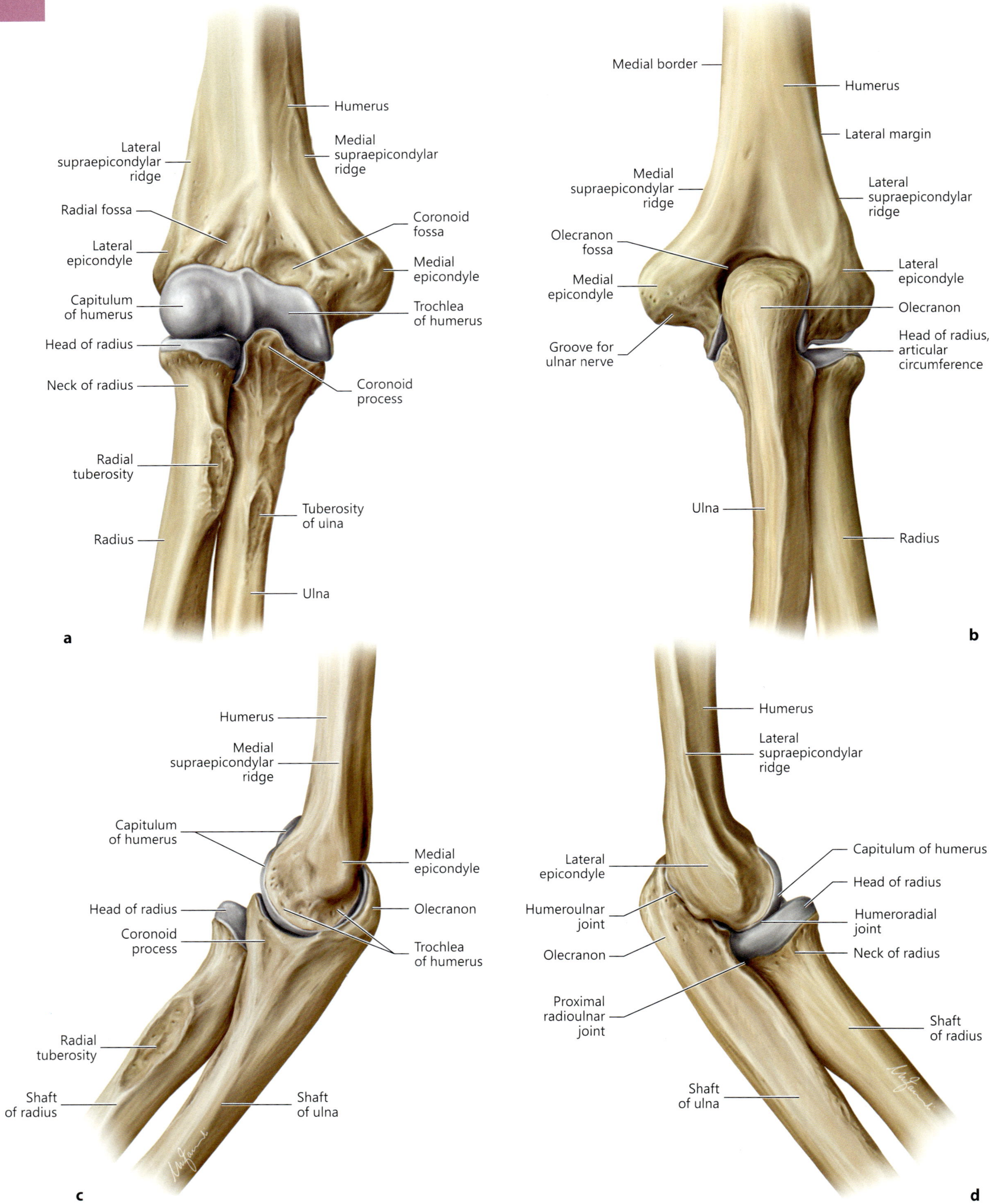

Joint of the right **elbow**. Anterior (**a**), posterior (**b**), medial (**c**) and lateral (**d**) views. Here you can see the relationships between the humerus, radius, and ulna, without the presence of the relative reinforcing ligaments.

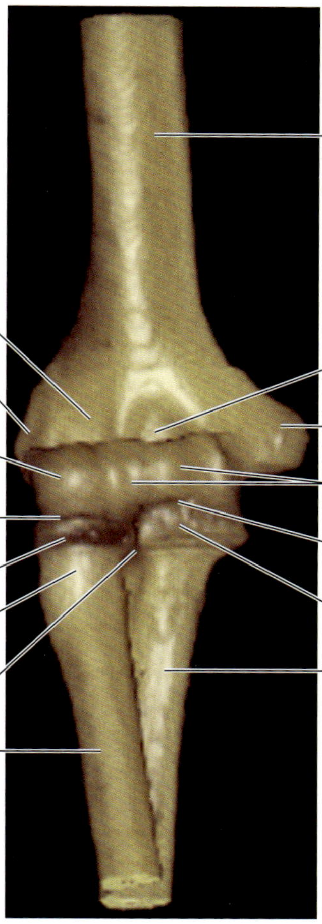

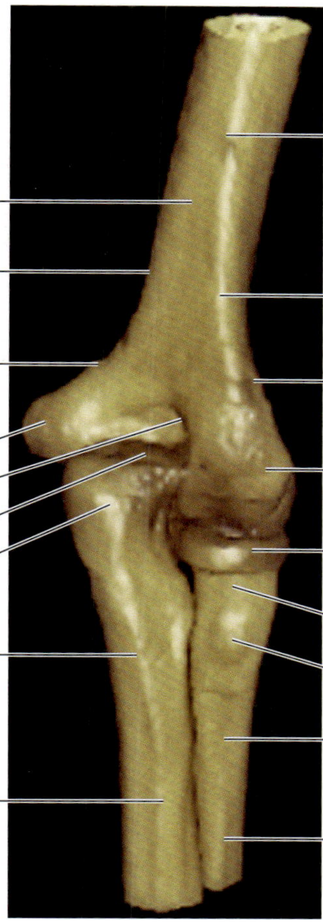

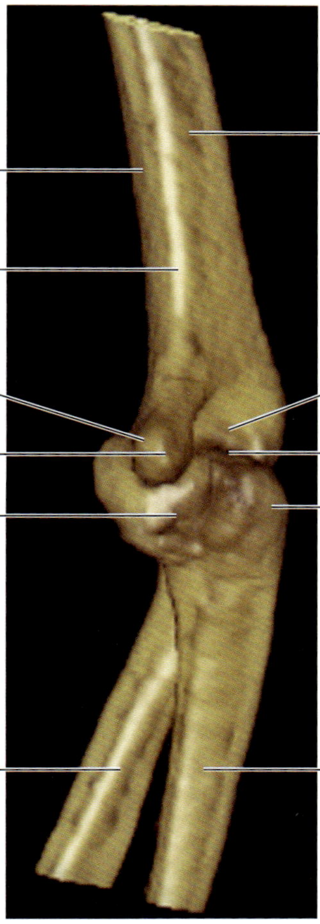

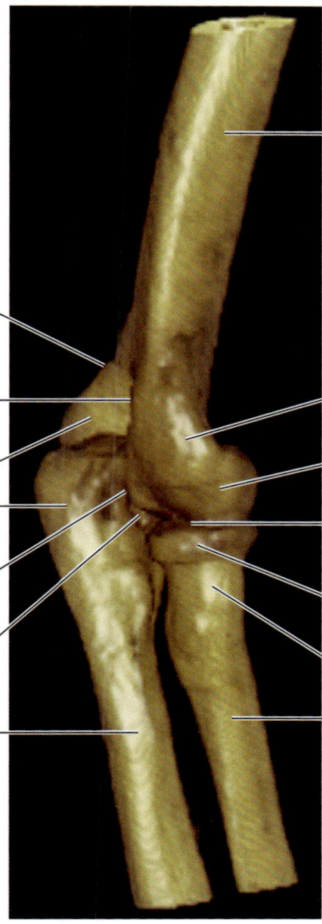

Three-dimensional images of a living human body, obtained by CT scan using a 3D volume rendering technique, of the relationships between the humerus, radius and ulna in the **elbow joint**. Anterior (**a**), posterior (**b**), medial (**c**) and lateral (**d**) views.

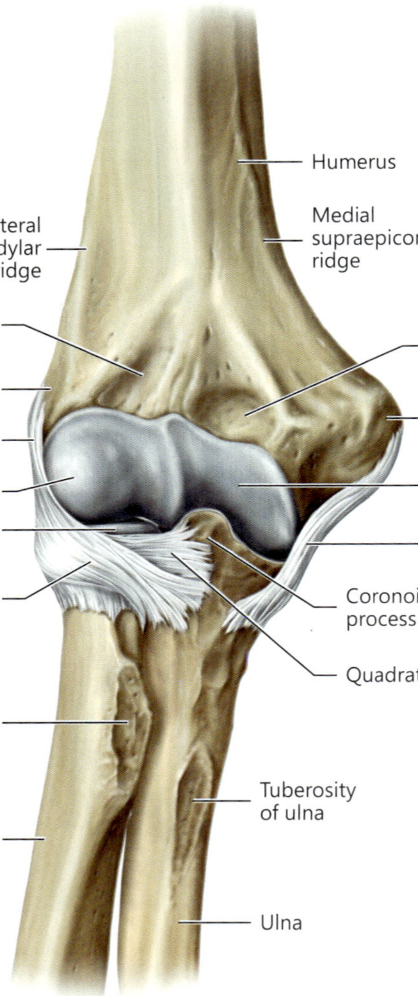

Right **elbow joint**, seen from the front. With the joint capsule removed, you can see the anular ligament of the radius and the radial and ulnar collateral ligaments.

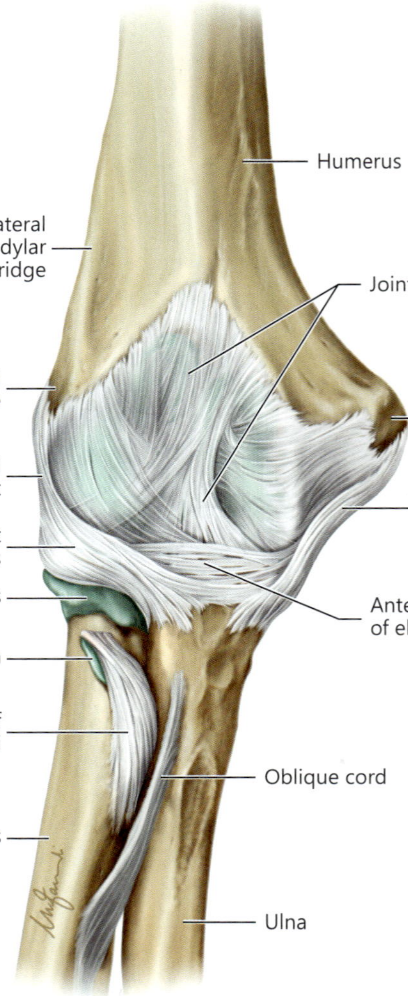

Right **elbow joint**. On the anterior surface the means of union of the humeroulnar, humeroradial and proximal radioulnar joints have been prepared.

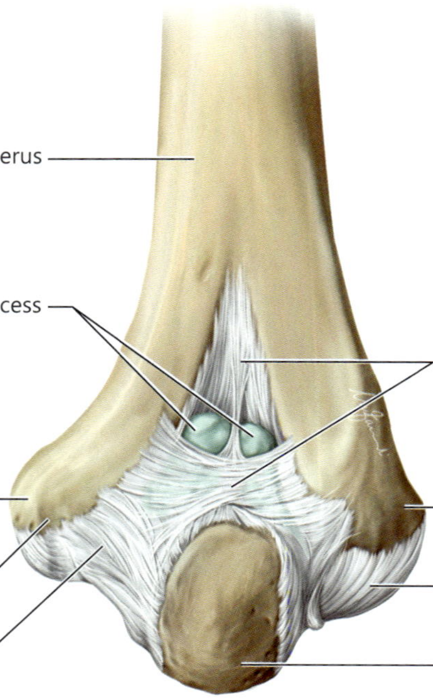

Right **elbow joint**, seen at the posterior surface, in the flexed position. Here you can see the dorsal part of the joint capsule.

Left **elbow joint**, lateral view. With the joint capsule removed the anular ligament of the radius and radial collateral ligament are visible.

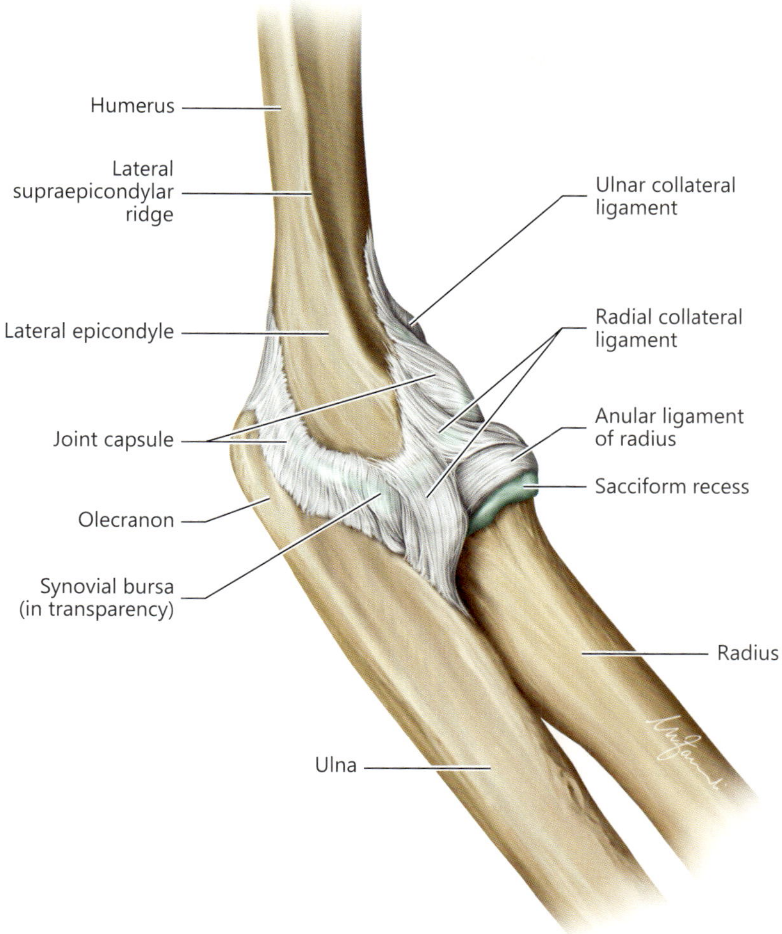

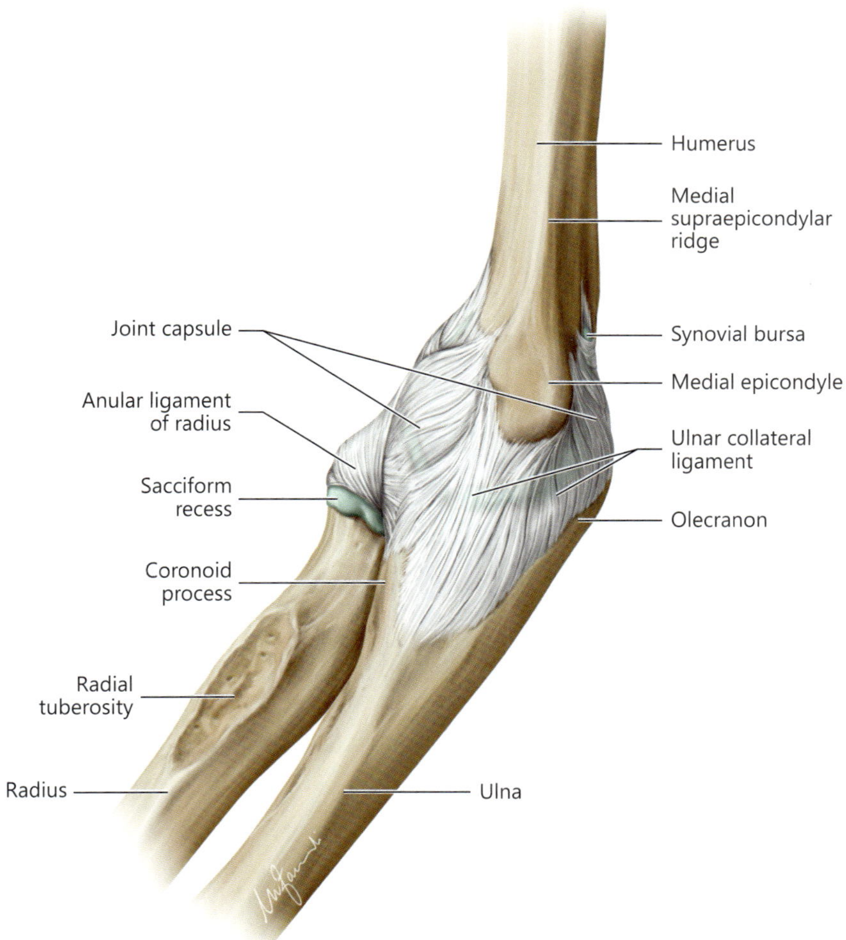

Right **elbow joint**, medial view. The joint capsule is in situ with its ligaments.

Right **elbow joint**. The sagittal section shows the extent of the articular cavity at the level of the humeroulnar joint.

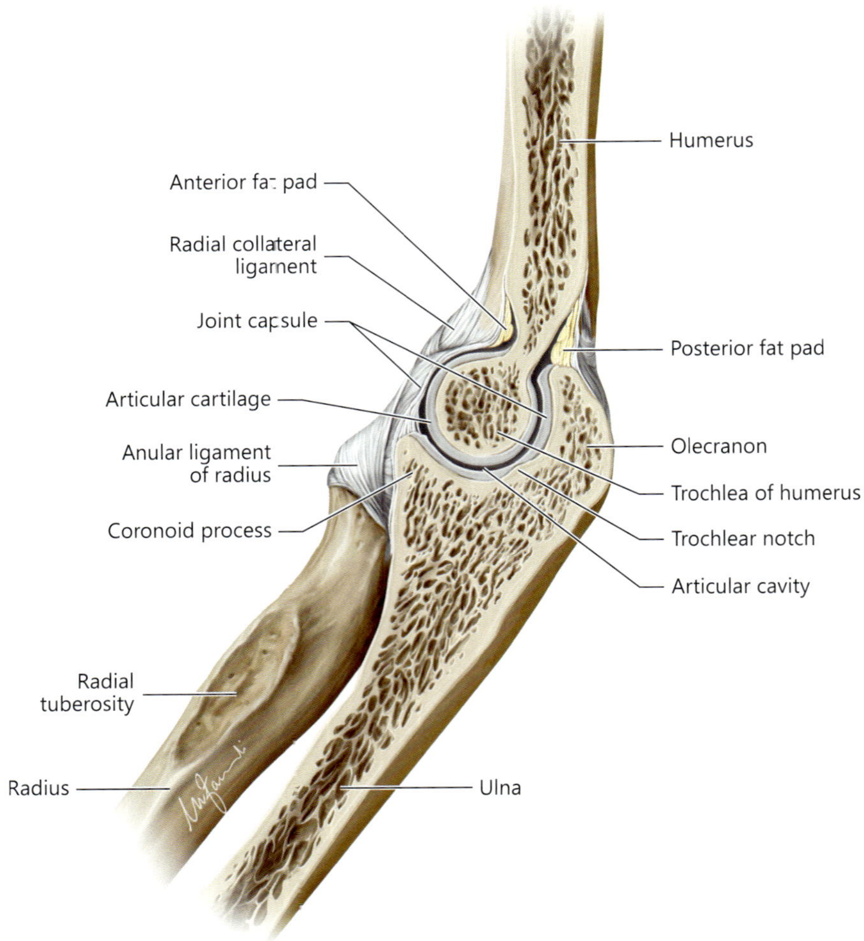

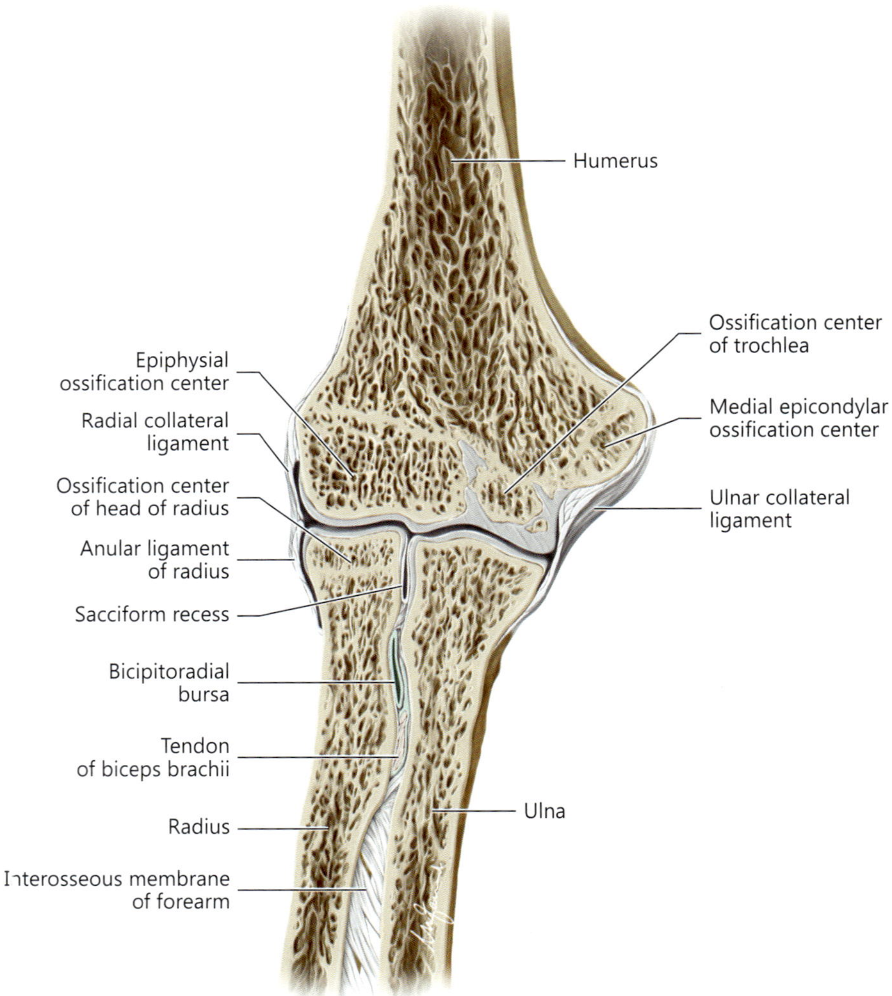

Right **elbow joint**. The frontal section, seen from the front, illustrates the extent of the articular cavity.

Proximal **articular surfaces** of the **radius** and **ulna**.

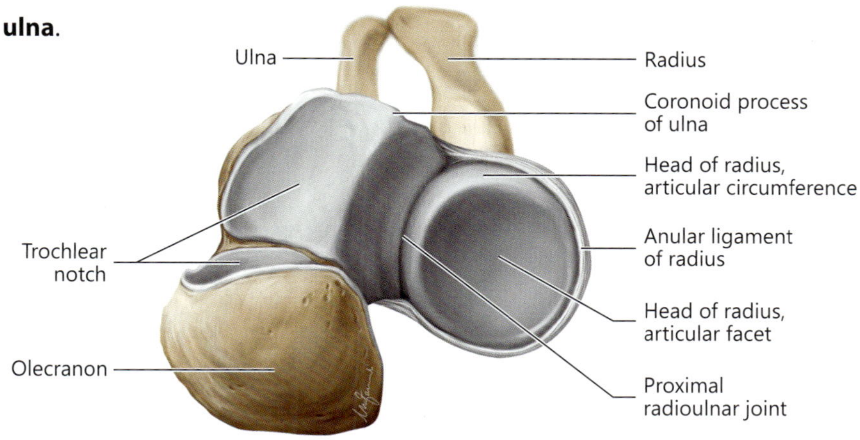

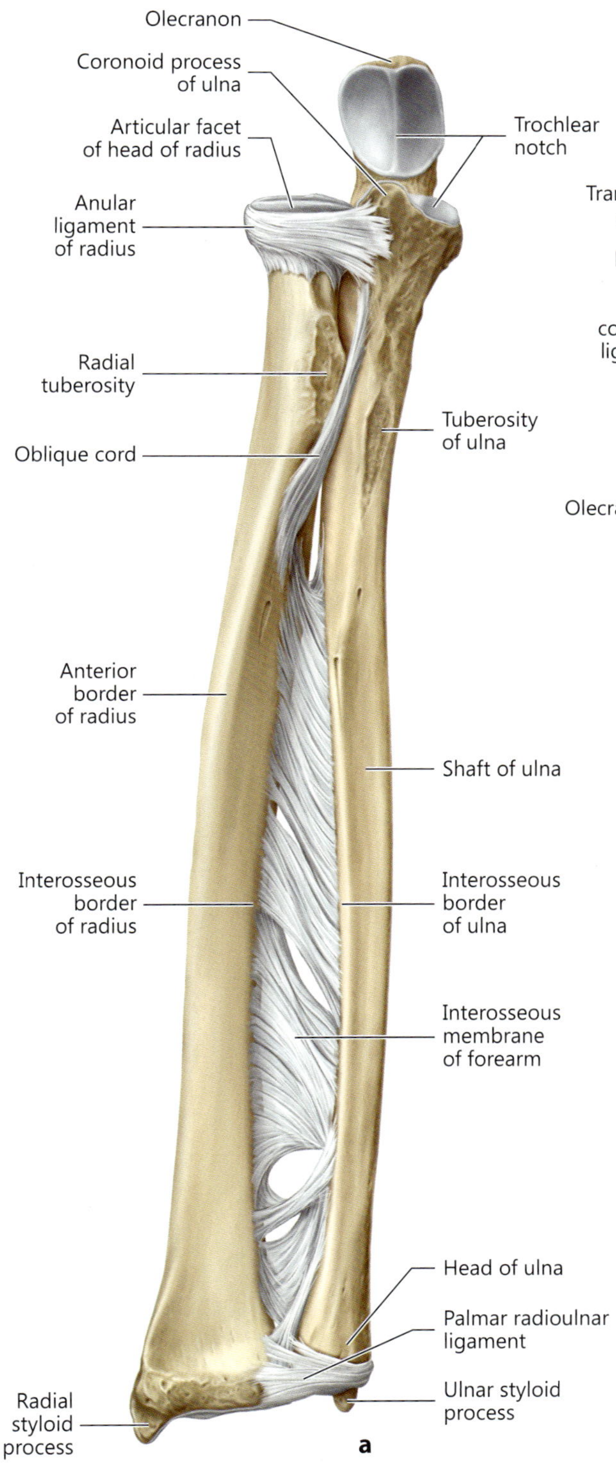

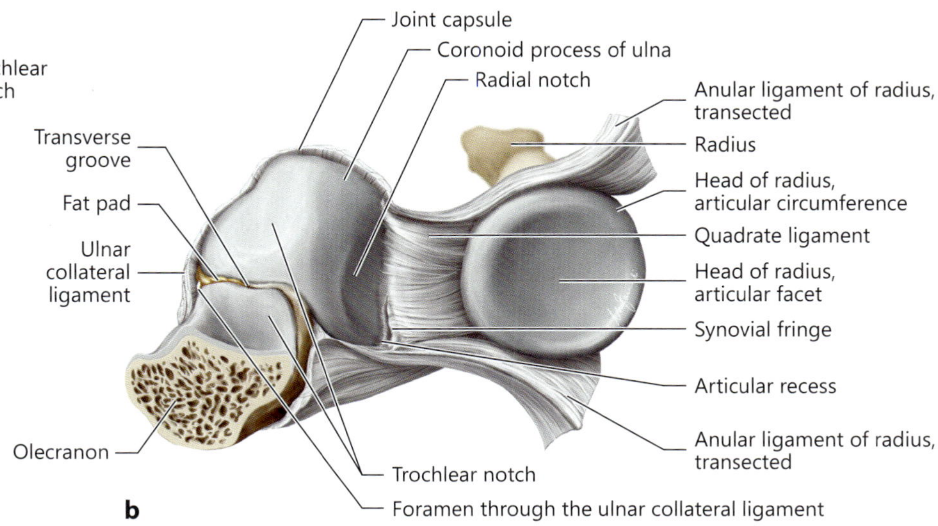

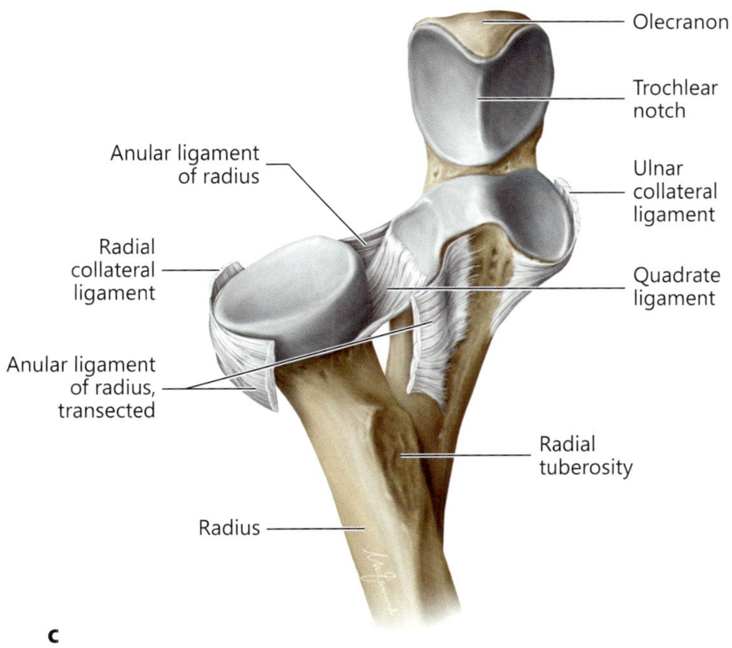

Articular structures between the right **radius and ulna. a**, Anterior view; **b**, proximal view; **c**, anterosuperior view. Here the proximal and distal radioulnar joints are depicted, along with the interosseous membrane, which serves as a means of union between the radial and ulnar diaphyses.

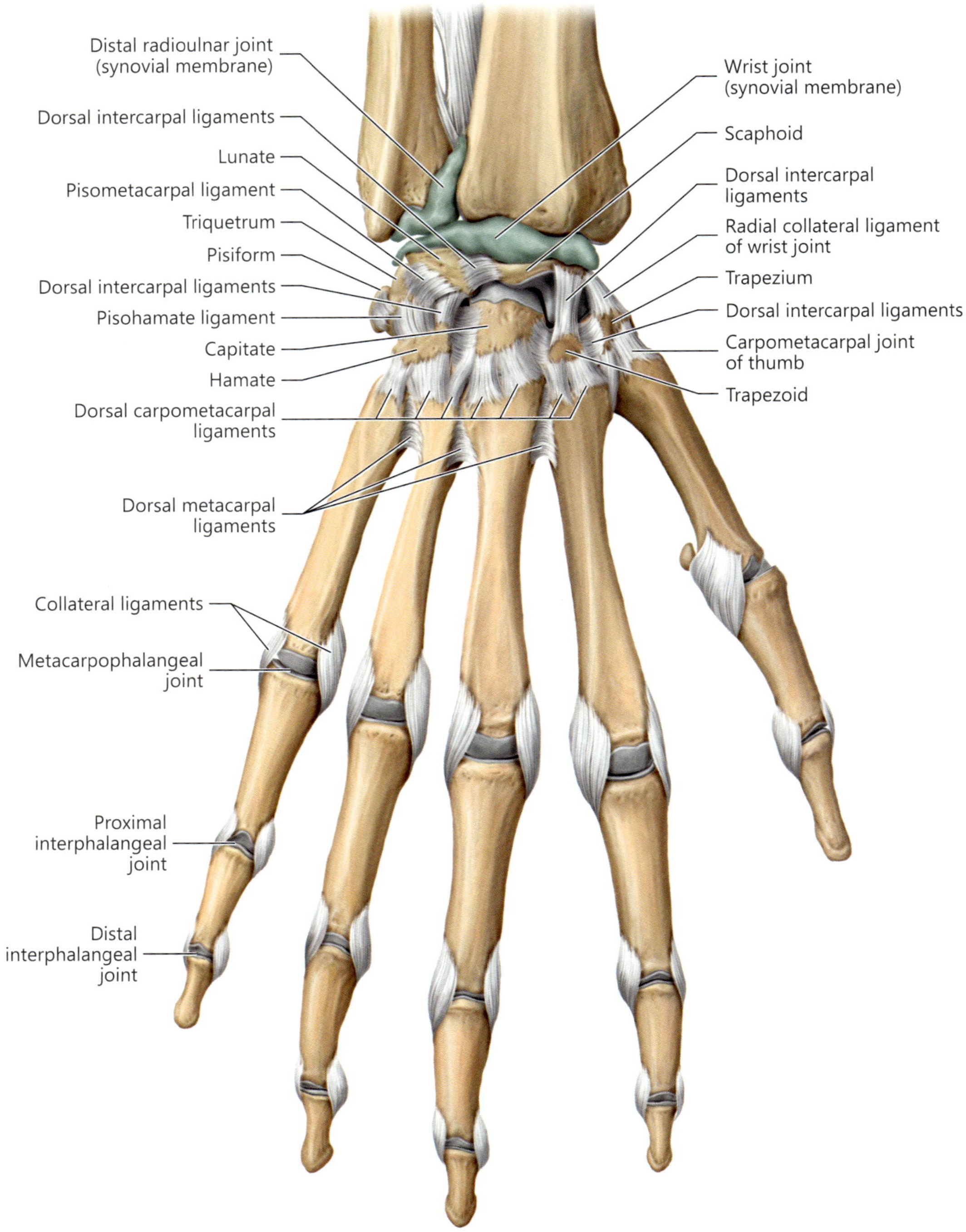

Joints of the right **wrist and hand**, posterior or dorsal view.

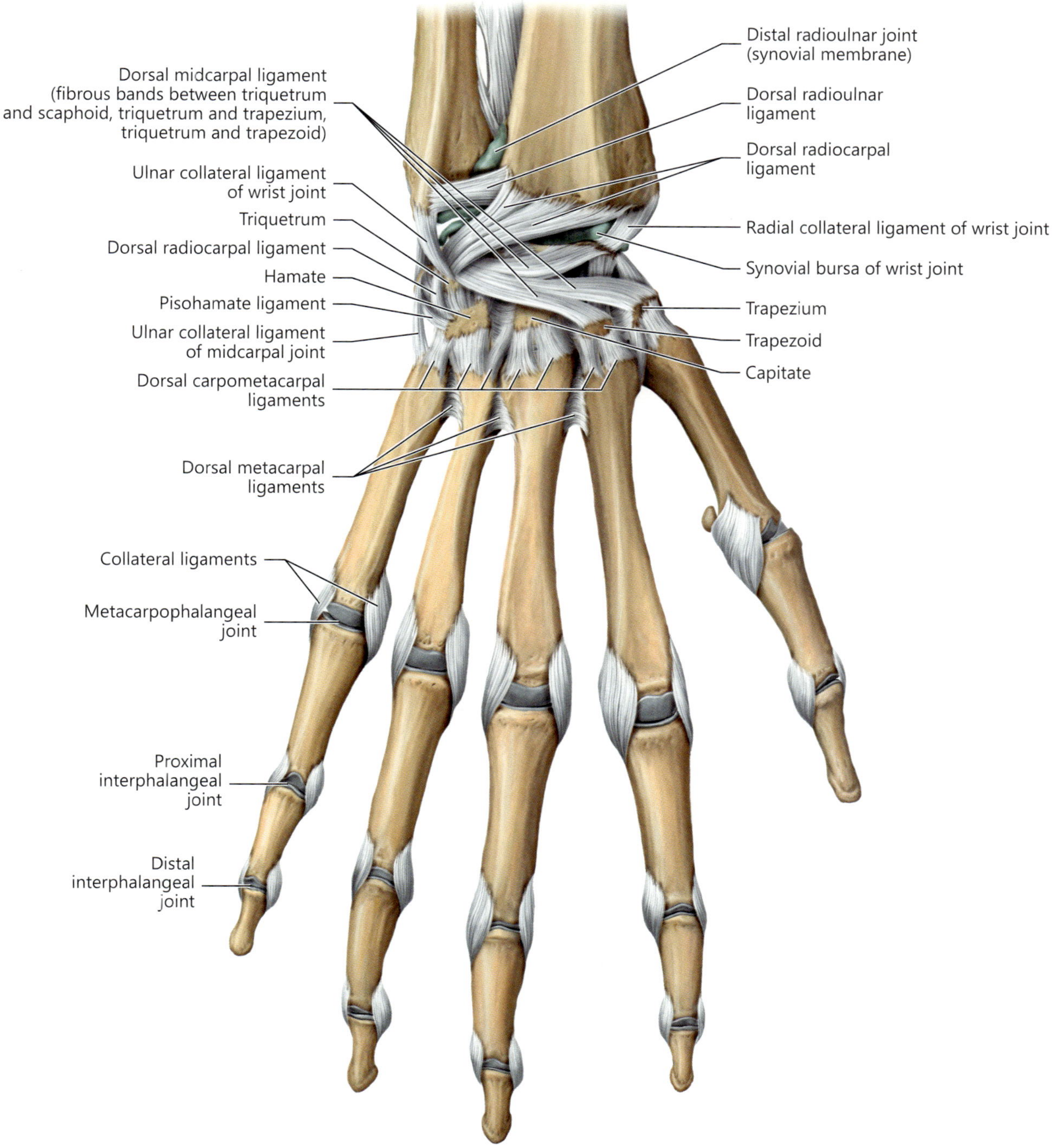

Ligaments of the joints of the right **wrist and hand**, posterior or dorsal view.

Means of union of the right **carpal joints**, palmar (or volar) view.

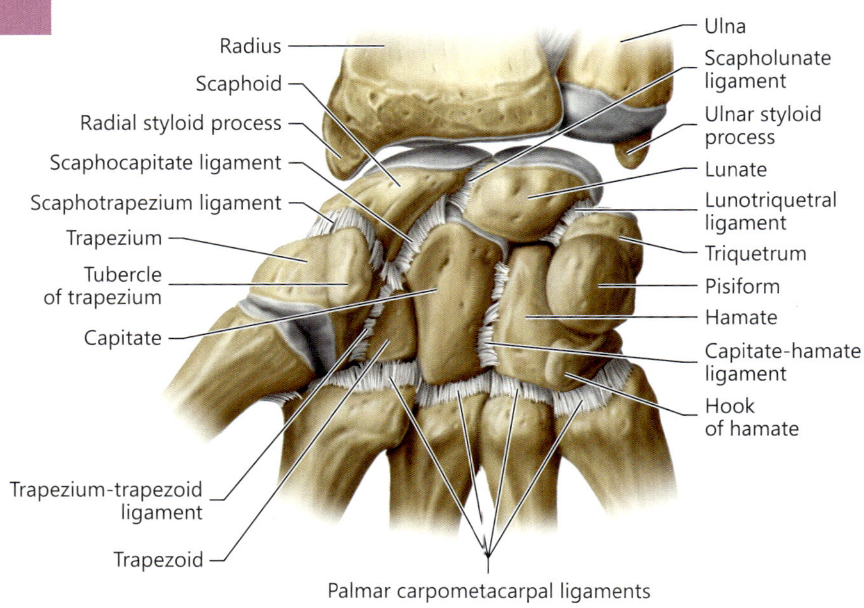

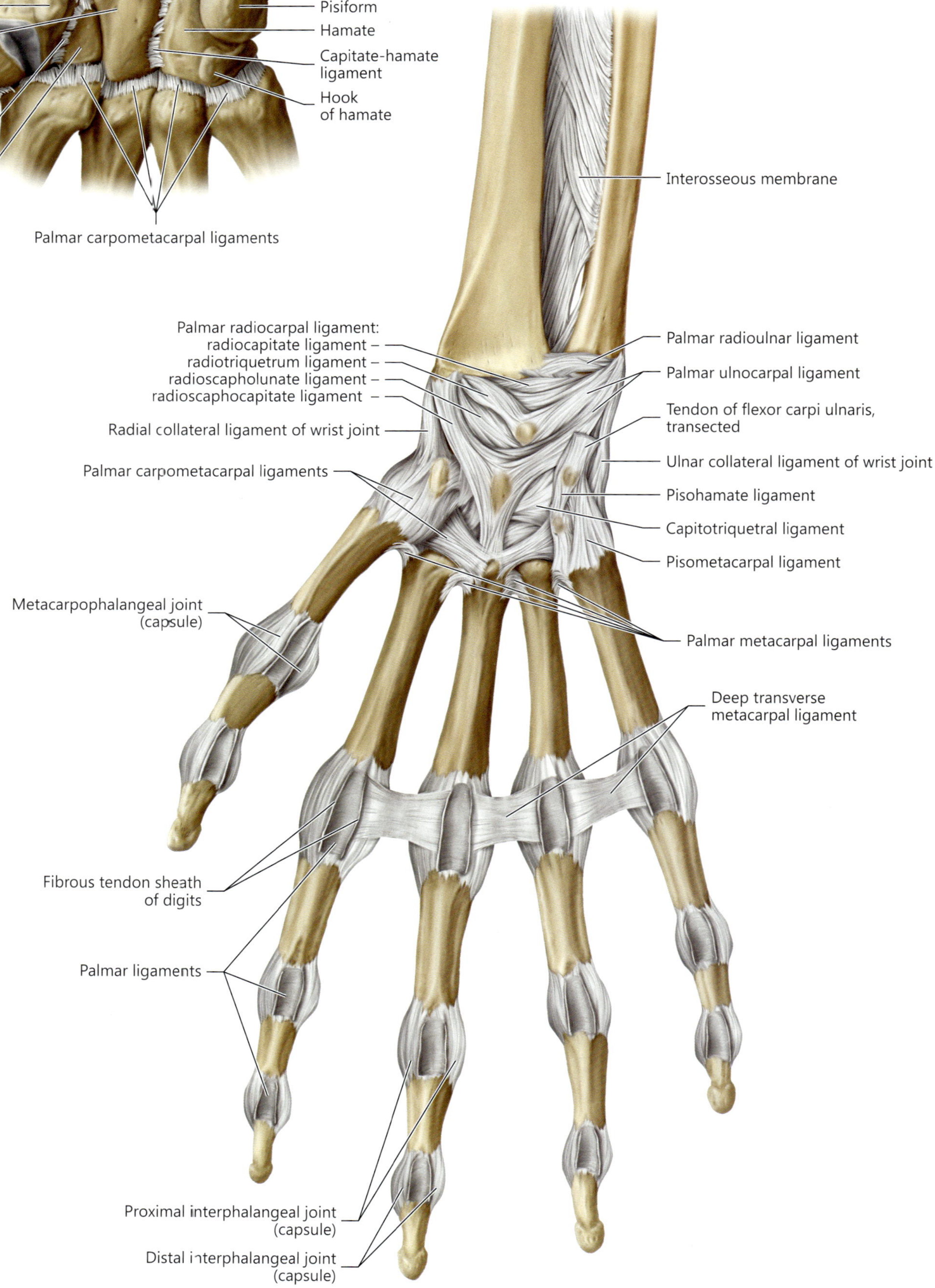

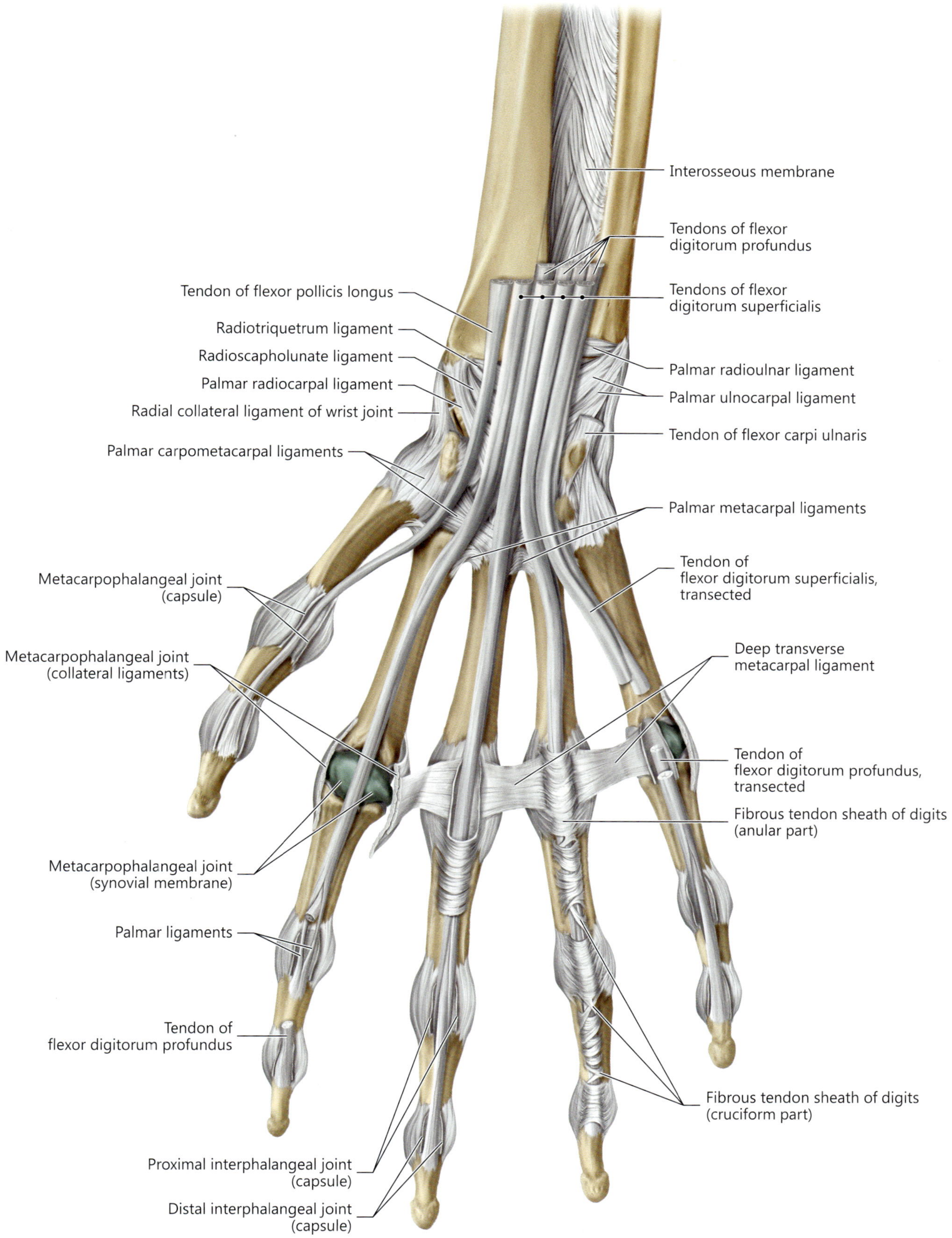

Tendons of the **superficial and deep flexor muscles of the fingers** which overlap with the ligamentous structures of the right wrist, palmar (or volar) view.

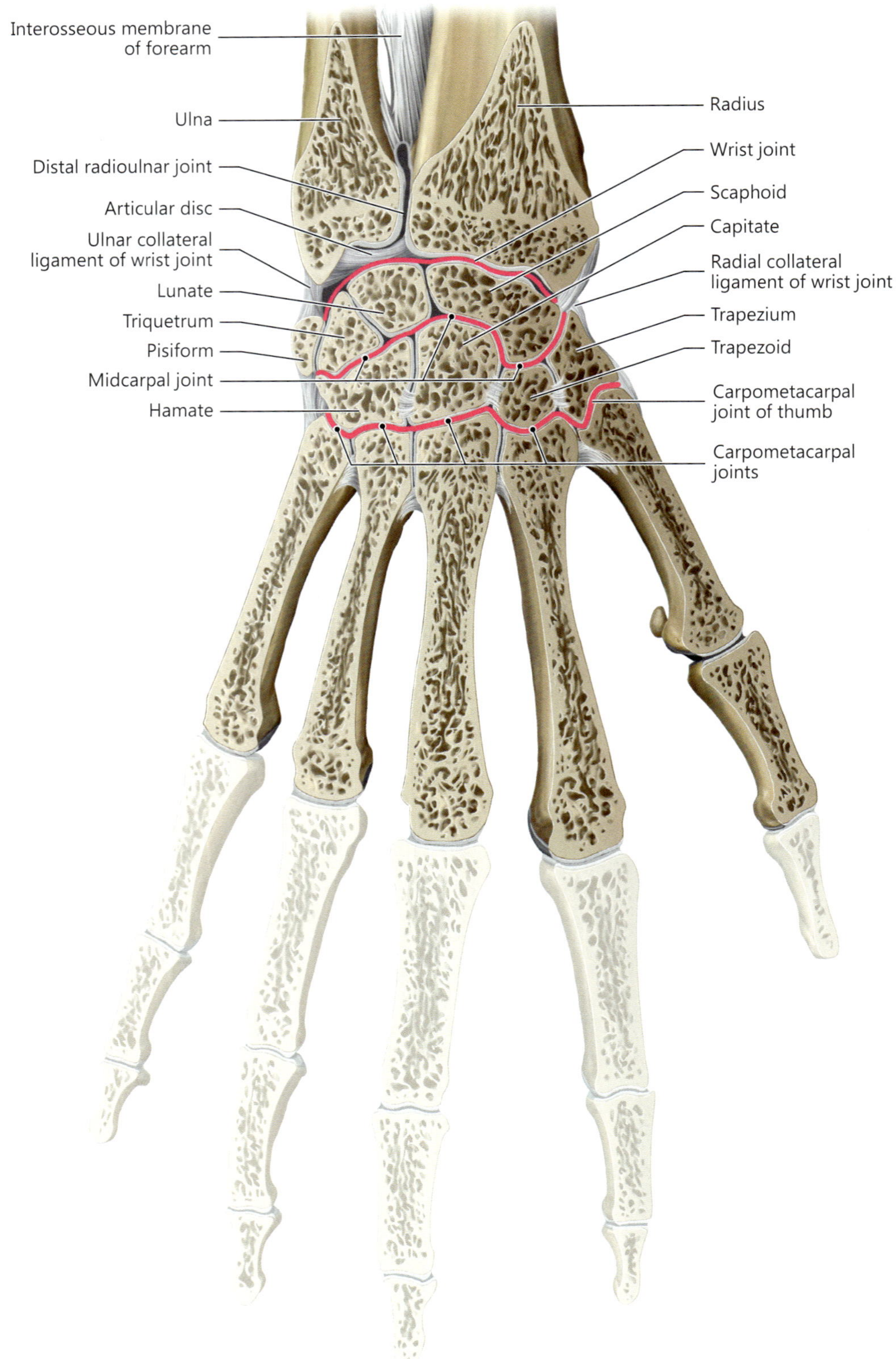

Joints of the right **hand**. The frontal plane of the section passes through the distal part of the forearm and the proximal part of the hand. It illustrates the complex system of articular cavities which is delimited at the level of the wrist, midcarpal, and carpometacarpal joints. The section is seen from the dorsal side.

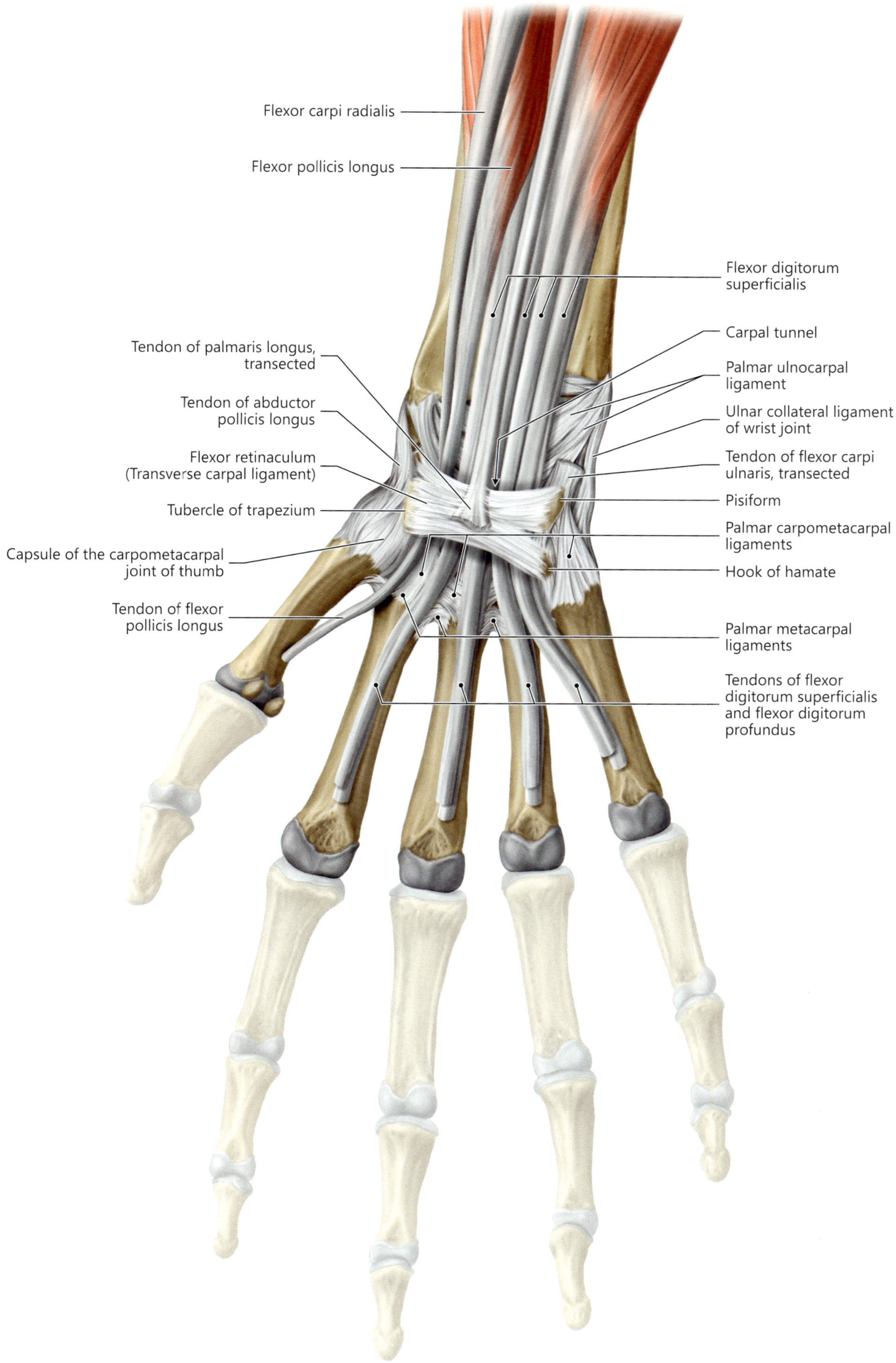

Means of union of the right **carpal joints**, volar view. Here you can see the flexor retinaculum delimiting the carpal tunnel.

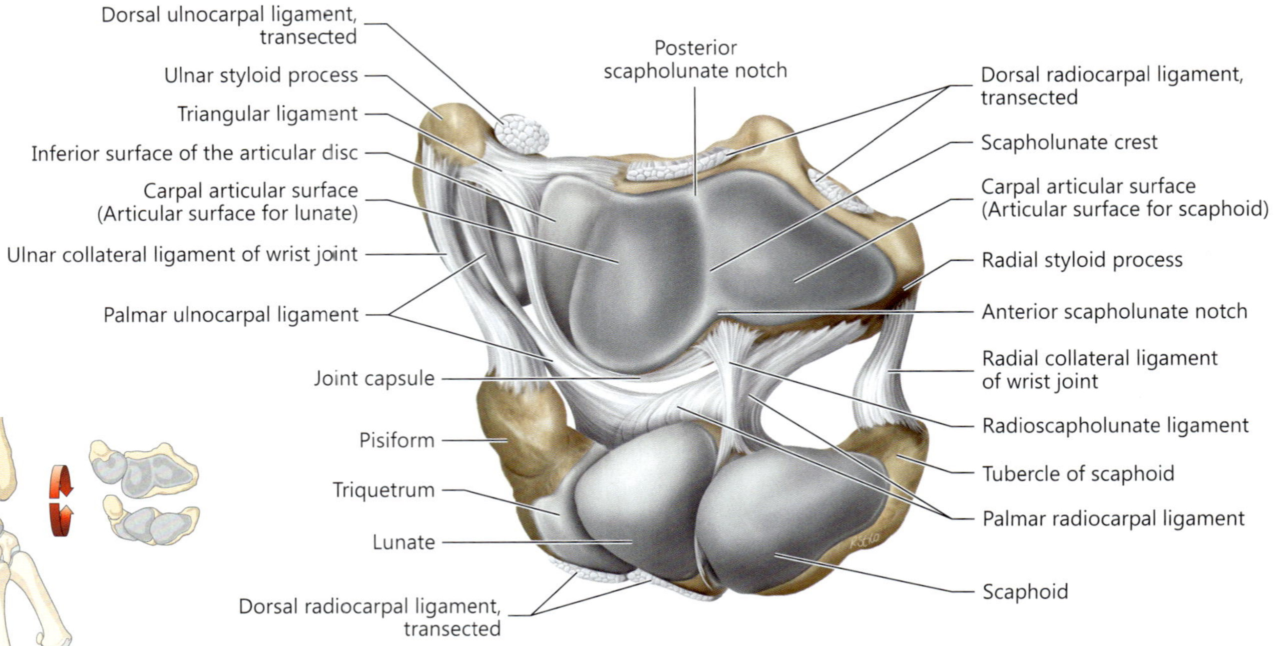

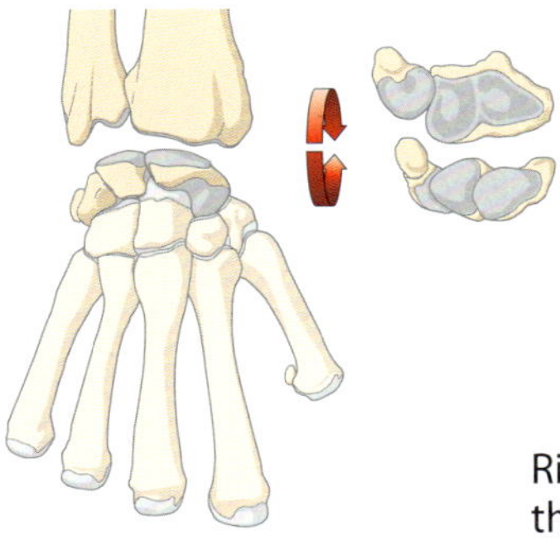

Right **wrist joint**. The radial surface has the shape of a glenoid cavity, while that of the carpal surface has the form of a condyle.

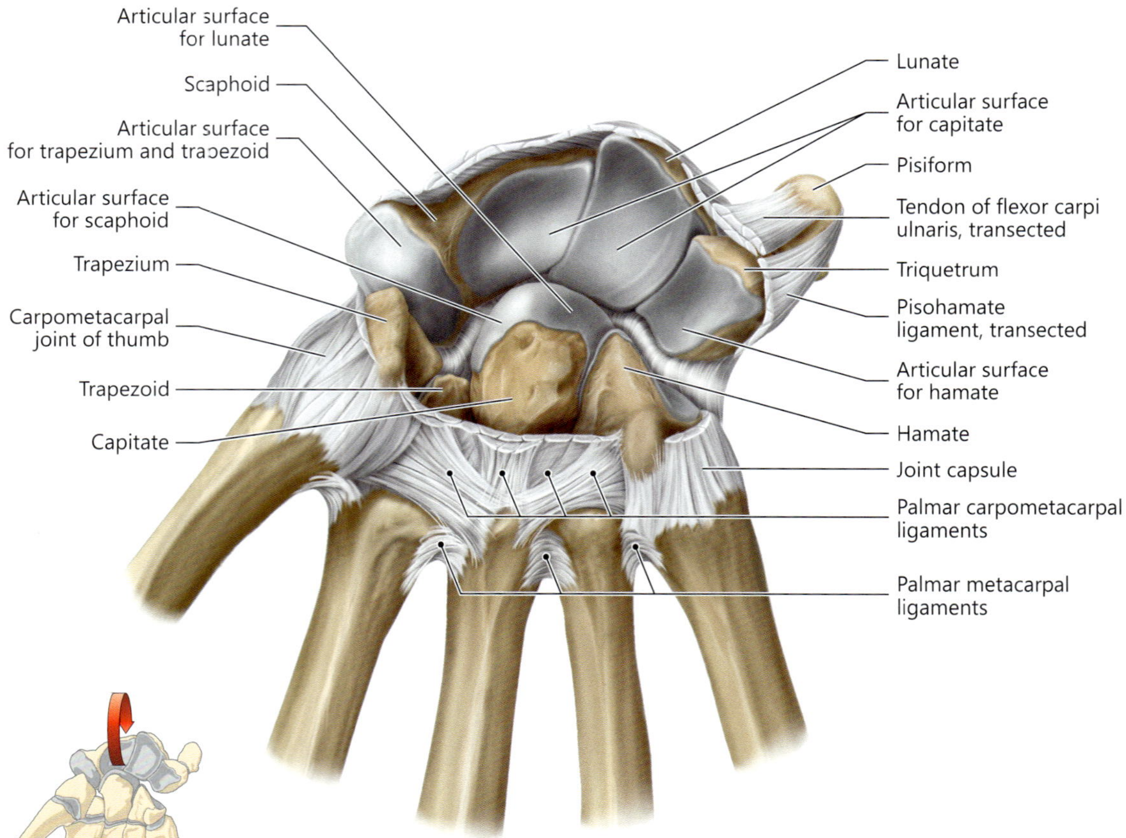

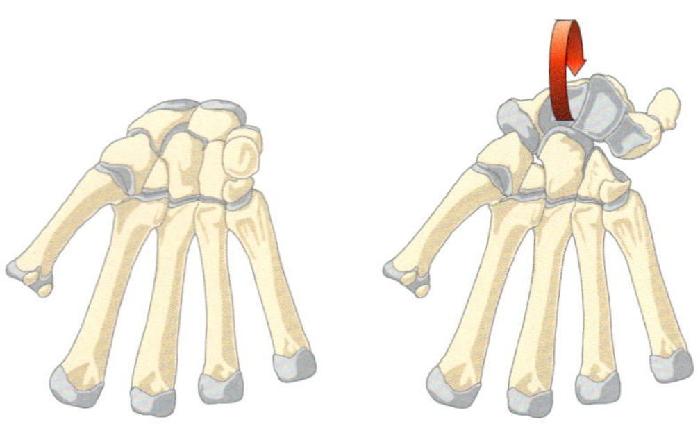

Right **midcarpal joint**. Articular surface given by the bones of the proximal and distal carpal row.

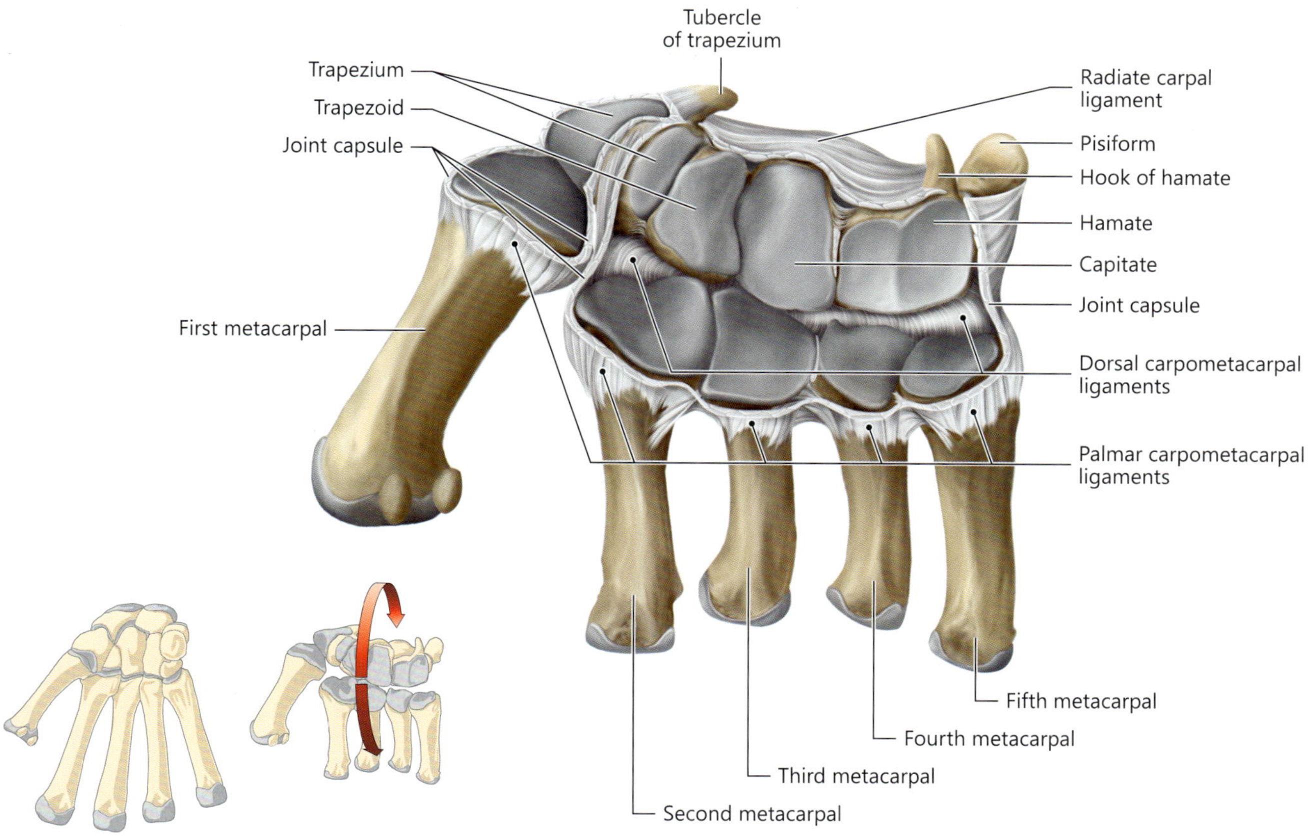

Right **carpometacarpal joint**. Here you can see the turned articular surface, which is formed by the bones of the distal carpal row and by the proximal extremities of the metacarpals.

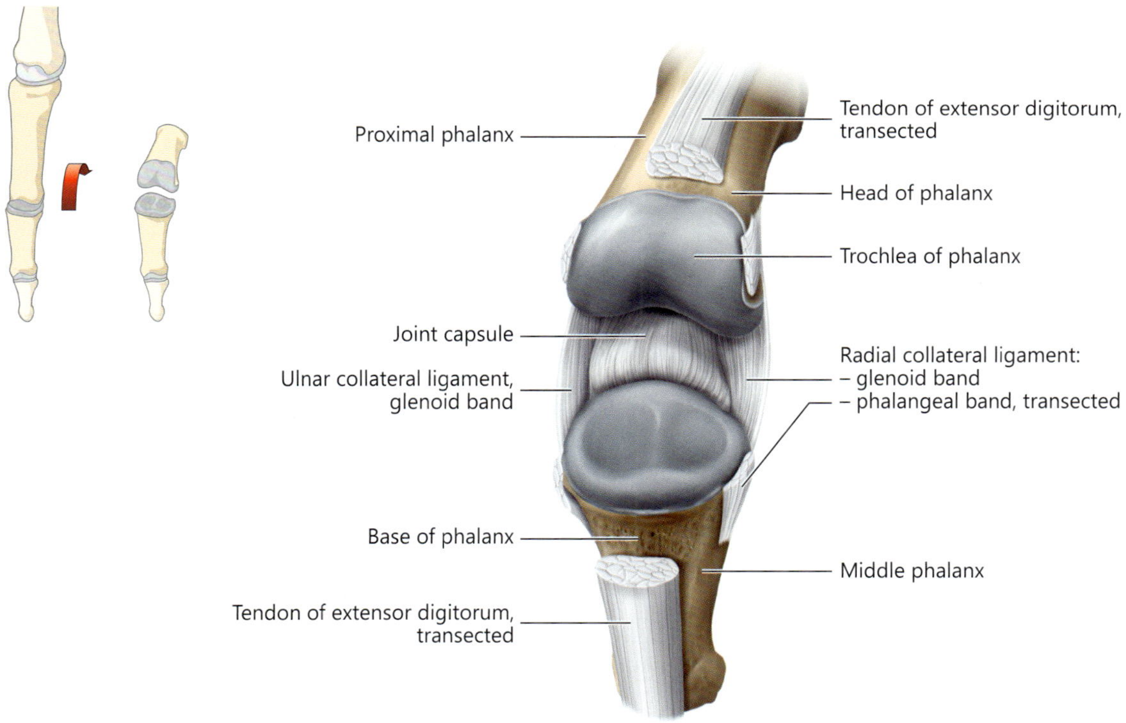

Proximal **interphalangeal joint**. The trochlea of the proximal phalanx has been turned over to show the base of the middle phalanx.

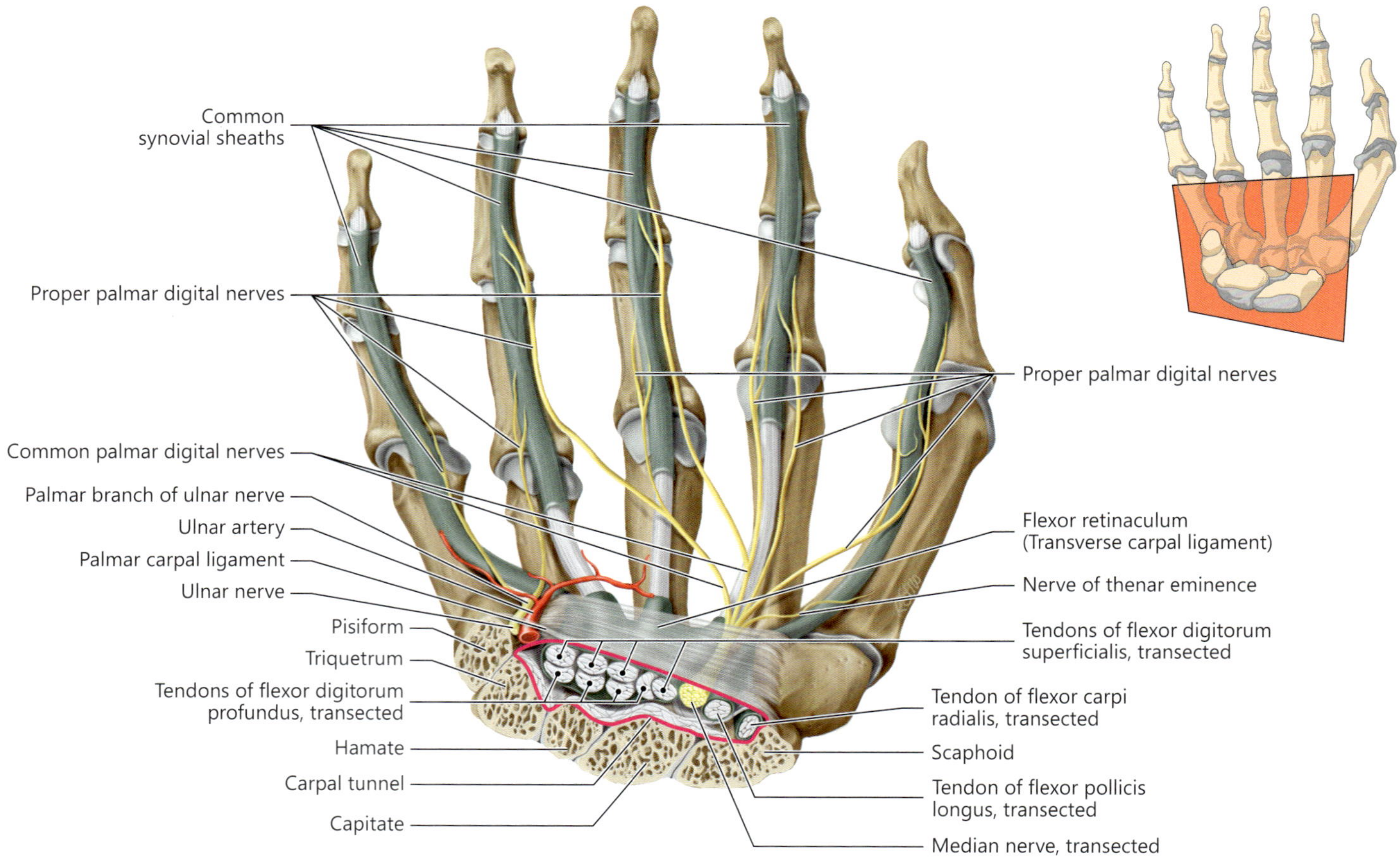

a

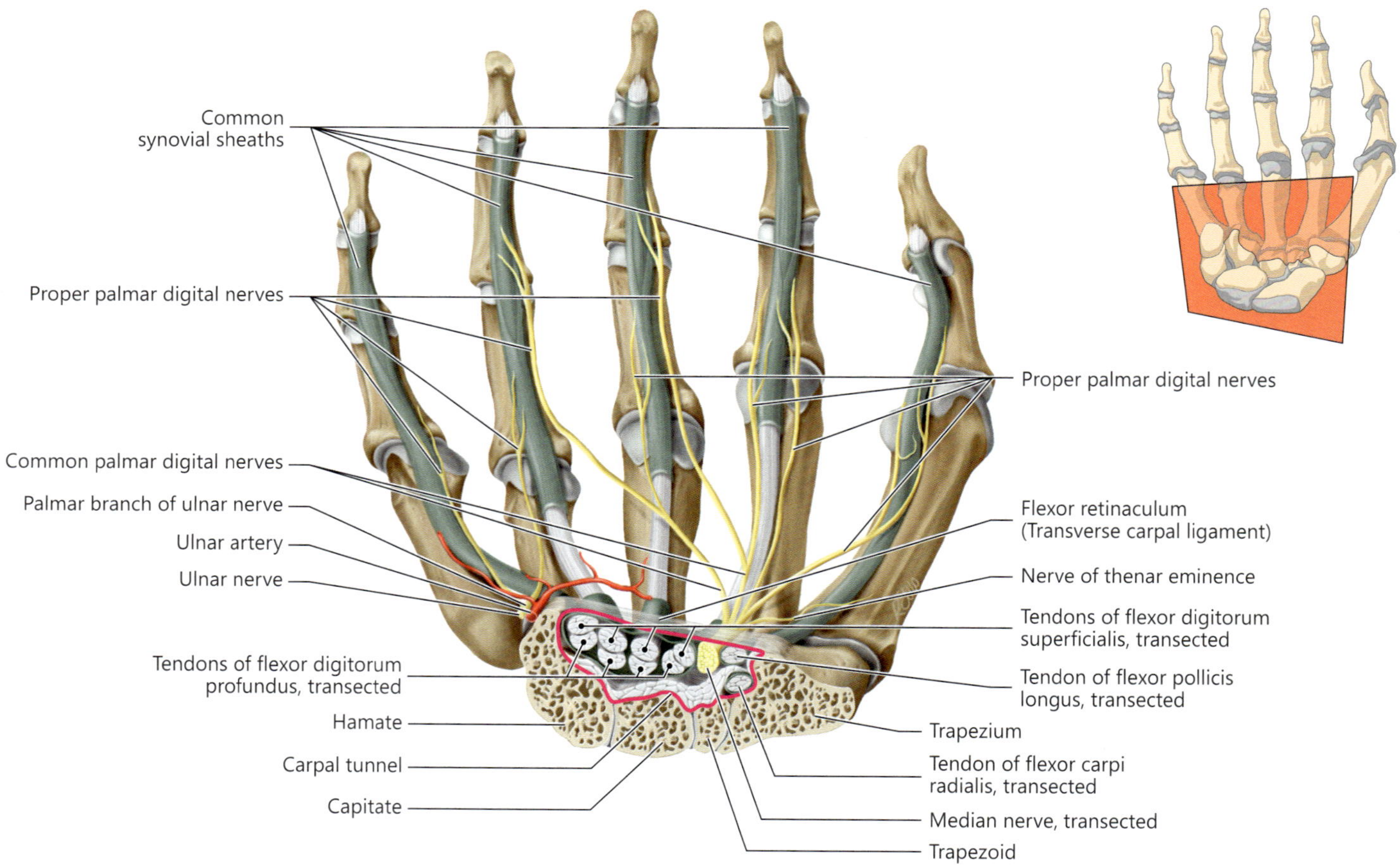

b

Section of the **carpal bones**. The flexor retinaculum is laterally fixed to the tubercles of the scaphoid and trapezium, and medially fixed to the pisiform and the hook of the hamate, thereby forming, together with the groove formed by the concavity of the carpal bones, the carpal tunnel (indicated with a red line). The palmar carpal ligament is a thickening of the antebrachial fascia on the anterior surface of the wrist, superficially and proximally with respect to the flexor retinaculum (transverse carpal ligament). **a**, Section conducted at the level of the proximal row; **b**, section conducted at the level of the distal row.

Metacarpophalangeal and **interphalangeal joints** on the right side, view from the radial side.

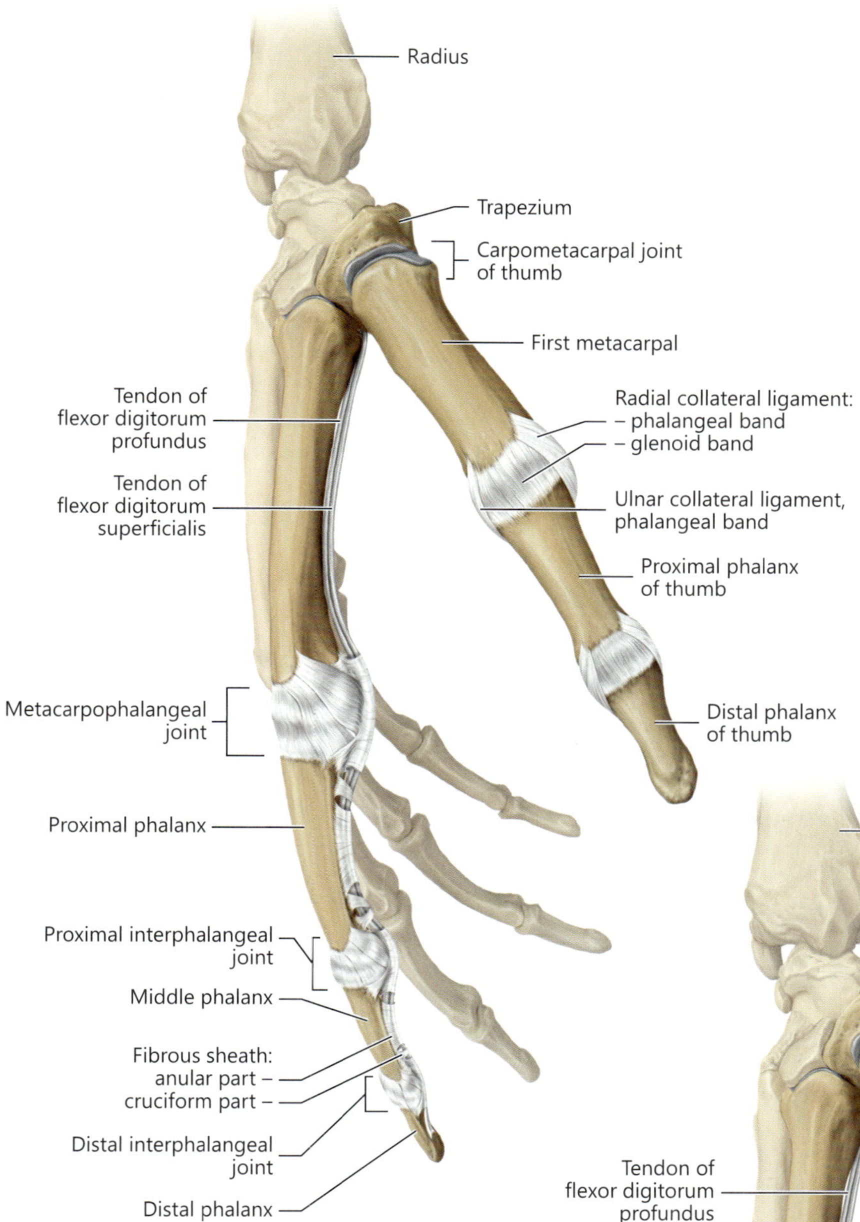

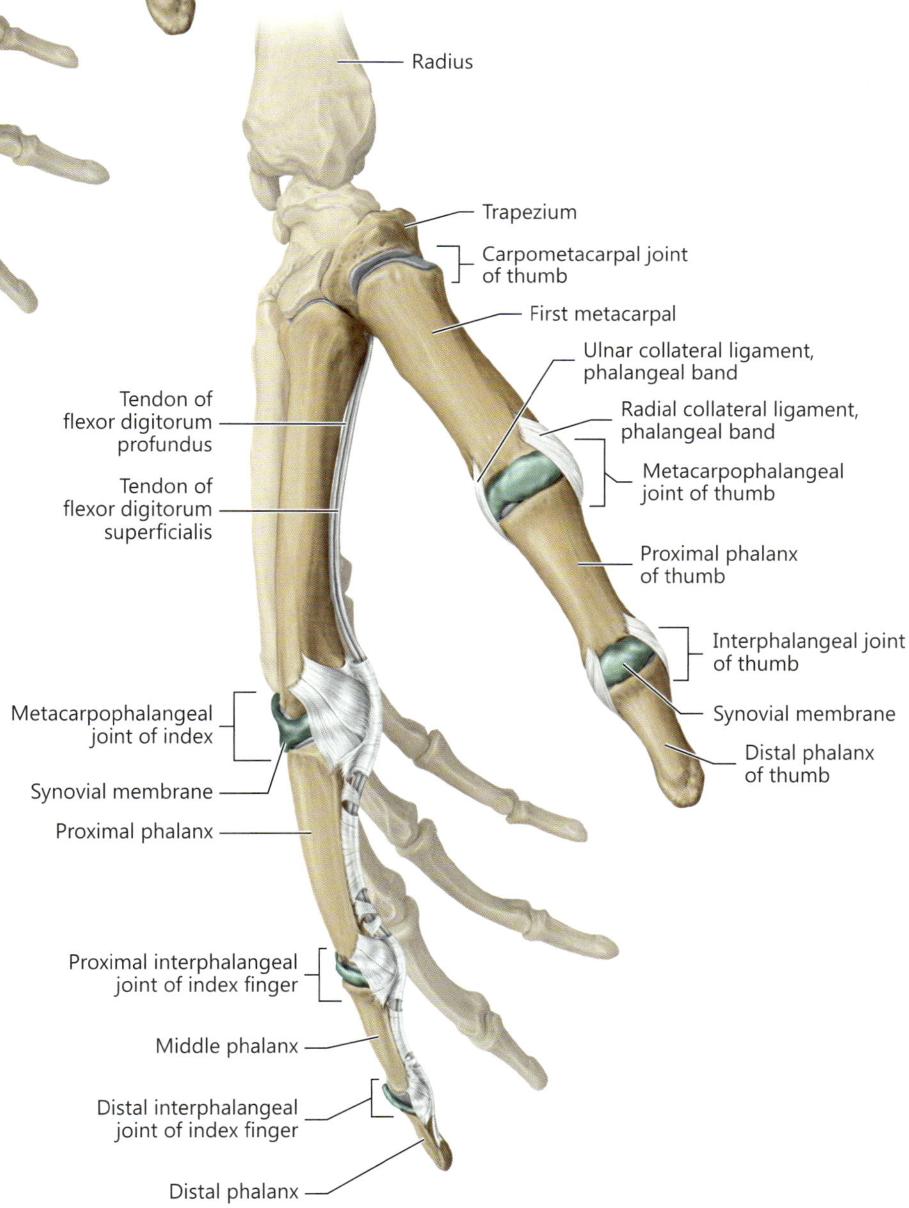

MOVEMENTS OF THE SHOULDER AND PECTORAL GIRDLE

Movements of the clavicle

Protraction and retraction of the clavicle (anteroposterior movements on the sagittal plane).
Elevation and depression of the clavicle (craniocaudal movements of low amplitude on the frontal plane).
Rotation around the longitudinal axis of the clavicle.
Total: 3 degrees of freedom (6 types of movement).

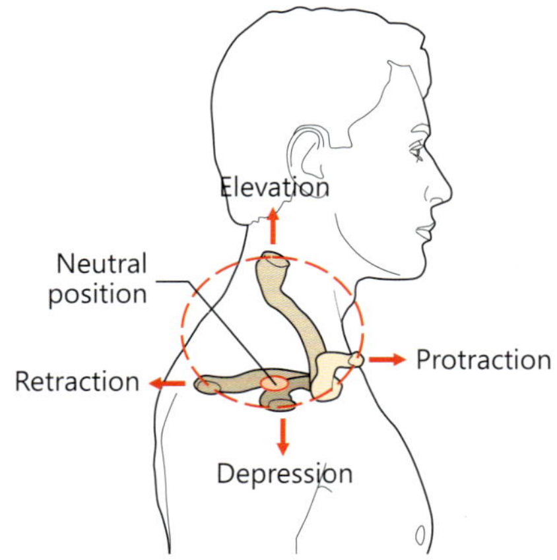

Movements of the sternoclavicular joint

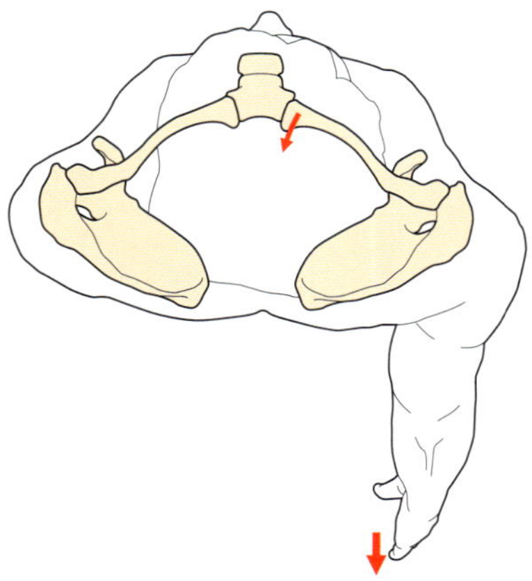

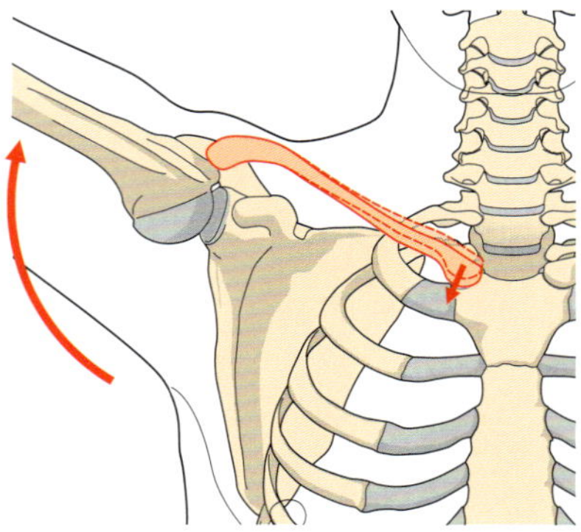

Protraction and retraction of the clavicle (anteroposterior movements on the sagittal plane):

- protraction (or antepulsion or anteposition) of about 3 cm
- retraction (or retropulsion or retroposition) of about 3 cm

Elevation and depression of the clavicle (craniocaudal movements on the frontal plane):

- elevation of about 10 cm:
  - the sternal end of the clavicle slides down, but the 1st rib, located just below the clavicle, limits its range of motion and therefore the elevation of the clavicle itself.
  - it is always accompanied by a certain degree of rotation and a consequent minimum protraction. The rotation occurs around the axis of the clavicle and is about 30°
- depression of about 3 cm:
  - the sternal end of the clavicle slides up
  - it is always accompanied by a certain degree of rotation and a consequent minimum retraction.

Rotation around the longitudinal axis of the clavicle of about 30°

Total: 3 degrees of freedom (6 types of movement)

Movements of the scapula

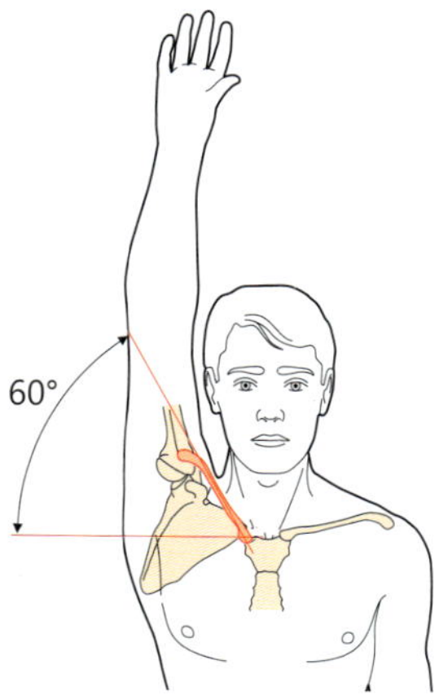

- The scapula (by means of the acromioclavicular joint) moves in synchrony with each movement of the clavicle:
  - if the various positions reached by the acromial end of the clavicle during the maximum amplitude of the clavicle's movements were joined together, the result would be an oval which is higher than it is wide, corresponding to an elevation of about 60°, a depression of 5°, and a protraction and retraction of 30° of the acromial end of the clavicle

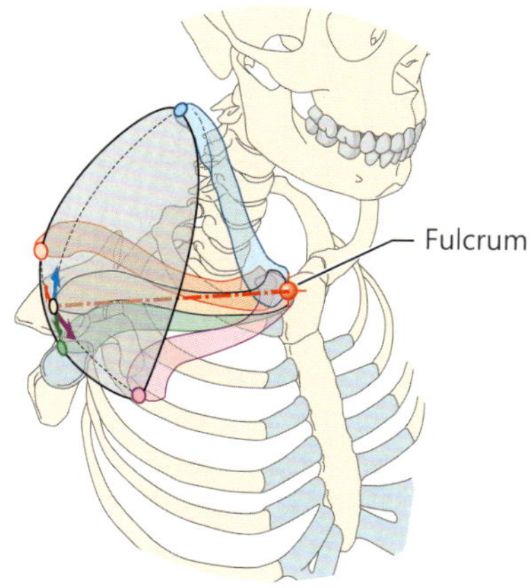

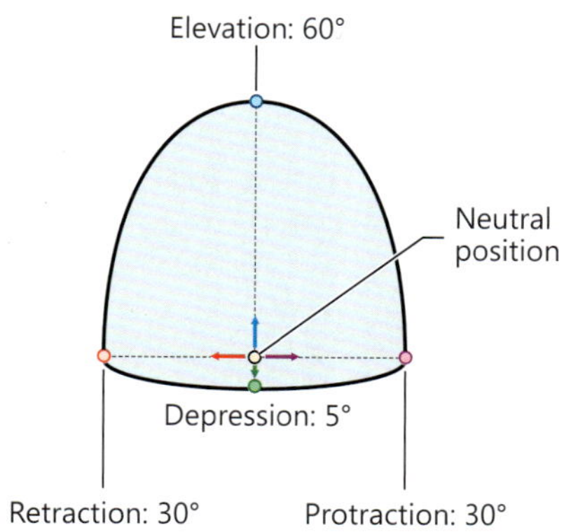

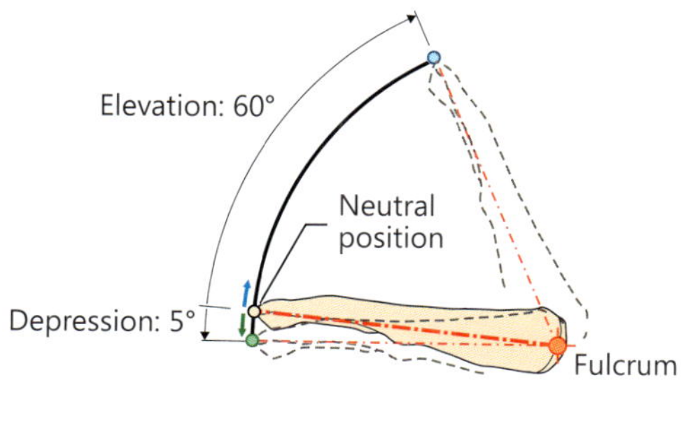

Frontal view

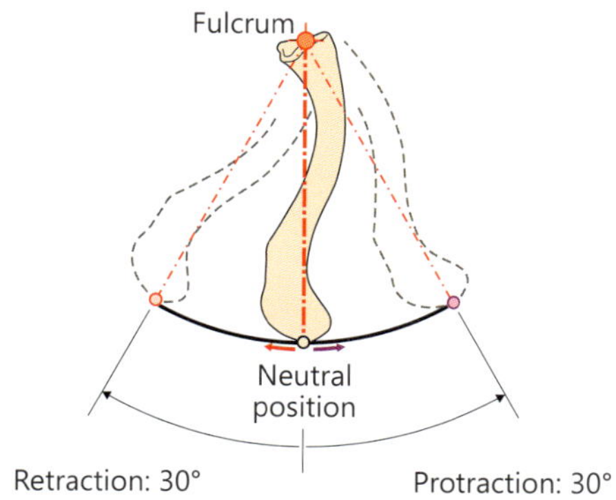

Sagittal view

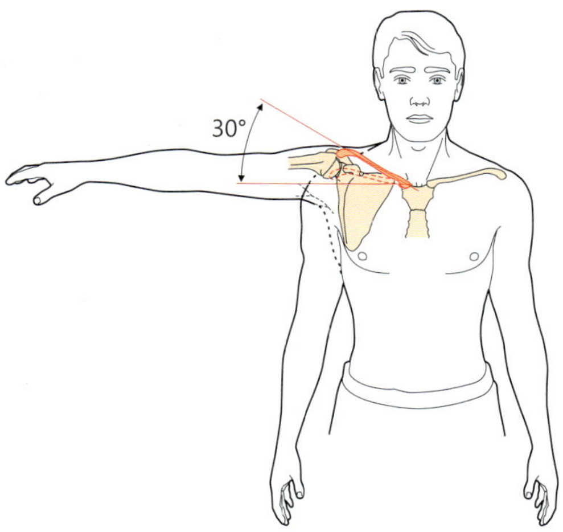

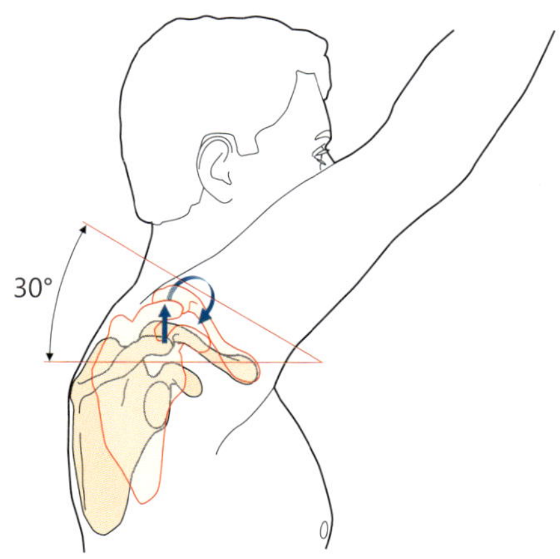

- Both clavicular joints aid in the movements of the scapula (the acromioclavicular joint to a lesser extent):
  - elevation of the scapula corresponds to elevation of the clavicle
  - depression of the scapula corresponds to depression of the clavicle
  - abduction of the scapula corresponds to protraction of the clavicle
  - adduction of the scapula corresponds to retraction of the clavicle
- Movements of the clavicle are combined with those of the scapula and humerus during abduction and flexion movements of the arm, thereby causing the backward and lateral displacement of the glenoid cavity:
  - external rotation of the scapula
  - simultaneous 30° elevation of the clavicle
  - rotation of the clavicle around its longitudinal axis (beings before the clavicle has reached an elevation of 30° and its amplitude is most likely less than 45°)
  - elevation of the acromial end of the clavicle from 30° to 60°.

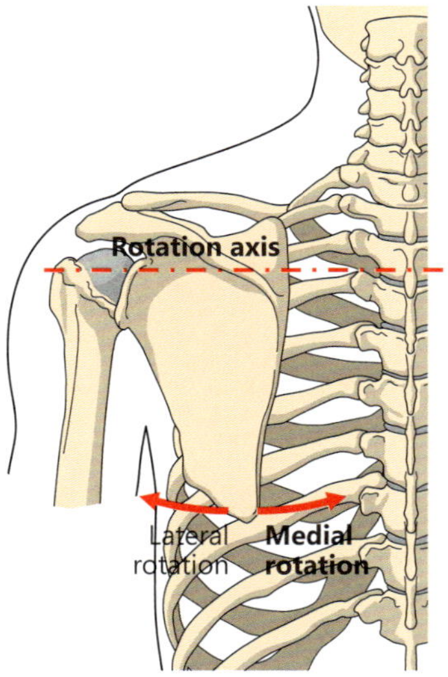

External rotation (extrarotation, lateral rotation)

- Lateral oscillation of the scapula
- The rotation axis, orthogonal to the plane of the scapula and passing just below the spine, moves inferiorly during movement
- The amplitude of the movement is around 60°
- The inferior angle of the scapula covers a distance of about 10 cm in the lateral direction.

It is the most important movement of the scapula, occurring both during shoulder abduction (with lateromedial elevation, in the side image), as well as during flexion (with anterosuperior elevation).

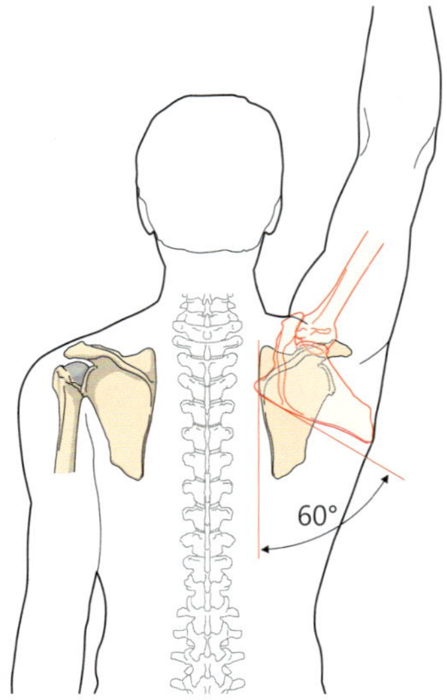

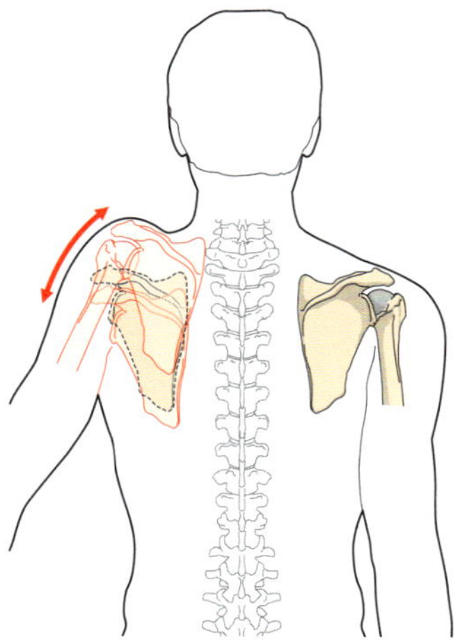

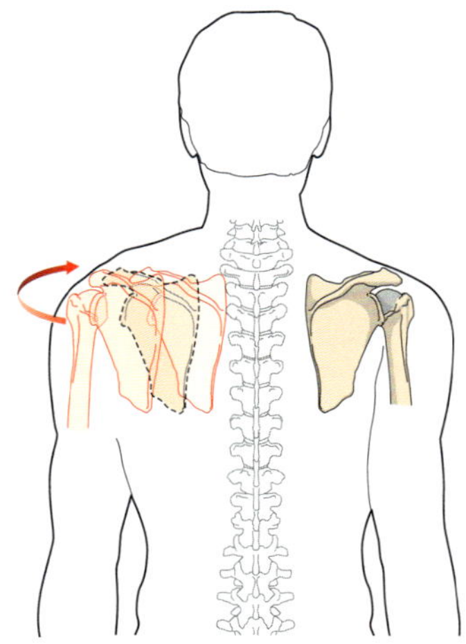

Elevation

- Upward movement of the scapula
- Elevation of about 10 cm.

Depression

- Downward movement of the scapula
- Depression of about 3 cm.

Adduction

- Approach of the medial border of the scapula to the vertebral column
- Corresponding retraction of the clavicle
- The angle between the clavicle and the scapula increases.

Abduction

- Distancing of the medial border of the scapula from the vertebral column
- Corresponding protraction of the clavicle
- The angle between the clavicle and the scapula is reduced.

Movements of the glenohumeral joint

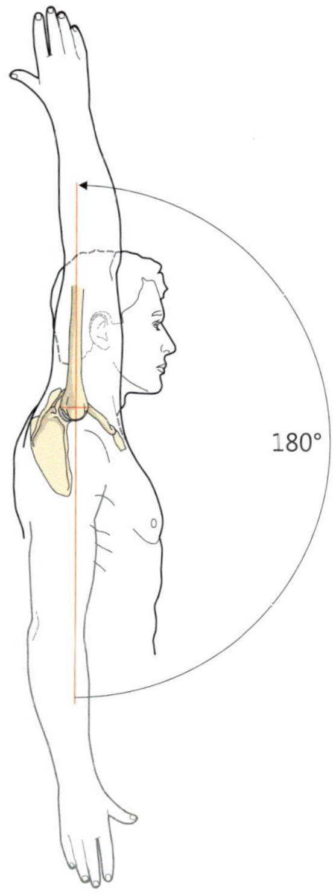

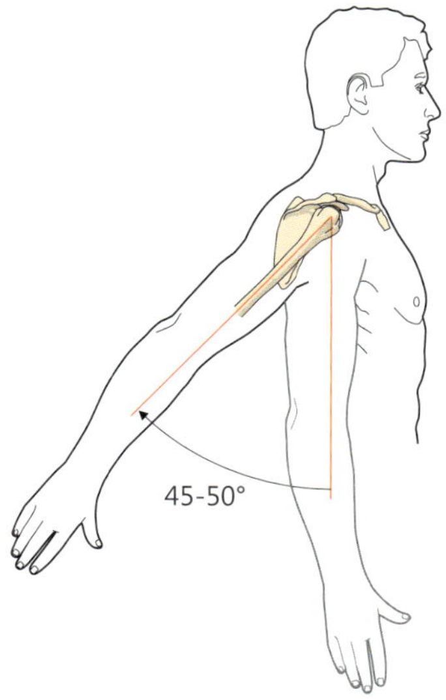

Flexion
- Occurs on the sagittal plane around the transverse (horizontal) axis
- Amplitude of about 180°
- Is accompanied by anterosuperior elevation.

Extension
- Occurs on the sagittal plane around the transverse (horizontal) axis
- Amplitude of about 45-50°.

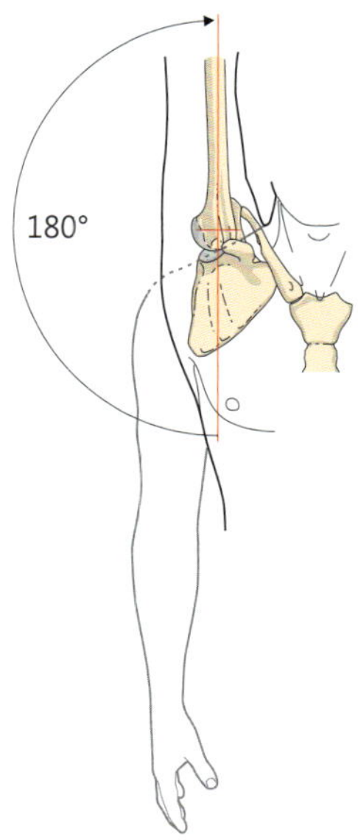

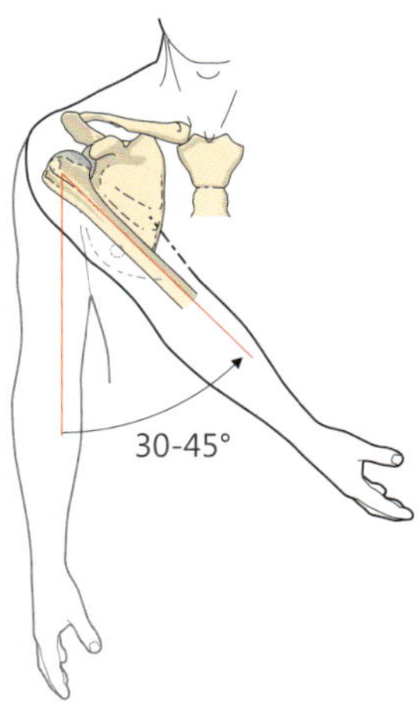

Abduction
- Occurs on the frontal plane around the sagittal axis
- Amplitude of 180°
- Is accompanied by lateromedial elevation.

Adduction
- Occurs on the frontal plane around the sagittal axis
- Absolute:
  - when performed from the anatomical position, preceded by a slight flexion of the trunk
  - amplitude of 30-45°
- Relative:
  - when performed from any abduction position.

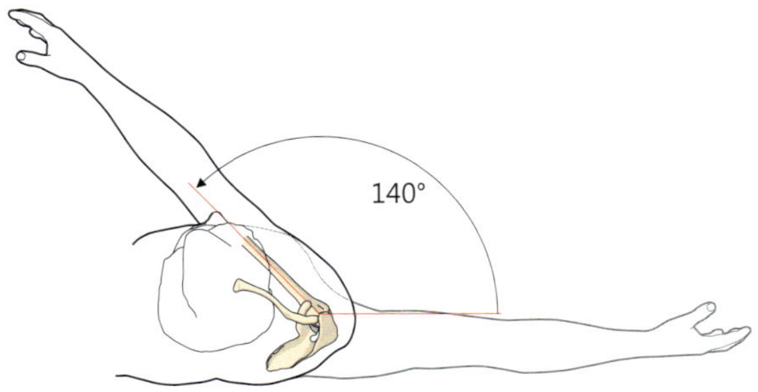

Internal rotation
- On the horizontal plane
- The starting position is an abduction of 90°
- Amplitude of 140°.

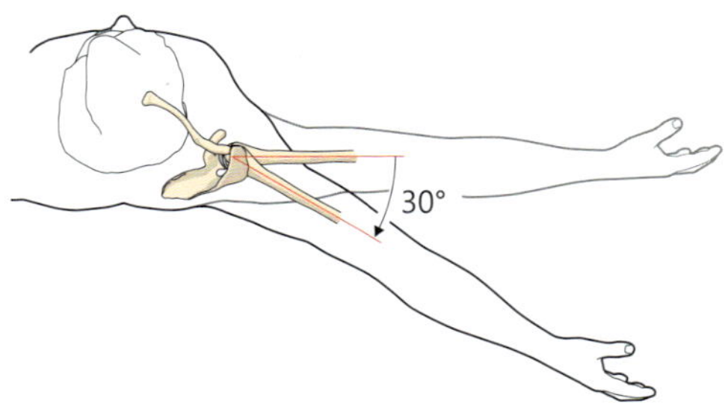

External rotation
- On the horizontal plane
- The starting position is an abduction of 90°
- Amplitude of 30°.

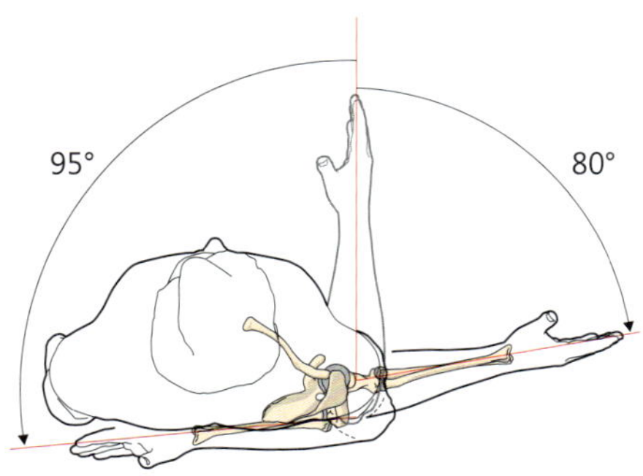

Rotation on the axis of the humeral diaphysis
- Glenohumeral joint movement around the longitudinal axis of the humeral diaphysis
- Starting position:
  - shoulder in a resting position (0°)
  - knee flexed at 90°
  - hand in any position (the forearm pronosupination component is excluded from the movement)
- External:
  - amplitude of 80°
- Internal:
  - the arm is brought behind the trunk
  - amplitude of 95°.

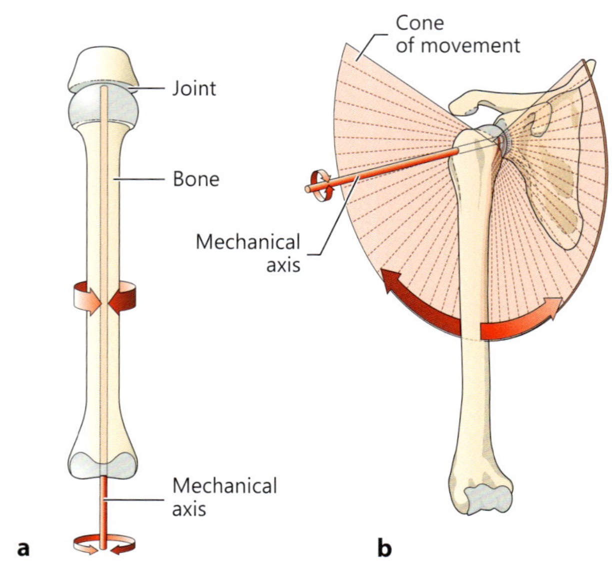

Circumduction
- Typical movement of all joints which have at least two degrees of freedom (flexion-extension and abduction-adduction)
- Maximum amplitude at the level of the shoulder.

MOVEMENTS OF THE ELBOW JOINT

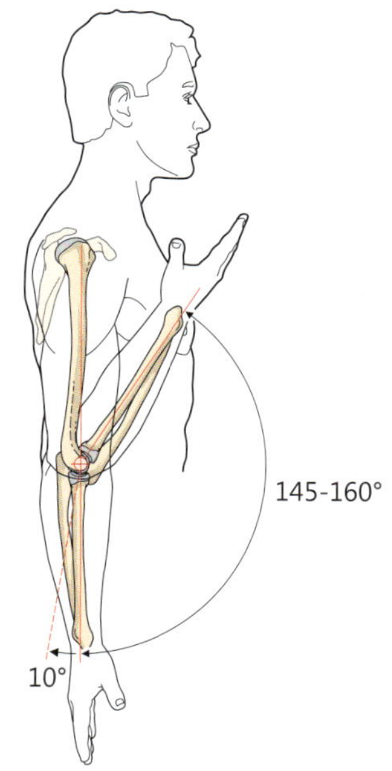

Flexion of the elbow
The reference position is given by the upper limb along the side, with the palm of the hand facing forward, and the movement brings the anterior surface of the forearm to that of the arm.

Extension of the elbow
- Relative – return of the elbow to the starting position from any position of flexion
- Absolute (hyperextension) (starting from the reference position).

Amplitude of the flexion-extension movement
The amplitude of flexion movement is:
- 145° (active movement)
- 160° (passive movement)

The range of motion during extension:
- corresponds to the amplitude of the reference position (achieved as a result of the distancing of the anterior surface of the forearm from that of the arm with the return to the starting position), equal to 175-180°
- absolute range is about 10° (with an interval of 5-15°), which must be added to the 180° of the reference position.

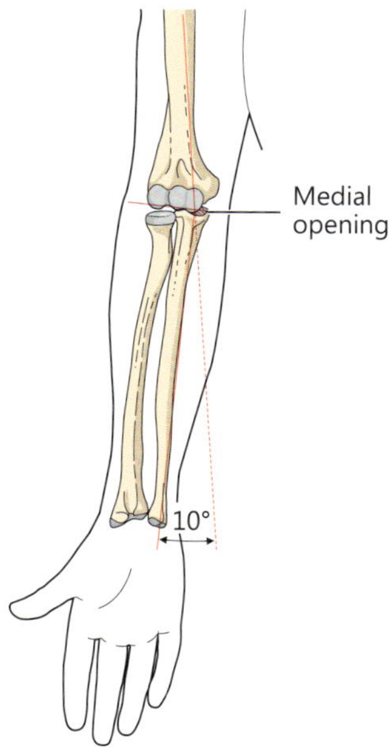

Axis
The flexion-extension movement occurs:
- on the sagittal plane
- around an oblique axis (which, below the lateral and medial epicondyles, crosses the condyle and the trochlea of the humerus).

Factors limiting flexion-extension movement
Active flexion is limited by:
- contact between the soft tissues of the anterior regions of the arm and forearm
- muscular contraction
- tension of the fibers of the anconeus and the triceps brachii.

Passive flexion is limited by:
- contact between the head of the radius and coronoid process of the ulna, respectively, with the radial and coronoid fossae of the humerus
- tension of the fibers of the flexor muscles of the forearm:
  - biceps brachii
  - brachialis
  - brachioradialis
- tension of the posterior capsuloligamentous system.

Extension is limited by:
- ulnar and radial collateral ligaments
- tension of the anterior capsuloligamentous system
- contact between the beak of the olecranon and the olecranon fossa of the humerus.

Pronosupination of the proximal and distal radioulnar joints

During the movement of pronosupination:

- the articular circumference of the head of the radius rotates on the osteofibrous ring formed by the radial notch of the ulna and the anular ligament
- the condyle of the humerus rotates on the articular fossa of the head of the radius
- the articular circumference of the head of the ulna rotates on the ulnar notch of the radius
- there is movement of the articular disc of the distal radioulnar joint.

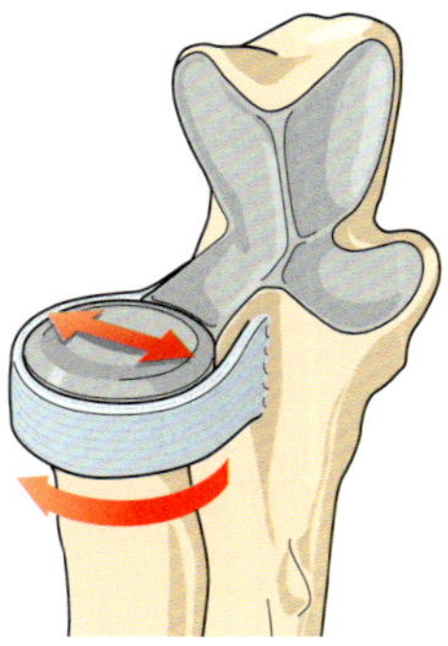

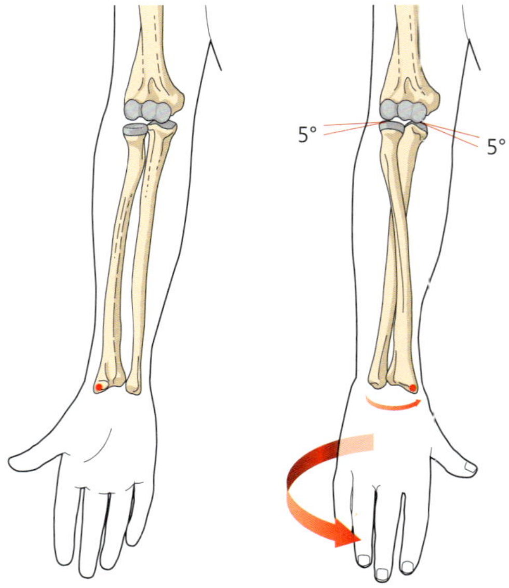

Pronation

- The radius rotates around the ulna
- The distal third of the ulna moves laterally
- At the end of the movement, the radius is positioned obliquely in relation to the ulna.

Supination

- The distal third of the ulna shows a slight medial movement
- At the end of the movement, the radius and ulna are parallel.

Amplitude of the pronosupination movement

The amplitude of the pronation movement is 85° (the movement is stopped due to the contact produced when the radius and ulna overlap).

The amplitude of the supination movement is 90°.

The global movement of pronosupination can reach, thanks to the involvement of the shoulder joint, an amplitude of:

- 360° when performed with the forearm extended and the arm aligned with the trunk or abducted at 90°
- 270° when performed with the forearm extended or the arm flexed at 90°
- 180° when performed with the forearm extended and the arm at the maximum lateromedial elevation.

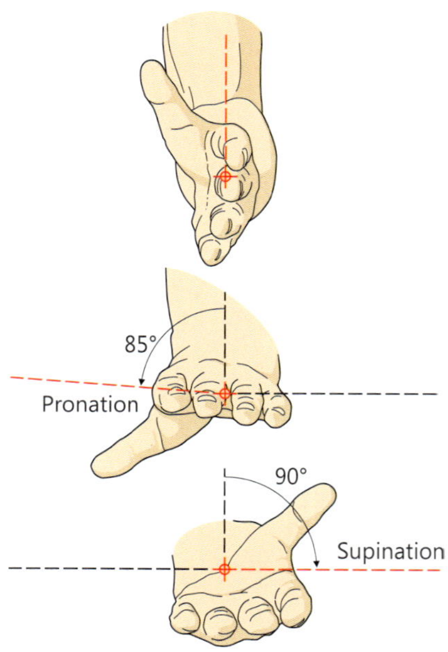

Axis
Pronosupination movement occurs around a vertical axis (which passes proximally through the condyle of the humerus and the center of the head of the radius and distally through the ulnar styloid process).

Factors limiting pronosupination movement
Pronation is limited by:
- contact between the radius and the ulna
- inferior posterior radioulnar ligament
- tension of the posterior capsuloligamentous system
- supinator muscles
- flexor digitorum profundus and flexor pollicis longus.

Supination is limited by:
- interosseous membrane
- oblique cord (Weitbrecht's cord)
- inferior anterior radioulnar ligament
- quadrate ligament
- tension of the anterior capsuloligamentous system
- pronator muscles.

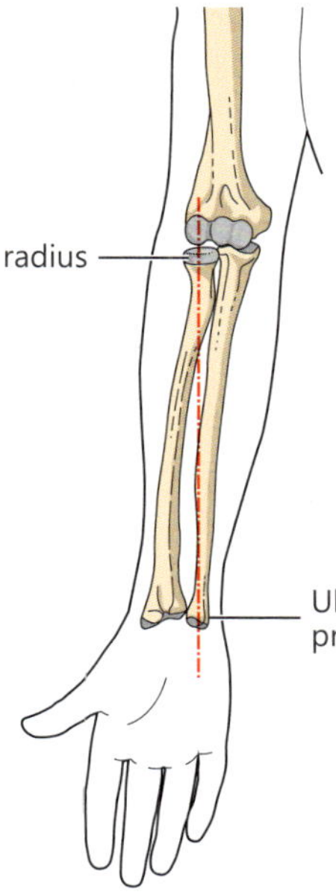

MOVEMENTS OF THE WRIST

The wrist and midcarpal joints, although anatomically distinct, form a functional unit.

Flexion (palmar flexion) and extension (dorsal flexion) of the wrist

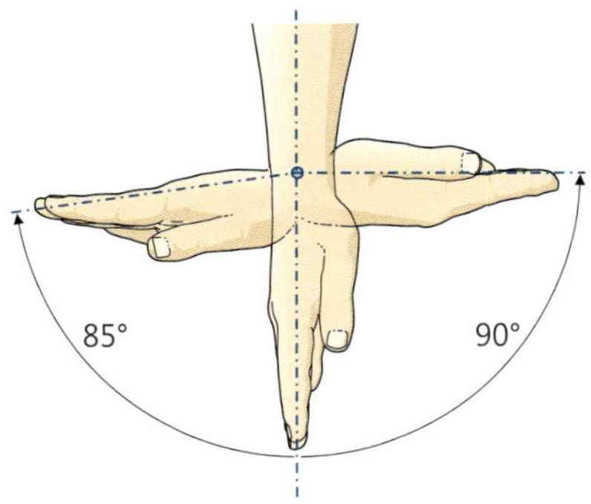

Amplitude:
- 85-90° → palmar flexion (50° – wrist joint; 35-40° – midcarpal joint)
- 85° → dorsal flexion (35° – wrist joint; 50° – midcarpal joint).

Maximum amplitude of the flexion movement:
- palmar, can be obtained by extending the digits of the hand
- dorsal, can be obtained by flexing the digits of the hand.

Rotation axis around which the flexion-extension movements occur:
- line joining the radial and ulnar styloid processes.

Radial abduction (true abduction) and ulnar abduction (adduction) of the wrist

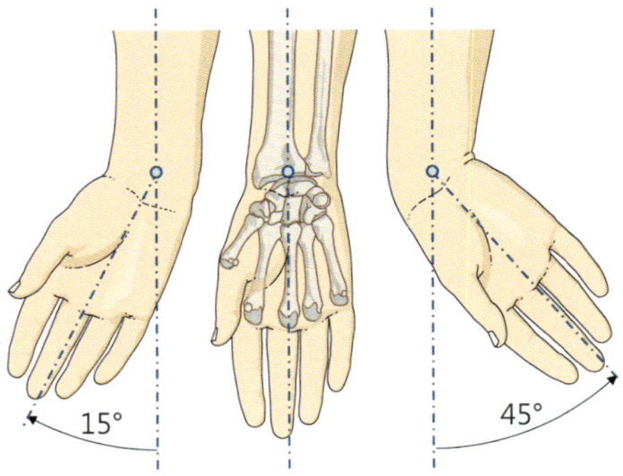

Amplitude:
- 15° → radial abduction
- 45° → ulnar abduction

In total 60° of abduction of which 37° related to the wrist joint and 23° to the midcarpal joint; this last is the sum of 15° of ulnar abduction and 8° of radial abduction.

The lower amplitude of radial abduction is due to blockade of the scaphoid under the radial styloid process, which has a greater distal protrusion with respect to the ulnar styloid process. The lower distal protrusion of the latter and the characteristics of the shape and position of the pisiform and triquetrum bones, which are more distant from the ulnar styloid process, make the ulnar abduction wider.

Rotation axis around which the abduction movements occur:
- sagittal line which passes from the most distal point of the axis of the forearm through the capitate and lunate.

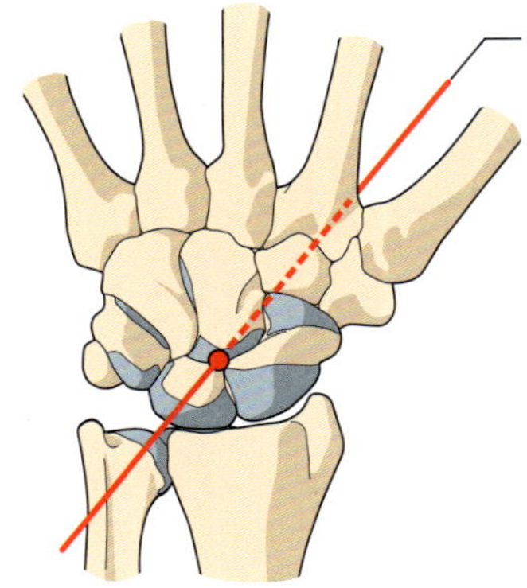

MOVEMENTS OF THE FINGERS

Movements of the 1st finger
The following joints primarily contribute to movements of the thumb:
- carpometacarpal (or trapeziometacarpal) joint of the thumb
- metacarpophalangeal joint of the thumb
- interphalangeal joint of the thumb.

Movements of the carpometacarpal joint of the 1st finger

Rotation axes
The rotation axes are oblique and the flexion-extension and abduction-adduction movements do not occur exactly in the sagittal and frontal planes. The flexion-extension movements occur in an oblique frontal plane around an oblique sagittal rotation axis. The abduction-adduction movements occur in an oblique sagittal plane around an oblique frontal rotation axis. The center of the flexion-extension axis passes through the trapezium, while the center of the abduction-adduction axis passes through the base of the first metacarpal.

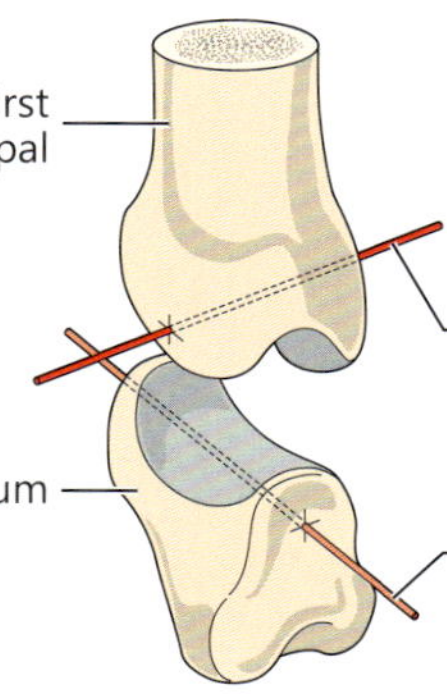

Saddle joints of the thumb.

Evaluation
The first metacarpal moves around a rectangular reference triad constituted by the transverse (**T**), frontal (**F**) and sagittal (**S**) planes. From a reference position in which the first metacarpal is in contact with the second metacarpal (approximately in the **F** plane), the flexion is the forward movement of the first metacarpal (arrow 1) and the extension is the reverse movement, while the abduction moves away the first metacarpal from the second metacarpal (arrow 2) and the adduction is the reverse movement.

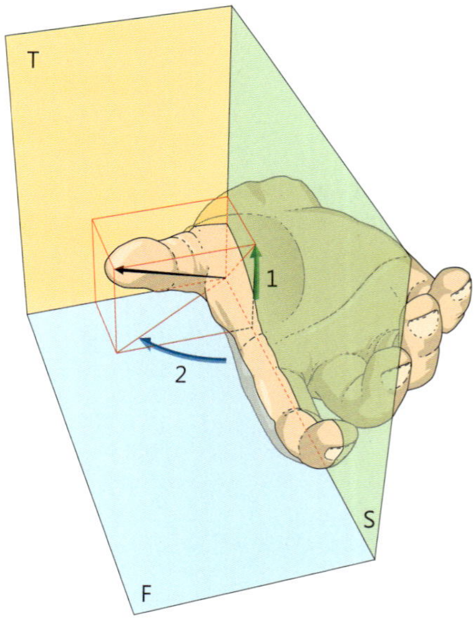

Carpometacarpal flexion-extension of the 1st finger

Flexion

- Reference position — Anatomical position with the thumb in contact with the 2nd metacarpal and the proximal phalanx of the 2nd finger
- Rotation axis — Oblique sagittal, with an angle of 15° with respect to the non-oblique sagittal axis, and orthogonal with respect to the axis of abduction-adduction
- Movement — Forward, upward, and inward (with rotation of the metacarpus around its own axis)
- Amplitude — 15°

Extension

- Movement — Backward, upward, and outward
- Amplitude — 40°

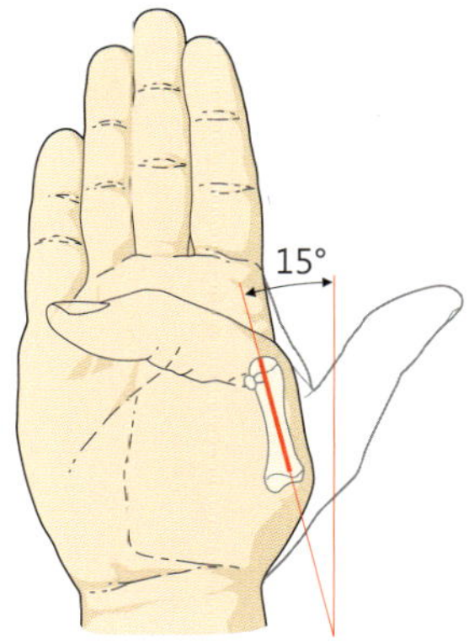

Carpometacarpal flexion of the thumb (volar view of the hand).

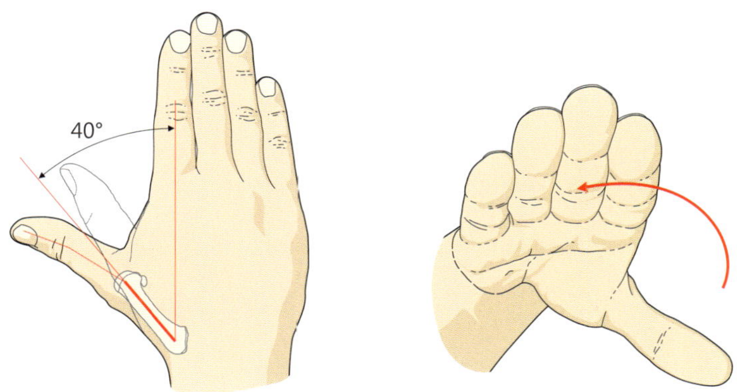

On the left, carpometacarpal extension of the thumb (dorsal view of the hand). On the right, carpometacarpal extension of the thumb (frontal view of the hand). The arrow indicates the direction of flexion.

Carpometacarpal abduction-adduction of the 1st finger

Abduction

- Reference position — The same as with flexion-extension movements
- Rotation axis — Oblique frontal, with an angle of approximately 30° with respect to the non-oblique frontal axis, and orthogonal with respect to the axis of flexion-extension
- Movement — Forward, upward, and inward
- Amplitude — 40-50° (70° passively)

Adduction

- Movement — Return to the reference position from abduction

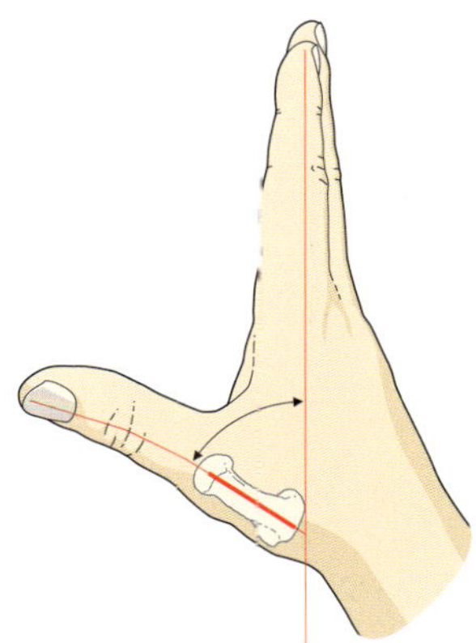

Carpometacarpal abduction of the thumb (right hand seen from the side of the thumb).

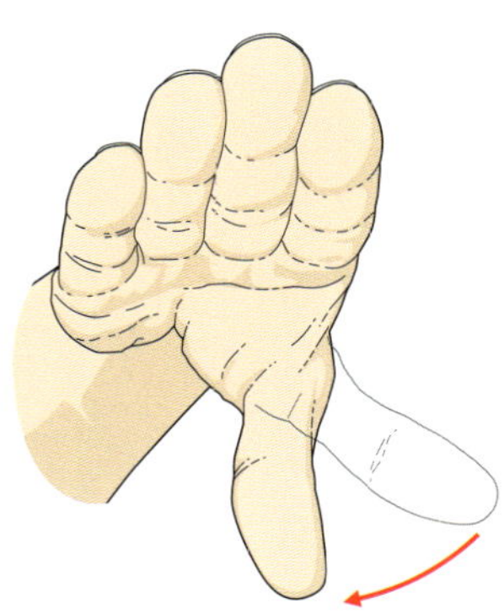

The downward arrow indicates the carpometacarpal abduction of the thumb (frontal view of the hand).

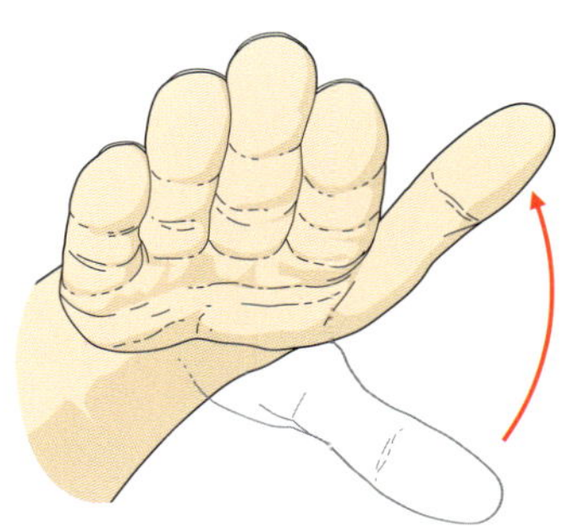

The upward arrow indicates the carpometacarpal adduction, or rather, hyperadduction of the thumb (frontal view of the hand).

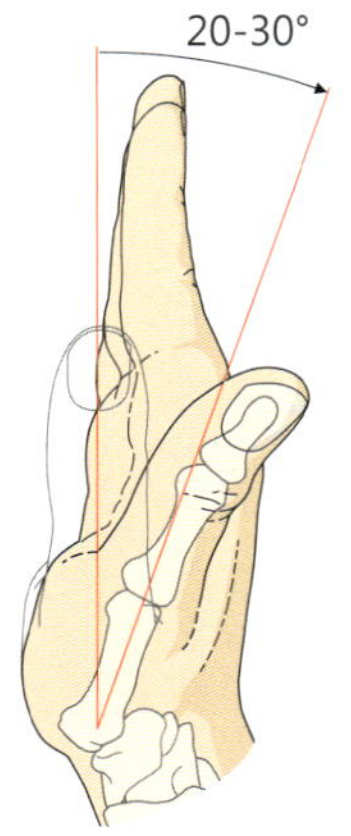

Hyperadduction
- Movement Adduction from the reference position
- Amplitude 20-30°

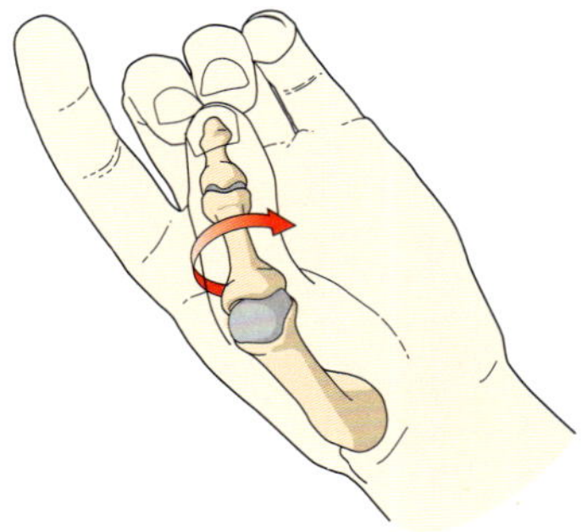

Metacarpophalangeal abduction-adduction of the 1st finger

- Amplitude Almost nothing

The opposition movements of the thumb to the other fingers (allowed by the carpometacarpal joint of the thumb) are based, at the level of the metacarpophalangeal joint of the thumb, on the internal and external rotation (active or passive) of the proximal phalanx with respect to the 1st metacarpal (axial rotation), which occurs by opposing the pulp of the thumb to the distal phalanx of the ring finger and leveraging on this.

Movements of the metacarpophalangeal joint of the 1st finger

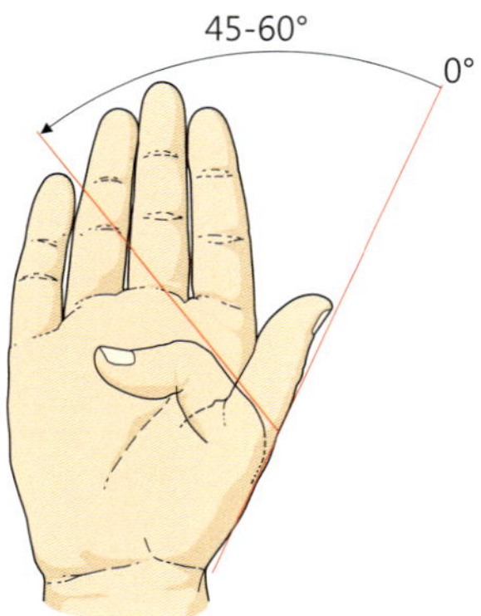

Metacarpophalangeal flexion-extension of the 1st finger

Flexion
- Amplitude 45-60°

Extension
- Amplitude 0°

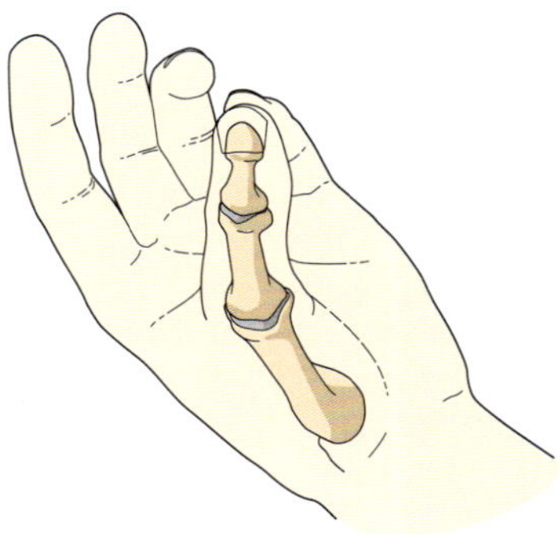

The carpometacarpal circumduction movement of the thumb, derived from a combination of abduction-extension and adduction-flexion movements, allows for the opposition and retroposition movements of the thumb to the other finger (contact between the thumb and the other fingers of the hand), as is characteristic of the human hand.

Movements of the 2nd, 3rd, 4th and 5th fingers

Metacarpophalangeal flexion-extension of the 2nd, 3rd, 4th, and 5th fingers

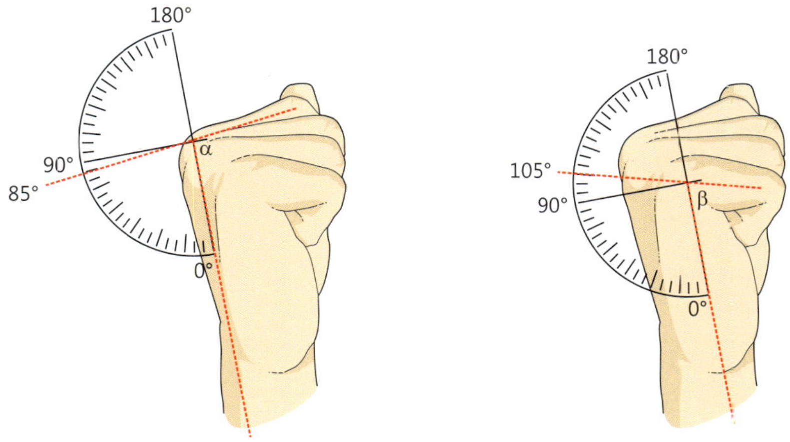

If you close your hand and make a fist, you can observe a significant difference between the angular amplitude of the metacarpophalangeal joint of the index finger (0-85°) and that of the little finger (0-105°).

Flexion

- Movement — Nearing of the proximal phalanx to the palm of the hand
- Amplitude — 90° (increases from the 2nd to the 5th finger)

Extension

- Movement — Return to the starting position

Hyperextension

- Active – about 20-30°
- Passive – up to 90° (in some subjects)

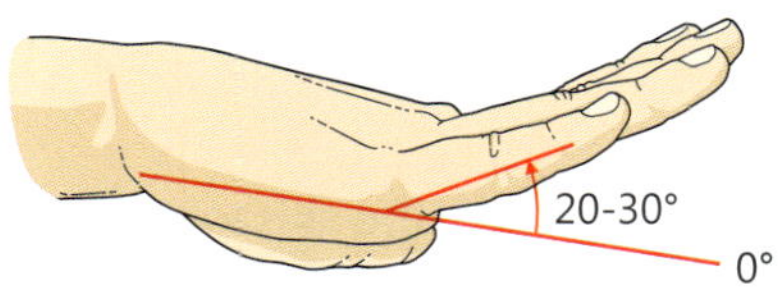

Active hyperextension of the proximal metacarpophalangeal joint.

Metacarpophalangeal abduction-adduction of the 2nd, 3rd, 4th, 5th fingers

Abduction

- Movement — Separation of the 2nd, 4th and 5th fingers from the 3rd finger (the 3rd finger refers to radial or ulnar abduction)
- Amplitude — 20-30° (maximum limit for the 3rd finger)

Adduction

- Movement — Return to the starting position

Movements of abduction and adduction are limited or non-existent if the proximal phalanges of the 2nd, 3rd, 4th and 5th fingers are flexed.

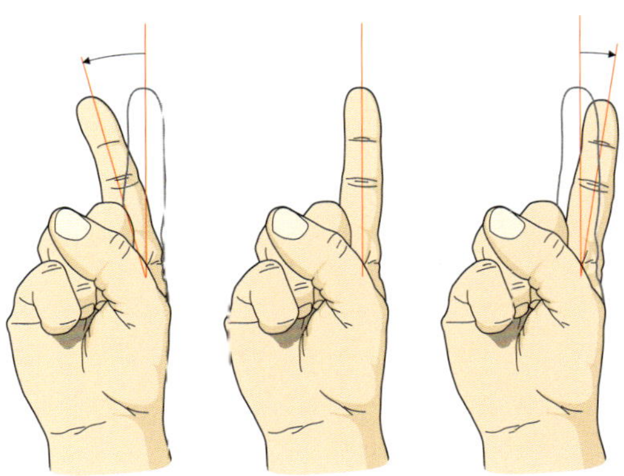

Adduction and abduction of the metacarpophalangeal joint of the 2nd finger.

Metacarpophalangeal axial rotation of the 2nd, 3rd, 4th, and 5th fingers

- Passive (up to 50-60°)

The passive external axial rotation of the 2nd finger is almost non-existent, unlike the internal rotation which is, on the contrary, quite wide (45°).

Movements of the interphalangeal joints of the 1st, 2nd, 3rd, 4th and 5th fingers

Flexion-extension of the interphalangeal joints of the 1st, 2nd, 3rd, 4th, and 5th fingers

Flexion

- Amplitude
  >90° (middle phalanx)
  60-70° (distal phalanx)
  Increases from the 2nd to the 5th finger
  Thumb → about 90° (active flexion)

Extension

- Amplitude
  Almost nothing, since the movement is represented by the return to the starting position and hyperextension is prevented by the accessory volar ligament.
  Distal phalanx → passive extension and modest active hyperextension
  Thumb → passive hyperextension of 30° and active hyperextension of about 5-10°.

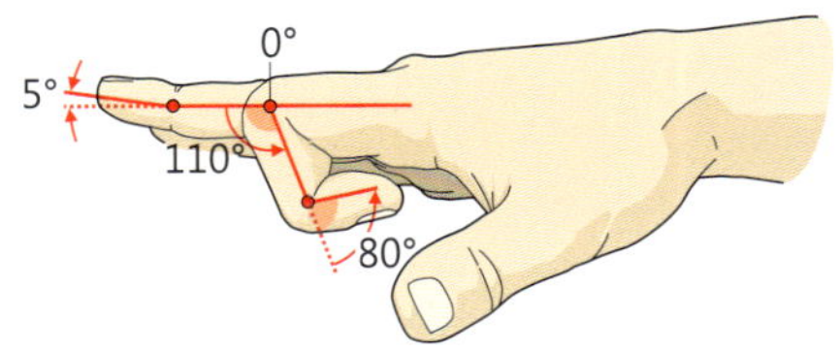

Axis of interphalangeal joints.

When the fingers are flexed, the axis joining the bases of the middle or distal phalanges (from the 2nd to 5th fingers) have a downward, oblique, and medial orientation. The axis joining the bases of the proximal phalanges (from the 3rd to the 5th fingers) has the same orientation as well. However, if you look at the tips of the 5th, 4th, and 3rd fingers (when in a position of flexion), you can see how they are oriented laterally downward, in the direction of the scaphoid. This arrangement of the fingers of the hand is of vital importance for the opposing movement of the thumb to the other fingers.

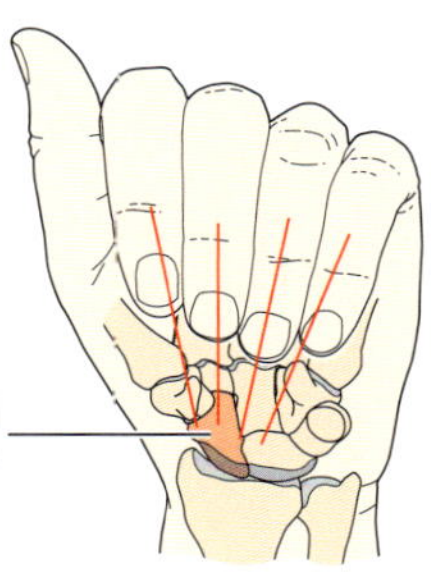

Orientation of the fingers, during flexion, towards the scaphoid.

Summary table of the joints and ligaments of the upper limb		
JOINT, LIGAMENT, OR MEMBRANE	***TYPE OF JOINT***	***BONES INVOLVED***
	JOINTS OF THE PECTORAL GIRDLE	
Coracoacromial ligament		
Superior transverse scapular ligament		
Inferior transverse scapular ligament (not always present)		
Acromioclavicular joint – acromioclavicular ligament – articular disc (not always present) – coracoclavicular ligament – – trapezoid ligament – – conoid ligament	Diarthrosis (plane joint)	Acromion of the scapula and acromial end of the clavicle
Sternoclavicular joint – articular disc – anterior sternoclavicular ligament – posterior sternoclavicular ligament – costoclavicular ligament – interclavicular ligament	Diarthrosis (saddle joint)	Sternal end of the clavicle, manubrium of the sternum, first costal cartilage
	JOINTS OF THE FREE PART OF THE UPPER LIMB	
Interosseous membrane of the forearm		
Oblique cord		
Glenohumeral joint (shoulder joint) – glenoid labrum – glenohumeral ligaments – coracohumeral ligament – transverse humeral ligament	Diarthrosis (ball and socket joint)	Glenoid cavity of the scapula and head of the humerus
Elbow joint – humeroulnar joint – humeroradial joint – proximal radioulnar joint – ulnar collateral ligament – radial collateral ligament – anular ligament of the radius – quadrate ligament – sacciform recess	 Diarthrosis (hinge joint or trochlea) Diarthrosis (condylar joint) Diarthrosis (pivot or trochoid joint)	 Trochlea of the humerus and trochlear notch of the ulna Condyle of the humerus and articular fossa of the head of the radius Articular circumference of the head of the radius and radial notch of the ulna
Distal radioulnar joint – articular disc – sacciform recess	Diarthrosis (pivot or trochoid joint)	Articular circumference of the head of the ulna and the ulnar notch of the radius
	JOINTS OF THE HAND	
Wrist joint (radiocarpal joint) – dorsal radiocarpal ligament – palmar radiocarpal ligament – dorsal ulnocarpal ligament – palmar ulnocarpal ligament – ulnar collateral ligament – radial collateral ligament	Diarthrosis (condylar joint)	Carpal articular surface of the radius and the articular disc of the distal radioulnar joint with the scaphoid, lunate, and triquetrum
Carpal joints (intercarpal joints) – midcarpal joint – radiate carpal ligament – dorsal intercarpal ligaments – palmar intercarpal ligaments – interosseous intercarpal ligaments – pisiform joint – – pisohamate ligament – – pisometacarpal ligament	Diarthroses (plane joints) Diarthrosis (plane and condylar joints) Diarthrosis (plane joint)	Opposing surfaces of the scaphoid, lunate, and triquetrum; opposing surfaces of the trapezium, trapezoid, capitate, and hamate Scaphoid with the trapezium and trapezoid (plane joint); scaphoid, lunate and triquetum with the capitate and hamate (condylar joint) Pisiform and triquetrum
Carpal tunnel		
Ulnar canal (Guyon's canal)		

continued

Summary table of the joints and ligaments of the upper limb *(continued)*

JOINT, LIGAMENT, OR MEMBRANE	*TYPE OF JOINT*	*BONES INVOLVED*
Carpometacarpal joints	Diarthroses (plane joints)	Base of the second metacarpal with the trapezoid and capitate; base of the third metacarpal with the capitate; base of the fourth metacarpal with the capitate and the hamate; base of the fifth metacarpal with the hamate
– dorsal carpometacarpal ligaments		
– palmar carpometacarpal ligaments		
– carpometacarpal joint of the thumb	Diarthrosis (saddle joint)	Base of the first metacarpal and the trapezium
Intermetacarpal joints	Diarthroses (plane joints)	Between the bases of the second, third, fourth, and fifth metacarpals
– dorsal metacarpal ligaments		
– palmar metacarpal ligaments		
– interosseous metacarpal ligaments		
– interosseous metacarpal spaces		
Metacarpophalangeal joints	Diarthroses (condylar joints)	Heads of the metacarpals and bases of the proximal phalanges
– collateral ligaments		
– plantar ligaments		
– deep transverse metacarpal ligament		
Interphalangeal joints of the hand	Diarthroses (hinge joints or trochleae)	Between the heads of the proximal phalanges and bases of the middle phalanges, heads of the middle phalanges and bases of the distal phalanges, head of the proximal phalanx and the base of the distal phalanx (thumb)
– collateral ligaments		
– plantar ligaments		

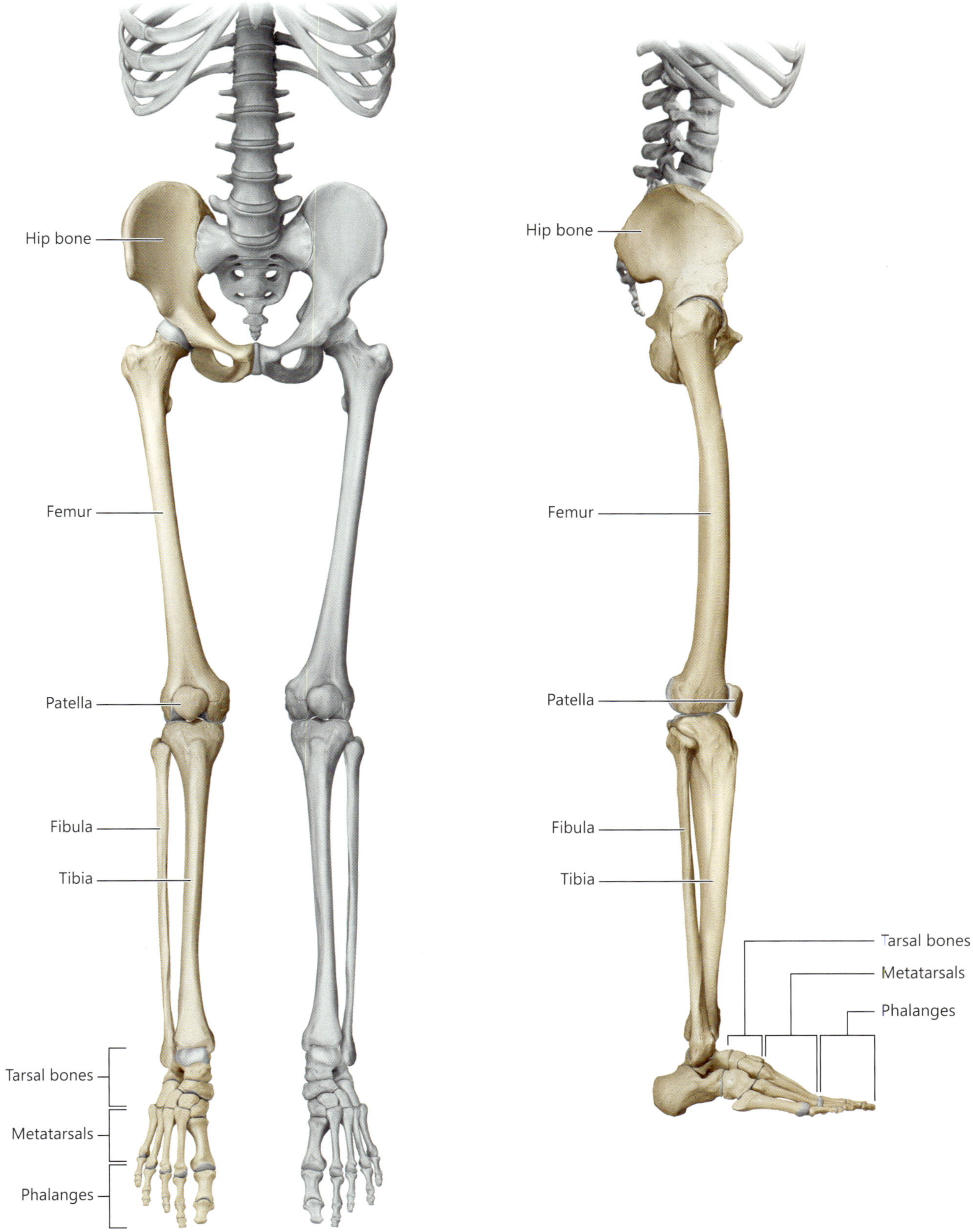

Osseous configuration of the **lower limb**. Anterior view (on the left) and lateral view (on the right).

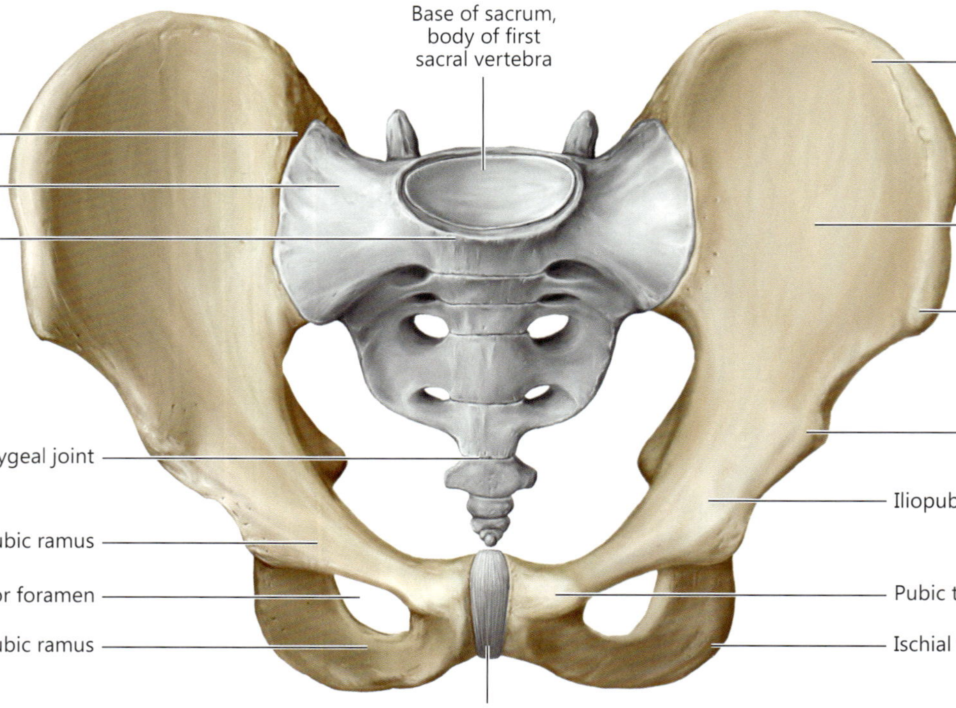

a

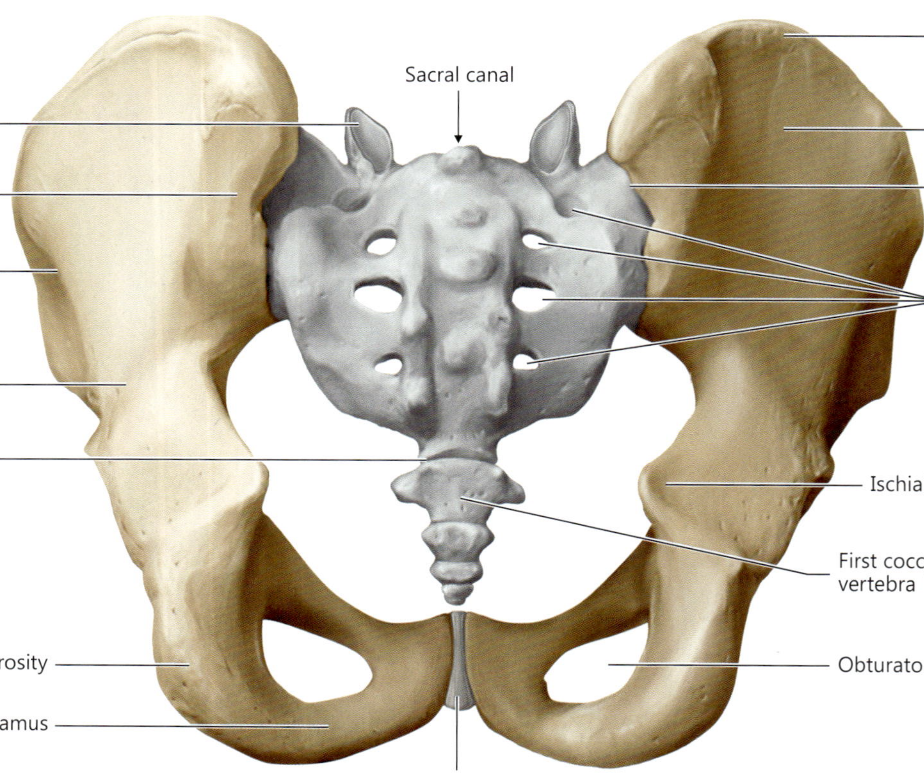

b

Male pelvis. The pelvic girdle is formed by the two bones of the hip articulated with one another and jointed with the sacrum through the sacroiliac joint. Anterosuperior (**a**) and posteroinferior (**b**) views. The hip bones make up part of the lower limb.

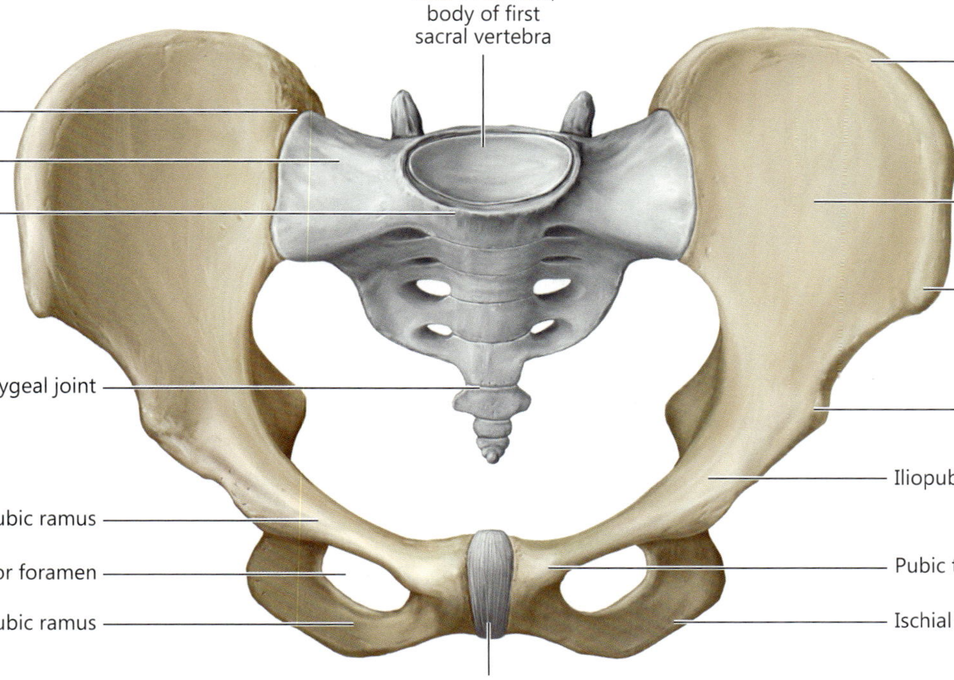

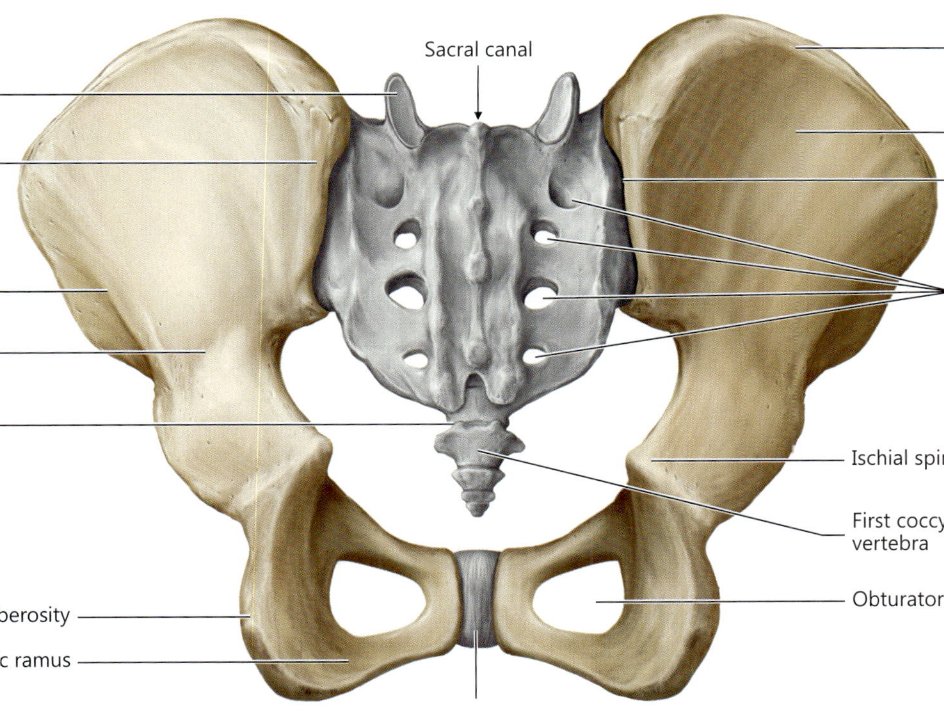

Female pelvis. The pelvic girdle is formed by the two bones of the hip articulated with one another and jointed with the sacrum through the sacroiliac joint. Anterosuperior (**a**) and posteroinferior (**b**) views. The hip bones make up part of the lower limb.

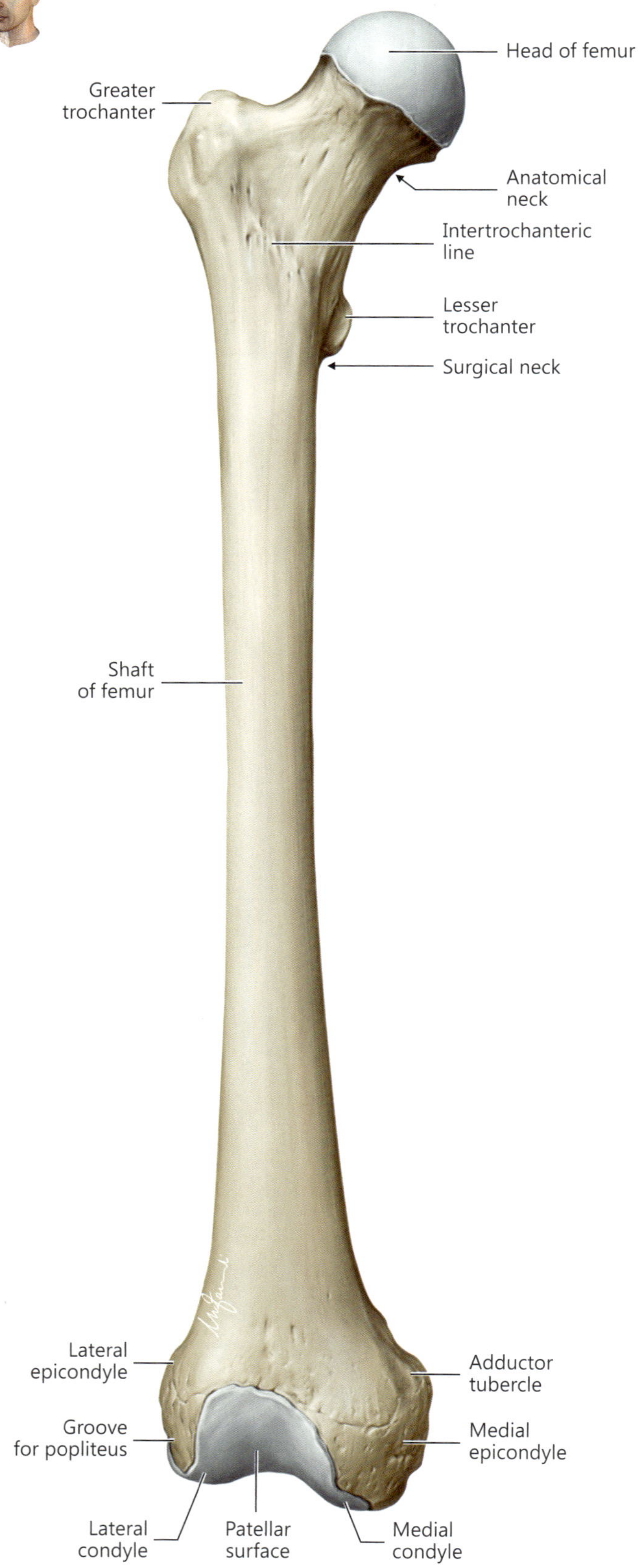

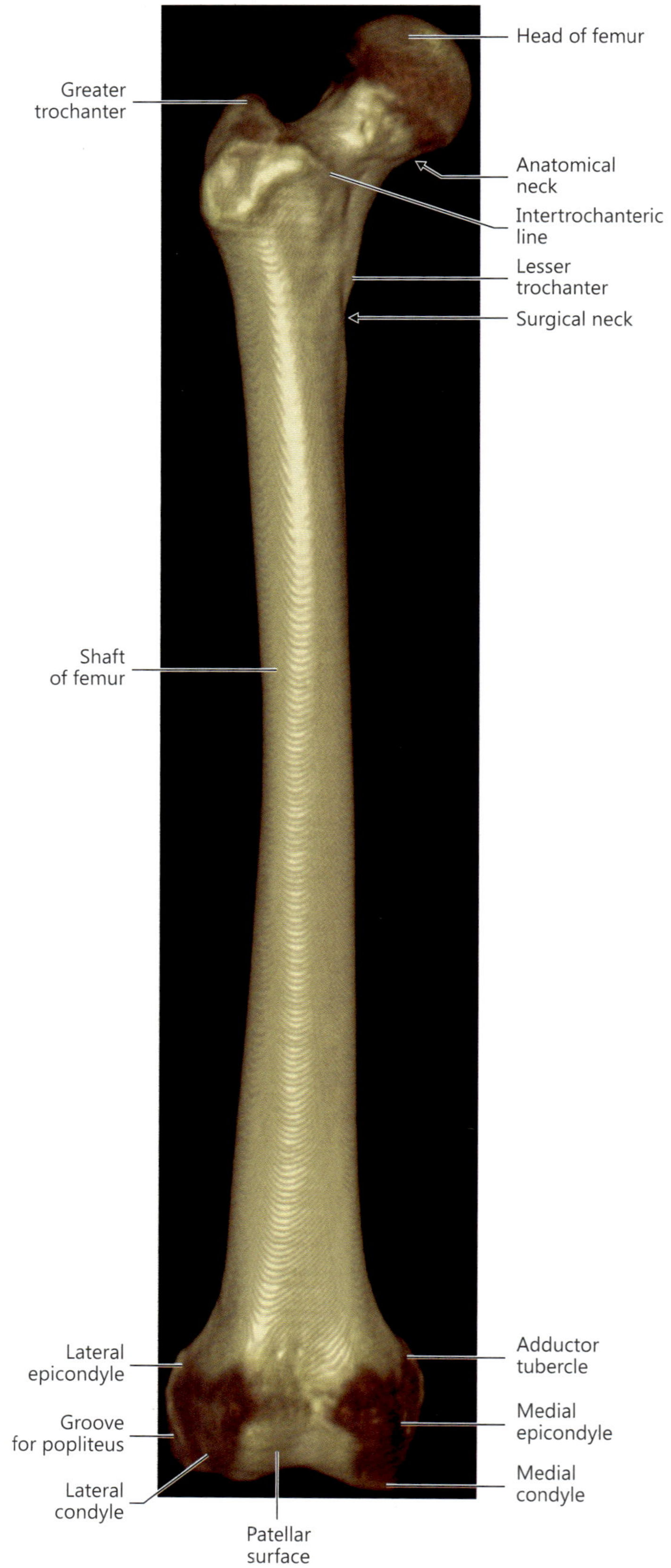

Femur. The femur constitutes the skeletal structure of the thigh. On the left, view of the anterior surface of the right femur. On the right, three-dimensional image of a living human body, obtained by CT scan using a 3D volume rendering technique, of the right femur.

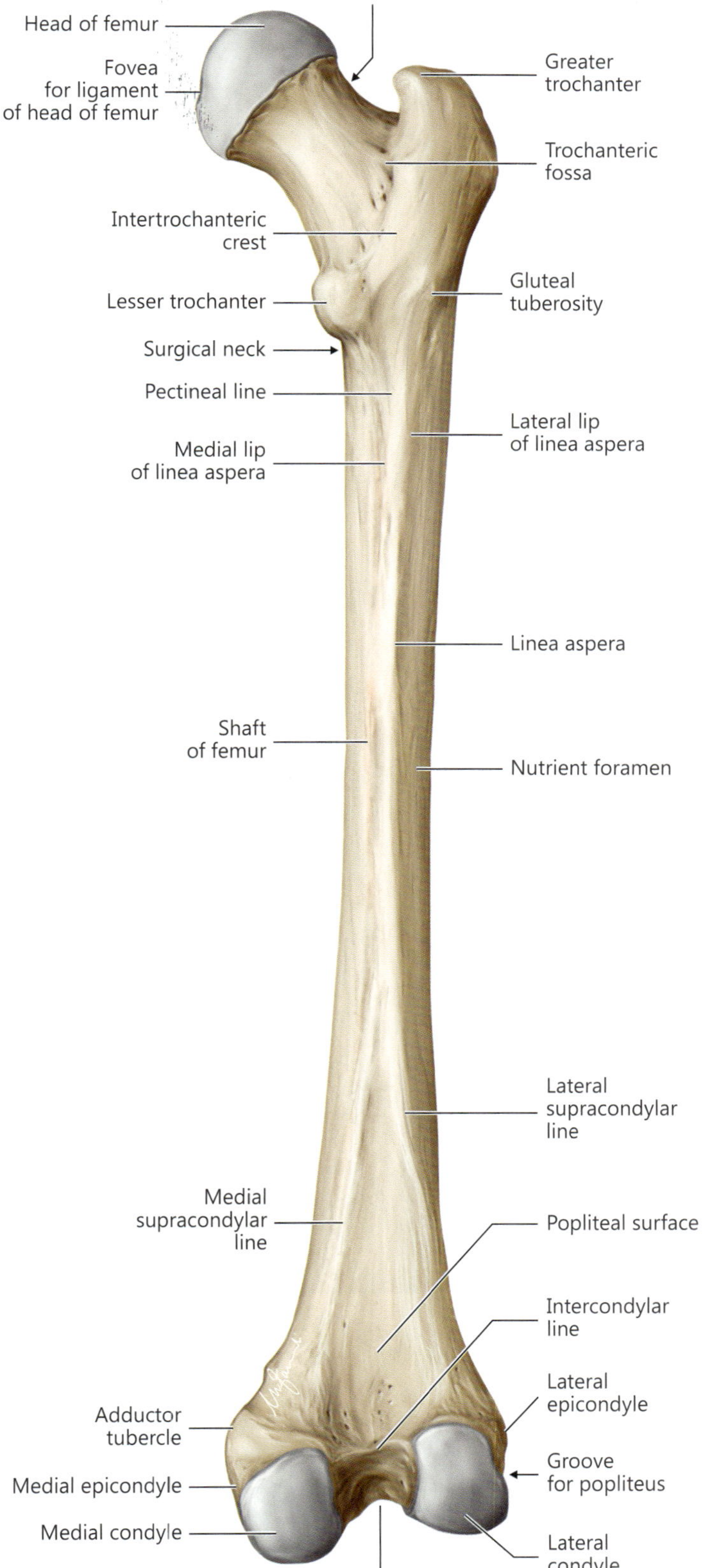

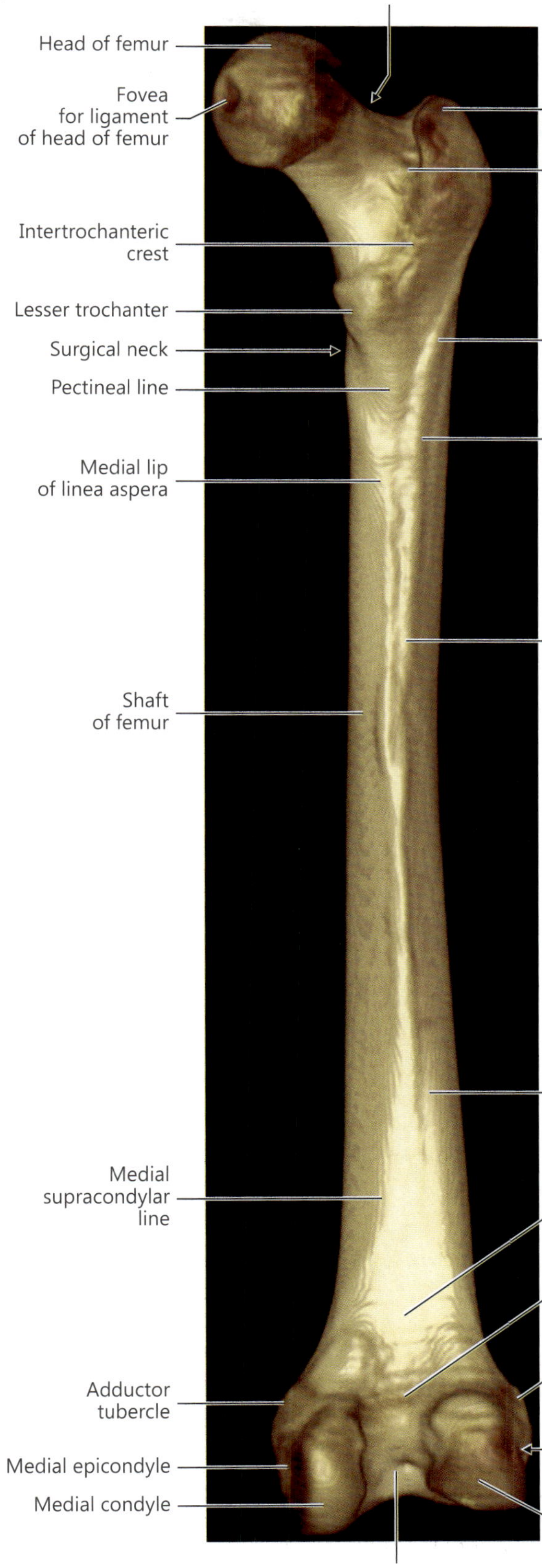

Femur. The femur constitutes the skeletal structure of the thigh. On the left, view of the posterior surface of the right femur. On the right, three-dimensional image of a living human body, obtained by CT scan using a 3D volume rendering technique, of the right femur.

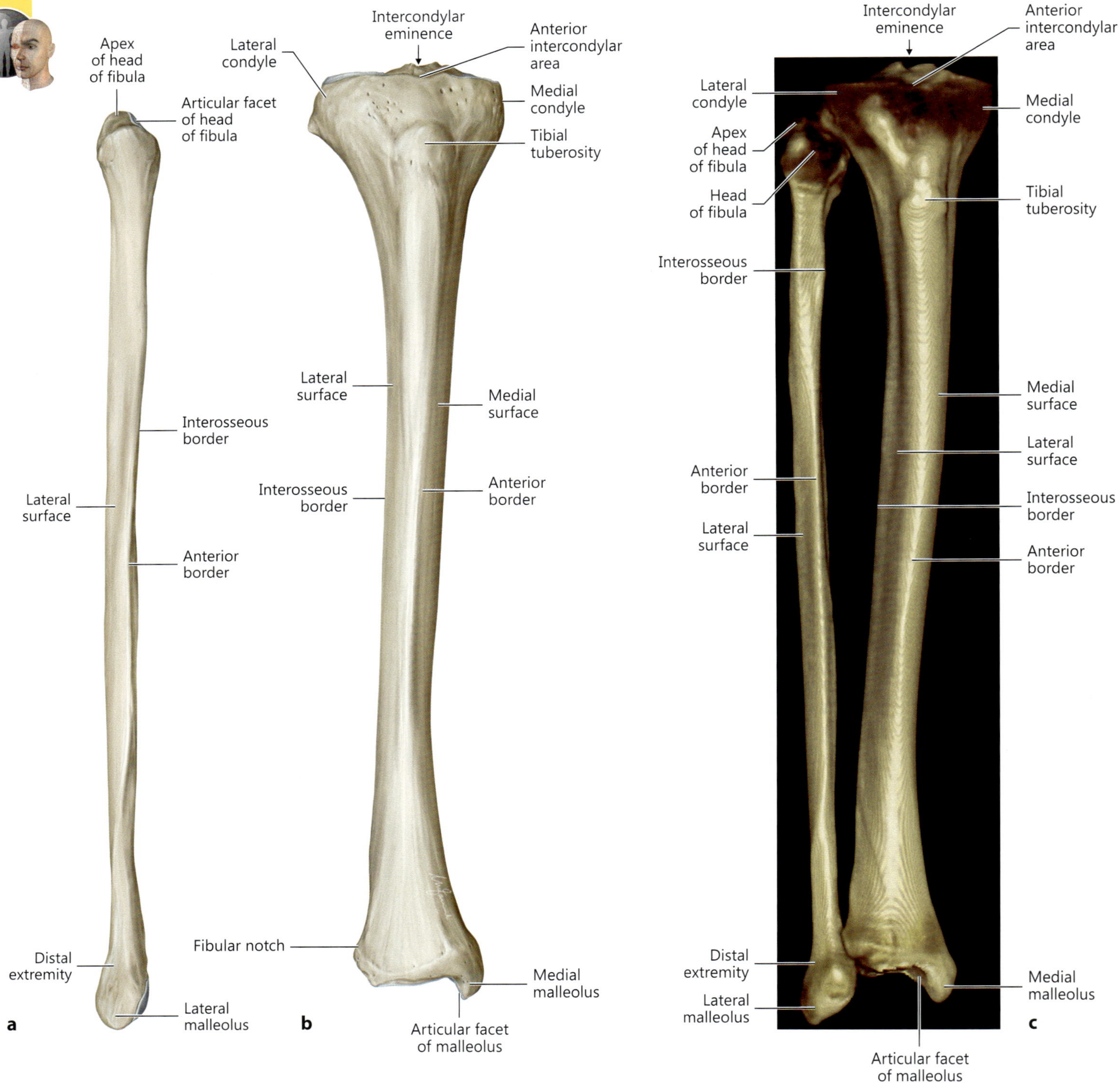

Right **fibula** (**a**) and **tibia** (**b**), anterior view. **c**, Three-dimensional image of a living human body, obtained by CT scan using a 3D volume rendering technique, showing the tibia and fibula, anterior view, in relation to one another.

➤

At the top of the next page you can see the right **tibia** (**a**) and **fibula** (**b**), posterior view. **c**, Three-dimensional image of a living human body, obtained by CT scan using a 3D volume rendering technique, showing the tibia and fibula, posterior view, in relation to one another.

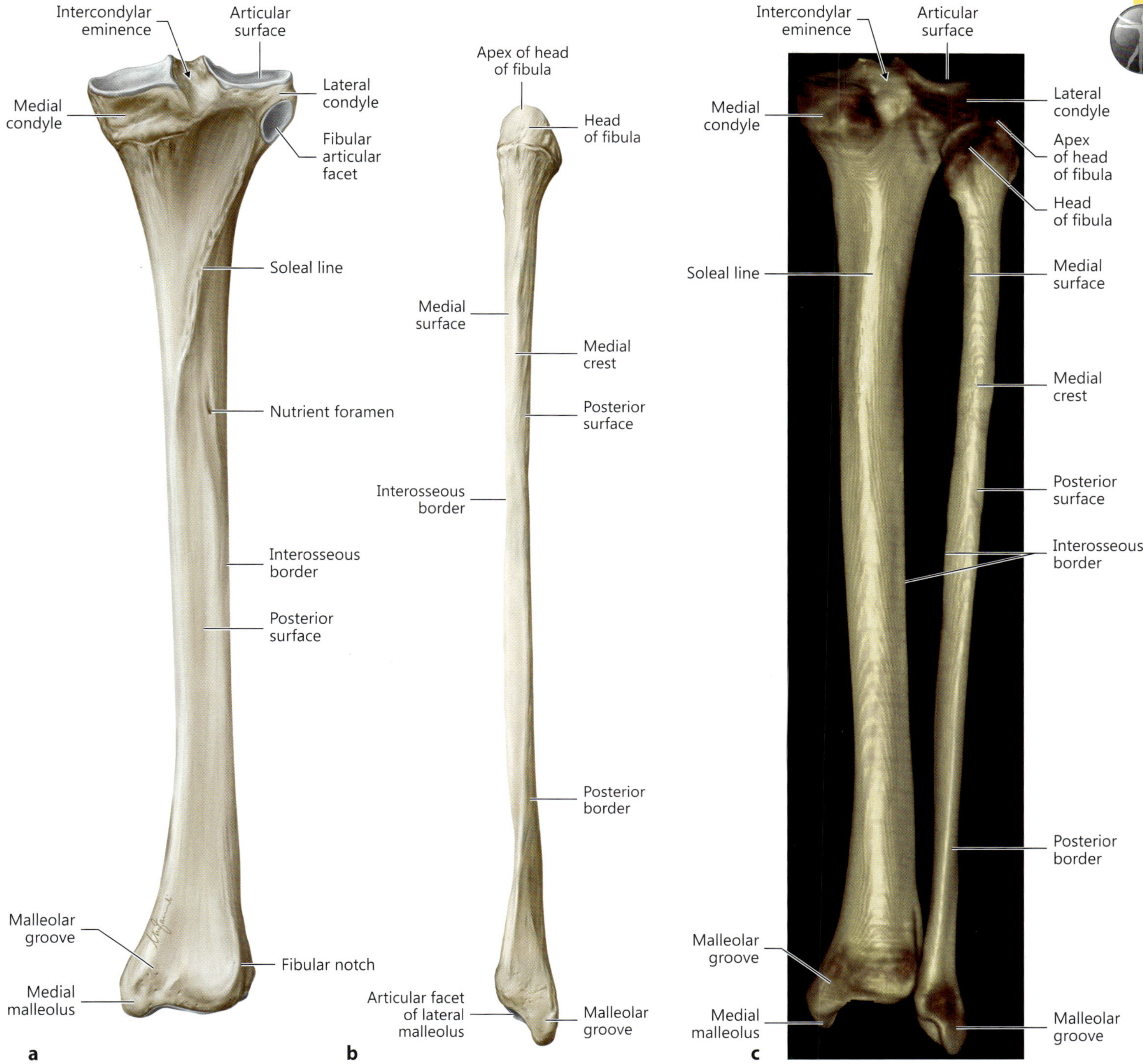

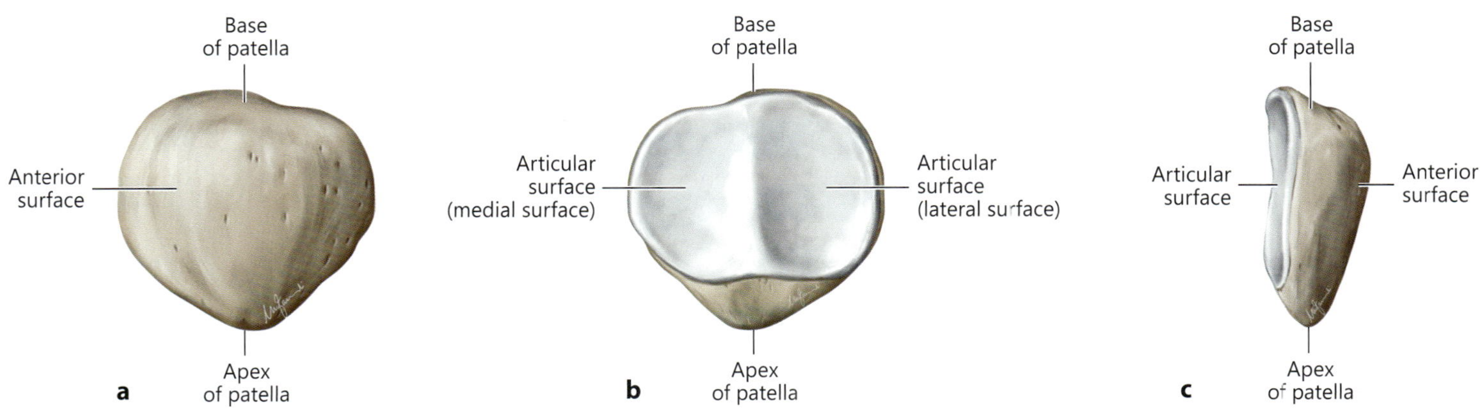

Right **patella**. Anterior (**a**), posterior (**b**) and lateral (**c**) views.

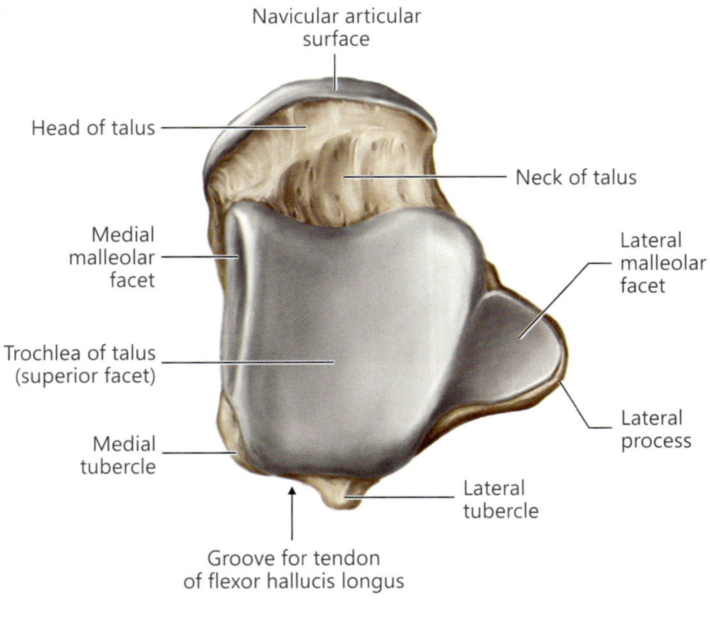

Talus, right foot, dorsal surface. Between the head and the trochlea there is a narrowing in the neck of the talus. On the sides of the trochlea two articular surfaces have been outlined: the lateral and medial malleolar facets.

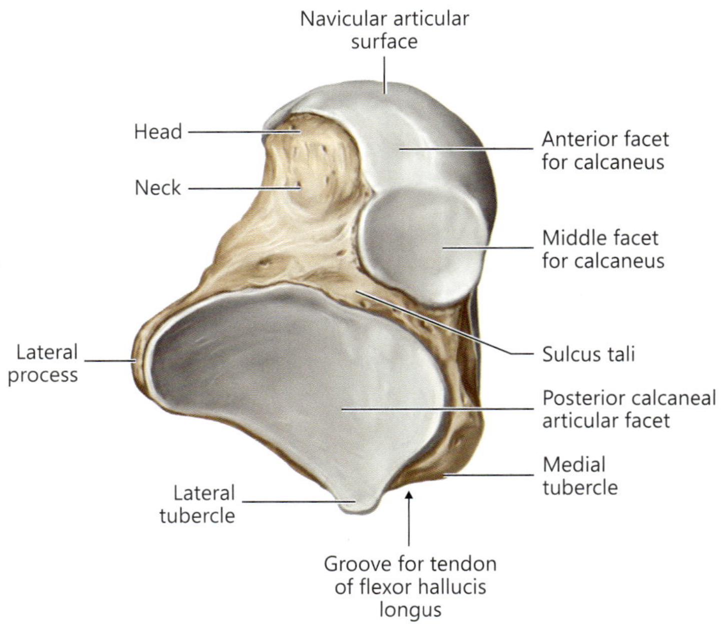

Talus, right foot, plantar surface. Much of this surface is occupied by the articular surfaces for the calcaneus. The sulcus tali is involved in the delimitation of the tarsal sinus.

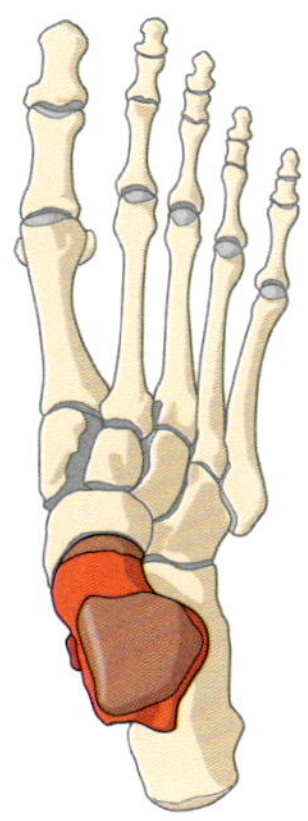

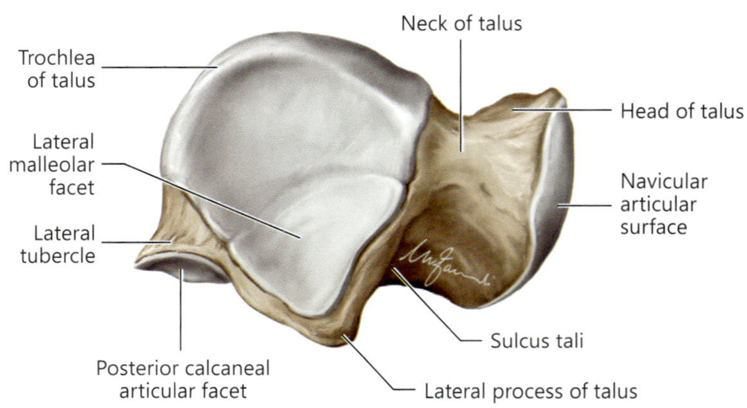

Talus, right foot, lateral surface. You can see the neck of the talus and, in the inferior and posterior part of the bone, the lateral tubercle that externally delimits the groove for the tendon of flexor hallucis longus.

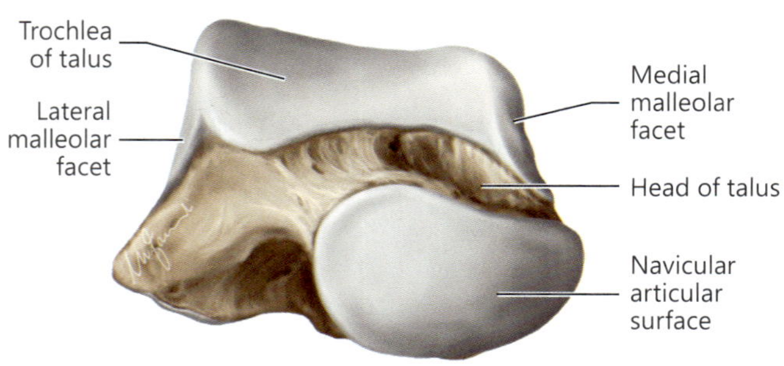

Talus, right foot, anterior surface. Here you can see the trochlea with the medial and lateral malleolar facets and the articular surface of the navicular.

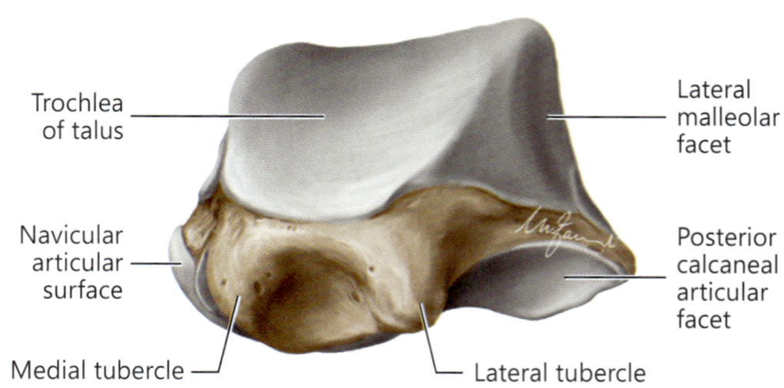

Talus, right foot, posterior surface. Here you can see the trochlea with the medial and lateral malleolar facets and the posterior calcaneal articular facet. Between the medial and lateral tubercles of the posterior process you can see the groove for the tendon of flexor hallucis longus.

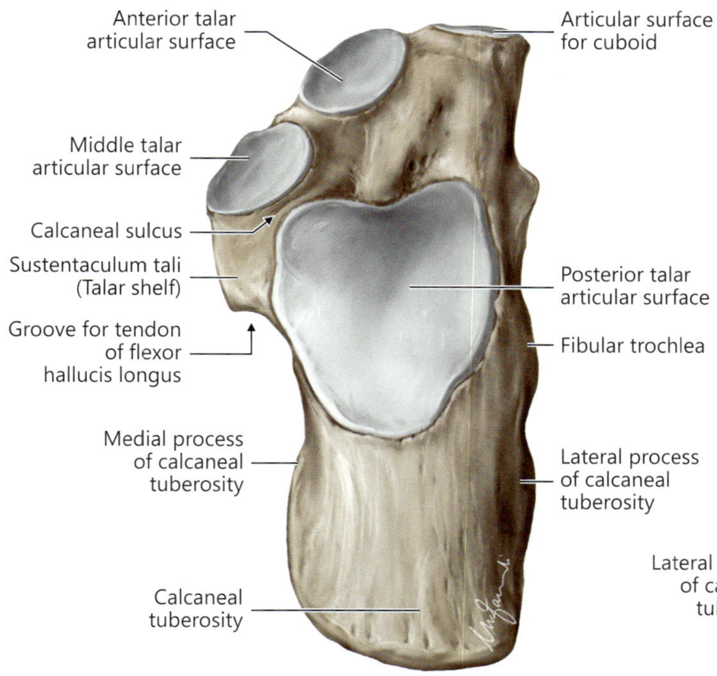

Calcaneus, right foot, dorsal surface. The posterior part of the bone rises into the robust posterior tuberosity.

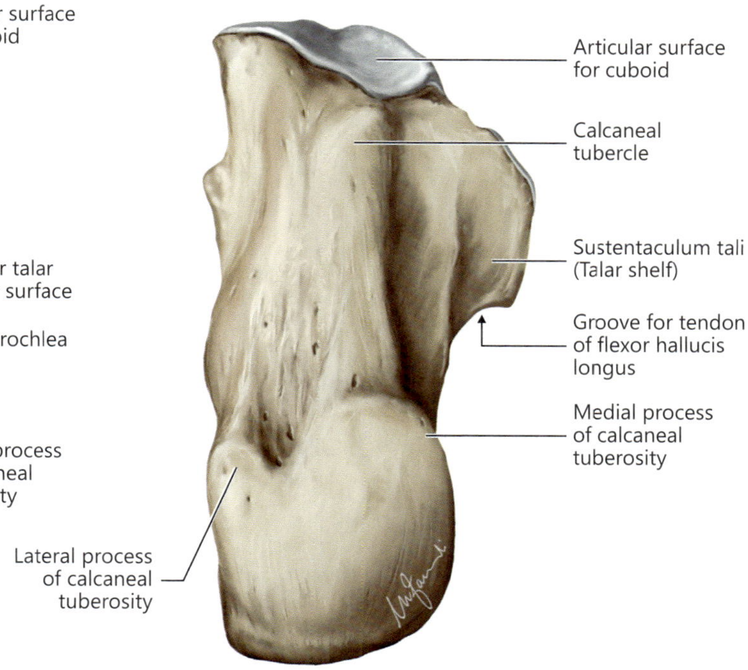

Calcaneus, right foot, inferior surface.

Calcaneus, right foot, lateral surface. The three articular surfaces for the talus are located in the anterior and superior part of the bone. ➤

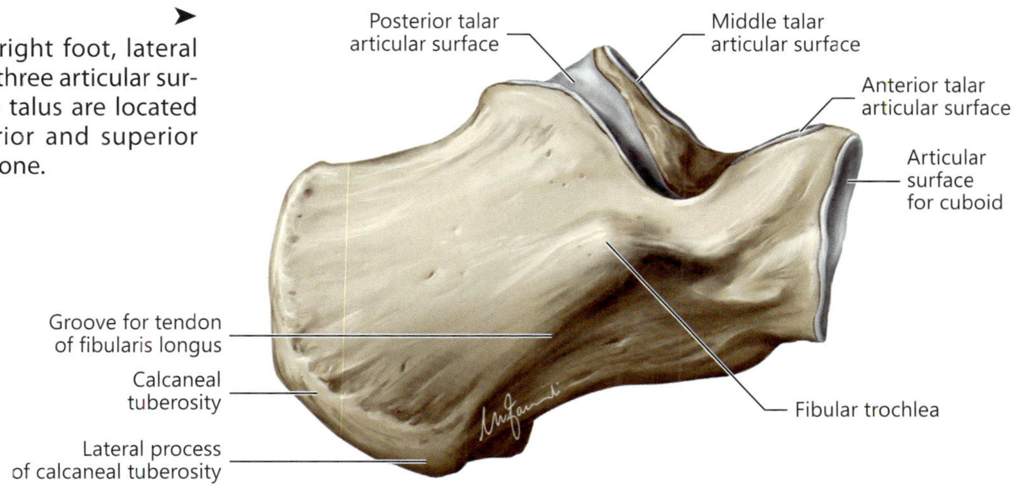

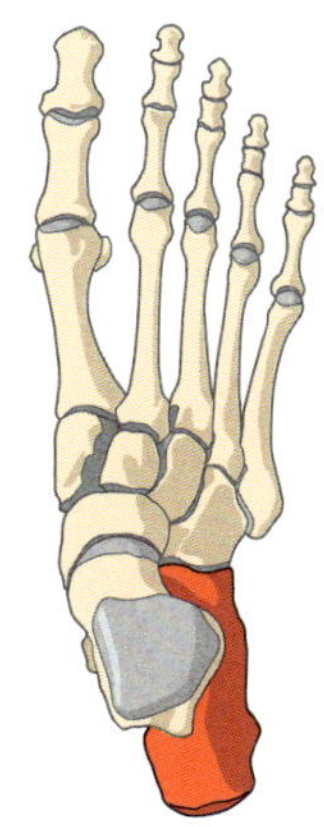

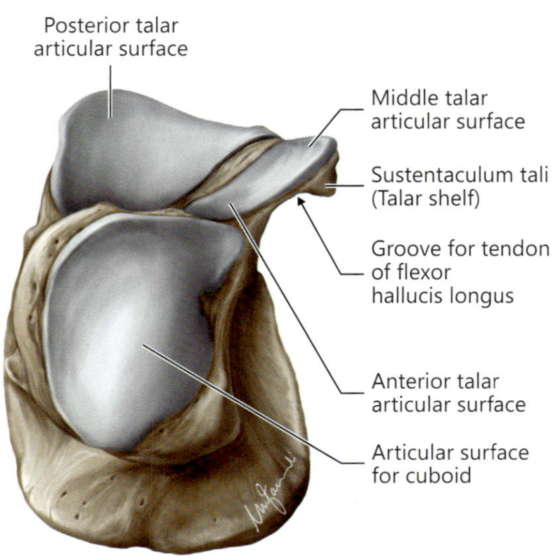

Calcaneus, right foot, anterior surface.

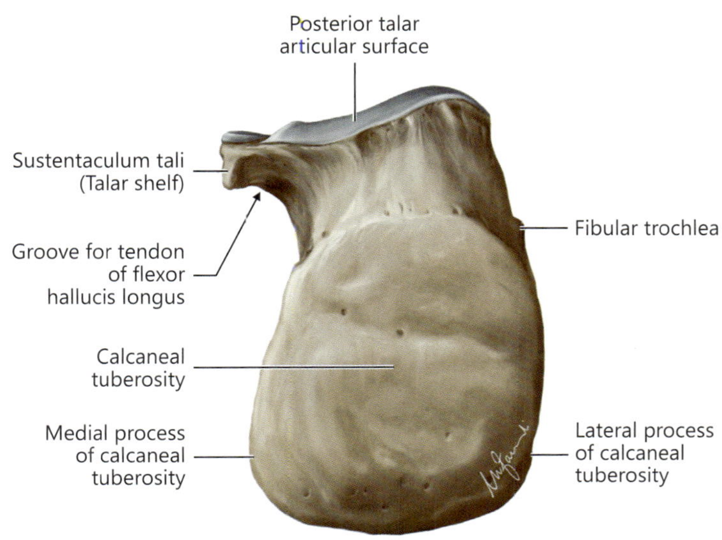

Calcaneus, right foot, posterior surface.

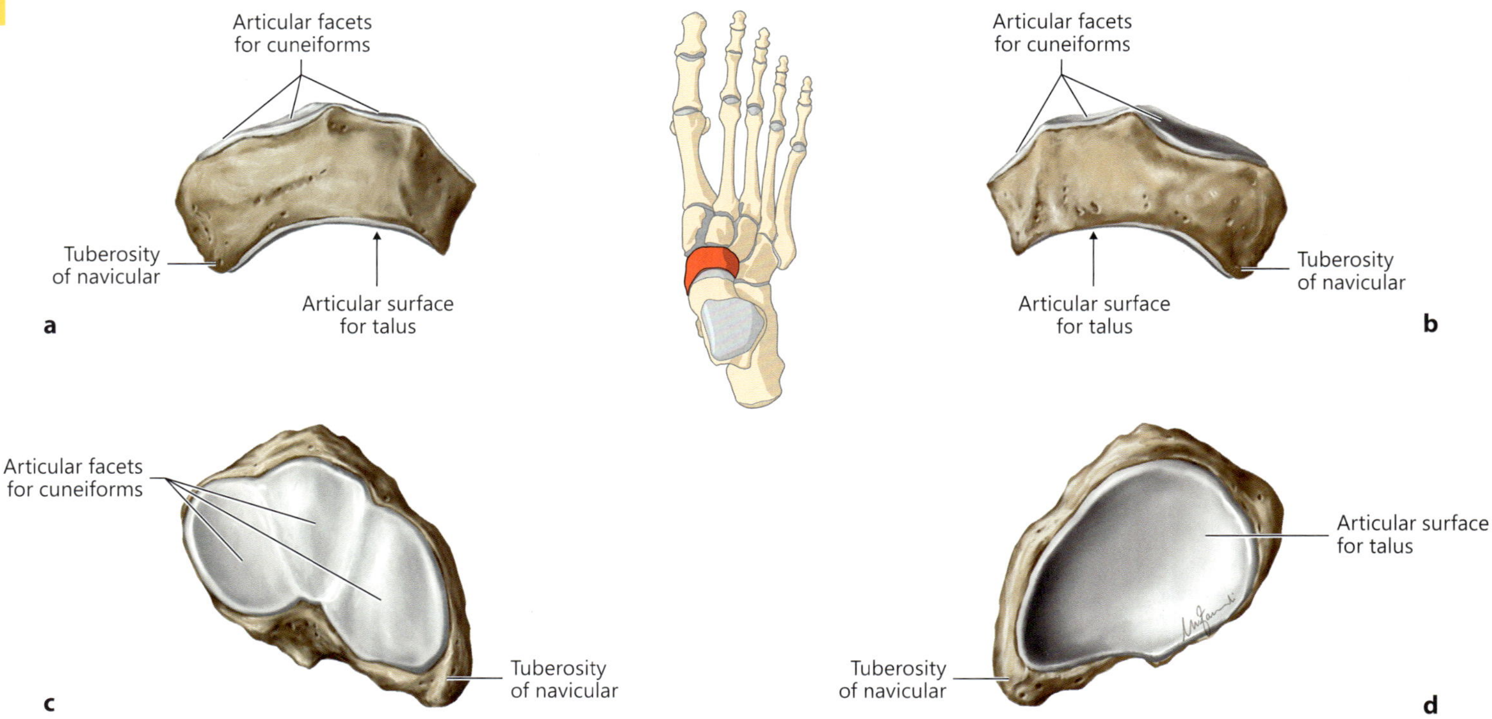

Navicular (or **scaphoid**), right foot. **a**, Dorsal surface; **b**, plantar surface; **c**, distal surface; **d**, proximal surface. Most of the surface of the bone is occupied by the articular surfaces for the talus and the three cuneiforms.

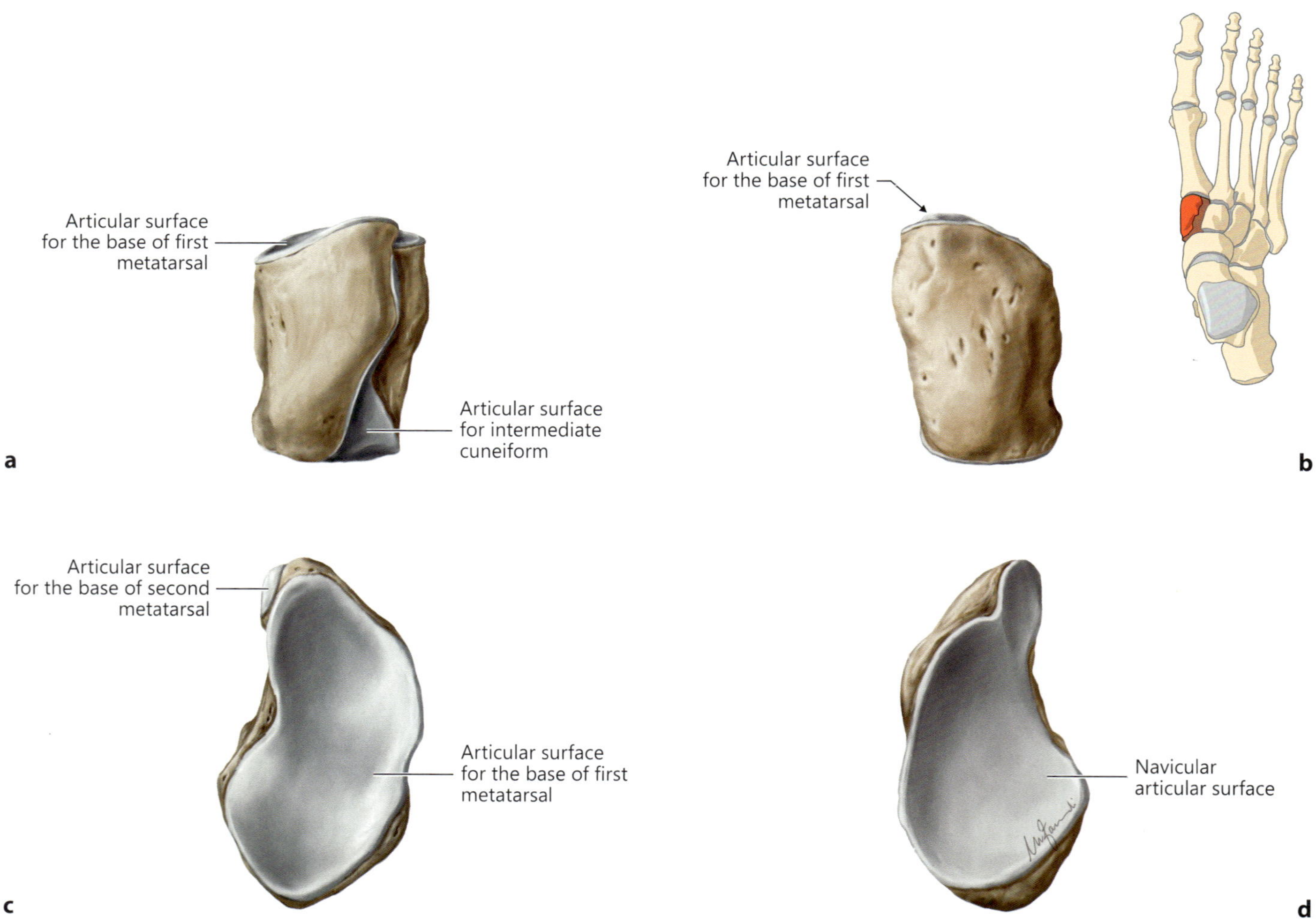

Medial cuneiform, right foot. **a**, Dorsal surface; **b**, plantar surface; **c**, distal surface; **d**, proximal surface.

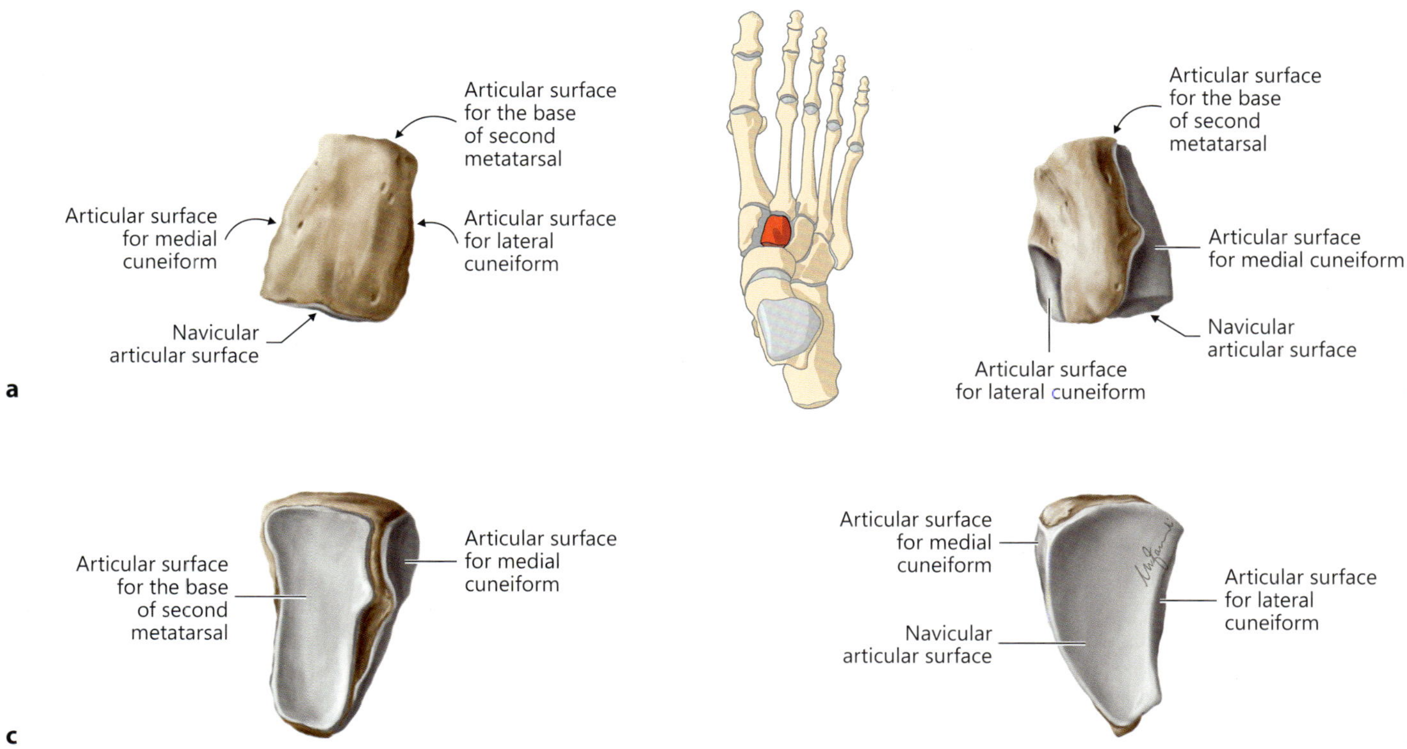

Intermediate cuneiform, right foot. **a**, Dorsal surface; **b**, plantar surface; **c**, distal surface; **d**, proximal surface.

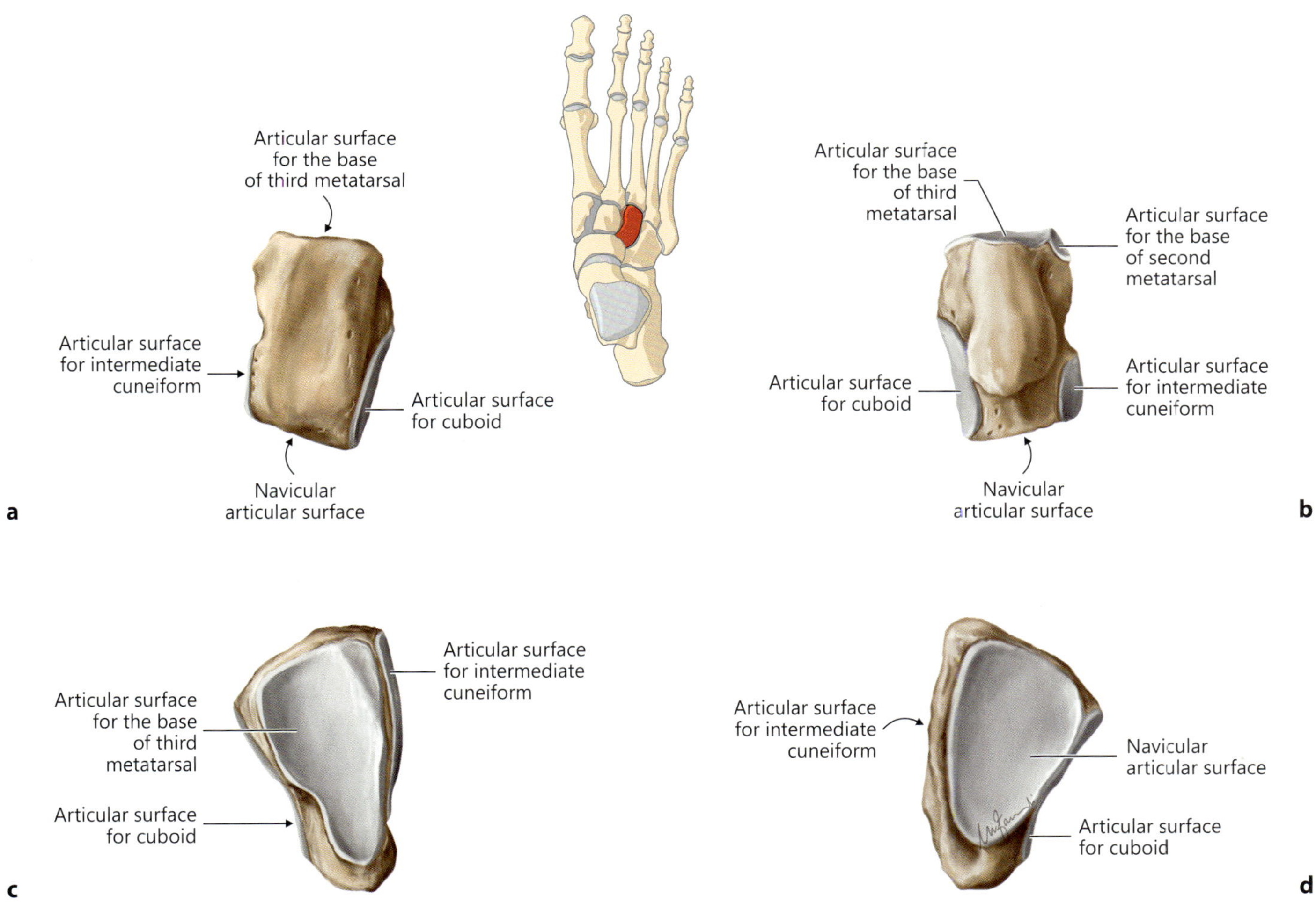

Lateral cuneiform, right foot. **a**, Dorsal surface; **b**, plantar surface; **c**, distal surface; **d**, proximal surface.

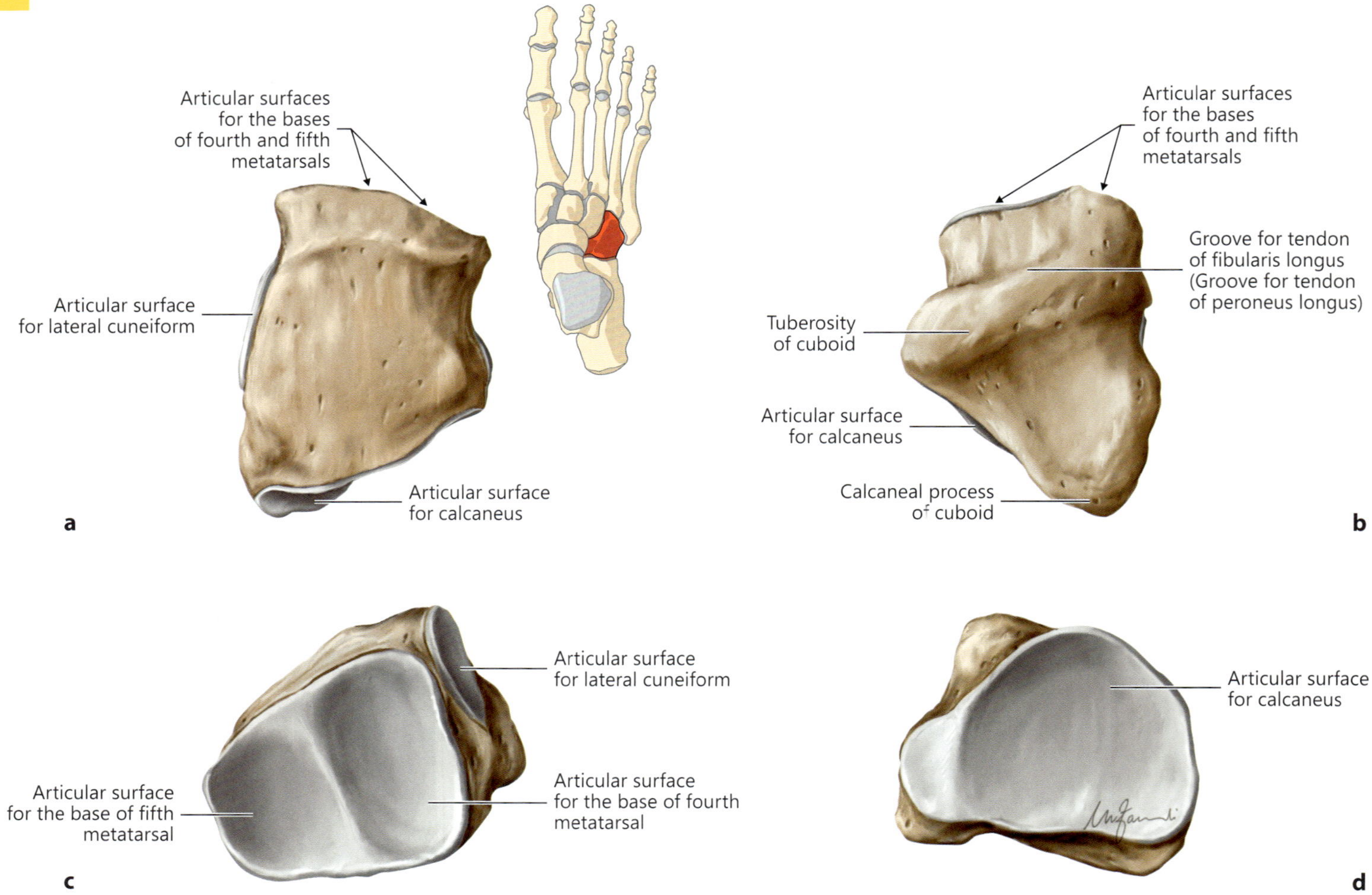

Cuboid, right foot. **a**, Dorsal surface; **b**, plantar surface; **c**, distal surface; **d**, proximal surface.

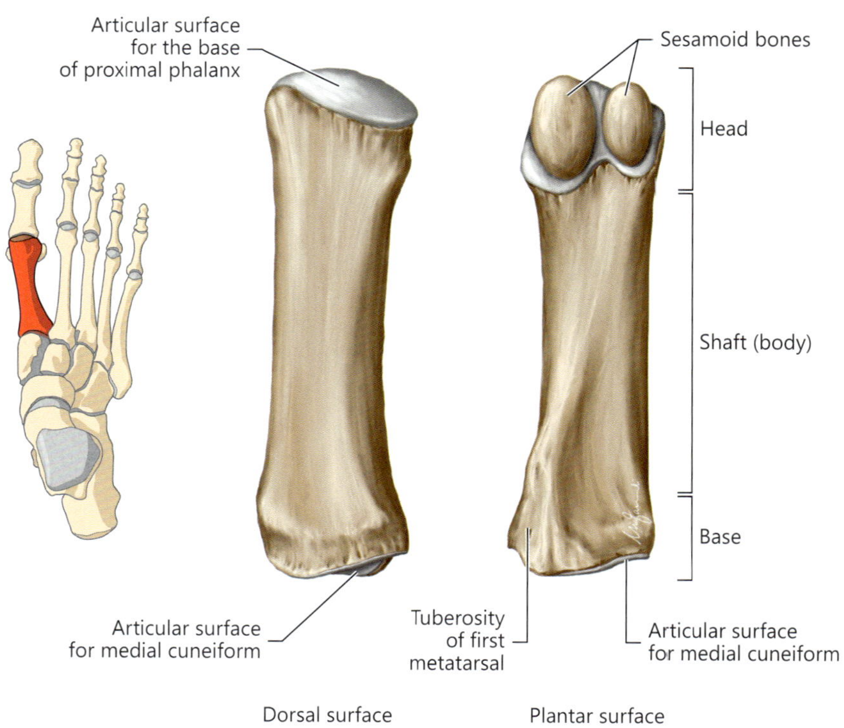

First metatarsal (I), right foot.

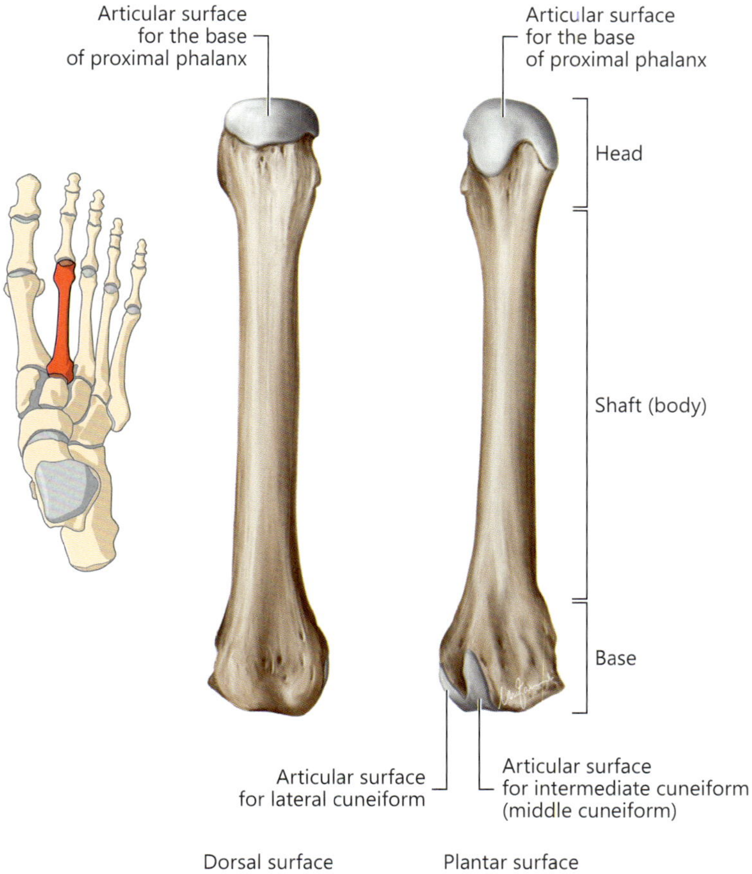

Second metatarsal (II), right foot.

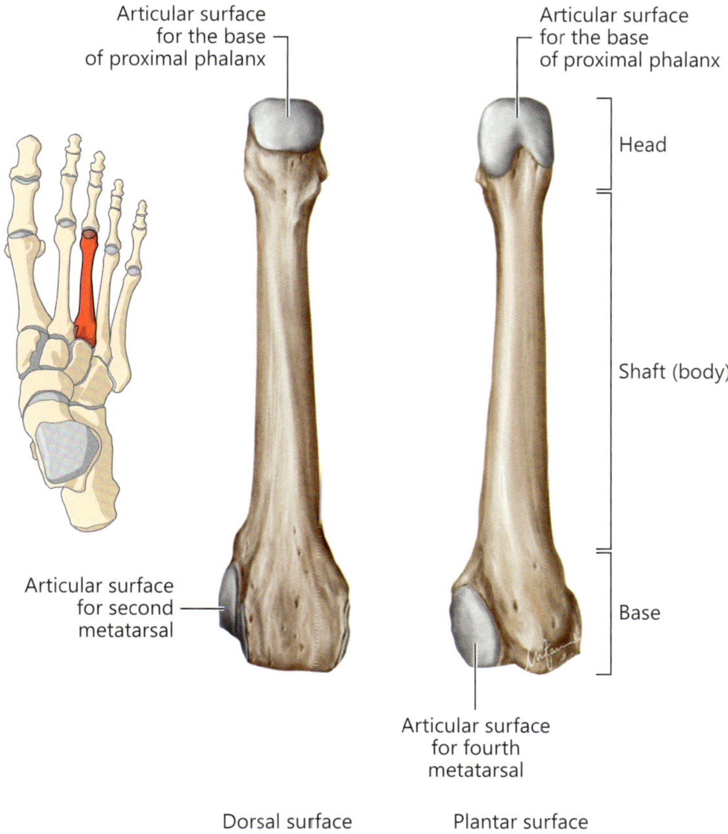

Third metatarsal (III), right foot.

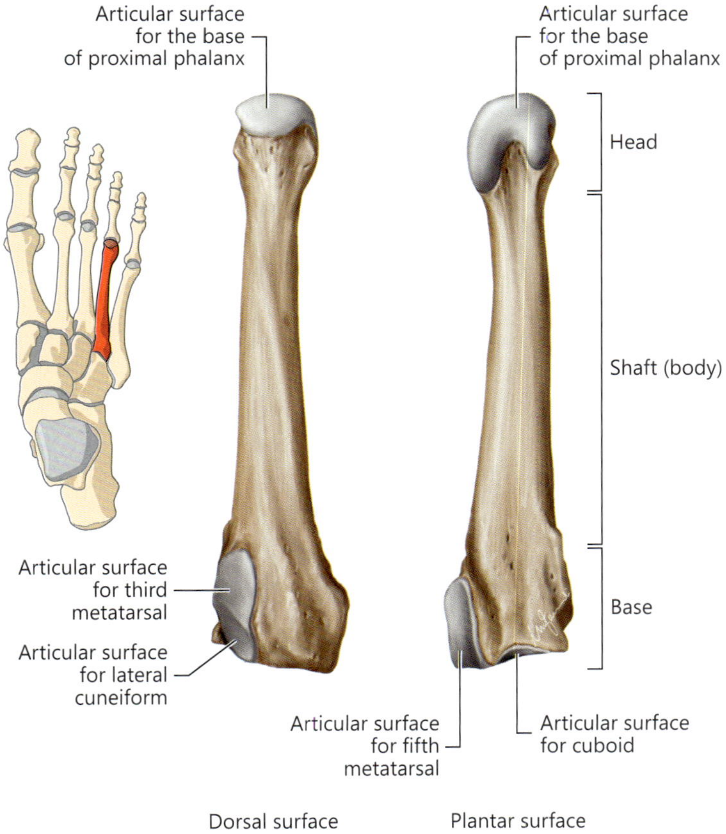

Fourth metatarsal (IV), right foot.

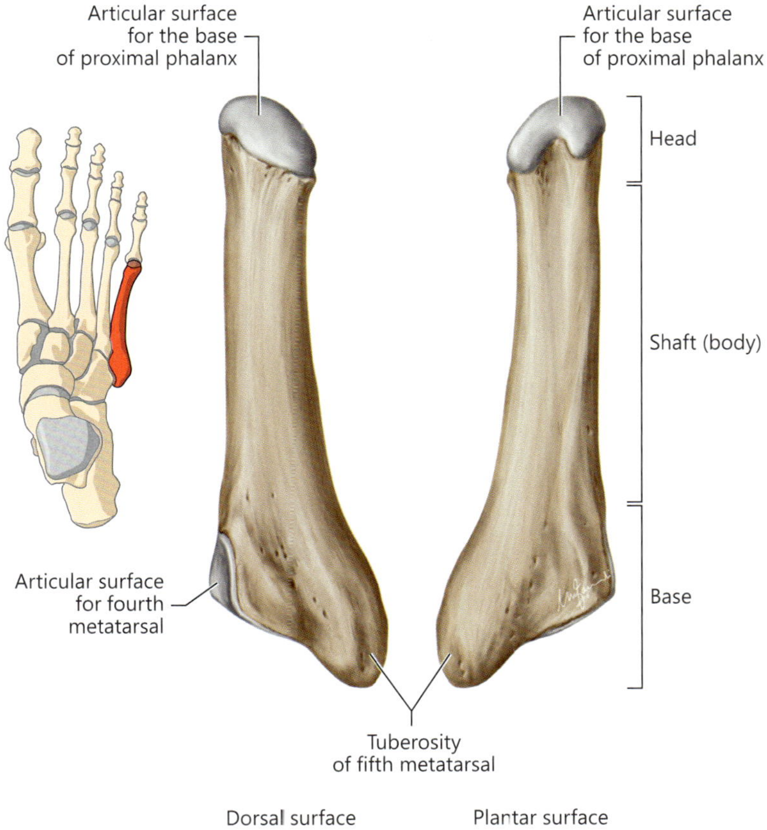

Fifth metatarsal (V), right foot.

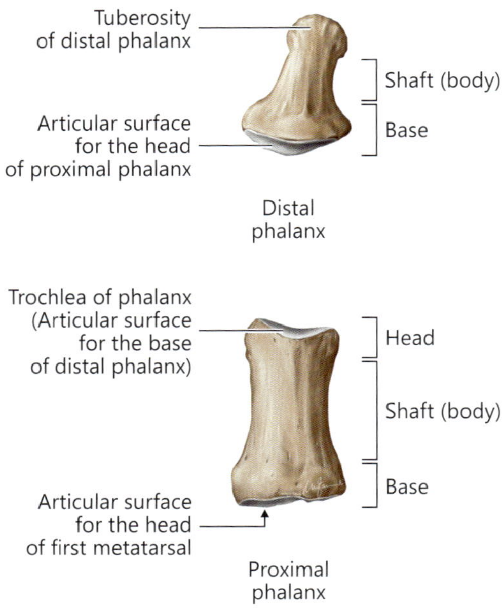

Proximal and **distal phalanges of the great toe** (dorsal surface), right foot.

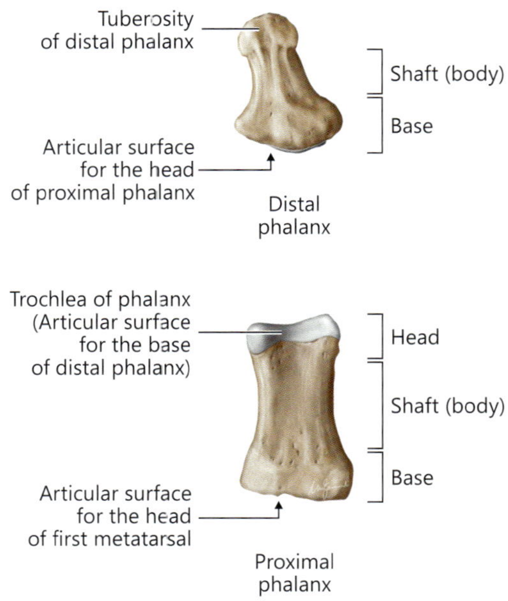

Proximal and **distal phalanges of the great toe** (plantar surface), right foot.

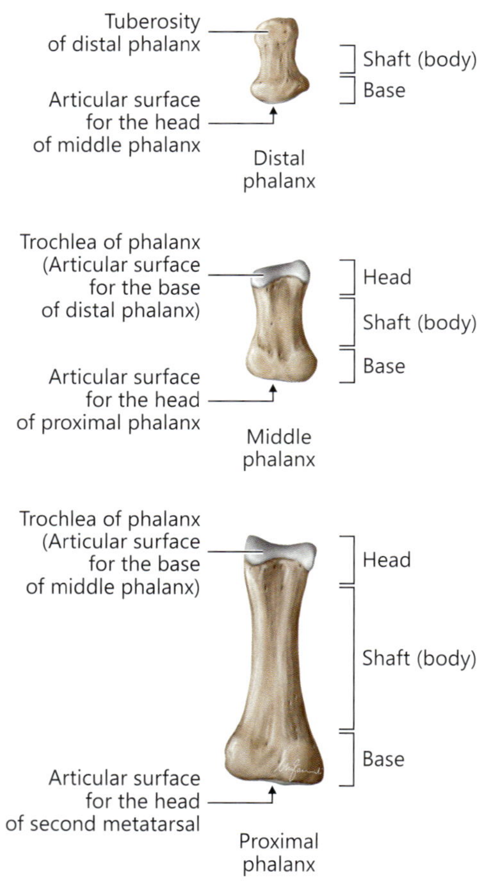

Proximal, **middle** and **distal phalanges of the second toe** (dorsal surface), right foot.

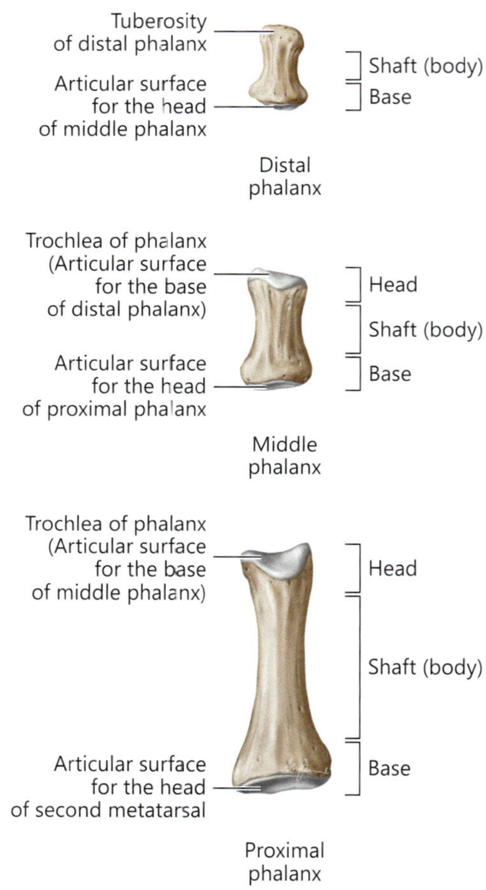

Proximal, **middle** and **distal phalanges of the second toe** (plantar surface), right foot.

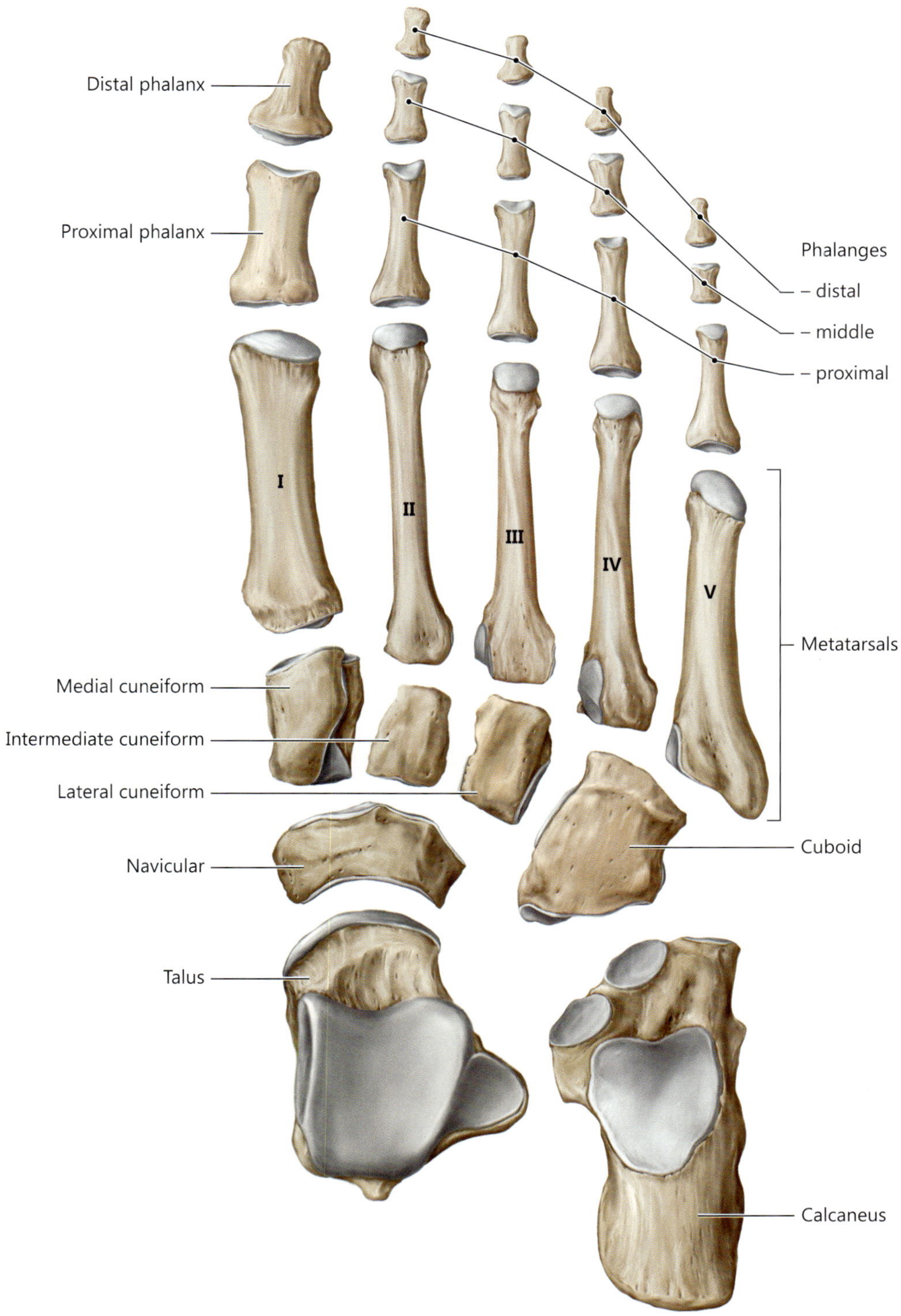

Right **foot**, dorsal view. The bones have been separated to reveal the articular surfaces. Here the calcaneus has been moved from its normal position towards the right to show the articular surfaces with the distal bones (navicular and cuboid).

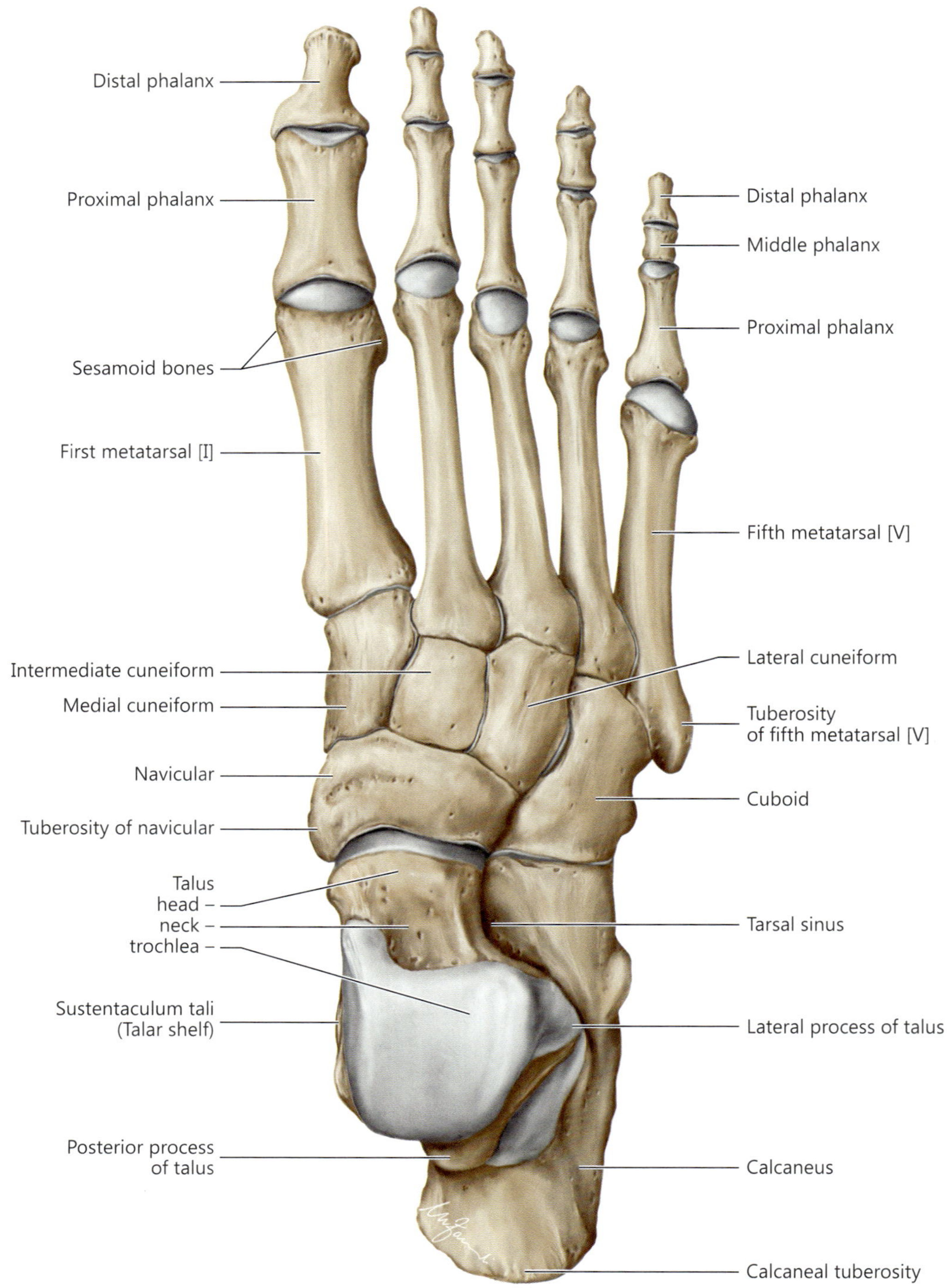

Right **foot**, skeleton seen from the dorsal surface. At the posterior and lateral part of the tarsus the tarsal sinus is delimited between the talus and the calcaneus.

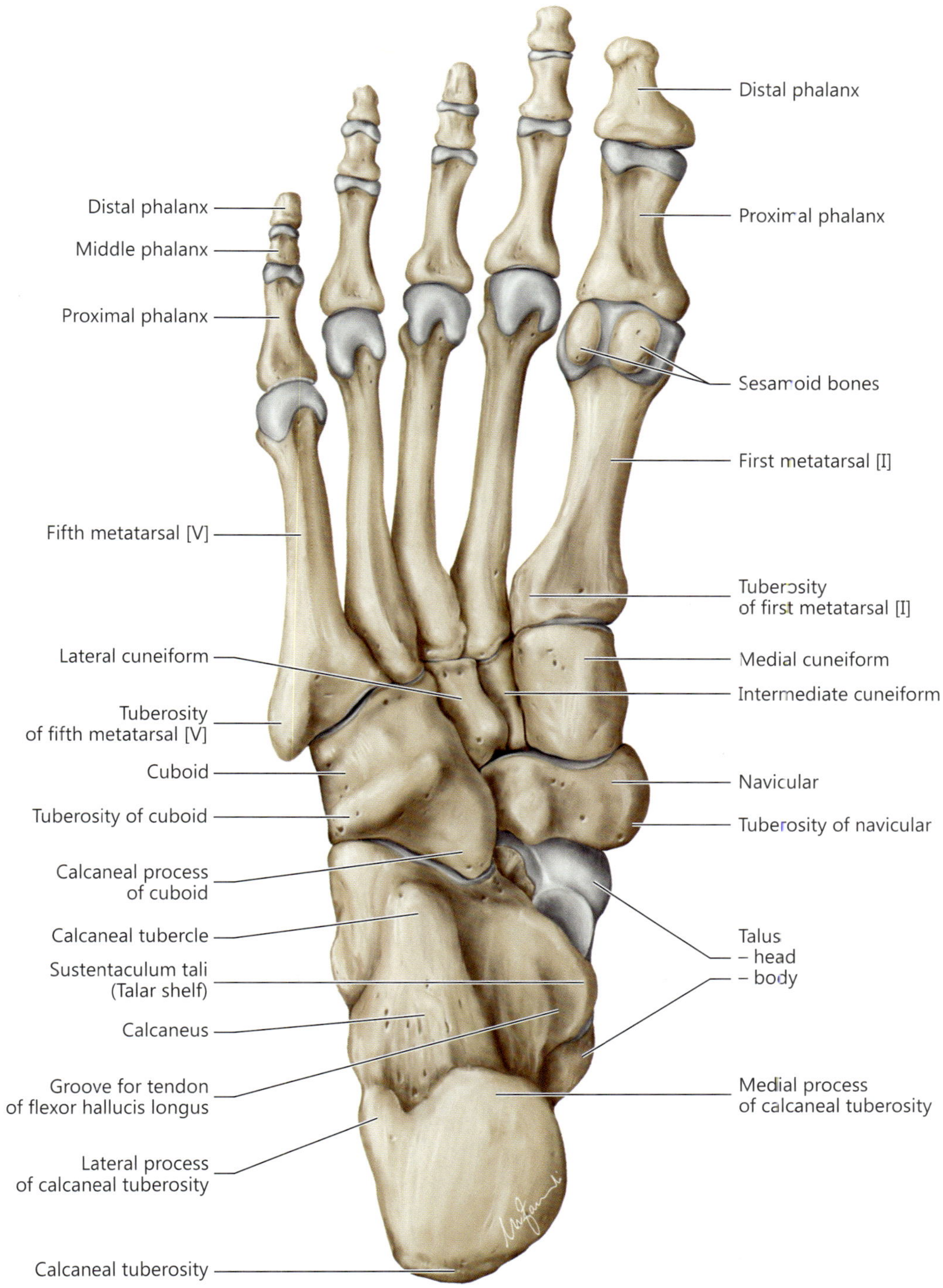

Right **foot**, skeleton seen from the plantar surface. In relation to the distal extremity of the 1st metatarsal and in proximity to the metatarsophalangeal joint there are two sesamoid bones.

Right **foot**, skeleton, lateral view.

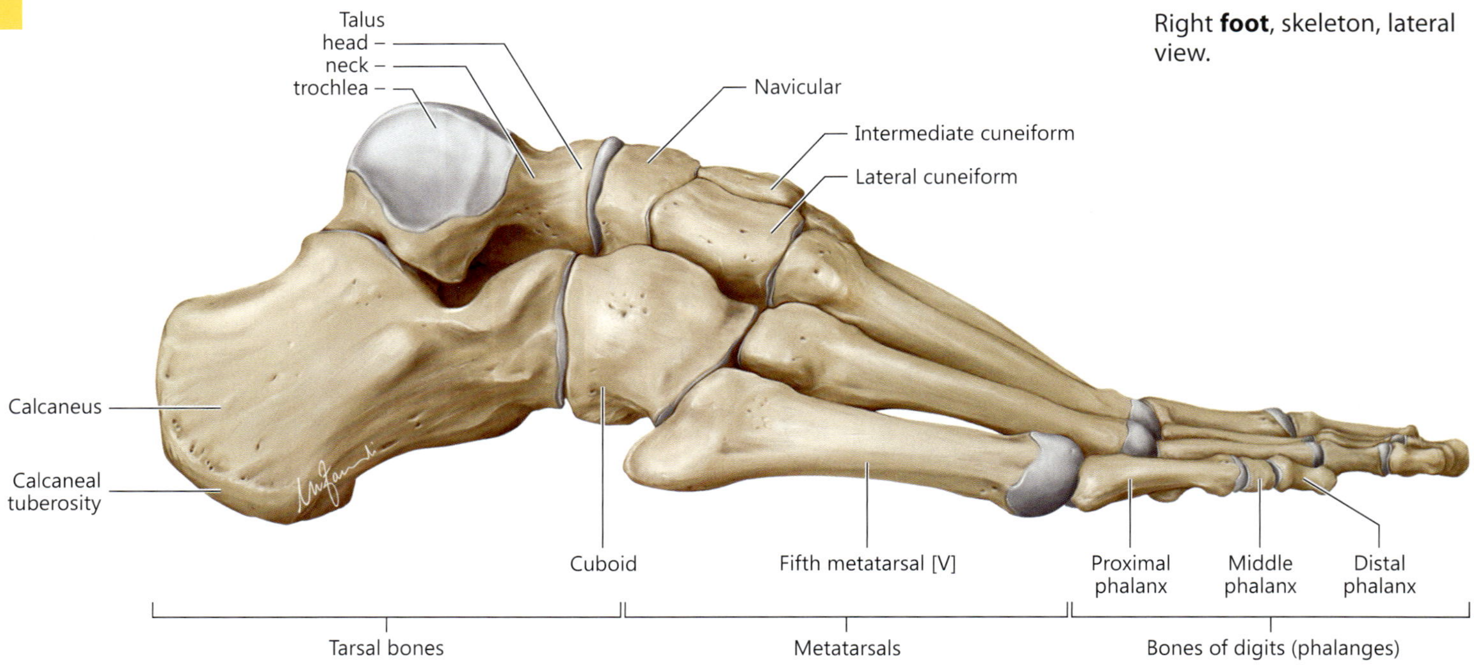

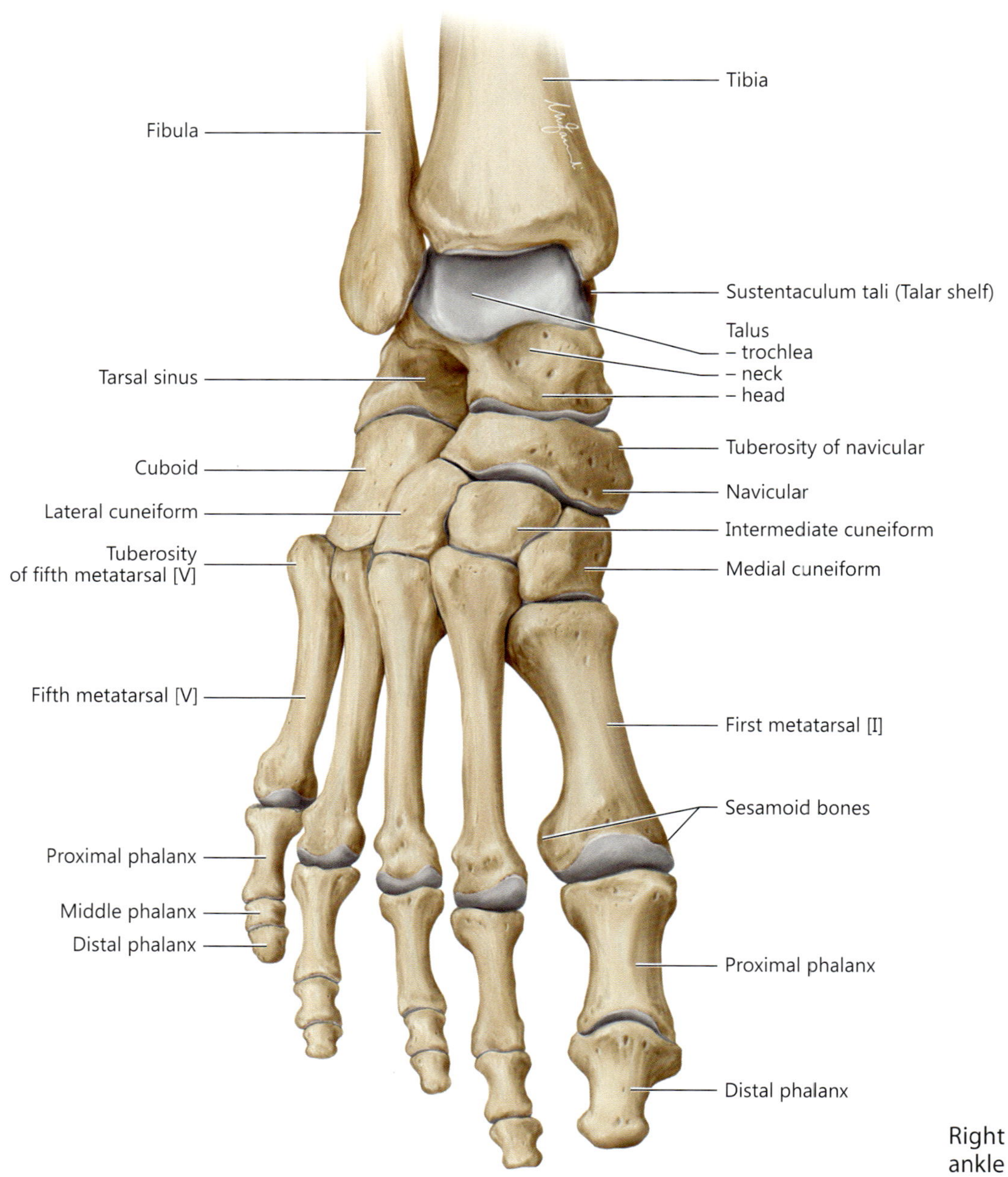

Right **foot**, bones and the ankle joint, anterior view.

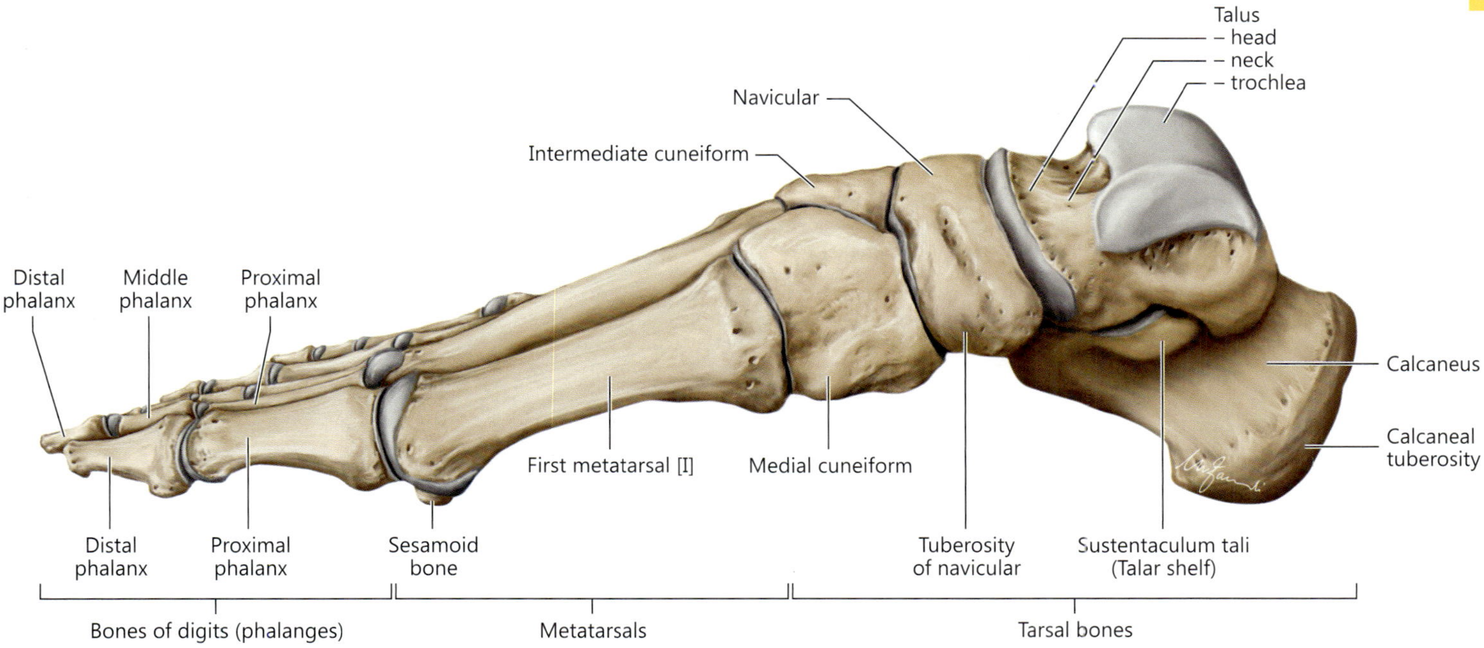

Right **foot**, skeleton, medial view. The deep medial part of longitudinal arch of foot, formed by the tarsal and metatarsal bones, has its keystone at the level of the navicular.

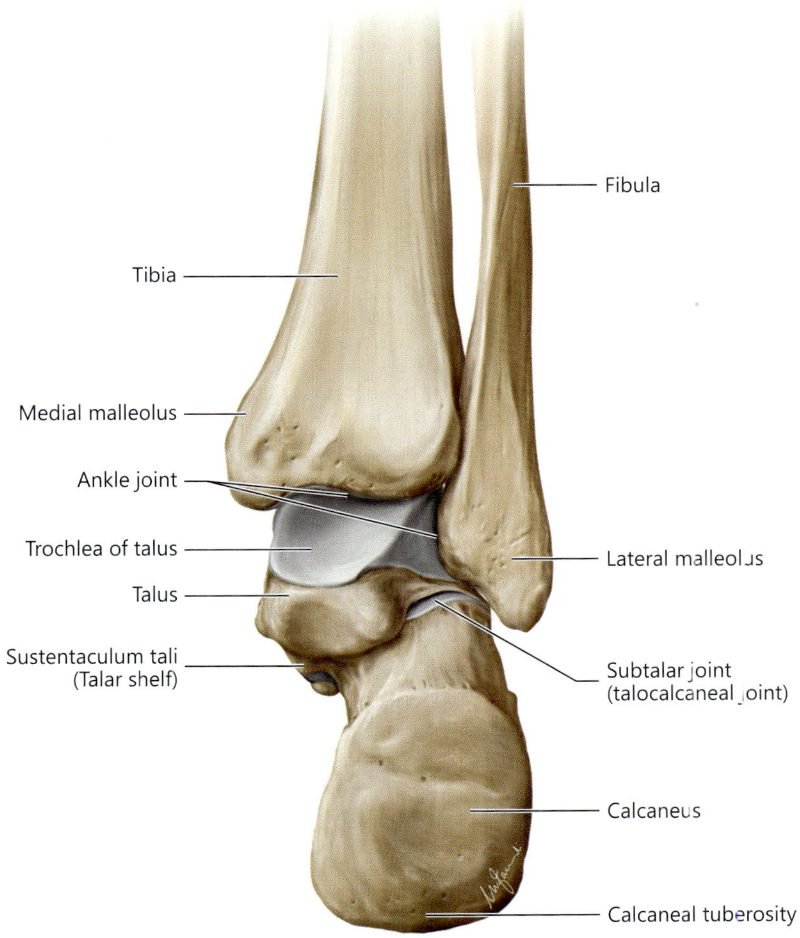

Right **foot**, bones and the ankle joint, posterior view.

Summary table of the bones and relative joints of the lower limb			
BONE	***LOCATION***	***DESCRIPTION***	***JOINTS***
Astragalus	☞ Talus		
Calcaneus	Foot	The largest tarsal bone, situated at the posterior part of the foot, has a cuboidal shape	Talus, cuboid
Cuboid	Foot	Pyramid-shaped bone at the lateral part of the foot, proximal with respect to the fourth and fifth metatarsals	Calcaneus, lateral cuneiform, fourth and fifth metatarsals, navicular
Fabella	Knee	Sesamoid bone (formed from fibrocartilage) on the lateral head of the gastrocnemius, behind the lateral condyle of the femur	Femur
Femur	Thigh	The longest and heaviest bone of the body, situated between the hip and the knee	Hip bone, patella, tibia
Fibula	Leg	Lateral bone of the leg	Tibia, talus
Hip	Hip	Large, wide, irregularly shaped bone, which constitutes the majority of the pelvis; it consists of three parts: ilium, ischium and pubis	Femur, sacrum, with the contralateral hip at the level of the pubic symphysis
Intermediate cuneiform second cuneiform	Foot	Wedge-shaped bone; the smallest of the three cuneiforms, located between the medial and lateral cuneiforms	Navicular, intermediate cuneiform, cuboid; second, third and fourth metatarsals
Lateral cuneiform external cuneiform third cuneiform	Foot	Cuneiform bone of intermediate size with respect to the other two, located at the center of the frontal row of the tarsal bones	Navicular, intermediate cuneiform, cuboid; second, third and fourth metatarsals
Medial cuneiform internal cuneiform first cuneiform	Foot	The largest of the three cuneiforms, situated at the medial side of the foot between the navicular and the base of the first metatarsal	Navicular, intermediate cuneiform, first and second metatarsals
Metatarsals [I-V]	Foot	Five thin bones of the foot, each consisting of a body and two extremities (head and base); they are numbered from the side of the great toe	Distal bones of the tarsus; the bases are articulated to one another, the heads with the corresponding proximal phalanges
Navicular scaphoid	Foot	Situated at the medial part of the tarsus, between the talus and the cuneiforms	Talus, three cuneiforms, occasionally with the cuboid
Patella	Knee	Flat, rounded, triangular bone (sesamoid), located anteriorly to the knee joint	Femur
Phalanges	Foot	Small, long bones in number of two in the great toe and three in the other toes	Proximal phalanges with the corresponding metatarsals and with the middle phalanges; middle phalanges with the distal and proximal phalanges; distal phalanges with the middle phalanges
Sesamoid bones	Limb	Small, round bones contained within the thick parts of certain tendons; some, constant, include the bones of the tendons of the quadriceps femoris, the flexor hallucis brevis, the fibularis longus, the tibialis anterior, and the psoas major; the patella is the sesamoid bone with the greatest volume	
Talus	Ankle	Second tarsal bone in terms of volume; supports the tibia and rests on the calcaneus	Tibia, fibula, calcaneus, navicular
Tibia	Leg	Located on the medial side of the leg between the joints of the ankle and the knee; second longest bone in the body	Superiorly with the femur and the fibula; inferiorly with the fibula and the talus

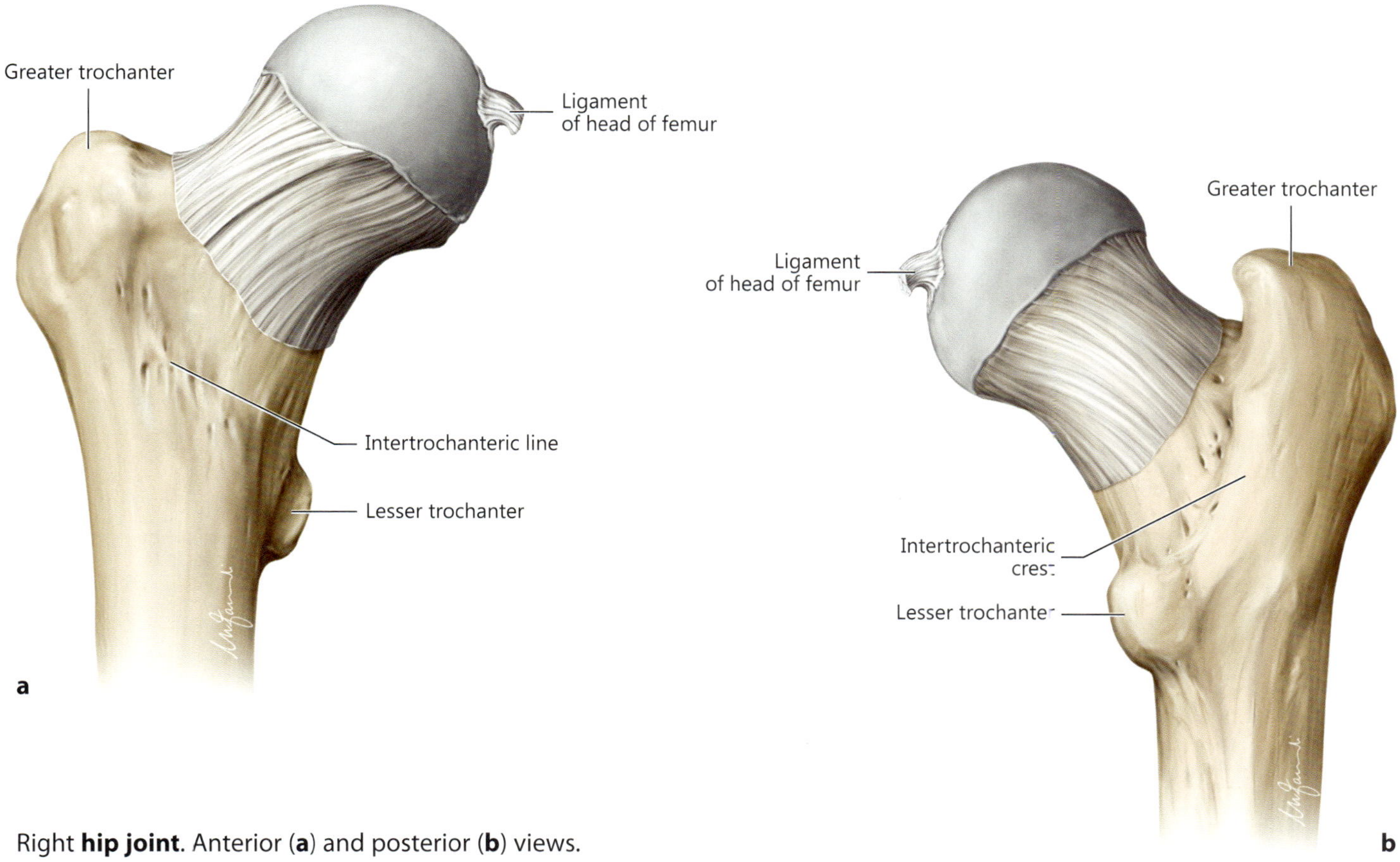

Right **hip joint**. Anterior (**a**) and posterior (**b**) views.

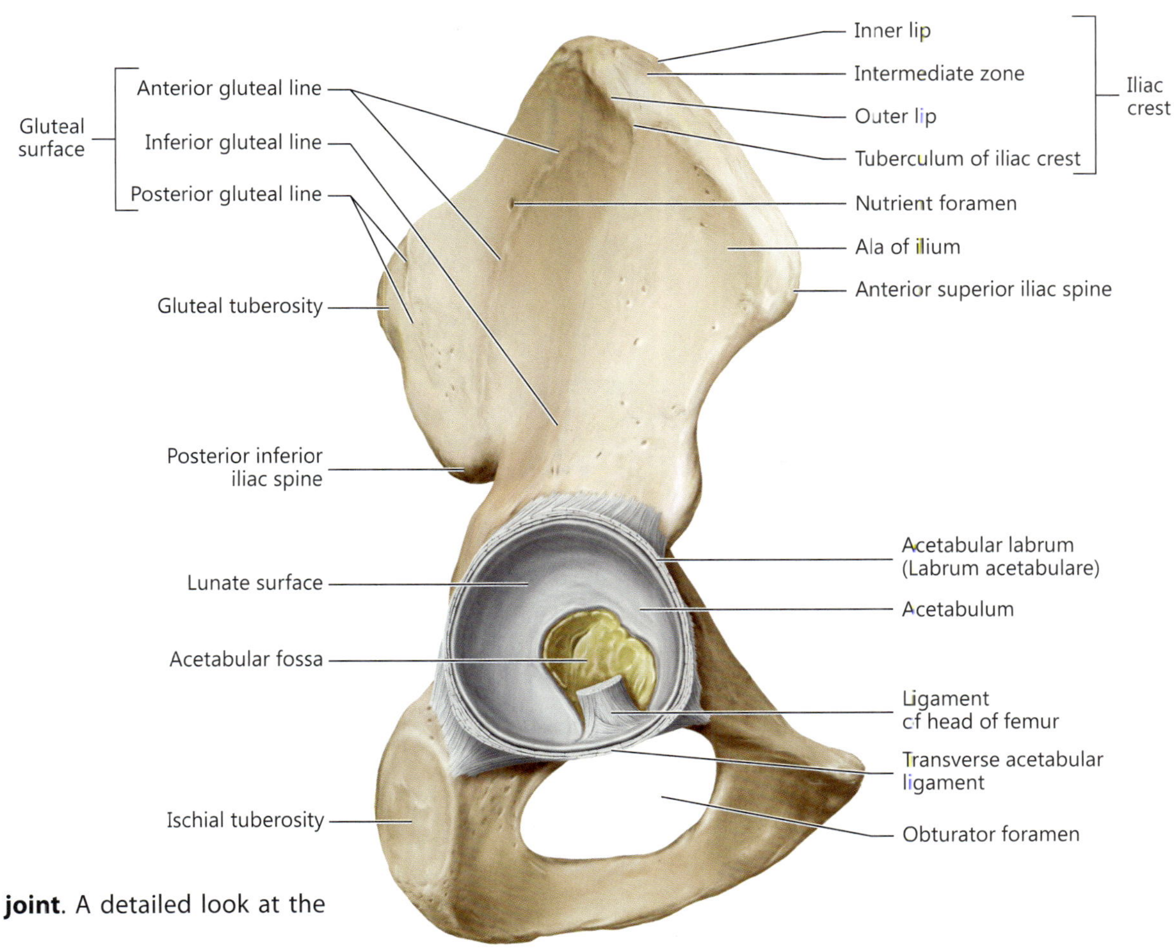

Right **hip joint**. A detailed look at the hip bone.

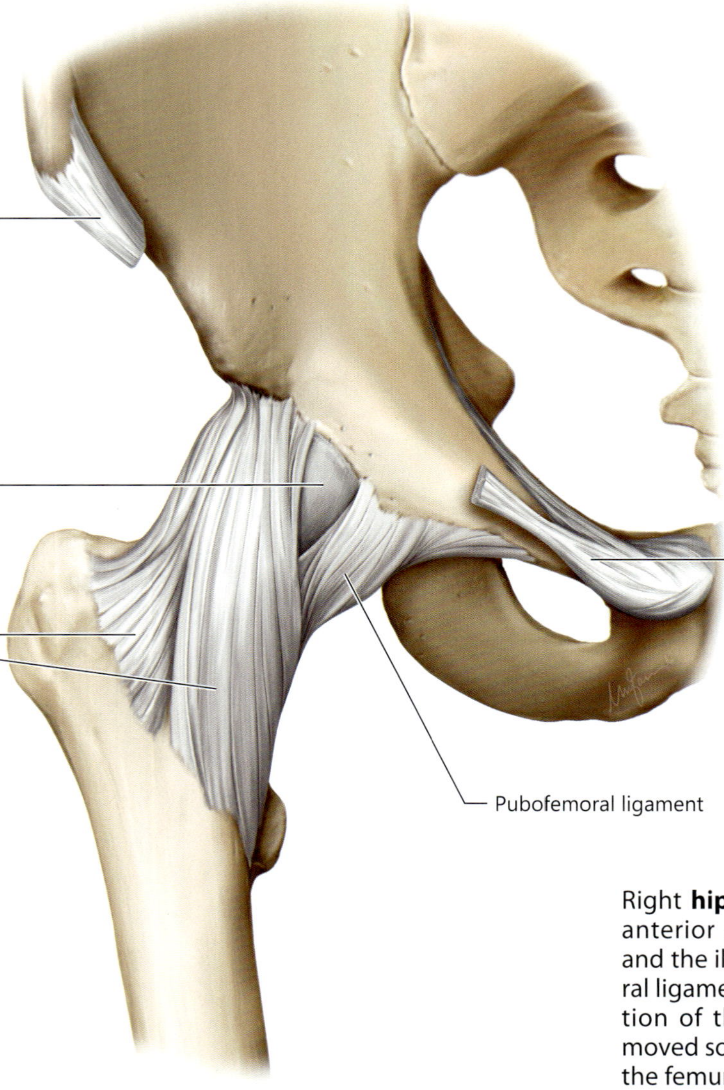

Right **hip joint**, anterior view. The anterior means of reinforcement and the iliofemoral and pubofemoral ligaments are shown here; a section of the capsule has been removed so as to uncover the head of the femur.

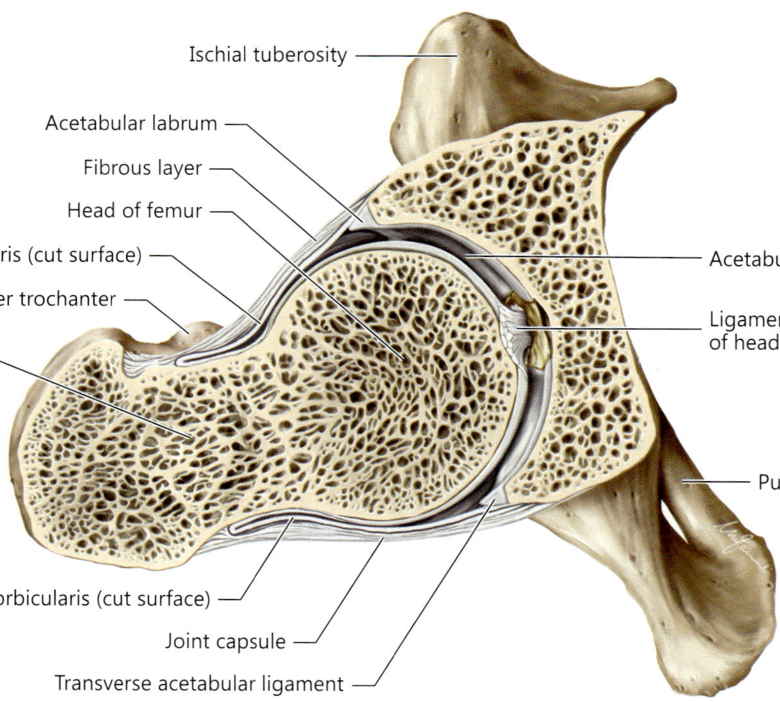

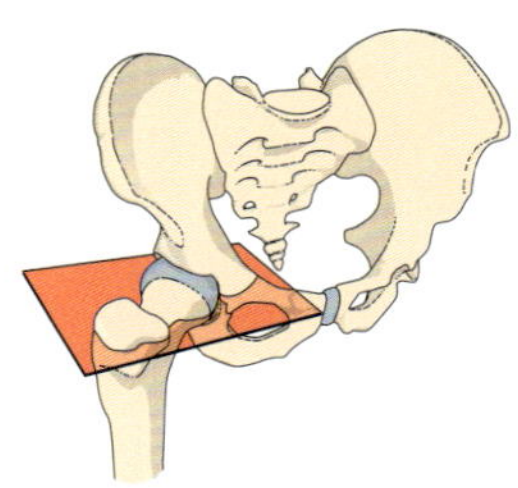

Right **hip joint**. The limits of the cavity are shown here in a section carried out along a plane parallel to the axis of the anatomical neck of the femur. In the cavity you can see the ligament of the head of the femur.

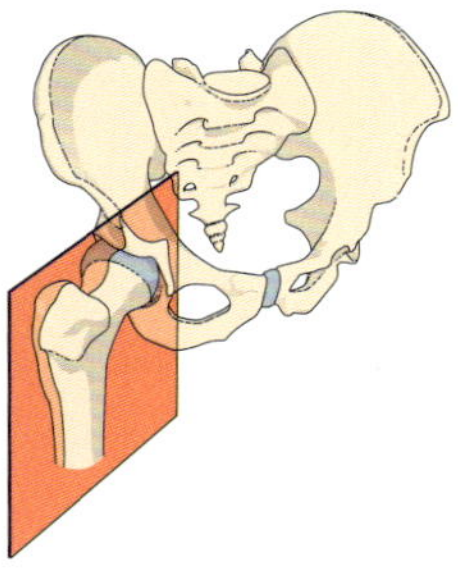

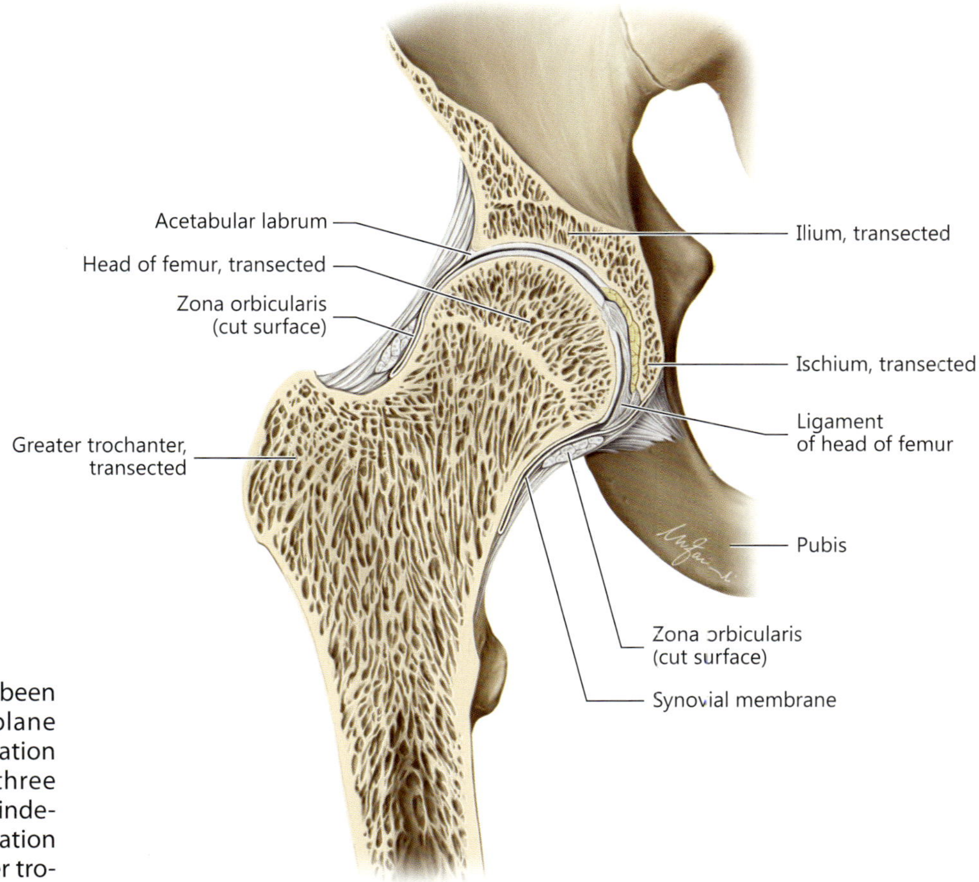

Right **hip joint**. The joint has been transected along a frontal plane showing the general conformation of the articular cavity. The three parts of the hip bone are still independent, as well as the ossification centers of the head and greater trochanter of the femur.

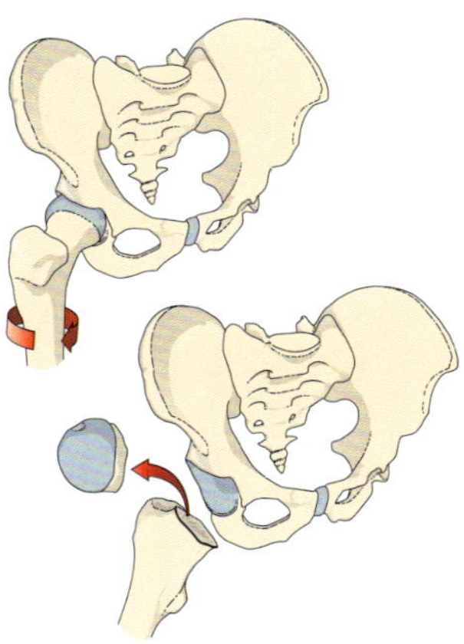

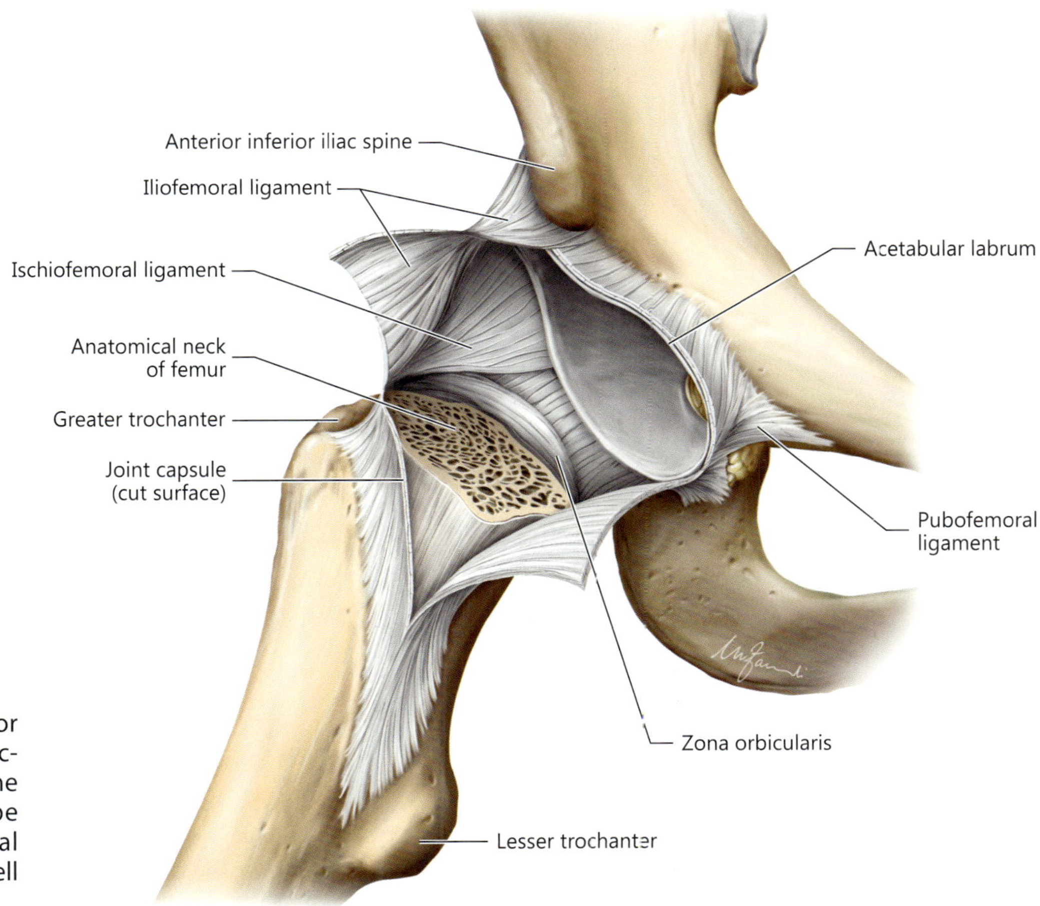

Right **hip joint**. After the anterior opening of the capsule and resection of the head of the femur, the means of reinforcement can be seen, along with the iliofemoral and pubofemoral ligaments, as well as the zona orbicularis.

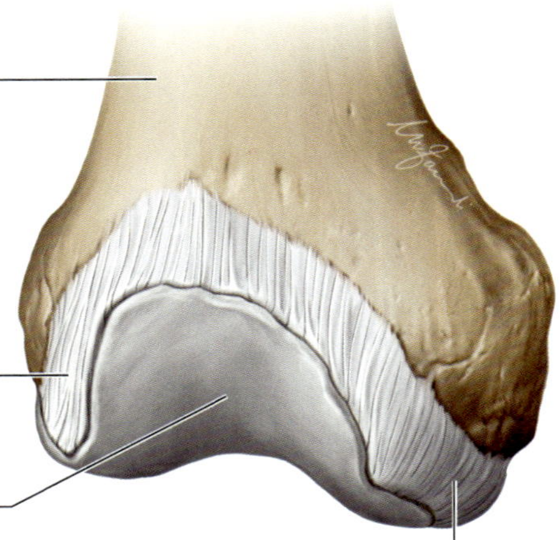

Right **knee** joint, seen from the front. The patellar surface of the femur is shown, which continues to the articular surface of the condyles.

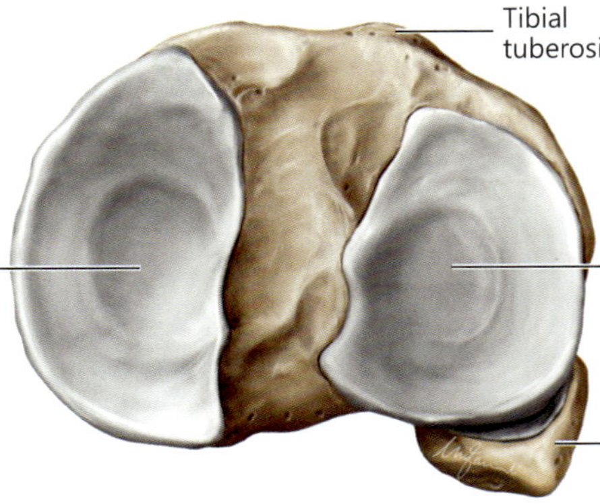

Right **knee** joint. Articular surface of the tibial condyles, seen from above, after removal of the menisci; the glenoid cavities appear as depressions on the surface of the tibial condyles.

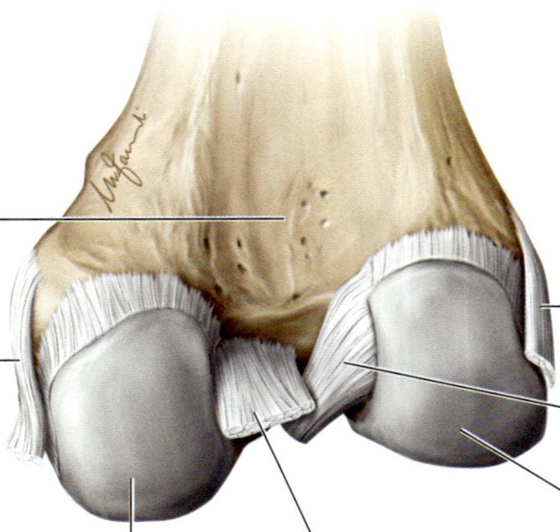

Right **knee** joint. Articular surfaces of the femoral condyles, seen from the posterior surface. The proximal insertions of the cruciate ligaments are also represented, as well as the collateral ligaments.

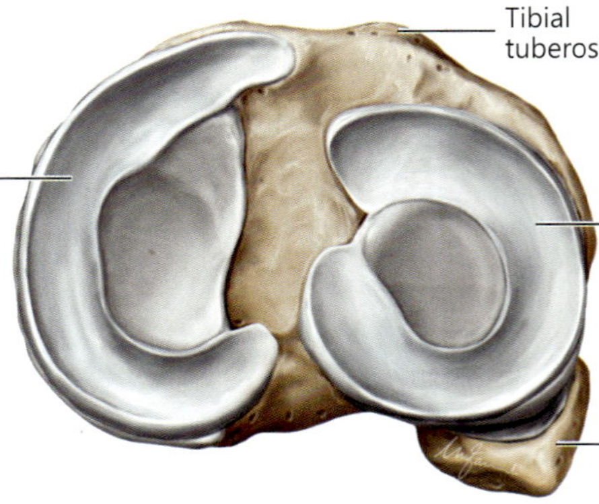

Right **knee** joint. Articular surface of the tibial condyles, seen from above; this surface is modified in its configuration due to the presence of the menisci, medial and lateral, which attach at the intercondylar region of the tibia.

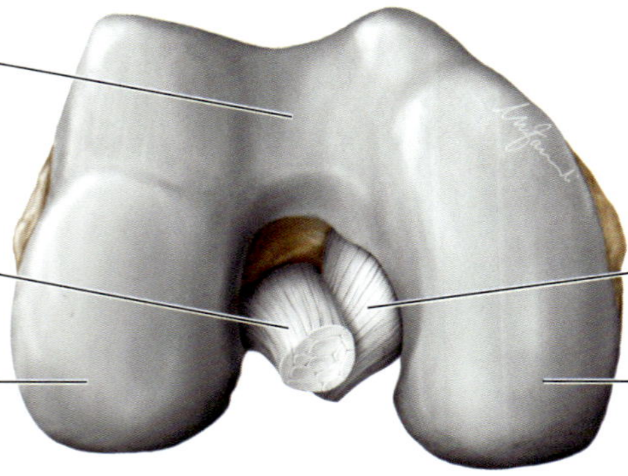

Right **knee** joint. The articular surface of the femur, the condyles, and the patellar surface are seen here from below. The insertion of the cruciate ligaments at the intercondylar fossa is also represented.

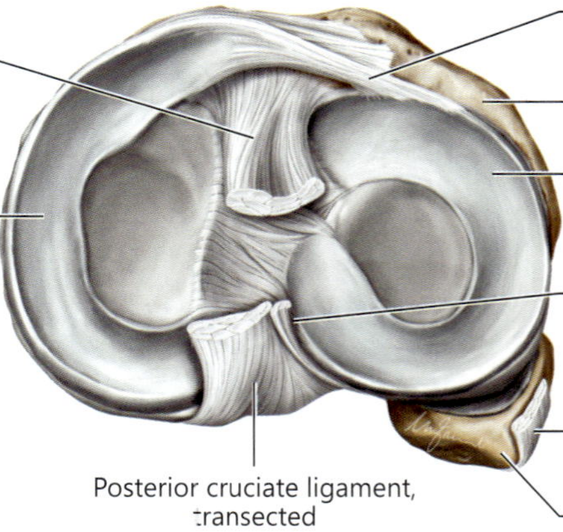

Right **knee** joint. The articular surface of the femur, the condyles, and the patellar surface are seen here from above. The insertion of the cruciate ligaments at the intercondylar fossa is visible here.

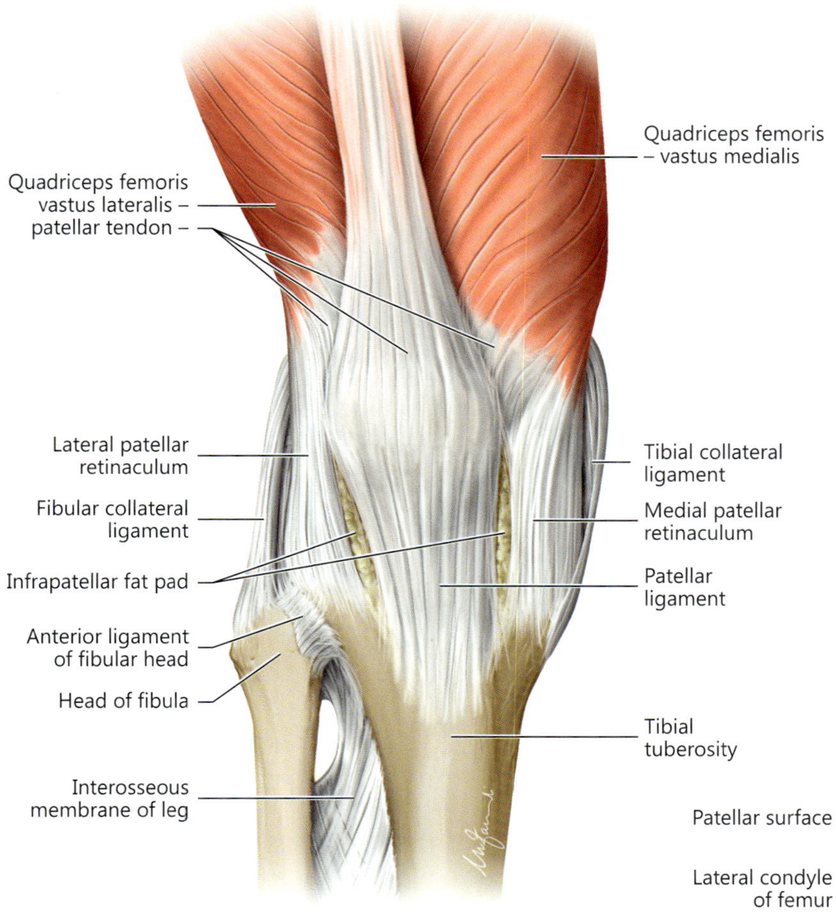

➤ Right **knee**. The means of union of the various joints are shown here from the front.

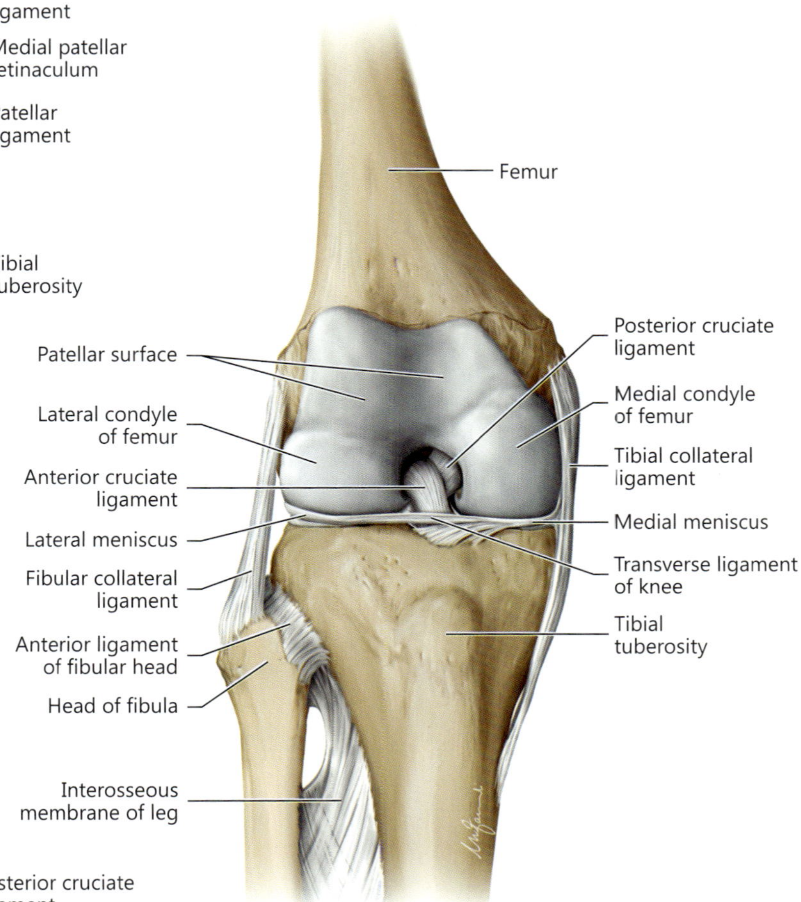

Right **knee**. The joint is seen during slight flexion. With the joint capsule removed, you can see the numerous means of union arranged between the femur, the tibia, and the fibula.

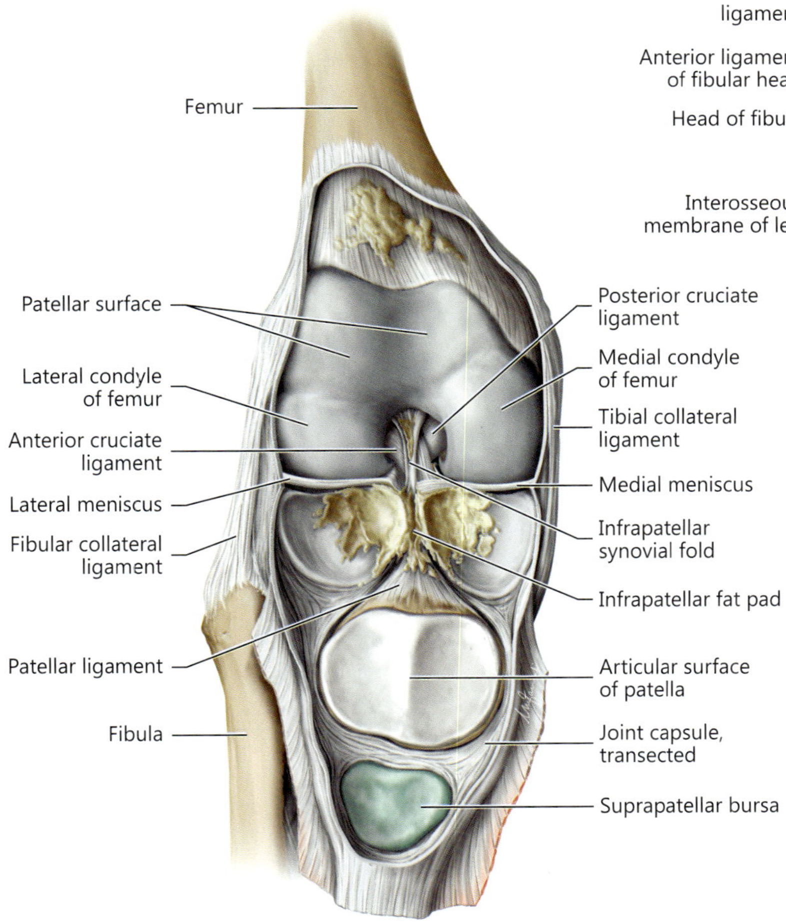

➤ Right **knee**. The anterior wall of the joint capsule has been opened by dissecting the tendon of the quadriceps muscle above the patella. The tendon and patella were then folded down to show the structures of the synovial capsule.

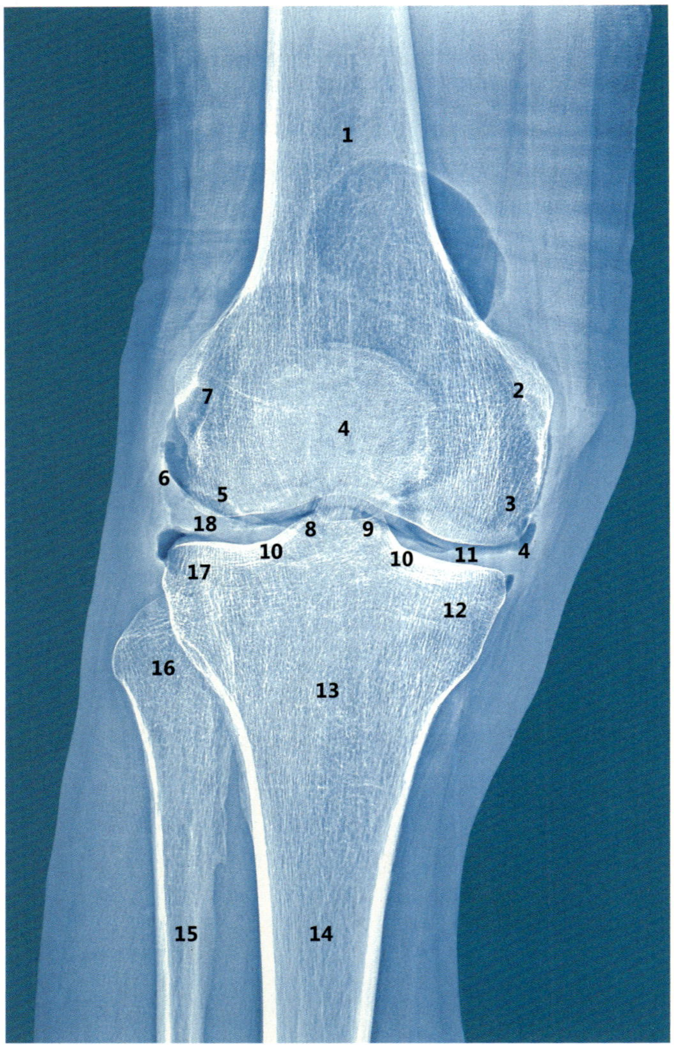

Right **knee**. Xeroradiographic image of the joint in an anteroposterior projection. Note the limits of the articular cavity which are evident with introduction of air into the joint. Between the two articular heads you can see the menisci with their connections to the joint capsule.

1, Shaft of femur
2, Medial epicondyle
3, Medial condyle of femur
4, Patella
5, Lateral condyle of femur
6, Joint capsule
7, Lateral epicondyle
8, Lateral intercondylar tubercle
9, Medial intercondylar tubercle
10, Superior articular surface of tibia
11, Medial meniscus
12, Medial condyle of tibia
13, Tibial tuberosity
14, Shaft of tibia
15, Shaft of fibula
16, Head of fibula
17, Lateral condyle of tibia
18, Lateral meniscus

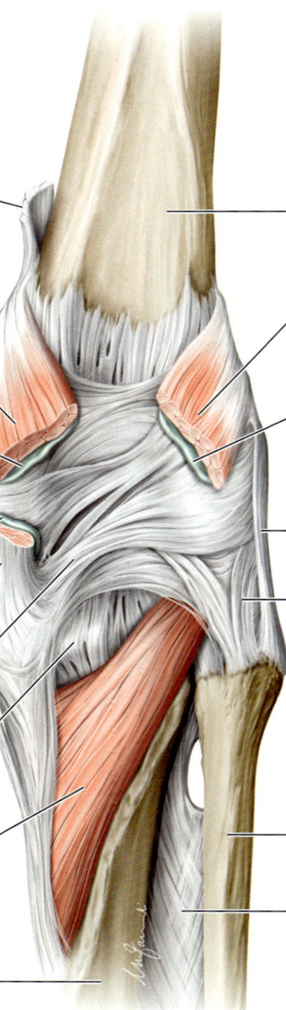

Right **knee**. Here the posterior surface of the joint capsule has been presented, on which you can see a few ligaments and reinforcing structures (oblique and arcuate popliteal ligaments). The abundance of muscle insertions on the joint capsule or nearby help to strengthen (posteriorly and laterally) the fibrous sleeve that holds the articular heads together.

Right **knee** seen posteriorly. By removing the joint capsule you can see the posterior contour of the menisci and some of the articular ligaments.

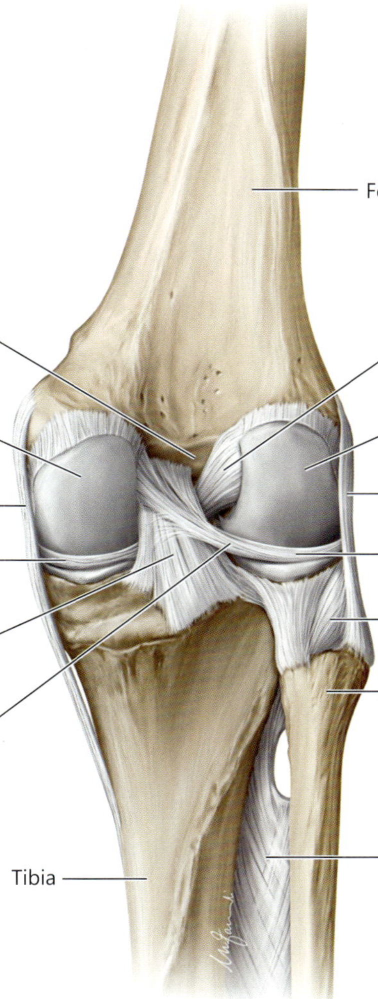

Three-dimensional image of a living human body, obtained by CT scan using a 3D volume rendering technique, showing the right **knee**, posterior view.

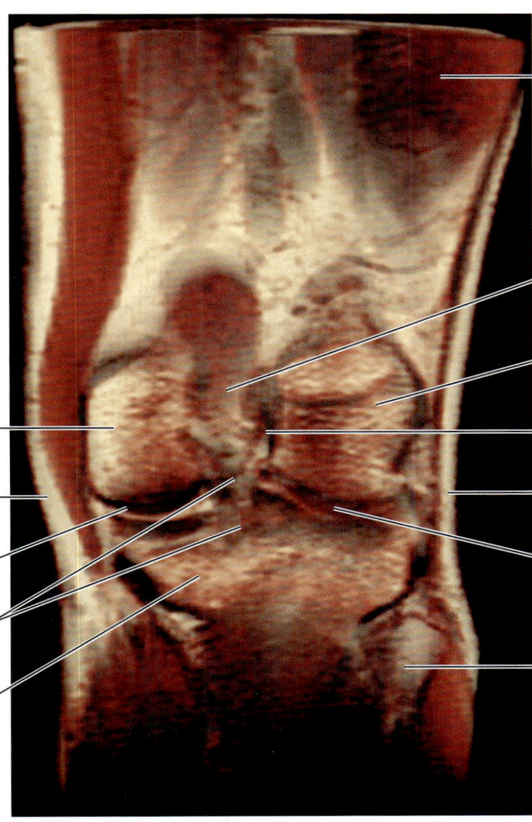

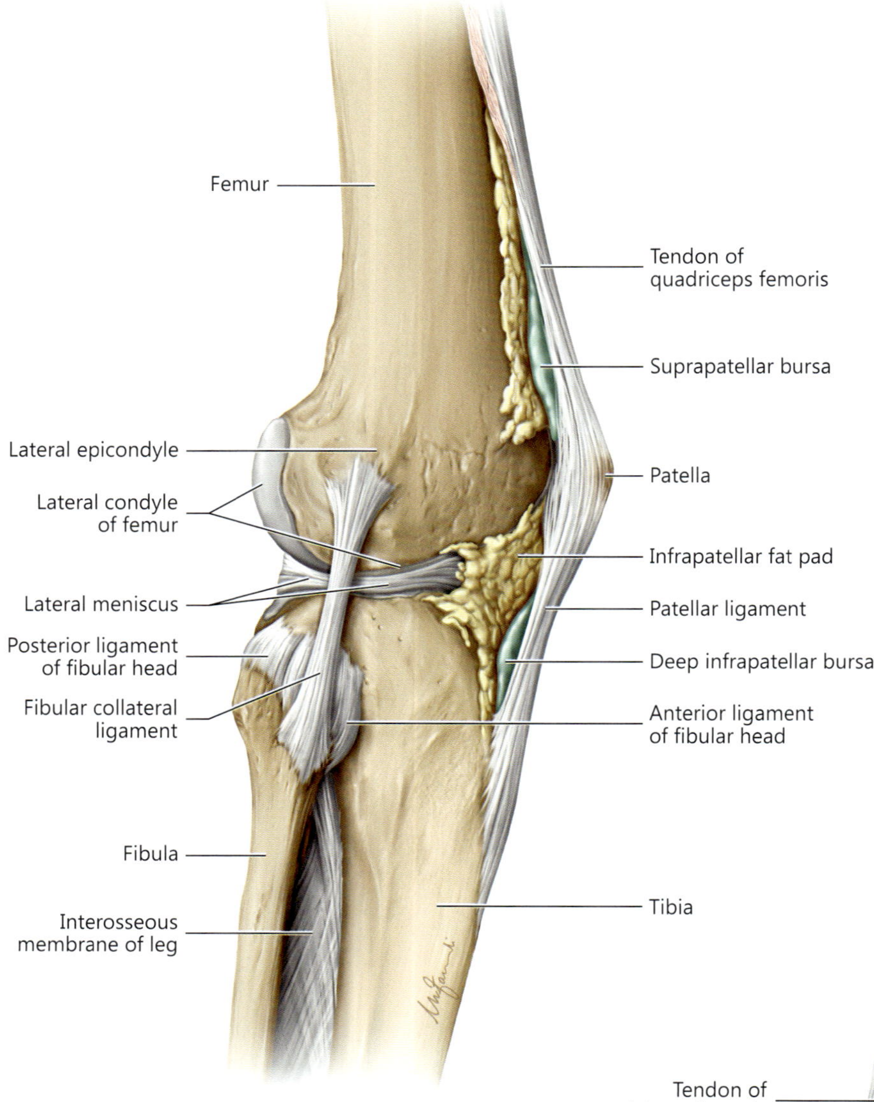

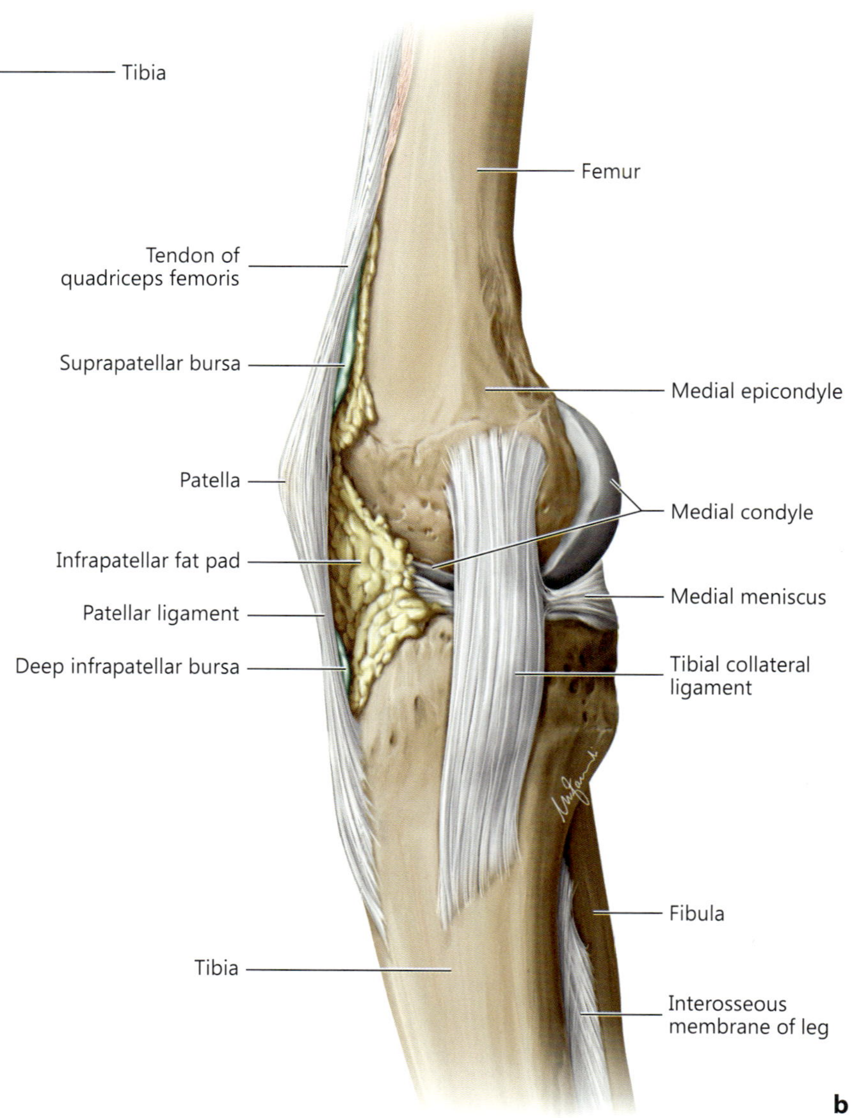

Right **knee joint**: ligaments and menisci. **a**, Lateral view; **b**, medial view.

Right **knee** seen laterally. The section has been done along a paramedian sagittal plane, allowing you to see the limits of the articular cavity with some of the anterior and posterior recesses. You can also see the suprapatellar and infrapatellar bursae, a section of the lateral meniscus, and the alar folds of the synovial capsule. Metaphyseal lines are still present in the femur and tibia at the union between the diaphysis and epiphysis.

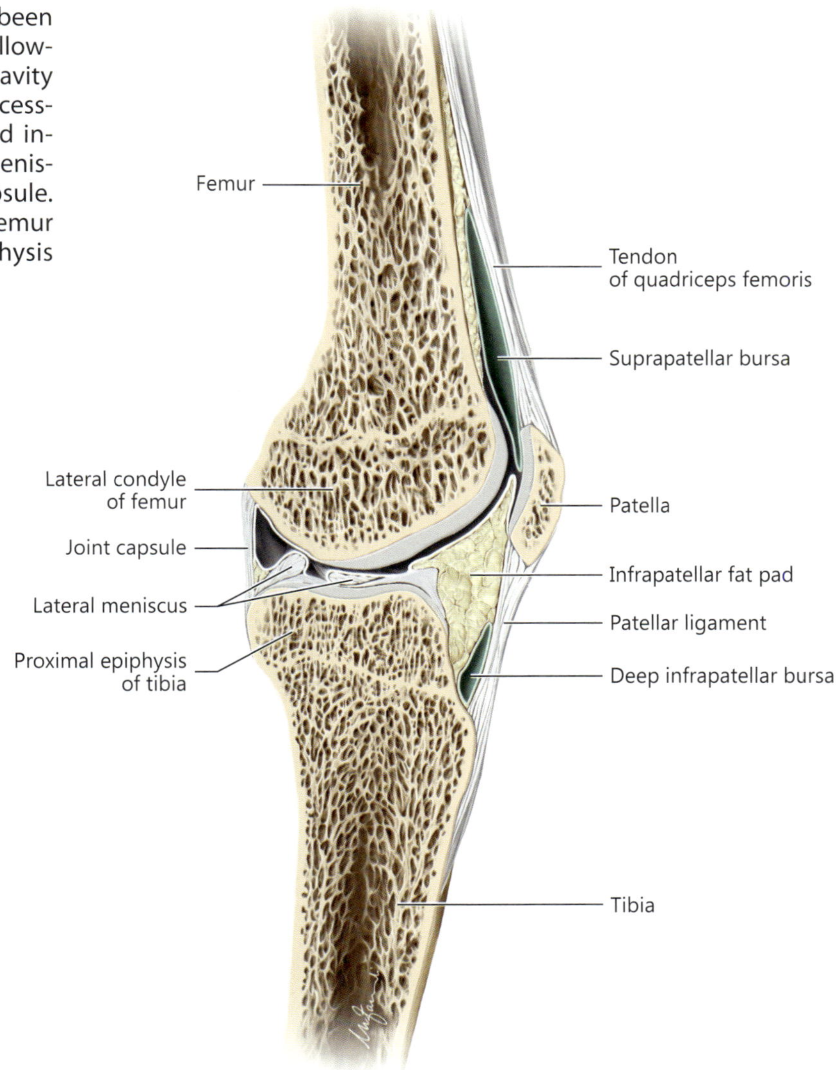

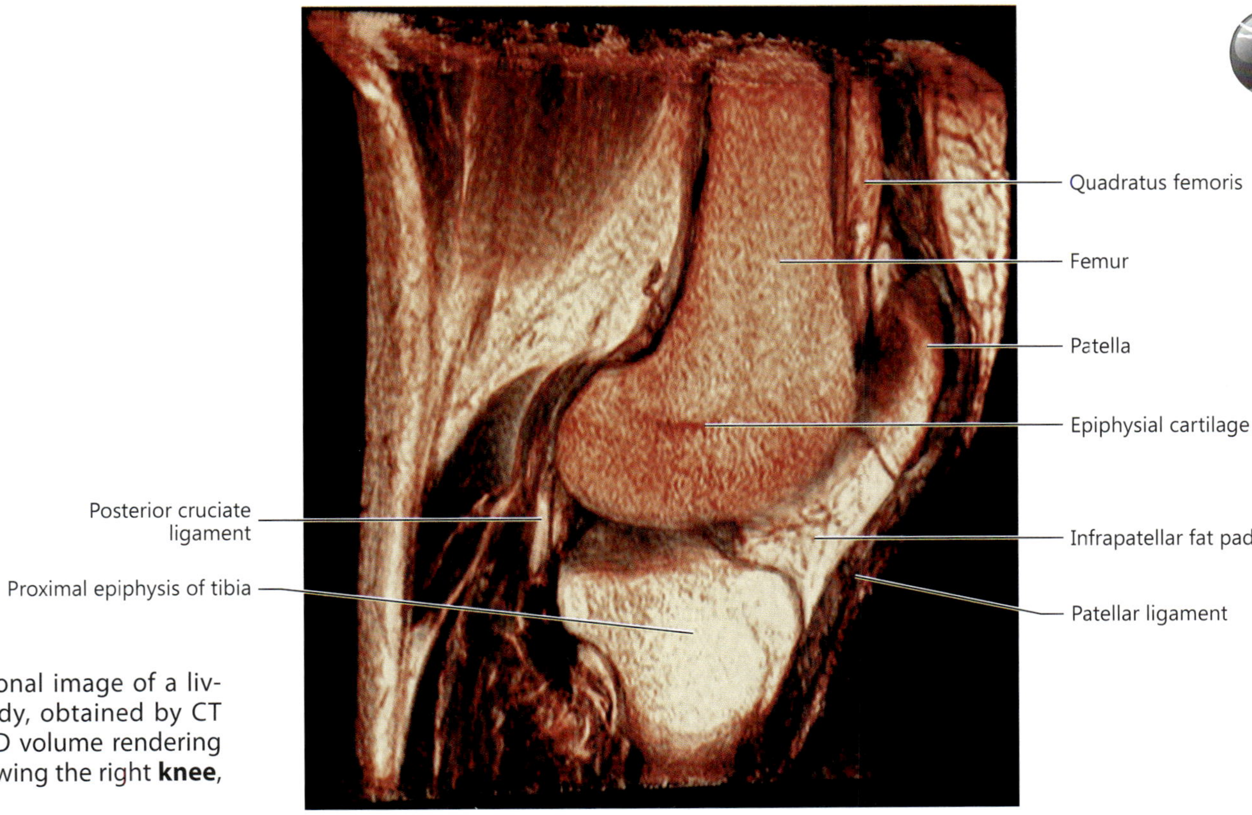

Three-dimensional image of a living human body, obtained by CT scan using a 3D volume rendering technique, showing the right **knee**, lateral view.

Right **knee**. Xeroradiographic image of the joint in a lateral position and during light flexion. Note the narrow contact between the femoral and tibial condyles. The opaque lines above and below the patella correspond to the quadriceps tendon.

1, Shaft of femur
2, Popliteal surface
3, Medial epicondyle
4, Medial condyle of femur
5, Lateral condyle of femur
6, Intercondylar eminence
7, Head of fibula
8, Shaft of fibula
9, Tibial tuberosity
10, Shaft of tibia
11, Medial condyle of tibia
12, Superior articular surface of tibia
13, Patella
14, Patellar surface of femur

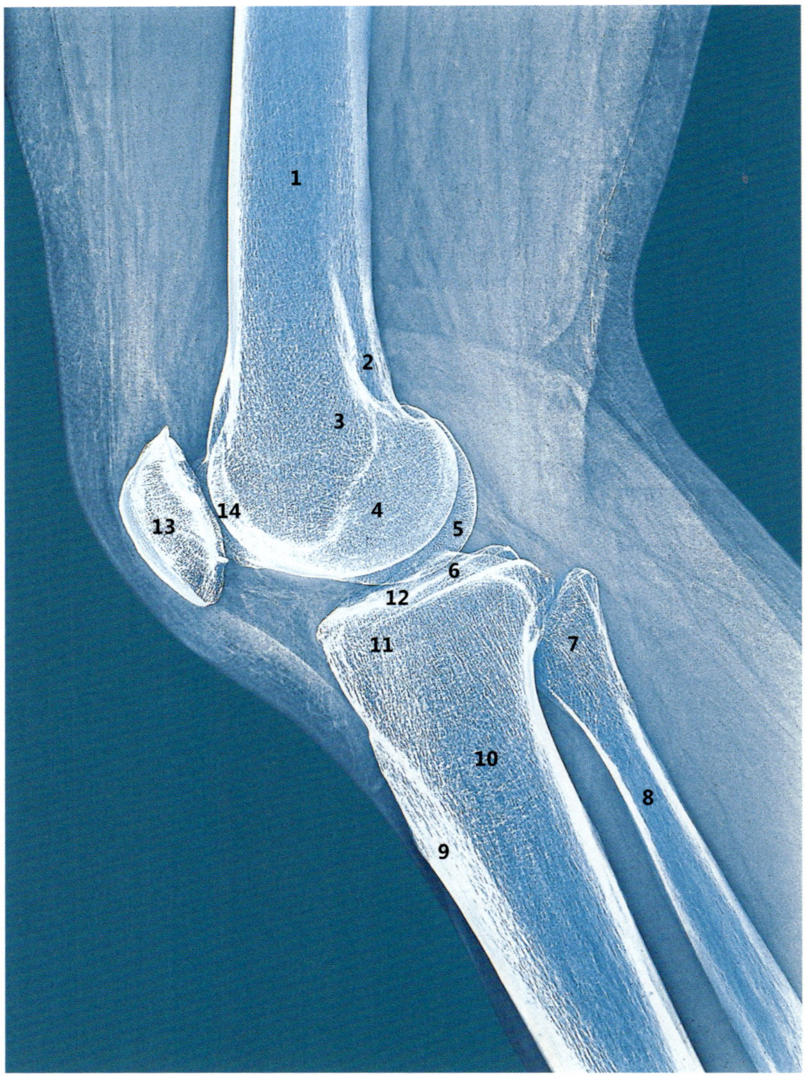

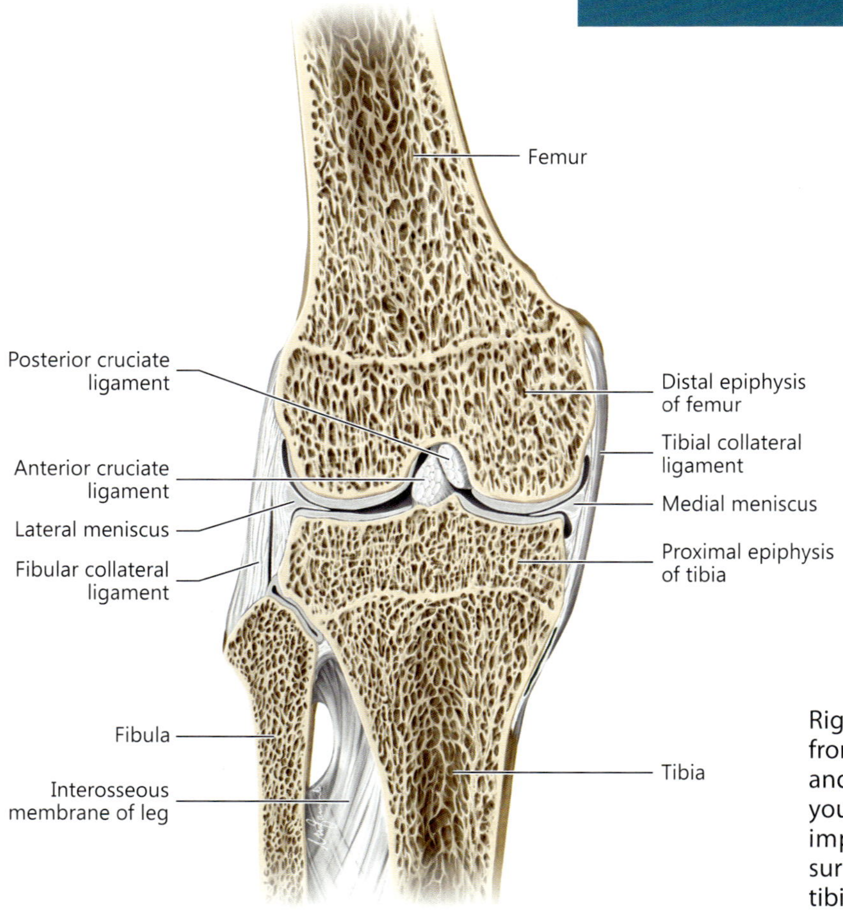

Right **knee**. Section along the frontal plane, seen from the front. Note the limits of the joint capsule and the relationship between the articular heads; you can also see the role that the menisci play in improving the concordance between opposing surfaces. Diaphysis and epiphysis of the femur and tibia are still separated by metaphyseal lines.

Right **knee**. The articular cavity has been stretched by the injection of a semisolid medium. Subsequent removal of the joint capsule allows you to obtain a mold of the cavity which expands mainly upward, to the back, and even more to the front of the femur.

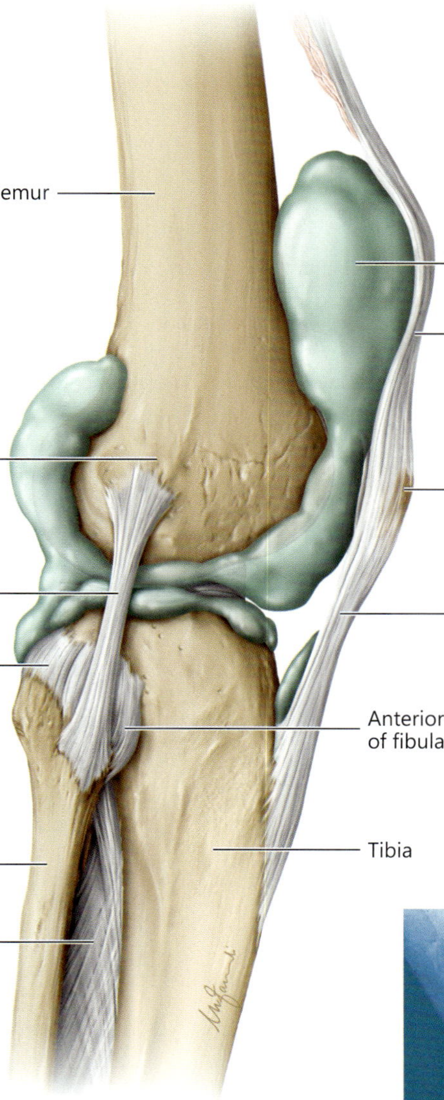

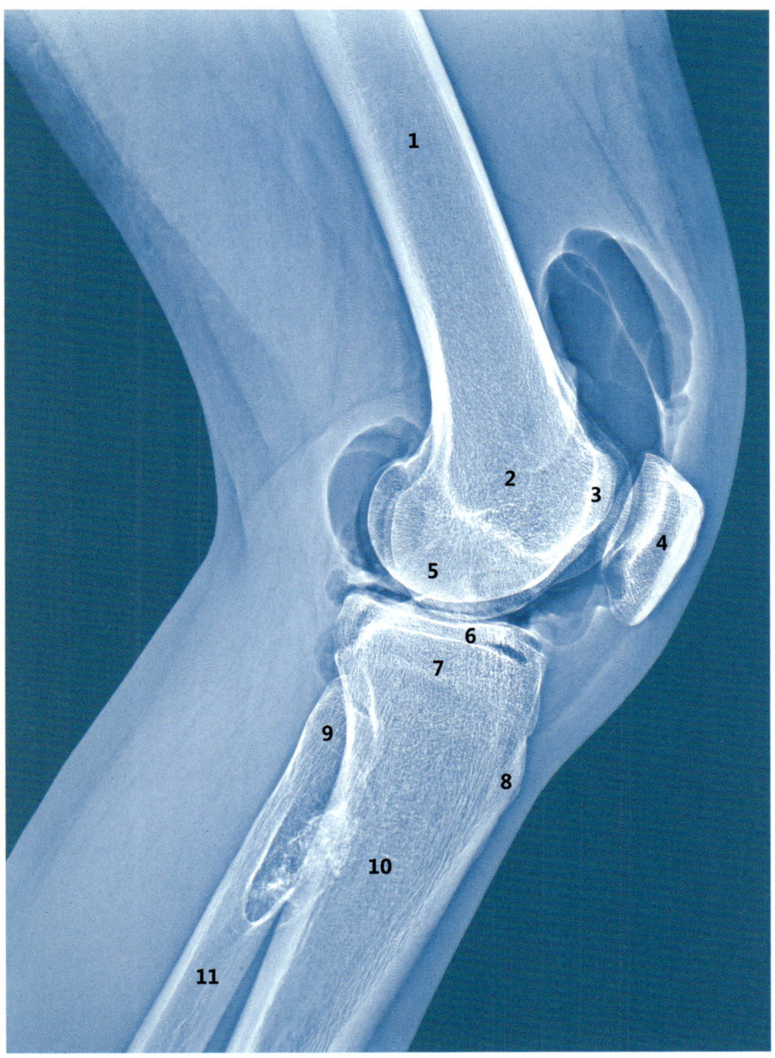

Right **knee**. Xeroradiographic image of the joint seen in a lateral projection. Introduction of air into the joint makes it possible to identify the limits of the articular cavity, and offers a very similar image to that which was prepared in the previous figure.

1, Shaft of femur
2, Lateral epicondyle
3, Patellar surface of femur
4, Patella
5, Lateral condyle of femur
6, Superior articular surface of tibia
7, Lateral condyle of tibia
8, Tibial tuberosity
9, Head of fibula
10, Shaft of tibia
11, Shaft of fibula

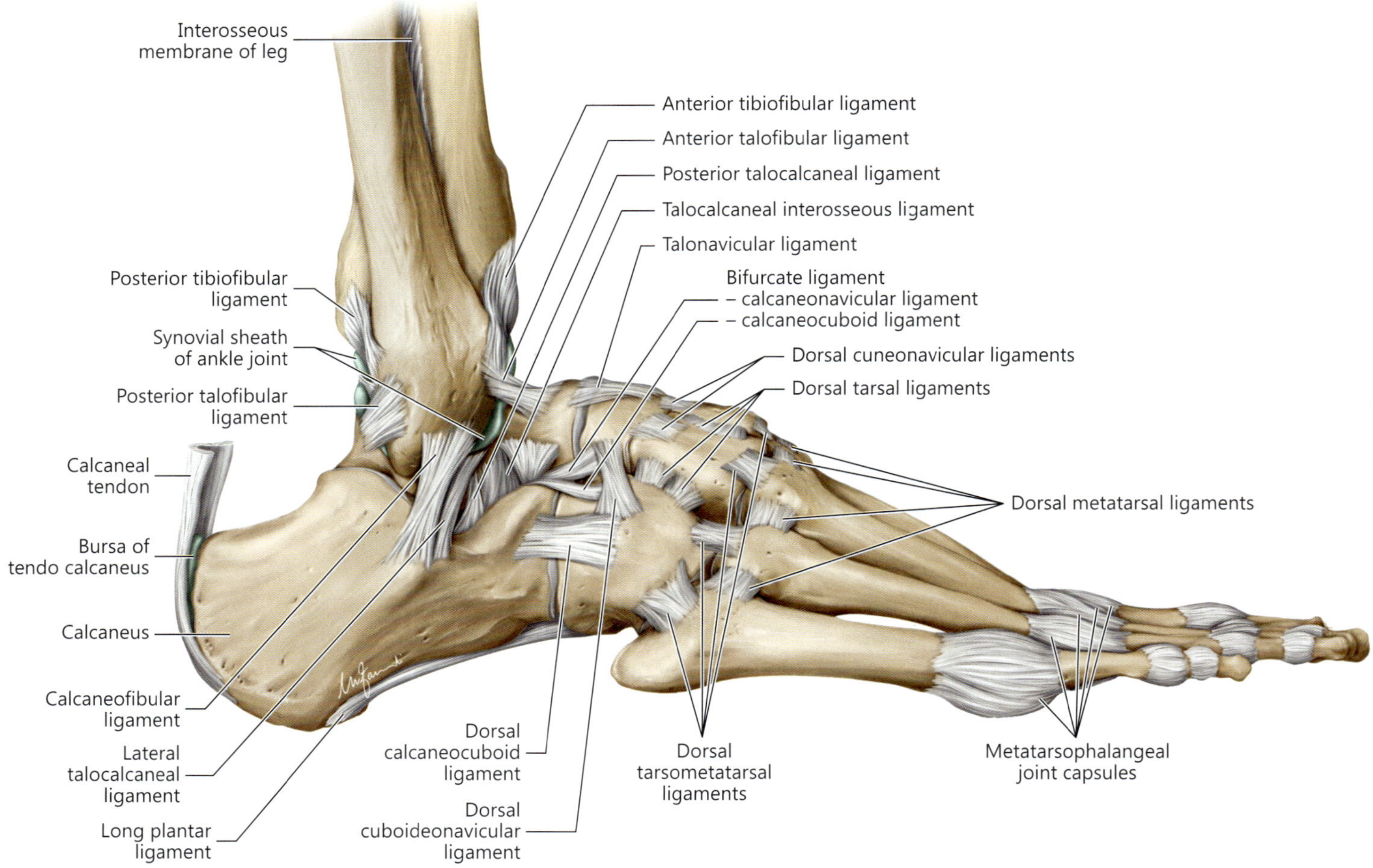

Ankle and tarsal joints on the right side. The means of union of the various joints are shown here laterally.

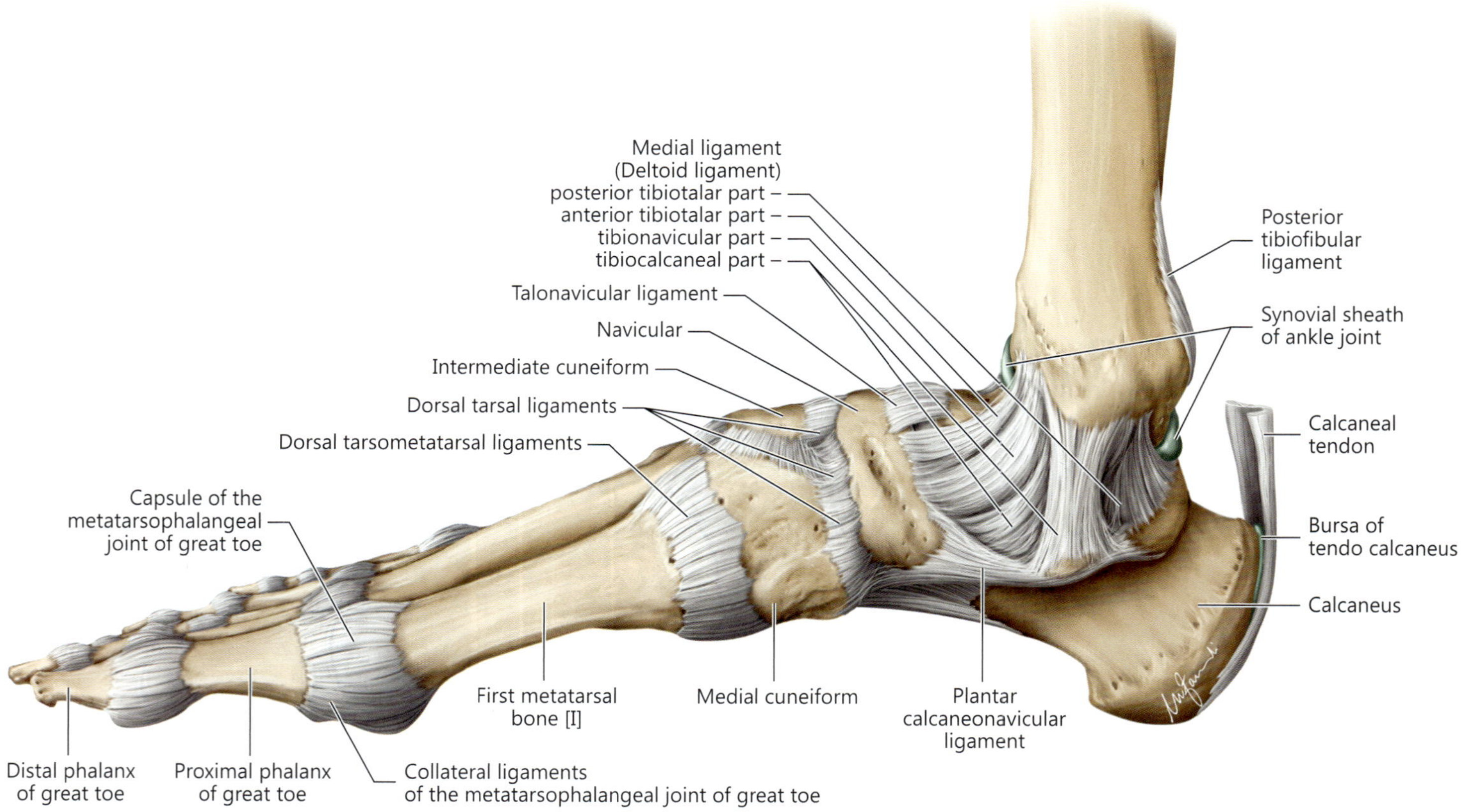

Ankle and tarsal joints on the right side. The means of union of the various joints are shown here medially.

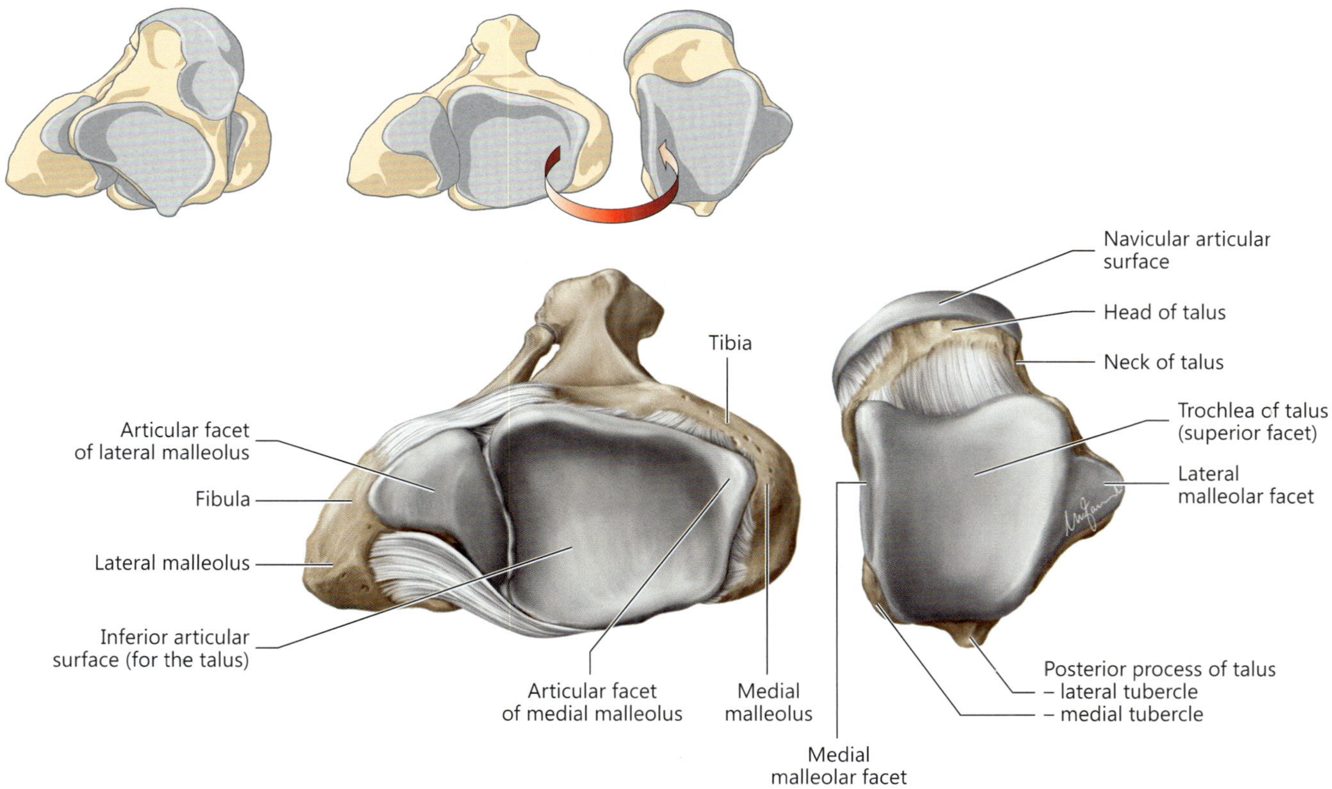

Right **ankle joint**. Configuration of the proximal articular surface.

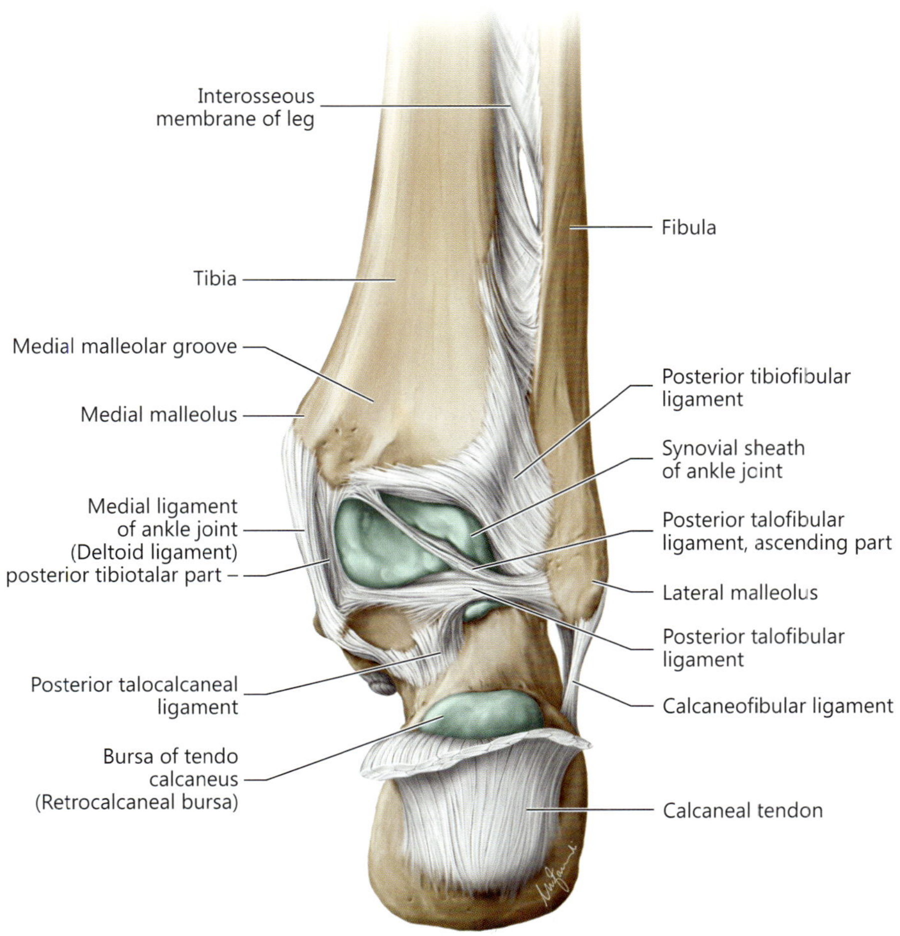

Ankle and subtalar (**talocalcaneal**) **joints** on the right side, seen from behind, with the foot in a state of flexion. The joint capsule has been removed to show the reinforcing ligaments and part of the articular surface of the talus.

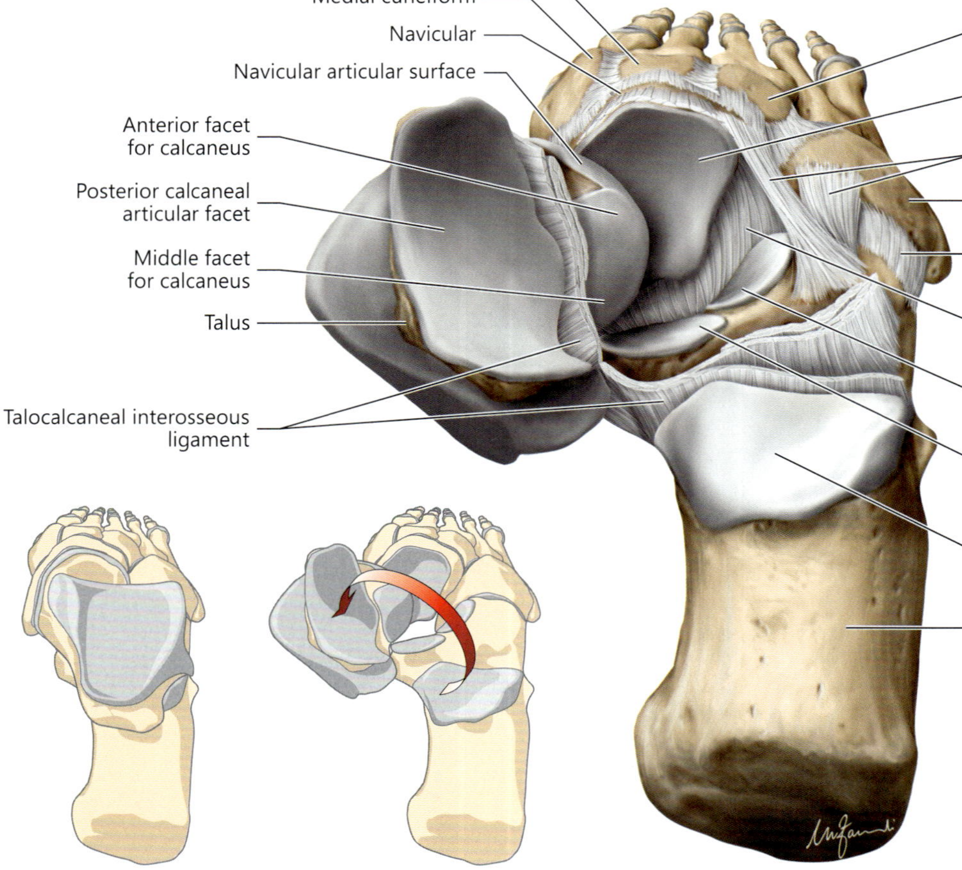

Right **subtalar and talocalcaneonavicular joints**. The talus has been turned to show the articular surfaces of the calcaneus and navicular (or scaphoid).

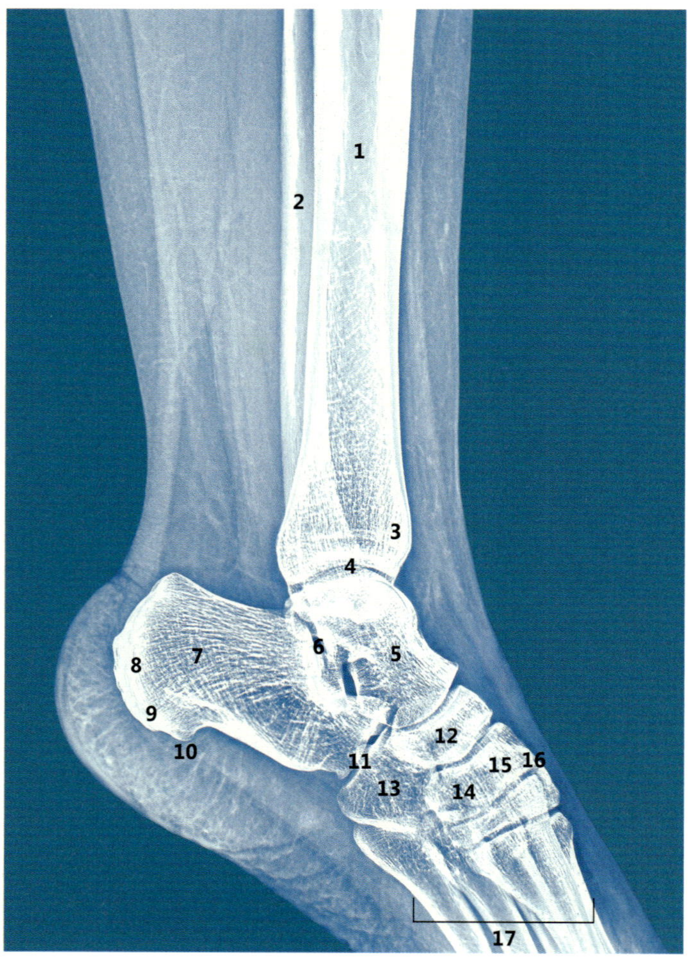

Xeroradiographic image of the **distal portion of the leg** and the **proximal portion of the foot** (ankle and part of the metatarsus). The trajectorial organization of the tarsal bones appears to be particularly pronounced in the calcaneus. The structure of the long bones can be detected in the metatarsals. The demonstration of the soft parts allows you to see the calcaneal insertion of the calcaneal tendon and the connective septa of the subcutaneous layer in the sole of the foot.

1, Shaft of tibia
2, Shaft of fibula
3, Medial malleolus
4, Ankle joint
5, Talus
6, Subtalar joint
7, Calcaneus
8, Calcaneal tuberosity
9, Lateral process of calcaneal tuberosity
10, Medial process of calcaneal tuberosity
11, Talocalcaneonavicular joint
12, Navicular
13, Cuboid
14, Lateral cuneiform
15, Intermediate cuneiform
16, Medial cuneiform
17, Metatarsals

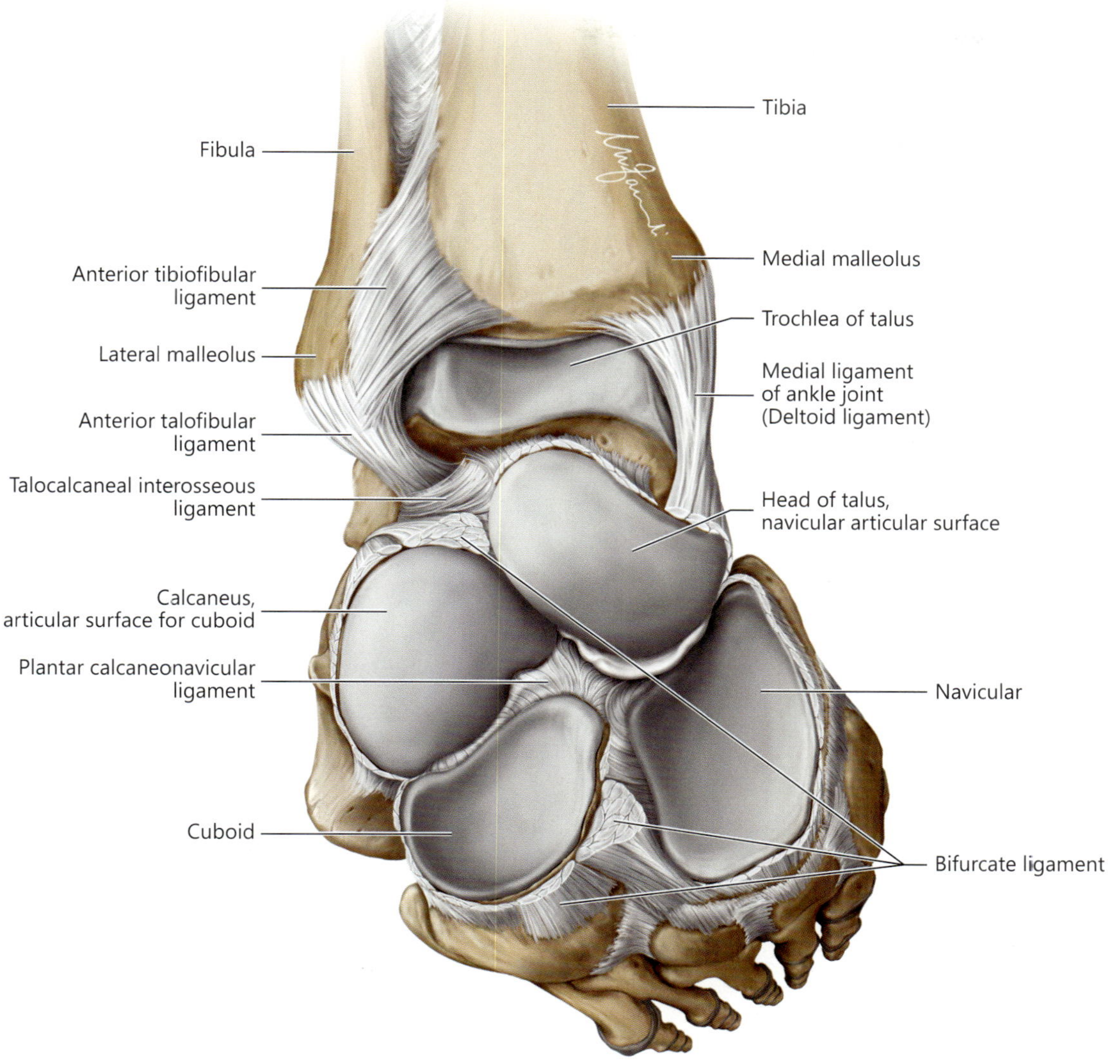

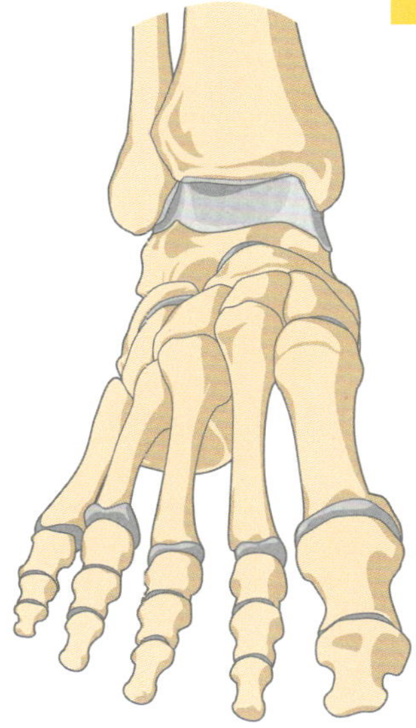

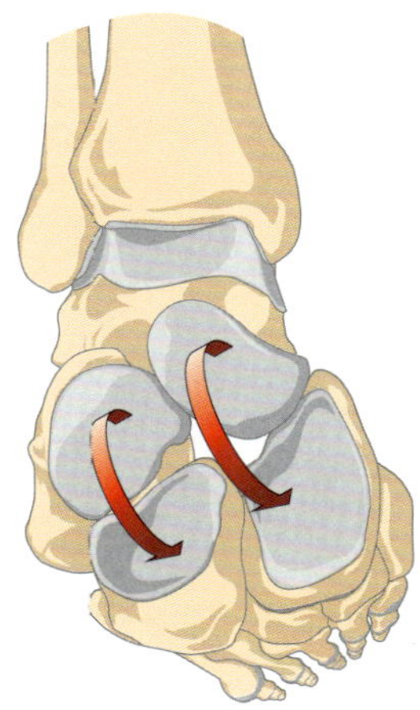

Right **transverse tarsal joint**, seen from above. The joint capsule has been removed and the foot has been brought into hyperextension, dissecting the bifurcated ligament. Thus you can see the opposing articular surfaces of the talus and of the calcaneus for the navicular and the cuboid, respectively.

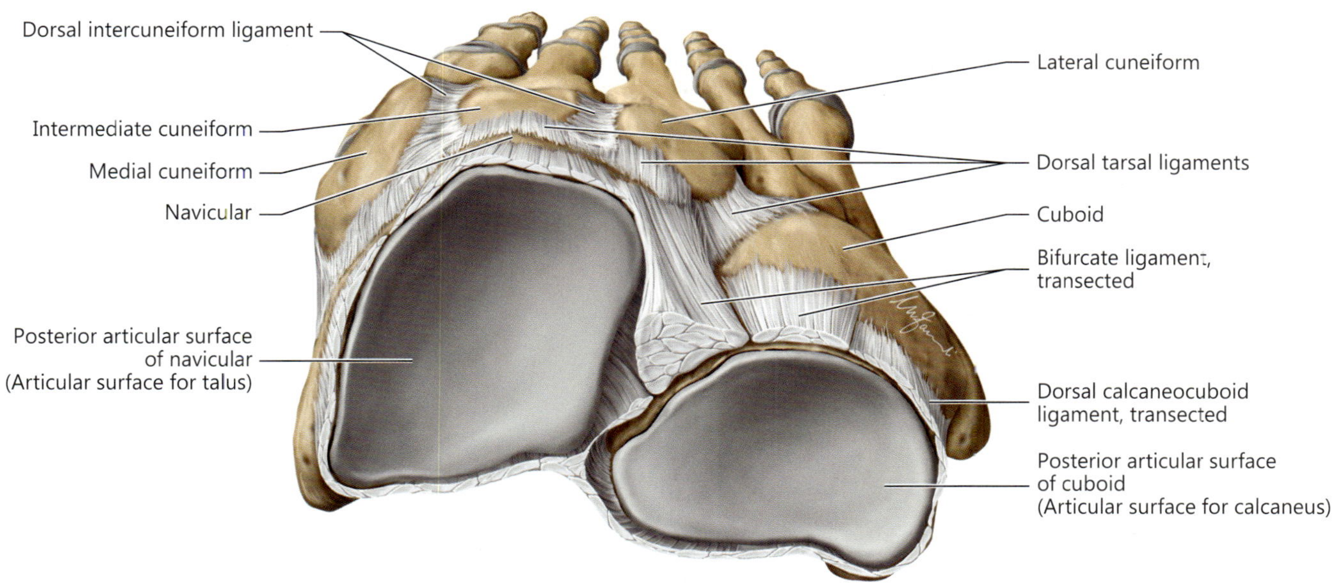

Right **transverse tarsal joint**. Articular surfaces of the navicular and cuboid for the talus and the calcaneus, respectively. The black line marks the limits of insertion of the joint capsule.

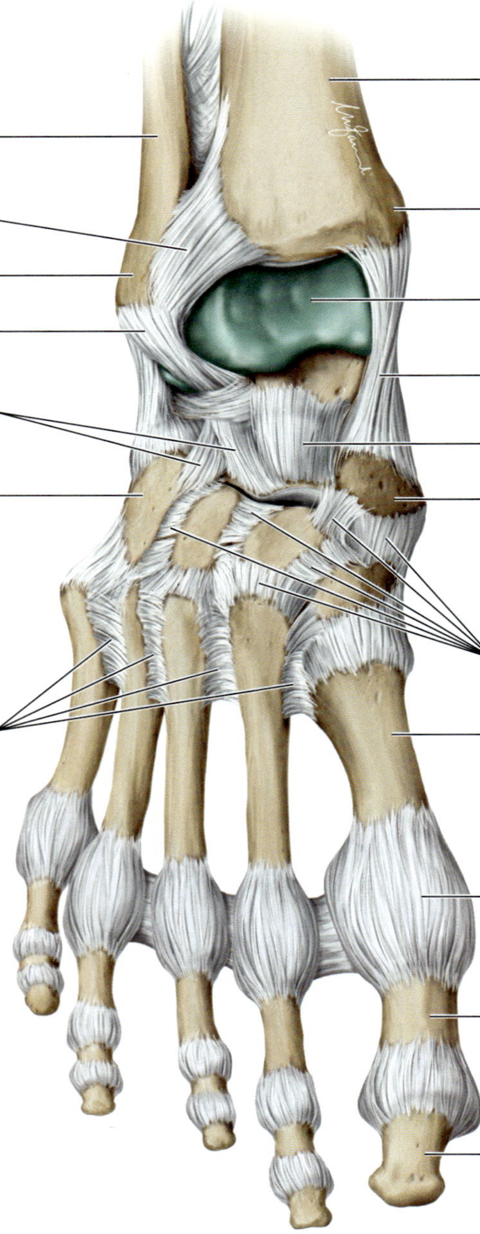

Ankle joint during plantar flexion, anterior view with the tarsal ligaments.

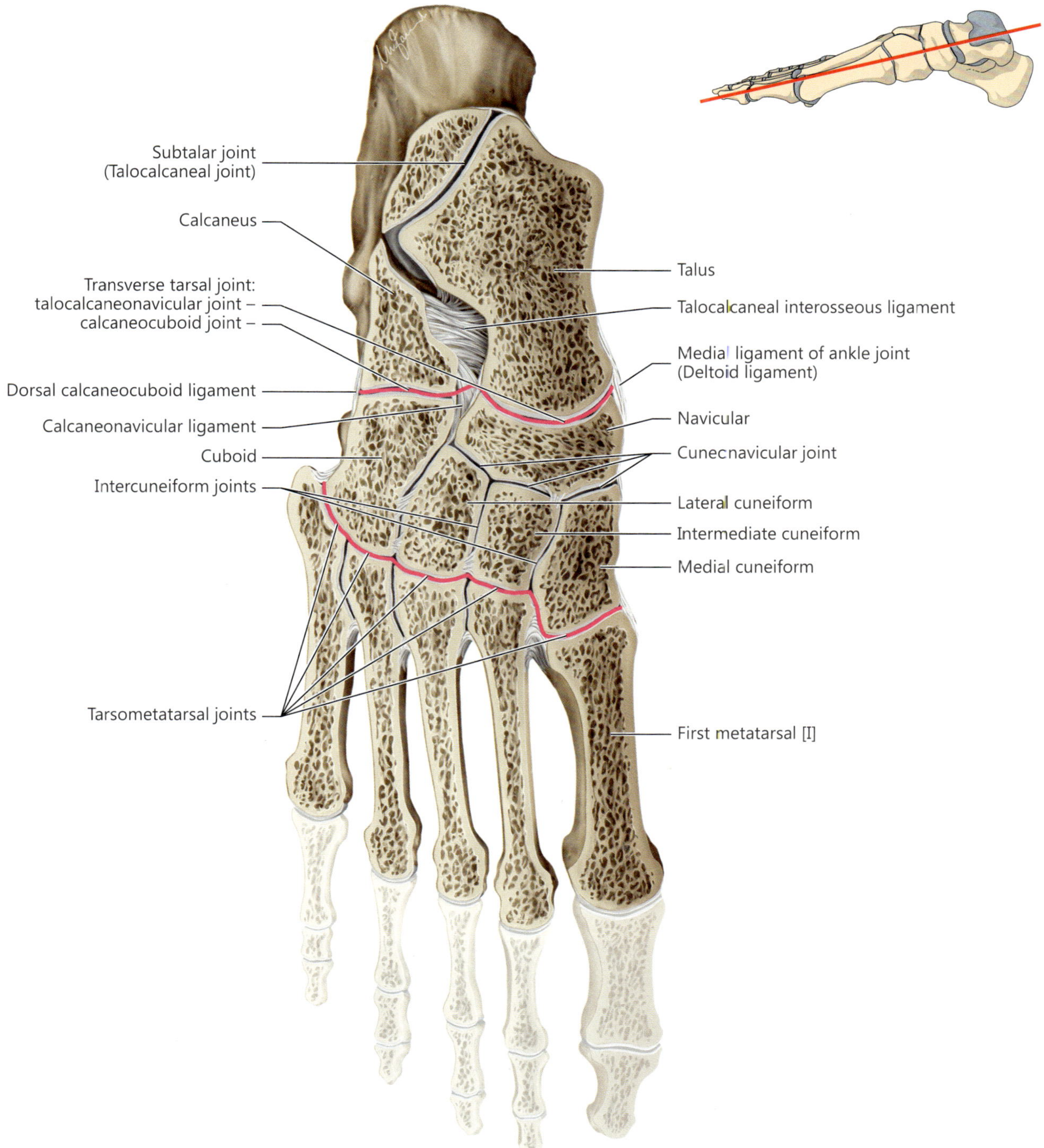

Right **tarsal joints**. The horizontal section plane, seen from above, shows the tarsal and tarsometatarsal joints. The two red lines, proximal and distal, indicate the transverse tarsal joint (Chopart's joint) and tarsometatarsal joints, respectively.

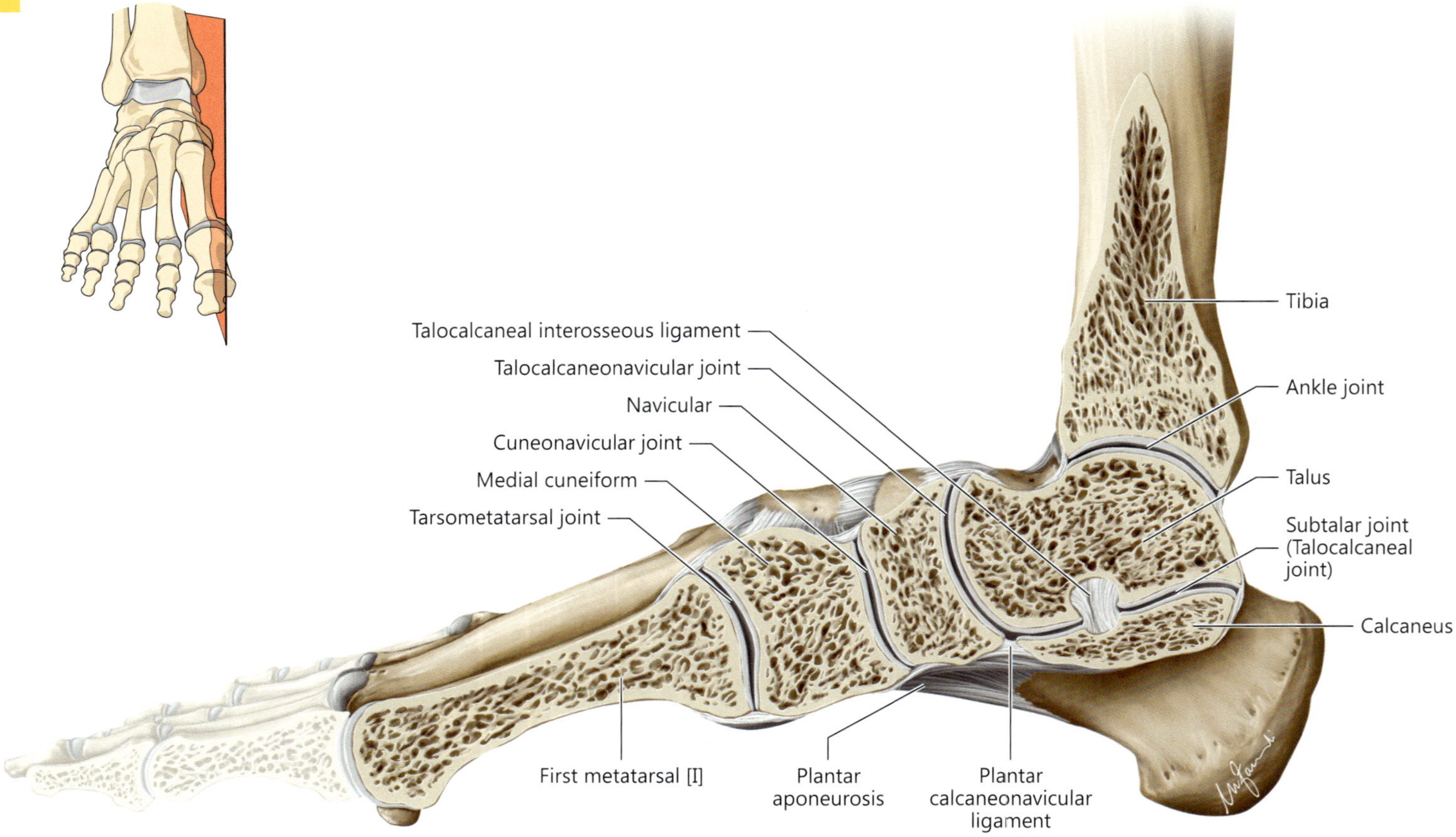

Right **transverse tarsal joint**. The articular relations between the tarsal bones are illustrated in a sagittal section of the foot, seen from the medial side.

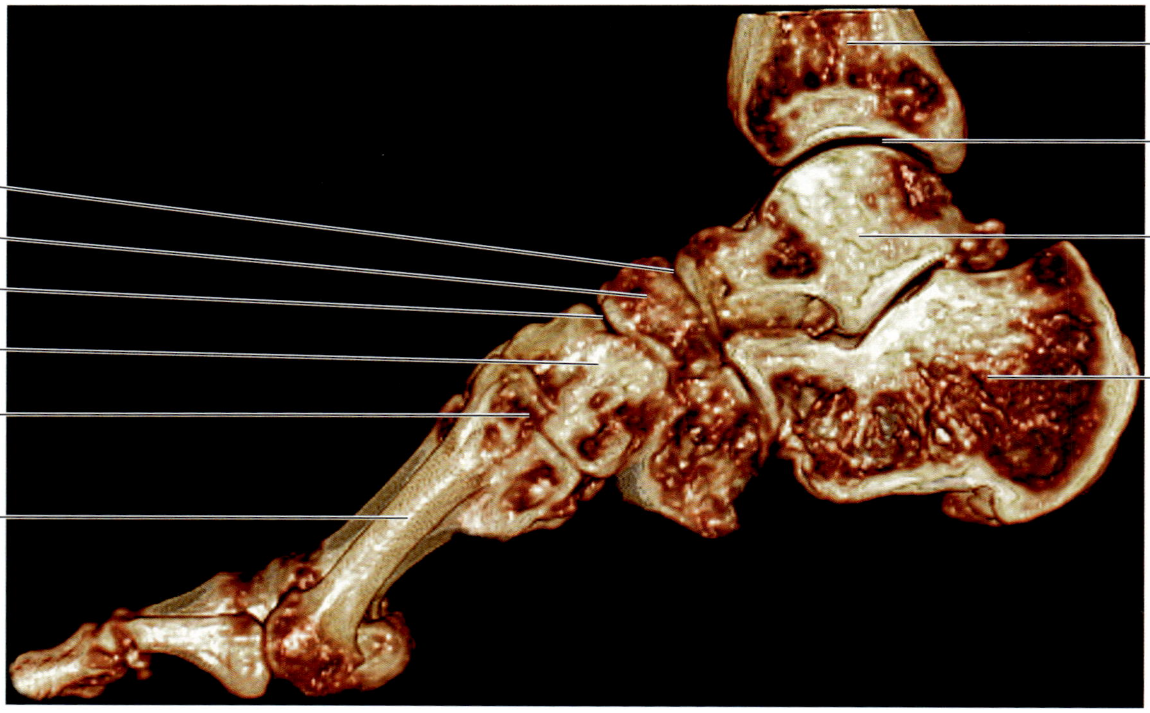

Three-dimensional image of a living human body, obtained by CT scan using a 3D volume rendering technique, showing the right **transverse tarsal joint** in a sagittal section of the right foot, seen from the medial side.

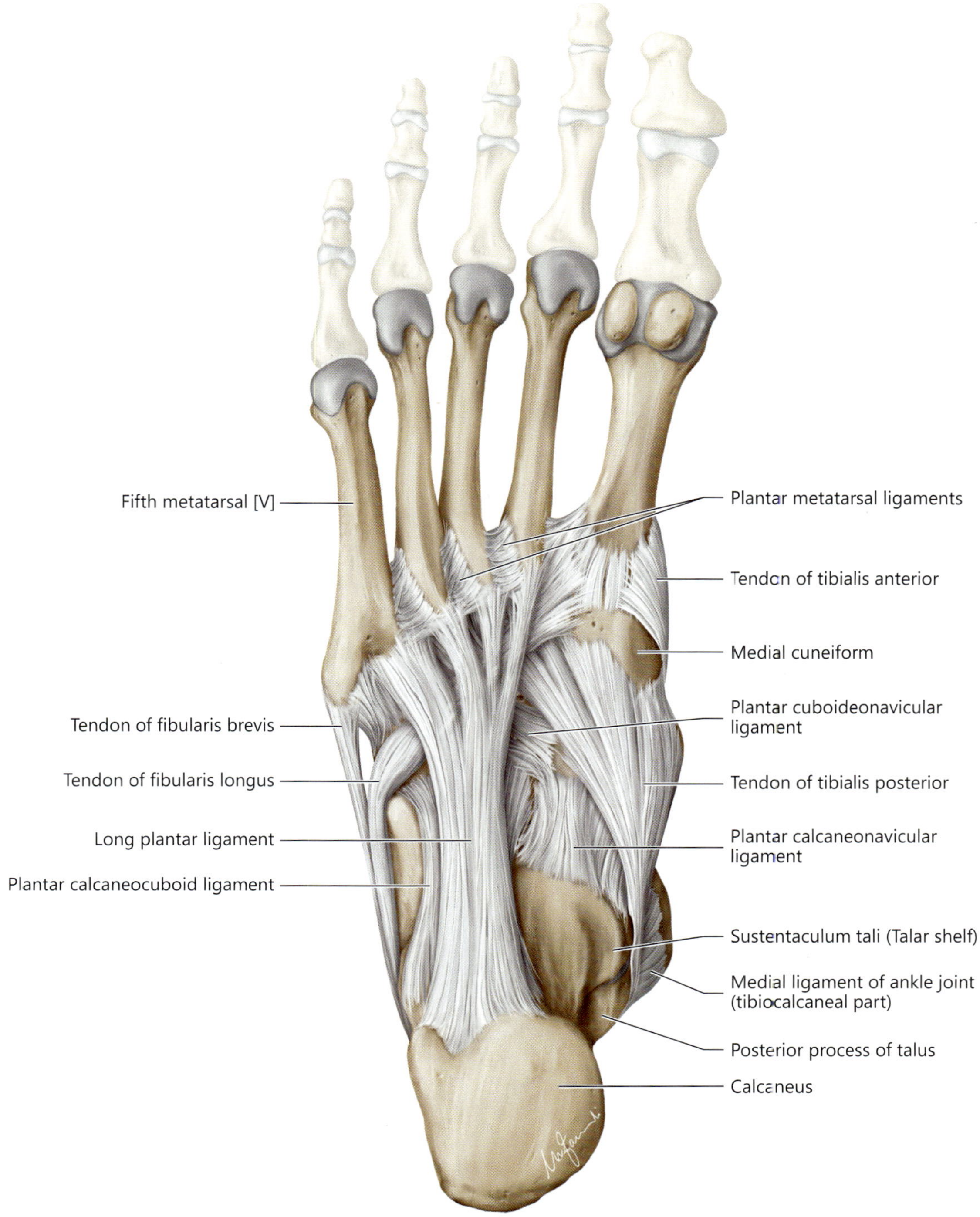

Ligaments of the sole of the right **foot**. Here the means of reinforcement of the tarsometatarsal and intermetatarsal joints are shown. The means of union of the joints, the tendons of the leg muscles, and the intrinsic muscles of the foot form a support structure on the sole of the foot which is strong and robust in terms of its various components and which helps to maintain the arched structure of the skeletal system.

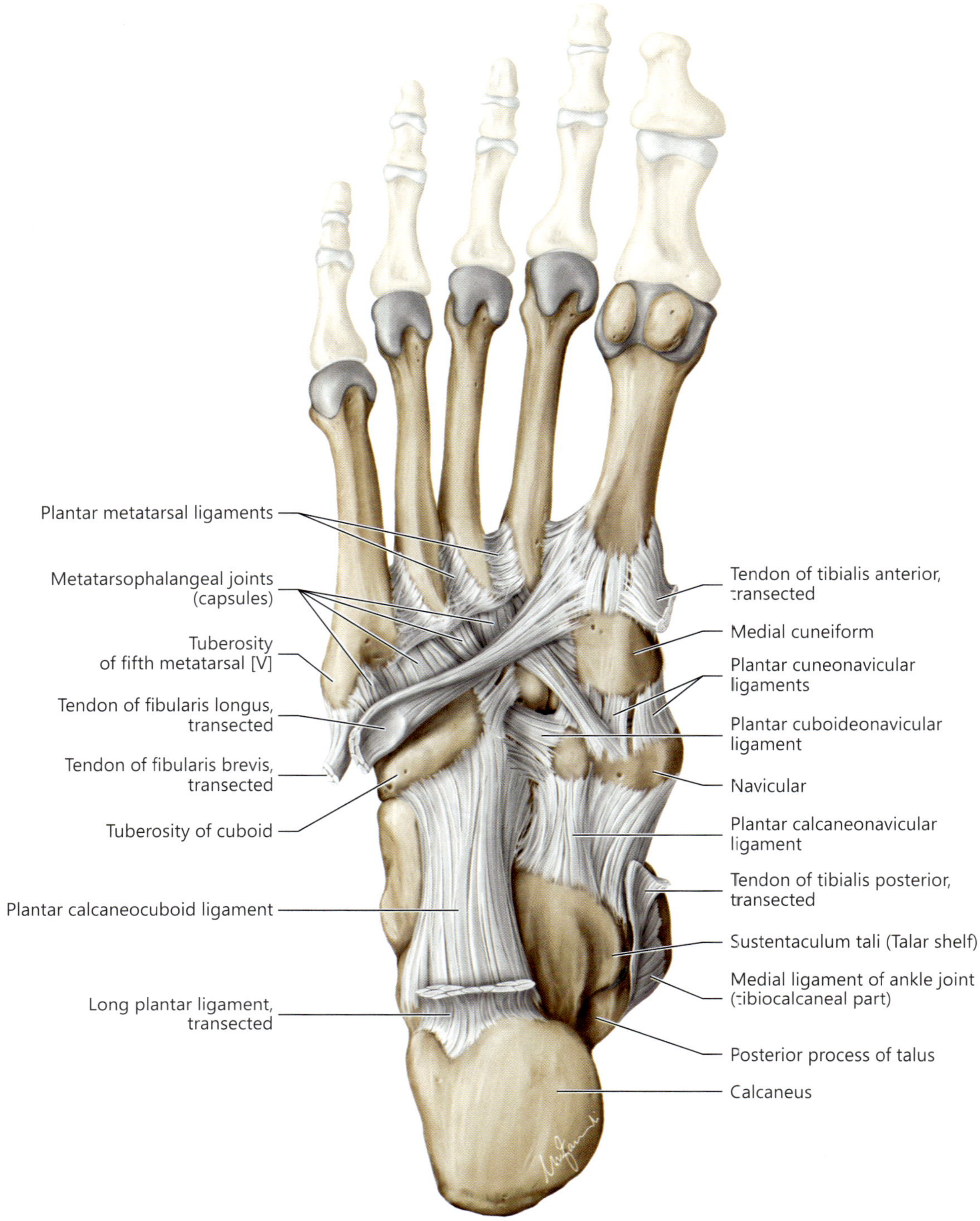

Ligaments of the sole of the right **foot**. The tendon of the fibularis longus has been removed from its articular contact point, raised, and turned forward. Here the means of reinforcement of the tarsometatarsal and intermetatarsal joints are also depicted. The means of union of the joints, the tendons of the leg muscles, and the intrinsic muscles of the foot form a support structure on the sole of the foot which is strong and robust in terms of its various components and which helps to maintain the arched structure of the skeletal system.

MOVEMENTS OF THE HIP

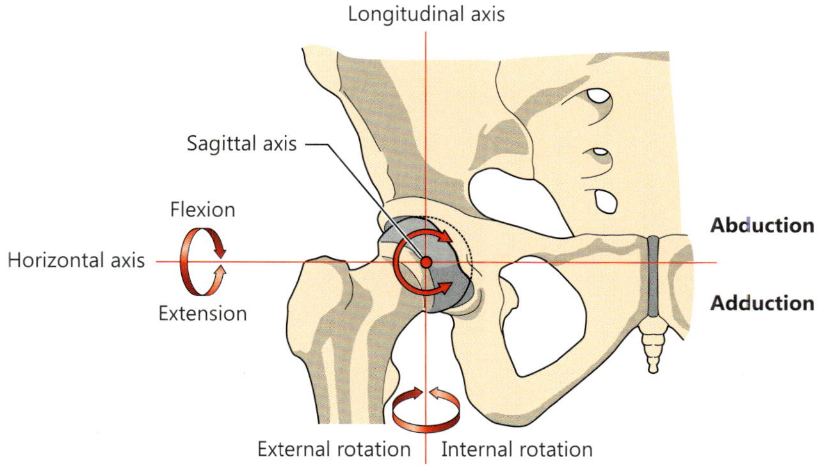

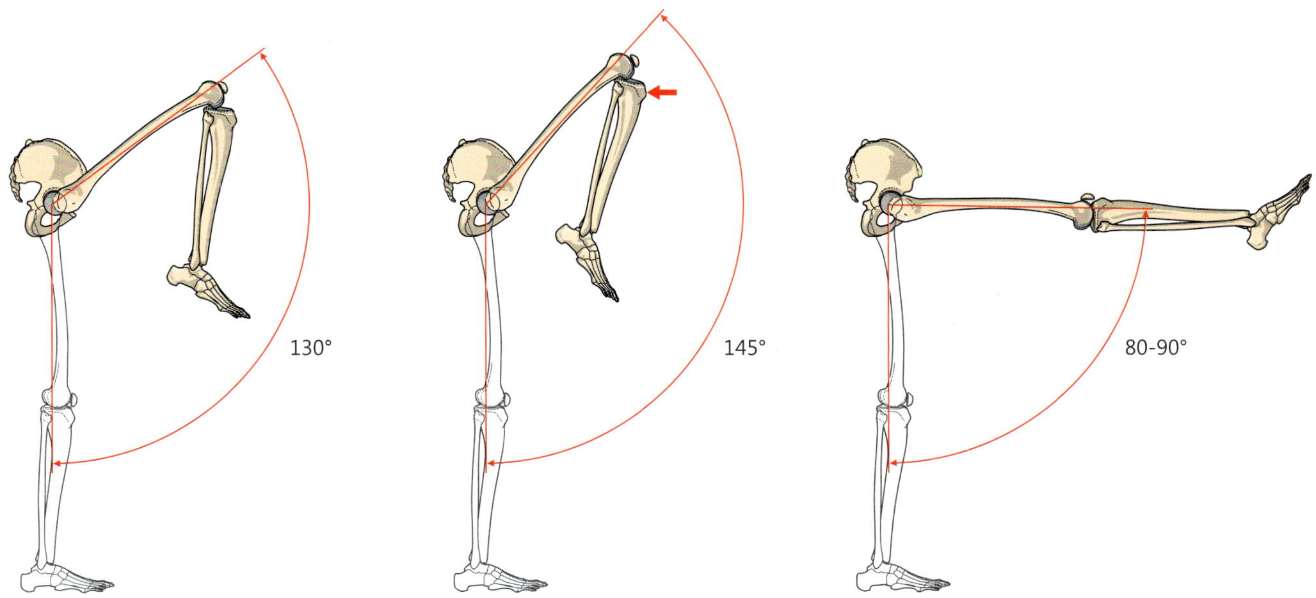

Flexion of the hip

- Occurs on the sagittal plane around a transverse or horizontal axis
- When upright and in the supine position it has an amplitude of:
  - 130°, with the knee flexed (actively performed)
  - 145°, with the knee flexed (passively performed)
  - 80-90°, with the knee extended (actively performed); the reduction in amplitude is due to the tension of the ischiocrural muscles (and the paravertebral muscles)

Limitations

- Tension of the ischiocrural muscle fibers
- Accentuation of lumbar lordosis

In the supine position, flexion of the hip is facilitated by the straightening of the lumbar lordosis which results in retroversion of the pelvis. This is also favored by flexion of the knee.

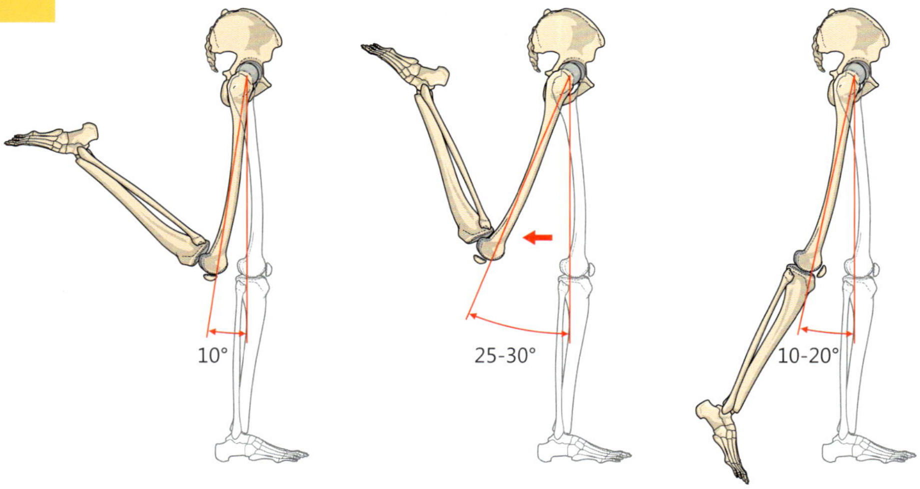

Extension of the hip
- Occurs on the sagittal plane around a transverse or horizontal axis
- Amplitude of:
  - 10°, with the knee flexed (actively performed)
  - 25-30°, with the knee flexed (passively performed)
  - 10-20°, with the knee extended.

Limitations
- Ligaments:
  - iliofemoral
  - ischiofemoral
  - pubofemoral
- Joint capsule (anterior part)
- Tension of the iliopsoas fibers.

Extension movements of the hip with greater amplitude require compensation movements, primarily the anteversion of the pelvis and an accentuation of the lumbar lordosis.

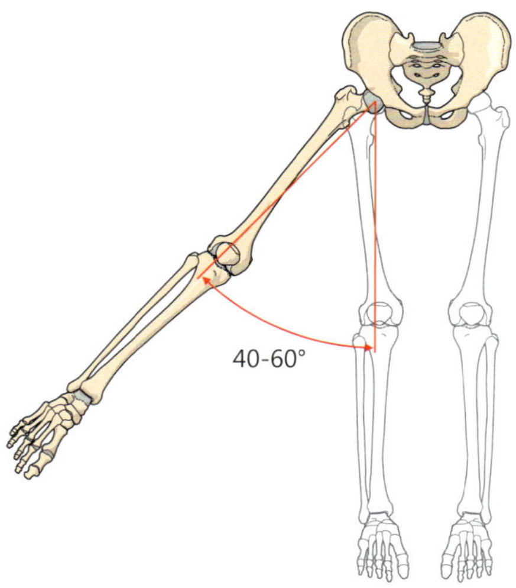

Abduction of the hip
- Occurs on the frontal plane around a sagittal axis
- Amplitude of 40-60°.

Limitations
- Ligaments:
  - iliofemoral
  - ischiofemoral
  - pubofemoral
- Joint capsule (inferior part)
- Tension of the fibers of the adductor muscles of the hip.

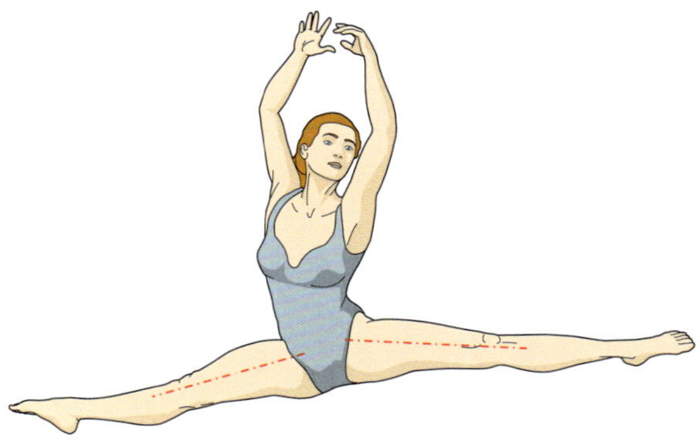

Extension of the hip is facilitated by flexion of the contralateral thigh, as is evident from the sagittal split.

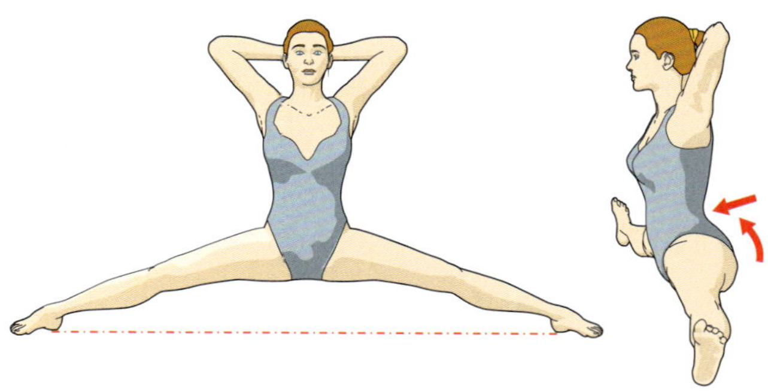

The abduction movement of the hip can reach values of 70° when the thigh is flexed 60° and externally rotated, and 90° in bilateral abduction performed by gymnasts (front split). In the latter case, compensational movements occur, represented by the anteversion of the pelvis and by an accentuation of lumbar lordosis, which reduce the tension of the ligaments of the hip joint.

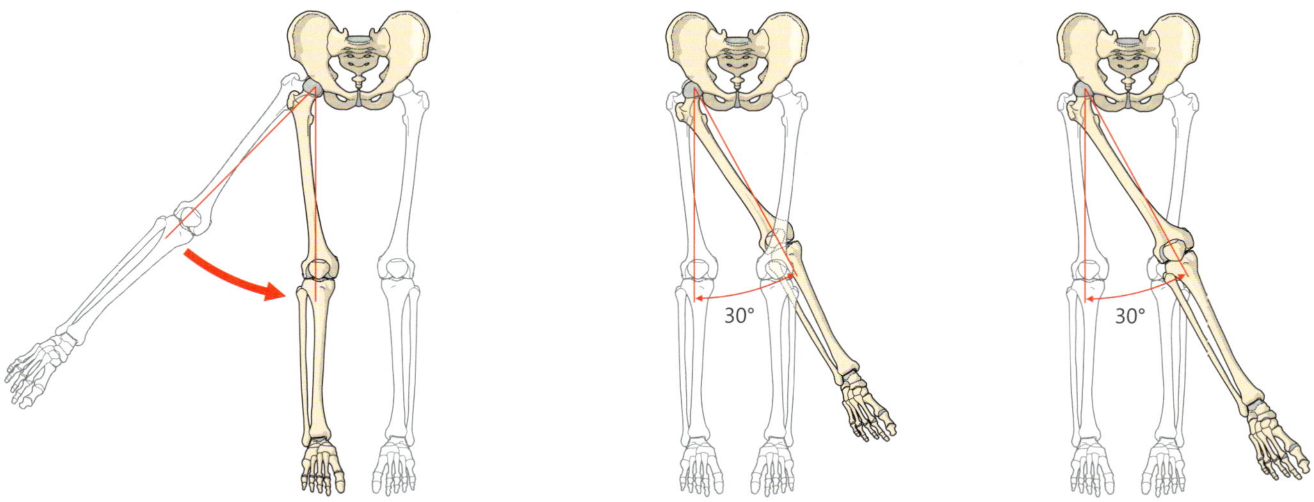

Adduction of the hip
- Occurs on the frontal plane around a sagittal axis
- Relative:
  - return from any position of abduction to the median line of symmetry
- Absolute:
  - is accompanied by a movement of:
    - flexion of the hip (forward movement of the contralateral limb)
    - extension of the hip (backward movement of the contralateral limb)
  - amplitude of: 30°.

Limitations
- Ligaments:
  - iliofemoral
  - ischiofemoral
- Joint capsule (superior part)
- Tension of the fibers of the abductor muscles of the hip.

A common example of a hip position during adduction is represented by the seated position with the legs crossed, in which the hip is also flexed and externally rotated (this is also the most unstable position of the hip joint).

External rotation (or supination) of the hip
- Occurs on the transverse plane around a longitudinal axis represented by the mechanical axis of the lower limb (a line vertically connecting the center of the head of the femur with the centers of the base of the tibia and foot)
- From an intermediate position between external and internal rotation (neutral position)
- Amplitude of:
  - 30°, with the knee extended:
    - in the upright position (including external rotation of the foot of 10-15°)
    - in the supine position (including external rotation of the foot of 10-15°)
  - 60°:
    - with the knee extended and hip flexed while standing or in the supine position
    - with the knee flexed (the greater amplitude of the movement of hip rotation is due to the additional movements derived from the knee in flexion, a position in which all the ligaments of the knee joint are released):
      - in the seated position
      - in the prone position.

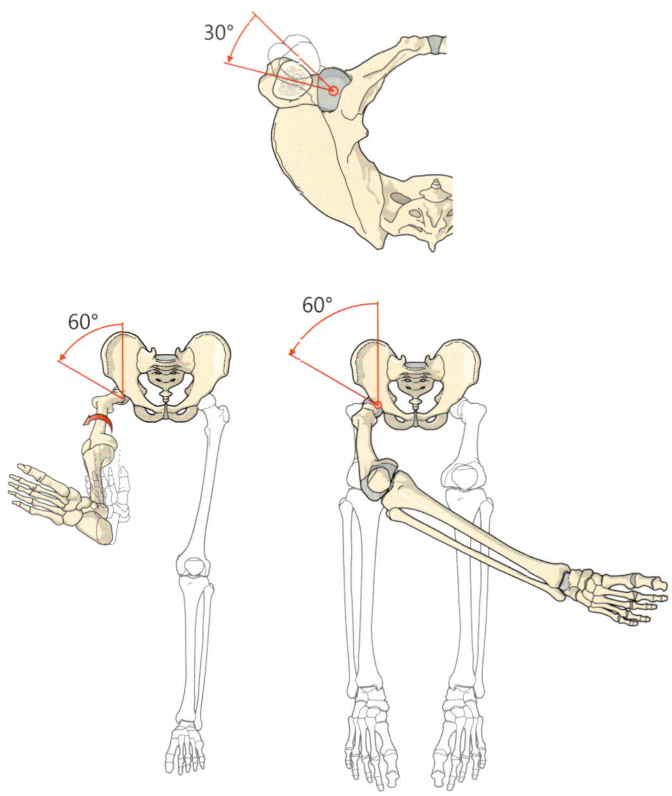

Limitations
- Ligaments:
  - iliofemoral
  - pubofemoral
- Joint capsule (anterior part)
- Tension of the fibers of the internal rotator muscles of the hip.

The range of motion of the external rotation of the hip is increased by flexion of the knee, which may be combined with rotational movements.

Internal rotation (or pronation) of the hip

- Occurs on the transverse plane around the mechanical axis of the lower limb
- From an intermediate position between external and internal rotation (neutral position)
- Amplitude of:
  - 25°, with the knee extended:
    - in the upright position
    - in the supine position
  - 50°, with the knee extended and thigh flexed while standing or in the supine position
  - 30°, with the knee flexed:
    - in the seated position
    - in the prone position.

Limitations

- ischiofemoral ligament
- Joint capsule (posterior part)
- Tension of the fibers of the external rotator muscles of the hip.

The range of motion of the internal rotation of the hip is increased by flexion of the knee, which may be combined with rotational movements.

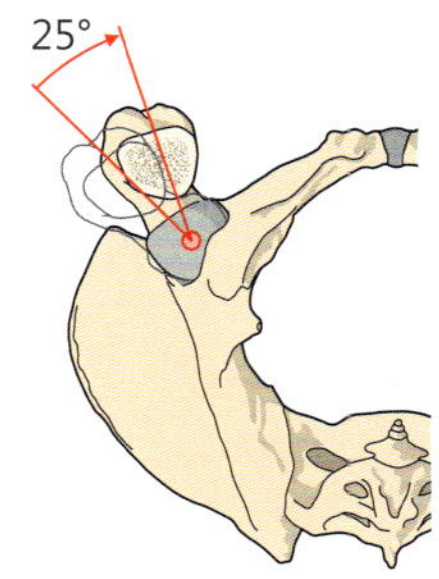

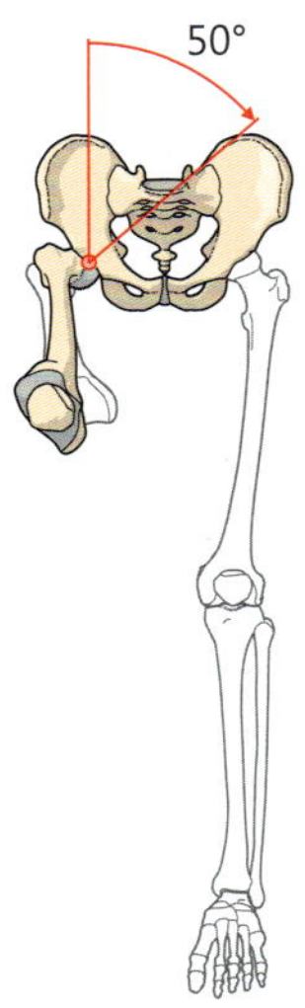

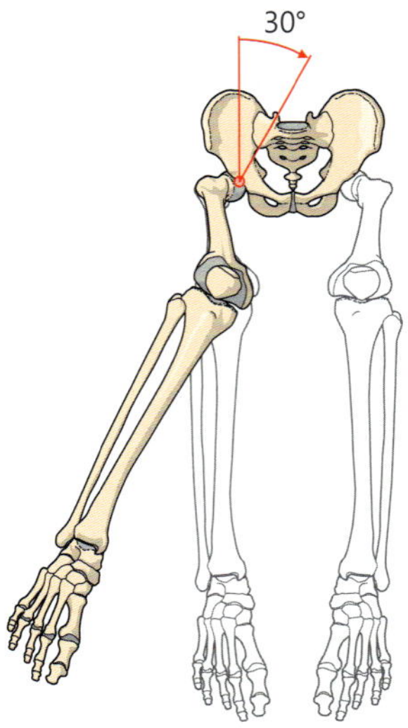

Circumduction of the hip

Definition: the set of the following types of movement:

- flexion – abduction – external rotation of the thigh
- internal rotation – adduction – extension of the thigh

which follow one another around the three axes.

The depiction of these successive movements is given by a cone with a vertex at the center of the head of the femur and base described by the different maximum amplitude of each movement.

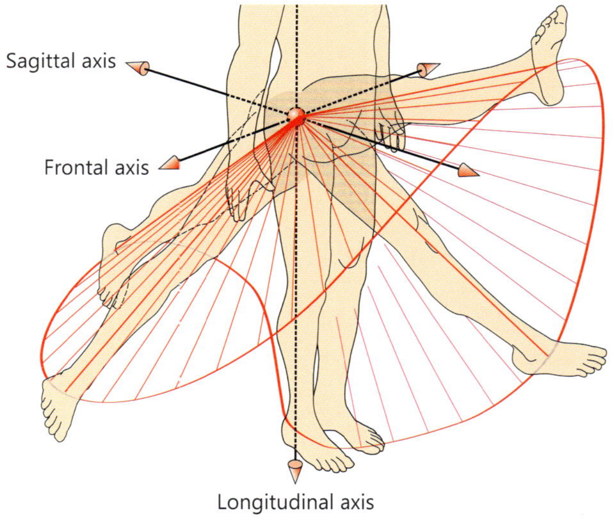

MOVEMENT OF THE MENISCI

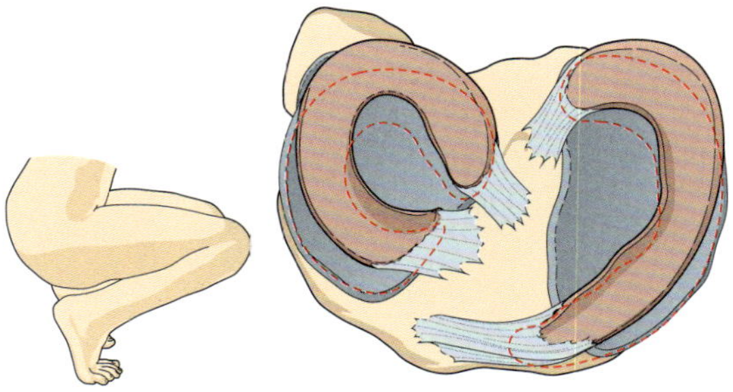

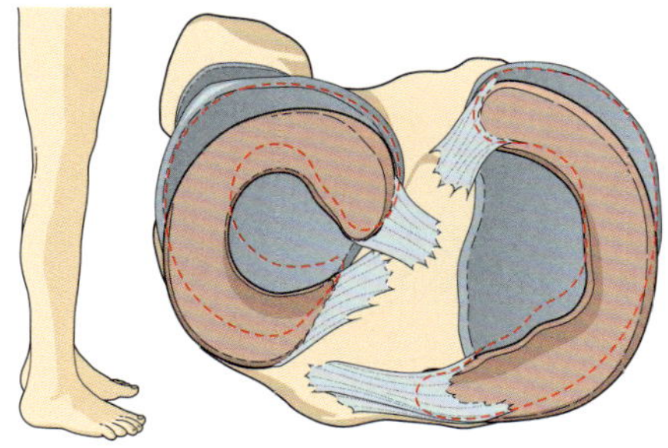

Flexion of the knee (combined action of forward sliding and backward rotation of the femoral condyles on the tibial surfaces):

- **posterior** movement of the menisci (medial and lateral) (this is actually a deformation of the menisci, since they are fixed to the glenoid cavity of the tibial condyles through the anterior and posterior horns while the remaining part of the meniscus is free):
  - is facilitated by the semimembranous and posterior cruciate ligament (medial meniscus)
  - is facilitated by the popliteus (lateral meniscus)

Extension of the knee (combined action of backward sliding and forward rotation of the femoral condyles on the tibial surfaces):

- **anterior** movement of the menisci (medial and lateral)
  - is facilitated by the alar ligaments of the patella (due to the forward movement of the patella itself)

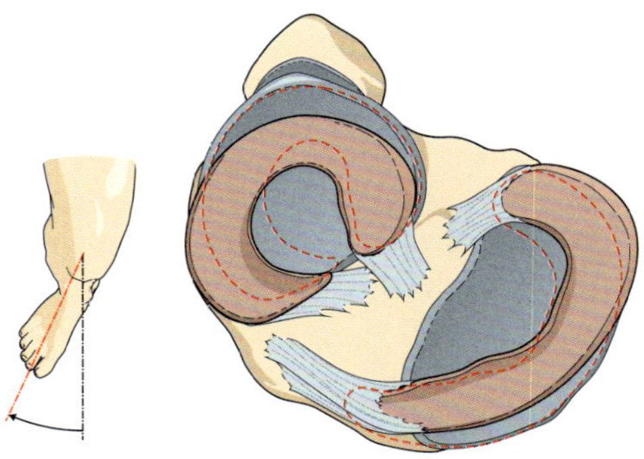

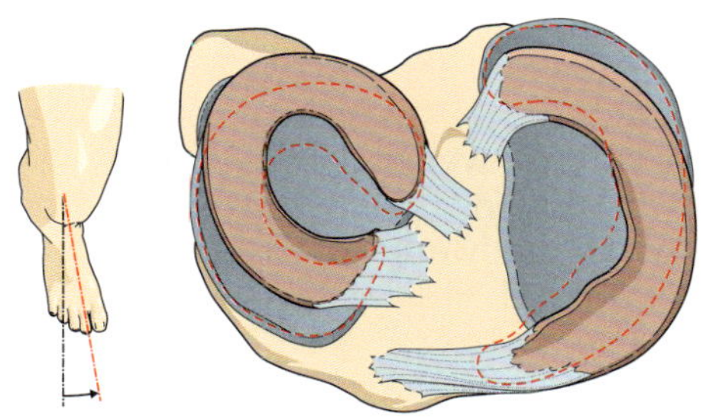

Internal rotation of the knee (external rotation of the tibia which corresponds to the internal rotation of the femoral condyles on the tibial surfaces)

- movement:
  - anterior movement of the lateral meniscus
  - posterior movement of the medial meniscus

External rotation of the knee (internal rotation of the tibia which corresponds to the external rotation of the femoral condyles on the tibial surfaces)

- movement:
  - posterior movement of the lateral meniscus
  - anterior movement of the medial meniscus

FUNCTION OF THE CRUCIATE LIGAMENTS

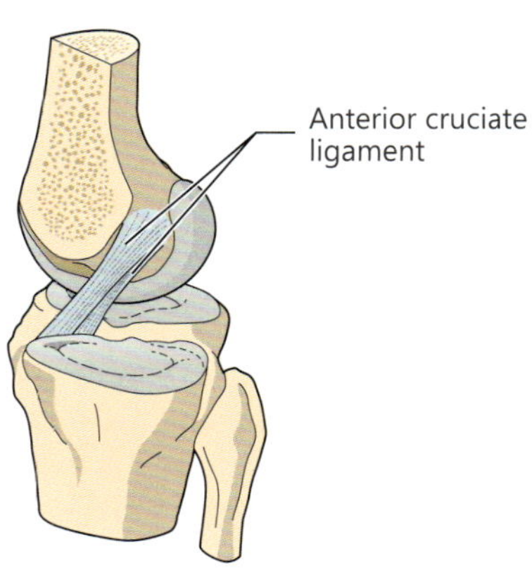

Right knee extended: medial view.
Function of the anterior cruciate ligament:

- stabilizes the knee anteroposteriorly
- limits:
  - anterior movement of the tibia on the femur
  - rotation of the knee

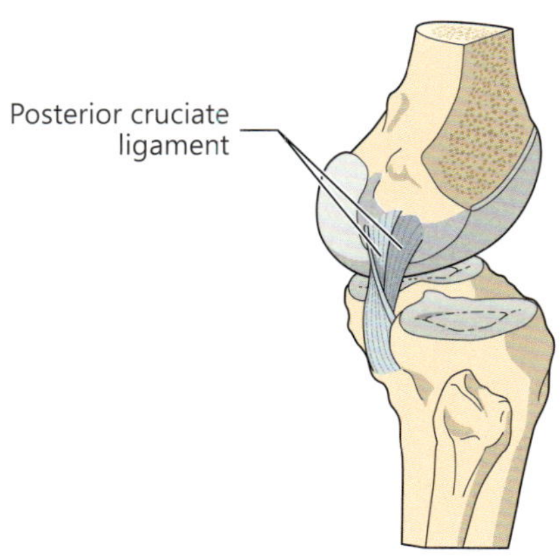

Right knee extended: lateral view.
Function of the posterior cruciate ligament:

- stabilizes the knee anteroposteriorly
- limits:
  - posterior movement of the tibia on the femur
  - rotation of the knee

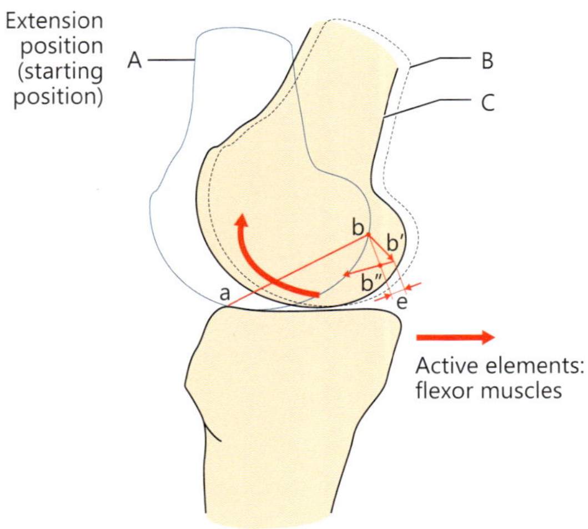

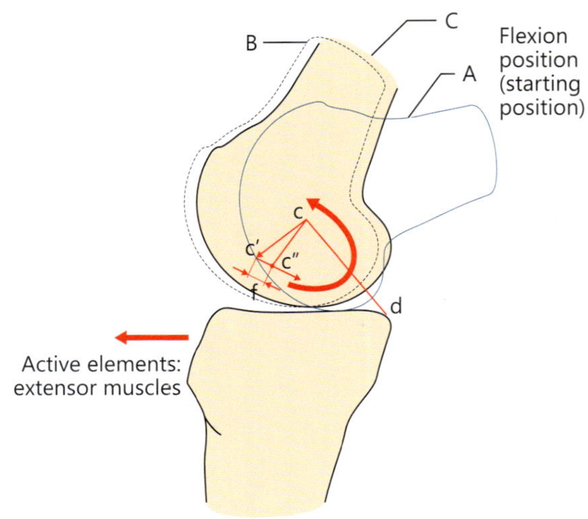

The cruciate ligaments, passive factors, control the sliding and rotational movements of the femoral condyles on the tibial condyles, which occur during flexion and extension of the knee. More precisely:

- during flexion there is an combined action of forward sliding and backward rotation of the lateral condyle of the femur – guided by the anterior cruciate ligament. If the lateral condyle of the femur rotates backward without slipping forward, point b, corresponding to the insertion of the anterior cruciate ligament to the medial surface of the condyle, would move along the bb' trajectory, corresponding to position B of the condyle. Following the forward sliding of the condyle, the insertion of the anterior cruciate ligament, which moves in a circle with a at the center and a radius of ab, instead finds itself at b'', corresponding to position C of the condyle, a more anterior position (lenght e) in comparison with position B.
- during extension there is an opposite action (backward sliding and forward rotation of the medial condyle of the femur) – guided by the posterior cruciate ligament. If the medial condyle of the femur rotates forward without slipping back, point c, corresponding to the insertion of the posterior cruciate ligament to the lateral surface of the condyle, would move along the cc' trajectory, corresponding to position B of the condyle. Following the backward sliding of the condyle, the insertion of the posterior cruciate ligament, which moves in a circle with d at the center and a radius of dc, instead finds itself at c'', corresponding to position C of the condyle, a more posterior position (lenght f) in comparison with position B.

MOVEMENTS OF THE PATELLA

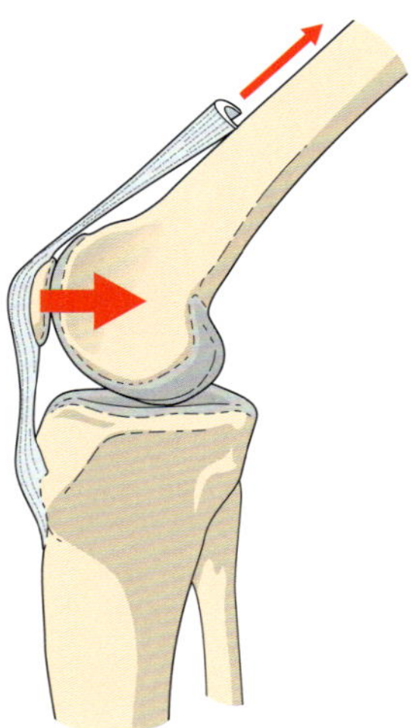

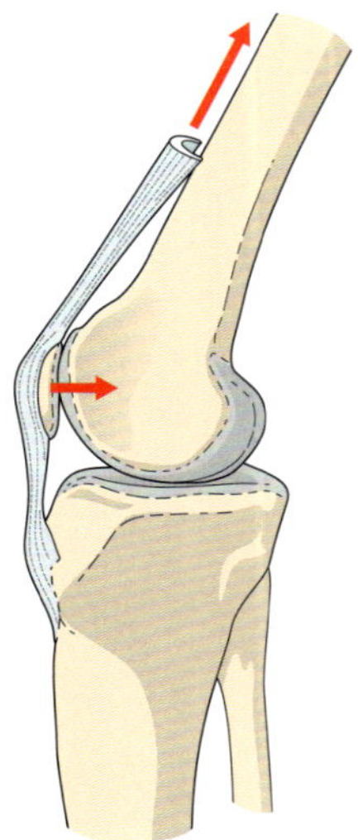

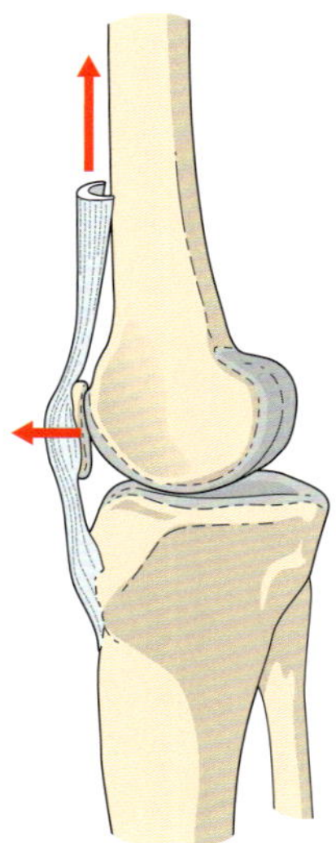

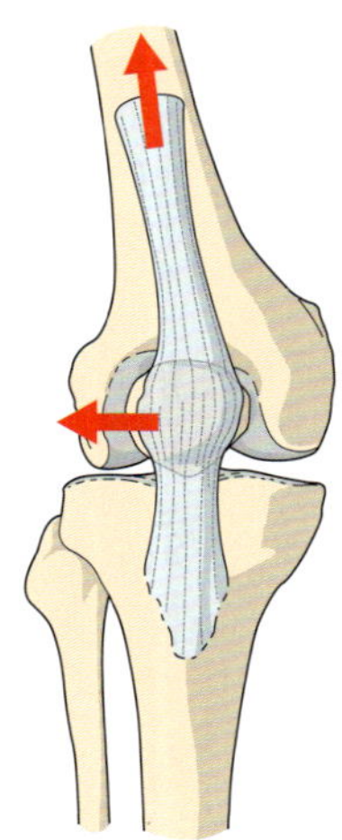

Flexion of the knee
Movement of the patella in a posteroinferior direction, upward, and anterosuperiorly, downward, from a vertical position corresponding to 0° of flexion to a horizontal position (posterior surface of the patella oriented downward and backward) in maximum flexion. This movement is equal to about twice the length of the patella and is made possible by the presence of the lateral patellar and subquadricipital capsular recesses.

Extension of the knee
Vertical position of the patella.

Hyperextension of the knee
Movement of the patella in a lateral direction, to a small degree (due to a decreased tension of the fibers of the quadriceps femoris, within which tendon the patella is contained, and to an increased tension in the patellar ligament, which is a continuation of the tendon of the quadriceps femoris, with an increased external angle of the knee, which becomes equal to or greater than 180°).

Internal rotation of the knee
Movement of the patella in a lateral direction produced by an increase in tension of the patellar ligament, caused by the external rotation of the thigh which accompanies the internal rotation of the leg.

External rotation of the knee
Movement of the patella in a medial direction produced by an increase in tension of the patellar ligament, caused by the internal rotation of the thigh which accompanies the external rotation of the leg.

ARCHES OF THE FOOT

- Longitudinal:
  - internal
  - external
- Transverse

The internal (medial) longitudinal arch is formed by:
- talus
- navicular
- 1st, 2nd, and 3rd cuneiforms
- 1st, 2nd, and 3rd metatarsals

The external (lateral) longitudinal arch is formed by:
- calcaneus
- cuboid
- 4th and 5th metatarsals

The transverse arch (orthogonal to the longitudinal arches) is formed by:
- 1st, 2nd, 3rd cuneiforms and cuboid
- 1st, 2nd, 3rd, 4th and 5th metatarsals (metatarsal arch at the level of the heads)

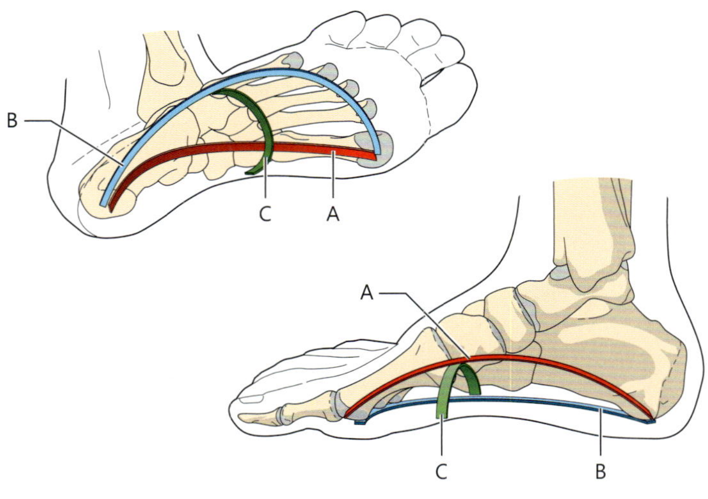

Outline of the three plantar arches. A, Internal longitudinal arch; B, external longitudinal arch; C, transverse arch.

HALLUX VALGUS

Outward deviation of the great toe, with superposition over the other toes.

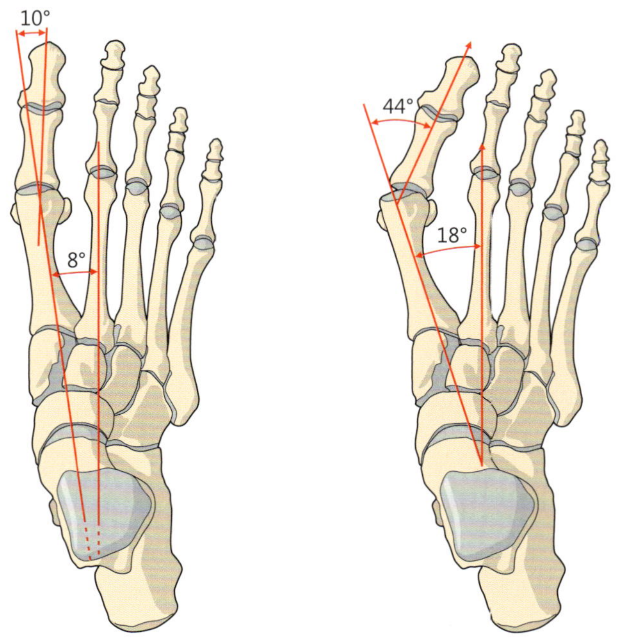

MOVEMENTS OF THE JOINTS OF THE ANKLE AND FOOT

Ankle joint

Flexion-extension of the ankle
- Flexion of the ankle (of the foot):
  - movement of the dorsum of the foot towards the tibia (dorsal flexion or dorsiflexion of the foot)
  - amplitude: 30-35°
- Extension of the ankle (of the foot):
  - movement of the foot in the opposite direction relative to the tibia (ventral flexion of the foot)
  - amplitude: 40-50°

Axis of movement
- Frontal axis passing through the medial (tibial) and lateral (fibular) malleoli; there is actually no connection between this axis and the frontal plane, since the fibular malleolus is located further back and further down in relation to the tibial malleolus, so during dorsiflexion of the foot partial movements of supination, abduction and inversion of the foot occur in succession

Reference position
- Angle of 90° between the axis of the foot and the axis of the leg.

During flexion-extension movements of the foot, the talus moves together with the calcaneus, forming a single body.

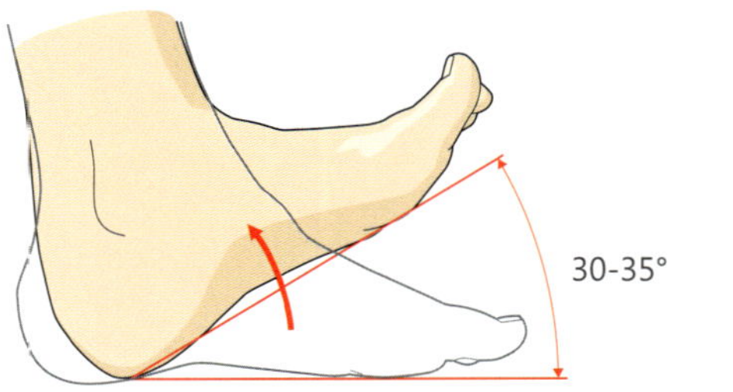

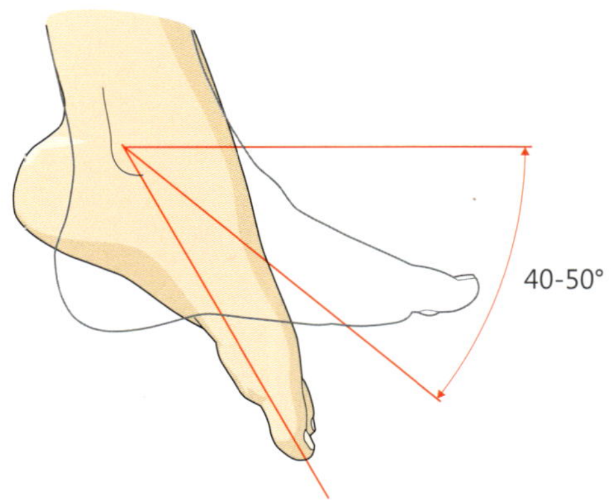

Subtalar (talocalcaneal) joint

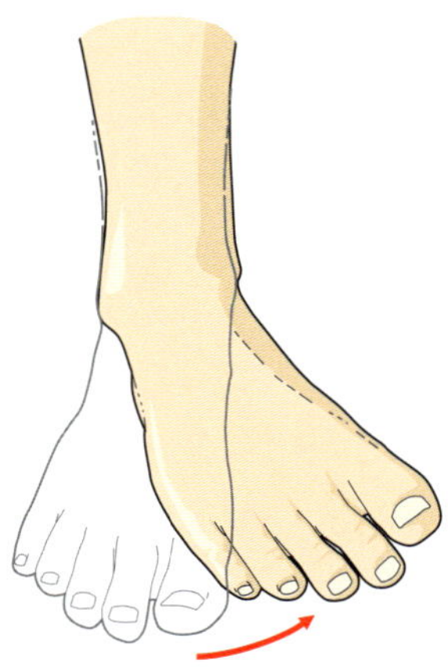

Adduction of the foot
The tip of the foot is brought inward (toward the median plane of symmetry).

The total amplitude of the abduction-adduction movements of the foot varies from 35° to 45°.

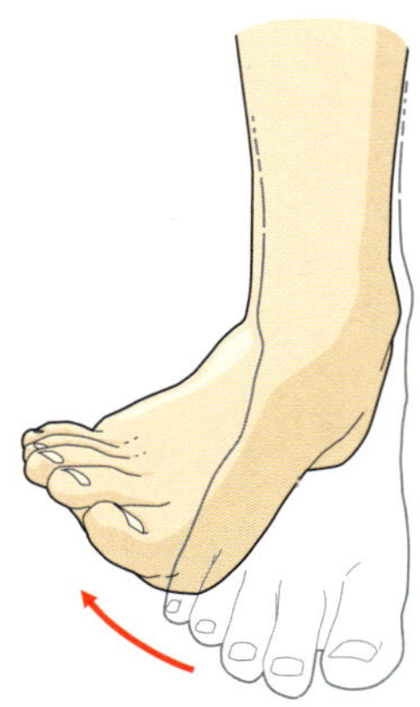

Abduction of the foot
The tip of the foot is brought outward (away from the median plane of symmetry).

The total amplitude of the abduction-adduction movements of the foot varies from 35° to 45°.

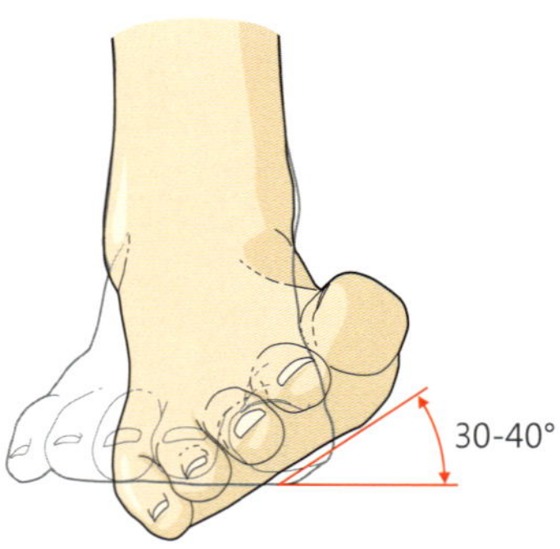

Supination of the foot
Rotation of the foot with inward exposure of the sole of the foot.

The amplitude of the supination movement of the foot is about 30-40°.

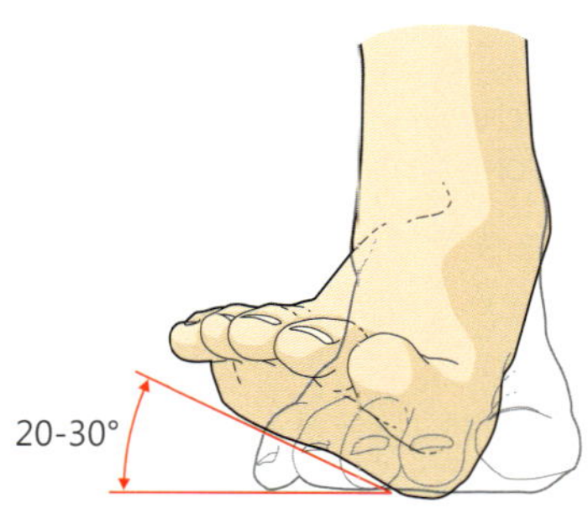

Pronation of the foot
Rotation of the foot with outward exposure of the sole of the foot.

The amplitude of the pronation movement of the foot is 20-30°.

The movements of adduction, abduction, supination and pronation, in relation to the mode of organization of the joints of the foot, do not occur individually, but are associated with one another and with the dorsal and plantar flexion of the foot, giving rise to movements of inversion and eversion of the foot.

Inversion and eversion of the foot

Inversion
Compound movements of adduction, supination, and plantar flexion of the foot which lead to the inward exposure of the sole of the foot.

Eversion
Compound movements of abduction, pronation, and dorsal flexion of the foot which lead to the outward exposure of the sole of the foot.

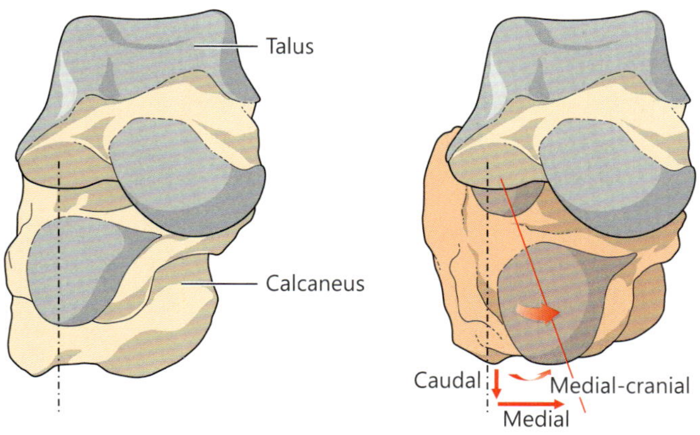

Neutral position Inversion

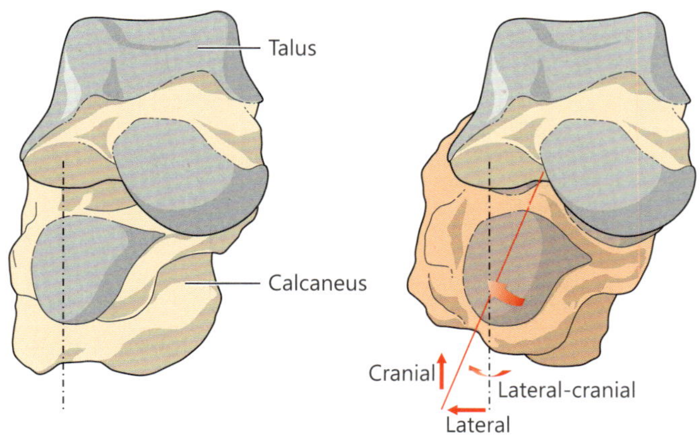

Neutral position Eversion

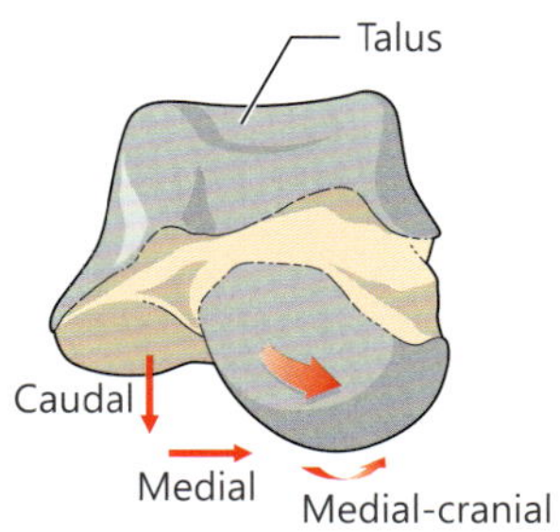

Sliding of the navicular in relation to the talus during inversion.

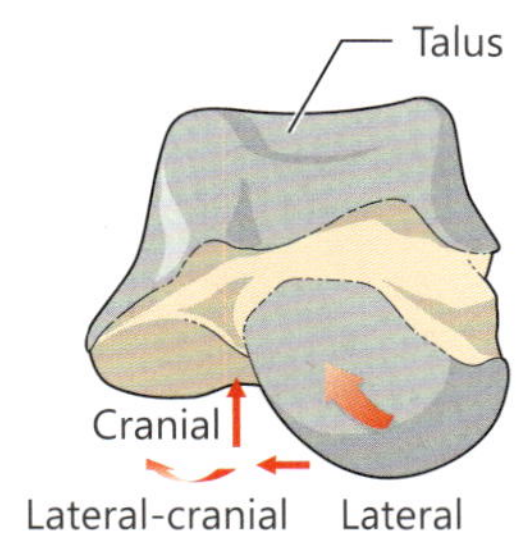

Sliding of the navicular in relation to the talus during eversion.

Axis of movement

- Between the posterior calcaneal tuberosity and the surface of the trochlea of the talus in the anteromedial position (this is an indicative axis, since individual movements occur, each around its own axis, which are added up together)
- Oblique direction from back to front, from lateral to medial, and from bottom upwards.

During eversion and inversion of the foot, the talus is fixed between the tibia and the fibula, allowing the calcaneus to move with the rest of the foot.

Summary table of the joints and ligaments of the lower limb

JOINT, LIGAMENT, OR MEMBRANE	TYPE OF JOINT	BONES INVOLVED
	JOINTS OF THE PELVIC GIRDLE	
Obturator canal		
Obturator membrane		
Pubic symphysis – interpubic disc – superior pubic ligament – inferior pubic ligament	Synarthrosis (syndesmosis or symphysis)	Between the two bones of the hip (pubic bones)
Sacroiliac joint – anterior sacroiliac ligament – interosseous sacroiliac ligament – posterior sacroiliac ligament – sacrotuberous ligament – – falciform process – sacrospinous ligament – greater sciatic foramen – lesser sciatic foramen	Synarthrosis (syndesmosis or symphysis or amphiarthrosis) or diarthrosis (plane joint)	Hip bone (iliac bone) and sacrum
	JOINTS OF THE FREE PART OF THE LOWER LIMB	
Hip joint – zona orbicularis – iliofemoral ligament – – transverse part – – descending part – ischiofemoral ligament – pubofemoral ligament – acetabular labrum – transverse acetabular ligament – ligament of the head of the femur	Diarthrosis (ball and socket joint)	Between the acetabulum of the hip bone and the head of the femur
Knee joint – lateral meniscus – – anterior meniscofemoral ligament – – posterior meniscofemoral ligament – medial meniscus – transverse ligament of the knee – anterior cruciate ligament – posterior cruciate ligament – infrapatellar synovial fold – – alar folds – fibular collateral ligament – tibial collateral ligament – oblique popliteal ligament – arcuate popliteal ligament – patellar ligament – medial patellar retinaculum – lateral patellar retinaculum – infrapatellar fat pad	Diarthrosis (hinge joint or trochlea)	Femur (condyles and patellar surface), tibia (superior articular surface) and patella (articular surface)
Tibiofibular joint – anterior ligament of the fibular head – posterior ligament of the fibular head	Diarthrosis (plane joint)	Tibia and fibula
Tibiofibular syndesmosis – interosseous membrane of the leg – anterior tibiofibular ligament – posterior tibiofibular ligament	Synarthrosis (syndesmosis)	Tibia and fibula
	JOINTS OF THE FOOT	
Ankle joint – medial ligament (deltoid ligament) – – tibionavicular part – – tibiocalcaneal part – – anterior tibiotalar part – – posterior tibiotalar part – lateral ligament – – anterior talofibular ligament – – posterior talofibular ligament – – calcaneofibular ligament	Diarthrosis (hinge joint or trochlea)	Tibia and fibula (tibiofibular mortise) and talus (trochlea)
Subtalar joint (talocalcaneal joint) – lateral talocalcaneal ligament – medial talocalcaneal ligament – posterior talocalcaneal ligament	Diarthrosis (pivot or trochoid joint)	Talus (posterior calcaneal articular facet) and calcaneus (posterior talar articular surface)

continued

Summary table of the joints and ligaments of the lower limb		(continued)
JOINT, LIGAMENT, OR MEMBRANE	***TYPE OF JOINT***	***BONES INVOLVED***
Transverse tarsal joint – talocalcaneonavicular joint – – plantar calcaneonavicular ligament – calcaneocuboid joint	 Diarthrosis (ball and socket joint) Diarthrosis (saddle joint)	 Between the head of the talus, the navicular (articular surface of the talus) and the calcaneus (anterior and middle talar articular surfaces) Calcaneus (articular surface for the cuboid) and cuboid
Cuboideonavicular joint (not always present)	Diarthrosis (plane joint)	Navicular and cuboid
Cuneonavicular joint	Diarthrosis (three plane joints)	Navicular and cuneiforms (medial, intermediate and lateral)
Intercuneiform joints	Diarthroses (two plane joints)	Between opposing surfaces of the three cuneiforms
Tarsal ligaments – tarsal interosseous ligaments – – talocalcaneal interosseous ligament – – cuneocuboid interosseous ligament – – intercuneiform interosseous ligaments – dorsal tarsal ligaments – – talonavicular ligament – – dorsal intercuneiform ligaments – – dorsal cuneocuboid ligament – – dorsal cuboideonavicular ligament – – bifurcate ligament – – – calcaneonavicular ligament – – – calcaneocuboid ligament – dorsal cuneonavicular ligaments – – dorsal calcaneocuboid ligament – plantar tarsal ligaments – – long plantar ligament – – plantar calcaneocuboid ligament – – plantar calcaneonavicular ligament – – plantar cuneonavicular ligaments – – plantar cuboideonavicular ligament – – plantar intercuneiform ligaments – – plantar cuneocuboid ligament		
Tarsometatarsal joints – dorsal tarsometatarsal ligaments – plantar tarsometatarsal ligaments – cuneometatarsal interosseous ligaments	Diarthroses (plane joints)	Between the medial cuneiform and the first metatarsal, the medial cuneiform and the second metatarsal, the intermediate cuneiform and the second metatarsal, the lateral cuneiform and the second metatarsal, the lateral cuneiform and the third metatarsal, the cuboid and the fourth metatarsal, the cuboid and the fifth metatarsal
Intermetatarsal joints – metatarsal interosseous ligaments – dorsal metatarsal ligaments – plantar metatarsal ligaments – intermetatarsal spaces	Capsular extensions of the tarsometatarsal joints between the bases of the metatarsals, with the exclusion of the first two joined solely by an interosseous ligament	Between the bases of the metatarsals
Metatarsophalangeal joints – collateral ligaments – plantar ligaments – deep transverse metatarsal ligament	Diarthroses (condylar joint)	Heads of the metatarsals and bases of the proximal phalanges
Interphalangeal joints of the foot – collateral ligaments – plantar ligaments	Diarthroses (hinge joint or trochlea)	Between the heads of the proximal phalanges and bases of the middle phalanges, heads of the middle phalanges and bases of the distal phalanges, the head of the proximal phalanx and the base of the distal phalanx (great toe)

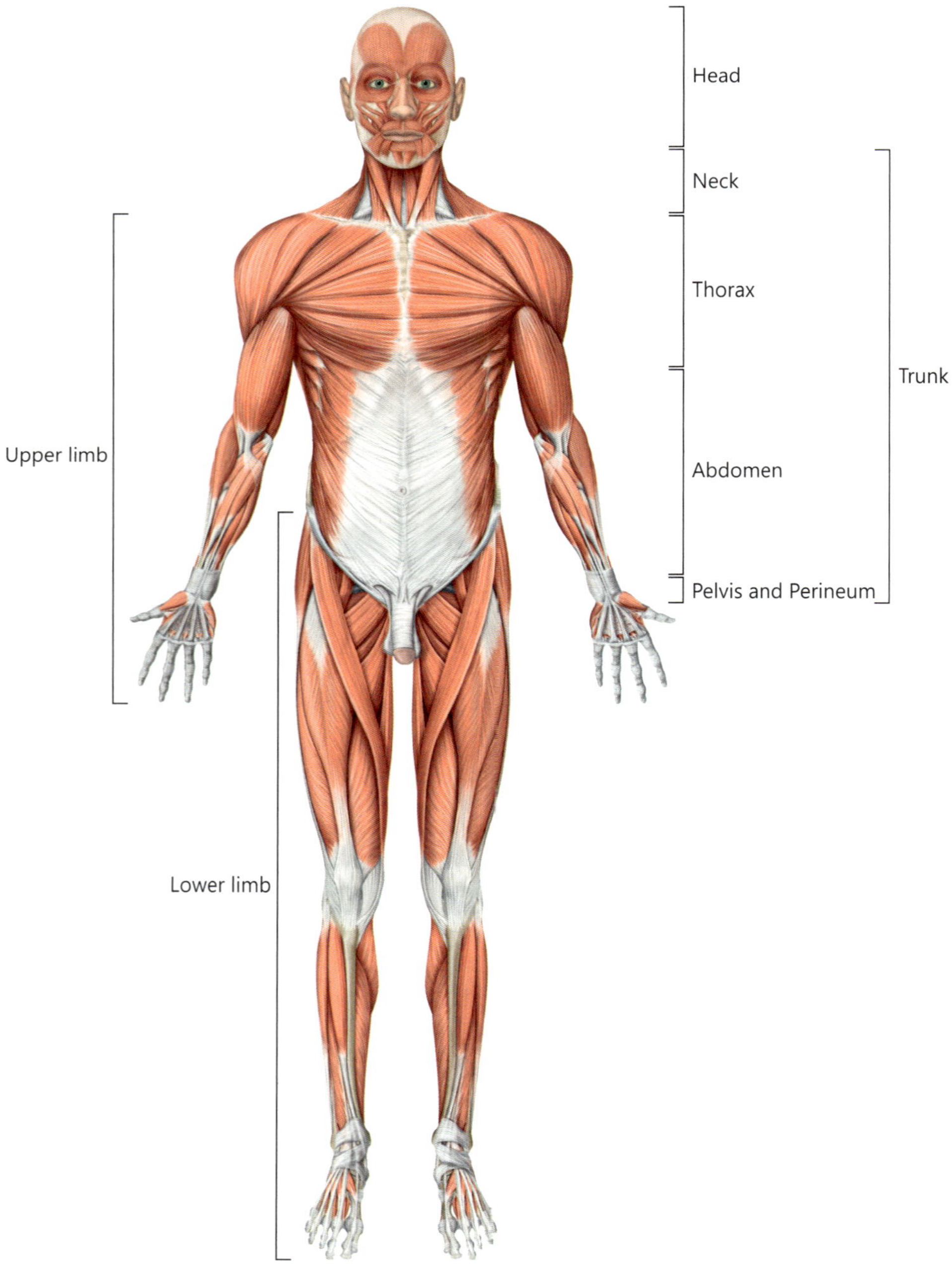

HEAD
Extraocular muscles
– inferior oblique
– inferior rectus
– lateral rectus
– levator palpebrae superioris
– medial rectus
– superior oblique
– superior rectus
Facial muscles
– auricularis anterior
– auricularis posterior
– auricularis superior
– buccinator
– corrugator supercilii
– depressor anguli oris
– depressor labii inferioris
– depressor septi nasi

continued

HEAD (continued)
– depressor supercilii
– epicranius
– – occipitofrontalis
– – temporoparietalis
– levator anguli oris
– levator labii superioris
– levator labii superioris alaeque nasi
– mentalis
– nasalis
– orbicularis oculi
– orbicularis oris
– procerus
– risorius
– transversus menti
– zygomaticus major
– zygomaticus minor
Longissimus capitis (*mainly in the neck*)

continued

HEAD (continued)
Masticatory muscles
– lateral pterygoid
– masseter
– medial pterygoid
– temporalis
Muscles of auditory ossicles
– stapedius
– tensor tympani
Muscles of soft palate and fauces
– levator veli palatini
– musculus uvulae
– palatoglossus
– palatopharyngeus
– tensor veli palatini
Muscles of tongue
– genioglossus
– hyoglossus
– – ceratoglossus
– – chondroglossus
– inferior longitudinal muscle
– palatoglossus
– styloglossus
– superior longitudinal muscle
– transverse muscle
– vertical muscle
Prevertebral muscles (*mainly in the neck*)
Rectus capitis lateralis (*mainly in the neck*)
Semispinalis capitis (*mainly in the neck*)
Spinalis capitis (*mainly in the neck*)
Splenius capitis (*mainly in the neck*)
Suboccipital muscles (*mainly in the neck*)

NECK
Infrahyoid muscles
– omohyoid
– sternohyoid
– sternothyroid
– thyrohyoid
Muscles of back proper
– cervical intertransversarii
– erector spinae
– – iliocostalis cervicis
– – longissimus capitis
– – longissimus cervicis
– – spinalis capitis
– – spinalis cervicis
– interspinales cervicis
– spinotransversales
– – splenius capitis
– – splenius cervicis
– transversospinales
– – multifidus cervicis
– – rotatores cervicis
– – semispinalis capitis
– – semispinalis cervicis
Platysma
Prevertebral muscles
– longus capitis
– longus colli
– rectus capitis anterior
Rectus capitis lateralis
Scalenus anterior
Scalenus medius
Scalenus posterior
Sternocleidomastoid
Suboccipital muscles
– obliquus capitis inferior
– obliquus capitis superior
– rectus capitis posterior major
– rectus capitis posterior minor
Suprahyoid muscles
– digastric
– geniohyoid
– mylohyoid
– stylohyoid
Trapezius (*mainly in the thorax*)

THORAX
Diaphragm
External intercostal muscles
Innermost intercostal muscles
Internal intercostal muscles
Levatores costarum
Muscles of back
– latissimus dorsi
– levator scapulae
– muscles proper
– – erector spinae
– – – iliocostalis lumborum, thoracic part
– – – longissimus thoracis
– – – spinalis thoracis
– – interspinales thoracis
– – thoracic intertransversarii
– – transversospinales
– – – multifidus thoracis
– – – rotatores thoracis
– – – semispinalis thoracis
– rhomboid major
– rhomboid minor
– serratus posterior inferior
– serratus posterior superior
– trapezius
Pectoralis major
Pectoralis minor
Serratus anterior
Sternalis
Subclavius
Subcostales
Transversus thoracis

ABDOMEN
Cremaster ♂
External oblique of abdomen
Iliopsoas
– iliacus (*mainly in the pelvis*)
– psoas major
Internal oblique of abdomen
Latissimus dorsi (*mainly in the thorax*)
Muscles of back proper
– erector spinae
– – iliocostalis lumborum, lumbar part
– interspinales lumborum
– intertransversarii lumborum
– transversospinales
– – multifidus lumborum
– – rotatores lumborum
Psoas minor
Pyramidalis
Quadratus lumborum
Rectus abdominis
Transversus abdominis

PELVIS
Gemellus inferior (*mainly in the lower limb*)
Gemellus superior (*mainly in the lower limb*)
Gluteus maximus (*mainly in the lower limb*)
Gluteus medius (*mainly in the lower limb*)
Gluteus minimus (*mainly in the lower limb*)
Iliopsoas
– iliacus
– psoas major (*mainly in the abdomen*)
Ischiococcygeus
Obturator externus (*mainly in the lower limb*)
Obturator internus

continued

PELVIS (continued)
Pelvic diaphragm
– levator ani
– – iliococcygeus
– – pubococcygeus
– – – puboanalis
– – – puboperineales
– – – puboprostaticus ♂
– – – pubovaginalis ♀
– – puborectalis
Piriformis
Psoas minor (*mainly in the abdomen*)
Quadratus femoris (*mainly in the lower limb*)
Tensor fasciae latae (*mainly in the lower limb*)

PERINEUM
Bulbospongiosus
External anal sphincter
Ischiocavernosus
Muscles of urogenital triangle
– deep transverse perineal muscle ♂
– external urethral sphincter ♂
– urogenital sphincter ♀
– – compressor urethrae ♀
– – external urethral sphincter ♀
– – sphincter urethrovaginalis ♀
Superficial transverse perineal muscle

UPPER LIMB
Abductor digiti minimi
Abductor pollicis brevis
Abductor pollicis longus
Adductor pollicis
Anconeus
Articularis cubiti
Biceps brachii
Brachialis
Brachioradialis
Coracobrachialis
Deltoid
Dorsal interossei
Extensor carpi radialis brevis
Extensor carpi radialis longus
Extensor carpi ulnaris
Extensor digiti minimi
Extensor digitorum
Extensor indicis
Extensor pollicis brevis
Extensor pollicis longus
Flexor carpi radialis
Flexor carpi ulnaris
Flexor digiti minimi brevis
Flexor digitorum profundus
Flexor digitorum superficialis
Flexor pollicis brevis
Flexor pollicis longus
Infraspinatus
Latissimus dorsi (*mainly in the thorax*)
Levator scapulae (*mainly in the thorax*)
Lumbricals
Opponens digiti minimi
Opponens pollicis
Palmar interossei
Palmaris brevis
Palmaris longus
Pectoralis major (*mainly in the thorax*)

continued

UPPER LIMB (continued)
Pectoralis minor (*mainly in the thorax*)
Pronator quadratus
Pronator teres
Rhomboid major (*mainly in the thorax*)
Rhomboid minor (*mainly in the thorax*)
Serratus anterior (*mainly in the thorax*)
Subclavius (*mainly in the thorax*)
Subscapularis
Supinator
Supraspinatus
Teres major
Teres minor
Trapezius (*mainly in the thorax*)
Triceps brachii

LOWER LIMB
Abductor digiti minimi
Abductor hallucis
Adductor brevis
Adductor hallucis
Adductor longus
Adductor magnus
Adductor minimus
Articularis genus
Biceps femoris
Dorsal interossei
Extensor digitorum brevis
Extensor digitorum longus
Extensor hallucis brevis
Extensor hallucis longus
Fibularis brevis
Fibularis longus
Fibularis tertius
Flexor digiti minimi brevis
Flexor digitorum brevis
Flexor digitorum longus
Flexor hallucis brevis
Flexor hallucis longus
Gemellus inferior
Gemellus superior
Gluteus maximus
Gluteus medius
Gluteus minimus
Gracilis
Iliopsoas
– iliacus (*mainly in the pelvis*)
– psoas major (*mainly in the abdomen*)
Lumbricals
Obturator externus
Obturator internus (*mainly in the pelvis*)
Pectineus
Piriformis (*mainly in the pelvis*)
Plantar interossei
Plantaris
Popliteus
Psoas minor (*mainly in the abdomen*)
Quadratus femoris
Quadratus plantae
Quadriceps femoris
– rectus femoris
– vastus intermedius
– vastus lateralis
– vastus medialis
Sartorius
Semimembranosus
Semitendinosus
Tensor fasciae latae
Tibialis anterior
Tibialis posterior
Triceps surae
– gastrocnemius
– soleus

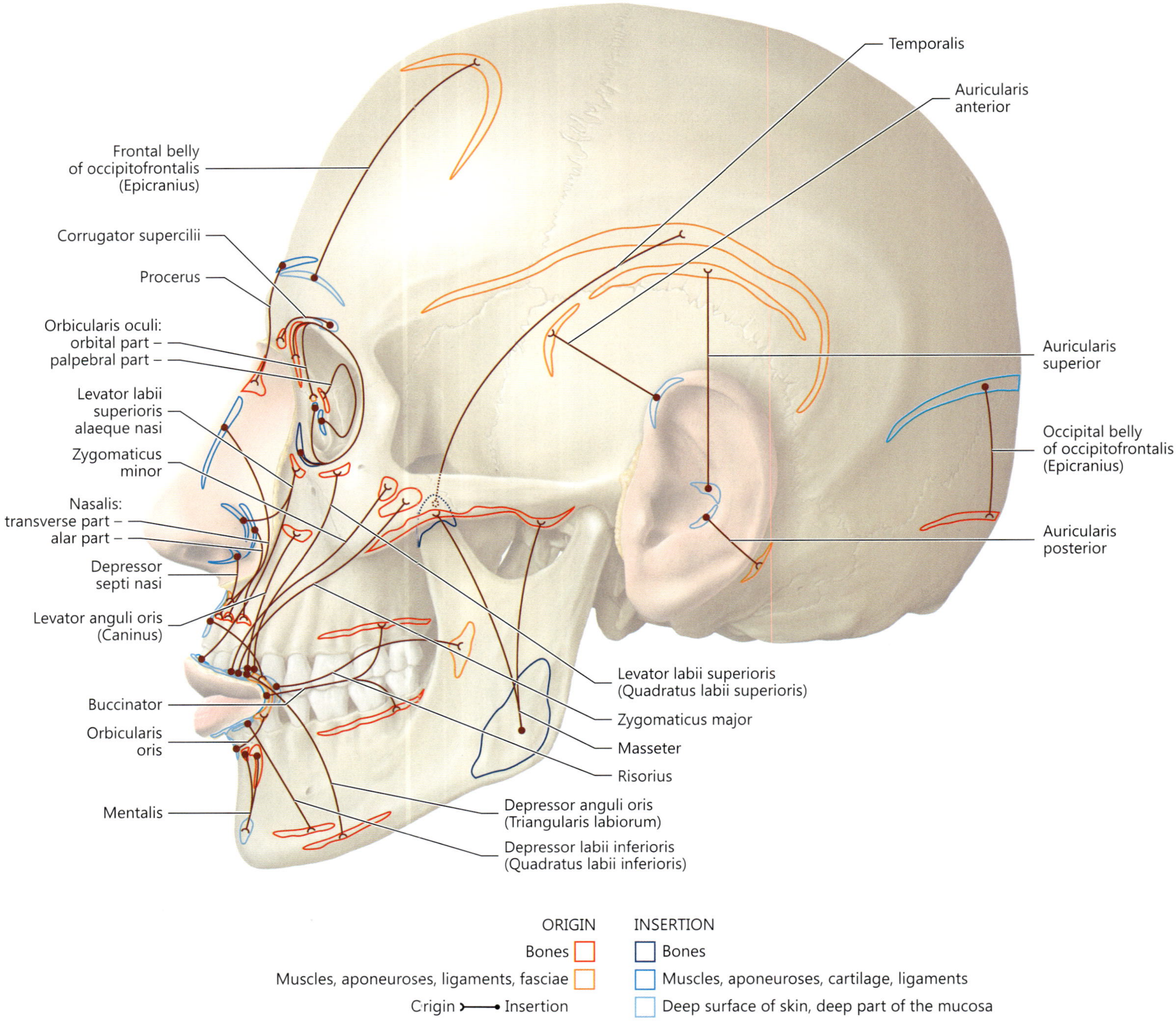

Cranium, left lateral view. The origins and insertions of the **muscles of the head** are shown.

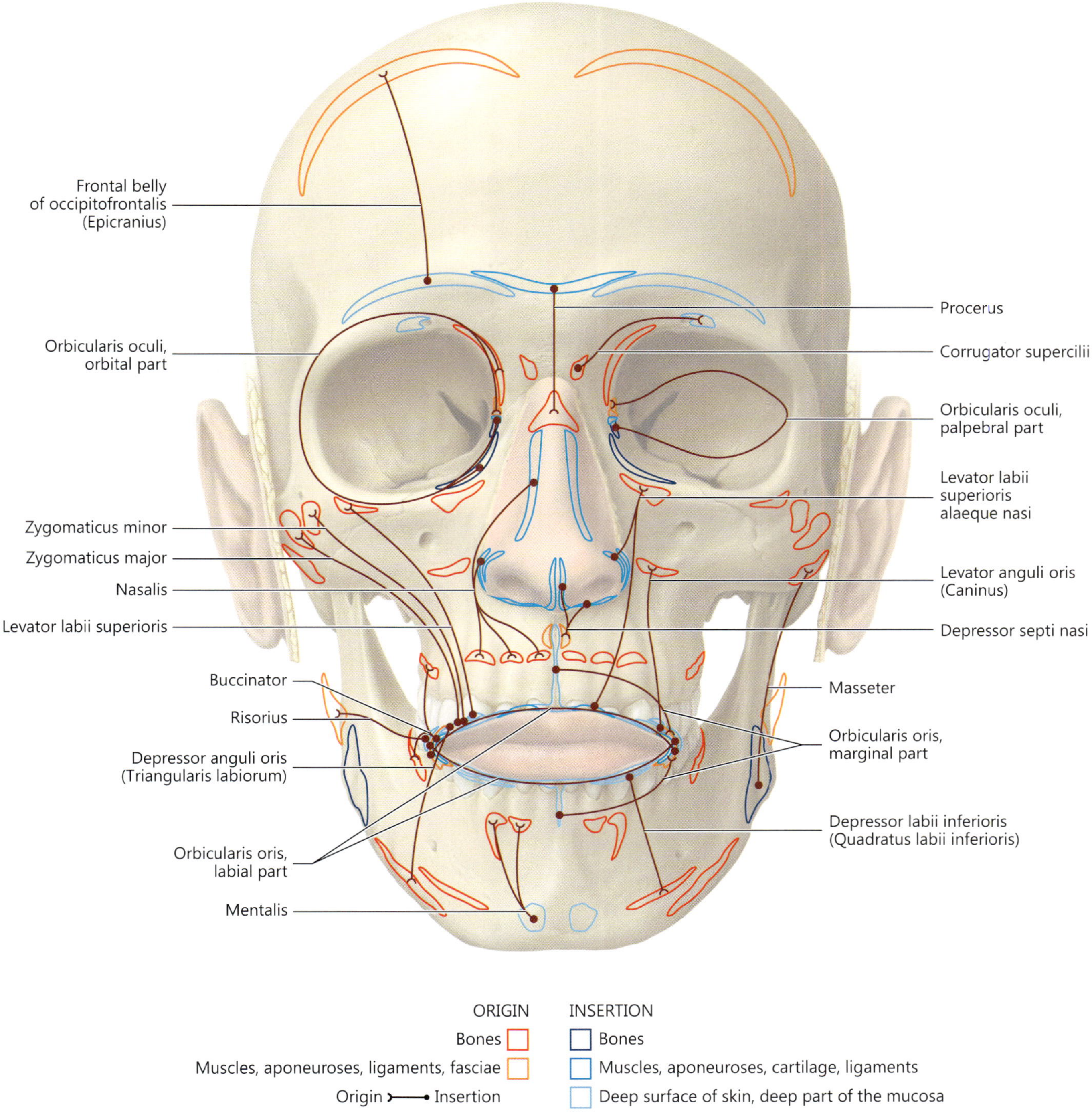

Cranium, frontal view. The origins and insertions of the **muscles of the head** are shown.

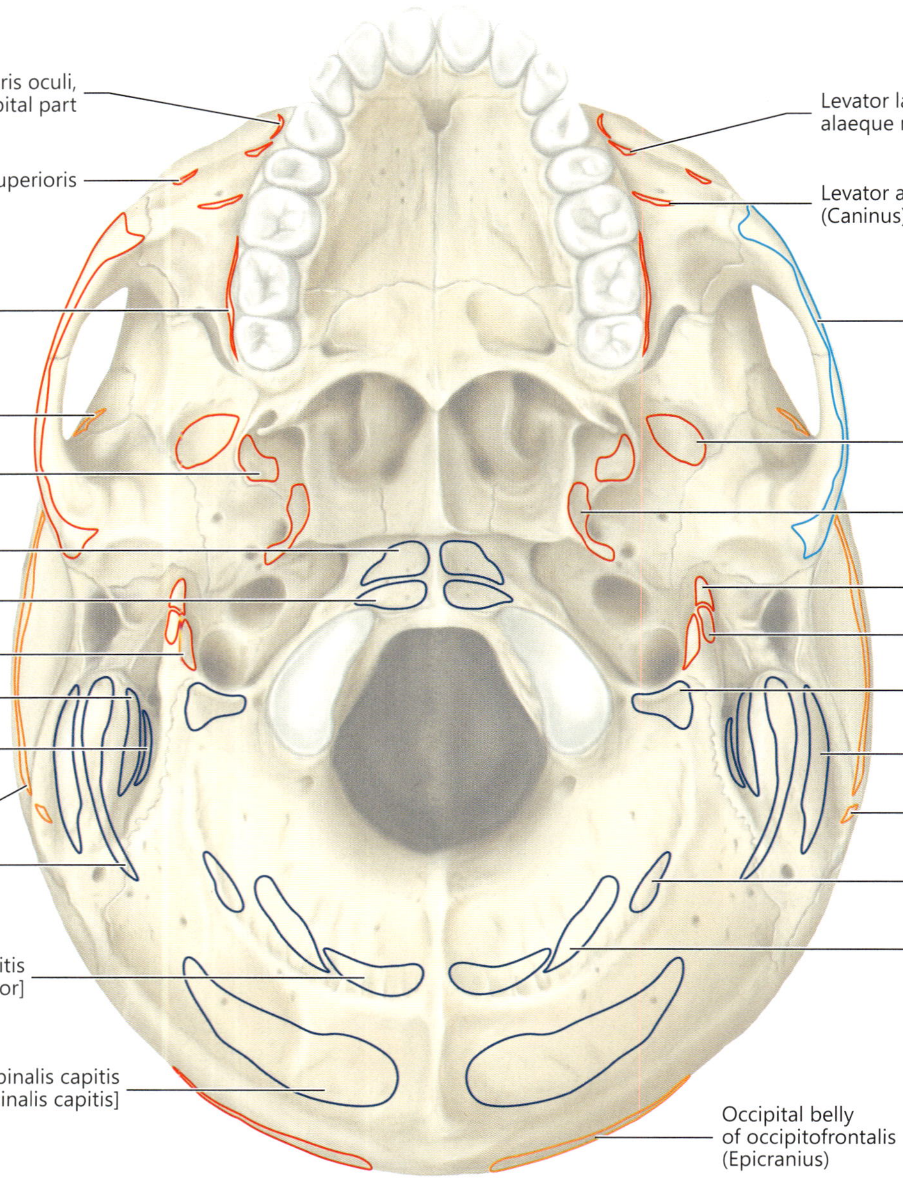

ORIGIN	INSERTION
Bones	Bones
Muscles, aponeuroses, ligaments, fasciae	Muscles, aponeuroses, cartilage, ligaments
Origin ›——• Insertion	Deep surface of skin, deep part of the mucosa

External surface of the cranial base. The origins and insertions of the **muscles of the head** from or to the cranial base are shown. Muscles with extracranial origin or insertion are shown in square brackets.

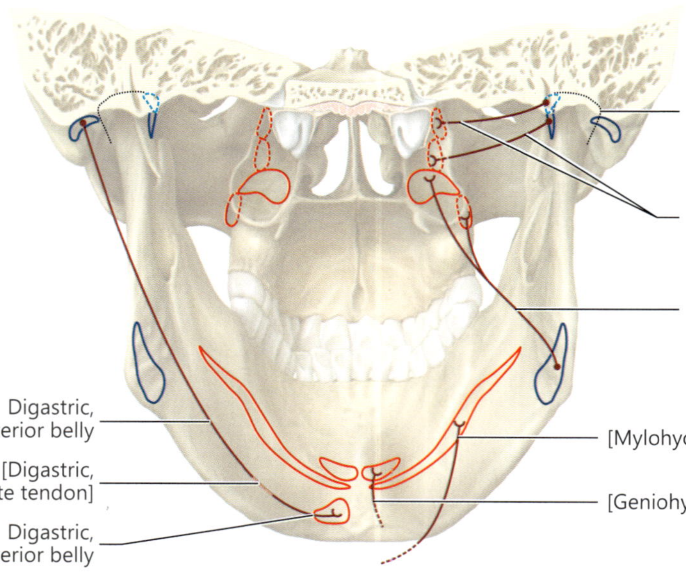

Origin and insertion of the **muscles of the head** and **neck** to the posterior surface of the body of the mandible and to the medial surface of the two rami of the mandible. Muscles with extracranial origin or insertion are shown in square brackets.

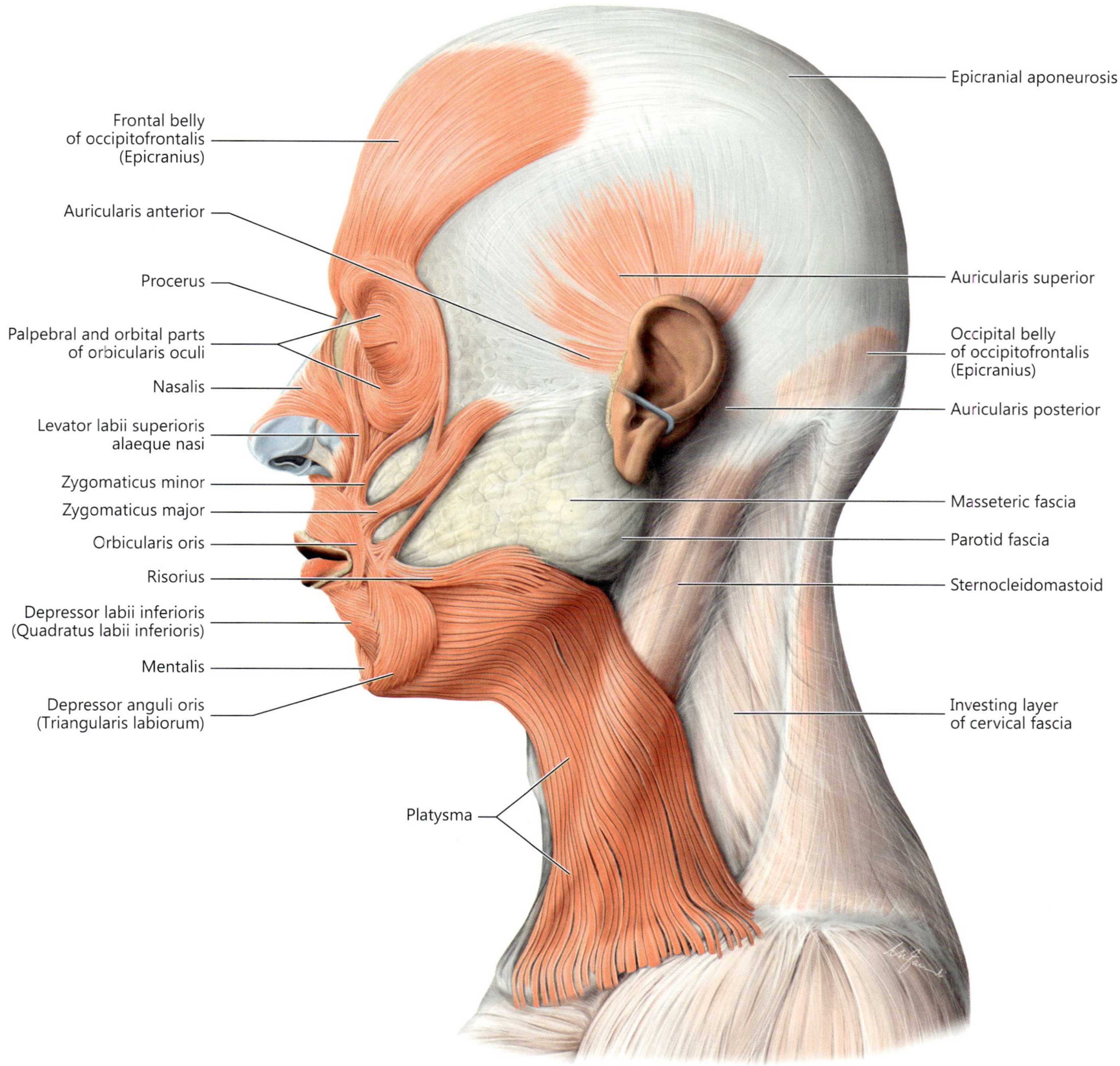

Muscles of the head and face (**mimic muscles**). Superficial layer, left lateral view.

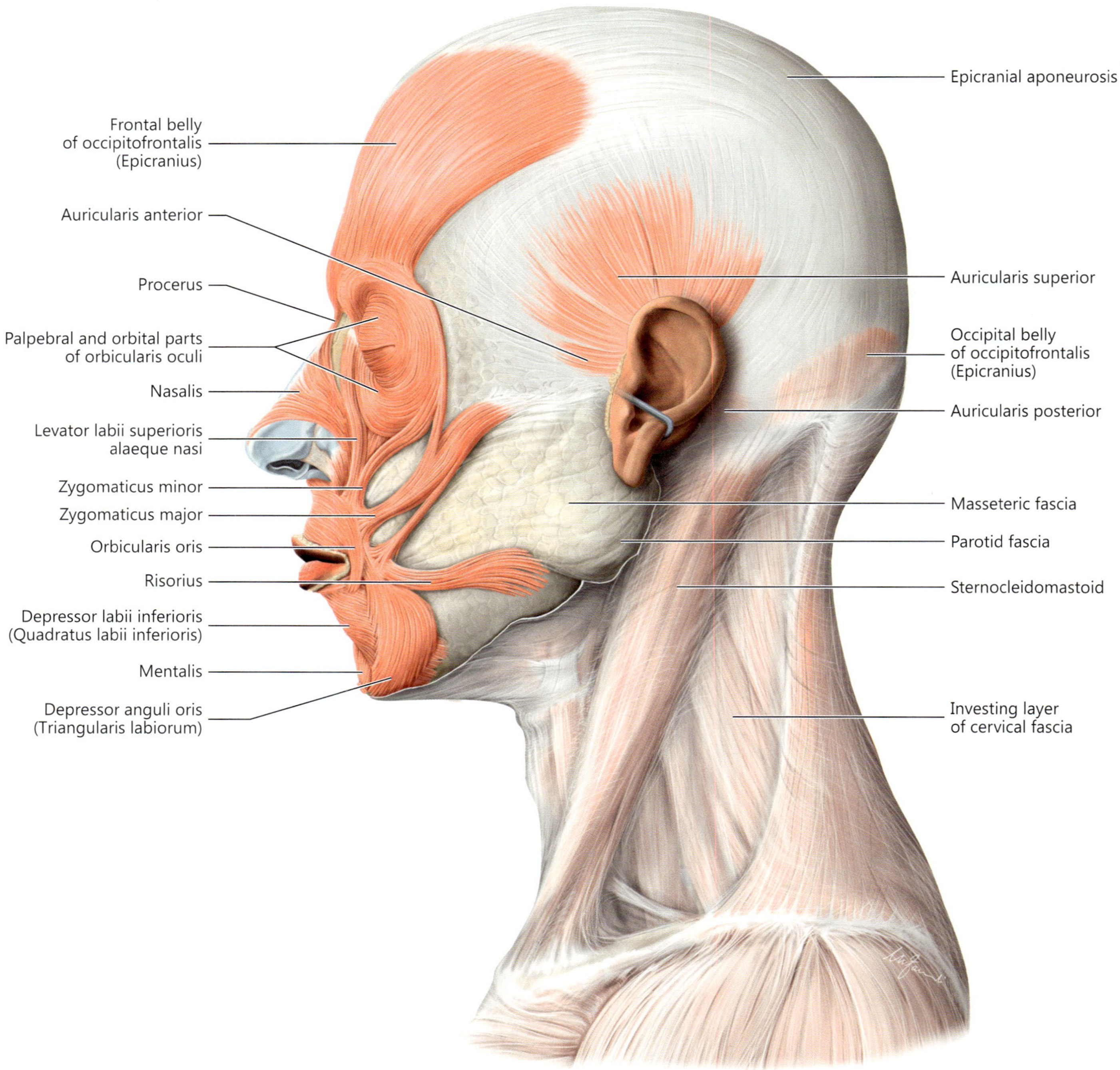

Superficial muscles of the head, left lateral view, and corresponding fasciae after removal of the platysma.

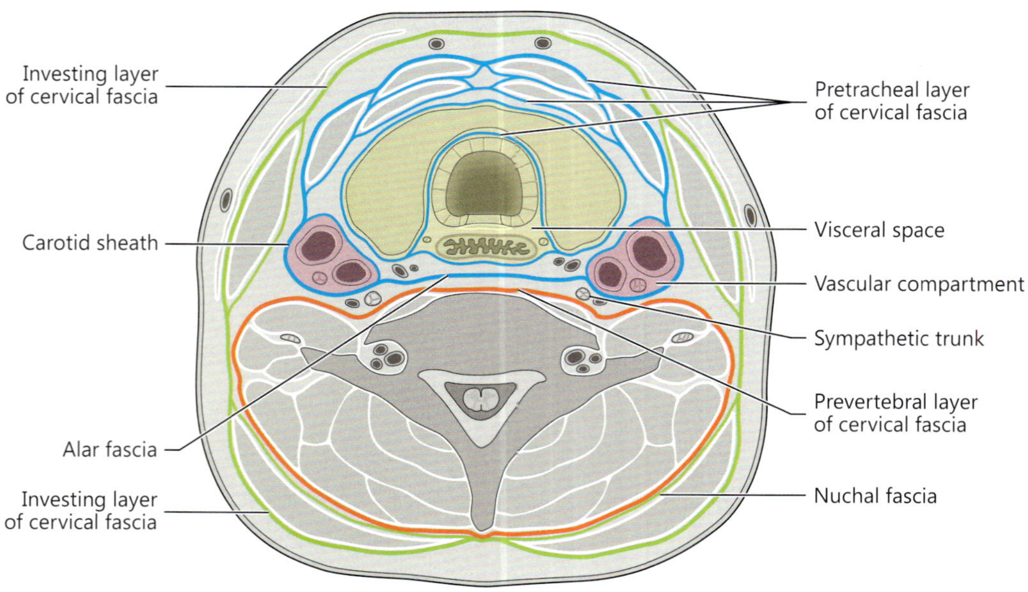

Fasciae and **spaces of the neck**.

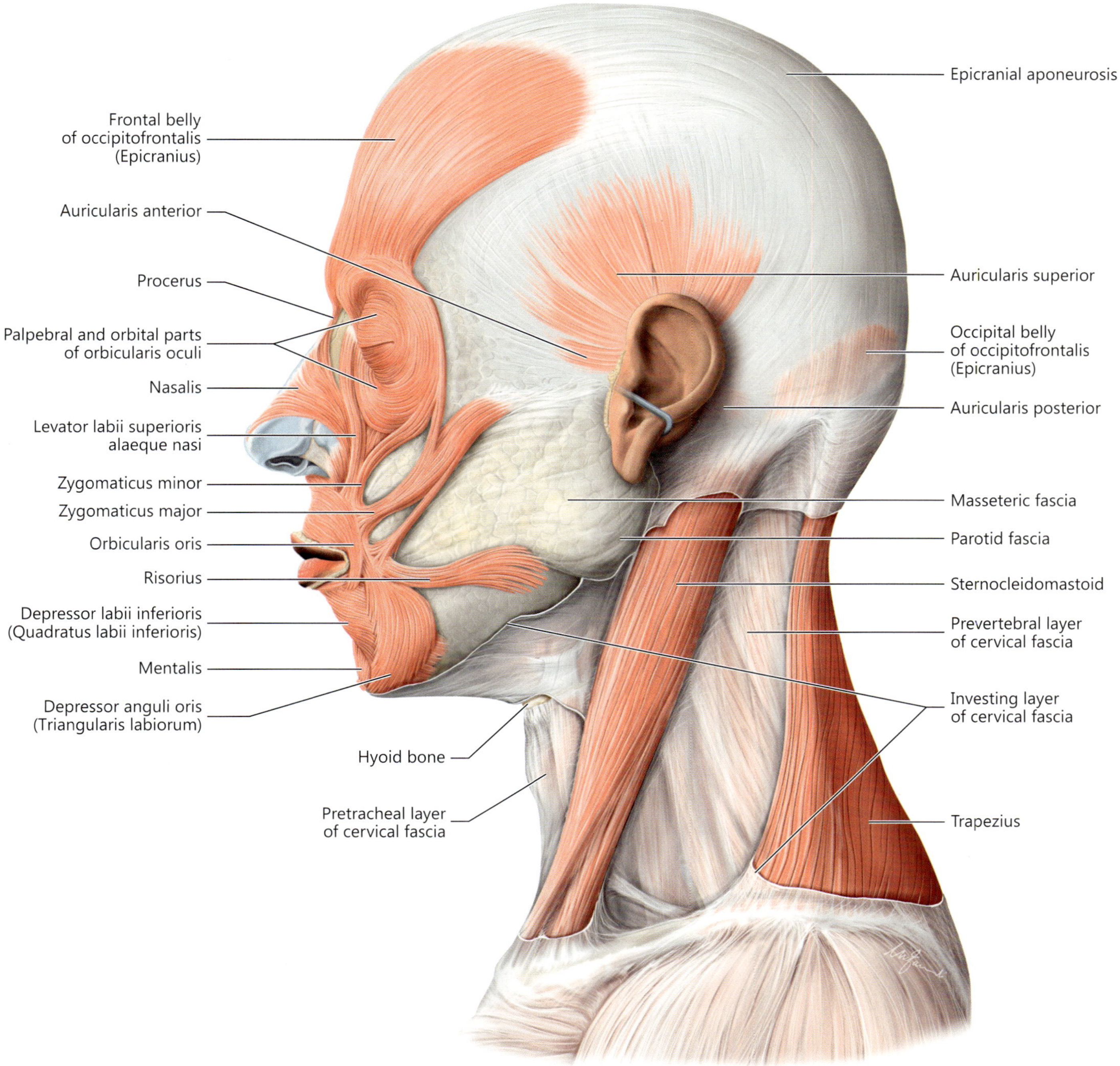

Superficial muscles of the head, left lateral view, after removal of the platysma and superficial fasciae.

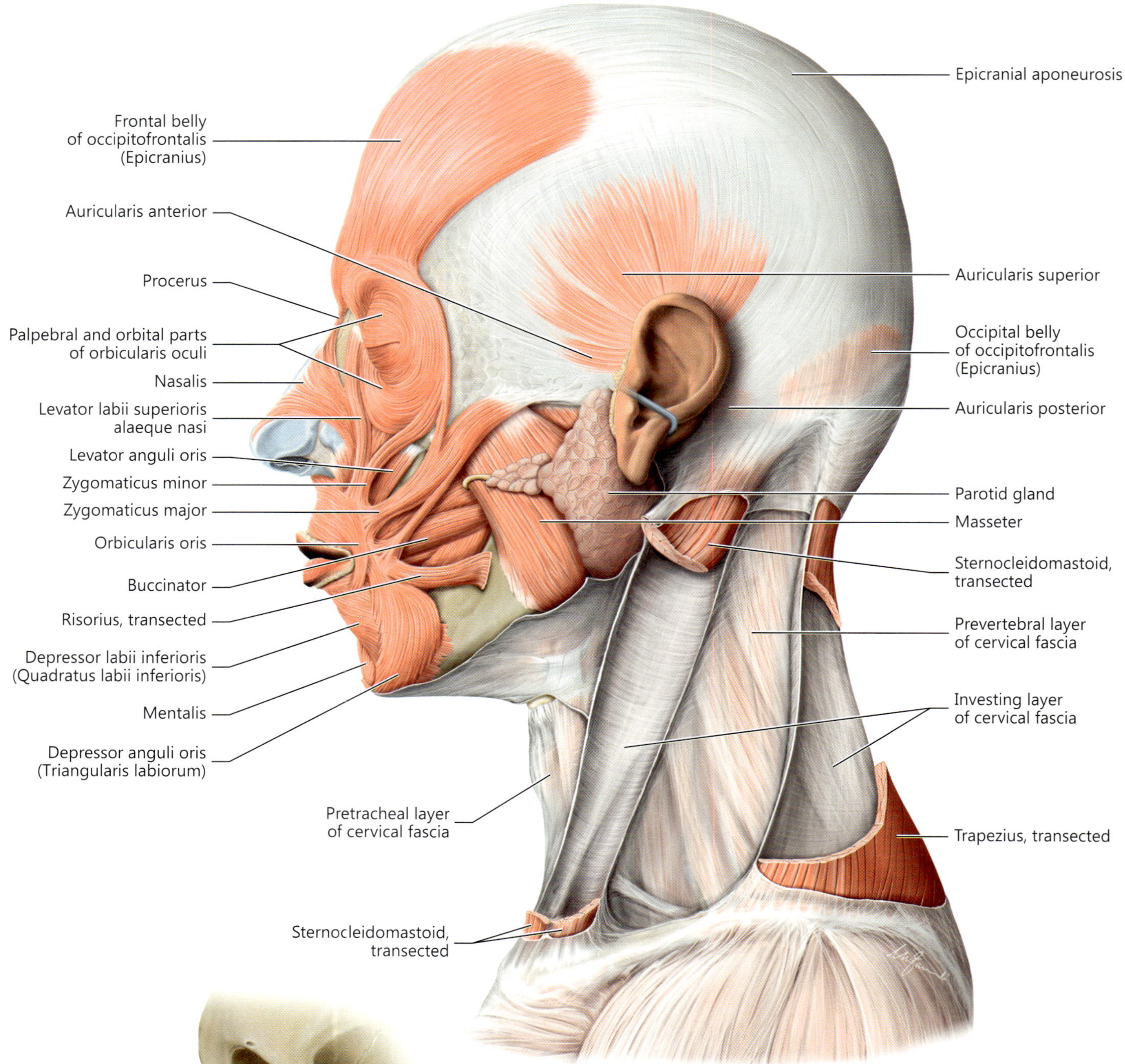

Muscles of the head, left lateral view, after removal of the platysma, sternocleidomastoid, trapezius and superficial fasciae.

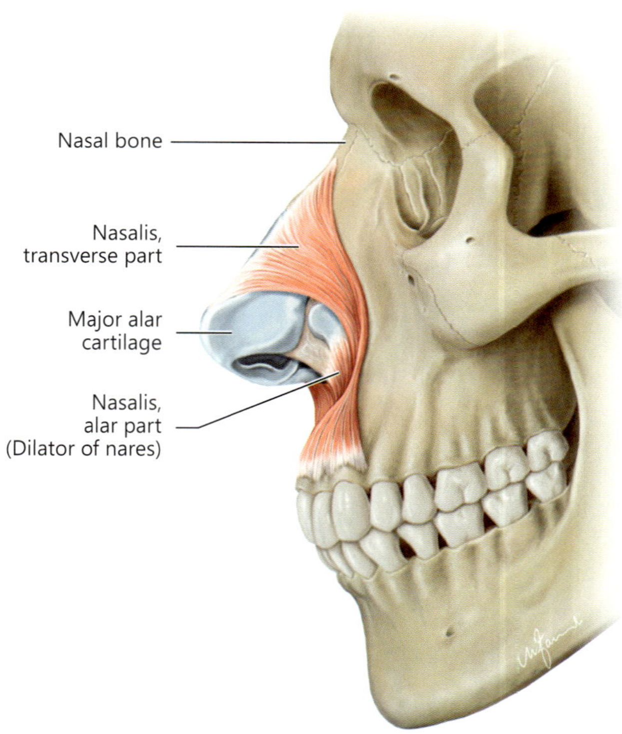

Muscles of the nose.

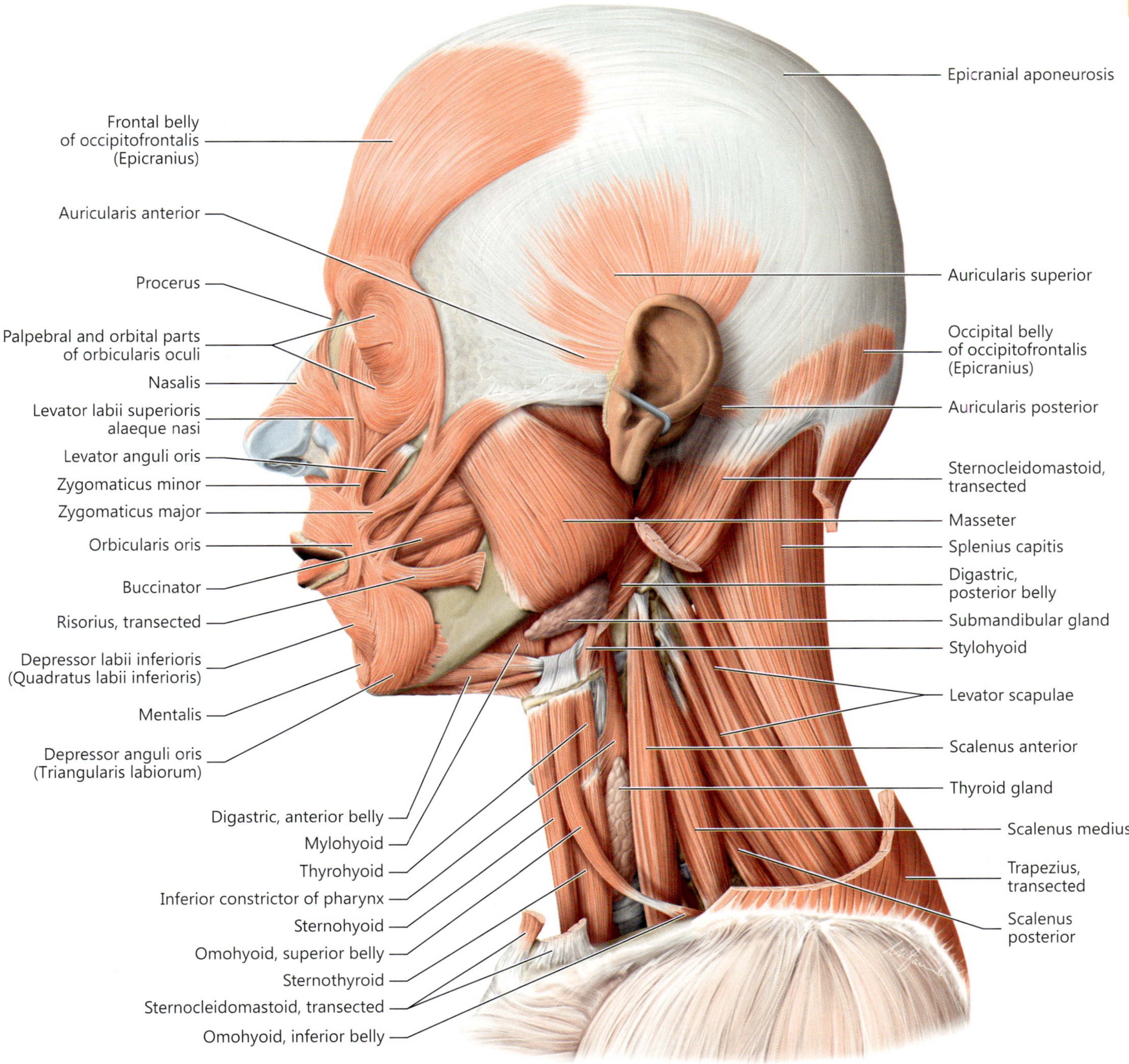

Preparation of the **mimic muscles** and the **anterior and lateral muscles of the neck**, left lateral view. Some superficial mimic muscles and the sternocleidomastoid and trapezius have been transected to expose the mimic muscles of the deeper layers.

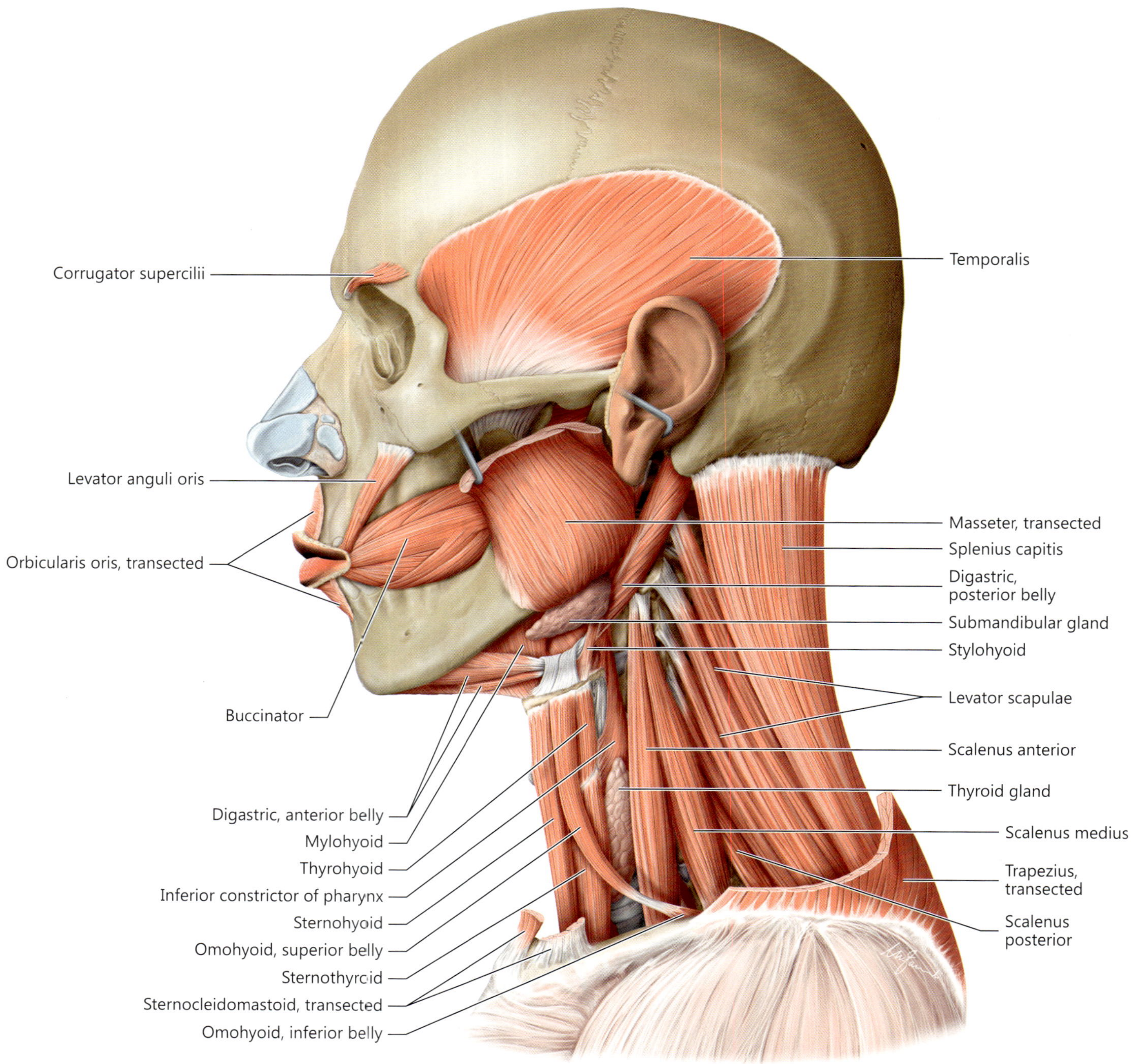

Deep muscles of the head and **anterior and lateral muscles of the neck**, left lateral view, after removal of the superficial muscles.

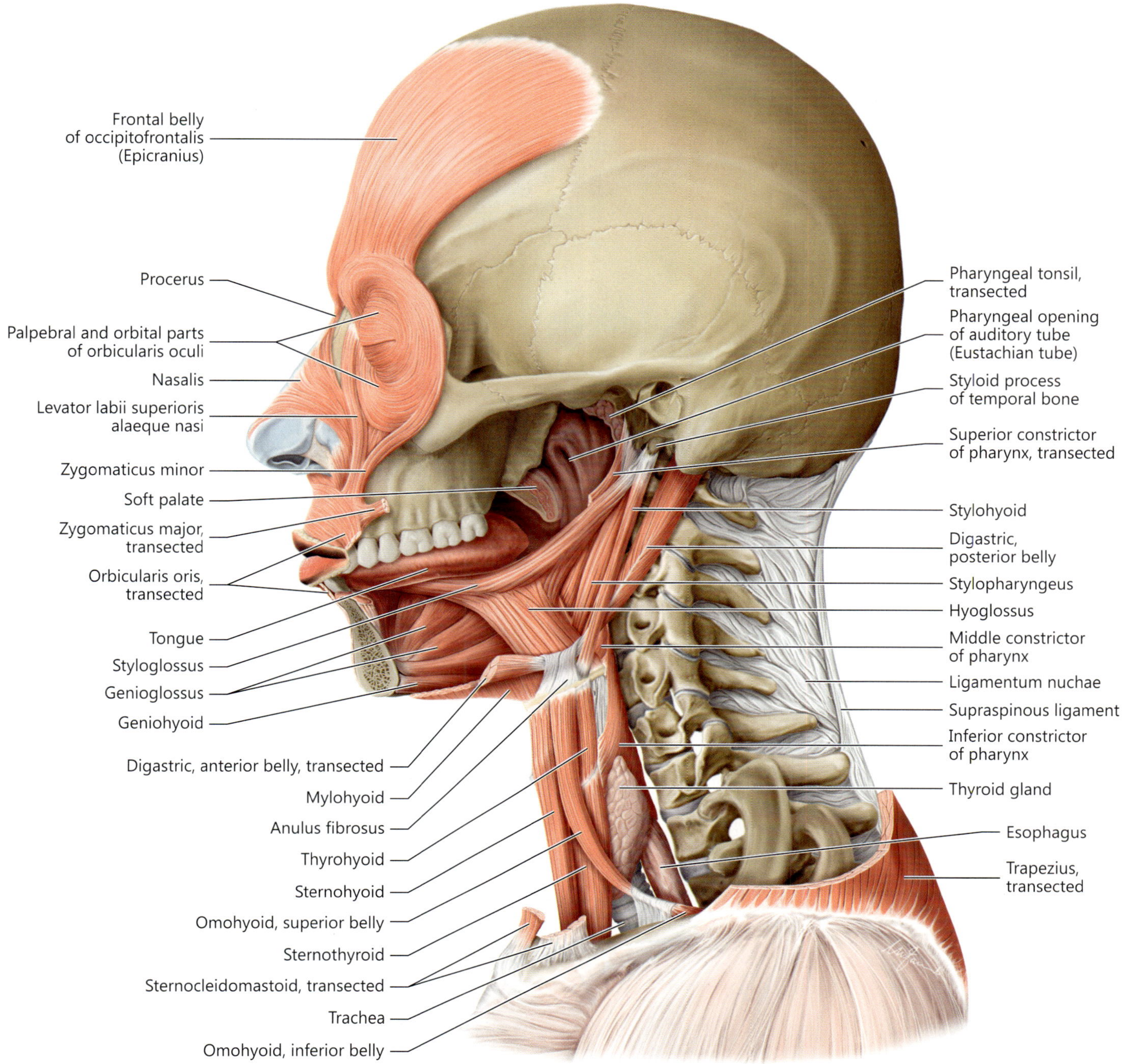

Muscles of the oral cavity and **infrahyoids**, right medial view, after resection of the left mandibular arcade. The infrahyoid muscles are shown.

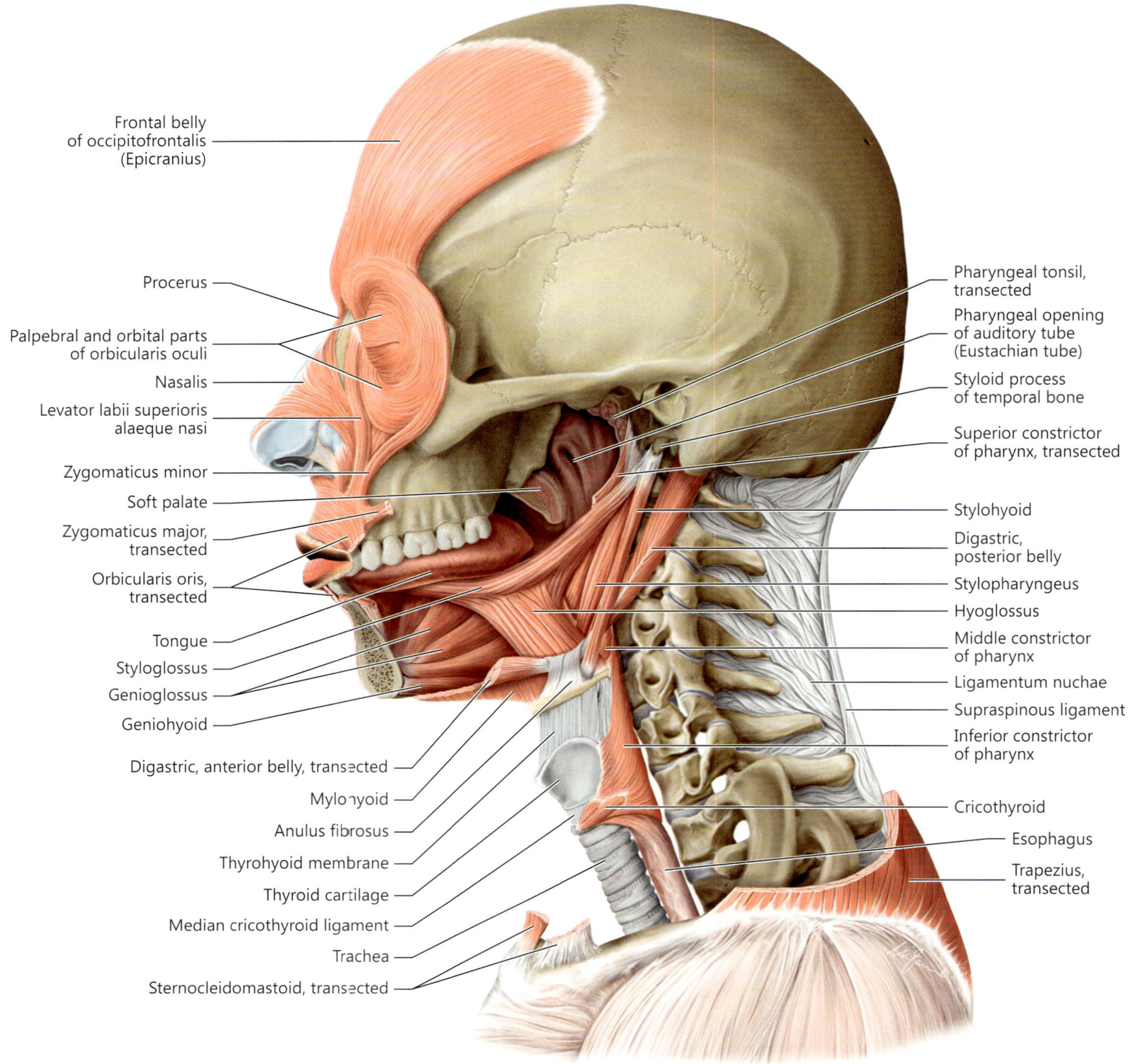

Muscles of the oral cavity, right medial view, after resection of the left mandibular arcade. The infrahyoid muscles have been transected, highlighting the pharynx, esophagus, larynx, and trachea.

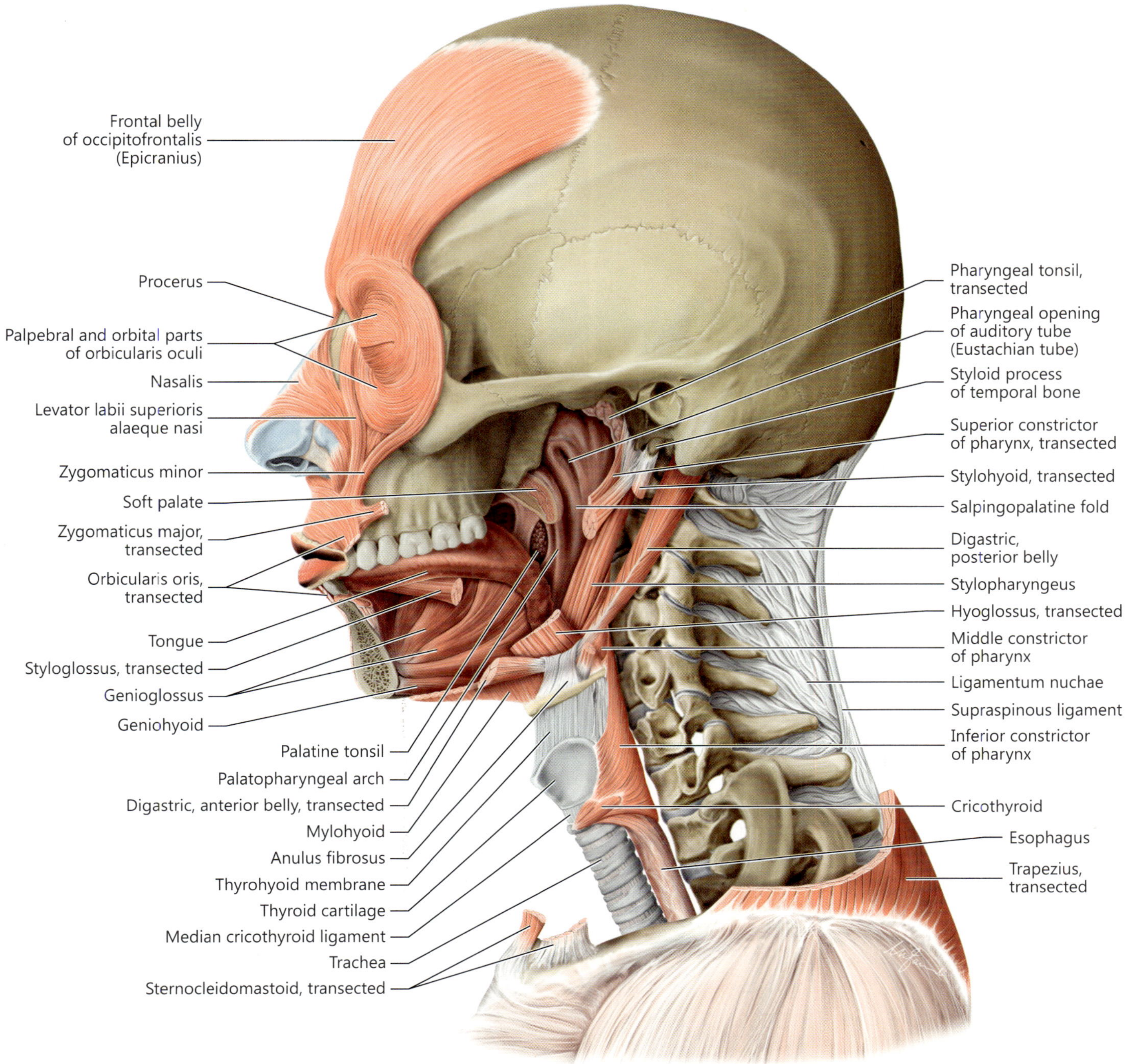

Muscles of the oral cavity, right medial view, after resection of the left mandibular arcade. The hyoglossus, styloglossus, and stylohyoid have been transected to illustrate the deep muscles of the oral cavity.

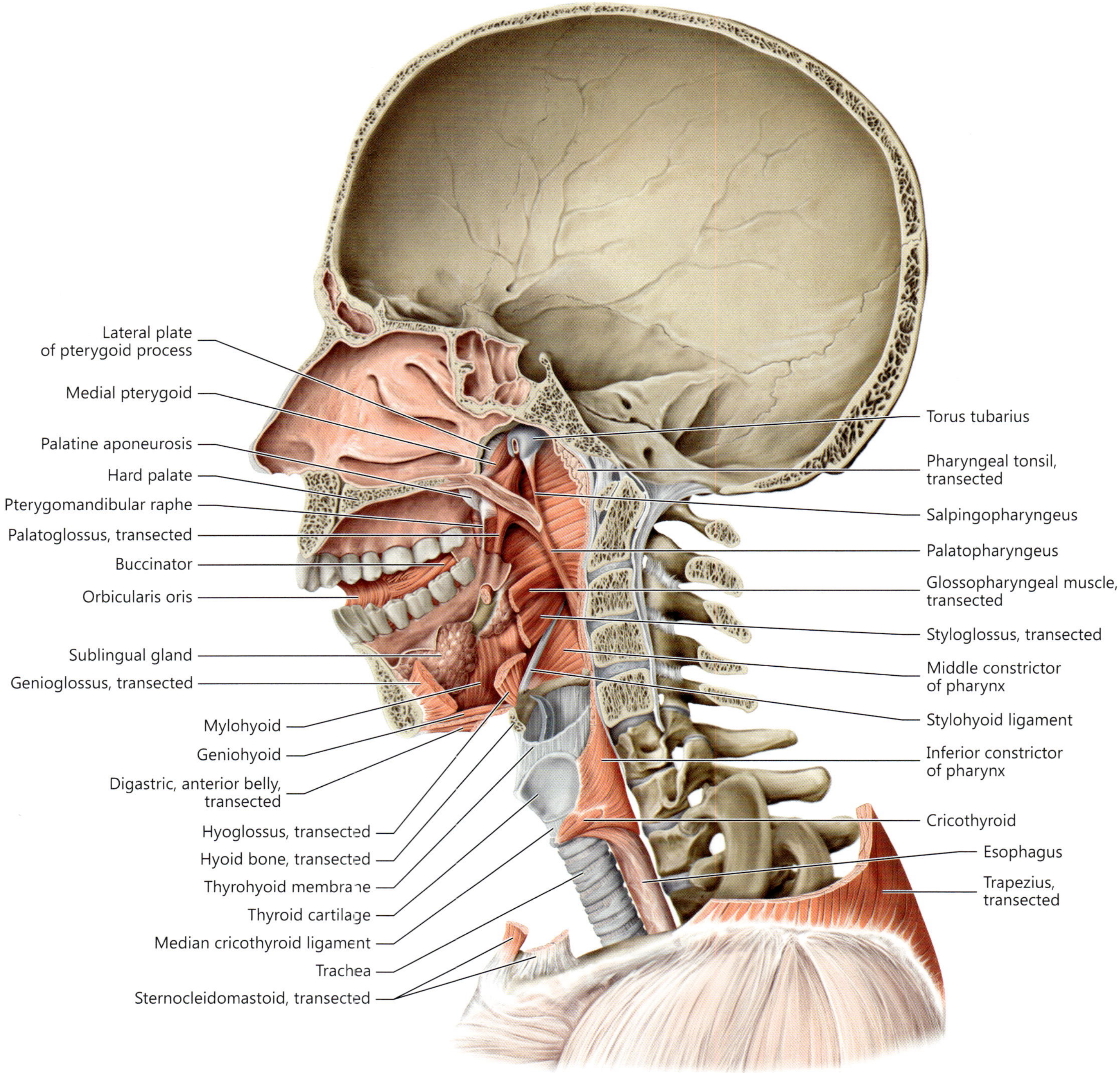

Muscles of the oral cavity. Section of the cranium at the level of the median line; right medial view showing the relationships between the palate and the muscles of the oral cavity.

Extrinsic muscles of the tongue, seen externally from below. The preparation of the surface layer of the suprahyoid region illustrates some of the muscles which extend from the hyoid bone to the tongue, as well as their relationships with the other suprahyoid muscles.

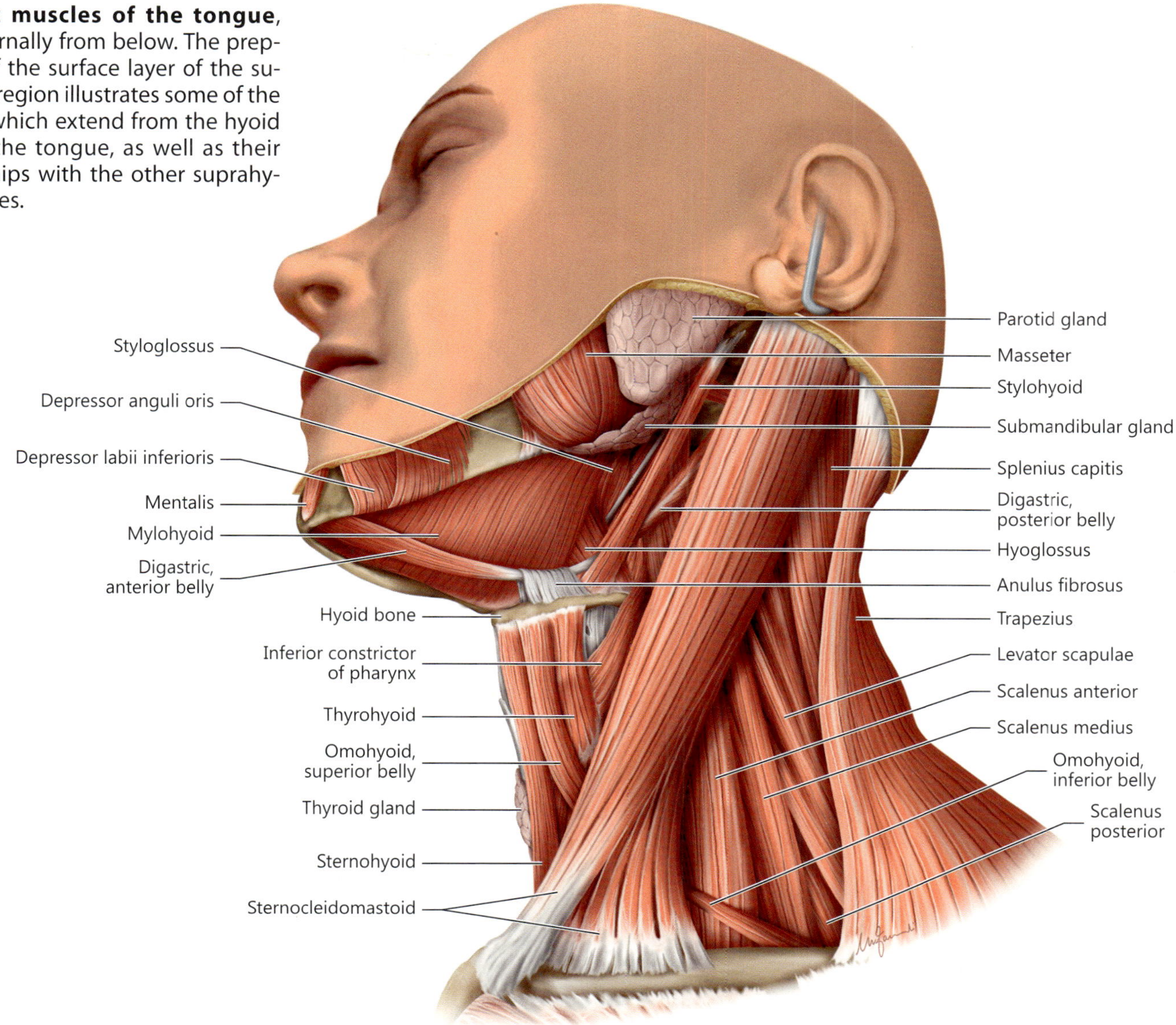

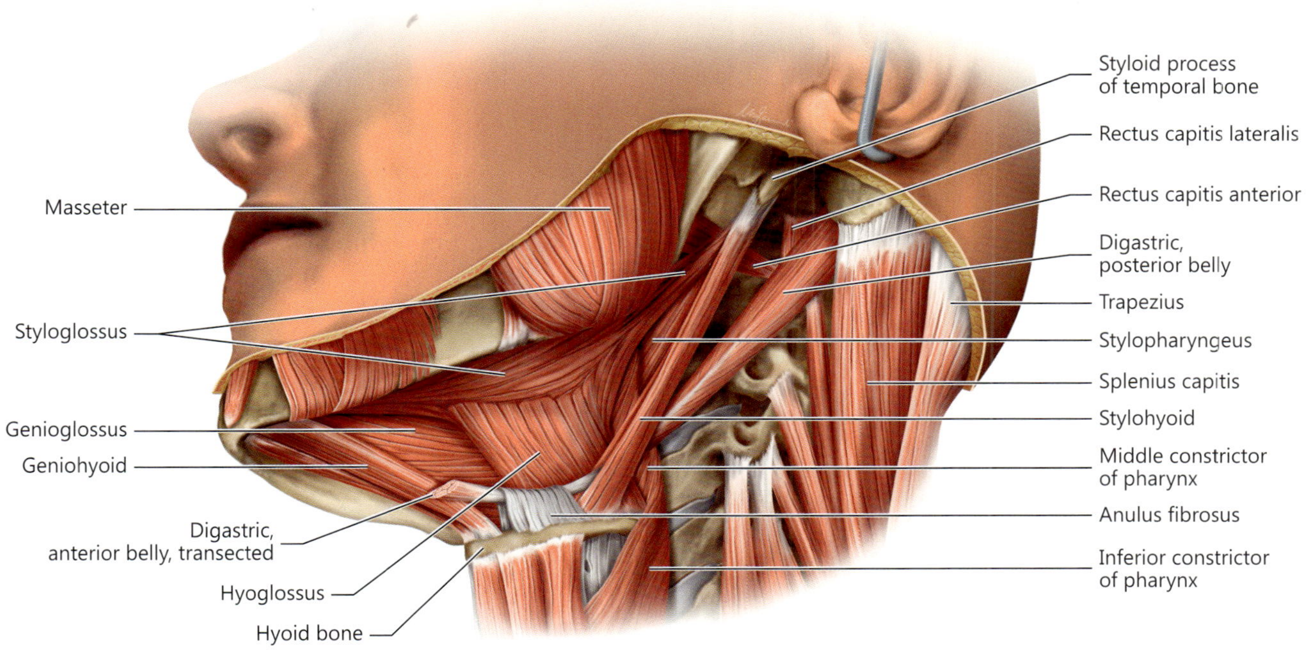

Extrinsic muscles of the tongue. The preparation of the deep layer of the suprahyoid region, in addition to the extrinsic muscles originating from the hyoid bone, also shows the muscles coming off of the styloid process and mandible.

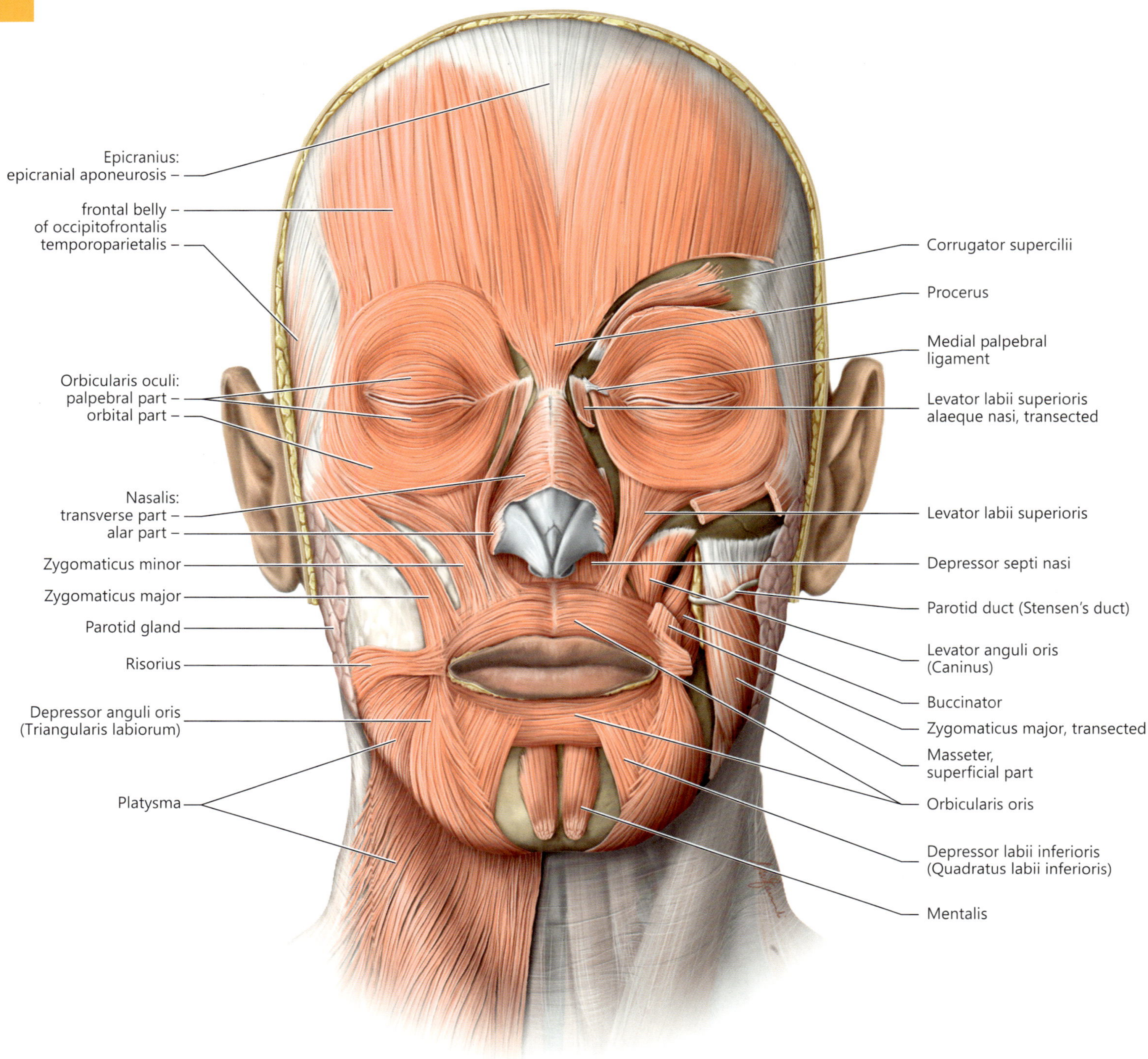

Facial muscles, anterior view. On the right surface layer (mimic muscles), on the left deep layer (mimic and masticatory muscles).

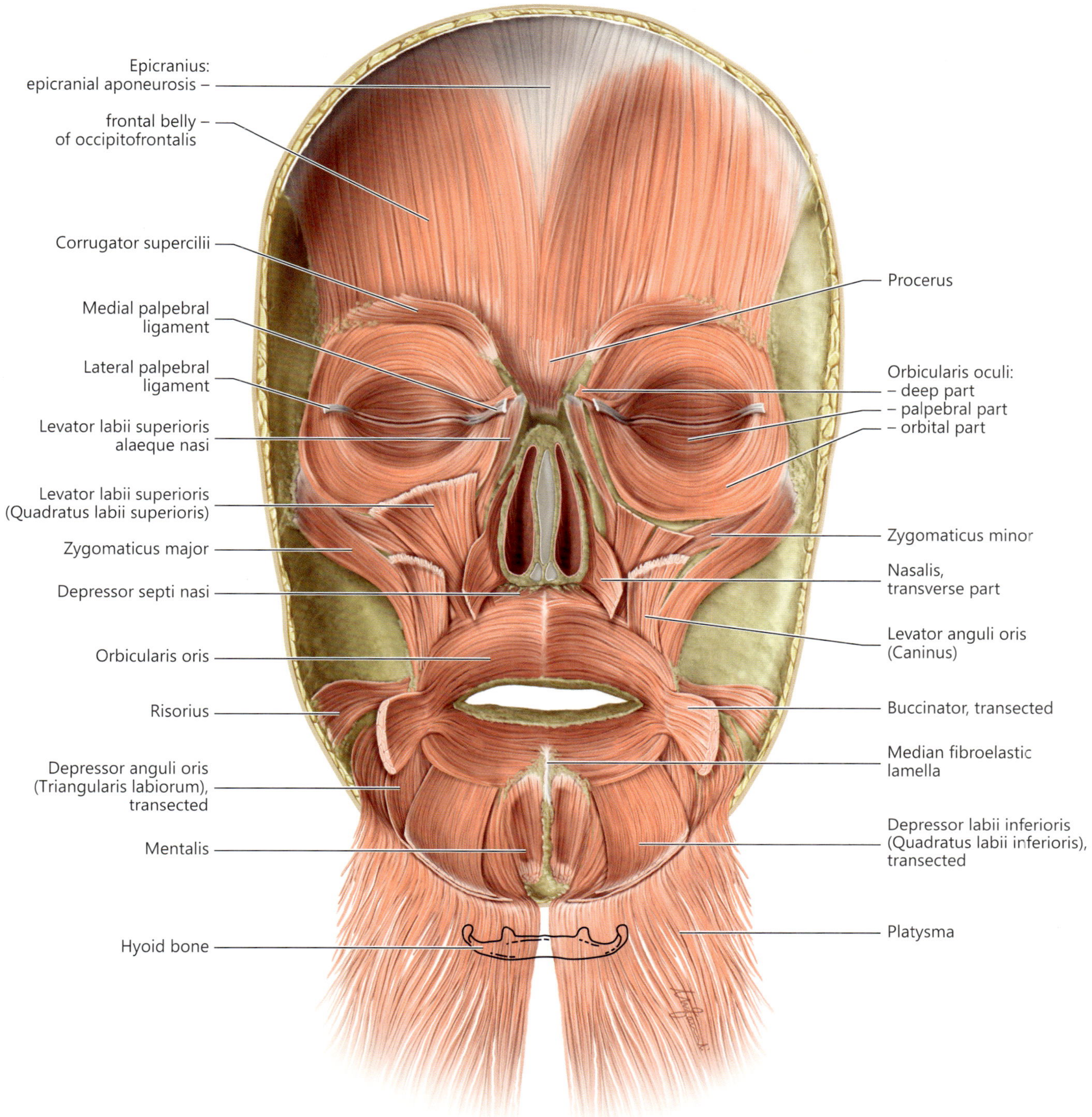

Facial muscles, anterior view from inside.

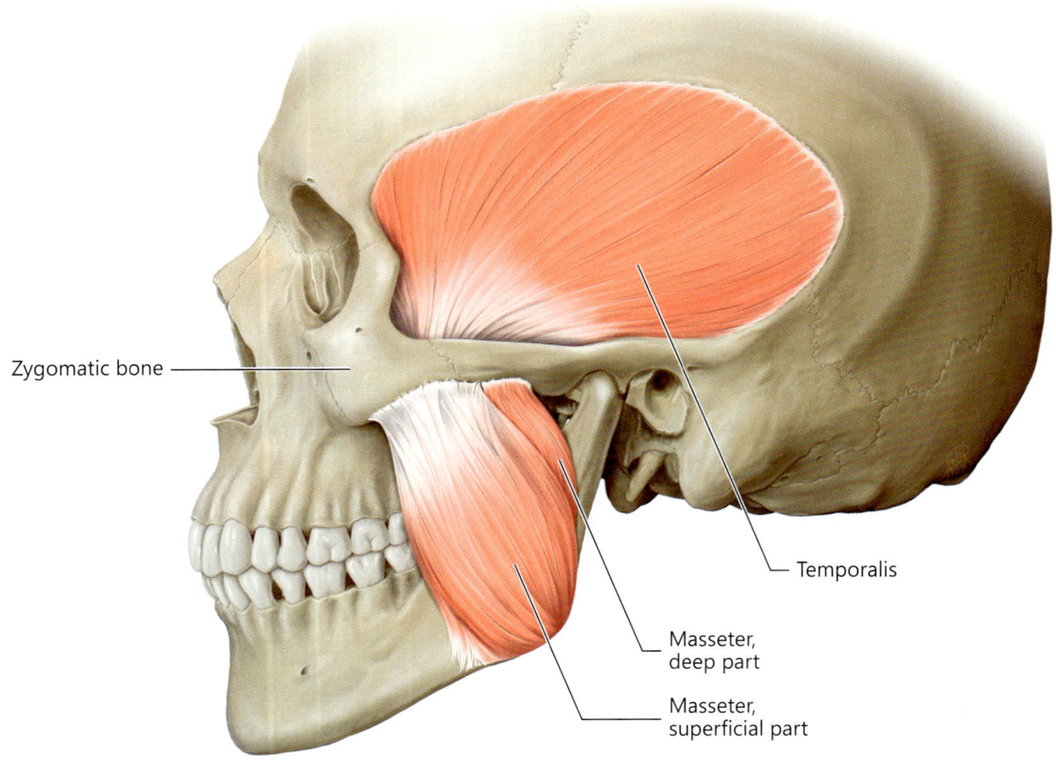

External masticatory muscles of the mandible. The masseter and temporalis have the function of raising the mandible.

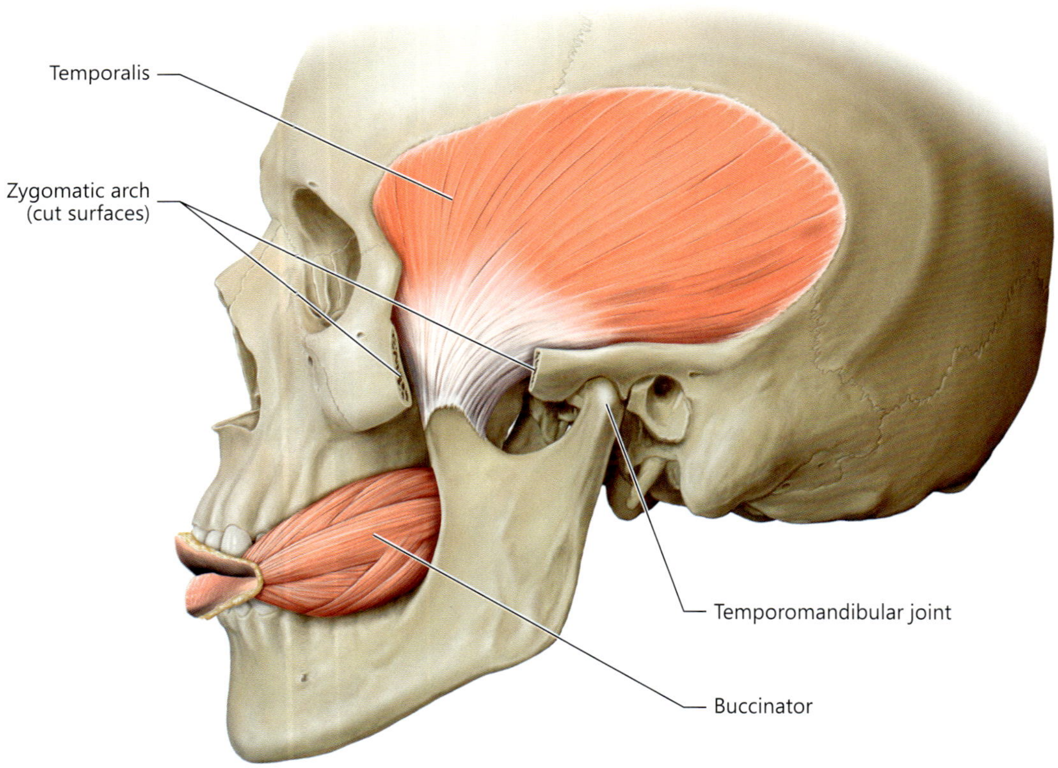

Preparation of the **masticatory muscles**. With the removal of the zygomatic arch is highlighted the insertion of the temporalis to the coronoid process of the mandible. The buccinator has also been prepared.

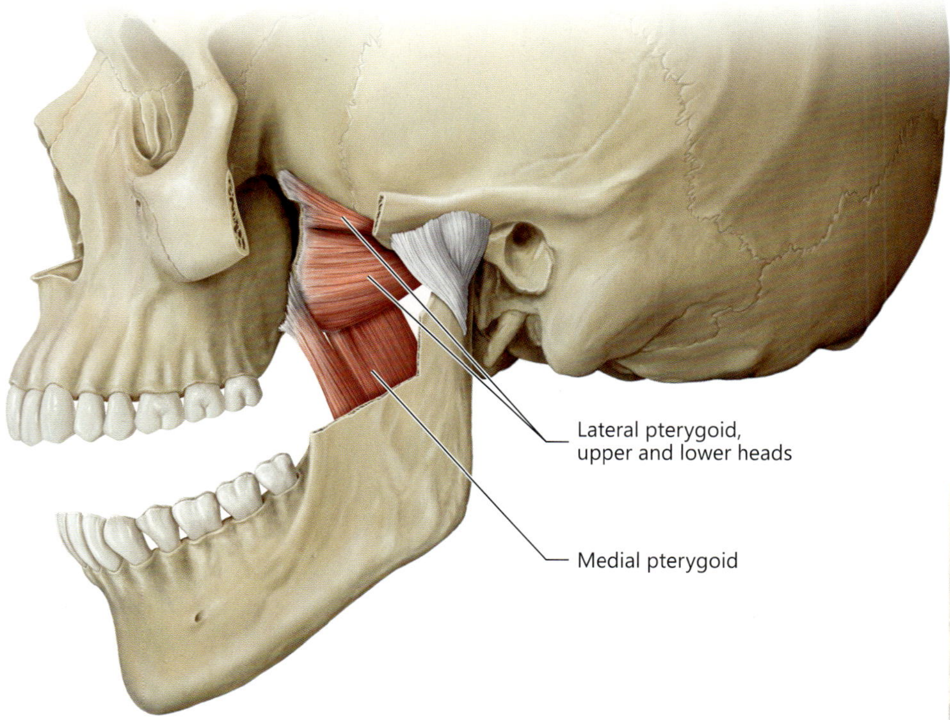

The superior and inferior heads of the **lateral pterygoid**. The coronoid process and part of the ramus of the mandible have been transected to show the superior and inferior heads of the lateral pterygoid. The superior head contracts when the mandible is raised, inhibiting the backward movement of the mandibular condyle.

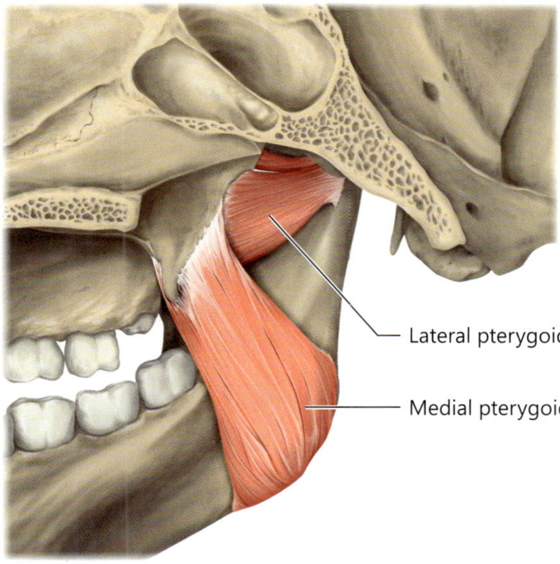

➤
Internal elevator muscles of the mandible. Lateral and medial pterygoids.

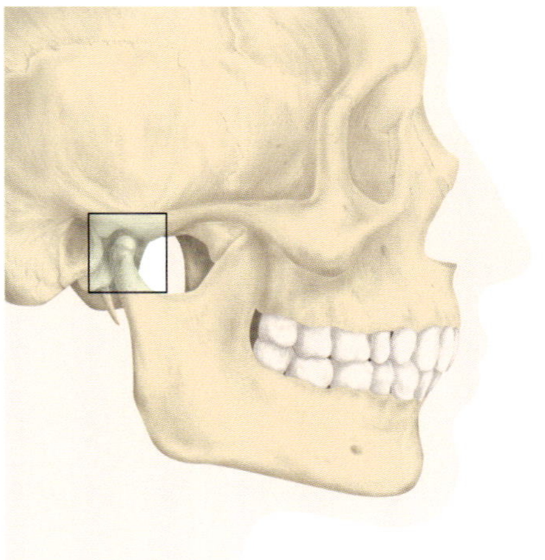

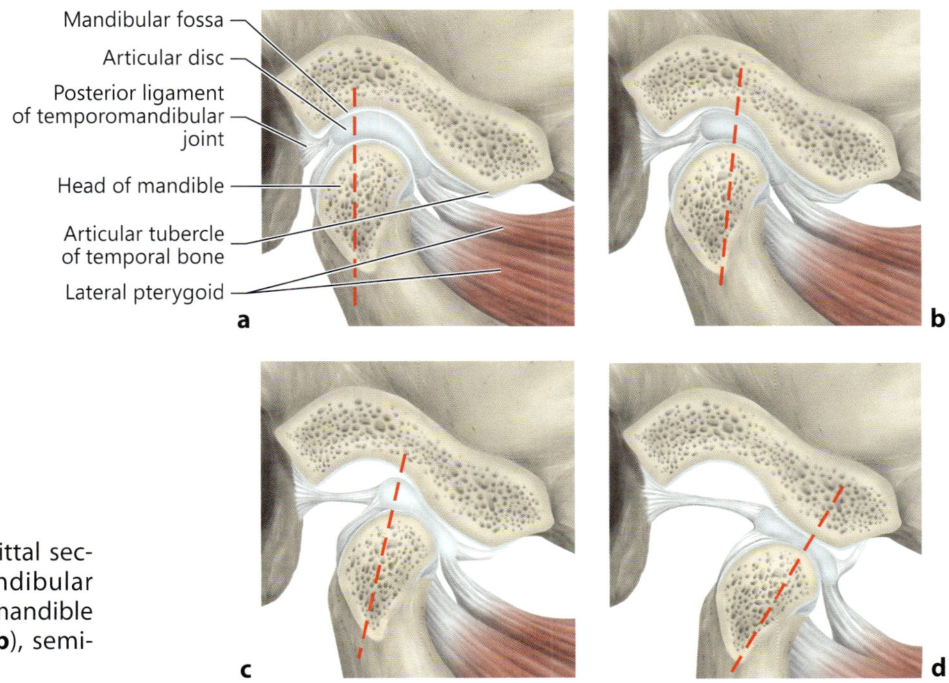

Combined mandibular movements. Sagittal sections passing through the temporomandibular joint in four different conditions: with the mandible raised (**a**), at the beginning of opening (**b**), semi-open (**c**), and fully open (**d**).

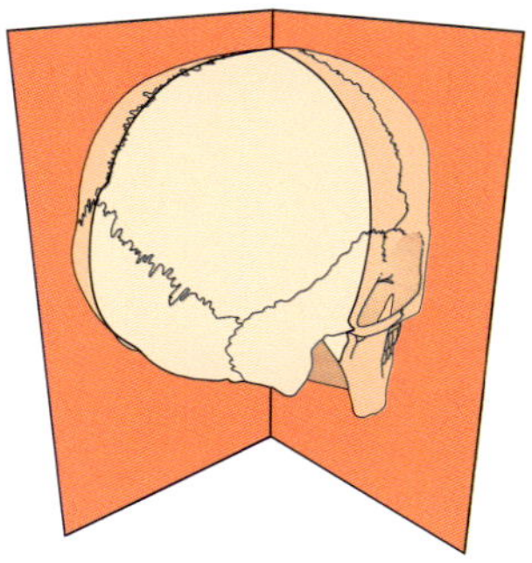

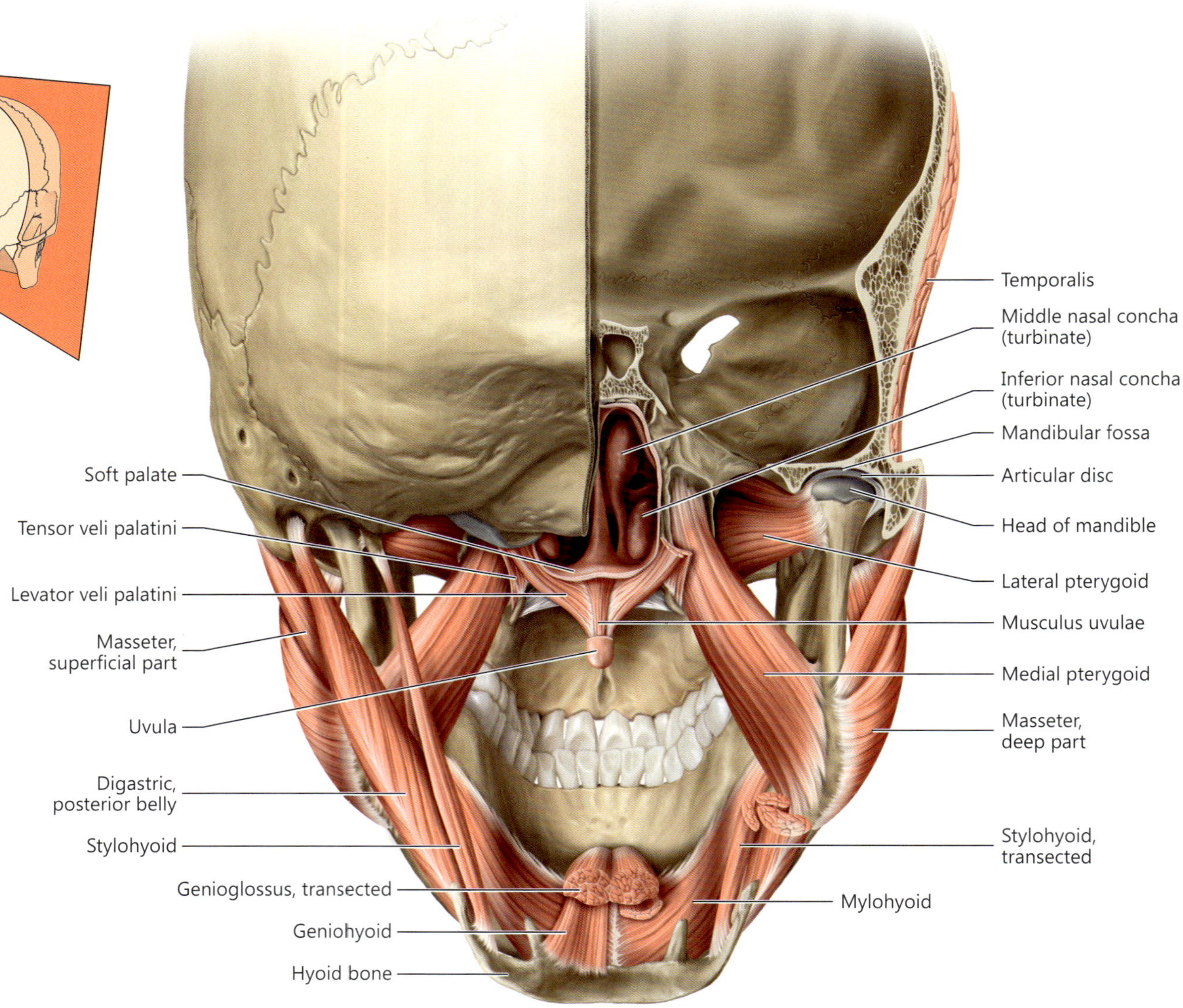

Muscles of the oral cavity, posterior view.

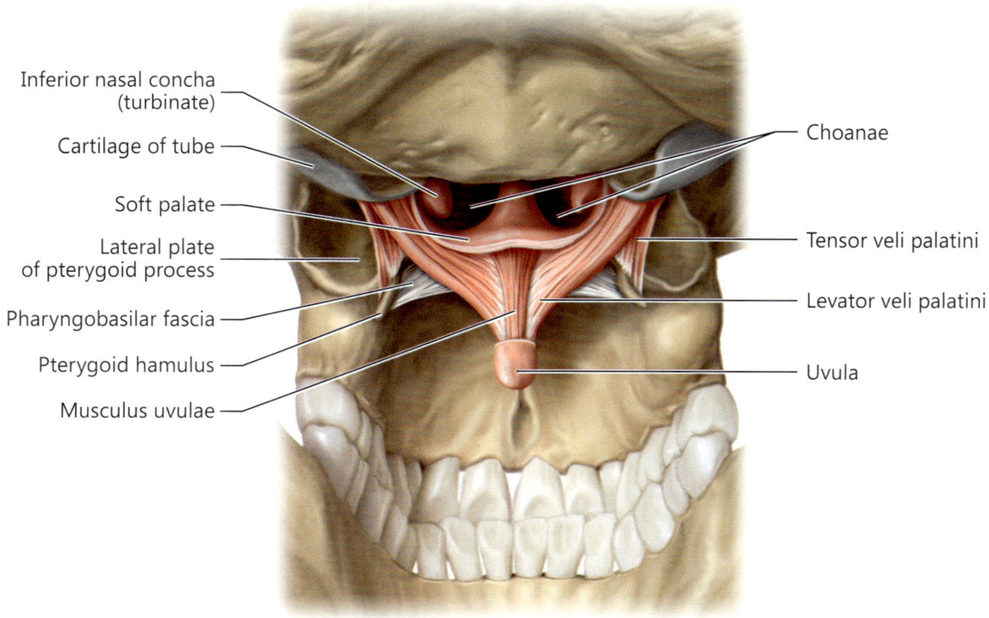

Muscles of the soft palate, posterosuperior view.

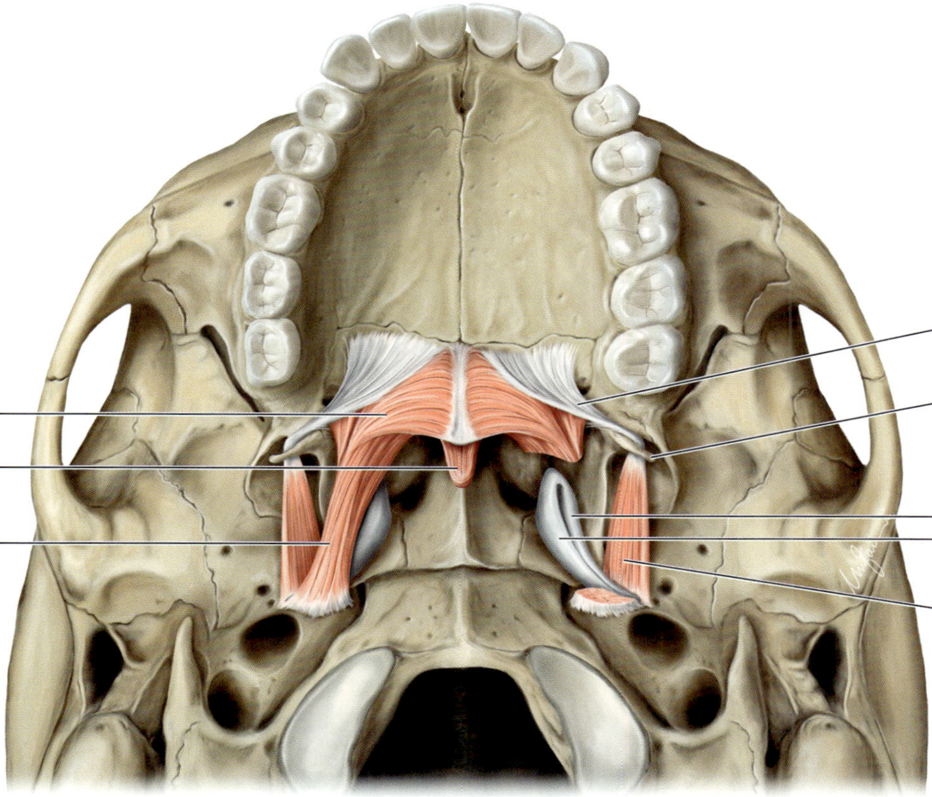

Muscles of the soft palate, inferior view.

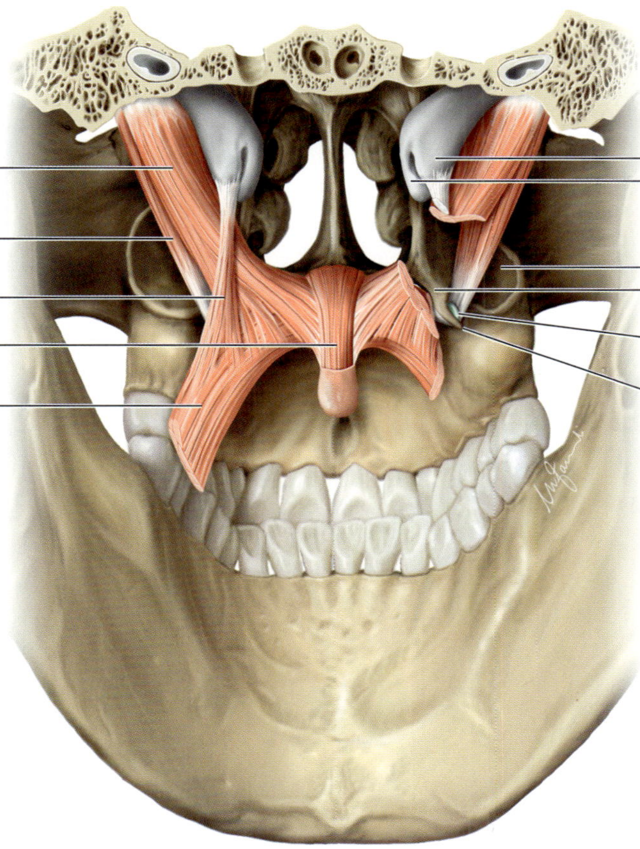

Muscles of the soft palate, posterior view.

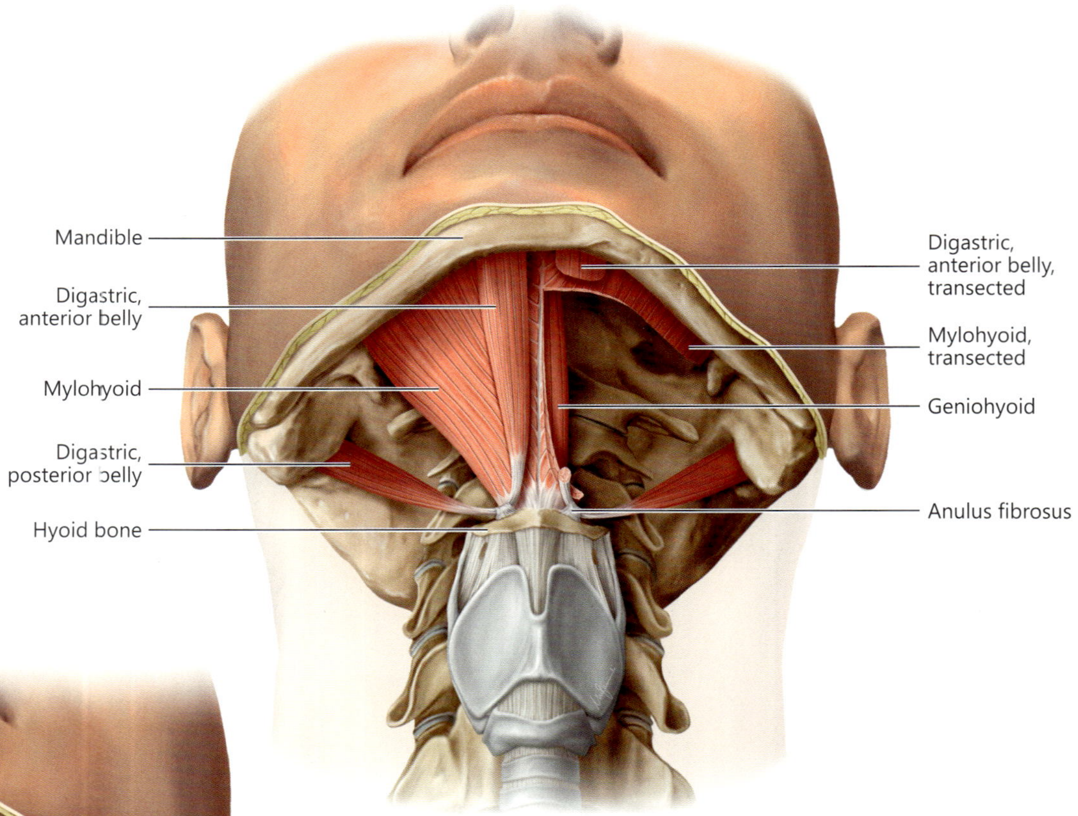

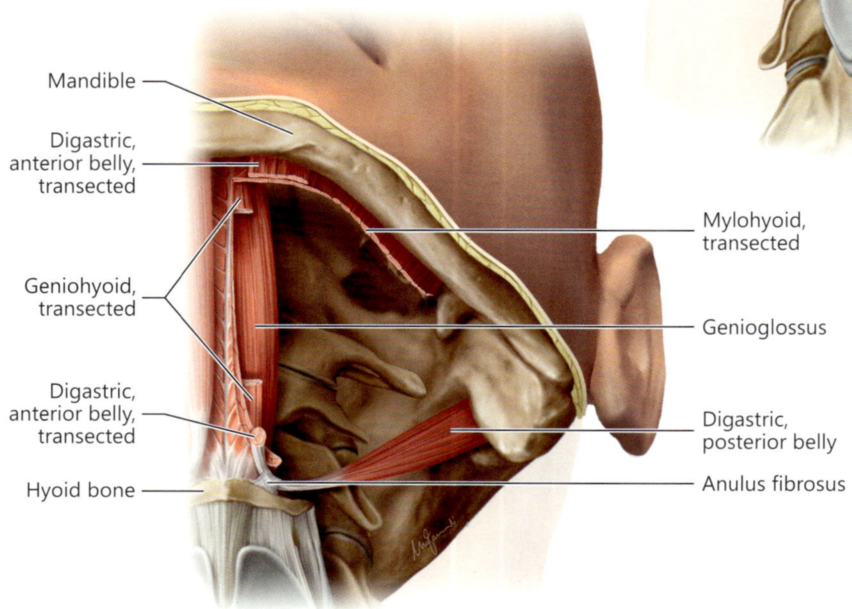

Muscles of the floor of the mouth (**mandibular depressors**). External view from below, with the next section of the geniohyoid (in detail).

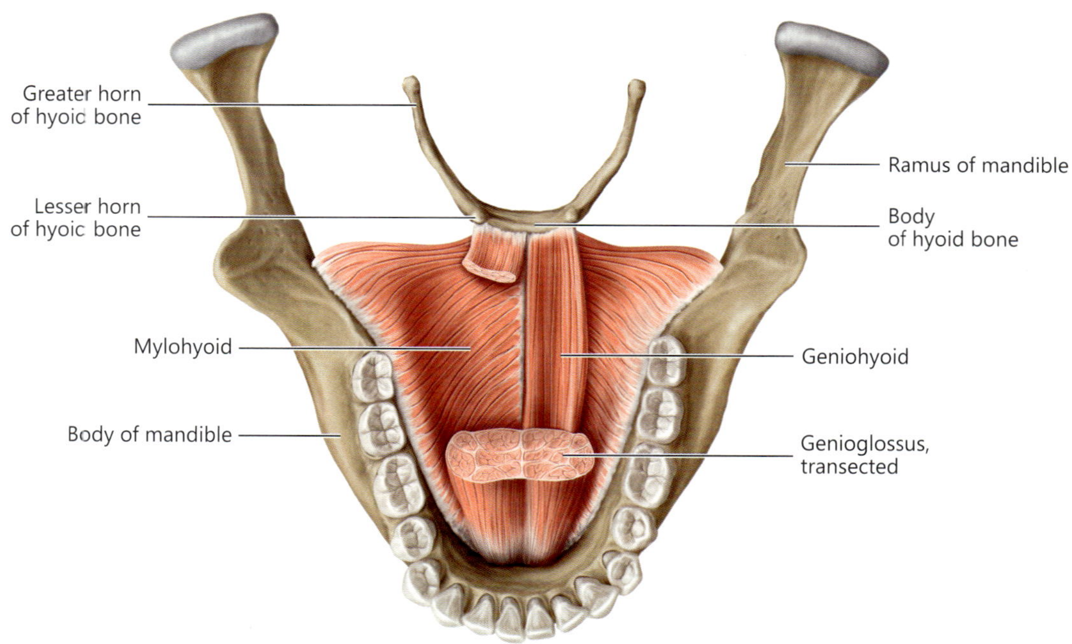

Muscles of the floor of the mouth, internal view from above.

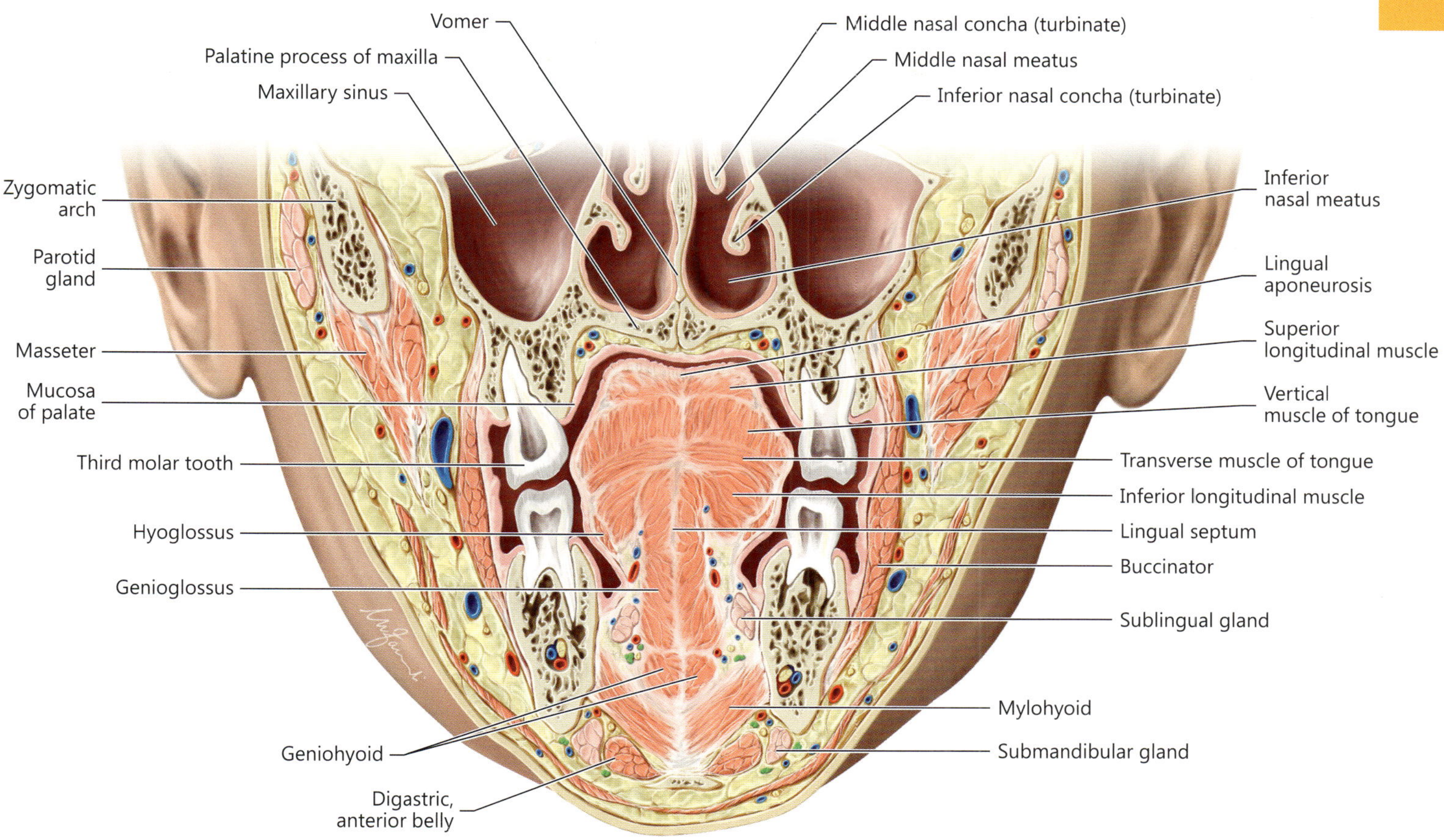

Muscles of the tongue and the floor of the mouth, frontal section at the level of the third molar.

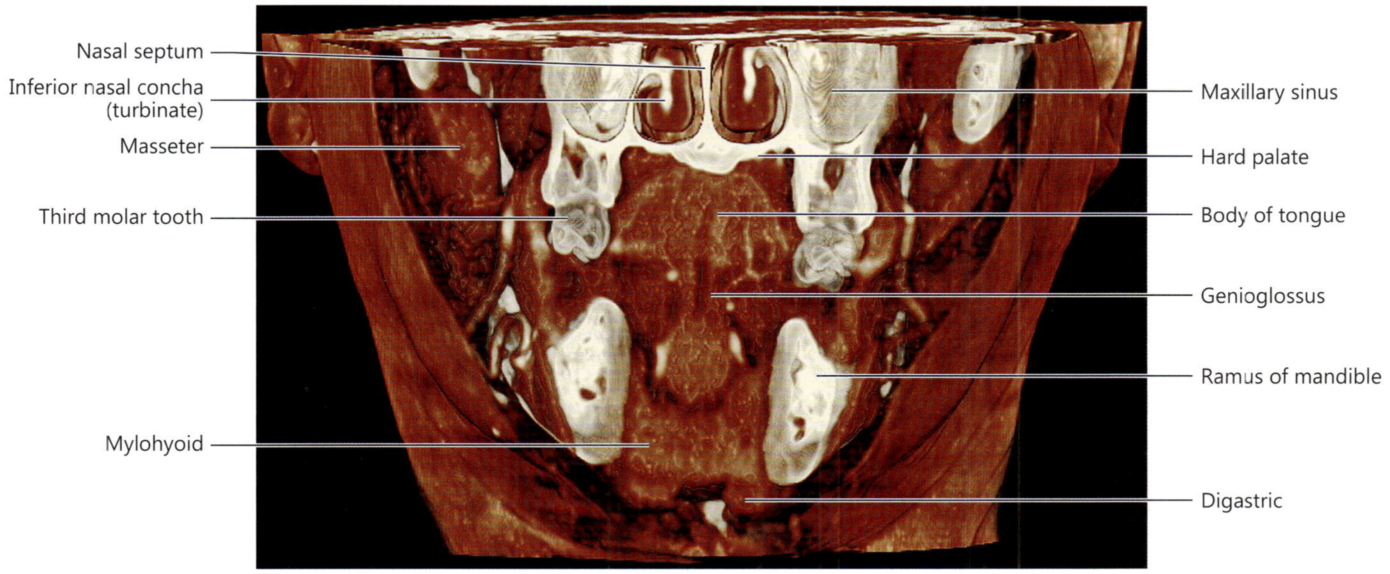

Three-dimensional image of a living human body, obtained using a 3D volume rendering technique, showing the **tongue** and the **muscles of the floor of the mouth**, frontal section at the level of the third molar.

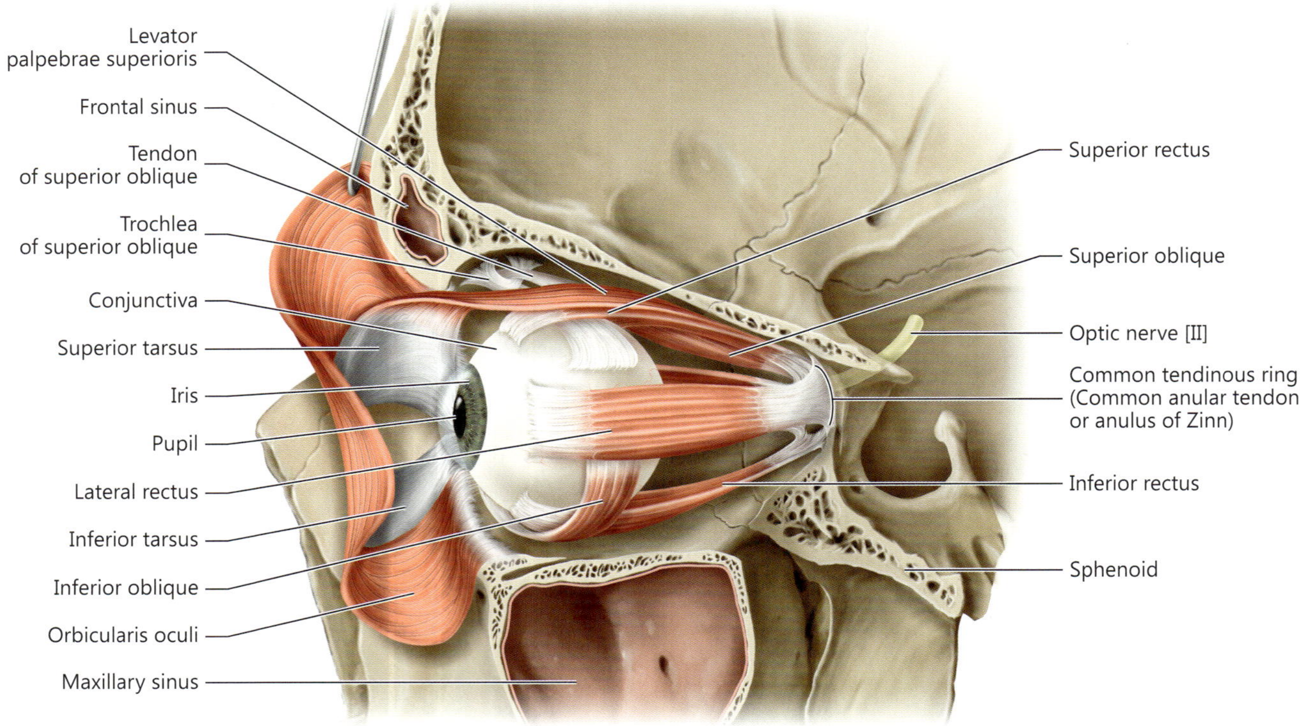

Extraocular muscles. Left eye, lateral view after removal of the lateral wall of the orbit.

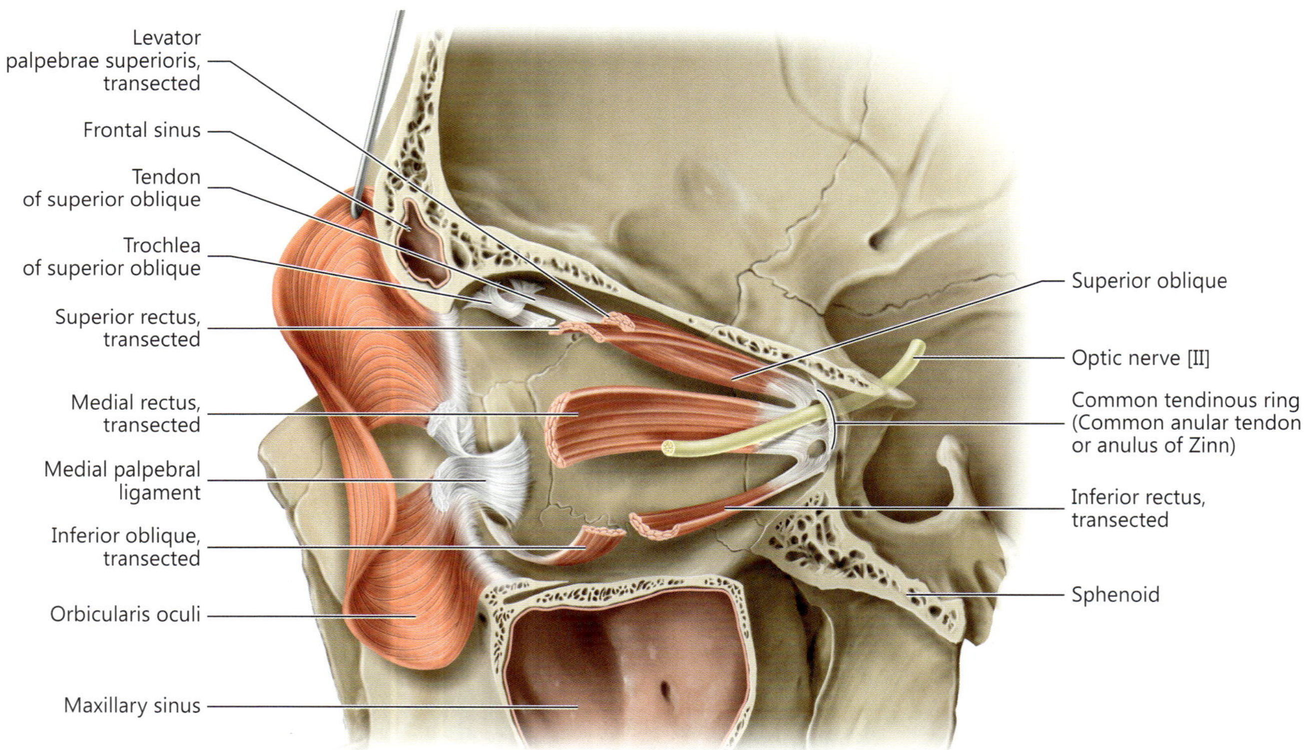

Extraocular muscles after removal of the lateral wall of the orbit and eyeball.

Extraocular muscles and **common tendinous ring**. View from above. The common tendinous ring surrounds the optic canal and the medial extremity of the superior orbital fissure.

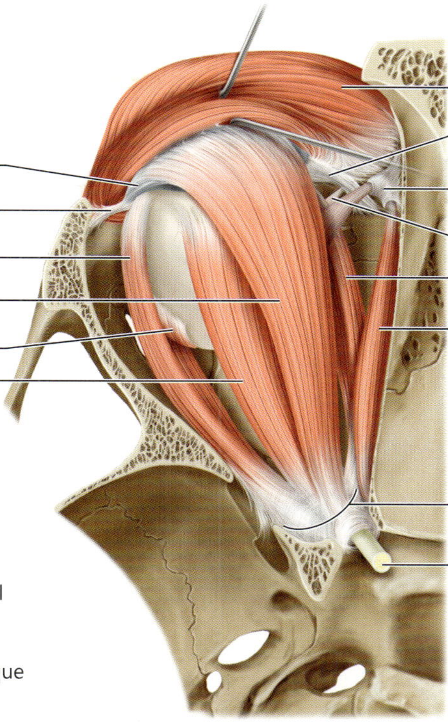

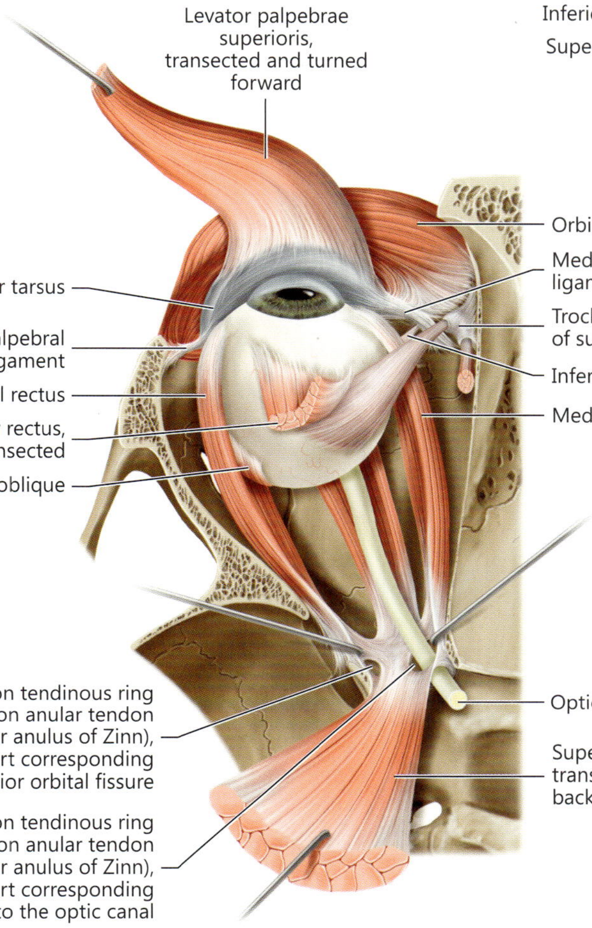

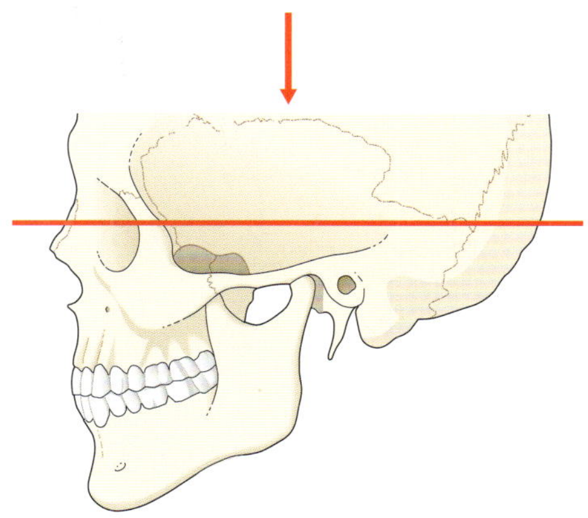

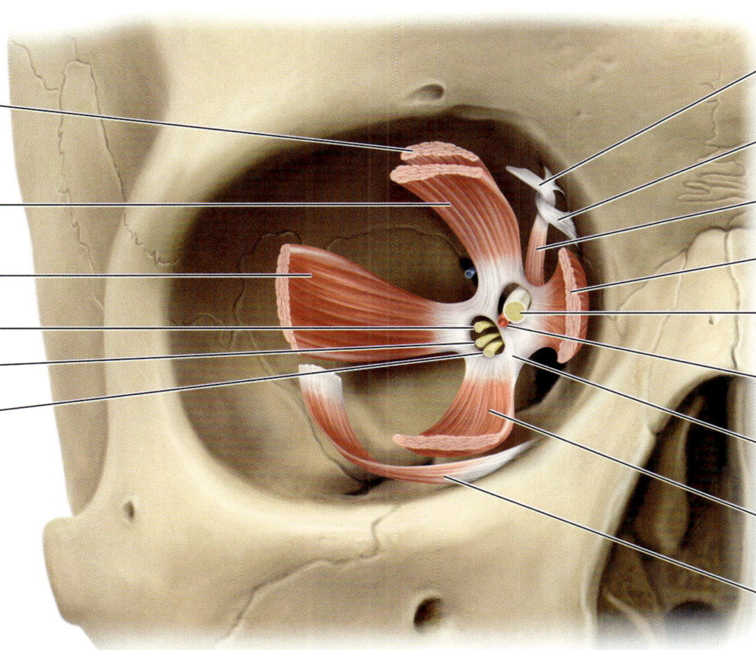

The common tendinous ring of the straight muscles of the eyeball. Anterior view. In addition to the optic nerve, the common tendinous ring is also crossed by the oculomotor, nasociliary, and abducens nerves.

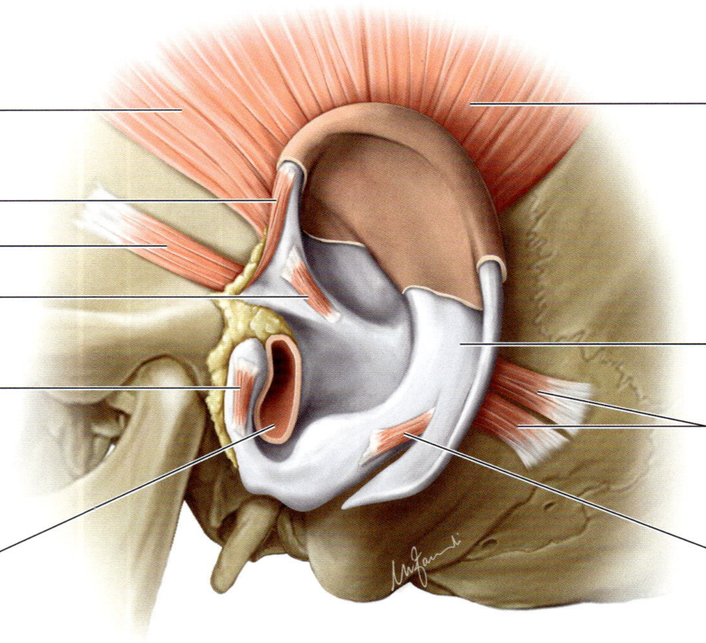

Left ear, external lateral view. Here beneath the skin, the muscles and cartilage of the auricle are shown, in particular the auricularis, helicis, tragicus, and antitragicus muscles.

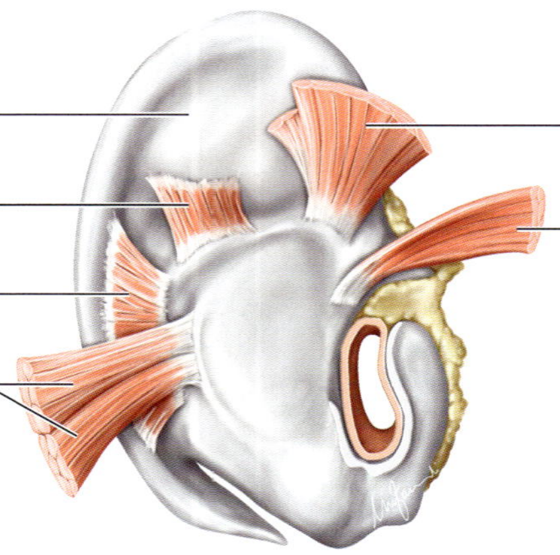

Left ear, medial view. The oblique and transverse muscles and cartilage of the auricle are shown.

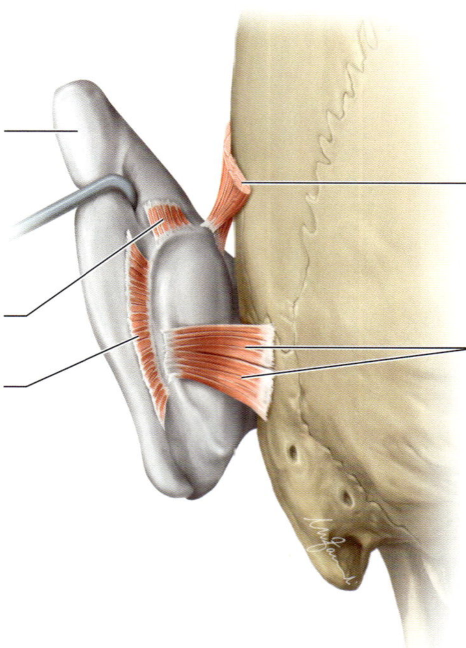

Left ear, posterior view of the cartilage and muscles of the auricle.

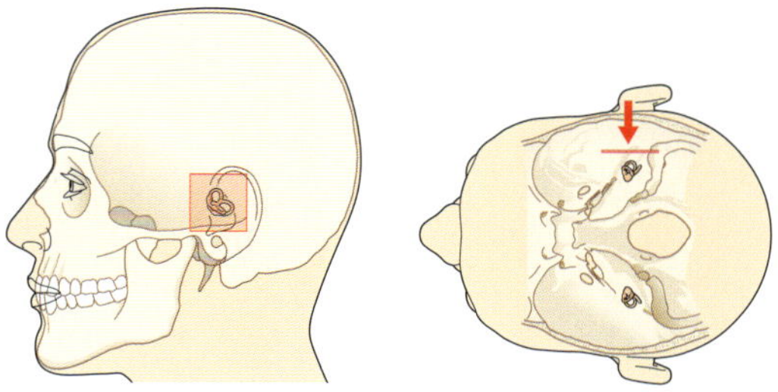

Transparency representation of the **bony labyrinth cavity** in relation to the medial wall of the tympanic cavity. The lateral wall of the vestibule corresponds to the oval window.

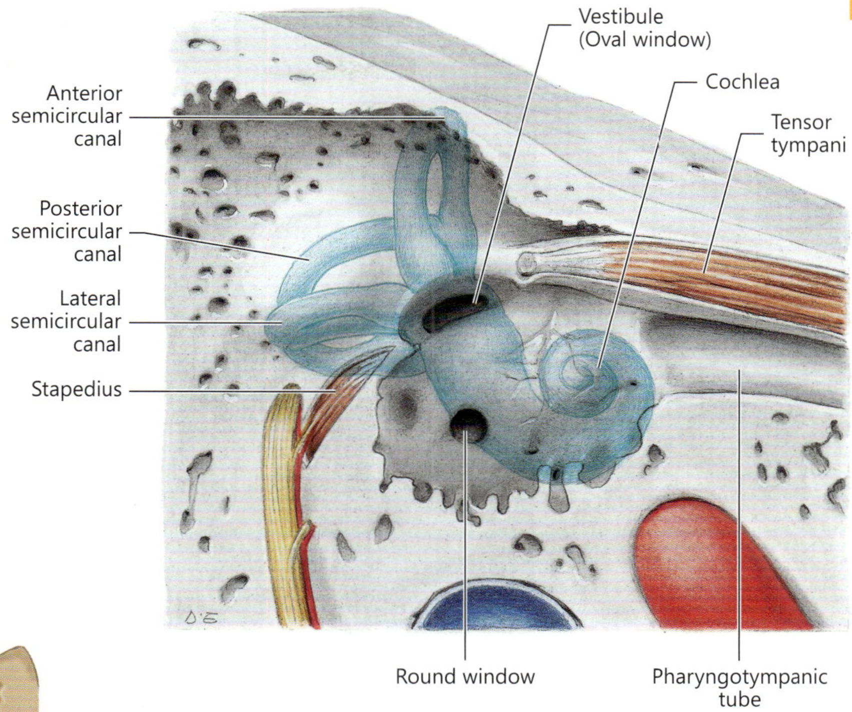

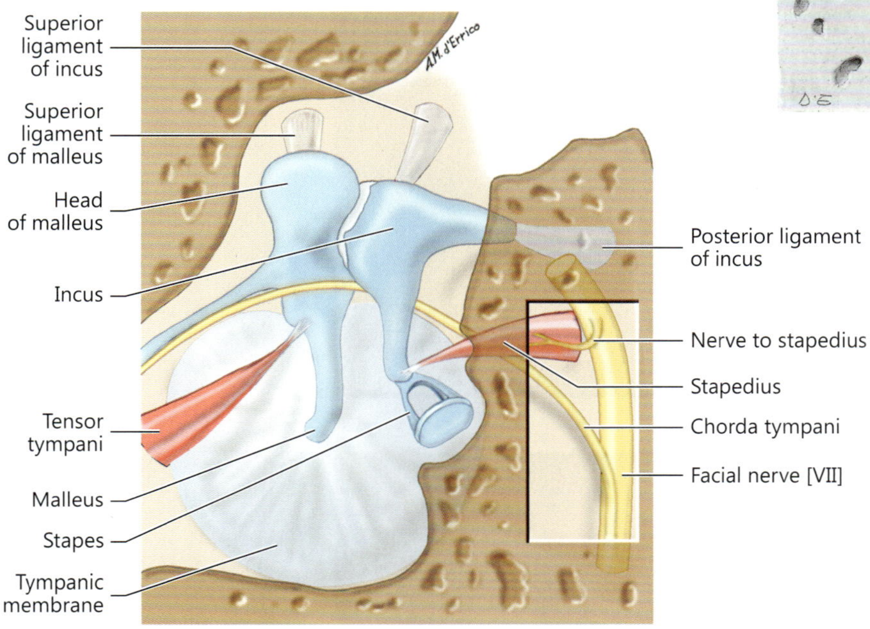

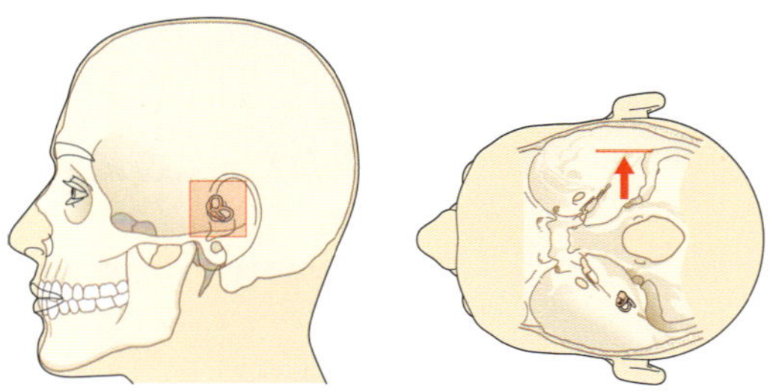

Schematic representation of **tympanic** cavity, seen from behind.

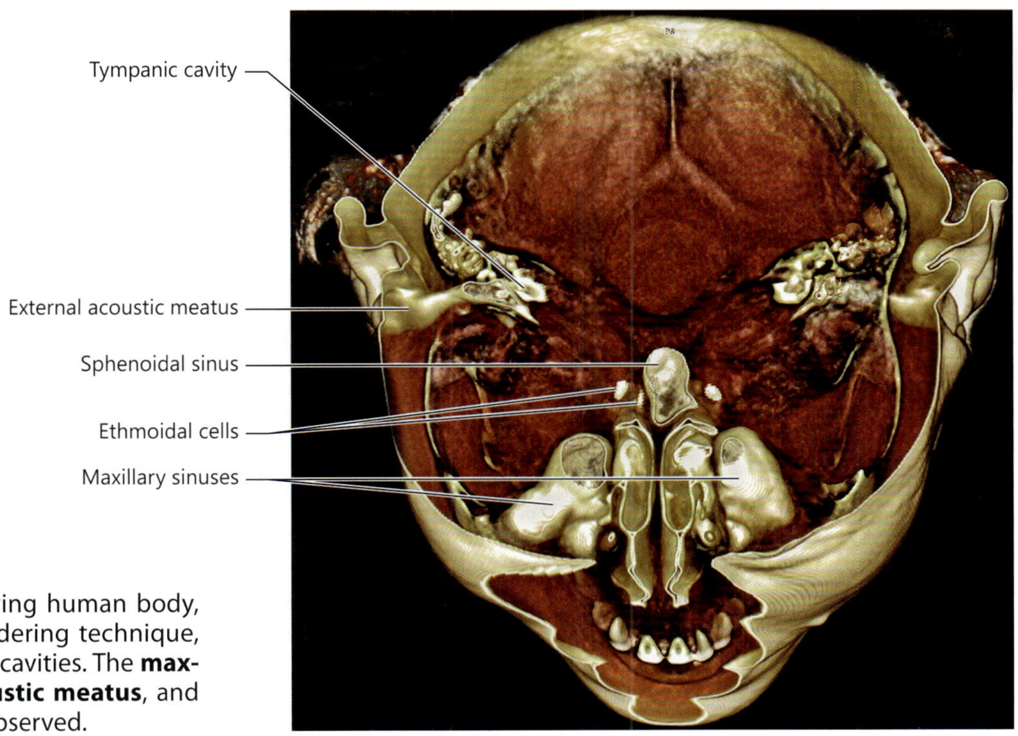

Three-dimensional image of a living human body, obtained using a 3D volume rendering technique, showing the limits of the air-filled cavities. The **maxillary sinuses**, the **external acoustic meatus**, and part of the **tympanic cavity** are observed.

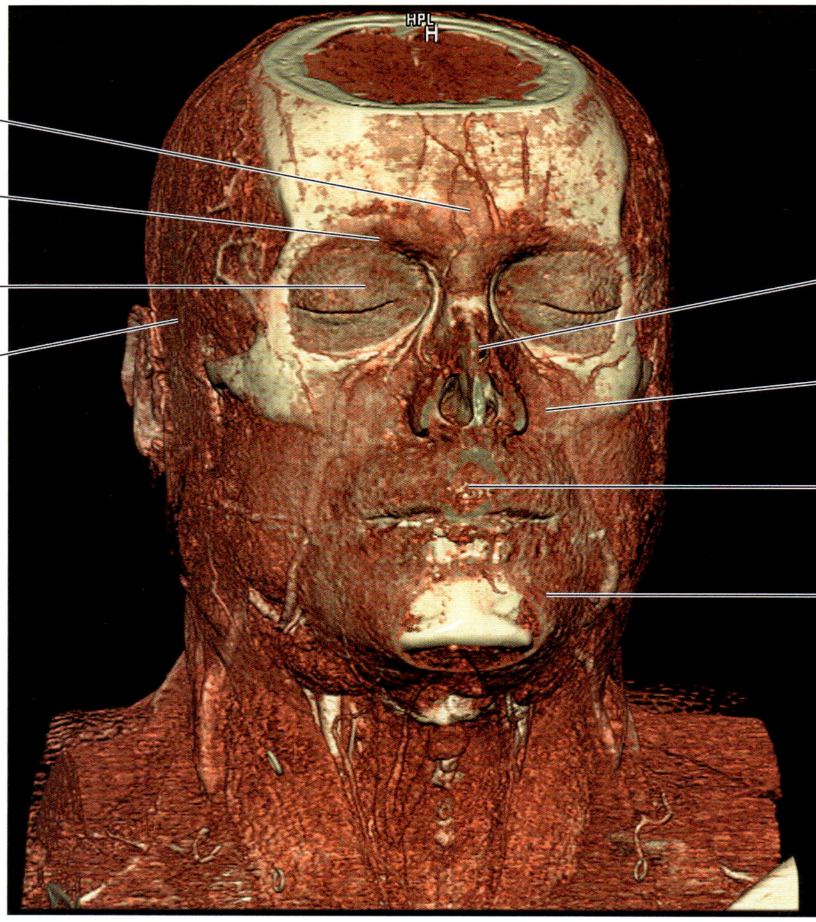

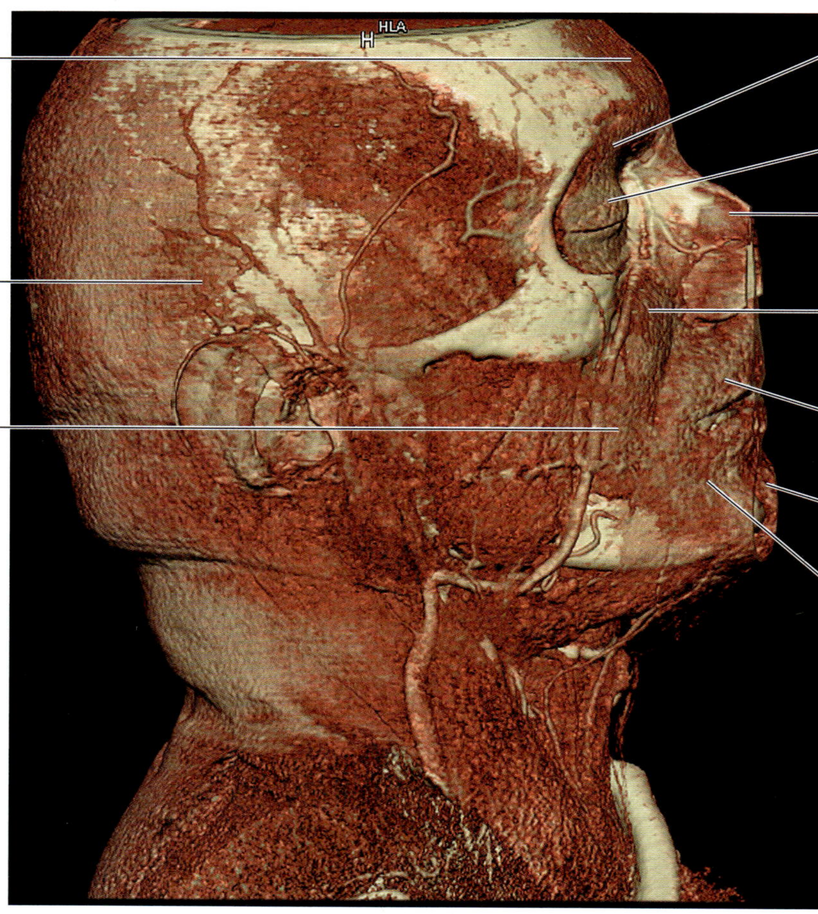

Three-dimensional images of a living human body, obtained using a 3D volume rendering technique, showing the **mimic muscles** of the head.

Three-dimensional image of a living human body, obtained using a 3D volume rendering technique, showing the **musculature of the head**.

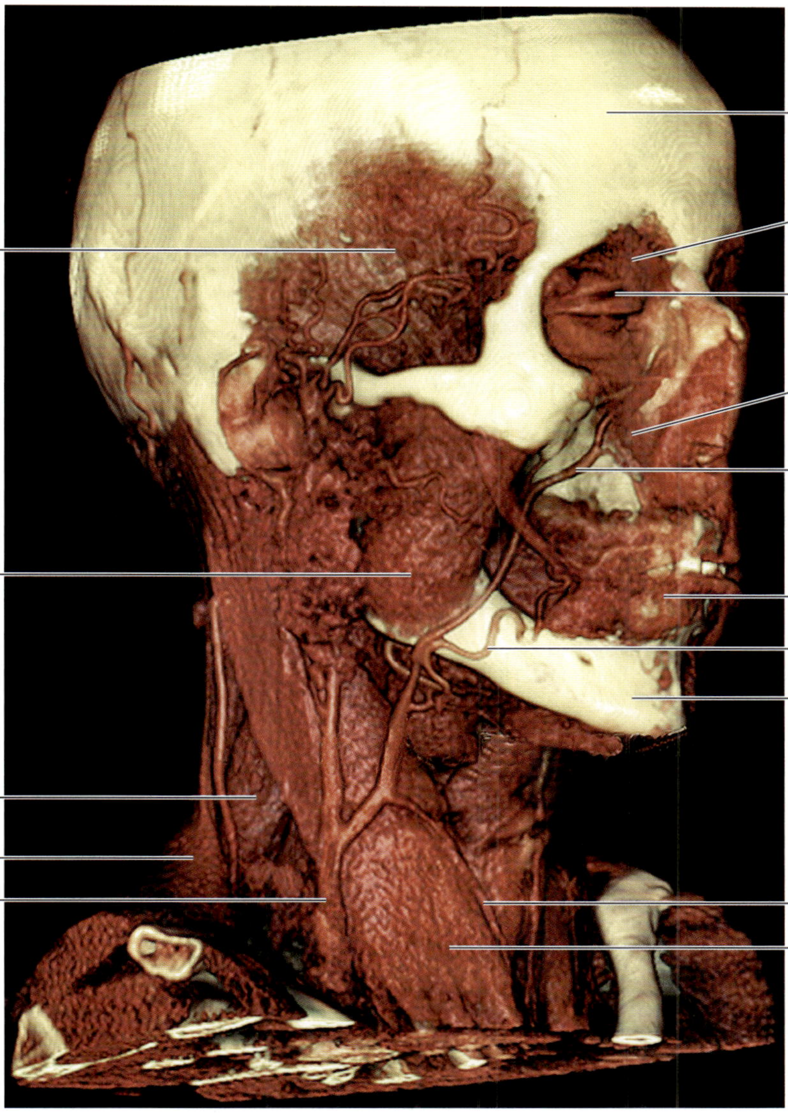

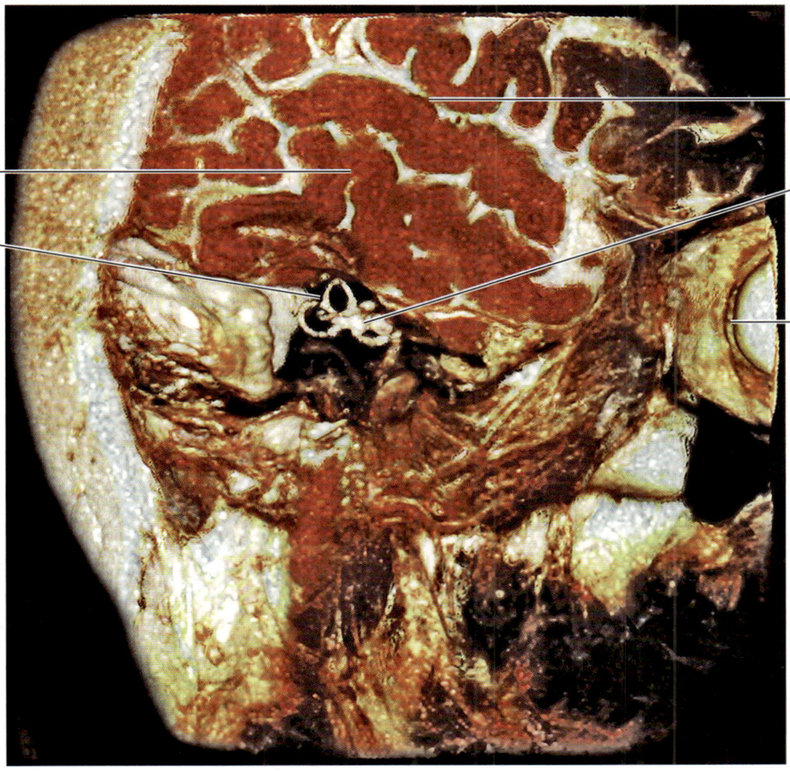

Three-dimensional image of a living human body, obtained using a 3D volume rendering technique, showing the parasagittal section tangent to the **cerebral hemispheres** and to the **eyeball**. The parameters have been modulated in such as way so as to partially strip away the bone tissue of the petrous part of temporal bone and to show the **membranous labyrinth**.

Origin and insertion of the **muscles of the face** (mimic muscles), including the masseter, sternocleidomastoid and platysma.

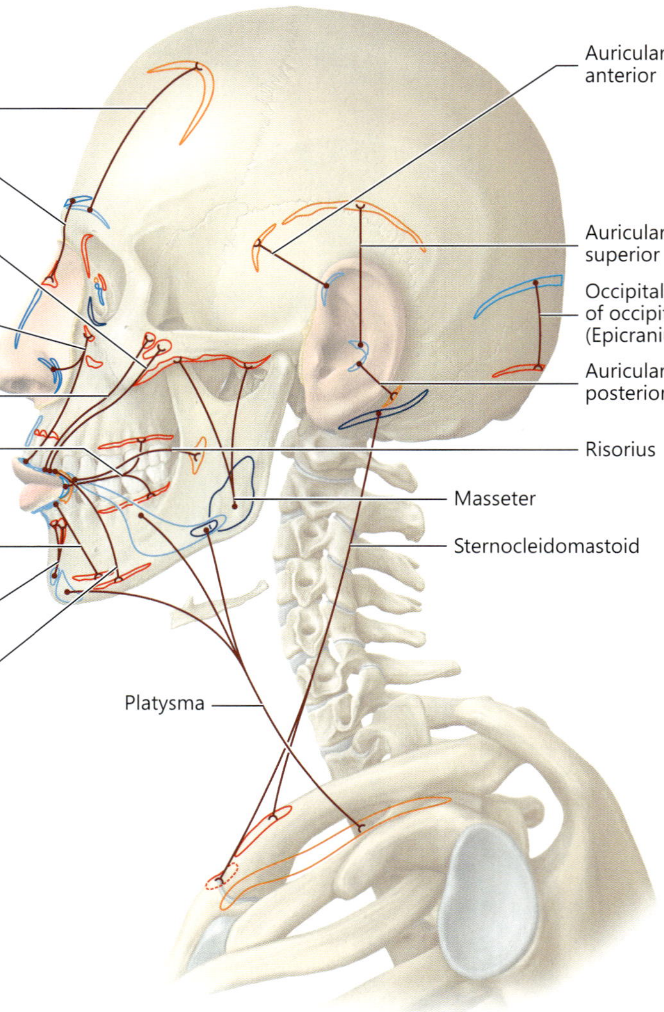

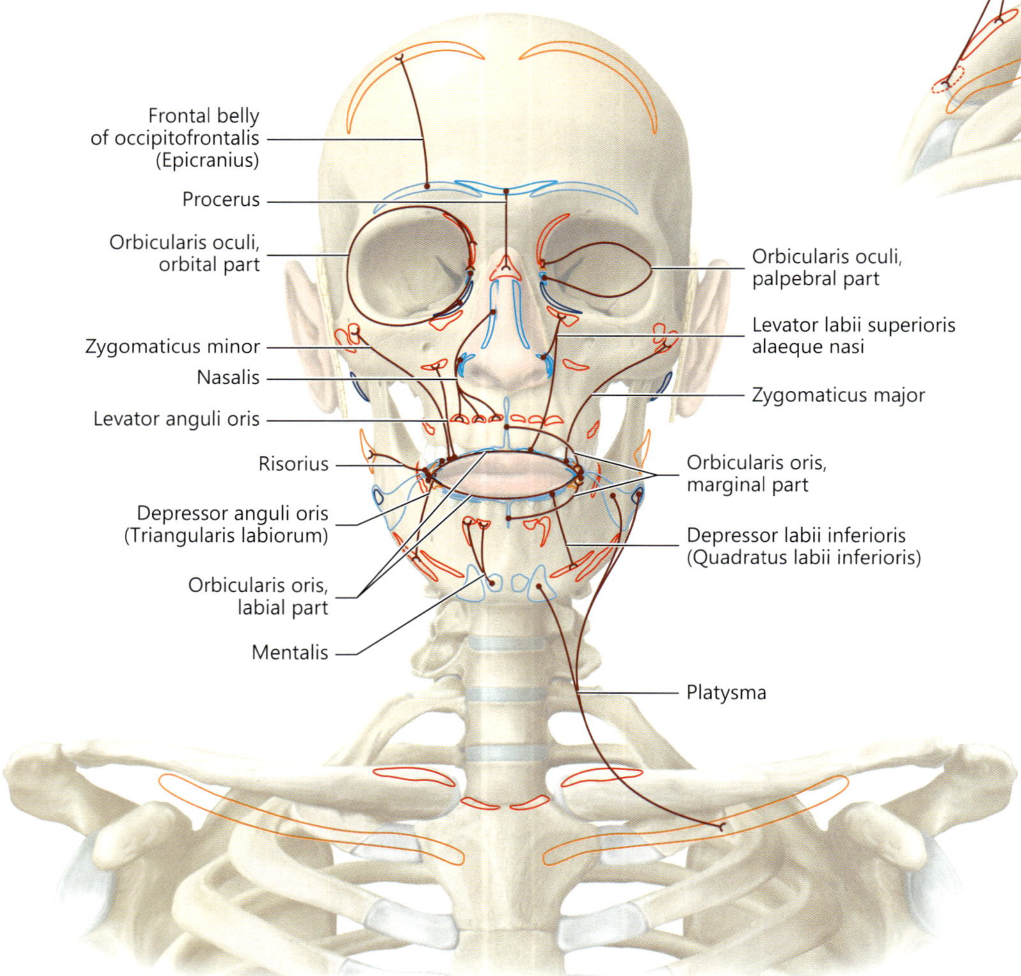

ORIGIN	INSERTION
Bones	Bones
Muscles, aponeuroses, ligaments, fasciae	Muscles, aponeuroses, cartilage, ligaments
Origin ⟶ Insertion	Deep surface of skin, deep part of the mucosa

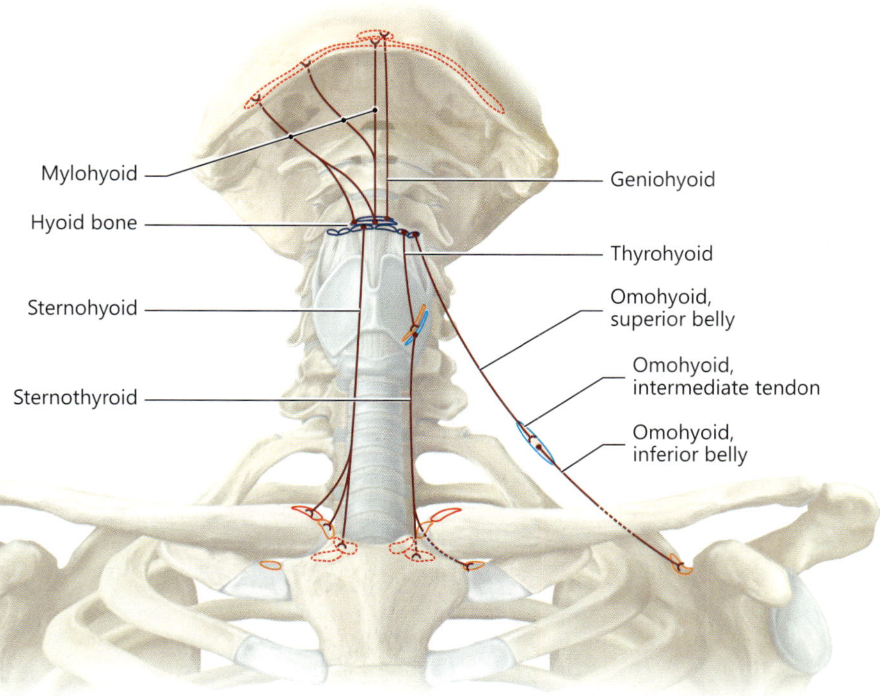

ORIGIN

Bones

Muscles, aponeuroses, ligaments, fasciae

Origin ⤚—• Insertion

INSERTION

Bones

Muscles, aponeuroses, cartilage, ligaments

Deep surface of skin, deep part of the mucosa

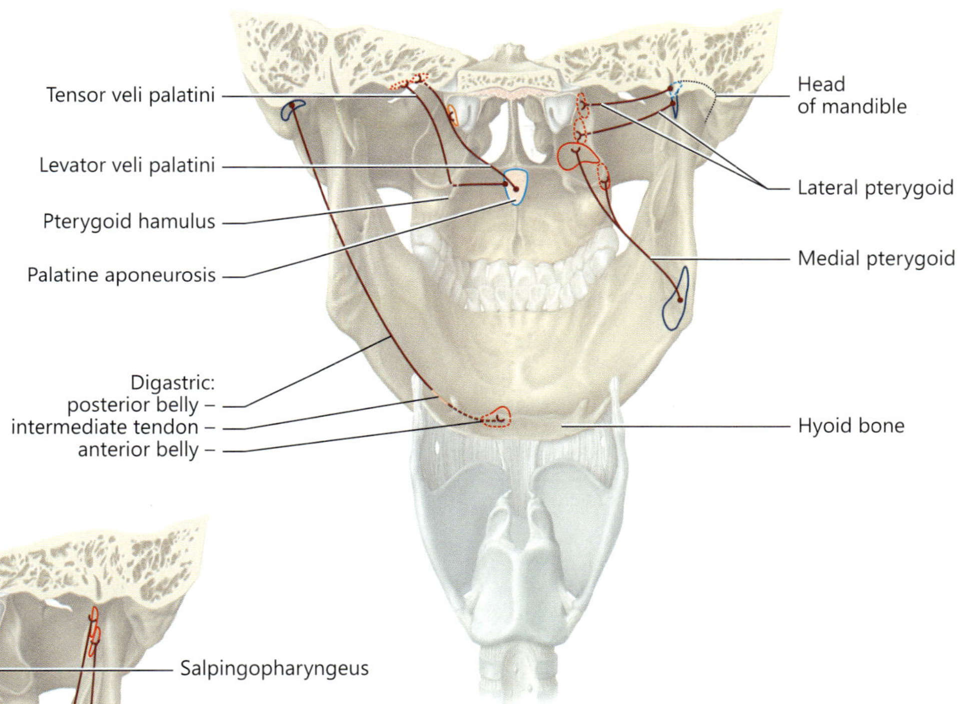

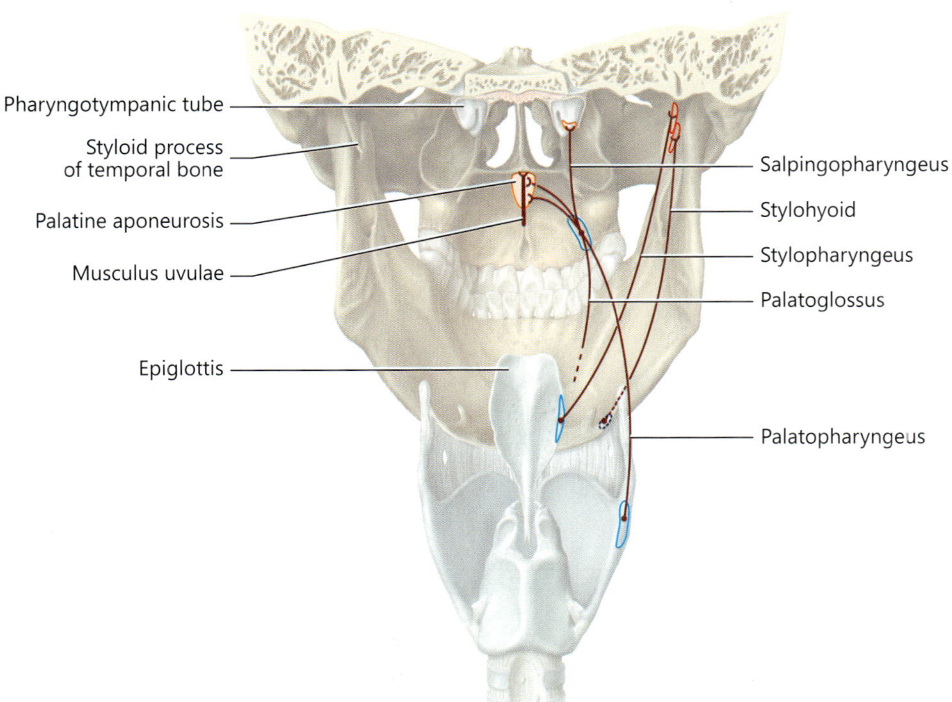

Origin and insertion of the **suprahyoid**, **infrahyoid**, and **medial** and **lateral pterygoid muscles**, as well as those of the **soft palate** and **fauces**, the **salpingopharyngeus**, and the **stylopharyngeus**.

Mimic muscles of the head					
MUSCLE	**ORIGIN**	**INSERTION**	**INNERVATION**	**ACTION**	**OBSERVATIONS**
Epicranius					
Epicranial aponeurosis or galea aponeurotica (origin of the frontal and occipital bellies of the occipitofrontalis and of the auricularis superior)			Facial nerve [VII]	Raising of the eyebrows and movement of the scalp forward and backward	The epicranial aponeurosis connects the occipitofrontalis and temporoparietalis, which in turn compose the muscular and tendinous layer of the scalp
Occipitofrontalis			Facial nerve [VII]		Wide and thick, quadrilateral in shape, two-bellied muscle
– Frontal belly (frontal muscle)	Anterior margin of the epicranial aponeurosis	Deep surface of the skin over the interciliary part and the superciliary arch of the frontal bone	Facial nerve [VII]	Protraction of the epicranial aponeurosis and raising of the skin of the eyebrows, with the appearance of transverse wrinkles on the forehead	
– Occipital belly (occipital muscle)	Posterior margin of the epicranial aponeurosis	External two-thirds of the superior nuchal line and the mastoid process	Facial nerve [VII]	Retraction of the epicranial aponeurosis	
Temporoparietalis	Skin of the temporal region and temporal fascia above the auricular cartilage	Lateral margin of the epicranial aponeurosis	Facial nerve [VII]	Scalp tension with retraction of the skin of the temporal region	Laminar (very thin)
Transverse occipital	External occipital protuberance and superior nuchal line	Highly variable (most often on the tendon of the sternocleidomastoid or on the posterior extremity of the auricularis posterior)	Facial nerve [VII]	Retraction of the epicranial aponeurosis	Supernumerary
Auriculares muscles					
Antitragicus	External surface of the antitragus	Tail of the helix and the antihelix	Facial nerve [VII]	Minimal forward movement of the antihelix	Rudimentary, quadrilateral in shape and more intense color
Auricularis anterior	Anterior part of the temporal fascia	The fibers converge backward toward the spine of the helix and the anterior margin of the concha of auricle	Facial nerve [VII]	Minimal forward movement of the auricle	Rudimentary
Auricularis posterior	Mastoid process	Above the oblique ridge of the eminentia conchae of the auricle	Facial nerve [VII]	Minimal backward movement of the auricle	Rudimentary
Auricularis superior	Epicranial aponeurosis and the temporal fascia	Cartilage of the auricle (superior part of the helix)	Facial nerve [VII]	Minimal raising of the auricle	Rudimentary
Helicis major	Superior and posterior parts of the cartilage of the helix	Spine of the helix	Facial nerve [VII]	Dilation of the external acoustic meatus	Tight vertical band
Helicis minor	Base of the crus of the helix	Anterior surface of the crus of the helix in its upward curvature above the tragus	Facial nerve [VII]	Minimal variation in the shape of the auricle at the level of the anterior margin of the cartilage	Oblique fascicle
Oblique muscle of auricle	Superior and posterior parts of the eminentia conchae	Cranial surface of the eminentia fossae triangularis	Facial nerve [VII]	Minimal variation in the shape of the auricle at the level of the superior margin of the cartilage	Rudimentary; consists of one or two small fascicles
Tragicus (muscle of the tragus)	Base of the tragus	Apex of the tragus	Facial nerve [VII]	Tragus tension with minimum dilation of the external acoustic meatus	Short and plane vertical band
Transverse muscle of auricle	Internal surface of the auricle (cranial surface) and eminentia conchae	Eminentia scaphae or groove of the helix	Facial nerve [VII]	Minimal retraction of the helix	Very short transversal fibers, partly tendineous and partly muscular

continued

Mimic muscles of the head *(continued)*

MUSCLE	ORIGIN	INSERTION	INNERVATION	ACTION	OBSERVATIONS
Circumorbital and palpebral muscles					
Corrugator supercilii	Medial extremity of the superciliary arch of the frontal bone	Deep surface of the skin of the eyebrow	Facial nerve [VII]	Medial depression of the eyebrow. Furrowing of the eyebrows, with corrugation of the forehead, produced by the simultaneous contraction of the two muscles	
Depressor supercilii	Nasal part of the frontal bone and fibers of the orbicularis oculi	Deep surface of the skin at the center of the eyebrow	Facial nerve [VII]	Depression and furrowing of the eyebrow	Also described as part of the orbicularis oculi
Orbicularis oculi			Facial nerve [VII]		
– Palpebral part	Posterior lacrimal crest and medial palpebral ligament	Lateral palpebral ligament	Facial nerve [VII]	As a sphincter. Closing of the palpebral fissure	Located in the tarsi of the eyelids. Besides, it contains striated muscle fibers. The tensor tarsi (or Horner's muscle) represents an accessory fascicle
– Orbital part	Medial wall of the orbital cavity and frontal process of the maxilla	Orbital tubercle of the zygmatic bone and near the origin	Facial nerve [VII]	Perpendicular corrugation of the skin of the forehead near the orbit	Circumscribes the edge of the orbit (concentric fibers). Forms a nearly complete ring around the orbit
– Deep part or lacrimal part	Deep surface of the orbital part	Posterior lacrimal crest	Facial nerve [VII]	Compression of the lacrimal sac	
Orbital muscle (orbital muscle of Müller)	Infraorbital groove and inferior orbital fissure	Periosteum of the orbit	Facial nerve [VII]	Slight protraction of the eyeball	Rudimentary; constituted by smooth muscle tissue
Muscles of the nose					
Depressor septi nasi	Incisive foramen of the maxilla	Nasal septum and ala of the nose	Facial nerve [VII]	Dilation of the naris and lowering of the ala of the nose	
Dilator of nares	Margin of the piriform opening of the maxilla posteriorly to the major alar cartilage	Skin that invests the inferior margin of the major alar cartilage	Facial nerve [VII]	Dilation of the naris in transversal sense	Also described as the alar part of the nasalis
Levator labii superioris alaeque nasi	Superior part of the frontal process of the maxilla	Deep surface of the skin of the upper lip and of the major alar cartilage	Facial nerve [VII]	Raising of the ala of the nose and the upper lip, with dilation of the naris	Small, thin and ribbon muscle
Nasalis			Facial nerve [VII]		
– Alar part	☞ Dilator of nares				
– Transverse part	Dorsum of the nose	Skin at the level of the ala of the nose and of the depressor septi nasi		Posteriorly reduces the width of the nasal cavity; anteriorly lifts the skin of the ala of the nose	
Procerus	Fascia attached to the periosteum of the lower part of the nasal bone	Skin of the glabella between the eyebrows	Facial nerve [VII]	Corrugation of the skin of the dorsum of the nose and lowering of the medial angle of the palpebral fissure	
Muscles of the mouth and lips					
Buccinator	Angle of the mouth, alveolar process of the maxilla and the alveolar part of the mandible, pterygoid hamulus, pterygomandibular raphe	Deep surface of the labial mucosa (some fibers intersect, entering the structure of the lips and taking part in the formation of the orbicularis oris)	Facial nerve [VII]	Retraction of the labial commissures or angles of the mouth, with compression of the cheeks, produced by the simultaneous contraction of the two muscles	Laminar; accessory muscle of mastication that constitutes the mass of the cheeks

continued

Mimic muscles of the head					*(continued)*
MUSCLE	***ORIGIN***	***INSERTION***	***INNERVATION***	***ACTION***	***OBSERVATIONS***
Depressor anguli oris (triangularis labiorum)	Middle third of the oblique line of the mandible	Skin of the labial commissure	Facial nerve [VII]	Lowering of the labial commissure or angle of the mouth. Protraction of the labial commissures following the simultaneous contraction of the two muscles	It characterizes the sad facial expression
Depressor labii inferioris (quadratus labii inferioris)	Oblique line of the mandible between the symphysis and the mental foramen	Deep surface of the skin of the lower lip	Facial nerve [VII]	Lowering, in the lateral part, of the lower lip and its eversion	Quadrilateral form
Levator labii superioris (quadratus labii superioris)	Frontal process of the maxilla, inferior margin of the orbit above the infraorbital foramen, lateral surface of the zygomatic bone	Deep surface of the skin of the upper lip and margin of the naris	Facial nerve [VII]	Raising of the upper lip and dilation of the naris	Plain form and elongated
Incisive muscle of the lower lip	Fibers of the orbicularis oris	Labial commissure	Facial nerve [VII]	Hollowing of the oral vestibule	Part of the orbicularis oris
Incisive muscle of the upper lip	Fibers of the orbicularis oris	Labial commissure	Facial nerve [VII]	Hollowing of the oral vestibule	Part of the orbicularis oris
Levator anguli oris (caninus)	Canine fossa of the maxilla (below the infraorbital foramen)	Deep surface of the skin and mucosa of the labial commissures	Facial nerve [VII]	Raising, in the medial part, of the labial commissure or angle of the mouth	
Mentalis	Incisive foramen of the mandible, on both sides of the median line, between the superior part of the mandibular symphysis and the mental protuberance	*Proximal*: alveolar yokes of the canine and inferior incisive teeth *Distal*: deep surface of the skin of the chin	Facial nerve [VII]	Raising of the mental protuberance with corrugation of the skin of the chin, and of the lower lip, with its protrusion and eversion	Paired, small and conoidal shape
Orbicularis oris			Facial nerve [VII]	Sphincter of the oral opening. Puckering, closure of the lips and their contraction	Forms two external half-rings, superior and inferior, which starting from the labial commissures, join together at the level of the median part of each lip
– Labial part	Mimic muscles surrounding the mouth	Deep surface of the skin and mucosa of the lips	Facial nerve [VII]		
– Marginal part	Mimic muscles surrounding the mouth	Deep surface of the skin and mucosa of the lips	Facial nerve [VII]		
Risorius	Zygomatic arch, parotid fascia, masseter fascia, fascia of the platysma	Skin and mucosa towards the labial commissure	Facial nerve [VII]	Retraction of the labial commissure or angle of the mouth. Increase in the transverse diameter of the mouth (smile expression) following the simultaneous contraction of the two muscles	Triangular shape, situated on each side of the face
Transversus menti	Below the mental tubercle of the mandible through interdigitations common to both muscles	Merger with the depressor anguli oris near the inferior margin of the ramus of the mandible	Facial nerve [VII]	Lowering of the lower lip with corrugation of the skin of the chin	Quadrilateral form
Zygomaticus major	Zygomatic arch	Deep surface of the skin and mucosa of the labial commissure	Facial nerve [VII]	Lateral raising of the labial commissure or angle of the mouth	
Zygomaticus minor	Inferior part of the lateral surface of the zygmatic bone	Deep surface of the skin of the upper lip	Facial nerve [VII]	Lateral raising of the median portion of the upper lip, with accentuation of the nasolabial sulcus	

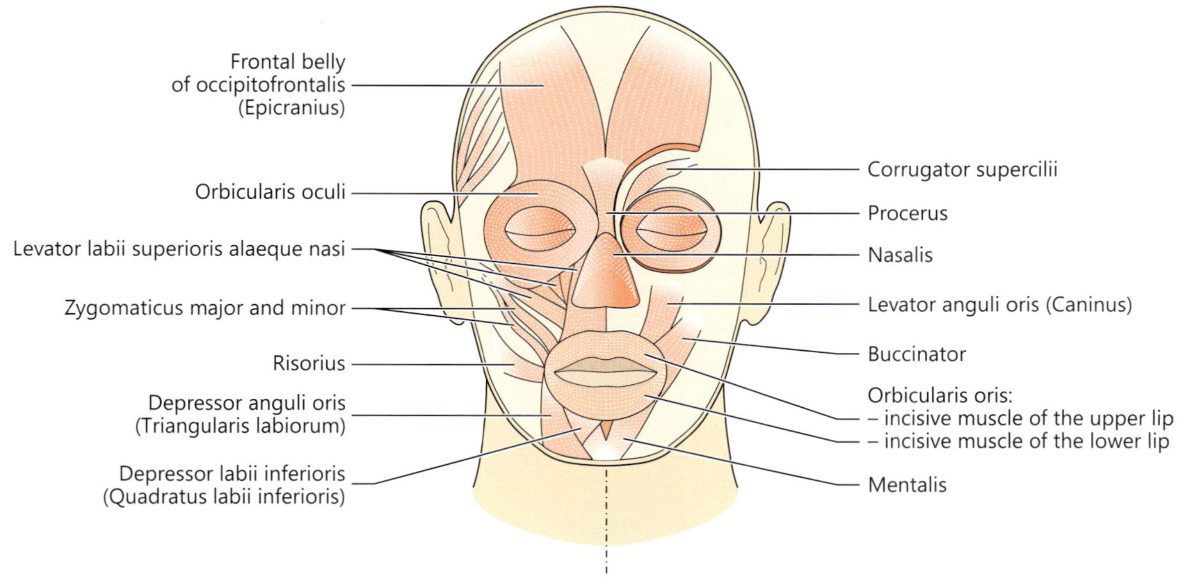

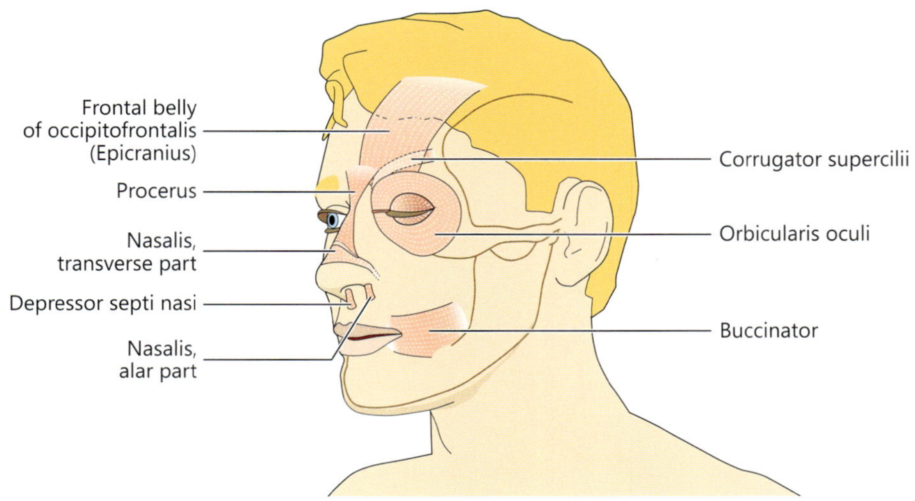

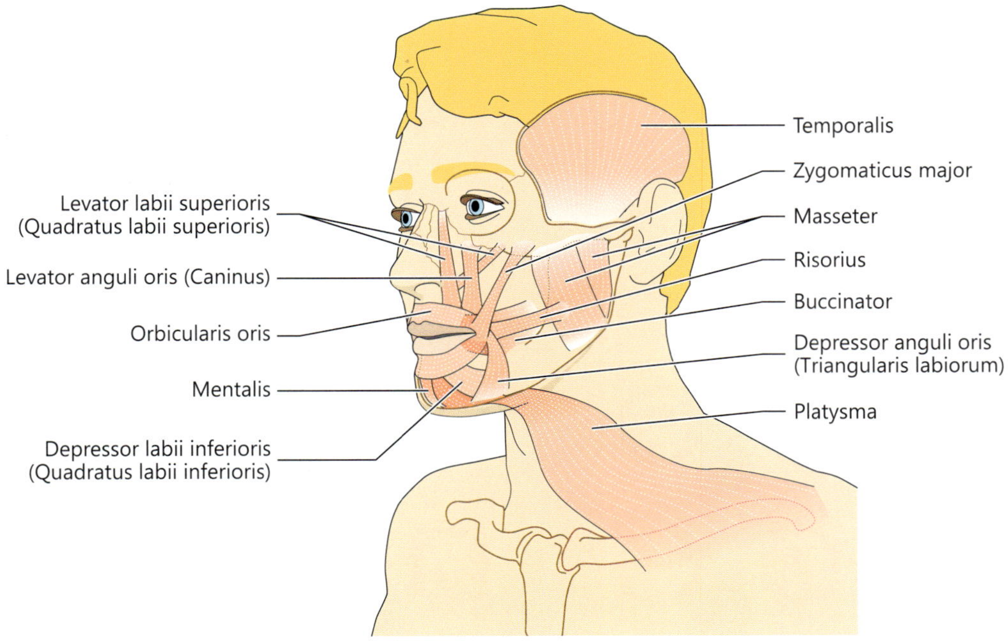

Mimic muscles.

Extraocular muscles*				
MUSCLE	***ORIGIN***	***INSERTION***	***INNERVATION***	***ACTION***
Inferior oblique	Floor of the orbital cavity (orbital surface of the maxilla) under the lacrimal fossa	Corneoscleral junction between the insertions of the lateral rectus and superior rectus	Oculomotor nerve [III]	Superior and lateral rotation of the eyeball
Inferior rectus	Lower segment of the common tendinous ring (anulus of Zinn)	Inferior part of the sclera just behind the corneoscleral junction	Oculomotor nerve [III]	Inferior and slightly medial rotation of the eyeball
Lateral rectus	Lateral margin of the common tendinous ring (anulus of Zinn)	Lateral part of the sclera just behind the corneoscleral junction	Abducens nerve [VI]	Lateral rotation of the eyeball
Levator palpebrae superioris**	Inferior surface of the lesser wing of sphenoid above and in front of the optic canal	Superior tarsus, skin of the upper eyelid and superior conjunctival fornix through a wide aponeurosis	Oculomotor nerve [III]	Raising of the superior eyelid. The palpebral part of the orbicularis oculi acts as its agonist
Medial rectus	Medial segment of the common tendinous ring (anulus of Zinn)	Medial part of the sclera just behind the corneoscleral junction	Oculomotor nerve [III]	Medial rotation of the eyeball
Superior oblique	Lesser wing of the sphenoid above the optic canal	After passing through the trochlea, the muscle fibers change direction and insert into the upper part of the sclera, deeply into the insertion of the superior rectus	Trochlear nerve [IV]	Inferior and lateral rotation of the eyeball
Superior rectus	Superior margin of the common tendinous ring (anulus of Zinn)	Superior part of the sclera just behind the corneoscleral junction	Oculomotor nerve [III]	Superior and slightly medial rotation of the eyeball

* In addition to the extraocular muscles there are also three intrinsic muscles: the ciliary, the sphincter pupillae, and the dilator pupillae, which are not taken into account in this part of the Atlas regarding the musculoskeletal system because they consist of smooth muscle fibers. Among the accessory organs of the eye there are the superior and inferior tarsal muscles, also composed of smooth muscle fibers, which form the muscular layer of the eyelids.

** From the underlying surface of the levator palpebrae superioris, a thin lamina of smooth muscle tissue, the superior tarsal muscle (Müller's), reaches the superior margin of the superior tarsus. This accessory muscle, also named as muscular tendon of the levator palpebrae superioris, is innervated by the internal carotid plexus of the sympathetic trunk.

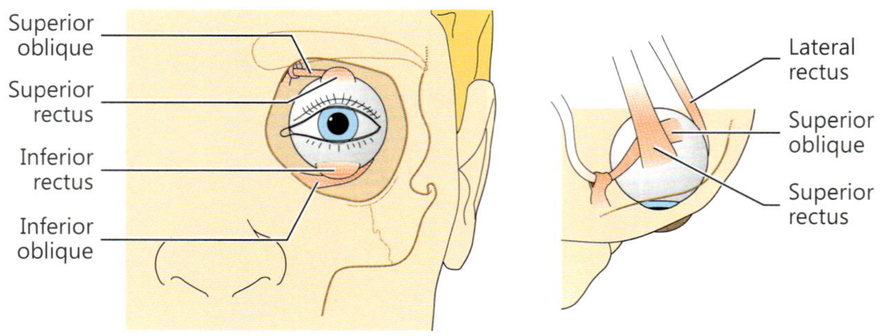

Frontal and superior aspect of the **left eye**.

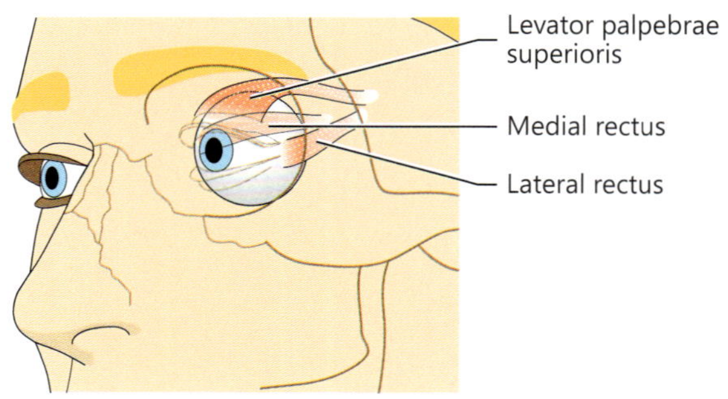

Lateral aspect of the **left eye**.

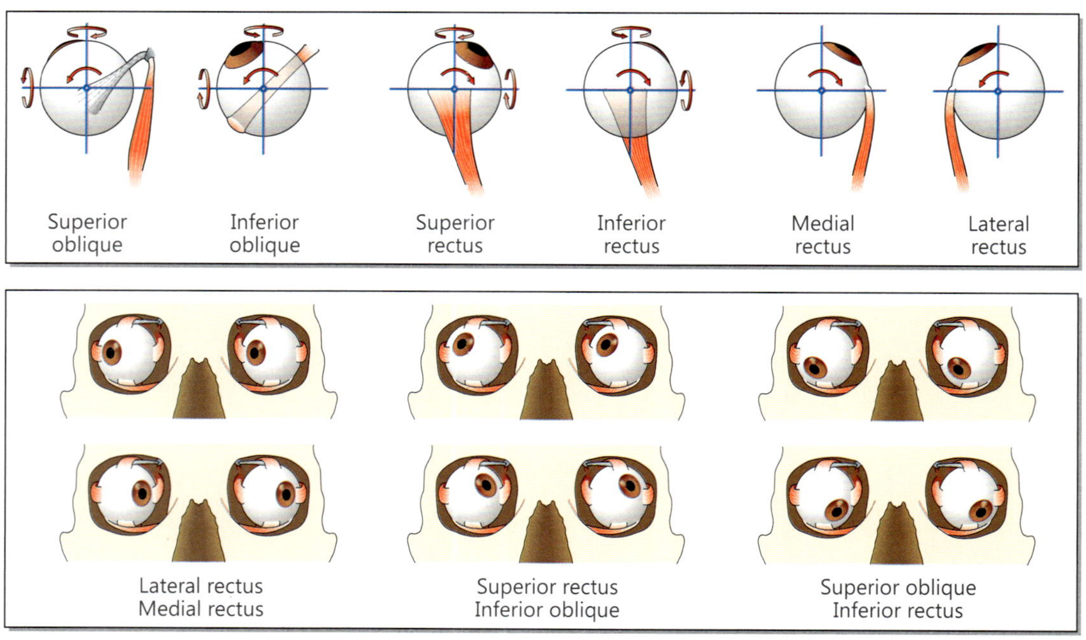

Schematic representation of the **movements of the eyeball** which are determined by the action of the extrinsic muscles.

Muscles of the auditory ossicles					
MUSCLE	**ORIGIN**	**INSERTION**	**INNERVATION**	**ACTION**	**OBSERVATIONS**
Stapedius	Pyramidal eminence of the mastoid wall of the tympanic cavity	Posterior part of the head of the stapes	Facial nerve [VII]	Retraction of the head of the stapes resulting in a slight tilt of the base of the stapes and a concomitant reduction in the tension of the tympanic membrane	The smallest skeletal muscle of the human body
Tensor tympani	Bony canal above the pharyngotympanic tube, cartilaginous part of the pharyngotympanic tube, and the greater wing of the sphenoid	Root of the handle of the malleus	Facial nerve [VII]	Medial movement of the handle of the malleus resulting in an increase in the tension of the tympanic membrane	

Functional anatomy

The auditory ossicles is a very efficient system for transferring sounds from the external ear to the internal ear. The tympanic membrane, when vibrating, transmits the movement to the malleus, which in turn transmits it to the incus, and from the incus to the stapes. If the tympanic membrane moves medially (towards the tympanic cavity), the same occurs for the stapes which moves deeper in the oval window; when the tympanic membrane moves laterally (towards the external acoustic meatus), the stapes follows. The movements of the stapes are transmitted from its base to the perilymph of the vestibule. The axis of the swinging movements of the auditory ossicles is a direct horizontal line in the anteroposterior direction from the petrotympanic fissure to the fossa of the incus, passing through the anterior process of the malleus and the short apophysis of the incus. Sound waves also enter the internal ear by vibration of the air contained in the tympanic cavity or by vibrations in the entire bone formation surrounding the cochlea.

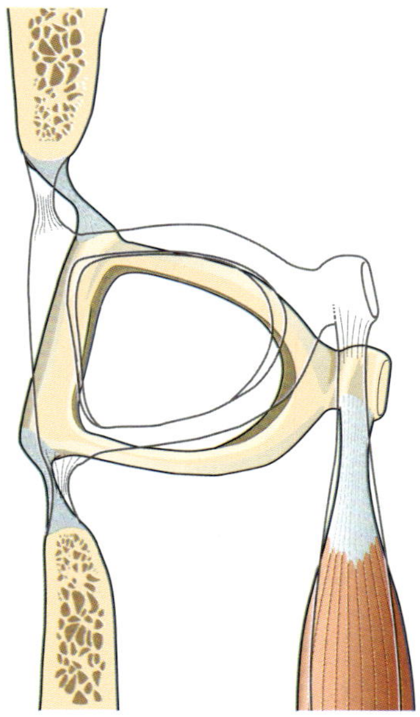

Mechanism of action of the **stapedius**.

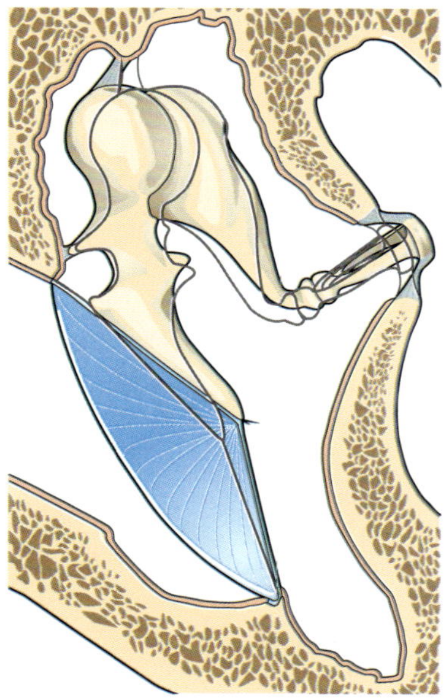

Transmission mechanisms mediated by the **auditory ossicles**.

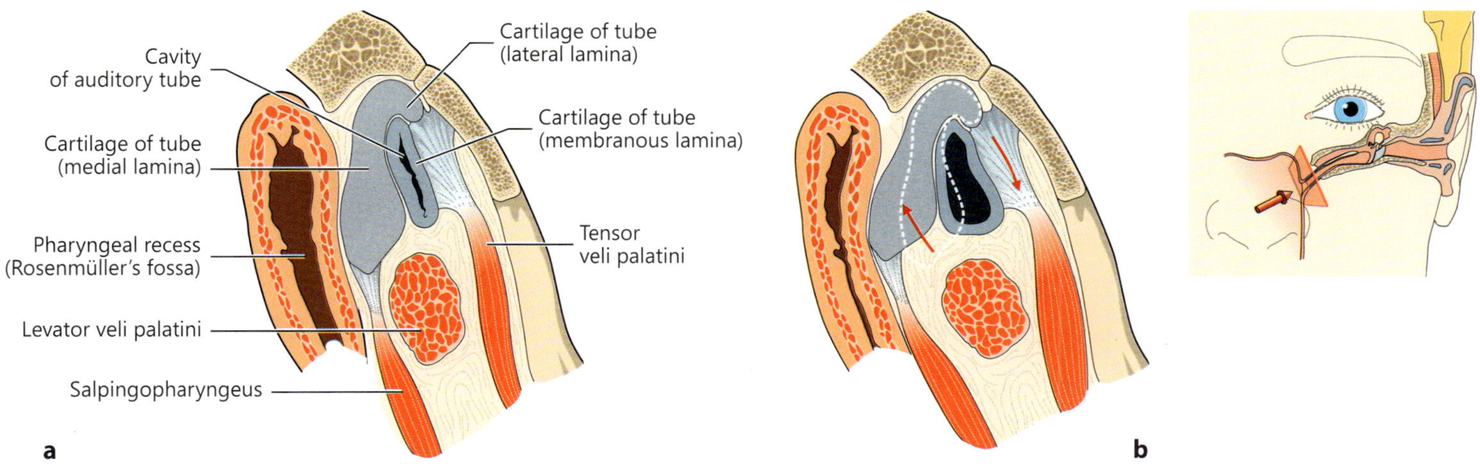

a, Fibrocartilaginous part of the **pharyngotympanic tube** in an oblique section conducted near the pharyngeal opening. It shows the characteristic crooked shape of the cartilage of the tube. **b**, Fibrocartilaginous part of the **pharyngotympanic tube** in conditions of patency determined by the contraction of the tensor veli palatini.

Masticatory muscles				
MUSCLE	**ORIGIN**	**INSERTION**	**INNERVATION**	**ACTION**
Masseter	*Superficial part*: zygomatic process of the maxilla; anterior two thirds of the zygomatic arch *Intermediate part*: medial surface of the anterior two thirds of the zygomatic arch; inferior margin of the posterior third of the ramus of the mandible *Deep part*: deep surface of the zygomatic arch	*Superficial part*: angle and lower half of the lateral surface of the ramus of the mandible *Intermediate part*: central part of the ramus of the mandible *Deep part*: superior part of the ramus of the mandible; coronoid process of the mandible	Trigeminal nerve [V]	Vigorous elevation of the mandible; modest deviation of the mandible to one side and to the other; protrusion and retrusion of the mandible; due to the oblique orientation of its fibers it can move the mandible in inferoposterior direction
Lateral pterygoid	*Superior head*: inferior part and lateral surface of the greater wing of the sphenoid *Inferior head*: lateral lamina of the pterygoid process	Pterygoid fovea on the internal surface of the neck of the mandible; articular disc and joint capsule of the temporomandibular joint	Trigeminal nerve [V]	Protrusion of the mandible (along with the medial pterygoid); lowering of the mandible (action concentrated with the digastric, mylohyoid and geniohyoid); control of the posterior movement of the articular disc of the temporomandibular joint and of the mandibular condyle when closing the mouth; deviation of the mandible to one side and to the other; during mastication it stabilizes the side in action
Medial pterygoid	Medial lamina of the pterygoid process; pyramidal process of the palatine bone; maxillary tuberosity	Posteroinferior part of the medial surfaces of the ramus and angle of the mandible	Trigeminal nerve [V]	Elevation of the mandible: protrusion of the mandible (along with the lateral pterygoid); lateral deviation of the mandible; during mastication it stabilizes the side in action
Temporalis	Temporal fossa; deep lamina of the temporal fascia	Coronoid process of the mandible; anterior margin of the ramus of the mandible almost until the last molar tooth	Trigeminal nerve [V]	Elevation of the mandible; deviation of the mandible to one side and to the other; retrusion of the mandible mediated by the action of the posterior fascicles

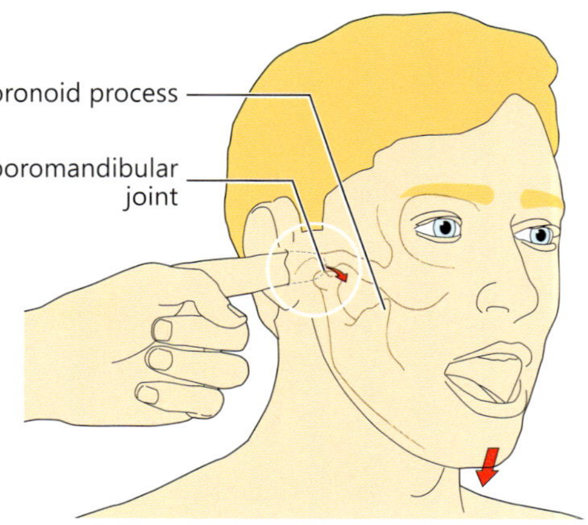

Palpation of the **temporomandibular joint**.

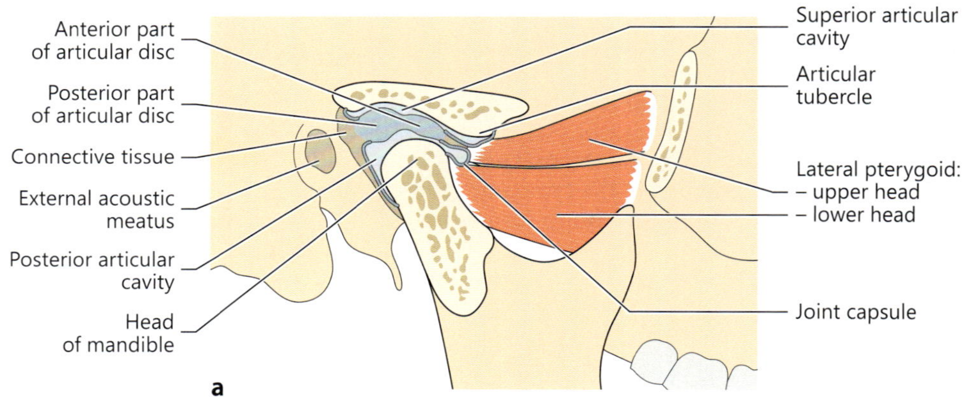

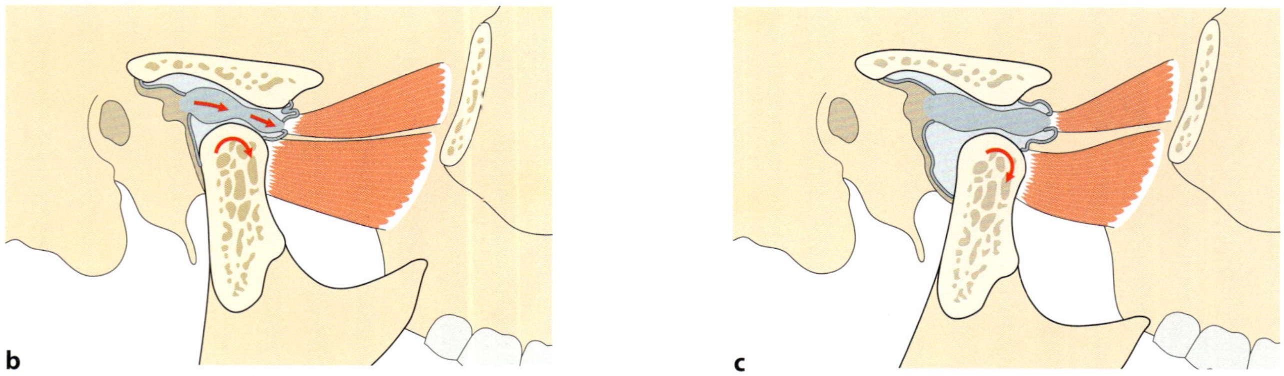

Position of the **temporomandibular joint**. **a**, Mouth closed; **b**, while opening the mouth; **c**, when the mouth is fully open.

Muscles of the tongue				
MUSCLE	**ORIGIN**	**INSERTION**	**INNERVATION**	**ACTION**
Extrinsic muscles				
Ceratoglossus (part of the hyoglossus)	Greater horn of the hyoid bone	Lateral margin of the tongue	Hypoglossal nerve [XII]	Retraction and lowering of the tongue
Chondroglossus (part of the hyoglossus)	Lesser horn of the hyoid bone	Posterior extremity of the lateral margin of the body of the tongue	Hypoglossal nerve [XII]	Lowering and retraction of the tongue
Genioglossus	Superior mental spine	Apex and dorsum of the tongue, superior margin of the body of the hyoid bone and the base of the epiglottis	Hypoglossal nerve [XII]	Lowering and retraction of the apex of the tongue. Protrusion of the tongue and protraction of the hyoid bone
Hyoglossus (constituted by the ceratoglossus, chondroglossus and basioglossus, to which refer the origin and insertion reported here)	Superior margin of the body of the hyoid bone	Lateral margin, lingual septum and apex of the tongue	Hypoglossal nerve [XII]	Retraction and lowering of the tongue
Palatoglossus (glossopalatine)	Inferior surface of the soft palate	Dorsum and lateral margin of the tongue	Accessory nerve [XI] through the pharyngeal plexus	Raising of the tongue and narrowing of the fauces
Styloglossus	Styloid process of the temporal bone and stylomandibular ligament	Longitudinal part: side of the tongue near the dorsal surface Oblique part: above the hyoglossus muscle	Hypoglossal nerve [XII]	Raising and retraction of the tongue
Intrinsic muscles				
Inferior longitudinal muscle of the tongue	Lamina propria of the mucosa of the base of the tongue	Submucosa of the inferior surface of the body of the tongue	Hypoglossal nerve [XII]	Shortening of the tongue with the retraction and lowering of the apex
Superior longitudinal muscle of the tongue	Deeply to the lamina propria of the mucosa of the tongue	Deeply to the lamina propria of the mucosa of the tongue	Hypoglossal nerve [XII]	Shortening of the tongue which assumes a superior concave shape
Transverse muscle of the tongue	Lingual septum	Submucosa of the margins of the tongue	Hypoglossal nerve [XII]	Reduction of the transverse diameter of the tongue
Vertical muscle of the tongue	Deep surface of the lamina propria of the dorsum	Submucosa of the inferior surface of the tongue	Hypoglossal nerve [XII]	Flattening of the tongue aiding in the processes of deglutition, mastication and speech

Muscles of the soft palate and fauces				
MUSCLE	**ORIGIN**	**INSERTION**	**INNERVATION**	**ACTION**
Levator veli palatini	Apex of the petrous part of the temporal bone and the cartilaginous part of the pharyngotympanic tube	Palatine aponeurosis	Glossopharyngeal nerve [IX] through the pharyngeal plexus	Raising of the soft palate
Musculus uvulae	Palatine aponeurosis and posterior nasal spine of the palatine bone	Mucosa of the uvula	Pharyngeal branch of the vagus nerve [X] through the pharyngeal plexus	Raising the uvula
Palatoglossus ☞ Palatoglossus between the Muscles of the tongue				
Palatopharyngeus (pharyngopalatine)				
– anterior fascicle	Soft palate	Posterior wall of the thyroid cartilage	Pharyngeal branch of the vagus nerve [X] (perhaps with fibers from the accessory nerve, XI) through the pharyngeal plexus	Raising of the pharynx, closure of the nasopharynx and restriction of the fauces
– posterior fascicle (palatopharyngeal sphincter)	Soft palate	Pharyngeal wall	Pharyngeal branch of the vagus nerve [X] (perhaps with fibers from the accessory nerve, XI) through the pharyngeal plexus	Raising of the pharynx, closure of the nasopharynx and restriction of the fauces
Tensor veli palatini	Spine of the sphenoid, pterygoid process, cartilage and membrane of the pharyngotympanic tube	Median line of the palatine aponeurosis, wall of the pharyngotympanic tube	Nerve to the tensor veli palatini, branch of the mandibular nerve (Vc or V3)	Raising of the palate

Muscles of the floor of the mouth

MUSCLE	ORIGIN	INSERTION	INNERVATION	ACTION
Digastric	Digastric fossa of the body of the mandible	Mastoid notch	Facial [VII] and glossopharyngeal [IX] (posterior belly) nerves Mylohyoid nerve (anterior belly)	Raising and retraction of the hyoid bone or retraction of the head (posterior belly) Lowering of the mandible or raising of the hyoid bone (anterior belly) Raising of the hyoid bone following the simultaneous contraction of the two muscles
Geniohyoid	Inferior mental spine	Middle part of the anterior surface of the hyoid bone	Hypoglossal nerve [XII]	Raising of the hyoid bone or lowering of the mandible
Mylohyoid	Internal surface of the mandible from the mental spine to the mylohyoid line	Anterior surface of the hyoid bone The more medial fibers come together on the median line forming the mylohyoid raphe	Mylohyoid nerve (branch of the inferior alveolar nerve, branch of the mandibular nerve, Vc or V3)	Raising of the floor of the mouth during deglutition Lowering of the mandible or raising of the hyoid bone
Stylohyoid	Styloid process of the temporal bone	Lateral extremity of the body of the hyoid bone	Facial nerve [VII]	Raising of the hyoid bone

Muscles of the pharynx

MUSCLE	ORIGIN	INSERTION	INNERVATION	ACTION	OBSERVATIONS
Inferior constrictor			Accessory nerve [XI] (cranial root) through the pharyngeal plexus		
– Thyropharyngeal part	Oblique line and inferior horn of the thyroid cartilage	Pharyngeal raphe		Restriction of the lower part of the pharynx during deglutition	The pharyngeal raphe inserts above the pharyngeal tubercle of the basilar part of the occipital bone
– Cricopharyngeal part	Lateral surface of the cricoid cartilage	Pharyngeal raphe		Principal component of the superior esophageal sphincter (consistent muscular tone, relaxation and rapid contraction during deglutition)	
Middle constrictor	Greater and lesser horns of the hyoid bone, stylohyoid ligament	Pharyngeal raphe	Accessory nerve [XI] (cranial root) through the pharyngeal plexus	Restriction of the medial part of the pharynx during deglutition	
Salpingopharyngeus	Cartilage of the pharyngotympanic tube	Pharyngeal wall (the muscle joins with the palatopharyngeal), thyroid cartilage	Accessory nerve [XI] (cranial root) through the pharyngeal plexus	Raising of the pharynx. Opening of the pharyngotympanic tube during deglutition (along with the tensor veli palatini)	
Stylopharyngeus	Styloid process of the temporal bone	Some fibers join with those of the constrictor muscles of the pharynx, others join with the fibers of the palatopharyngeus fitting into the thyroid cartilage	Glossopharyngeal nerve [IX]	Raising of the pharynx and of the larynx	
Superior constrictor	Pterygoid hamulus, medial lamina of the pterygoid process pterygomandibular raphe, mylohyoid line of the mandible	Pharyngeal raphe	Accessory nerve [XI] (cranial root) through the pharyngeal plexus	Restriction of the upper part of the pharynx during deglutition	

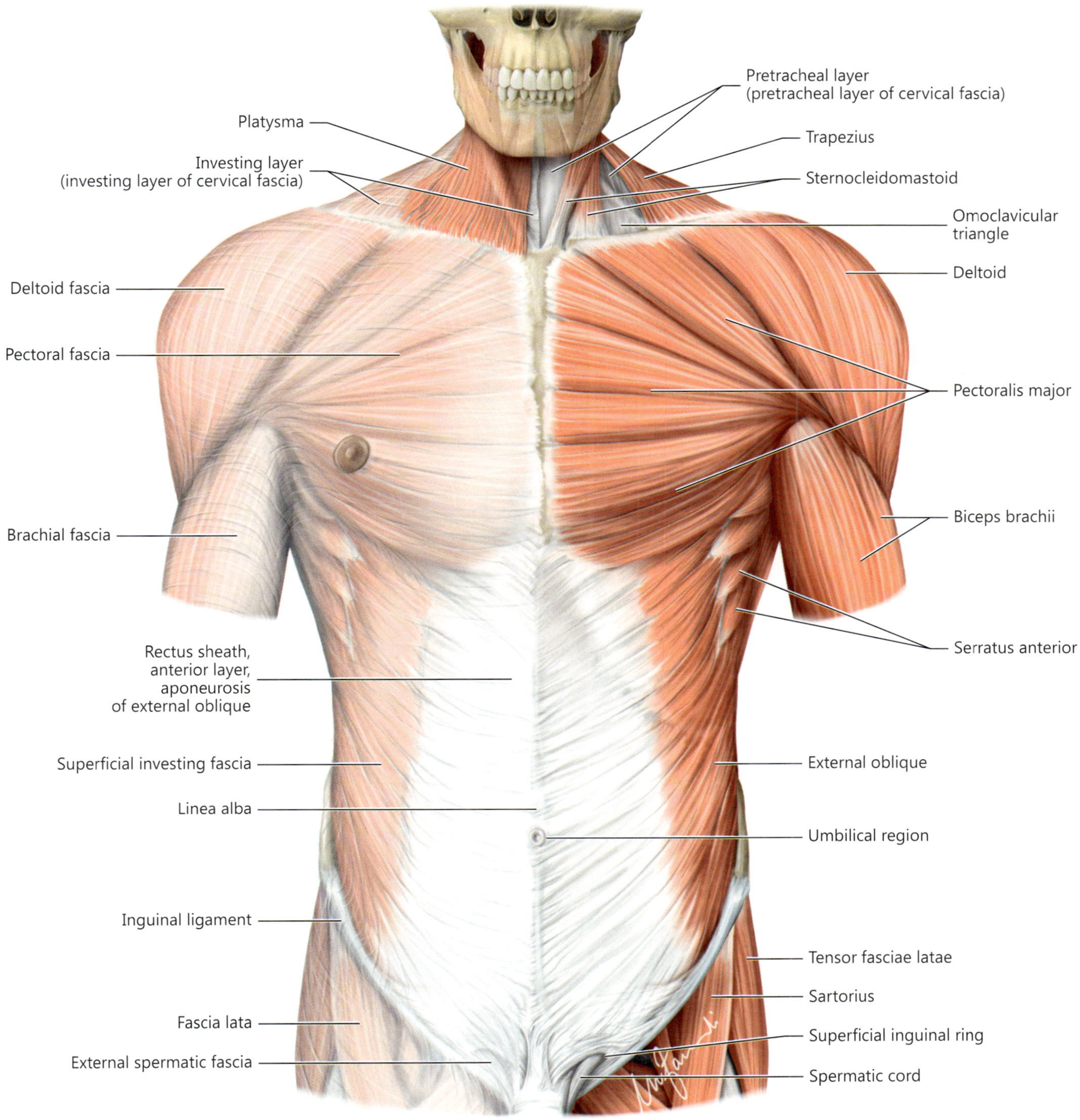

Preparation illustrating the stratigraphic sequence of the **muscles of the anterior wall of the trunk**. On the right, the investing fasciae have been left in place; on the left, they have been removed. The linea alba, between the borders of the rectus abdominis muscles, is formed by the intersection along the midline of the aponeurosis of the external oblique, internal oblique, and transversus abdominis.

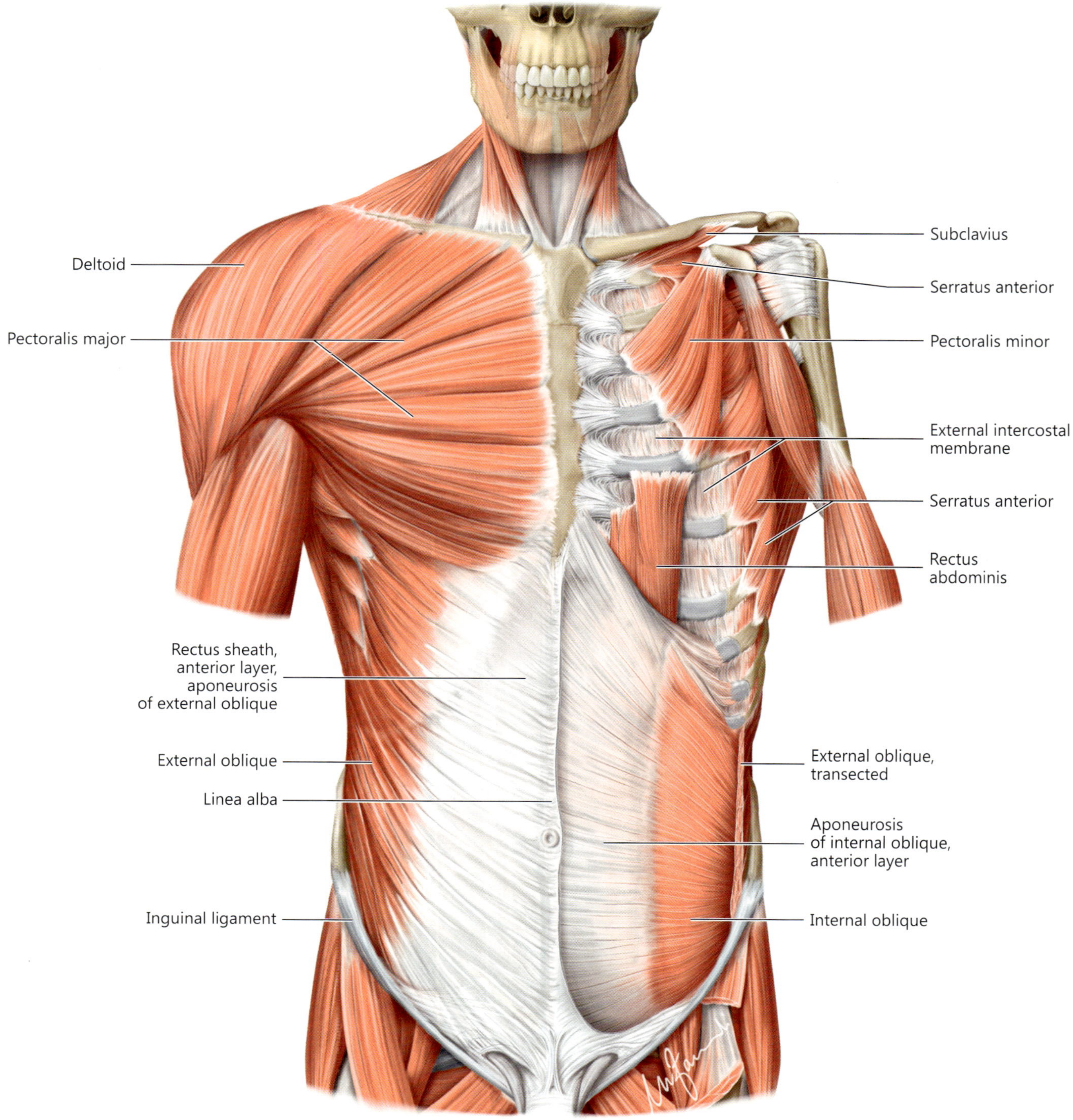

Preparation illustrating the stratigraphic sequence of the **muscles of the anterior wall of the trunk**. The right side shows the surface layer. On the left, the trapezius and deltoid have been removed, revealing the deeper muscles below. The linea alba, between the borders of the rectus abdominis muscles, is formed by the intersection along the midline of the aponeurosis of the external oblique, internal oblique, and transversus abdominis.

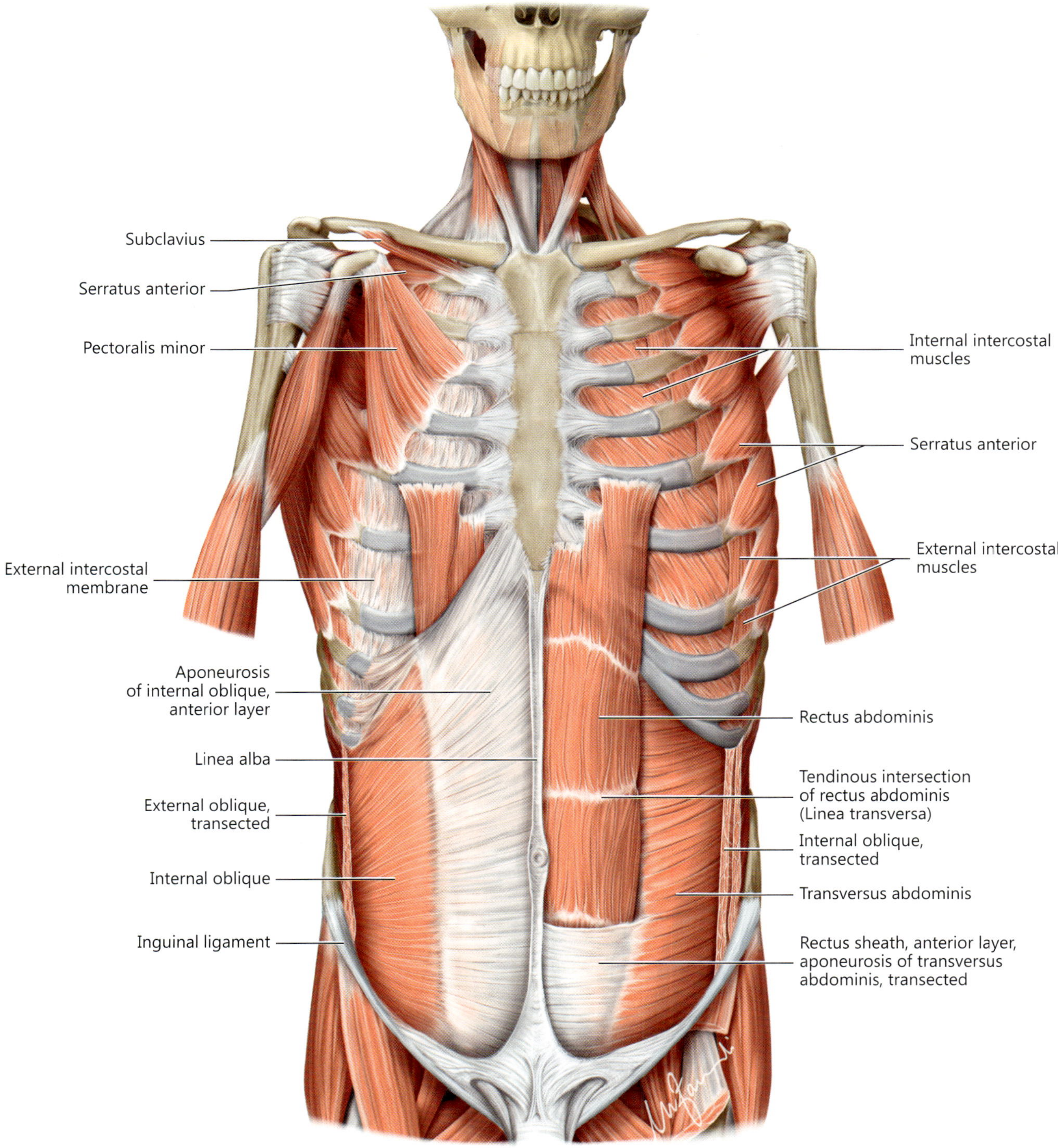

Preparation illustrating the stratigraphic sequence of the **muscles of the anterior wall of the trunk**. On the left, the anterior thoracoappendicular muscles and the internal oblique have been removed. On the right, the trapezius, deltoid, pectoralis major, biceps brachii, and external oblique have been removed. The anterior layer of the rectus sheath is formed in the upper three-quarters by the aponeurosis of the external oblique and the anterior splitting layer of the aponeurosis of the internal oblique, and in the lower quarter also by the aponeurosis of the transversus abdominis. On the left, a dissection of the aponeurosis of the internal oblique reveals the formation of the anterior layer of the rectus sheath by the aponeurosis of the transversus abdominis in the lower quarter.

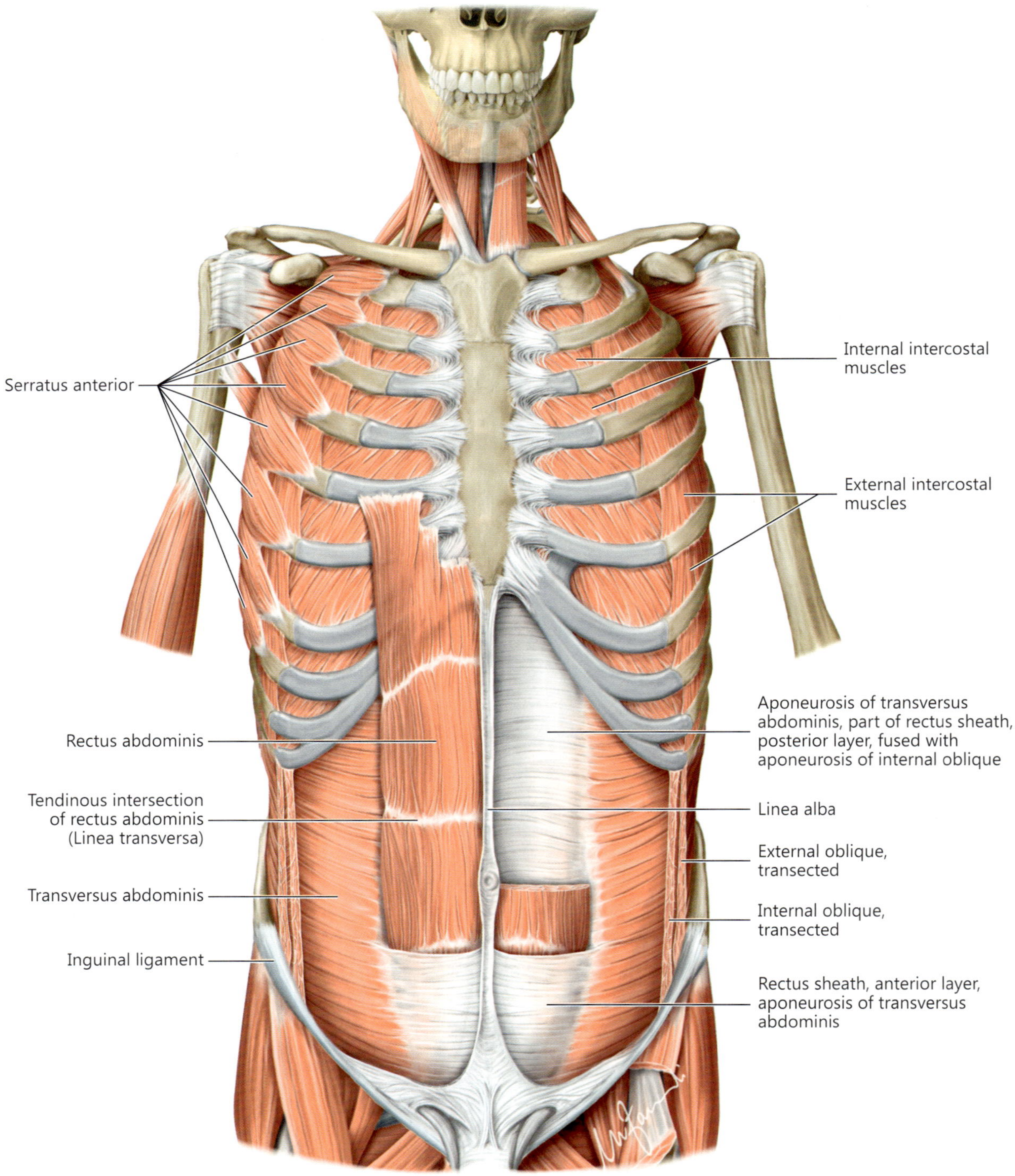

Preparation illustrating the stratigraphic sequence of the **muscles of the anterior wall of the trunk**. On the left, the serratus anterior has been removed and the rectus abdominis has been transected. Comparison with the previous and following pictures highlights the organization of the rectus sheath, which surrounds the muscle. The anterior layer of the sheath is formed in the upper three-quarters by the aponeurosis of the external oblique and the anterior splitting layer of the aponeurosis of the internal oblique, and in the lower quarter also by the aponeurosis of the transversus abdominis. The posterior layer of the sheath is formed in the upper three-quarters by the posterior splitting layer of the aponeurosis of the internal oblique, of the aponeurosis of the transversus abdominis, and of the transversalis fascia; in the lower quarter solely by the transversalis fascia.

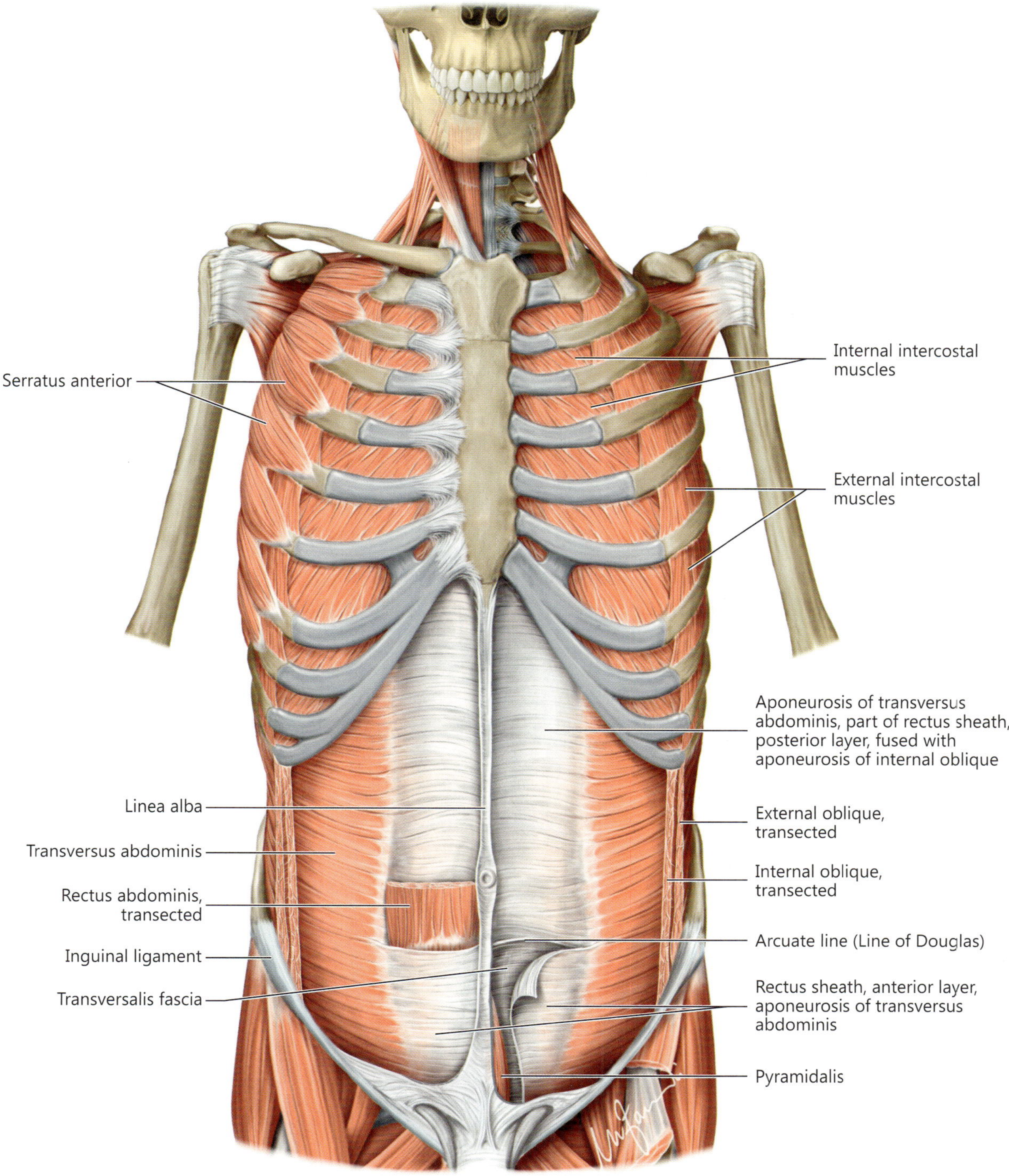

Preparation illustrating the stratigraphic sequence of the **muscles of the anterior wall of the trunk**. On the left, the rectus abdominis has been removed. The posterior layer of the rectus sheath is formed in the upper three-quarters by the posterior splitting layer of the aponeurosis of the internal oblique, of the aponeurosis of the transversus abdominis, and of the transversalis fascia; in the lower quarter solely by the transversalis fascia. The transition point between the two corresponds to an inferiorly concave line, the arcuate line.

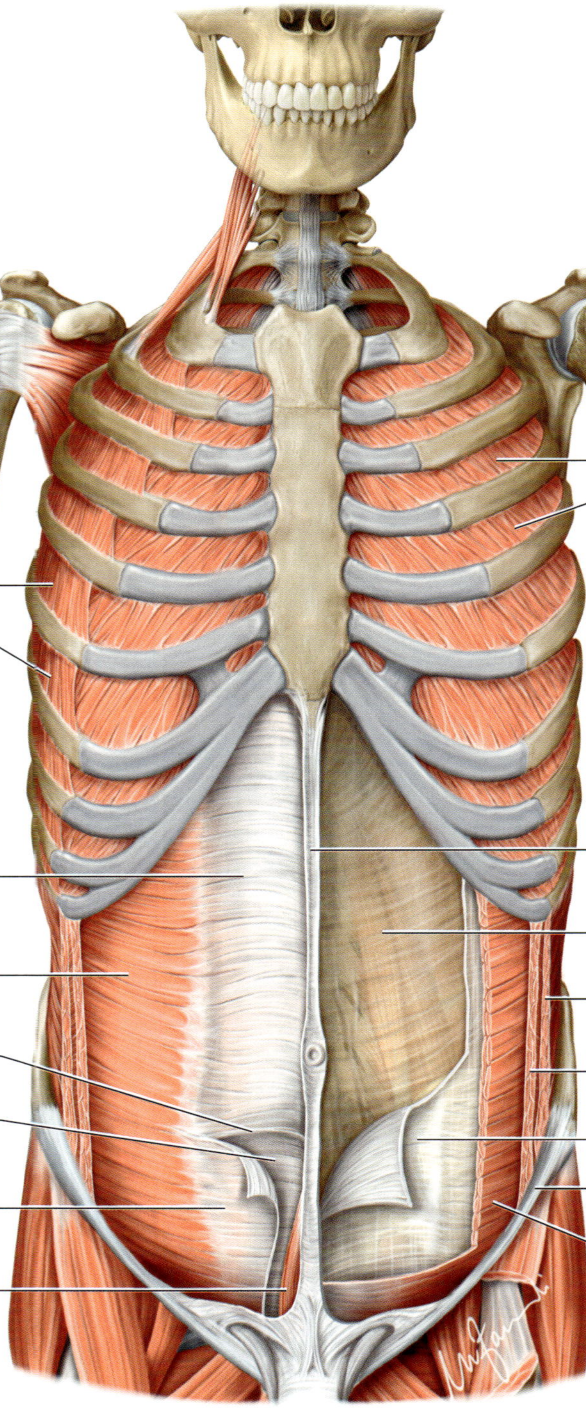

Preparation illustrating the stratigraphic sequence of the **muscles of the anterior wall of the trunk**. On the right, the rectus abdominis has been removed, leaving in place the posterior layer of its sheath. In the lower quarter, below the arcuate line, this layer is formed solely by the transversalis fascia. On the left, also the posterior layer of the sheath has been transected in the upper three-quarters, exposing the parietal peritoneum.

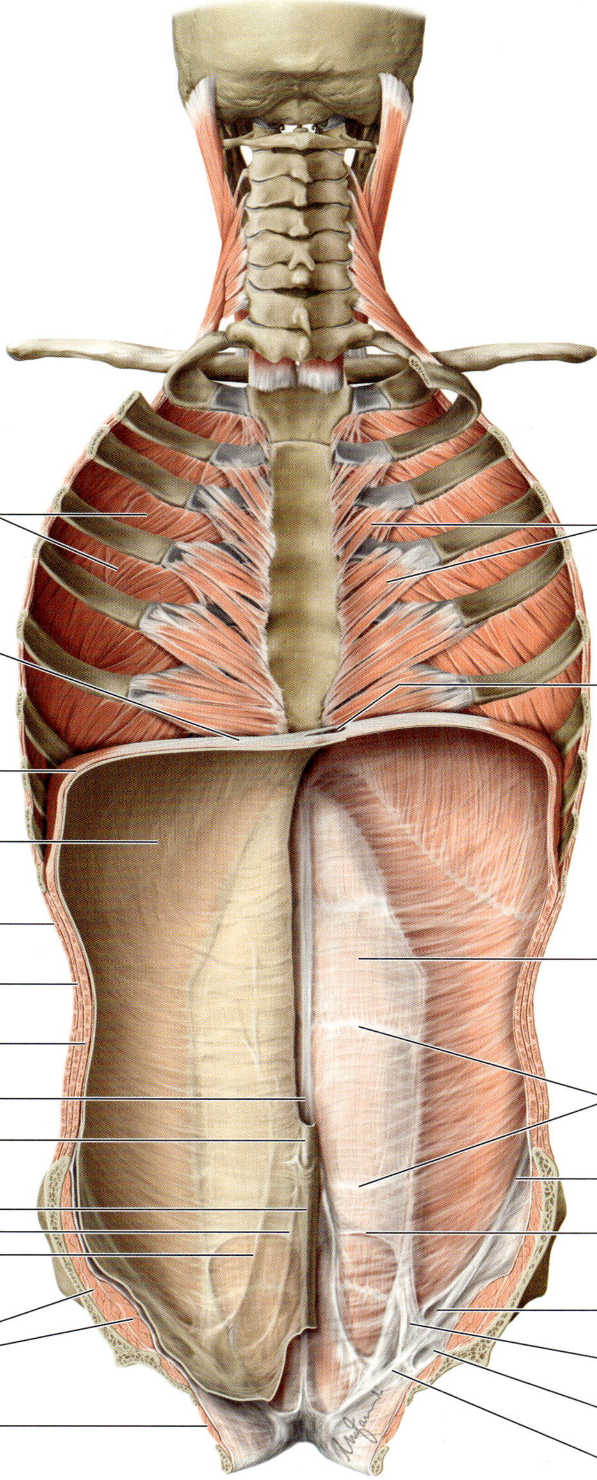

Preparation illustrating the **muscles of the anterior wall of the abdomen**. Posterior view from the inside of the thoracic and abdominal cavities. Above the arcuate line, the posterior layer of the rectus sheath is formed, from its lower depths to the surface, by the transversalis fascia, the aponeurosis of the transversus abdominis, and the poster or splitting layer of the aponeurosis of the internal oblique; below the arcuate line, solely by the transversalis fascia.

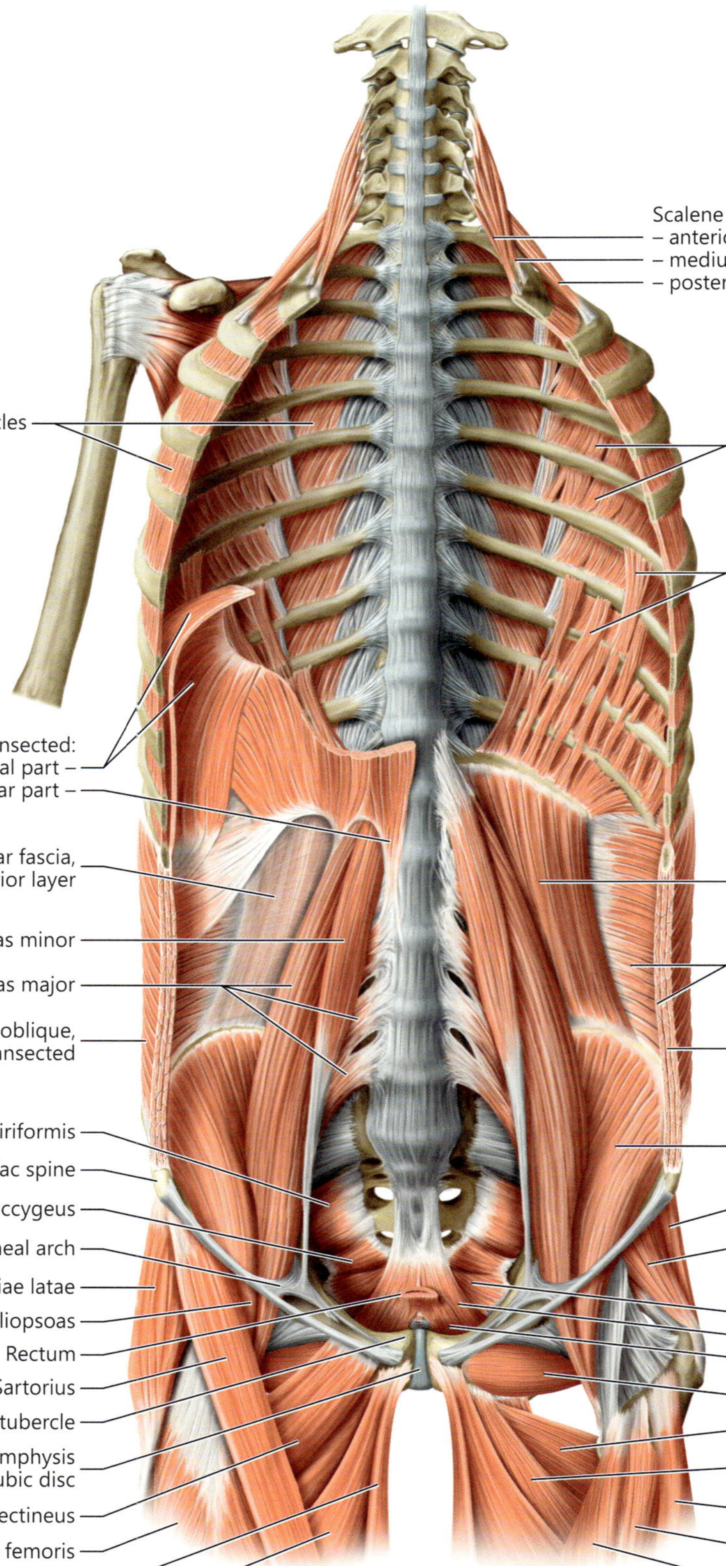

Here you can see the **muscles of the posterior portion of the trunk**. Observe how the iliacus and psoas major protrude from the pelvis through the muscular space, inserting on the lesser trochanter of the femur. On the left side, the diaphragm has been transected in order to view the proximal insertion of the psoas muscles. On the right side, the quadratus lumborum is covered by the anterior layer of the thoracolumbar fascia. On the lateral border of the muscle, this layer merges with the middle layer (☞ page 403) of the same fascia, and with the deep lamina of the posterior layer of the thoracolumbar fascia (☞ page 404), forming the posterior insertion of the internal oblique and transversus abdominis, laterally to the paravertebral muscles and above the iliac crest.

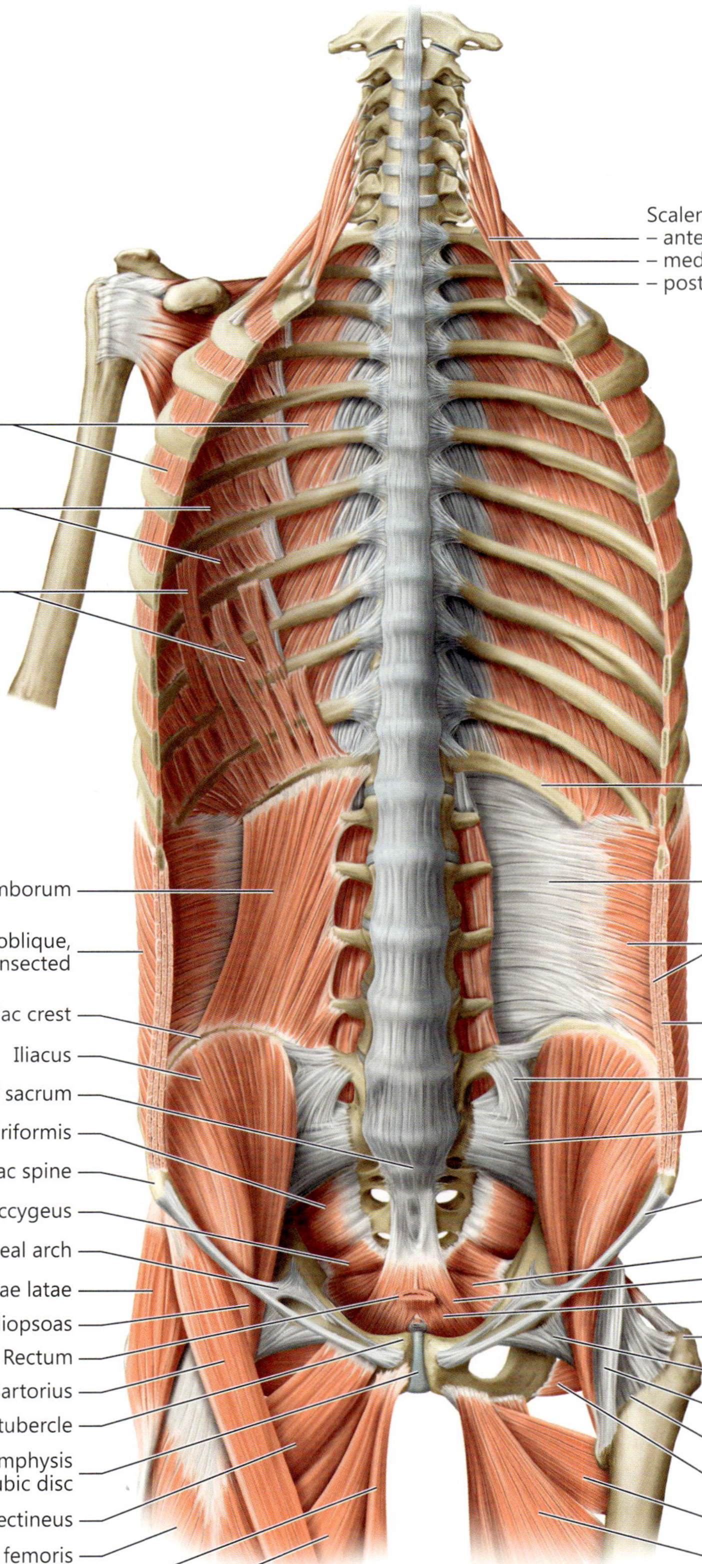

Here you can see the **muscles of the posterior portion of the trunk**. Both the left and right side correspond to the deeper layers of the previous image. On the left, the removal of the quadratus lumborum highlights the middle layer of the thoracolumbar fascia. This layer merges with the anterior layer (☞ page 402) of the same fascia, and with the deep lamina of the posterior layer of the thoracolumbar fascia (☞ page 404), to form the posterior insertion of the internal oblique and transversus abdominis, laterally to the paravertebral muscles and above the iliac crest.

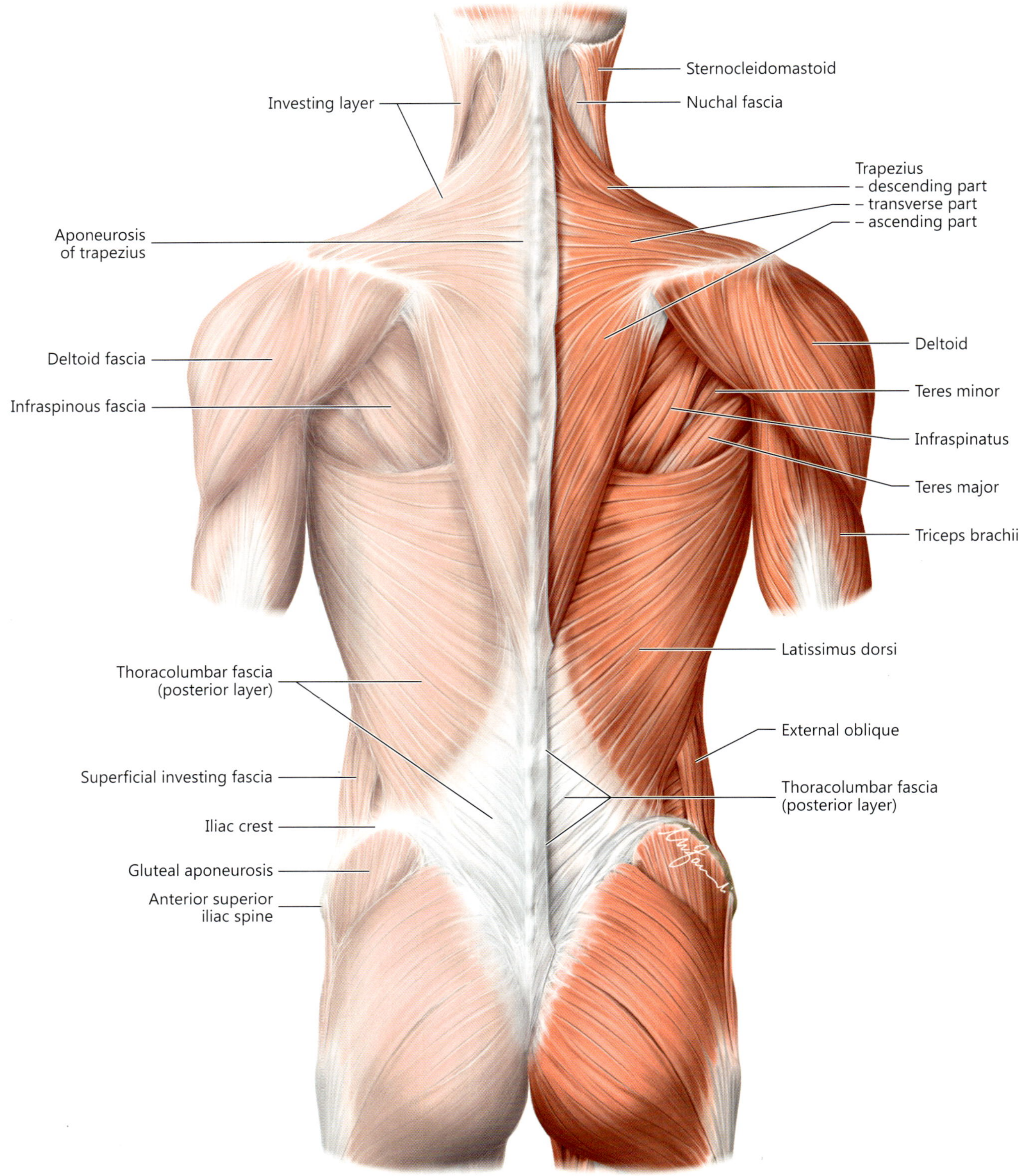

Preparation illustrating the stratigraphic sequence of the **muscles of the posterior wall of the trunk**. On the left, the more superficial investing fasciae have been left in place. On the right, we have removed the superficial lamina of the posterior layer of the thoracolumbar fascia, which follows downwards the nuchal fascia and covers the erector spinae. At L IV-L V, the posterior layer of the thoracolumbar fascia splits into a superficial lamina (shown here on the left side) and a deep lamina, which connect together at the level of L IV-L V. Many of the fibers of this superficial lamina derive from the aponeurosis of the latissimus dorsi.

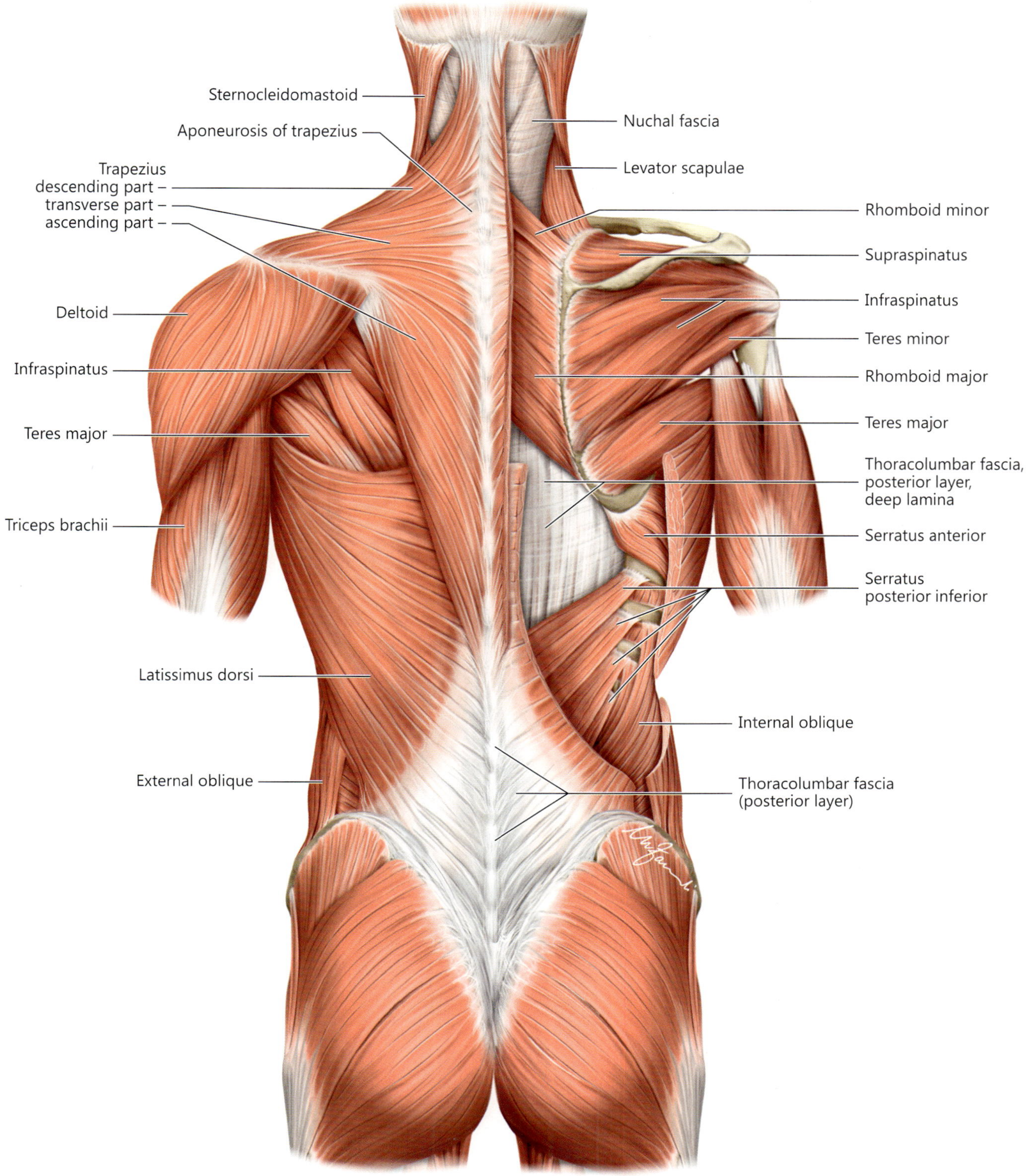

Preparation illustrating the stratigraphic sequence of the **muscles** and **fasciae of the posterior wall of the trunk.** On the right, the deltoid has been removed and the trapezius, latissimus dorsi, and external oblique have been partially transected to show the muscles of the underlying plane and part of the thoracolumbar fascia (deep lamina of the posterior layer). This lamina separates the layer of superficial muscles (shown in the image) from the deeper muscles. The posterior layer of the thoracolumbar fascia is thinner at the top, and gradually thickens toward the bottom. It extends laterally up to the costal angle. At the bottom, the deep lamina of the posterior layer of the thoracolumbar fascia runs deep to the serratus posterior inferior and is reinforced by the insertions of this muscle and those of the aponeurosis of the latissimus dorsi.

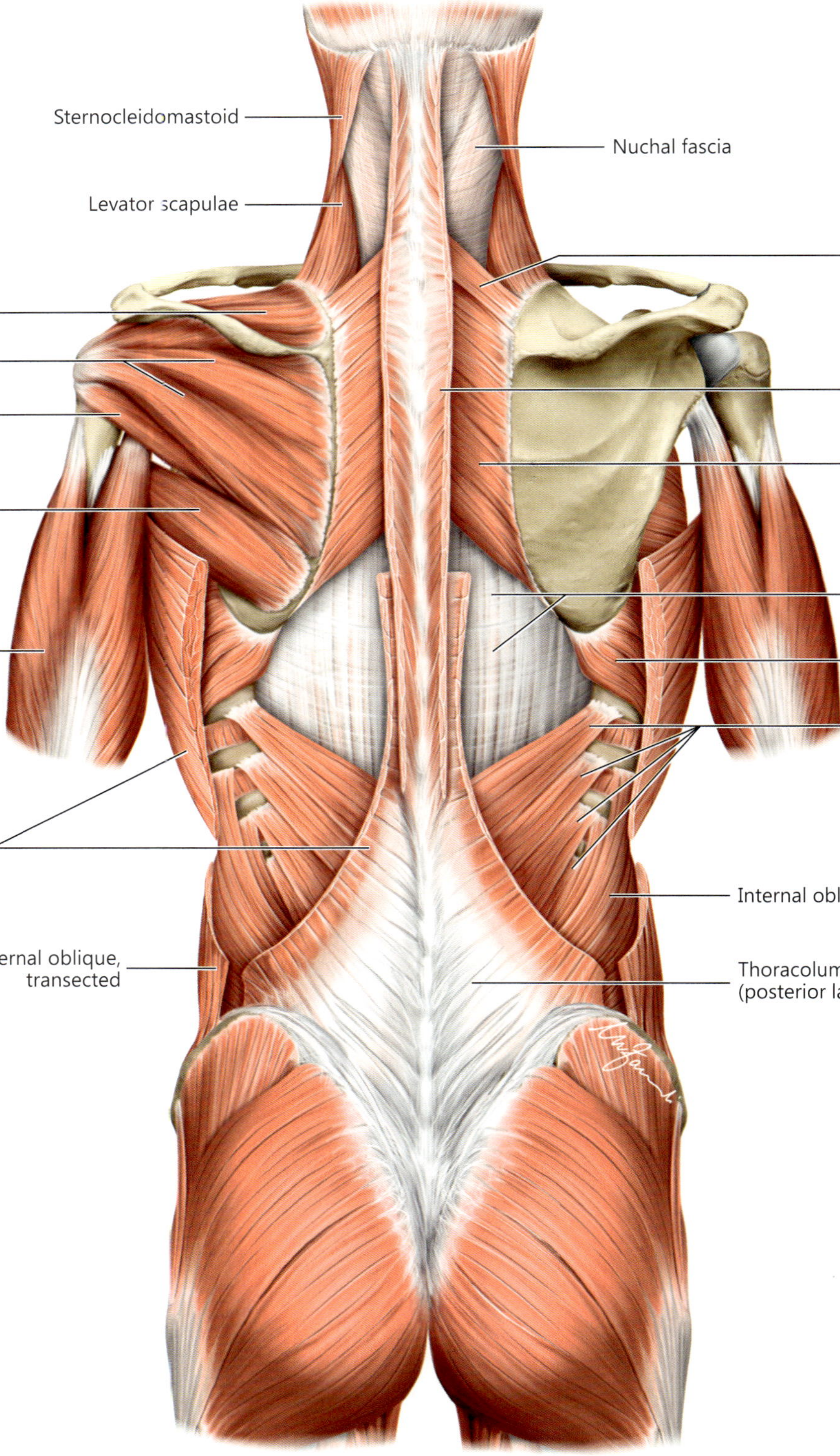

Preparation illustrating the stratigraphic sequence of the **muscles** and **fasciae of the posterior wall of the trunk**. The left side corresponds to the description of the right side of the previous image. On the right, the posterior muscles of the scapula have been removed (supraspinatus, infraspinatus, teres minor and teres major).

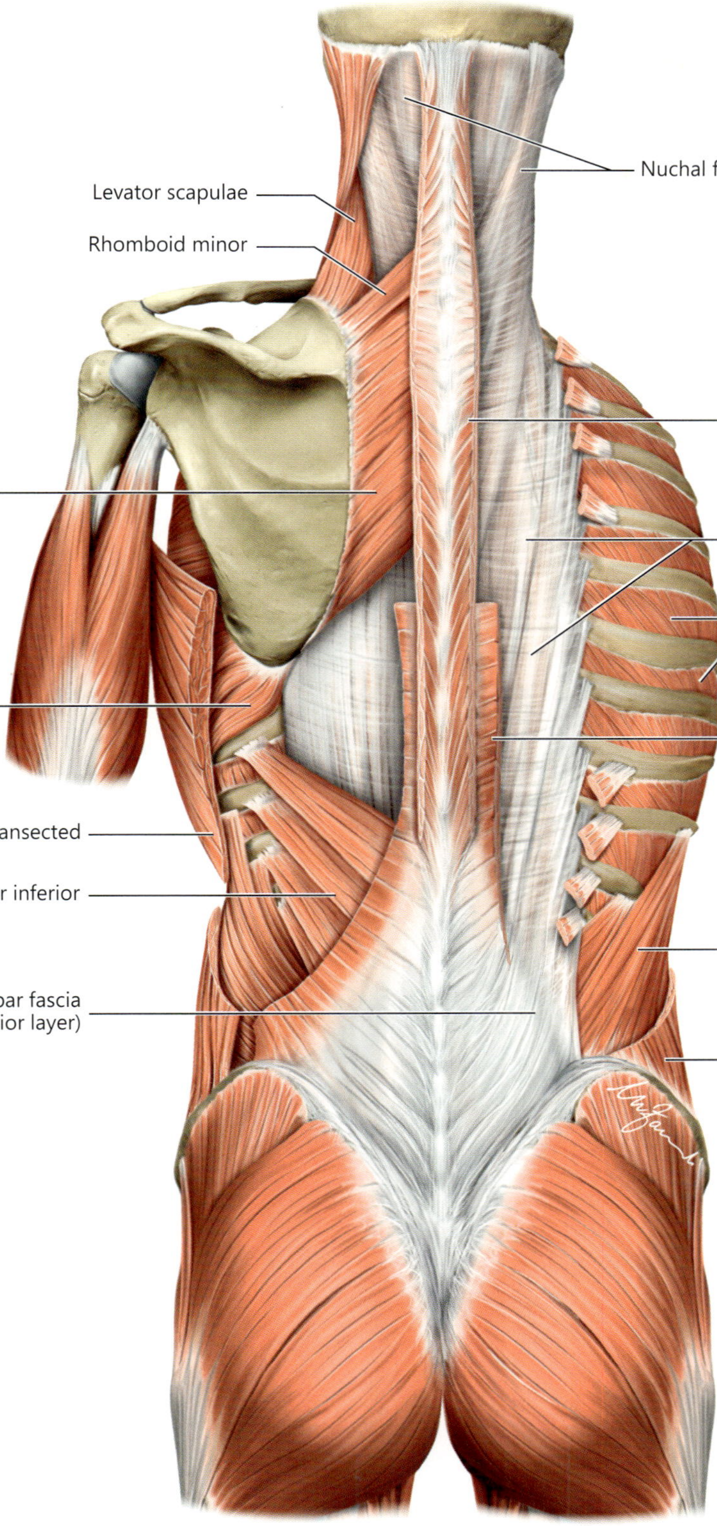

Preparation illustrating the stratigraphic sequence of the **muscles** and **fasciae of the posterior wall of the trunk**. The left side corresponds to the deepest layer seen in the previous figure. On the right, the semispinalis capitis, splenius capitis, levator scapulae, rhomboid major and rhomboid minor have been removed, and the trapezius, latissimus dorsi and serratus posterior inferior have been transected. This reveals the muscles of the deep layer of the posterior wall of the trunk, which are covered by the deep lamina of the posterior layer of the thoracolumbar fascia (longissimus thoracis, spinalis, iliocostalis thoracis, and iliocostalis lumborum).

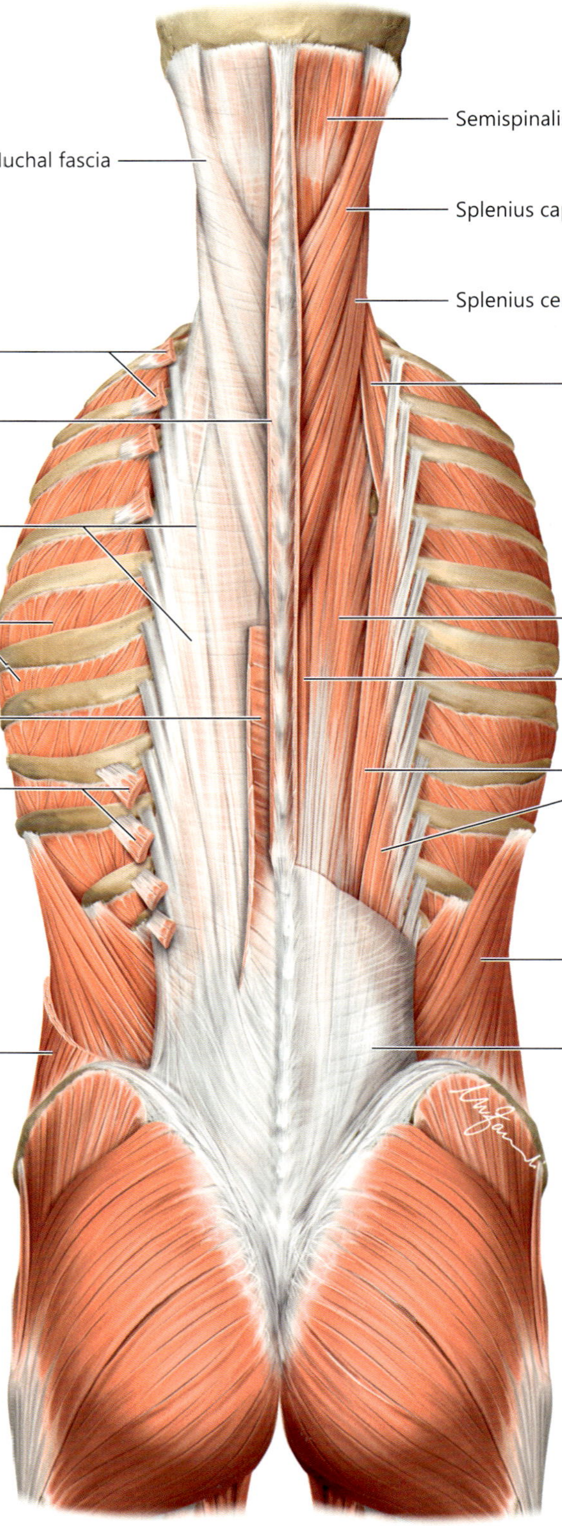

Preparation illustrating the stratigraphic sequence of the **muscles** and **fasciae of the posterior wall of the trunk**. The left side corresponds to the deepest layer seen in the previous figure. On the right, the superior part of the thoracolumbar fascia has been removed, exposing the muscles of the deep layer it covers. The erector spinae consists of three muscle bundles: the iliocostalis (lateral), the longissimus (intermediate), and the spinalis (medial). The iliocostalis is, in turn, formed by a lumbar part and a thoracic part. The longissimus consists of three parts: the longissimus capitis, longissimus cervicis, and longissimus thoracis.

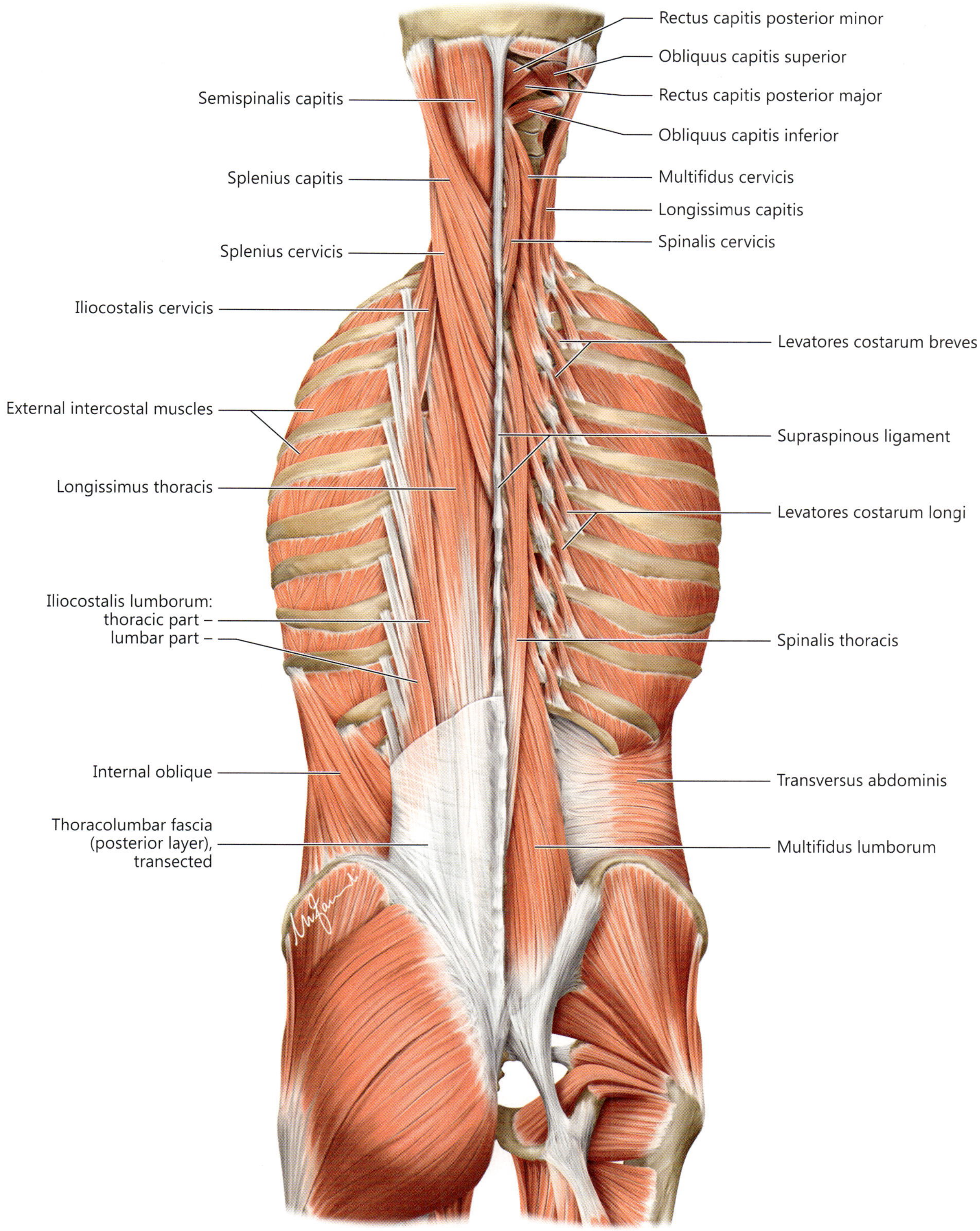

Preparation illustrating the stratigraphic sequence of the **muscles** and **fasciae of the posterior wall of the trunk**. The left side corresponds to the deepest layer seen in the previous figure. On the right, the semispinalis capitis, splenius capitis, splenius cervicis, and iliocostalis cervicis have been removed, as well as the lateral (iliocostalis) and intermediate (longissimus) parts of the erector spinae, exposing the transversospinales: multifidus, semispinalis, and below these, the rotator muscles. The multifidus muscles each consist of three parts (cervical, thoracic, and lumbar), as the semispinalis muscles (of the trunk, neck, and head).

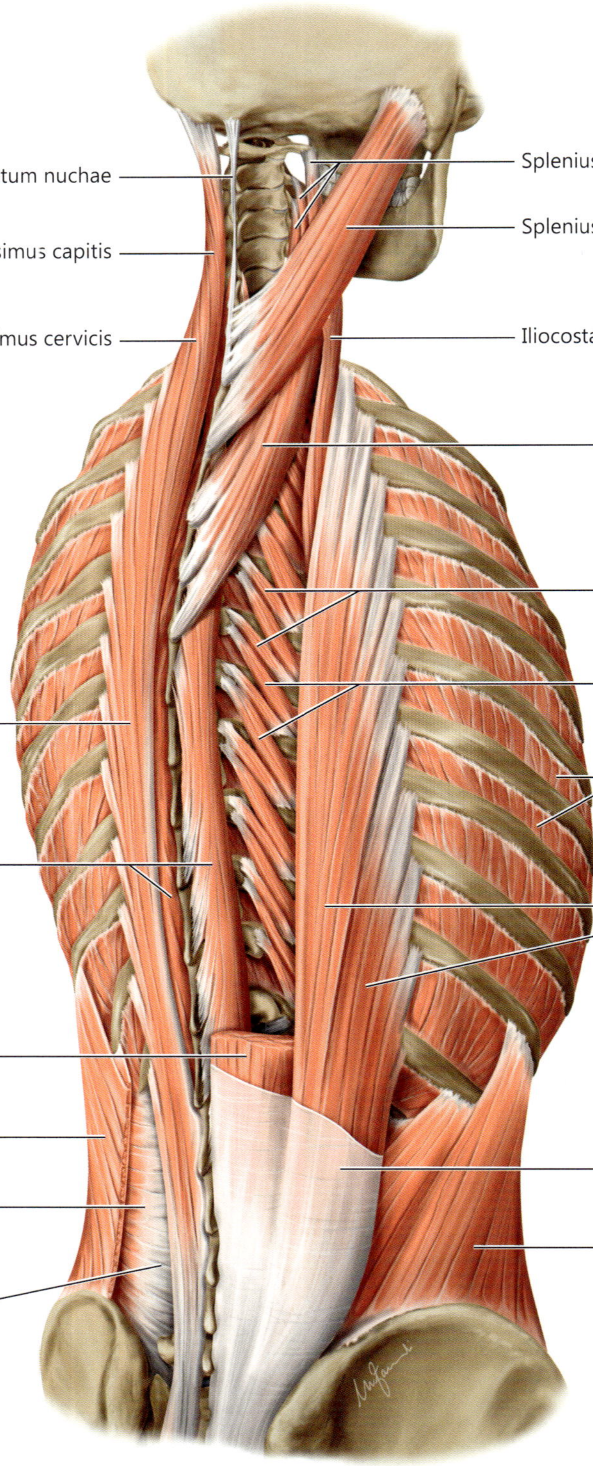

Preparation illustrating the stratigraphic sequence of the **muscles of the posterior wall of the trunk**, right posterolateral view. On the right, the longissimus thoracis has been removed to show the points of origin and insertion of the levatores costarum, longi and breves, as well as the path of the splenius cervicis and spinalis.

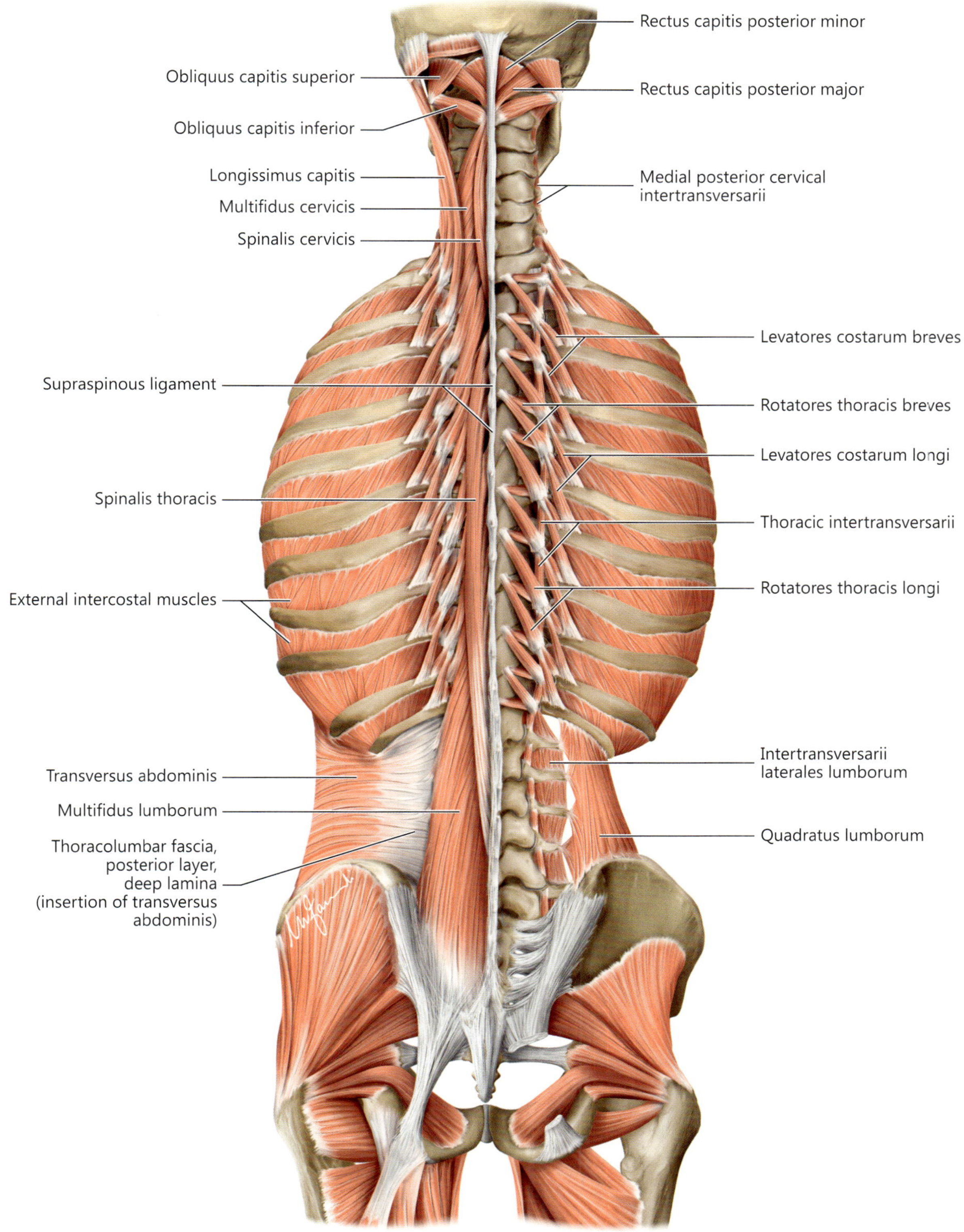

Preparation illustrating the points of origin and insertion of the **transversospinales**, **suboccipital** and **external intercostal muscles**.

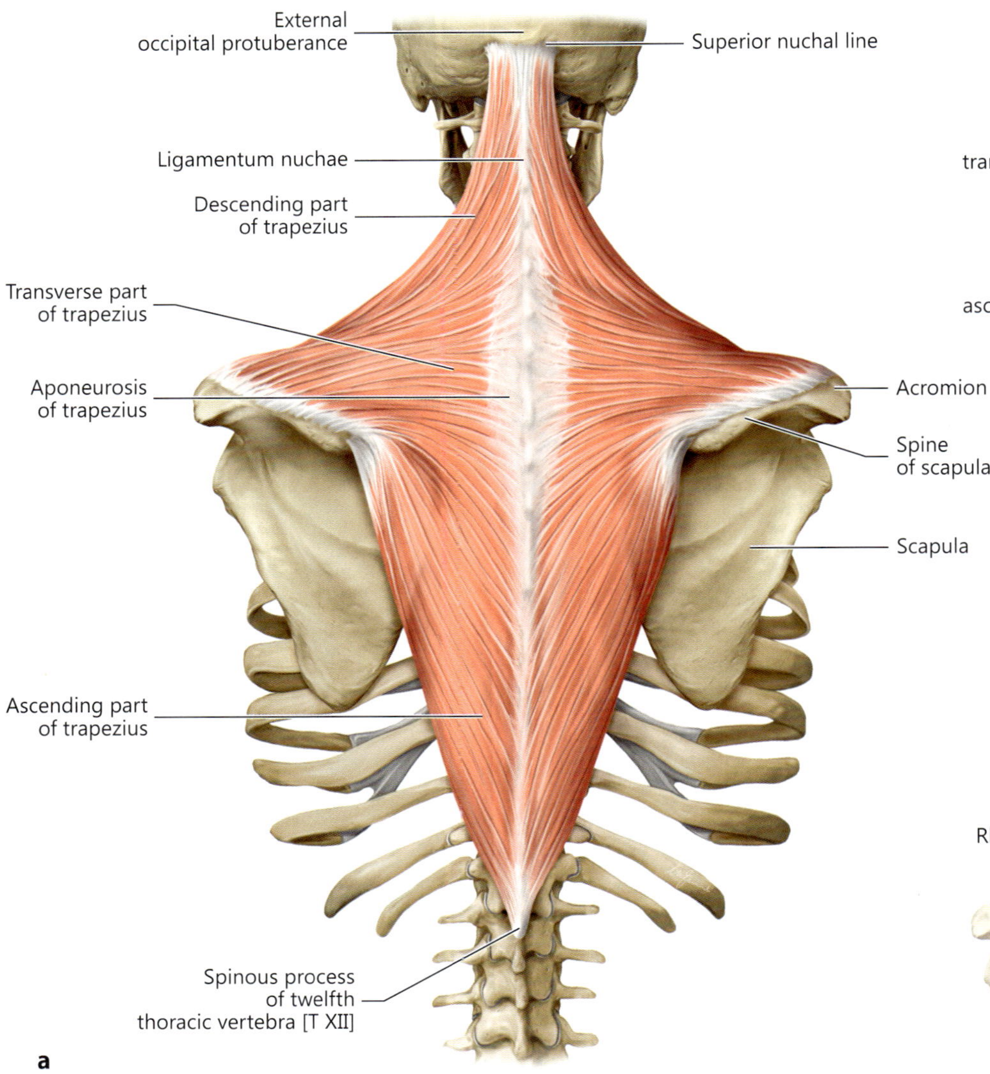

a

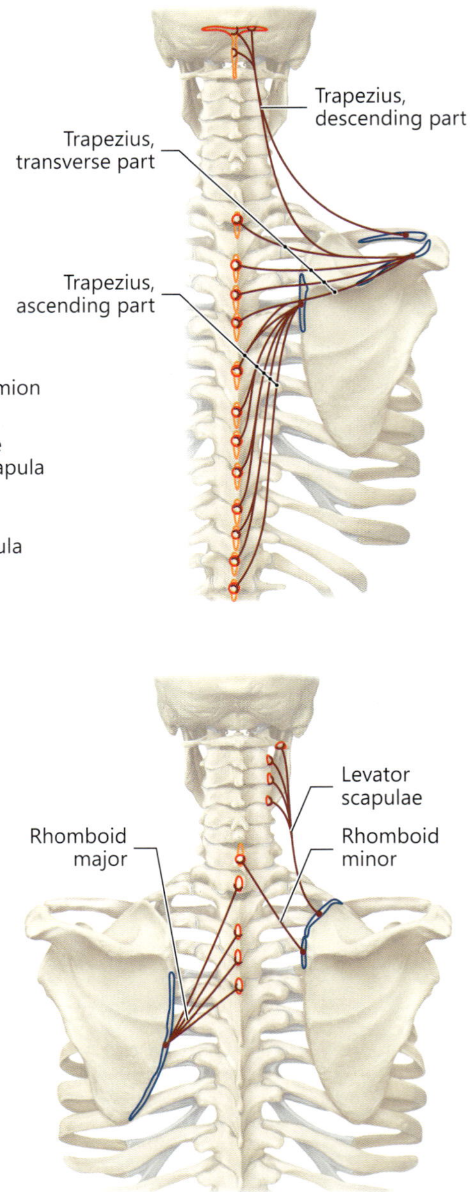

c

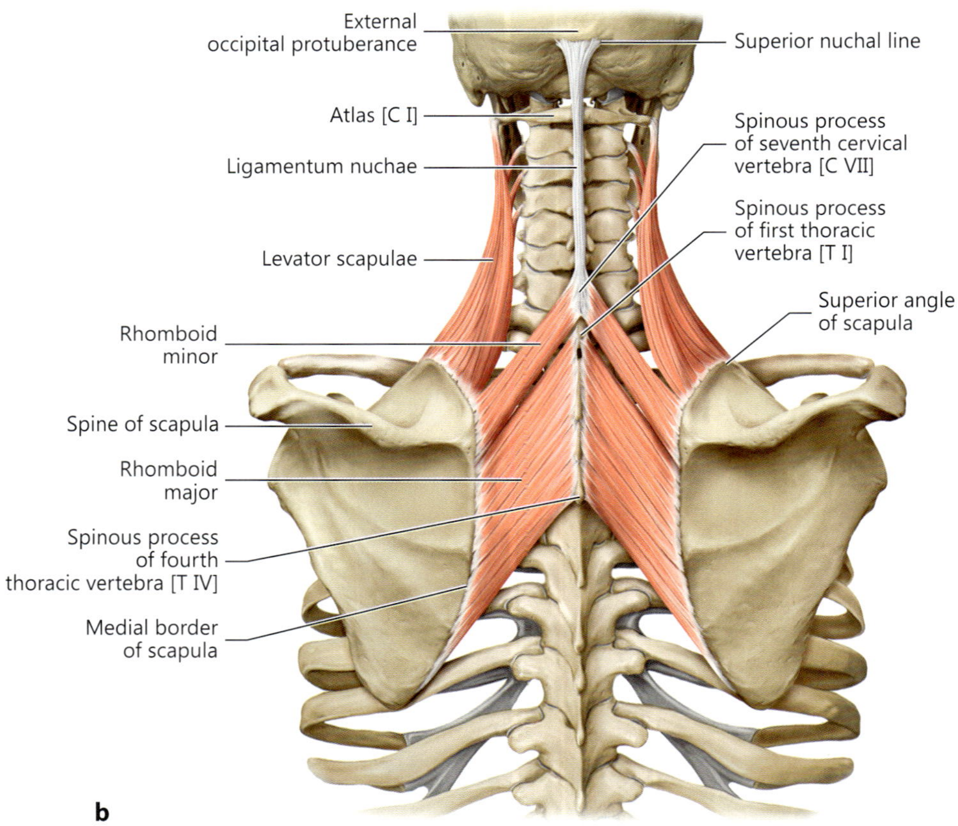

b

ORIGIN
Bones
Ligaments
Origin ⟶ Insertion
INSERTION
Bones

Preparation illustrating the points of origin and insertion of the **trapezius** (**a**) and the **levator scapulae**, **rhomboid major** and **rhomboid minor** (**b**).

c, Schematic representation of the origin and insertions of the trapezius, levator scapulae, rhomboid major, and rhomboid minor (spinoappendicular muscles or muscles of the superficial layer of the muscles of the back, along with the latissimus dorsi). The rhomboids and levator scapulae are located deep to the trapezius and latissimus dorsi.

Preparation illustrating the points of origin and insertion of the **latissimus dorsi** and **teres major**.

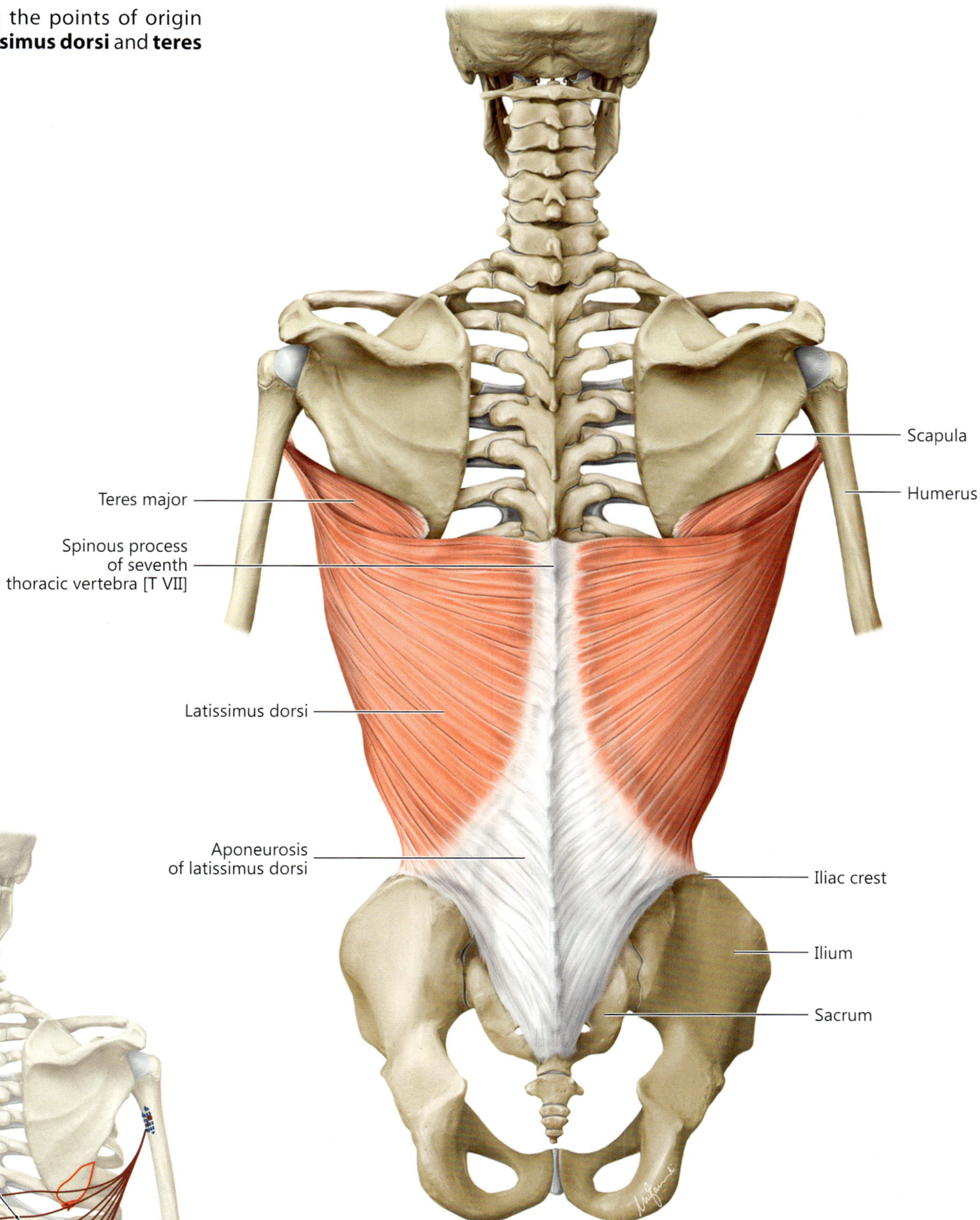

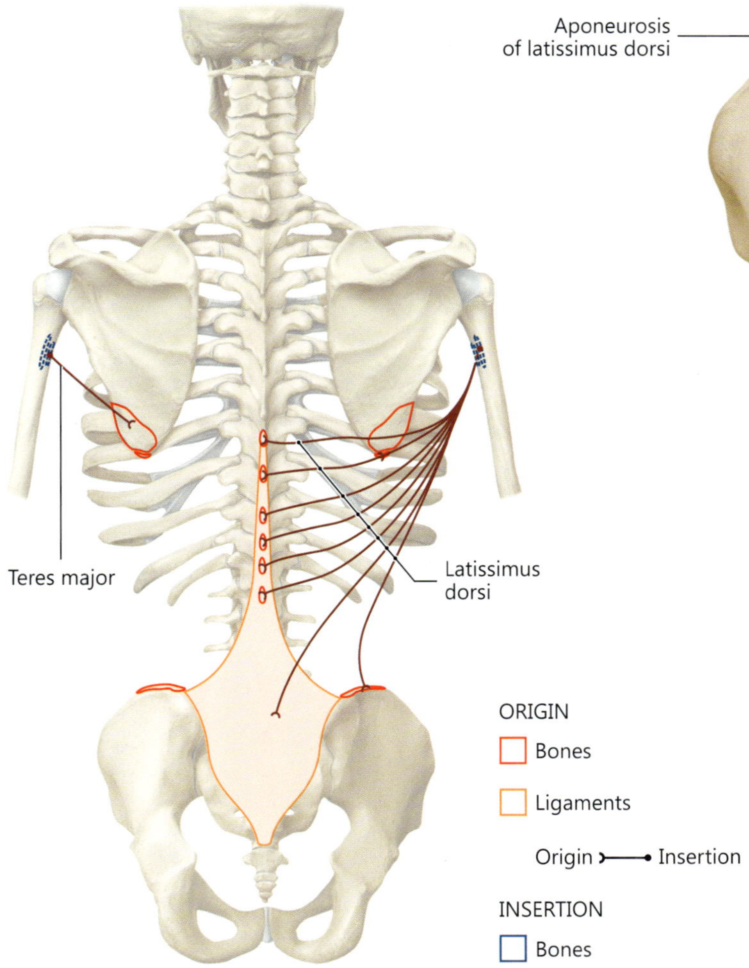

➤ Schematic representation of the origin and insertions of the **latissimus dorsi** (spinoappendicular muscle or muscle of the superficial layer of the muscles of the back) and **teres major** (muscle of the upper limb). The latissimus dorsi originates in the posterior layer of the thoracolumbar fascia from the spinous processes of the last thoracic and lumbar vertebrae, from the median sacral crest, and from the outer lip of the iliac crest, and inserts on the crest of the lesser tubercle of the humerus. The teres major originates from the inferior angle of the scapula and the infraspinous fossa, and inserts on the crest of the lesser tubercle and on the intertubercular sulcus of the humerus.

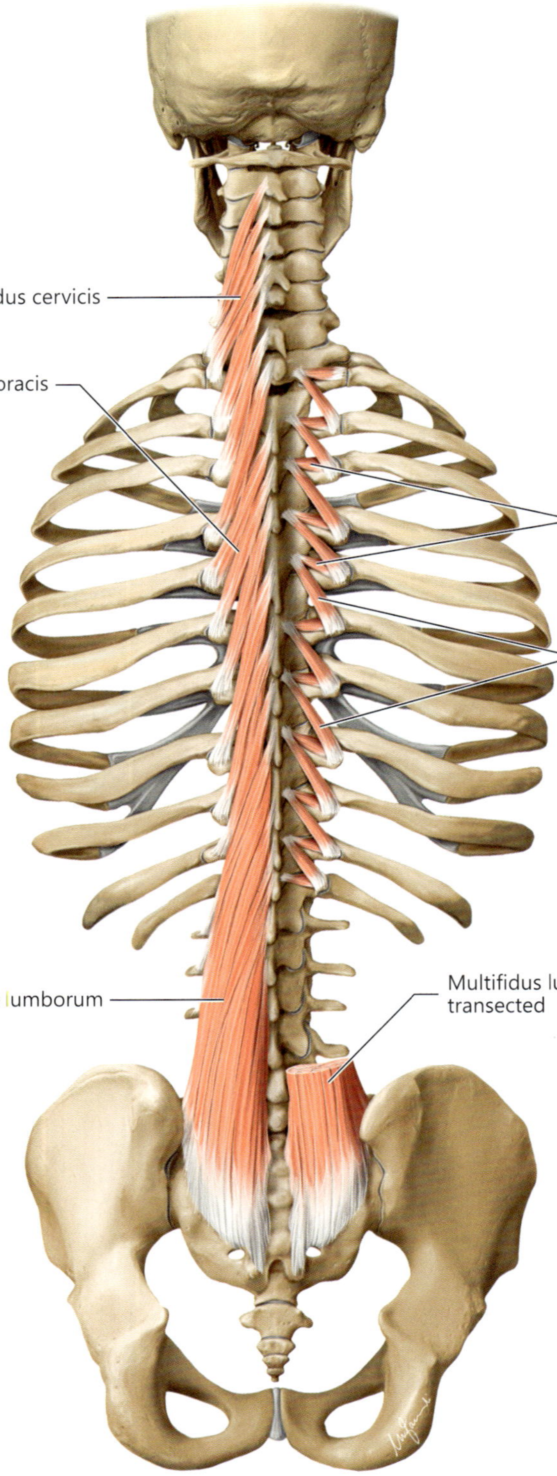

Preparation illustrating the points of origin and insertion of the **rotatores breves** and **longi of the thorax** and the **multifidus cervicis**, **thoracis**, and **lumborum** (along with the semispinalis this muscle group forms part of the transversospinales).

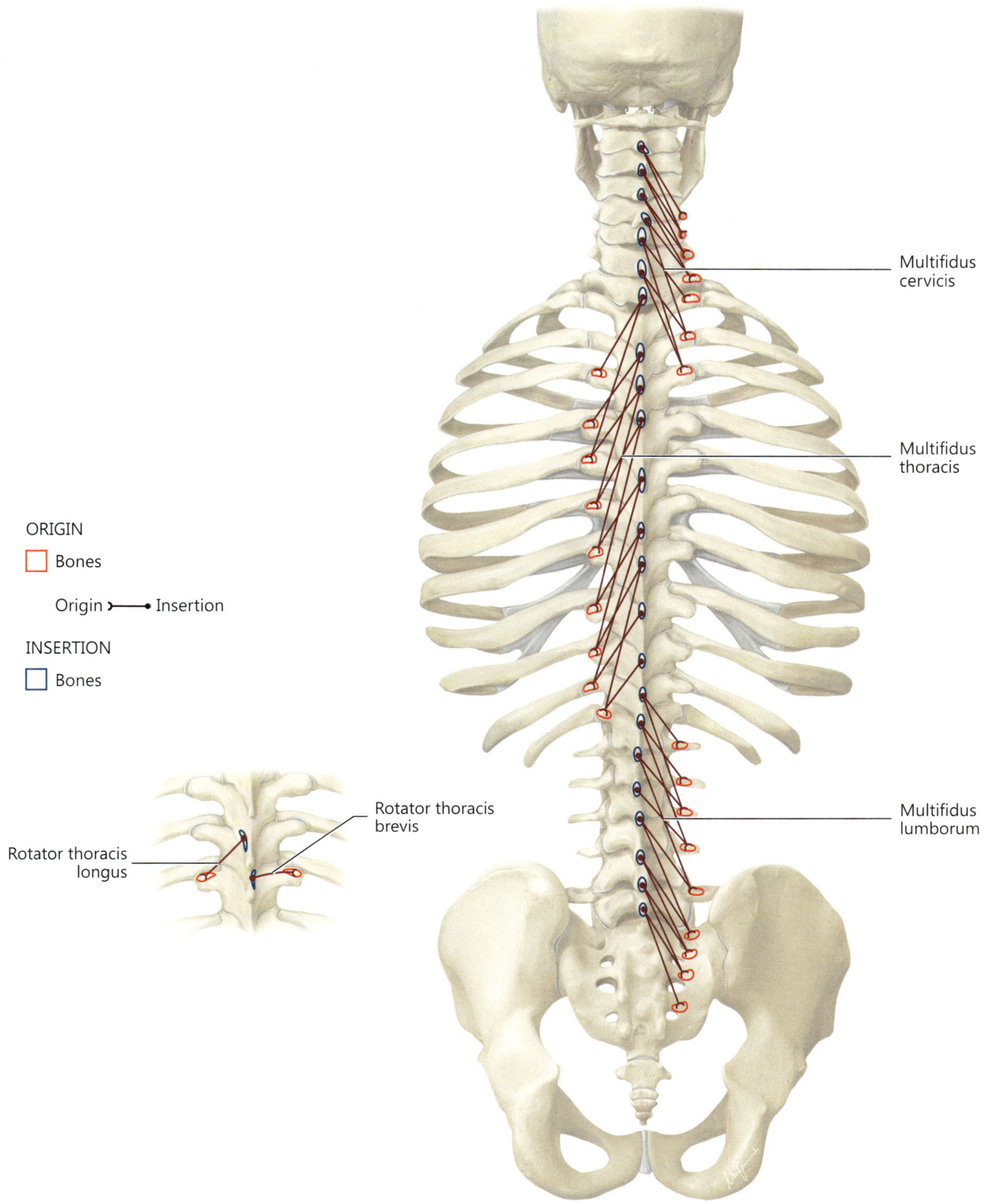

Schematic representation of the origins and insertions of the multifidus and rotator muscles (which, along with the semispinalis, form part of the transversospinales). These muscles are discussed here in the order they present themselves, from the surface downward. The **multifidus cervicis** extends from the articular processes of C IV-C VII to the posterior surface of the laminae and spinous processes of C II-C VI; the **multifidus thoracis** extends from the transverse processes of T I-T XII to the posterior surface of the laminae and the base of the spinous processes of C VII and T I-T XI; the **multifidus lumborum** extends from the medial sacral crests, from the posterior superior iliac spine, and from the mammillary processes of the lumbar vertebrae, to the posterior surface of the laminae and the base of the spinous processes of T X-T XII and L I-L V. The **rotatores thoracis** (clearly present, in contrast to the rotatores cervicis and lumborum) extend from the superior and posterior portion of the transverse process of one thoracic vertebra to the inferior border and lateral surface of the lamina of the overlying thoracic vertebra (rotatores breves), and from the transverse process of one thoracic vertebra to the base of the spinous process of the thoracic vertebra next to the overlying vertebra (rotatores longi).

Preparation illustrating the points of origin and insertion of the **spino-transversales**, **splenius capitis** and **splenius cervicis**, **levatores costarum**, both **longi** and **breves**, and **quadratus lumborum**. Note the relationship between the splenius cervicis and the levatores costarum.

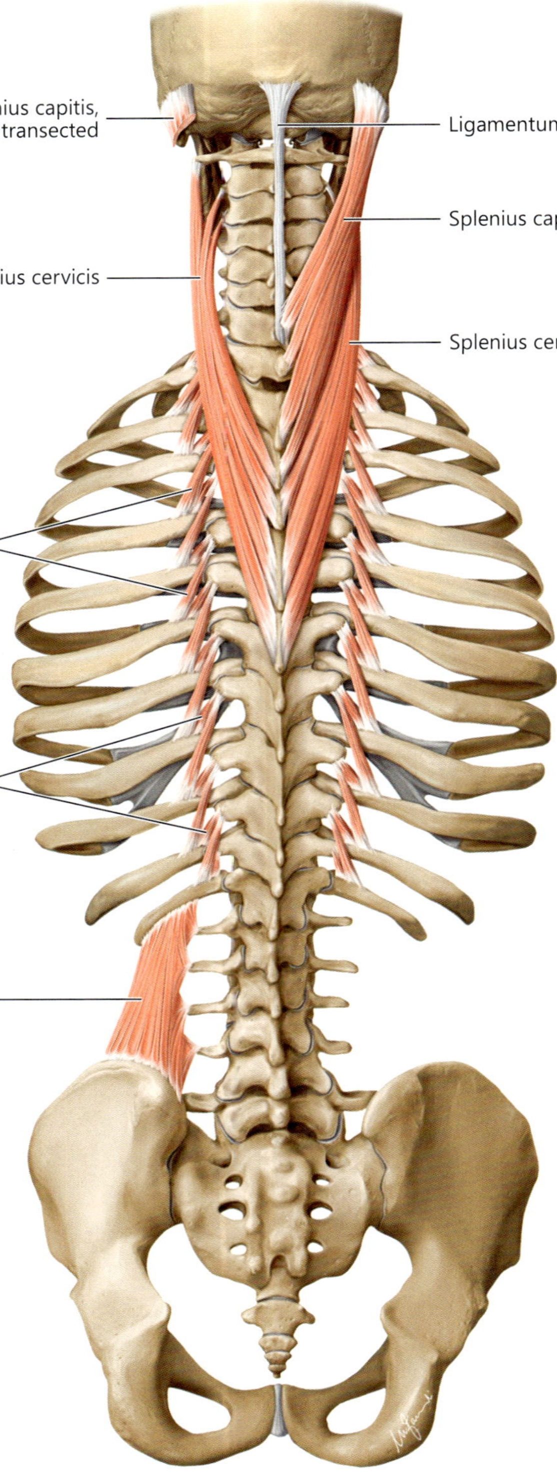

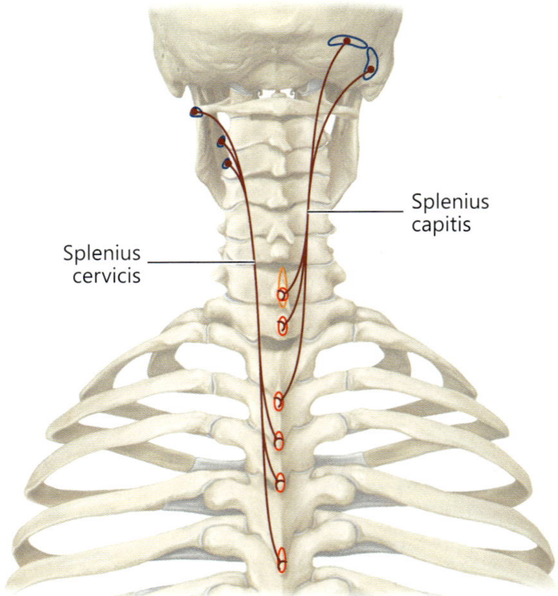

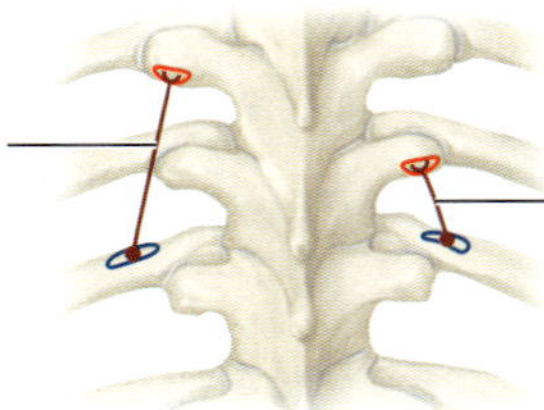

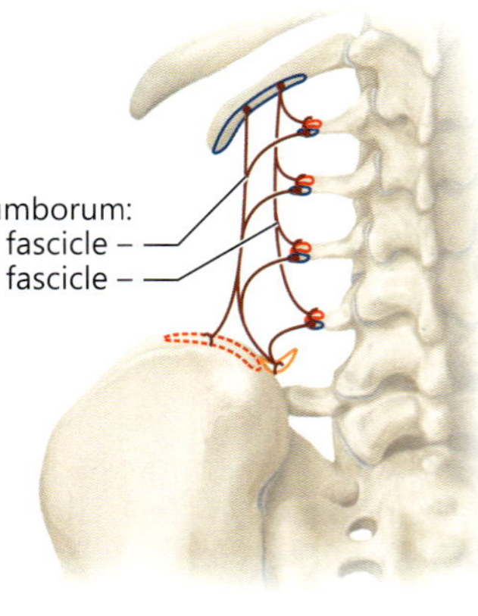

➤

Schematic representation of the origin and insertions of the levatores costarum (breves and longi), splenius (capitis and cervicis), and quadratus lumborum. The **splenius capitis** originates from the ligamentum nuchae and spinous processes of the seventh cervical vertebra and first two thoracic vertebrae, and inserts on the superior nuchal line and the mastoid process of the temporal bone. The **splenius cervicis** originates from the spinous processes of the third, fourth, and fifth thoracic vertebrae and inserts on the transverse processes of the first three cervical vertebrae. The **levatores costarum** originate from the transverse processes of the seventh cervical vertebra and of the first eleven thoracic vertebrae, and insert on the superior border and external surface of the underlying rib (levatores breves), or of the successive rib (levatores longi in the inferior part of the thorax). The posterior fascicle of the **quadratus lumborum** originates from the inner lip of the iliac crest, inserting on the costal processes of the first four lumbar vertebrae and on the inferior border of the twelfth rib. The anterior fascicle originates from the costal processes of the second, third, and fourth lumbar vertebrae, inserting on the twelfth rib.

ORIGIN

☐ Bones

☐ Ligaments

Origin ➤—• Insertion

INSERTION

☐ Bones

Preparation illustrating the points of origin and insertion of the **semispinalis** (capitis, cervicis, and thoracis) and **serratus posterior** (superior and inferior).

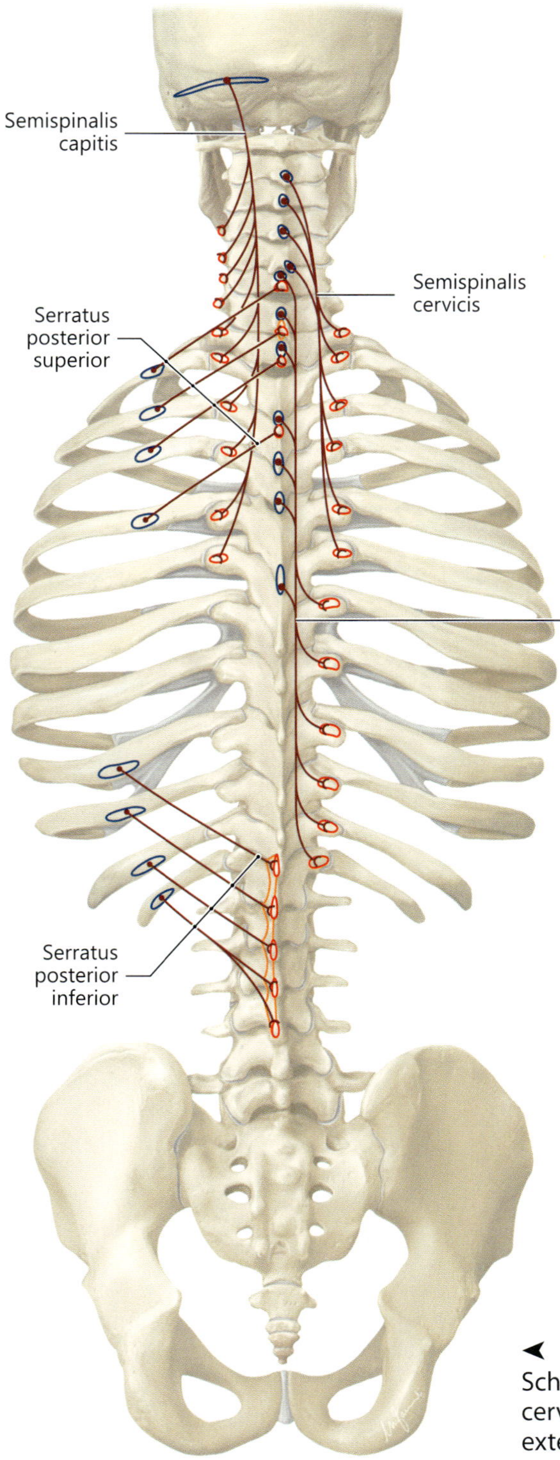

ORIGIN

Bones

Ligaments

Origin ➤—• Insertion

INSERTION

Bones

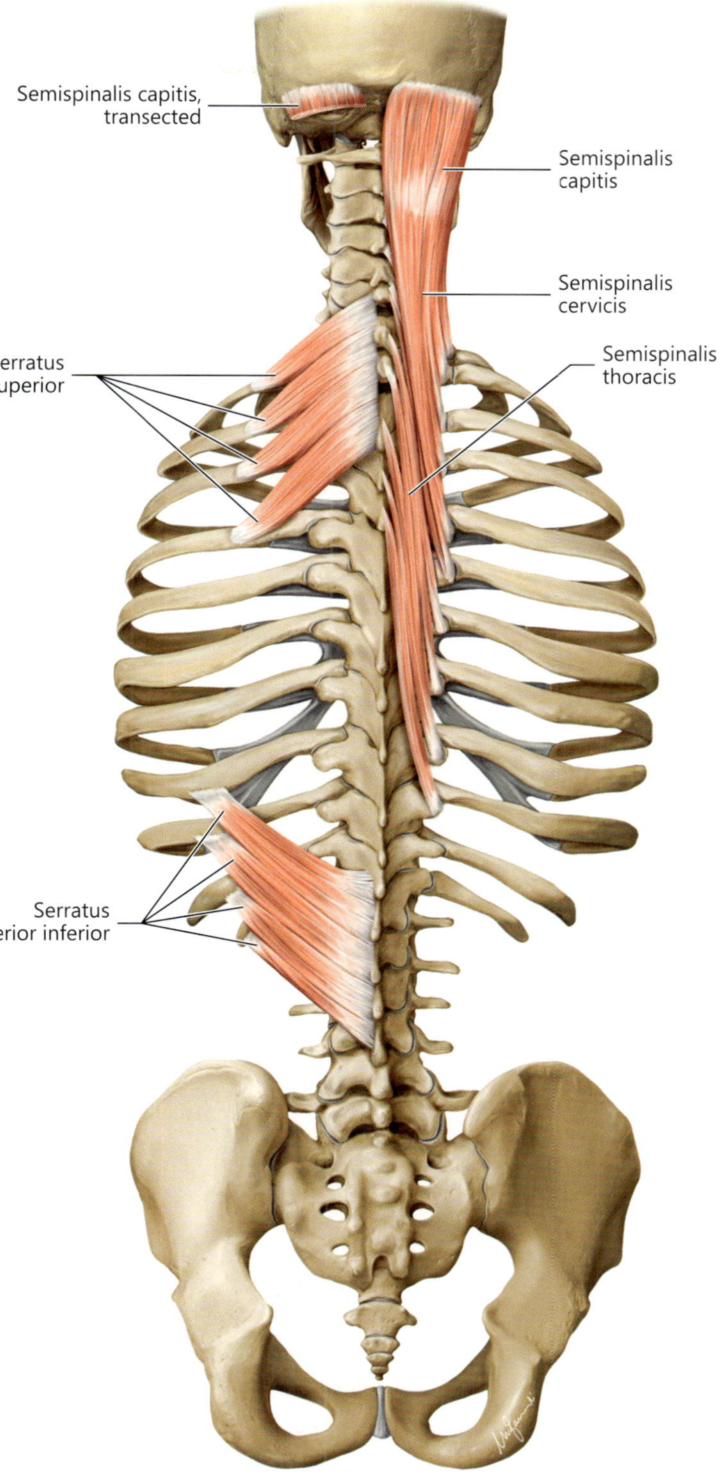

➤ Schematic representation of the origin and insertions of the semispinalis (capitis, cervicis, and thoracis) and serratus posterior muscles. The **semispinalis capitis** extends from the transverse processes of C IV-C VII and T I-T VI to the squamous part of the occipital bone, between the superior and inferior nuchal lines; the **semispinalis cervicis** extends from the transverse processes of T I-T VI to the spinous processes of C II-C V; the **semispinalis thoracis** extends from the transverse processes of T X-T XII to the spinous processes of C VI-C VII and T I-T VI. The **serratus posterior superior** originates from the spinous processes of the last two cervical vertebrae and of the first two thoracic vertebrae, and inserts on the external surface of the second through fifth ribs. The **serratus posterior inferior** originates from the spinous processes of the last two thoracic vertebrae and of the first three lumbar vertebrae, and inserts on the external surface of the last four ribs.

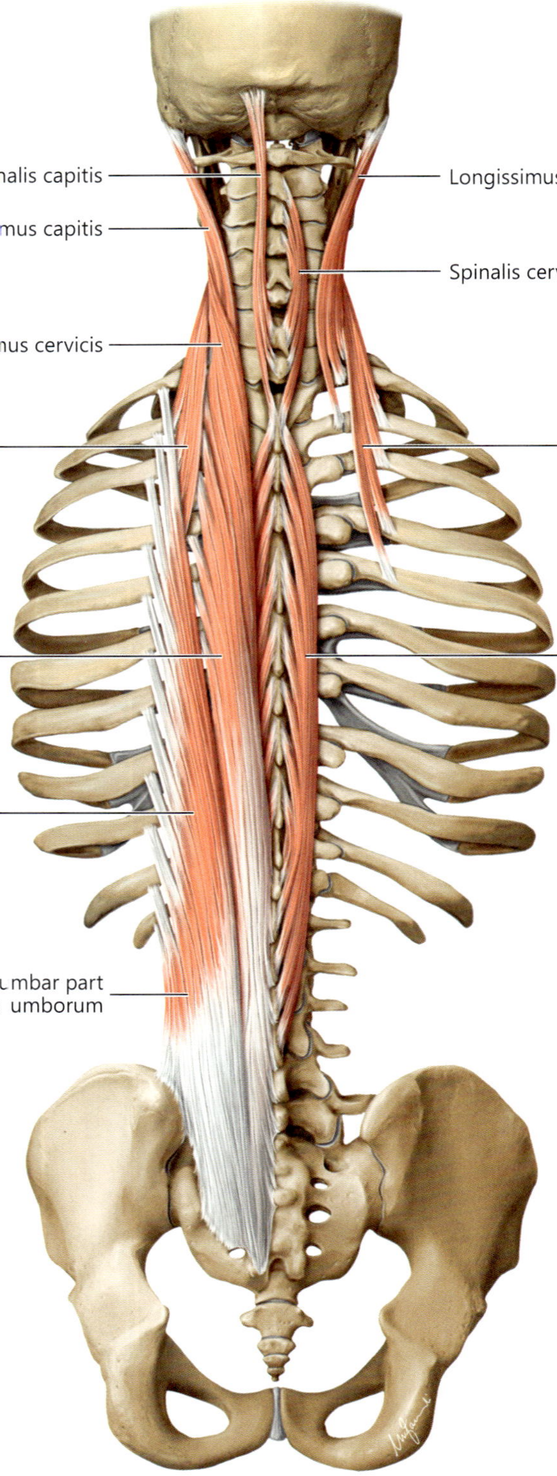

Preparation illustrating the points of origin and insertion of the **erector spinae**. On the right side, the iliocostalis lumborum, longissimus thoracis, and spinalis capitis have been removed to show the points of origin and insertion of the spinalis cervicis and thoracis.

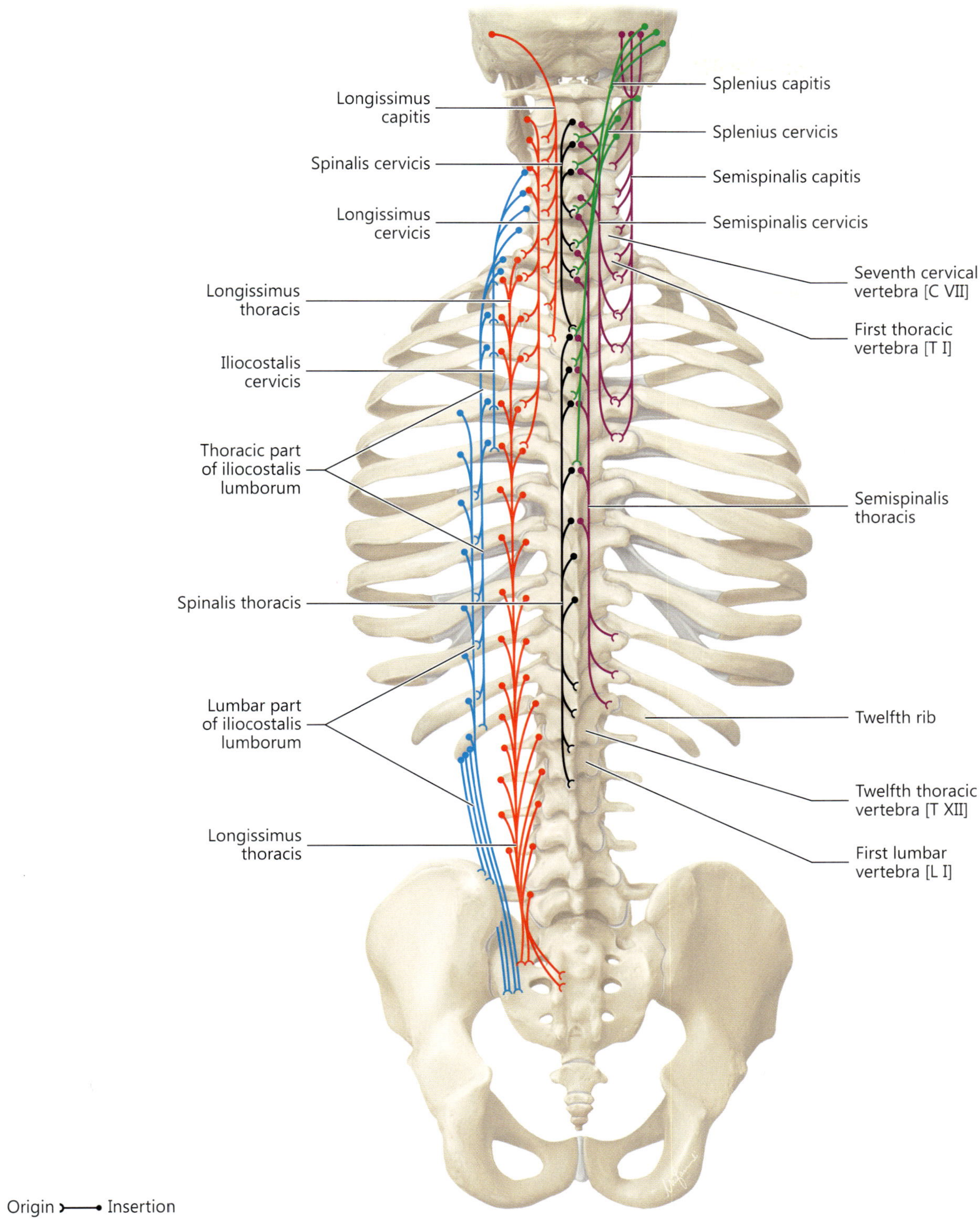

Schematic representation of the origin and insertion of the **paravertebral muscles** in the superficial and intermediate layers.

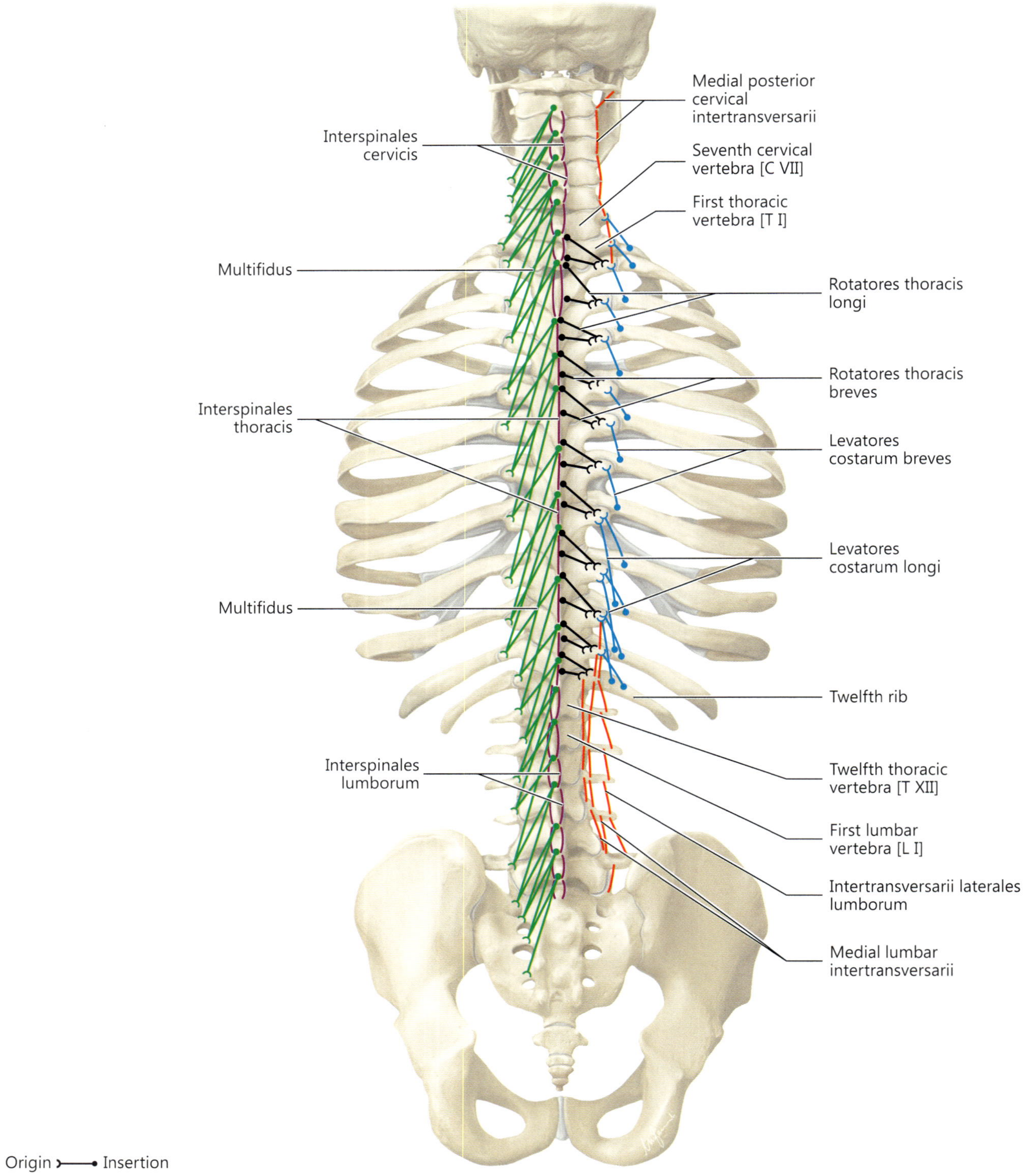

Schematic representation of the origin and insertion of the **paravertebral muscles** in the intermediate and deep layers.

Lateral wall of the **thorax** and **abdomen**. The upper limb has been removed, and the latissimus dorsi and pectoralis major have been partially transected to display the **pectoralis minor**, **serratus anterior**, and **external oblique**.

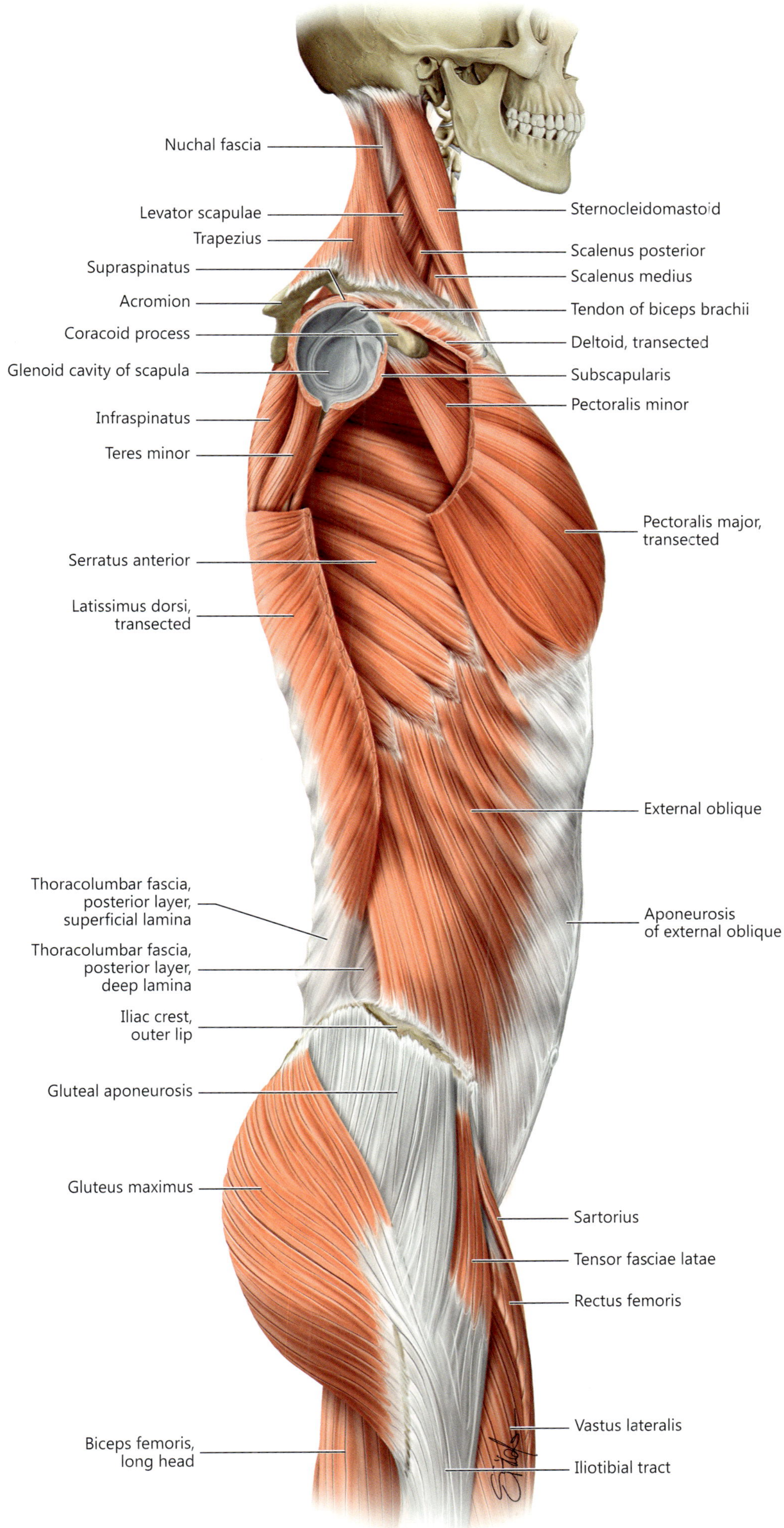

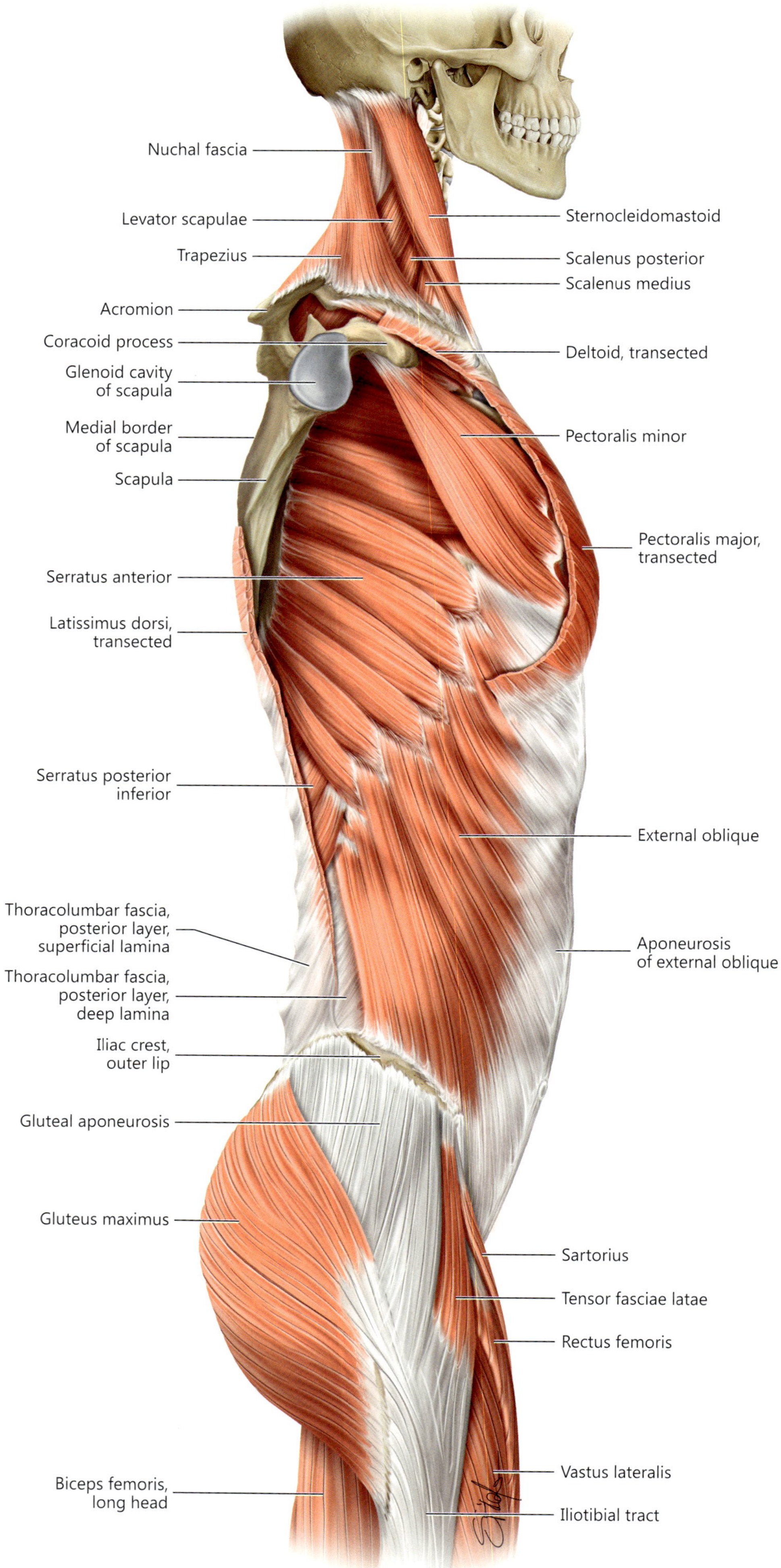

Extensive resection of the **latissimus dorsi** and **pectoralis major** highlights the **pectoralis minor**, located deep to the pectoralis major, and **serratus anterior**, found on the lateral surface of the rib cage. The **external oblique**, along with the **internal oblique** over which it is superimposed, and the **transversus abdominis**, which lies deep to the **internal oblique**, form the lateral wall of the abdomen.

After removal of the pectoral girdle, latissimus dorsi, and serratus anterior, and with extensive resection of the pectoralis minor, you can see the **external intercostal muscles** and the **external oblique**, with the former occupying the intercostal spaces from the vertebral column to the union of the ribs with the costal cartilages. The external oblique extends from the rib cage to the pelvis, moving from top to bottom and back to front.

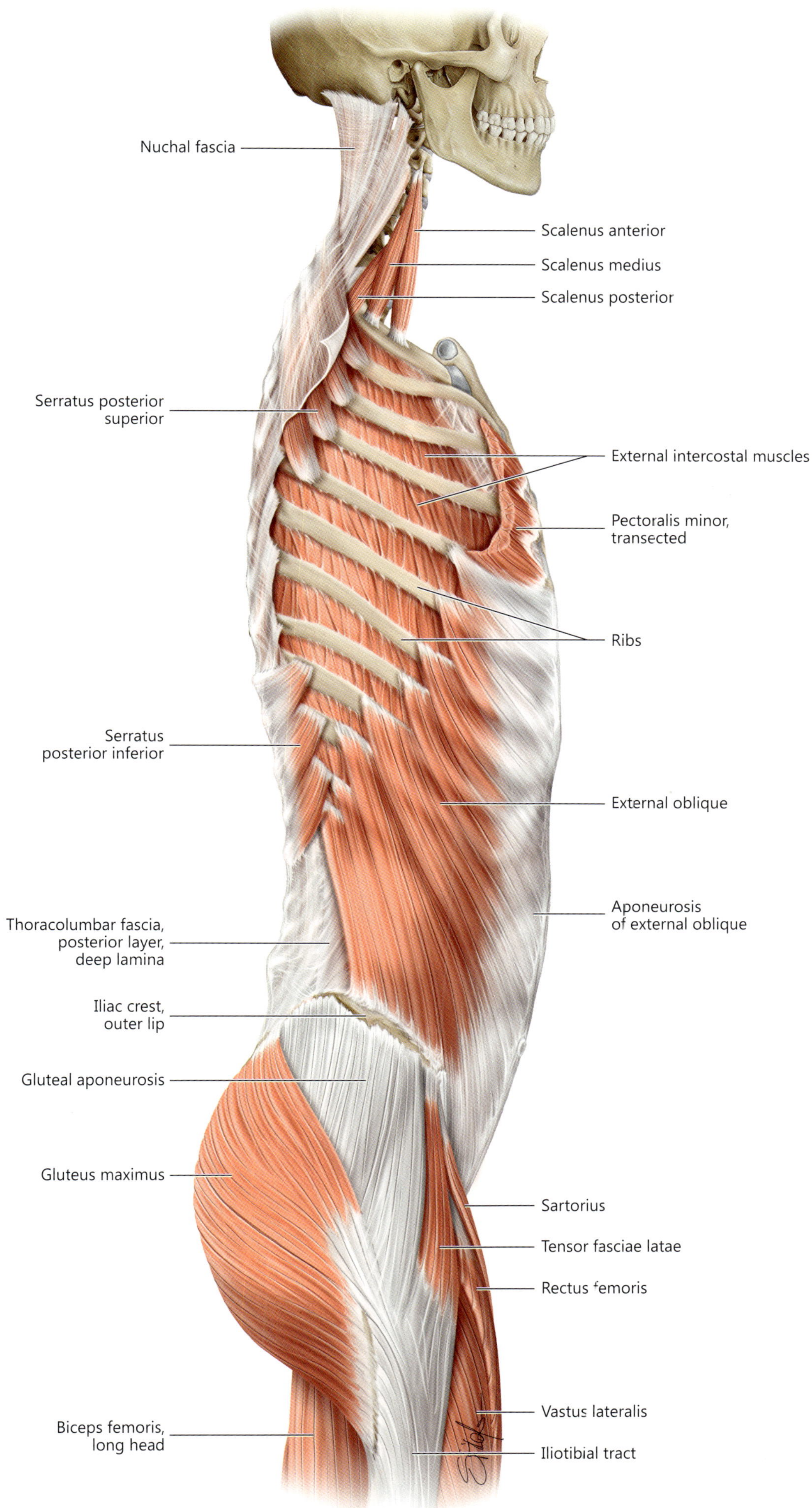

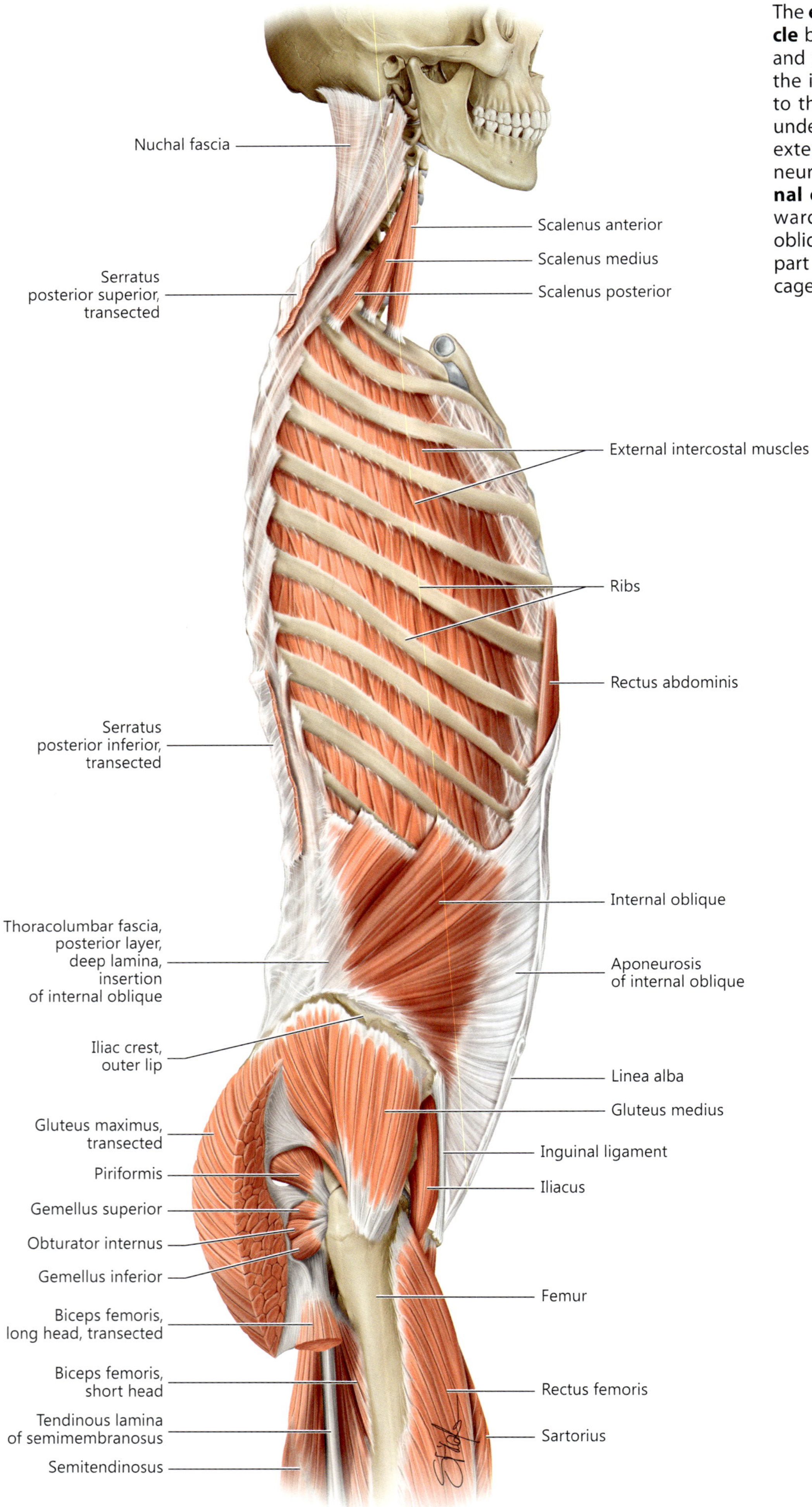

The **external intercostal muscle** bundles have a downward and forward orientation, from the inferior border of the ribs to the superior border of the underlying rib. Removal of the external oblique and its aponeurosis highlights the **internal oblique**, which goes upward and forward along an oblique path from the inferior part of the abdomen to the rib cage.

The **internal intercostal muscles**, the bundles of which are oriented downward and backward, extend from the sternum to the costal angle. Removal of the internal oblique and anterior splitting layer of its aponeurosis reveals the **rectus abdominis** anteriorly, which lacks the anterior layer of its sheath in the upper three-quarters. The **transversus abdominis** is also visible laterally.

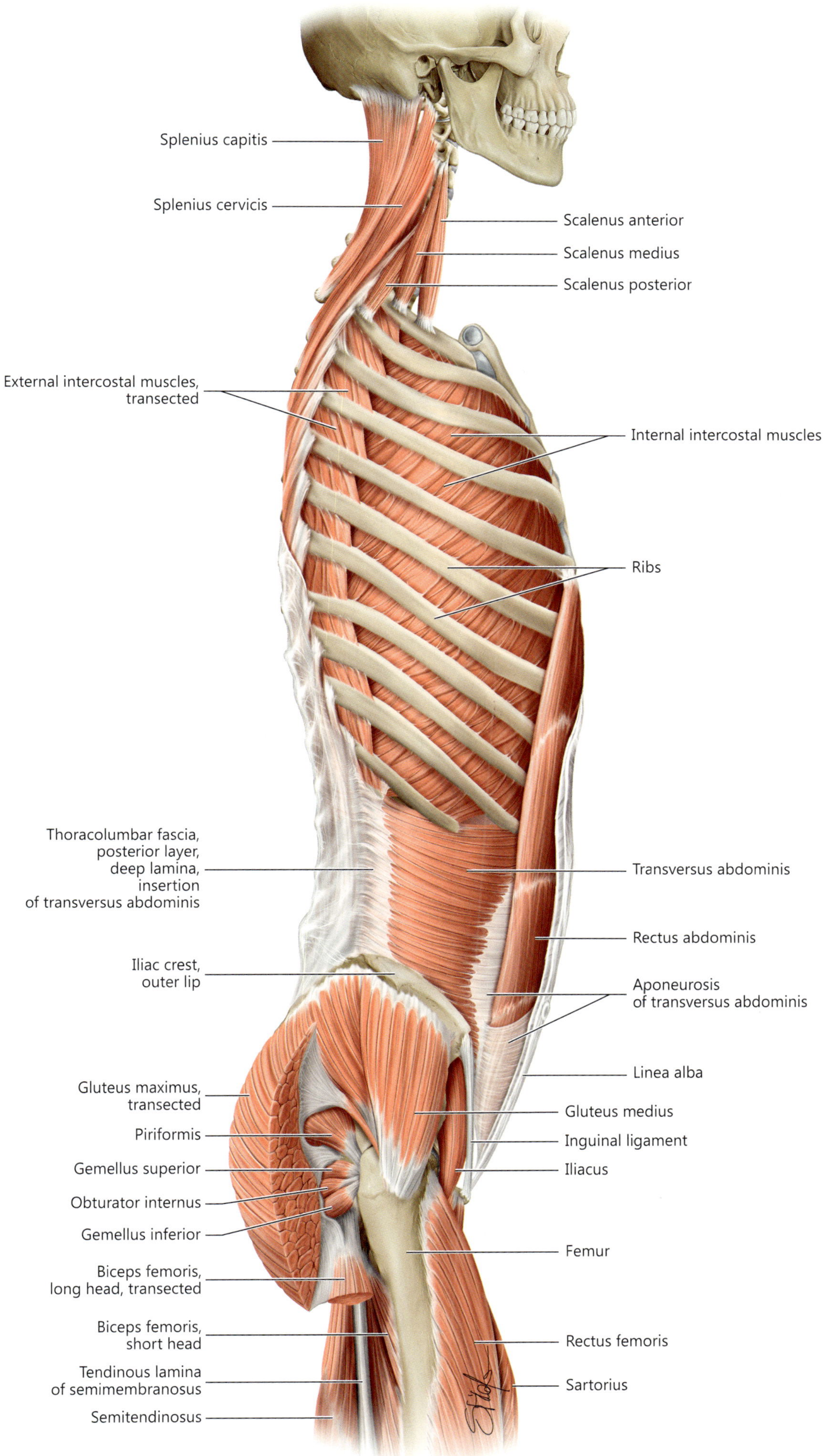

Resection of the intercostal muscles makes it possible to view the general form of the rib cage in a lateral projection. The **diaphragm** extends into the rib cage up to the level of the fourth rib, on the right, but during forced inspiration is lowered by about two intercostal spaces. The resection of the transversus abdominis, on the right, highlights the posterior muscles of the abdomen.

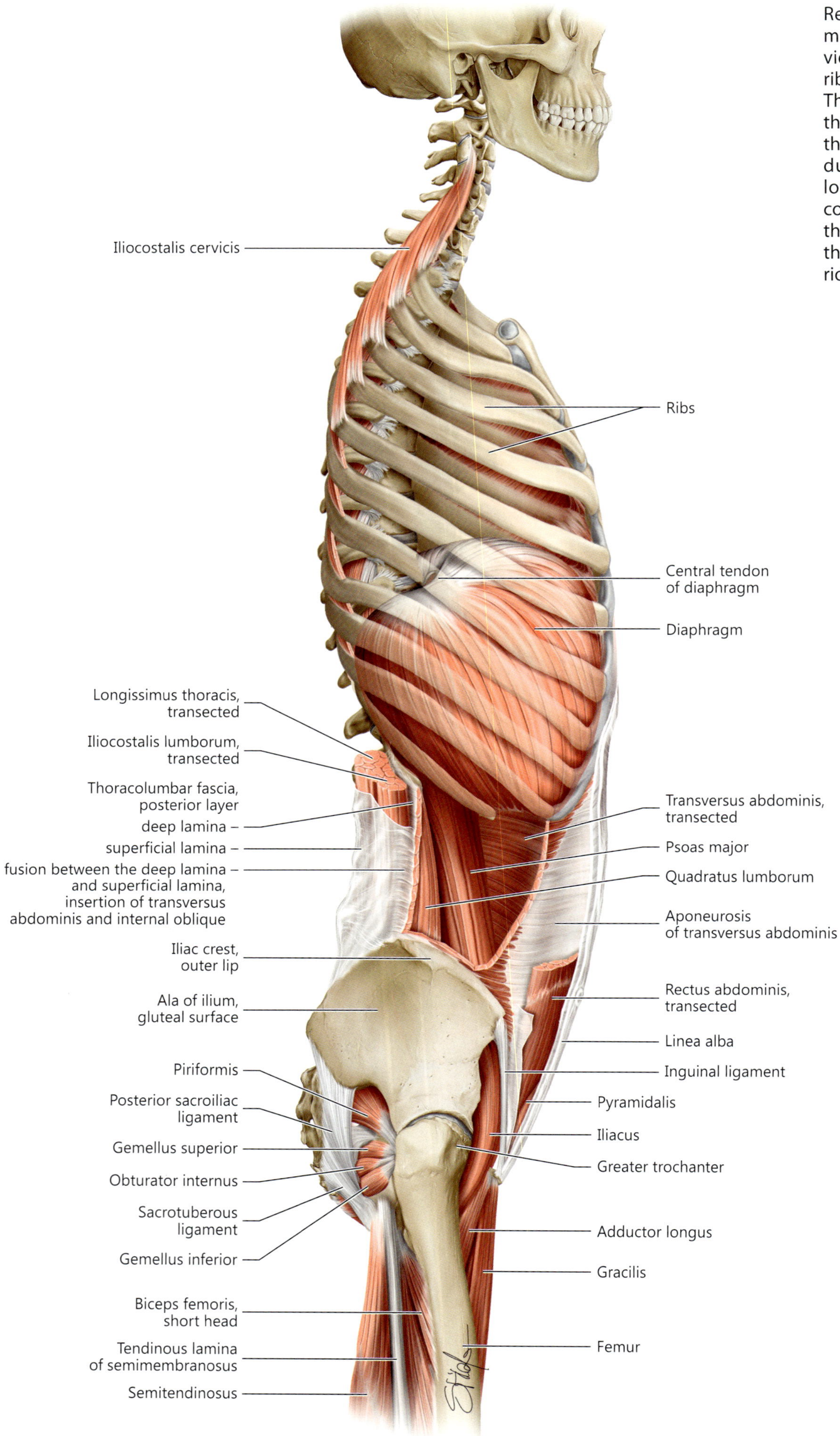

Posterior muscles of the abdomen: the **quadratus lumborum** extends from the twelfth rib to the iliac crest; the **psoas major**, which originates from the twelfth thoracic vertebra and the first four lumbar vertebrae, turns downward until it reaches the iliac fossa, where it is juxtaposed with the iliacus. The bundles of the two muscles pass below the inguinal ligament occupying the muscular space.

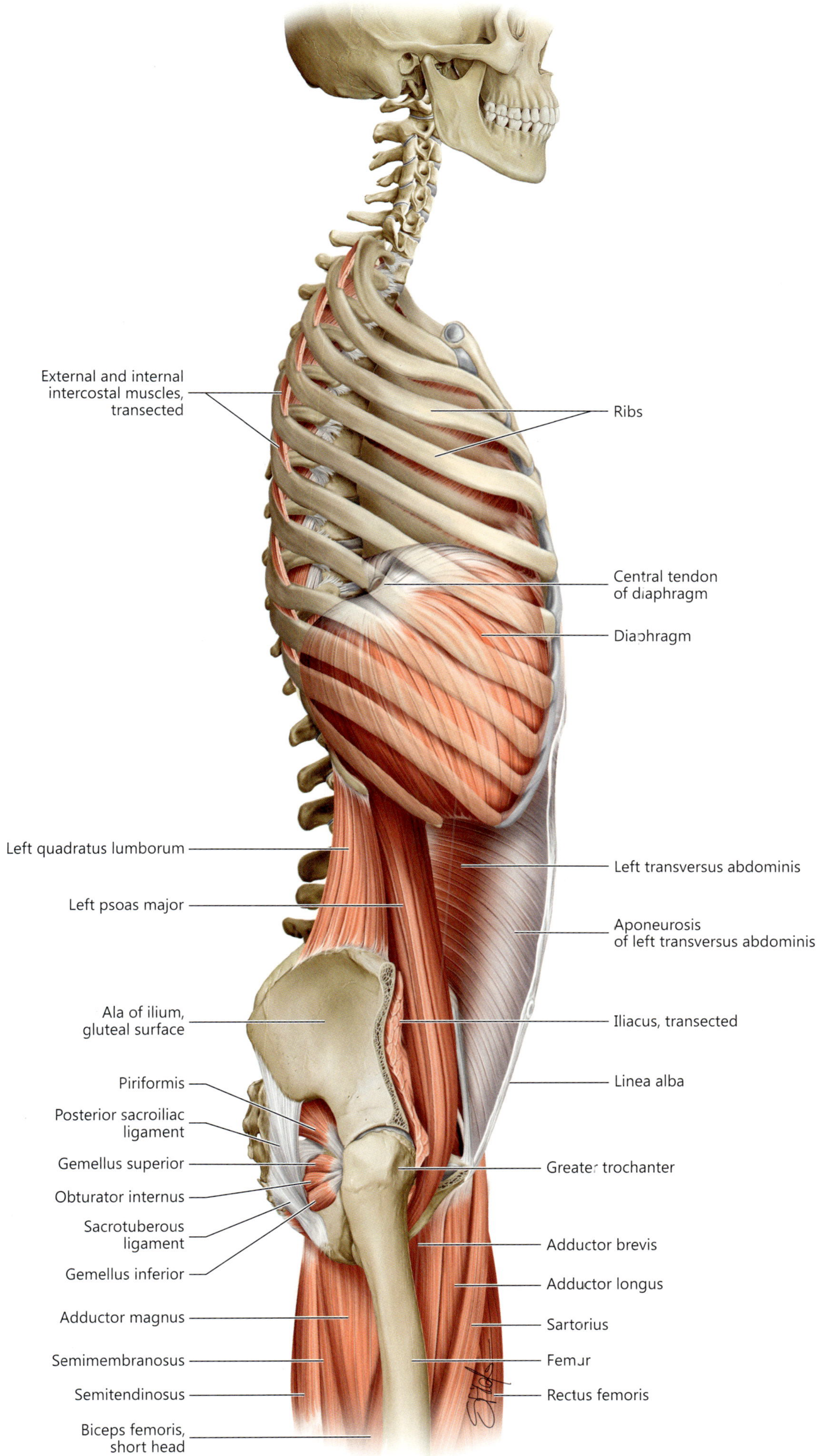

Removal of the hip bone, on the right, reveals the muscles of the pelvic wall. The **pirifor-mis** forms part of the postero-lateral wall of the lesser pelvis, while the **obturator internus** forms part of the anterolateral wall.

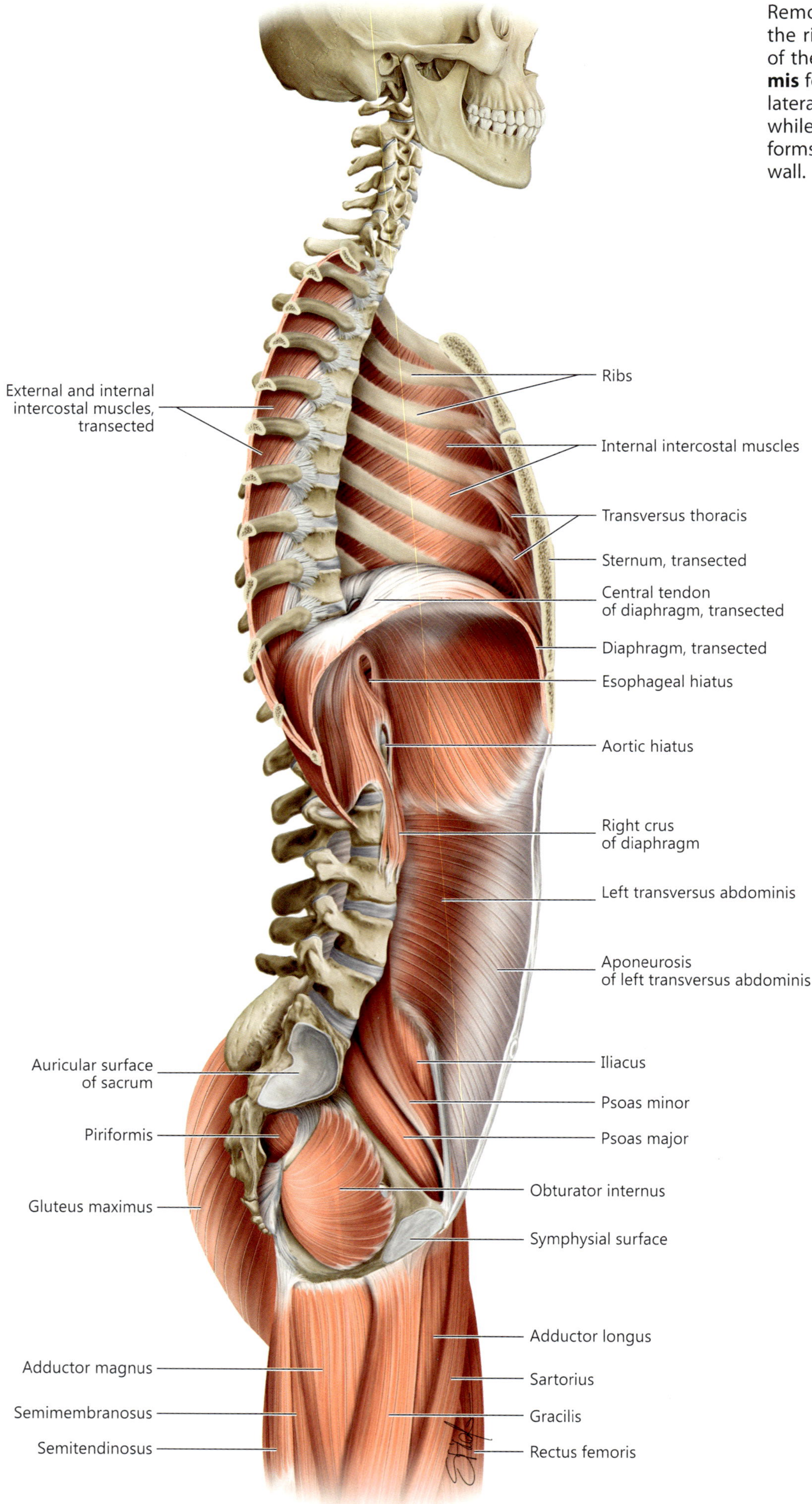

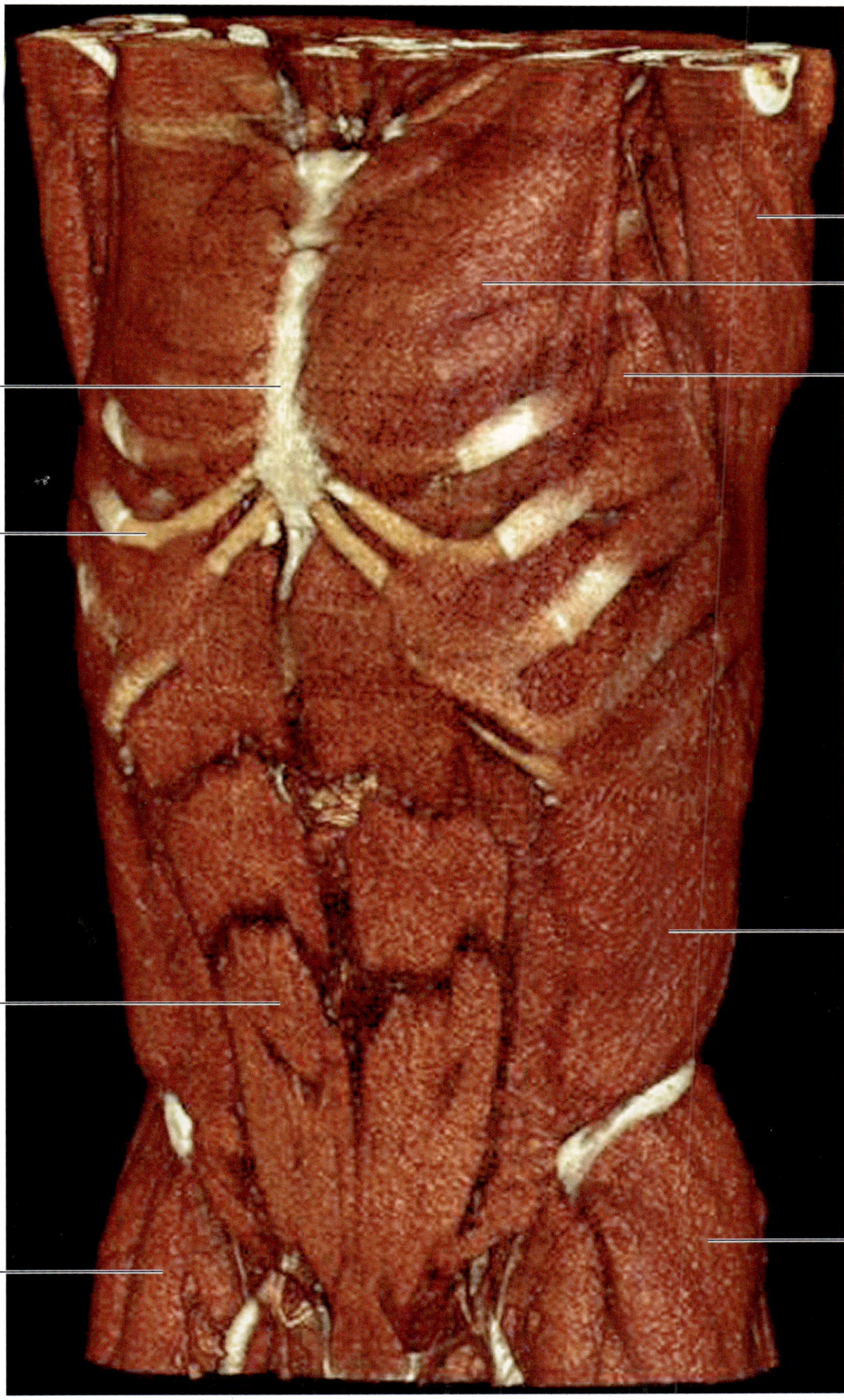

Three-dimensional image of a living human body, obtained by CT scan using a 3D volume rendering technique, showing the **muscles of the anterior wall of the trunk**. Left anterior view.

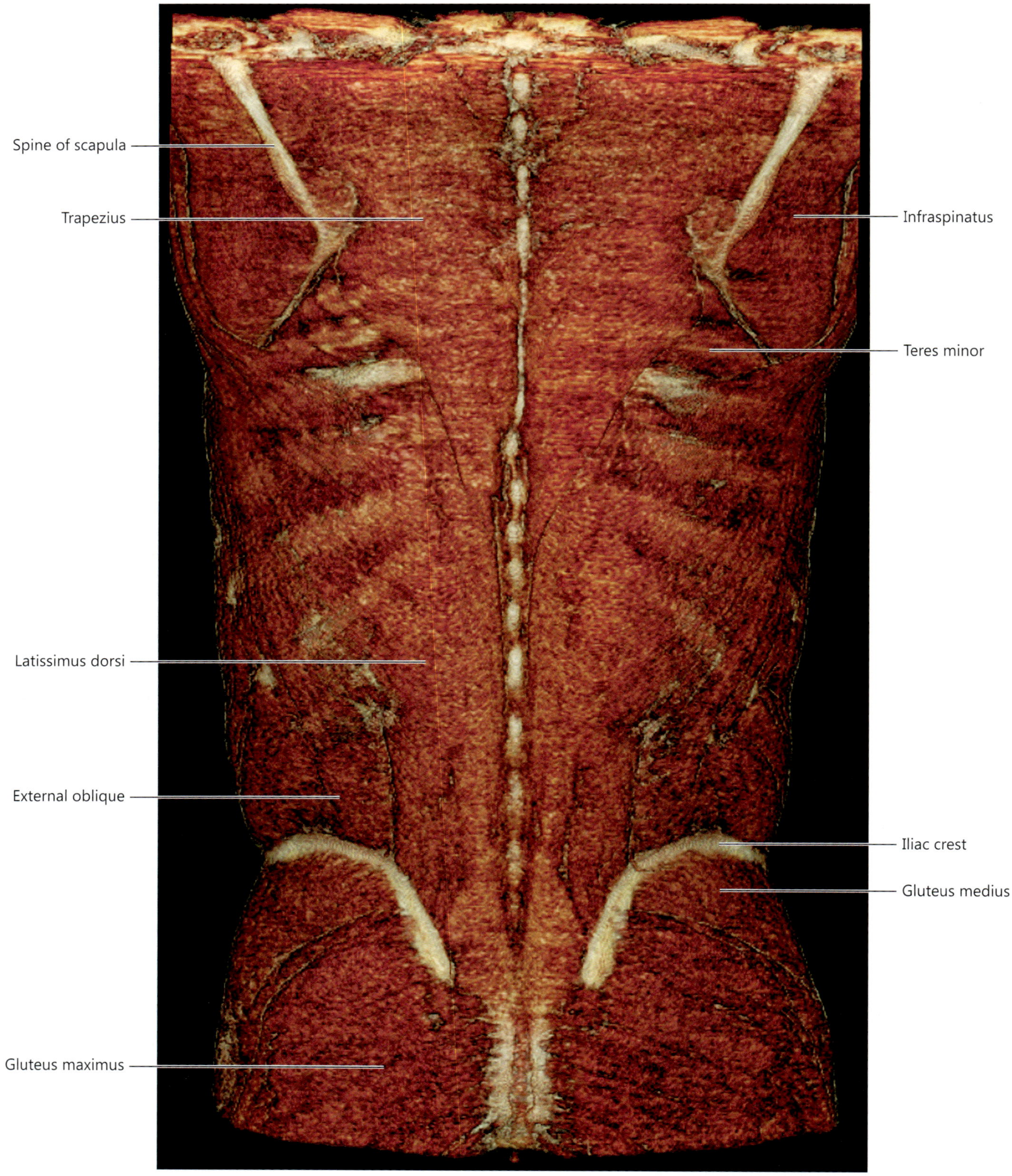

Three-dimensional image of a living human body, obtained by CT scan using a 3D volume rendering technique, showing the **superficial layer of the muscles of the posterior wall of the trunk**.

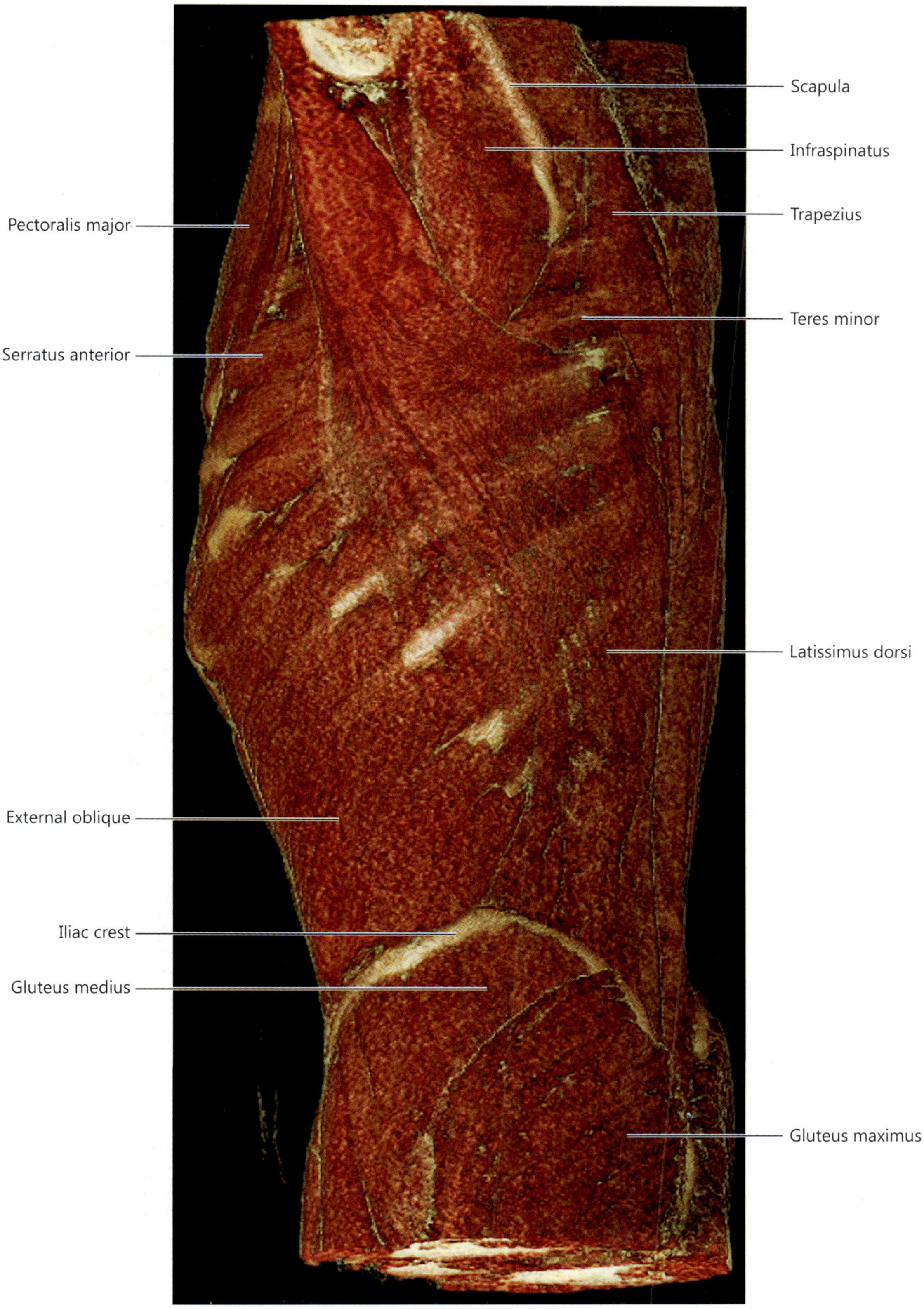

Three-dimensional image of a living human body, obtained by CT scan using a 3D volume rendering technique, showing the **muscles of the lateral wall of the trunk**.

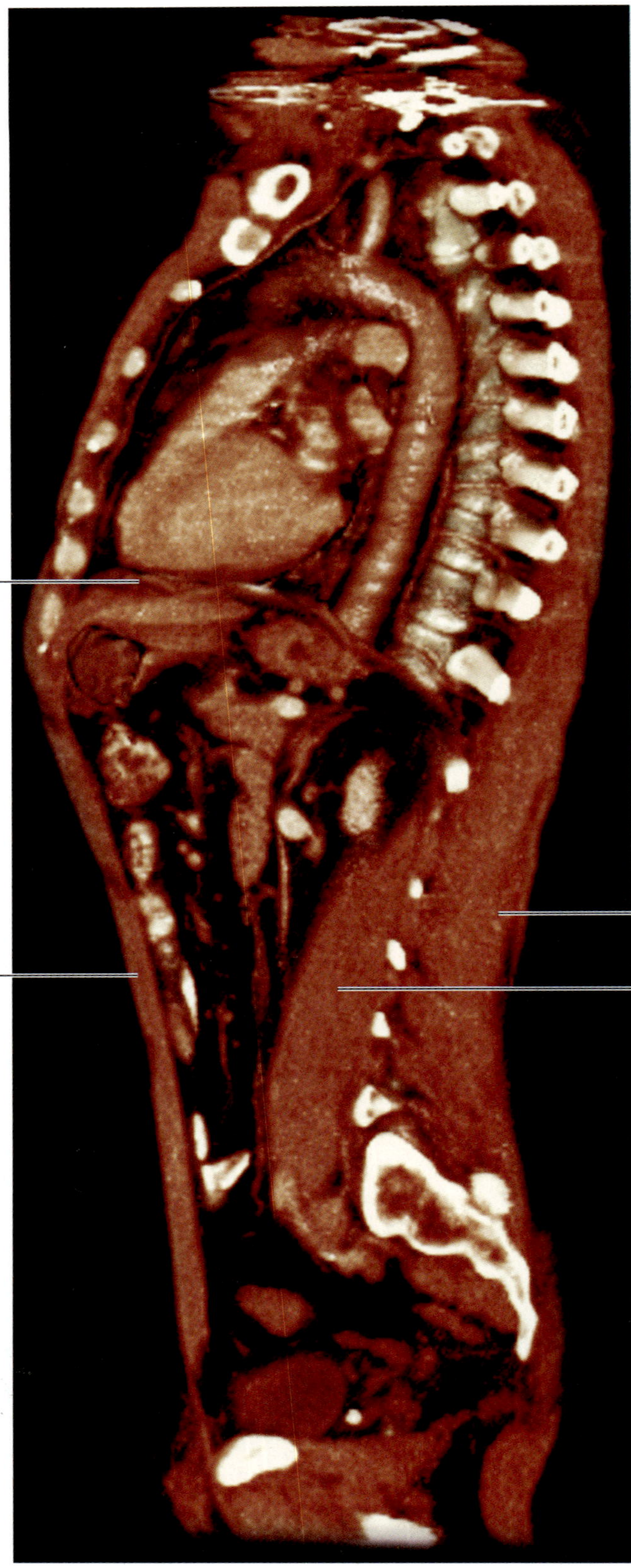

Three-dimensional image of a living human body, obtained by CT scan using a 3D volume rendering technique, showing the **anterior** and **posterior muscles of the trunk** in a paramedian sagittal view.

Origin, insertion, action, and innervation of the muscles of the back (extrinsic and intrinsic muscles of the vertebral column)

MUSCLE	ORIGIN	INSERTION	INNERVATION	ACTION
	SPINOCOSTAL AND SPINOAPPENDICULAR MUSCLES (FLAT EXTRINSIC MUSCLES OF THE VERTEBRAL COLUMN)			
Serratus posterior inferior	Spinous processes of the last two thoracic vertebrae and of the first two or three lumbar vertebrae, and corresponding supraspinous ligament	Inferior border of the last four ribs, just past the costal angle	Intercostal nerves [T9-T12]	Outward and downward traction of the ribs, in contrast to the depression of the central tendon of the diaphragm (expiratory muscle)
Serratus posterior superior	Inferior part of the ligamentum nuchae, spinous processes of the seventh cervical vertebra and of the first two or three thoracic vertebrae, and corresponding supraspinous ligament	Superior border of the second, third, fourth, and fifth ribs, just past the costal angle	Intercostal nerves [T2-T5]	Elevation of the ribs (inspiratory muscle)
Levator scapulae	Posterior tubercles of the transverse processes of the first four or five cervical vertebrae	Medial border of the scapula, superior to the origin of the spine of the scapula	Third and fourth cervical nerves [C3, C4] and dorsal nerve of the scapula [C5]	Elevation and medial movement of the scapula
Latissimus dorsi	Spinous processes of the last six thoracic vertebrae, thoracolumbar fascia, and iliac crest	Crest of the lesser tubercle of the humerus	Thoracodorsal nerve [C6-C8]	Adduction, extension, and internal rotation of the shoulder; with a fixed point on the humerus, elevation of the trunk and ribs
Rhomboid major	Spinous processes from the second to the fifth thoracic vertebrae	Medial border of the scapula	Dorsal nerve of the scapula [C4, C5]	Adduction, internal rotation, and elevation of the scapula
Rhomboid minor	Spinous processes of the seventh cervical vertebra and of the first thoracic vertebra, and the inferior part of the ligamentum nuchae	Medial border of the scapula, near the origin of the spine of the scapula	Dorsal nerve of the scapula [C4, C5]	Adduction, internal rotation, and elevation of the scapula
Trapezius	*Descending part*: superior nuchal line, external occipital protuberance, and ligamentum nuchae	Lateral third of the posterior border of the clavicle	Spinal root of the accessory nerve [XI] and branches of the cervical plexus [C3, C4]	Elevation of the scapula, with a fixed point on the column
	Transverse part: spinous processes of the seventh cervical vertebra and of the first three thoracic vertebrae and corresponding interspinous ligaments	Medial border of the acromion and posterior border of the spine of the scapula	Spinal root of the accessory nerve [XI] and branches of the cervical plexus [C3, C4]	Medial movement of the scapula, with a fixed point on the column
	Ascending part: spinous processes of the thoracic vertebrae (from the third to the twelfth), and corresponding interspinous ligament	Medial border of the scapula at the point of origin of the spine	Spinal root of the accessory nerve [XI] and branches of the cervical plexus [C3, C4]	Depression of the scapula, with a fixed point on the column
				Joint action of all three parts, with a fixed point on the scapula: ipsilateral inclination of the head and elevation of the trunk (by a joint action of the muscles on both sides)
	MUSCLES OF THE BACK PROPER OR SPINODORSAL MUSCLES (LONG INTRINSIC MUSCLES OF THE VERTEBRAL COLUMN)			
Erector spinae (sacrospinal)	Long muscle extending from the sacrum to the nape, with a wide and thick tendon inserting on the median sacral crest, the spinous processes of the lumbar vertebrae and the last two thoracic vertebrae, and the posterior part of the iliac crest, where it is possible to distinguish three parts or columns: the iliocostalis (lateral division), the longissimus (intermediate division), and the spinalis (medial division)			
– **iliocostalis**			Posterior rami of the spinal nerves (from C4 to L3)	
– – **cervicis** (☞ Table: Extensor muscles of the head and neck)	Costal angles from the third to the sixth rib	Posterior tubercles of the transverse processes from C IV to C VI		Extension and ipsilateral inclination of the neck
– – **lumborum**				
– – – thoracic part	Superior border of the costal angles from the sixth to the twelfth rib	Superior border of the costal angles from the first to the sixth rib; posterior surface of the transverse process of C VII		Extension and ipsilateral inclination of the trunk
– – – lumbar part	Outer lip of the iliac crest, posterior layer of the thoracolumbar fascia	Last eight ribs, near the costal angle		Extension and ipsilateral inclination of the trunk

continued

Origin, insertion, action, and innervation of the muscles of the back (extrinsic and intrinsic muscles of the vertebral column)				*(continued)*
MUSCLE	***ORIGIN***	***INSERTION***	***INNERVATION***	***ACTION***
– **longissimus**			Posterior rami of the spinal nerves (from C4 to L3)	
– – **capitis** (☞ Table: Extensor muscles of the head and neck)	Transverse processes from T I to T V and the articular processes from C III to C VII	Mastoid process of the temporal bone		Extension of the head and neck and ipsilateral inclination of the neck
– – **cervicis** (☞ Table: Extensor muscles of the head and neck)	Transverse processes from T I to T V	Posterior tubercles of the transverse processes from C II to C VI		Extension and ipsilateral inclination of the neck
– – **thoracis**	Posterior surfaces of the transverse and accessory processes from L I to L V; middle layer of the thoracolumbar fascia	Apex of the transverse processes from T I to T XII, and between the tubercles and costal angles of the last nine or ten ribs		Extension and ipsilateral inclination of the trunk
– **spinalis**			Posterior rami of the spinal nerves (from C4 to L3)	
– – **capitis** (☞ Table: Extensor muscles of the head and neck)	Inferior part of the ligamentum nuchae and spinous processes of C VII, T I, and T II	Squamous part of the occipital bone between the superior and inferior nuchal lines, merging with the semispinalis capitis		Extension of the head and neck
– – **cervicis** (☞ Table: Extensor muscles of the head and neck)	Inferior part of the ligamentum nuchae and spinous processes of C VII, T I, and T II	Spinous processes from C I to C III		Extension and lateral inclination of the neck
– – **thoracis**	Spinous processes of T XI, T XII, L I, and L II	Spinous processes from T I to T IV or T VIII		Extension and lateral inclination of the trunk
Interspinales	Short muscles between the spinous processes of contiguous vertebrae to the sides of the interspinous ligaments in the cervical, thoracic, and lumbar tracts of the vertebral column		Posterior rami of the spinal nerves (from C3 to L5)	Extension of the trunk (posture control)
Intertransversarii	Short muscles between the transverse processes of contiguous vertebrae in the cervical, thoracic, and lumbar tracts of the vertebral column		Anterior [C1-C8] and posterior (C1 to L5) rami of the spinal nerves	Lateral inclination of the trunk (posture control)
Spinotransversales				
– **splenius**			Posterior rami of the cervical nerves [C2-C4]	
– – **capitis** (☞ Table: Extensor muscles of the head and neck)	Inferior half of the ligamentum nuchae, spinous processes of C VII and from T I to T IV and corresponding supraspinous ligament	Mastoid process of the temporal bone, occipital bone below the lateral third of the superior nuchal line		Extension of the head and neck
– – **cervicis** (☞ Table: Extensor muscles of the head and neck)	Spinous processes from T III to T VI	Posterior tubercles of the transverse processes from C I to C III		Extension and ipsilateral rotation of the neck
Transversospinales				
– **multifidus muscles [multifidus cervicis** (☞ Table: Extensor muscles of the head and neck), **multifidus thoracis, multifidus lumborum]**	Posterior surface of the sacrum; aponeurosis of the erector spinae; posterior superior iliac spine; posterior sacroiliac ligament; transverse processes from C IV to L V	Spinous processes from one to four of the overlying vertebrae	Posterior rami of the spinal nerves (from C1 to L5)	Extension and contralateral rotation of the trunk (posture control)
– **rotator muscles [rotatores cervicis breves and longi** (☞ Table: Extensor muscles of the head and neck), **rotatores thoracis, rotatores lumborum]**	Superior and posterior surface of the transverse processes of the cervical, thoracic, and lumbar vertebrae	Inferior and lateral surfaces of the lamina of the overlying vertebra (breves) or subsequent vertebra (longi) in the cervical, thoracic, and lumbar tracts of the vertebral column	Posterior rami of the spinal nerves (from C1 to L5)	Extension and contralateral rotation of the trunk (posture control)

continued

Origin, insertion, action, and innervation of the muscles of the back (extrinsic and intrinsic muscles of the vertebral column)				*(continued)*
MUSCLE	***ORIGIN***	***INSERTION***	***INNERVATION***	***ACTION***
– **semispinalis**			Posterior rami of the spinal nerves (from C1 to L5)	
– – **capitis** (☞ Table: Extensor muscles of the head and neck)	Apex of the transverse processes of C VII and from T I to T VII, and the articular processes from C IV to C VI	Medial part of the region between the superior and inferior nuchal lines of the occipital bone		Extension and contralateral rotation of the head
– – **cervicis** (☞ Table: Extensor muscles of the head and neck)	Transverse processes from T I to T VI	Spinous processes from C I to C V		Extension and contralateral rotation of the neck
– – **thoracis**	Transverse processes from T VI to T X	Spinous processes of C VI and C VII and from T I to T IV		Extension and contralateral rotation of the trunk

For the muscles of the thorax ☞ pg. 484

Fasciae and fascial structures of the trunk				
STRUCTURE	***POSITION***	***ORGANIZATION***	***INSERTION***	***CONTINUATION***
		NECK		
Investing layer of cervical fascia (superficial layer of the cervical fascia)	Immediately below the subcutaneous tissue, invests all of the neck	It splits laterally surrounding the sternocleidomastoid, posteriorly surrounding the trapezius and the sternum above, delimiting the suprasternal space	At the top: the inferior margin of the mandible, the mastoid process of the temporal bone, and the superior nuchal line, up to the external occipital protuberance; at the bottom: the manubrium of the sternum, the anterior border of the clavicle, the acromion, and the spine of the scapula; anteriorly: at the midline, joins with the pretracheal layer of cervical fascia; at the level of the anterior border of the trapezius, gives rise to an extension which passes in front of the scalene muscles and attaches to the anterior tubercles of the transverse processes of the cervical vertebrae	Superiorly with the fasciae of the face; inferiorly with the pectoral and deltoid fasciae; posteriorly with the nuchal fascia; medially with the prevertebral layer of cervical fascia; anteriorly with the pretracheal layer of cervical fascia
Pretracheal layer of cervical fascia	Deep to the investing layer of cervical fascia, extending from the hyoid bone above to the superior opening of the thorax below	Trapezoidal in shape, with a smaller base at the top and a larger base at the bottom, completely enveloping the infrahyoid muscles to terminate laterally after splitting to surround the omohyoid; inferiorly, behind the manubrium of the sternum, it creates a split that descends into the mediastinum all the way to the pericardium	At the top: the hyoid bone; at the bottom: the posterior surface of the sternum and the posterior border of the clavicle; it forms the dense connective tissue surrounding the neurovascular bundle of the neck, and adheres to the subclavian and internal jugular veins, keeping them open, as well as to the carotid sheath; laterally, after wrapping around the omohyoid, it gives rise to a posterior extension that merges with the prevertebral layer of cervical fascia	Anteriorly with the investing layer of cervical fascia; posteriorly with the prevertebral layer of cervical fascia
Prevertebral layer of cervical fascia	Extends between the cranial base and the first thoracic vertebrae	Anteriorly to the vertebral bodies and the prevertebral and scalene muscles; posteriorly to the pharynx, esophagus, and neurovascular bundle of the neck	At the top: the inferior surface of the basilar part of the occipital bone; laterally, the anterior tubercles of the transverse processes of the cervical vertebrae	Anteriorly with the pretracheal layer of cervical fascia, laterally with the investing layer of cervical fascia; inferiorly, with the endothoracic fascia
Carotid sheath	Dense connective tissue surrounding the neurovascular bundle of the neck			
Nuchal fascia	Between the trapezius and splenius capitis	It splits inferiorly to wrap around the serratus posterior superior	Superiorly, at the squamous part of the occipital bone; medially, at the ligamentum nuchae	Laterally with the investing layer of cervical fascia; inferiorly with the posterior layer of the thoracolumbar fascia

continued

Fasciae and fascial structures of the trunk				(continued)
STRUCTURE	**POSITION**	**ORGANIZATION**	**INSERTION**	**CONTINUATION**
		THORAX		
Diaphragmatic fascia	Inferior surface of the diaphragm	Invests the inferior surface of the diaphragm, separating it from the parietal peritoneum		Inferiorly with the transversalis fascia
Clavipectoral fascia	Robust lamina located posteriorly to the clavicular head of the pectoralis major, where it covers the section between the pectoralis minor and subclavius	Splits to wrap around the subclavius and pectoralis minor	Superiorly, to the clavicle, anteriorly and posteriorly to the subclavian groove; laterally to the coracoid process	Posteriorly, with the posterior extension of the pretracheal layer of cervical fascia that merges with the prevertebral layer of cervical fascia; medially, with the thoracic fascia; at the level of the superior border of the pectoralis minor, with the pectoral fascia, and inferiorly, with the axillary fascia
Endothoracic fascia (parietal fascia of thorax)	Inside the rib cage	Invests the inside of the rib cage, separating the internal intercostal muscles, subcostales, and transversus thoracis, as well as the diaphragm, from the parietal pleura		Superiorly, with the prevertebral layer of cervical fascia
Pectoral fascia (retromammary)	Thin lamina investing the pectoralis major, with extensions between its bundles	Anteriorly, has a relationship with the mammary gland through the retromammary space	Superiorly, to the clavicle; anteriorly, at the midline, to the anterior surface of the sternum	Laterally, between the inferior borders of the pectoralis major and latissimus dorsi, with the axillary fascia; inferiorly, with the rectus sheath; deeply, with the thoracic fascia; at the level of the superior border of the pectoralis minor, with the clavipectoral fascia
Thoracic fascia (external)	Externally to the rib cage	External covering (epimysium) of the external intercostal muscles		Superficially, with the pectoral fascia; laterally, with the clavipectoral fascia
Thoracolumbar fascia	Back and lumbar region	Splits into three layers:		
	Invests the erector spinae	– posterior (superficial) splits into two laminae: superficial and deep (of the posterior layer)	Inferiorly, at the outer lip of the iliac crest; medially, at the transverse processes of the thoracic vertebrae and costal processes of the lumbar vertebrae, and at the median sacral crest; laterally, at the costal angles	Superiorly, with the nuchal fascia; between the last rib and the iliac crest, with the anterior layer, inferolaterally to the fascia of the gluteus maximus
	Anteriorly invests the quadratus lumborum	– anterior (deep) or quadratus lumborum fascia, limited at the lumbar region	Superiorly, at the twelfth rib; inferiorly, at the inner lip of the iliac crest; medially, at the costal processes of the first four lumbar vertebrae	Between the last rib and the iliac crest, with the posterior layer
	Posteriorly invests the quadratus lumborum	– middle: merges laterally with the posterior layer to form a raphe from which the internal oblique and transversus abdominis originate	Extremity of the costal processes of the lumbar vertebrae	

continued

Fasciae and fascial structures of the trunk *(continued)*

STRUCTURE	POSITION	ORGANIZATION	INSERTION	CONTINUATION
ABDOMEN				
Umbilical ring ☞ Linea alba				
Erector spinae aponeurosis	Lumbar region	Tendinous lamina of the muscle origin, merged with the thoracolumbar fascia	Spinous processes of the lumbar vertebrae, dorsal surface of the sacrum, posterior part of the iliac crest	
Aponeurosis of the external oblique ☞ Rectus sheath; Inguinal ligament; Inguinal canal				
Aponeurosis of the internal oblique ☞ Rectus sheath				
Aponeurosis of the transversus abdominis ☞ Rectus sheath				
Iliopectineal arch ☞ Iliac fascia; Inguinal ligament				
Inguinal canal	Extends between the inferior borders and aponeurosis of the external oblique, internal oblique, and transversus abdominis and the inguinal ligament, into which runs the spermatic cord (♂) or the round ligament of the uterus (♀)	Openings: – superficial inguinal ring: – – lateral crus: fibers of the aponeurosis of the external oblique which reach the pubic tubercle from the anterior superior iliac spine – – medial crus: fibers of the aponeurosis of the external oblique which extend from the anterior superior iliac spine to the pubic symphysis (the lateral border of the ring, between the two crura, is formed by the intercrural fibers, and the medial border by the reflected ligament, consisting of the aponeurosis of the contralateral external oblique, which extends past the midline) – deep inguinal ring, medially delimited by the interfoveolar ligament Walls: – superior wall: inferior border of the internal oblique and transversus abdominis – inferior wall: medial half of the inguinal ligament – anterior wall: aponeurosis of the external oblique adhering to the anterior border of the inguinal ligament – posterior wall: transversalis fascia that extends to the posterior border of the inguinal ligament		
Parietal abdominal fascia (endoabdominal) ☞ Investing abdominal fascia; Transversalis fascia; Iliac fascia; Fascia of individual organs (extraperitoneal)				
Visceral abdominal fascia	Below the visceral peritoneum, including the fascia of the intraperitoneal organs, which consists of the adventitia			
Investing abdominal fascia	Located deeply in relation to the subcutaneous tissue, invests the muscles of the abdomen	Divided into three fasciae:		
	Invests the outer surface of the external oblique	– superficial: at the bottom, forms the suspensory ligament of the penis (♂) or clitoris (♀)	Medially merges with the rectus sheath	Inferiorly, with the external spermatic fascia, which invests the spermatic cord
	Between the external oblique and internal oblique	– intermediate		
	Between the internal oblique and transversus abdominis	– deep		Inferiorly, with the perineal fascia

continued

Fasciae and fascial structures of the trunk				*(continued)*
STRUCTURE	***POSITION***	***ORGANIZATION***	***INSERTION***	***CONTINUATION***
Extraperitoneal fascia	Layer of connective tissue, between the transversalis fascia and the parietal peritoneum, from which the extraperitoneal ligaments originate			
Iliac fascia (iliopsoas fascia)	Invests the iliopsoas and gives rise to the iliopectineal arch, between the inferior surface of the inguinal ligament and the iliopubic ramus, which separates the vascular space from the muscular space	Iliac portion (from the inguinal ligament) and a portion of the psoas muscle (from the arcuate line)		Superiorly with the transversalis fascia, inferiorly with the obturator fascia
Umbilical fascia ☞ Transversalis fascia				
Fascia of the extraperitoneal organs	Formed by the adventitia			
Thoracolumbar fascia ☞ THORAX				
Transversalis fascia	Invests the internal surface of the transversus abdominis, separating it from the parietal peritoneum. Between the inguinal ligament and the superior pubic ramus, it cuts off communication between the abdominal cavity and the vascular space, forming the femoral septum. At the umbilical level, it forms the umbilical ring, and the thickest part located below the umbilical region is known as the umbilical fascia. The internal surface of the fascia, above the inguinal ligament, is reinforced: laterally, by the interfoveolar ligament; medially, and from the inside out, by the conjoint tendon (inguinal falx) and the reflected ligament (formed by the aponeurosis of the contralateral external oblique); and at the level of the posterior border of the inguinal ligament, by the Iliopubic tract (triangularly shaped, with a base corresponding to the inguinal ligament and an apex that reaches the arcuate line). The lateral border of the interfoveolar ligament is concave, and medially delimits the deep inguinal ring	Quadrangular lamina with four borders:		Deeply, with the extraperitoneal fascia
		– superior	At the iliac crest and the posterior border of the inguinal ligament	With the diaphragmatic fascia
		– inferior		With the iliac and pelvic fascia
		– anterior: contributes to the formation of the posterior layer of the rectus sheath		
		– posterior: invests the anterior fascia of the psoas major		

continued

Fasciae and fascial structures of the trunk				*(continued)*
STRUCTURE	***POSITION***	***ORGANIZATION***	***INSERTION***	***CONTINUATION***
Rectus sheath	Encloses the rectus abdominis	Formed by the aponeurosis of the external oblique, internal oblique, and transversus abdominis, as well as by the transversalis fascia. It is divided into two layers or laminae: – anterior: formed in the upper three-quarters by the aponeurosis of the external oblique and by the anterior splitting layer of the aponeurosis of the internal oblique; formed in the lower quarter by the aponeurosis of the transversus abdominis – posterior: formed in the upper three-quarters by the posterior splitting layer of the aponeurosis of the internal oblique, the aponeurosis of the transversus abdominis, and the transversalis fascia; formed in the lower quarter solely by the transversalis fascia (the point of passage between the two layers corresponds to the inferiorly concave arcuate line, at which point the upper three-quarters of the posterior layer come to an end)		
Muscular space ☞ Inguinal ligament				
Vascular space ☞ Inguinal ligament				
Inguinal ligament	Fibrous cord between the anterior superior iliac spine and the pubic tubercle, formed by the inferior fibers of the aponeurosis of the external oblique	Surfaces: – superior: concave (at its medial half, receives the spermatic cord ♂ or round ligament of the uterus ♀) – inferior: adheres laterally to the iliopsoas fascia contained in the muscular space along with the femoral nerve, and medially to the adventitia of the femoral vessels, which are contained in the vascular space (delimited medially by the lacunar ligament, which extends backward from the medial part of the inguinal ligament to insert posteriorly at the pectineal line; the two spaces are separated by the iliopectineal arch) – anterior border: aponeurosis of the external oblique – posterior border: Iliopubic tract of the transversalis fascia		
Interfoveolar ligament ☞ Transversalis fascia				
Lacunar ligament ☞ Inguinal ligament				
Medial umbilical ligament ☞ PELVIS				
Median umbilical ligament ☞ PELVIS				
Reflected ligament ☞ Inguinal canal; Transversalis fascia				
Linea alba	Fibrous raphe between the rectus abdominis muscles; extends from the xiphoid process of the sternum to the superior border of the pubic symphysis	Formed by the intersection, on the midline, of the aponeurosis of the external oblique, internal oblique, and transversus abdominis; interrupted at the umbilical level, where the transversalis fascia forms the umbilical ring		
Arcuate line ☞ Rectus sheath				
Femoral septum ☞ Transversalis fascia				
Conjoint tendon (inguinal falx) ☞ Transversalis fascia				
Iliopubic tract ☞ Transversalis fascia				

continued

Fasciae and fascial structures of the trunk *(continued)*

STRUCTURE	POSITION	ORGANIZATION	INSERTION	CONTINUATION
		PELVIS		
Tendinous arch of the levator ani	Arching line from the pubis to the ischial spine, from which the levator ani originates	Thickening of the obturator fascia	Posterior surface of the superior pubic ramus and ischial spine	
Obturator canal ☞ Obturator membrane				
Extraperitoneal fascia	Connective tissue that can support blood vessels and nerves or serve as the point of origin for ligaments	Contains smooth muscle fibers, and can be connected to the skeletal muscle fibers through the epimysium		
– **cardinal ligament**	Thickening of connective tissue, directed transversally from the lateral wall of the pelvis to the lateral border of the uterus and vagina	Parametrium at the level of the supravaginal part of the cervix		
– **uterosacral ligament** (rectouterine)	Between the uterine cervix and rectum	Reinforcing ligament		
– **round ligament of the uterus**	Fibrous cord extending from the superolateral angle of the uterus to the anterolateral wall of the pelvis, then extending toward the internal inguinal ring, entering the inguinal canal, from which it exits and ends at the fat pad of the labium majus	A vestige of the caudal gonadal fold, enveloped in the inguinal canal by a connective tissue sheath derived from the transversalis fascia		
– **vesicouterine ligament** (pubocervical)	Between the bladder and the cervix	Bundles of loose connective tissue		
Parietal pelvic fascia (endopelvic fascia)	Invests the pelvic wall	Covers the levator ani, ischiococcygeus, piriformis, and obturator internus, and, anteriorly, the deep transverse perineal muscle		With the transversalis fascia
– **piriformis fascia**	Invests the piriformis		Sacrum, near the anterior sacral foramina	
– **obturator fascia**	Invests the obturator internus	Forms an arch under the obturator nerves and vessels, completing the obturator canal	Posterior surface of the superior pubic ramus, pubic arch, and sacrotuberous ligament	With the superior fascia of the pelvic diaphragm and with the iliac fascia
– – **tendinous arch of the pelvic fascia**	Thickening of the tendinous arch extending between the pubic symphysis and the levator ani, continuing posteriorly to the ischial spine	Structure connecting to the pelvic wall	Pubic symphysis and ischial spine	
– **presacral fascia**	Area of connective tissue located anteriorly to the sacrum, between the visceral pelvic fascia of the posterior wall of the rectum and the superior fascia of the pelvic diaphragm			
– **fascia of individual organs**	Covers the muscles of the pelvis			
– **rectosacral fascia** (♂)	Between the rectal ampulla and sacrum			
– **superior fascia of the pelvic diaphragm**	Investing fascia of the levator ani and ischiococcygeus	Invests the superior surface of the pelvic diaphragm		With the obturator fascia
– – **lateral ligament of the rectum** (lateral rectal)	Between the rectum and the pelvic wall	Posterior lamina of the endopelvic fascia		

continued

Fasciae and fascial structures of the trunk				*(continued)*
STRUCTURE	**POSITION**	**ORGANIZATION**	**INSERTION**	**CONTINUATION**
– – **lateral ligament of the bladder** (lateral vesical)	Between the bladder and pelvic wall	Anterior lamina of the endopelvic fascia		
– – – false	From the bladder to the pelvic wall			
– – – true	From the lateral wall of the bladder to the tendinous arch of the pelvic fascia			
– – **posterior ligament of the bladder** (posterior vesical) (false ligament) (♂)	To the side of the bladder, around the rectum up to the anterior surface of the sacrum			
– – **puboprostatic ligament** (lateral) (♂)	Lateral fascial thickening to the pubic symphysis, between the prostate, bladder, and pelvic wall		Tendinous arch of the pelvic fascia	
– – **lateral pubovesical ligament** (♀)	Fascial thickening that extends laterally to the pubic symphysis, between the bladder and the pelvic wall		Tendinous arch of the pelvic fascia	
– – **medial pubovesical ligament** (♀)	Fascial thickening that extends anteromedially between the bladder and the pubic symphysis			
– – **pubovesical ligament** (medial puboprostatic) (♂)	Fascial thickening between the anterior wall of the prostate, the bladder, and the pubic symphysis	Contains smooth muscle fibers (puboprostaticus)		
– – **superior vesical ligament** (false ligament)	Covers the median umbilical ligament (urachus)			
– – – lateral	Covers the medial umbilical ligament			
Pelvic fascia ☞ Visceral pelvic fascia; Parietal pelvic fascia; Extraperitoneal fascia				
Visceral pelvic fascia	Invests the pelvic organs	Particularly thick around the bladder and rectal ampulla		With the transversalis fascia
– **fascia of the individual organs**	Formed by the adventitia			
– **rectoprostatic fascia** (rectovesical septum or ligament) (♂)	Septum between the rectum and prostate or bladder (extending from the rectovesical pouch to the perineal body)	Contains smooth muscle fibers	Perineal body	
– **rectovaginal fascia** (rectovaginal septum) (♀)	Septum between the rectum and the vagina (extending from the rectouterine pouch to the perineal body)	Contains smooth muscle fibers	Perineal body	
Medial umbilical ligament (umbilical artery, occluded part)	From the umbilical scar (under the peritoneum) to the lateral surface of the bladder	Continues with the patent tract of the umbilical artery		
Median umbilical ligament (urachus)	From the apex of the bladder to the umbilical region (running along the posterior surface of the anterior abdominal wall)	Fibrous vestige of the urachus		

continued

Fasciae and fascial structures of the trunk				*(continued)*
STRUCTURE	***POSITION***	***ORGANIZATION***	***INSERTION***	***CONTINUATION***
		PERINEUM		
Perineal body	Layer of elastic connective tissue between the rectum and vagina (♀), and the rectum and urethra (♂)	Formed by the superior and inferior fasciae of the pelvic diaphragm and by the tendons of the levator ani, the deep transverse perineal muscle, the bulbospongiosus, and the external anal sphincter, with smooth muscle fibers originating from the bladder and rectum, connected to the capsule of the prostate (♂) and to the vagina (♀)		
Inferior fascia of the pelvic diaphragm	Investing fascia of the levator ani and ischiococcygeus	Invests the inferior surface of the pelvic diaphragm	Ischiopubic rami	
Perineal fascia (superficial investing fascia of the perineum or deep perineal fascia)	Inferior boundary of the superficial perineal pouch	Covers the superficial transverse perineal muscle, the ischiocavernosus, and the bulbospongiosus. Merges anteriorly with the suspensory ligament of the penis or clitoris	Ischiopubic rami	With the deep investing abdominal fascia
Anococcygeal body	Band of connective tissue between the anus and the coccyx	Contains smooth muscle fibers		
Transverse perineal ligament	Thickened anterior border of the perineal membrane	Fibrous expansion of the anterior border of the deep transverse perineal muscle		
Perineal membrane	Underlying fascia of the deep transverse perineal muscle	Forms a triangle with a forward apex, toward the triangular ligament of the pubis, and a backward base toward the rectum	Ischiopubic rami and perineal body	
Perineal raphe	From the anus, extending forward along the perineum	In continuity with the raphe of the scrotum (♂)		

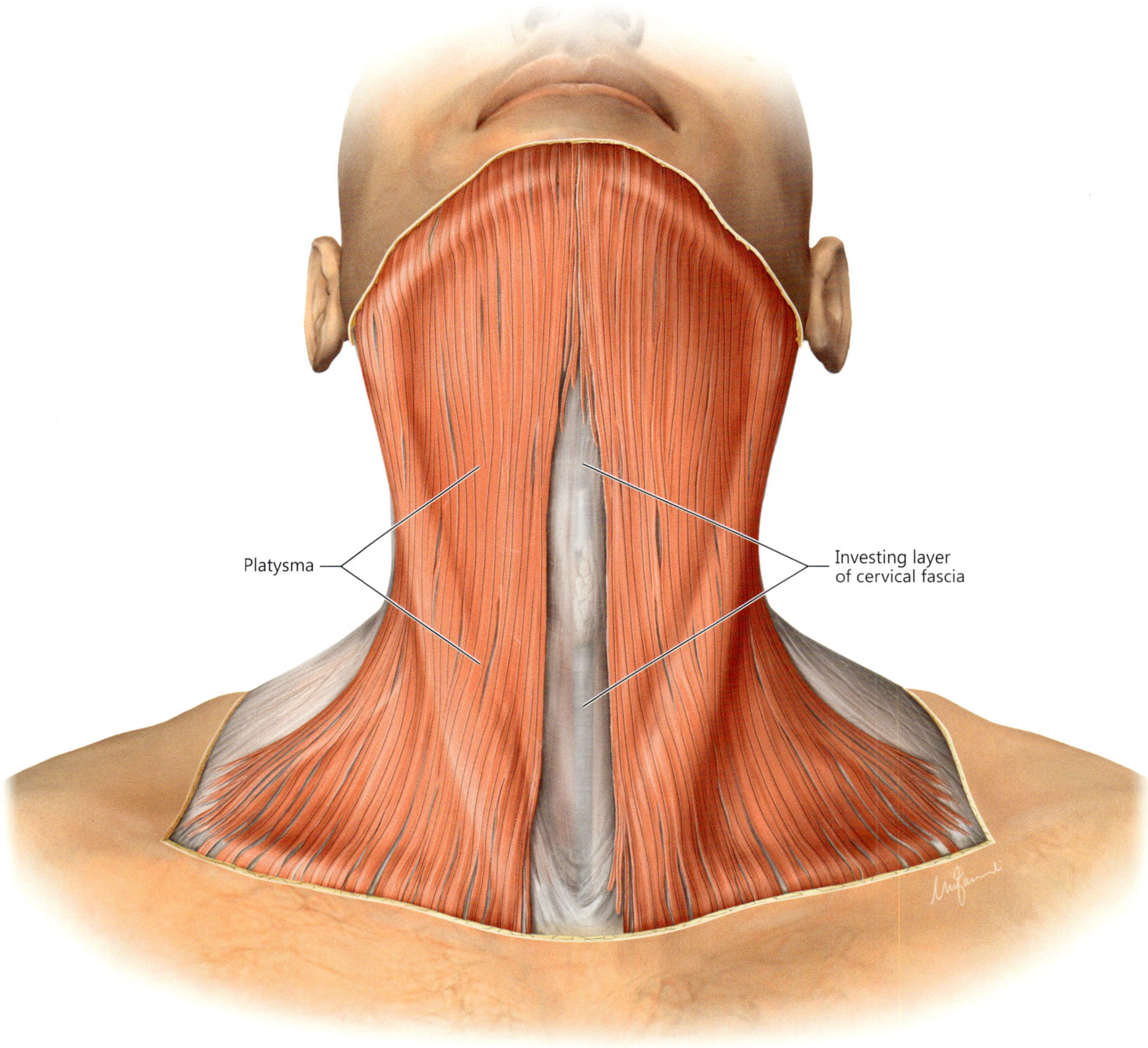

Platysma. Lateral muscle of the neck in the thickness of the subcutaneous tissue, superficially to the investing layer of cervical fascia.

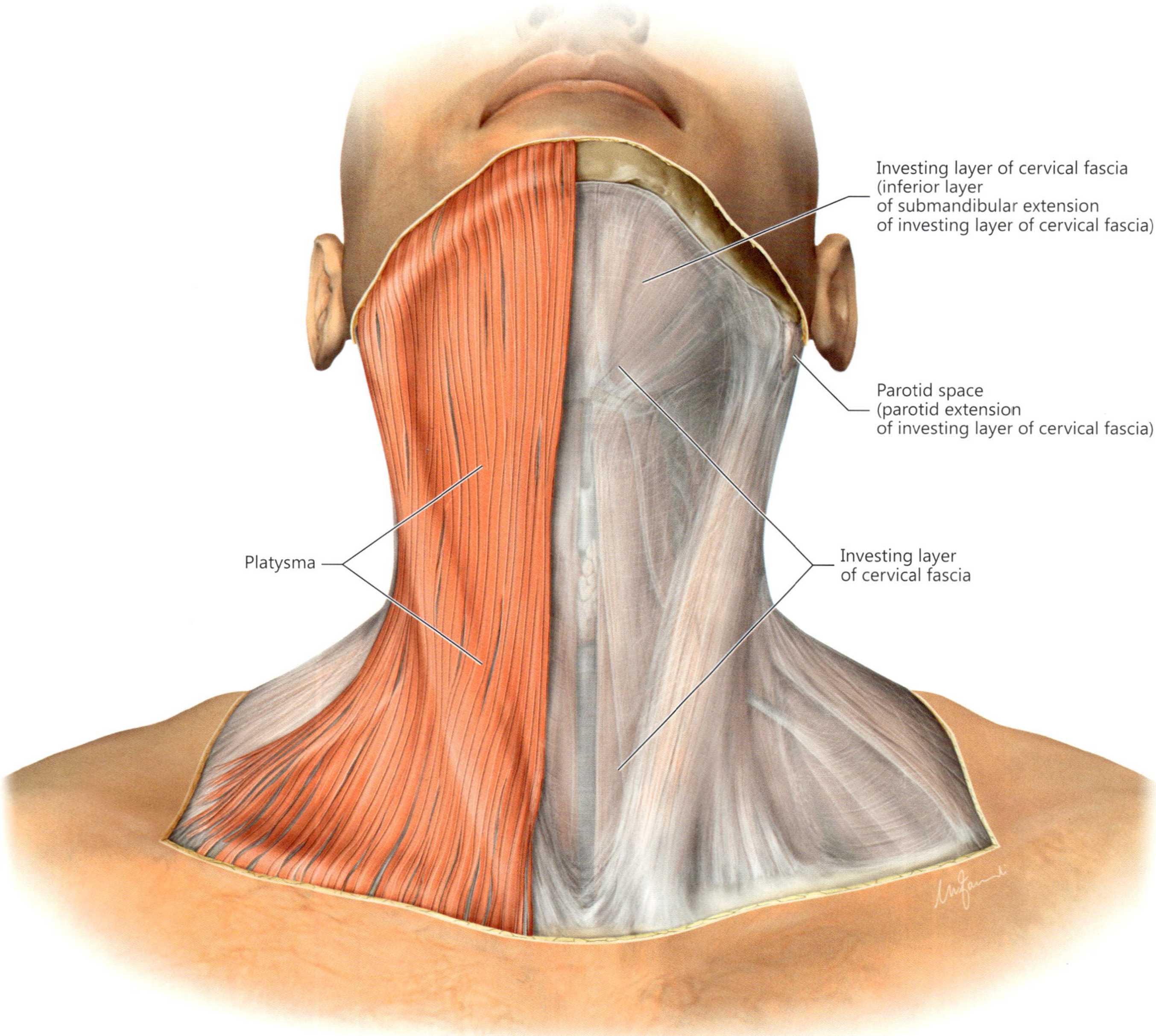

The **investing layer of cervical fascia**, which covers the entire neck, is located just below the subcutaneous tissue, or hypodermis, and platysma. It continues superiorly with the masseteric and parotid fasciae (anterolateral surface of the head) and with the nuchal fascia (posterior surface of the head); it continues inferiorly with the pectoral and deltoid fasciae (anteriorly) and the thoracolumbar fascia (posteriorly).

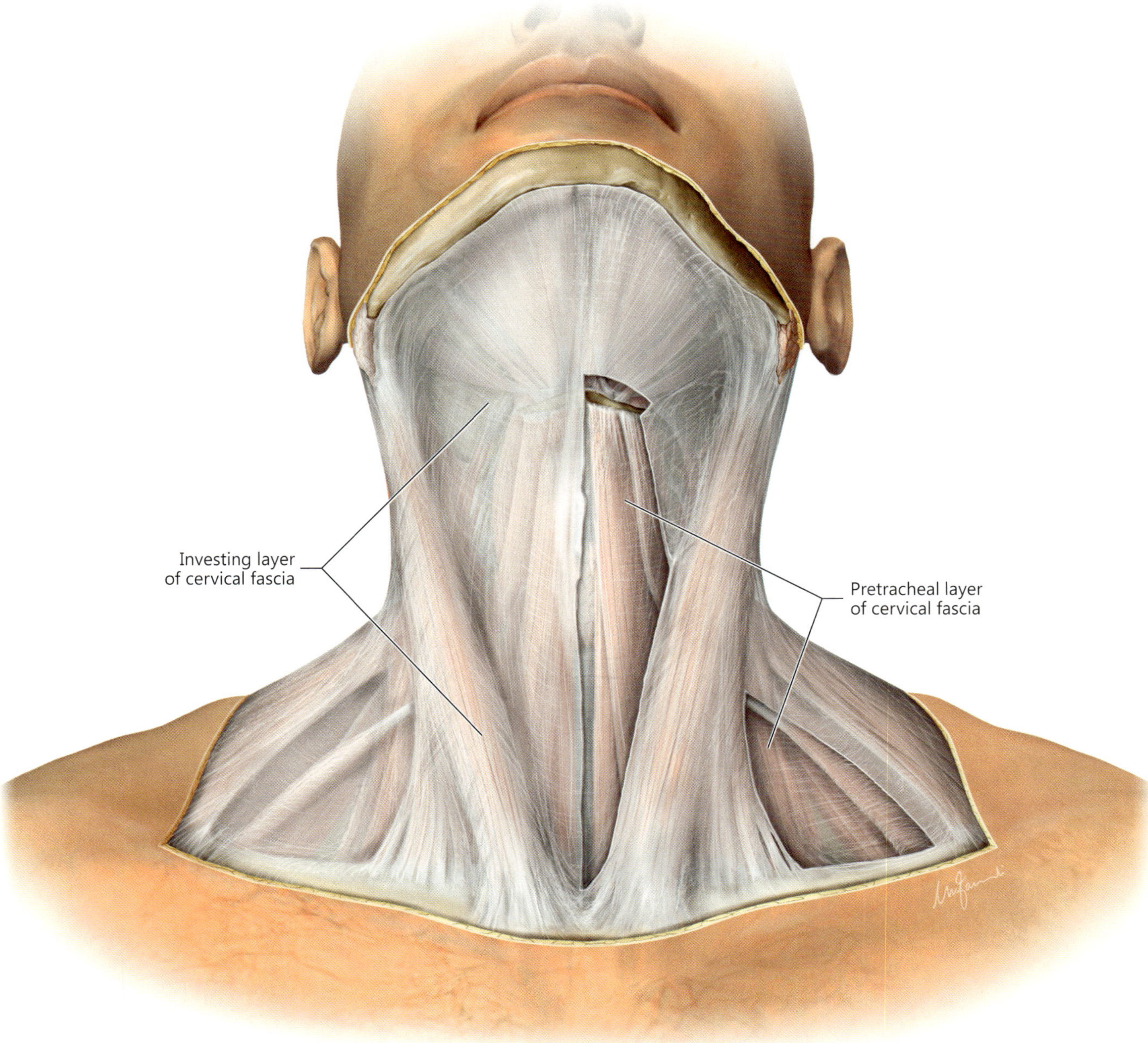

The **pretracheal layer of cervical fascia**, deeply located in relation to the investing layer of cervical fascia, has a trapezoidal shape with the smaller base at the top, at the level of hyoid bone, and the larger base at the bottom, at the superior opening of the thorax.

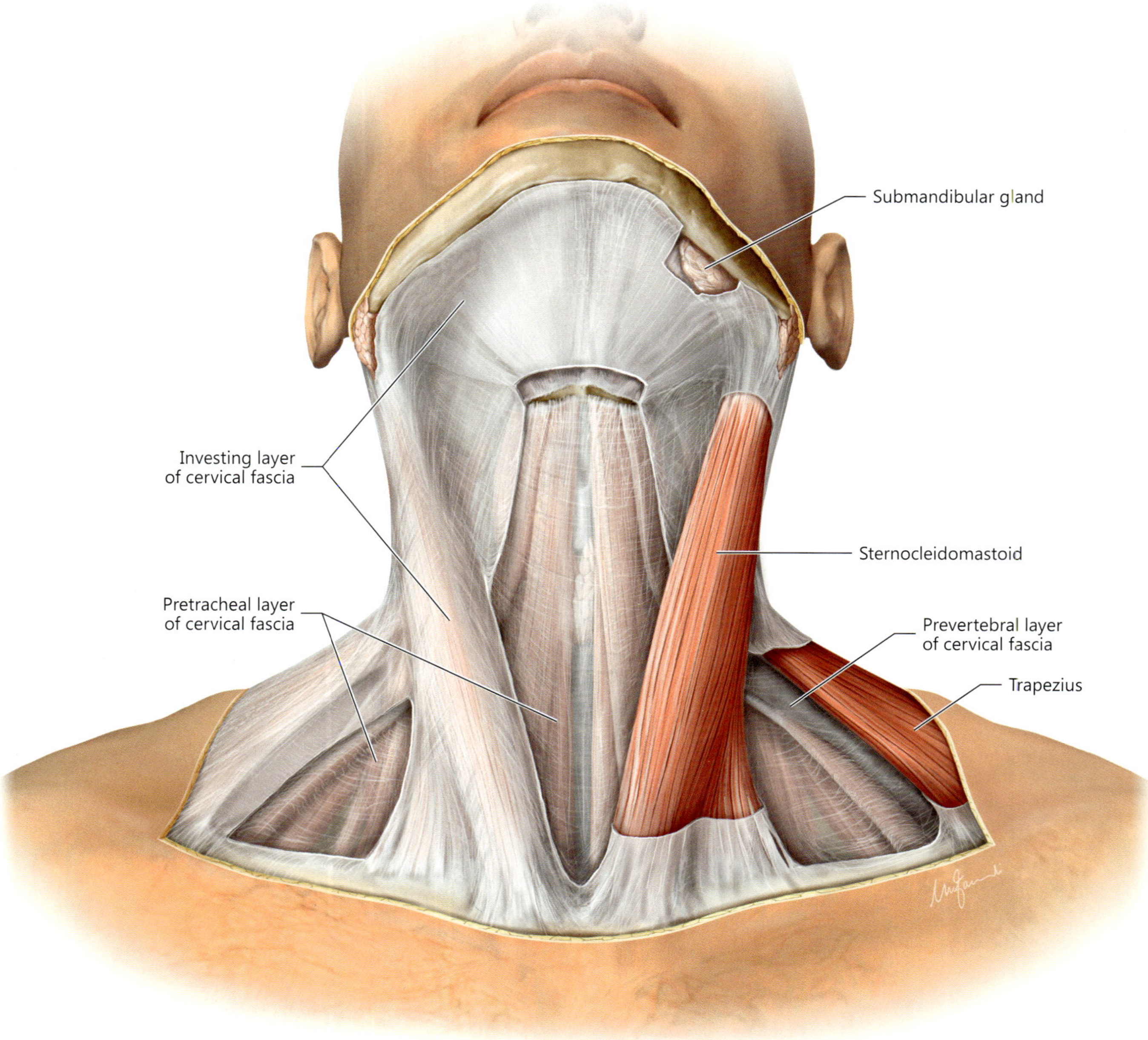

The **prevertebral layer of cervical fascia**, which extends from the cranial base to the first thoracic vertebrae, is located anteriorly to the vertebral bodies and the prevertebral and scalene muscles.

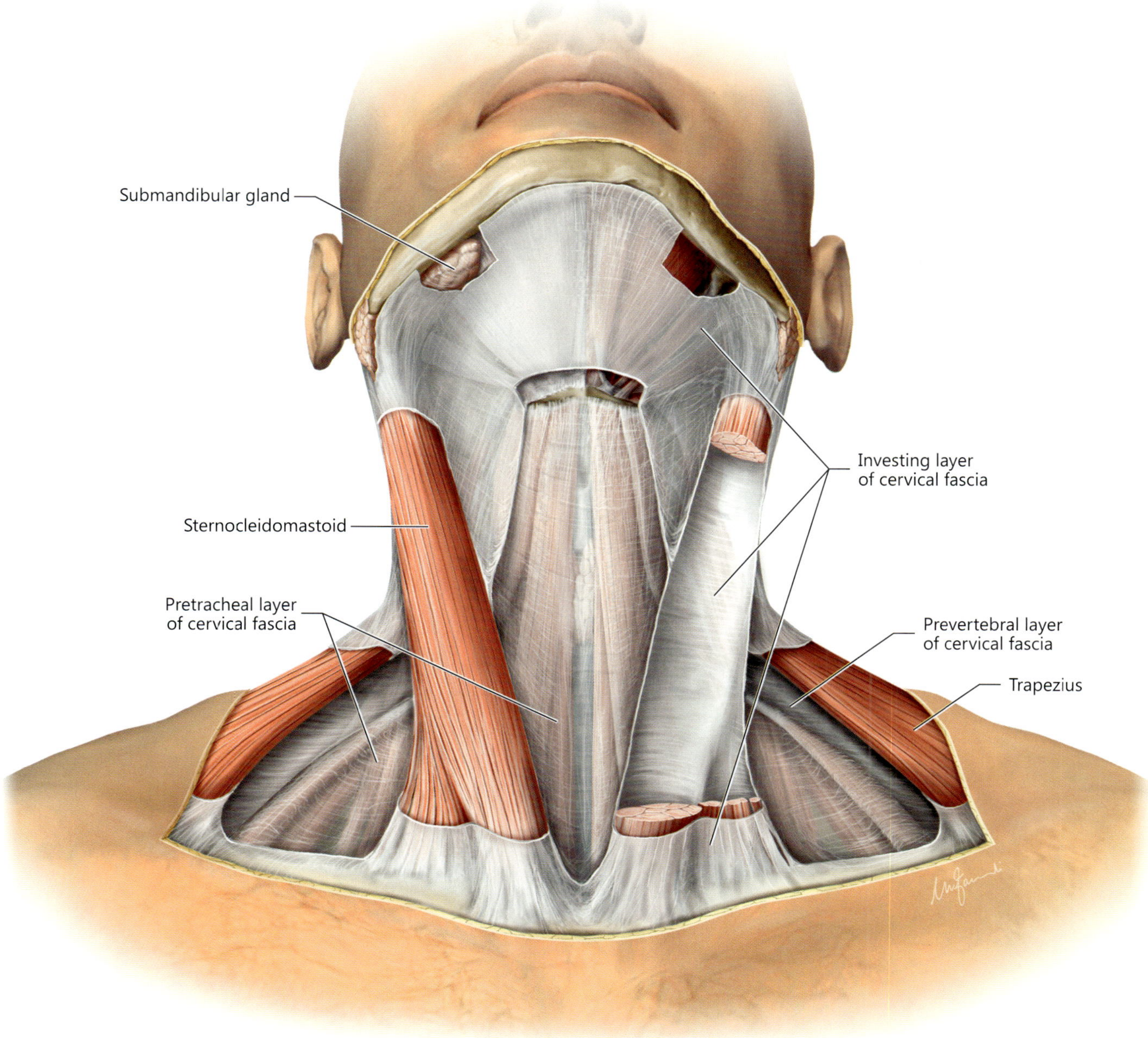

The **sternocleidomastoid** is enveloped in a splitting of the investing layer of cervical fascia, which also surrounds the trapezius.

Suprahyoid and Infrahyoid Muscles, and Cervical Fascia

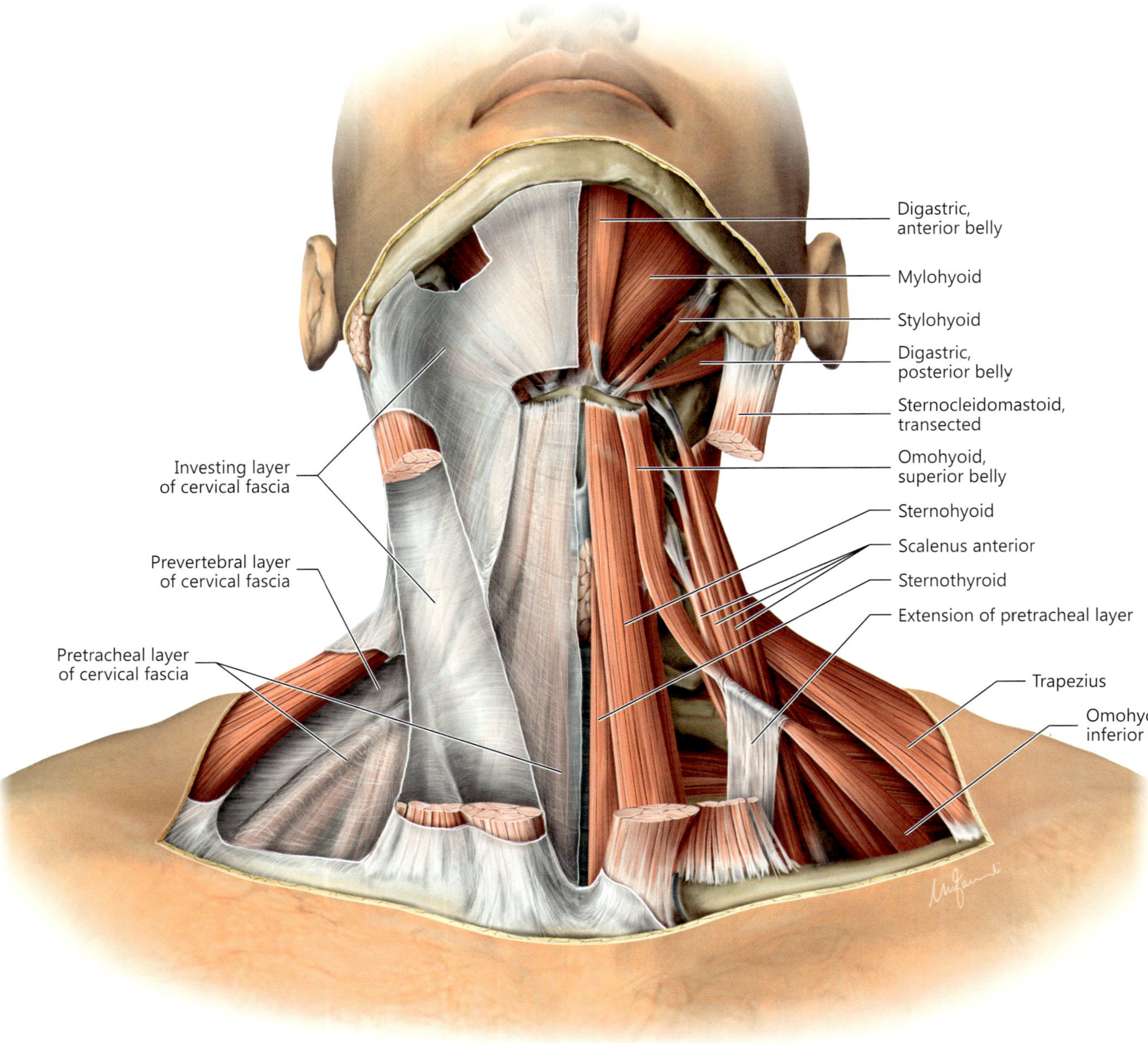

Suprahyoid and **infrahyoid** muscles. A split of the pretracheal layer of cervical fascia surrounds the infrahyoid muscles and ends laterally folding around the omohyoid. Its posterior extension (extension of pretracheal layer) moves from the omohyoid to the prevertebral layer of cervical fascia and merges with it.

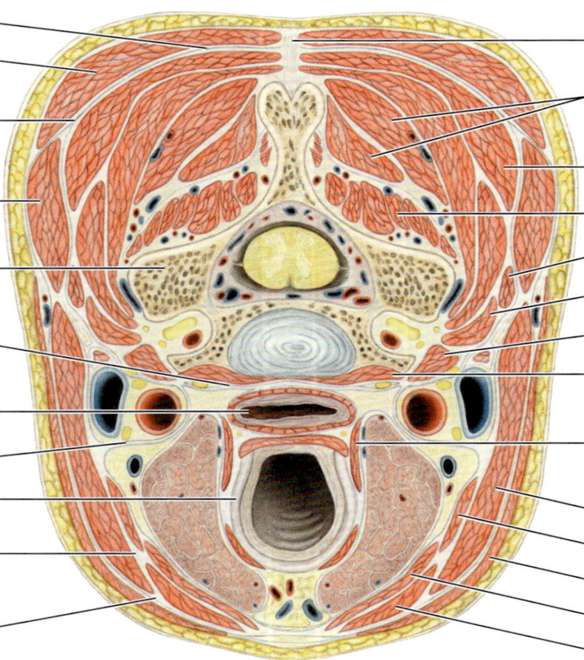

Fasciae of the neck. Transversal section of the neck at the level of the sixth cervical vertebra, schematic representation of their organization.

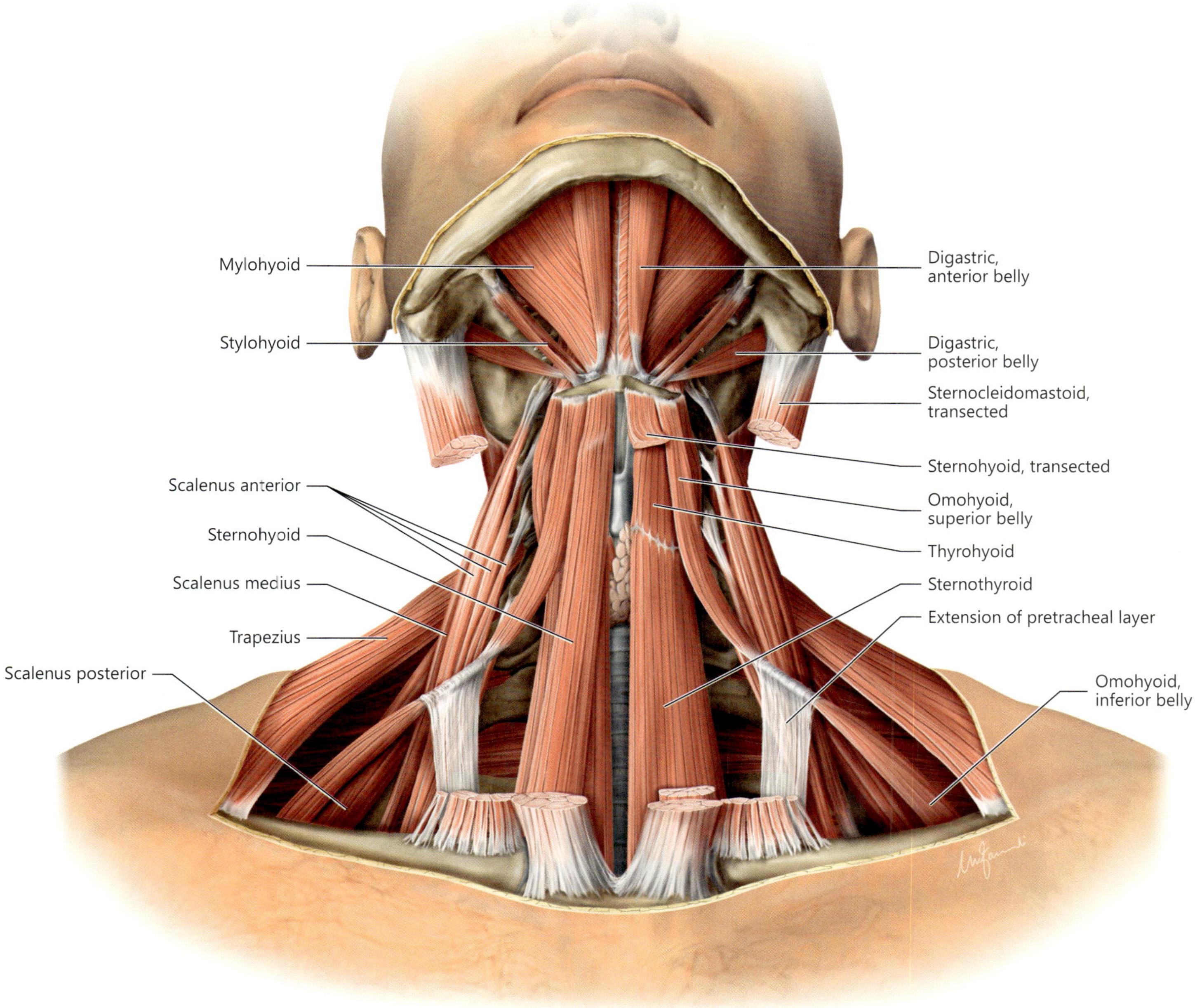

Preparation of the **muscles of the neck**, anterior projection. The sternocleidomastoid muscles on both sides have been transected and mostly removed. On the left, the sternohyoid has been transected and largely removed.

Suprahyoid, Infrahyoid, Scalene, and Prevertebral Muscles

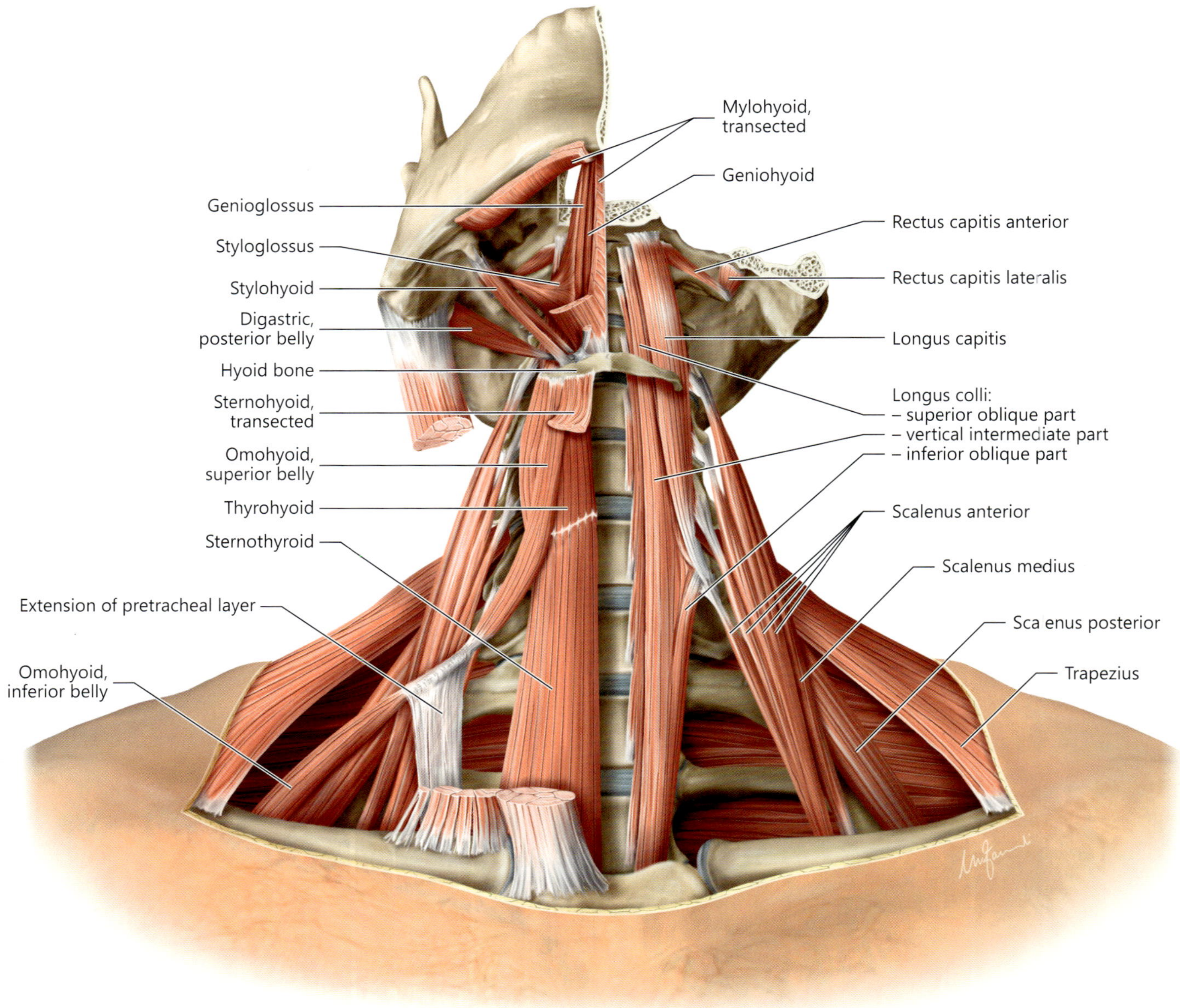

The three **scalene muscles** (**scalenus anterior**, **medius** and **posterior**) are located deeply in relation to the sternocleidomastoid, laterally in relation to the cervical column, and they connect the cervical vertebrae to the first two ribs.

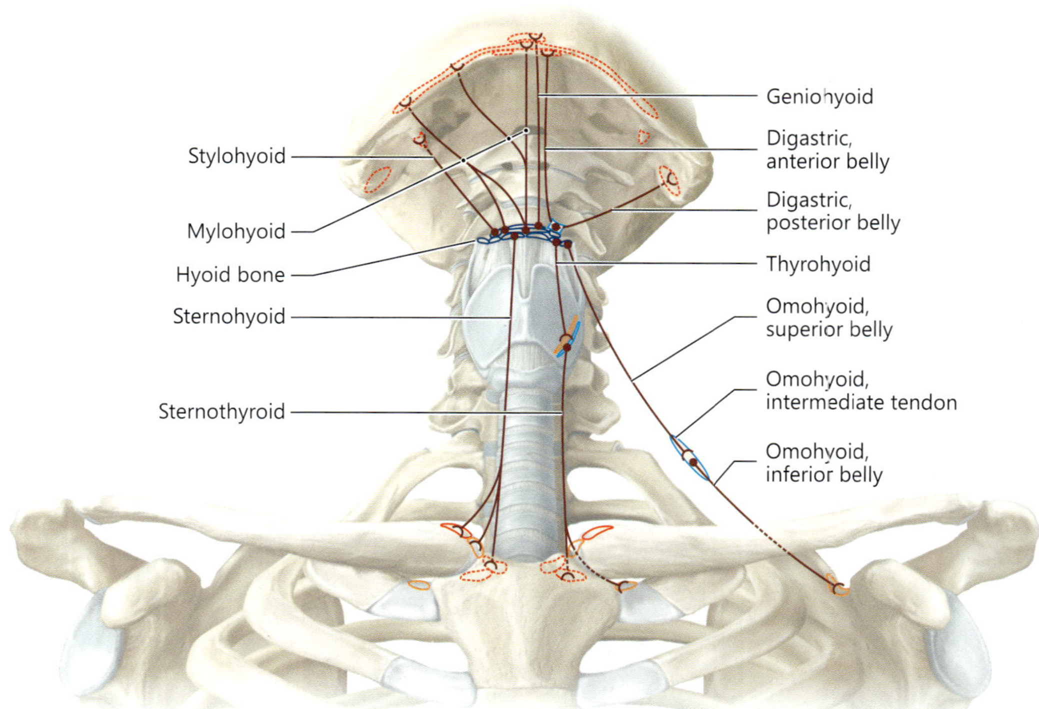

ORIGIN

- Bones
- Cartilage, joint capsule

Origin ⟶ Insertion

INSERTION

- Bones
- Cartilage, tendon

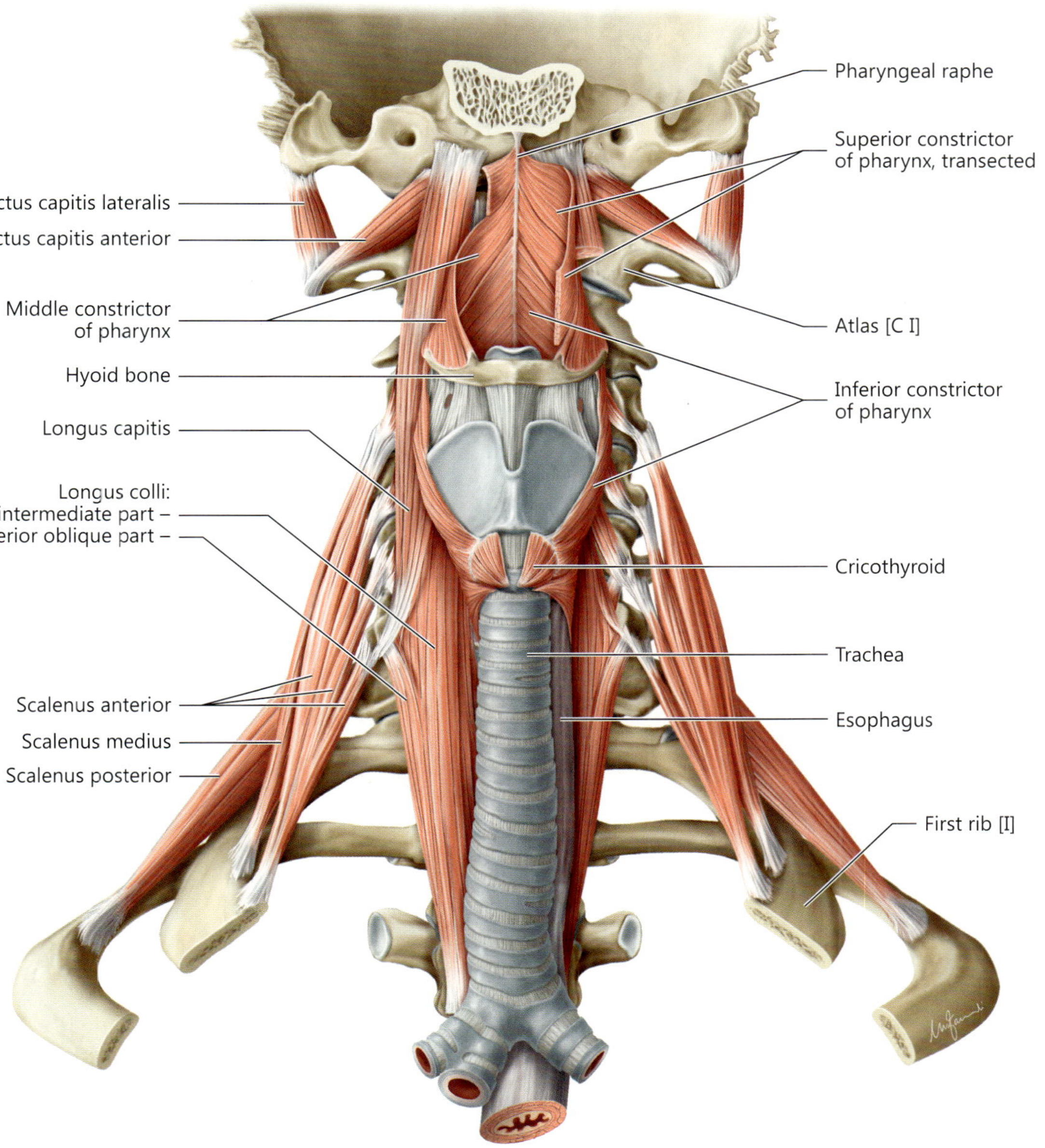

Preparation of the **prevertebral muscles**, anterior view. These are long muscles that are adjacent to the ventrolateral surface of the vertebral column, covered by the prevertebral layer of cervical fascia; at the side of the prevertebral muscles, the anterior, middle, and posterior scalene muscles have been prepared. The prevertebral muscles extend between the cranial base and the cervical column, while the scalene muscles extend from the vertebral column to the upper part of the thorax.

➤

Schematic representation of the origin and insertion of the **suprahyoid** and **infrahyoid muscles**. The suprahyoid muscles include the **digastric**, **stylohyoid**, **mylohyoid** and **geniohyoid**. The anterior belly of the digastric originates from the digastric fossa of the body of the mandible and, extending inferiorly and posteriorly, continues into the intermediate tendon, attached to the lateral side of the body of the hyoid bone. From the intermediate tendon, the posterior belly moves superiorly and posteriorly, inserting on the mastoid notch of the temporal bone. The stylohyoid, originating from the styloid process of the temporal bone, inserts on the lateral extremity of the body of the hyoid bone. The mylohyoid, which originates from the internal surface of the mandible, between the mental spines (superior and inferior) and the mylohyoid line, inserts on the anterior surface of the body of the hyoid bone. The geniohyoid, which originates from the mental spines, attaches to the anterior surface of the hyoid bone. The infrahyoid muscles include the **sternohyoid**, **omohyoid**, **sternothyroid**, and **thyrohyoid**. The sternohyoid originates from the internal surface of the manubrium of the sternum, the sternoclavicular joint capsule and the medial end of the clavicle, and inserts on the inferior border or the hyoid bone. The inferior belly of the omohyoid, from its origin at the suprascapular notch, extends superiorly and medially reaching the intermediate tendon. From the intermediate tendon, the superior belly, also moving superiorly and medially, inserts on the inferior border of the hyoid bone. The sternothyroid originates from the posterior surface of the sternum and from the first costal cartilage and inserts on the oblique line of the thyroid cartilage, where the thyrohyoid originates, which inserts on the inferior border of the body of the hyoid bone.

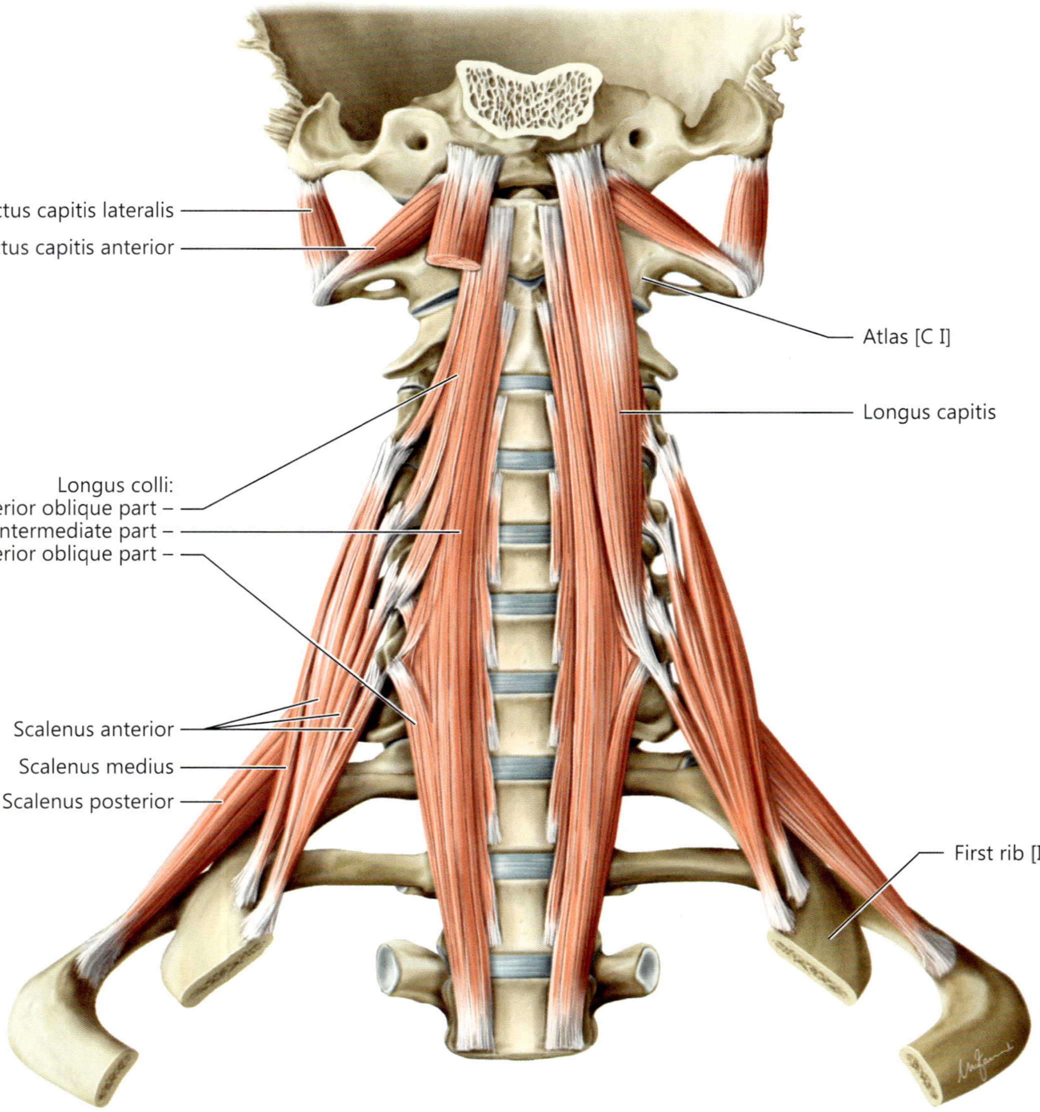

The **prevertebral muscles** (rectus capitis anterior, longus capitis, and longus colli) are located deep within the neck, in front of the cervical vertebral column, and they are covered by the prevertebral layer of cervical fascia, as are the scalene muscles.

ORIGIN

Bones

Origin → Insertion

INSERTION

Bones

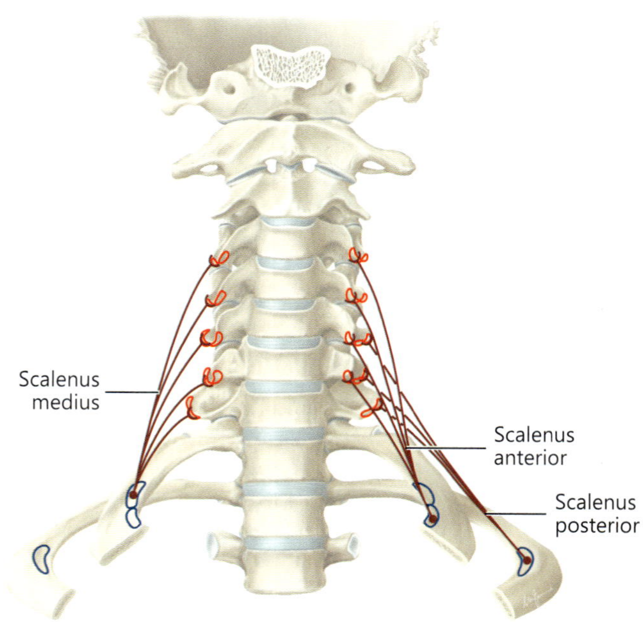

Schematic representation of the origin and insertion of the **scalene muscles**. The scalene muscles originate from the cervical vertebrae and insert on the first rib (scalenus anterior and medius) and on the second rib (scalenus posterior).

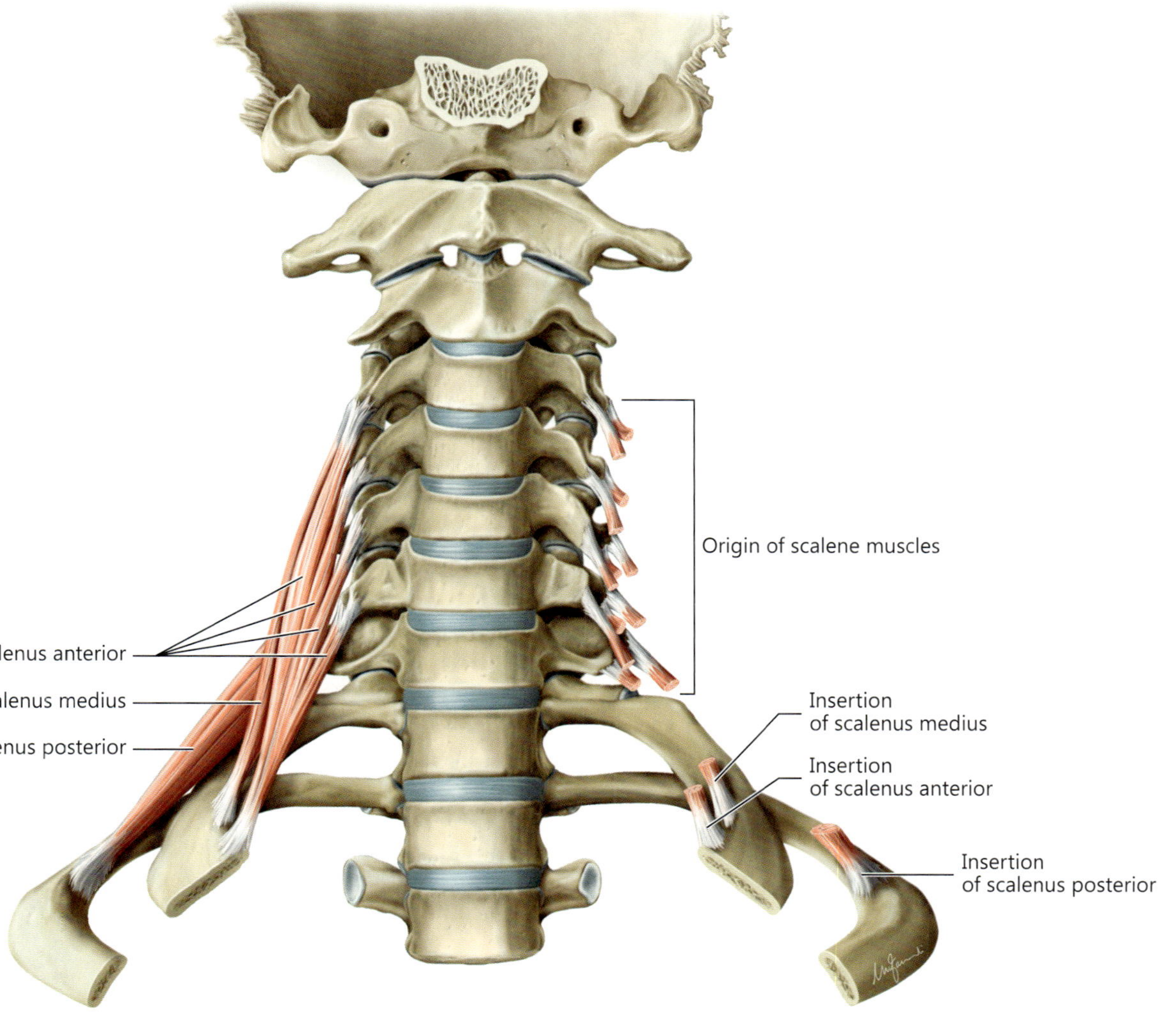

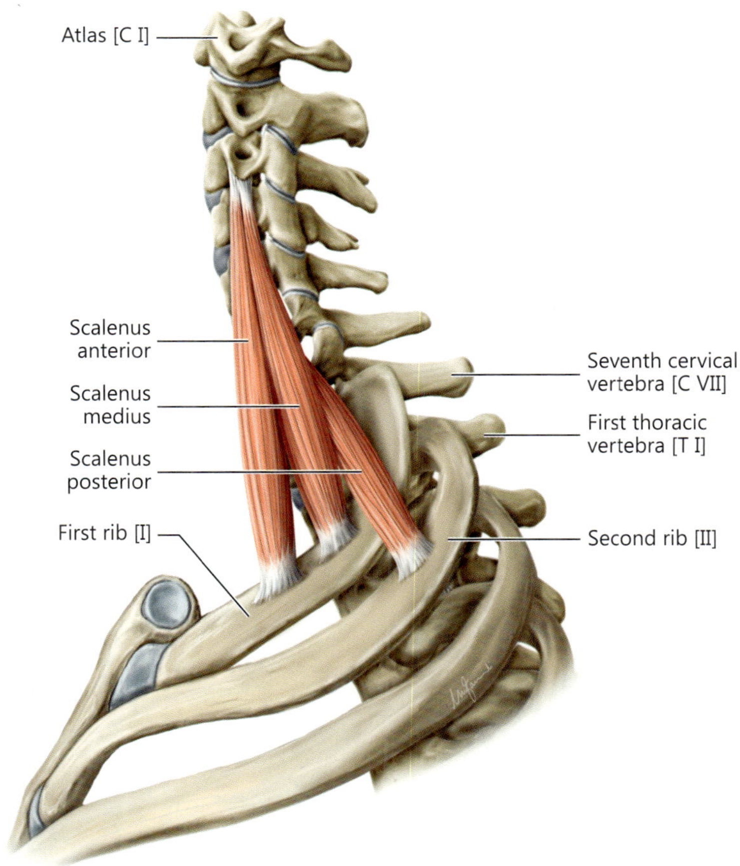

Scalene muscles. In the image above the muscles on the left have been resected to show the insertions on the cervical vertebrae and on the ribs. At the bottom of the preceding page the origins and insertions are schematically represented. At the bottom of this page, left lateral view.

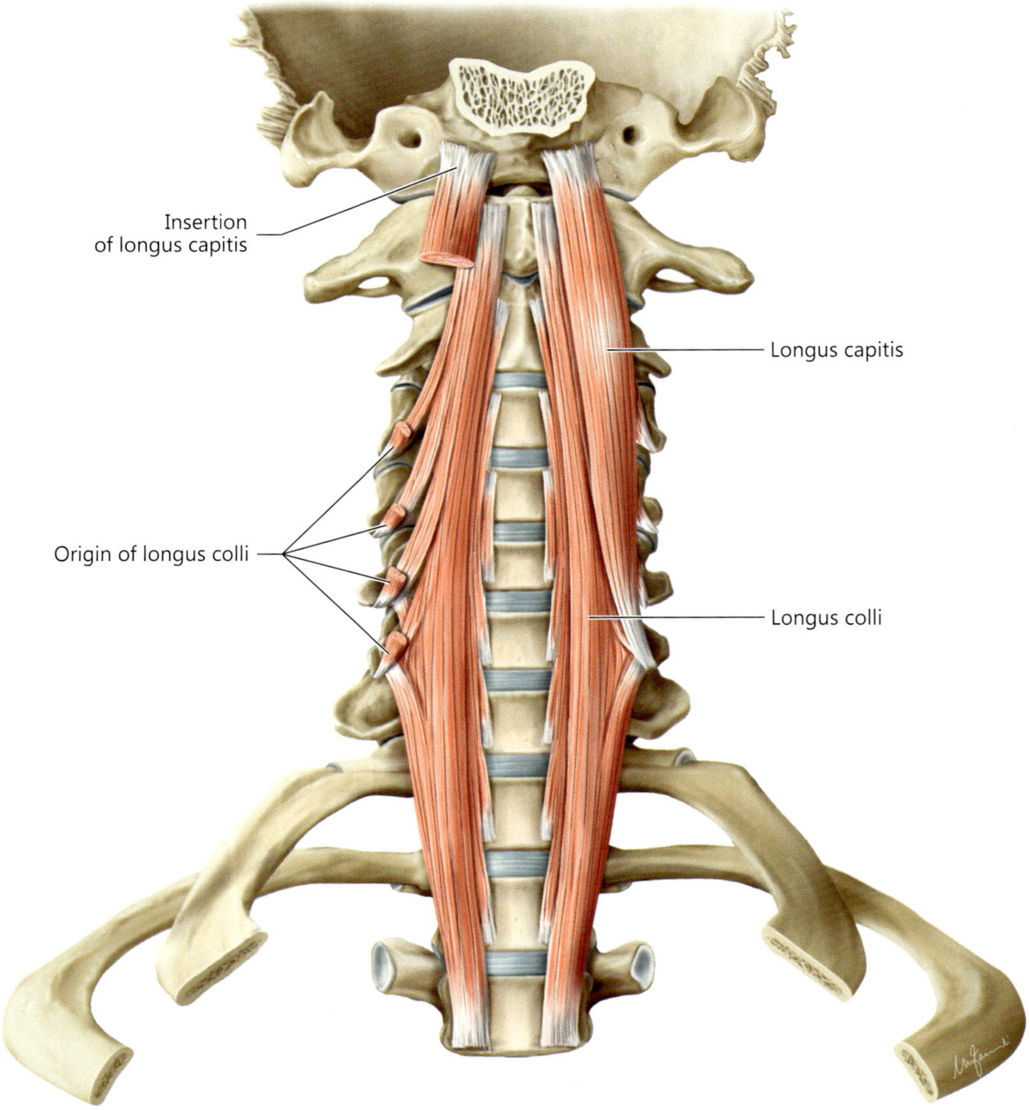

Longus colli and **longus capitis**. The longus colli, on the right side, has been resected to show its origin on the cervical vertebrae.

ORIGIN
Bones

Origin Insertion

INSERTION
Bones

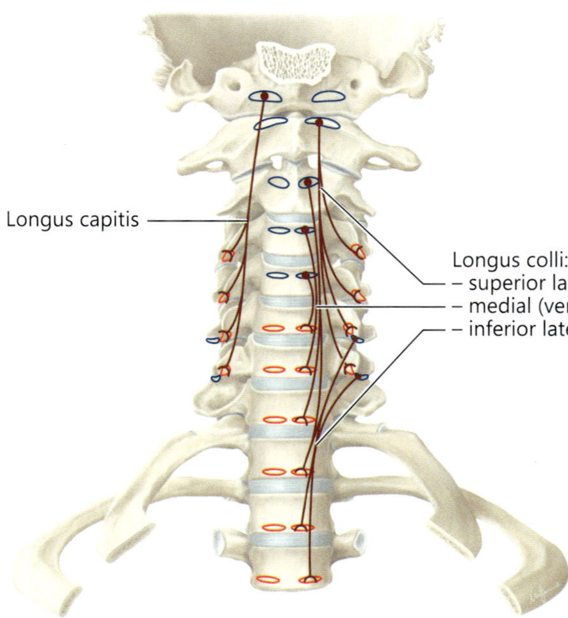

Schematic representation of the origin and insertion of the **longus colli** and **longus capitis**. The longus colli joins the first three thoracic vertebrae and the last three cervical vertebrae to the first cervical vertebrae; the longus capitis connects the occipital bone to the cervical vertebrae.

Infrahyoid muscles, anterior view. The sternohyoid and omohyoid are located on a more superficial plane, while the sternothyroid and thyrohyoid are located more deeply.

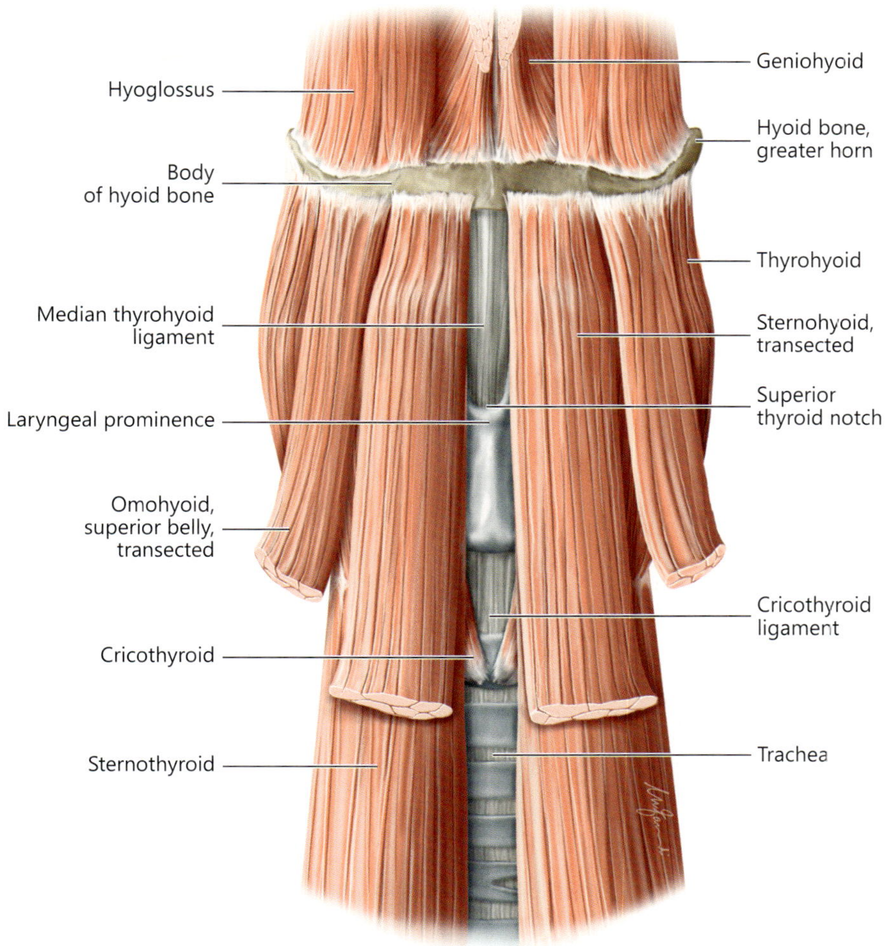

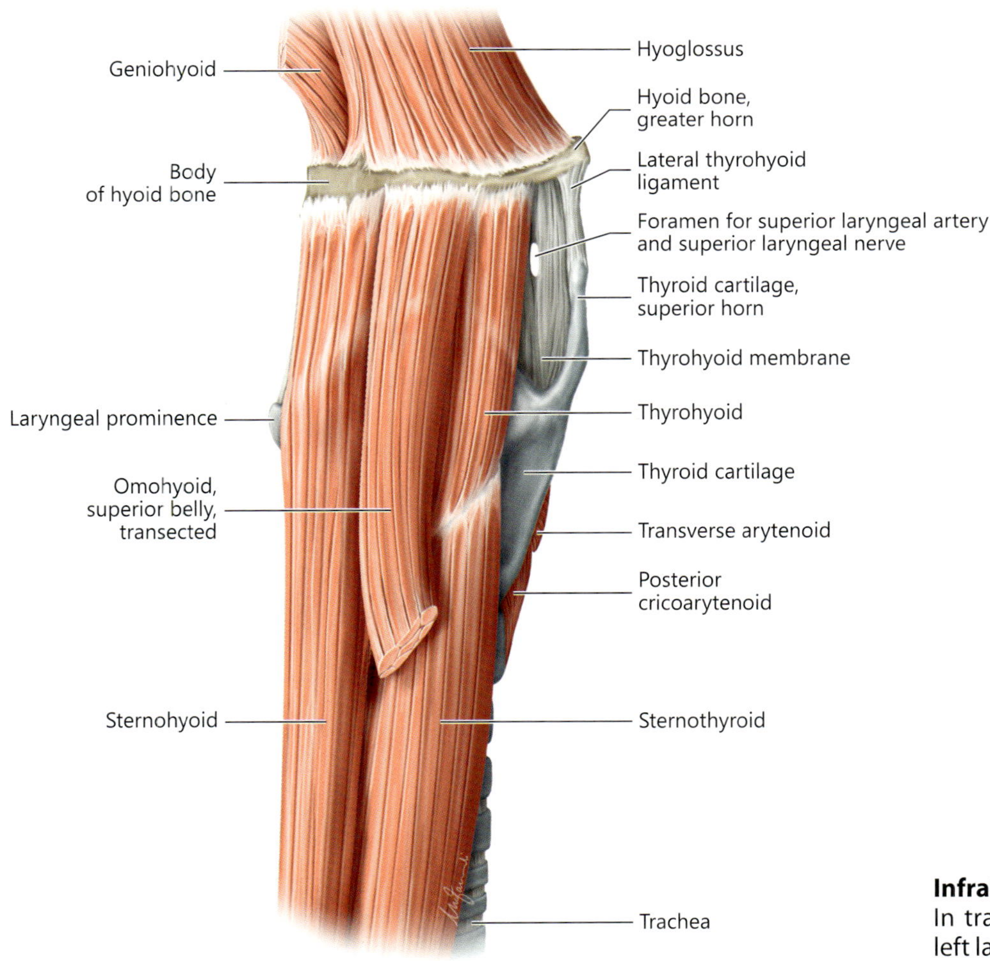

Infrahyoid muscles, lateral view. In transparency, you can see the left lamina of the thyroid cartilage.

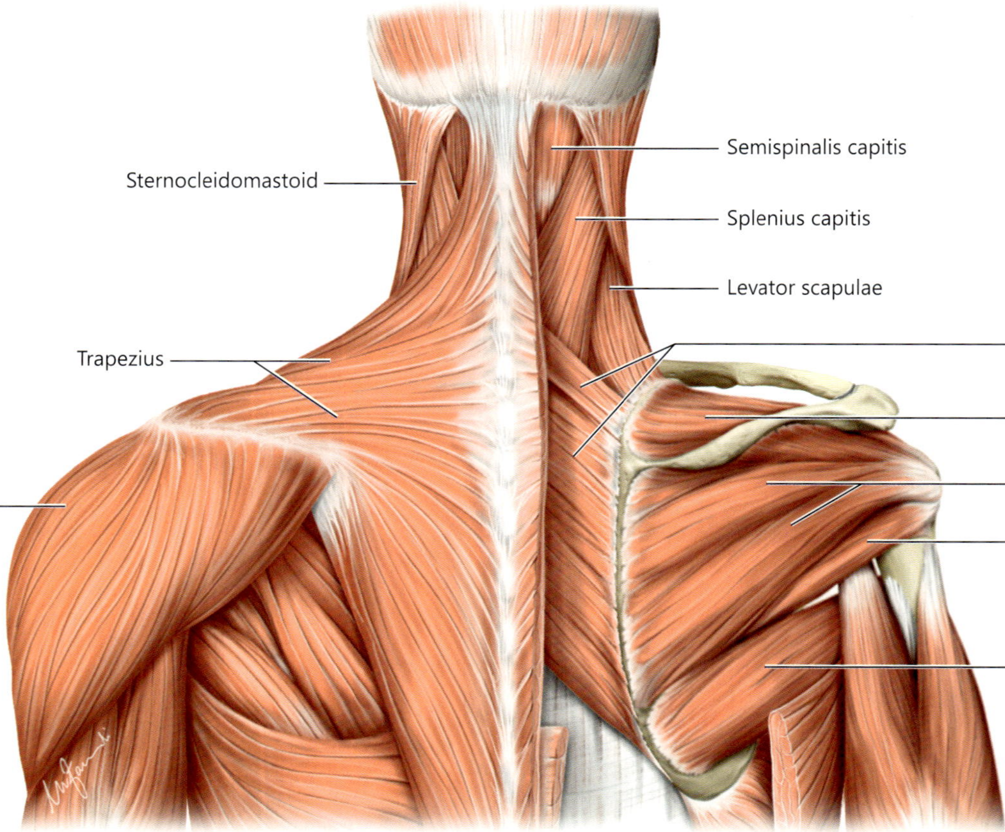

The **muscles of the back** are arranged in three layers. On the superficial plane are the spinoappendicular muscles (of which the trapezius, rhomboid major and minor, and levator scapulae are represented here) which, originating from the vertebral column, insert on the upper limb.

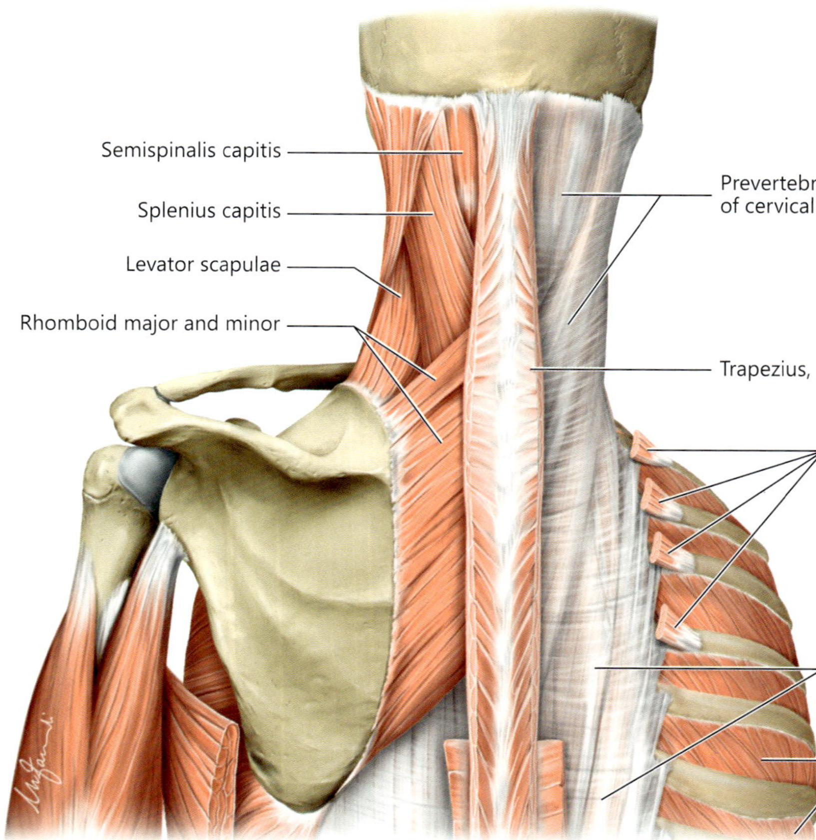

The **prevertebral layer of cervical fascia** originates from the basilar part of the occipital bone and is fixed laterally to the anterior tubercles, in the cervical vertebrae, where it continues with a deep extension of the investing layer of cervical fascia.

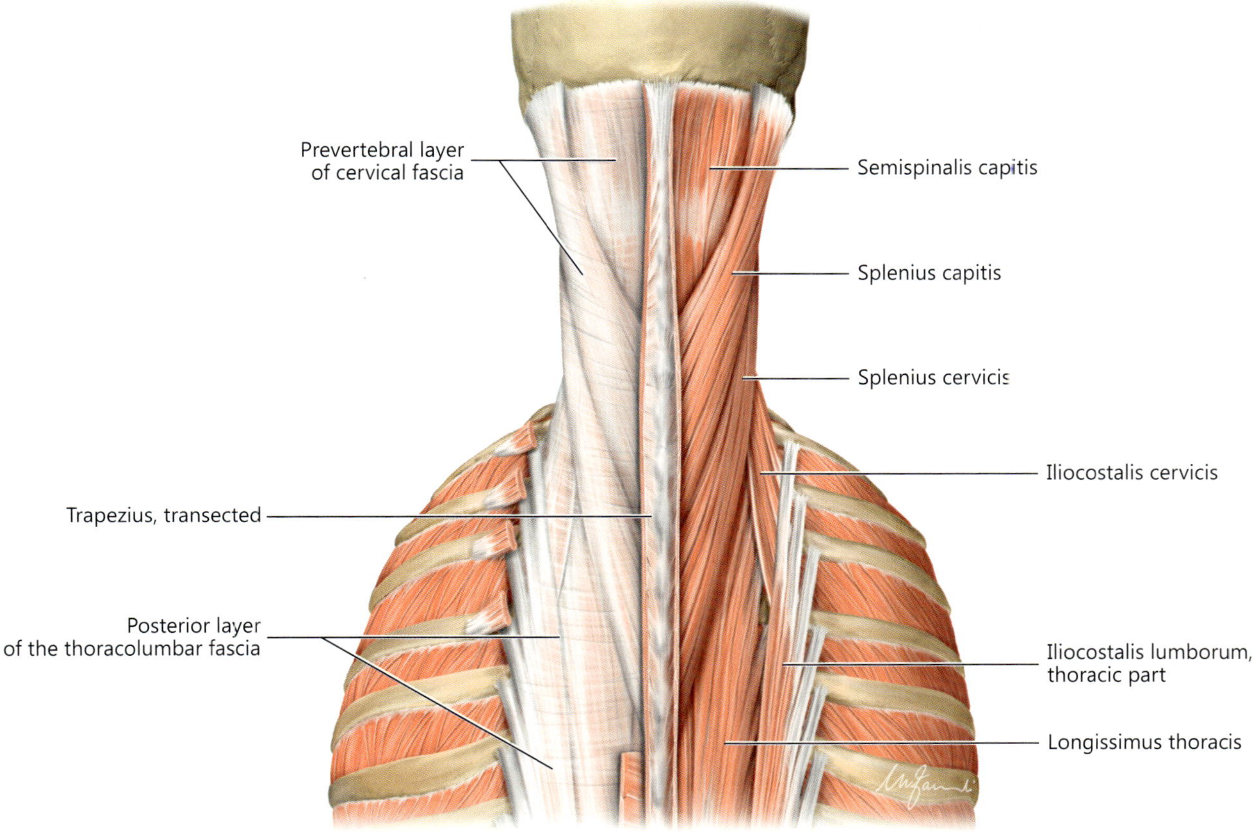

Spinotransversales (splenius capitis and splenius cervicis), **erector spinae** (of which the iliocostalis cervicis is represented here), and the **transversospinales** (of which the semispinalis capitis is represented here).

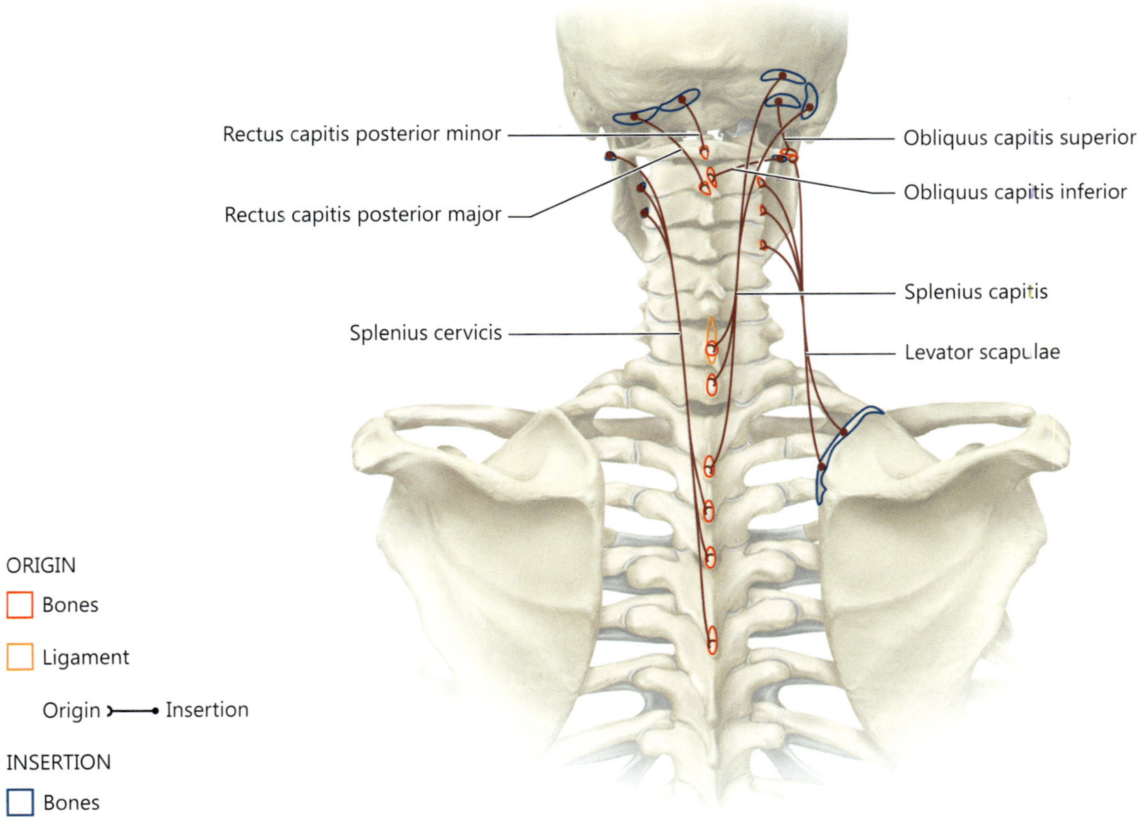

Schematic representation of the origin and insertion of the **suboccipital muscles**, the **splenius capitis** and the **splenius cervicis**, as well as the **levator scapulae**.

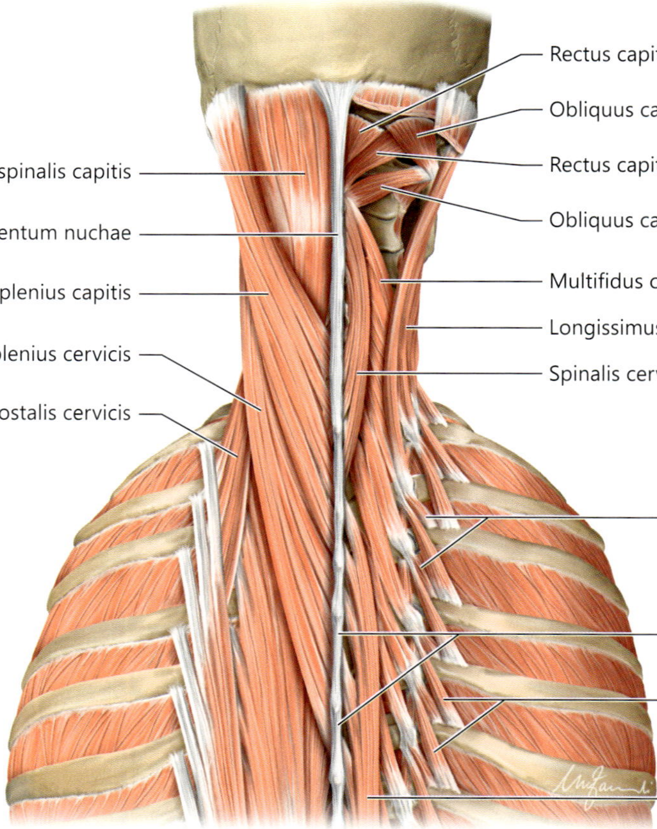

On the right, **suboccipital muscles**, **transversospinales** (multifidus cervicis), and **erector spinae** (longissimus capitis and spinalis cervicis); on the left, on a more superficial plane, **transversospinales** (semispinalis capitis), **spinotransversales** (splenius capitis and splenius cervicis), and **erector spinae** (iliocostalis cervicis).

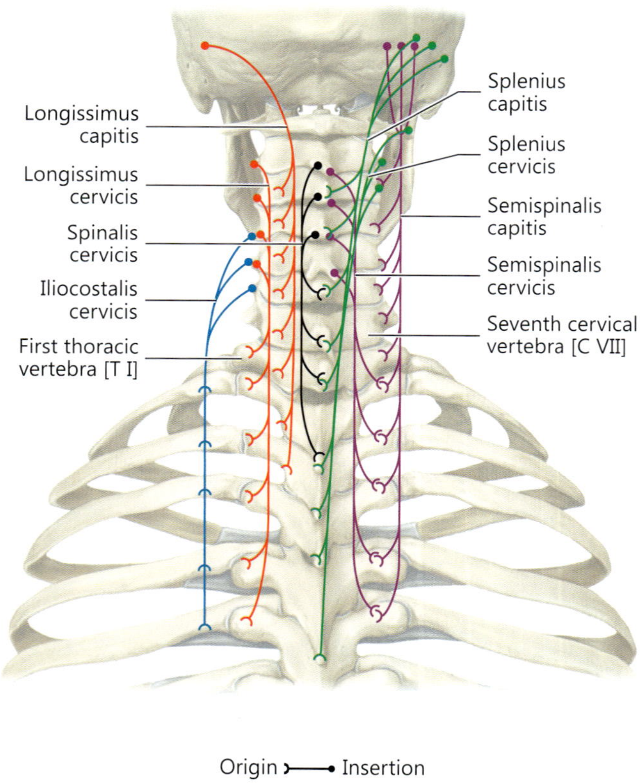

Schematic representation of the origins and insertions of the **spinotransversales** (splenius capitis and splenius cervicis), **erector spinae** (for the iliocostal parts of the neck, **longissimus capitis and longissimus cervicis**, **spinalis cervicis**), and **transversospinales** (for the semispinalis capitis and semispinalis cervicis).

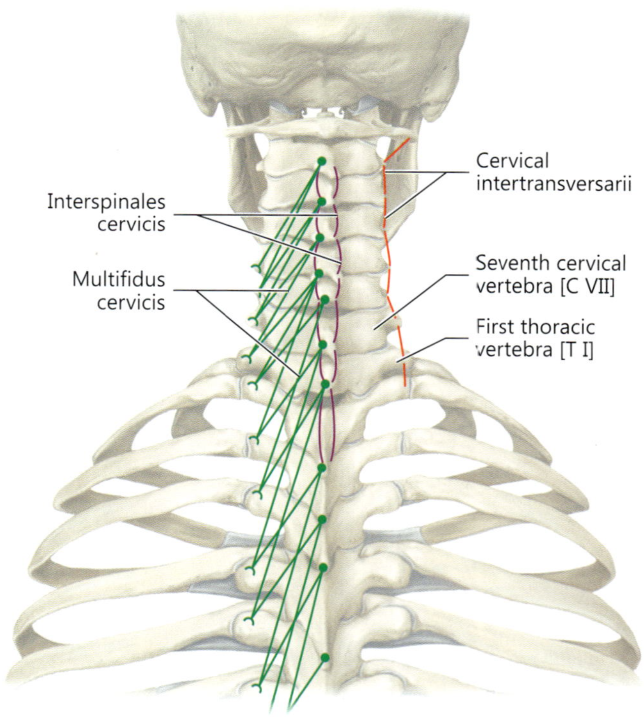

Schematic representation of the origins and insertions of the **interspinales cervicis** and of the **cervical intertransversarii** as well as the **multifidus cervicis**. The interspinales and intertransversarii are small fascicles tense between the spinous processes and between the transverse processes of contiguous vertebrae, respectively. The multifidus is formed by small fascicles stretched between and overlying the transverse processes and spinous processes of contiguous vertebrae. In the case of these muscles there is no distinction between origin and insertion.

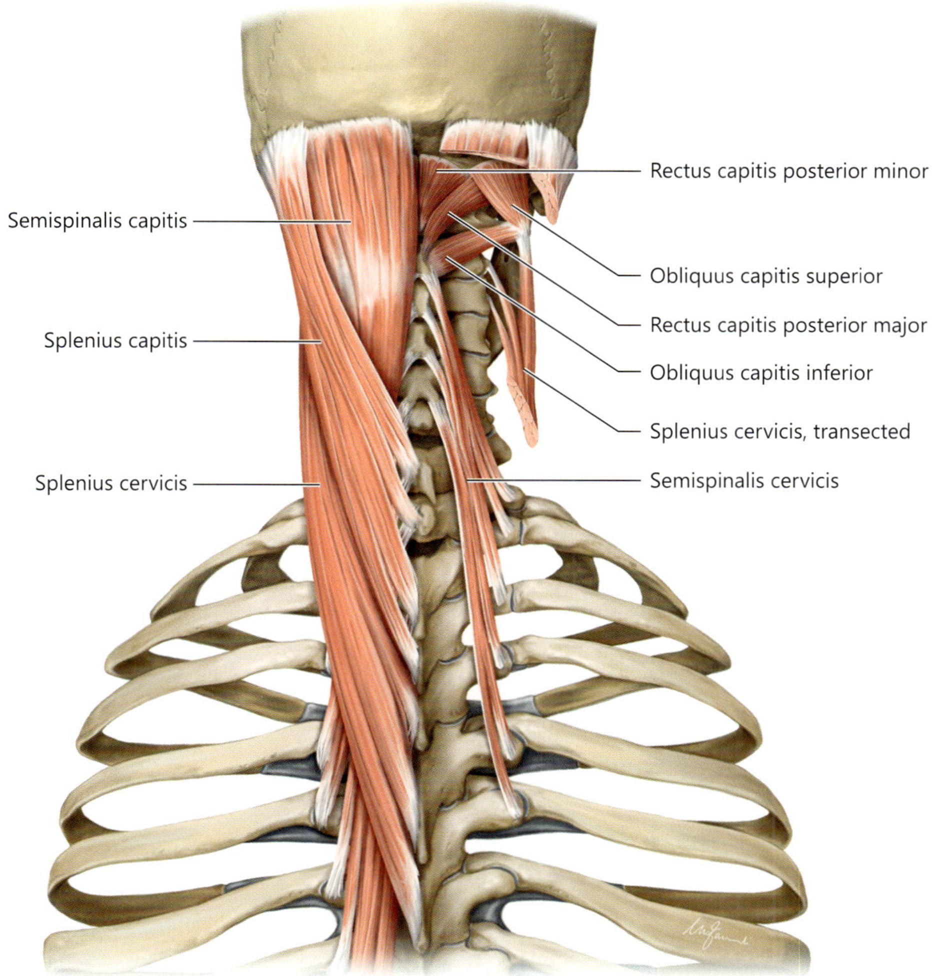

Splenius capitis and **splenius cervicis** and **semispinalis capitis** on the left; on the right, the muscles have been transected to show the deepest layer of the underlying muscles (**suboccipital muscles** and **semispinalis cervicis**).

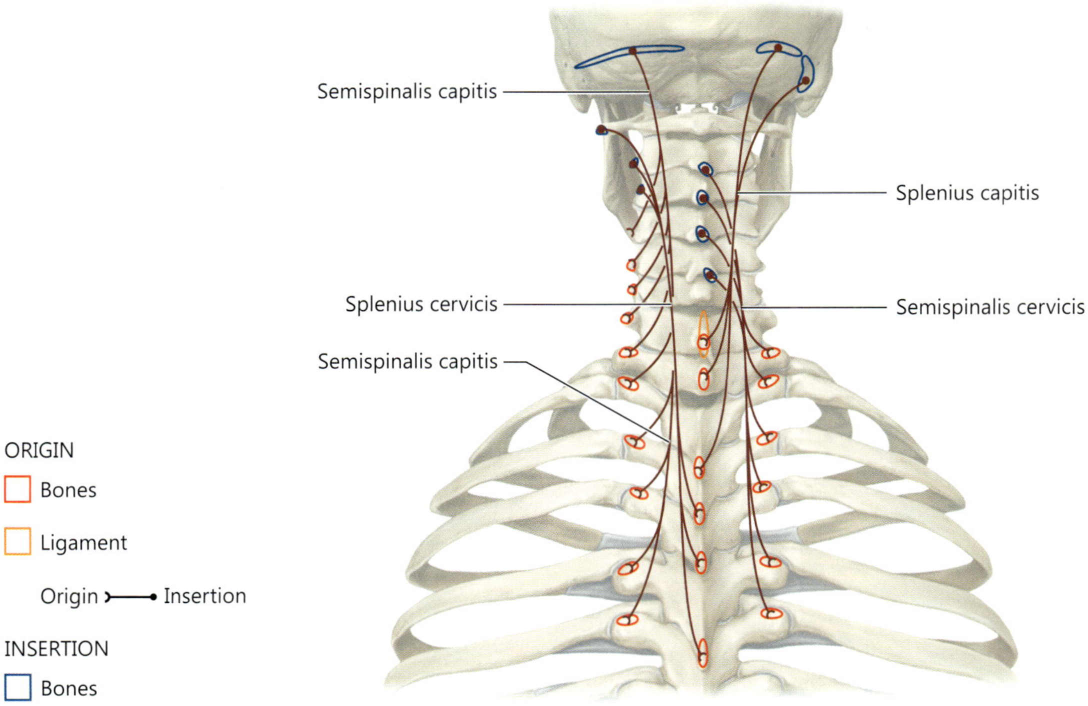

Schematic representation of the origins and insertions of the **transversospinales** (for the semispinalis capitis and semispinalis cervicis) and the **spinotransversales** (splenius capitis and splenius cervicis).

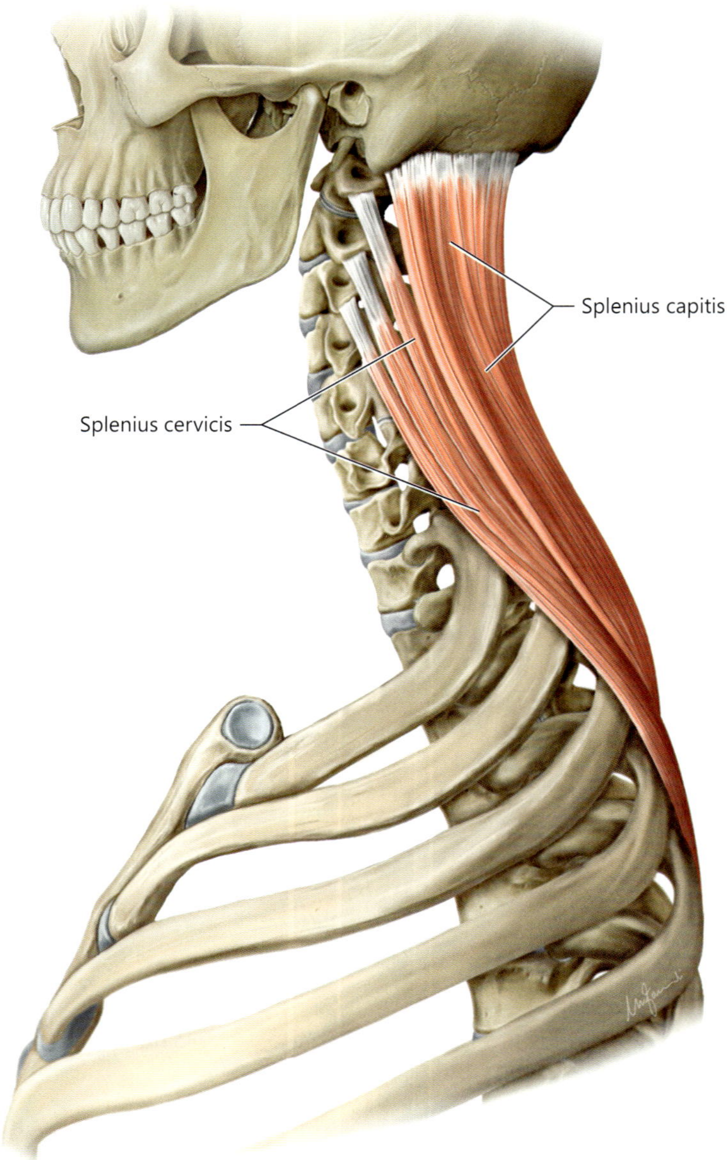

Splenius capitis and **splenius cervicis**, lateral view.

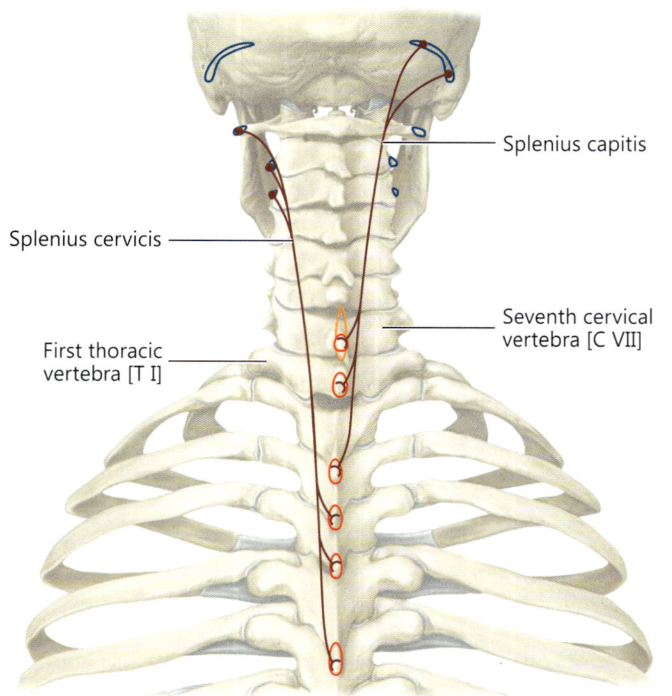

ORIGIN

Bones

Ligament

Origin ⟶ Insertion

INSERTION

Bones

Origins and insertions of the **spinotransversales**. The splenius capitis originates from the ligamentum nuchae, the spinous processes of the seventh cervical vertebra and of the first two thoracic vertebrae, and inserts on the superior nuchal line and the mastoid process of the temporal bone. The splenius cervicis originates from the spinous processes of the third, fourth, and fifth thoracic vertebrae and inserts on the transverse processes of the first three cervical vertebrae.

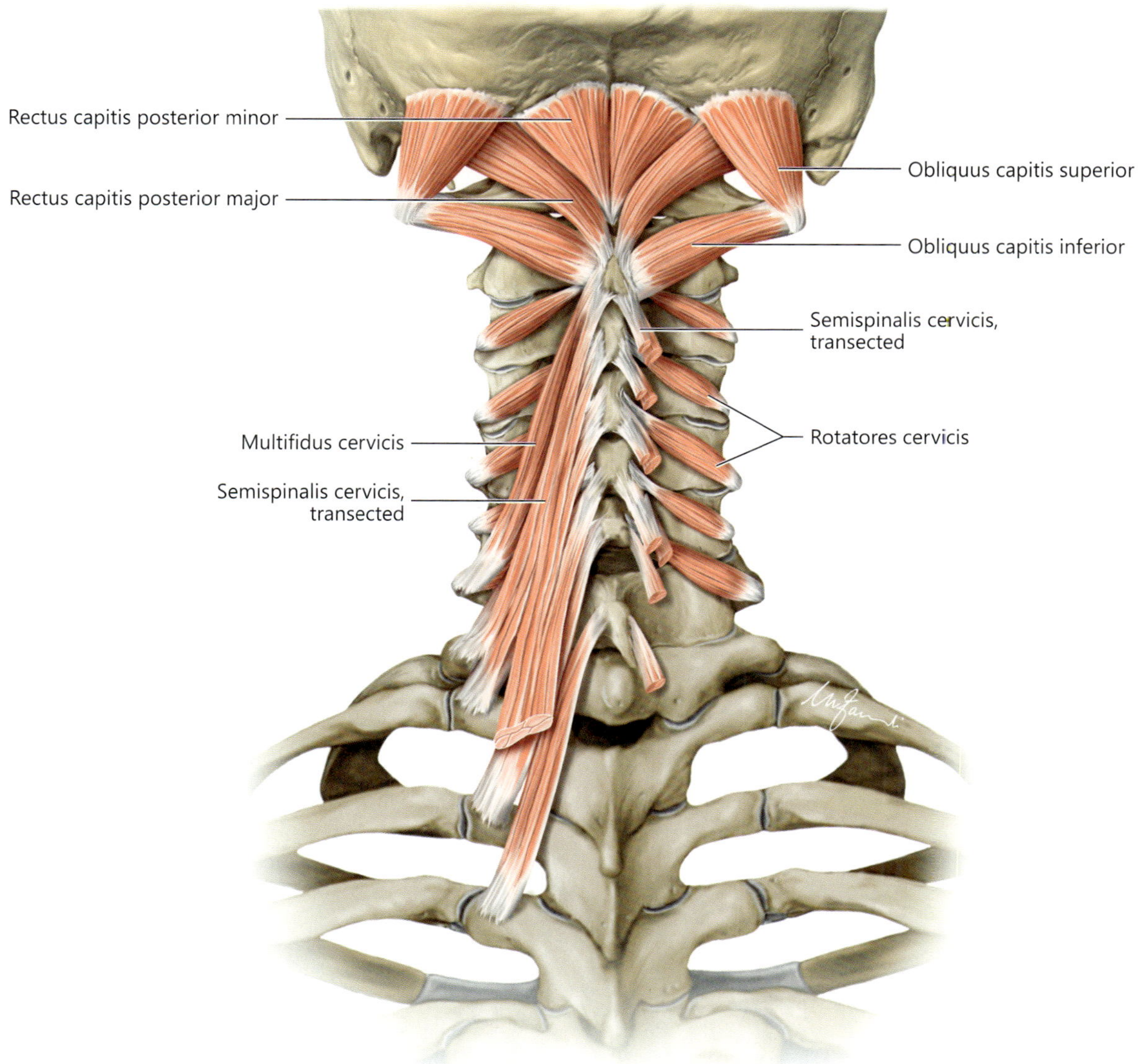

Intersection formed by the **suboccipital muscles** and the **transversospinales** (multifidus and rotatores).

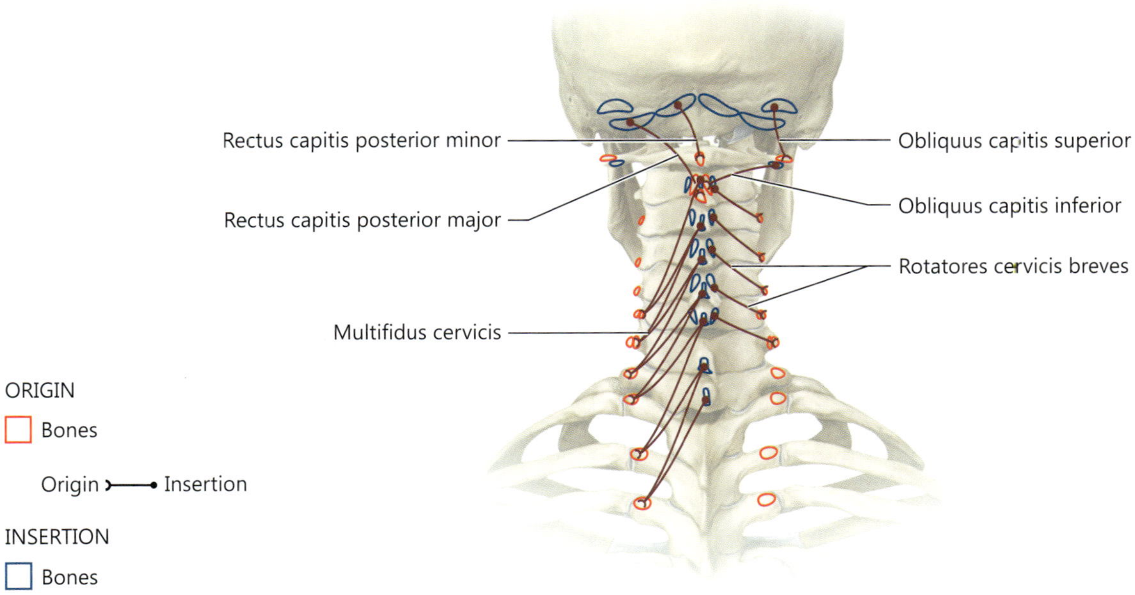

Schematic representation of the origins and insertions of the **suboccipital muscles** and **transversospinales** (for the multifidus cervicis and rotatores cervicis).

Muscles of Soft Palate and Fauces

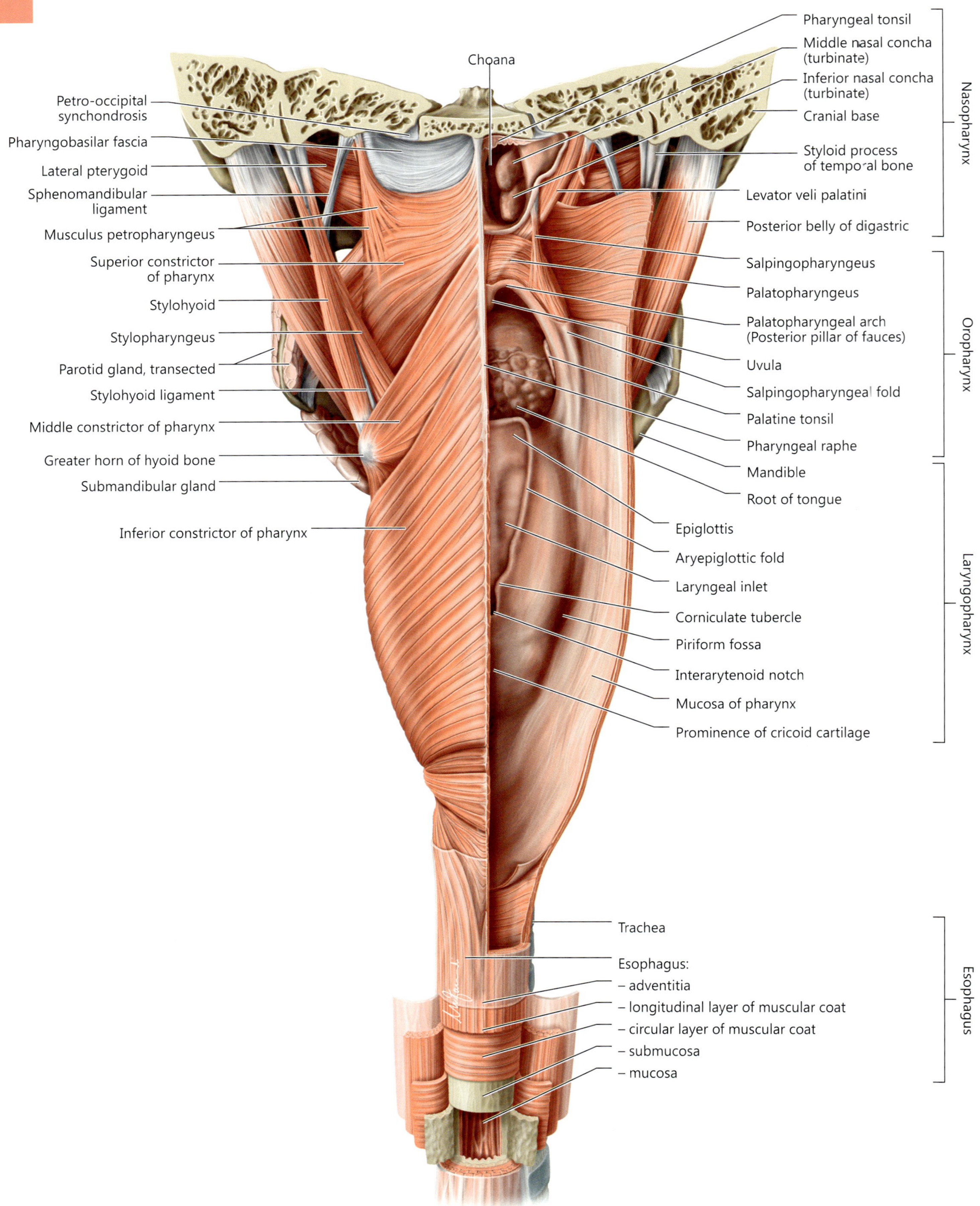

Muscles of the soft palate and **pharyngeal muscles**, posterior view. On the right, the inside of the pharynx after resection of the posterior wall.

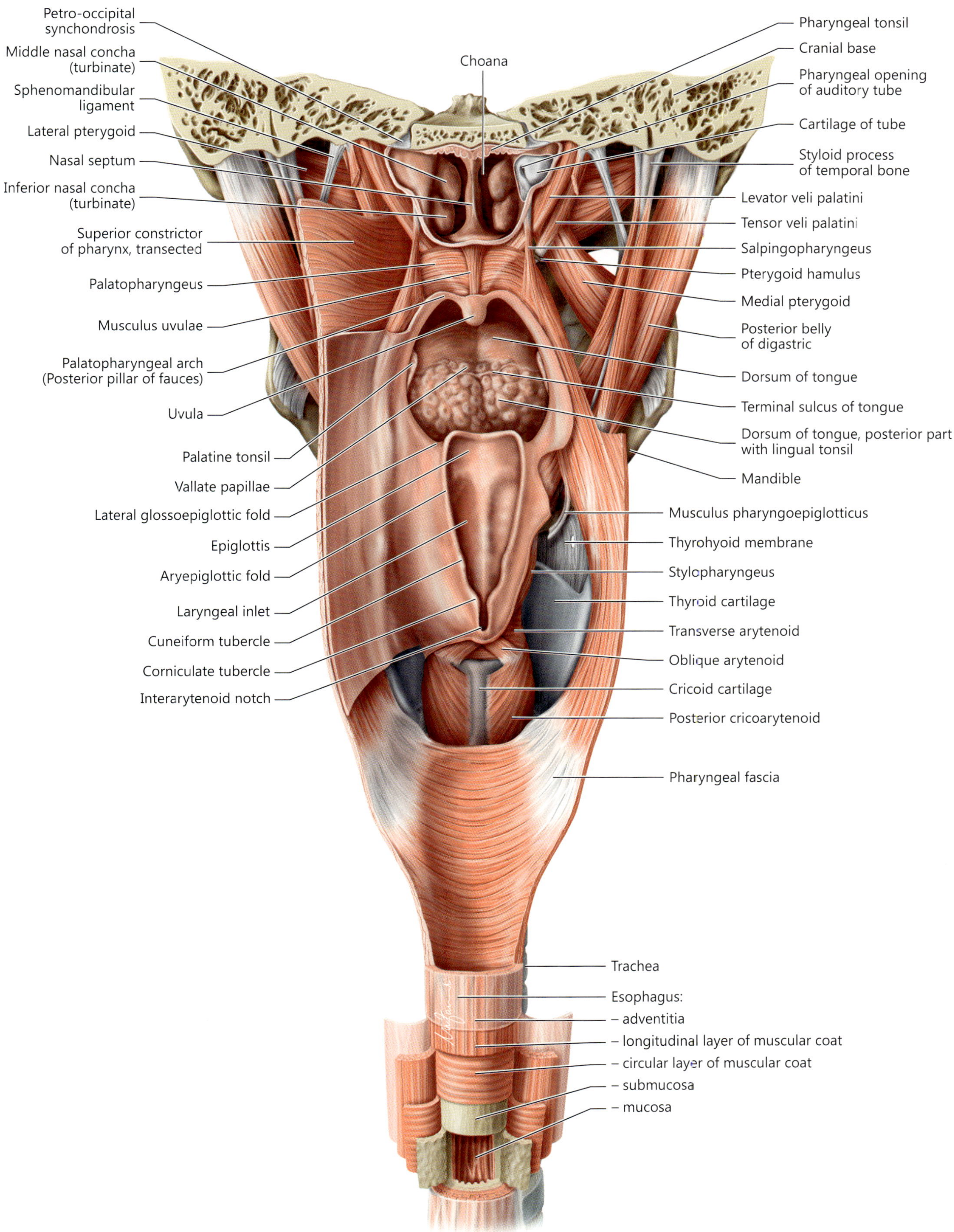

Muscles of the soft palate, **pharyngeal** and **laryngeal muscles**, posterior view after removal of the anterolateral wall of the pharynx.

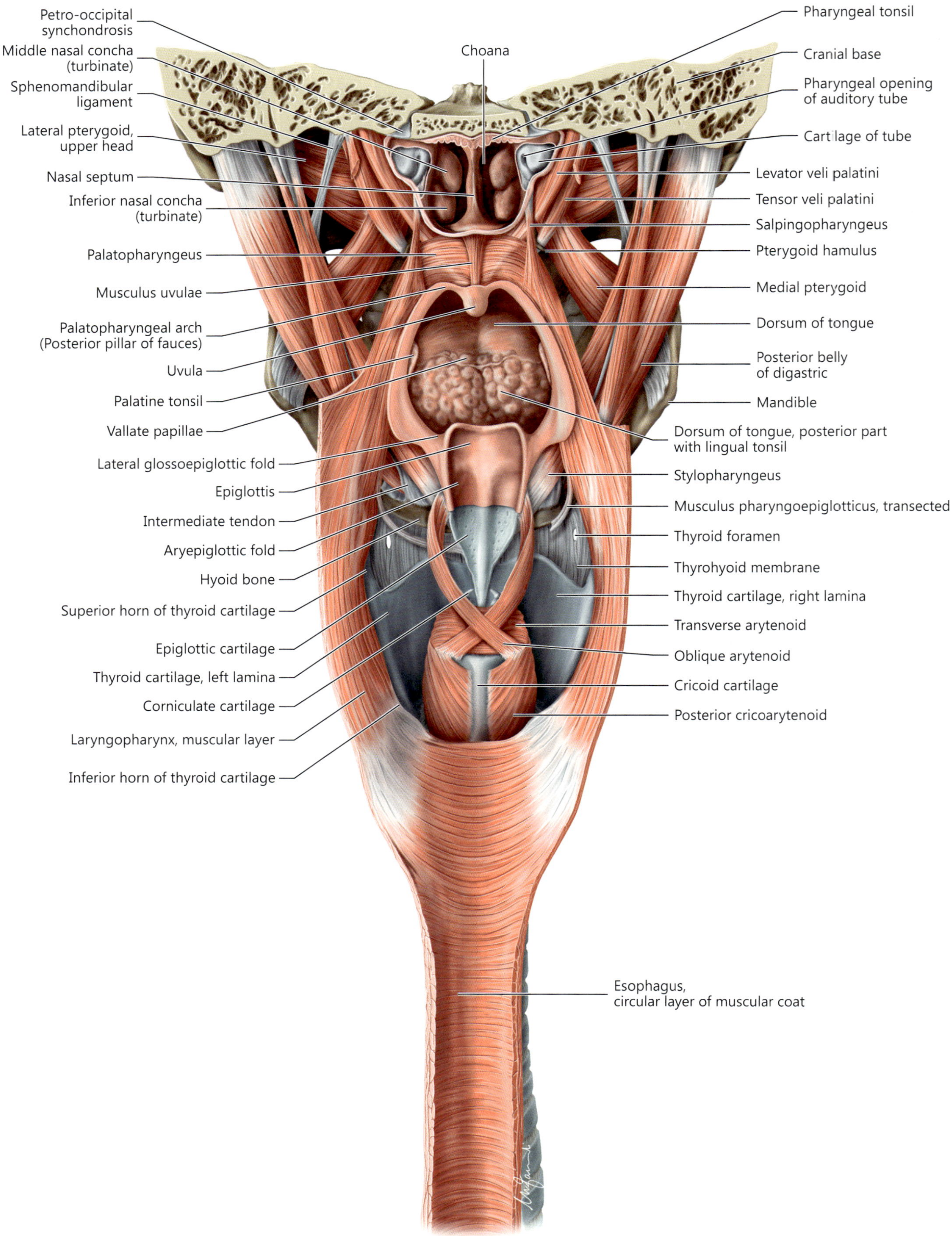

Muscles of the soft palate and **fauces**, posterior view. The pharynx has been mostly removed in order to focus on the larynx.

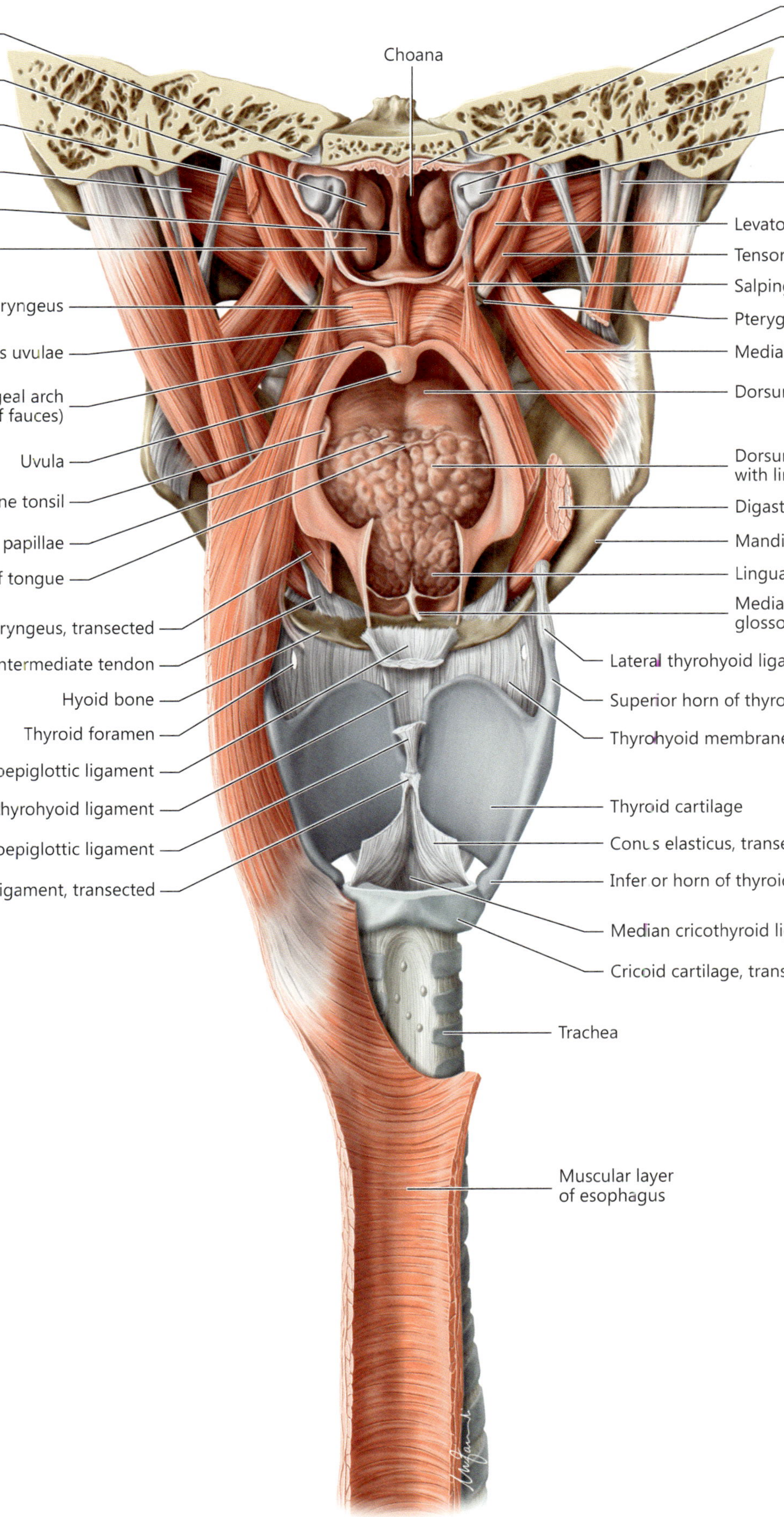

Muscles of the soft palate and **fauces**, posterior view. The pharynx has been mostly removed in order to focus on the larynx.

Muscles of Soft Palate and Fauces

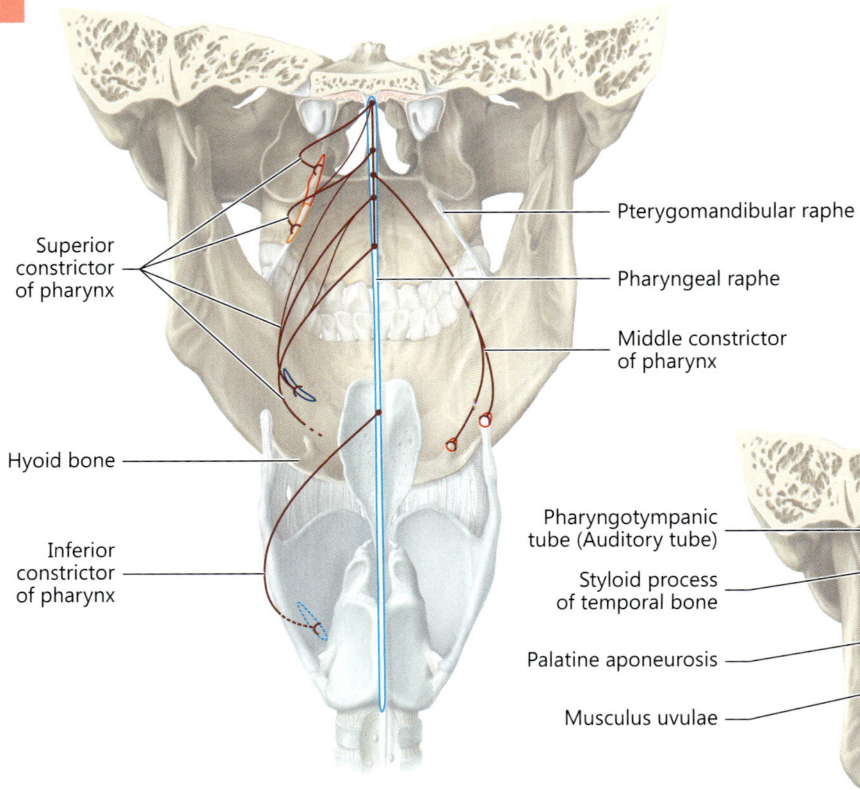

ORIGIN

- Bones
- Aponeuroses, cartilage, raphe

Origin ⟶ Insertion

INSERTION

- Bones
- Raphe, cartilage, aponeuroses, septum, tendon, capsule and articular disc

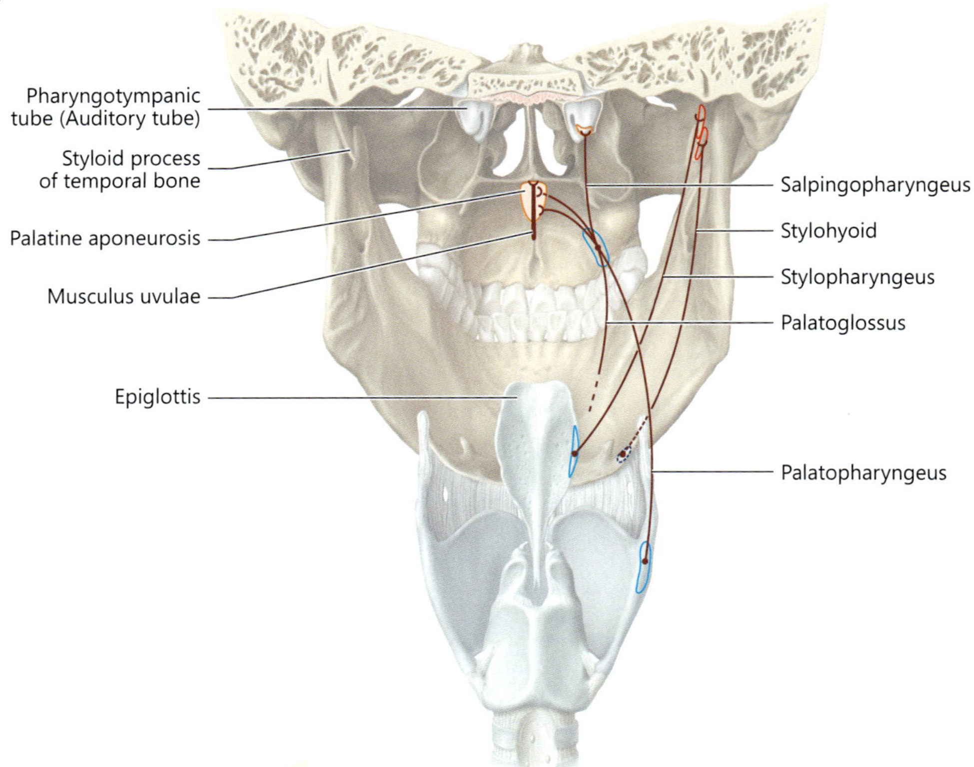

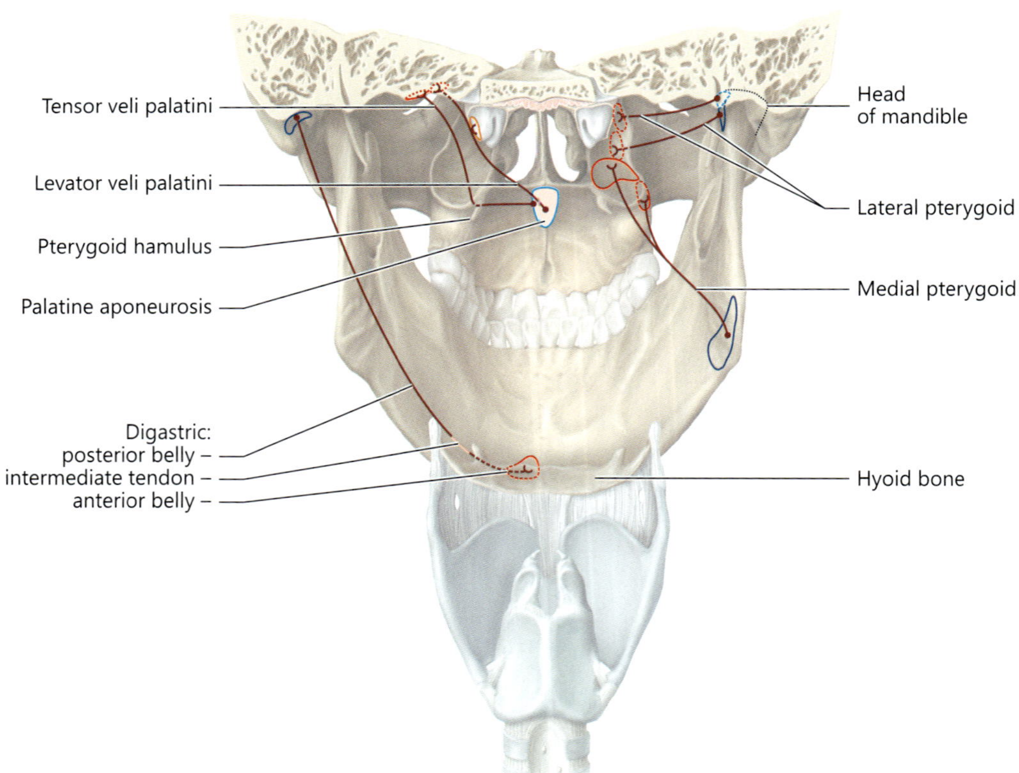

Schematic representation of the origin and insertions of the **muscles of the soft palate** and the **pharyngeal muscles**. The muscles are treated within the muscles of the head with the exception of the stylohyoid (flexor muscle of the head and neck). The palatoglossus inserts on the back and lateral border of the tongue (dotted line).

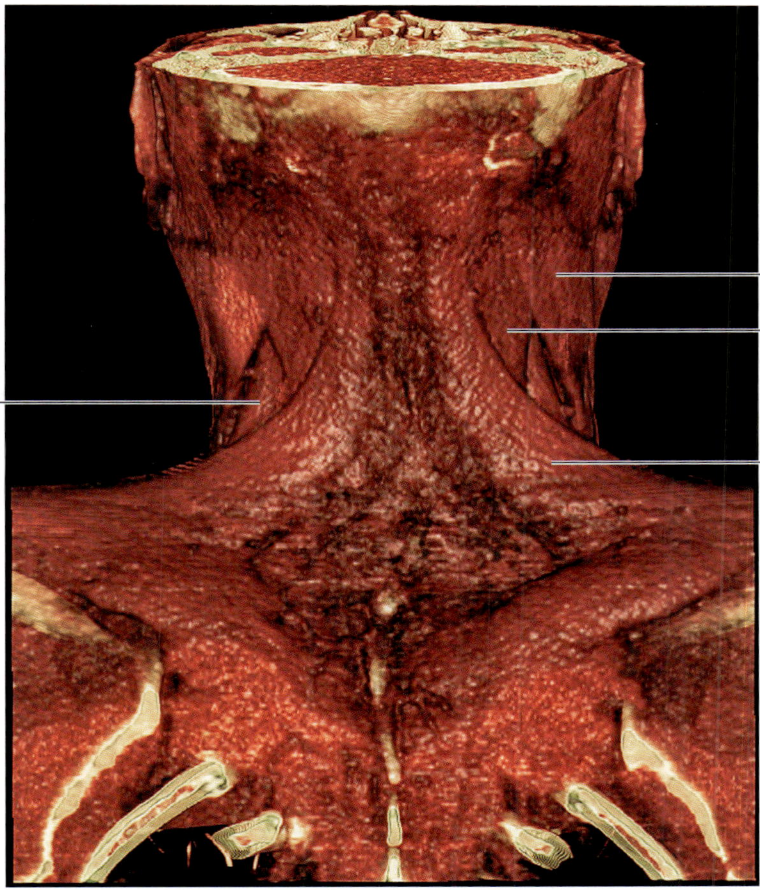

Three-dimensional image of a living human body, obtained by CT scan using a 3D volume rendering technique, showing some of the **posterior muscular structures of the neck**.

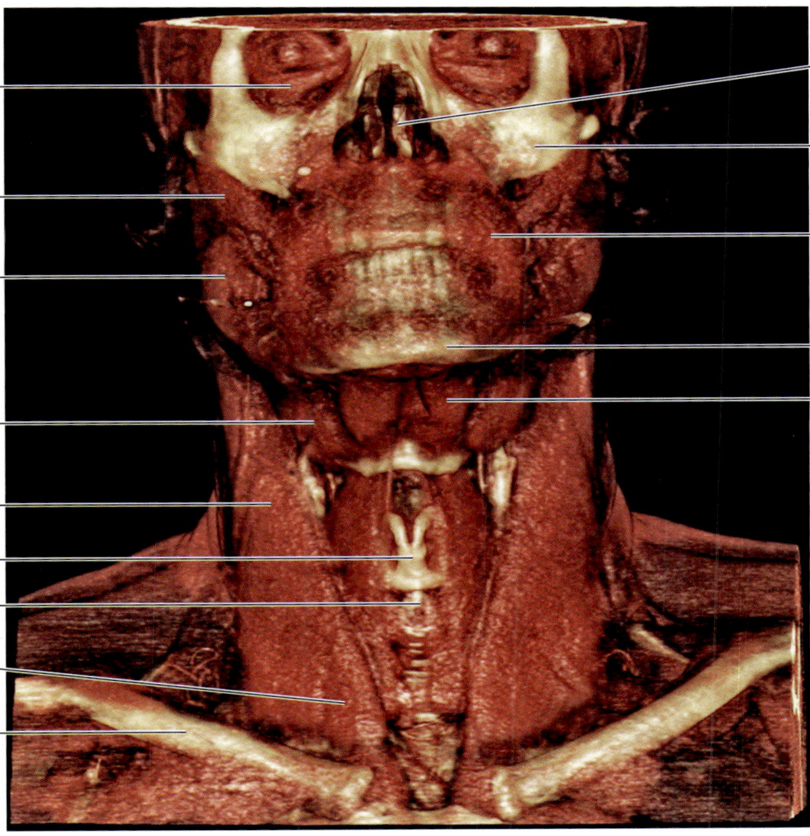

Three-dimensional image of a living human body, obtained by CT scan using a 3D volume rendering technique, showing, in a frontal projection, the **muscles of the neck** and **face**.

Muscles of the neck			
REGION	**FUNCTIONAL GROUP**	**MUSCLE**	**ANATOMICAL CLASSIFICATION**
Anterior	Levator muscles of the hyoid bone	Digastric	Suprahyoid muscle
	Levator muscles of the hyoid bone	Geniohyoid	Suprahyoid muscle
	Levator muscles of the hyoid bone	Mylohyoid	Suprahyoid muscle
	Depressor muscles of the hyoid bone	Omohyoid	Infrahyoid muscle
	Depressor muscles of the hyoid bone	Sternohyoid	Infrahyoid muscle
	Depressor muscles of the hyoid bone	Sternothyroid	Infrahyoid muscle
	Levator muscles of the hyoid bone	Stylohyoid	Suprahyoid muscle
	Depressor muscles of the hyoid bone	Thyrohyoid	Infrahyoid muscle
Lateral	Tensor muscle of the superficial fascia (used as donor in the reconstruction of mimic muscles)	Platysma	Muscle of the neck
	Lateral inclinator muscles of the head and neck	Rectus capitis lateralis	Suboccipital muscle
	Lateral inclinator muscles of the neck Flexor muscles of the neck Rotator muscles of the neck	Scalenus anterior	Muscle of the neck
	Lateral inclinator muscles of the head and neck Rotator muscles of the head and neck Flexor muscles of the neck	Scalenus medius	Muscle of the neck
	Lateral inclinator muscles of the neck Inclinator muscles of the neck Rotator muscles of the neck	Scalenus posterior	Muscle of the neck
	Extensor muscles of the head and neck Lateral inclinator muscles of the head and neck Rotator muscles of the head and neck	Sternocleidomastoid	Muscle of the neck
Prevertebral	Flexor muscles of the head and neck	Longus colli	Muscle of the neck
	Flexor muscles of the head and neck	Longus capitis	Muscle of the neck
	Flexor muscles of the head and neck	Rectus capitis anterior	Suboccipital muscle
Posterior (nuchal)			
– spinohumeral group	Extensor muscles of the head and neck Lateral extensor muscles of the head and neck Rotator muscles of the head and neck	Levator scapulae	Muscle of the back
	Extensor muscles of the head and neck Lateral inclinator muscles of the head and neck Rotator muscles of the head and neck	Trapezius	Muscle of the back
– spinodorsal group	Extensor muscles of the trunk	Iliocostalis cervicis	Muscle of the back proper Erector spinae
	Flexor muscles of the head and neck Lateral inclinator muscles of the head and neck Lateral rotator muscles of the head and neck	Anterior cervical intertransversarii	Muscles of the back
	Lateral inclinator muscles of the head and neck Extensor muscles of the head and neck Rotator muscles of the head and neck	Lateral posterior cervical intertransversarii	Muscles of the back
	Extensor muscles of the head and neck Lateral inclinator muscles of the head and neck Rotator muscles of the head and neck	Longissimus cervicis	Muscle of the back proper Erector spinae
	Extensor muscles of the head and neck Lateral inclinator muscles of the head and neck Rotator muscles of the head and neck	Longissimus capitis	Muscle of the back proper Erector spinae
	Extensor muscles of the head and neck Lateral inclinator muscles of the head and neck Rotator muscles of the head and neck	Spinalis cervicis	Muscle of the back proper Erector spinae
	Extensor muscles of the head and neck Lateral inclinator muscles of the head and neck Rotator muscles of the head and neck	Spinalis capitis	Muscle of the back proper Erector spinae
	Extensor muscles of the head and neck Lateral inclinator muscles of the head and neck Rotator muscles of the head and neck	Splenius cervicis	Muscle of the back proper Spinotransversalis
	Extensor muscles of the head and neck Lateral inclinator muscles of the head and neck Rotator muscles of the head and neck	Splenius capitis	Muscle of the back proper Spinotransversalis

continued

Muscles of the neck *(continued)*

REGION	FUNCTIONAL GROUP	MUSCLE	ANATOMICAL CLASSIFICATION
– vertebro-occipital group	Extensor muscles of the head and neck Lateral inclinator muscles of the head and neck Rotator muscles of the head and neck	Rectus capitis posterior major	Suboccipital muscle
	Extensor muscles of the trunk	Interspinales cervicis	Muscles of the back proper Interspinales
	Lateral inclinator muscles of the head and neck Extensor muscles of the head and neck Rotator muscles of the head and neck	Medial posterior cervical intertransversarii	Muscles of the back proper Intertransversarii
	Extensor muscles of the head and neck Lateral inclinator muscles of the head and neck Rotator muscles of the head and neck	Multifidus cervicis	Muscles of the back proper Transversospinalis
	Extensor muscles of the head and neck Lateral inclinator muscles of the head and neck Rotator muscles of the head and neck	Obliquus capitis inferior	Suboccipital muscle
	Extensor muscles of the head and neck Lateral inclinator muscles of the head and neck Rotator muscles of the head and neck	Obliquus capitis superior	Suboccipital muscle
	Extensor muscles of the head and neck Lateral inclinator muscles of the head and neck Rotator muscles of the head and neck	Rectus capitis posterior minor	Suboccipital muscle
	Extensor muscles of the head and neck Lateral inclinator muscles of the head and neck Rotator muscles of the head and neck	Rotatores cervicis	Muscles of the back proper Transversospinales
	Extensor muscles of the head and neck Lateral inclinator muscles of the head and neck Rotator muscles of the head and neck	Semispinalis cervicis	Muscles of the back proper Transversospinalis
	Extensor muscles of the head and neck Lateral inclinator muscles of the head and neck Rotator muscles of the head and neck	Semispinalis capitis	Muscles of the back proper Transversospinalis

Extensor muscles of the head and neck				
MUSCLE	***ORIGIN***	***INSERTIONS***	***INNERVATION***	***ACTION***
Muscles with an extensor component				
Levator scapulae	Transverse processes of the first four cervical vertebrae	Medial border of the scapula between the superior angle and the origin of the spine	Third and fourth cervical nerve [C3-C4]; dorsal scapular nerve [C5]	Elevation and medial rotation of the scapula
Superior fibers of the trapezius	Medial third of the superior nuchal line; external occipital protuberance; ligamentum nuchae	Posterior border of the lateral third of the clavicle	Spinal root of the accessory nerve [XI] and branches of the cervical plexus [C3-C4]	Extension of the head and neck
Sternocleidomastoid	*Sternal head*: superior surface of the manubrium of the sternum *Clavicular head*: superior surface of the lateral third of the clavicle	Lateral surface of the mastoid process; lateral half of the superior nuchal line	Spinal root of the accessory nerve [XI] and branches of the cervical plexus (C2, C3, and at times C4)	Extension of the neck; flexion of the neck; contralateral rotation of the neck; ipsilateral inclination of the neck
Extensor muscles proper				
Rectus capitis posterior major	Spinous process of the axis	Lateral part of the inferior nuchal line and osseous area immediately below	Posterior ramus of the first cervical nerve [C1]	Extension of the head and ipsilateral rotation of the head
Obliquus capitis inferior	Lateral surface of the spinous process and the adjacent superior surface of the lamina of the axis	Inferior and posterior surfaces of the transverse process of the atlas	Posterior ramus of the first cervical nerve [C1]	Ipsilateral rotation of the head
Obliquus capitis superior	Superior surface of the transverse process of the atlas	Occipital bone between the superior and inferior nuchal lines, lateral to the semispinalis capitis	Posterior ramus of the first cervical nerve [C1]	Extension of the head and its ipsilateral inclination
Rectus capitis posterior minor	Posterior tubercle of the atlas	Medial part of the inferior nuchal line and osseous area between this and the foramen magnum	Posterior ramus of the first cervical nerve [C1]	Extension of the head
Splenius cervicis	Spinous processes from T III to T VI	Posterior tubercles of the transverse processes of the three superior cervical vertebrae	Posterior rami of the cervical nerves [C2-C4]	Extension of the neck; ipsilateral rotation of the neck
Splenius capitis	Inferior half of the ligamentum nuchae; spinous processes of C VII and from T I to T IV and corresponding supraspinous ligaments	Mastoid process of the temporal bone; occipital bone below the lateral third of the superior nuchal line	Posterior rami of the cervical nerves [C2-C4]	Extension of the neck; ipsilateral rotation of the neck
Common extensor muscles				
Iliocostalis cervicis	Costal angles from the 3rd to the 6th ribs	Posterior tubercles of the transverse processes from C IV to C VI	Posterior rami of the cervical nerves [C4-C8]	Extension of the neck; lateral inclination of the neck (ipsilateral)
Interspinales cervicis	Short muscles between the spinous processes of contiguous vertebrae to the sides of the interspinous ligament in the cervical, thoracic, and lumbar tracts		Posterior rami of the cervical nerves [C1-C8]	Extension of the trunk (postural control)
Anterior, medial posterior, and lateral posterior cervical intertransversarii	Short muscles between the transverse processes of contiguous vertebrae in the cervical, thoracic, and lumbar tracts		Anterior and posterior rami of the cervical nerves [C1-C8]	Extension of the trunk; lateral inclination of the trunk (postural control)
Longissimus cervicis	Transverse processes from T I to T V	Posterior tubercles of the transverse processes from C II to C VI	Posterior rami of the cervical nerves [C4-C8]	Extension and lateral inclination of the neck (ipsilateral)
Longissimus capitis (trachelomastoid)	Transverse processes from T I to T V; articular processes from C III to C VII	Mastoid process of the temporal bone	Posterior rami of the cervical nerves [C4-C8]	Extension of the head and neck, rotation of the head and lateral inclination of the neck (ipsilateral)
Spinalis cervicis	Inferior surface of the ligamentum nuchae, spinous processes of C VII, T I and T II	Spinous processes from C I to C III	Posterior rami of the cervical nerves [C4-C8]	Extension of the neck and lateral inclination of the neck
Spinalis capitis	Spinous process of C VII	Close to the midline between the superior and inferior nuchal lines of the occipital bone	Posterior rami of the cervical nerves [C4-C8]	Extension of the head and neck

continued

Extensor muscles of the head and neck *(continued)*

MUSCLE	ORIGIN	INSERTIONS	INNERVATION	ACTION
Spinous transverse complex of the neck				
Multifidus cervicis	Posterior surface of the sacrum; aponeurosis of the erector spinae; posterior superior iliac spine; posterior sacroiliac ligament; transverse processes from C IV to L V	To the spinous processes of the overlying vertebrae (from one to four)	Posterior rami of the cervical nerves [C1-C8]	Extension of the trunk; contralateral rotation (postural control)
Rotatores cervicis	Superior and posterior surface of the transverse processes of the cervical, thoracic and lumbar vertebrae	Inferior and lateral surface of the lamina of the overlying vertebra in the cervical, thoracic and lumbar tracts	Posterior rami of the cervical nerves [C1-C8]	Extension and contralateral rotation of the trunk (postural control)
Semispinalis cervicis	Transverse processes from T I to T VI	Spinous processes from C I to C V	Posterior rami of the cervical nerves [C1-C8]	Extension of the neck; contralateral rotation of the neck
Semispinalis capitis	Apices of the transverse processes of C VII and from T I to T VII; articular processes from C IV to C VI	Medial part of the region between the superior and inferior nuchal lines of the occipital bone	Posterior rami of the cervical nerves [C1-C8]	Extension of the head; contralateral rotation of the head

Functional aspects of the extensor muscles of the head and neck

- **Muscles with an extensor component**
  - The trapezius has an extensor action, contracting bilaterally, through the superior and middle fascicles (descending and transverse fibers), maintaining its insertion on the clavicle and spine of the scapula as a fixed point
  - The sternocleidomastoid extends the head by contracting bilaterally when the mastoid insertion is located posteriorly to the origin on the clavicle and on the sternum, rather than anteriorly
  - The levator scapulae extends the neck by contracting bilaterally with a fixed point on the scapula; in unilateral contraction with a fixed point on the scapula it tilts and rotates the neck ipsilaterally.

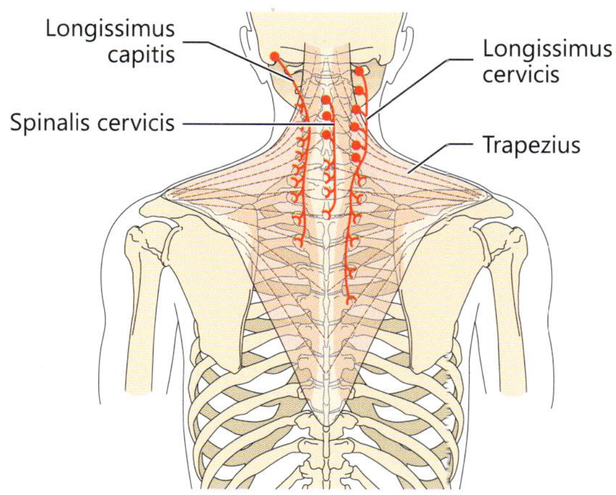

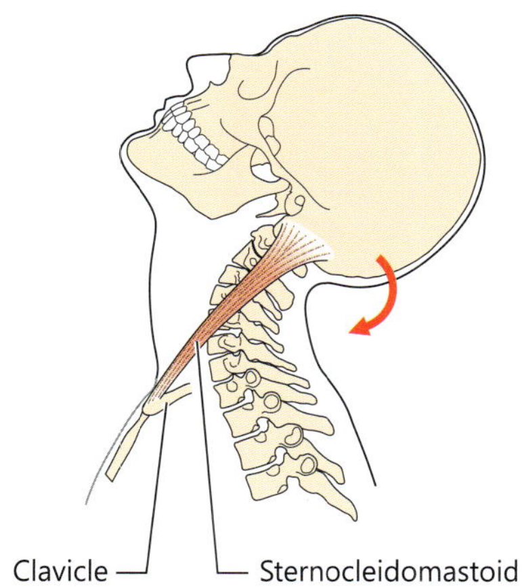

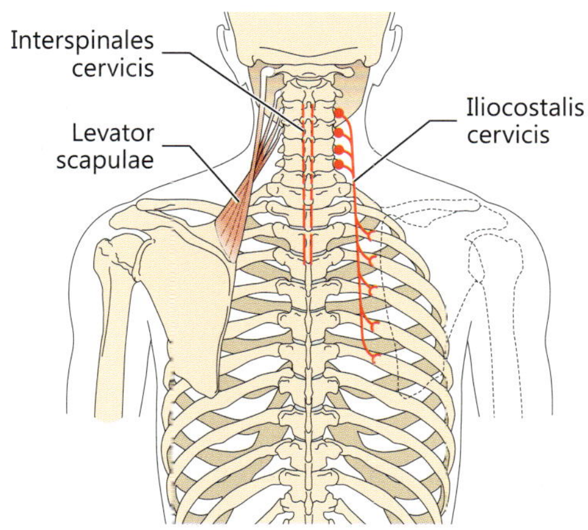

- **Extensor muscles proper of the head and neck**
  - Present only at this site
  - The rectus capitis posterior minor, rectus capitis posterior major, obliquus capitis superior, and obliquus capitis inferior extend the head by contracting bilaterally
  - The unilateral contraction of these muscles tilts the head laterally.

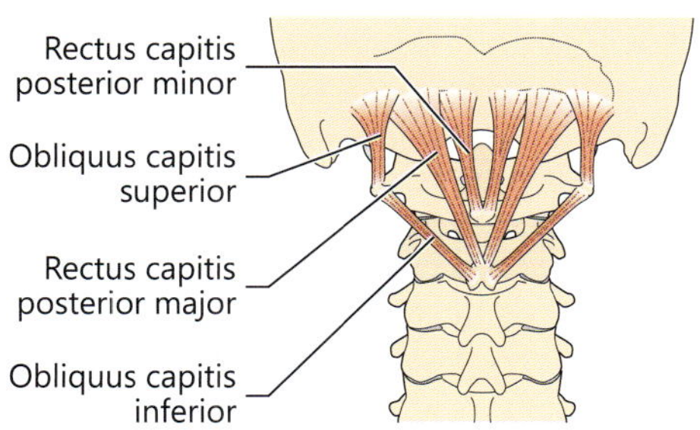

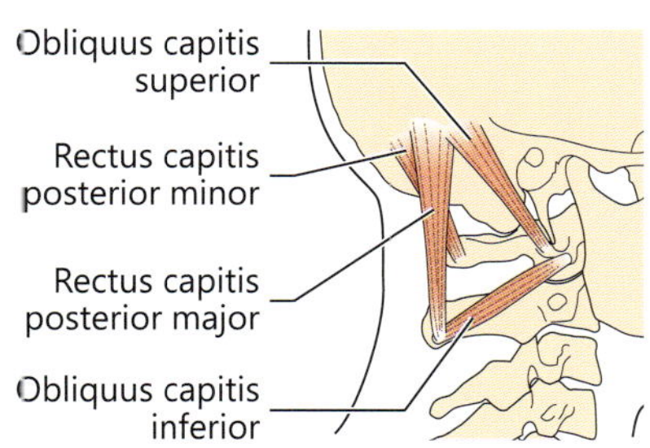

- The rectus capitis posterior major and rectus capitis posterior minor (which replace the interspinales between the occipital bone and the atlas, and between the atlas and the axis), and the obliquus capitis inferior rotate the head ipsilaterally
- The obliquus capitis superior (which replaces the intertransversarius between C II and C I) rotates the head contralaterally
- The splenius capitis and splenius cervicis extend the head and neck by contracting bilaterally, while their unilateral contraction rotates the neck and head ipsilaterally.

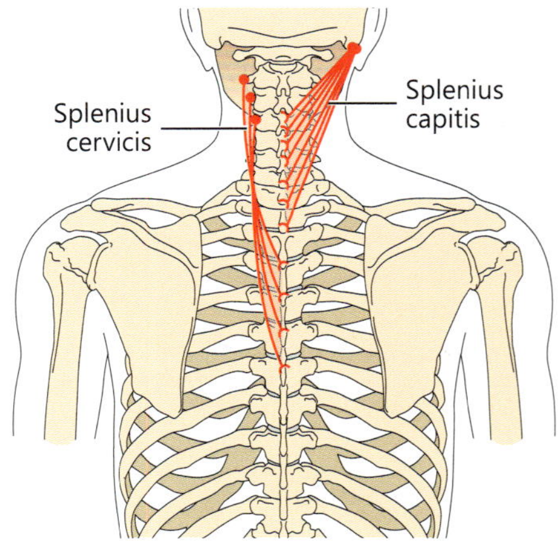

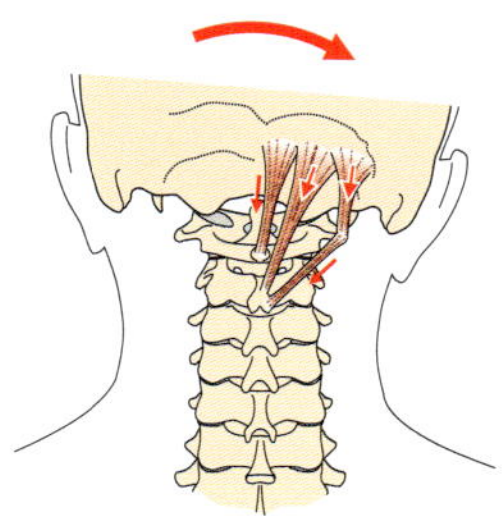

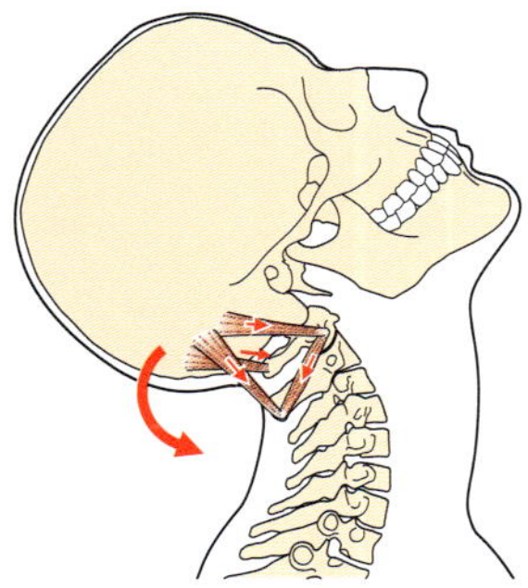

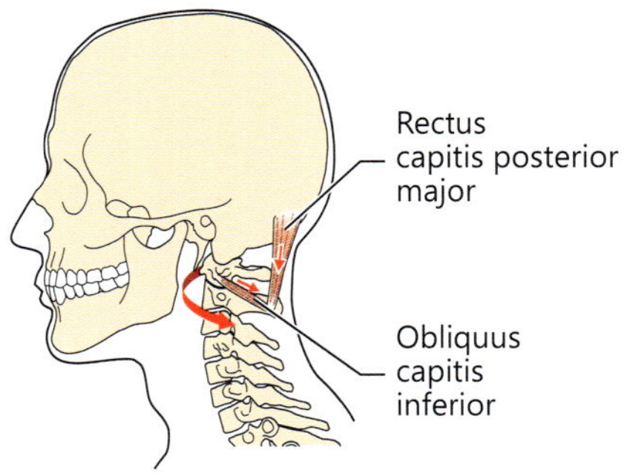

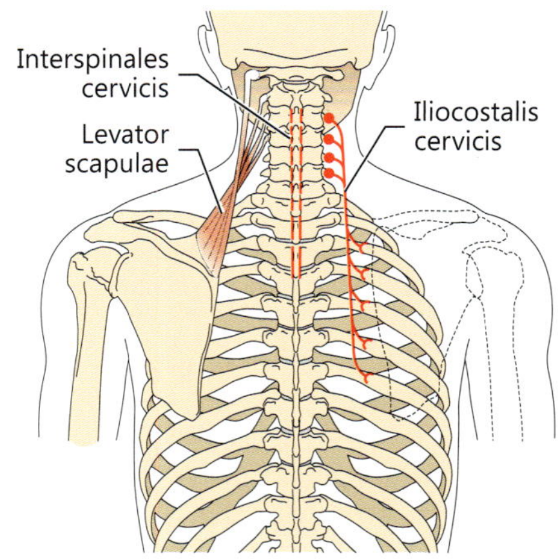

- **Common extensor muscles of the head and neck**
  - Also present throughout the vertebral column
  - Extensors when they contract bilaterally
  - Their unilateral contraction contributes to the lateral inclination of the head and neck with the exception of the transversospinal complex of the neck and the longissimus capitis (the first rotates the head and neck contralaterally, the second ipsilaterally).

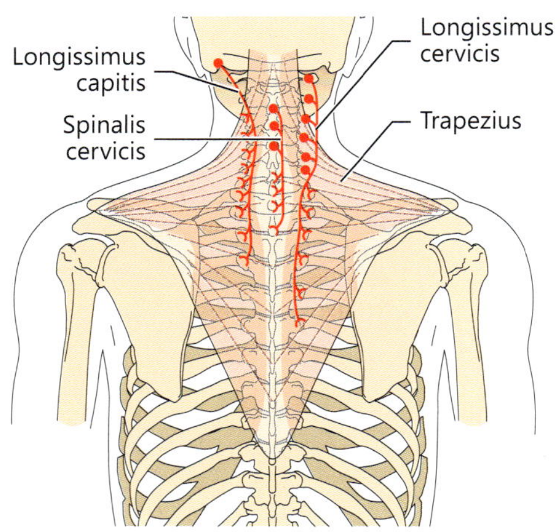

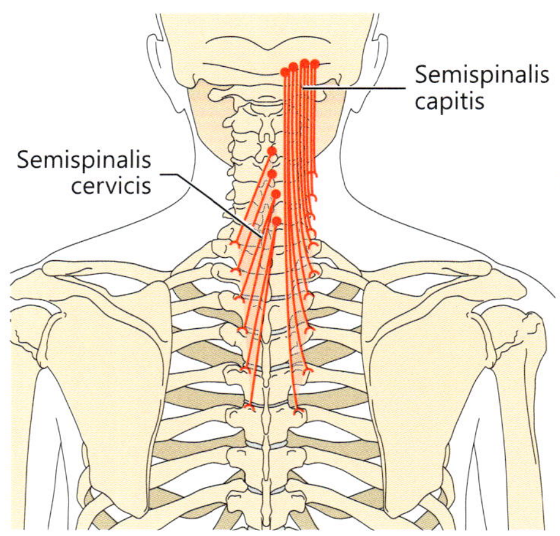

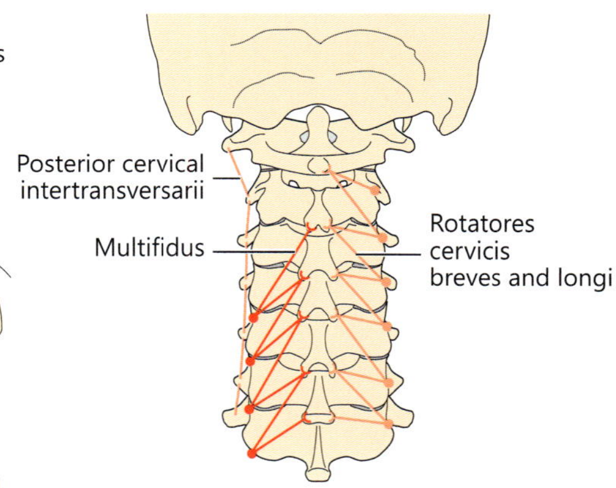

- **Functional applications of the extensor muscles of the head and neck**
  - Their bilateral contraction causes an extension of the neck by eliminating the actions of lateral inclination and rotation which occur with unilateral contraction
  - In the case of a forced extension, they contract after the articular excursion motion has been completed, or when the extension is performed against resistance
    - the actions performed with the upper limbs above the head require the contraction of these muscles to the maximum range of the articular motion, as happens, for example, when you retrieve a book from a high shelf
    - other activities that require the contraction to the maximum range of articular excursion are represented, for example, by drinking from a glass or looking down a bowling alley when the ball is released from one's hand
  - They contract and work against resistance when raising the head from a supporting surface in the prone position and they control the flexion of the neck when the head is lowered again toward the supporting surface
  - They contract during activities in which the head is tilted forward, such as reading or writing
  - The activity of these muscles ceases when the neck is fully flexed and the position of the head is maintained by the tension of the ligamentum nuchae.

Flexor muscles of the head and neck				
MUSCLE	**ORIGIN**	**INSERTIONS**	**INNERVATION**	**ACTION**
Suprahyoid				
Digastric	Digastric fossa of the body of the mandible	Mastoid notch	Facial [VII] and glossopharyngeal [IX] nerves (posterior belly) Nerve to mylohyoid (Vc or V3) (anterior belly)	Elevation and retraction of the hyoid bone or retraction of the head (posterior belly) Depression of the mandible or elevation of the hyoid bone (anterior belly) Elevation of the hyoid bone following simultaneous contraction of the two muscles
Geniohyoid	Inferior mental spine	Medial part of the anterior surface of the hyoid bone	Branch of the first cervical spinal nerve (C1), which branch off from the hypoglossal nerve	Elevation of the hyoid bone or depression of the mandible
Mylohyoid	Internal surface of the mandible from the mental spines to the mylohyoid line	Anterior surface of the hyoid bone The most medial fibers come together at the midline forming the mylohyoid raphe	Nerve to mylohyoid (branch of the inferior alveolar nerve, branch of the mandibular nerve, Vc or V3)	Elevation of the floor of the mouth during deglutition Depression of the mandible or elevation of the hyoid bone
Stylohyoid	Styloid process of the temporal bone	Lateral extremity the body of the hyoid bone	Facial nerve [VII]	Elevation of the hyoid bone
Infrahyoid				
Omohyoid	Superior border of the scapula close to the scapular notch; superior transverse scapular ligament	A band of prevertebral layer of cervical fascia holds the intermediate portion of the muscle inferiorly, in the direction of the clavicle and of the first rib; changing the direction thereof at this level; inferior border of the body of the hyoid bone	Ansa cervicalis (branch of C1 to the superior belly through superior root; branches of C1, C2 and C3 to the inferior belly)	Depression of the hyoid bone
Sternohyoid	Posterior surface of the medial extremity of the clavicle; posterior sternoclavicular ligament; superior and posterior surface of the manubrium of the sternum	Inferior surface of the body of the hyoid bone	Ansa cervicalis (branches of C1, C2 and C3)	Depression of the hyoid bone
Sternothyroid	Posterior surface of the manubrium of the sternum below the origin of the sternohyoid and border of the first costal cartilage	Oblique line on the lamina of the thyroid cartilage	Ansa cervicalis (branches of C1, C2 and C3)	Depression of the larynx
Thyrohyoid	Oblique line on the lamina of the thyroid cartilage	Inferior border of the greater horn and the adjacent part of the body of the hyoid bone	Thyrohyoid branch of the first cervical spinal nerve (C1), which branch off from the hypoglossal nerve	Depression of the hyoid bone; elevation of the larynx
Prevertebral				
Longus cervicis	*Inferior oblique part*: anterior surface of the vertebral bodies from T I to T III *Superior oblique part*: anterior tubercles of the transverse processes from C III to C V *Vertical part*: anterior surface of the vertebral bodies from T I to T III and from C V to C VII	*Inferior oblique part*: anterior tubercles of the transverse processes of C V and C VI *Superior oblique part*: anterolateral surfaces of the anterior tubercle of the atlas *Vertical part*: anterior surface of the vertebral bodies from C II to C IV	Anterior rami of the cervical nerves [C2-C6]	Flexion of the neck; contralateral rotation of the neck (inferior oblique fibers); lateral inclination of the neck (oblique fibers)
Longus capitis	Anterior tubercles of the transverse processes from C III to C VI	Inferior surface of the basilar part of the occipital bone	Anterior rami of the cervical nerves [C1-C3]	Flexion of the head
Rectus capitis anterior	Anterior surface of the lateral mass of the atlas; root of the transverse process of the atlas	Inferior surface of the basilar part of the occipital bone anteriorly to the occipital condyle	Anterior rami of the cervical nerves [C1, C2]	Flexion of the head
Rectus capitis lateralis	Superior surface of the transverse process of the atlas	Inferior surface of the jugular process of the occipital bone	Anterior rami of the cervical nerves [C1, C2]	Ipsilateral inclination and support of the head

continued

Flexor muscles of the head and neck				(continued)
MUSCLE	***ORIGIN***	***INSERTIONS***	***INNERVATION***	***ACTION***
Scalene muscles				
Scalenus anterior	Anterior tubercles of the transverse processes from C III to C VI	Tubercle of the scalene muscle on the internal border of the 1st rib and ridge on the superior surface of the rib anteriorly to the groove for the subclavian artery	Anterior rami of the cervical nerves [C4-C6]	Flexion of the neck and its ipsilateral inclination; contralateral rotation of the neck
Scalenus medius	Transverse process of the axis; anterior surface of the posterior tubercles of the transverse processes from C III to C VII	Superior surface of the 1st rib between the costal tubercle and the groove for the subclavian artery	Anterior rami of the cervical nerves [C3-C8]	Ipsilateral inclination of the neck
Scalenus posterior	Posterior tubercles of the transverse processes from C IV to C VI	Lateral surface of the 2nd rib	Anterior rami of the cervical nerves [C6-C8]	Ipsilateral inclination of the neck
Sternocleido-mastoid	*Sternal head*: superior surface of the manubrium of the sternum *Clavicular head*: superior surface of the lateral third of the clavicle	Lateral surface of the mastoid process; lateral half of the superior nuchal line	Spinal root of the accessory nerve [XI] and branches of the cervical plexus (C2, C3, and at times C4)	Extension of the neck; flexion of the neck; contralateral rotation of the neck; ipsilateral inclination of the neck

- **Functional aspects of the flexor muscles of the head and neck**
  - Flexion of the head and neck requires less muscular strength than extension due to the tendency of the head to fall forward
  - The suprahyoid muscles flex the head and neck with a fixed point (origin) on the hyoid bone and insertion on the mandible and other bones of the cranium
  - The infrahyoid muscles flex the head and neck with a fixed point (origin) on the bones and cartilages of the neck and trunk, and insertions on the hyoid bone
  - Flexion of the head and neck by the suprahyoid and infrahyoid muscles is only possible with the mouth closed (thanks to the action of the masticatory muscles which act as fixator muscles)
  - The prevertebral muscles (longus capitis, longus colli, rectus capitis anterior, rectus capitis lateralis) flex the head and neck by contracting bilaterally; when they contract unilaterally, all four muscles tilt the neck ipsilaterally; additionally the longus capitis and rectus capitis anterior rotate the head contralaterally and ipsilaterally, respectively
  - The scalene muscles are primarily inspiratory muscles; by contracting bilaterally and acting synergistically they flex the head; when they instead contract unilaterally they tilt the head laterally
  - The sternocleidomastoid is a secondary inspiratory muscle; the rectus capitis anterior and rectus capitis lateralis contract and bring the mastoid insertion of the muscle anteriorly to the clavicular and sternal origin, thus allowing for them to have a flexor action on the head and neck.

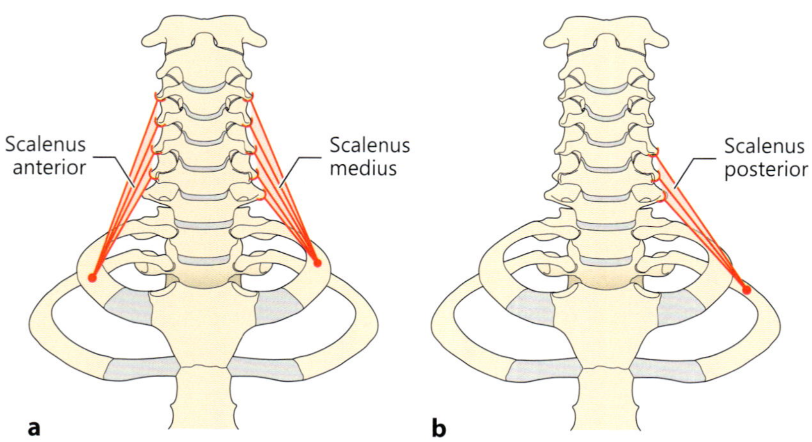

Scalene muscles. **a**, Scalenus anterior, scalenus medius; **b**, scalenus posterior.

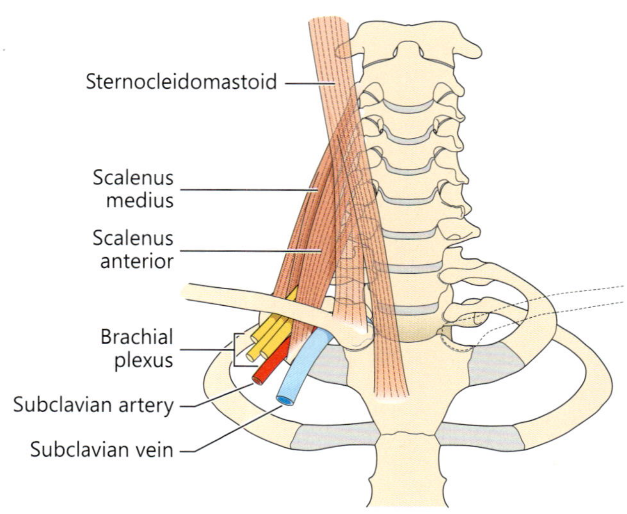

Interscalene triangle.

Action of the sternocleidomastoid. Anterior stabilization.

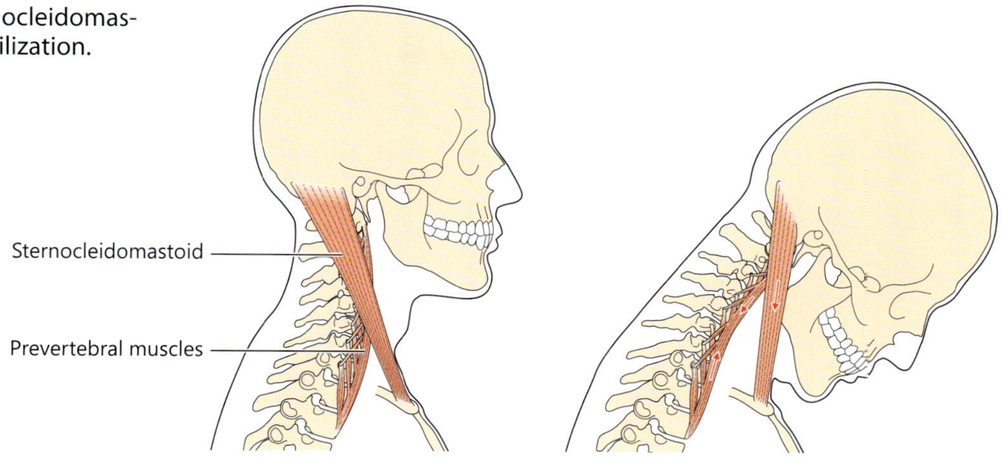

- **Functional applications of the flexor muscles of the head and neck**
  - They contract when the movement occurs against resistance, which can be represented by the weight of the head itself
  - They flex the head and neck and maintain this position when the head is raised from a supporting surface in a supine position, such as, for example, when you get out of bed
  - They control the extension of the neck when the head lowers back toward the supporting surface, as occurs when you stretch out in a supine position
  - In the upright position the flexor muscles of the head and neck contract only with complete, forced flexion
    - this combined action of the flexor muscles occurs, for example, when, looking down, you loosen or fasten the first button of your shirt, or when you fasten a necklace around your neck
  - When you eat, the bilateral contraction of the sternocleidomastoid muscles brings the head forward and helps the longus colli to flex the cervical column
  - Electromyographic studies have shown a moderate bilateral activity of the longus colli and sternocleidomastoid in maintaining the head's forward position.

Suprahyoid and infrahyoid muscles and the hyoid bone

The hyoid bone is located just below the mandible and has the following characteristics:
- body with four appendices – greater and lesser horns
- they attach to the suprahyoid and infrahyoid muscles.

The suprahyoid and infrahyoid muscles may be considered as a muscular sling with the hyoid bone as its pivot. Deglutition, mastication, and speech are the main functions of the suprahyoid and infrahyoid muscles, which are accomplished by acting on the hyoid bone, the mandible, and the thyroid cartilage. These muscles can also flex the cervical column when the masseter and temporalis contract to close the mouth, thus ensuring a correct closure of the mouth while the masticatory muscles stabilize the temporomandibular joint.

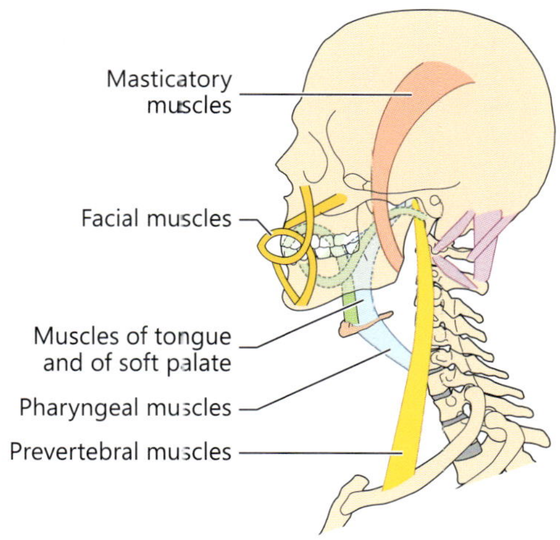

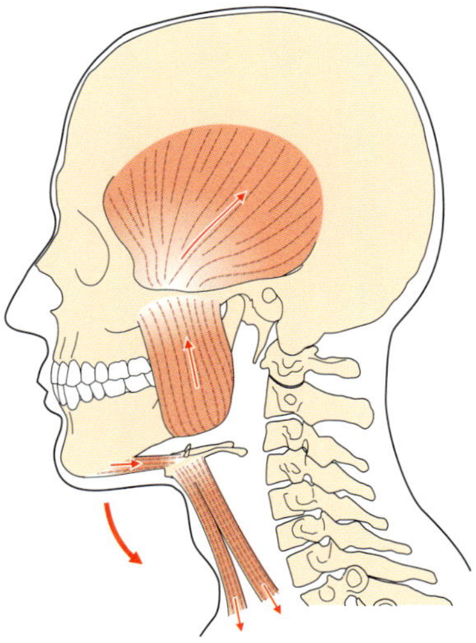

Flexion of the cervical column caused by the suprahyoid and infrahyoid muscles during closing of the mouth.

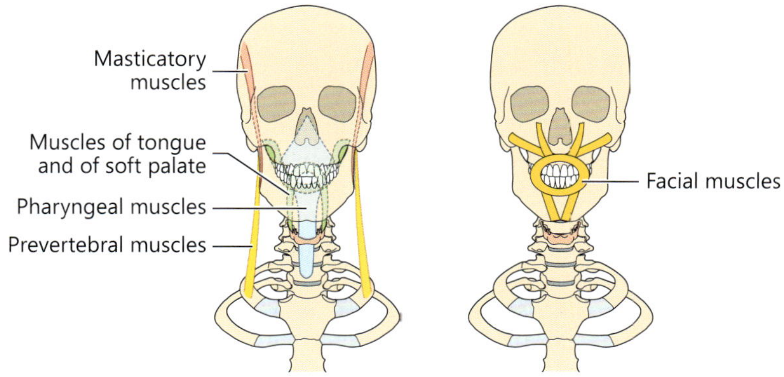

Arrangement of the suprahyoid and infrahyoid muscles in vertical rows.

Lateral inclination of the head and neck

It is produced by unilateral contraction of the muscles:
- levator scapulae (with a fixed point on the scapula)
- rectus capitis posterior minor
- rectus capitis posterior major
- obliquus capitis superior
- obliquus capitis inferior
- iliocostalis cervicis
- longissimus cervicis
- spinalis capitis
- spinalis cervicis
- semispinalis capitis
- semispinalis cervicis
- cervical intertransversarii
- interspinales cervicis
- longus capitis
- longus colli
- rectus capitis anterior
- rectus capitis lateralis
- scalene muscles.

Functional applications

- The unilateral contraction of the extensor muscles and of many flexor muscles of the head and neck tilts the head laterally and neck to the same side
- From a functional point of view, these muscles contract homolaterally to:
  - tilt the cervical column laterally
  - position and control the inclination of the head so as to allow one to correctly view objects placed on different planes
  - regulate the position of the head and facilitate corporal realignment in an upright position starting from a decubitus or upside down position
  - maintain the position of the head in the transition from the lateral decubitus position to the seated position, such as, for example, when you get out of bed.

Rotation of the head and neck

Ipsilateral rotation
It is produced by unilateral contraction of the muscles:
- rectus capitis posterior minor
- rectus capitis posterior major
- obliquus capitis inferior
- splenius capitis
- splenius cervicis
- longissimus capitis
- levator scapulae
- rectus capitis lateralis.

The most powerful ipsilateral rotator muscles are the splenius capitis and splenius cervicis.

Contralateral rotation
This is produced by the unilateral contraction of the following muscles:
- obliquus capitis superior
- semispinalis cervicis
- semispinalis capitis
- multifidus cervicis
- rotatores cervicis
- descending part (superior part) of the trapezius
- sternocleidomastoid
- longus capitis.

The most powerful contralateral rotator muscle is the sternocleidomastoid.

Functional applications

- The primary function of the rotator muscles of the head and neck consists of rotating the head and neck in order to look from one side to the other, for example when you look over your shoulder when backing up your car or when you hit the ball after having followed its trajectory while playing tennis
- In the rolling motion from the supine position to the lateral decubitus or prone position the movement of the trunk may begin with the rotation of the head and neck in the desired direction
- The rotator muscles of the head and neck contract when you express refusal with your head

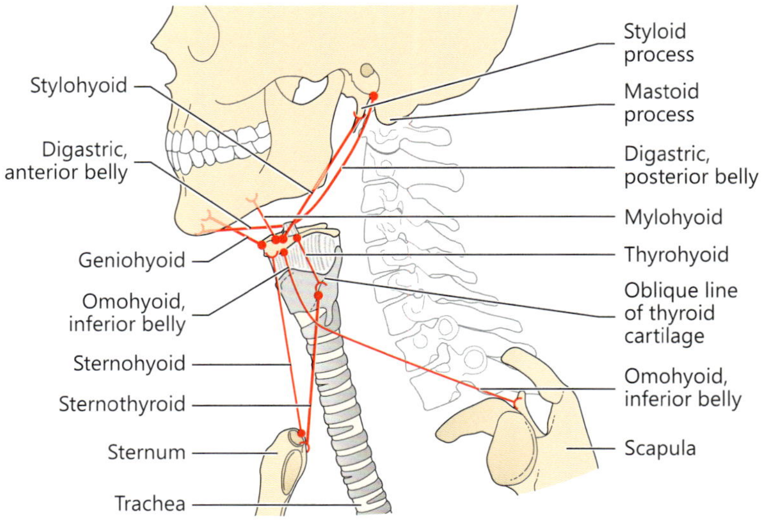

Suprahyoid and infrahyoid muscles: origins and insertions.

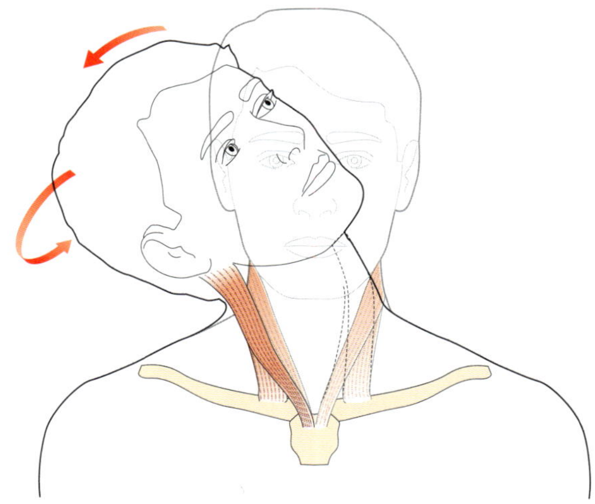

Action of the sternocleidomastoid: homolateral inclination, contralateral rotation.

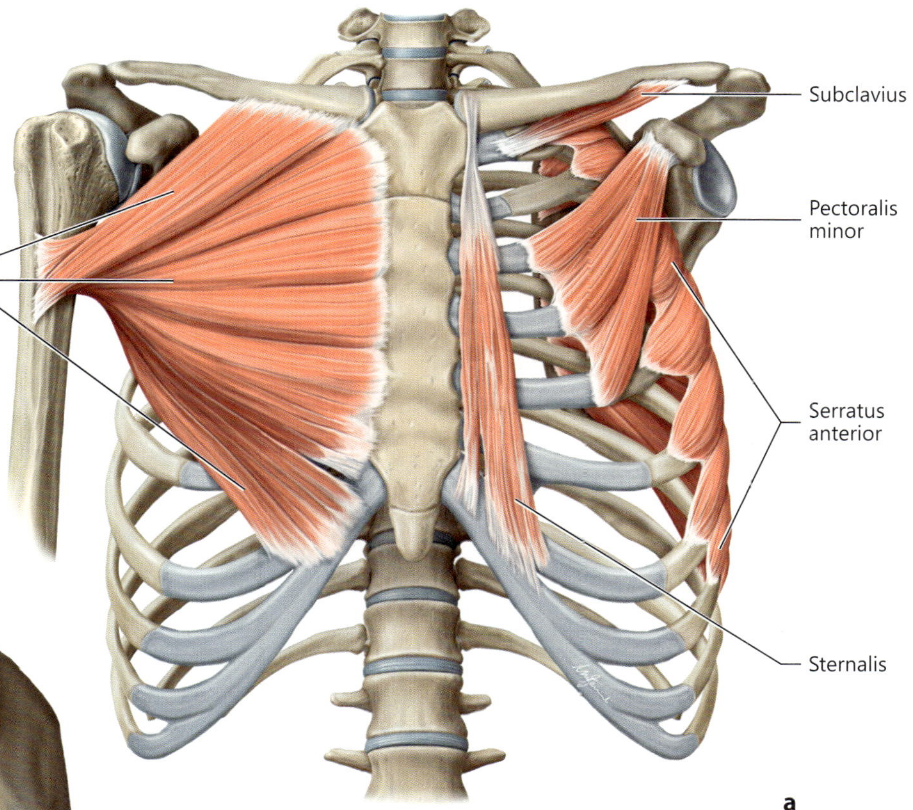

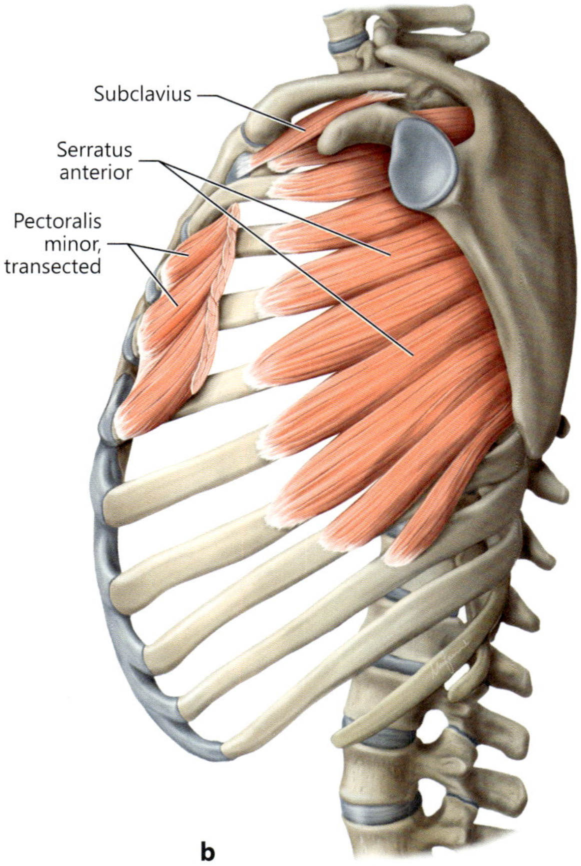

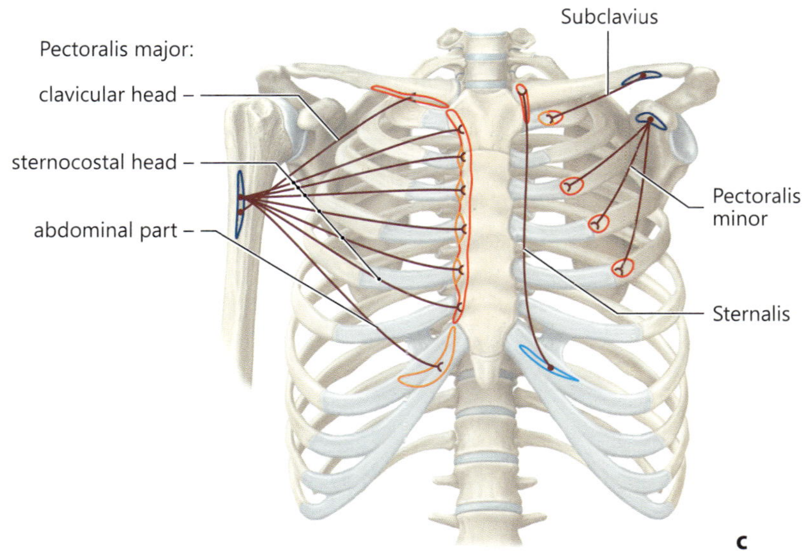

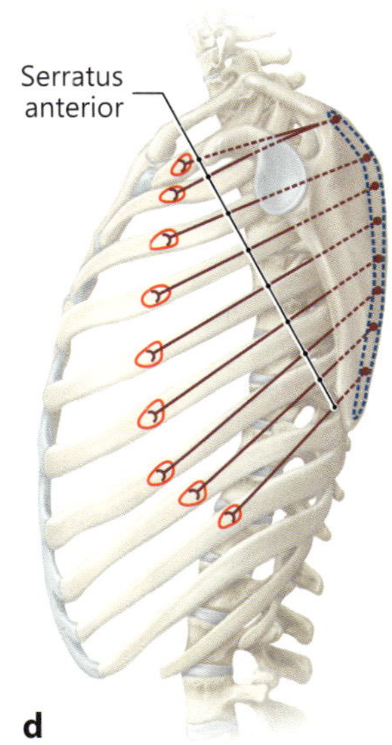

Preparation of the **muscles of the anterior and lateral wall of the thorax**, anterior and lateral views. **a**, On the left, the pectoralis major has been removed to uncover the pectoralis minor. **b**, The pectoralis minor has been transected to highlight the serratus anterior. **c-d**, Schematic representation of the origins and insertions of the pectoralis major, pectoralis minor, and serratus anterior. The **pectoralis major** consists of three heads: the clavicular head, the sternocostal head, and the abdominal part, which insert at the crest of the greater tubercle of the humerus. The **pectoralis minor** originates from the second, third, and fourth ribs, to insert on the coracoid process of the scapula. The **serratus anterior** originates from the lateral surface of the first ten ribs, to insert on the medial border of the scapula. The **sternalis** (inconstant) consists of a small muscular band located superficially and parallel to the costal border of the sternum between the clavicle and the last costal cartilages.

ORIGIN
- Bones
- Aponeuroses, cartilage, raphe

Origin ⟶ Insertion

INSERTION
- Bones
- Raphe, cartilage, aponeuroses, septum, tendon, capsule and articular disc

Preparation of the **intercostal muscles**. **a**, Anterior view of the rib cage; **b**, posterior view. In **a** you can see the opposite oblique course of the bundles in the external and internal intercostal muscles. Schematic representation of the origin and insertions of the **internal** (**c**) and **external** (**d**) **intercostal muscles**. The former extend from the medial lip of the costal groove to the superior border of the underlying cartilage and rib; the latter extend from the inferior border of the ribs to the superior border of the underlying rib.

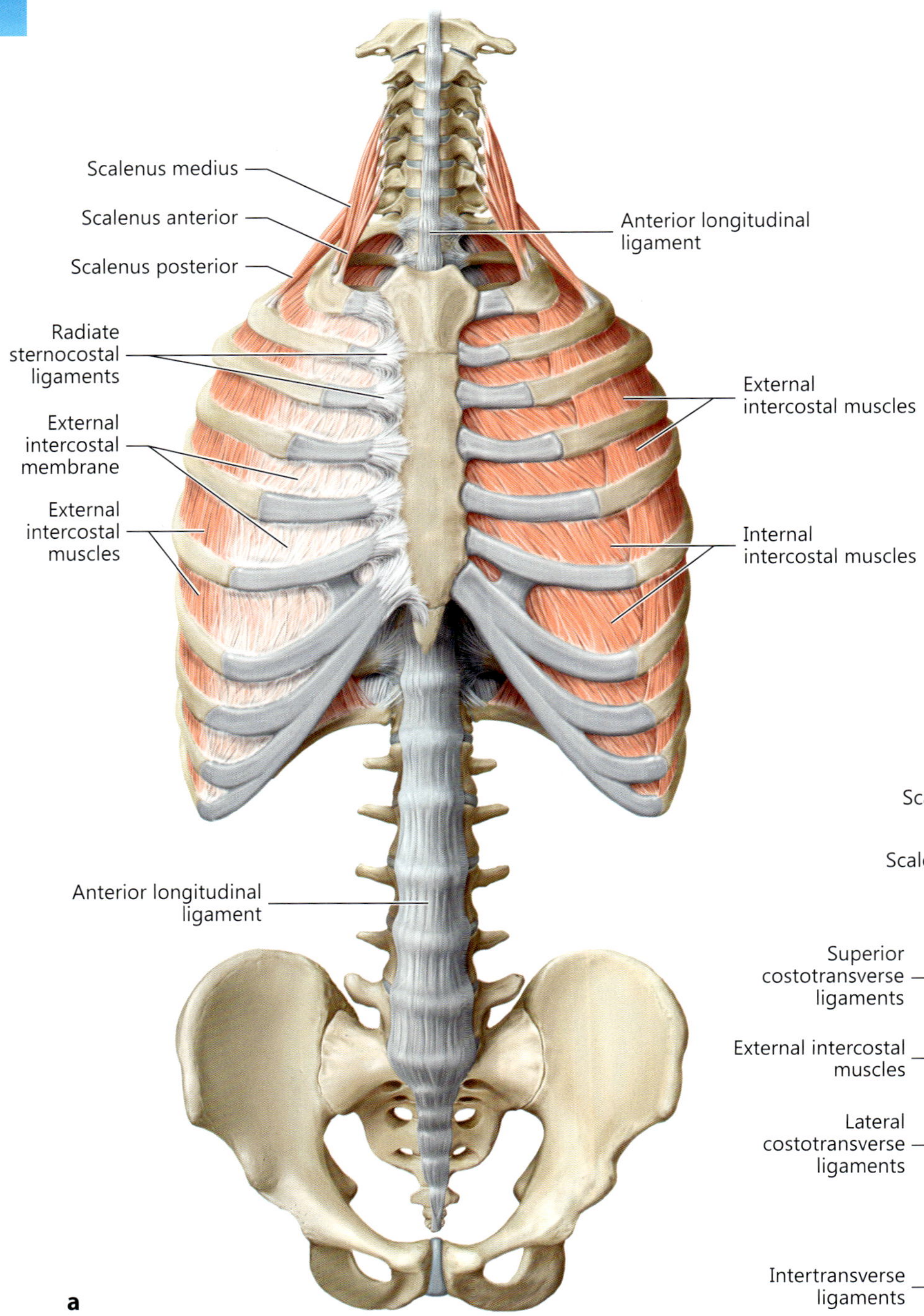

a

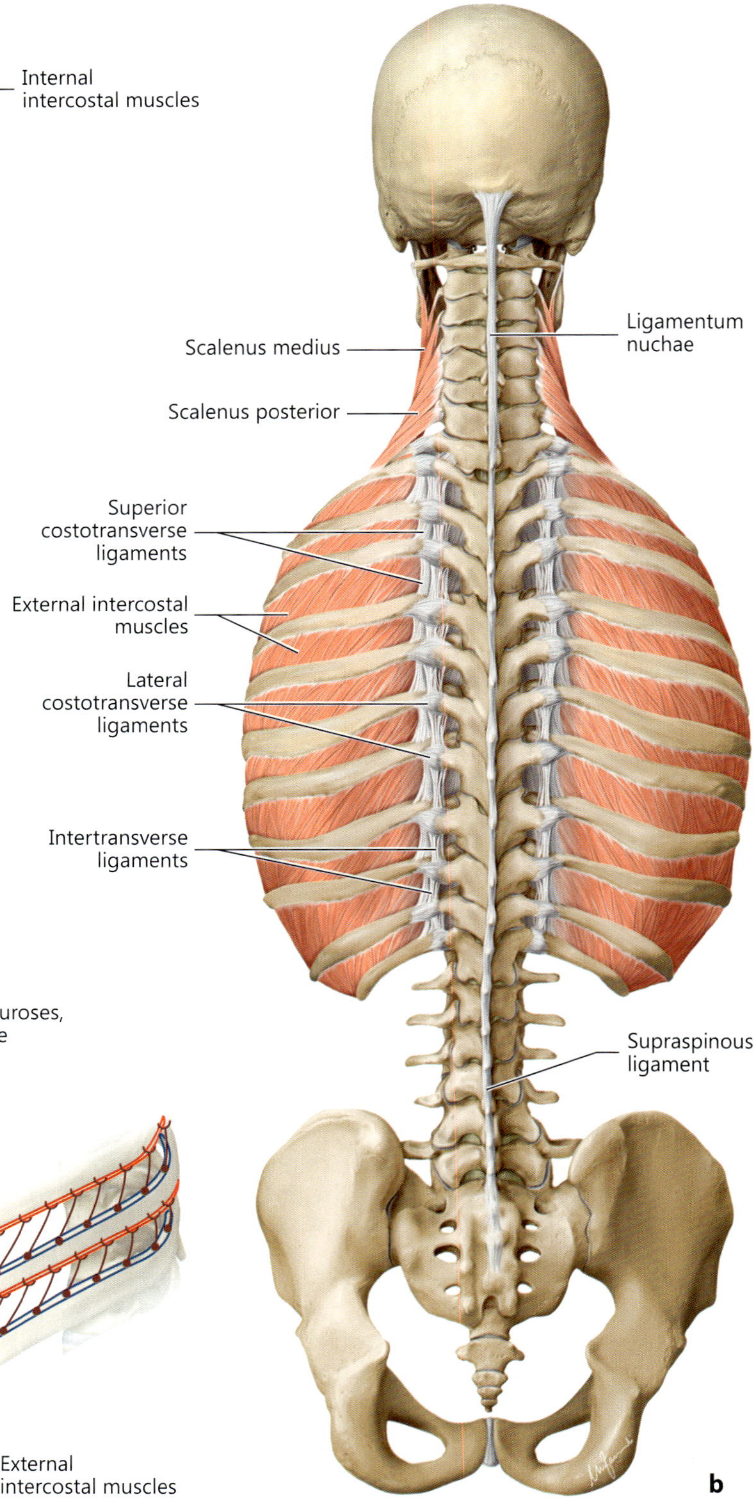

b

ORIGIN

- Bones
- Aponeuroses, cartilage, raphe

Origin ⟩—• Insertion

INSERTION

- Bones
- Raphe, cartilage, aponeuroses, septum, tendon, capsule and articular disc

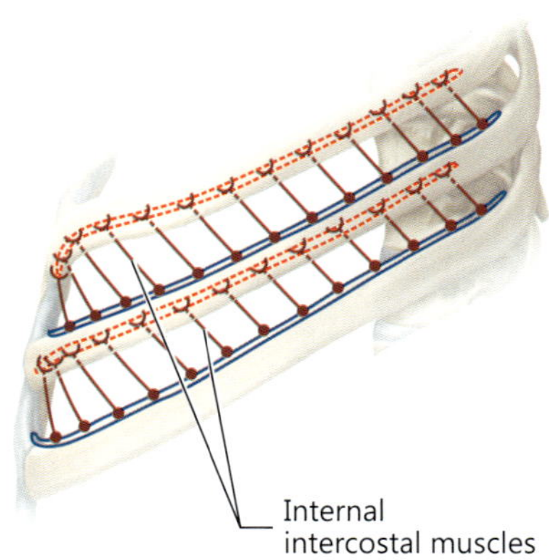

c

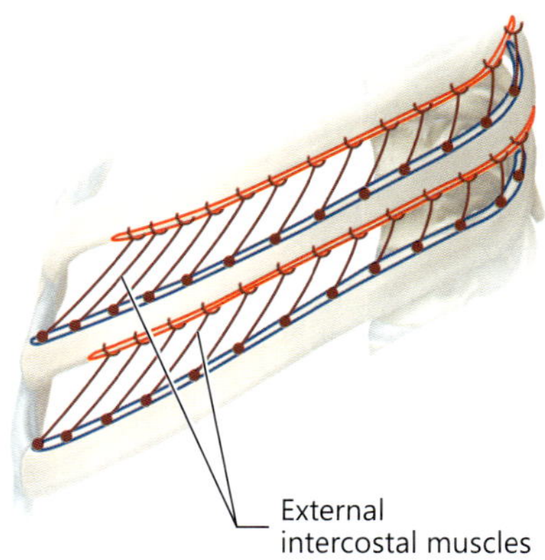

d

Muscles of the rib cage. **a**, Partially open thoracic wall, after removal of the left anterolateral half; **b**, right anterolateral half, including the sternum, internal view.
c, **d**, Schematic representation of the origins and insertions of the subcostales, transversus thoracis, innermost intercostal muscles, and sternalis. The **subcostales** extend from the internal surface of one rib to the internal surface of the underlying rib or the next rib; the **transversus thoracis**, from the posterior surface of the body and of the xiphoid process of the sternum, to the internal surface and inferior border of the costal cartilages from the second to the sixth ribs. The **innermost intercostal muscles** extend from the inferior border of the rib to the superior border of the underlying rib. The **sternalis**, which is not always present, is formed by one or more superficial vertical muscle fascicles that extend from the inferior costal cartilages, to insert into the top part of the sternum or superior costal cartilages.
Schematic representation of the origin and insertions of the **scalene muscles** ☞ pg. 454, **internal** and **external intercostal muscles** ☞ pg. 480.

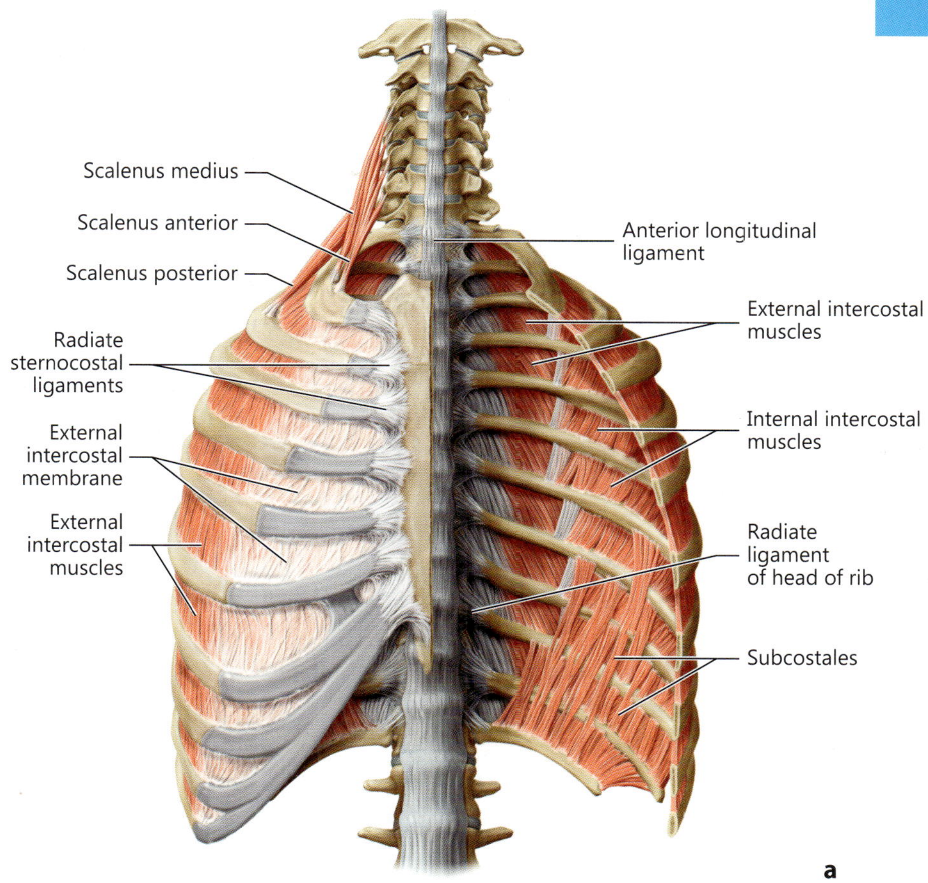

a

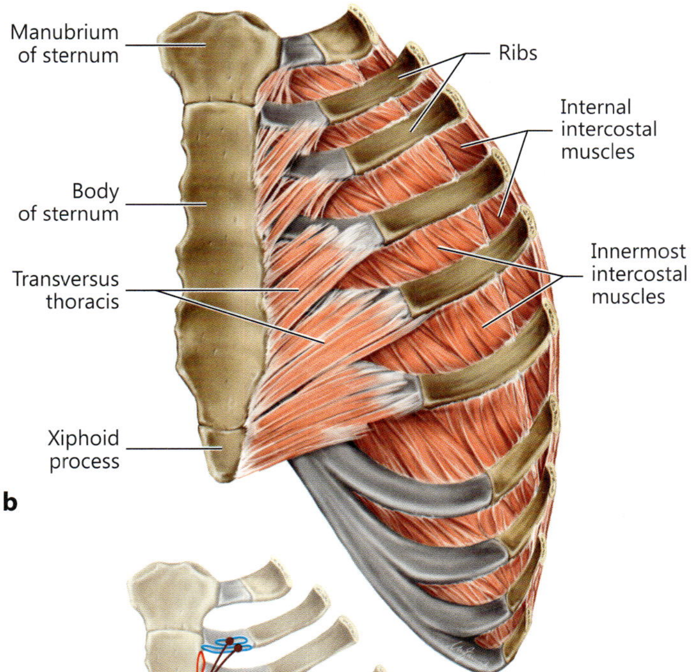

b

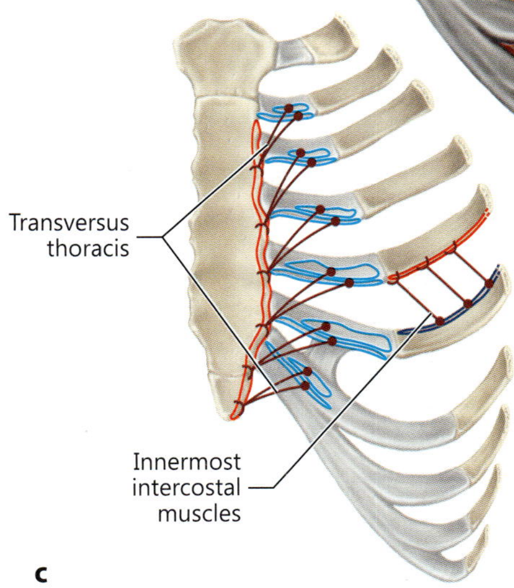

c

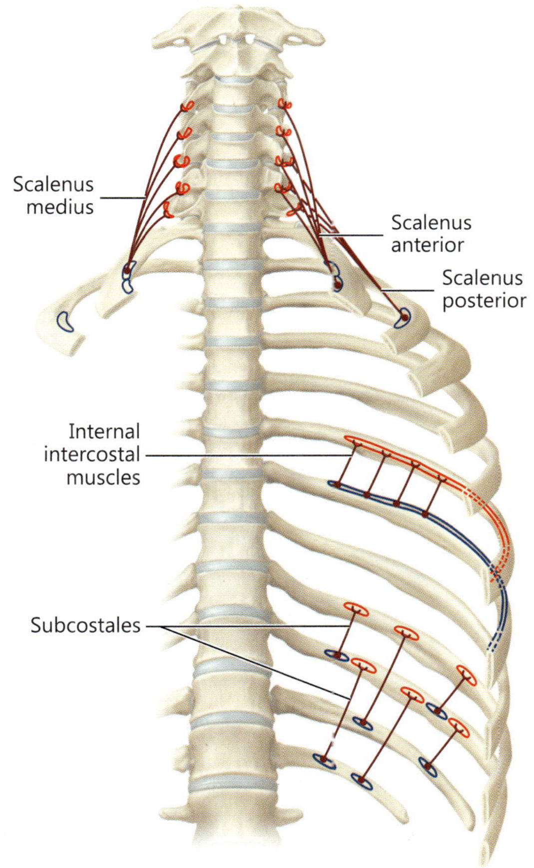

d

ORIGIN
Bones
Aponeuroses, cartilage, raphe
Origin ⟶ Insertion

INSERTION
Bones
Raphe, cartilage, aponeuroses, septum, tendon, capsule and articular disc

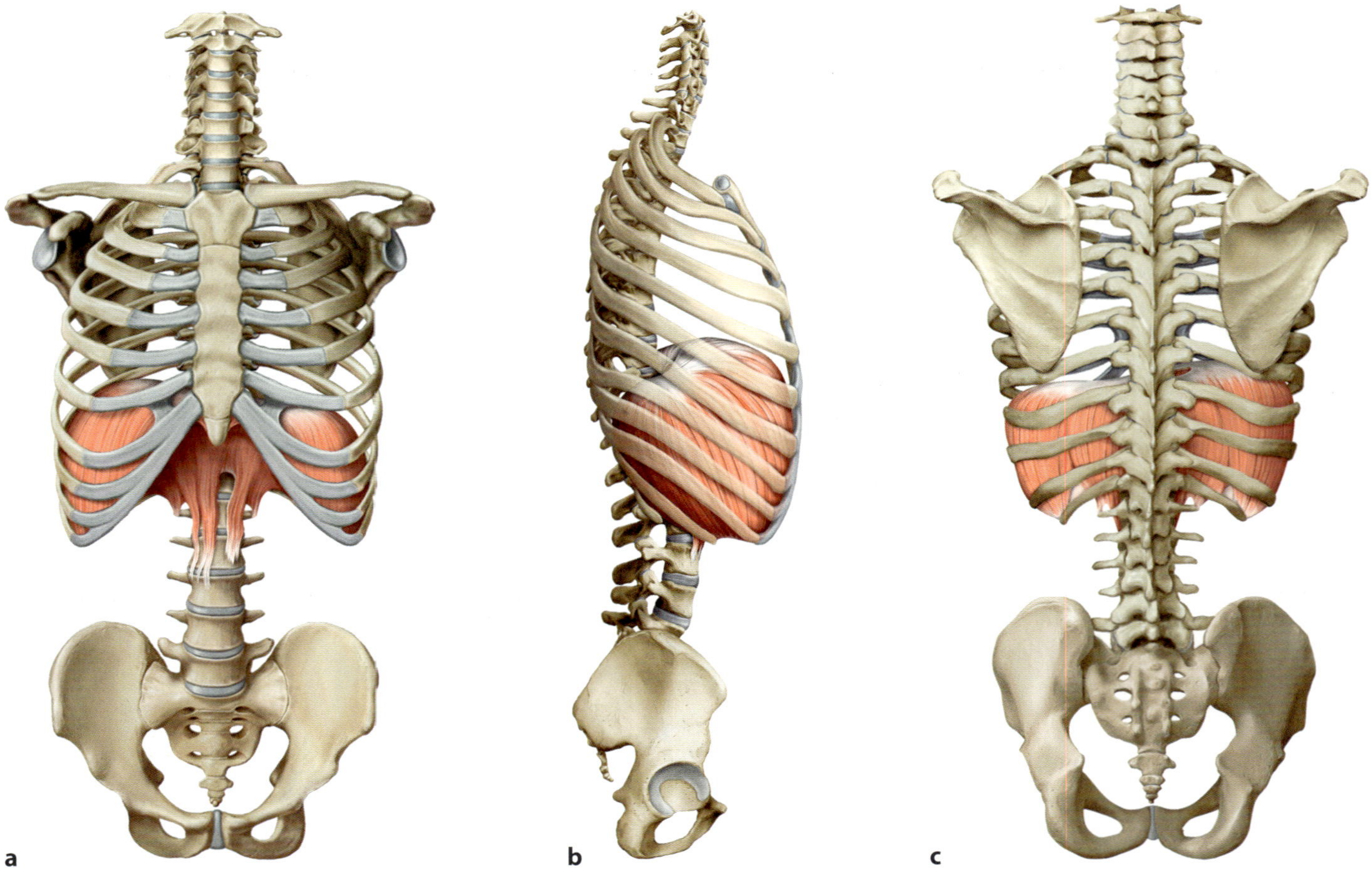

Diaphragm. **a**, Anterior view; **b**, right lateral view; **c**, posterior view.

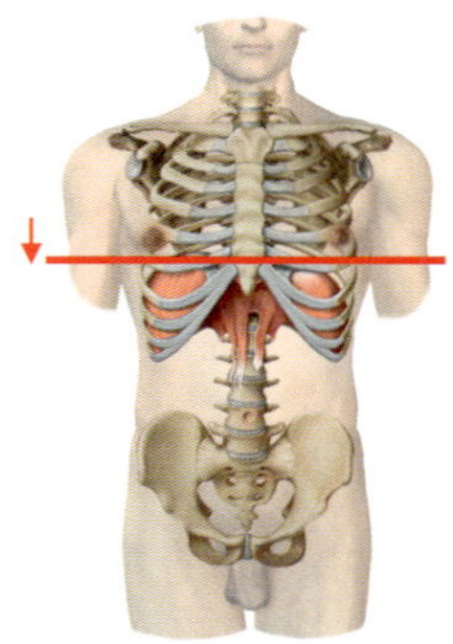

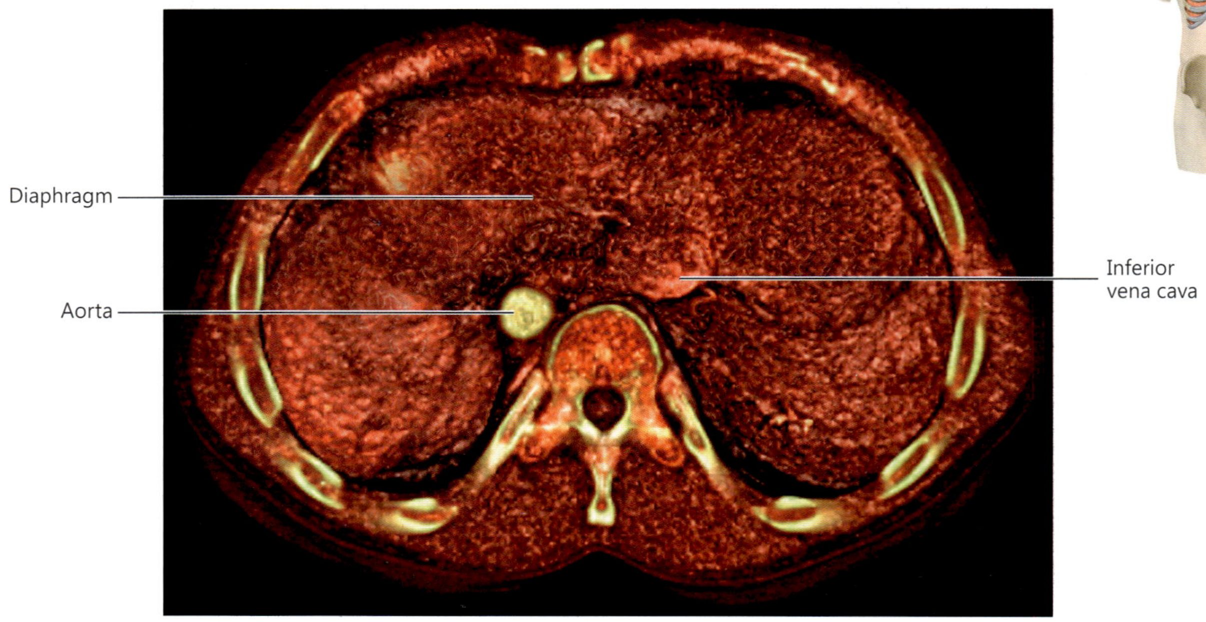

Three-dimensional image of a living human body, obtained by CT scan using a 3D volume rendering technique, showing the **transverse section of the trunk** obtained by a plane passing through the **diaphragm**.

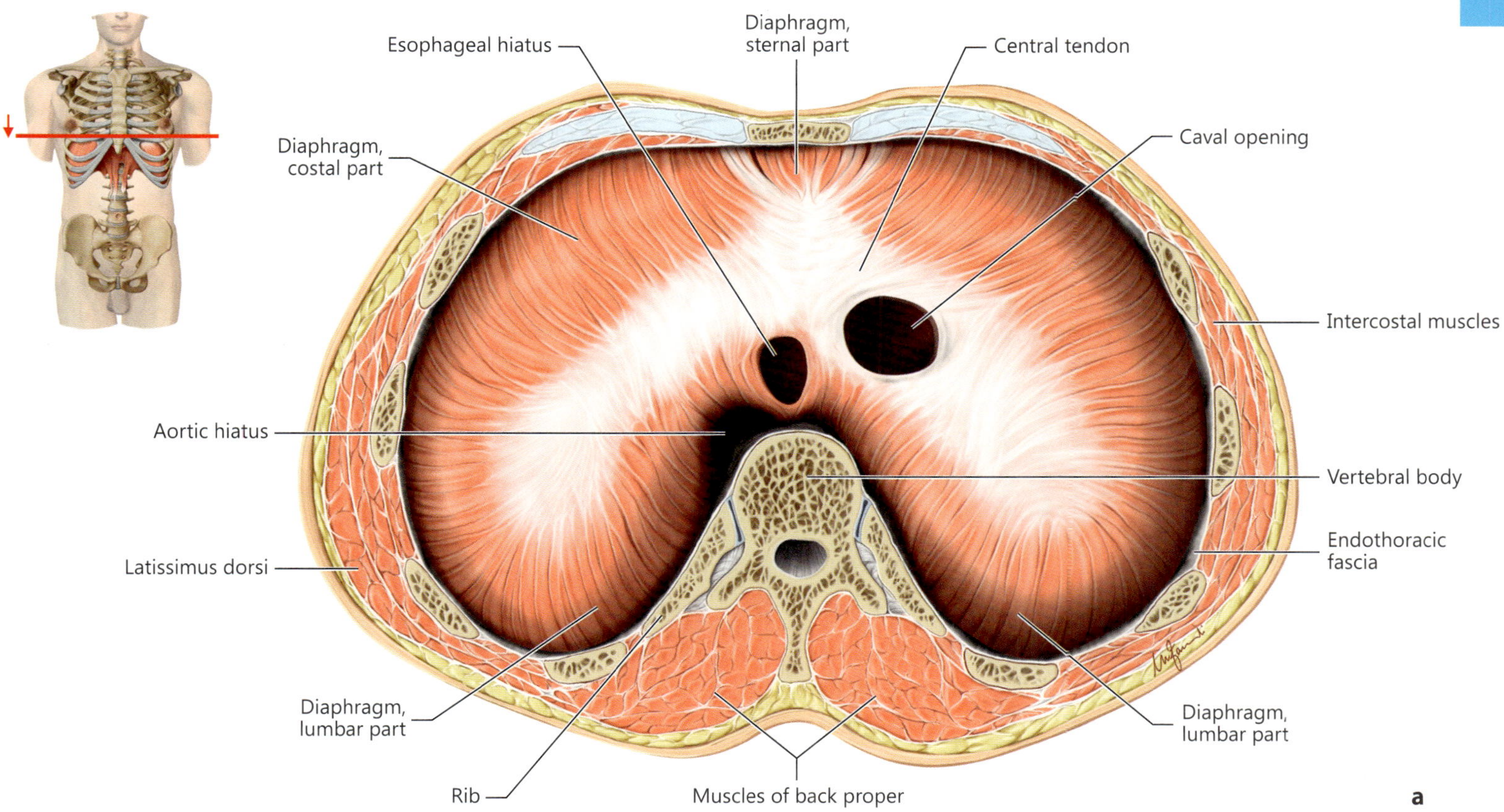

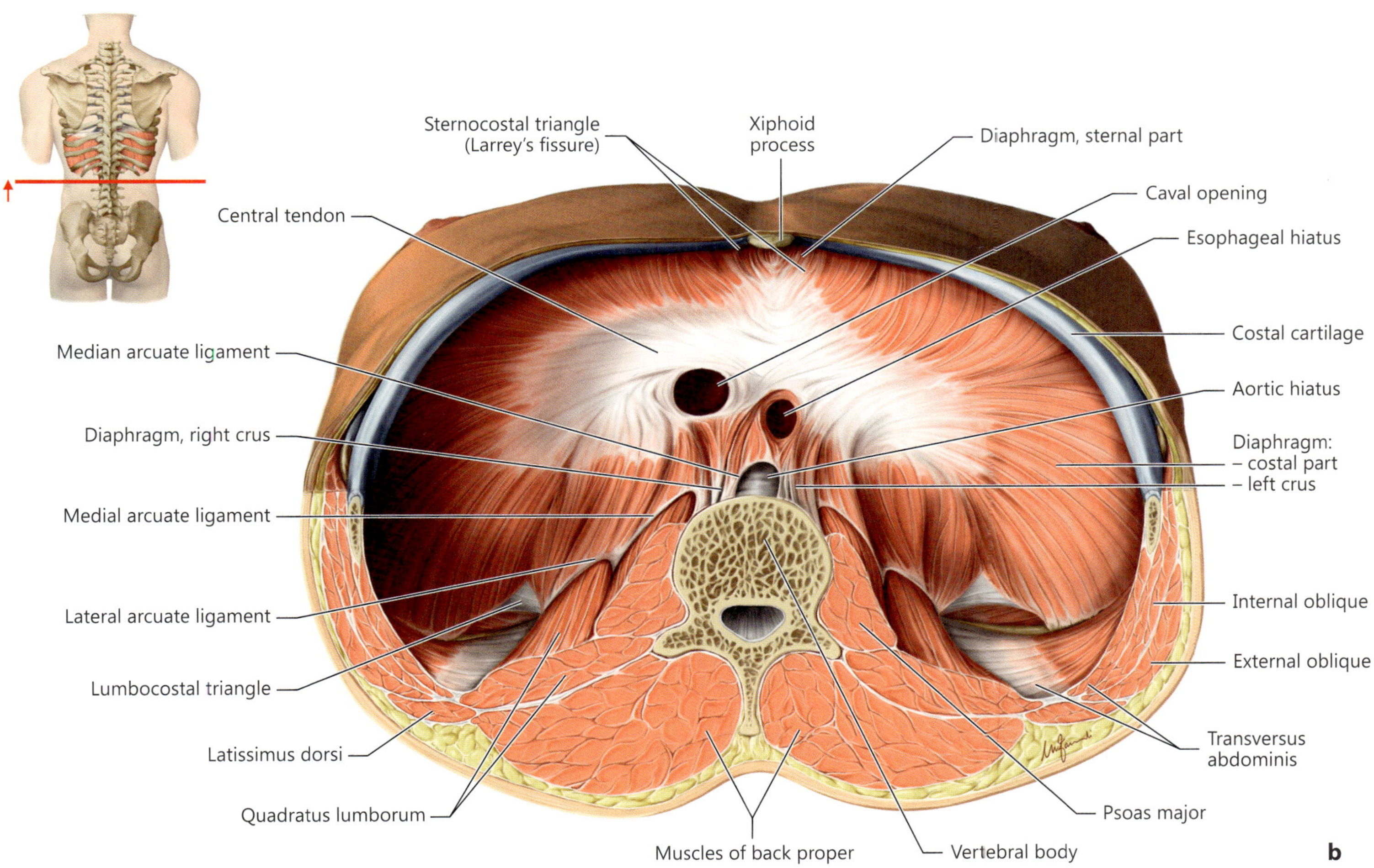

Diaphragm. **a**, Superior surface; **b**, inferior surface.

Origin, insertion, action, and innervation of the muscles of the thorax				
MUSCLE	***ORIGIN***	***INSERTIONS***	***INNERVATION***	***ACTION***
Serratus anterior	Lateral surface of the first ten ribs	Medial (vertebral) border of the scapula	Long thoracic nerve [C5-C7]	Forward and lateral traction of the scapula and its rotation when the arm is raised. When the scapula is fixed, elevation of the ribs (inspiratory muscle)
Diaphragm	Sternal part: posterior surface of the xiphoid process Costal part: internal surface of the last six costal cartilages and of the adjacent ribs on both sides Lumbar part: medial and lateral arcuate ligaments and lumbar vertebrae through two crura (right and left)	Central tendon of the diaphragm	Phrenic nerve [C3-C5] (the peripheral part of the muscle receives sensory fibers from the last six or seven intercostal nerves, T6-T12)	The initial contraction of the muscle pushes the central tendon downwards, which, once the movement has ceased, becomes a fixed point for the successive contraction of the fibers, with elevation of the second to tenth ribs and increase of the transverse and anteroposterior diameters of the thorax. This increase in thoracic capacity, which occurs during inspiration, confers the role of primary inspiratory muscle upon the diaphragm
Levatores costarum	Transverse processes of the seventh cervical vertebra and first eleven thoracic vertebrae	Superior border and external surface of the underlying rib between the tubercle and the costal angle	Posterior rami of the thoracic nerves [T1-T12]	Elevation of the ribs (inspiratory muscles)
External intercostal	Inferior border of the ribs	Superior border of the underlying rib	Intercostal nerves [T1-T12]	Contraction of the ribs (inspiratory muscles)
Internal intercostal	Medial lip of costal groove	Superior border of the underlying rib and cartilage	Intercostal nerves [T1-T12]	Contraction of the ribs (expiratory muscles)
Innermost intercostal	Inferior border of the ribs	Superior border of the underlying rib	Intercostal nerves [T1-T12]	Contraction of the ribs (expiratory muscles)
Pectoralis major	Clavicular head: medial half of the anterior border of the clavicle Sternocostal head: anterior surface of the sternum and from the second to the sixth costal cartilage Abdominal part: superior part of the anterior layer of the rectus sheath	Crest of the greater tubercle of the humerus	Medial and lateral pectoral nerves (C5-C8 and T1)	Flexion, adduction, and internal rotation of the shoulder. With a fixed point on the humerus, elevation of the trunk (inspiratory muscle)
Pectoralis minor	Superior border and external surface of the third, fourth and fifth ribs (often of the second, third, and fourth ribs)	Coracoid process of the scapula	Medial and lateral pectoral nerves (C5-C8 and T1)	Downward traction of the scapula. With a fixed point on the scapula, elevation of the ribs (inspiratory muscle)
Subcostales	Internal surface of the rib near the costal angle	Internal surface of the underlying rib or the one next to it	Intercostal nerves [T1-T12]	Depression of the ribs (expiratory muscles)
Sternalis	Small muscular bundle, the presence of which is inconstant, located superficially to the sternocostal head of the pectoralis major (parallel to the border of the sternum)			
Sternocostal	☞ Transversus thoracis			
Subclavius	Superior surface of the first rib and adjacent costal cartilage	Inferior surface of the clavicle (subclavian groove)	Subclavian nerve [C5, C6]	Depression of the clavicle
Transversus thoracis	Posterior surface of the xiphoid body and process of the sternum	Internal surface and inferior border of the costal cartilages, (from the second to the sixth)	Intercostal nerves [T2-T6]	Downward traction of the costal cartilages (expiratory muscle)

For the intrinsic muscles of the vertebral column ☞ pg. 435-437

Note: Other muscles located in the thorax or with an origin or insertion in the thorax
The **pleuroesophageus** and **trachealis** are bundles of smooth muscle fibers that extend, respectively, from the left mediastinal pleura to the muscular layer of the esophagus, and as a bridge between the free ends of the C-shaped tracheal cartilages.
The **scalene muscles**, which insert at the first two ribs. and the **sternocleidomastoid**, which originates from the sternum and clavicle, are considered among the muscles of the neck (☞ pg. 452).
The **serratus posterior superior** and **inferior** (**spinocostal muscles**), which insert respectively at the second to fifth ribs and at the last four ribs, originating at the thorax level from the spinous processes T I and T II, and T XI and T XII, and the **latissimus dorsi** and **trapezius** (**spinoappendicular muscles**), originating at the thorax level, respectively, from the spinous processes from T VI or T VII to T XII and from T I to T XII, which line the thorax posterolaterally and posterosuperiorly, are considered among the muscles of the back (☞ pg. 412-413, 417).
The **quadratus lumborum**, which inserts at the twelfth rib, is considered among the muscles of the abdomen (☞ pg. 488, 497).

Rectus abdominis and **pyramidalis**. **a**, Anterior view. The **rectus abdominis** appears as a thick layer extending from the rib cage to the pelvis. The **pyramidalis** is located on the lower portion of the abdominal wall, superficial to the rectus abdominis. **b**, Schematic representation of the origin and insertion of the **rectus abdominis**: from the external surface of the fifth, sixth, and seventh costal cartilages, and from the xiphoid process of the sternum to the superior pubic ramus, between the pubic tubercle and symphysis. The latter is the origin of the **pyramidalis** which ascends upward to insert on the linea alba.

ORIGIN
Bones
Aponeuroses, cartilage, raphe
Origin ⟶ Insertion
INSERTION
Bones
Raphe, cartilage, aponeuroses, septum, tendon, capsule and articular disc

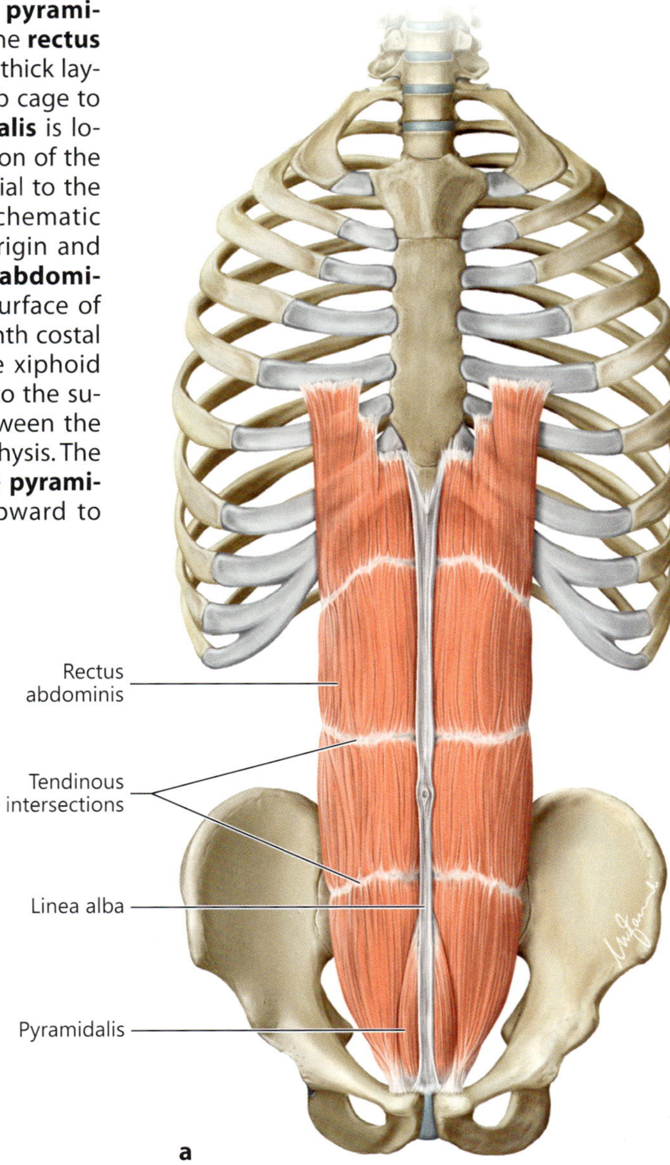

a

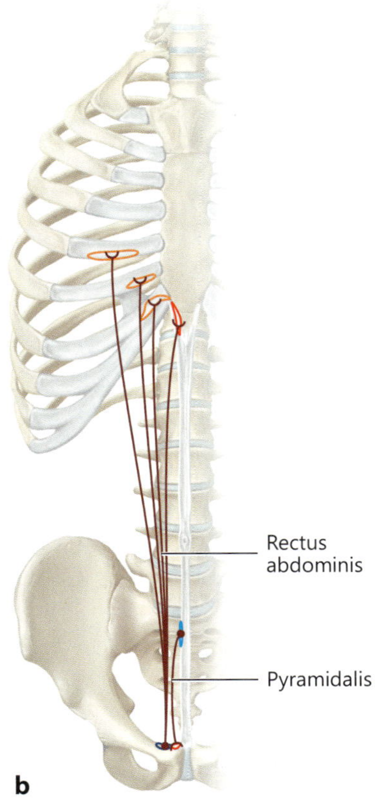

b

Iliopsoas. a, Anterior view. The muscle consists of two parts, the **psoas major** and the **iliacus**. The two muscles unite distally to insert at the lesser trochanter of the femur. The psoas minor is positioned anterior to the psoas major and is not consistently present. **b**, Schematic representation of the origin and insertions of the **iliopsoas** and **psoas minor**. The psoas major extends from the transverse processes, the vertebral bodies and intervertebral discs of the lumbar vertebrae, and from the twelfth thoracic vertebra to the lesser trochanter of the femur. The iliacus extends from the iliac fossa (upper two-thirds), from the iliac crest (inner lip), from the anterior sacroiliac and iliolumbar ligaments, and from the upper part of the lateral surface of the sacrum, to the tendon (lateral surface) of the psoas major and to the lesser trochanter of the femur. The psoas minor extends from the last thoracic vertebra and the first lumbar vertebra to the iliopubic ramus and the iliac fascia.

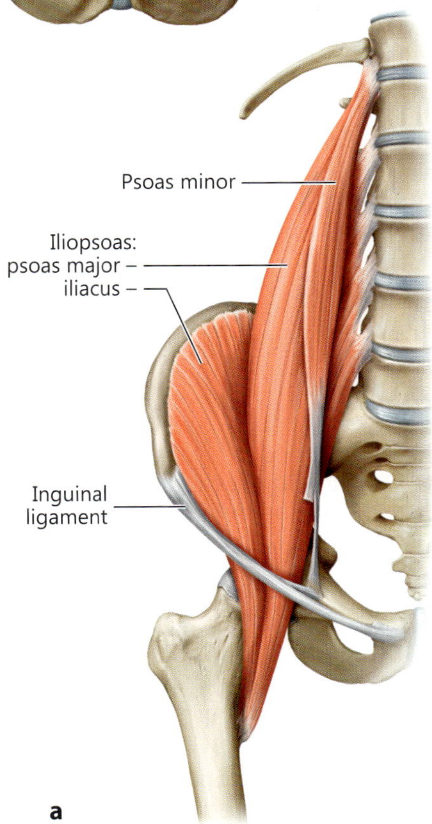

a

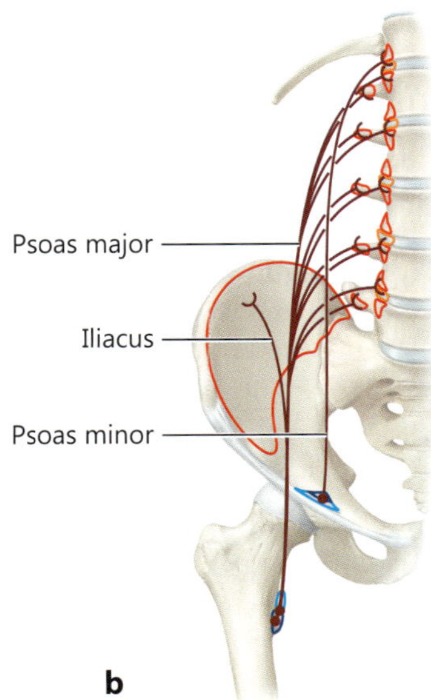

b

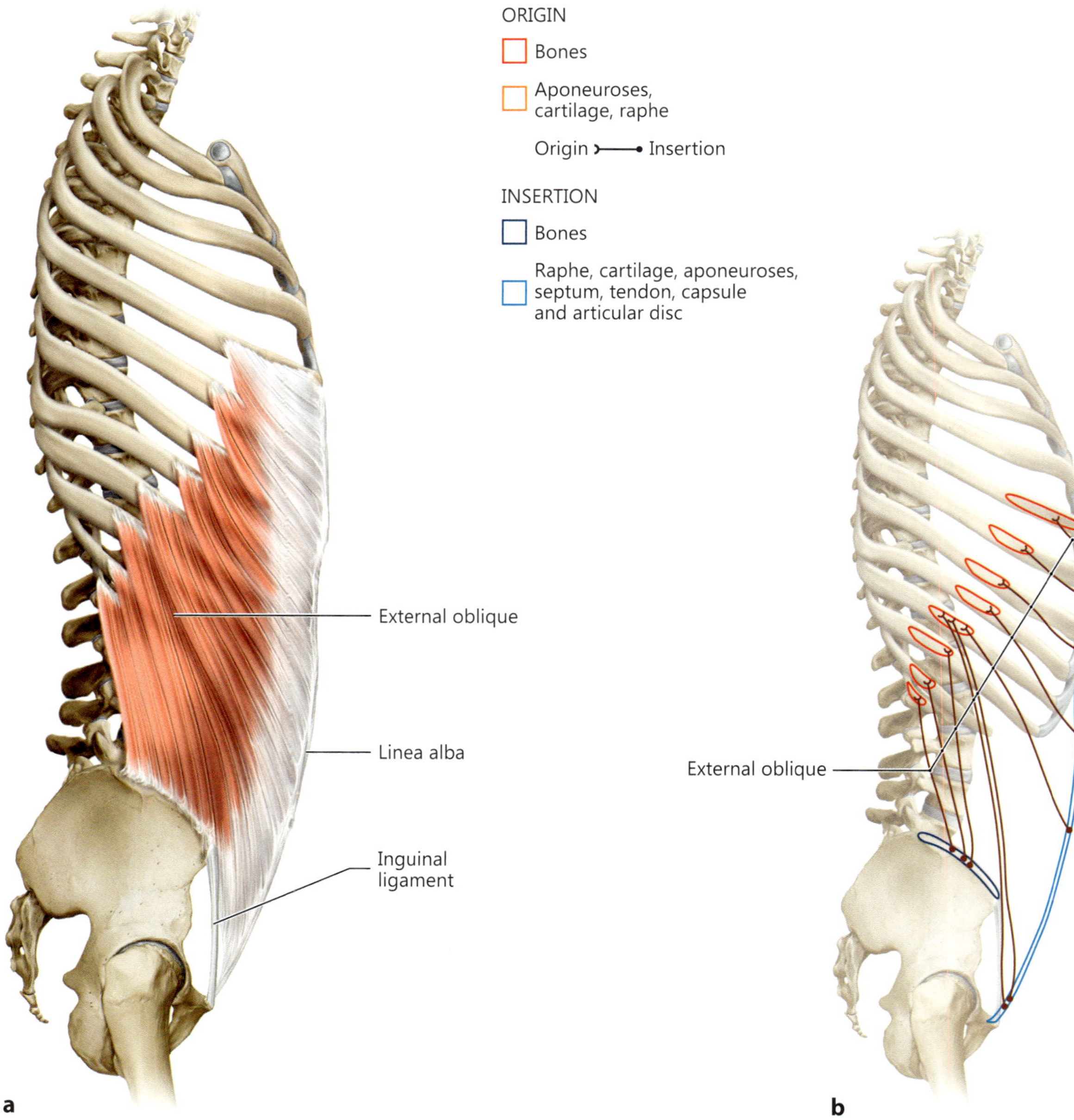

External oblique. a, The muscle extends from the rib cage to the pelvis, moving from top to bottom and back to front. The fibers of its aponeurosis help to form the anterior layer of the rectus sheath, and they contribute to the formation of the **linea alba** by intersecting with the fibers on the opposite side. The lower fibers form the inguinal ligament, stretched between the anterior superior iliac spine and the pubic tubercle. **b**, Schematic representation of the origin and insertion of the external oblique: from the external surface of the last eight ribs to the outer lip of the iliac crest.

➤

Internal oblique. c, Located deeply in relation to the external oblique, it extends from the inferior part of the abdomen to the rib cage, moving from bottom to top and back to front. The anterior splitting layer of its aponeurosis helps to form the anterior layer of the rectus sheath, and after intersecting with that of the opposite side it contributes to the formation of the **linea alba**. The posterior splitting layer contributes to the formation of the posterior layer of the rectus sheath (in the upper three-quarters). **d**, Schematic representation of the origin and insertion of the internal oblique: from the inguinal ligament, the anterior superior iliac spine, the intermediate line, and the thoracolumbar fascia, to the last three costal cartilages.

Transversus abdominis. e, Located deeply in relation to the internal oblique, it is arranged transversely with posteroanterior direction of the fibers. Its aponeurosis contributes to the formation of the linea alba, as well as the posterior layer (in the upper three-quarters) and the anterior layer (in the lower quarter) of the rectus sheath. The transversalis fascia completes the formation of the posterior layer of the sheath. **f**, Schematic representation of the origin of the transversus abdominis: from the last six costal cartilages, the thoracolumbar fascia, the iliac crest, and the inguinal ligament.

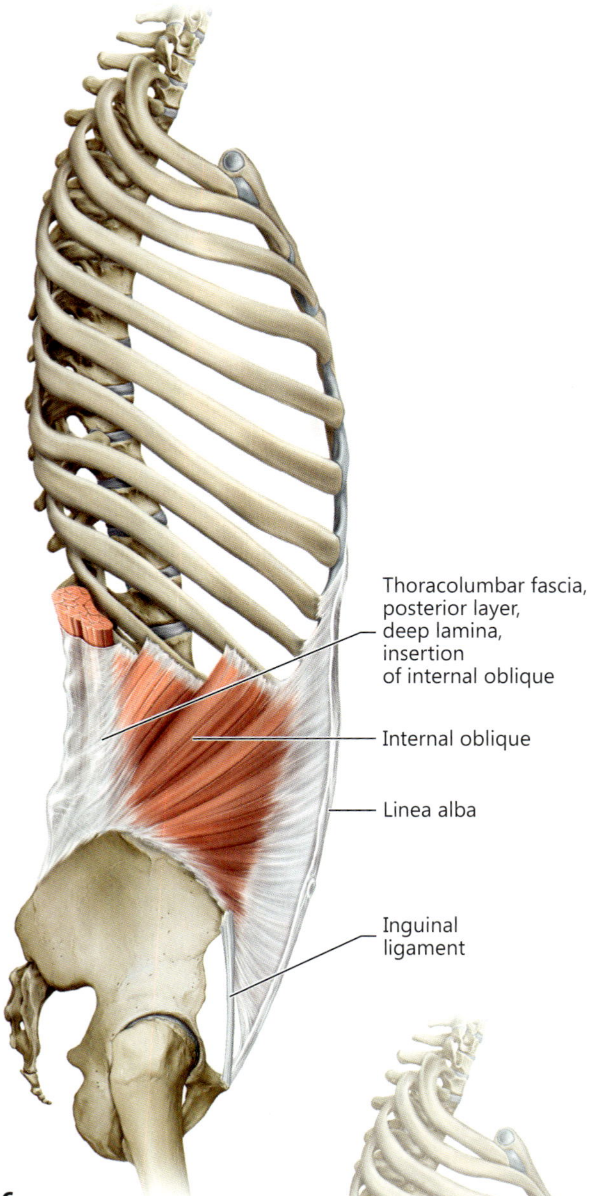

c

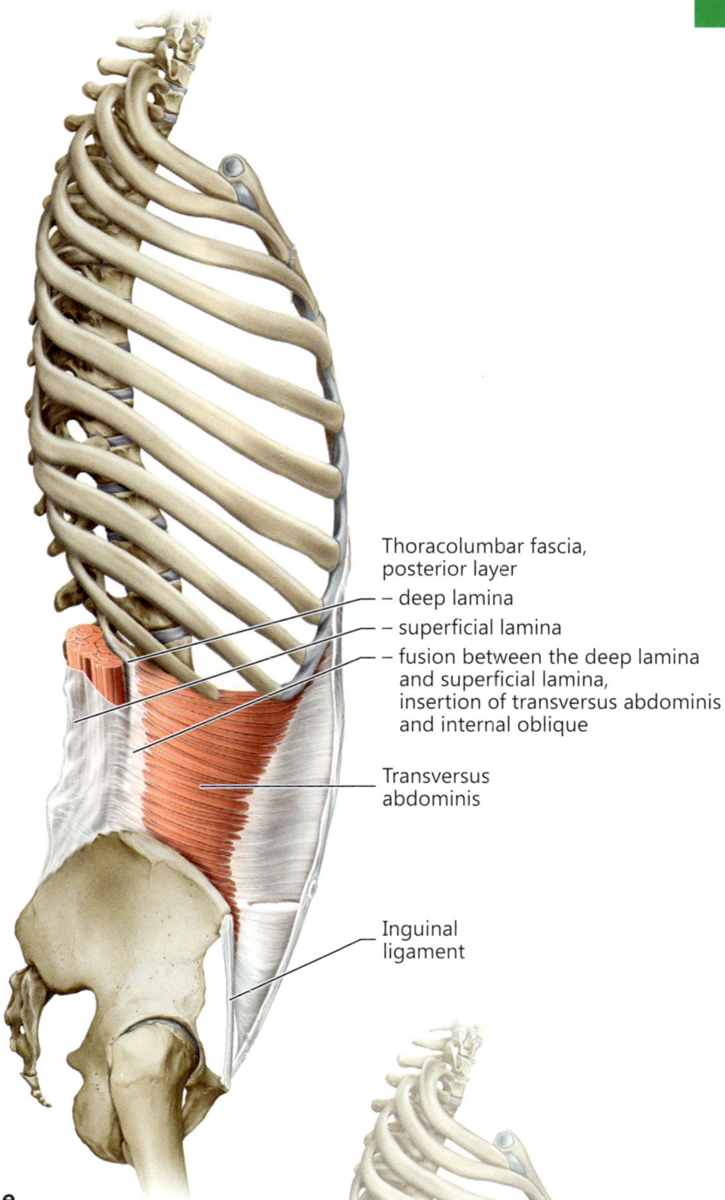

e

ORIGIN
- Bones
- Aponeuroses, cartilage, raphe

Origin ⟶ Insertion

INSERTION
- Bones
- Raphe, cartilage, aponeuroses, septum, tendon, capsule and articular disc

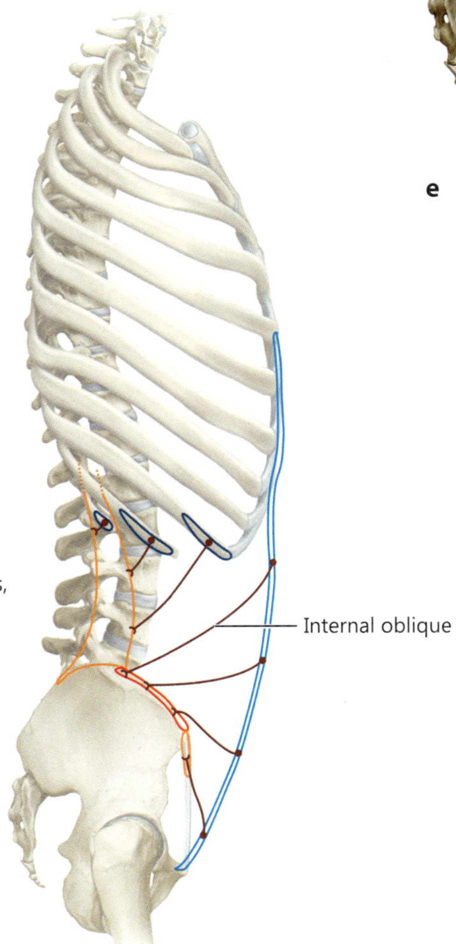

d

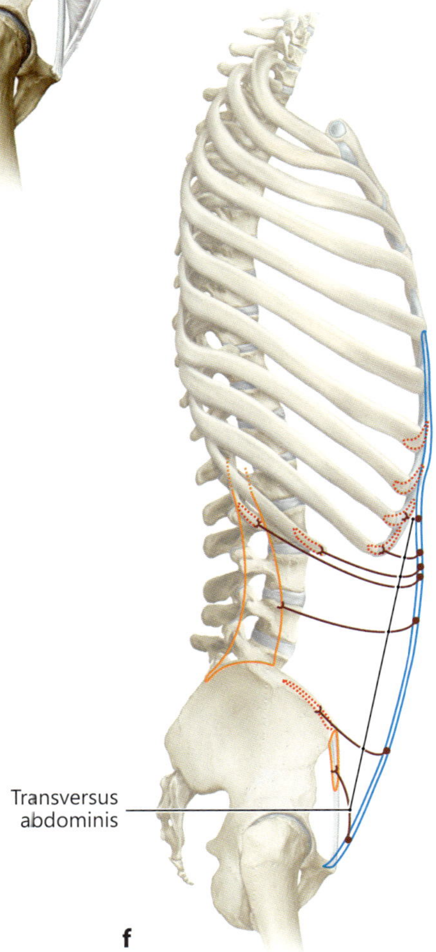

f

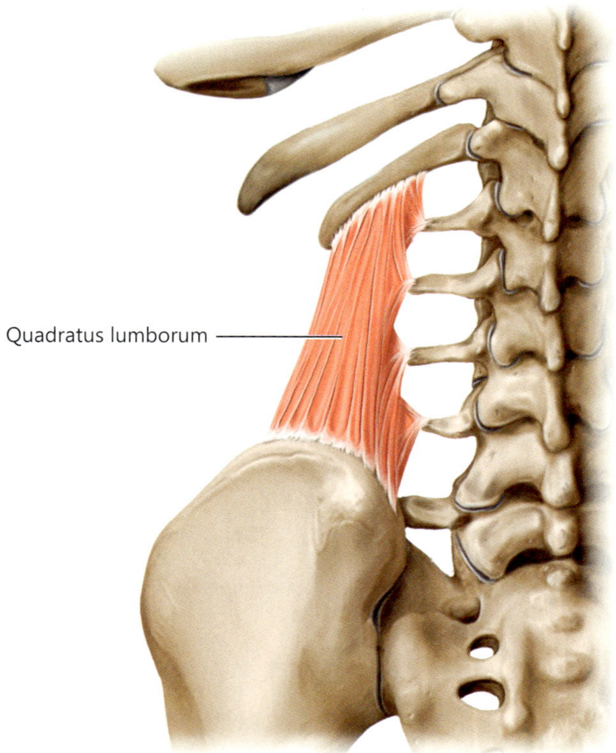

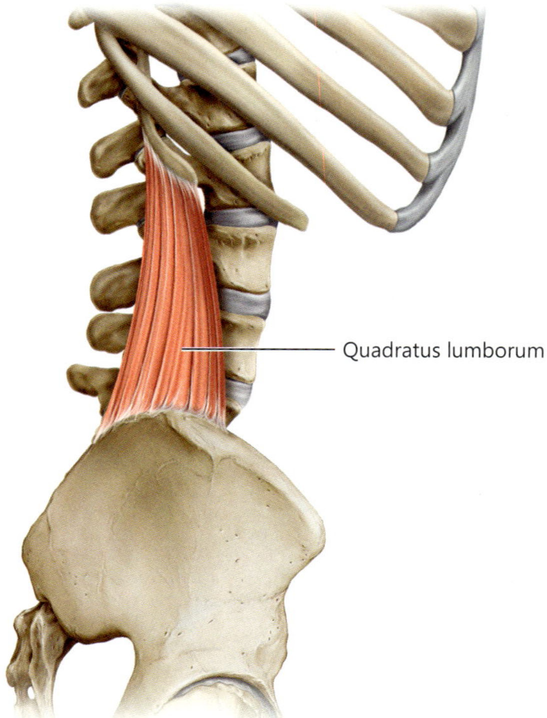

Preparation showing the points of origin and insertion of the **quadratus lumborum**.

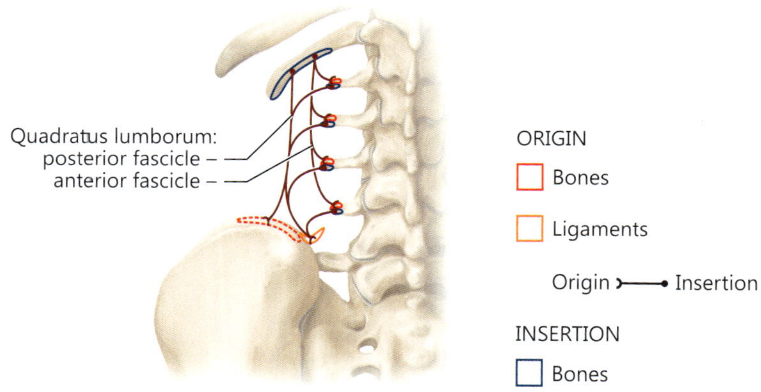

Schematic representation of the origin and insertions of the **quadratus lumborum**. The posterior fascicle originates from the inner lip of the iliac crest, and inserts on the costal processes of the first four lumbar vertebrae and on the inferior border of the twelfth rib. The anterior fascicle originates from the costal processes of the second, third, and fourth lumbar vertebrae, and inserts on the twelfth rib.

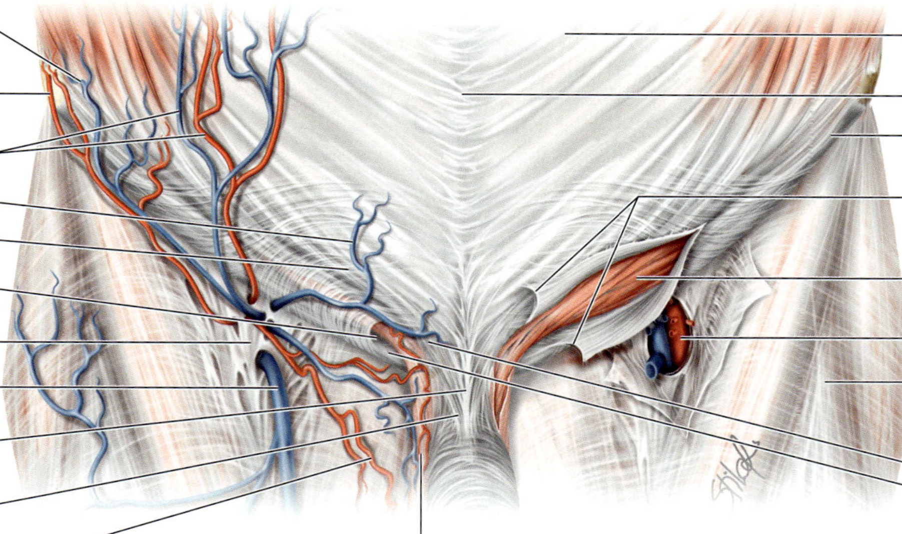

Inguinal region and inferior part of the anterior abdominal wall. To the left is a section of the anterior wall of the **inguinal canal**, consisting of the aponeurosis of the external oblique which adheres to the anterior border of the inguinal ligament, highlighting the cremaster, which covers the inguinal and scrotal portions of the spermatic cord in males. In females, the cremaster is represented by a few fibers, which descend to the round ligament of the uterus. The cremasteric fascia (or external spermatic fascia) covers the scrotal portion of the spermatic cord, detaching from the aponeurosis of the external oblique at the external inguinal ring. The suspensory and fundiform ligaments form the suspensory system of the penis in males (or maintain the position of the angle of the clitoris in females). The first originates from the pubic symphysis, the second from the linea alba, and both join with the fascia of penis.

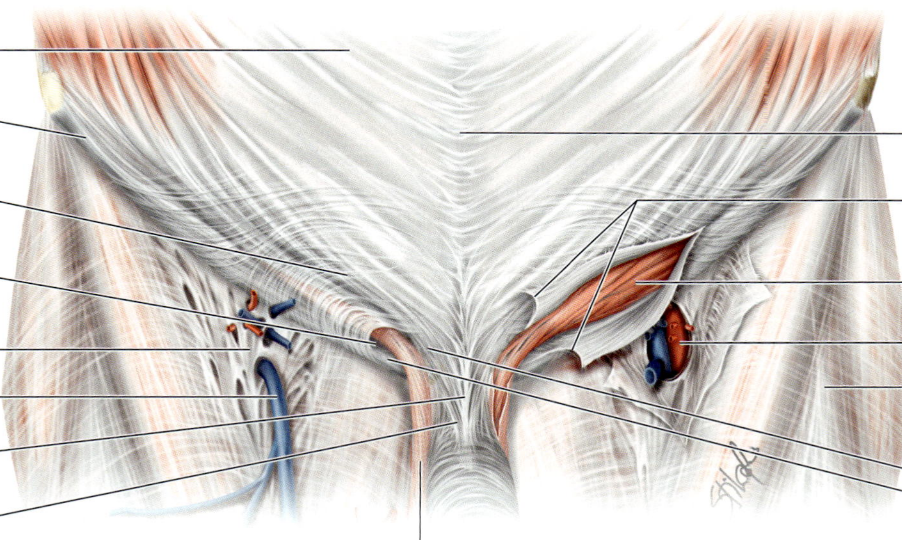

In males, the superficial **inguinal ring** is crossed by the spermatic cord, which in the scrotal portion is also covered externally by the cremasteric fascia; in females it is crossed by the round ligament of the uterus. The fibers of the aponeurosis of the external oblique, which originate from the anterior superior iliac spine, extend toward the pubic tubercle forming the lateral crus of the ring, while those that reach the pubic symphysis form the medial crus. The suspensory and fundiform ligaments of the penis, shown here, are replaced in females by the suspensory and fundiform ligaments of the clitoris. The cremaster is formed by the muscular fascicles loosely arranged along the spermatic cord in males, and in females consists of a few fibers that descend into the round ligament of the uterus.

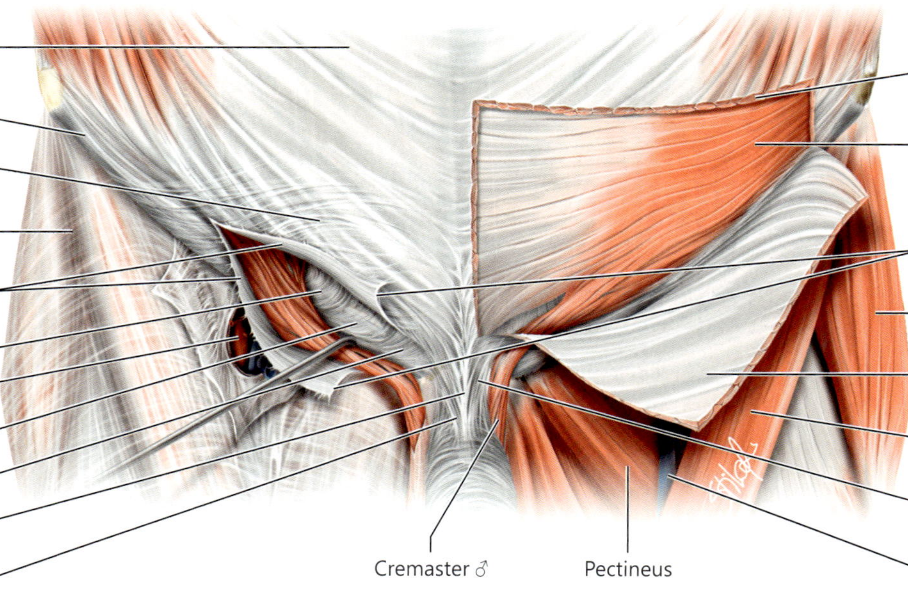

The **superior wall of the inguinal canal** corresponds to the inferior border of the internal oblique and transversus abdominis, and the inferior wall corresponds to the medial half of the inguinal ligament, which includes the spermatic cord in males and the round ligament of the uterus in females. The posterior wall is formed by the transversalis fascia, which attaches to the posterior border of the inguinal ligament. At this level, the transversalis fascia is reinforced laterally by the interfoveolar ligament, medially and from inside to outside by the conjoint tendon (or inguinal falx), and by the reflected ligament. The suspensory and fundiform ligaments of the penis are replaced in females by the suspensory and fundiform ligaments of the clitoris. The cremaster is represented in females by few fibers which descend into the round ligament of the uterus.

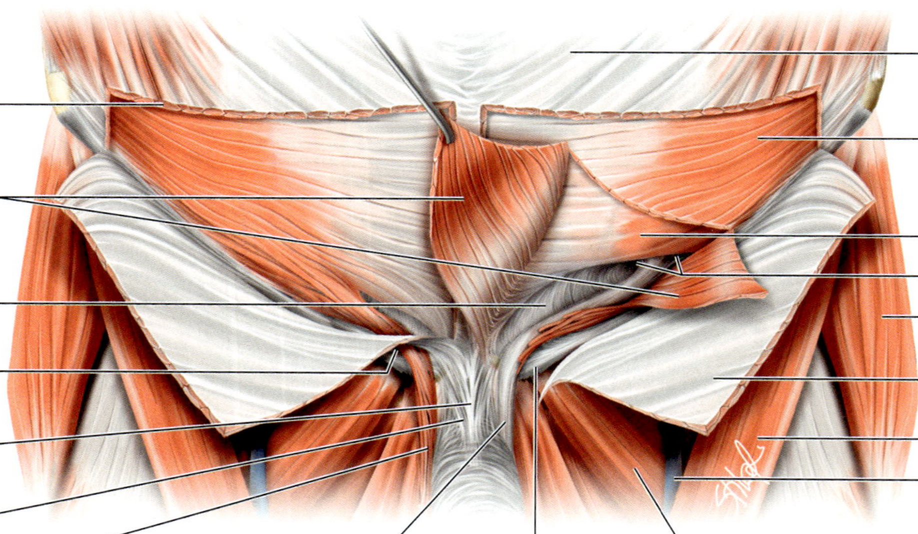

The **deep inguinal ring** corresponds with the point where the parietal peritoneum and the transversalis fascia invaginate to accommodate the descent of the testis into the scrotum. In adults, this gives way to the spermatic cord in males or the round ligament in females. The suspensory and fundiform ligaments of the penis are replaced in females by the suspensory and fundiform ligaments of the clitoris. The cremaster is represented in females by few fibers which descend into the round ligament of the uterus.

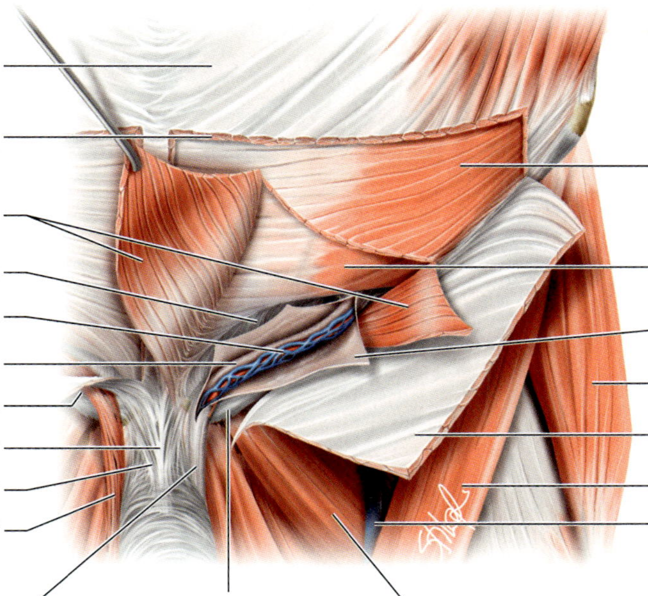

The internal spermatic fascia, or tunica vaginalis communis, which derives from the transversalis fascia and forms the innermost lining of the spermatic cord, has been transected in the inguinal canal, highlighting the **pampiniform plexus** and the **ductus deferens**. In males, the pampiniform plexus is formed by the testicular veins and forms one of the main components of the spermatic cord, moving anteriorly to the ductus deferens. In females, the pampiniform plexus is formed by the ovarian veins, which emerge from the ovaries and run into the mesovarium and suspensory ligament of the ovary. The suspensory and fundiform ligaments of the penis are replaced in females by the suspensory and fundiform ligaments of the clitoris. The cremaster is represented in females by a few fibers which descend into the round ligament of the uterus.

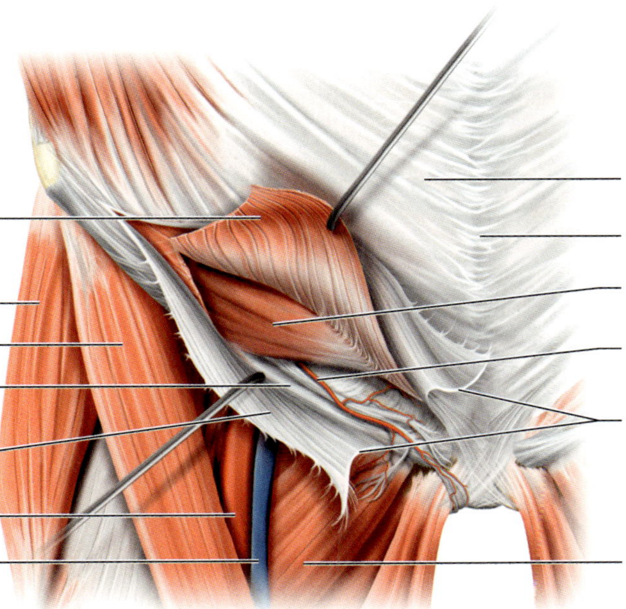

In females, the inguinal canal is crossed by the **round ligament of the uterus**, surrounded by a connective sheath derived from the transversalis fascia. After exiting the inguinal canal through the superficial inguinal ring, the round ligament of the uterus unravels into the fat pad of the labium majus. In males, the inguinal canal is crossed by the spermatic cord, on which the fascicles of the cremaster are loosely arranged, covered by the cremasteric fascia. The artery of the round ligament of the uterus runs with the round ligament of the uterus in the inguinal canal. In males, it is replaced by the cremasteric artery, which runs along with the spermatic cord into the inguinal canal and vascularizes the cremaster and the coverings of the spermatic cord.

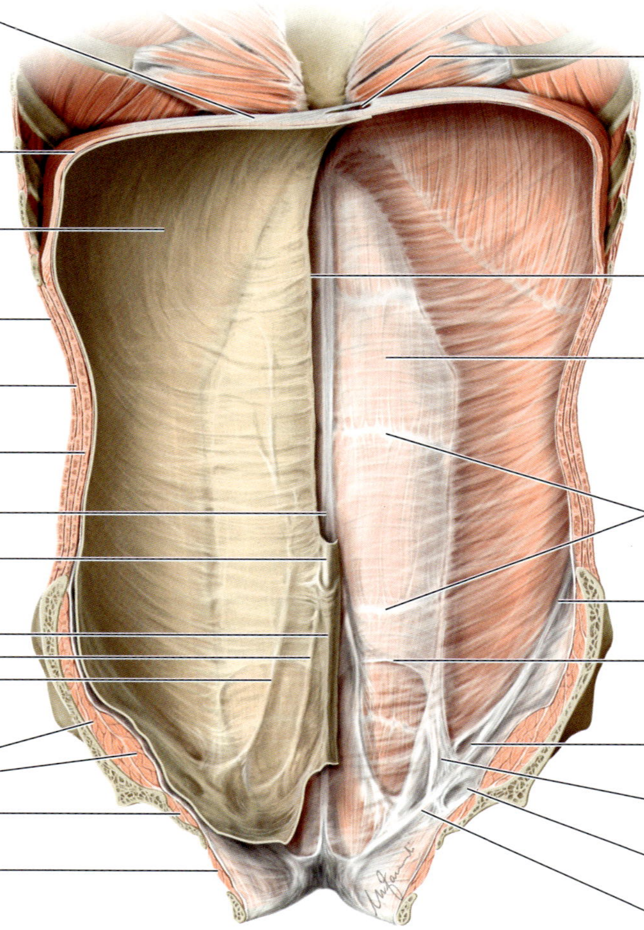

Central tendon of diaphragm, transected
Diaphragm, transected
Parietal peritoneum
External oblique, transected
Internal oblique, transected
Transversus abdominis, transected
Linea alba
Median umbilical ligament
Umbilical fold
median –
medial –
lateral –
Iliopsoas, transected
iliacus –
psoas major –
Iliacus, transected
Obturator internus, transected
Opening of inferior vena cava
Falciform ligament
Rectus sheath, posterior layer, transversalis fascia
Tendinous intersections of rectus abdominis
Inguinal ligament
Arcuate line (Line of Douglas)
Deep inguinal ring
Interfoveolar ligament
Iliopectineal arch
Lacunar ligament

a

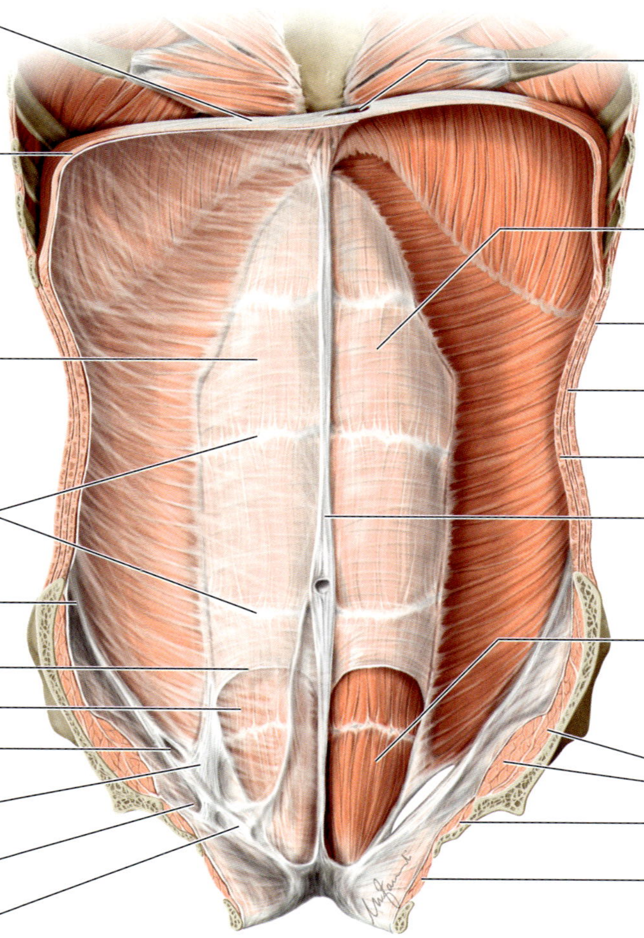

Central tendon of diaphragm, transected
Diaphragm, transected
Rectus sheath, posterior layer, lined internally by transversalis fascia
Tendinous intersections of rectus abdominis
Inguinal ligament
Arcuate line (Line of Douglas)
Transversalis fascia
Deep inguinal ring
Interfoveolar ligament
Iliopectineal arch
Lacunar ligament
Opening of inferior vena cava
Rectus sheath, posterior layer, aponeurosis of transversus abdominis
External oblique, transected
Internal oblique, transected
Transversus abdominis, transected
Linea alba
Rectus abdominis
Iliopsoas, transected
– iliacus
– psoas major
Iliacus, transected
Obturator internus, transected

b

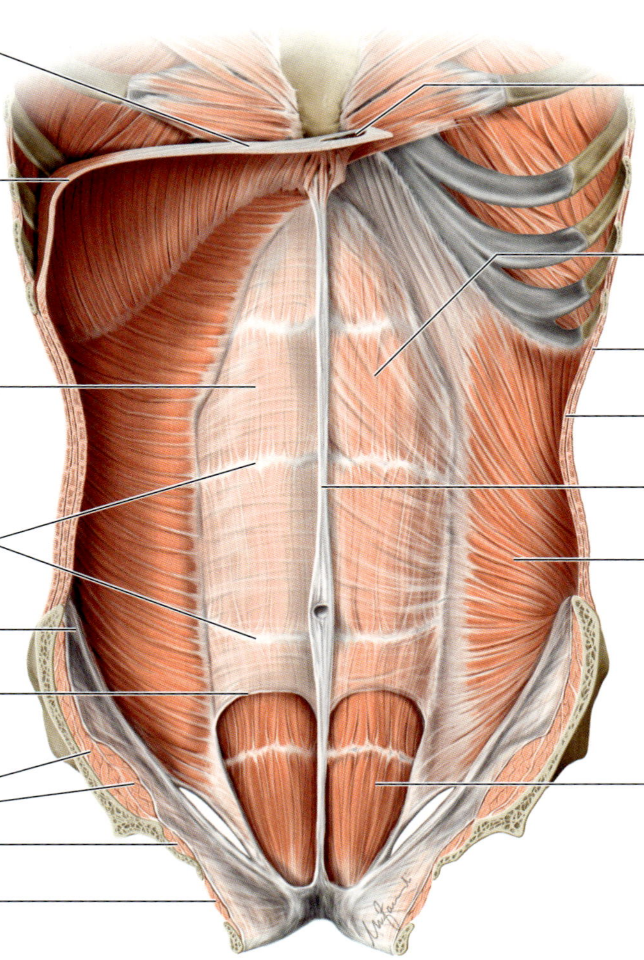

c

a, Posterior, or internal, view of the **anterior abdominal wall**. In left side of the wall, the anterior parietal peritoneum appears raised in the median (residual urachus), medial (umbilical ligament) and lateral (inferior epigastric artery) umbilical folds. Between the median and medial folds, between the medial and lateral folds and laterally to the latter, there are the supravesical fossa, medial inguinal fossa (also known as the inguinal triangle), and lateral inguinal fossa, respectively, which are all potential sites for the formation of hernias. In the right side of the wall, the anterior parietal peritoneum has been removed to highlight the rectus sheath. This sheath is formed by two layers, of which the posterior is shown here, consisting, from inside outwards, of the transversalis fascia, the aponeurosis of the transversus abdominis, and the posterior splitting layer of the aponeurosis of the internal oblique in the upper three-quarters, and only of the transversalis fascia in the lower quarter. From the inside moving outwards, the anterior layer is formed in the upper three-quarters by the anterior splitting layer of the aponeurosis of the internal oblique and the aponeurosis of the external oblique, and in the lower quarter also by the aponeurosis of the transversus abdominis.

b, Proceeding stratigraphically towards the outside and after removing the anterior parietal peritoneum, one can see the transversalis fascia to the left, and, after removing the transversalis fascia, the aponeurosis of the transversus abdominis becomes visible on the right. These contribute to the formation of the posterior layer of the rectus sheath, which in the lower quarter, beneath the arcuate line, consists of only the transversalis fascia.

c, On the right, with the aponeurosis of the transversus abdominis removed, one can see the posterior splitting layer of the aponeurosis of the internal oblique, which aids the formation of the posterior layer of the rectus sheath in the upper three-quarters, as well as the aponeurosis of the transversus abdominis, which can still be seen on the left.

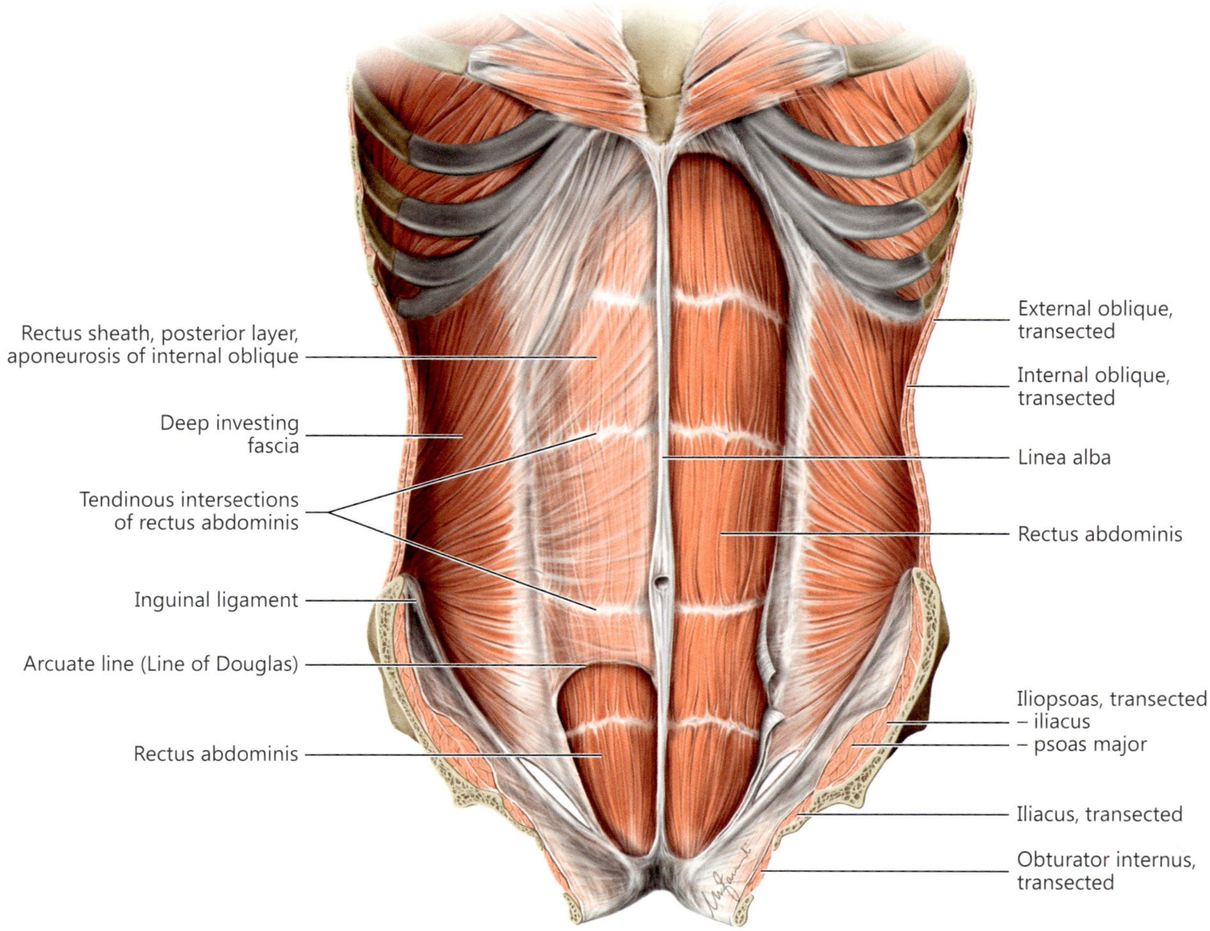

Rectus sheath, posterior layer, aponeurosis of internal oblique
Deep investing fascia
Tendinous intersections of rectus abdominis
Inguinal ligament
Arcuate line (Line of Douglas)
Rectus abdominis
External oblique, transected
Internal oblique, transected
Linea alba
Rectus abdominis
Iliopsoas, transected
– iliacus
– psoas major
Iliacus, transected
Obturator internus, transected

d

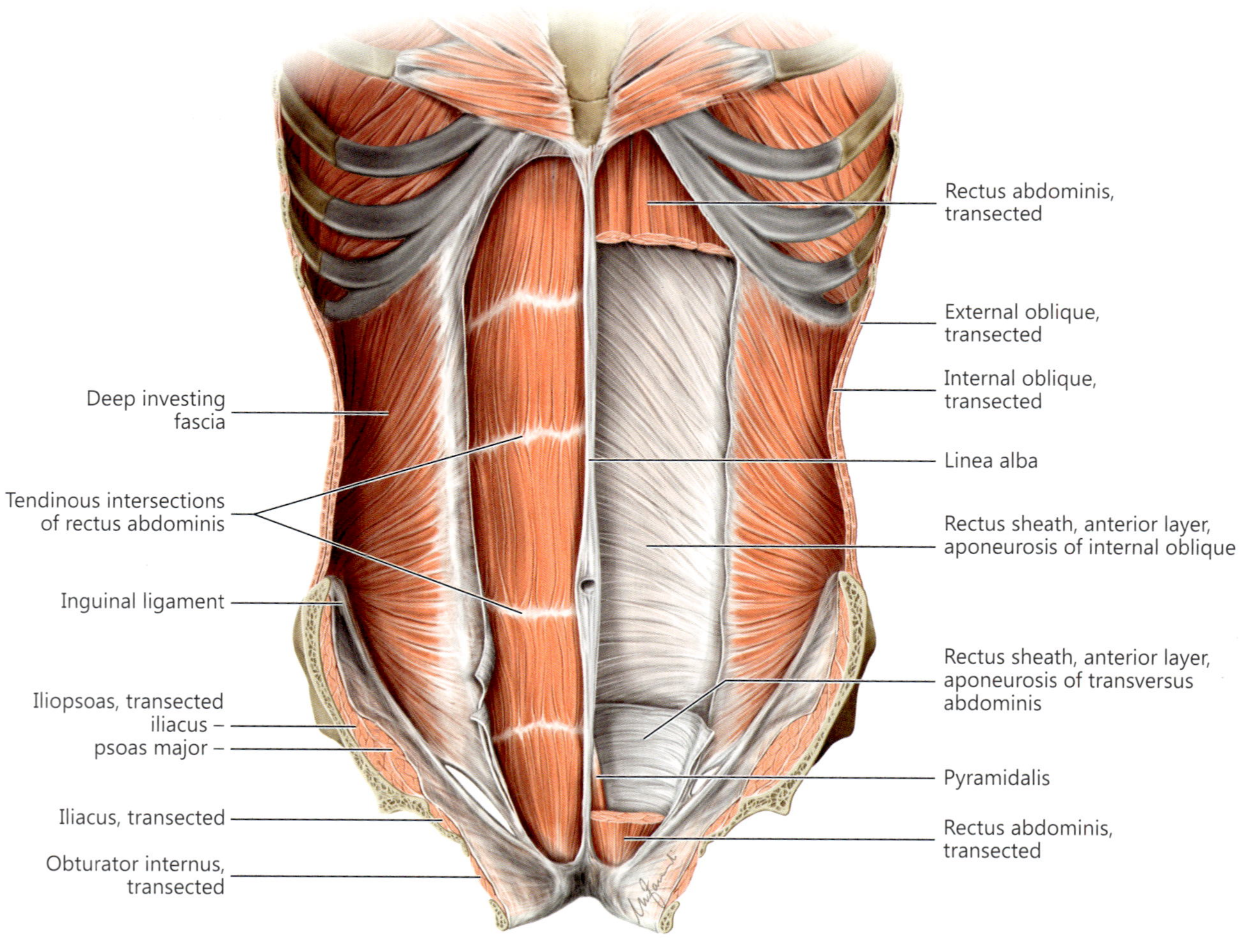

Deep investing fascia
Tendinous intersections of rectus abdominis
Inguinal ligament
Iliopsoas, transected
iliacus –
psoas major –
Iliacus, transected
Obturator internus, transected
Rectus abdominis, transected
External oblique, transected
Internal oblique, transected
Linea alba
Rectus sheath, anterior layer, aponeurosis of internal oblique
Rectus sheath, anterior layer, aponeurosis of transversus abdominis
Pyramidalis
Rectus abdominis, transected

e

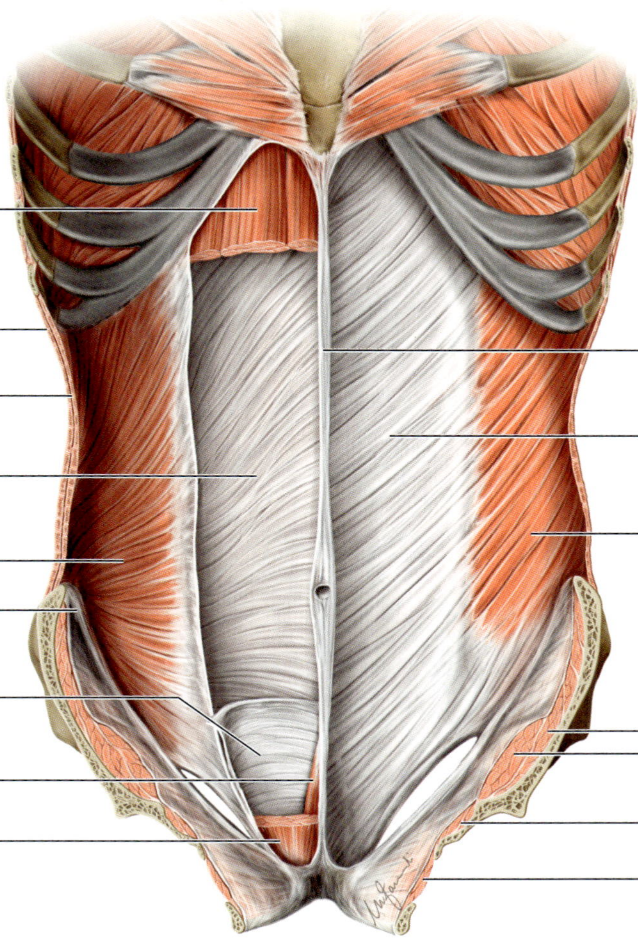

f

d, Moving from the inside out, the rectus abdominis is now visible on the right also in the upper three-quarters, after removal of the posterior splitting layer of the aponeurosis of the internal oblique. Note on the left the deep investing fascia of the abdomen, located between the internal oblique and transversus abdominis, the latter of which has been already removed in the previous picture.

e, Approaching the surface, on the right the rectus abdominis has been transected, allowing you to observe the anterior layer of the rectus sheath, formed in the upper three-quarters by the anterior splitting layer of the aponeurosis of the internal oblique (visible here) and the aponeurosis of the external oblique, and in the lower quarter also by the aponeurosis of the transversus abdominis (visible here).

f, Finally, on the right, the anterior splitting layer of the aponeurosis of the internal oblique and the aponeurosis of the transversus abdominis have been removed to highlight the aponeurosis of the external oblique. Again, on the right, note the intermediate investing fascia of the abdomen, located between the external and internal oblique, which has been removed.

Transverse sections of the **anterior abdominal wall** performed above the umbilical region (**a**), at the level of the umbilical region (**b**), below the arcuate line of the rectus sheath (**c**), and in proximity of the pubis (**d**).

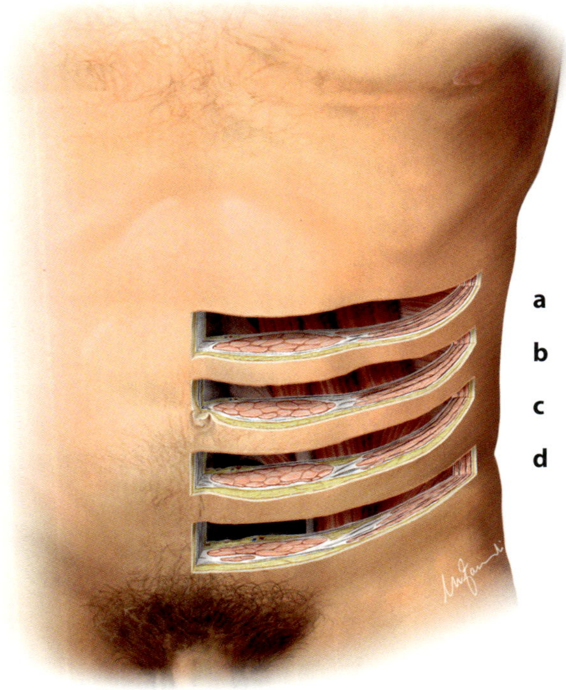

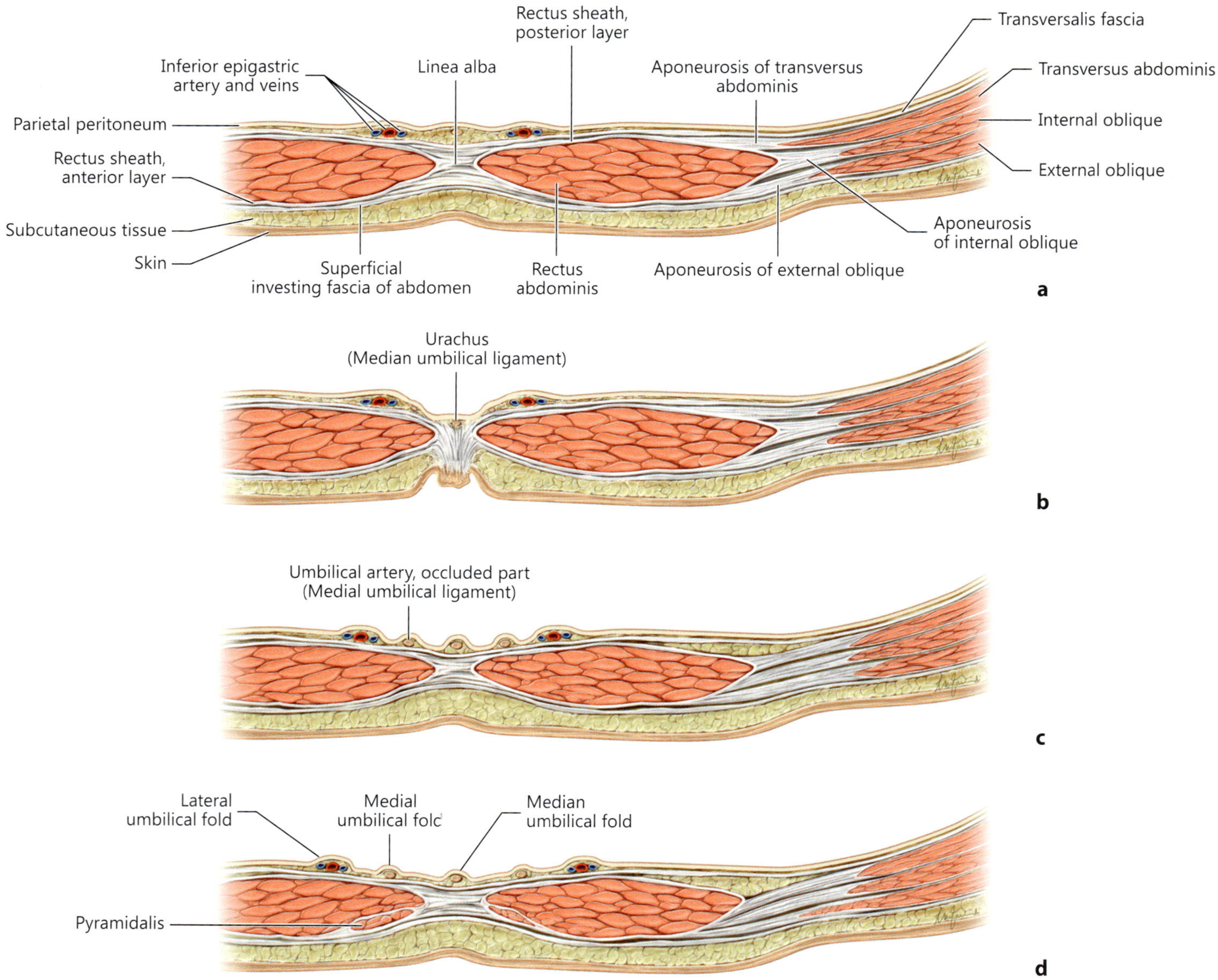

Origin, insertion, action, and innervation of the muscles of the abdomen					
MUSCLE	*ORIGIN*	*INSERTIONS*	*INNERVATION*	*ACTION*	*OBSERVATIONS*
Cremaster	In males, a few fascicles of the internal oblique join with fascicles originating from the pubic tubercle, which are arranged around the spermatic cord and the testicle		Genital branch of the genitofemoral nerve [L1, L2]	Lifting of the testicle	
Iliopsoas					
– psoas major	Anterior surface of the transverse processes of all the lumbar vertebrae, lateral surface of the intervertebral bodies and discs of the twelfth thoracic vertebra and of all the lumbar vertebrae	Lesser trochanter of the femur	Branches of the lumbar plexus (primarily L1, L2, with some contribution from L3)	Forward flexion of the trunk (by a joint action of the muscles on both sides)	
– psoas minor	Bodies of the last thoracic vertebra and of the first lumbar vertebra, and interposed intervertebral disc	Iliopubic ramus and iliac fascia	Branches of the lumbar plexus [L1]	Flexion of the trunk and tension of the iliac fascia	
– iliacus	Upper two-thirds of the iliac fossa, the inner lip of the iliac crest, the anterior sacroiliac and iliolumbar ligaments, and the superior part of the lateral surface of the sacrum	Lateral surface of the tendon of the psoas major, lesser trochanter of the femur	Branches of the lumbar plexus [L2, L3]	Forward flexion of the hip (with a fixed point on the pelvis); flexion and homolateral inclination of the trunk (with a fixed point on the femur)	
External oblique	External and inferior surface of the last eight ribs	Outer lip of the iliac crest, aponeurosis of the external oblique	Intercostal nerves, iliohypogastric nerve, and ilioinguinal nerve (T7-T12, and to a lesser degree L1)	Contralateral rotation of the trunk (by the action of the muscle on just one side); forward flexion of the trunk (by a joint action of the muscles on both sides)	The inferior fibers of the aponeurosis form the inguinal ligament and the lateral and medial crus of the superficial inguinal ring; the more medial fibers form the rectus sheath, as well as the linea alba by intersecting with those of the opposite side
Internal oblique	Lateral third of the inguinal ligament, anterior superior iliac spine, intermediate line of the iliac crest, and thoracolumbar fascia	Inferior border of the last three costal cartilages, aponeurosis of the internal oblique	Intercostal nerves, iliohypogastric nerve, and ilioinguinal nerve (T7-T12, and to a much lesser degree L1)	Homolateral rotation of the trunk (by the action of the muscle on just one side); forward flexion of the trunk (by a joint action of the muscles on both sides)	The aponeurosis, which forms the rectus sheath and the linea alba, joins with its inferior border to the aponeurosis of the transversus abdominis, forming the inguinal falx
Pyramidalis	Superior pubic ramus	Linea alba, about halfway	Subcostal nerve (last intercostal nerve, T12)	Tension of the linea alba	
Quadratus lumborum	Posterior fascicle: inner lip of the iliac crest Anterior fascicle: costal processes of the second, third, and fourth lumbar vertebrae	Costal processes of the first four lumbar vertebrae and the inferior border of the twelfth rib Twelfth rib	Subcostal nerve (last intercostal nerve and anterior rami of the first two or three lumbar nerves, T12, L1, L2 or L1-L3)	Depression of the twelfth rib and lateral inclination of the lumbar tract of the vertebral column	
Rectus abdominis	Xiphoid process and external surface of the fifth, sixth, and seventh costal cartilages	Superior pubic ramus	Last intercostal nerves and iliohypogastric nerve (T6-T12 or T7-T12)	Forward flexion of the trunk (by a joint action of the muscles on both sides)	The belly of the muscle is interrupted by three or four transverse tendinous intersections
Transversus abdominis	Internal surface of the last six costal cartilages, thoracolumbar fascia, inner lip of the iliac crest, and lateral half of the inguinal ligament	Aponeurosis of the transversus abdominis	Intercostal nerves, iliohypogastric nerve, and ilioinguinal nerve (T7-T12, L1)	Depression of the ribs (expiratory muscle) and increase of the abdominal pressure, along with the external oblique and internal oblique	The aponeurosis, which forms the rectus sheath and the linea alba, joins inferiorly the aponeurosis of the internal oblique, forming the inguinal falx

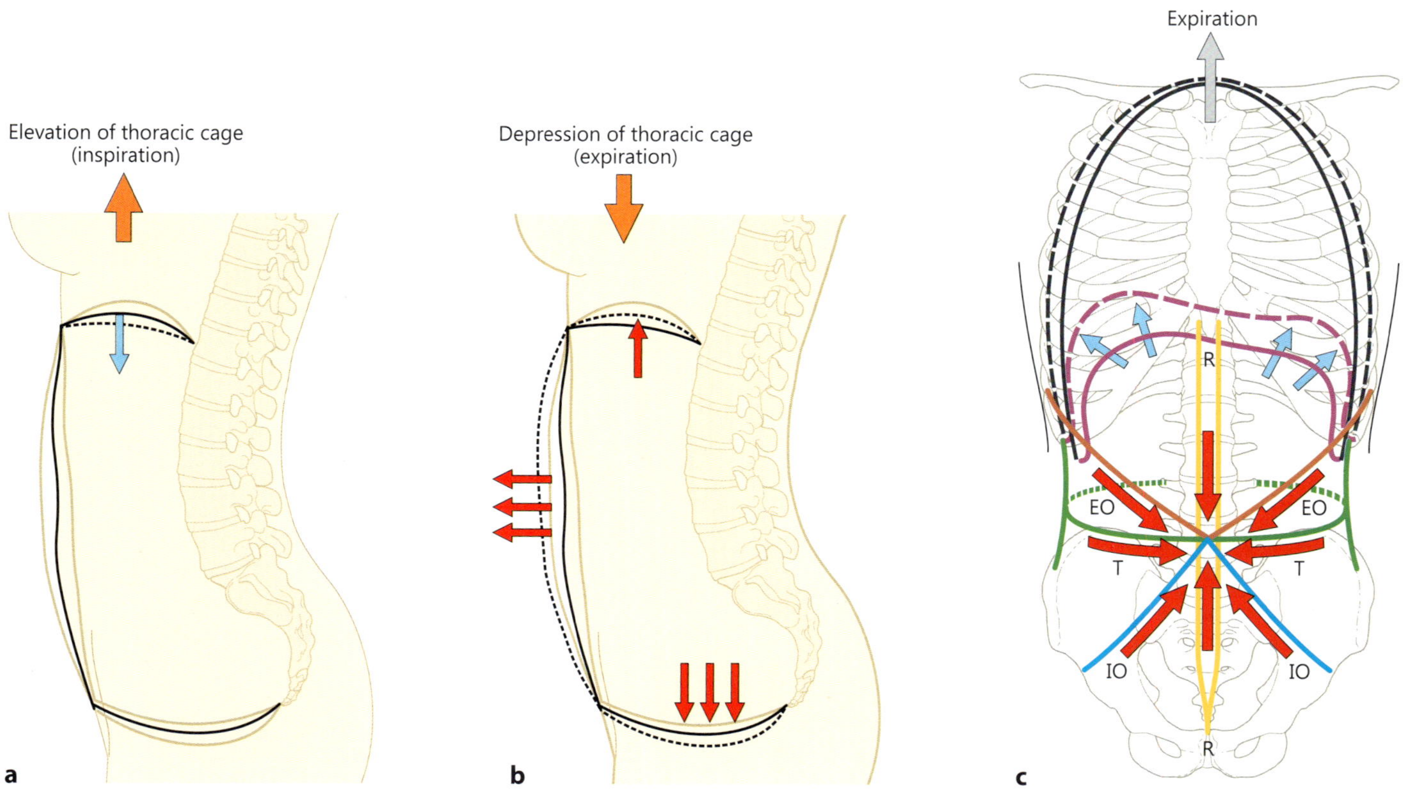

Elevation of the rib cage during the inspiratory phase preceding the forced expiration associated with blowing one's nose, coughing, or sneezing when stress urinary incontinence is present. The continuous line indicates the starting position, the dotted line shows the final position. The arrows indicate the associated movements of the abdominal wall, diaphragm, and pelvic floor during the inspiration (blue arrow) preceding forced expiration (red arrows). **a**, With the elevation of the rib cage required by the inspiratory effort, before coughing for example, the abdominal wall tends to be stretched and pushed slightly backward, even if the diaphragm descends. This does not happen with the occurrence of stress urinary incontinence, in which, to the contrary, there may be a protrusion of the abdominal wall. **b**, With expiratory effort, the rib cage is lowered, the abdominal wall is pushed outward, and the muscles of the pelvic floor move inferiorly. **c**, During inspiration, the abdominal belt formed by the vertical (rectus abdominis, R), oblique (external oblique, EO, and internal oblique, IO), and transverse systems (transversus abdominis, T) contracts (the red arrows outline the contraction of the abdominal wall) and antagonizes the action of the diaphragm, which in the meantime relaxes, resulting in the depression of the ribs and thus of the inferior opening of the thorax (from the dotted line to the continuous line). The increase in intra-abdominal pressure pushes the abdominal organs upwards, raising (blue arrows) the central tendon of the diaphragm (red line), resulting in the closure of the costodiaphragmatic recesses.

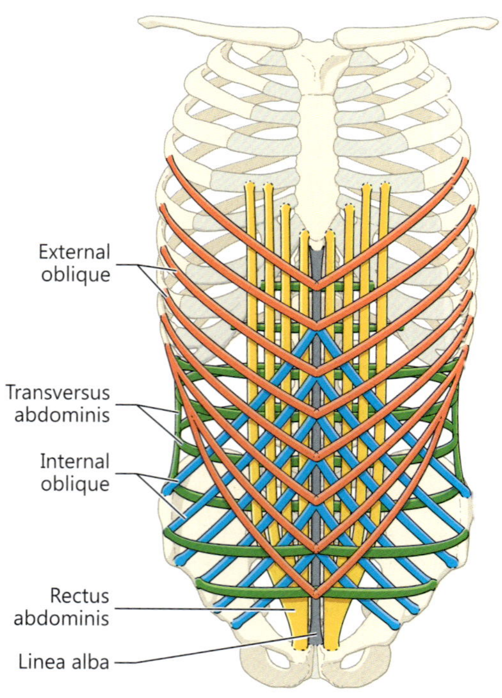

➤ The anterolateral abdominal muscles – rectus, external and internal oblique, and transversus – together form a functional unit comparable to a belt, consisting of three contraction systems of the abdominal wall: vertical, oblique and transverse. During inspiration, the abdominal belt tonically contains the abdominal organs, preventing the upward and downward push caused by the descent of the central tendon after contraction of the diaphragm. Additionally, the subsequent cessation of said descent caused by this belt fixes the central tendon, allowing the ribs (rib cage) elevation due to the final action of the diaphragm.

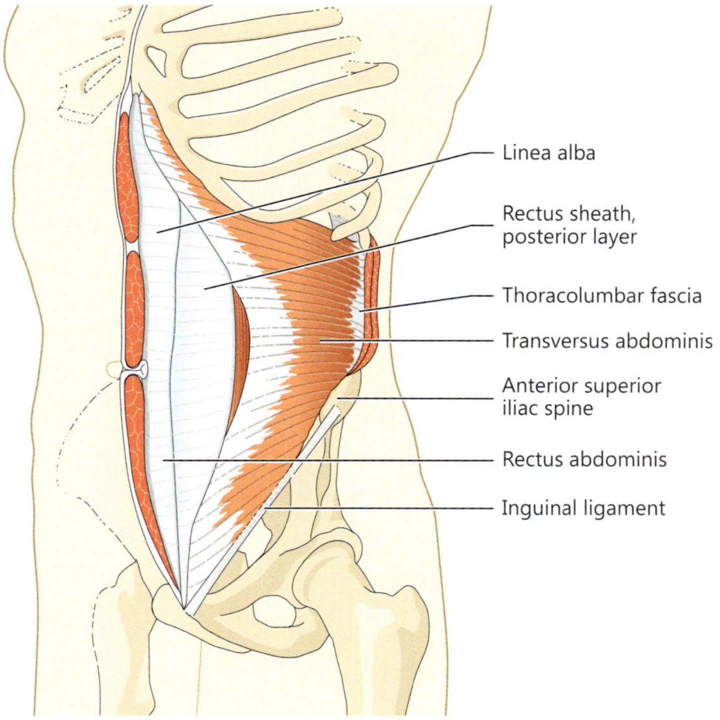

Location of the transversus abdominis.

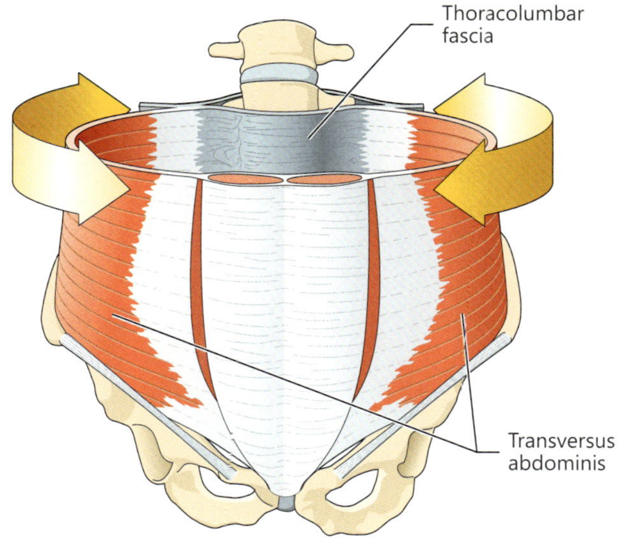

Reduction in the circumference of the abdomen during the recruitment of the transversus abdominis, with tensioning of the thoracolumbar fascia and increase in the intra-abdominal pressure. The reduction in the abdominal circumference is caused by the retraction of the anterior abdominal wall, due to the orientation of the muscle fibers in relation to the linea alba, formed by the aponeurosis of the transversus abdominis, external oblique, internal oblique, and rectus abdominis. The transversus abdominis originates from both the thoracolumbar fascia and the internal surface of the last six costal cartilages, the inner lip of the iliac crest, and the lateral half of the inguinal ligament.

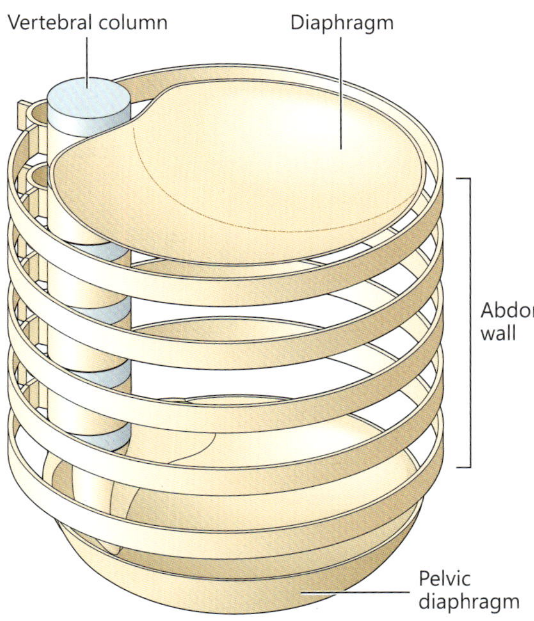

Model of the abdominal cavity as a "cell" or "box" comprised of the pelvic diaphragm, the respiratory diaphragm, the vertebral column, and the abdominal wall. The perfect organization of the parts makes it possible to prevent and manage excessive pressure on the various components.

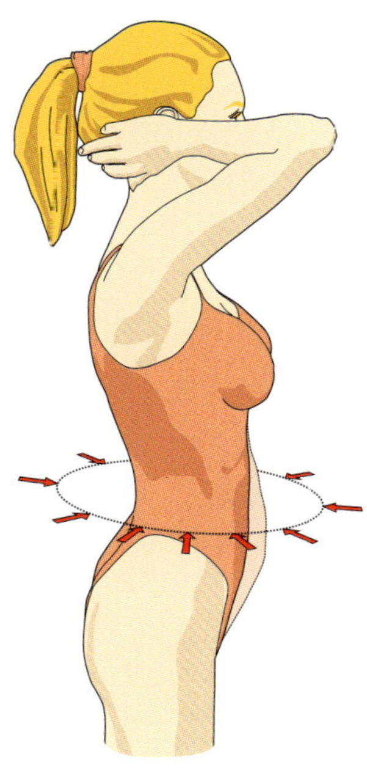

Hollowing mechanism for contraction of the transversus abdominis. The traction or inward push of the anterior abdominal wall determines a greater activation (recruitment) of the transversus abdominis, the action of which is similar to that of a tightened corset, given its posterior origin from the thoracolumbar fascia, between the twelfth rib and the iliac crest, and the anterior insertion at the iliac crest.

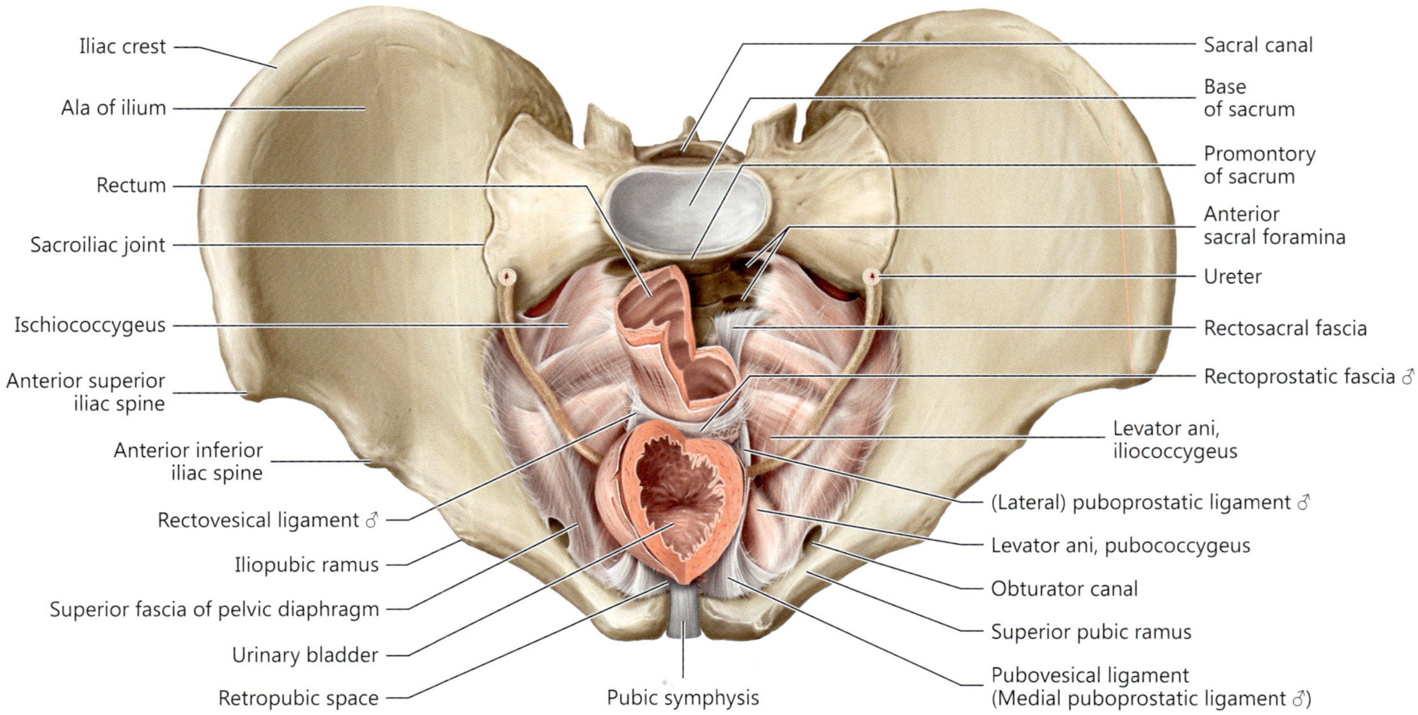

Male pelvis: appendages. Anterosuperior view. On the right, the rectum and the ureter have been transected transversely at the pelvic inlet, while the bladder appears in the oblique sagittal section. On the left, the rectum and the bladder have been transected at the level of the pelvic diaphragm (pelvic floor).

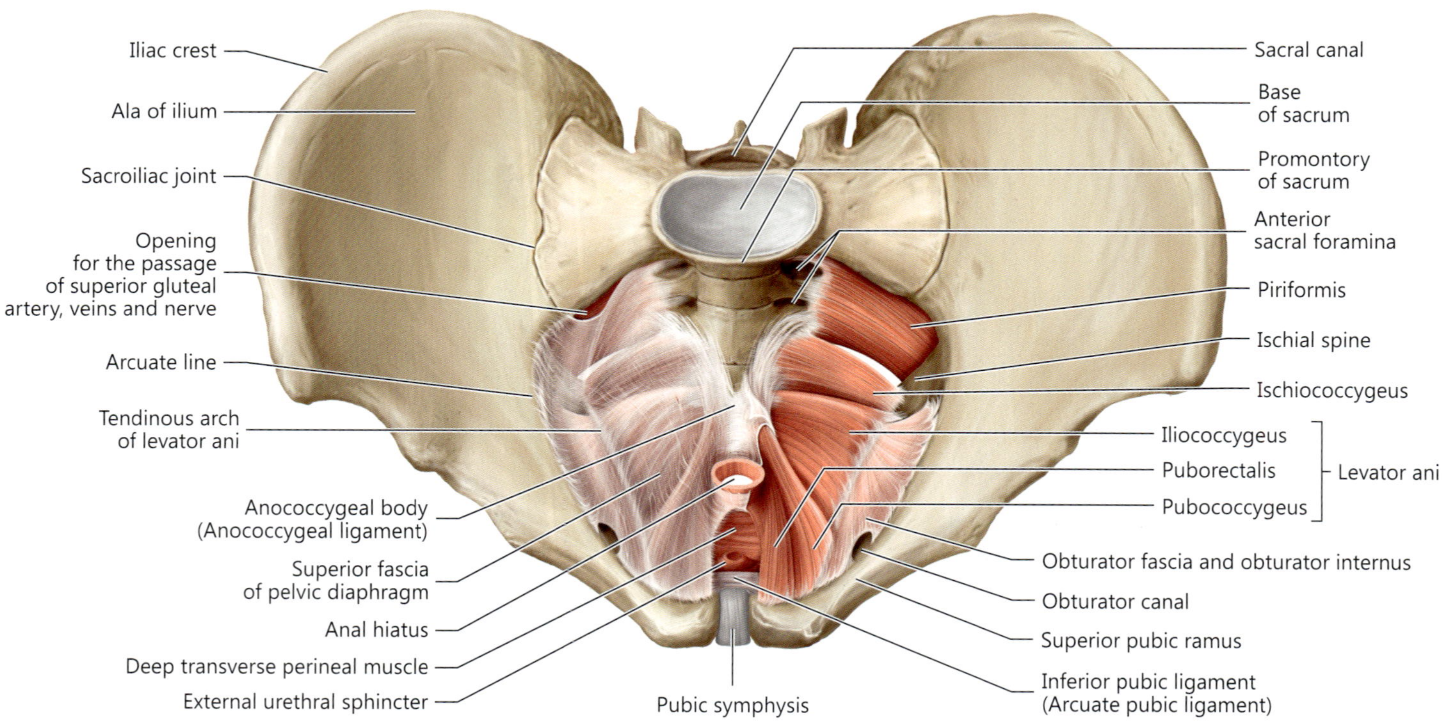

Male pelvis: walls and floor. Anterosuperior view. On the right, you can see the superior fascia of the pelvic diaphragm, depending on the parietal pelvic fascia, which also includes the piriformis fascia and the obturator fascia. On the left, with the superior fascia of the pelvic diaphragm removed, you can see the levator ani.

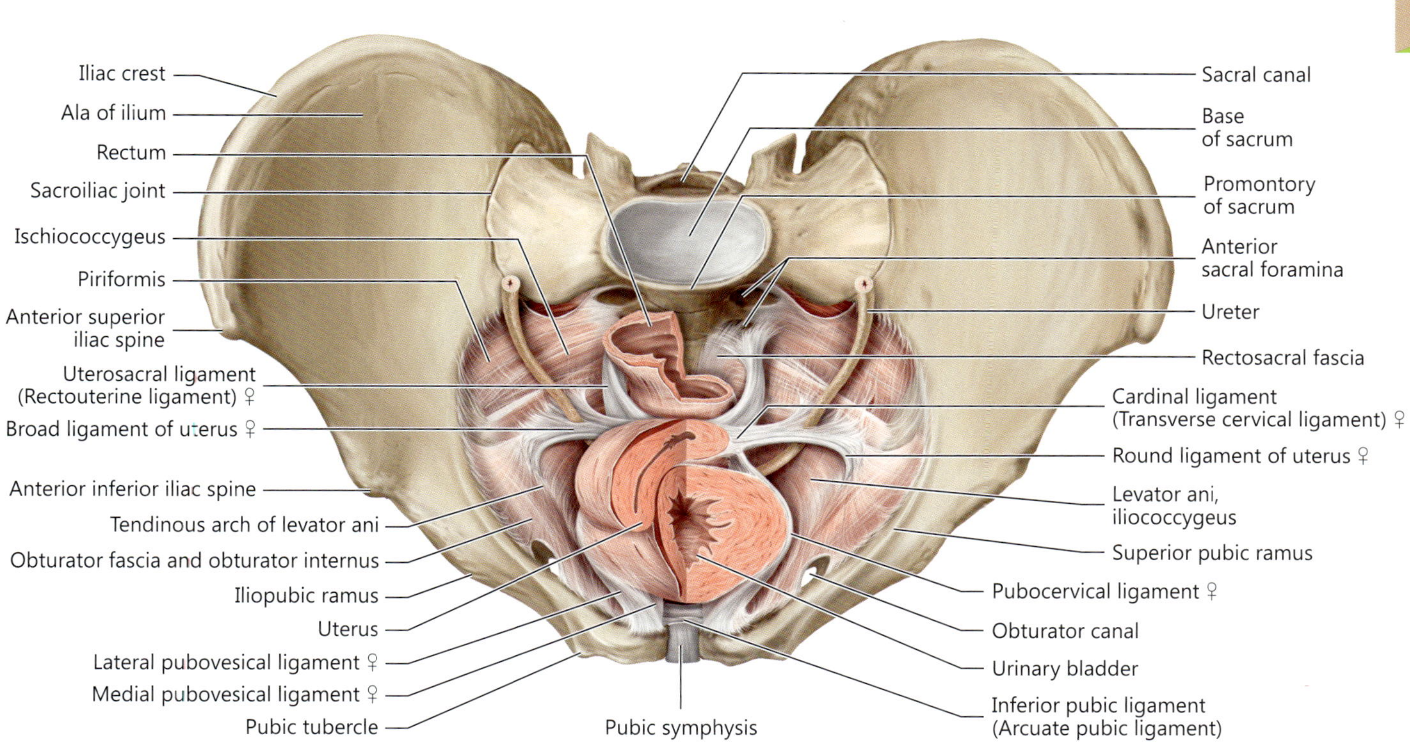

Female pelvis: appendages. Anterosuperior view. On the right, the rectum and ureter have been transected transversely at the pelvic inlet, while the uterus and bladder appear in the oblique sagittal section. On the left, the rectum, bladder, and uterus have been transected at the level of the pelvic diaphragm (pelvic floor).

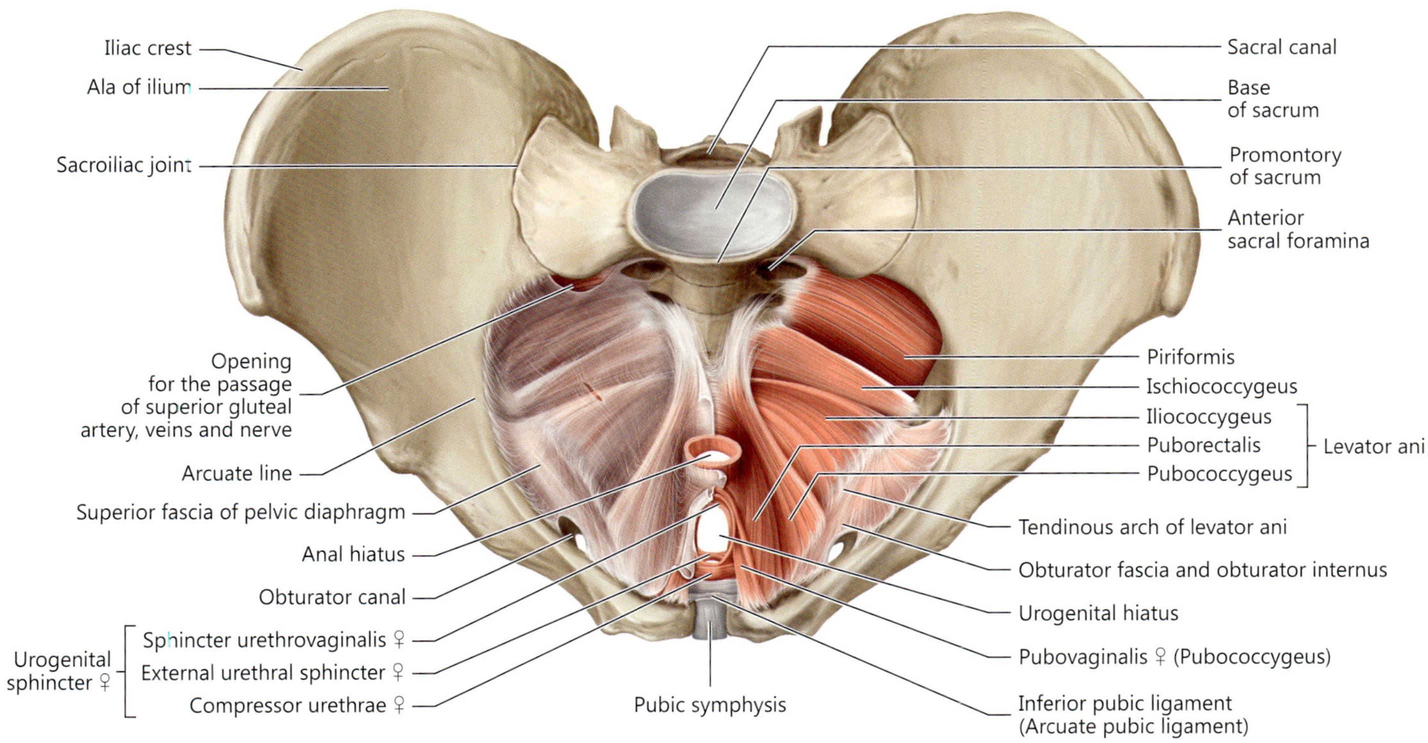

Female pelvis: walls and floor. Anterosuperior view. On the right, you can see the superior fascia of the pelvic diaphragm, depending on the parietal pelvic fascia, which also includes the piriformis fascia and the obturator fascia. On the left, with the superior fascia of the pelvic diaphragm removed, you can see the levator ani.

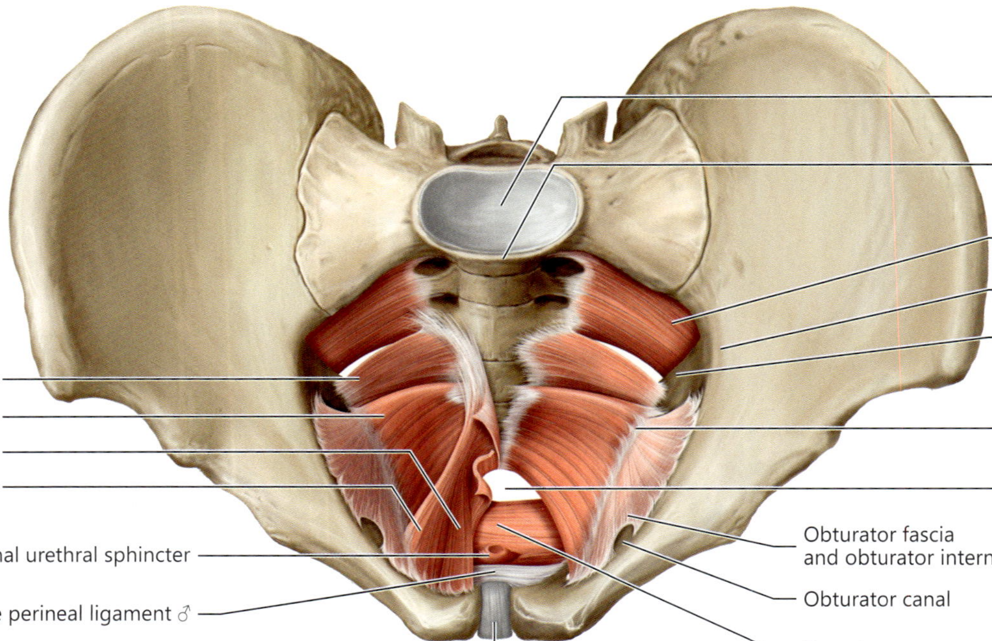

Male pelvis: pelvic diaphragm. Anterosuperior view. On the right, with the superior fascia of the pelvic diaphragm removed, you can see the **levator ani**, which has an extensive origin from the posterior surface of the superior pubic ramus, from the symphysis to the ischial spine, connecting to the tendinous arch of the levator ani. The lateral fibers of the anterior fascicles insert on the lateral border of the coccyx and the last sacral vertebra (**pubococcygeus**), while the medial fibers intersect posteriorly to the rectum (**puborectalis**) with the contralateral fibers to form, along the midline, the anococcygeal body or ligament. The posterior fascicles (**iliococcygeus**) end at the lateral border of the coccyx and at the anococcygeal ligament. Medial fibers of the anterior fascicles also make up the **puboanalis**, **puboperinealis**, and **puboprostaticus**. The first inserts at the external surface of the anal canal, while the others insert at the perineal body. On the left, removal of the puborectalis and pubococcygeus highlights the external urethral sphincter and the deep transverse perineal muscle.

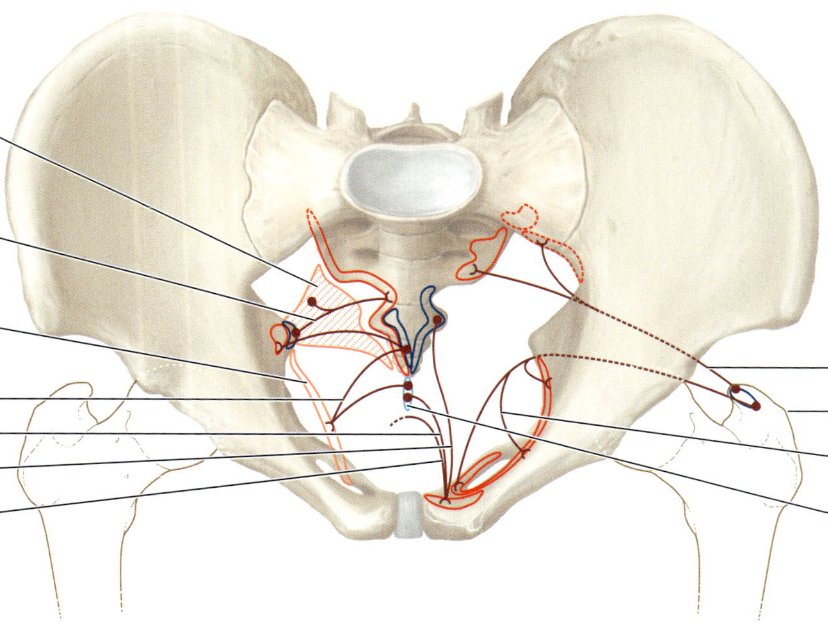

Male pelvis: schematic representation of the origin and insertion of the **muscles of the pelvic diaphragm** (**levator ani** and **ischiococcygeus**) and the **muscles of the pelvic wall** (**obturator internus** and **piriformis**). The **ischiococcygeus** originates from the lateral border of the last three sacral vertebrae and from the coccyx, and ends with a tapered extremity on the ischial spine. Being located on the pelvic surface of the sacrospinous ligament, the two can become fused together. The **obturator internus** joins the internal surface of the hip bone (obturator foramen, bone surface between this and the greater sciatic notch, obturator membrane) to the femur (trochanteric fossa). The **piriformis** originate from the pelvic surface of the sacrum, laterally to the second, third, and fourth sacral foramina, and from the border of the greater sciatic notch, to insert on the upper extremity of the internal surface of the greater trochanter of the femur.

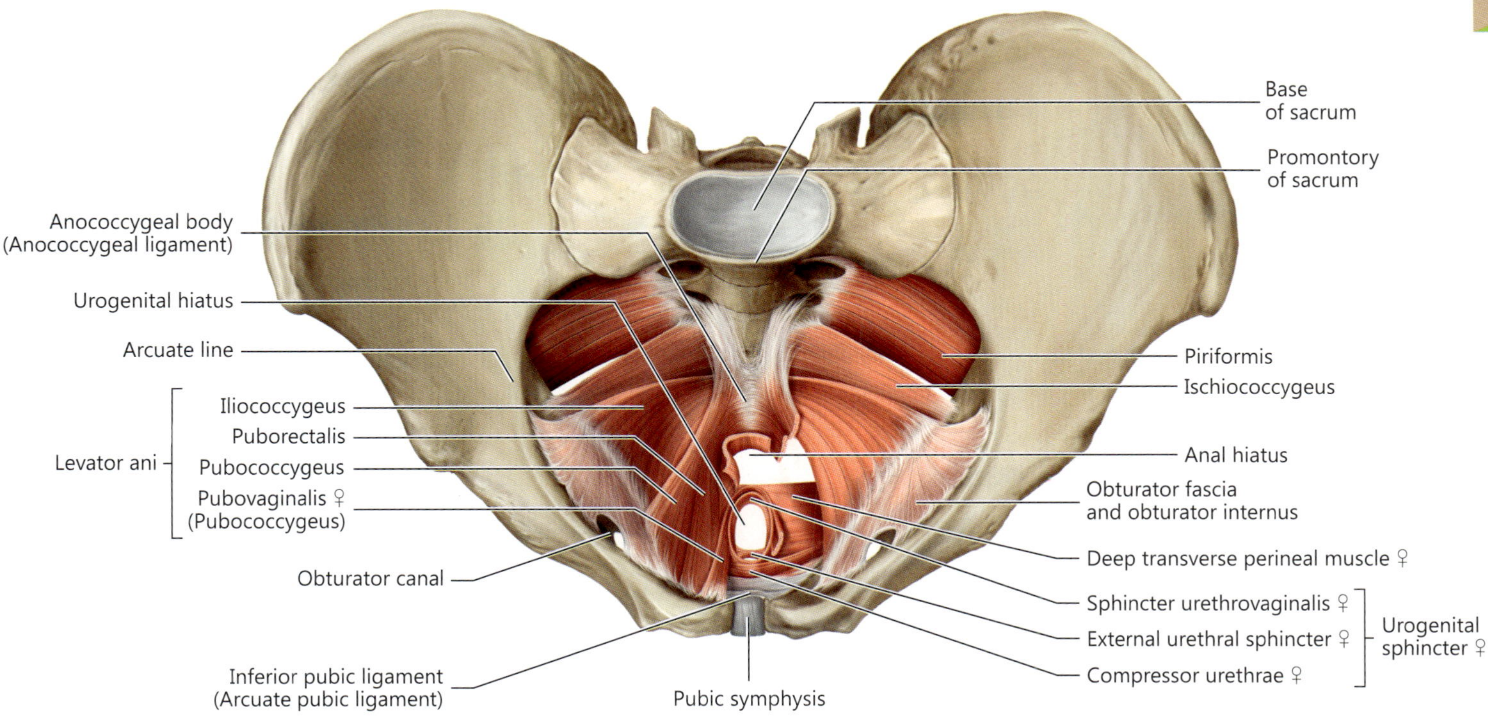

Female pelvis: pelvic diaphragm. Anterosuperior view. On the right, with the superior fascia of the pelvic diaphragm removed, you can see the **levator ani**, which has an extensive origin from the posterior surface of the superior pubic ramus, from the symphysis to the ischial spine, connecting to the tendinous arch of the levator ani. The lateral fibers of the anterior fascicles insert on the lateral border of the coccyx and the last sacral vertebra (**pubococcygeus**), while the medial fibers intersect posteriorly to the rectum (**puborectalis**) with the contralateral fibers to form, along the midline, the anococcygeal body or ligament. The posterior fascicles (**iliococcygeus**) end at the lateral border of the coccyx and at the anococcygeal ligament. Medial fibers of the anterior fascicles also make up the **puboanalis**, **puboperinealis**, and **pubovaginalis**. The first inserts at the external surface of the anal canal, the second at the perineal body, and the third joins with its contralateral counterpart surrounding the posterior wall of the vagina. On the left, the removal of the puborectalis and pubococcygeus highlights the urogenital sphincter and the smooth muscle adjacent to it in the deep perineal pouch.

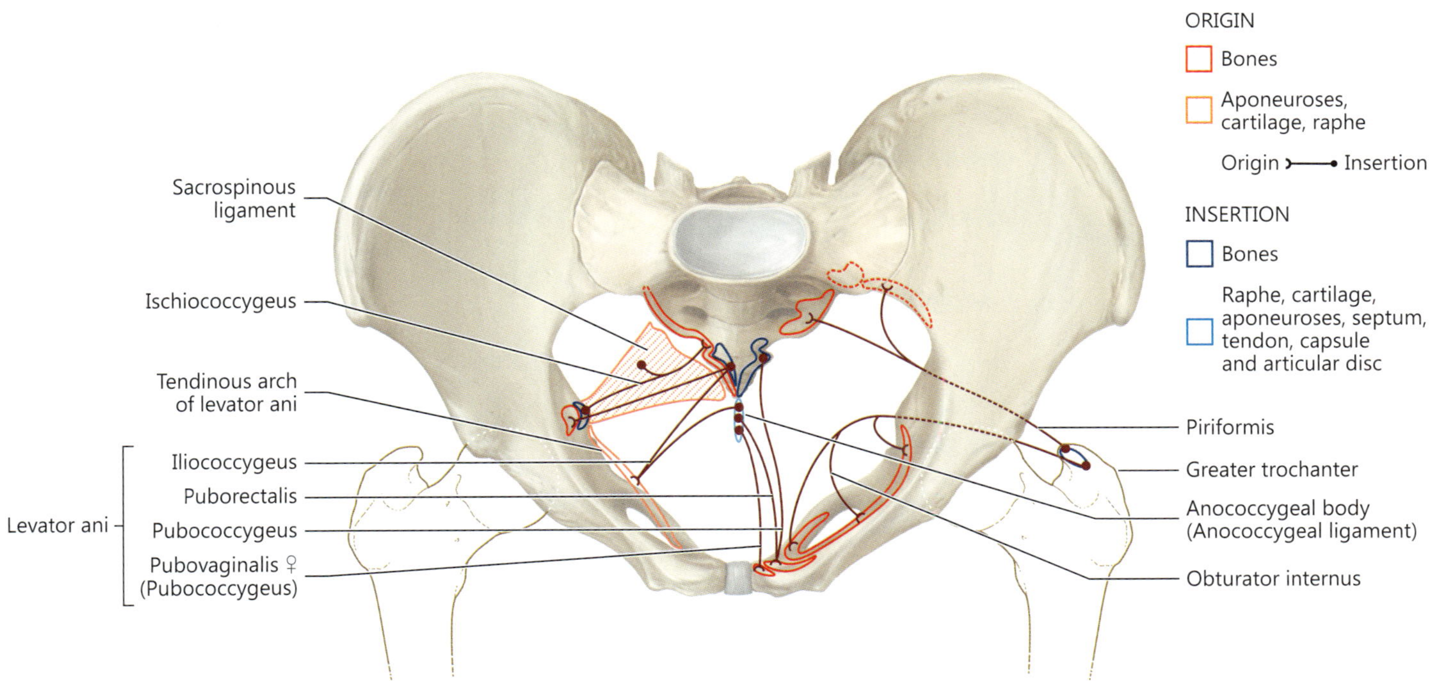

Female pelvis: schematic representation of the origin and insertion of the **muscles of the pelvic diaphragm** (**levator ani** and **ischiococcygeus**) and the **muscles of the pelvic wall** (**internal obturator** and **piriformis**). The **ischiococcygeus** originates from the lateral border of the last three sacral vertebrae and from the coccyx, and ends with a tapered extremity on the ischial spine. Being located on the pelvic surface of the sacrospinous ligament, the two can become fused together. The **obturator internus** joins the internal surface of the hip bone (obturator foramen, bone surface between this and the greater sciatic notch, obturator membrane) to the femur (trochanteric fossa). The **piriformis** originate from the pelvic surface of the sacrum, laterally to the second, third, and fourth sacral foramina, and from the border of the greater sciatic notch, to insert on the upper extremity of the internal surface of the greater trochanter of the femur.

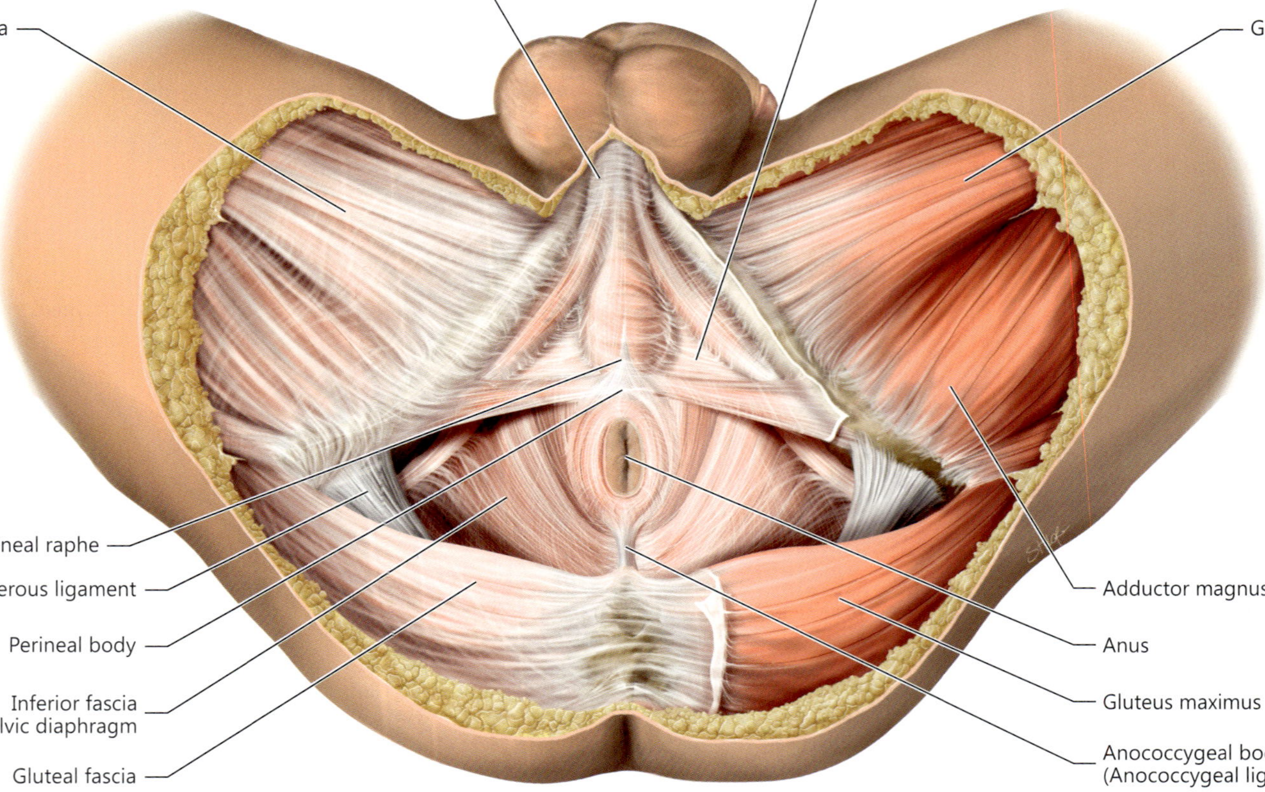

Male perineum. Inferior view. On the right, you can see the fascia lata and gluteal fascia, which have been removed on the left to show the gracilis, adductor magnus, and gluteus maximus.

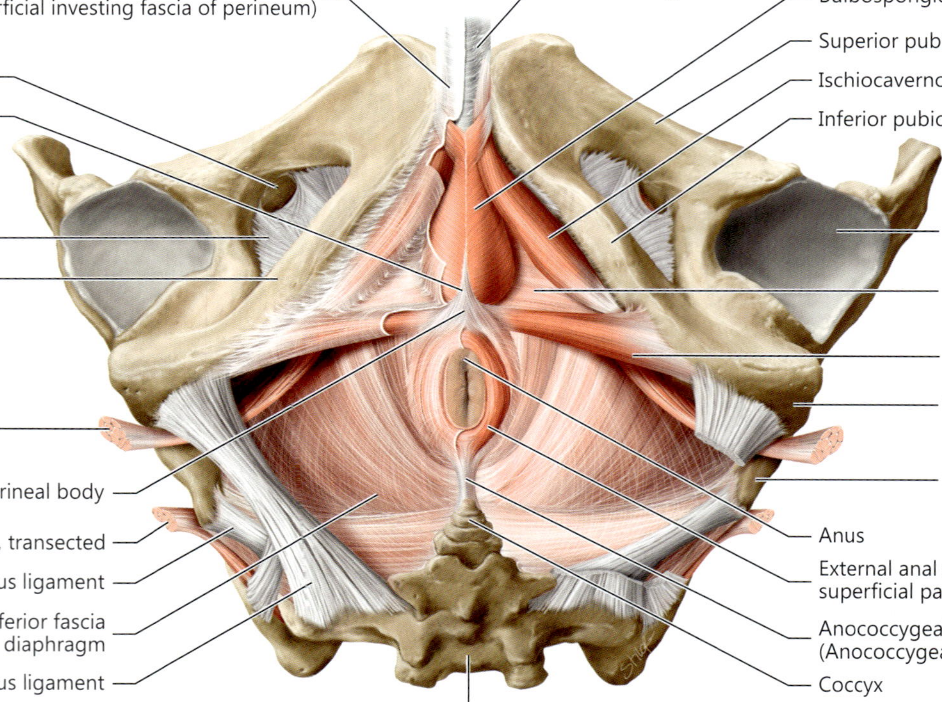

Male urogenital triangle (diaphragm) and **anal triangle**. In males, the urogenital triangle contains the bulb of the urethra and the insertions of the penis. On the right, you can see the perineal fascia, which superficially covers the perineum and has been removed on the left to highlight the bulbospongiosus, ischiocavernosus, and superficial transverse perineal muscle.

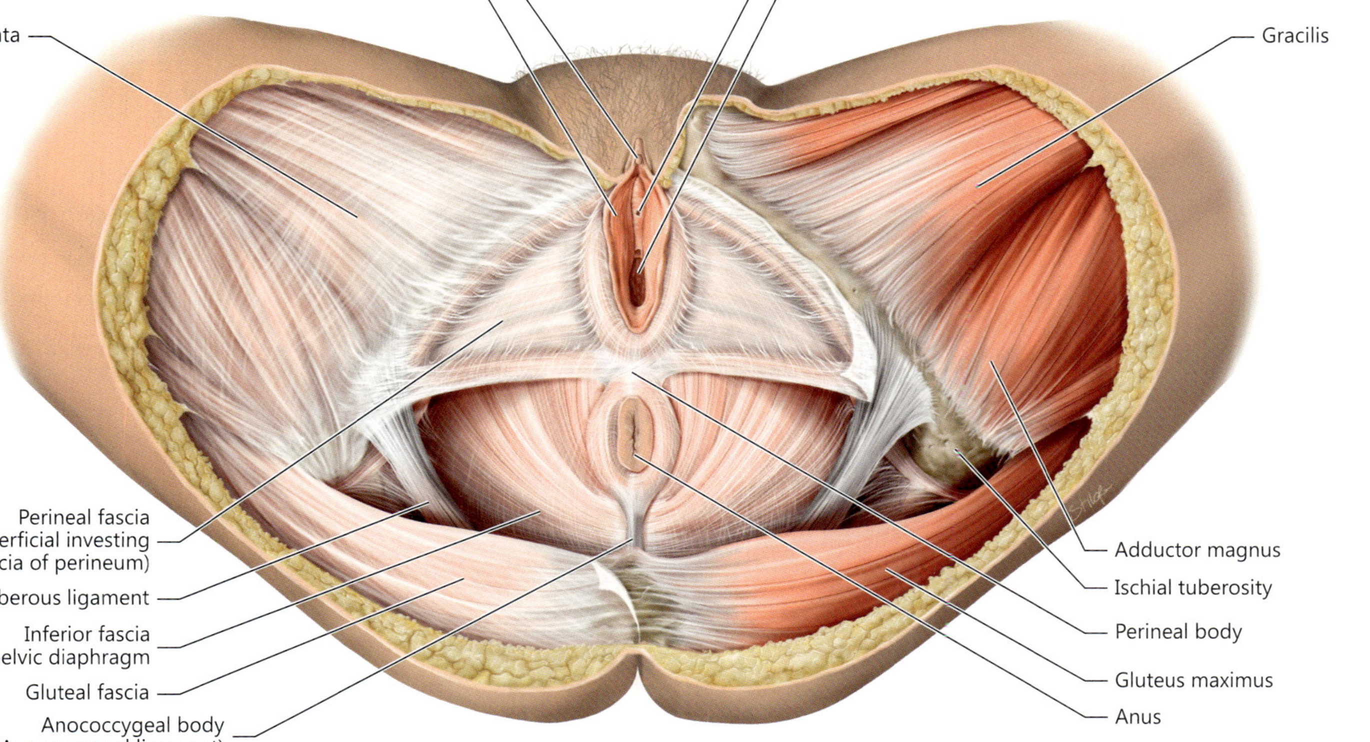

Female perineum. Inferior view. On the right, the fascia lata and gluteal fascia are visible, which have been removed on the left to show the gracilis, adductor magnus, and gluteus maximus.

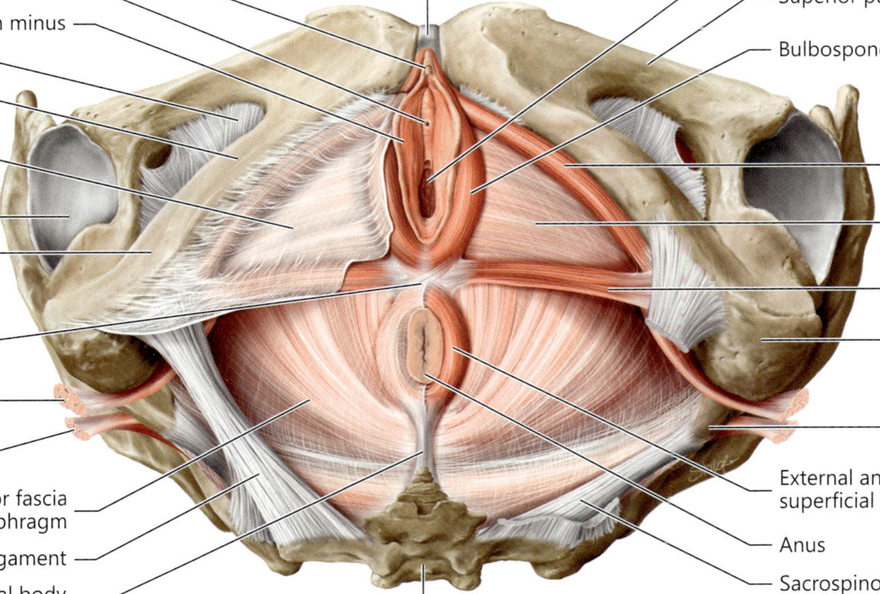

Female urogenital triangle (diaphragm) and **anal triangle**. In females, the urogenital triangle contains the mons pubis, labia majora, labia minora, clitoris, and vaginal and urethral orifices. On the right, you can see the perineal fascia, which superficially covers the perineum and has been removed on the left to highlight the bulbospongiosus, ischiocavernosus, and superficial transverse perineal muscle.

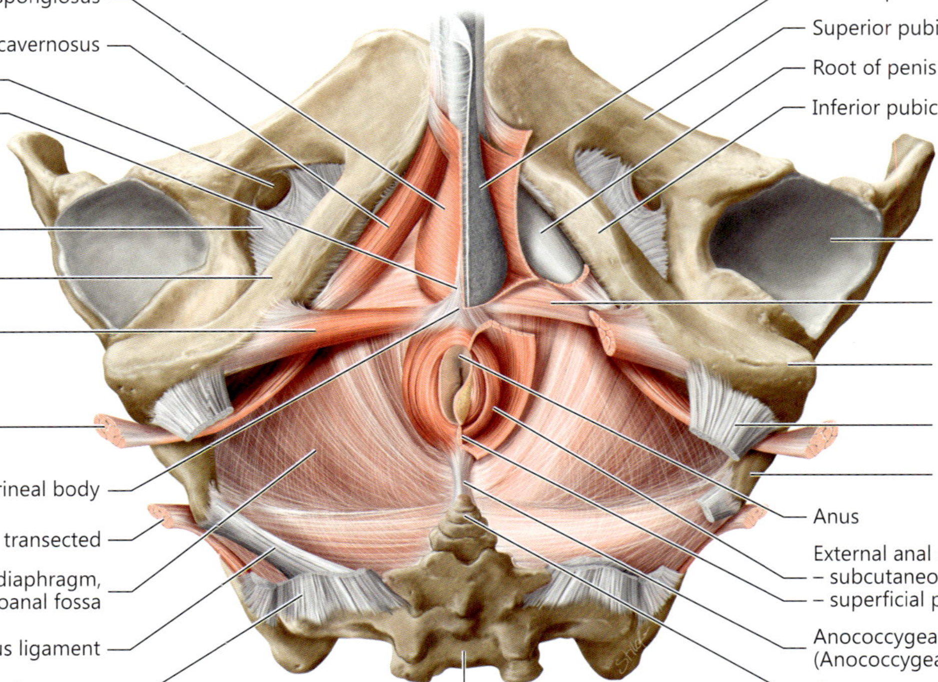

Male perineum. On the right, in the urogenital diaphragm or triangle you can see the bulbospongiosus and ischiocavernosus, transected on the left to highlight the bulb and root of the penis. On the right, in the anal triangle you can see the external anal sphincter, the upper part of which has been transected and turned to the left to show the subcutaneous part of the muscle.

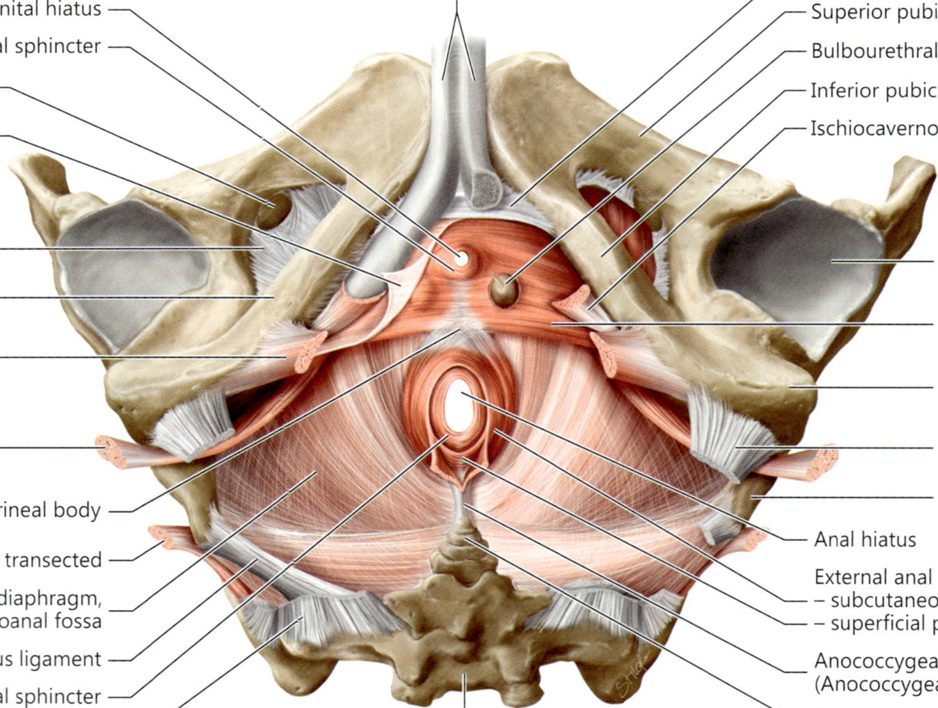

Male perineum. On the right, in the urogenital diaphragm or triangle you can see the corpus cavernosum of the penis, transected on the left, where the perineal membrane has been removed to show the deep transverse perineal muscle and the external urethral sphincter. In the anal triangle, the external anal sphincter has been transected to highlight the internal anal sphincter and the anal hiatus.

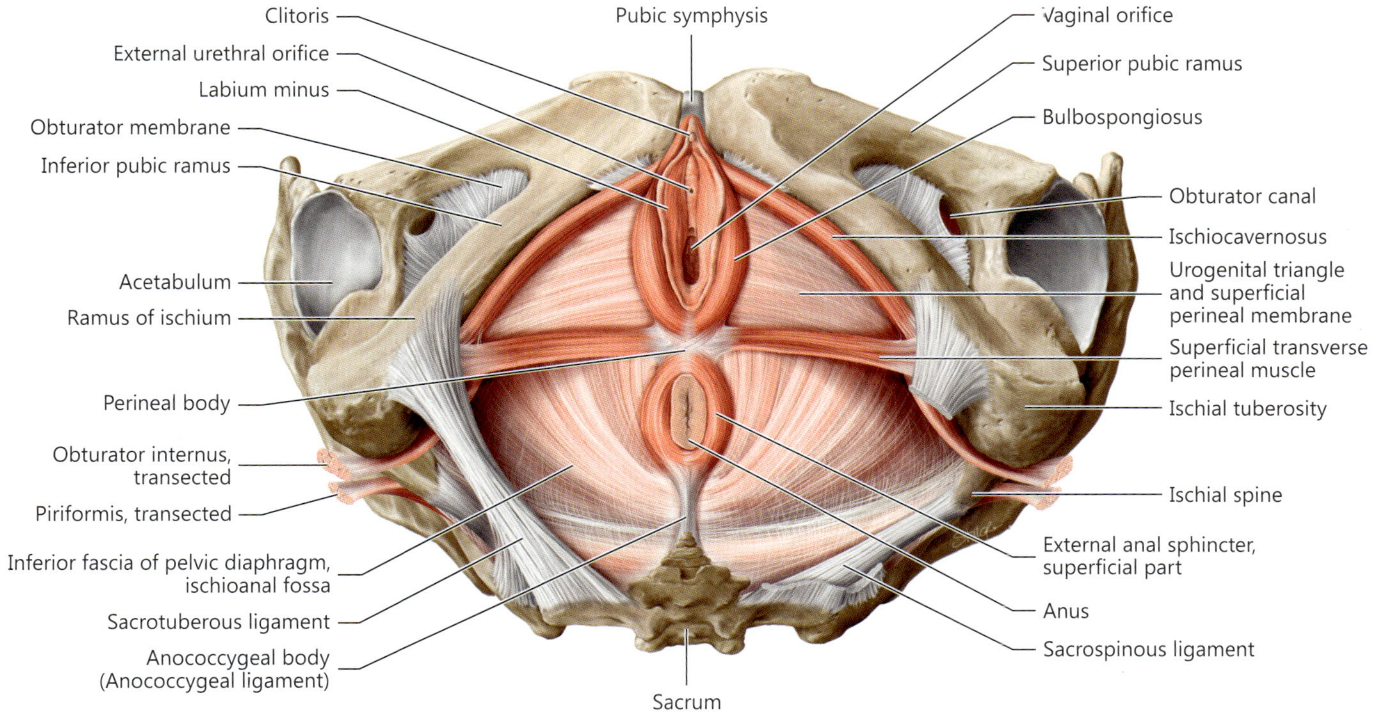

Female perineum. In the urogenital diaphragm or triangle you can see the labia majora, labia minora, clitoris, and vaginal and urethral orifices. The perineal membrane is also visible, extending above the deep perineal pouch. In the anal triangle you can see the inferior fascia of the pelvic diaphragm and the external anal sphincter, and on the right, the sacrotuberous ligament, which has been transected on the left to show the sacrospinous ligament and the ischial spine.

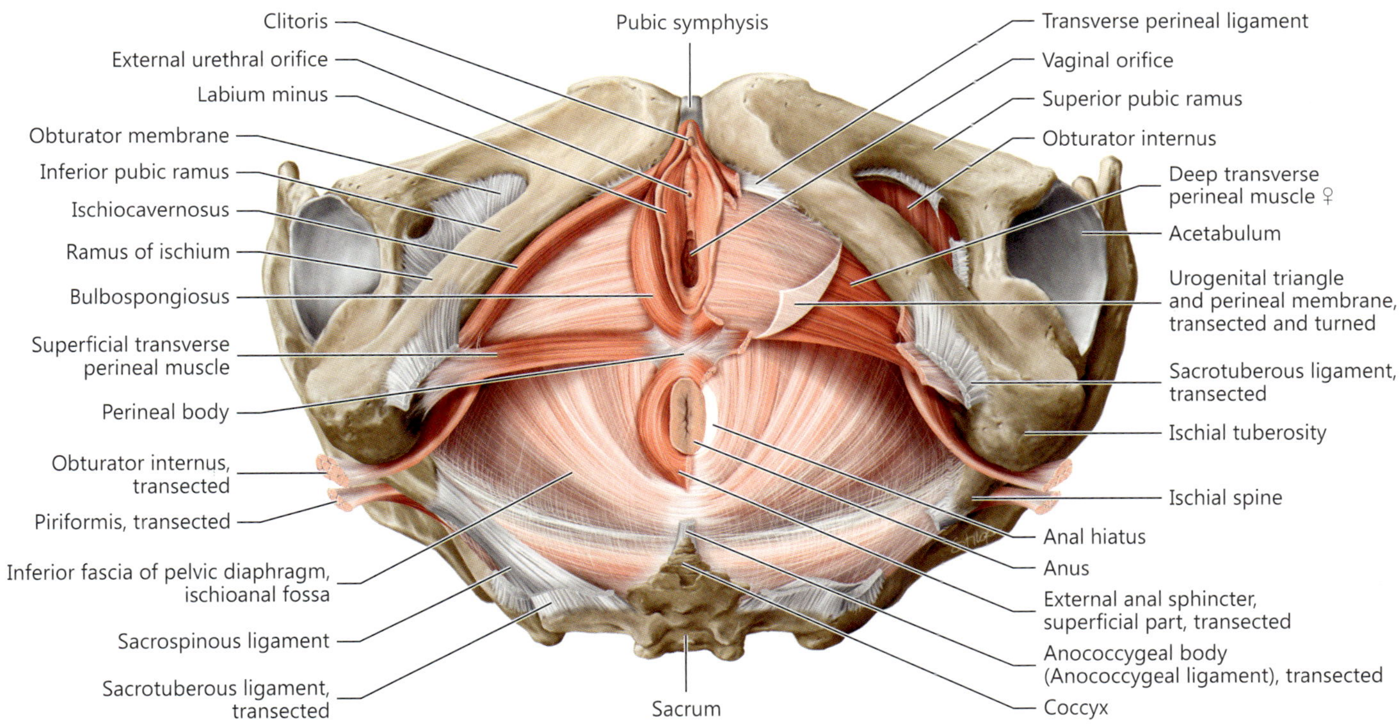

Female perineum: with respect to the top image, on the left side the perineal membrane has been transected and turned to highlight the deep transverse perineal muscle, while on the right side the sacrotuberous ligament has been transected. On the left, the removal of the external anal sphincter highlights the anal hiatus.

In women, the perineal musculature is relatively undeveloped. The deep transverse perineal muscle, sometimes simply referred to as smooth muscle, is mainly composed of smooth muscle fibers inserting into the connective tissue. Behind this is the thin superficial transverse perineal muscle. Together, these two muscles do not form an actual muscle layer, and as such, the classical concept of the urogenital diaphragm has recently been subject to reevaluation.

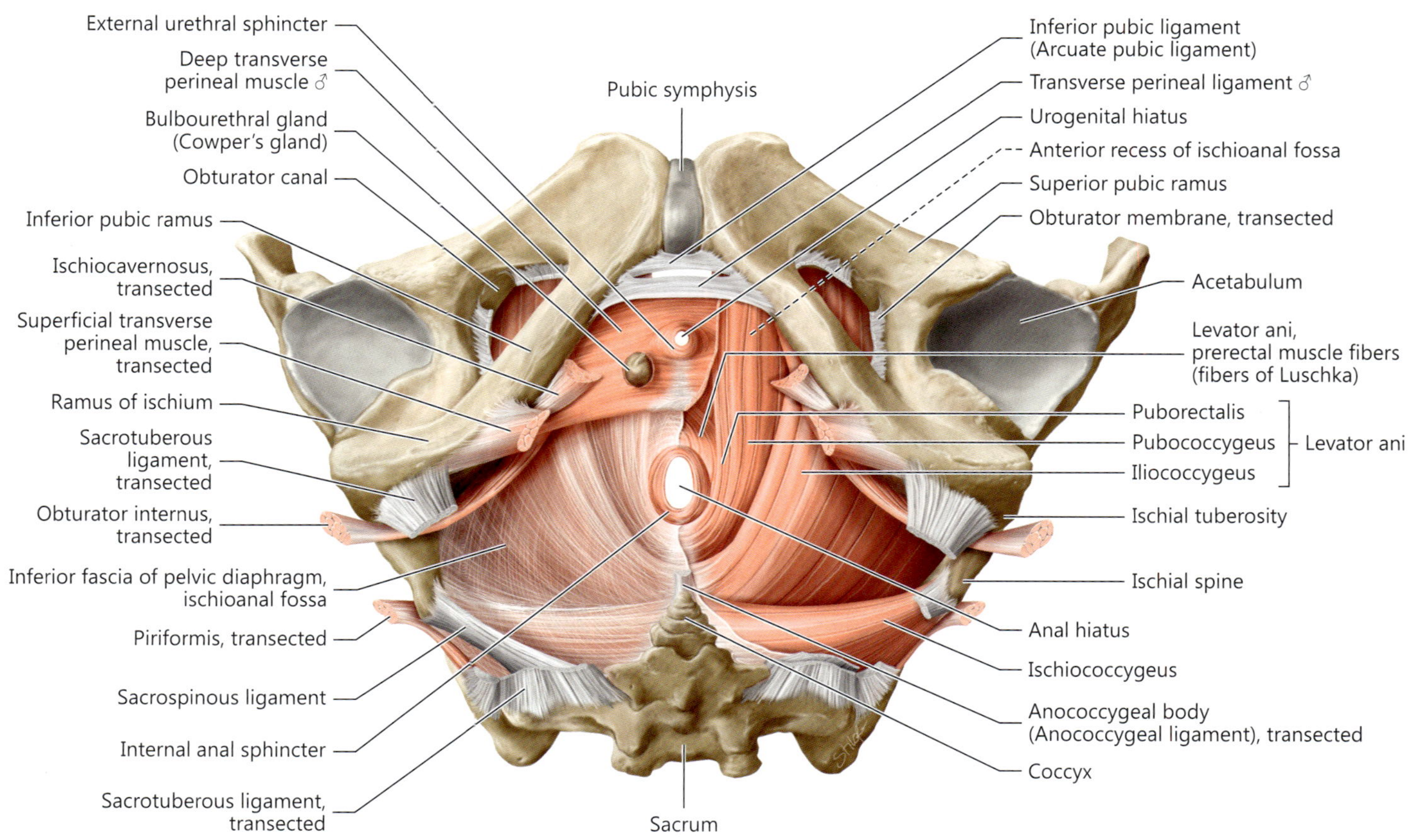

Male perineum and **pelvic diaphragm**. Inferior view. On the right, the urogenital and anal triangles are visible following the dissection of the ischiocavernosus and superficial transverse perineal muscle. On the left, the removal of the inferior fascia of the pelvic diaphragm and of the deep transverse perineal muscle highlights the levator ani.

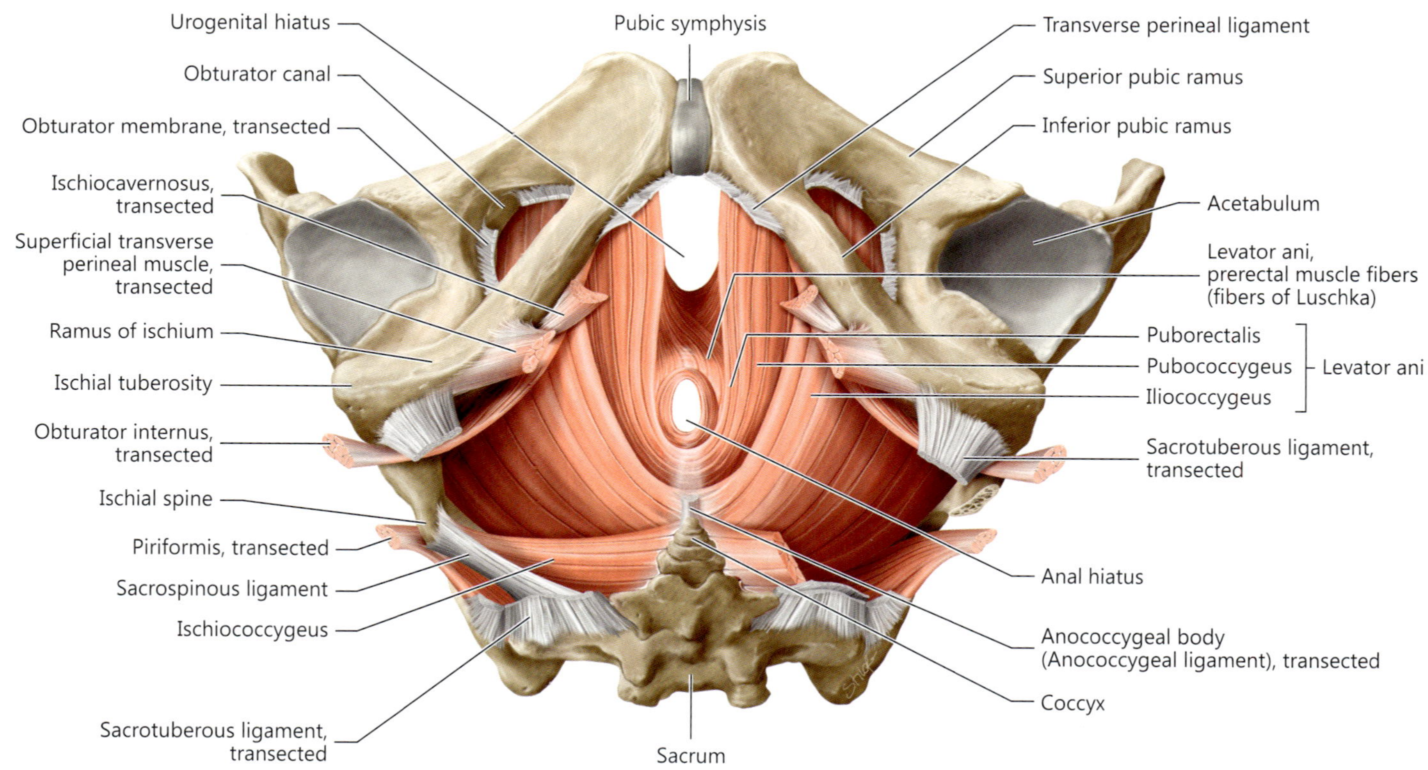

Male pelvic diaphragm. Inferior view. With respect to the top image, the inferior fascia of the pelvic diaphragm and the deep transverse perineal muscle have also been removed on the right, highlighting the levator ani. On the left, the ischiococcygeus has also been transected.

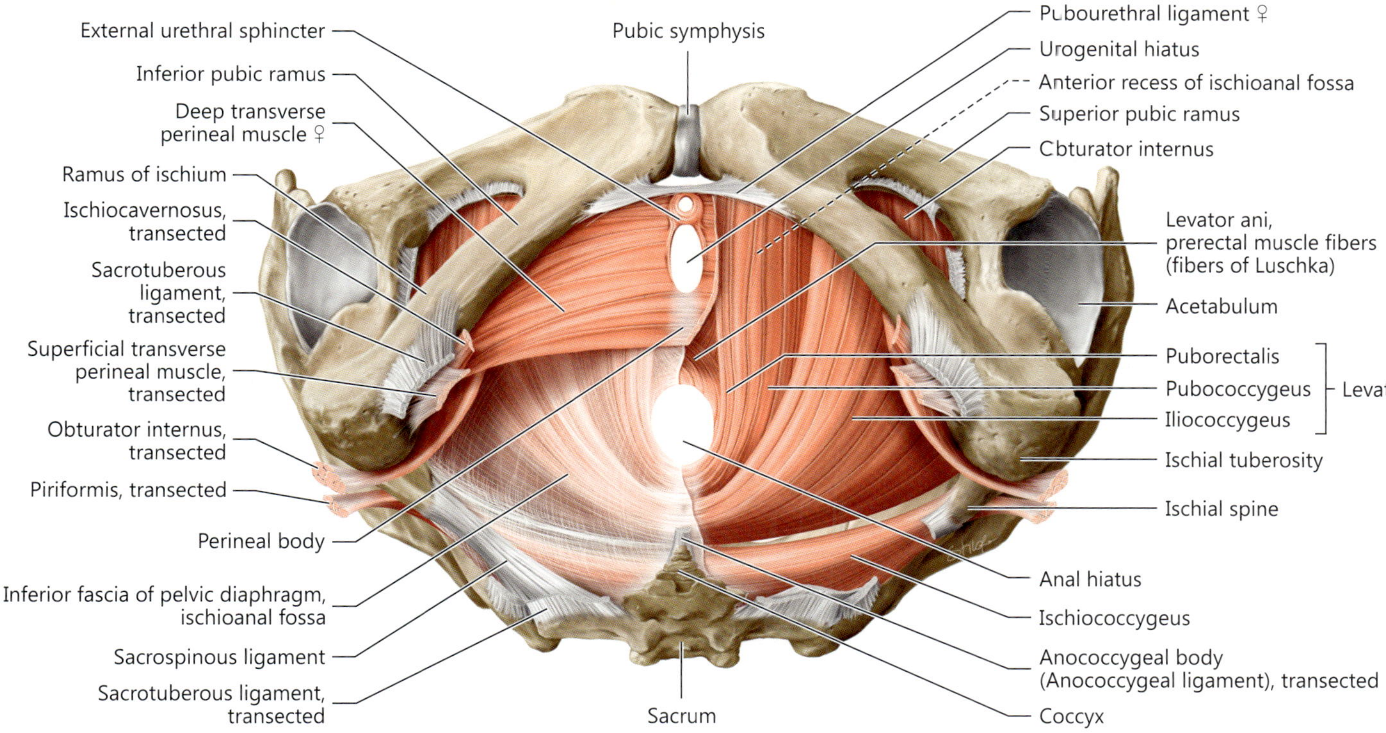

Female perineum and **pelvic diaphragm**. Inferior view. On the right, the urogenital and anal triangles are visible following the dissection of the ischiocavernosus and superficial transverse perineal muscle. On the left, the removal of the inferior fascia of the pelvic diaphragm and the smooth muscle adjacent to the urogenital sphincter highlights the levator ani.

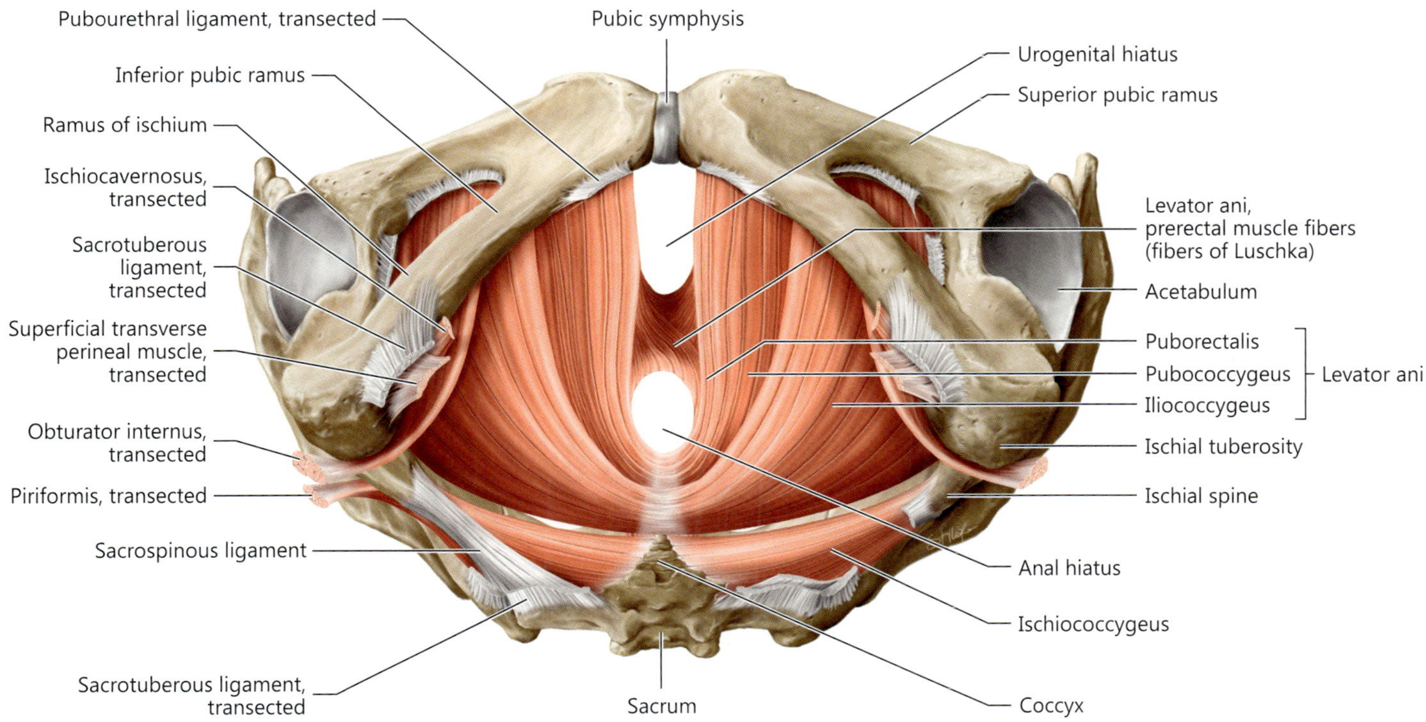

Female pelvic diaphragm. Inferior view. With respect to the top image, the inferior fascia of the pelvic diaphragm and the smooth muscle adjacent to the urogenital sphincter have also been removed on the right, highlighting the levator ani. On the left, the sacrospinous ligament has been transected.

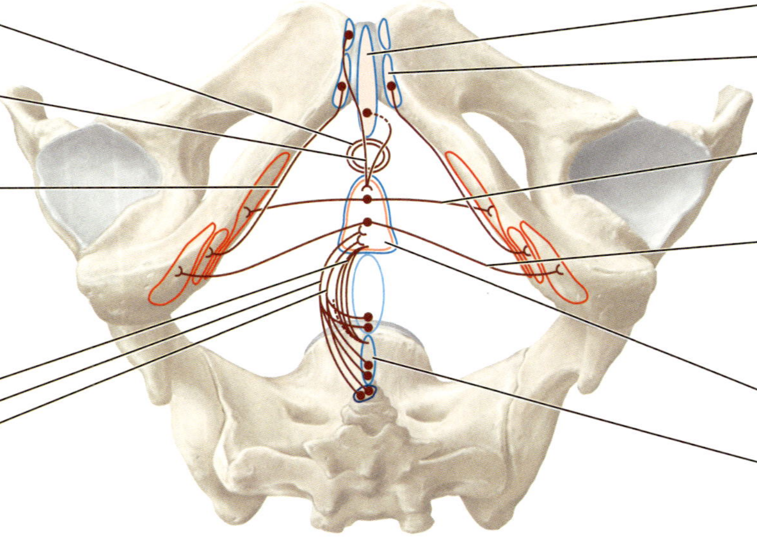

ORIGIN

Bones

Aponeuroses, cartilage, raphe

Origin ⟩—• Insertion

INSERTION

Bones

Raphe, cartilage, aponeuroses, septum, tendon, capsule and articular disc

Male perineum: schematic representation of the origin and insertions of the external urethral sphincter, bulbospongiosus, ischiocavernosus, superficial and deep transverse perineal muscles, and external anal sphincter. The **external urethral sphincter** originates from the internal surface of the inferior pubic ramus and from the anterior surface of the prostate, to insert on the perineal body and on the posterior surface of the prostate. The **bulbospongiosus** and **ischiocavernosus** have their origins, respectively, from the perineal raphe and the perineal body, and from the ischial tuberosity; the former inserts on the corpus spongiosum of the penis, and the latter on the tunica albuginea of the corpora cavernosa of the penis, below the pubic symphysis. The **superficial** and **deep transverse perineal muscles** extend from the internal surface of the ischial tuberosity, in the case of the former, and from the medial surface of the ramus of the ischium, in the case of the latter, to the perineal body. The deep part of the **external anal sphincter** consists of fibers common to those of the puborectalis that extend anteriorly and posteriorly to the anal canal, intersecting with those of the opposite side. The subcutaneous part extends from the perineal raphe to the deep surface of the dermis, while the superficial part extends from the perineal body, reaching the tip of the coccyx through the anococcygeal ligament.

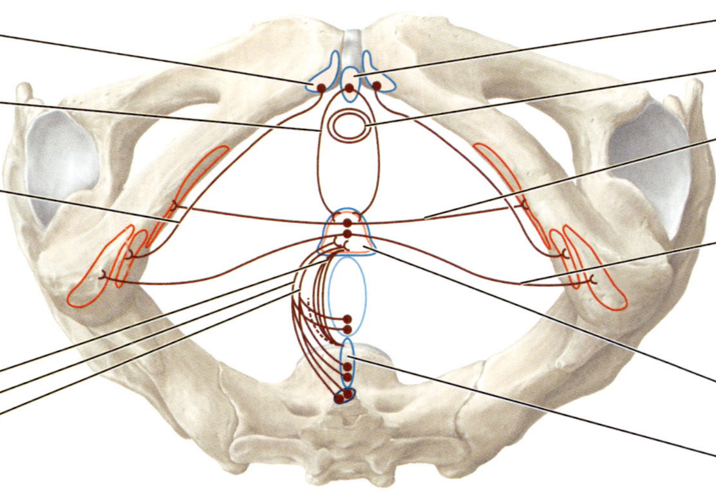

Female perineum. Schematic representation of the origin and insertions of the external urethral sphincter, bulbospongiosus, ischiocavernosus, superficial and deep transverse perineal muscles, and external anal sphincter. The **external urethral sphincter** originates from the internal surface of the inferior pubic ramus and inserts on the anterior wall of the vagina. The **bulbospongiosus** and **ischiocavernosus** have their origins, respectively, from the perineal raphe and the perineal body, and from the ischial tuberosity. The former inserts on the crus of the clitoris, and the latter below the pubic symphysis on the tunica albuginea of the corpora cavernosa of the clitoris. The **superficial** and **deep transverse perineal muscles** extend from the internal surface of the ischial tuberosity, in the case of the former, and from the medial surface of the ramus of the ischium, in the case of the latter, to the perineal body. The deep part of the **external anal sphincter** consists of fibers connected to those of the puborectalis that extend anteriorly and posteriorly to the anal canal, intersecting with those of the opposite side. The subcutaneous part extends from the perineal raphe to the deep surface of the dermis, while the superficial part extends from the perineal body, reaching the tip of the coccyx through the anococcygeal ligament.

Right lateral view of the **pelvis**. The piriformis forms part of the posterolateral wall of the lesser pelvis, and leaves the pelvis through the greater sciatic foramen.

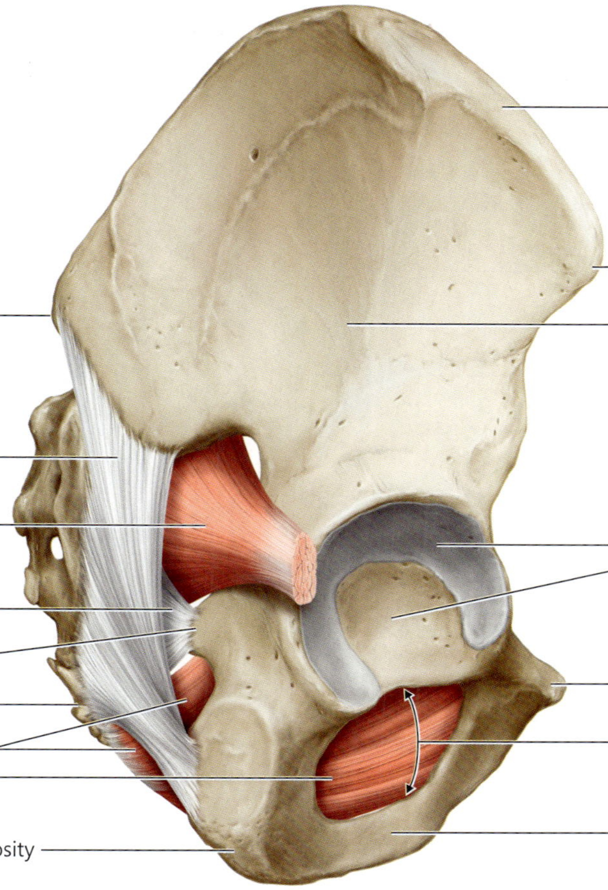

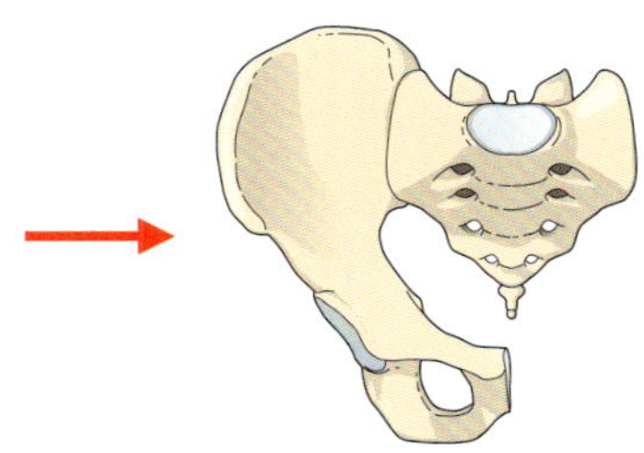

Right lateral view of the **pelvis**. The resection of the ischium and of the sacrotuberous and sacrospinous ligaments shows the ischiococcygeus and levator ani.

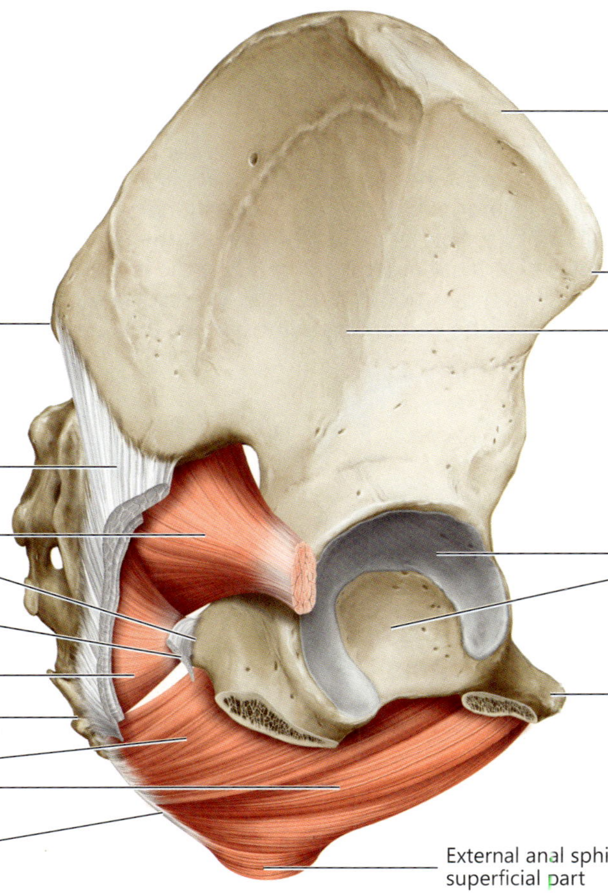

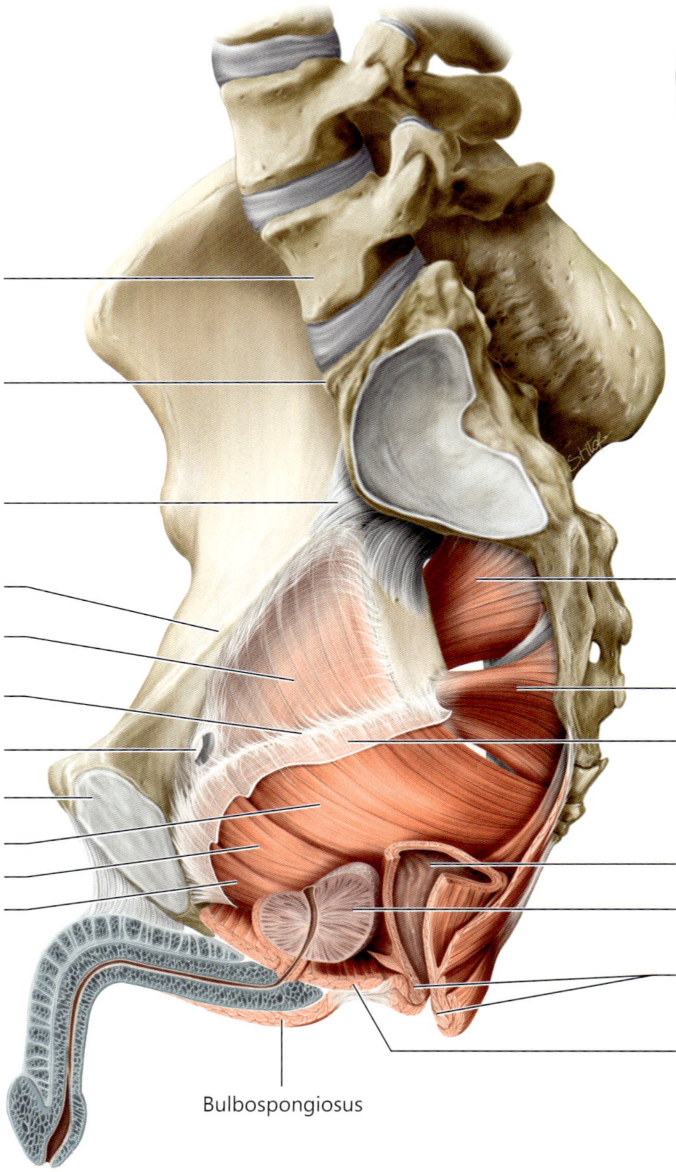

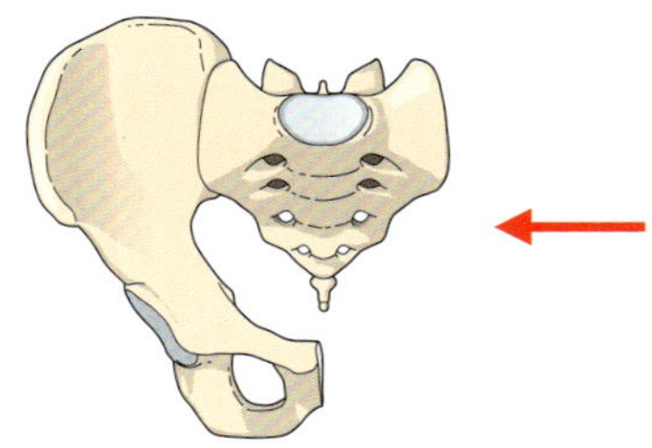

Male pelvis and **perineum**. Median sagittal section. The parietal pelvic fascia on the medial surface of the obturator internus forms the obturator fascia, connected to the posterior part of the arcuate line of the ilium. Behind the obturator canal, its thickened portion, the tendinous arch of the levator ani, is the site of insertion for this muscle, which is a large lamina of variable thickness inserted on the internal surface of the lesser pelvis, forming a large portion of its floor (pelvic diaphragm). The external anal sphincter is an ovoid formation of striated muscle surrounding the inferior part of the anal canal. The deep transverse perineal muscle, which in females is replaced by smooth muscle tissue associated to the urogenital sphincter, forms a partial lamina extending from the medial surface of the ischiopubic ramus above the deep perineal pouch.

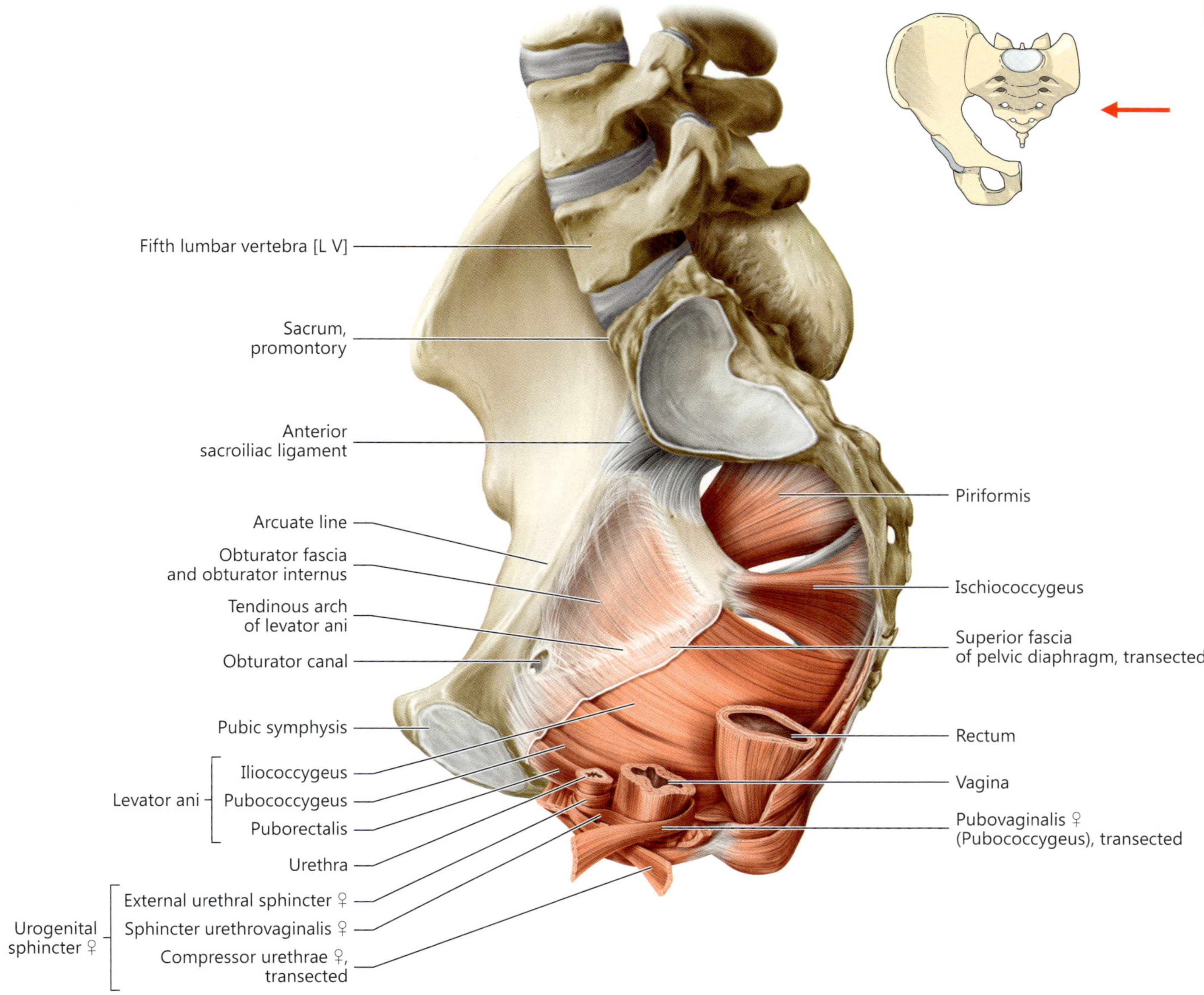

Female pelvis and **perineum**. Median sagittal section. The urogenital sphincter, formed by the external urethral sphincter, sphincter urethrovaginalis, and compressor urethrae, is only found in females. The external urethral sphincter surrounds the middle and lower thirds of the urethra. The sphincter urethrovaginalis, originating from the perineal body, is formed by fibers that extend forward on both sides of the vagina and urethra, to join with the contralateral fibers anteriorly to the urethra. The compressor urethrae, which originates from the ischiopubic rami, is formed by the fibers that extend forward to join with the contralateral fibers anteriorly to the urethra, below the external urethral sphincter.

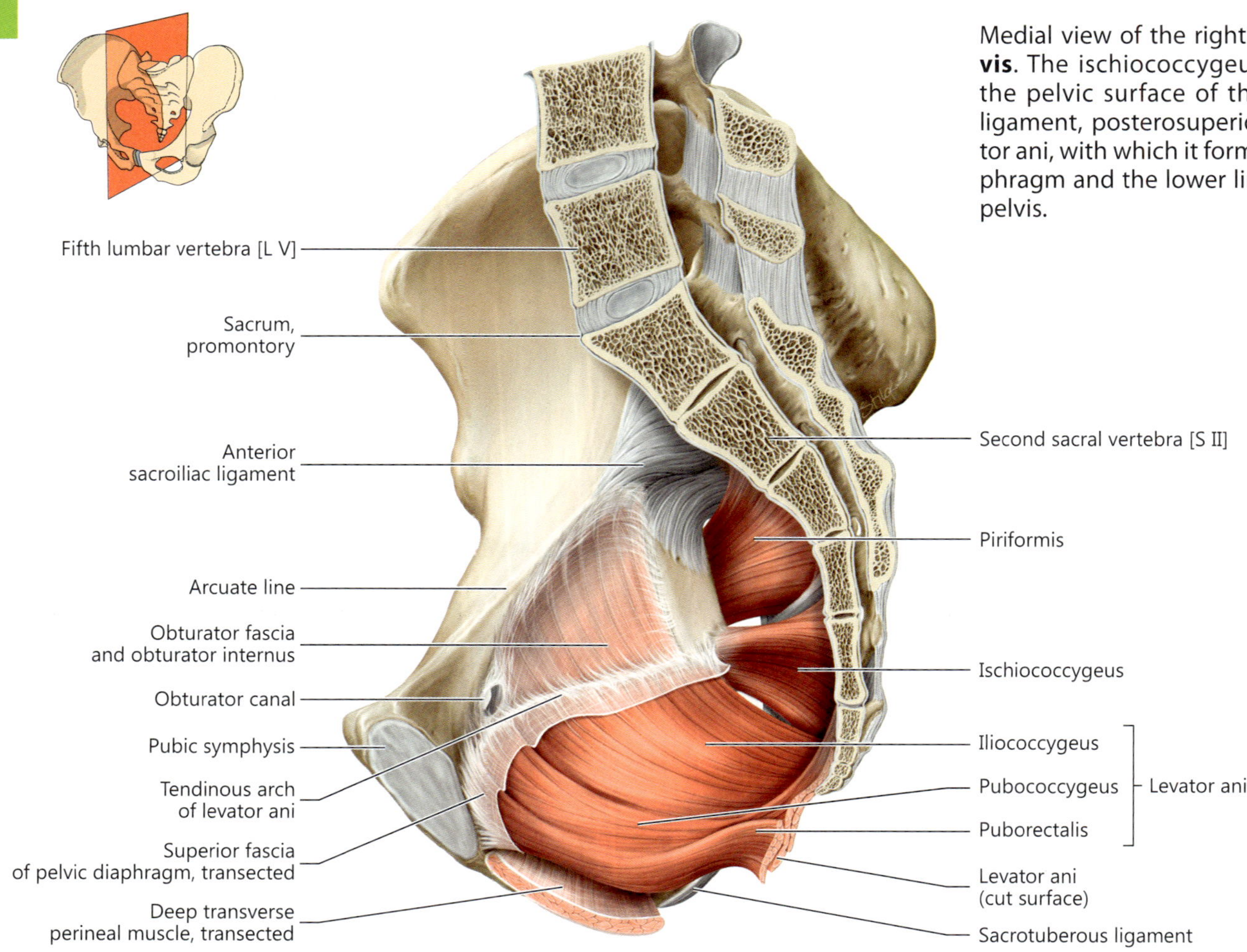

Medial view of the right **male hemipelvis**. The ischiococcygeus is located on the pelvic surface of the sacrospinous ligament, posterosuperiorly to the levator ani, with which it forms the pelvic diaphragm and the lower limit of the lesser pelvis.

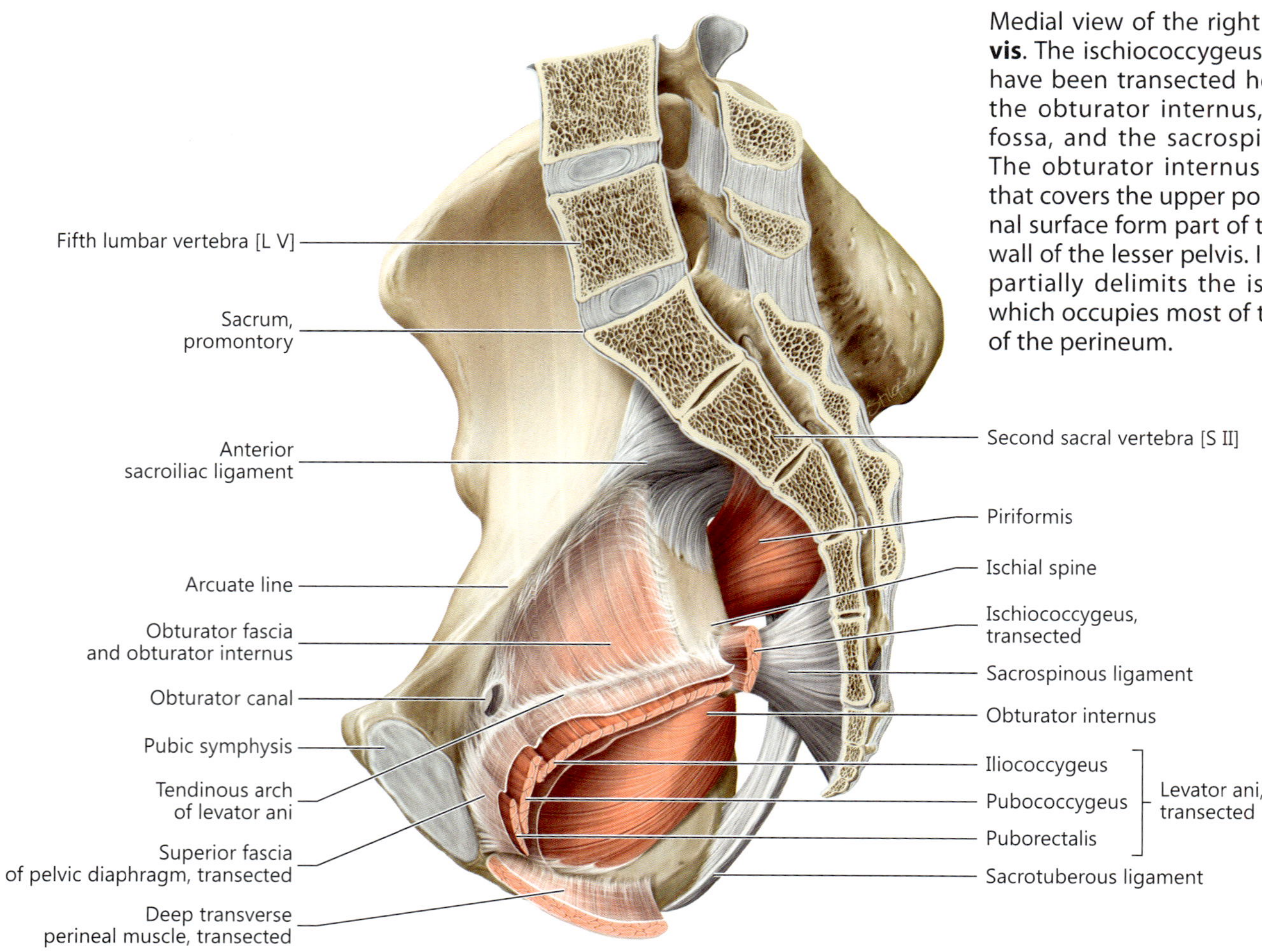

Medial view of the right **male hemipelvis**. The ischiococcygeus and levator ani have been transected here to highlight the obturator internus, the ischioanal fossa, and the sacrospinous ligament. The obturator internus and the fascia that covers the upper portion of its internal surface form part of the anterolateral wall of the lesser pelvis. Its lower portion partially delimits the ischioanal fossa, which occupies most of the anal triangle of the perineum.

ORIGIN

Bones

Aponeuroses, cartilage, raphe

Origin ⟶ Insertion

INSERTION

Bones

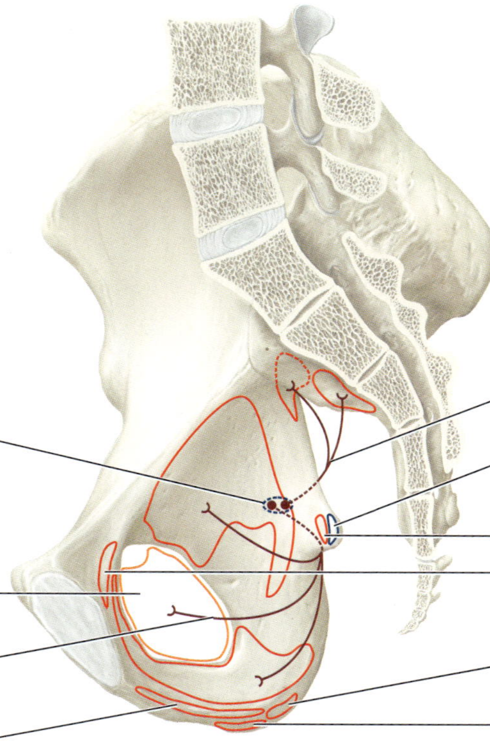

Schematic representation of the origin and insertions of the piriformis, obturator internus, ischiococcygeus, iliococcygeus, and pubococcygeus (parts of the levator ani), superficial and deep transverse perineal muscles, and ischiocavernosus. The **piriformis** originates from the pelvic surface of the sacrum, laterally to the second, third, and fourth sacral foramina, and from the border of the greater sciatic notch, to insert on the upper extremity of the internal surface of the greater trochanter of the femur. The **obturator internus** joins the internal surface of the hip bone (obturator foramen, the bone surface between this and the greater sciatic notch, and obturator membrane) to the femur (trochanteric fossa). The **ischiococcygeus** originates from the lateral border of the last three sacral vertebrae and from the coccyx, and ends with a tapered extremity on the ischial spine. The **iliococcygeus** originates from the tendinous arch of the levator ani and from the ischial spine, ending at the lateral border of the coccyx and the anococcygeal body or ligament. The **pubococcygeus**, from the posterior surface of the superior pubic ramus, on the side of the symphysis, extends to the lateral border of the coccyx and last sacral vertebra. The **superficial** and **deep transverse perineal muscles** extend from the internal surface of the ischial tuberosity, in the case of the former, and from the medial surface of the ramus of the ischium, in the case of the latter, to the perineal body. The **ischiocavernosus** originates from the ischial tuberosity and inserts below the pubic symphysis on the tunica albuginea of the corpora cavernosa of the penis (♂) or clitoris (♀).

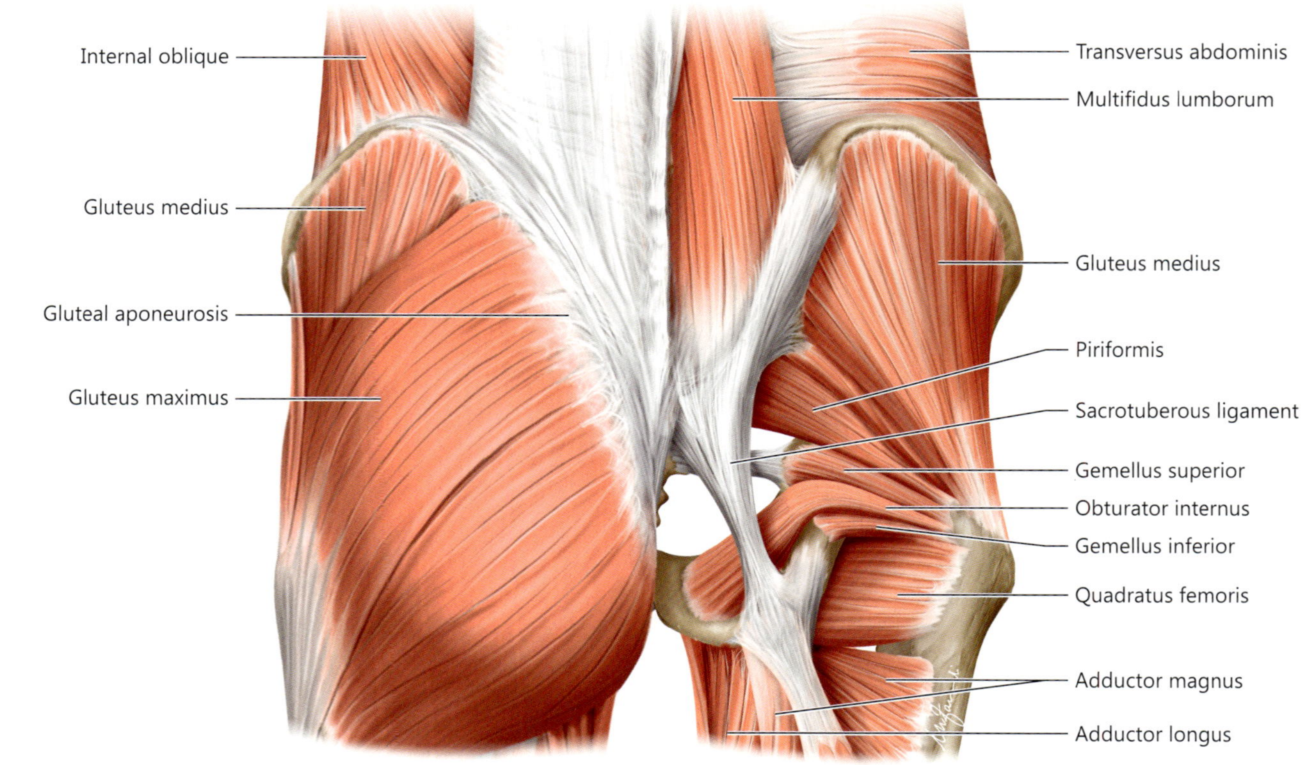

Gluteal region. On the left, the superficial plane, following removal of the gluteal fascia. On the right, after resection of the gluteus maximus, you can see the gluteus medius on an intermediate plane, and more deeply the piriformis, obturator externus and internus, gemellus superior and inferior, and quadratus femoris.

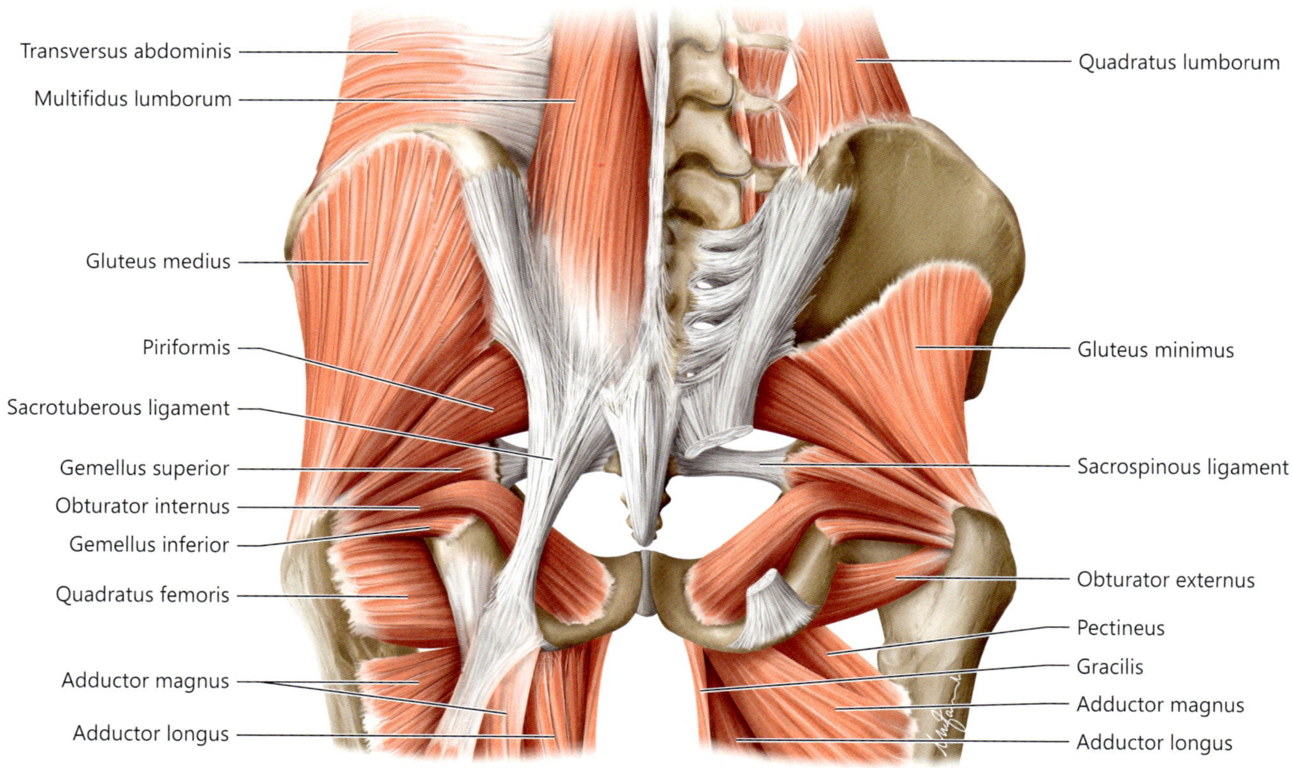

Gluteal region. On the left, you can see the dissection plane visible on the right in the figure above. The fascicles of the piriformis exit the lesser pelvis through the greater sciatic foramen. On the right, after resection of the gluteus medius, you can see the gluteus minimus. Removal of the quadratus femoris highlights the obturator externus.

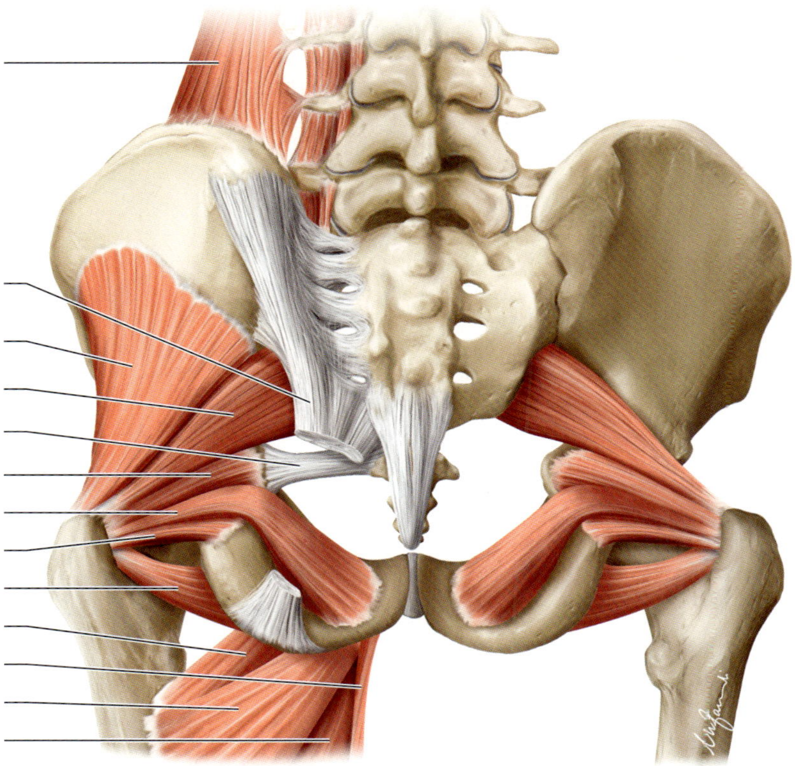

Gluteal region. On the left, you can see the dissection plane visible on the right in the figure at the bottom of the previous page, after further dissection of the sacrotuberous ligament. You can clearly see the obturator internus, which forms part of the anterolateral wall of the lesser pelvis.

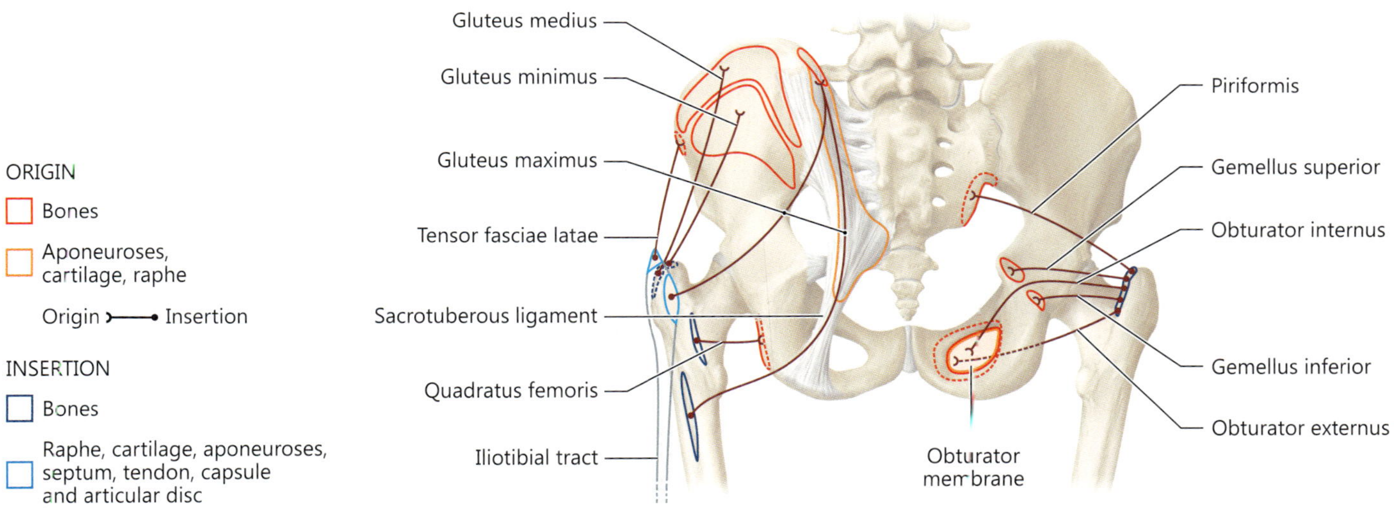

Schematic representation of the origin and insertions of the gluteal muscles, tensor fasciae latae, quadratus femoris, piriformis, gemellus superior and inferior, and obturator internus and externus. The **gluteus maximus**, from the iliac crest, posterior gluteal line, lateral surfaces of the sacrum and coccyx, and sacrotuberous ligament, extends to the gluteal tuberosity of the femur and the iliotibial tract; the **gluteus medius**, from the gluteal fascia, iliac crest, gluteal surface of the ilium between the anterior and posterior gluteal lines, and from the anterior superior iliac spine, inserts on the external surface of the greater trochanter; the **gluteus minimus**, from the gluteal surface of the ilium between the anterior and posterior gluteal lines and from the iliac crest, inserts on the anterior border of the greater trochanter. The **tensor fasciae latae**, which originates from the anterior superior iliac spine and from the anterior extremity of the iliac crest, joins with its tendon to the iliotibial tract. The **quadratus femoris** originates from the ischial tuberosity and inserts on the intertrochanteric crest. The **piriformis**, from the pelvic surface of the sacrum, laterally to the anterior sacral foramina, and from the border of the greater sciatic notch, extends to the upper extremity of the internal surface of the greater trochanter. The **gemellus superior** and **inferior** extend from the ischial spine, in the case of the former, and from the ischial tuberosity, in the case of the latter, to the trochanteric fossa. The **obturator externus**, which originates from the external border of the obturator foramen and from the obturator membrane, and the **obturator internus**, which originates from the internal border of the obturator foramen, from the obturator membrane, and from the bone surface between the obturator foramen and the greater sciatic notch, insert on the trochanteric fossa.

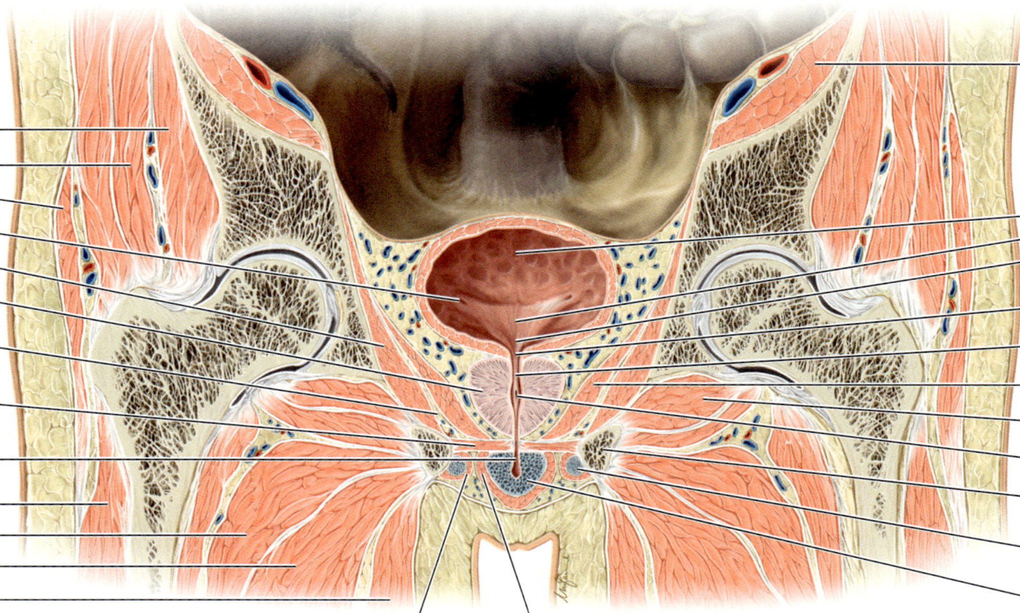

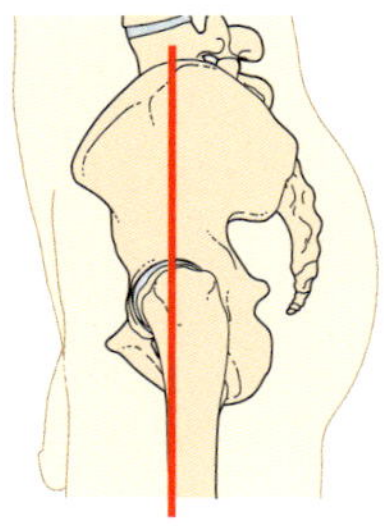

Male pelvis and perineum. Anterior view. Frontal section through the lesser pelvis, on the plane of the urethra and urinary bladder. The bladder and prostate are contained anteriorly within the lesser pelvis; the intermediate part of the urethra (membranous urethra), the bulb of the penis, and the initial part of the corpora cavernosa within the anterior perineum (urogenital diaphragm).

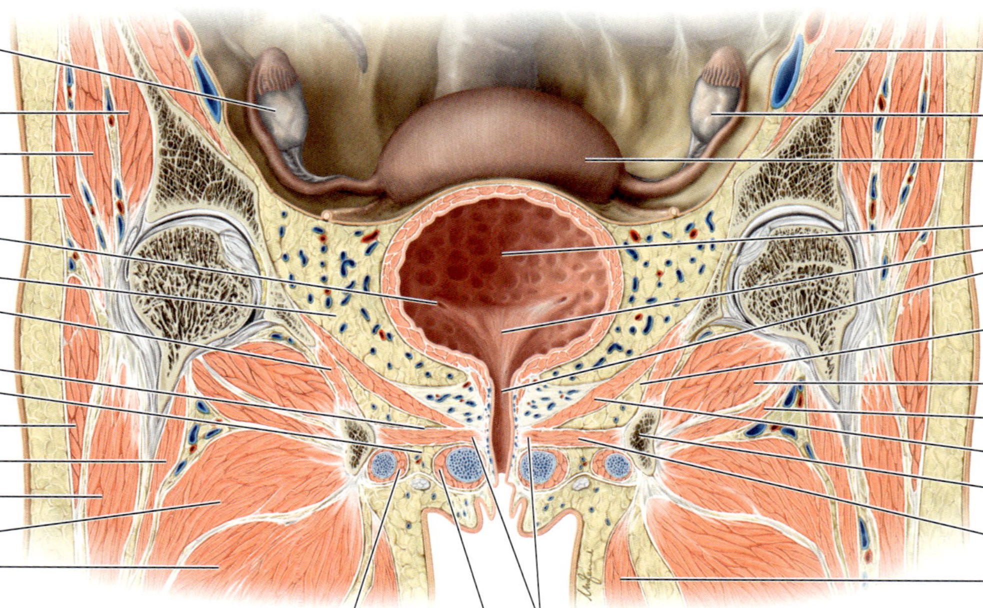

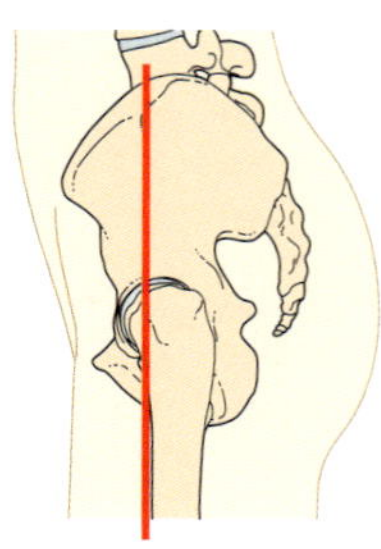

Female pelvis and perineum. Anterior view. Frontal section through the lesser pelvis on the plane of the urethra and urinary bladder. In females, the bladder appears more developed transversely, and is flatter in the anteroposterior direction. The female urethra passes through the pelvic diaphragm and, after a short path in the anterior perineum, opens into the vestibule of the vagina.

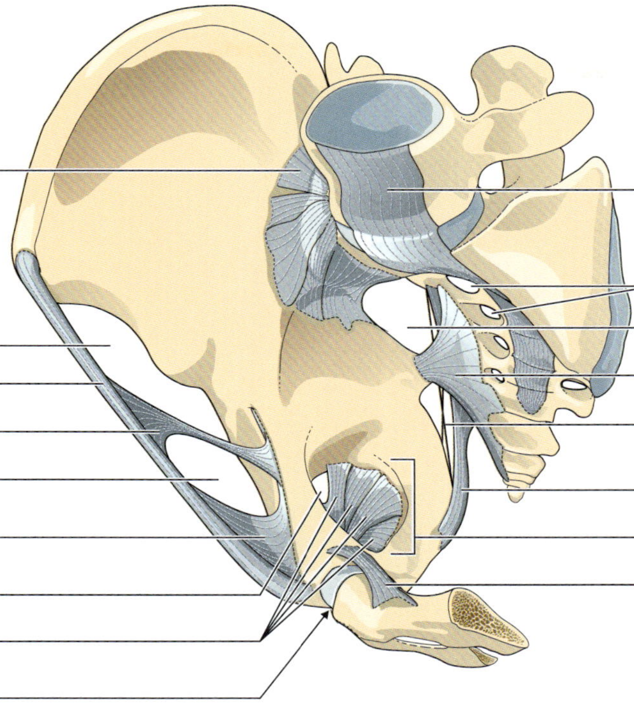

Greater and **lesser sciatic foramen**. The sciatic and pudendal nerves, and the gluteal vessels and nerves, exit the pelvis through the greater sciatic foramen, to reach the gluteal region. The pudendal nerve and the internal pudendal vessels enter the perineum from the gluteal region through the lesser sciatic foramen, crossing the ischioanal fossa.

Ligaments and foramina of the pelvis	
Ligaments – inferior pubic (arcuate pubic) – anterior longitudinal – anterior sacroiliac – sacrospinous – sacrotuberous	 Between the inferior pubic tubercles Anterior surface of the vertebral bodies and partially of the intervertebral discs from the occipital bone to the pelvic surface of the sacrum From the pelvic surface of the sacrum to the medial part of the iliac fossa Between the lateral border of the sacrum and coccyx and the ischial spine From the posterior superior and posterior inferior iliac spines and from the lateral border of the sacrum and coccyx to the ischial tuberosity
Foramina – greater sciatic notch – vascular space – muscular space – obturator – lesser sciatic notch – anterior sacral	 Deep notch above the ischial spine, the formation of which also involves the body of the ilium, converted into the greater sciatic foramen by the sacrospinous ligament[a] Delimited by the inferior surface of the inguinal ligament, the lacunar ligament, and the iliopectineal arch[b] Delimited by the inferior surface of the inguinal ligament, the iliopectineal arch, and the ilium[c] Below the acetabulum, delimited superoanteriorly by the pubis and posteroinferiorly by the ischium, closed off by the obturator membrane except in the obturator canal (obturator groove of the superior pubic ramus forming a canal by an interruption of the obturator membrane)[d] Notch below the ischial spine, converted into the lesser sciatic foramen by the sacrotuberous and sacrospinous ligaments[e] Pelvic surface of the sacrum, laterally to the transverse lines. Communicating with the sacral canal, they provide passage to the anterior rami of the sacral nerves

[a] Crossed by the piriformis, sciatic nerve, superior gluteal artery and veins, inferior gluteal artery, veins, and nerve, internal pudendal artery and vein, pudendal nerve (exiting from the pelvis), nerves to the obturator internus, nerve to the quadratus femoris, and posterior cutaneous nerve of thigh
[b] Crossed by the femoral artery and vein, and the lymphatic vessels
[c] Crossed by the femoral nerve and iliopsoas
[d] Crossed by the obturator artery, veins, and nerve
[e] Crossed by the tendon of the obturator internus, nerve to the obturator internus, internal pudendal artery and vein, and pudendal nerve (entering the perineum)

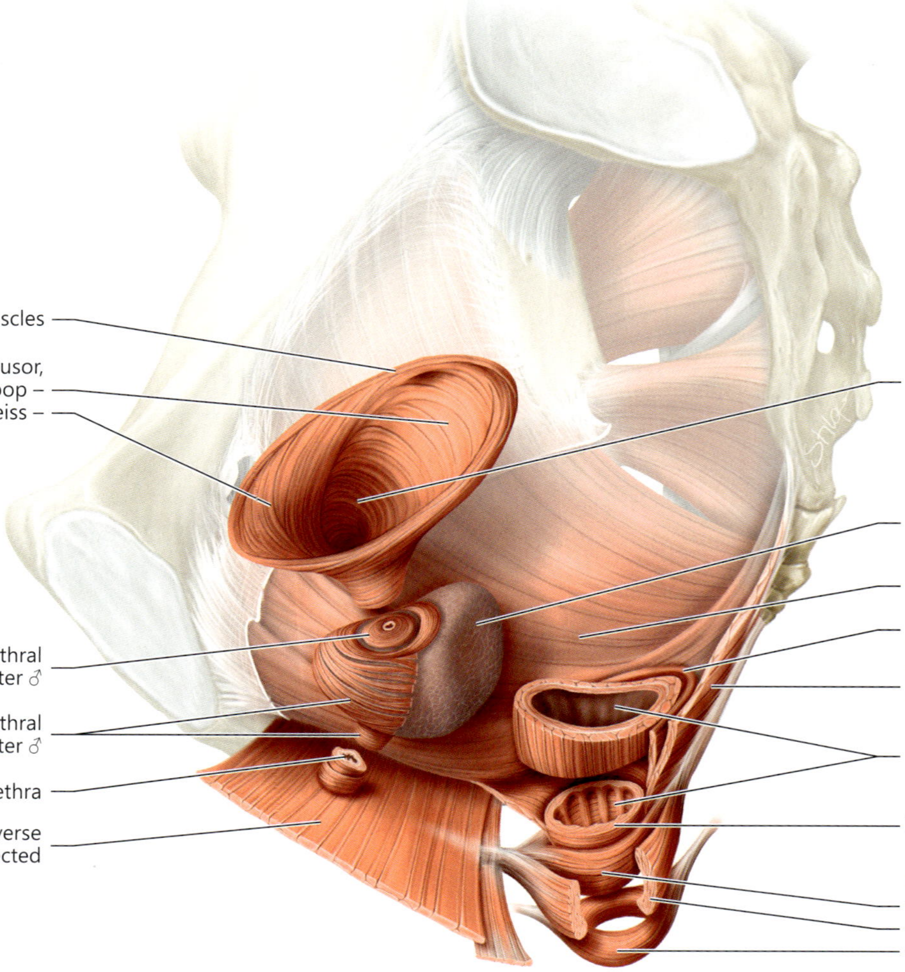

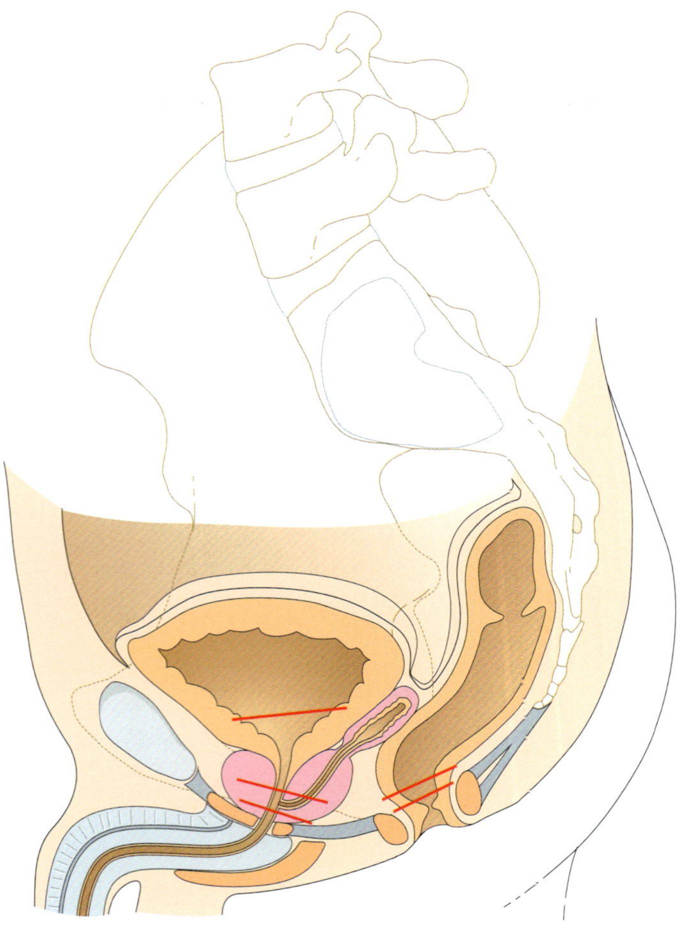

Sphincter muscles of the pelvis and perineum in males. At the level of the neck of the bladder of the **detrusor** (smooth muscle) (**sphincter of the bladder**), the muscular coat is characterized by three layers (external longitudinal, intermediate circular, and internal longitudinal); it is possible to distinguish fascicles arranged in loops (posterior loop and loop of Heiss). The **trigonal muscles** (**superficial** and **deep**), or trigonal ring (muscular layer of the trigone of the bladder), are also involved in the formation of the sphincter of the bladder. The **internal urethral sphincter** (smooth muscle), also known as the **supracollicular** or **preprostatic sphincter**, is formed by the circular layer of the muscular coat of the intramural (preprostatic) part of the urethra. The **external urethral sphincter** (striated muscle) is located in the intermediate part of the urethra (membranous urethra) as an external layer of the muscular coat. In the anal canal, at the level of the sinuses, the thick circular (internal) layer of the muscular coat is crossed obliquely by the fascicles coming from the external longitudinal layer, which are arranged more internally in a longitudinal direction (thus forming a third layer). Fascicles of the levator ani also extend to the external longitudinal layer. The **internal sphincter** (smooth muscle), surrounded externally by the **external sphincter** (striated muscle), which consists of a deep part, a superficial part, and a subcutaneous part, constitutes the sphincter system of the anus. The red lines indicate the levels of transverse sections through the detrusor, the internal urethral sphincter ♂, the intermediate part of the male urethra (membranous urethra), the anal canal, and the internal anal sphincter.

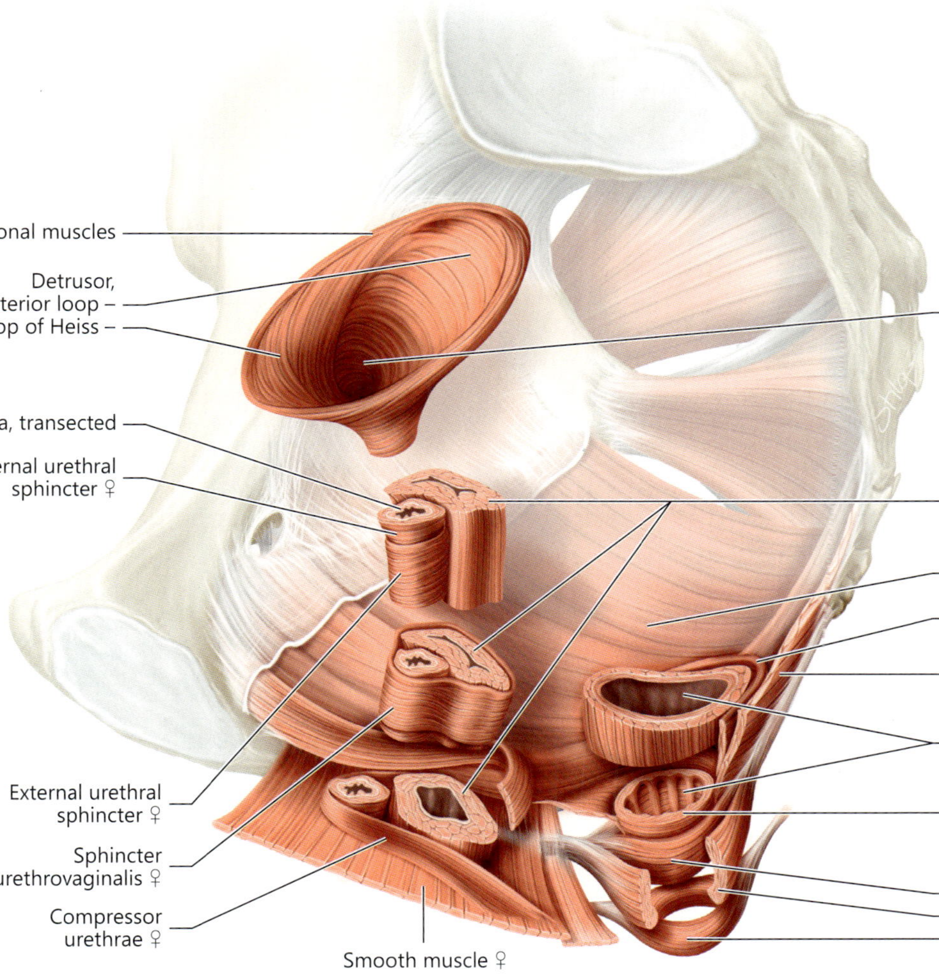

Sphincter muscles of the pelvis and perineum in females. The sphincter muscles of the bladder and anus (internal and external) do not present substantial differences in females as compared to males (☞ see the description on the previous page). The only exception is the **external anal sphincter**, which may be shorter in women in the anterior portion, and may also have a slightly thinner wall. At the urethral level, in females there are the **urogenital sphincter**, formed by three striated sphincter muscles – external urethral sphincter, sphincter urethrovaginalis, and compressor urethrae – and the **internal urethral sphincter**, formed by the circular layer of the muscular coat of the urethra. The **external urethral sphincter** (striated muscle) surrounds the urethra beyond its middle third portion. The **compressor urethrae** originates from the ischiopubic ramus and connects anteriorly to the urethra with the contralateral muscle, forming a flattened area before the urethra. The **sphincter urethrovaginalis** originates from the perineal body, and, after extending to the side of the vagina and urethra, connects with the contralateral muscle to form the flattened area anterior to the urethra. The red lines indicate the levels of transverse sections through the detrusor, the internal urethral sphincter ♀ and the vagina, the sphincter urethrovaginalis and the vagina, the anal canal, and the internal anal sphincter.

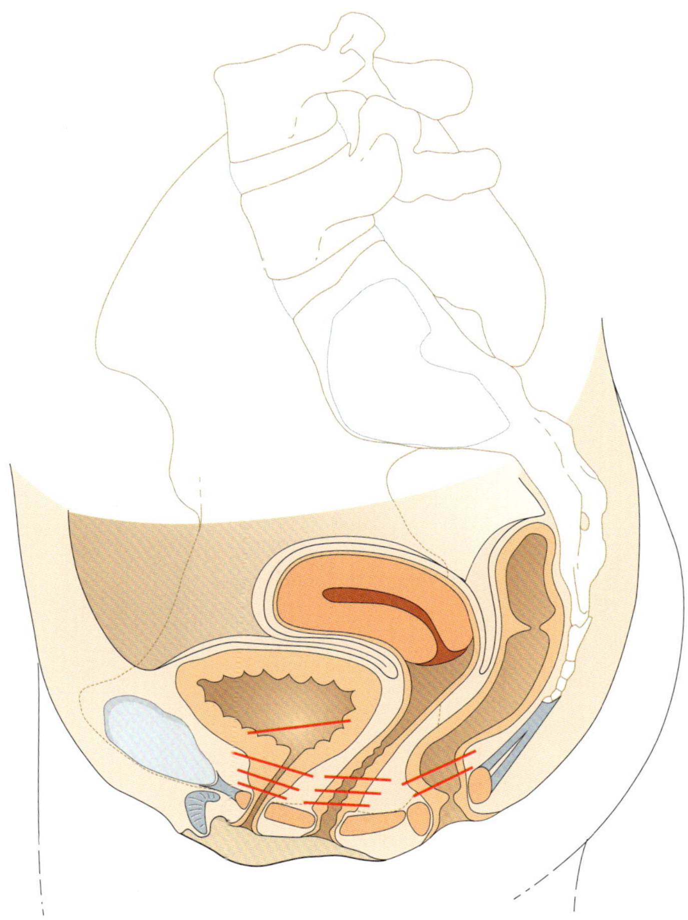

Sphincter muscles of the pelvis and perineum		
Bladder	Muscular coat (smooth muscles) – detrusor – – bladder neck part (sphincter of the bladder or internal sphincter of the bladder) – – – internal longitudinal layer (plexiform) – – – circular layer (intermediate) – – – external longitudinal layer – – – posterior loop – – – loop of Heiss (detrusor loop) – trigonal muscle (trigonal ring) – – superficial trigonal muscle – – deep trigonal muscle	
Urethra	*Male* Intramural (preprostatic) part of urethra – muscular coat (smooth muscles) – – circular layer (internal or smooth urethral sphincter or supracollicular sphincter or preprostatic sphincter) Intermediate part of urethra (membranous urethra) – external urethral sphincter (striated sphincter or rhabdosphincter) (striated muscles)	*Female* Muscular coat (smooth muscles) – circular layer (external) (internal or smooth urethral sphincter) Urogenital sphincter (striated muscles) – external urethral sphincter (striated sphincter or rhabdosphincter) – sphincter urethrovaginalis – compressor urethrae
Anal canal	Muscular coat (smooth muscles) – circular layer (internal) (internal or smooth anal sphincter)* – longitudinal layer (external)* External anal sphincter – subcutaneous part – superficial part – deep part	

* The circular layer, which is thicker at the level of the anal sinuses, is crossed obliquely by the fascicles that extend from the external longitudinal layer to the internal layer, thus forming three layers, two of which are longitudinal (external and internal), while one is circular (intermediate). Additionally, striated fascicles of the levator ani extend to the external longitudinal layer, intersecting with the smooth fascicles.

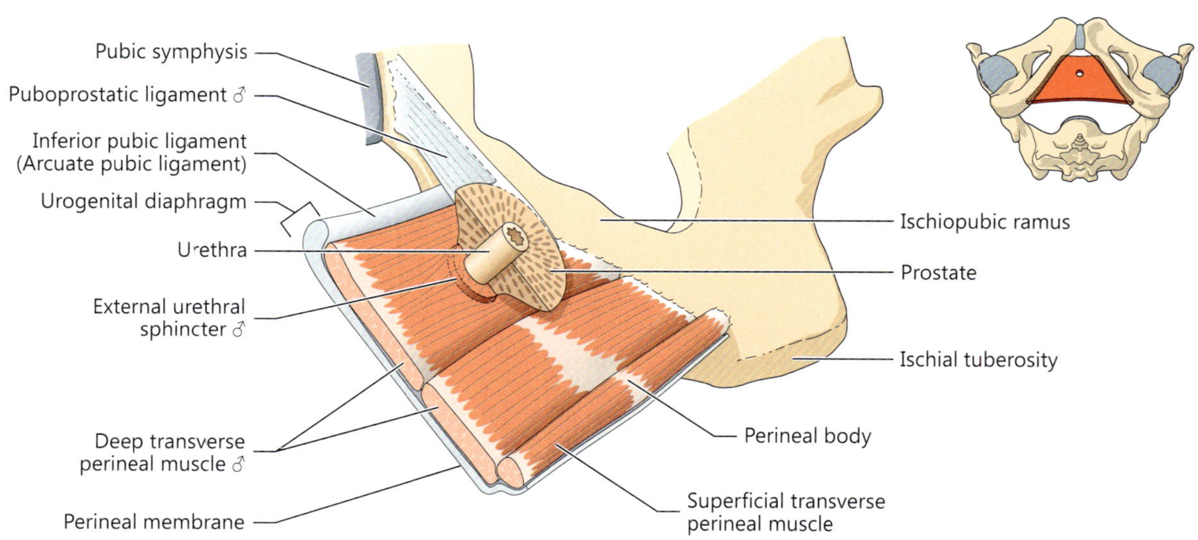

a

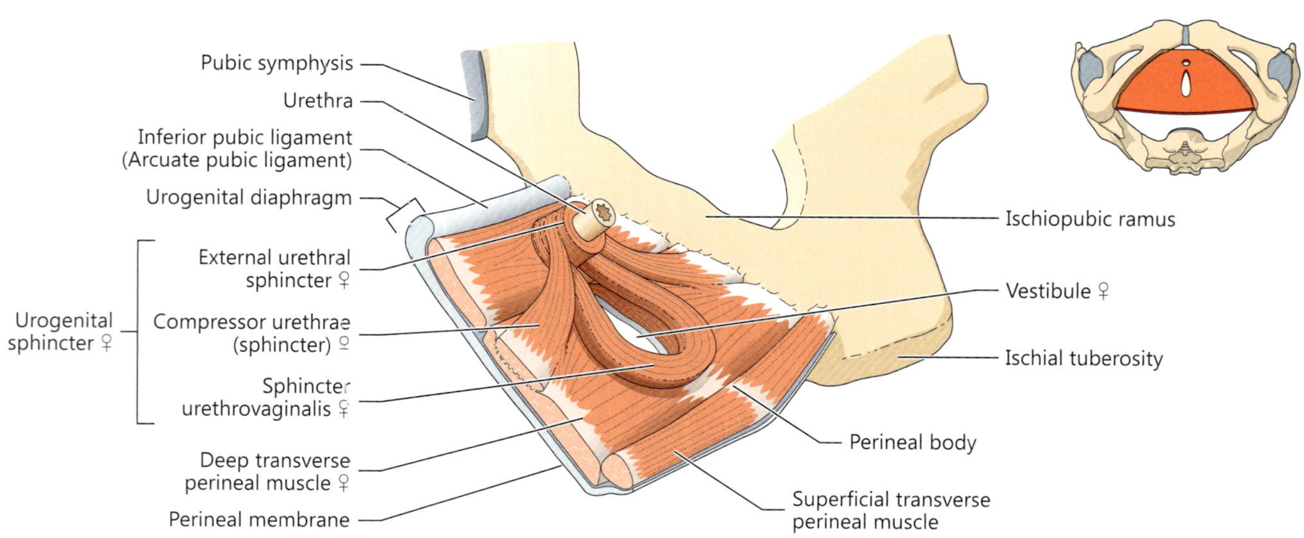

b

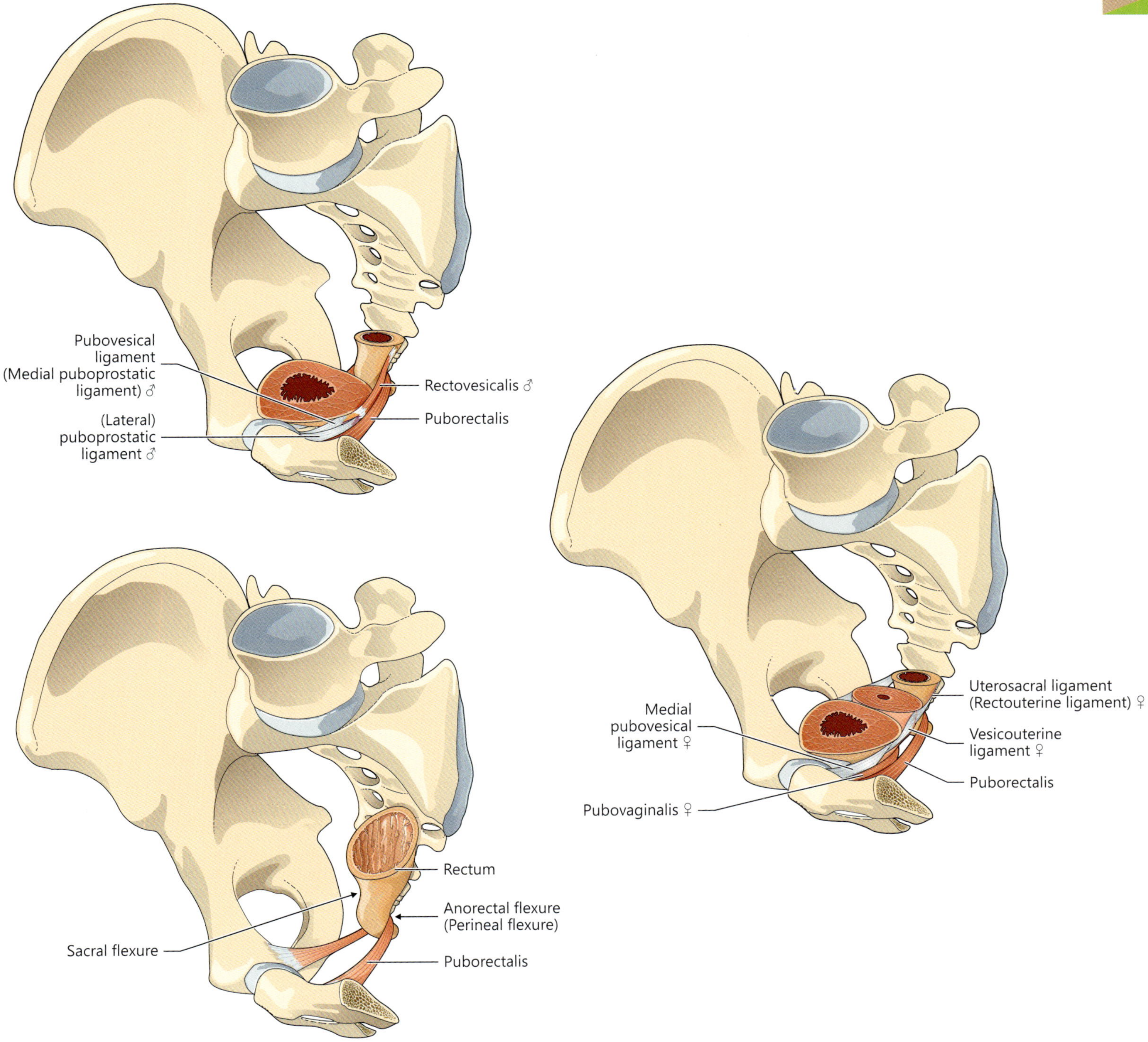

Constrictor muscles of the rectum (puborectalis) and **of the vagina** (pubovaginalis) and **means of attachment in relation to the pelvic fascia**. The rectovesicalis or rectovesical ligament is a fibromuscular connection (smooth muscle fibers) from the base of the bladder and prostate to the lateral surfaces of the rectum.

◄
a, The intermediate part of the urethra in males, while crossing the urogenital diaphragm, is surrounded by a ring of striated muscle fibers (**external urethral sphincter**).
b, The **striated muscles in relation to the urethra and vagina** of females form a continuous mass of muscle tissue known as the urogenital sphincter, which consists of two parts: one surrounds the urethra (external urethral sphincter), and the other is arranged around the urethra and vagina (sphincter urethrovaginalis). These two parts are complemented by fibers that attach to the ischiopubic rami, forming the compressor urethrae. From the fundus of the bladder, inside the pelvis, the urogenital sphincter extends into the perineum, around the vagina, passing through the urogenital hiatus of the pelvic diaphragm. At the level of the urogenital diaphragm (or triangle), there is no superior fascia to close the deep perineal pouch or to form the floor of the urogenital hiatus.

Muscles of the pelvis					
MUSCLE	**ORIGIN**	**INSERTIONS**	**INNERVATION**	**ACTION**	**OBSERVATIONS**
INTERNAL MUSCLES OF THE HIP					
Iliopsoas (☞ table Muscles of the Abdomen)					
– **psoas major**	Anterior surfaces of the transverse processes of all the lumbar vertebrae; lateral surfaces of the intervertebral bodies and discs of the twelfth thoracic vertebra and of all the lumbar vertebrae	Lesser trochanter of the femur	Branches of the lumbar plexus (primarily L1, L2, with some contribution from L3)	Forward flexion of the trunk (by a joint action of the muscles on both sides)	
– **iliacus**	Upper two-thirds of the iliac fossa; inner lip of the iliac crest, anterior sacroiliac and iliolumbar ligaments; upper part of the lateral surface of the sacrum	Lateral surface of the tendon of the psoas major, lesser trochanter of the femur	Muscular branches of the femoral nerve [L2, L3]	Forward flexion of the hip (with a fixed point on the pelvis); flexion and homolateral inclination of the trunk (with a fixed point on the femur)	
Psoas minor	Bodies of the last thoracic vertebra and of the first lumbar vertebra and interposed intervertebral disc	Iliopubic ramus and iliac fascia	Branches of the lumbar plexus [L1]	Flexion of the trunk and tension of the iliac fascia	
EXTERNAL MUSCLES OF THE HIP					
Muscles that insert on the external side of the pelvis					
Tensor fasciae latae	Anterior superior iliac spine and anterior extremity of the iliac crest	Lateral condyle of the tibia, into which inserts the iliotibial tract, formed by the union of the muscular tendon with the fascia lata	Superior gluteal nerve [L4, L5, S1]	Tension of the fascia lata, contributing to maintenance of the upright position; flexion and abduction of the hip, extension of the knee	
Gluteus maximus	Iliac crest, posterior gluteal line, lateral surfaces of the sacrum and coccyx, and sacrotuberous ligament	Gluteal tuberosity of the femur and iliotibial tract	Inferior gluteal nerve [L5, S1, S2]	Extension and external rotation of the hip. With a fixed point on the femur, extension of the pelvis, thus contributing to maintenance of the upright position and deambulation	
Gluteus medius	Gluteal fascia, iliac crest, gluteal surface of the ilium between the anterior and posterior gluteal lines, anterior superior iliac spine	External surface of the greater trochanter	Superior gluteal nerve [L4, L5, S1]	Abduction of the hip and its internal rotation, by the action of the anterior muscle fascicles; external rotation by the action of the posterior fascicles. With a fixed point on the femur, extension and lateral inclination of the pelvis. The contraction of both muscles contributes to maintenance of the upright position	
Gluteus minimus	Gluteal surface of the ilium between the anterior and inferior gluteal lines, iliac crest	Anterior border of the greater trochanter	Superior gluteal nerve [L4, L5, S1]	Abduction and internal rotation of the hip. With a fixed point on the femur, extension and lateral inclination of the pelvis. The contraction of both muscles contributes to maintenance of the upright position	
Gemellus inferior	Ischial tuberosity	Trochanteric fossa through the tendon of the obturator internus	Nerve to the quadratus femoris [L4, L5, S1]	External rotation and stabilization of the hip joint	
Gemellus superior	Ischial spine	Trochanteric fossa through the tendon of the obturator internus	Nerve to the obturator internus [L5, S1, S2]	External rotation and stabilization of the hip joint	
Obturator externus	External border of the obturator foramen and obturator membrane	Trochanteric fossa	Posterior branch of the obturator nerve [L3, L4]	External rotation and stabilization of the hip joint	

continued

Muscles of the pelvis (continued)

MUSCLE	ORIGIN	INSERTIONS	INNERVATION	ACTION	OBSERVATIONS
Quadratus femoris	Ischial tuberosity	Intertrochanteric crest	Nerve to the quadratus femoris [L5, S1]	External rotation and stabilization of the hip joint	
Muscles that insert on the internal side of the pelvis (muscles of the pelvic wall)					
Obturator internus	Internal border of the obturator foramen, obturator membrane, and the bone surface between the obturator foramen and greater sciatic notch	Trochanteric fossa	Nerve to the obturator internus [L5, S1]	External rotation and stabilization of the hip joint	
Piriformis	Pelvic surface of the sacrum, laterally to the anterior sacral foramina, border of the greater sciatic notch	Upper extremity of the internal surface of the greater trochanter	Nerve to the piriformis [L5, S1, S2]	Abduction, external rotation, and stabilization of the hip joint	
PELVIC DIAPHRAGM (PELVIC FLOOR)					
Levator ani	Laterally from the posterior surface of the superior pubic ramus, on the side of the symphysis, to the ischial spine, and from the tendinous arch of the levator ani	Lateral border of the coccyx, last sacral vertebra, and anococcygeal body or ligament	Pudendal plexus [S2-S4]	Elevation of the pelvic floor	The tendinous arch of the levator ani is a fibrous cord that thickens the obturator fascia, extending from the anterosuperior boundary to the posteroinferior boundary of the obturator foramen
– **iliococcygeus**	Tendinous arch of the levator ani and ischial spine (posterior fascicles of the levator ani)	Lateral border of the coccyx and anococcygeal body or ligament		Constriction of the rectum	
– **pubococcygeus**	Posterior surface of the superior pubic ramus, on the side of the symphysis (anterior fascicles of the levator ani)	Lateral border of the coccyx and last sacral vertebra (lateral fascicles of the pubococcygeus)		Lateral compression of the organs crossing the pelvic floor	
– – **puboanalis**	Posterior surface of the superior pubic ramus, on the side of the symphysis (medial fascicles of the pubococcygeus)	External surface of the anal canal		Constriction of the anus	
– – **puboperinealis**	Posterior surface of the superior pubic ramus, on the side of the symphysis (medial fascicles of the pubococcygeus)	Perineal body		Constriction of the perineal body	
– – **puboprostaticus** (levator prostatae) ♂	Posterior surface of the superior pubic ramus, on the side of the symphysis (medial fascicles of the pubococcygeus)	Perineal body		Constriction of the prostate	
– – **pubovaginalis** (constrictor of the vagina) ♀	Posterior surface of the superior pubic ramus, on the side of the symphysis (medial fascicles of the pubococcygeus)	Fascicles of the pubococcygeus surrounding the posterior wall of the vagina, interposing with the contralateral ones		Constriction of the vagina	Muscular sling that contracts to strongly put the anterior wall of the vagina in contact with the posterior wall
– **puborectalis**	Posterior surface of the superior pubic ramus, on the side of the symphysis (anterior fascicles of the levator ani)	Posteriorly to the rectum, the medial fibers intertwine with the contralateral fibers to form the anococcygeal body (or ligament) along the midline		Constriction of the rectum	
Ischiococcygeus	Lateral border of the last three sacral vertebrae and of the coccyx, and adjacent portion of the sacrospinous ligament	Ischial spine	Pudendal plexus [S3, S4]	Constriction of the rectum	

Muscles of the perineum					
MUSCLE	**ORIGIN**	**INSERTIONS**	**INNERVATION**	**ACTION**	**OBSERVATIONS**
External anal sphincter			Inferior anal (rectal) nerves and small motor branches of the pudendal nerve [S2-S4]	Muscle tone normally keeps the last portion of the rectum and anal canal closed	Cylindrical striated muscle, with an oval profile in the anteroposterior direction, surrounding the lowest part of the anal canal
– deep part (superior fibers)	Upper fibers that join laterally with the lower fibers of the puborectalis (which represents the middle-most part of the levator ani)	Anteriorly to the anal canal, a few fibers intertwine with those of the opposite side, merging with the fibers of the superficial transverse perineal muscle; posteriorly to the anal canal, a few fibers insert at the anococcygeal ligament or body			
– subcutaneous part (inferior fibers that extend under the inferior border of the internal anal sphincter)	Perineal raphe	Deep surface of the dermis			The submucosa separates these fibers from the epithelium of the inferior part of the anal canal
– superficial part (intermediate fibers that surround the inferior part of the internal anal sphincter, a few of which intersect anteriorly and posteriorly)	Perineal body	Tip of the coccyx through the anococcygeal ligament			
Of the urogenital triangle (**trigone** or **diaphragm**)			Branches of the pudendal nerve, branches of the sacral plexus and pelvic splanchnic nerves [S2-S4]		
– **urogenital sphincter** ♀					
– – **compressor urethrae** ♀	Medial surface of the ischiopubic ramus	The muscle fibers join with the contralateral fibers, forming a flat tendinous raphe, which has a band shape and passes anteriorly to the urethra		Stretching and compression of the urethra and vagina	
– – **external urethral sphincter** (**striated**) ♀	Internal surface of the inferior pubic ramus	Anterior wall of the vagina		Forced closure of the urethra	Sleeve-like covering of the intermediate part of urethra (membranous urethra), partially surrounding also the vagina
– – **sphincter urethrovaginalis** ♀	Perineal body	The muscle fibers join with the contralateral fibers, forming an aponeurotic sling on the anterior wall of the urethra		Stretching and compression of the urethra and vagina	

continued

Muscles of the perineum *(continued)*

MUSCLE	ORIGIN	INSERTIONS	INNERVATION	ACTION	OBSERVATIONS
– **external urethral sphincter (striated)** ♂	Internal surface of the inferior pubic ramus, and anterior surface of the prostate	Perineal body and posterior surface of the prostate		Forced closure of the urethra and involvement in the mechanisms of ejaculation	Sleeve-like covering of the intermediate part of urethra (membranous urethra), partially surrounding also the prostate
– **deep transverse perineal muscle**	Medial surface of the ramus of the ischium	Perineal body (the muscle fibers join with the contralateral fibers)		Increased resistance of the pelvic diaphragm (floor), for stabilization; tension of the perineal body, limiting its movements	
Ischiocavernosus	Internal surface of the ischial tuberosity	Under the pubic symphysis, on the tunica albuginea of the corpora cavernosa of the penis (♂) or clitoris (♀)	Perineal branches of the pudendal nerve [S2-S4]	Compression of the root of the corpora cavernosa, contributing to the erection of the penis (♂) or clitoris (♀)	
Bulbospongiosus	Perineal raphe and perineal body	Dorsal surface of the corpus spongiosum of the penis, where the muscle fibers intertwine with the contralateral fibers (♂); crus of the clitoris (♀)	Perineal branches of the pudendal nerve [S2-S4]	Compression of the urethra, facilitating the discharge of its content; compression of the corpora cavernosa and corpus spongiosum of the penis and of the deep dorsal vein of the penis, facilitating erections (♂); restriction of the vaginal orifice and compression of the bulb of the vestibule, contributing to the erection of the clitoris (♀)	
Superficial transverse perineal muscle	Internal surface of the ischial tuberosity	Perineal body (the muscle fibers join with the contralateral fibers)	Perineal branches of the pudendal nerve [S2-S4]	Tension of the perineal body (by a joint action of the two muscles)	

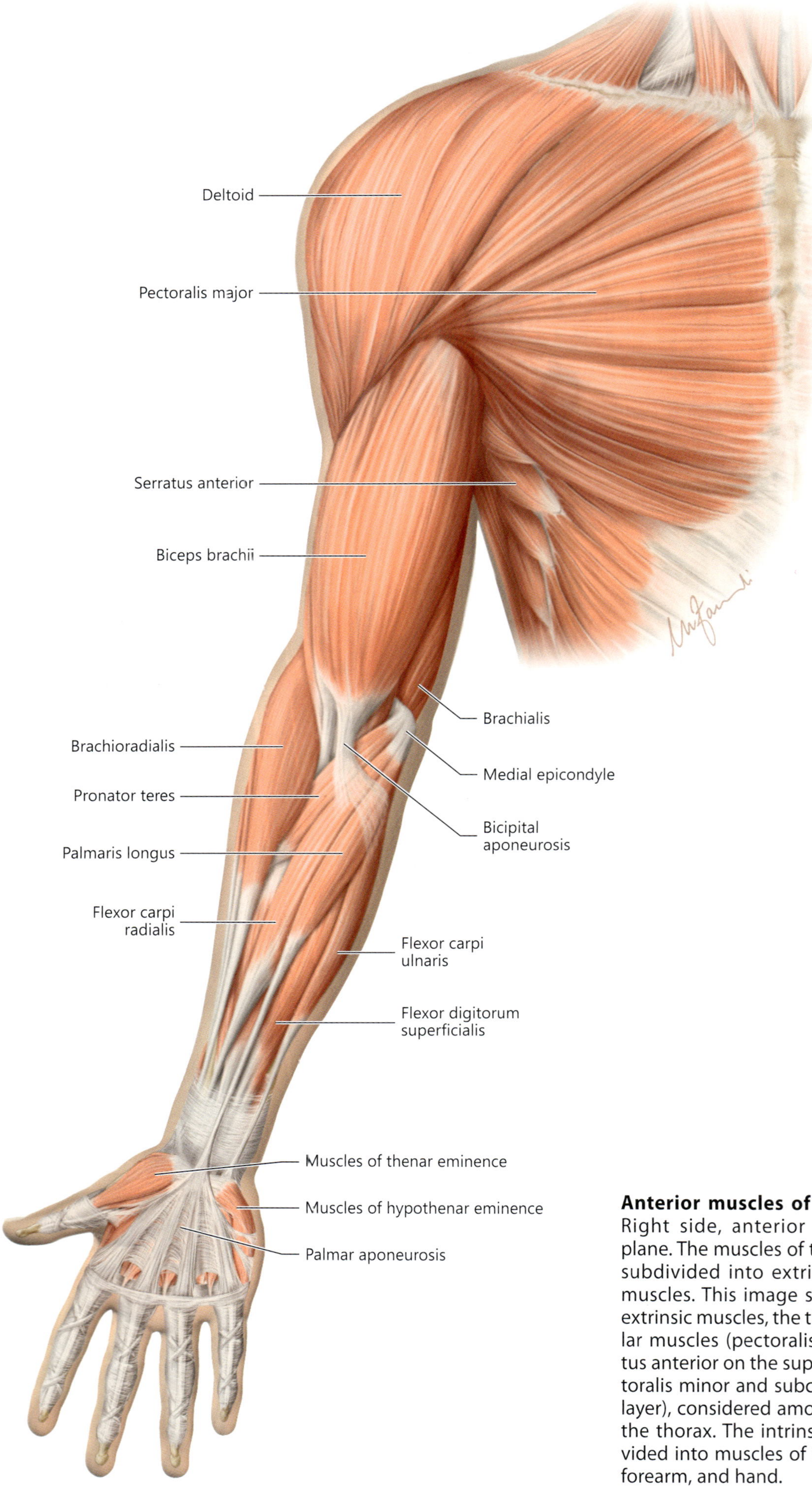

Anterior muscles of the upper limb. Right side, anterior view, superficial plane. The muscles of the upper limb are subdivided into extrinsic and intrinsic muscles. This image shows, among the extrinsic muscles, the thoracoappendicular muscles (pectoralis major and serratus anterior on the superficial plane; pectoralis minor and subclavius in the deep layer), considered among the muscles of the thorax. The intrinsic muscles are divided into muscles of the shoulder, arm, forearm, and hand.

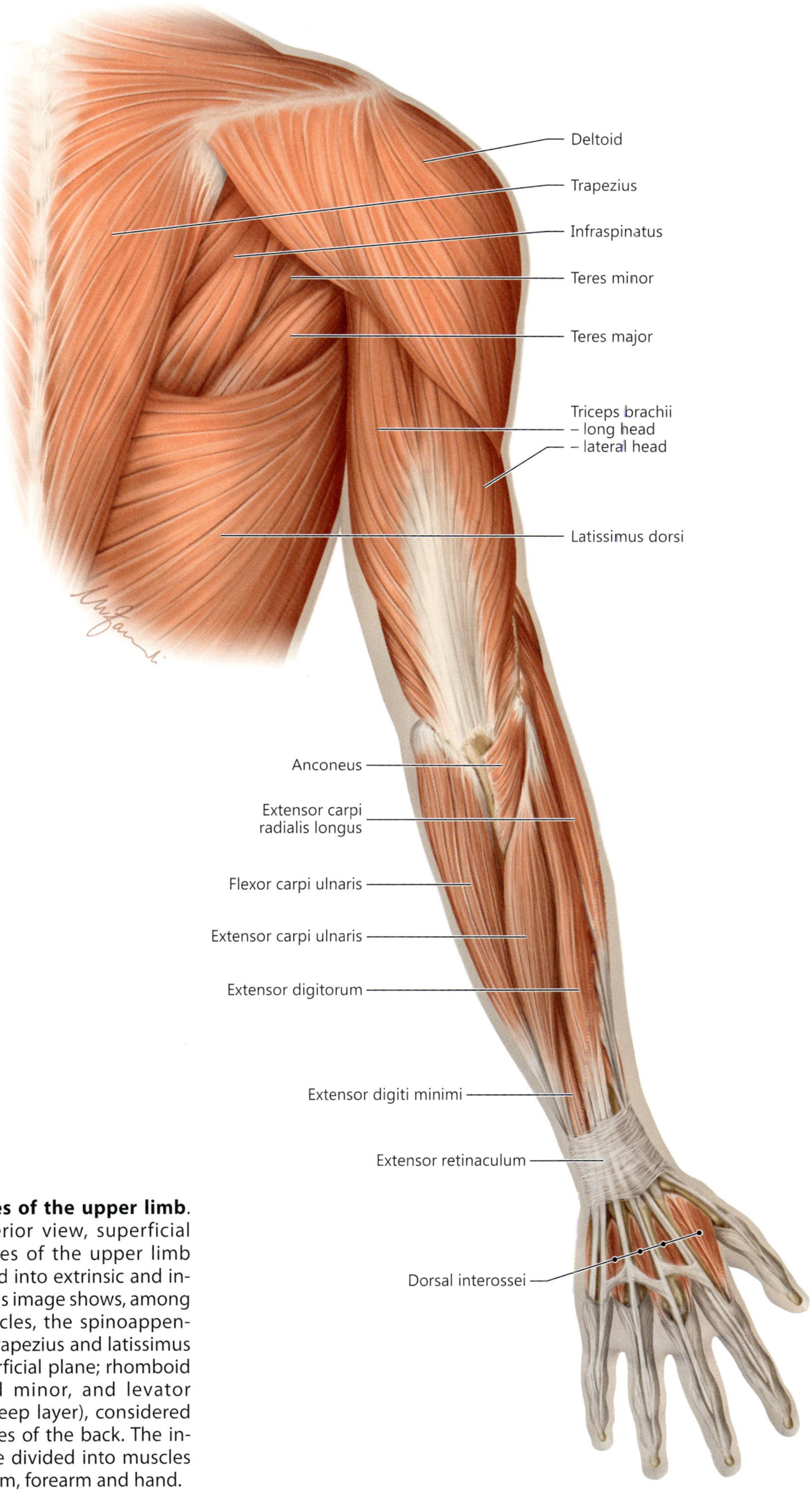

Posterior muscles of the upper limb. Right side, posterior view, superficial plane. The muscles of the upper limb may be subdivided into extrinsic and intrinsic muscles. This image shows, among the extrinsic muscles, the spinoappendicular muscles (trapezius and latissimus dorsi on the superficial plane; rhomboid major, rhomboid minor, and levator scapulae in the deep layer), considered among the muscles of the back. The intrinsic muscles are divided into muscles of the shoulder, arm, forearm and hand.

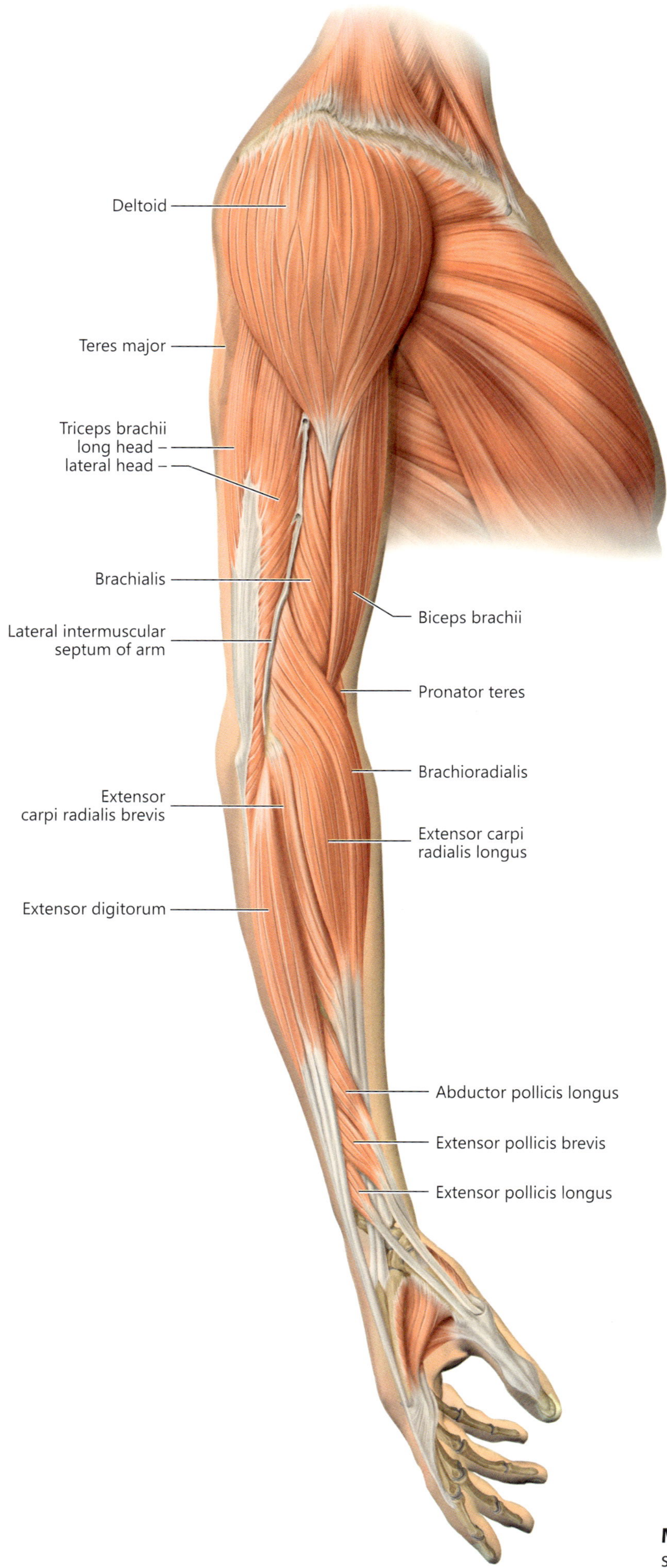

Muscles of the upper limb. Right side, lateral view, superficial plane.

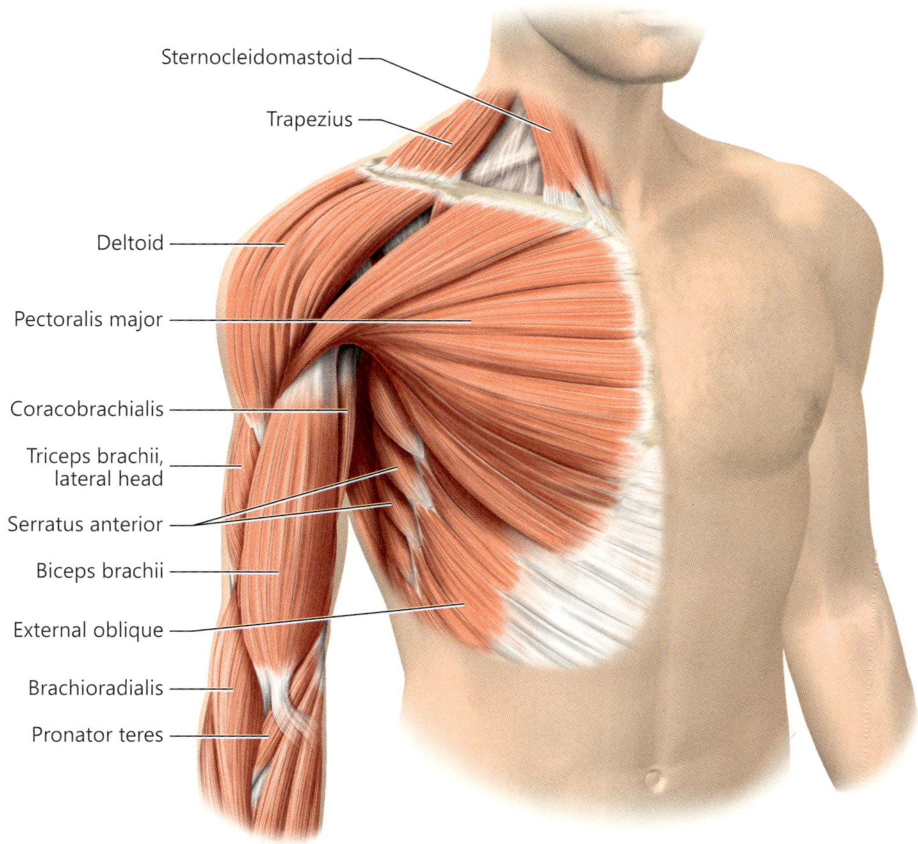

Schematic representation showing the **muscles of the anterior wall of the trunk**, the **shoulder**, and the **arm**, after the removal of the associated cutaneous and subcutaneous layers.

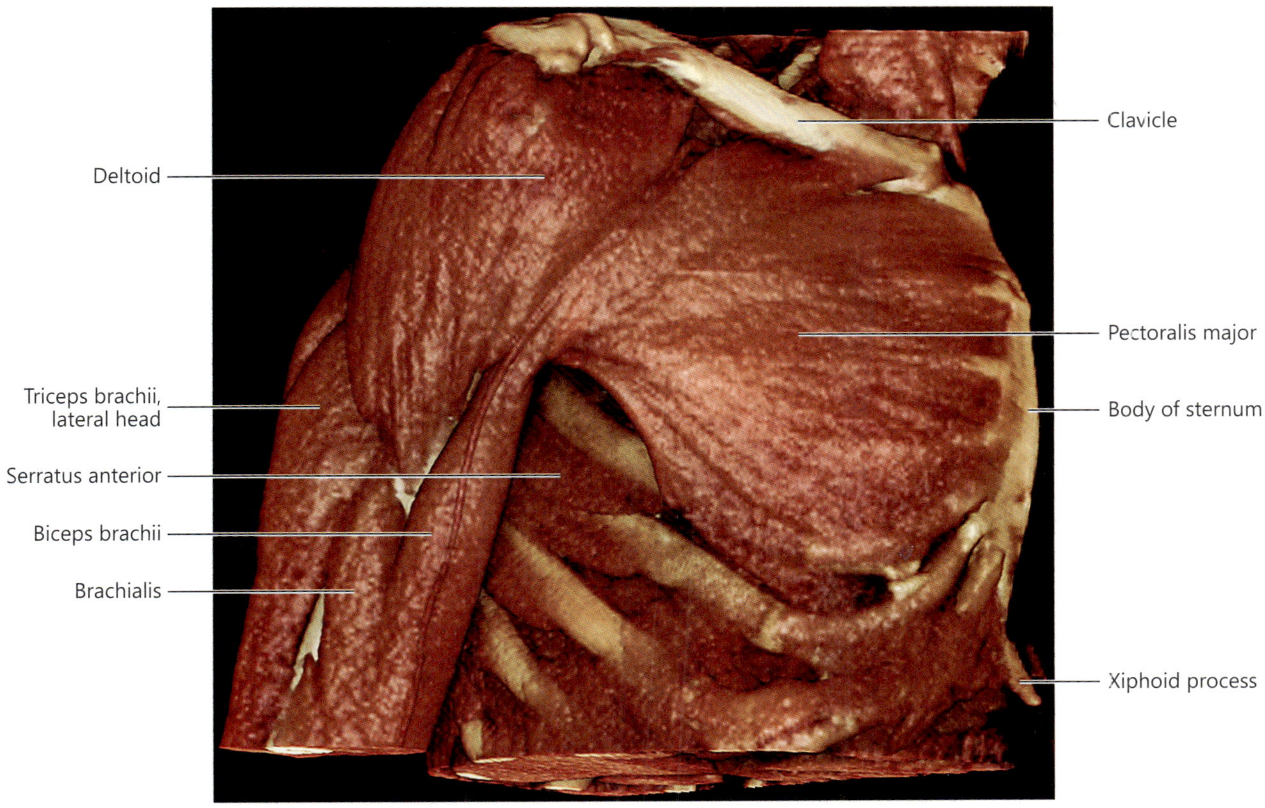

Three-dimensional image of a living human body, obtained by CT scan using a 3D volume rendering technique, showing the **muscles of the anterior wall of the trunk**, the **shoulder**, and the **arm**.

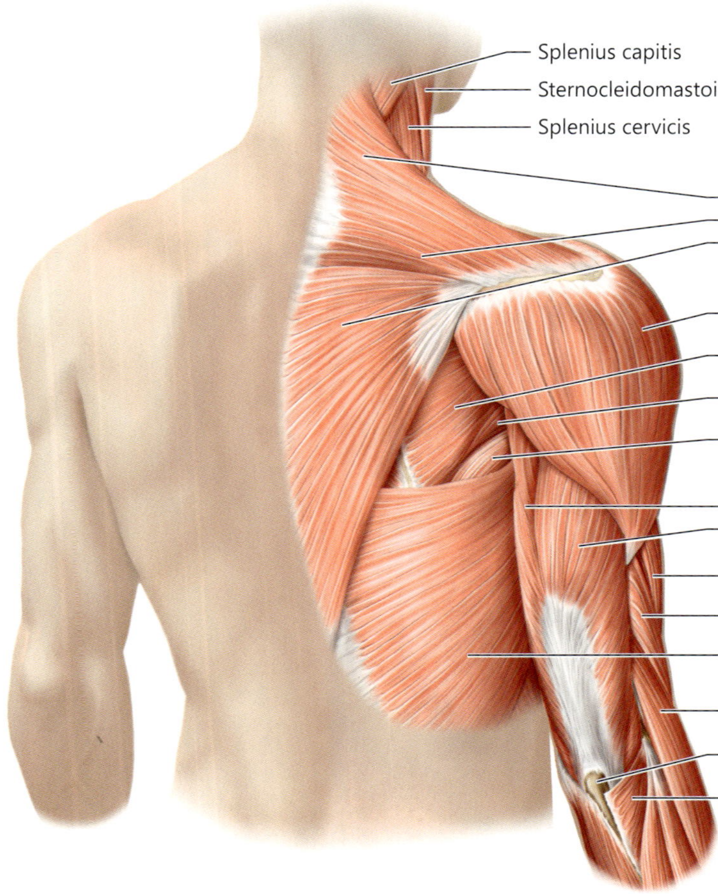

Schematic representation showing the **muscles of the posterior wall of the trunk**, the **shoulder**, and the **arm**, after the removal of the associated cutaneous and subcutaneous layers.

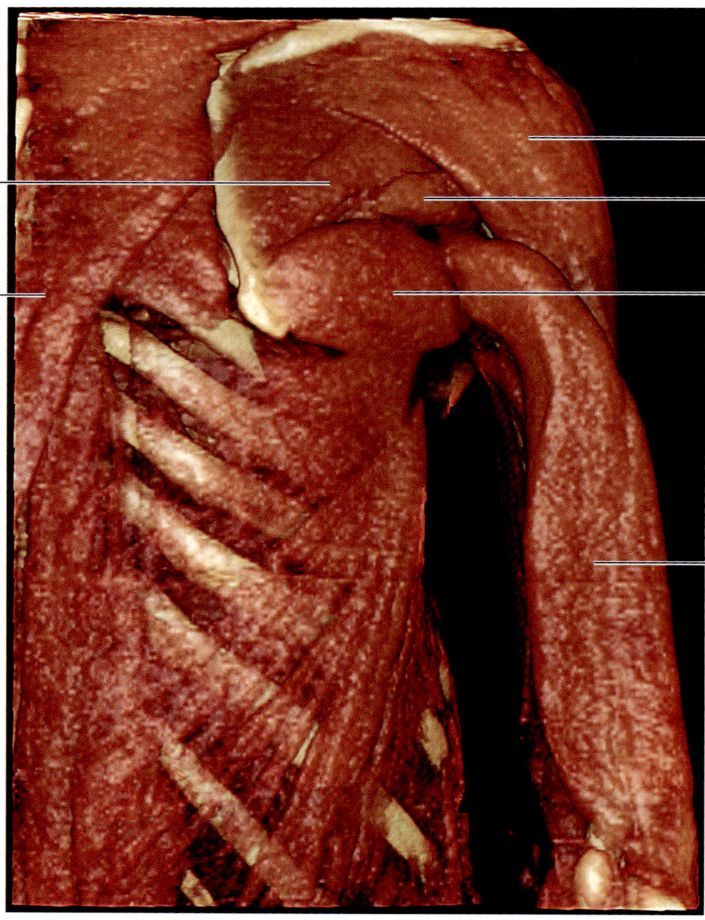

Three-dimensional image of a living human body, obtained by CT scan using a 3D volume rendering technique, showing the **muscles of the posterior wall of the trunk**, the **shoulder**, and the **arm**.

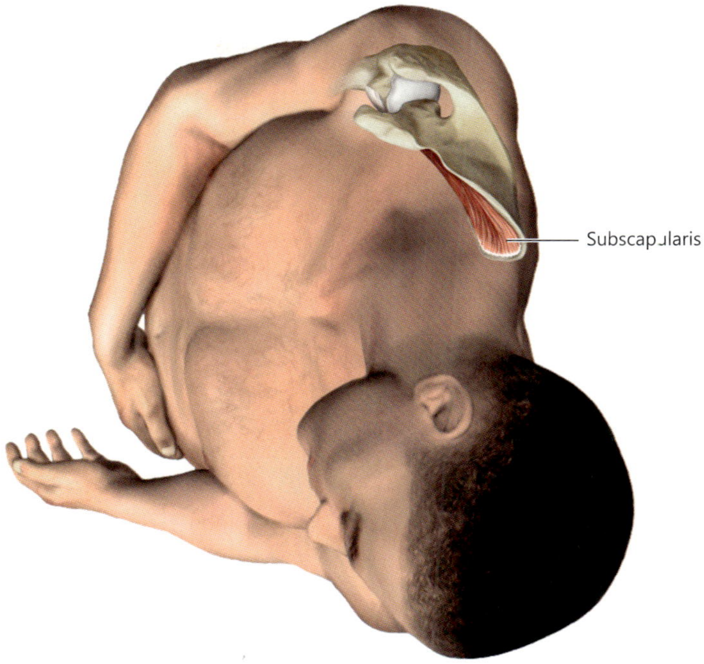

Schematic representation showing the position of the **subscapularis** viewed from above.

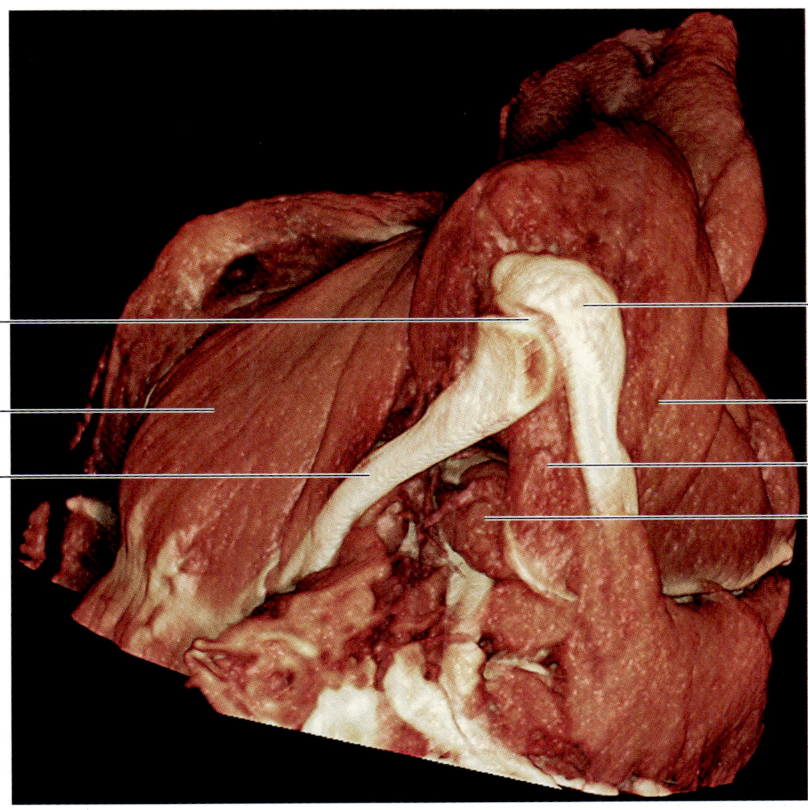

Three-dimensional image of a living human body, obtained by CT scan using a 3D volume rendering technique, showing the right **shoulder muscles** viewed from above.

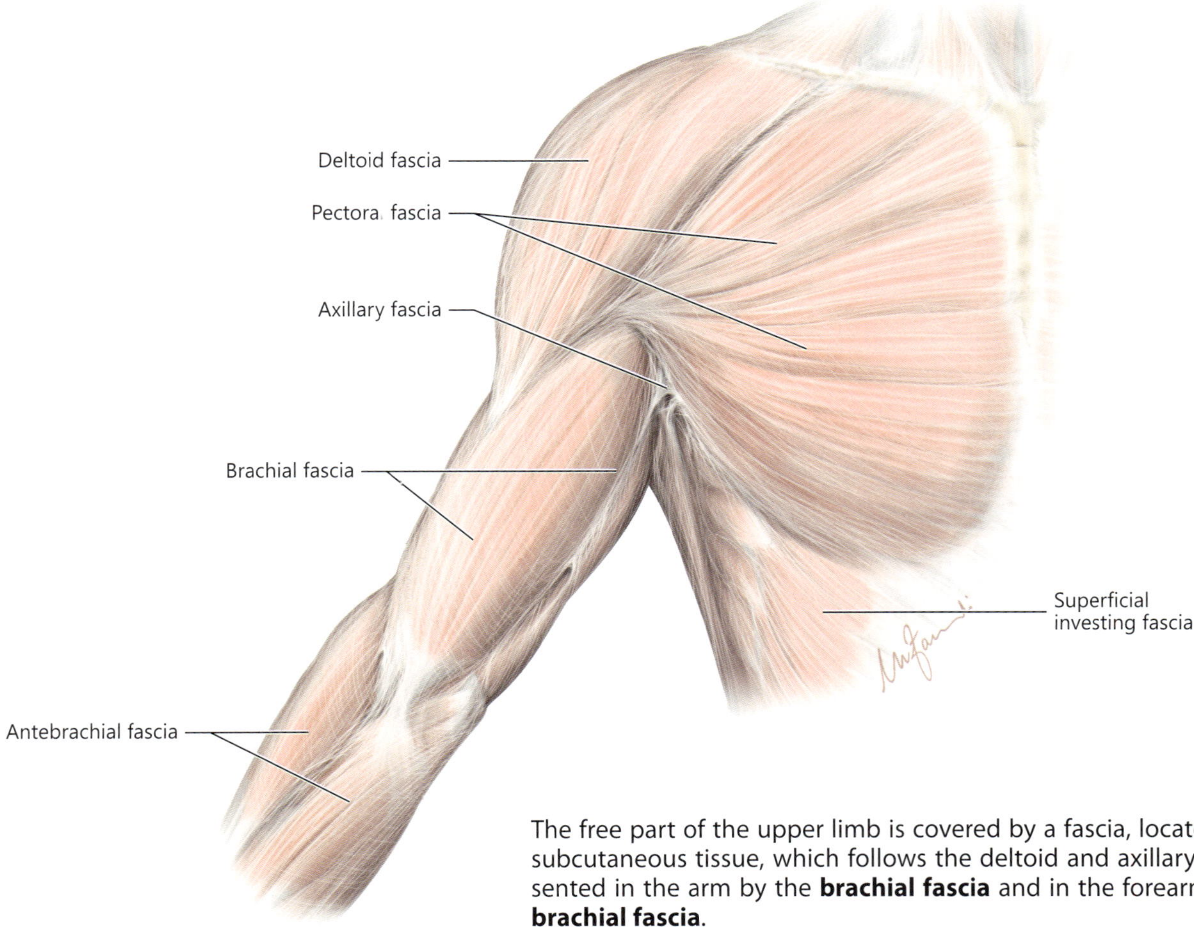

The free part of the upper limb is covered by a fascia, located deep to the subcutaneous tissue, which follows the deltoid and axillary fasciae, represented in the arm by the **brachial fascia** and in the forearm by the **antebrachial fascia**.

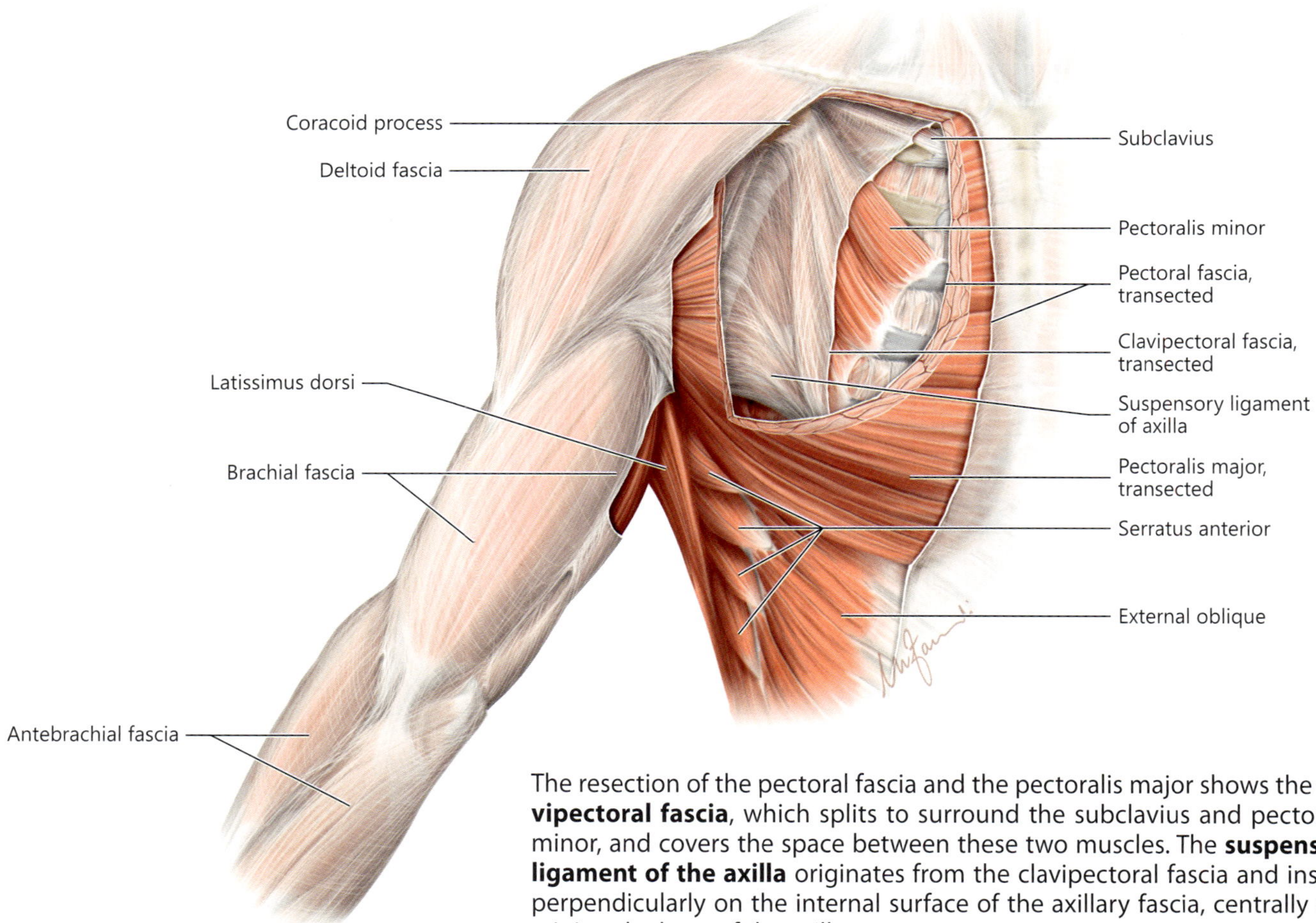

The resection of the pectoral fascia and the pectoralis major shows the **clavipectoral fascia**, which splits to surround the subclavius and pectoralis minor, and covers the space between these two muscles. The **suspensory ligament of the axilla** originates from the clavipectoral fascia and inserts perpendicularly on the internal surface of the axillary fascia, centrally sustaining the base of the axilla.

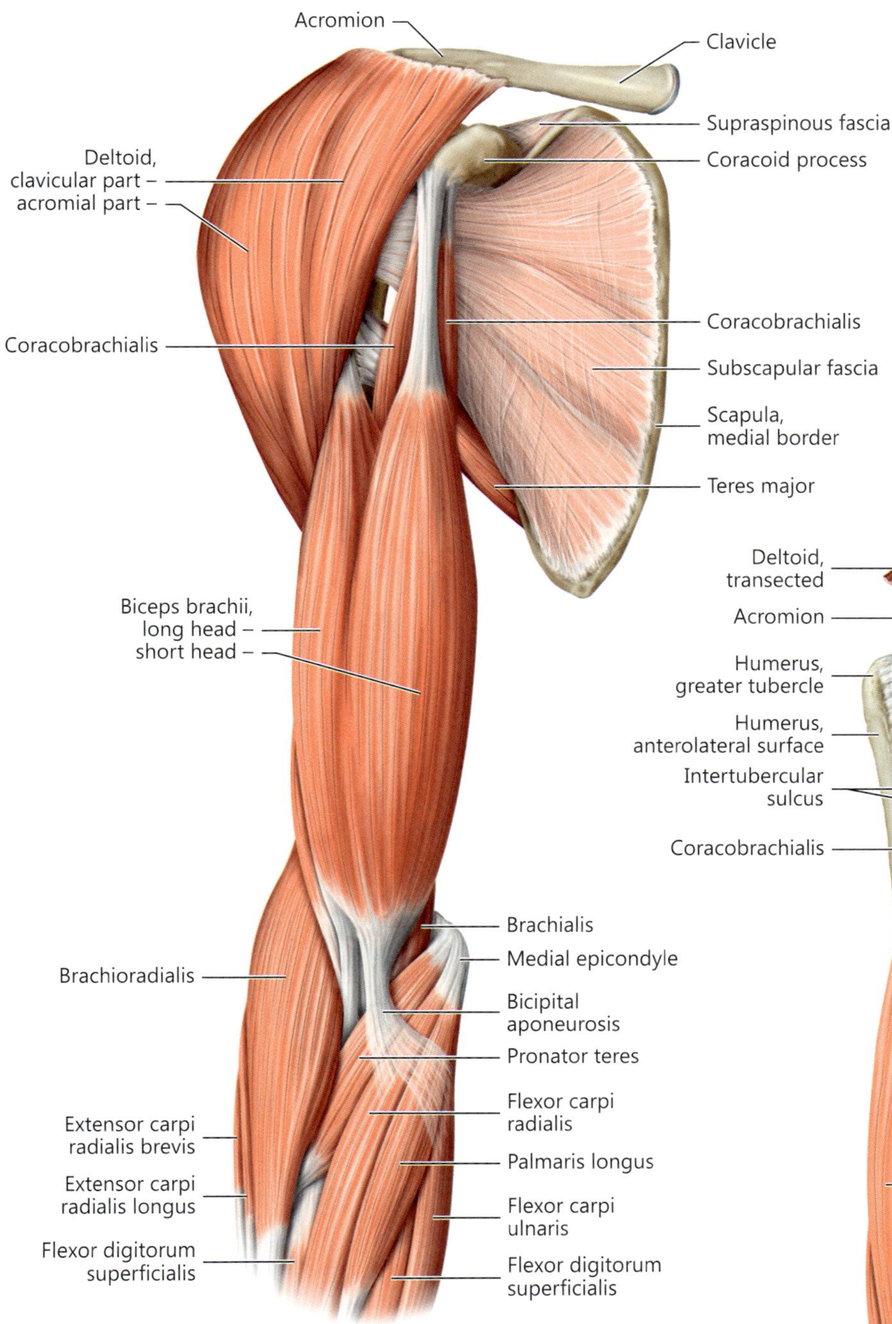

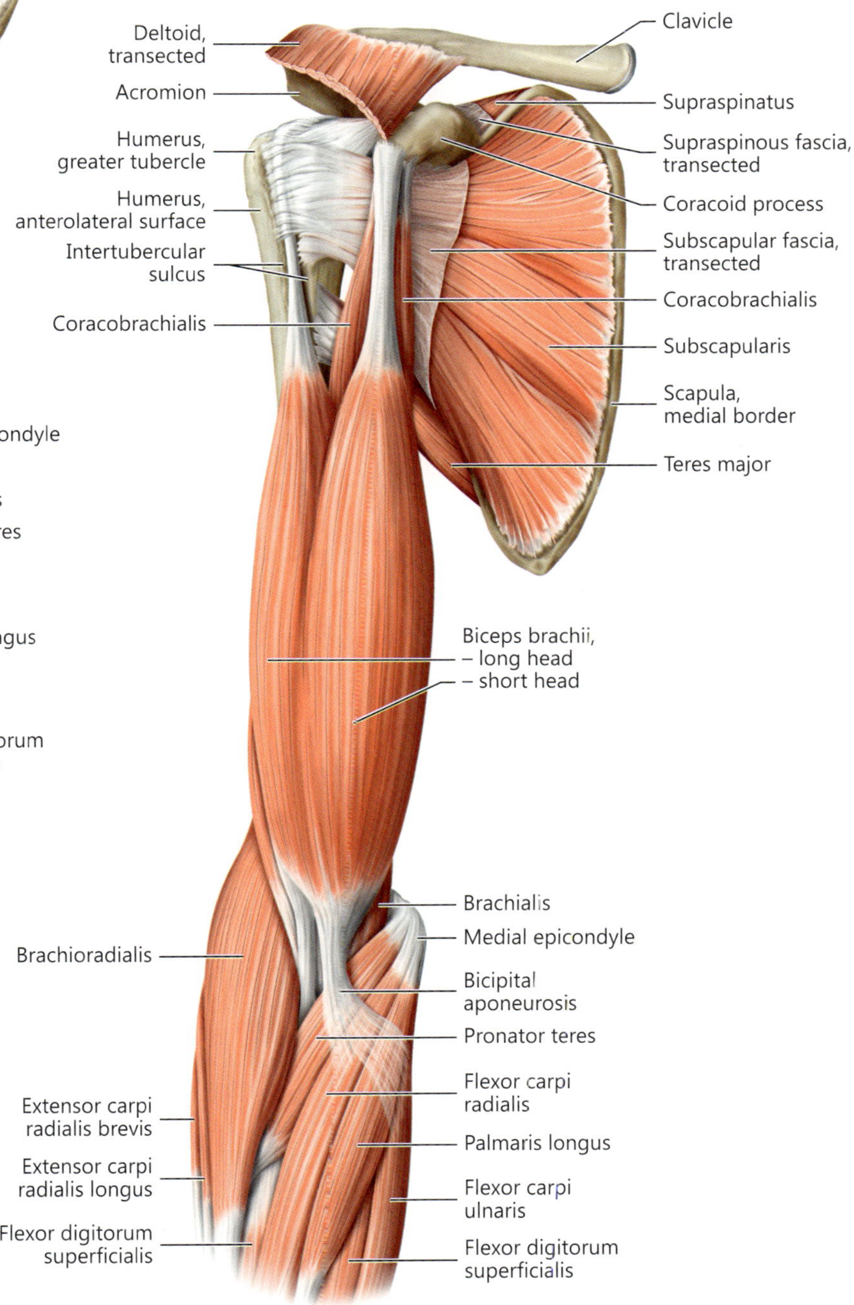

Anterior muscles of the shoulder, **the arm**, and **the forearm**. Right side, anterior view. The resection of the deltoid reveals the origin of the long head of the biceps brachii, with the long tendon that runs into the glenohumeral joint and then into the intertubercular sulcus (bicipital groove) of the humerus.

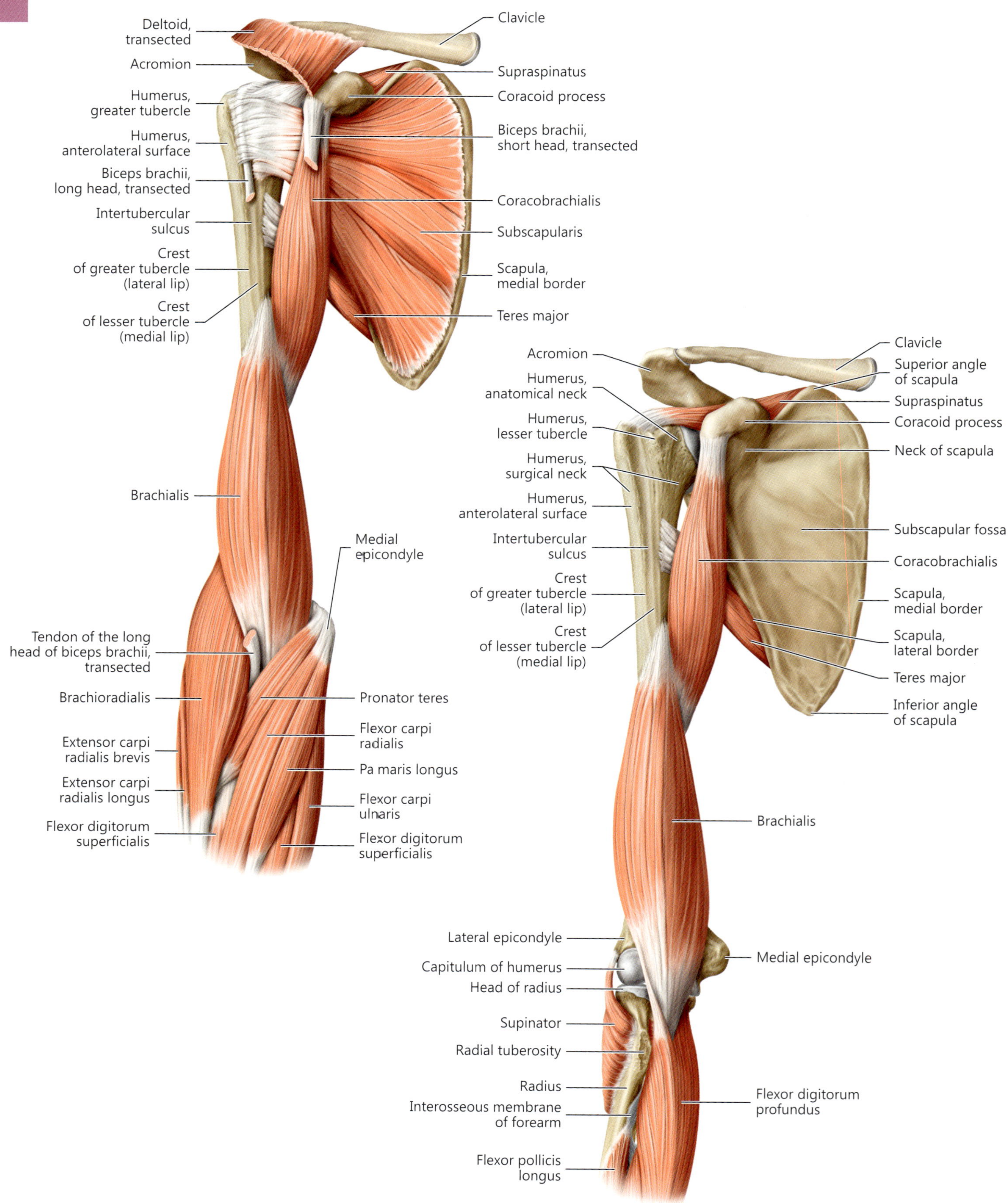

Anterior muscles of the shoulder, **the arm**, and **the forearm**. Right side, anterior view. The removal of the biceps brachii reveals the coracobrachialis and brachialis, located more deeply. In the forearm, the resection of the anterior muscles of the most superficial layer (pronator teres, flexor carpi radialis, palmaris longus, and flexor carpi ulnaris) and of the flexor digitorum superficialis (which constitutes the second layer) shows the muscles of the third layer: the flexor digitorum profundus and the flexor pollicis longus. The resection of the brachioradialis shows the supinator, located in the deep layer of the posterior muscles of the forearm.

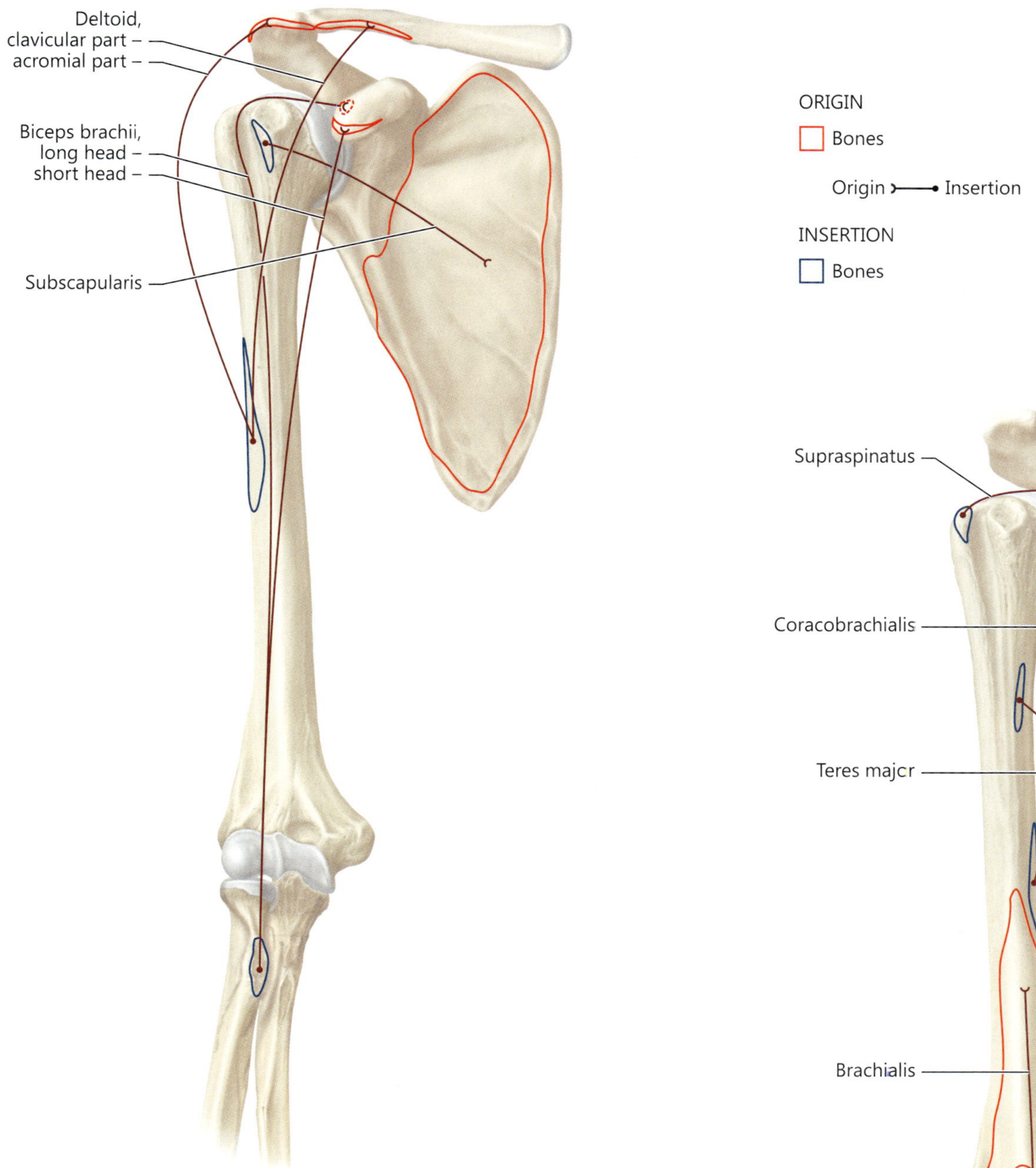

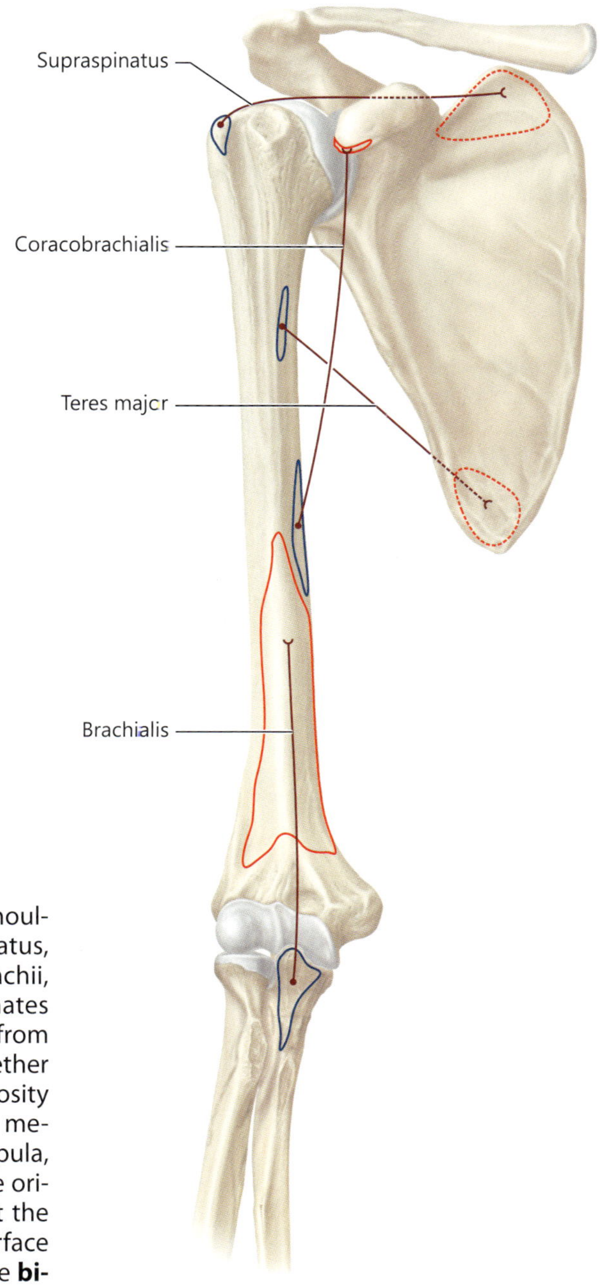

Schematic representation of the origin and insertions of the muscles of the shoulder (clavicular and acromial parts of the deltoid, as well as the supraspinatus, subscapularis, and teres major) and of the muscles of the arm (biceps brachii, brachialis, and coracobrachialis). The clavicular part of the **deltoid** originates from the lateral third of the anterior border of the clavicle, the acromial part from the apex and the lateral border of the acromion. The muscular fascicles, together with the fascicles of the spinal part of the muscle, insert on the deltoid tuberosity of the shaft (body) of the humerus. The **supraspinatus** originates from the medial two-thirds of the supraspinous fossa of the posterior surface of the scapula, and inserts on the superior facet of the greater tubercle of the humerus. The origin of the **subscapularis** is from the subscapular fossa, and its insertion at the lesser tubercle of the humerus. The **teres major** runs from the posterior surface of the scapula to the posterior lip of the bicipital groove of the humerus. The **biceps brachii** has its origin with both a long and a short head: the first from the supraglenoid tubercle of the scapula, the second from the apex of the coracoid process of the scapula. In the middle of the arm, the two heads unite into a belly that continues to a tendon inserting at the radial tuberosity. The **coracobrachialis** runs from the coracoid process of the scapula to the middle third of the anteromedial surface of the shaft of the humerus, while the **brachialis** originates from the anteromedial and anterolateral surfaces of the humerus and inserts at the tuberosity of the ulna.

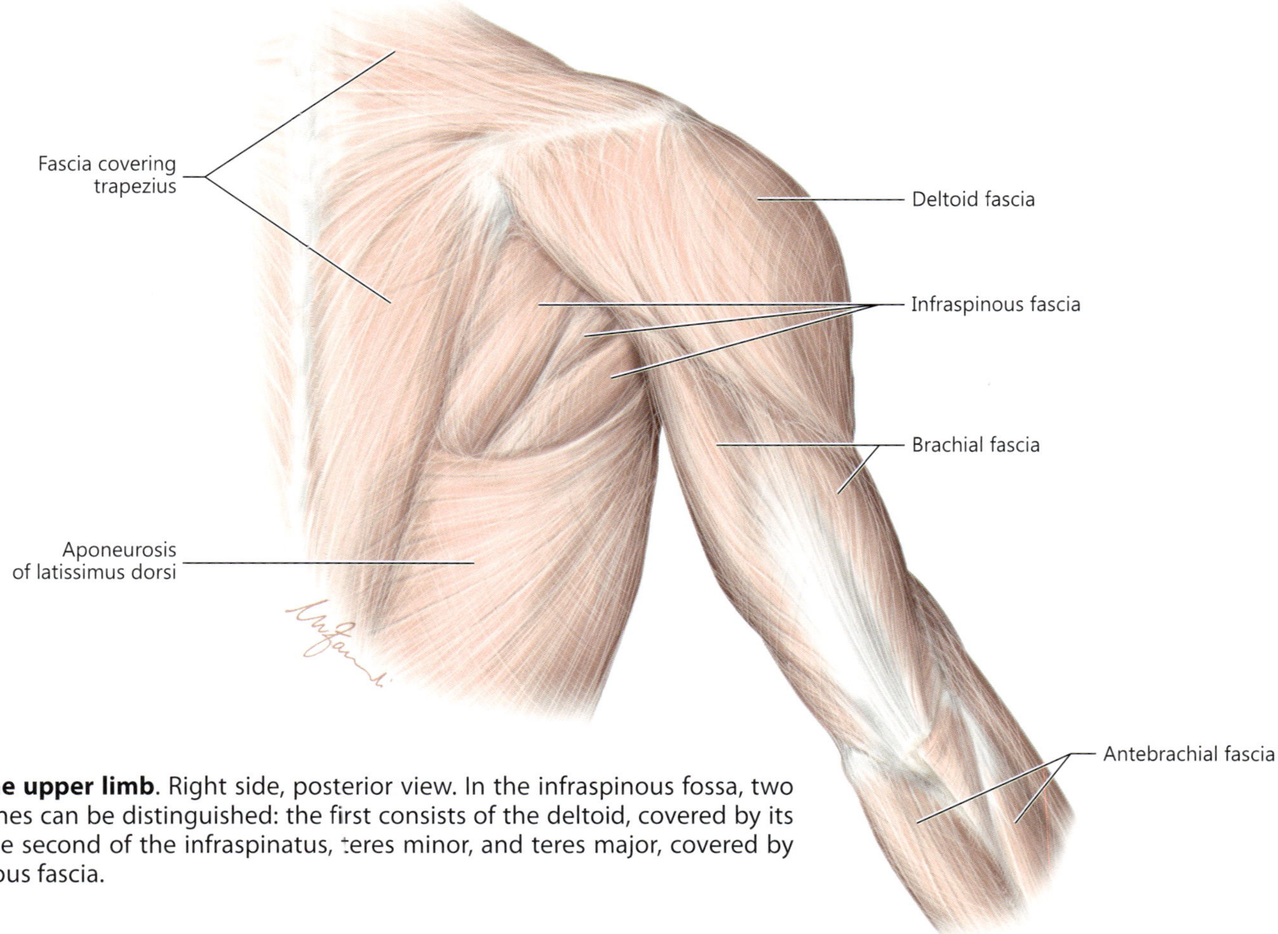

Fasciae of the upper limb. Right side, posterior view. In the infraspinous fossa, two muscular planes can be distinguished: the first consists of the deltoid, covered by its fascia, and the second of the infraspinatus, teres minor, and teres major, covered by the infraspinous fascia.

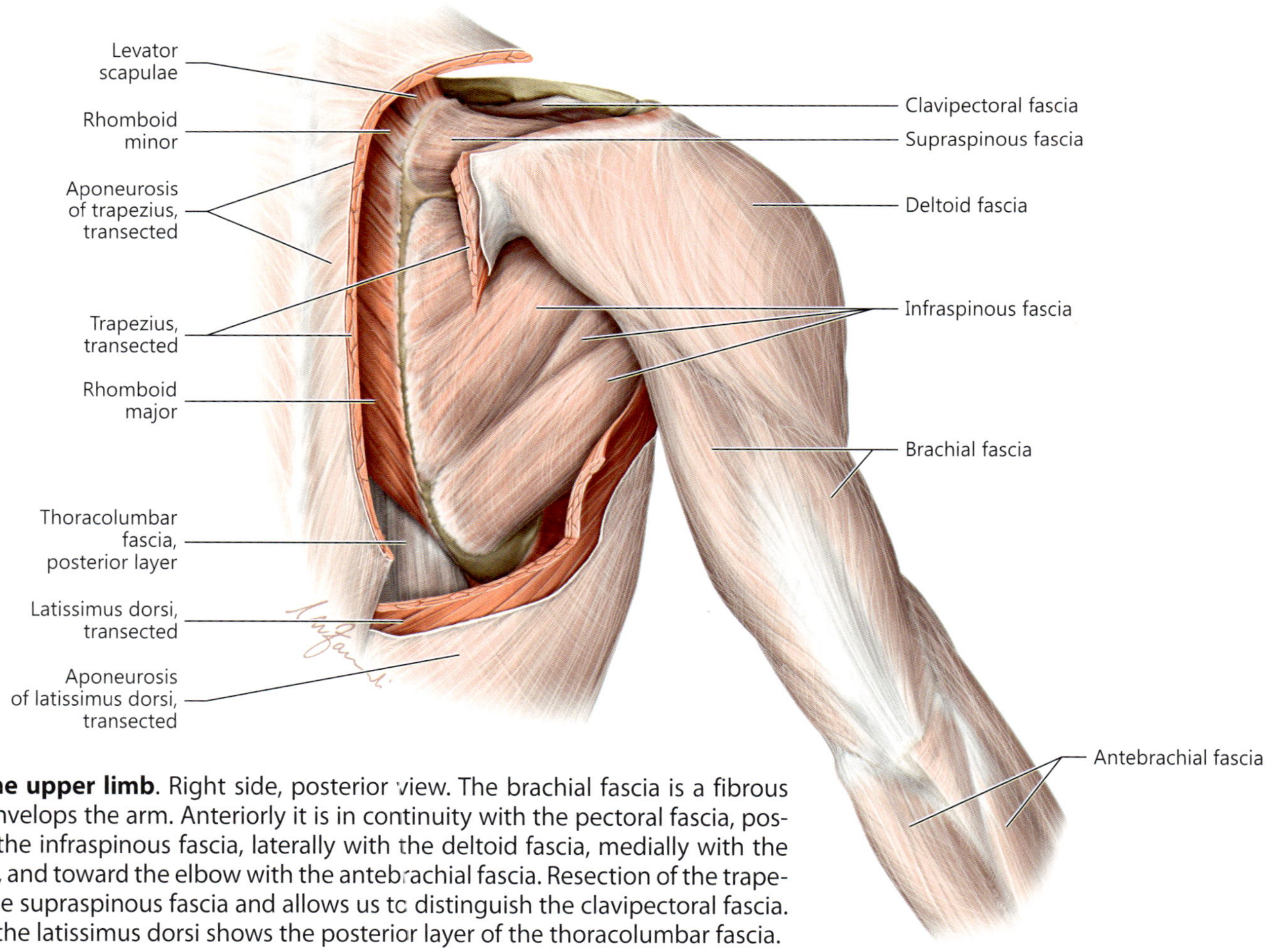

Fasciae of the upper limb. Right side, posterior view. The brachial fascia is a fibrous sleeve that envelops the arm. Anteriorly it is in continuity with the pectoral fascia, posteriorly with the infraspinous fascia, laterally with the deltoid fascia, medially with the axillary fascia, and toward the elbow with the antebrachial fascia. Resection of the trapezius shows the supraspinous fascia and allows us to distinguish the clavipectoral fascia. Resection of the latissimus dorsi shows the posterior layer of the thoracolumbar fascia.

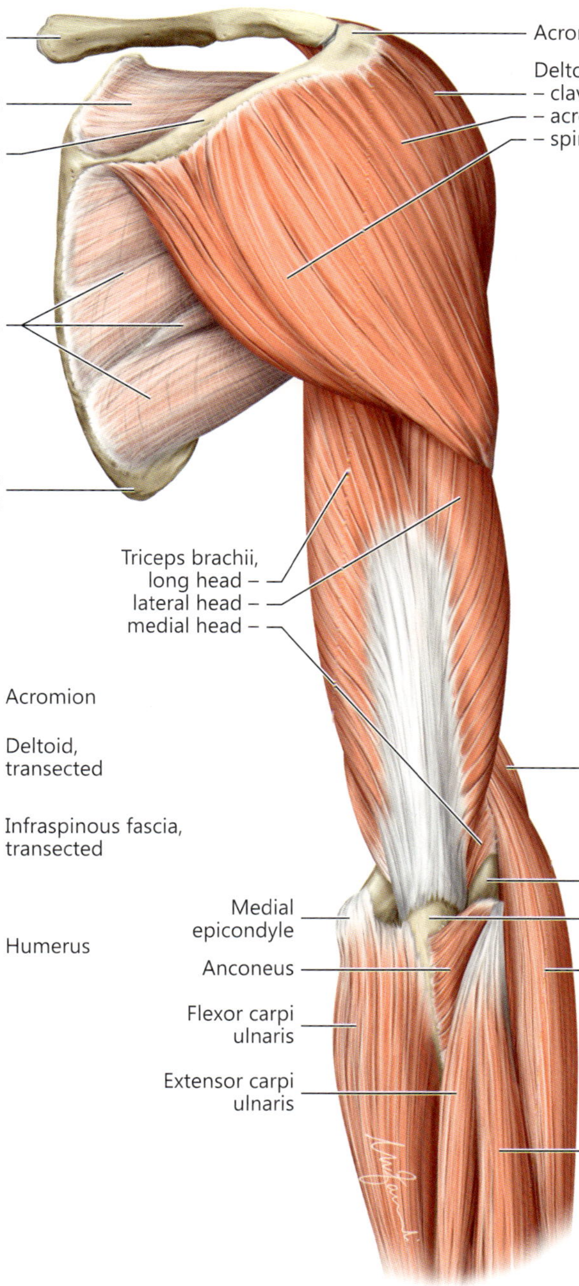

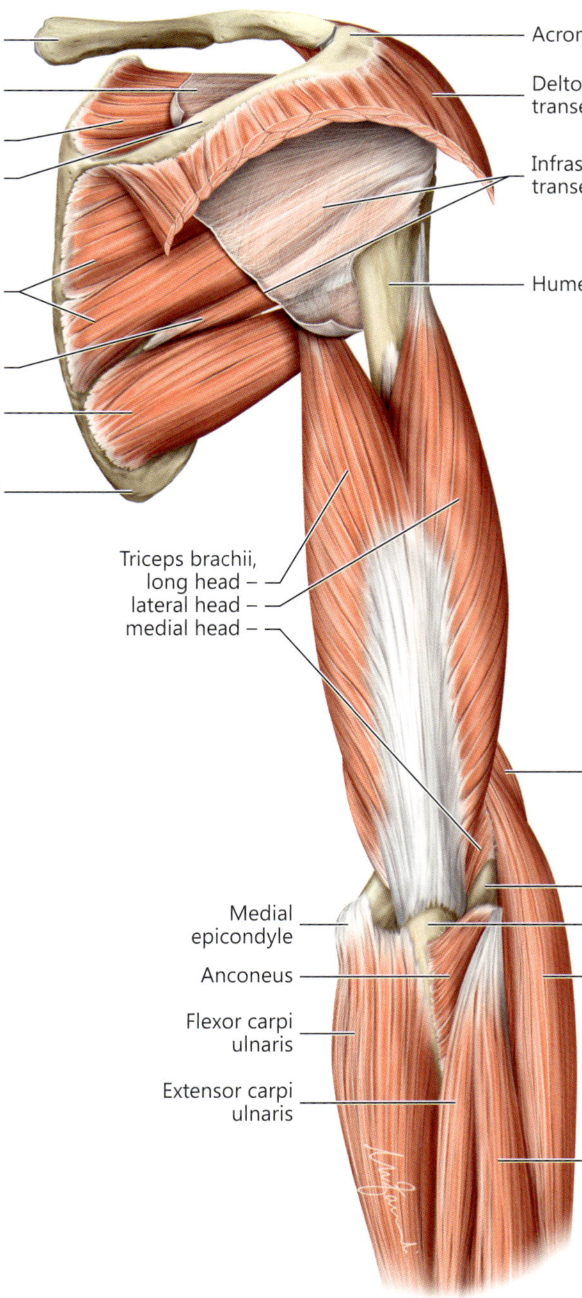

Posterior muscles of the shoulder, the arm, and **the forearm**. Right side, posterior view. Above: middle plane after removal of the rhomboid minor, rhomboid major, serratus anterior, and levator scapulae. Below: middle plane after resection of the deltoid.

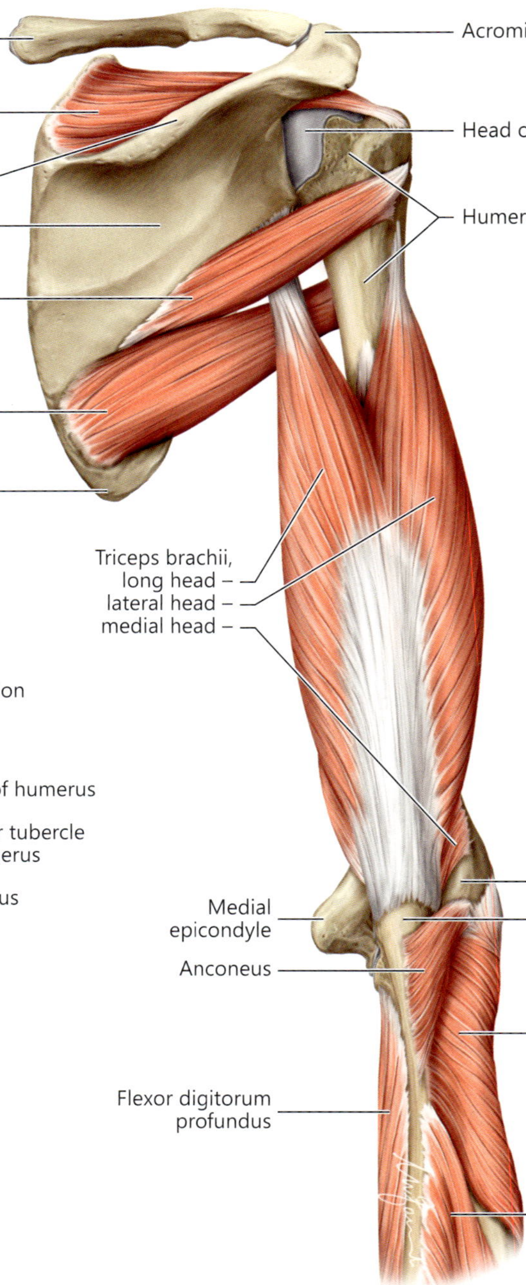

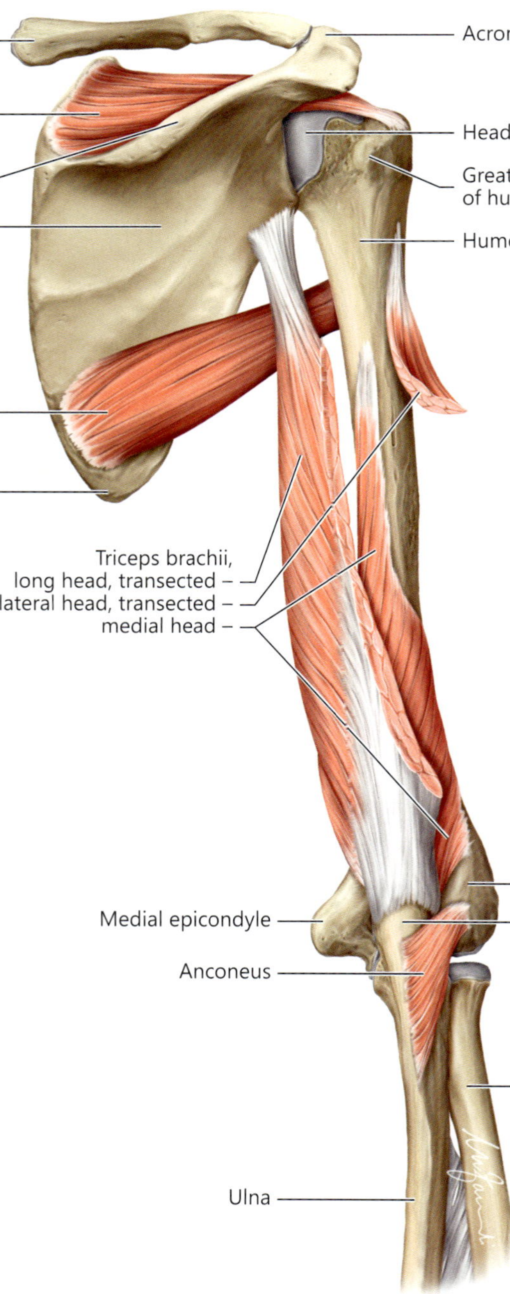

Posterior muscles of the shoulder, **the arm**, and **the forearm**. Right side, posterior view. Above: the removal of the infraspinatus shows the teres minor. Resection of the superficial plane of the posterior muscles in the forearm shows, in the deep plane, the supinator and abductor pollicis longus. Below, deep plane after removal of the teres minor and partial resection of the triceps brachii. Resection of the supinator and abductor pollicis longus in the forearm shows the skeletal plane.

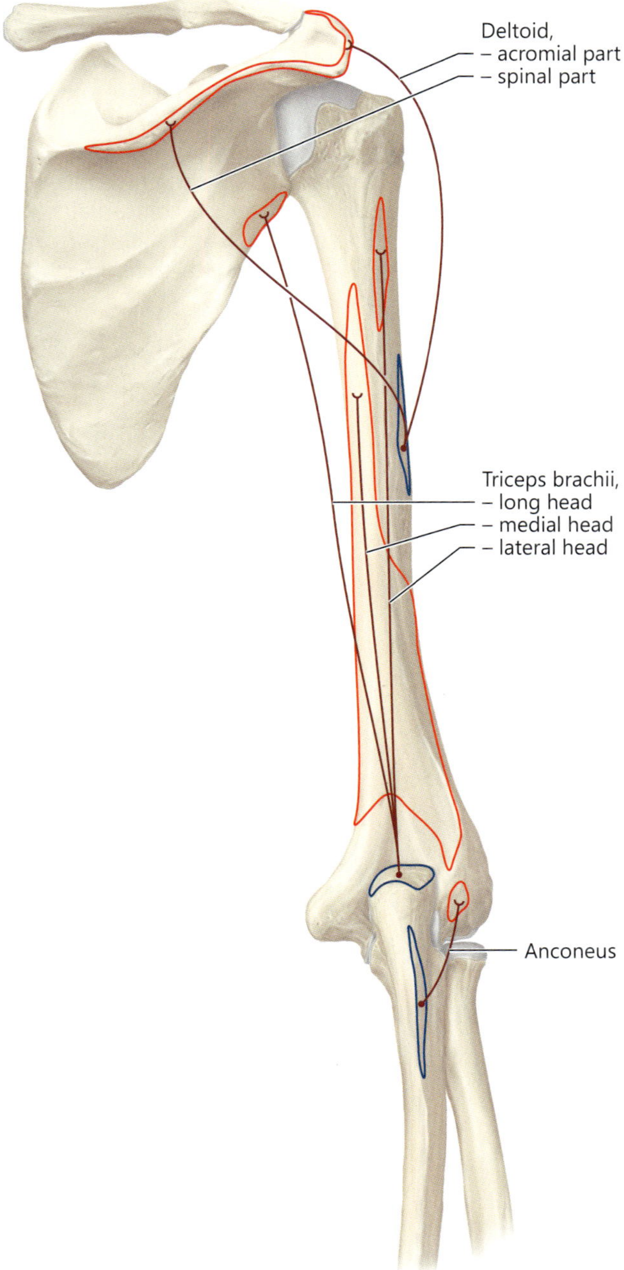

ORIGIN
Bones

Origin ➤—• Insertion

INSERTION
Bones

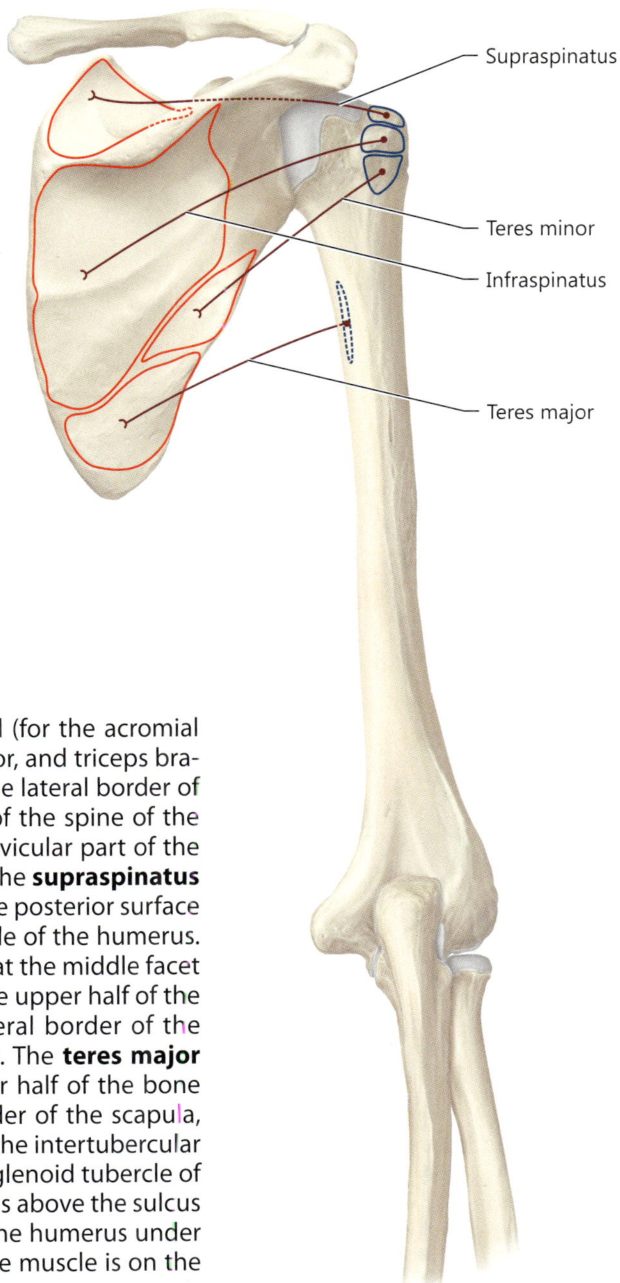

Schematic representation of the origin and insertions of the deltoid (for the acromial and spinal parts), supraspinatus, infraspinatus, teres minor, teres major, and triceps brachii. The acromial part of the **deltoid** originates from the apex and the lateral border of the acromion; the spinal part originates from the posterior border of the spine of the scapula. The muscular fascicles, together with the fascicles of the clavicular part of the muscle, insert on the deltoid tuberosity of the shaft of the humerus. The **supraspinatus** originates from the medial two-thirds of the supraspinatus fossa of the posterior surface of the scapula, and inserts at the superior facet of the greater tubercle of the humerus. The **infraspinatus** originates from the infraspinous fossa and inserts at the middle facet of the greater tubercle of the humerus. The **teres minor** runs from the upper half of the bone surface located between the infraspinous fossa and the lateral border of the scapula to the inferior facet of the greater tubercle of the humerus. The **teres major** originates from the inferior angle of the scapula and from the lower half of the bone surface located between the infraspinous fossa and the lateral border of the scapula, and inserts on the crest of the lesser tubercle of the humerus and on the intertubercular sulcus. The long head of the **triceps brachii** originates from the infraglenoid tubercle of the scapula, the lateral head from the posterior surface of the humerus above the sulcus for the radial nerve, and the medial head from the same surface of the humerus under the sulcus for the radial nerve. The insertion of the three heads of the muscle is on the olecranon of the ulna.

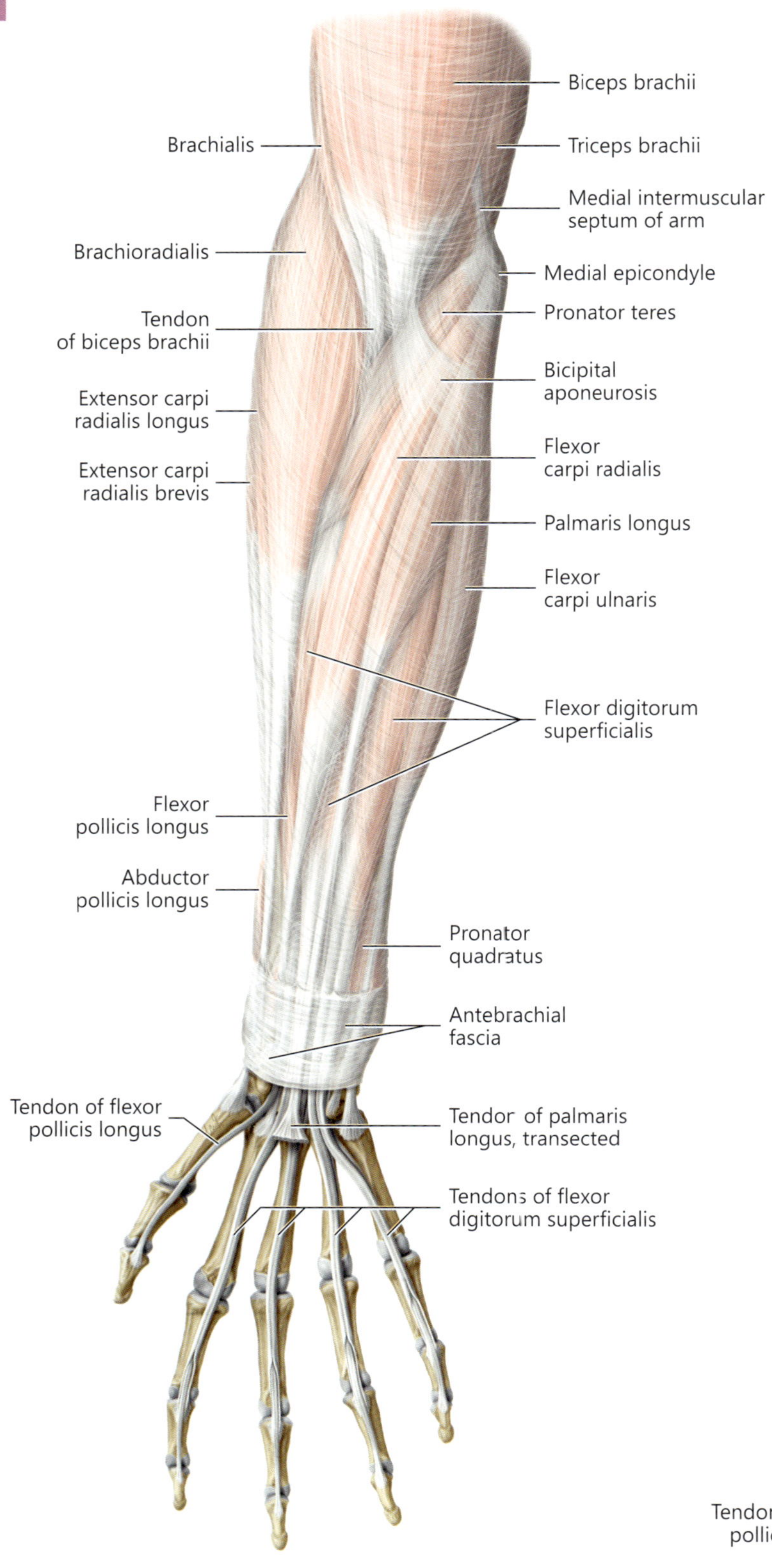

Anterior and **lateral muscles of the forearm**. Right side, anterior view. The first layer (superficial) of the anterior muscles, in a lateromedial direction, is formed by the pronator teres, flexor carpi radialis, palmaris longus, and flexor carpi ulnaris. The lateral muscles, in an anteroposterior direction, comprise the brachioradialis, extensor carpi radialis longus, and extensor carpi radialis brevis. The muscles are covered by the antebrachial fascia.

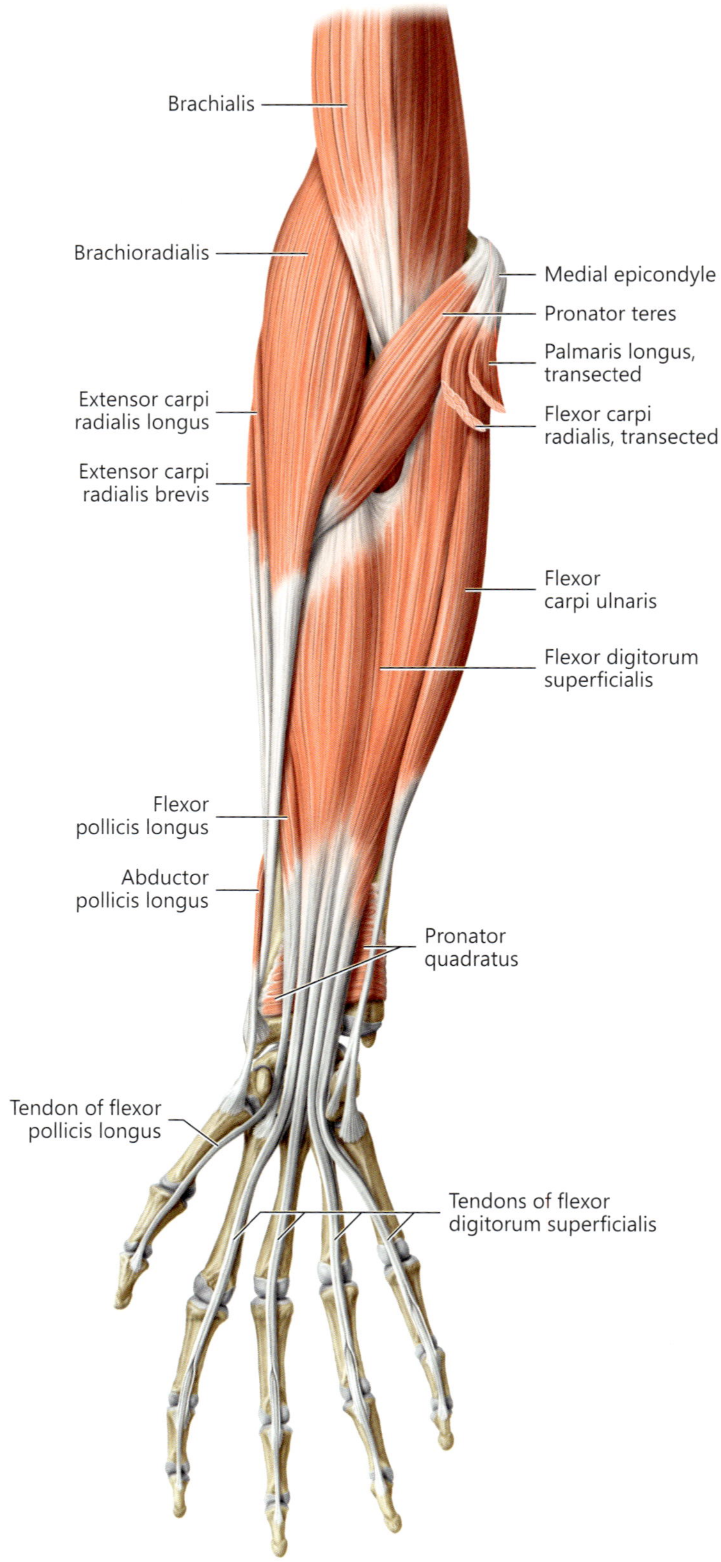

Anterior and **lateral muscles of the forearm**. Right side, anterior view. The removal of the palmaris longus and flexor carpi radialis, belonging to the first layer of anterior muscles of the forearm, shows the flexor digitorum superficialis, which constitutes the second layer of the anterior muscles of the forearm.

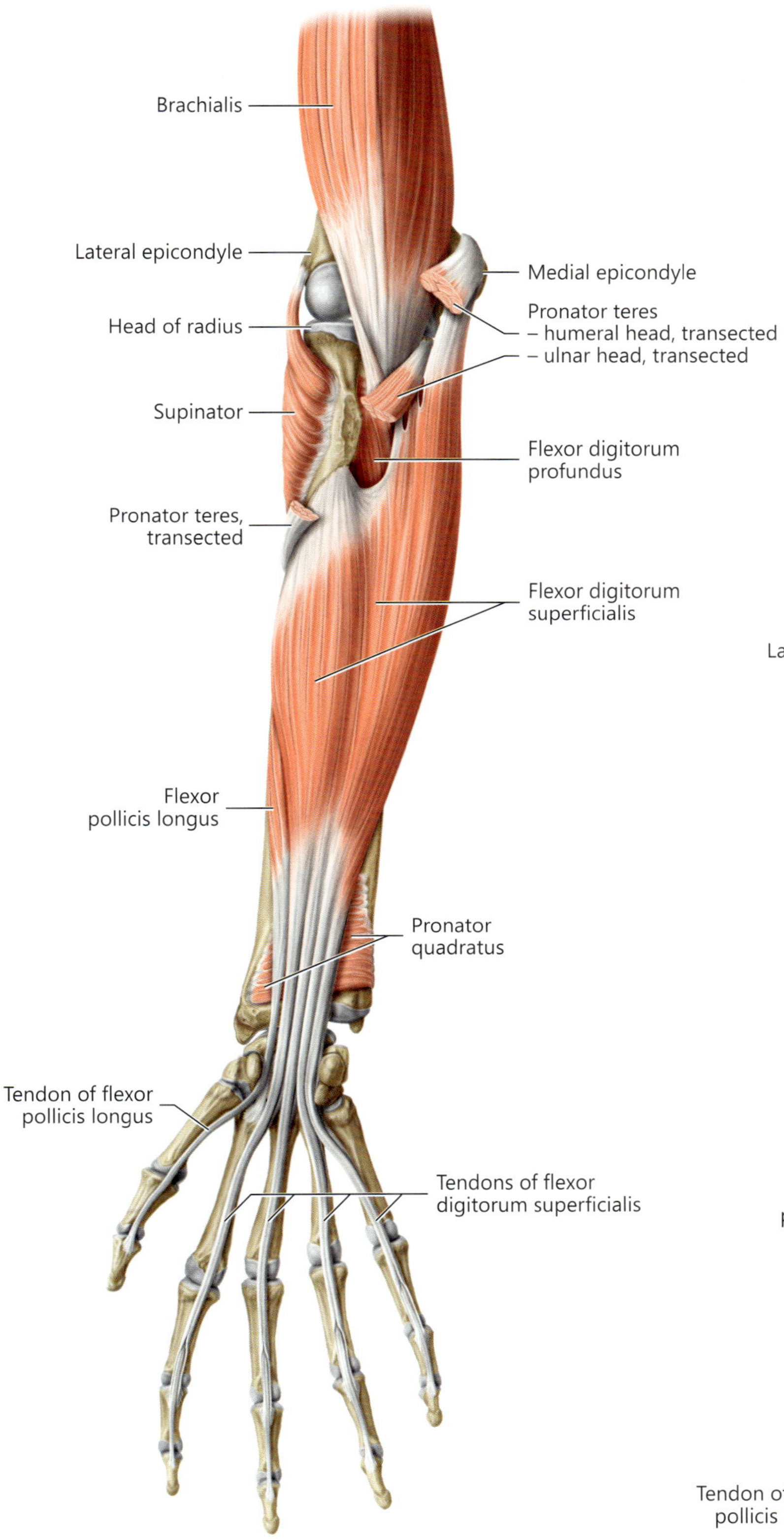

Anterior and **lateral muscles of the forearm**. Right side, anterior view. The removal of the superficial layer of the anterior muscles of the forearm, which have their common origin from the medial epicondyle of the humerus, exposes the second layer, made up of the flexor digitorum superficialis. Also visible, after the removal of the lateral muscles of the forearm, is the supinator in the deep plane of the posterior muscles of the forearm.

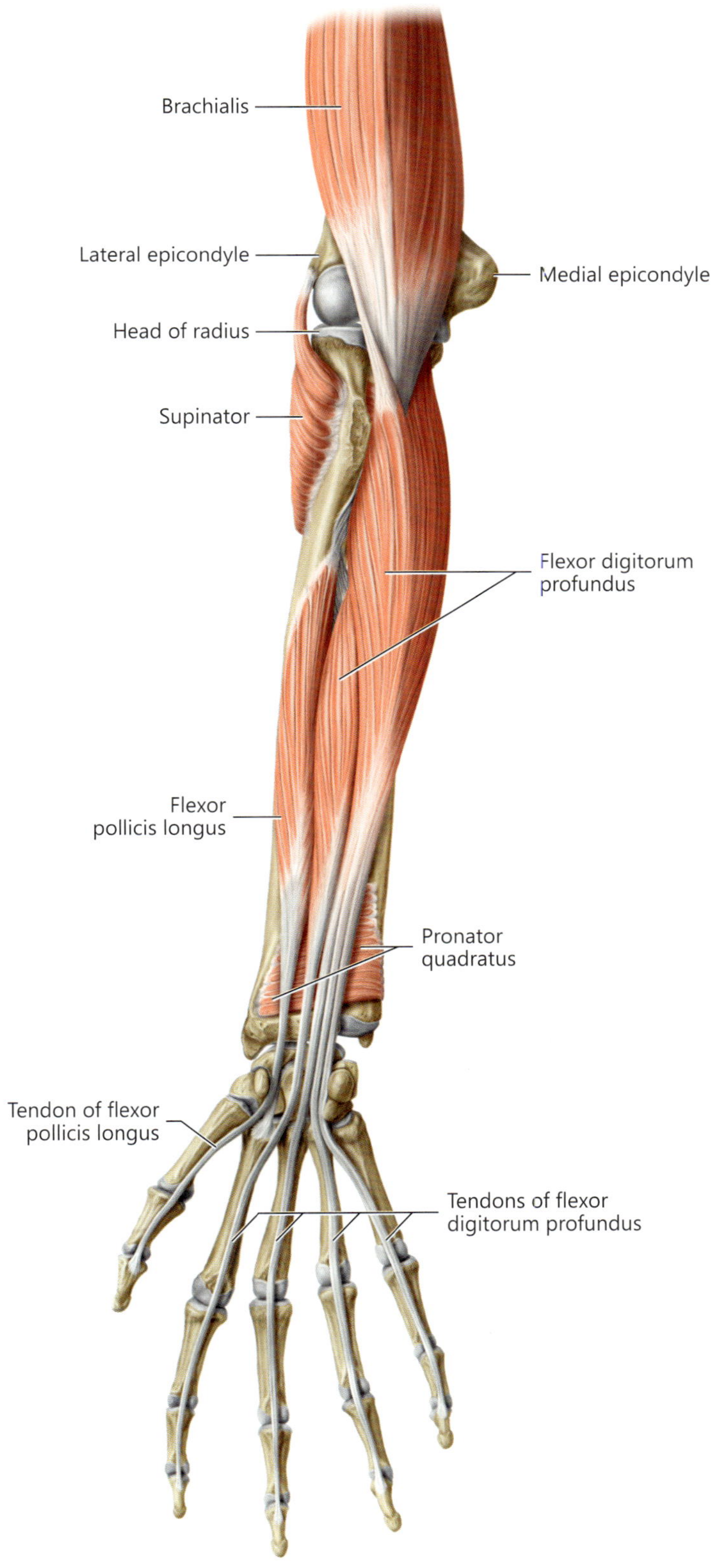

Anterior and **lateral muscles of the forearm**. Right side, anterior view. The removal of the flexor digitorum superficialis shows the flexor digitorum profundus and the flexor pollicis longus, which constitute the third layer of the anterior muscles of the forearm. Also visible is the pronator quadratus, which itself forms the fourth layer of the anterior muscles of the forearm.

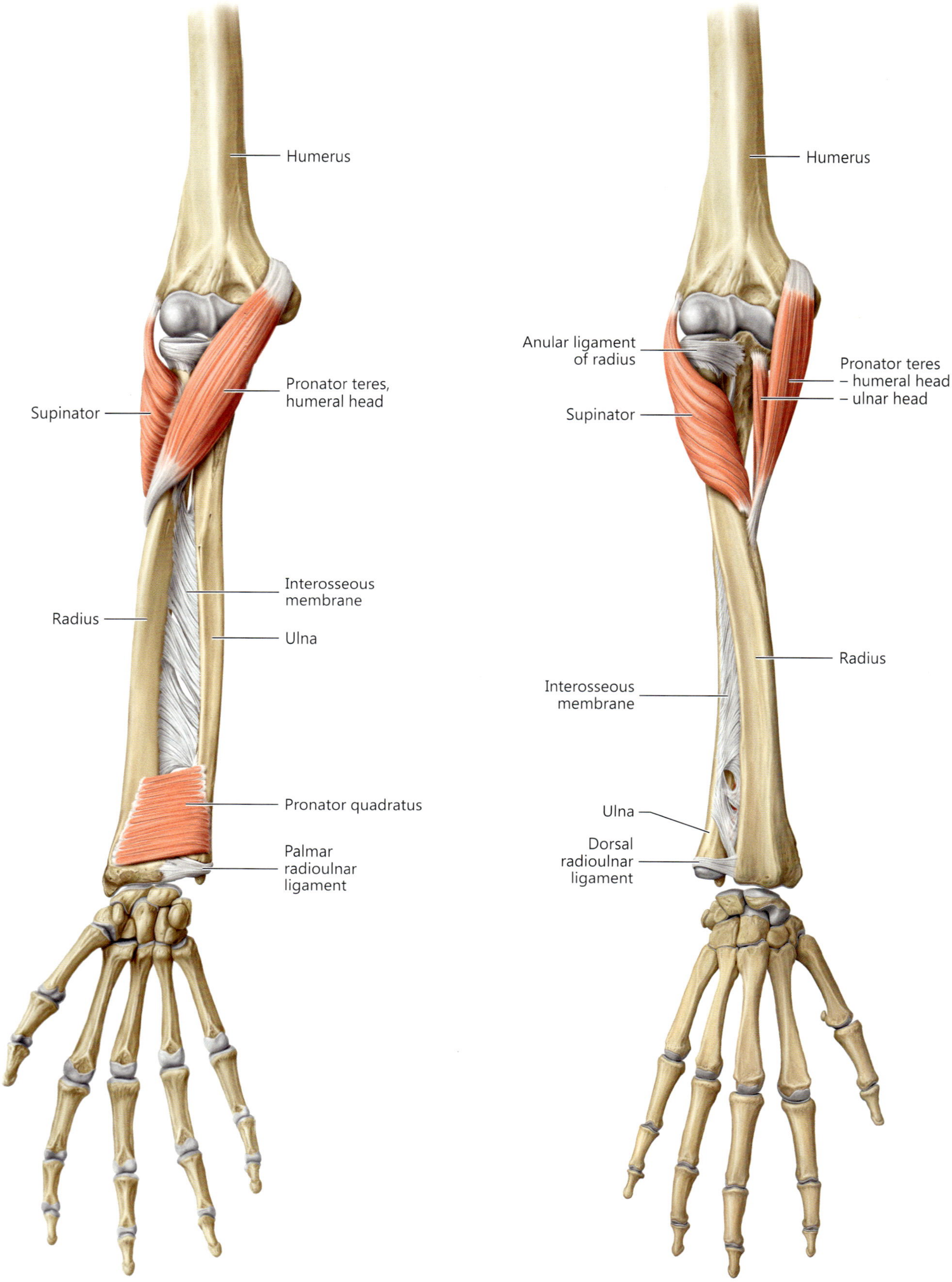

Right hand in supine (left) and **prone position** (right). The supinator, located in the deep plane of the posterior muscles of the forearm, connects the humerus to the radius. The pronator teres, the most lateral of the superficial layer of anterior muscles of the forearm, originates with two heads, humeral and ulnar, which insert on the lateral surface of the radius. The pronator quadratus forms the fourth layer of the anterior muscles of the forearm, originating from the anterior border of the ulna and inserting at the anterior border of the radius.

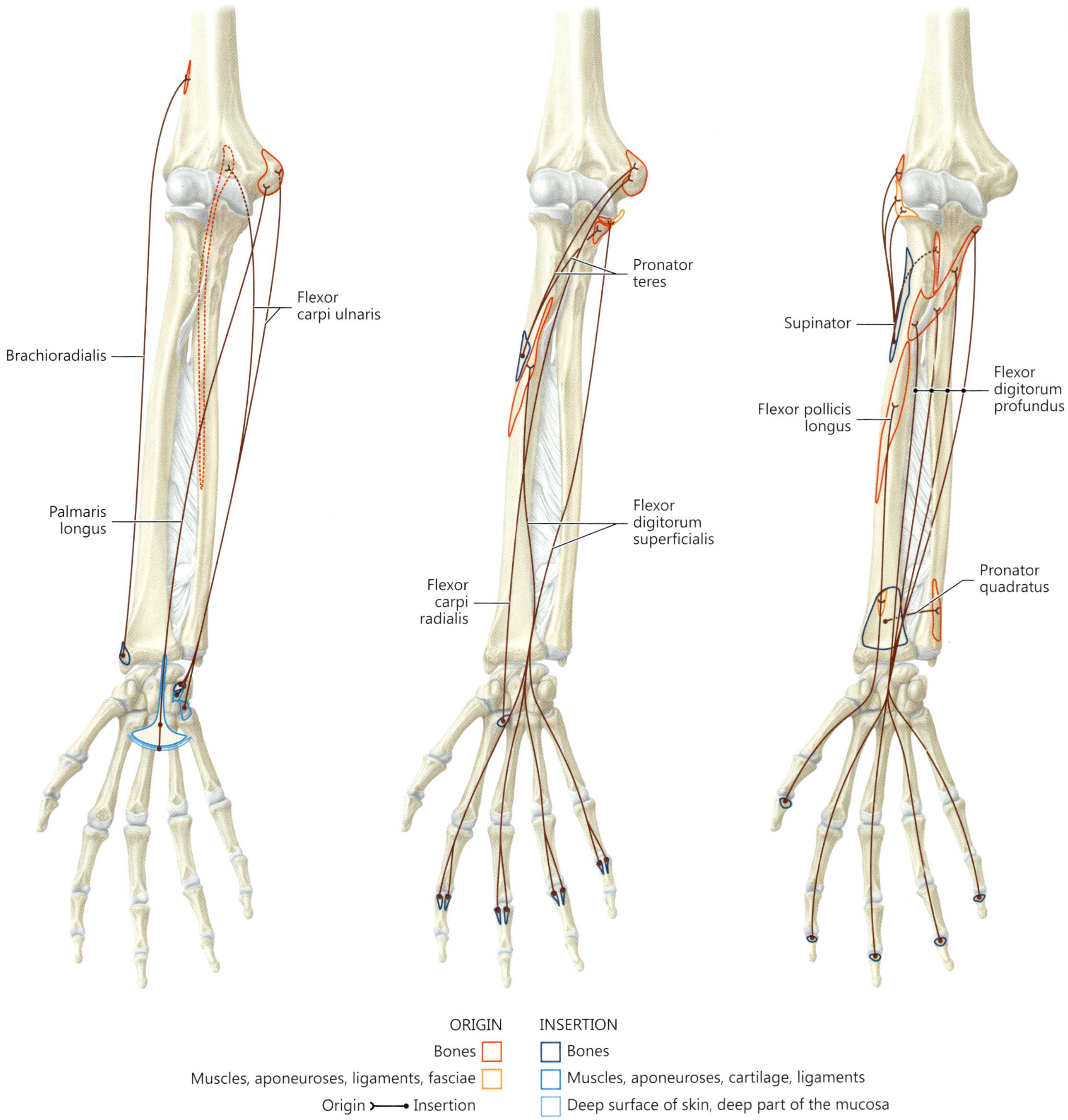

Schematic representation of the origin and insertions of the anterior muscles of the forearm, the extensor carpi radialis longus, and the supinator. The **pronator teres** extends from the medial epicondyle and from the coronoid process to the lateral surface of the radius. The **flexor carpi radialis** reaches, from the medial epicondyle, the second metacarpal. The **palmaris longus** runs from the medial epicondyle to the palmar aponeurosis. The **flexor carpi ulnaris** runs from the medial epicondyle, the olecranon, and the posterior border of the ulna, to the pisiform and, through the pisohamate and pisometacarpal ligaments, to the hamate and the fifth metacarpal. The **flexor digitorum superficialis** originates from the medial epicondyle, the ulnar collateral ligament, the coronoid process, and the anterior border of the radius. Its four tendons divide into two splits that insert on the middle phalanges of the last four fingers. The **flexor digitorum profundus** originates from the superior two-thirds of the anterior and medial surfaces of the ulna, the interosseous membrane of the forearm and the interosseous border of radius. Its four tendons, after crossing the splitting of the tendons of the flexor digitorum superficialis, insert on the distal phalanges of the last four fingers. The **flexor pollicis longus** extends from the proximal two-thirds of the anterior surface of the radius, the interosseous membrane and the coronoid process of the ulna to the distal phalanx of the thumb. The **pronator quadratus** is located between the anterior border of the ulna and the anterior border of the radius. The **brachioradialis** from the lateral supraepicondylar ridge and the lateral epicondyle inserts on the radial styloid process. The **supinator** reaches the anterior border and the lateral surface of the radius from the lateral epicondyle, the radial collateral and anular ligaments and from the supinator crest.

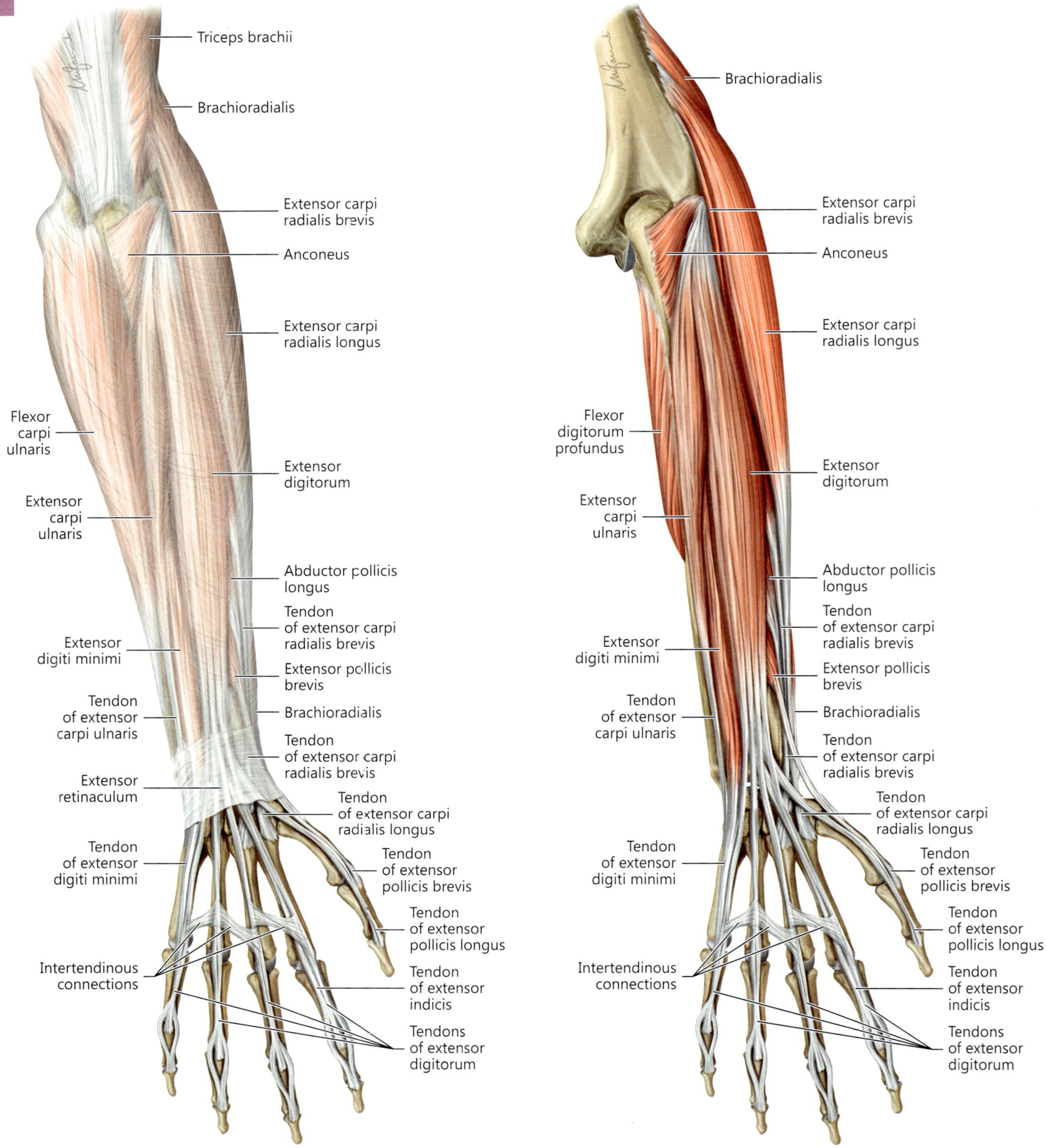

Lateral and **posterior muscles of the forearm**. Right side, posterior view. The lateral muscles, in an anteroposterior direction, are made up of the brachioradialis, extensor carpi radialis longus, and extensor carpi radialis brevis. The posterior muscles are distributed in two superimposed layers: superficial (extensor digitorum, extensor digiti minimi, and extensor carpi ulnaris) and deep (supinator, abductor pollicis longus, extensor pollicis longus and brevis, and extensor indicis). The anconeus is a posterior muscle of the arm in continuity with the inferolateral part of the medial head of the triceps brachii, which has been removed in the image on the right, along with the flexor carpi ulnaris, to show the brachioradialis and flexor digitorum profundus, respectively. In the figure on the left, the muscles are covered by the antebrachial fascia.

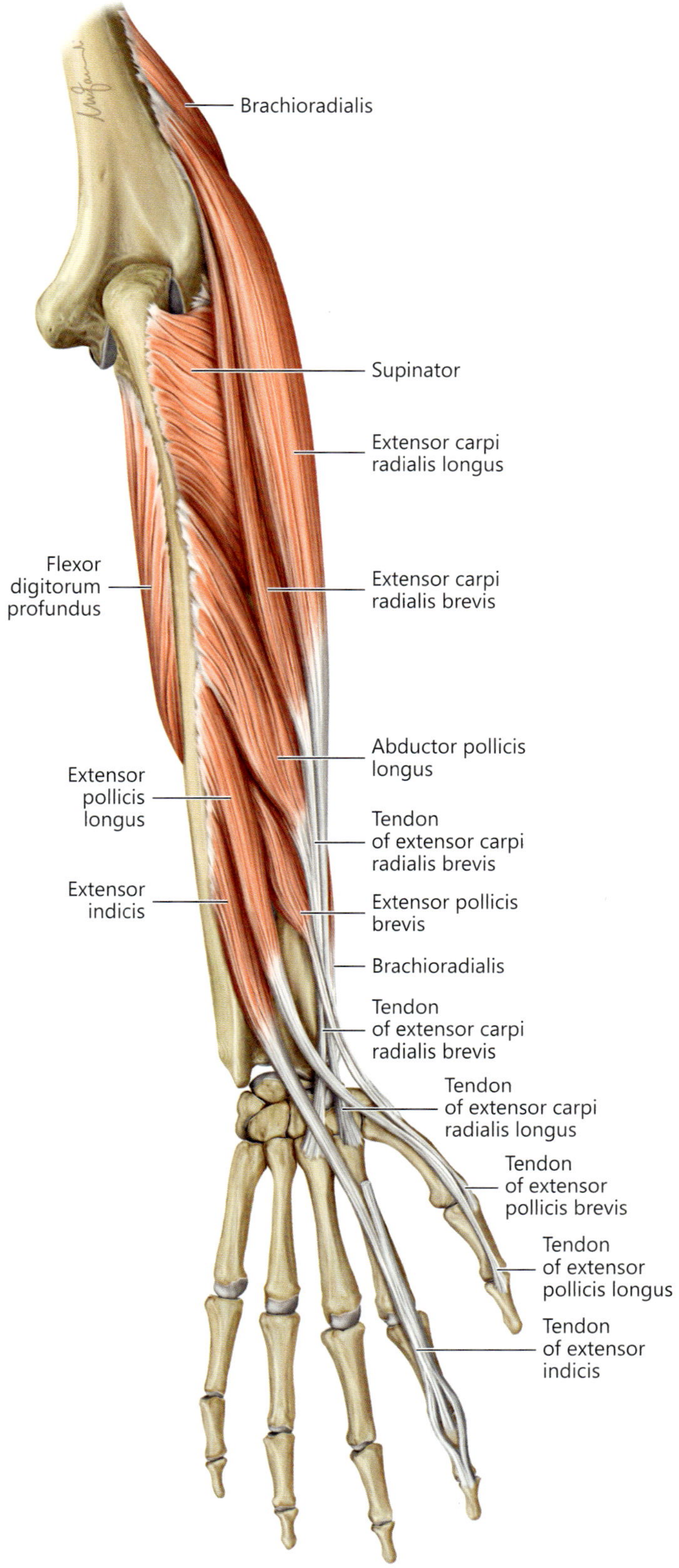

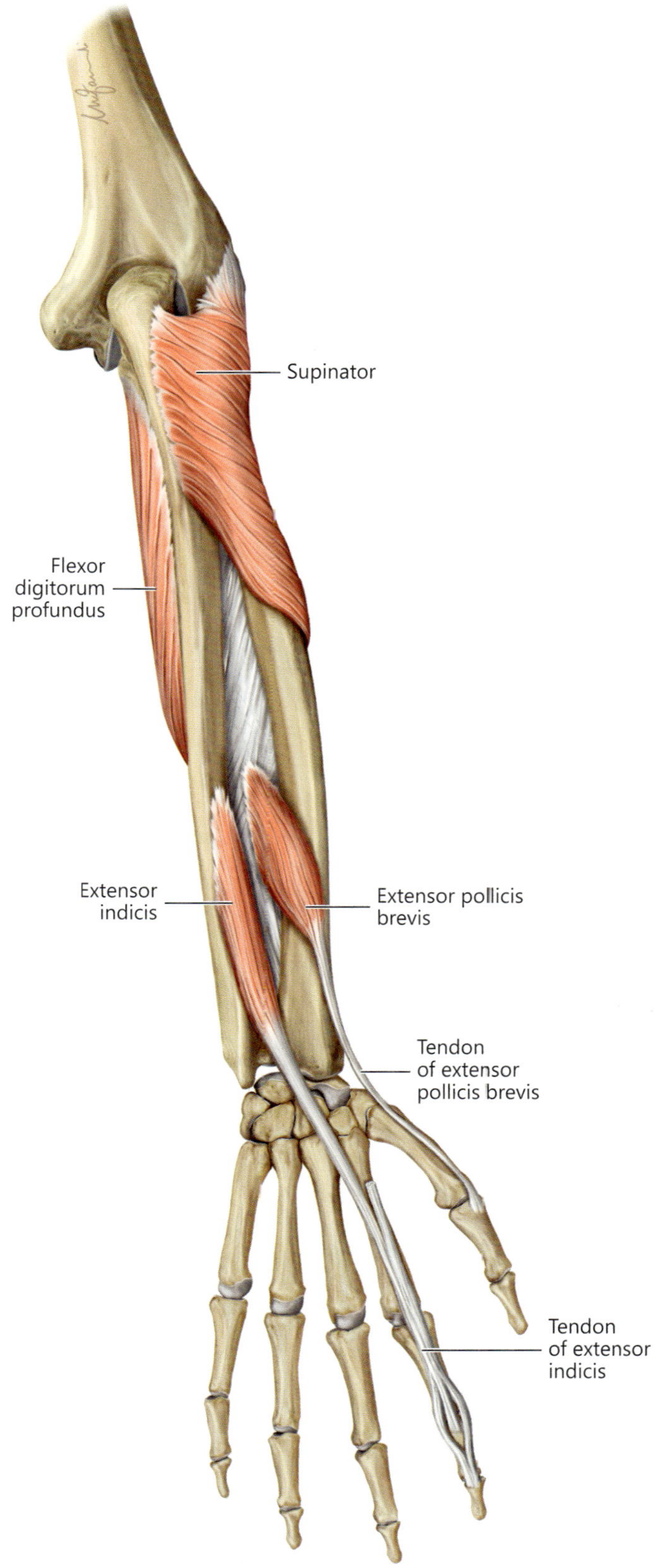

Lateral and **posterior muscles of the forearm**. Right side, posterior view. In the image on the left, the resection of the extensor digitorum, extensor digiti minimi, and extensor carpi ulnaris, which form the superficial layer of the posterior muscles of the forearm, reveals the muscles of the deep layer: supinator, abductor pollicis longus, extensor pollicis brevis, extensor pollicis longus, and extensor indicis. In the image on the right, the removal of the lateral muscles of the forearm renders the supinator more visible.

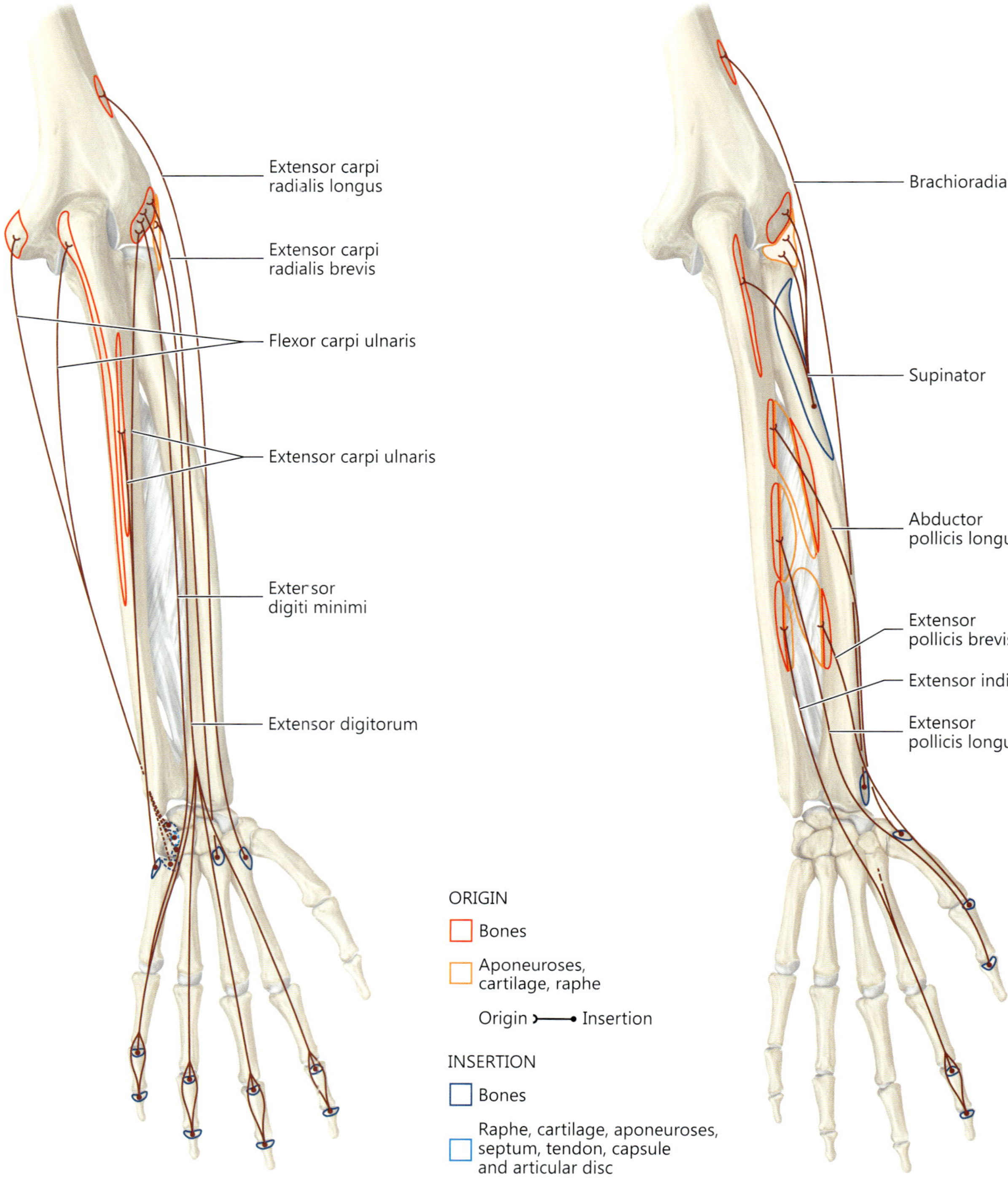

Schematic representation of the origin and insertions: the **extensor carpi radialis longus** originates from the lateral supracondylar border of the humerus and inserts at the base of the second metacarpal. The **extensor carpi radialis brevis** extends from the lateral epicondyle and from the radial collateral ligament to the base of the third metacarpal. The **extensor carpi ulnaris** originates from the lateral epicondyle and from the posterior border of the ulna, and inserts on the base of the fifth metacarpal. The **extensor digitorum** originates from the lateral epicondyle. Its four tendons divide into three splits: the middle split inserts on the dorsal surface of the base of the middle phalanx; the lateral and medial splits join to insert on the base of the distal phalanx of the second, third, fourth, and fifth finger. The **extensor pollicis brevis** originates from the posterior surface of the radius and from the interosseous membrane, and inserts at the base of the proximal phalanx of the thumb. The **extensor pollicis longus** extends from the posterolateral surface of the ulna and from the interosseous membrane to the base of the distal phalanx of the thumb. The **extensor indicis** originates from the posterior surface of the ulna and from the interosseous membrane, and connects to the ulnar side of the tendon for the index of the extensor digitorum. The **extensor digiti minimi** extends from the lateral epicondyle to the dorsal digital expansion of the fifth finger. The **flexor carpi ulnaris** originates with the humeral head from the medial epicondyle, and with the ulnar head from the medial border of the olecranon and the posterior border of the ulna. The two heads are fixed to the pisiform, the hamate, the base of the fifth metacarpal, and to the flexor retinaculum. The **brachioradialis** runs from the lateral border of the humerus and the lateral intermuscular septum to the radial styloid process. The **supinator** originates from the lateral epicondyle, from the radial collateral and anular ligaments, and from the supinator crest on the shaft (body) of the ulna, and inserts at the proximal third of the anterior and lateral surfaces of the radius. The **abductor pollicis longus** extends from the posterior surface of the ulna, of the radius, and of the interosseous membrane to the trapezius and to the base of the first metacarpal.

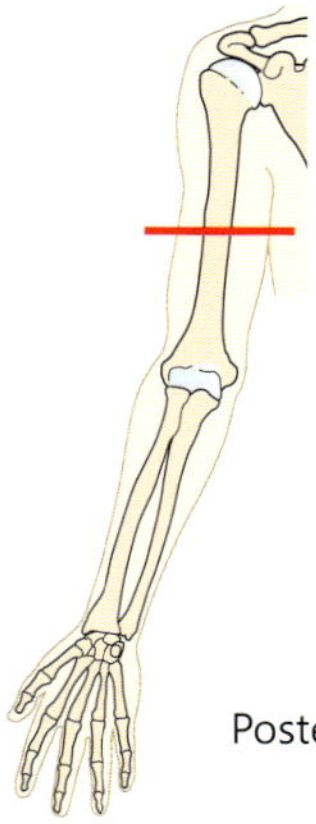

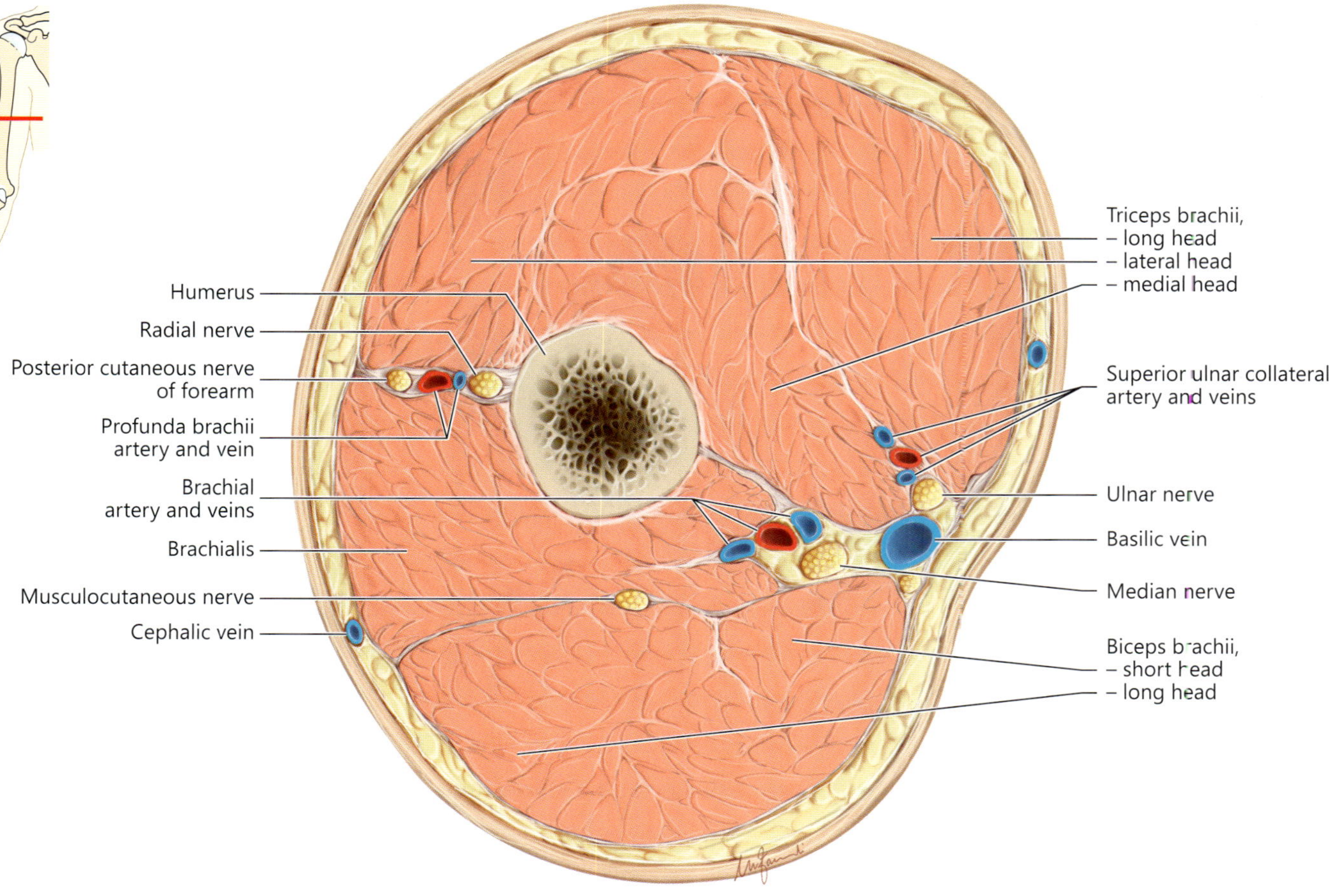

Schematic representation showing a transverse section of the arm in correspondence to the middle of the **humeral diaphysis**.

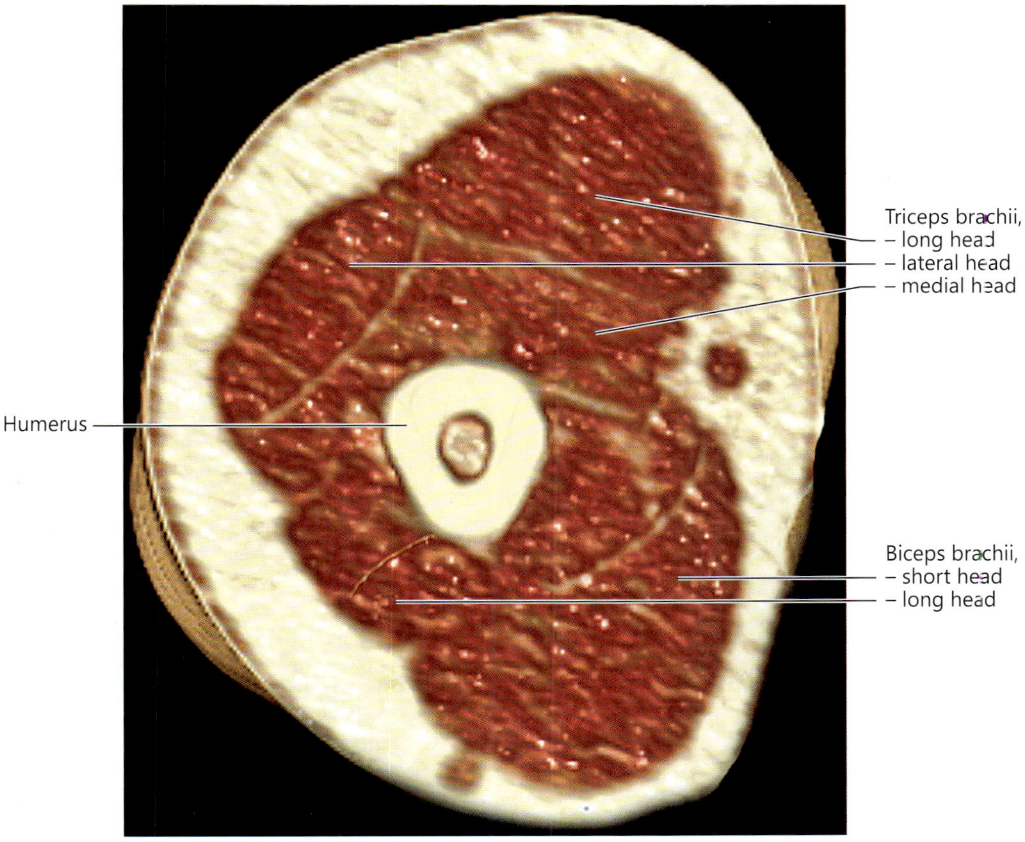

Three-dimensional image of a living human body, obtained by CT scan using a 3D volume rendering technique, showing the **muscles of the arm** in a transverse section in correspondence to the middle of the humeral diaphysis.

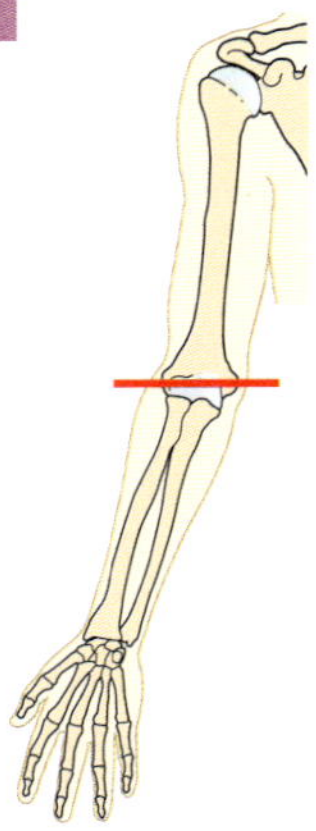

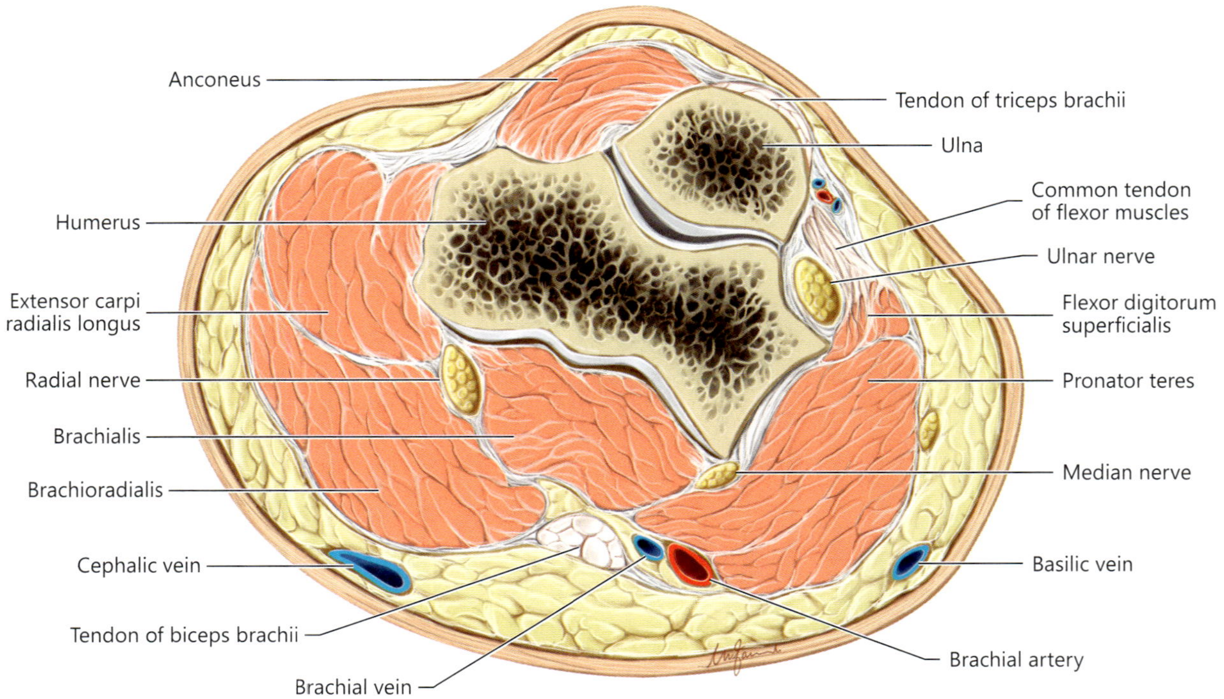

Transverse section at the **elbow joint**.

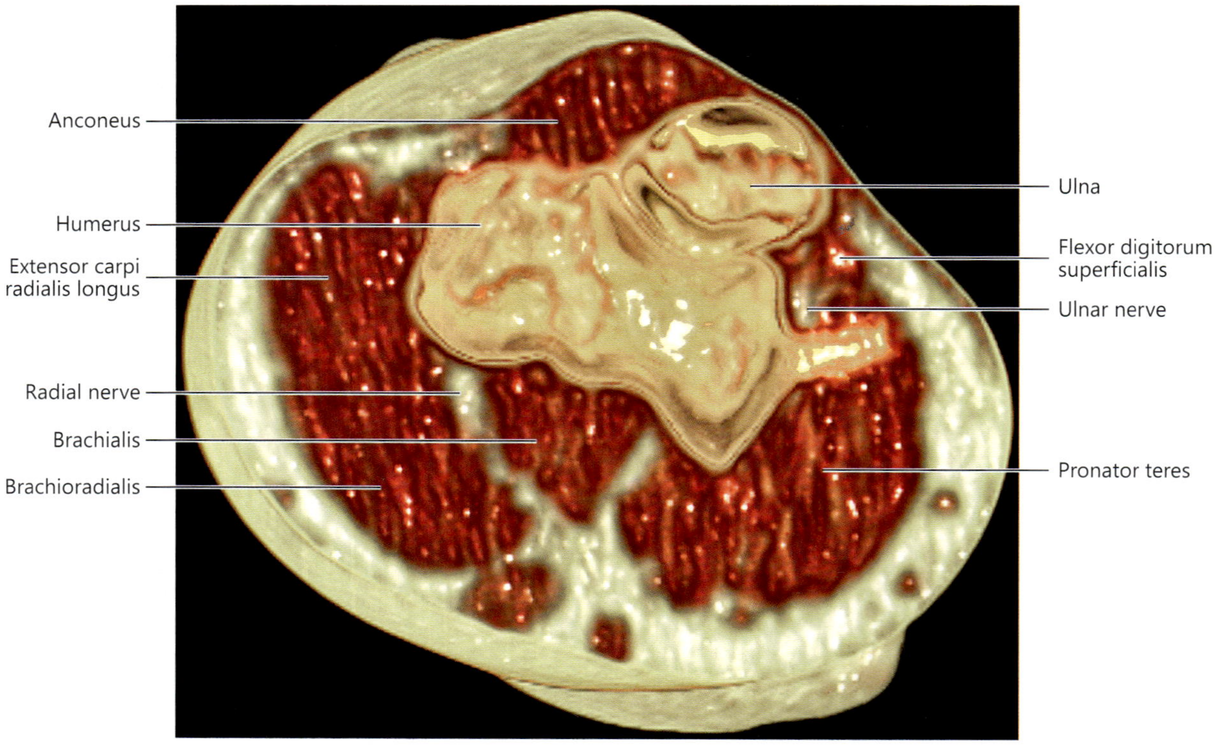

Three-dimensional image of a living human body, obtained by CT scan using a 3D volume rendering technique, showing the **muscles of the** right **forearm** in a transverse section in correspondence to the elbow joint.

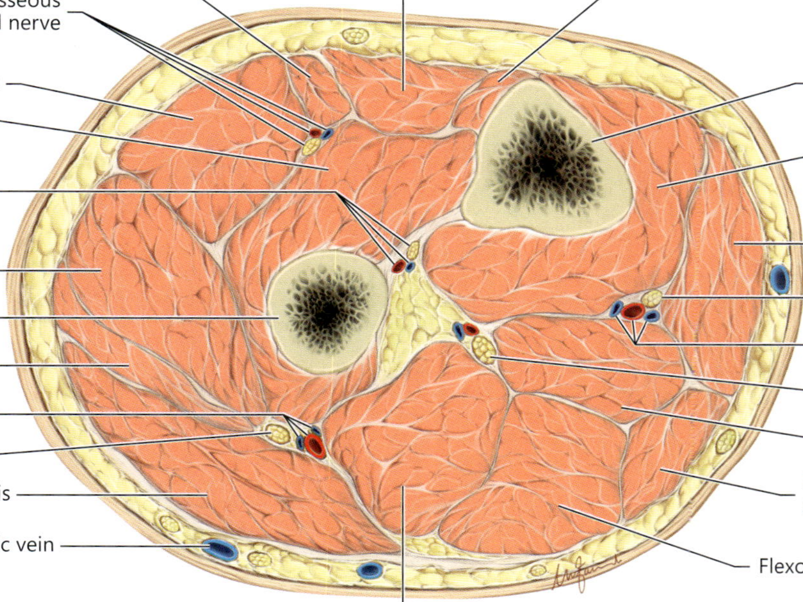

Transverse section at the proximal third of the forearm. The **anterior forearm muscles**, with prevalent flexor action, are enclosed in the anterior compartment; in the posterior compartment are the **posterior muscles**, with principally extensor action; in the anterolateral part of the forearm are the **lateral muscles**, brachioradialis, and extensor carpi radialis longus and brevis.

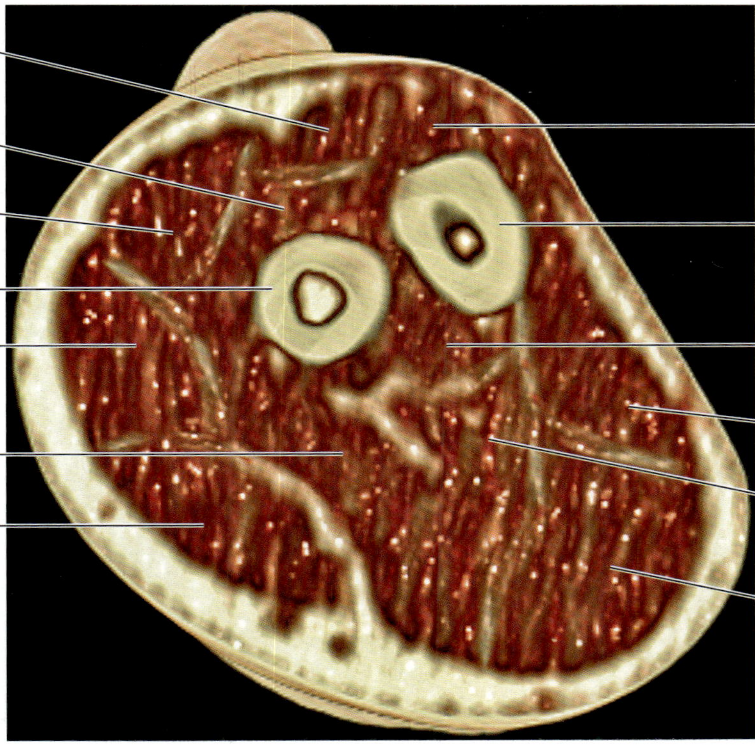

Three-dimensional image of a living human body, obtained by CT scan using a 3D volume rendering technique, showing the **muscles of the** right **forearm** in a transverse section at the proximal third.

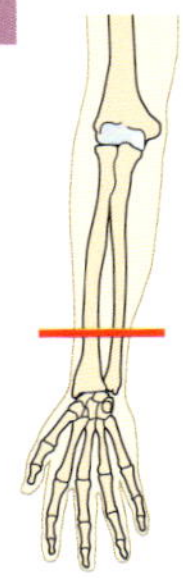

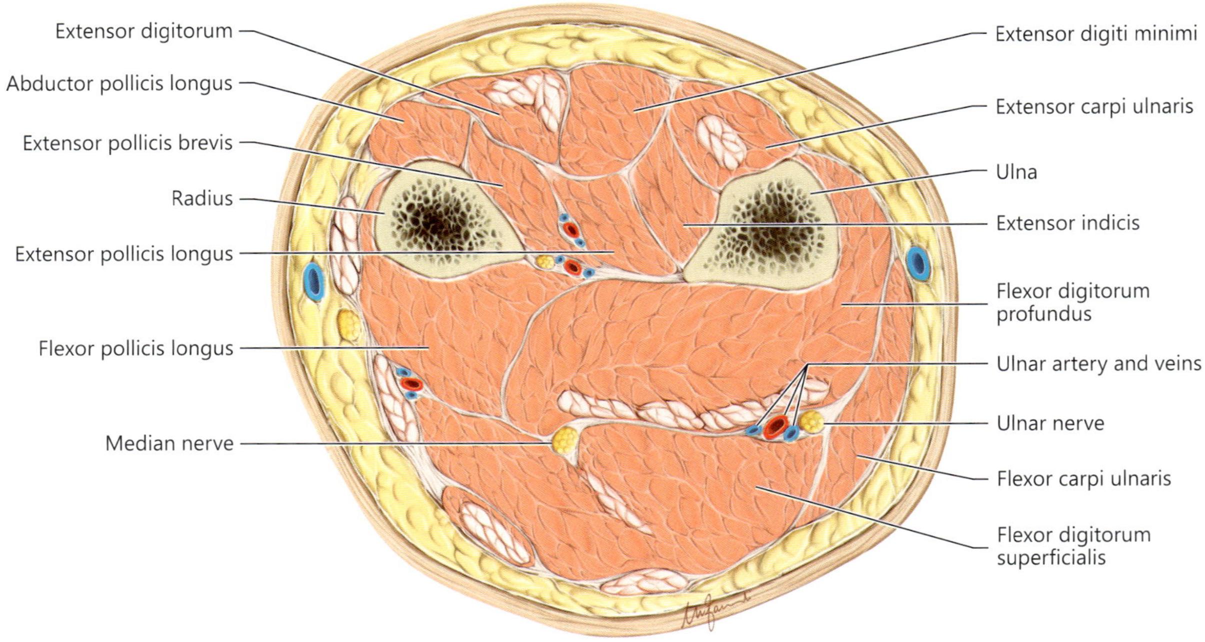

Transverse section at the distal third of the forearm. The **posterior muscles of the forearm** are located on two superimposed planes. The most superficial one, in a lateromedial direction, is formed by the extensor digitorum, the extensor digiti minimi, and the extensor carpi ulnaris. The deep layer is made up of the supinator, abductor pollicis longus, and extensor pollicis brevis, pollicis longus, and indicis.

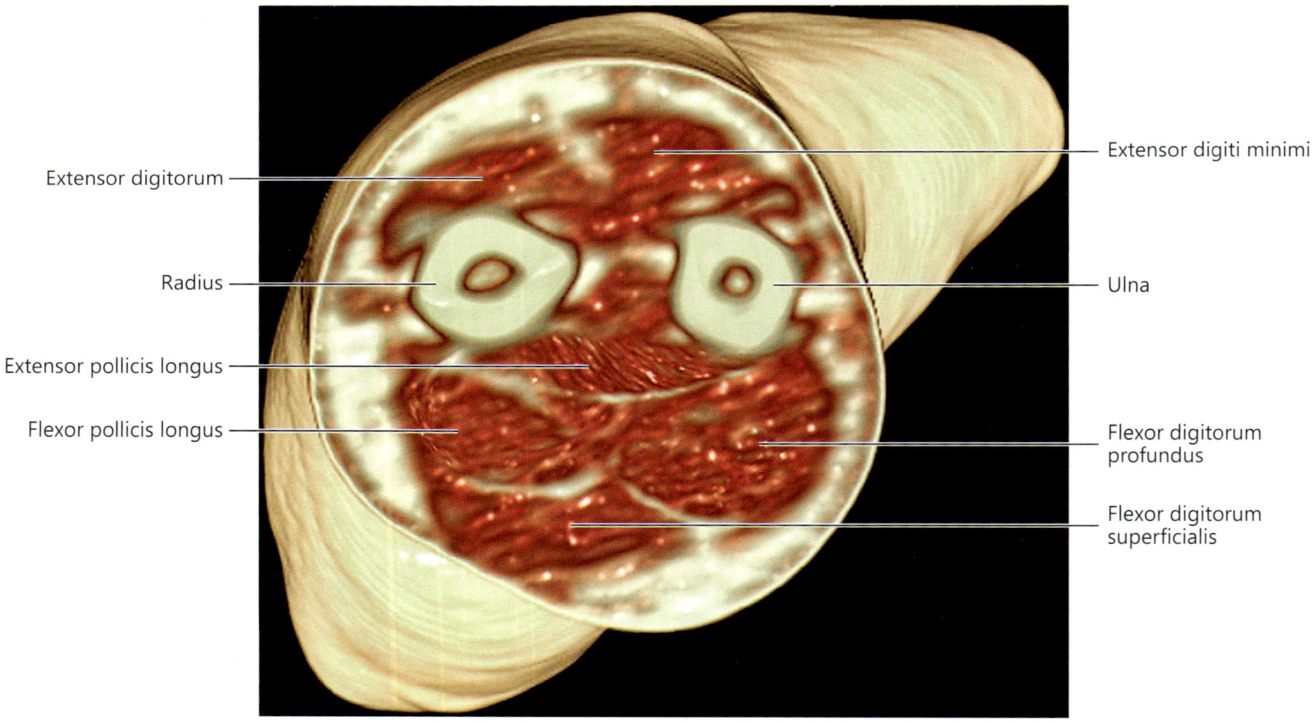

Three-dimensional image of a living human body, obtained by CT scan using a 3D volume rendering technique, showing the **muscles of the** right **forearm** in a transverse section at the distal third. The flexor digitorum superficialis constitutes the second layer of the anterior muscles of the forearm; the flexor digitorum profundus and flexor pollicis longus form the third layer.

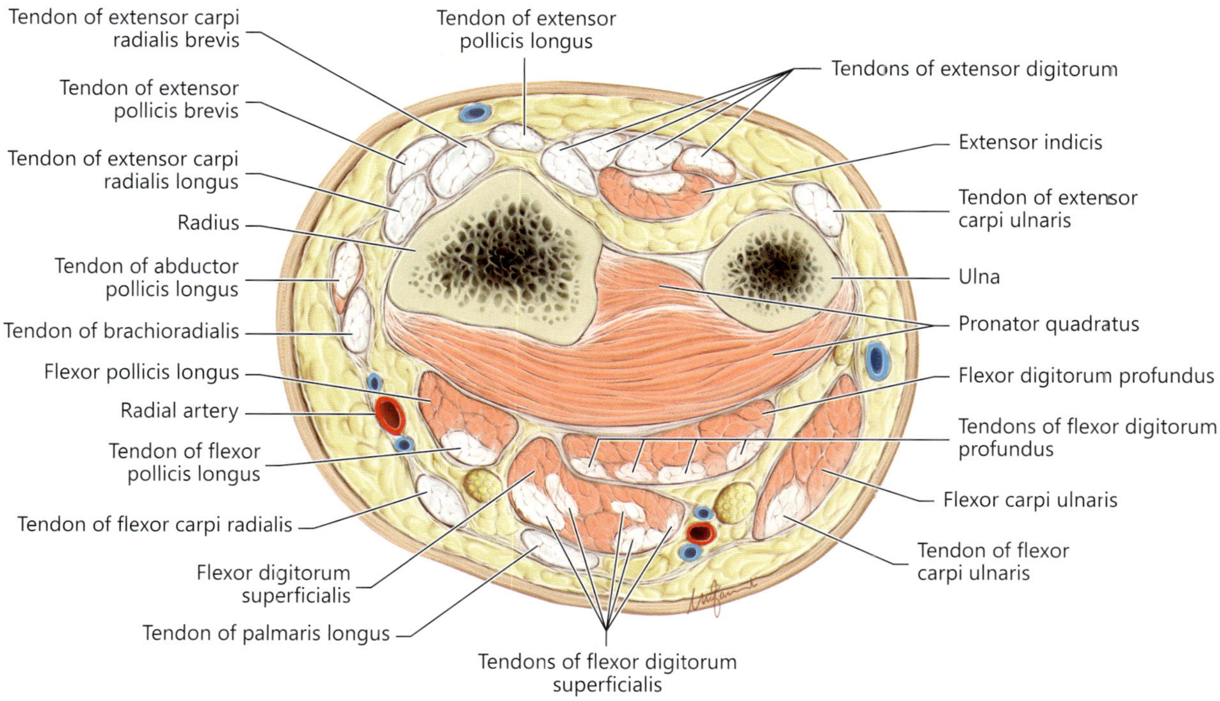

Transverse section at the distal end of the forearm. We see the **tendons of the flexor muscles**: digitorum superficialis and profundus, carpi radialis and pollicis longus before their entry into the carpal tunnel, and those of the **extensor muscles** and **abductor pollicis longus** before their passage into the canals of the extensor retinaculum.

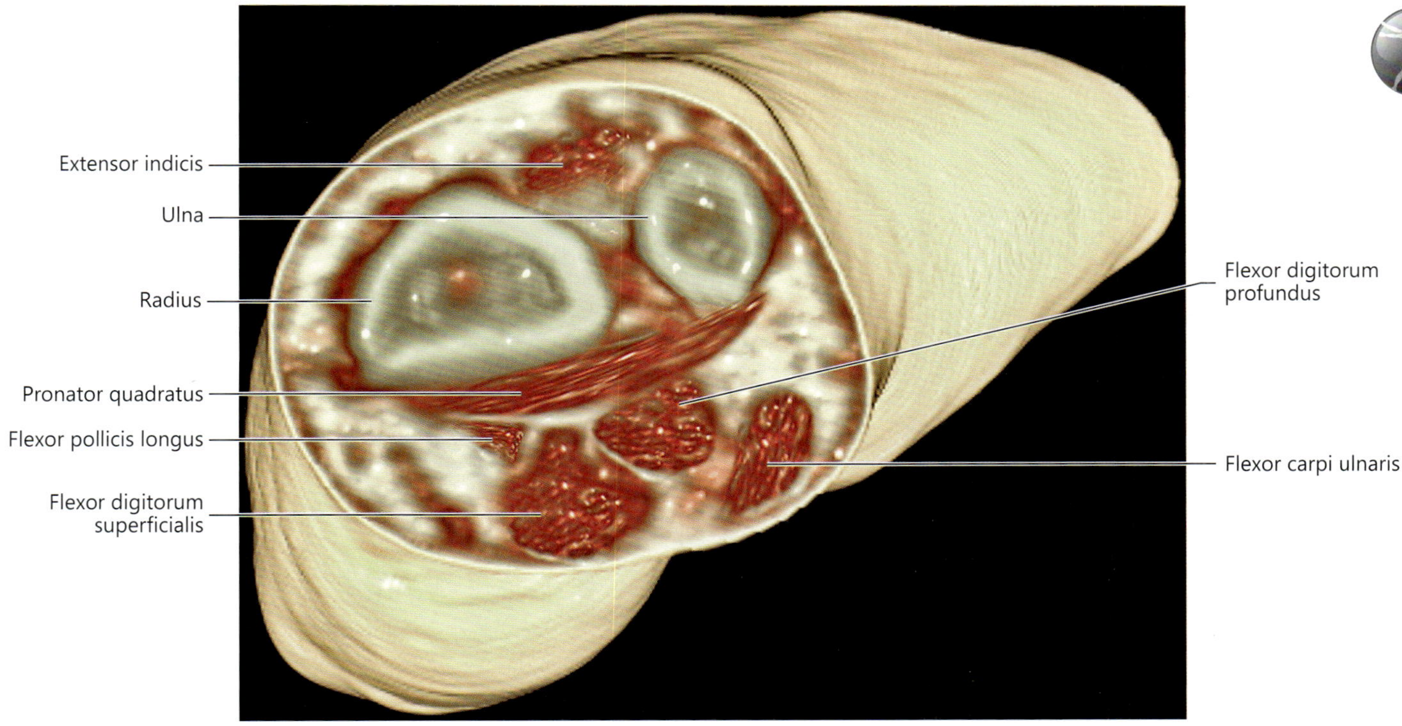

Three-dimensional image of a living human body, obtained by CT scan using a 3D volume rendering technique, showing the **muscles of the** right **forearm** in a transverse section at the distal end. The pronator quadratus, located in the distal part of the forearm near the interosseous membrane, forms the fourth layer of forearm extensor muscles.

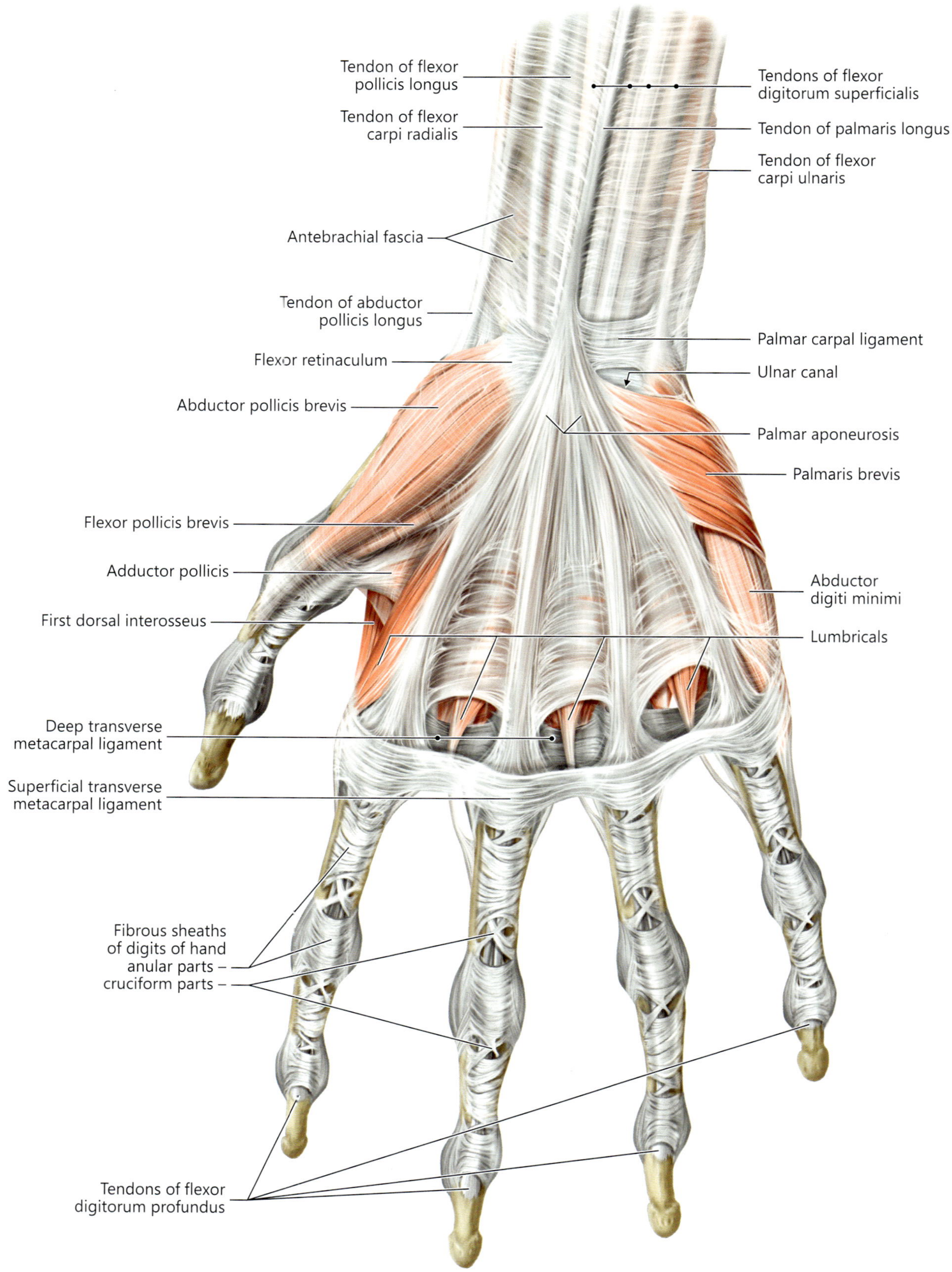

Fascial structures of the right **hand**, anterior view. The palmar aponeurosis, into which the antebrachial fascia continues, adheres deeply to the flexor retinaculum (transverse carpal ligament), which, together with the sulcus formed by the concavity of the carpal bones, forms the carpal tunnel. The superficial transverse metacarpal ligament (natatory ligament) is in close contact with the subcutaneous tissue. Fibrous sheaths cover the synovial sheaths of the digits. The abductor pollicis brevis is covered by the fascia of the thenar eminence, the abductor digiti minimi is covered by the fascia of the hypothenar eminence.

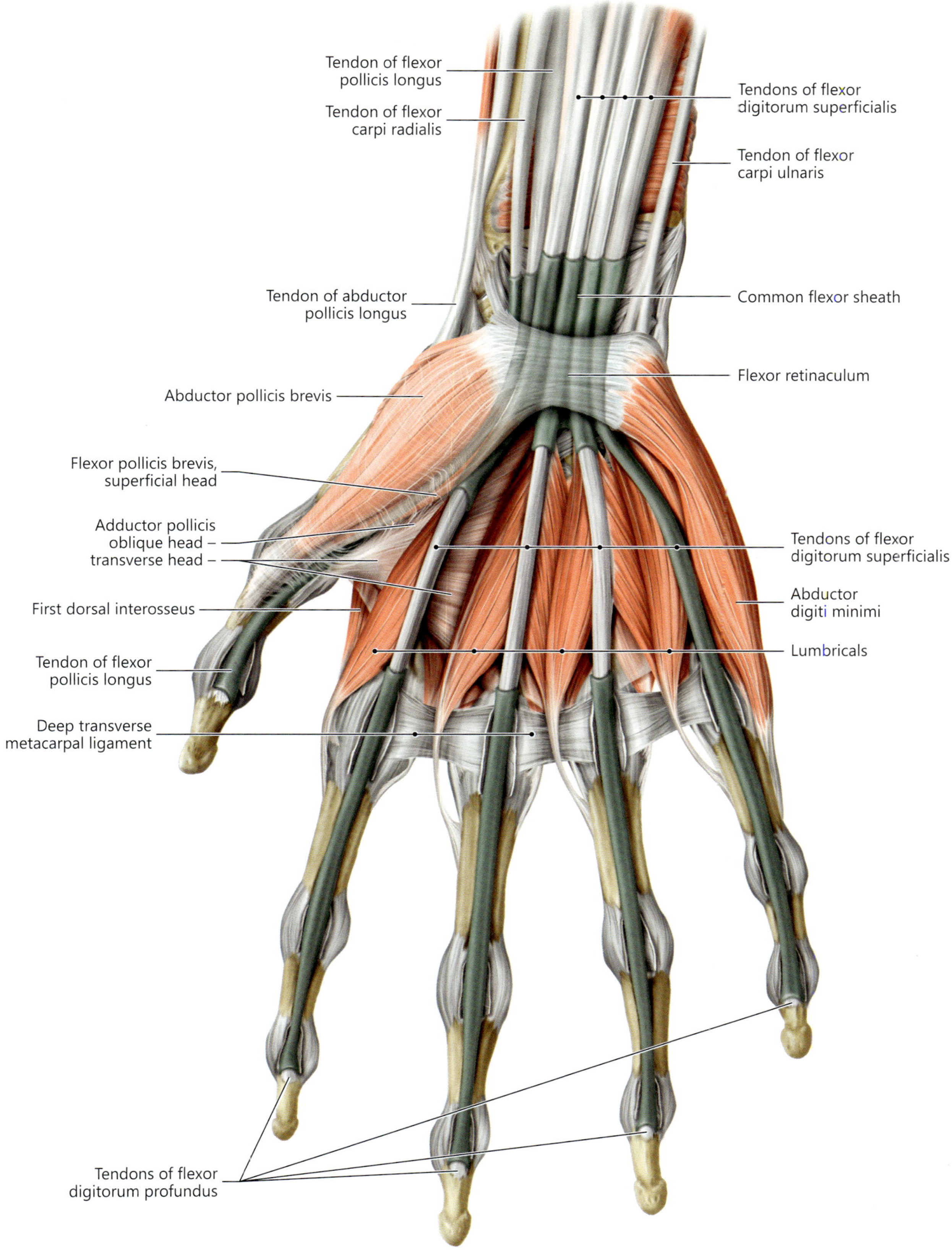

Right hand, anterior view. The **carpal tunnel** (**carpal canal**), between the flexor retinaculum (thickening of the antebrachial fascia) and the concavity of the carpal bones, is subdivided into two parts, lateral and medial, by a septum that is attached to the trapezoid and the capitate. In the lateral part run the tendons of the flexor carpi radialis and flexor pollicis longus, each covered by its own tendinous sheath. In the medial part run the tendons of the flexor digitorum superficialis and profundus, covered by a common tendinous sheath. The abductor pollicis brevis and adductor pollicis are covered by the fascia of the thenar eminence.

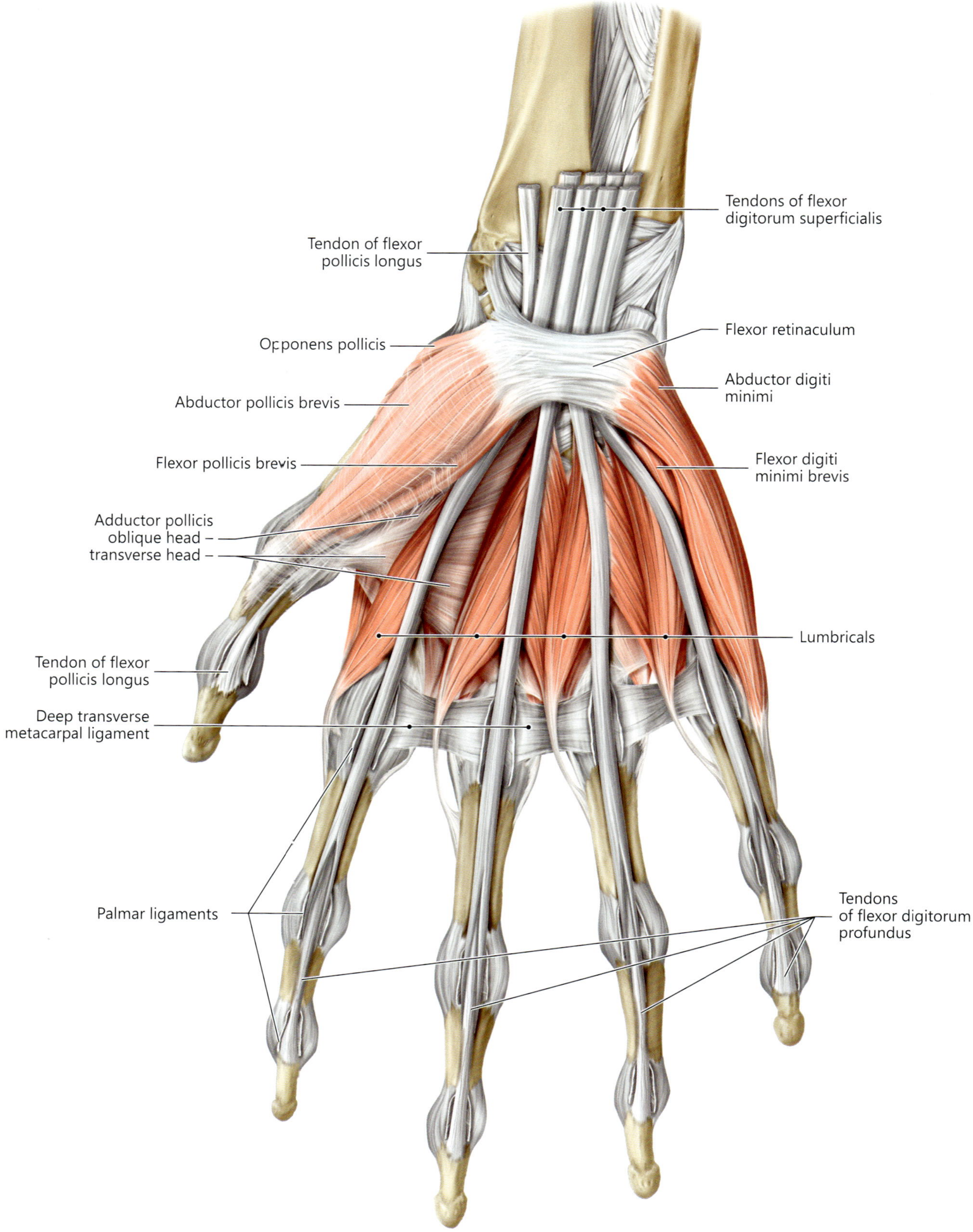

Muscles of the right **hand**, anterior view. The muscles of the thenar eminence are four muscles intended to the thumb: abductor pollicis brevis, opponens pollicis, flexor pollicis brevis, and adductor pollicis. The four lumbricals are located among the tendons of the flexor digitorum profundus. The muscles of the hypothenar eminence visible in this image are the abductor digiti minimi, located superficially and medially, and the flexor digiti minimi brevis, located superficially and laterally. The abductor pollicis brevis and adductor pollicis are covered by the fascia of the thenar eminence.

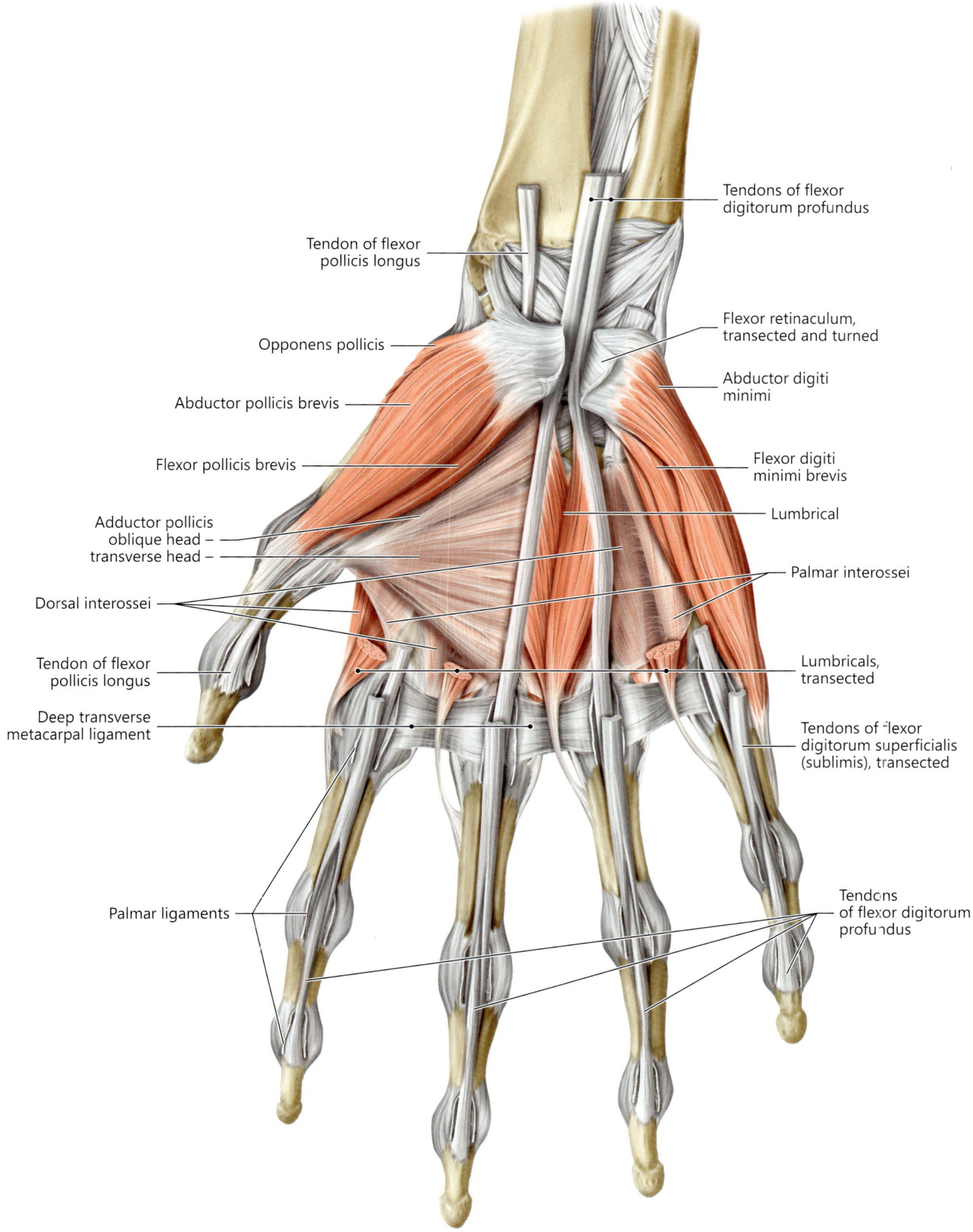

Muscles of the right **hand**, anterior view. The resection of the first, second, and fourth lumbricals and of the tendons of the flexor digitorum superficialis and profundus shows the adductor pollicis and some palmar and dorsal interossei. The adductor pollicis is covered by the fascia of the thenar eminence. The interossei are covered by the deep palmar fascia.

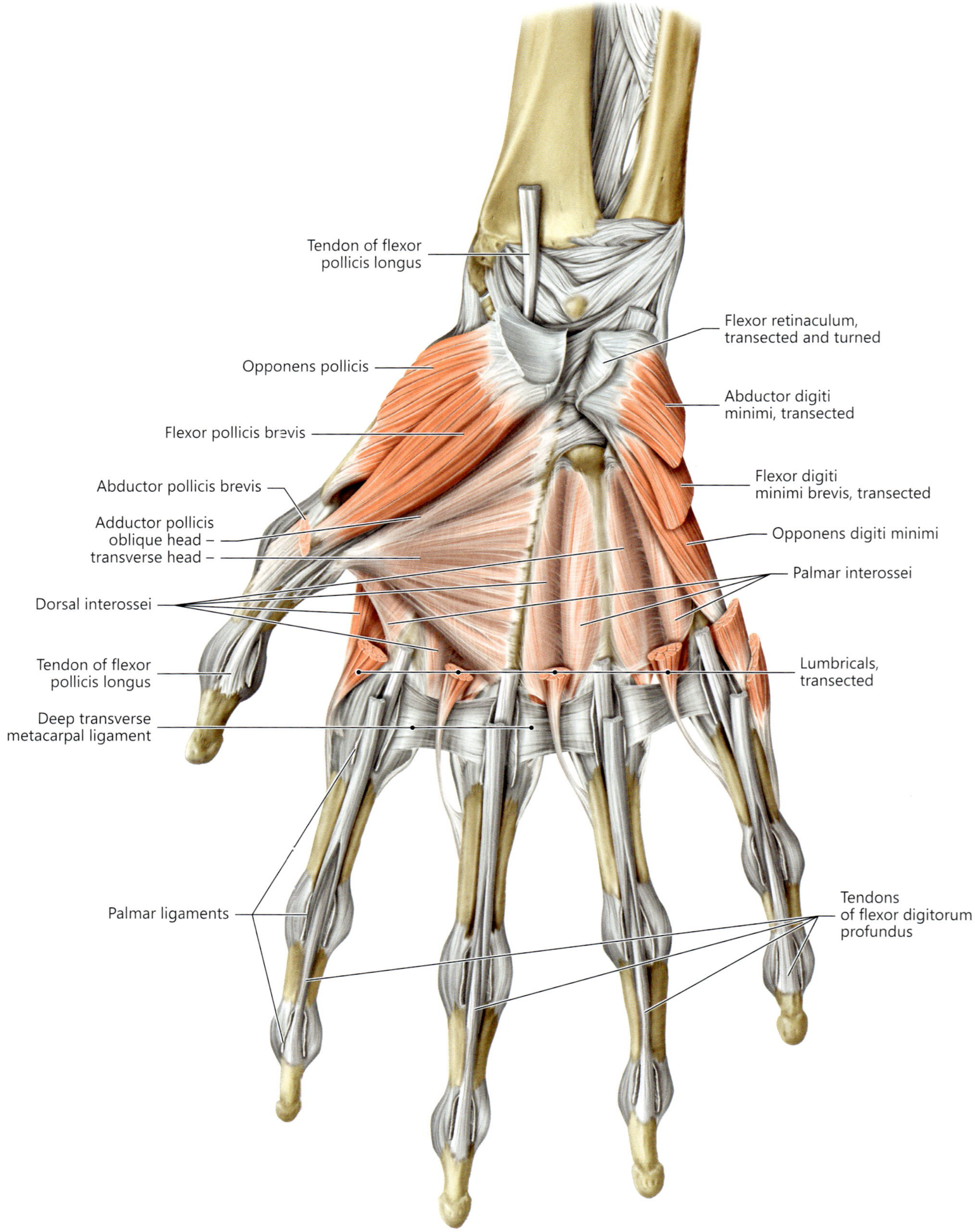

Muscles of the right **hand**, anterior view. The resection of the abductor pollicis brevis shows the superficial and deep heads of the flexor pollicis brevis. The removal of the third lumbrical highlights the palmar and dorsal interossei. The resection of the abductor digiti minimi and flexor digiti minimi brevis shows the opponens digiti minimi, located deep within the hypothenar eminence. The adductor pollicis is covered by the fascia of the thenar eminence. The interossei are covered by the deep palmar fascia.

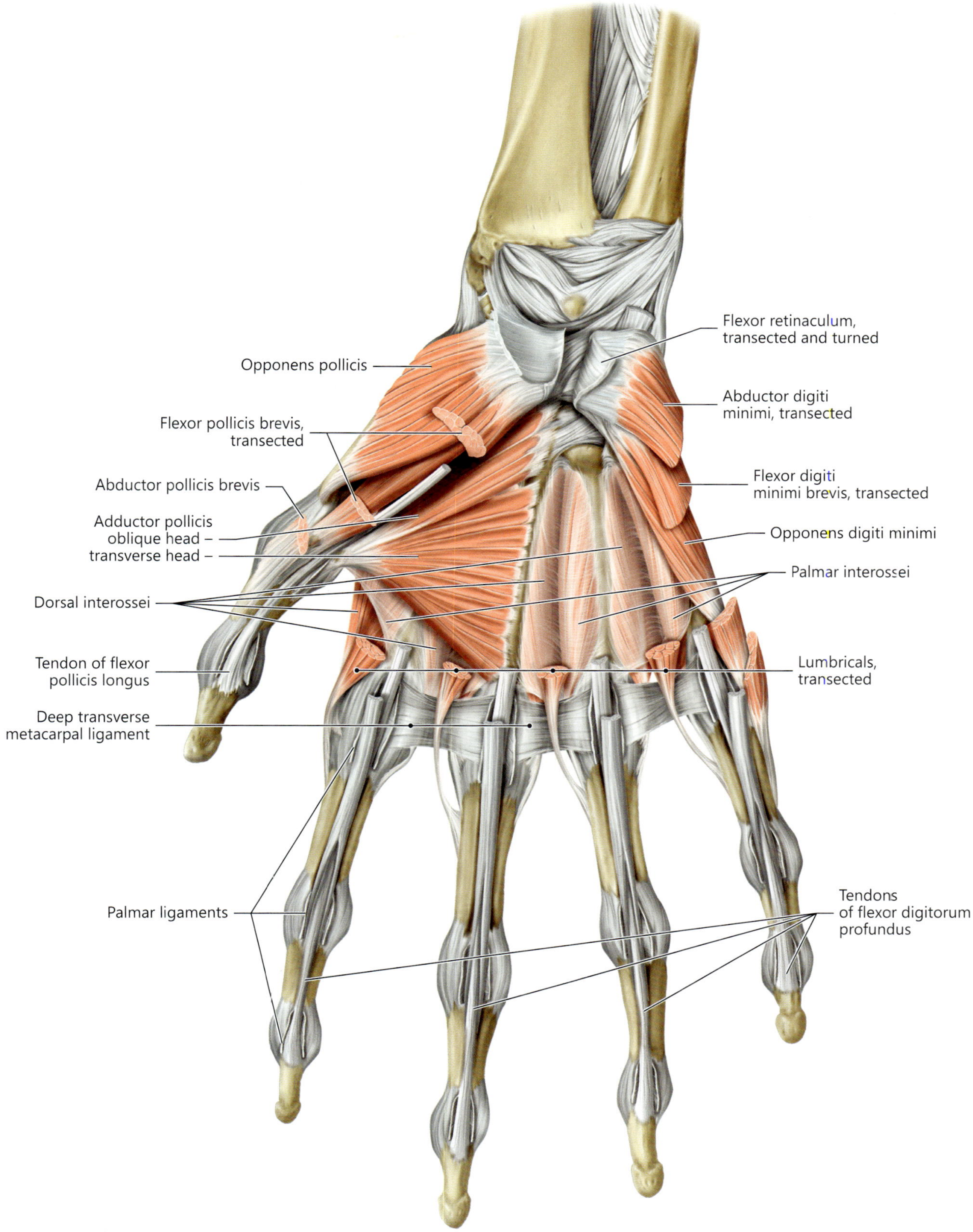

Muscles of the right **hand**, anterior view. With regards to the previous image, the superficial and deep heads of the flexor pollicis brevis have been transected; between thems runs the tendon of the flexor pollicis longus, which appears here partially transected. The interossei are covered by the deep palmar fascia.

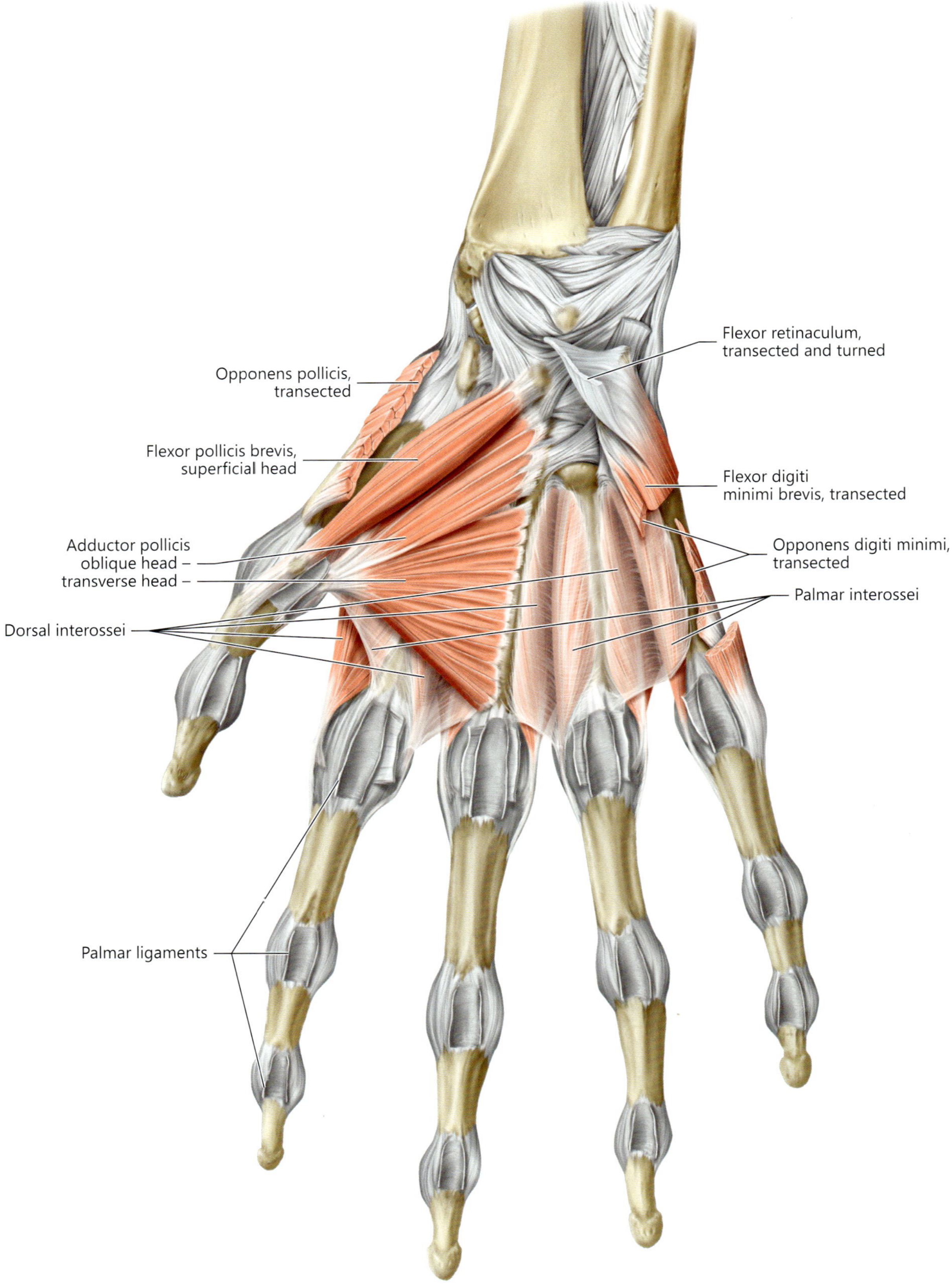

Muscles of the right **hand**, anterior view. The resection of the opponens digiti minimi shows the fourth palmar interosseous. In this image, the opponens pollicis has been transected as well, leaving intact, instead, the superficial head of the flexor pollicis brevis. The interossei are covered by the deep palmar fascia.

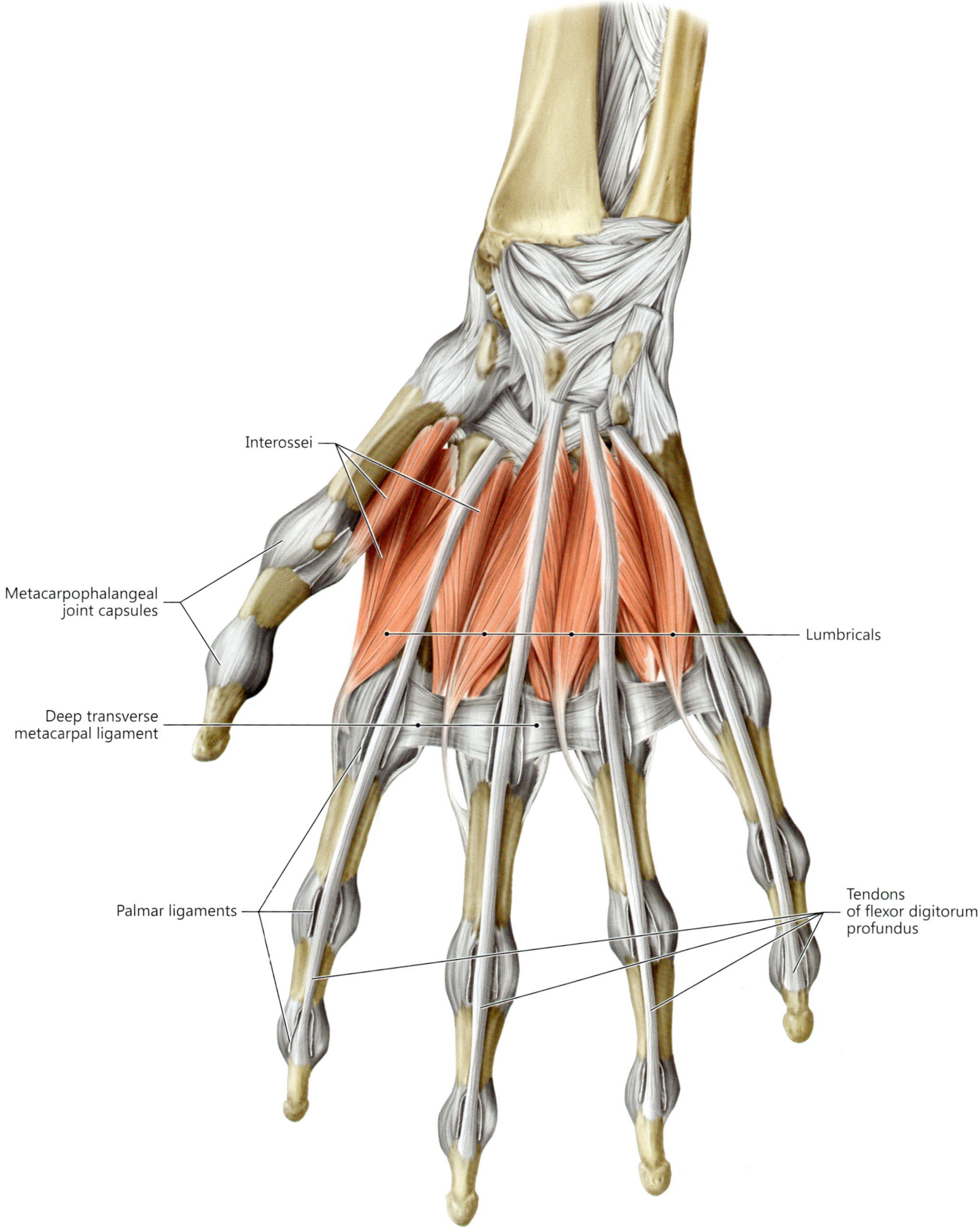

Muscles of the right **hand**, anterior view. The palmar muscles are located within the middle part of the palm and are constituted by the lumbricals and palmar and dorsal interossei. The lumbricals (1st-4th) originate from the lateral border of the tendons of the flexor digitorum profundus. After reaching the lateral contour of the 2nd, 3rd, 4th, and 5th metacarpophalangeal joints, they run on the dorsal surface of the proximal phalanx, uniting with a small tendon to the tendons of the extensor digitorum.

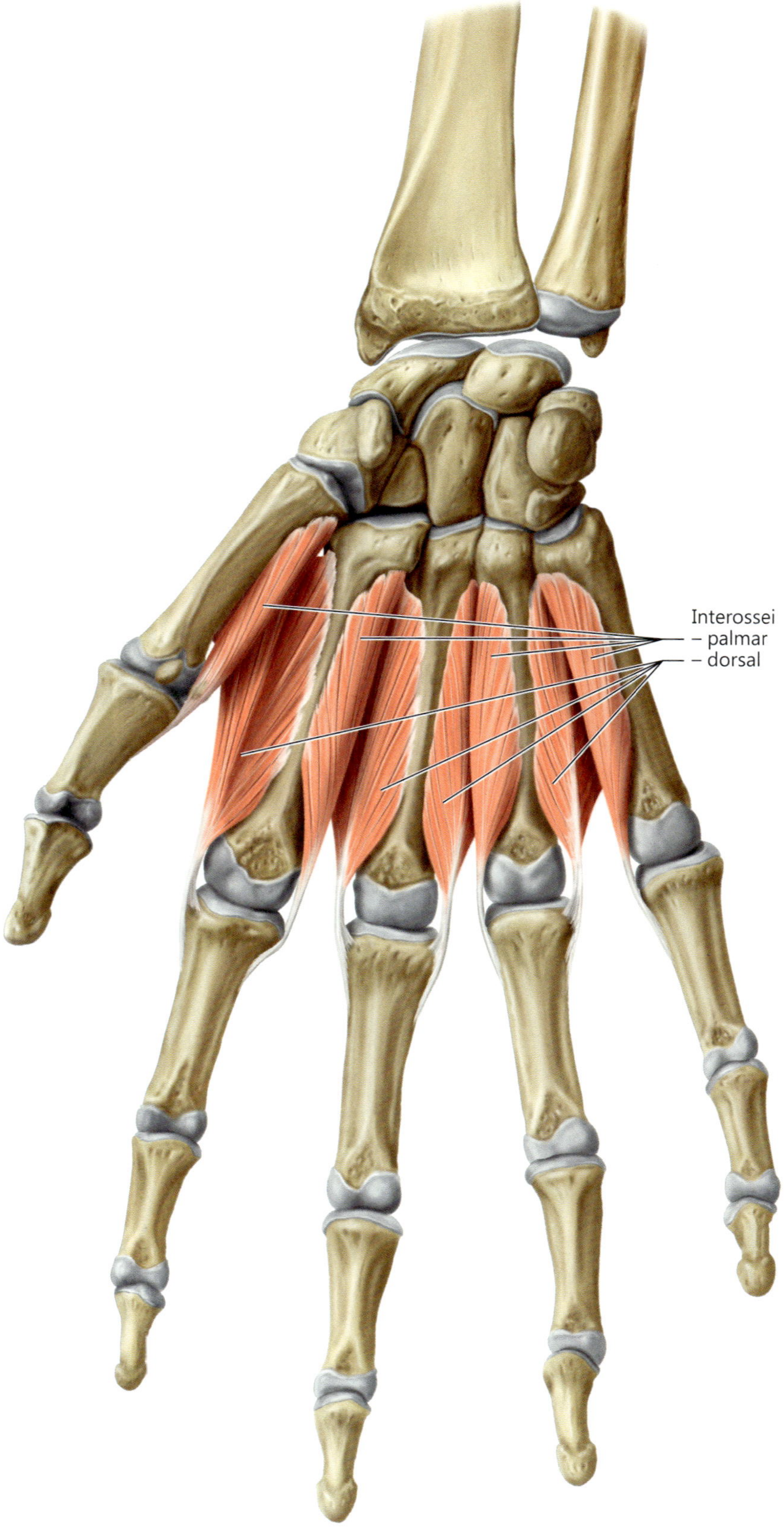

Muscles of the right **hand**, anterior view. The removal of the lumbricals shows the palmar (1st-4th) and dorsal (1st-4th) interossei located in the spaces between the metacarpals, the latter being dorsal to the former. The first palmar interosseous, in the first intermetacarpal space, is often rudimentary or missing, given the presence of the adductor pollicis.

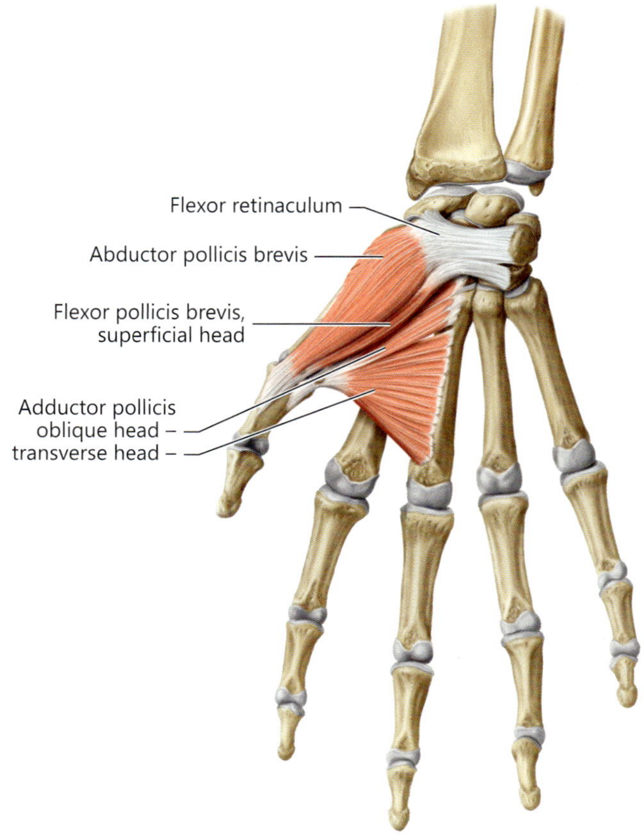

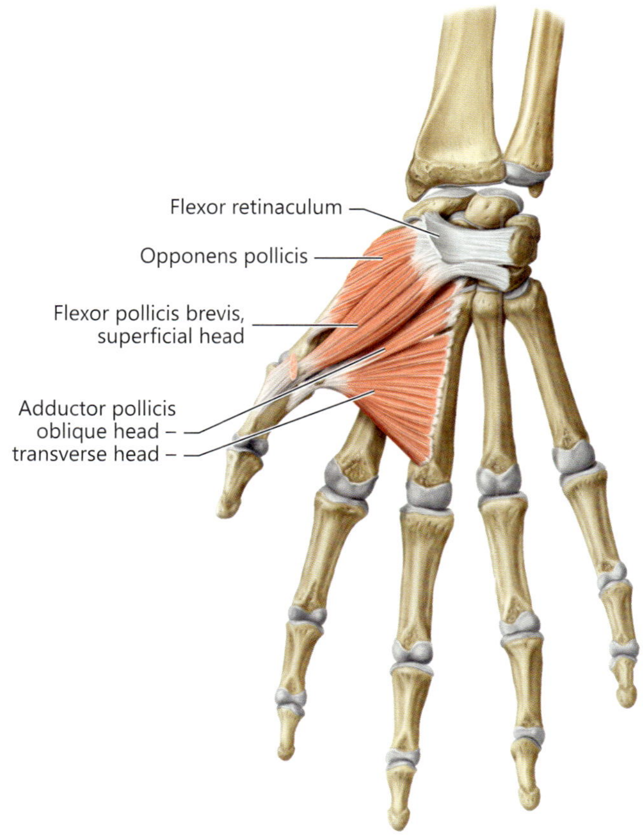

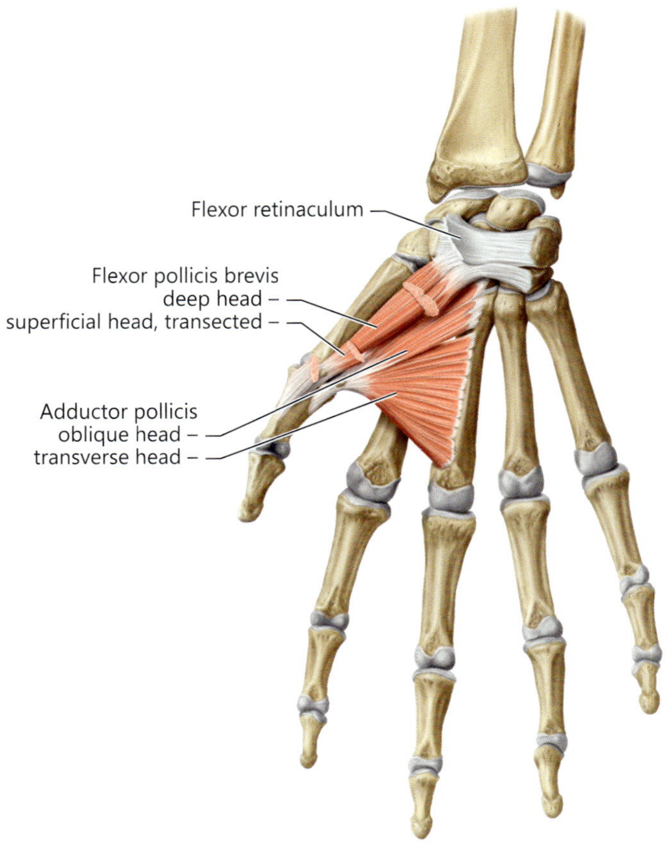

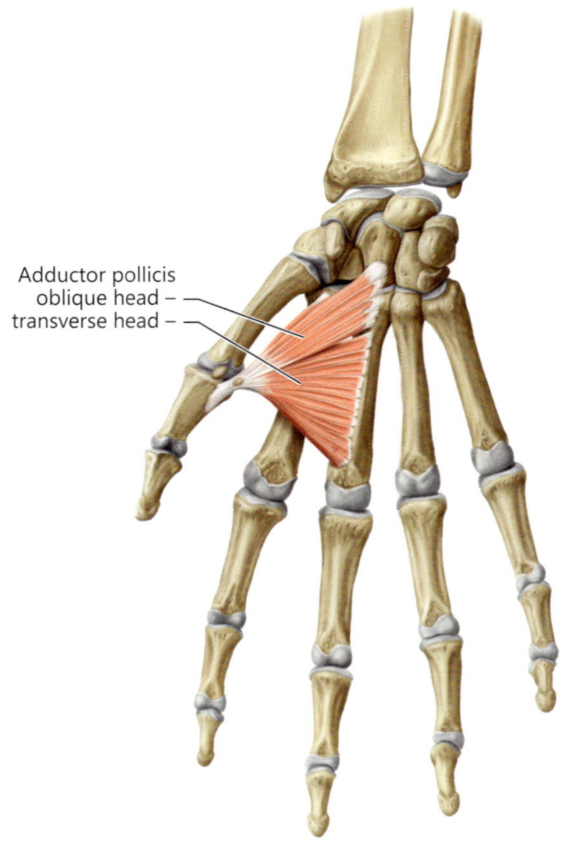

Right hand, anterior view. **Muscles of the thenar eminence**, from the surface to the deep part. The abductor pollicis brevis is the most superficial; the opponens pollicis is a triangular muscle located deep to the abductor brevis; the flexor pollicis brevis is a two-headed muscle located deep to the abductor brevis; the adductor pollicis, which is two-headed and triangular as well, is the deepest.

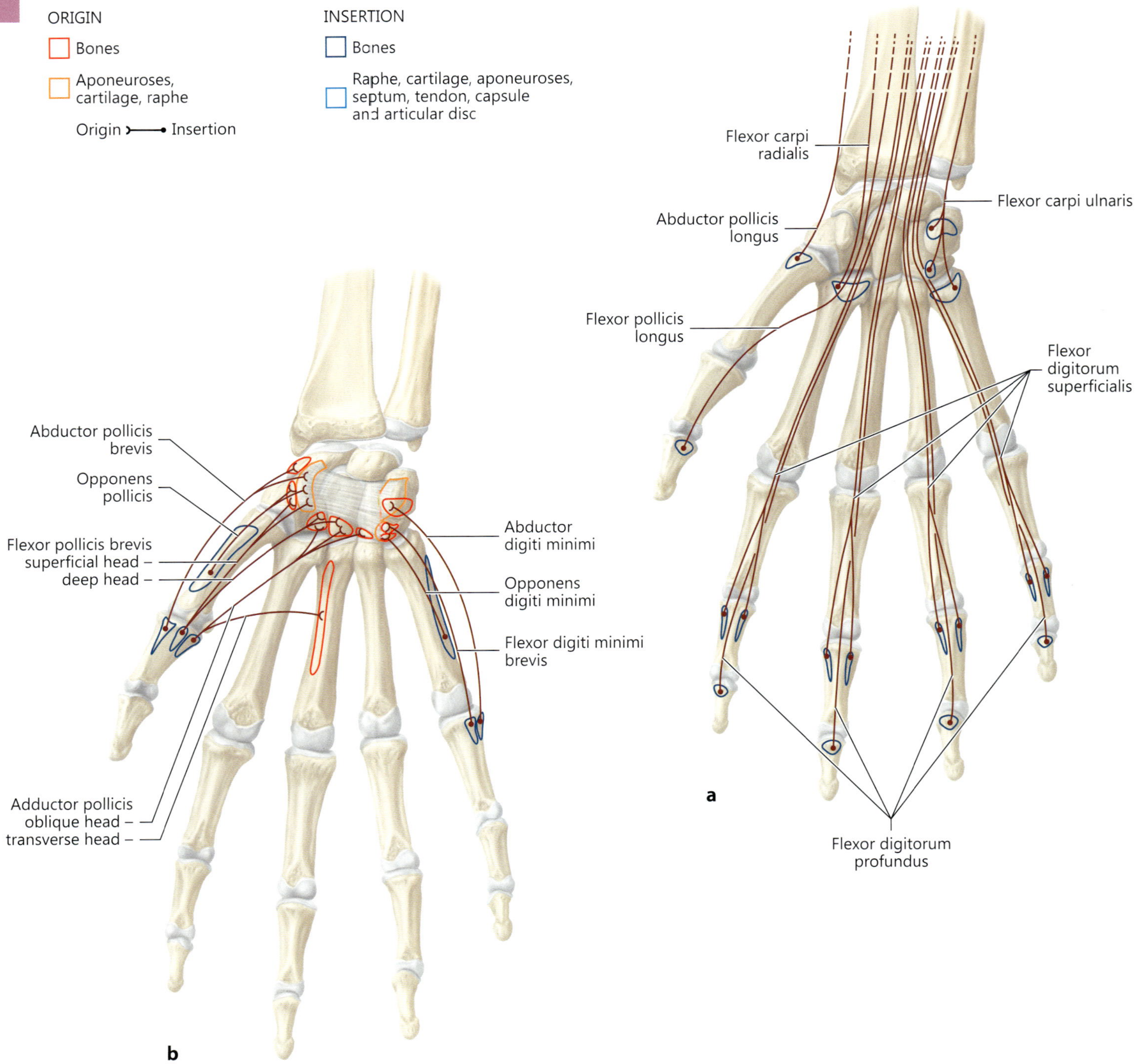

Schematic representation of the insertions of the flexor muscles of the wrist and fingers, abductor pollicis longus and flexor pollicis longus, and the origin and insertions of the intrinsic muscles of the hand: abductor, adductor and opponens, lumbricals, and palmar and dorsal interossei.

a, The **flexor carpi radialis** inserts on the second and third metacarpals, the **flexor carpi ulnaris** on the pisiform, the hamate, and the fifth metacarpal. The **flexor digitorum superficialis** and **profundus** in the middle of the forearm divide into four bellies, each of which continues into a tendon. The tendons of the flexor digitorum superficialis, after running into the carpal tunnel, divide, near the proximal phalanges, in two splits that insert into the lateral part of the anterior surface of the middle phalanges of the last four fingers. Those of the flexor digitorum profundus, after running into the carpal tunnel and crossing the split of the tendons of the flexor digitorum superficialis, insert into the volar surface of the bases of the distal phalanges of the last four fingers. The **abductor pollicis longus** inserts on the trapezium and the first metacarpal, and the **flexor pollicis longus** inserts on the distal phalanx of the thumb.

b, The **abductor pollicis brevis** runs from the flexor retinaculum and the tubercle of the scaphoid to the lateral surface of the base of the proximal phalanx of the thumb; the **opponens pollicis** from the flexor retinaculum and from the tubercle of the trapezium reaches the first metacarpal; the **flexor pollicis brevis** from the flexor retinaculum and the tubercle of the trapezium (superficial head), the capitate and the trapezoid (deep head), extends to the base of the proximal phalanx of the thumb; the **adductor pollicis** from the trapezoid, the capitate, the hamate (oblique head) and the margin of the third metacarpal (transverse head) runs to the medial surface of the base of the proximal phalanx of the thumb. The **abductor digiti minimi** from the pisiform, the pisoham-

ORIGIN

Bones

Aponeuroses, cartilage, raphe

Origin ⟶ Insertion

INSERTION

Bones

Raphe, cartilage, aponeuroses, septum, tendon, capsule and articular disc

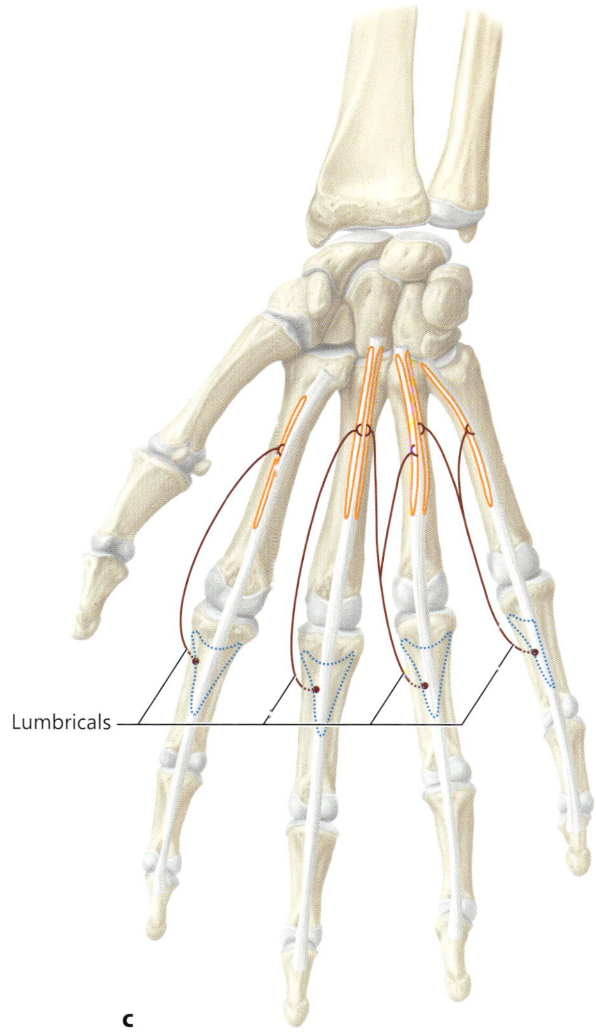

c

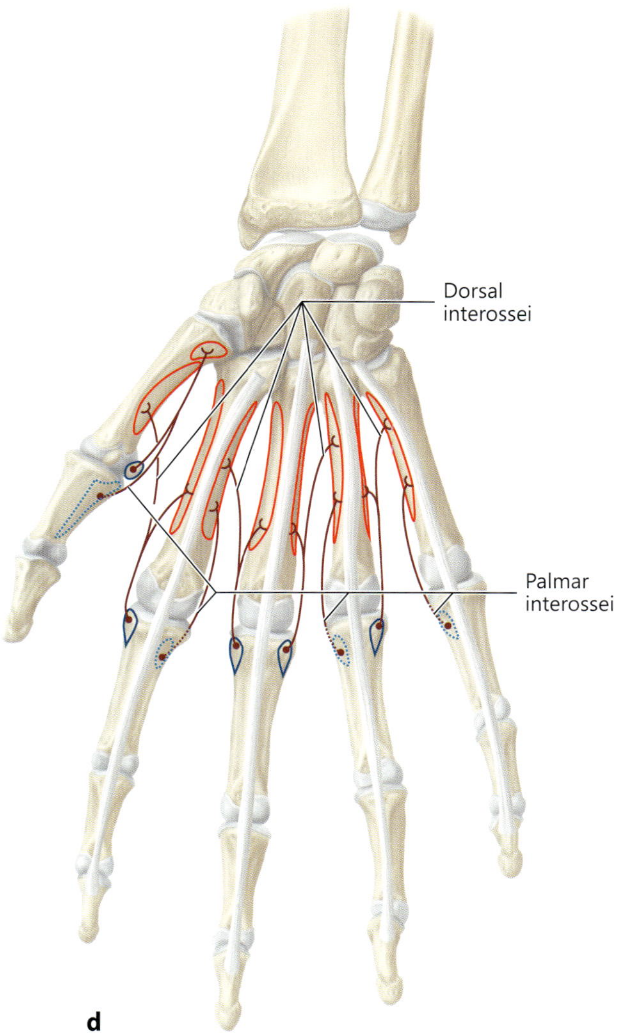

d

ate ligament, and the tendon of the flexor carpi ulnaris reaches the proximal phalanx of the little finger; the **flexor digiti minimi brevis** from the hamate and from the flexor retinaculum extends to the proximal phalanx of the little finger; the **opponens digiti minimi** from the hamate and from the flexor retinaculum reaches the fifth metacarpal.

c, The **lumbricals** originate from the tendons of the flexor digitorum profundus (the first and second lumbricals originate from the radial side and from the volar surface of the tendons that insert at the distal phalanges of the second and third finger; the third lumbrical from the adjacent sides of the tendons that insert at the distal phalanges of the third and fourth fingers; the fourth lumbrical from the adjacent sides of the tendons that insert at the distal phalanges of the fourth and fifth finger) and insert at the radial border of the dorsal digital expansion of the corresponding fingers (last four).

d, The **palmar interossei** are three or four muscles originating from the metacarpals, except for the third metacarpal, and insert at the dorsal digital expansion of the thumb, index, ring and little fingers. This first palmar interosseous, if present, also inserts on the proximal phalanx of the thumb and on the sesamoid bone located on the ulnar side of the base of the proximal phalanx of the thumb; the third or fourth palmar interosseous (according to whether the first one is present) also on the radial side of the base of the proximal phalanx of the little finger. The **dorsal interossei** originate from the adjacent surface of the first and second metacarpals (the first), second and third metacarpals (the second), third and fourth metacarpals (the third), and fourth and fifth metacarpals (the fourth), and insert on the dorsal digital expansions of the index, middle, and ring fingers. The first dorsal interosseous also inserts on the radial side of the base of the proximal phalanx of the index; the second and the third respectively on the radial side and on the ulnar side of the base of the proximal phalanx of the middle finger; the fourth on the ulnar side of the base of the proximal phalanx of the ring finger.

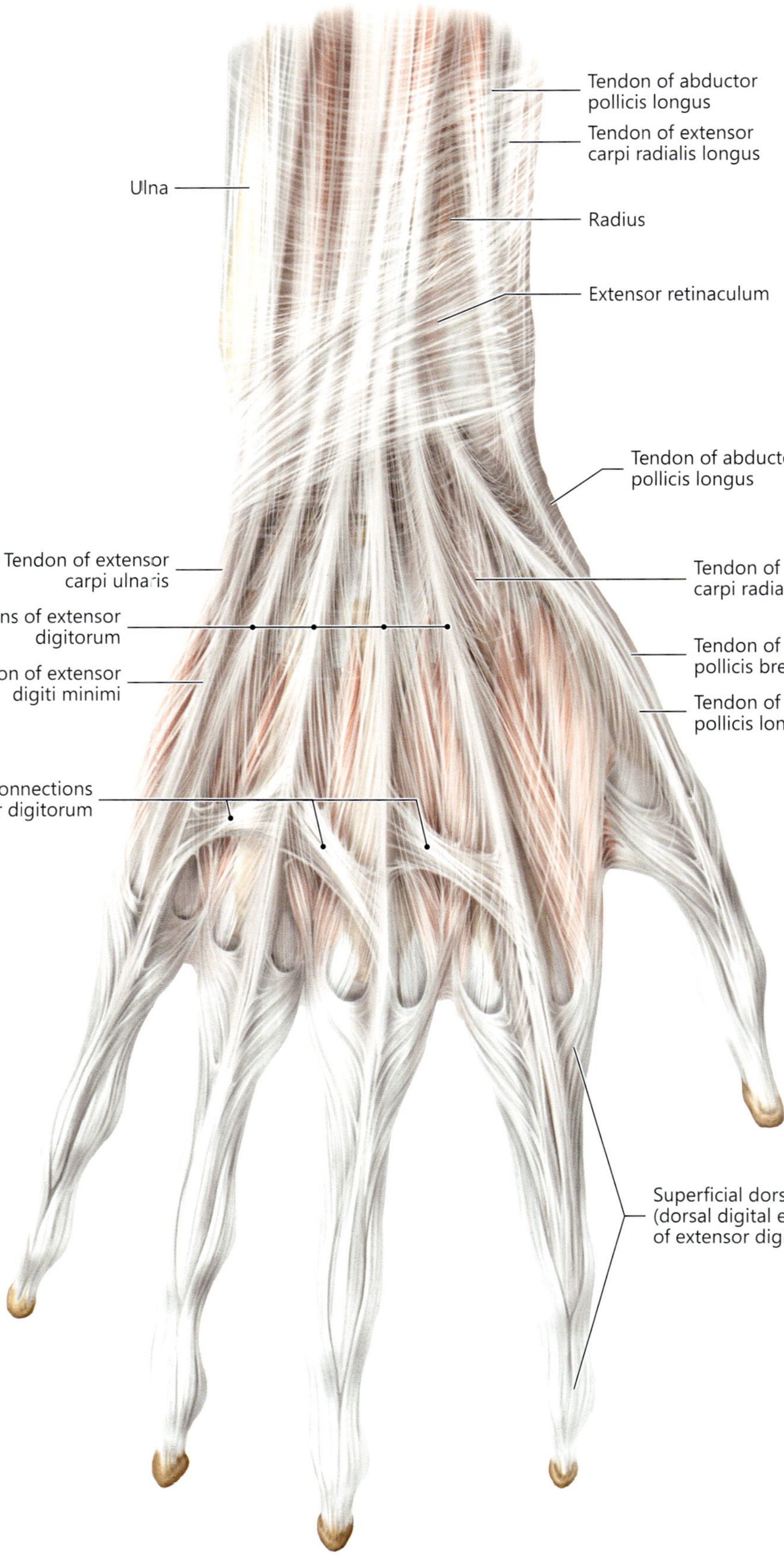

Fascial structures of the right **hand**, posterior view. We see the superficial dorsal fascia, the extension of the extensor retinaculum.

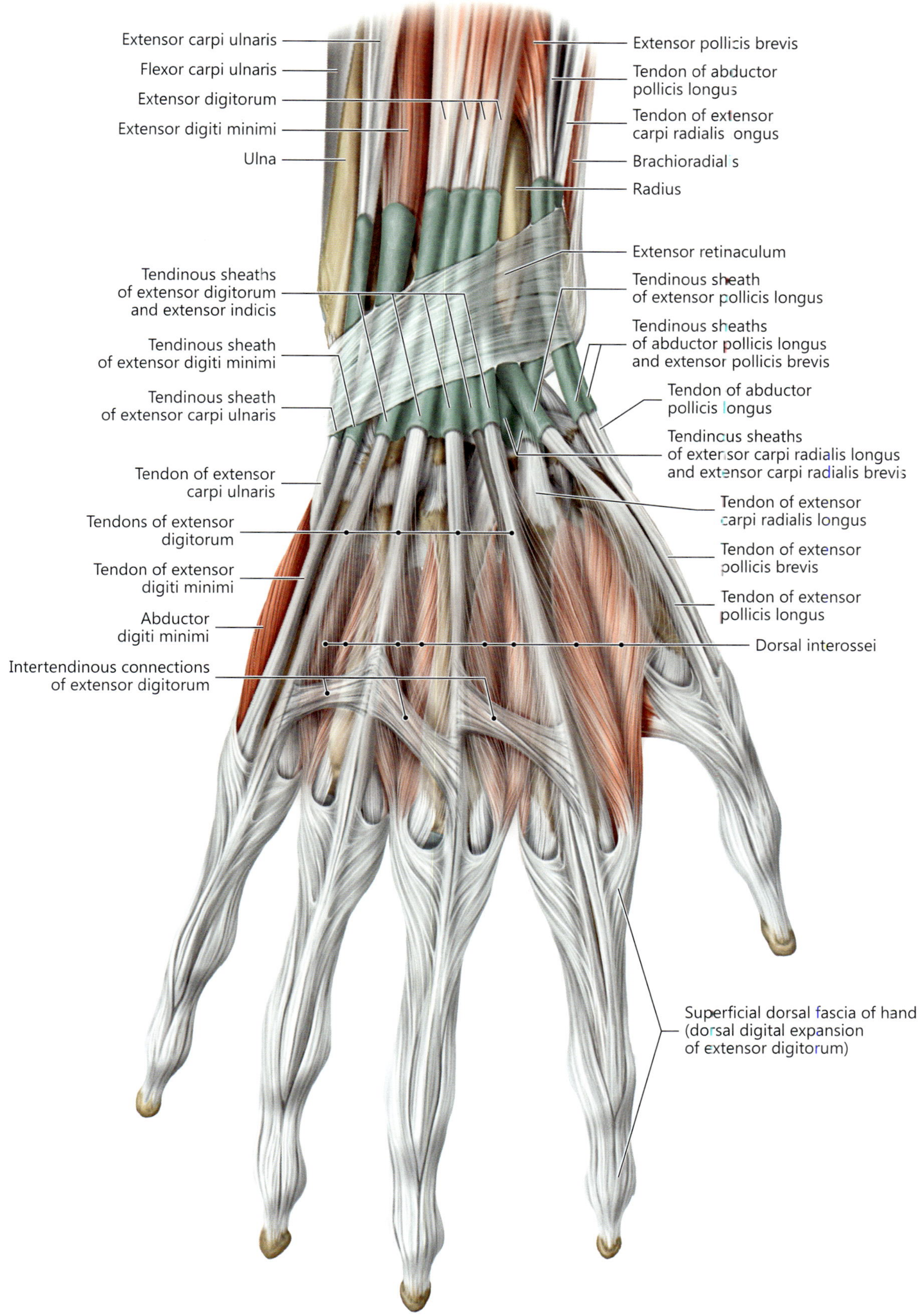

Right hand, posterior view. **Dorsal carpal synovial tendinous sheaths**, which envelop the tendons of the extensor muscles in the canals located deep to the extensor retinaculum. Inwardly, a fine layer of the dorsal fascia (deep dorsal fascia) is located between the tendons of the extensor digitorum longus and the dorsal interossei.

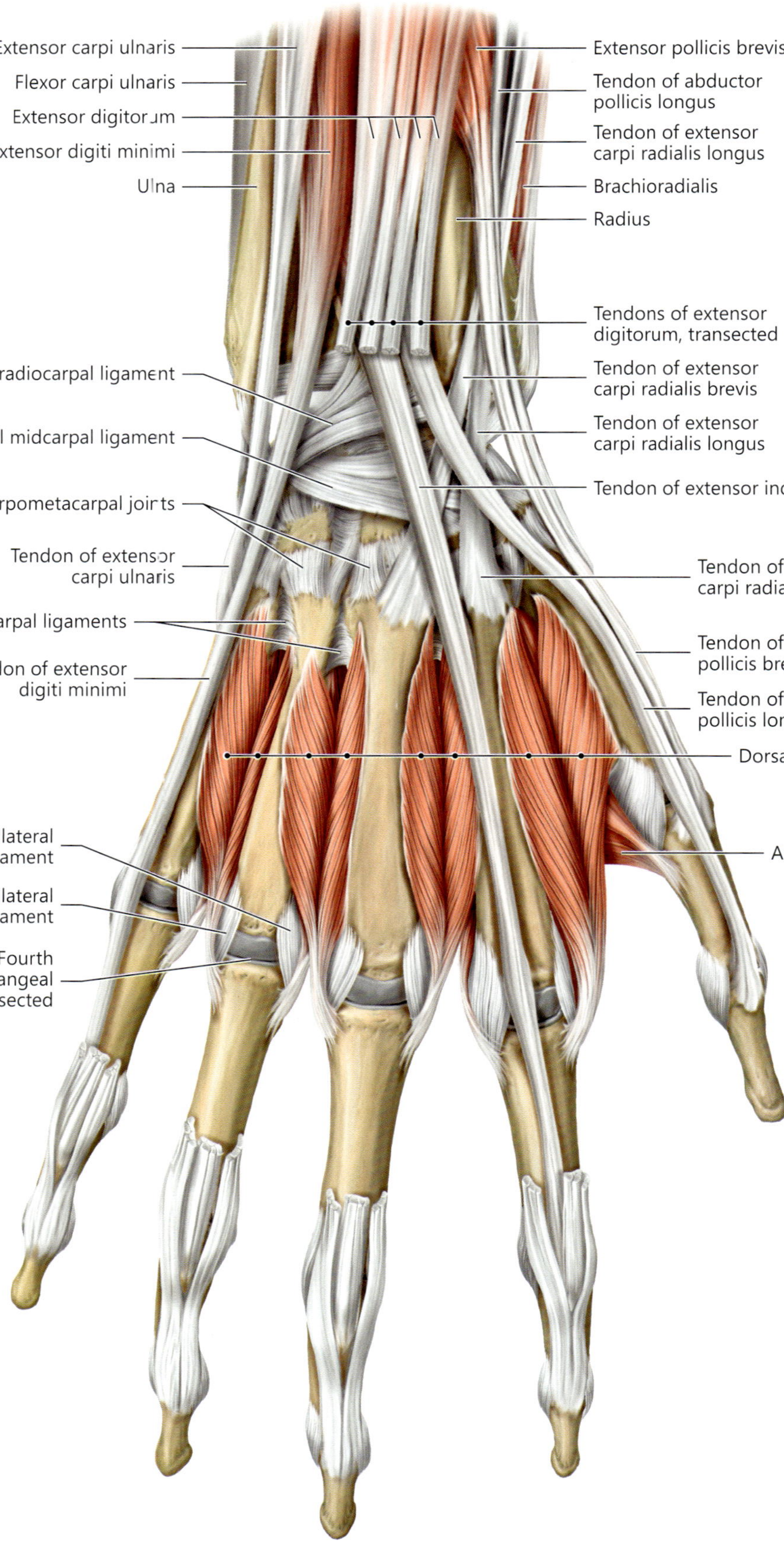

Muscles of the right **hand**, posterior view. After sectioning the tendons of the extensor digitorum, the tendons of the extensor indicis and pollicis longus appear. We can also see the tendons of the other extensor muscles: extensor carpi ulnaris, digiti minimi, carpi radialis brevis and longus, and pollicis brevis. The tendon of the extensor pollicis longus with the tendons of the abductor longus and extensor pollicis brevis delimits, on the lateral border of the dorsum of the hand, a triangular depression called the "anatomical snuffbox".

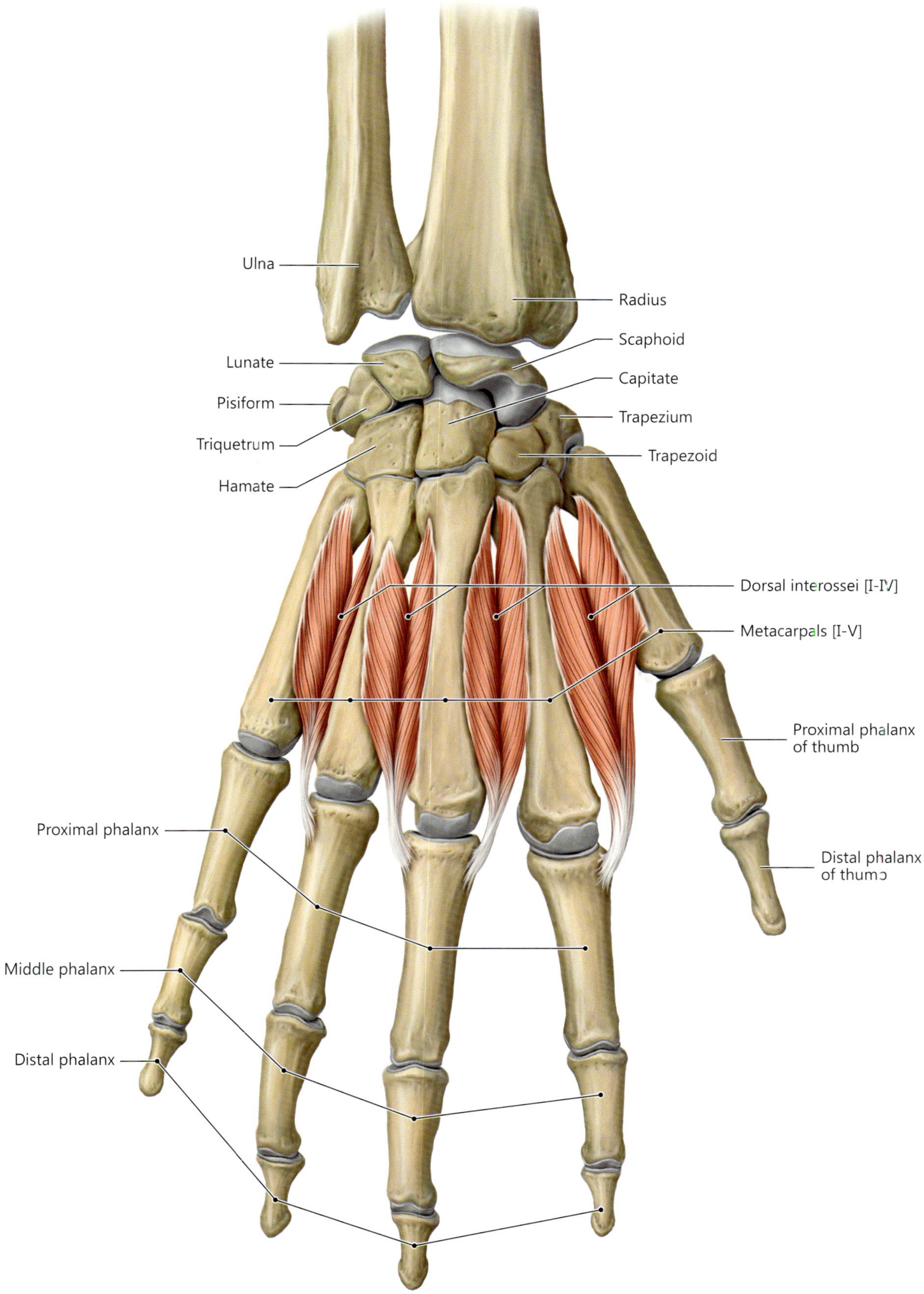

Muscles of the right **hand**, posterior view. The dorsal interossei are four flattened muscles located in the interosseous space between the metacarpals, where they occupy the two halves of the corresponding interosseous space, inserting on the two sides of the metacarpals that delimit each space.

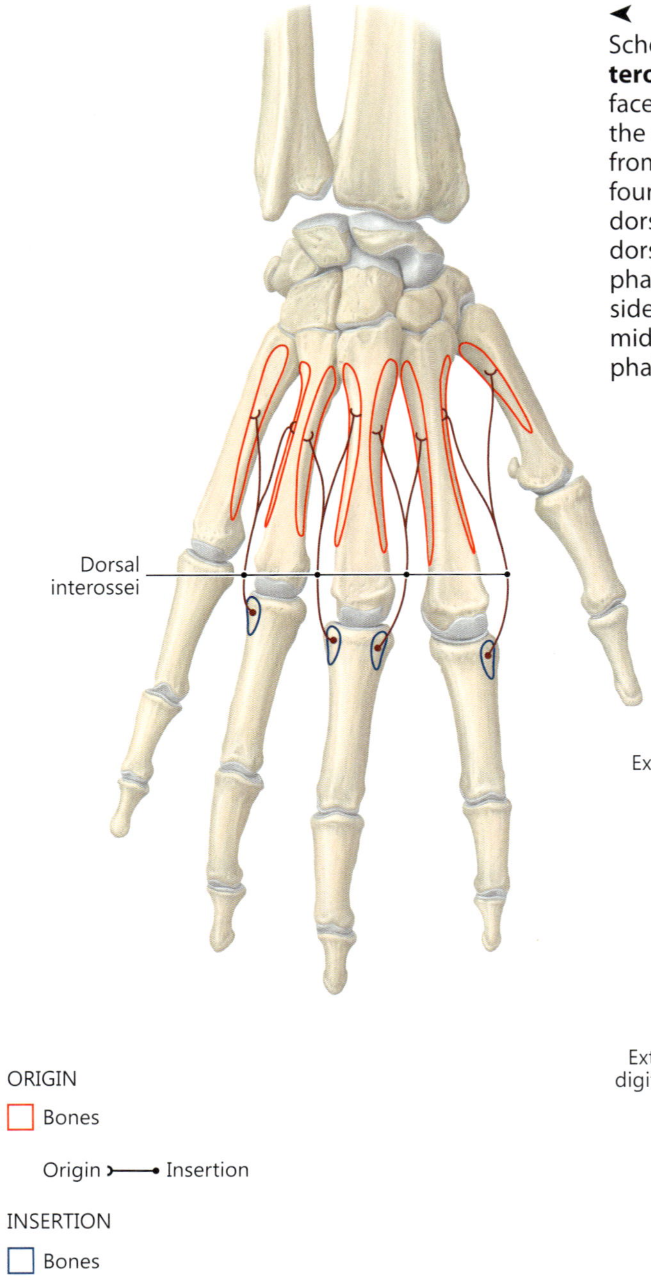

◄ Schematic representation of the origin and insertions of the **dorsal interossei**. The first dorsal interosseous originates from the adjacent surface of the first and the second metacarpals; the second originates from the adjacent surface of the second and the third metacarpals; the third from that of the third and fourth metacarpals; the fourth from that of the fourth and fifth metacarpals. In addition to the common insertion at the dorsal digital expansion of the index, middle, and ring fingers, the first dorsal interosseous inserts on the radial side of the base of the proximal phalanx of the index; the second and the third respectively on the radial side and on the ulnar side of the base of the proximal phalanx of the middle finger; the fourth on the ulnar side of the base of the proximal phalanx of the ring finger.

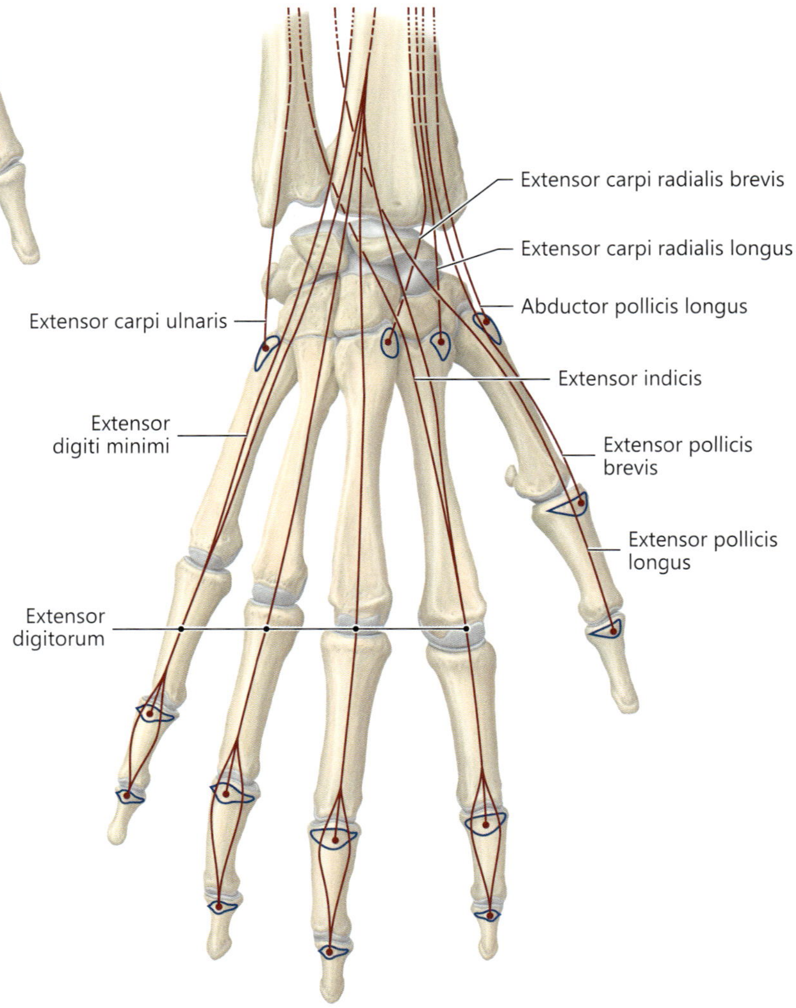

Schematic representation of the insertions of the extensor muscles of the wrist and fingers, and the abductor pollicis longus. The **extensor carpi radialis longus** originates from the lateral supraepicondylar ridge and from the lateral epicondyle. The **extensor carpi radialis brevis** originates from the lateral epicondyle and from the radial collateral ligament. The first inserts on the second metacarpal; the second inserts on the third metacarpal. The **extensor carpi ulnaris** runs from the lateral epicondyle and from the posterior surface of the ulna to the tubercle on the ulnar surface of the base of the fifth metacarpal. The **extensor digitorum** originates from the lateral epicondyle. Just below the halfway point of the forearm, the muscular belly proceeds into four tendons that run into the fourth canal of the extensor retinaculum. After reaching the proximal phalanx of the last four digits, the tendons divide into three splits: the middle split inserts at the dorsal surface of the base of the middle phalanx, the lateral and medial splits join to insert at the base of the distal phalanx of the second, third, fourth, and fifth finger. The **extensor pollicis longus** reaches the dorsal surface of the base of the distal phalanx of the thumb from the posterior and lateral surfaces of the ulna and from the interosseous membrane; the **extensor pollicis brevis** extends from the posterior surface of the radius and from the interosseous membrane to the dorsal surface of the base of the proximal phalanx of the thumb. The **extensor indicis** originates from the posterior surface of the ulna and from the interosseous membrane and connects to the ulnar side of the tendon for the index of the extensor digitorum. The **extensor digiti minimi** originates from the lateral epicondyle and attaches to the dorsal digital expansion of the little finger. The **abductor pollicis longus** reaches the trapezium and the radial surface of the base of the first metacarpal from the posterior surface of the ulna and the radius and from the interosseous membrane.

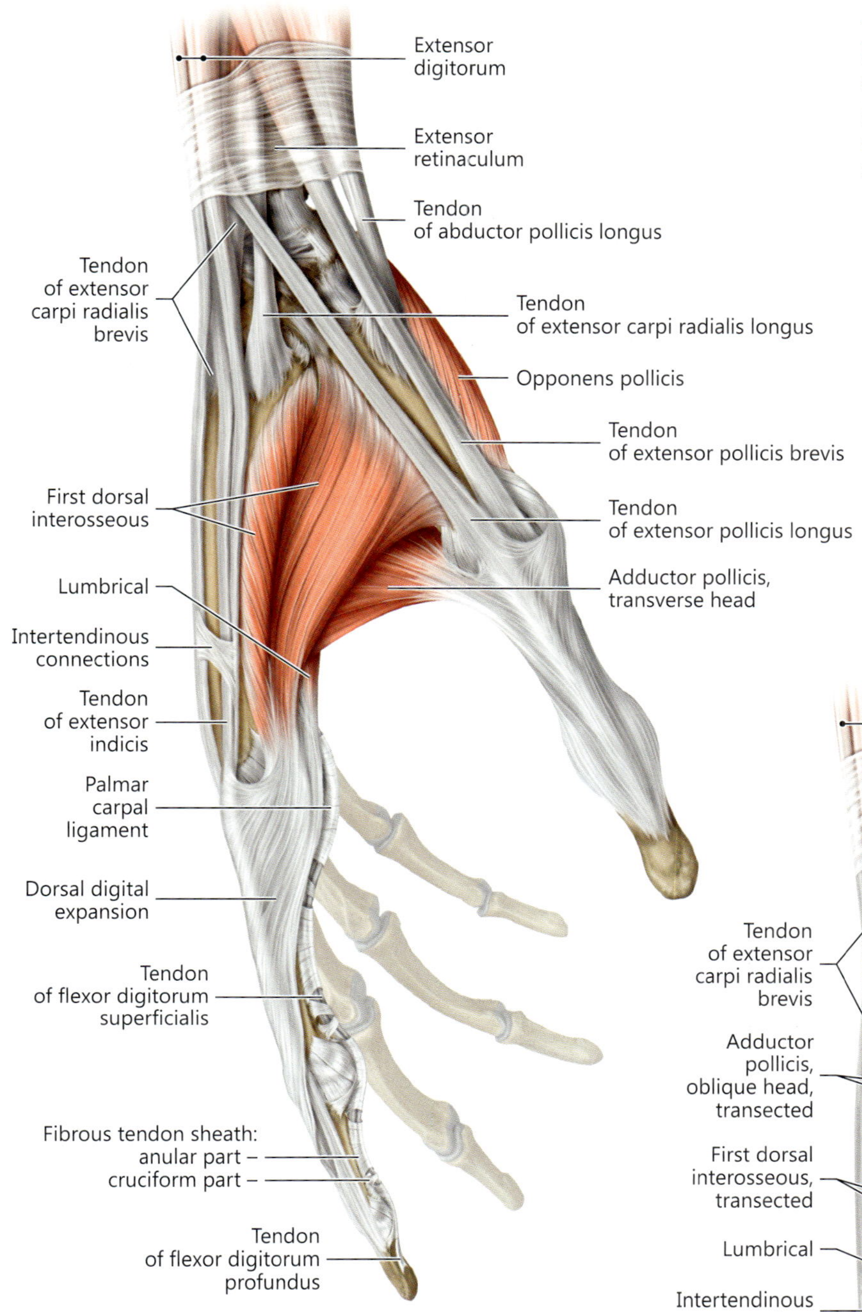

➤ Right **hand**, radial view. We see the tendons of the extensor carpi radialis longus and brevis, extensor pollicis longus and brevis, the muscles of the thenar eminence, opponens pollicis and adductor pollicis, and the first dorsal interosseous.

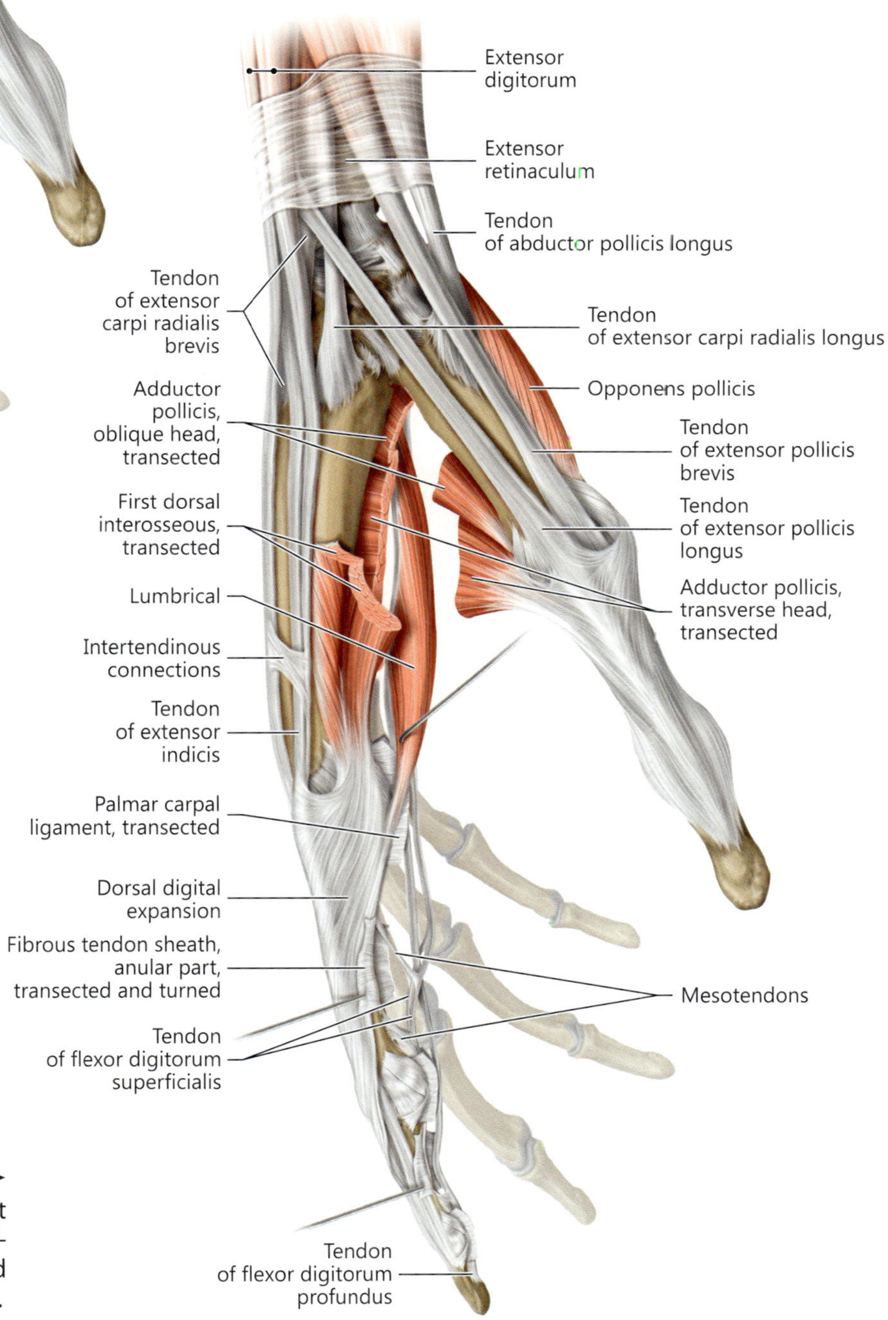

➤ Right **hand**, radial view. After resection of the first dorsal interosseous and of the two heads of the adductor pollicis, we see the first lumbrical (lowered to show its origin on the flexor profundus tendon).

Transverse section of the right upper limb at the **wrist**. The distal surface of the section has been turned upwards and backwards. We see the tendons of the extensor and flexor longi muscles and the corresponding synovial sheaths.

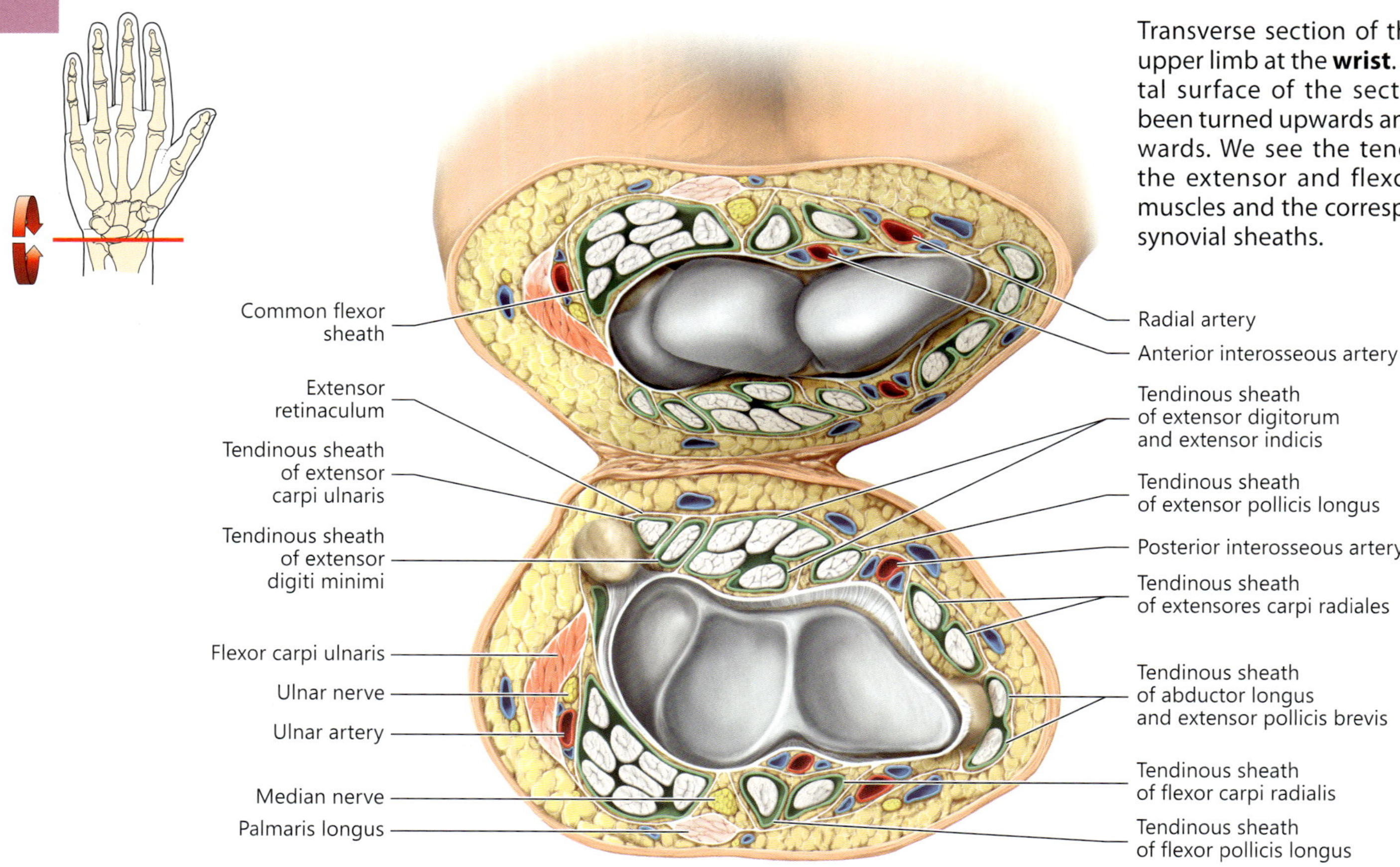

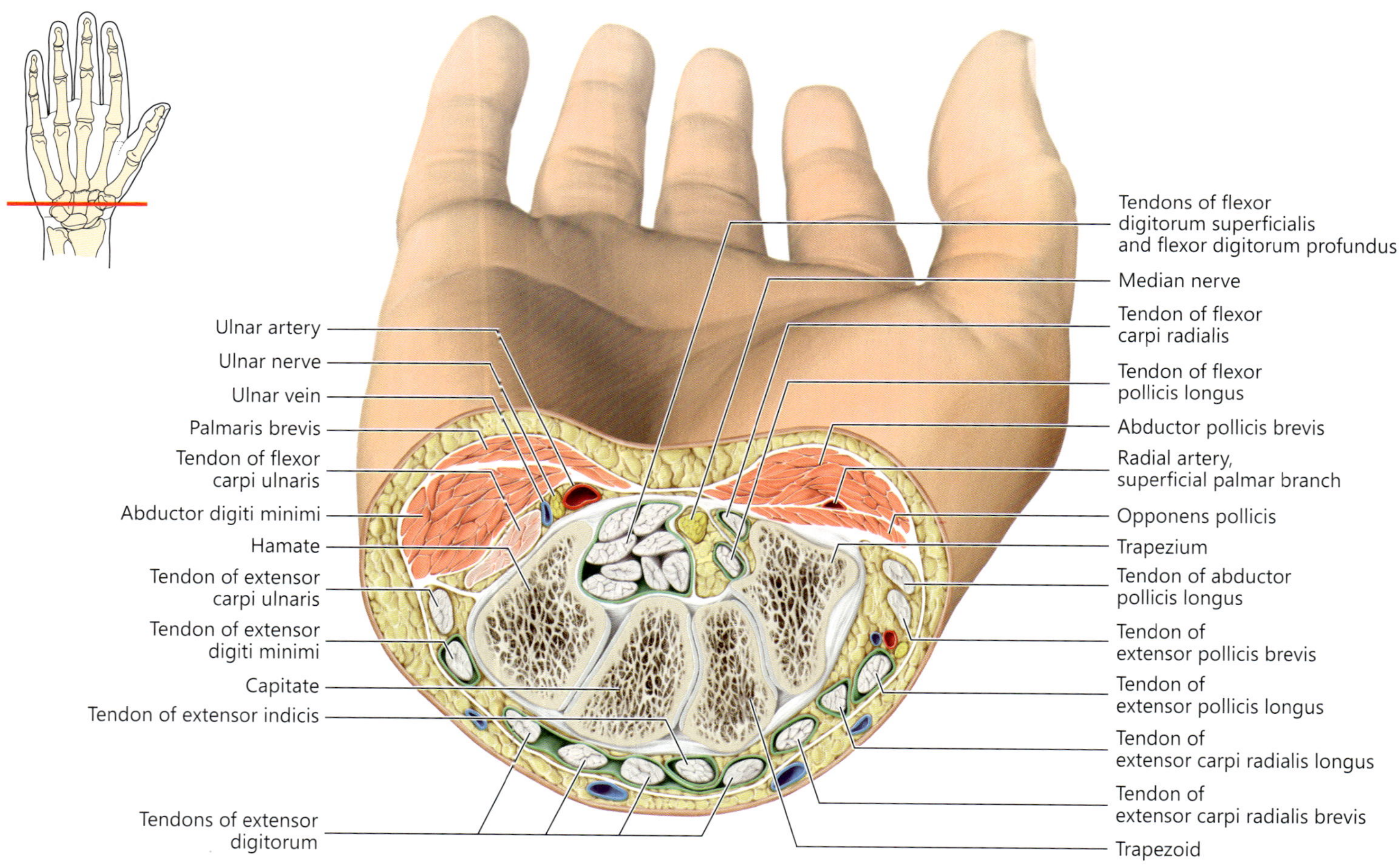

Transverse section from the right **hand** at the bones of the distal row of the wrist. We see the tendons of the extensor muscles on the dorsal surface of the wrist and those of the flexor digitorum muscles in the middle space between the palmar aponeurosis and the interosseous plane.

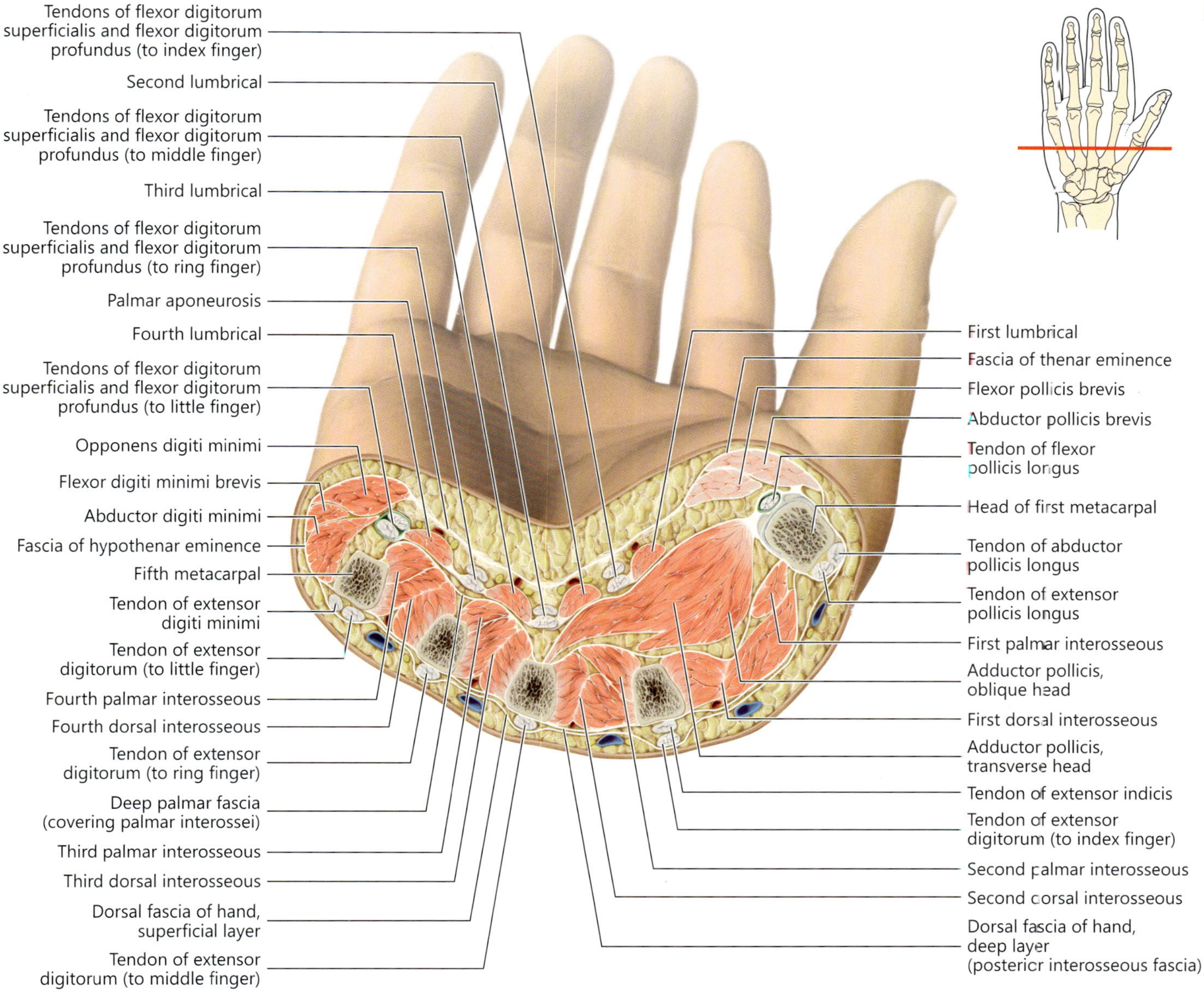

Transverse section of the right **hand** at the metacarpals. We see the three subfascial spaces of the palm of the hand, containing the muscles of the thenar eminence (lateral space), the tendons of the flexor digitorum muscles and the lumbricals (middle space), and the muscles of the hypothenar eminence (medial space).

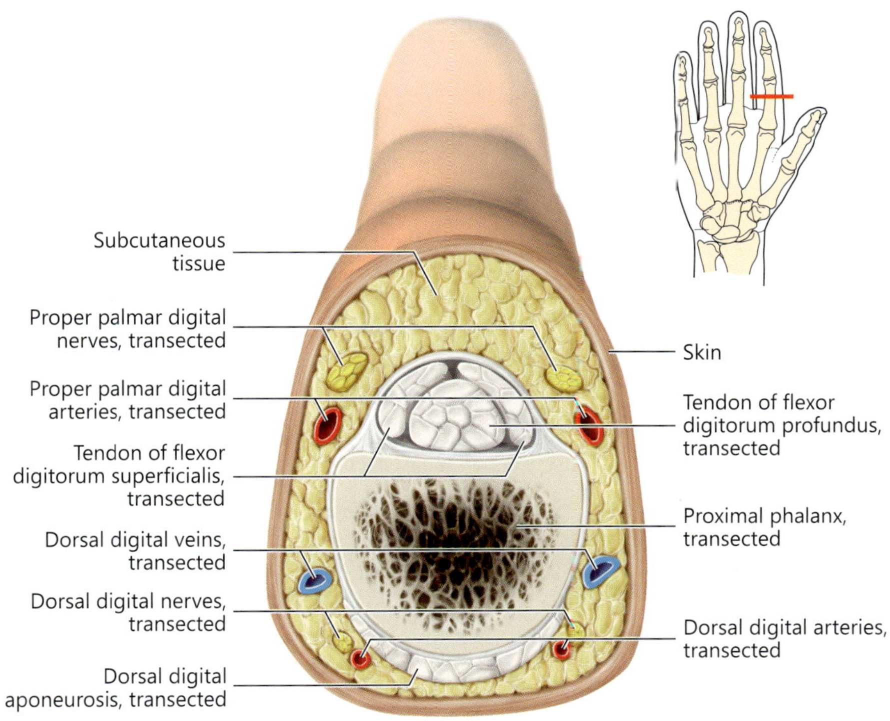

Transverse section of the second finger of the right hand at the proximal **phalanx**. The tendons of the flexor digitorum superficialis, near the proximal phalanx of the second, third, and fourth fingers, divide into two splits that insert in the lateral parts of the palmar surface of the middle phalanx. The tendon of the flexor digitorum profundus passes through the splitting of the tendon of the flexor digitorum superficialis and reaches the distal phalanx.

MUSCULAR COAPTATION OF THE SHOULDER
The muscles of the shoulder play an extremely important role, beyond just movements to include the limitation of some types of movements, intervening with a coapting action.
In fact, the periarticular muscles with a transverse direction, which are true active ligaments of the joint, ensure the coaptation of the joint surfaces, constraining the humeral head in the glenoid cavity.
When the long head of the biceps brachii contracts, the tendon inserted at the supraglenoid tubercle turns the head of the humerus inwards.
At the same time, the longitudinal muscles of the arm and the pectoral girdle impede the humeral head from dislocating below the glenoid cavity due to the traction of a weight held in the hand or solely due to the weight of the upper limb.
However, these muscles do not actively intervene except in the case of large weights (supported by the upper limb); the fundamental role of support in normal conditions is carried out not only by the coracohumeral ligament, but mainly by the lower part of the glenohumeral joint capsule.
Conversely, the upward luxation of the head, due to excessive contraction of these longitudinal muscles, is impeded and stopped by the acromiocoracoid vault, thickened by the insertion of the infraspinatus. If this is destroyed, the humeral head strikes directly upon the inferior surface of the acromion and the coracoacromial ligament, causing pain in scapulohumeral periarthritis or rotator cuff tear syndrome.

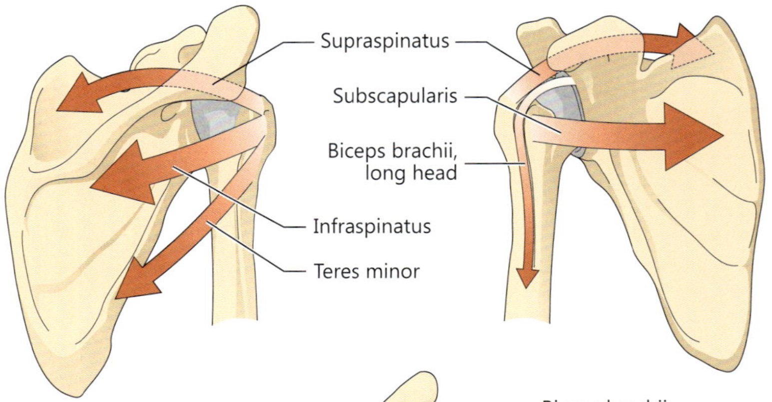

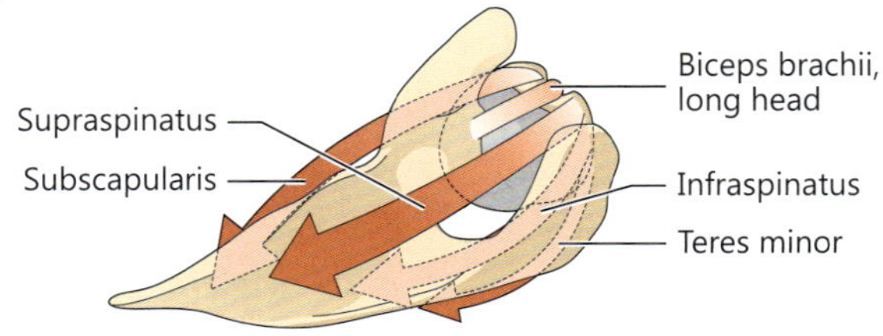

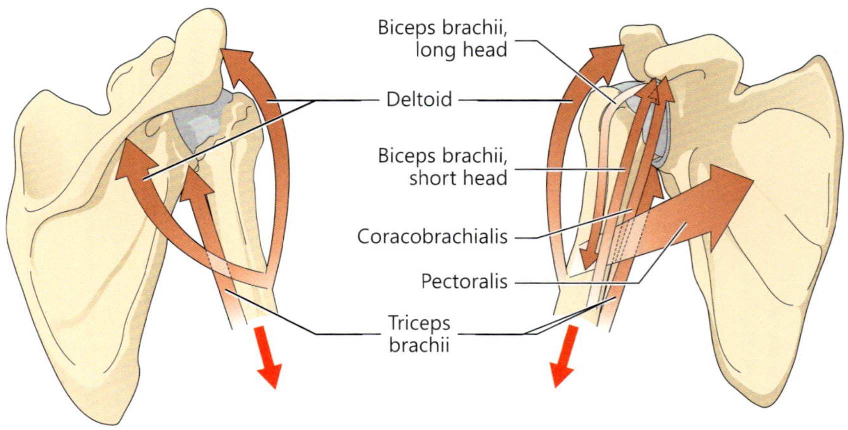

ROTATOR CUFF
Formed by these muscles:
- supraspinatus
- infraspinatus
- teres minor
- subscapularis

These muscles act synergistically (agonists), lowering the head of the humerus and bringing it closer to the glenoid cavity. The head of the humerus is thus fixed in the glenoid cavity. In reality, there is a reciprocal movement of more or less contemporary approaching of the two joint heads that leads to the centering of the head of the humerus in the glenoid cavity.

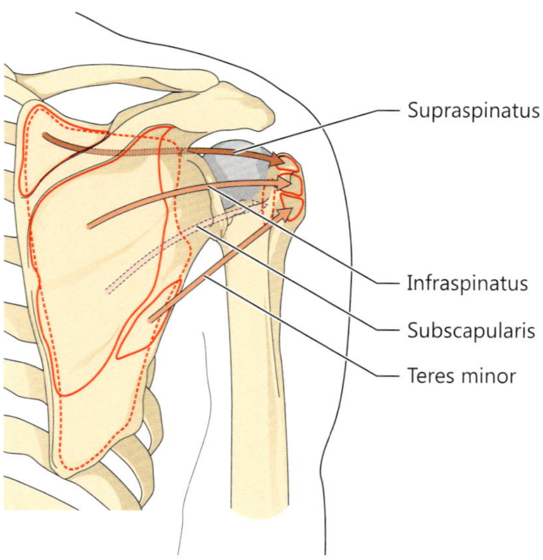

Synergistic action of the rotator cuff and the deltoid
The acromial part of the deltoid (principal shoulder abductor muscle) presents, in the initial phase of the abduction movement, a vector of force directed upwards, which the rotator muscles oppose. Their action results in a slight lowering of the head of the humerus, which antagonizes the action of the vector of force of the deltoid upwards, avoiding the compression of the head of the humerus against the roof of the glenoid cavity and favoring instead its accommodation.

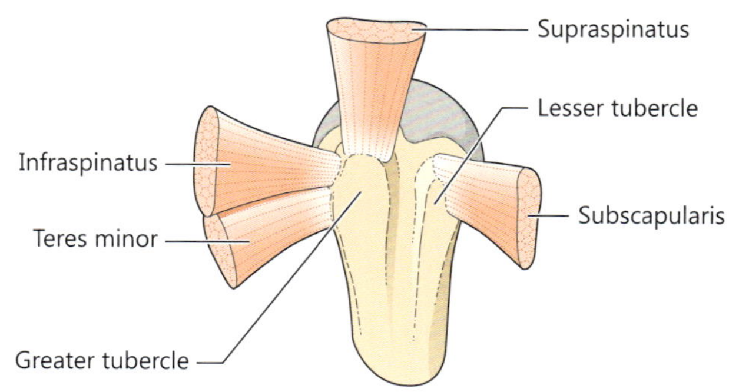

Extensor carpi radialis longus
Also an abductor of the hand.

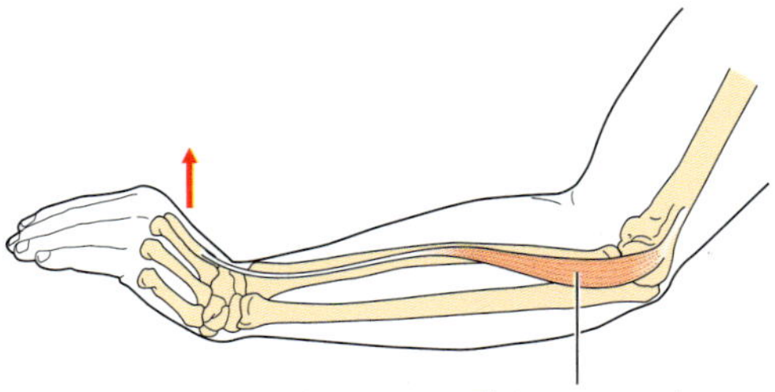

Extensor carpi radialis brevis
Extension and abduction of the hand.

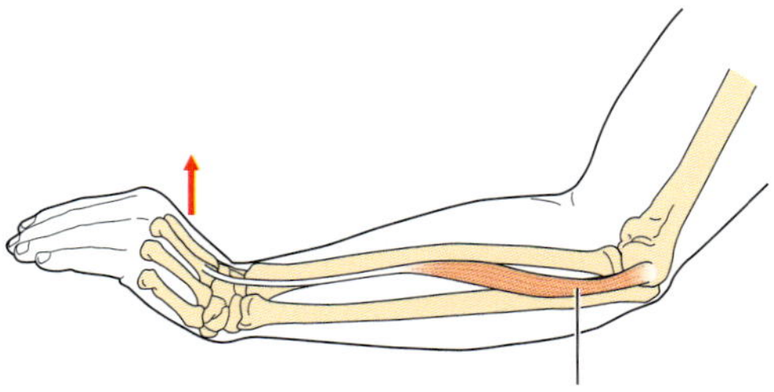

Extensor indicis
Extension of the index finger.

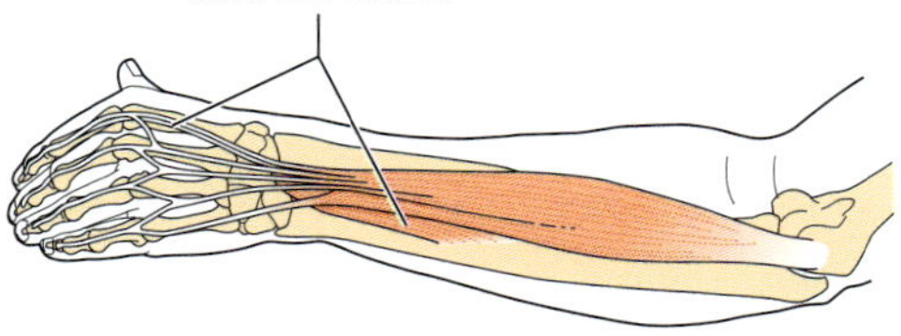

Extensor pollicis longus
Extension of the thumb.

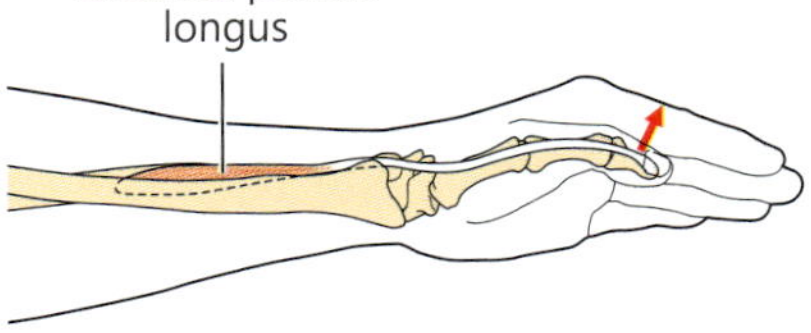

Extensor digiti minimi
Also an adductor of the hand.
Extension of the little finger.

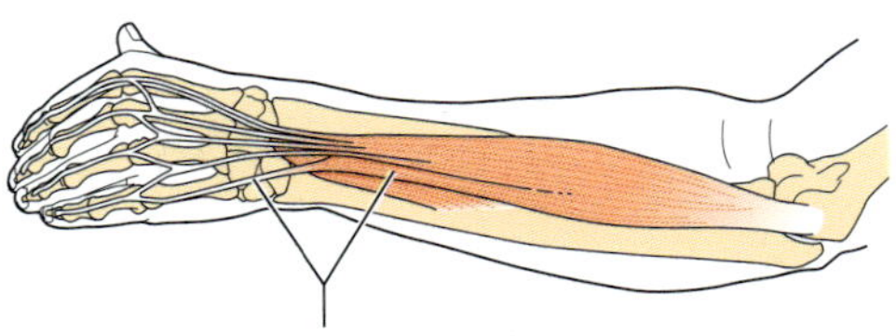

Abductor (or radial abductor) muscles
(the action of the muscles previously discussed is no further described)
(from greatest to least muscular force)

- Extensor carpi radialis longus
- Extensor carpi radialis brevis
- Abductor pollicis longus
- Extensor pollicis longus
- Flexor carpi radialis
- Flexor pollicis longus
- Extensor pollicis brevis

Extensor pollicis brevis
Extension and abduction of the thumb.

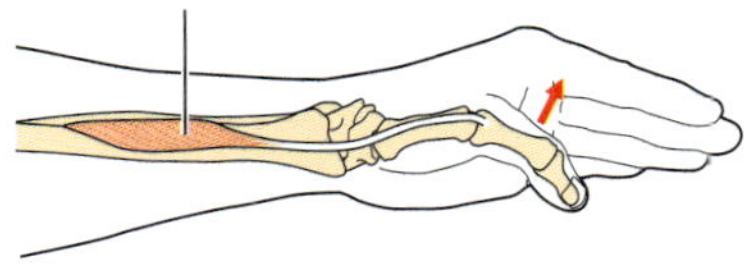

Adductor (or ulnar abductor) muscles
(the action of the muscles previously discussed is no further described)
(from greatest to least muscular force)

- Extensor carpi ulnaris
- Flexor carpi ulnaris
- Extensor digitorum
- Extensor digiti minimi

Extensor carpi ulnaris
Extensor action (dorsal flexion) at the wrist joint and flexor action (palmar flexion) at the midcarpal joint.

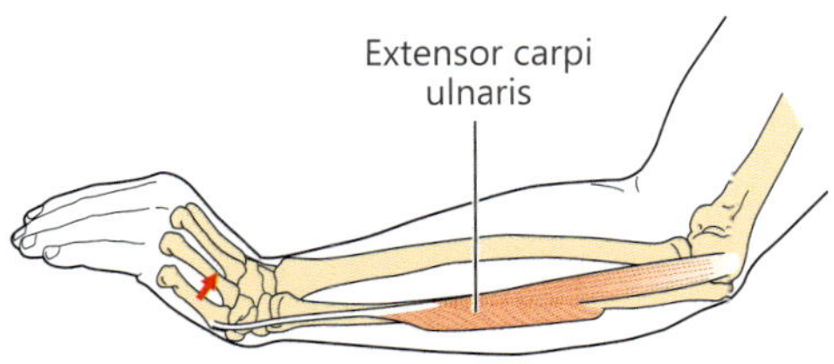

Intrinsic muscles of the hand

- Thenar eminence
- Hypothenar eminence
- Lumbricals
- Interossei:
  - palmar
  - dorsal

Muscles of the thenar eminence
Bulging in the palm of the hand laterally and in continuity with the thumb. Allow opposition movements of the thumb.

- Flexor pollicis brevis
- Abductor pollicis brevis
- Abductor pollicis
- Opponens pollicis

- *Flexor pollicis brevis*
  Flexion and adduction of the proximal phalanx of the thumb and the first metacarpal, bringing the thumb upwards, forwards and medially, in an intermediate position between flexion and adduction.

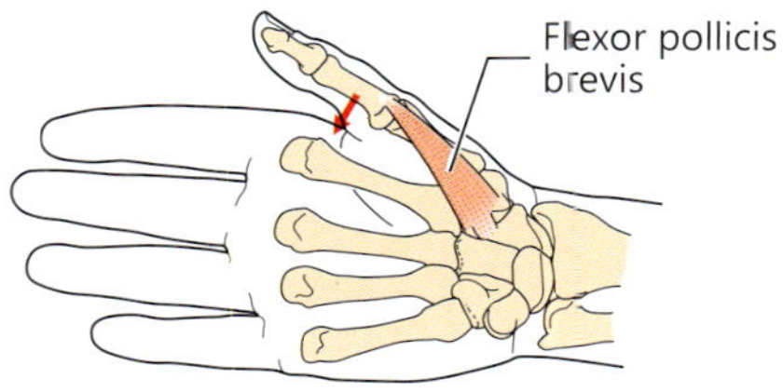

- *Abductor pollicis brevis*
  Abduction of the proximal phalanx of the thumb and the first metacarpal (which rotates internally), bringing the thumb outwards, forwards, and upwards (the movement occurs principally on the sagittal plane), in position of flexion.

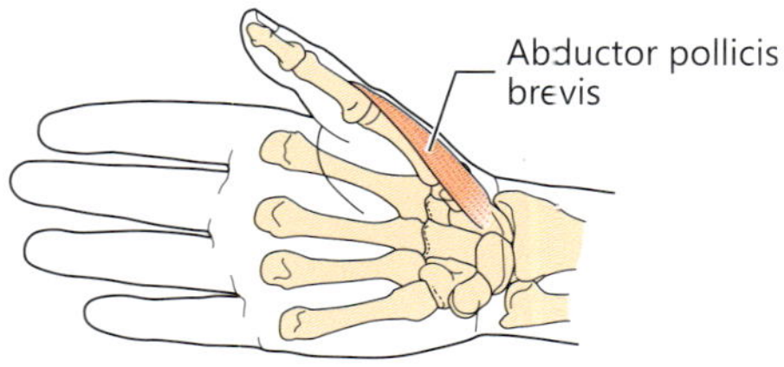

- *Adductor pollicis*
  Adduction and flexion of the proximal phalanx of the thumb and the first metacarpal, bringing the thumb forwards and inwards, contributing to the opposition movement of the thumb.

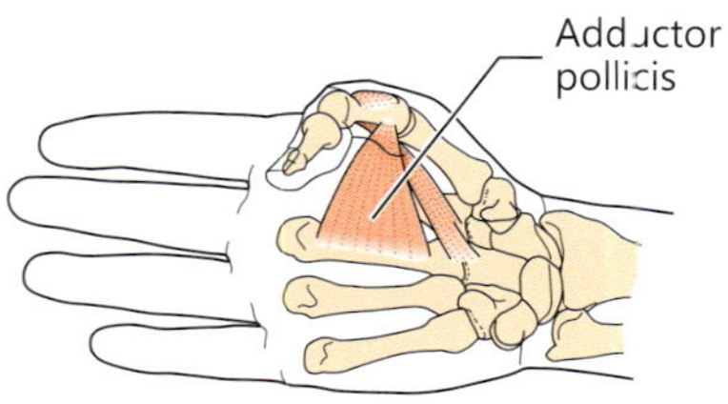

- *Opponens pollicis*
  Movement of the first metacarpal forwards and upwards and then inwards and backwards, in position of flexion-adduction or circumduction, through which the distal phalanx of the thumb opposes the other fingers.

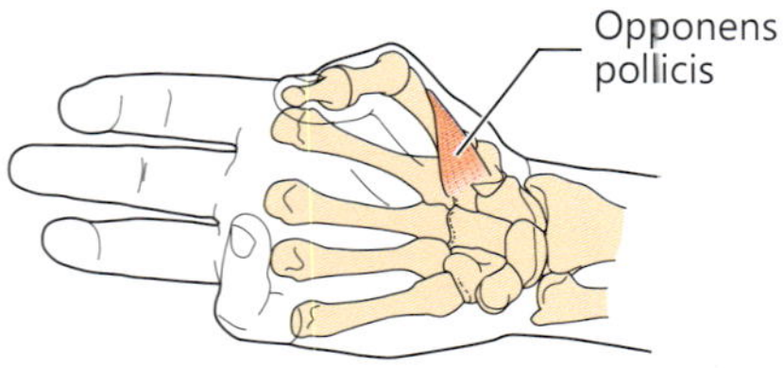

Muscles of the hypothenar eminence
Bulging in the palm of the hand in medial and proximal location. Allow opposition movements of the little finger.

- Abductor digiti minimi
- Flexor digiti minimi brevis
- Opponens digiti minimi

- *Abductor digiti minimi*
  Abduction in an ulnar direction and flexion of the proximal phalanx of the little finger.

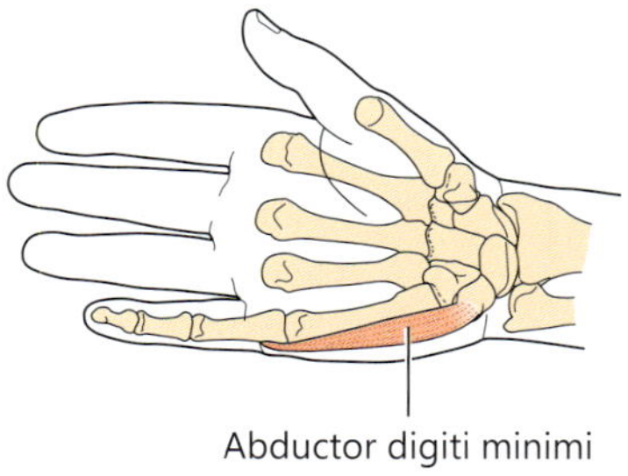

- *Flexor digiti minimi brevis*
  Flexion of the proximal phalanx of the little finger, contributing to the opposition movement of the little finger to the thumb.

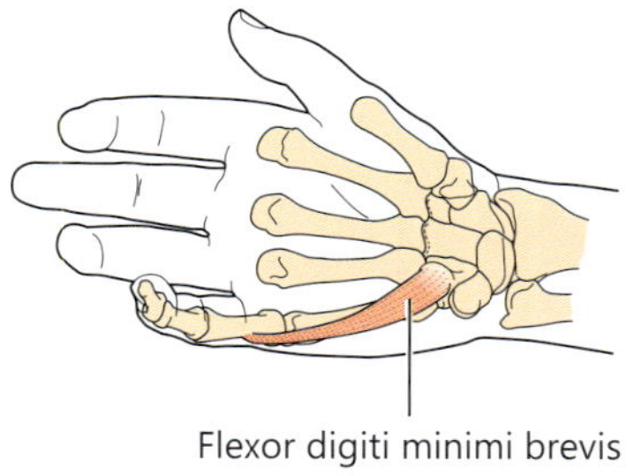

- *Opponens digiti minimi*
  Flexion and radial rotation of the fifth metacarpal, contributing to the opposition movement of the little finger to the thumb.

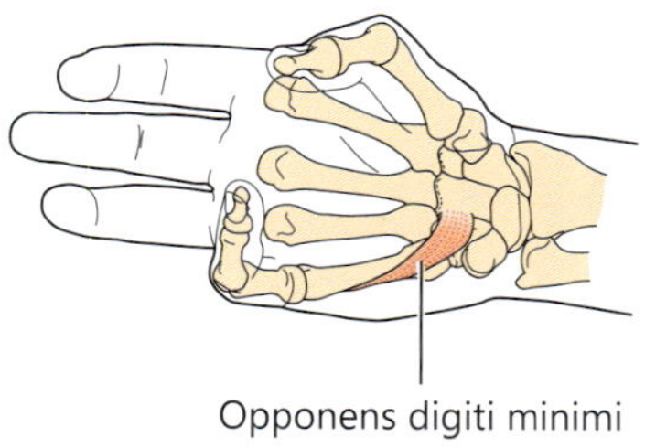

Lumbricals
Four in number, they flex the proximal phalanx and extend the middle and distal phalanges of the last four fingers.

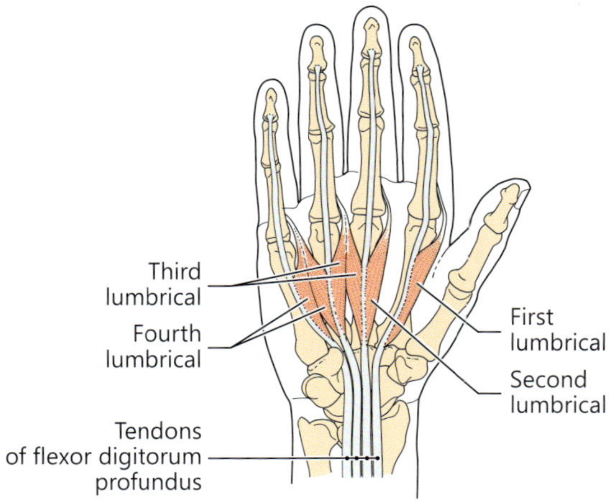

Interossei

- Palmar. Three or four in number:
  - bring index, ring, and little finger toward middle finger (adduction of fingers)
  - flex the proximal phalanx and extend the middle and distal phalanges of the index, ring, and little fingers.

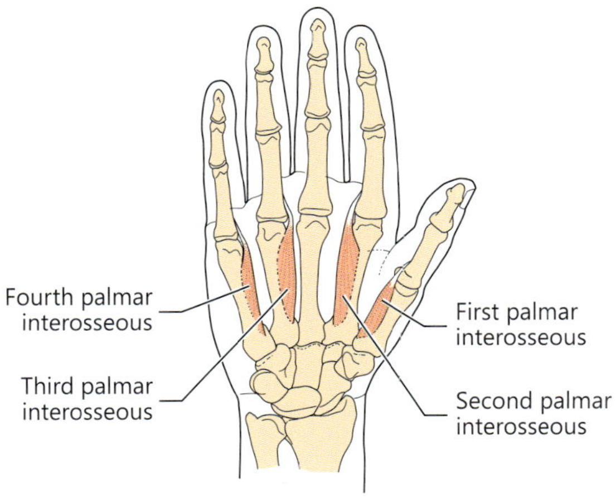

- Dorsal. Four in number:
  - separate the index and ring finger from the middle finger (abduction of the fingers)
  - abduct the middle finger in ulnar or radial direction
  - flex the proximal phalanx and extend the middle and distal phalanges of the last four fingers.

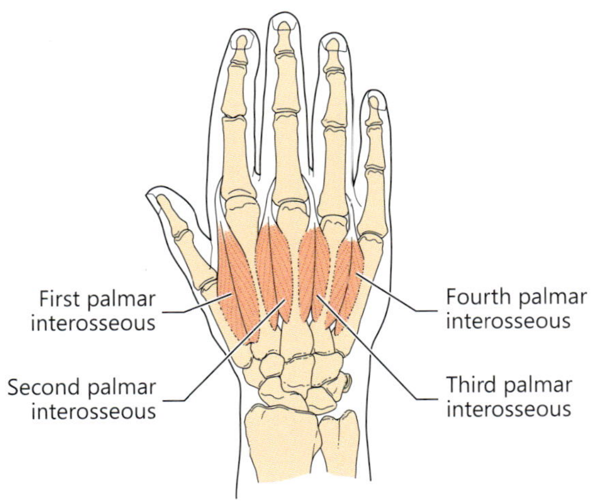

PREHENSION AND GRIP FUNCTIONS

Thumb

The thumb occupies a very particular function and position in the hand, because it is indispensable to the formation of the so-called thumb-finger pinch with each of the other fingers, particularly the index, and also to the constitution of the force grip with the other four fingers. It may thus take part in actions associated with the grips concerning the hand itself. Without the thumb, the hand loses the larger part of its functional possibilities.

Opposition of the thumb
Opposition is the essential movement of the thumb, which allows the pulp of the thumb to be brought into contact with the pulp of one of the other fingers in order to create a thumb-finger pinch. There is, therefore, not a single opposition, but a whole range of oppositions that carry out a great variety of grips and actions, following the number of fingers involved and their method of association. Without the thumb, the hand loses nearly all of its functional value to such a degree that complex surgical operations are necessary in order to reconstruct the thumb using the remaining elements; these are called, in fact, finger pollicization operations.

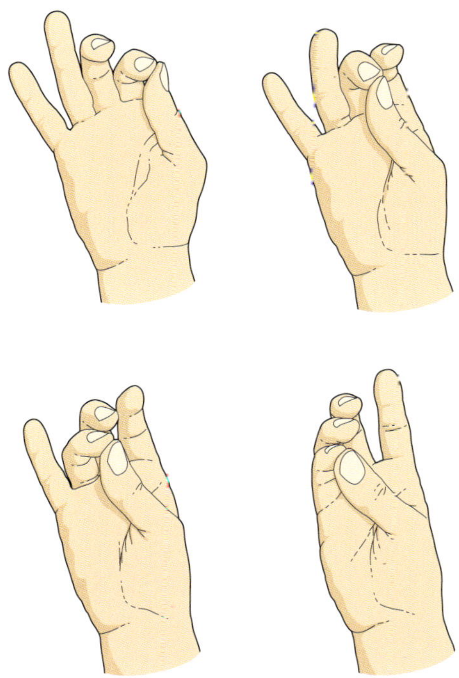

Prehension

The total of the functions brought into play when an object is grasped with the hands. These are comprised of:

- intentionality
- continuous sensory control
- gripping mechanism.

Grip

Mechanical manual component of prehension.
Grips may be classified as:

- power grip
- precision grip and fine manipulation.

Power grip

The object is grasped tightly in a vice formed by the flexed fingers and the palm of the hand, to which extra pressure produced by the thumb may be added.

Four phases are distinguished:

- opening of the hand
- placement of the fingers
- movement of the fingers towards the object
- realization of the grip.

Precision grip and fine manipulation

The object is held between the volar surface of one or more fingers and the thumb in opposition, stabilized and kept still in order to be moved via the actions of the proximal part of the upper limb.

The object may be manipulated while it is held in the hand (fine manipulation or precision dexterity).

In the precision grip and fine manipulation, three phases are recognized:

- opening of the hand
- placement of the fingers
- movement of the fingers towards the object.

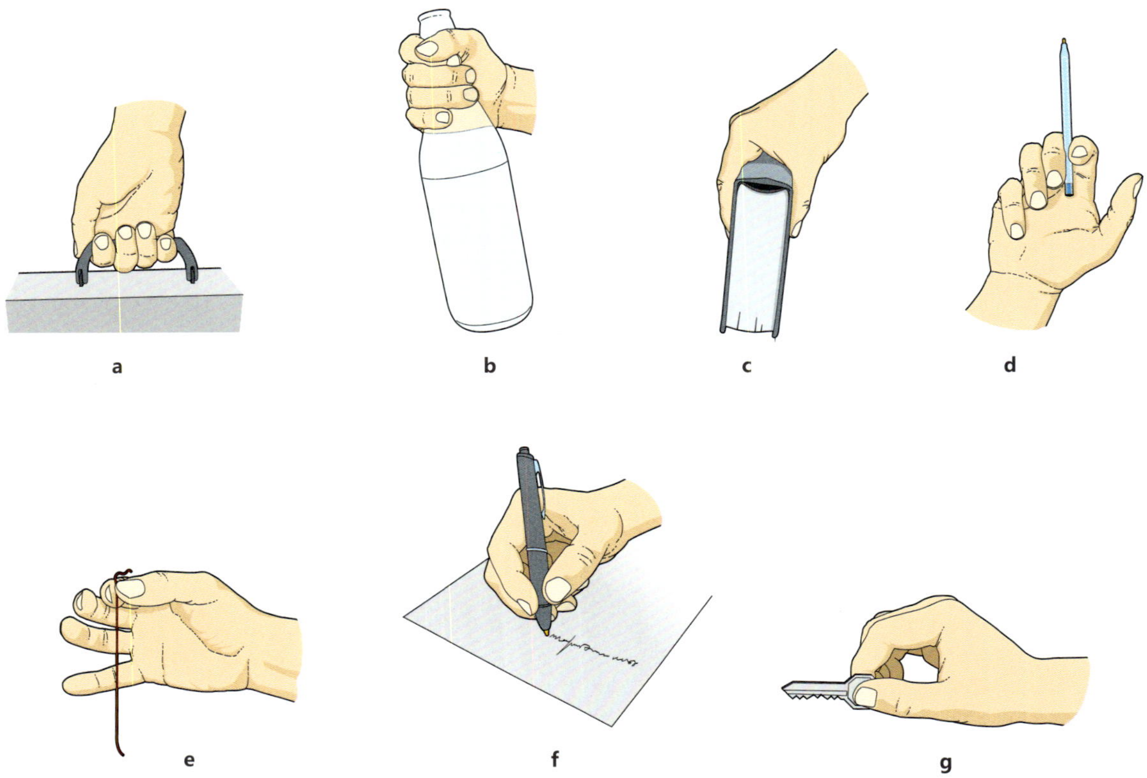

a, Hook grip (or by digital-palmar opposition); **b**, full hand palmar grip; **c**, vice grip; **d**, Interdigital pinch grip (or laterolateral interdigital grip); **e**, superior pinch (or grip by terminal opposition); **f**, grip by subterminal opposition; **g**, inferior grip (or grip by subterminolateral opposition or lateral pinch).

Extrinsic muscles of the upper limb				
MUSCLE	***ORIGIN***	***INSERTION***	***INNERVATION***	***ACTION***
		THORACOAPPENDICULAR MUSCLES		
Pectoralis major	*Clavicular head*: anterior border of the medial third of the clavicle *Sternocostal head*: medial half of the anterior surface of the sternum; cartilages of the first six or seven ribs *Abdominal part*: aponeurosis of the external oblique	*All three parts*: crest of the greater tubercle of the humerus (lateral lip of the intertubercular sulcus)	Medial and lateral pectoral nerves [C5-C8, T1]	Horizontal adduction and internal rotation of the shoulder. With fixed point on the humerus, elevation of the trunk (accessory inspiratory muscle)
Pectoralis minor	External surfaces of the ribs from the second to the fourth or from the third to the fifth in proximity to the costal cartilages; fascia of the corresponding external intercostal muscles	Medial border and superior surface of coracoid process of the scapula	Medial and lateral pectoral nerves [C5-C8, T1]	Forward displacement and medial rotation of the scapula; with fixed point on the scapula, elevation of the ribs (inspiratory muscle)
Subclavius	Medial and superior surfaces of the first rib	Inferior surface of the middle third and the lateral third of the clavicle (subclavian groove)	Subclavian nerve [C5, C6]	Depression and protraction of the clavicle
Serratus anterior	External surfaces and superior borders of the first eight, nine, or ten ribs; fascia which covers the corresponding intercostal muscles	Costal surface of the medial border of the scapula, comprising the superior angle and the inferior angle	Long thoracic nerve [C5-C7]	Abduction and external rotation of the scapula; with fixed point on the scapula, elevation of the ribs (inspiratory muscle)
	SPINOAPPENDICULAR MUSCLES (SUPERFICIAL PLANE OF THE MUSCLES OF THE BACK)			
Superficial layer				
Trapezius			Spinal root of the accessory nerve [XI] and anterior rami of the third and fourth cervical nerve [C3, C4]	With fixed point on the scapula, homolateral inclination of the head following contraction of the entire muscle; elevation of the trunk, as in climbing following bilateral contraction
	Descending part: medial third of the superior nuchal line, external occipital protuberance, ligamentum nuchae	Posterior border of the lateral third of the clavicle		Elevation of the scapula with external rotation (abduction)
	Transverse part: spinous processes from T I to T V and corresponding supraspinous ligament	Medial border of the acromion and superior border of the spine of the scapula		Adduction of the scapula
	Ascending part: spinous processes from T VI to T XII and corresponding supraspinous ligament	Tubercle at the apex of the triangular surface in correspondence to the medial extremity of the spine of the scapula		Depression of the scapula with internal rotation (adduction)
Latissimus dorsi	Posterior membrane of the thoracolumbar fascia, which originates from the spinous processes of the lumbar and sacral vertebrae; corresponding supraspinous ligament; posterior surface of the iliac crest; spinous processes of the last six thoracic vertebrae anteriorly to the insertion of the trapezius; last three or four ribs; inferior angle of the scapula	Bottom of the intertubercular sulcus of the humerus	Thoracodorsal nerve [C6-C8]	Extension, adduction, and internal rotation of the shoulder. With fixed point on the humerus, elevation of the trunk and the ribs
Deep layer				
Rhomboid minor	Lower part of ligamentum nuchae; spinous processes of C VII and T I and corresponding supraspinous ligament	Base of the smooth triangular region in correspondence to the root of the spine of the scapula	Dorsal scapular nerve [C4, C5]	Adduction and medial rotation of the scapula

continued

Extrinsic muscles of the upper limb				*(continued)*
MUSCLE	***ORIGIN***	***INSERTION***	***INNERVATION***	***ACTION***
Rhomboid major	Spinous processes from T II to T V and corresponding supraspinous ligament	Medial border of the scapula between the root of the spine and the inferior angle of the scapula	Dorsal scapular nerve [C4, C5]	Adduction and medial rotation of the scapula
Levator scapulae	Transverse processes of the first four cervical vertebrae	Medial border of the scapula between the superior angle and the origin of the spine	Dorsal scapular nerve [C3-C5]	Elevation and medial rotation of the scapula

Intrinsic muscles of the upper limb				
MUSCLE	***ORIGIN***	***INSERTION***	***INNERVATION***	***ACTION***
MUSCLES OF THE SHOULDER				
Deltoid			Axillary nerve [C5, C6]	Abduction of the shoulder up to 90° following contraction of the entire muscle; with fixed point on the humerus, elevation of the trunk, as in climbing
	Clavicular part: anterior border of the lateral third of the clavicle	*All three parts*: deltoid tuberosity on the lateral surface of the humeral diaphysis		Flexion and internal rotation of the shoulder
	Acromial part: lateral border and superior surface of the acromion			Abduction of the shoulder
	Spinal part: lower lip of the crest of the spine of the scapula			Extension and external rotation of the shoulder
Subscapularis	Medial two-thirds of the subscapular fossa	Lesser tubercle of the humerus; anterior part of the capsule of the shoulder joint	Superior and inferior subscapular nerves [C5, C6]	Internal rotation of the shoulder
Supraspinatus	Medial two-thirds of the supraspinous fossa	Superior facet of the greater tubercle of the humerus	Suprascapular nerve [C5, C6]	Abduction of the shoulder
Infraspinatus	Medial two-thirds of the infraspinous fossa	Middle facet of the greater tubercle of the humerus	Suprascapular nerve [C5, C6]	External rotation of the shoulder
Teres minor	Upper two-thirds of the lateral surface of the dorsal surface of the scapula near the lateral border	Inferior facet of the greater tubercle of the humerus	Axillary nerve [C5, C6]	External rotation of the shoulder
Teres major	Posterior surface of the inferior angle of the scapula	Crest of the lesser tubercle (medial lip of the intertubercular sulcus of the humerus)	Inferior subscapular nerve [C5, C6]	Extension and internal rotation of the shoulder
MUSCLES OF THE ARM				
Anterior muscles (in the anterior compartment of the arm)				
Biceps brachii	*Short head*: apex of the coracoid process of the scapula *Long head*: supraglenoid tubercle of the scapula	*Both heads*: radial tuberosity and antebrachial fascia (via the bicipital aponeurosis)	Musculocutaneous nerve [C5, C6]	Flexion, abduction, adduction, and rotation of the shoulder; flexion and supination of the elbow
Coracobrachialis (located deep and medially to the short head of the biceps brachii)	Coracoid process of the scapula, medially to the short head of the biceps brachii	Middle third of the anteromedial surface of the humerus	Musculocutaneous nerve [C5-C7]	Flexion and adduction of the shoulder
Brachialis (located deep to the biceps brachii)	Distal half of the anteromedial and anterolateral surfaces of the humerus	Tuberosity of the ulna	Musculocutaneous nerve [C5, C6] and radial nerve [C7]	Flexion of the elbow

continued

Intrinsic muscles of the upper limb				*(continued)*
MUSCLE	***ORIGIN***	***INSERTION***	***INNERVATION***	***ACTION***
Posterior muscles (in the posterior compartment of the arm)				
Triceps brachii	*Long head*: infraglenoid tubercle of the scapula *Lateral head*: posterior surface of the humerus between the radial groove and the insertion of the teres minor; lateral intermuscular septum of the arm *Medial head*: posterior surface of the humerus below the radial groove, between the trochlea of the humerus and the insertion of the teres minor; medial and lateral intermuscular septa of the arm	*All three heads*: apex, posterior surface and borders of the olecranon; some fibers proceed distally merging with the antebrachial fascia	Radial nerve [C6-C8]	Extension and adduction of the shoulder; extension of the elbow
Anconeus (in continuity with the inferolateral part of the medial head of the triceps brachii)	Posterior part of the lateral epicondyle of the humerus and radial collateral ligament of the elbow joint	Posterior and lateral surfaces of the olecranon	Radial nerve [C6-C8]	Extension of the elbow
MUSCLES OF THE FOREARM				
Anterior muscles (arranged in four superimposed layers, from the surface to within)				
First layer (from lateral to medial)				
Pronator teres	*Humeral head*: anterior surface of the medial epicondyle of the humerus; medial border of the humerus; brachial fascia *Ulnar head*: medial surface of the coronoid process of the ulna	*Both heads*: middle part of the lateral surface of the radius	Median nerve [C6, C7]	Pronation of the elbow
Flexor carpi radialis	Common origin of the flexor muscles from the medial epicondyle of the humerus	Volar surface of the base of the second metacarpal	Median nerve [C6, C7]	Flexion of the wrist and (radial) abduction of the hand (with contribution to the pronation of the elbow)
Palmaris longus	Common origin of the flexor muscles from the medial epicondyle of the humerus	Distal palmar surface of the flexor retinaculum; palmar aponeurosis; skin and fascia of the palm of the hand located distally and in the interdigital spaces	Median nerve [C7, C8]	Flexion of the wrist and anchoring of the fascia and the skin of the palm of the hand (with contribution to the flexion and pronation of the elbow)
Flexor carpi ulnaris	*Humeral head*: common origin of the flexor muscles from the medial epicondyle of the humerus *Ulnar head*: medial border of the olecranon and upper two-thirds of the posterior border of the ulna	*Both heads*: pisiform; hamate (pisohamate ligament), base of the fifth metacarpal (pisometacarpal ligament)	Ulnar nerve [C7, C8, T1]	Flexion of the wrist and adduction of the hand
Second layer				
Flexor digitorum superficialis	*Humeroulnar head*: common origin of the flexor muscles from the medial epicondyle of the humerus, ulnar collateral ligament of the elbow joint and medial surface of the coronoid process of the ulna *Radial head*: anterior border of the radius, from the radial tuberosity to the insertion of the pronator teres	At the mid-point of the forearm the muscle divides into four bellies, which each continues in a tendon. These, after passing into the carpal tunnel, near the proximal phalanges divide into two splits that insert at the lateral parts of the anterior surface of the middle phalanges of the second, third, fourth, and fifth fingers	Median nerve [C8, T1]	Flexion of the proximal and medial phalanges of the fingers

continued

Intrinsic muscles of the upper limb				*(continued)*
MUSCLE	***ORIGIN***	***INSERTION***	***INNERVATION***	***ACTION***
Third layer				
Flexor digitorum profundus	Upper three-quarters of the anterior and medial surface of the ulna; medial surface of the coronoid process of the ulna; upper three-quarters of the posterior border of the ulna; anterior surface of the lateral half of the interosseous membrane	At the mid-point of the forearm the muscle is divided into four bellies, which each continues in a tendon. These, after passing into the carpal tunnel and crossing the split of the tendons of the flexor digitorum superficialis, insert at the volar surface of the bases of the distal phalanges of the second, third, fourth, and fifth fingers	The lateral part of the muscle (muscular bellies for the middle and index finger) is innervated by the anterior interosseous nerve, branch of the median nerve [C7, C8]. The medial part of the muscle (muscular bellies for the little and ring fingers) is innervated by the ulnar nerve [C7, C8, T1]	Interphalangeal flexion of the fingers (above all of the distal phalanges)
Flexor pollicis longus	Anterior surface of the radius between the radial tuberosity and the pronator quadratus; anterior surface of the lateral half of the interosseous membrane; lateral surface of the coronoid process of the ulna	Volar surface of the base of the distal phalanx of the thumb	Anterior interosseous nerve, branch of the median nerve [C7, C8]	Flexion of the first metacarpal and of the proximal and distal phalanges of the thumb (with contribution to the radial abduction and flexion of the wrist)
Fourth layer				
Pronator quadratus	Distal quarter of the anterior surface and of the anterior border of the ulna	Distal quarter of the anterior surface and of the anterior border of the radius	Anterior interosseous nerve, branch of the median nerve [C7, C8]	Pronation of the forearm
Lateral muscles				
Brachioradialis (located more anteriorly)	Proximal two-thirds of the lateral border of the humerus; lateral intermuscular septum of the arm	Radial styloid process	Radial nerve [C5, C6]	Flexion of the elbow
Extensor carpi radialis longus (located posteriorly to the brachioradialis)	Lower third of the lateral supraepicondylar ridge of the humerus; common origin of the extensor muscles from the lateral epicondy e of the humerus	Dorsal surface of the base of the second metacarpal	Radial nerve [C6, C7]	Extension and radial abduction of the hand (with contribution to the flexion of the elbow)
Extensor carpi radialis brevis (situated more posteriorly)	Common origin of the extensor muscles from the lateral epicondyle of the humerus; radial collateral ligament of the elbow joint	Dorsal surface of the base of the third metacarpal	Posterior interosseous nerve, branch of the radial nerve [C7, C8]	Extension and radial abduction of the hand (with contribution to the flexion of the elbow)
Posterior muscles				
Superficial plane				
Extensor digitorum (located more laterally)	Common origin of the extensor muscles from the lateral epicondyle of the humerus	Slightly below the halfway point of the forearm, the muscular belly proceeds into four tendons that pass into the fourth canal of the extensor retinaculum. Having reached the proximal phalanx of the last four fingers, the tendons divide into three splits: the middle one inserts on the dorsal surface of the base of the middle phalanx, the lateral and medial splits join together to insert at the base of the distal phalanx of the second, third, fourth, and fifth fingers	Posterior interosseous nerve, branch of the radial nerve [C7, C8]	Extension and abduction of the fingers (with contribution to the extension and adduction of the wrist)

continued

Intrinsic muscles of the upper limb				*(continued)*
MUSCLE	***ORIGIN***	***INSERTION***	***INNERVATION***	***ACTION***
Extensor digiti minimi (located medially to the extensor digitorum)	Common origin of the extensor muscles from the lateral epicondyle of the humerus	Dorsal digital expansion* of the fifth finger	Posterior interosseous nerve, branch of the radial nerve [C7, C8]	Extension of the little finger (with contribution to the extension and adduction of the wrist)
Extensor carpi ulnaris (located more medially)	Common origin of the extensor muscles from the lateral epicondyle of the humerus; posterior border of the ulna	Tubercle on the ulnar surface of the base of the fifth metacarpal	Posterior interosseous nerve, branch of the radial nerve [C7, C8]	Extension and adduction (ulnar abduction) of the hand
Deep plane				
Supinator	Lateral epicondyle of the humerus; radial collateral ligament of the elbow joint; anular ligament of the radius; supinator crest on the shaft (body) of the ulna	Proximal third of the anterior and lateral surfaces of the radius	Posterior interosseous nerve, branch of the radial nerve [C6, C7]	Supination of the elbow
Abductor pollicis longus (located more laterally)	Posterior surface of the ulnar diaphysis distal to the insertion of the anconeus; posterior surface of the radial diaphysis distal to the insertion of the supinator; posterior surface of the interosseous membrane	Trapezium and radial surface of the base of the first metacarpal	Posterior interosseous nerve, branch of the radial nerve [C7-C8]	Radial abduction of the first metacarpal (with contribution to the supination of the elbow and to the flexion and radial abduction of the wrist)
Extensor pollicis brevis (located medially to the abductor pollicis longus)	Posterior surface of the radius below the abductor pollicis longus and posterior surface of the interosseous membrane	Dorsal surface of the base of the proximal phalanx of the thumb	Posterior interosseous nerve, branch of the radial nerve [C7, C8]	Extension and adduction of the proximal phalanx of the thumb
Extensor pollicis longus (located between the extensor pollicis brevis and extensor indicis)	Middle third of the posterolateral surface of the ulna and posterior surface of the interosseous membrane	Dorsal surface of the base of the distal phalanx of the thumb	Posterior interosseous nerve, branch of the radial nerve [C7, C8]	Extension of the thumb (with contribution to the extension of the wrist)
Extensor indicis (located more medially)	Posterior surface of the ulna distal to the origin of the extensor pollicis longus; posterior surface of the interosseous membrane	Joins the ulnar side of the tendon for the index finger of the extensor digitorum at the head of the second metacarpal	Posterior interosseous nerve, branch of the radial nerve [C7, C8]	Extension of the index finger (with contribution to the extension of the wrist)
		MUSCLES OF THE HAND		
Muscles of the thenar eminence				
Abductor pollicis brevis (located more superficially)	Flexor retinaculum; tubercles of the scaphoid and the trapezium; tendon of the abductor pollicis longus	Radial surface of the base of the proximal phalanx of the thumb; dorsal digital expansion* of the thumb	Median nerve [C8, T1]	Radial abduction of the proximal phalanx of the thumb and the first metacarpal (movement is made outwards, forwards, and upwards)
Opponens pollicis (located deep to the abductor pollicis brevis)	Flexor retinaculum; tubercle of the trapezium	Lateral surface and lateral side of the volar surface of the first metacarpal	Median nerve [C8-T1]	Opposition of the thumb (abduction, flexion, and internal rotation of the first metacarpal)
Flexor pollicis brevis (located deep to the abductor pollicis brevis)	*Superficial head*: flexor retinaculum and tubercle of the trapezium *Deep head*: capitate, trapezoid and palmar intercarpal ligaments	From the union of the two heads originates a short tendon that inserts on the radial side of the base of the proximal phalanx of the thumb	The superficial head is innervated by the median nerve [C8, T1], the deep head by the deep branch of the ulnar nerve [C8, T1]	Flexion of the proximal phalanx of the thumb and of the first metacarpal (movement is made upwards, forwards, and medially)

continued

Intrinsic muscles of the upper limb				*(continued)*
MUSCLE	***ORIGIN***	***INSERTION***	***INNERVATION***	***ACTION***
Adductor pollicis (located more deeply)	*Oblique head*: capitate and volar surfaces of the bases of the second and third metacarpals *Transverse head*: distal two-thirds of the volar surface of the diaphysis of the third metacarpal	Ulnar surface of the base of the proximal phalanx of the thumb; dorsal digital expansion* of the thumb	Deep branch of the ulnar nerve [C8, T1]	Adduction and flexion of the proximal phalanx of the thumb and the first metacarpal
Muscles of the hypothenar eminence				
Palmaris brevis (rudimentary cutaneous muscle)	Medial border of palmar aponeurosis	Deep layer of the skin of the medial border of the hypothenar eminence	Superficial branch of the ulnar nerve [C8, T1]	Corrugation of the skin of the hypothenar eminence
Abductor digiti minimi (located superficially and medially)	Pisiform; pisohamate ligament; tendon of the flexor carpi ulnaris	Ulnar surface of the base of the proximal phalanx of the little finger; dorsal digital expansion* of the little finger	Deep branch of the ulnar nerve [C8, T1]	Abduction of the little finger (with contribution to the flexion of the proximal phalanx of the thumb)
Flexor digiti minimi brevis (located superficially and laterally)	Hamate; flexor retinaculum	Ulnar surface of the base of the proximal phalanx of the little finger	Deep branch of the ulnar nerve [C8, T1]	Flexion of the proximal phalanx of the little finger
Opponens digiti minimi (located deeply)	Hamate; flexor retinaculum	Ulnar surface and adjacent volar surface of the fifth metacarpal	Deep branch of the ulnar nerve [C8, T1]	Opposition of the little finger (flexion and internal rotation of the fifth metacarpal)
Palmar muscles				
Lumbricals (located among the tendons of the flexor digitorum profundus)	Tendons of the flexor digitorum profundus: – *first and second lumbricals*: origin from radial sides and volar surfaces of the tendons that insert at the distal phalanges of the second and third fingers – *third lumbrical*: origin from the adjacent sides of the tendons that insert at the distal phalanges of the third and fourth fingers – *fourth lumbrical*: origin from the adjacent sides of the tendons that insert at the distal phalanges of the fourth and fifth fingers	Radial border of the dorsal digital expansion* of the corresponding fingers (index, middle, ring, and little fingers)	The first and second lumbricals are innervated by the median nerve [C8, T1], the third and fourth by the deep branch of the ulnar nerve [C8, T1]. The third is often innervated by the median nerve [C8, T1], instead of the ulnar nerve. Sometimes, the first and the second are innervated by the deep branch of the ulnar nerve [C8, T1], instead of the median nerve	Flexion of the proximal phalanges and extension of the middle and distal phalanges of the last four fingers
Interossei – palmar (three or four muscles located in the interosseous spaces between the metacarpals)	*First palmar interosseous* (not always present): origin from the ulnar side of the base of the first metacarpal *Second palmar interosseous*: origin from the ulnar side of the volar surface of the second metacarpal *Third palmar interosseous*: origin from the radial side of the volar surface of the fourth metacarpal *Fourth palmar interosseous*: origin from the radial side of the volar surface of the fifth metacarpal	Common insertion to the dorsal digital expansions* of the thumb, index, ring, and little fingers. The first palmar interosseous also inserts on the proximal phalanx of the thumb and on the sesamoid bone located on the ulnar side of the base of the proximal phalanx of the thumb, and the fourth also inserts on the radial side of the base of the proximal phalanx of the little finger	Deep branch of the ulnar nerve [C8, T1]	(Radial) abduction of the little finger and ring finger (bringing closer to the middle finger) and ulnar abduction (adduction) of the index finger (bringing closer to the middle finger). Flexion of the proximal phalanges and extension of the middle and distal phalanges of the index, ring and little finger

continued

Intrinsic muscles of the upper limb				(continued)
MUSCLE	ORIGIN	INSERTION	INNERVATION	ACTION
– dorsal (located in the interosseous spaces between the metacarpals, dorsal to the palmar interossei)	*First dorsal interosseous*: origin from the adjacent surfaces of the first and second metacarpals *Second dorsal interosseous*: origin from the adjacent surfaces of the second and third metacarpals *Third dorsal interosseous*: origin from the adjacent surfaces of the third and fourth metacarpals *Fourth dorsal interosseous*: origin from the adjacent surfaces of the fourth and fifth metacarpals	Common insertion to the dorsal digital expansions* of the index, middle, and ring fingers. The first dorsal interosseous inserts on the radial side of the base of the proximal phalanx of the index finger, the second and third respectively on the radial side and on the ulnar side of the base of the proximal phalanx of the middle finger, the fourth on the ulnar side of the base of the proximal phalanx of the ring finger	Deep branch of the ulnar nerve [C8, T1]	(Radial) abduction of the index finger and adduction (ulnar abduction) of the ring and little fingers (separation from the middle finger). Flexion of the proximal phalanges and extension of the middle and distal phalanges of the last four fingers

* Prolongation of the dorsal fascia of the hand onto the dorsal surface of the fingers (dorsal digital expansion of the extensor digitorum).

Fasciae and fascial structures of the upper limb

STRUCTURE	POSITION	ORGANIZATION	INSERTION	CONTINUATION
Axillary fascia	Base of the axilla	Supported centrally by the perpendicular insertion of the suspensory ligament of the axilla, anchors the skin to the floor of the axilla	Stretched as a bridge between the inferior border of the pectoralis minor anteriorly and the inferior border of the latissimus dorsi posteriorly	Clavipectoral and brachial fasciae
Suspensory ligament of the axilla	Axillary cavity (produced by the tension of this ligament)	Connection between clavipectoral fascia and axillary fascia	At right angle on the internal surface of the axillary fascia	
Deltoid fascia	Over the deltoid		Clavicle, acromion, and spine of the scapula	Pectoral, infraspinous, and brachial fasciae
Infraspinous fascia	Covers the infraspinatus, teres minor, and teres major		Borders of the infraspinous fossa	Deltoid fascia
Supraspinous fascia	Covers the supraspinatus		Scapula	
Subscapular fascia	Covers the subscapularis		Circumference of the subscapular fossa	
Brachial fascia (deep fascia of the arm)	Arm	Fine, loose sheath that forms a complete sleeve around the muscles of the arm, sending septa through the muscles	Epicondyle of the humerus and olecranon of the ulna	Axillary, deltoid, pectoral, and antebrachial fasciae
Medial intermuscular septum of the arm	Arm	Dependency of the brachial fascia: originates from its deep surface and subdivides the arm into an anterior compartment (or of the flexor muscles of the arm) and a posterior one (or of the extensor muscles of the arm)	From the medial border of the intertubercular sulcus of the humerus, along the medial supraepicondylar ridge, to the medial epicondyle	
Lateral intermuscular septum of the arm	Arm	Dependency of the brachial fascia: originates from its deep surface and subdivides the arm into an anterior compartment (or of the flexor muscles of the arm) and a posterior one (or of the extensor muscles of the arm)	From the lateral border of the intertubercular sulcus of the humerus, along the lateral supraepicondylar ridge, to the lateral epicondyle	
Antebrachial fascia (deep fascia of the forearm)	Forearm	Compact sheath that envelops, collectively and individually, the muscles of the forearm, sending septa through the muscles. Along with the interosseous membrane, it separates, in the forearm, the anterior compartment (of the flexor muscles of the forearm, superficial and deep) from the posterior one (of the extensor muscles of the forearm) and from the proximolateral one, which includes the brachioradialis, extensor carpi radialis longus and brevis	Olecranon and posterior border of the ulna	Brachial and dorsal fasciae of the hand and palmar aponeurosis
Palmar carpal ligament	Wrist	Thickening of the antebrachial fascia anteriorly on the wrist, superficially and proximally to the flexor retinaculum with which it delimits the ulnar canal	Between the pisiform and the hamate	Extensor retinaculum
Dorsal fascia of the hand	Dorsum of the hand	Extension of the extensor retinaculum (superficial dorsal fascia of the hand). Deeply, a fine layer of this fascia (deep dorsal fascia of the hand or posterior interosseous fascia) is located between the tendons of the extensor digitorum longus and the dorsal interossei	Metacarpals at the ulnar and radial borders of the dorsum of the hand	Antebrachial fascia (extensor retinaculum). Distally it continues as a dorsal aponeurosis of the fingers (dorsal digital expansion or aponeurosis)

continued

Fasciae and fascial structures of the upper limb *(continued)*

STRUCTURE	POSITION	ORGANIZATION	INSERTION	CONTINUATION
Extensor retinaculum (dorsal carpal ligament)	At the wrist level	Dorsal thickening of the antebrachial fascia formed by connective fibers distributed transversally. The deep septa identify six canals traversed by the tendons of the extensor muscles within the synovial sheaths: – first canal: sheath of the tendons of the abductor longus and extensor pollicis brevis – second canal: sheath of the tendons of the extensor carpi radialis longus and extensor carpi radialis brevis (sheath of the tendons of the extensor carpi radialis) – third canal: sheath of the tendon of the extensor pollicis longus – fourth canal: sheath of the tendons of the extensor digitorum and extensor indicis – fifth canal: sheath of the tendon of the extensor digiti minimi – sixth canal: sheath of the tendon of the extensor carpi ulnaris	Laterally at the lateral border of the radius, medially at the triquetrum and at the pisiform, in its path across the wrist at crests on the posterior surface of the radius	Palmar carpal ligament
Superficial transverse metacarpal ligament (natatory ligament)	Roots of the fingers	Thin ligament, in close contact with the subcutaneous tissue, to which longitudinal fascicles coming from the palmar aponeurosis radiate		
Palmar aponeurosis	Palm of the hand	Triangular in shape with proximal apex and distal base, it is formed by longitudinal fascicles that end at the metacarpophalangeal joints and by transverse fascicles arranged in correspondence to the base		Antebrachial fascia. It is reinforced by the tendon of the palmaris longus and adheres deeply to the flexor retinaculum
Palmar fascia	Palm of the hand	Extended continuously over the entire palmar region; can be subdivided into three parts: fascia of the thenar eminence (radial part which covers the muscles of the thenar eminence), fascia of the hypothenar eminence (ulnar part which covers the muscles of the hypothenar eminence), and the palmar fascia proper, or palmar aponeurosis (intermediate part), from the deep surface of which two fibrous septa originate which divide the palm of the hand into three subfascial spaces: lateral space, containing the muscles of the thenar eminence; medial space, occupied by the muscles of the hypothenar eminence, except for the palmaris brevis placed in the subcutaneous tissue; and middle space, in which lie the tendons of the flexor digitorum muscles and the lumbricals		
Deep palmar fascia	Palm of the hand	Covers the palmar interossei		Palmar aponeurosis
Flexor retinaculum (transverse carpal ligament)	Wrist	Ventral thickening of the antebrachial fascia, formed by connective fibers distributed transversally, which, together with the sulcus formed by the concavity of the carpal bones, form the carpal tunnel	Laterally to the tubercles of the scaphoid and the trapezium; medially to the pisiform and to the hook of the hamate	
Tendinous chiasm	Fingers	Crossing of the tendons of the flexor digitorum profundus with those of the flexor digitorum superficialis in the gap created by the decussation of the fibers of the tendons of the flexor digitorum superficialis		

continued

Fasciae and fascial structures of the upper limb				*(continued)*
STRUCTURE	***POSITION***	***ORGANIZATION***	***INSERTION***	***CONTINUATION***
Carpal tunnel	Wrist	Between the flexor retinaculum and the concavity of the carpal bones, it is subdivided into two parts, lateral and medial, by a septum that is fixed to the trapezoid and the capitate from the deep surface of the retinaculum – lateral part: sheaths of the tendons of the flexor carpi radialis and flexor pollicis longus – medial part: common sheath of the tendons of the flexor digitorum superficialis and profundus (common sheath of the tendons of the flexor muscles)		
Ulnar canal (Guyon's canal or ulnar tunnel)	Wrist	Space included between the flexor retinaculum, deeply, and the palmar carpal ligament, superficially, crossed by the ulnar artery and nerve		
Fibrous sheaths of digits of hand	Fingers	Cover the synovial sheaths of the hand fingers and are made up of a cruciform part and an anular part		Fasciae of the hand

Tendinous sheaths of the upper limb	
Intertubercular (synovial) sheath	Tendons of the long head of the biceps brachii
Carpal (synovial) sheaths	
– palmar	Tendons of the flexor muscles in the carpal tunnel
– dorsal	Tendons of the extensor muscles in the canals situated deep to the extensor retinaculum
Fibrous sheaths of digits of hand	☞ Fasciae and fascial structures of the upper limb
Synovial sheaths of digits of hand	
– vincula tendinum	Folds of the synovial membrane which give passage to the blood vessels that reach the tendons
– – breves	Triangular bands, two in number for each finger: one connects the deep surface of the tendon of the flexor digitorum superficialis, near the insertion, to the proximal interphalangeal joint and to the head of the proximal phalanx; the other connects the deep surface of the tendon of the flexor digitorum profundus, near the insertion, to the distal interphalangeal joint and to the head of the middle phalanx
– – longi	Filiform fascicles: the fascicle of the tendon of the flexor digitorum superficialis, usually double, affixes from the deep surface of the tendon to the proximal extremity of the proximal phalanx and to the metacarpophalangeal joint; the fascicle of the tendon of the flexor digitorum profundus originates from the deep surface of the tendon, after the crossing in the tendinous chiasm, and runs to the adjacent surface of the tendon of the flexor digitorum superficialis

Muscles of the hip, **the thigh**, and **the leg**. Right side, anterior view.

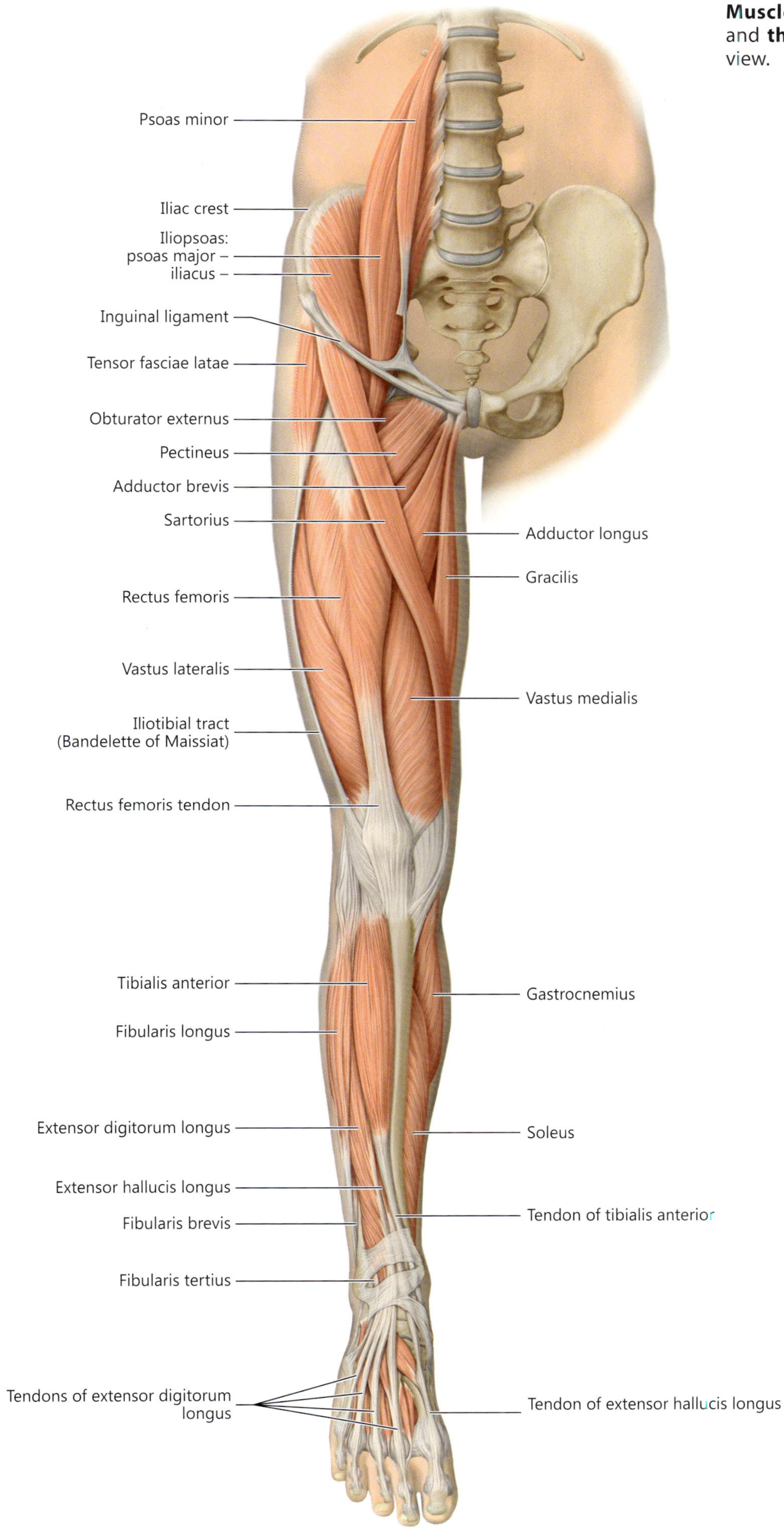

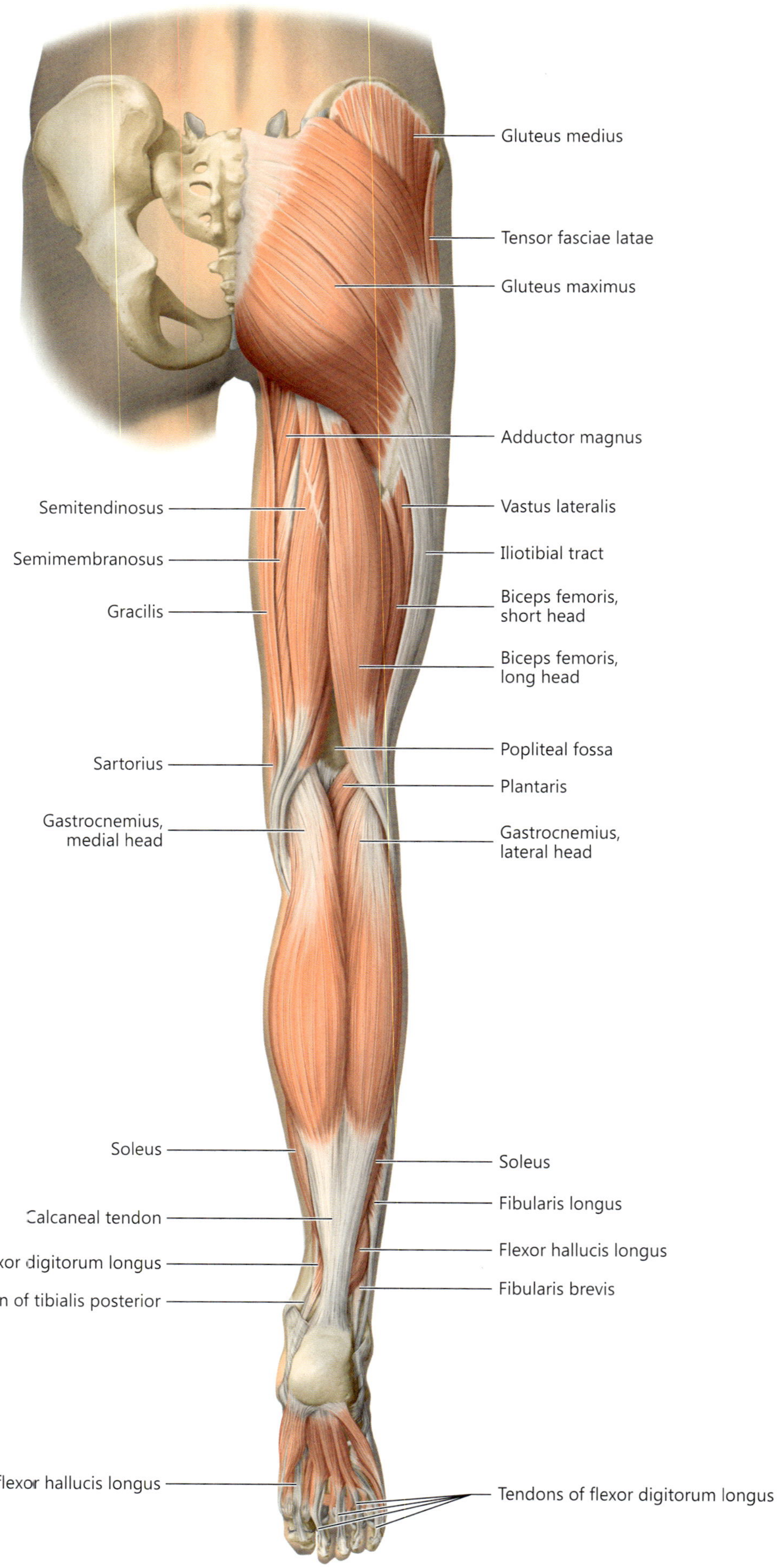

Muscles of the hip, **the thigh**, and **the leg**. Right side, posterior view.

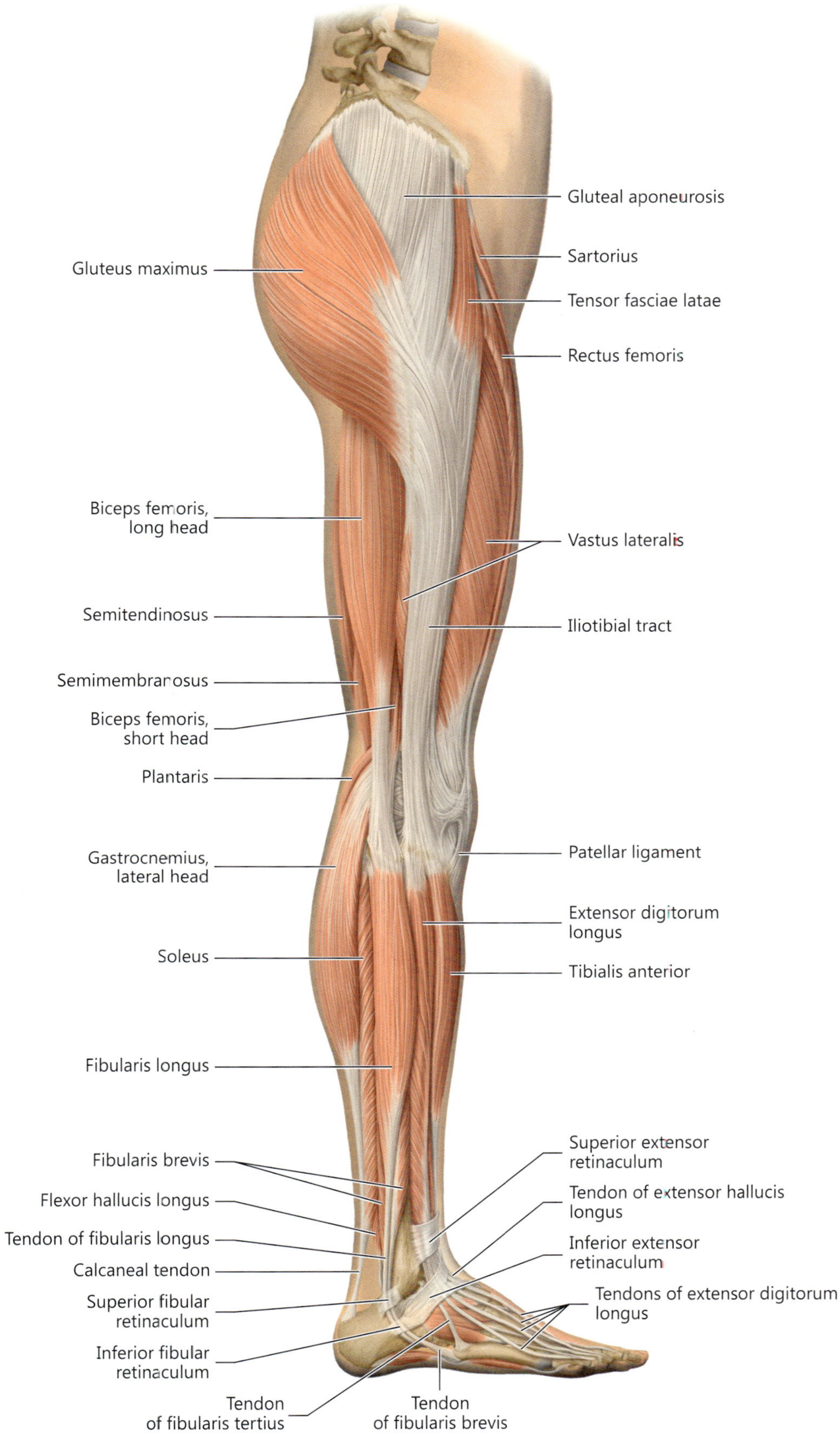

Muscles of the hip, **the thigh**, and **the leg**. Right side, lateral view.

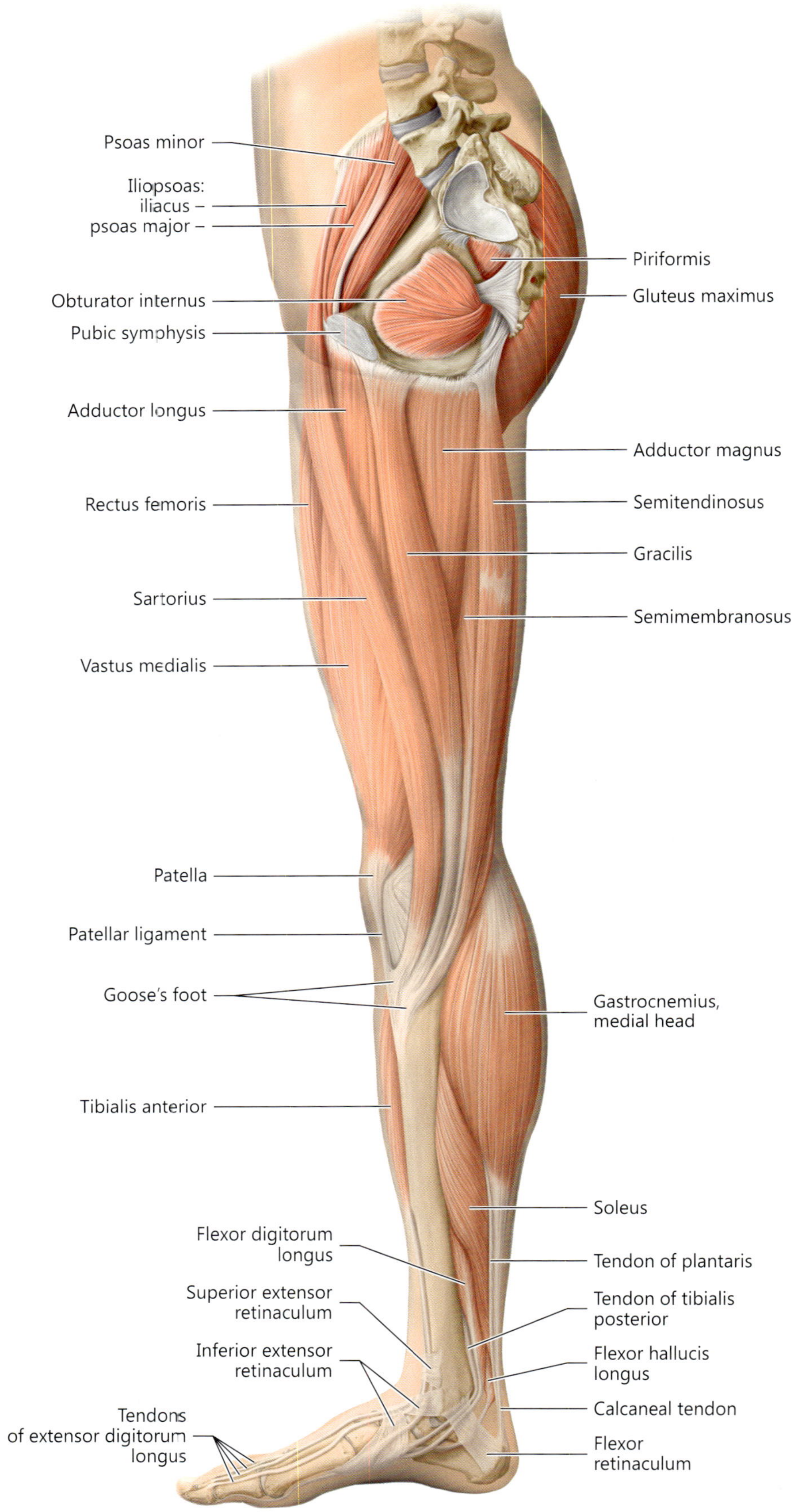

Muscles of the hip, **the thigh**, and **the leg**. Right side, medial view.

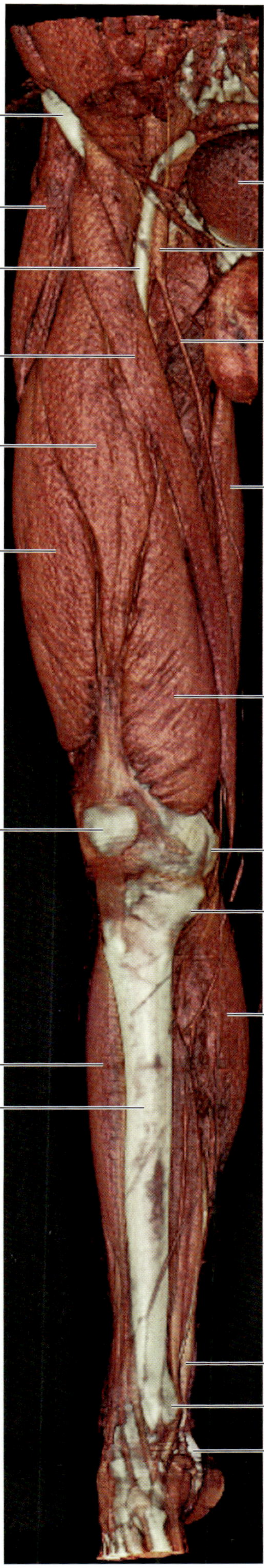

Three-dimensional image of a living human body, obtained by CT scan using a 3D volume rendering technique, showing the superficial plane of the **muscles of the thigh** and **of the leg** viewed from an anteromedial projection.

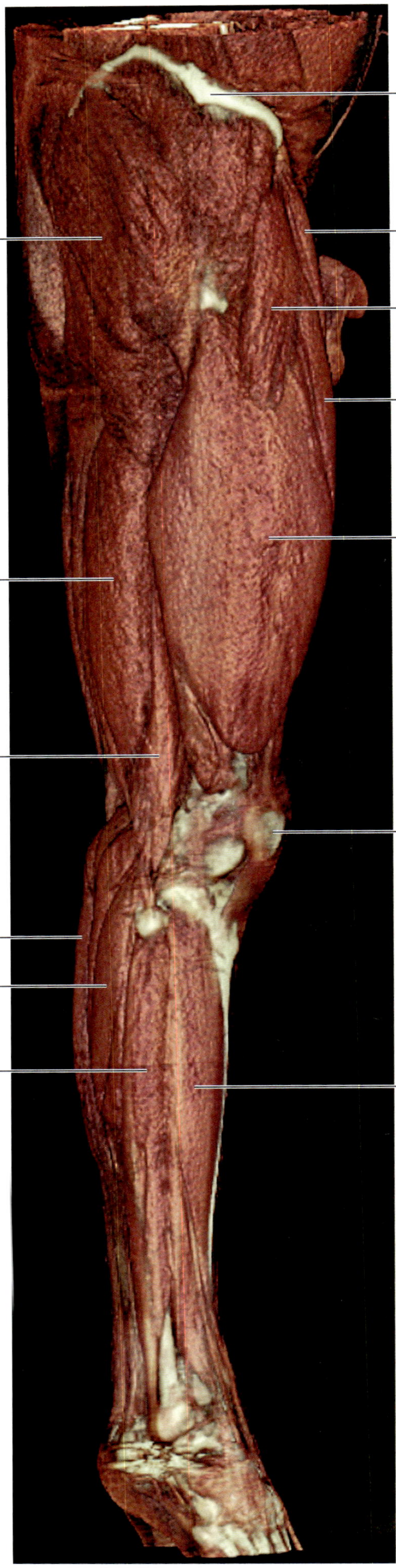

Three-dimensional image of a living human body, obtained by CT scan using a 3D volume rendering technique, showing the **muscles of the thigh** and **of the leg** in a lateral projection.

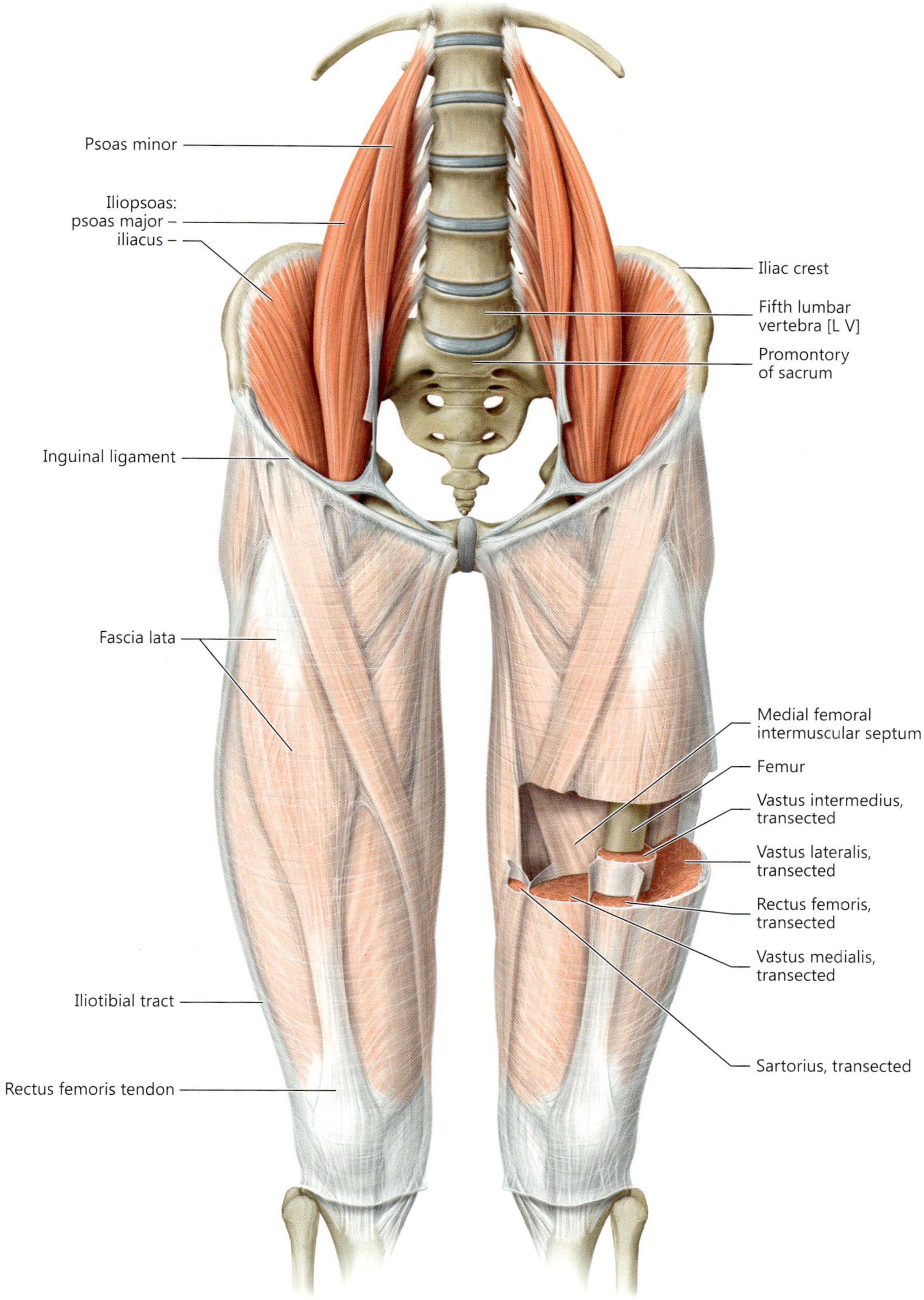

Fasciae and **fascial structures of the thigh**, anterior view. On the left, we see the fascia lata or femoral fascia, which surrounds the thigh like a sleeve, thickening laterally on the iliotibial tract. It further accommodates, in one of its splits, the sartorius and the tensor fasciae latae. On the right, we see the intermuscular septa, which, originating from the deep surface of the fascia lata, run toward the femur, where they insert on the surface of the periosteum. This osteofascial septum (formed by the two intermuscular septa and the femur) separates two muscular compartments of the thigh, one anterior and one posterior. The anterior compartment of the thigh contains the anterior muscles of the thigh, while the posterior compartment is occupied by the posterior and medial muscles of the thigh. Unlike the anterior and posterior muscle groups of the thigh, which, as stated above, occupy different osteofascial spaces, the (medial) adductor muscles of the thigh are not contained in a compartment delimited by fascial planes. However, references are often found to a medial compartment that could contain the adductor magnus, adductor longus, and pectineus.

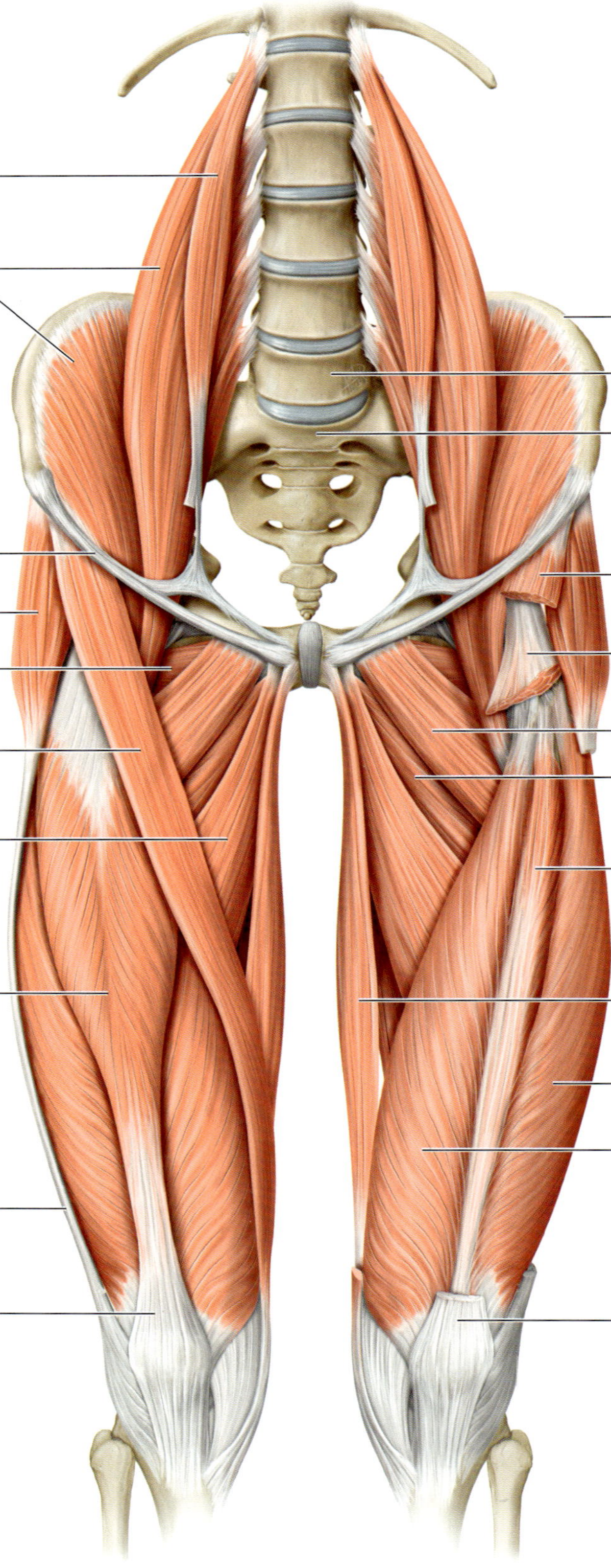

Internal muscles of the hip and **anterior muscles of the thigh**. On the left, we see the internal muscles of the hip (psoas major and iliacus, iliopsoas and psoas minor), the anterior superficial muscles of the thigh (sartorius and rectus femoris), and the superficial and intermediate layers of the medial muscles of the thigh (gracilis, pectineus, adductor longus, and adductor brevis). On the right, the sartorius and rectus femoris have been transected, revealing the other anterior muscles of the thigh: vastus medialis, vastus lateralis, and vastus intermedius, which, along with the rectus femoris, constitute the quadriceps femoris.

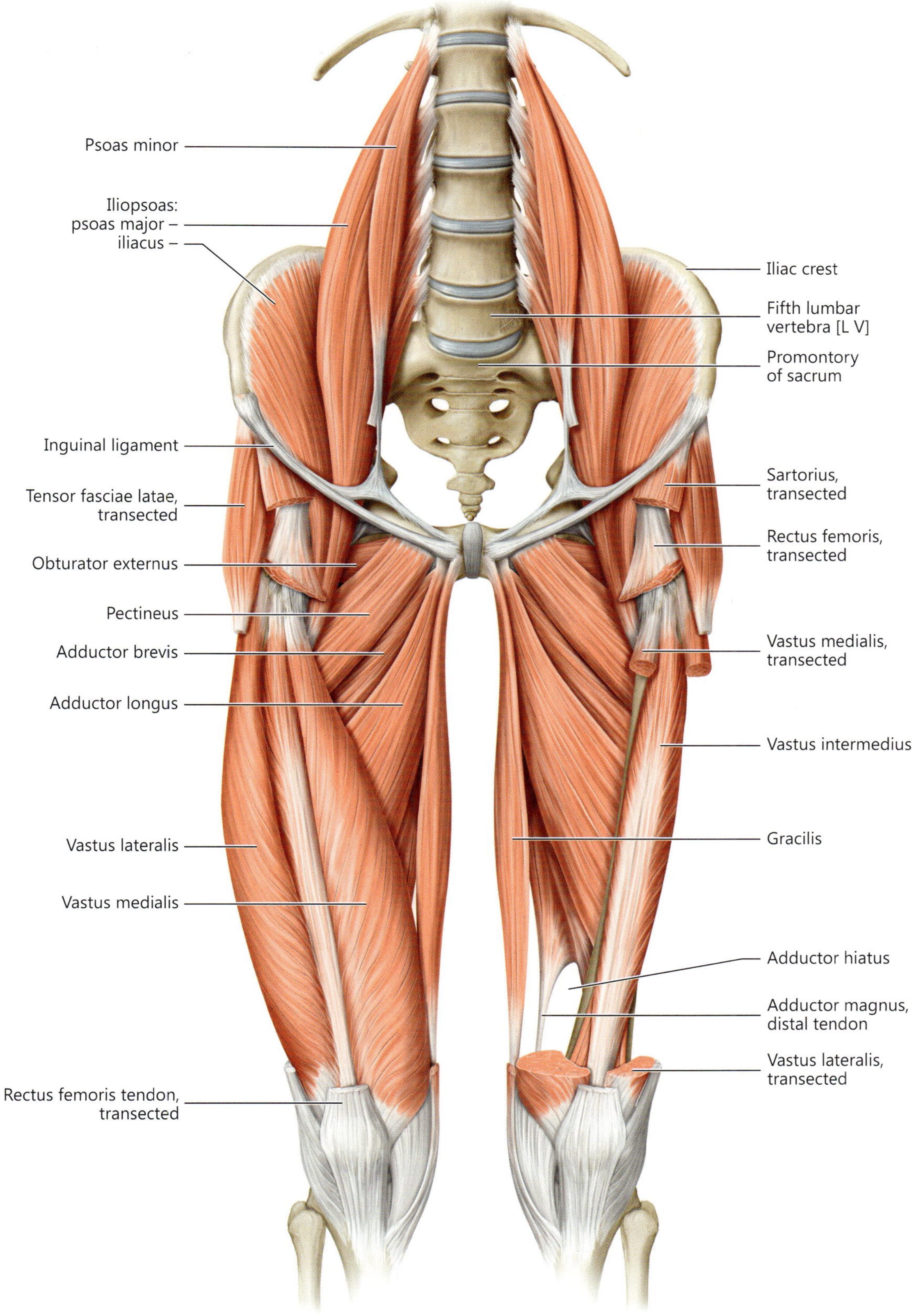

Internal muscles of the hip and **anterior muscles of the thigh**. Deeper plane than the previous figure. On the right, in addition to the sartorius and rectus femoris (more superficial), the vastus medialis and vastus lateralis are also transected, exposing more extensively the medial muscles of the thigh: gracilis, pectineus, adductor longus and brevis, and giving a glimpse of the adductor magnus. The former muscles form the superficial and intermediate medial muscular planes, and the adductor magnus forms the deep layer.

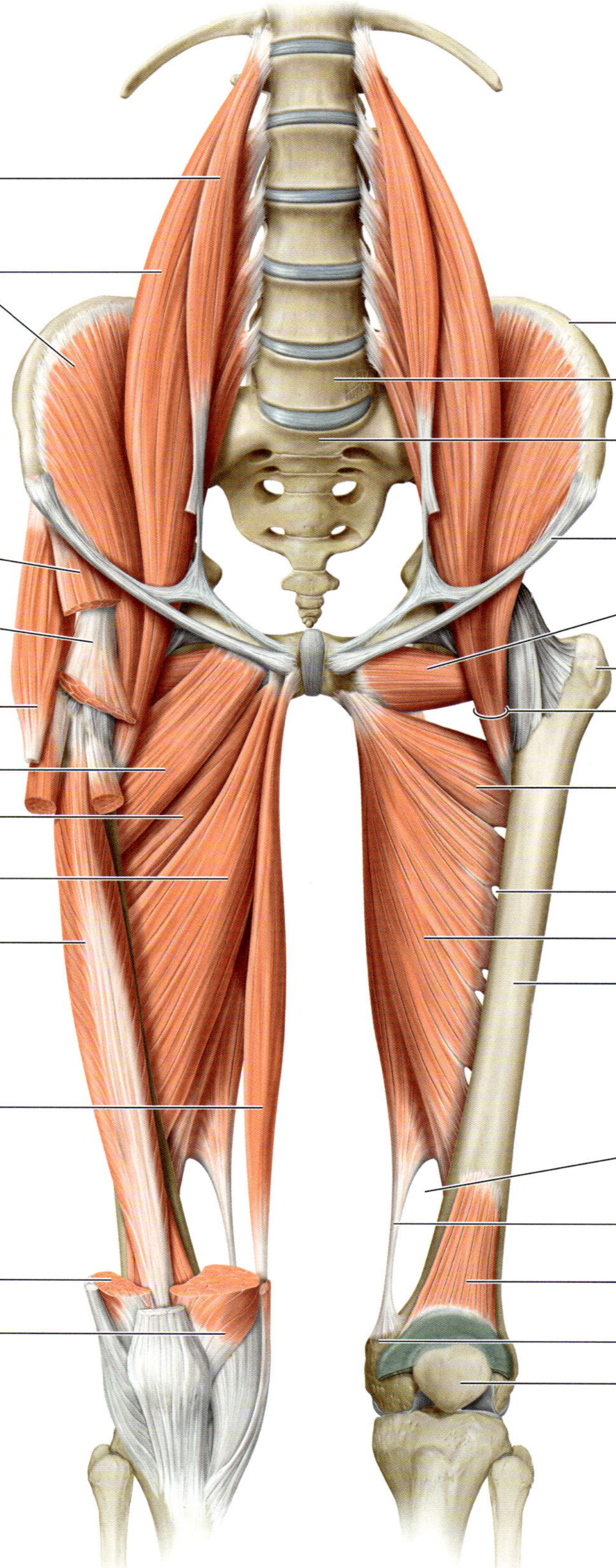

Internal muscles of the hip and **anterior muscles of the thigh**. Deeper plane than the previous figure. On the right, the superficial and intermediate layers of the medial muscles of the thigh have been removed; these are formed by the gracilis, pectineus, adductor longus, and adductor brevis, thus revealing the adductor magnus. Its superior and middle fascicles insert at the medial lip of the linea aspera of the femur, delimiting several orifices that give passage to the perforating vessels. Between the middle and inferior fascicles, the adductor hiatus is formed (ring of the adductor magnus), crossed by femoral vessels.

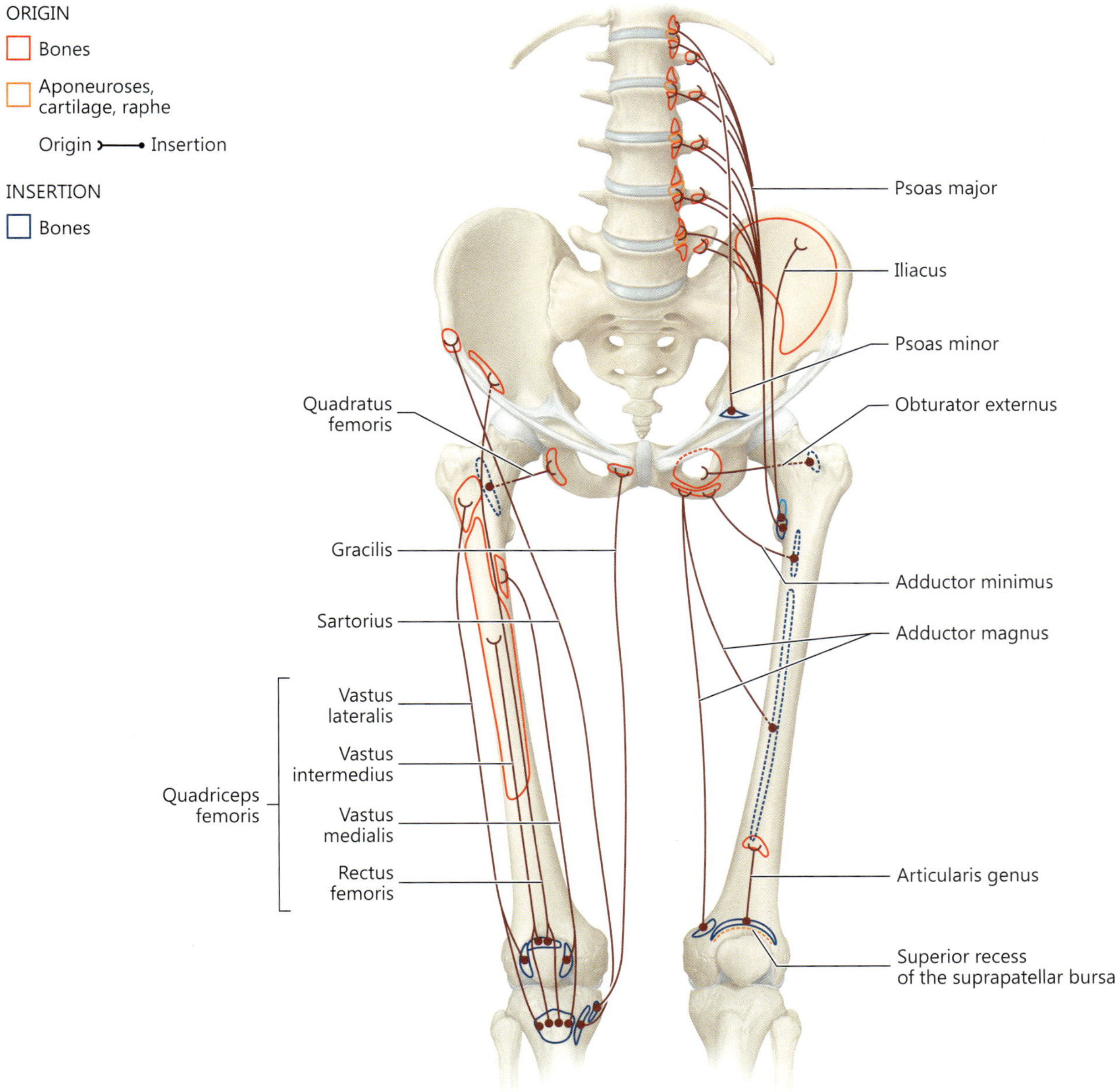

Schematic representation of the origin and insertions of the muscles of the hip and the thigh, anterior view. The **psoas minor** originates from the bodies of the last thoracic vertebra and of the first lumbar vertebra, and from the corresponding intervertebral disc, and inserts on the iliopubic ramus and on the iliac fascia. The **psoas major** originates from the transverse processes of the lumbar vertebrae and from the intervertebral bodies and discs of the twelfth thoracic vertebra and of all the lumbar vertebrae, and inserts on the lesser trochanter of the femur. The **iliacus** runs from the upper two-thirds of the iliac fossa, from the inner lip of the iliac crest, from the anterior sacroiliac and iliolumbar ligaments, and from the lateral part of the sacrum, to the tendon of the psoas major and to the lesser trochanter of the femur. The **obturator externus** originates from the external border of the obturator foramen and from the obturator membrane and inserts on the trochanteric fossa. The **adductor minimus** extends from the inferior pubic ramus to the superior third of the medial lip of the linea aspera. The **adductor magnus** originates from the ischiopubic ramus and from the ischial tuberosity, and reaches the medial lip of the linea aspera of the femur, the medial supracondylar line, and the adductor tubercle. The **articularis genus** extends from the anterior surface of the shaft of the femur to the superior recess of the suprapatellar bursa. The origin of the **quadratus femoris** is from the ischial tuberosity, and its insertion is on the intertrochanteric crest. The **gracilis** originates from the anterior surface of the ischiopubic ramus and from the lower half of the body of the pubis, and inserts on the medial surface of the tibia. The **sartorius** extends from the anterior superior iliac spine and from the upper half of the underlying notch to the upper part of the medial surface of the tibia. From the upper part of the intertrochanteric line and from the anterior and inferior borders of the greater trochanter, the **vastus lateralis** reaches the lateral border and the base of the patella and the tibial tuberosity through the tendon of the quadriceps femoris. The **vastus intermedius** runs from the upper half of the lateral lip of the linea aspera and from the upper two-thirds of the shaft of the femur to the base of the patella and to the tibial tuberosity through the tendon of the quadriceps femoris. The **vastus medialis** originates from the lower part of the intertrochanteric line, from the medial lip of the linea aspera, and from the medial supracondylar line, and inserts on the medial border of the patella and on the tibial tuberosity through the tendon of the quadriceps femoris. The **rectus femoris** originates with two tendons: straight head, from the anterior inferior iliac spine, and reflected head, from the supra-acetabular groove, and inserts on the base of the patella and on the tibial tuberosity through the tendon of the quadriceps femoris.

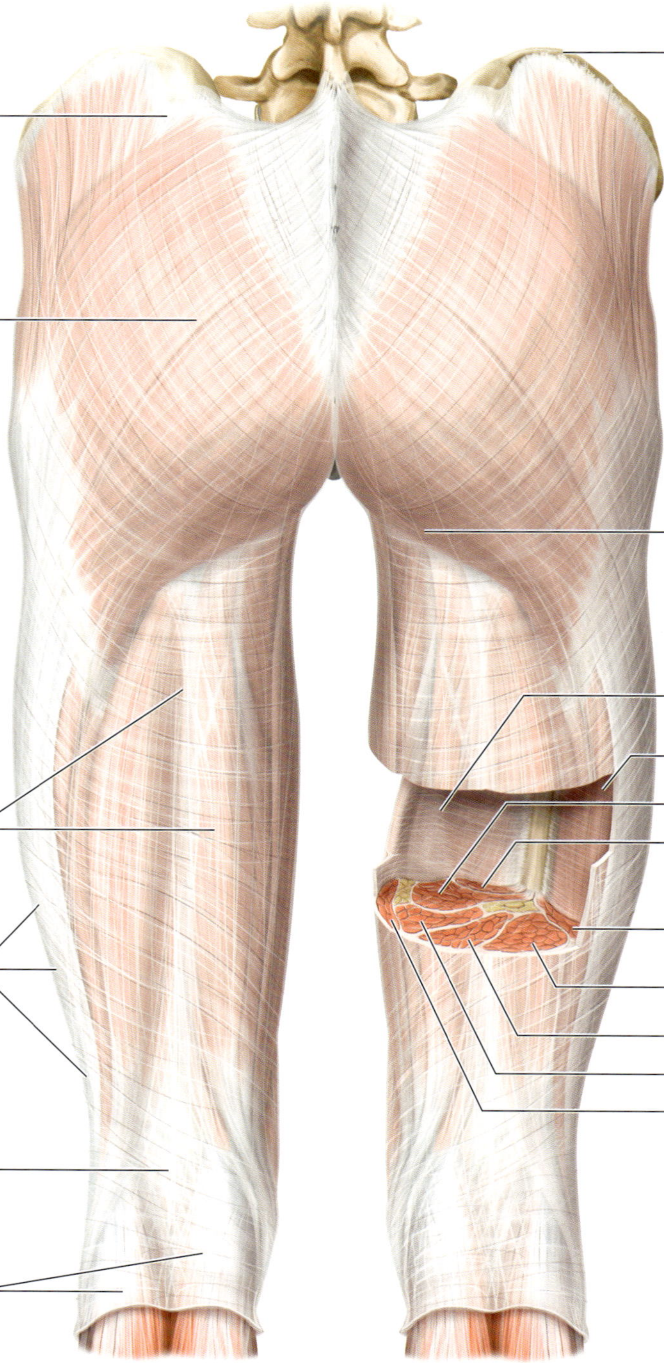

Fasciae and **fascial structures of the thigh**, posterior view. On the left, we see the fascia lata or femoral fascia, which surrounds the thigh like a sleeve, thickening laterally on the iliotibial tract. On the right, the section conducted in the middle of the thigh shows the contents of the posterior compartment of the thigh, limited by the osteofascial structure made up of the fascia lata, its two dependencies represented by the medial and lateral intermuscular septa of the thigh, and the femur. We see, in cross-section, the posterior muscles of the thigh: biceps femoris, semitendinosus and semimembranosus, and the intermediate and deep layers of the medial muscles, made up respectively of the adductor longus and adductor magnus.

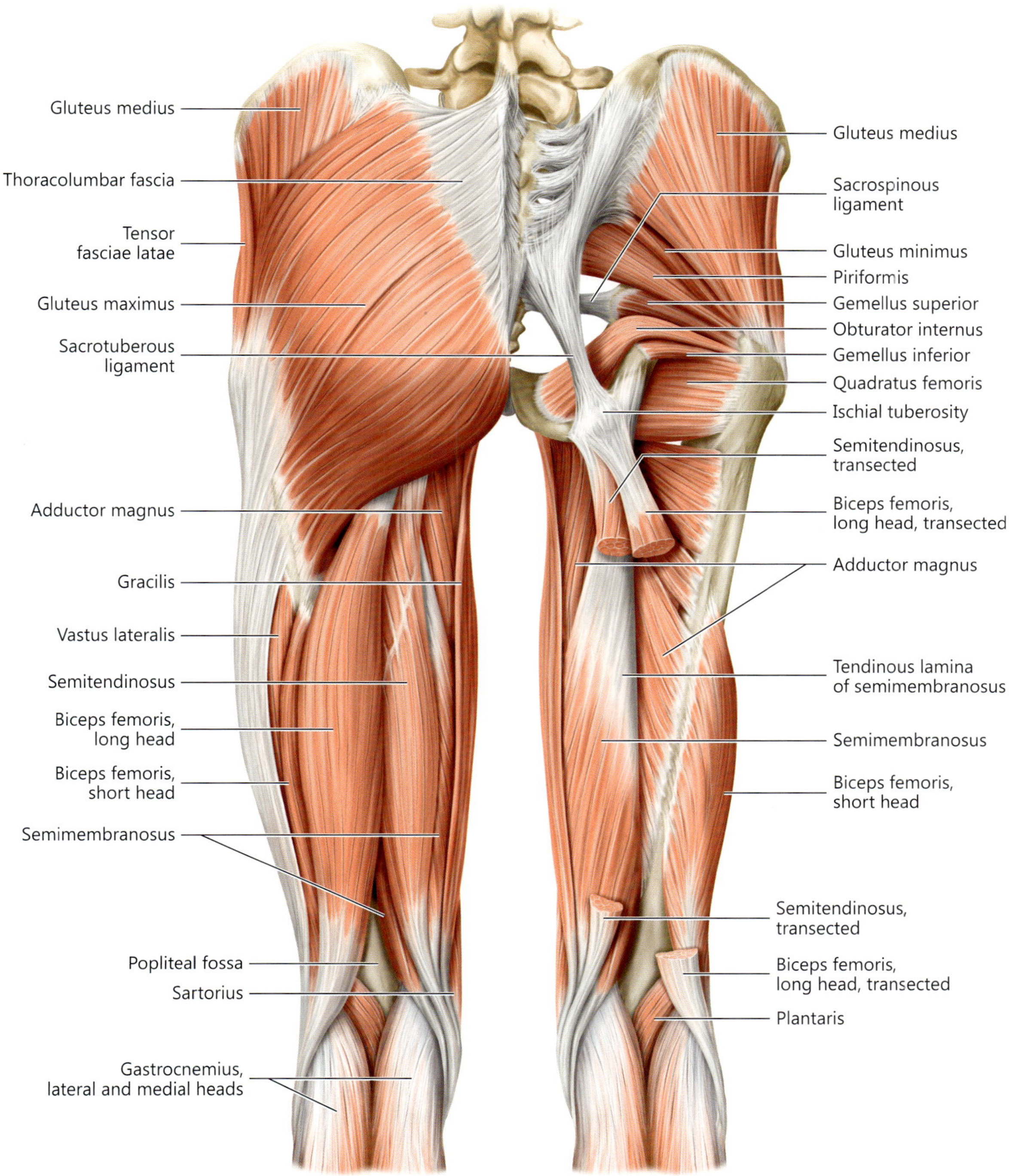

External muscles of the hip and **posterior muscles of the thigh**. On the left, we see the superficial layer of the external muscles of the hip, formed by the tensor fasciae latae and gluteus maximus, and, on the right, after removal of the gluteus maximus, the gluteus medius, in an intermediate position, deep to the gluteus maximus. The posterior compartment of the thigh is occupied by the semitendinosus and long head of the biceps femoris, visible superficially on the left, and the semimembranosus and short head of the biceps femoris, visible extensively on the right after removal of the semitendinosus and the long head of the biceps femoris.

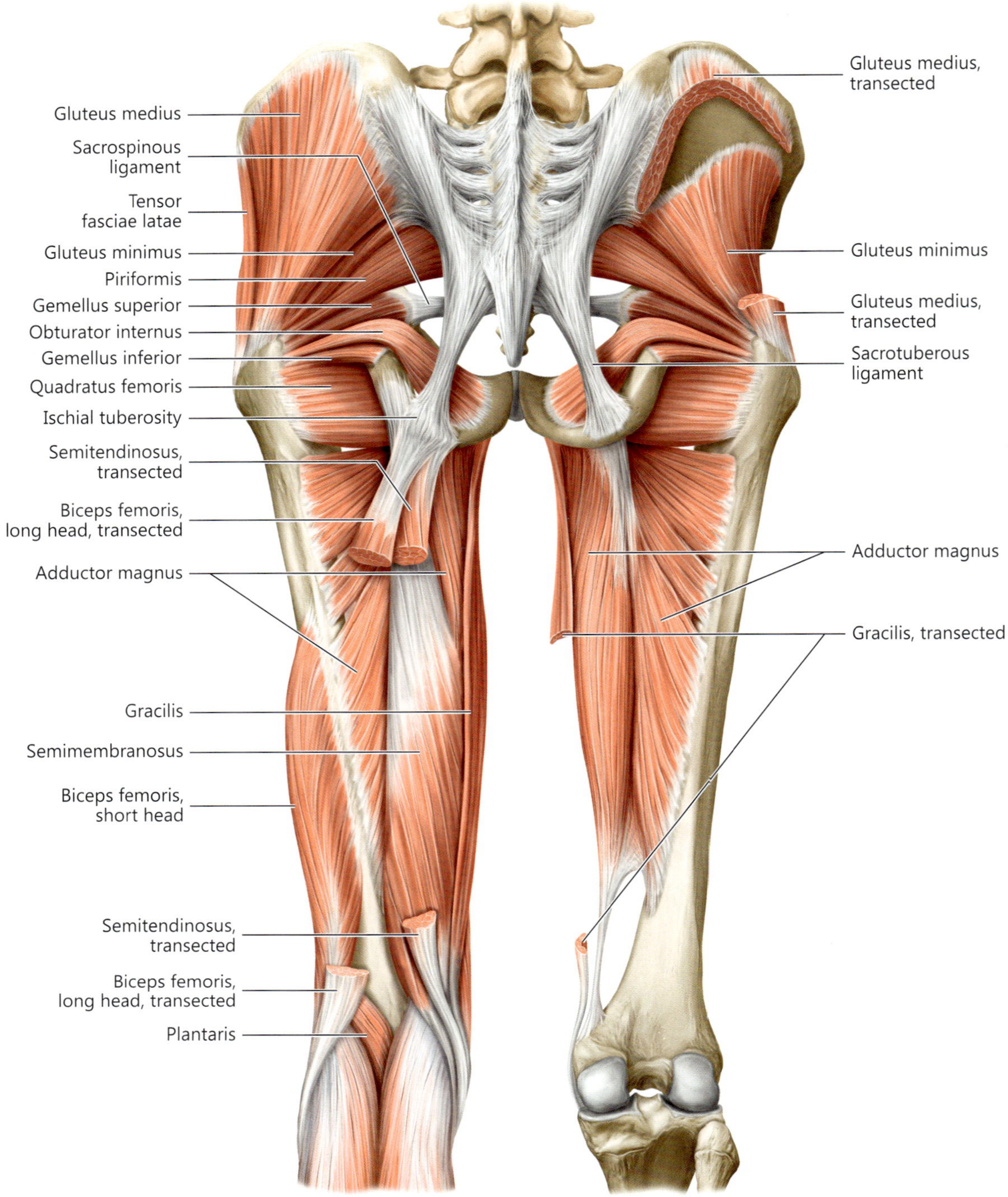

External muscles of the hip and **posterior muscles of the thigh**. On the left, with the gluteus maximus removed, we see the gluteus medius and the deep layer of the external muscles of the hip, made up of the piriformis, obturator internus and obturator externus (here not visible), gemellus superior and inferior, quadratus femoris and gluteus minimus. The gluteus minimus appears widely on the right after cross-section of the gluteus medius. Also on the right, the removal of the posterior muscles of the thigh and the sectioning of the gracilis reveals the adductor magnus in the medial compartment of the thigh.

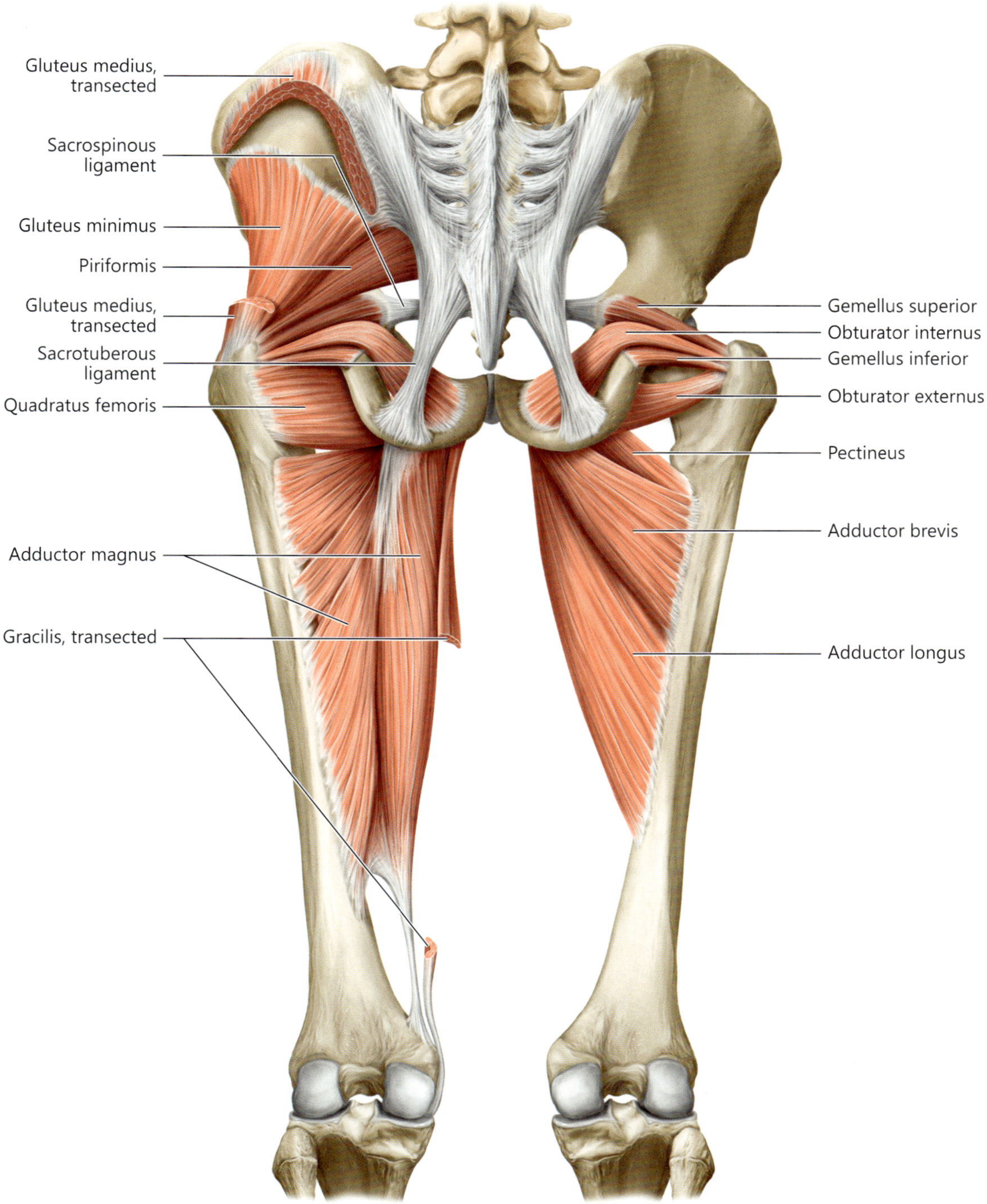

External muscles of the hip and **medial muscles of the thigh**, posterior view. On the left, after resection of the gluteus maximus and gluteus medius and removal of the posterior muscles of the thigh, appear, respectively, the external muscles of the hip located more deeply, except for the obturator externus, visible on the right after removal of the quadratus femoris, and the adductor magnus, which forms the deep layer of the medial muscles of the thigh. Its removal reveals, on the right, the adductor brevis, which forms the intermediate layer, and the adductor longus and pectineus, which, along with the gracilis visible on the left where it has been transected to better show the adductor magnus, form the superficial layer of the medial muscles of the thigh.

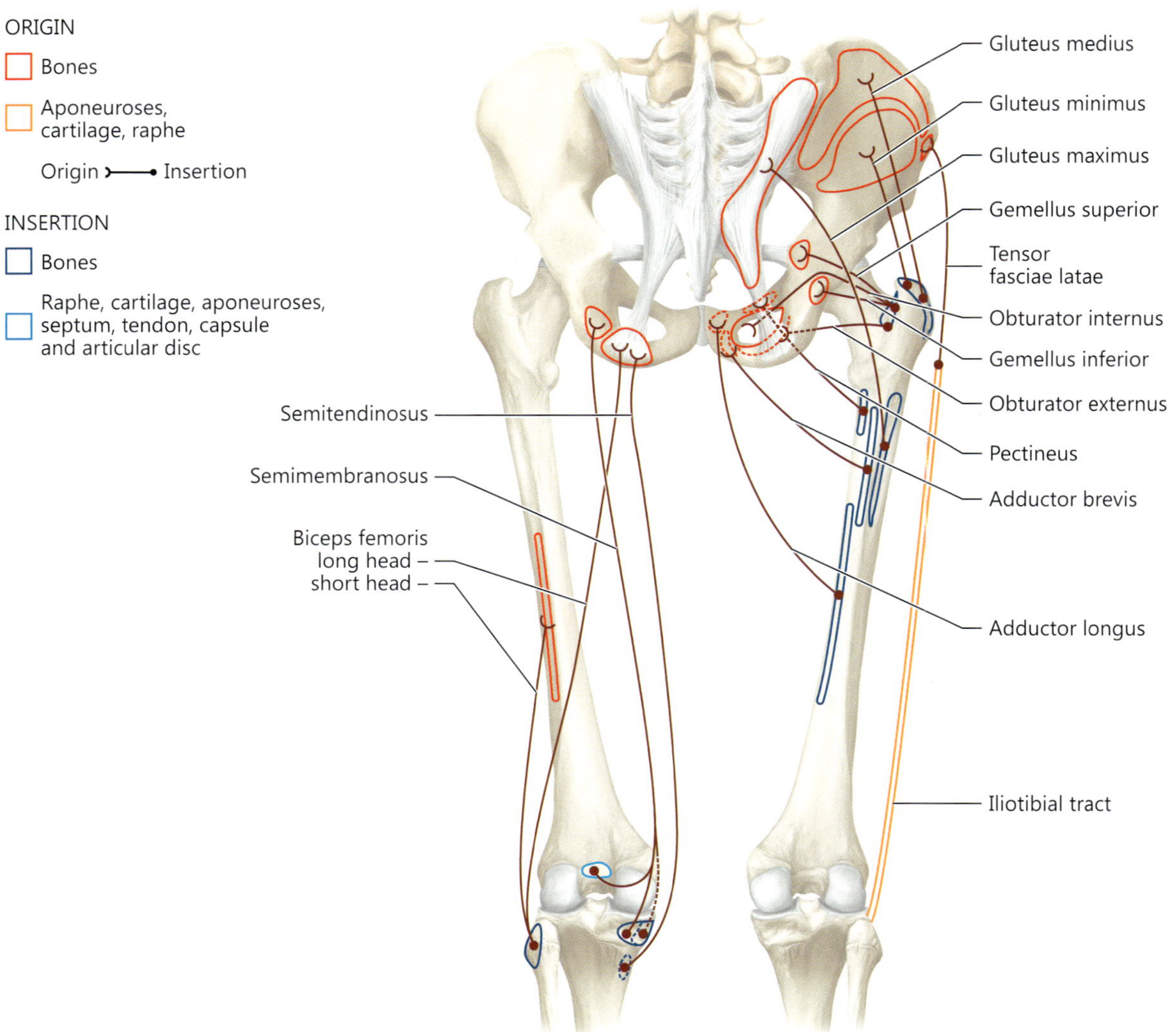

Schematic representation of the origin and insertions of the muscles of the hip and the thigh, posterior view. The **gluteus medius** originates from the gluteal fascia, the iliac crest, the gluteal surface of the ilium, and the anterior superior iliac spine, and inserts on the external surface of the greater trochanter. The **gluteus minimus** extends from the gluteal surface of the ilium and from the iliac crest to the anterior border of the greater trochanter. From the iliac crest, the posterior gluteal line, the lateral surfaces of the sacrum and the coccyx, and from the sacrotuberous ligament, the **gluteus maximus** runs to the gluteal tuberosity of the femur. The **gemellus superior** originates from the ischial spine and inserts on the trochanteric fossa through the tendon of the obturator internus. The **tensor fasciae latae** extends from the anterior superior iliac spine and from the iliac crest to the lateral condyle of the tibia, on which the iliotibial tract inserts, formed by the union of the tendon of the muscle with the fascia lata. The **obturator internus** originates from the internal border of the obturator foramen, from the obturator membrane, and from the bone surface between the obturator foramen and the greater sciatic notch, and inserts on the trochanteric fossa. From the ischial tuberosity, the **gemellus inferior** affixes to the trochanteric fossa through the tendon of the obturator internus. The **obturator externus** originates from the external border of the obturator foramen and from the obturator membrane, and inserts on the trochanteric fossa. The **pectineus** extends from the pecten pubis (pectineal line of the pubis) to the pectineal line of the femur. The **adductor brevis** originates from the external surface of the body of the pubis and from the inferior pubic ramus, and inserts between the lesser trochanter and the linea aspera and on the upper part of the linea aspera. The origin of the **adductor longus** is from the superior pubic ramus, and its insertion is on the medial lip of the linea aspera in the middle third of the shaft of the femur. The **semitendinosus** originates from the inferomedial sulcus on the superior surface of the ischial tuberosity and inserts on the medial surface of the tibia. The **semimembranosus** runs from the superolateral surface of the ischial tuberosity to the medial condyle of the tibia. The distal insertion tendon is divided into three parts: the straight tendon, which inserts on the posterior surface of the medial condyle of the tibia; the oblique popliteal ligament (or recurrent tendon), that reinforces the posterior part of the capsule of the knee joint; the reflected tendon, that ends anteriorly to the medial condyle of the tibia. The **biceps femoris** originates with two heads: long, from the inferomedial sulcus on the superior surface of the ischial tuberosity and from the inferior part of the sacrotuberous ligament, and short, from the lateral lip of the linea aspera and from the lateral supracondylar line, and inserts on the head of the fibula, on the lateral condyle of the tibia, and on the fibular collateral ligament.

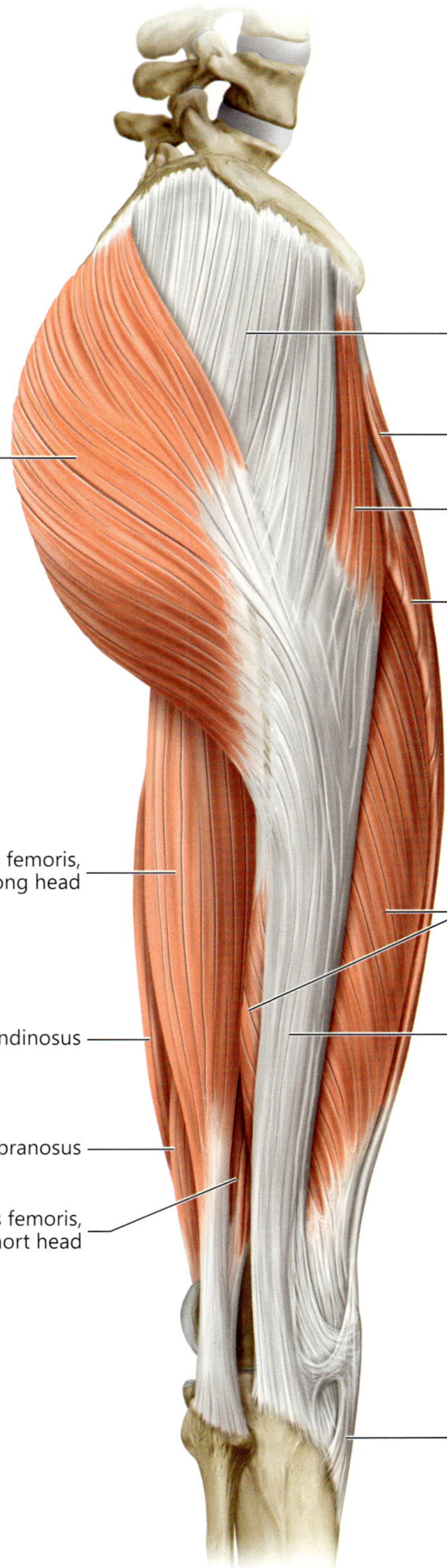

External muscles of the hip and **posterior muscles of the thigh**. Right side, lateral view. The external muscles of the hip are located in the gluteal region and are distributed in several layers, the most superficial of which is visible here, formed by the tensor fasciae latae and gluteus maximus. The posterior muscles of the thigh are represented by the biceps femoris, semitendinosus, and semimembranosus. On the lateral surface of the thigh, the fascia lata thickens forming the iliotibial tract, the superior extremity of which opens into two layers that include and anchor the tensor fasciae latae.

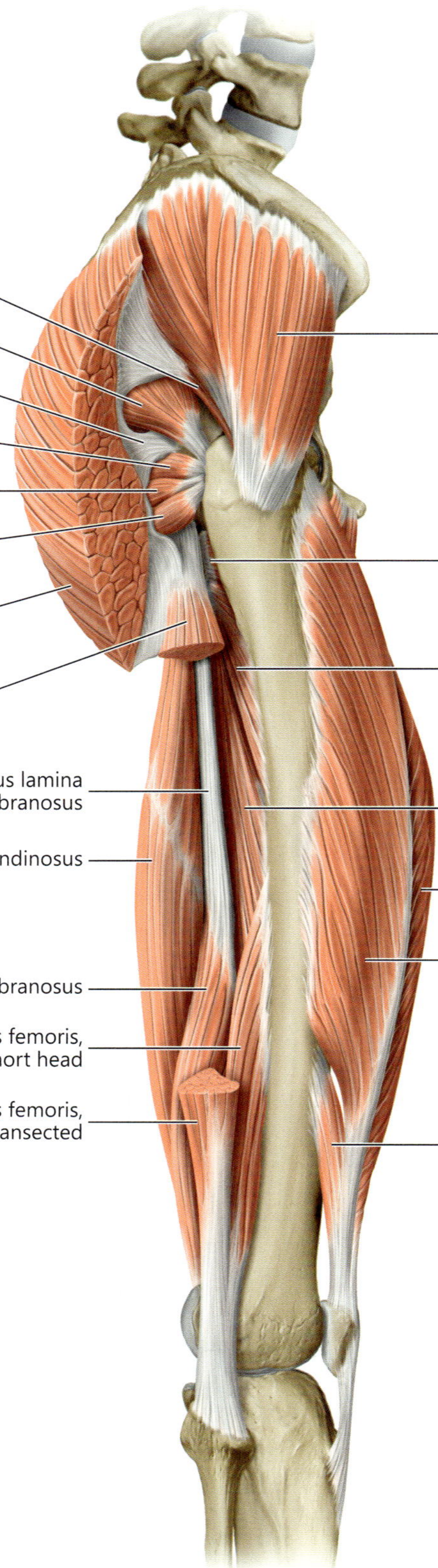

External muscles of the hip and **posterior muscles of the thigh**. Right side, lateral view. With the most superficial external muscles of the hip (represented by the tensor fasciae latae and gluteus maximus) transected, the gluteus medius appears in an intermediate location, and near their insertions on the femur, the more deeply located muscles: the gluteus minimus, piriformis, gemellus superior, obturator internus, gemellus inferior, and quadratus femoris. The resection of the long head of the biceps femoris reveals the other two posterior muscles of the thigh: the semitendinosus and the semimembranosus. The latter is thus named because in its proximal third it is formed by a wide tendinous lamina, seen here from the side.

ORIGIN

Bones

Origin ➤—• Insertion

INSERTION

Bones

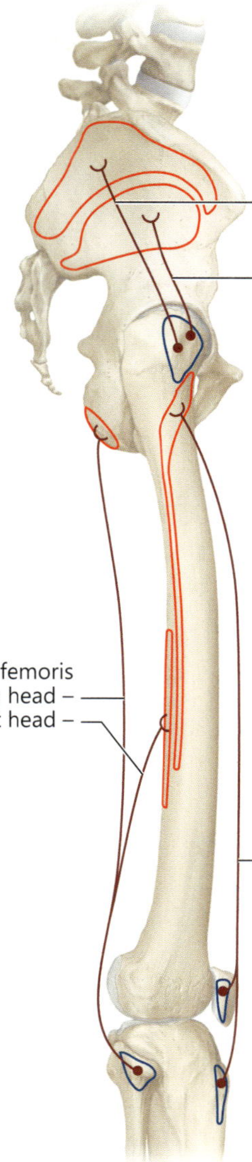

Schematic representation of the origin and insertions of the gluteus medius, gluteus minimus, biceps femoris, and vastus lateralis. The **gluteus medius** originates from the gluteal fascia, the iliac crest, the gluteal surface of the ilium between the anterior and posterior gluteal lines, and the anterior superior iliac spine, and inserts on the external surface of the greater trochanter. The **gluteus minimus** extends from the gluteal surface of the ilium between the anterior and inferior gluteal lines and from the iliac crest to the anterior border of the greater trochanter. The **biceps femoris** originates with two heads: long, from the inferomedial sulcus on the superior surface of the ischial tuberosity and from the inferior part of the sacrotuberous ligament, and short, from the lateral lip of the linea aspera and from the lateral supracondylar line, and inserts on the head of the fibula, the lateral condyle of the tibia and the fibular collateral ligament. From the upper part of the intertrochanteric line and from the anterior and inferior borders of the greater trochanter, the **vastus lateralis** reaches the lateral border and the base of the patella and the tibial tuberosity through the tendon of the quadriceps femoris.

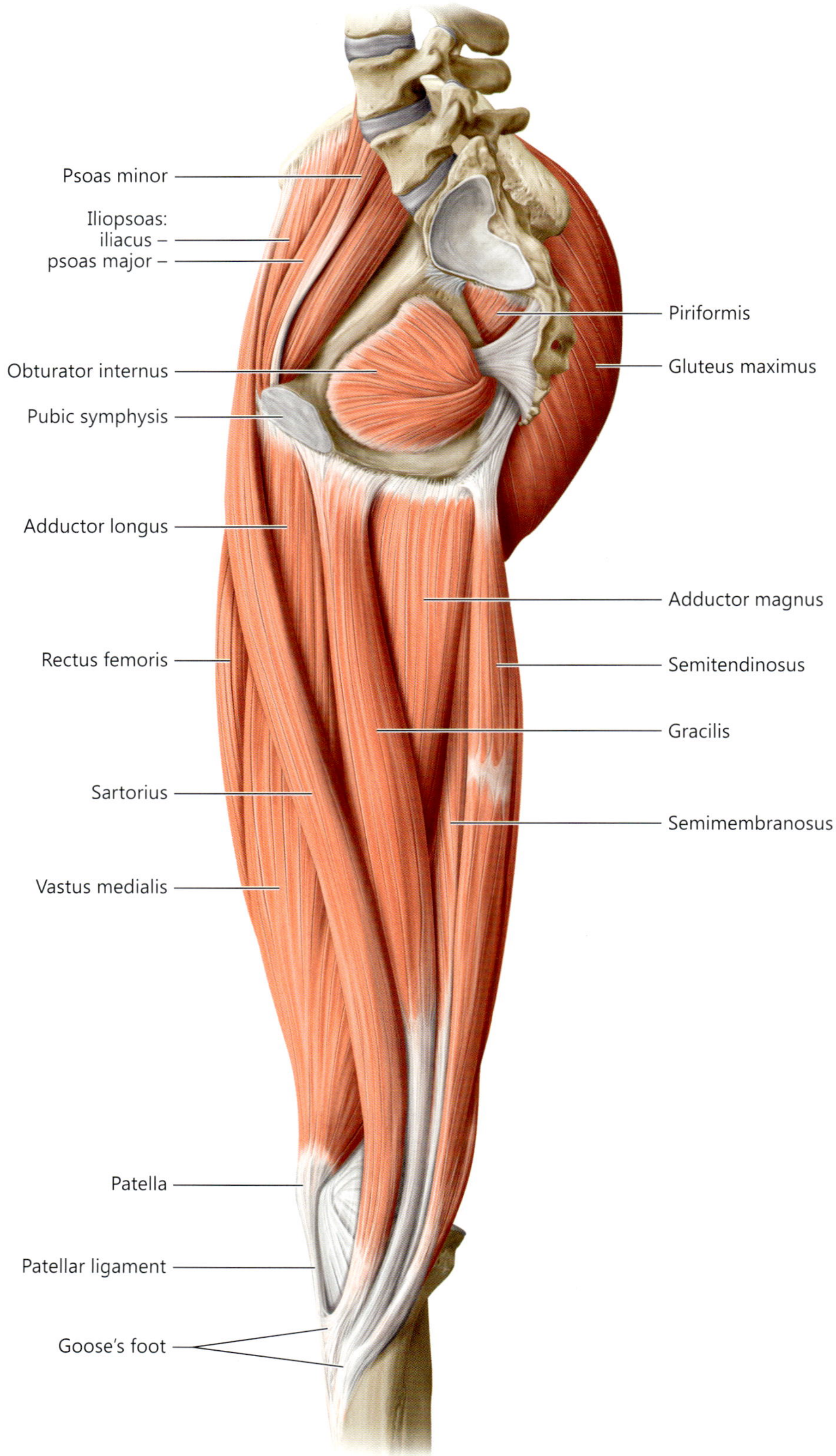

Muscles of the hip and **of the thigh**. Right side, medial view. Median sagittal section. The sartorius, the most superficial of the anterior muscles of the thigh, inserts on the superior extremity of the medial surface of the tibia with a tendon joined with fibrous expansions to the tendons of the gracilis and semitendinosus, forming a structure that, due to its appearance, is called goose's foot. The adductor magnus, along with the adductor minimus (not visible here), forms the deep layer of the medial muscles of the thigh.

ORIGIN

Bones

Origin ⟶ Insertion

INSERTION

Bones

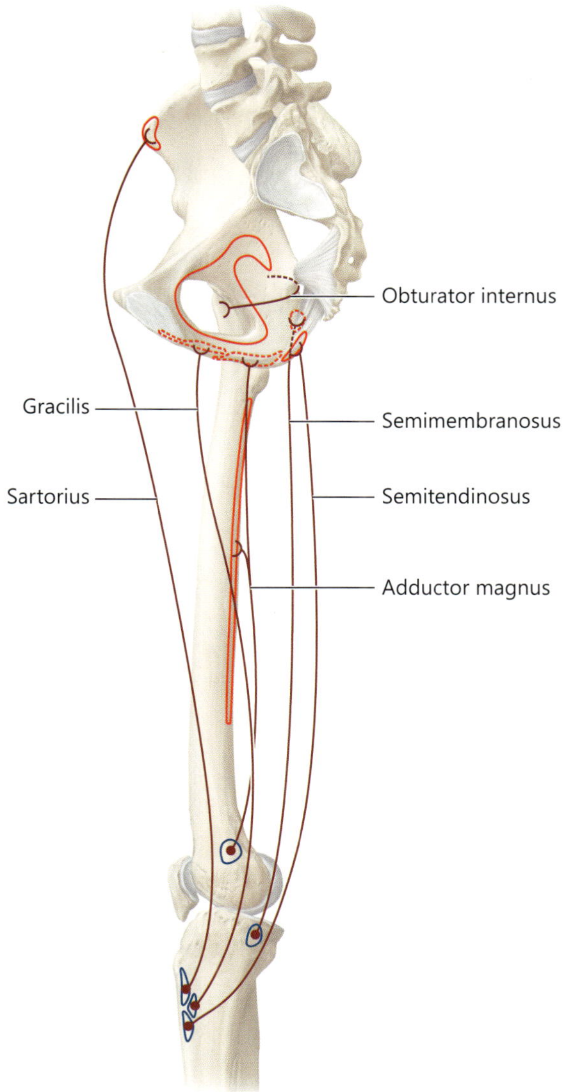

Schematic representation of the origin and insertions of the gracilis, sartorius, semitendinosus, adductor magnus, obturator internus, and semimembranosus. The **gracilis** originates from the anterior surface of the ischiopubic ramus and from the lower half of the body of the pubis, and inserts on the upper part of the medial surface of the tibia. The **sartorius** extends from the anterior superior iliac spine and from the upper half of the underlying notch to the upper part of the medial surface of the tibia. The **obturator internus** originates from the internal border of the obturator foramen, from the obturator membrane, and from the bone surface between the obturator foramen and the greater sciatic notch, and inserts on the trochanteric fossa. The **semimembranosus** runs from the superolateral surface of the ischial tuberosity to the posterior surface of the medial condyle of the tibia. The **semitendinosus** originates from the inferomedial sulcus on the superior surface of the ischial tuberosity, and inserts on the proximal part of the medial surface of the tibia. From the anterior surface of the ischiopubic ramus and from the inferolateral surface of the ischial tuberosity, the **adductor magnus** runs to the medial lip of the linea aspera of the femur, to the medial supracondylar line, and to the adductor tubercle.

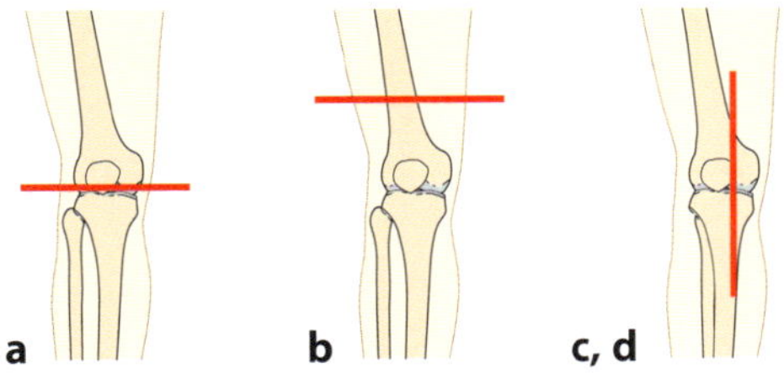

a b c, d

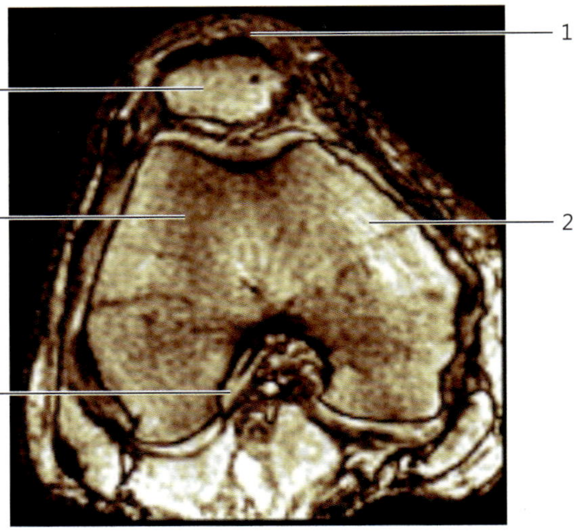

a

Computerized three-dimensional reconstructions using a direct volume rendering technique from axial nuclear magnetic resonance scans (**a**, **b**) and from sagittal nuclear magnetic resonance scans, gray scale (**c**) and color and opacity mapping (**d**).
1, Patellar ligament
2, medial condyle of femur
3, anterior cruciate ligament
4, lateral condyle of femur
5, patella
6, vastus lateralis
7, femur
8, biceps femoris
9, semimembranosus
10, vastus medialis
11, articular cartilage
12, infrapatellar fat pad (of Hoffa)
13, medial meniscus

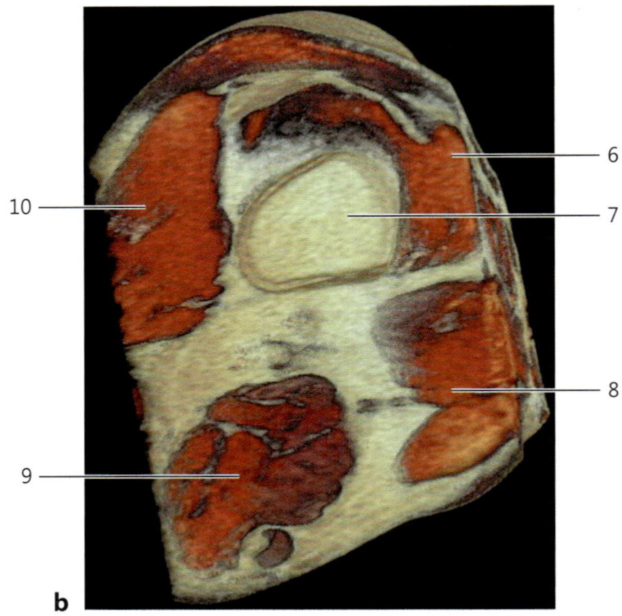

b

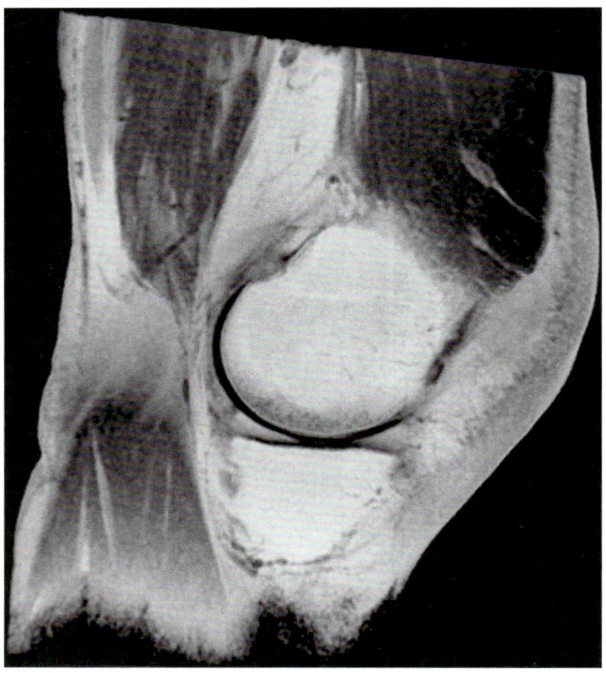

c

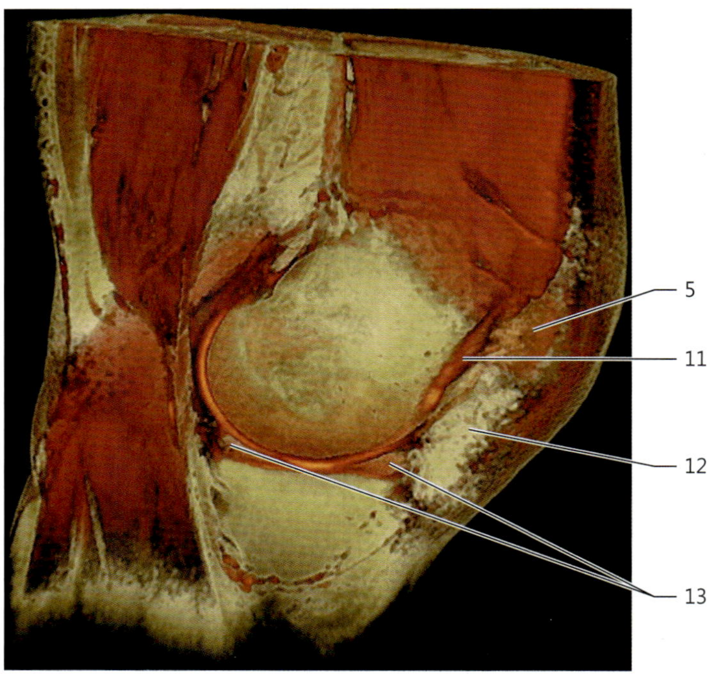

d

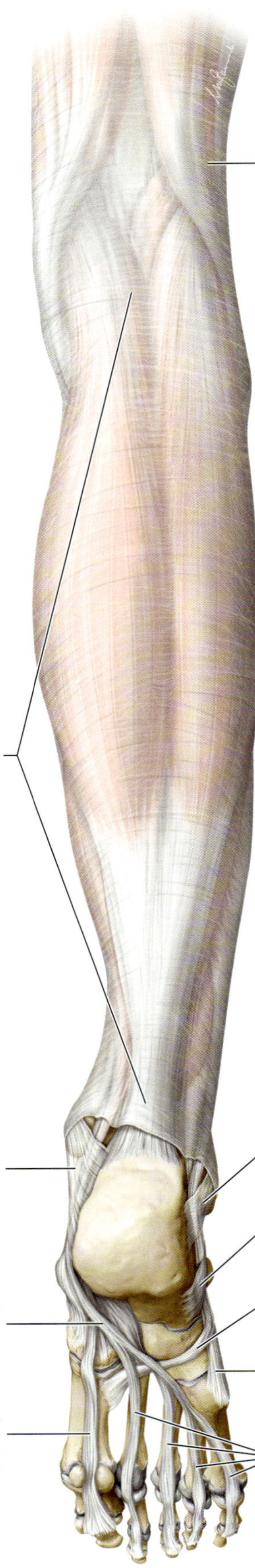

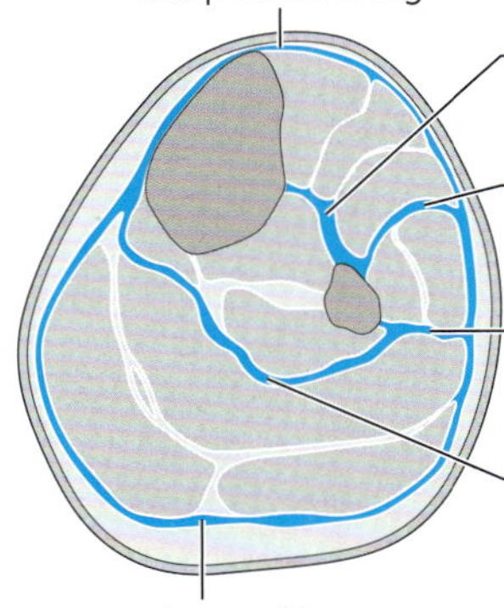

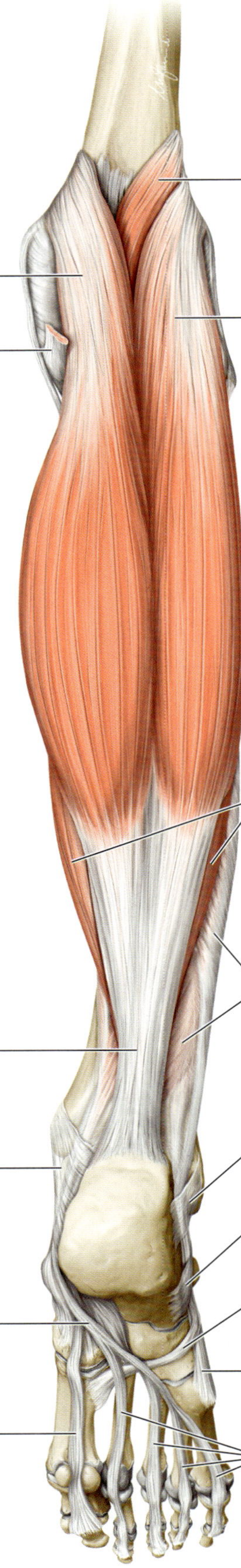

Posterior muscles of the right **leg**. On the left, fasciae and fascial structures of the leg and ankle. The deep fascia of the leg, which follows the fascia lata, thickens at the ankle, forming, laterally and medially, respectively the fibular retinacula and the flexor retinaculum, which keep in place the tendons of the lateral and deep posterior muscles of the leg that reach the foot. On the center, fascial compartments of the leg. On the right, superficial layer of the posterior muscles of the leg, formed by the triceps surae (gastrocnemius and soleus) and plantaris.

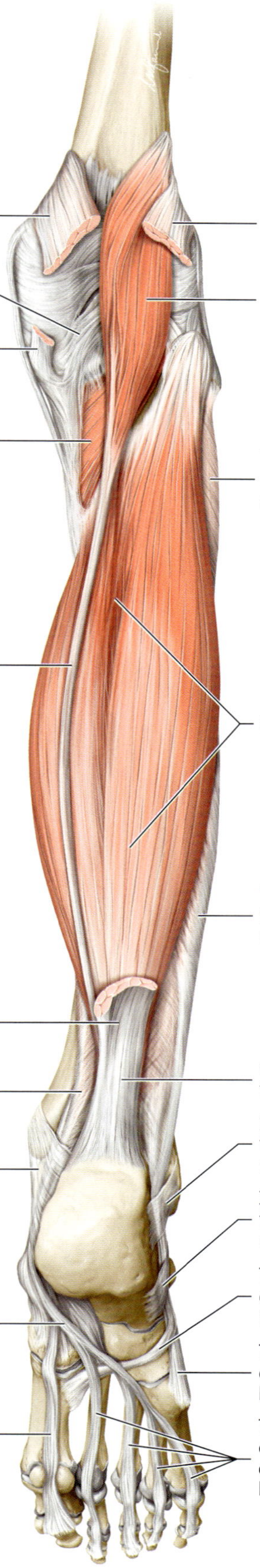

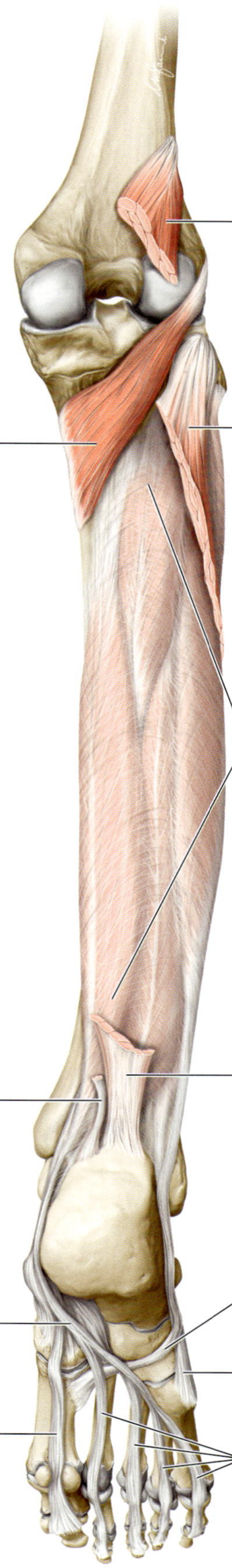

Posterior muscles of the right **leg**. On the left, the removal of the gastrocnemius, located superficially, reveals the soleus (which, together with the gastrocnemius, constitutes the triceps surae), and the plantaris, located more deeply. On the right, with the soleus and plantaris transected, the transverse intermuscular septum and the popliteus appear, located deeply within the popliteal fossa. The septum originates at the tendinous arch of the soleus, and extends from the medial border of the tibia to the posterior border of the fibula, separating the superficial posterior muscles of the calf from the deep ones.

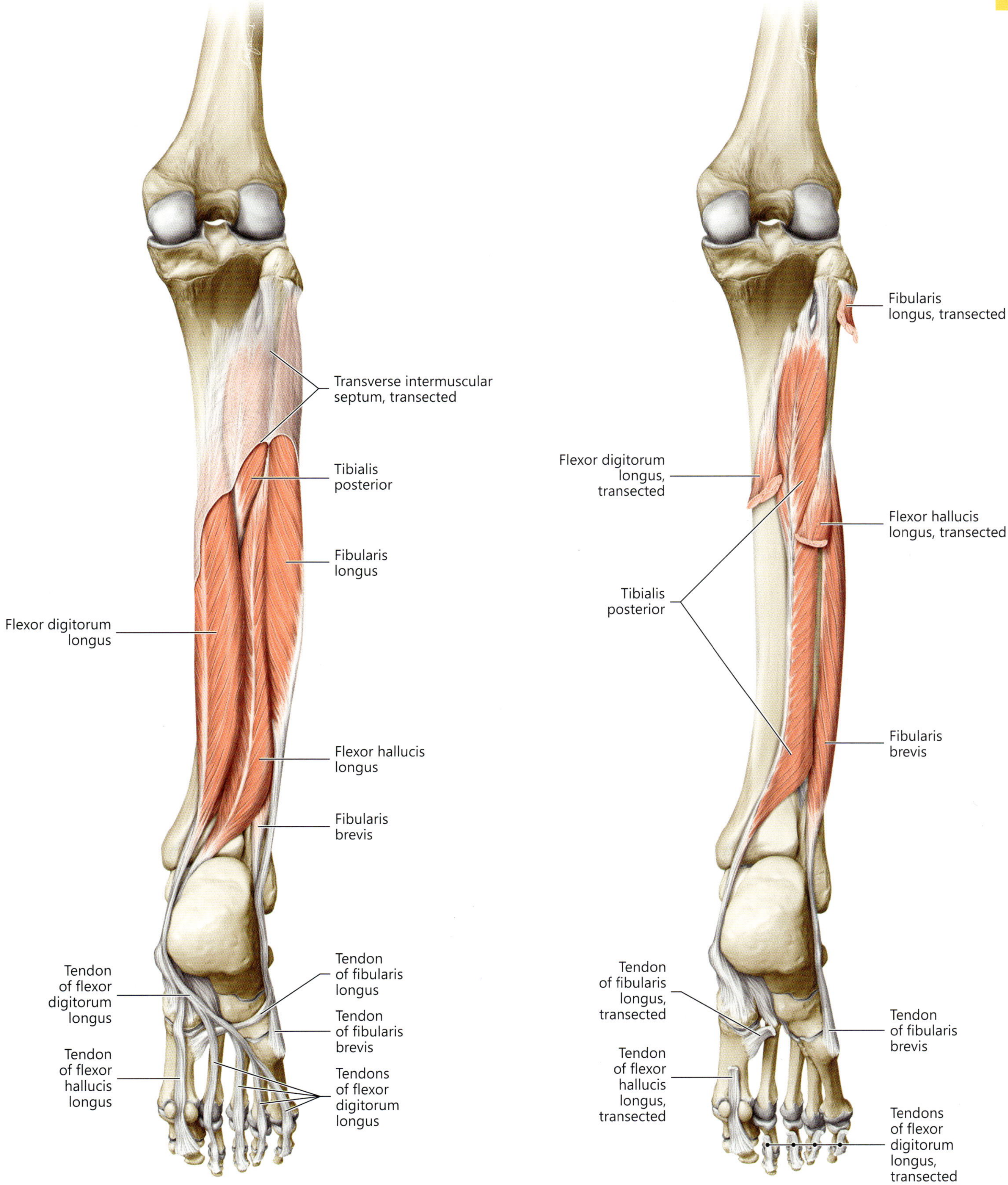

Posterior muscles of the right **leg**. On the left, the popliteus has been removed, and the transverse intermuscular septum or posterior intermuscular septum of the leg has been transected and removed (☞ see figure on page 624) in order to show the deep layer of the posterior muscles of the leg: tibialis posterior, flexor digitorum longus, and flexor hallucis longus. Also visible is the fibularis brevis, housed in the lateral compartment. On the right, the removal of the flexor digitorum longus, flexor hallucis longus, and fibularis longus reveals, from medially to laterally, the posterior surface of the tibia, the tibialis posterior, and the fibularis brevis.

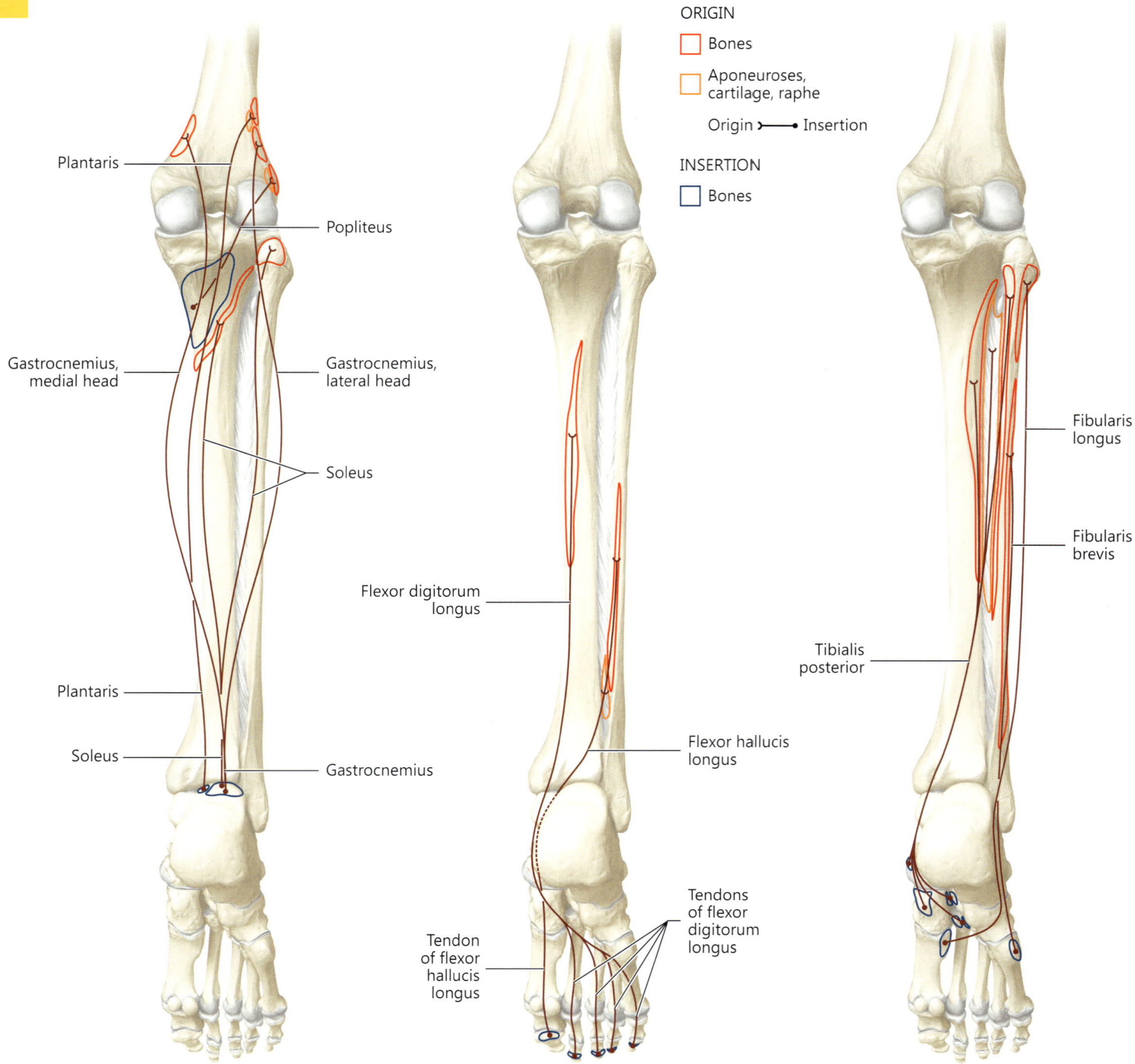

Schematic representation of the origin and insertions of the plantaris, popliteus, gastrocnemius, soleus, flexor digitorum longus, flexor hallucis longus, tibialis posterior, fibularis longus, and fibularis brevis. The **plantaris** extends from the lateral condyle of the femur to the medial surface of the calcaneus. The **popliteus** originates from the posterior surface of the lateral epicondyle of the femur and from the capsule of the knee joint, and inserts on the posterior surface of the tibia. The **gastrocnemius** originates with two heads, the medial and lateral heads, from the condyles of the femur, and reaches the calcaneus through the calcaneal tendon. The **soleus** runs from the posterior surface of the head and the proximal quarter of the shaft of the fibula, from the soleal line and from the middle third of the medial border of the tibia to the calcaneus through the calcaneal tendon. The **flexor digitorum longus** originates from the posterior surface of the tibia and inserts on the plantar surface of the base of the distal phalanges of the last four toes. The **flexor hallucis longus** originates from the posterior surface of the fibula and from the interosseous membranes and is affixed onto the plantar surface of the base of the distal phalange of the great toe. The **tibialis posterior** extends from the upper two-thirds of the posterior and lateral surfaces of the tibia, from the posterior surface of the interosseous membrane and from the medial surface of the proximal two-thirds of the fibula to the tuberosity of the navicular and to the plantar surface of the cuneiforms. The **fibularis longus** originates from the head and the upper two-thirds of the lateral surface of the fibula, and it inserts on the lateral surface of the base of the first metatarsal and on the medial and intermediate cuneiforms. The **fibularis brevis** originates from the lower two-thirds of the lateral surface of the fibula and attaches to the tuberosity of the fifth metatarsal.

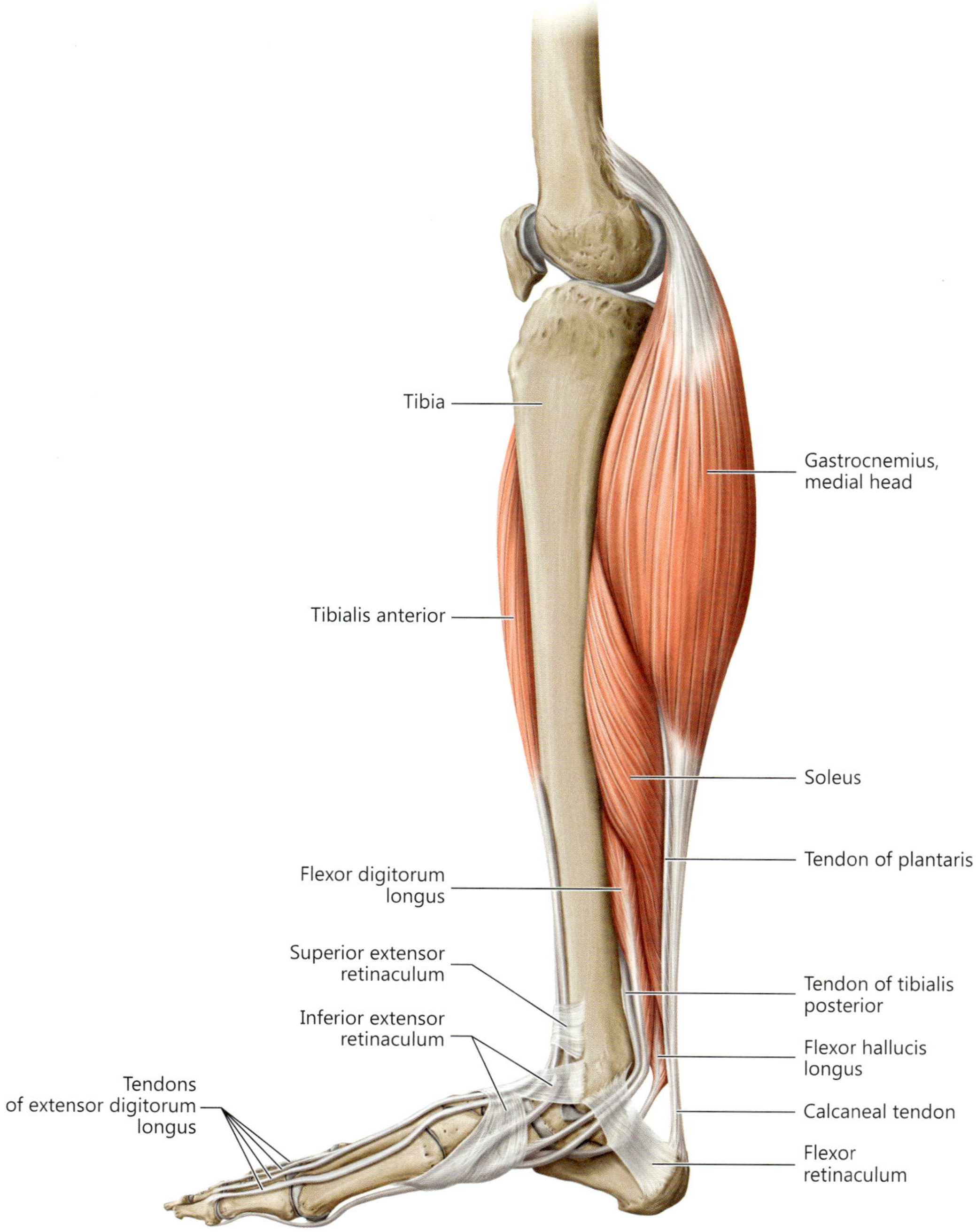

Muscles of the right **leg**, medial view. The posterior muscles of the leg are arranged in two superimposed layers. The superficial layer is formed by the gastrocnemius (here only the medial head is seen), soleus and plantaris, the tendon of which, long and thin, reaches the medial border of the calcaneal tendon and inserts in the medial part of the posterior surface of the calcaneus. The deep layer is made up of the popliteus (not visible here), flexor hallucis longus, tibialis posterior (which is the deepest), and flexor digitorum longus (which is the most medial). Deep to the flexor retinaculum, stretched between the anterior border of the medial malleolus and the medial surface of the calcaneus, run the tendons of the deep posterior muscles of the leg that reach the foot.

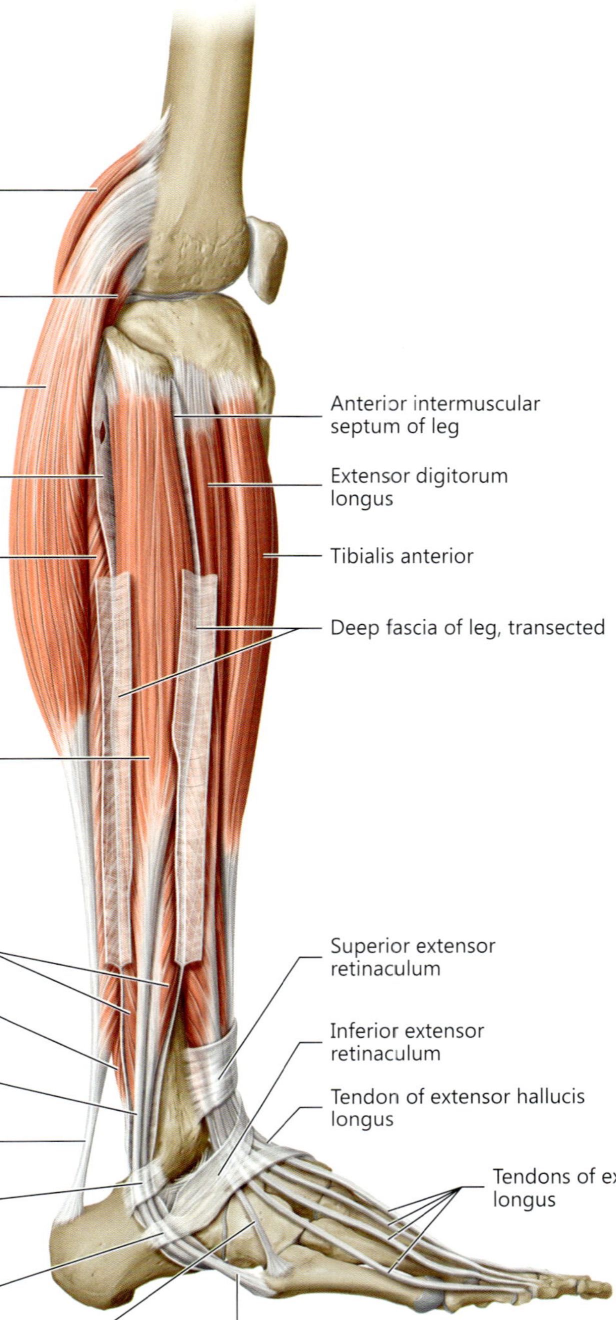

Lateral muscles of the right **leg**. From the deep surface of the deep fascia of the leg originate the anterior and posterior intermuscular septa of the leg, which insert on the anterior and lateral borders of the fibula, delineating the lateral compartment of the leg, or the fibular compartment of the leg. The fibularis longus is located superficially to the fibularis brevis. Deep to the fibular retinacula, stretched between the lateral malleolus and the lateral surface of the calcaneus, run the tendons of the two muscles inside of a common sheath.

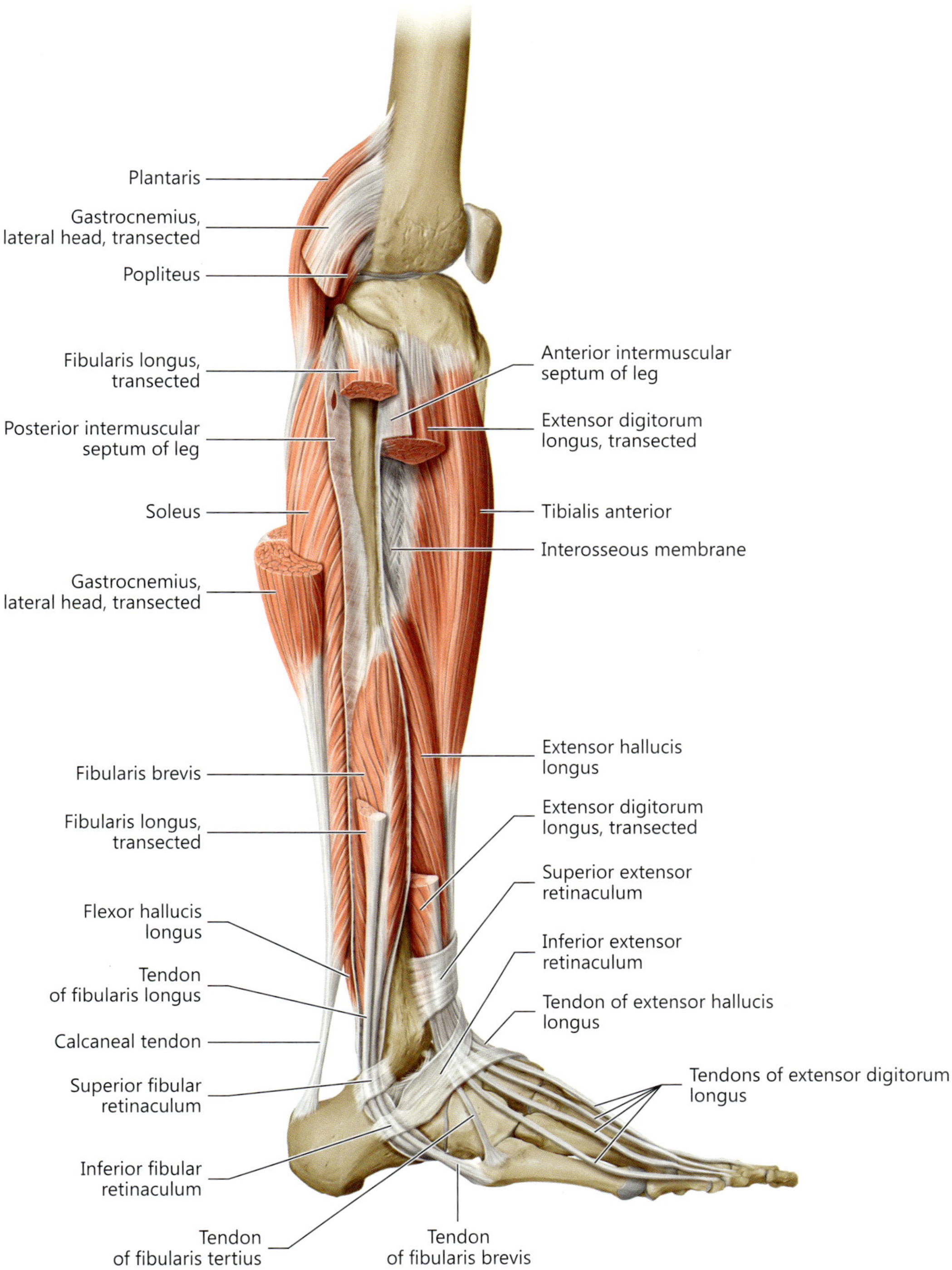

Muscles of the leg. Right side, lateral view. The fibularis longus in the lateral compartment, the gastrocnemius in the posterior compartment, and the extensor digitorum longus in the anterior compartment have been transected. In this way, in an antero-posterior direction, the extensor hallucis longus, fibularis brevis, and soleus are revealed. The fibularis brevis is located deeply, between the tibialis anterior and the extensor digitorum longus. The extensor retinacula, anteriorly at the level of the ankle, keep the position of the tendons of the anterior muscles of the leg that reach the foot.

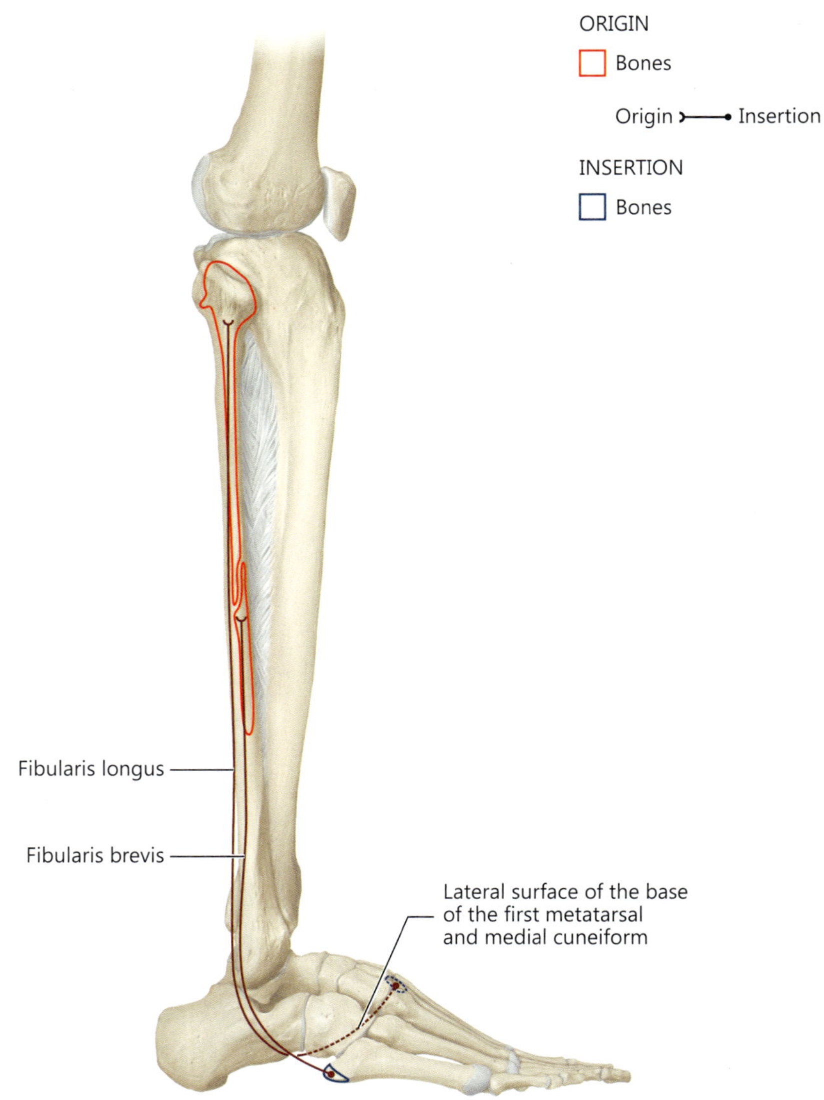

Schematic representation of the origin and insertions of the fibularis longus and fibularis brevis. The **fibularis longus** originates from the head and the upper two-thirds of the lateral surface of the fibula, and sometimes with few fibers from the lateral condyle of the tibia. It inserts on the lateral surface of the base of the first metatarsal and on the lateral surface of the medial cuneiform. The **fibularis brevis** originates from the lower two-thirds of the lateral surface of the fibula, and attaches to the tuberosity of the fifth metatarsal.

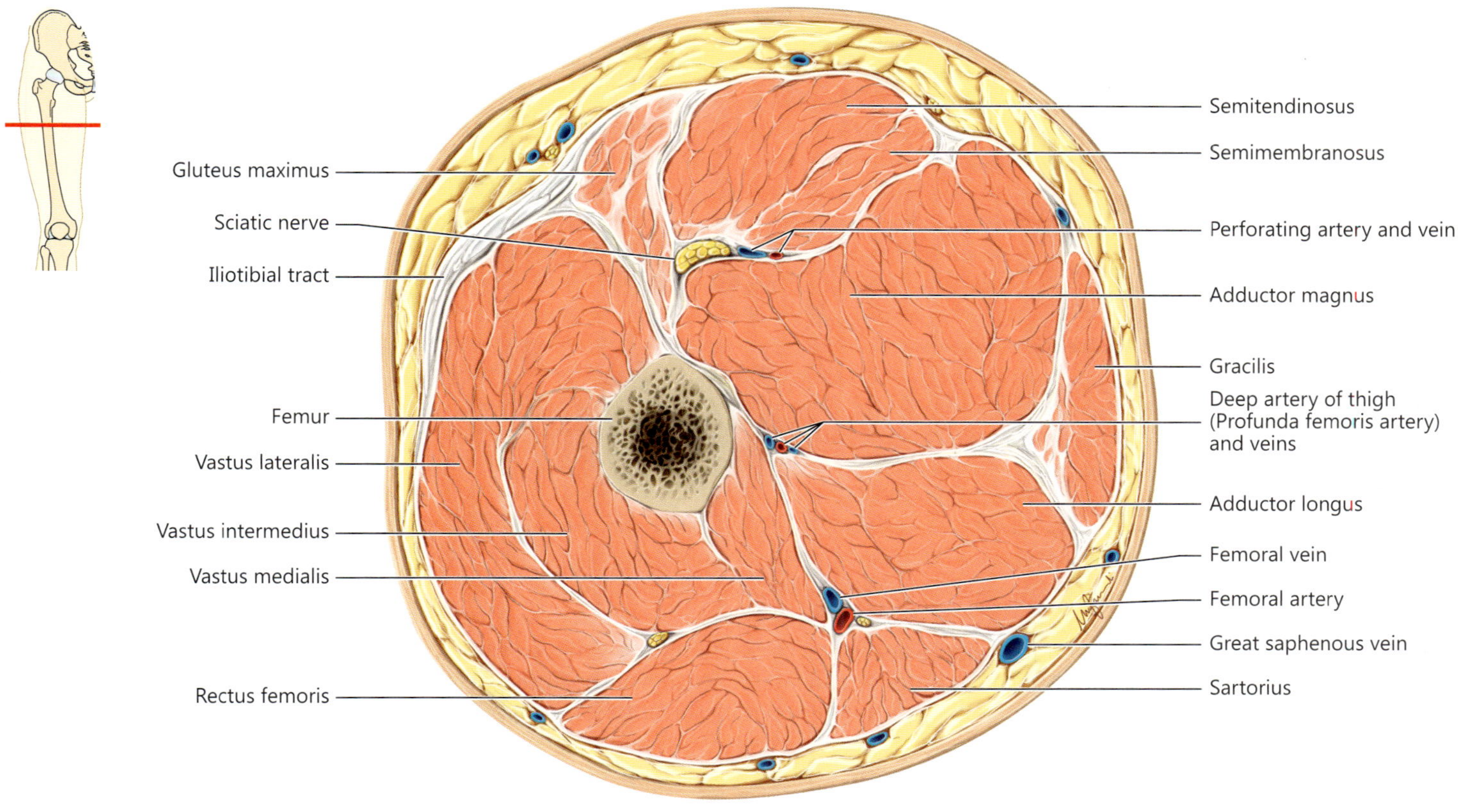

Transverse section of the **thigh** between the proximal and middle thirds. Right side, viewed from above. The anterior compartment of the thigh is occupied by the quadriceps femoris (rectus femoris and vasti); the medial compartment by the adductor muscles; the posterior compartment by the semitendinosus, semimembranosus, and biceps femoris (the latter not visible at this level). The sartorius is contained in a split of the fascia lata.

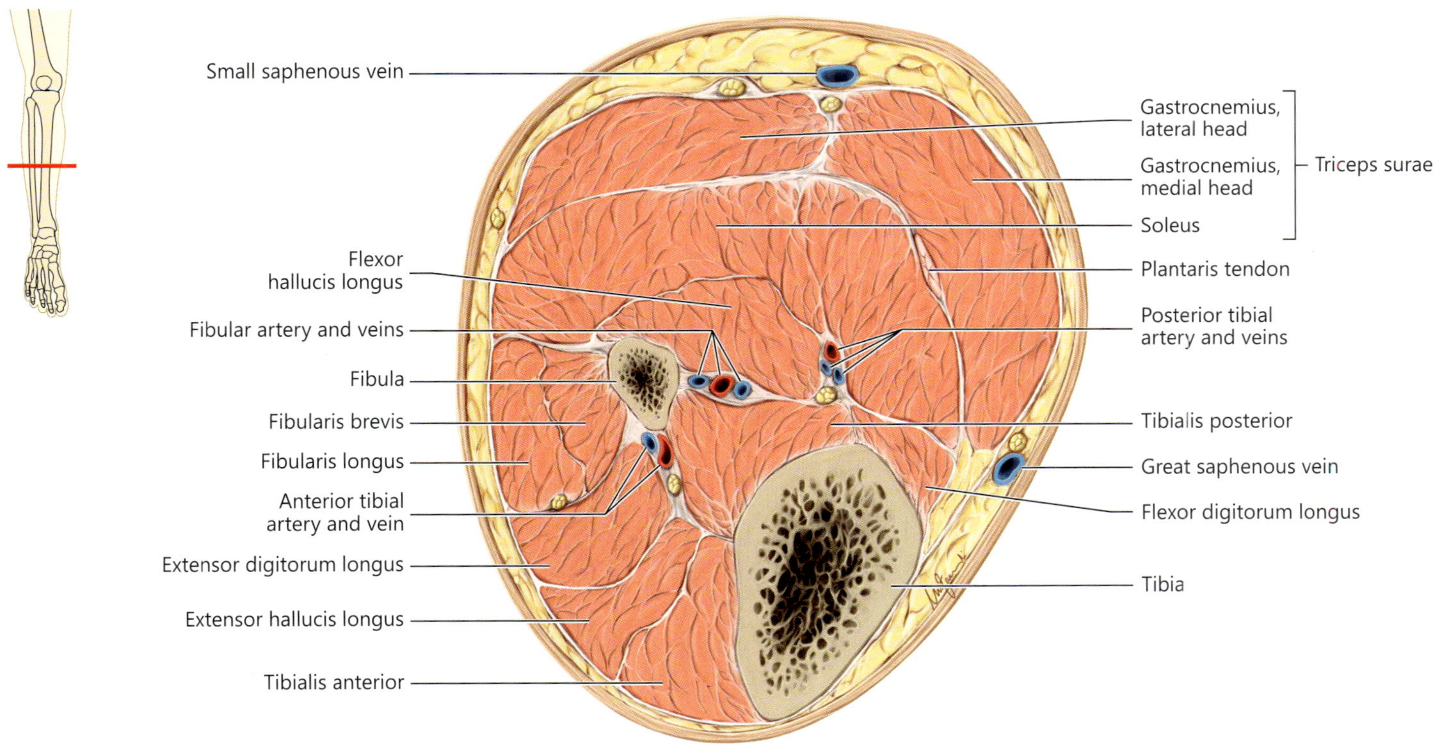

Transverse section of the **leg** in the middle of the diaphyses of the tibia and the fibula. Right side, viewed from above. The anterior compartment of the leg contains the tibialis anterior, extensor hallucis longus, extensor digitorum longus, and the fibularis tertius (the latter not visible at this level). The fibularis longus and brevis occupy the lateral compartment of the leg (fibular compartment of the leg). The posterior muscles of the leg comprise, superficially, the triceps surae (gastrocnemius and soleus) and plantaris, and deeply, the flexor hallucis longus, tibialis posterior, flexor digitorum longus, and popliteus (the latter not visible at this level).

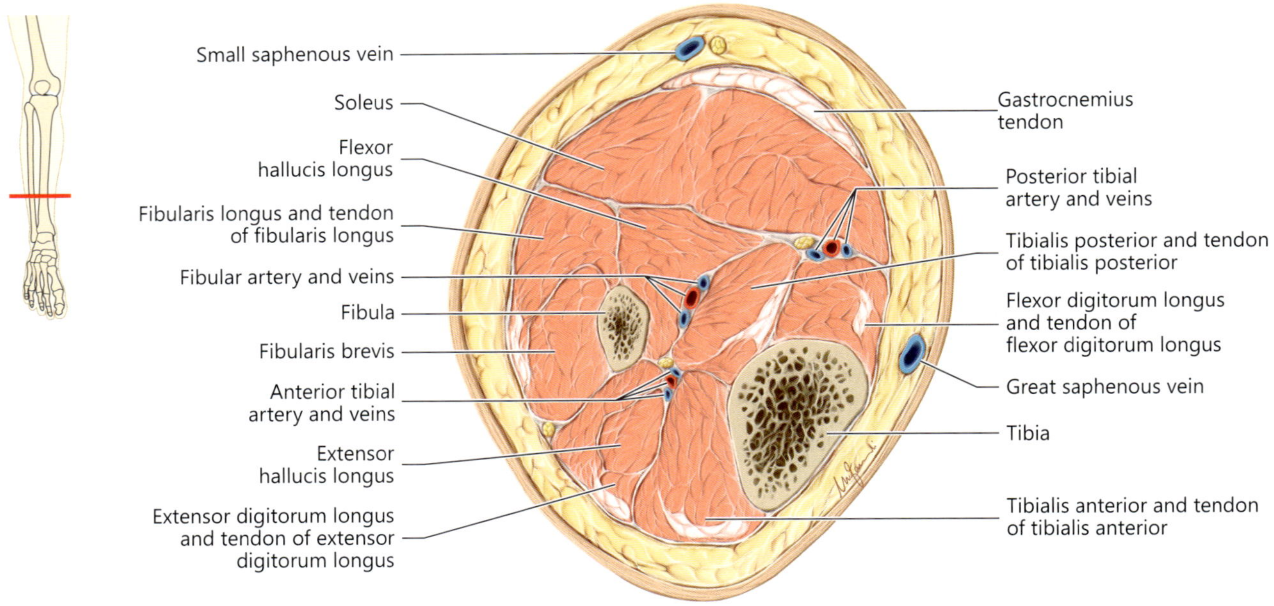

Transverse section of the **leg** between the middle and distal thirds. Right side, viewed from above. The anterior and posterior intermuscular septa of the leg, which insert at the anterior and lateral borders of the fibula, delineate the fibular compartment of the leg. The skeleton of the leg and the interosseous membrane separate the anterior compartment of the leg from the posterior one.

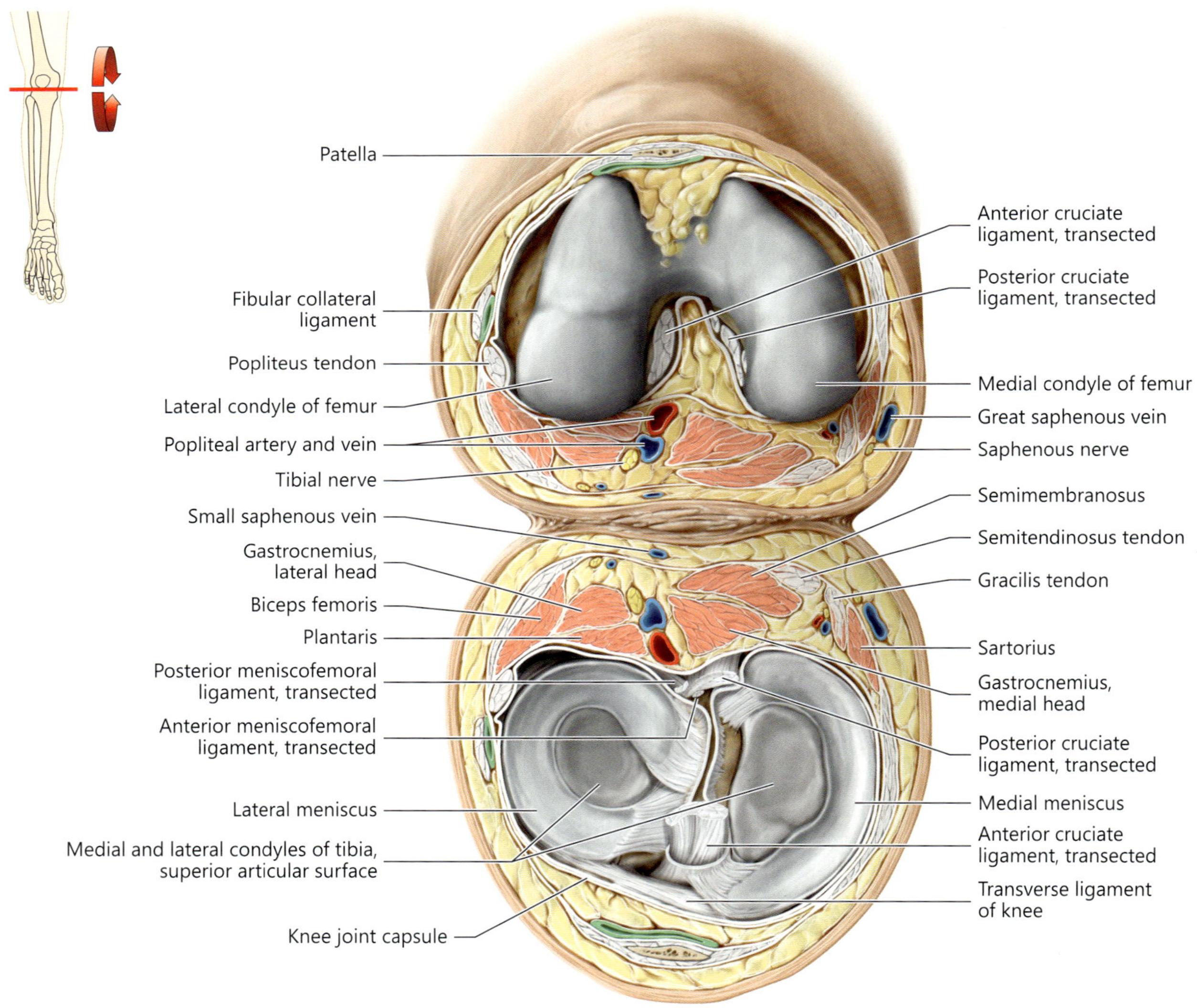

Right **knee**. The joint has been opened with a transverse section that has separated the articular surface of the femur from that of the tibia. Distal view (from below) of the condyles of the femur, and proximal view (from above) of those of the tibia, after turning the surface of the inferior section downwards and forwards.

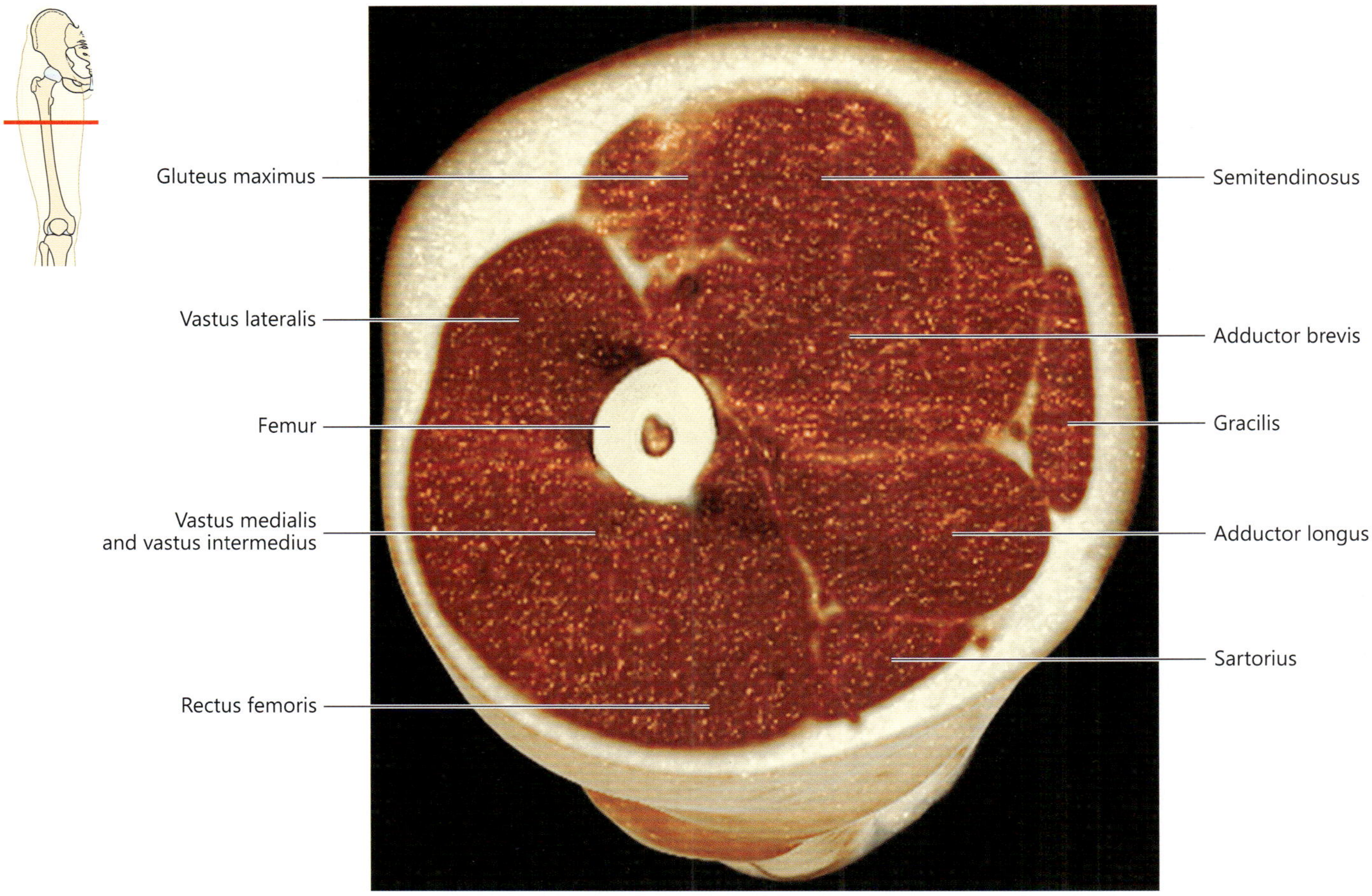

Three-dimensional image of a living human body, obtained by CT scan using a 3D volume rendering technique, showing the **thigh** in a transverse section between the proximal and middle thirds.

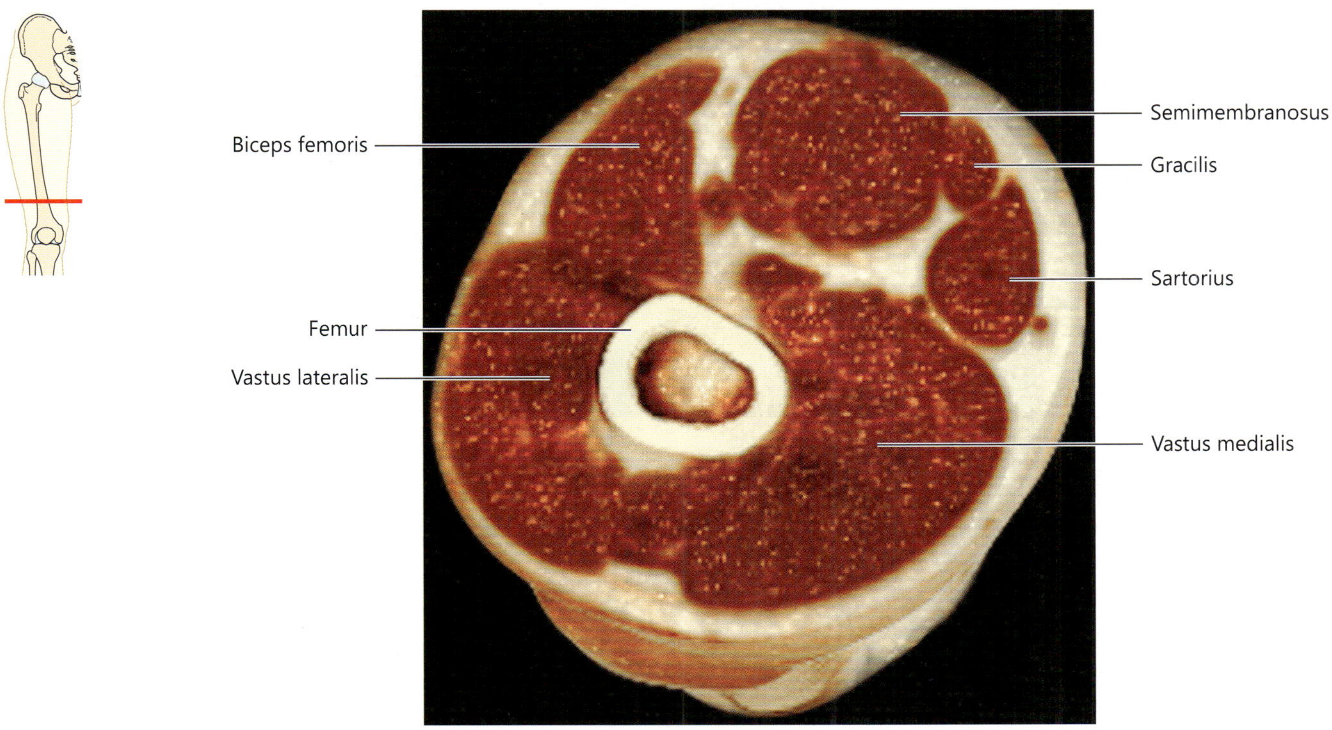

Three-dimensional image of a living human body, obtained by CT scan using a 3D volume rendering technique, showing the **thigh** in a transverse section in correspondence to the distal third, slightly above the knee joint.

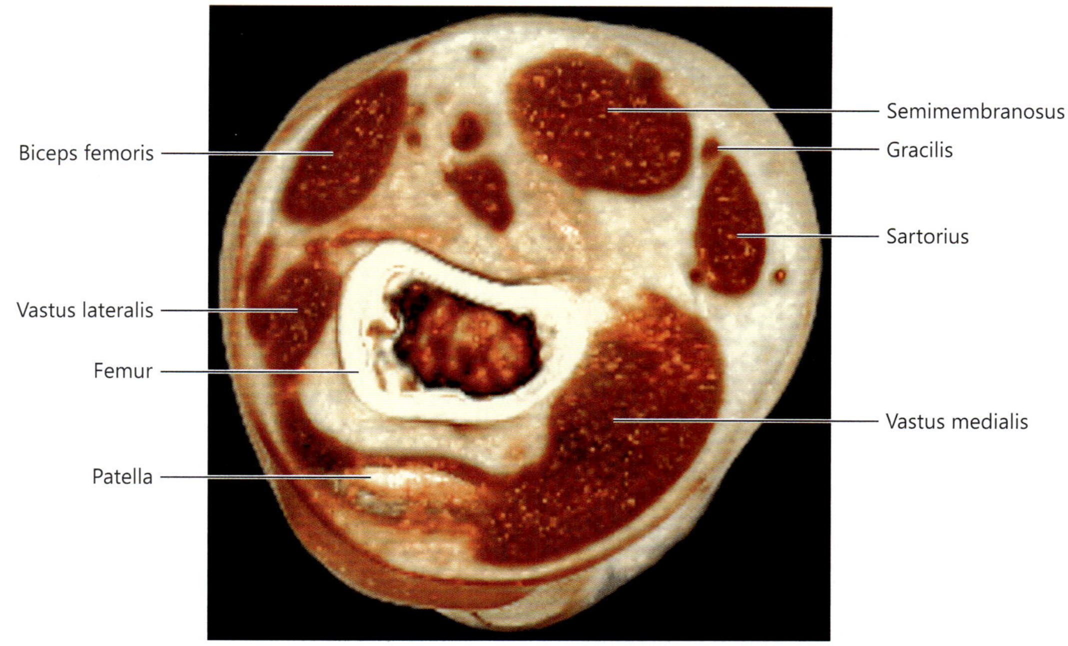

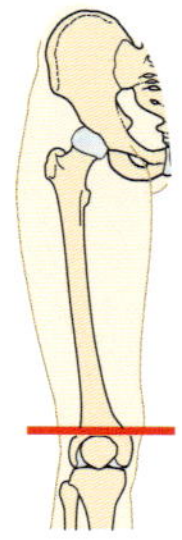

Three-dimensional image of a living human body, obtained by CT scan using a 3D volume rendering technique, showing the **thigh** in a transverse section in correspondence to the distal third.

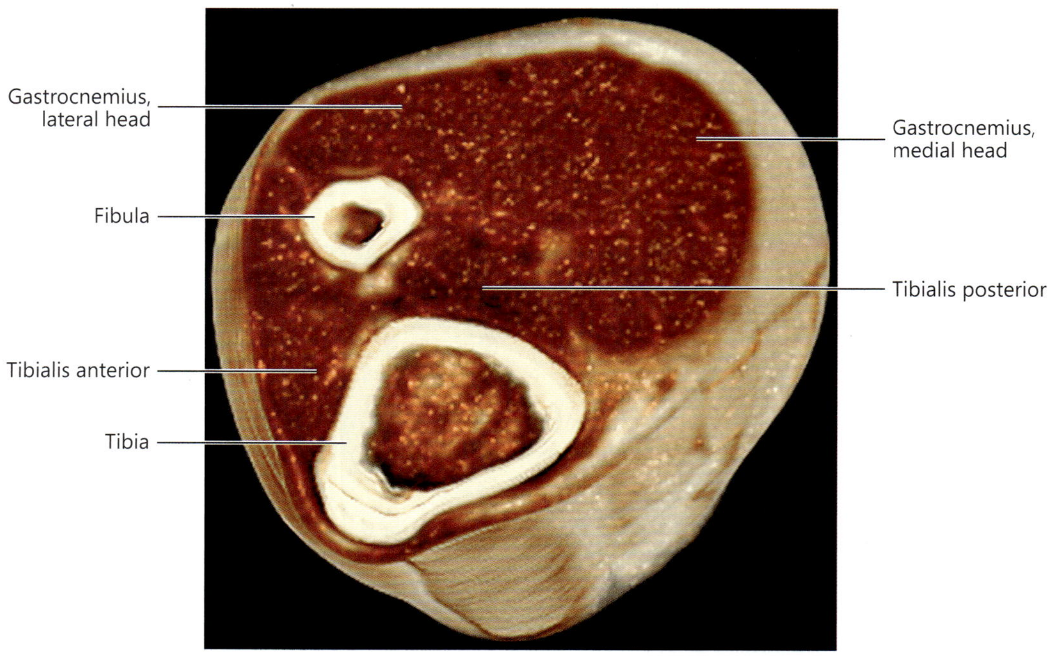

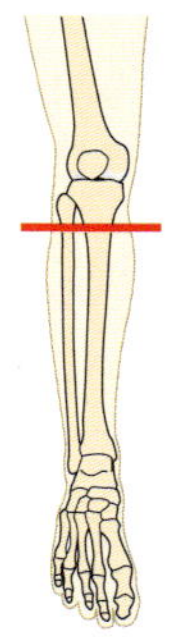

Three-dimensional image of a living human body, obtained by CT scan using a 3D volume rendering technique, showing the **leg** in a transverse section in correspondence to the proximal third.

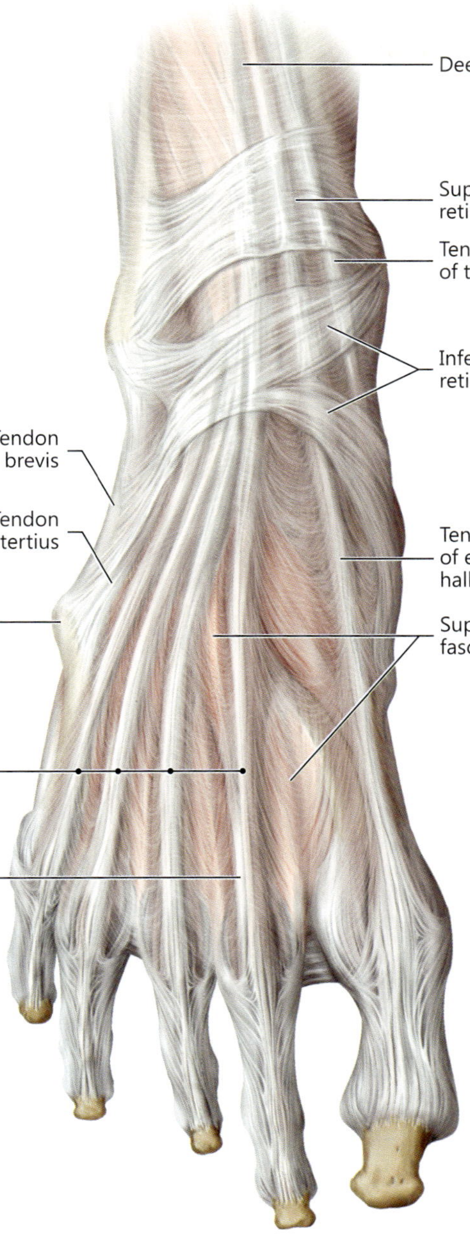

Dorsum of the right **foot**. The tendon of the extensor digitorum longus, after passing deep to the superior and inferior extensor retinaculum, divides into four tendons for the last four toes, which diverge on the dorsum of the foot. The tendons of the tibialis anterior and extensor hallucis longus run deep to the extensor retinacula as well, while those of the fibular muscles are held in position by the fibular retinacula.

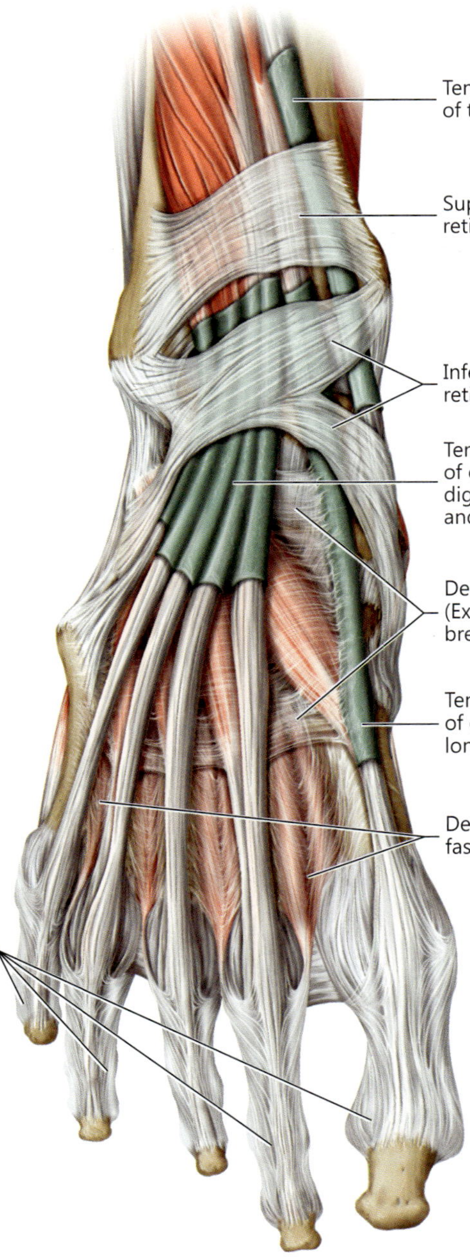

Dorsum of the right **foot**. From the deep surface of the extensor retinacula, fibrous septa originate, which insert at the skeleton and identify three osteofibrous canals where the tendons of the anterior muscles of the leg run into three synovial sheaths: one for the tibialis anterior, one for the extensor hallucis longus, one for the extensor digitorum longus and the fibularis tertius.

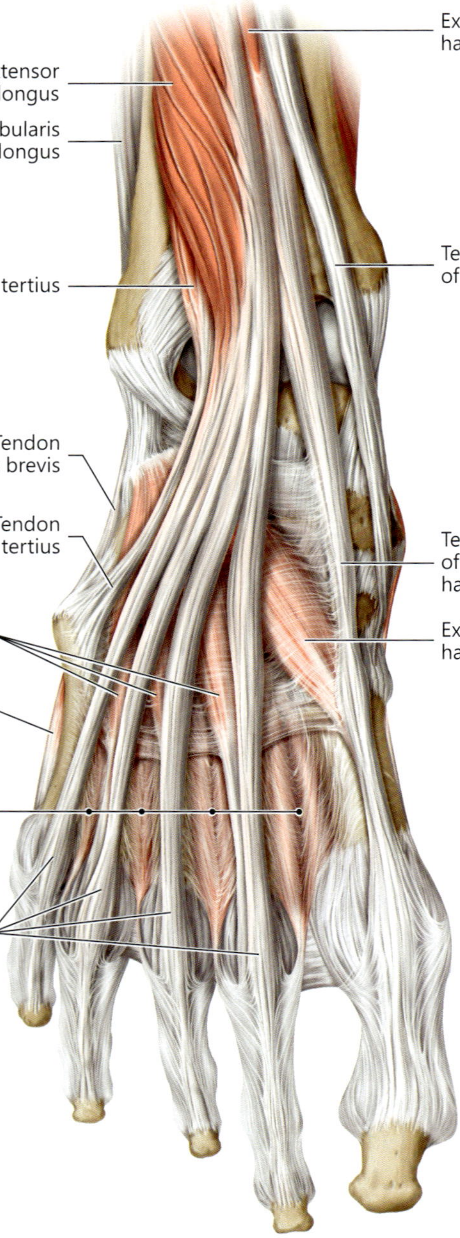

➤ **Dorsum of the** right **foot**. The sheaths of the tendons of the anterior muscles of the leg and the extensor retinacula have been removed. Each of the tendons of the extensor digitorum longus divides into three splits, the middle of which inserts at the base of the middle phalanx, while the other two attach to the base of the distal phalanx.

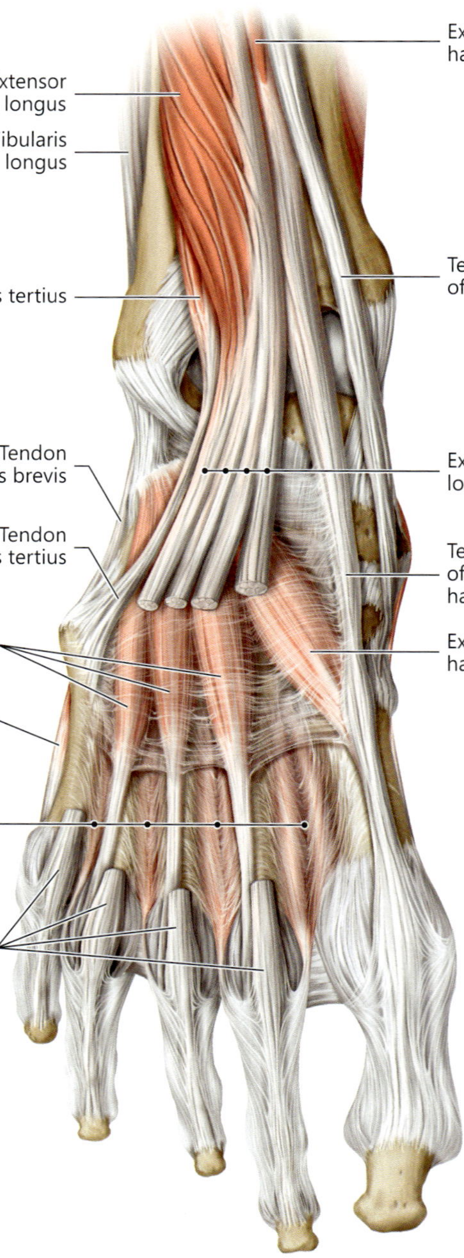

➤ **Dorsum of the** right **foot**. With the tendons of the extensor digitorum longus transected, the muscles of the dorsum of the foot, the extensor digitorum brevis and the extensor hallucis brevis, here still covered by the deep dorsal fascia of the foot that also covers the dorsal interossei, become visible.

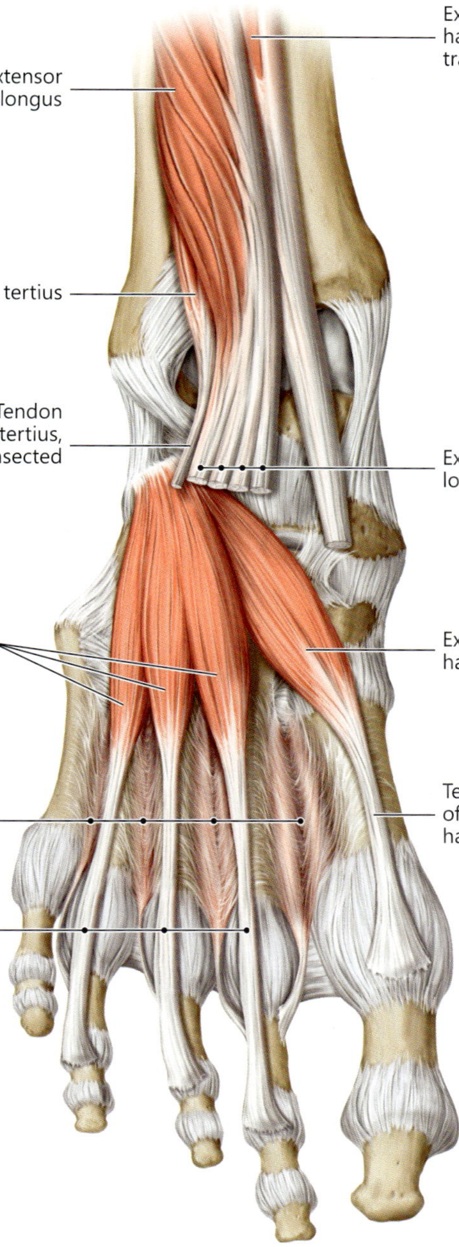

➤ **Dorsum of the** right **foot**. With the deep dorsal fascia of the foot that covers the extensor digitorum brevis and extensor hallucis brevis transected, we see the belly of the extensor digitorum brevis. This muscle is divided into three heads, which end with tendons reaching the second, third, and fourth toes, uniting to the lateral surface of the tendons of the extensor digitorum longus.

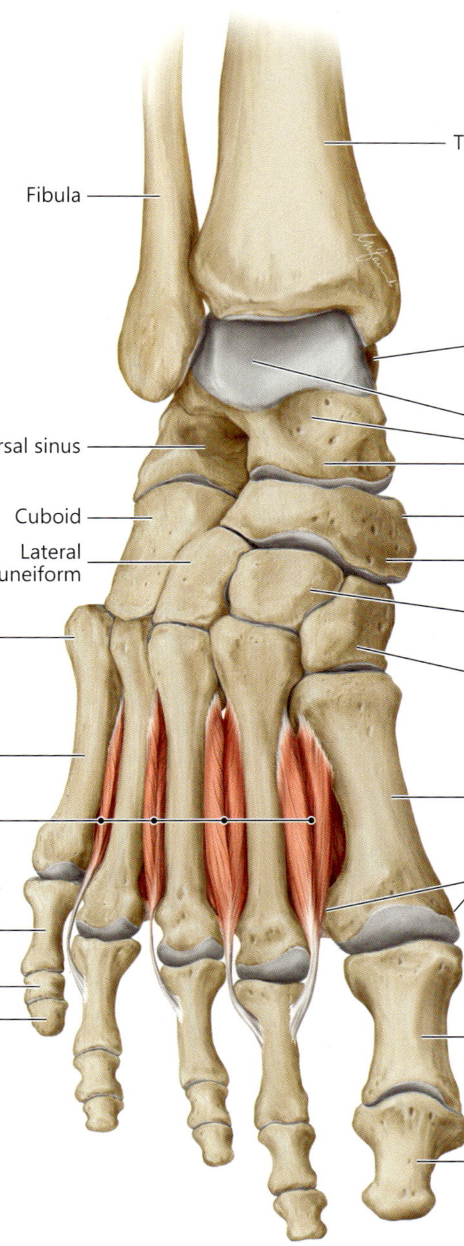

➤ **Dorsum of the** right **foot**. The four dorsal interossei, here without the deep dorsal fascia that covers them, each originate with two heads from the facing surfaces of the metatarsals. The first and the second dorsal interossei insert at the second toe; the third and the fourth, respectively, at the third and fourth toes.

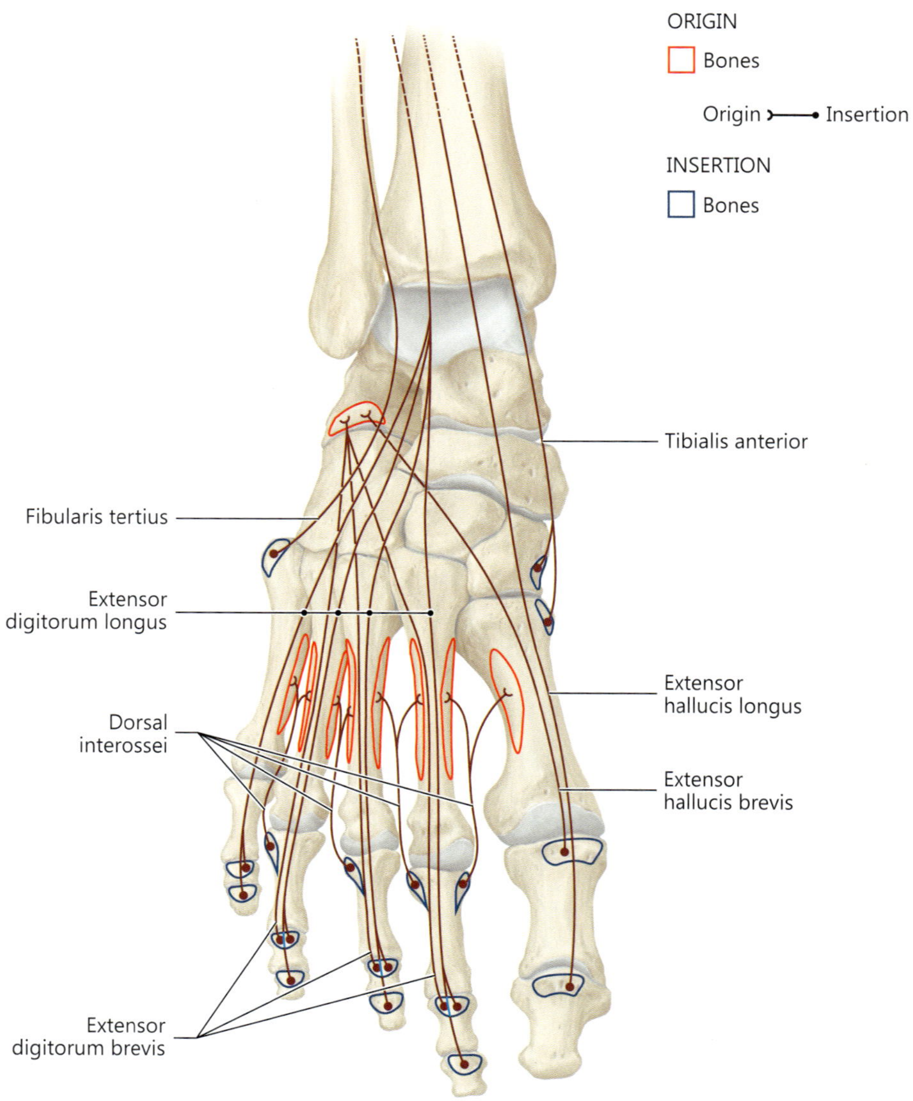

Schematic representation of the insertions of the tibialis anterior, extensor digitorum longus, extensor hallucis longus, and fibularis tertius, and of the origin and insertions of the extensor digitorum brevis, extensor hallucis brevis, and dorsal interossei. The **tibialis anterior** inserts on the medial and inferior surfaces of the medial cuneiform and on the medial surface of the base of the first metatarsal, the **extensor digitorum longus** on the dorsal surface of the base of the middle and distal phalanges of the last four toes, the **extensor hallucis longus** on the dorsal surface of the base of the distal phalanx of the great toe, the **fibularis tertius** on the dorsal surface of the base of the fifth metatarsal. The **extensor digitorum brevis** originates from the distal part of the superolateral surface of the calcaneus, immediately in front of the tarsal sinus. Its three tendons join the lateral surface of the tendons of the extensor digitorum longus for the second, third, and fourth toes. The **extensor hallucis brevis** has a common origin with the extensor digitorum brevis, and inserts on the dorsal surface of the base of the proximal phalanx of the great toe. The **dorsal interossei** originate from the adjacent sides of the metatarsals: the first runs to the medial surface of the base of the proximal phalanx of the second toe, while the second, third, and fourth reach, respectively, the lateral surfaces of the bases of the proximal phalanges of the second, third, and fourth toes.

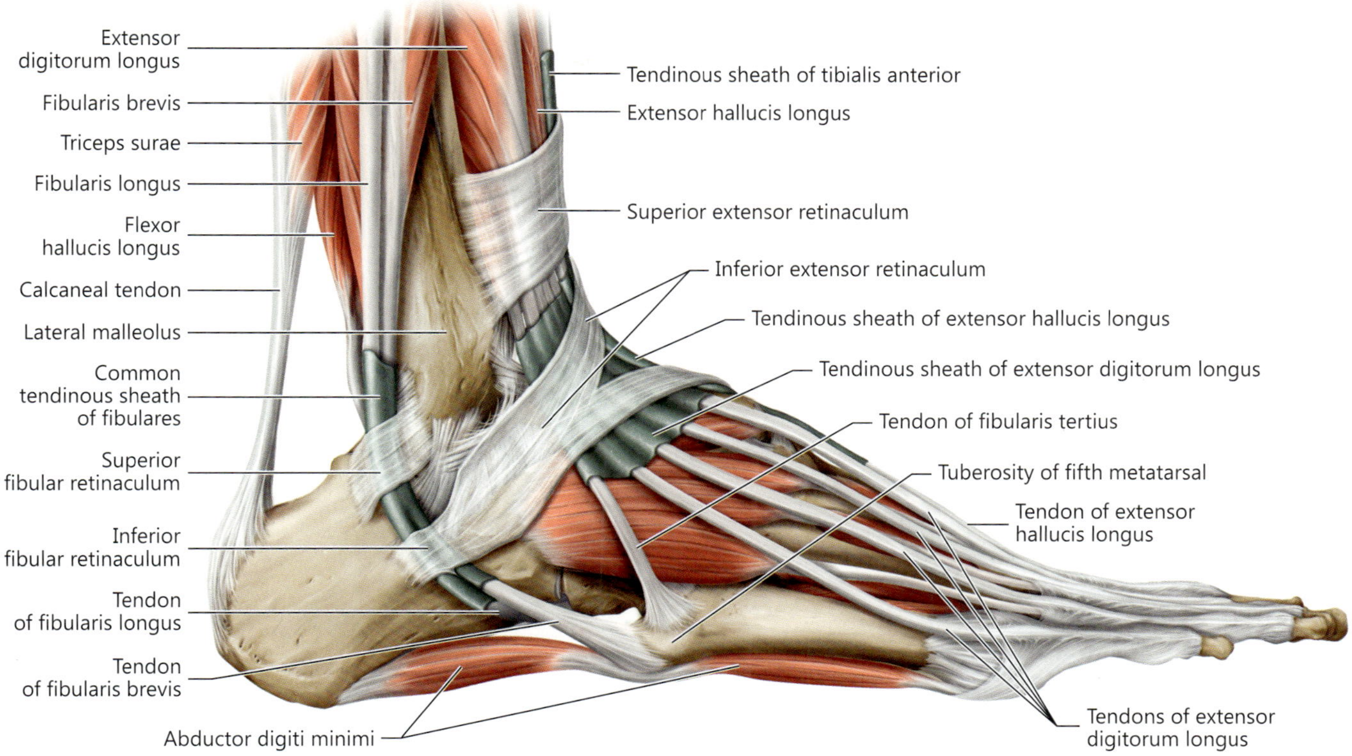

Right foot, lateral view. **Anterior tarsal** and **fibular tendinous sheaths** and **anterior** and **lateral fascial structures** at the level of the ankle. The tendons of the anterior muscles of the leg run inside of three synovial sheaths: one for the tibialis anterior, one for the extensor hallucis longus, and one for the extensor digitorum longus and fibularis tertius. The three sheaths are located deep to the extensor retinacula. The tendons of the fibularis longus and fibularis brevis run inside of a common sheath deep to the fibular retinacula.

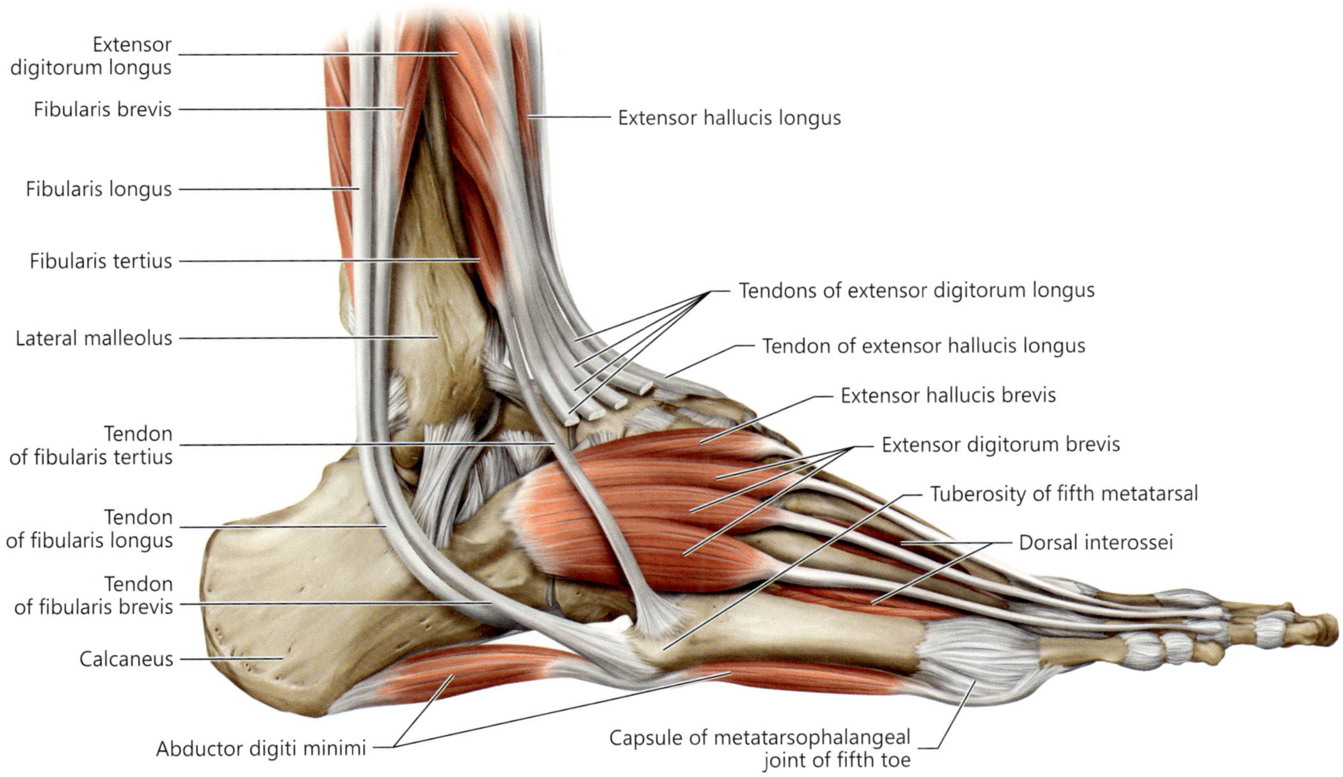

Muscles of the right **foot**, lateral view, after removal of the extensor and fibular retinacula and of the sheaths of the corresponding tendons, and the resection of the tendons of the extensor digitorum longus. The muscles of the dorsum of the foot are shown: the extensor digitorum brevis and the extensor hallucis brevis. The abductor digiti minimi is the most superficial of the lateral muscles of the sole of the foot. The tendons of the fibularis longus and fibularis brevis run together behind the lateral malleolus and then diverge, running respectively into the sole and along the lateral border of the foot.

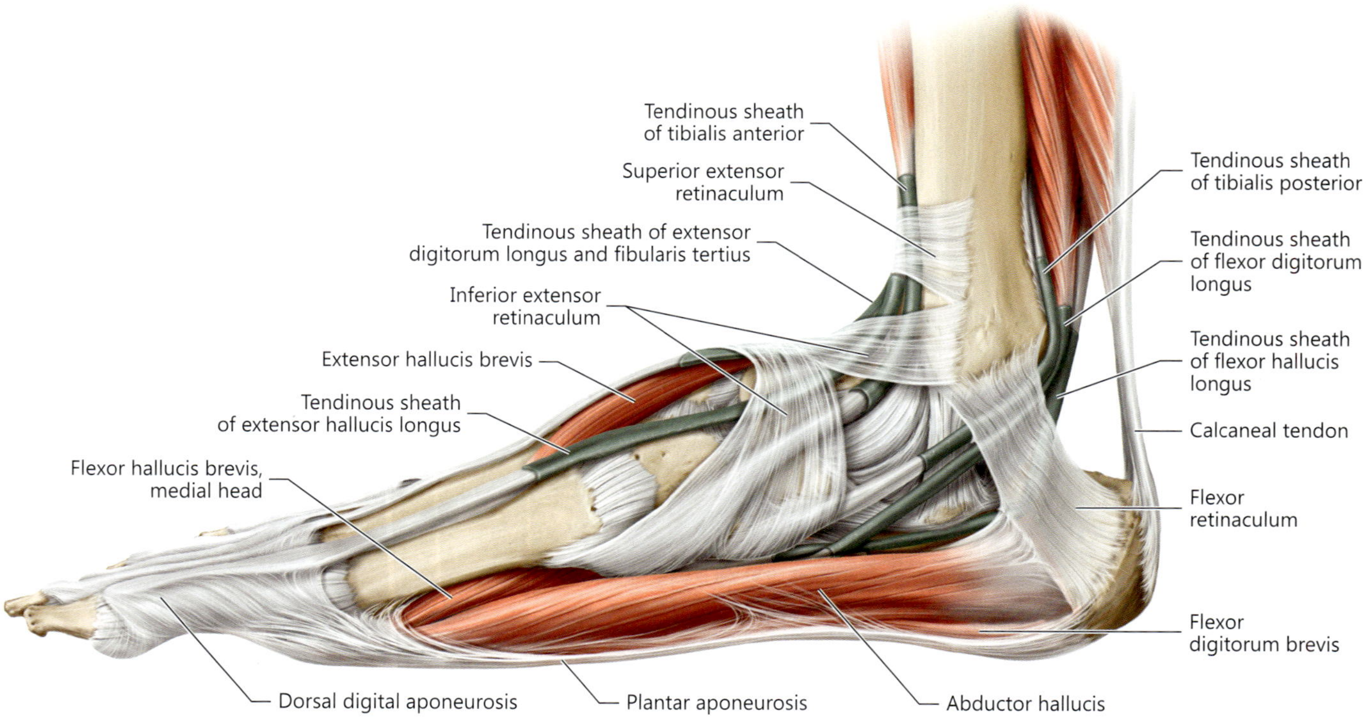

Right foot, medial view. **Anterior tarsal** and **tibial tendinous sheaths** and **anterior** and **medial fascial structures** at the level of the ankle. Deep to the extensor retinacula run the tendons of the anterior muscles of the leg, within three synovial sheaths: one for the tibialis anterior, one for the extensor hallucis longus, and one for the extensor digitorum longus and fibularis tertius. Deep to the flexor retinaculum run the tendons of the deep posterior muscles of the leg, within three synovial sheaths for the tibialis posterior, the flexor hallucis longus, and the flexor digitorum longus.

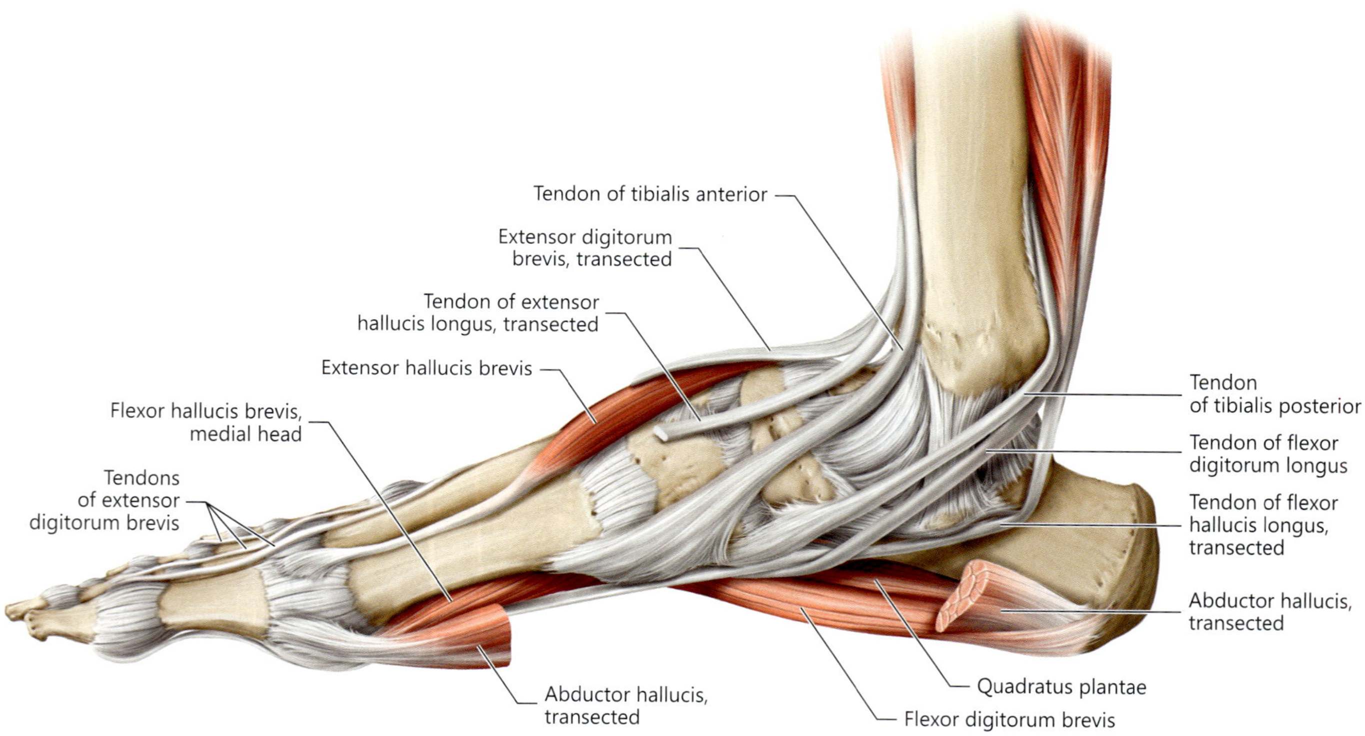

Muscles of the right **foot**, medial view. We see the tendons of the posterior muscles of the leg, located in the deep layer: the tibialis posterior, flexor digitorum longus, and flexor hallucis longus; of the anterior muscles of the leg: tibialis anterior and extensor hallucis longus; of the muscles of the dorsum of the foot: extensor digitorum brevis and extensor hallucis brevis. The resection of the abductor hallucis, the most superficial of the medial muscles of the sole of the foot, shows the flexor hallucis brevis and the muscles of the intermediate region of the sole of the foot: flexor digitorum brevis (more superficial) and quadratus plantae (on the intermediate plane).

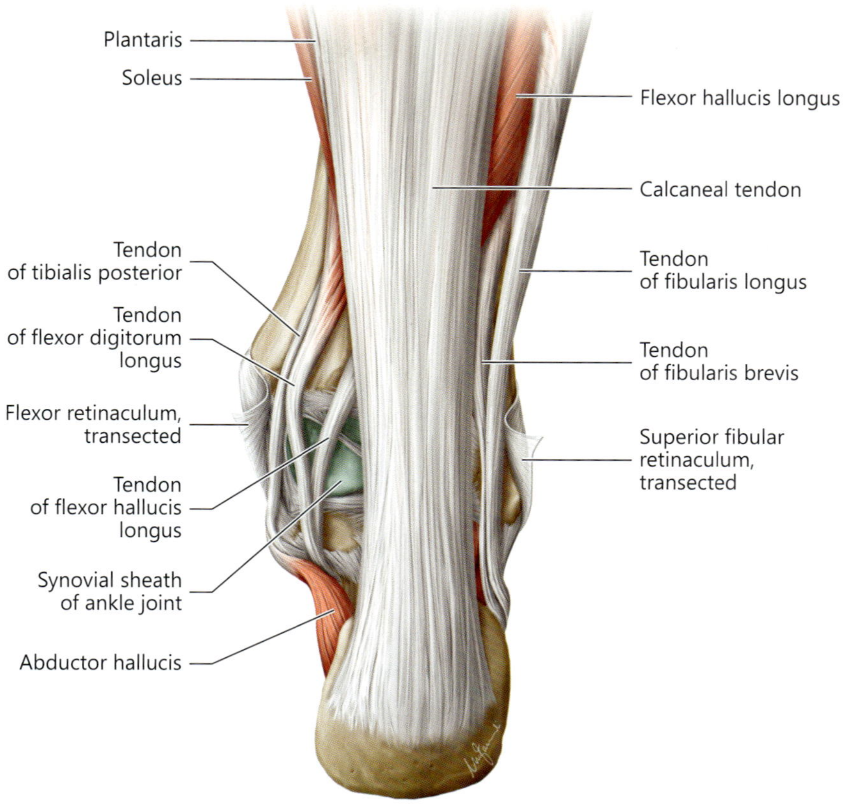

Muscles of the right **foot**, posterior view. The calcaneal tendon (Achilles tendon), formed by the union of the aponeuroses of the gastrocnemius and soleus, inserts at the calcaneal tuberosity. The tendons of the fibularis longus and fibularis brevis run together behind the lateral malleolus, deep to the fibular retinacula. The tendons of the deep posterior muscles of the leg, tibialis posterior, flexor hallucis longus, and flexor digitorum longus run behind the medial malleolus, deep to the flexor retinaculum.

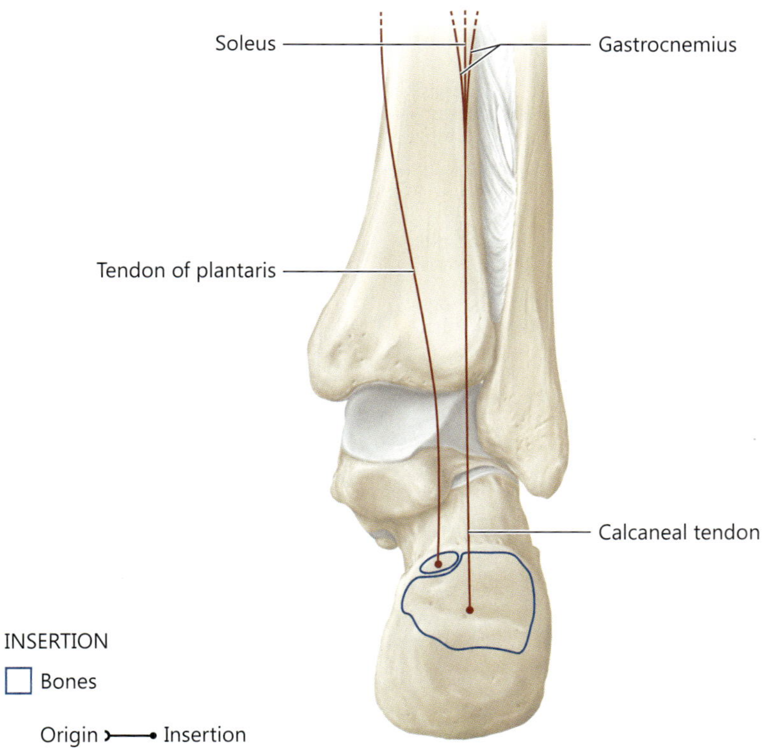

Schematic representation of the insertions of the gastrocnemius, soleus, and plantaris. The **gastrocnemius** and **soleus** insert on the calcaneal tuberosity through the calcaneal tendon, into which their aponeuroses continue. The tendon of the **plantaris** runs between the gastrocnemius and soleus, and after reaching the medial border of the calcaneal tendon, inserts on the medial part of the posterior surface of the calcaneus.

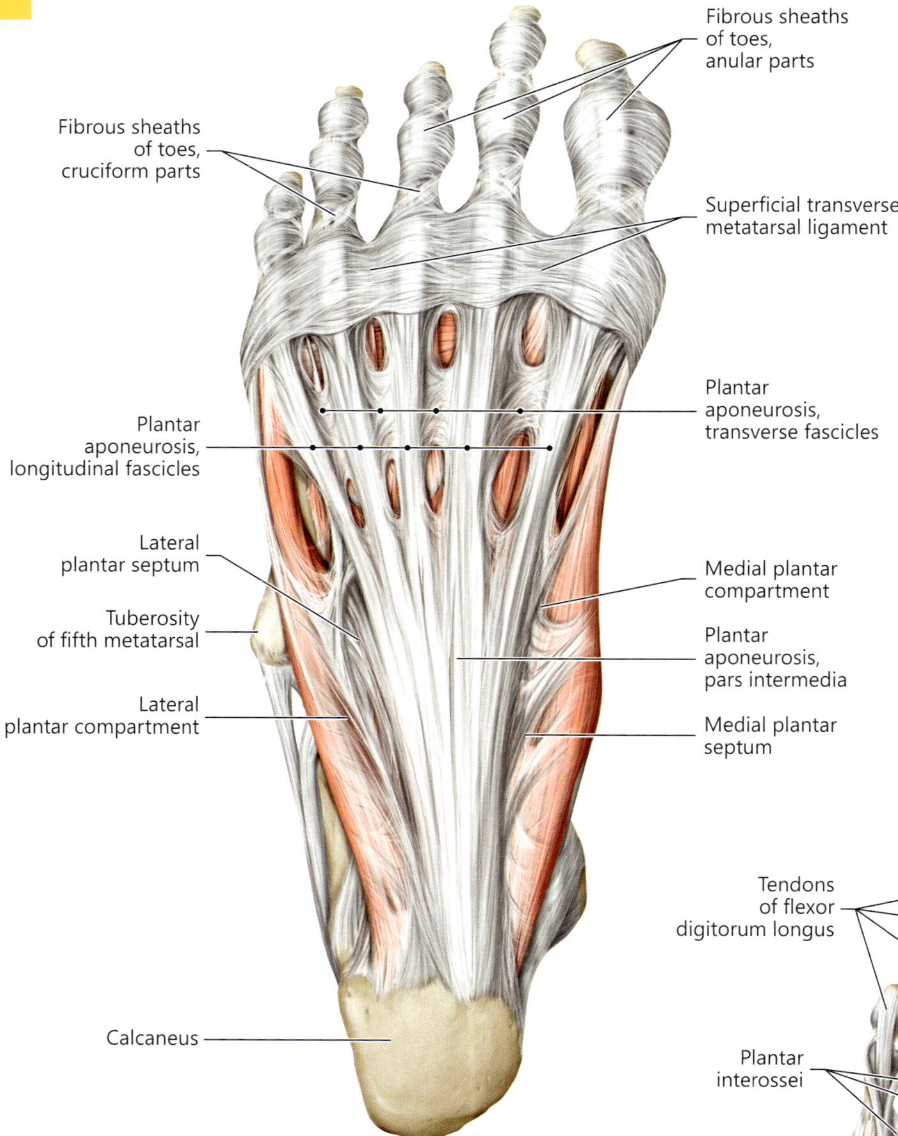

Muscles of the sole of the right **foot**. The plantar aponeurosis is formed by longitudinal fascicles and transverse fascicles that join them together. From its deep surface, the medial and lateral plantar septa originate, which delineate three osteofibrous compartments in which the medial, intermediate, and lateral muscles of the sole of the foot are contained.

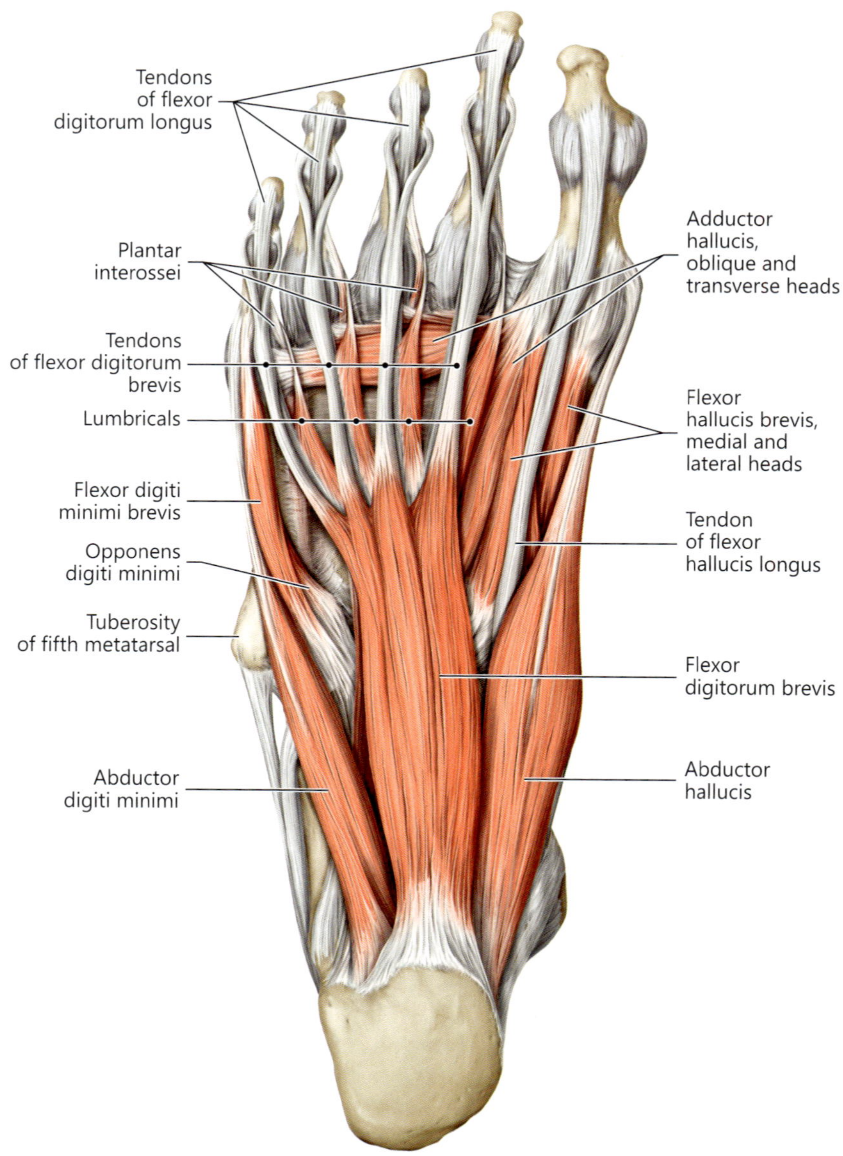

Muscles of the sole of the right **foot**. With the plantar aponeurosis transected, the medial and lateral muscles of the sole of the foot and the first layer, more superficial, of the muscles of the intermediate region appear. The medial muscles run to the great toe: the abductor, flexor brevis, and adductor. The lateral muscles act upon the little toe: abductor, opponens, and flexor brevis. The flexor digitorum brevis forms the superficial layer of the intermediate muscles.

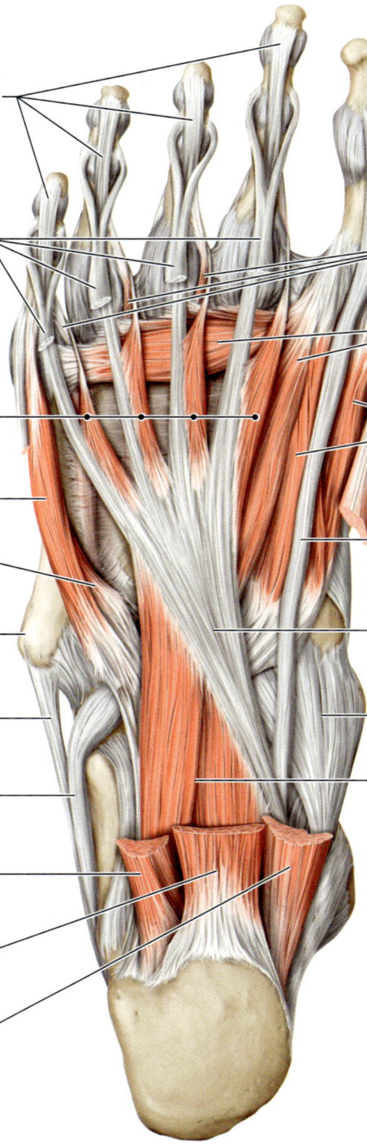

➤ **Muscles of the sole of the** right **foot**. After removal of the flexor digitorum brevis, proceeding from the surface to within, in the intermediate region of the sole there appears a second muscular plane, formed by the quadratus plantae and the lumbricals. In the intermediate compartment run the tendons of the flexor digitorum longus and flexor hallucis longus as well.

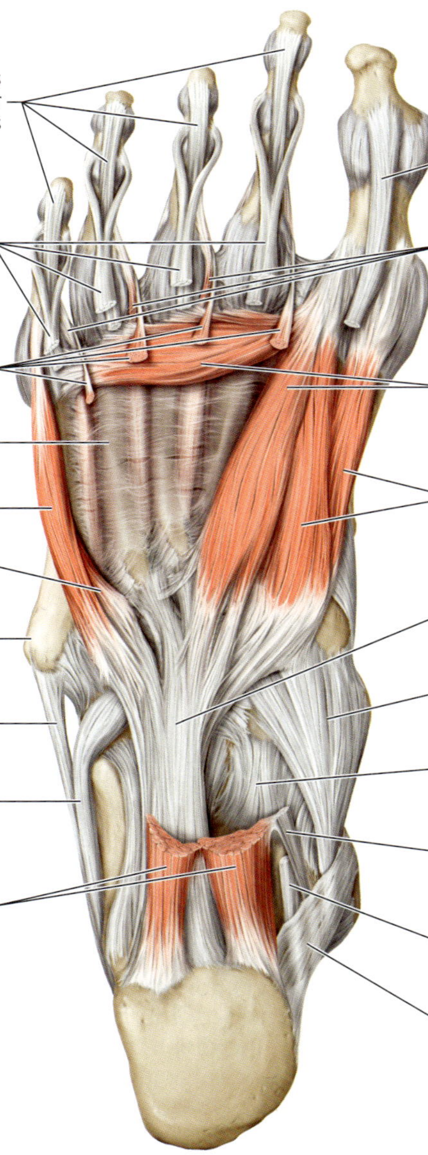

➤ **Muscles of the sole of the** right **foot**. The removal of the tendons of the flexor digitorum longus and flexor hallucis longus, the quadratus plantae, and the lumbricals shows the long plantar ligament and the deep plantar fascia that covers the plantar interossei, which, together with the dorsal interossei, constitute the deepest layer of the muscles of the intermediate region of the sole of the foot.

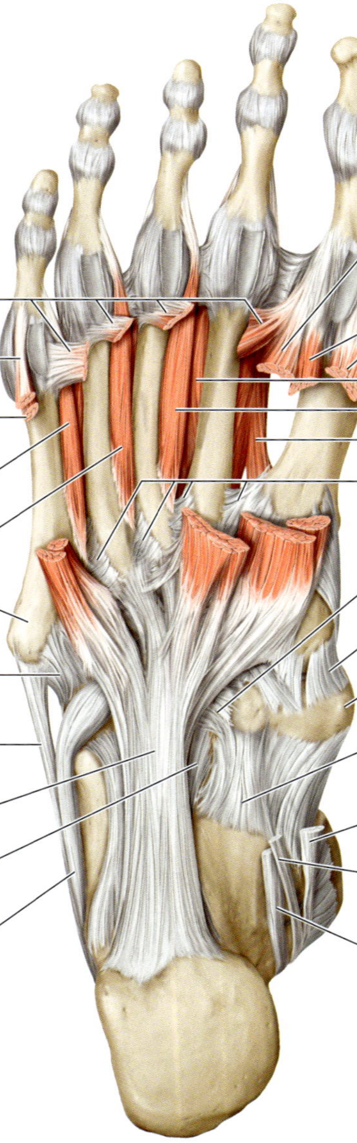

◄ **Muscles of the sole of the** right **foot**. With the deep medial muscles (flexor hallucis brevis and adductor hallucis) and deep lateral muscles (flexor digiti minimi brevis and opponens digiti minimi) from the sole of the foot and the deep plantar fascia transected, we see the plantar and dorsal interossei, which form the third layer, the deepest of the muscles of the intermediate region of the sole of the foot.

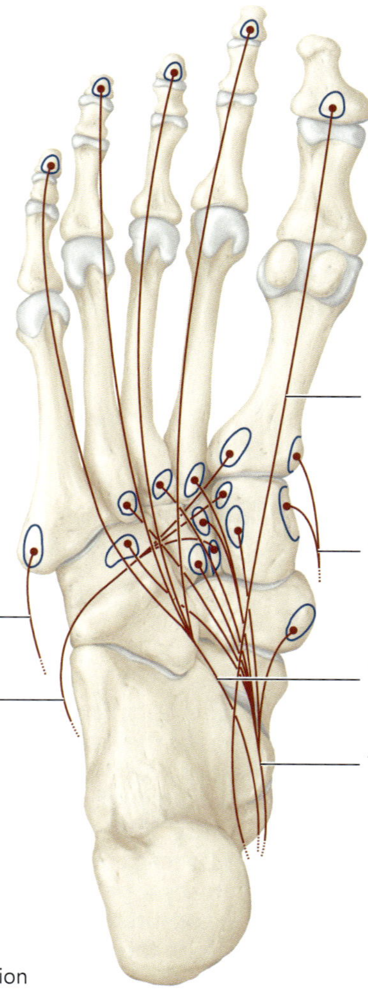

► Schematic representation of the insertions of the fibularis longus and brevis, flexor hallucis longus, flexor digitorum longus, tibialis anterior, and tibialis posterior. The **fibularis longus** inserts on the lateral surface of the base of the first metatarsal and on the medial and intermediate cuneiforms; the **fibularis brevis** on the tuberosity of the fifth metatarsal. The **flexor hallucis longus** inserts on the plantar surface of the base of the distal phalanx of the great toe, while the **flexor digitorum longus** inserts on the plantar surface of the base of the distal phalanges of the other four toes. The **tibialis anterior** inserts on the medial and inferior surfaces of the medial cuneiform and on the medial surface of the base of the first metatarsal; the **tibialis posterior** on the tuberosity of the navicular, with tendinous expansions directed to the plantar surface of the three cuneiforms.

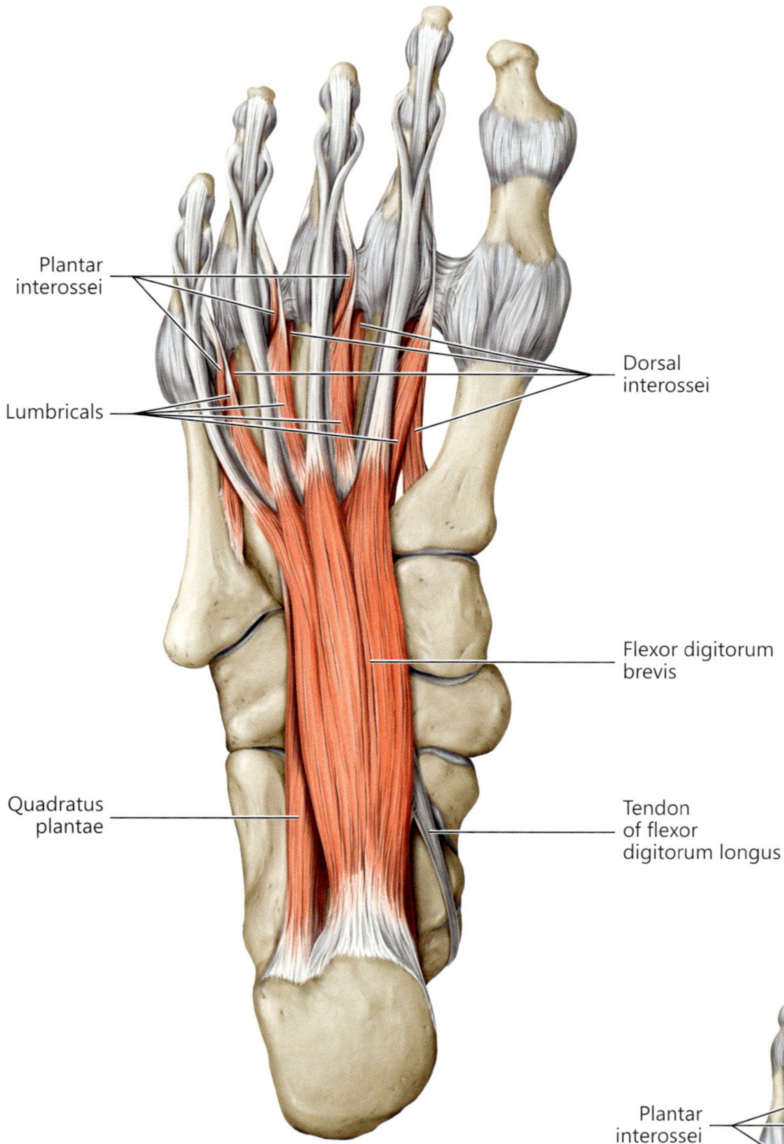

➤ **Muscles of the sole of the** right **foot**. The muscles of the intermediate region of the sole of the foot are arranged in three superimposed planes, deep to the plantar aponeurosis. The first layer, the most superficial, is formed by the flexor digitorum brevis; the second, intermediate, is formed by the quadratus plantae and the lumbricals; the third, the deepest, by the dorsal and plantar interossei.

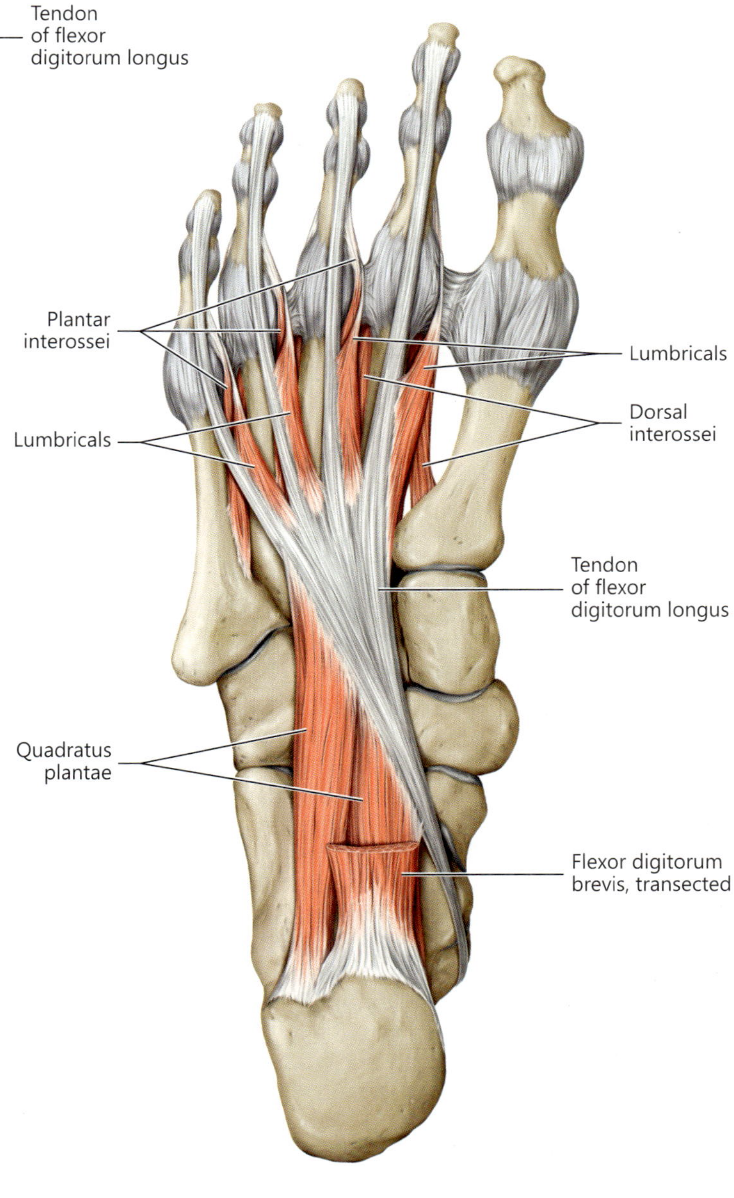

➤ **Muscles of the sole of the** right **foot**. With the flexor digitorum brevis, which constitutes the most superficial muscular layer of the intermediate region of the sole of the foot, transected, we see the tendon of the flexor digitorum longus, which divides into four terminal tendons, these too housed in the intermediate compartment of the foot, and the quadratus plantae, which, along with the lumbricals, forms the second muscular plane of the intermediate region of the sole of the foot.

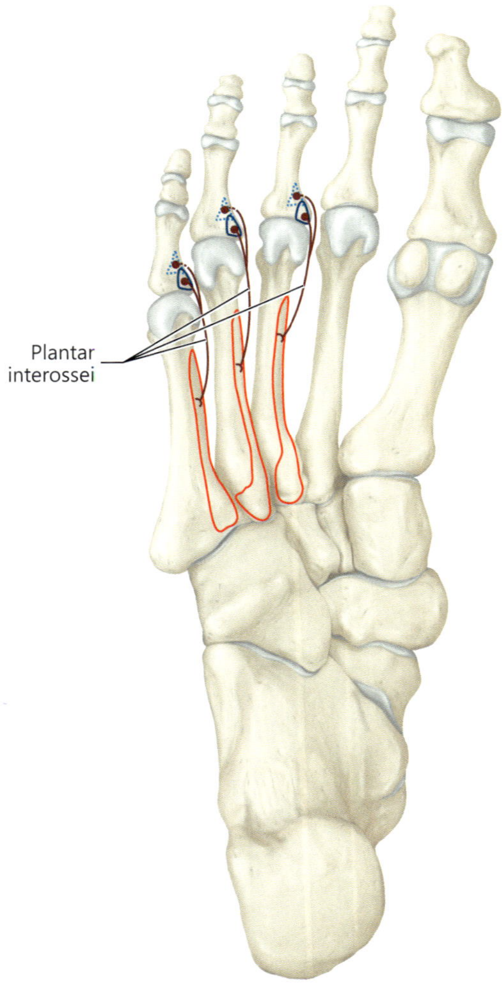

◄ Schematic representation of the origin and insertions of the plantar interossei. The **plantar interossei** originate from the bases and the medial surfaces of the third, fourth, and fifth metatarsals and insert on the medial surfaces of the bases of the proximal phalanges of the third, fourth, and little toes and on the dorsal digital expansion or aponeurosis.

ORIGIN

- Bones
- Aponeuroses, cartilage, raphe

Origin ≻——• Insertion

INSERTION

- Bones
- Raphe, cartilage, aponeuroses, septum, tendon, capsule and articular disc

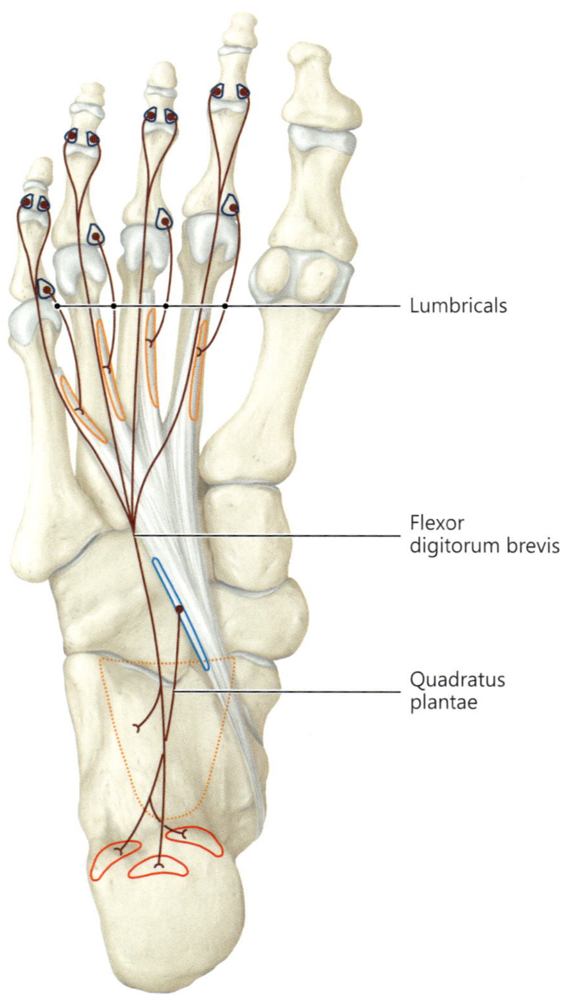

► Schematic representation of the origin and insertions of the lumbricals, quadratus plantae, and flexor digitorum brevis. The **first lumbrical** originates from the medial surface of the tendon for the second toe of the flexor digitorum longus; the **second**, the **third**, and the **fourth lumbricals** from the adjacent sides of the tendons of the flexor digitorum longus. The insertion of the **lumbricals** is on the medial surface of the dorsal digital expansion or aponeurosis on the proximal phalanges of the last four toes. The **quadratus plantae** originates from the lateral and medial borders of the inferior surface of the calcaneus and inserts laterally to the tendon of the flexor digitorum longus before its division into four terminal tendons. The **flexor digitorum brevis** extends from the medial process of the calcaneal tuberosity and from the plantar aponeurosis to the medial and lateral surfaces of the middle phalanges of the last four toes.

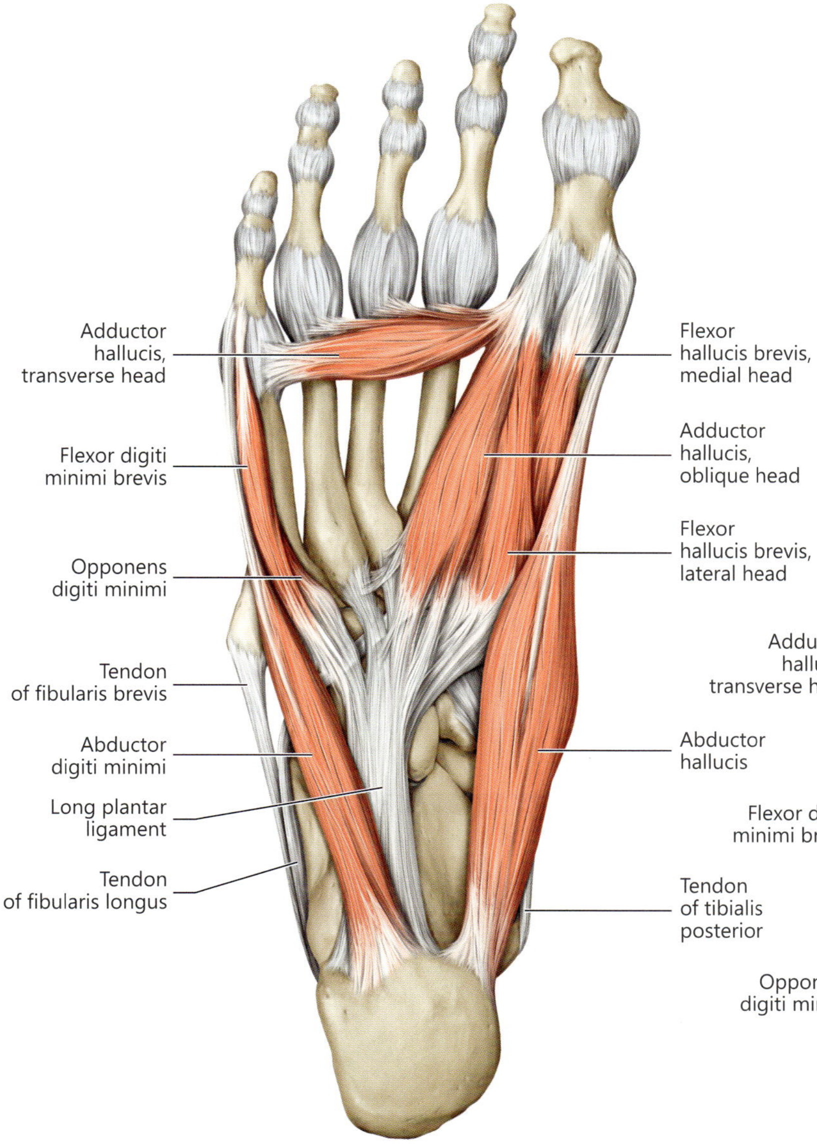

◄ **Medial** and **lateral muscles of the sole of the** right **foot**. The abductor hallucis is the most superficial of the medial muscles, while the flexor hallucis brevis and adductor hallucis are located more deeply. The most superficial of the lateral muscles is the abductor digiti minimi, deep to which are the flexor digiti minimi brevis and the opponens digiti minimi.

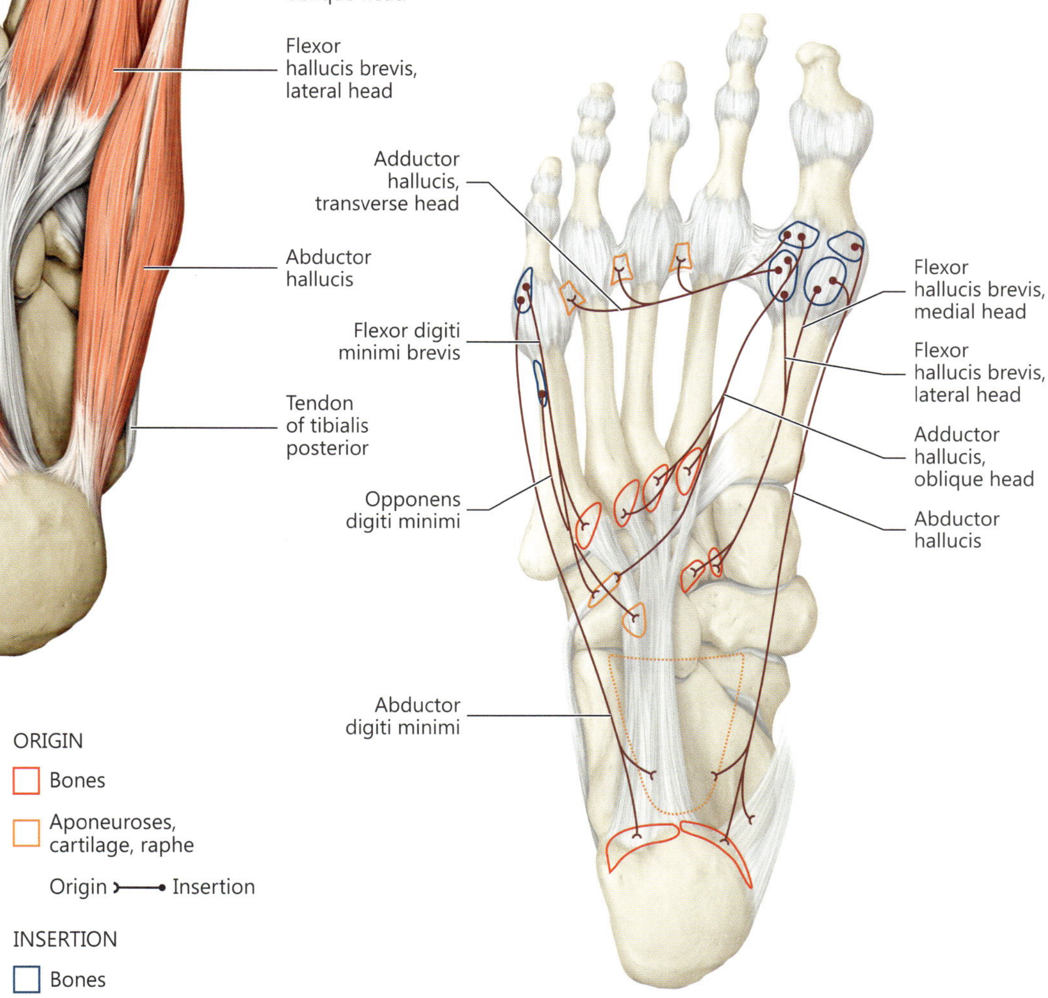

Schematic representation of the origin and insertions of the medial and lateral muscles of the sole of the foot. The **abductor hallucis** originates from the medial process of the calcaneal tuberosity, from the flexor retinaculum and from the plantar aponeurosis, and inserts on the medial surface of the base of the proximal phalanx of the great toe. The origin of the **flexor hallucis brevis** is from the plantar surface of the medial and intermediate cuneiforms with fascicles that divide into a medial head, which inserts on the medial sesamoid bone of the metatarsophalangeal joint of the great toe, and into a lateral head, which partially merges with the adductor hallucis and inserts on the lateral sesamoid bone of the metatarsophalangeal joint of the great toe. The **adductor hallucis** is formed by two heads, an oblique one and a transverse one. The first originates from the bases of the second, third, and fourth metatarsals and from the sheath of the fibularis longus, the second from the plantar ligaments of the metatarsophalangeal joints of the third, fourth, and little toes and from the deep transverse metatarsal ligament of the same toes. Both heads insert on the lateral sesamoid bone of the first metatarsophalangeal joint and on the base of the proximal phalanx of the great toe. The **abductor digiti minimi** originates from the medial and lateral processes of the calcaneal tuberosity, from the part of the bone between the two processes and from the plantar aponeurosis, and inserts on the lateral surface of the base of the proximal phalanx of the little toe. The **flexor digiti minimi brevis** extends from the medial part of the plantar surface of the base of the fifth metatarsal and from the sheath of the fibularis longus to the lateral surface of the base of the proximal phalanx of the little toe. The **opponens digiti minimi** originates from the long plantar ligament and inserts on the lateral border of the fifth metatarsal.

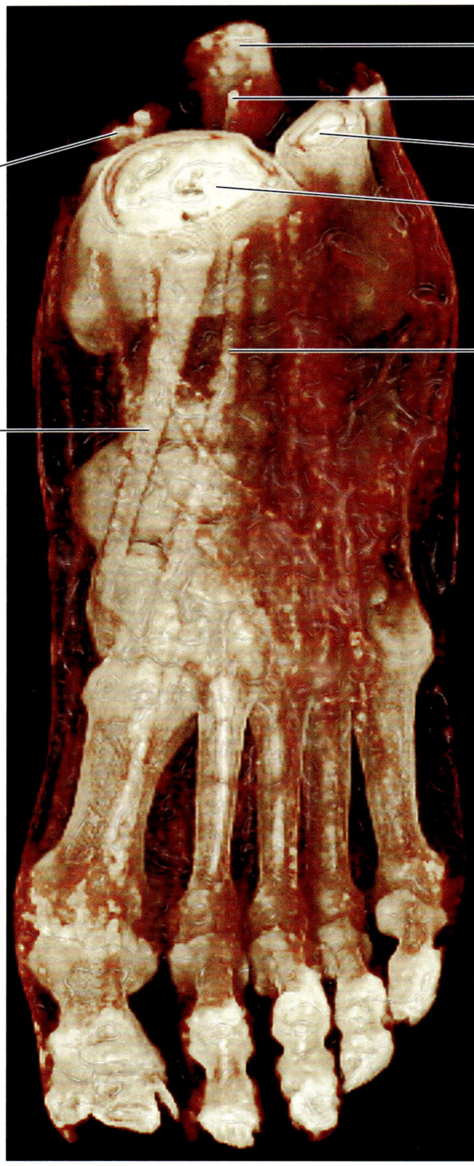

Three-dimensional image of a living human body, obtained by CT scan using a 3D volume rendering technique, showing the **foot** from a dorsal view.

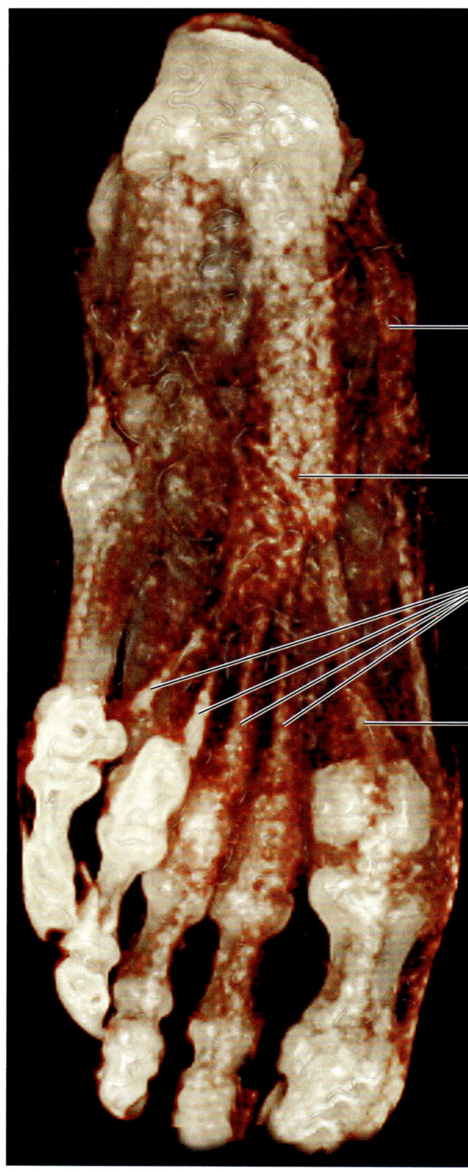

Three-dimensional image of a living human body, obtained by CT scan using a 3D volume rendering technique, showing the **foot** from a plantar view.

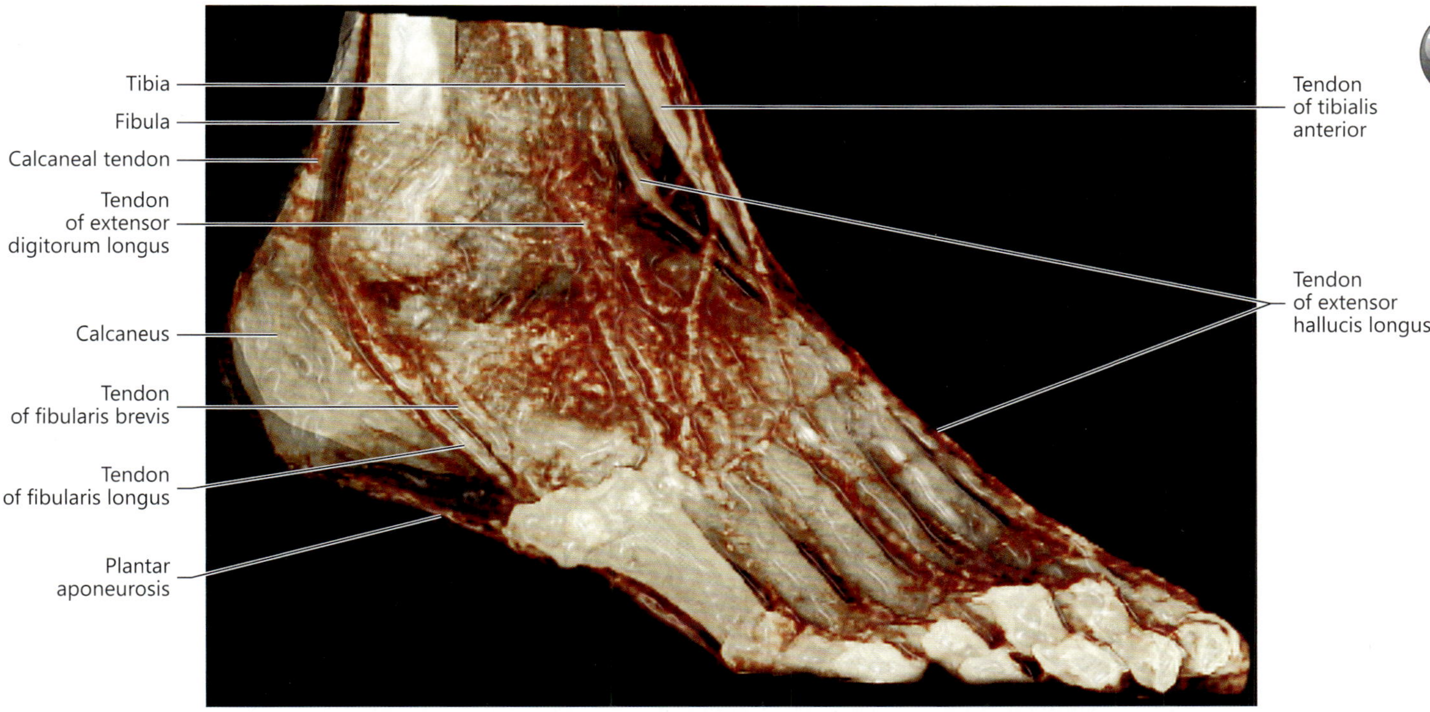

Three-dimensional image of a living human body, obtained by CT scan using a 3D volume rendering technique, showing the **foot** from a lateral view.

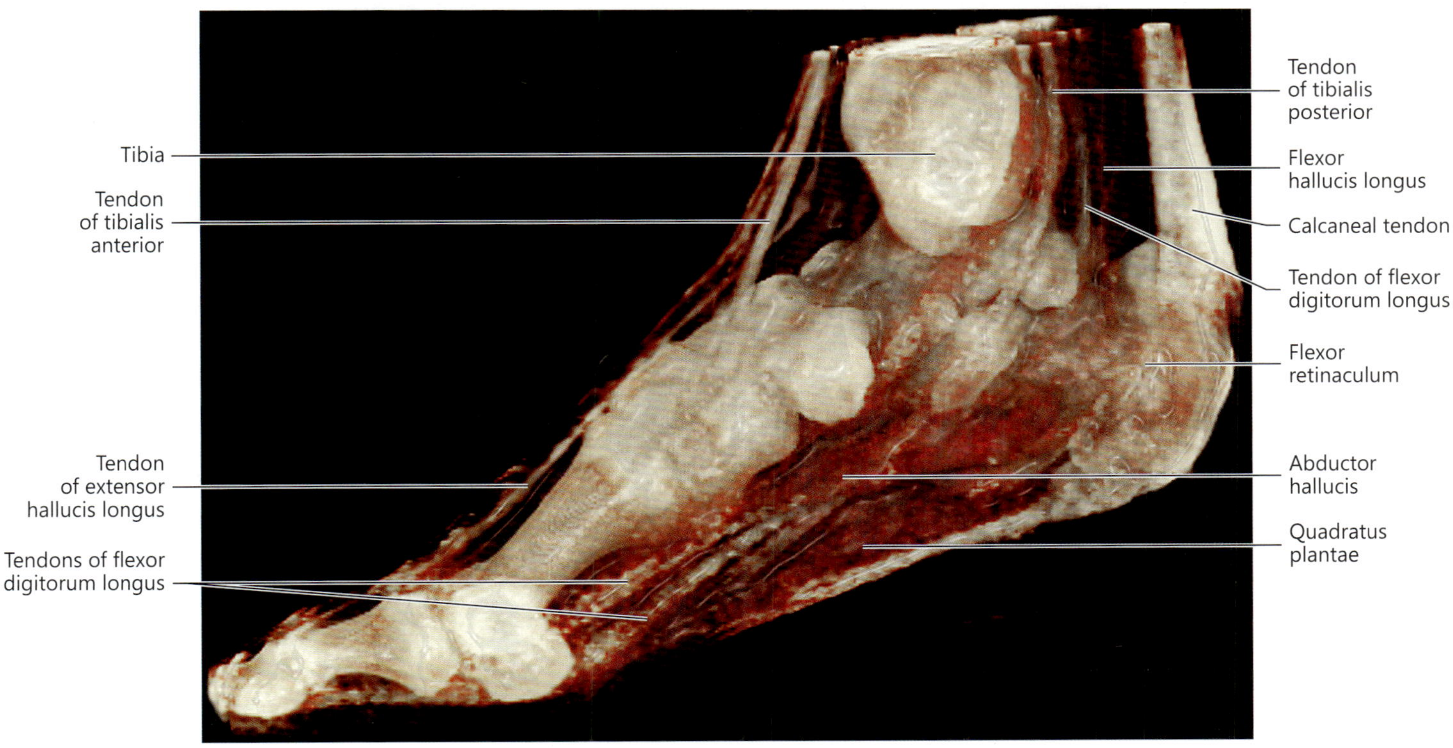

Three-dimensional image of a living human body, obtained by CT scan using a 3D volume rendering technique, showing the **foot** from a medial view.

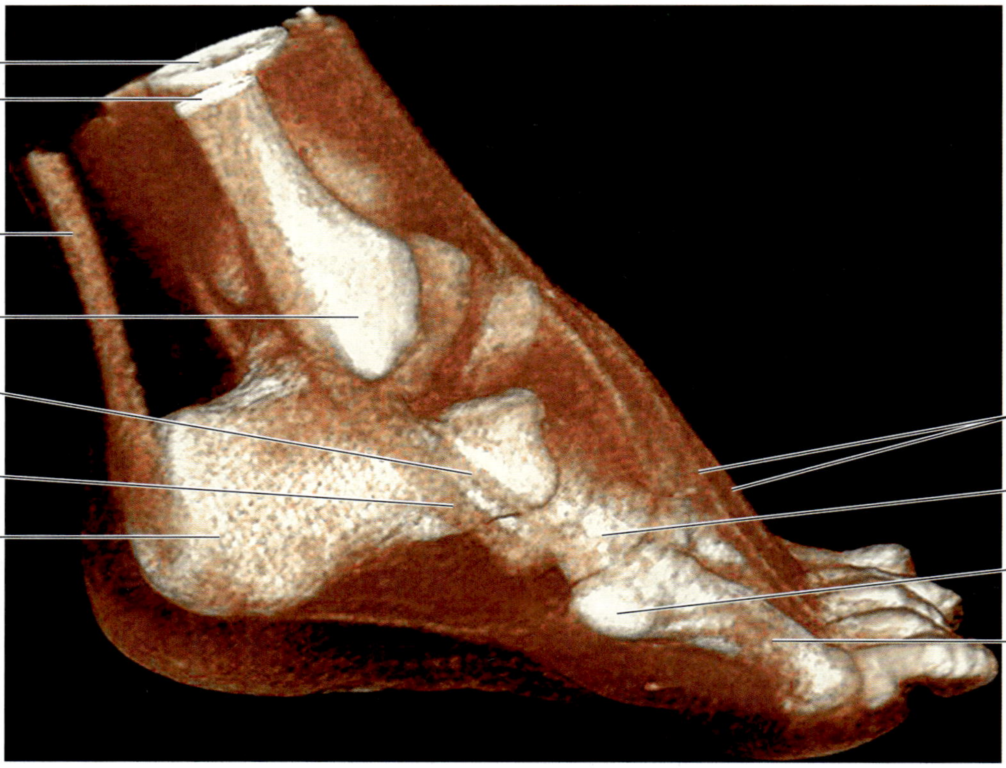

Three-dimensional image of a living human body, obtained by CT scan using a 3D volume rendering technique, showing the **tendons of the muscles of the leg and dorsum of the** right **foot** in a lateral projection.

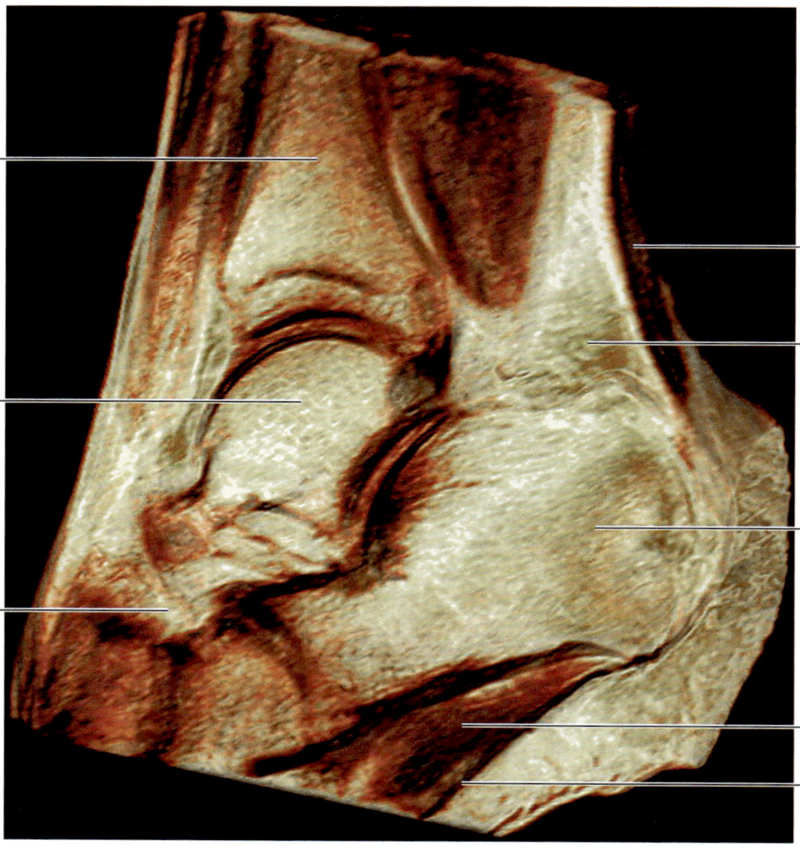

Three-dimensional image of a living human body, obtained by CT scan using a 3D volume rendering technique, showing the **insertion of the calcaneal tendon** on the calcaneal tuberosity and the ankle, subtalar, and talocalcaneonavicular joints in a medial projection of the right foot.

MUSCULAR COAPTATION OF THE HIP JOINT

The muscles play a very important role in keeping the articular surfaces in contact. It is possible to notice how there is a certain equilibrium in their respective roles. In fact, in front of the joint, the muscles are less numerous (red arrow A), but the ligaments, indicated by the black arrow, are robust; posteriorly, on the other hand, the inverse happens: the muscles (red arrow B) are predominant over the ligaments. It should be noted as well that the action of the ligaments varies according to the position of the hip: in a reference position or in extension, the ligaments are stretched, creating effective coaptation; under flexion, on the contrary, the ligaments are released and the head is no longer pressed against the acetabulum. The flexion position of the hip thus entails the release of the ligaments, a position of instability for the joint. When adduction is added, as in the seated position crossing the legs, a slight bump in the direction of the axis of the femur is sufficient to create posterior luxation, with or without breakage of the posterior border of the acetabulum, a very common occurrence that is seen in impacts against the dashboard in automobile accidents.

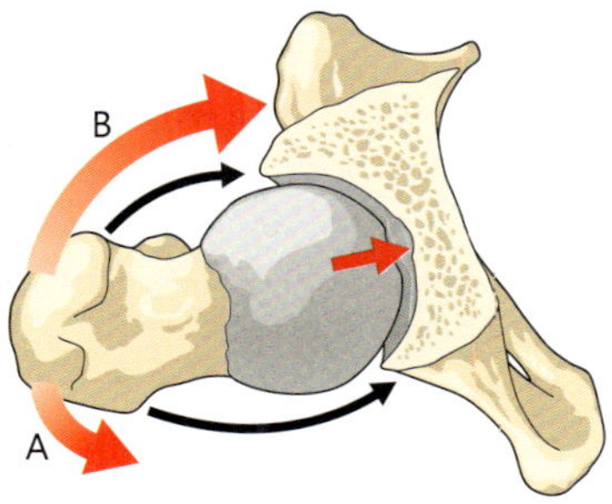

TRANSVERSE BALANCE OF THE PELVIS

When the pelvis is under bilateral support, its transverse balance is ensured by the simultaneous and bilateral action of the adductor and abductor muscles. When these opposing actions are balanced, as shown in figure **a**, the pelvis is stable in a symmetrical position, as for example in the "at attention" position. If the abductors prevail on one side, while on the other side the adductors are predominant, as in figure **b**, the pelvis moves laterally towards the side of the adductors; without the intervention of new muscular balance there is a lateral fall.

On the other hand, when the pelvis has unilateral support, as shown in figure **c**, the transverse balance is ensured exclusively by the action of the abductors on the side of the support. Thus, stimulated by the body weight W applied to the center of gravity, the pelvis tends to tilt around the bearing hip. Thus, the pelvic girdle may be compared to a type-one lever, where the point of support is represented by the bearing hip O, the resistance by the body weight W applied at the center of gravity G and the power by the force of the gluteus medius GF applied onto the external iliac fossa. In order for the line passing through the center of the hips to remain horizontal during unilateral support, the force of the gluteus medius must be sufficient to balance the weight of the body, considering the inequalities of lever arms OE and OG. In this balance of the pelvis, the gluteus minimus and medius are strongly supported by the tensor fasciae latae TFL.

If one of these muscles becomes insufficient, the action of gravity is no longer counterbalanced and the pelvis inclines to the opposite side with an angle that widens with the severity of the paralysis.

The stability of the pelvis created by the gluteus medius and minimus and the tensor fasciae latae is indispensable during normal gait. In fact, under unilateral support, the line of the pelvis, represented by the bisiliac line, remains horizontal and practically parallel to the line of the shoulders. If the muscles are paralyzed on the side of the support, the pelvis tilts on the other side, which could provoke a fall if the trunk were not to move as a block towards the side of the support with an inverse tilt to the line of the shoulders. This characteristic attitude under unilateral support associated with the tilting of the pelvis on the opposite side and the displacement of the superior part of the trunk is the Duchenne-Trendelenburg sign, a sign of paralysis or insufficiency of the gluteus minimus and medius.

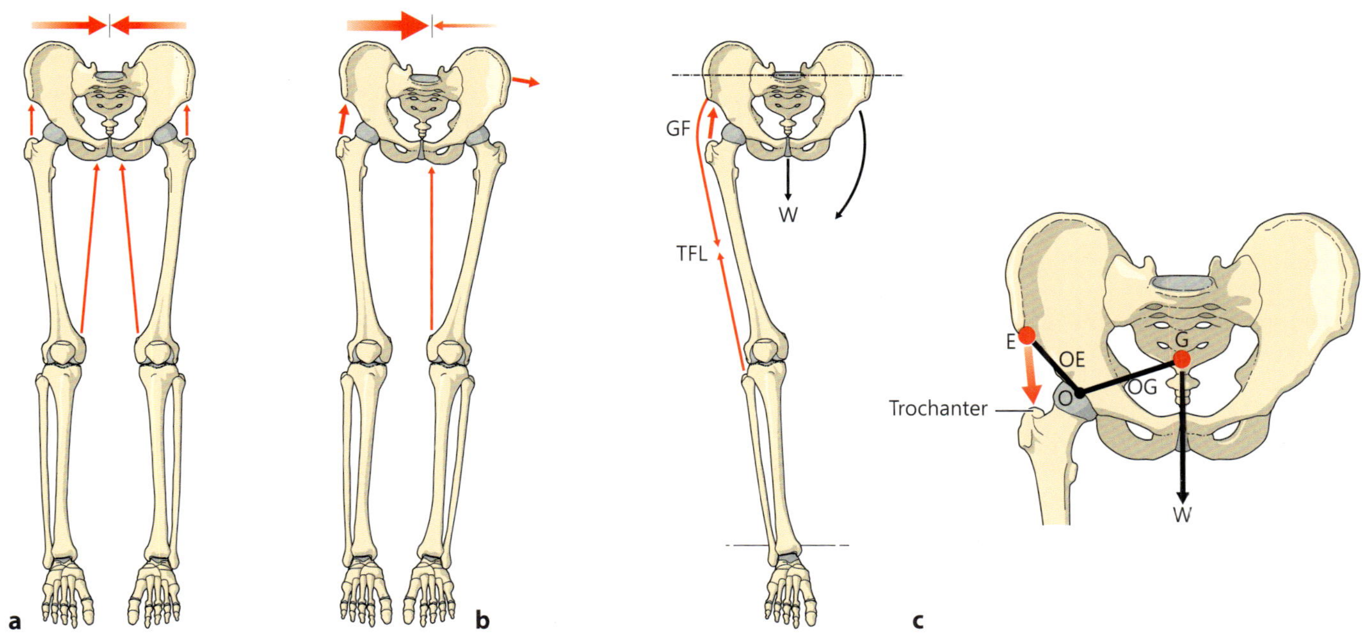

FUNCTIONAL ASPECTS OF THE MUSCLES OF THE HIP

FLEXION OF THE HIP

- The contraction of the flexor muscles of the hip is exemplified by activities such as:
  - putting on pants
  - climbing a ladder
  - climbing stairs
  - entering or exiting a bathtub
- The hip flexor muscles contract along with the muscles of the abdomen in order to lift the trunk when moving from a supine to a sitting position
- The iliopsoas:
  - is the only flexor muscle capable of flexion of the hip beyond 90° in a seated position, for example in activities such as:
    - sitting with lower limbs crossed
    - bringing the foot onto the contralateral thigh in seated position
    - putting on a sock
  - controls the movements of the trunk and the pelvis when leaning backwards in a seated position to look backwards
  - lowers the trunk when lying supine from the sitting position
- The adductor muscles contribute to the flexion of the hip when it is extended
- The gracilis flexes the hip above all in the initial phase of movement and when the knee is extended and not flexed
- The intervention of the adductor muscles and the gracilis in the movement of the flexion of the hip is exemplified by activities such as:
  - kicking a ball
  - swimming with fins
- The intervention of the sartorius as a flexor, abductor, and external rotator muscle of the hip and as flexor of the knee is exemplified by the action of bringing a foot onto the contralateral thigh in a seated position.

Flexor muscles of the hip

(from greatest to least muscular force)

- Iliopsoas
- Rectus femoris
- Sartorius
- Tensor fasciae latae
- Pectineus
- Adductor longus
- Adductor brevis
- Gracilis.

Action of the flexor muscles of the hip

Iliopsoas

- Principal hip flexor muscle
- Adducts and rotates the hip externally
- Inclines the vertebral column laterally (with fixed point on the distal insertion)
- Flexes the trunk forwards by action of both muscles (with fixed point on the distal insertion)
- Favors the stability of the hip joint (together with the iliofemoral ligament)
- Acts constantly during walking (with maximum work of the iliacus during the oscillation phase of walking).

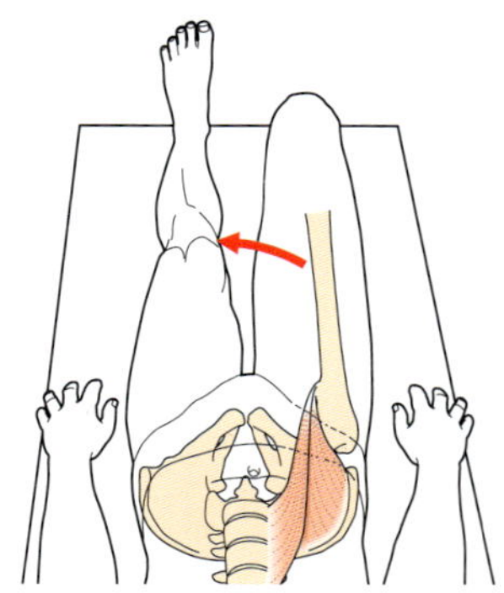

Rectus femoris

- Second hip flexor muscle
- Extends the knee (together with the vastus medialis, vastus lateralis, and vastus intermedius, with which it constitutes the quadriceps femoris)
- Facilitates the anteversion of the pelvis (with fixed point on the tibia).

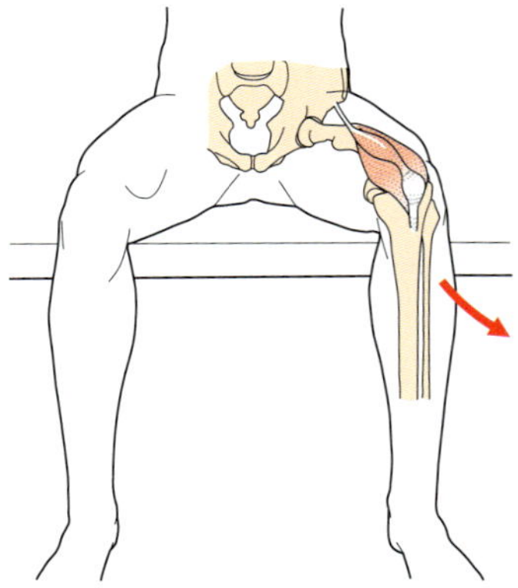

Sartorius

- Third hip flexor muscle
- Adducts and rotates the hip externally
- Facilitates the anteversion of the pelvis (contraction of both muscles with fixed point on the tibia)
- Flexes and internally rotates the knee.

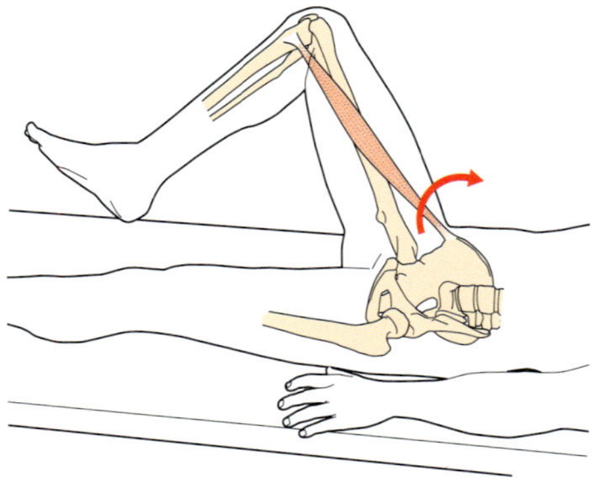

Tensor fasciae latae
- Flexion, abduction, and internal rotation of the hip
- Flexion of the knee and external rotation of the flexed knee.

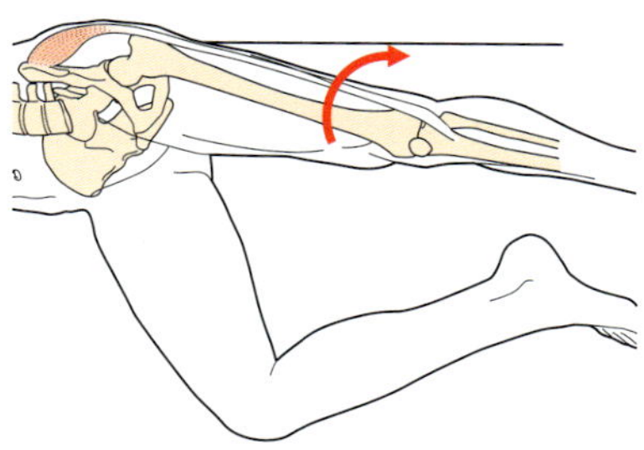

Pectineus
- Flexion and adduction of the hip.

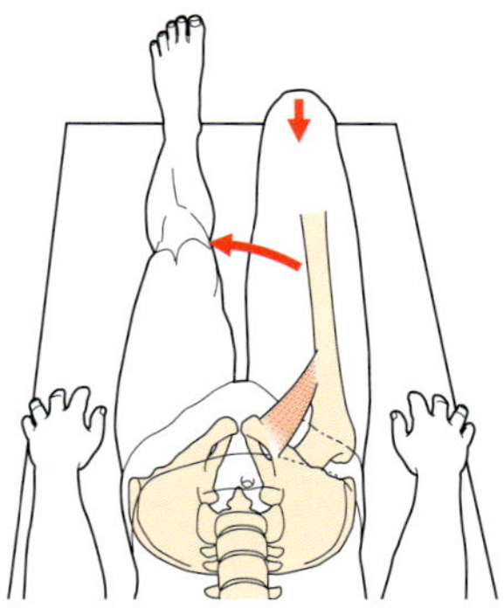

Adductor longus
- Weak flexor of the hip
- Principal action of adduction and external rotation of the hip.

Adductor brevis
- Weak flexor of the hip
- Principal action of adduction and external rotation of the hip.

Gracilis
- Weak flexor of the hip
- Adduction of the hip when the knee is extended
- Internal rotation of the hip
- Flexion of the knee.

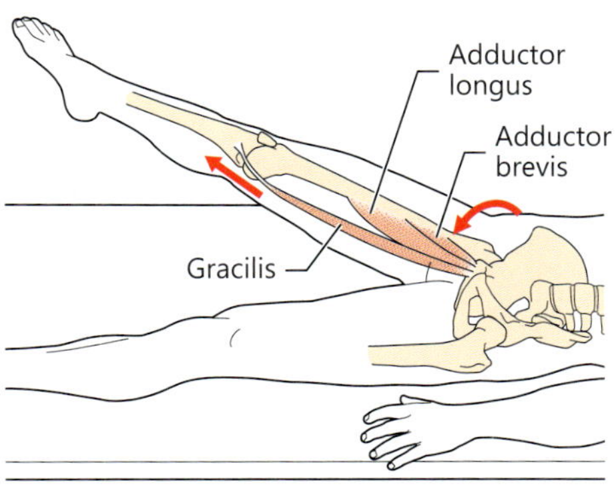

EXTENSION OF THE HIP
- The contribution of the extensor muscles of the hip to the activities of daily life is in part determined by the position of the hip and the strength required for its extension
- The position of the knee between 0 and 90° of flexion does not influence the extension of the hip
- The action of the hip extensor muscles is exemplified by activities in which the body is lifted, such as:
  - rising from a seated position
  - climbing stairs
  - jumping
- The hip extensor muscles:
  - contract during lifting activities carried out with hips and knees flexed
  - control the anteversion of the pelvis when one flexes forwards in a seated or standing position
  - produce retropulsion of the pelvis when sitting or getting up
  - are active in the pushing phase of skating
- The hamstring or ischiocrural muscles (semimembranosus, semitendinosus, and biceps femoris):
  - initiate the hip extension movement
  - extend the hip while standing upright
- The gluteus maximus:
  - contracts when the thigh is moved from position in hyperextension or when extension is carried out against resistance
  - contracts when bending forward, for example to change gears while driving or to look into a low shelf
  - extends the hip during upright standing in the case of increase of resistance
- The adductor magnus extends the hip when it is flexed between 0 and 90°.

Extensor muscles of the hip
(from greatest to least muscular force)
- Gluteus maximus
- Gluteus medius
- Gluteus minimus
- Semimembranosus
- Semitendinosus
- Adductor magnus (and adductor minimus)
- Biceps femoris (long head).

Action of the extensor muscles of the hip

Gluteus maximus
- Principal action of extension and external rotation of the hip. Its extensor action is greater when the thigh is flexed between 45 and 90°
- It is both abductor (insertion of the fibers to the iliotibial tract) and adductor of the thigh (insertion of fibers to the gluteal tuberosity of the femur)
- Retroversion of the pelvis (action of both muscles with fixed distal point).

The gluteus maximus is principally involved in activities such as rising from a seat, walking uphill, climbing stairs, and its action is very often used in sports (cycling, starting blocks in racing).

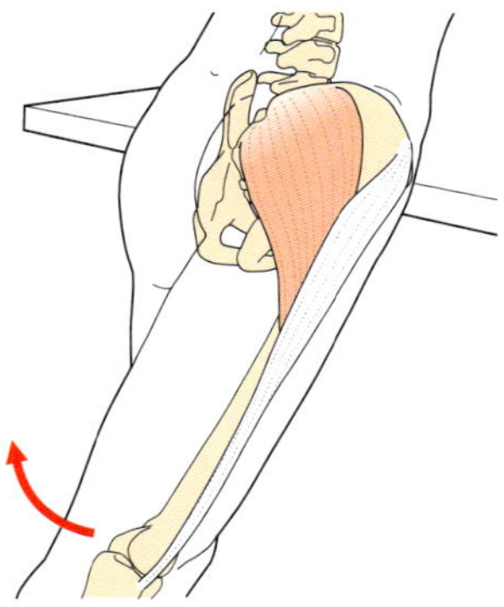

Gluteus medius
- Extension and external rotation of the hip (posterior fibers)
- Flexion and internal rotation of the hip (anterior fibers)
- Principal action of abduction of the hip.

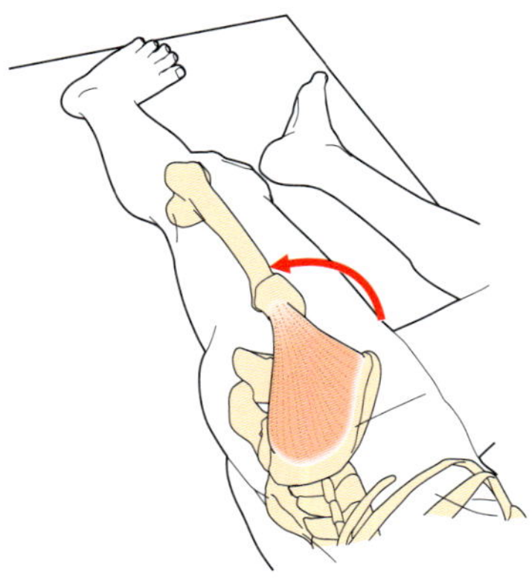

Gluteus minimus
- Extension of the hip (posterior fibers)
- Flexion of the hip (anterior fibers)
- Principal action of abduction of the hip.

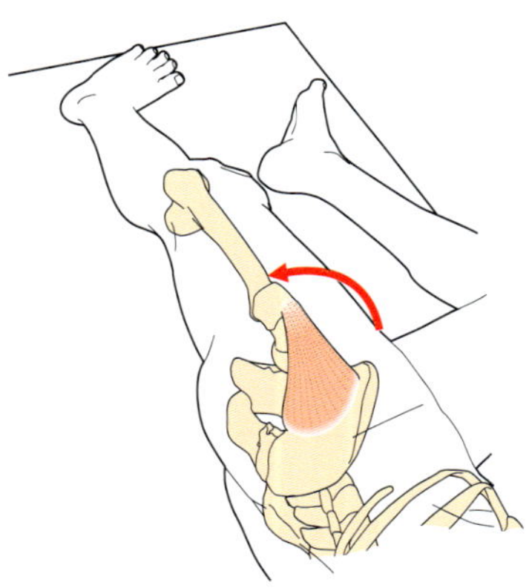

Semimembranosus
- Extension of the hip
- Principal action of flexion and internal rotation of the knee.

Semitendinosus
- Extension of the hip
- Principal action of flexion and internal rotation of the knee.

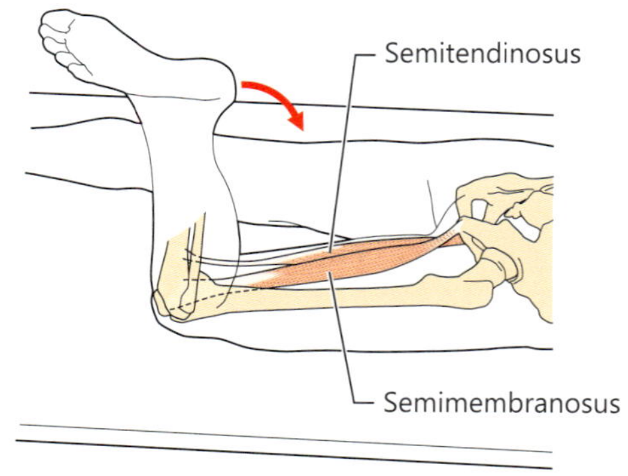

Adductor magnus (and adductor minimus)
- Extension of the hip
- Principal action of adduction of the hip
- External rotation of the hip (fibers that insert at the linea aspera)
- Internal rotation of the hip (fibers that insert at the adductor tubercle) rotated externally and flexed.

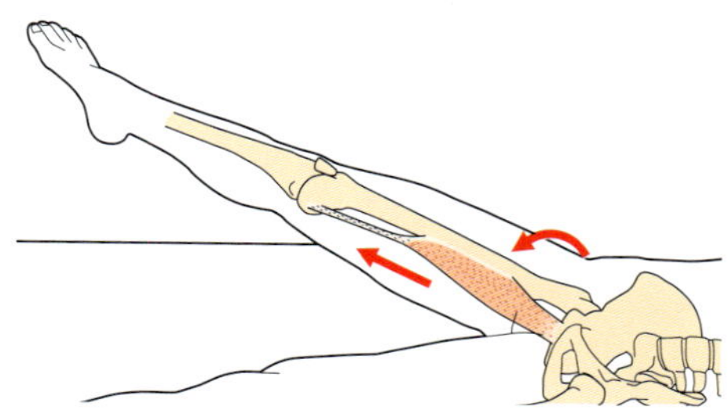

Biceps femoris (long head)
- Extension of the thigh
- Flexion and external rotation of the knee.

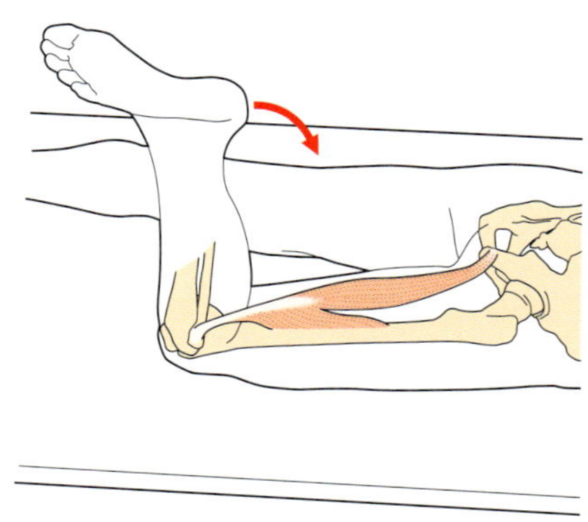

ABDUCTION OF THE HIP

- The principal function of the hip abductor muscles is to keep the pelvis in a horizontal position in single-foot support
- When the foot is lifted from the ground, the pelvis, not supported on the side of oscillation, tends to drop due to the weight of the head, upper limbs, trunk, and lifted lower limb, which makes it rotate around the head of the femur of the supporting limb
- The hip abductor muscles of the supporting side, contracting with inverted origins and insertions, drag the iliac crest of the same side downwards, causing the rotation of the pelvis around the head of the femur and elevating the non-supporting side
- The action of keeping the pelvis in a horizontal position during single-foot support by the hip abductor muscles occurs during walking, running, or when kicking a ball
- During the single-foot support phase, the contraction of the hip abductor muscles may not be necessary to maintain the pelvis in a horizontal position, if the weight of the trunk is shifted onto the supporting foot in such a way that the barycenter of the head, arms, and trunk pass through the hip joint of the supporting side
- The hip abductor muscles can contract bilaterally in activities that require the abduction of the non-supporting lower limb, such as when mounting a bicycle
- The upper fibers of the gluteus maximus contribute to the abduction of the hip when an increase in strength is required.

Abductor muscles of the hip
(the action of the muscles previously discussed is no further described)
(from greatest to least muscular force)

- Gluteus medius
- Tensor fasciae latae
- Gluteus maximus
- Gluteus minimus
- Piriformis
- Obturator internus
- Sartorius.

Action of the abductor muscles of the hip

Piriformis

- Abduction of the flexed hip
- External rotation of the extended hip.

Obturator internus

- Abduction of the hip in seated (flexed) position with knee extended
- Principal action of external rotation of the hip.

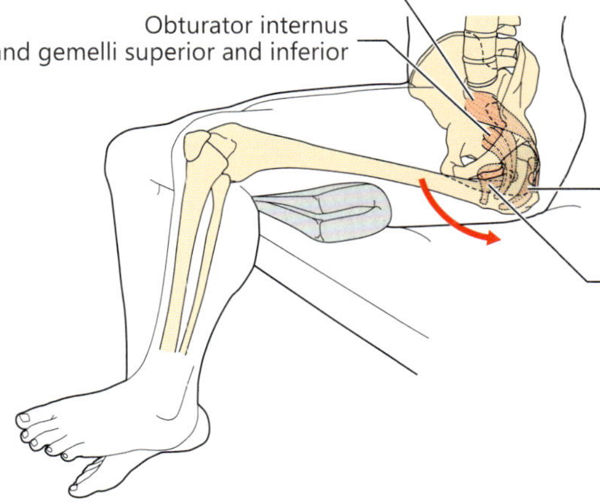

ADDUCTION OF THE HIP

- The action of the hip adductor muscles is exemplified by activities such as:
  - climbing a rope
  - kicking a ball, crossing the body anteriorly with the foot
  - horseback riding
- In various activities, the hip adductor muscles act principally as postural muscles rather than as agonists of the adduction movement.

Adductor muscles of the hip
(the action of the muscles previously discussed is no further described)
(from greatest to least muscular force)

- Adductor magnus (and adductor minimus)
- Adductor longus
- Adductor brevis
- Gluteus maximus
- Gracilis
- Pectineus
- Quadratus femoris
- Obturator externus.

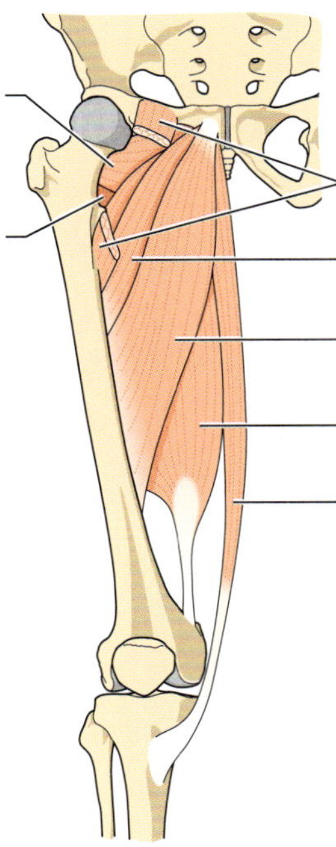

Action of the adductor muscles of the hip

Quadratus femoris
- Weak adductor
- Principal action of external rotation of the hip
- Facilitates the retroversion of the pelvis (contraction of both muscles).

Obturator externus
- Weak adductor
- Principal action of external rotation of the hip.

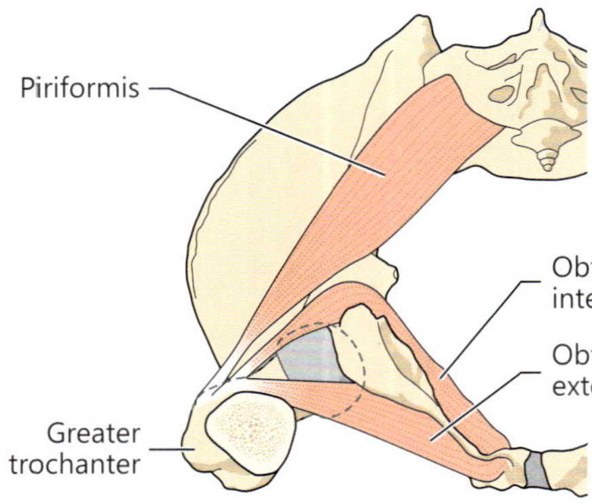

INTERNAL ROTATION OF THE HIP
- The internal rotator muscles are active during walking or when, in the upright position, rotation on one foot is carried out
- The semimembranosus and semitendinosus perform internal rotation of the thigh when the thigh is extended.

Internal rotator muscles of the hip
(the action of the muscles previously discussed is no further described)
(from greatest to least muscular force)
- Gluteus medius
- Gluteus minimus
- Tensor fasciae latae
- Adductor magnus
- Pectineus
- Gracilis.

EXTERNAL ROTATION OF THE HIP
- The piriformis and obturator internus work most effectively as external rotators of the hip when the hip is extended, becoming less valid when the hip is flexed
- The gluteus maximus and biceps femoris also rotate the hip externally when it is extended
- Activities that require the contraction of the external rotator muscles are exemplified by:
  - mounting a bicycle
  - sitting, bringing the foot onto the contralateral thigh in order to tie the shoe.

External rotator muscles of the hip
(the action of the muscles previously discussed is no further described)
(from greatest to least muscular force)
- Gluteus maximus
- Quadratus femoris
- Obturator internus
- Gluteus medius
- Gluteus minimus
- Iliopsoas
- Obturator externus
- Adductor longus
- Adductor brevis
- Adductor magnus
- Piriformis
- Sartorius
- Gemellus superior
- Gemellus inferior
- Biceps femoris.

Action of the external rotator muscles of the hip

Gemellus superior
- External rotation of the hip (sole action).

Gemellus inferior
- External rotation of the hip (sole action).

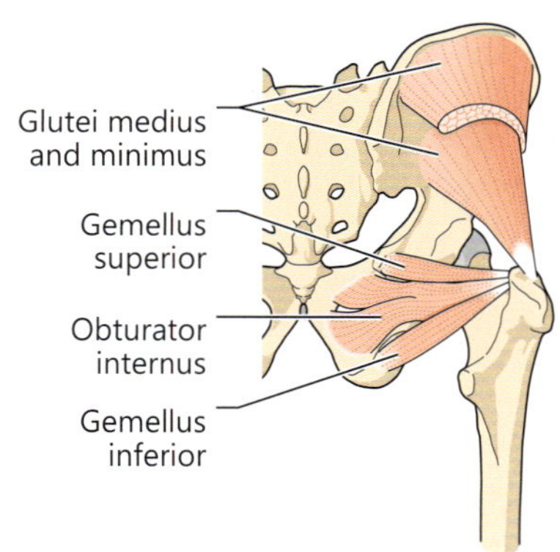

FUNCTIONAL ASPECTS OF THE MUSCLES OF THE KNEE

FLEXION OF THE KNEE

- The popliteus and the short head of the biceps femoris are the only monoarticular flexor muscles of the knee (they move only the leg):
  - the action of the popliteus as flexor of the knee is negligible
  - its principal function consists of rotating the leg internally when it begins to be flexed, thus unblocking the knee joint
  - when loads are transported while walking downhill, the activity of the popliteus is increased in order to stabilize the knee in the central part of the foot's support phase
  - when a squatting position is assumed (to lift, for instance, an object from the ground), the popliteus contracts in order to avoid the forward movement of the femur relative to the tibia
- The majority of the knee flexor muscles is made up of biarticular muscles (capable of moving the thigh or the foot as well):
  - the gastrocnemius flexes the leg and controls its hyperextension
- The action of the knee flexor muscles is exemplified by the seated position in which the ankle is supported on the contralateral thigh or by the position with legs crossed at the level of the ankles
- The knee flexor muscles contract as well when the flexion of the knee is forced at the end of the excursion, as happens for example when a sock is put on
- During upright standing, the knee flexor muscles contract, for example, when the sole of the foot is examined
- During walking or running, the knee flexor muscles contract eccentrically in order to decelerate the leg that is extended to take a step forward
- Climbing stairs does not require the contraction of the knee flexor muscles, since the flexion of the knee is carried out passively following the active flexion of the hip
- The knee flexor muscles also have rotatory action on the leg (with the exception of the gastrocnemius):
  - the biceps femoris rotates the leg externally
  - the other flexor muscles rotate the leg internally
  - they initiate and control the rotation of the leg in activities such as running and turning
  - they are also active during squatting and kneeling when the upper limbs and the trunk cause movement of the leg on the ground.

Flexor muscles of the knee
(from greatest to least muscular force)

- Biceps femoris
- Sartorius
- Semimembranosus
- Semitendinosus
- Gracilis
- Popliteus
- Gastrocnemius.

Action of the flexor muscles of the knee

Biceps femoris

- Flexion of the knee
- External rotation of the knee
- Participates in the extension of the hip (long head).

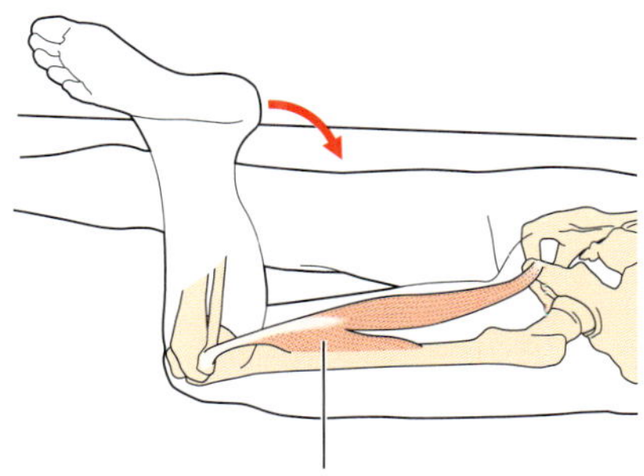

The biceps femoris, together with the semimembranosus and semitendinosus, forms the hamstring muscle group, also called ischiocrural muscles due to their origin from the ischial tuberosity and their tibial insertion. They:

- extend the hip
- flex the knee
- rotate the knee
- participate in the retroversion of the pelvis.

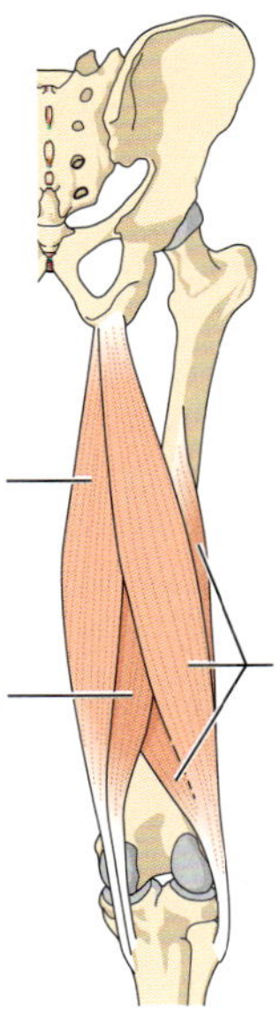

Sartorius

- Flexion and internal rotation of the knee
- Flexion of the hip
- Abduction of the hip
- External rotation of the hip.

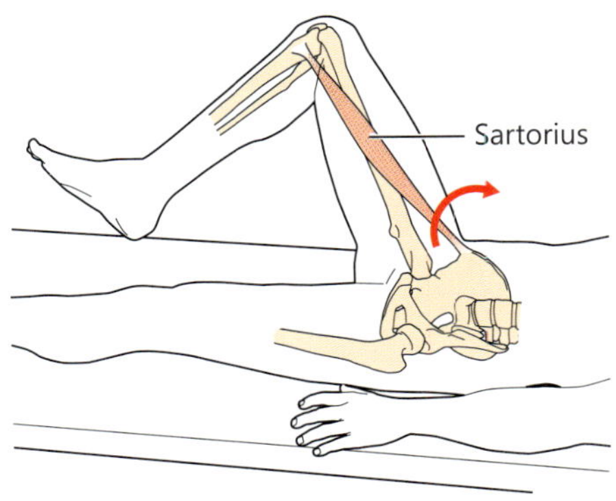

Semimembranosus
- Flexion of the knee
- Internal rotation of the knee
- Participates in the extension of the hip.

Semitendinosus
- Flexion of the knee
- Internal rotation of the knee
- Participates in the extension of the hip.

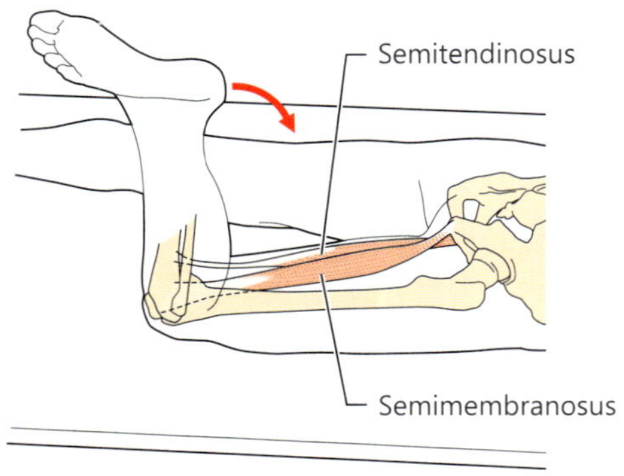

Gracilis
- Participates in the flexion and internal rotation of the knee
- Weak flexor of the hip
- Adduction and internal rotation of the hip when the knee is extended.

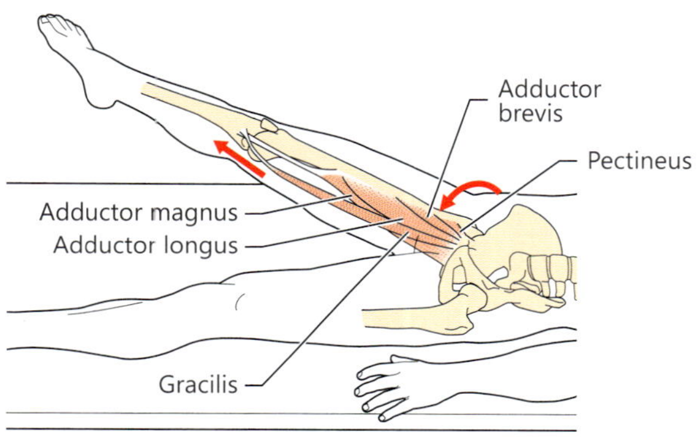

The insertion of the sartorius, gracilis, and semitendinosus occurs through a common tendon called goose's foot due to its shape, which inserts on the upper part of the medial surface of the tibia.

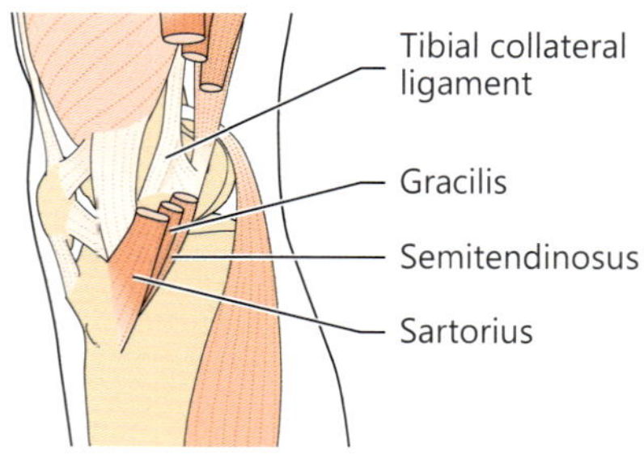

Popliteus
- Flexion of the knee
- Internal rotation of the knee:
  - during the initial flexion of the knee, due to the course of its fibers directed from above to below and from the outside to the inside on the posterior surface of the leg (from the lateral condyle of the femur to the popliteal line on the posterior surface of the tibia), frees the knee (leg) from a potential external rotation; during this action, it stabilizes the knee on the sagittal and transverse planes, assisting the action of the posterior cruciate ligament and avoiding forward slippage of the femoral condyles.

Gastrocnemius
(medial and lateral heads)
- Strong flexor of the knee, above all in the initial phase, when the knee is extended
- Stabilizes the knee in relation to the tibia (above all in the presence of wide extensor movements of the leg such as those made during running and jumping)
- Plantar flexion of the foot (together with the soleus, with which it forms the triceps surae).

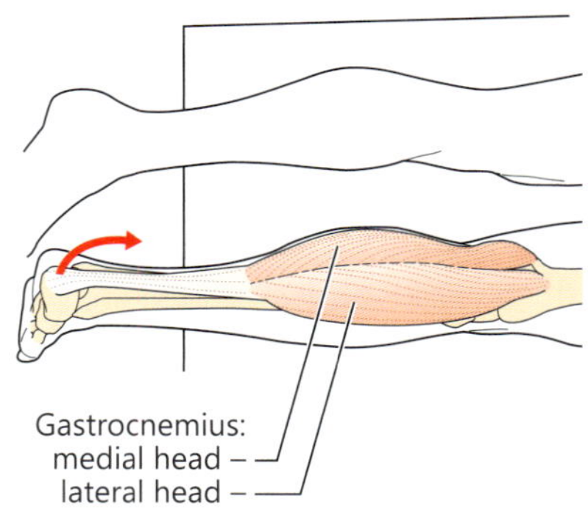

EXTENSION OF THE KNEE
- The rectus femoris acts upon the thigh and the leg, and is more effective as an extensor of the knee when the hip is extended and its fibers are stretched:
  - the thigh must be stabilized in order for the rectus femoris to fully act as an extensor of the knee
- The vastus medialis contracts along with the vastus lateralis and vastus intermedius during the extension of the knee for the entire range of the joint's movement:
  - the inferior oblique fibers of the vastus medialis act in the final phase of the extension of the knee in order to prevent the displacement of the patella in a lateral direction (the patella is pulled medially)

Two principle patterns of movement are carried out with the lower limb during the execution of the principal activities of daily life:

- the first is comprised of the flexion of the hip, the flexion of the knee and the dorsal flexion of the foot:
  - in this pattern of movement, the knee extensor muscles contract in an eccentric way in order to control its flexion (an example is offered by activities during which the foot is fixed to the ground and the body is lowered, as in squatting to raise an object, sitting on a chair, or descending stairs)
- the second is comprised of the extension of the hip, extension of the knee and plantar flexion of the foot:
  - in this pattern of movement, the knee extensor muscles contract concentrically in order to extend the leg (examples are activities such as getting up out of a chair, jumping, climbing stairs, and getting up to exit the bathtub)

- When the foot is not supported, the extensor muscles of the knee contract if the extension is carried out against resistance, which may be created by the weight of the limb as well
- Examples of functional activities that require the contraction of the extensor muscles of the knee are kicking a ball, putting on pants, swimming the breaststroke.

Extensor muscles of the knee

- Quadriceps femoris
  - Vastus lateralis
  - Vastus intermedius
  - Vastus medialis
  - Rectus femoris.

Action of the extensor muscles of the knee

Quadriceps femoris

- The only extensor muscle of the knee (the extensor action of some fibers of the tensor fasciae latae being negligible)
- Greater extensor work of the vastus medialis when the hip is flexed and the knee is rotated internally
- Greater extensor work of the vastus lateralis when the hip is flexed and the knee is rotated externally
- Greater force of the vastus medialis during extensions against resistance
- Hip flexion action by the rectus femoris, more efficient when the knee is extended.

Overall, the work of the quadriceps femoris is more than three times greater than the work of the knee flexor muscles.

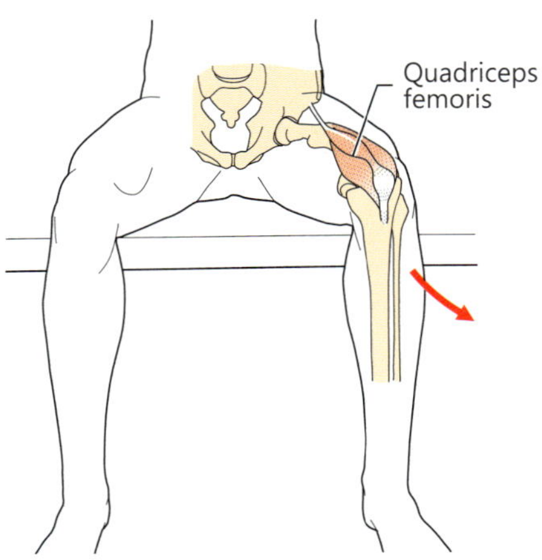

ROTATION OF THE KNEE

The flexor muscles of the knee are also rotators of the knee and may be divided into two groups: internal and external rotators, in relation to their insertion on the outside or inside of the vertical axis of rotation of the knee. The external rotator muscles, contracting, bring backwards the external part of the lateral condyle of the tibia, thus causing the tip of the foot to rotate externally. Conversely, the internal rotator muscles, with the exception of the popliteus, pull backwards the internal part of the medial condyle of the tibia, causing the tip of the foot to rotate internally. They also restrain external rotation when the knee is flexed. The popliteus, the only monoarticular internal rotator muscle, originating externally from the lateral epicondyle of the femur and inserting on the posterior surface of the tibia, acts as internal rotator during the flexion of the knee, when its fossula of origin is moved upwards or forwards.

External rotator muscles of the knee

(the action of the muscles previously discussed is no further described)
(from greatest to least muscular force)

- Biceps femoris
- Tensor fasciae latae
- Vastus lateralis.

Action of the external rotator muscles of the knee

Tensor fasciae latae

The tensor fasciae latae acts as an external rotator of the knee only when the knee is flexed.

Vastus lateralis

The vastus lateralis actually stabilizes a position of internal rotation of the knee in flexion when returning from the flexed position to the extended position.

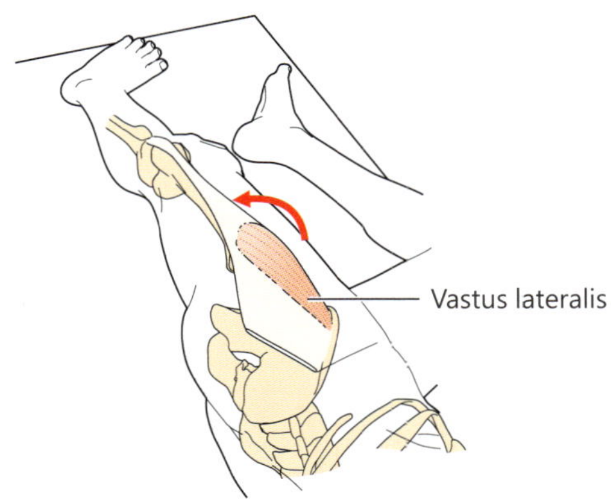

Internal rotator muscles of the knee

(the action of the muscles previously discussed is no further described)
(from greatest to least muscular force)

- Semimembranosus
- Semitendinosus
- Gracilis
- Sartorius
- Popliteus.

The work of the external rotator muscles of the knee is approximately equal to the work of the internal rotator muscles and much less than that of the knee extensor muscles (quadriceps femoris).

FUNCTIONAL ASPECTS OF THE MUSCLES OF THE ANKLE AND THE FOOT

PLANTAR FLEXION OF THE FOOT

- The gastrocnemius acts as a plantar flexor of the foot:
  - when the foot is already in a position of plantar flexion
  - when strong contraction is required
  - when the knee is extended
- the soleus is principally active during plantar flexion of the foot:
  - when the foot is flexed dorsally
  - when minimal contraction is required
- the tibialis posterior, flexor hallucis longus, flexor digitorum longus and fibularis longus and brevis act as accessory plantar flexors
- The actions of the plantar flexor muscles are exemplified by activities in which the foot is in plantar flexion against resistance, such as:
  - rising on toes
  - jumping
  - pressing the accelerator of an automobile
- The plantar flexor muscles of the foot:
  - contract also when the movement is forced at extreme degrees of plantar flexion, such as in the action of putting on a sock
  - control the dorsal flexion of the foot during descent of stairs when the foot is fixed
  - contract in the pushing phase of walking, realizing their greatest effort during deambulation.

Plantar flexor muscles of the foot
(from greatest to least muscular force)

- Triceps surae: gastrocnemius and soleus
- Flexor hallucis longus
- Fibularis longus
- Fibularis brevis
- Plantaris
- Flexor digitorum longus
- Tibialis posterior.

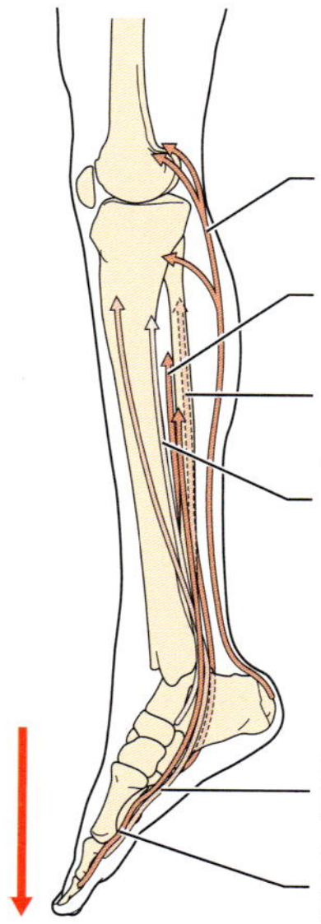

Actions of the plantar flexor muscles of the foot

Gastrocnemius

- Strong plantar flexor of the foot (the action of the muscle is favored by the extension of the leg)
- Supination of the foot
- Strong flexors of the knee, above all in the initial phase of movement, when the knee is extended
- Affixes the knee to the tibia
- Supports the longitudinal arches of the foot.

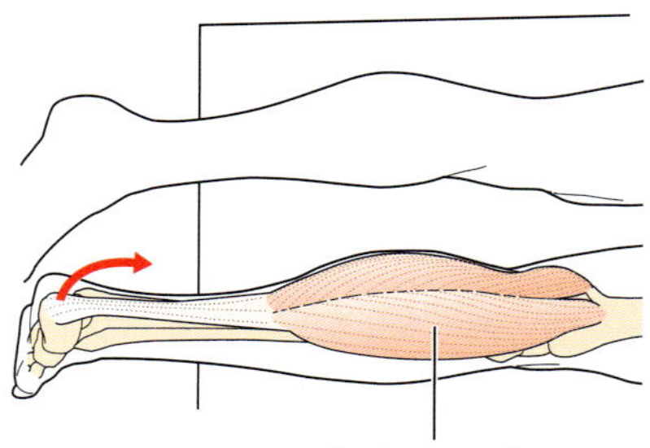

Soleus

- Plantar flexor of the foot
- Abduction of the foot
- Supination of the foot
- Supports the longitudinal arches of the foot.

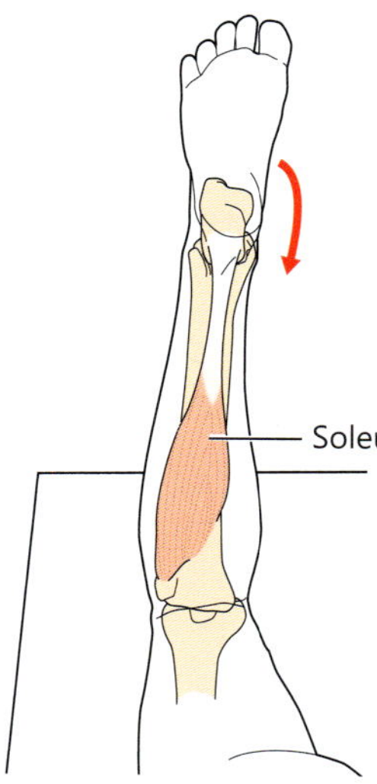

The gastrocnemius and soleus, which together form the triceps surae, carry out a common action of plantar flexion of the foot when particular muscular force is required, and also play an important postural role.

Flexor hallucis longus
- Plantar flexion of the great toe
- Supination of the foot
- Contributes to the plantar flexion of the foot
- Supports the longitudinal arches of the foot.

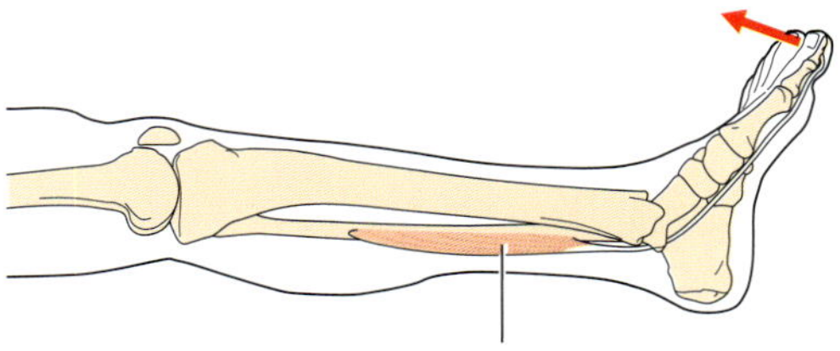

Fibularis longus
- Pronation of the foot
- Contributes to the plantar flexion of the foot
- Supports the transverse arch of the foot.

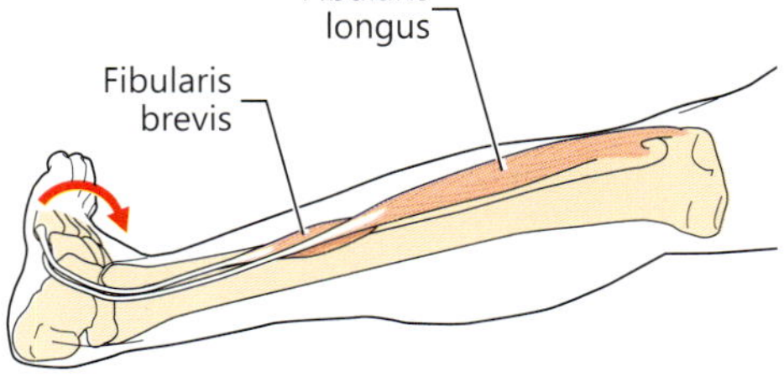

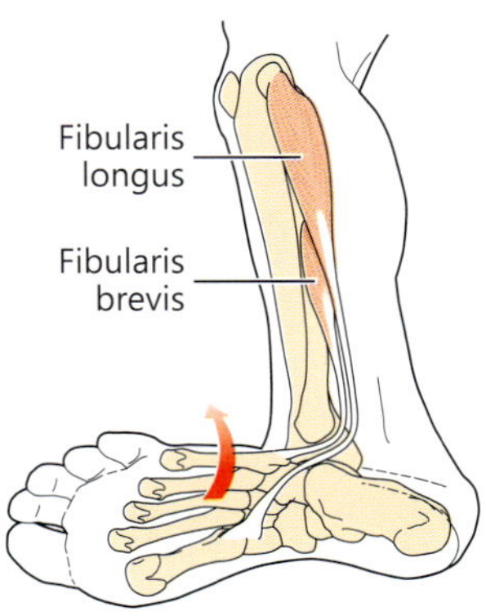

Fibularis brevis
- Action similar to that of the fibularis longus.

Plantaris
- Modest plantar flexion of the foot
- Modest supination of the foot.

Flexor digitorum longus
- Flexion of the last four toes
- Contributes to the plantar flexion of the foot
- Contributes to the supination of the foot
- Supports the longitudinal arches of the foot.

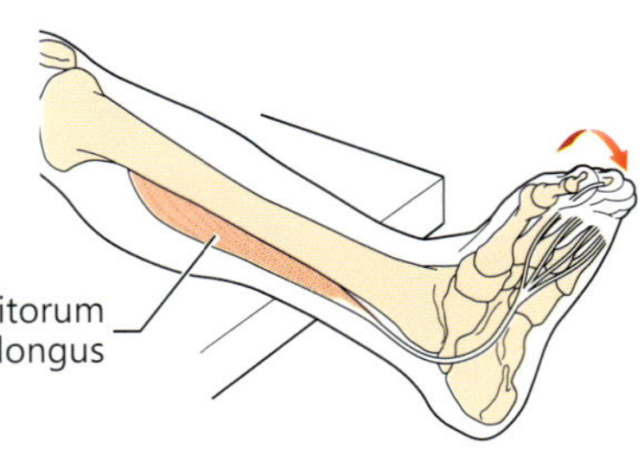

Tibialis posterior
- Plantar flexion of the foot
- Adduction of the foot
- Supination of the foot
- Supports the medial longitudinal and transversal arches of the foot.

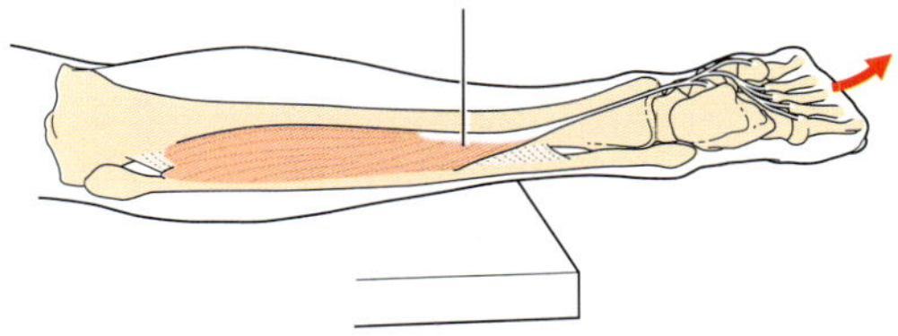

DORSAL FLEXION OF THE FOOT
- The tibialis anterior, which initiates the dorsal flexion of the foot, and the extensor hallucis longus are dorsal flexor muscles stronger than the extensor digitorum longus and fibularis tertius
- The dorsal flexor muscles of the foot:
  - keep the foot under dorsal flexion during activities such as cutting the toenails or tying shoes
  - control the plantar flexion of the foot when it is lowered against the ground, as happens when the foot is slowly pressed against the ground or in response to the load during the cycle of a footstep
  - stabilize the tibia on the ankle when rising from the seated position to a standing position with the feet supported on the floor
  - are active in almost all phases of walking (support of the heel, median support phase until the lifting of the foot and in nearly all of the oscillatory phase).

Dorsal flexor muscles of the foot
(from greatest to least muscular force)
- Tibialis anterior
- Extensor digitorum longus
- Extensor hallucis longus
- Fibularis tertius.

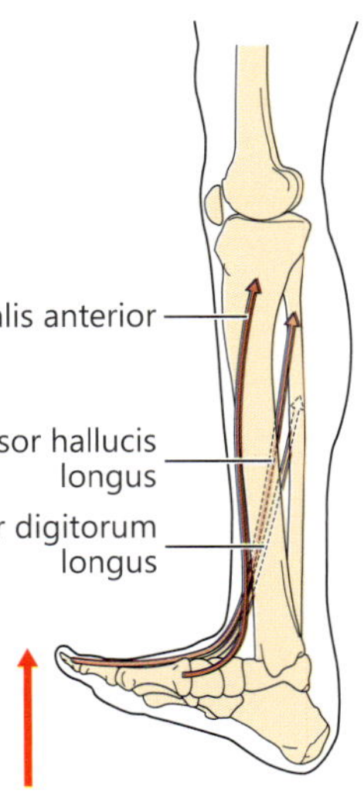

Action of the dorsal flexor muscles of the foot

Tibialis anterior
- Principal dorsal flexor muscle of the foot
- Adduction of the foot
- Supination of the foot
- Contributes to the support of the longitudinal arches of the foot.

The action of the tibialis anterior is markedly greater than that of the other three dorsal flexor muscles of the foot, both for its muscular dimensions and its function, which involves the entire foot. With a fixed point on the foot, it brings the anterior surface of the leg towards the dorsum of the foot, an action that occurs particularly during fast walking.

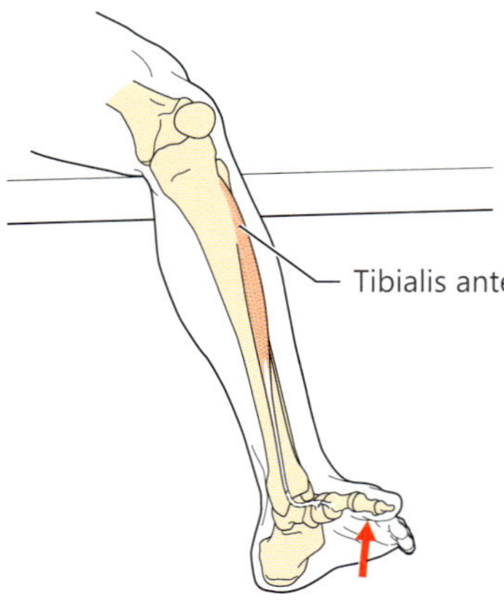

Extensor digitorum longus
- Dorsal flexion of the foot (in agonist synergy with the tibialis anterior)
- Abduction of the foot (in antagonist synergy with the tibialis anterior)
- Pronation of the foot (in antagonist synergy with the tibialis anterior)
- With fixed point on the foot, brings the anterior surface of the leg towards the dorsum of the foot.

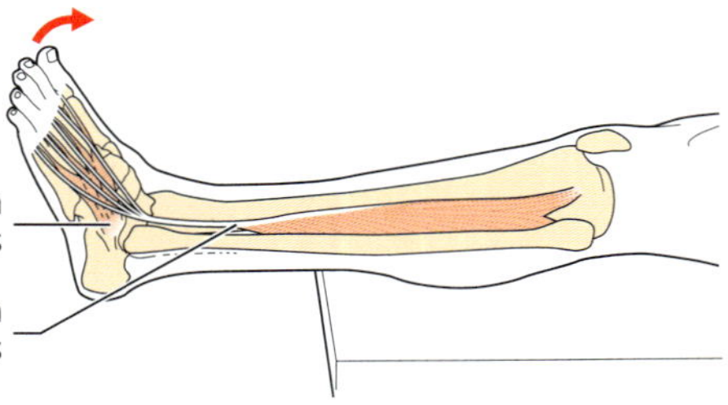

Extensor hallucis longus
- Dorsal flexion of the great toe
- Contributes to the dorsal flexion of the foot
- With fixed point on the foot, brings the anterior surface of the leg towards the dorsum of the foot.

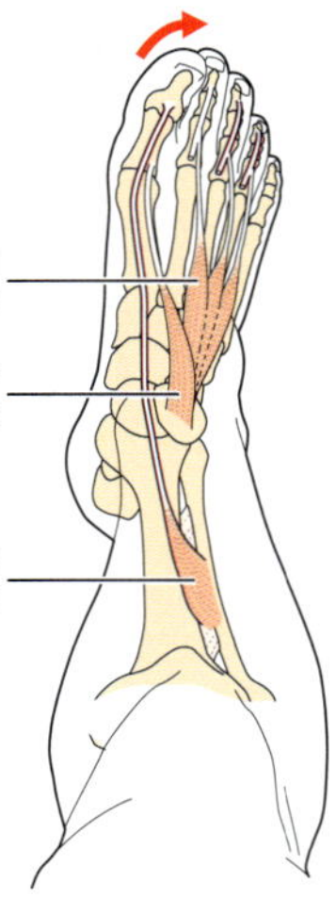

Fibularis tertius
- Contributes to the dorsal flexion of the foot
- Abduction of the foot (action similar to that of the fibularis brevis and longus, and of the extensor digitorum longus)
- Pronation of the foot (action similar to that of the fibularis brevis and longus, and of the extensor digitorum longus).

The fibularis tertius may be considered as a fifth terminal tendon of the extensor digitorum longus, inserted dorsally to the base of the fifth metatarsal.

INVERSION OF THE FOOT
(adduction + supination + plantar flexion of the foot)
- The tibialis posterior is the principal invertor muscle of the foot
- The gastrocnemius and soleus cause the inversion of the calcaneus during the plantar flexion of the foot
- The tibialis anterior and the flexor muscles of the last four toes act as weak invertors of the foot from the position of eversion to the intermediate position
- The tibialis anterior acts only as an invertor muscle when the medial border of the foot is lifted during the inversion movement
- The action of the invertor muscles of the foot is exemplified by activities such as bringing the foot onto the contralateral thigh in a sitting position in order to look at the sole, or walking on bumpy terrain, contributing in this last situation to the stabilization of the foot.

Invertor muscles of the foot
(the action of the muscles previously discussed is no further described)
(from greatest to least muscular force)
- Triceps surae (gastrocnemius and soleus)
- Tibialis posterior
- Flexor hallucis longus
- Flexor digitorum longus
- Tibialis anterior
- Plantaris.

Contributions to the inversion movement of the foot:
- triceps surae (greater contribution) – plantar flexion and supination
- tibialis anterior – adduction and supination
- tibialis posterior, flexor hallucis longus and digitorum longus, plantaris – plantar flexion and supination.

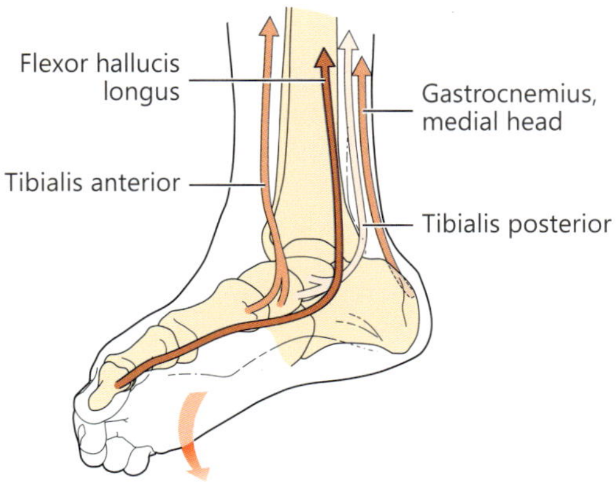

EVERSION OF THE FOOT
(abduction + pronation + dorsal flexion of the foot)
- The fibularis longus and brevis, assisted by the extensor digitorum longus and by the fibularis tertius, are responsible for the eversion of the foot:
  - the action of these muscles is carried out, as well, during walking on bumpy terrain.

Evertor muscles of the foot
(the action of muscles previously discussed is no further described)
(from greatest to least muscular force)
- Fibularis longus
- Fibularis brevis
- Extensor digitorum longus
- Fibularis tertius.

Contributions to the eversion movement of the foot:
- fibularis longus and brevis – abduction and pronation
- extensor digitorum longus and fibularis tertius – dorsal flexion, abduction, and pronation of the foot.

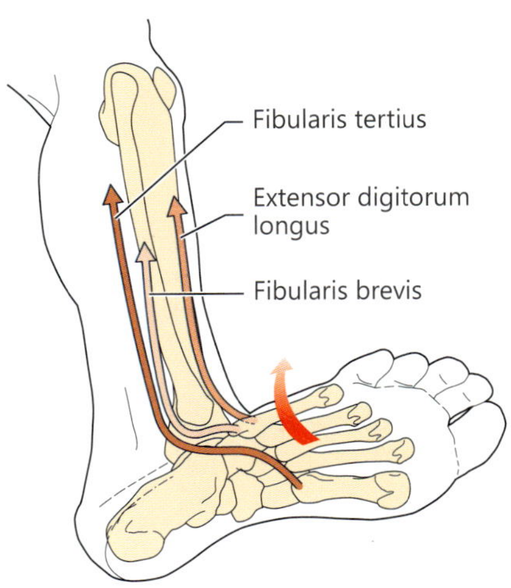

FLEXION OF THE TOES
- The flexor muscles of the great toe are comprised of:
  - flexor hallucis longus
  - flexor hallucis brevis
  - abductor hallucis, which flexes the proximal phalanx and extends the distal phalanx of the great toe
- The flexor digitorum longus and brevis flex the last four toes
- The flexor and abductor digiti minimi also contribute to the flexion of the little toe
- The flexor muscles of the toes:
  - act in such a way that the great toe presses firmly on the ground and the other toes hold on the ground, favoring the maintenance of balance during the unilateral support of the foot or support on the tiptoes
  - contract eccentrically in order to control the passive extension of the toes, which occurs when one lowers oneself to pick up an object from the ground, or during walking.

Flexor muscles of the toes
(the action of the muscles previously discussed is no further described)
(from greatest to least muscular force)
- Flexor digitorum longus
- Flexor hallucis longus
- Flexor digitorum brevis
- Flexor hallucis brevis
- Quadratus plantae
- Lumbricals
- Flexor digiti minimi brevis
- Opponens digiti minimi.

Action of the flexor muscles of the toes

Flexor digitorum brevis
- Flexes the middle phalanges of the last four toes.

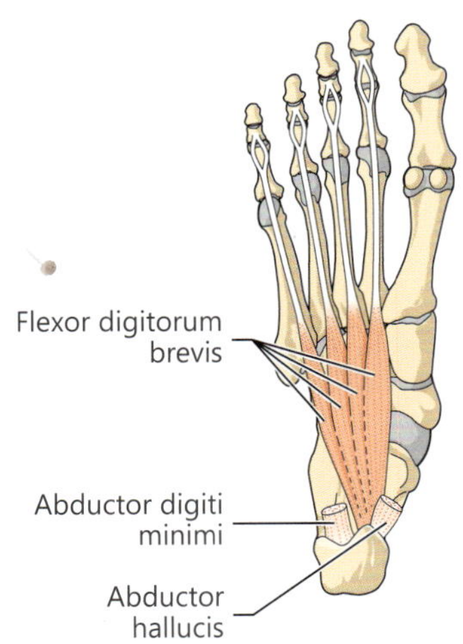

Flexor hallucis brevis
- Flexes the proximal phalanx of the great toe.

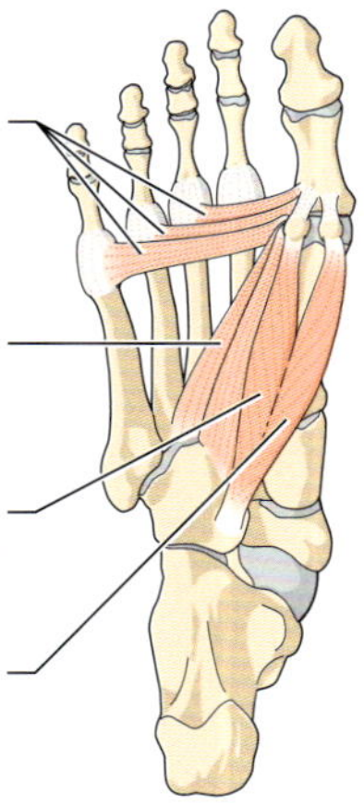

Quadratus plantae
- Increases the flexor action of the middle phalanges of the last four toes carried out by the flexor digitorum longus.

Lumbricals
- Flex the first phalanx and extend the middle and distal phalanges of the last four toes.

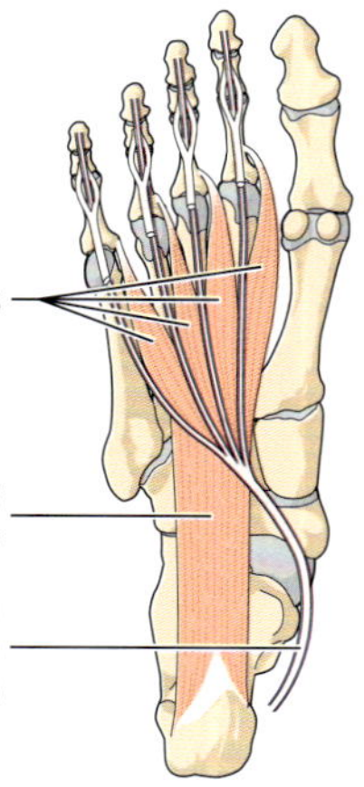

Flexor digiti minimi brevis
- Flexes the little toe plantarly.

Opponens digiti minimi
- Flexes the fifth metatarsal
- Contributes to supporting the longitudinal arches of the foot.

EXTENSION OF THE TOES
The extensor system of the foot is analogous to that of the hand.
- The extensor muscles of the great toe are comprised of:
  - the extensor hallucis longus
  - the extensor hallucis brevis
  - the abductor hallucis
- The extensor digitorum longus and brevis extend the proximal phalanges of the last four toes:
  - the extensor digitorum brevis does not reach the little toe; however, fibers coming from the abductor digiti minimi and the flexor digiti minimi brevis attach to the dorsal digital expansion of the little toe, aiding in its extension
- The lumbricals and interossei flex the proximal phalanges and simultaneously extend the middle and distal phalanges of the last four toes
- The extensor muscles of the toes contract during walking and climbing stairs:
  - the action of these muscles is exemplified by the extension of the toes when cutting the toenails.

Extensor muscles of the toes
- Extensor digitorum brevis
- Extensor hallucis brevis.

Action of the extensor muscles of the toes

Extensor digitorum brevis
- Extends the second, third, and fourth toes.

Extensor hallucis brevis
- Extends the great toe.

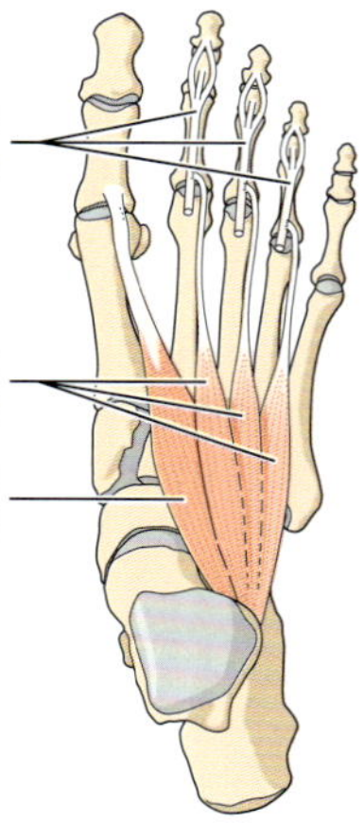

ABDUCTION AND ADDUCTION OF THE TOES
These movements occur only at the level of the metatarsophalangeal joints, and thus only involve the proximal phalanges of the toes.

Abductor muscles of the toes
(from greatest to least muscular force)
- Abductor hallucis
- Abductor digiti minimi
- Dorsal interossei.

Adductor muscles of the toes
(from greatest to least muscular force)
- Adductor hallucis
- Plantar interossei.

Action of the abductor and adductor muscles of the toes

Abductor hallucis
- Abducts and slightly flexes the great toe.

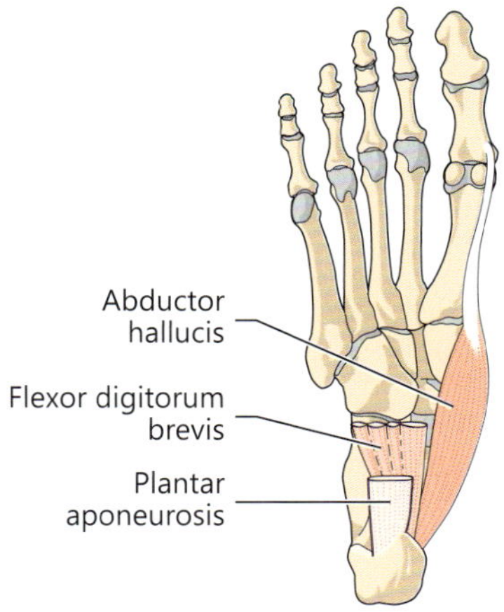

Adductor digiti minimi
- Abducts and flexes the little toe.

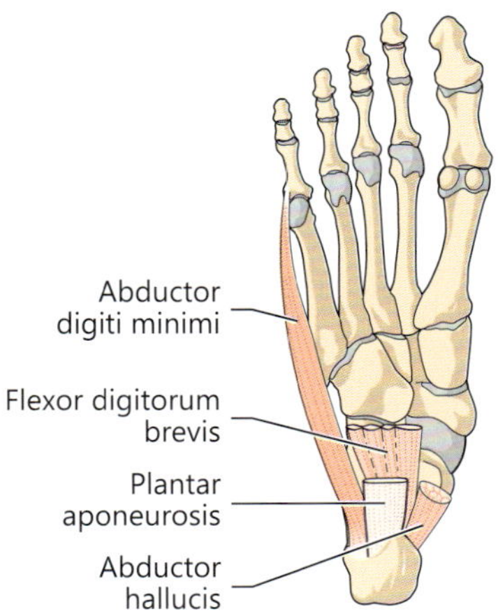

Dorsal interossei
- Second, third, and fourth dorsal interossei → abduct (separate) the second, third, and fourth toes from the great toe
- First dorsal interosseous → adducts (brings closer) the second toe to the great toe
- Flex the proximal phalanges and extend the distal ones.

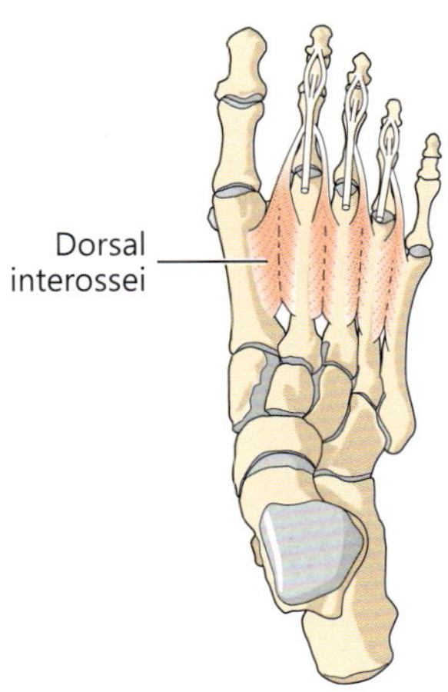

Adductor hallucis
- Moves the great toe in a medial direction and flexes its proximal phalanx
- Stretches the longitudinal arches of the foot.

Plantar interossei
- Adduct (bring closer) the second, fourth, and little toes to the third toe
- Flex the proximal phalanges and extend the distal ones (action in common with the dorsal interossei).

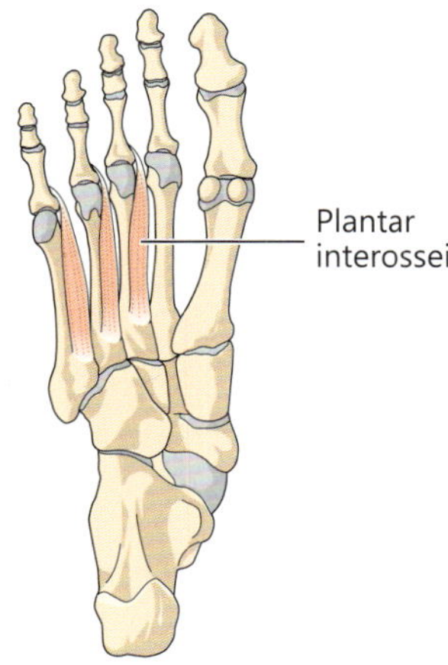

INTRINSIC MUSCLES OF THE FOOT AND SUPPORT OF THE ARCHES OF THE FOOT
- A function of the intrinsic muscles of the foot is to stabilize the foot during its forward thrust
- The principal intrinsic muscles of the foot (abductor hallucis, flexor hallucis brevis, flexor digitorum brevis, and abductor digiti minimi) constitute the most important muscular support of the arches of the foot, exerting a strong flexion of the forefoot and helping to stabilize the transverse tarsal joint
- The intrinsic muscles of the foot contract in order to stabilize the foot when the weight is transferred onto the forefoot while one lifts onto the tiptoes or while climbing and descending stairs
- The triceps surae, fibularis longus and fibularis brevis, tibialis anterior and tibialis posterior contract along with the intrinsic muscles of the foot in order to stiffen the foot in activities such as running and climbing.

MUSCLE	ORIGIN	INSERTIONS	INNERVATION	ACTION	OBSERVATIONS
Muscles of the lower limb					
MUSCLES OF THE HIP					
Internal muscles of the hip					
Iliopsoas (☞ table Muscles of the abdomen)					
– **psoas major**	Anterior surface of the transverse processes of L I-L V, lateral surface of the intervertebral bodies and discs of T XII, L I-L V	Lesser trochanter of the femur	Branches of the lumbar plexus (principally L1, L2, with some contribution of L3)	Flexion, adduction, and external rotation of the hip (with fixed point on the column and on the pelvis); forward flexion of trunk (for joint action of the muscles of both sides); homolateral inclination of the trunk (with fixed point on the femur)	
– **iliacus**	Upper two-thirds of the iliac fossa, inner lip of the iliac crest, anterior sacroiliac and iliolumbar ligaments, upper part of the lateral surface of the sacrum	Lateral surface of the tendon of the psoas major, lesser trochanter of the femur	Branches of the lumbar plexus [L2, L3]	Flexion, adduction, and external rotation of the hip (with fixed point on the pelvis); flexion and homolateral inclination of the trunk (with fixed point on femur)	
Psoas minor	Bodies of T XII and L I and interposed intervertebral disc	Iliopubic ramus and iliac fascia	Branches of the lumbar plexus [L1]	Flexion of the trunk and tension of the iliac fascia	
External muscles of the hip					
Superficial					
Gluteus maximus	Iliac crest, posterior gluteal line, lateral surfaces of the sacrum and coccyx, sacrotuberous ligament	Gluteal tuberosity of the femur	Inferior gluteal nerve [L5, S1, S2]	Extension and external rotation of the hip; abduction (superior fascicles) and adduction (inferior fascicles) of the hip; with fixed point on the femur, extension and retroversion of the pelvis, thus contributing to maintenance of the upright position and deambulation	
Tensor fasciae latae	Anterior superior iliac spine and anterior extremity of the iliac crest	Lateral condyle of the tibia, to which is attached by union of the tendons of the muscle with the fascia lata, forming the iliotibial tract	Superior gluteal nerve [L4, L5, S1]	Flexion, abduction, and internal rotation of the hip; tension of the fascia lata, contributing in this way to maintenance of the upright position, flexion and extension of the knee, via the iliotibial tract, respectively in the last 20° of extension and after the first 20° of flexion of the knee	
Intermediate					
Gluteus medius	Gluteal fascia, iliac crest, gluteal surface of the ilium between the anterior and posterior gluteal lines, anterior superior iliac spine	External surface of the greater trochanter	Superior gluteal nerve [L4, L5, S1]	Abduction of the hip; external rotation and extension of the hip (by the action of the posterior fascicles); internal rotation and flexion of the hip (by the action of the anterior fascicles)	
Deep					
Gemellus inferior	Ischial tuberosity	Trochanteric fossa joining the tendon of the obturator internus	Nerve to the quadratus femoris [L4, L5, S1]	External rotation and stabilization of the hip	
Gemellus superior	Ischial spine	Trochanteric fossa joining the tendon of the obturator internus	Nerve to the obturator internus [L5, S1, S2]	External rotation and stabilization of the hip	

continued

Muscles of the lower limb					(continued)
MUSCLE	***ORIGIN***	***INSERTIONS***	***INNERVATION***	***ACTION***	***OBSERVATIONS***
Obturator externus	External border of obturator foramen and obturator membrane	Trochanteric fossa	Posterior branch of the obturator nerve [L3, L4]	External rotation, stabilization and adduction of the hip	
Obturator internus	Internal border of the obturator foramen, obturator membrane, and bone surface between the obturator foramen and the greater sciatic notch	Trochanteric fossa	Nerve to the obturator internus [L5, S1]	External rotation and stabilization of the hip (abduction of the hip from a seated position with knee extended)	
Gluteus minimus	Gluteal surface of the ilium between the anterior and inferior gluteal lines, iliac crest	Anterior border of the greater trochanter	Superior gluteal nerve [L4, L5, S1]	Abduction and internal rotation of the hip; with fixed point on the femur, extension and lateral inclination of the pelvis. The contraction of both muscles contributes to maintenance of the upright position	
Piriformis	Pelvic surface of the sacrum, laterally to the anterior sacral foramina, border of the greater sciatic notch	Upper extremity of the internal surface of the greater trochanter	Nerve to the piriformis [L5, S1, S2]	Abduction of the hip (from position of flexion); stabilization and external rotation of the hip (from position of extension)	
Quadratus femoris	Ischial tuberosity	Intertrochanteric crest	Nerve to the quadratus femoris [L5, S1]	External rotation, stabilization and adduction of the hip	
MUSCLES OF THE THIGH					
Anterior					
Articularis genus	Anterior surface of the inferior part of the shaft of the femur	Superior recess of the suprapatellar bursa	Femoral nerve [L2-L4]	Traction in proximal direction of the suprapatellar bursa, which is thus brought away from the patella during the extension of the knee	Made up of several deep fascicles of the vastus intermedius
Quadriceps femoris					
– **rectus femoris**	Straight head: anterior surface of the anterior inferior iliac spine Reflected head: supra-acetabular groove and capsule of the hip joint	Base of the patella and tibial tuberosity through the tendon of the quadriceps femoris	Femoral nerve [L2-L4]	Extension of the knee; flexion of the hip	The inferior fibers of the muscle, originating mostly from the tendon of the adductor magnus and running nearly horizontally with insertion on the medial border of the patella, are sometimes defined as the vastus medialis obliquus
– **vastus intermedius**	Upper half of the lateral lip of the linea aspera, upper two-thirds of the anterior and lateral surface of the shaft of the femur	Base of the patella and tibial tuberosity through the tendon of the quadriceps femoris	Femoral nerve [L2-L4]	Extension of the knee	
– **vastus lateralis**	Upper part of the intertrochanteric line (spiral line of the femur), anterior and inferior borders of the greater trochanter, tendon of the gluteus maximus	Lateral border and base of the patella, tibial tuberosity through the tendon of the quadriceps femoris	Femoral nerve [L2-L4]	Extension of the knee; external rotation of the knee from a position of internal rotation and flexion	
– **vastus medialis**	Lower part of the intertrochanteric line (spiral line of the femur), medial lip of the linea aspera, proximal part of the medial supracondylar line, tendons of the adductor longus and adductor magnus, medial intermuscular septum of the thigh	Medial border of the patella, tibial tuberosity through the tendon of the quadriceps femoris	Femoral nerve [L2-L4]	Extension of the knee; internal rotation of the knee from a position of external rotation and flexion	

continued

Muscles of the lower limb					*(continued)*
MUSCLE	***ORIGIN***	***INSERTIONS***	***INNERVATION***	***ACTION***	***OBSERVATIONS***
Sartorius	Anterior superior iliac spine and upper half of the underlying notch	Upper part of the medial surface of the tibia (in front of the gracilis and semitendinosus)	Femoral nerve [L2, L3]	Flexion, abduction and external rotation of the hip; flexion and internal rotation of the knee; anteversion of the pelvis (with fixed point on the tibia)	The tendon of insertion, joined with fibrous expansions to the tendons of the gracilis and semitendinosus, constitute the goose's foot
Medial (adductors)					
Superficial					
Adductor longus	Superior pubic ramus, medially to the origin of the pectineus, between the pubic symphysis and tubercle	Medial lip of the linea aspera in the middle third of the shaft of the femur	Obturator nerve [L1-L4]	Adduction, external rotation and flexion of the hip	
Gracilis	Anterior surface of the ischiopubic ramus and lower half of the body of the pubis	Upper part of the medial surface of the tibia (between the sartorius and semitendinosus)	Obturator nerve [L2-L4]	Adduction of the hip (with knee extended); internal rotation and flexion of the hip; flexion of the knee	
Pectineus	Pecten pubis (pectineal line of the pubis), between the iliopubic ramus and the pubic tubercle	Pectineal line of the femur, between the lesser trochanter and the linea aspera	Femoral nerve [L2, L3]	Flexion and adduction of the hip	
Intermediate					
Adductor brevis	External surface of the body of the pubis and inferior pubic ramus, between the gracilis and obturator externus	Between the lesser trochanter and the linea aspera and on the upper part of the linea aspera	Obturator nerve [L2-L4]	Adduction; external rotation and flexion of the hip	
Deep					
Adductor minimus	Inferior pubic ramus	Upper third of the medial lip of the linea aspera	Obturator nerve [L2]	Adduction and external rotation of the hip	Considered also as superior part of the adductor magnus
Adductor magnus	Anterior surface of the ischiopubic ramus and inferolateral surface of the ischial tuberosity	Medial lip of the linea aspera of the femur, medial supracondylar line and adductor tubercle	Obturator and tibial nerves [L3-L5]	Adduction, extension, and external rotation of the hip; internal rotation of the hip (from position of external rotation and flexion)	
Posterior					
Biceps femoris	Long head: inferomedial sulcus on the superior surface of the ischial tuberosity; inferior part of the sacrotuberous ligament Short head: lateral lip of the linea aspera and lateral supracondylar line	Head of the fibula, lateral condyle of the tibia, fibular collateral ligament	Sciatic nerve (tibial nerve) [L5, S1, S2] Sciatic nerve (common fibular nerve) [L5, S1, S2]	Flexion of the knee and external rotation of the knee from a position of flexion	The semimembranosus, semitendinosus and biceps femoris unite the inferior part of the ischium and the femur to the bones of the leg (tibia and fibula), for which reason they are generically defined as ischiocrural muscles
Semimembranosus	Superolateral surface of the ischial tuberosity	Medial condyle of the tibia (posteriorly and anteriorly) and capsule of the knee joint through the oblique popliteal ligament	Sciatic nerve (tibial nerve) [L5, S1, S2]	Flexion of the knee and internal rotation of the knee from a position of flexion	
Semitendinosus	Inferomedial sulcus on the superior surface of the ischial tuberosity	Proximal part of the medial surface of the tibia	Sciatic nerve (tibial nerve) [L5, S1, S2]	Flexion of the knee and internal rotation of the knee from a position of flexion	

continued

Muscles of the lower limb					(continued)
MUSCLE	***ORIGIN***	***INSERTIONS***	***INNERVATION***	***ACTION***	***OBSERVATIONS***
		MUSCLES OF THE LEG			
Anterior					
Extensor hallucis longus	Middle two-quarters of the medial surface of the fibula, anterior surface of the interosseous membrane	Dorsal surface of the base of the proximal phalanx and that of the distal phalanx of the great toe	Deep fibular nerve [L4, L5, S1]	Interphalangeal extension of the great toe	
Extensor digitorum longus	Lateral condyle of the tibia, proximal three-quarters of the medial surface of the fibula, anterior surface of the interosseous membrane	Dorsal surface of the base of the middle phalanx of the last four toes; dorsal surface of the base of the distal phalanx of the last four toes	Deep fibular nerve [L5, S1]	Extension of the metatarsophalangeal and interphalangeal joints of the last four toes	
Fibularis tertius	Lower third of the medial surface of the fibula and corresponding part of the interosseous membrane	Dorsal surface of the base of the fifth metatarsal	Deep fibular nerve [L4, L5, S1]	Dorsal flexion and eversion of the foot	
Tibialis anterior	Lateral condyle of the tibia, proximal half of the lateral surface of the shaft of the tibia; anterior surface of the interosseous membrane	Medial and inferior surface of the medial cuneiform; medial surface of the base of the first metatarsal	Deep fibular nerve [L4, L5, S1]	Dorsal flexion, inversion and adduction of the foot	
Lateral					
Fibularis brevis	Lower two-thirds of the lateral surface of the fibula	Tuberosity of the fifth metatarsal	Superficial fibular nerve [L4, L5, S1]	Plantar flexion and eversion of the foot	
Fibularis longus	Head and upper two-thirds of the lateral surface of the fibula; some fibers originate from the lateral condyle of the tibia	Lateral surface of the base of the first metatarsal and medial cuneiform	Superficial fibular nerve [L4, L5, S1]	Plantar flexion and eversion of the foot	
Posterior					
Superficial					
Plantaris	Lateral condyle of the femur in posterior, superior, and medial location	Medial part of the posterior surface of the calcaneus	Tibial nerve [L4, L5, S1]	Plantar flexion and inversion of the foot	
Triceps surae					
– gastrocnemius	Medial head: proximal and posterior surfaces of the medial condyle of the femur posteriorly to the adductor tubercle Lateral head: lateral and posterior surfaces of the lateral condyle of the femur; lower part of the lateral supracondylar line	Calcaneal tuberosity, through the calcaneal tendon (Achilles tendon)	Tibial nerve [S1, S2]	Plantar flexion, inversion and abduction of the foot; flexion of the knee	
– soleus	Posterior surface of the head and proximal quarter of the shaft of the fibula, soleal line, middle third of the medial border of the tibia	Calcaneal tuberosity, through the calcaneal tendon (Achilles tendon)	Tibial nerve [S1, S2]	Plantar flexion and inversion of the foot	

continued

Muscles of the lower limb					(continued)
MUSCLE	ORIGIN	INSERTIONS	INNERVATION	ACTION	OBSERVATIONS
Deep					
Flexor hallucis longus	Distal two-thirds of the posterior surface of the fibula, posterior surface of the interosseous membrane	Plantar surface of the base of the distal phalanx of the great toe	Tibial nerve [S1-S3]	Interphalangeal plantar flexion of the great toe	
Flexor digitorum longus	Posterior surface of the middle three-fifths of the tibia, below the soleal line	Plantar surface of the bases of the distal phalanges of the last four toes	Tibial nerve [L5, S1, S2]	Distal interphalangeal plantar flexion of the last four toes	
Popliteus	Posterior surface of the lateral epicondyle of the femur, capsule of the knee joint	Posterior surface of the tibia, superiorly to the soleal line	Tibial nerve [L4, L5, S1]	Flexion and internal rotation of the knee	
Tibialis posterior	Upper two-thirds of the posterior and lateral surfaces of the tibia below the soleal line, posterior surface of the interosseous membrane, medial surface of the proximal two-thirds of the fibula	Tuberosity of the navicular, with tendinous expansions directed to the plantar surface of the three cuneiforms	Tibial nerve [L5, S1]	Plantar flexion, inversion and adduction of the foot	
MUSCLES OF THE FOOT					
Muscles of the dorsum of the foot					
Extensor hallucis brevis	Common origin with the extensor digitorum brevis	Dorsal surface of the base of the proximal phalanx of the great toe	Deep fibular nerve [L5, S1]	Extension and adduction of the great toe	Considered as well as the medial part of the extensor digitorum brevis
Extensor digitorum brevis	Distal part of the superolateral surface of the calcaneus, immediately in front of the tarsal sinus	Tendons that join the lateral surface of the tendons of the extensor digitorum longus for the second, third, and fourth toes	Deep fibular nerve [L5, S1]	Extension of the phalanges of the second, third, and fourth toes	
Muscles of the sole of the foot					
Medial					
Abductor hallucis	Medial process of the calcaneal tuberosity, flexor retinaculum, plantar aponeurosis	Medial surface of the base of the proximal phalanx of the great toe	Medial plantar nerve [S1, S2]	Abduction and plantar flexion of the great toe	
Adductor hallucis	Oblique head: bases of the second, third, and fourth metatarsals; sheath of the fibularis longus Transverse head: plantar ligaments of the metatarsophalangeal joints of the third, fourth, and little toes and deep transverse metatarsal ligament of the same toes	Both heads: lateral sesamoid bone of the first metatarsophalangeal joint and base of the proximal phalanx of the great toe	Lateral plantar nerve [S2, S3]	Adduction of the great toe	
Flexor hallucis brevis	Plantar surface of the medial and intermediate cuneiforms	The muscular fascicles divide into a medial head that inserts at the medial sesamoid bone of the metatarsophalangeal joint of the great toe, and into a lateral head that partially merges with the adductor hallucis, and inserts at the lateral sesamoid bone of the metatarsophalangeal joint of the great toe	Medial plantar nerve [L5, S1]	Metatarsophalangeal plantar flexion of the great toe; adduction and abduction of the great toe, through, respectively, the lateral and medial heads of the muscle	

continued

Muscles of the lower limb					(continued)
MUSCLE	**ORIGIN**	**INSERTIONS**	**INNERVATION**	**ACTION**	**OBSERVATIONS**
Lateral					
Abductor digiti minimi	Medial and lateral processes of the calcaneal tuberosity and bone between the two processes, plantar aponeurosis	Lateral surface of the base of the proximal phalanx of the little toe	Lateral plantar nerve [S1-S3]	Abduction and flexion of the little toe	
Flexor digiti minimi brevis	Medial part of the plantar surface of the base of the fifth metatarsal, sheath of the fibularis longus	Lateral surface of the base of the proximal phalanx of the little toe	Lateral plantar nerve [S1, S2]	Metatarsophalangeal plantar flexion of the little toe	
Opponens digiti minimi	Long plantar ligament	Lateral border of the fifth metatarsal	Lateral plantar nerve [S2, S3]	Adduction and flexion of the little toe	Muscle not always present
Intermediate					
Flexor digitorum brevis	Medial process of the calcaneal tuberosity, plantar aponeurosis	Medial and lateral surfaces of the middle phalanges of the last four toes	Medial plantar nerve [S1, S2]	Proximal interphalangeal plantar flexion of the last four toes	
Dorsal interossei	Adjacent sides of the metatarsals	First dorsal interosseous: medial surface of the base of the proximal phalanx of the second toe; from the second to the fourth dorsal interossei: lateral surfaces of the bases of the proximal phalanges of the second, third, and fourth toes; dorsal digital expansion or aponeurosis	Lateral plantar nerve [S2, S3]	Abduction of the second, third, and fourth toes and flexion of the metatarsophalangeal joints	
Plantar interossei	Bases and medial surfaces of the third, fourth, and fifth metatarsals	Medial surfaces of the bases of the proximal phalanges of the third, fourth, and little toes; dorsal digital expansion or aponeurosis	Lateral plantar nerve [S2, S3]	Adduction of the third, fourth, and little toes and flexion of the metatarsophalangeal joints	
Lumbricals	First lumbrical: medial surface of the tendon for the second toe of the flexor digitorum longus; from the second to the fourth lumbricals: adjacent sides of the tendons of the flexor digitorum longus	Medial surface of the dorsal digital expansion or aponeurosis on the proximal phalanges of the last four toes	First lumbrical: medial plantar nerve [S2, S3]; from the second to the fourth lumbricals: lateral plantar nerve [S2, S3]	Flexion of the proximal phalanges and extension of the middle and distal phalanges of the last four toes	
Quadratus plantae	Lateral and medial borders of the inferior surface of the calcaneus	Laterally to the tendon of the flexor digitorum longus before its division into four terminal tendons	Lateral plantar nerve [S1, S2]	Flexion of the middle phalanges of the last four toes	

Fasciae and fascial structures of the lower limb

STRUCTURE	POSITION	ORGANIZATION	INSERTION	CONTINUATION
Gluteal aponeurosis	Gluteal region	Extends in front of the gluteus medius, and at the anterosuperior or superolateral border of the gluteus maximus it divides into three layers, superficial and middle, which form a sheath around the gluteus maximus, and deep, which, in addition to covering the gluteus medius, covers the other muscles located deeply	Superiorly and anteriorly at the iliac crest and at the posterior border of the tensor fasciae latae	Distally with the gluteal fascia
Gluteal fascia	Gluteal region	Fascia lata that covers the gluteus maximus		Proximally with the gluteal aponeurosis, distally with the fascia lata
Fascia lata (femoral fascia)	Thigh	Surrounds the thigh like a sleeve, thickening laterally in the iliotibial tract and being perforated anteriorly at the level of the femoral triangle. From its deep surface originate the medial and lateral intermuscular septa of the thigh. Accommodates, in one of its splits, the sartorius and the tensor fasciae latae	Superiorly and posteriorly on the posterior portion of the sacrum and coccyx, laterally on the external border of the iliac crest, anteriorly on the inguinal ligament and the superior pubic ramus, medially on the inferior pubic ramus, the ischial ramus, the ischial tuberosity, and the inferior border of the sacrotuberous ligament, inferiorly on the condyles of the femur and of the tibia and on the head of the fibula	Proximally with the gluteal fascia and the perineal fascia, distally with the deep fascia of the leg
Medial and lateral intermuscular septa of the thigh	Thigh	Originating from the deep surface of the fascia lata, they identify an anterior compartment of the thigh, which contains the anterior muscles of the thigh, and a posterior compartment, which contains the posterior and medial muscles of the thigh, with the exception of the sartorius and the tensor fasciae latae	Respectively, on the medial and lateral lips of the linea aspera of the femur	
Iliotibial tract	Lateral surface of the thigh	Thickening of the fascia lata over the lateral surface of the thigh. The superior extremity opens into two layers that include and anchor the tensor fasciae latae	Gerdy's tubercle on the anterolateral part of the lateral condyle of the tibia	
Femoral triangle (inguinofemoral Scarpa's triangle)	Triangular region in the proximal portion of the anterior surface of the thigh	Delimited superiorly by the inguinal ligament, inferolaterally by the sartorius and inferomedially by the adductor longus, it is superficially covered by the fascia lata, also called in this location the inguinofemoral fascia, which in the central part of the triangle takes the name of the cribriform fascia. The apex of the triangle is located where the sartorius stands above the adductor longus, while the deep muscular plane is made up laterally by the iliacus and psoas major and medially by the pectineus and adductor longus. It contains the femoral vessels in the deepest part (femoral canal), the femoral nerve, laterally to the femoral artery, lymph nodes, and adipose tissue		
Inguinofemoral fascia	Inguinofemoral region	Between the subcutaneous tissue and the muscular layer, it may be subdivided into three portions: lateral (from the tensor fasciae latae to the medial border of the sartorius), medial (corresponding to the gracilis and adductor longus), and intermediate (between the aforementioned ones), called the cribriform fascia	Proximally, at the anterior border of the inguinal ligament	Gluteal fascia, perineal fascia, and fascia lata

continued

Fasciae and fascial structures of the lower limb				*(continued)*
STRUCTURE	***POSITION***	***ORGANIZATION***	***INSERTION***	***CONTINUATION***
Cribriform fascia	Central part of the femoral triangle	Part of the fascia lata pierced with many holes that serve as a passage for arteries, veins, lymphatic vessels, and nerves, which from the subcutaneous tissue descend into the subfascial layer or vice versa, the most important of which is the one crossed by the saphenous vein (saphenous opening)		
Femoral canal	Femoral triangle	Fibrous canal, depending on the fascia lata, containing the femoral vessels (femoral artery in the lateral third, femoral vein in the intermediate third, lymphatic vessels and lymph nodes in the medial third). The anterior wall is formed by the cribriform fascia, the posterolateral and posteromedial walls by the deep layer of the fascia lata, the proximal extremity by the femoral ring and the distal extremity by the saphenous opening		
Saphenous opening	Cribriform fascia (3-4 centimeters distally to the inguinal ligament)	Medial border corresponding to the fascia that covers the pectineus, lateral border formed by a medially concave lunate bundle (fascial thickening), called falciform margin		
Falciform margin	Saphenous opening	Lateral border of the saphenous opening with two extremities or horns, the proximal of which (superior horn) is close to the inguinal ligament, while the distal one (inferior horn) runs medially, passing under the arch of the great saphenous vein to end laterally to the vein, on the fascia of the pectineus		
Iliopectineal arch	Part of the iliac fascia that extends obliquely from the inguinal ligament to the iliopubic ramus	Divides the space between the inguinal ligament and the anterior border of the hip bone into two orifices: lateral, which connect the internal iliac fossa with the inguinofemoral region and is filled by the iliopsoas (muscular space), which carries the femoral nerve on its medial side; and medial, which connects the inguinofemoral triangle with the abdominal cavity (vascular space or femoral ring)		
Femoral ring (vascular space)	Orifice located medially to the iliopectineal arch, which connects the inguinofemoral triangle with the abdominal cavity	Serves as a passage for the femoral vessels (femoral artery, which occupies its lateral part, applied against the iliopectineal arch, and femoral vein, located just medially to the artery)		
Femoral septum	Femoral ring	Part of the transverse fascia that, medially to the femoral vein, descends up to the pecten pubis, closing the space between the femoral vessels and the lacunar ligament		
Muscular space ☞ Iliopectineal arch				
Vascular space ☞ Iliopectineal arch				
Anteromedial intermuscular septum (vastoadductor or subsartorial fascia)	Triangular lamina that extends from the inferomedial border of the adductor magnus to the vastus medialis	Covered by the sartorius, It forms the anteromedial boundary of the adductor canal		

continued

Fasciae and fascial structures of the lower limb *(continued)*

STRUCTURE	POSITION	ORGANIZATION	INSERTION	CONTINUATION
Adductor hiatus (ring of the adductor magnus)	Adductor canal	Between the middle muscular fascicles of the adductor magnus, which insert at the medial lip of the linea aspera of the femur, and the inferior fascicles, which insert at the adductor tubercle, it gives passage to the femoral vessels that pass through the popliteal fossa		
Adductor canal	Distal two-thirds of the medial part of the thigh	Intermuscular canal that begins at the apex of the femoral triangle and extends up to the insertion of the tendon of the adductor magnus. Delimited anterolaterally by the vastus medialis, posteromedially by the adductor longus, and distally to the adductor longus, by the adductor magnus, it contains the femoral artery and vein and the saphenous nerve		
Deep fascia of the leg	Leg	Receives lateral expansions from the tendon of the biceps femoris and medial expansions from the tendons of the sartorius, gracilis, semitendinosus, and semimembranosus, and merges with the periosteum on the subcutaneous surfaces of the tibia, the head of the fibula, and the lateral malleolus. From its deep surface originate the anterior and posterior intermuscular septa of the leg	Border of the patella, patellar ligament, tibial tuberosity, condyles of the tibia, and head of the fibula	Superiorly with the fascia lata, inferiorly with the extensor and flexor retinacula
Transverse intermuscular septum (posterior intermuscular septum of the leg)	Leg	Originates at the tendinous arch of the soleus and extends transversally from the medial border of the tibia to the posterior border of the fibula, separating the superficial posterior muscles of the calf from the deep muscles		Popliteal fascia, flexor retinaculum and superior fibular retinaculum
Anterior and posterior intermuscular septa	Leg	Originating from the deep surface of the deep fascia of the leg, they insert at the anterior and lateral borders of the fibula and identify a fibular compartment. Along with the skeleton of the leg and the interosseous membrane, they separate the compartment of the anterior muscles of the leg from that of the posterior muscles		
Popliteal fascia	Popliteal fossa	Depending on the deep fascia of the leg that covers the popliteal fossa, reinforced by transverse fibers and often perforated by the small saphenous vein and by the sural nerve		
Tendinous arch of the soleus	Leg	Fibrous arch stretched between the head of the fibula and the soleal line, behind the interosseous membrane, with which it delimits the ring of the soleus, where the popliteal vessels and the tibial nerve run	Head of the fibula, and the soleal line on the posterior surface of the tibia	
Superior extensor retinaculum (transverse ligament of the leg)	Ankle	Ventral thickening of the deep fascia of the leg formed by transverse fibers, from the deep surface of which fibrous septa originate that insert on the skeleton, delineating three osteofibrous canals where the tendons of the anterior muscles of the leg run	Anterior border of the tibia and the fibula	Deep fascia of the leg

continued

Fasciae and fascial structures of the lower limb				*(continued)*
STRUCTURE	***POSITION***	***ORGANIZATION***	***INSERTION***	***CONTINUATION***
Inferior extensor retinaculum (cruciate ligament of the leg)	Ankle	Transverse Y-shaped band, with the stem of the Y positioned laterally and the upper and lower arms medially, from the deep surface of which fibrous septa originate that insert on the skeleton, delineating three osteofibrous canals where the tendons of the anterior muscles of the leg run	Superior surface of the calcaneus, calcaneal sulcus, medial malleolus, medial cuneiform and navicular	Deep fascia of the leg and superficial dorsal fascia of the foot
Flexor retinaculum (laciniate ligament)	Ankle	Medial thickening of the deep fascia of the leg, from the deep surface of which fibrous septa originate that insert on the skeleton, delineating three osteofibrous canals where the tendons of the posterior muscles of the leg run	Medial malleolus and medial process of the calcaneal tuberosity	Deep fascia of the leg and plantar aponeurosis
Superior and inferior fibular retinacula	Ankle	Lateral thickenings of the deep fascia of the leg, deep to which run the tendons of the fibularis longus and fibularis brevis	Lateral malleolus and lateral surface of the calcaneus	Deep fascia of the leg
Superficial dorsal fascia of the foot	Foot	Covers the tendons of the extensor muscles and distally extends on the metatarsals and on the phalanges as the dorsal digital aponeurosis	Lateral and medial to the skeleton of the foot	Inferior extensor retinaculum and, corresponding to the borders of the foot, with the plantar aponeurosis
Deep dorsal fascia of the foot	Foot	Located deep to the tendons of the extensor muscles, covers the dorsal interossei and the skeleton of the foot		
Plantar aponeurosis (plantar fascia)	Foot	Divided into three parts (medial, intermediate, and lateral), covers the sole of the foot with longitudinal fascicles that from the calcaneus reach the metatarsophalangeal joints, and transverse fascicles that unite the longitudinal fascicles. From the deep surface, plantar septa originate that identify three osteofibrous compartments containing the muscles of the sole of the foot	Medial process of the calcaneal tuberosity	Flexor retinaculum and, corresponding to the borders of the foot, with the superficial dorsal fascia of the foot
Deep plantar fascia (plantar interosseous fascia)	Metatarsals and interosseous spaces	Not very differentiated and difficult to isolate, It resembles a layer of loose connective tissue rather than a fascia	Lateral border of the first metatarsal and medial border of the fifth metatarsal	Superficial transverse metatarsal ligament
Superficial transverse metatarsal ligament	Heads of the metatarsals	Plantar interdigital ligament that, distal to the heads of the metatarsals, merges with the deep plantar fascia		Deep plantar fascia

Tendinous sheaths of the lower limb	
(Synovial) sheaths in the three osteofibrous canals of the extensor retinacula (anterior tarsal)	Tendon of the tibialis anterior Tendon of the extensor hallucis longus Tendons of the extensor digitorum longus and fibularis tertius
(Synovial) sheaths in the three osteofibrous canals of the flexor retinaculum (tibial tarsal)	Tendon of the tibialis posterior Tendon of the flexor hallucis longus Tendons of the flexor digitorum longus
Common (synovial) sheath deep to the fibular retinacula (fibular tarsal)	Tendons of the fibularis longus and fibularis brevis
Plantar (synovial) sheath (fibular tarsal)	Tendon of the fibularis longus
Sheaths of the toes – synovial sheaths of the toes – – vincula tendinum – fibrous sheaths of the toes (anular and cruciform parts)	 Tendons of the flexor digitorum longus and brevis Folds of the synovial membrane, which give passage to blood vessels that reach the tendons Cover the synovial sheaths of the toes

Index

This index provides that single entry words are organized under the relative structures (i.e. Artery(ies), Bone(s), Muscle(s), etc.). For example, the single entry word "Humerus" is under "Bone(s)" or the single entry word "Latissimus dorsi" is under "Muscle(s)". This concerns also structures like ducts, prominence, tracts, etc. Organs and regional structures (i.e. "Colon") have their main entries and subentries. Page numbers with "t" refer to tables.

D

G

M

N

R

FOUR STEPS TO ACCESS DIGITAL CONTENTS

1	CONNECT	Go to the website indicated on the label that includes the access code
2	REGISTER	Register at the Publisher Portal (first time only) to receive a username and password
3	ACCESS	Enter your username and password to access the restricted contents
4	TYPE THE CODE	On the web page dedicated to the work, type in the personal access code hidden under the protection of the label applied to this page

After the first access, digital contents will be available on the website entering your username and password.

The **online access** grants **personal license** limited to a **single user** for each code.
The **access is limited to the individual user** and is not a library or institutional access license.

Sharing of passwords and/or code is not allowed; any misuse of the personal code will invalidate and lock it. The access cannot be shared and will expire according as indicated in the license agreement subscribed at the first access. Further details may be provided upon acceptance of the license agreement. The use of the code is subject to acceptance of the conditions. Any violation or attempted violation of code protection will not be accepted.

To remove the protection use a coin to scrape with light pressure

EE5893003862
www.evirtualcampus.com/en

Hardware and software requirements:
PC with Windows, Macintosh or Linux,
latest generation of internet browser such as Internet Explorer (version 9 onwards), Firefox, Chrome etc.; Internet connection.

Technical Help Desk: by email at support@ediermes.com

Contributors
English Edition

Editors

Giuseppe Anastasi	University of Messina, Italy
Eugenio Gaudio	University of Roma (Sapienza), Italy
Carlo Tacchetti	University Vita-Salute San Raffaele, Milan, Italy

Editor English Edition

Estomih Mtui	Weill Cornell Medicine, New York, USA

Coauthors

Paolo Castano†	University Vita-Salute San Raffaele, Milan, Italy
Sergio Castorina	University of Catania, Italy
Ottavio Cremona	University Vita-Salute San Raffaele, Milan, Italy
Raffaele De Caro	University of Padova, Italy
Guido Macchiarelli	University of l'Aquila, Italy
Mario Rende	University of Perugia, Italy
Domenico Ribatti	University of Bari, Italy
Chiarella Sforza	University of Milan, Italy

Other contributors

Demetrio Milardi, University of Messina, Italy, *3D volume rendering images*

Web contents

Estomih Mtui, Weill Cornell Medicine, New York, USA
Angelo Favaloro, University of Messina, Italy
Antonio Esposito, University Vita-Salute San Raffaele, Milan, Italy
Maurizio Vertemati, Department of Biomedical and Clinical Sciences "L. Sacco" University of Milan, Italy

Giuseppe Anastasi
Eugenio Gaudio
Carlo Tacchetti

Estomih Mtui
Editor English Edition

HUMAN ANATOMY
Atlas

Volume 2

HEAD AND NECK
THORAX

edi·ermes

HUMAN ANATOMY - Atlas - Volume 2 - Giuseppe Anastasi, Eugenio Gaudio, Carlo Tacchetti (eds)
Estomih Mtui (ed English edition)

ISBN 978-88-7051-589-3

A book is the final product of a very complex series of operations that requires numerous tests on texts and images. It is almost impossible to publish a book with no errors. We will be grateful to those who find them and notify us. For enquiries or suggestions about this volume, please use the following address:

External relations - Edi.Ermes srl - Viale Enrico Forlanini, 65 - 20134 Milan (Italy)
Phone +39.02.70.21.121 - Fax +39.02.70.21.12.83

Drawings, Edi.Ermes Archive/Andrea Rossi Raccagni, Marco Fanuli, Raffaella Stilo, Maria Acquaviva, Anna Maria D'Errico, Ettore Orlandi

Translated and revised by Rosemary Allpress

Printed in November 2017 by Faenza Printing Industries SpA
for Edi.Ermes - viale Enrico Forlanini, 65 - 20134 Milan, Italy
http://www.ediermes.com - Phone +39.02.70.21.121 - Fax +39.02.70.21.12.83

Foreword

The **Human Anatomy – Atlas** is in my opinion the finest Atlas of Human Anatomy that has ever been written. The main factors that make it so unique include the quality of both the anatomical and radiological images and the way they have been organized.

Artists and radiologists working in collaboration with Edi.Ermes in Milan have created a very high quality 3-D reconstructed images from CT scans and MRI's, used in the Atlas. This approach will give the student and the reader a very clear and accurate understanding of the three dimensional presentation of each structure covered in the Atlas.

All the images are accompanied by a very detailed information related to the anatomy and where appropriate clinical correlates have also been included. This approach will assist the student to relate each structure studied with a clinical condition.

In addition, the authorship of this Atlas has organized the images so that they present the flow of information needed by the student to build a clear image of the human anatomy.

An emphasis on cross sectional anatomy and lymphatic drainage has been made in this Atlas by including several illustrations and radiological images.

Each chapter of the Atlas covering both the anatomy and neuroanatomy contains an introduction, dissection movies and interactive slides used in building the anatomical regions with a stratigraphic approach.

It has been a great pleasure to collaborate with Edi.Ermes in preparing the English edition of this Atlas. I wish to thank in particular Raffaele Grandi the CEO of Edi.Ermes for allowing me to be part of the development team for this Atlas of Human Anatomy.

Estomih Mtui, MD
Editor English Edition

Professor of Clinical Anatomy in Radiology
Chief, Division of Anatomy
Department of Radiology
Weill Cornell Medicine
New York, USA

Contents Plan

Volume 1

Volume 2

Volume 3

Table of Contents

HEAD AND NECK

THORAX

1

HEAD AND NECK

Petros (Athens, 1928 - Milan, 2014), *Talete*, oil on canvas, cm 150 × 200, 1990

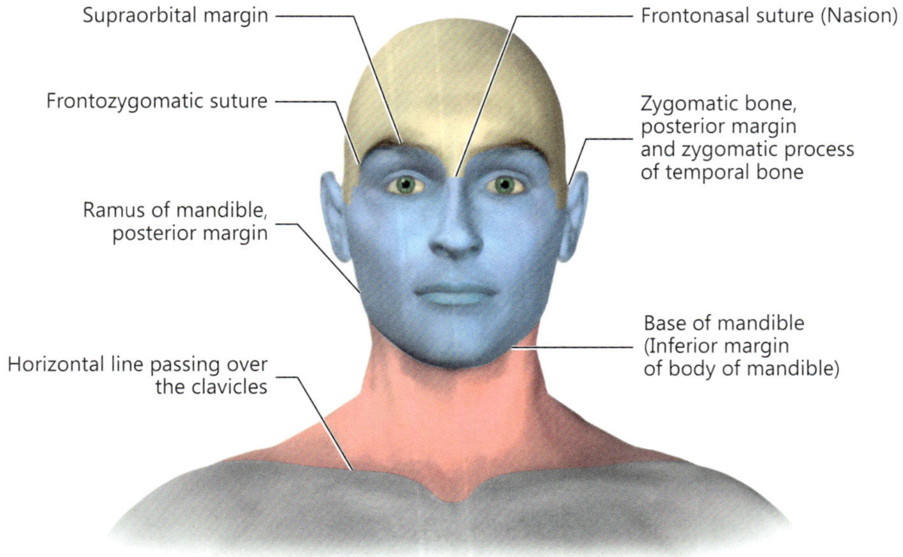

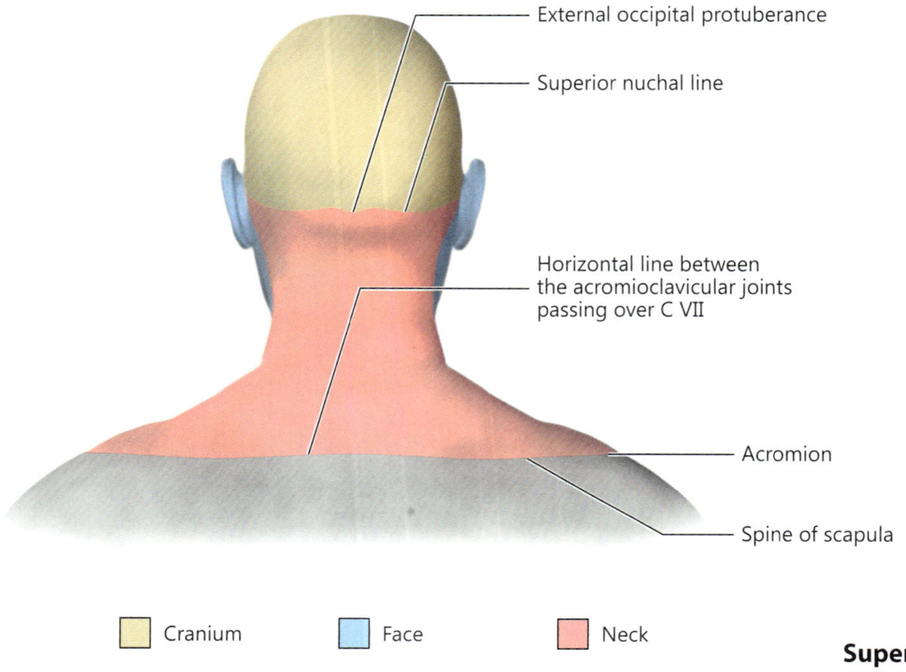

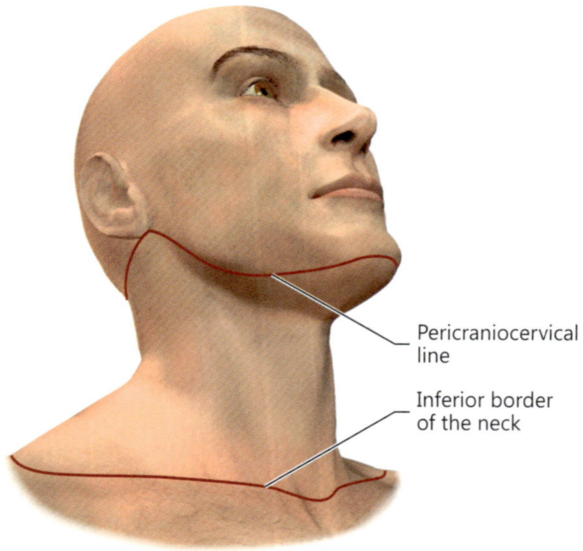

Superficial boundaries of the head and **neck**. The **head** is composed of the **cranium** (*yellow*) and **face** (*blue*), and is separated from the neck by the pericraniocervical line. This follows the inferior margin of the body of the mandible until its angle, then the posterior margin of the ramus, the temporomandibular joint, the edge of the mastoid process and the superior nuchal line, until it reaches the external occipital protuberance.
The **neck** (*pink*) lies between the pericraniocervical line (which separates it from the head) and the inferior border of the neck (which separates it from the thorax). The inferior border of the neck begins at the jugular notch of the sternum, and from there follows, on either side, a line that passes over the clavicles and acromion to reach the spinous process of the seventh cervical vertebra.

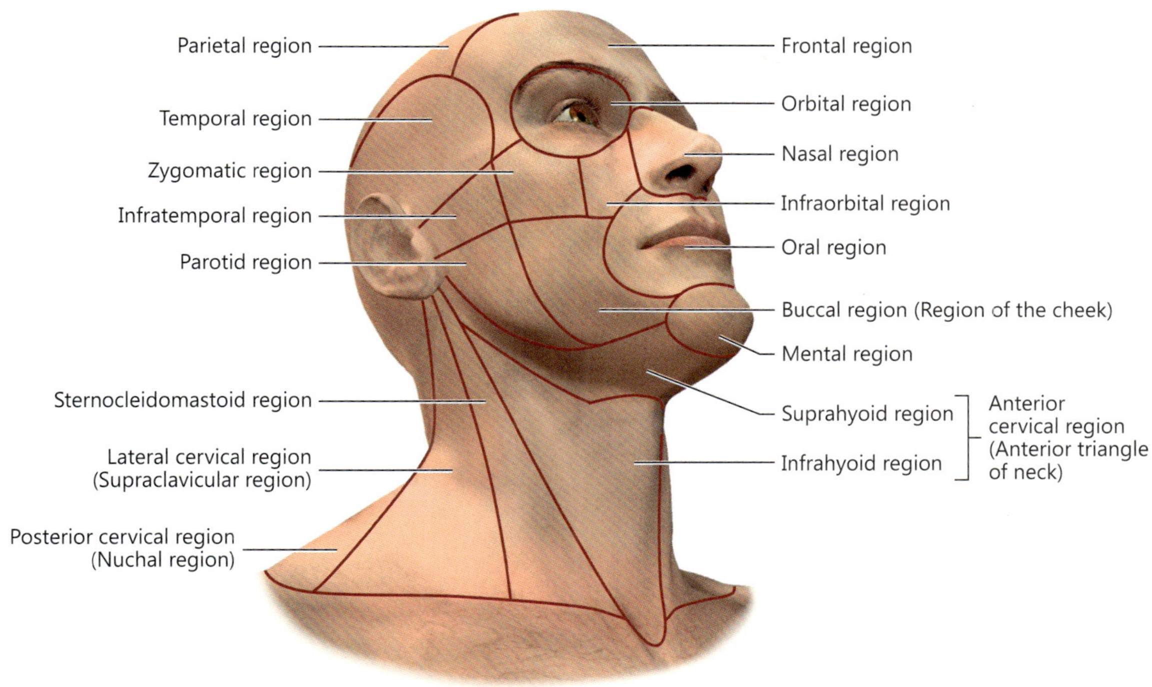

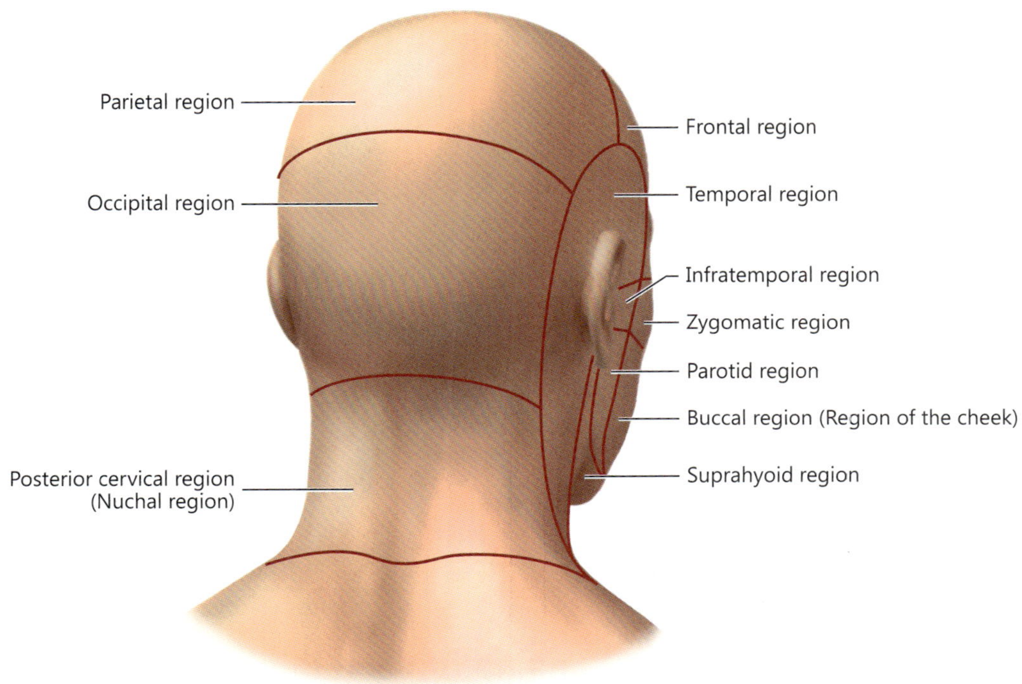

Regional subdivision of the head and the neck.

REGIONS OF THE HEAD	*REGIONS OF THE NECK*
Facial region – buccal region (cheek region) – infraorbital region – mental region (chin region) – nasal region – oral region (mouth region) – orbital region – parotid region – zygomatic region Frontal region] Occipitofrontal region Occipital region] Parietal region Temporal region Infratemporal region	Anterior cervical region (Anterior triangle of neck) – suprahyoid region – infrahyoid region – visceral space (Visceral compartment of neck) Sternocleidomastoid region or carotid region – vascular compartment (Carotid compartment) Lateral cervical region (Supraclavicular region) – supraclavicular compartment Posterior cervical region (Nuchal region)

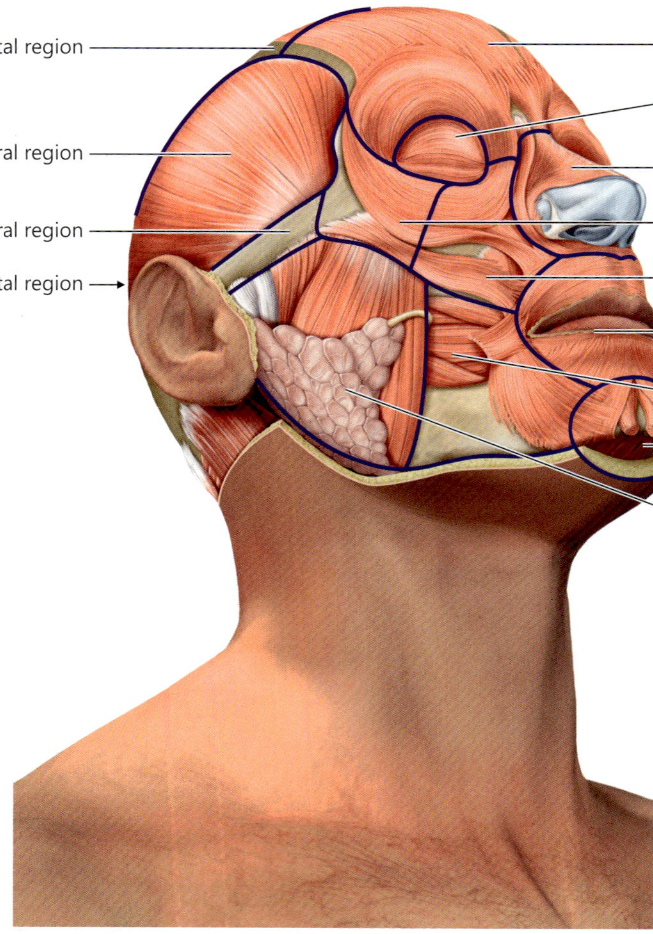

Regions of the head: right anterior oblique view. The frontal region corresponds to the area of the forehead that extends under the coronal suture. Adjacent to it is the parietal region, situated in correspondence to the parietal bone, while the temporal region is situated in correspondence to the squamous part of the temporal bone. The infratemporal region (or fossa) is covered by the zygomatic arch. Posteriorly, the occipital region (here not visible) lies over the occipital bone. The facial region is composed of the anterior fascial regions of the head: i.e. nasal, oral, mental, orbital (around the eye), infraorbital (to the side of the nose), buccal (to the side of the oral region), zygomatic (in correspondence to the zygomatic bone) and parotid (containing the masseter and the parotid gland).

Regions of the head: subdivision and boundaries

REGIONS OF THE HEAD	*BOUNDARIES OR LOCATION*
Facial	
– buccal (cheek region)	Zygomatic bone (superiorly), base of the mandible (inferiorly), nasolabial and nasogenian sulci (medially) and anterior border of the masseter (laterally). Corresponds to the outer surface of the buccinator
– infraorbital	Infraorbital margin (superiorly), sulcus delimiting the ala of nose (medially) and nasolabial sulcus (inferiorly)
– mental (chin region)	Mentolabial sulcus (superiorly) and base of the mandible (inferiorly)
– nasal	Root of nose (superiorly); line from medial angle of the eye to the ala of nose (laterally)
– oral (mouth region)	Nasolabial sulcus (superiorly) and mentolabial sulcus (inferiorly)
– orbital	Suprapalpebral sulcus (superiorly) and infrapalpebral sulcus (inferiorly)
– parotid (parotid-masseteric)	Zygomatic arch (superiorly), base of the mandible (inferiorly), anterior border of the masseter (anteriorly), anterior border of the sternocleidomastoid and external acoustic meatus (posteriorly)
– zygomatic	Corresponds to the zygomatic bone
Frontal	Edge of the frontal bone: coronal suture (superiorly), supraorbital margins of the frontal bone and frontonasal suture (inferiorly), sphenofrontal and frontozygomatic sutures (laterally)
Infratemporal	Corresponds to the zygomatic arch
Occipital	Edge of the occipital bone: lambdoid (superiorly) and occipitomastoid (inferiorly) sutures
Parietal	Edge of the parietal bone: sagittal suture (superiorly), coronal suture (anteriorly), squamous and parietomastoid sutures (inferiorly), lambdoid suture (posteriorly)
Temporal	Zygomatic bone, zygomatic process of the frontal bone and temporal crest (anteriorly), superior temporal line up to the asterion (superiorly and posteriorly); horizontal line passing through the zygomatic arch (inferiorly)

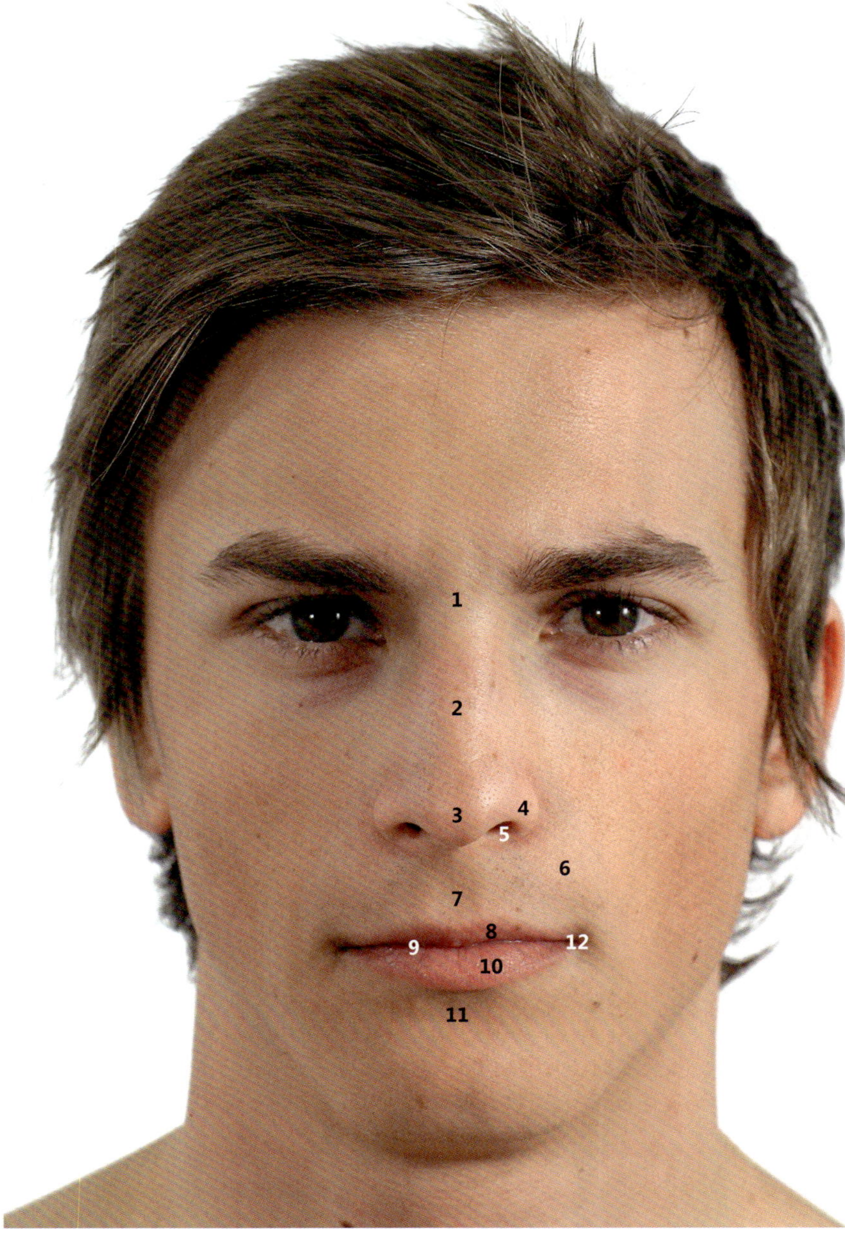

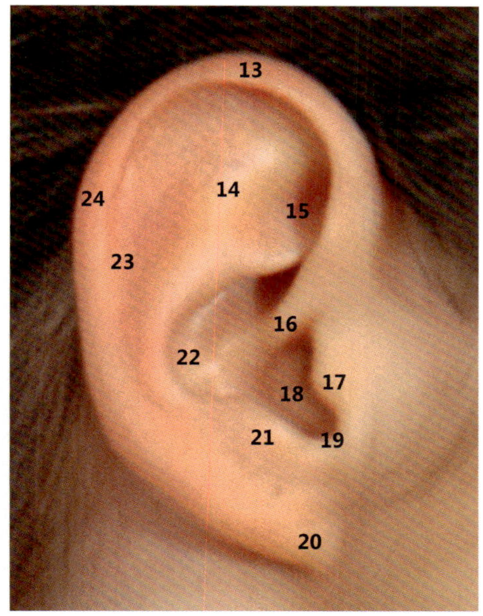

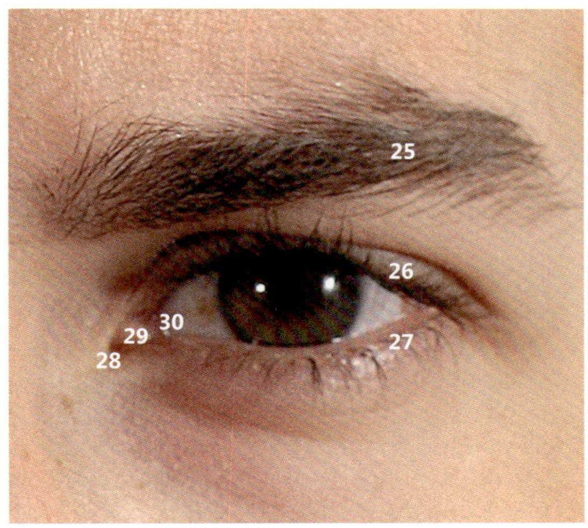

	Nose:
1	– root
2	– dorsum
3	– apex
4	– ala
5	– naris
	Lips:
6	– nasolabial sulcus
7	– philtrum
8	– upper lip
9	– oral fissure
10	– lower lip
11	– mentolabial sulcus
12	– angle of mouth

	Auricle (external ear):
13	– helix
14	– antihelix
15	– triangular fossa
16	– crus of helix
17	– tragus
18	– external acoustic meatus
19	– intertragic incisure
20	– lobule of auricle
21	– antitragus
22	– concha of auricle
23	– groove of helix
24	– auricular tubercle

	Eye:
25	– eyebrow (or superciliary arch)
26	– superior eyelid
27	– inferior eyelid
28	– lacrimal caruncle
29	– plica semilunaris
30	– lacus lacrimalis

General structure of the **external nose**, **eye**, **lips** and **external ear** in relation to the surrounding frontal regions.

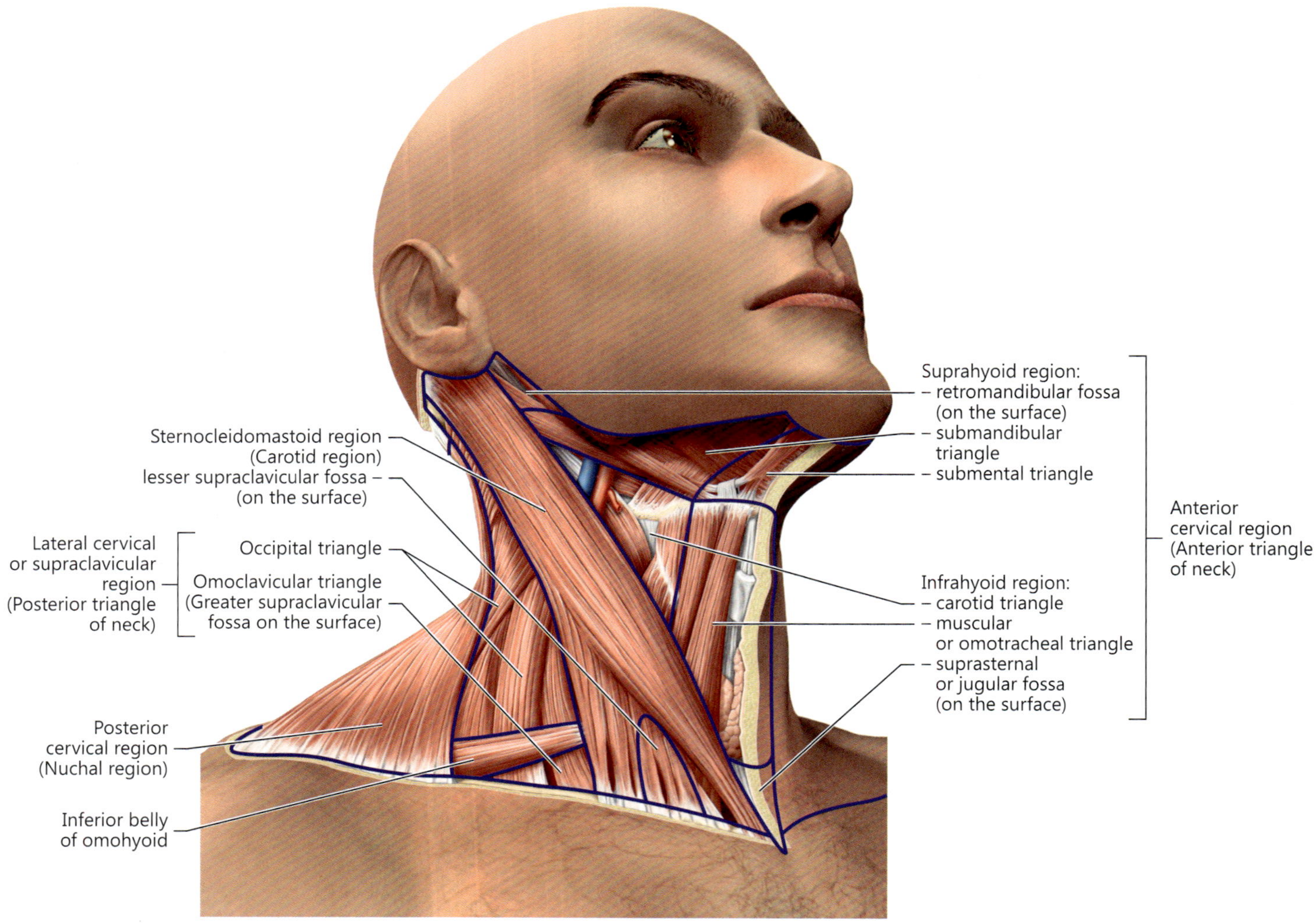

Regions of the neck: right lateral view. The sternocleidomastoid divides the anterior cervical region, or anterior triangle of the neck, with its base above (mandible) and apex below, from the lateral one, or posterior triangle of the neck, with its base below (clavicle) and apex above. The posterior cervical region lies within the anterior borders of the two trapezius muscles, which separate it from the posterior triangles of the neck. The anterior triangle of the neck is subdivided into smaller triangles: the submandibular, submental, carotid and muscular triangles. The posterior triangle is subdivided into the occipital and omoclavicular triangles.

Regions of the neck: subdivision and boundaries

REGIONS OF THE NECK	*BOUNDARIES OF THE REGIONS AND OF THE CORRESPONDING CONNECTIVE SPACES*	*TRIANGLES AND FOSSAE*	
Anterior cervical region:	*Superior*: base of the mandible *Lateral*: anterior border of the sternocleidomastoid *Medial*: anterior median line. The apex of the triangle corresponds to the jugular notch, the base to the inferior margin of the mandible and to a line that unites the angle of the mandible with the mastoid process	Anterior triangle of neck:	
Suprahyoid region	*Superior*: base of the mandible *Inferior*: horizontal line passing through the hyoid bone *Lateral*: anterior border of the sternocleidomastoid	– submandibular triangle	
		– retromandibular fossa	The retromandibular fossa, bordered by the ramus of the mandible, posterior belly of the digastric and by a split of the investing layer of cervical fascia between the angle of the mandible and anterior border of the sternocleidomastoid, contains the deep part of the parotid gland
		– submental triangle	

continued

Regions of the neck: subdivision and boundaries *(continued)*

REGIONS OF THE NECK	*BOUNDARIES OF THE REGIONS AND OF THE CORRESPONDING CONNECTIVE SPACES*	*TRIANGLES AND FOSSAE*	
Infrahyoid region	*Superior*: horizontal line passing through the hyoid bone *Inferior*: jugular notch of the sternum *Lateral*: anterior border of the sternocleidomastoid. Deeply, it is bounded by the anterior fascia of the visceral sheath	– carotid triangle – muscular (or omotracheal) triangle – – suprasternal or jugular fossa	The suprasternal fossa is a superficial recess situated in correspondence to the jugular notch, bordered laterally by the sternal heads of the sternocleidomastoids
Visceral space (visceral compartment of neck)	*Anterior*: suprahyoid and infrahyoid regions *Posterior*: anterior surface of the cervical column covered by the prevertebral layer of cervical fascia *Lateral*: border with the vascular (carotid) compartment *Superior*: cranial base *Inferior*: mediastinum		
Sternocleidomastoid region or carotid region	Anterior and posterior borders of the sternocleidomastoid	Lesser supraclavicular fossa	The lesser supraclavicular fossa is a superficial recess situated between the clavicular and sternal heads of the sternocleidomastoid
Sternocleidomastoid compartment	Splitting of the investing layer of cervical fascia in which the sternocleidomastoid is enclosed		
Vascular compartment (carotid compartment)	*Anterolateral*: sternocleidomastoid region *Posterior*: prevertebral muscles and transverse processes of the cervical vertebrae *Medial*: border with the visceral space		
Lateral cervical region (supraclavicular region)	*Anterior*: posterior border of the sternocleidomastoid *Posterior*: anterior border of the trapezius *Superior*: intersection of the two sternocleidomastoids and the trapezius *Inferior*: tract of clavicle between the origin of the sternocleidomastoid and the insertion of the trapezius	Posterior triangle of neck: – occipital triangle – omoclavicular triangle (subclavian triangle) – – greater supraclavicular fossa	The greater supraclavicular fossa, situated between the posterior border of the sternocleidomastoid, the middle third of the clavicle and the inferior belly of the omohyoid, is the area of superficial recess corresponding to the omoclavicular triangle
Supraclavicular compartment	*Lateral*: skin, subcutaneous tissue, investing layer of cervical fascia and, in the lower part, pretracheal layer of cervical fascia and omohyoid *Anteromedial*: scalenus anterior *Posterior*: transverse processes of the cervical vertebrae covered, above, by the levator scapulae and splenius cervicis and, below, by the scalenus medius and scalenus posterior		
Posterior cervical region or nuchal region	*Inferior*: horizontal line that unites the acromion of the scapulae crossing the seventh cervical vertebra or vertebra prominens *Superior*: superior nuchal line and external occipital protuberance *Lateral*: anterior borders of the trapezius muscles		

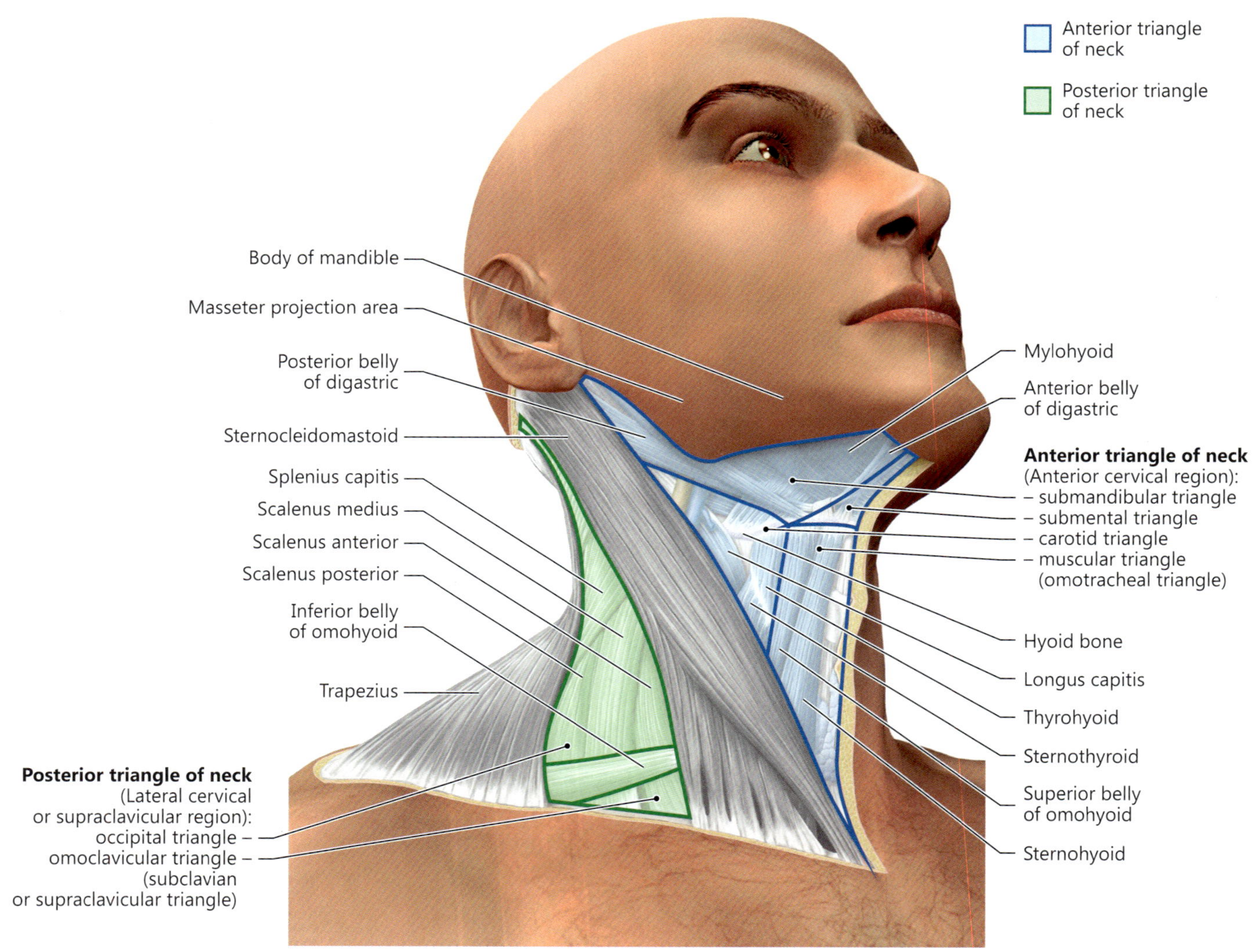

Anterior and **lateral cervical regions** and their respective **triangles**.

Triangles of the neck: subdivision, boundaries and contents

NAME	BOUNDARIES	SUBDIVISION	CONTENT
Anterior triangle (anterior cervical region)	*Anterior*: median line of the neck; *posterior*: anterior border of the sternocleidomastoid; *superior* (base of the triangle): inferior margin of the base of the mandible and its projection to the mastoid process; *apex*: manubrium of the sternum. Covered by: skin, subcutaneous tissue (superficial fascia of the skin), platysma and investing layer of cervical fascia	Divided by the anterior and posterior bellies of the digastric and by the superior belly of the omohyoid into the submental, submandibular, carotid and muscular triangles	*Bones*: hyoid bone *Muscles*: suprahyoid muscles: mylohyoid, geniohyoid, stylohyoid, digastric; infrahyoid muscles: sternohyoid, omohyoid, sternothyroid, thyrohyoid *Other formations*: ☞ submental triangle, submandibular triangle, carotid triangle and muscular triangle
– submental triangle	Single triangle between the anterior bellies of the digastric muscles; *apex*: chin; *base*: body of the hyoid bone; *floor*: both mylohyoid muscles		*Veins*: small veins uniting to form the anterior jugular vein *Lymph nodes*: submental

continued

Triangles of the neck: subdivision, boundaries and contents			(continued)
NAME	***BOUNDARIES***	***SUBDIVISION***	***CONTENT***
– submandibular triangle (digastric triangle)	*Superior*: base of the mandible and its projection to the mastoid process; *posteroinferior*: posterior belly of the digastric and the stylohyoid; *anteroinferior*: anterior belly of the digastric; *floor*: mylohyoid and hyoglossus. Covered by: skin, subcutaneous tissue (superficial fascia of the skin), platysma and investing layer of cervical fascia		*Muscles*: styloglossus, stylopharyngeus *Arteries*: facial, submental, external carotid, internal carotid, mylohyoid branch of the inferior alveolar artery *Veins*: facial, internal jugular, small veins uniting to form the anterior jugular vein *Lymph nodes*: submandibular, submental *Nerves*: mylohyoid, glossopharyngeal [IX], vagus [X], branches of facial nerve and transverse cervical nerve *Organs*: inferior part of the parotid gland, submandibular gland
– carotid triangle	*Posterior*: sternocleidomastoid; *anteroinferior*: superior belly of the omohyoid; *superior*: stylohyoid and posterior belly of the digastric; anterior angle and adjacent floor of the hyoid bone; *floor*: parts of the thyrohyoid, hyoglossus, middle constrictor of pharynx and inferior constrictor of pharynx. Covered by: skin, subcutaneous tissue (superficial fascia of the skin), platysma and investing layer of cervical fascia		*Arteries*: superior part of the common carotid artery and its division into the external carotid and internal carotid arteries, branches of the external carotid artery (superior thyroid, lingual, facial, occipital, ascending pharyngeal arteries), sternocleidomastoid branches of the occipital artery *Veins*: internal jugular, superior thyroid, lingual, facial, ascending pharyngeal and, sometimes, occipital *Lymph nodes*: deep lateral cervical *Nerves*: vagus [X], accessory [XI], hypoglossal [XII], superior root of ansa cervicalis, external and internal branches of the superior laryngeal nerve, branches of the facial nerve and transverse cervical nerve (within the investing layer of cervical fascia)
– muscular triangle (omotracheal triangle)	*Anterior*: median line of the neck from the hyoid bone to the sternum; *inferoposterior*: anterior border of the sternocleidomastoid; *posterosuperior*: superior belly of the omohyoid		*Muscles*: infrahyoid muscles: omohyoid, sternohyoid, sternothyroid and thyrohyoid
Posterior triangle (lateral cervical or supraclavicular region)	*Anterior*: posterior border of the sternocleidomastoid; *posterior*: anterior border of the trapezius; *inferior*: middle third of the clavicle; *apex*: between insertions to the occipital bone of the sternocleidomastoid and the trapezius and, sometimes, also the semispinalis capitis; *roof*: investing layer of cervical fascia; *floor*: prevertebral layer of cervical fascia superimposing the splenius capitis, levator scapulae and scalene muscles. Covered by: skin, subcutaneous tissue (superficial fascia of the skin), platysma, and investing layer of cervical fascia	Divided by the inferior belly of the omohyoid into the occipital triangle, above, and the omoclavicular triangle, below	*Arteries*: transverse cervical, suprascapular, occipital, third part of the subclavian artery *Veins*: transverse cervical veins, suprascapular veins, external jugular, subclavian *Lymph nodes*: accessory (of the accessory nerve) *Nerves*: accessory [XI], phrenic, cervical and brachial plexuses, cutaneous branches of the cervical plexus (lesser occipital, great auricular, transverse cervical and supraclavicular nerves)
– occipital triangle	Same boundaries as the posterior triangle except for the *inferior border*: inferior belly of the omohyoid; *floor*: splenius capitis, levator scapulae, scalenus medius and scalenus posterior. Covered by: skin, subcutaneous tissue (superficial fascia of the skin), platysma (inferiorly), investing layer of cervical fascia		*Arteries*: transverse cervical, occipital *Veins*: transverse cervical *Lymph nodes*: accessory (of the accessory nerve) *Nerves*: accessory [XI], cutaneous and muscular branches of the cervical plexus (supraclavicular nerves, phrenic nerve), superior part of the brachial plexus
– omoclavicular triangle or subclavian triangle (supraclavicular triangle)	Same boundaries as the posterior triangle except for the *superior border*: omohyoid; *floor*: first rib, scalenus medius, first digitation of the serratus anterior. Covered by: skin, subcutaneous tissue (superficial fascia of the skin), platysma, investing layer of cervical fascia	Corresponds to the inferior part of the greater supraclavicular fossa	*Arteries*: suprascapular, transverse cervical, third part of the subclavian artery *Veins*: subclavian, suprascapular, transverse cervical, external jugular *Lymph nodes*: supraclavicular *Nerves*: supraclavicular nerves, subclavian nerve, trunks of the brachial plexus

Sagittal section of the neck: the boundaries of the **fasciae** are indicated by colored lines.

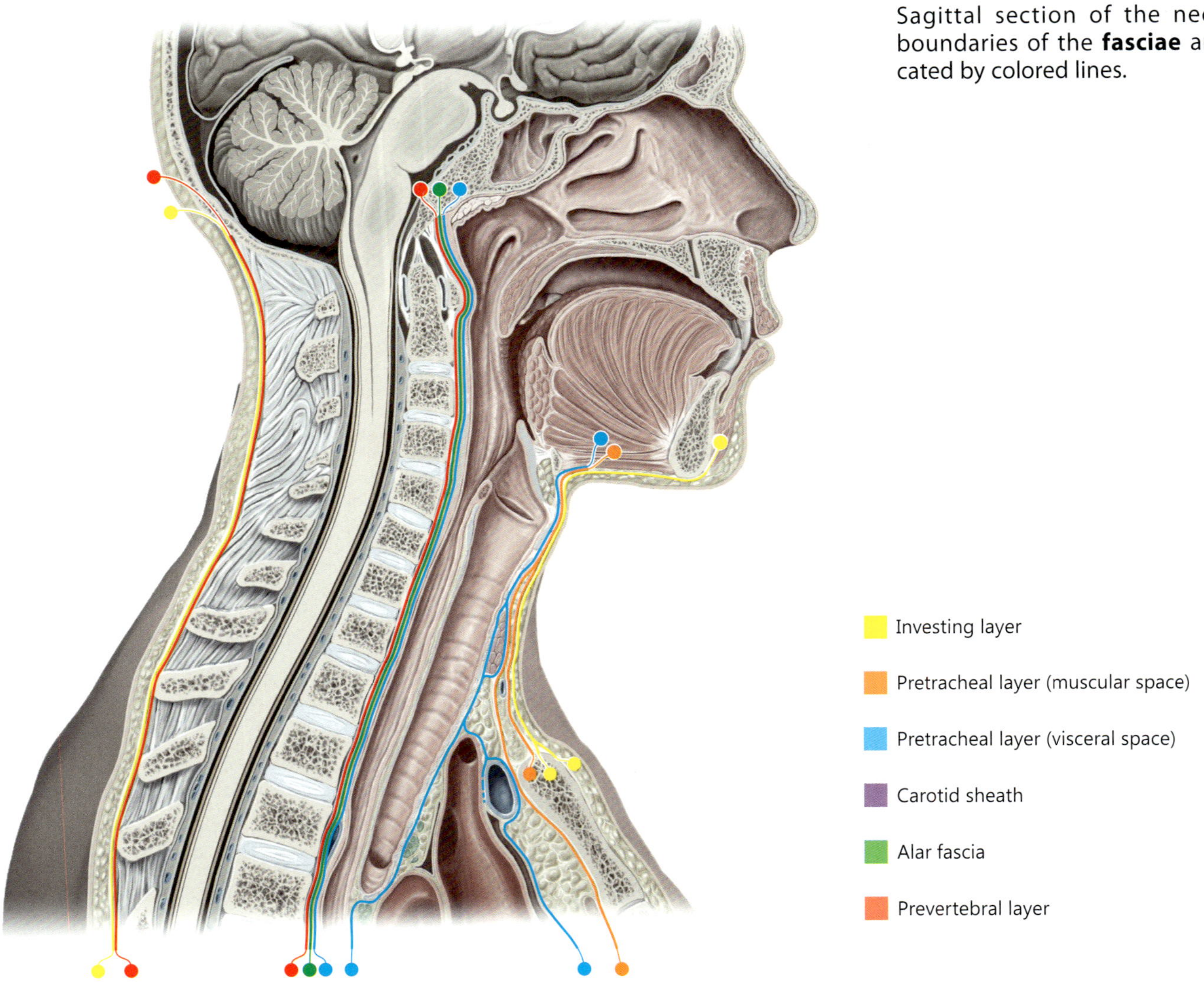

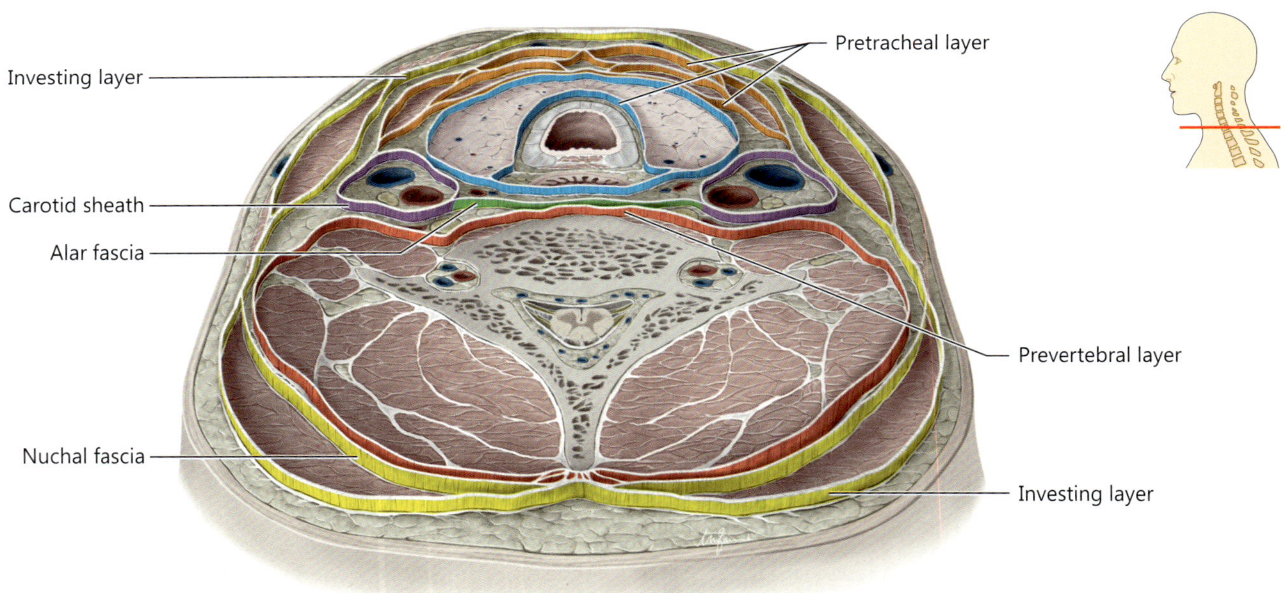

Transverse section of the neck through the 7th cervical vertebra, schematic representation showing its subdivisions by the **cervical** and **nuchal fasciae**.

Sagittal section of the neck: the **compartments** are indicated by different colors.

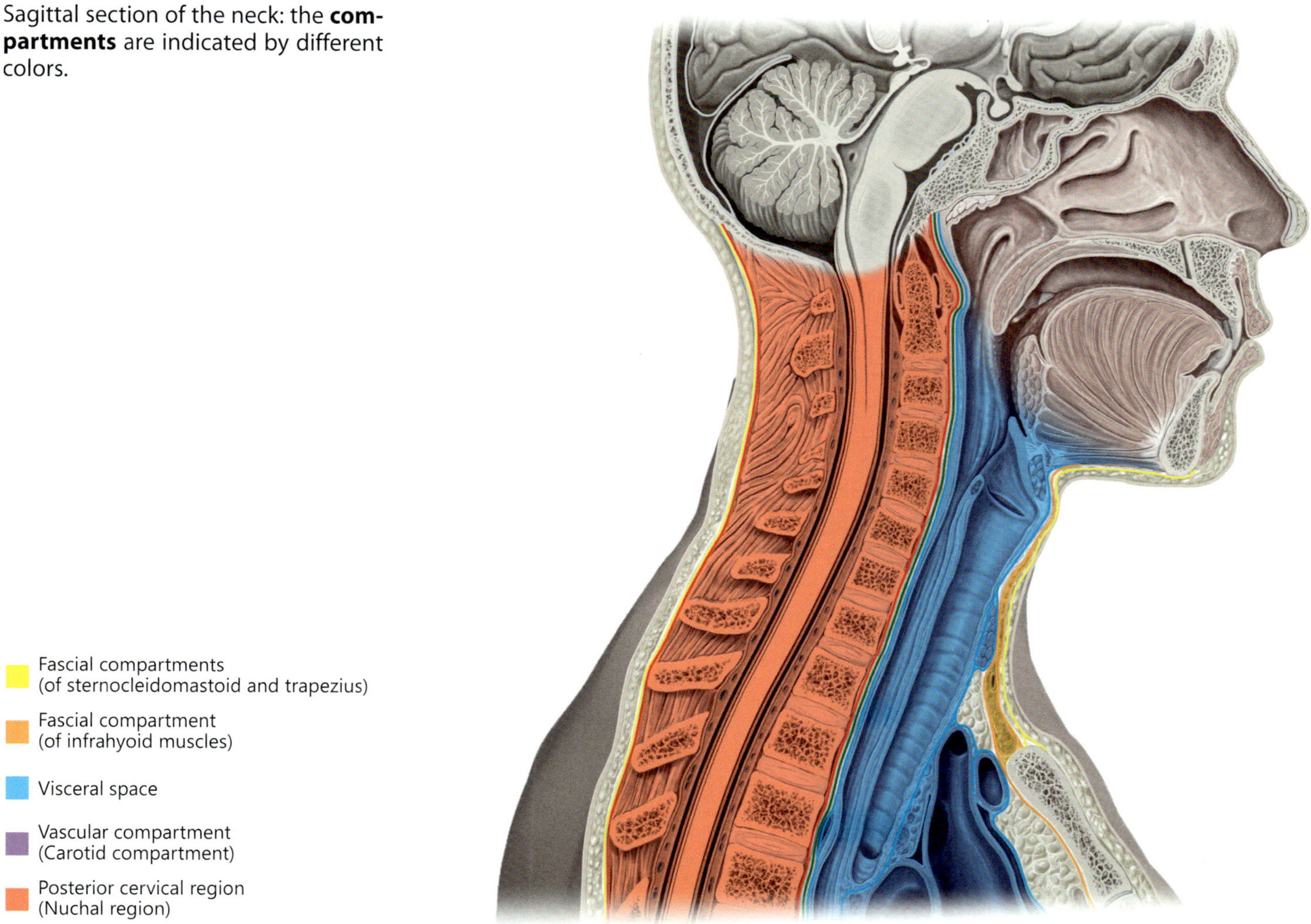

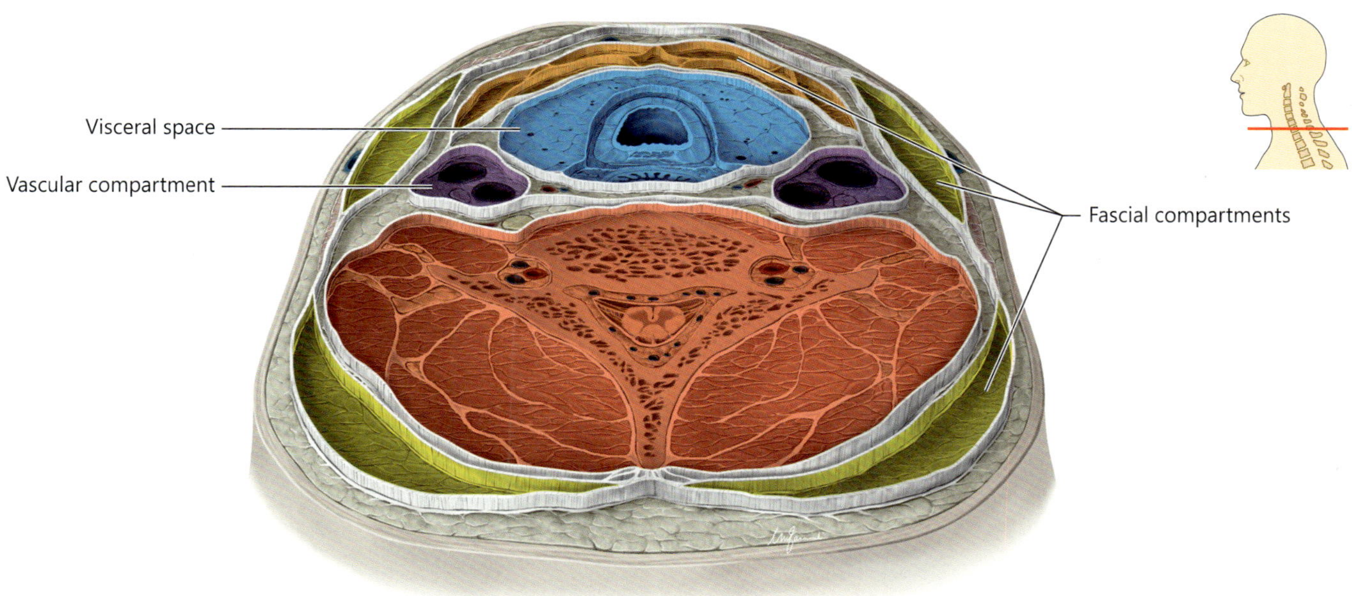

Transverse section of the neck through the seventh cervical vertebra, showing: the **visceral space** or **visceral compartment of the neck**, bounded, anteriorly, by the pretracheal layer of cervical fascia investing the infrahyoid muscles and, posteriorly, by the same layer on the posterior wall of the esophagus and around the trachea; the **vascular compartment** within the carotid sheath, which collects the neurovascular bundle of the neck; the **fascial compartments** formed by splits of the investing layer of cervical fascia to enclose the sternocleidomastoid and trapezius muscles; the **platysma**, which occupies the subcutaneous tissue, between the superficial fascia of skin and the investing layer of cervical fascia; the **fascial compartments** formed by splits of the pretracheal layer that envelop the infrahyoid muscles.

Compartments of the head: boundaries and contents			
NAME	*BOUNDARIES*	*SUBDIVISION*	*CONTENT*
Anterior subparotid space (paratonsillar space)	Between the medial pterygoid and lateral part of the pharynx		*Organs*: pharyngeal tonsil *Other formations*: adipose tissue
Cerebellar compartment	☞ Rhombencephalon cavity		
Cerebral compartment	*Superior and lateral*: calvaria *Inferior*: anterior and middle cranial fossae and tentorium cerebelli	Divided into two halves, right and left, by the falx cerebri	Cerebral hemispheres
Hypophyseal compartment	Between the diaphragma sellae and the sella turcica		*Organs*: pituitary gland
Lateral pharyngeal space	☞ Parapharyngeal space		
Lateropharyngeal space	☞ Parapharyngeal space		
Leptomeningeal spaces	Located between the arachnoid mater and the pia mater, they are described as cisterns: the cerebellomedullary cisterns, chiasmatic cistern, interpeduncular cistern, quadrigeminal cistern (or cistern of great cerebral vein), and the cistern of lateral cerebral fossa		
Masseteric compartment	*Lateral*: masseteric fascia *Medial*: ramus of the mandible		*Muscles*: masseter *Arteries*: masseteric *Nerves*: masseteric
Masticator space	Investing layer of cervical fascia surrounding, laterally, the masseter and, medially, the pterygoid muscles		*Bones*: mandible *Muscles*: temporalis, masseter, medial pterygoid and lateral pterygoid *Arteries*: maxillary *Veins and plexuses*: maxillary vein, pterygoid plexus *Nerves*: inferior alveolar and lingual *Other formations*: buccal fat pad
Maxillopharyngeal space	☞ Parapharyngeal space		
Meckel's cavity	Splitting of the dura mater that covers the trigeminal impression on the anterior surface of the petrous part of the temporal bone		*Nerves and ganglia*: trigeminal ganglion, trigeminal nerve [V], ophthalmic nerve [V_1], maxillary nerve [V_2] and mandibular nerve [V_3]
Parapharyngeal space (lateral pharyngeal, pharyngomaxillary or maxillopharyngeal space)	*Anterior*: pterygomandibular raphe *Posterior*: prevertebral layer of cervical fascia *Medial*: buccopharyngeal fascia and, above the mandible, investing (or superficial) layer of cervical fascia *Lateral*: medial pterygoid and parotid gland *Superior*: petrous part of the temporal bone and sphenoid *Inferior*: hyoid bone	Divided by the styloid process of the temporal bone into two parts: retrostyloid (retrostyloid space) and prestyloid (prestyloid space)	Retrostyloid part: *Arteries*: internal carotid, ascending pharyngeal and ascending palatine *Veins*: internal jugular *Nerves and trunks*: glossopharyngeal [IX], vagus [X], accessory [XI] and hypoglossal [XII] nerves, sympathetic trunk (within the prevertebral layer of cervical fascia) Prestyloid part: In addition to the formations already present in the retrostyloid part, with the exclusion of the sympathetic trunk, since the deep part of the parotid gland projects onto the lateral surface of the prestyloid space: *Arteries*: external carotid with its terminal branches, superficial temporal and maxillary *Veins*: retromandibular *Nerves*: facial [VII] and auriculotemporal
Paratonsillar space	☞ Anterior subparotid space		
Parotid compartment	Investing layer of cervical fascia through septa of connective tissue running from the parotid fascia into the parotid gland		*Arteries*: external carotid, superficial temporal, maxillary, posterior auricular, tympanic, transverse facial *Veins*: retromandibular *Lymph nodes*: deep parotid *Nerves and plexuses*: facial and auriculotemporal nerves, parotid plexus *Organs*: parotid gland and parotid duct
Peripharyngeal space	☞ Parapharyngeal space or Retropharyngeal space		
Pharyngomaxillary space	☞ Parapharyngeal space		
Prefascial compartment	Fascial sheath of the eyeball		Eyeball
Prestyloid space	☞ Parapharyngeal space		

continued

Compartments of the head: boundaries and contents			(continued)
NAME	**BOUNDARIES**	**SUBDIVISION**	**CONTENT**
Retrofascial compartment	Fascial sheath of the eyeball and walls of the orbit		*Muscles*: extraocular muscles *Arteries*: ophthalmic, lacrimal, central retinal, supraorbital, short and long posterior ciliary arteries, muscular arteries, anterior and posterior ethmoidal arteries, medial palpebral arteries, dorsal nasal, supratrochlear *Veins*: superior and inferior ophthalmic veins, central retinal *Nerves and ganglia*: optic [II], oculomotor [III], trochlear [IV], abducens [VI], lacrimal, nasociliary, frontal, zygomatic and zygomaticofacial and zygomaticotemporal branches of the zygomatic nerve, long ciliary nerves, anterior and posterior ethmoidal nerves, infratrochlear, supratrochlear, supraorbital, ciliary ganglion *Other formations*: retrobulbar fat
Retropharyngeal space (retrovisceral space)	*Anterior*: pretracheal layer of cervical fascia surrounding the pharynx (buccopharyngeal fascia) *Posterior*: alar fascia (part of the deep cervical fascia or prevertebral layer of cervical fascia)		*Lymph nodes*: retropharyngeal
Retrostyloid space	☞ Parapharyngeal space		
Retrovisceral space	☞ Retropharyngeal space		
Rhombencephalon cavity (cerebellar compartment)	*Superior*: tentorium cerebelli *Inferior*: cerebellar fossae *Anterior*: clivus and petrous part of the temporal bone		Rhombencephalon (cerebellum, pons and myelencephalon)
Subdural space	Virtual space between the dura mater and the arachnoid mater		
Sublingual compartment	*Superior*: mucous membrane of the mouth *Anterolateral*: internal surface of the body of the mandible *Posteromedial*: hyoglossus *Inferior*: mylohyoid In front: *Anterior*: mandibular symphysis *Posteroinferior*: genioglossus *Superior*: mucous membrane of the mouth		*Arteries*: sublingual *Veins*: sublingual *Nerves*: hypoglossal [XII] and lingual *Organs*: sublingual gland, prolongation of the submandibular gland and submandibular duct *Other formations*: adipose tissue
Temporal compartment	*Lateral*: investing layer of cervical fascia fixed to the infratemporal crest of the sphenoid, and zygomatic bone *Medial*: periosteum of the temporal bone		*Muscles*: temporalis *Arteries*: superficial temporal, middle temporal, anterior and posterior deep temporal arteries *Veins*: superficial temporal veins and deep temporal veins *Nerves*: auriculotemporal and deep temporal nerves *Other formations*: adipose tissue
Tonsillar compartment	*Superior*: soft palate *Inferior*: horizontal line prolonging the lateral border of the tongue *Anterior*: palatoglossal arch *Posterior*: palatopharyngeal arch *Lateral*: buccopharyngeal fascia		*Organs*: palatine tonsil

Compartments of the neck: boundaries and contents			
NAME	*BOUNDARIES*	*SUBDIVISION*	*CONTENT*
Carotid compartment	☞ Vascular compartment		
Carotid space	☞ Vascular compartment		
"Danger" space	Between the two layers of the prevertebral layer of cervical fascia (between the alar fascia and the prevertebral layer)		Areolar connective tissue
Lateral vascular space	☞ Vascular compartment		
Pre-epiglottic space	*Superior*: glossoepiglottic folds *Inferior*: insertion of the epiglottic cartilage on the thyroid cartilage *Anterior*: thyrohyoid membrane *Posterior*: epiglottic cartilage		Loose connective tissue
Prevertebral space	*Anterior*: prevertebral layer *Posterior*: vertebral column		*Muscles*: erector spinae, prevertebral, scalene muscles *Arteries*: vertebral *Veins*: vertebral *Nerves and plexuses*: phrenic nerve, brachial plexus
Previsceral space	☞ Visceral space		
Retroesophageal space (retrovisceral)	*Anterior*: posterior wall of the esophagus *Posterior*: prevertebral layer *Inferior*: at the level of the second-fourth thoracic vertebra (where the prevertebral layer fuses with the connective tissue on the posterior surface of the esophagus)		Areolar connective tissue
Retrovisceral space	☞ Retroesophageal space		
Sternocleidomastoid compartment	Splitting of the investing layer of cervical fascia		*Muscles*: sternocleidomastoid
Submandibular compartment	*Anterior and lateral*: mandible *Superior*: mucous membrane of the tongue *Posteroinferior*: hyoid bone *Inferior*: investing layer of cervical fascia		*Arteries*: facial, with its glandular branches, and submental *Veins*: facial and lingual *Nerves*: hypoglossal [XII], mylohyoid and lingual *Lymph nodes*: submental and submandibular *Organs*: submandibular gland and submandibular duct
Supraclavicular compartment	*Lateral*: skin, subcutaneous tissue and investing layer of cervical fascia; pretracheal layer of cervical fascia and omohyoid inferiorly *Anteromedial*: scalenus anterior *Posterior*: transverse processes of the cervical vertebrae covered, in the upper part, by the levator scapulae and splenius cervicis and, in the lower part, by the scalenus medius and scalenus posterior *Apex*: convergence towards the top of the sternocleidomastoid, trapezius, splenius cervicis and levator scapulae *Base*: clavicle and superior border of the scapula	The base of the compartment is subdivided into two portions, lateral and medial, by the superior surface of the first rib	*Arteries*: subclavian and origin of its branches: vertebral and internal thoracic artery, and thyrocervical and costocervical trunks *Veins*: subclavian *Lymph nodes and ducts*: supraclavicular nodes, thoracic duct and right lymphatic duct *Nerves, plexuses and ganglia*: accessory [XI] and phrenic nerves, brachial plexus, cervicothoracic (stellate) ganglion *Other formations*: cervical pleura
Suprasternal space	Splitting of the investing layer of cervical arsch		*Veins*: anterior jugular veins with jugular venous arch
Vascular compartment (carotid compartment, carotid space or lateral vascular space)	Located within the carotid sheath, to whose formation all three layers of the cervical fascia (investing, pretracheal and prevertebral) contribute		*Arteries*: common carotid, external carotid and internal carotid *Veins*: internal jugular *Lymph nodes*: deep lateral cervical *Nerves, branches and plexuses*: vagus nerve [X], carotid branch of glossopharyngeal nerve [IX], common carotid plexus and internal carotid plexus, hypoglossal nerve [XII] *Other formations*: adipose tissue

continued

Compartments of the neck: boundaries and contents			(*continued*)
NAME	***BOUNDARIES***	***SUBDIVISION***	***CONTENT***
Visceral compartment of neck	☞ Visceral space		
Visceral space (previsceral space or visceral compartment of neck)	*Anterior*: pretracheal layer of cervical fascia and infrahyoid muscles *Posterior*: pretracheal layer around the trachea and adhering to the posterior wall of the esophagus *Superior*: insertions of the infrahyoid muscles on the thyroid cartilage of the larynx	Anterior cervical space: space lying between the fascia covering the infrahyoid muscles and the perithyroid fascia, with the posterolateral border consisting of the carotid sheath and sternocleidomastoid fascia Perithyroid "danger" space: interstitium between the internal surface of the perithyroid sheath and the thyroid gland capsule	*Organs*: pharynx, esophagus, larynx, trachea, thyroid and parathyroid glands

Schematic representation of the branches of the **aorta**, showing the arch and its branches involved in the vascularization of the superior part of the thoracic cage, neck and head.

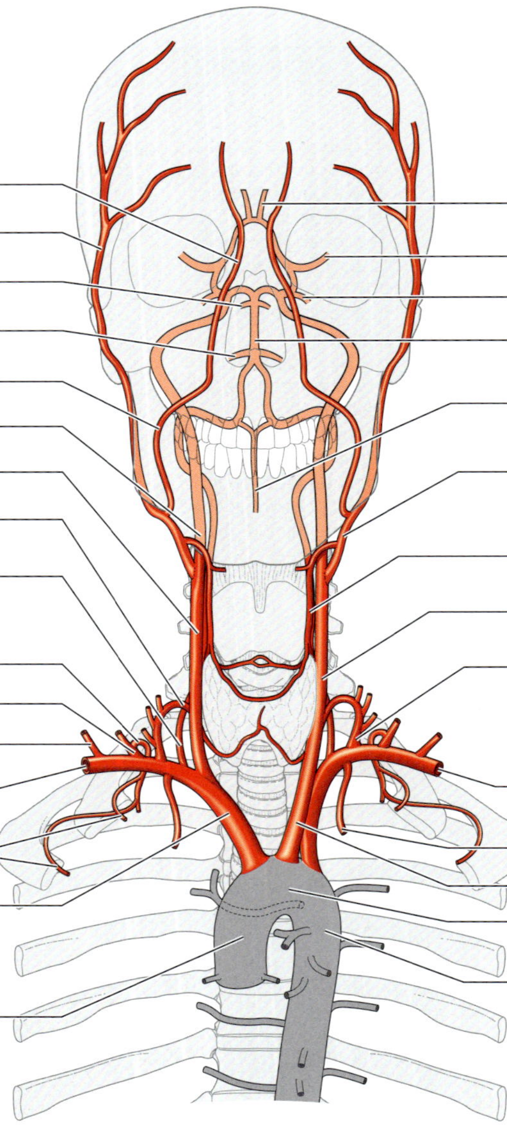

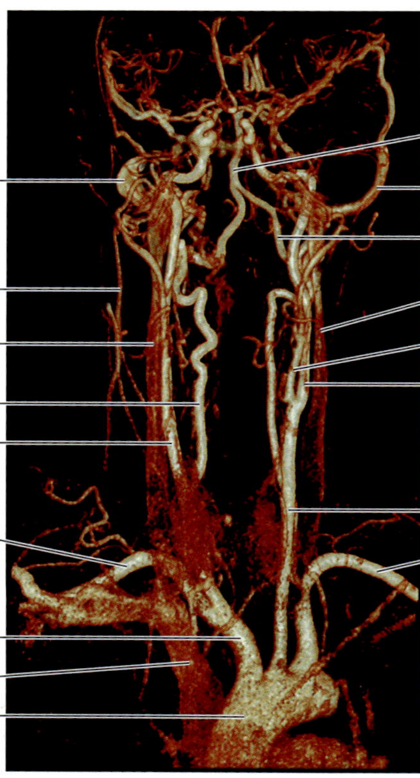

Three-dimensional image of a living human body, obtained by contrast-enhanced NMR using a 3D volume rendering technique, showing the **arteries** and **veins of the head** and **neck**.

Arteries of the Head and Neck

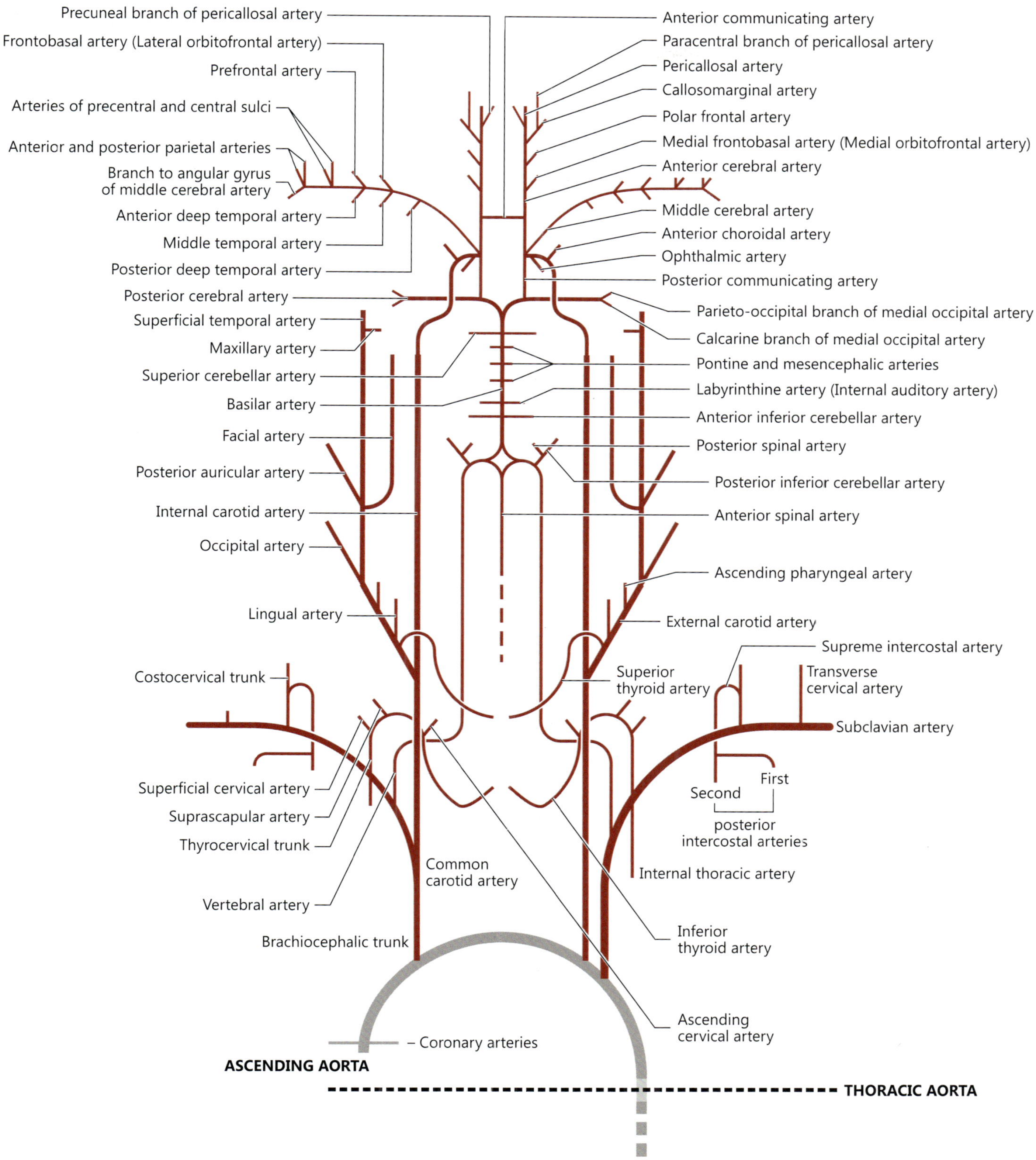

Schematic representation of the arterial blood supply to the head and neck highlighting, in particular, the ramifications of the external carotid, internal carotid, vertebral and basilar arteries.
The **external carotid artery** has as its collateral branches the superior thyroid, ascending pharyngeal, lingual, facial, occipital and posterior auricular arteries; its terminal branches are the superficial temporal and internal maxillary arteries.
The **internal carotid artery** has as its collateral branches the caroticotympanic arteries, the artery of pterygoid canal, the cavernous branches (not indicated here) in the carotid canal, and the ophthalmic artery in the cranial cavity; its main terminal branches are the anterior choroidal, anterior cerebral, middle cerebral and posterior communicating arteries.
The **vertebral artery** has as its collateral branches the spinal and muscular branches in the cervical part and meningeal and medullary branches in the intracranial part (not indicated here), and the posterior inferior cerebellar artery and anterior and posterior spinal arteries. At the point of confluence of the two vertebral arteries begins the **basilar artery**, which has as its collateral branches the pontine and mesencephalic arteries, the labyrinthine artery, the anterior inferior cerebellar and superior cerebellar arteries, and as its terminal branches the posterior cerebral arteries. Details of the blood supply to the brain are reported in the chapter on the nervous system.

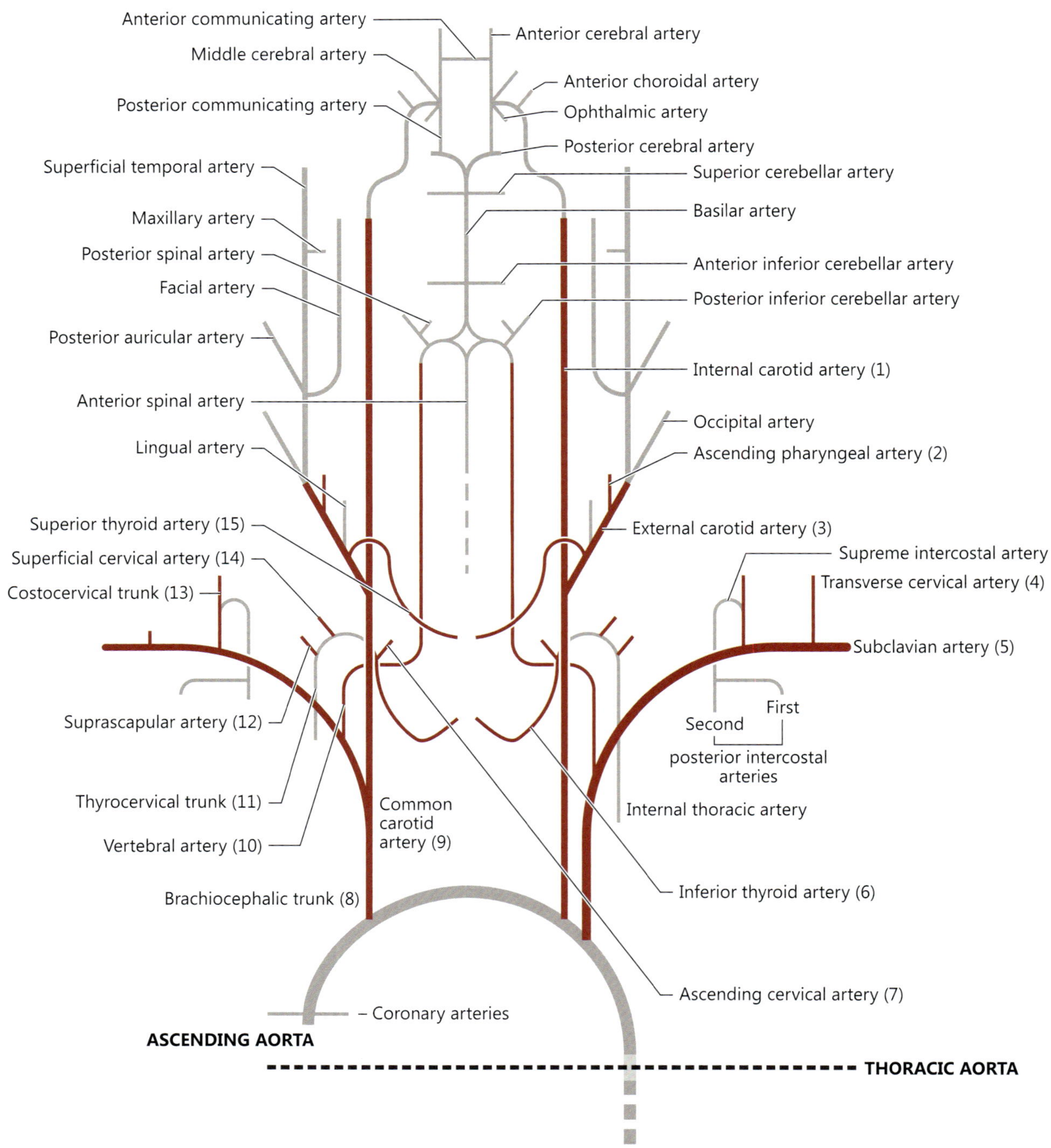

Schematic representation of the arterial blood supply to the head and neck.
This page focuses on the role of the subclavian arteries involved together with some of their collateral branches: the **vertebral arteries**, from whose confluence originates the **basilar artery**, the internal thoracic artery, the **thyrocervical** and **costocervical trunks**, and the **transverse cervical artery**. The relative territories of supply (indicated numerically) are described in the table alongside. In general, the **vertebral artery** supplies the semispinalis and suboccipital muscles, the cervical spinal cord, myelencephalon (or medulla oblongata or bulb), cerebellum and dura mater of the cerebellar fossa, while the **basilar artery** supplies the pons, internal ear, cerebellum, cortex of the temporal and occipital lobes, third ventricle, thalamus, hypothalamus and mesencephalon.

Main arteries of the neck: territories of blood supply		
	ARTERIES OF THE NECK	***TERRITORIES OF SUPPLY***
1	Internal carotid artery (cervical part)	Middle ear, orbital cavity, nasal cavity, brain, pituitary gland, trigeminal ganglion, meninges
2	Ascending pharyngeal artery	Pharynx, soft palate, longus capitis and longus colli, sympathetic trunk, palatine tonsil, pharyngotympanic tube, tympanic cavity, dura mater of the cerebellar fossa
3	External carotid artery	Anterior part of the face and neck, lateral part of the head, walls of the cranial cavity, dura mater, posterior part of the scalp
4	Transverse cervical artery	Posterior cervical, shoulder and back muscles
5	Subclavian artery	Territory of supply of the vertebral artery (☞ 10), anterior and superior thoracic wall, upper limb
6	Inferior thyroid artery	Infrahyoid muscles, longus colli, scalenus anterior and inferior constrictor of pharynx, cervical part of the esophagus, muscles and mucosa of the larynx, inferior part of the pharynx, trachea, posteroinferior part of the thyroid gland
7	Ascending cervical artery	Sternocleidomastoid, omohyoid, trapezius, longus capitis, phrenic nerve, spinal cord
8	Brachiocephalic trunk (at the level of the superior border of the right sternoclavicular joint where it divides into the right common carotid and subclavian arteries)	Head and neck, upper limb, superior part of the thorax
9	Common carotid artery	Head and neck
10	Vertebral artery	Semispinalis capitis, suboccipital muscles, prevertebral muscles and muscles of back proper, vertebral canal, meninges and spinal cord of the cervical tract, myelencephalon, cerebellum, dura mater of the cerebellar fossa, pons, mesencephalon, thalamus, hypothalamus, third ventricle, temporal and occipital lobes, internal ear
11	Thyrocervical trunk	Territories of supply of the inferior thyroid artery (☞ 6), ascending cervical artery (☞ 7), suprascapular artery (☞ 12) and transverse cervical artery (☞ 4)
12	Suprascapular artery	Clavicle, scapula, sternocleidomastoid, subclavius, supraspinatus and subscapularis, acromioclavicular and glenohumeral joints, suprascapular region, skin over the acromion
13	Costocervical trunk	First two intercostal spaces, semispinalis capitis and semispinalis cervicis, spinal cord
14	Superficial cervical artery	Trapezius
15	Superior thyroid artery	Thyroid gland, infrahyoid muscles and sternocleidomastoid, muscles and mucosa of the larynx

N.B.: for a general overview of the arteries and their territories of supply, see the table summarizing the arteries of the neck.

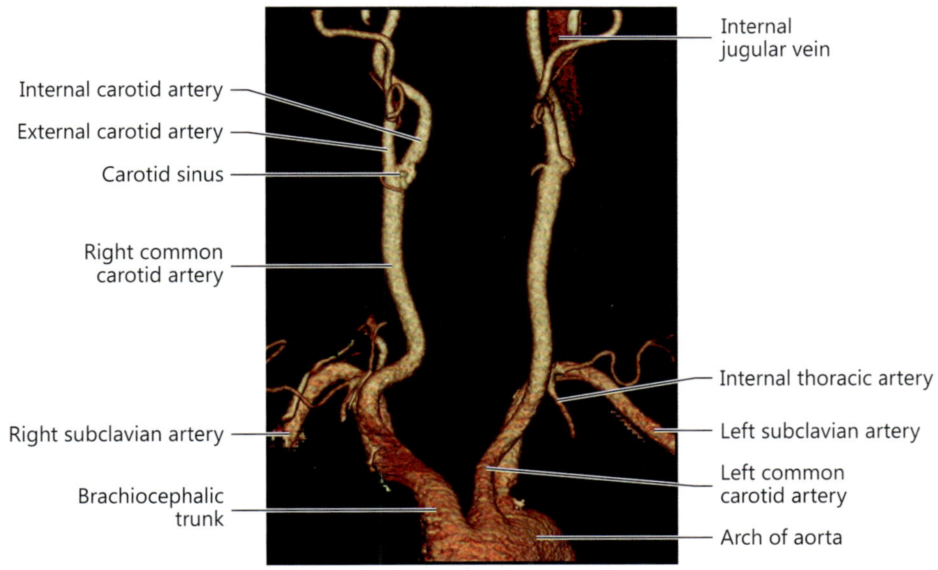

Three-dimensional image of a living human body, obtained by CT scan using a 3D volume rendering technique, showing the ramification and distribution of the **vessels that originate from the arch of the aorta**. The convex face of the arch gives rise to the brachiocephalic trunk, the left common carotid artery and left subclavian artery. The brachiocephalic trunk subdivides into the right common carotid and the right subclavian arteries. The common carotid arteries are the main arteries supplying blood to the head and neck and terminate with subdivision into the external carotid and internal carotid arteries.

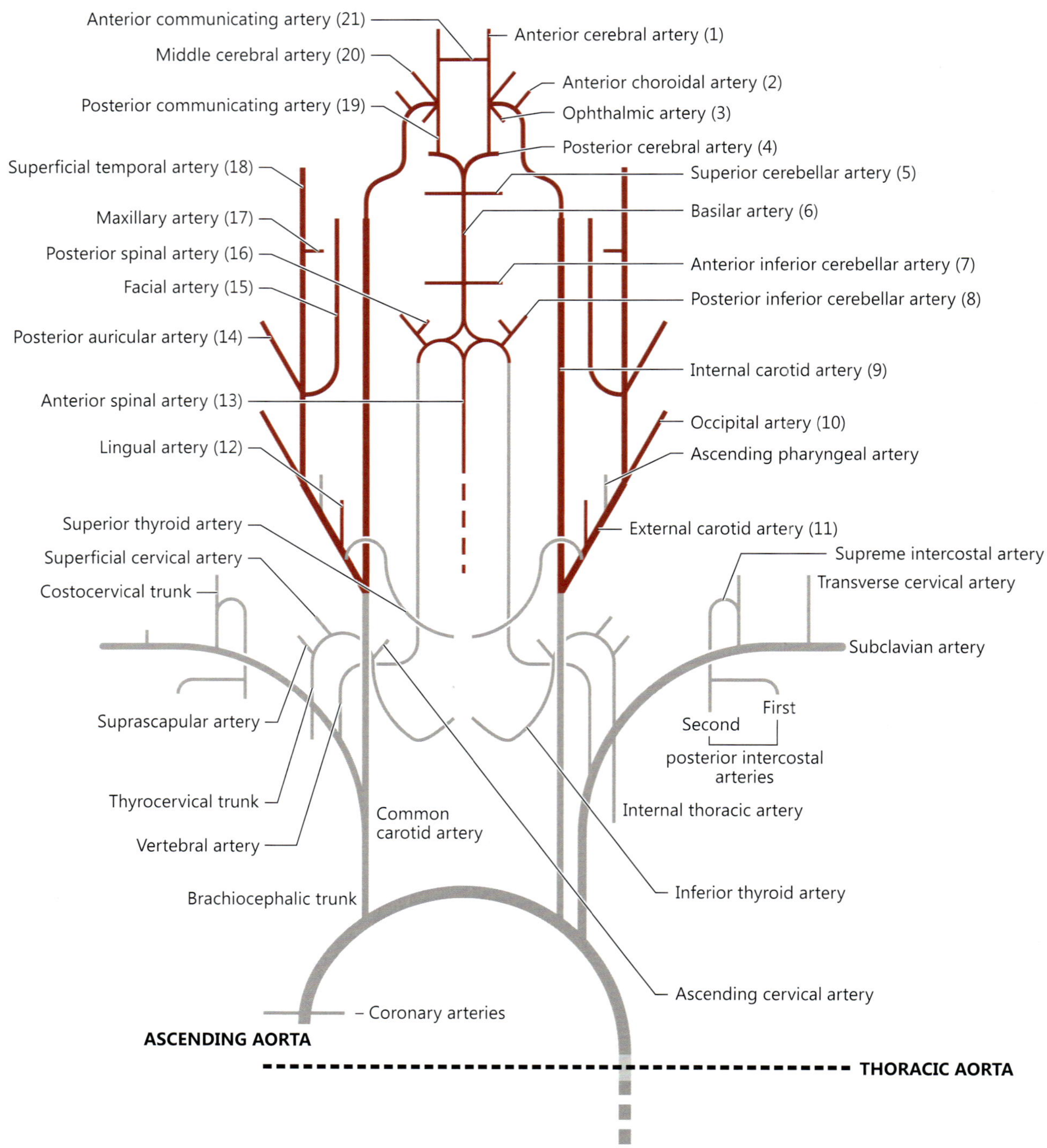

Schematic representation of the arterial blood supply to the head.
This page focuses on the role of the external and internal carotid arteries involved. The relative territories of blood supply (indicated numerically) are described in the table alongside. In general, the **external carotid artery** supplies the neck, face and the walls of the cranium, while the **internal carotid artery** supplies the brain and intracranial formations of the orbital cavity and nasal cavities. The circulation upstream of the cerebral arterial circle, or circle of Willis, is described in the chapter on the nervous system.

Main arteries of the head: territories of blood supply		
	ARTERIES OF THE HEAD	TERRITORIES OF SUPPLY
1	Anterior cerebral artery	Hypothalamus, lamina terminalis, anterior perforated substance, anterior commissure, column of the fornix, caudate nucleus, putamen, pallidum, internal capsule, choroid plexuses, lateral ventricle, corpus striatum, corpus callosum, medial part of the frontal and parietal lobes, straight gyrus, olfactory bulb
2	Anterior choroidal artery	Choroid plexus of the lateral ventricle, optic tract, optic radiation, cerebral peduncle, hippocampus, fimbria of hippocampus, lateral geniculate body, caudate nucleus, red nucleus, pallidum, amygdaloid body, hypothalamus, tuber cinereum, substantia nigra, posterior limb of the internal capsule
3	Ophthalmic artery	Orbital cavity, eyeball, extraocular muscles, eyelids, dura mater of the anterior cranial fossa, vault and lateral walls of the nasal cavities
4	Posterior cerebral artery	Thalamus, walls of the third ventricle, hypothalamus, globus pallidus, cerebral peduncle, mesencephalon, tectal plate, epiphysis, medial and lateral geniculate bodies, uncus, parahippocampal gyrus, medial occipitotemporal gyrus, lateral occipitotemporal gyrus, temporal and occipital lobes, choroid plexus of the lateral ventricle and choroid plexus of the third ventricle
5	Superior cerebellar artery	Superior surface of the cerebellum, vermis of cerebellum [I-X], pons, epiphysis, superior medullary velum, choroid membrane of the third ventricle
6	Basilar artery	Internal ear, pons, cerebellum, mesencephalon, thalamus, hypothalamus, epiphysis, third ventricle, cerebral ventricles, cortex of the temporal and occipital lobes
7	Anterior inferior cerebellar artery	Anterolateral region of the inferior surface of the hemisphere of cerebellum, choroid plexus of the fourth ventricle
8	Posterior inferior cerebellar artery	Hemisphere of cerebellum [H II - H X], vermis of cerebellum [I-X], choroid plexus of the fourth ventricle
9	Internal carotid artery	Brain, pituitary gland, orbit, nasal cavity, trigeminal ganglion, meninges, middle ear
10	Occipital artery	Lateral and posterior cervical muscles, auricle, mastoid cells, dura mater of the posterior cranial fossa, diploë, muscles and skin of the occipital region, scalp
11	External carotid artery	Anterior parts of the face and neck and lateral part of the head, walls of the cranial cavity, dura mater, posterior part of the scalp
12	Lingual artery	Muscles and mucous membrane of the tongue, sublingual gland, gingivae, lingual and palatine tonsils, epiglottis, suprahyoid muscles
13	Anterior spinal artery	Anterior two-thirds of the spinal cord
14	Posterior auricular artery	Middle ear, mastoid cells, auricle, digastric, stylohyoid, sternocleidomastoid, skin of the occipital region, parotid gland
15	Facial artery	Soft palate, palatine tonsil, submandibular gland and labial glands, skin and muscles of the chin, lips and external nose, ala and dorsum of the nose
16	Posterior spinal artery	Myelencephalon, spinal cord, pia mater
17	Maxillary artery	Tympanic cavity, external acoustic meatus, dental arcades, upper lip, chin, dura mater of the middle cranial fossa, of the walls and the calvaria, trigeminal ganglion, temporalis, masseter, buccinator, extraocular muscles, lacrimal sac, lacrimal gland, palatine tonsil, palate, nasopharynx, pharyngotympanic tube, nasal cavity and paranasal sinuses
18	Superficial temporal artery	Masseter and temporalis, frontal belly of occipitofrontalis, orbicularis oculi, external acoustic meatus, auricle, skin of the face, scalp, parotid gland, temporomandibular joint
19	Posterior communicating artery	Base of the telencephalon between the infundibulum and optic tract, optic chiasm, hippocampus, cerebral peduncle, internal capsule, anterior third of the thalamus, third ventricle, anterior perforated substance
20	Middle cerebral artery	Lentiform nucleus, internal capsule, caudate nucleus, corpus striatum, insula, frontal lobe, parietal lobe, temporal lobe, precentral gyrus, postcentral gyrus, inferior frontal gyrus, superior temporal gyrus, middle temporal gyrus, supramarginal gyrus and angular gyrus
21	Anterior communicating artery	Cerebral cortex, orbitomesial giri

N.B.: for a general overview of the arteries and their territories of supply, see the table summarizing the arteries of the head.

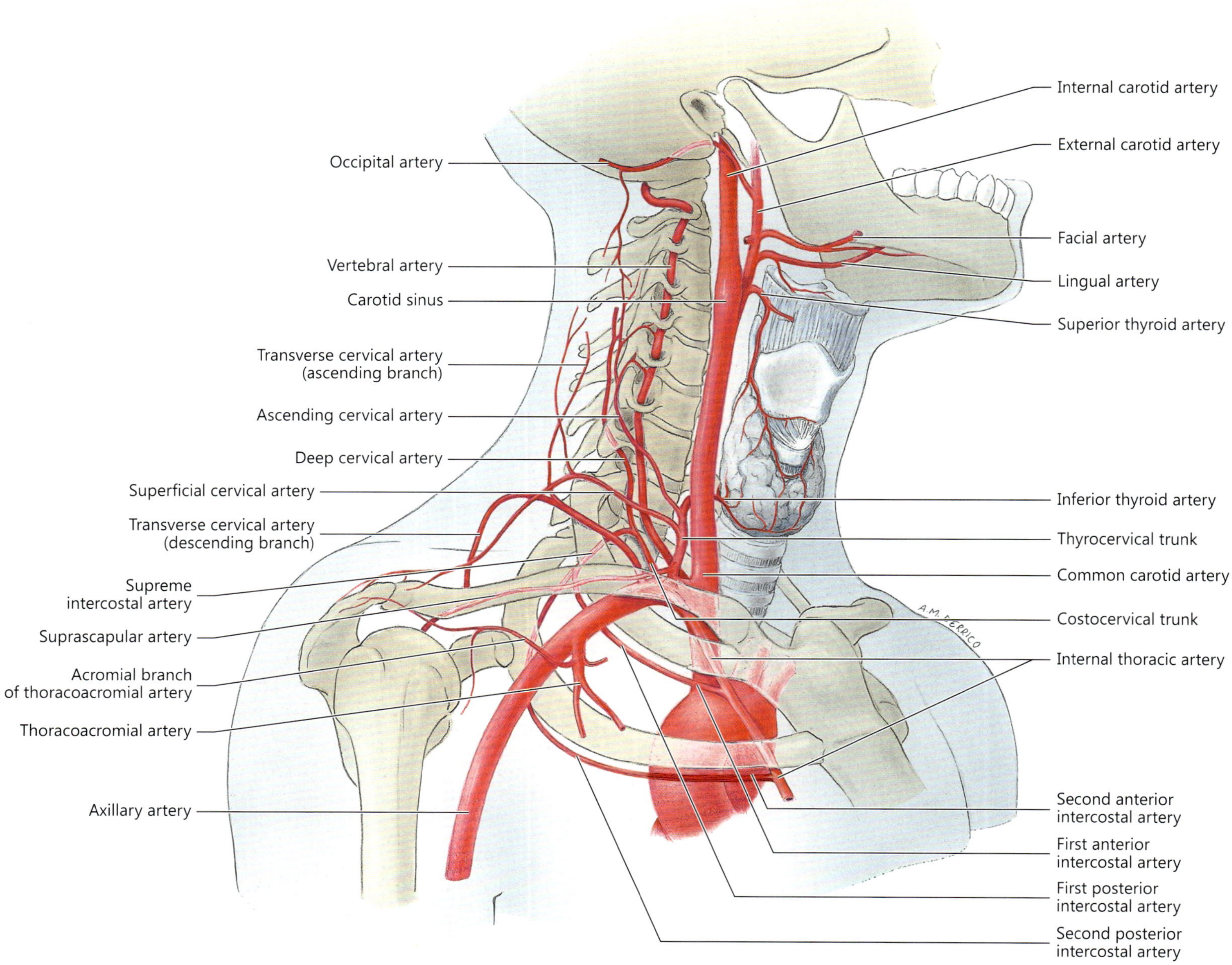

Right subclavian artery and its branches. Note the course of the **vertebral artery** that, on reaching the 6th cervical vertebra, ascends the neck passing through the foramina transversaria; inferiorly, the **supreme intercostal artery** supplies branches for the 1st and 2nd intercostal spaces, and further on anastomoses with the **internal thoracic artery**.

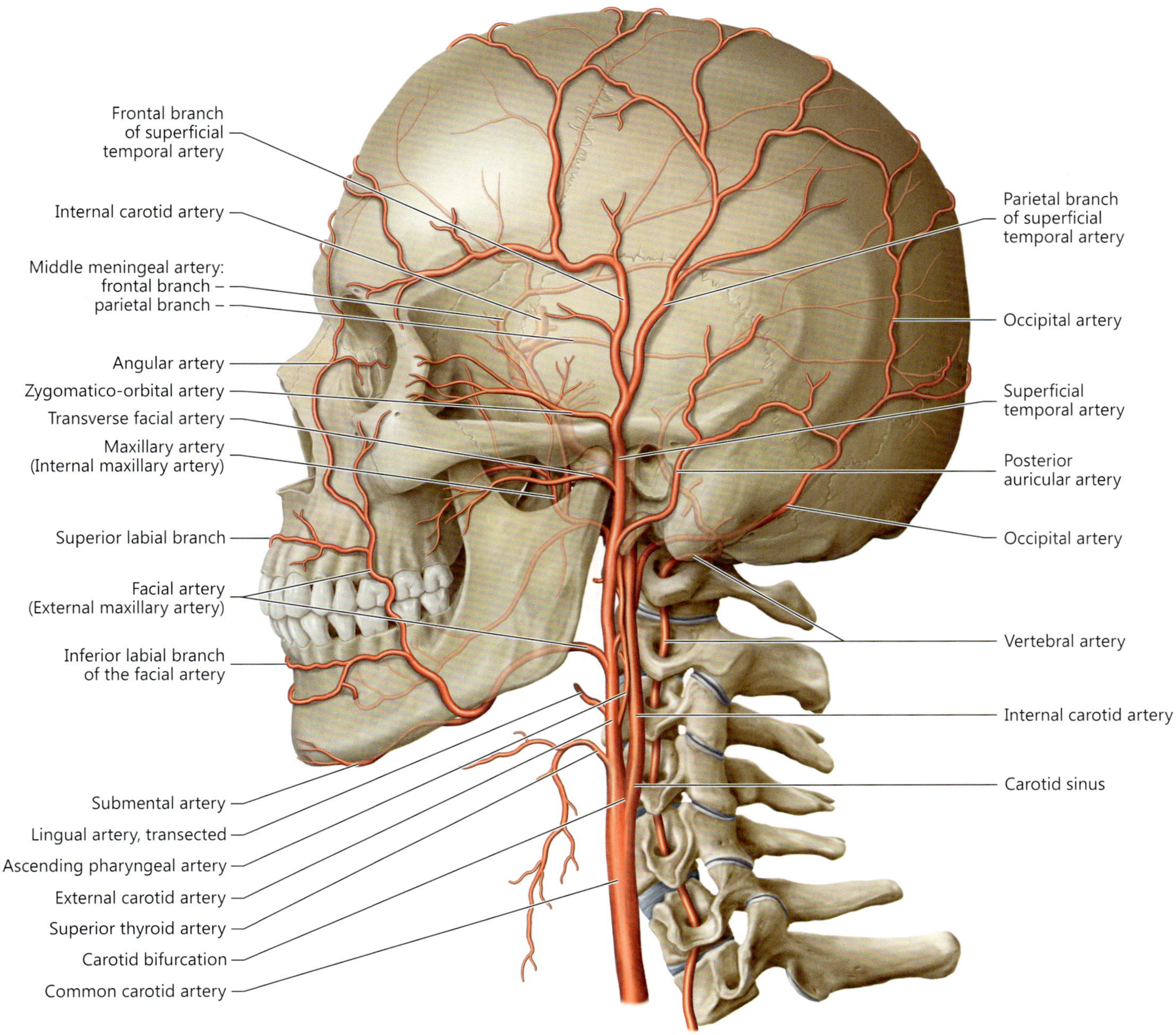

Course of the **external carotid artery** and relative collateral and terminal branches. Also visible are the internal carotid artery and vertebral artery and, in transparency, some of their intracranial branches.

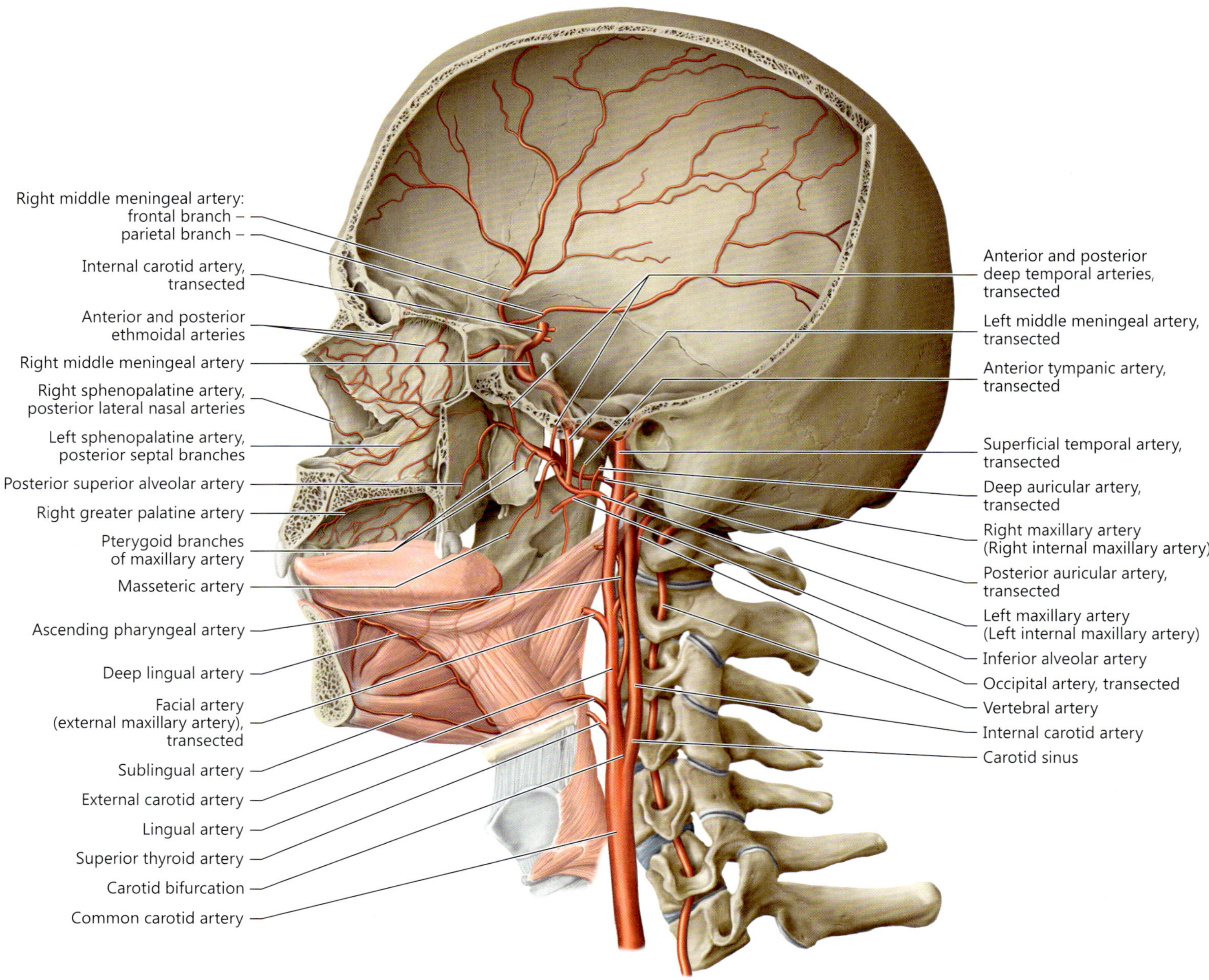

The collateral and terminal branches of the **external carotid artery** have important relations with the cranium and with the muscles of the mouth and tongue.

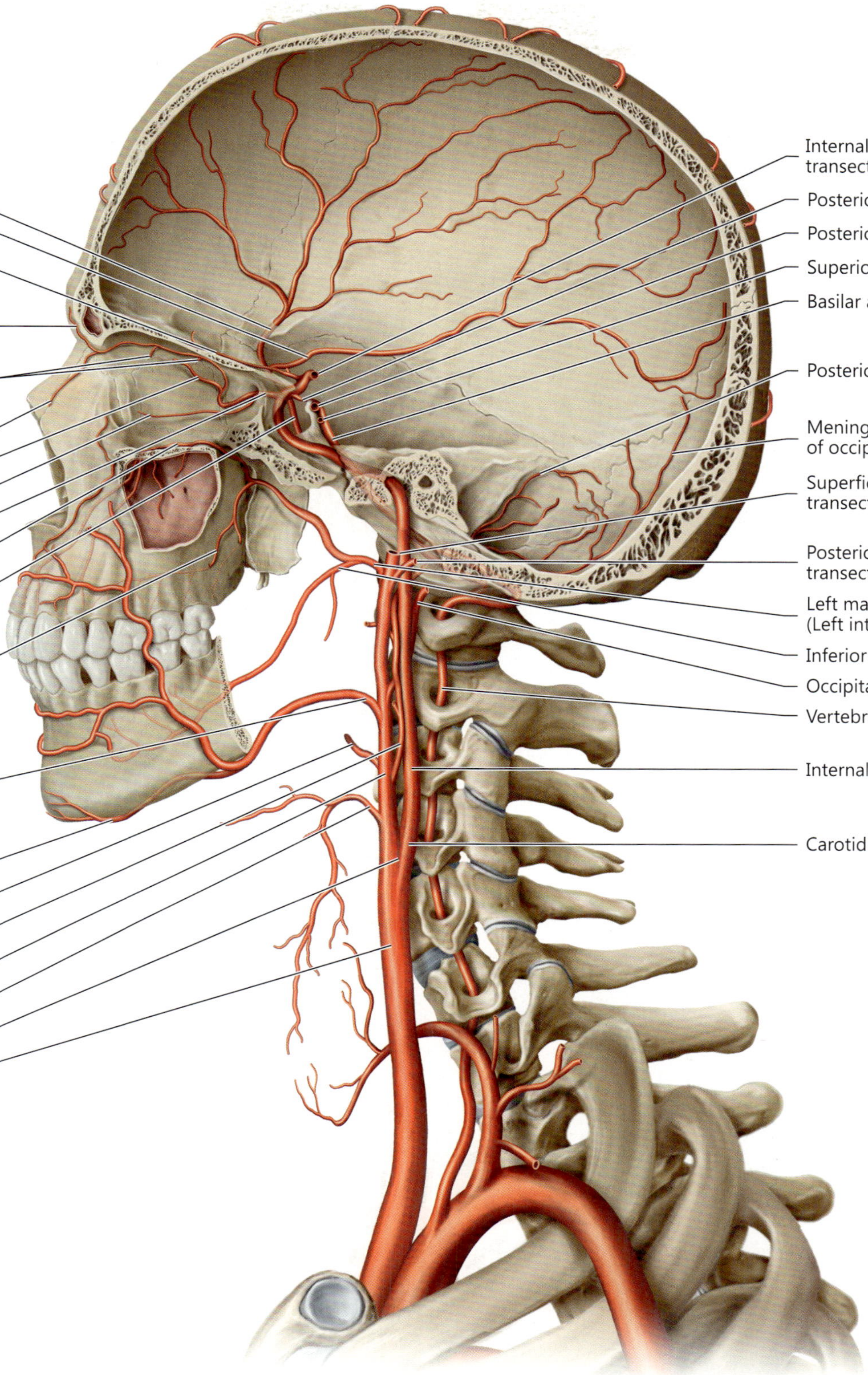

Course and branches of the **external carotid artery** and **maxillary artery** (in the three portions: mandibular, pterygoid and pterygopalatine). With respect to the previous figure, large parts of the soft tissues, facial skeleton and calvaria have been resected and removed. Highlighted are the right middle meningeal artery, with its intracranial course, and the meningeal branches of the posterior cranial fossa: the posterior meningeal artery, branch of the ascending pharyngeal artery, and the meningeal branch of the occipital artery. Further detail on the course of the maxillary (or internal maxillary) artery can be found on page 141.

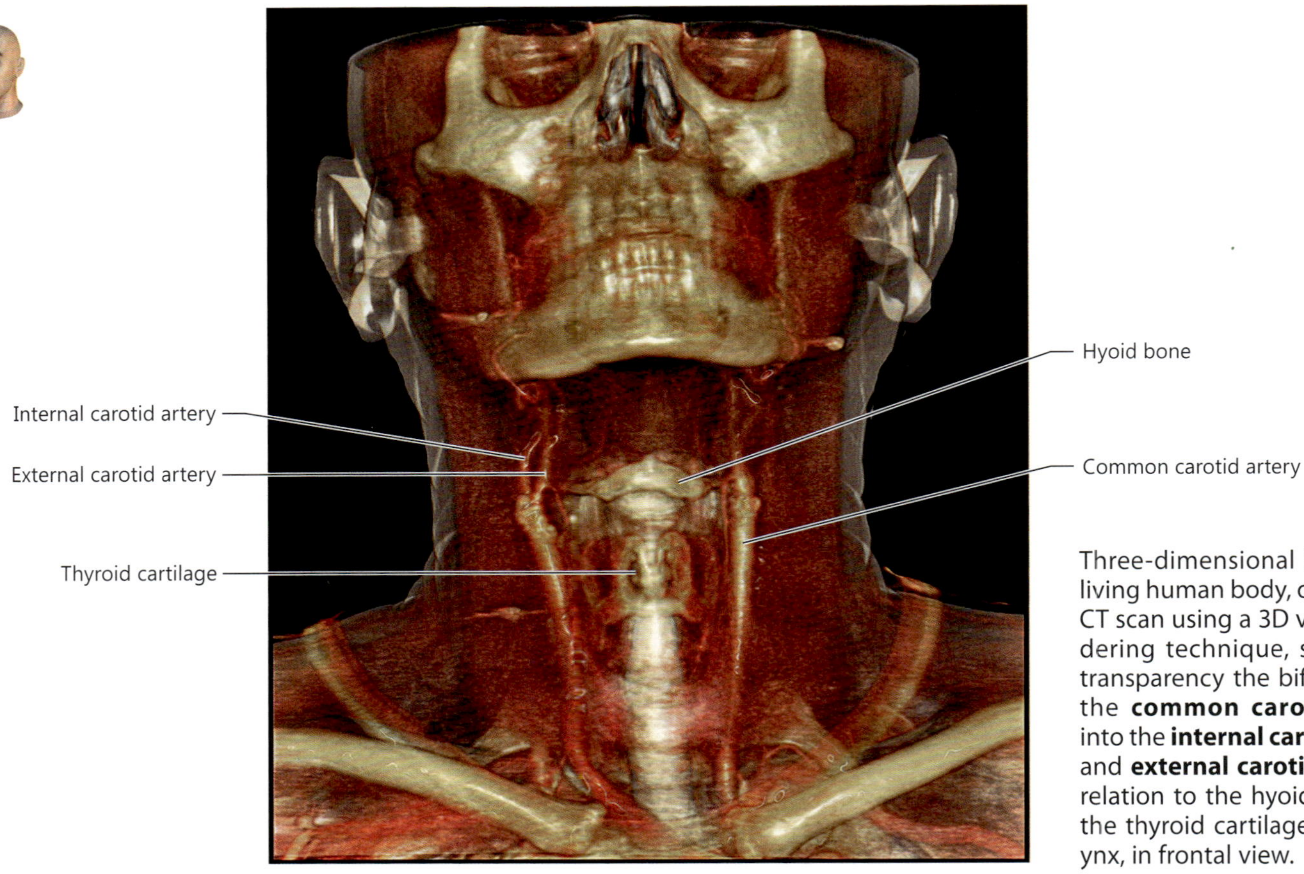

Three-dimensional image of a living human body, obtained by CT scan using a 3D volume rendering technique, showing in transparency the bifurcation of the **common carotid artery** into the **internal carotid artery** and **external carotid artery** in relation to the hyoid bone and the thyroid cartilage of the larynx, in frontal view.

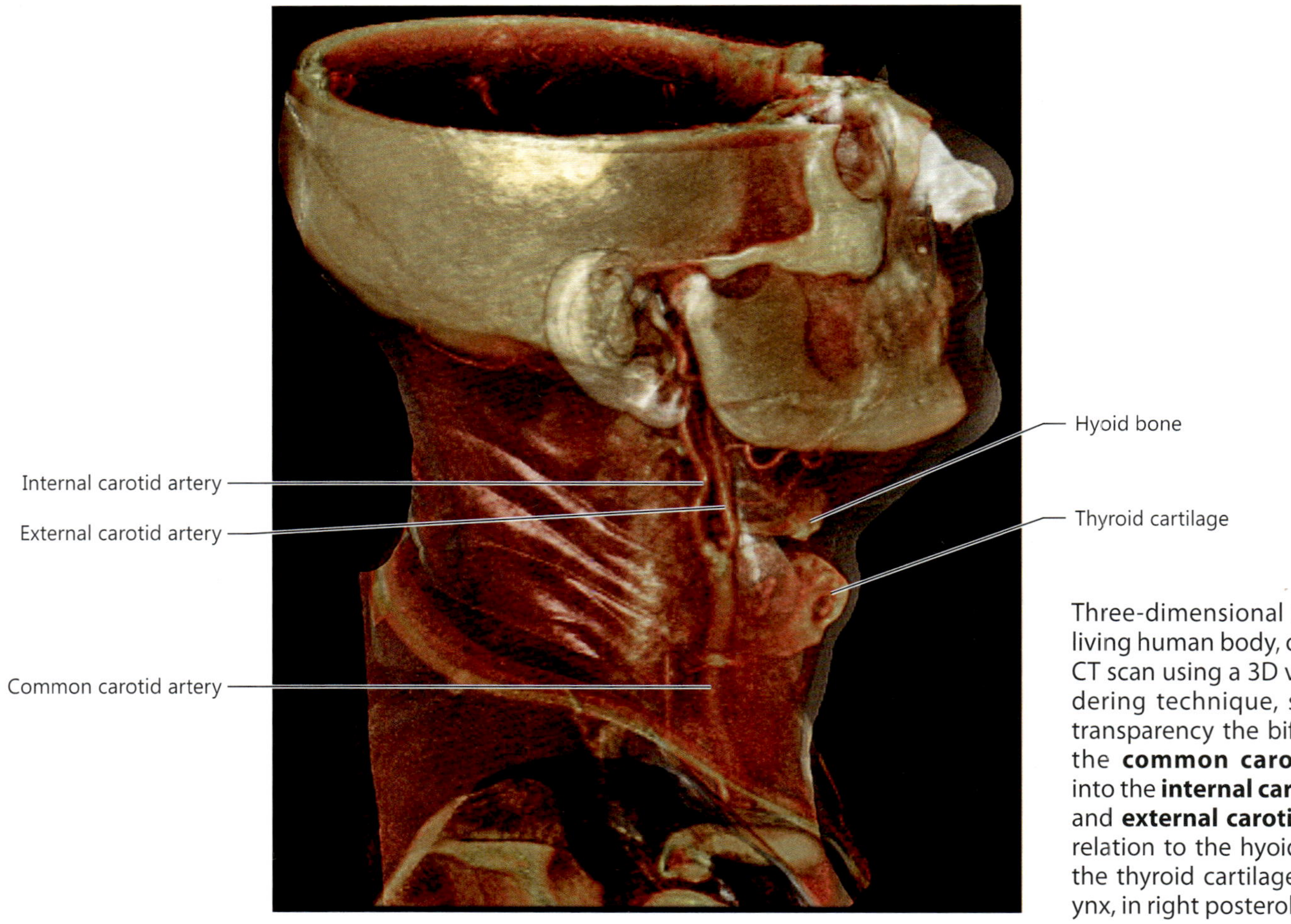

Three-dimensional image of a living human body, obtained by CT scan using a 3D volume rendering technique, showing in transparency the bifurcation of the **common carotid artery** into the **internal carotid artery** and **external carotid artery** in relation to the hyoid bone and the thyroid cartilage of the larynx, in right posterolateral view.

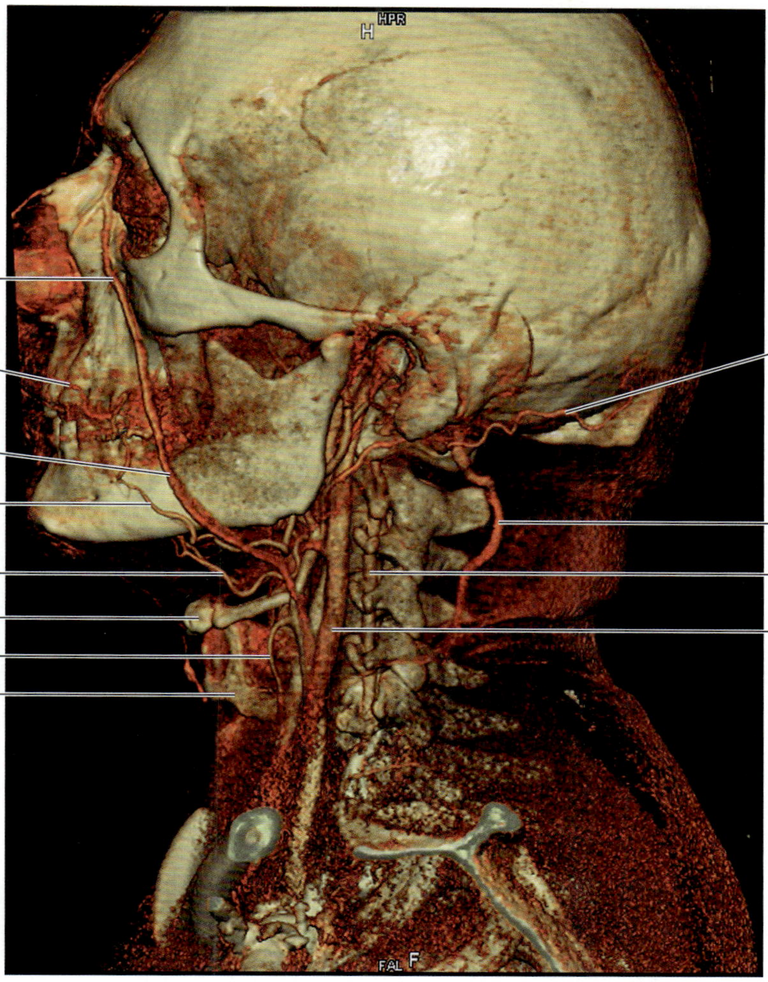

Three-dimensional images of a living human body, obtained by CT scan using a 3D volume rendering technique. Above: **collateral branches of the external carotid artery** and course of the **vertebral artery**. Below: **arteries of the neck** in relation to the skeleton of the neck and head, viewed on a deeper plane with respect to the image above, for which reason the external maxillary artery in its tangential course to the mandible is not visible, thus making it possible to see the internal maxillary artery.

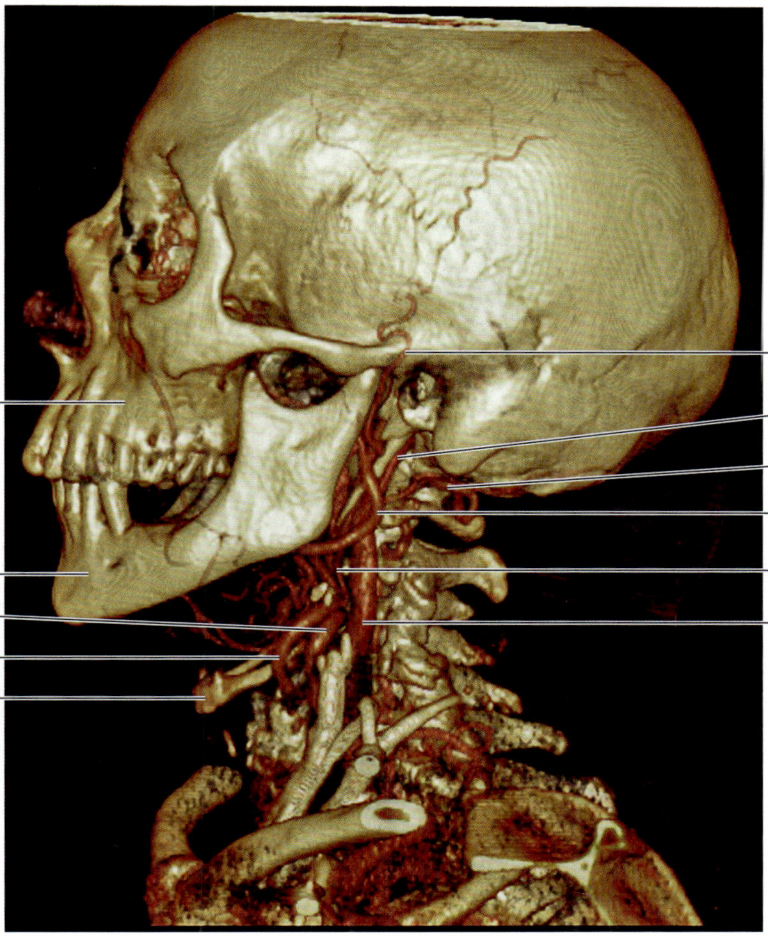

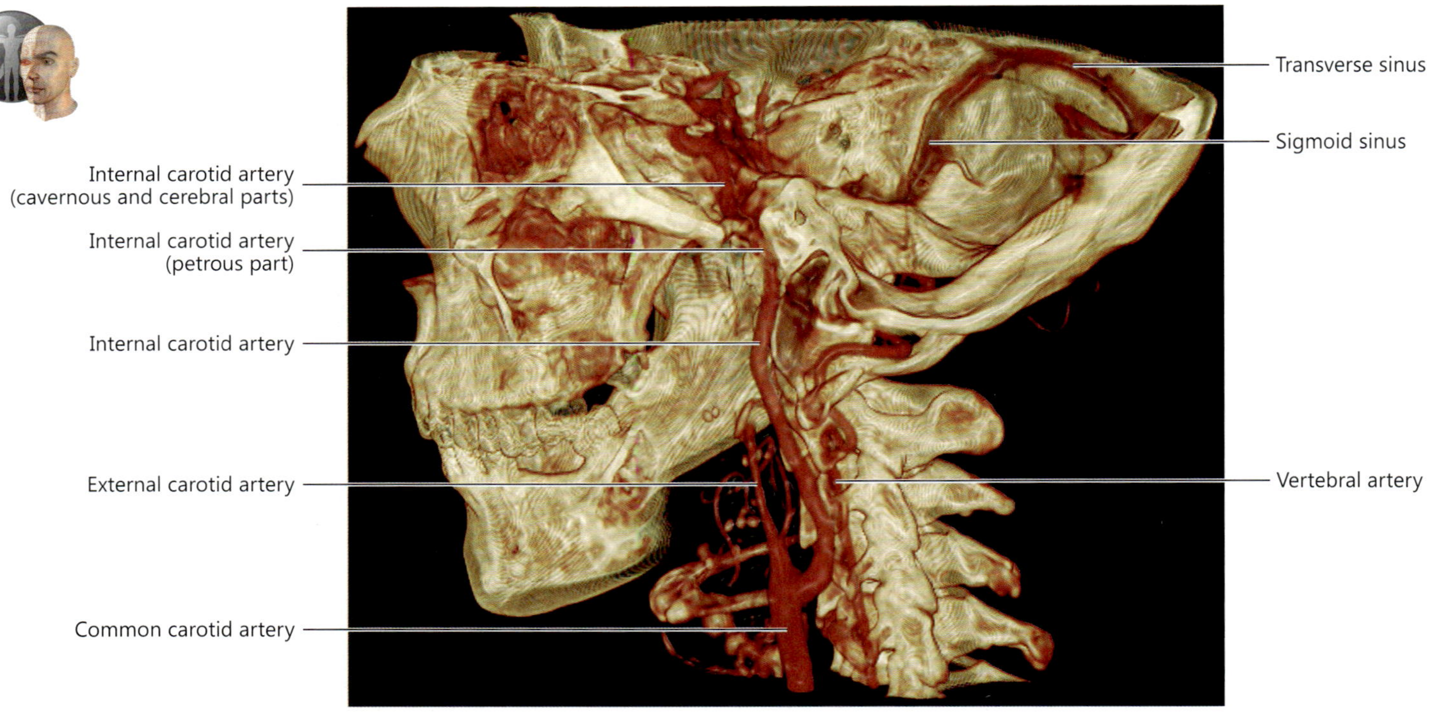

Three-dimensional image of a living human body, obtained by contrast-enhanced CT scan using a 3D volume rendering technique, showing the course of the **internal carotid artery**, in relation to the skeleton, from the bifurcation to the intracranial part.

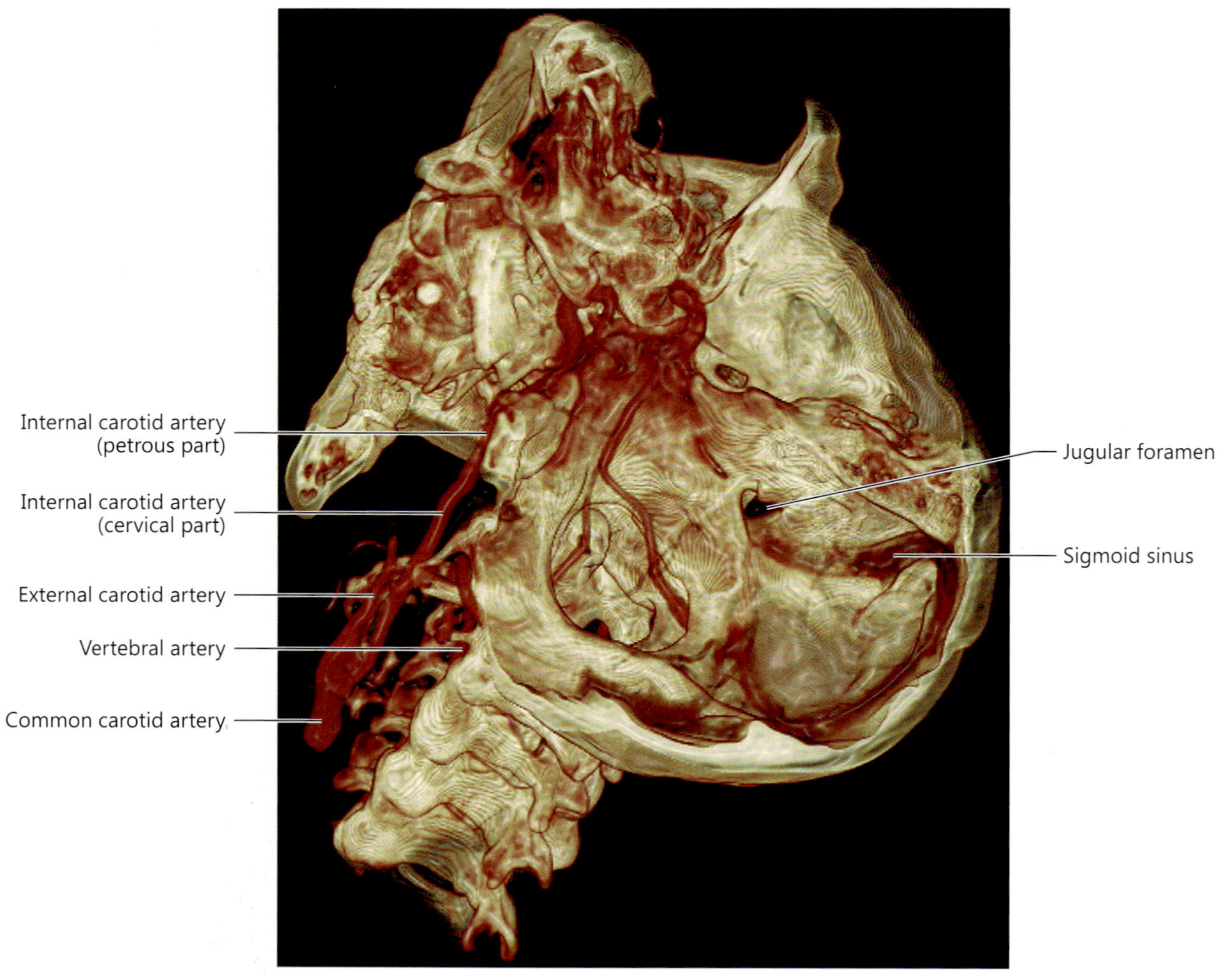

Three-dimensional image of a living human body, obtained by contrast-enhanced CT scan using a 3D volume rendering technique, showing the course of the **internal carotid artery**, in relation to the skeleton, from the bifurcation to the intracranial part, viewed from above and posterolaterally.

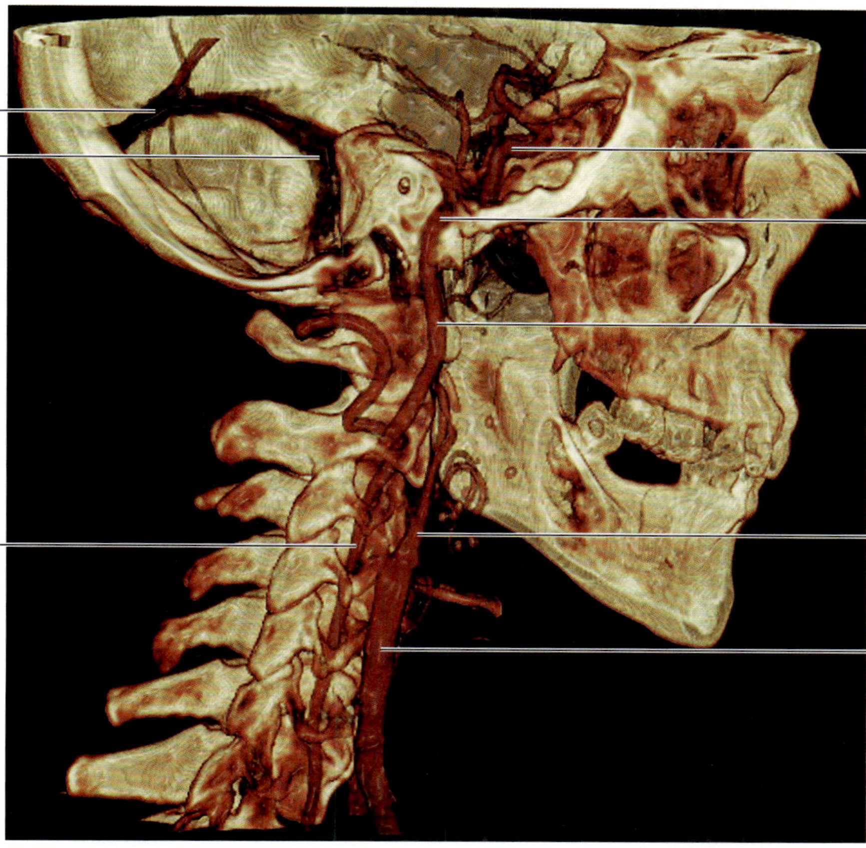

Three-dimensional image of a living human body, obtained by contrast-enhanced CT scan using a 3D volume rendering technique, showing the course of the **internal carotid artery**, in relation to the skeleton, from the point proximal to the entrance into the carotid canal to its end.

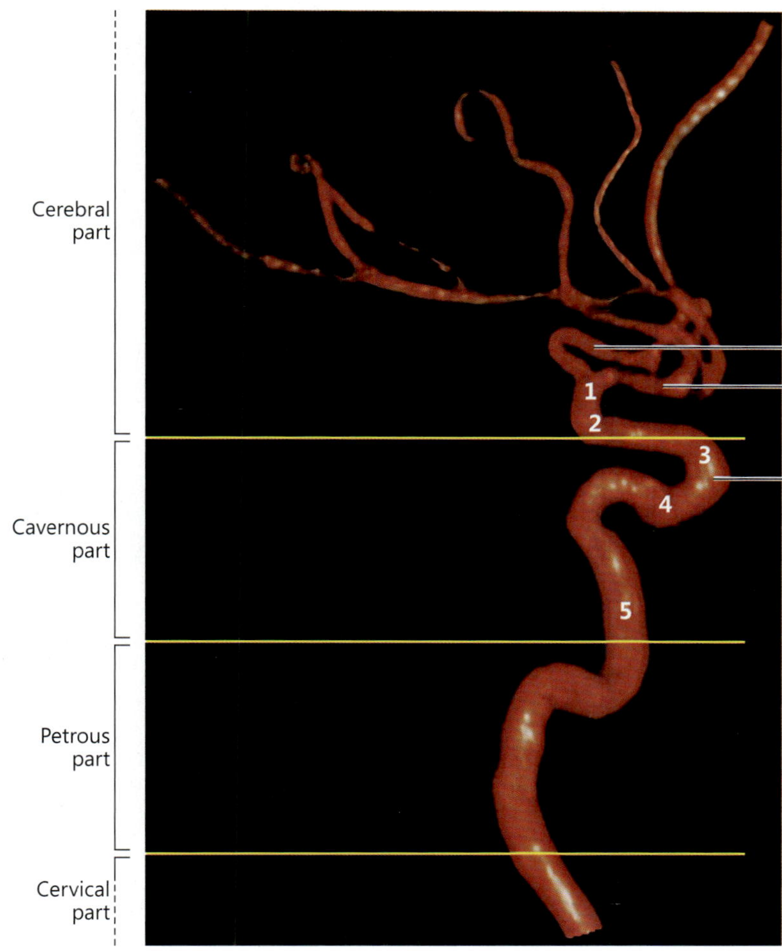

Three-dimensional image of a living human body, obtained by contrast-enhanced CT scan using a 3D volume rendering technique, showing the course of the **internal carotid artery**. The intracranial part subdivides into the cavernous part and cerebral part and, successively, the two parts further subdivide into five segments, according to Fisher's scheme:

1. communicating or distal supraclinoid tract
2. ophthalmic or proximal supraclinoid tract
3. clinoid tract
4. cavernous tract
5. precavernous tract

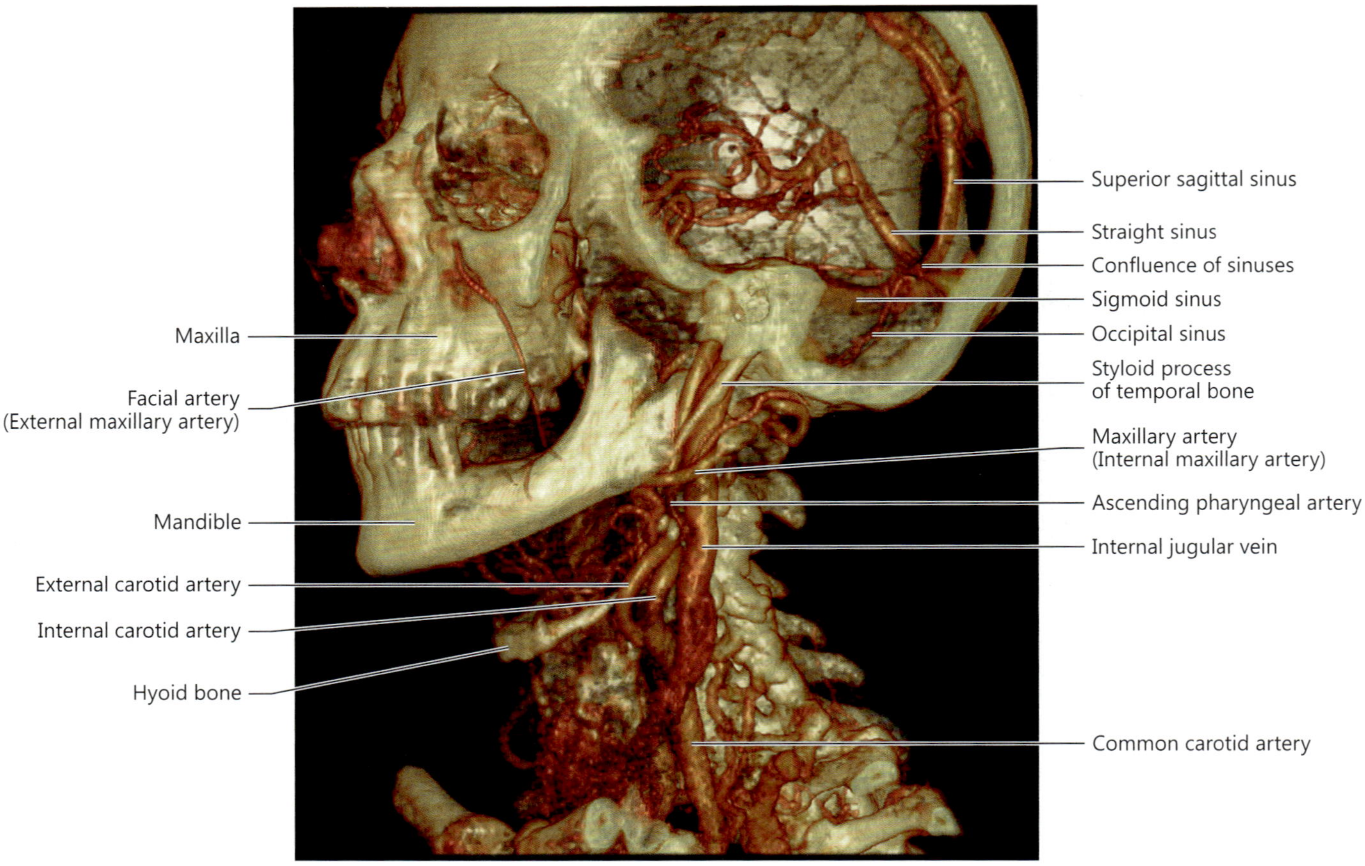

Three-dimensional image of a living human body, obtained by CT scan using a 3D volume rendering technique, showing the vessels passing through the neck in relation to the bone structures, in left anterolateral view. With the left cranial wall removed, one can observe the principal **venous sinuses** within the **cranial cavity**.

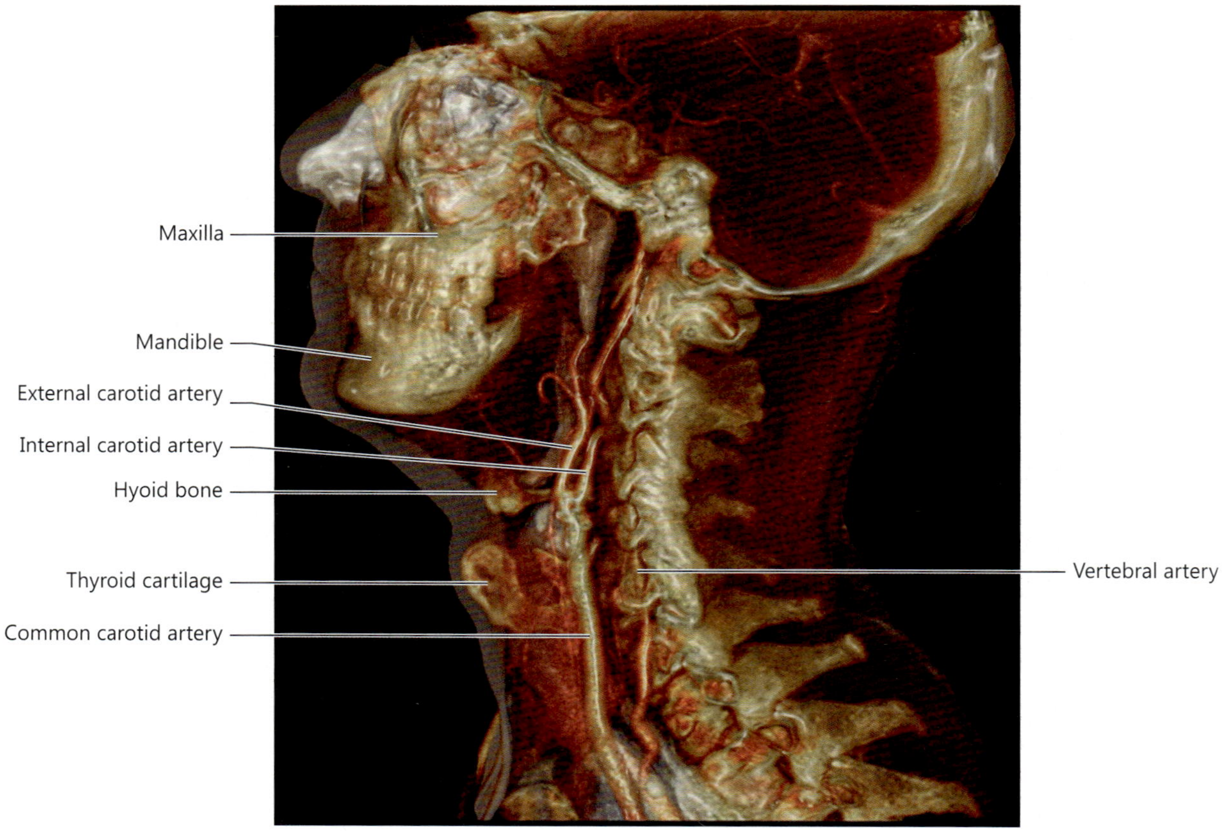

Three-dimensional image of a living human body, obtained by CT scan using a 3D volume rendering technique, showing the passage of the **vessels within the neck** in relation to the bone structures, in left lateral view.

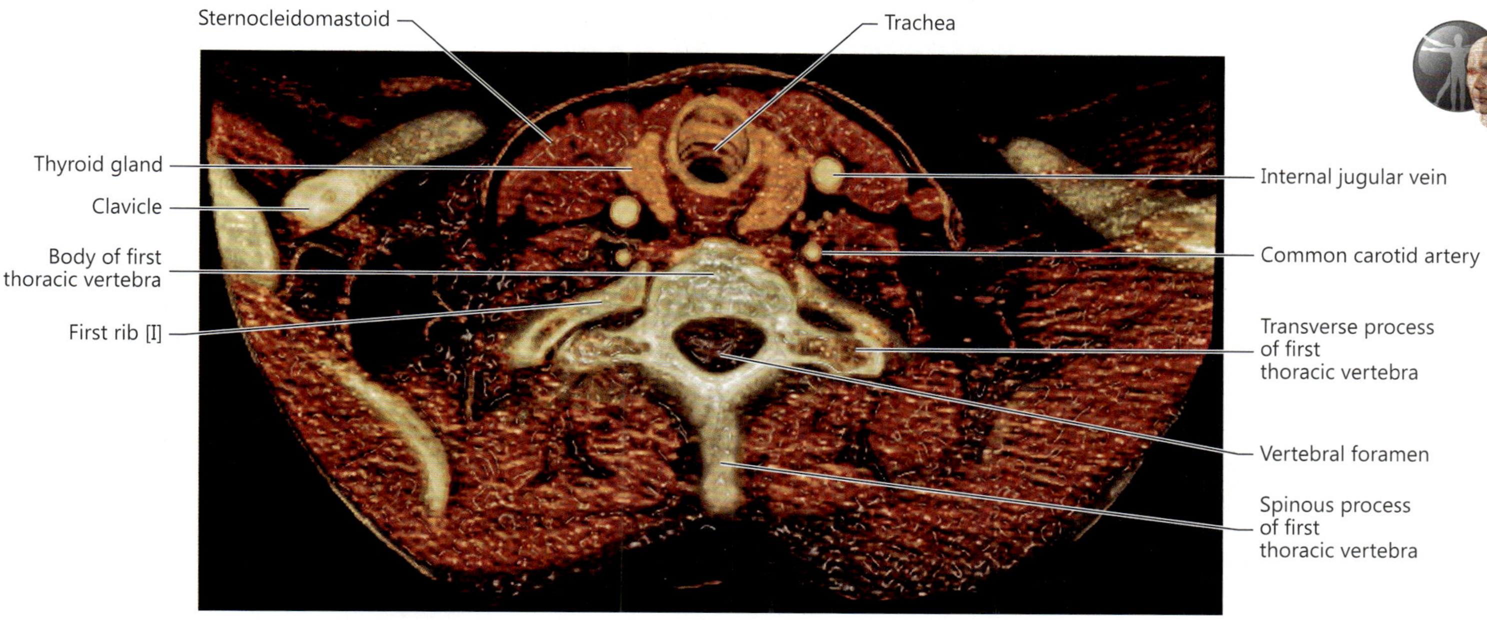

Three-dimensional image of a living human body, obtained by CT scan using a 3D volume rendering technique, showing a transverse section of the **neck** at the **level of the first thoracic vertebra**.

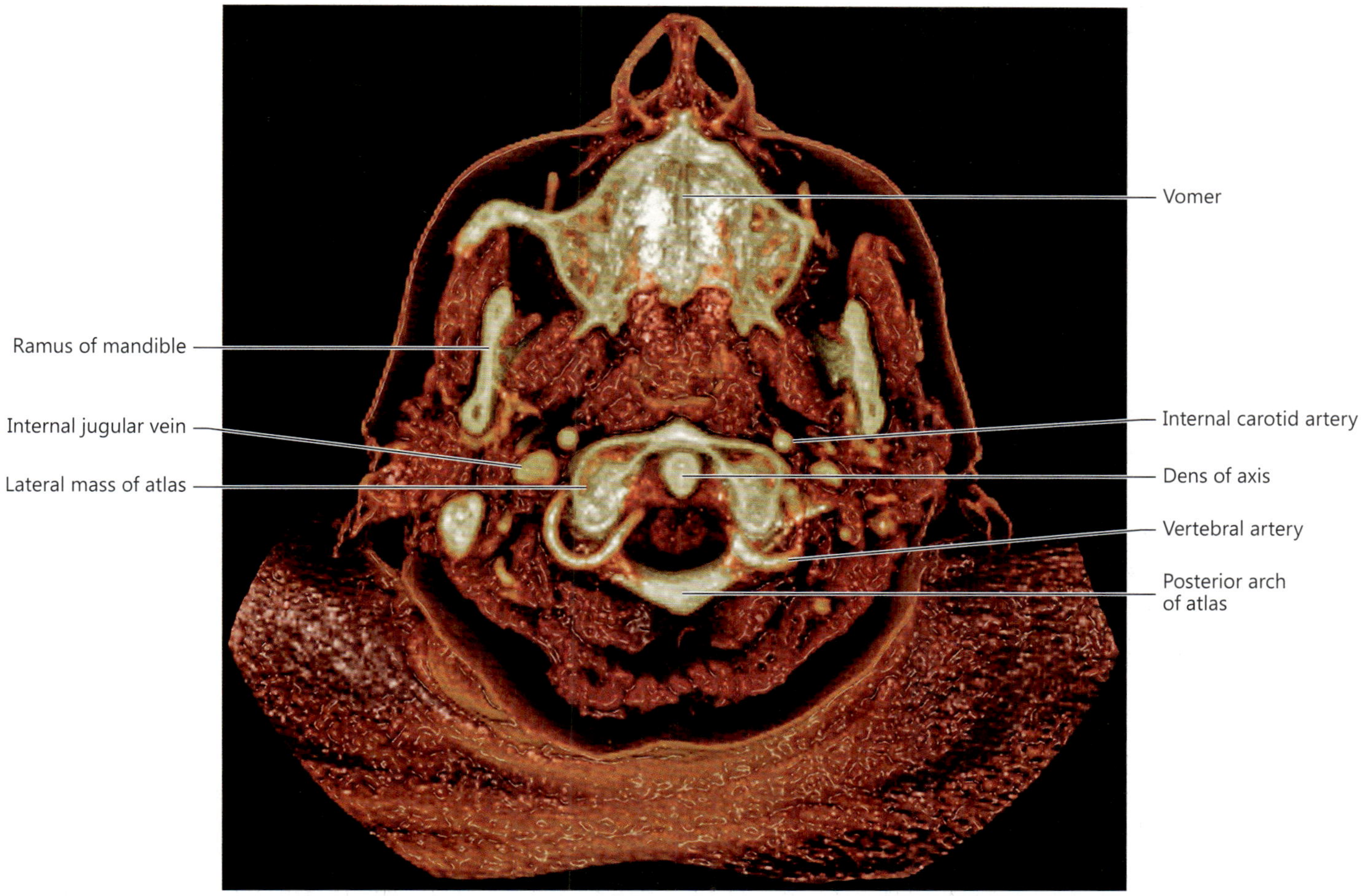

Three-dimensional image of a living human body, obtained by CT scan using a 3D volume rendering technique, showing a transverse section of the **neck** at the **level of the atlas-axis joint**.

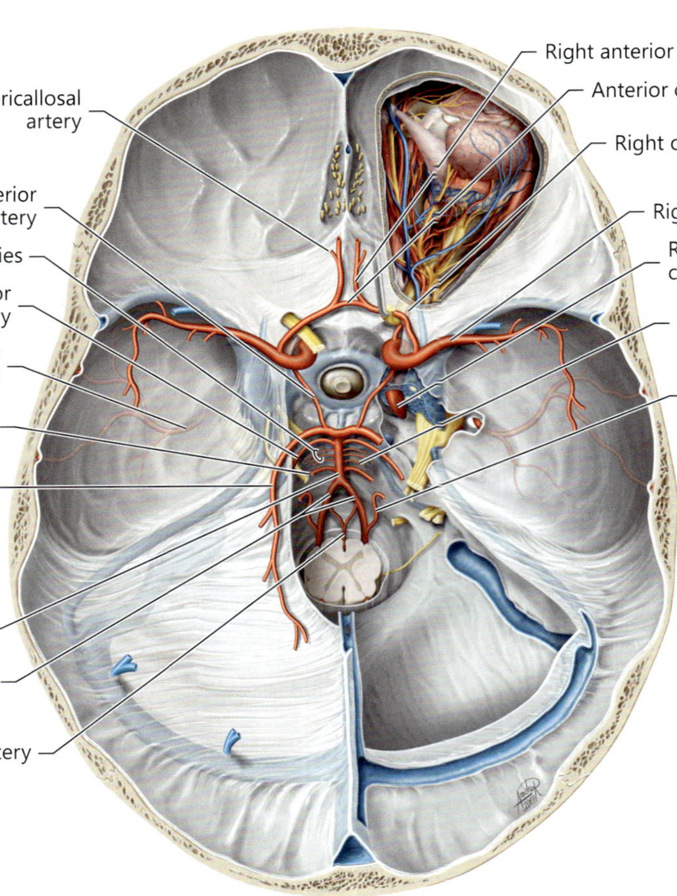

The **circle of Willis** (cerebral arterial circle) and intracranial branches of the **internal carotid artery**, **vertebral artery** and **basilar artery**.
Below: the circle of Willis is shown as it looks in a three-dimensional image of a living human body, obtained by CT scan using a 3D volume rendering technique.

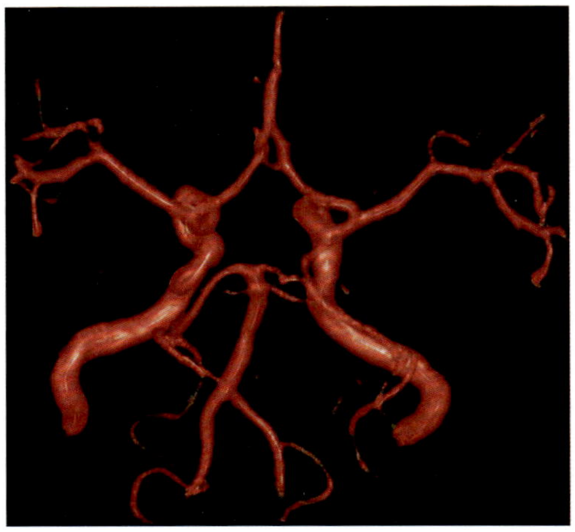

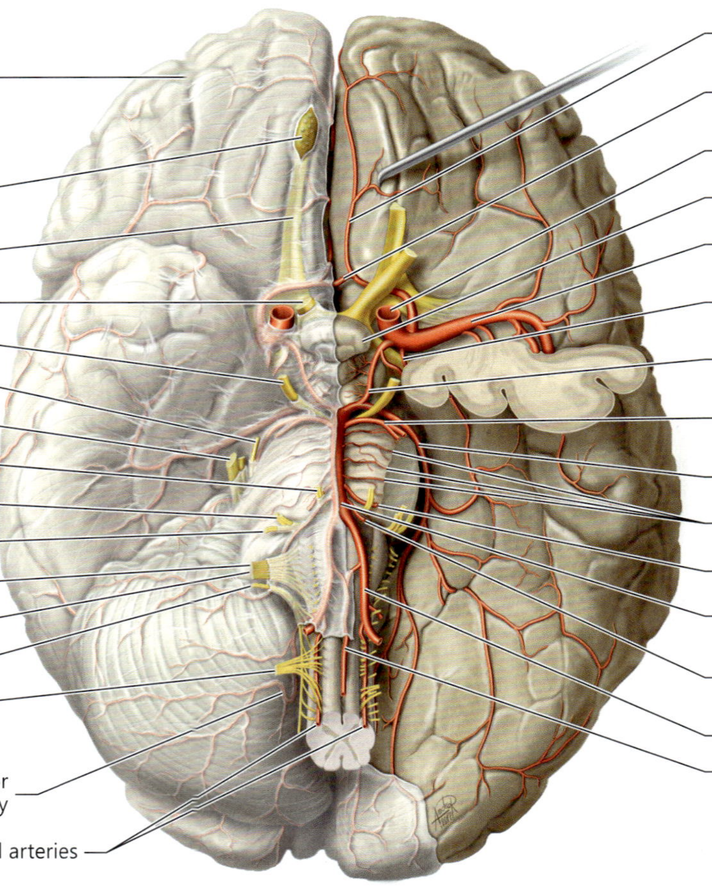

Arteries of the base of the brain: the left cerebellar hemisphere and part of the left temporal lobe have been removed to show the middle and posterior cerebral arteries. On the left side of the figure, the apparent origins of the cranial nerves are indicated.

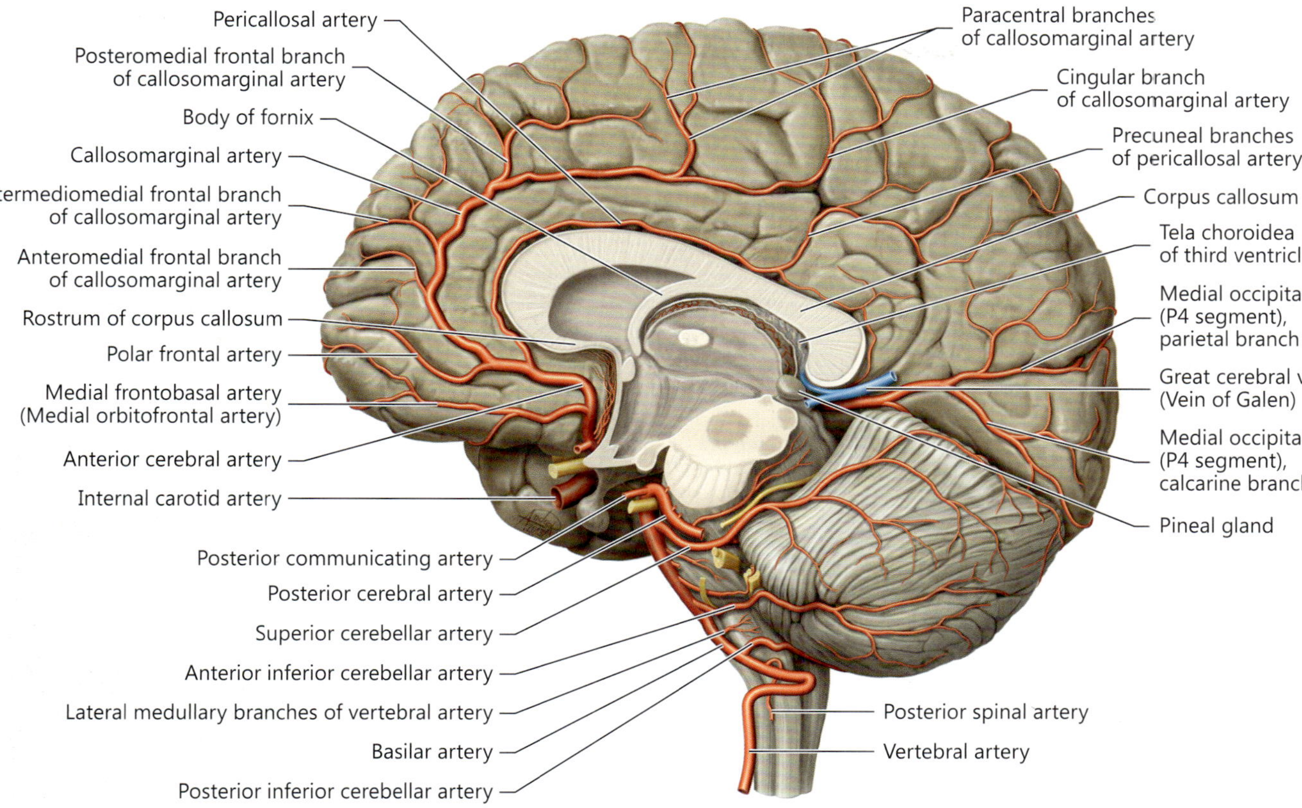

Anterior and **posterior cerebral arteries**. Medial view of the right cerebral hemisphere.

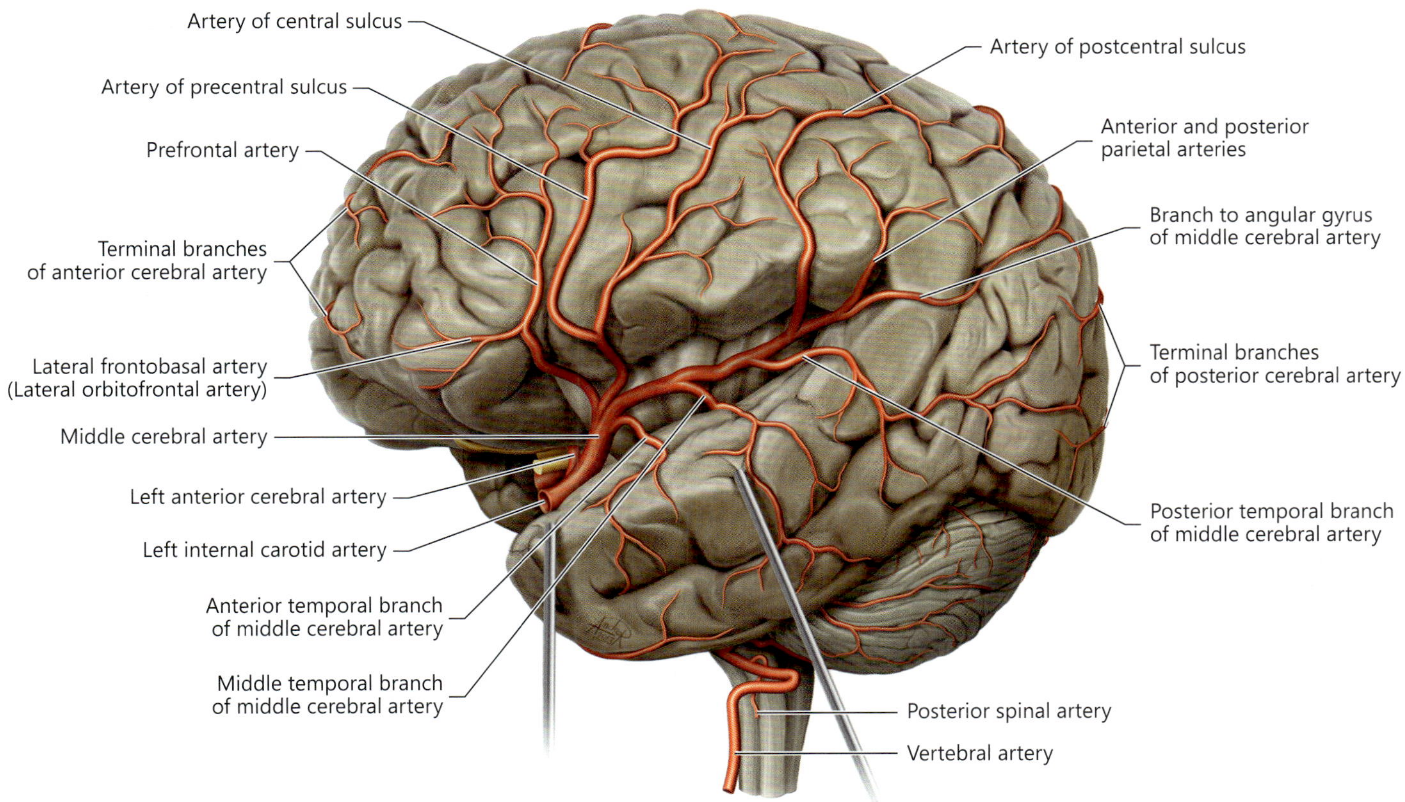

Middle cerebral artery. Lateral view of the left cerebral hemisphere. The lips of the lateral sulcus on the superolateral face of the hemisphere have been parted to show the insular part of the middle cerebral artery.

Summary table of the arteries of the head			
ARTERY	***ORIGIN***	***BRANCHES***	***TERRITORIES OF SUPPLY***
angular	Facial		Ala of nose, lacrimal sac
anterior cerebral	Internal carotid (at the medial end of the lateral sulcus of telencephalon)	*Precommunicating part (A1 segment)*: anteromedial central arteries; anterior communicating artery; *postcommunicating part (A2 segment)*: distal medial striate, medial frontobasal (orbitofrontal), polar frontal, callosomarginal and pericallosal arteries	Hypothalamus, lamina terminalis, anterior perforated substance, anterior commissure, column of the fornix, caudate nucleus, putamen, pallidum, internal capsule, choroid plexuses, lateral ventricle, corpus striatum, corpus callosum, medial part of the frontal and parietal lobes, straight gyrus, olfactory bulb
anterior choroidal	Internal carotid (cerebral part)	Choroidal branches to the lateral ventricle, choroidal branches to the third ventricle, branches to the anterior perforated substance, to the optic chiasm, to the optic tract, to the lateral geniculate body, to the genu of the internal capsule, to the posterior limb of the internal capsule, to the retrolentiform limb of the internal capsule, to the globus pallidus, to the tail of the caudate nucleus, to the hippocampus, to the uncus, to the amygdaloid body, to the tuber cinereum, to the hypothalamic nuclei, to the thalamic nuclei, to the substantia nigra, to the red nucleus, to the crus cerebri	Choroid plexus of the lateral ventricle, optic tract, optic radiation, cerebral peduncle, hippocampus, fimbria of hippocampus, lateral geniculate body, caudate nucleus, red nucleus, pallidum, amygdaloid body, hypothalamus, tuber cinereum, substantia nigra, posterior limb of the internal capsule
anterior ciliary	Ophthalmic or lacrimal (muscular branches)		Conjunctiva, iris
anterior communicating	Anterior cerebral (anastomosis between the anterior cerebral arteries)	Anteromedial central arteries	Anterior perforated substance
anterior conjunctival	Ophthalmic (muscular branches or anterior ciliary arteries)		Conjunctiva
anterior deep temporal	Maxillary		Temporalis
anterior ethmoidal	Ophthalmic	Anterior meningeal, anterior septal, anterior lateral nasal	Anterior and middle ethmoidal cells, frontal sinus, dura mater, nasal cavity
anterior inferior cerebellar	Basilar	Labyrinthine artery	Anterolateral region of the posterior surface of the hemisphere of cerebellum, choroid plexus of the fourth ventricle
anterior parietal	Middle cerebral (superior terminal branches)		Parietal lobe
anterior perforating	Precommunicating part of the anterior cerebral artery (anteromedial central branches)		Anterior perforated substance
anterior spinal	Vertebral (intracranial part)		Anterior two-thirds of the spinal cord
anterior superior alveolar (anterior dental)	Infraorbital		Incisor and canine teeth of the maxillary dental arcade, mucosa of the maxillary sinus
anterior temporal	Middle cerebral (sphenoid part)		Temporal gyri
anterior tympanic	Maxillary		Tympanic membrane, tympanic cavity, auditory ossicles
anterolateral central	Middle cerebral (sphenoid part)	Proximal and distal lateral striate arteries	Corpus striatum, lentiform nucleus
anteromedial central	Anterior cerebral (precommunicating part)	Proximal medial striate arteries, supraoptic artery, anterior perforating arteries, preoptic arteries	Anteromedial part of the corpus striatum and part of the thalamus
	Anterior communicating	Suprachiasmatic, median commissural and median callosal arteries	Anteromedial part of the corpus striatum and part of the thalamus

continued

Summary table of the arteries of the head *(continued)*

ARTERY	*ORIGIN*	*BRANCHES*	*TERRITORIES OF SUPPLY*
ascending palatine	Facial		Soft palate, palatine glands, pharyngotympanic tube
ascending pharyngeal	External carotid	Posterior meningeal artery; pharyngeal branches; inferior tympanic artery	Wall of the pharynx, soft palate, tonsils, ear, meninges, posterior cervical and head muscles
basilar	From the union of the two vertebral arteries	Anterior inferior cerebellar artery, pontine arteries, mesencephalic arteries, superior cerebellar artery, posterior cerebral artery	Internal ear, pons, cerebellum, mesencephalon, thalamus, hypothalamus, epiphysis, third ventricle, cerebral ventricles, cortex of the temporal and occipital lobes
buccal (buccinator)	Maxillary		Buccinator, mucosa of the maxillary gingivae, mucosa and skin of the cheeks
callosomarginal	Anterior cerebral (postcommunicating part)	Anteromedial frontal, intermediomedial frontal, posteromedial frontal, cingular, paracentral branches	Paracentral lobule, parts of the cingulate gyrus
caroticotympanic	Internal carotid (petrous part)		Tympanic cavity
central retinal	Ophthalmic	Extraocular part; intraocular part	Retina
collicular (quadrigeminal)	Posterior cerebral (precommunicating part)		Tectal plate
deep auricular	Maxillary		Skin of the external acoustic meatus, external surface of the tympanic membrane, temporomandibular joint
deep lingual (or ranine)	Lingual (terminal part)		Mucosa and intrinsic muscles of the tongue
descending palatine	Maxillary	Greater palatine artery and lesser palatine arteries; pharyngeal branch	Soft palate, hard palate, palatine tonsil, gingivae
distal medial striate	Anterior cerebral (postcommunicating part)		Anterior part of the caudate nucleus, putamen, anterior limb of the internal capsule
dorsal lingual	Lingual		Mucosa of the posterior part of the tongue; palatoglossal arch, lingual tonsil, epiglottis
dorsal nasal (external nasal)	Ophthalmic		Skin of the nose, lacrimal sac
episcleral	Ophthalmic (muscular branches or anterior ciliary arteries)		Iris, ciliary body
external carotid	Common carotid	*Anterior tract*: superior thyroid, lingual and facial arteries; *posterior tract*: occipital and posterior auricular arteries; *medial tract*: ascending pharyngeal artery; *terminal tract*: superficial temporal and maxillary (or internal maxillary) arteries	Anterior parts of the face and neck and lateral part of the head, walls of the cranial cavity, dura mater, posterior part of the scalp
facial (external maxillary)	External carotid	*Cervical part*: ascending palatine artery, tonsillar branch; glandular branches; submental artery; *facial part*: inferior and superior labial branches	Soft palate, palatine tonsil, submandibular gland and labial glands, skin and muscles of the chin, of the lips and external nose, ala and dorsum of the nose
greater palatine	Descending palatine		Hard palate, gingivae, palatine glands
hyaloid (generally disappears in the ninth month of intrauterine life)	Central retinal		Vitreous body, lens
inferior alveolar (inferior dental or mandibular)	Maxillary	Dental, peridental, mental, mylohyoid	Mandible, mandibular dental arcade, gingiva, mucous membrane of the mouth, of the lower lip and cheek, mylohyoid
inferior hypophysial	Internal carotid (cavernous part)		Neurohypophysis
inferior labial	Facial		Labial glands, mucosa and muscles of the lower lip

continued

Summary table of the arteries of the head			(*continued*)
ARTERY	***ORIGIN***	***BRANCHES***	***TERRITORIES OF SUPPLY***
inferior tympanic	Ascending pharyngeal		Medial wall of the tympanic cavity
infraorbital	Maxillary	Anterior superior alveolar arteries; dental branches, peridental branches	Orbit, maxilla, maxillary sinus, maxillary dental arcade, inferior eyelid, extraocular muscles, cheeks, ala of nose
insular	Middle cerebral (insular part)		Insula
internal carotid	Common carotid	*Cervical part*: carotid sinus; *petrous part*: caroticotympanic arteries, artery of the pterygoid canal (vidian artery); *cavernous part*: tentorial basal, tentorial marginal, meningeal, cavernous, inferior hypophysial artery; branches to the trigeminal ganglion, branches to nerves of the internal carotid artery; *cerebral part*: ophthalmic, superior hypophysial, anterior cerebral, middle cerebral, posterior communicating, anterior choroidal and uncal arteries; clivus branches, meningeal branch; carotid syphon	Brain, pituitary gland, orbit, nasal cavity, trigeminal ganglion, meninges, middle ear
labyrinthine (internal auditory)	Anterior inferior cerebellar (or from the basilar artery)		Internal ear
lacrimal	Ophthalmic (near the optic canal)	Lateral palpebral arteries; anastomotic branch with the middle meningeal artery	Lacrimal gland, conjunctiva, superior rectus and lateral rectus (extraocular muscles), cheeks, ciliary processes, eyelids
lateral frontobasal (orbitofrontal)	Middle cerebral (superior terminal branches)		Lateral half of the inferior surface of the frontal lobe
lateral nasal (of the ala of nose)	Facial		Ala of nose
lateral occipital (P3 segment of posterior cerebral artery)	Posterior cerebral	Anterior temporal, intermediate (middle) temporal, posterior temporal	Anterior, medial, intermediate and posterior parts of the temporal lobe
lateral palpebral	Lacrimal		Eyelids, conjunctiva
lesser palatine	Descending palatine		Soft palate, palatine tonsil
lingual	External carotid	Suprahyoid branch; dorsal lingual branches; sublingual and deep lingual arteries	Muscles and mucous membrane of the tongue, sublingual gland, gingivae, lingual and palatine tonsils, epiglottis, suprahyoid muscles
long posterior ciliary (2 in number)	Ophthalmic		Iris, ciliary body
mammillary	Posterior communicating		Mammillary bodies
masseteric	Maxillary (or internal maxillary)		Masseter
maxillary (internal maxillary or deep facial)	External carotid	*Mandibular part*: deep auricular, anterior tympanic, inferior alveolar and middle meningeal arteries; *pterygoid part*: pterygomeningeal, masseteric, anterior deep temporal, posterior deep temporal and buccal arteries; *pterygopalatine part*: posterior superior alveolar artery, infraorbital artery, descending palatine artery, artery of pterygoid canal, sphenopalatine artery	Tympanic cavity, external acoustic meatus, dental arcades, upper lip, chin, dura mater of the middle cranial fossa, of the walls and the calvaria, trigeminal ganglion, temporalis, masseter, buccinator, extraocular muscles, lacrimal sac, lacrimal gland, palatine tonsil, palate, nasopharynx, pharyngotympanic tube, nasal cavity and paranasal sinuses
medial frontobasal (orbitofrontal)	Anterior cerebral (postcommunicating part)		Medial half of the inferior surface of the frontal lobe
medial occipital (P4 segment of posterior cerebral artery)	Posterior cerebral	Dorsal branch to the corpus callosum, parietal, parieto-occipital, calcarine, occipitotemporal	Dorsum of the corpus callosum, precuneus, cuneus, lingual gyrus, posterior part of the lateral surface of the occipital lobe
medial palpebral	Ophthalmic (near the trochlea of the superior oblique)	Posterior conjunctival arteries; inferior palpebral arch, superior palpebral arch	Eyelids, conjunctiva, nasolacrimal duct

continued

Summary table of the arteries of the head *(continued)*

ARTERY	ORIGIN	BRANCHES	TERRITORIES OF SUPPLY
median callosal	Anterior communicating (anteromedial central branches)		Lamina terminalis and rostrum of corpus callosum
median commissural	Anterior communicating (anteromedial central branches)		Supraoptic commissure and optic chiasm
mesencephalic	Basilar		Area of the mesencephalon located medially to the reticular part of the substantia nigra
middle cerebral	Internal carotid	*Sphenoid part (horizontal part or M1 segment)*: anterolateral central arteries; uncal, polar temporal and anterior temporal arteries; *insular part (M2 segment)*: insular arteries; *inferior terminal branches (inferior cortical or M2 segment)*: anterior temporal, intermediate temporal, posterior temporal, temporo-occipital, branch to the angular gyrus; *superior terminal branches (superior cortical or M2 segment)*: lateral frontobasal (orbitofrontal) artery, prefrontal artery, artery of precentral sulcus, artery of central sulcus, artery of postcentral sulcus, anterior parietal artery, posterior parietal artery	Lentiform nucleus, internal capsule, caudate nucleus, corpus striatum, insula, frontal lobe, parietal lobe, temporal lobe, precentral gyrus, postcentral gyrus, inferior frontal gyrus, superior temporal gyrus, middle temporal gyrus, supramarginal gyrus and angular gyrus
middle meningeal (sphenospinous or pterygomeningeal)	Maxillary (or internal maxillary)	Accessory, frontal, orbital, parietal, petrosal; superior tympanic artery; anastomotic branch with the lacrimal artery	Cranium, dura mater, tensor tympani, trigeminal ganglion, orbit
middle temporal	Superficial temporal (immediately above the zygomatic arch)		Temporalis
muscular	Ophthalmic	Anterior ciliary, anterior conjunctival and episcleral arteries	Extraocular muscles
occipital	External carotid	Mastoid, auricular, sternocleidomastoid, occipital, meningeal, descending	Lateral and posterior cervical muscles, auricle, mastoid cells, dura mater of the posterior cranial fossa, diploë, muscles and skin of the occipital region, scalp
of central sulcus	Middle cerebral (superior terminal branches)		Central sulcus
of postcentral sulcus	Middle cerebral (superior terminal branches)		Postcentral gyrus
of precentral sulcus	Middle cerebral (superior terminal branches)		Precentral gyrus
of pterygoid canal (vidian)	Maxillary Internal carotid (petrous part)	Pharyngeal	Sphenoidal sinus Mucosa of the superior part of the nasopharynx, pharyngotympanic tube and tympanic cavity
of tuber cinereum	Posterior communicating	Medial, lateral	Tuber cinereum
ophthalmic	Internal carotid	*Orbital part*: lacrimal artery; recurrent meningeal; supraorbital, posterior ethmoidal, anterior ethmoidal, supratrochlear, medial palpebral and dorsal nasal arteries; *ocular part*: central retinal, short posterior ciliary, long posterior ciliary and muscular arteries	Orbital cavity, eyeball, extraocular muscles, eyelids, dura mater of the anterior cranial fossa, vault and lateral walls of the nasal cavities
pericallosal	Anterior cerebral (postcommunicating part)	Precuneal, parieto-occipital	Corpus callosum, precuneus, medial and superolateral surfaces of the posterior parietal lobe up to the paracentral lobule
polar frontal	Anterior cerebral (postcommunicating part)		Frontal lobe
polar temporal	Middle cerebral (sphenoid part)		Temporal lobe
pontine	Basilar	Medial, lateral	Pons and adjacent areas

continued

Summary table of the arteries of the head *(continued)*

ARTERY	ORIGIN	BRANCHES	TERRITORIES OF SUPPLY
posterior auricular	External carotid	Stylomastoid, posterior tympanic, auricular and occipital arteries; parotid branch	Middle ear, mastoid cells, auricle, digastric, stylohyoid, sternocleidomastoid, skin of the occipital region, parotid gland
posterior cerebral	Terminal bifurcation of the basilar artery	*Precommunicating part (P1 segment)*: posteromedial central, short circumferential arteries; thalamoperforating and collicular arteries; *postcommunicating part (P2 segment)*: posterolateral central; thalamogeniculate artery; posterior medial choroidal, posterior lateral choroidal, peduncular; *lateral occipital (P3 segment)*: anterior temporal, intermediate temporal, posterior temporal; *medial occipital (P4 segment)*: dorsal branch to the corpus callosum, parietal, parieto-occipital, calcarine, occipitotemporal	Thalamus, walls of third ventricle, hypothalamus, globus pallidus, cerebral peduncle, mesencephalon, tectal plate, epiphysis, medial and lateral geniculate bodies, uncus, parahippocampal gyrus, medial occipitotemporal gyrus, lateral occipitotemporal gyrus, temporal and occipital lobes, choroid plexus of the lateral ventricle and choroid plexus of the third ventricle
posterior communicating	Internal carotid (cerebral part); links the internal carotid artery with the posterior cerebral artery	Posteromedial central arteries; chiasmatic; arteries of tuber cinereum, thalamotuberal artery; hypothalamic; mammillary arteries; branch to the oculomotor nerve	Base of the telencephalon between the infundibulum and optic tract, optic chiasm, hippocampus, cerebral peduncle, internal capsule, anterior third of the thalamus, third ventricle
posterior conjunctival	Medial palpebral arteries		Conjunctiva
posterior deep temporal	Maxillary	Pterygoid	Temporalis, medial pterygoid and lateral pterygoid
posterior ethmoidal	Ophthalmic		Posterior ethmoidal cells, dura mater, nasal cavity
posterior inferior cerebellar	Vertebral (intracranial part) or basilar	Posterior spinal artery (can originate from the vertebral artery); cerebellar tonsillar branch, choroidal branch to the fourth ventricle	Hemisphere of cerebellum [H II - H X], vermis of cerebellum [I-X], choroid plexus of the fourth ventricle
posterior lateral nasal	Sphenopalatine		Posterior part of the nasal conchae and lateral wall of the nose
posterior meningeal	Ascending pharyngeal		Bones and dura mater of the posterior cranial fossa
posterior parietal	Middle cerebral (superior terminal branches)		Parietal lobe
posterior spinal	Posterior inferior cerebellar or vertebral (intracranial part)		Myelencephalon, spinal cord, pia mater
posterior superior alveolar (posterior dental)	Maxillary	Dental, peridental	Premolar and molar teeth of the maxillary dental arcade, mucosa of the maxillary sinus, gingivae
posterior tympanic	Posterior auricular	Mastoid, stapedial branch	Tympanic cavity, posterior part of the tympanic membrane
posterolateral central	Posterior cerebral (postcommunicating part)		Posterolateral part of the mesencephalon
posteromedial central	Posterior communicating Posterior cerebral (precommunicating part)	Anterior and posterior branches Anterior and posterior branches	Posteromedial part of the mesencephalon Posteromedial part of the mesencephalon
prefrontal	Middle cerebral (superior terminal branches)		Frontal lobe
preoptic	Precommunicating part of the anterior cerebral artery (anteromedial central branches)		Preoptic area
proximal medial striate	Precommunicating part of the anterior cerebral artery (anteromedial central branches)		Inferior part of the frontal lobe, thalamus, corpus striatum
pterygomeningeal	Maxillary		Trigeminal ganglion, dura mater, internal table of the calvaria
short circumferential	Posterior cerebral (precommunicating part)		Posterior part of the thalamus

continued

Summary table of the arteries of the head *(continued)*

ARTERY	*ORIGIN*	*BRANCHES*	*TERRITORIES OF SUPPLY*
short posterior ciliary (from 6 to 12 in number)	Ophthalmic or one of its branches		Choroid and ciliary processes of the eyeball
sphenopalatine (nasopalatine)	Maxillary	Posterior lateral nasal arteries; posterior septal branches	Frontal sinus, maxillary sinus, sphenoidal sinus, ethmoidal cells, nasal septum
stylomastoid	Posterior auricular		Tympanic cavity, stapes, stapedius, mastoid cells, semicircular canals
sublingual	Lingual		Sublingual gland, mylohyoid, geniohyoid and genioglossus, mucous membrane of the mouth and of the gingivae
submental	Facial		Mylohyoid, digastric and platysma, sublingual gland, skin of the chin, lips
superficial temporal	External carotid	Parotid; transverse facial artery; anterior auricular; zygomatico-orbital and middle temporal arteries; frontal, parietal	Masseter and temporalis, frontal belly of the occipitofrontalis, orbicularis oculi, external acoustic meatus, auricle, skin of the face, scalp, parotid gland, temporomandibular joint
superior cerebellar	Basilar (near its termination)	Medial: superior vermian, lateral	Superior surface of cerebellum, vermis of cerebellum [I-X], pons, epiphysis, superior medullary velum, choroid membrane of the third ventricle
superior hypophysial	Internal carotid (cerebral part)		Pars tuberalis of the adenohypophysis and infundibulum of the neurohypophysis, median eminence
superior labial	Facial	Nasal septal branch	Upper lip, nasal septum, ala of nose
superior tympanic	Middle meningeal		Tympanic cavity, tensor tympani
suprachiasmatic	Anterior communicating (anteromedial central branches)		Supraoptic recess and hypothalamic area
supraoptic	Precommunicating part of the anterior cerebral artery (anteromedial central branches)		Superior part of the optic nerve and of the optic chiasm
supraorbital	Ophthalmic (where it crosses the optic nerve)	Diploic	Skin, muscles and pericranium of the forehead, superior rectus of the eyeball, levator palpebrae, diploë
supratrochlear (frontal)	Ophthalmic		Skin, muscles and pericranium of the forehead
thalamogeniculate	Posterior cerebral (postcommunicating part)		Thalamus, geniculate bodies
thalamoperforating	Posterior cerebral (precommunicating part)		Thalamic nuclei
thalamotuberal	Posterior communicating		Thalamus and adjacent structures
transverse facial	Superficial temporal (in the tract that runs in the parotid gland)		Parotid g and and parotid duct, masseter, skin of the face
uncal	Internal carotid (cerebral part) or, sometimes, middle cerebral (sphenoid part)		Uncus
vertebral	Subclavian	*Intracranial part*: meningeal; posterior inferior cerebellar and anterior spinal arteries; medial medullary, lateral medullary	Bodies of the vertebrae, deep muscles of the neck, falx cerebelli, spinal cord, cerebellum, brainstem
zygomatico-orbital	Superficial temporal (occasionally middle temporal)		Lateral part of the orbit, orbicularis oculi

Arteries of the brain
- anterior choroidal
- anterior cerebral
- middle cerebral
- posterior communicating
- cerebral arterial circle
- posterior cerebral

Cerebral arterial circle
– internal carotid
– – anterior cerebral
– – – anterior communicating
– – middle cerebral
– – posterior communicating
– basilar
– – posterior cerebral

Summary table of the arteries of the neck

ARTERY	*ORIGIN*	*BRANCHES*	*TERRITORIES OF SUPPLY*
anterior spinal	Vertebral		Anterior two-thirds of the spinal cord
ascending cervical	Thyrocervical trunk (or superior thyroid artery)	Spinal	Sternocleidomastoid, omohyoid, trapezius, longus capitis, phrenic nerve, spinal cord
ascending palatine	Facial		Muscles and mucosa of the soft palate
ascending pharyngeal (pharyngomeningeal)	External carotid	Pharyngeal branches; inferior tympanic artery, posterior meningeal artery	Pharynx, soft palate, longus capitis and longus colli, sympathetic trunk, palatine tonsil, pharyngotympanic tube, tympanic cavity, dura mater of the cerebellar fossa
brachiocephalic trunk	Arch of aorta	Thyroid ima artery (inconstant)	Head and neck, upper limb, superior part of the thorax
common carotid	*To the right*: bifurcation of the brachiocephalic trunk; *to the left*: most elevated part of the arch of aorta	External carotid and internal carotid arteries	Head and neck
costocervical trunk	Subclavian	Supreme intercostal and deep cervical arteries	First two intercostal spaces, semispinalis capitis and semispinalis cervicis, spinal cord
cricothyroid	Superior thyroid		Median cricothyroid ligament
deep cervical	Costocervical trunk		Spinal cord, semispinalis capitis and semispinalis cervicis
descending palatine	Maxillary	Greater palatine artery, lesser palatine arteries; pharyngeal branch	Soft palate, palatine tonsil
– pharyngeal branch	Descending palatine		Mucosa of the vault of the nasal cavities and pharynx, sphenoidal sinus, pharyngotympanic tube
dorsal scapular (deep branch)	Trasversa del collo (o tronco tireocervicale)		Muscles of the shoulder and the back
esophageal (esophageal branches, from 4 to 5 in number)	Thoracic aorta; inferior thyroid artery; left gastric artery		Esophagus
external carotid	Common carotid	*Anterior tract*: superior thyroid artery, lingual artery, facial artery; *posterior tract*: occipital artery, posterior auricular artery; *medial tract*: ascending pharyngeal artery; *terminal tract*: superficial temporal artery, maxillary artery	Anterior part of the face and neck, lateral part of the head, walls of the cranial cavity, dura mater, posterior part of the scalp
inferior laryngeal	Inferior thyroid		Muscles and mucosa of the larynx
inferior thyroid	Thyrocervical trunk	Inferior laryngeal artery; tracheal, esophageal, pharyngeal and glandular branches	Infrahyoid muscles, longus colli, scalenus anterior and inferior constrictor of pharynx, cervical part of the esophagus, muscles and mucosa of the larynx, inferior part of the pharynx, trachea, posteroinferior part of the thyroid gland
infrahyoid	Superior thyroid		Thyrohyoid
internal carotid	Common carotid	*Cervical part*: carotid sinus; *petrous part*: caroticotympanic arteries, artery of the pterygoid canal; *cavernous part*: inferior hypophysial artery; tentorial basal, tentorial marginal, meningeal, cavernous branches, branch to the trigeminal ganglion, branches to nerves of the internal carotid artery; *cerebral part*: ophthalmic, superior hypophysial, posterior communicating, anterior choroidal and uncal arteries; clivus branches, meningeal branch; anterior cerebral and middle cerebral arteries	Middle ear, orbital cavity, nasal cavity, brain, pituitary gland, trigeminal ganglion, meninges

continued

Summary table of the arteries of the neck *(continued)*

ARTERY	ORIGIN	BRANCHES	TERRITORIES OF SUPPLY
internal thoracic (internal mammary)	First part of the subclavian artery	Musculophrenic, superior epigastric and pericardiacophrenic arteries; mediastinal, thymic, bronchial, tracheal, sternal, perforating, lateral costal, anterior intercostal	Thoracic wall (the internal thoracic artery is listed in this table as it begins in the neck, but it belongs to the thorax in terms of the branches it gives rise to and its area of supply)
mylohyoid	Inferior alveolar		Mylohyoid
occipital	External carotid	Mastoid, sternocleidomastoid, auricular, meningeal, descending, occipital	Dura mater, diploë, mastoid cells, muscles of neck and scalp
posterior spinal	Posterior inferior cerebellar (or vertebral)		Myelencephalon, spinal cord, pia mater
sternocleidomastoid	Superior thyroid		Sternocleidomastoid
subclavian	*To the right*: brachiocephalic trunk; *to the left*: arch of aorta	Vertebral artery; thyrocervical trunk; internal thoracic artery; costocervical trunk; becomes the axillary artery in correspondence to the external margin of the first rib	Same territory of supply as the vertebral artery, anterior and superior thoracic wall, upper limb
submental	Facial		Skin and muscles of the chin and the lower lip
superficial cervical	Transverse cervical (or thyrocervical trunk)	Ascending, descending	Trapezius
superior laryngeal	Superior thyroid (or external carotid)		Muscles, mucosa and cartilages of the larynx
superior thyroid	External carotid	Infrahyoid arteries, sternocleidomastoid, superior laryngeal and cricothyroid arteries; anterior, posterior and lateral glandular branches	Thyroid gland, infrahyoid muscles and sternocleidomastoid, muscles and mucosa of the larynx
suprahyoid	Lingual		Genioglossus and geniohyoid
suprascapular (transverse scapular)	Thyrocervical	Acromial	Clavicle, scapula, sternocleidomastoid, subclavius, supraspinatus and subscapularis, acromioclavicular and glenohumeral joints, suprascapular region, skin over the acromion
supreme intercostal (at its origin from the costocervical trunk)	Costocervical trunk	First and second posterior intercostal arteries	Thoracic wall, first and second intercostal spaces
thyrocervical trunk	First part of the subclavian artery	Inferior thyroid, ascending cervical, suprascapular, transverse cervical and dorsal scapular arteries	Same areas of supply as the inferior thyroid, ascending cervical, suprascapular and transverse cervical arteries
thyroid ima (inconstant)	Arch of aorta or brachiocephalic trunk		Thyroid gland, thymus
tonsillar	Facial or ascending palatine		Pharyngeal tonsil and posterior part of the tongue
transverse cervical	Thyrocervical trunk (or from the third part of the subclavian artery)	Superficial (superficial cervical artery), deep (dorsal scapular artery)	Posterior cervical, shoulder and back muscles
vertebral (prevertebral, cervical and atlantic part)	Subclavian	Spinal (radicular and segmental medullary artery) and muscular branches, meningeal, medial and lateral medullary branches; posterior inferior cerebellar artery (cerebellar tonsillar branch and choroidal branch to the fourth ventricle; posterior spinal artery); anterior spinal artery	Semispinalis capitis, suboccipital muscles, prevertebral muscles and muscles of back proper, vertebral canal, meninges and spinal cord of the cervical tract, myelencephalon, cerebellum, dura mater of the cerebellar fossa, pons, mesencephalon, thalamus, hypothalamus, third ventricle, temporal and occipital lobes, internal ear
– muscular branches	Vertebral		Prevertebral muscles and muscles of back proper
– spinal branches (spinal or vertebromedullary arteries)	Vertebral	Radicular, segmental medullary	Vertebral canal, meninges and spinal cord of the cervical tract

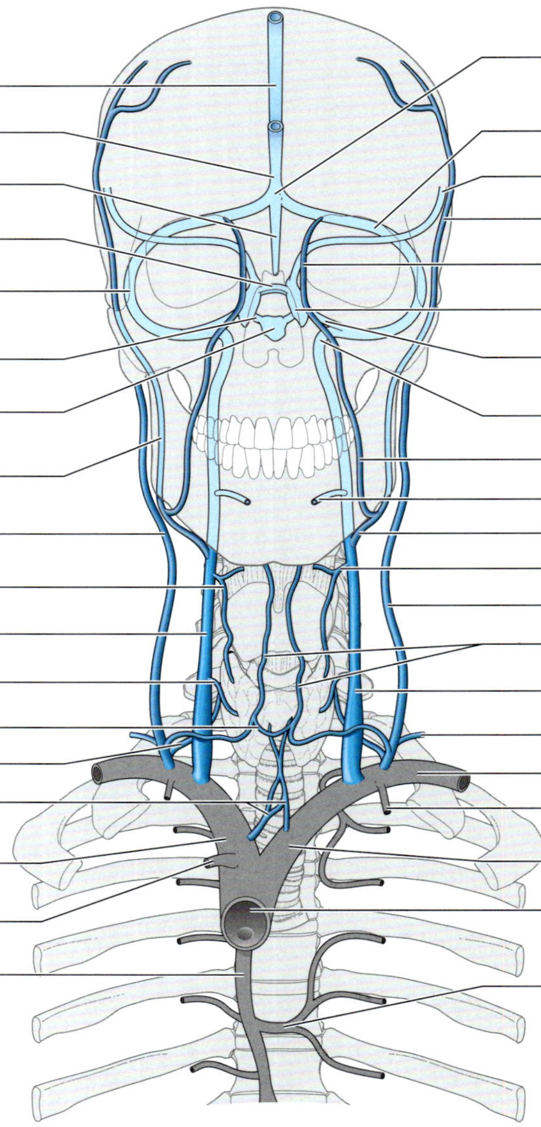

Schematic representation of the **dural venous sinuses** and **veins of the neck** in anterior projection. The right and left transverse sinuses continue with the sigmoid sinuses and then with the internal jugular veins, through jugular foramina of the external surface of the cranial base. Tributaries of the internal jugular vein in the cervical tract are: the common facial vein, which originates at the angle of the mandible from the confluence of the (anterior) facial vein and anterior division of the retromandibular vein; the lingual and pharyngeal veins, and the superior and middle thyroid veins. The external jugular vein originates at the level of the angle of the mandible from the posterior auricular vein and the posterior division of the retromandibular vein. The anterior jugular veins, which originate at the level of the hyoid bone from the confluence of the veins that drain the submandibular region, anastomose under the thyroid gland through the jugular venous arch, then continue in a lateral direction opening into the external jugular veins.

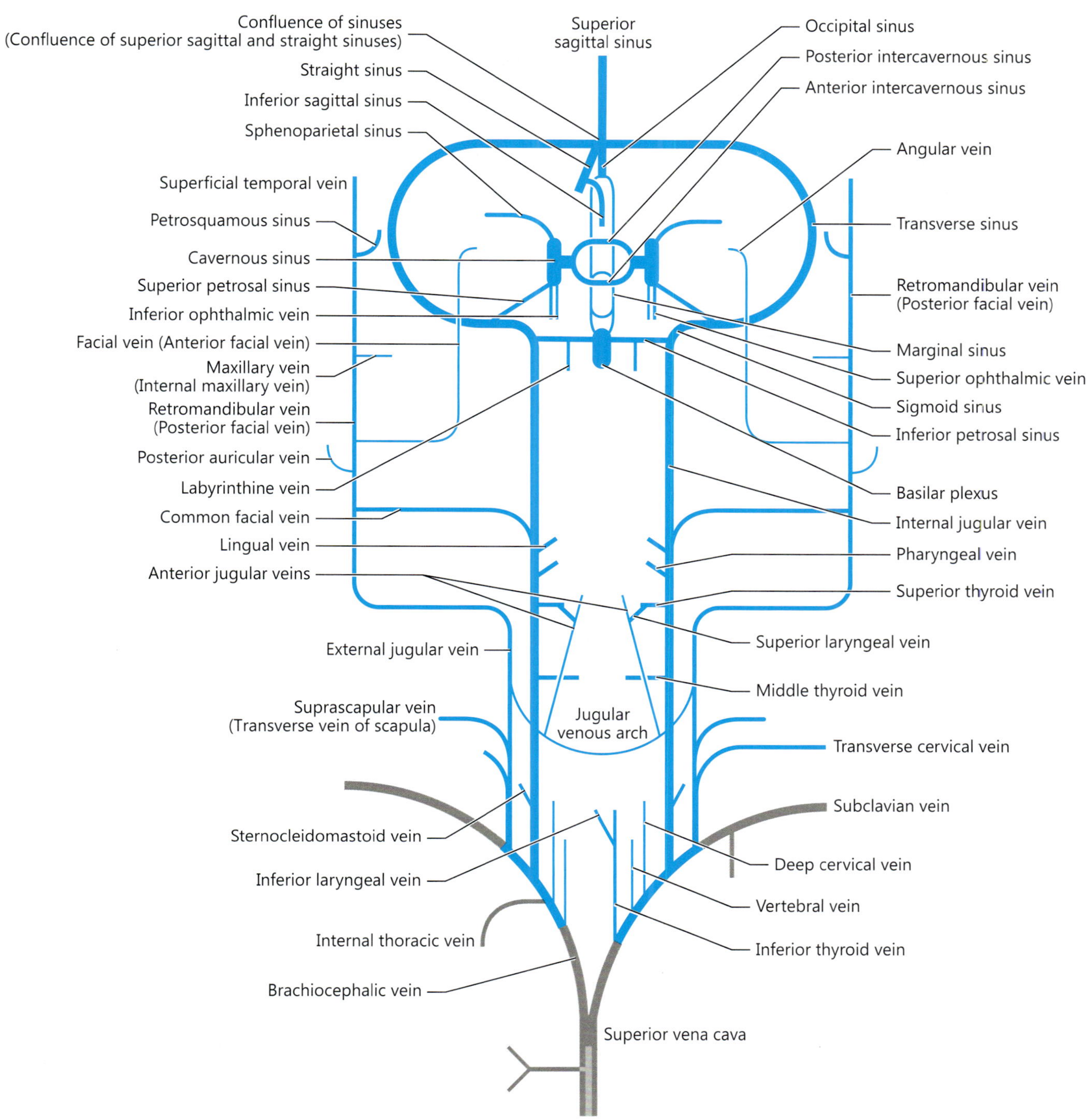

Schematic representation of the **external** and **internal jugular vein tributaries** in the neck. The suprascapular and anterior jugular veins are tributaries of the external jugular vein. Branches that enter the cervical tract of the internal jugular vein are: the common facial, lingual, sternocleidomastoid, pharyngeal, superior thyroid and middle thyroid veins. From the confluence of the internal jugular and subclavian veins originate the brachiocephalic veins. The inferior thyroid veins and the vertebral veins terminate in the brachiocephalic veins (or innominate veins). The external jugular vein terminates in the subclavian vein, which receives also the transverse cervical vein. The latter can, however, unite with the suprascapular vein in a common trunk terminating in correspondence to the terminal tract of the external jugular vein. The jugular venous arch is an anastomotic trunk that unites the two anterior jugular veins.

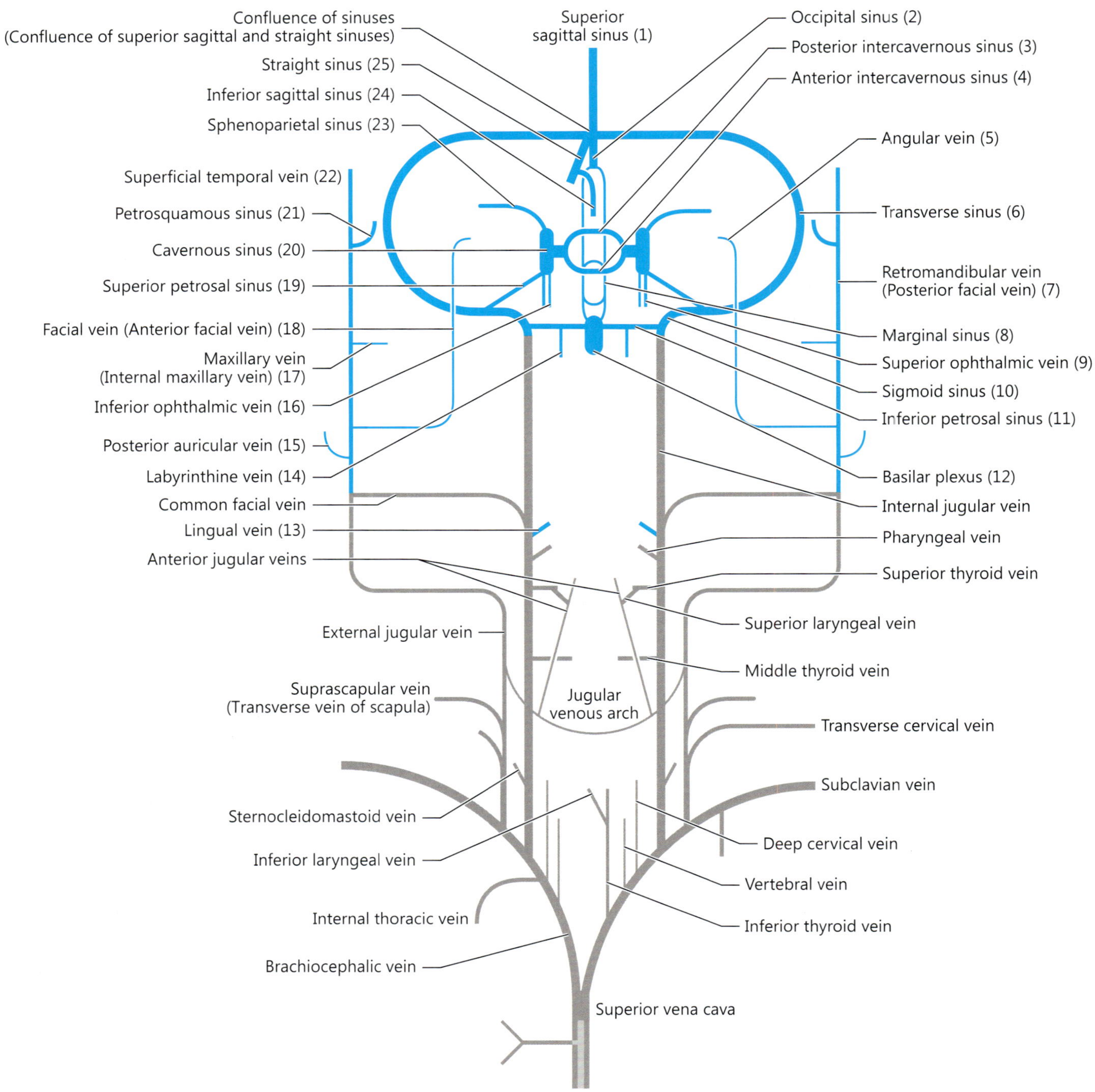

Schematic representation of the **roots of the internal jugular vein**. The roots of the internal jugular vein are composed of the **dural venous sinuses** and the veins that enter it. The internal jugular vein originates at the level of the posterolateral part of the jugular foramen as a continuation of the sigmoid sinus in which the transverse sinus continues, into which drain the diploic, meningeal, cerebral veins and those of the visual and auditory organs. The branches that open into the cranial tract of the internal jugular vein are the inferior petrosal sinus, the vein of the cochlear aqueduct and the emissary veins of the internal carotid venous plexus. The sequence of graphic representations follows the course of blood flow.

➤

Three-dimensional image of a living human body, obtained by contrast-enhanced CT scan using a 3D volume rendering technique, showing the **dural venous sinuses** viewed from above after almost complete removal of the superior sagittal sinus.

Venous sinuses and main veins of the head: territories of drainage		
	VENOUS SINUSES AND VEINS OF THE HEAD	***TERRITORIES OF DRAINAGE***
1	Superior sagittal sinus	Medial and superolateral surfaces of the cerebral hemispheres, diploë, meninges
2	Occipital sinus	Dura mater, diploë, cerebellum
3	Posterior intercavernous sinus	Pituitary gland
4	Anterior intercavernous sinus	Pituitary gland
5	Angular vein	Frontal region
6	Transverse sinus	Inferior surfaces, medial surfaces and inferior halves of the lateral surfaces of the cerebral hemispheres, cerebellum, labyrinth
7	Retromandibular vein (Posterior facial vein)	Parotid gland, temporomandibular joint, masseter, temporalis, medial pterygoid, lateral pterygoid and frontal belly of occipitofrontalis, skin of the face, of the auricle, of the external acoustic meatus and of the frontal and occipital regions
8	Marginal sinus	Posterior cranial fossa
9	Superior ophthalmic vein	Eyeball, extraocular muscles and eyelids
10	Sigmoid sinus	Inferior surfaces, medial surfaces and inferior halves of the lateral surfaces of the cerebral hemispheres, cerebellum, labyrinth
11	Inferior petrosal sinus	Meninges, cerebellum, pons, myelencephalon, labyrinth
12	Basilar plexus	Occipital bone, pons and myelencephalon
13	Lingual vein	Tongue
14	Labyrinthine vein	Internal ear
15	Posterior auricular vein	Posterior part of the ear
16	Inferior ophthalmic vein	Inferior rectus and inferior oblique (extraocular muscles), lacrimal sac, eyelids
17	Maxillary vein (Internal maxillary vein)	Pterygoid plexus (structures of the infratemporal fossa)
18	Facial vein (Anterior facial vein)	Superficial structures of the face
19	Superior petrosal sinus	Inferior surfaces, parts of the medial surfaces inferior and posterior to the inferior frontal sulcus, and inferior halves of the lateral surfaces of the cerebral hemispheres, cerebellum, pons, tympanic cavity, vestibular labyrinth
20	Cavernous sinus	Eyeball and organs contained within the orbital cavity, cerebral hemispheres
21	Petrosquamous sinus	Meninges
22	Superficial temporal vein	Forehead and scalp
23	Sphenoparietal sinus	Diploë, meninges, inferior, medial surfaces and inferior half of the lateral surfaces of the cerebral hemispheres, corpus callosum
24	Inferior sagittal sinus	Falx cerebri (or cerebral falx), corpus callosum
25	Straight sinus	Basal nuclei, walls of the lateral ventricles, central part of the cerebral hemispheres, tentorium cerebelli

N.B.: for veins of the brain and a general overview of the veins of the head and their territories of drainage, see the table summarizing the veins of the head.

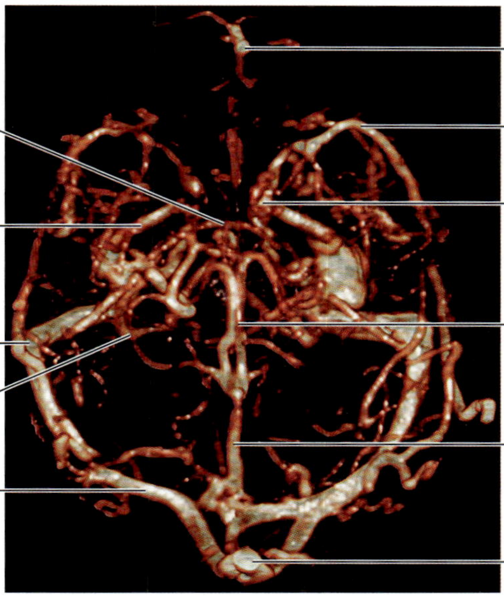

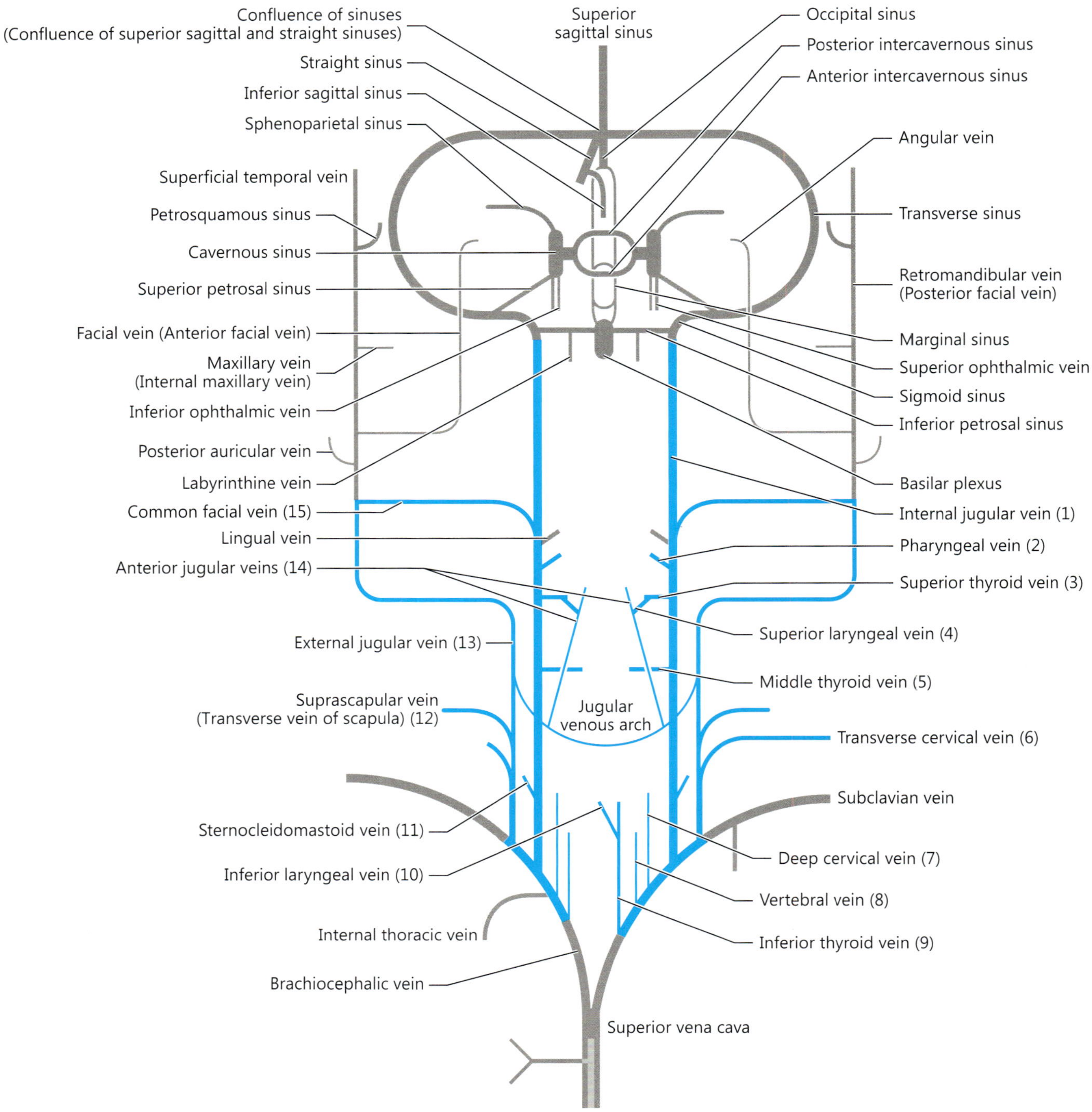

Schematic representation of the **internal jugular** and **external jugular venous trees.** The internal jugular vein originates as a continuation of the sigmoid sinus, which is a continuation at full caliber of the transverse sinus, into which drain the diploic, meningeal, cerebral veins and those of the visual and auditory organs. Tributary branches of the internal jugular vein are the inferior petrosal sinus, the vein of the cochlear aqueduct, and the emissary veins of the internal carotid venous plexus in the cranial tract, and the common facial, lingual, pharyngeal, superior thyroid and middle thyroid veins in the cervical tract. The external jugular vein originates from the confluence of the retromandibular (posterior facial) vein and posterior auricular vein, and has as its main tributaries the suprascapular, transverse cervical and anterior jugular veins. The external jugular vein flows into the subclavian vein within a short distance of the confluence of the internal jugular vein with the subclavian vein.

Main veins of the neck: territories of drainage		
	VEINS OF THE NECK	TERRITORIES OF DRAINAGE
1	Internal jugular vein	Cranium, cranial dura mater, brain, organs of sight and hearing, structures of the face and of the deep regions of the neck
2	Pharyngeal vein	Pharyngeal plexus
3	Superior thyroid vein	Superior part of the thyroid gland, larynx
4	Superior laryngeal vein	Glands, muscles and mucosa of the larynx
5	Middle thyroid vein	Inferior part of the thyroid gland
6	Transverse cervical vein	Trapezius and posterior cervical muscles
7	Deep cervical vein	Deep posterior cervical muscles
8	Vertebral vein	Internal and external vertebral venous plexuses
9	Inferior thyroid vein	Unpaired thyroid plexus
10	Inferior laryngeal vein	Muscles and mucosa of the larynx
11	Sternocleidomastoid vein	Sternocleidomastoid
12	Suprascapular vein (Transverse vein of scapula)	Clavicle, scapula, sternocleidomastoid, subclavius, supraspinatus and subscapularis, acromioclavicular and glenohumeral joints, suprascapular region, skin over the acromion
13	External jugular vein	Superficial territories of the posterior region of the neck, integuments and superficial muscles of the anterior region of the neck, integuments and superficial muscles of the occipital region and the nuchal region
14	Anterior jugular vein	Integuments and superficial muscles of the anterior region of the neck
15	Common facial vein	Territories of drainage of the facial and retromandibular veins

N.B.: for a general overview of the veins of the neck and their territories of drainage, see the table summarizing the veins of the neck.

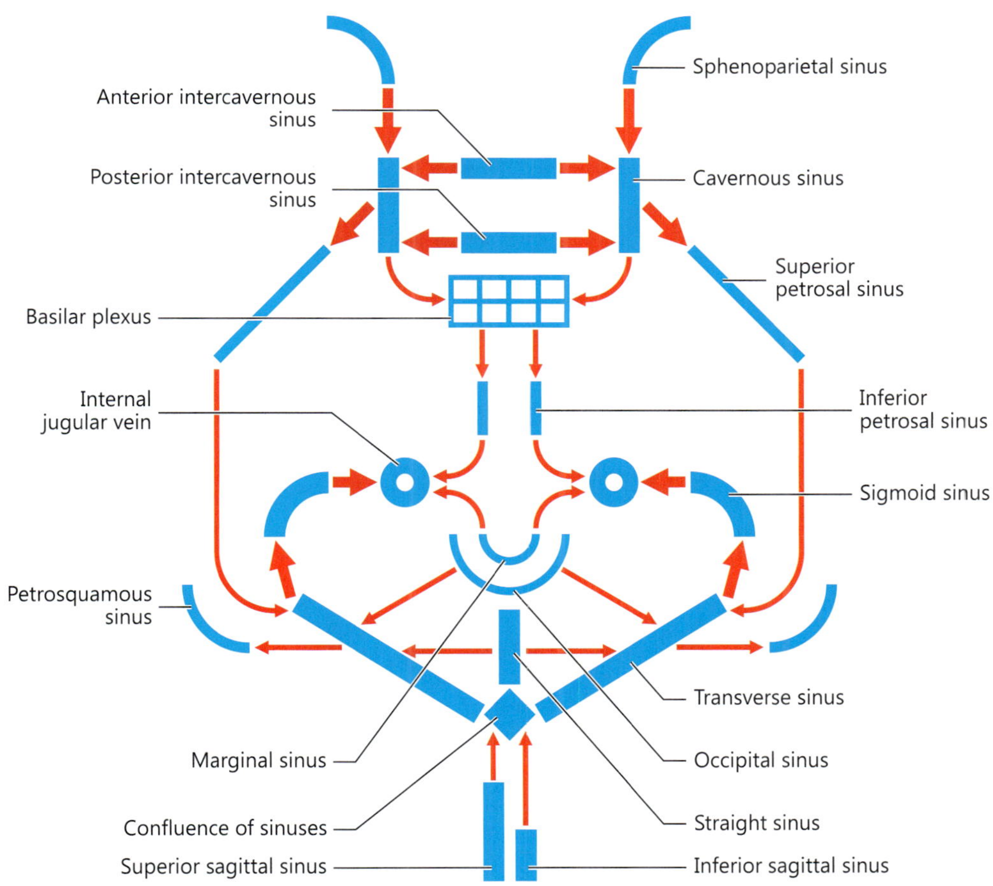

Schematic representation showing the origins and terminations of the **dural venous sinuses**. Arrows indicate the direction of drainage. The sphenoparietal sinus originates from the superior sagittal sinus, though actually today it is considered to be formed by the sinus of the lesser wing of the sphenoid and the parietal part of the anterior branch of the middle meningeal vein. The superior sagittal sinus originates from the confluence of veins in the foramen cecum. The inferior sagittal sinus begins in a blind end. The petrosquamous sinus has two drainage canals, into the retromandibular vein and into the pterygoid plexus, and it generally disappears in the last three months of intrauterine life.

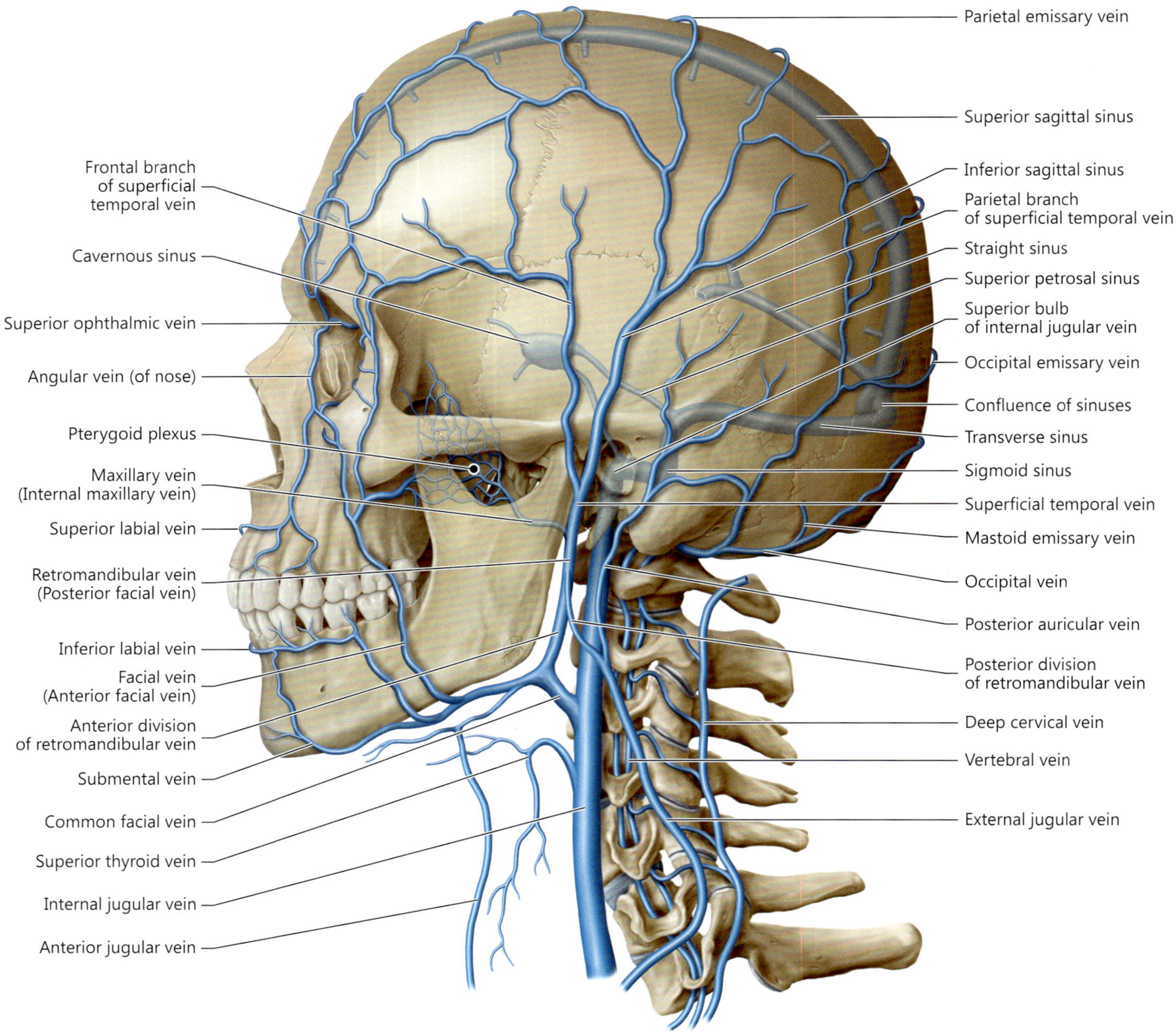

Superficial and **deep circulation** of the veins of the head and neck, in left lateral view. Looking through into the interior of the cranium, some dural venous sinuses are visible.

Superficial and Deep Venous Drainage

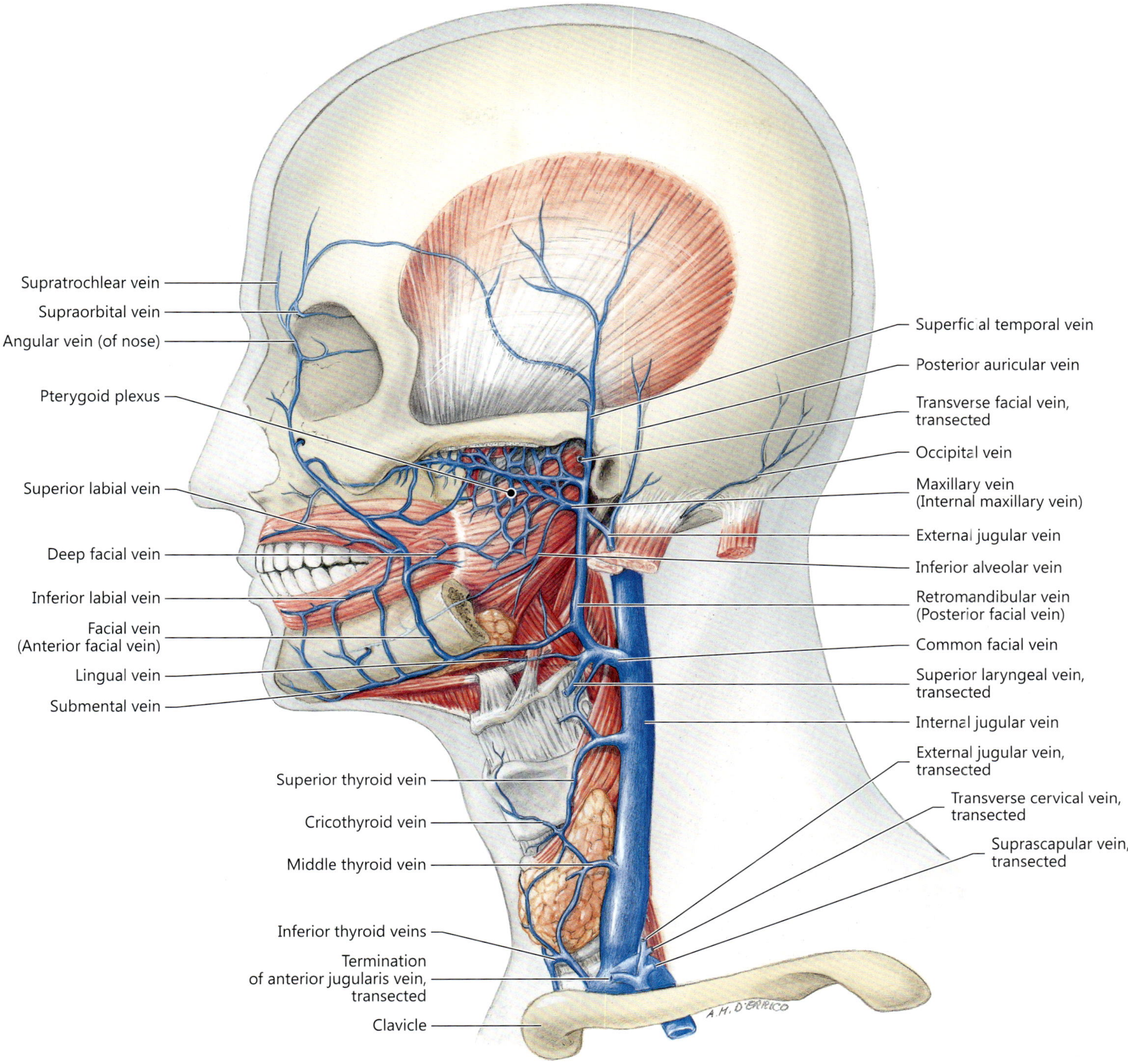

Illustration of the **internal jugular vein**, left side, in the cervical region, and its venous tributaries. The pterygoid venous plexus, drained by the internal maxillary vein, can be seen.

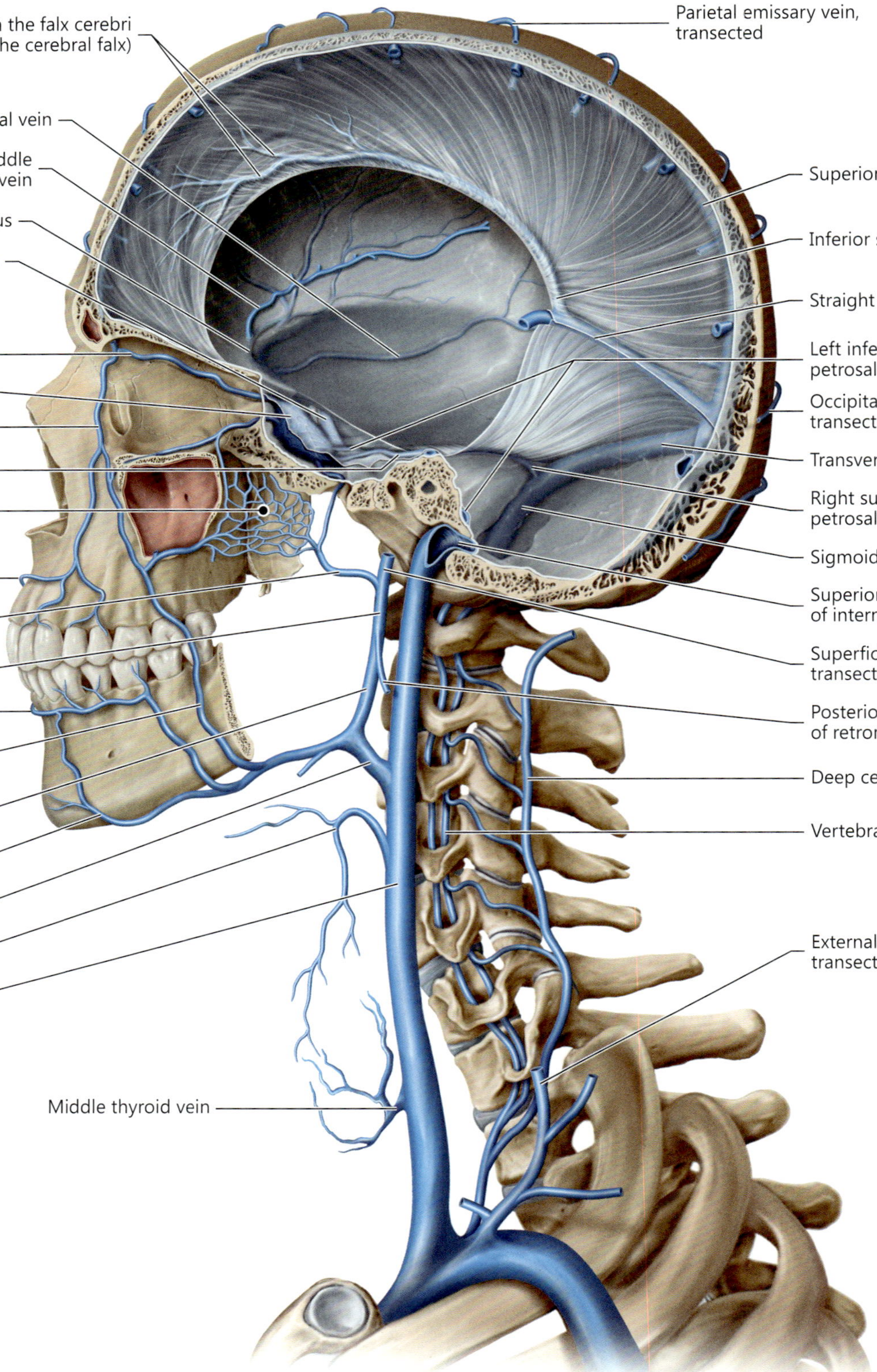

Internal jugular vein showing in detail its facial and head tributaries. Inside the cranium are the principal **dural venous sinuses** from whose confluence it originates.

Three-dimensional image of a living human body, obtained by CT scan using a 3D volume rendering technique, showing the **internal jugular vein** and **dural venous sinuses** in relation to the skeleton and contiguous vascular structures, in lateral view.

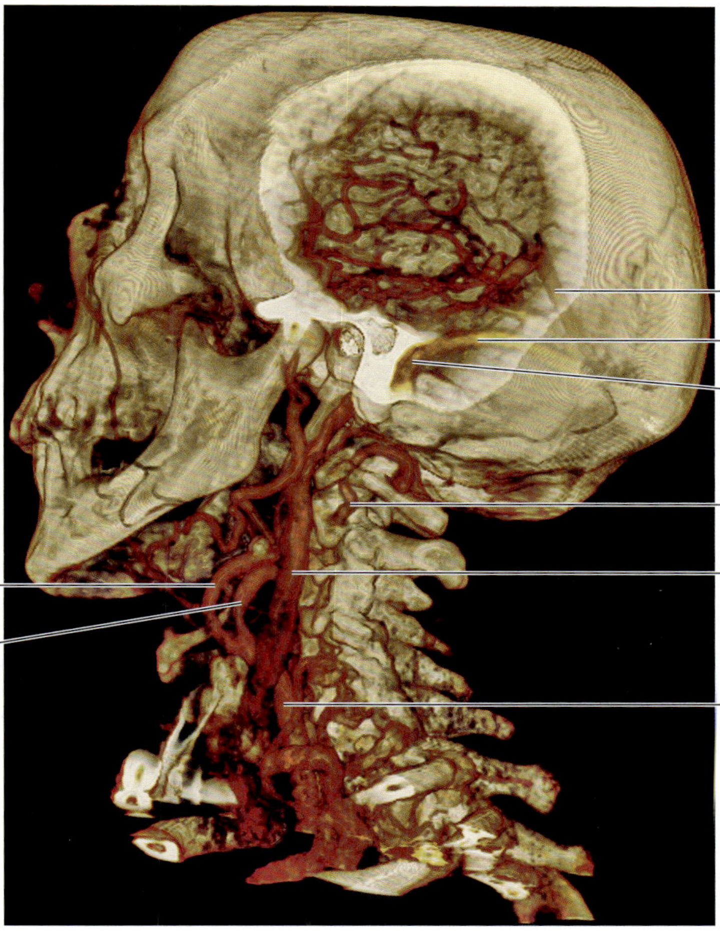

Three-dimensional image of a living human body, obtained by CT scan using a 3D volume rendering technique, showing the **internal jugular vein** and **carotid arteries** in relation to the skeleton, in posterolateral view.

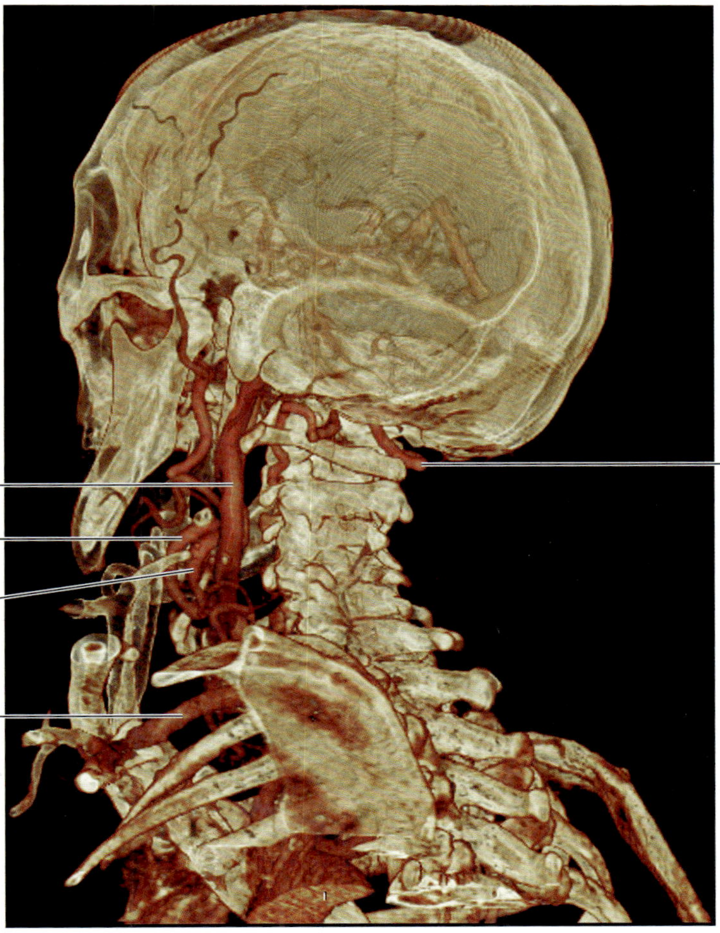

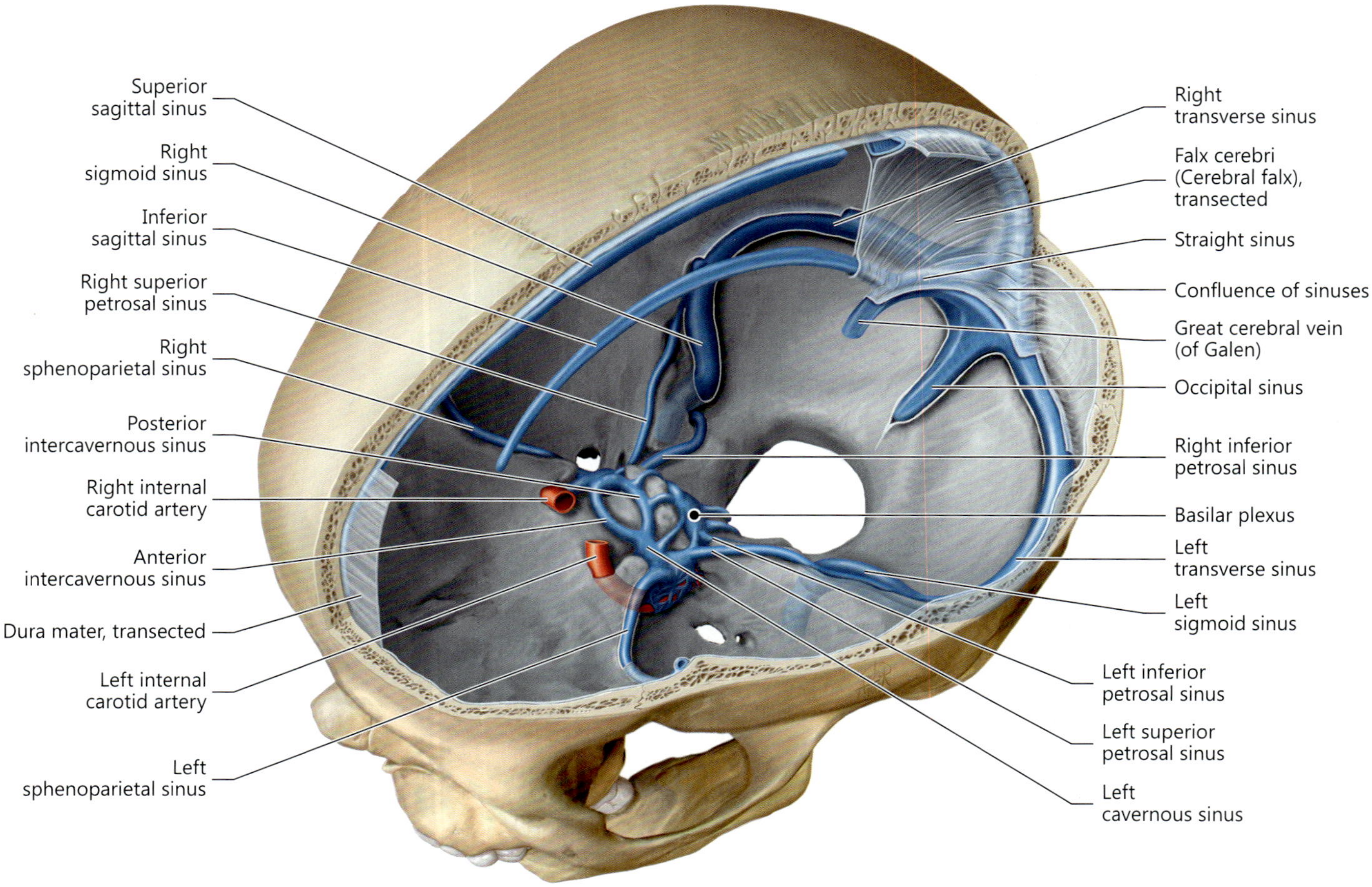

Schematic representation showing the structure and course of the **dural venous sinuses** also in relation to the bones forming the cranium.

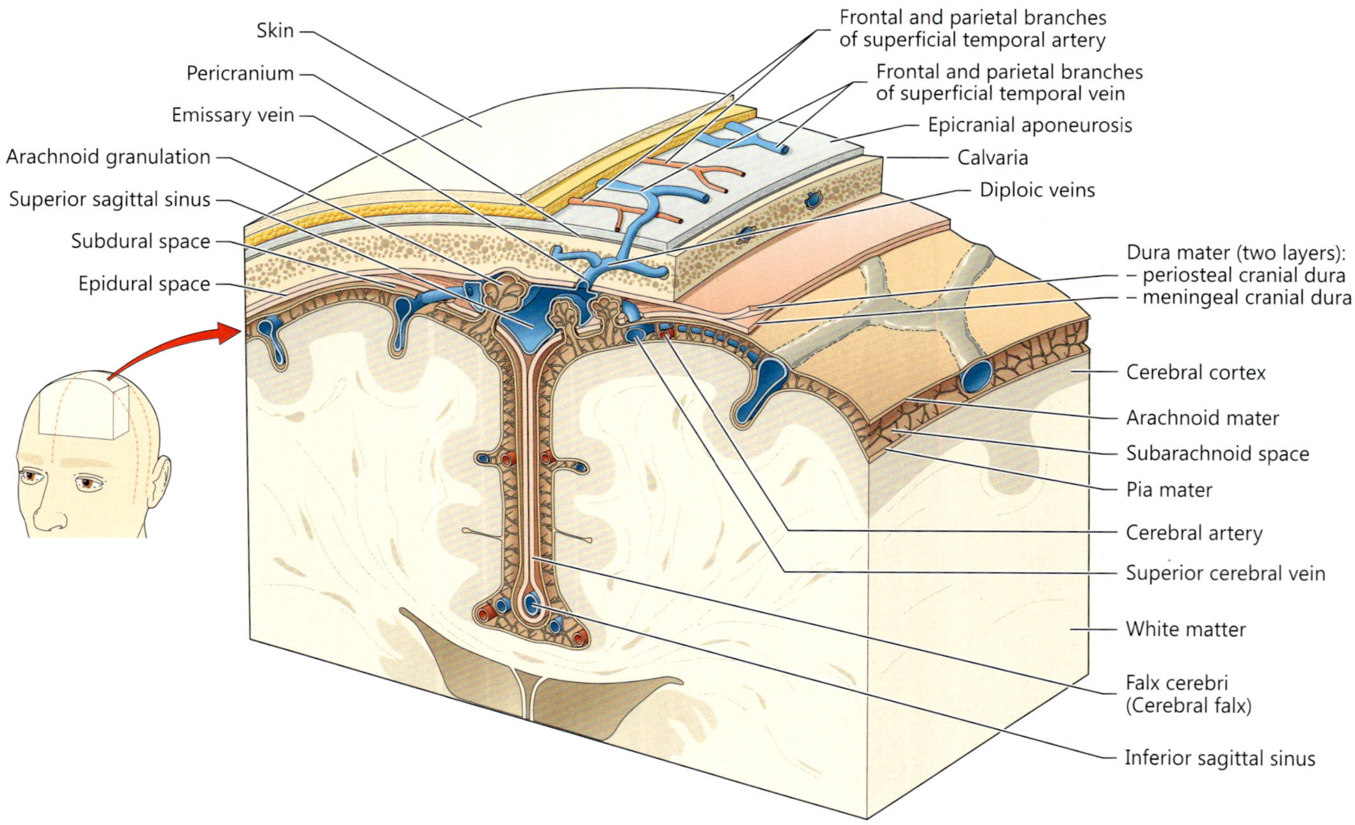

Position of the **superior sagittal sinus** in a frontal section of the cranium. Note the triangular form of the sinus, which receives, on the left, a cerebral vein and, on the right, diploic veins, in the venous lake that it forms.

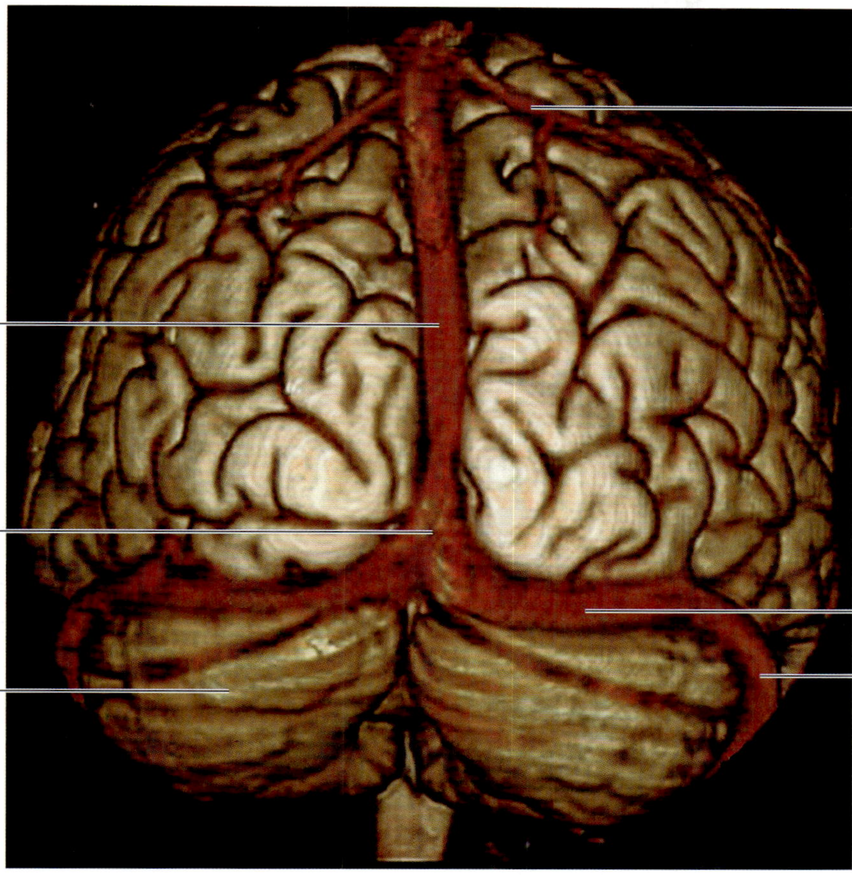

Three-dimensional image of a living human body, posterior view, obtained by CT scan using a 3D volume rendering technique, showing some **vessels** and **venous sinuses** inside the cranium, in relation to the brain hemispheres and cerebellum.

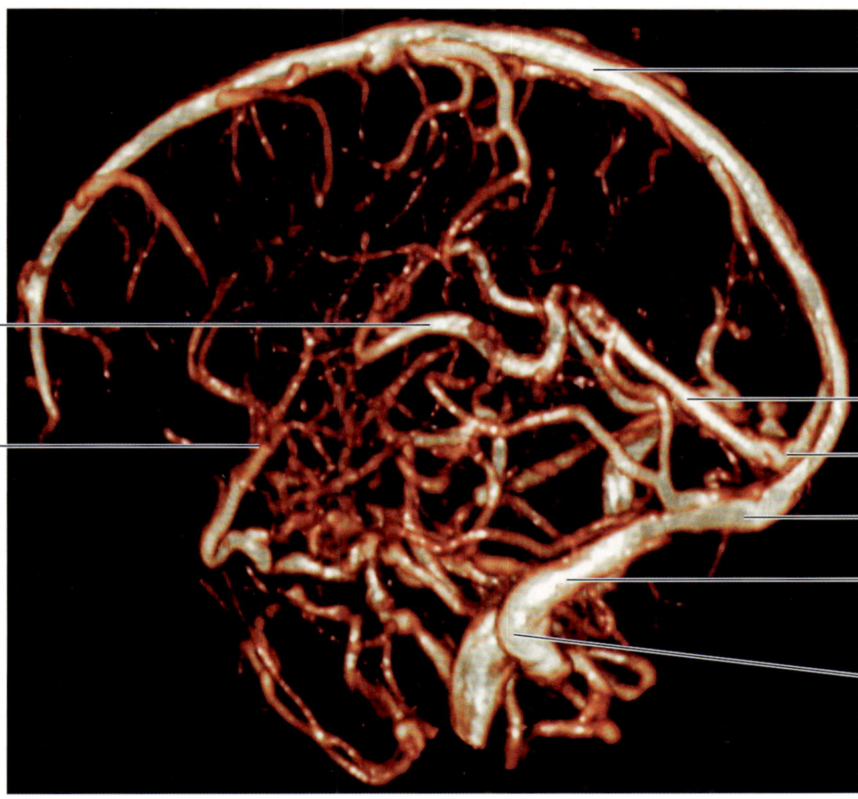

Three-dimensional image of a living human body, sagittal view, obtained by CT scan using a 3D volume rendering technique, showing the **dural venous sinuses** in left lateral view.

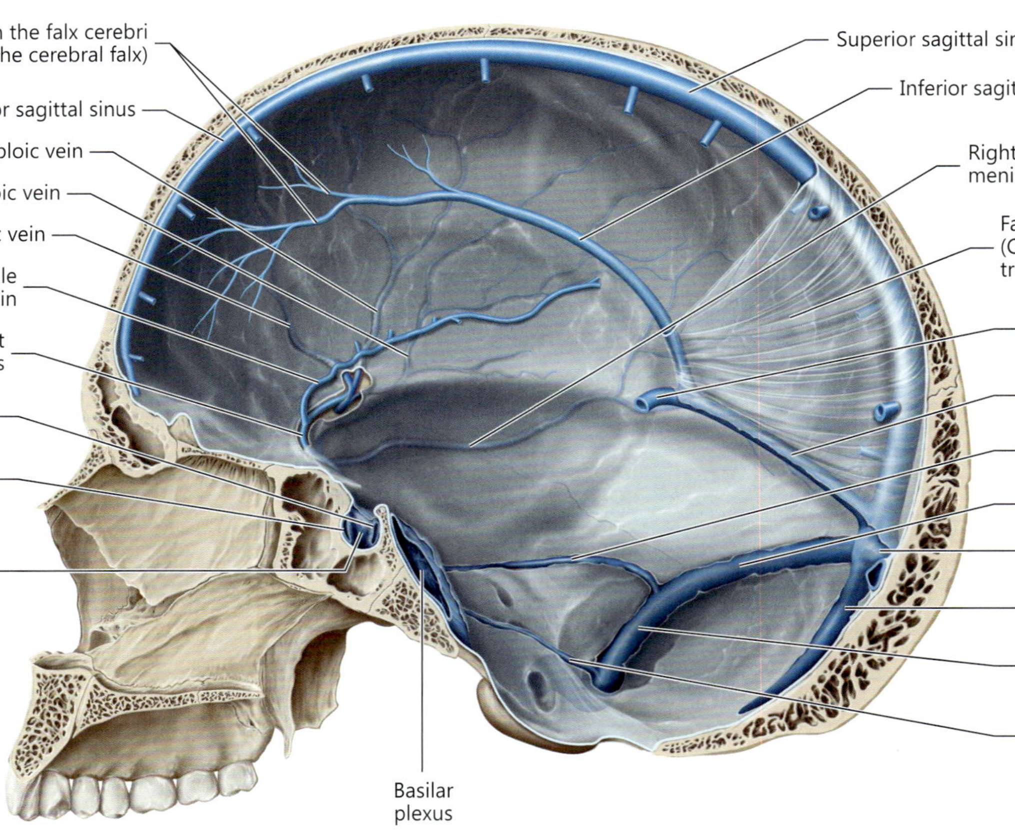

Schematic representation of the **dural venous sinuses** in a sagittal section of the cranium immediately to the right of the median plane; the cranial cavity has been emptied leaving visible part (the left surface) of the falx cerebri (cerebral falx).

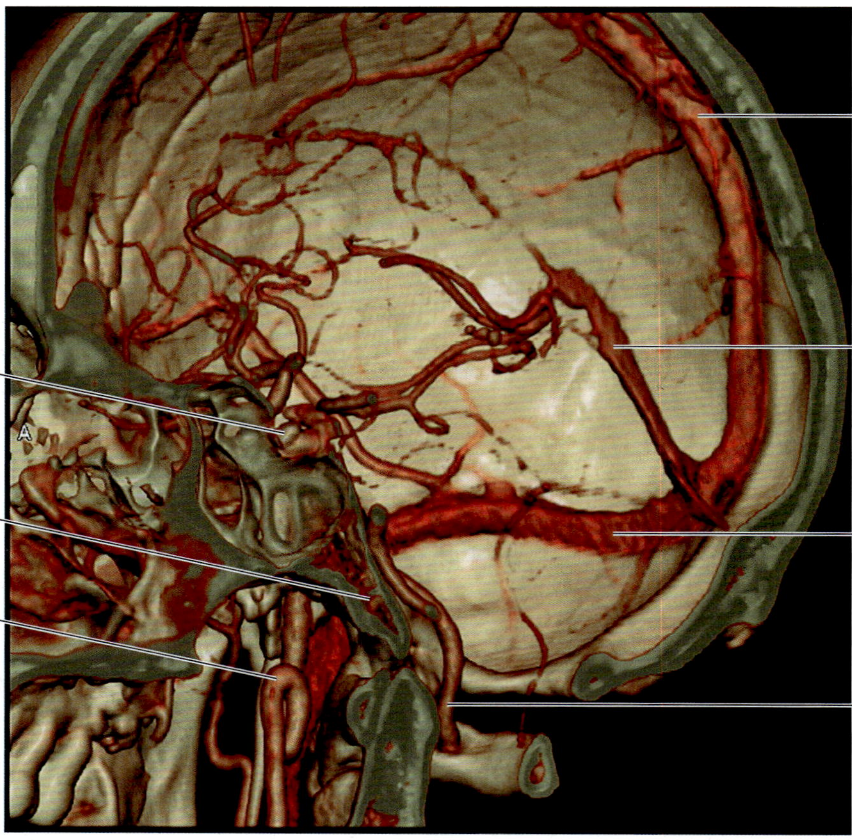

Three-dimensional image of a living human body, obtained by CT scan using a 3D volume rendering technique, showing in left posterolateral view two of the three **arterial sources of the neurocranium** and some **venous sinuses**.

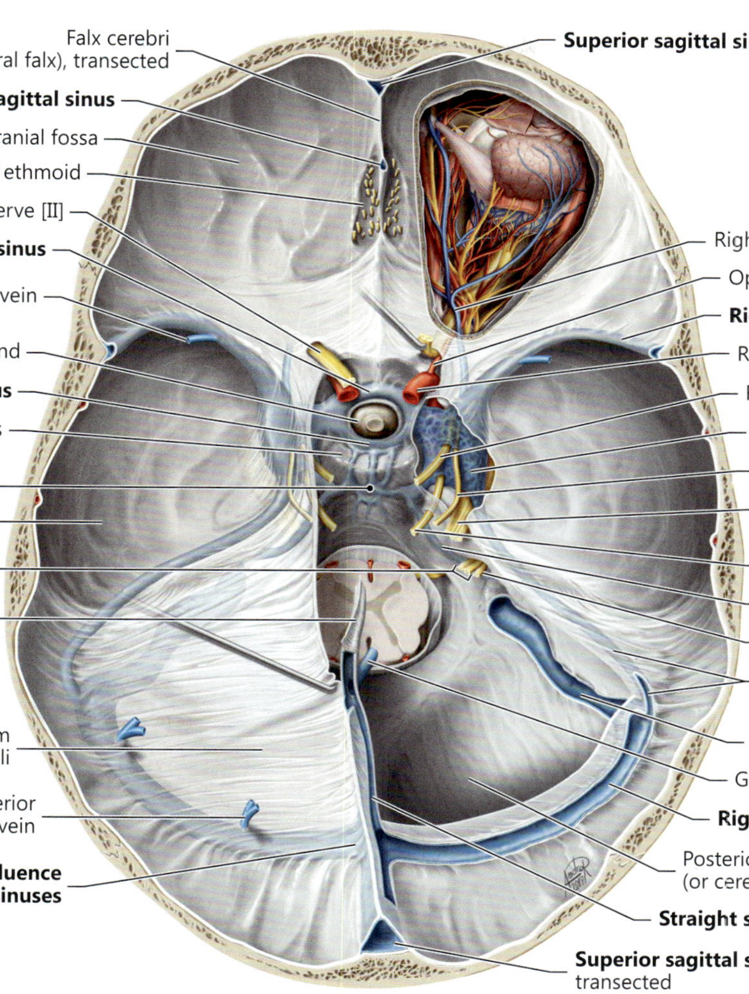

View, looking down from above, of some **dural venous sinuses** on the cranial base. With the aid of resection and demolition, one can see directly the left cavernous sinus and left superior ophthalmic vein.

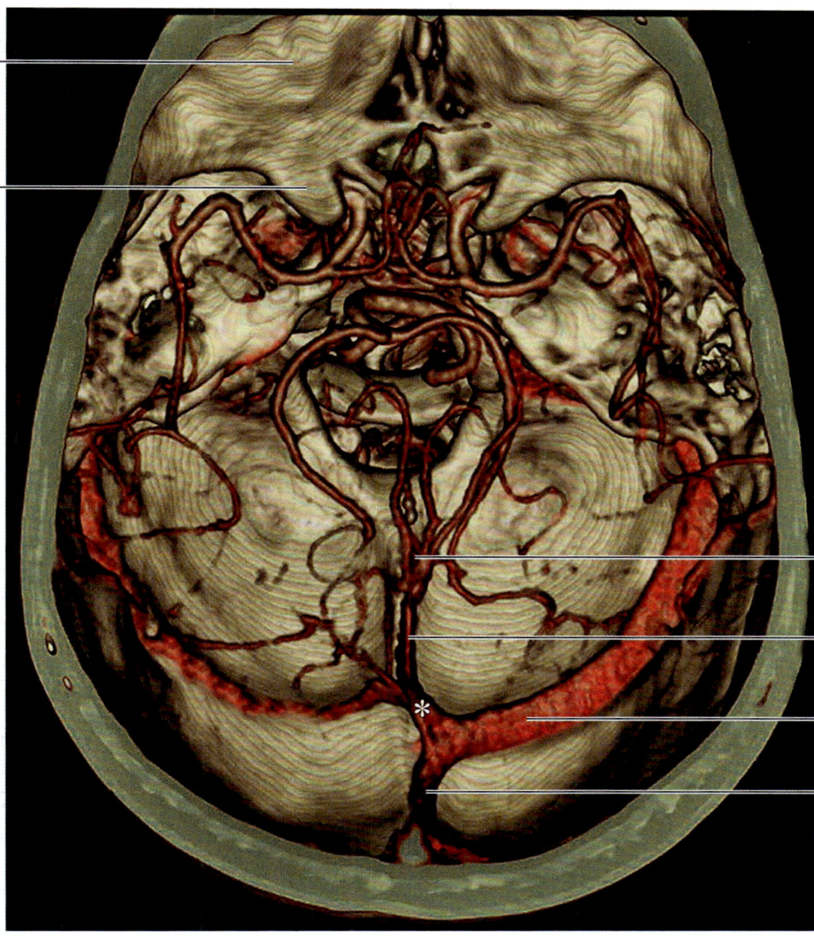

Three-dimensional image of a living human body, obtained by CT scan using a 3D volume rendering technique, showing some **vessels** and **venous sinuses** located principally in the posterior cranial fossa. In particular, an anatomic variant can be observed; in contrast to what is seen in the image above, the superior sagittal sinus drains first and predominantly into the right transverse sinus and only secondarily, through a communicating tract (*), into the left transverse sinus. It is important to remember that this variant does not correspond to any pathology.

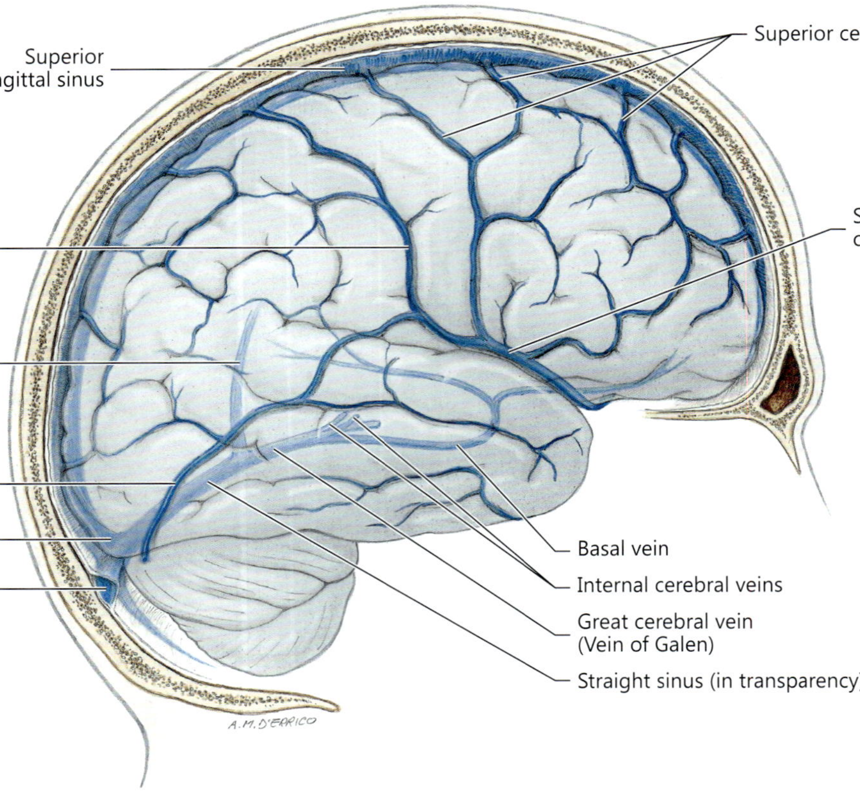

Illustration showing the **cerebral veins** after resection and removal of the right half of the cranium. One can see, on the surface of the right hemisphere of the telencephalon, the superior cerebral veins and, in the lateral sulcus, the middle cerebral vein. This vein, in its continuation as the superior anastomotic vein, connects the superior sagittal sinus, the sphenoparietal sinus and the cavernous sinus. In transparency, the internal cerebral veins.

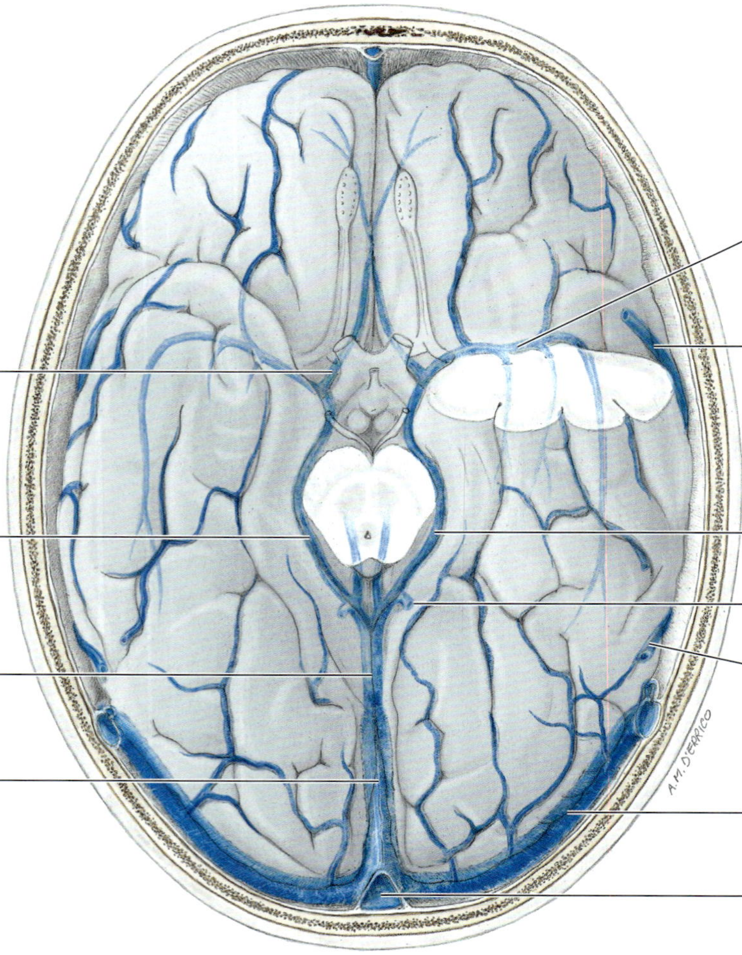

Cerebral veins on the basal surface of the brain. Resection and removal of part of the left temporal lobe permits one to see directly the insular vein. Note the great cerebral vein (vein of Galen), formed by the union of the two internal cerebral veins, into which drain the deep cerebral veins.

Summary table of the dural venous sinuses					
NAME	**ORIGIN**	**LOCATION OR COURSE**	**DRAINAGE AND TRIBUTARIES**	**CONTINUATION OR TERMINATION**	**CONNECTIONS**
Anterior intercavernous sinus (anterior coronary sinus)		Between the cavernous sinuses of the two sides	Pituitary gland; small sinuses from the pituitary gland		
Basilar plexus (unpaired)	Clivus of the occipital bone (between the internal and external layers of the dura mater)		Occipital veins, pontine veins, veins of the medulla oblongata		Intracranial: cavernous, inferior and, sometimes, superior petrosal sinuses Extracranial: internal vertebral venous plexuses
Cavernous sinus (paired)		Laterally to the sella turcica of the sphenoid, from the medial extremity of the superior orbital fissure at the apex of the petrous part of the temporal bone	Eyeball and organs contained in the orbital cavity, cerebral hemispheres; superior and inferior ophthalmic veins, central retinal vein, inferior cerebral veins and superficial middle cerebral vein, sphenoparietal sinus, superior anastomotic vein	Drainage into the superior and inferior petrosal sinuses, into the basilar, internal carotid and pterygoid plexuses, into the facial vein through the superior ophthalmic vein	Anterior and posterior intercavernous sinuses (between the cavernous sinuses of the two sides), into which drain small sinuses from the pituitary gland, and basilar plexus
Confluence of sinuses (torcula Herophili, sometimes missing)	Confluence of the superior sagittal and straight sinuses	Internal occipital protuberance	Superior and occipital sagittal sinuses	Gives origin to the transverse sinuses	
Inferior petrosal sinus (paired)	Posterior extremity of the cavernous sinus	Groove for the inferior petrosal sinus of the temporal bone, medial part of the petro-occipital synchondrosis up to the jugular foramen	Meninges, cerebellum, pons, myelencephalon, labyrinth; meningeal, cerebellar, pontine veins, veins of medulla oblongata, labyrinthine vein and, at times, condylar emissary veins, petro-occipital sinus and pharyngeal veins	Superior bulb of the internal jugular vein or, sometimes, in the suboccipital venous plexus through a vein in the hypoglossal canal	Intracranial: basilar plexus
Inferior sagittal sinus (unpaired)	In a blind end	Posterior half or two-thirds of the free margin of the falx cerebri (cerebral falx)	Falx cerebri (cerebral falx), corpus callosum; posterior vein of the corpus callosum and small veins from the falx cerebri	Ends in the straight sinus	
Marginal sinus (unpaired)		Surrounds the foramen magnum	Posterior cranial fossa	Sigmoid sinus or superior bulb of the internal jugular vein	Intracranial: basilar plexus and occipital sinus Extracranial: internal vertebral venous plexuses and deep cervical vein
Occipital sinus (usually unpaired, sometimes paired or missing)	Near the foramen magnum, from many small canals, one of which is a joint canal with the termination of the sigmoid sinus	Adherent margin of the falx cerebelli (cerebellar falx), bifurcation into two branches that follow the margins of the foramen magnum	Dura mater, diploë, cerebellum; veins from the tentorium cerebelli, diploic veins and cerebellar veins	Ends in the confluence of sinuses or in one of the transverse sinuses or in the straight sinus	Intracranial: superior sagittal and sigmoid sinuses Extracranial: internal vertebral venous plexuses
Petrosquamous sinus		In a sulcus or canal along the junction between the squamous and petrous parts of the temporal bone	Meninges	Transverse sinus or retromandibular vein	Extracranial: retromandibular vein
Posterior intercavernous sinus (posterior coronary sinus)		Between the cavernous sinuses of the two sides	Pituitary gland; small sinuses from the pituitary gland		

continued

Summary table of the dural venous sinuses *(continued)*

NAME	ORIGIN	LOCATION OR COURSE	DRAINAGE AND TRIBUTARIES	CONTINUATION OR TERMINATION	CONNECTIONS
Sigmoid sinus (paired)	Continuation of the transverse sinus	Sulcus on the mastoid process of the temporal bone, jugular process of the occipital bone, jugular foramen	Inferior and medial surfaces, and inferior halves of the lateral surfaces of the cerebral hemispheres, cerebellum, labyrinth	Internal jugular vein	Mastoid and condylar emissary veins, occipital sinus
Sphenoparietal sinus (paired)	Superior sagittal sinus	Sulcus or canal in the parietal bone, along the posterior side of the coronal suture, sulcus on the lesser wing of the sphenoid, along its posterior margin	Diploë, meninges, inferior and medial surfaces, and inferior halves of the lateral surfaces of the cerebral hemispheres, corpus callosum; diploic veins, in particular, the anterior temporal diploic vein, middle meningeal veins, anterior cerebral veins and superficial middle cerebral vein, and vein from the temporal lobe	Anterior part of the cavernous sinus	
Straight sinus (unpaired)	Continuation of the inferior sagittal sinus	Attachment of the falx cerebri to the tentorium cerebelli	Basal nuclei, walls of the lateral ventricles, central part of the cerebral hemispheres, tentorium cerebelli; great cerebral vein, superior veins of cerebellar hemisphere and veins from the tentorium cerebelli, occipital sinus	Ends in the confluence of sinuses or in the transverse sinus	
Superior petrosal sinus (paired)	Posterior part of the cavernous sinus	Groove for the superior petrosal sinus of the temporal bone, adherent margin of the tentorium cerebelli	Inferior surfaces, parts of the medial surfaces inferiorly and posteriorly to the inferior frontal sulcus, and inferior halves of the lateral surfaces of the cerebral hemispheres, cerebellum, pons, tympanic cavity, vestibular labyrinth; inferior cerebral, cerebellar, pontine, tympanic veins, vein of the vestibular aqueduct and, sometimes, superior anastomotic vein	Ends in the transverse sinus in the tract where this continues in the sigmoid sinus	Intracranial: at times, inferior petrosal sinus and basilar plexus
Superior sagittal sinus (unpaired)	Near the crista galli, immediately behind the foramen cecum of the frontal bone, through confluence of the veins of the foramen cecum	Adherent convex margin of the falx cerebri, sulcus on the internal surface of the frontal bone, adjacent margins of the parietal bones, squamous part of the occipital bone, internal occipital protuberance	Medial and superolateral surfaces of the cerebral hemispheres, diploë, meninges; superior cerebral veins	Ends in the confluence of sinuses or in the right transverse sinus	Intracranial: lateral lacunae in the dura mater, which drain the meningeal and diploic veins, left transverse and occipital sinuses Extracranial: parietal emissary vein
Transverse sinus (paired)	Conjunction of the bifurcating branches of the superior sagittal and straight sinuses or confluence of sinuses At times, right transverse sinus: direct continuation of the superior sagittal sinus; left transverse sinus: direct continuation of the straight sinus	Internal occipital protuberance, groove for the transverse sinus of the parietal bone (squamous part) and the occipital bone (squamous part) in the adherent margin of the tentorium cerebelli, groove for the transverse sinus of the occipital bone (lateral part), posterior part of the jugular foramen	Inferior and medial surfaces, and inferior halves of the lateral surfaces of the cerebral hemispheres, cerebellum, labyrinth; inferior cerebral veins, inferior veins of cerebellar hemisphere, labyrinthine, diploic and inferior anastomotic veins, superior petrosal sinus and occipital sinus	Continuation in the sigmoid sinus	Intracranial: mastoid and occipital emissary veins

Summary table of the veins of the head			
VEIN	**ORIGIN**	**DRAINAGE**	**CONFLUENCE**
angular	From the union of the supratrochlear and supraorbital veins in correspondence to the medial angle of the eye and root of nose	Frontal region; receives the superior and inferior palpebral veins and external nasal veins	Facial vein
anterior auricular	Anteriorly to the external ear	Anterior surface of the external ear	Pterygoid plexus or superficial temporal veins
anterior cerebral	Superior surface of the corpus callosum	Anterior perforated substance, lamina terminalis, rostrum of corpus callosum, septum pellucidum	Basal vein
anterior ciliary	Ciliary body	External part of the ciliary muscle; receive the scleral venous sinus and the scleral veins	Superior ophthalmic vein
anterior temporal diploic	Diploë of the lateral part of the frontal bone	Frontal bone	Sphenoparietal sinus, deep temporal vein or middle meningeal vein
basal (of Rosenthal)	Anterior perforated substance	Rostral parts of the cingulate gyrus, corpus callosum, orbital surface of the frontal lobe, insula, part of the ventral striatum; receives the anterior cerebral veins; the deep middle cerebral vein, of which the insular veins are tributaries; the inferior thalamostriate veins; the vein of the olfactory gyrus; the inferior ventricular vein; the inferior choroid vein and the peduncular veins	Internal cerebral veins
central retinal	Capillaries of the retina	Retina	Superior ophthalmic vein or cavernous sinus
cerebellar (include the superior and inferior veins of cerebellar hemisphere and precentral cerebellar vein, superior and inferior veins of vermis, petrosal vein)	Cerebellum	Cerebellum	Transverse, superior petrosal, occipital and straight sinuses, great cerebral vein
ciliary	Ciliary body	Sclera	Superior ophthalmic vein
cochlear	Osseous spiral lamina and basal lamina	Cochlea	Labyrinthine veins
common facial	Formal term for the inferior trunk of the facial vein (sometimes united with the retromandibular vein) that empties into the internal jugular vein		
common modiolar	Modiolus	Cochlea	Vein of the cochlear aqueduct
condylar emissary	Condylar canal	Sigmoid sinus	Suboccipital venous plexus
conjunctival	Eye	Bulbar conjunctiva	Episcleral veins
deep cerebral (two in number; include the basal vein and the great cerebral vein)	Corpus callosum	Lateral ventricle, third ventricle, basal nuclei, semioval center (white matter between the cortex and subcortical nuclei), thalamus	Straight sinus
deep facial	Between the buccinator and masseter	Buccinator, zygomaticus major and zygomaticus minor, masseter	Facial vein
deep lingual vein (ranine)	Tongue	Apex and deep part of the tongue	Lingual vein
deep middle cerebral (deep Sylvian)	Inferior part of the lateral sulcus	Insula and insular gyri	Basal vein
deep temporal	Deep to the anterior part of the temporalis	Temporal region	Pterygoid plexus
dorsal lingual	Tongue	Posterior parte of the tongue	Lingual vein
emissary of foramen lacerum (two or three in number)	Foramen lacerum	Cavernous sinus	Pterygoid plexus
emissary of foramen of Vesalius	Foramen of Vesalius (when present)	Cavernous sinus	Pterygoid plexus
episcleral	Sclera, near the corneoscleral junction	Medial angle of the eye, sclera, conjunctiva; receive the palpebral and conjunctival veins	Superior ophthalmic vein or anterior ciliary veins

continued

Summary table of the veins of the head *(continued)*

VEIN	ORIGIN	DRAINAGE	CONFLUENCE
ethmoidal	Ethmoidal cells	Ethmoidal cells, frontal sinus, dura mater, walls of the nasal cavity, nasal septum	Superior ophthalmic vein
external jugular	Angle of the mandible through confluence of the retromandibular and posterior auricular veins	Superficial territories of the posterior region of the neck, integuments and superficial muscles of the anterior region of the neck, integuments and superficial muscles of the occipital region and the nuchal region; receives the anterior jugular, suprascapular, transverse cervical and, sometimes, occipital veins	Internal jugular vein or subclavian vein
external nasal (several in number)	Ala of nose	External part of the nose	Angular vein or facial vein
external palatine	Palate	Soft palate and palatine tonsil	Facial vein
facial (anterior facial)	Continuation of the angular vein	Superficial structures of the face; receives the superior and inferior labial veins, deep facial, parotid, external palatine and submental veins	Common facial vein or internal jugular vein
frontal diploic	Diploë of the median part of the frontal bone	Frontal bone	Supraorbital vein and superior sagittal sinus
great cerebral (of Galen)	Around the posterior part of the corpus callosum	Thalamus, hypothalamus, basal nuclei and mesencephalon; receives the internal cerebral veins, of which the superior choroid vein is a tributary; superior thalamostriate vein, of which the anterior and posterior veins of the septum pellucidum are tributaries, medial vein of the lateral ventricle, lateral vein of the lateral ventricle, of which the veins of the caudate nucleus are tributaries; lateral direct veins and posterior vein of the corpus callosum	Anterior extremity of the straight sinus
inferior alveolar (inferior dental)	Mandibular canal	Mandible and mandibular dental arcade	Pterygoid plexus
inferior anastomotic (Labbé's vein)	Temporal lobe	Superficial middle cerebral vein	Transverse sinus
inferior cerebral	Inferior surface of the cerebral hemispheres	Inferior surface of the cerebral hemispheres; receive the vein of the uncus and the orbital and temporal veins	Superior sagittal, cavernous, superior petrosal, transverse and straight sinuses
inferior choroid	Lateral ventricle	Inferior part of the choroid plexus of the lateral ventricle	Basal vein
inferior labial	Lower lip	Labial glands, mucosa and muscles of the lower lip	Facial vein
inferior ophthalmic	Anterior part of the floor and medial wall of the orbit	Inferior rectus and inferior oblique (extraocular muscles), lacrimal sac, eyelids	Pterygoid plexus and cavernous sinus
inferior palpebral	Inferior eyelid	Inferior eyelid	Facial vein
inferior thalamostriate (inferior striate)	Corpus striatum	Perforated substance	Basal vein
inferior vein of vermis	Behind the cerebellar tonsil	Vermis and cerebral hemispheres	Transverse sinus or straight sinus
inferior veins of cerebellar hemisphere	Inferior surface of the cerebellum	Inferior surface of the cerebellum	Transverse, superior petrosal and occipital sinuses
internal cerebral (deep cerebral veins of Galen; two in number)	Interventricular foramen	Deep parts of the cerebral hemispheres; receive the superior choroid vein	Great cerebral vein
internal jugular	At the level of the posterolateral part of the jugular foramen as a continuation of the sigmoid sinus which is the continuation of the transverse sinus	Cranium, cranial dura mater, brain, organs of sight and hearing, facial and deep neck structures; in the cranial tract: receives the inferior petrosal sinus, the vein of the cochlear aqueduct and the emissary veins of the internal carotid venous plexus; in the cervical tract: receives the facial, sternocleidomastoid, lingual, pharyngeal from the pharyngeal plexus, superior thyroid, middle thyroid and superior laryngeal veins	Unites with the subclavian vein to form the brachiocephalic vein (innominate vein)

continued

Summary table of the veins of the head			(*continued*)
VEIN	***ORIGIN***	***DRAINAGE***	***CONFLUENCE***
labyrinthine (internal auditory)	Internal ear	Internal ear	Inferior petrosal sinus or transverse sinus
lacrimal	Orbit	Lacrimal gland, eyelids, conjunctiva	Superior ophthalmic vein
lingual	Tongue	Tongue; receives the dorsal lingual veins, vena comitans of hypoglossal nerve, sublingual vein and deep lingual vein	Internal jugular vein or inferior part of the facial vein
mastoid emissary (mastoid)	Mastoid foramen	Transverse sinus	Posterior auricular vein or occipital vein
maxillary	Between the head of the mandible and the sphenomandibular ligament	Pterygoid plexus	Retromandibular vein
meningeal	Plexiform vessels In the dura mater	Cranial dura mater	Efferent vessels in the external layer of the dura mater that link with the lacunae of the dura mater sinuses
middle meningeal	Lateral lacunae of the superior sagittal sinus	Dura mater, internal surface of the cranium, trigeminal ganglion	Pterygoid plexus
middle temporal	Lateral angle of the eye	Temporalis	Retromandibular vein
nasofrontal	Anterior medial part of the orbit	Frontal belly of the occipitofrontalis, medial angle of the eye and root of nose	Superior ophthalmic vein
occipital	Plexus on the posterior part of the cranium	Occipital region	Vertebral vein and deep cervical vein or external jugular vein through the posterior auricular vein or internal jugular vein (following the course of the occipital artery)
occipital diploic (largest of the four diploic veins)	Diploë of the occipital bone	Occipital bone	Occipital vein, transverse sinus or occipital emissary vein
occipital emissary	Occipital protuberance	Confluence of sinuses and occipital diploic vein	Occipital vein
of brainstem (include the pontomesencephalic, interpeduncular, intercollicular, lateral mesencephalic, and pontine veins, and veins of the medulla oblongata, of the lateral recess of the fourth ventricle, and of the cerebellomedullary cistern)	Brainstem	Brainstem	Basal and petrosal veins, inferior petrosal sinus and marginal sinus
of cochlear aqueduct (of cochlear canaliculus)	Perilymphatic space	Cochlea	Internal jugular vein
of medulla oblongata (of the myelencephalon; includes the anteromedian, anterolateral, transverse, dorsal and posteromedian medullary veins)	Myelencephalon	Myelencephalon	Anterior spinal vein and petrosal vein
of pterygoid canal (vidian)	Pterygoid canal	Sphenoidal sinus	Pterygoid plexus
of vestibular aqueduct	Vestibular aqueduct	Internal ear	Superior petrosal sinus
parietal emissary	Parietal foramen	Scalp	Superior sagittal sinus
parotid (anterior and posterior)	Parotid gland	Superficial and deep parts of the parotid gland, parotid duct	Pterygoid plexus (anterior veins), facial vein (posterior veins)
petrosal	Pons	Myelencephalon, pons and cerebellum	Superior petrosal sinus

continued

Summary table of the veins of the head *(continued)*

VEIN	ORIGIN	DRAINAGE	CONFLUENCE
pontine (include the anteromedian, anterolateral, transverse and lateral pontine veins)	Pons	Pons	Petrosal vein
posterior auricular	Plexus behind the ear	Posterior part of the ear	Forms the external jugular vein In connection with the posterior division of the retromandibular vein
posterior temporal diploic	Diploë of the parietal bone	Parietal bone	Transverse sinus, posterior auricular vein or mastoid emissary vein
precentral cerebellar	Precentral fissure	Vermis and superior part of the hemisphere of the cerebellum	Great cerebral vein
pterygoid plexus	Infratemporal fossa	Structures of the infratemporal fossa; receives the middle meningeal veins, deep temporal veins, vein of the pterygoid canal, anterior auricular veins, parotid veins, articular veins, tympanic veins, stylomastoid vein, superior and inferior alveolar veins	Retromandibular vein or maxillary veins
retromandibular (posterior facial vein)	Anteriorly to the auricle	Parotid gland, temporomandibular joint, masseter, temporalis, medial pterygoid, lateral pterygoid and frontal belly of occipitofrontalis, skin of the face, of the auricle, of the external acoustic meatus and the frontal and occipital regions; receives the superficial and middle temporal veins, transverse facial vein and maxillary veins and drains the pterygoid plexus	Common facial vein through the anterior division that unites with the facial vein; external jugular vein through the posterior division that unites with the posterior auricular vein
stylomastoid	Stylomastoid foramen	Mastoid cells, tympanic cavity, semicircular canals	Pterygoid plexus
sublingual	Under the tongue	Sublingual gland, mylohyoid, mucous membrane of the mouth, gingivae, alveolar process of the mandible	Lingual vein
submental	Under the base of the mandible	Submandibular gland; mylohyoid, digastric and platysma	Facial vein
superficial cerebral (include the superior and inferior cerebral veins and the superficial middle cerebral vein)	Surface of the cerebral hemispheres	Superior, lateral, medial and inferior surfaces of the cerebral hemispheres	Superior sagittal, cavernous, superior petrosal, transverse, straight and sphenoparietal sinuses
superficial middle cerebral (superficial Sylvian)	Lateral surface of the cerebral hemisphere	Lateral surface of the cerebral hemisphere; receives the superior anastomotic vein and inferior anastomotic vein	Superior petrosal and sphenoparietal sinuses
superficial temporal	Plexus at the apex and on the lateral surface of the cranium	Forehead and scalp	Retromandibular vein
superior anastomotic (of Trolard)	Lateral sulcus	Superficial middle cerebral vein	Superior sagittal sinus
superior cerebral (8-12 in number)	Cerebral cortex	Superior, lateral, medial surfaces of the cerebral hemispheres; receive the prefrontal, frontal, parietal, temporal and occipital veins	Superior sagittal sinus
superior choroid	Lateral ventricle	Lateral ventricle and choroid plexus of the lateral ventricle, corpus callosum	Internal cerebral veins
superior labial	Upper lip	Upper lip, nose, nasal septum	Facial vein
superior ophthalmic	Internal angle of the orbit	Eyeball, extraocular muscles and eyelids; receives the nasofrontal vein; ethmoidal veins; lacrimal vein; vorticose veins; ciliary veins; anterior ciliary veins, of which the scleral venous sinus and the scleral veins are tributaries; central retinal vein; episcleral veins, of which the palpebral and conjunctival veins are tributaries	Cavernous sinus

continued

Summary table of the veins of the head *(continued)*

VEIN	ORIGIN	DRAINAGE	CONFLUENCE
superior palpebral	Superior eyelid	Superior eyelid	Facial vein or angular vein and superior ophthalmic vein
superior thalamostriate (superior or terminal striate)	Deep part of the telencephalon	Corpus striatum, thalamus and corpus callosum; receives the anterior and posterior veins of the septum pellucidum, medial and lateral veins of the lateral ventricle	Great cerebral vein or internal cerebral vein
superior vein of vermis	Culmen	Culmen	Great cerebral vein
superior veins of cerebellar hemisphere	Superior surface of the vermis	Superior surface of the cerebellum	Straight sinus and internal cerebral veins
supraorbital	Forehead	Frontal belly of occipitofrontalis	Angular vein
supratrochlear (generally two in number)	Forehead	Forehead, near the median line	Angular vein
transverse facial	Cheek	Zygomatic region	Retromandibular vein or superficial temporal veins
tympanic	Middle ear	Tympanic cavity, tympanic membrane, mastoid cells, pharyngotympanic tube	Pterygoid plexus
vestibular	Internal ear	Saccule and utricle	Labyrinthine veins and vein of the vestibular aqueduct
vorticose (generally 4 or 5 in number)	Eyeball, halfway between the lamina cribrosa of the sclera and the corneoscleral junction	Choroid	Superior ophthalmic vein or inferior ophthalmic vein

N.B.: for dural venous sinuses, see the table summarizing the venous sinuses.

Cerebral veins
- superficial cerebral
- deep cerebral
- of brainstem
- cerebellar

Summary table of the veins of the neck

VEIN	ORIGIN	DRAINAGE	CONFLUENCE
anterior jugular	At the level of the hyoid bone (inferiorly anastomoses with the contralateral vein through the jugular venous arch)	Integuments and superficial muscles of the anterior region of the neck	External jugular vein
ascending cervical (anterior vertebral)	Anteriorly to the bodies and transverse processes of the cervical vertebrae, from the anterior external vertebral venous plexus	Anterior external vertebral venous plexus, prevertebral and scalene muscles	Vertebral vein
brachiocephalic (brachiocephalic venous trunk or innominate vein)	Behind the sternoclavicular joint through confluence of the internal jugular vein and subclavian vein	Head, neck, upper limb, parts of the thoracic walls and some thoracic organs	The two right and left brachiocephalic veins unite to form the superior vena cava
common facial (formal term for the inferior trunk of the facial vein that empties into the internal jugular vein)	At the level of the angle of the mandible through confluence of the facial and retromandibular veins	Territories of drainage of the facial and retromandibular veins	Internal jugular vein
deep cervical (posterior jugular)	Superior part of the posterior cervical region	Deep posterior cervical muscles; drains the suboccipital venous plexus	Brachiocephalic vein, at the point of union between the internal jugular vein and subclavian vein

continued

Summary table of the veins of the neck *(continued)*

VEIN	ORIGIN	DRAINAGE	CONFLUENCE
external jugular	Angle of the mandible through confluence of the retromandibular and posterior auricular veins	Superficial territories of the posterior region of the neck, integuments and superficial muscles of the anterior region of the neck, integuments and superficial muscles of the occipital region and the nuchal region; receives the anterior jugular, suprascapular, transverse cervical and, sometimes, occipital veins	Internal jugular vein or subclavian vein
inferior laryngeal	Posterior part of the larynx	Muscles and mucosa of the larynx	Brachiocephalic vein
inferior thyroid	Anterior surface of the isthmus of thyroid gland	Unpaired thyroid plexus	Brachiocephalic vein
internal jugular	At the level of the posterolateral part of the jugular foramen as a continuation of the transverse sinus	Cranium, cranial dura mater, brain, organs of sight and hearing, facial and deep neck structures; in the cranial tract: receives the inferior petrosal sinus, the vein of the cochlear aqueduct and the emissary veins of the internal carotid venous plexus; in the cervical tract: receives the facial, sternocleidomastoid, lingual, pharyngeal from the pharyngeal plexus, superior thyroid, middle thyroid and superior laryngeal veins	Unites with the subclavian vein to form the brachiocephalic vein
middle thyroid	Lateral surface of the thyroid gland	Inferior part of the thyroid gland; receives tributary veins from the trachea and larynx	Internal jugular vein
occipital	Posterior venous plexus of the scalp	Integuments and superficial muscles of the occipital region and the nuchal region	External jugular vein or vertebral vein
pharyngeal (several in number)	External surface of the pharynx	Pharyngeal plexus	Internal jugular vein
posterior external jugular (posterior subcutaneous vein of neck)	Occipital region	Superficial territories of the posterior region of the neck	External jugular vein
retromandibular (posterior facial)	Anterior and posterior dividing branches that unite, respectively, with the facial vein to form the common facial vein, and with the posterior auricular vein to form the external jugular vein	Territory of drainage corresponding to the territory of supply of the superficial temporal artery and part of that of the maxillary artery	Internal and external jugular veins
sternocleidomastoid	Sternocleidomastoid	Sternocleidomastoid	Internal jugular vein or superior thyroid vein
subclavian	Medial point of the clavicle	Superior regions of the thorax, upper limb, some structures of the head and neck; receives the external jugular vein	At the level of the sternoclavicular joint unites with the internal jugular vein to form the brachiocephalic vein
superior laryngeal	Larynx	Glands, muscles and mucosa of the larynx	Superior thyroid vein or internal jugular vein
superior thyroid	Superior part of the lobe of the thyroid gland	Superior part of the thyroid gland; receives the superior laryngeal and cricothyroid veins and, sometimes, sternocleidomastoid vein	Internal jugular vein
suprascapular (transverse vein of scapula)	Superior border of the scapula; often unites in a common trunk with the transverse cervical veins	Clavicle, scapula, sternocleidomastoid, subclavius, supraspinatus and subscapularis, acromioclavicular and glenohumeral joints, suprascapular region, skin over the acromion	External jugular vein
transverse cervical	Lateral cervical region; they often unite in a common trunk with the suprascapular vein	Trapezius and posterior cervical muscles	External jugular vein or, sometimes, subclavian vein
vertebral	Between the occipital bone and the atlas, from the internal vertebral venous plexuses	Internal and external vertebral venous plexuses; receives the intervertebral vein	Brachiocephalic vein

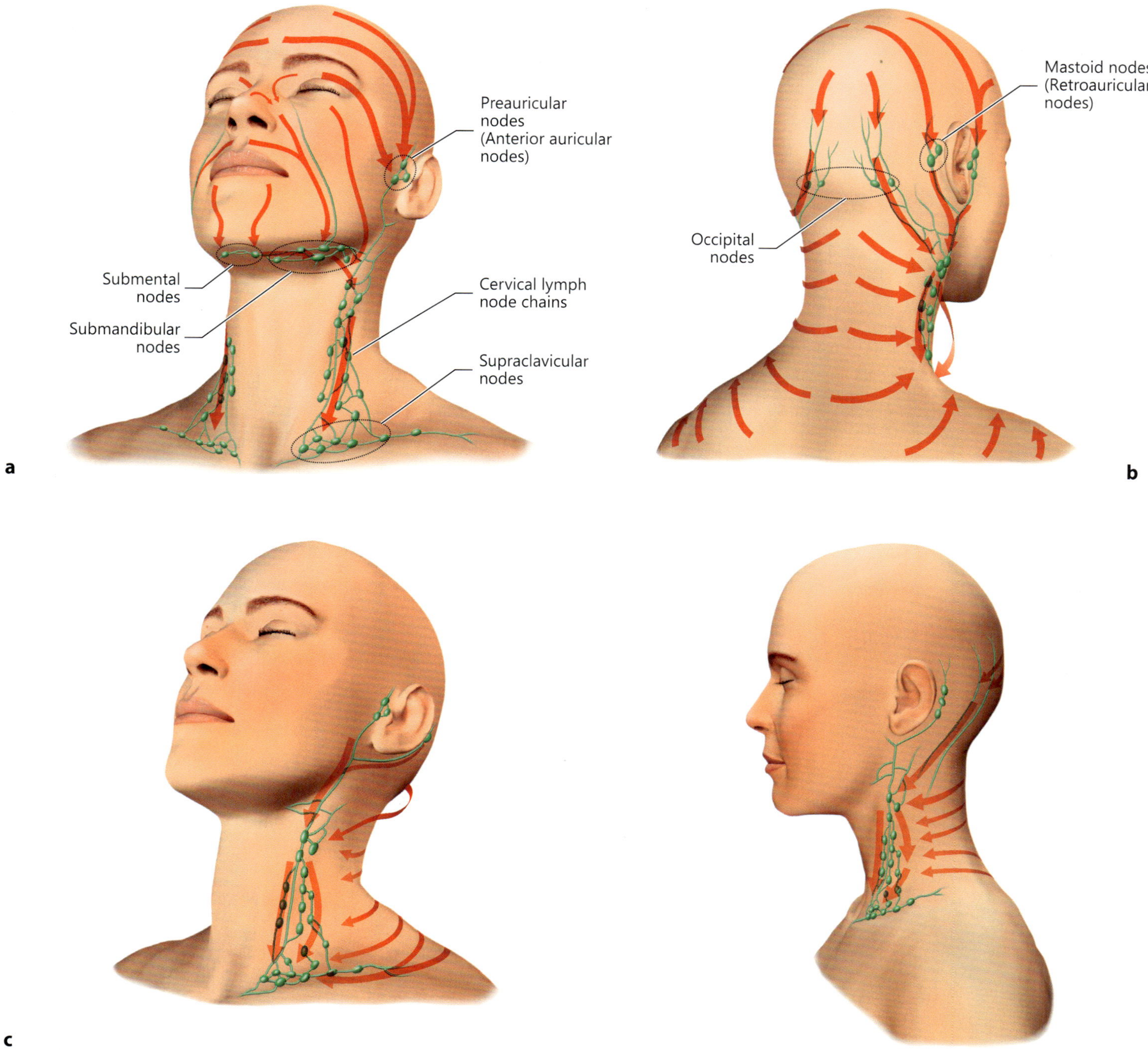

Lymph nodes and **lymphatic drainage of the head** and **neck**. Frontal (**a**), posterior (**b**) and lateral (**c**) views.

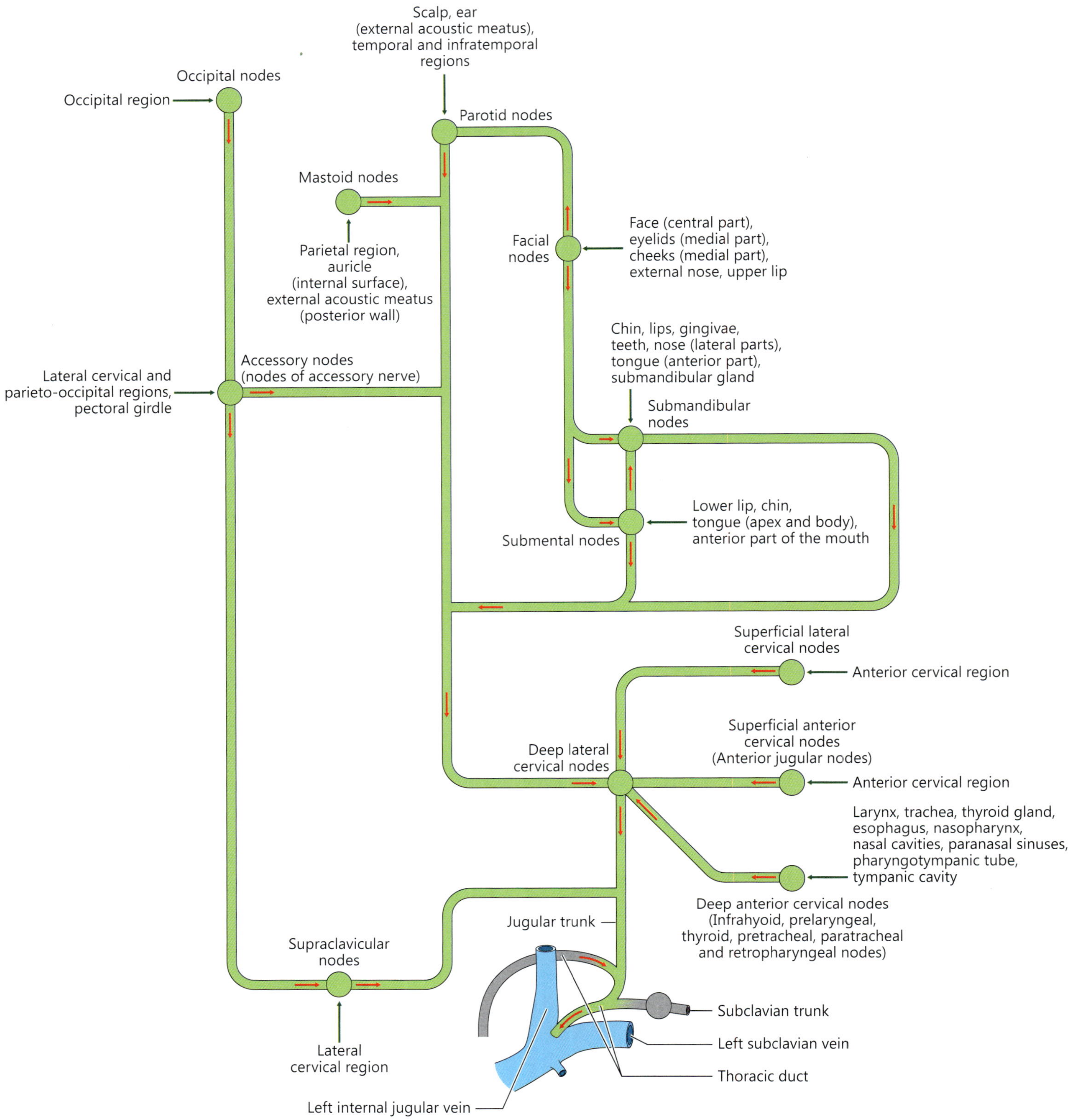

Schematic representation of the **lymphatic groups** of the **head**. Arrows indicate the direction of lymphatic flow. The principal lymphatic ducts (jugular, subclavian, bronchomediastinal, thoracic) reach the homolateral union of the subclavian and internal jugular veins into which they discharge either directly or through a common lymphatic duct.

Distribution of **lymphatic groups of the head** and **neck** in horizontal and vertical chains (dotted lines). There are two horizontal chains, superior and inferior, and four vertical chains, situated at superficial and deep levels.

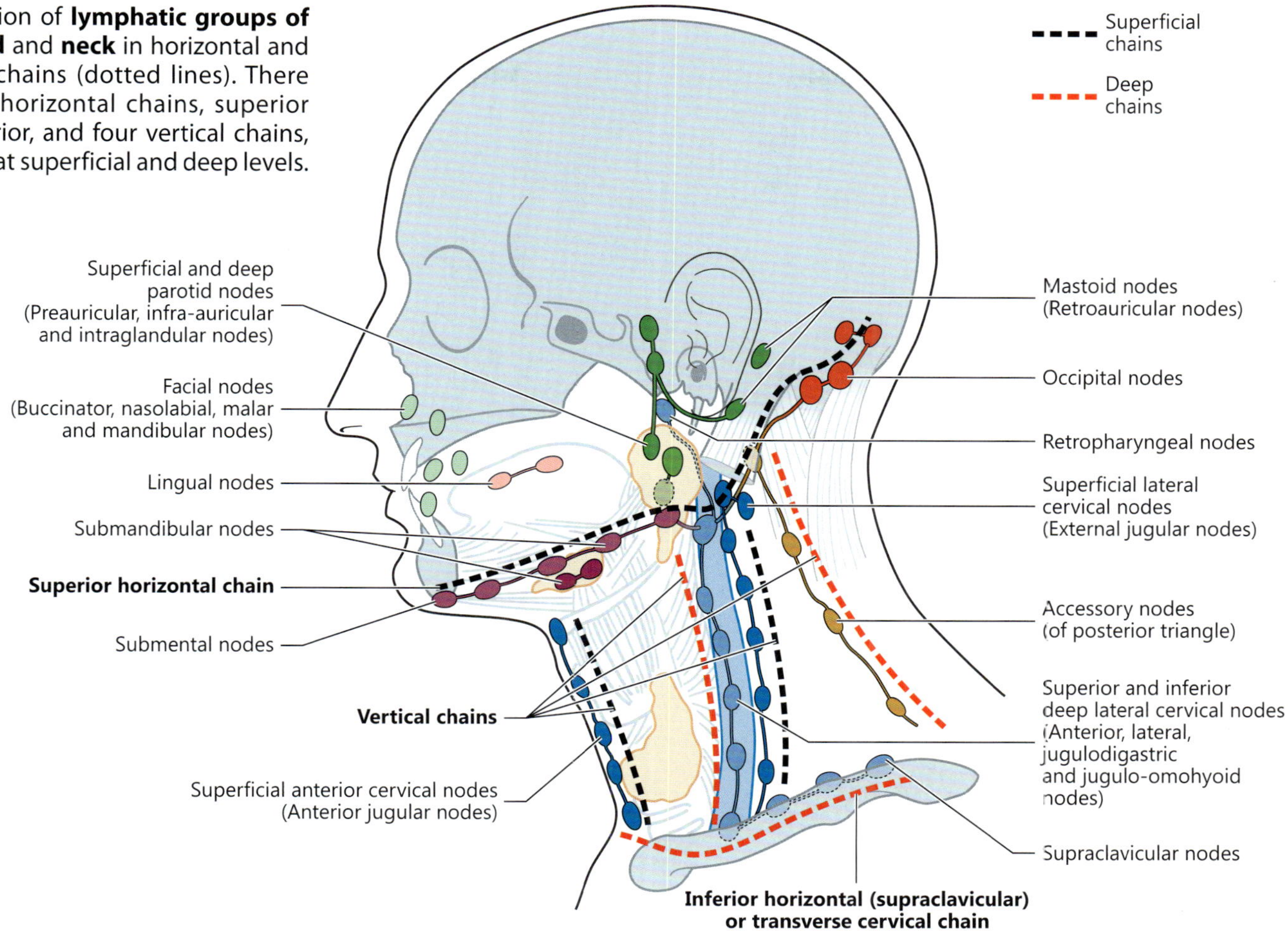

Schematic representation of **deep lymphatic groups of the head** and **neck**. The different groups, with their relative tributaries and flow pathways, are differentiated by color. Arrows indicate the direction of lymphatic flow. All lymph nodes in the region lead to the deep cervical nodes, from where the jugular trunk originates.

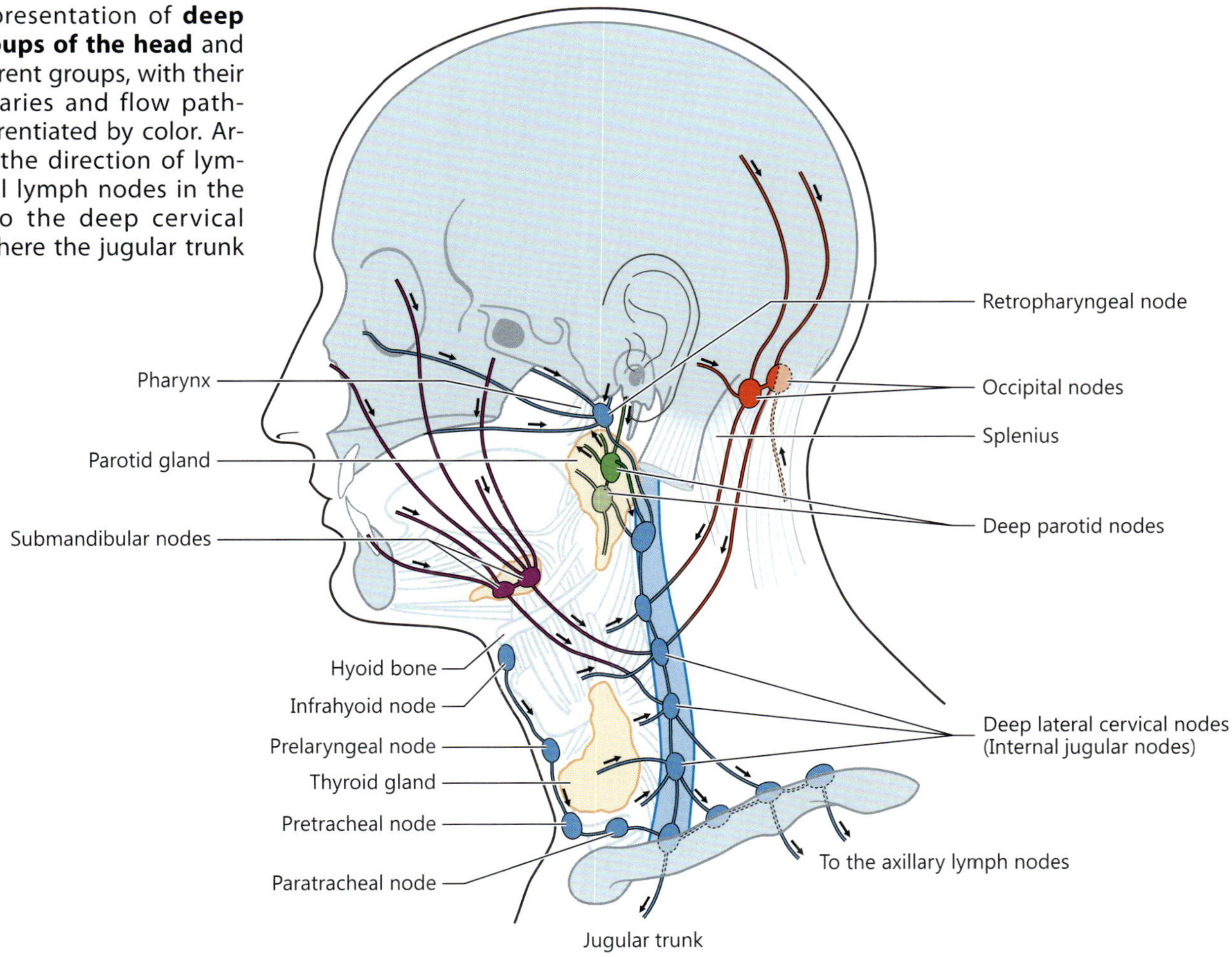

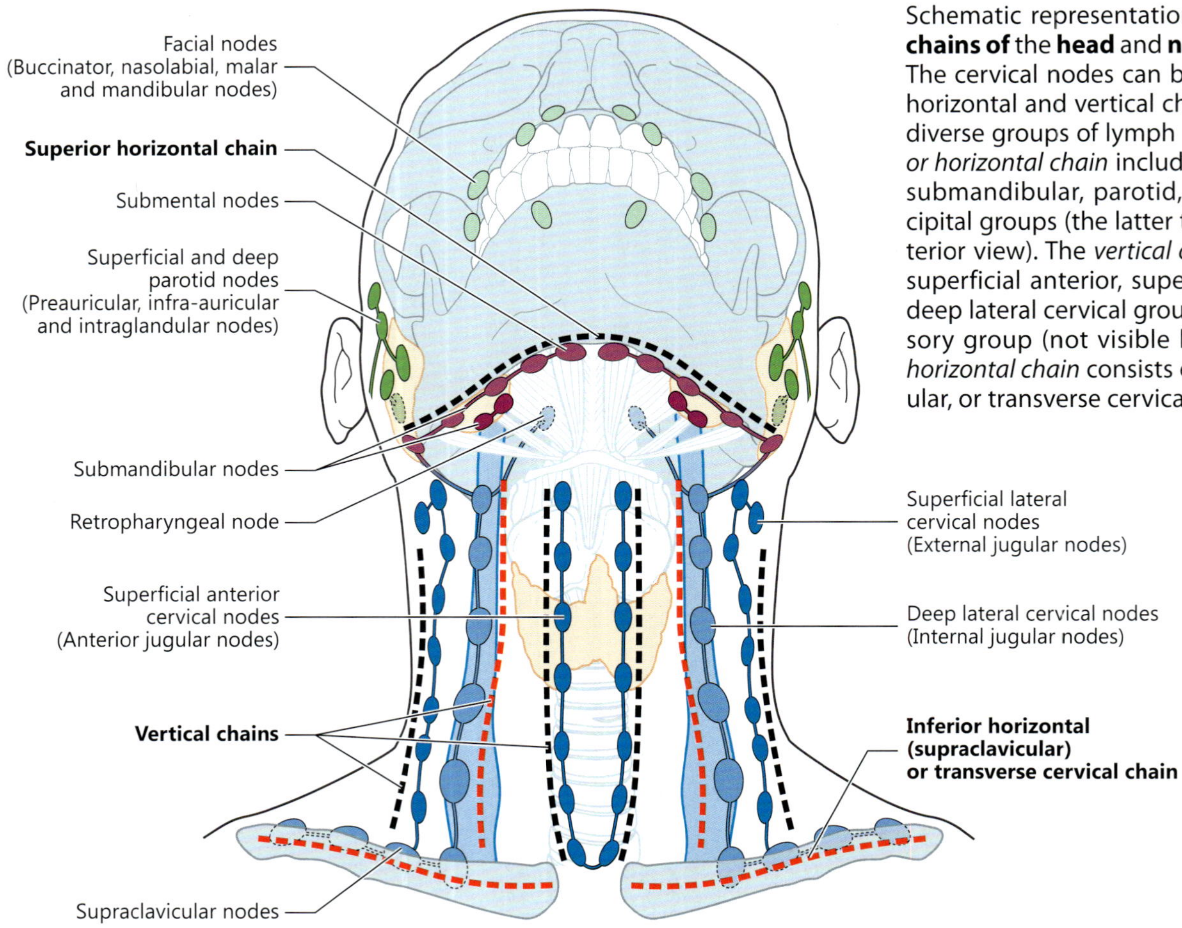

Schematic representation of **lymph node chains of** the **head** and **neck**, anterior view. The cervical nodes can be subdivided into horizontal and vertical chains consisting of diverse groups of lymph nodes. The *superior horizontal chain* includes the submental, submandibular, parotid, mastoid and occipital groups (the latter two visible in posterior view). The *vertical chains* include the superficial anterior, superficial lateral, and deep lateral cervical groups, and the accessory group (not visible here). The *inferior horizontal chain* consists of the supraclavicular, or transverse cervical, group.

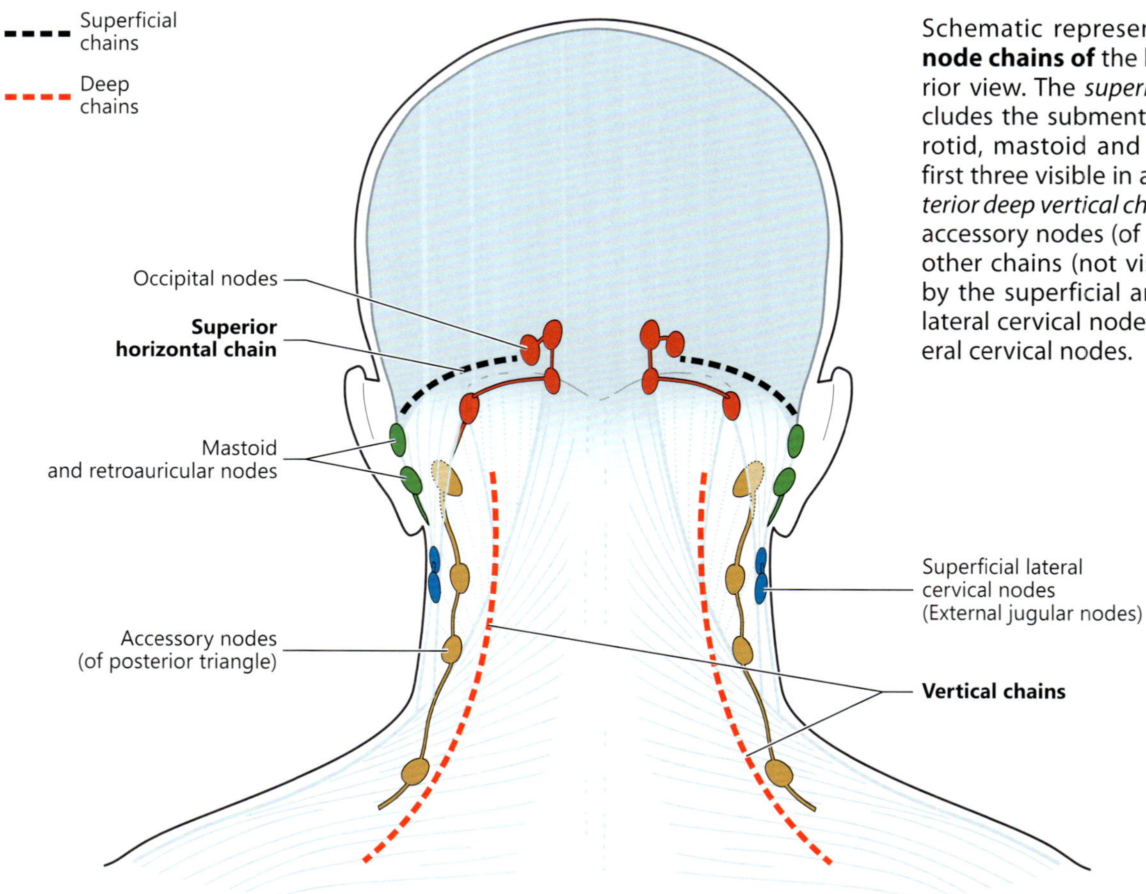

Schematic representation of the **lymph node chains of** the **head** and **neck**, posterior view. The *superior horizontal chain* includes the submental, submandibular, parotid, mastoid and occipital groups (the first three visible in anterior view). The *posterior deep vertical chain* is composed of the accessory nodes (of posterior triangle); the other chains (not visible here) are formed by the superficial anterior and superficial lateral cervical nodes, and by the deep lateral cervical nodes.

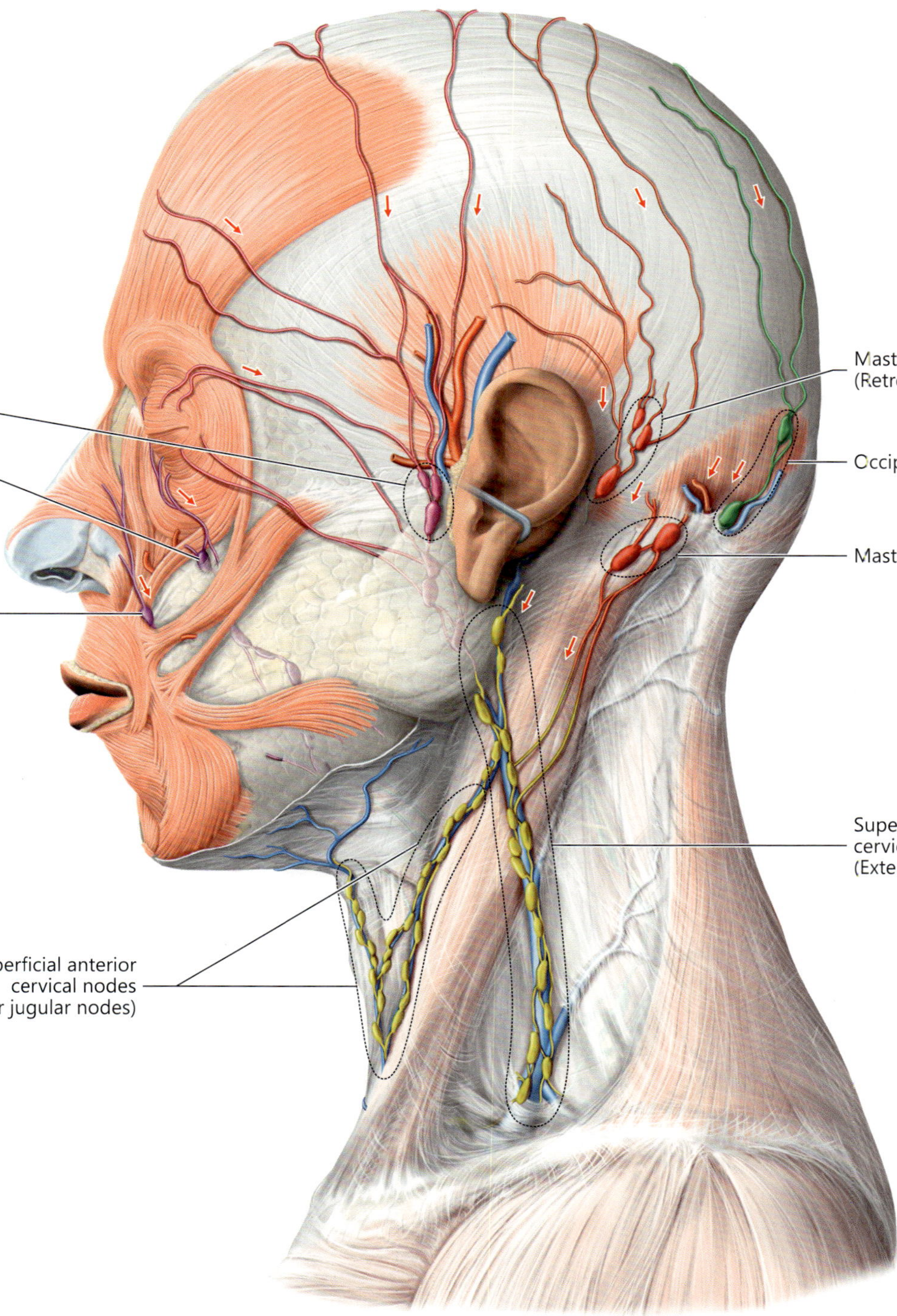

Lymph nodes of the **head** and **neck**. The cervicocephalic nodes are situated at superficial and deep levels. The **superficial parotid nodes** are located anteriorly to the tragus, above the parotid fascia. The **mastoid nodes**, which include also the **retroauricular nodes**, are located superficially to the origin of the sternocleidomastoid. The **occipital nodes** are situated superficially to the origin of the descending part of the trapezius. **Cervical nodes** are found also along the external jugular vein, superficially with respect to the sternocleidomastoid (**external jugular nodes**), and along the anterior jugular vein (**anterior jugular nodes**). The **malar node**, located superficially near the zygomaticus minor, and the **nasolabial node**, also superficial, located deep to the nasolabial sulcus, belong to the lymph nodes of the face (**facial nodes**).

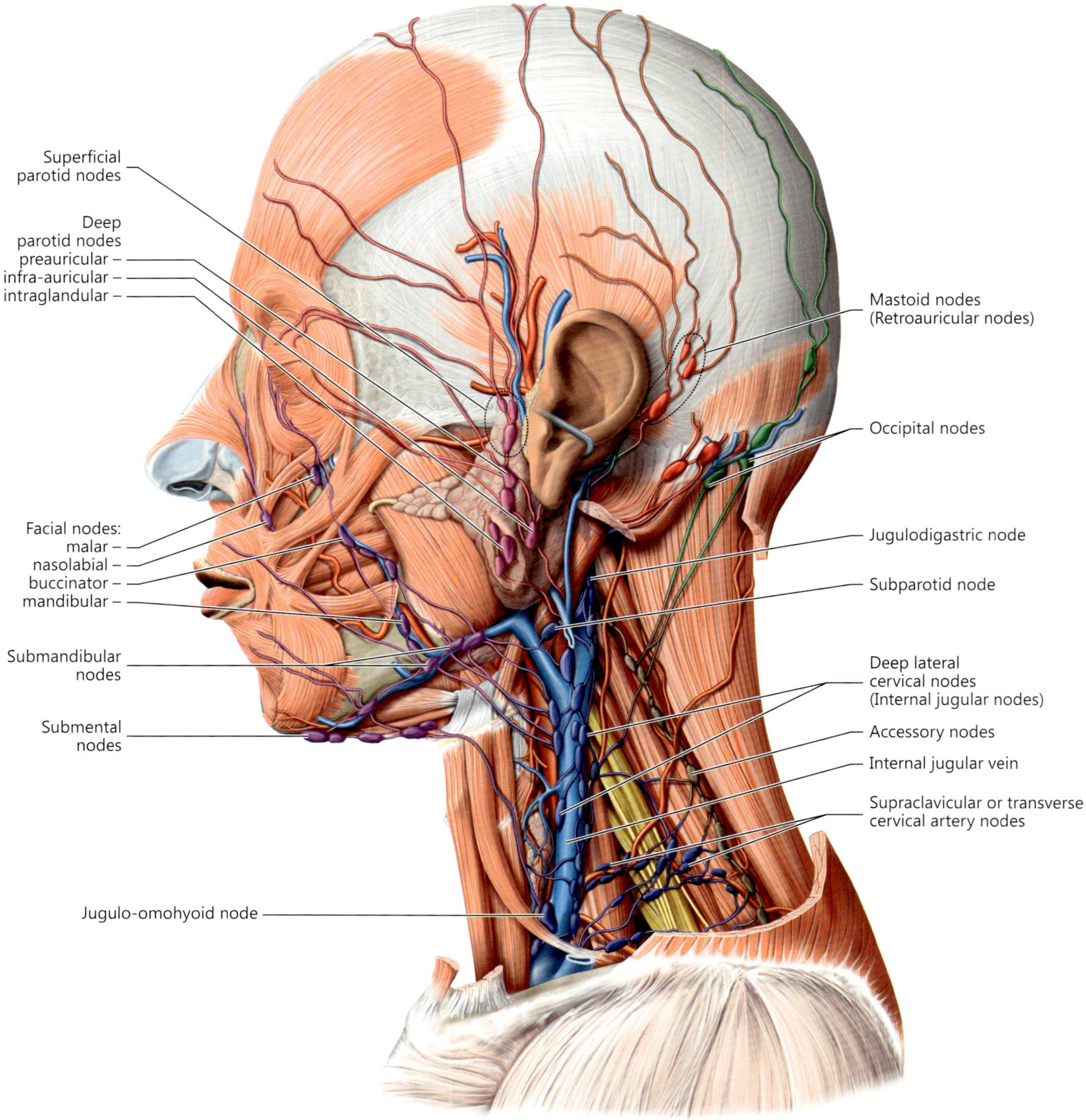

Lymph nodes of the **head** and **neck**. The **deep parotid nodes** are located beneath the parotid fascia at the preauricular and infra-auricular sites and inside the parotid gland (**intraglandular nodes**). The lymph nodes of the face (**facial nodes**) include, besides the **malar** and **nasolabial nodes**, also the **buccinator node**, on the surface of the buccinator, and **mandibular node**, situated in front of the masseter. In the submandibular triangle, at a deeper level with respect to the investing layer of cervical fascia, are the **submandibular nodes**, located on the superior face of the submandibular gland. The **submental nodes** are located on the inferior surface of the mylohyoid, between the anterior bellies of the digastric muscles. Located along the course of the internal jugular vein are the **deep lateral cervical nodes** (**internal jugular nodes**), as well as the **jugulodigastric node**, in the triangle bounded by the posterior belly of the digastric, the internal jugular vein and the facial vein, and **jugulo-omohyoid node**, situated at the level of the intermediate tendon of the omohyoid. The **accessory nodes**, along the anterior border of the trapezius, follow the path of the accessory nerve and from the upper extremity of the internal jugular nodal chain travel downward and posteriorly until they are below the trapezius. The **supraclavicular nodes** or **transverse cervical artery nodes** follow the path of the transverse cervical artery.

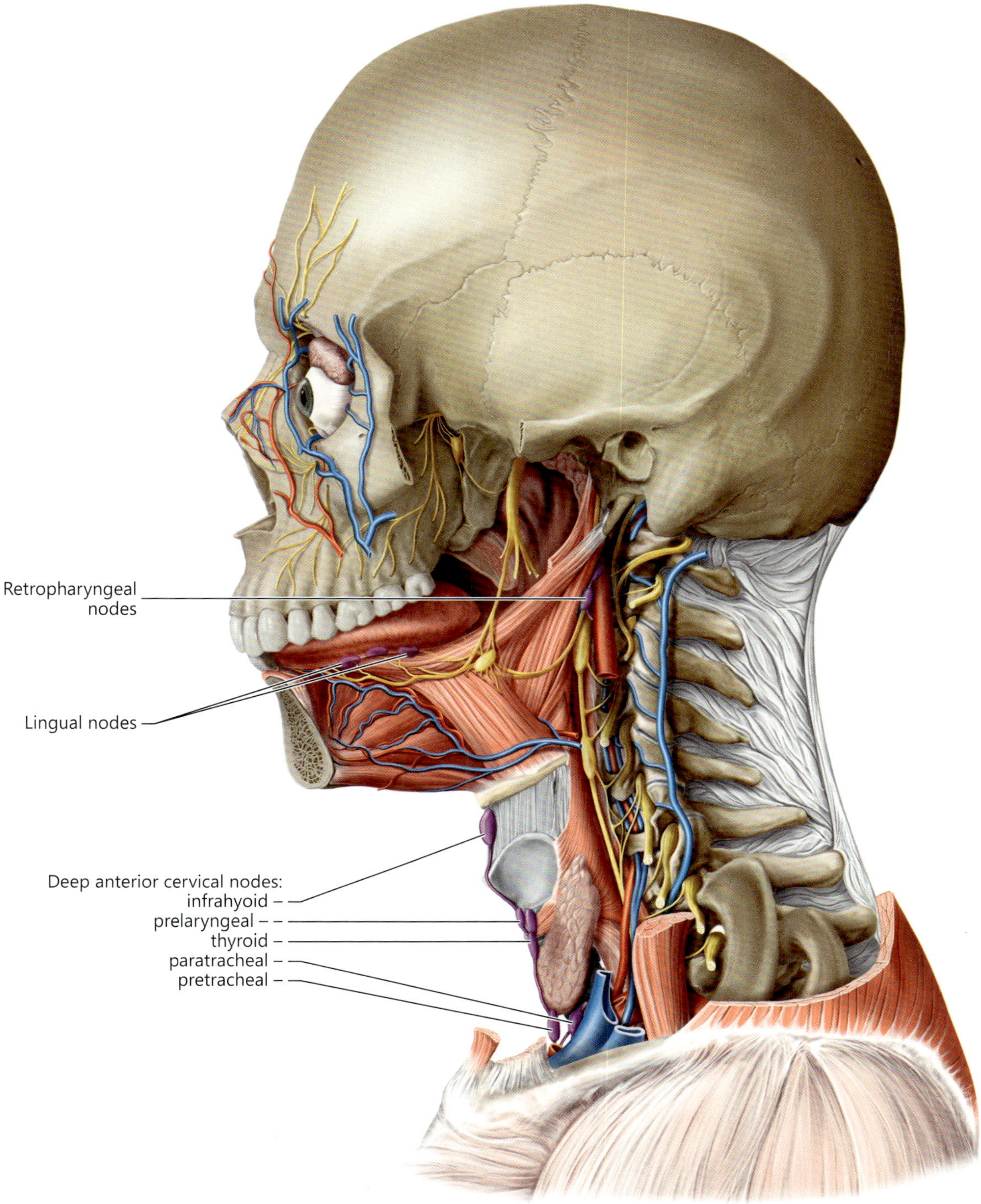

Lymph nodes of the **head** and **neck**. The deep anterior cervical nodes comprise the **infrahyoid nodes**, which include the **prelaryngeal nodes**, and the **thyroid**, **pretracheal** and **paratracheal nodes**, together defined also as anterior visceral nodes. The **retropharyngeal nodes** (these, too, are deep anterior cervical nodes) are situated above, between the posterolateral wall of the pharynx and the prevertebral layer of cervical fascia, on the adventitia that covers the posterior part of the nasopharynx. The **lingual nodes**, when present, lie on the external surface of the hyoglossus and between the two genioglossus muscles or along the lingual vein.

Summary table of the lymph nodes of the head

LYMPH NODE	GROUP	SITE	AFFERENTS	EFFERENTS
Deep parotid	Superior horizontal chain	Deep	Scalp, external acoustic meatus, temporal and infratemporal regions	Nodes of the internal jugular vein
– **infra-auricular**				
– **intraglandular**				
– **preauricular** (anterior auricular)	Superior horizontal chain	Deep	Anterior part of the auricle and anterior wall of the external acoustic meatus	Nodes of the internal jugular vein
Facial		Superficial	Cheeks, lips	Superficial parotid nodes
– **buccinator**				
– **malar**				
– **mandibular** (supramandibular)				
– **nasolabial**				
Lingual (inconstant)		Deep	Inferior surface and lateral margin of the tongue, medial anterior two-thirds of the dorsum of the tongue	Superior deep cervical nodes
Mastoid (retroauricular)	Superior horizontal chain	Superficial	Parietal region, internal surface of the auricle, posterior wall of the external acoustic meatus	Nodes of the internal jugular vein
Occipital	Superior horizontal chain	Superficial	Nuchal region	Nodes of the spinal accessory nerve
Submandibular	Superior horizontal chain and level I	Superficial	Chin, lips, gingivae, teeth, lateral parts of the nose, middle third of the tongue, submandibular gland, periocular area	Nodes of the internal jugular vein and contralateral submandibular nodes
Submental	Superior horizontal chain and level I	Superficial	Lower lip, chin, apex and body of the tongue, anterior part of the mouth	Retromandibular and jugulo-omohyoid nodes
Subparotid		Deep	Scalp, external acoustic meatus, temporal and infratemporal regions	Nodes of the internal jugular vein
Superficial parotid	Superior horizontal chain	Superficial	Scalp, external acoustic meatus, periocular area, nasal, upper frontal, temporal and anterior auricular regions	Nodes of the internal jugular vein

Summary table of the lymph nodes of the neck

LYMPH NODE	GROUP	SITE	AFFERENTS	EFFERENTS
Accessory (nodes of posterior triangle or nodes of spinal accessory nerve)	Vertical chain, level V and posterior triangle	Deep	Lateral region of the neck and parieto-occipital region, pectoral girdle, posterior third of the tongue	Anastomosis with the deep cervical nodes and inferior horizontal chain
Anterior cervical		Superficial and deep		
– deep		Deep	Larynx, trachea, thyroid gland, esophagus	Nodes of the internal jugular vein
– – infrahyoid				
– – – prelaryngeal	Anterior visceral and level VI			
– – paratracheal	Anterior visceral			
– – pretracheal	Anterior visceral and level VI			
– – thyroid	Anterior visceral and level VI			
– superficial (anterior jugular or of anterior jugular vein)	Vertical chain	Superficial	Anterior region of the neck	Transverse cervical artery nodes and nodes of the internal jugular vein
Lateral cervical	Vertical chain	Superficial and deep		
– deep (of internal jugular vein or internal, anterior and posterolateral jugular)	Posterior triangle	Deep	Nodes of the external jugular vein and retropharyngeal nodes, nasal cavities, pharynx, esophagus, larynx, trachea, submental region, mandible, middle third of the tongue, nuchal region	Jugular trunk
– – superior	Superior cervical chain and level II			
– – – lateral				
– – – anterior				
– – – jugulodigastric (subdigastric)		Deep	Palatine tonsil, retromolar triangle, posterior third of the tongue	Jugular trunk
– – middle	Middle cervical chain and level III			
– – inferior	Inferior cervical chain and level IV			
– – – laterale				
– – – anteriori				
– – – jugulo-omohyoid		Deep	Anterior part of the tongue	Jugular trunk
– superficial (of external jugular vein or external jugular)		Superficial	Lateral region of the neck, periocular area, nasal regions	Transverse cervical artery nodes and nodes of the internal jugular vein
Retropharyngeal		Deep	Nasopharynx, nasal cavities, paranasal sinuses, pharyngotympanic tube, tympanic cavity	Transverse cervical artery nodes and nodes of the internal jugular vein
Supraclavicular (transverse cervical artery nodes or nodes of transverse cervical chain)	Inferior horizontal (supraclavicular) chain, level V and posterior triangle	Deep	Collectors from the paraspinal and subclavicular nodes, lymph of the mammary region and of the upper limb	Subclavian nodes (apical axillary)

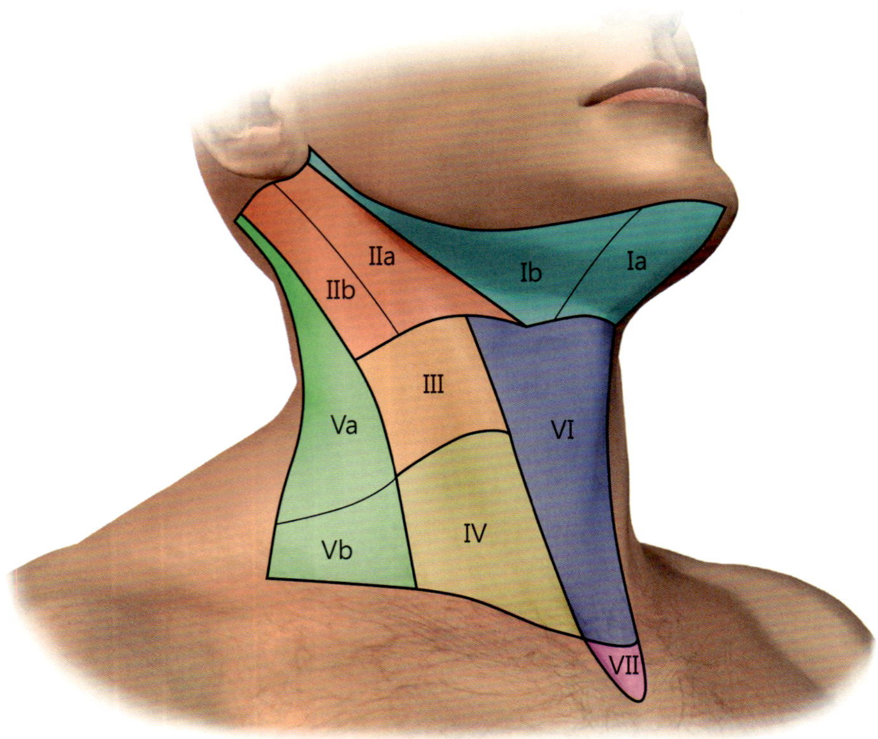

Simplified classification of the clinically palpable **lymphatic groups** of **head** and **neck**. The seven levels of subdivision identified are listed in the table below.

Levels of subdivision of the lymphatic groups of the head and neck

LEVEL	*GROUP*	*CHAIN*
I	Submental and submandibular	Superior horizontal
a	submental	
b	submandibular	
II	Superior deep lateral cervical, including the jugulodigastric node	Superior cervical (upper part of the internal jugular chain or vertical chain of superficial and deep lateral cervical nodes)
a	superior deep lateral cervical	
b	jugulodigastric	
III	Middle deep lateral cervical, including the jugulo-omohyoid node	Middle cervical (middle part of the internal jugular chain or vertical chain of superficial and deep lateral cervical nodes)
IV	Inferior deep lateral cervical	Inferior cervical (lower part of the internal jugular chain or vertical chain of superficial and deep lateral cervical nodes)
V	Accessory (or of posterior triangle), supraclavicular (or transverse cervical artery)	Vertical of the accessory nodes and inferior horizontal supraclavicular or transverse cervical
a	accessory (or of posterior triangle)	
b	supraclavicular (or transverse cervical artery)	
VI	Deep anterior cervical: thyroid, prelaryngeal and pretracheal nodes	
VII	Paratracheal and juxtaesophageal	

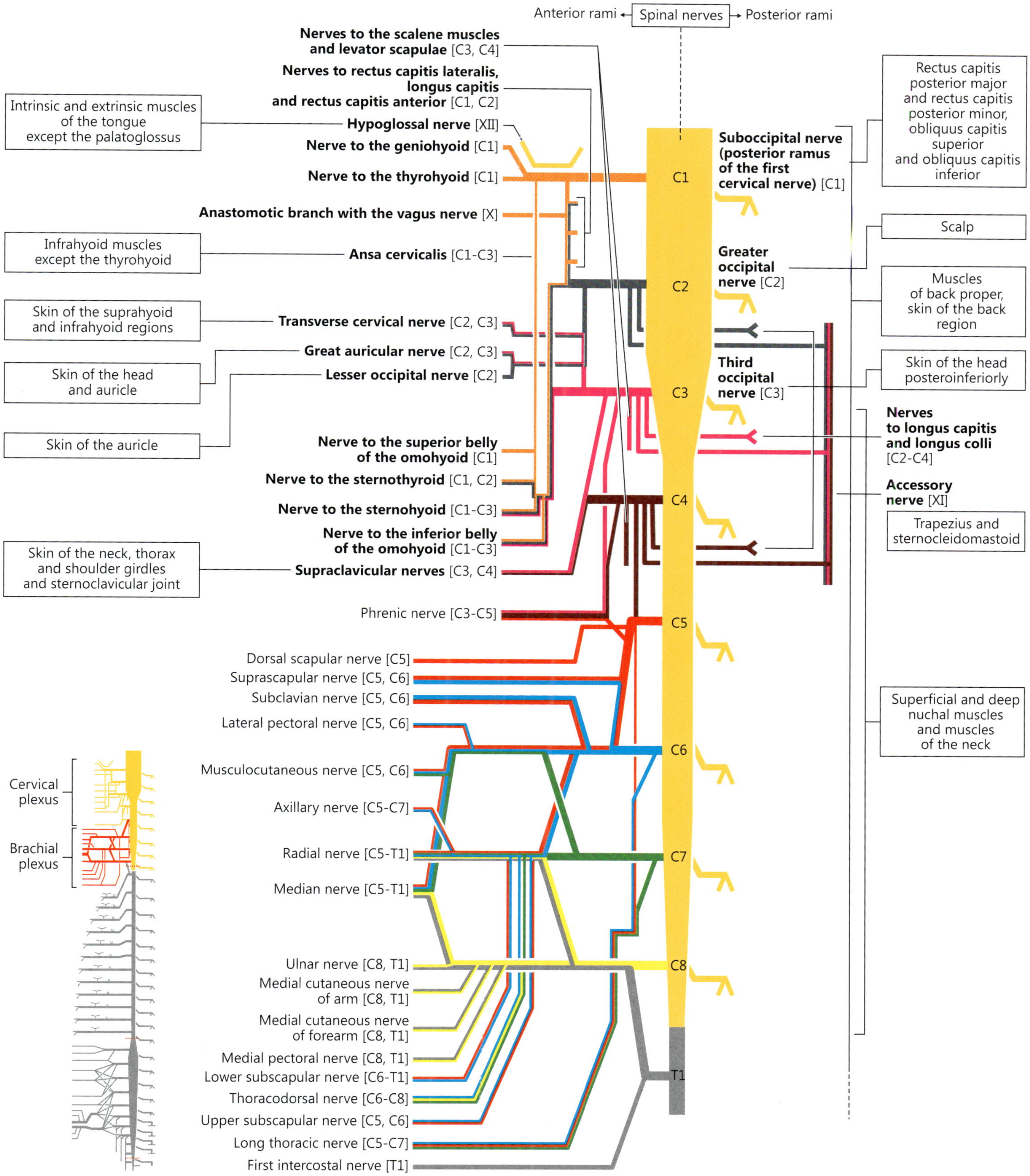

Schematic representation of the **cervical** and **brachial plexuses**. On the left-hand side of the image, at the top, the cervical plexus is represented. It is formed by the anterior rami of the first four cervical nerves. Note that each anterior ramus, with the exception of the first cervical nerve [C1], divides into an ascending and descending branch, which unite with the corresponding branches of contiguous nerves to constitute the cervical loops; the descending branch of the fourth cervical nerve [C4] anastomoses with the anterior ramus of the fifth cervical nerve [C5], which is part of the brachial plexus. The lower part of the diagram, on the left, shows the brachial plexus. It is composed of the anterior rami of the fifth, sixth, seventh and eighth cervical nerves [C5, C6, C7, C8] and of the first thoracic nerve [T1] (whose anterior ramus, which unites with the anterior ramus of C8, is indicated here in *gray* since it originates in the thorax). On the right-hand side of the diagram, the dorsal branches of the cervical nerves are represented.

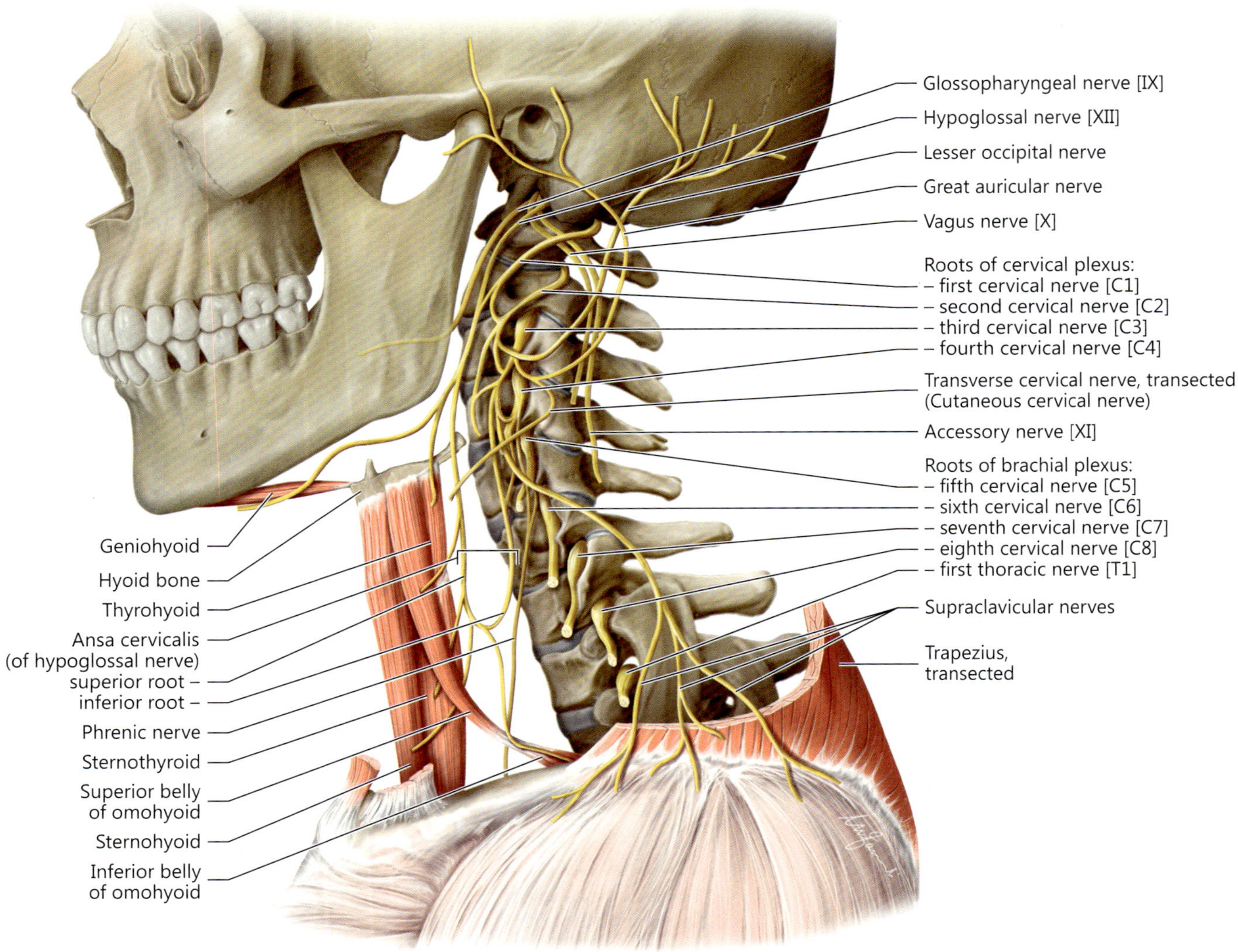

Figure showing the **cervical plexus** in relation to the skeleton of the cervical column. The relationship of the cervical plexus to the glossopharyngeal [IX], vagus [X], accessory [XI] and hypoglossal [XII] nerves can also be seen.

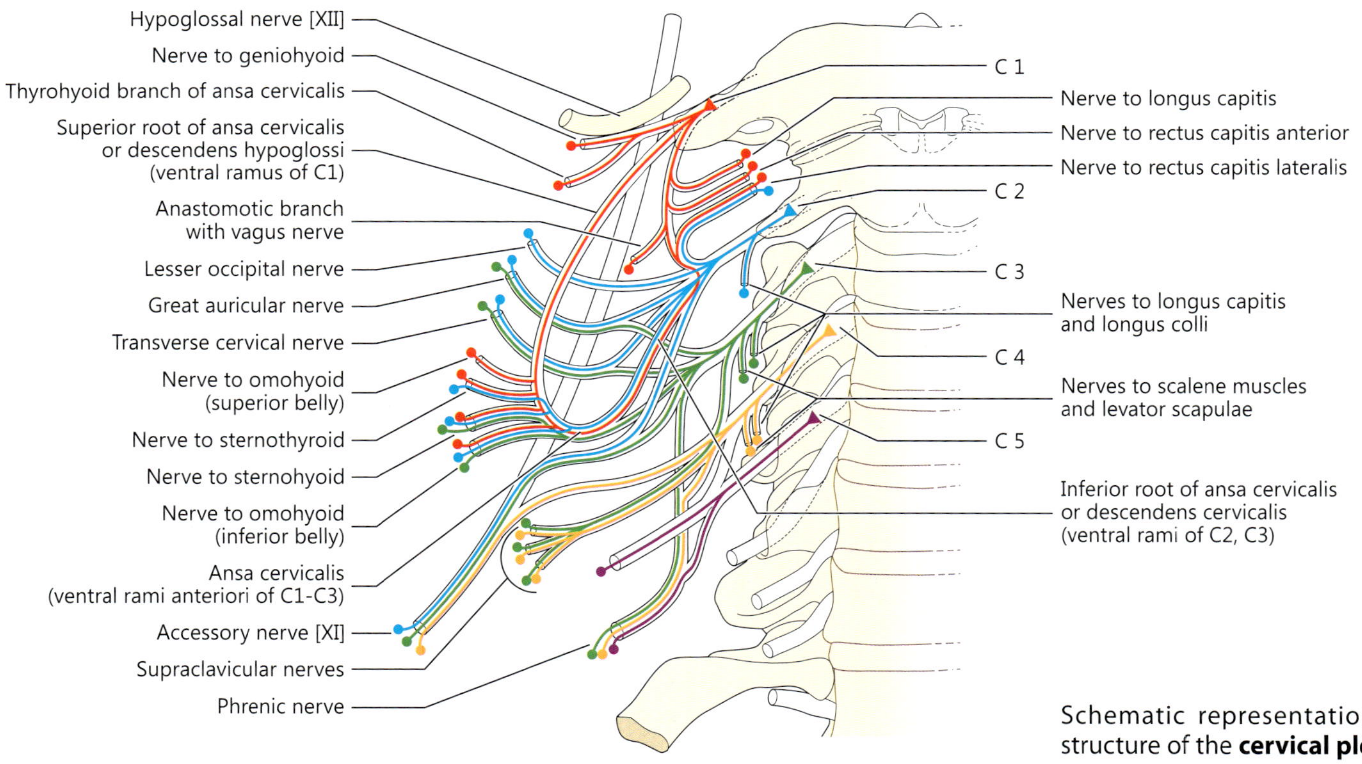

Schematic representation of the structure of the **cervical plexus**.

Figure showing the **brachial plexus** in relation to the skeleton of the vertebral column.

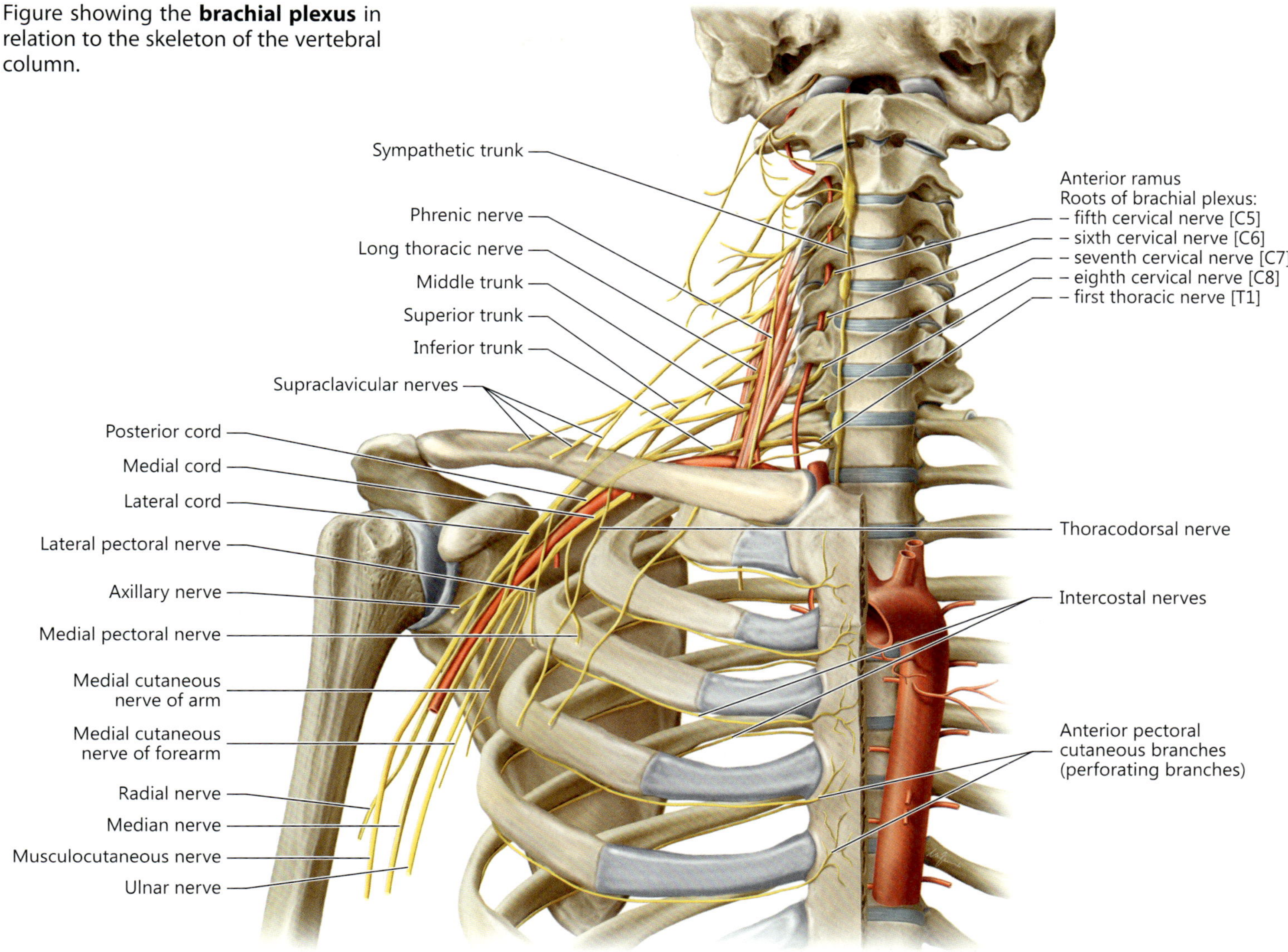

Schematic representation of the structure of the **brachial plexus**. One can see the position of the anterior branches (ant. b.), with the formation of the trunks (superior, middle and inferior) and cords, and of the posterior branches (post. b.) and terminal branches (term. b.). The phrenic nerve [C3, C4, C5] belongs to the cervical plexus; the first intercostal nerve to the thoracic nerves (anterior ramus of first thoracic nerve).

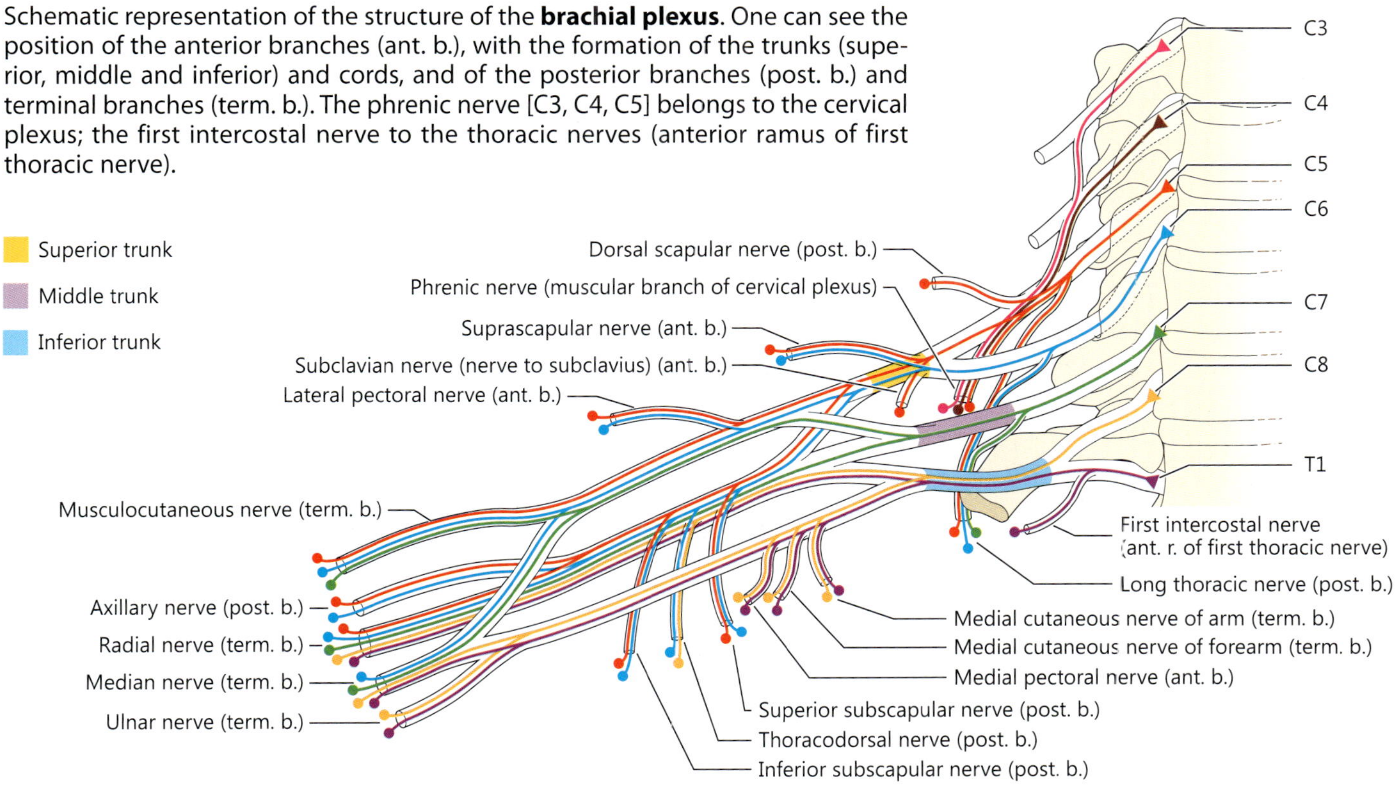

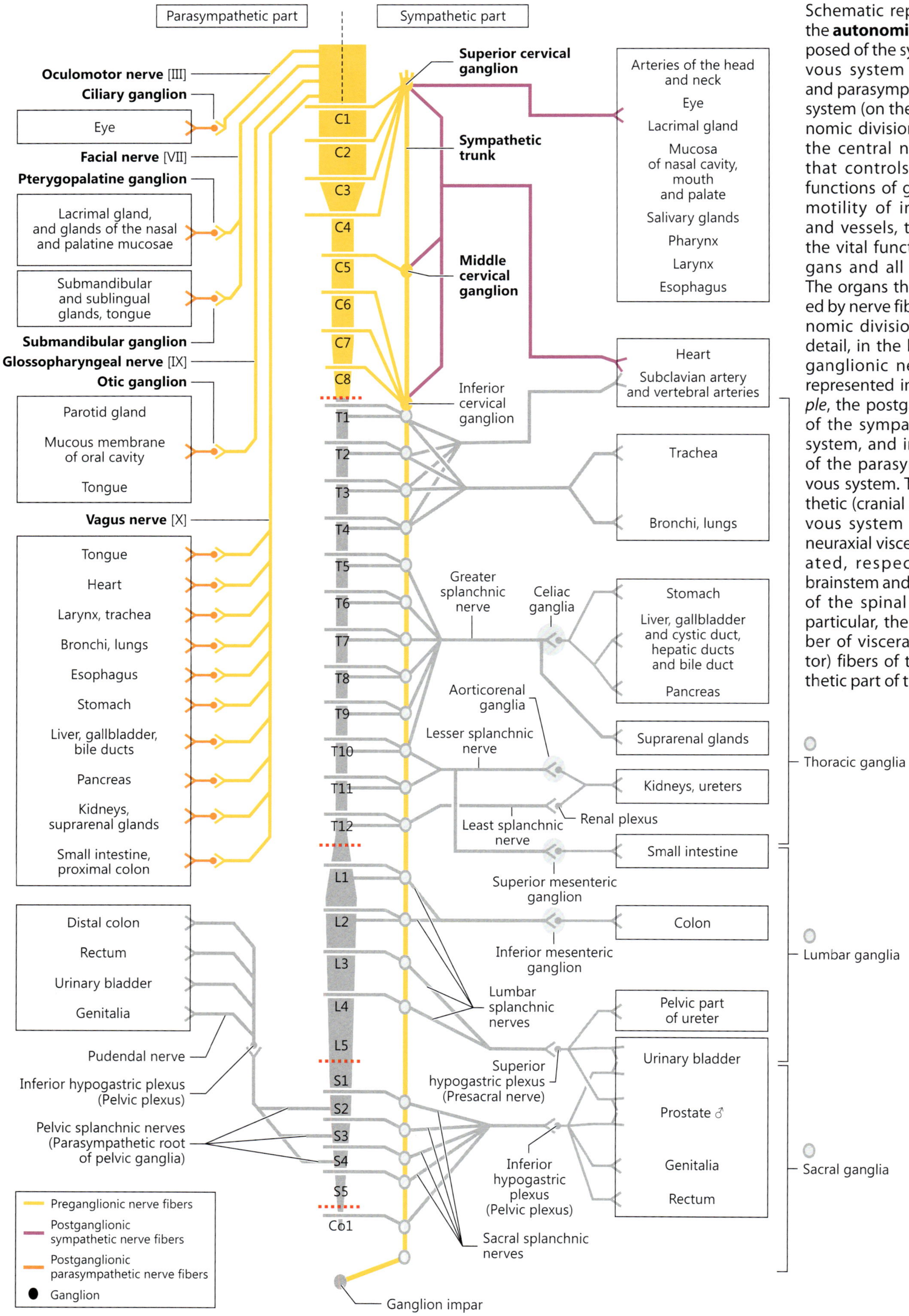

Schematic representation of the **autonomic division** composed of the sympathetic nervous system (on the right) and parasympathetic nervous system (on the left). The autonomic division is the part of the central nervous system that controls the secretory functions of glands and wall motility of internal organs and vessels, thus regulating the vital functions of the organs and all body systems. The organs that are innervated by nerve fibers of the autonomic division are listed, in detail, in the boxes. The preganglionic nerve fibers are represented in *yellow*; in *purple*, the postganglionic fibers of the sympathetic nervous system, and in *orange* those of the parasympathetic nervous system. The parasympathetic (cranial and sacral) nervous system comprises the neuraxial visceral centers situated, respectively, in the brainstem and sacral segment of the spinal cord. Note, in particular, the elevated number of visceral efferent (motor) fibers of the parasympathetic part of the vagus nerve.

Organization of the **autonomic division** in the head region. While the parasympathetic fibers travel inside the cranial nerves III, VII, IX and X, the sympathetic fibers penetrate into the cranium as plexuses that accompany the vertebral and carotid (internal and external) arteries. They can link to cranial nerves and cross parasympathetic ganglia without being interrupted.

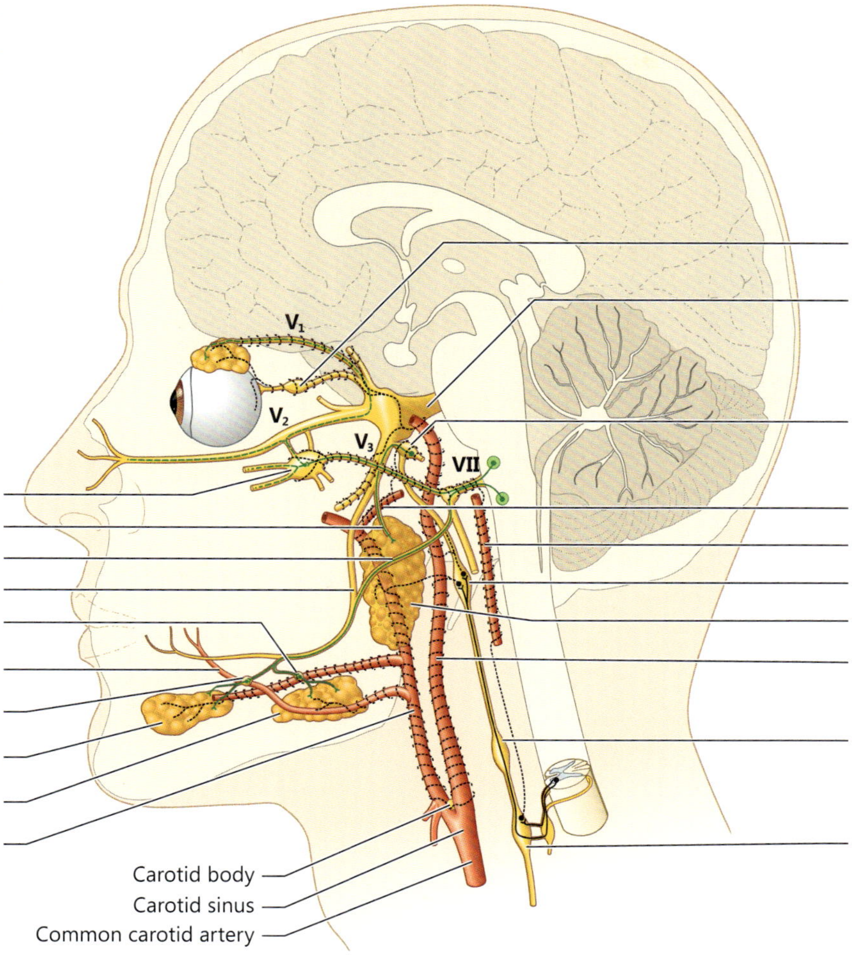

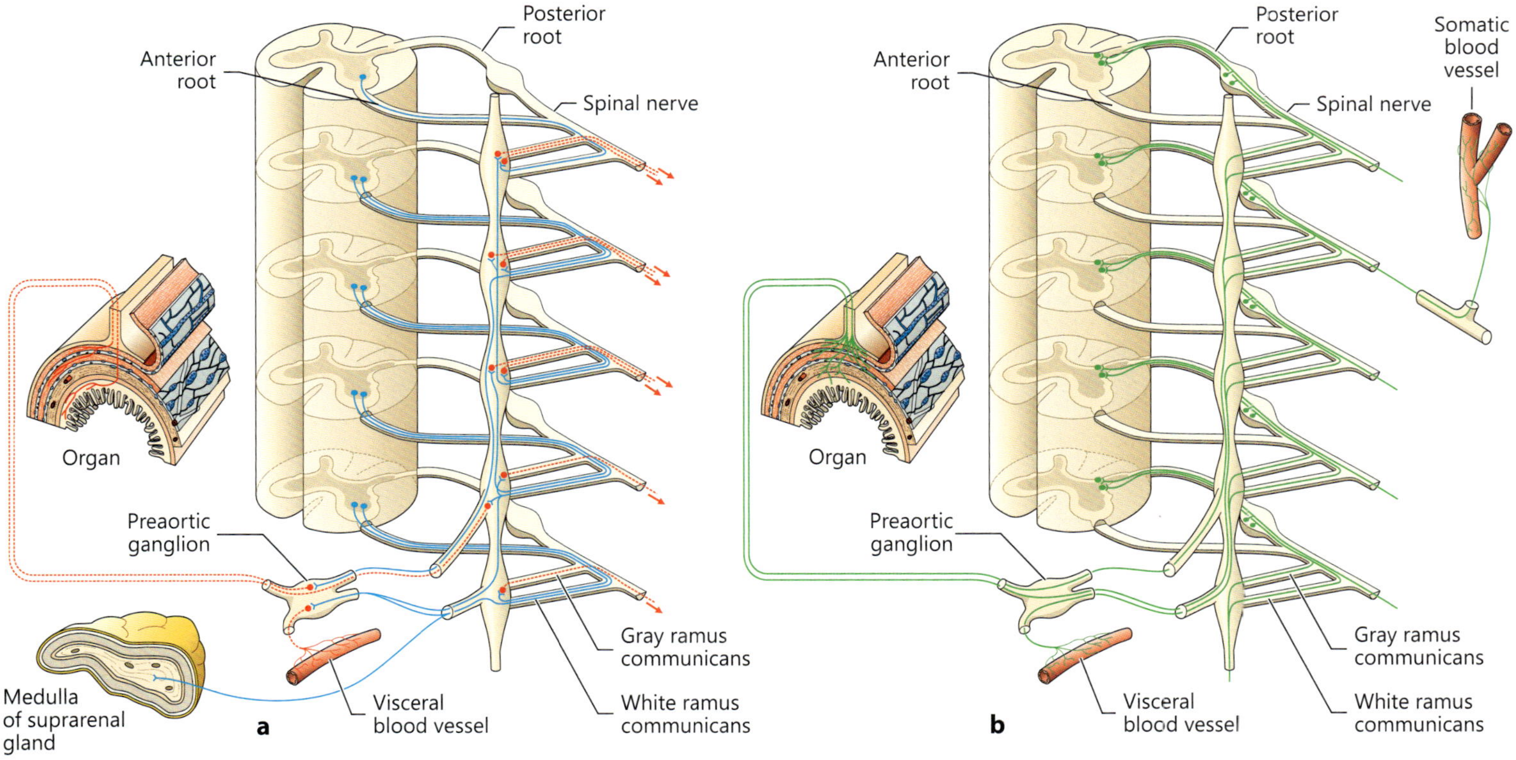

a, Peripheral distribution of the **sympathetic visceral motor fibers**. The different pathways that the preganglionic fibers (*blue*) of medullary origin can follow are indicated, as well as the possible peripheral paths of the postganglionic fibers (*red*) towards vessels and organs. **b**, Peripheral distribution of the **sympathetic visceral sensory fibers**. Localized in the spinal ganglion, together with the somatic sensory neurons, visceral sensory neurons (*green*) are distributed to the periphery with the spinal nerves or, through the trunk of the sympathetic system, with the autonomic nerves.

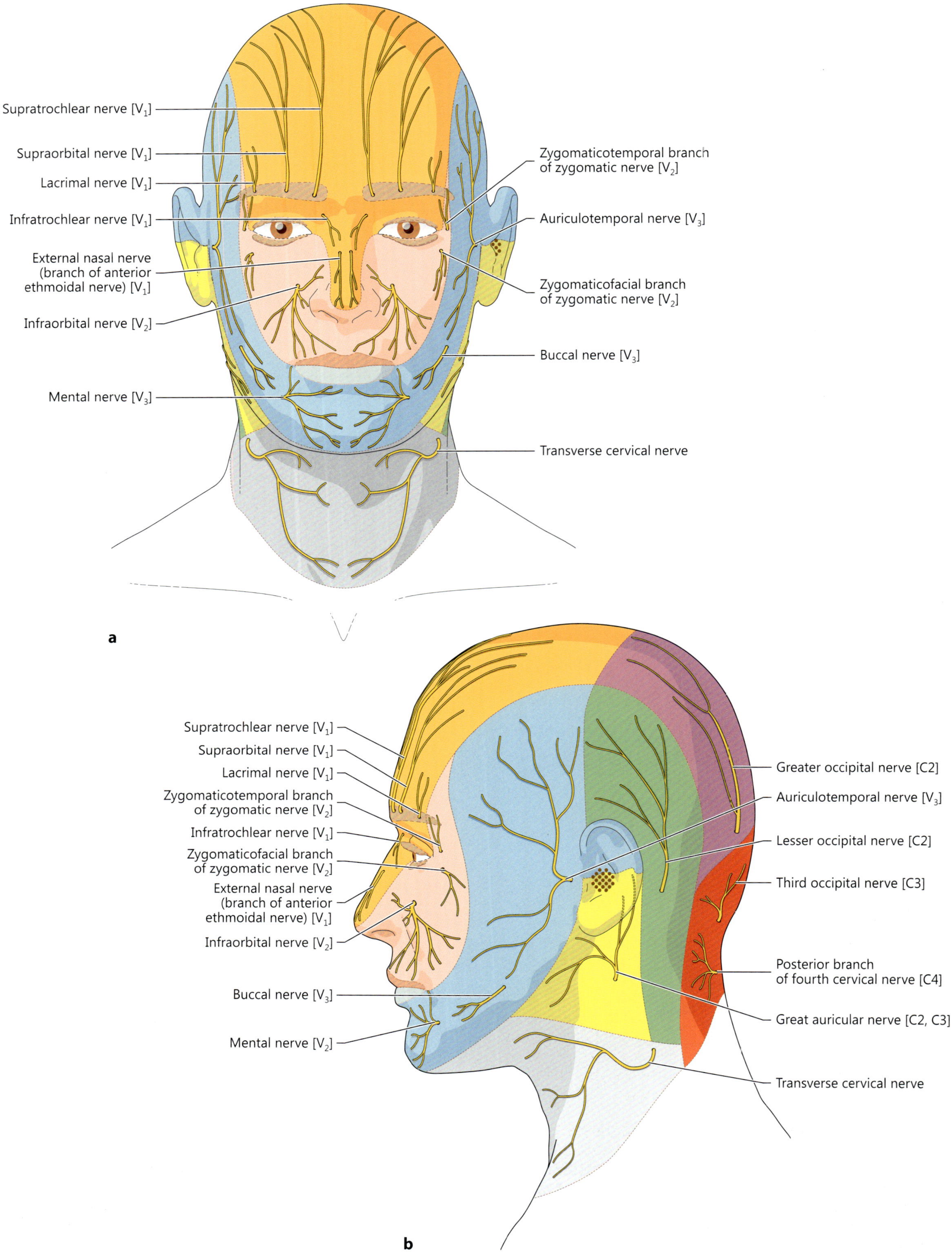
Supratrochlear nerve [V1]
Supraorbital nerve [V1]
Lacrimal nerve [V1]
Infratrochlear nerve [V1]
External nasal nerve (branch of anterior ethmoidal nerve) [V1]
Infraorbital nerve [V2]
Mental nerve [V3]
Zygomaticotemporal branch of zygomatic nerve [V2]
Auriculotemporal nerve [V3]
Zygomaticofacial branch of zygomatic nerve [V2]
Buccal nerve [V3]
Transverse cervical nerve
a
Supratrochlear nerve [V1]
Supraorbital nerve [V1]
Lacrimal nerve [V1]
Zygomaticotemporal branch of zygomatic nerve [V2]
Infratrochlear nerve [V1]
Zygomaticofacial branch of zygomatic nerve [V2]
External nasal nerve (branch of anterior ethmoidal nerve) [V1]
Infraorbital nerve [V2]
Buccal nerve [V3]
Mental nerve [V2]
Greater occipital nerve [C2]
Auriculotemporal nerve [V3]
Lesser occipital nerve [C2]
Third occipital nerve [C3]
Posterior branch of fourth cervical nerve [C4]
Great auricular nerve [C2, C3]
Transverse cervical nerve
b

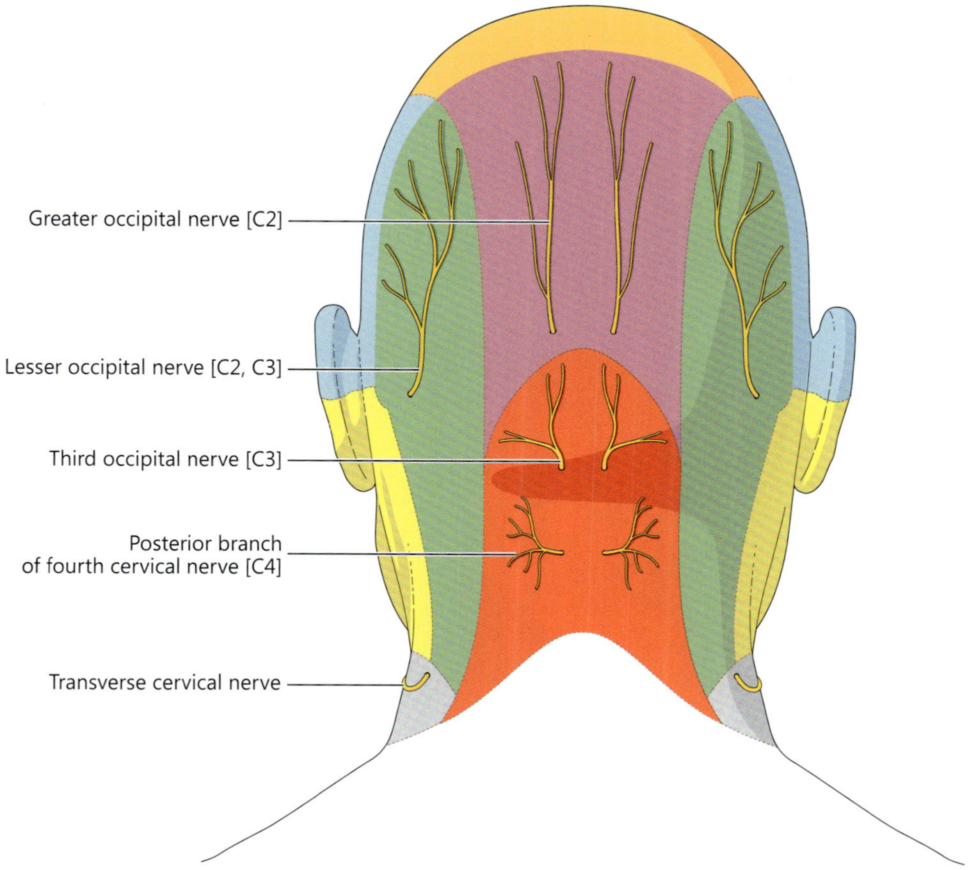

a, **Cutaneous territories of innervation of the head** and **neck**, anterior view. Indicated in *orange*, *pink* and *blue* are the areas supplied, respectively, by the ophthalmic, maxillary and mandibular nerves. In *yellow*, is the territory of distribution of the great auricular nerve (anterior rami of C2 and C3 via the cervical plexus). The dotted area, representing the skin of the concha of auricle, receives sensory fibers from the facial [VII], glossopharyngeal [IX] and vagus [X] nerves. The transverse cervical nerve innervates the skin in the suprahyoid and infrahyoid regions and lateral cervical region (in *gray*).

b, **Cutaneous territories of innervation of the head** and **neck**, left lateral view. The trigeminal nerve supplies, with its three branches, the whole face and the anterior half of the cranium. Indicated in *orange*, *pink* and *blue* are the areas supplied, respectively, by the ophthalmic, maxillary and mandibular nerves. In *yellow* is the territory of distribution of the great auricular nerve (anterior rami of C2 and C3 via the cervical plexus). The dotted area, representing the skin of the concha of auricle, receives sensory fibers from the facial [VII], glossopharyngeal [IX] and vagus [X] nerves. The transverse cervical nerve innervates the skin in the suprahyoid and infrahyoid regions and in the lateral cervical region (in *gray*).

c, **Cutaneous territories of innervation of the head** and **neck**, posterior view. Indicated in *purple*, *green* and *red* are the areas supplied by the posterior and anterior rami of the cervical nerves: respectively, the greater occipital nerve (posterior ramus of C2); the lesser occipital nerve (anterior ramus of C2 via the cervical plexus); and the third occipital nerve (posterior ramus of C3) and the posterior ramus of C4. Indicated in *yellow* is the territory of distribution of the great auricular nerve (anterior rami of C2 and C3 via the cervical plexus). The transverse cervical nerve innervates the skin in the suprahyoid and infrahyoid regions and in the lateral cervical region (in *gray*). Indicated in *orange* is the area supplied by the ophthalmic nerve via the supratrochlear and supraorbital nerves.

External surface of the cranial base, showing **the passage of vessels** and **nerves through the foramina**.

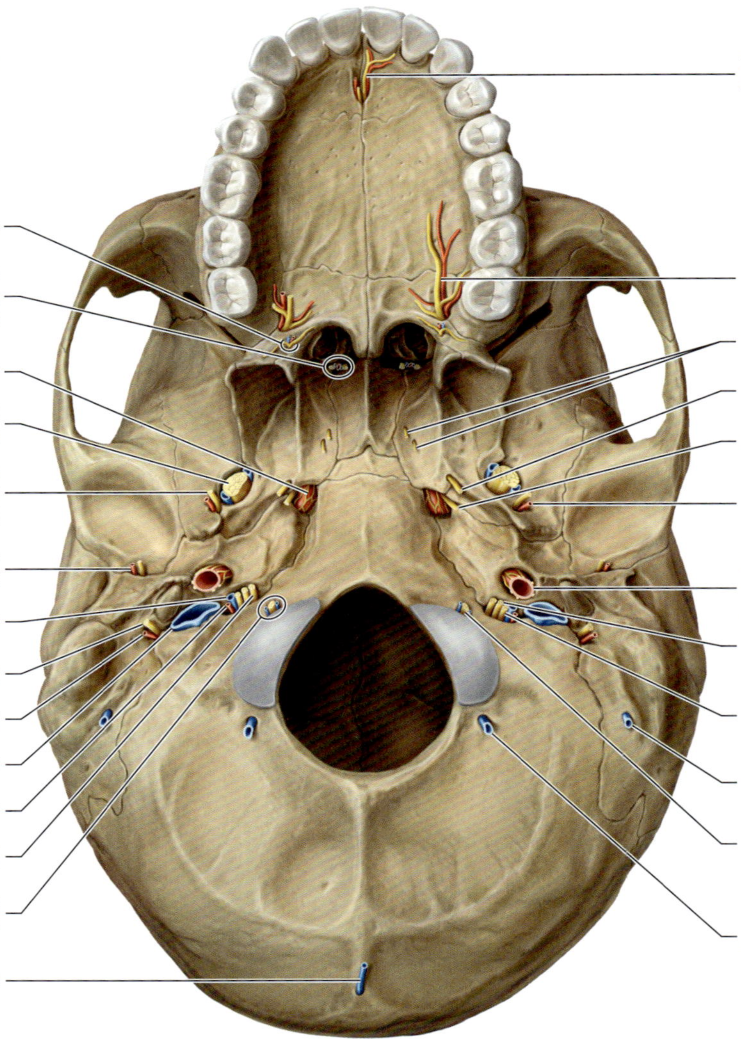

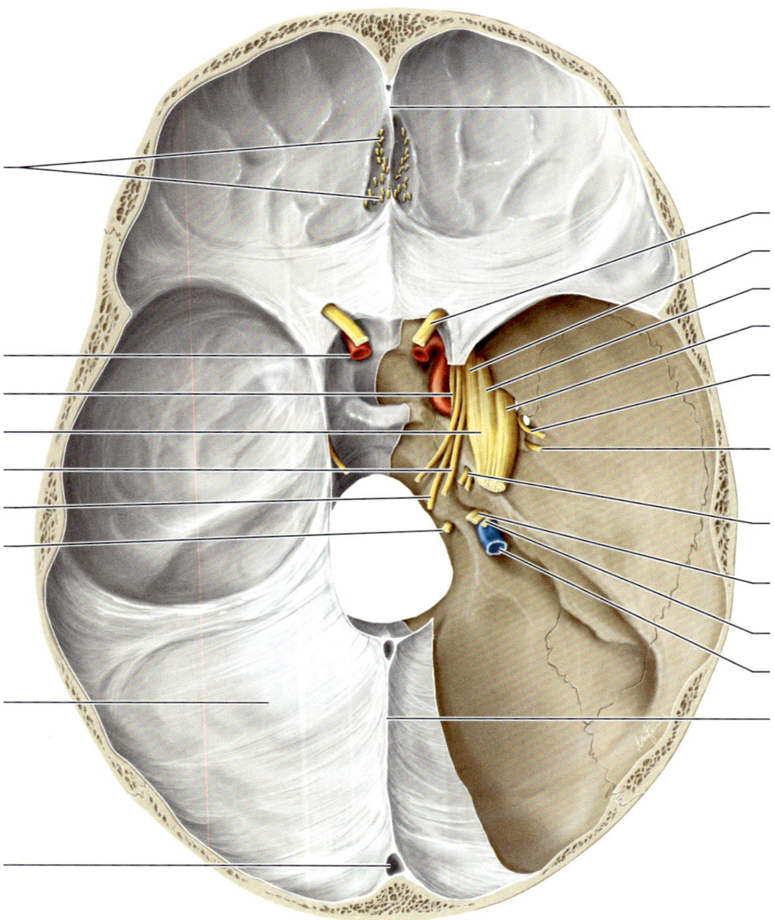

Internal surface of the cranial base, showing the **points of exit of cranial nerves** through the bony foramina.

Canals and canaliculi		
CANAL OR CANALICULUS	***BONE***	***FORMATIONS THAT PASS THROUGH IT***
Alveolar canals	Maxilla	Posterior superior alveolar artery and vein; posterior superior alveolar branches of the superior alveolar nerves
Canal for auditory tube	Temporal bone	Pharyngotympanic tube (auditory tube)
Canal for tensor tympani	Temporal bone	Tensor tympani
Canaliculus for chorda tympani (petrotympanic fissure)	Temporal bone	Chorda tympani
Caroticotympanic canaliculi	Temporal bone	Caroticotympanic arteries and nerves
Carotid canal	Temporal bone	Internal carotid artery; internal carotid venous plexus; internal carotid nerve plexus
Cochlear canaliculus	Temporal bone	Perilymph
Condylar canal	Occipital bone	Condylar emissary vein
Diploic canals	Diploë	Diploic veins
Facial canal	Temporal bone	Facial nerve [VII]
Greater palatine or pterygopalatine canal	Maxilla and palatine bone	Greater palatine and lesser palatine arteries and veins, greater palatine nerve and lesser palatine nerves
Hypoglossal canal	Occipital bone	Meningeal branch of the ascending pharyngeal artery; emissary vein from the sigmoid sinus to the internal jugular vein; hypoglossal nerve [XII]; meningeal branch of the hypoglossal nerve
Incisive canals	Maxilla	Sphenopalatine artery and veins; nasopalatine nerve
Infraorbital canal	Maxilla	Infraorbital and anterior superior alveolar arteries; infraorbital nerve and middle superior and anterior superior alveolar branches of the superior alveolar nerves
Lesser palatine canals	Palatine bone	Lesser palatine arteries, veins and nerves
Mandibular canal	Mandible	Inferior alveolar artery, vein and nerve
Mastoid canaliculus	Temporal bone	Auricular branch of the vagus nerve (at entry)
Musculotubal canal	Temporal bone	Septum of musculotubal canal; canal for auditory tube and canal for tensor tympani
Nasolacrimal canal	Inferior nasal concha, maxilla and lacrimal bone	Nasolacrimal duct
Optic canal	Sphenoid	Ophthalmic artery, central retinal vein; optic nerve [II], autonomic nerve fibers
Palatovaginal or pharyngeal canal	Sphenoid and palatine bone	Pharyngeal branch of the artery of pterygoid canal and of the descending palatine artery; branches of pterygopalatine ganglion and pharyngeal nerve
Pterygoid or vidian canal	Sphenoid	Artery, vein and nerve of the pterygoid canal (mixed sympathetic and parasympathetic nerve)
Tympanic canaliculus	Temporal bone	Tympanic nerve (from the glossopharyngeal nerve [IX])
Vestibular canaliculus	Temporal bone	Vestibular aqueduct containing the endolymphatic sac
Vomerorostral or vomerobasilar canal	Sphenoid and vomer	Small veins
Vomerovaginal canal	Sphenoid and, at times, vomer	Branch of the sphenopalatine artery; branches of the pterygopalatine ganglion

Fissures		
FISSURE	***BONE***	***FORMATIONS THAT PASS THROUGH IT***
Inferior orbital fissure	Sphenoid and maxilla	Infraorbital artery and vein; inferior ophthalmic vein; infraorbital and zygomatic nerves from the maxillary nerve [V_2]; branches of pterygopalatine ganglion
Petrotympanic fissure	Temporal bone	Chorda tympani through a small patent portion, the canaliculus for chorda tympani
Superior orbital fissure	Sphenoid	Superior ophthalmic vein; oculomotor nerve [III], trochlear nerve [IV], lacrimal, frontal and nasociliary nerves (from the ophthalmic nerve, V_1), abducens nerve [VI]
Tympanomastoid fissure	Temporal bone	Auricular branch of vagus nerve (at exit)

Foramina

FORAMEN	BONE	FORMATIONS THAT PASS THROUGH IT
Anterior and posterior incisive foramina	Maxilla	Sphenopalatine artery and vein; left and right nasopalatine nerves
Anterior ethmoidal foramen	Ethmoid	Anterior ethmoidal artery; ethmoidal veins; anterior ethmoidal nerve
Cribriform foramina	Ethmoid	Olfactory nerves [I] (from the nasal cavity to the anterior cranial fossa)
Foramen cecum	Frontal bone	Pharyngeal veins, pterygoid venous plexus, vein from the mucosa of the nasal cavity to the superior sagittal sinus
Foramen lacerum anterior	Sphenoid and temporal bone	Meningeal branch of the ascending pharyngeal artery; artery, vein and nerve of the pterygoid canal; internal carotid artery and greater petrosal nerve (they pass through it only in part)
Foramen magnum or great occipital foramen	Occipital bone	Continuation of the spinal cord in the myelencephalon; vertebral arteries and veins, and anterior and posterior spinal arteries and veins; spinal root of the accessory nerve
Foramen ovale	Sphenoid	Accessory and (sometimes) frontal branches of the middle meningeal artery; emissary vein from the cavernous sinus to the pterygoid plexus; mandibular nerve [V_3]; accessory meningeal nerve; lesser petrosal nerve (sometimes)
Foramen petrosum (not always present)	Temporal bone	Lesser petrosal nerve
Foramen rotundum	Sphenoid	Maxillary nerve [V_2]
Foramen spinosum	Sphenoid	Parietal branch of the middle meningeal artery; middle meningeal veins; nervus spinosus (meningeal branch of the mandibular nerve)
Frontal notch/foramen	Frontal bone	Supratrochlear artery and veins; medial branch of the supraorbital nerve
Greater palatine foramen	Palatine bone	Greater palatine artery, vein and nerve
Infraorbital foramen	Maxilla	Infraorbital artery, vein and nerve
Jugular foramen or foramen lacerum posterior	Temporal bone	Inferior petrosal sinus (anteriorly); meningeal branch of the ascending pharyngeal artery and glossopharyngeal [IX], vagus [X], accessory [XI] nerves (in the center); occipital artery and veins, sigmoid sinus which continues in the internal jugular vein and occipital emissary vein (posteriorly)
Lesser palatine foramina	Palatine bone and maxilla	Lesser palatine arteries, veins and nerves
Mandibular foramen	Mandible	Inferior alveolar artery, vein and nerve
Mastoid foramen	Temporal bone	Mastoid emissary vein
Mental foramen	Mandible	Mental branches of the inferior alveolar artery and vein; mental nerve
Optic foramen	Sphenoid	Optic nerve [II] and ophthalmic artery
Parietal foramen	Parietal bone	Parietal emissary vein
Posterior ethmoidal foramen	Ethmoid	Posterior ethmoidal artery; ethmoidal veins; posterior ethmoidal nerve
Sphenoidal emissary foramen (not always present)	Sphenoid	Emissary vein from the pterygoid plexus to the cavernous sinus (vein of Vesalius)
Sphenopalatine foramen	Palatine bone	Branches of the maxillary nerve to the pterygopalatine ganglion (sphenopalatine nerves); sphenopalatine artery and vein
Stylomastoid foramen	Temporal bone	Stylomastoid branch of the posterior auricular artery and stylomastoid vein; facial nerve [VII]
Supraorbital notch/foramen	Frontal bone	Supraorbital artery, vein and nerve
Zygomatico-orbital foramen	Zygomatic bone	Zygomaticofacial and zygomaticotemporal branches of the zygomatic nerve
Zygomaticofacial foramen	Zygomatic bone	Zygomaticofacial branch of the zygomatic nerve
Zygomaticotemporal foramen	Zygomatic bone	Zygomaticotemporal branch of the zygomatic nerve

Hiatuses

HIATUS	BONE	FORMATIONS THAT PASS THROUGH IT
Hiatus for greater petrosal nerve	Temporal bone	Petrosal branch of the middle meningeal artery; greater petrosal nerve
Hiatus for lesser petrosal nerve	Temporal bone	Lesser petrosal nerve

Meatuses

MEATUS	BONE	FORMATIONS THAT PASS THROUGH IT
External acoustic meatus	Temporal bone	–
Internal acoustic meatus	Temporal bone	Labyrinthine arteries and veins; facial [VII], vestibulocochlear [VIII] and intermediate nerves

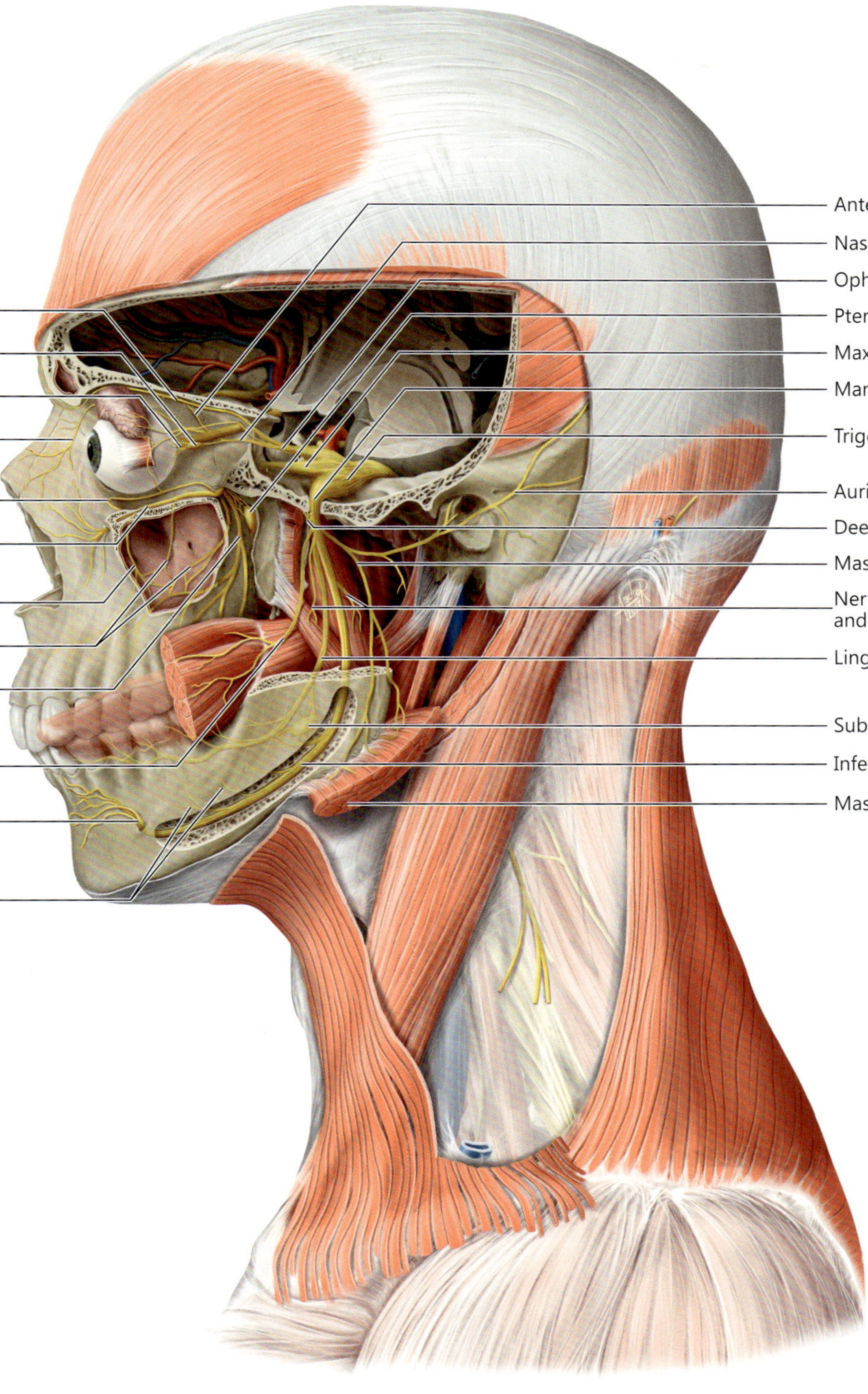

The **trigeminal nerve** [V] divides into three branches: the ophthalmic [V_1], maxillary [V_2] and mandibular [V_3] nerves. The first, a somatic sensory branch, innervates the forehead, calvaria and part of the nose, the eyeball, conjunctiva and part of the mucosa of the nasal cavity. The second branch, also somatic sensory, innervates an extended area of the face and mucosa of the nasal cavities; in addition, it supplies the mucosa of the superior part of the oral cavity and of the superior dento-alveolar arch. The third, a somatic motor and somatic sensory branch, innervates, with its somatic motor component, the masticatory muscles, tensor tympani, tensor veli palatini, anterior belly of the digastric, and mylohyoid, and, with its somatic sensory component, the inferior part of the face, part of the oral mucosa, the inferior dento-alveolar arch, most of the temporal region, part of the auricle, the tympanic membrane, and the masticatory muscles for proprioceptive sensitivity. The three branches, with the exception of the somatic motor component of the mandibular nerve, originate from the convex margin of the trigeminal ganglion or semilunar ganglion.

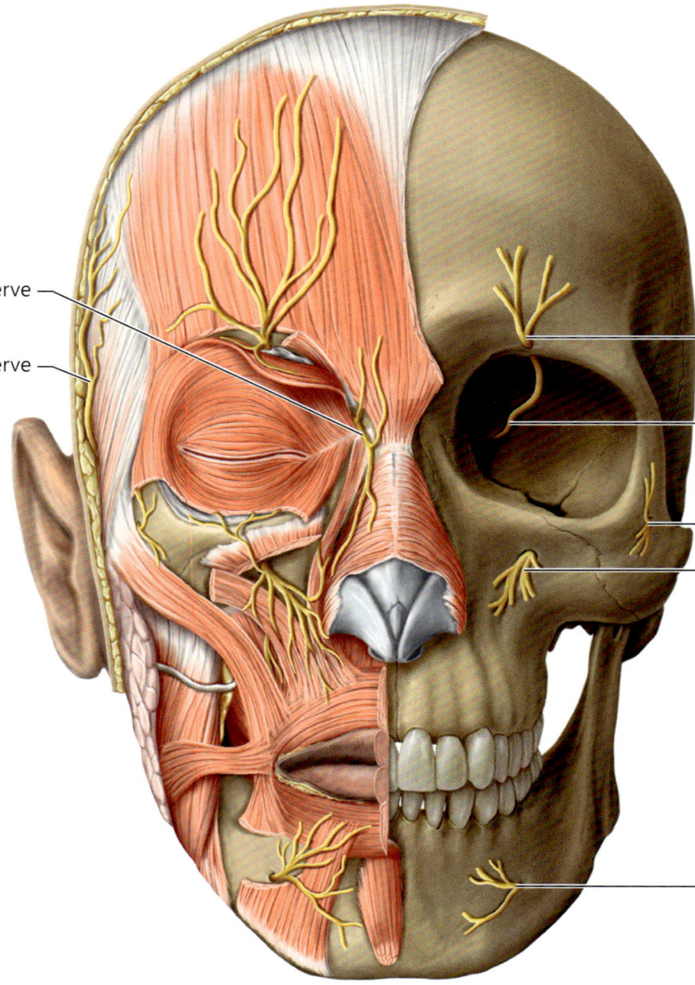

In the left hemicranium, one can see the supraorbital nerve, a branch of the frontal nerve which in turn is a terminal branch of the ophthalmic nerve (**first branch of the trigeminal nerve**), the zygomatic and infraorbital nerves, branches of the maxillary nerve (**second branch of the trigeminal nerve**), and the mental nerve, a branch of the inferior alveolar nerve, in its turn a dividing branch of the mandibular nerve (**third branch of the trigeminal nerve**). They emerge, respectively, from the supraorbital, zygomaticofacial, infraorbital and mental foramina. In the right hemicranium, one can see, in addition to the auriculotemporal nerve, the muscular and, hence also, cutaneous territories pertaining to these nerves. Also seen is the infratrochlear nerve, one of the terminal branches of the nasociliary nerve, which is the medial branch of division of the ophthalmic nerve.

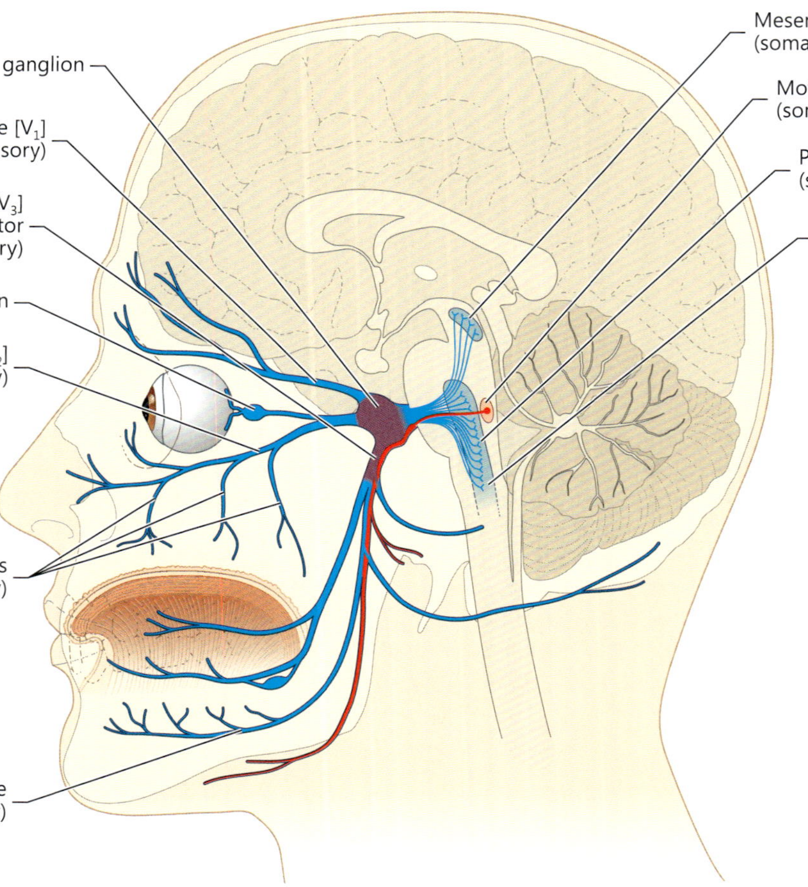

The somatic sensory fibers of the **trigeminal nerve** originate in the trigeminal ganglion. The ganglionic protoneurons send their central projection to the principal and spinal nuclei, and in small part also to the mesencephalic nucleus, and their peripheral projection to the three nerve branches: ophthalmic, maxillary and mandibular. The mesencephalic nucleus receives central projections of ganglionic protoneurons that gather principally the proprioceptive sensitivity of muscles. The mesencephalic nucleus is also composed of protoneurons that receive the mechanoreceptive sensitivity from the desmodentium of the superior and inferior dental arcades. The somatic motor component originates from the motor neurons of the motor nucleus, also known as the masticatory nucleus.

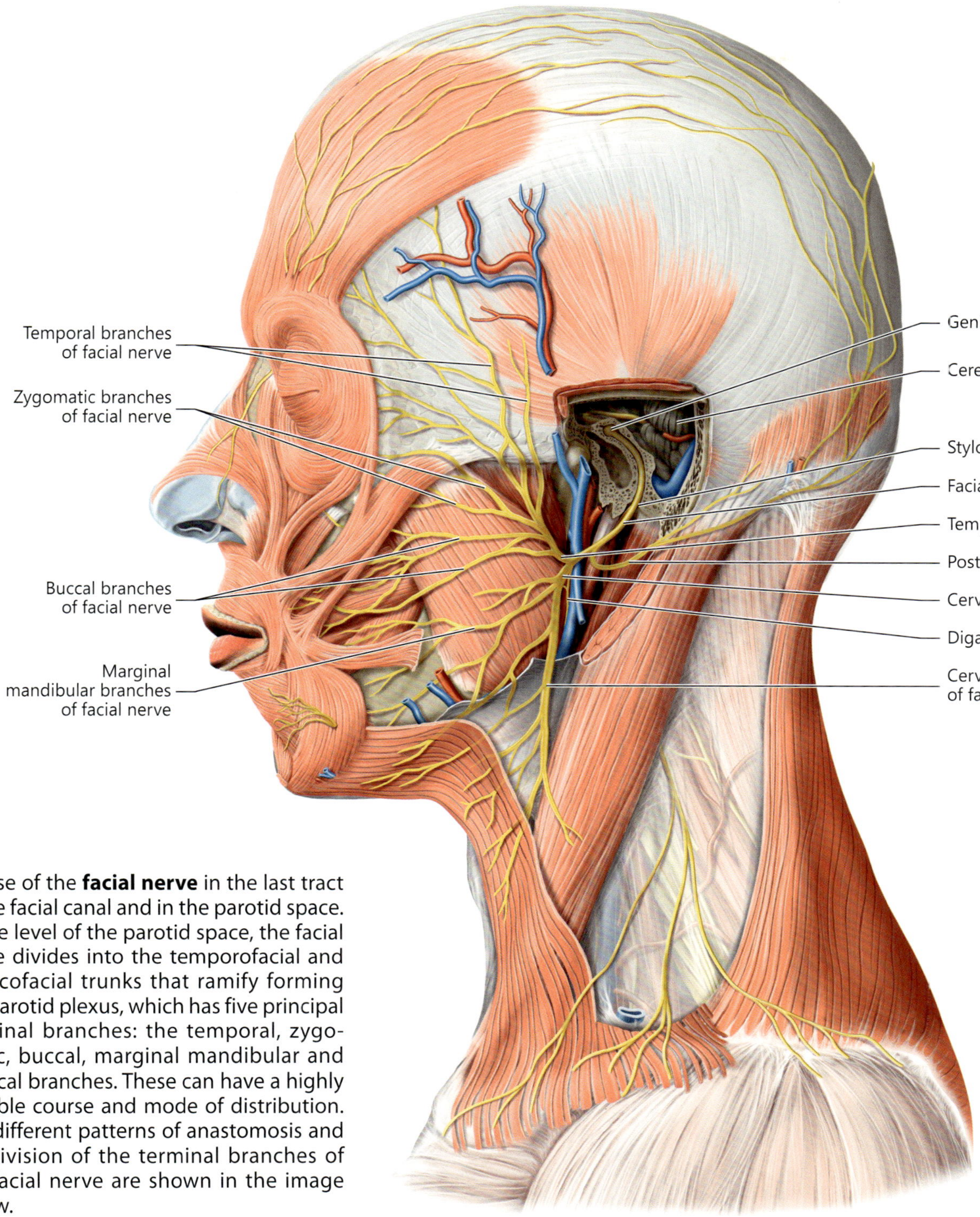

Course of the **facial nerve** in the last tract of the facial canal and in the parotid space. At the level of the parotid space, the facial nerve divides into the temporofacial and cervicofacial trunks that ramify forming the parotid plexus, which has five principal terminal branches: the temporal, zygomatic, buccal, marginal mandibular and cervical branches. These can have a highly variable course and mode of distribution. The different patterns of anastomosis and subdivision of the terminal branches of the facial nerve are shown in the image below.

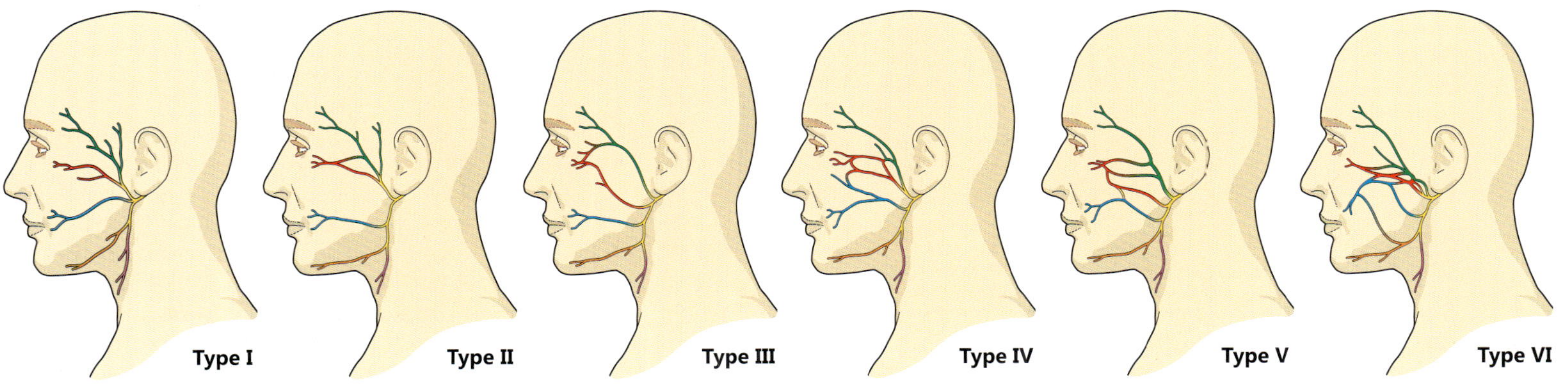

Modified from Berkovits BKB, Moxham BJ 2002, Head and Neck Anatomy. London: Martin Dunitz.

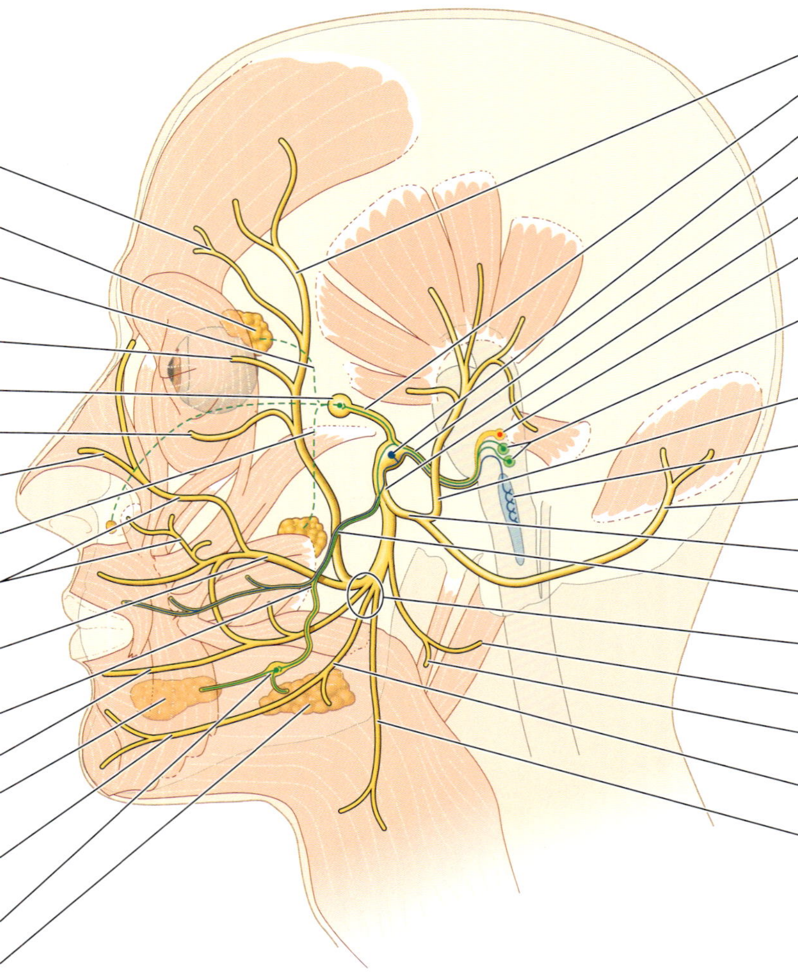

Schematic representation of the **facial nerve** and its branches. The somatic motor component appears clearly predominant with respect to the visceral motor (in *green*) and special visceral sensory (gustatory, in *blue*) components.

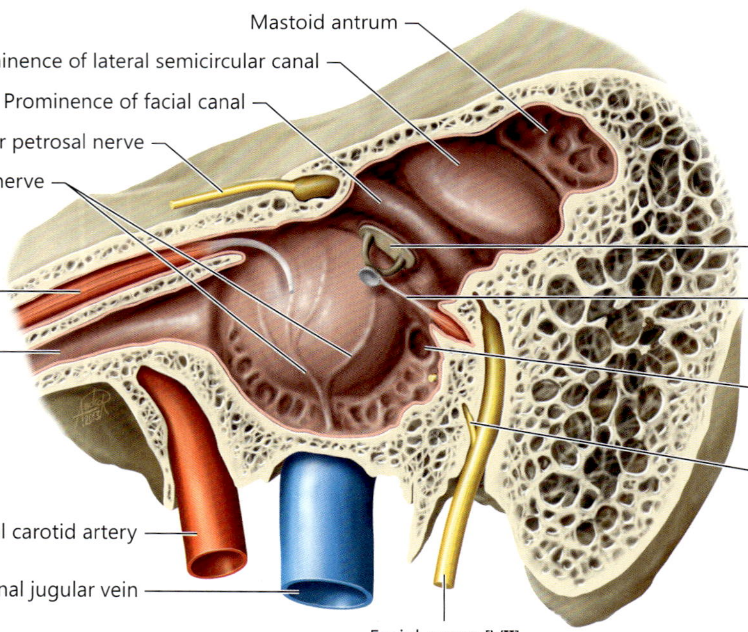

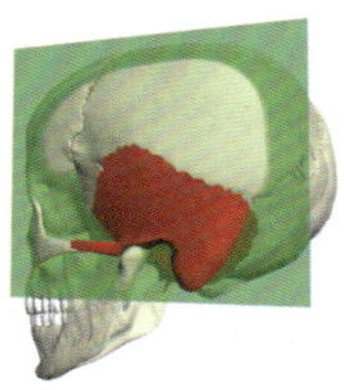

Tympanic cavity, opened, and viewed from the exterior; one can see the **intrapetrous course of the facial nerve** and the facial canal in relation to the internal carotid artery and internal jugular vein. The facial nerve exits from the cranium in front of the mastoid process, passing through the stylomastoid foramen, which, in this diagram, has been partially removed. One can also see the relationship of the facial nerve, in its intrapetrous journey and near the point where it emerges from the cranium, with the bone cavities of the mastoid antrum.

Summary table of the nerves of the head			
NERVE	**ORIGIN**	**RAMIFICATIONS**	**DISTRIBUTION**
abducens, VI cranial nerve	Emerges from the brainstem in correspondence to inferior border of the pons	Filaments	Lateral rectus of the eyeball
anterior ampullary (superior ampullary)	Vestibular nerve [VIII]	Filaments	Ampulla of the anterior semicircular duct
anterior auricular (generally two in number)	Auriculotemporal nerve	Filaments	Skin of the anterosuperior part of the external ear (principally of the helix and tragus)
anterior ethmoidal	Nasociliary nerve	Internal, lateral and medial nasal branches, and external nasal branch	Mucosa of the anterior ethmoidal cells and the nasal cavities
– **internal nasal branches**	Anterior ethmoidal nerve	Filaments	Mucosa of the vault of the nasal cavities
– **external nasal nerve**	Anterior ethmoidal nerve	Filaments	Skin of the ala of nose, vestibule and dorsum of the nose, including the apex of the nose
auriculotemporal	Mandibular nerve [V_3]	Parotid and superficial temporal branches, branches to the tympanic membrane, communicating branches with the facial nerve; nerve to external acoustic meatus and anterior auricular nerves	Skin anterior to the auricle and of the posterior two-thirds of the temporal region; skin of the tragus and adjacent to the helix; skin of the roof of the external acoustic meatus and of the superior part of the tympanic membrane; temporomandibular joint
buccal (buccinator)	Mandibular nerve [V_3]	Collateral branches and anastomotic branches with the buccal branches of the facial nerve	Skin and oral mucosa of the cheek (above and deep to the anterior part of the buccinator); vestibular gingiva adjacent to second and third molar teeth
caroticotympanic	Superior cervical ganglion and tympanic plexus	Filaments	Internal ear, pharyngotympanic tube
chorda tympani	Intermediate nerve [VII] and lingual nerve	Filaments	Anterior two-thirds of the tongue, submandibular and sublingual glands
cochlear [VIII] (acoustic)	Vestibulocochlear nerve [VIII]	Filaments	Spiral organ (of Corti)
cranial (encephalic)	12 nerve pairs: olfactory [I]; optic [II]; oculomotor [III]; trochlear [IV]; trigeminal [V]; abducens [VI]; facial [VII]; vestibulocochlear [VIII]; glossopharyngeal [IX]; vagus [X]; accessory [XI]; hypoglossal [XII]		
deep petrosal	Internal carotid plexus	Unites with the greater petrosal nerve to form the nerve of pterygoid canal	Glands of the nasal and palatine mucosae, lacrimal gland
deep temporal (generally two in number)	Mandibular nerve [V_3]	Filaments	Temporalis
external carotid	Superior cervical ganglion	Filaments	External carotid artery and its branches, sublingual and submandibular glands
facial, VII cranial nerve (facial nerve proper)	Emerges from the brainstem in correspondence to the inferior border of the pons	Nerve to stapedius and posterior auricular nerve: occipital and auricular branches; digastric branch, stylohyoid branch, communicating branch with the hypoglossal nerve; parotid plexus: temporal, zygomatic, buccal, marginal mandibular and cervical branches	Facial muscles, muscles of the scalp and auricle; buccinator, platysma, stapedius, stylohyoid and posterior belly of the digastric
frontal	Ophthalmic nerve [V_1]	Supraorbital and supratrochlear nerves	Conjunctiva, skin of the superior eyelid and the forehead, corrugator supercilii and frontal belly of the occipitofrontalis, scalp and frontal sinus
glossopharyngeal, IX cranial nerve	Emerges from the superior part of the myelencephalon	Tympanic nerve: tympanic plexus, tubal branch and caroticotympanic nerves; pharyngeal, tonsillar and lingual branches, stylopharyngeal branch and carotid branch; communicating branches with the auricular branch of vagus nerve, with the meningeal branch, with the auriculotemporal nerve and with the chorda tympani; lesser petrosal nerve	Mucosa of the middle ear and pharyngotympanic tube, tongue, pharynx, fauces, palatine tonsil, baroreceptors of the carotid sinus, stylopharyngeus, parotid gland

continued

Summary table of the nerves of the head *(continued)*

NERVE	ORIGIN	RAMIFICATIONS	DISTRIBUTION
great auricular	Cervical plexus (from the anterior rami of C2 and C3)	Posterior and anterior rami and anastomotic branches with the facial nerve	Skin above the angle of the mandible and inferior to the lobule of the auricle; skin of the auricle; parotid fascia
greater occipital	Posterior ramus of C2	Terminal branches and anastomotic branches with the posterior rami of C1 and C3	Scalp of the occipital region; obliquus capitis inferior, semispinalis capitis, longissimus capitis, splenius capitis, splenius cervicis and trapezius
greater palatine (anterior palatine)	Maxillary nerve [V_2]	Posterior inferior nasal nerves	Gingivae, mucosa, glands and taste buds of the palate
– **posterior inferior nasal nerves**	Greater palatine nerve	Filaments	Mucosa of the nasal conchae
greater petrosal (greater superficial petrosal)	Intermediate nerve [VII]	Unites with the sympathetic root of pterygopalatine ganglion to form the nerve of pterygoid canal	Nasal and palatine glands, lacrimal gland
hypoglossal, XII cranial nerve	Emerges from the anterolateral sulcus of the myelencephalon, between the pyramid and inferior olive	Lingual and anastomotic branches with the cervical plexus	Intrinsic and extrinsic muscles of the tongue
inferior alveolar (inferior dental)	Mandibular nerve [V_3]	Nerve to mylohyoid, inferior dental plexus: inferior dental branches and inferior gingival branches, mental nerve	Mylohyoid and anterior belly of the digastric, periodontium and pulp of the mandibular teeth, gingiva of the alveolar arch of the mandible, skin of the chin, mucosa of the lower lip
infraorbital	Maxillary nerve [V_2]	Middle and anterior branches of the superior alveolar nerves, palpebral branches, nasal branches and superior labial branches	Maxillary sinus; premolar teeth, canine teeth and maxillary incisors; skin and palpebral conjunctiva of the inferior eyelid; skin of the cheek, lateral part of the nose and anteroinferior to the nasal septum; skin and oral mucosa of the upper lip
– **inferior palpebral branches**	Infraorbital	Filaments	Inferior eyelid
infratrochlear	Nasociliary nerve	Palpebral branches	Skin lateral to the root of the nose; skin and palpebral conjunctiva adjacent to the medial angle of the eye, to the lacrimal sac and to the lacrimal caruncle
– **palpebral branches**	Infratrochlear	Filaments	Superior eyelid
intermediate [VII] (glossopalatine or of Wrisberg)	Emerges from the brainstem in correspondence to the inferior border of the pons	Greater petrosal nerve and chorda tympani; communicating branches with the tympanic plexus and the vagus nerve	Gustatory papillae of the anterior two-thirds of the tongue, nasal and soft palate glands, submandibular, sublingual and lacrimal glands, skin of the external acoustic meatus and the mastoid process
internal carotid	Superior cervical ganglion	Filaments	Internal carotid artery and its branches, dilator pupillae, salivary, lacrimal and pineal glands
lacrimal	Ophthalmic nerve [VI]	Terminal branches	Lacrimal gland (postganglionic secretory fibers originating in the pterygopalatine ganglion that reach the lacrimal nerve via the zygomaticotemporal branch of the zygomatic nerve); small area of the skin and palpebral conjunctiva of the lateral third of the superior eyelid
lateral ampullary	Vestibular nerve [VIII]	Filaments	Ampulla of the lateral semicircular duct
lesser occipital	Cervical plexus (from the anterior rami of C2 and C3)	Filaments	Scalp posterior to the auricle; skin of the occipital and mastoid regions

continued

Summary table of the nerves of the head			*(continued)*
NERVE	***ORIGIN***	***RAMIFICATIONS***	***DISTRIBUTION***
lesser palatine	Maxillary nerve [V_2]	Tonsillar branches	Soft palate, uvula, palatine tonsil
lesser petrosal	Glossopharyngeal nerve [IX]	Ganglionic, filaments	Parotid and minor salivary glands
lingual (small hypoglossal)	Mandibular nerve [V_3]	Branches to the isthmus of fauces, communicating branches with the hypoglossal nerve, lingual and ganglionic branches (to the submandibular ganglion and sublingual ganglion); chorda tympani and sublingual nerve	Mucosa of anterior two-thirds of the tongue, floor of the mouth, gingivae and submandibular and sublingual glands
long ciliary (2 or 3 in number)	Nasociliary nerve	Filaments	Iris and cornea
mandibular [V_3] (inferior maxillary)	Trigeminal nerve [V]	Nerve to medial pterygoid and nerve to lateral pterygoid, nerve to tensor veli palatini and nerve to tensor tympani, masseteric nerve; deep temporal, auriculotemporal, buccal, lingual, inferior alveolar and mental nerves; meningeal branch and branch to otic ganglion	Masticatory muscles, tensor tympani, tensor veli palatini, mylohyoid and anterior belly of the digastric; mandible, gingiva and mandibular teeth; anterior two-thirds of the tongue, cheek, inferior part of the face; meninges, temporomandibular joint, skin of the temporal region, external ear
– **meningeal branch** (nervus spinosus)	Mandibular nerve [V_3]	Filaments	Meninges
masseteric	Mandibular nerve [V_3]	Filaments	Masseter, temporomandibular joint
maxillary [V_2] (superior maxillary)	Trigeminal nerve [V]	Nasopalatine, pharyngeal, greater palatine, lesser palatine, superior alveolar, zygomatic and infraorbital nerves; meningeal, ganglionic (to pterygopalatine ganglion), orbital, and lateral and medial posterior superior nasal branches	Skin of the middle part of the face, nose, inferior eyelids and upper lip; gingiva and maxillary teeth; tonsils and floor of the mouth, soft palate, maxillary sinus, mucosa of the nasopharynx
– **meningeal branch** (middle meningeal branch)	Maxillary nerve [V_2]	Filaments	Dura mater of the middle cranial fossa
– **ganglionic branches** (to pterygopalatine ganglion) (pterygopalatine or sphenopalatine nerves or nerves of Bock)	Maxillary nerve [V_2]	Postganglionic nerve fibers	Glands of the nasal and palatine mucosae, lacrimal gland
– **posterior superior nasal branches**	Maxillary nerve [V_2]	Filaments	Vault of the pharynx, lateral wall of the nasal cavities, mucosa of the nasal septum, glands of the nasal mucosa
mental	Mandibular nerve [V_3]	Mental, labial and gingival branches	Skin of the chin; skin and oral mucosa of the lower lip
nasociliary (nasal)	Ophthalmic nerve [V_1]	Communicating branches with the ciliary ganglion, long ciliary nerves	Mucosa of the nasal cavity, frontal sinus; iris, cornea, conjunctiva, lacrimal sac, skin of the eyelids and ala of nose
nasopalatine (of Scarpa)	Maxillary nerve [V_2]	Filaments	Anterior part of the hard palate
oculomotor, III cranial nerve (common oculomotor)	Mesencephalon in correspondence to the medial surface of the cerebral peduncle	Superior, inferior and ganglionic branches (to the ciliary ganglion)	Extraocular muscles: superior rectus, levator palpebrae superioris, medial rectus, inferior rectus and inferior oblique; intrinsic muscles of the eyeball: sphincter pupillae, ciliary muscle
of pterygoid canal (vidian)	Formed by the union of the greater petrosal and deep petrosal nerves	Filaments	Glands of the nasal and palatine mucosae, lacrimal gland

continued

Summary table of the nerves of the head *(continued)*

NERVE	*ORIGIN*	*RAMIFICATIONS*	*DISTRIBUTION*
olfactory, I cranial nerve	Olfactory part of the nasal mucosa	Filaments	Olfactory bulb
ophthalmic [V_1]	Trigeminal nerve [V]	Lacrimal, frontal, nasociliary, anterior and posterior ethmoidal and infratrochlear nerves; recurrent meningeal branch or tentorial nerve	Dura mater, eyeball, conjunctiva, lacrimal gland; mucous membrane of the nose and the paranasal sinuses; skin of the forehead, the eyelids and the nose
– **recurrent meningeal branch** (tentorial nerve or of Arnold)	Ophthalmic nerve [V_1]	Filaments	Dura mater bordering the cavernous sinus and forming the tentorium cerebelli
optic, II cranial nerve	Ganglionic layer of the retina	Filaments	Lateral geniculate body
posterior ampullary (inferior ampullary)	Vestibular nerve [VIII]	Filaments	Ampulla of the posterior semicircular duct
posterior auricular	Facial nerve [VII]	Auricular and occipital branches	Auricularis posterior and occipital belly of the occipitofrontalis, skin of the external ear
posterior ethmoidal	Nasociliary nerve	Anterior meningeal branch	Mucosa of the posterior ethmoidal cells and the paranasal sinuses; meninges
– **anterior meningeal branch**	Posterior ethmoidal nerve	Filaments	Meninges
saccular	Vestibular nerve [VIII]	Filaments	Saccule
short ciliary (from 6 to 10 in number)	Ciliary ganglion	Filaments	Ciliary body, iris and cornea
sublingual	Lingual nerve	Filaments	Sublingual gland and mucosa of the floor of the mouth
superior alveolar (superior dental)	Maxillary nerve [V_2]	Anterior and posterior superior alveolar branches, middle superior alveolar branch	Maxillary sinus, periodontium and pulp of the maxillary teeth, gingiva of the alveolar arch of the maxilla
– **anterior superior alveolar branches** (anterior superior dental)	Superior alveolar nerves	Superior dental plexus: superior dental branches and superior gingival branches	Maxillary sinus, periodontium and pulp of the maxillary teeth, gingiva of the alveolar arch of the maxilla, floor of the nose
– **posterior superior alveolar branches** (posterior superior dental)	Superior alveolar nerves	Filaments	Maxillary sinus, periodontium and pulp of the maxillary teeth, gingiva of the alveolar arch of the maxilla
– **middle superior alveolar branch** (middle superior dental)	Superior alveolar nerves	Filaments	Maxillary sinus, periodontium and pulp of the maxillary teeth, gingiva of the alveolar arch of the maxilla
supraorbital	Frontal nerve	Medial and lateral branches	Skin and palpebral conjunctiva of the middle third of the superior eyelid; skin and periosteum of the anterolateral part of the forehead and scalp up to the vertex of the calvaria
supratrochlear	Frontal nerve	Lateral and medial descending branches	Skin and palpebral conjunctiva of the medial third of the superior eyelid; skin and periosteum of the anterolateral part of the forehead
terminal	Associated to the olfactory nerve [I]	Filaments	Inferior and anterior parts of the nasal septum
third occipital	Posterior ramus of C3	Filaments	Scalp of the inferior occipital and suboccipital regions
to external acoustic meatus	Auriculotemporal nerve	Filaments	External acoustic meatus

continued

Summary table of the nerves of the head (continued)

NERVE	ORIGIN	RAMIFICATIONS	DISTRIBUTION
to lateral pterygoid	Mandibular nerve [V_3]	Filaments	Lateral pterygoid
to medial pterygoid	Mandibular nerve [V_3]	Filaments	Medial pterygoid
to stapedius	Facial nerve [VII]	Filaments	Stapedius
to tensor tympani	Mandibular nerve [V_3]	Filaments	Tensor tympani
to tensor veli palatini	Mandibular nerve [V_3]	Filaments	Tensor veli palatini
trigeminal, V cranial nerve (largest of the cranial nerves)	Emerges from the brainstem in correspondence to the inferior surface of the pons	Two roots (motor and sensory) that reach the trigeminal ganglion near the apex of the petrous part of the temporal bone, from which arise the ophthalmic [V_1], maxillary [V_2] and mandibular [V_3] branches or nerves	Skin of the face, masticatory muscles, teeth, mouth and nasal cavity
trochlear, IV cranial nerve (smallest of the cranial nerves)	Emerges from the mesencephalon immediately behind the inferior colliculus	Filaments	Superior oblique of the eyeball
tympanic (Jacobson's nerve)	Glossopharyngeal nerve [IX]	Tympanic plexus: tubal branch and caroticotympanic nerves	Tympanic cavity and membrane, mastoid cells, pharyngotympanic tube, parotid gland
utricular	Vestibular nerve [VIII]	Filaments	Utricle
utriculoampullary	Vestibular nerve [VIII]	Filaments	Macula of utricle and ampulla of the lateral and anterior semicircular ducts
vagus, X cranial nerve (pneumogastric)	Emerges from the posterior lateral sulcus of the myelencephalon	Meningeal, auricular, communicating with the glossopharyngeal nerve, pharyngeal, superior and inferior cervical cardiac, thoracic cardiac, bronchial and renal branches; anterior (anterior gastric and hepatic branches) and posterior (posterior gastric and celiac branches) vagal trunks; superior laryngeal nerve (external and internal branches) and recurrent laryngeal nerve (tracheal, esophageal and pharyngeal branches)	Heart, wall of the aorta and of its major branches; glands and smooth muscle of the respiratory tract, of the digestive tract (except for its terminal part) and of part of the urinary system; striated muscle of the pharynx, of the larynx and of the proximal part of the esophagus; carotid sinus, carotid body; mucosa of the larynx, pharynx, trachea, esophagus and other thoracic and abdominal organs; taste buds of the mucosa of the epiglottis and oropharynx; skin of the auricle, of the external acoustic meatus and of the lateral surface of the tympanic membrane
– **meningeal branch**	Vagus nerve [X]	Filaments	Meninges
vertebral	Inferior cervical ganglion	Filaments	Vertebral artery and its extracranial and intracranial branches
vestibular [VIII]	Vestibulocochlear nerve [VIII]	Utricular, utriculoampullary, anterior ampullary, lateral ampullary, posterior ampullary and saccular nerves	Macula of the utricle and the saccule and ampulla of the semicircular ducts
vestibulocochlear, VIII cranial nerve (statoacoustic)	Emerges from the brainstem between the pons and the myelencephalon, formed by the union of the vestibular and cochlear nerves	Vestibular (medial), cochlear (lateral)	Membranous labyrinth of the internal ear
vomeronasal	Present in the nasal septum of the fetus but disappears before birth		
zygomatic	Maxillary nerve [V_2]	Zygomaticotemporal and zygomaticofacial	Skin of the temple and the zygomatic region
– **zygomaticofacial branch**	Zygomatic nerve	Filaments	Skin on the protuberant part of the cheek
– **zygomaticotemporal branch**	Zygomatic nerve	Branch to the lacrimal nerve	Anterior part of the glabrous skin of the temporal fossa

N.B.: the accessory nerve (XI cranial nerve), given that it emerges at the level of the first 4-5 cervical neuromeres and supplies the superior part of the trapezius and sternocleidomastoid, is reported in the table summarizing nerves of the neck.

Summary table of the nerves of the neck

NERVE	ORIGIN	RAMIFICATIONS	DISTRIBUTION
accessory (spinal accessory), XI cranial nerve	*Cranial root*: lateral face of the myelencephalon *Spinal root*: first five cervical segments of the spinal cord	Internal and external branches	*Cranial root*: via the pharyngeal branches of the vagus nerve, to the pharyngeal muscles and muscles of the palate, with the exception of the tensor veli palatini (innervated by the nerve to medial pterygoid, a branch of the mandibular nerve, third branch of the trigeminal nerve), and, via the recurrent laryngeal nerve, to the thyroarytenoid and lateral cricoarytenoid *Spinal root*: sternocleidomastoid and trapezius
ansa cervicalis (ansa of hypoglossal nerve)	*Superior root*: anastomotic branch with the hypoglossal nerve arising from the anterior ramus of the first cervical nerve *Inferior root* (or descendens cervicalis): anastomotic branches with the hypoglossal nerve arising from anterior rami of the second and third cervical nerves that flow into a common trunk which joins the superior root	Filaments	Omohyoid, sternohyoid and sternothyroid
brachial plexus	Anterior rami from the fifth to eighth cervical nerves and of the first thoracic nerve	Trunks (primary trunks) and cords (secondary trunks)	Upper limb (sensory and motor innervation)
cervical (8 pairs of spinal nerves)	Cervical segments of the spinal cord	Anterior and posterior rami	Cervical and brachial plexuses (anterior rami), skin and superficial and deep muscles in the posterior part of the neck (posterior rami)
cervical plexus	Anterior rami from the first to fourth cervical nerves	*Cutaneous branches*: lesser occipital, great auricular, transverse cervical and supraclavicular nerves; *muscular branches*: nerves to rectus capitis anterior and rectus capitis lateralis, longus capitis and longus colli, geniohyoid, thyrohyoid, omohyoid, sternohyoid, sternocleidomastoid, trapezius, levator scapulae, scalenus medius, phrenic nerve; *anastomotic branches*: gray rami communicantes with the sympathetic trunk, anastomotic branches with the hypoglossal nerve (ansa cervicalis)	Muscles and skin of the neck and the superior part of the back; diaphragm
glossopharyngeal, IX cranial nerve	Posterior lateral sulcus of the myelencephalon (sulcus of mixed nerves)	Tympanic, carotid, pharyngeal, stylopharyngeal tonsillar, lingual branches	Tongue and pharynx, fauces, palatine tonsil, baroreceptors of the carotid sinus, stylopharyngeus
jugular	Superior cervical ganglion	Filaments	Superior ganglion of the vagus nerve and inferior ganglion of the glossopharyngeal nerve
recurrent laryngeal (inferior laryngeal)	Vagus nerve	Pharyngeal, esophageal and tracheal branches	All intrinsic muscles of the larynx, excepting the cricothyroid
suboccipital	Posterior rami of the first cervical nerve	Filaments	Suboccipital muscles
superior laryngeal	Vagus nerve, near the inferior ganglion situated in correspondence to the jugular foramen	External and internal	Mucosa of the larynx, in particular of the epiglottis; inferior constrictor of pharynx and cricothyroid
superior, middle and inferior cervical ganglion	Cervical segment of the sympathetic trunk	Gray rami communicantes with the cervical nerves and white ramus communicans with the first thoracic nerve *Perivascular branches*: external carotid and vertebral nerves *Visceral branches*: laryngopharyngeal, superior, middle and inferior cervical cardiac nerves	Vessels and organs connected to the neck, spinal and meningeal vessels of the neck, cardiac plexus

continued

Summary table of the nerves of the neck			(continued)
NERVE	***ORIGIN***	***RAMIFICATIONS***	***DISTRIBUTION***
to geniohyoid	First cervical nerve via fibers traveling with the hypoglossal nerve	Filaments	Geniohyoid
to mylohyoid	Inferior alveolar nerve (branch of the mandibular nerve, third branch of the trigeminal nerve) just before entering the mandibular foramen	Filaments	Mylohyoid and inferior belly of the digastric
to omohyoid	Ansa cervicalis	Filaments	Omohyoid
to sternohyoid	Ansa cervicalis	Filaments	Sternohyoid
to sternothyroid	Ansa cervicalis	Filaments	Sternothyroid
to thyrohyoid (thyrohyoid branch of C1)	First cervical nerve	Filaments branching out via the hypoglossal nerve	Thyrohyoid
transverse cervical (cutaneous cervical)	Cervical plexus (from the anterior rami of C2 and C3)	Ascending and descending branches	Skin of the suprahyoic and infrahyoid regions and of the lateral cervical region
vagus, X cranial nerve	Posterior lateral sulcus of the myelencephalon (sulcus of mixed nerves)	Cervical collateral branches: meningeal, auricular, pharyngeal, superior and inferior cervical cardiac, superior laryngeal and recurrent laryngeal	In the neck: all muscles of the larynx, trachea, pharynx, esophagus
vertebral	Inferior cervical ganglion	Filaments	Spinal and meningeal vessels of the neck

N.B.: The posterior ramus of the fourth cervical nerve [C4], via its medial and lateral branches, innervates the skin of the posterior cervical region. The vagus, glossopharyngeal and recurrent laryngeal (a branch of the vagus) nerves ramify in the neck forming a network of pharyngeal branches, the pharyngeal plexus, which supplies the pharynx. The pharynx also receives laryngopharyngeal branches from the superior cervical ganglion of the sympathetic trunk.

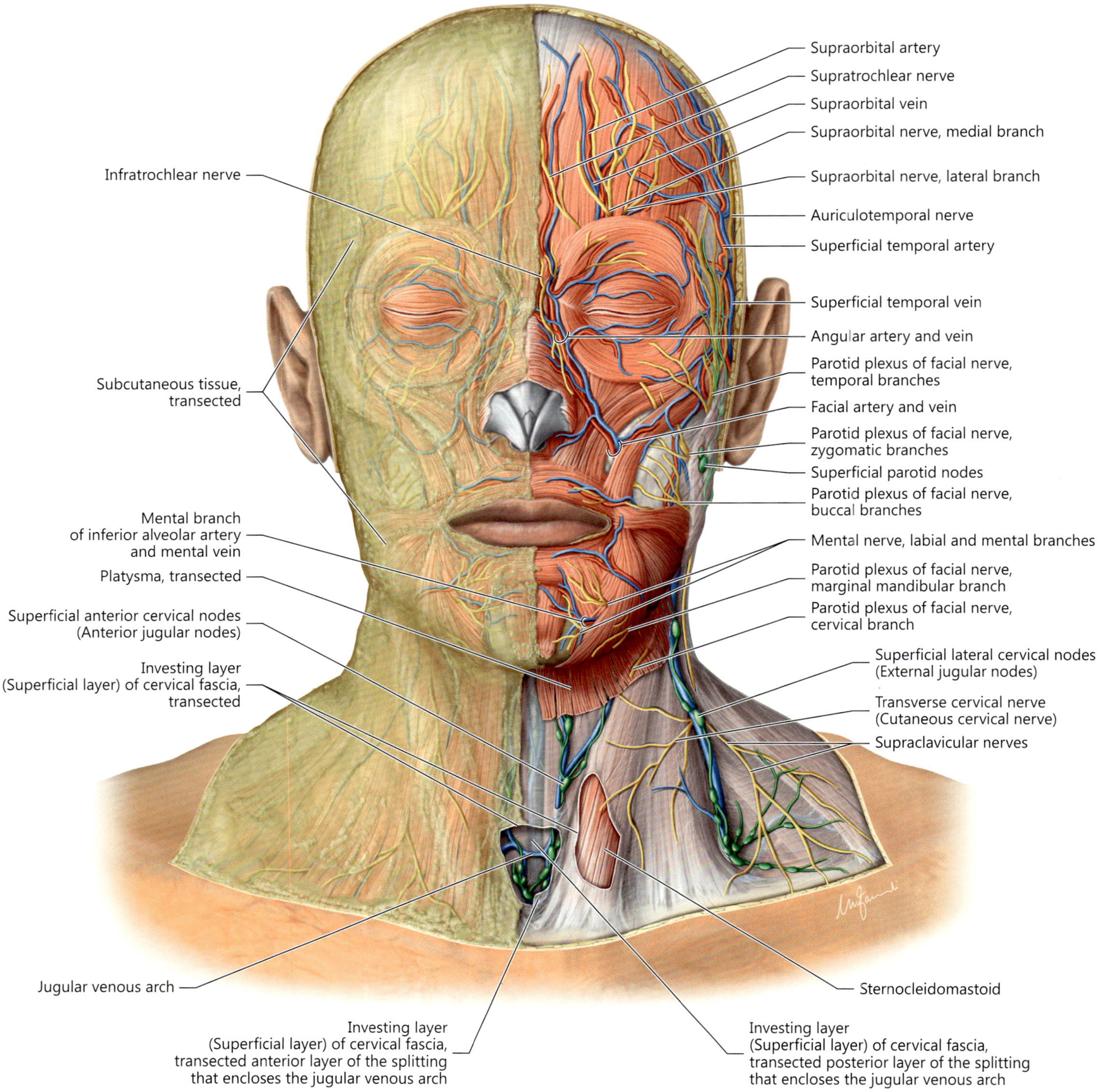

Head and **neck**, anterior view. On the left side of the head and neck, the skin and subcutaneous tissue have been removed and the platysma transected. One can see the vessels and nerves that supply the skin and muscles of the face, in particular the parotid plexus, which gives rise to the temporal, zygomatic, buccal, marginal mandibular and cervical branches of the facial nerve.

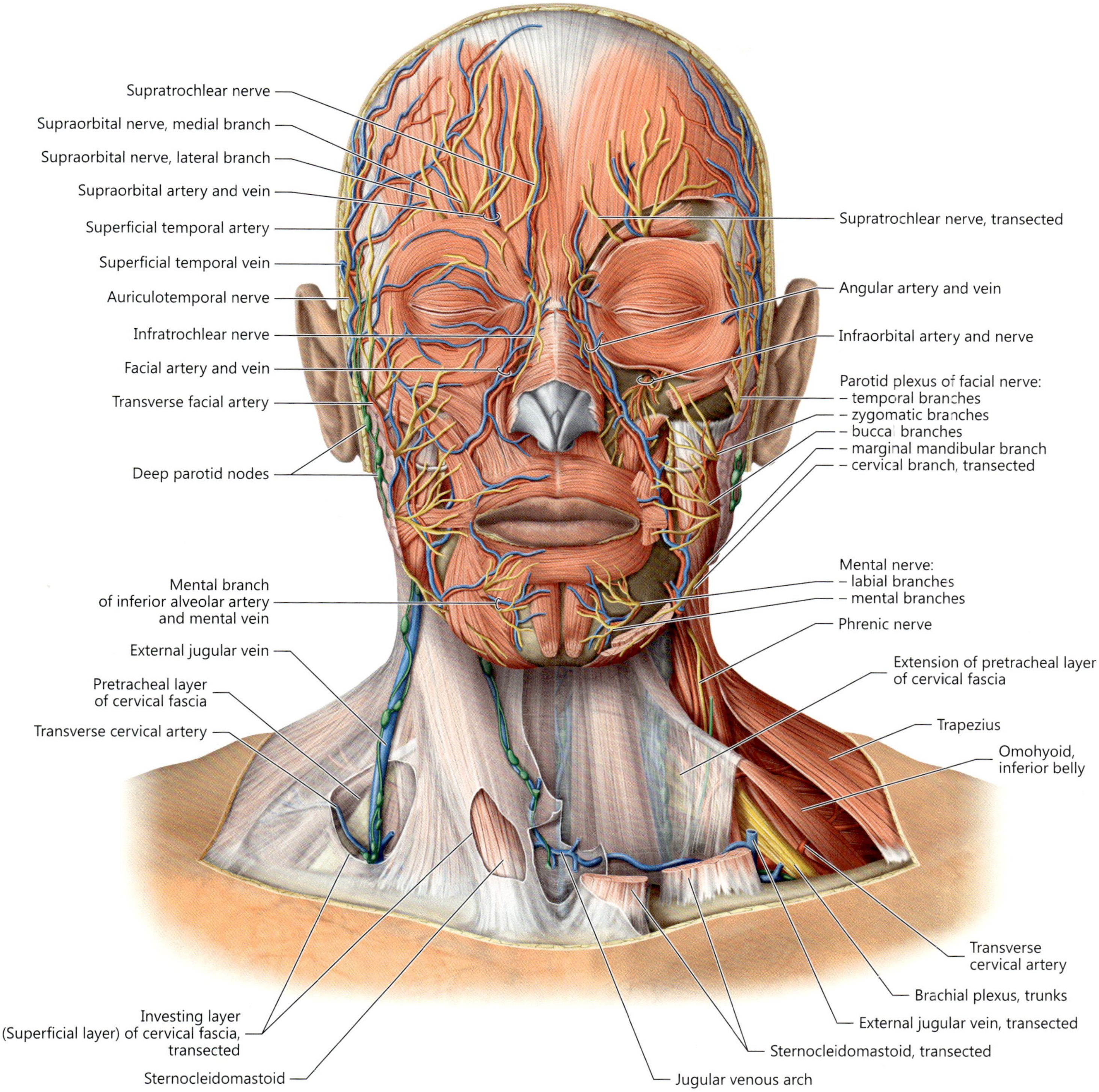

Head and **neck**, anterior view. On the right side of the head and neck, after the skin, subcutaneous tissue and platysma have been removed and the parotid fascia transected, one can see the parotid plexus of the facial nerve, the supratrochlear and supraorbital nerves, branches of the frontal nerve, and the auriculotemporal nerve, branch of the mandibular nerve. On the left side of the head, the orbicularis oculi, zygomaticus major and minor, levator labii superioris, levator anguli oris, depressor labii inferioris and depressor anguli oris have been transected. Besides the terminal branches of the parotid plexus of the facial nerve, one can see the facial artery and vein.

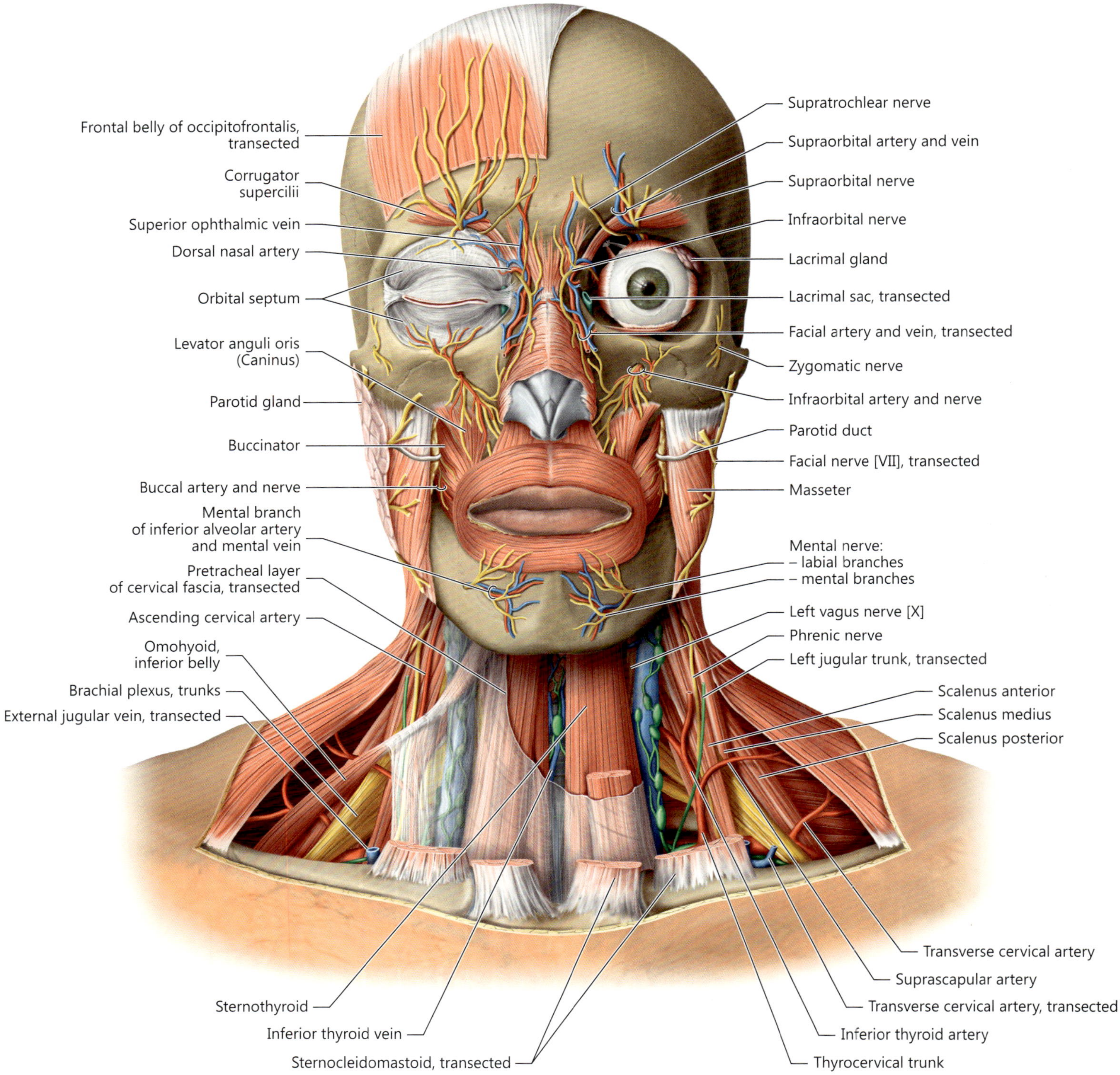

Head and **neck**, anterior view. On the right side of the cranium, the facial muscles have been transected to show the formations that emerge from the supraorbital, zygomaticofacial, infraorbital and mental foramina. On the left side, the eyeball is visible after removal of the orbital septum and tarsus of both eyelids, which together form a diaphragm separating the content of the orbit from the superficial layers of the eyelids.

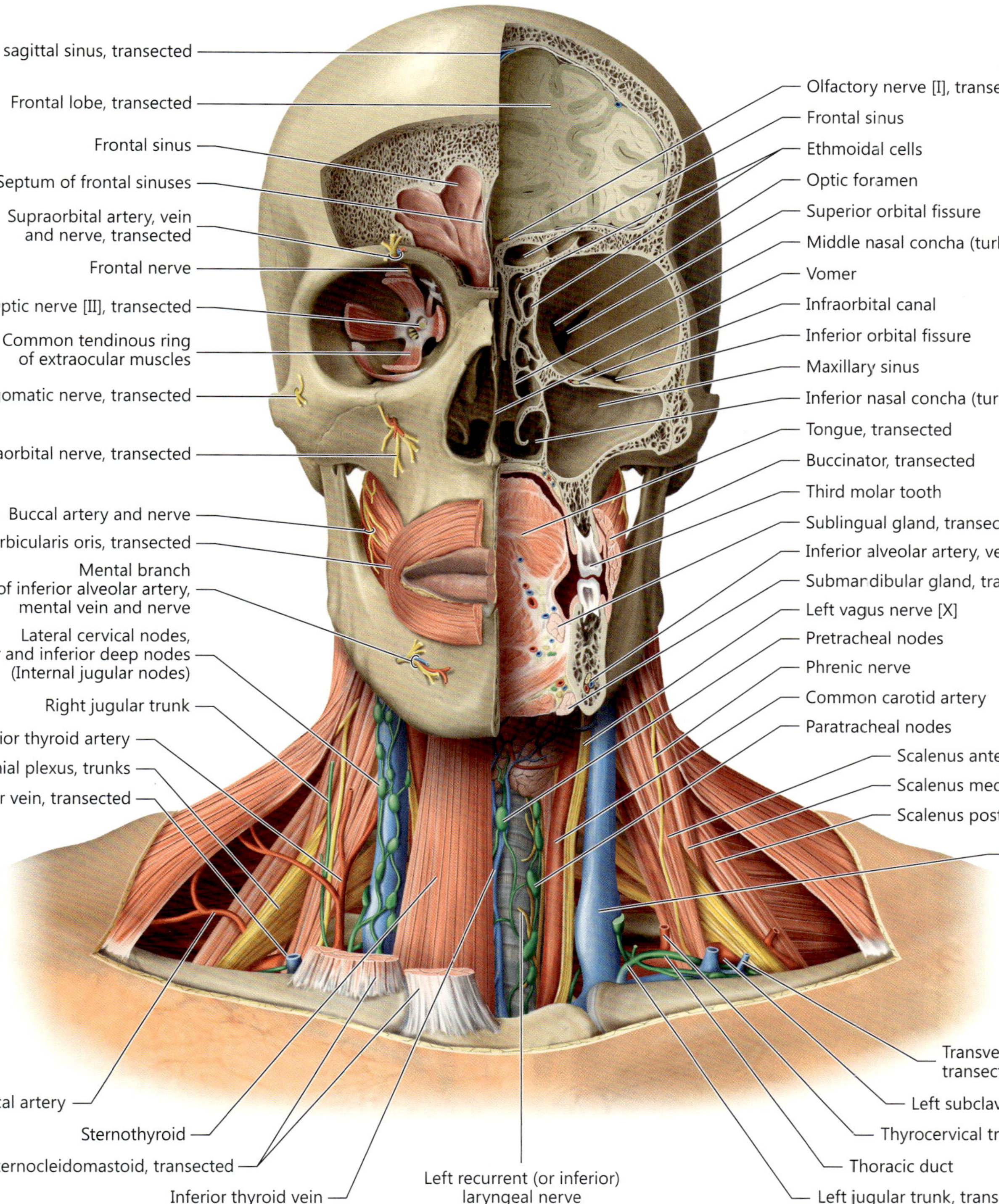

Head and **neck**, anterior view. In the right hemicranium, one can see the internal configuration of the frontal sinus after demolition of its anterior bony wall, as well as the tendinous ring of the extraocular muscles in the orbital cavity, the supraorbital, zygomaticofacial, infraorbital and mental foramina and the formations that emerge from them. In the left hemicranium, a frontal section at the level of the third molar tooth has been performed showing the ethmoidal cells, the muscles of the tongue and the floor of the mouth.

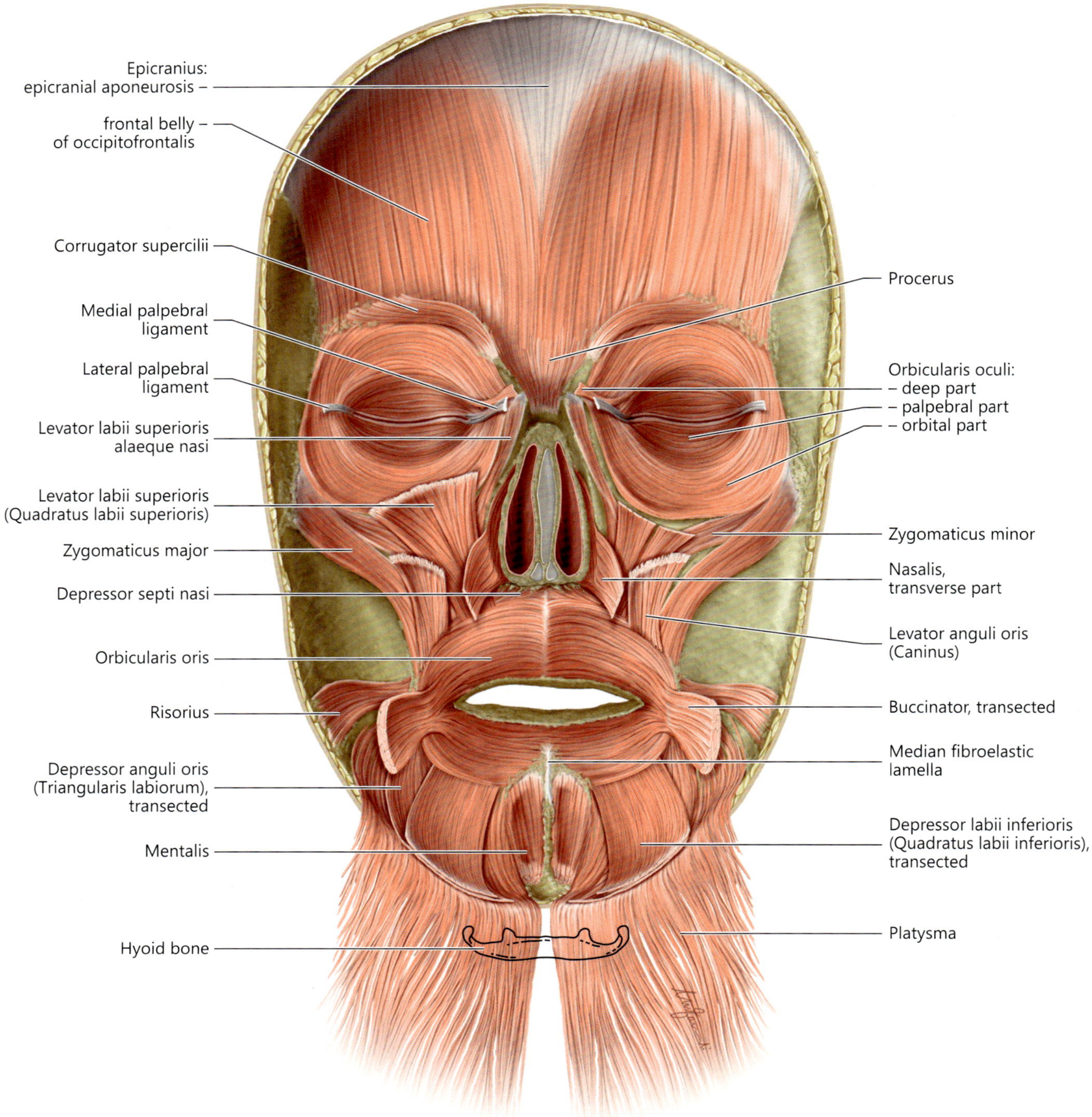

Facial muscles, view of the facial muscles and platysma, from the inside.

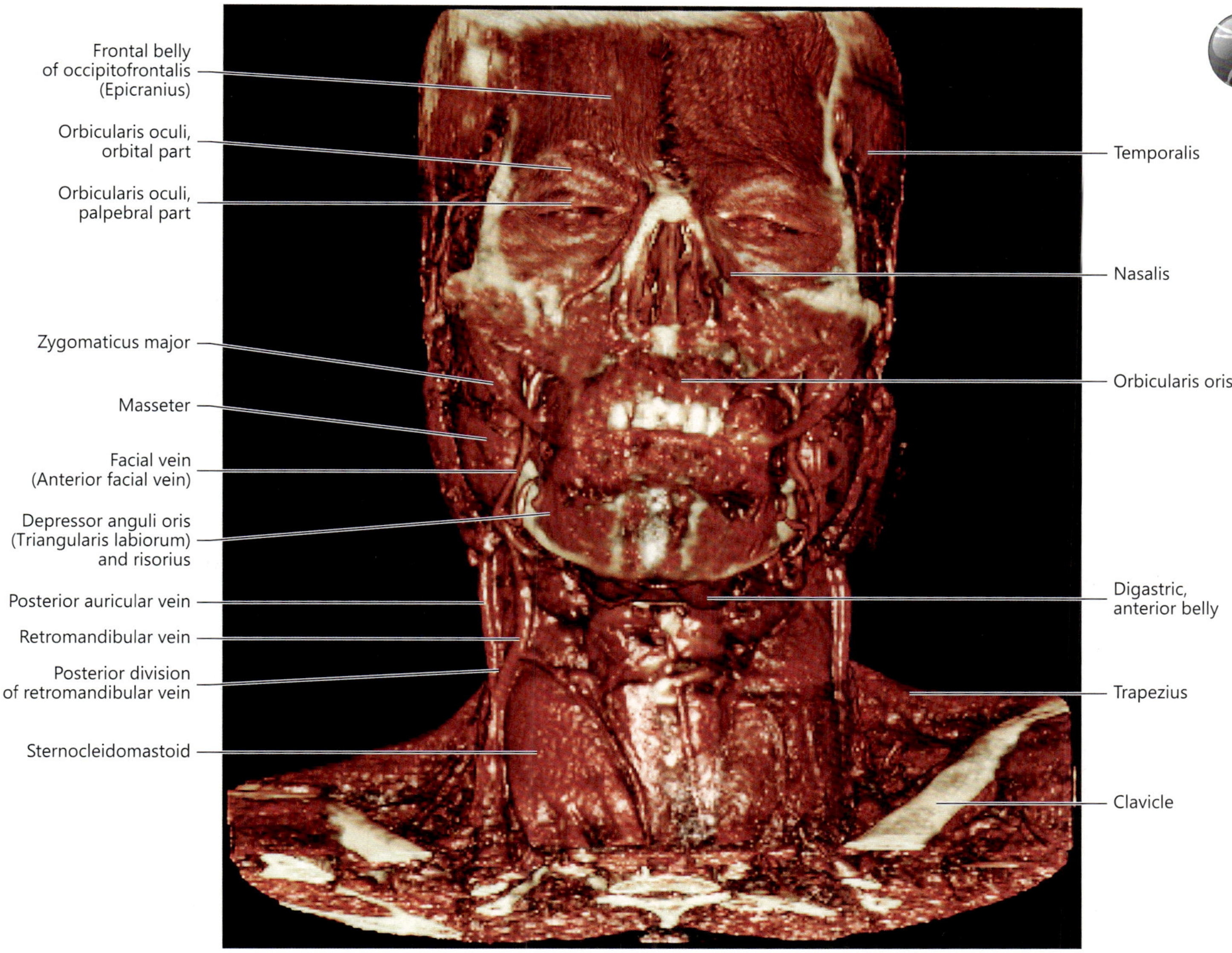

Three-dimensional image of a living human body, obtained by CT scan using a 3D volume rendering technique, showing the **cutaneous** (or **mimic**) **muscles** of the face and **masticatory muscles** of the superficial plane, anterior view. The mimic muscles constitute the superficial or subcutaneous muscle layer of the face, and their contraction causes the skin to stretch and folds to form, modifying the facial expression. The temporalis and masseter form the superficial layer of the masticatory muscles.

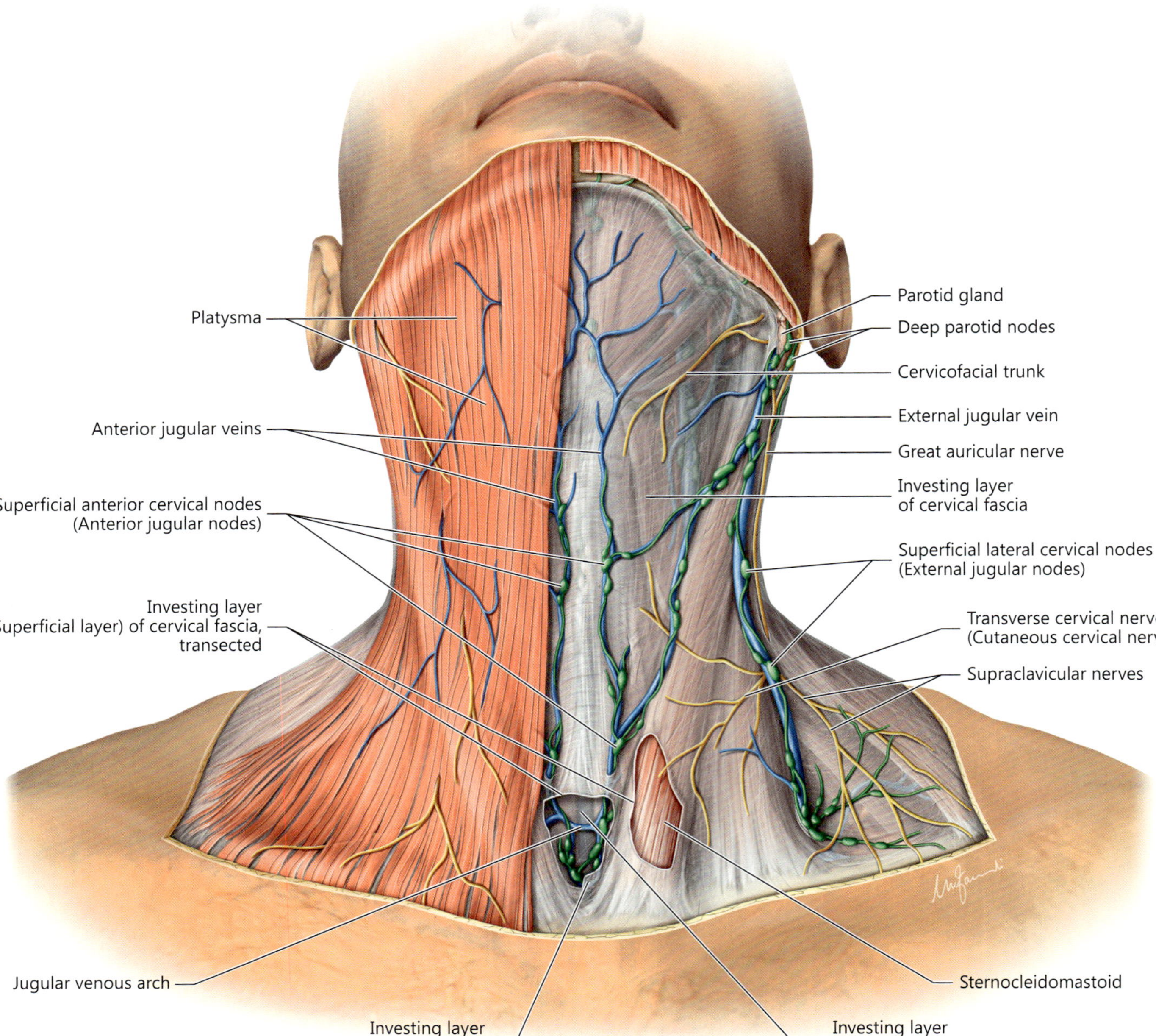

Anterior cervical region. In the right half of the neck, one can see, on a more superficial plane, the platysma, contained within the layer of subcutaneous tissue superficially to the investing layer (or superficial layer) of cervical fascia. On the left side, after removal of the platysma, one can see, superficially to the investing layer, the anterior jugular vein and the external jugular vein, with the chains of lymph nodes linked to the path they follow: the superficial anterior and superficial lateral cervical nodes (or anterior jugular nodes and external jugular nodes). In correspondence to the median line, the investing layer thickens and forms the linea alba of the neck, adhering to the underlying pretracheal layer of cervical fascia, or visceral fascia. Moving in a lateral direction, the investing layer splits to enclose the sternocleidomastoid. Posteriorly to the muscle, the fascia unites to form a single layer, only to split once again to enclose the trapezius. The fascia attaches to the sternum by splitting into two layers that go to the anterior and posterior borders of the manubrium, so delimiting the suprasternal space in which the jugular venous arch is enclosed.

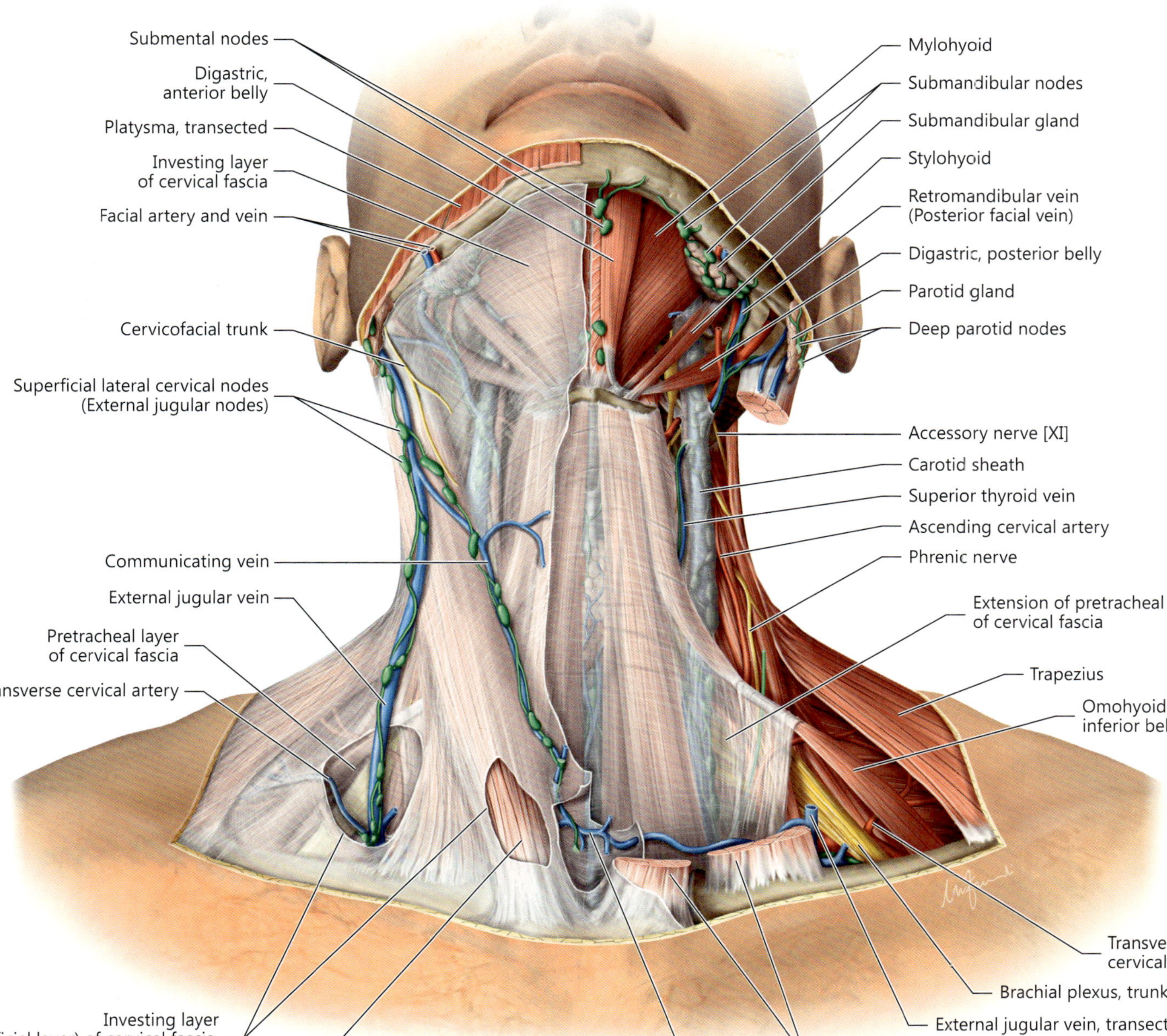

Veins of the neck, anterior view. In the right half of the neck, one can see the external jugular vein that runs down the lateral part of the neck to the middle of the clavicle where it goes deep, piercing the investing and pretracheal layers of cervical fascia, to empty into the subclavian vein at a short distance from its confluence with the internal jugular vein. On the left side, removal of the external jugular vein and investing layer reveals the pretracheal layer (visceral fascia or visceral middle layer of deep cervical fascia) between the hyoid bone and superior thoracic aperture. The pretracheal layer splits to enclose the infrahyoid muscles and, deeply, borders with the carotid sheath that encloses the neurovascular bundle of the neck. Laterally, the pretracheal layer terminates by splitting around the omohyoid. In correspondence to the superior thoracic aperture, it attaches to the posterior surface of the sternum, to the posterior border of the clavicle and adheres to the subclavian and internal jugular veins maintaining them open.

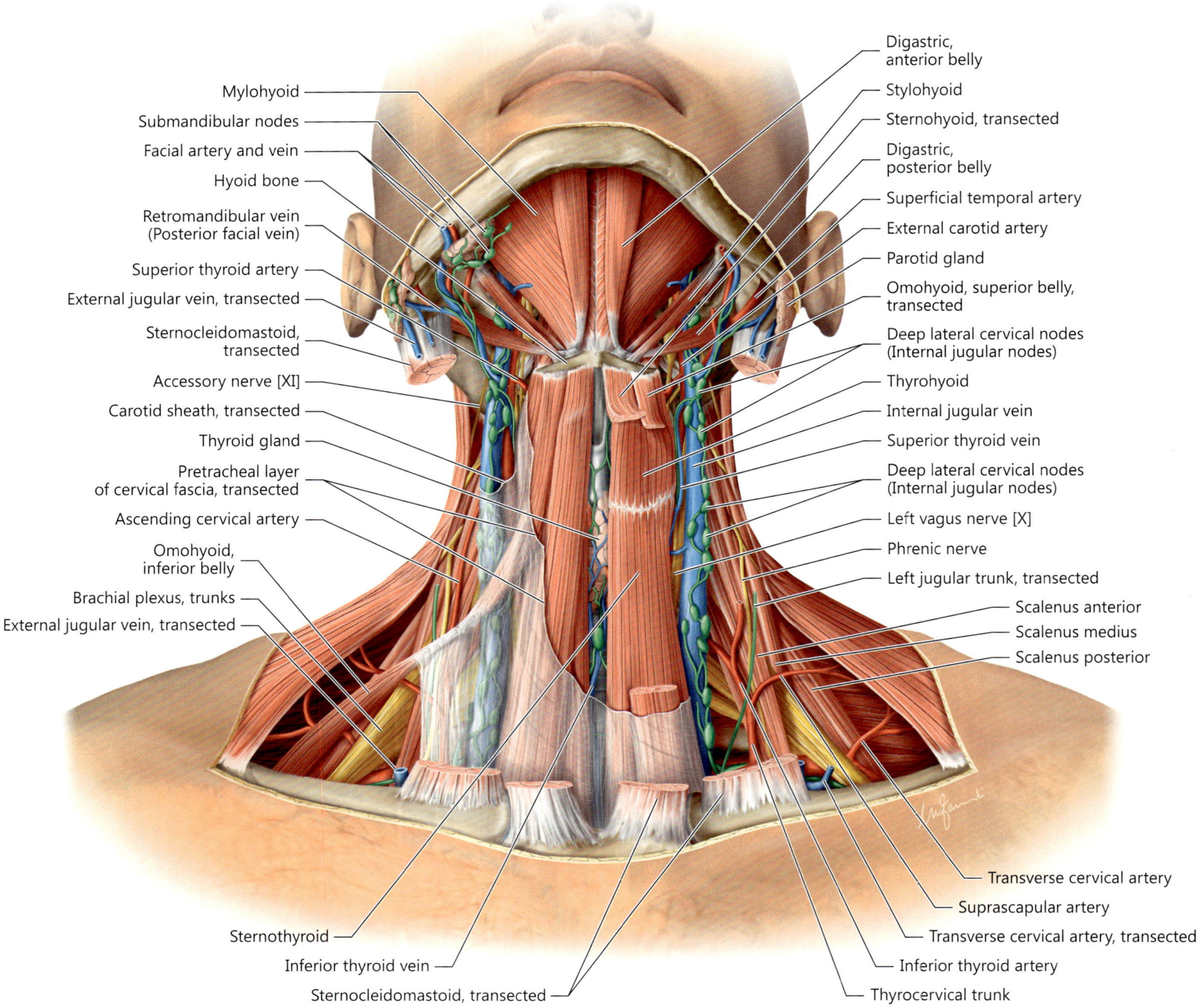

Neck, anterior view. Transection of the sternocleidomastoid, external jugular vein, investing layer and, in part, also pretracheal layer of cervical fascia enables one to observe, in the right half of the neck, the suprahyoid and infrahyoid muscles and the internal jugular vein inside the carotid sheath, which has been in part transected. On the left side, the sternohyoid and omohyoid, located on a superficial plane, have been transected and the carotid sheath removed, rendering visible the sternothyroid and thyrohyoid, located on a deeper plane, and the internal jugular vein with deep lateral cervical nodes located along the length of its course. One can also see the trunks of the brachial plexus above the clavicle, between the scalenus anterior and the scalenus medius and scalenus posterior, and the thyrocervical trunk which, arising from the subclavian artery near the medial border of the scalenus anterior, branches into the inferior thyroid and suprascapular arteries.

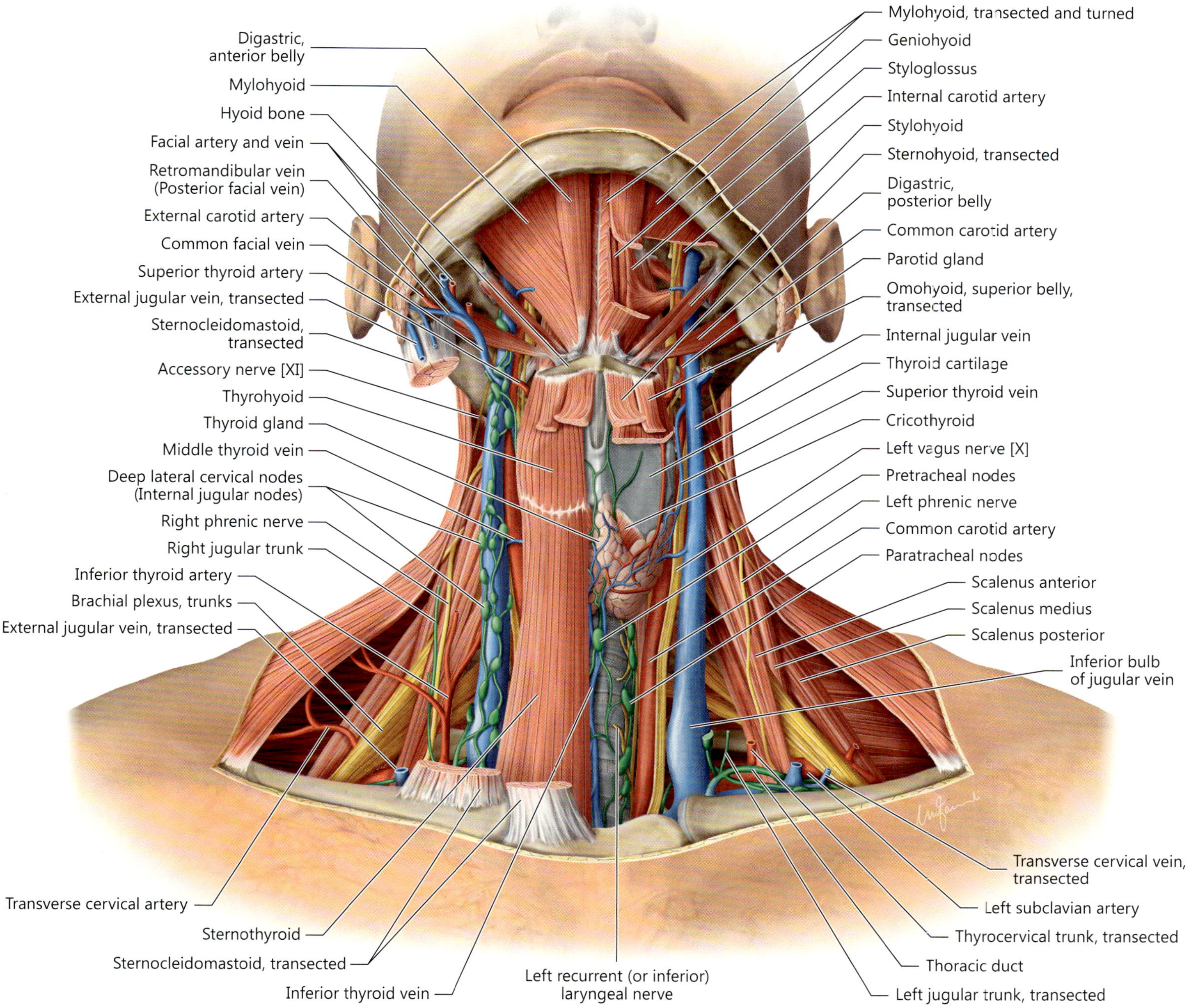

Neck, anterior view. In the right half of the neck, with the sternocleidomastoid, external jugular vein, carotid sheath and sternohyoid and omohyoid transected, it is possible to see the phrenic nerve lying on the anterior surface of the scalenus anterior, the internal jugular vein with lymph nodes present along the length of its course, the sternothyroid and thyrohyoid and the brachial plexus, between the scalenus anterior and the scalenus medius and scalenus posterior. On the left side, with the sternothyroid and thyrohyoid transected, one can see the thyroid cartilage of the larynx, the thyroid gland, the trachea with its pretracheal and paratracheal nodes, the common carotid artery and the vagus nerve. The neurovascular bundle of the neck is also visible, at the top, after transecting the anterior belly of the digastric and the mylohyoid.

Shown in detail, to the right, is the **jugular foramen** through which the glossopharyngeal [IX], vagus [X] and accessory [XI] nerves and the internal jugular vein pass.

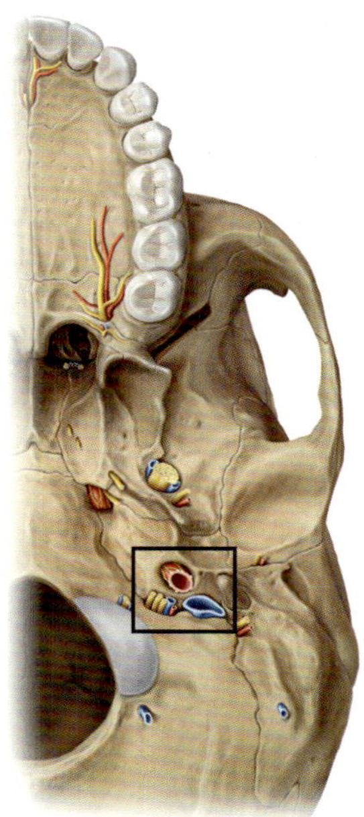

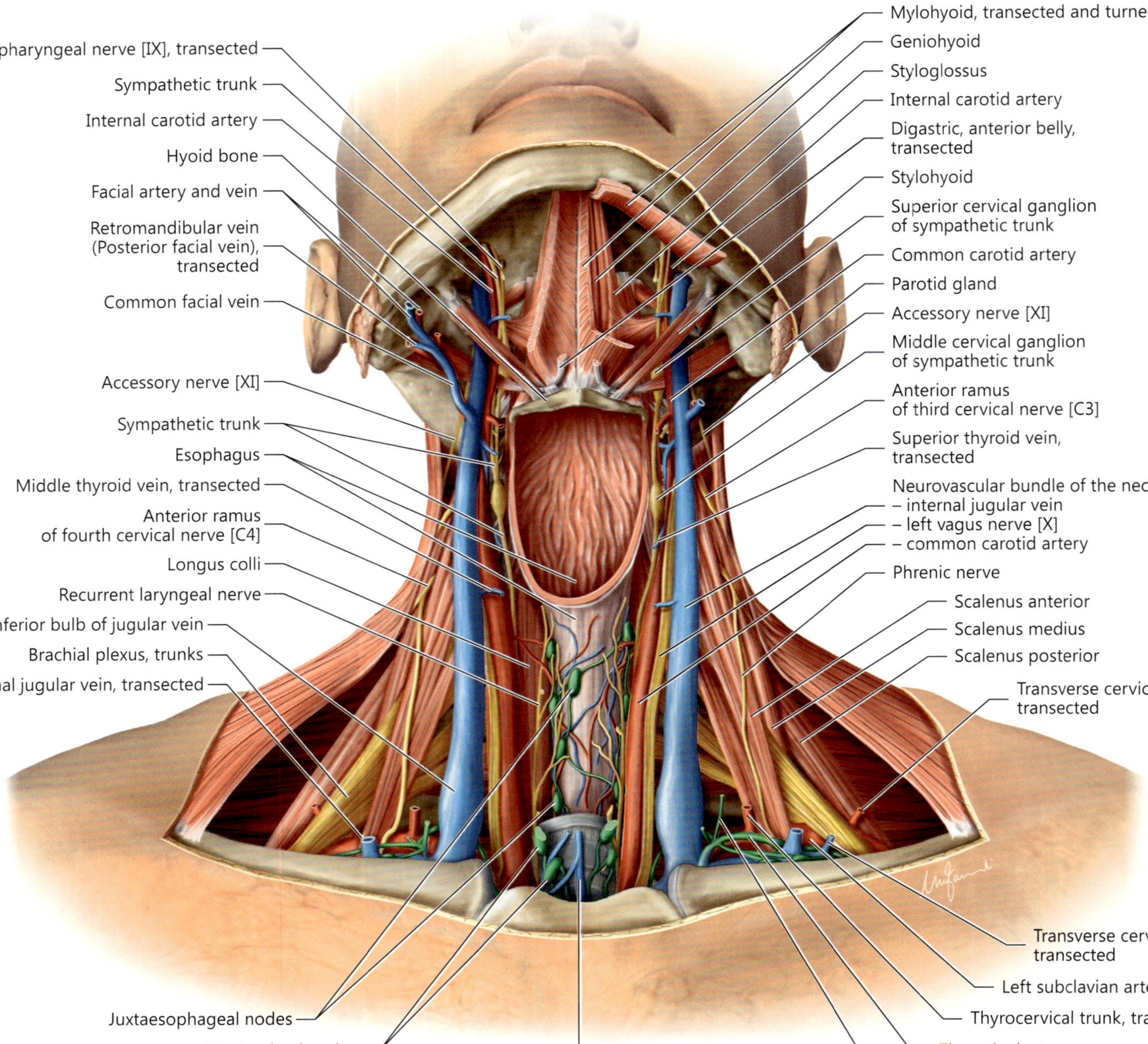

Neck, anterior view. The infrahyoid muscles, thyroid gland, larynx and lymph nodes of the internal jugular vein have been removed as well as the cervical part of the trachea, leaving exposed the posterior wall of the oropharynx and laryngopharynx. The neurovascular bundle of the neck is clearly visible, on the right and left, running anteriorly to the sympathetic trunk, which is attached to the prevertebral layer of cervical fascia (deep cervical fascia) that separates it from the longus capitis and longus colli. The vagus nerve lies in the dihedral angle, which is open posteriorly, formed by the internal carotid artery and internal jugular vein flanking each other. On the upper right side, removal of the suprahyoid muscles (with exception of the stylohyoid) makes it possible to see the common facial vein leading into the internal jugular vein. On the upper left side, the anterior belly of the digastric and the mylohyoid have been transected (the latter has also been turned), and the common facial vein has been transected.

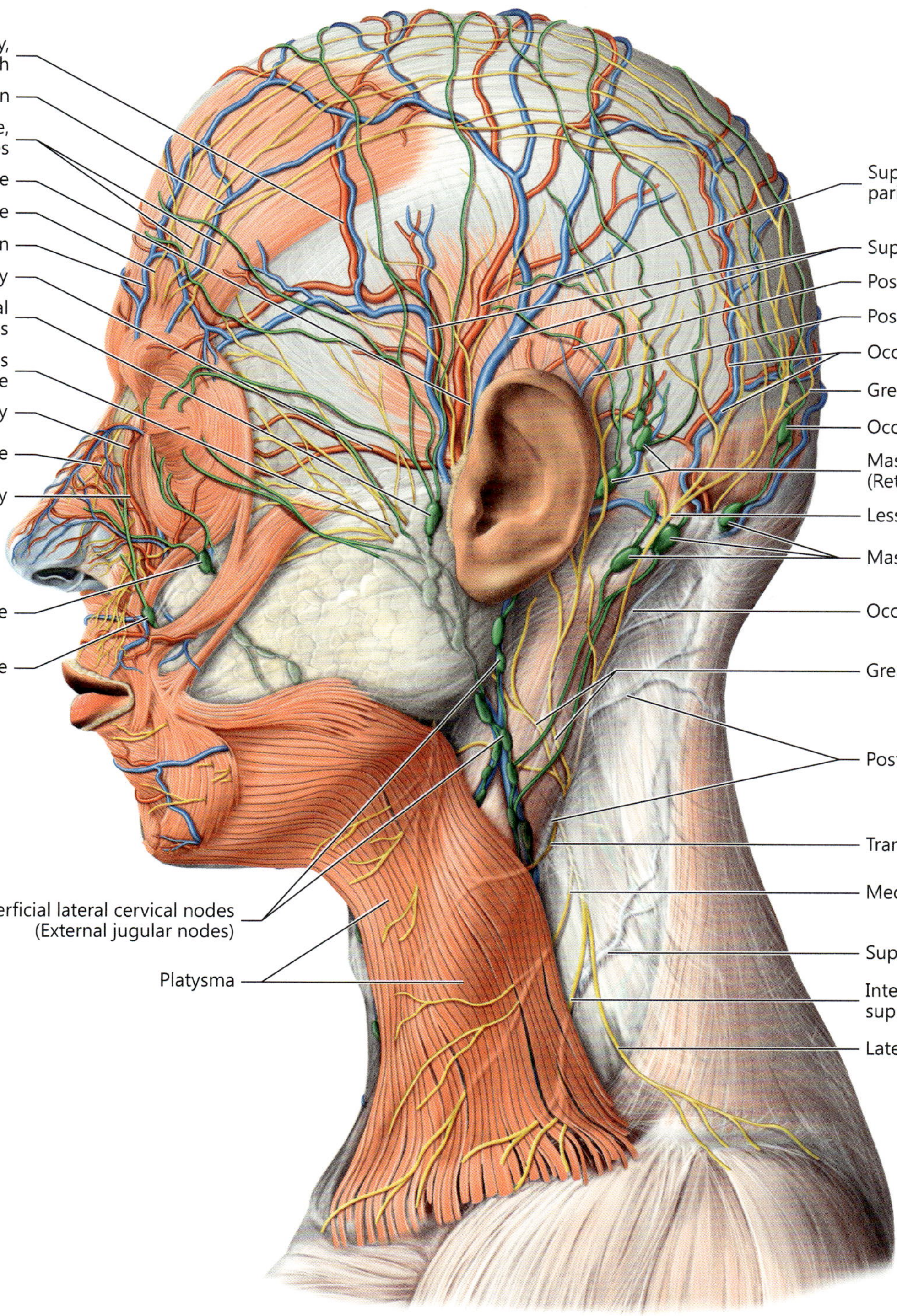

Head and **neck**, left lateral view. Superficial plane after removal of the skin and subcutaneous tissue. One can see the superficial temporal artery, terminal branch of the external carotid artery that emerges from the surface of the parotid gland and divides, in the temporal region, into a frontal and a parietal branch. The zygomatico-orbital artery, collateral branch of the superficial temporal artery, runs along the superior edge of the zygomatic arch to the lateral angle of the orbit. The occipital artery passes through the cranial insertions of the sternocleidomastoid and trapezius and climbs up on the epicranial aponeurosis. The angular artery, continuation of the facial artery, anastomoses with the dorsal nasal artery, branch of the ophthalmic artery. The superficial temporal vein originates from a subcutaneous venous plexus extending over a wide area of the scalp, and anastomoses with the contralateral vein and with the supratrochlear, supraorbital, posterior auricular and occipital veins.

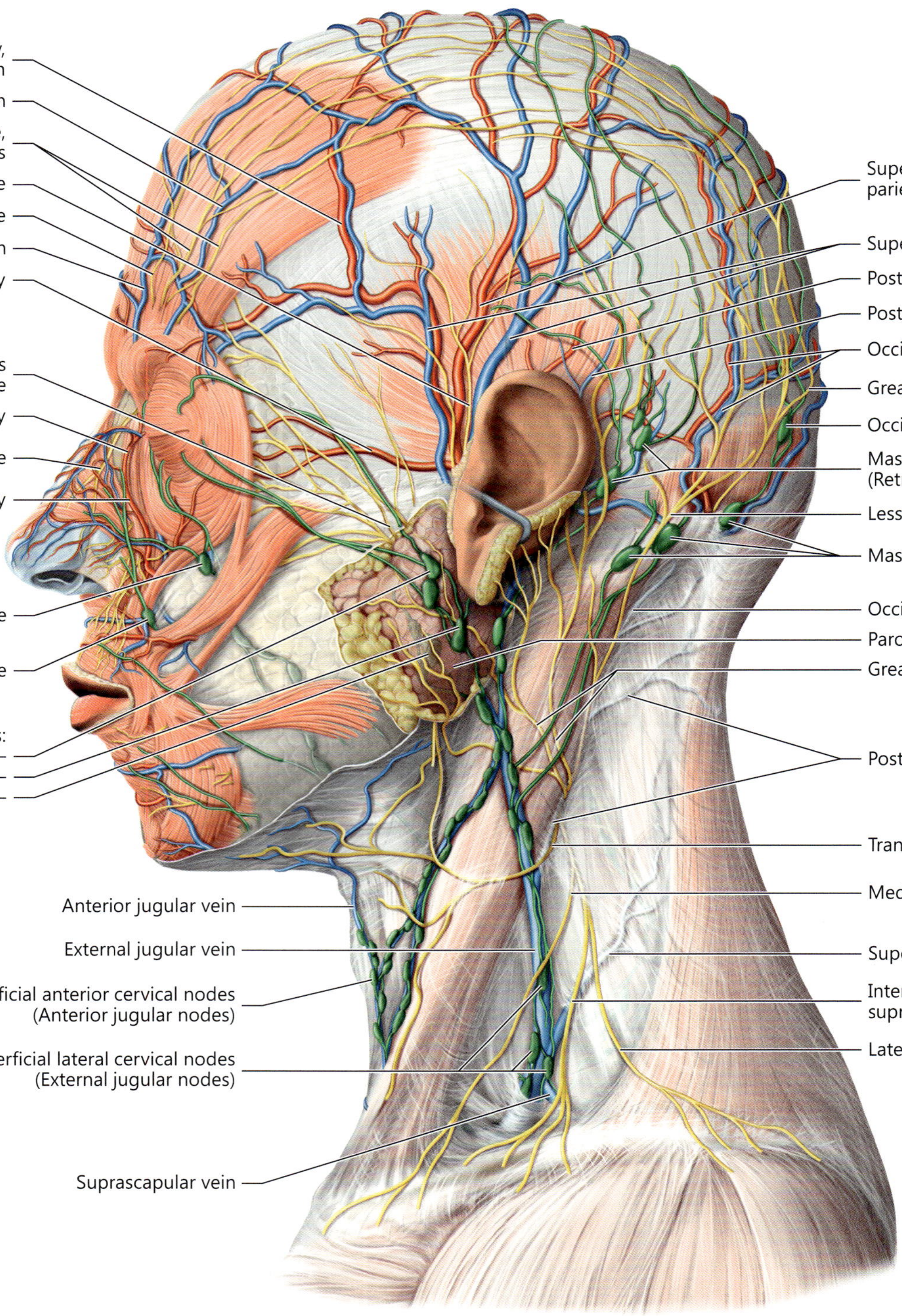

Head and **neck**, left lateral view. Superficial plane after removal of the skin, subcutaneous tissue and the platysma. The parotid fascia has also been partially transected. One can see, in particular, the external jugular and anterior jugular veins. Along the course of the two veins are the superficial anterior and superficial lateral cervical nodes (anterior jugular and external jugular nodes). One can also see the group of deep parotid nodes, the mastoid nodes and occipital nodes, the malar node and the nasolabial node. The great auricular nerve is the largest of the superficial branches of the cervical plexus. It winds around the posterior border of the sternocleidomastoid, then climbs back up its lateral surface and proceeds to the auricle, whose skin it innervates. The transverse cervical nerve, it too a cutaneous branch of the cervical plexus, turns around the posterior border of the sternocleidomastoid, then proceeds in a forward direction crossing the external jugular vein. At the level of the anterior border of the sternocleidomastoid, it innervates the suprahyoid and infrahyoid regions with superior and inferior branches. Other cutaneous branches of the cervical plexus are the supraclavicular nerves, which innervate the superior part of the thorax and the shoulder. The greater occipital nerve, a posterior ramus of C2, becomes subcutaneous in the occipital region where it divides into its terminal branches that constitute a dense plexus on the occipital belly of the occipitofrontalis and on the epicranial aponeurosis.

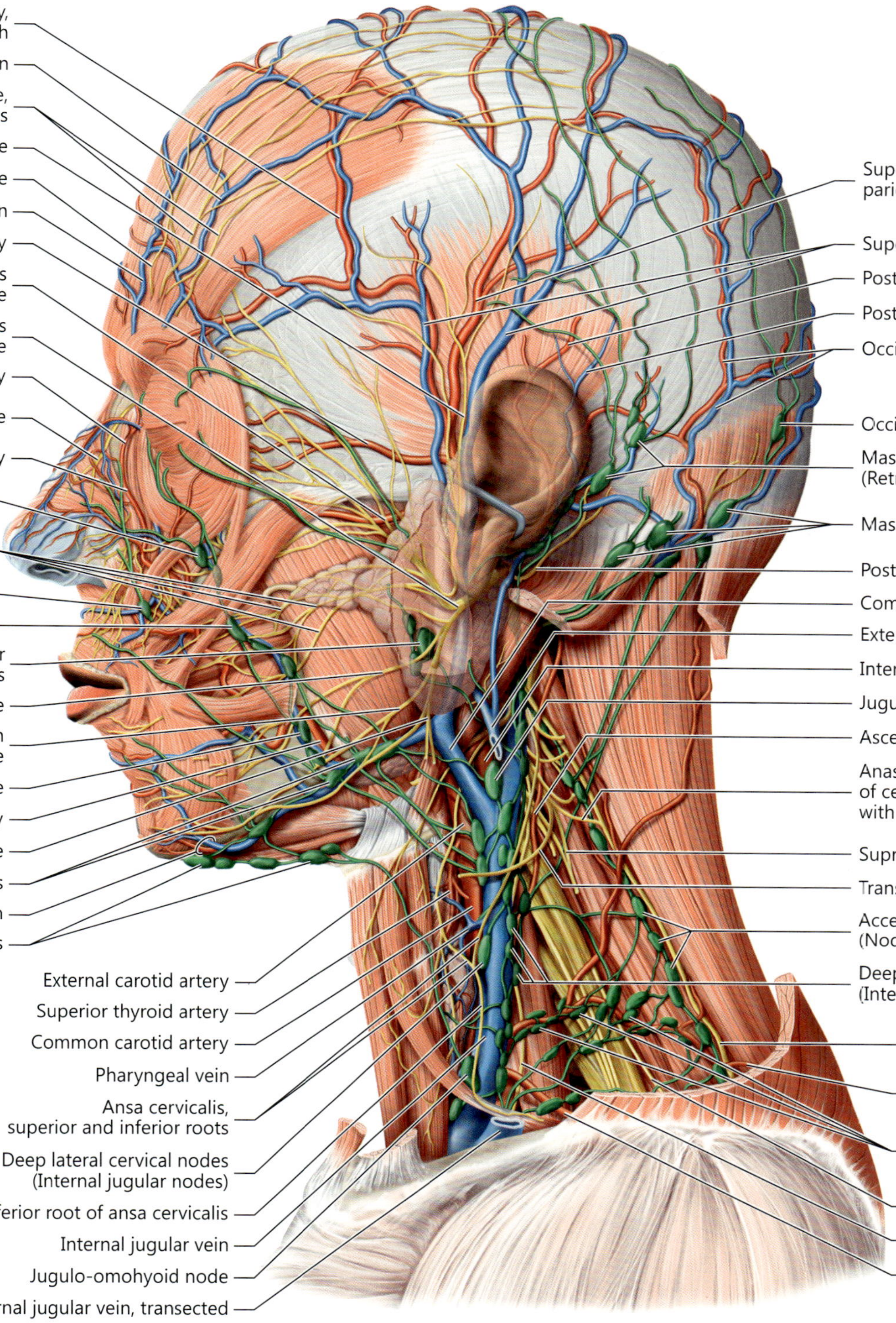

Head and **neck**, left lateral view. Deep plane. The sternocleidomastoid and trapezius and the parotid fascia have been transected. One can see, in particular, the internal jugular vein. The external jugular vein has been transected at its beginning and end. An opening has been made in the parotid gland to show the parotid plexus of the facial nerve and the auriculotemporal nerve that passes right through the gland. From the parotid plexus emerge the temporal, zygomatic, buccal, marginal mandibular and cervical branches of the facial nerve. The posterior auricular nerve originates from the facial nerve after it exits from the stylomastoid foramen. The auriculotemporal nerve, a branch of the mandibular nerve, anastomoses during its passage through the parotid gland with the parotid plexus of the facial nerve, then heads towards the exterior, between the external acoustic meatus and superficial temporal vessels, and proceeds vertically upwards in the temporal region. The chain of deep lateral cervical nodes along the internal jugular vein, that of the accessory nodes along the path of the accessory nerve, and that of the supraclavicular nodes along the transverse cervical artery constitute the posterior triangle. Clearly visible also are the facial, malar, nasolabial, buccinator and mandibular nodes, and the submandibular and submental nodes.

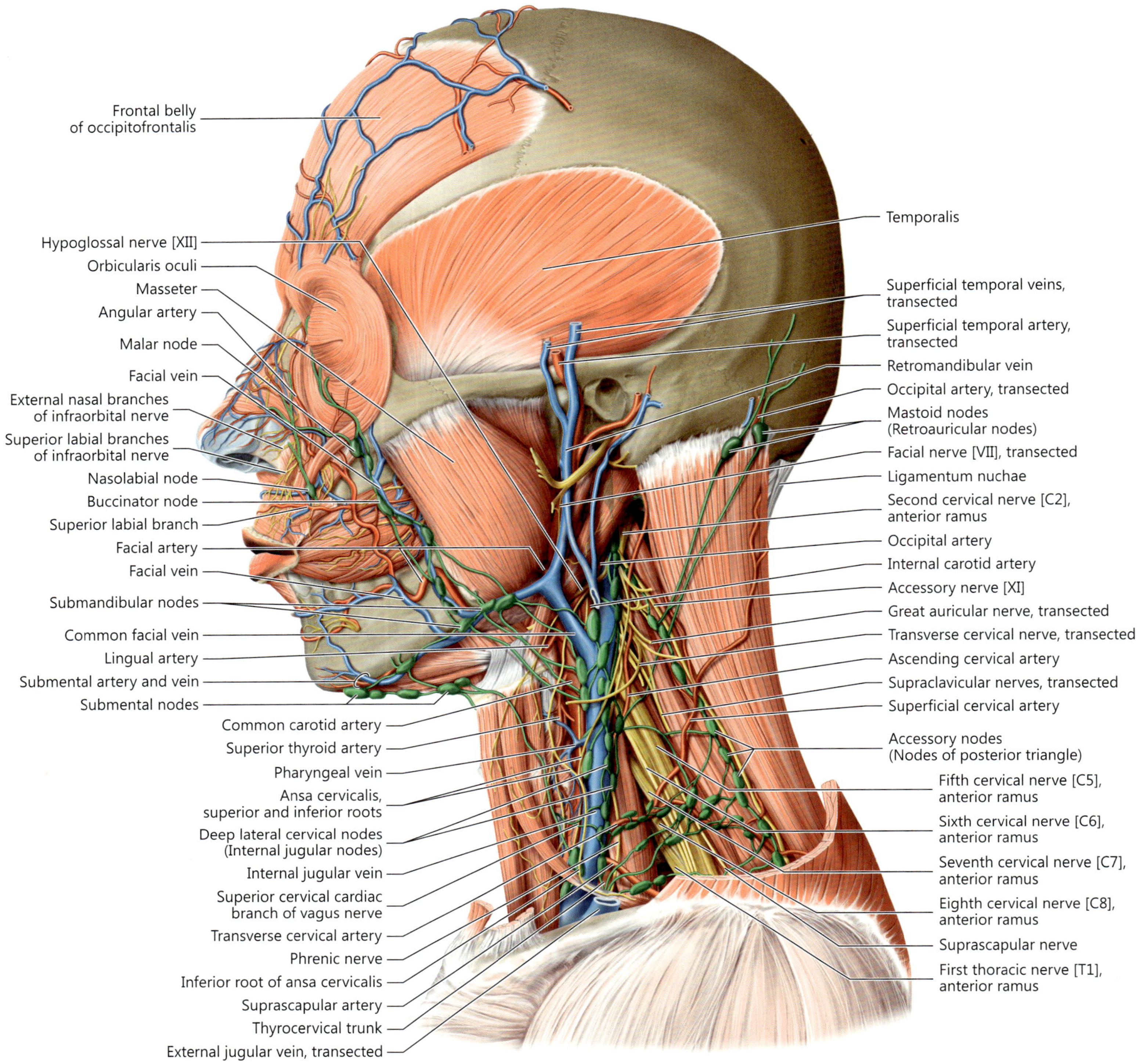

Head and **neck**, left lateral view. Deep plane. The sternocleidomastoid and trapezius, parotid gland, some facial muscles, epicranial aponeurosis and the occipital belly of the occipitofrontalis have been transected. One can see, in addition to the internal jugular vein, the facial vein and trunks of the brachial plexus, formed by the anterior rami of the fifth, sixth, seventh and eighth cervical and first thoracic nerves. The superior root of the ansa cervicalis, containing fibers from the first cervical nerve, proceeds downwards after leaving the hypoglossal nerve, in front of, or inside, the carotid sheath (which here has been removed), and unites with the inferior root of the ansa cervicalis, coming from the second and third cervical nerves. The ansa cervicalis, also known as ansa of hypoglossal nerve, innervates the omohyoid, sternohyoid and sternothyroid.

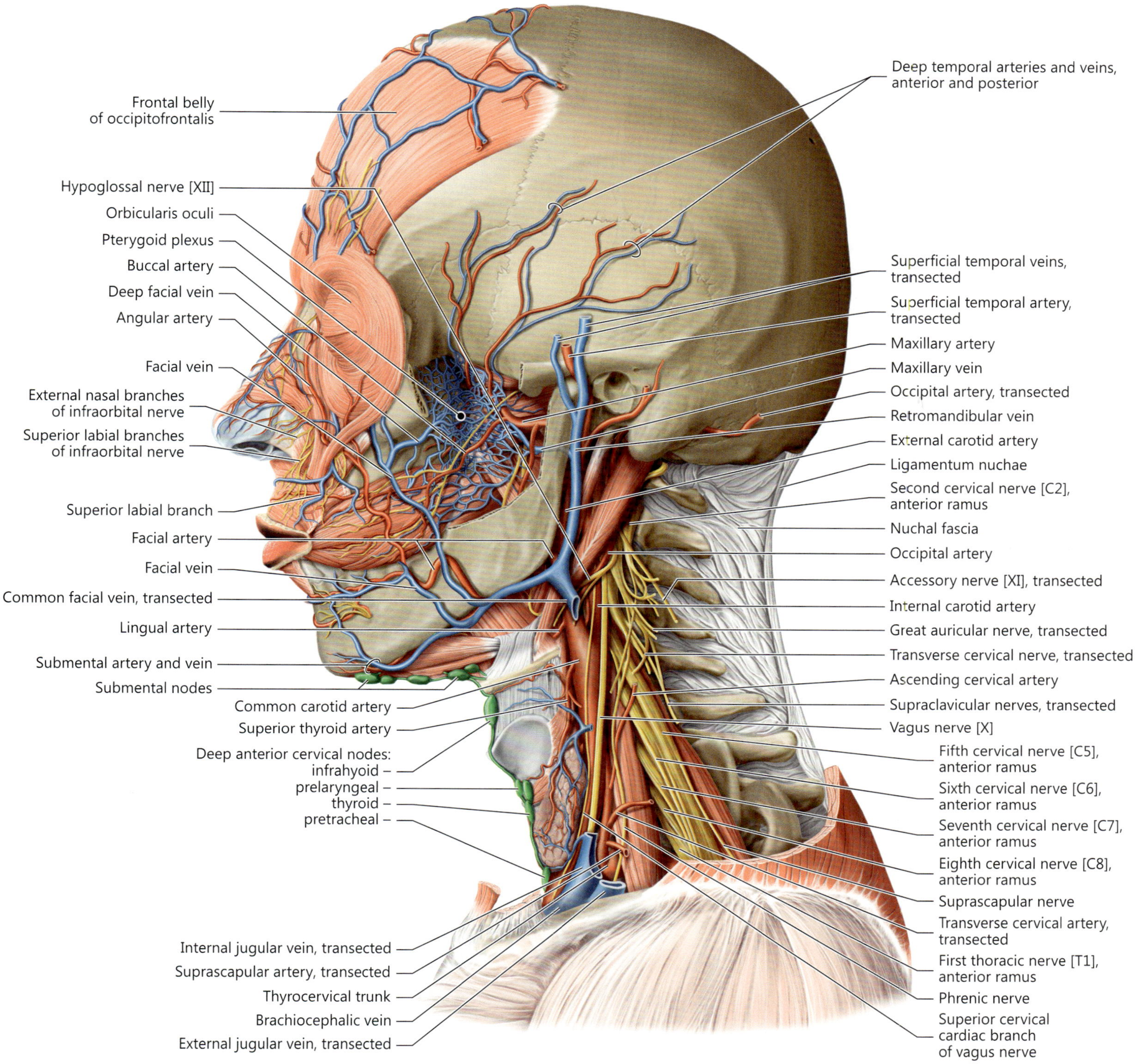

Head and **neck**, left lateral view. Deep plane. The sternocleidomastoid and trapezius, parotid gland, some facial muscles, epicranial aponeurosis, occipital belly of the occipitofrontalis, and the infrahyoid muscles, masseter, temporalis and splenius capitis have been transected. The removal of the internal jugular vein reveals the common carotid artery and the vagus nerve. Also visible are the thyroid cartilage and thyroid gland. The vagus nerve is located in the dihedral angle, which is open posteriorly, formed by the common carotid artery and internal jugular vein. One can see the group of deep anterior cervical nodes, or anterior visceral nodes, that drain the larynx, trachea and thyroid gland. In the infratemporal fossa the pterygoid plexus can be seen, situated in part between the temporalis and lateral pterygoid and partly between the lateral and medial pterygoid. From the confluence of the veins of the plexus originates the maxillary vein, which joins the superficial temporal veins forming the retromandibular vein. The pterygoid plexus communicates with the facial vein via the deep facial vein.

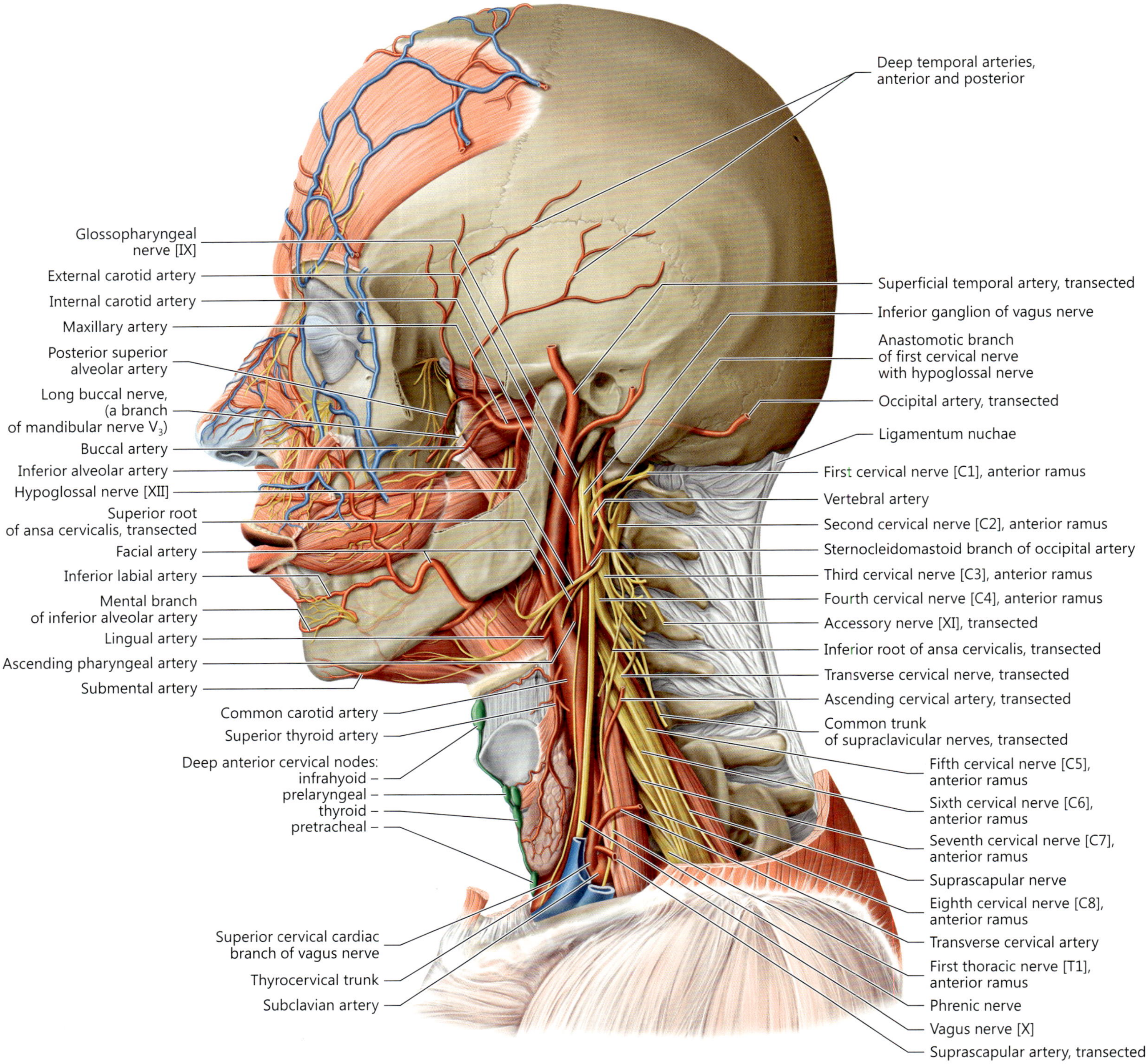

Head and **neck**, left lateral view. Deep plane. Almost the entire venous system has been removed. One can see, in addition to the brachial plexus, the cervical plexus, formed by the anterior rami of the first, second, third and fourth cervical nerves. The phrenic nerve, the largest and longest branch arising from the cervical plexus, runs obliquely downwards and medially, on the anterior surface of the scalenus anterior. In the parapharyngeal space, note the hypoglossal nerve, which crosses the internal carotid artery and vagus nerve forming a loop around the sternocleidomastoid branch of the occipital artery. The maxillary artery, the largest terminal branch of the external carotid artery, proceeds medially to the mandibular condyle, passes through the infratemporal fossa and, heading superficially (as shown here) or deeply to the lower head of the lateral pterygoid, terminates in the pterygopalatine fossa.

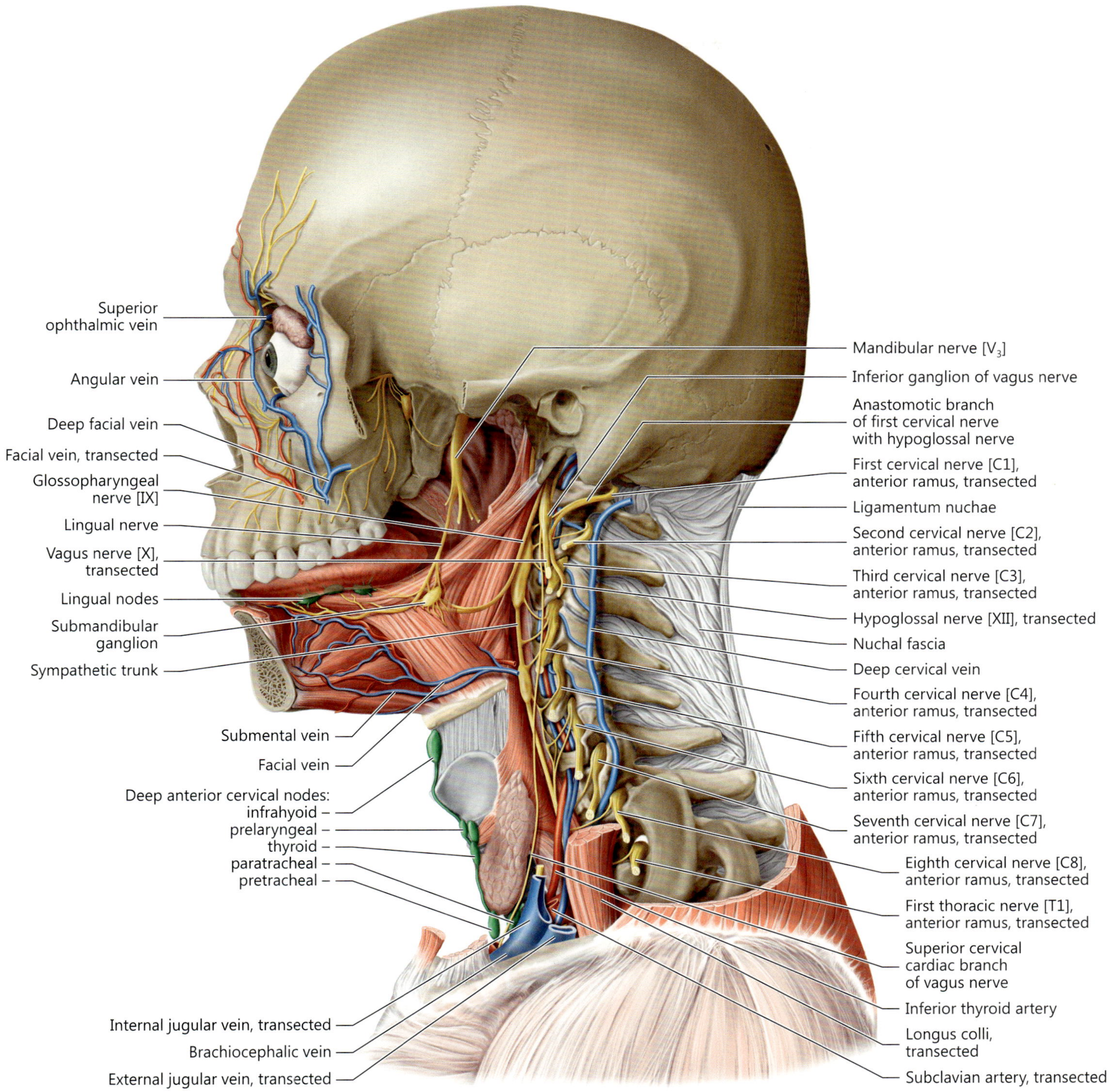

Head and **neck**, left lateral view. Deep plane. The neurovascular bundle of the neck and anterior rami of the cervical nerves have been transected, showing the sympathetic trunk. Removal of the left hemi-mandible reveals the lingual nerve and submandibular ganglion. The cervical segment of the sympathetic trunk extends from the cranial base to the superior thoracic aperture, attached to the prevertebral layer of cervical fascia, which separates it from the transverse processes of the cervical vertebrae and from the longus capitis and longus colli. Note the lingual nerve, a branch of the mandibular nerve, which makes an anterosuperiorly concave curve in a downward and forward direction. Also the glossopharyngeal nerve [IX] proceeds obliquely downwards and forwards making an anterosuperiorly concave curve. The vertebral vein originates between the occipital bone and atlas, and after descending through the foramina transversaria of the first six cervical vertebrae it heads anteriorly to empty into the brachiocephalic vein.

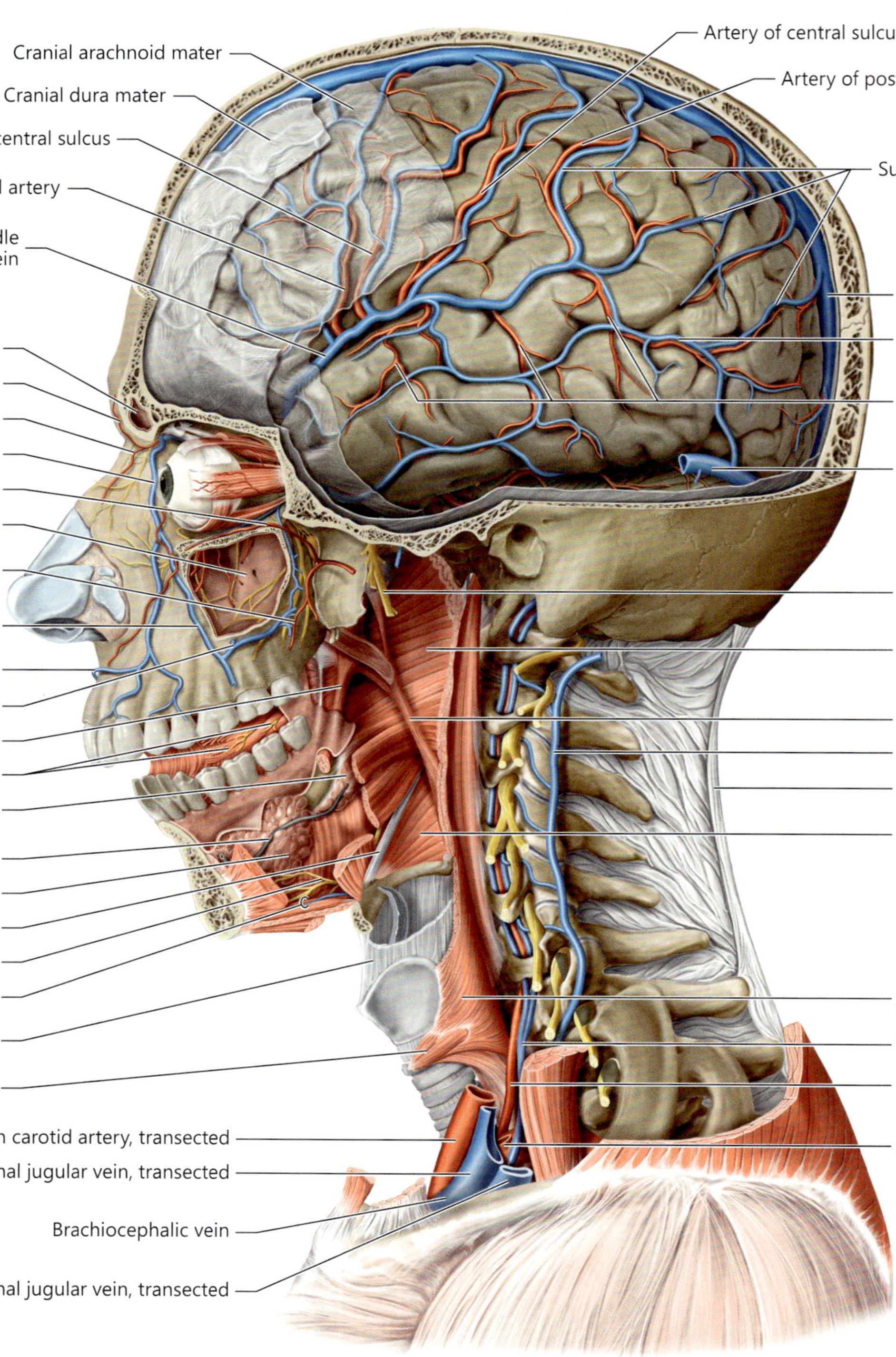

Head and **neck**, left lateral view. With most of the left hemicranium removed, one can see the superficial cerebral veins and branches of the middle cerebral artery on the surface of the left cerebral hemisphere. In the face, a parasagittal transection has been carried out making it possible to see both the orbital cavity and the interior of the maxillary sinus; one can also see the angular vein, which originates superficially from the medial angle of the eye and, running obliquely downward laterally to the root of the nose, as a facial vein, receives the deep facial vein from the pterygoid plexus. In the inferior part of the mouth, instead, the left hemi-mandible and respective soft tissues, including the entire tongue, have been removed, so as to render clearly visible the sublingual gland with its neurovascular relations. In the neck, a paramedian sagittal section reveals the deep cervical vein that originates inferiorly to the mastoid process and descends posteriorly to the transverse processes of the cervical vertebrae. It is possible, moreover, to see the vertebral vein, which originates between the occipital bone and atlas, and descends through the foramina transversaria of the first six cervical vertebrae, in a lateral position with respect to the vertebral artery. In the anterior part of the neck, a more external paramedian section allows one to see the laryngotracheal canal and its relationship to the left common carotid artery and left internal jugular vein.

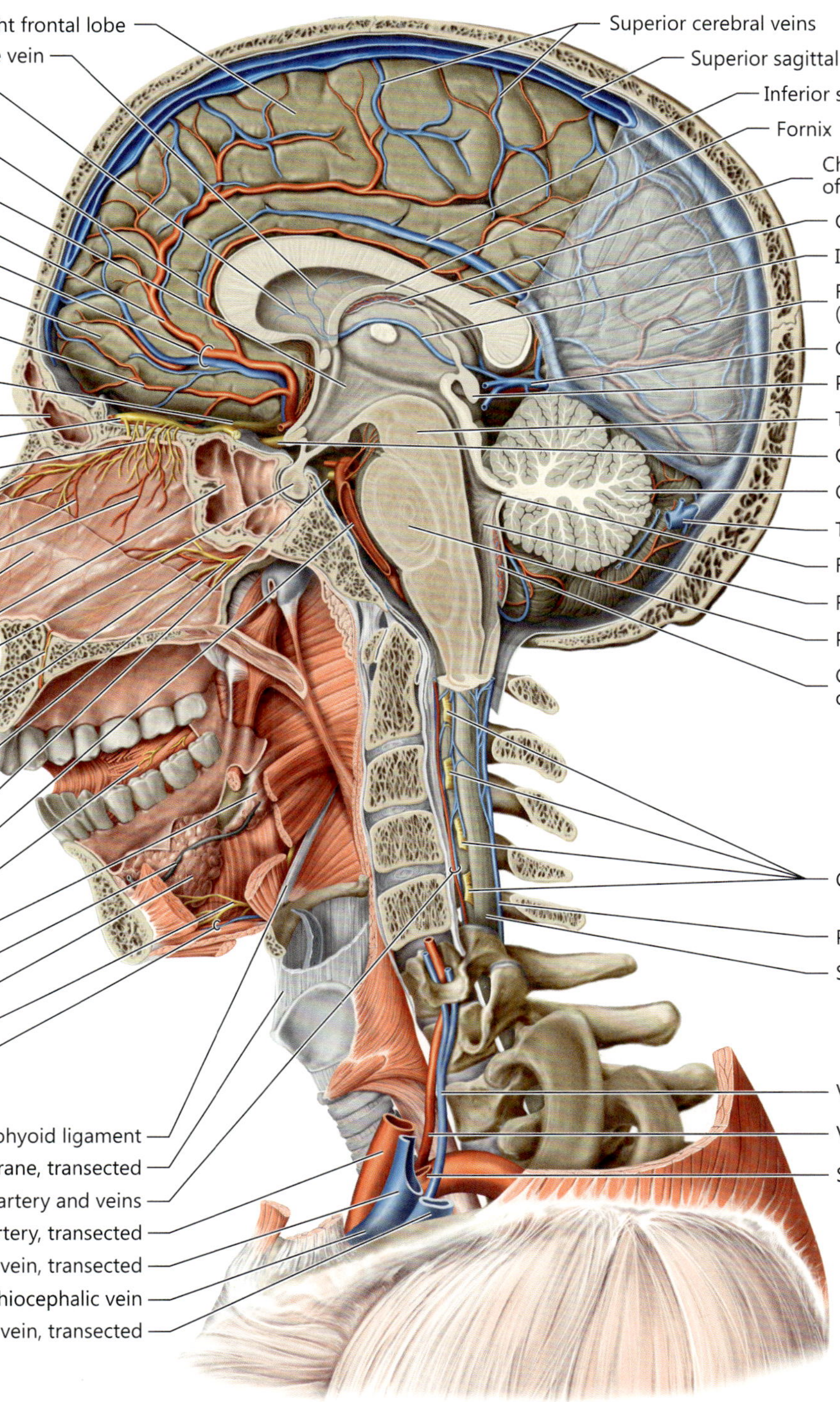

Sagittal section of head and **neck**, left lateral view, showing the contents of the cranial cavity and vertebral canal, the right nasal cavity, oral cavity, and large vessels transected at the base of the neck. The cerebral veins have an independent course with respect to the arteries and are more numerous. They can be classified into: superficial cerebral veins (of which the superior cerebral veins are visible), numbering 8-10 per hemisphere, that drain the medial and superolateral surfaces of the hemispheres and empty mostly into the superior sagittal sinus, and deep cerebral veins (of which here can be seen the internal cerebral vein and the great cerebral vein, or vein of Galen). The two internal cerebral veins, right and left, run anteroposteriorly between the two layers of the choroid membrane and terminate by uniting to form the great cerebral vein, a median unpaired venous trunk, approximately 1 cm long, situated in the transverse fissure of the brain, that runs between the cerebellum and splenium of corpus callosum terminating in the straight sinus. Below the hyoid bone, the transection plane becomes parasagittal allowing one to see, from the left, the larynx and trachea, and, lower down, the origin and partial course of the vertebral artery and vein.

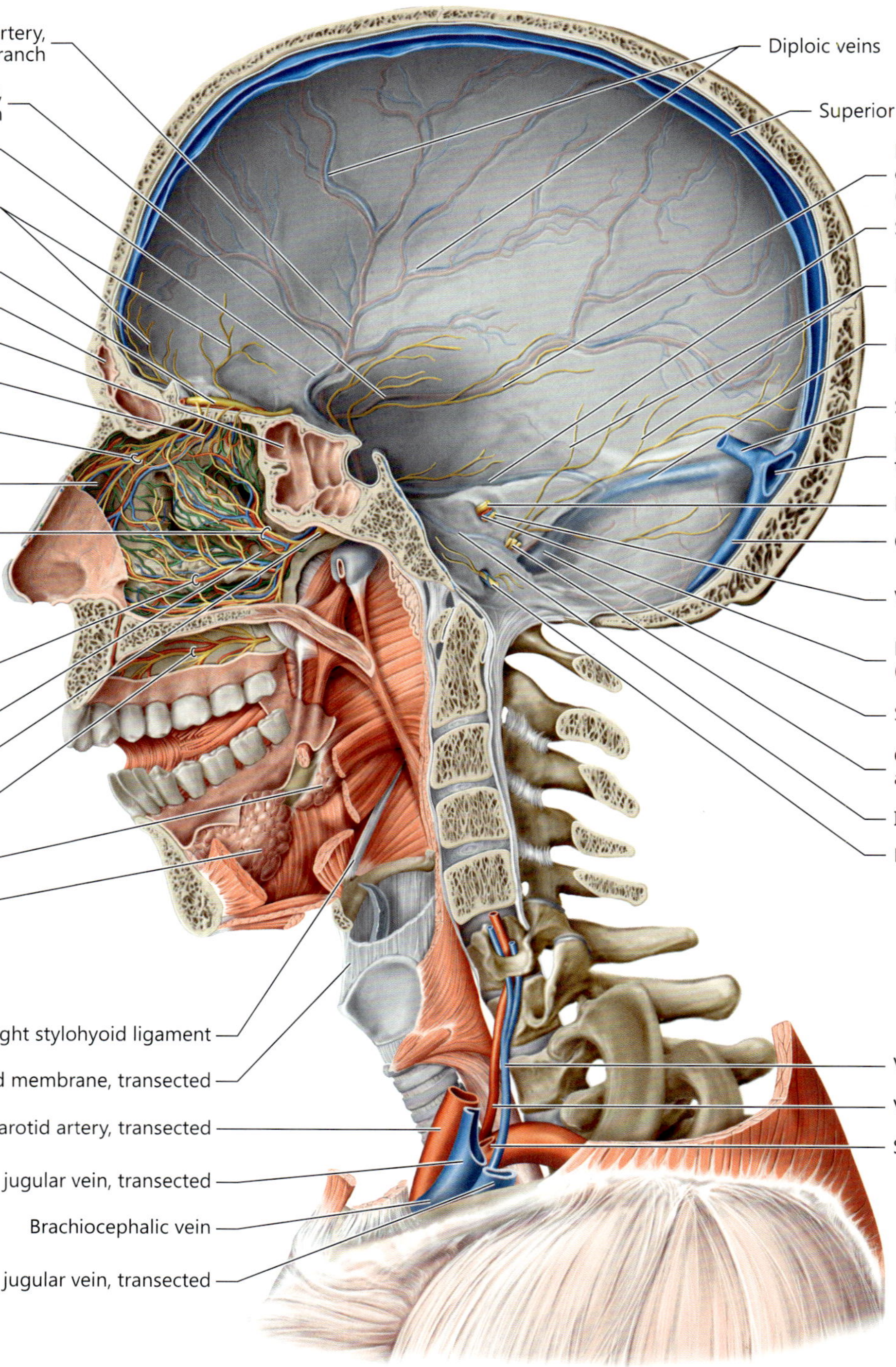

Head and **neck**, left lateral view. With the left hemicranium and brain removed, one can see the lateral wall of the cranium from the inside. The middle meningeal artery, which originates from the maxillary artery, penetrates into the middle cranial fossa through the foramen spinosum. The middle meningeal vein, a satellite of the middle meningeal artery, runs between the two layers of cranial dura mater and can be considered homologous to the dural venous sinuses. After removal of a large part of the nasal septum, it is also possible to see the lateral wall of the right nasal cavity. The posterior lateral nasal arteries originate from the sphenopalatine artery, which penetrates into the nasal cavity through the sphenopalatine foramen together with the nasopalatine nerve. The greater palatine artery emerges through the greater palatine foramen on the vault of the palate, heads forward, enters the incisive canal and anastomoses with a branch of the sphenopalatine artery. The anterior ethmoidal artery, a branch of the ophthalmic artery, passes through the anterior ethmoidal foramen together with the anterior ethmoidal nerve. Below the hyoid bone, the transection plane becomes parasagittal allowing one to see, from the left, the larynx and trachea, and, further down, the origin and partial course of the vertebral artery and vein.

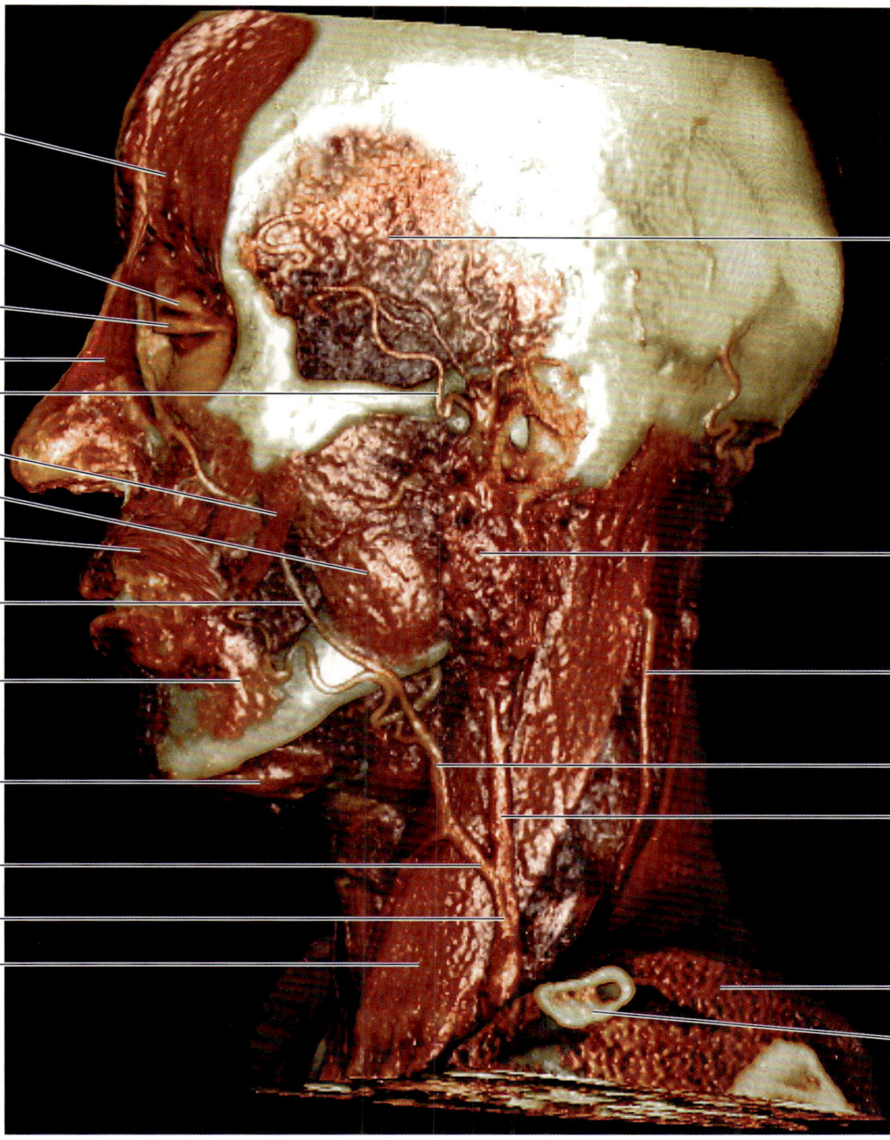

Three-dimensional image of a living human body, obtained using a 3D volume rendering technique, of the head and neck in left lateral view, showing the **cutaneous** (or **mimic**) **muscles**, the **temporalis** and **masseter** (**masticatory muscles**), the **sternocleidomastoid** and the **parotid gland**, whose superficial part or external surface, triangular in form, is visible. The anterior margin of the superficial part of the gland borders with the masseter, and is superimposed on it to some extent; the posterior margin borders with the sternocleidomastoid; the base, superior, corresponds to the cartilaginous part of the external acoustic meatus and to the zygomatic arch; the apex, inferior, is near the submandibular gland.

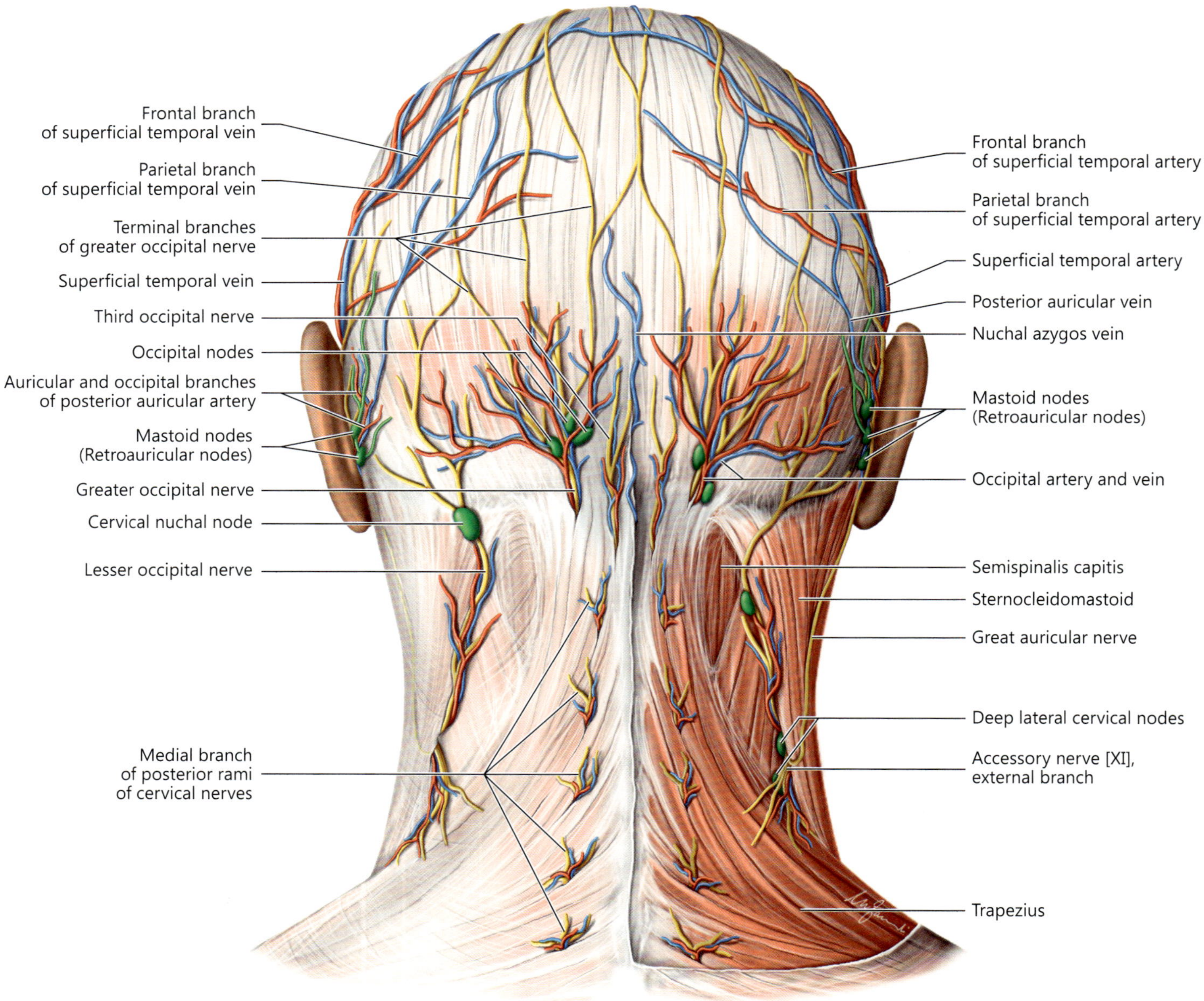

Head and **neck**, posterior view. In the left half, with the skin and subcutaneous tissue removed, one can observe in particular the occipital nodes and the mastoid or retroauricular nodes. In the right half, in addition to the skin and subcutaneous tissue, the investing layer of cervical fascia has been transected. One can see in particular the great auricular nerve, the largest of the superficial branches of the cervical plexus, which proceeds upwards on the lateral surface of the sternocleidomastoid.

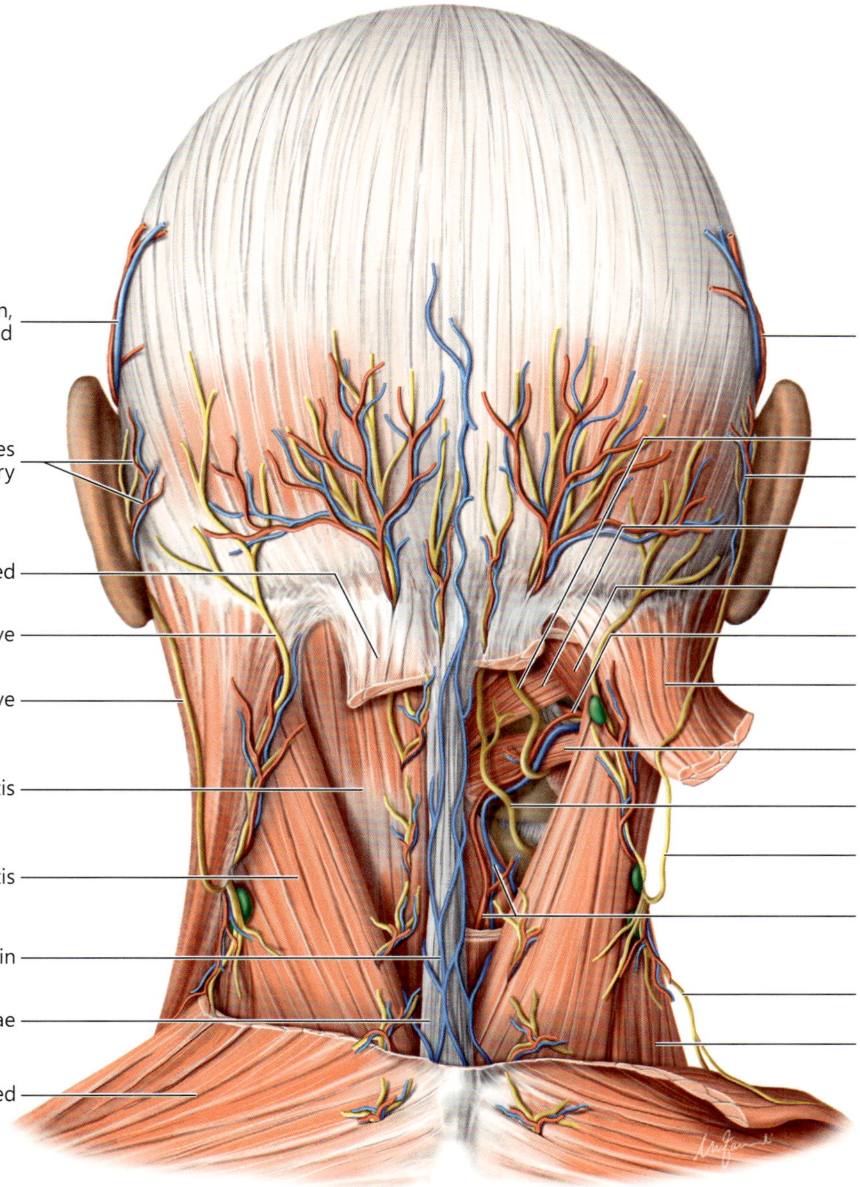

Head and **neck**, posterior view. In the left half, after removal of the skin, subcutaneous tissue and investing layer of cervical fascia, the trapezius and the nuchal fascia that separates it from the splenius capitis have been transected, revealing the semispinalis capitis. The lesser occipital nerve, a cutaneous branch of the cervical plexus, climbs up into the head along the posterior border of the sternocleidomastoid. In the right half, also the semispinalis capitis has been removed to expose the suboccipital muscles, and the sternocleidomastoid has been transected. One can see the vertebral artery and the greater occipital and third occipital nerves, posterior rami of C2 and C3, the second and third cervical nerves.

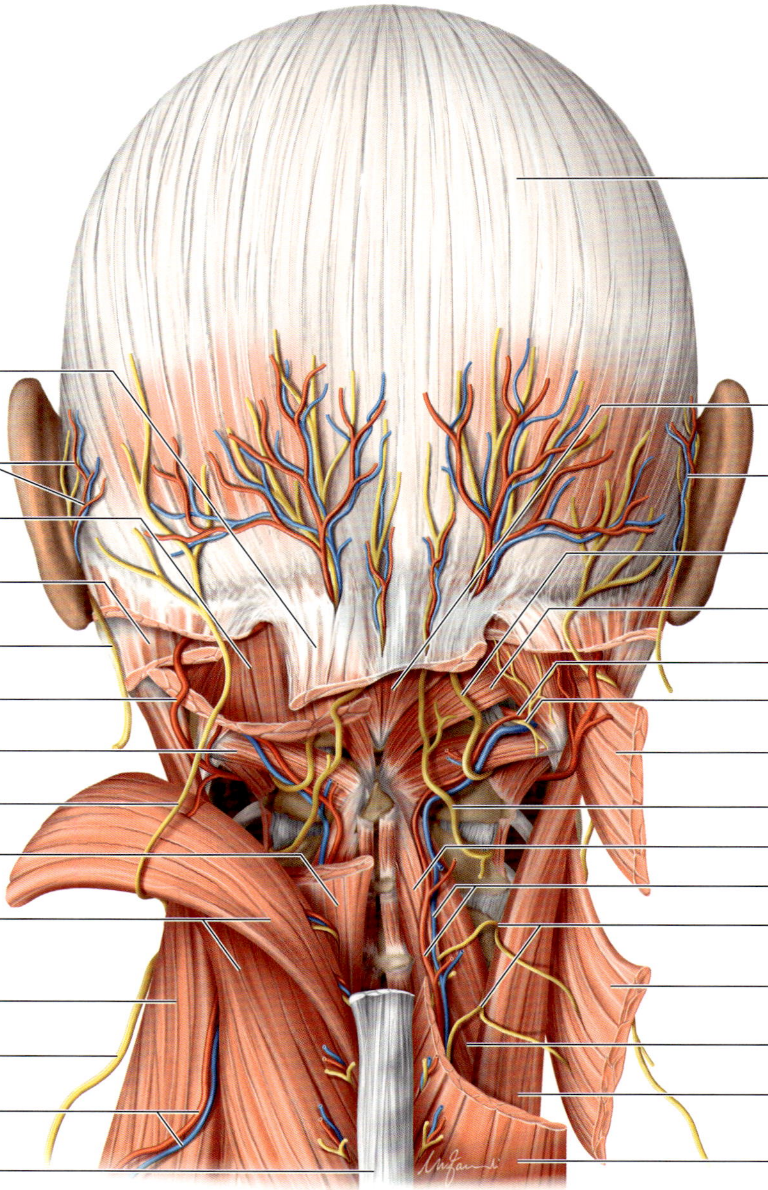

Head and **neck**, posterior view. In the left half, in addition to the trapezius, sternocleidomastoid and semispinalis capitis, the splenius capitis has been transected and turned, showing the occipital artery. On the median plane, the ligamentum nuchae has been transected. In the right half, also the levator scapulae has been transected, exposing the longissimus capitis and semispinalis cervicis. In the interstitium delimited medially by the rectus capitis posterior major and laterally by the obliquus capitis superior and obliquus capitis inferior, one can see the suboccipital nerve, a posterior ramus of C1, the first cervical nerve.

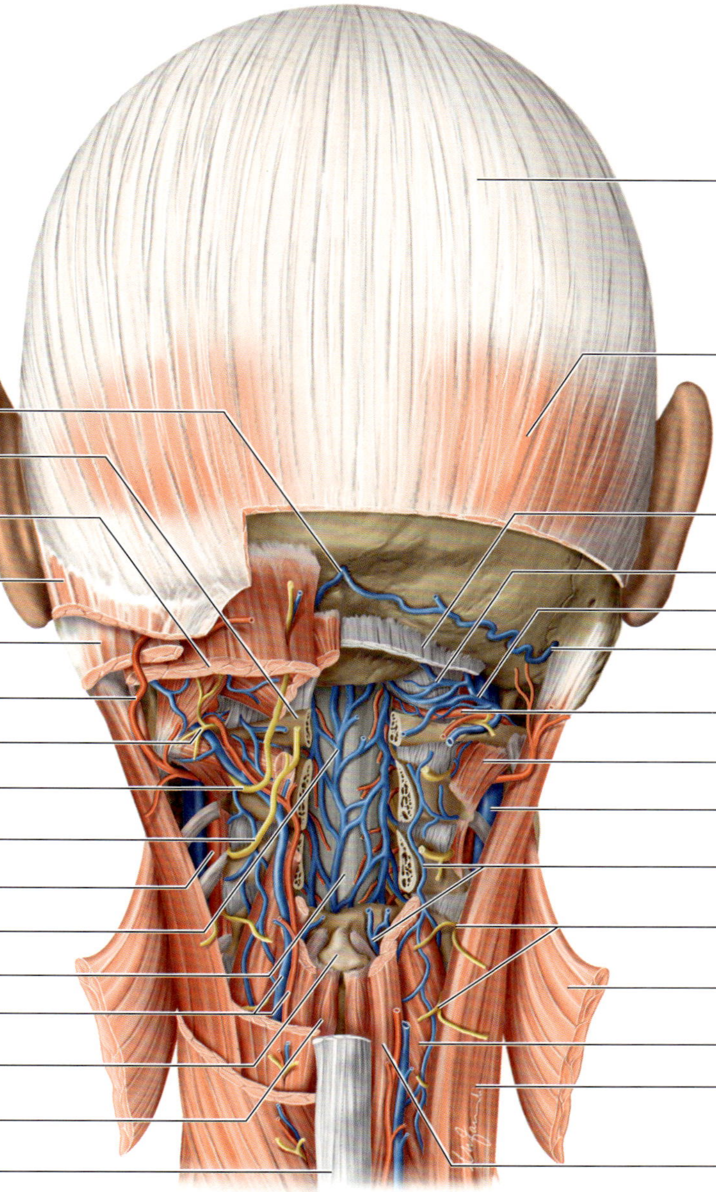

Head and **neck**, posterior view. In the left half, the trapezius, sternocleidomastoid, semispinalis capitis, semispinalis cervicis, splenius capitis and levator scapulae have been removed. On the median plane, the ligamentum nuchae has been transected and, to the right and left, the arches of the first three cervical vertebrae have been transected, revealing the posterior external vertebral venous plexus. In the right half, the insertion of the occipital belly of the occipitofrontalis and the atlanto-occipital membrane have also been transected, showing the suboccipital venous plexus, the occipital, condylar and mastoid emissary veins, and the atlantic part of the vertebral artery.

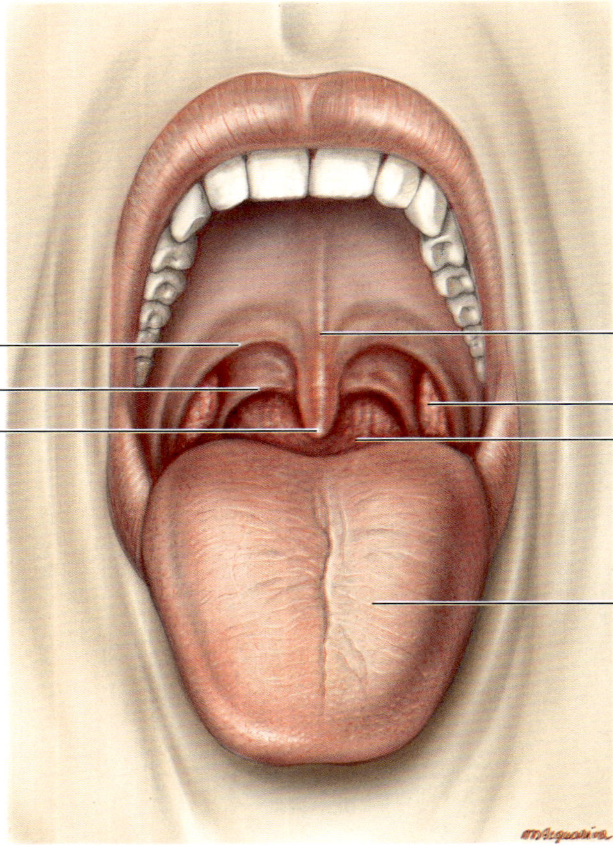

Oral cavity and isthmus of the fauces viewed from the front.

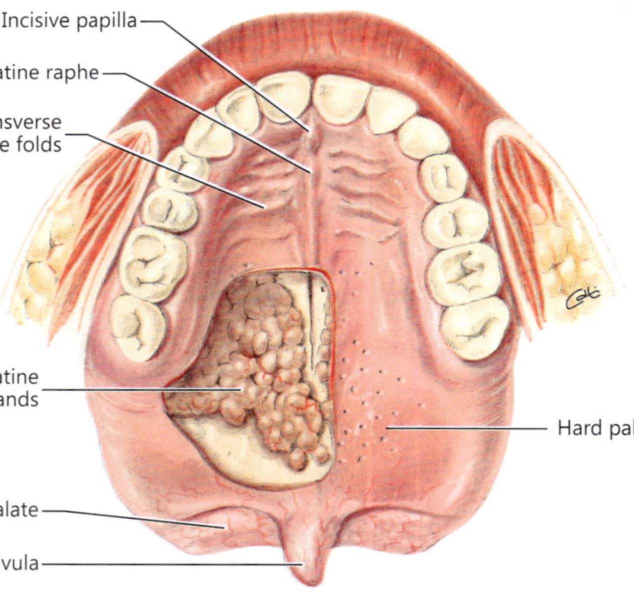

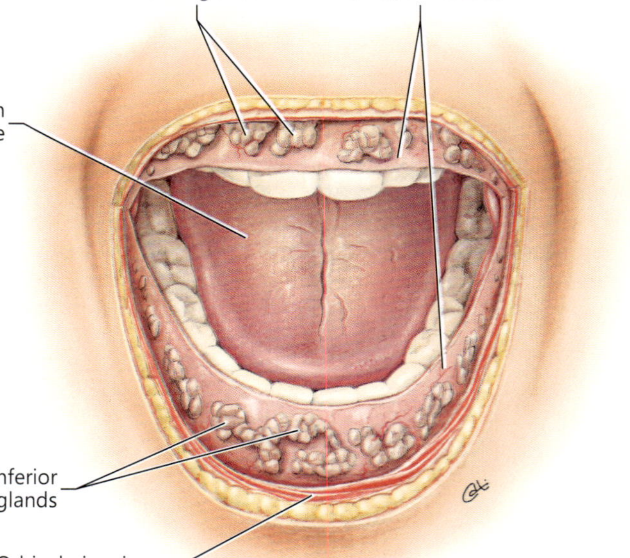

Palate. In the anterior third of the hard palate, the mucous membrane displays the transverse palatine folds and, in median position, the palatine raphe. Posteriorly, with the mucous membrane removed, one can see the lobules of the palatine glands, which extend also to the soft palate.

After the skin and muscle layer have been removed, the lower labial mucosa turned and the upper labial mucosa pulled downwards, one can see the distribution of lobules of the **labial glands** that form a glandular layer deep to the lamina propria of labial mucosa. The glandular lobules appear less densely concentrated near the commissures.

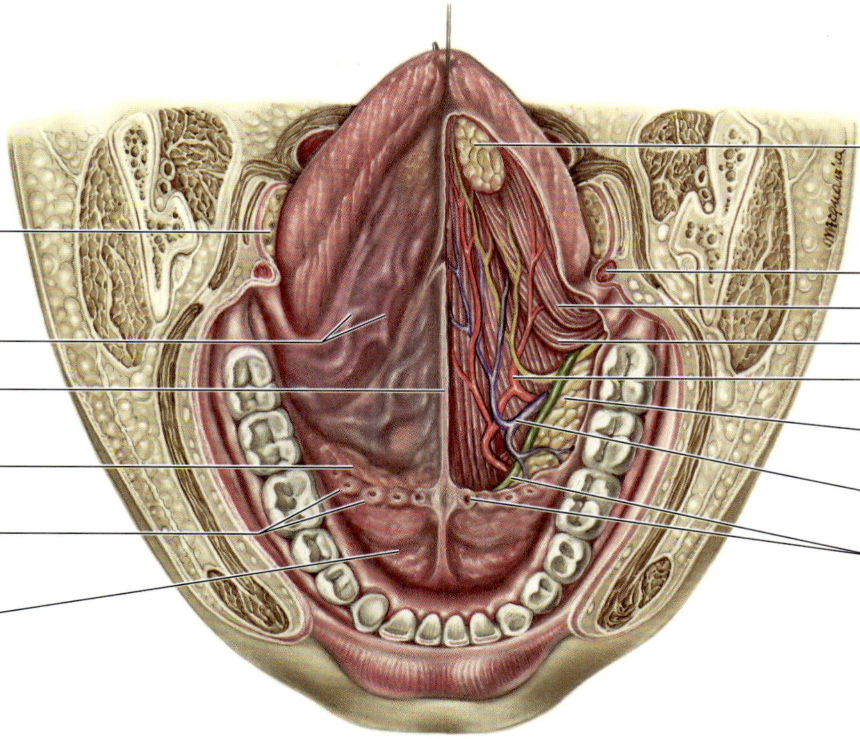

Sublingual fold. Transverse section of the head passing through the oral fissure. The tongue raised backwards shows the sublingual fold. On the right side, where the mucosa is intact, the veins of the submucosa can be seen showing through. On the left, the mucosa has been removed, and one can see affluents of the lingual vein and branches of the deep lingual artery. In addition, in the sublingual fold, numerous orifices where the submandibular and sublingual glands open into the sulcus are visible.

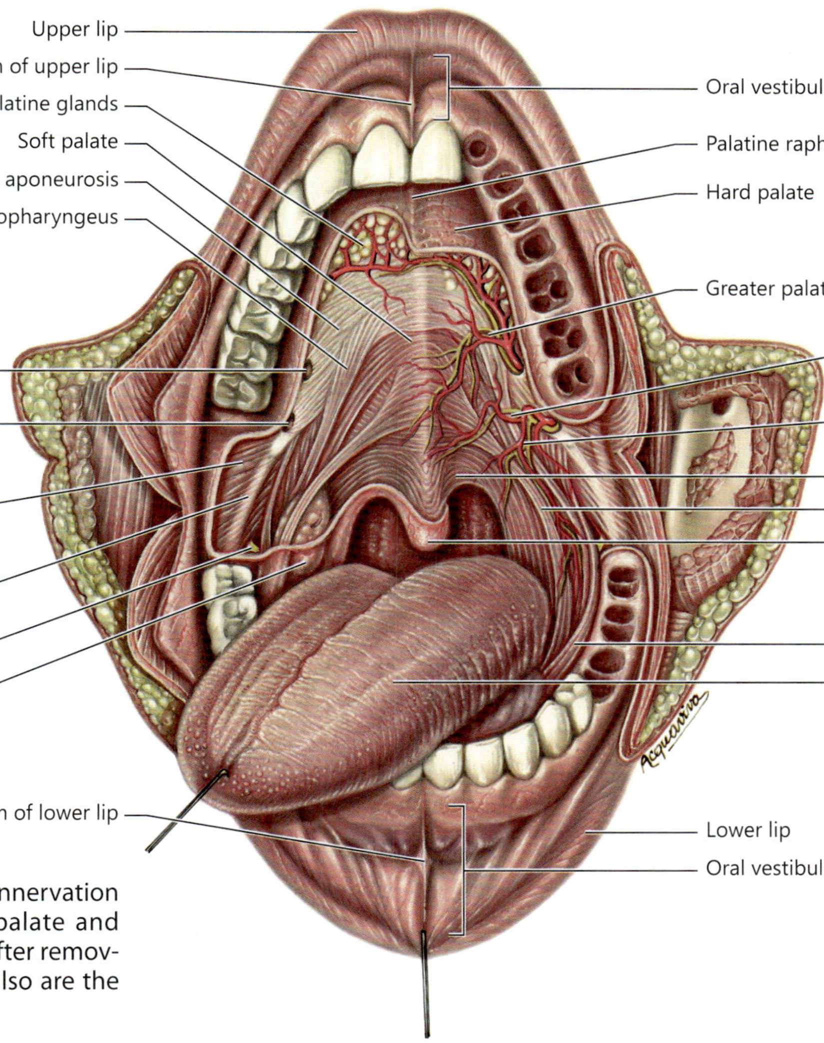

Arterial blood supply and innervation of the **hard palate**, soft palate and palatoglossal arch, visible after removal of the mucosa. Evident also are the muscles of the **soft palate**.

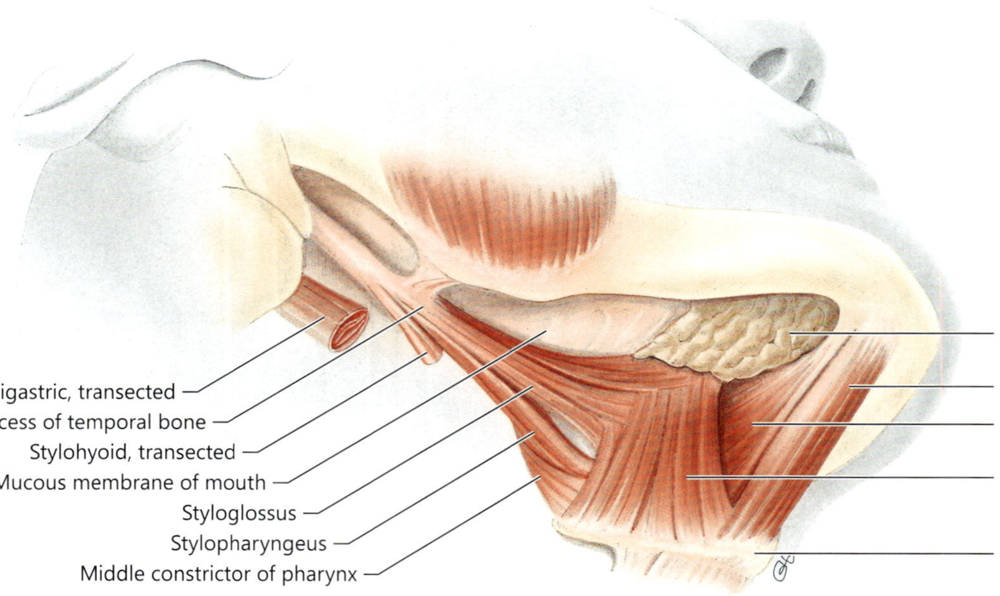

Diagram showing the **extrinsic muscles of the tongue**. This view of the deep layer of the suprahyoid region shows, alongside the extrinsic muscles that originate from the hyoid bone, also the muscles that originate from the styloid process and mandible.

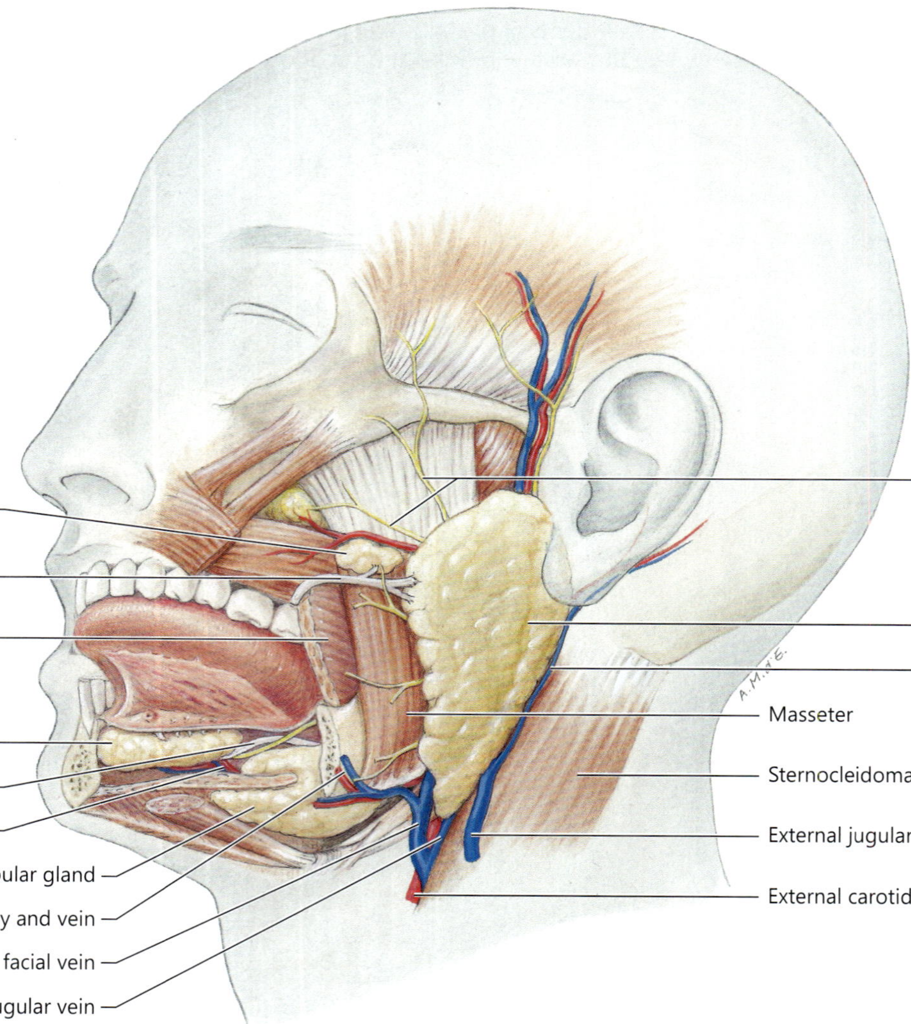

Diagram showing the location of the left **submandibular** and **sublingual glands** in relation to the floor of the mouth. The submandibular duct is flanked by the anterior prolongation of the homonymous gland, which, running above the mylohyoid, borders directly with the sublingual gland.

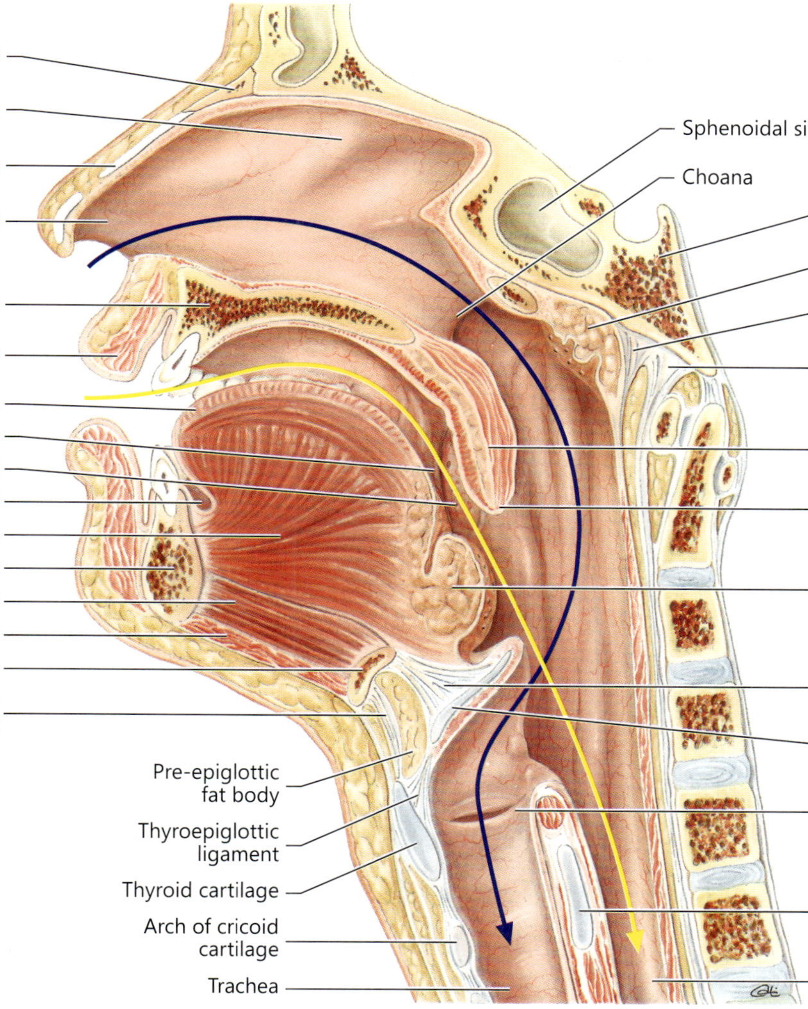

Diagram illustrating the **crossing**, at the level of the pharynx, **of the airways** (in *blue*) and **digestive tract** (in *yellow*). The place of crossing is at the point of the laryngeal inlet, in correspondence to the epiglottis.

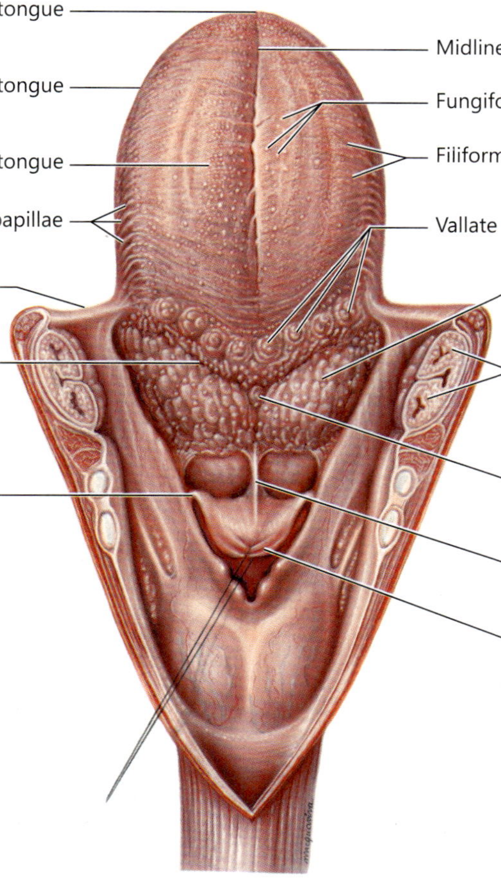

Dorsal surface of the tongue. Note the different external configuration of the body of the tongue with respect to the root. The *terminal sulcus* represents the boundary between the two regions. The tongue here is viewed from behind, after ample demolition of the posterior wall of the pharynx.

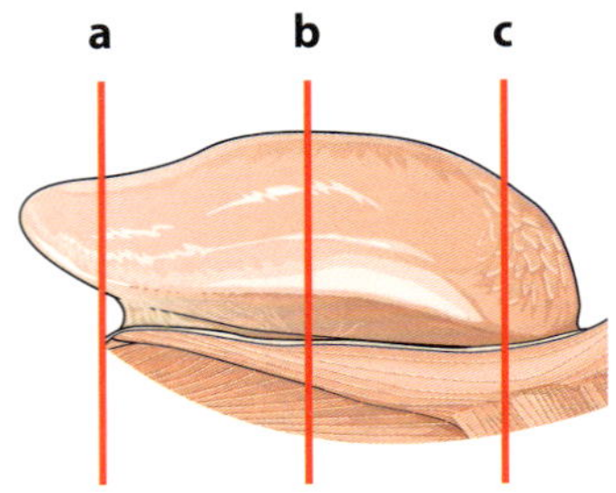

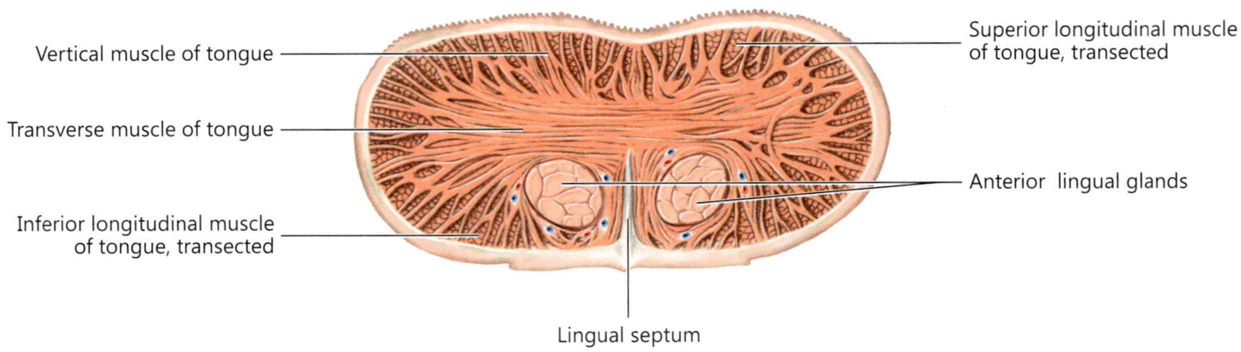

a

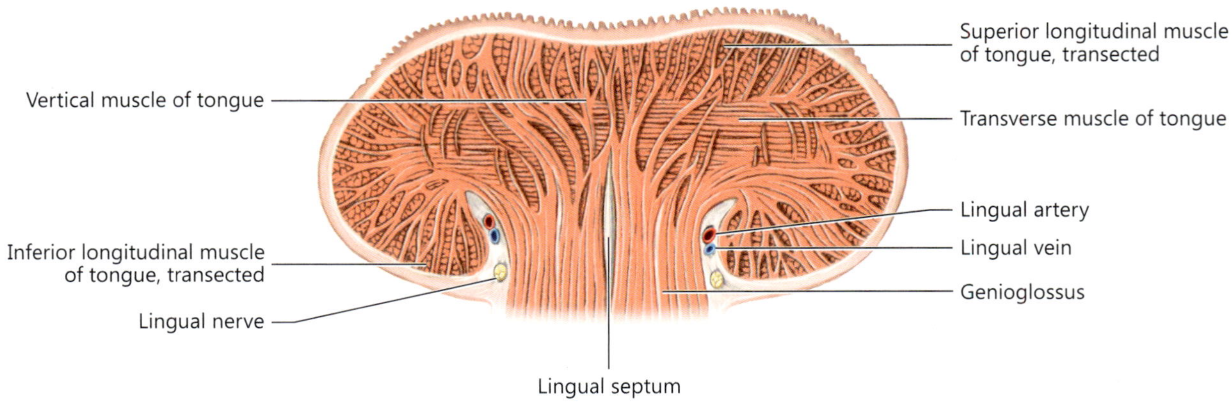

b

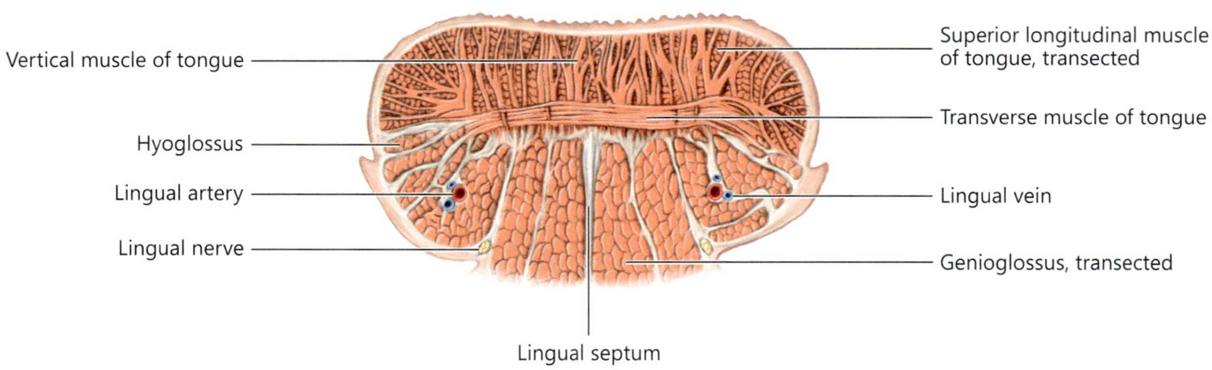

c

Structure of the tongue in frontal sections conducted at the level of: the body (**a**) and root of the tongue (**b** and **c**; see scheme above). The organ has a prevalently fibromuscular constitution and is covered, on the surface, by the mucous membrane. The tongue, as a whole entity, contains glands (here the anterior lingual glands are represented).

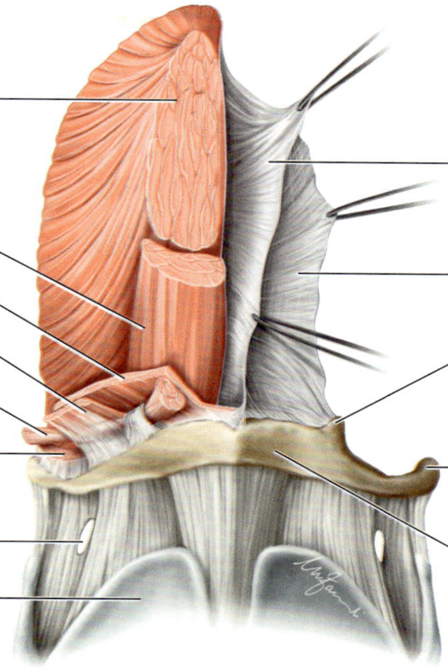

Inferior surface and **root of the tongue**. This view highlights the **fibrous skeleton of the tongue**, on the left side, formed by the lingual septum and hyoglossal membrane. The *hyoglossal membrane* is a broad thin lamina, located on an almost frontal plane, that inserts on the body of the hyoid bone, between the lesser horns, and extends into the root of the tongue. The *lingual septum* has a crescent-shaped form and extends, starting from the inferior surface of the hyoglossal membrane and hyoid bone, on a median sagittal plane between the two genioglossus muscles.

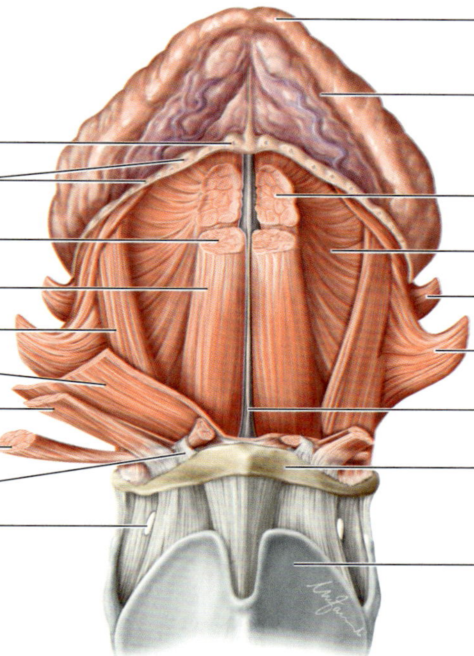

Inferior surface and **root of the tongue**. One can see the extrinsic muscles of the tongue. These include the genioglossus and hyoglossus, and the palatoglossus and styloglossus, which here have been transected. The floor of the mouth is formed by the mylohyoid and geniohyoid, in part transected, classified as part of the suprahyoid muscles. The fimbriated folds run from front to back and from the interior to exterior in the lateral parts of the inferior surface of the tongue.

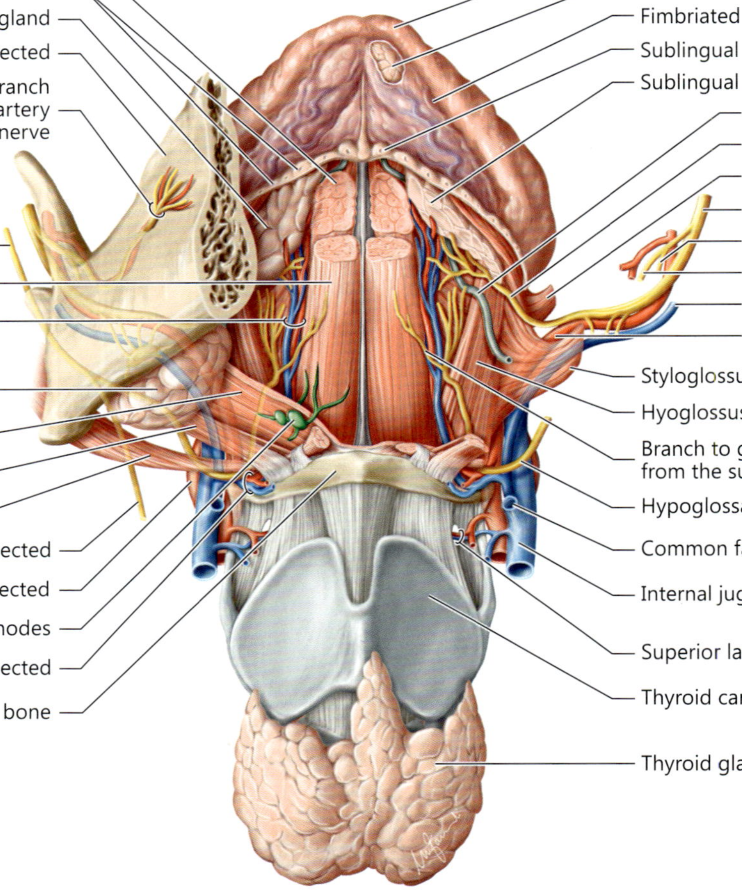

Floor of the mouth, anterior view. The tongue has been raised displaying its inferior surface. On the right, the body of the mandible has been transected in part and one can see the genioglossus and geniohyoid, transected at the point of their insertions into the superior and inferior mental, or genial, spines. One can see the sublingual and submandibular glands, the lingual artery and vein and, visible through the body of the mandible, the hypoglossal nerve. In the left half, with the mandible removed and the mylohyoid, stylohyoid and posterior belly of the digastric transected, one can see the lingual nerve, a branch of the mandibular nerve, the sublingual duct, and the retromandibular, common facial and internal jugular veins, transected. At the bottom, beneath the right and left laminae of the thyroid cartilage, one can see the thyroid gland.

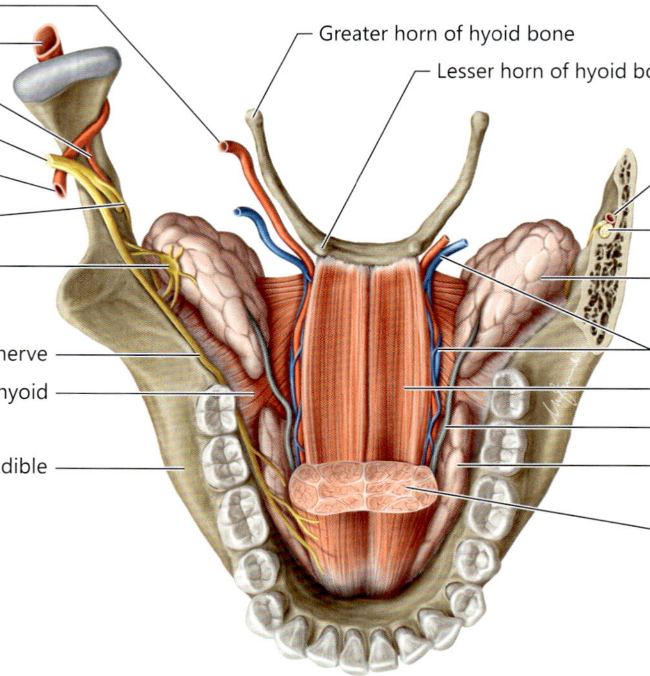

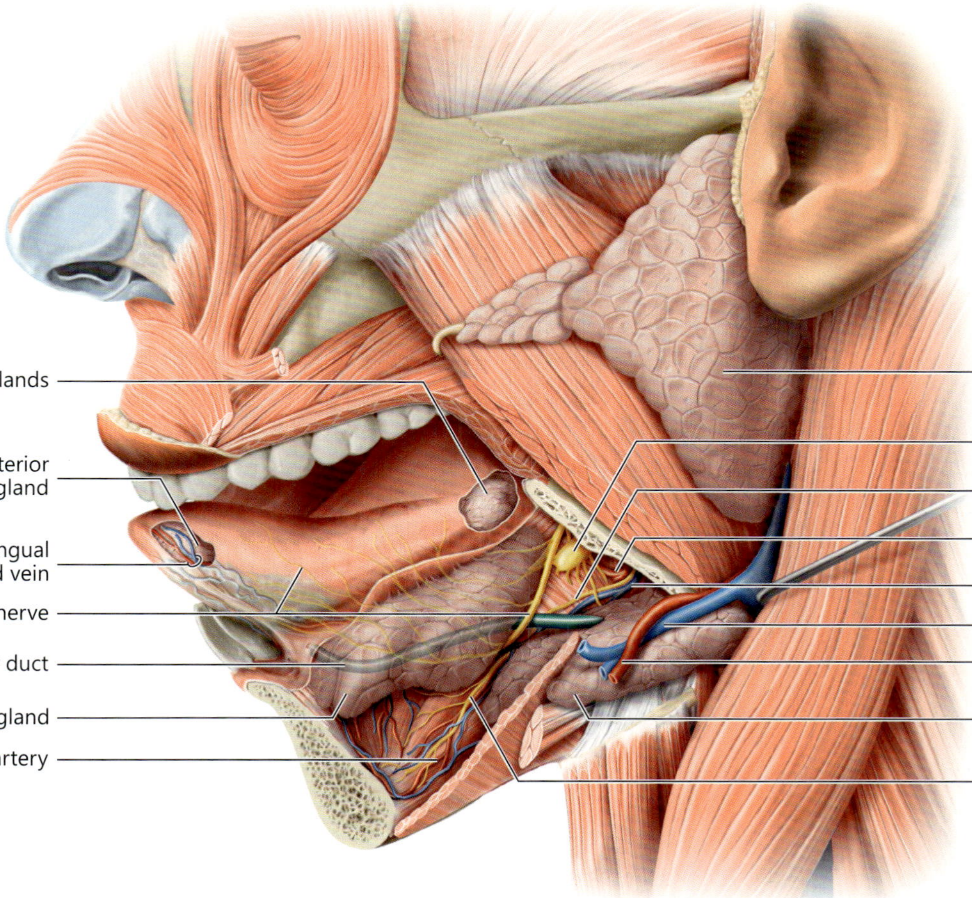

Left lateral view of the **oral cavity** after removal of part of the left body of the mandible and of the mylohyoid, stylohyoid, hyoglossus, genioglossus, digastric, buccinator, and masseter. One can see the major salivary glands (submandibular and sublingual glands) and some lingual glands. Also the parotid gland is visible in the parotid space. The lingual nerve, a branch of the mandibular nerve [V_3], enters the tongue on the lateral surface of the hyoglossus, in correspondence to the upper part of the submandibular space. Passing between the hyoglossus and the posterior border of the mylohyoid, it reaches the sublingual space, crossing below and in a lateromedial direction the submandibular duct. It then runs between the hyoglossus and genioglossus, located medially, and the sublingual gland, situated laterally. The hypoglossal nerve [XII], after reaching the submandibular space, passes into the interstitium between the hyoglossus and mylohyoid, deep to the submandibular gland and its excretory duct, anastomosing with the lingual nerve situated superiorly.

◄
Submandibular and **sublingual glands**, superior view after removal of the tongue. The submandibular ganglion receives afferent branches from the lingual nerve, a branch of the mandibular nerve. These are preganglionic fibers that arise from the superior salivatory nucleus, emerge from the brainstem with the intermediate nerve of the facial nerve, pass into the chorda tympani and unite with the lingual nerve along its course in the infratemporal fossa. Some postganglionic fibers travel directly to the submandibular gland, others return to the lingual nerve, which distributes them to the sublingual gland and to the minor salivary glands.

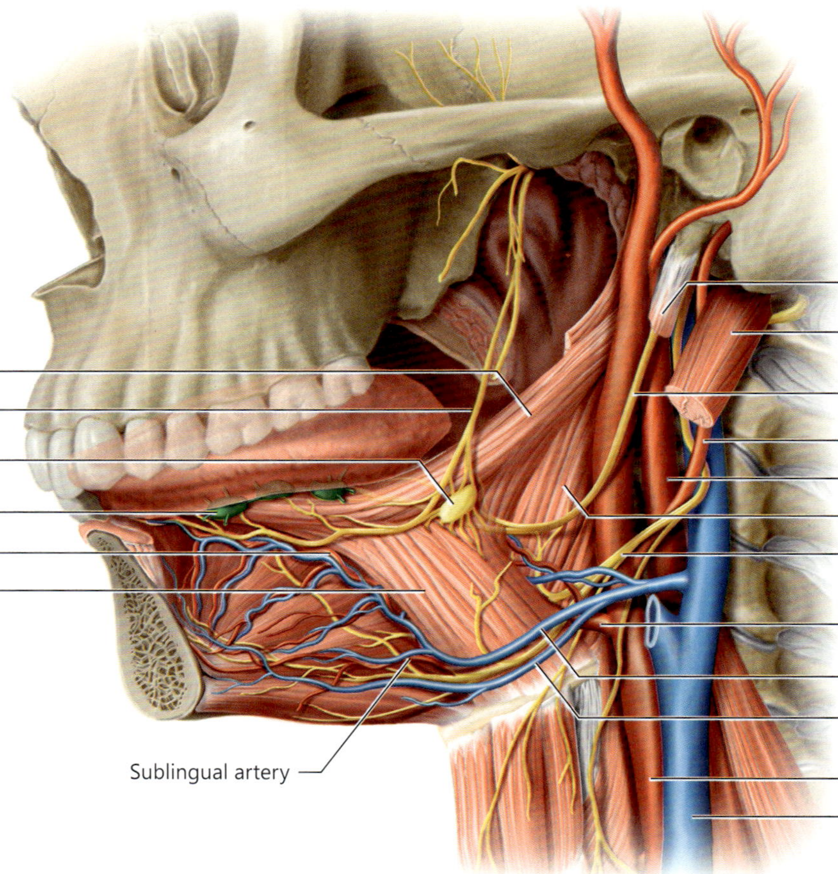

Vessels and nerves of the tongue. Left lateral view after removal of the left hemi-mandible, mylohyoid, stylohyoid, genioglossus, digastric, buccinator and masseter, and of the salivary glands. The submandibular ganglion receives afferent branches from the lingual nerve, a branch of the mandibular nerve [V_3]. Some efferent branches of the ganglion travel back to the lingual nerve, through which they are distributed. The glossopharyngeal nerve [IX], after emerging from the jugular foramen, follows an oblique path downwards and forwards, passing between the internal carotid artery and the internal jugular vein, then between the stylopharyngeus and styloglossus and, finally, attached to the stylopharyngeus. The hypoglossal nerve [XII], after its initial course behind the styloid muscles and neurovascular bundle of the neck, passes between the internal carotid artery and internal jugular vein, deep to the stylohyoid and posterior belly of the digastric, then winds around the occipital artery and, further down, superficially, crosses the lingual artery.

a, Distribution of **papillae on the surface of the tongue**. **b**, Three-dimensional graphic reconstruction of the filiform, vallate, foliate (situated at the edges of the body of the tongue and vestigial in man), and fungiform papillae. The filiform papillae contain only mechanoreceptors and respond to stimuli of a mechanical nature; the others contain also taste buds and, hence, respond also to chemical stimuli. **c**, Each taste bud is made up of a variety of taste sensory cells (organ of taste), each of which is specialized in mediating a certain type of flavor (in the image shown, the color of the cells corresponds to a specific flavor) in such a way that each taste bud perceives all the flavors. The apical surface of the taste cells forms microvilli, also termed taste hairs, which enter into contact, through the taste pore, with the chemical stimuli. At the basal extremity, the taste sensory cells in the posterior third of the tongue receive peripheral projections of protoneurons of the inferior or petrosal ganglion of the glossopharyngeal nerve [IX], whose central projections reach the nucleus of the solitary tract. The taste sensory cells of the anterior two-thirds of the tongue receive instead peripheral projections of the protoneurons of the geniculate ganglion of the intermediate nerve of the facial nerve [VII], whose central projections also reach the nucleus of the solitary tract. **d**, Recent functional and molecular studies have demonstrated that, in contrast to what is commonly held, the sensitivity to different modes of taste is not spatially distributed on the tongue, i.e. there is not a specific map of flavors. Each part of the tongue, in particular the zones marked by different colors, is sensitive to all flavors, as described above in **c** in regard to taste buds.

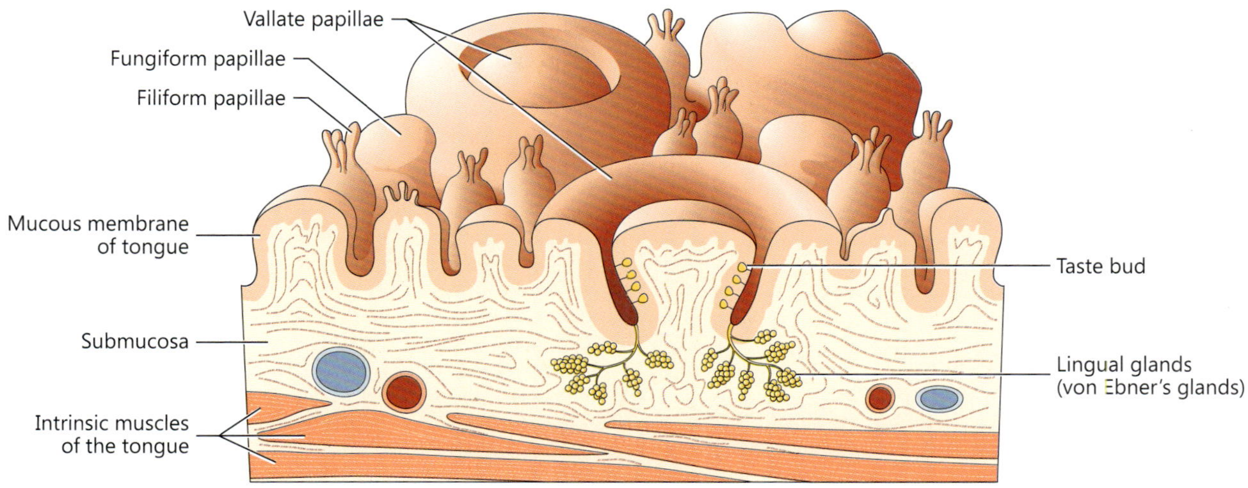

The **mucous membrane of the tongue** has a velvet-like texture owing to the presence of diverse types of papillae. The filiform papillae are features of the lamina propria of the mucosa which present at their apex fine secondary evaginations of a filamentous aspect. The fungiform papillae, also offshoots of the lamina propria, are mushroom-shaped with a narrow base and swollen apex. The vallate papillae, also termed circumvallate papillae, do not protrude from the lamina propria but are cylindrical features of connective tissue that have a circular sulcus or dip. The taste buds are contained in the epithelium covering the sides of the sulcus.

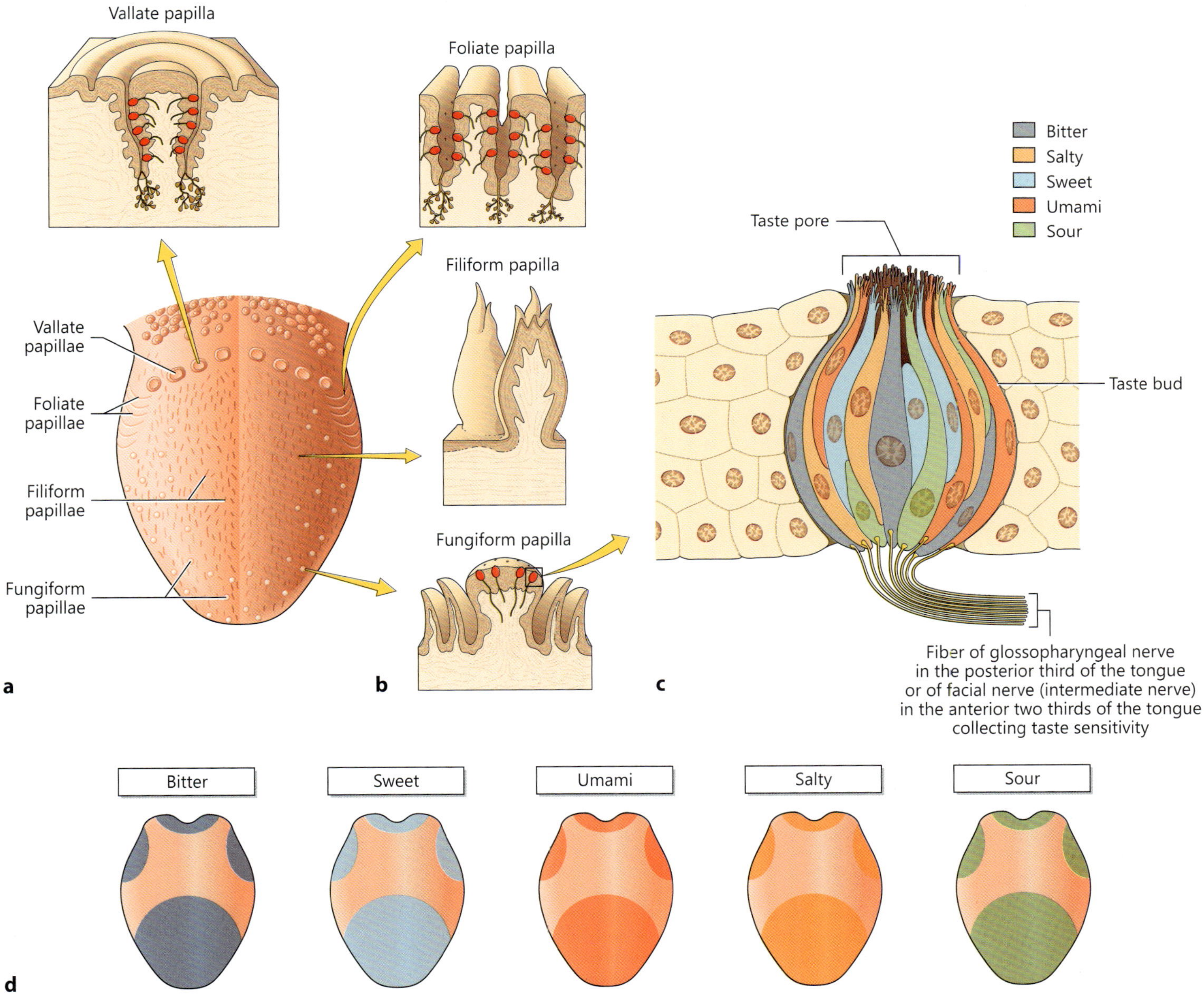

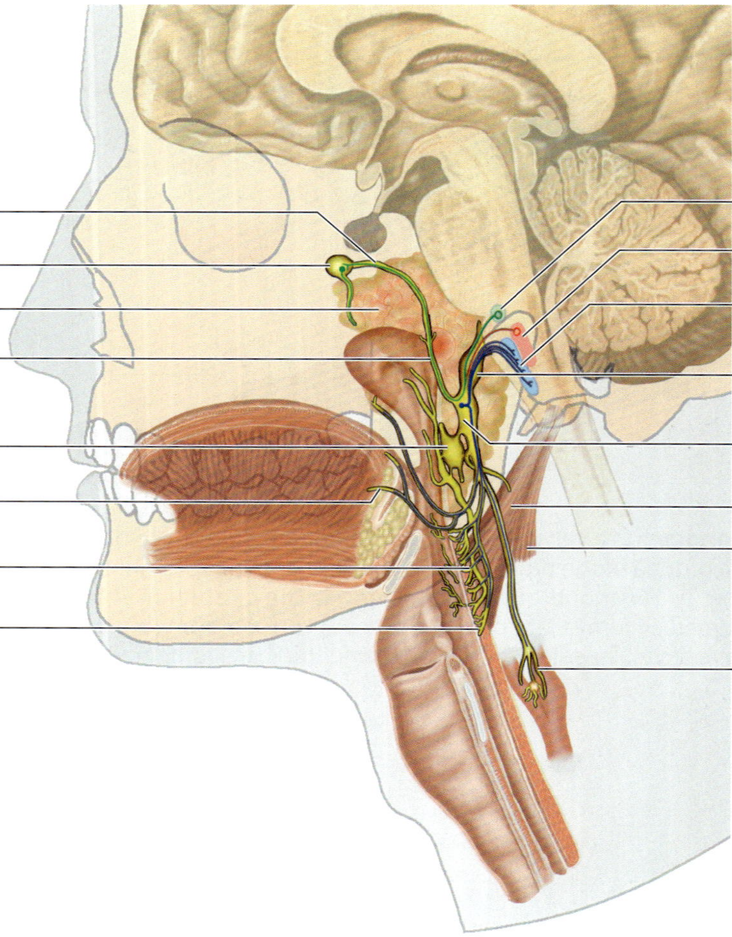
Lesser petrosal nerve
Otic ganglion
Parotid gland
Tympanic nerve
Superior cervical ganglion
Lingual branches of glossopharyngeal nerve
Pharyngeal branches of glossopharyngeal nerve
Tonsillar branches of glossopharyngeal nerve
Inferior salivatory nucleus
Nucleus ambiguus
Nucleus of solitary tract
Superior ganglion of glossopharyngeal nerve
Inferior ganglion of glossopharyngeal nerve
Styloglossus
Stylopharyngeus, transected
Carotid branch of glossopharyngeal nerve

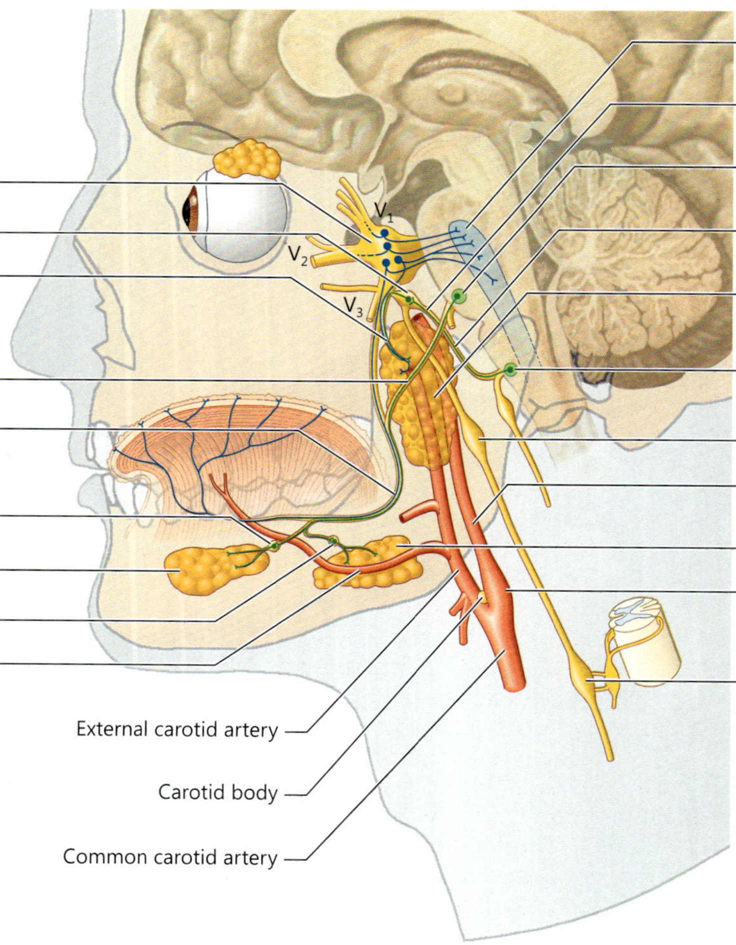
Trigeminal ganglion
Otic ganglion
Auriculotemporal nerve
Chorda tympani
Lingual nerve
Sublingual ganglion (inconstant)
Sublingual gland
Submandibular ganglion
Lingual artery
External carotid artery
Carotid body
Common carotid artery
Main mesencephalic nucleus and spinal nucleus of trigeminal nerve
Lesser petrosal nerve
Superior salivatory nucleus
Tympanic nerve (Jacobson's nerve)
Parotid gland
Inferior salivatory nucleus
Superior cervical ganglion
Internal carotid artery
Submandibular gland
Carotid sinus
Inferior cervical ganglion
V_1
V_2
V_3

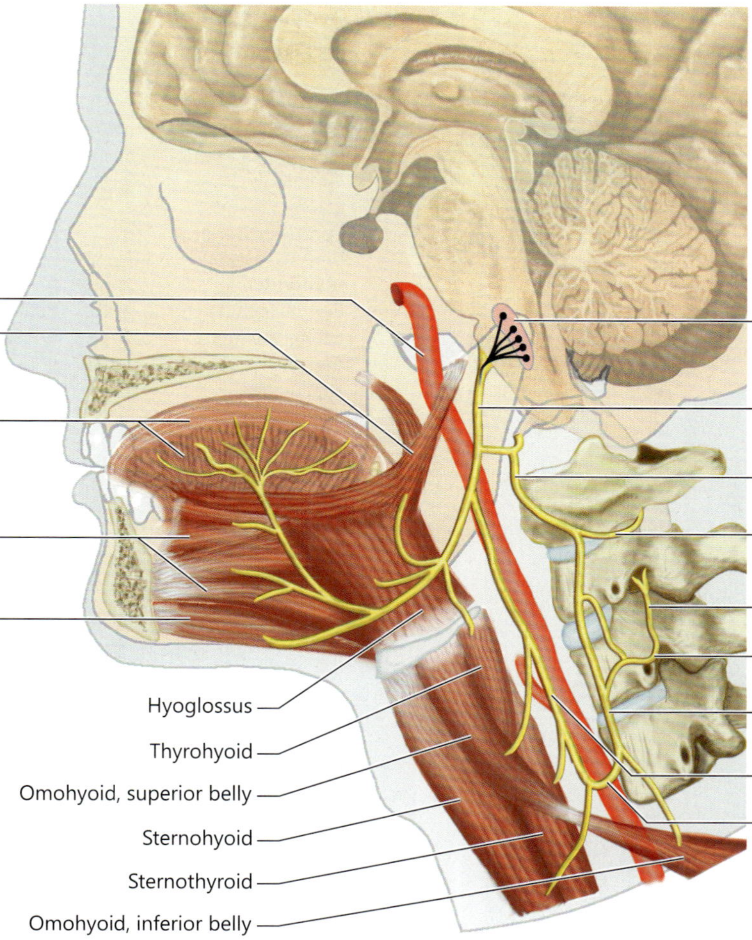

Course and distribution of the **hypoglossal nerve** [XII]. The nerve, which is exclusively somatic motor, supplies the muscles of the tongue and also the supra- and infrahyoid muscles. These latter are actually innervated by the anterior rami of the first cervical nerves, through the ansa cervicalis. The superior root of the ansa cervicalis, also known as the descending branch of the hypoglossal nerve, contains only fibers coming from the first cervical nerve [C1]. The inferior root of the ansa cervicalis, also known as the descending cervical nerve, is formed of fibers coming from the second [C2] and third [C3] cervical nerves. The two roots unite to form the ansa cervicalis, known also as ansa of the hypoglossal nerve, whose branches innervate the sternohyoid, sternothyroid and inferior belly of the omohyoid. The superior belly of the omohyoid is innervated by a branch of the superior root of the ansa cervicalis. The thyrohyoid is innervated by a branch containing fibers coming from the first cervical nerve [C1] that separates from the hypoglossal nerve after the latter has given rise to the superior root of the ansa cervicalis.

➤

At the top of the previous page, course and branches of the **glossopharyngeal nerve** [IX]. Its somatic motor component supplies the stylopharyngeus. The somatic sensory fibers (general somatic sense) receive sensory input from a limited zone of the auricle. The motor or visceral effector component reaches the parotid and minor salivary glands. Visceral sensory fibers conduct special visceral stimuli (specific taste sense) from the taste buds of the posterior third of the tongue and visceral stimuli (visceral general sense) from the chemoreceptors of the carotid body, from the baroreceptors of the carotid sinus, and from the mucosa of the tympanic cavity, pharyngotympanic tube, nasopharynx, isthmus of fauces, palatine tonsil and the posterior third of the tongue.

At the bottom of the previous page, diagram showing the connections of the **otic**, **submandibular** and **sublingual ganglia**. The three ganglia receive parasympathetic fibers (*green*) from the superior salivatory nucleus (of the facial nerve) and inferior salivatory nucleus (of the glossopharyngeal nerve). The otic ganglion receives preganglionic fibers coming from the inferior salivatory nucleus that exit from the brainstem with the glossopharyngeal nerve [IX] and pass into the tympanic nerve and then into the lesser petrosal nerve. Most of the postganglionic fibers enter into the auriculotemporal nerve and reach the parotid gland; others go to the minor salivary glands by means of peripheral branches of the mandibular nerve [V_3]. Some postganglionic vasodilator fibers go to peripheral innervation zones of the mandibular nerve [V_3]. The submandibular and sublingual (inconstant) ganglia receive preganglionic fibers coming from the superior salivatory nucleus, which emerge from the brainstem with the intermediate nerve, a component of the facial nerve [VII], pass into the chorda tympani and then unite with the lingual nerve, a branch of the mandibular nerve [V_3]. Some postganglionic fibers supply the submandibular gland, others reach the lingual nerve and from there supply the sublingual and minor salivary glands.

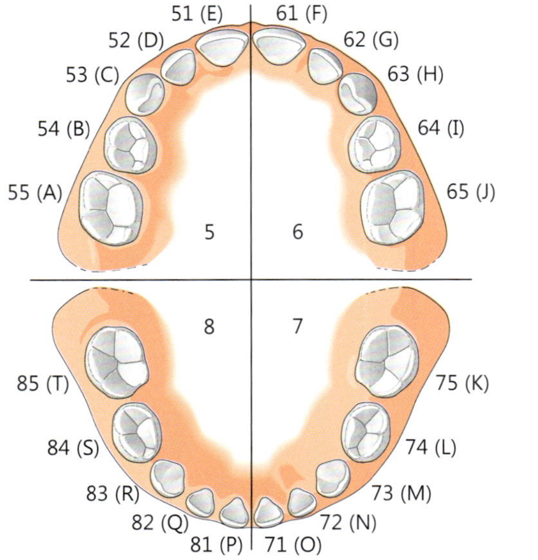

a

Deciduous teeth		
	ERUPTION	***EXFOLIATION***
Maxilla		
Second molars	23-31 months	10-12 years
First molars	14-18 months	9-11 years
Canines	17-23 months	9-12 years
Lateral incisors	10-16 months	7-8 years
Central incisors	6-10 months	6-7 years
Mandible		
Central incisors	8-12 months	6-7 years
Lateral incisors	9-13 months	7-8 years
Canines	16-22 months	10-12 years
First molars	13-19 months	9-11 years
Second molars	25-33 months	10-12 years

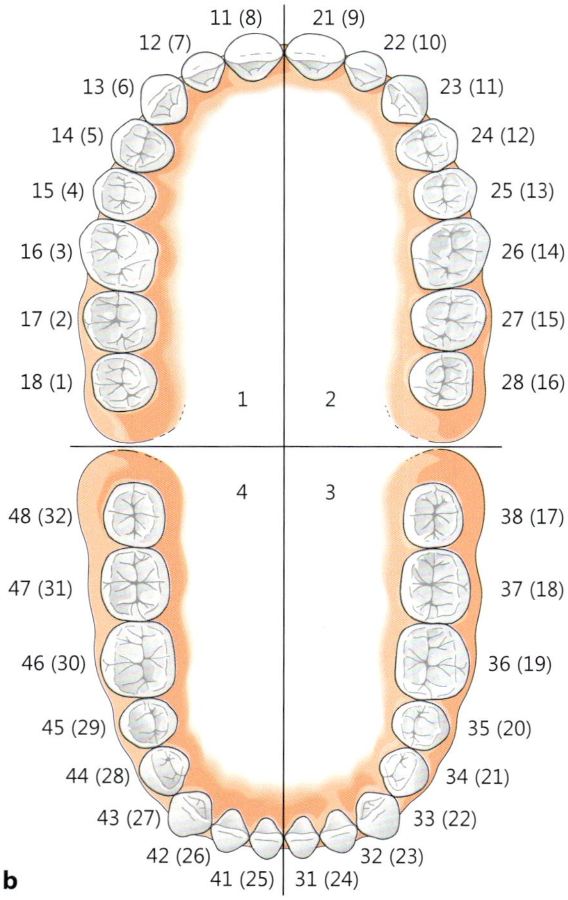

b

Permanent teeth		
	ERUPTION	***ROOT COMPLETION***
Maxilla		
First molars	6-7 years	9-10 years
Central incisors	7-8 years	10 years
Lateral incisors	8-9 years	11 years
First premolars	10-11 years	12-13 years
Second premolars	10-12 years	12-14 years
Canines	11-12 years	13-15 years
Second molars	12-13 years	14-16 years
Third molars	17-21 years	18-25 years
Mandible		
First molars	6-7 years	9-10 years
Central incisors	6-7 years	9-11 years
Lateral incisors	7-8 years	10 years
Canines	9-10 years	12-14 years
First premolars	10-12 years	12-13 years
Second premolars	11-12 years	13-14 years
Second molars	11-13 years	14-15 years
Third molars	17-21 years	18-25 years

Dentition. a, numbering and representation of **deciduous teeth** according to the Fédération Dentaire International (FDI) and American Dental Association (ADA; indicated in brackets). In the FDI numbering, each tooth is classified by two digits: the first indicates the hemiarcade, the second, the type of tooth. The hemiarcade is indicated by the number 5, 6, 7, or 8 (in order, right and left maxillary hemiarcades and left and right mandibular hemiarcades), while the tooth (central incisor, lateral incisor, canine, 1st molar, 2nd molar) is indicated by a number from 1 to 5. In the ADA system of numbering, deciduous teeth are represented by the letters of the English alphabet from A to T starting from the upper right 2nd molar and proceeding left to the upper left 2nd molar, then continuing inferiorly from the lower left 2nd molar to the lower right 2nd molar; **b**, numbering of **permanent teeth** according to the Fédération Dentaire International (FDI) and American Dental Association (ADA; indicated in brackets). In the FDI numbering, each tooth is classified by two digits: the first indicates the hemiarcade, the second, the type of tooth. The hemiarcade is indicated by the number 1, 2, 3, or 4 (in order, right and left maxillary hemiarcades, and left and right mandibular hemiarcades), while the tooth (central incisor, lateral incisor, canine, 1st premolar, 2nd premolar, 1st molar, 2nd molar, 3rd molar) is indicated by a number from 1 to 8. In the ADA system of numbering, permanent teeth are numbered from 1 to 32 starting from the upper right 3rd molar and proceeding left to the upper left 3rd molar, then continuing inferiorly from the lower left 3rd molar to the lower right 3rd molar.

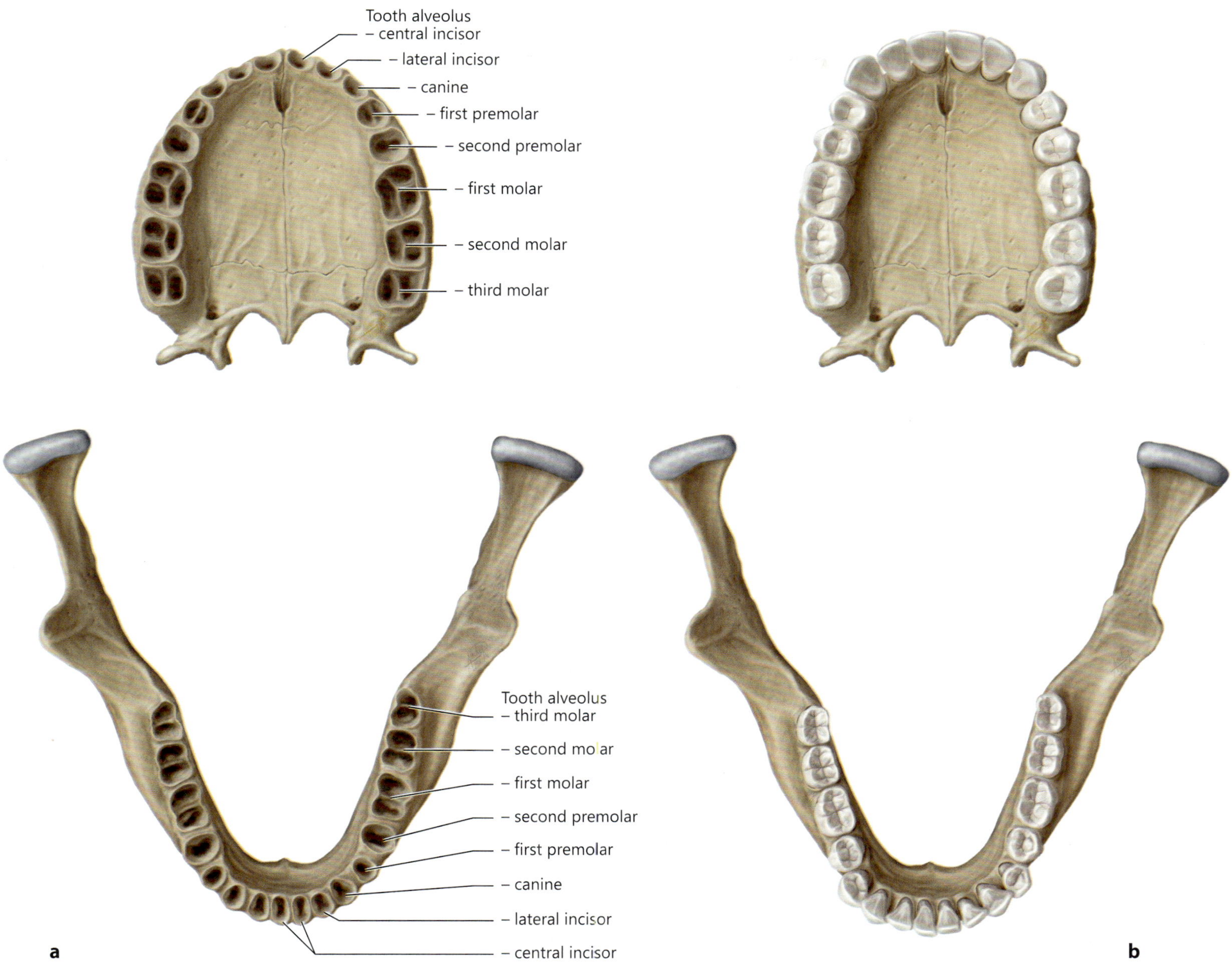

Maxillary, or upper (above), **and mandibular**, or lower (below), **alveolar arches** (**a**) **and dental arcades** (**b**), without and with dentition, viewed respectively from below and from above. Monoradicular teeth are held in dental alveoli with a single cavity, pluriradicular teeth in alveoli with a cavity subdivided by septa.

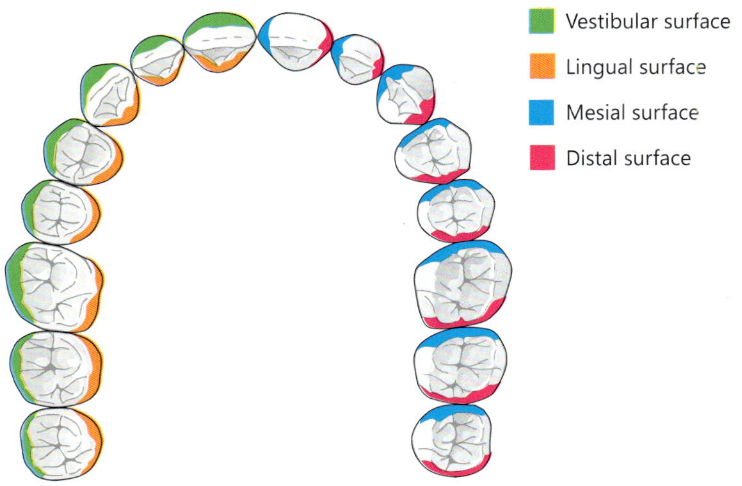

Terminology of position and surfaces of the teeth	
Vestibular	Facing towards the oral vestibule *Vestibular surface*: the external surface of the teeth
Lingual	Facing towards the tongue *Lingual surface*: internal surface of the teeth (many use the term "palatine" for the corresponding surfaces of the superior teeth)
Mesial	Facing the median line *Mesial surface*: surface of the tooth turned towards the median line; faces the tooth that precedes it
Distal	Facing in the opposite direction with respect to the median line *Distal surface*: surface of the tooth furthest from the median line; faces the tooth that follows it

The mesial and distal surfaces are indicated as *proximal surfaces*.

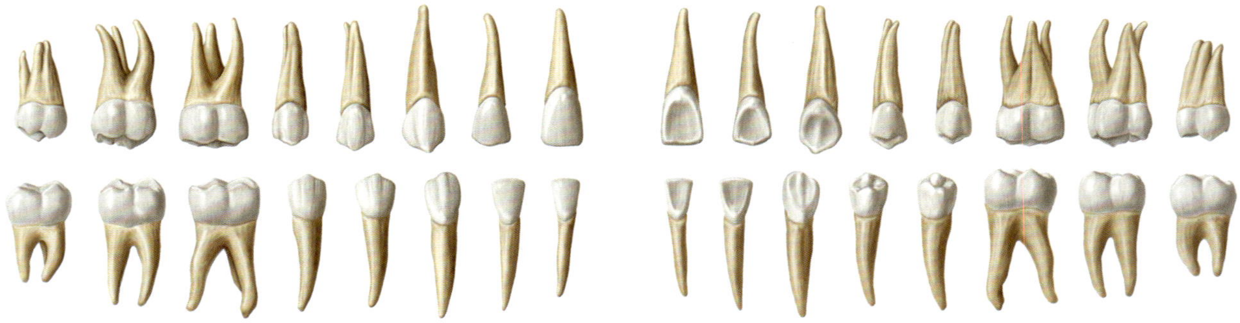

Dentition. **a**, **Permanent teeth** of the right **upper hemiarcade**, viewed from the vestibular surface and from the lingual surface. The premolars present a slight bifidity of the roots; the molars are triradicular; **b**, **permanent teeth** of the right **lower hemiarcade** viewed from the vestibular surface and from the lingual surface. The premolars are monoradicular, while the molars present two roots.

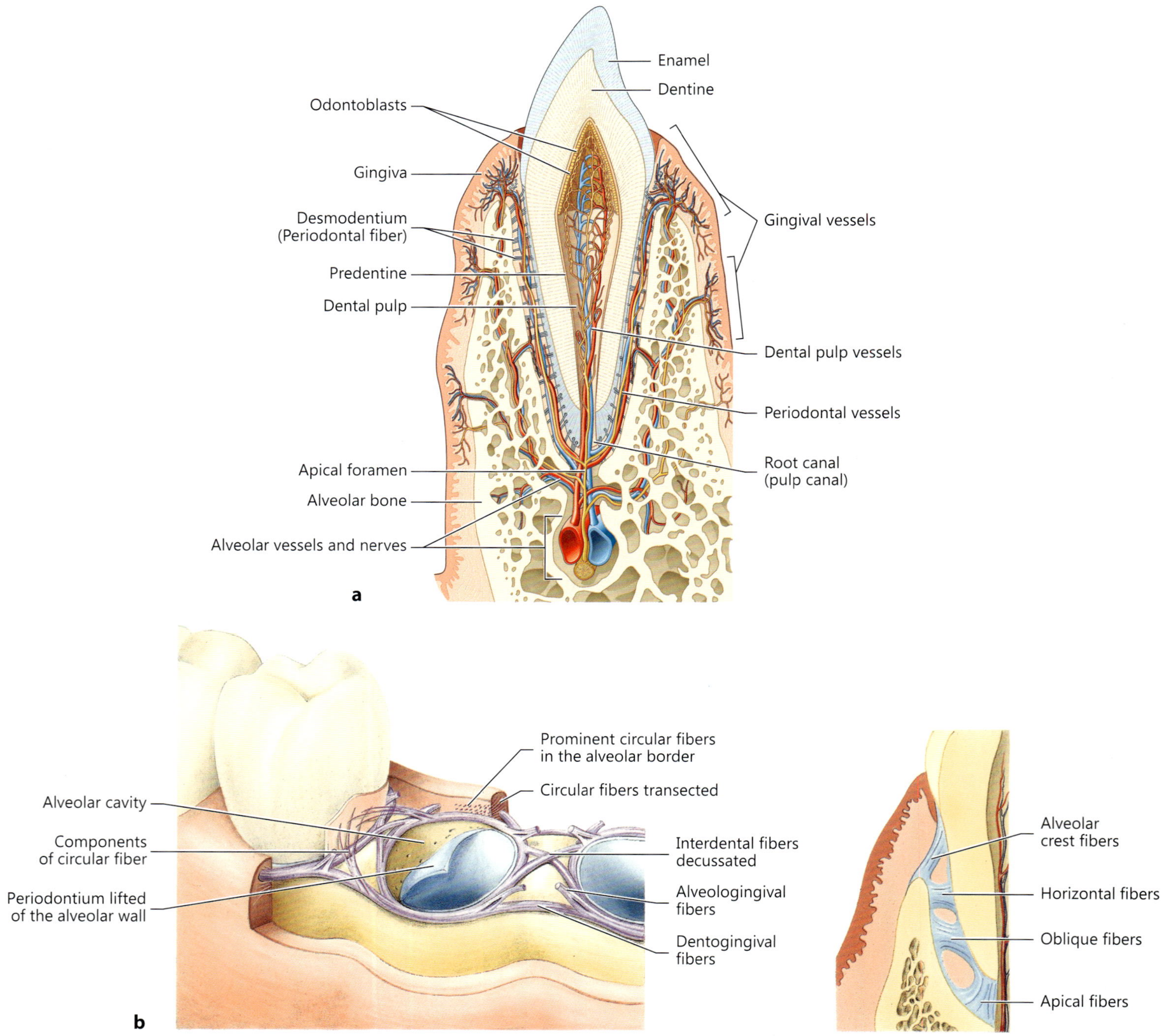

Dentition. **a**, Irroration of the oral tissues. The blood vessel supply and innervation of the desmodentium and dental pulp are shown; **b**, general structure of the periodontium (system of fixation of the tooth).

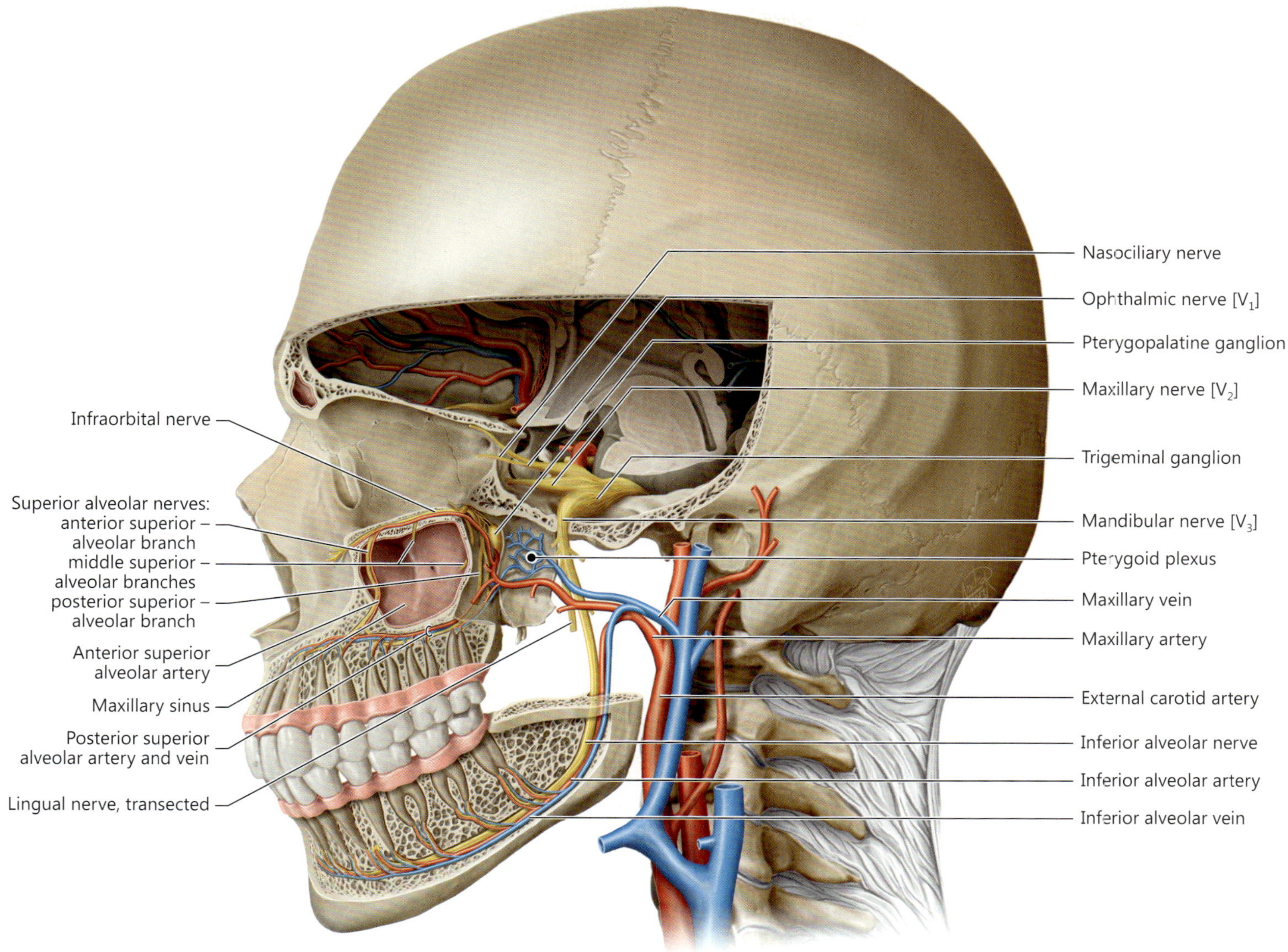

Viscerocranium transected at different levels, left lateral view. The left lateral wall of the neurocranium has also been partially removed to show the trigeminal or semilunar ganglion in the middle cranial fossa, near the apex of the petrous part of the temporal bone. The **mandibular nerve** [V_3] exits from the middle cranial fossa through the foramen ovale to reach the upper part of the infratemporal fossa where it divides into two large trunks, anterolateral and posteromedial. The **inferior alveolar nerve** arises from the posteromedial trunk and heads downwards, forwards and laterally describing an anterosuperiorly concave arch. After reaching the mandibular foramen on the medial surface of the mandible, it enters the mandibular canal together with the inferior alveolar artery, a branch of the **maxillary artery**. The posterior superior alveolar branch of the superior alveolar nerves arises from the **maxillary nerve** [V_2], whereas the anterior and middle superior alveolar branches originate from the infraorbital nerve, a terminal branch of the maxillary nerve [V_2], while it travels through the infraorbital canal. These branches pass in canaliculi excavated in the wall of the maxillary sinus, through which they reach the alveolar process of the maxilla. The posterior superior alveolar artery, a branch of the maxillary artery, descends forwards down on the maxillary tuberosity, dividing into small branches that penetrate into the alveolar canals of the maxilla. The anterior superior alveolar arteries, of which one single branch is represented here, arise from the infraorbital artery, a branch of the maxillary artery, within the orbital cavity, and penetrate into the incisive canals of the maxilla.

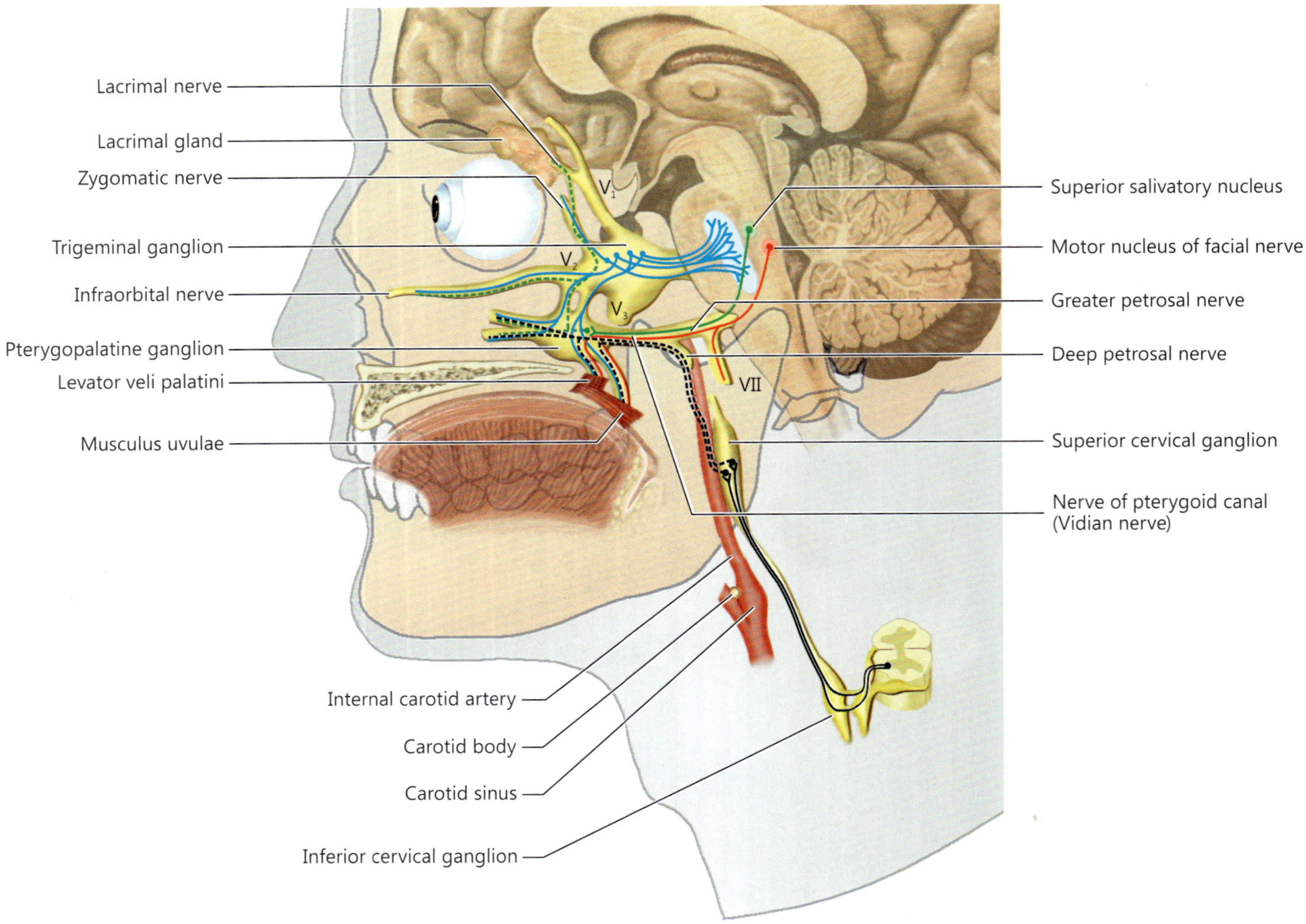

Connections of the **pterygopalatine ganglion** (or sphenopalatine ganglion). Motor and somatosensory fibers (*red* and *blue*) and sympathetic fibers (*black*) pass through the pterygopalatine ganglion, but only the parasympathetic fibers (*green*) originating from the superior salivatory nucleus sinapse there.

Diagram showing the course and distribution of the **trigeminal nerve** ☞ pag. 86.

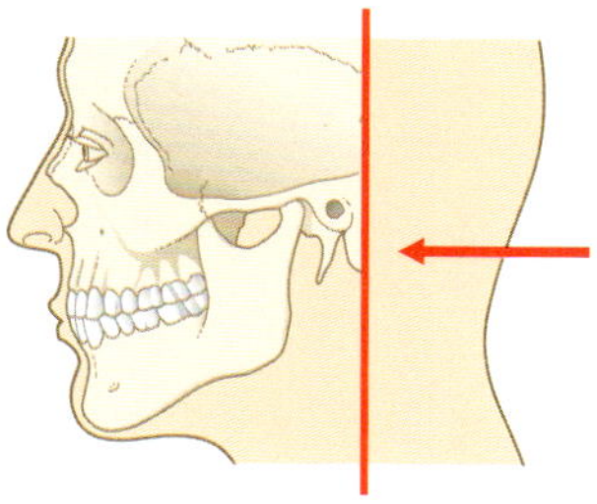

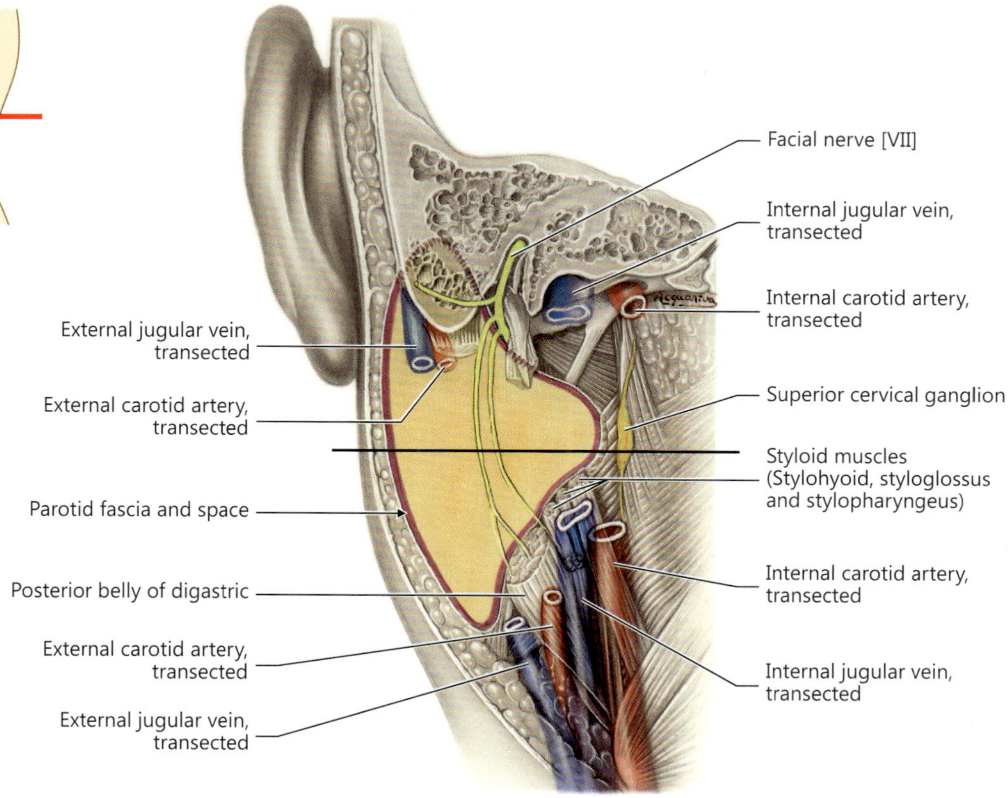

Superior and inferior, superficial and deep boundaries of the **parotid space** (area in *yellow*) in frontal section. The principal muscles, blood vessels and nerve formations related to the parotid space and the gland contained within it are represented. The horizontal line indicates the plane of transection that is illustrated in the following figure.

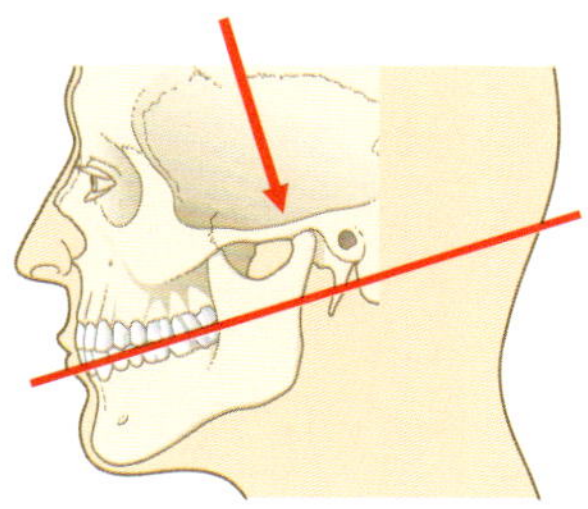

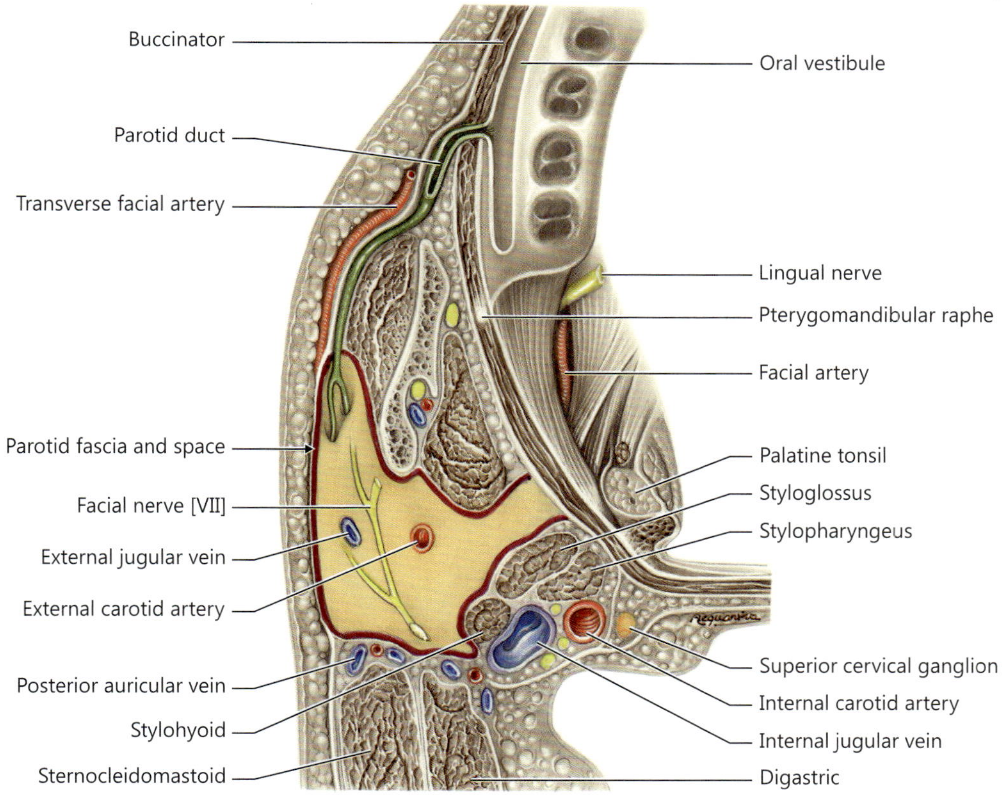

Transverse section of the **parotid space** (area in *yellow*) passing through the plane indicated in the preceding figure. Besides the superficial and deep relations of the parotid gland, also its anterior and posterior relations are illustrated.

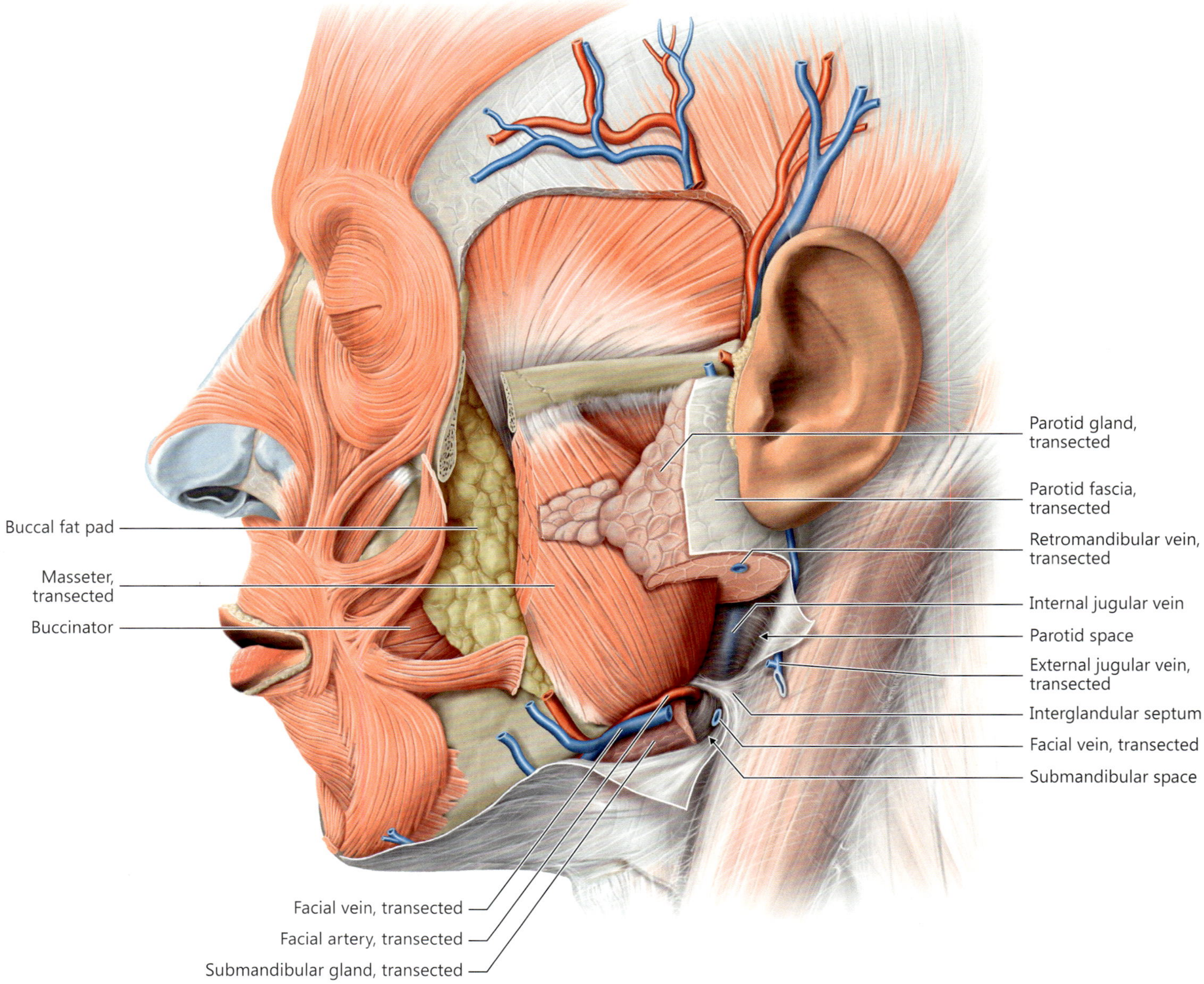

Head, left lateral view. After removing the skin and subcutaneous tissue, the zygomatic arch, masseter, parotid fascia and parotid gland have been transected. In the thickness of the posterior part of the cheek is the buccal fat pad, which occupies a space delimited laterally by the medial surface of the masseter and medially by the lateral surface of the buccinator. The parotid and submandibular glands are divided by a fibrous lamina formed by the fusion of the superficial and deep layers of the parotid fascia: the interglandular septum.

➤

Schematic representation showing the path of the **maxillary artery** in the **infratemporal fossa**. The maxillary artery is the largest terminal branch of the external carotid artery. After its origin behind the neck of the mandible within the parotid gland, it passes medially to the neck of the mandible and crosses the infratemporal fossa. This part of the artery, known as the pterygoid portion, can run either along the lateral surface of the lateral pterygoid, between the lower head of the lateral pterygoid and inferior part of the temporalis (in approximately 60% of cases), or along its medial surface, between the lower head of the lateral pterygoid and the medial pterygoid (in approximately 40% of cases), and it terminates in the pterygopalatine fossa.

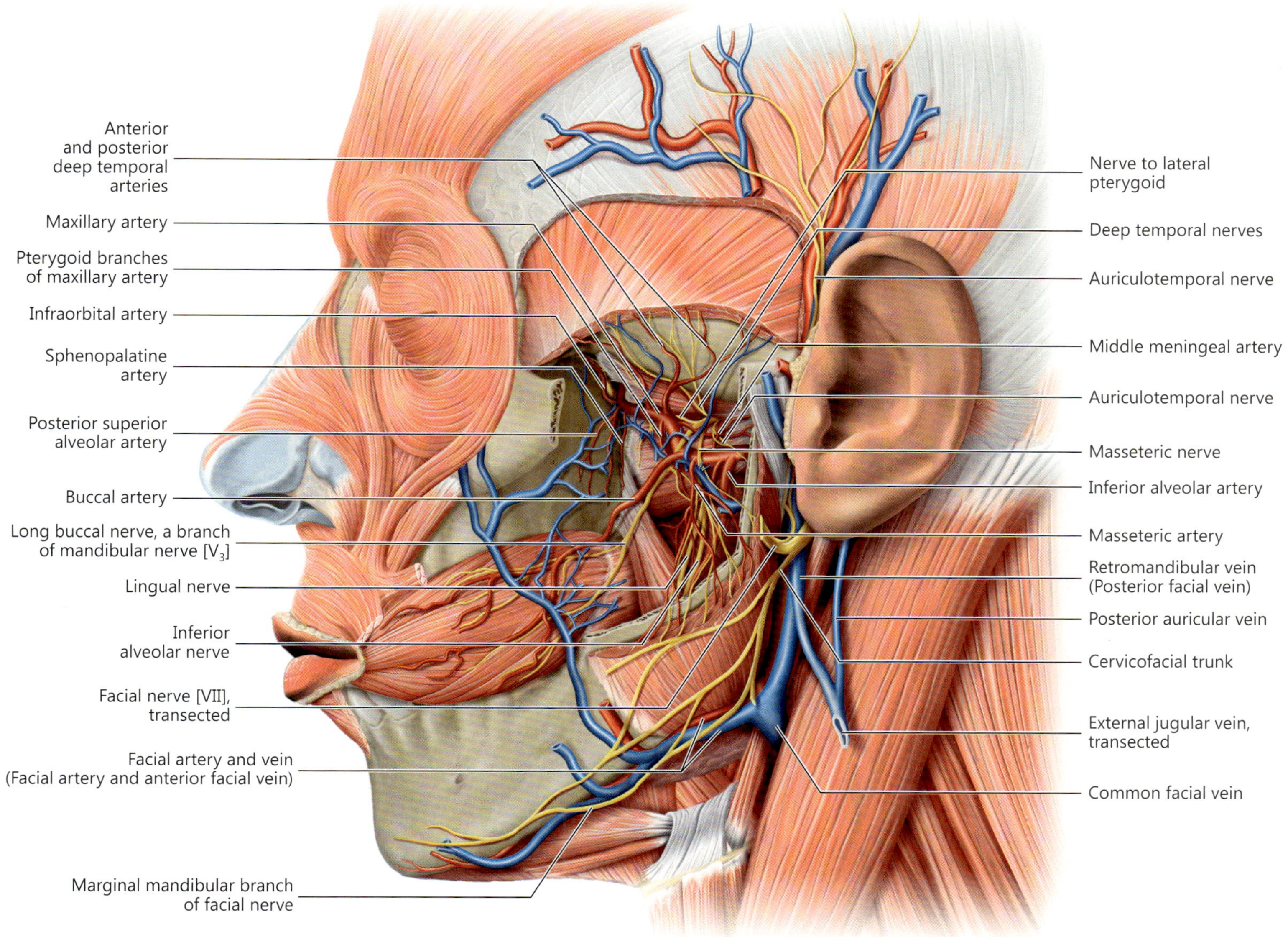

Infratemporal fossa, deep plane. Left lateral view after removal of the zygomatic arch and part of the temporalis. The maxillary artery, largest terminal branch of the external carotid artery, originates behind the neck of the mandible inside the parotid gland, passes medially to the neck of the mandible and crosses the infratemporal fossa heading laterally (as shown in the image and in the diagram below on the right), or medially (diagram below on the left) to the lower head of the lateral pterygoid, and terminates in the pterygopalatine fossa. In the infratemporal fossa one can see the inferior alveolar, masseteric, buccal, and anterior and posterior deep temporal arteries and the pterygoid branches of the maxillary artery, while in the pterygopalatine fossa one can see the posterior superior alveolar artery and infraorbital artery and the terminal branch of the maxillary artery, represented by the sphenopalatine artery.

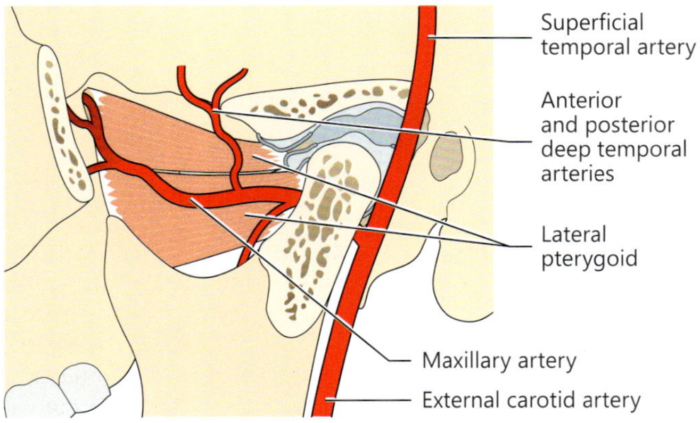

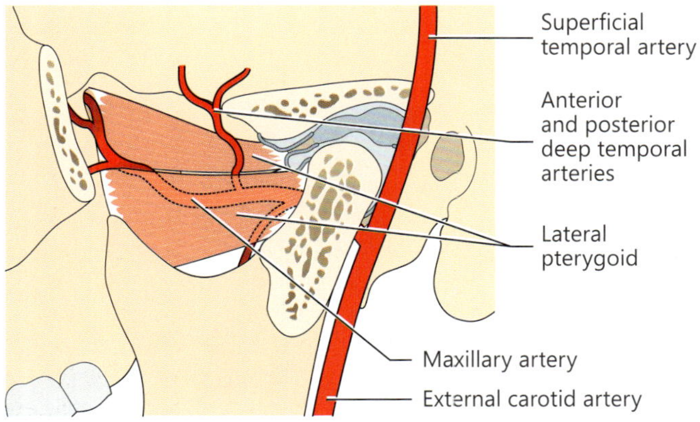

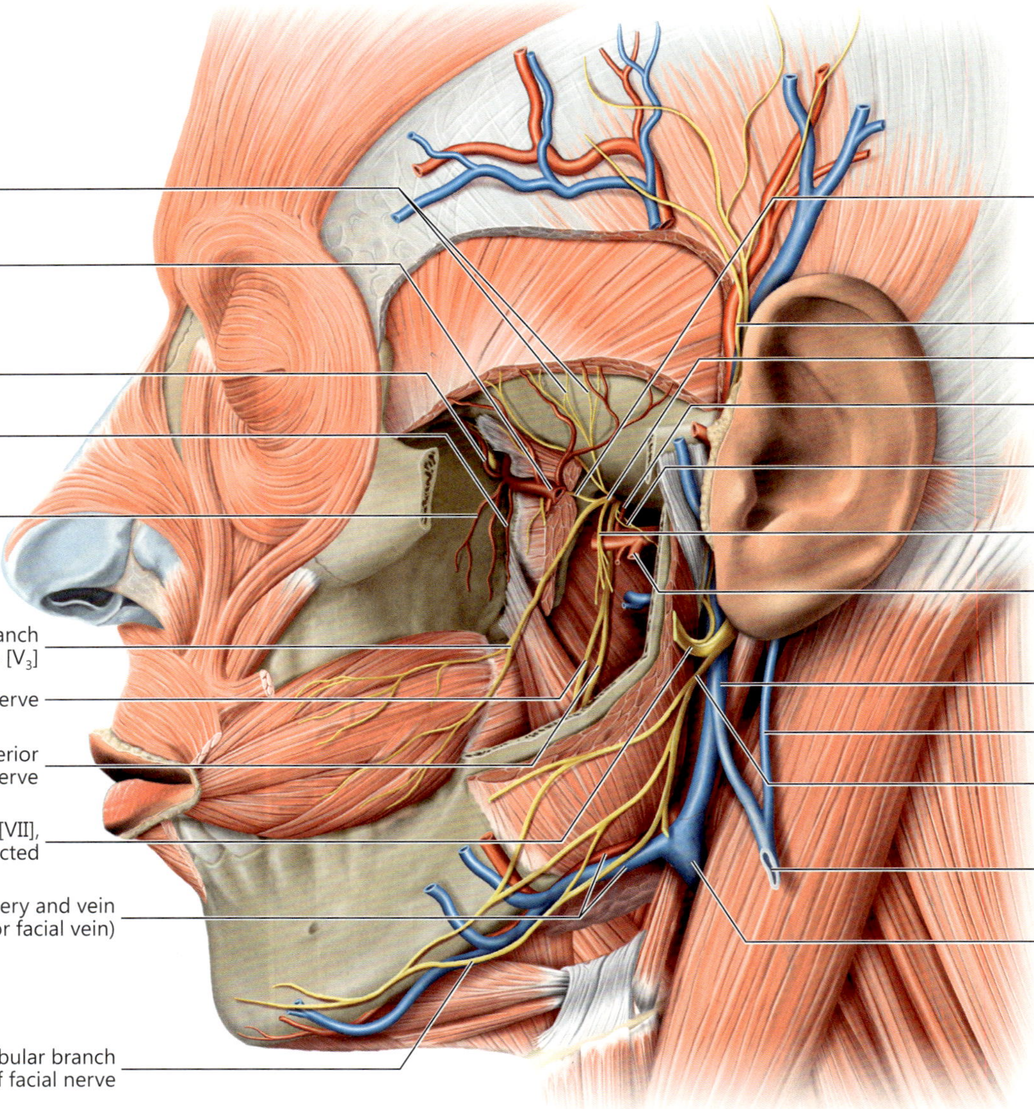

Infratemporal fossa, deep plane. Left lateral view after removal of the zygomatic arch, part of the temporalis and upper and lower heads of the lateral pterygoid. The mandibular nerve [V_3] arrives in the upper part of the infratemporal fossa after passing through the foramen ovale and here divides into two trunks, anterolateral and posteromedial. From the former arise the deep temporal, buccal, masseteric and lateral pterygoid nerves, from the latter the lingual, inferior alveolar, auriculotemporal and medial pterygoid nerves. The auriculotemporal nerve separates from the posterior trunk of the mandibular nerve with two roots that encircle the middle meningeal artery.

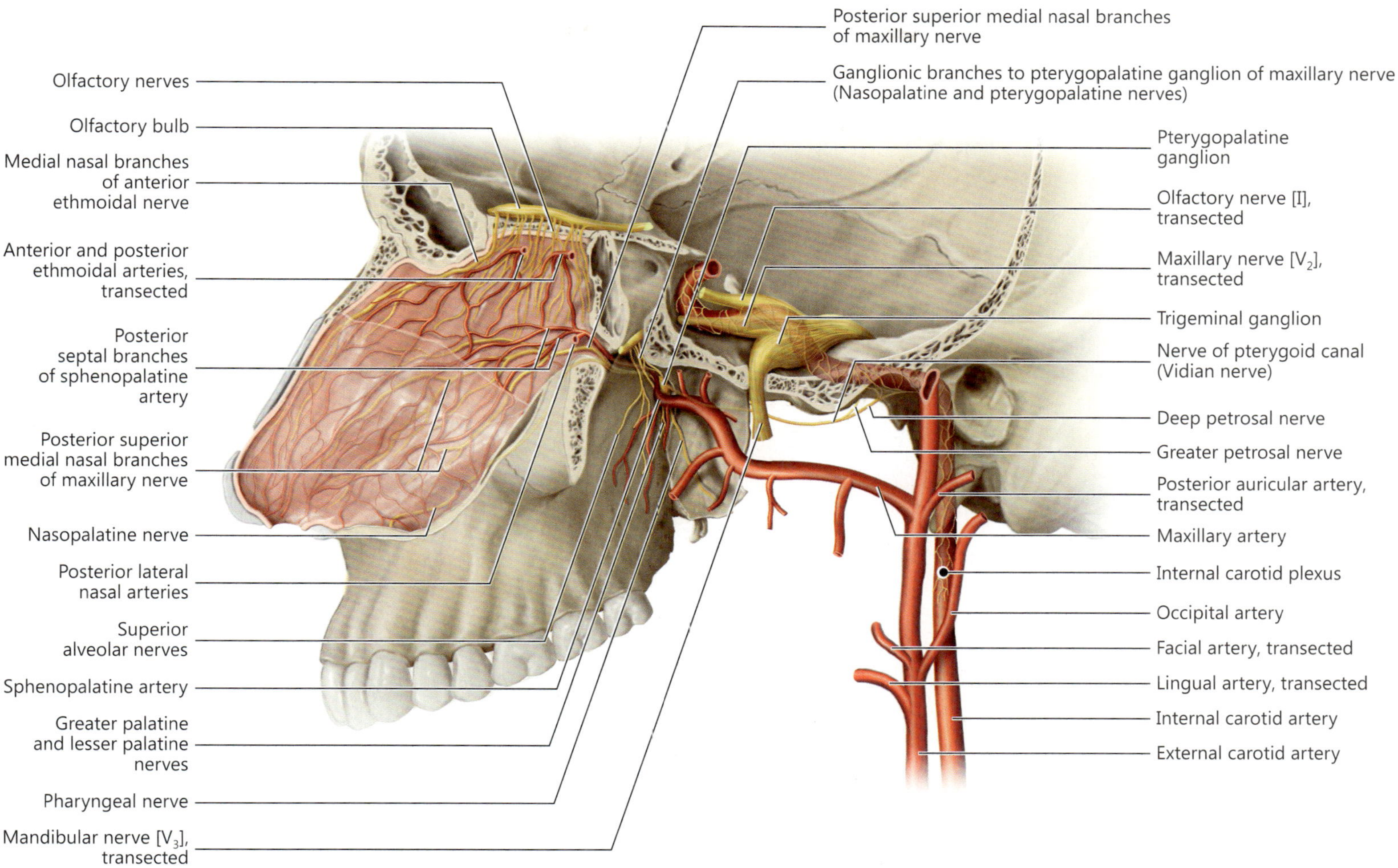

Left lateral view of the medial wall of the left **nasal cavity** after removal of its lateral wall. One can see the nasal and palatine branches of the maxillary nerve [V_2]. Also the medial wall of the pterygopalatine fossa has been removed in order to show the pterygopalatine (or sphenopalatine) ganglion, which receives the nerve of the pterygoid canal (or Vidian nerve), into which the greater petrosal nerve, a branch of the facial nerve [VII], flows as well as the deep petrosal nerve, coming from the internal carotid plexus. To it travel also the ganglionic branches to the pterygopalatine ganglion of the maxillary nerve (or the nasopalatine or pterygopalatine nerves). Efferent branches of the pterygopalatine ganglion are the pharyngeal, greater palatine and lesser palatine nerves and the posterior superior nasal branches of the maxillary nerve, of which the largest is the nasopalatine nerve. One can also see the trigeminal or semilunar ganglion, located in the middle cranial fossa near the apex of the petrous part of the temporal bone.

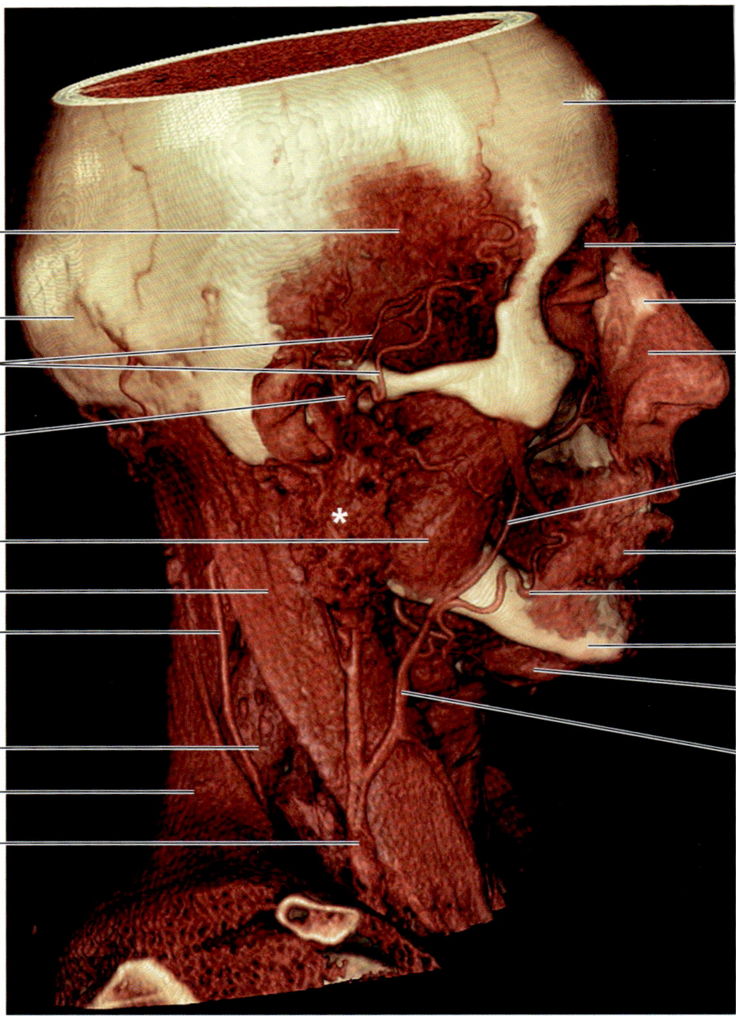

Three-dimensional image of a living human body, obtained using a 3D volume rendering technique, showing some of the **bony structures of the cranium** and muscle structures of the neck. The parotid space is indicated with an asterisk and reported in the figure below.

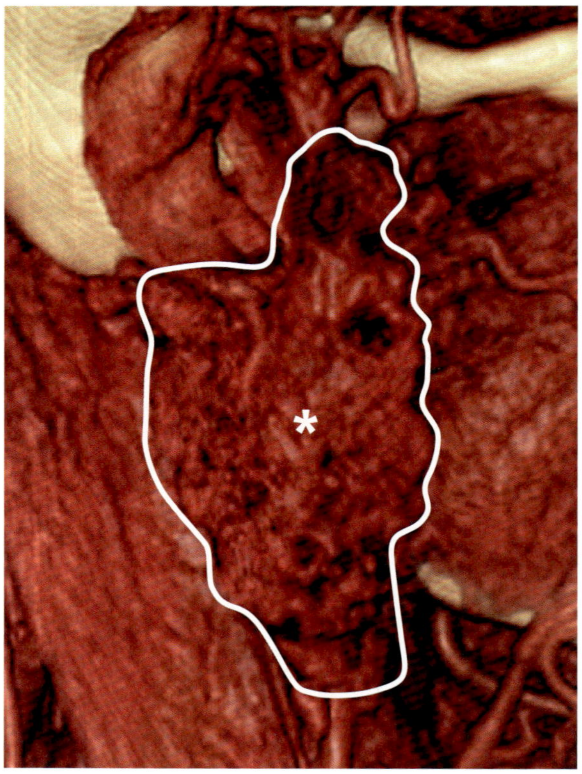

Close-up detail of the **parotid space**, showing its boundaries on the skin plane, i.e. below the auricle and external acoustic meatus, behind the ramus of the mandible and in front of the sternocleidomastoid.

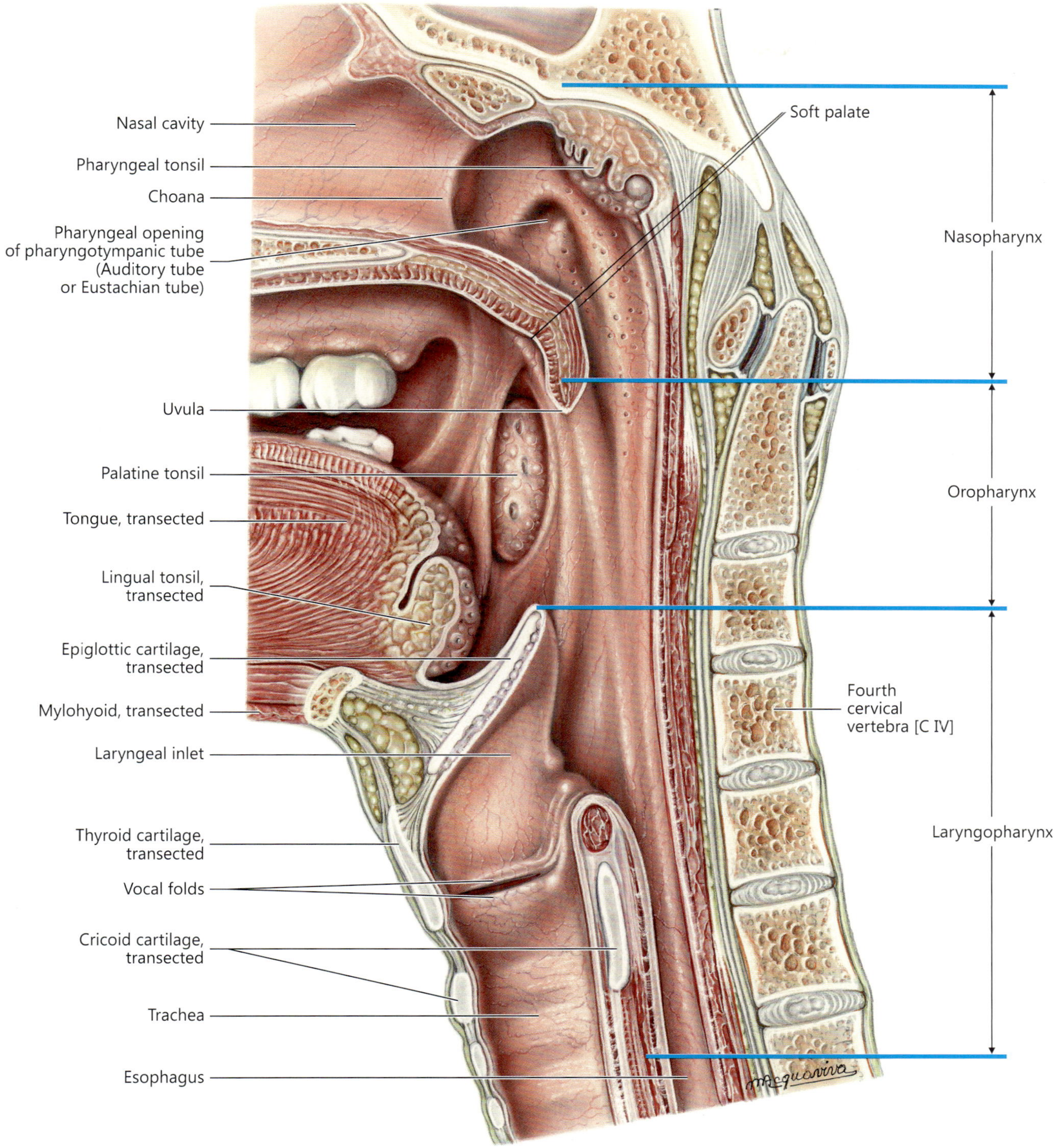

Sagittal section of the head and neck showing the lateral wall of the **isthmus of the fauces**. The palatine tonsil is enclosed in the space between the palatoglossal arch (anteriorly) and the palatopharyngeal arch (posteriorly). Also visible are the lateral wall of the pharynx, with the pharyngeal opening of the pharyngotympanic tube (auditory tube), the initial part of the esophagus and, in front of it, the larynx and initial part of the trachea.

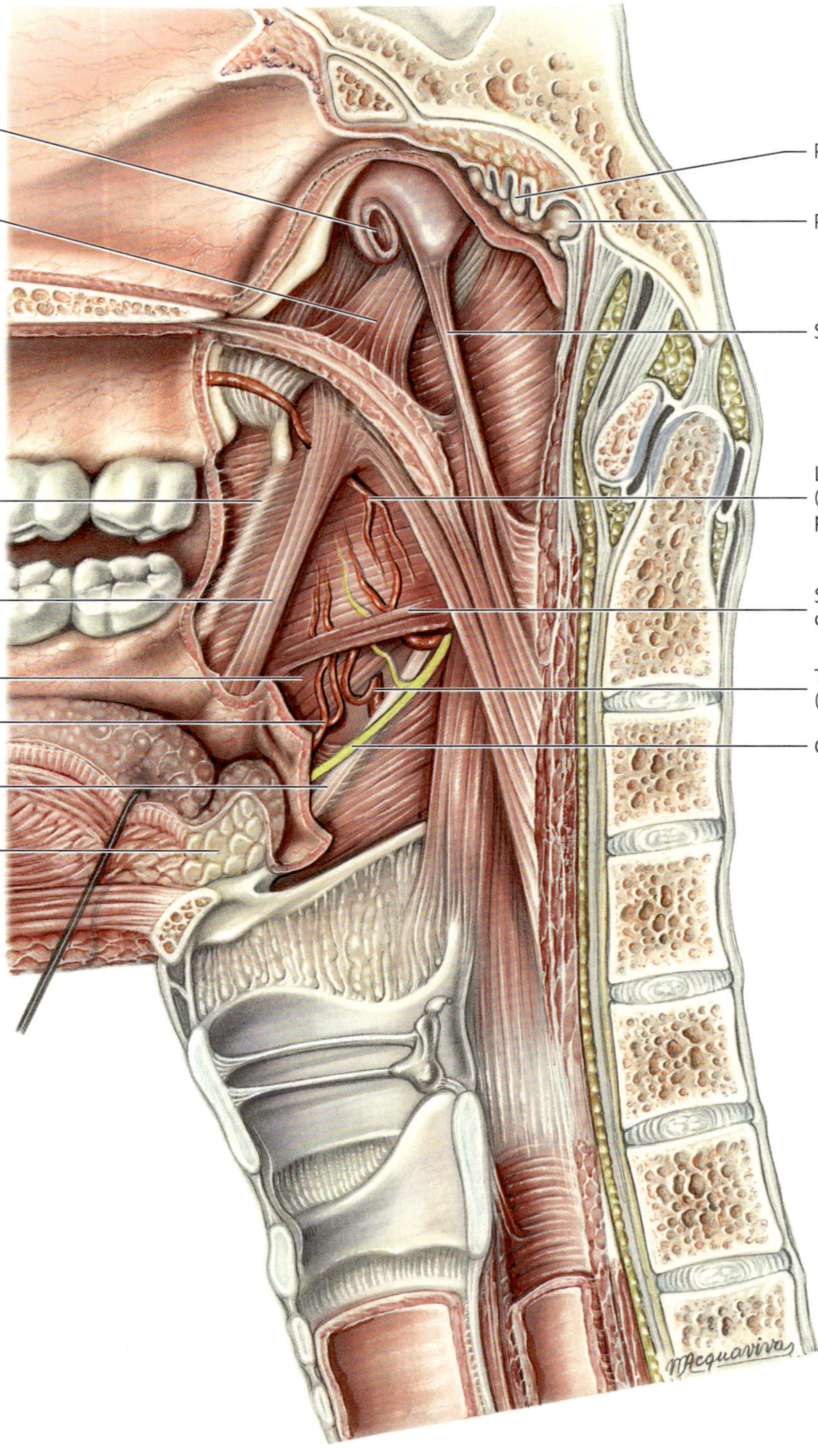

Lateral wall of the **isthmus of the fauces**, after removing the mucosa of the palate and palatine tonsil. Some of the muscles and deep nerves related to the palatine tonsil can be seen, as well as the muscular plane that supports the organ. Also highlighted is the musculature of the lateral wall of the pharynx.

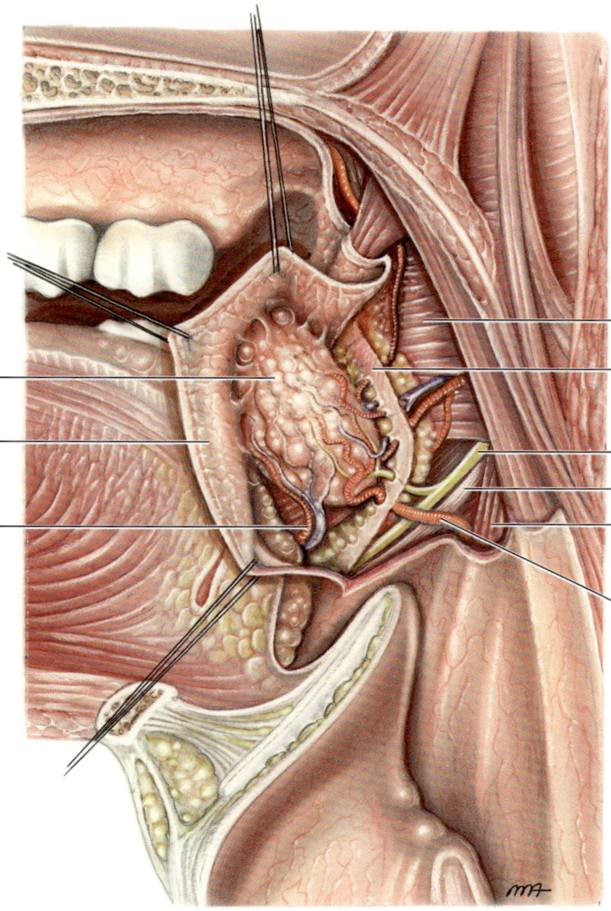

After section of the peritonsillar mucosa and its detachment with respect to the deep planes, one has a view of the neurovascular peduncles of the **palatine tonsil**.

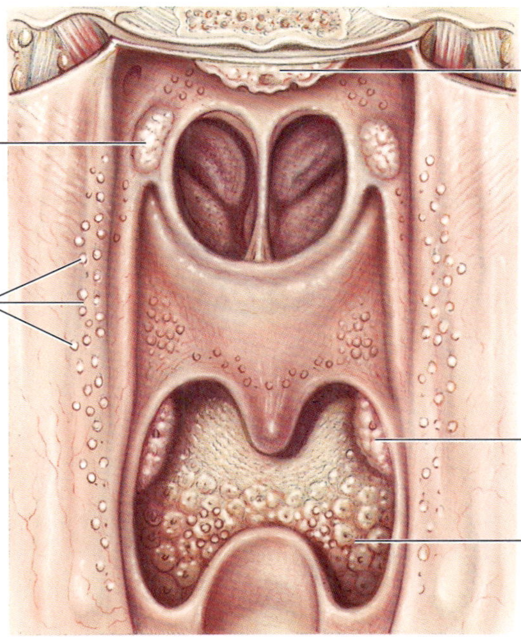

Pharyngeal lymphoid ring (Waldeyer's ring). Ring-like arrangement of lymphoid tissue in the pharynx. The ring is formed by the pharyngeal tonsil, tubal tonsils, palatine tonsils and lingual tonsil.

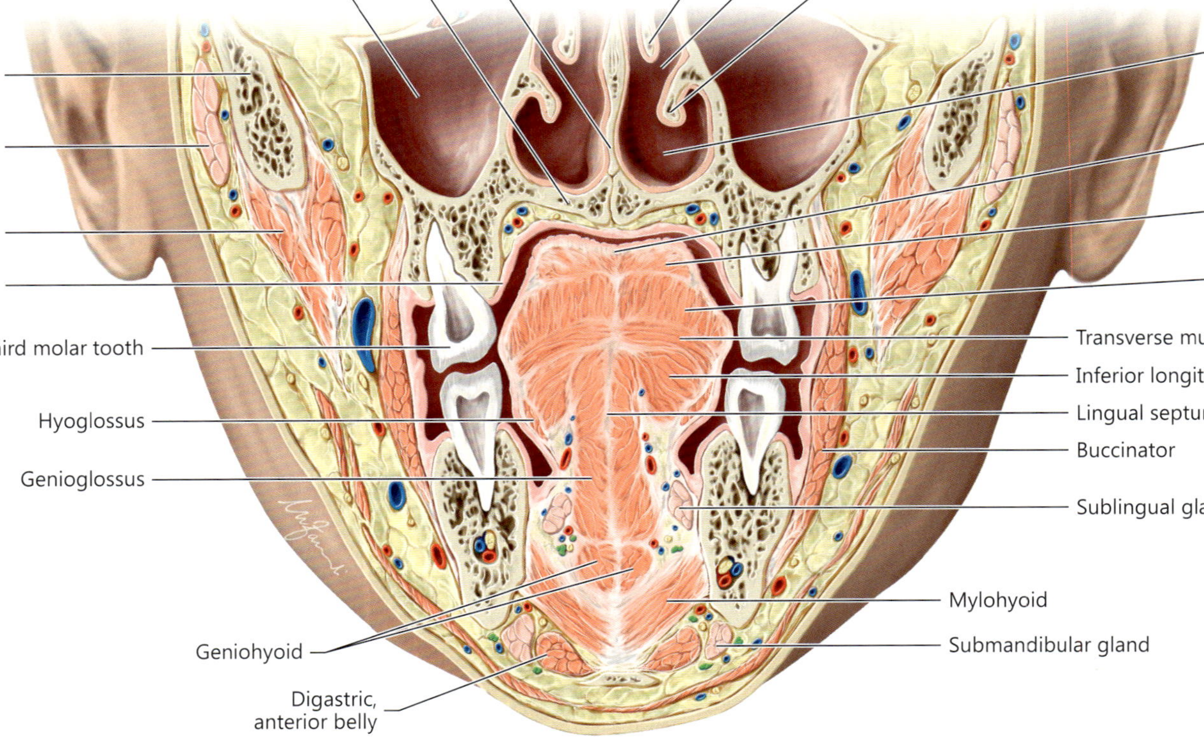

Muscles of the tongue and floor of the mouth, frontal section at the level of the third molar.

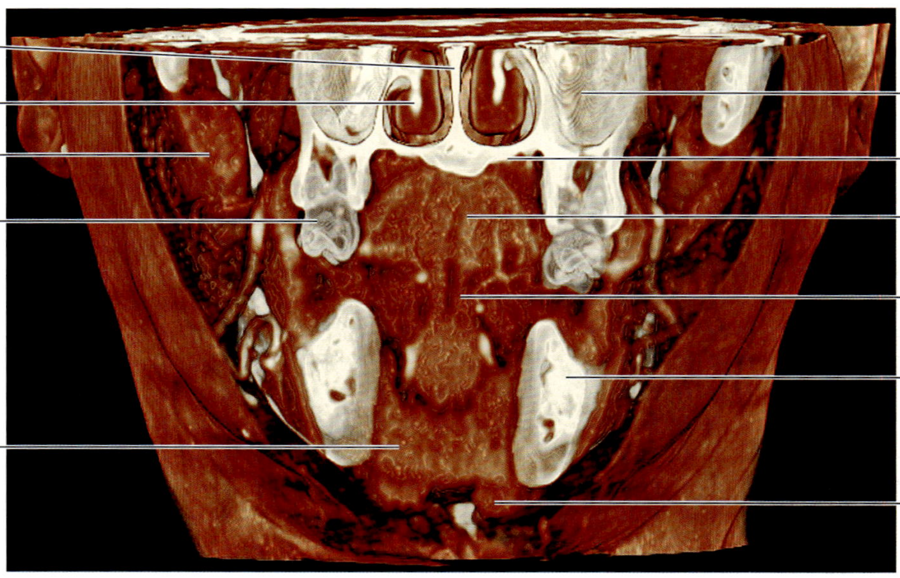

Three-dimensional image of a living human body, obtained using a 3D volume rendering technique, showing the **tongue** and **muscles of the floor of the mouth**, frontal section at the level of the third molar.

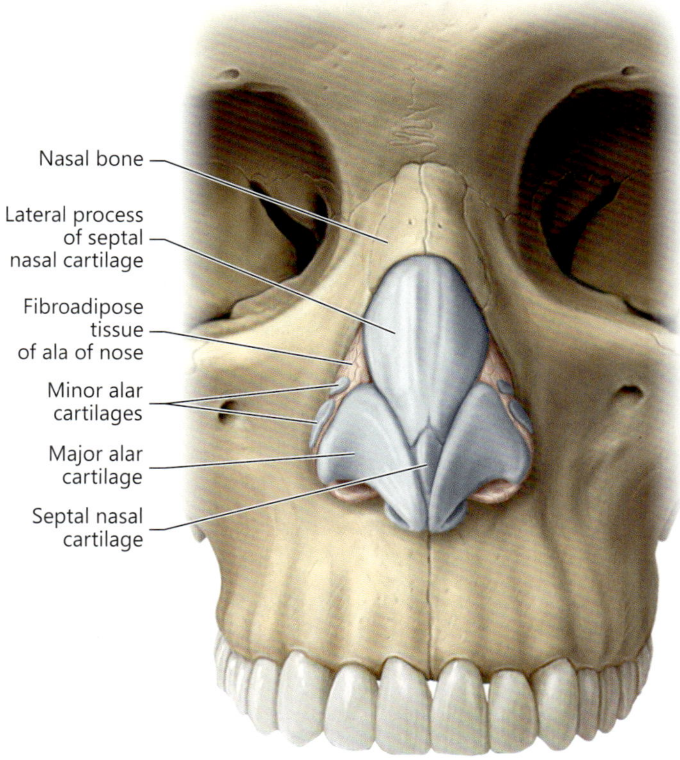

Constitution of the **external nose**, viewed in anterior projection. The major and minor alar cartilages and septal nasal cartilage form the skeleton of the mobile part of the nasal pyramid.

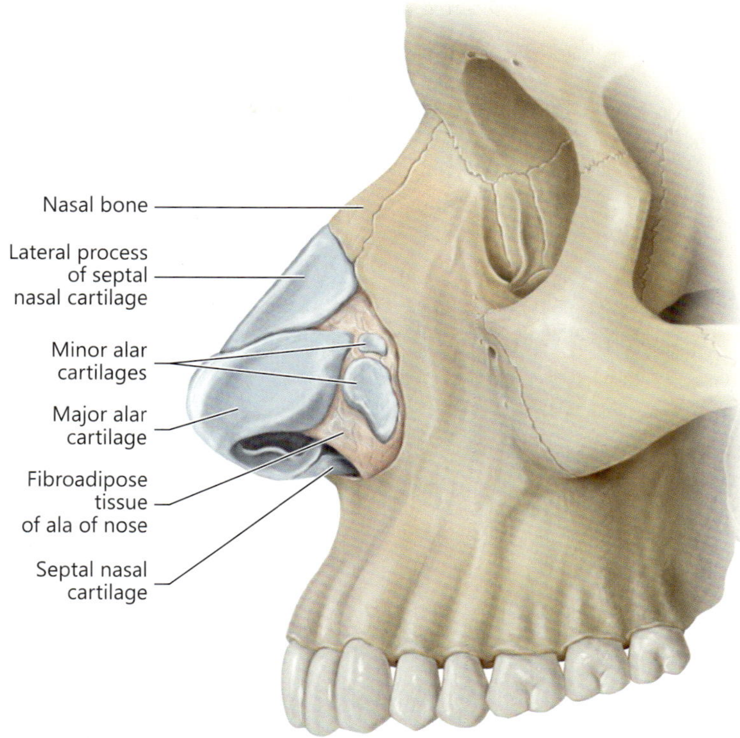

Constitution of the **external nose**, viewed in left lateral projection. The nasal pyramid is formed of bone at the apex, while the intermediate and basal parts have a cartilaginous constitution.

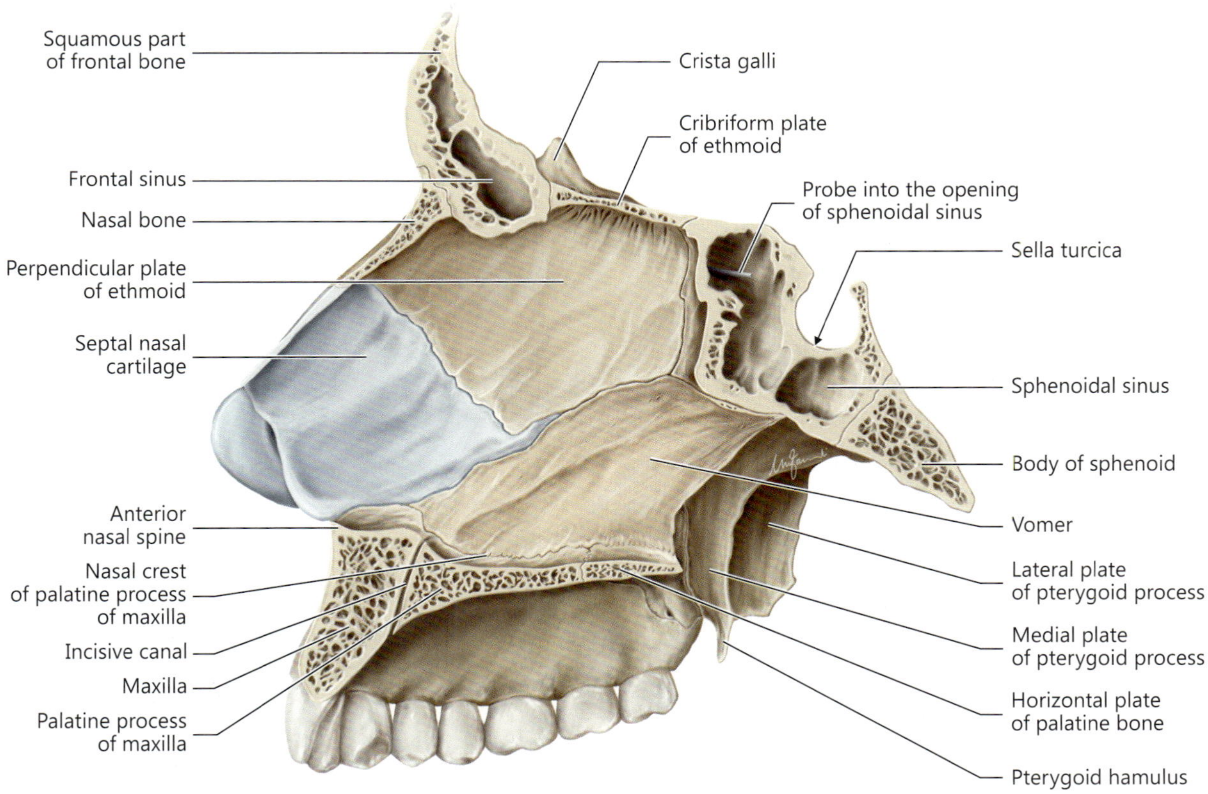

Constitution of the **nasal septum**, viewed in left lateral projection. One can see the two bony components of the septum, the vomer and perpendicular plate of the ethmoid, which unite forming an angle open to the front into which the septal nasal cartilage fits.

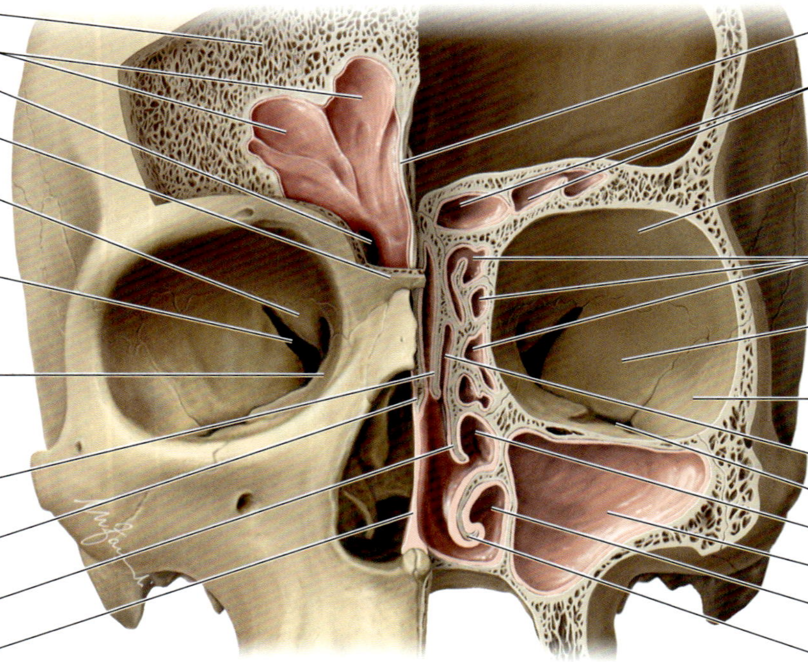

Nasal cavity, **orbital cavity**, and **paranasal sinuses** (anterior view). On the right, the squamous part of the frontal bone has been transected, displaying the frontal sinus. On the left, a frontal section has been performed passing through the orbital surface of the frontal bone to display the ethmoidal cells, nasal meatuses and maxillary sinus. The three nasal conchae protrude from the lateral wall of the nasal cavity: from top to bottom, the superior, middle and inferior nasal conchae, which constitute the roof of the respective meatuses. Respiratory mucosa lines both sides of the conchae and, through the communicating orifices, also the paranasal sinuses. The ethmoidal cells, contained prevalently in the ethmoid, open for the most part into the middle nasal meatus, some into the superior nasal meatus.

For further detail ☞ pages 170-171 of volume 1.

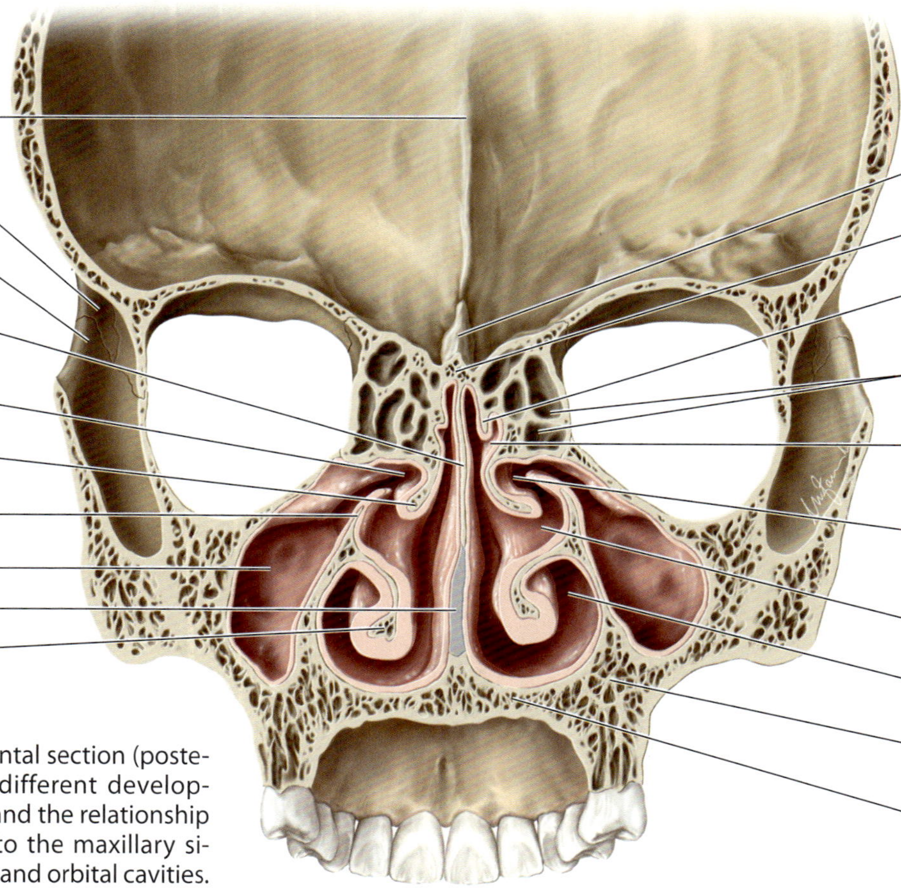

Nasal cavities in a frontal section (posterior view). Note the different development of the conchae and the relationship of the nasal cavities to the maxillary sinuses, ethmoidal cells and orbital cavities.

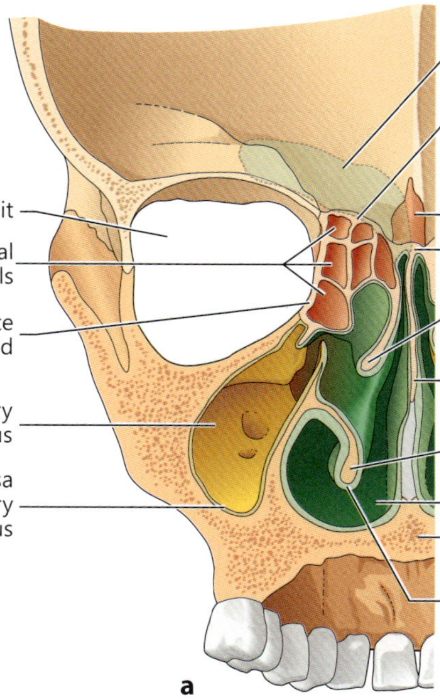

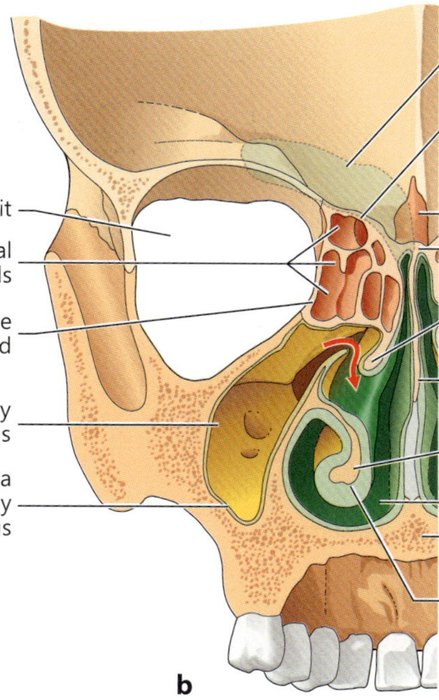

Schematic representation showing frontal sections of the **nasal cavity**. In **a**, the nasal cavity has been transected passing through the ethmoidal infundibulum. The transection in **b** highlights the entry of the maxillary sinus into the nasal cavity (arrow).

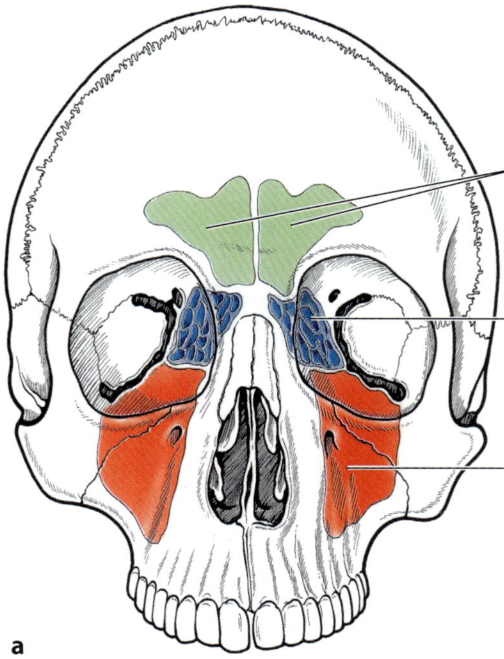

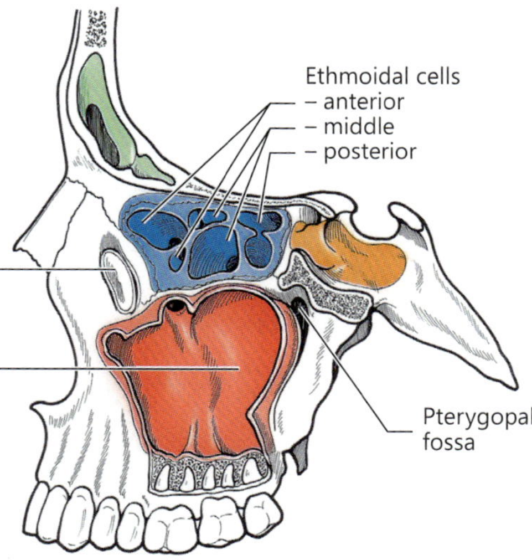

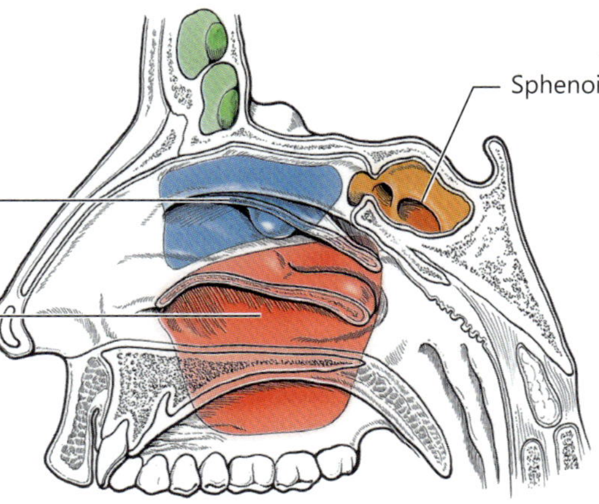

General conformation of the **paranasal sinuses**, in transparent view, in relation to the craniofacial skeleton. In **a**, anterior projection, in **b** and **c**, respectively, lateral and medial projections. The **maxillary sinuses** are visible in the three-dimensional image of a living human body on page 381 of volume 1.

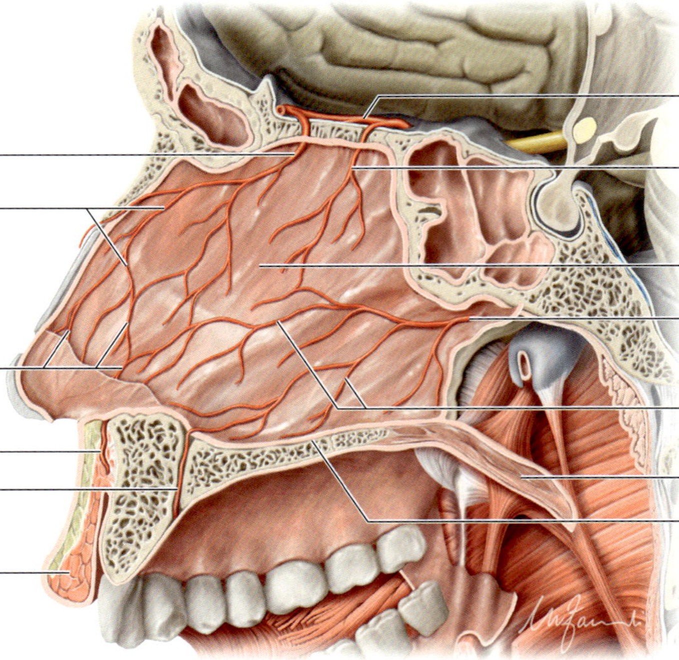

Paramedian sagittal section of the left **nasal cavity**. One can see the arterial branches that supply the nasal septum. The anterior and posterior ethmoidal arteries, branches of the ophthalmic artery, supply nasal branches that cross the cribriform plate. Anterior septal branches of the anterior ethmoidal artery anastomose with posterior septal branches of the sphenopalatine artery, a branch of the maxillary artery, which penetrates into the nasal cavity through the sphenopalatine foramen. Also the superior labial branch of the facial artery supplies a branch to the nasal septum. The greater palatine artery, a branch of the descending palatine artery, itself a branch of the maxillary artery, after emerging on the palatal vault, heads forward and enters the incisive canal and anastomoses with a branch of the sphenopalatine artery. The Kiesselbach's area, in the anteroinferior part of the nasal septum, corresponds to the vascular plexus formed by anastomoses of the anterior ethmoidal, sphenopalatine, greater palatine and superior labial arteries.

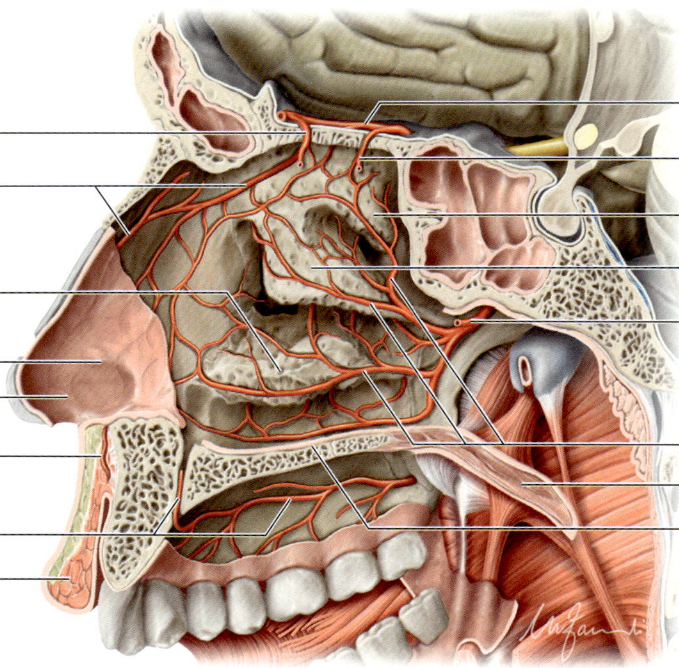

Sagittal section of the right **nasal cavity**, left lateral view. One can see the arterial branches that supply the lateral wall of the nasal cavity: the anterior lateral nasal branches, which stem from the anterior ethmoidal artery, a branch of the ophthalmic artery, and the posterior lateral nasal arteries, branches of the sphenopalatine artery, which anastomose with the ethmoidal arteries and with nasal branches of the greater palatine artery.

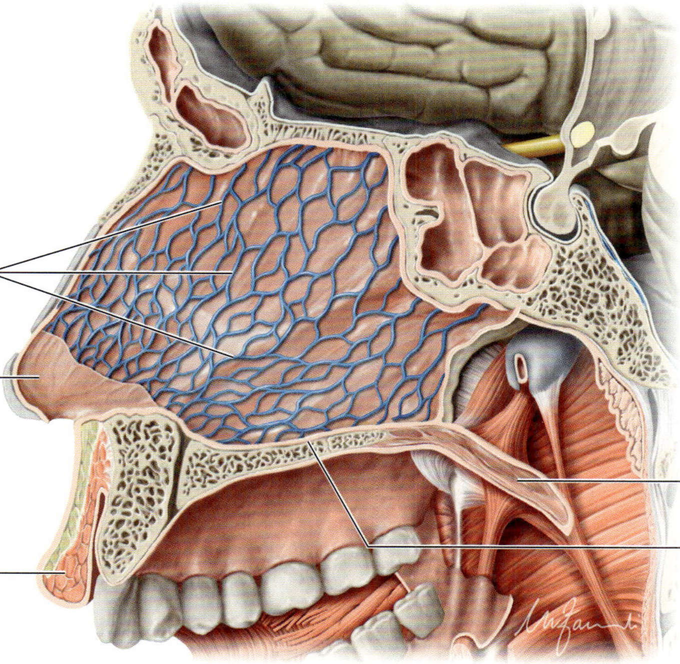

Paramedian sagittal section of the left **nasal cavity**, showing the superficial venous plexus in the lamina propria of the respiratory region at the level of the nasal septum.

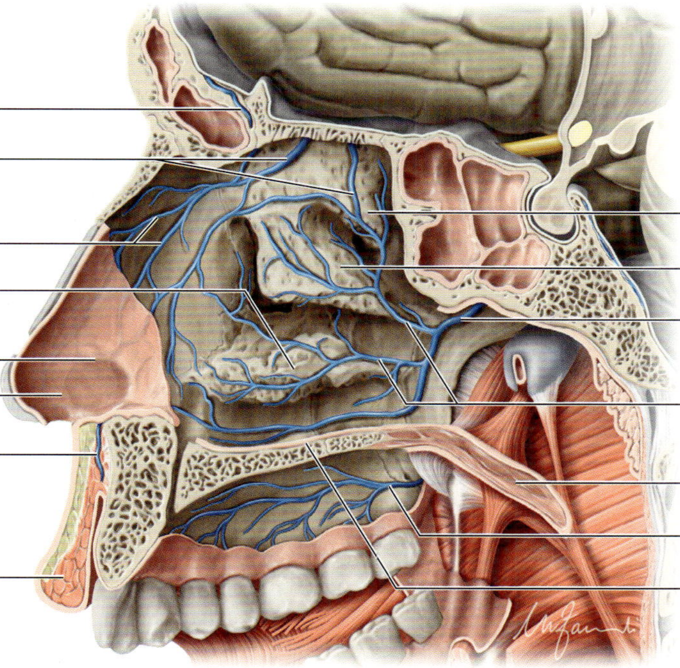

Sagittal section of the right **nasal cavity**, left lateral view. The veins of the nasal cavity terminate: posteriorly, in the sphenopalatine vein, which drains into the pterygoid plexus; superiorly, in the ethmoidal veins, which drain into the superior ophthalmic vein; and anteriorly, in the superior labial vein, an affluent branch of the facial vein. The presence in man of the vein of the foramen cecum remains a controversial issue. The vein should provide intracranial drainage of the nasal mucosa into the superior sagittal sinus through one of the frontal veins.

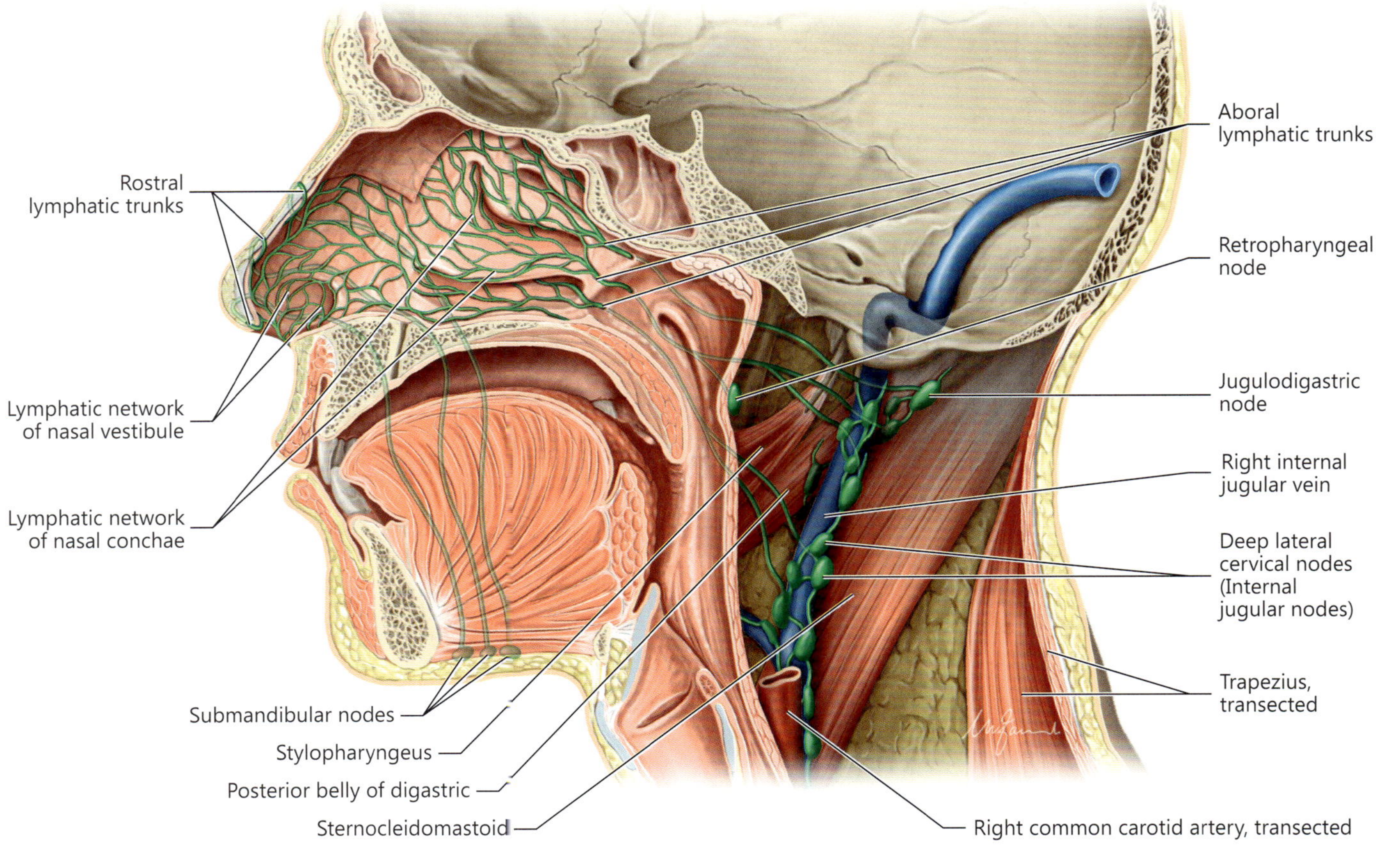

Sagittal section of the right **nasal cavity** (anteriorly) and oblique section on a plane passing through the mastoid process of the temporal bone and basilar part of the occipital bone (posteriorly). One can see the lymphatic network of the mucosa of the nasal cavity, the efferent trunks of which drain into the submandibular, retropharyngeal, deep lateral cervical (internal jugular) and jugulodigastric nodes.

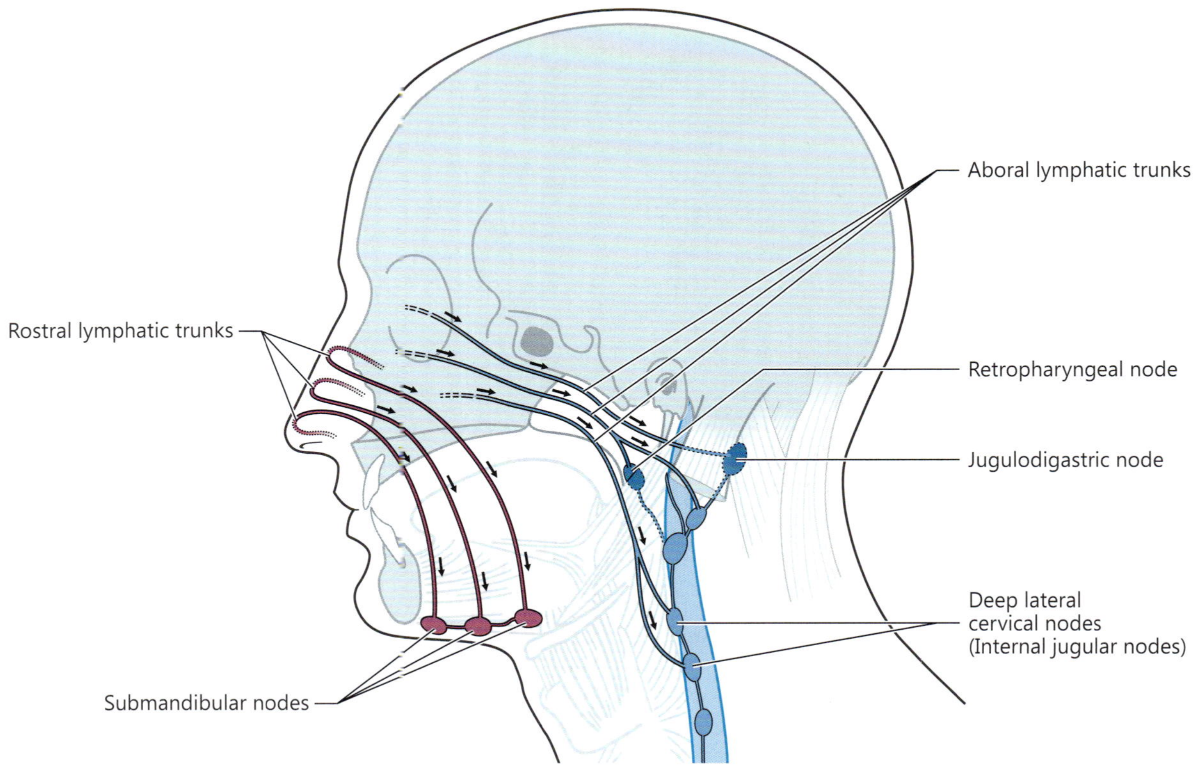

Schematic representation showing the **pathways of lymphatic drainage** of the mucosa of the nasal cavity to the submandibular, retropharyngeal, deep lateral cervica (internal jugular) and jugulodigastric nodes.

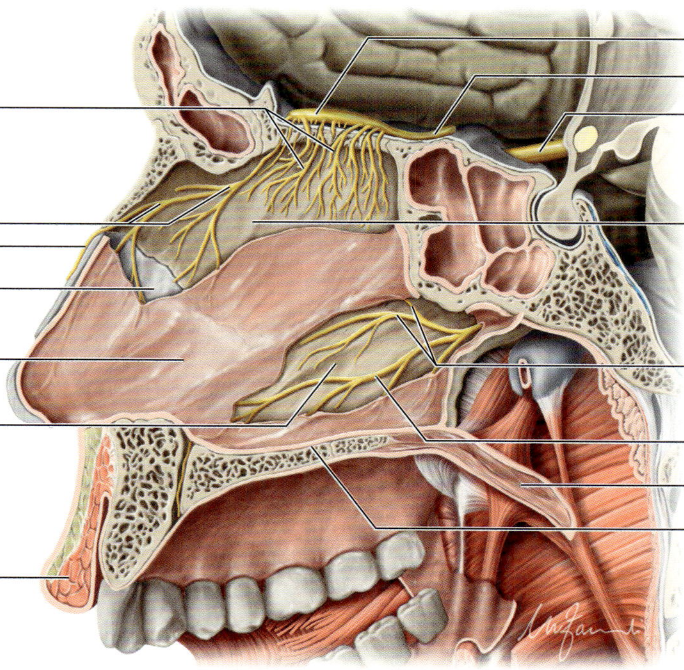

Paramedian sagittal section of the left **nasal cavity**, left lateral view. The anterior ethmoidal nerve, a branch of the nasociliary nerve, itself a branch of the ophthalmic nerve [V_1], supplies internal medial nasal branches, while posterior superior medial nasal branches are supplied by the maxillary nerve [V_2]. The nasopalatine nerve, a branch of the maxillary nerve [V_2], following the sulcus of the vomer running obliquely forwards and downwards, descends down the wall of the nasal septum reaching the incisive canal, through which it enters the oral cavity. The olfactory nerve [I] is composed of filaments which, crossing the cribriform foramina of the ethmoid, penetrate into the inferior surface of the olfactory bulb, situated in the anterior cranial fossa, above the cribriform plate of the ethmoid.

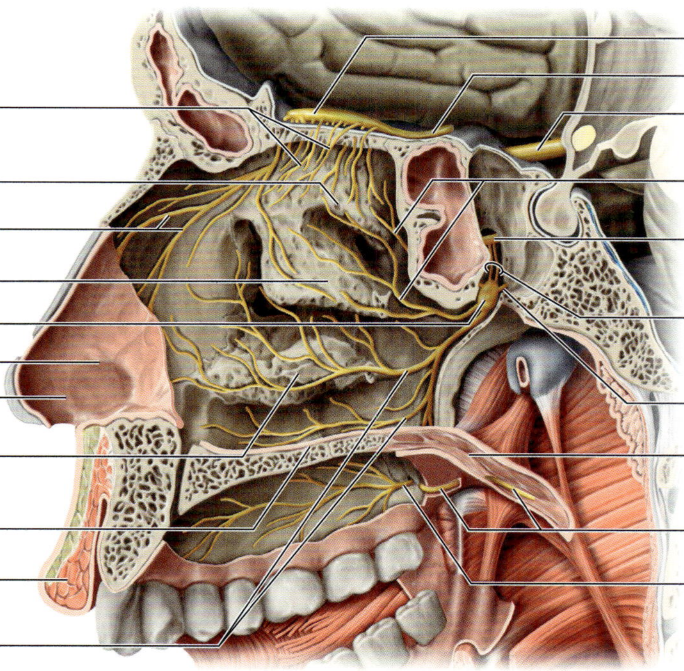

Sagittal section of the right **nasal cavity**, left lateral view. From the anterior ethmoidal nerve, a branch of the ophthalmic nerve [V_1], lateral nasal branches head to the vault and lateral wall of the nasal cavity. The medial surface of the pterygopalatine ganglion gives off numerous fibers that pass into the nasal cavity through the sphenopalatine foramen. The lateral group consists of posterior superior lateral nasal branches of the maxillary nerve [V_2] that go to the lateral wall of the nasal cavity, which is reached also by the posterior inferior nasal branches of the greater palatine nerve, which originates from the inferior part of the pterygopalatine ganglion and descends through the greater palatine canal to the palate. Also the lesser palatine nerves originate from the inferior part of the pterygopalatine ganglion, and pass through the lesser palatine canals from which they emerge to head behind, through the lesser palatine foramina, to the soft palate.

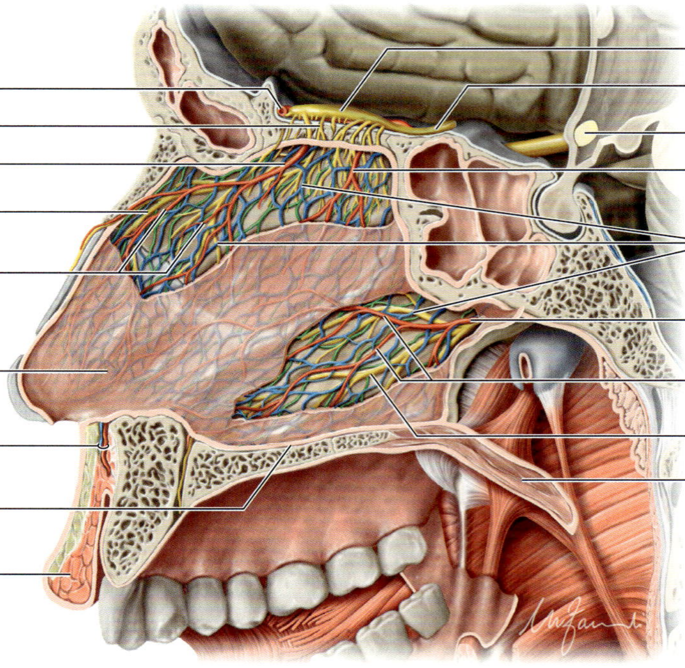

Paramedian sagittal section of the left **nasal cavity**. The olfactory region and part of the respiratory region of the mucosa of the nasal cavity have been transected to display the particularly rich **vascular network** of the **nasal mucosa**, which in some zones such as the middle part of the nasal septum is highly developed.

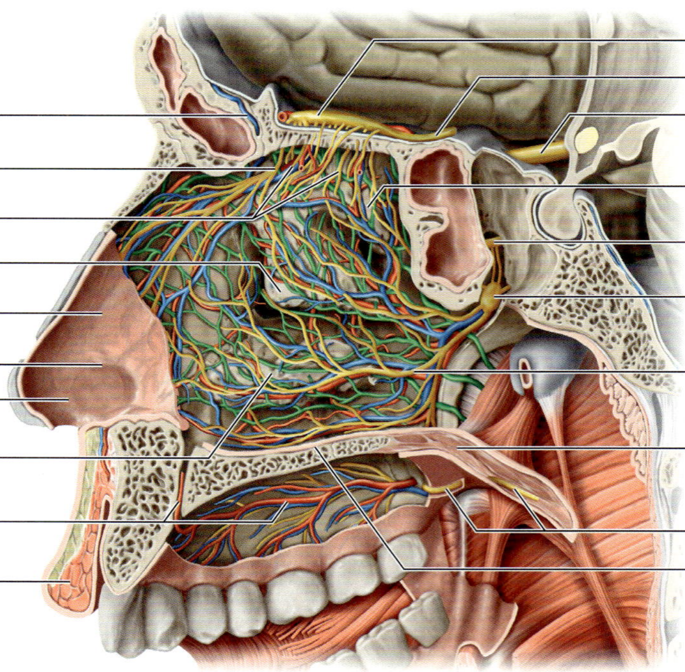

Sagittal section of the right **nasal cavity**, left lateral view. The vascular system is particularly well developed, especially in certain zones such as the inferior nasal meatus. The arteries, starting from deep branches situated near the periosteum, run perpendicularly to the surface of the respiratory region forming an intermediate plexus in the lamina propria and ramifying into a dense subepithelial capillary network. In correspondence to the middle and inferior nasal conchae, there is a **superficial venous plexus** that gives the lamina propria a cavernous tissue-like appearance due to the presence of large venous caverns partitioned by fine fibrous septa.

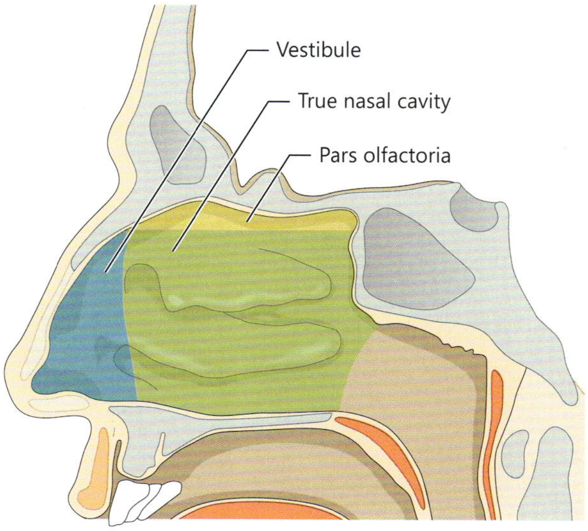

Subdivision of the **nasal cavity**.

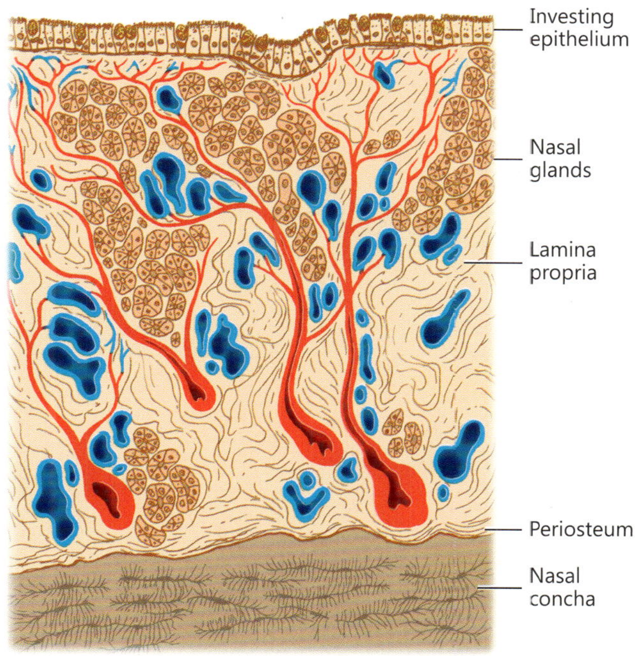

Schematic representation showing the structure of the **respiratory region**. The rich vascular supply is indicated in *red* (arterial part) and *blue* (venous part). There is direct continuity between the two systems through arteriovenous anastomoses.

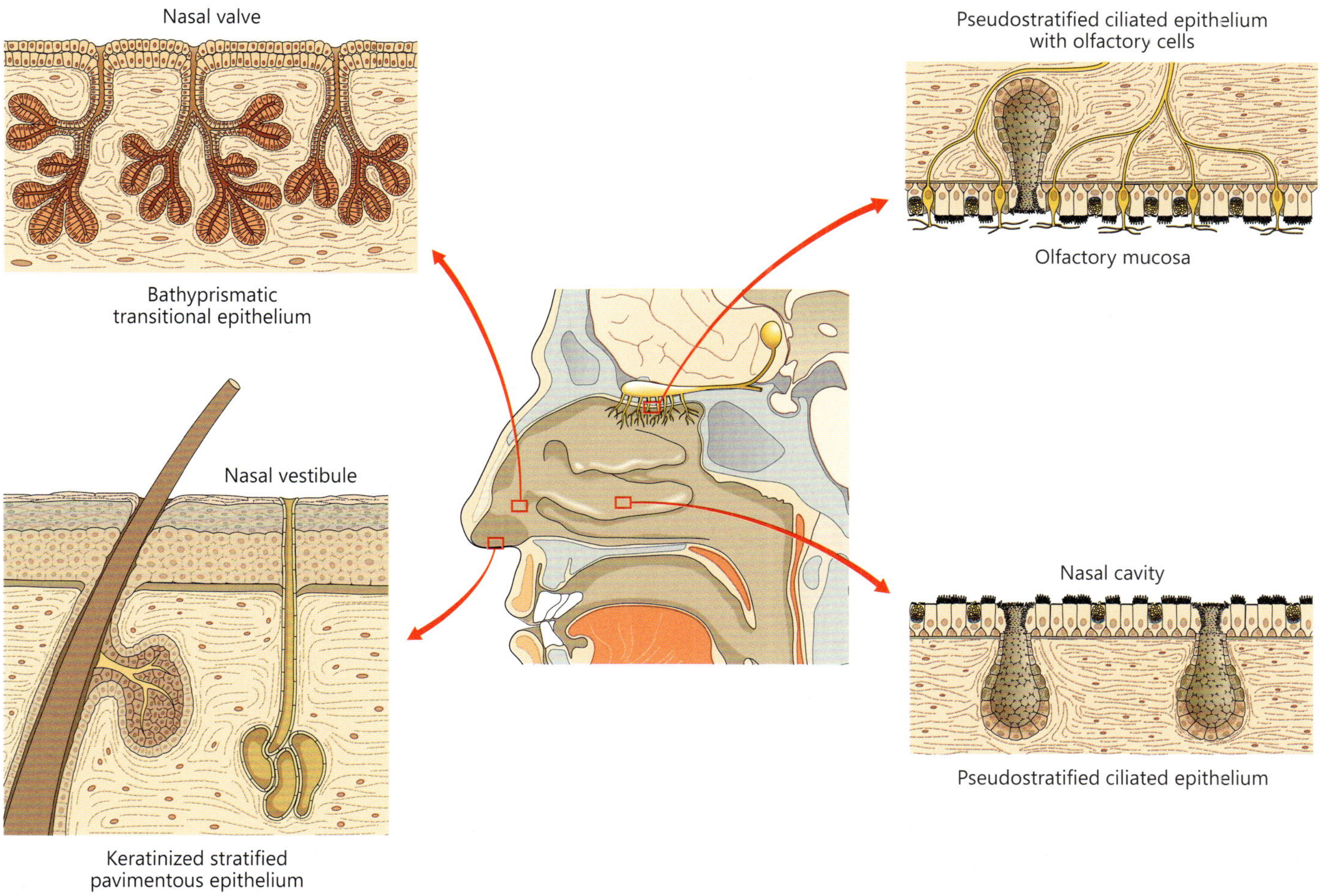

Nasal mucosa: histological variations according to the nasal cavity zone.

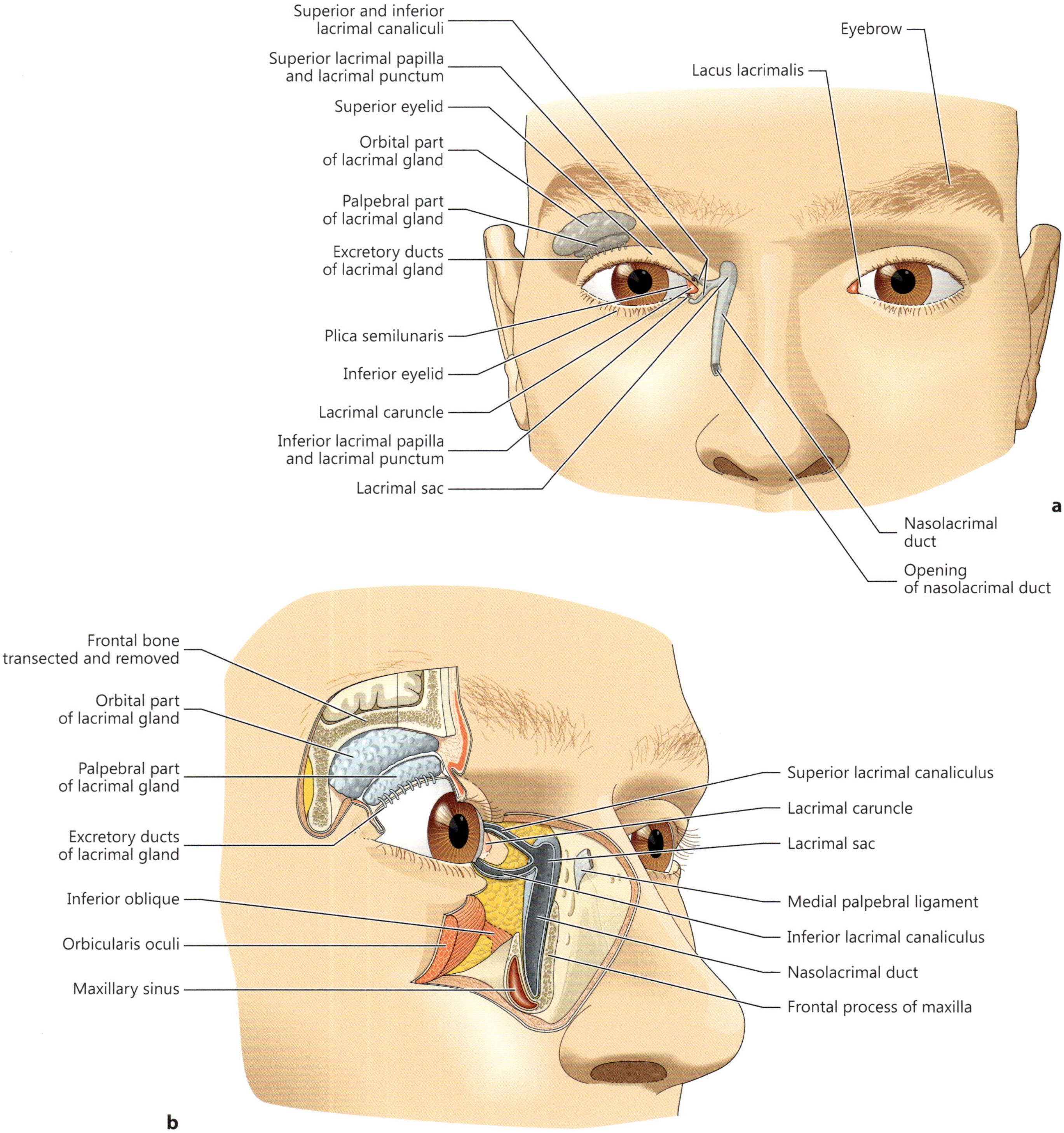

a, Eye and accessory structures as seen through the palpebral fissure. In transparent view, the formations of the protective apparatus of the eyeball. **b**, Organs of the lacrimal apparatus.

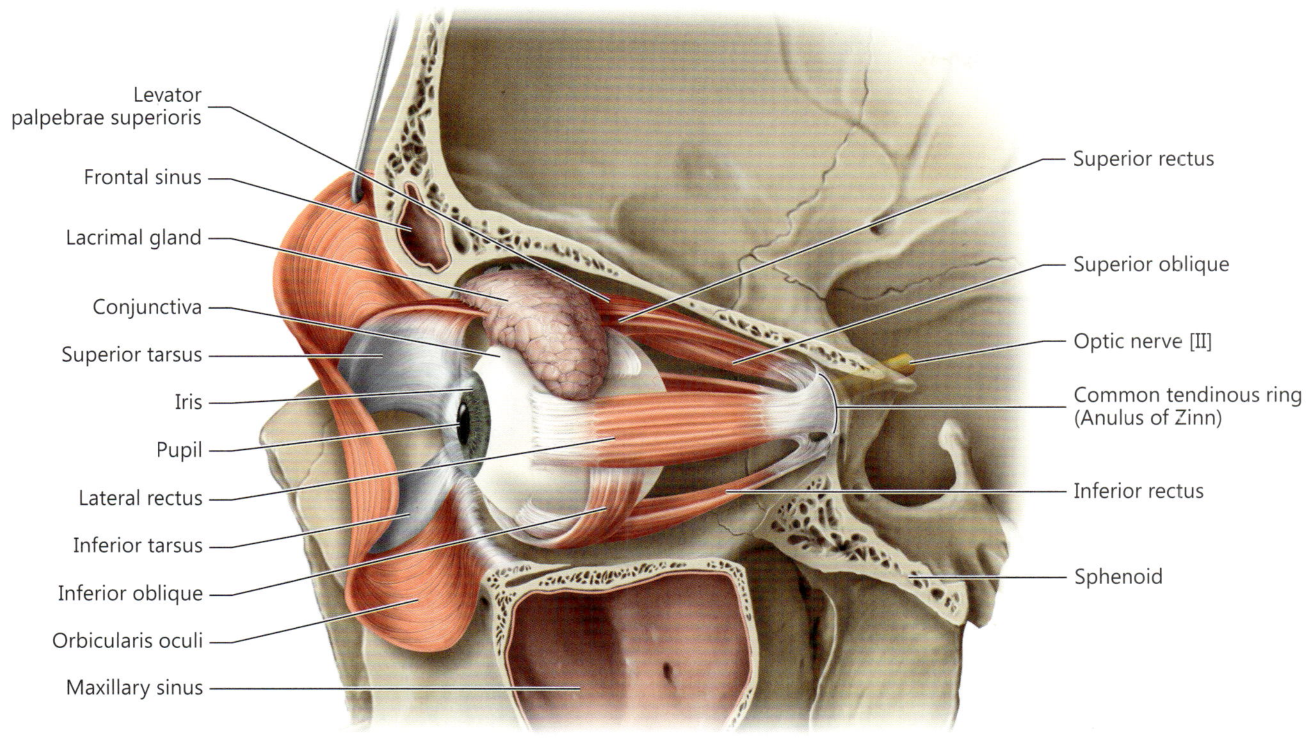

Extraocular muscles. Left eye, lateral view after removal of the lateral wall of the orbit.

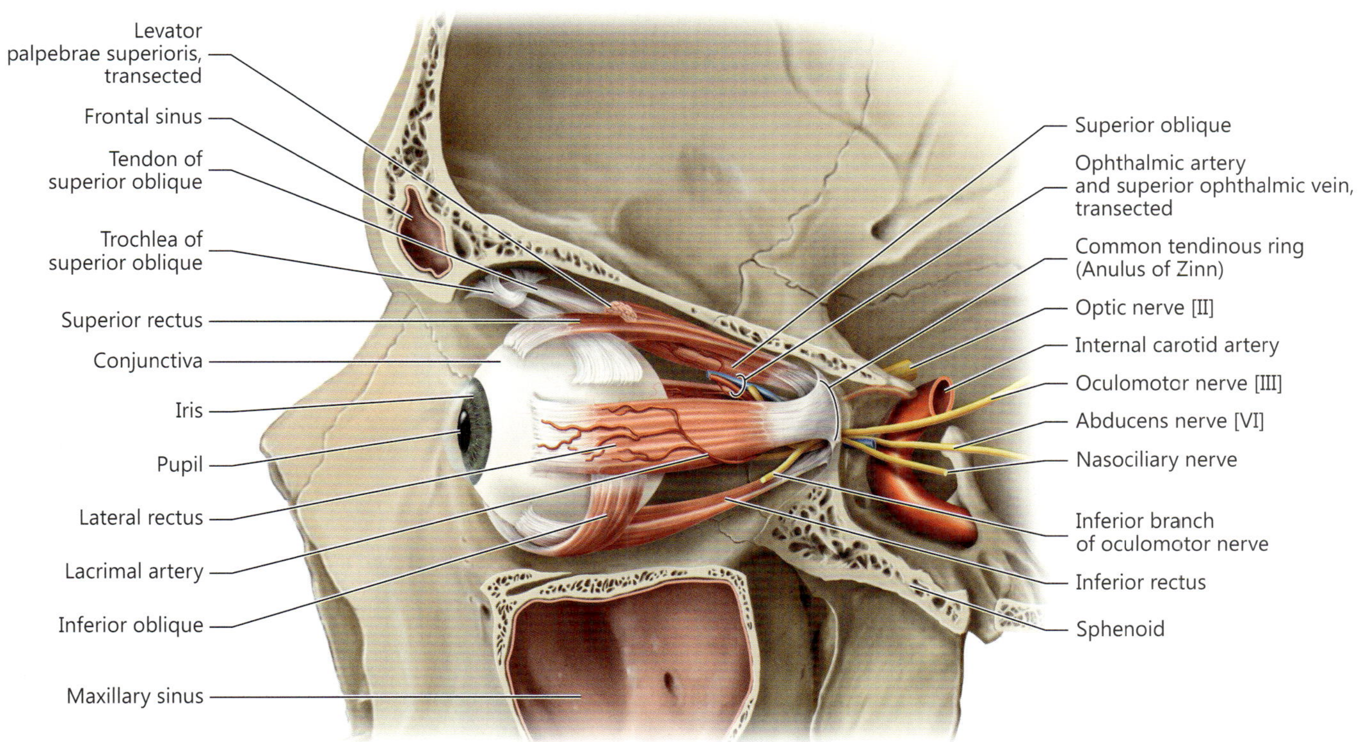

Common tendinous ring (**anulus of Zinn**). The tendinous ring is the origin of the rectus muscles and, partially, of the levator palpebrae superioris and superior oblique. Through its superior compartment run the optic nerve [II] and ophthalmic artery, while through its inferior compartment run the oculomotor nerve [III], superior ophthalmic vein, abducens nerve [VI] and the nasociliary nerve, a branch of the ophthalmic nerve [V_1].

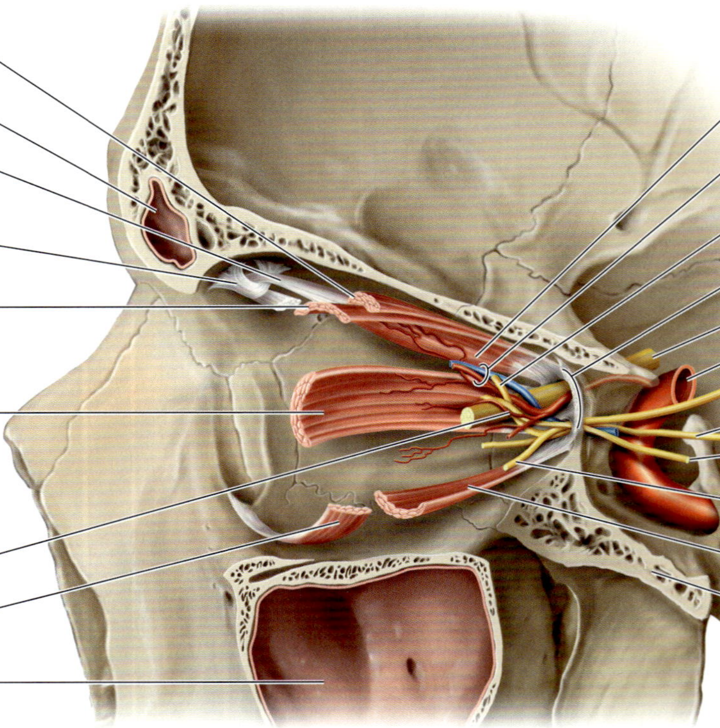

Extraocular muscles of the left eye, lateral view after removal of the lateral wall of the orbit, lateral rectus and eyeball.

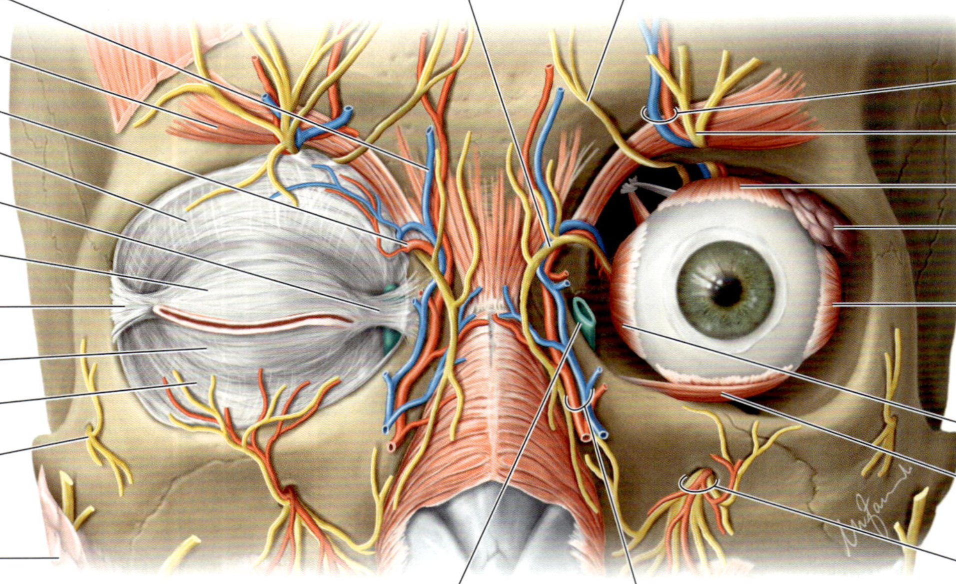

In the right eye, the orbicularis oculi has been transected, showing the orbitopalpebral part of the orbital contents, in front of the **orbital septum**. In the left eye, also the orbital septum has been transected, revealing the orbitobulbar part behind the septum. The **supratrochlear nerve**, a branch of the frontal nerve, itself a branch of the ophthalmic nerve [V_1], exits from the orbital cavity above the trochlea of the **superior oblique**. The **supraorbital nerve**, which is the other branch of the frontal nerve, exits from the orbital cavity together with the **supraorbital artery**, a branch of the ophthalmic artery, and the **supraorbital vein**, which gives rise to the angular vein, passing through the supraorbital notch or foramen of the frontal bone. The **infraorbital nerve**, which represents the last tract of the maxillary nerve [V_2], exits from the infraorbital canal through the infraorbital foramen, together with the **infraorbital artery**, a branch of the maxillary artery.

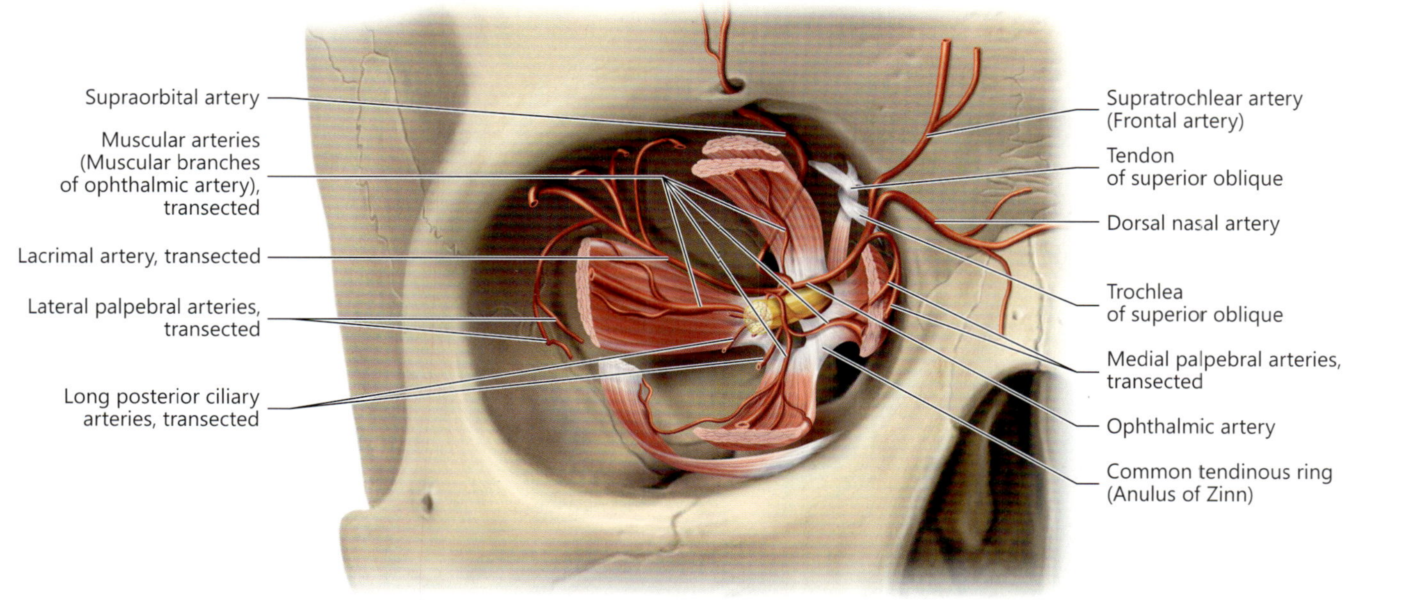

Arteries of the orbit, anterior view after removal of the eyeball. At the level of the medial angle of the eye, the **ophthalmic artery** divides into its terminal branches, i.e. the **medial palpebral arteries**, **dorsal nasal artery** and **supratrochlear** (or frontal) **artery**. The **lateral palpebral arteries** are terminal branches of the lacrimal artery, which is a branch of the ophthalmic artery that arises just after its entrance in the orbital cavity and runs along the superior border of the lateral rectus to supply the lacrimal gland. The **supraorbital artery** originates from the tract of the ophthalmic artery where it crosses the optic nerve, and passes between the superior rectus and levator palpebrae superioris, in the vault of the orbital cavity, exiting through the supraorbital notch or foramen. The **muscular arteries** are branches of the ophthalmic artery that supply the extraocular muscles, accompanying the branches of the oculomotor nerve [III]. The **long posterior ciliary arteries** originate from the tract of the ophthalmic artery that runs under the optic nerve.

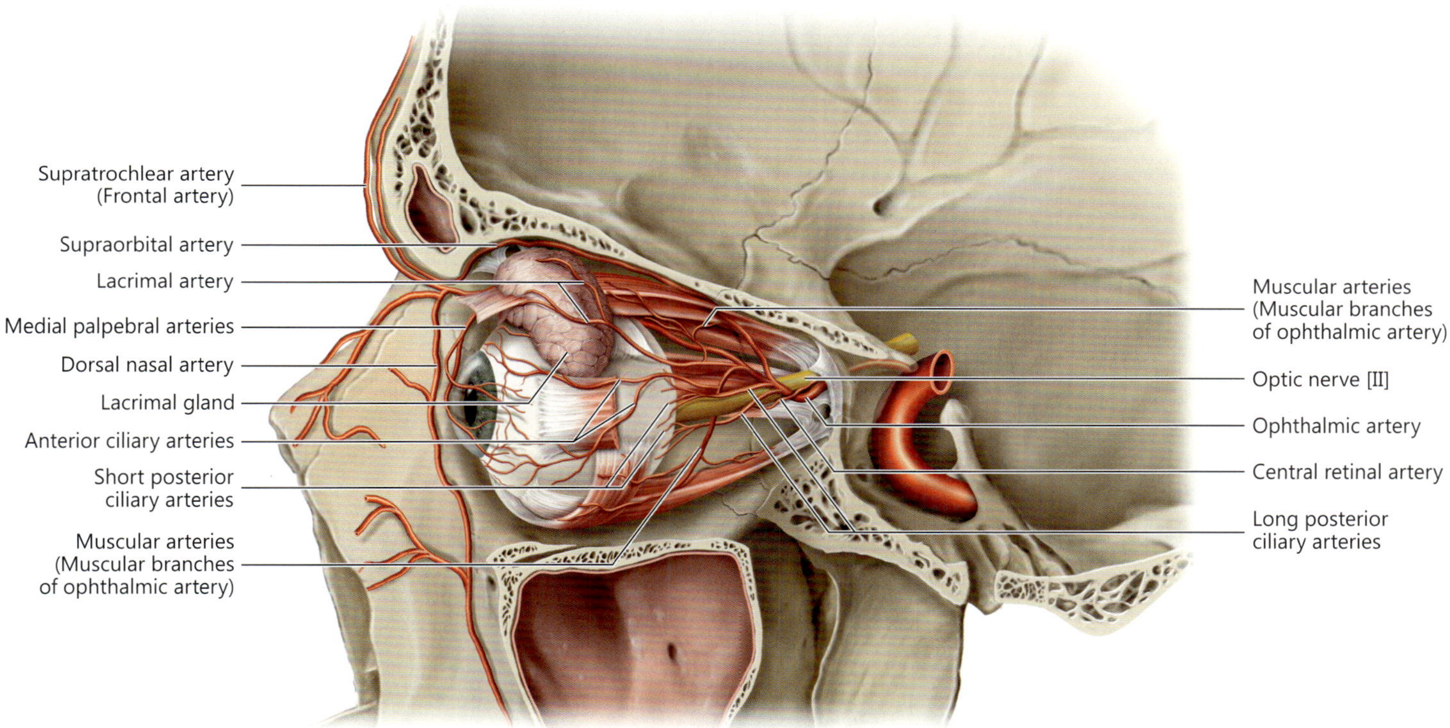

Lateral view of the left **orbital cavity** after removal of the lateral wall. The **central retinal artery**, first and smallest collateral branch of the ophthalmic artery, runs in a lateral position with respect to the optic nerve, and then penetrates its inferomedial surface. The **lacrimal artery** runs along the superior border of the lateral rectus. The **long posterior ciliary arteries** originate from the tract of the ophthalmic artery laterally to the optic nerve and branch into the **short posterior ciliary arteries**. The **muscular arteries** travel to the orbital muscles accompanying the branches of the oculomotor nerve [III] and give rise to the **anterior ciliary arteries**. The **supraorbital artery** passes between the superior rectus and levator palpebrae superioris and exits from the orbit through the supraorbital foramen. Terminal branches of the ophthalmic artery are the **medial palpebral**, **dorsal nasal** and **supratrochlear** (or frontal) **arteries**.

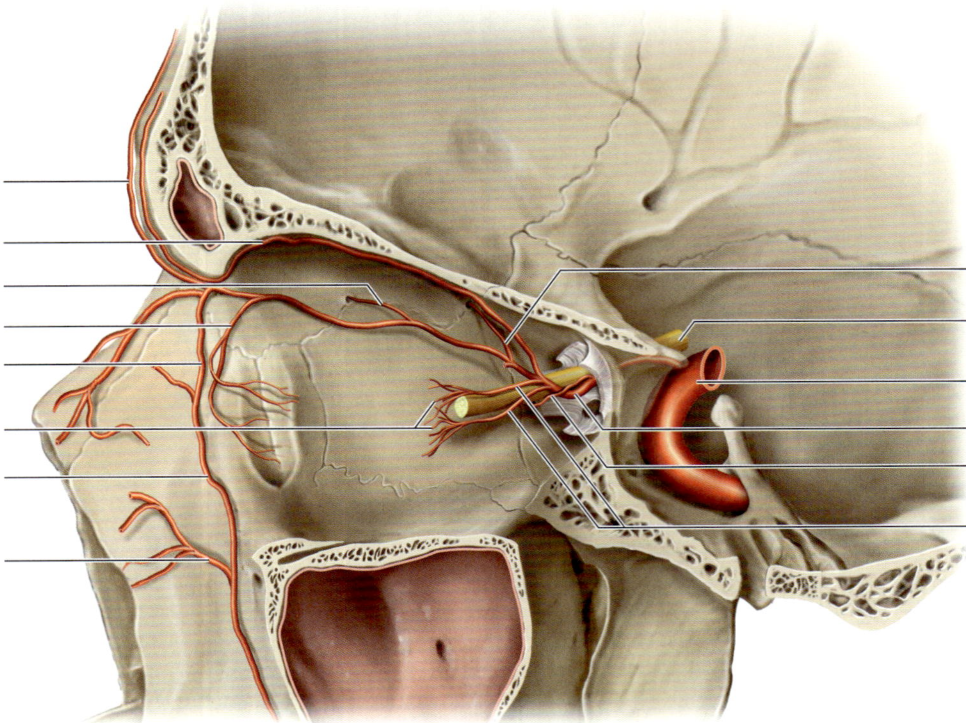

Medial wall of the **orbit**. Paramedian sagittal section, left lateral view, after removal of the eyeball and accessory visual structures. The **internal carotid artery** enters the cranial cavity through the internal opening of the carotid canal of the temporal bone and, turning upwards, makes an S-shaped curve to enter the cavernous sinus, here removed. At a point corresponding to the anterior clinoid process of the lesser wing of the sphenoid, it gives off the **ophthalmic artery** which passes through the optic canal and enters the orbital cavity running laterally beneath the optic nerve [II]. The **central retinal**, **long** and **short posterior ciliary**, **supraorbital**, and the **posterior** and **anterior ethmoidal arteries** enter, respectively, the posterior and anterior ethmoidal foramina; the medial palpebral, supratrochlear (or frontal) and dorsal nasal arteries are terminal branches of the ophthalmic artery.

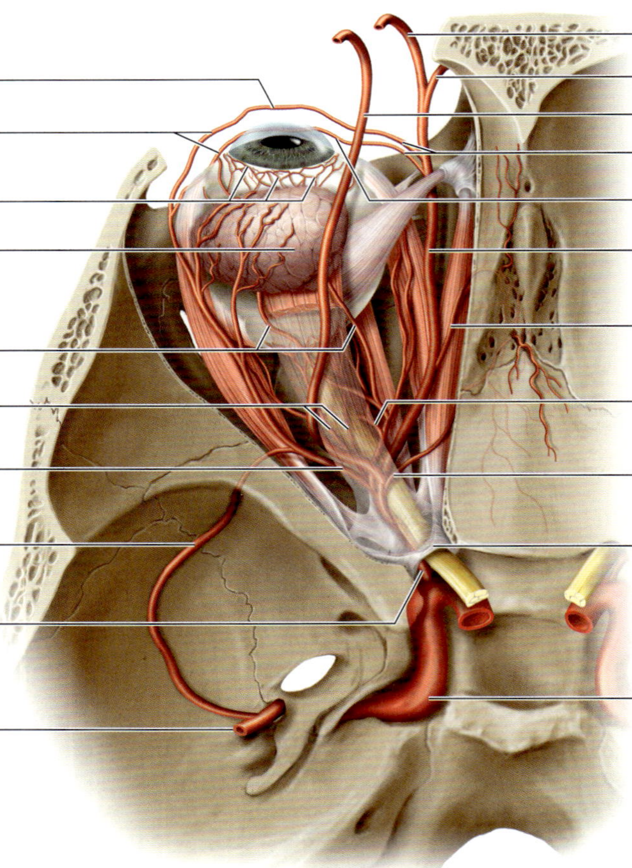

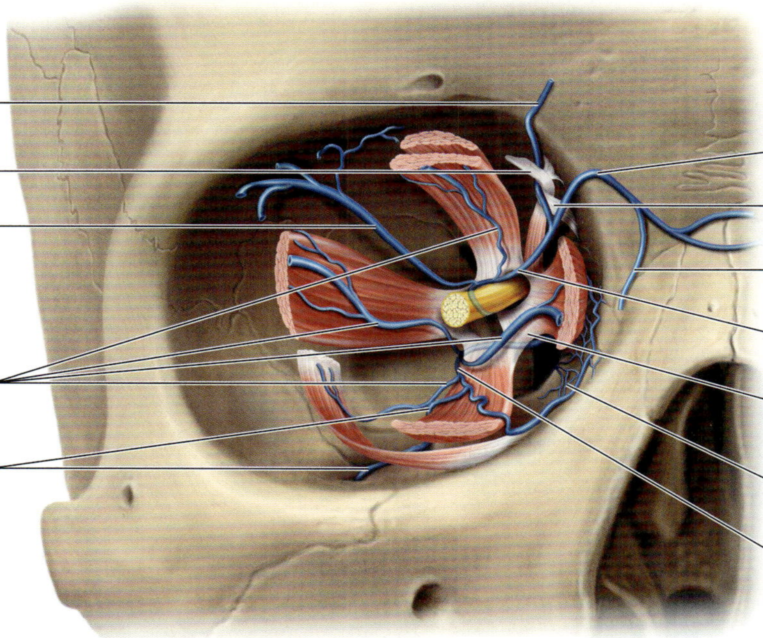

Veins of the orbit, anterior view. The **superior ophthalmic vein**, originating from the confluence of the **nasofrontal vein** and a venous root connected to the supraorbital vein, runs in the superior part of the orbital cavity. Its course corresponds more or less to that of the ophthalmic artery. The **inferior ophthalmic vein**, originating from a venous network connected to the venules of the lacrimal sac, runs in the inferior part of the orbit, above the inferior rectus. Among the affluent branches of the ophthalmic veins are the **muscular veins**, satellites of the muscular arteries or muscular branches of the ophthalmic artery. The **lacrimal vein**, a satellite of the lacrimal artery, empties into the superior ophthalmic vein.

➤

Transverse section of the cranium passing just below the roof of the **orbital cavity**, superior view. The **ophthalmic artery**, an offshoot of the internal carotid artery after leaving the cavernous sinus, supplies the orbital cavity and its contents, the eyelids, the dura mater of the anterior cranial fossa, and the vault and lateral walls of the nasal cavities. The **lacrimal artery** runs along the superior border of the lateral rectus and gives rise to the recurrent meningeal branch, which passes through the lateral part of the superior orbital fissure and anastomoses with the middle meningeal artery. The **lateral palpebral arteries**, terminal branches of the lacrimal artery, anastomose with the **medial palpebral arteries**, terminal branches of the ophthalmic artery, forming the superior and inferior **palpebral arches**. The **long posterior ciliary arteries** branch into **short posterior ciliary arteries**, while the **anterior ciliary arteries** originate from the muscular arteries, supplying the muscles of the orbit. The **major circulus arteriosus of the iris** is formed by anastomoses of division branches of the long posterior ciliary arteries with branches of the anterior ciliary arteries and between the branches of the long posterior ciliary arteries themselves.

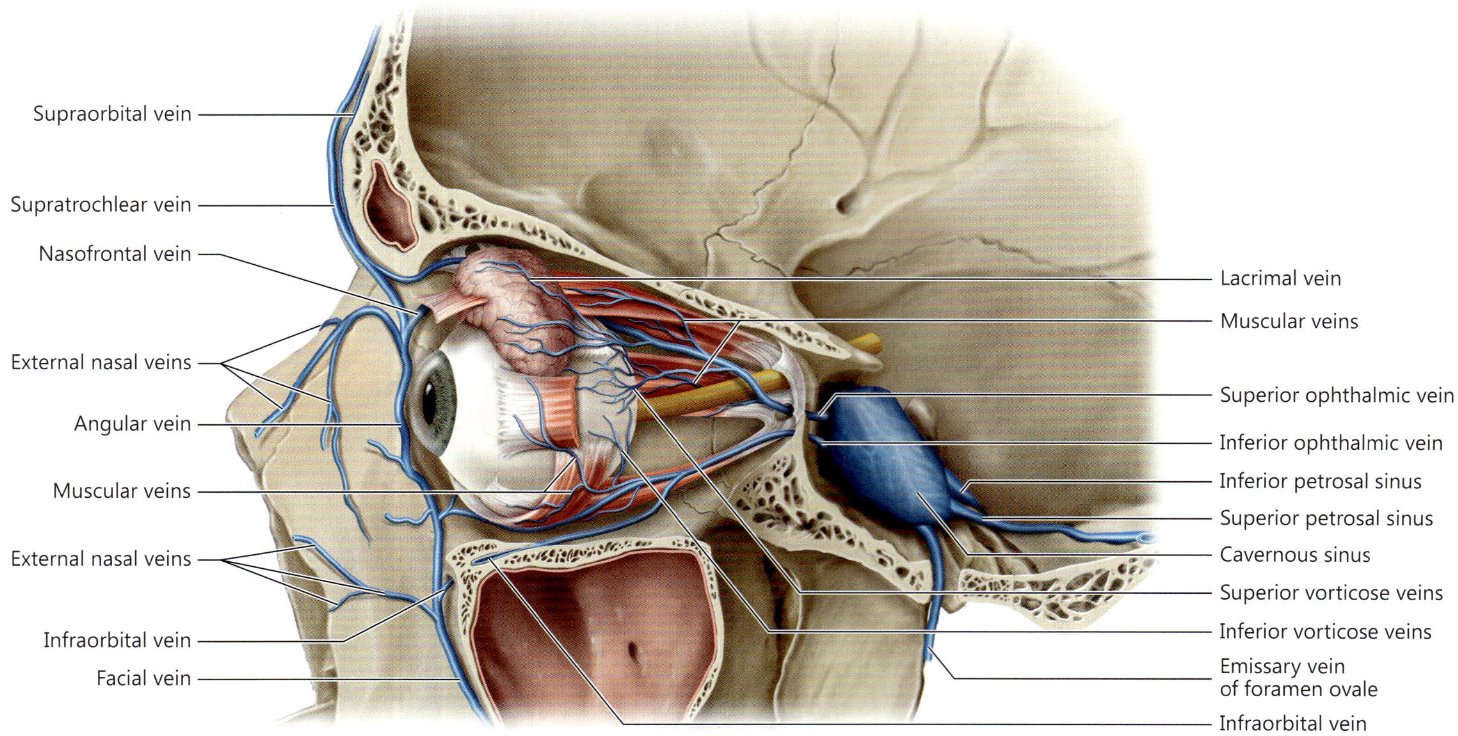

Lateral view of the left **orbital cavity** after removal of the lateral wall. The **superior ophthalmic vein** arises from the confluence of the nasofrontal vein with a venous root connected to the supraorbital vein and, after passing through the superior orbital fissure, empties into the cavernous sinus. Among its affluent branches are the lacrimal vein, muscular veins and superior vorticose veins. The **inferior ophthalmic vein** originates in correspondence to the medial inferior angle of the orbit, runs along the floor of the orbit above the inferior rectus and, after crossing the superior orbital fissure, terminates in the cavernous sinus. Tributaries of it are the muscular veins and inferior vorticose veins. The **angular vein** originates superficially at the medial angle of the eye from the confluence of the frontal, supraorbital and nasofrontal veins, runs obliquely downwards and finally heads in a posterolateral direction constituting the **facial vein**.

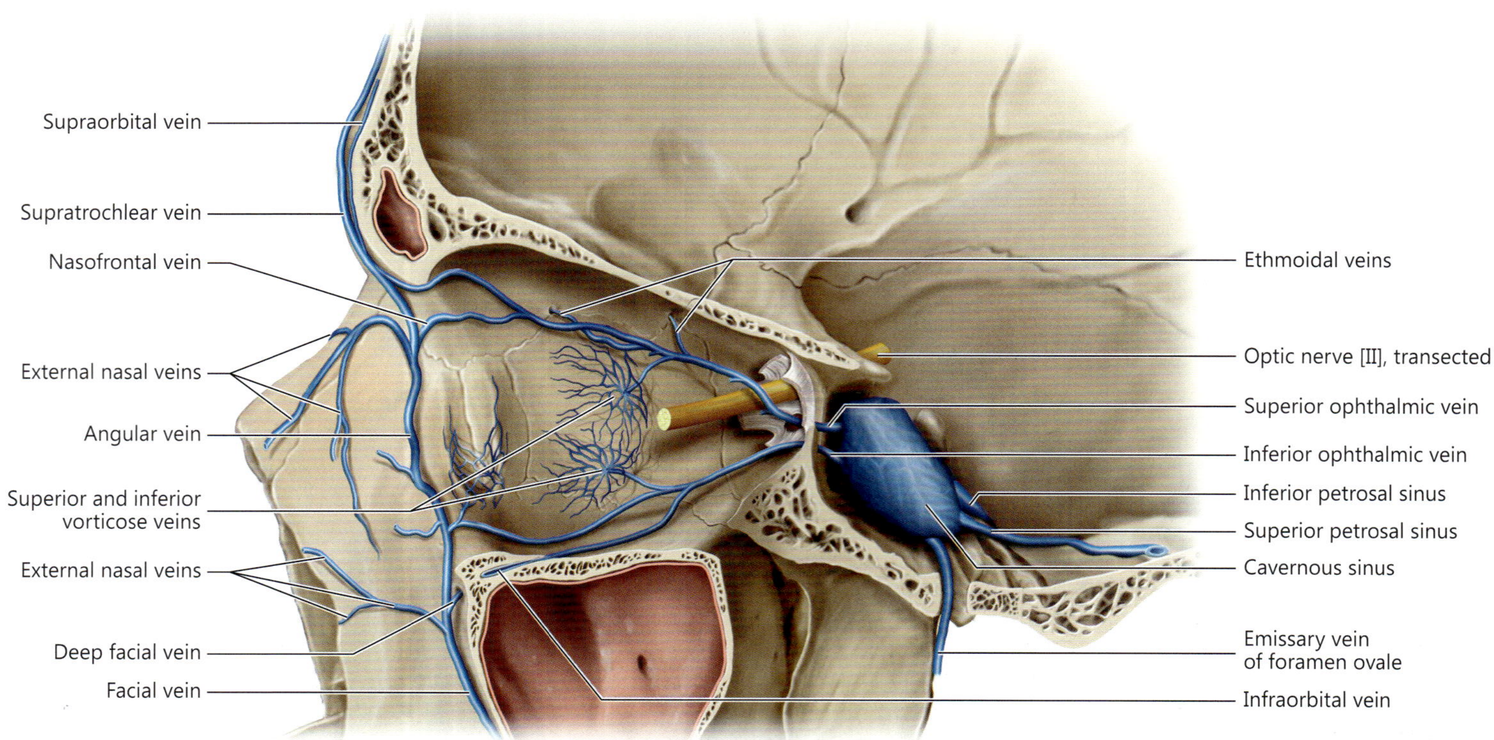

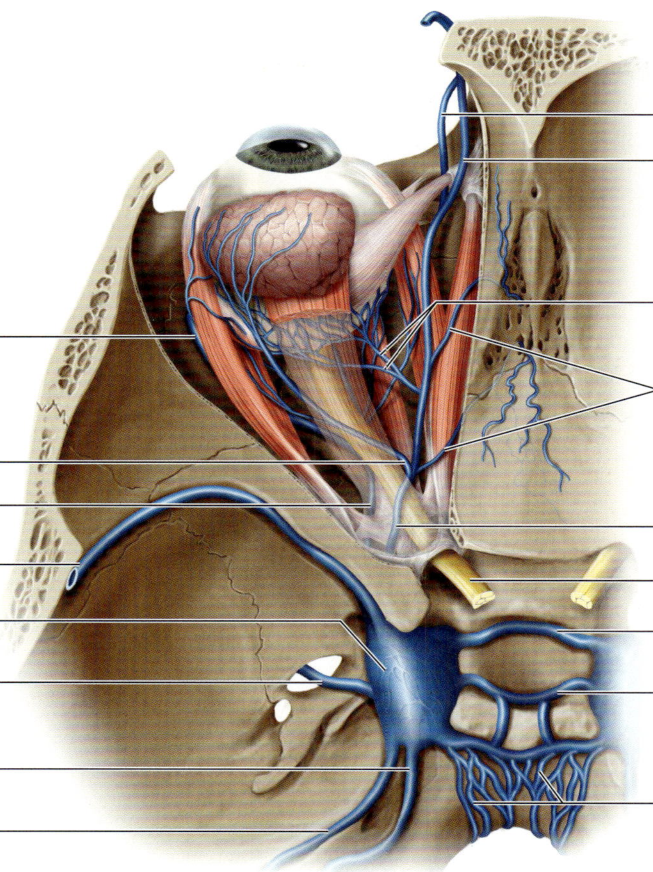

Transverse section of the cranium passing just under the roof of the **orbital cavity**, superior view. The **veins of the orbit** are shown. The **superior ophthalmic vein**, the largest vein of the organ of sight, originates from the confluence of the nasofrontal and supraorbital veins and empties into the anterior extremity of the cavernous sinus. Its affluents are the **ethmoidal veins**, **lacrimal vein** and **superior vorticose veins**. The **cavernous sinus**, extending from the medial extremity of the superior orbital fissure to the apex of the petrous part of the temporal bone, receives also the **inferior ophthalmic vein**. The cavernous sinuses of the two sides are in communication with each other by means of the **intercavernous sinuses** and drain into the **superior** and **inferior petrosal sinuses**, into the basilar plexus and the pterygoid plexus via the **emissary vein of the foramen ovale** or a venous network of the foramen ovale. The **infraorbital vein**, satellite of the infraorbital artery, runs through the infraorbital canal and groove, exits from the orbital cavity through the inferior orbital fissure and drains into the pterygoid plexus.

➤

Medial wall of the **orbit**. Paramedian sagittal section, left lateral view, after removal of the eyeball and accessory visual structures. The **cavernous sinus** is situated laterally to the sella turcica of the sphenoid, between the medial extremity of the superior orbital fissure and the apex of the petrous part of the temporal bone. Running through its cavity are numerous trabeculae and blood and nerve vessels. The internal carotid artery, here not shown, passes through the cavernous sinus in a posteroanterior direction. The **superior ophthalmic vein**, whose course in the superior part of the orbital cavity corresponds more or less to that of the ophthalmic artery, originates from the confluence of the nasofrontal vein and a venous root connected to the supraorbital vein. After passing through the superior orbital fissure, it empties into the anterior extremity of the cavernous sinus, which receives also the **inferior ophthalmic vein**, which arises in correspondence to the inferior and medial angle of the orbit from a venous network connected to the venules of the lacrimal sac and lower eyelid.

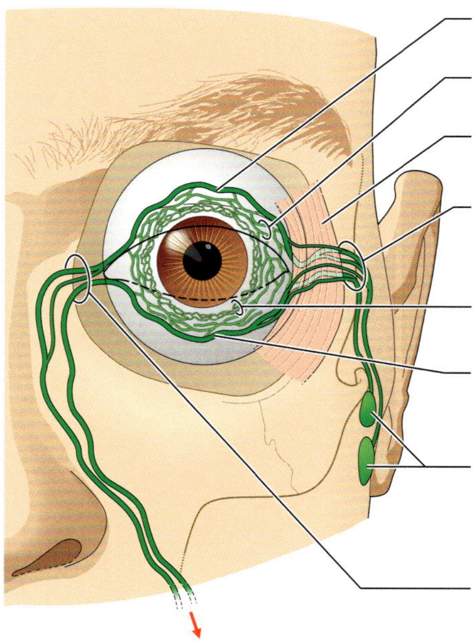

The **lymphatic capillaries of the bulbar conjunctiva** constitute a pericorneal lymphatic ring around the corneoscleral junction, approximately 1 mm wide, in continuity with a radial lymphatic network disposed around the cornea. This network flows into two collectors, a superior and an inferior one, which proceed into the medial and lateral lymphatic collectors, at the angles of the eye. The medial group converges towards the root of the nose, follows the facial vein and reaches the submandibular nodes. The lateral group drains into the preauricular and infra-auricular deep parotid nodes.

The **superficial lymphatic network of the eyelids**, localized in the skin, and the deep or conjunctival lymphatic network, situated in correspondence to the palpebral conjunctiva and bulbar conjunctiva, drain through medial and lateral lymphatic collectors, which constitute the paths of lymphatic drainage for the eyelids, into, respectively, the submandibular nodes and the preauricular and infra-auricular deep parotid nodes.

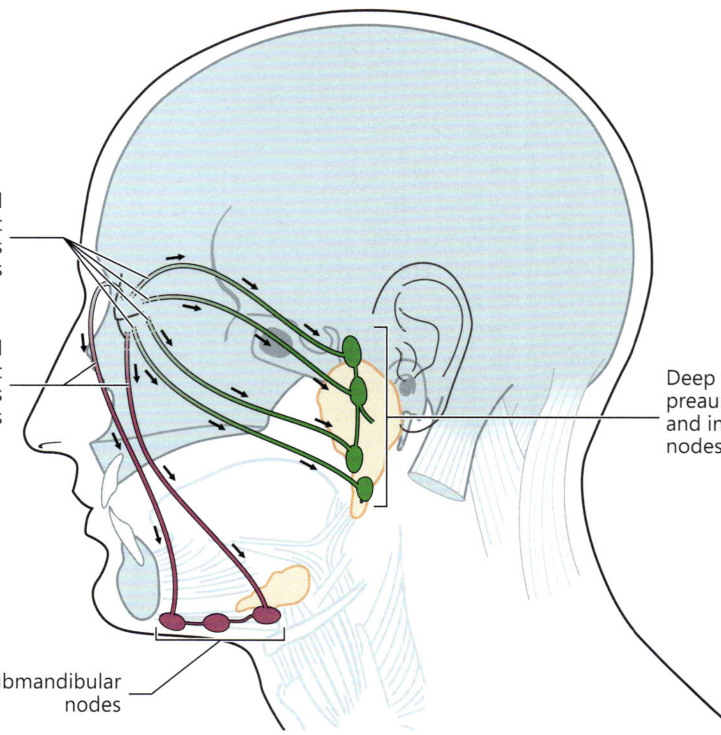

Lateral view of the left **orbital cavity** after removal of the lateral wall. At the boundary with the superior orbital fissure, the **ophthalmic nerve** [V_1] divides into the **nasociliary**, **frontal** and **lacrimal nerves**. The lacrimal nerve, through a communicating branch with the zygomaticotemporal nerve, a branch of the zygomatic nerve, itself a branch of the maxillary nerve [V_2], receives postganglionic fibers to the lacrimal gland arising from the pterygopalatine ganglion. The long, or sensory, root of the **ciliary ganglion** is a collateral branch of the nasociliary nerve whose fibers, passing through the ciliary ganglion without being interrupted, follow the short ciliary nerves up to the eyeball. The ciliary ganglion has two other roots: a short, or parasympathetic, one and a sympathetic one. The short root is a collateral branch of the oculomotor nerve [III] containing preganglionic fibers from the accessory nuclei of the oculomotor nerve (nucleus of Edinger-Westphal). The postganglionic fibers supply the sphincter pupillae and ciliary muscle. The sympathetic root comes directly from the internal carotid plexus and contains postganglionic vasomotor fibers arising from the superior cervical ganglion. The long root of the ciliary ganglion also contains postganglionic fibers from the superior cervical ganglion that, through the internal carotid plexus, anastomose with the ophthalmic nerve [V_1] on their way to the dilator pupillae via the short ciliary nerves.

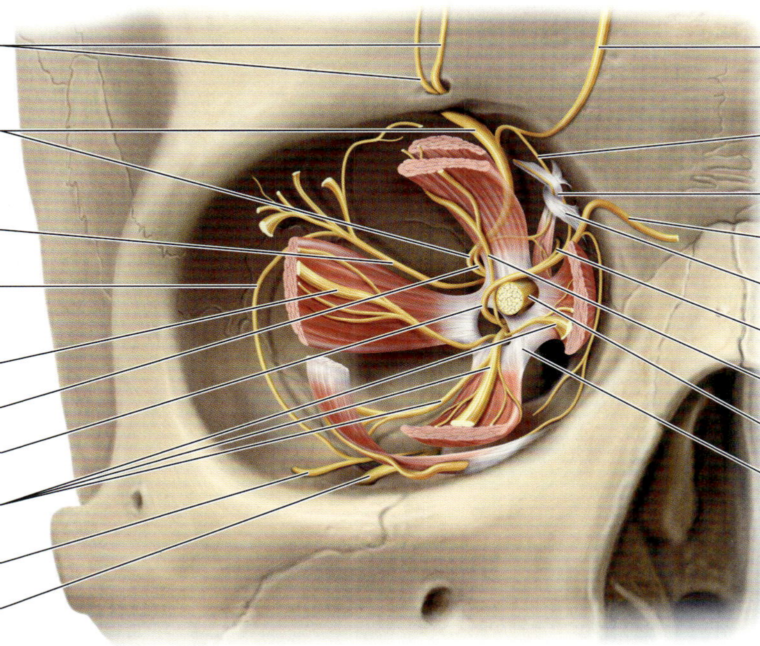

Nerves of the orbit, anterior view. The **oculomotor nerve** [III], at its arrival in correspondence to the posterior clinoid process, runs in the superior lateral wall of the cavernous sinus, and at the level of the superior orbital fissure divides into two branches, superior and inferior, that enter the orbital cavity through the common tendinous ring. The superior branch innervates the superior rectus and levator palpebrae superioris, while the inferior branch innervates the medial rectus, inferior rectus and inferior oblique. Also the **abducens nerve** [VI], which innervates the lateral rectus, and the **nasociliary nerve**, a branch of the ophthalmic nerve [V_1], pass through the ring. The abducens nerve [VI], after its course in the cavernous sinus laterally to the internal carotid artery, enters the orbital cavity through the superior orbital fissure and passes inside the tendinous ring in a position between the superior and inferior branches of the oculomotor nerve and inferolaterally to the nasociliary nerve. The nasociliary nerve, which has branched off from the ophthalmic nerve [V_1] before entering the orbital cavity, passes through the common tendinous ring and continues in a medial direction crossing over the ophthalmic artery and the optic nerve [II]. The **trochlear nerve** [IV] passes instead above the ring, together with the **lacrimal** and **frontal nerves**, branches of the ophthalmic nerve, to supply the superior oblique. The **supratrochlear** and **supraorbital nerves** are the two terminal branches of the frontal nerve, while the **infratrochlear** and **anterior ethmoidal nerves** are terminal branches of the nasociliary nerve.

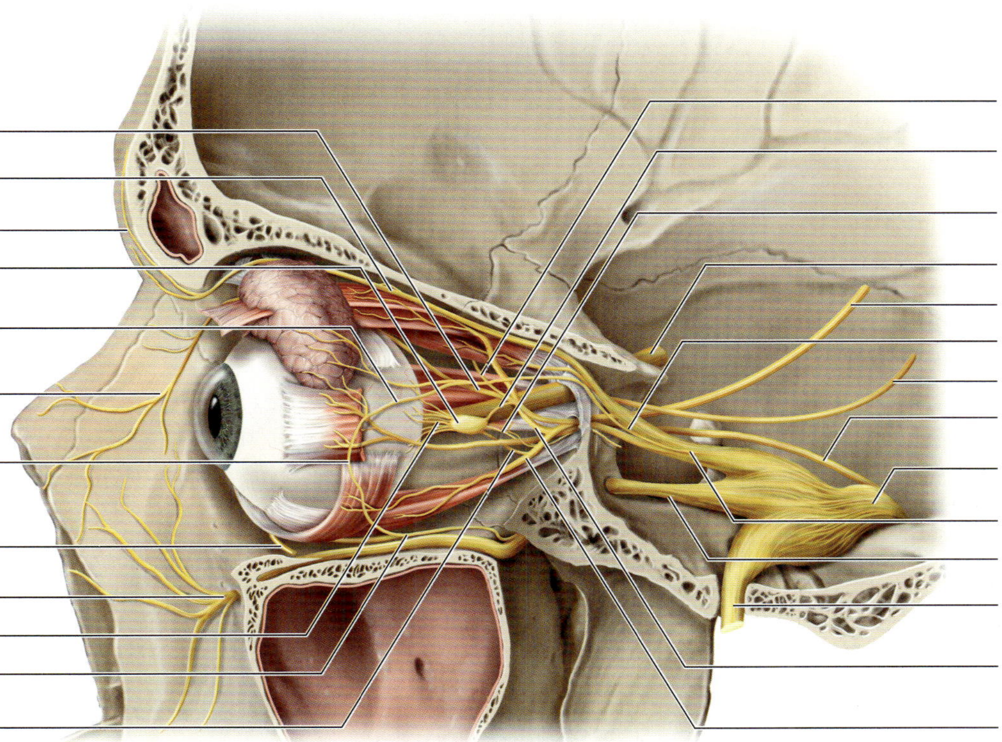

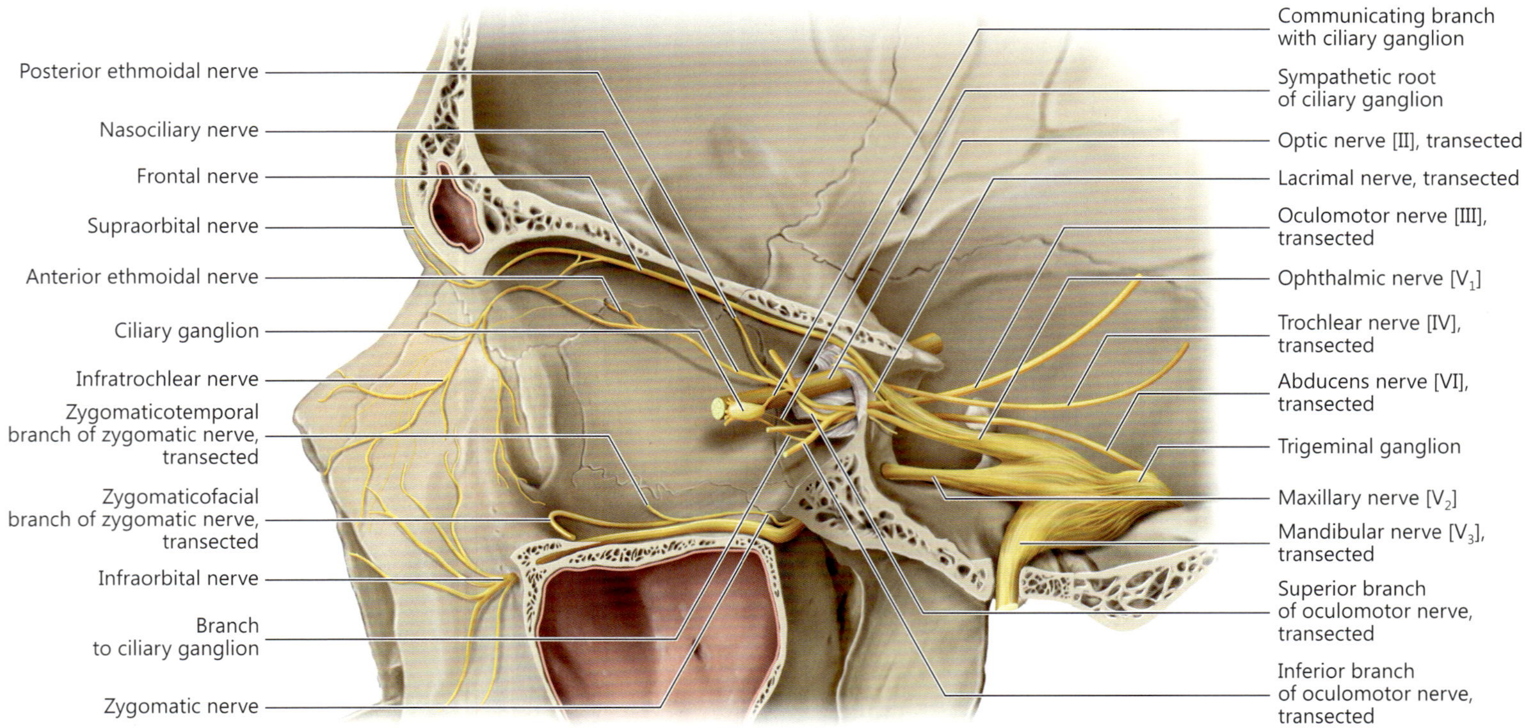

Medial wall of the **orbit**. The **nasociliary nerve**, a medial division of the ophthalmic nerve [V_1], enters the orbital cavity passing through the common tendinous ring (anulus of Zinn) and continues in a medial direction with two terminal branches, the anterior ethmoidal and infratrochlear nerves. The **oculomotor nerve** [III] divides into two branches, superior and inferior, that pass through the common tendinous ring, together with the abducens nerve [VI]. The **trochlear nerve** [IV] passes into the orbital cavity above the common tendinous ring, together with the frontal and lacrimal nerves, intermediate and lateral divisions of the ophthalmic nerve [V_1]. The **maxillary nerve** [V_2] enters the orbital cavity through the inferior orbital fissure and continues its path forward on the floor of the orbit and in the infraorbital canal, as the infraorbital nerve. The **zygomatic nerve**, a branch of the maxillary nerve, divides off from the maxillary nerve at the border with the inferior orbital fissure and heads laterally forwards dividing into two terminal branches, the zygomaticotemporal and zygomaticofacial nerves.

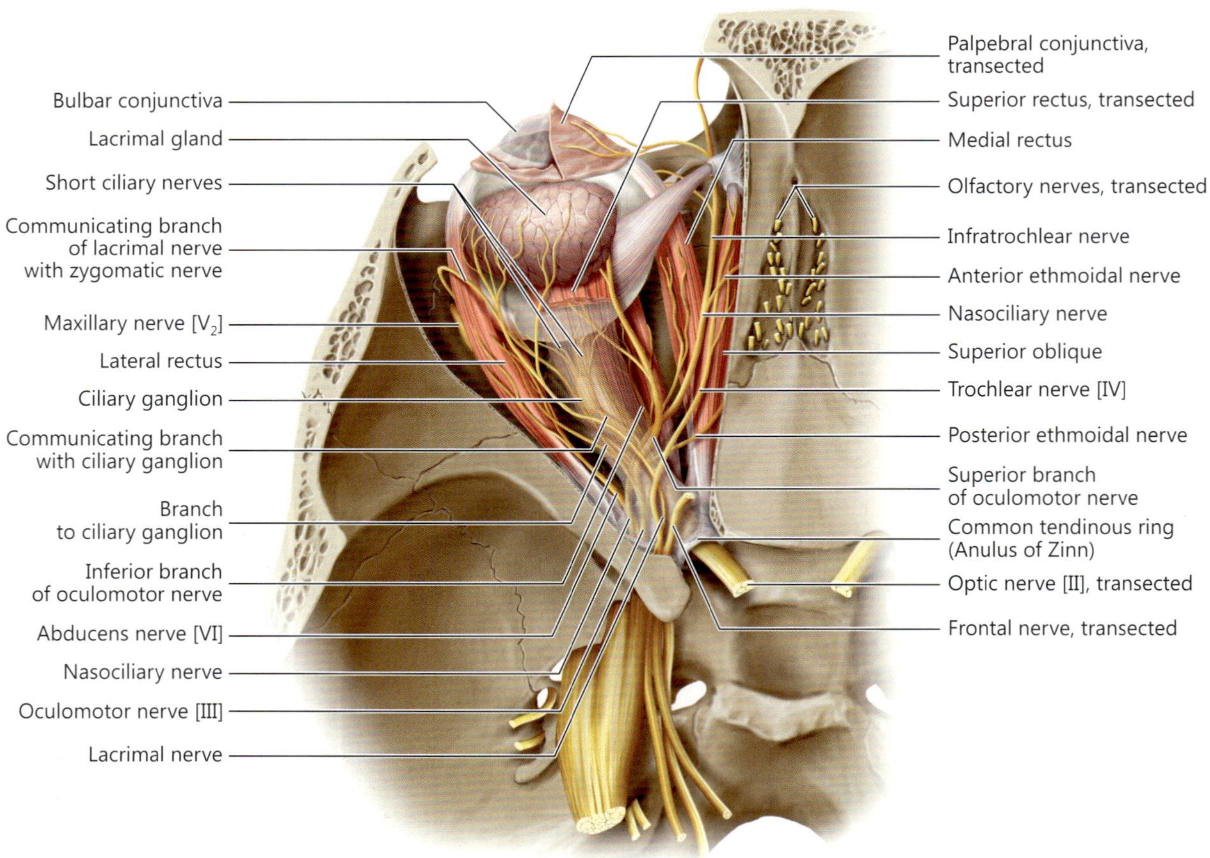

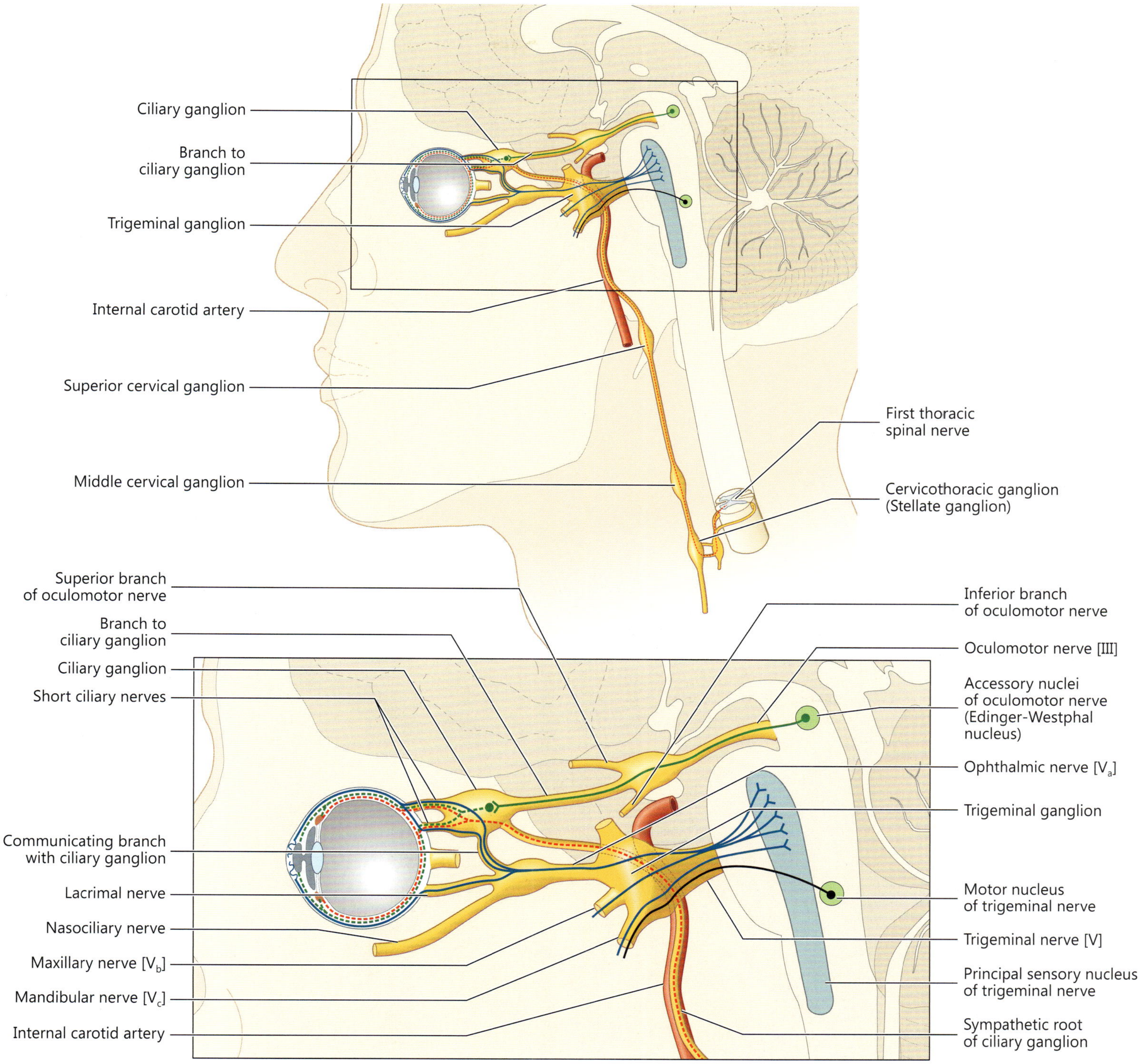

Diagram showing the connections of the **ciliary ganglion** (in the box). The only fibers that form synapses in the ganglion are the parasympathetic ones (*green*); sympathetic fibers (*red* dotted line) and sensory fibers (*blue*) pass through the ganglion without sinapsing there. The three types of fiber constitute, respectively, the parasympathetic root, sympathetic root, and sensory root of the ciliary ganglion.

◄
Transverse section of the cranium passing just below the roof of the **orbital cavity**, superior view. One can see the nerves that enter the orbital cavity, through the optic canal and superior compartment of the common tendinous ring: the **optic nerve** [II]; through the superior orbital fissure, inside the common tendinous ring in its inferior part, the superior and inferior branches of the **oculomotor nerve** [III], and the **abducens nerve** [VI] and **nasociliary nerve**, a branch of the ophthalmic nerve [V_1]; outside the common tendinous ring, the **trochlear nerve** [IV] and **frontal** and **lacrimal nerves**, branches of the ophthalmic nerve [V_1]. The common tendinous ring (anulus of Zinn), funnel-shaped with the superior part corresponding to the optic foramen, which is the opening of the optic canal, is the point of origin of the rectus muscles of the eyeball and, partially, of the levator palpebrae superioris and superior oblique.

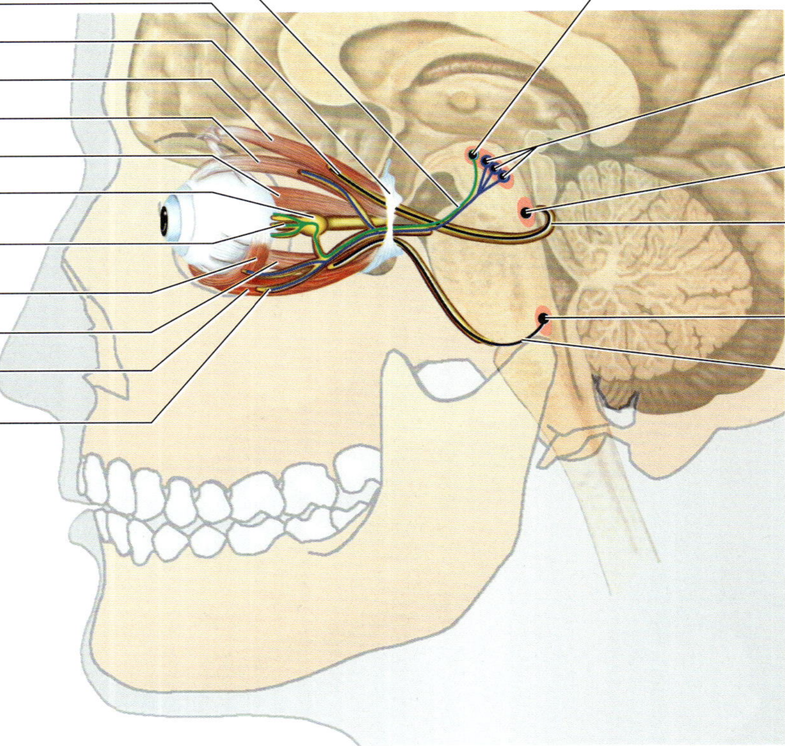

Diagram showing the origin and distribution of the **oculomotor**, **trochlear** and **abducens nerves**. Only the oculomotor nerve has a visceral motor component (*green*).

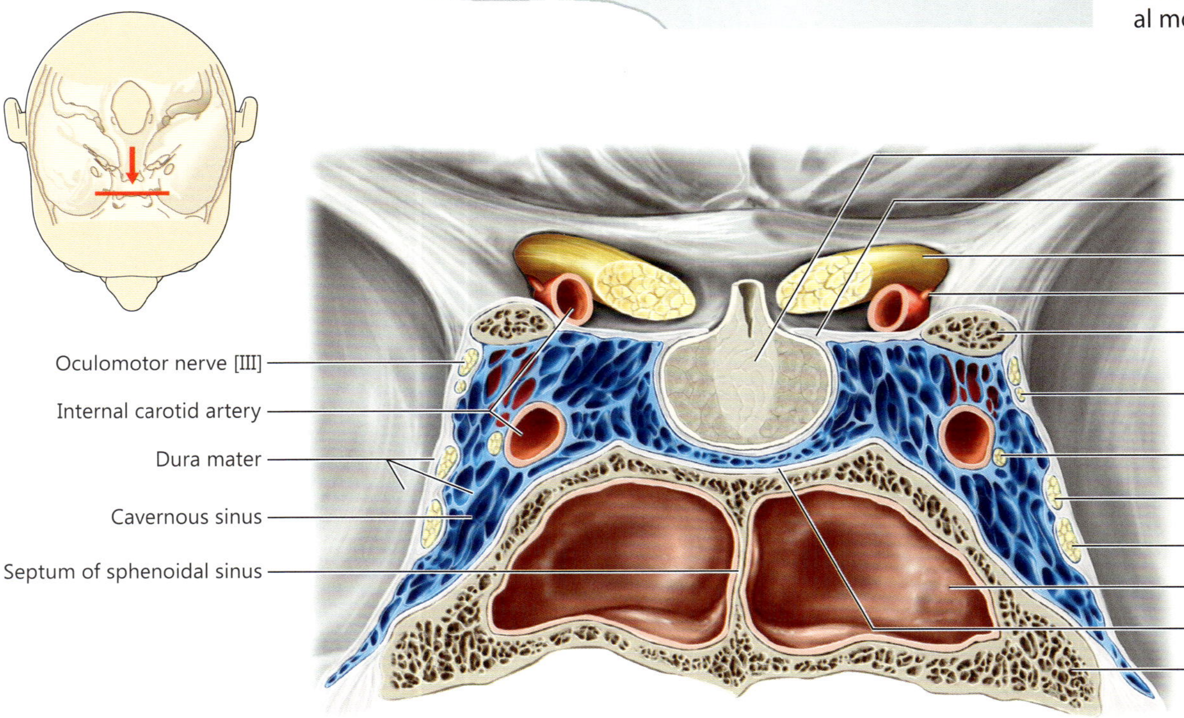

Frontal section of the head passing through the **anterior clinoid processes** of the sphenoid. The **cavernous sinus**, paired, is situated lateral to the sella turcica of the sphenoid, which contains the pituitary gland. Running through its cavity are numerous trabeculae or fibrous bands. The **internal carotid artery** passes through the cavernous sinus in a posteroanterior direction, and the **abducens nerve** [VI] runs inferolaterally to it. Within the lateral wall of the cavernous sinus run the **oculomotor** [III], **trochlear** [IV], **ophthalmic** [V_1] and **maxillary** [V_2] **nerves**.

➤

Medial wall of the **orbit**. Paramedian sagittal section, left lateral view after removal of the eyeball and accessory visual structures. The **vorticose veins**, in general four, one for each quadrant of the eyeball, subdivided into superior and inferior (here one can see the medial ones), drain blood from the choroid as well as from the ciliary body and iris, and are tributaries of the **superior** and **inferior ophthalmic veins**, which receive also the **anterior ciliary veins**, draining blood from the anterior part of the ciliary body and aqueous humor from the scleral venous sinus, and the (posterior) **ciliary veins**, that collect blood coming from the posterior part of the sclera.

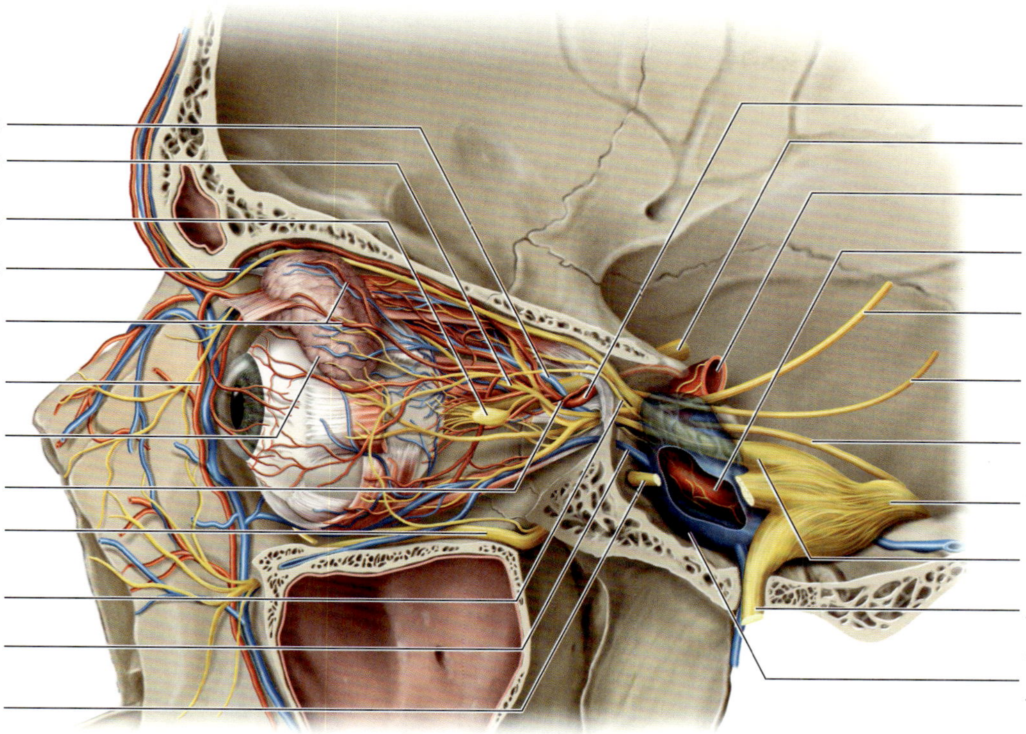

Lateral view of the left **orbital cavity** and middle cranial fossa after removal of the lateral wall. The lateral wall of the cavernous sinus has been transected in part to display its contents. The **oculomotor nerve** [III], in correspondence to the posterior clinoid process of the sphenoid, pierces the internal layer of the dura mater penetrating into the superior lateral wall of the cavernous sinus, within which run also the **trochlear nerve** [IV], beneath the oculomotor nerve [III], the **ophthalmic nerve** [V_1], which runs obliquely, from bottom to top, crossing under the trochlear [IV] and oculomotor [III] nerves, and, in the lowest region, also the **maxillary nerve** [V_2]. The **internal carotid artery** enters the cavernous sinus with an S-shaped curve and at its exit crosses laterally the **optic nerve** and passes through the arachnoid mater so coming to an end. Inside the cavernous sinus, the internal carotid artery is surrounded by sympathetic nerve fibers constituting the cavernous plexus, formed by the **internal carotid nerve**, a perivascular branch of the cervical segment of the sympathetic trunk that accompanies the internal carotid artery and its intracranial branches.

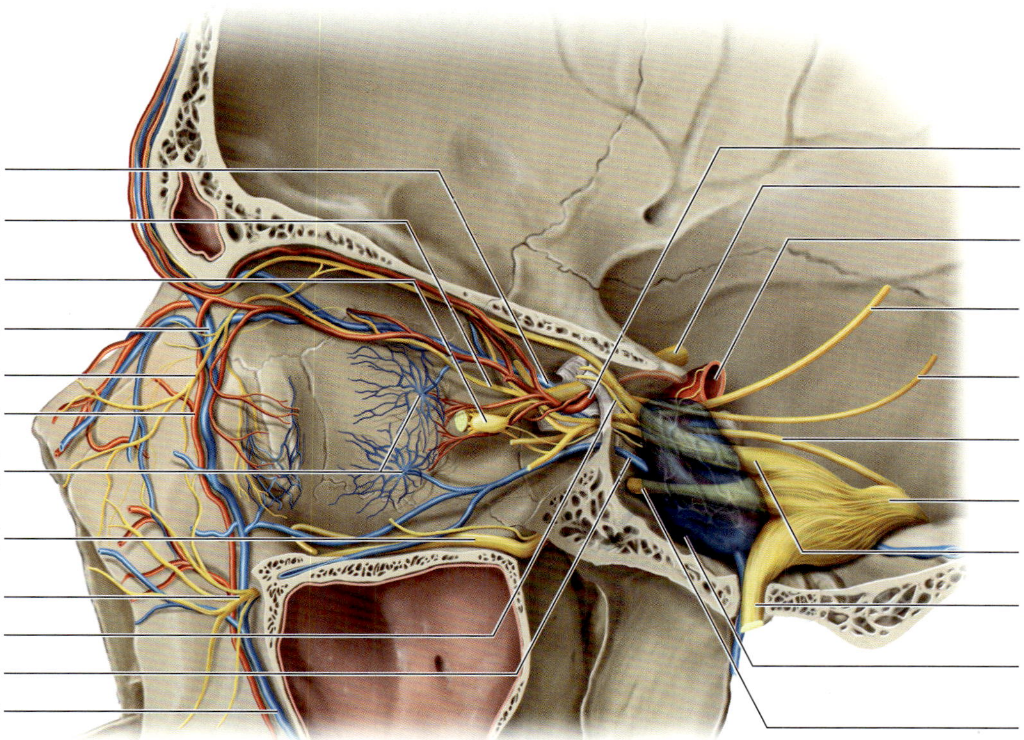

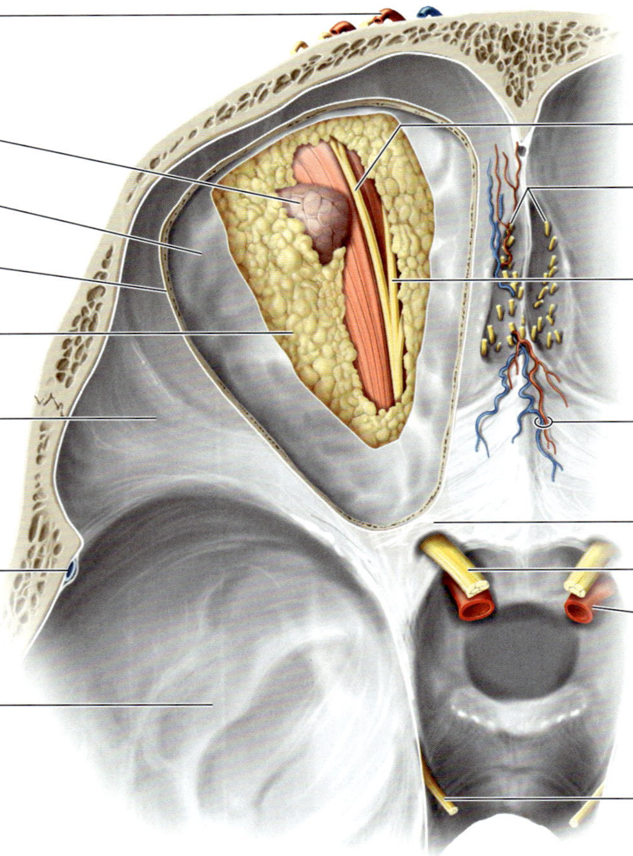

Transverse section of the cranium passing just below the roof of the **orbital cavity**, superior view. One can see the retrobulbar fat that fills the spaces of the orbital cavity located between the periosteum, which covers its internal surface, called the periorbita, and the organs contained in it.

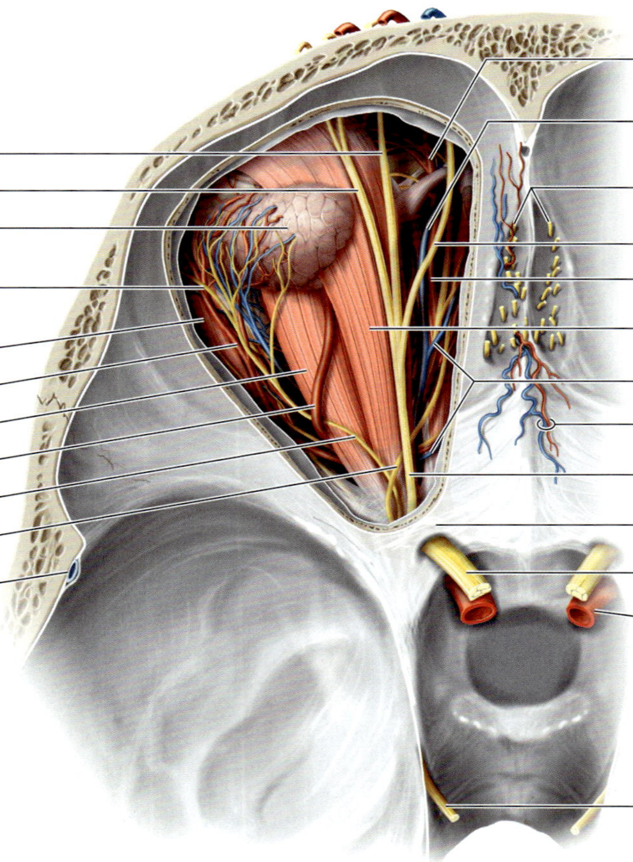

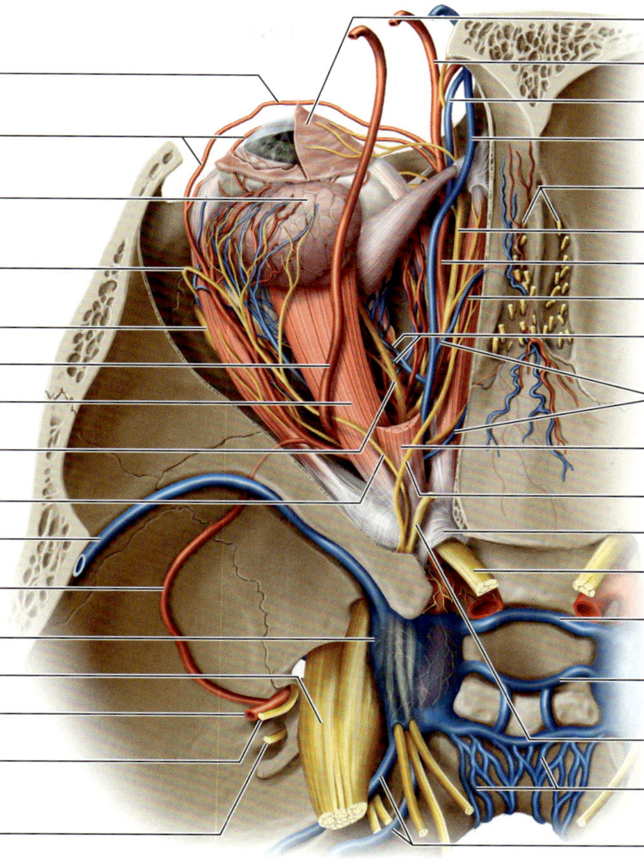

Transverse section of the cranium passing just below the roof of the **orbital cavity**, superior view. The levator palpebrae superioris has been transected to show the contents of the medial part of the orbit. One can see the three nerves that enter the orbital cavity passing above and lateral to the common tendinous ring: the **trochlear nerve** [IV], that crosses from above the tendon of the levator palpebrae superioris, the **frontal nerve**, an intermediate terminal branch of the ophthalmic nerve [V_1], and the **lacrimal nerve**, a lateral terminal branch of the ophthalmic nerve.

◄
Transverse section of the cranium passing just below the roof of the **orbital cavity**, superior view. The **frontal nerve**, an intermediate terminal branch of the ophthalmic nerve [V_1], runs on the vault of the orbital cavity in direct contact with the periosteum, above the levator palpebrae superioris, and divides into two branches whose paths diverge slightly: the **supratrochlear** and **supraorbital nerves**. The **lacrimal nerve**, a lateral terminal branch of the ophthalmic nerve, runs in the external part of the orbital vault, near the periosteum, following the superior border of the lateral rectus together with the **lacrimal artery**, and advancing forwards it reaches the lacrimal gland, in the vicinity of which it receives a **communicating branch with the zygomatic nerve**, a branch of the maxillary nerve [V_2], containing postganglionic fibers from the pterygopalatine ganglion coming to innervate the lacrimal gland.

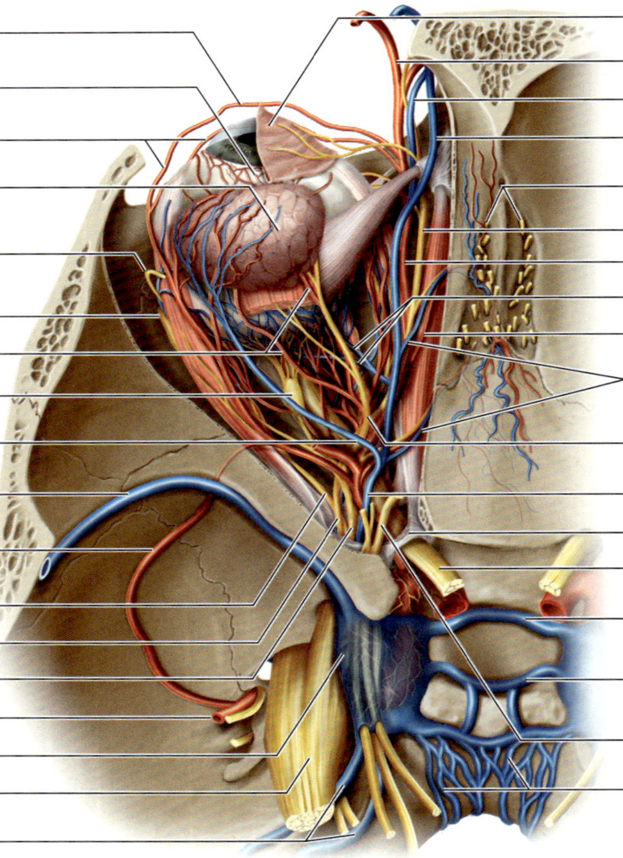

Transverse section of the cranium passing just below the roof of the **orbital cavity**, superior view. One can see the contents of the orbit, after removal of the superior rectus and levator palpebrae superioris, and the **vascular and nerve formations** that penetrate into the **orbital cavity** through the superior orbital fissure: the **trigeminal ganglion** in the middle cranial fossa, near the apex of the petrous part of the temporal bone; the **cavernous sinus**, and the formations contained within it, laterally to the sella turcica of the sphenoid; the **basilar plexus** on the clivus of the occipital bone; and the **olfactory nerves** that pass through the small holes of the cribriform plate of the ethmoid and penetrate into the anterior cranial fossa.

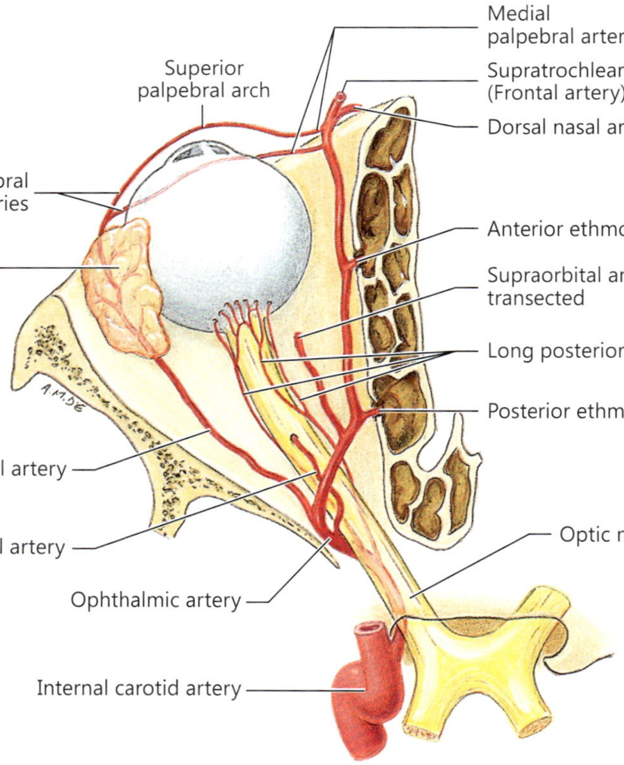

Left **ophthalmic artery** and its branches viewed from above, after demolition of the orbit and removal of part of its contents.

Three-dimensional image of a living human body, obtained by CT scan using a 3D volume rendering technique, showing the **extraocular muscles of the right eyeball** viewed laterally.

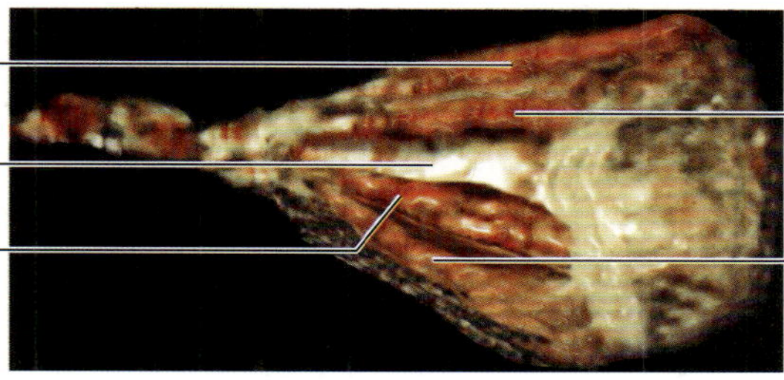

Three-dimensional image of a living human body, obtained by CT scan using a 3D volume rendering technique, showing the **extraocular muscles of the eyeballs** viewed superiorly.

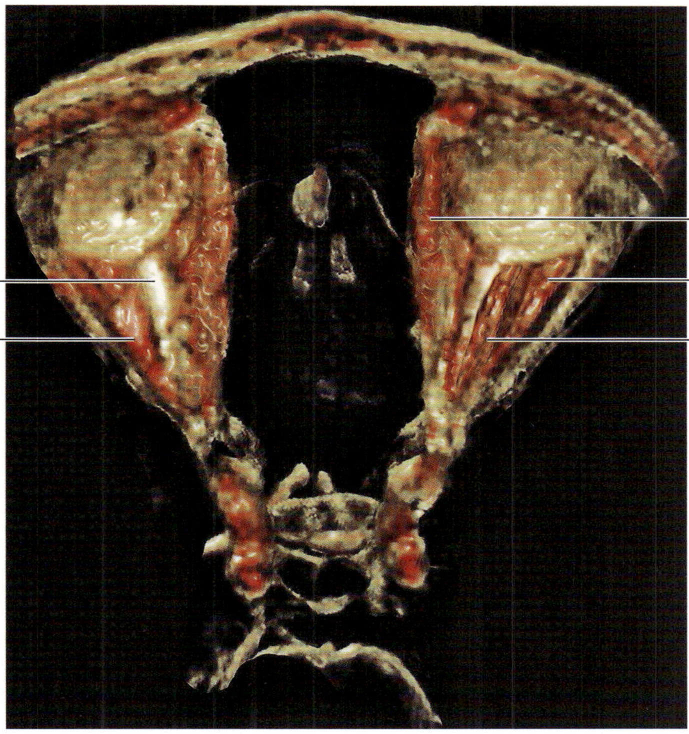

Extraocular muscles*

MUSCLE	ORIGIN	INSERTION	INNERVATION	ACTION
Inferior oblique	Floor of the orbital cavity (orbital surface of the maxilla) under the lacrimal fossa	Corneoscleral junction between the insertions of the lateral rectus and superior rectus	Oculomotor nerve [III]	Superior and lateral rotation of the eyeball
Inferior rectus	Inferior segment of the common tendinous ring (Anulus of Zinn)	Inferior part of the sclera just behind the corneoscleral junction	Oculomotor nerve [III]	Inferior and slightly medial rotation of the eyeball
Lateral rectus	Lateral border of the common tendinous ring (Anulus of Zinn)	Lateral part of the sclera just behind the corneoscleral junction	Abducens nerve [VI]	Lateral rotation of the eyeball
Levator palpebrae superioris	Roof of the orbital cavity (orbital part of the frontal bone)	Superior tarsus, skin of the superior eyelid, and superior conjunctival fornix	Oculomotor nerve [III]	Raising of the superior eyelid
Medial rectus	Medial segment of the common tendinous ring (Anulus of Zinn)	Medial part of the sclera just behind the corneoscleral junction	Oculomotor nerve [III]	Medial rotation of the eyeball
Superior oblique	Lesser wing of the sphenoid above the optic canal	After passing through the trochlea the muscle fibers change direction and insert on the superior part of the sclera, deep to the insertion of the superior rectus	Trochlear nerve [IV]	Inferior and lateral rotation of the eyeball
Superior rectus	Superior border of the common tendinous ring (Anulus of Zinn)	Superior part of the sclera just behind the corneoscleral junction	Oculomotor nerve [III]	Superior and slightly medial rotation of the eyeball

* In addition to the extraocular muscles there are three intrinsic muscles: the ciliary muscle, sphincter pupillae and dilator pupillae, not considered in this section of the Atlas dedicated to the musculoskeletal system since they are formed of smooth muscle fibers. Among the accessory visual structures are the superior tarsal and inferior tarsal muscles, also composed of smooth muscle fiber, that form the muscle layer of the eyelids.

Eye movements ☞ page 390 of volume 1.

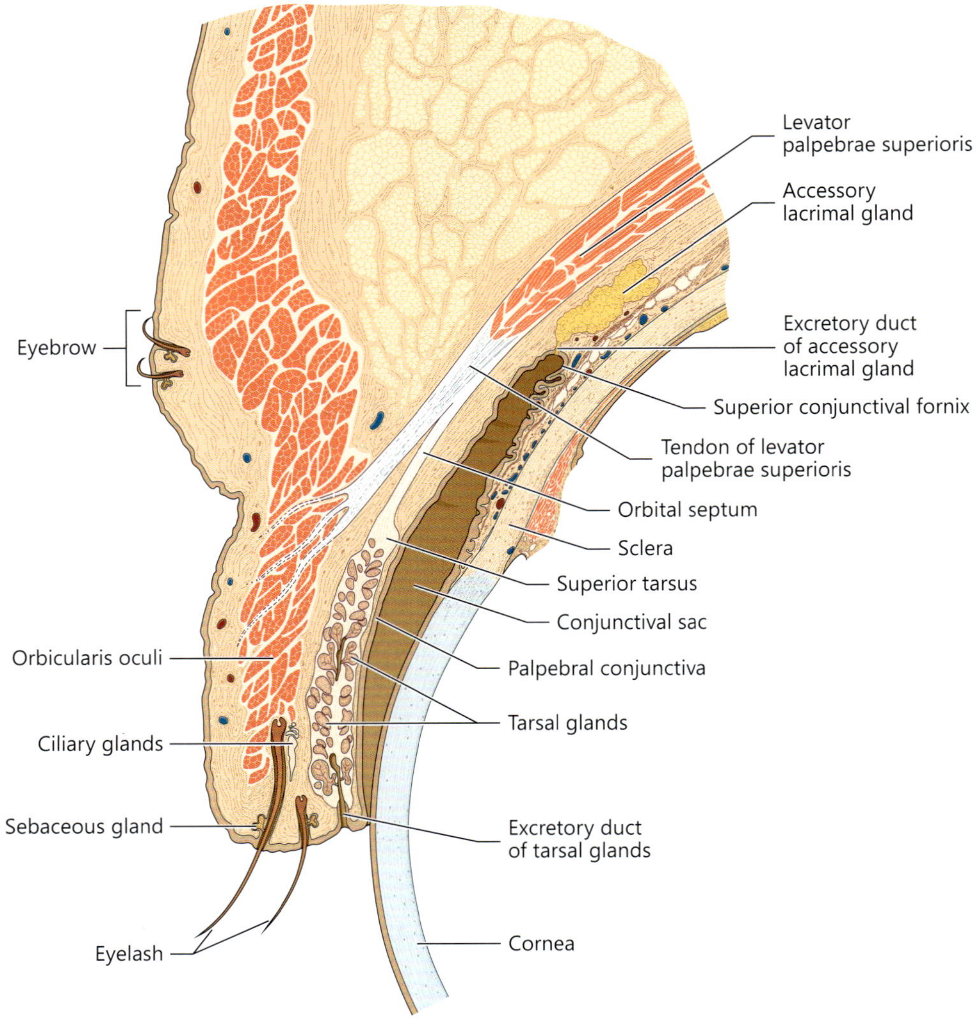

Superior eyelid, sagittal section. The diagram shows the relationship between the various formations of connective tissue, muscles and glands.

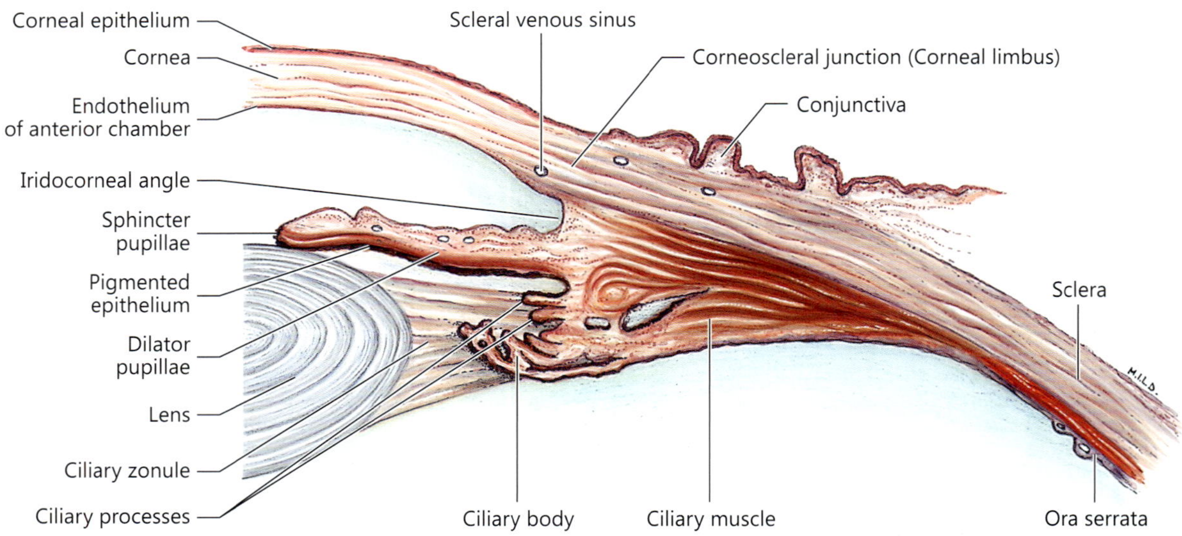

Anterior segment of the eye viewed in a **meridian section of the eyeball**. One can see the relationships of the sclera and cornea with the conjunctiva, ciliary processes and iris.

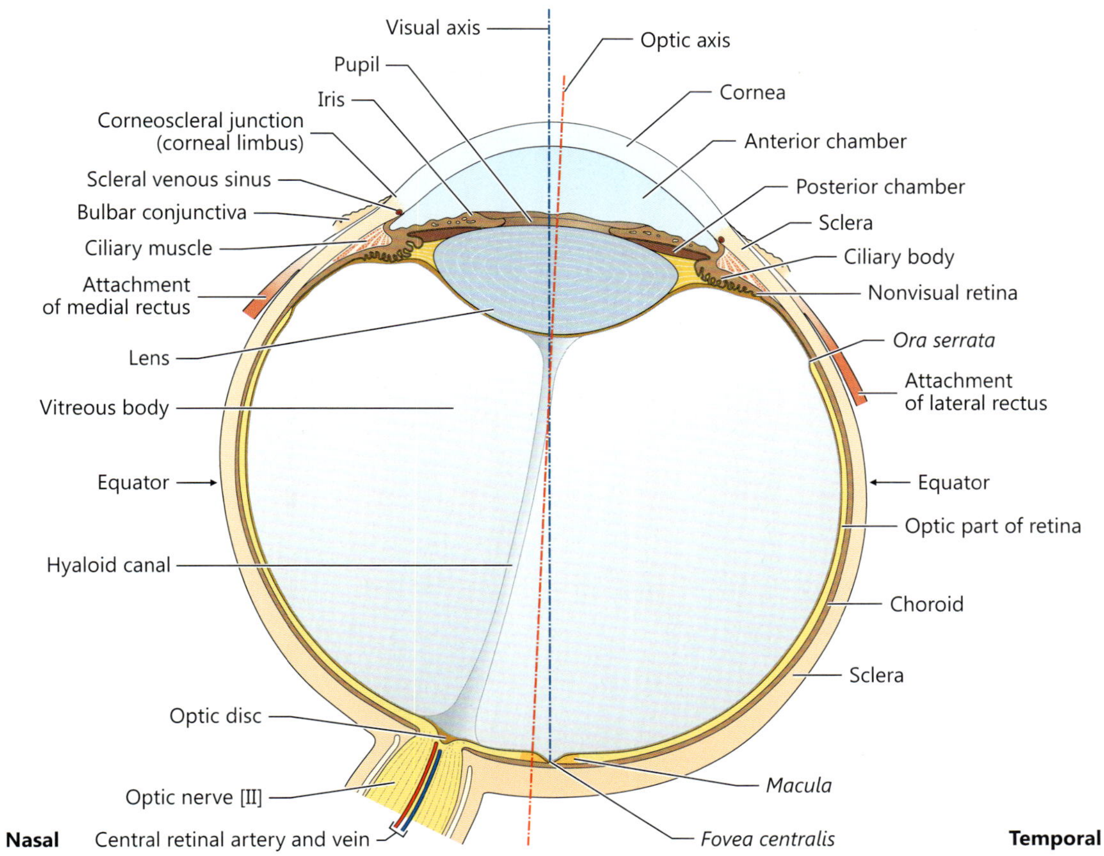

Schematic representation of the general structure of the right **eyeball** transected on the horizontal plane. The aqueous humor is shown in *blue*. The sclera is continuous with the cornea. The tendons of the extraocular muscles (such as the lateral rectus and medial rectus shown in the figure) insert into the sclera. The optic axis or external axis of the eyeball (*red*) and the visual axis (*blue*) are indicated.

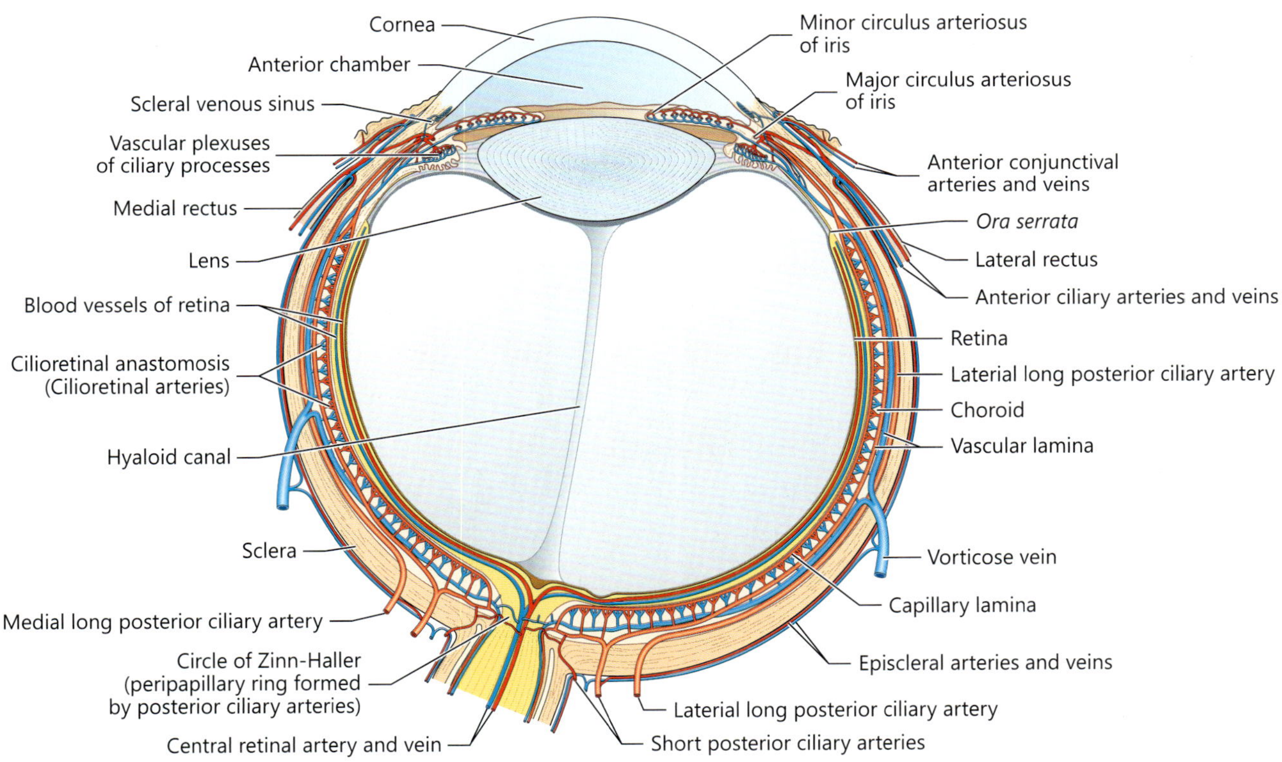

Diagram showing the **blood circulation of the eyeball**. The principal sources of arterial blood for the layers, or tunics, of the eyeball are the central retinal artery and the ciliary arteries. These vessels cross the wall of the eyeball in correspondence, respectively, to the optic nerve and the corneoscleral junction (or corneal limbus) (anterior ciliary arteries). The vorticose veins, which cross the wall of the eyeball itself, emerge in the equatorial region of the eyeball.

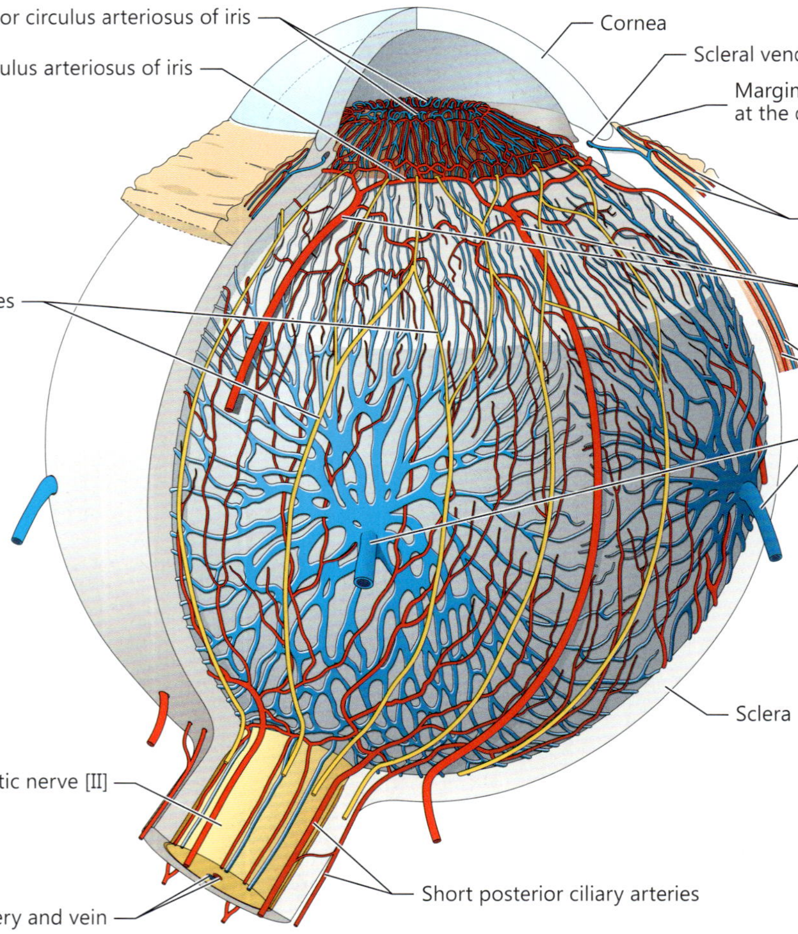

Blood vessels of the eyeball.

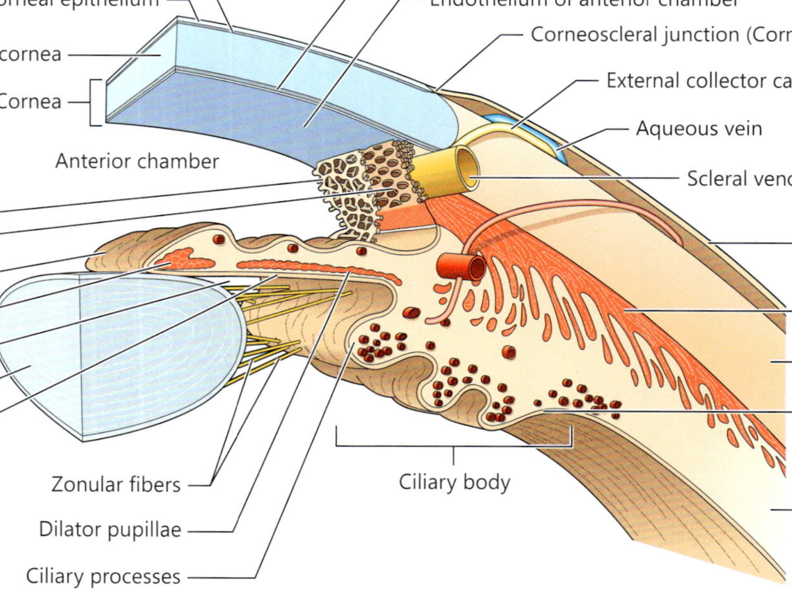

Eyeball, schematic representation of a horizontal section of the anterior part. The fibrous layer of the eyeball is visible, with the sclera, corneoscleral junction and cornea characterized by a 5-layer structure, and the vascular layer, with the iris, ciliary body and choroid, and the anterior and posterior chambers. The lens is shown suspended by fibers that constitute the zonule of Zinn (zonular fibers); these fibers, separating from the ciliary body where they originate, move to the equator of the lens where they insert.

Note: The aqueous vein, a tributary of the anterior ciliary vein, receives aqueous humor from the scleral venous sinus. The external collector canals transport aqueous humor from the scleral venous sinus to the scleral veins.

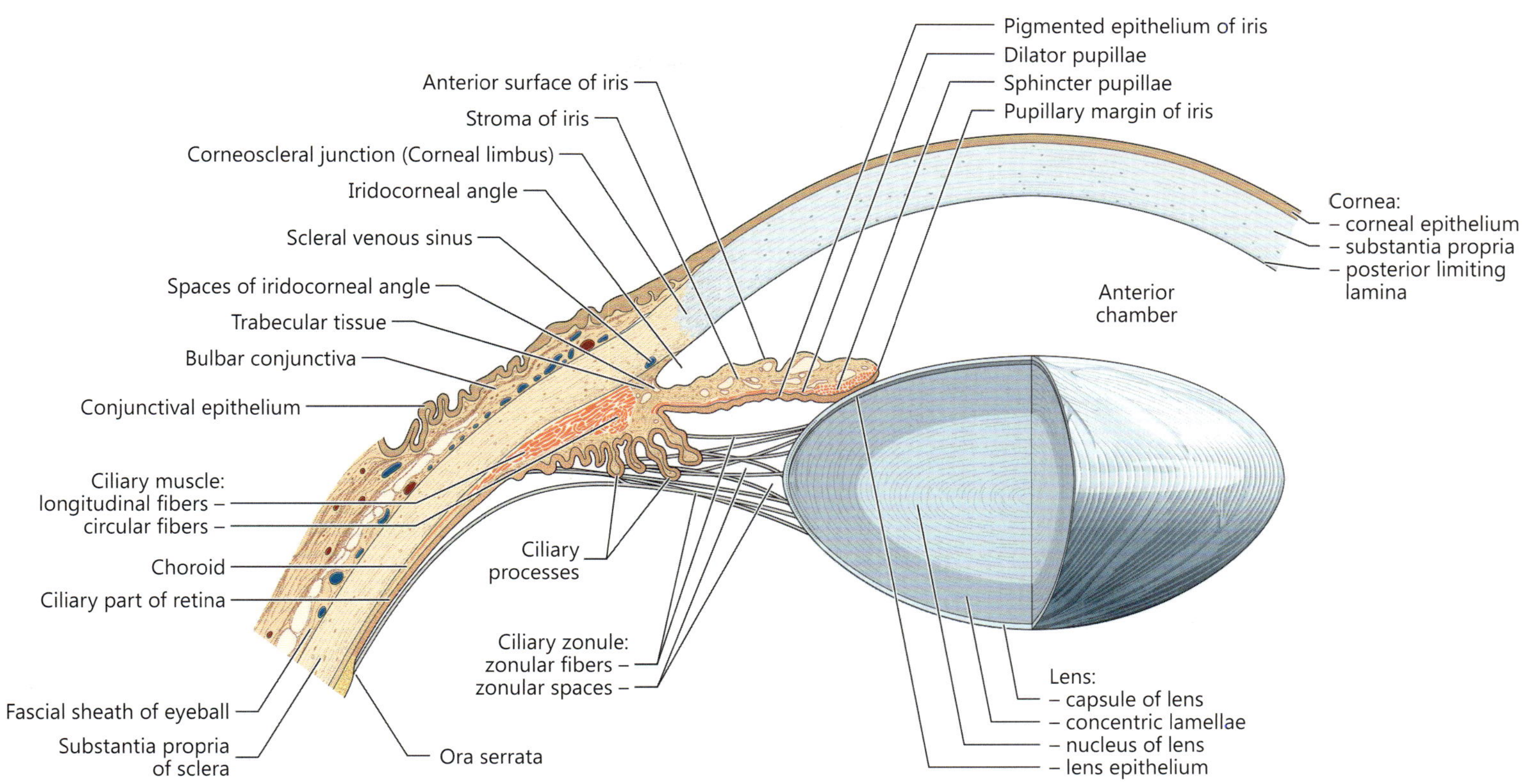

Eye, structure of the anterior (iridociliary) part of the vascular layer.

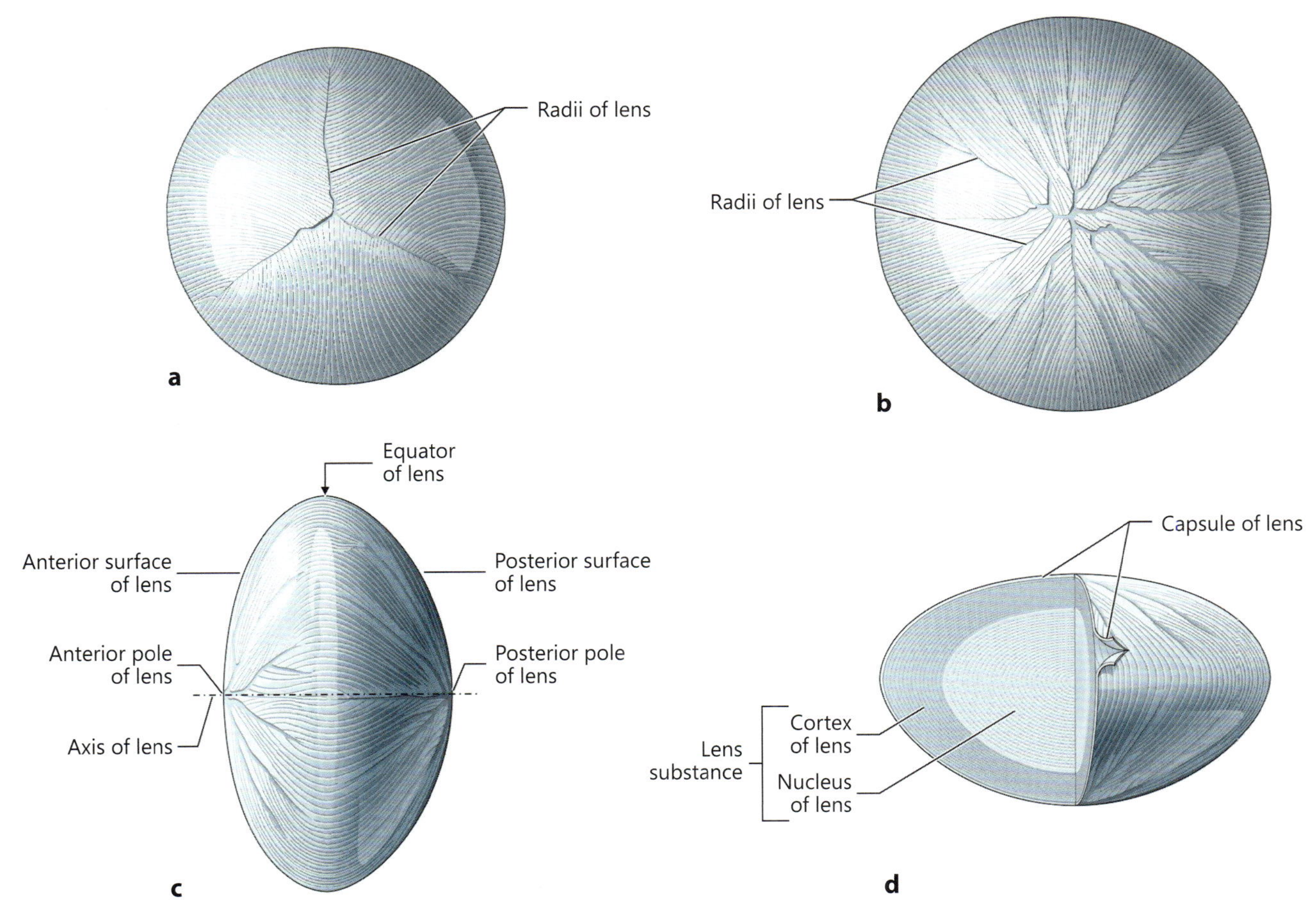

Lens, external and internal conformation. **a**, Anterior surface of the lens, in the child, with a triradiate star figure. **b**, Anterior surface of the lens, in the adult, with a multiradiate star figure. **c**, Lateral projection. **d**, Equatorial section, in which the nucleus of the lens is clearly visible.

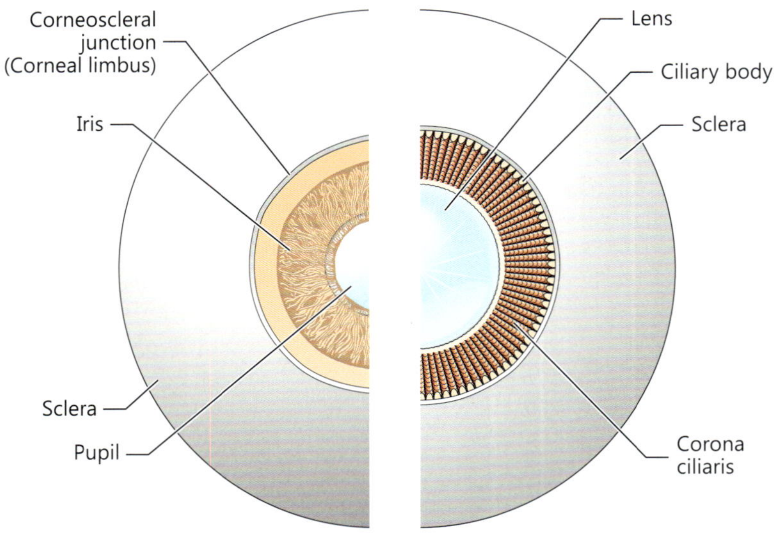

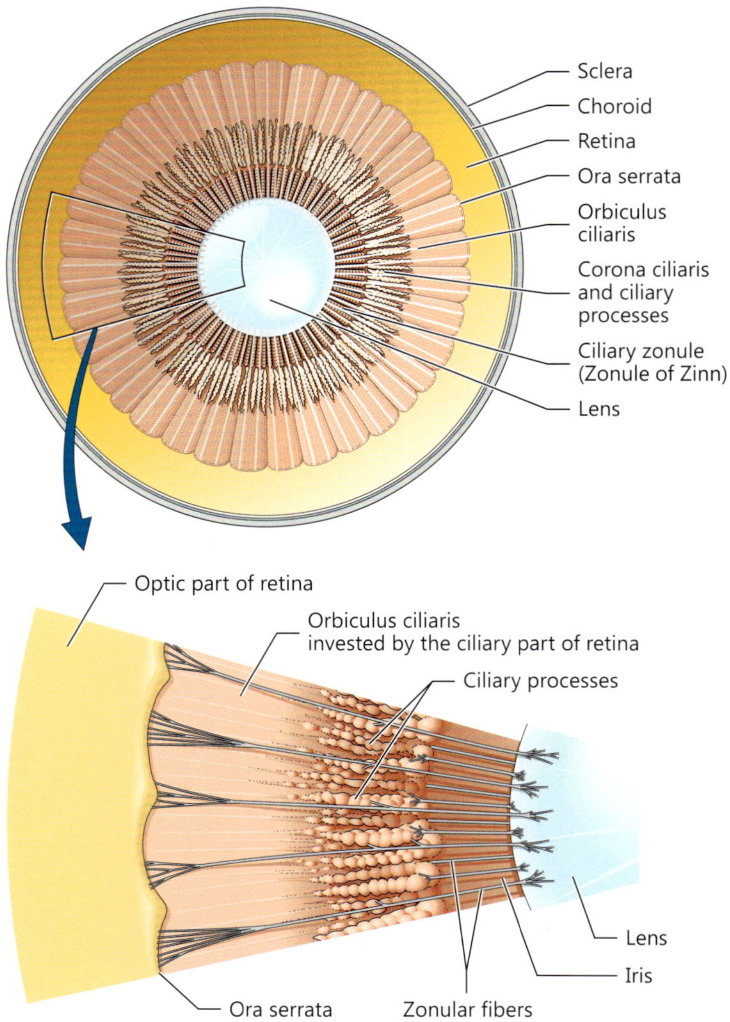

Eyeball, frontal section. **a**, The cornea (on the left) and half of the iris (on the right) have been removed in order to display the pupil, bounded by the iris (on the left), and the lens with its suspensory apparatus (on the right). **b**, Anterior segment viewed posteriorly.

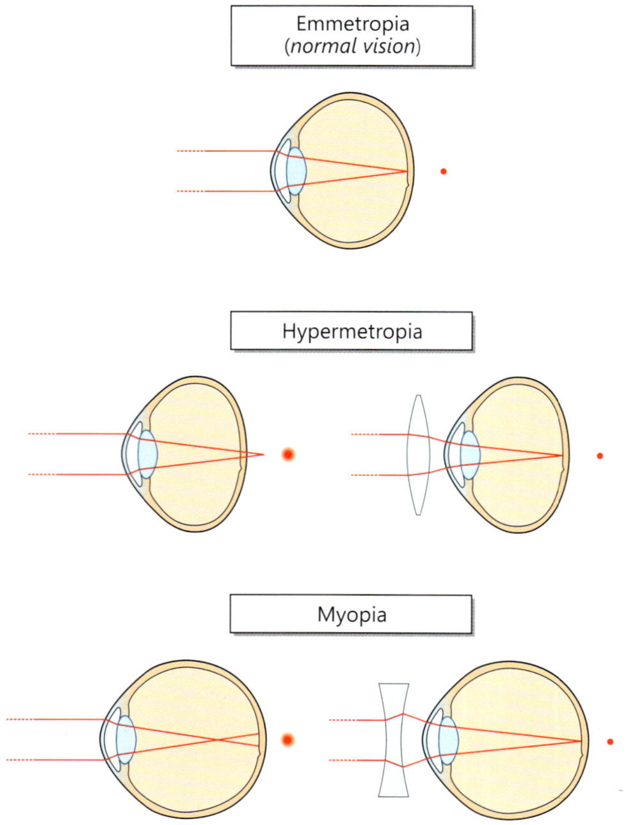

In an **emmetropic eye** accommodated for an object at infinity, the focus of the dioptric system of the eye is located on the retina, where the parallel light rays (coming from infinity) converge to form a focused image. In **hypermetropia**, the dioptric power of the eye is not sufficient to make the parallel rays converge on the retina (generally because the anteroposterior length of the eyeball is reduced with respect to the norm). On the contrary, in the **myopic eye** the excessive length of the eyeball makes the rays converge in front of the retina. In both cases, therefore, in the eye accommodated for infinity, the images that form on the retina are out of focus. Hypermetropia can be corrected with convergent lenses (and myopia with divergent lenses) whose dioptric power, added to (or subtracted from) that of the eye's dioptric system, brings the image into focus on the retina. Convergent lenses are also called positive (their dioptric power adds to that of the eye), while divergent lenses are termed negative (their dioptric power subtracts from that of the eye).

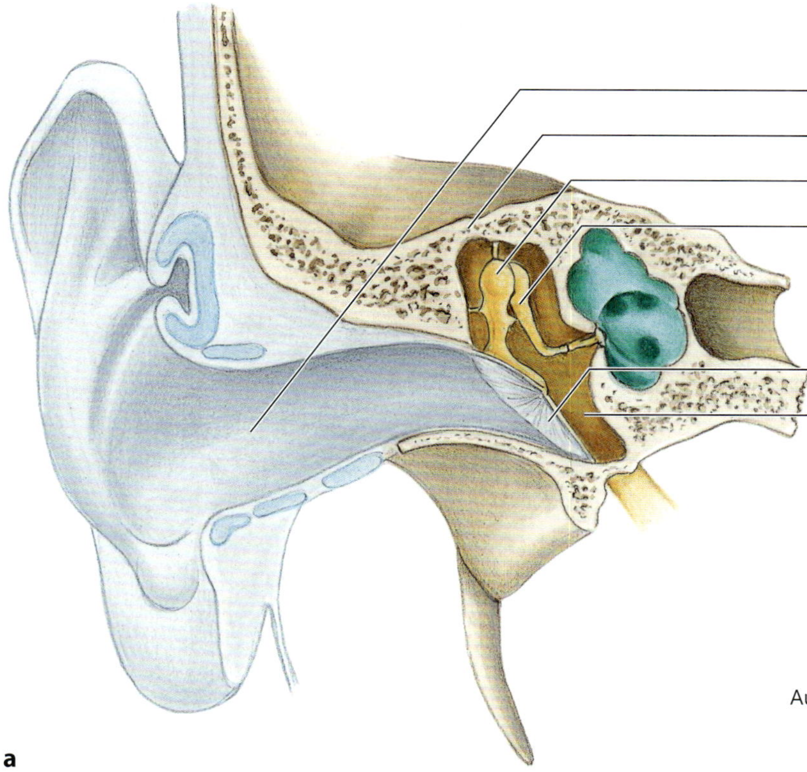

a

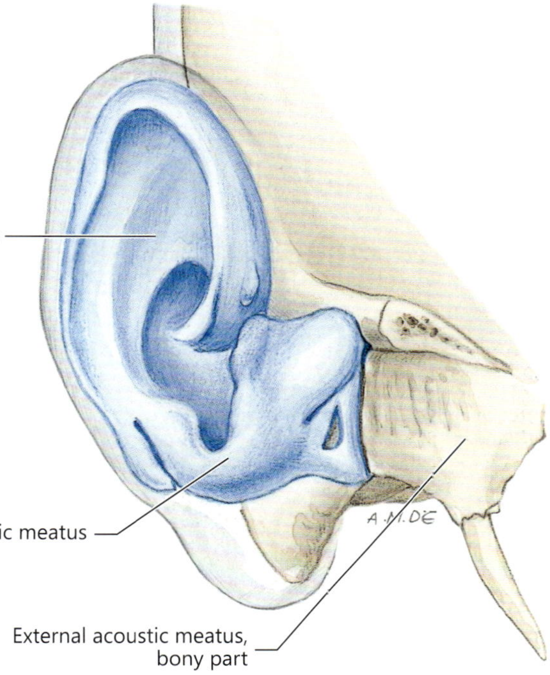

b

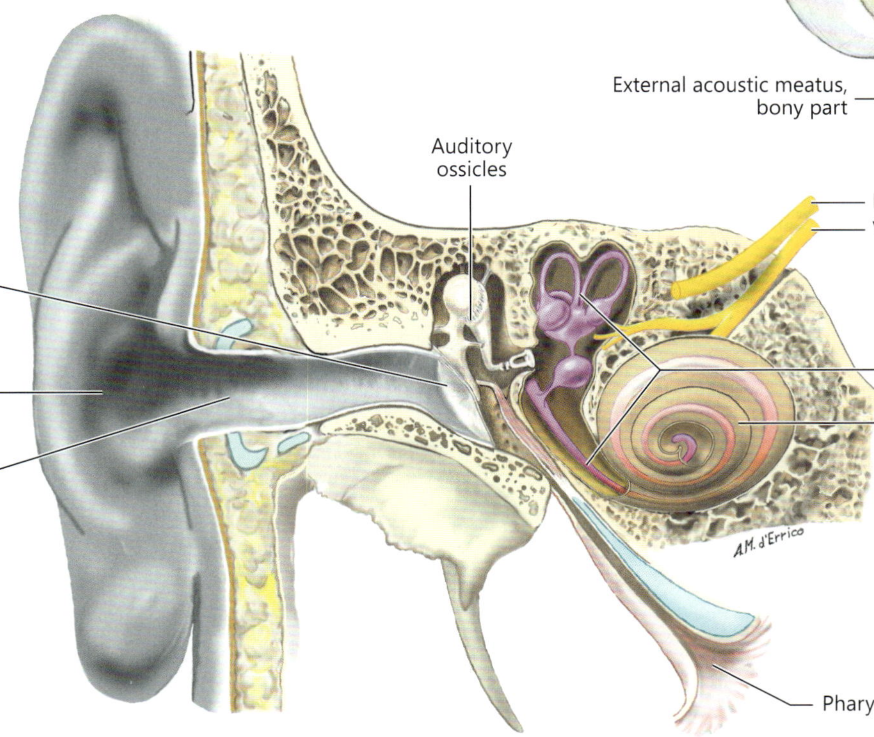

c

Structure of the **auditory system**. **a**, In *blue*, the **external ear**; in *yellow*, the **middle ear**; in *green*, the **internal ear**. **b**, Indicated in *blue* are the **cartilaginous parts of the external ear**: the auricle and the cartilaginous external acoustic meatus, which proceeds into a segment consisting of a bony part. **c**, **Internal ear**: vestibulocochlear organ. Indicated in *purple* are the **vestibular** and **cochlear labyrinths**.

For a three-dimensional image illustrating the membranous labyrinth, see volume 1, page 383.

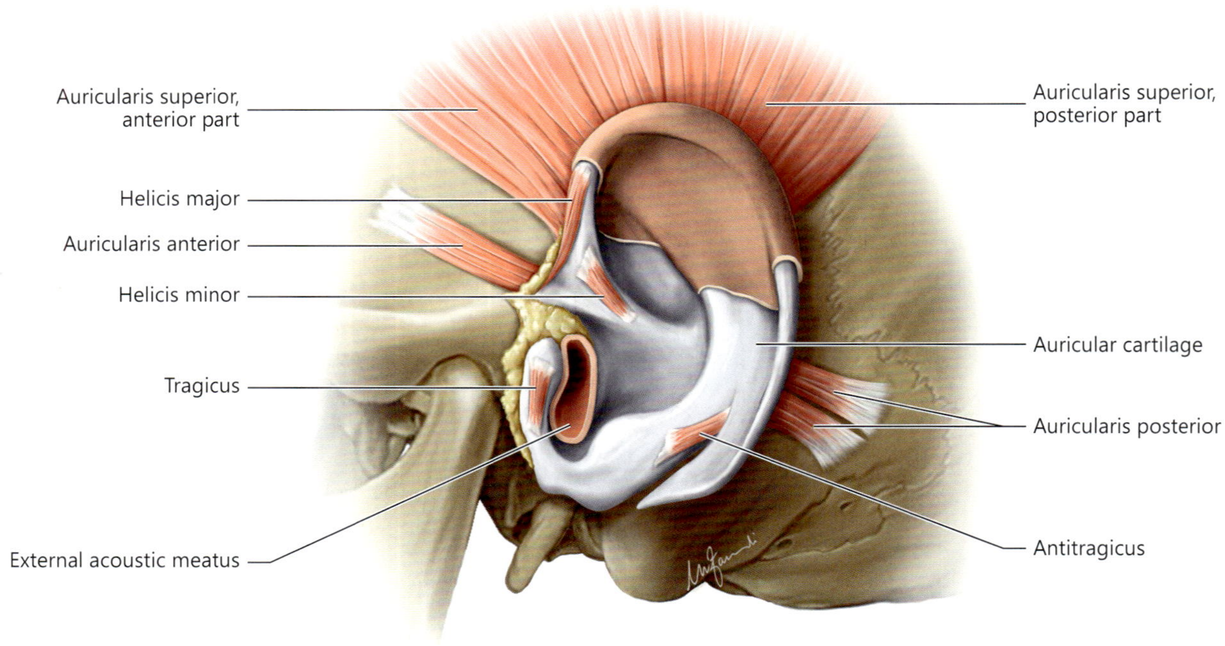

Left ear, external lateral view. Visible beneath the skin are the muscles and cartilage of the **auricle**, in particular the auricularis muscles, helicis muscles, tragicus and antitragicus.

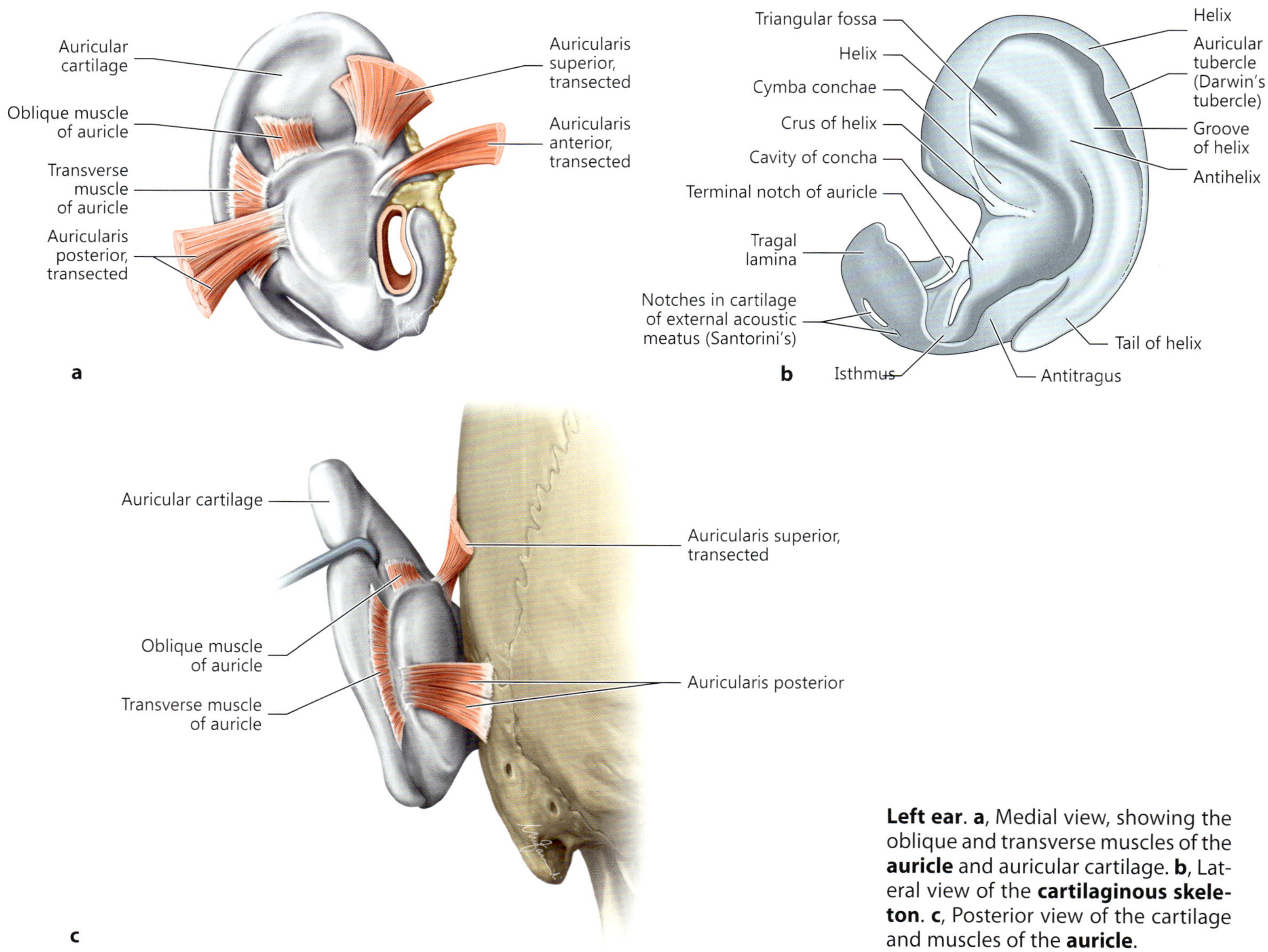

Left ear. **a**, Medial view, showing the oblique and transverse muscles of the **auricle** and auricular cartilage. **b**, Lateral view of the **cartilaginous skeleton**. **c**, Posterior view of the cartilage and muscles of the **auricle**.

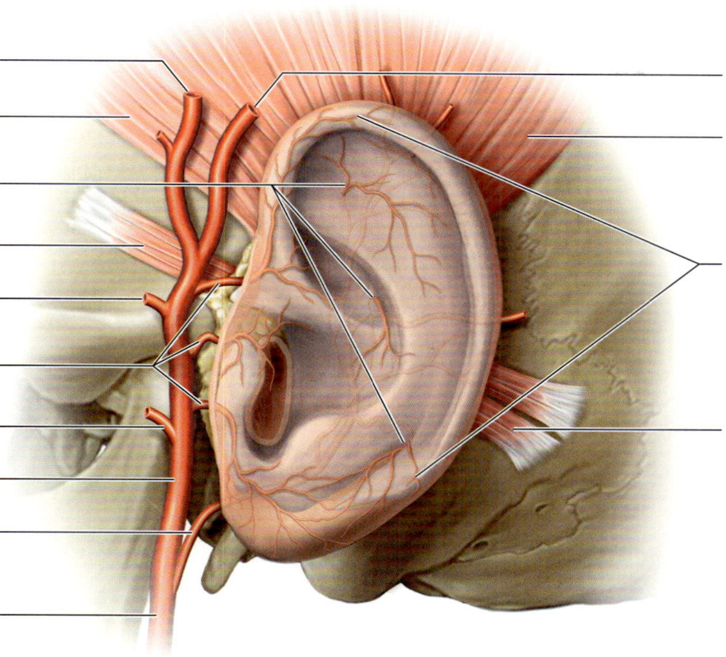

Arteries of the auricle, lateral view of the left external ear. Anterior auricular branches of the **superficial temporal artery**, a branch of the external carotid artery, supply the anterior half of the lateral surface of the auricle. The medial surface and posterior half of the lateral surface of the auricle are supplied by the **posterior auricular artery**, a branch of the external carotid artery. Anastomoses are abundant among the different arterial branches supplying the auricle.

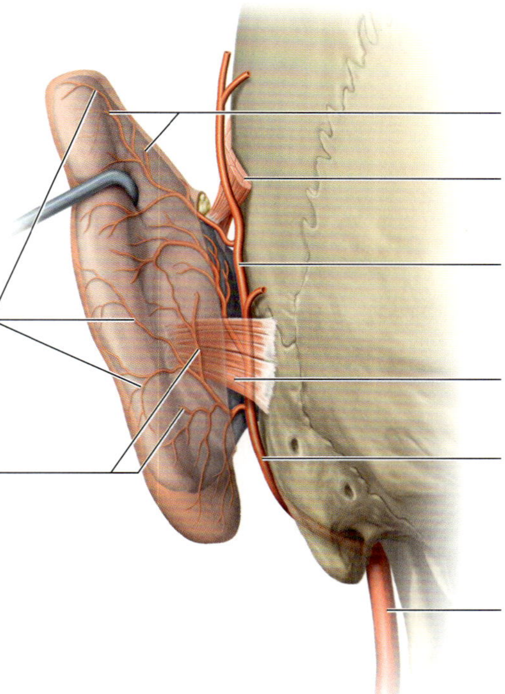

Arteries of the auricle, posterior view of the left external ear. The **posterior auricular artery**, a branch arising from the external carotid artery, ascends to the groove between the auricle and mastoid process of the temporal bone where it continues as an **auricular branch of the posterior auricular artery**, supplying the territory of the medial surface and posterior half of lateral surface of the auricle.

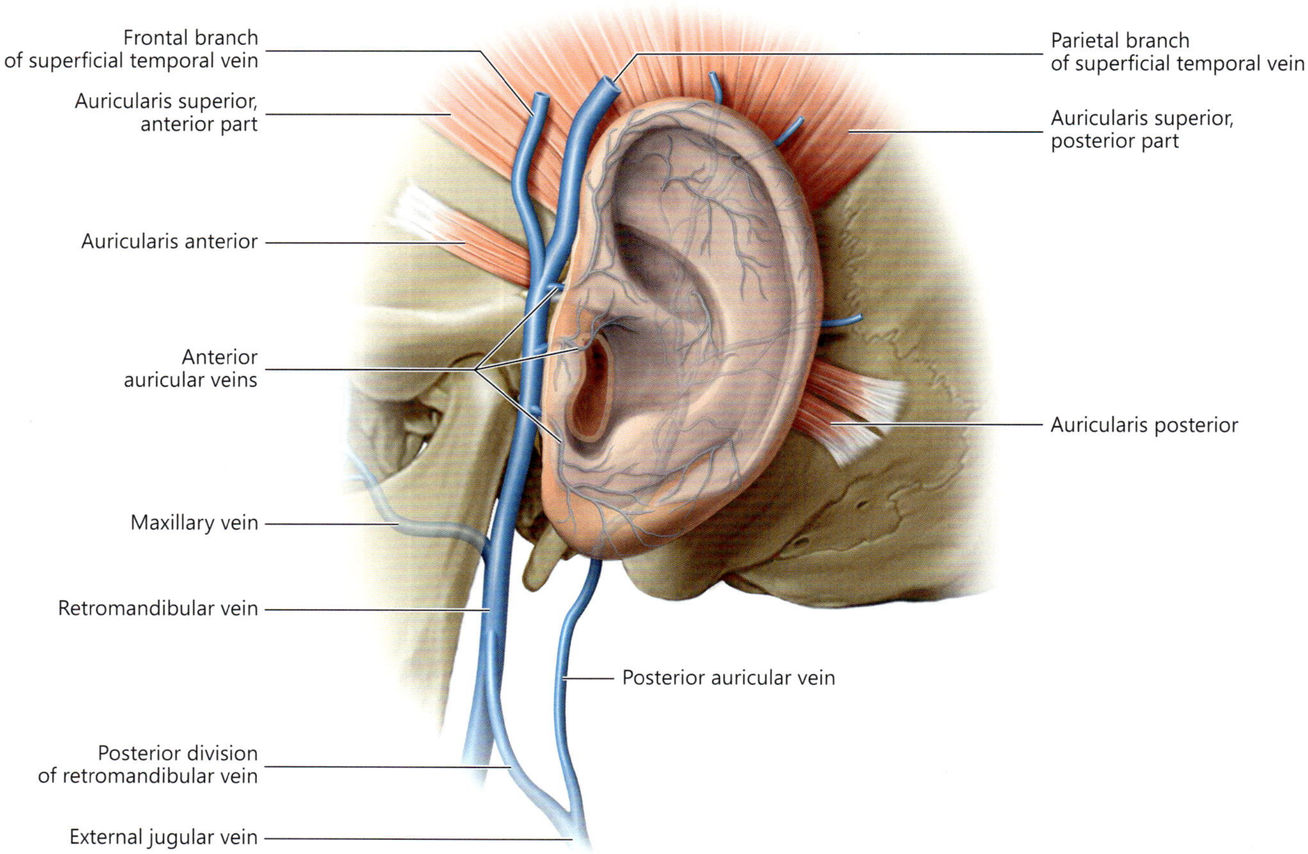

Veins of the auricle, lateral view of the left external ear. The **anterior auricular veins** drain into the **superficial temporal vein**, which unites with the **maxillary vein** forming the **retromandibular vein** or **posterior facial vein**.

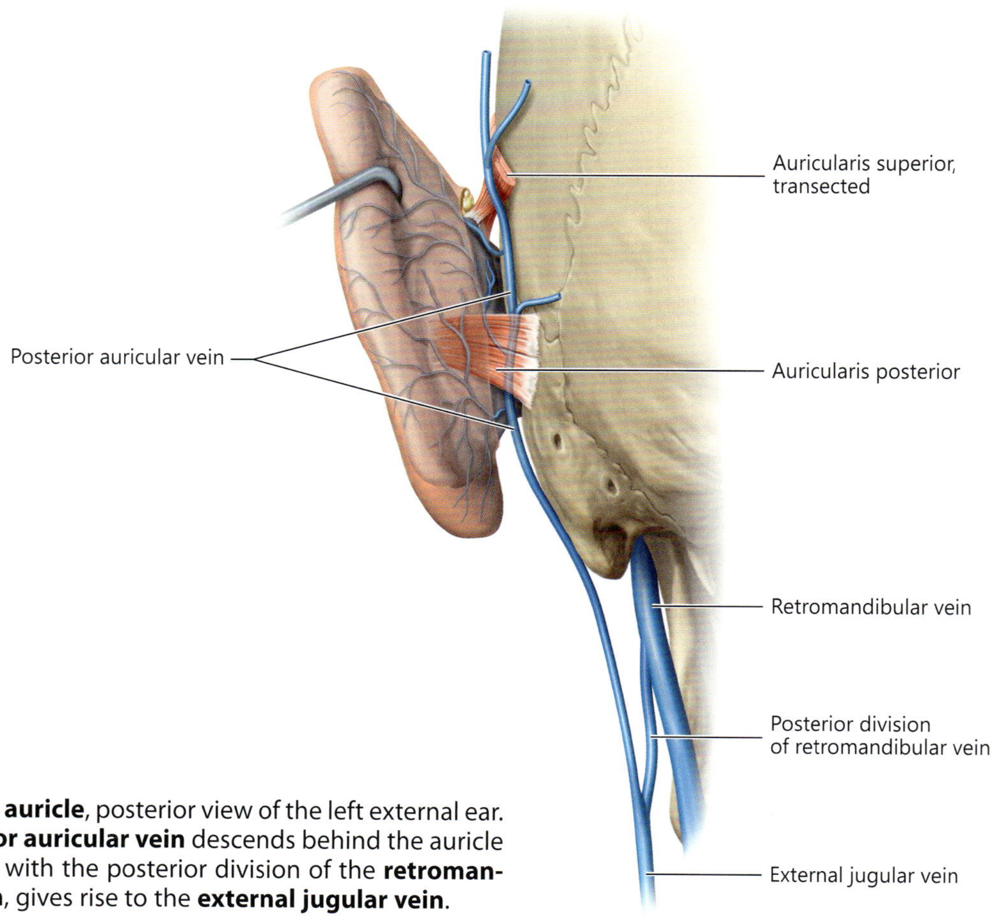

Veins of the auricle, posterior view of the left external ear. The **posterior auricular vein** descends behind the auricle and, uniting with the posterior division of the **retromandibular vein**, gives rise to the **external jugular vein**.

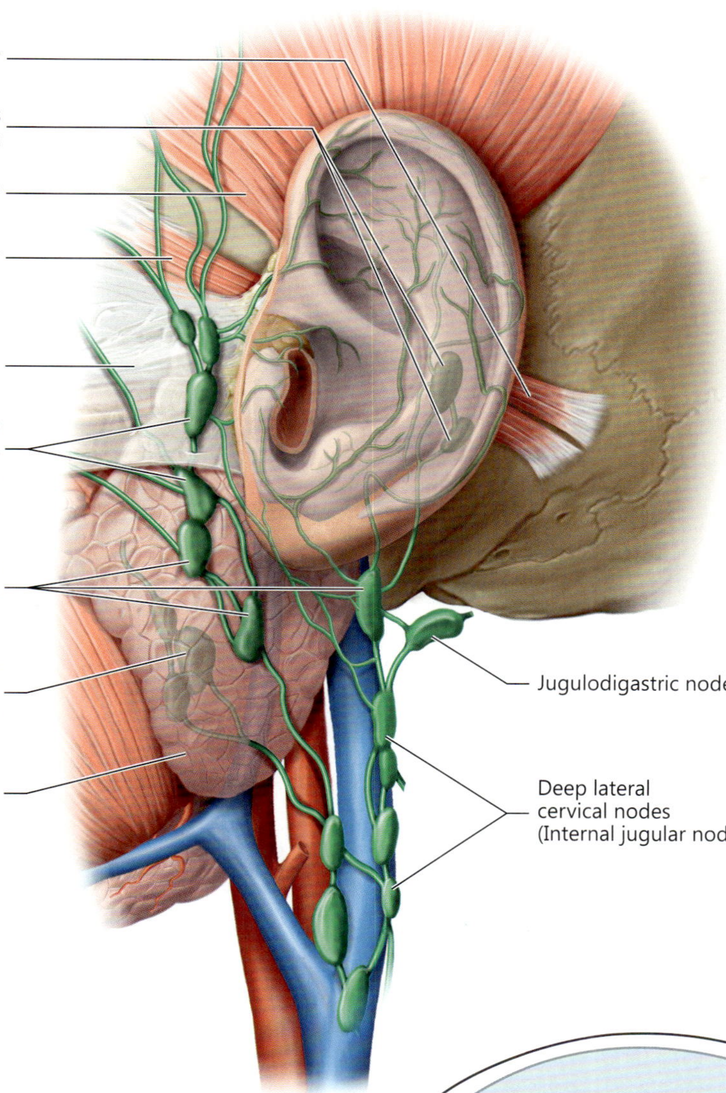

Lymphatic drainage of the auricle. The lymphatic network of the medial surface of the auricle drains into an **infra-auricular node**, belonging to the **group of deep parotid nodes**, situated in contact with the internal jugular vein, in front of the sternocleidomastoid. The lymphatic network of the lateral surface anterior to the concha of the auricle drains into the **preauricular superficial** and **deep parotid nodes**, above and below the concha into the **mastoid nodes**, while at the level of the ear lobe it drains into the **infra-auricular deep parotid nodes**.

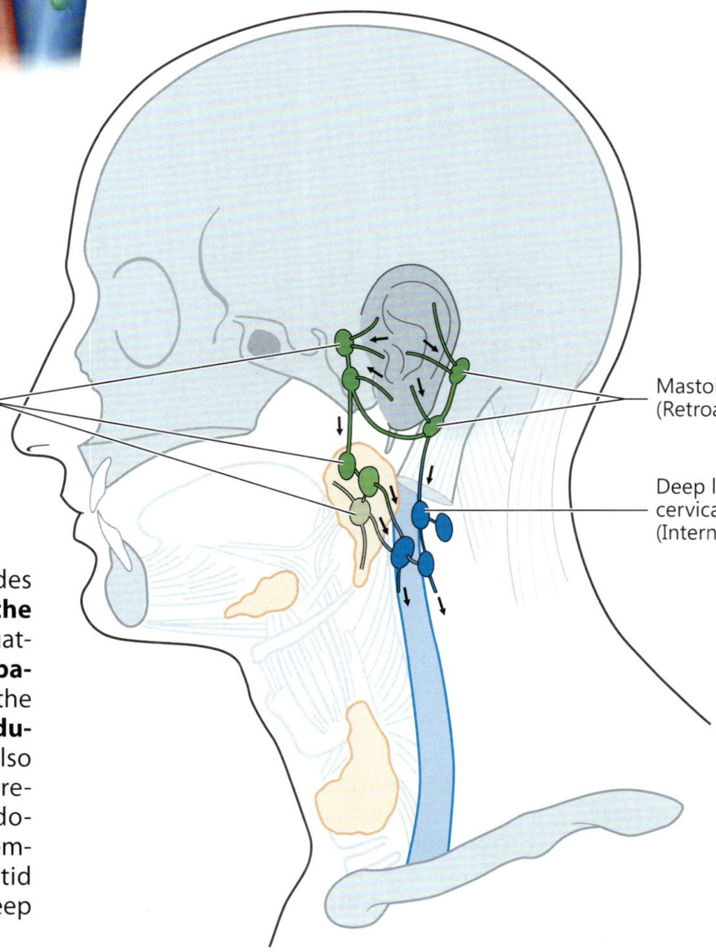

Schematic representation of the lymph nodes associated to the **lymphatic networks of the auricle**. The **superficial parotid nodes**, situated above the parotid fascia, and the **deep parotid nodes**, subfascial, are subdivided into the **preauricular**, **infra-auricular** and **intraglandular nodes**. The **mastoid nodes**, including also the retroauricular nodes, are located in correspondence to the origin of the sternocleidomastoid from the mastoid process of the temporal bone. The superficial and deep parotid nodes and mastoid nodes drain into the deep lateral cervical, or internal jugular, nodes.

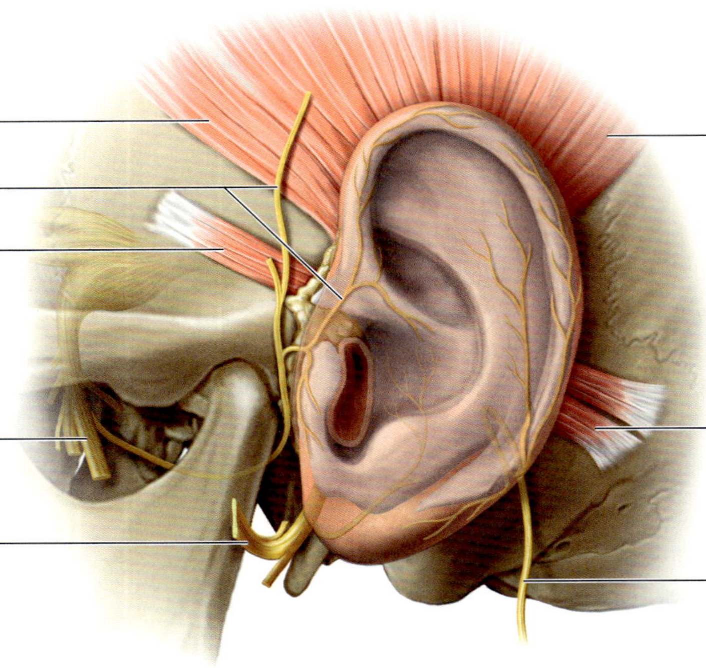

The **auriculotemporal nerve**, a branch of the mandibular nerve [V_3], after passing completely through the parotid gland, exchanging anastomoses with the parotid plexus of the facial nerve, runs upwards between the external acoustic meatus and superficial temporal vessels, supplying the anterior part of the auricle and skin of the temporal region.

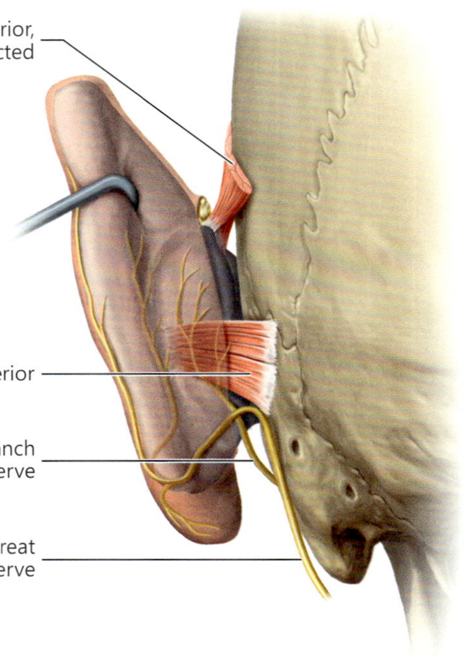

The **great auricular nerve** [C2, C3] is a cutaneous branch of the cervical plexus that runs behind and below the sternocleidomastoid and then winds around the posterior margin and ascends on its lateral surface up to the auricle, whose skin it innervates. Along its course it gives off sensory branches that anastomose with the facial nerve [VII] and cutaneous branches for the parotid region.

Schematic representation showing the **innervation of the auricle**. In addition to the **auriculotemporal** and **great auricular nerves**, also the **facial nerve** [VII], with the auricular branch of the posterior auricular nerve, innervates the auricle through fibers that reach, in particular, the concha and initial part of the external acoustic meatus. The skin of the concha of the auricle is innervated not only by the facial nerve but also by the **vagus** [X] and **glossopharyngeal** [IX] **nerves**, through the auricular branch of the vagus nerve and the communicating branch of the glossopharyngeal nerve with the auricular branch of the vagus nerve.

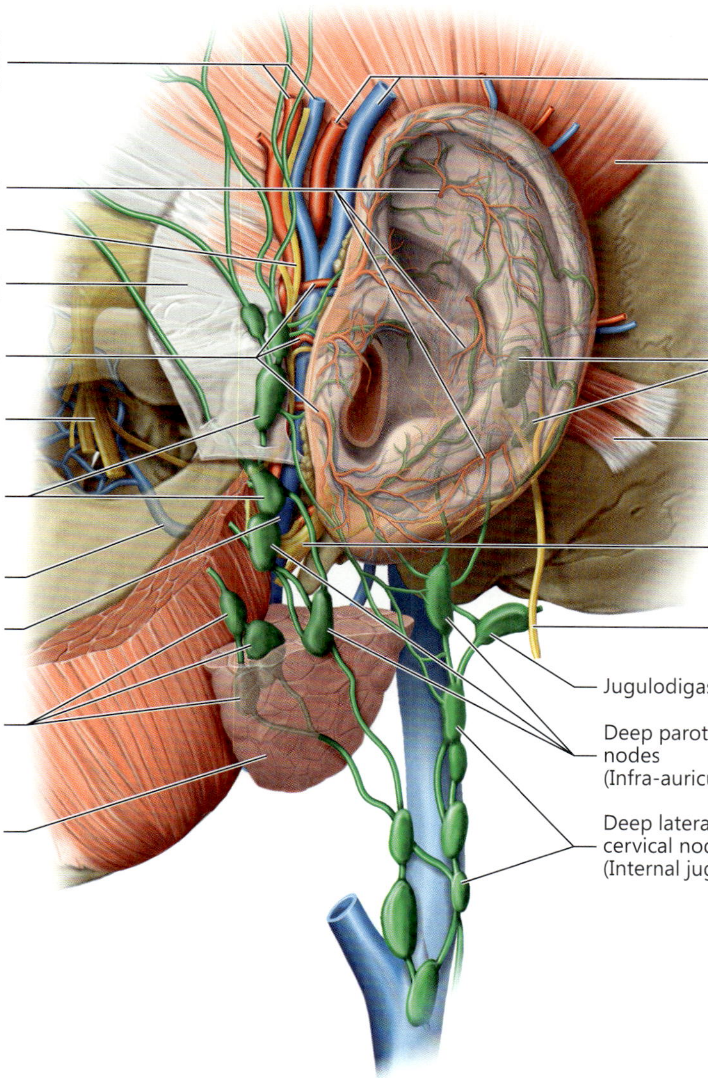

Vascularization, innervation and lymphatic drainage of the **external ear**, left lateral view of the lateral surface of auricle. The arterial supply of the auricle is provided principally by the superficial temporal and posterior auricular arteries; the veins are satellites of the arteries, and lymphatic drainage is towards the parotid and mastoid nodes. Sensory innervation is provided by the great auricular and auriculotemporal nerves, the latter a branch of the mandibular nerve [V_3], with contributions from the facial [VII], vagus [X] and glossopharyngeal [IX] nerves.

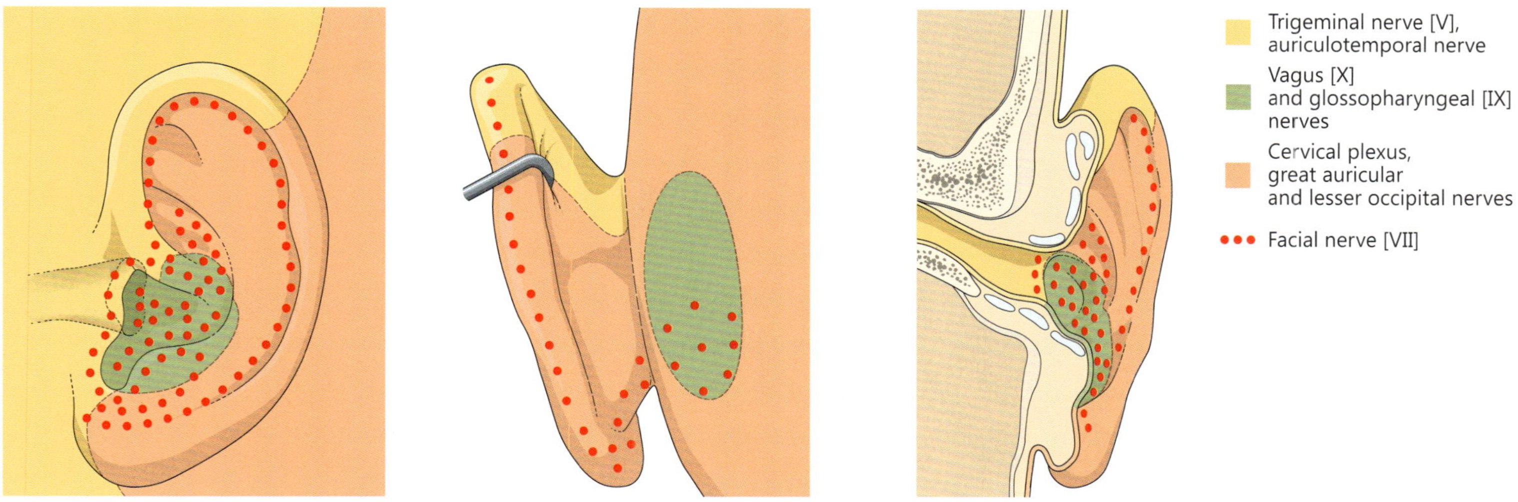

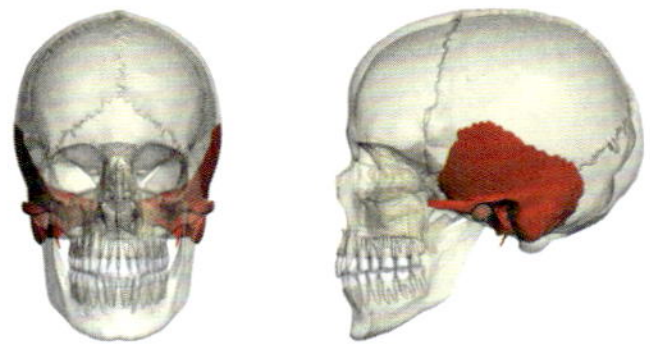

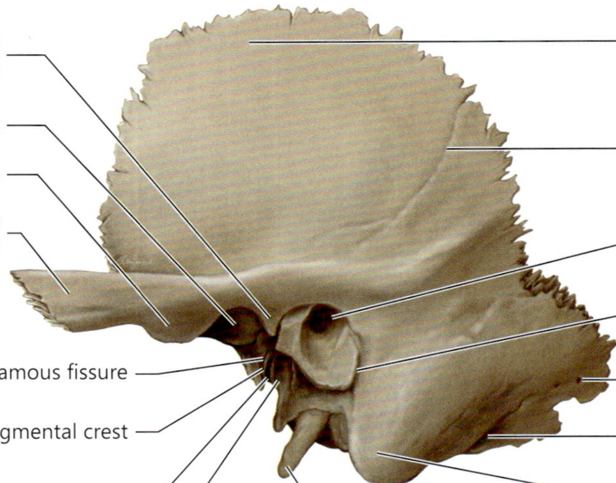

Left **temporal bone**, lateral surface of the extracranial surface.

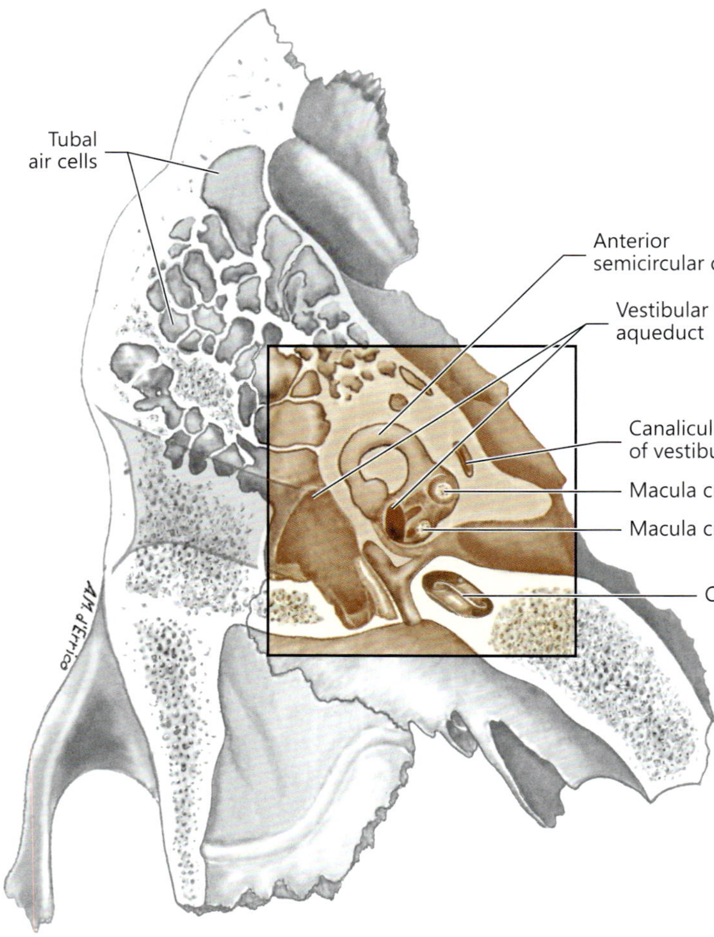

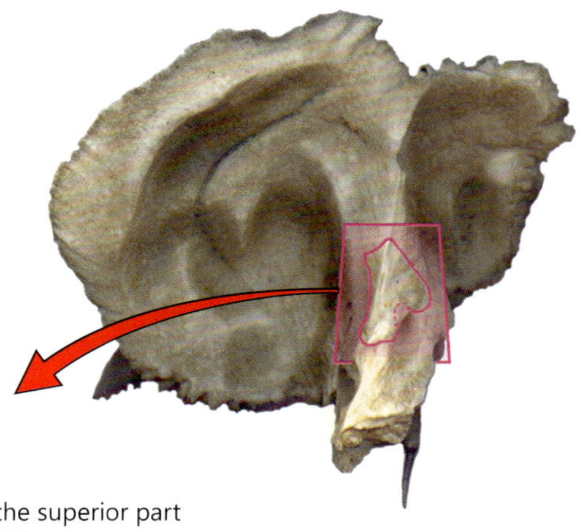

Transverse section of the **petrous part of the temporal bone** at the level of the **vestibule**. The medial wall has a cribriform appearance due to the passage of filaments of the utriculoampullary nerve (through the macula cribrosa superior) and saccular nerve (through the macula cribrosa media), and of branches of the vestibular nerve [VIII].

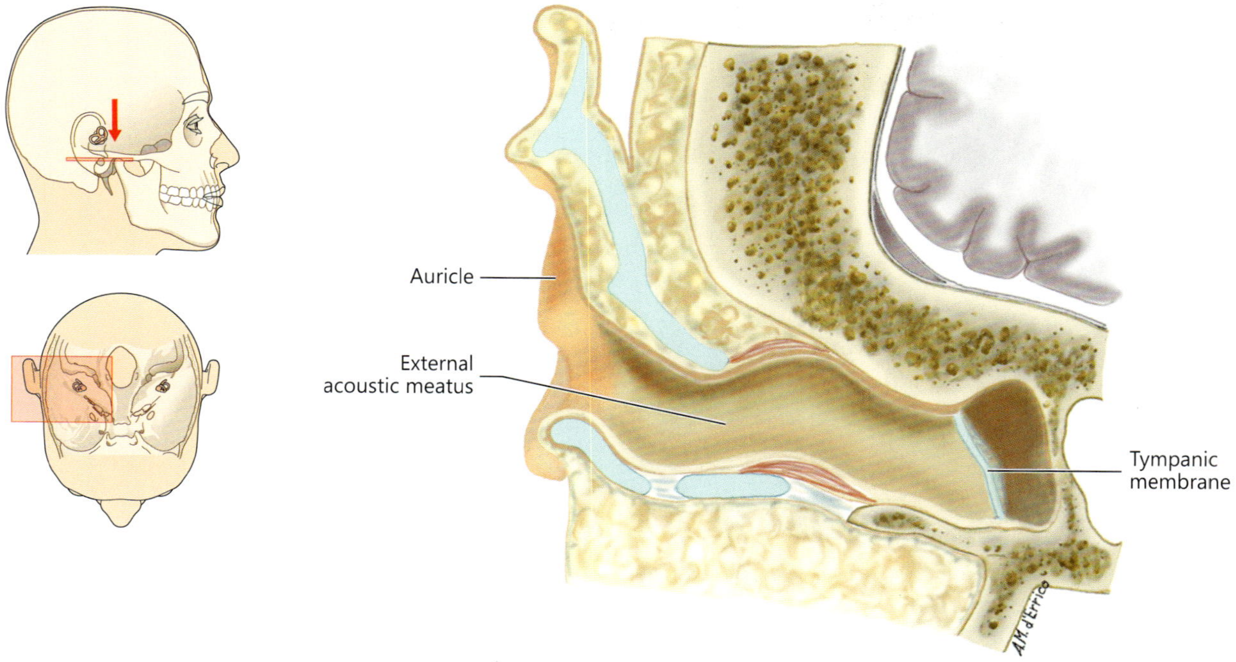

External acoustic meatus of the right ear in transverse section.

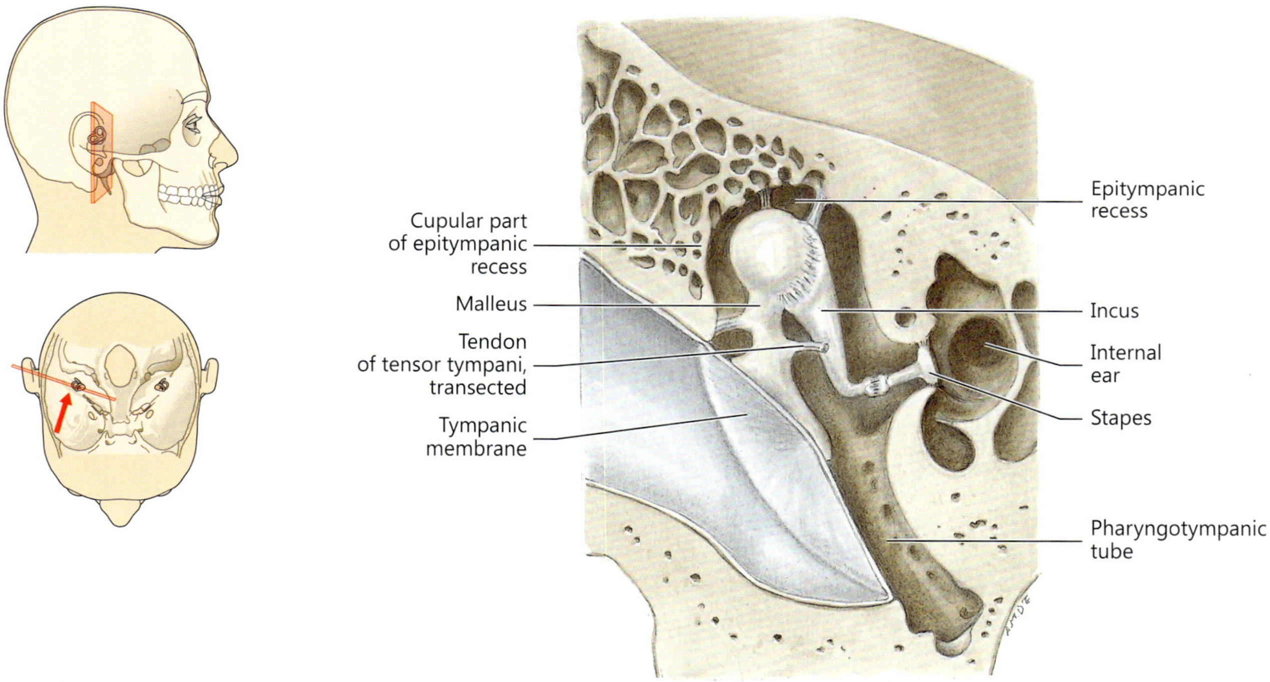

Right middle ear viewed from in front, in an oblique frontal section. The **tympanic cavity** is occupied in part by the ossicles.

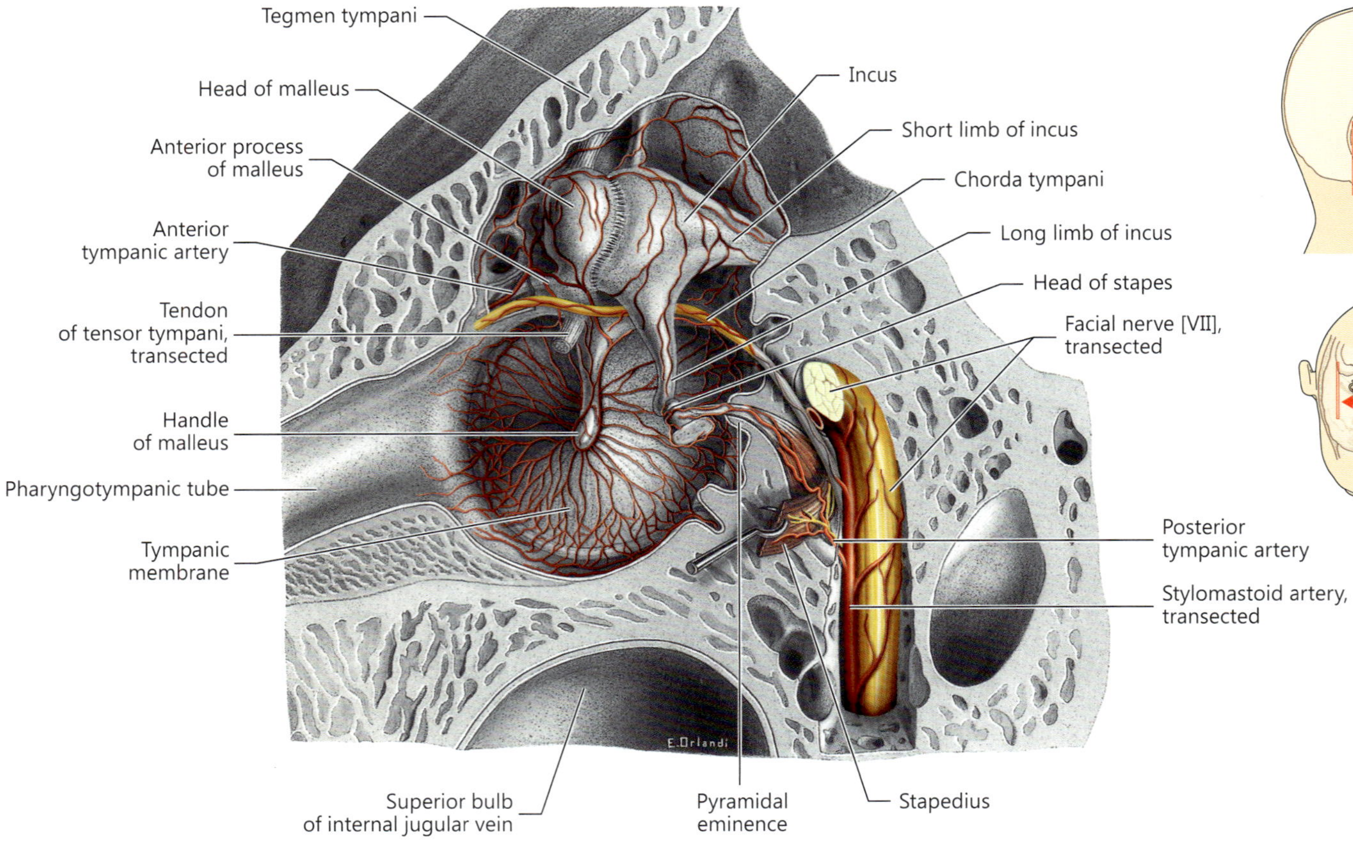

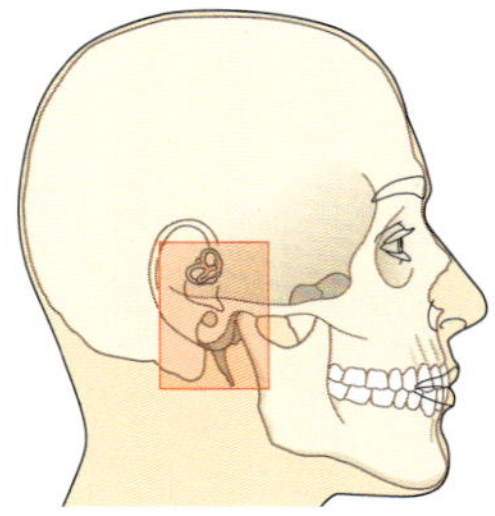

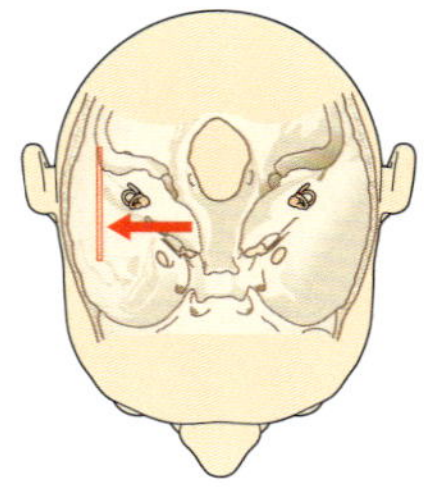

Right tympanic cavity transected sagittally: lateral half of the section viewed from the interior. Blood is supplied to the mucous surface of the tympanic membrane by branches of the **anterior tympanic artery**, a branch of the maxillary artery, and of the **stylomastoid artery**, a branch of the posterior auricular artery.

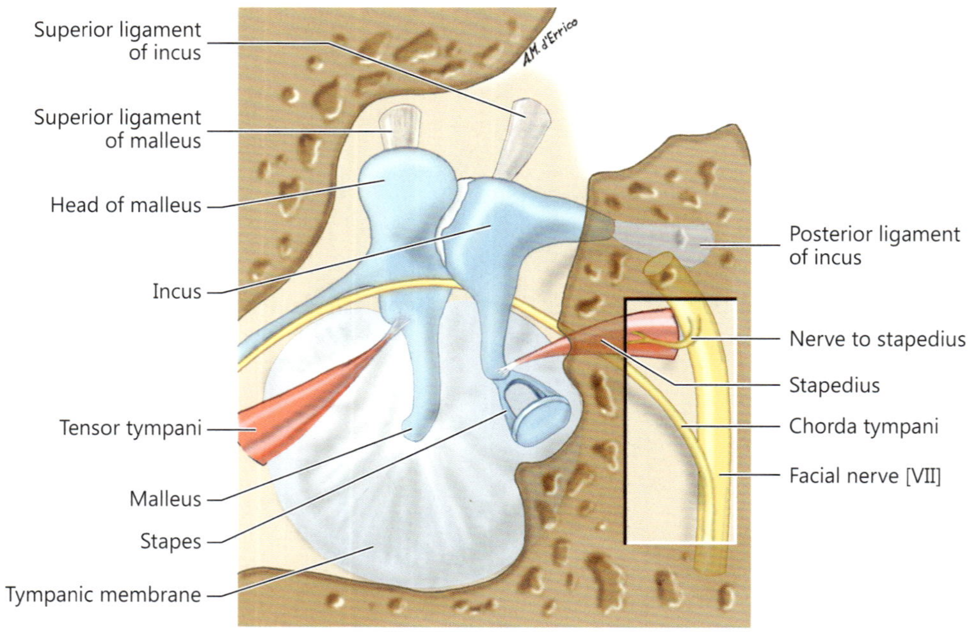

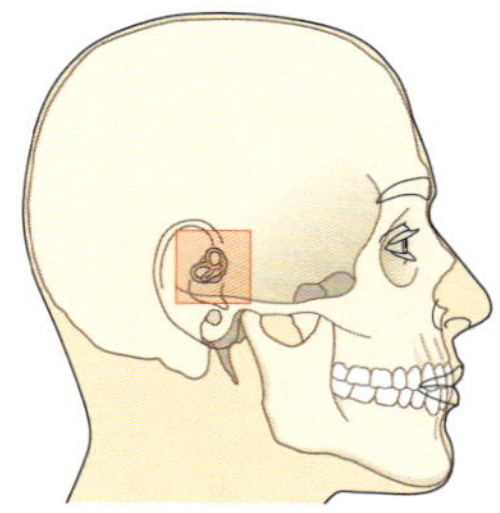

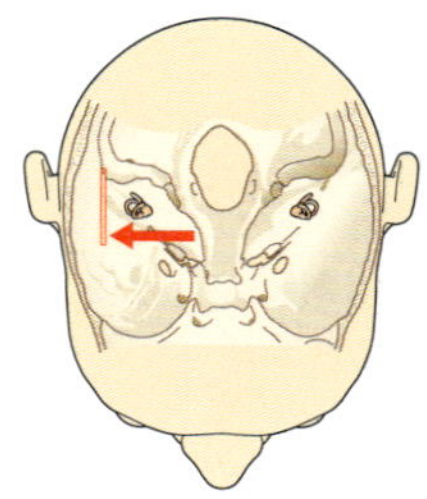

Schematic representation showing the right **tympanic cavity** viewed from the interior.

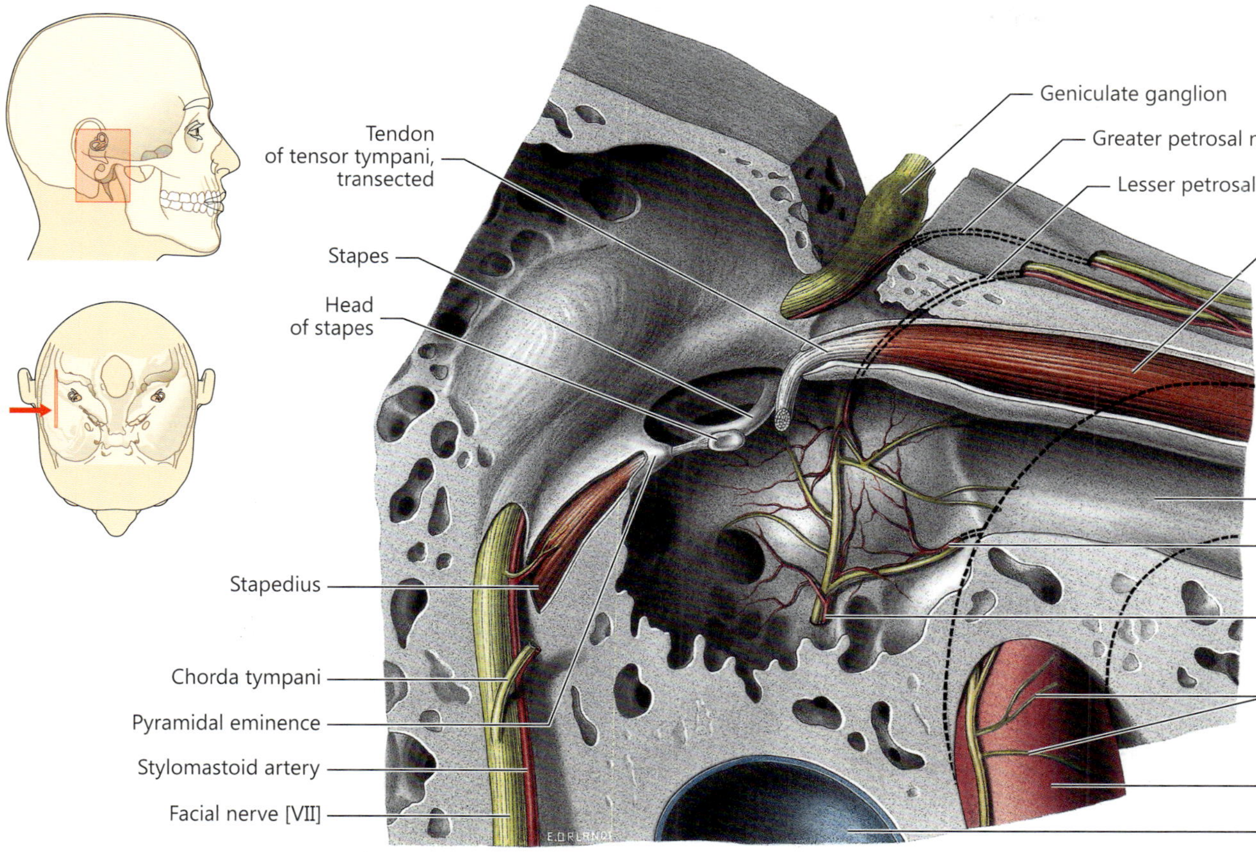

Arterial blood supply of the right **tympanic cavity**. The medial part of the cavity is viewed in a sagittal section of the middle ear. One can see the distribution of the various arterial branches that supply the mucosa of the tympanic cavity.

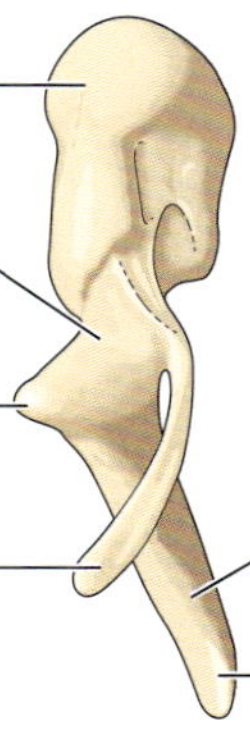

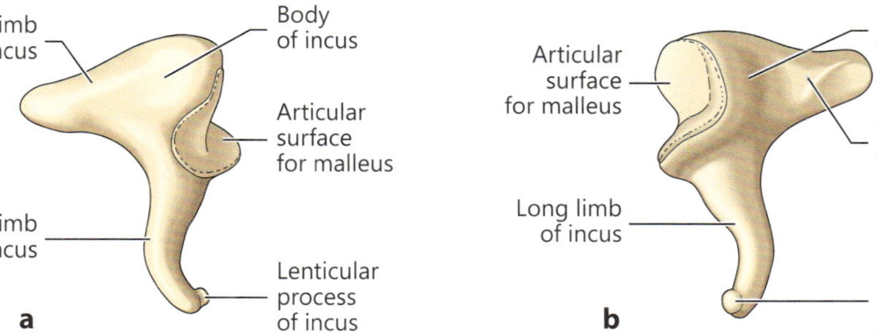

Incus of the auditory ossicles of the right middle ear, lateral (**a**) and medial (**b**) views.

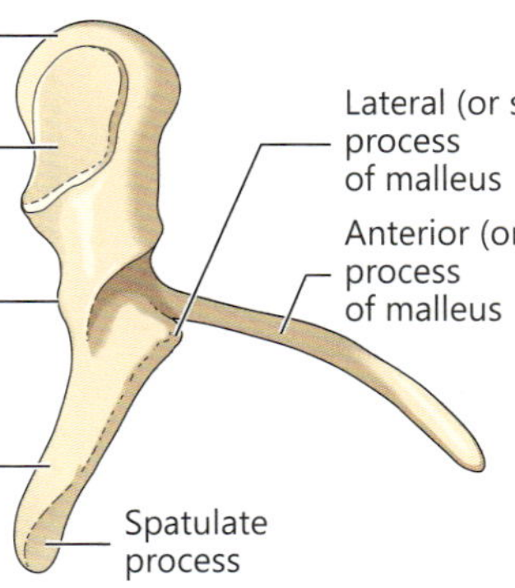

Malleus of the auditory ossicles of the right middle ear, anterior (**a**) and lateral (**b**) views.

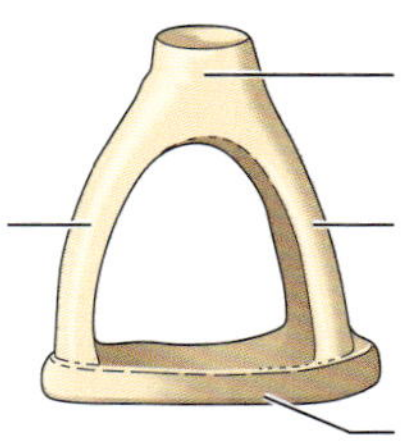

Stapes of the auditory ossicles of the right middle ear, viewed from above.

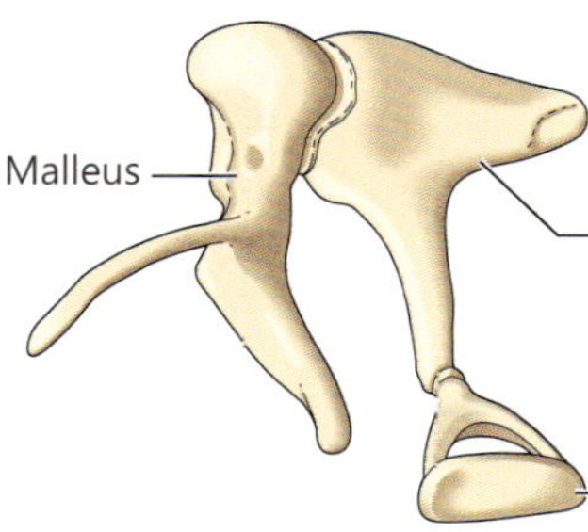

Auditory ossicles of the right middle ear, medial view. One can see the relationship between the three ossicles when they are placed in reciprocal articulation.

Muscles of the auditory ossicles					
MUSCLE	**ORIGIN**	**INSERTION**	**INNERVATION**	**ACTION**	**OBSERVATIONS**
Stapedius	Pyramidal eminence of the mastoid wall of the tympanic cavity	Posterior part of the head of the stapes	Facial nerve [VII]	Retracts the head of the stapes with a consequent slight inclination of its base. Function: reduces tension of the tympanic membrane	Smallest skeletal muscle in the human body
Tensor tympani	Bony canal above the pharyngotympanic tube, cartilaginous part of the pharyngotympanic tube and greater wing of the sphenoid	Root of the handle of the malleus	Facial nerve [VII]	Pulls the handle of the malleus medially. Function: increases tension of the tympanic membrane	

For the functional anatomy of the mechanisms of sound transmission ☞ page 391, volume 1

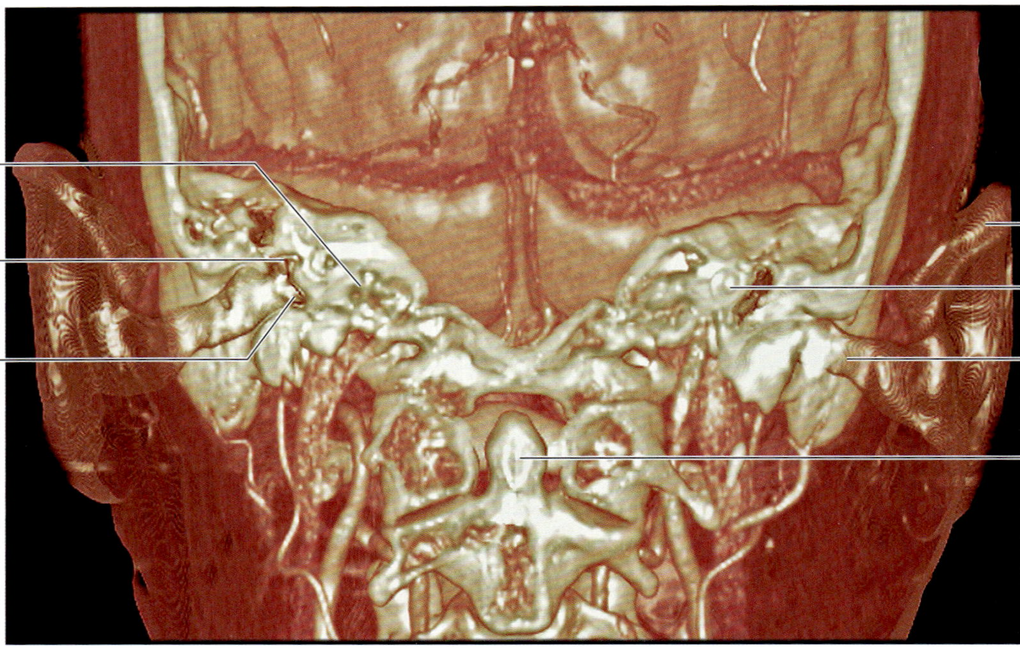

Three-dimensional image of a living human body, obtained by CT scan using a 3D volume rendering technique, showing, in a coronal section of the head viewed posteriorly, the structures of the **external**, **middle** and **internal ear**.

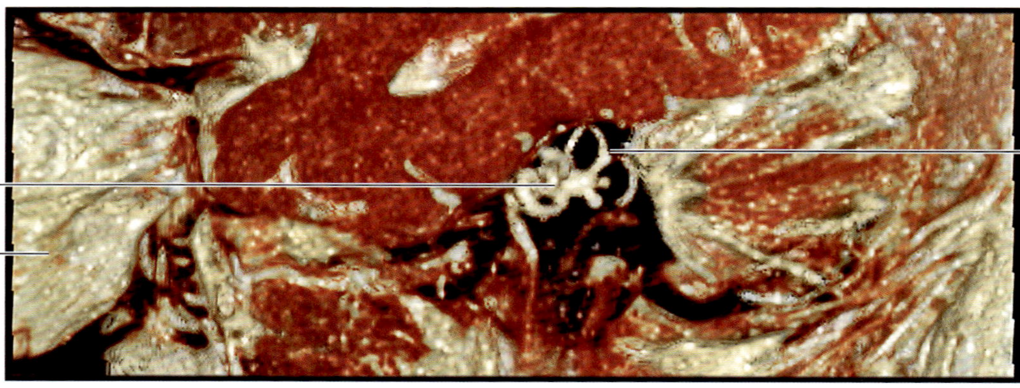

Three-dimensional image of a living human body, obtained by CT scan using a 3D volume rendering technique showing, in a tangential parasagittal section lateral to the eyeball, the **semicircular ducts** and **cochlear duct**.

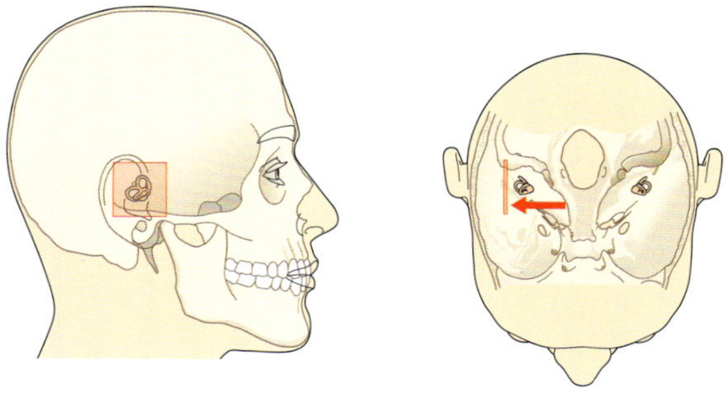

Transparent diagram showing the **bony labyrinth cavity** of the right ear in relation to the medial wall of the tympanic cavity. The lateral wall of the vestibule corresponds to the oval window.

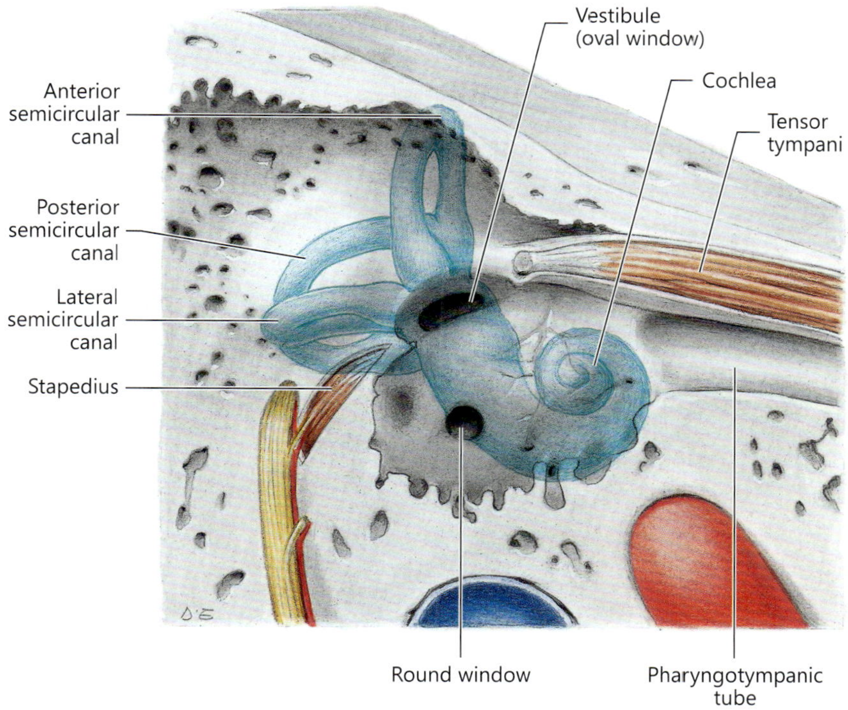

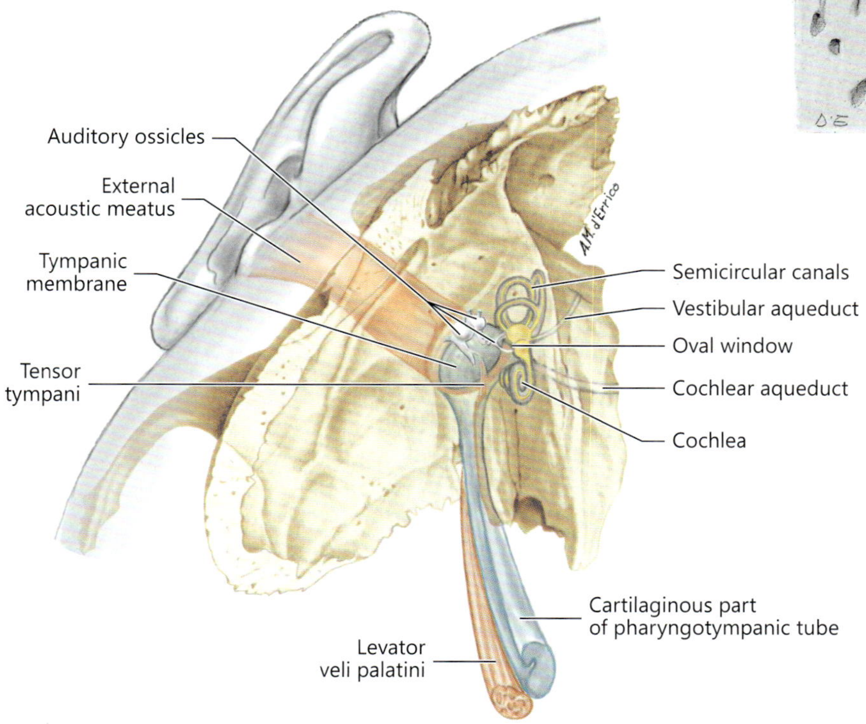

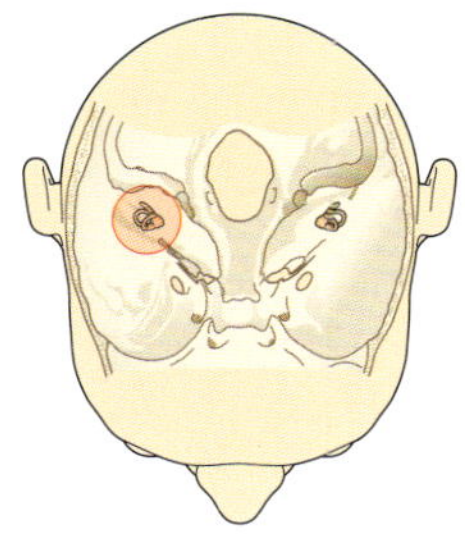

Components of the bony labyrinth of the right ear seen in relation to the **middle ear**, in superomedial view.

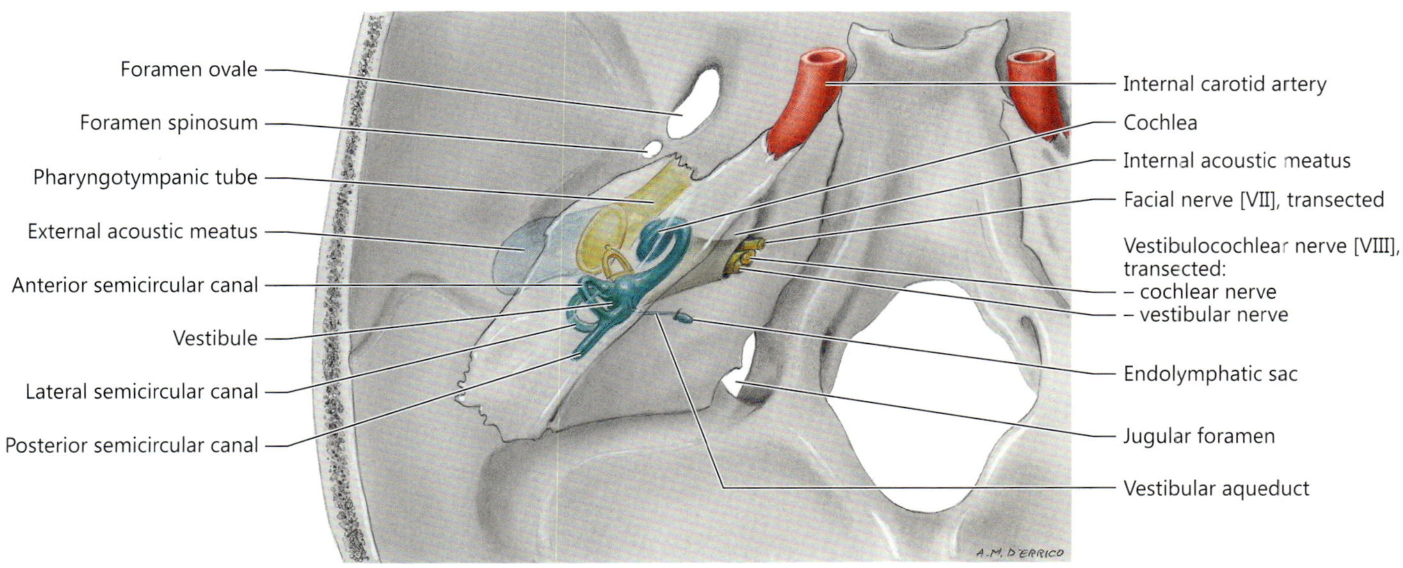

Topography of the anterior and posterior formations of the **internal ear** and their relationships with the other parts of the auditory system. Medially to the bony labyrinth one finds the internal acoustic meatus, which provides a passage for the vestibulocochlear [VIII] and facial [VII] nerves. For a three-dimensional image, see volume 1, page 381.

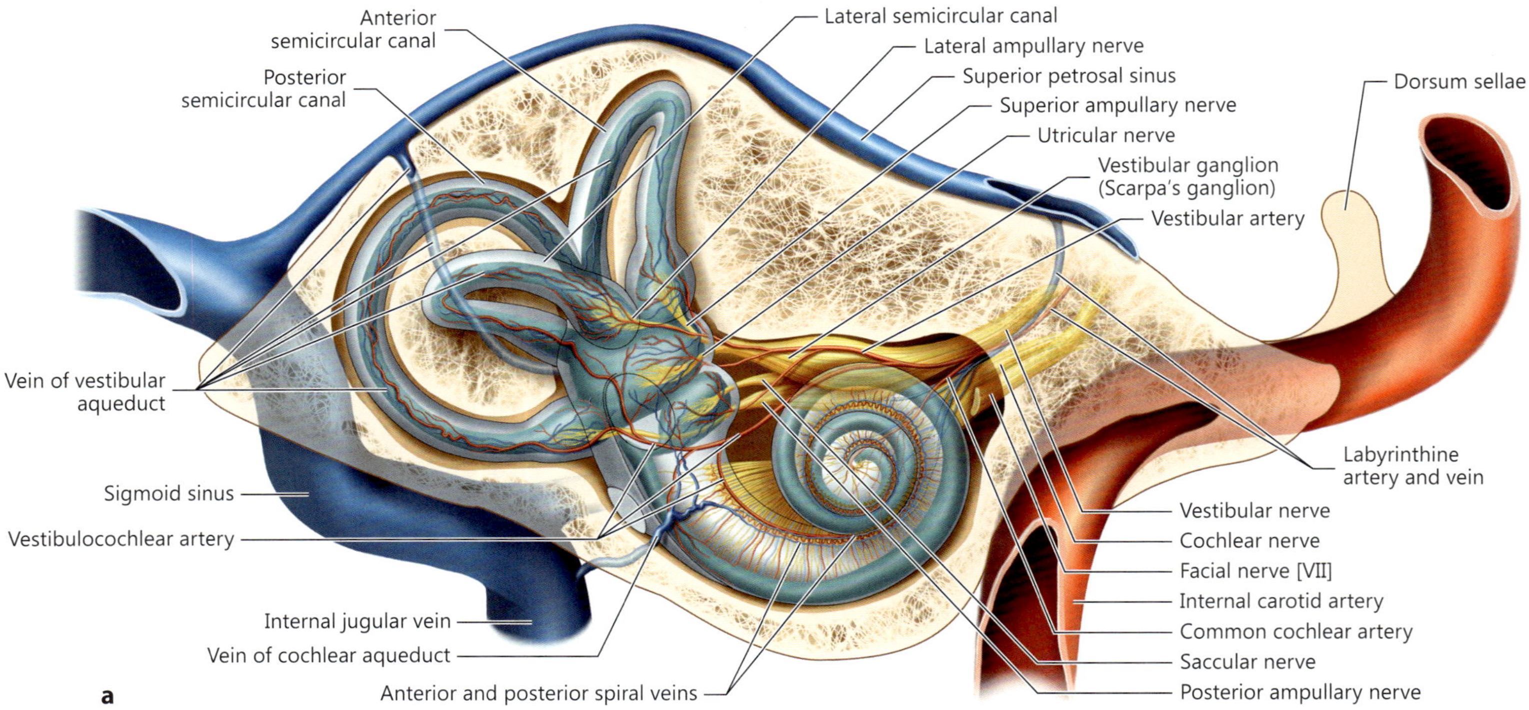

a

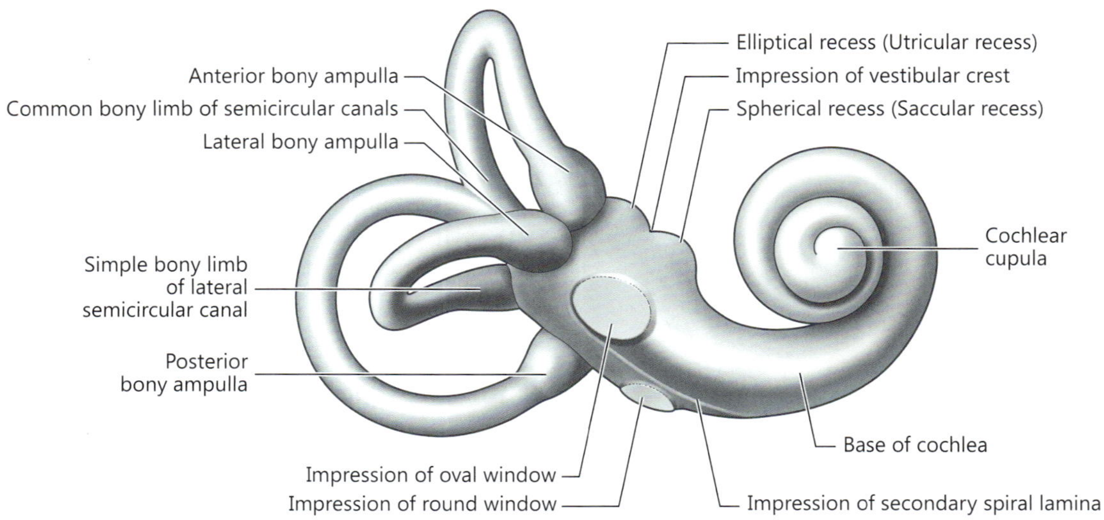

b

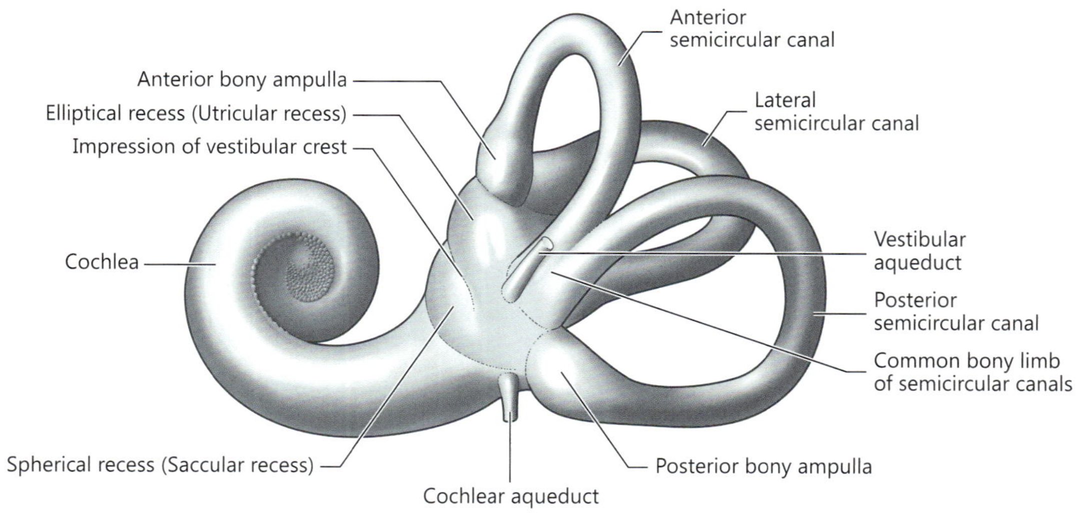

c

Right **bony labyrinth**. **a**, View from the exterior. Part of the bony structure has been demolished to display the vestibular and cochlear labyrinths. **b**, Lateral view. The cochlea, vestibule and semicircular canals follow one another in an anteroposterior succession. **c**, Medial view, showing the position of the vestibular and cochlear aqueducts.

Vestibular and cochlear labyrinths of the right **internal ear**. **a**, View from the external side, **b**, view from the internal side.

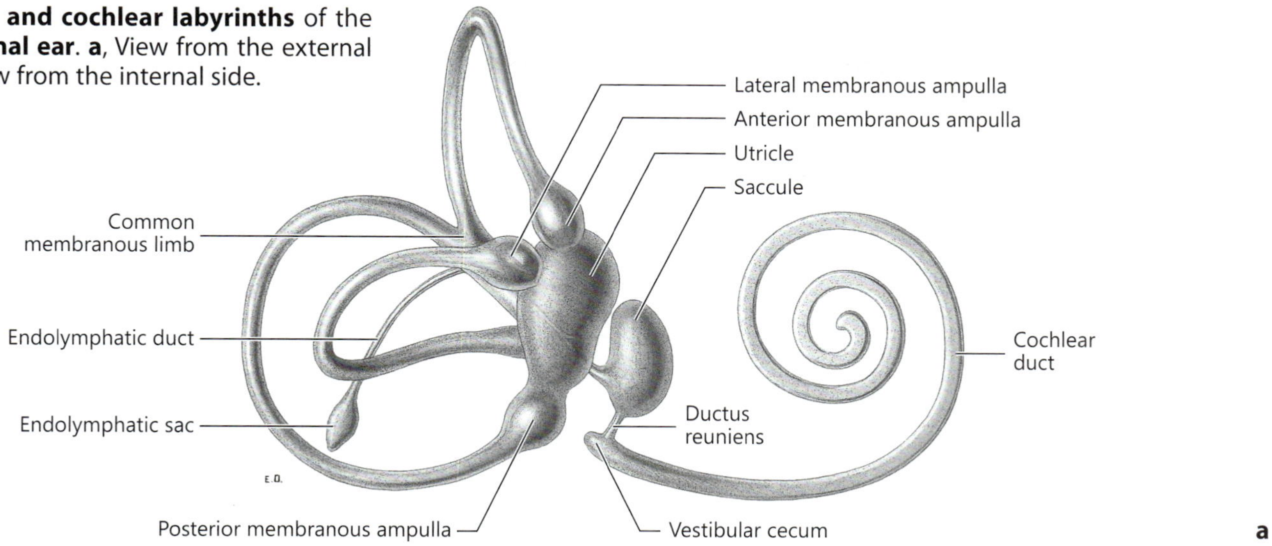

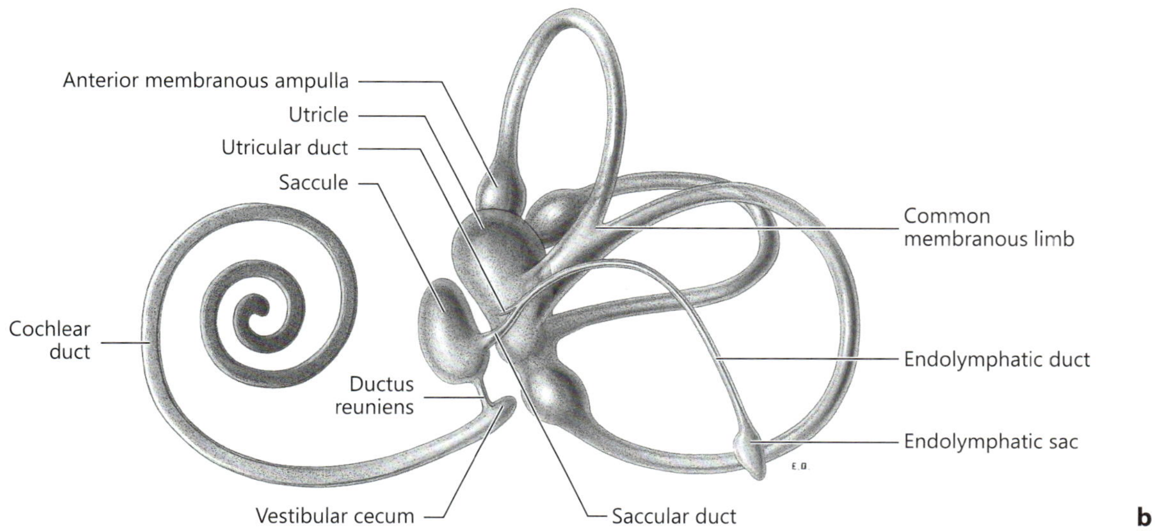

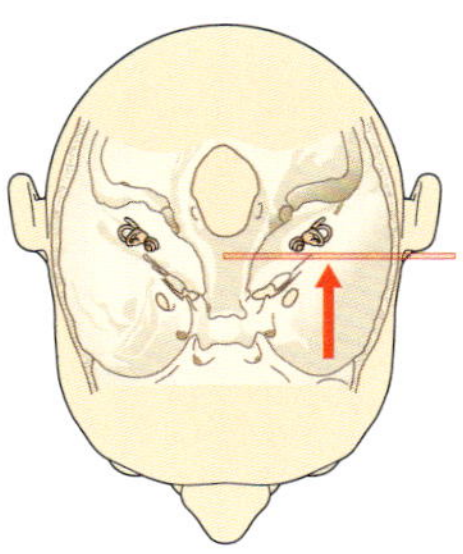

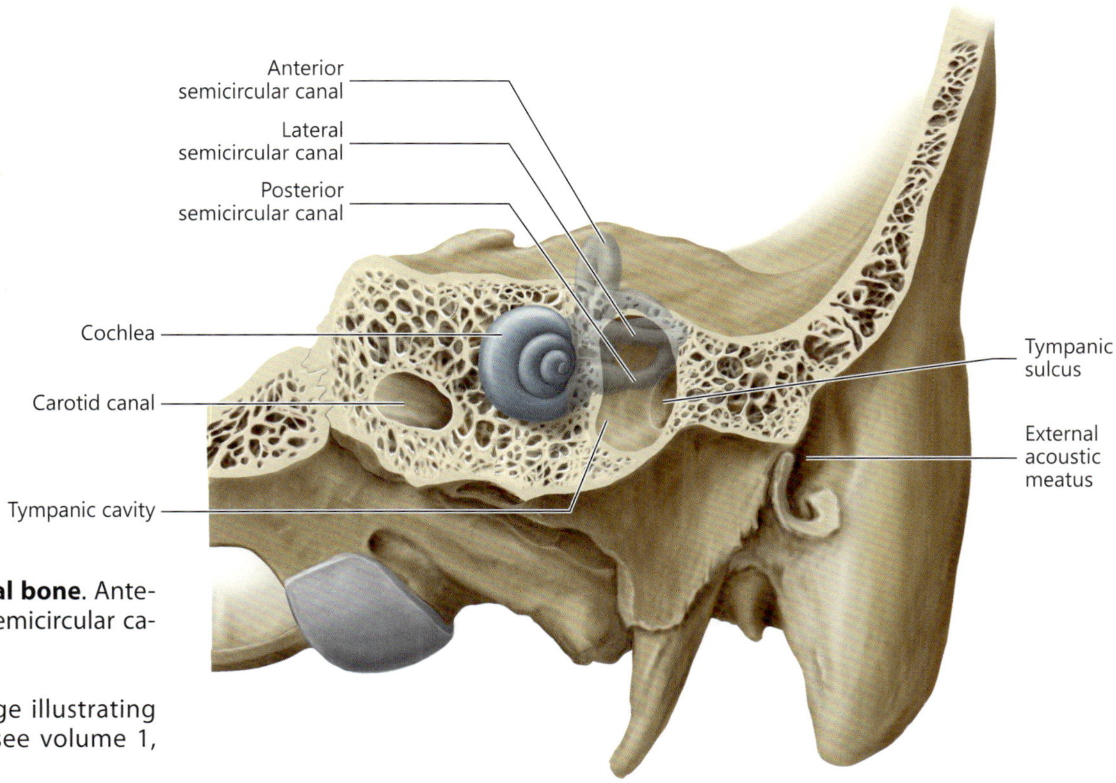

Frontal section of the **temporal bone**. Anterior view of the cochlea and semicircular canals, in transparency.

For a three-dimensional image illustrating the membranous labyrinth, see volume 1, page 383.

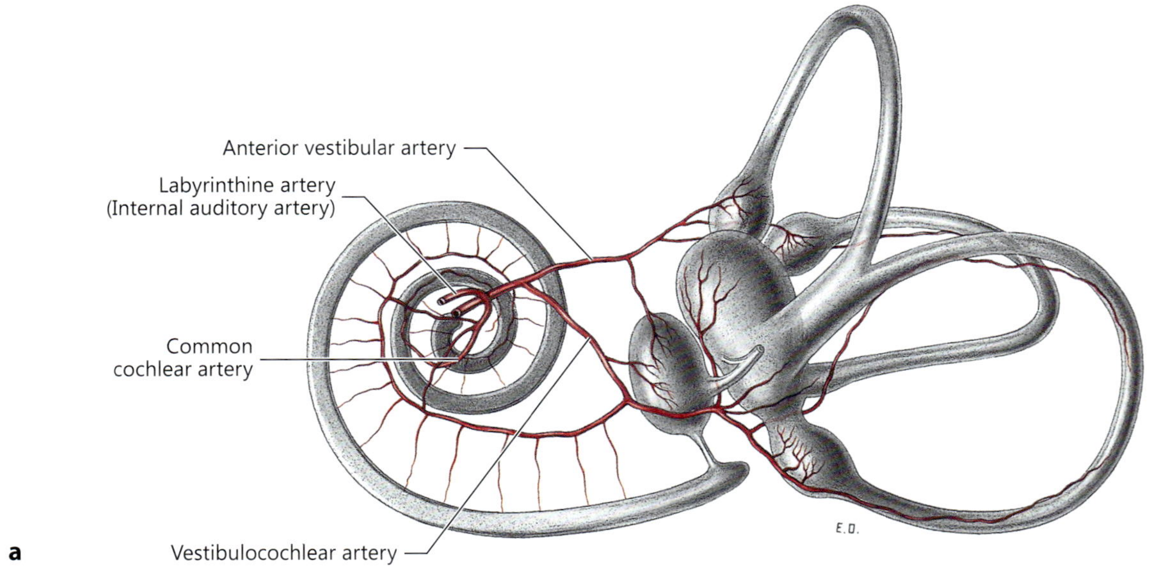

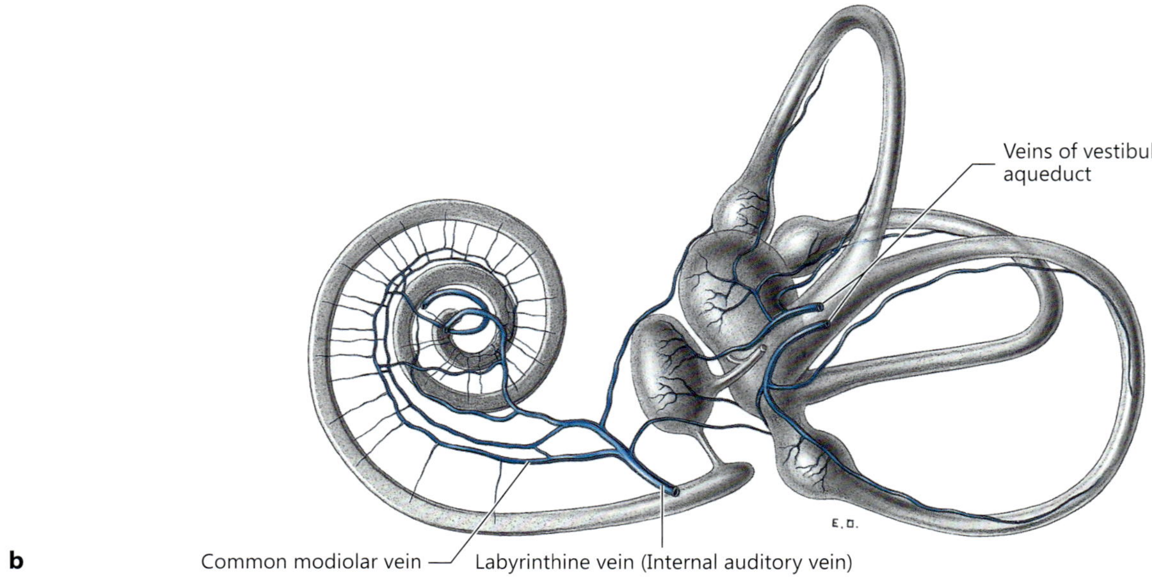

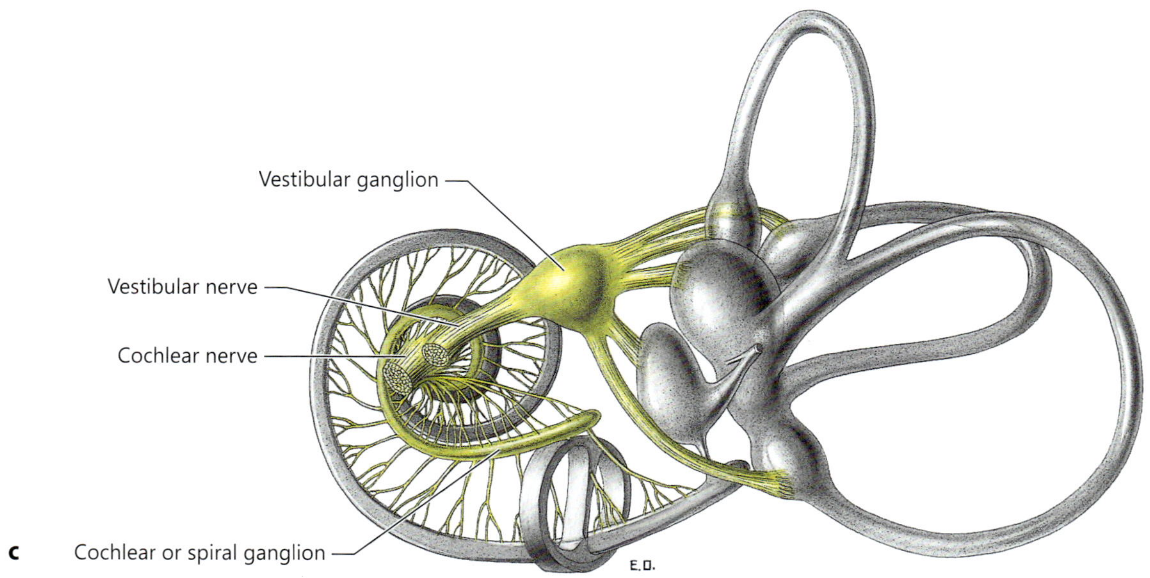

a, Arterial blood supply of the **vestibular** and **cochlear labyrinths**. **b**, Venous drainage of the **vestibular** and **cochlear labyrinths**. **c**, Territory of innervation of the vestibulocochlear nerve [VIII]. Peripheral branches of the vestibular and of the cochlear nerves are shown at their point of departure from, respectively, the vestibular ganglion (or Scarpa's ganglion) and the spiral ganglion (or Corti's ganglion).

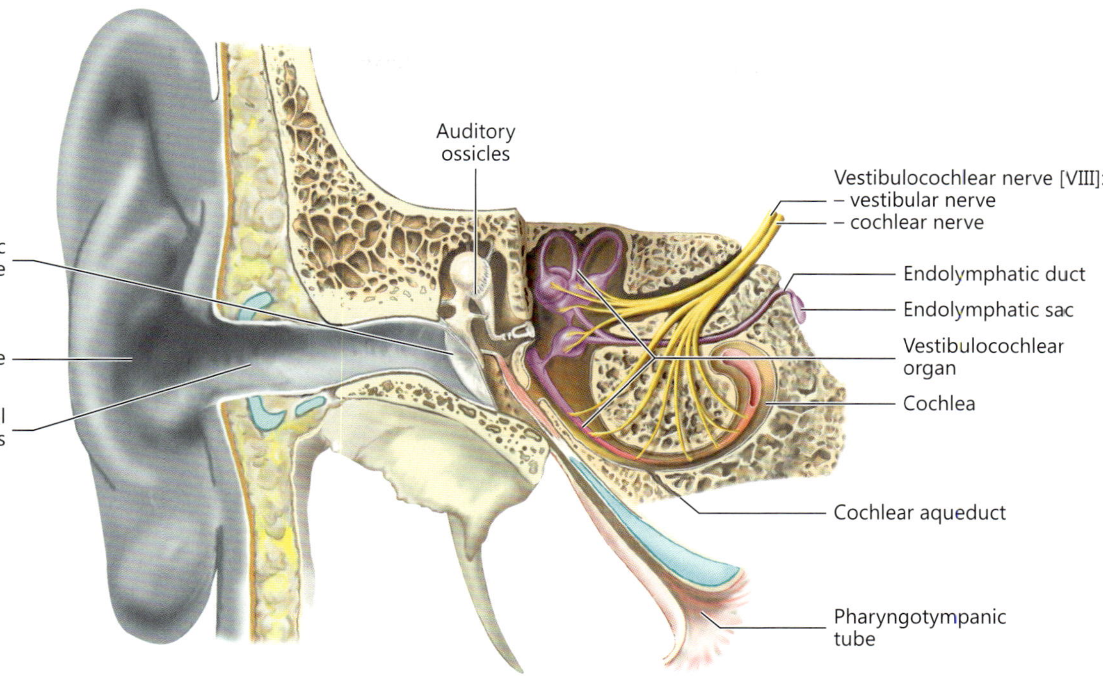

a

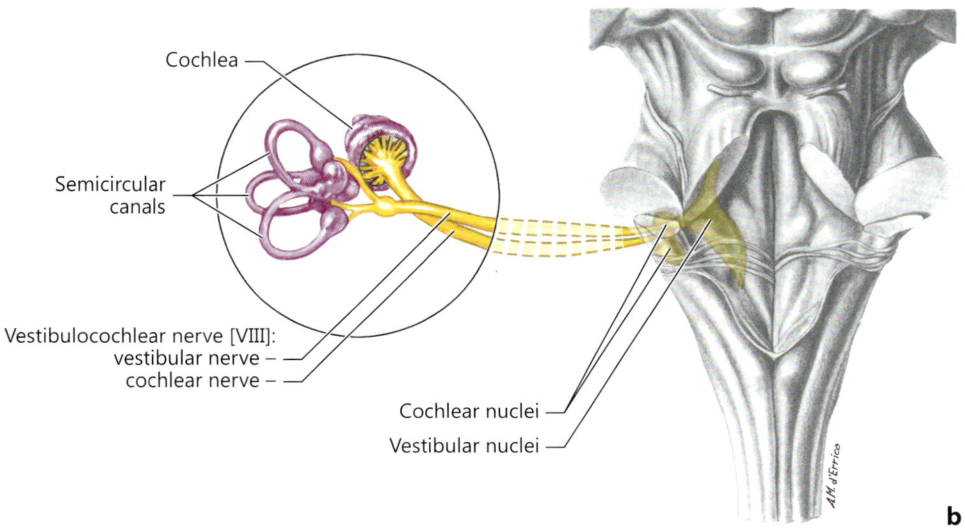

b

a, Peripheral distribution of the **vestibulocochlear nerve** [VIII]. Branches of the nerve penetrate into the labyrinth through the fundus of the internal acoustic meatus (☞ page 142, volume 1) and are distributed to the macula receptors of the utricle and saccule, to the ampullary crests of the semicircular ducts (vestibular nerve), and to the acoustic receptors of the cochlear duct (cochlear nerve). **b**, The site of the terminal nuclei of the **vestibular** and **cochlear nerves** in the bulbopontine tegmentum is indicated. The two nerves originate from their respective ganglia: the vestibular ganglion (Scarpa's ganglion), and the cochlear, or spiral, ganglion (Corti's ganglion).

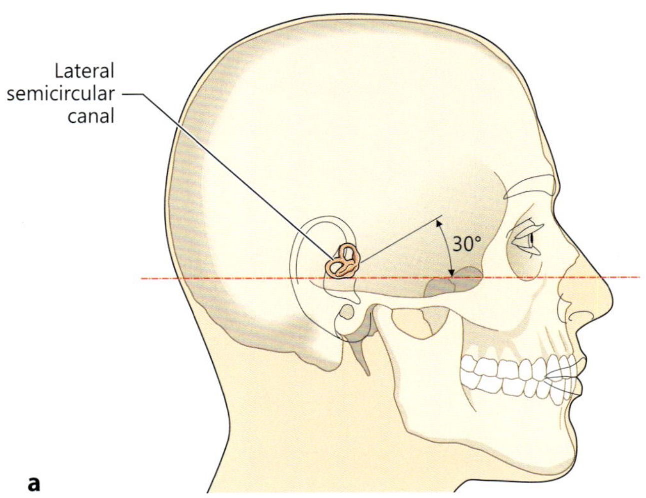

a

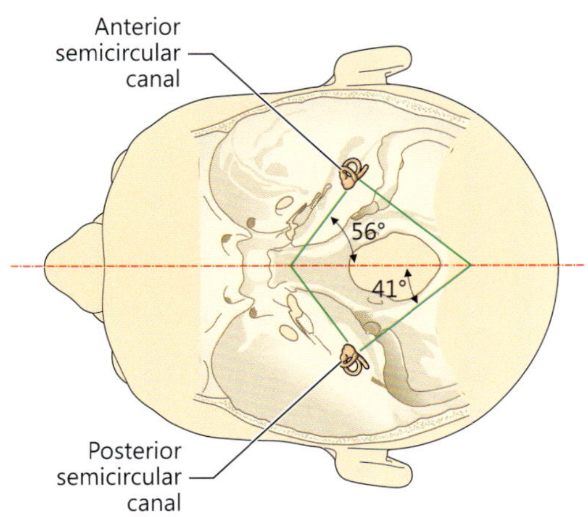

b

Orientation and angles formed by the **semicircular canals**. The anterior, posterior and lateral semicircular canals are located on the two sides in an orthogonal position (at right angles) to each other. The lateral semicircular canal is almost on a horizontal plane, inclined 30° forward and toward the top (**a**); the other canals are vertical, and oriented with respect to the sagittal plane at 56° (posterior canal) and 41° (anterior canal) (**b**).

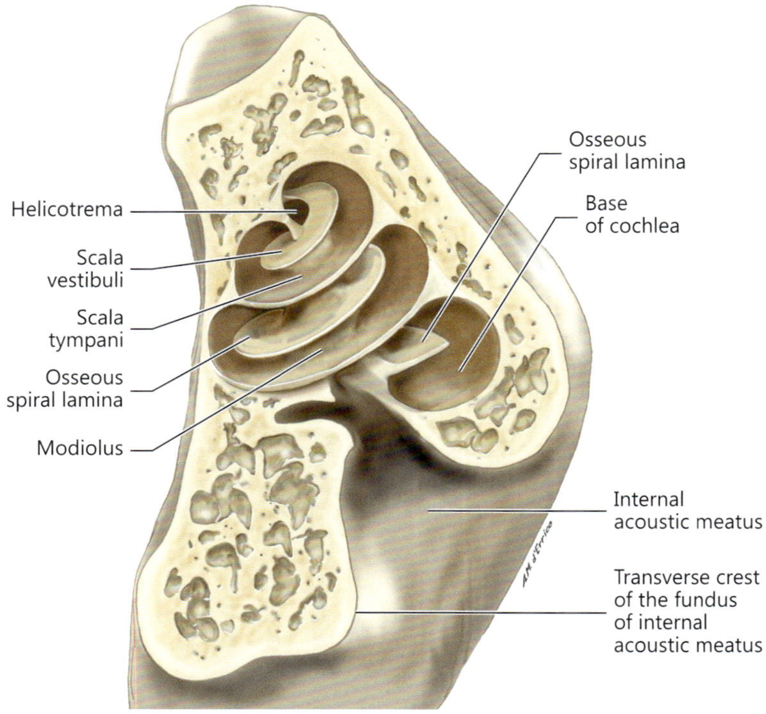

Cochlea, partially open, of the bony labyrinth of the right internal ear. The compact bone tissue of the wall is bounded by spongy bone of the petrous part of the temporal bone.

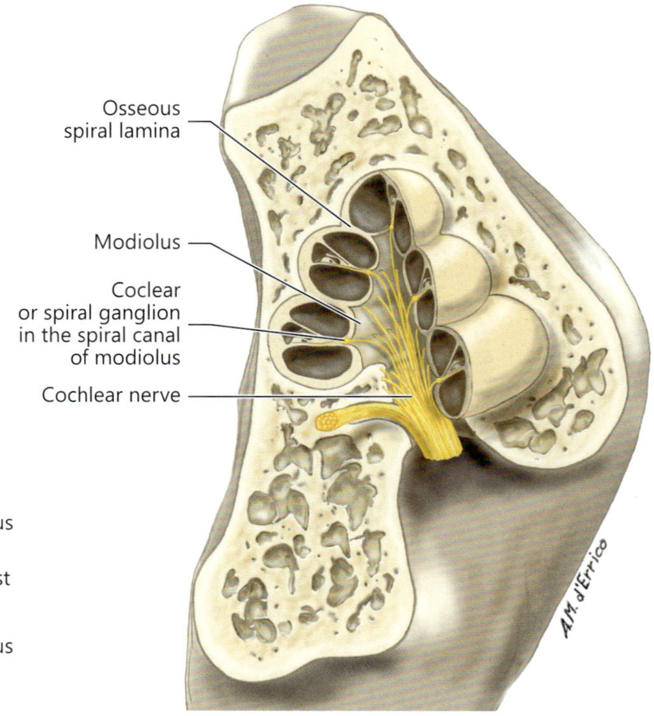

Section of the **cochlea** illustrating the modes of termination of the **cochlear nerve** and the position of the **cochlear**, or spiral, **ganglion** (Corti's ganglion) in the spiral canal of the modiolus.

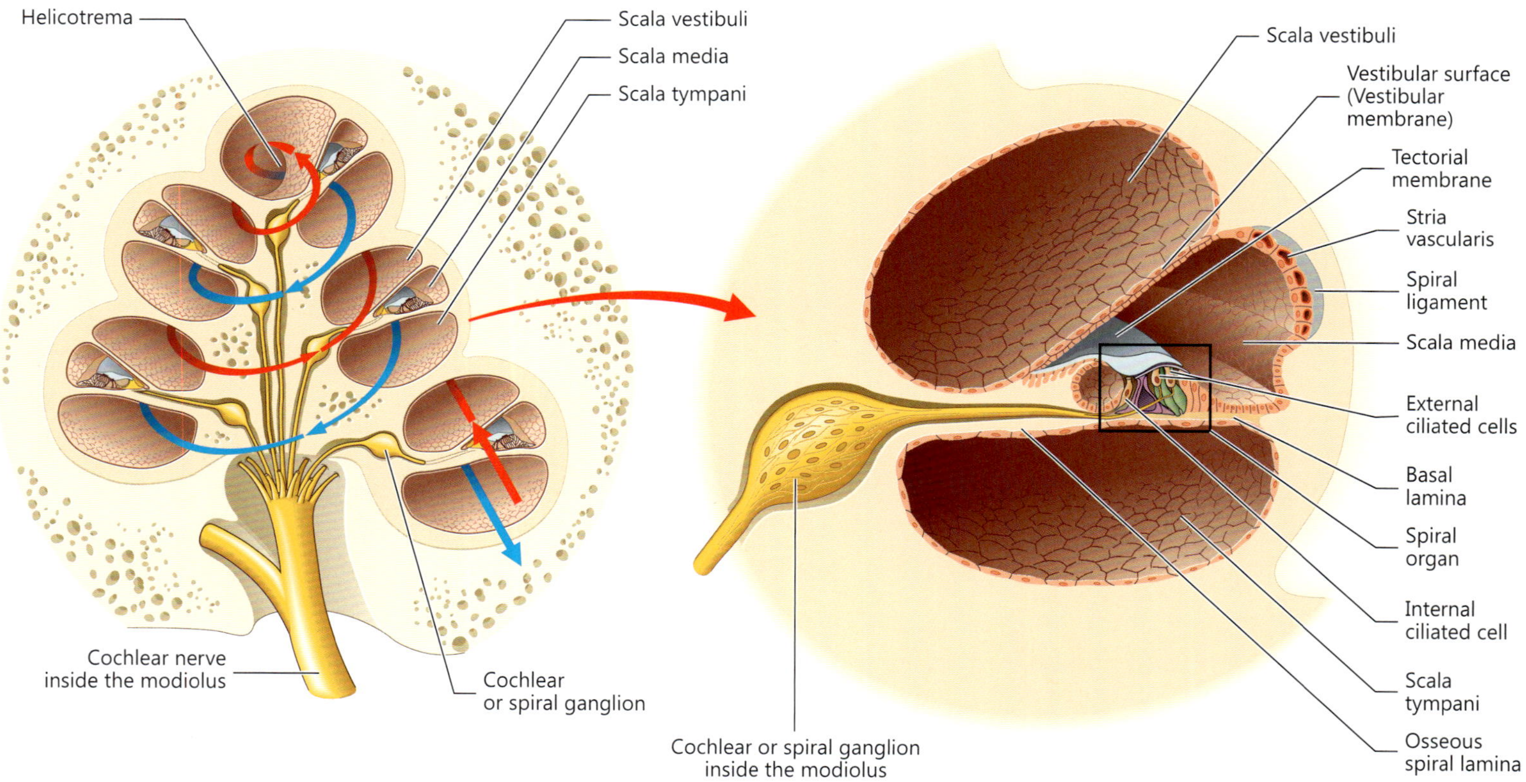

Structure of the **internal ear**. The cochlea is a spiraliform canaliculus that completes two and a half spirals over a length of approximately 35 mm. In its interior, the structures of the cochlear labyrinth demarcate three ducts, or scalae, that run parallel for the whole length. In the enlargement, the relative disposition of the three ducts is illustrated: the scala vestibuli, scala tympani and scala media. The scala media, which has a triangular cross-section, is bordered by three membranes: the basal lamina, on which the spiral organ (or organ of Corti) is found, separates it from the scala tympani; the vestibular surface (vestibular, or Reissner's, membrane) separates it from the scala vestibuli; the stria vascularis forms the boundary with the external surface of the cochlea. Lastly, the scala tympani and scala vestibuli continue, one into the other, through the helicotrema, situated at the apex of the cochlear duct.

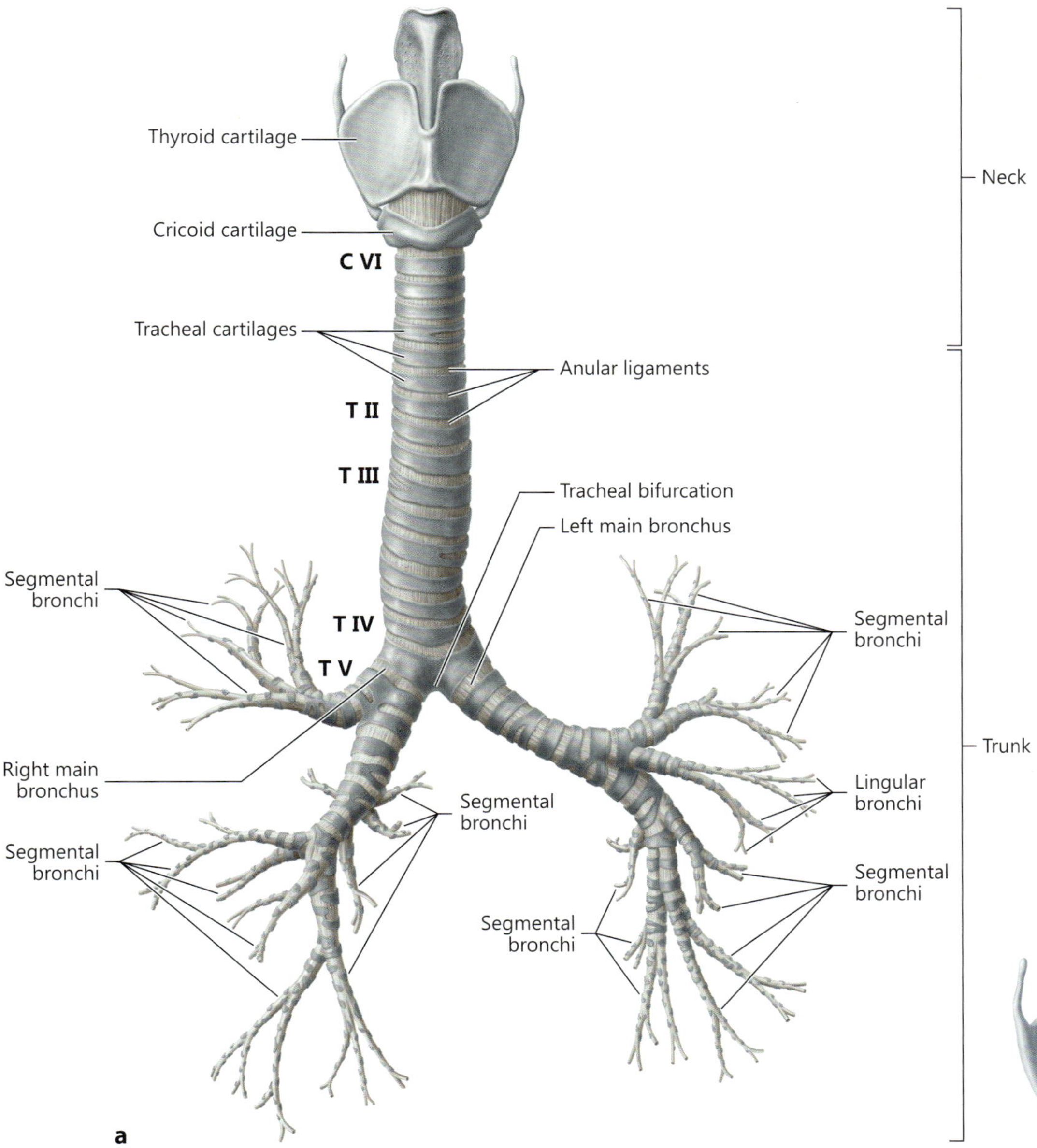

a

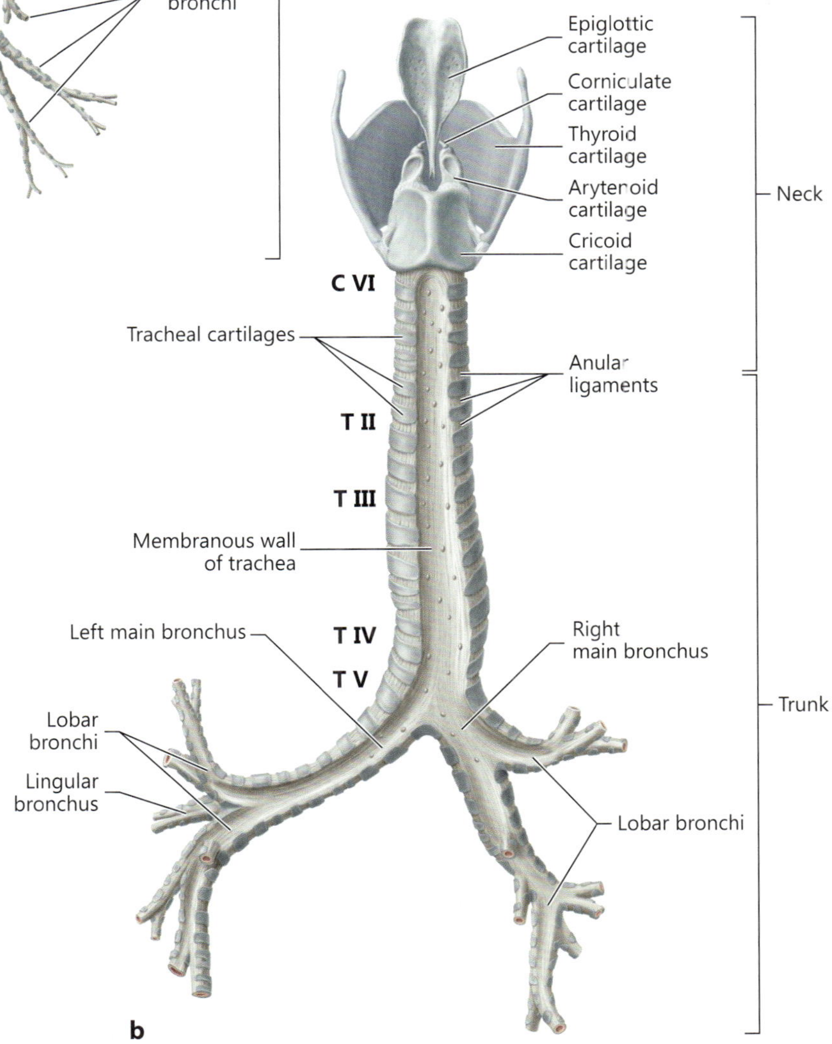

b

a, Anterior projection of the **trachea** and bronchial subdivisions. The trachea extends from the cricoid cartilage of the larynx, at the level of C VI, to the superior margin of T V, at which point it divides into the two main bronchi. The right main bronchus has a more vertical path than the left. **b**, Posterior projection of the **trachea**. The tracheal rings are incomplete posteriorly and continuity of the wall is provided by the membranous part of the trachea.
For more detail on the trachea, see chapter on the thorax.

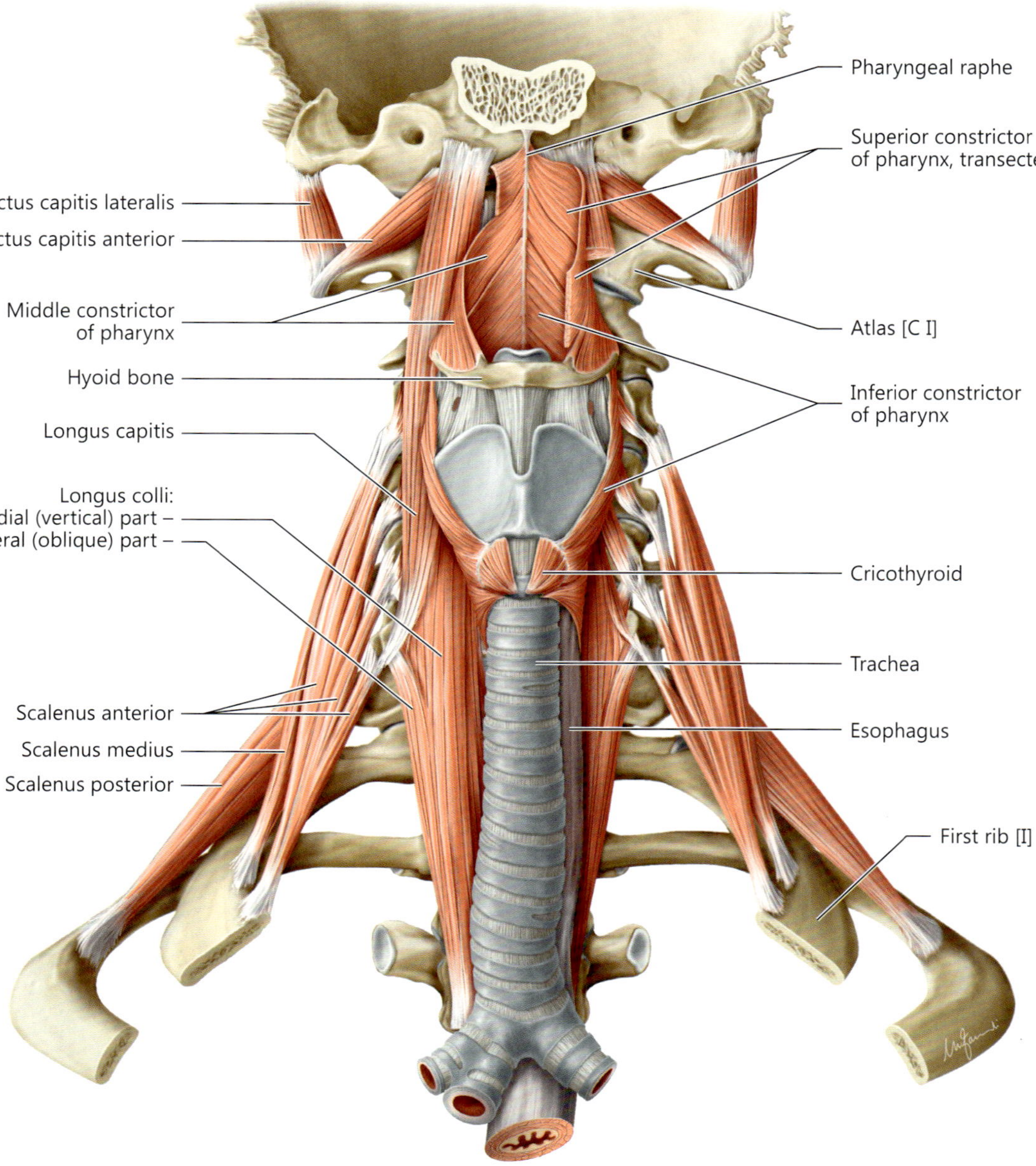

Diagram illustrating the relations of the pharynx, trachea and esophagus with the **prevertebral muscles** of the neck, viewed from the front. The prevertebral muscles of the neck are long muscles that are attached to the ventrolateral surface of the column, and covered by the prevertebral layer of cervical fascia; at the side of the prevertebral muscles are displayed the scalene muscles (scalenus anterior, scalenus medius and scalenus posterior). The prevertebral muscles stretch from the cranial base to the cervical column, whereas the scalene muscles go from the cervical column to the superior part of the thorax.
For a further description of the pharynx, see pages 145-147.

Infrahyoid muscles in relation to the larynx, anterior view. The sternohyoid and omohyoid are located on a more superficial plane, the sternothyroid and thyrohyoid are situated more deeply in direct contact with the larynx.

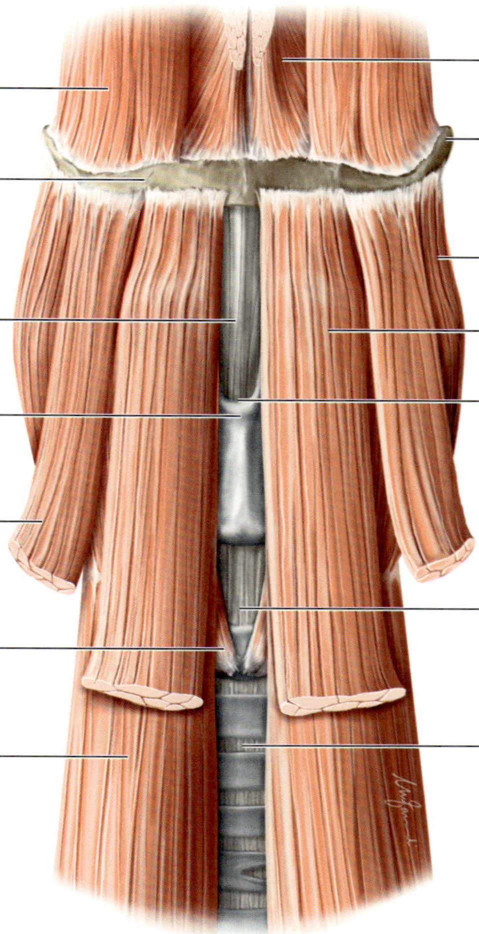

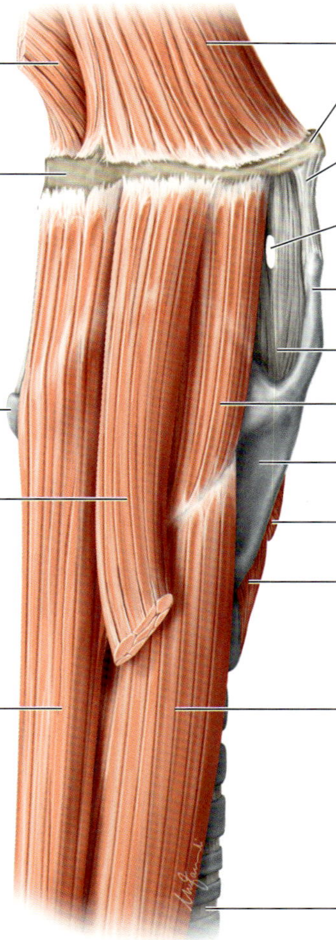

Infrahyoid muscles, left lateral view. The laryngeal prominence, or Adam's apple, is the most protuberant part of the angle formed on the median line by the two laminae, right and left, of the thyroid cartilage. The sternothyroid is situated deeply to the sternohyoid and unites the sternum to the larynx. The thyrohyoid is the continuation upwards of the sternothyroid and, situated deeply to the sternohyoid, it extends to the hyoid bone.

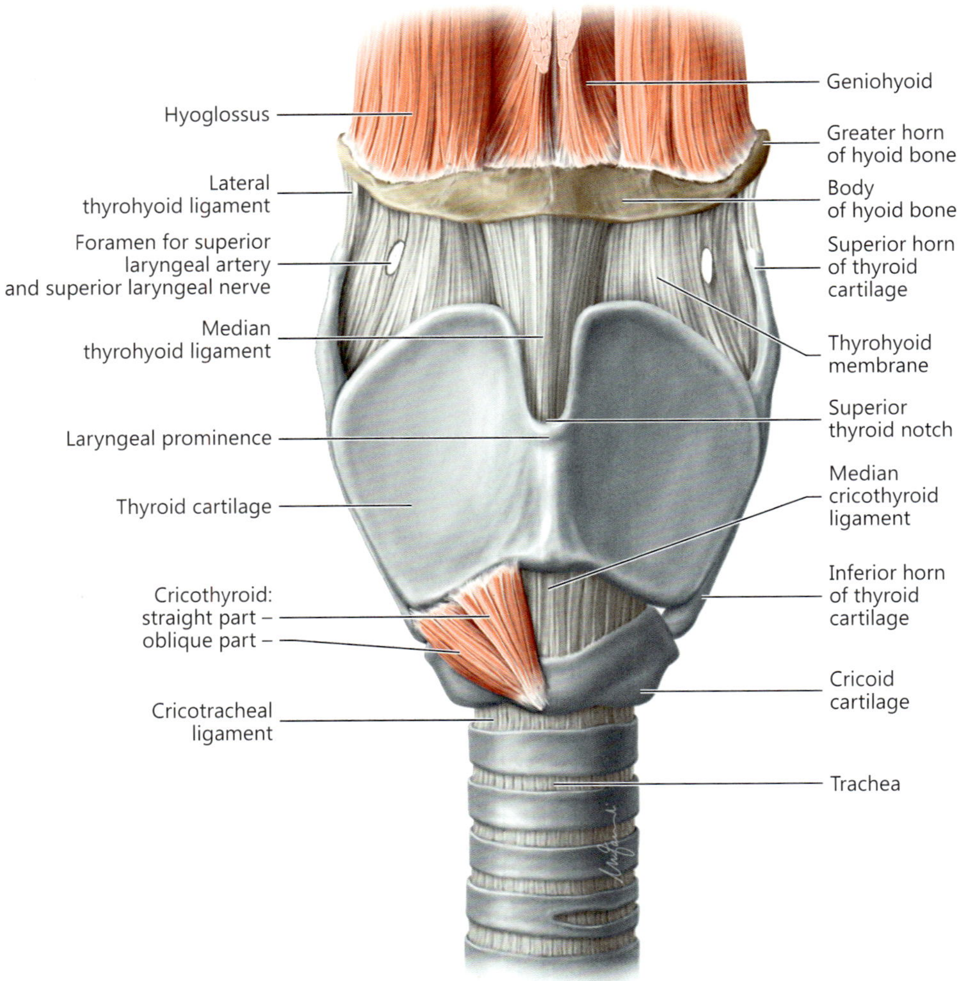

After removal of the infrahyoid muscles, one can see the **thyroid cartilage** of the larynx with the ligaments that connect it to the hyoid bone (**thyrohyoid membrane**, **median** and **lateral thyrohyoid ligaments**), and the **median cricothyroid ligament** of the cricothyroid joint. When the **cricothyroid** contracts it tends to elongate the vocal ligament, and is hence also called the tensor muscle of the vocal cords.

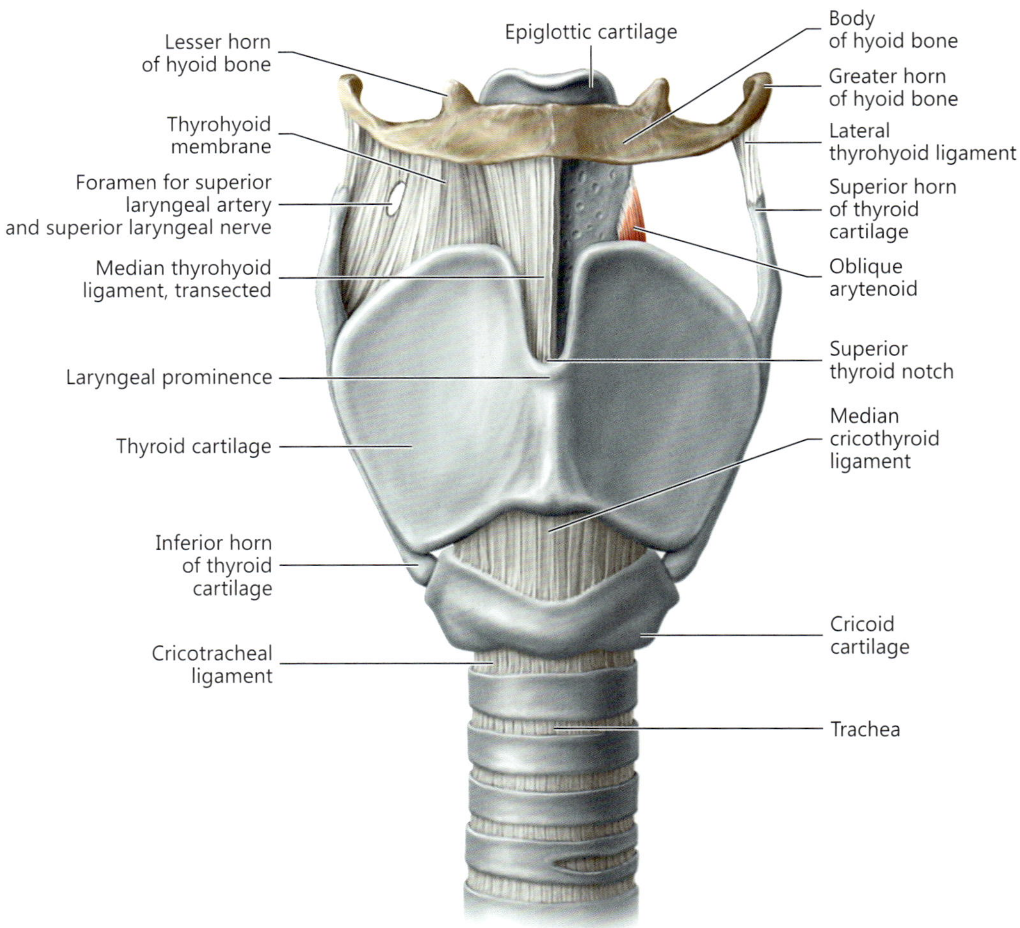

External configuration of the larynx, anterior view. One can see the relation of the organ to the hyoid bone, superiorly, and to the trachea, inferiorly. In the left part of the larynx, the thyrohyoid membrane and median thyrohyoid ligament have been removed so as to display the **oblique arytenoid**.

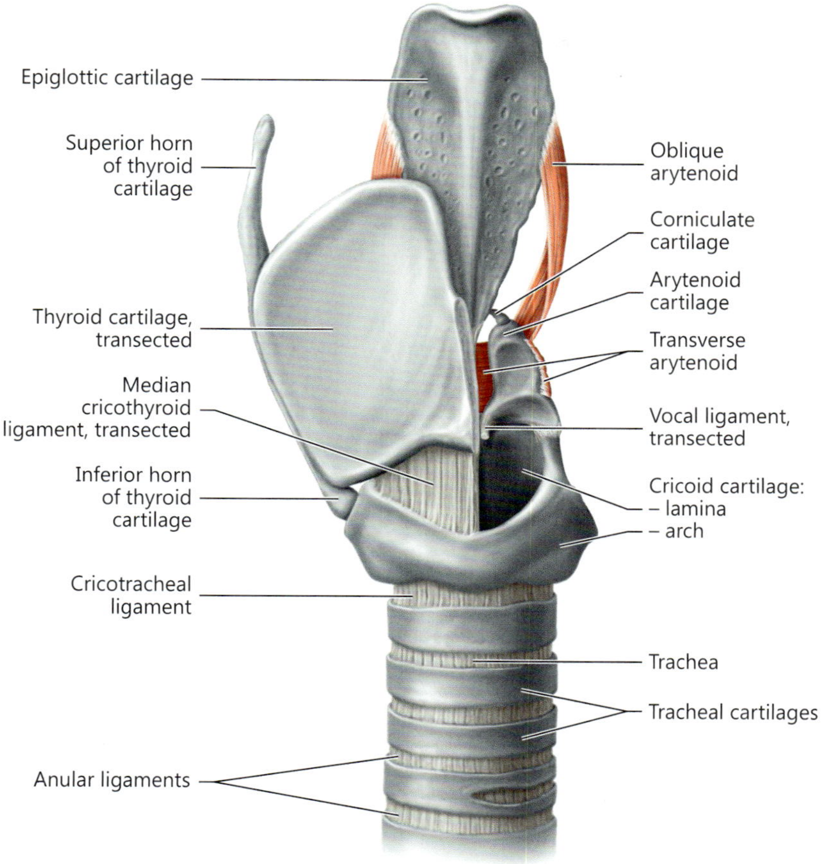

Larynx, anterior view. In the right half of the larynx, the thyroid cartilage and median cricothyroid ligament can be seen; in the left half, they have been transected to expose the arytenoid and corniculate cartilages, the **oblique and transverse arytenoid**, and the arch of the cricoid cartilage.

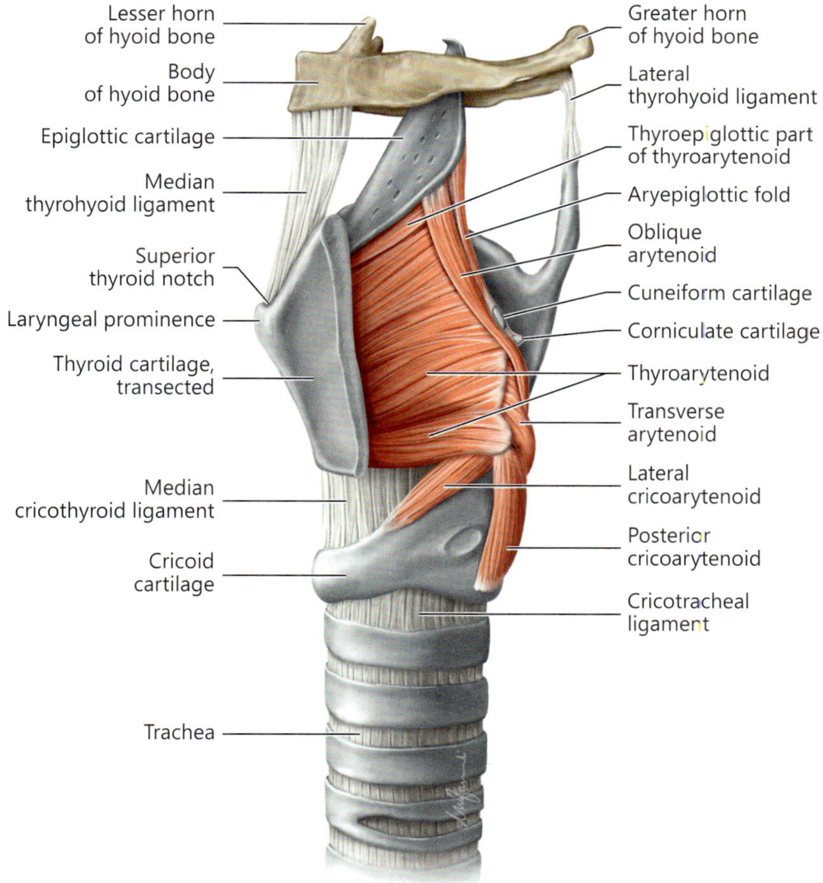

Muscles of the larynx, left lateral view, after demolition of the left lamina of the thyroid cartilage. The thyroarytenoid (above) and lateral cricoarytenoid (below) form part of the lateral wall of the larynx.

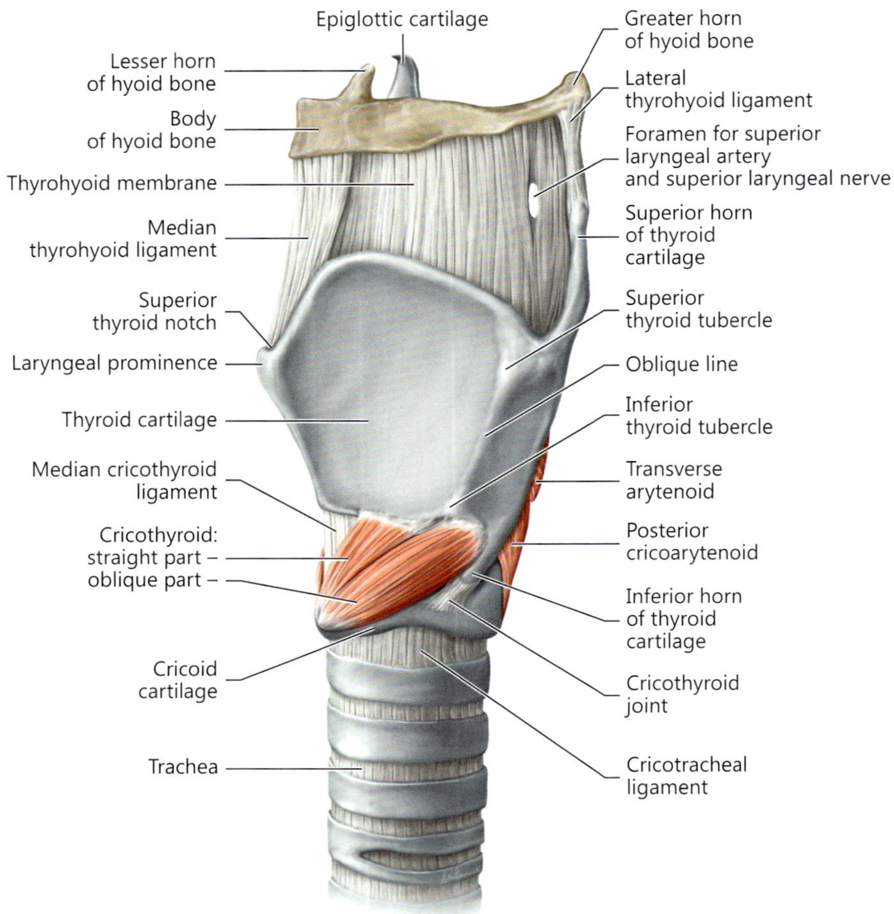

Cricothyroid, left lateral view. One can distinguish the two parts of the muscle: the straight part (medial) and the oblique part (lateral).

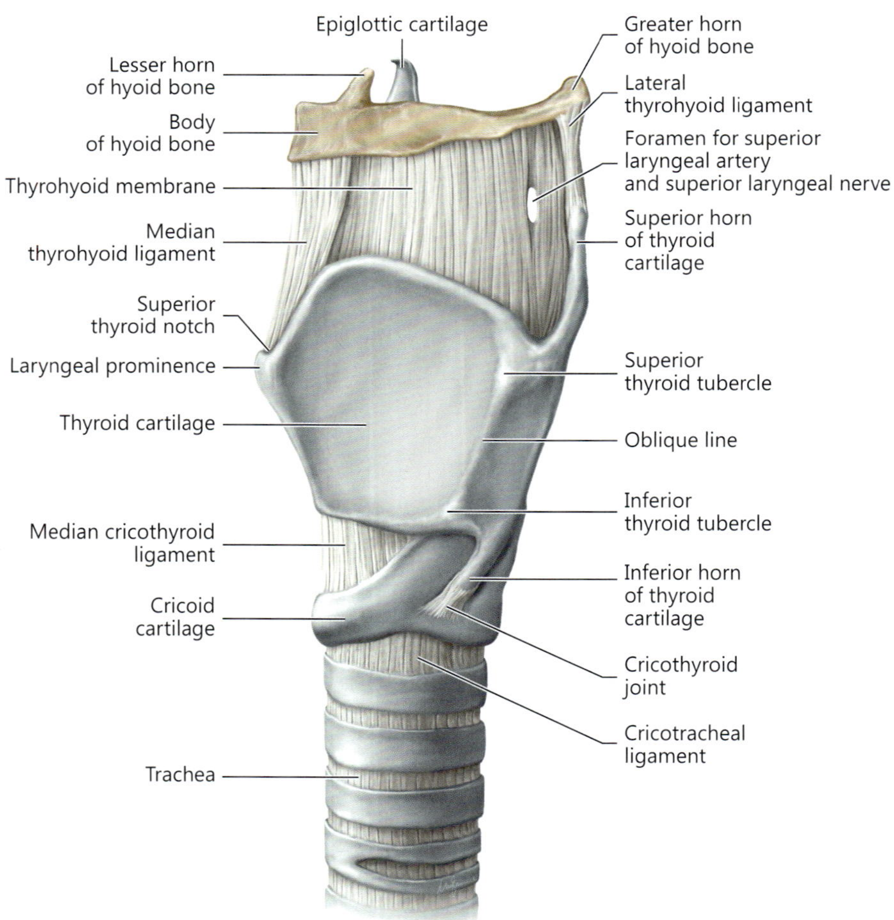

Principal means of union of the **laryngeal cartilages**, left lateral view. One can also see the connections and relations of the larynx with the hyoid bone (above) and trachea (below).

Larynx and **trachea**, posterior view. The arytenoid cartilages provide attachment for all the intrinsic laryngeal muscles except the cricothyroid muscles. Here one can see the **posterior cricoarytenoid muscles**, the **oblique arytenoid** (or aryepiglottic) **muscles** with their aryepiglottic parts, and the **transverse arytenoid**.

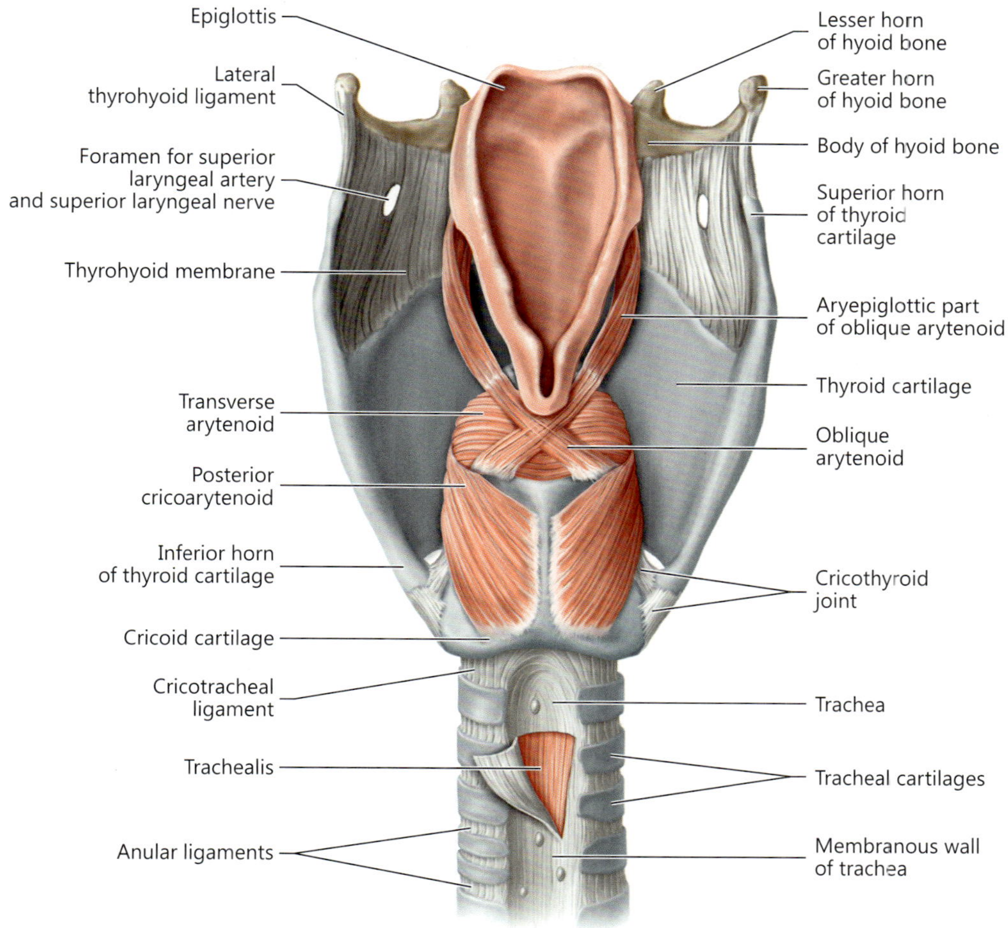

Larynx and **trachea**, posterior view. One can see the unpaired cartilages of the larynx: the thyroid, cricoid, and epiglottic cartilages. The corniculate cartilages, hook-like in form, surmount the apices of the arytenoid cartilages, which have the form of a triangular pyramid whose posterior concave surface corresponds to the **transverse arytenoid**, stretching between the lateral borders of the two arytenoid cartilages. The **posterior cricoarytenoid muscles** are situated behind the cricoid cartilage, below the mucosa of the pharynx. The cervical part of the trachea includes the first 5 to 6 tracheal rings. The tracheal rings are incomplete posteriorly, where they are continuous with the membranous wall of the trachea.

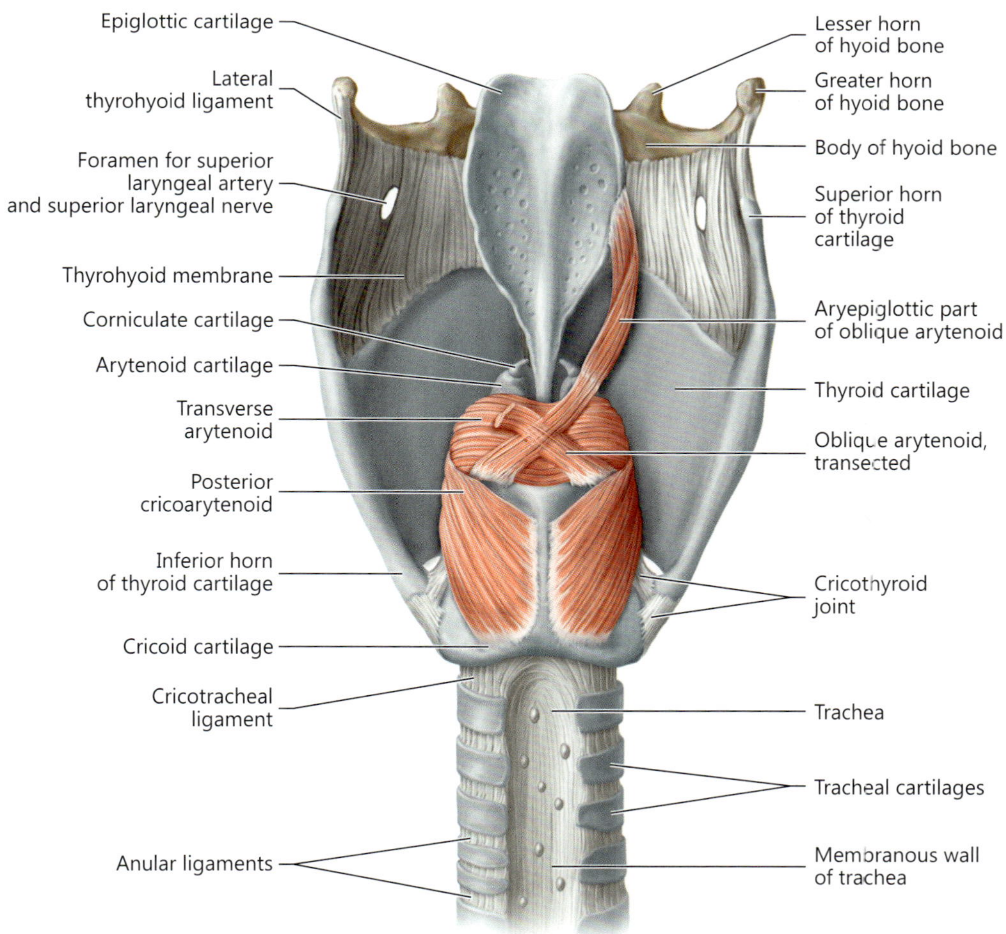

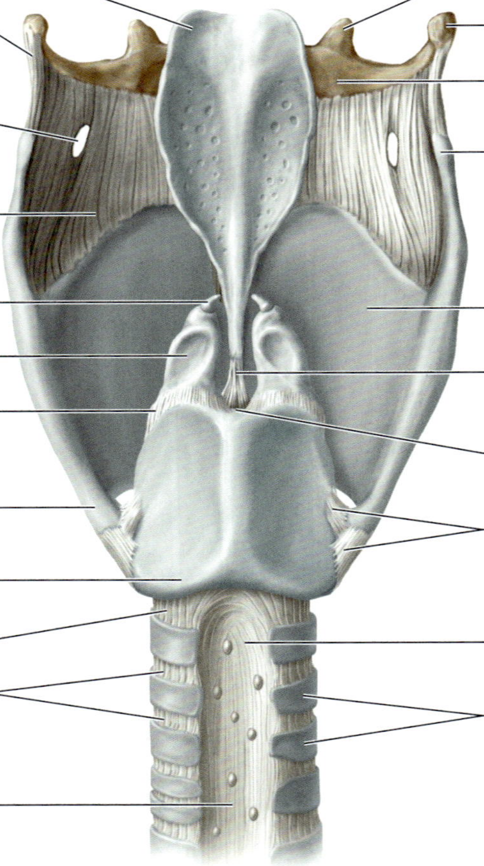

Larynx and **trachea**, posterior view. After removal of the intrinsic muscles of the larynx, one can see the laryngeal cartilages: the epiglottic cartilage, which unites with the thyroid cartilage through the thyroepiglottic ligament; the thyroid cartilage, which articulates with the cricoid cartilage through the inferior horns; the arytenoid cartilages, which sit on the arytenoid articular surfaces of the lamina of the cricoid cartilage; the corniculate cartilages, which are connected to the apices of the arytenoid cartilages; and the cricoid cartilage which, besides articulating with the thyroid and arytenoid cartilages, is united by means of the cricotracheal ligament to the first tracheal ring.

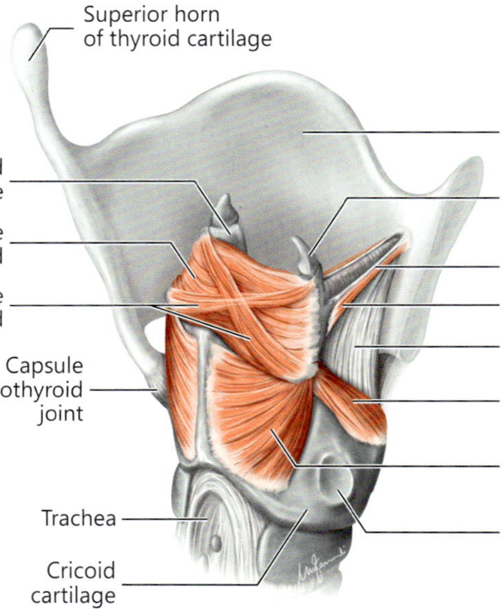

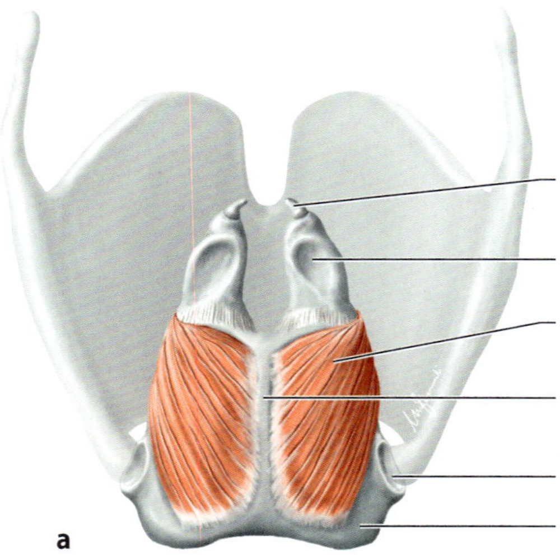

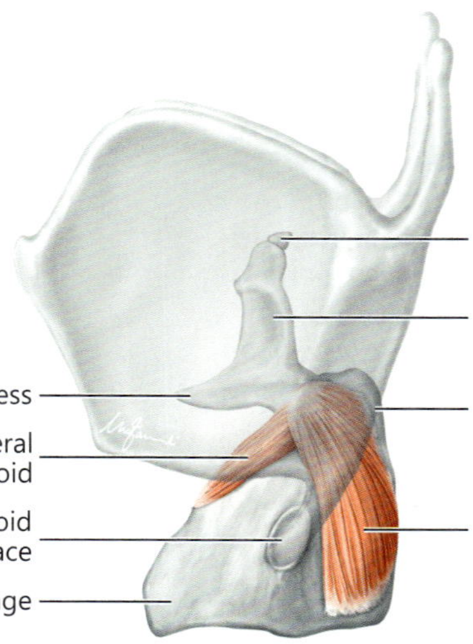

Cricoarytenoid muscles. **a**, Posterior view of the larynx showing the posterior cricoarytenoid. **b**, Posterolateral projection. The epiglottic cartilage has been removed and the right lamina of the thyroid cartilage transected. One can see the conus elasticus and the vocal ligament, which together with the vocalis constitutes the vocal fold, or true vocal cord, stretching from the internal surface of the thyroid cartilage, immediately under the notch, to the vocal process of the arytenoid cartilage. **c**, The left lateral projection enables one also to see the characteristic position of the lateral cricoarytenoid.

Larynx, view looking down from behind. The thyroarytenoid has been transected in the right half to expose the vocalis. Also visible are the straight and oblique parts of the cricothyroid and the lateral cricoarytenoid. Posteriorly, one can see the transverse arytenoid, covered by the oblique arytenoid muscles, and, behind the cricoid cartilage, the posterior cricoarytenoid muscles.

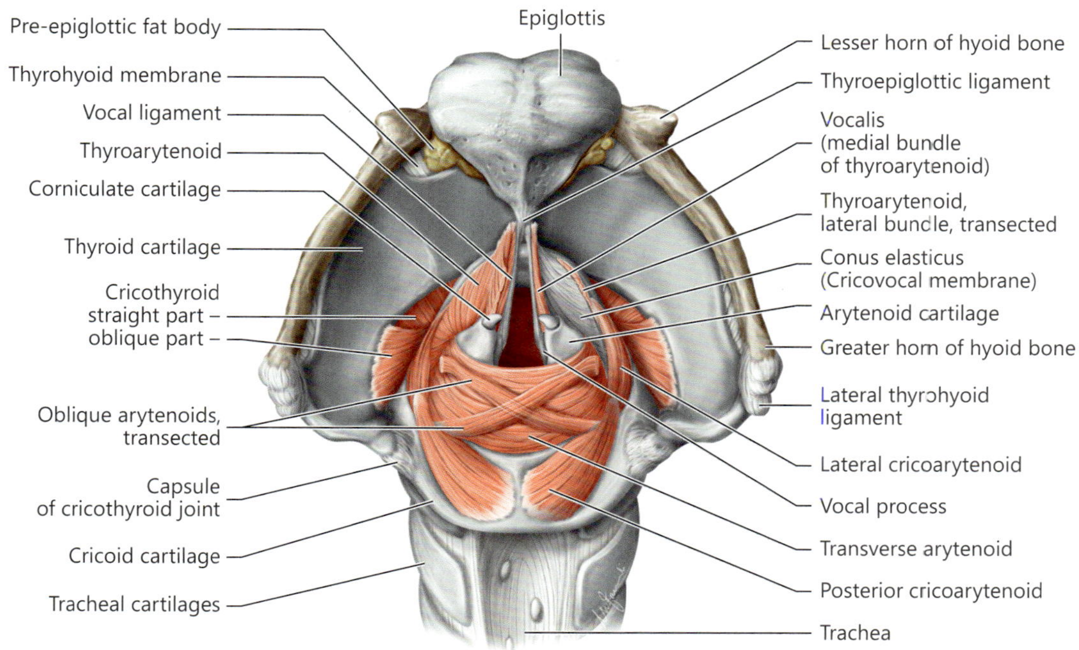

Larynx, view looking down from behind. After removal of the posterior cricoarytenoid muscles, transverse arytenoid and oblique arytenoid muscles, one can see the lamina of the cricoid cartilage and the arytenoid cartilages, whose vocal processes with their apices continue in the vocal ligaments.

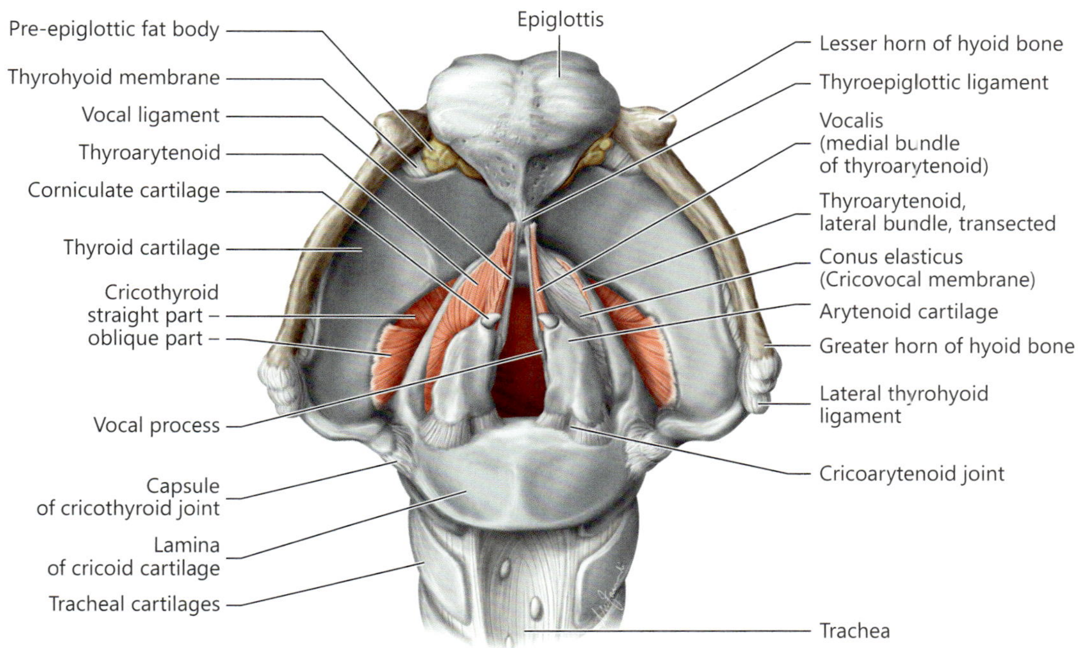

Larynx, superior view. With the muscles of the larynx removed, one can clearly see the conus elasticus, which, inferiorly, inserts on the superior circumference of the ring of the cricoid cartilage, while, at the top, it narrows and flattens into two laminae that terminate with the vocal ligament. Also visible are the vocal and muscular processes at the base of the arytenoid cartilages.

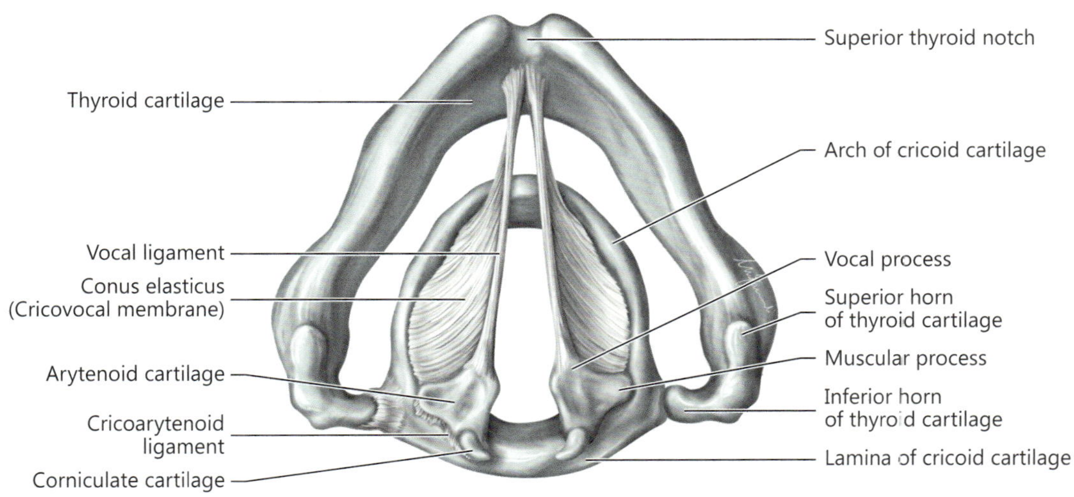

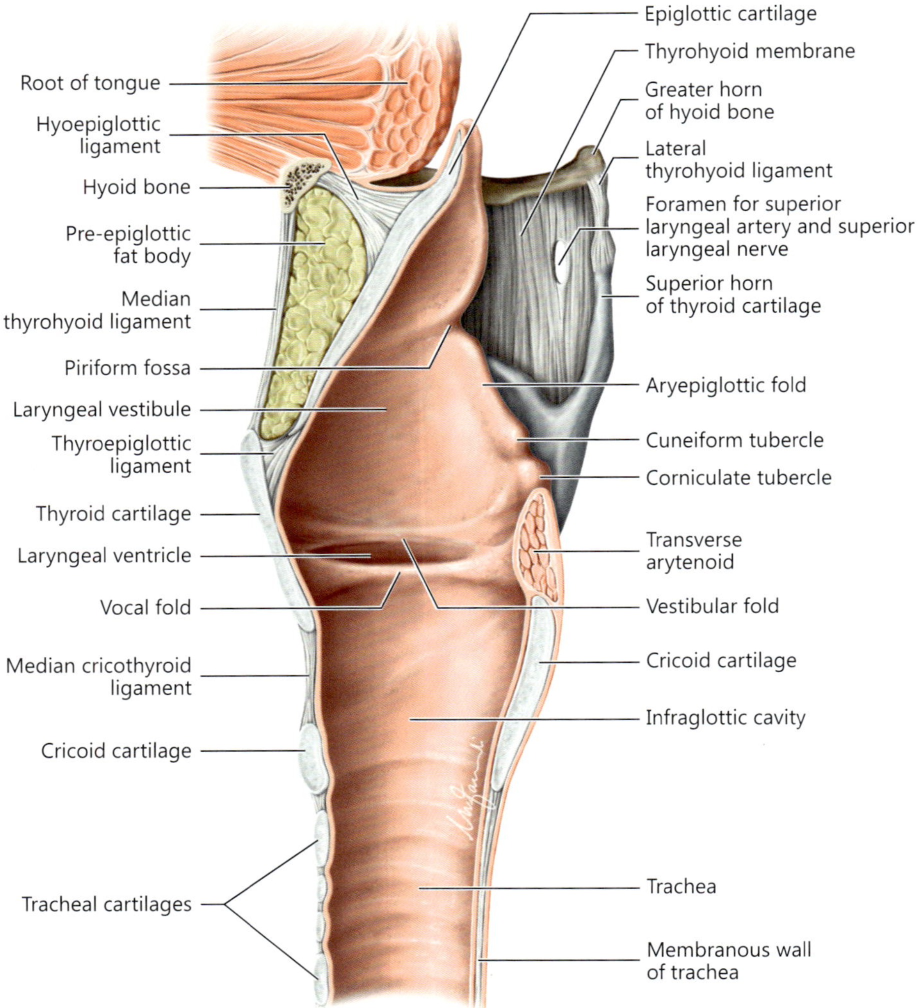

Median sagittal section of the **larynx**. One can see the laryngeal cavity, whose intermediate portion presents a sagittal fissure due to the presence, on each side, of two horizontal anatomical features oriented anteroposteriorly: the *vestibular folds*, above, and the *vocal folds*, below.

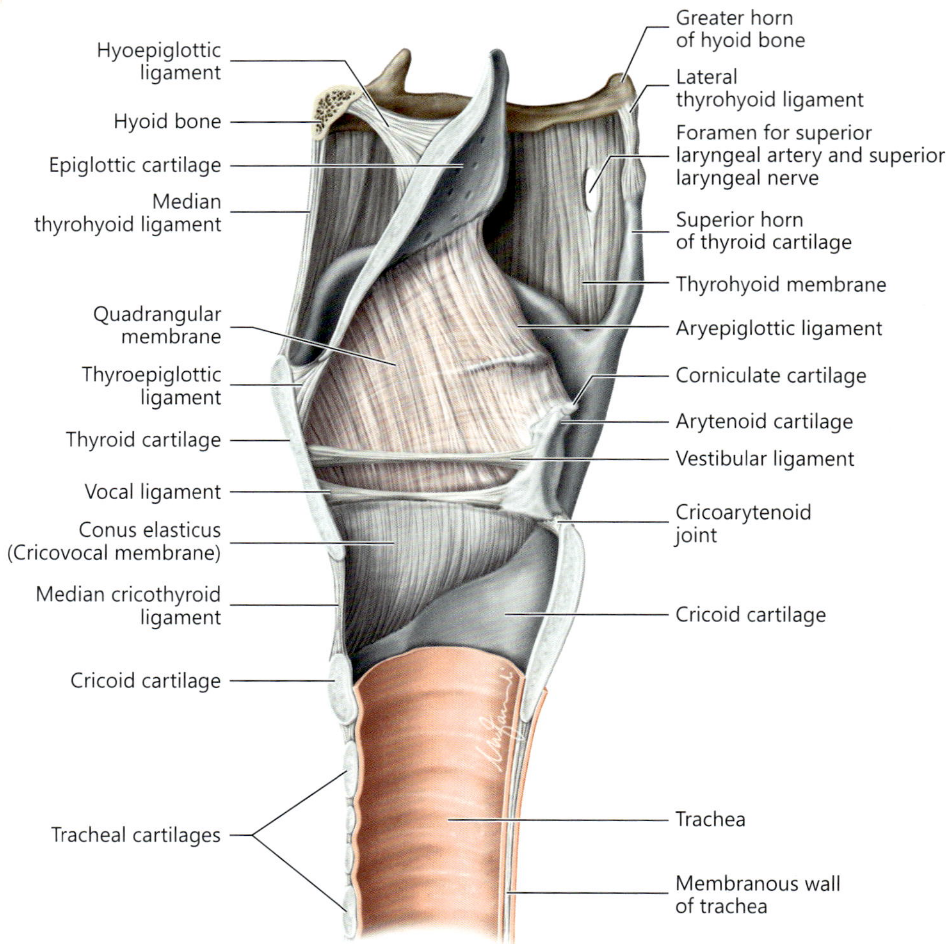

Median sagittal section of the **larynx**. One can see, in particular: the vestibular ligament, which represents the scaffold supporting the vestibular fold, or false vocal cord; the conus elasticus, whose free edge, thickened, is the vocal ligament, which together with the vocalis constitutes the vocal fold, or true vocal cord; and the quadrangular membrane, whose posterosuperior margin, free and thickened, forms the aryepiglottic ligament, while its inferior one, also thickened, constitutes the vestibular ligament.

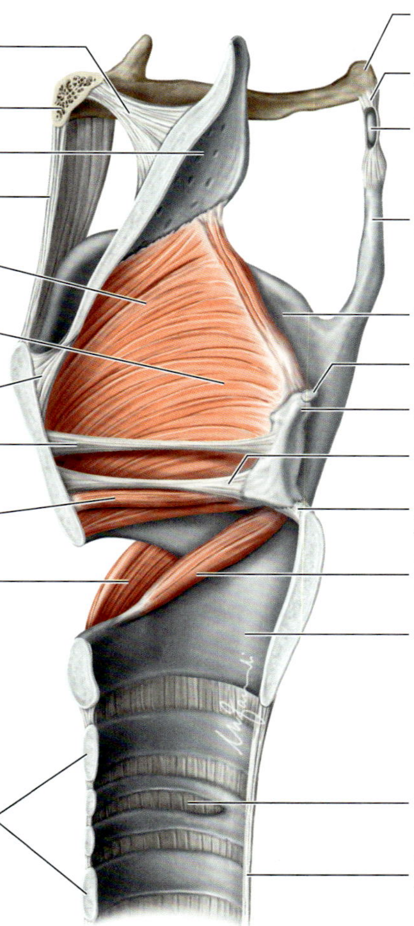

Median sagittal section of the **larynx**. The following muscles can be seen: the vocalis, the isometric contraction of which places the vocal ligament under tension (the vocal ligament and vocalis together constitute the vocal fold, or true vocal cord); the lateral cricoarytenoid; the cricothyroid, also known as the tensor muscle of the vocal cords because, contracting, it tenses and elongates the vocal ligament; and the thyroarytenoid, of which one can distinguish, besides the thyroepiglottic part or thyroepiglottic muscle, also a lateral bundle and a medial bundle or, as otherwise known (the latter), the vocalis.

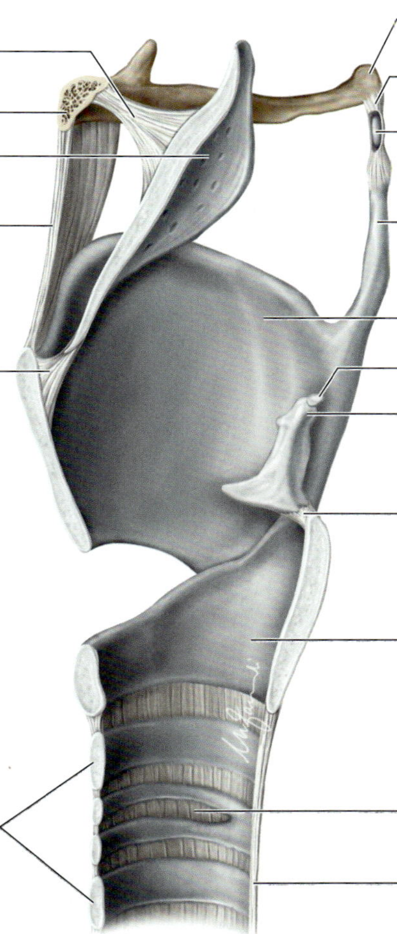

Median sagittal section of the **larynx**. One can distinguish the various cartilaginous elements articulated one to another for contiguity and united at a distance by means of ligaments. Here are shown the thyroepiglottic ligament and the hyoepiglottic ligament, which connects the epiglottic cartilage of the larynx to the hyoid bone, contributing to the return of the epiglottis after swallowing.

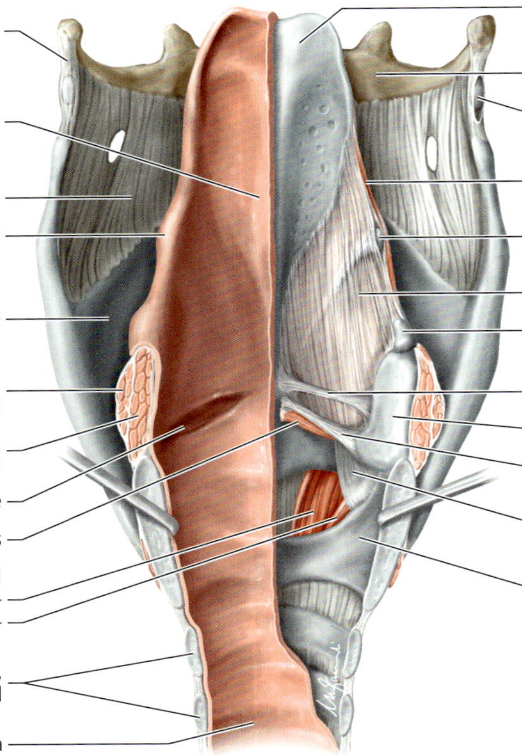

Larynx, posterior view after transecting the arytenoid muscles and cricoid cartilage. In the left half, one can see the laryngeal cavity covered by the mucosa of the larynx with, in succession from top to bottom, the laryngeal inlet, laryngeal vestibule, laryngeal ventricle, glottis and infraglottic cavity. In the right half, with the mucosa removed, the quadrangular membrane with the vestibular ligament is visible as well as the conus elasticus with the vocal ligament.

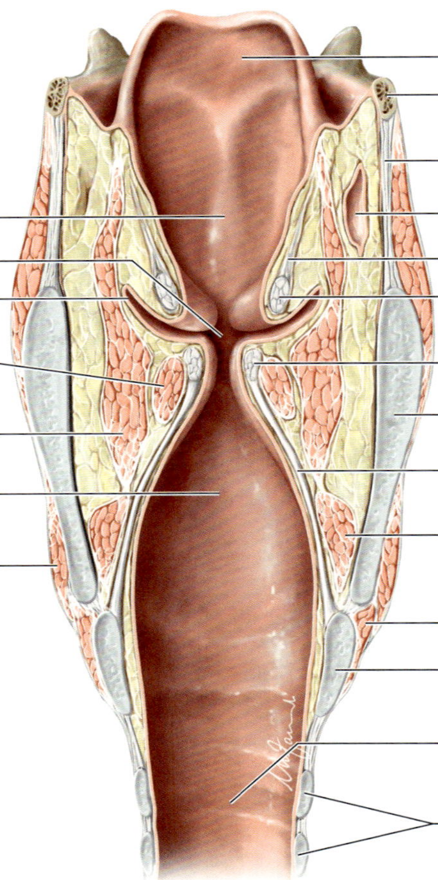

Frontal section of the **larynx**, posterior view of the anterior part. One can see the laryngeal vestibule, between the laryngeal inlet and the rima vestibuli, situated among the vestibular folds, which contain in their thickness the vestibular ligament. The vocal folds are situated below the vestibular folds and delimit the rima glottidis, the narrowest point of the laryngeal cavity. They contain in their anterior part the vocal ligament and the vocalis. The infraglottic cavity extends between the glottis and a plane passing through the inferior margin of the cricoid cartilage, narrowing as it descends.

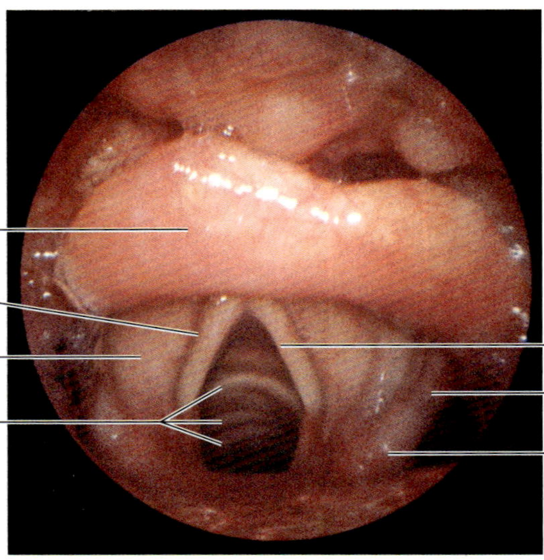

Rima glottidis observed with the laryngoscope. The vocal folds (true vocal cords) are found below the vestibular folds (ventricular folds or false vocal cords). The free edge of each true vocal cord protrudes toward the laryngeal cavity and delimits, with that of the other side, the rima glottidis, triangular in shape with its apex in front, corresponding to the apex of the thyroid cartilage, and its base behind, between the arytenoid cartilages.

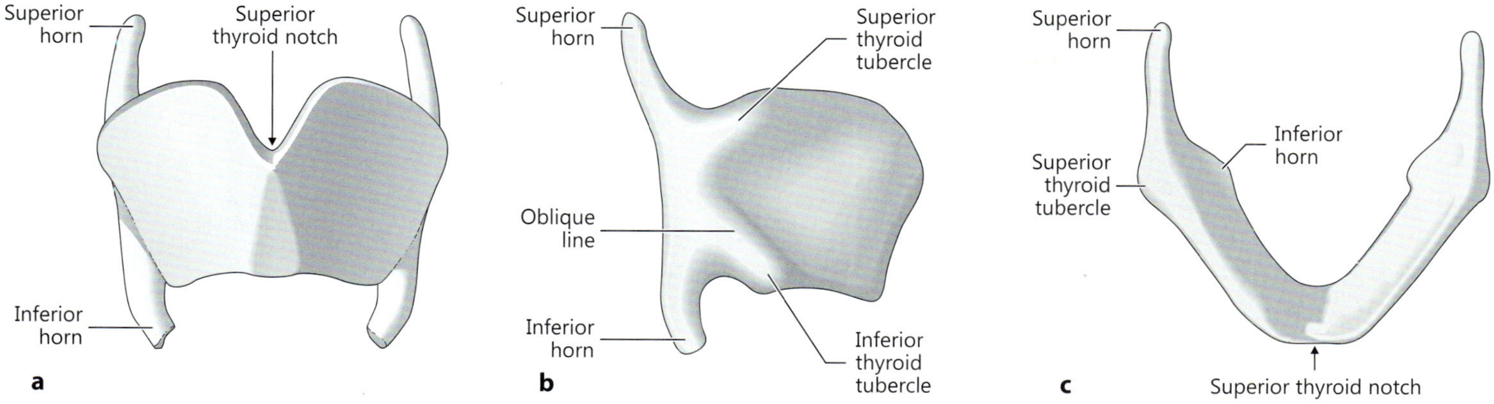

Thyroid cartilage in the male, viewed in anterior (**a**), lateral (**b**) and superior (**c**) projection.

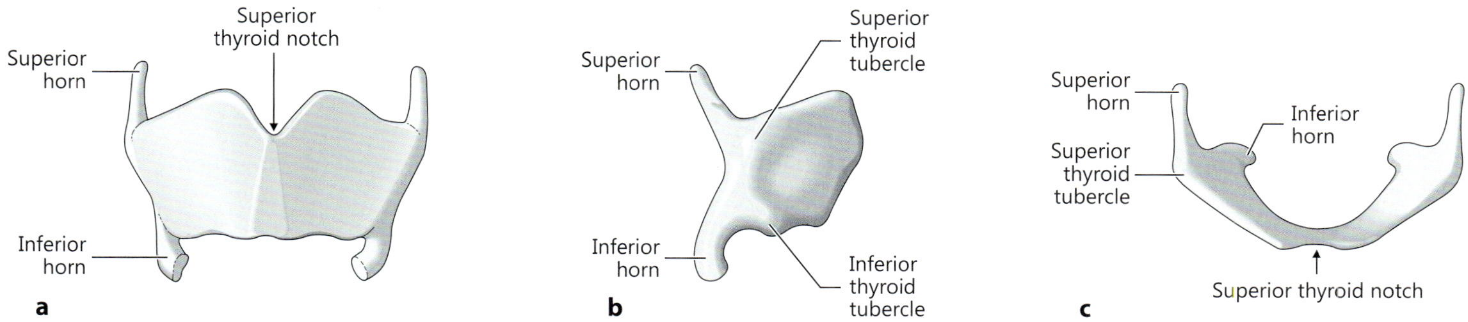

Thyroid cartilage in the female, in anterior (**a**), lateral (**b**) and superior (**c**) projection. Compared to the male larynx, numerous differences can be seen in the conformation of the thyroid cartilage. In particular, the angle formed by the two laminae of the thyroid cartilage, right and left, at the point where they meet in the laryngeal prominence is much wider in the female than in the male.

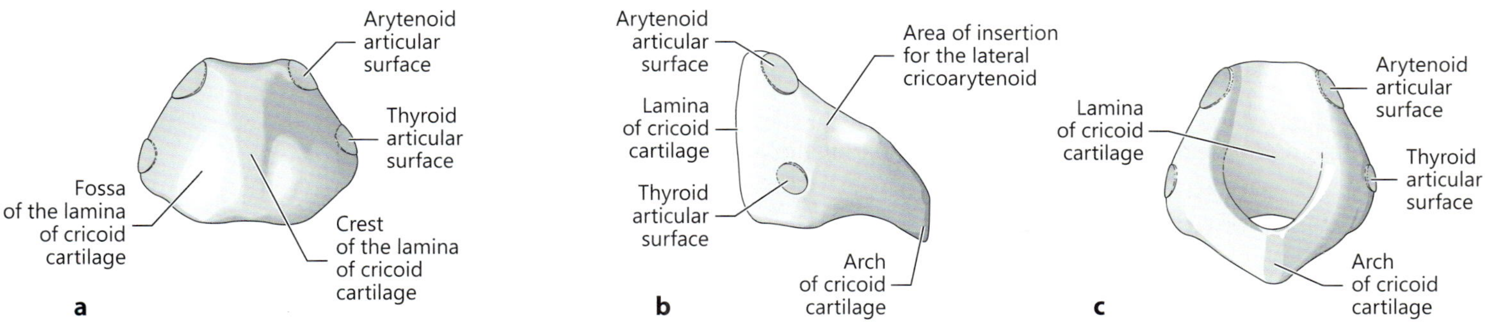

Cricoid cartilage in the male, viewed in dorsal (**a**), lateral (**b**) and anterior (**c**) projection. The areas indicated in *darker gray* correspond to the articular surfaces.

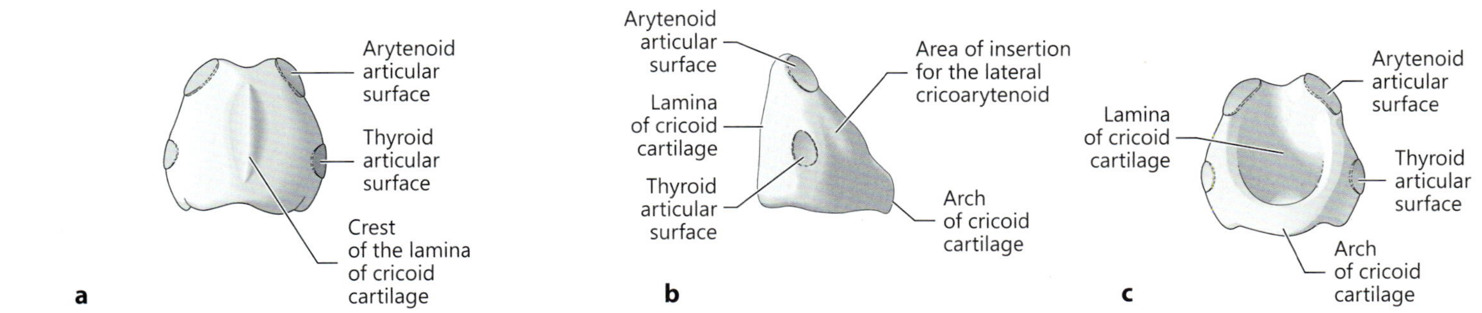

Cricoid cartilage in the female, viewed in dorsal (**a**), lateral (**b**) and anterior (**c**) projection. The cavity demarcated by the cartilaginous ring is more rounded compared to the male larynx.

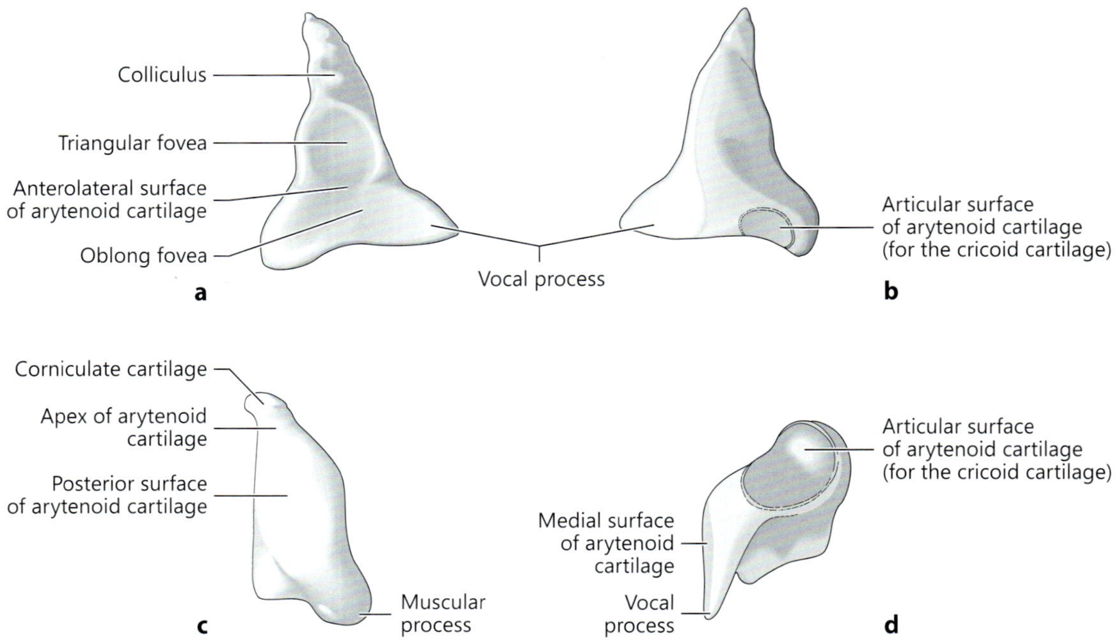

Right arytenoid cartilage of a male, in lateral (**a**), medial (**b**), posterior (**c**) and inferior (**d**) projection.

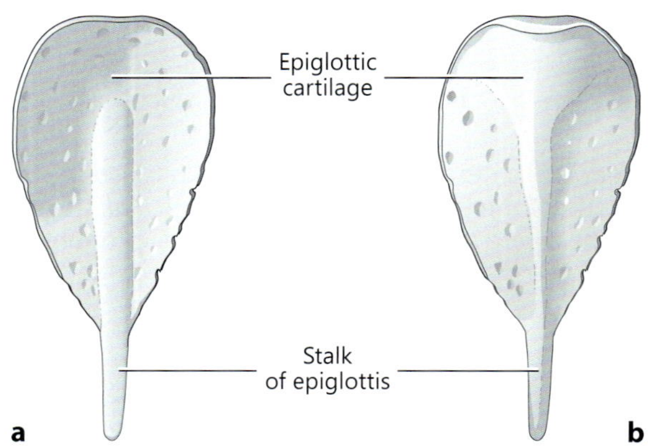

Conformation of the **epiglottic cartilage** and characteristics of its dorsal (**a**) and ventral (**b**) surfaces. The dorsal surface of the cartilage is perforated by small pits that lodge glandular lobules.

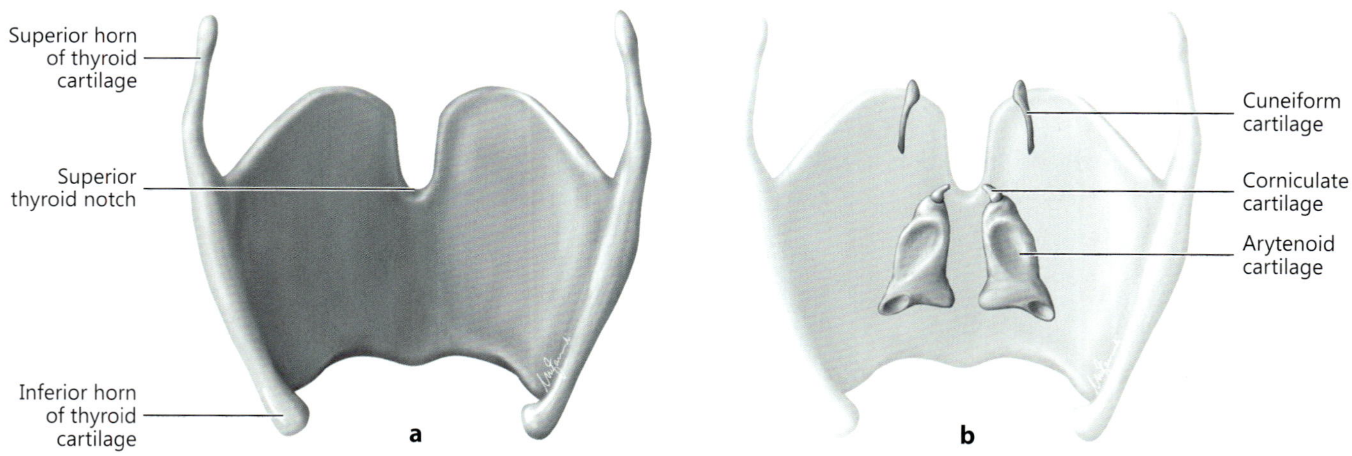

Cartilages of the larynx. **a**, Posterior view of the thyroid cartilage, which is located in the anterior, lateral and superior part of the larynx. **b**, The arytenoid cartilages, connected to which are the corniculate cartilages, are located in the superior and posterior part of the larynx. The cuneiform cartilages, parallel to the anterior margin of the arytenoid cartilages, are considered to be detached fragments of the epiglottic cartilage that run in the thickness of the quadrangular membrane. The posterior pole of the cuneiform cartilage, facing backwards and upwards, is responsible for the formation of the cuneiform tubercle that can be seen in the aryepiglottic fold.

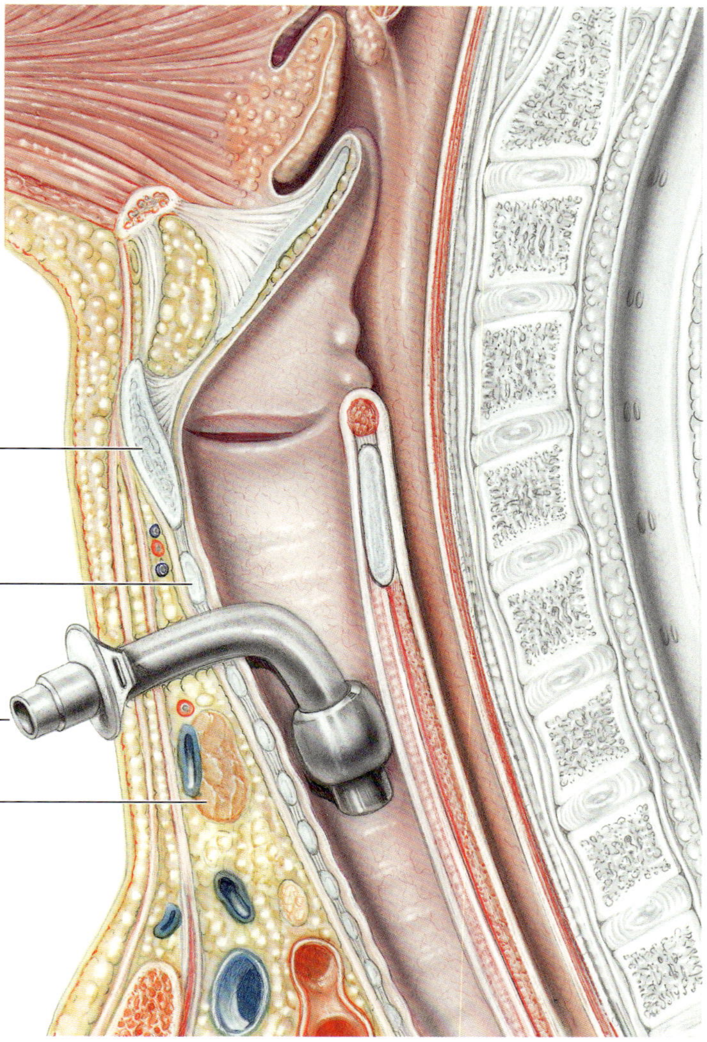

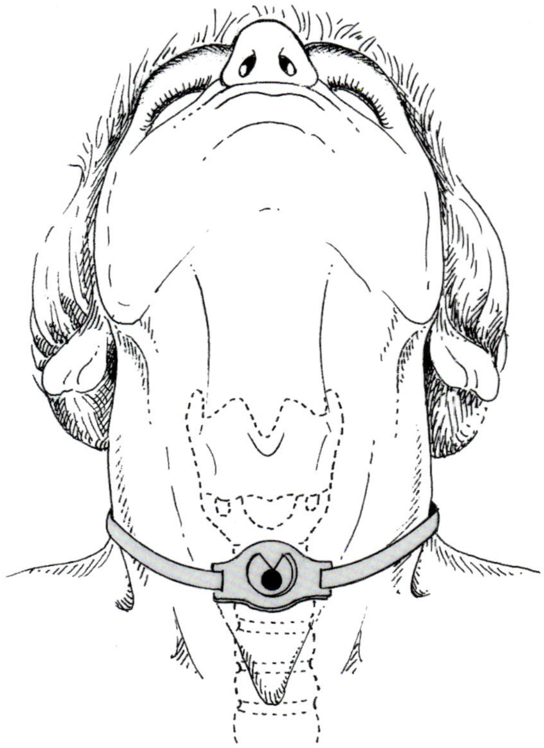

A **tracheotomy** consists of an incision of the trachea made in the neck, to connect the trachea itself with the exterior. Through the incision a cannula is introduced which allows air to reach the lungs.

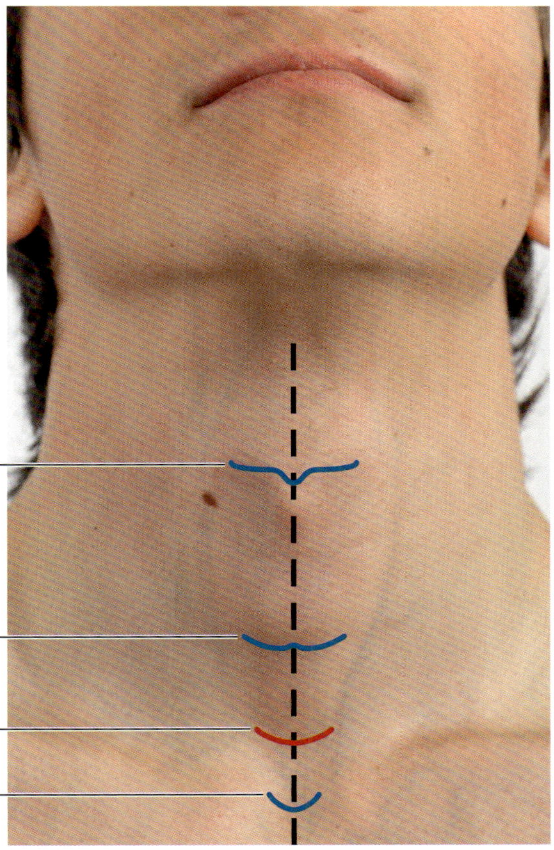

Landmarks for the tracheotomy. By palpation one can identify the two laminae and the superior notch of the thyroid cartilage. The location of the jugular notch corresponds to the superior border of the manubrium of the sternum between the clavicular notches.

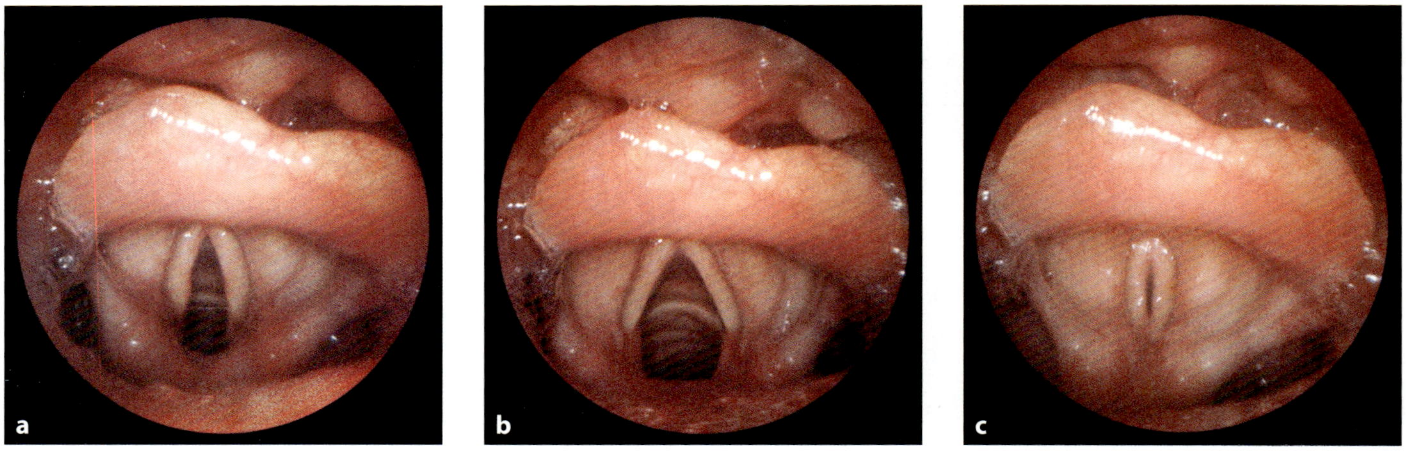

Rima glottidis observed with the laryngoscope. **a**, Open during respiration; **b**, open to maximum extent during a forced inspiration; **c**, closed during phonation.

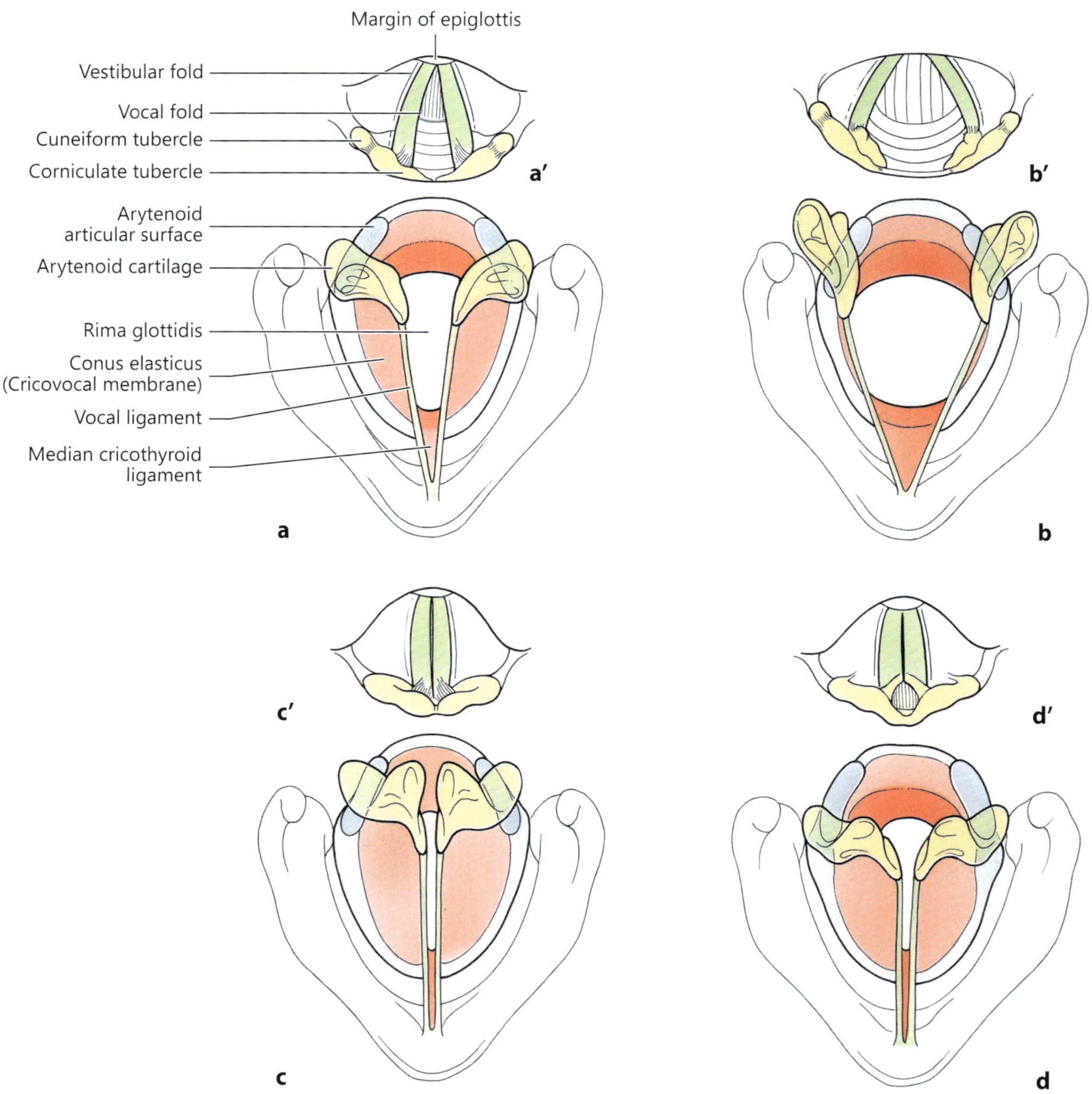

Characteristics of the **rima glottidis** and its variations in preparations of the laryngeal skeleton (**a**, **b**, **c**, **d**) and in laryngoscope images (**a′**, **b′**, **c′**, **d′**). **a-a′**, In vocal rest position; **b-b′**, during a normal phase of breathing; **c-c′**, during phonation with normal vocal tone; **d-d′**, during whisper phonation.

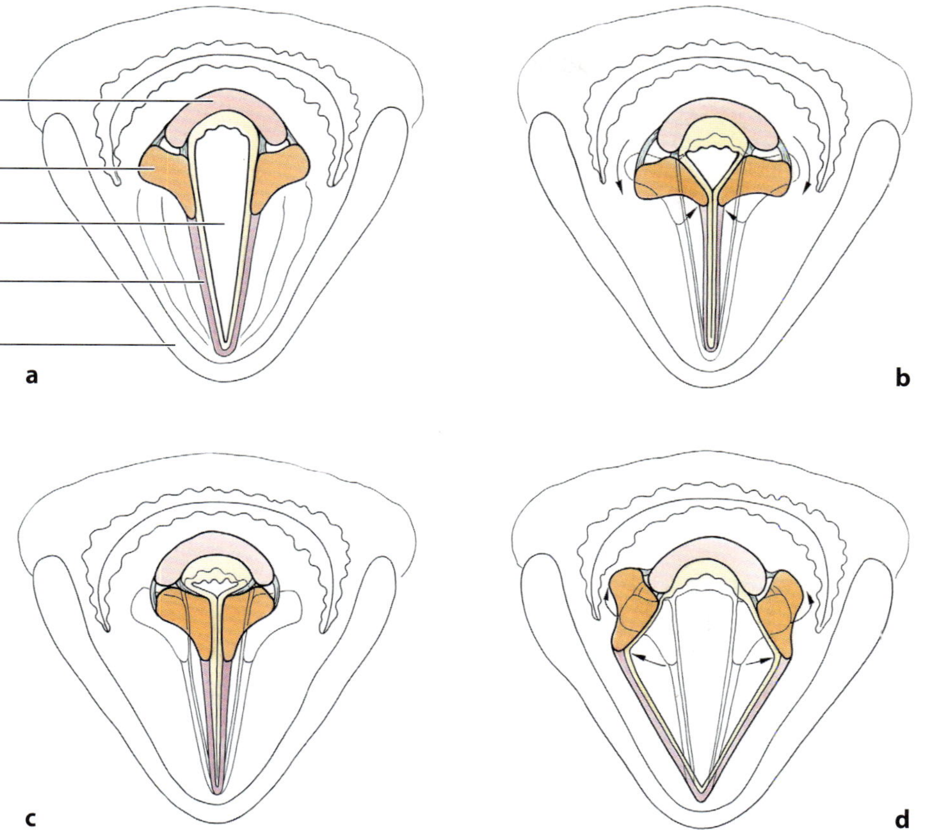

Position and movements of the **arytenoid cartilages** and variations of the rima glottidis. **a**, Position of vocal rest. **b**, Adduction of the vocal processes and tension of the vocal folds. **c**, Adduction of the arytenoid cartilages as a whole and tension of the vocal folds. **d**, Abduction of the vocal processes with consequent opening of the rima glottidis without tension of the vocal folds.

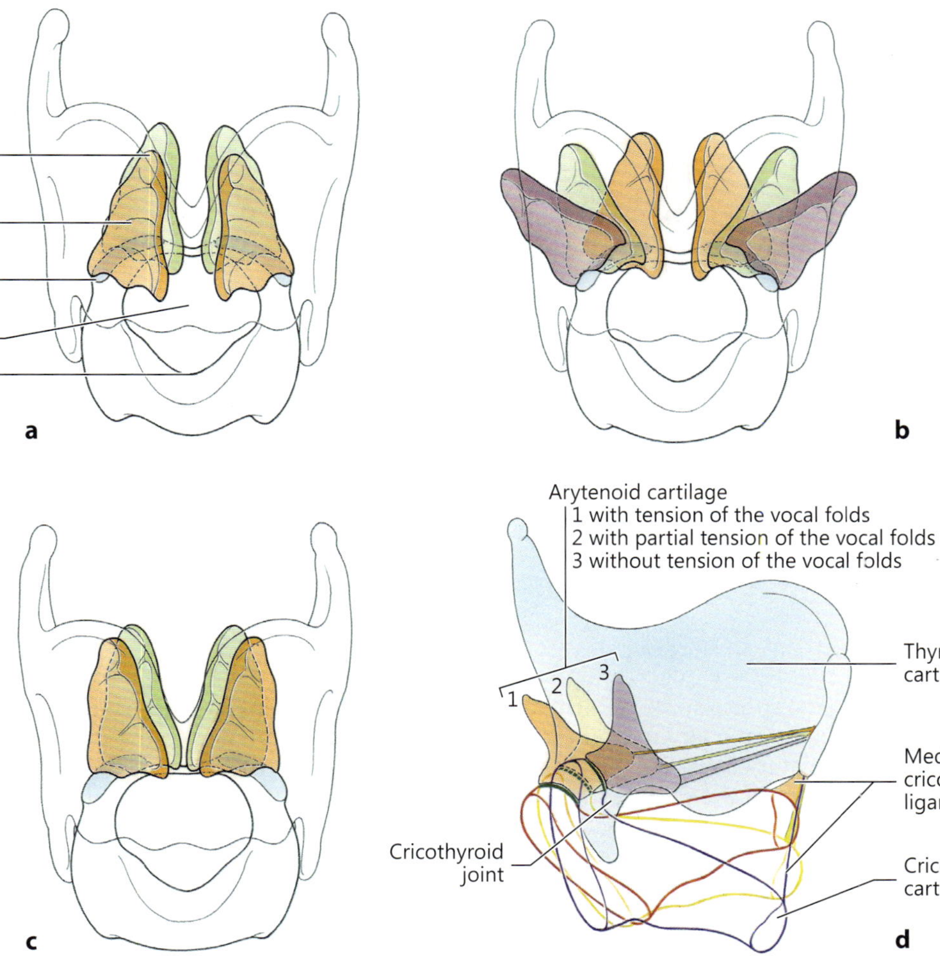

Diagram illustrating the movements of the **arytenoid cartilages** with respect to the **cricoid cartilage**. **a**, Movement in anteroposterior direction; **b**, movement in lateromedial direction with inclination; **c**, movement in lateromedial direction without inclination; **d**, relative movements of the arytenoid cartilages and cricoid cartilage in relation to the different length of the vocal folds.

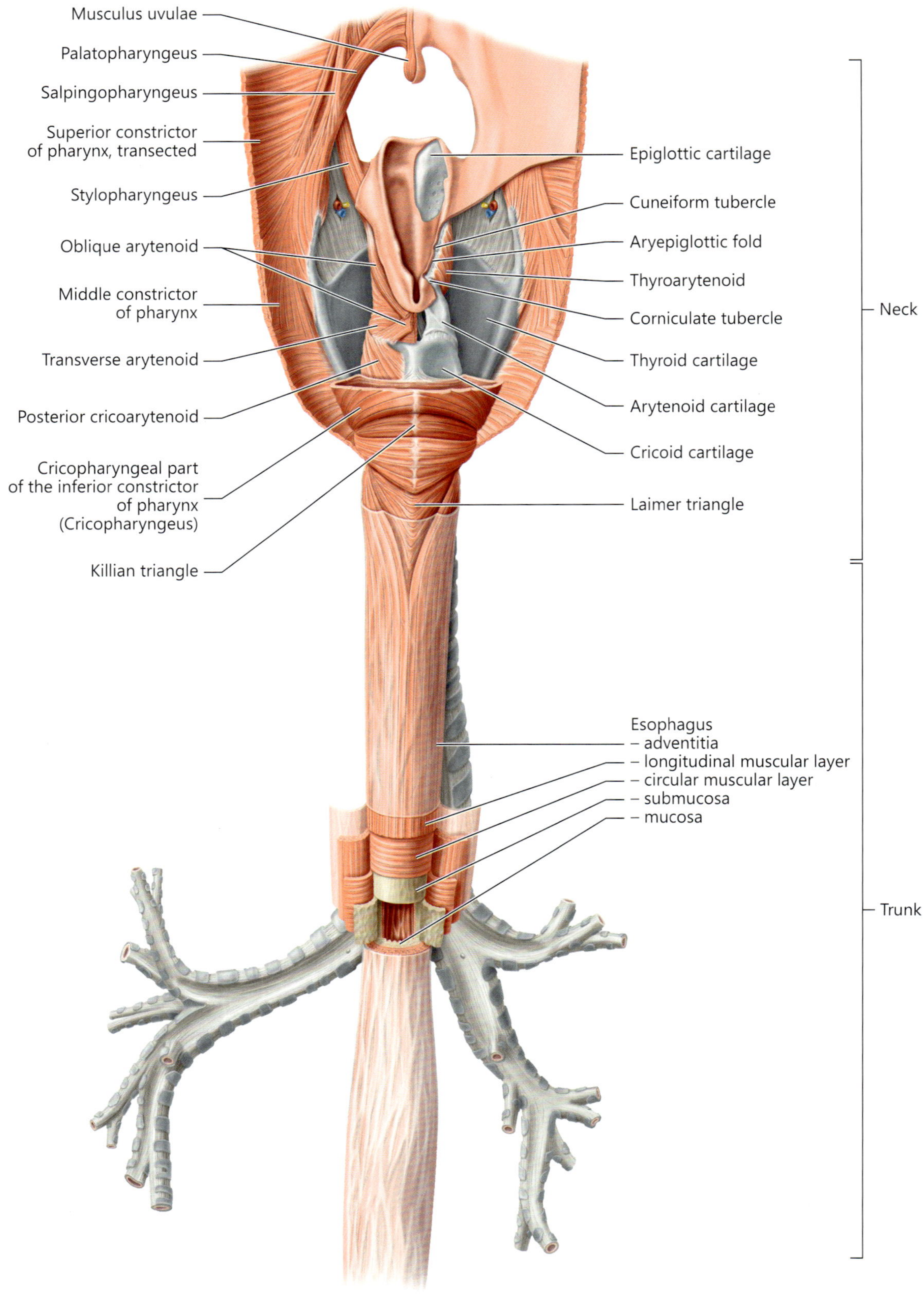

The **esophagus**, approximately 25 cm long, extends from C VI to T XI. It starts from the pharynx in the neck, at the level of the cricoid cartilage of the larynx, and proceeds into the thorax, running in the posterior mediastinum. It terminates in the abdominal cavity at the level of the cardia with the abdominal part of the esophagus.
The superior third of the esophagus presents a striated muscular layer, which plays a role in the swallowing process. In the inferior two-thirds, the muscular coat is composed of two layers of smooth muscle, an inner circular one and an outer longitudinal one (transected in the figure). The anterior surface of the esophagus adheres to the posterior surface of the trachea by means of small bundles of loose connective tissue and smooth muscle (tracheoesophageal muscle).

The structure of the esophagus is represented on page 314 (Thorax - Esophagus)

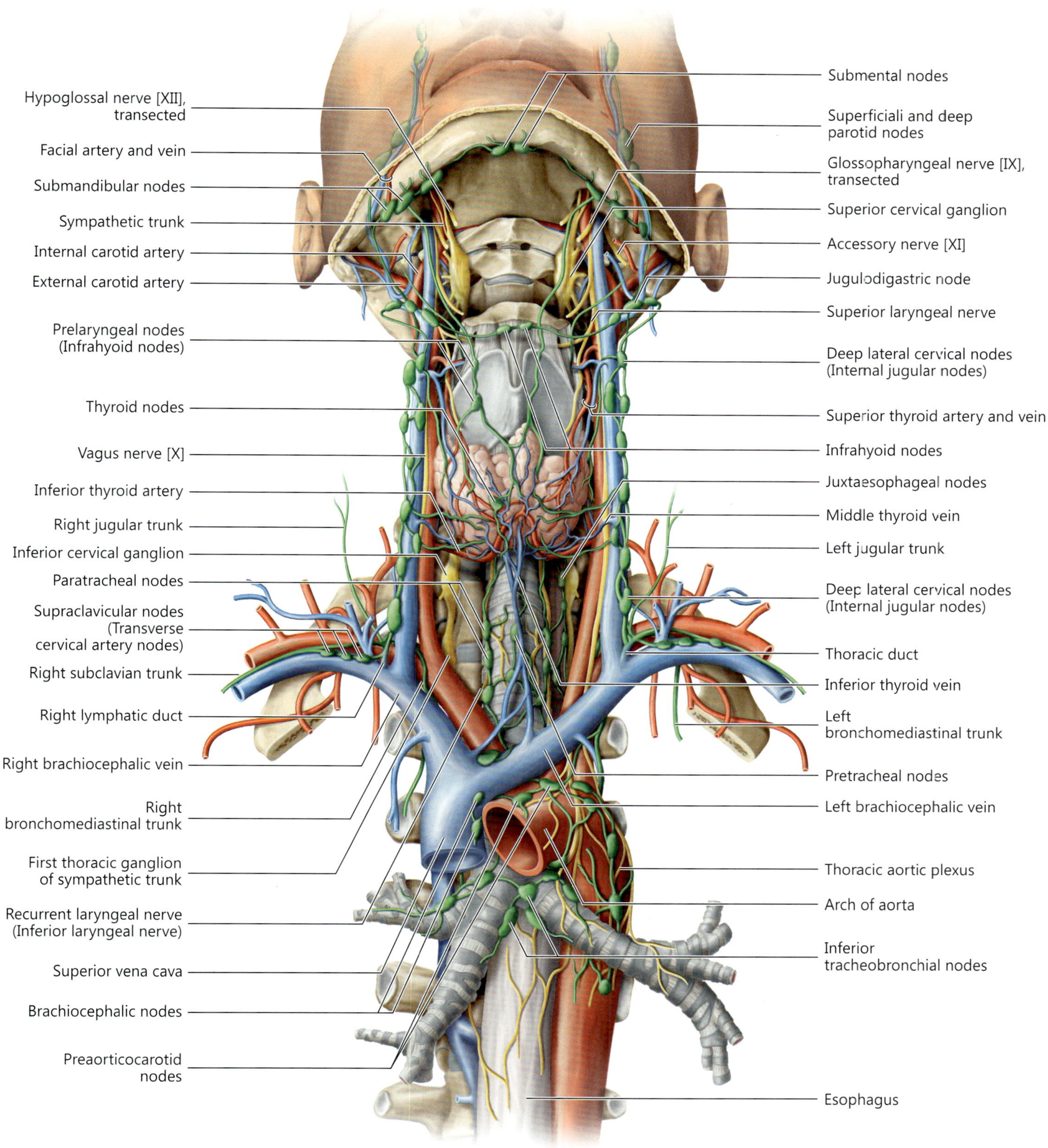

Neck, anterior view. The two common carotid arteries run vertically from the sternoclavicular joint to the point halfway between the angle of the mandible and mastoid process of the temporal bone. The two vessels are separated by a space occupied by the trachea at the base of the neck and, superiorly, by the thyroid gland, larynx and pharynx. The internal jugular veins descend in the neck externally and posteriorly to the internal carotid arteries and then laterally to the common carotid arteries.

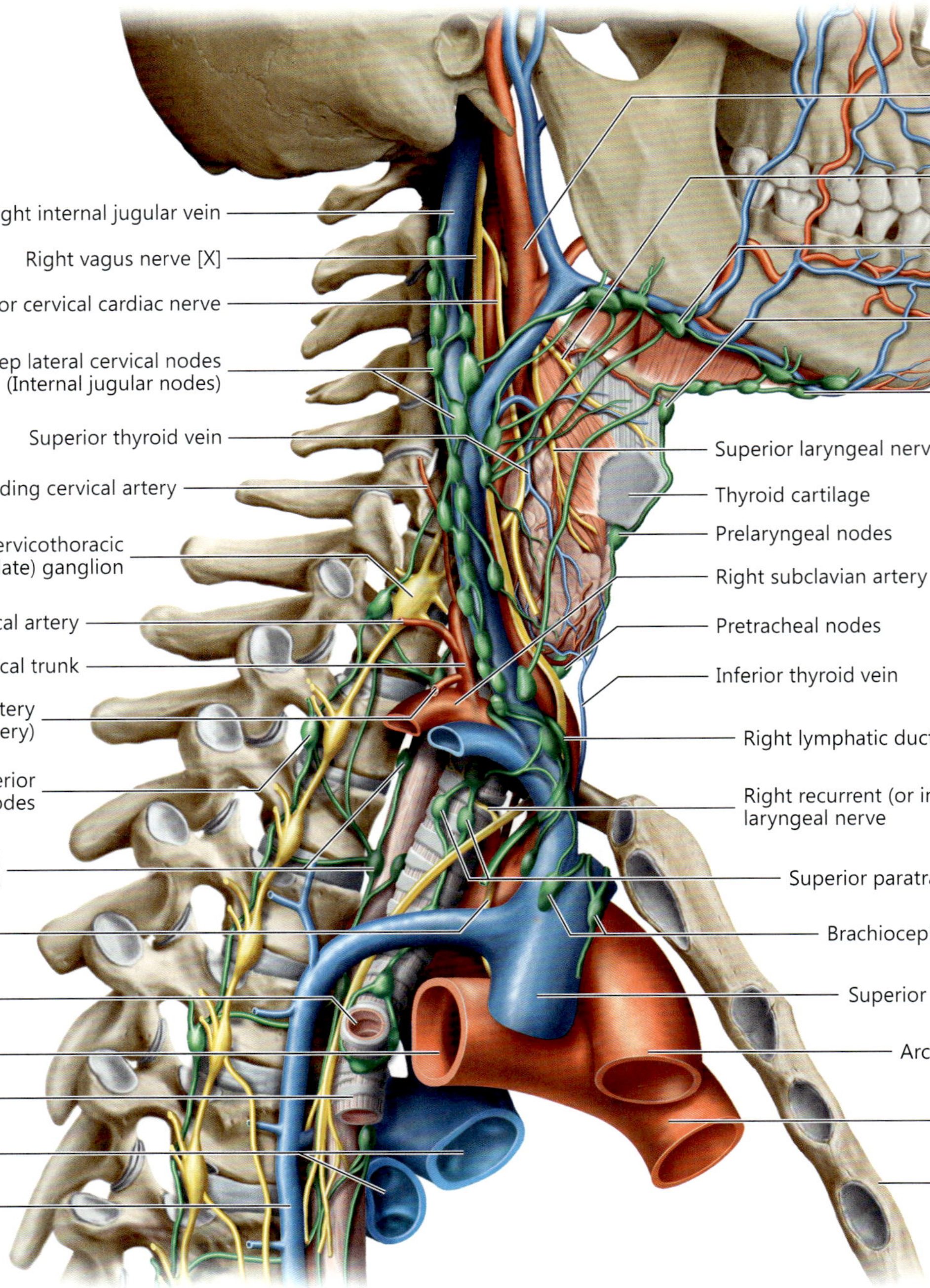

Vessels and nerves of the neck, right lateral view. One can see, in particular, the internal jugular vein, the vagus nerve [X] with the superior and the recurrent (or inferior) laryngeal nerves, the superior cervical cardiac nerve, which arises from the superior cervical ganglion of the sympathetic trunk, the cervicothoracic (or stellate) ganglion, the superior thyroid vein, an affluent of the internal jugular vein, and the inferior thyroid vein, an affluent of the brachiocephalic vein.

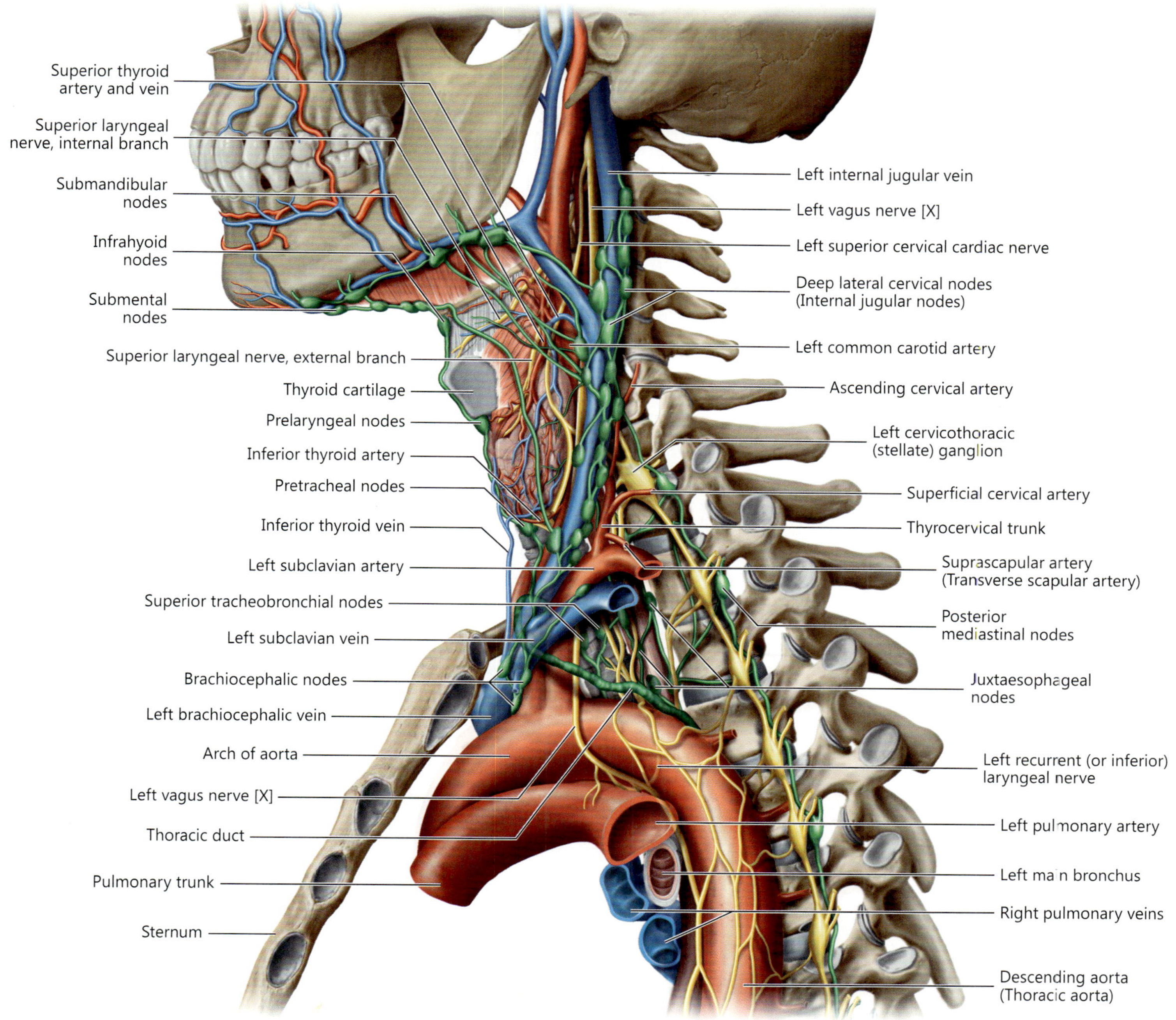

Vessels and nerves of the neck, left lateral view. One can see, in particular: the common carotid artery; the superior and inferior thyroid arteries, which stem, respectively, from the external carotid artery and thyrocervical trunk; the thyrocervical trunk, from which arise also the ascending cervical artery (which otherwise can originate from the inferior thyroid artery); the suprascapular (or transverse scapular) artery, and the superficial cervical artery (which can otherwise originate from the transverse cervical artery, itself a branch of either the thyrocervical trunk or the subclavian artery).

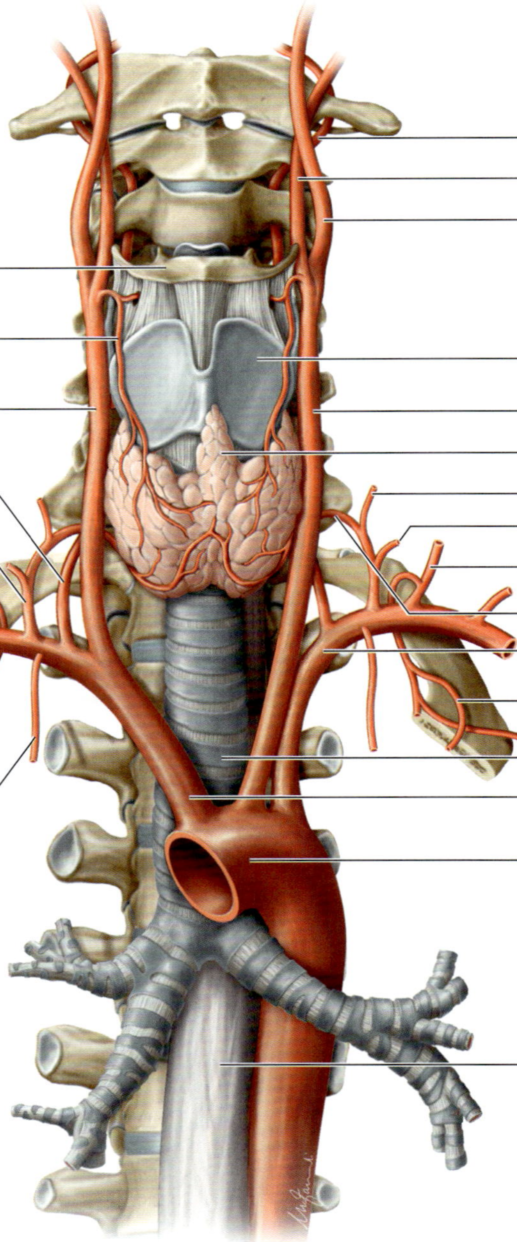

Arteries of the neck, anterior view. The **common carotid artery** originates on the right side from the brachiocephalic trunk, behind the sternoclavicular joint, while on the left it arises directly from the arch of the aorta. At the level of the superior border of the thyroid cartilage of the larynx, it divides into the internal and external carotid arteries. The **subclavian artery**, that originates on the right from the brachiocephalic trunk and on the left from the arch of the aorta, can be subdivided into three portions: the first, from its origin to the medial border of the scalenus anterior; the second, posterior to this muscle; and the third, from the lateral border of the scalenus anterior to the external border of the first rib. From the first part arise: superoposteriorly, the **vertebral artery**, which passes through the foramina transversaria of all cervical vertebrae except the seventh; inferiorly, the **internal thoracic artery**; and, anteriorly, the **thyrocervical trunk**, which divides into the **suprascapular** (or transverse scapular), **superficial cervical** and **inferior thyroid arteries**. From the second part of the subclavian artery (on the right) and from the first part (on the left), the **costocervical trunk** arises posteriorly, which then divides into the **supreme intercostal** and **deep cervical** (here not visible) **arteries**. The third portion of the subclavian artery gives rise to the **transverse cervical artery** and to the posterior scapular artery. In one-third of cases, the superficial cervical and posterior scapular arteries originate as superficial and deep branches of the transverse cervical artery. The **superior thyroid artery** originates anteriorly from the external carotid artery at the level of the greater horn of the hyoid bone.

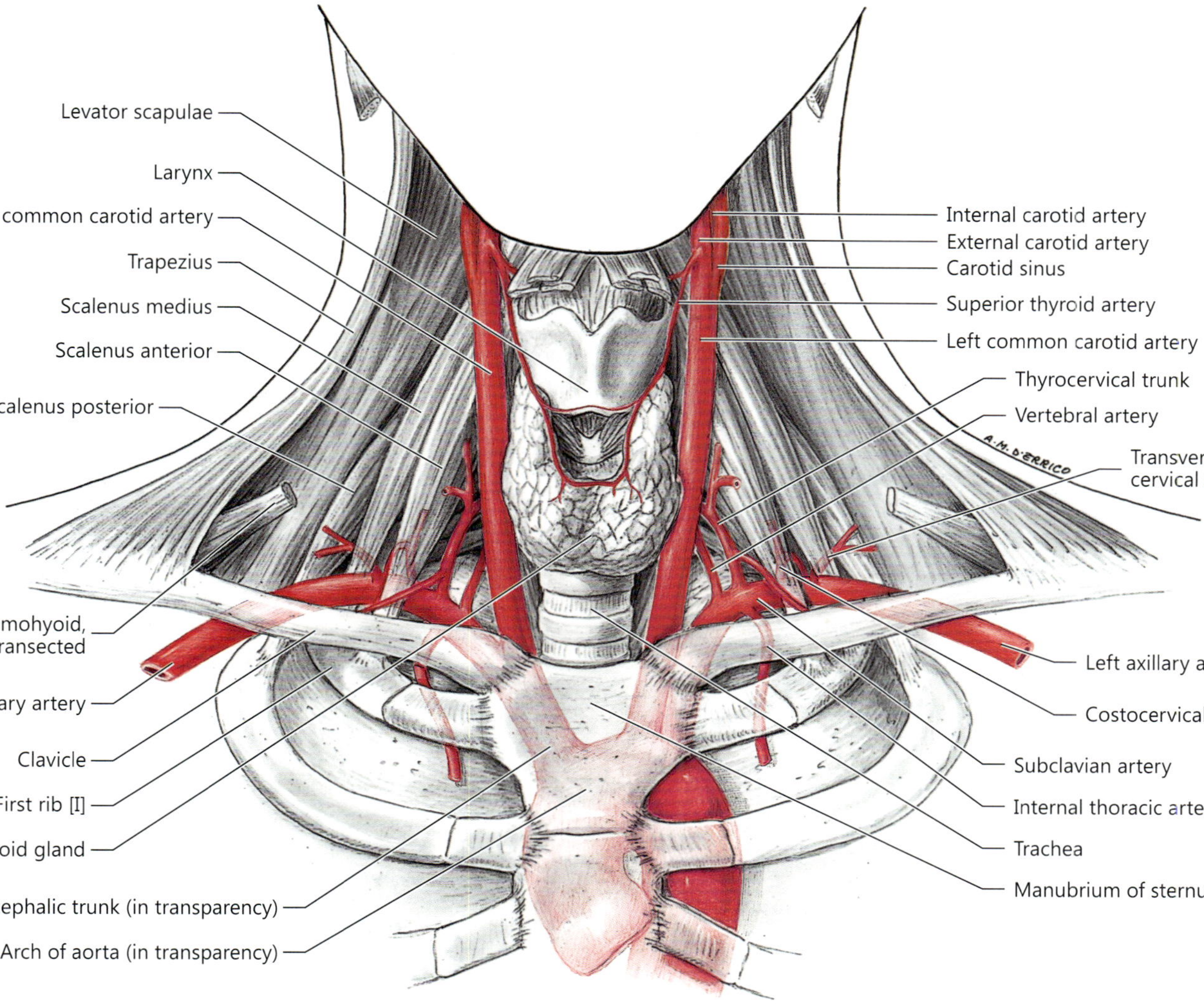

Branches of the **arch of the aorta**. To highlight the position of the arteries and their relations, the anterior, cutaneous and muscular planes of the thorax and neck have been amply demolished. The **brachiocephalic trunk** and the thoracic portion of the **common carotid** and left **subclavian arteries** are visible in transparency. One can see the collateral branches of the subclavian artery (resected at a short distance from their origin) and the **superior thyroid artery**, first collateral branch of the external carotid artery.

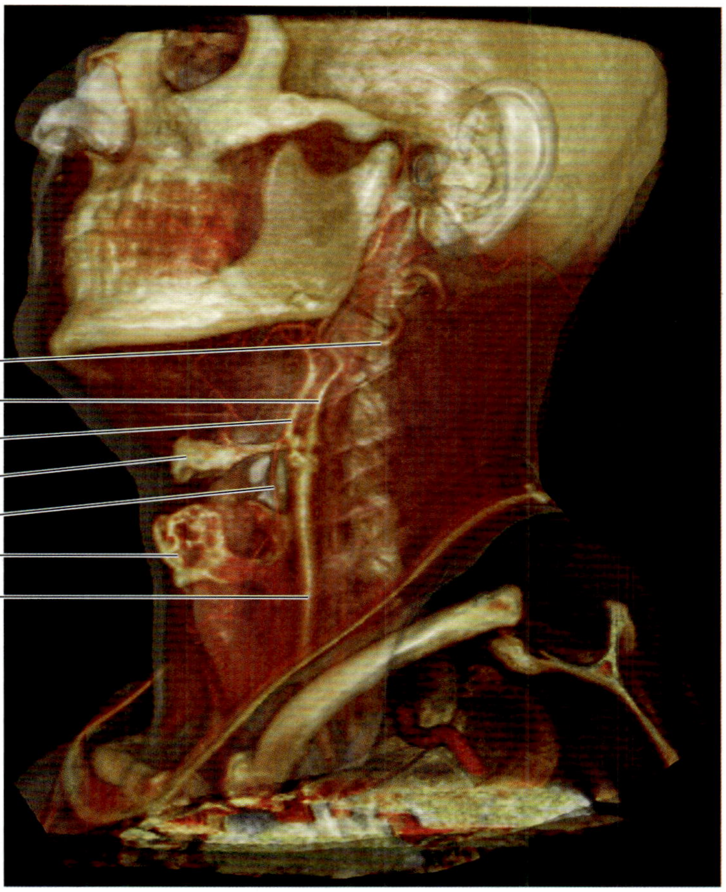

Three-dimensional image of a living human body, obtained by CT scan using a 3D volume rendering technique, showing the **vessels passing through the neck** in relation to the bony structures, in left lateral view.

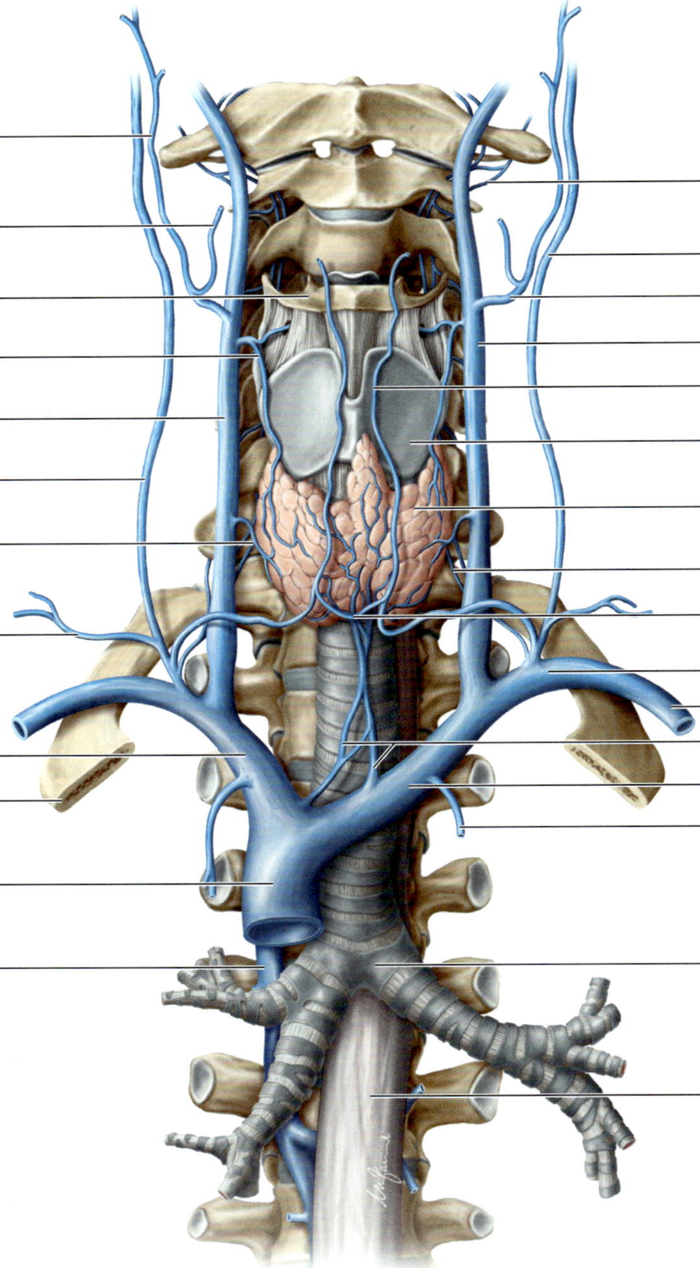

Veins of the neck, anterior view. The **internal jugular vein**, which drains blood from the cranium, brain, superficial parts of the face and much of the neck, descends in the neck and unites with the **subclavian vein** posteriorly to the sternal end of the clavicle, forming the **brachiocephalic vein**. The **external jugular vein** collects blood principally from the scalp and face and is formed by the union of the posterior division of the **retromandibular vein** (or **posterior facial vein**) with the **posterior auricular vein**. It descends in the lateral part of the neck and empties into the subclavian vein at a short distance from its confluence with the internal jugular vein. The **anterior jugular vein**, which drains blood from the skin and muscles of the anterior cervical region, descends anteriorly in the neck to its base and then heads laterally to unite with the termination of the external jugular vein or with the subclavian vein. The two anterior jugular veins are united, just above the manubrium of the sternum, by the jugular venous arch. The **common facial vein**, formed by the union of the anterior branch of the retromandibular vein with the facial vein (or anterior facial vein), is not always present. The facial vein can, in fact, drain directly into the internal jugular vein, while the retromandibular vein can connect to the external jugular vein, through an anastomotic branch, before uniting with the facial vein.

➤

Schematic representation showing the **superior vena cava** and large affluent veins. The drainage territories of the principal vessels are indicated. The superior vena cava collects venous blood from the territories of the body above the diaphragm, with the exception of the heart. It is formed by the confluence of the right and left brachiocephalic veins and the only affluent branch it receives during its course is the azygos vein. It has an almost vertical path and terminates in the posterosuperior part of the right atrium of the heart. The azygos, hemiazygos, accessory hemiazygos and, in part, subclavian and brachiocephalic veins drain territories of the thorax and of the upper limbs.

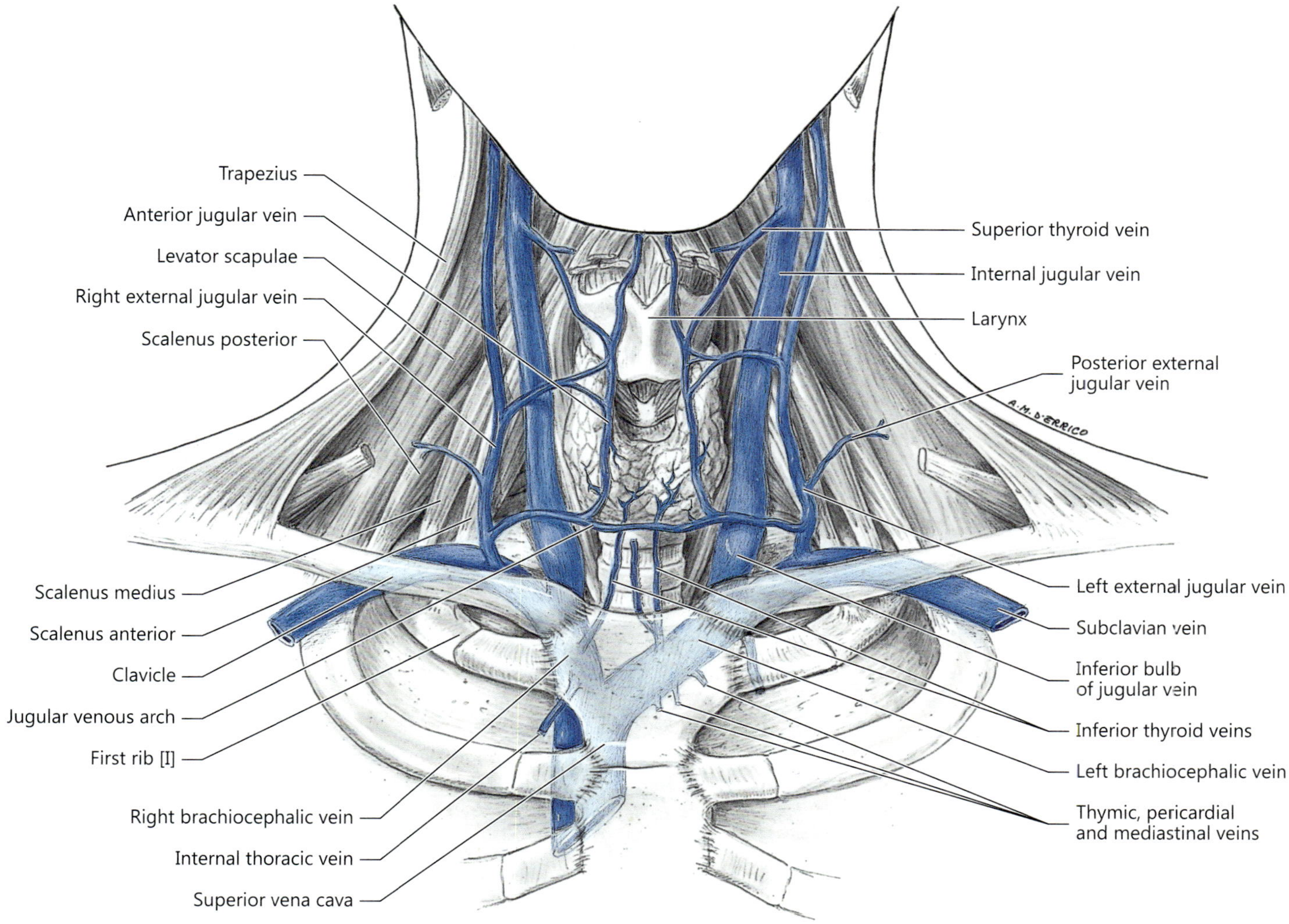

Diagram of the **internal jugular veins**, **subclavian veins** and of their affluent branches. Through the thoracic wall, in transparency, one can see the two **brachiocephalic veins**, right and left, from their origin to the point of their confluence into the **superior vena cava**.

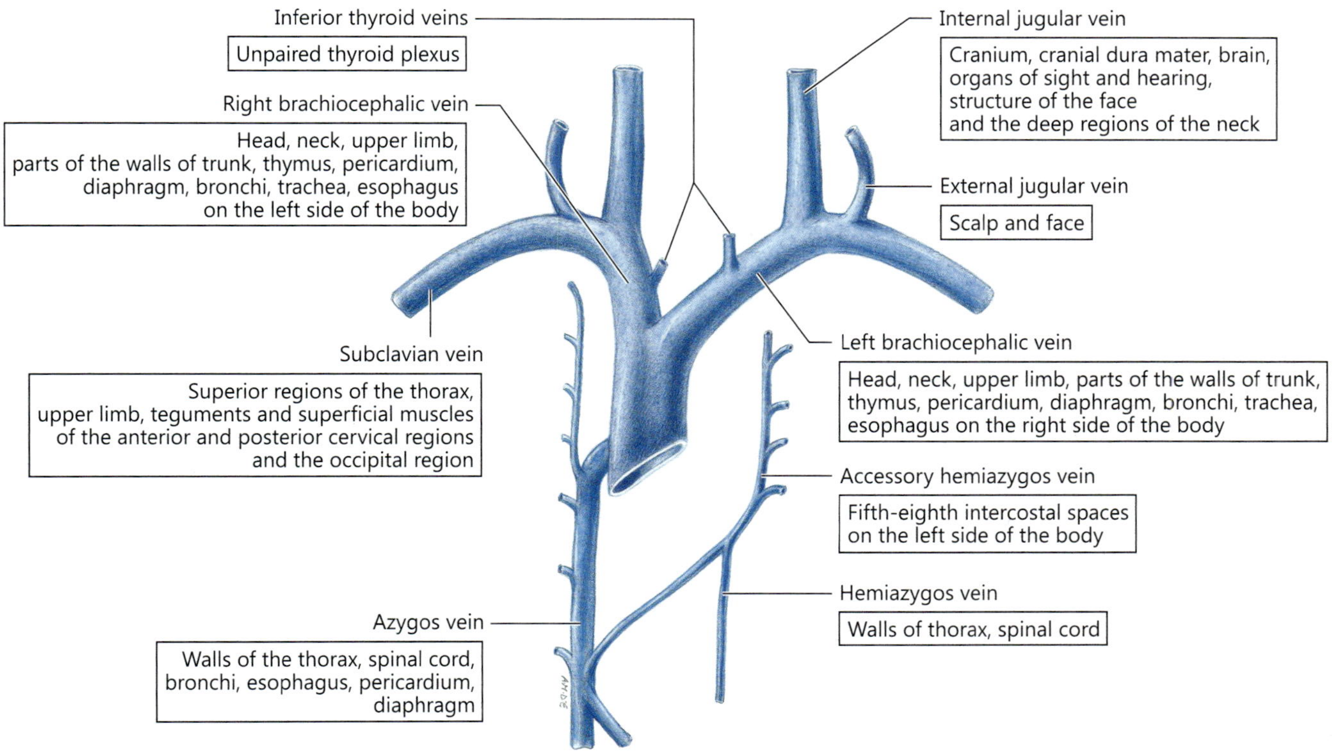

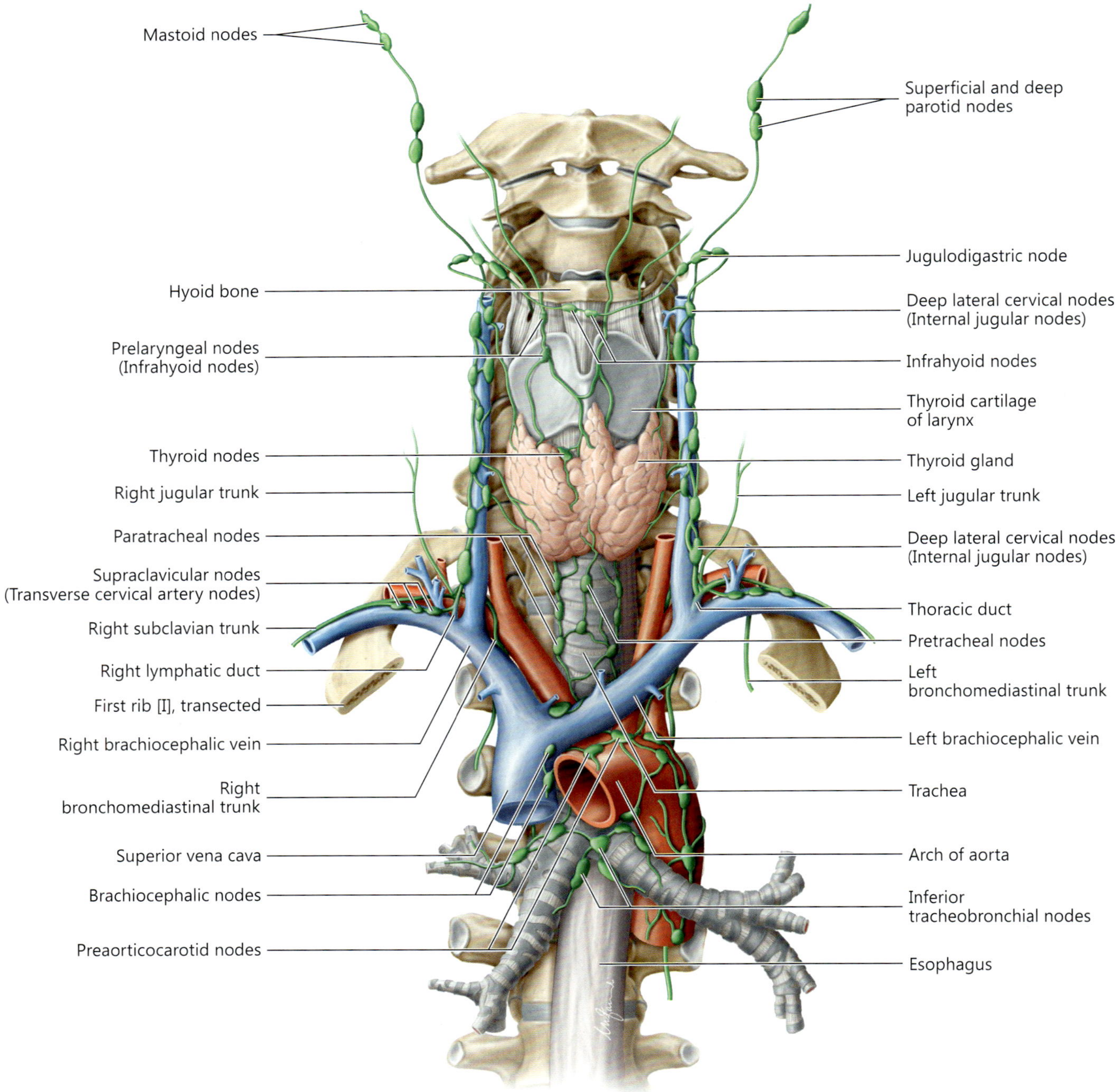

Deep lymph nodes of the neck, anterior view. The **deep anterior cervical nodes** (**anterior visceral nodes**) below the hyoid bone in relation with the larynx (**prelaryngeal infrahyoid nodes**), thyroid gland (**thyroid nodes**) and trachea (**pretracheal and paratracheal nodes**) are represented. Also represented are the **deep lateral cervical nodes** located along the course of the internal jugular vein (**internal jugular nodes**) as well as the **supraclavicular nodes** (**transverse cervical artery nodes**), situated in the supraclavicular fossa, which are related to the transverse cervical artery. The **brachiocephalic nodes**, situated in the superior mediastinum and related to the large vessels, the para-aortic nodes, situated anteriorly and laterally to the ascending aorta and arch of the aorta, and the **inferior tracheobronchial nodes**, situated in correspondence to the tracheal bifurcation, are classified as lymph nodes of the thorax.

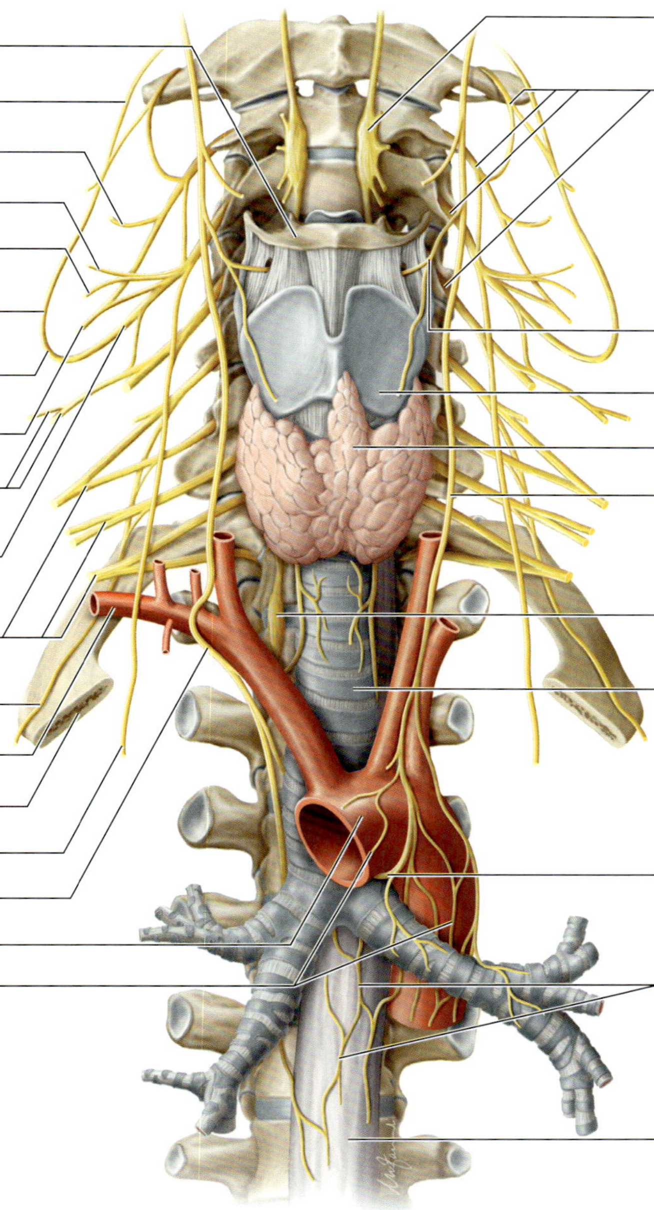

Nerves of the neck, anterior view. The anterior rami of the first four cervical nerves [C1-C4] divide into branches that unite forming anastomotic loops constituting the **cervical plexus**, situated in front of the transverse processes of the cervical vertebrae. The cutaneous branches of the cervical plexus are exclusively sensory nerves: the **lesser occipital**, **great auricular**, **transverse cervical** and **supraclavicular nerves**. The **ansa cervicalis** (or **ansa of the hypoglossal nerve**, cranial nerve XII) is formed by two roots: a superior one, derived from the union with the hypoglossal nerve of an anastomotic branch coming from the anterior ramus of C1, and an inferior one, derived from the union of anastomotic branches coming from the anterior rami of C2 and C3. The **phrenic nerve**, the largest branch of the cervical plexus, originates with a principal root from C4 and with two accessory, inconstant, roots from C3 and C5. The primary trunks of the **brachial plexus**, formed by the anterior rami of C5, C6, C7, C8 and T1, occupy the supraclavicular space, while the cords (secondary trunks) are contained in the axilla. The **vagus nerve** [X] gives rise to numerous branches in the neck, including the **recurrent** (or inferior) **laryngeal nerve** which on the right forms a loop around the subclavian artery, while on the left it loops around the arch of the aorta, as it ascends in the neck. The cervical segment of the sympathetic trunk, which extends from the cranial base to the superior thoracic aperture, includes three ganglia, one of which is visible here: the **superior cervical ganglion**, in front of the transverse processes of the second and third cervical vertebrae.

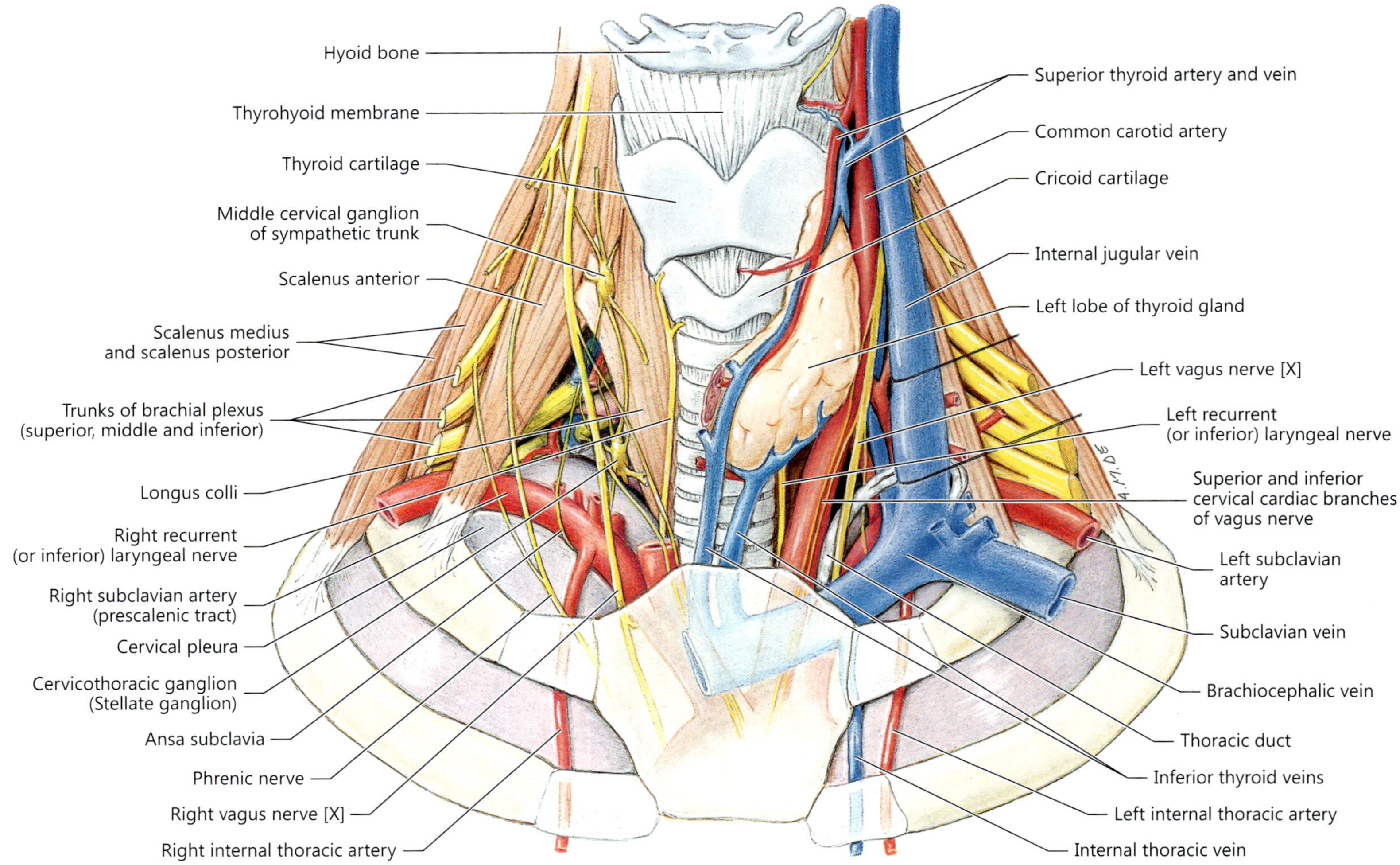

Schematic representation showing the course and relations of the **common carotid** and **subclavian arteries** and of the **subclavian** and **internal jugular veins**. The anterior thoracic wall and anterior cervical region have been amply demolished to highlight the organs and the vascular and nerve formations located behind.

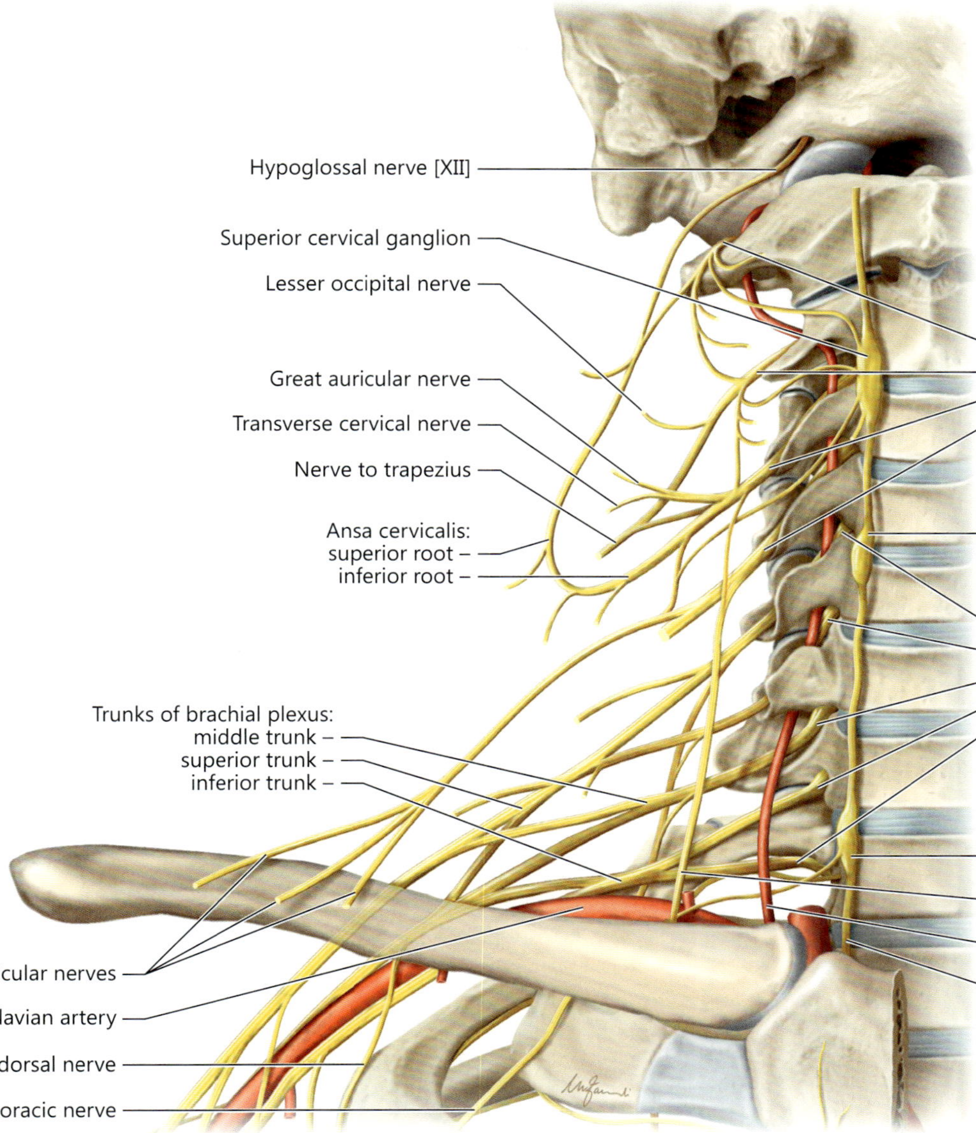

Cervical plexus, **trunks of the brachial plexus**, **ansa cervicalis** and **cervical segment of the sympathetic trunk**, anterior view. The cervical plexus is formed by the anterior rami of the first four cervical nerves that unite to form anastomotic loops, or arches, superimposed in a vertical direction. The anterior ramus of the first cervical nerve anastomoses with the hypoglossal nerve [XII]: part of the anastomotic fibers unite for a brief tract with the hypoglossal nerve, then separate as a descending branch that constitutes the superior root of the ansa cervicalis, or ansa of the hypoglossal nerve, whose inferior root (descending cervical nerve) is formed instead of branches originating from the anterior rami of the second and third cervical nerves which unite, continuing in the superior root. The brachial plexus, formed by anterior rami of the last four cervical nerves and first thoracic nerve, is characterized by anastomoses and subdivisions that form three (primary) trunks above the clavicle (supraclavicular part of the cervical nerves), from which arise the cords of the plexus (or secondary trunks).

For a lateral view and schematic representation of the cervical plexus ☞ page 76.
The brachial plexus is illustrated on page 77.

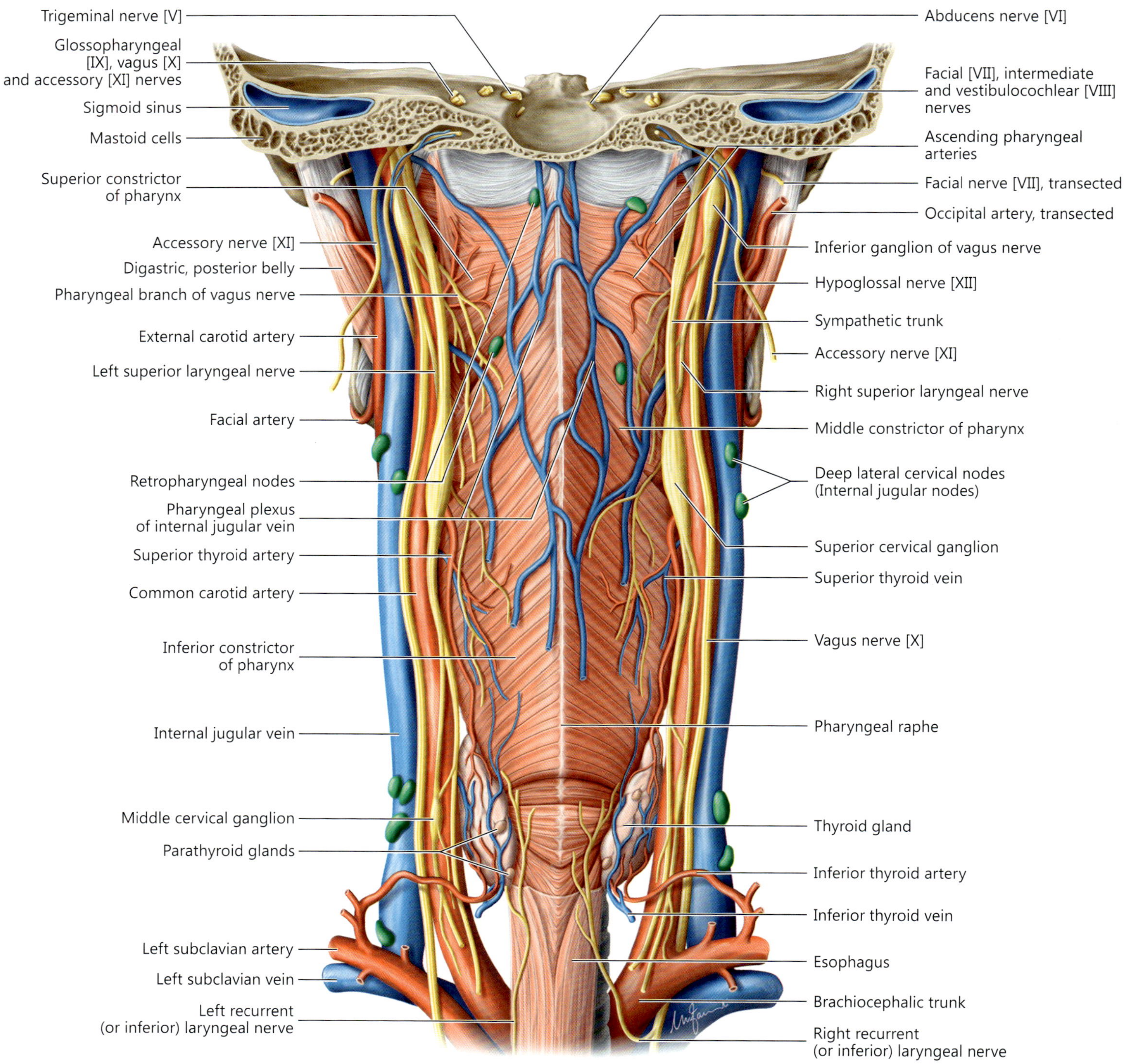

Vessels and nerves in the **parapharyngeal** (**lateral pharyngeal**) **space**, posterior view after removal of the vertebral column and muscles of the back. The **common carotid artery** ascends in the neck up to the superior border of the thyroid cartilage of the larynx where it divides into the **external and internal carotid arteries**. Together with the **internal jugular vein**, which descends in the neck along the lateral wall of the pharynx, at first posteriorly to the internal carotid artery and then laterally to the common carotid artery, and with the **vagus nerve**, in the posteriorly open dihedral angle that the two vessels form alongside each other, it constitutes the *neurovascular bundle of the neck* inside the carotid sheath, which here has been removed. The cervical segment of the sympathetic trunk extends from the cranial base to the superior thoracic aperture, posteriorly to the neurovascular bundle of the neck.

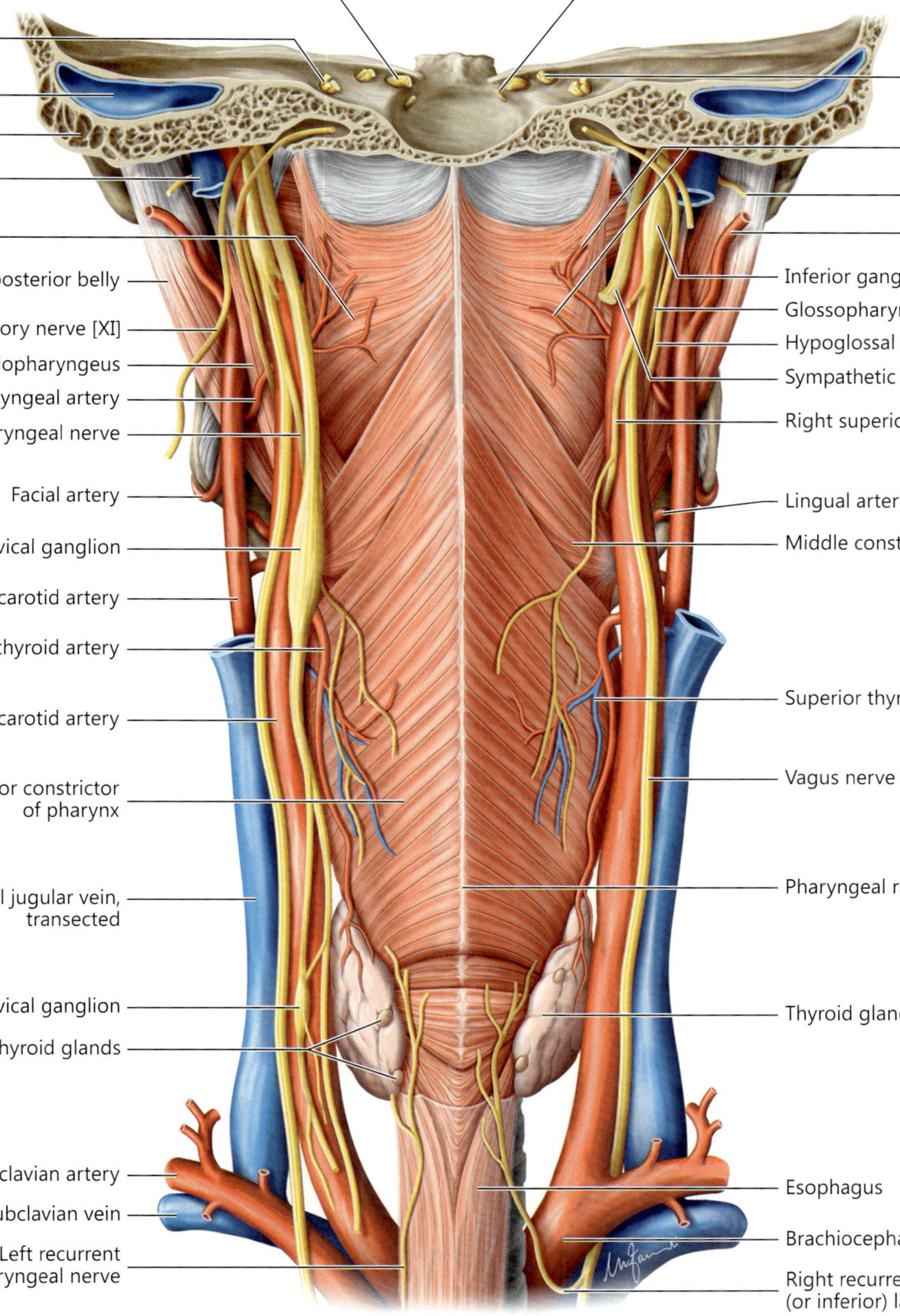

Vessels and nerves in the **parapharyngeal** (**lateral pharyngeal**) **space**, posterior view after removal of the vertebral column and muscles of the back. The internal jugular vein has been transected revealing the **external carotid artery**, and the origins of its collateral branches such as the **superior thyroid artery**, the **facial artery**, the **lingual artery**, the **ascending pharyngeal artery** and the **occipital artery**. The **superior laryngeal nerve** departs from the inferior, or nodose, ganglion of the vagus nerve or slightly below. The **recurrent laryngeal**, or inferior, **nerve** is so defined because of the loop that it makes, first descending into the thorax and then re-ascending in the neck to reach the larynx.

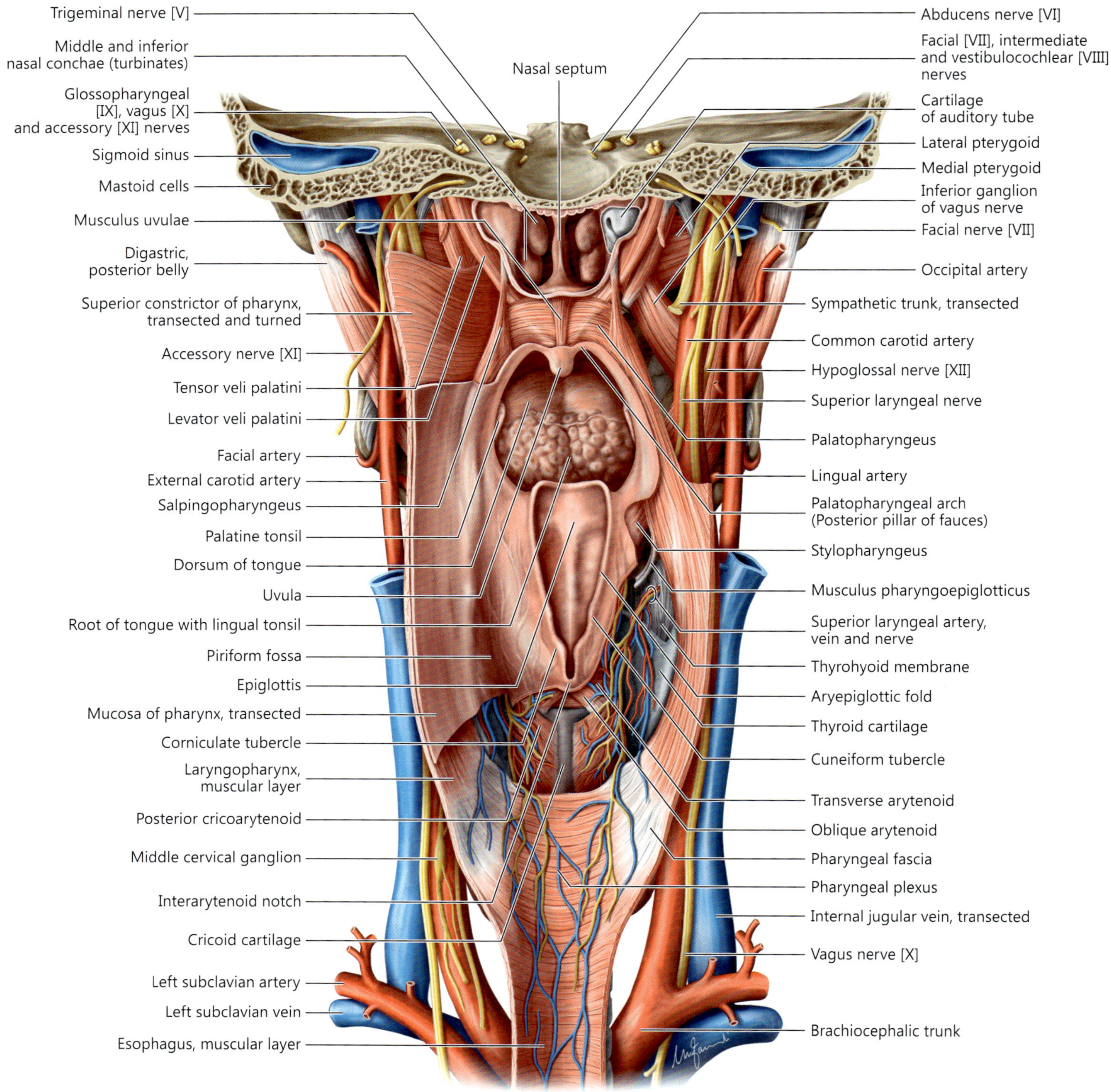

Posterior view of the **pharynx** after opening its posterior wall. One can see the anterior wall of the pharynx, which is complete only in correspondence to the posterior surface of the larynx. Above it, one finds, from bottom to top: the **laryngeal inlet**, an oval opening bounded anteriorly by the free border of the epiglottis and laterally and posteriorly by the aryepiglottic folds; the posterior surface of the **epiglottis**; the **isthmus of the fauces**, formed by the root of the tongue with the **lingual tonsil** on its dorsal surface, by the free border of the soft palate with the **uvula** and by the **palatoglossal and palatopharyngeal arches**; the posterior surface of the **soft palate** and the **choanae**, quadrangular openings bounded superiorly by the body of the sphenoid, laterally by the medial plates of the pterygoid processes of the sphenoid, medially by the posterior border of the vomer and inferiorly by the posterior border of the horizontal plate of the palatine bone.

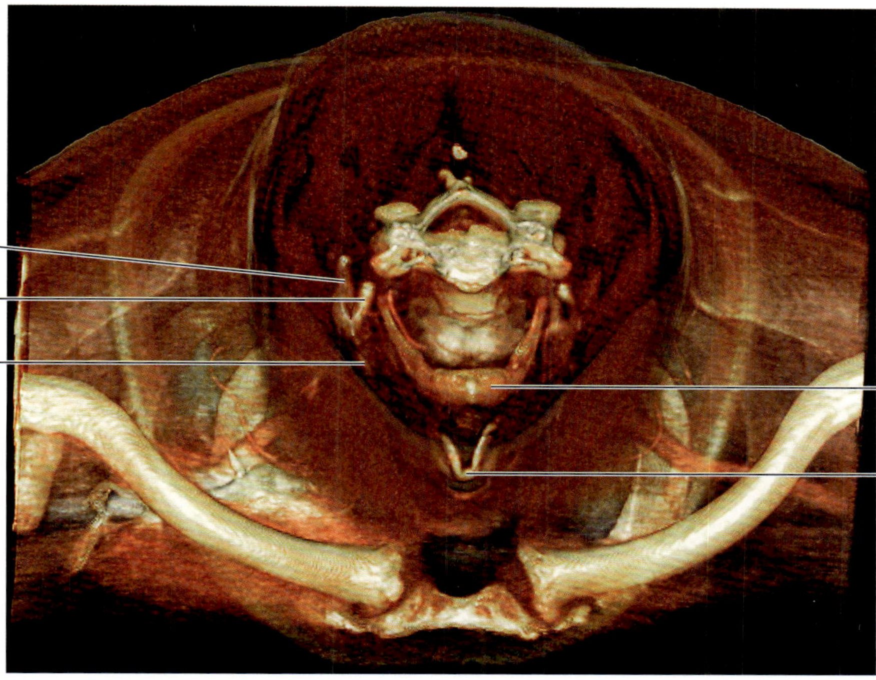

Three-dimensional image of a living human body, obtained by CT scan using a 3D volume rendering technique, showing the **vessels passing through the neck** in relation to the bony structures, in a transverse section immediately above the hyoid bone.

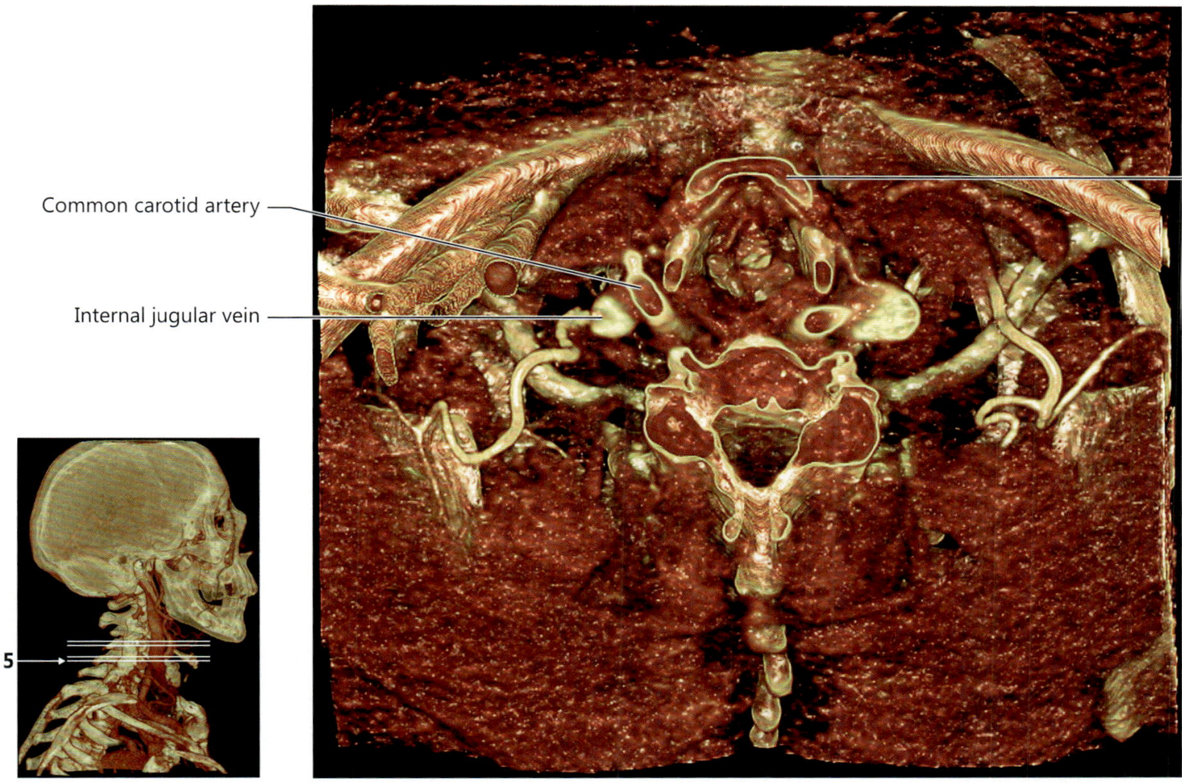

Three-dimensional image of a living human body, obtained by CT scan using a 3D volume rendering technique, showing the **vessels passing through the neck**, in a transverse section immediately below the inferior margin of the mandible before the bifurcation of the common carotid artery, on a plane (5) that passes through the fifth cervical vertebra.

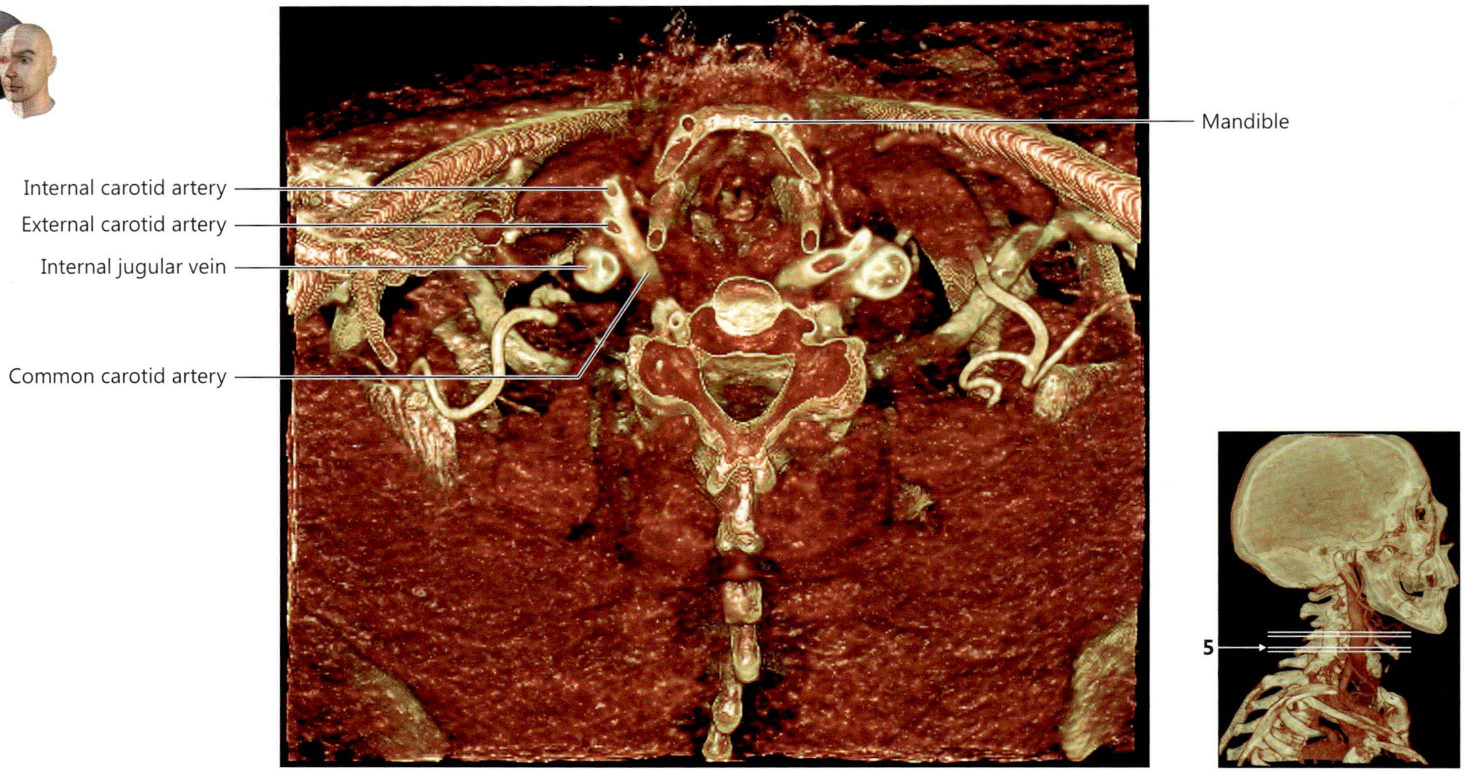

Three-dimensional image of a living human body, obtained by CT scan using a 3D volume rendering technique, showing the **vessels passing through the neck**, in a transverse section immediately above the bifurcation of the common carotid artery, on a plane (5) that passes through the fifth cervical vertebra.

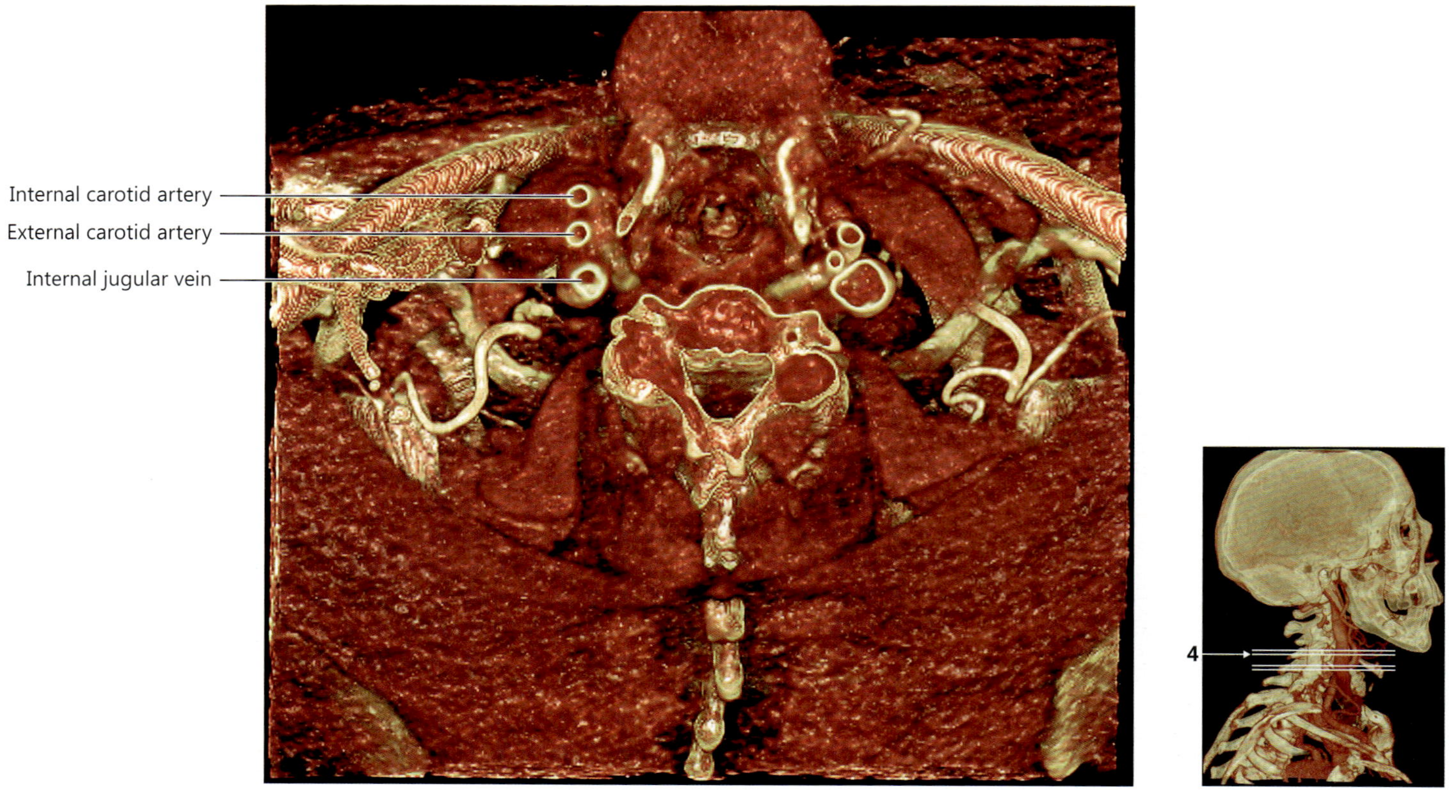

Three-dimensional image of a living human body, obtained by CT scan using a 3D volume rendering technique, showing the **vessels passing through the neck**, in a transverse section immediately above the bifurcation of the common carotid artery, on a plane (4) that passes through the fourth cervical vertebra.

THORAX

Petros (Athens, 1928 - Milan, 2014), *Pitagoro archi*, oil on canvas, cm 80 × 60, 1991

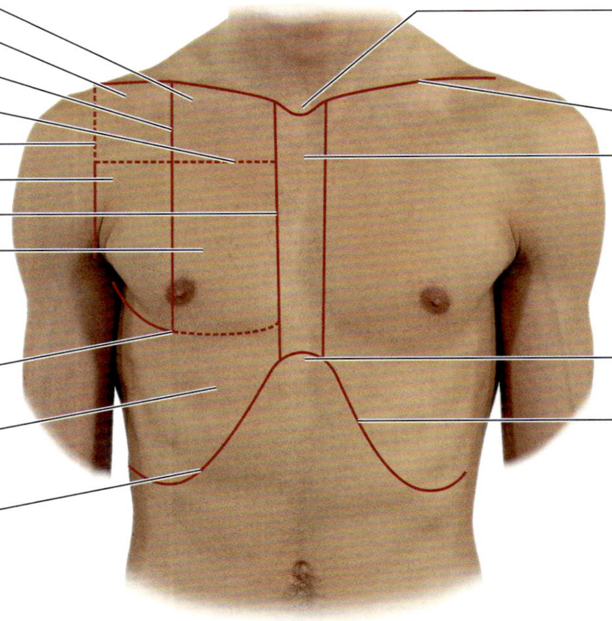

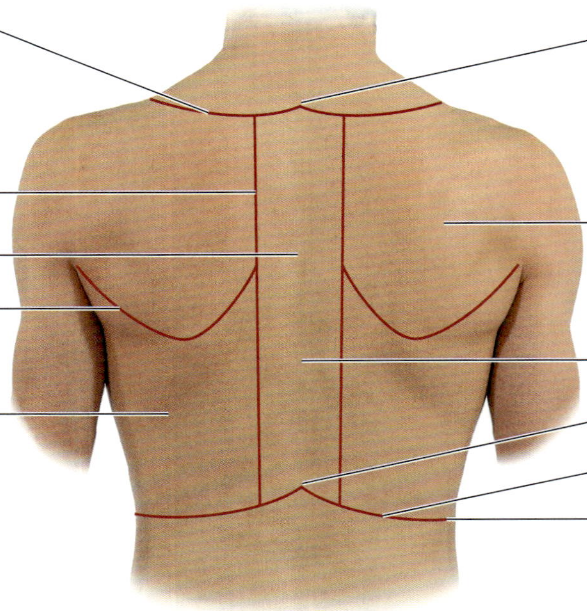

THORACIC REGIONS	***BOUNDARIES OR LOCATION***
Infraclavicular region and infraclavicular fossa (clavipectoral or deltopectoral triangle)	Infraclavicular fossa: depression of the skin between the clavicle (superiorly), deltoid (laterally) and pectoralis major (medially), to which the clavipectoral triangle corresponds at deep level. Superior, inferior, medial and lateral boundaries of the infraclavicular region: inferior border of the neck, line passing over the third rib, sternal line and anterior axillary line
Costal	Superficial area of the thorax between the inframammary sulcus and costal margin, including, posteriorly, the infrascapular region. Inferior boundary: thoracoabdominal line. Note: the diaphragmatic cupula is situated at a higher level than the thoracoabdominal line and some abdominal organs are located in a zone belonging superficially to the thorax, but deeply to the abdomen (hypochondrium)
Pectoral	Area overlying the pectoralis major
– inframammary	Area of the pectoral region underlying the mammary region. Inferior boundary: inframammary sulcus
– lateral pectoral	Area of the pectoral region between the midclavicular (medially) and anterior axillary (laterally) lines, including the anterior and medial walls of the axilla (pectoralis major, pectoralis minor and serratus anterior)
– mammary	Area of the pectoral region corresponding to the breast
Presternal (or sternal)	Area in front of the sternum, between the jugular notch (superiorly), xiphoid process (inferiorly) and sternal lines (laterally)
Vertebral (at the thoracic level)	Region of the back, at the thoracic level, between the spinous processes of the seventh cervical vertebra [C VII] (superiorly) and the twelfth thoracic vertebra [T XII] (inferiorly) and vertical lines passing through the medial border of the scapula or the angle of the rib
Infrascapular	Region of the back underlying the scapular region. Inferior and superior boundaries: twelfth rib [XII] and scapula
Interscapular	Region of the back between the two scapulae

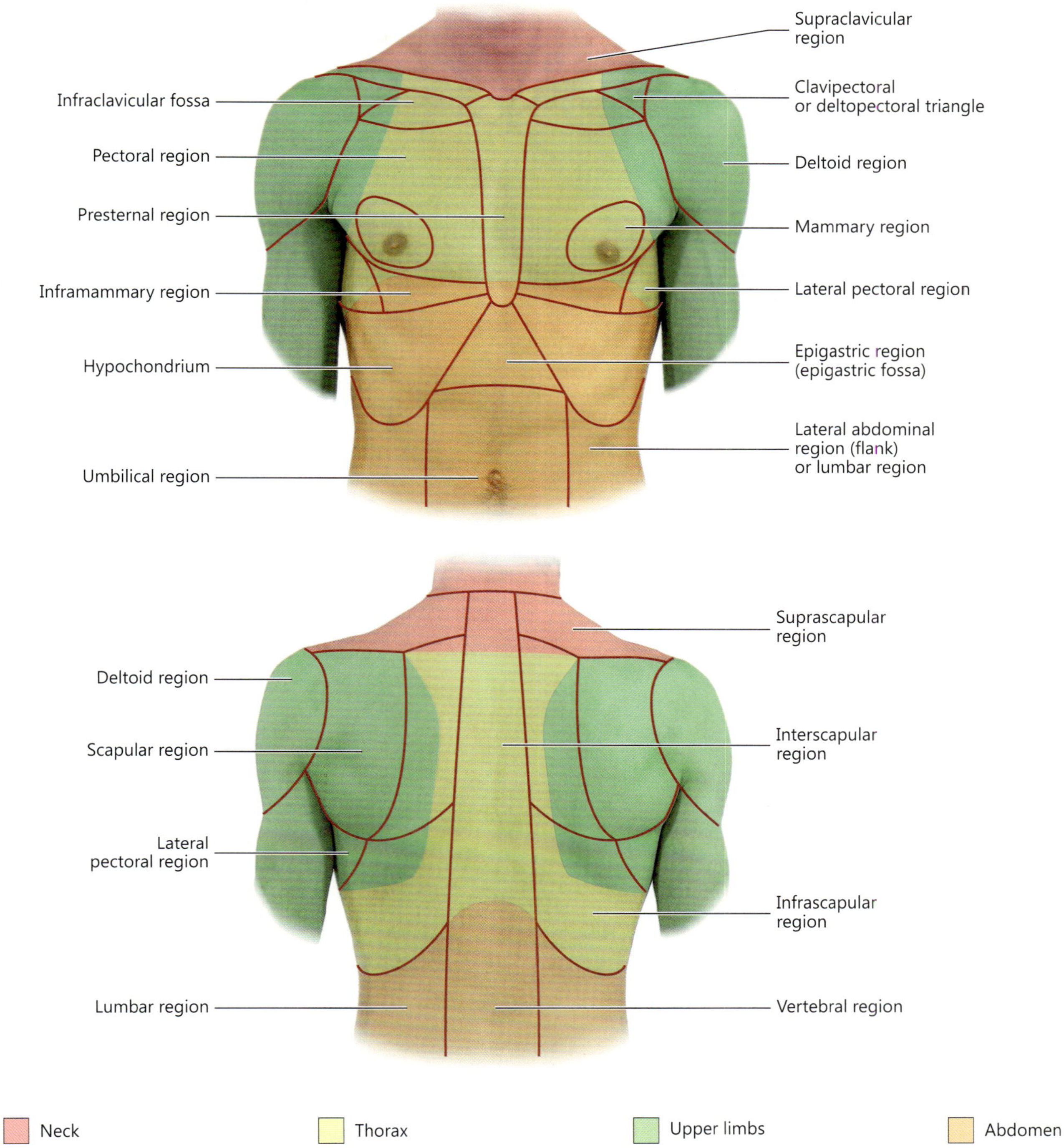

Projection of the thoracic cavity and regions of the thorax. The thoracic cavity has different boundaries from those of the wall. At the top it surpasses the inferior border of the neck reaching the supraclavicular fossae with the cervical pleurae and the apexes of the lung, while below it corresponds to the diaphragmatic cupula which is situated at a higher level than the thoracoabdominal line, so that some abdominal organs are located in a zone that, superficially, belongs to the thorax. Superficially this boundary corresponds approximately to the xiphoid line in front, while at the back it coincides with the thoracoabdominal line. The thorax is separated from the upper limb by a line that anteriorly passes medially to the glenohumeral joint and posteriorly along the medial border of the scapula.

◄

Boundaries of the thorax. The superficial boundaries of the thorax are represented by the **inferior border of the neck** – that separates the thorax from the neck – which starts from the jugular notch of the sternum and follows, on each side, the superior border of the clavicle and the acromion to reach the spinous process of the seventh cervical vertebra, and the **thoracoabdominal line** – that separates the thorax from the neck – which begins at the base of the xiphoid process of the sternum and follows, on each side, the costal margin and the inferior border of the twelfth rib to reach the spinous process of the twelfth thoracic vertebra.

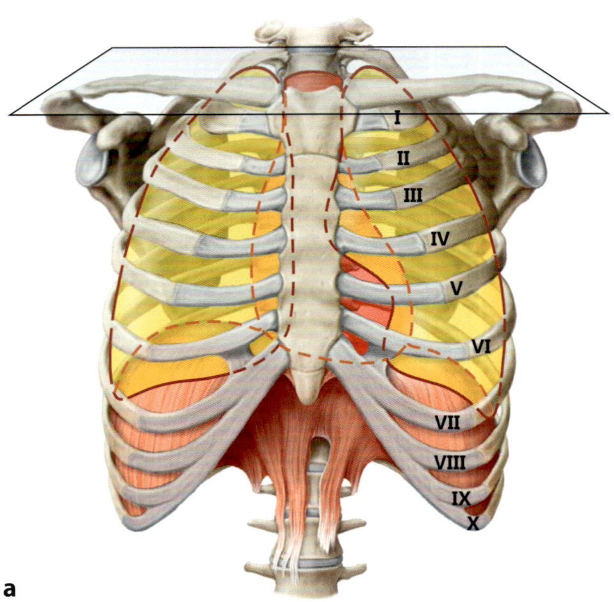

a

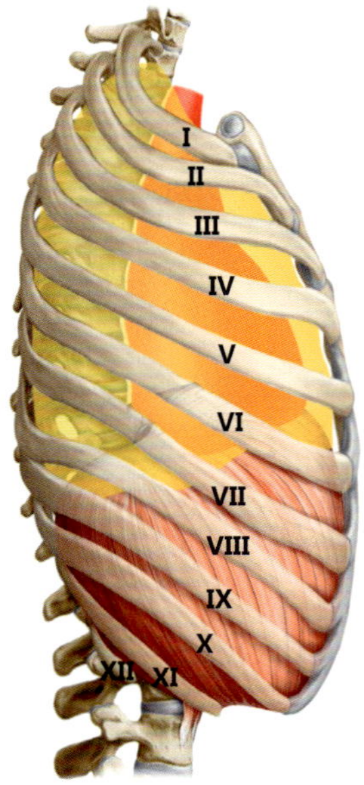

b

The **thoracic cavity** is contained within the thoracic cage. At the top, an imaginary transverse plane passing through the first rib separates it from the neck (**a**). At the bottom, the diaphragm separates it from the abdominal cavity (**a**, **b**). The concavity of the diaphragmatic cupula, convex at the top, encloses the superior part of the abdominal cavity (supramesocolic space and high part of the retroperitoneal space) and its contents. The thoracic cavity, thus, does not coincide with the boundaries of the thoracic cage (☞ Musculoskeletal System - General Anatomy) (**c**).

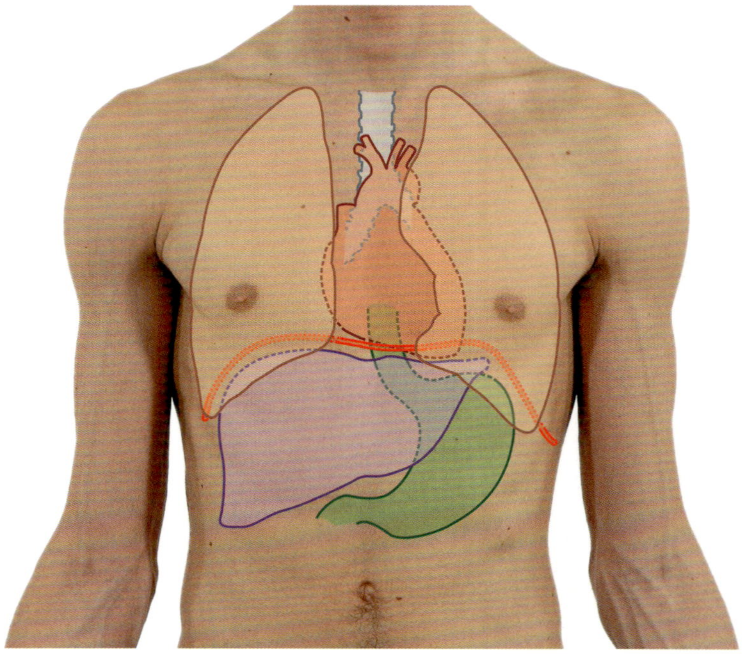

c

X-ray image of the **heart**, posteroanterior projection. In the thoracic cavity, two large lateral cavities can be detected, the pleural compartments, separated by a median space, the mediastinum. The pleural compartments house the lungs and the pleurae. The mediastinum contains numerous anatomical structures, the largest of which is the heart.

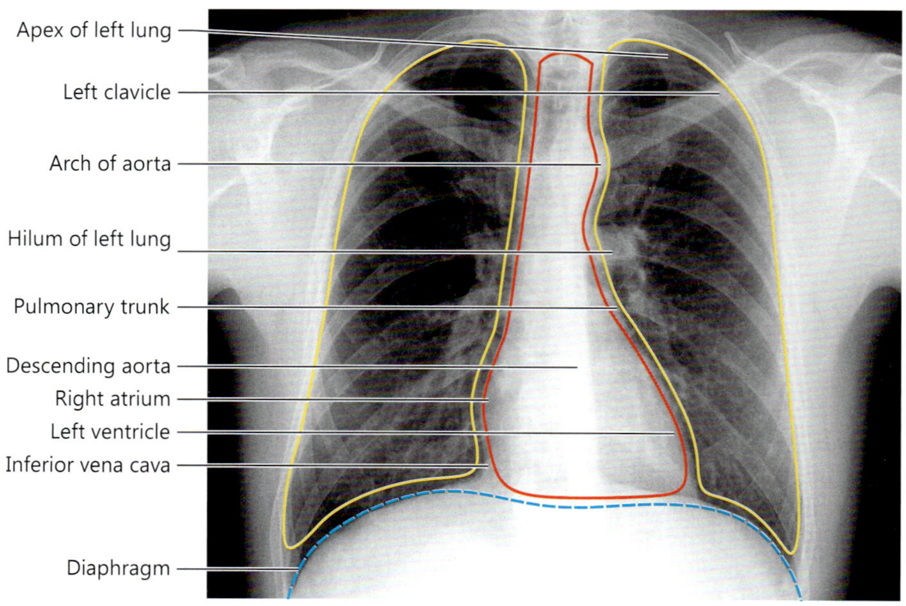

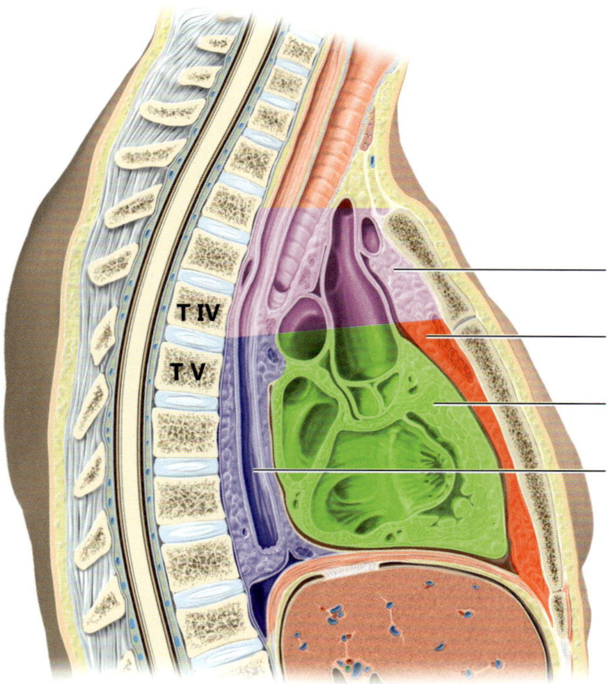

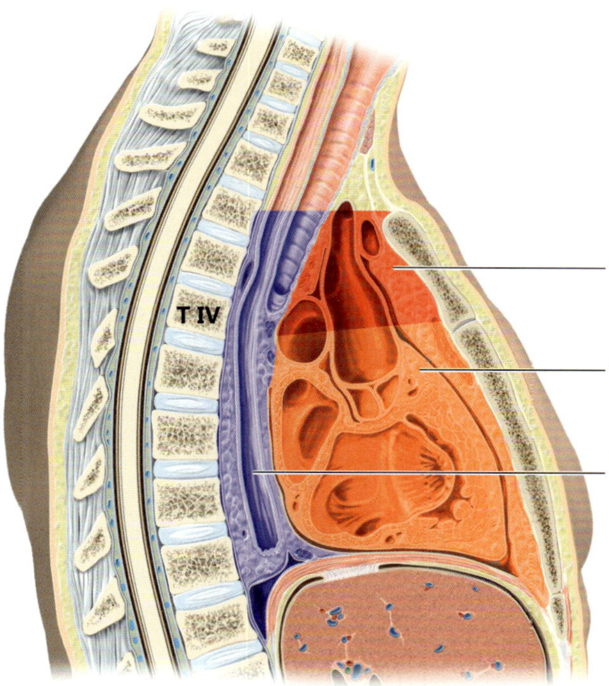

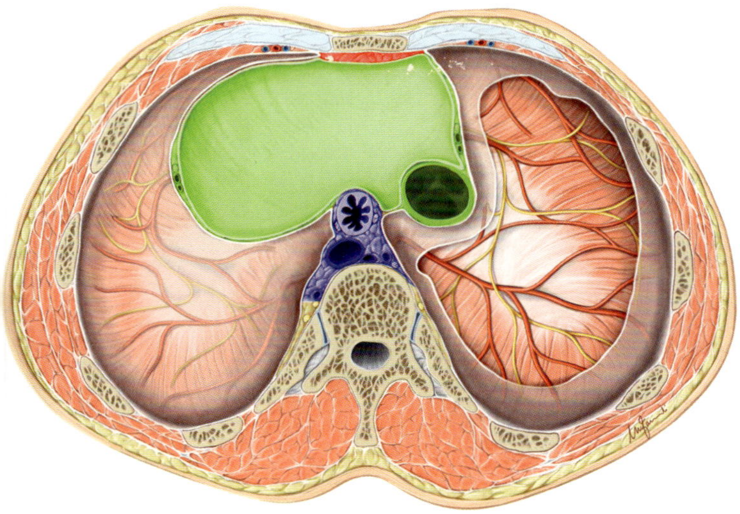

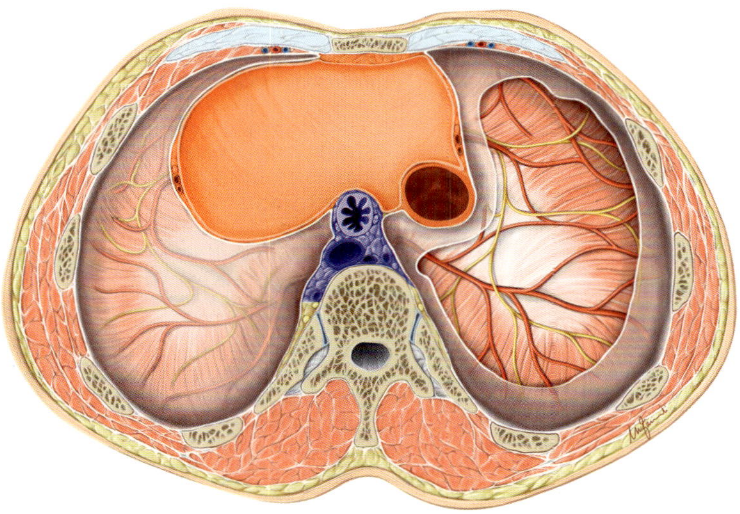

Subdivisions of the **mediastinum** according to the international anatomical nomenclature. The subdivisions are described in the table below. In the image above, the boundary between the superior and inferior mediastinum is represented by a plane passing through the intervertebral disc between T IV and T V and the sternal angle.

Subdivisions of the mediastinum according to the Latin school. The subdivision between anterior and posterior mediastinum is defined by the course of the posterior layer of the pretracheal layer of cervical fascia (☞ Head and Neck - Fasciae of the Neck), which extends downwards to the great vessels of the heart. The subdivision between superior and inferior mediastinum is limited to the anterior mediastinum.

SUBDIVISION	BOUNDARIES
Mediastinum	*Upper*: oblique plane passing through the superior border of the manubrium of sternum and first rib. *Lower*: central part of the diaphragm. *Front*: sternum and costal cartilages [I-VII]. *Back*: thoracic part of the vertebral column [T I-T XII]
Superior mediastinum	*Inferior boundary*: plane passing through intervertebral disc between T IV and T V and the sternal angle
Inferior mediastinum	*Inferior boundary*: diaphragm
• anterior	Between the sternum and the sternocostal surface of the pericardium
• middle	Between the sternocostal surface of the pericardium and a frontal plane passing through the posterior surface of the trachea and main bronchi and the posterior surface of the pericardium
• posterior	Between the posterior surfaces of the trachea, main bronchi and pericardium and the vertebral column (between T IV/T V and T XII, in according to the Anglo-saxon school, or T I-T XII, in according to the Latin school)

Contents of the mediastinum

DENOMINATION	*CONTENTS*
Superior mediastinum	• Arch of aorta • Branches of the arch of aorta (brachiocephalic trunk, left common carotid artery, left subclavian artery) • Supreme intercostal artery (right and left) • Pericardiacophrenic artery (right and left) • Internal thoracic artery (right and left) • Thoracic duct • Esophagus • Pretracheal layer of cervical fascia • Thoracic ganglia of sympathetic trunk (right and left) • Insertions of sternohyoid and sternothyroid (right and left) • Insertions of longus colli on the first three thoracic vertebrae [T I-T III], medial and inferior lateral bundles (right and left) • Brachiocephalic nodes • Posterior intercostal nodes (right and left) • Paratracheal nodes (right and left) • Preaorticocarotid nodes • Prevascular nodes • Superior tracheobronchial nodes (right and left) • Upper half of the superior vena cava • Phrenic nerves • Vagus nerves (right and left) • Right and left recurrent (or inferior) laryngeal nerves • Cardiac plexus • Thymus or residual thymic tissue and fibroadipose tissue • Trachea • Bronchomediastinal trunks • Accessory hemiazygos vein (first tract) • Supreme intercostal vein (right and left) • Brachiocephalic vein (right and left) • Pericardiacophrenic vein (right and left) • Internal thoracic vein (right and left)
Posterior mediastinum	• Thoracic aorta • Branches of the thoracic aorta: mediastinal branches, bronchial branches, esophageal branches, pericardial branches • Right Intercostal arteries from 4th to 12th • Thoracic duct • Esophagus • Thoracic ganglia of sympathetic trunk (right and left) • Pulmonary (triangular) ligaments (right and left) • Vertebropericardial ligaments • Posterior intercostal nodes • Posterior mediastinal nodes (juxtaesophageal nodes, interaorticoesophageal nodes, juxta-aortic nodes, posterior superior diaphragmatic nodes) • Splanchnic nerves • Esophageal plexus • Azygoesophageal and interaortoesophageal pleural recesses • Anterior and posterior vagal trunks • Azygos vein on right side • Hemiazygos vein and accessory hemiazygos vein (last tract) on left side
Middle mediastinum	• Ascending aorta • Bronchial branches of thoracic aorta (right and left) • Pericardial branches of thoracic aorta • Pericardiacophrenic artery and vein (right and left) • Pulmonary arteries (right and left) • Tracheal bifurcation • Main bronchi (right and left) • Heart • Pericardiacophrenic ligaments • Middle superior diaphragmatic nodes (diaphragmatic anterior mediastinal nodes) • Superior and inferior tracheobronchial nodes • Lower half of the superior vena cava • Phrenic nerves • Pericardium • Cardiac plexus • Pulmonary plexus • Pulmonary trunk and ligamentum arteriosum (ligamentum of Botallo) • Opening of azygos vein into the superior vena cava • Bronchial veins (right and left) • Superior and inferior pulmonary veins (right and left)
Anterior mediastinum	• Sternopericardial ligaments (superior and inferior) • Prepericardial nodes, lateral pericardial nodes and anterior superior diaphragmatic nodes • Costomediastinal pleural recesses • Loose connective tissue • Thymus or residual thymic tissue

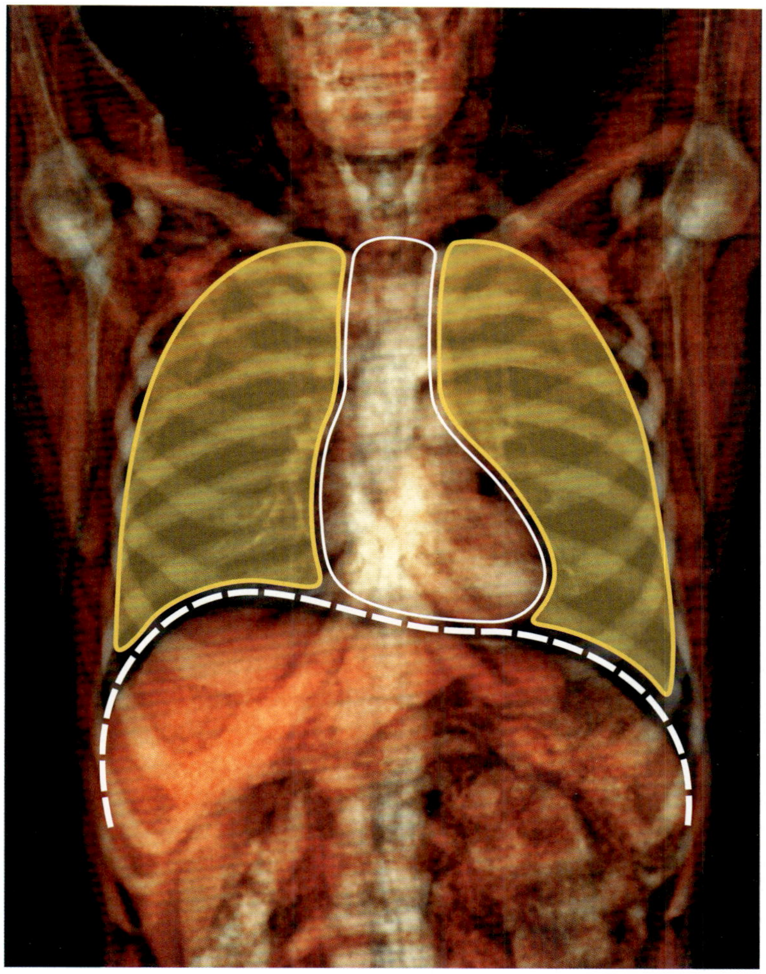

The lungs are housed in the **pleural compartments** inside the thoracic cavity (three-dimensional image of a living human body, obtained by CT scan using a 3D volume rendering technique).

Pleural compartments

DENOMINATION	BOUNDARIES	CONTENT	BLOOD SUPPLY	INNERVATION	LYMPHATIC DRAINAGE
Right and left pleural compartments	Superior: subclavian vessels, brachial plexus, scalenus anterior Inferior: diaphragm Medial: mediastinum Lateral: ribs and intercostal muscles	Right and left lungs	Lung: pulmonary artery, pulmonary veins; bronchial branches of thoracic aorta and bronchial veins	Lung: anterior and posterior pulmonary plexuses (branches of the vagus nerve and the sympathetic part)	Lung: superficial (subpleural) lymphatic plexus and deep lymphatic plexus (which accompanies the branches of the pulmonary and bronchial vessels) draining into the bronchopulmonary nodes (the deep lymphatic plexus, through the intrapulmonary nodes)
		Visceral pleura Pleural cavity	Visceral pleura: bronchial branches of thoracic aorta and pulmonary veins	Visceral pleura: pulmonary plexuses	Visceral pleura: subserosal lymphatic network connected to the lymphatic vessels which accompany the pulmonary veins and drain into the tracheobronchial nodes
		Parietal pleura	Parietal pleura: intercostal arteries (costal part of parietal pleura), superior phrenic arteries and inferior phrenic arteries (diaphragmatic part of parietal pleura), mediastinal branches of thoracic aorta (mediastinal part of parietal pleura), azygos vein and internal thoracic veins	Parietal pleura: intercostal nerves, vagus nerve, phrenic nerve, branches of the sympathetic part	Parietal pleura: intercommunicating lymphatic networks which drain into the parasternal, intercostal and diaphragmatic nodes

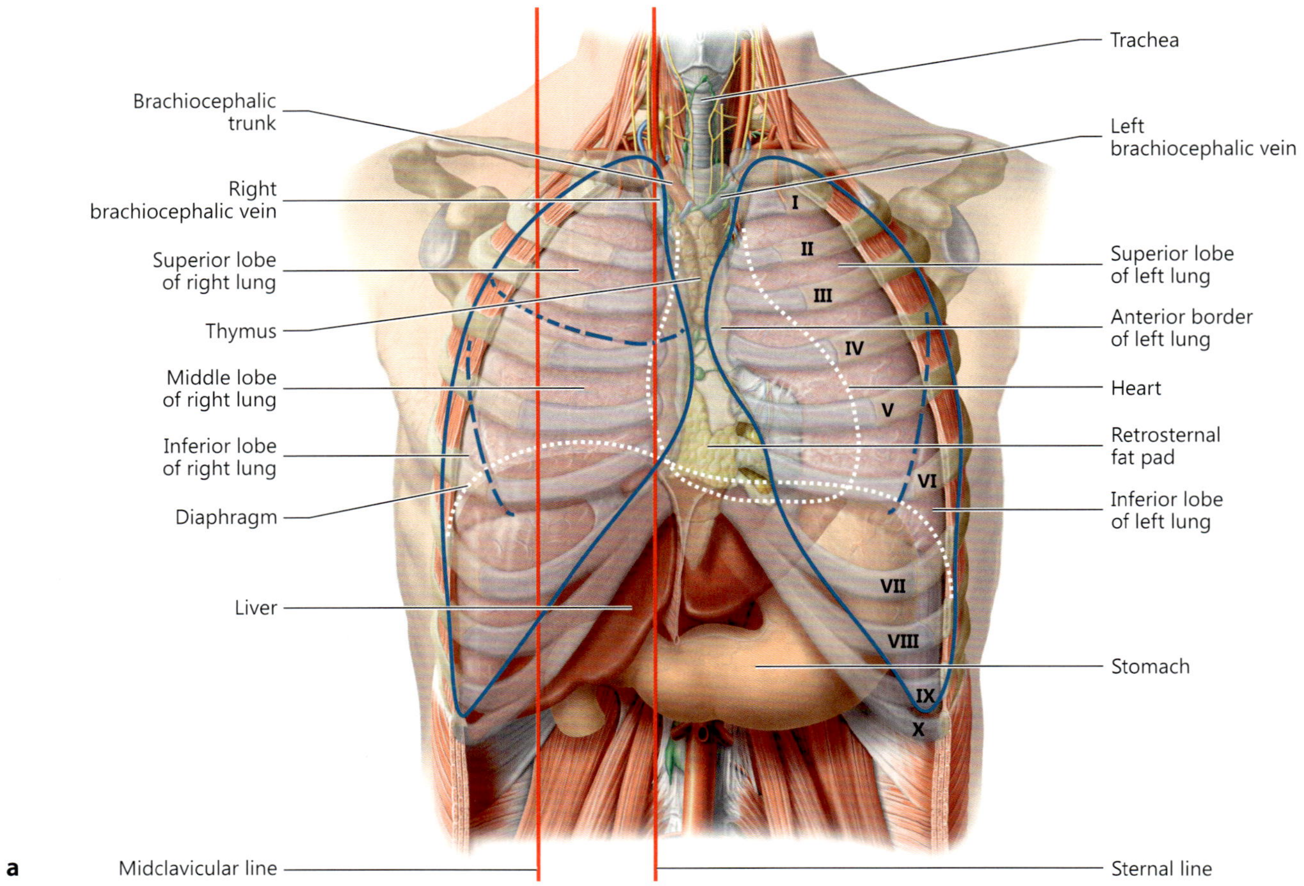

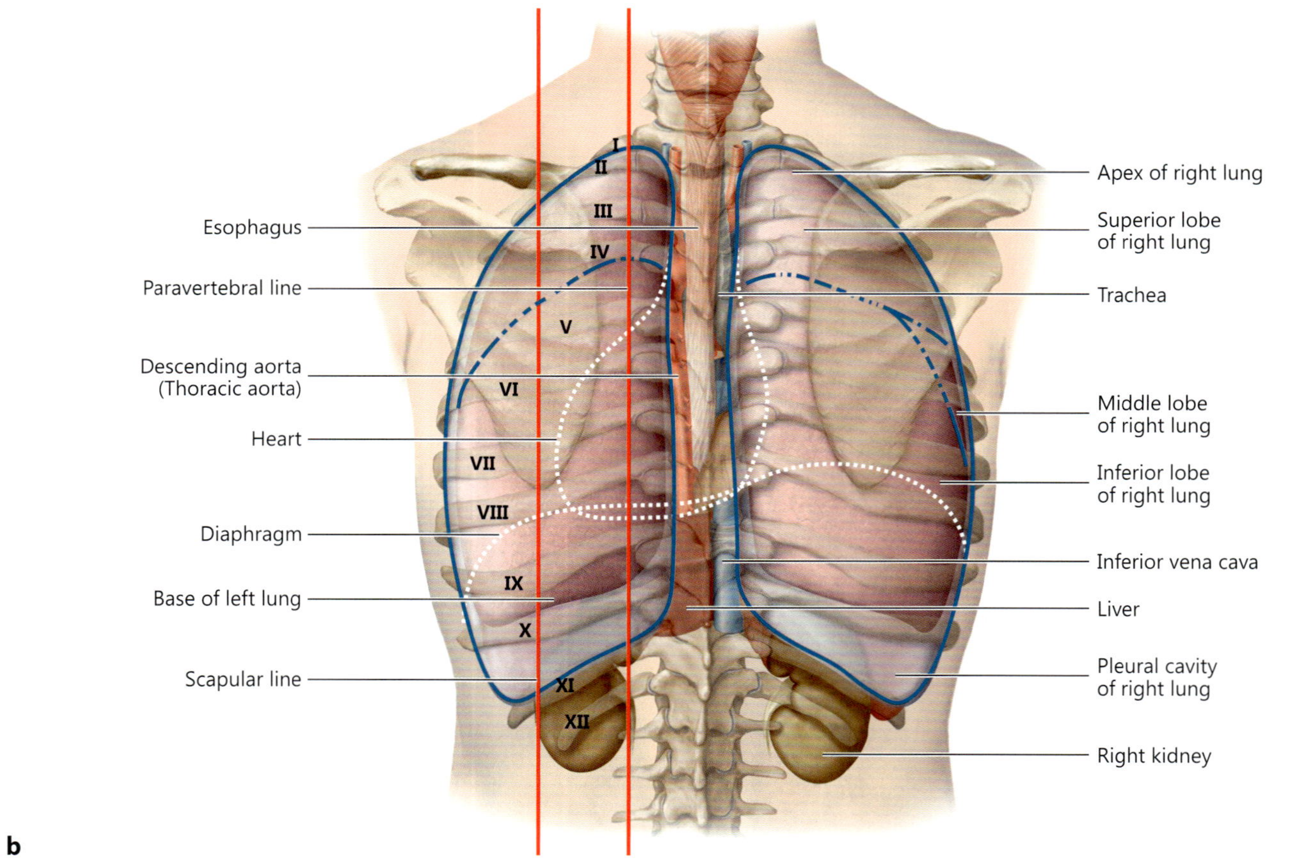

Projection onto the surface of the thoracic cage of the visceral and vascular formations of the **mediastinum** and of the **pleural compartments**. The *white* dotted lines define the contours of the diaphragm, heart and the great vessels. The *blue* lines define the boundaries of the parietal pleura. Within these boundaries one can see, in transparency, the lungs and the course of the pulmonary fissures (indicated by the dotted line). Also visible, in transparency, are some internal organs of the abdominal cavity on account of the latter's extension into the thoracic cage below the diaphragm. **a**, Anterior projection; **b**, posterior projection.

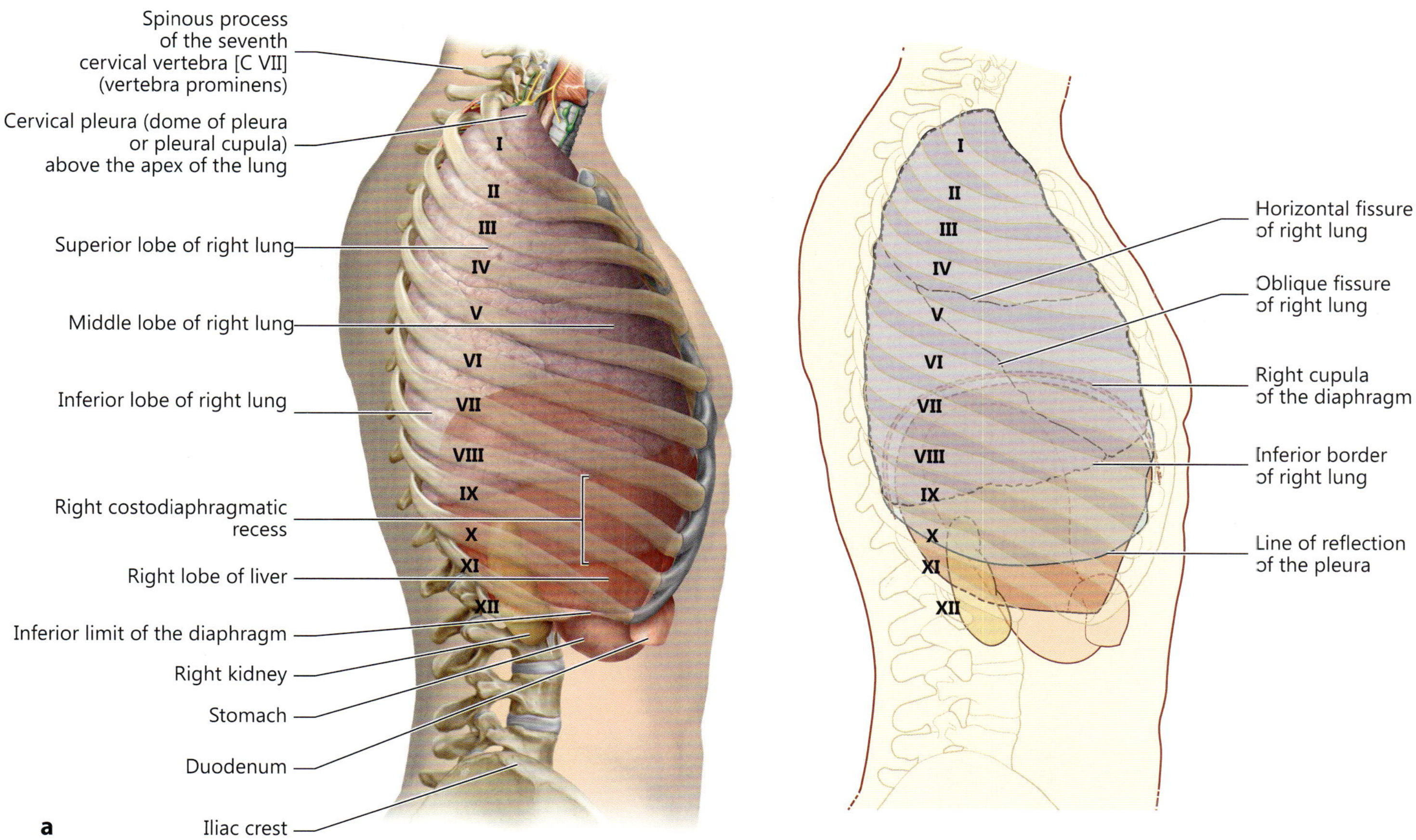

a

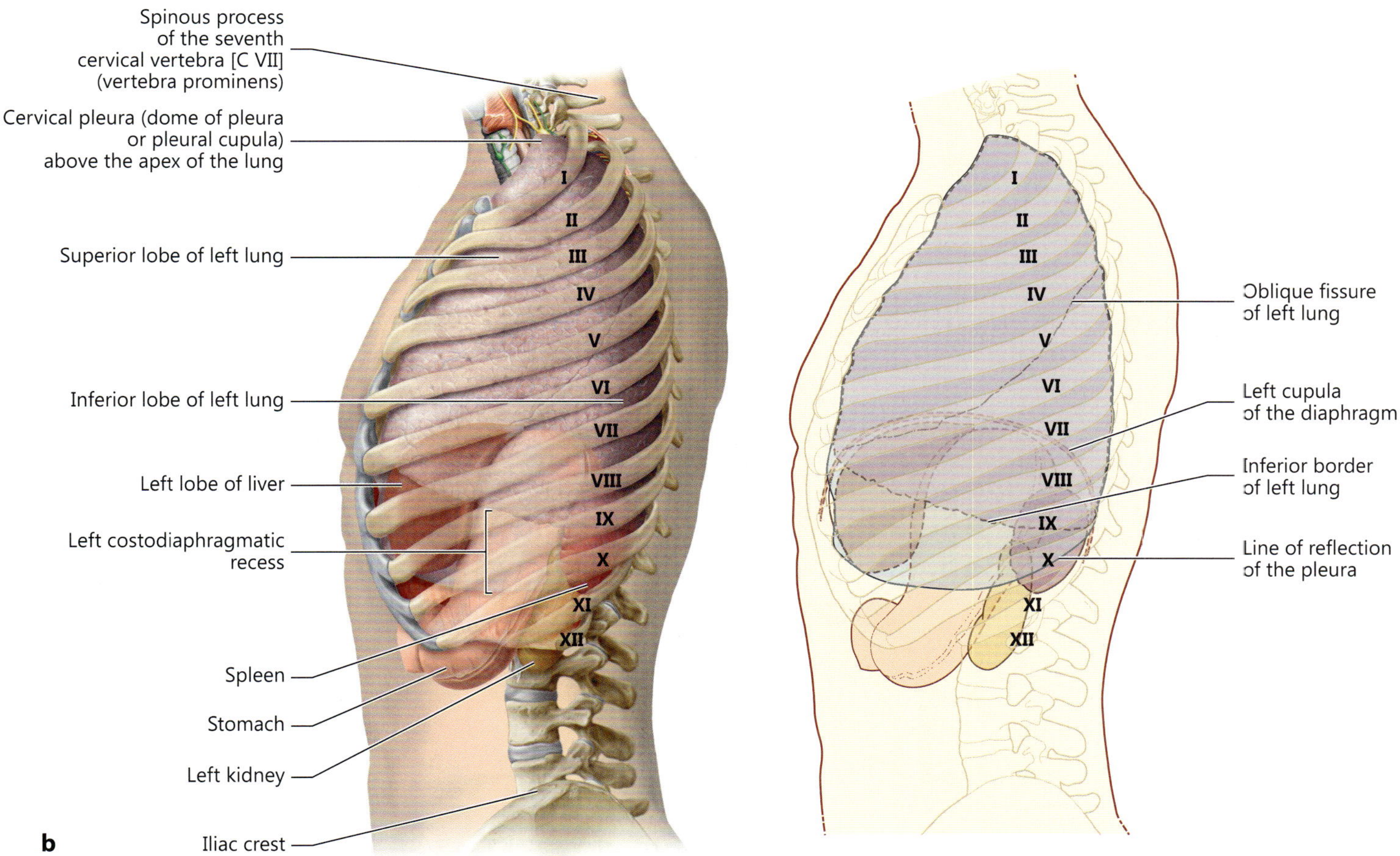

b

Projection onto the thoracic cage of the **lungs** and **supramesocolic** and **retroperitoneal organs** located in the thoracic cage. **a**, Right lateral projection. **b**, Left lateral projection.

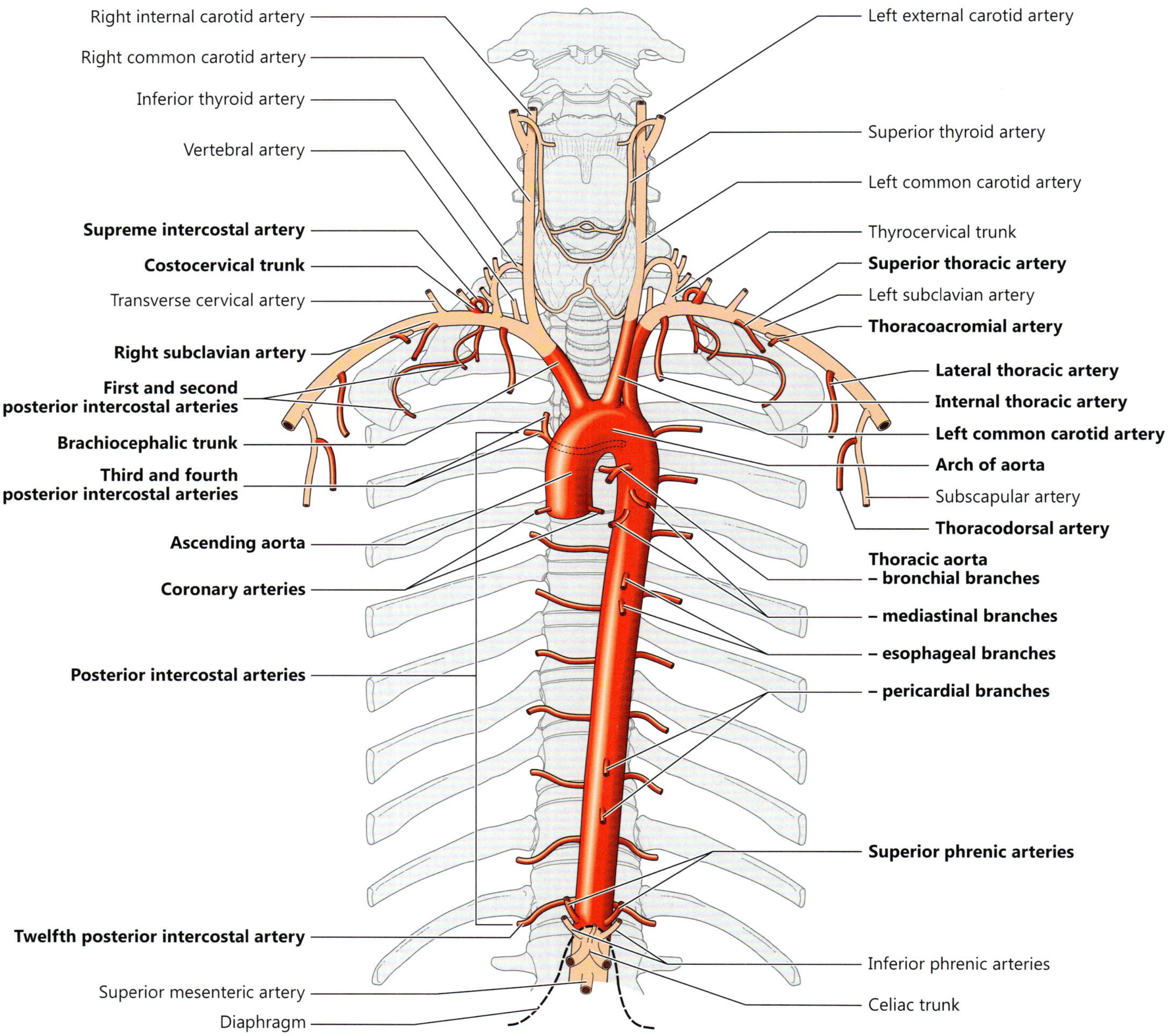

Schematic representation of the **arch of the aorta** and **thoracic aorta** with the arteries and branches involved in the visceral and parietal blood supply of the thorax.

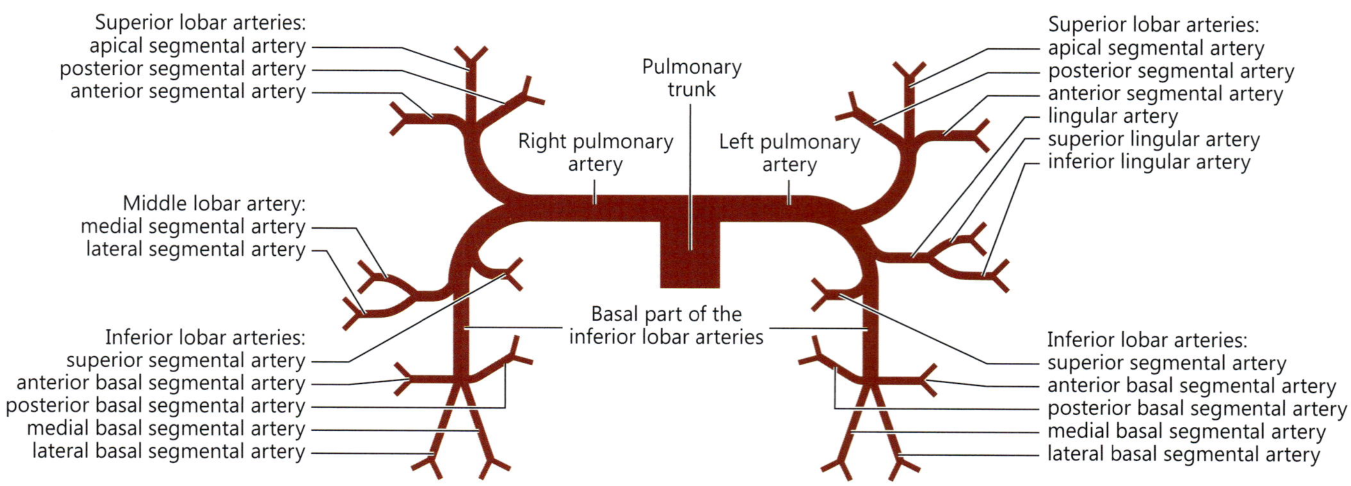

Schematic representation of the division of the **pulmonary trunk** into the **right and left pulmonary arteries**, which ramify attaching themselves to the bronchial branches and following their subdivisions.

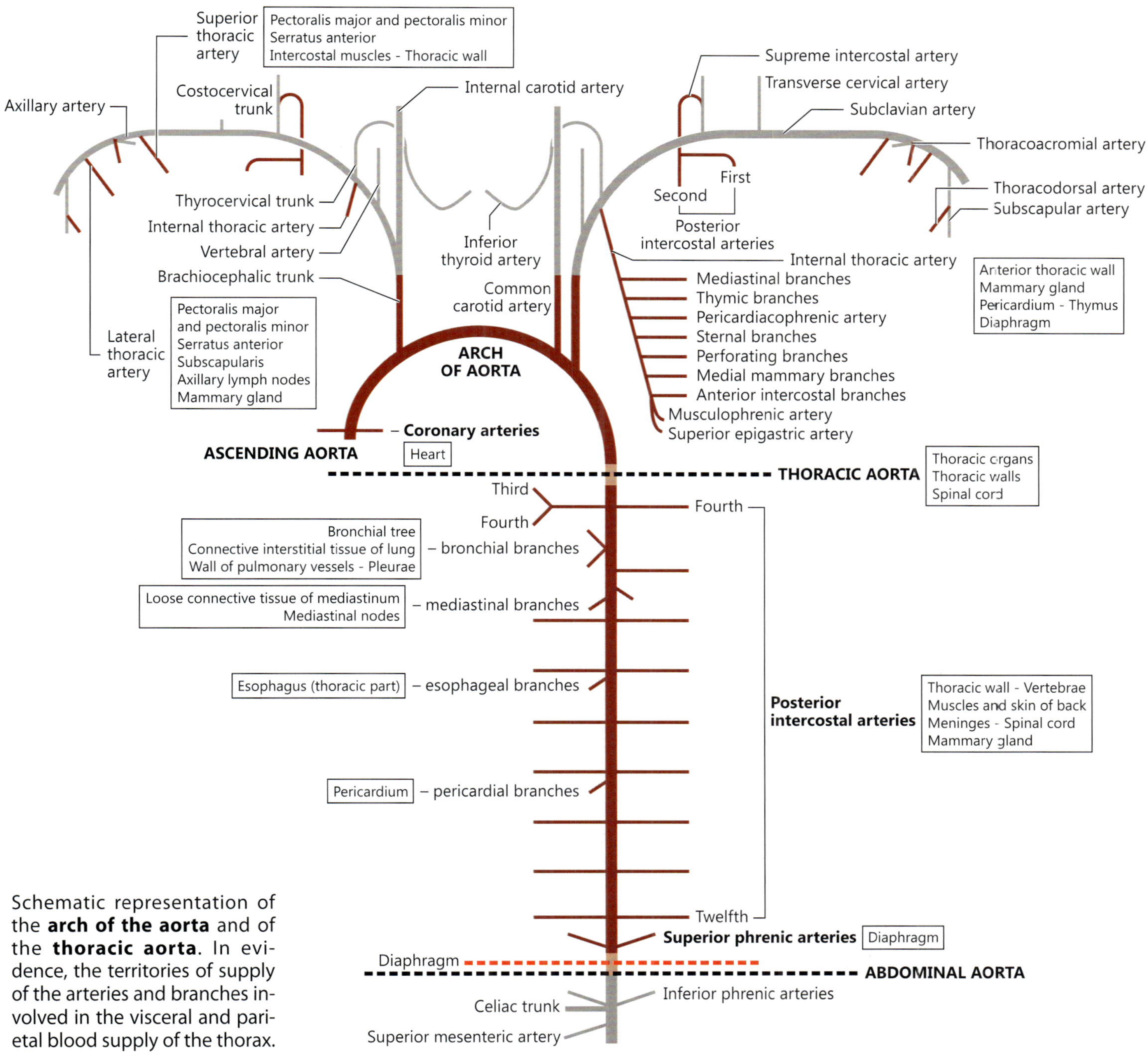

Schematic representation of the **arch of the aorta** and of the **thoracic aorta**. In evidence, the territories of supply of the arteries and branches involved in the visceral and parietal blood supply of the thorax.

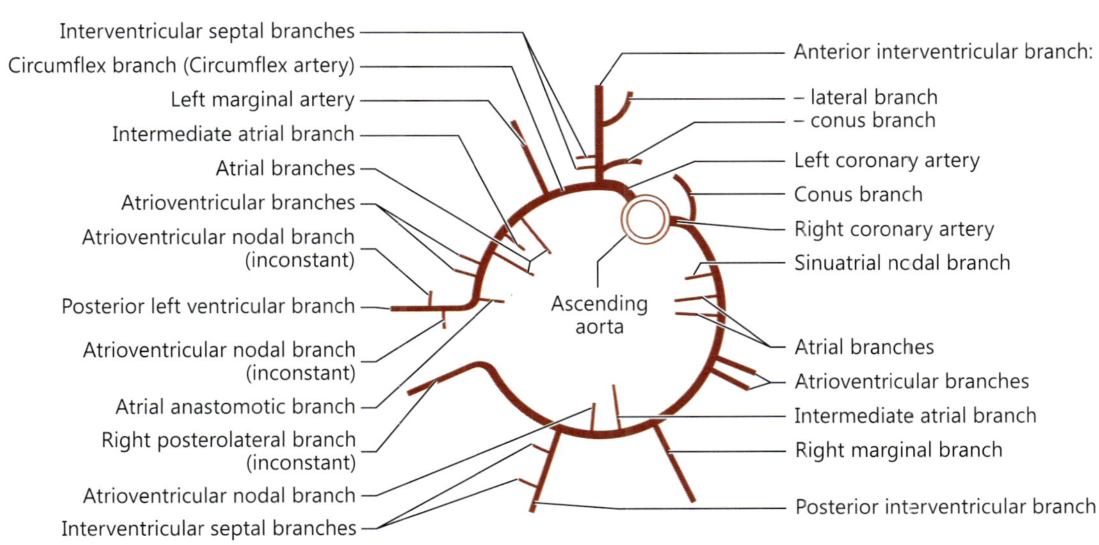

Schematic representation of the **coronary circulation**. The coronary arteries and their principal branches run on the external surface of the heart, in the coronary and interventricular sulci.

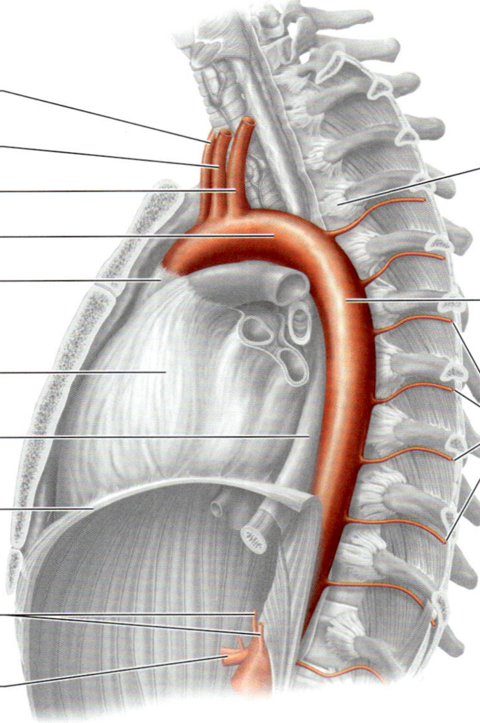

Course and relations of the **aorta** in the mediastinum and in the passageway between the **thoracic part** and **abdominal part**. The ascending aorta is partially covered by pericardium. The arch of the aorta can be seen in relation to the root of the left lung, and the thoracic aorta in relation to the esophagus. From the posterolateral surface of the thoracic aorta branch off the posterior intercostal arteries.

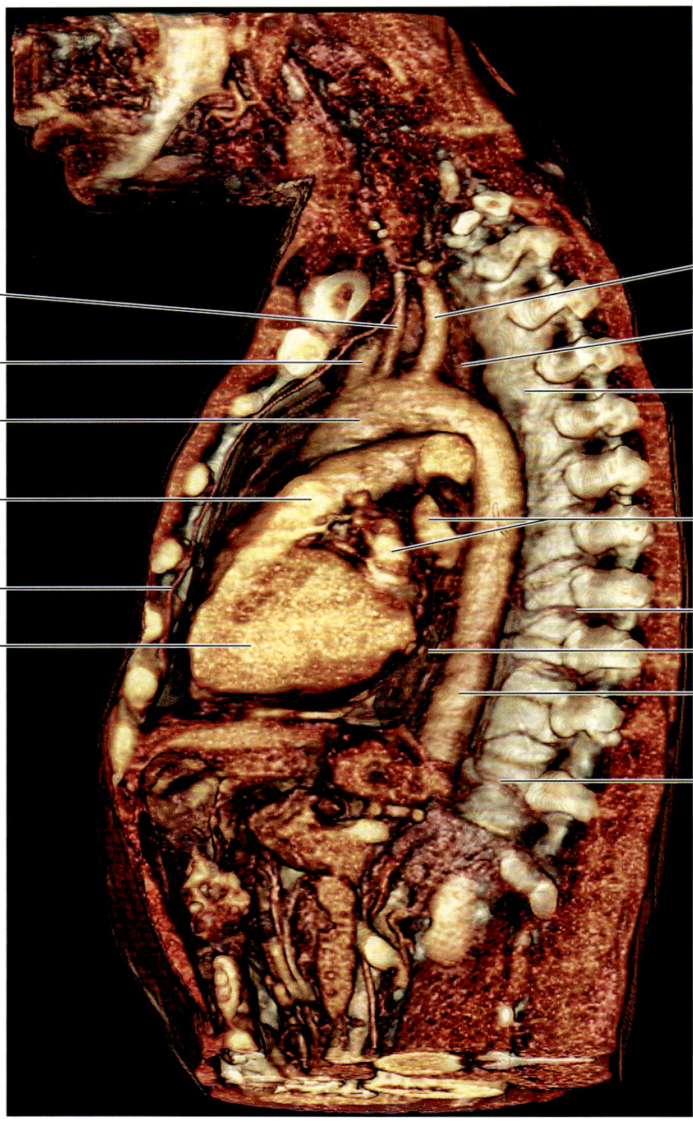

Three-dimensional image of a living human body, obtained by CT scan using a 3D volume rendering technique, showing the **thorax**, in sagittal section, at the level of the **aorta**.

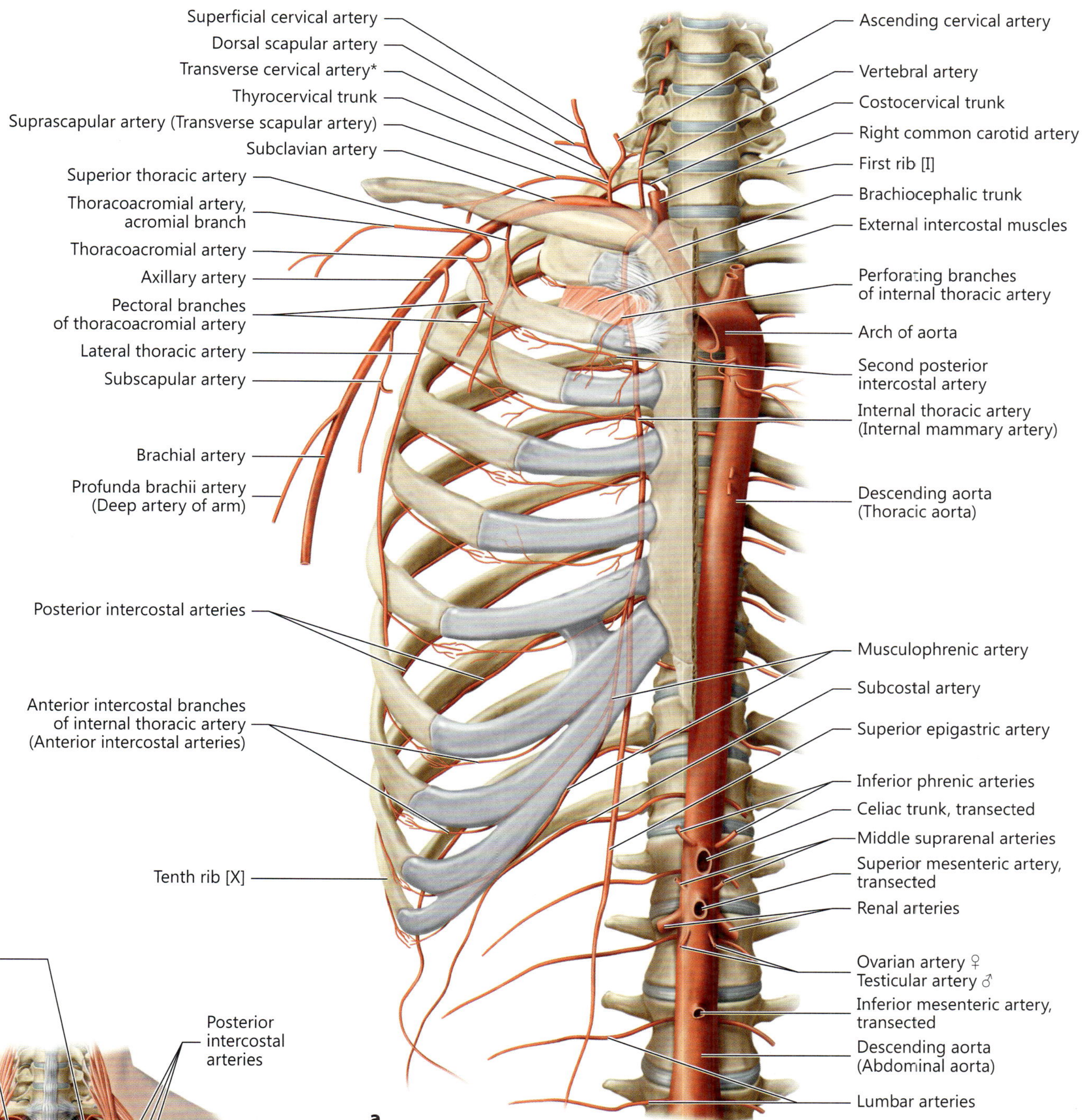

a

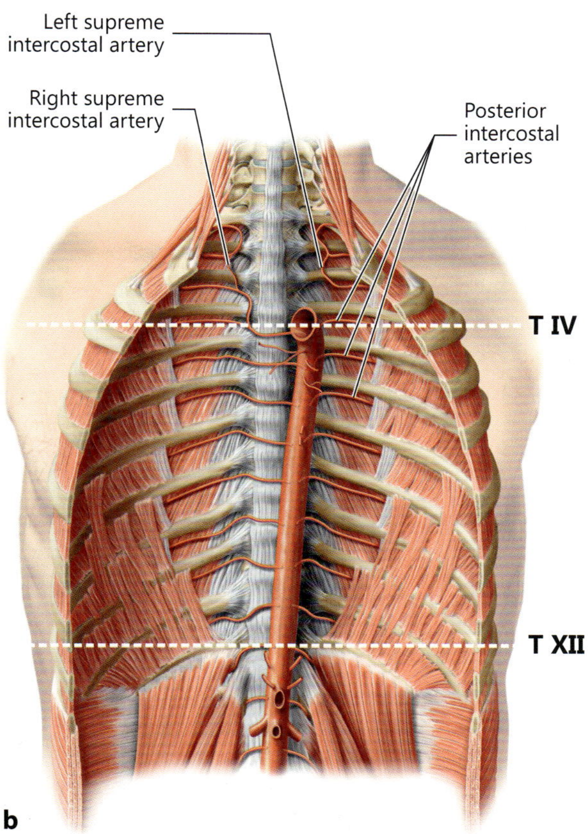

b

The **parietal blood supply of the thorax** has a metameric arrangement and receives branches of the descending aorta (thoracic aorta) and subclavian artery. The posterior intercostal arteries run in their respective intercostal spaces, against the superior border. The first two originate from the supreme intercostal artery (a branch of the costocervical trunk of the subclavian artery); the last nine pairs, from the descending aorta (thoracic aorta) (**a**, **b**). The subcostal artery completes the series, taking its position under the twelfth rib (**b**). The right intercostal arteries, which originate from the aorta, and the right subcostal artery pass in front of the vertebral column to reach their respective intercostal spaces, taking a longer route than the contralateral ones (**a**). Once past the angle, the posterior intercostal arteries proceed along the body of the ribs and anastomose with the corresponding anterior intercostal branches of the internal thoracic artery (a branch of the subclavian artery), or of its musculophrenic branch. The last two arteries anastomose instead with the superior epigastric artery (a branch of the internal thoracic artery) and with the first lumbar artery (a branch of the descending abdominal aorta). Also the superior thoracic, thoracoacromial and lateral thoracic branches of the axillary artery are part of the parietal circulation (**b**).

* Here it is represented the origin of the thyrocervical trunk and not directly of the subclavian artery.

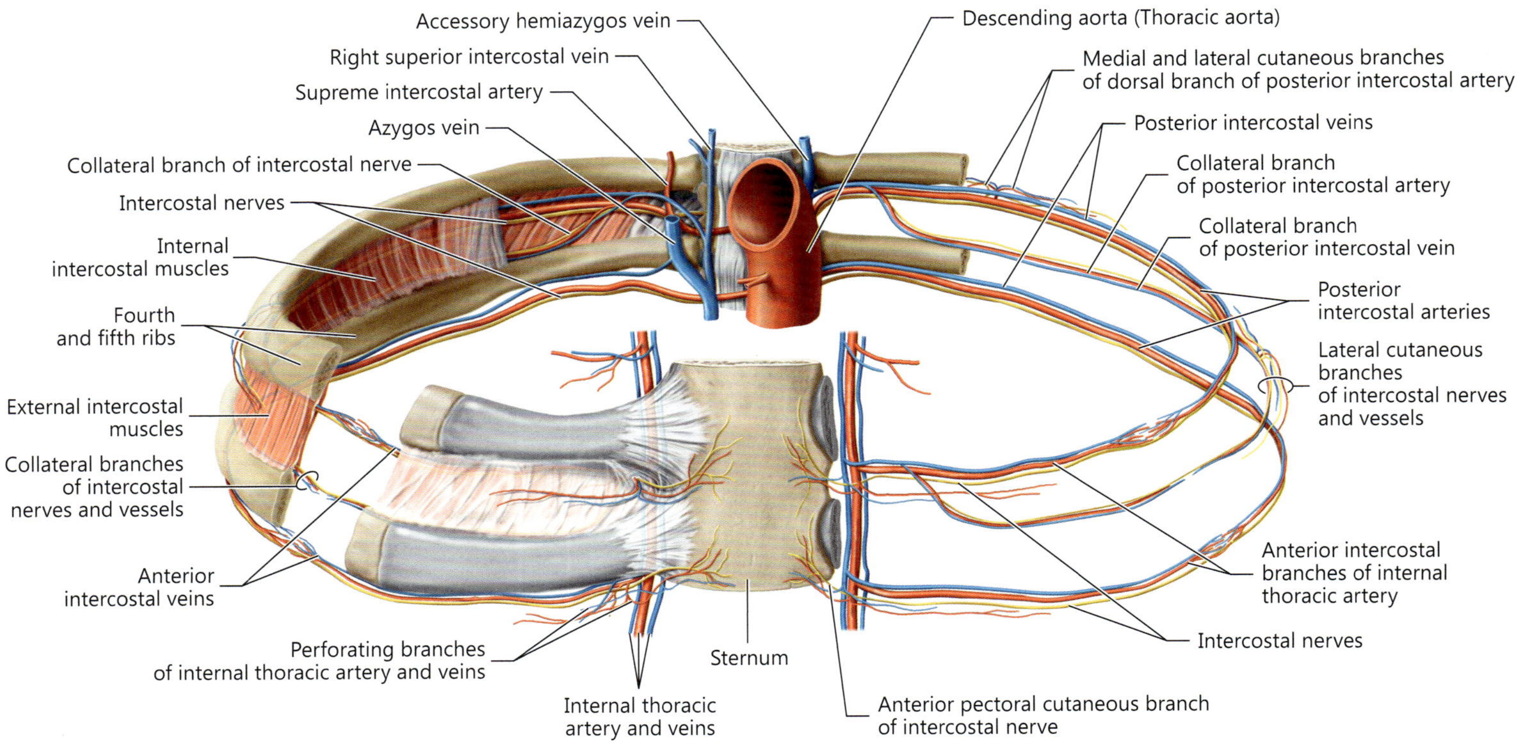

Pathway of the **intercostal blood vessels** and **nerves** at the level of T IV-T V.

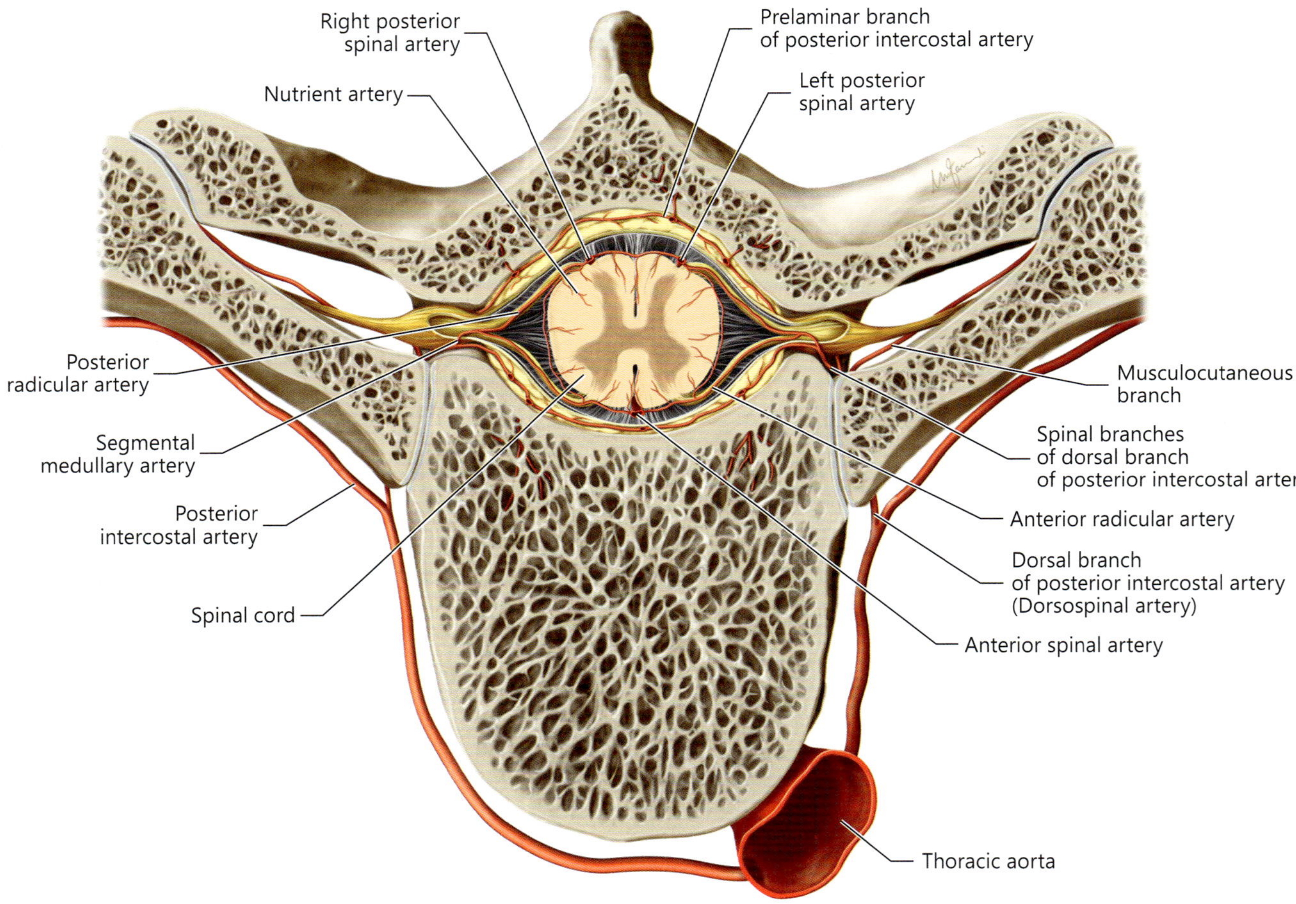

Transverse section of the dorsal vertebral column; representation of the **arterial circulation of the spinal cord**. The segmental branches of the spinal arteries reach the spinal cord after passing through the intervertebral foramina together with the medullary nerves. Each spinal artery is linked to those situated above and below it by longitudinal anastomoses, one anterior and two posterior, which extend from the interior of the neurocranium to the conus medullaris. From the anastomotic tracts and from the perimedullary network which they create arise the nutrient arteries of the spinal cord.

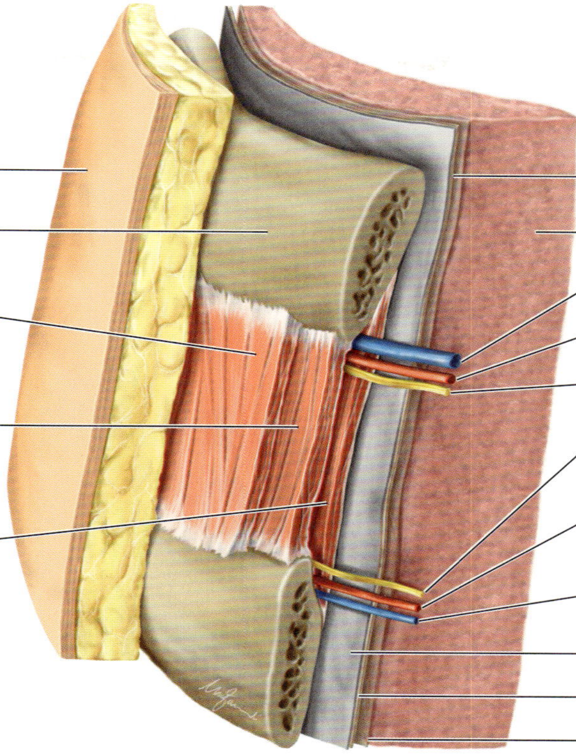

Dissection of the **thoracic wall** at the level of an intercostal space, displaying the passage of the intercostal blood vessels and nerves.

Images of the **internal wall** of the **thoracic cavity** obtained during a thoracoscopy. The endoscope is positioned, within the pleural cavity, between the parietal and visceral layers of pleura.

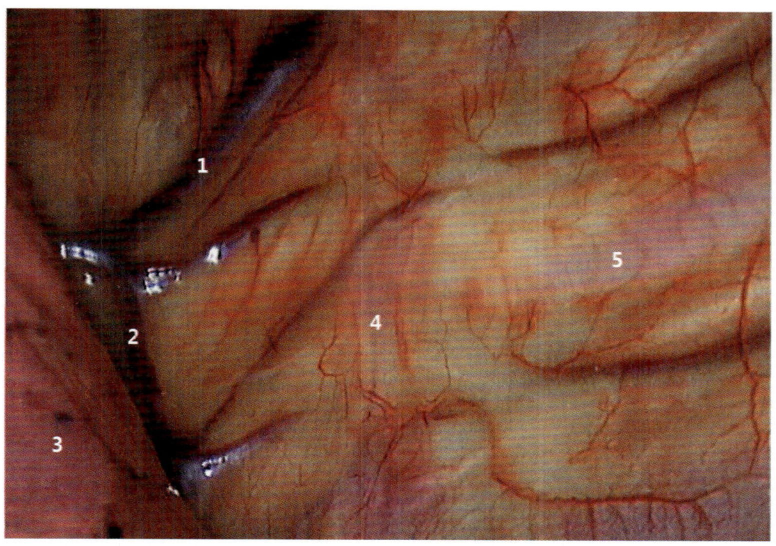

1, Posterior intercostal vein
2, Accessory hemiazygos vein
3, Lung invested by the visceral layer of pleura
4, Thoracic ganglia of sympathetic trunk covered by the parietal layer of pleura
5, Rib covered by the endothoracic fascia and parietal layer of pleura

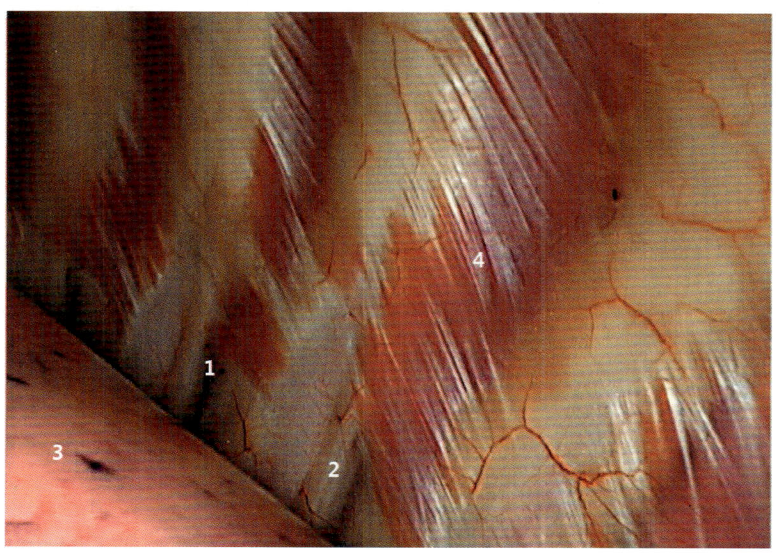

1, Intercostal vessels
2, Rib
3, Lung invested by the visceral layer of pleura
4, Internal intercostal muscles covered by the parietal layer of pleura and endothoracic fascia

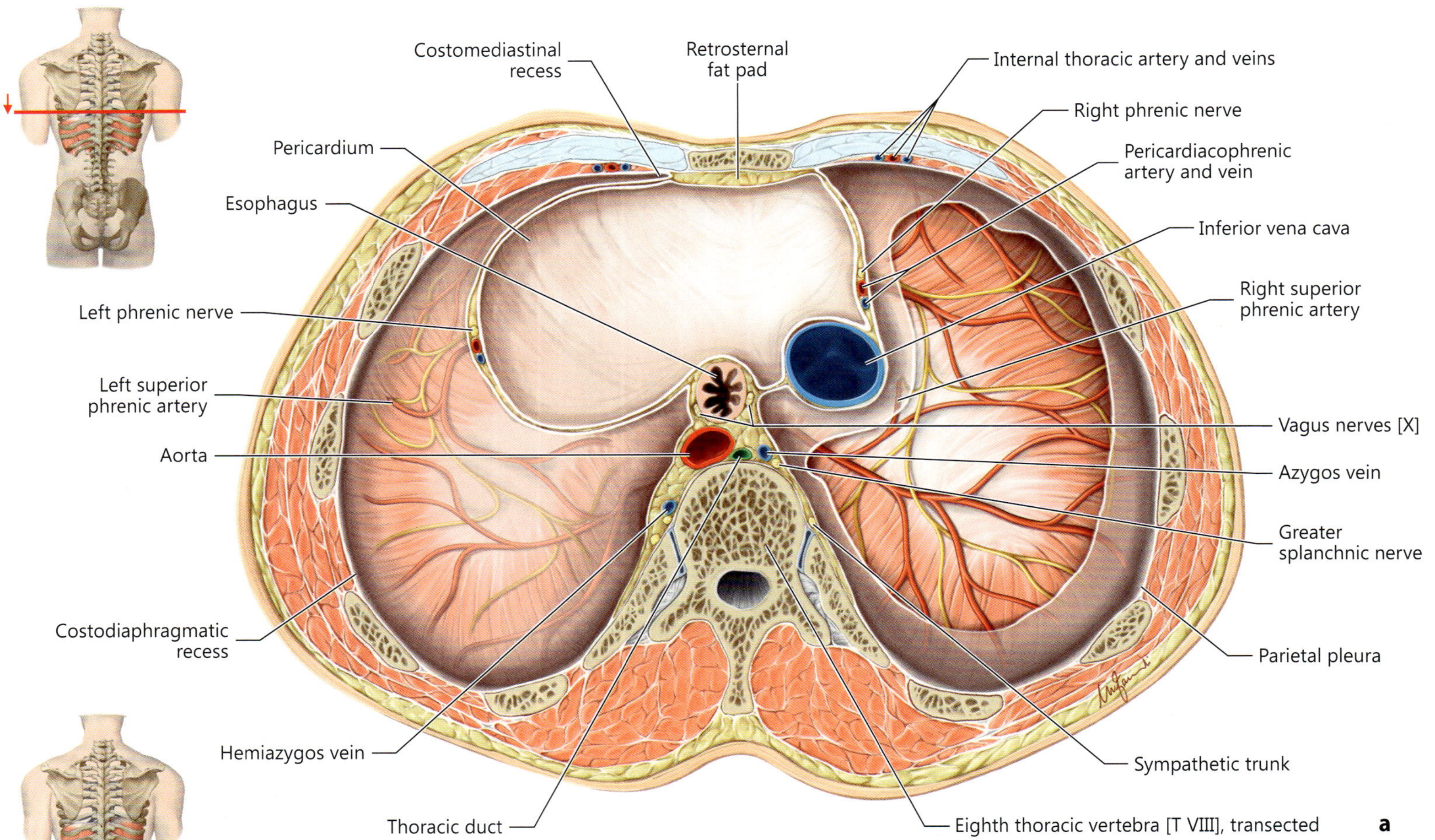

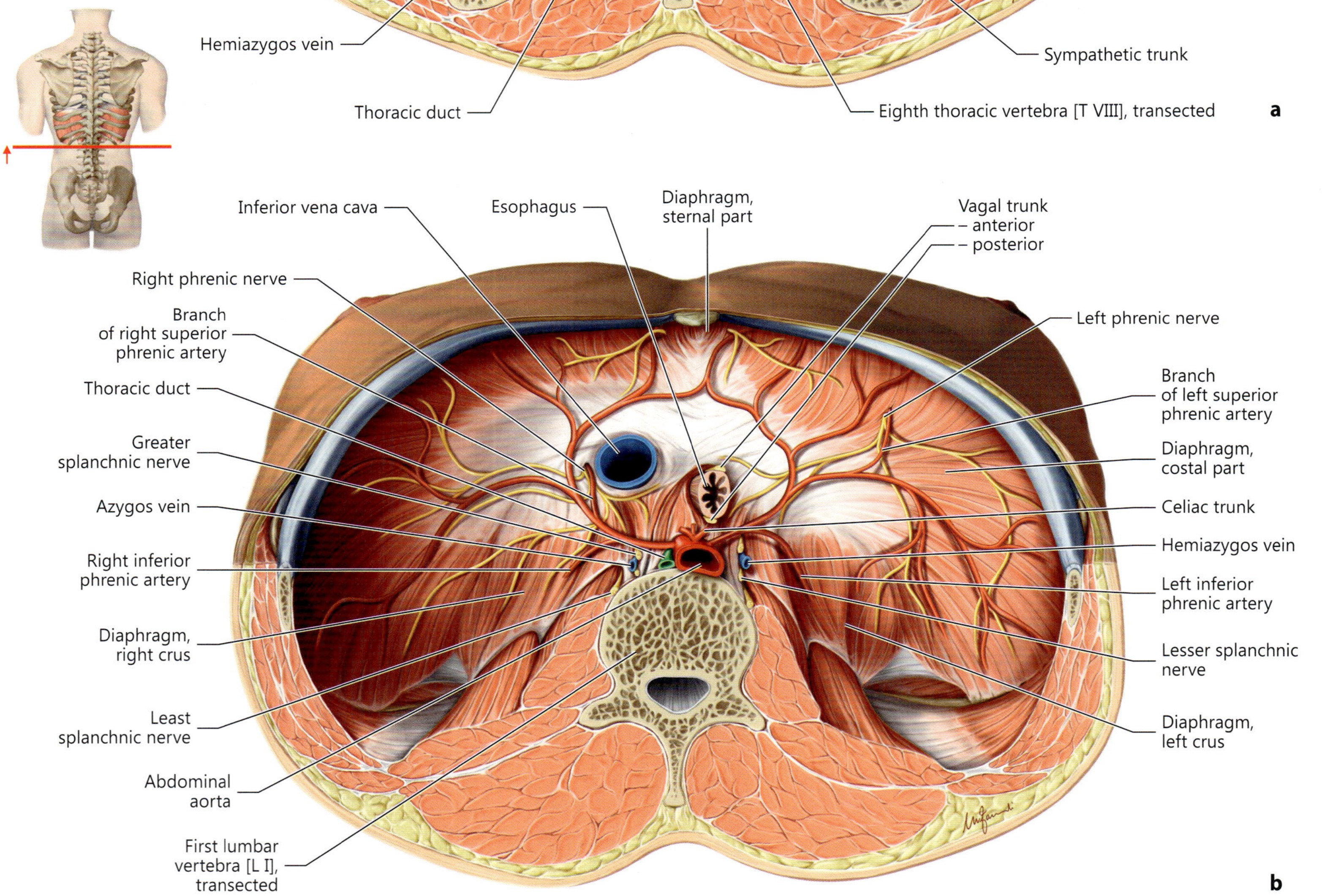

Superior (**a**) and inferior (**b**) view of the **diaphragm**. The parietal vascularization of the thorax is completed by the blood supply to the diaphragm carried out by the superior phrenic arteries (branches of the descending thoracic aorta) and the pericardiacophrenic artery (a branch of the internal thoracic artery) on the superior surface of the diaphragm (**a**) and by the inferior phrenic arteries (branches of the descending abdominal aorta) (**b**).

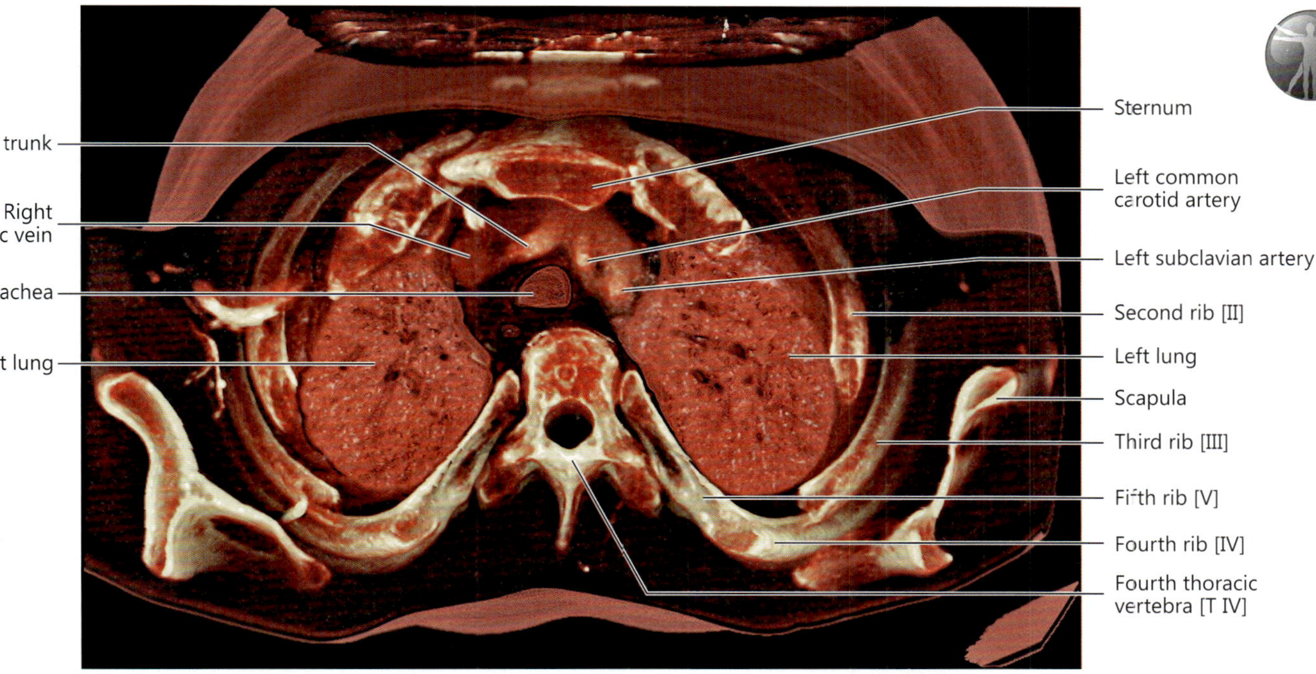

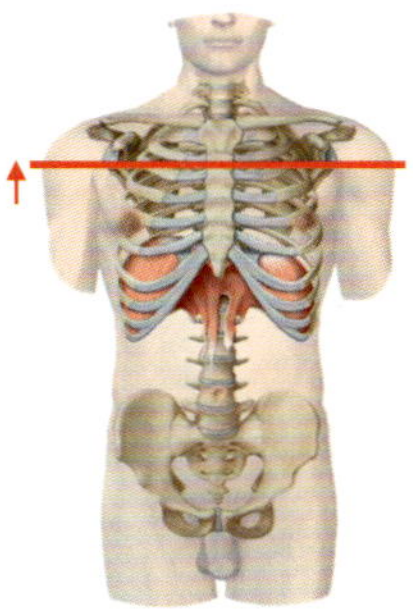

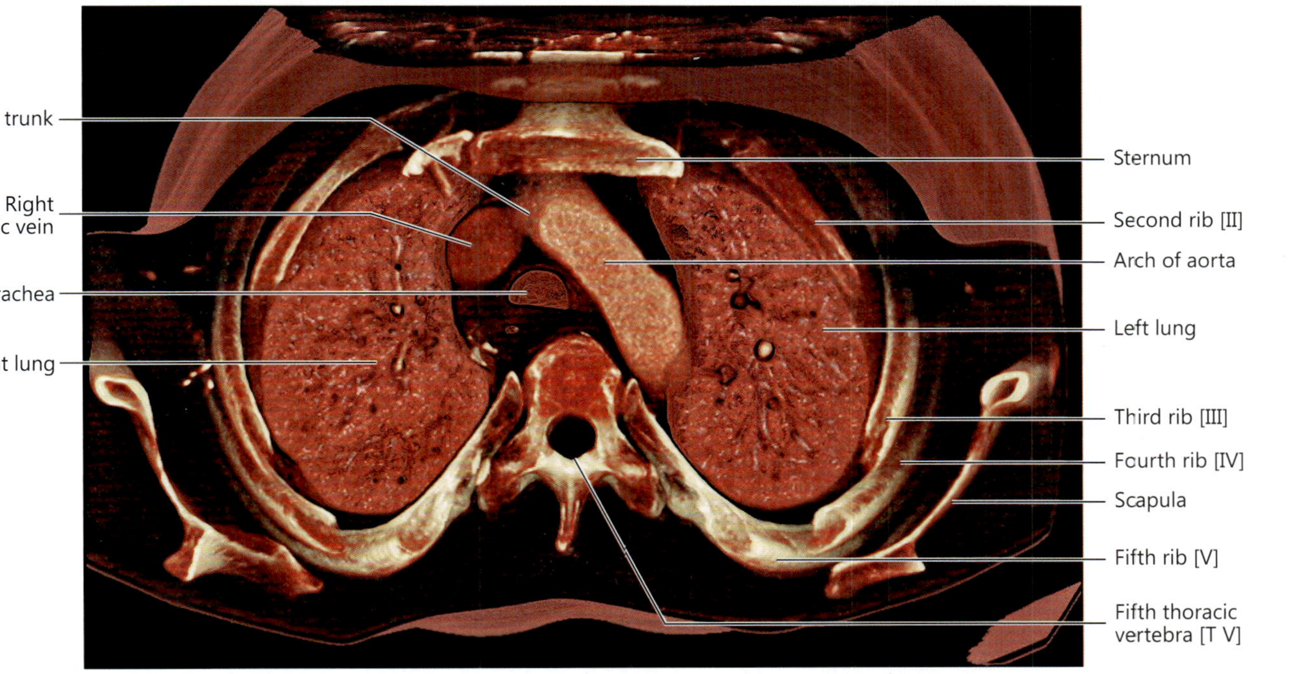

The thorax contains, besides the vessels of the large circle, also the vessels of the small circle of pulmonary circulation. In this series of three-dimensional images of a living human body, obtained by CT scan using a 3D volume rendering technique, featuring transverse sections at the level of the vertebrae T IV and T V, one can see the **pulmonary vessels** (arteries and veins) viewed from below, in radiological position.

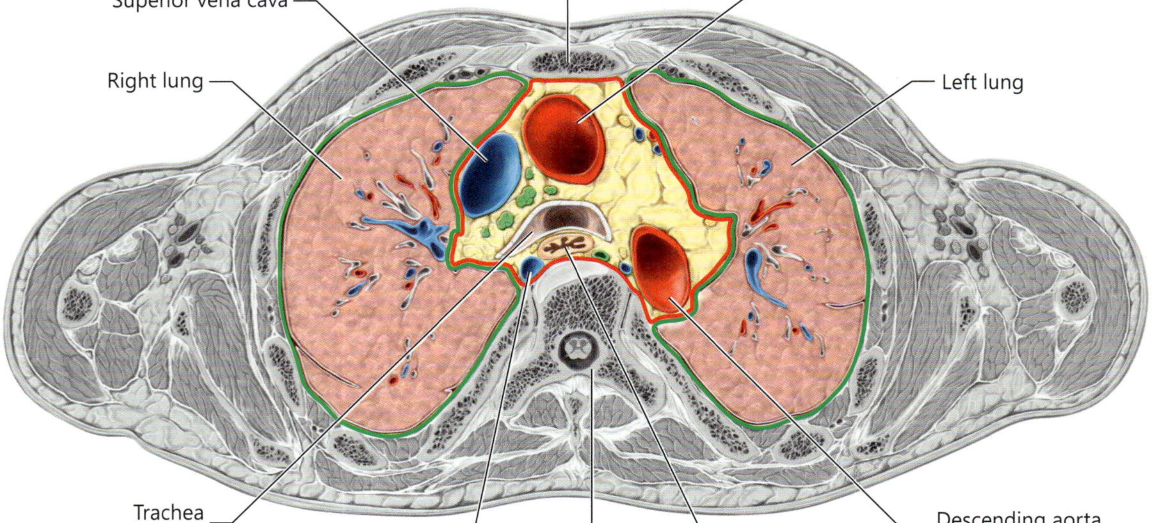
Sternum
Superior vena cava
Ascending aorta
Right lung
Left lung
Trachea
(bifurcation)
Azygos vein
Sixth thoracic
vertebra [T VI]
Esophagus
Descending aorta
(Thoracic aorta)

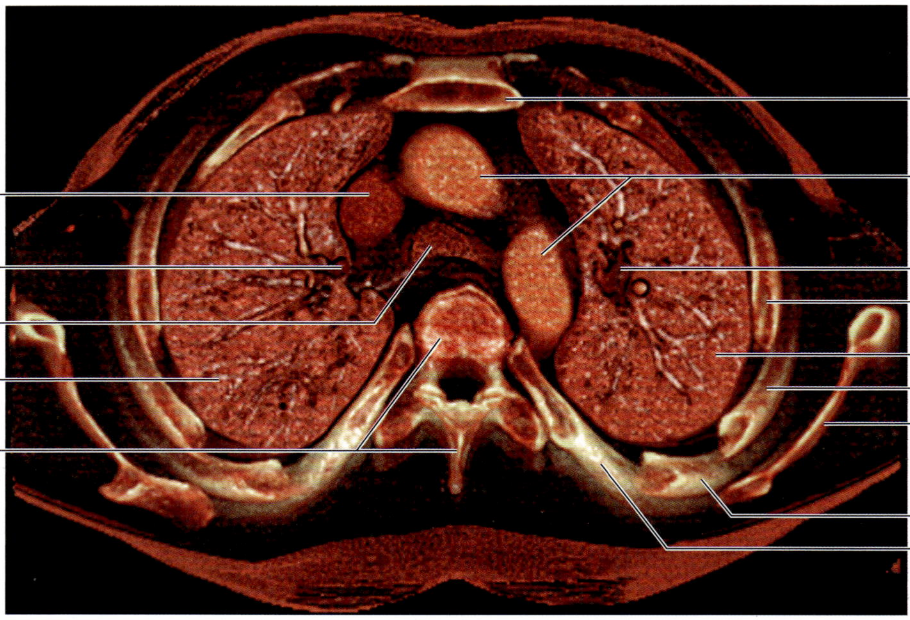
Sternum
Arch of aorta
Superior vena cava
Hilum of right lung
Hilum of left lung
Third rib [III]
Trachea
Left lung
Right lung
Fourth rib [IV]
Scapula
Sixth thoracic
vertebra [T VI]
Fifth rib [V]
Sixth rib [VI]

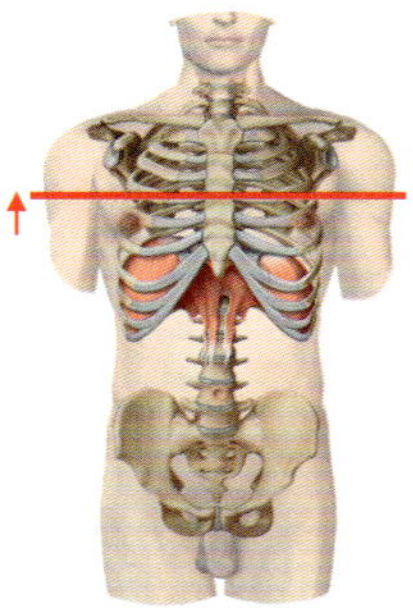

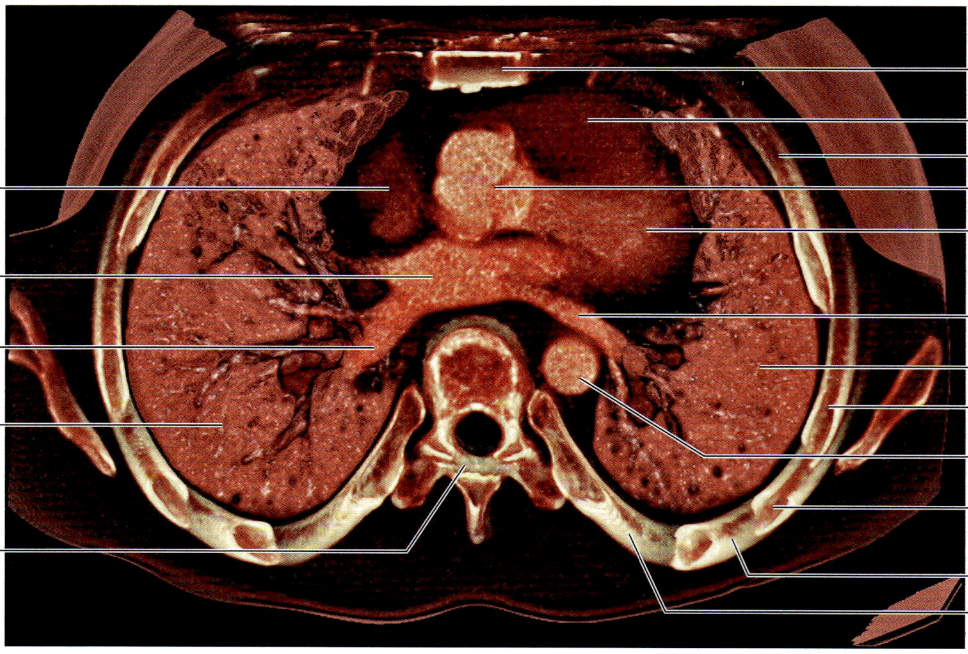
Sternum
Right ventricle
Fourth rib [IV]
Right atrium
Aortic valve
Left ventricle
Left atrium
Left inferior
pulmonary vein
Right inferior
pulmonary vein
Left lung
Fifth rib [V]
Right lung
Descending aorta
(Thoracic aorta)
Sixth rib [VI]
Seventh thoracic
vertebra [T VII]
Seventh rib [VII]
Eighth rib [VIII]

Summary table of the arteries of the thorax

ARTERY	ORIGIN	BRANCHES	TERRITORIES OF SUPPLY
anterior spinal (in the tract that runs along the anterior median fissure in the thoracic part of the spinal cord)	Vertebral artery		Anterior surface of the myelencephalon (medulla oblongata or bulb) spinal cord, filum terminale, meninges
aorta			See specific branches
ascending aorta	Left ventricle	Right and left coronary arteries	Sinuatrial node, atrioventricular node, right and left atria, interventricular septum and right and left ventricles
arch of aorta	Ascending aorta at the level of the superior border of the second right sternocostal joint	Brachiocephalic trunk, left common carotid artery, left subclavian artery; continues as the thoracic aorta	Thoracic organs, thoracic walls, spinal cord, head, neck, upper limb, thyroid gland and thymus, muscles of the arm and shoulder, brain
descending aorta			
– thoracic (thoracic aorta)	Arch of aorta at the level of the fourth thoracic vertebra	*Visceral*: pericardial, bronchial, esophageal and mediastinal branches; *parietal*: posterior intercostal, subcostal and superior phrenic arteries; continues as the abdominal aorta	Thoracic organs, thoracic walls, spinal cord
– – bronchial branches (bronchial arteries)	Thoracic aorta		Bronchi, lung parenchyma, tracheobronchial nodes, esophagus
– abdominal (abdominal aorta)	Descending aorta at the level of the aortic hiatus (in correspondence to the inferior border of T XII)		☞ Territories of supply of the arteries of the abdomen
brachiocephalic trunk	Arch of aorta	Right common carotid, right subclavian, thyroid ima (inconstant) arteries	Right side of the head, neck and upper limb, thyroid gland, thymus
costocervical trunk (at the origin of the thoracic tract of the subclavian artery, on the left)	Subclavian artery	Deep cervical and supreme intercostal arteries	Deep muscles of the neck, first and second intercostal spaces, spinal cord
first and second posterior intercostal	Supreme intercostal artery	Dorsal (or dorsospinal), medial and lateral cutaneous, spinal (or vertebromedullary artery), postcentral, prelaminar; posterior and anterior radicular arteries, segmental medullary artery; collateral (intercostal or supracostal), lateral cutaneous, lateral mammary	Superior part of the thoracic wall, vertebrae, spinal cord, meninges, muscles and skin of the back, mammary gland
internal thoracic	Subclavian artery	Mediastinal, thymic, bronchial (inconstant) and tracheal (inconstant); pericardiacophrenic artery; sternal, perforating, medial mammary, lateral costal (inconstant), anterior intercostal; musculophrenic and superior epigastric arteries	Anterior thoracic wall, diaphragm, pericardium, thymus, mammary gland
– anterior intercostal branches (anterior intercostal arteries)	Internal thoracic artery		First five or six intercostal spaces, pectoralis major and pectoralis minor

continued

◄
The thorax contains, besides the vessels of the large circle, also the vessels of the small circle of pulmonary circulation. On the page beside, two three-dimensional serial images of a living human body, obtained by CT scan using a 3D volume rendering technique, show transverse sections at the level of the vertebrae T VI and T VII, viewed from below, in radiological position. In the first, one can see the initial and final tracts of the arch of the aorta, and the position of the arch with respect to the trachea and its left bifurcation. In the second, the **pulmonary vessels** (arteries and veins) are visible, viewed from below, in radiological position. Also evident are the relations of these structures with the origin of the aorta (one can see the aortic valve at the level of the vertebra T VII) and the thoracic aorta.
At the top, a graphic representation, shown for comparative purposes, of a section at the level of the superior border of the vertebra T VI.

Summary table of the arteries of the thorax *(continued)*

ARTERY	ORIGIN	BRANCHES	TERRITORIES OF SUPPLY
lateral thoracic	Thoracoacromial, subscapular or axillary arteries	Lateral mammary	Pectoralis major and pectoralis minor, serratus anterior and subscapularis; axillary lymph nodes, mammary gland
left common carotid (in the thoracic tract)	Arch of aorta		Head and neck
left coronary	Ascending aorta, left aortic sinus (of Valsalva)	Anterior interventricular (or anterior interventricular or anterior descending artery), conus branch, lateral branch of anterior interventricular branch, interventricular septal, circumflex (or circumflex artery), atrial anastomotic, atrioventricular, left marginal, intermediate atrial, posterior left ventricular branch, sinuatrial nodal and atrioventricular nodal (inconstant), atrial	Interventricular septum, left atrium, left and right ventricles
left pulmonary	Pulmonary trunk	Superior lobar arteries: anterior and posterior apical segmental arteries and inferior and superior lingular arteries; inferior lobar arteries: superior segmental artery and anterior, lateral, medial and posterior basal segmental arteries	Left lung
musculophrenic	Internal thoracic artery		Diaphragm, seventh, eighth and ninth intercostal spaces, pericardium
pericardiacophrenic	Internal thoracic artery		Diaphragm, pericardium, pleura
posterior spinal (in the tract that runs along the posterolateral sulcus in the thoracic part of the spinal cord)	Posterior inferior cerebellar artery and vertebral artery		Myelencephalon (medulla oblongata or bulb), posterior part of the spinal cord, cauda equina, meninges, fourth ventricle
pulmonary trunk	Conus arteriosus	Right and left pulmonary arteries	Lungs
right coronary	Ascending aorta, right aortic sinus (of Valsalva)	Atrioventricular, conus branch, sinuatrial nodal, atrial, right marginal, intermediate atrial, posterior interventricular (or posterior interventricular or posterior descending artery), interventricular septal, atrioventricular nodal, right posterolateral (inconstant)	Sinuatrial node, atrioventricular node, right atrium, interventricular septum, right and left ventricles
right pulmonary	Pulmonary trunk	Superior lobar arteries: anterior and posterior apical segmental arteries; middle lobar artery: medial and lateral segmental arteries; inferior lobar arteries; superior segmental artery and anterior, lateral, medial and posterior basal segmental arteries	Right lung
subclavian (from the origin to the superior thoracic aperture, on the right side; thoracic tract, on the left)	*Right side*: brachiocephalic trunk; *left side*: arch of aorta	Vertebral artery, thyrocervical trunk, internal thoracic artery, costocervical trunk, dorsal scapular artery (originating more often from the thyrocervical trunk); becomes the axillary artery in correspondence to the outer border of the first rib	Neck, thoracic wall, muscles of the shoulder and the arm, spinal cord and brain
subcostal (in the tract that passes behind the lateral arcuate ligament)	Thoracic aorta	Dorsal and spinal	Superior abdominal wall; anastomoses with the superior epigastric, eleventh posterior intercostal and first lumbar arteries

continued

Summary table of the arteries of the thorax			(continued)
ARTERY	***ORIGIN***	***BRANCHES***	***TERRITORIES OF SUPPLY***
superior epigastric (in the part that runs between the sternal and costal insertions of the diaphragm)	Internal thoracic artery		Diaphragm, xiphoid process, rectus abdominis, liver, peritoneum; anastomoses with the inferior epigastric artery which creates a connection between the system of the subclavian artery and that of the external iliac artery
superior phrenic	Thoracic aorta		Diaphragm
superior thoracic	Axillary or thoracoacromial arteries		Pectoralis major and pectoralis minor, serratus anterior and intercostal muscles, thoracic wall
supreme intercostal	Costocervical trunk	First and second posterior intercostal arteries	First and second intercostal spaces
third to eleventh posterior intercostal	Thoracic aorta	Dorsal (or dorsospinal), medial and lateral cutaneous, spinal (or vertebromedullary artery), postcentral, prelaminar; posterior and anterior radicular arteries, segmental medullary artery; collateral (intercostal or supracostal), lateral cutaneous, lateral mammary, bronchial (or bronchial artery; can originate from the third right posterior intercostal artery)	Thoracic wall, vertebrae, spinal cord, meninges, muscles and skin of the back, mammary gland
thoracoacromial	Axillary artery	Acromial, clavicular, deltoid, pectoral	Pectoralis major and pectoralis minor, deltoid and subclavius, mammary gland, sternoclavicular joint, coracoid process, acromion
thoracodorsal	Subscapular artery		Subscapularis, latissimus dorsi, serratus anterior and intercostal muscles
thyroid ima	Arch of aorta or brachiocephalic trunk		Thyroid gland, thymus
vertebral (at its origin from the subclavian artery)	Subclavian	*Cervical part*: spinal and muscular branches; *intracranial part*: meningeal branches; posterior and anterior spinal arteries, posterior inferior cerebellar artery; medial and lateral medullary branches	Vertebral body, deep muscles of the neck, falx cerebelli, spinal cord, cerebellum, brainstem

ARTERY	
abdominal	☞ abdominal aorta
acromiothoracic	☞ thoracoacromial artery
dorsal spinal	☞ posterior spinal artery
innominate	☞ brachiocephalic trunk
internal mammary	☞ internal thoracic artery
long thoracic or **external mammary**	☞ lateral thoracic artery
twelfth thoracic	☞ subcostal artery
ventral spinal	☞ anterior spinal artery

The table alongside lists some arteries by their name normally used or eponym, with the corresponding reference to the official term.

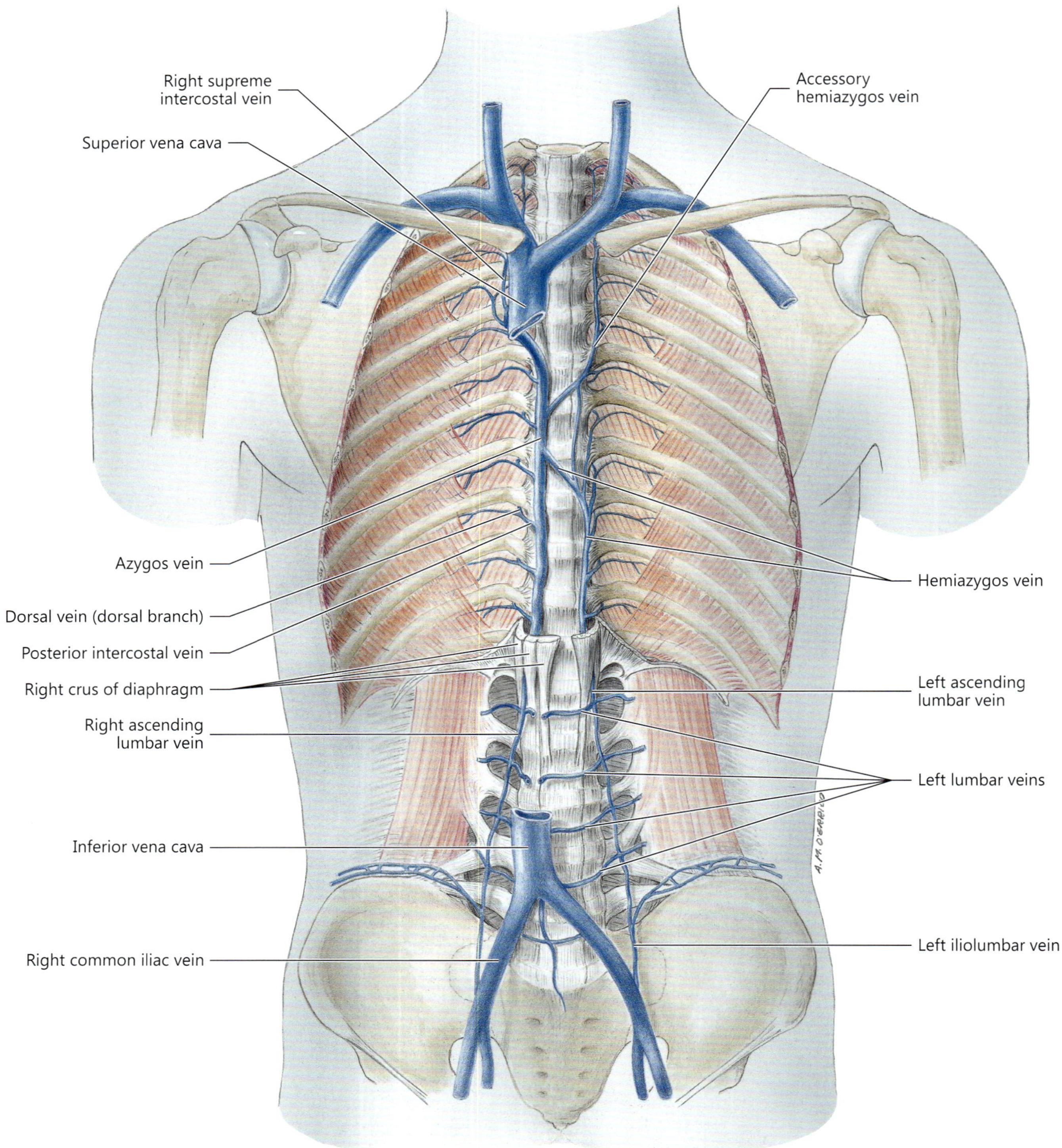

The **azygos**, **hemiazygos** and **accessory hemiazygos veins** viewed from the front after demolition of the thoracic and abdominal walls and removal of the splanchnic organs. Note the dorsospinal branches of the intercostal veins which drain the vertebral plexuses. The system of the azygos veins constitutes an important deep anastomotic system between the superior vena cava and the inferior vena cava through the ascending lumbar veins.

Schematic representation of the superior vena cava and of the major affluent veins ☞ page 222 (Neck - Veins).

➤

Schematic representation showing the affluent branches of the **superior vena cava**, **coronary sinus** and **veins of the pulmonary circulation**. The superior vena cava, which is formed by the confluence of the right and left brachiocephalic veins, after receiving refluent blood from the walls of the thorax, spinal cord and some structures of the posterior mediastinum through the azygos vein, reaches the right atrium of the heart, into which also the coronary sinus opens, near the opening of the inferior vena cava. The pulmonary veins conduct arterial blood from the lungs to the left atrium of the heart.

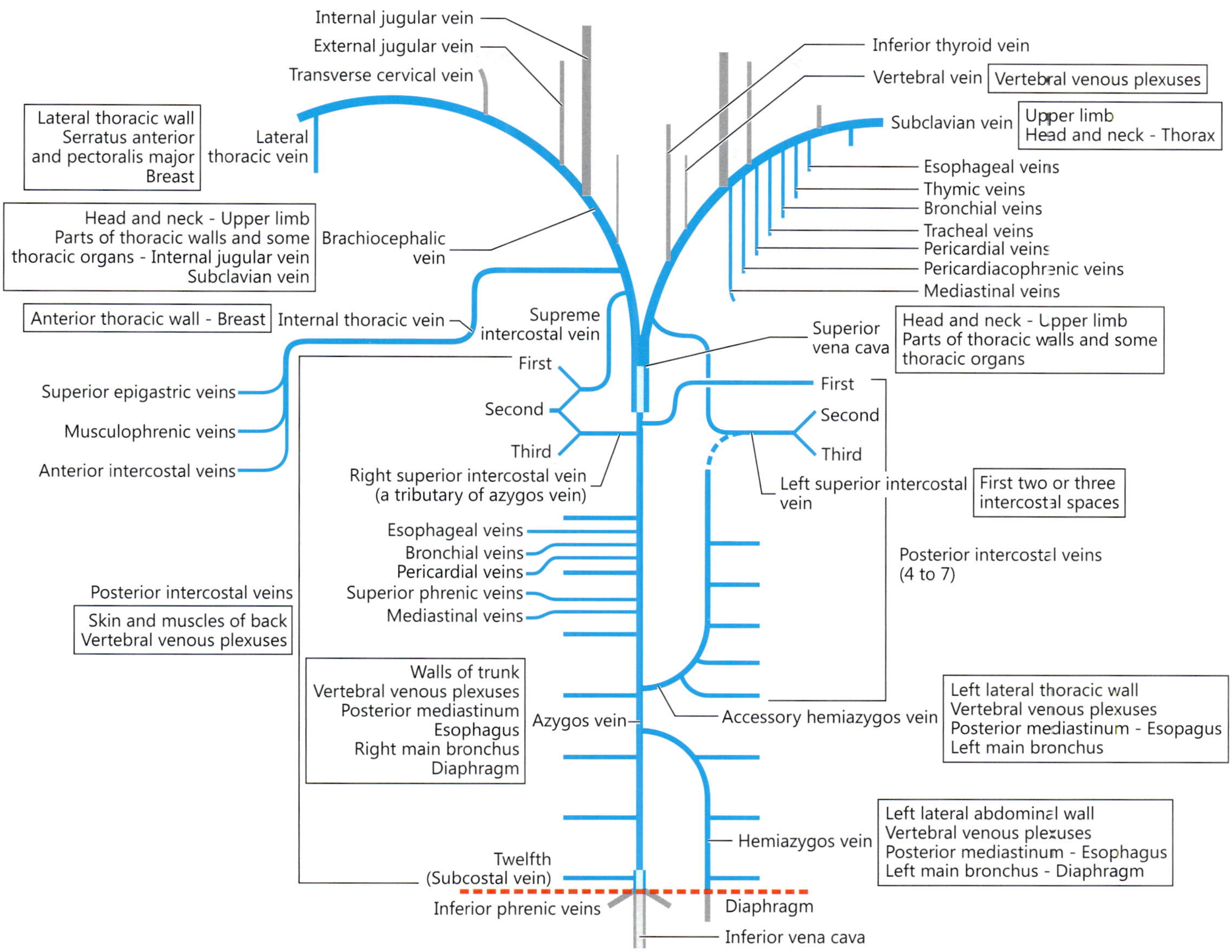
Internal jugular vein
External jugular vein
Transverse cervical vein
Lateral thoracic wall
Serratus anterior
and pectoralis major
Breast
Lateral
thoracic vein
Head and neck - Upper limb
Parts of thoracic walls and some
thoracic organs - Internal jugular vein
Subclavian vein
Brachiocephalic
vein
Anterior thoracic wall - Breast
Internal thoracic vein
Supreme
intercostal vein
First
Second
Third
Superior epigastric veins
Musculophrenic veins
Anterior intercostal veins
Right superior intercostal vein
(a tributary of azygos vein)
Esophageal veins
Bronchial veins
Pericardial veins
Superior phrenic veins
Mediastinal veins
Posterior intercostal veins
Skin and muscles of back
Vertebral venous plexuses
Walls of trunk
Vertebral venous plexuses
Posterior mediastinum
Esophagus
Right main bronchus
Diaphragm
Azygos vein
Twelfth
(Subcostal vein)
Inferior phrenic veins
Inferior thyroid vein
Vertebral vein
Vertebral venous plexuses
Subclavian vein
Upper limb
Head and neck - Thorax
Esophageal veins
Thymic veins
Bronchial veins
Tracheal veins
Pericardial veins
Pericardiacophrenic veins
Mediastinal veins
Superior
vena cava
Head and neck - Upper limb
Parts of thoracic walls and some
thoracic organs
First
Second
Third
Left superior intercostal
vein
First two or three
intercostal spaces
Posterior intercostal veins
(4 to 7)
Accessory hemiazygos vein
Left lateral thoracic wall
Vertebral venous plexuses
Posterior mediastinum - Esopagus
Left main bronchus
Hemiazygos vein
Left lateral abdominal wall
Vertebral venous plexuses
Posterior mediastinum - Esophagus
Left main bronchus - Diaphragm
Diaphragm
Inferior vena cava

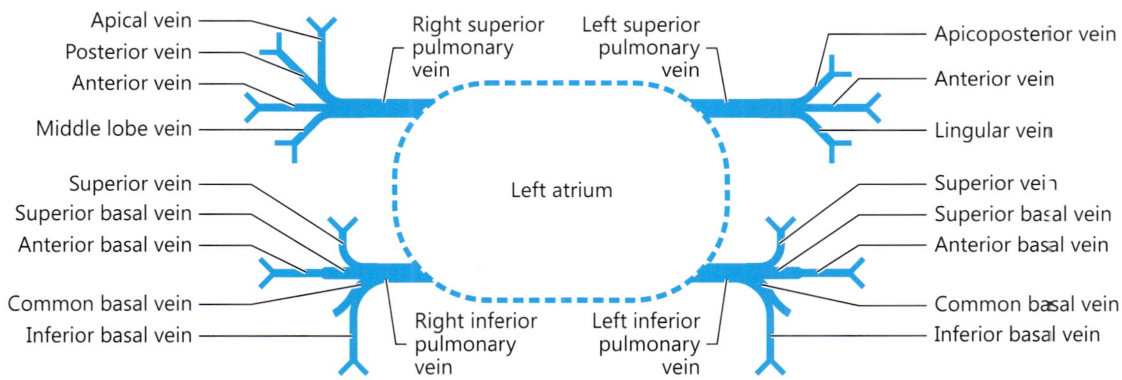
Apical vein
Posterior vein
Anterior vein
Middle lobe vein
Right superior
pulmonary
vein
Left superior
pulmonary
vein
Apicoposterior vein
Anterior vein
Lingular vein
Left atrium
Superior vein
Superior basal vein
Anterior basal vein
Common basal vein
Inferior basal vein
Right inferior
pulmonary
vein
Left inferior
pulmonary
vein
Superior vein
Superior basal vein
Anterior basal vein
Common basal vein
Inferior basal vein

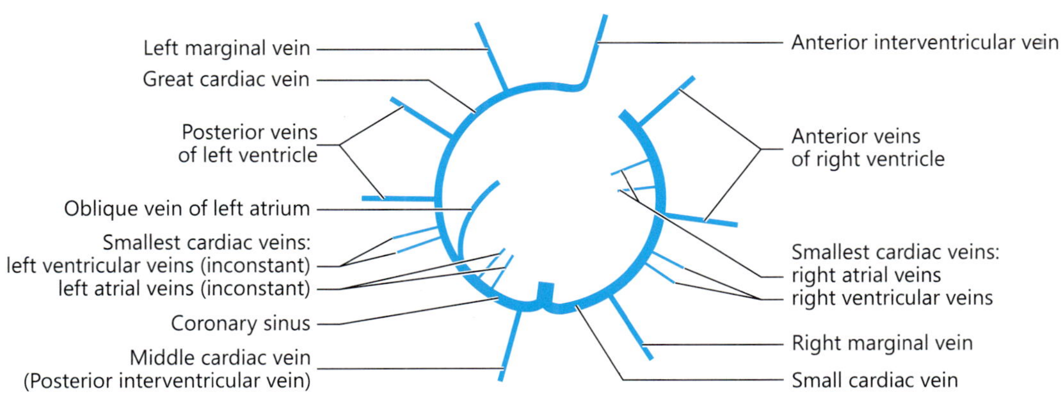
Left marginal vein
Great cardiac vein
Posterior veins
of left ventricle
Oblique vein of left atrium
Smallest cardiac veins:
left ventricular veins (inconstant)
left atrial veins (inconstant)
Coronary sinus
Middle cardiac vein
(Posterior interventricular vein)
Anterior interventricular vein
Anterior veins
of right ventricle
Smallest cardiac veins:
right atrial veins
right ventricular veins
Right marginal vein
Small cardiac vein

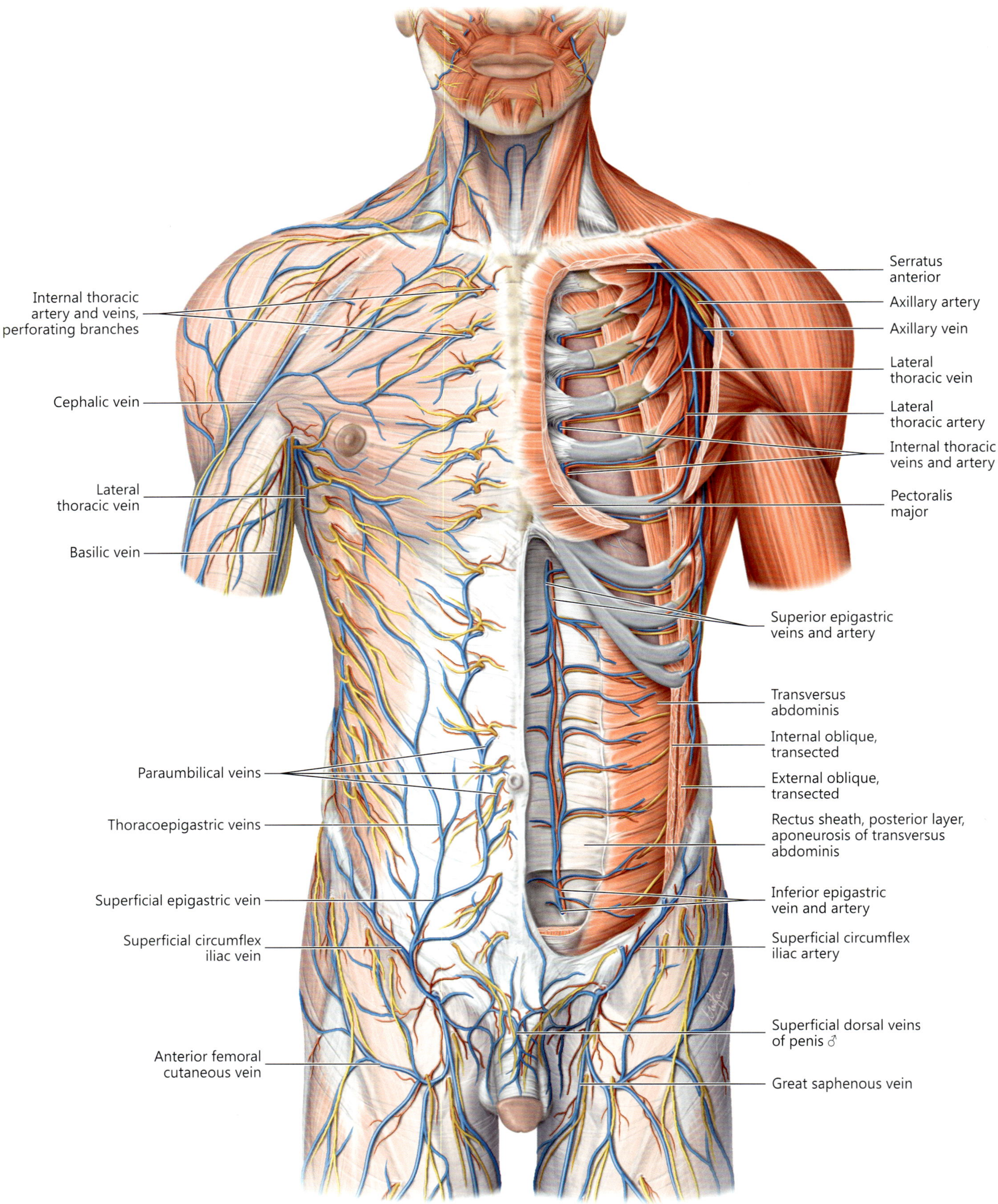

Parietal vascularization of the trunk: **cava-caval anastomoses**. The superficial and deep venous drainage of the thorax and of the abdomen are communicating. These anastomoses connect the circulation of the superior vena cava with that of the inferior vena cava. At superficial level, the lateral thoracic veins (affluents of the axillary veins) anastomose with the superficial epigastric veins (affluents of the femoral veins), through the thoracoepigastric veins. At deep level, in the context of the rectus sheath, the superior epigastric veins (affluents of the internal thoracic veins) anastomose with the inferior epigastric veins (affluents of the external iliac veins).

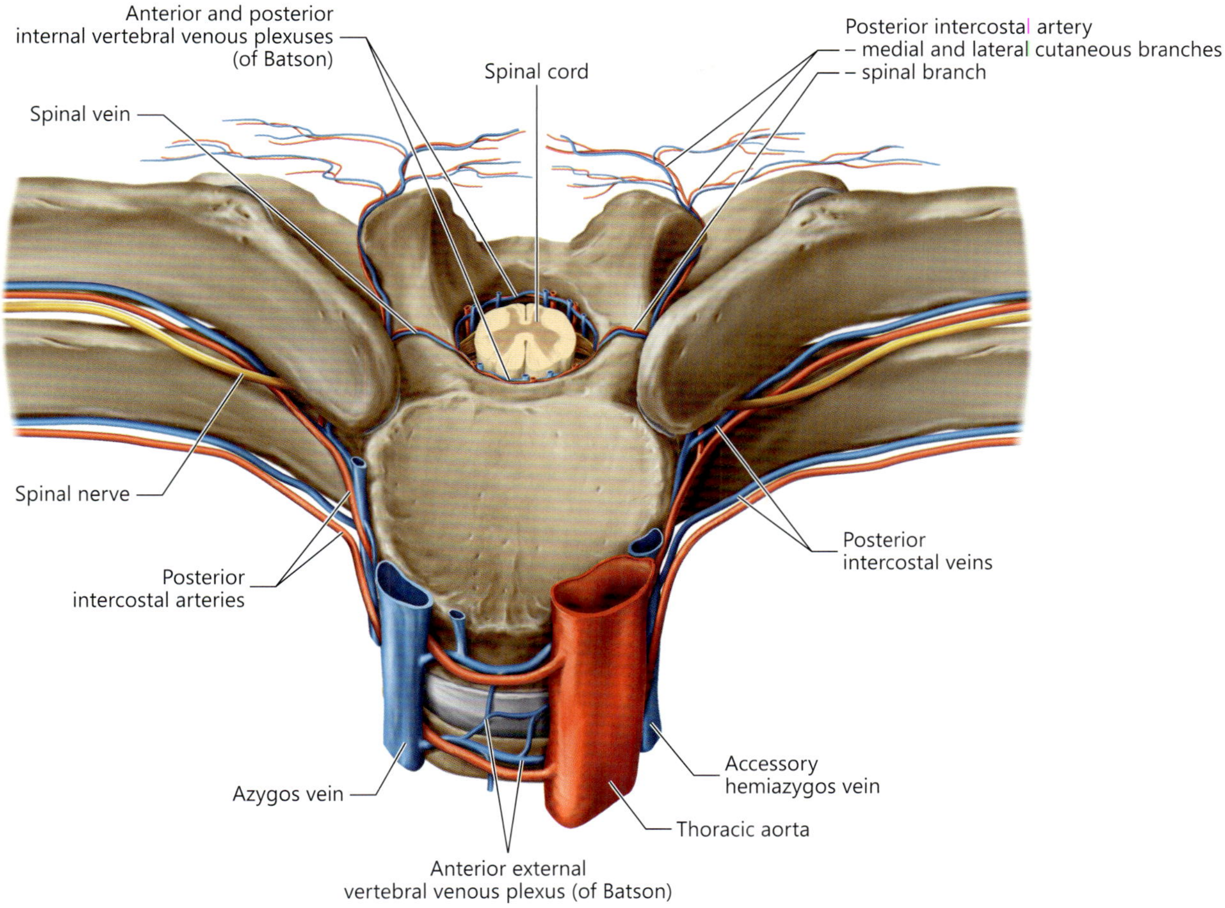

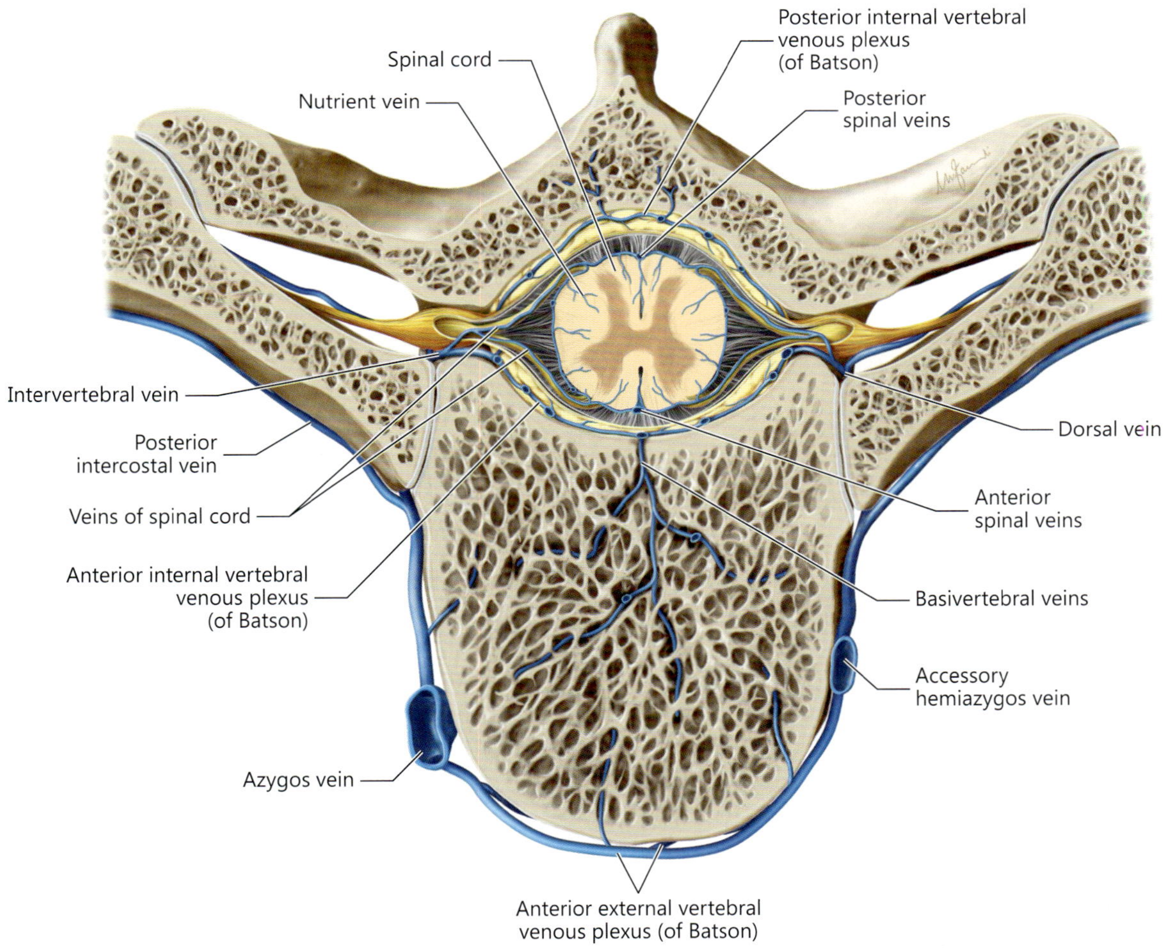

Transverse section of the dorsal vertebral column. Disposition of the **internal vertebral venous plexuses** of the epidural space. The plexuses extend longitudinally from the foramen magnum to the apex of the sacrum.

Summary table of the veins of the thorax

VEIN	*ORIGIN/LOCATION*	*DRAINAGE*	*CONFLUENCE*
accessory hemiazygos	Left side of T II-T III; crosses T VII-T VIII moving rightwards	From four to seven posterior intercostal veins	Azygos vein
anterior and posterior spinal	Longitudinal venous plexus along the surface of the spinal cord	Meninges and spinal cord	Anterior internal vertebral venous plexus
anterior intercostal (12 pairs)	Inferior border of each rib	Ribs and intercostal muscles	Internal thoracic and musculophrenic veins
anterior veins of right ventricle or anterior cardiac veins (three or four in number; one of these ascends along the right margin and is known as the small coronary)	Anterior or sternocostal surface of the heart	Ventral side of the right ventricle	Small cardiac vein or right atrium
azygos	Anterior surface of L I; right side of T IV; anterior arch, passes over the top of the root of the right lung; opening into the superior vena cava	Ascending lumbar, right subcostal, posterior intercostal, hemiazygos, esophageal, mediastinal, pericardial and right bronchial veins	Superior vena cava
basivertebral	Anterior internal vertebral venous plexus; spongy bone of vertebral body	Vertebrae	Anterior internal and external vertebral venous plexuses
brachiocephalic	Origin behind the sternoclavicular joint through the confluence of the internal jugular and subclavian veins; termination behind the first sternocostal joint	Head, neck, upper limb, parts of thoracic walls, thymus, pericardium, diaphragm, bronchi, trachea, esophagus	Merge with each other to form the superior vena cava
bronchial	They follow the ramifications of the bronchi	Main bronchi and roots of the lungs	*Right side*: azygos vein; *left side*: left superior intercostal vein, accessory hemiazygos vein or left brachiocephalic vein
coronary sinus	Posterior part of the coronary sulcus	Great, middle and small cardiac veins, posterior vein of left ventricle, oblique vein of left atrium	Right atrium, between the opening of inferior vena cava and the right atrioventricular orifice
esophageal (varying in number)	Along the esophagus	Esophagus	*Right side*: azygos vein; *left side*: hemiazygos, left gastric and inferior thyroid or left brachiocephalic veins on left side
great cardiac	Anterior interventricular sulcus; coronary sulcus on the left	Tributary veins from the left atrium and from both ventricles; anterior interventricular and left marginal veins	Left extremity of the coronary sinus
hemiazygos	Left side of T XII; crosses T VIII-T IX moving rightwards	Left ascending lumbar, four or five posterior caudal intercostal, left subcostal, esophageal and mediastinal veins	Azygos vein
inferior vena cava	Anterior right surface of L IV; Right side of the lumbar segment of the vertebral column; diaphragmatic opening of the inferior vena cava at the level of the fifth costal cartilage	Both common iliac veins; lumbar, renal, right testicular (♂), right ovarian (♀), suprarenal, inferior phrenic and hepatic veins	Right atrium of the heart
internal thoracic	From the confluence of the musculophrenic and superior epigastric veins	Besides the musculophrenic and superior epigastric veins, other affluent branches are the anterior intercostal and subcutaneous abdominal veins	Brachiocephalic vein
intervertebral (in the thoracic segment of the vertebral column)	Internal vertebral venous plexuses; passes through the intervertebral foramen accompanied by the roots and initial tract of the spinal nerve	Superficial planes of the posterior regions of the trunk, erector spinae, vertebrae, meninges and spinal cord; in part, also the external vertebral venous plexuses	Vertebral, posterior intercostal, lumbar and lateral sacral veins

continued

Summary table of the veins of the thorax			*(continued)*
VEIN	***ORIGIN/LOCATION***	***DRAINAGE***	***CONFLUENCE***
lateral thoracic	Lateral thoracic wall	Lateral thoracic wall; anastomoses, through the costoaxillary veins, with the posterior intercostal veins	Axillary vein
left inferior pulmonary	From the left lung to the heart	Inferior lobe of left lung; receives the superior vein and common, superior (which receives the anterior basal vein) and inferior basal veins	Left atrium
left superior intercostal	From the confluence of the 2nd-3rd or 2nd-4th left posterior intercostal vein; in front of the left costovertebral joints	2nd-3rd or 2nd-4th intercostal space; left bronchial vein and, occasionally, left superior phrenic vein	Left brachiocephalic vein or accessory hemiazygos vein
left superior pulmonary	From the left lung to the heart	Superior lobe of left lung; receives the apicoposterior, anterior and lingular veins	Left atrium
mediastinal	Mediastinum	Areolar tissue and lymph nodes of the anterior mediastinum, pericardium	Azygos and brachiocephalic veins or superior vena cava
middle cardiac or **posterior interventricular**	Posterior interventricular sulcus	Tributary veins from both ventricles	Right extremity of the coronary sinus
musculophrenic	From the confluence of the 6th-9th anterior intercostal vein they follow the internal border of the thoracic cage until the sixth costal cartilage	Diaphragm and inferior intercostal spaces	Internal thoracic veins
oblique vein of left atrium	Posterior wall of the left atrium	Posterior part of the left atrium	Coronary sinus
pericardiacophrenic	Satellites of the pericardiacophrenic artery and phrenic nerve; they ascend from the superior surface of the diaphragm between the pleura and pericardium	Diaphragm; tributary veins from the pericardium	Brachiocephalic vein
pericardial (varying in number)	Fibrous pericardium	Pericardium	Azygos and brachiocephalic veins or superior vena cava
posterior intercostal (12 pairs)	Inferior border of each rib, from the neck to approximately half of the body of the rib	Skin and muscles of the back, vertebrae, meninges and spinal cord; dorsal vein or branch, intervertebral vein and spinal vein or branch	*Right side*: azygos and right superior intercostal veins; *left side*: left brachiocephalic and hemiazygos or accessory hemiazygos veins
posterior vein of left ventricle	Diaphragmatic or inferior surface of the heart near the apex	Posterior surface of the left ventricle	Coronary sinus
right inferior pulmonary	From the right lung to the heart	Inferior lobe of right lung; receives the superior vein and common basal vein, where the superior basal vein (which receives the anterior basal vein) and inferior basal vein drain	Left atrium
right superior intercostal	From the confluence of the 2nd-3rd or 2nd-4th right posterior intercostal vein; in front of the right costovertebral joints	2nd-3rd or 2nd-4th intercostal space	Azygos vein
right superior pulmonary	From the right lung to the heart	Superior and middle lobes of right lung; receives the apical, anterior, posterior veins and middle lobe veins	Left atrium
small cardiac vein	Right margin of the heart; coronary sulcus on the right	Posterior part of right atrium and ventricle; right marginal vein and anterior veins of right ventricle or anterior cardiac veins	Right extremity of the coronary sinus or right atrium
smallest cardiac veins	Thickness of the myocardium	Muscular wall of the heart; right atrial and ventricular veins; left atrial and ventricular veins (inconstant)	Majority in the atria, some in the ventricles

continued

Summary table of the veins of the thorax *(continued)*

VEIN	ORIGIN/LOCATION	DRAINAGE	CONFLUENCE
subclavian	From the external border of the first rib to the sternal end of the clavicle	Continuation of the axillary vein; external jugular vein and anterior jugular vein (occasionally)	Unites with the internal jugular vein to form the brachiocephalic vein
subcostal	Satellite of the subcostal artery along the inferior border of twelfth rib	Anterior abdominal wall	Azygos vein
superior epigastric	Satellites of the superior epigastric artery	Territory corresponding to the supply territory of the superior epigastric artery	Unite with the musculophrenic veins to form the internal thoracic veins
superior phrenic	Superior surface of the diaphragm; they ascend along the lateral surface of the pericardium	Diaphragm	*Right side*: azygos vein or confluence of the two brachiocephalic veins; *left side*: left brachiocephalic vein
superior vena cava	Superior mediastinum, at the level of the first costal cartilage, from the confluence of the brachiocephalic veins; in correspondence to the third costal cartilage	Supradiaphragmatic part of body	Roof of the right atrium
supreme intercostal	From the confluence of 1st-2nd posterior intercostal vein; in front of the costovertebral joints	1st intercostal space	Brachiocephalic vein or superior intercostal vein or vertebral vein of the corresponding side
thoracoacromial	Satellite of the thoracoacromial artery	Acromion, coracoid process, sternoclavicular joint; tributary veins from the deltoid, subclavius, pectoralis minor and pectoralis major	Cephalic vein or axillary vein or subclavian vein
thymic (two or three in number)	Thymus	Parenchyma of thymus	Left brachiocephalic and inferior thyroid veins or, at times, superior phrenic veins
tracheal (varying in number)	Along the surface of the trachea	Wall of the trachea	Thyroid plexus and brachiocephalic vein or superior vena cava

VEIN	
acromiothoracic	☞ thoracoacromial vein
anterior mediastinal	☞ mediastinal veins
azygos minor inferior	☞ hemiazygos vein
azygos minor superior	☞ accessory hemiazygos vein
greater azygos	☞ azygos vein
inferior coronary	☞ middle cardiac vein or posterior interventricular vein
innominate of Vieussens	☞ anterior veins of right ventricle or anterior cardiac veins
internal mammary	☞ internal thoracic vein
left azygos	☞ hemiazygos vein
left coronary	☞ great cardiac vein
long thoracic	☞ lateral thoracic vein
Marshall's vein	☞ oblique vein of left atrium
right coronary	☞ small cardiac vein
small coronary or **of Galen**	☞ anterior veins of right ventricle or anterior cardiac veins
Thebesian	☞ smallest cardiac veins

The table alongside lists some veins by their name normally used or eponym, with the corresponding reference to the official term.

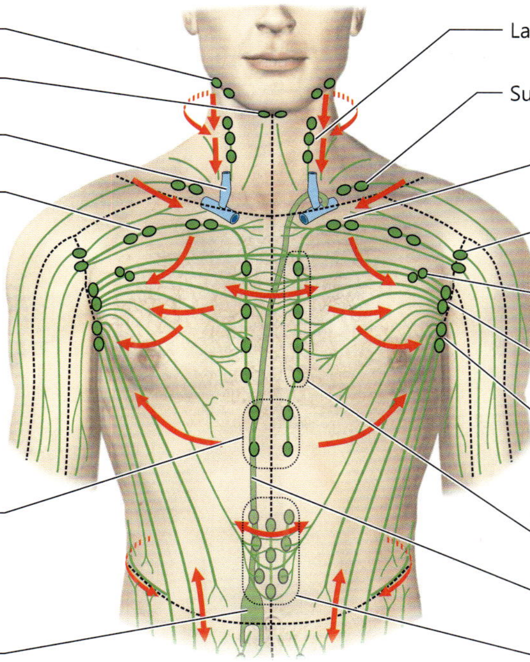

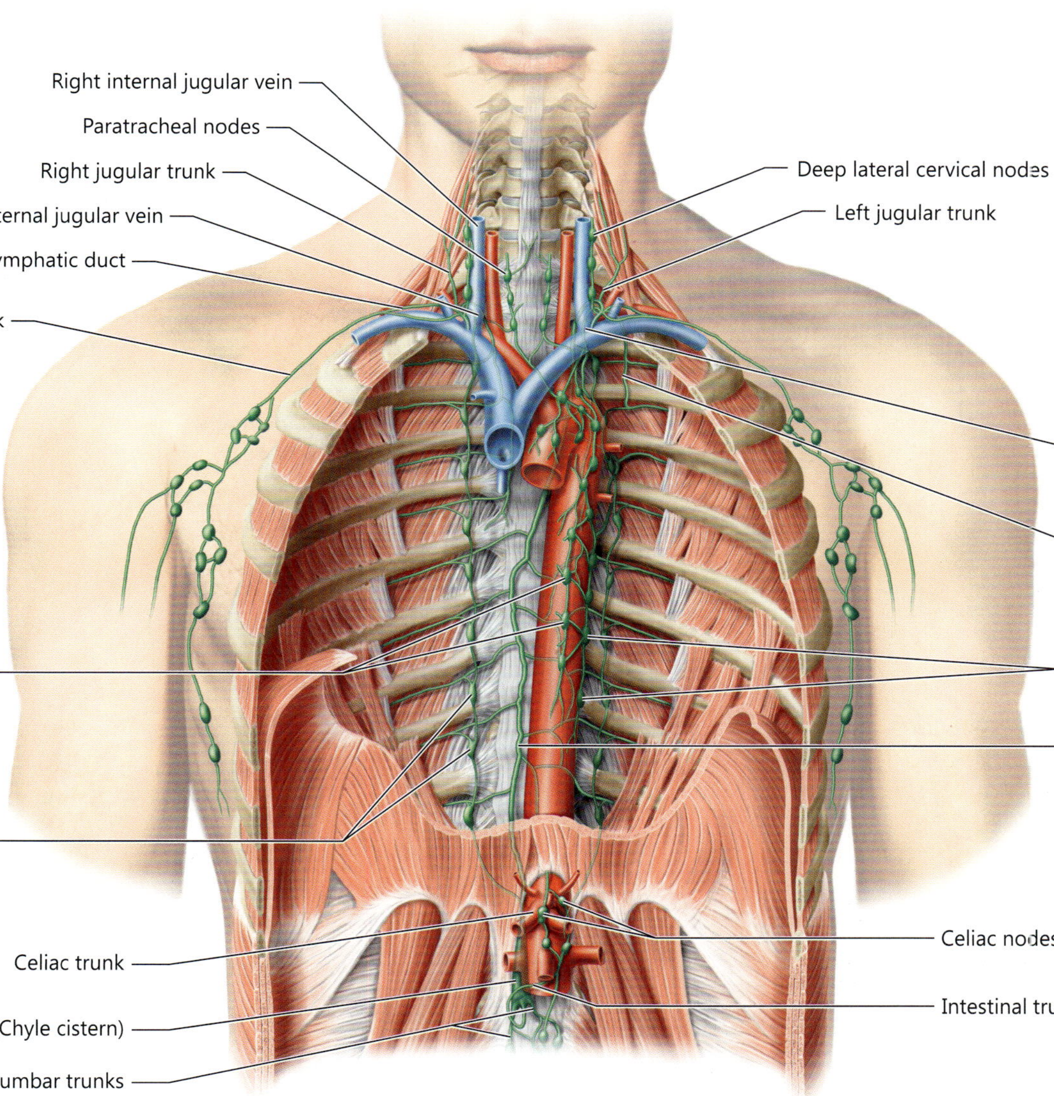

At the top, schematic representation showing the principal **superficial lymphatic vessels** and **lymphatic groups** of the thorax and the flow paths of lymphatic drainage. Below, disposition of the principal deep lymphatic ducts and lymphatic groups of the posterior mediastinum.

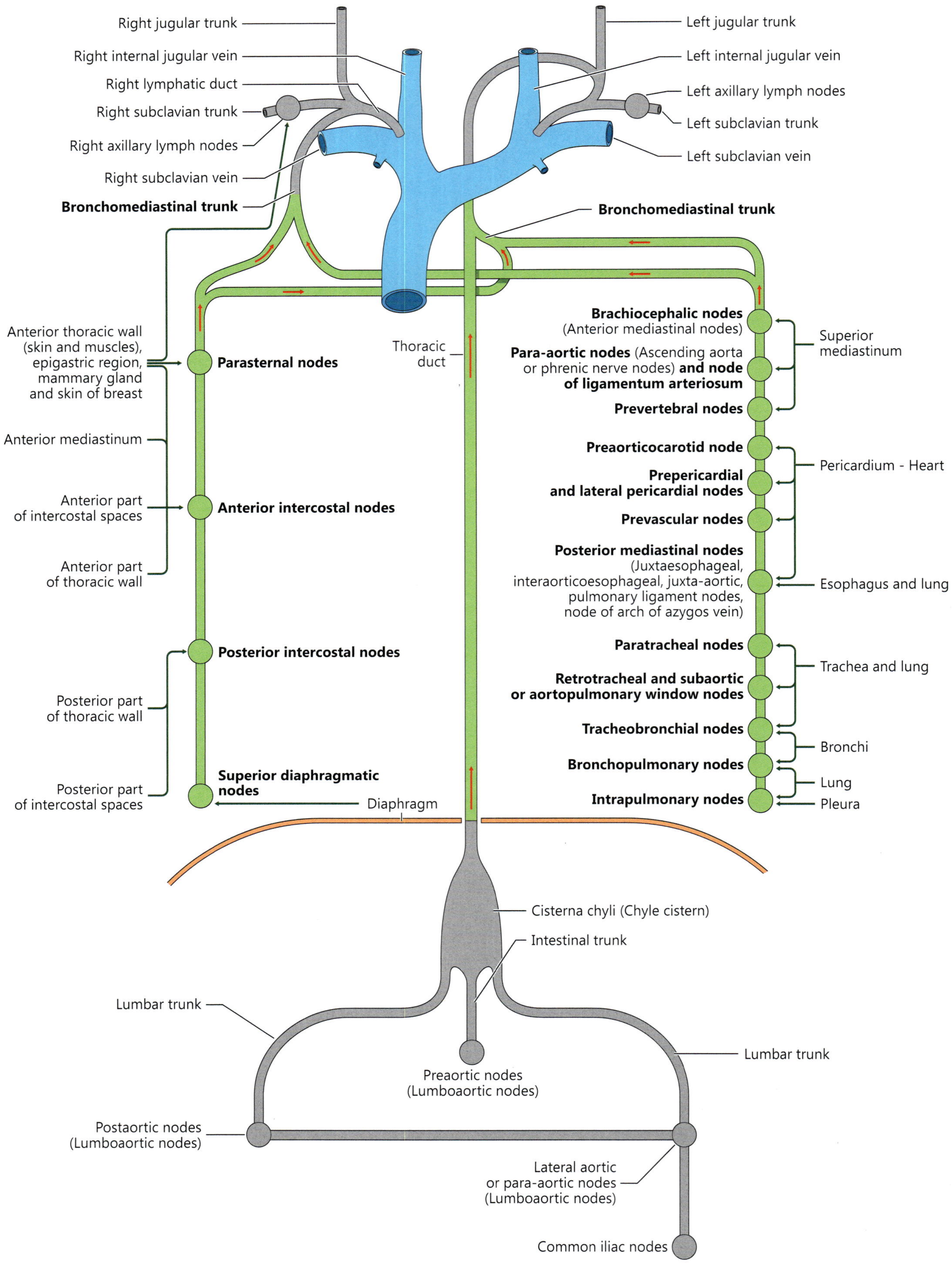

Schematic representation of the principal **lymphatic groups** and **lymphatic ducts**. The lymph nodes and ducts of the thorax and their visceral and parietal territories of drainage are highlighted.

Classification of the lymph nodes of the mediastinum

LYMPH NODE STATION OR GROUP	LYMPH NODES	LOCATION	REGION	AFFERENTS	EFFERENTS
1	**Superior mediastinal** (proper)	Above a horizontal line crossing the left brachiocephalic vein in correspondence to the trachea	Superior mediastinum	Thymus, thyroid gland, heart, pericardium, diaphragmatic pleura and mediastinal pleura, superior diaphragmatic nodes	Right and left bronchomediastinal trunks
2	**Superior paratracheal**	Below the above indicated boundary for the superior mediastinal nodes and above a horizontal line tangential to the superior border of the arch of aorta	Superior mediastinum	Lungs, bronchi, trachea, heart, juxtaesophageal nodes	Inferior paratracheal nodes and inferior cervical nodes
3	**Prevascular and retrotracheal**	Behind the trachea in front of the great vessels	Superior mediastinum	Thymus, thyroid gland, heart, pericardium, diaphragmatic pleura and mediastinal pleura, superior diaphragmatic nodes, superior lobe of left lung (to the prevascular nodes); lungs, bronchi, trachea, heart, juxtaesophageal nodes (to the retrotracheal nodes)	Right and left bronchomediastinal trunks
4	**Inferior paratracheal**	On each side of the median line crossing the trachea between a horizontal line tangential to the superior border of the arch of aorta and a horizontal line passing over the superior borders of the right and left superior lobar bronchi	Superior mediastinum	Lungs, bronchi, trachea, heart, juxtaesophageal nodes	Thoracic duct
5	**Subaortic** (or aortopulmonary window nodes, also known as Aortic)	In the aortopulmonary window laterally to the ligamentum arteriosum or to the aorta or left pulmonary artery	Mediastinum	Lungs, bronchi, trachea, heart, juxtaesophageal nodes	Thoracic duct
6	**Para-aortic** (ascending aorta nodes or phrenic nerve nodes, also known as Aortic)	In front of and laterally to the ascending aorta and arch of aorta or to the brachiocephalic trunk below a line tangential to the superior border of the arch of aorta	Mediastinum	Thymus, thyroid gland, heart, pericardium, diaphragmatic pleura and mediastinal pleura, superior diaphragmatic nodes	Right and left bronchomediastinal trunks
7	**Subcarinal**	Below the carina of trachea not associated to the right and left inferior lobar bronchi	Inferior mediastinum	Lungs, bronchi, trachea, heart, juxtaesophageal nodes	Paratracheal nodes, in particular on the right side
8	**Juxtaesophageal** (paraesophageal)	Adjacent to the wall of the esophagus to the right and left of the median line	Inferior mediastinum	Esophagus, pericardium, diaphragm, inferior diaphragmatic nodes, left lobe of liver	Thoracic duct, tracheobronchial, subcarinal, lateral aortic and celiac nodes
9	**Pulmonary ligament nodes**	Within the pulmonary ligament	Inferior mediastinum	Basal segments of the inferior lobes of lungs, esophagus	Thoracic duct, tracheobronchial, subcarinal, lateral aortic and celiac nodes

SUPERIOR MEDIASTINAL NODES	INFERIOR MEDIASTINAL NODES	ANTERIOR MEDIASTINAL NODES	POSTERIOR MEDIASTINAL NODES
Superior mediastinal nodes (proper) Superior paratracheal nodes Prevascular nodes Prevertebral nodes Inferior paratracheal nodes	Subcarinal nodes Juxtaesophageal nodes Pulmonary ligament nodes	Brachiocephalic nodes (or nodes of the transverse chain) Preaorticocarotid nodes Node of ligamentum arteriosum Prevascular nodes	Juxtaesophageal nodes Interaorticoesophageal nodes Juxta-aortic nodes Node of arch of azygos vein

Summary table of the lymphatic groups of the thorax

LYMPH NODES	*GROUP OR STATION*	*SITE*	*AFFERENTS*	*EFFERENTS*
			PARIETAL THORACIC LYMPH NODES	
Intercostal	Parietal	Deep	Intercostal collectors	Posterior mediastinal nodes, thoracic duct
Paramammary	Parietal	Superficial	Superficial mammary region	Axillary lymph nodes
Parasternal	Parietal	Deep	Collectors from the skin and muscles of the anterior thoracic wall, from the epigastric region, from the skin of the breast, from the anterior diaphragmatic nodes, and costal part of parietal pleura	Bronchomediastinal trunk
Superior diaphragmatic	Parietal	Deep	Collectors from the diaphragm and deep ascending from the liver	Mediastinal and parasternal nodes
			VISCERAL THORACIC LYMPH NODES	
Brachiocephalic (nodes of the transverse chain or anterosuperior mediastinal)	Anterior mediastinal	Deep	Thyroid gland, thymus, heart, pericardium, lateral diaphragmatic nodes	Bronchomediastinal trunk
Bronchopulmonary (pulmonary) – right – left (hilar or left hilar nodes)	Tracheobronchial	Deep	Lungs, visceral pleura and other thoracic organs	Tracheobronchial nodes (only some directly into the thoracic duct)
Interaorticoesophageal	Posterior mediastinal and Station 8	Deep	Mediastinal pleura and thoracic part of the esophagus, posterior pericardium	Tracheobronchial nodes (only some directly into the thoracic duct)
Intrapulmonary – interlobar – lobar – segmental – subsegmental		Deep	Lungs	Right thoracic duct or lymphatic duct
Juxta-aortic	Posterior mediastinal	Deep	Mediastinal pleura, thoracic part of the esophagus, posterior pericardium	Tracheobronchial nodes (only some directly into the thoracic duct)
Juxtaesophageal (paraesophageal)	Posterior mediastinal or inferior mediastinal and Station 8	Deep	Mediastinal pleura, thoracic part of the esophagus, posterior pericardium	Tracheobronchial nodes (only some directly into the thoracic duct)
Lateral pericardial		Deep	Thoracic organs	Right thoracic duct or lymphatic duct
Mediastinal – anterior (☞ Anterior mediastinal group) – – diaphragmatic – inferior (☞ Inferior mediastinal group) – posterior (☞ Posterior mediastinal group) – – diaphragmatic – superior (☞ Superior mediastinal group)	Station 1	Deep	Diaphragm, diaphragmatic pleura, liver Diaphragmatic pleura	Right thoracic duct or lymphatic duct Right thoracic duct or lymphatic duct
Node of arch of azygos vein	Posterior mediastinal	Deep	Thoracic organs	Right thoracic duct or lymphatic duct
Node of ligamentum arteriosum	Anterior mediastinal	Deep	Thoracic organs	Right thoracic duct or lymphatic duct
Para-aortic (ascending aorta or phrenic nerve nodes)	Station 6	Deep	Thoracic organs	Right thoracic duct or lymphatic duct
Paratracheal – superior – right and left inferior	Superior mediastinal Station 2 Tracheobronchial and Station 4	Deep	Trachea, bronchi, mediastinal pleura, posterior pericardium	Tracheobronchial nodes or thoracic duct
Preaorticocarotid	Anterior mediastinal	Deep	Left lung	Bronchomediastinal trunk

continued

Summary table of the lymphatic groups of the thorax				*(continued)*
LYMPH NODES	***GROUP OR STATION***	***SITE***	***AFFERENTS***	***EFFERENTS***
Prepericardial		Deep	Thoracic organs	Right thoracic duct or lymphatic duct
Prevascular	Anterior mediastinal or superior mediastinal and Station 3	Deep	Right lung	Nodes of the transverse chain
Prevertebral	Superior mediastinal	Deep	Thoracic organs	Right thoracic duct or lymphatic duct
Pulmonary ligament nodes	Inferior mediastinal and Station 9	Deep	Thoracic organs	Right thoracic duct or lymphatic duct
Retrotracheal	Station 3	Deep	Thoracic organs	Right thoracic duct or lymphatic duct
Subaortic (aortopulmonary window nodes)	Station 5	Deep	Thoracic organs	Right thoracic duct or lymphatic duct
Subcarinal	Station 7	Deep	Thoracic organs	Right thoracic duct or lymphatic duct
Tracheobronchial – right superior (hilar or right hilar nodes) and left superior – inferior	Tracheobronchial	Deep	Heart, trachea, bronchi, lungs, posterior mediastinal nodes	Bronchomediastinal trunk

Summary table of the lymphatic ducts involved in the drainage of the thorax			
LYMPH NODES	***ORIGIN***	***TERRITORIES OF DRAINAGE***	***OPENING***
Right and left bronchomediastinal trunks	Brachiocephalic and tracheobronchial lymphatic groups	*Right*: thoracic wall (deep), right lung, right bronchial tree, trachea, esophagus, right part of the heart, liver *Left*: thoracic wall (deep), left lung, left bronchial tree, trachea, most of the esophagus, left part of the heart	*Right*: right jugulosubclavian confluence or right lymphatic duct *Left*: left jugulosubclavian confluence or terminal part of the thoracic duct
Right and left subclavian trunks	Apical group of the axillary lymph nodes	Upper limb, part of the thoracic wall, most of the breast, anterosuperior (supraumbilical) part of the anterior abdominal wall	*Right*: right jugulosubclavian confluence or right lymphatic duct *Left*: left jugulosubclavian confluence or terminal part of the thoracic duct
Right lymphatic duct	Right jugular, subclavian and at times bronchomediastinal trunks	Territories of drainage of the right bronchomediastinal, subclavian and jugular trunks	Right jugulosubclavian confluence
Thoracic duct	Right and left lumbar trunks and intestinal trunk (cisterna chyli)	Territories of drainage of the right and left lumbar trunks, and of the left intestinal, bronchomediastinal, subclavian and jugular trunks	Left jugulosubclavian confluence

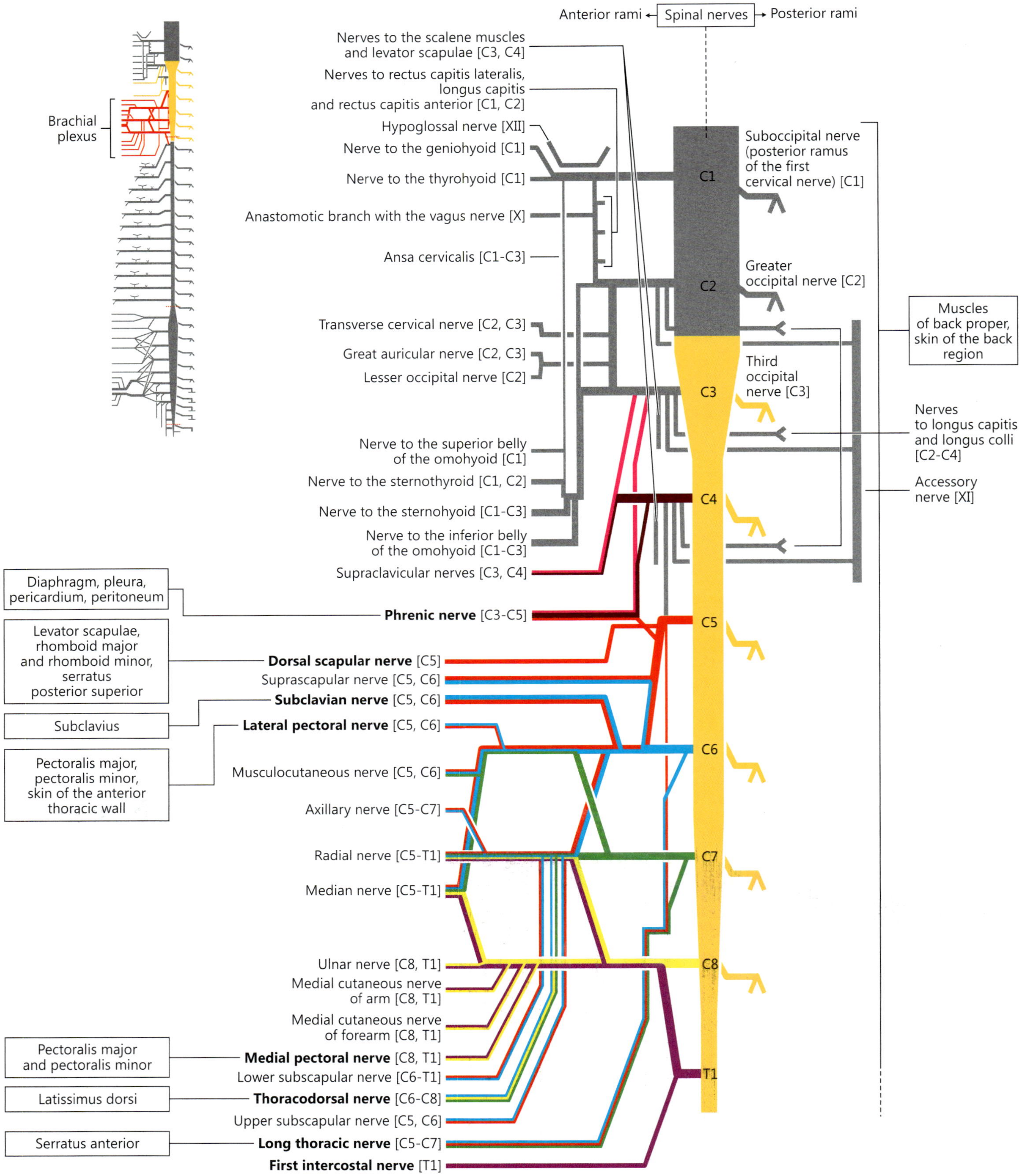

Schematic representation of the **cervical** and **brachial plexuses**. In the left part of the picture, the cervical plexus (a description of which is given in the chapter on the neck) is represented in *gray*. It is formed by the anterior rami of the first four cervical nerves. Note that each anterior ramus, with the exception of the 1st cervical nerve, divides into two branches, an ascending and a descending one, which unite with the corresponding rami of the contiguous nerves to construct the cervical loops; the descending branch of the 4th cervical nerve anastomoses with the anterior ramus of the 5th cervical nerve, which forms part of the brachial plexus. From the cervical plexus arise cutaneous and muscular anastomotic branches. The brachial plexus is represented in different colors. It is composed of the anterior rami of the 5th, 6th, 7th and 8th cervical nerves. From this plexus originate, besides the terminal branches, also collateral branches. In the right part of the picture, the dorsal branches of the cervical nerves are represented.

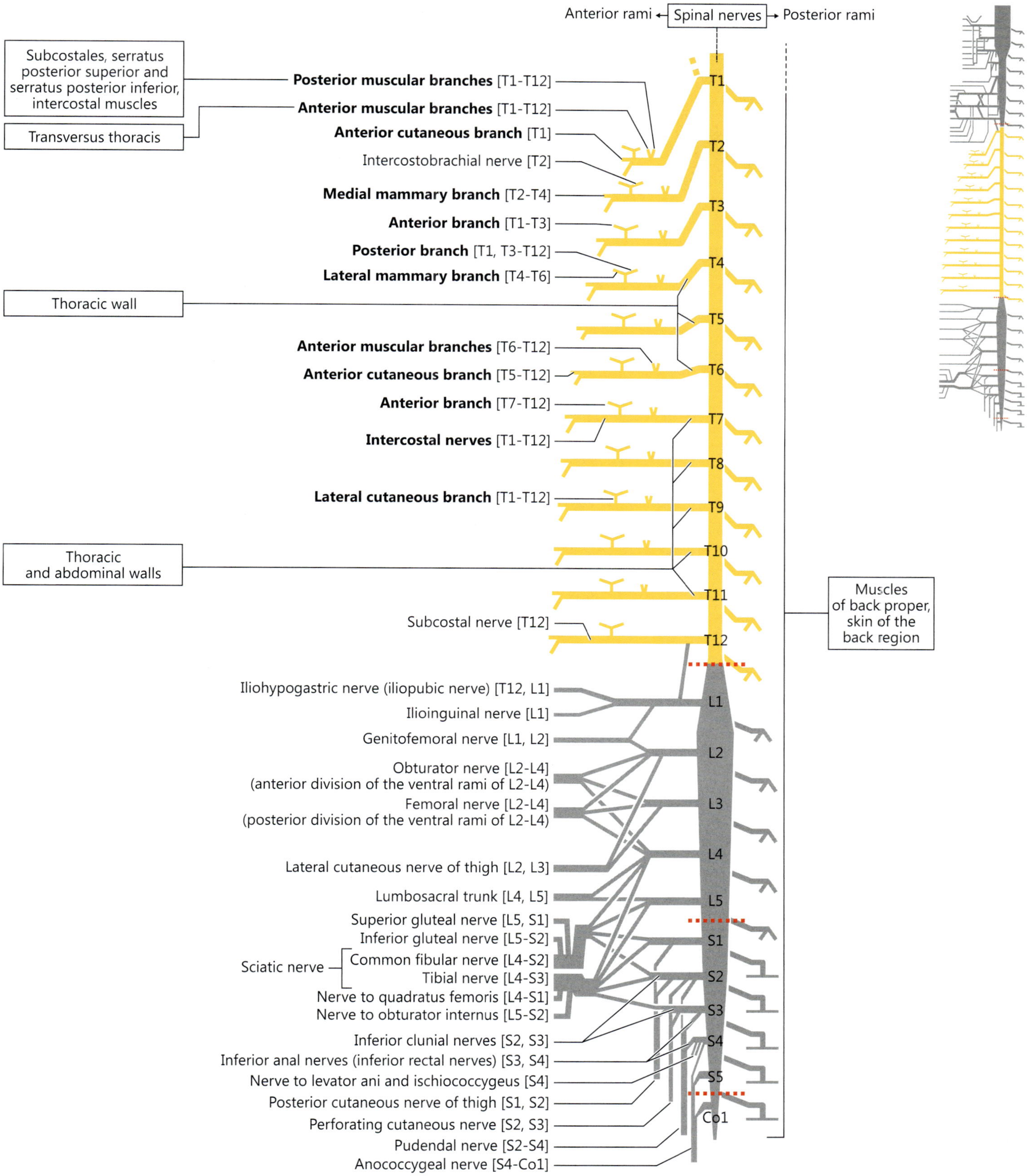

Schematic representation of the **intercostal nerves**, and of the **lumbar**, **sacral** and **coccygeal plexuses**. In the left part of the picture, the intercostal nerves are represented in *yellow*. Composed of the anterior rami of the twelve pairs of thoracic nerves, these are mixed nerves which provide cutaneous innervation to the thoracoabdominal wall and muscles of the thorax. The lumbar, sacral and coccygeal plexuses are represented in *gray*, and will be illustrated in the section on the abdomen. In the right part of the picture, the dorsal branches of the intercostal nerves are represented.

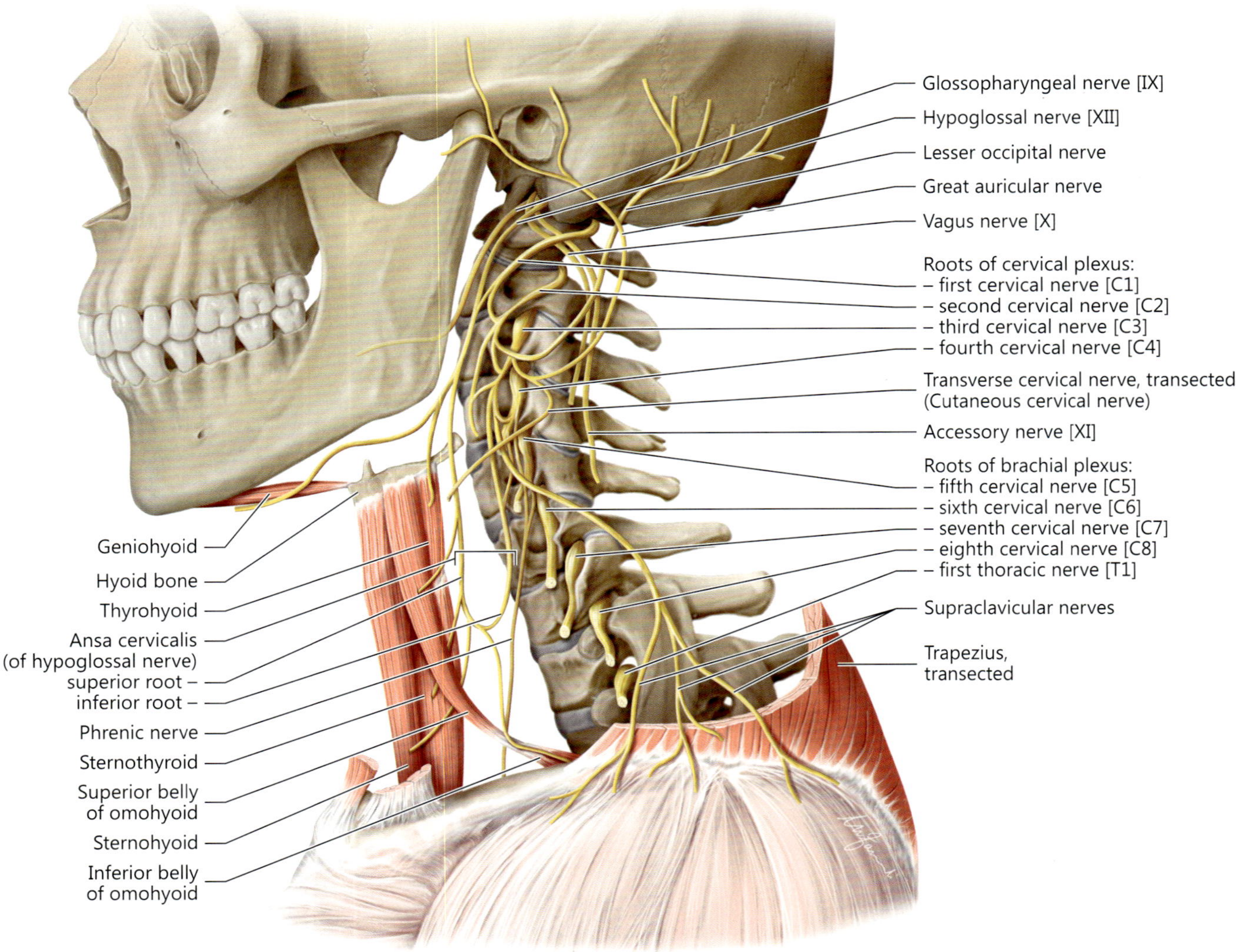

Illustration showing the **cervical plexus** in relation to the skeleton of the vertebral column.

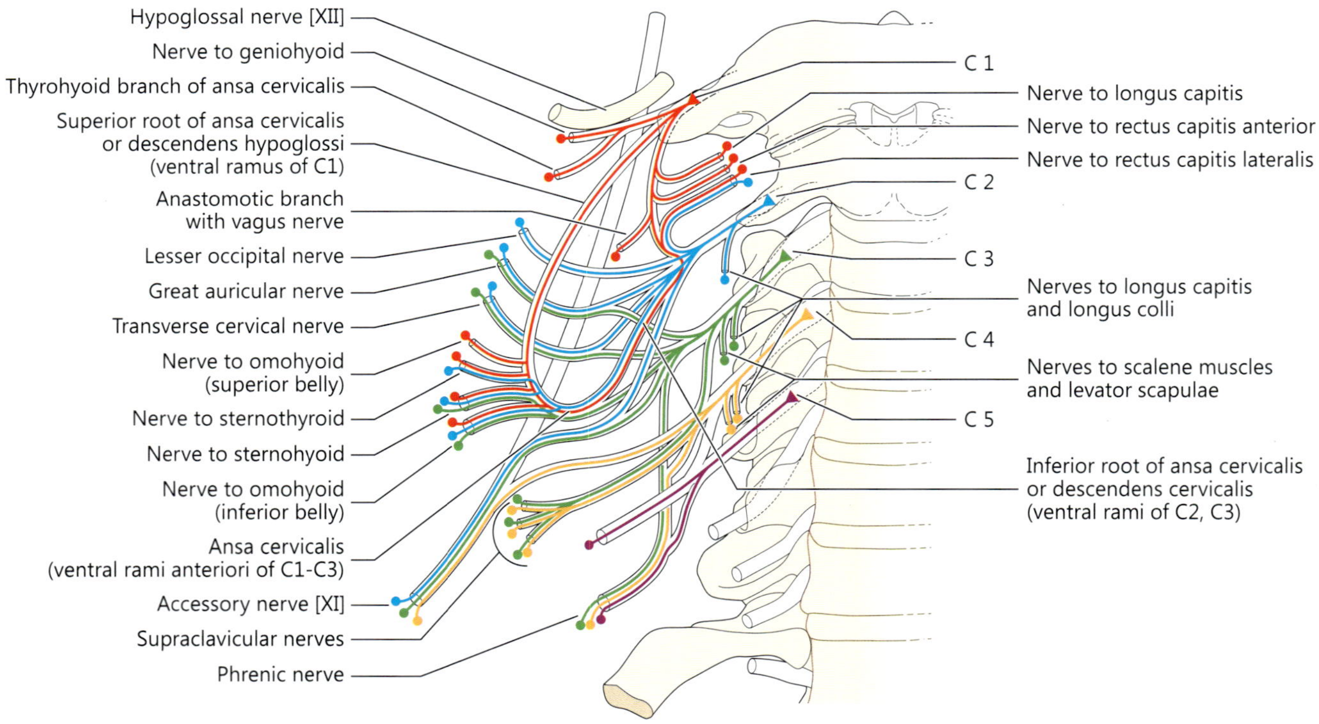

Schematic representation showing the structure of the **cervical plexus**.

Schematic representation showing the **brachial plexus** in relation to the skeleton of the vertebral column.

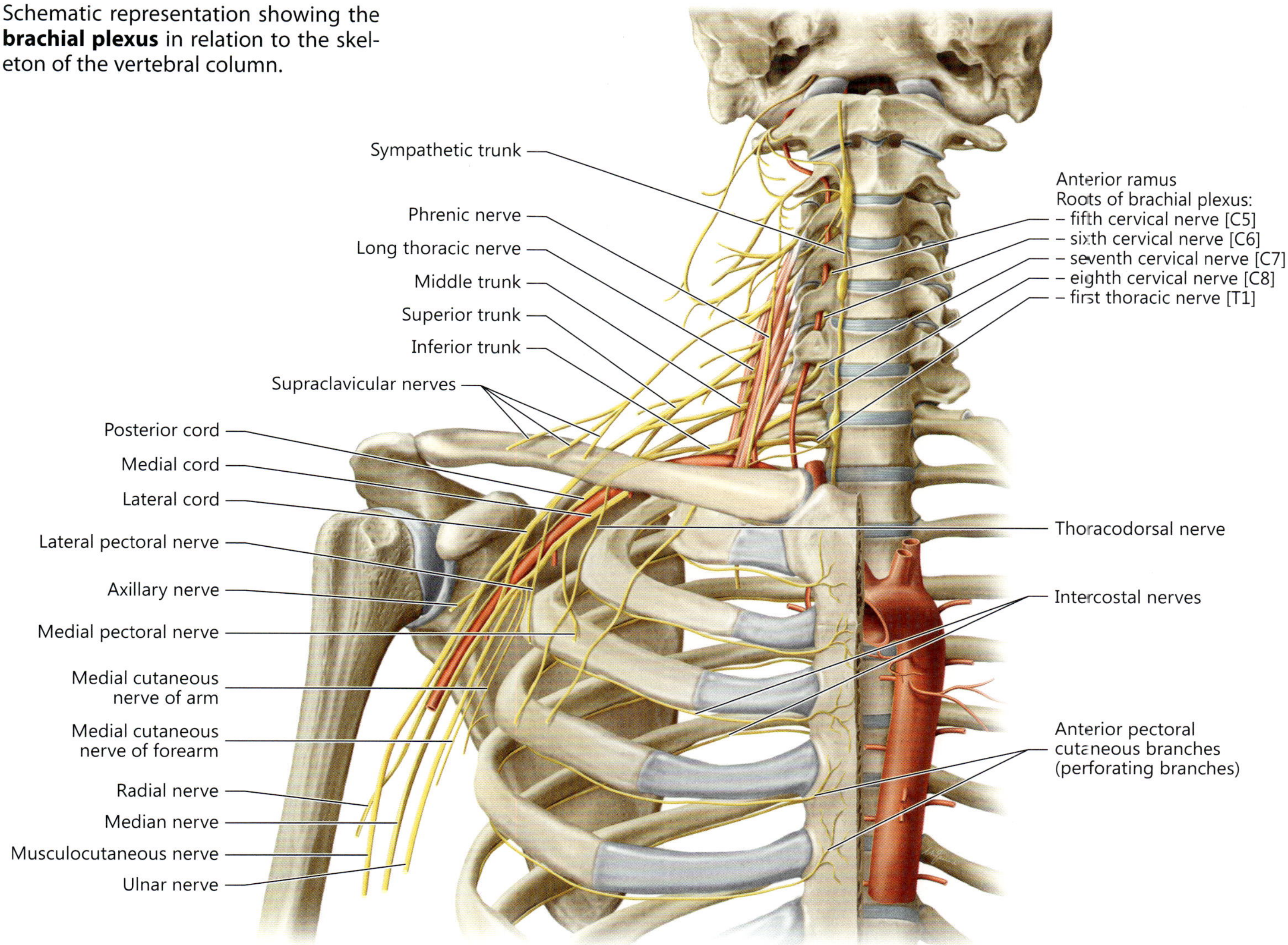

Schematic representation showing the structure of the **brachial plexus**. One can see the position of the anterior branches (ant. b.), with the formation of the trunks (superior, middle and inferior) and cords, and of the posterior (post. b.) and terminal branches (term. b.). The phrenic nerve [C3, C4, C5] belongs to the cervical plexus; the first intercostal nerve to the thoracic nerves (anterior ramus of the first thoracic nerve).

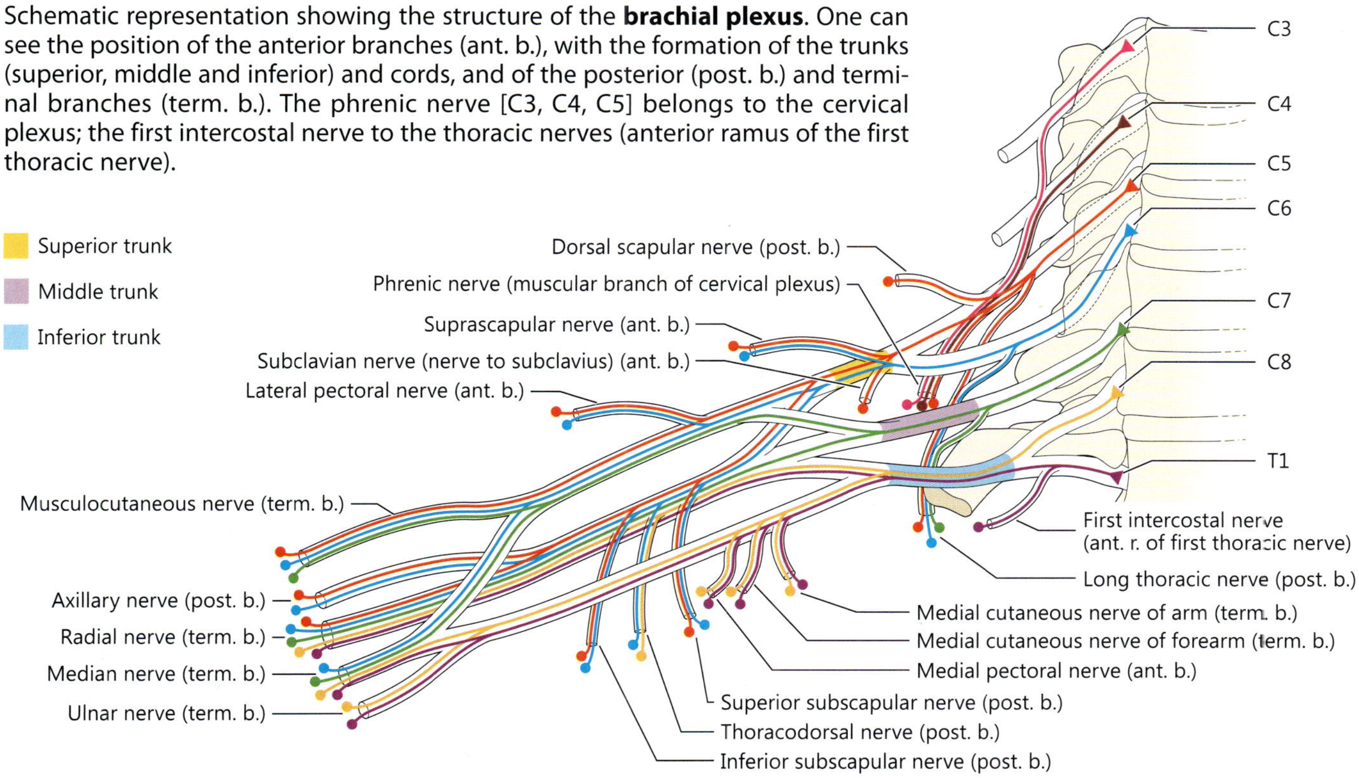

Course of the **posterior rami of the spinal nerves** which, heading dorsally, divide into a lateral branch and a medial branch. Note the particular disposition of the dorsal branches of the first three cervical rami, in particular of the third occipital nerve which ascends in the nuchal region.

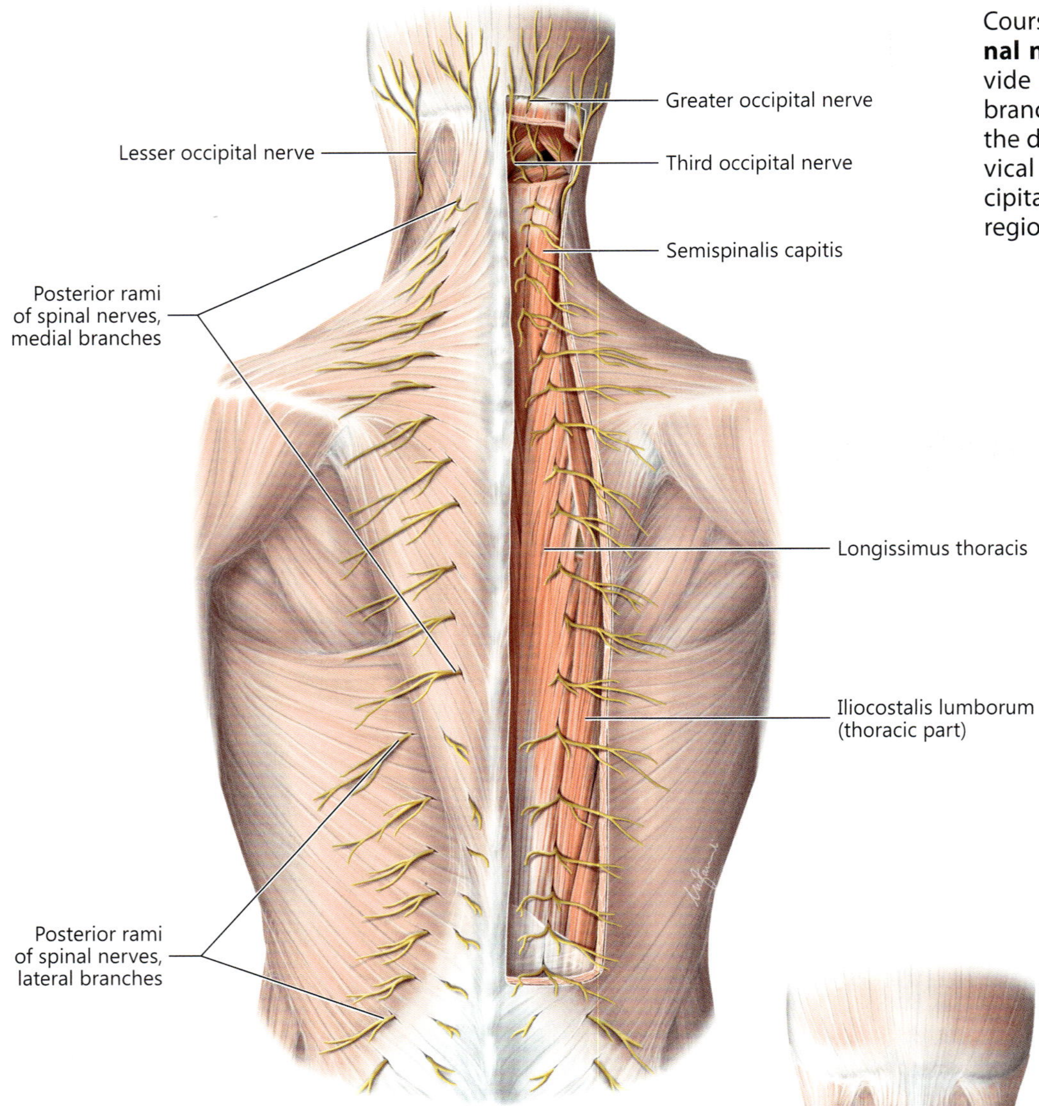

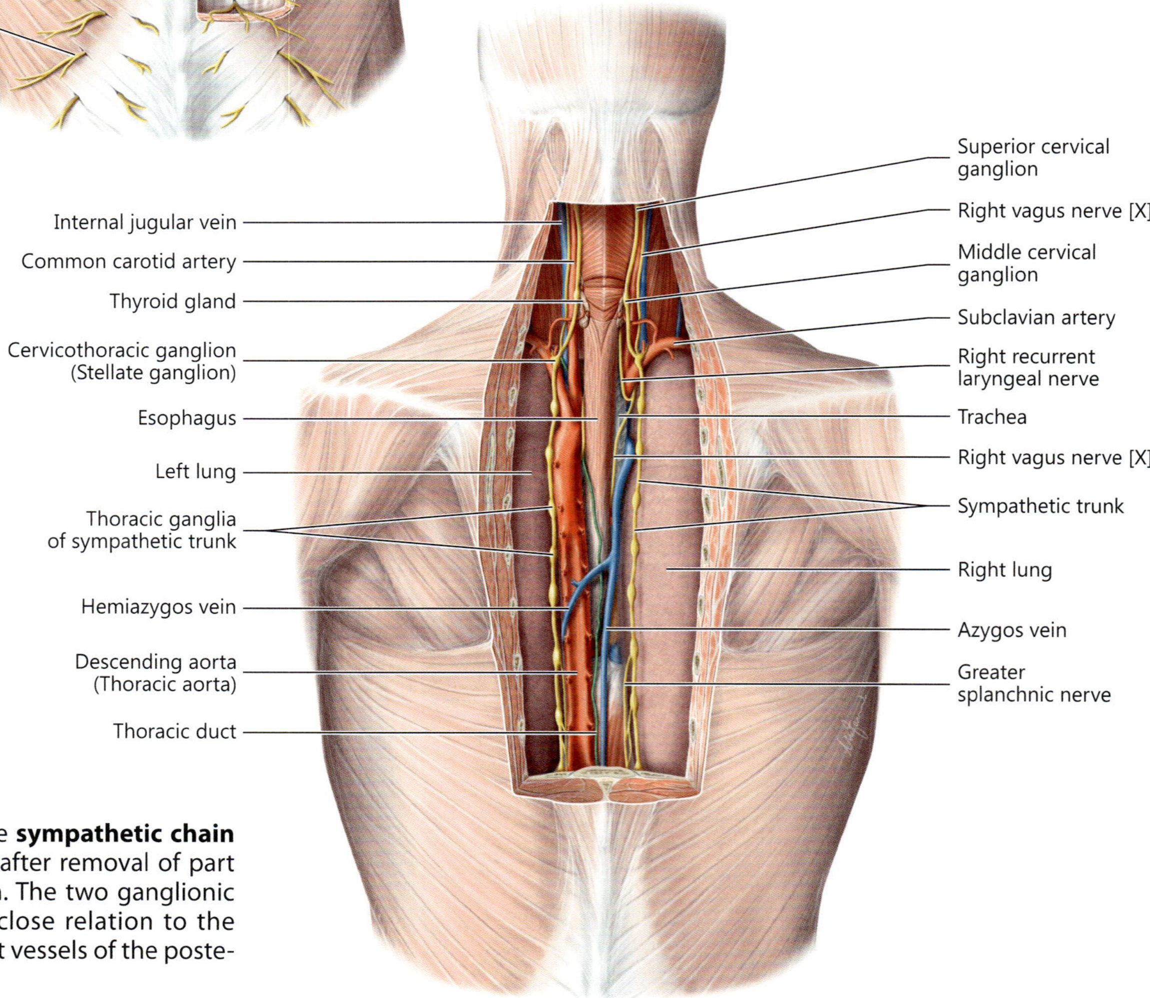

Thoracic segment of the **sympathetic chain** in posterior projection, after removal of part of the vertebral column. The two ganglionic chains are situated in close relation to the parietal pleura and great vessels of the posterior mediastinum.

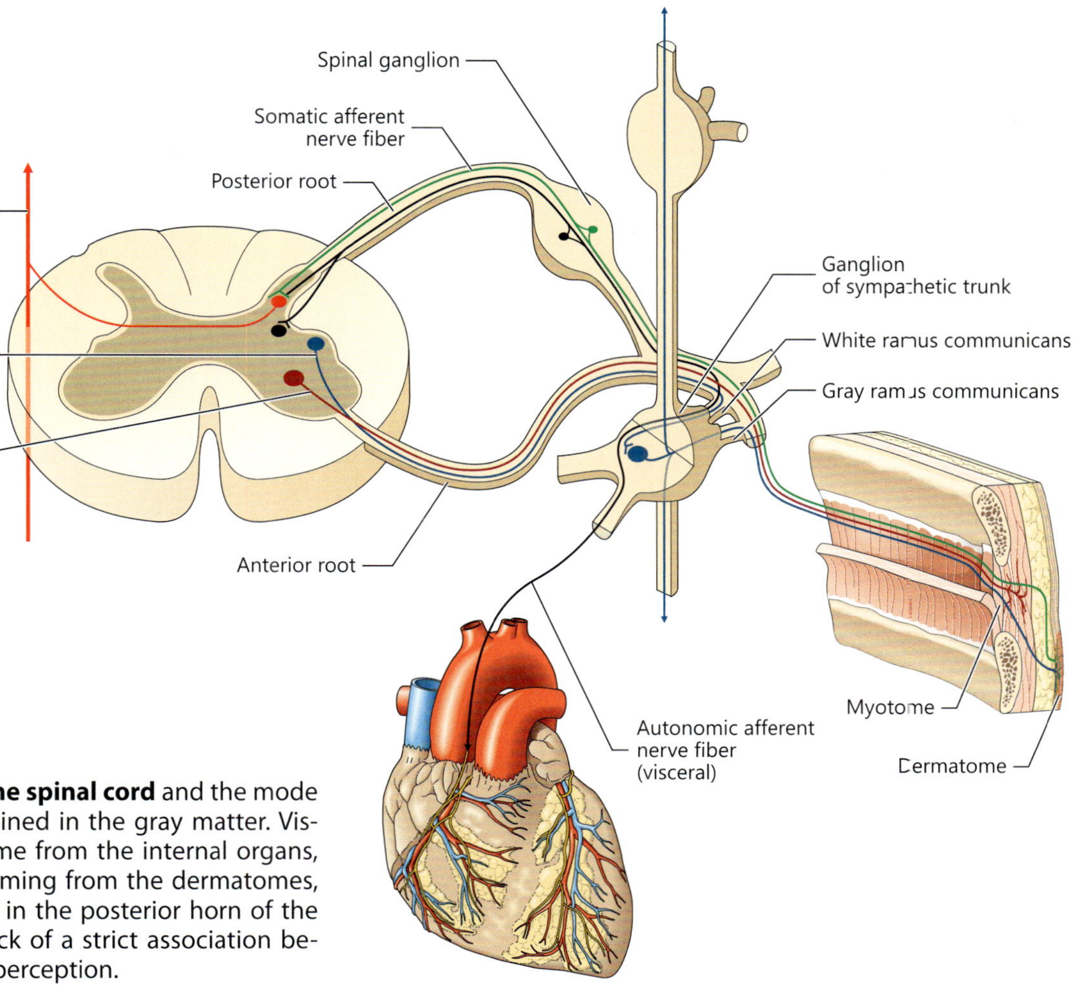

Diagram showing a **section of the spinal cord** and the mode of activity of the neurons contained in the gray matter. Visceral pain sensations, which come from the internal organs, and somatic pain sensations, coming from the dermatomes, terminate on the same neurons in the posterior horn of the spinal cord. This explains the lack of a strict association between the painful event and its perception.

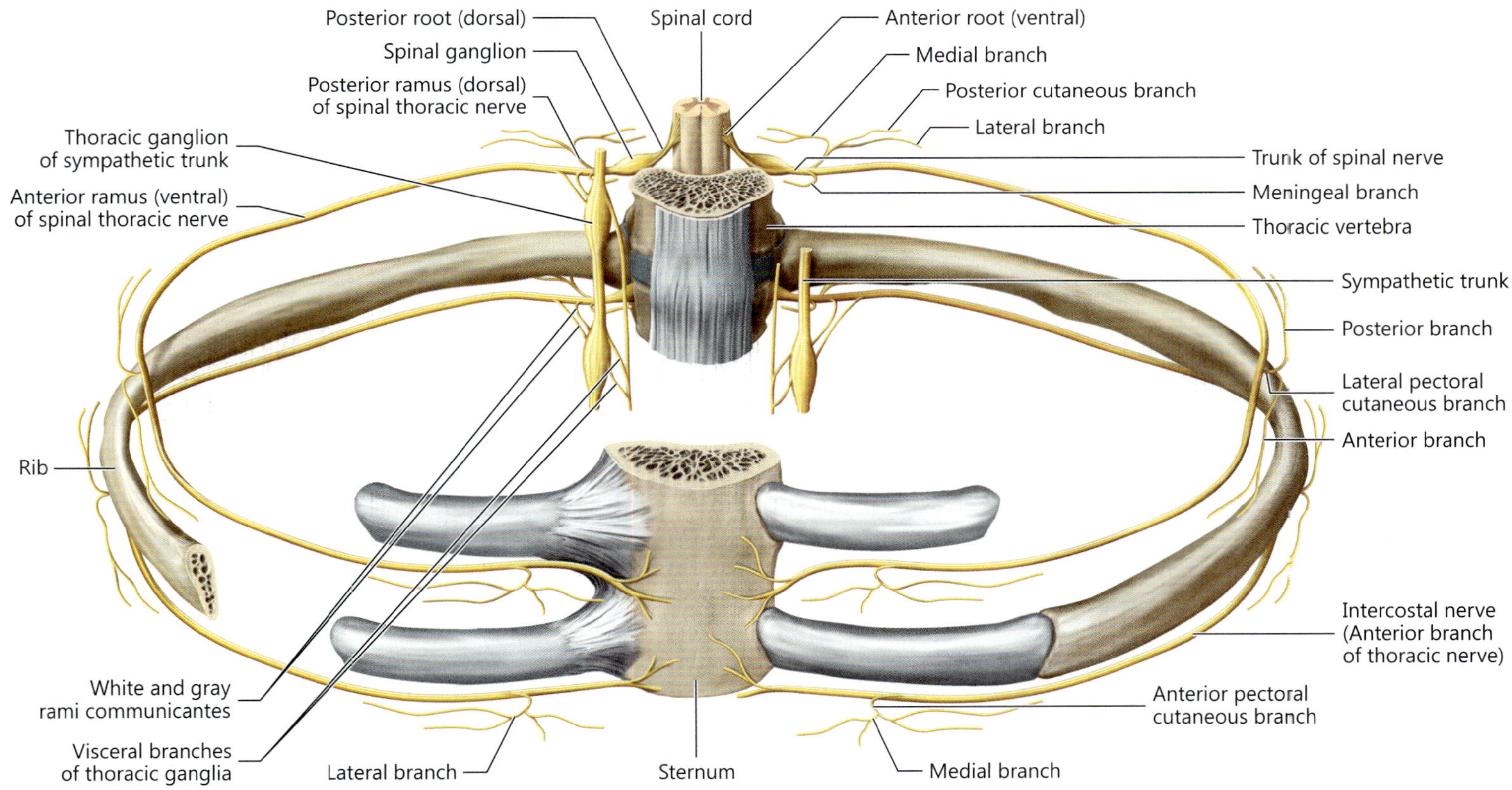

Schematic representation showing the course of the **thoracic nerves** in relation to the ribs and sternum. The anterior rami of the thoracic nerves (intercostal nerves) head forwards following the same course as the ribs and giving off anterior, lateral and medial cutaneous branches.

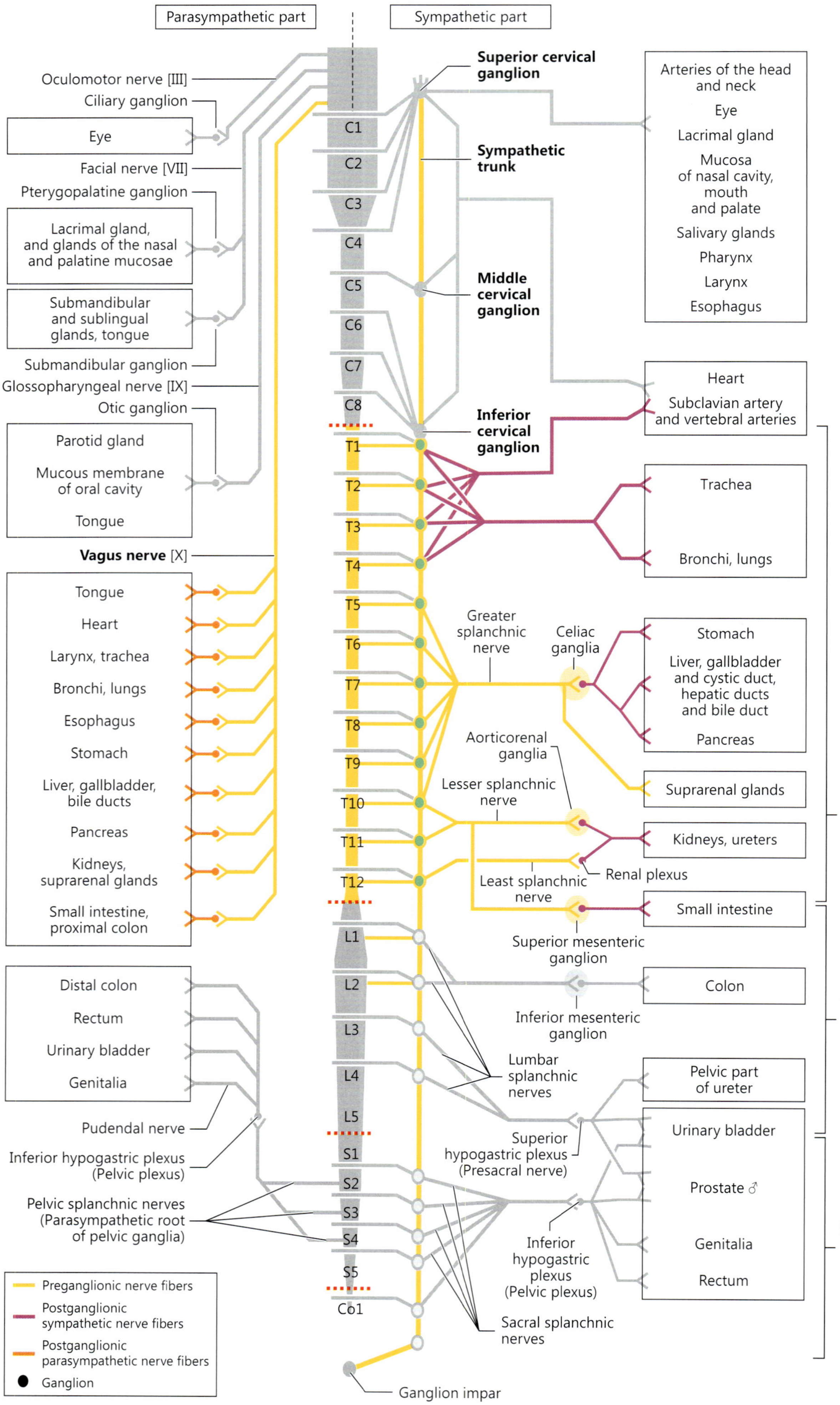

Schematic representation of the **autonomic division** composed of the sympathetic nervous system (on the right) and parasympathetic nervous system (on the left). The autonomic division is the part of the central nervous system that controls the secretory functions of glands and wall motility of internal organs and vessels, thus regulating the vital functions of the organs and all body systems. The organs that are innervated by nerve fibers of the autonomic division are listed, in detail, in the boxes. The preganglionic nerve fibers are represented in *yellow*; in *purple*, the postganglionic fibers of the sympathetic nervous system, and in *orange* those of the parasympathetic nervous system. The parasympathetic (cranial and sacral) nervous system comprises the neuraxial visceral centers situated, respectively, in the brainstem and sacral segment of the spinal cord. Note, in particular, the elevated number of visceral efferent (motor) fibers of the parasympathetic part of the vagus nerve.

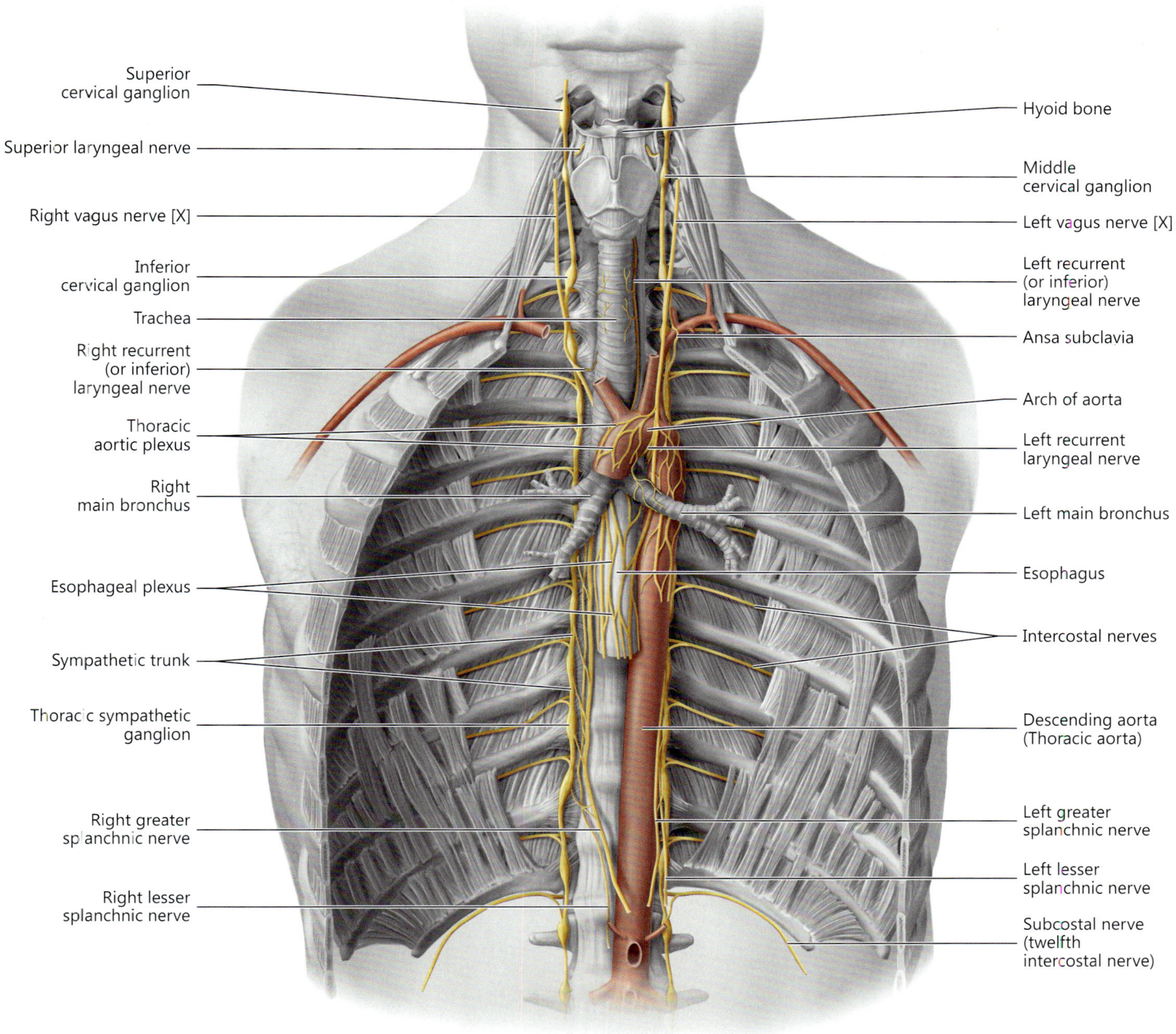

Disposition of the **sympathetic chain** in the cervical and thoracic segments.

Summary table of the nerves of the thorax. In simple bold, the nerves that originate from the spinal nerves; in bold underlined, the nerves of the sympathetic nervous system; in *italics*, the vagus nerve, whose visceral component is an afferent of the parasympathetic nervous system

NERVE(S)	*ORIGIN*	*RAMIFICATIONS*	*DISTRIBUTION*
accessory phrenic	Inconstant branch of the fifth cervical nerve which originates together with the subclavian nerve	Unites with the phrenic nerve	Diaphragm
ansa subclavia	Branch connecting the cervicothoracic ganglion with the middle cervical ganglion forming an ansa around the subclavian artery		
bronchial branches	From the vagus nerve [X], in the thorax	To the pulmonary plexus	Lungs
dorsal scapular (first posterior thoracic nerve)	Supraclavicular part of the brachial plexus, from the superior trunk [C4, C5]	Filaments	Levator scapulae, rhomboid major, rhomboid minor and serratus posterior superior
esophageal branches	From the thoracic segment of the sympathetic trunk		Esophagus
greater splanchnic	From the thoracic segment of the sympathetic trunk with four or five roots; runs in the thorax and abdomen	To the celiac plexus and celiac ganglia	Through the celiac plexus, to the other visceral plexuses of the abdominal part
inferior cervical cardiac	Inferior cervical ganglion and first thoracic ganglion (the two ganglia can fuse together to form the cervicothoracic ganglion or stellate ganglion)	To the cardiac plexus	Heart
intercostal (anterior rami of thoracic nerves from T I to T XI; the anterior ramus of T XII is the subcostal nerve)	Thoracic nerves	White and gray rami communicantes, lateral pectoral cutaneous branch, lateral mammary branch of the fourth, fifth and sixth intercostal nerve, anterior pectoral cutaneous branch, medial mammary branch of the second, third and fourth intercostal nerve, muscular branches, collateral branch, anterior and lateral abdominal cutaneous branches, intercostobrachial nerve	The first 2 nerves supply fibers to the upper limbs and thoracic wall; the 4 successive nerves to the thoracic wall; the 5 inferior nerves to the thoracic and abdominal walls; subcostales, serratus posterior superior, serratus posterior inferior external, internal and innermost intercostal muscles, transversus thoracis, abdominal muscles (rectus abdominis, external oblique, internal oblique, transversus abdominis, pyramidalis)
intercostobrachial (posterior branch of the lateral pectoral cutaneous branch of the second intercostal nerve)	Second intercostal nerve	Anastomoses with the medial and posterior cutaneous nerves of arm	Skin of the superior half of the medial and posterior surfaces of the arm
intermediate supraclavicular	Cervical plexus [C3, C4]	Filaments	Skin over the pectoralis major and deltoid
lateral pectoral (first anterior thoracic nerve)	Infraclavicular part of the brachial plexus, from the lateral cord [C5, C6]	Filaments	Pectoralis major
lateral supraclavicular	Cervical plexus [C3, C4]	Filaments	Skin over the superior and posterior parts of shoulder
lesser splanchnic	From the thoracic segment of the sympathetic trunk, in general with three roots; runs in the thorax and abdomen	To the celiac plexus and renal and suprarenal plexuses	Through the celiac plexus, to the other visceral plexuses of the abdominal part
long thoracic (second posterior thoracic nerve)	Supraclavicular part of the brachial plexus, from the superior and middle trunks [C5, C6, C7]	Filaments	Serratus anterior
medial pectoral (second anterior thoracic nerve)	Infraclavicular part of the brachial plexus, from the medial cord [C8, T I]	Filaments	Pectoralis minor and inferior part of pectoralis major
medial supraclavicular	Cervical plexus [C3, C4]	Filaments	Skin of the medial part of the infraclavicular region until the median line, sternoclavicular joint
middle cervical cardiac	Middle cervical ganglion	To the deep part of the cardiac plexus	Heart
phrenic	Cervical plexus, with a principal root from the fourth cervical nerve and two accessory, inconstant, roots from the third and fifth cervical nerves	Pericardial branch and phrenicoabdominal branches	Diaphragm, pericardium, pleura, great vessels passing through the diaphragm, peritoneum
posterior rami of thoracic nerves	Thoracic nerves	Medial and lateral branches	Skin of the vertebral region and muscles of back proper

continued

Summary table of the nerves of the thorax. In simple bold, the nerves that originate from the spinal nerves; in <u>bold underlined</u>, the nerves of the sympathetic nervous system; in *italics*, the vagus nerve, whose visceral component is an afferent of the parasympathetic nervous system *(continued)*

NERVE(S)	ORIGIN	RAMIFICATIONS	DISTRIBUTION
recurrent laryngeal (inferior laryngeal)	Vagus nerve [X]	Tracheal, esophageal and pharyngeal branches	Intrinsic muscles of the larynx with the exception of the cricothyroid, innervated by the superior laryngeal nerve; mucosa of larynx below the vocal folds, trachea, esophagus, inferior constrictor of pharynx
subclavian	Supraclavicular part of the brachial plexus, from the superior trunk [C5]	Filaments, anastomosis with the phrenic nerve	Subclavius, sternoclavicular joint
subcostal	Anterior ramus of the twelfth thoracic nerve	Lateral pectoral cutaneous branch, anastomotic branch with the lumbar plexus	Abdominal wall and skin of the gluteal region
superior and inferior cervical cardiac branches	From the vagus nerve [X], in the neck	To the cardiac plexus	Heart
<u>**superior cervical cardiac**</u>	Inferior part of the superior cervical ganglion	To the cardiac plexus	Heart
<u>**thoracic cardiac branches**</u>	From the second to the fifth thoracic ganglion	To the cardiac plexus	Heart
thoracic cardiac branches	From the vagus nerve [X], in the thorax	To the cardiac plexus	Heart
thoracic nerves [T I-T XII] (12 pairs of spinal nerves)	Thoracic segment of the spinal cord	Anterior and posterior rami	Thoracic and abdominal walls, skin of the buttocks
<u>**thoracic pulmonary branches**</u>	From the second to the fifth or sixth thoracic ganglion	To the pulmonary plexus	Lungs
<u>**thoracic segment of sympathetic trunk**</u>	Sympathetic trunk in the thorax; generally eleven thoracic ganglia, the first thoracic ganglion often being fused with the inferior cervical ganglion	White and gray rami communicantes with the thoracic nerves and perivascular and visceral peripheral branches (greater and lesser splanchnic nerves and esophageal branches)	Thoracic aortic, cardiac, pulmonary and celiac plexuses through the greater and lesser splanchnic nerves
thoracodorsal	Infraclavicular part of the brachial plexus, from the posterior cord [C6, C7, C8]	Filaments	Latissimus dorsi
vagus [X]	Myelencephalon (medulla oblongata or bulb)	Meningeal, auricular, communicating branch with glossopharyngeal nerve, pharyngeal, superior and inferior cervical cardiac branches and thoracic cardiac, bronchial, renal branches; anterior and posterior vagal trunks (anterior and posterior gastric branches, hepatic and celiac branches); superior and recurrent laryngeal nerves (tracheal, esophageal and pharyngeal branches)	Organs of the thorax and abdomen: heart, wall of the aorta and of its large branches; glands and smooth muscle of the airways, digestive tract (except for its terminal part) and of part of the urinary system; striated muscle of pharynx, larynx and proximal part of the esophagus; aortic sinus, aortic body, mucosa of the larynx, pharynx, trachea, esophagus and of the other thoracic and abdominal organs; taste buds in the mucosa of the epiglottis and in the oropharynx; skin of the auricle, of the external acoustic meatus and of the lateral surface of the tympanic membrane

NERVE	
anterior supraclavicular	☞ medial supraclavicular nerves
great cardiac	☞ middle cervical cardiac nerve
internal respiratory nerve of Bell	☞ phrenic nerve
long subscapular	☞ thoracodorsal nerve
middle supraclavicular	☞ intermediate supraclavicular nerves
pneumogastric	☞ vagus nerve [X]
posterior scapular	☞ dorsal scapular nerve
to pectoralis major	☞ lateral pectoral nerve
to pectoralis minor	☞ medial pectoral nerve
to serratus anterior	☞ long thoracic nerve

The table alongside lists some nerves by their name normally used or eponym, with the corresponding reference to the official term.

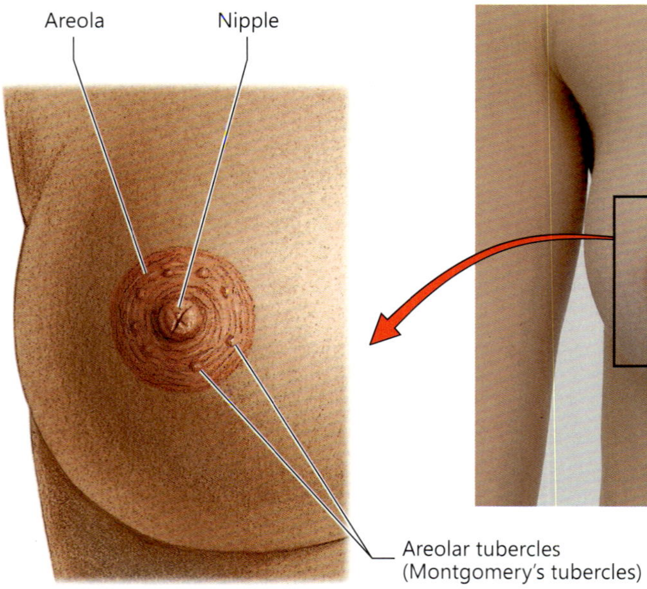

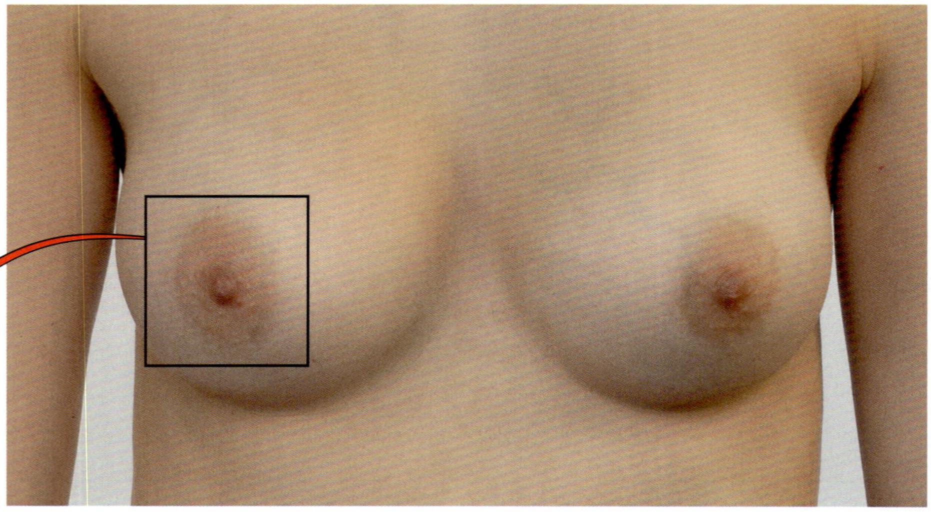

The area occupied by the **breast** of an adult woman extends from the third to the seventh rib on the thoracic cage (inframammary sulcus). The surface of the breast presents a hyperpigmented area (areola) surmounted by a raised zone, the nipple. At its apex open the outlets of the lactiferous ducts. The nipple is generally situated in correspondence to the fourth rib, lateral with respect to the midclavicular line. On the surface of the areola, the areolar tubercles (Montgomery's tubercles) are evident. These are sebaceous glands whose secretion has a lubricating effect and increases during lactation. The breast also contains the mammary gland.

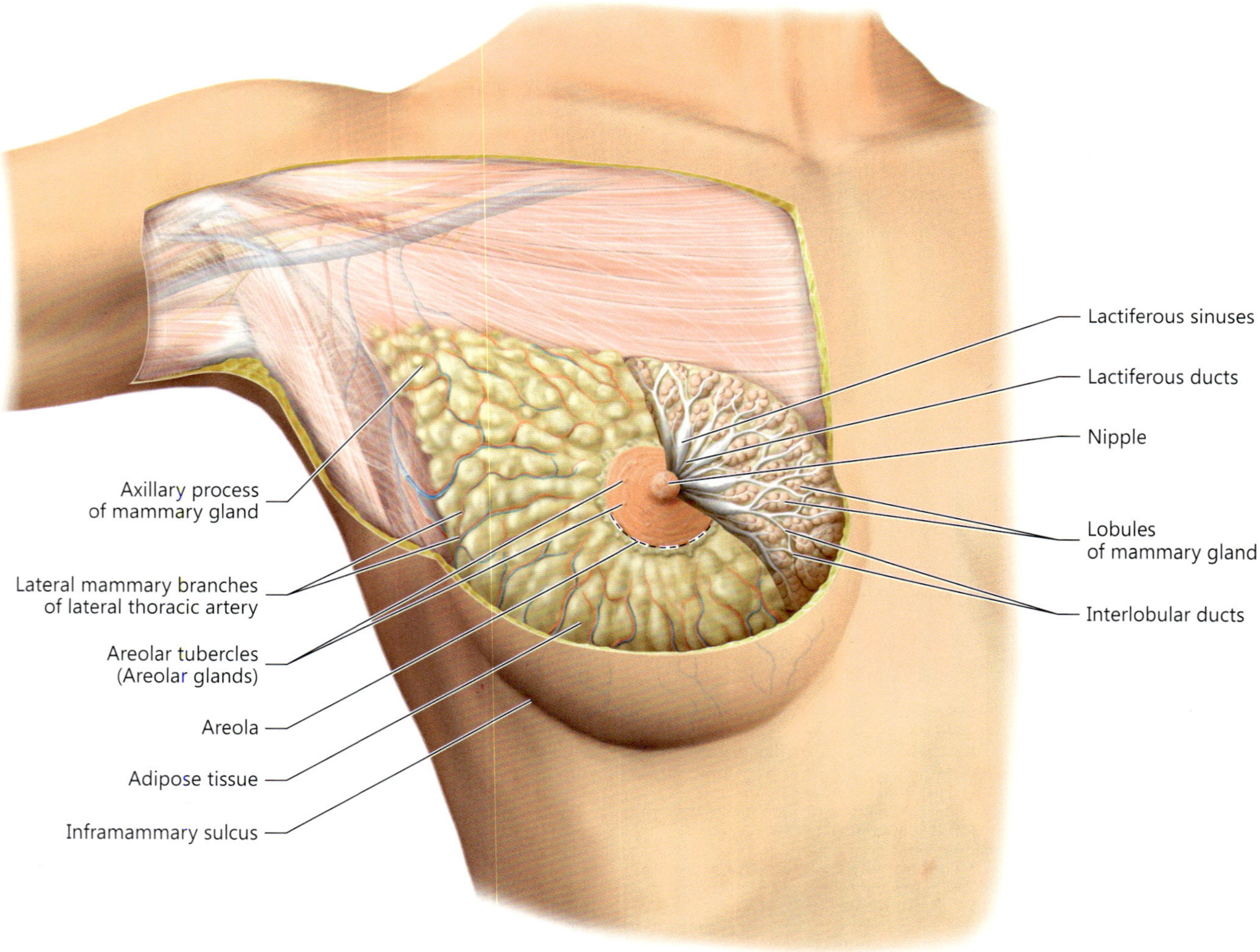

Representation of the **mammary gland** and of the investing layer of the pectoralis major. The medial part of the breast has been transected to show the lobular aspect of the gland, and the lactiferous sinuses and ducts. In the lateral part, one can see the abundant adipose tissue of the subcutaneous layer.

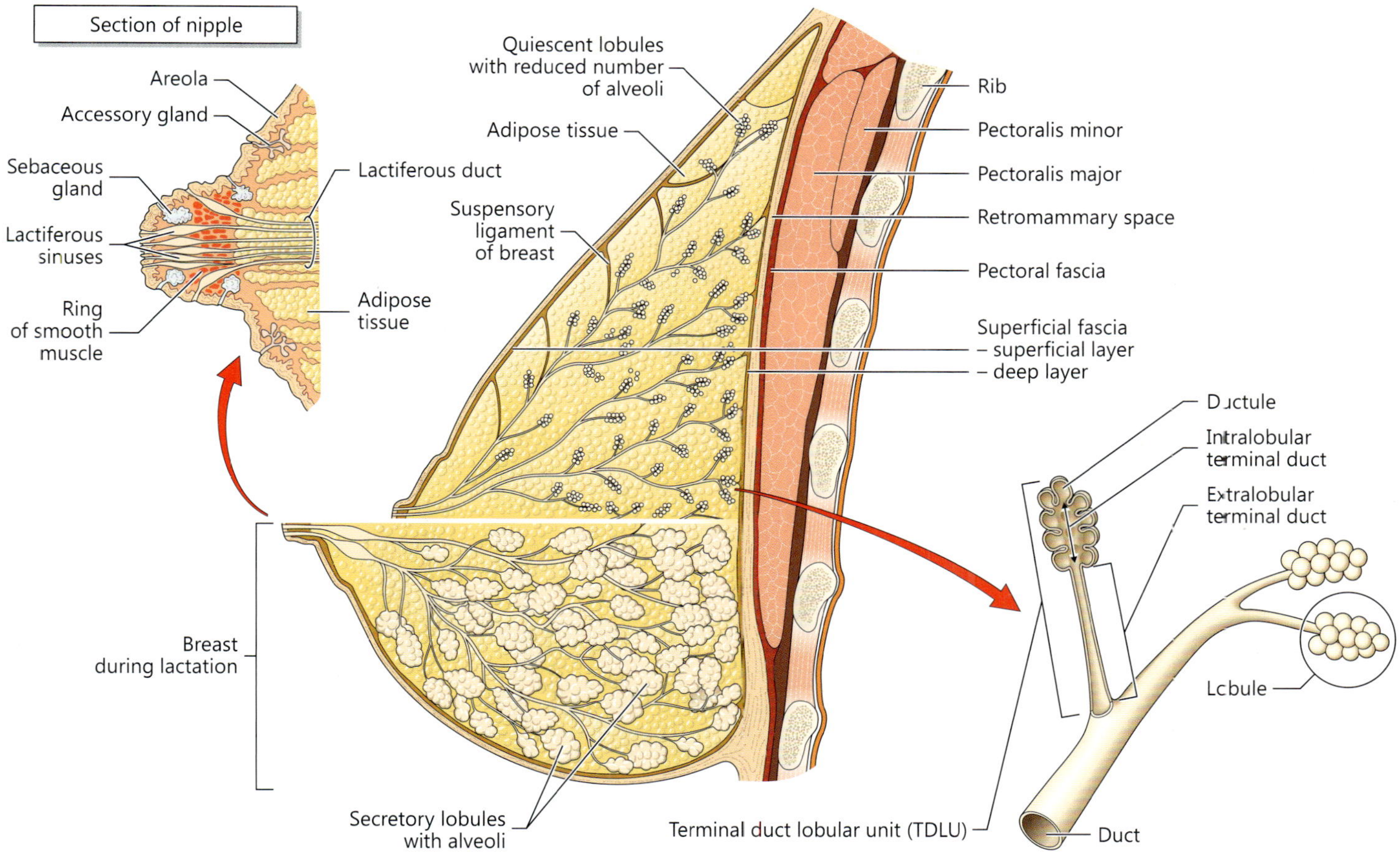

Schematic representation of a **sagittal section of the breast**. The upper part shows a breast in resting phase; the lower part a breast during lactation. The pectoral fascia splits into two layers, between which the mammary gland is located. The deep layer invests the muscle; the superficial layer extends in the subcutaneous layer and sends out fibrous septa which, extending into the thickness of the gland, constitute its stroma (Cooper's ligament or suspensory ligament of the breast). At the top, a close-up detail of a section of the nipple is presented, showing the lactiferous sinuses, the absence of adipose tissue and the presence of smooth muscle tissue. Below is shown the structure of the parenchyma of the gland during lactation: the interlobular ducts ramify into extralobular ducts which, terminating inside lobules, form the intralobular ducts. From these stem numerous ductules which drain the single alveoli of the gland.

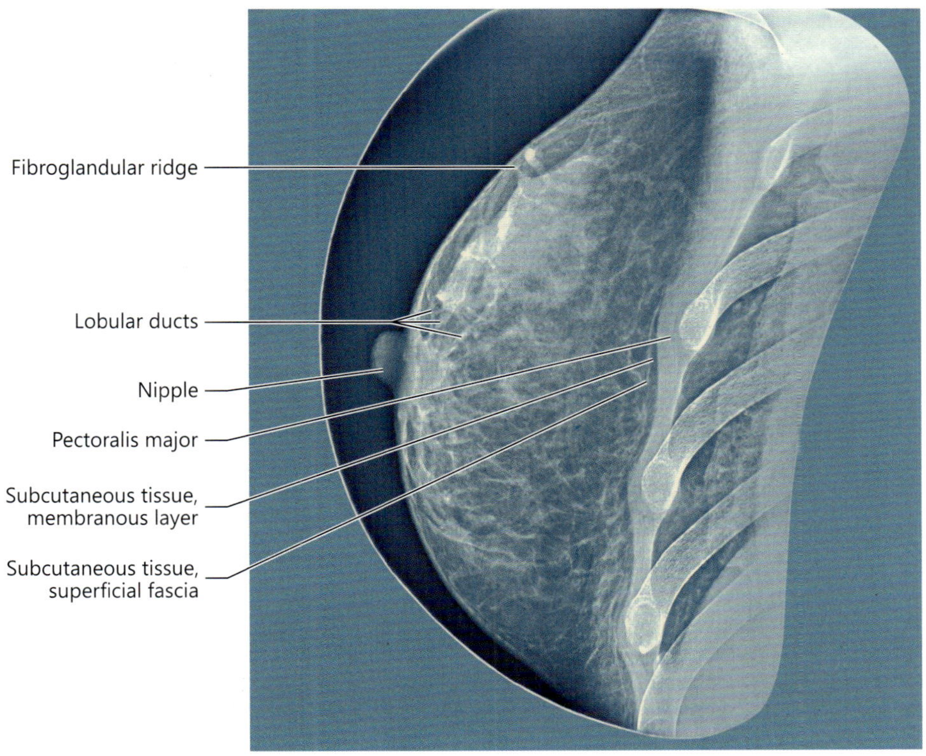

A **mammography** is a radiographic technique at low dosage that consents to visualize the connective tissue of the glandular stroma of the breast and of the lobular ducts. Alterations of this tissue can indicate the presence of neoplasias.

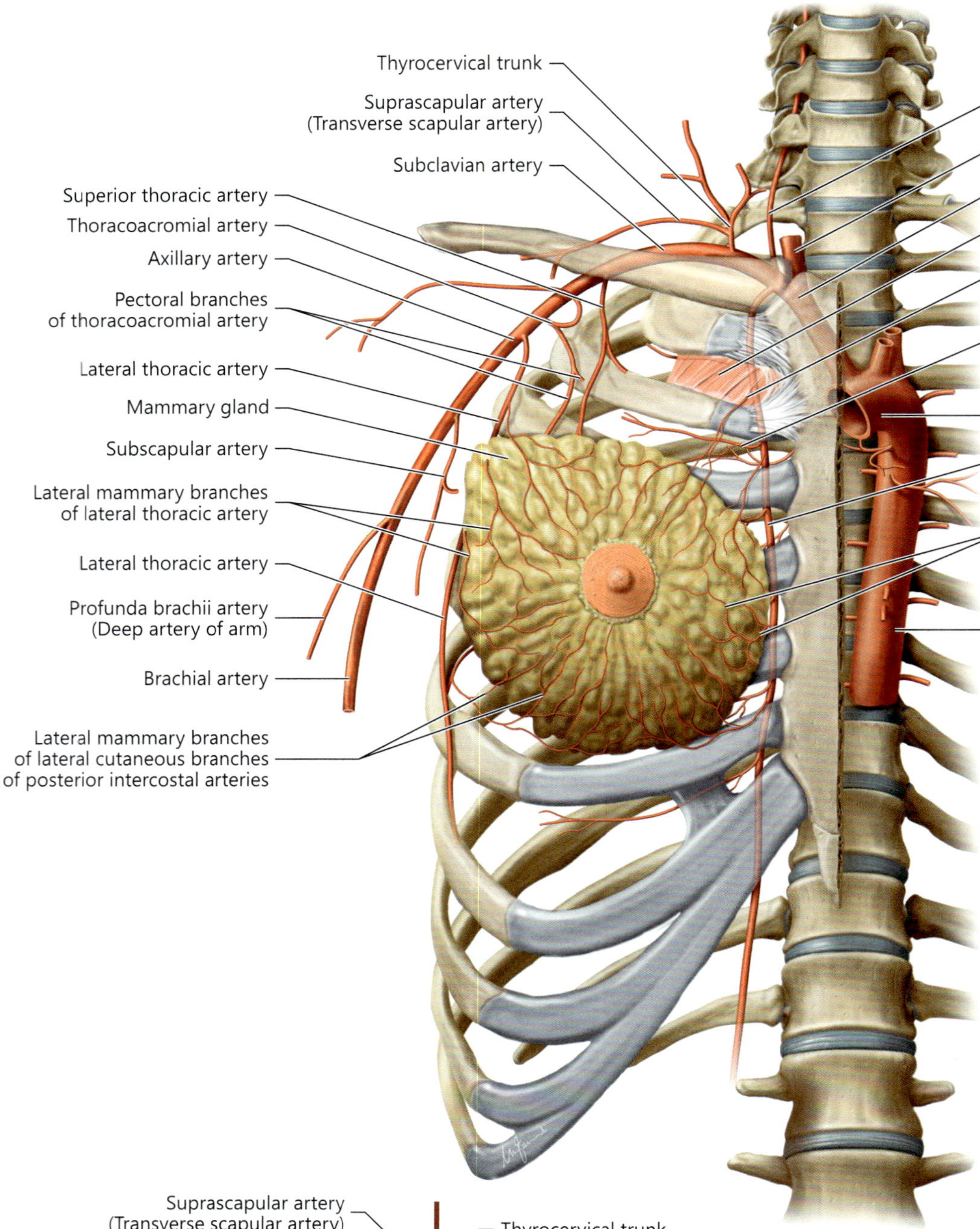

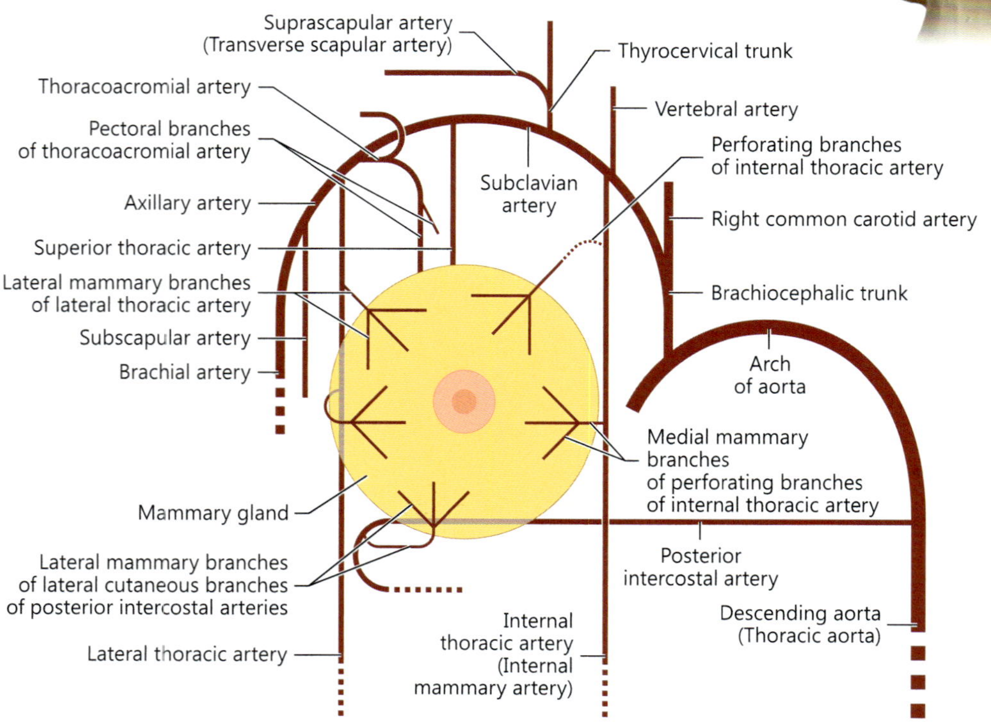

Arterial blood supply of the breast is carried out prevalently by branches of the subclavian artery (internal thoracic artery) and of the axillary artery (lateral thoracic artery). In lesser measure, also the posterior intercostal arteries (branches of the descending thoracic aorta) and the thoracoacromial artery (branch of the axillary artery) are involved.

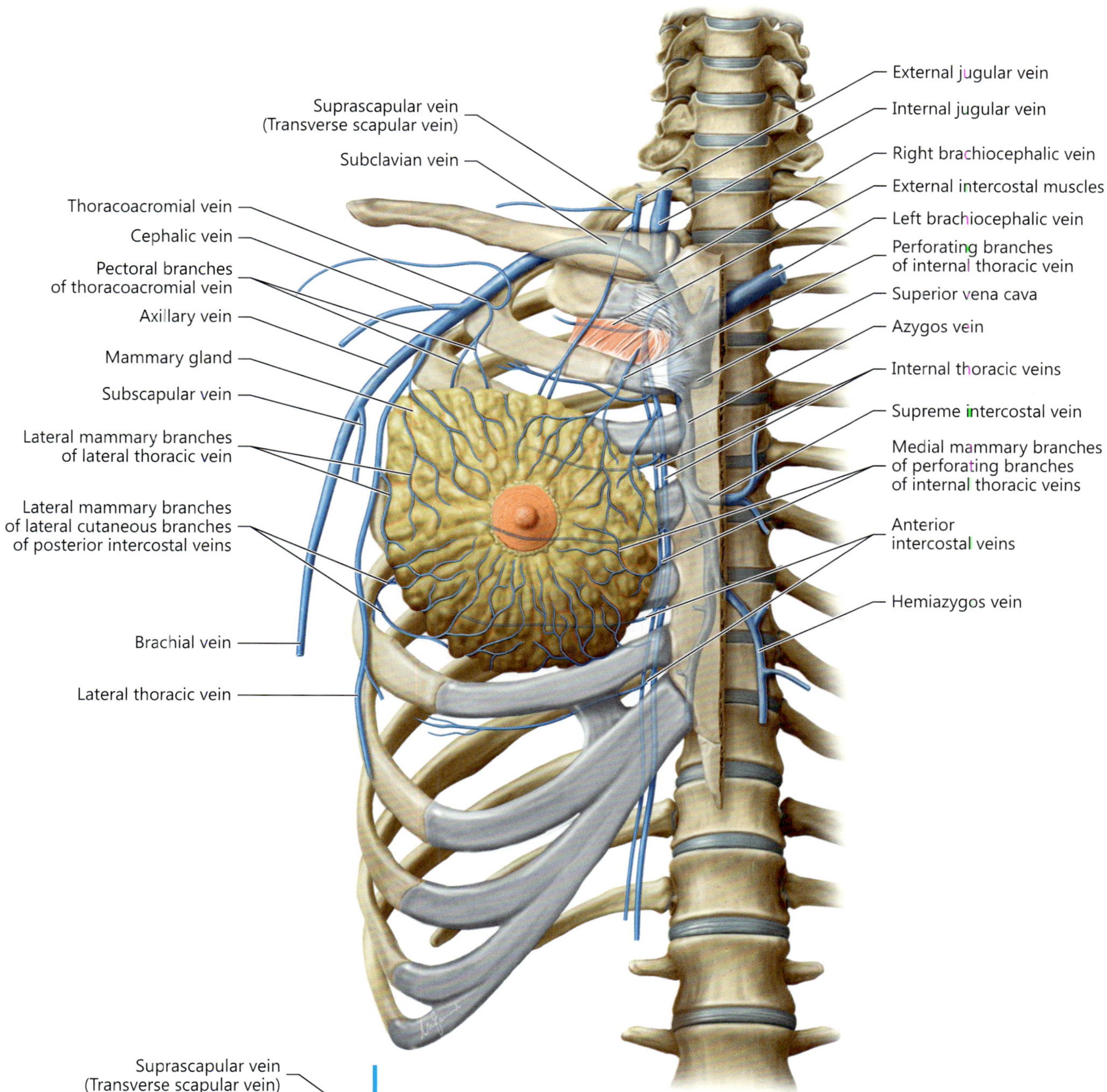

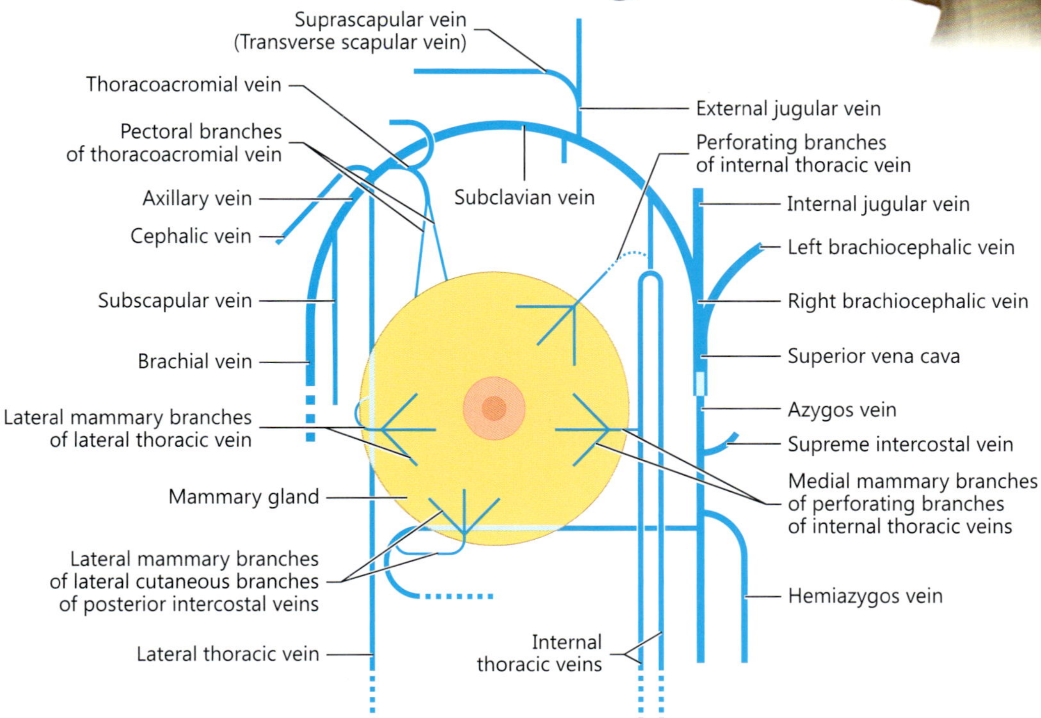

Venous drainage of the breast is carried out prevalently by satellite veins of the arteries involved in the breast.

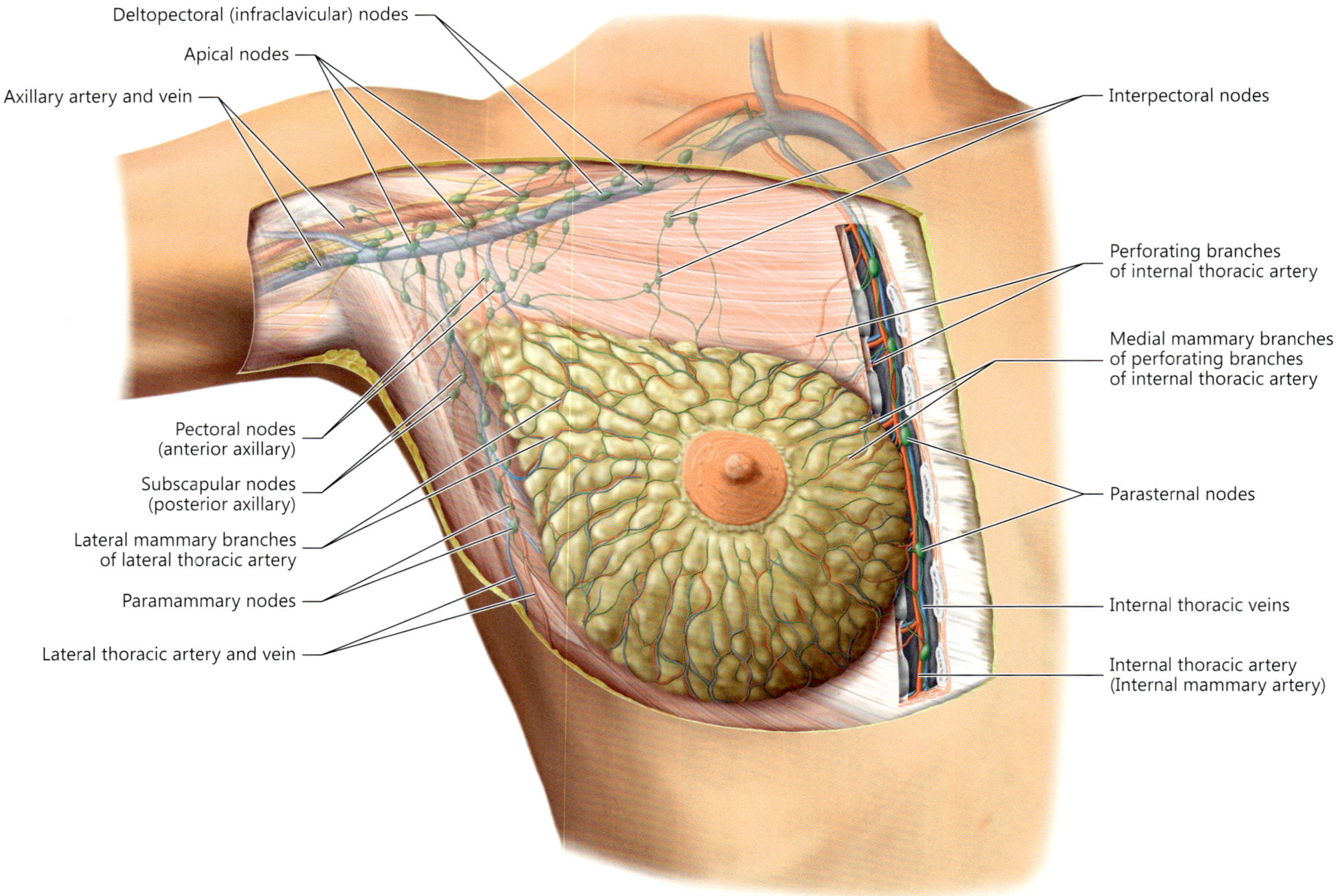

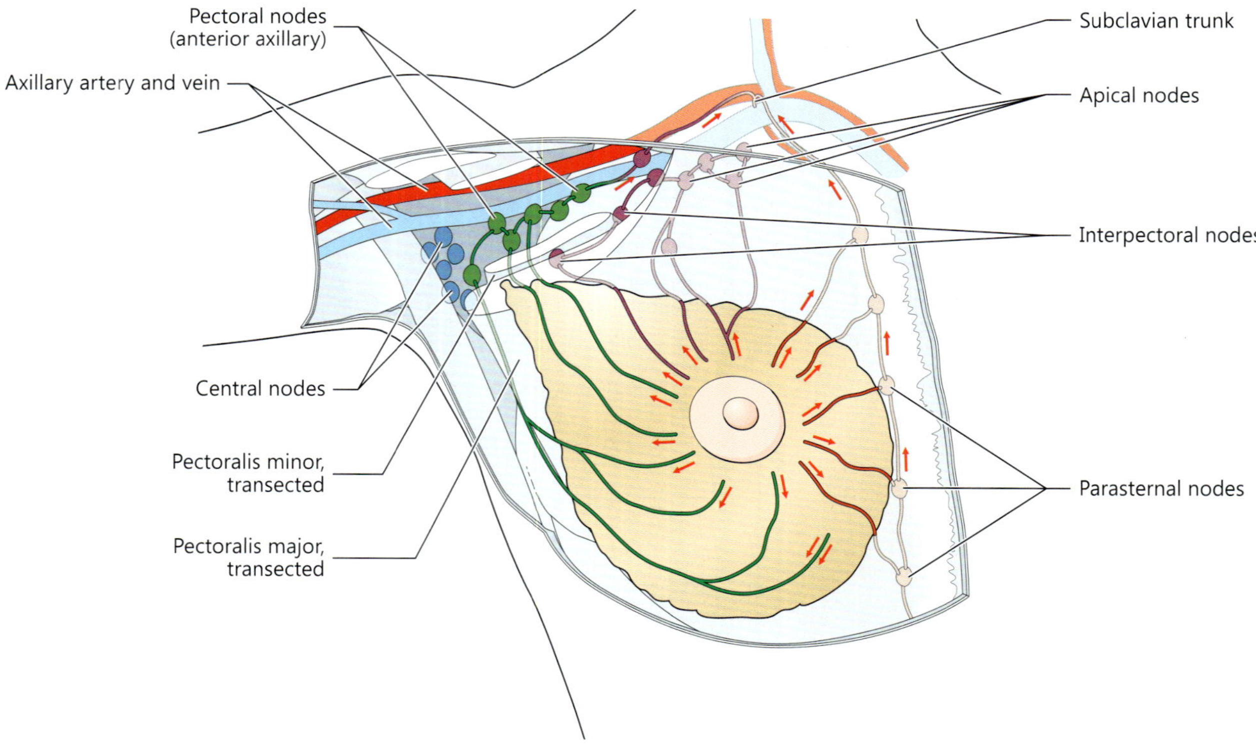

The **lymphatic drainage of the breast** follows the same path as that of the veins. Drainage of the lateral quadrants is carried out prevalently by the axillary lymphatic group. The superomedial quadrant is handled by the deltopectoral (infraclavicular) nodes of the axillary lymphatic group and by the parasternal nodes. The inferomedial quadrant is handled by the parasternal nodes.

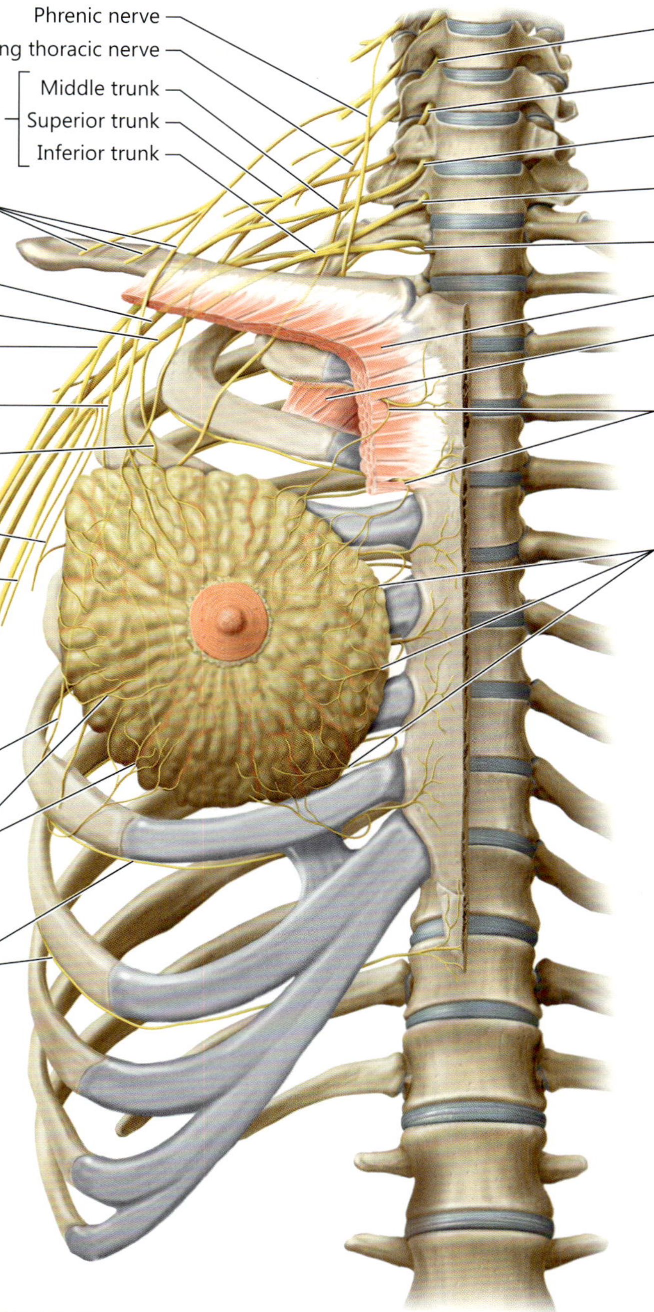

Illustration of the **brachial plexus** and branches that innervate the mammary gland. The lateral and medial mammary branches (collateral branches of the intercostal nerves) are distributed over the whole surface of the mammary gland.

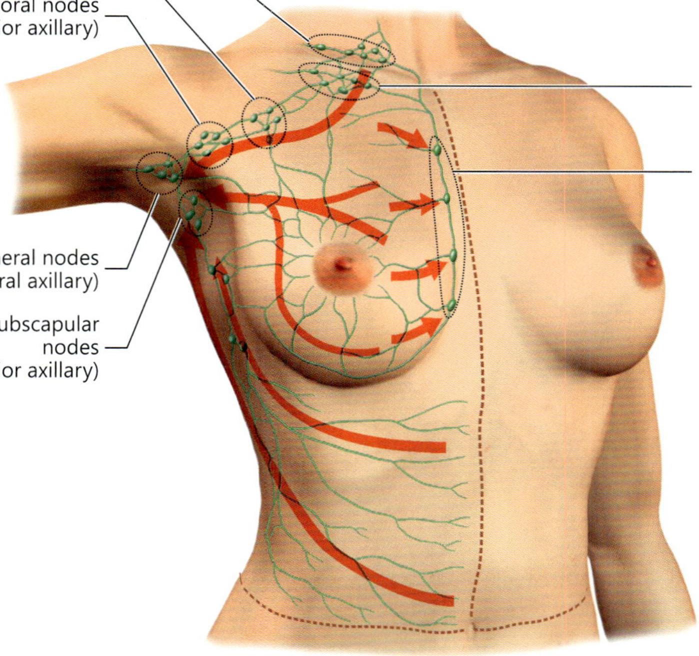

The physiologic **drainage of the breast** is essentially conducted in the direction of the axillary nodes and of the internal mammary lymphatic chain, situated at homolateral parasternal level. As an accessory channel, drainage can also be directed towards the homolateral supraclavicular nodes.

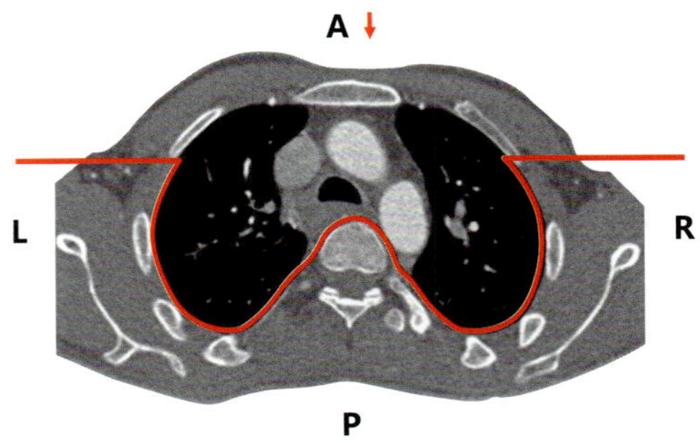

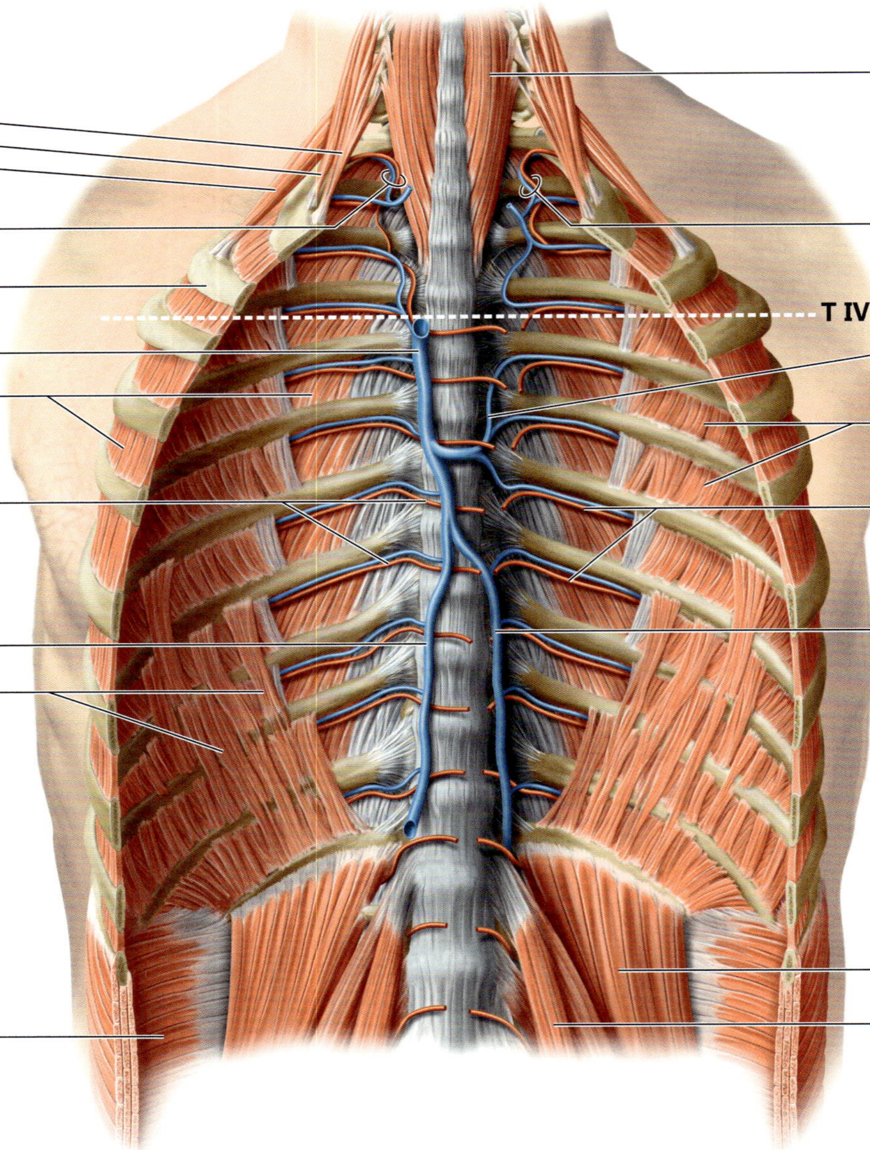

CONTENTS OF THE THORAX (in bold are the structures added to the posterior wall)
• Accessory hemiazygos vein • Azygos vein, on the right • Hemiazygos vein • Insertions of medial and lateral inferior parts of longi colli, T I-T III (right and left) • Posterior intercostal arteries from 4th to 12th (the right arteries pass in front of the vertebral column) • Supreme intercostal arteries (right and left) (1st, 2nd and 3rd posterior intercostal arteries) • Supreme intercostal veins (right and left) (intercostal veins from 1st to 2nd)

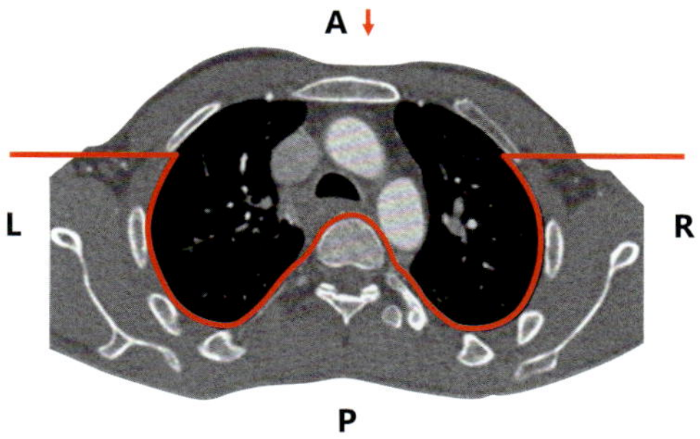

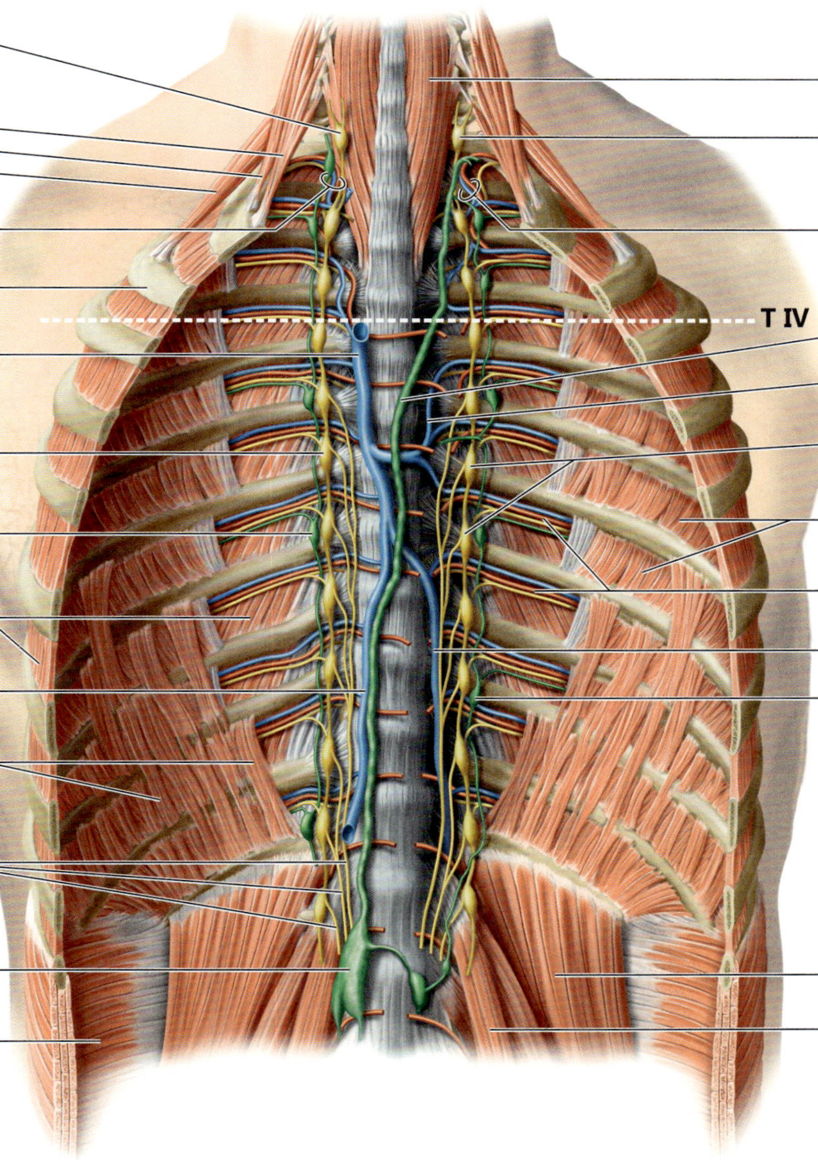

CONTENTS OF THE THORAX (in bold are the structures added with respect to the preceding table)
• Cervicothoracic (stellate) ganglion (right and left) • Greater, lesser and least splanchnic nerves (right and left) • Left intercostal nerves • Posterior intercostal nodes (right and left) • Thoracic duct • Thoracic ganglia of sympathetic trunk (right and left)

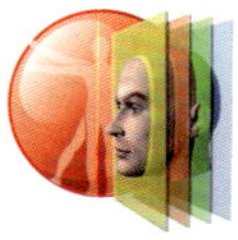

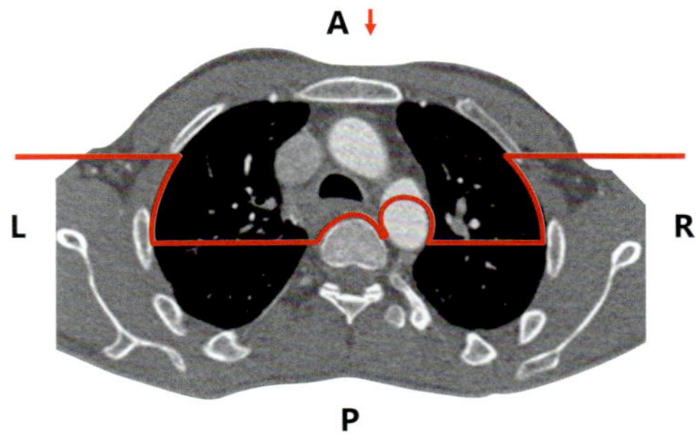

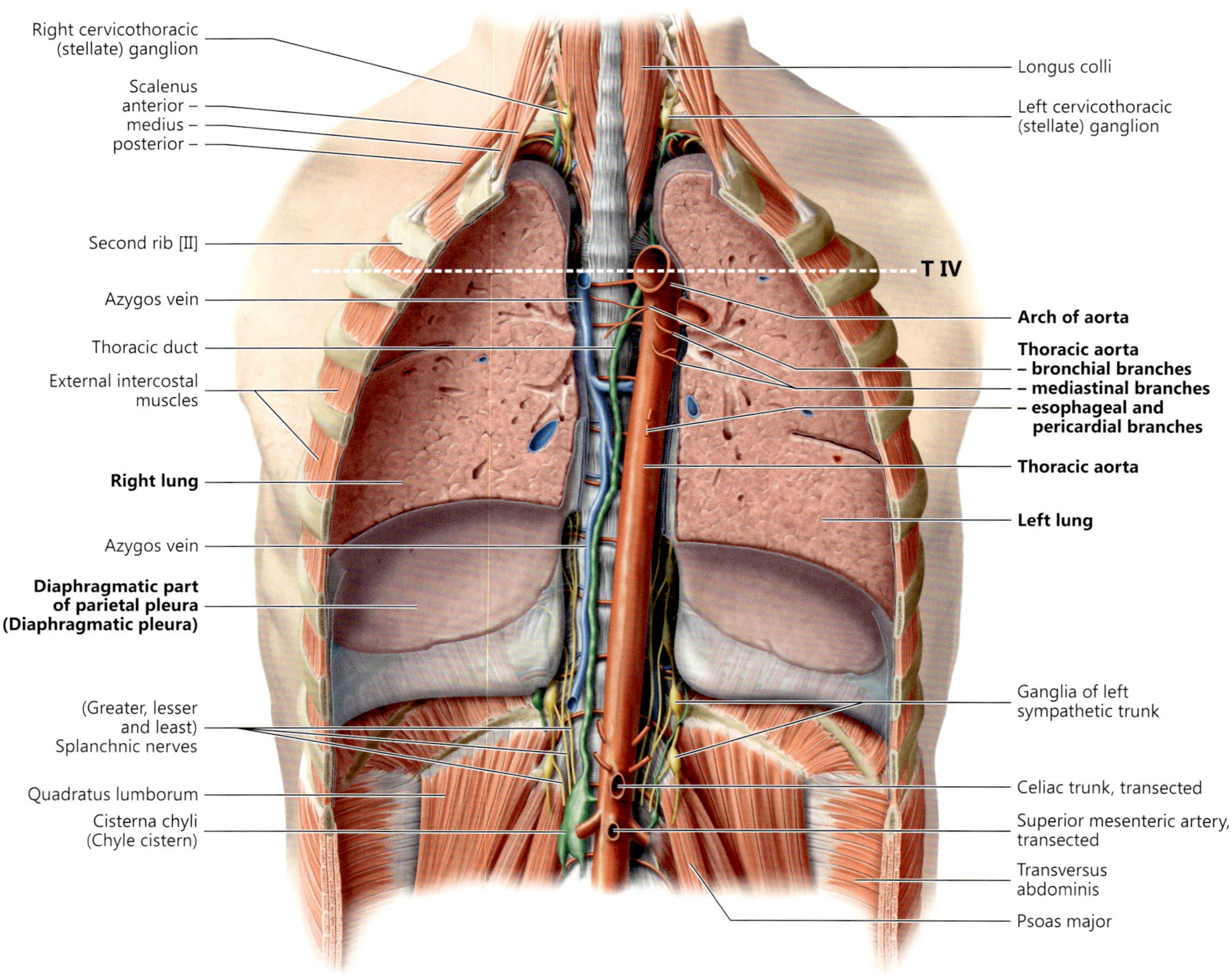

CONTENTS OF THE THORAX (in bold are the structures added with respect to the preceding table)
• Diaphragmatic pleura • Lungs (right and left) • Posterior third of arch of aorta • Thoracic aorta and its collateral branches (bronchial, mediastinal, esophageal and pericardial)

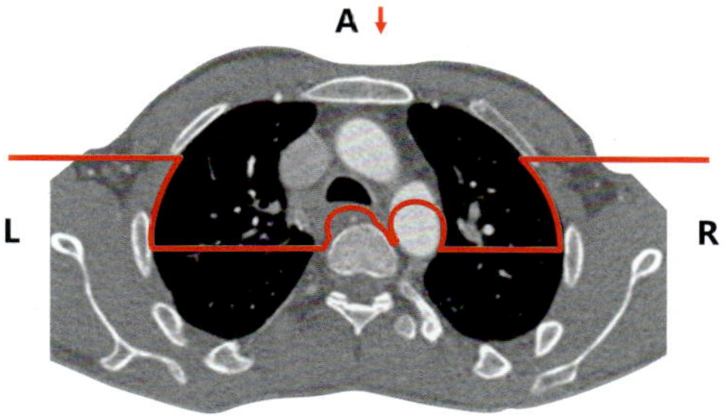

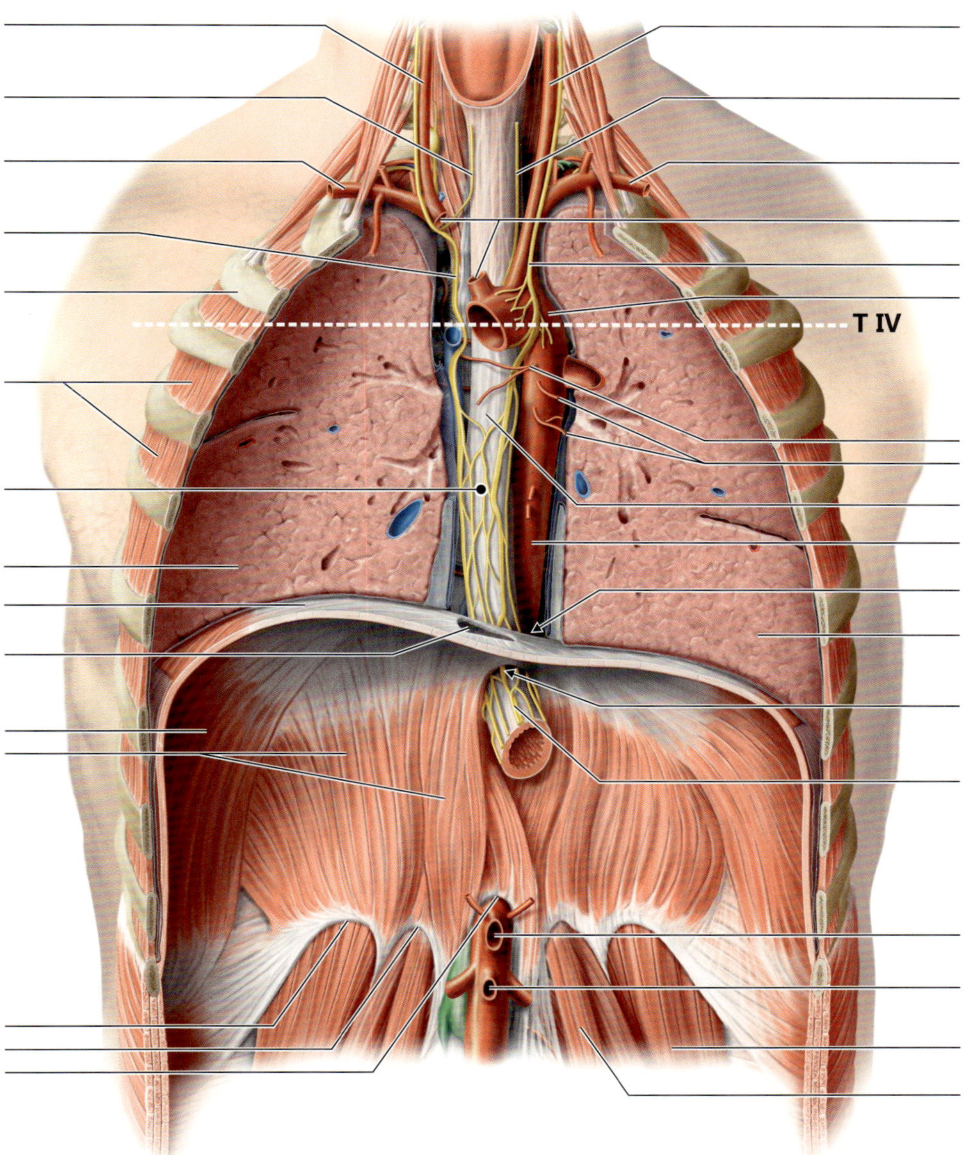

CONTENTS OF THE THORAX (in bold are the structures added with respect to the preceding table)
• Anterior vagal trunk • Branches of arch of aorta (left and right subclavian arteries, left and right common carotid arteries, brachiocephalic trunk) • Central tendon of diaphragm (opening of inferior vena cava, aortic hiatus, esophageal hiatus) • Diaphragm (costal and lumbar parts) • Esophageal plexus • Esophagus • Recurrent laryngeal nerves (inferior laryngeal nerves) (right and left) • Vagus nerves (right and left)

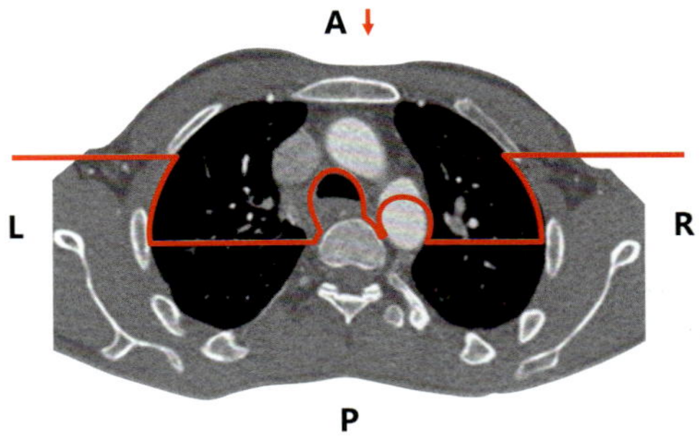

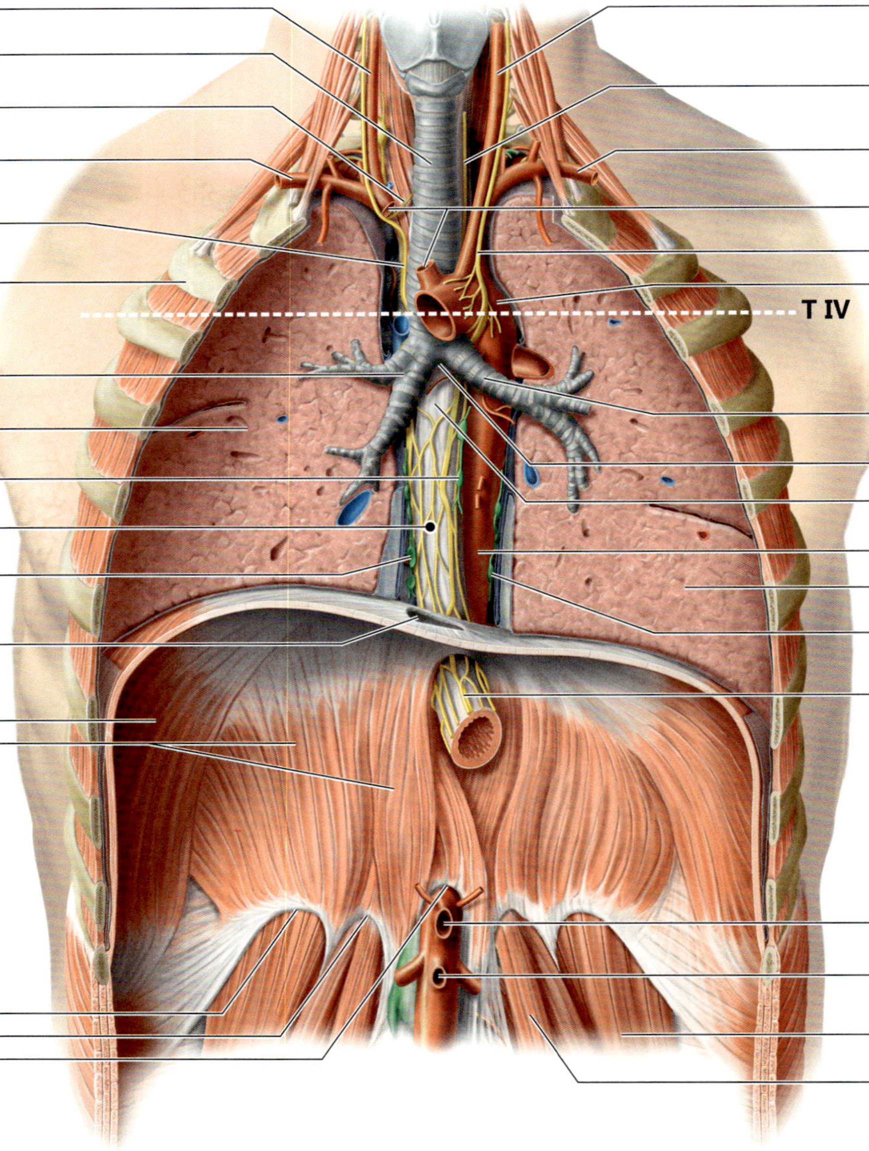

CONTENTS OF THE THORAX (in bold are the structures added with respect to the preceding table)
• Main bronchi (right and left) • Posterior mediastinal nodes (juxtaesophageal, juxta-aortic, interaorticoesophageal) • Trachea • Tracheal bifurcation

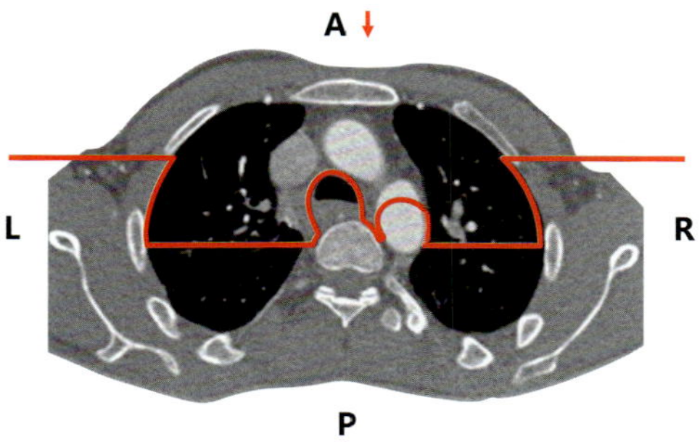

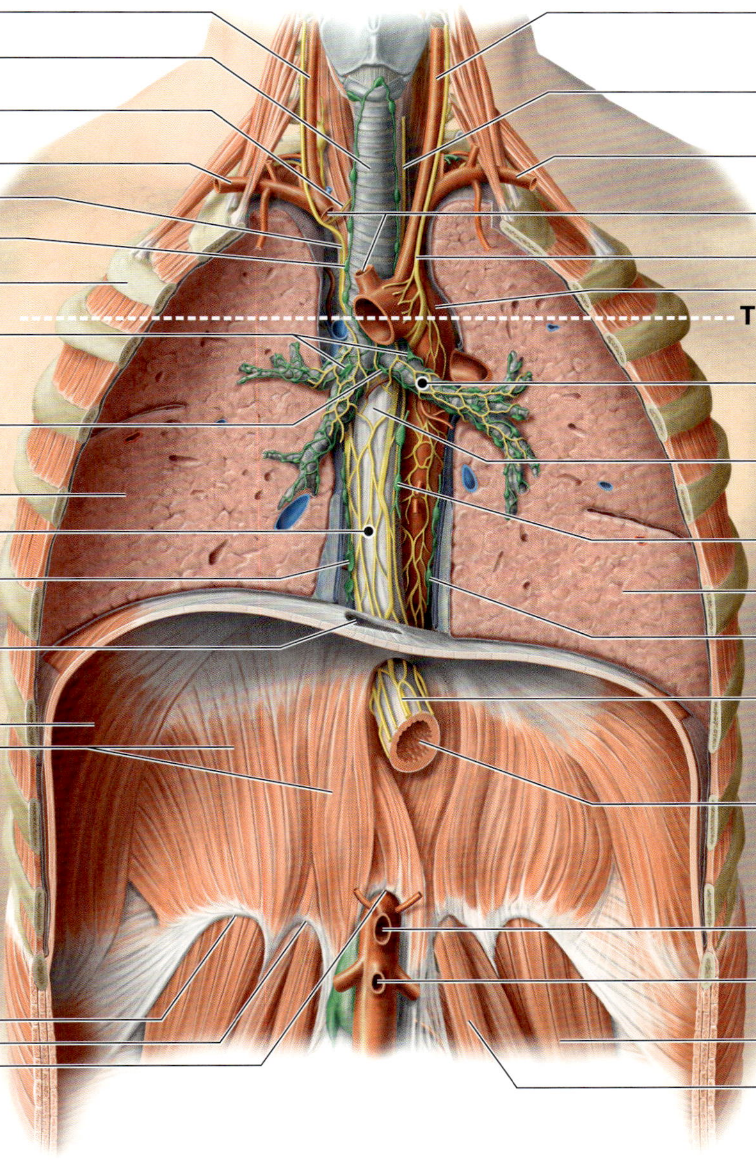

CONTENTS OF THE THORAX (in bold are the structures added with respect to the preceding table)
• Paratracheal nodes (right and left) • Pulmonary plexus • Superior and inferior tracheobronchial nodes

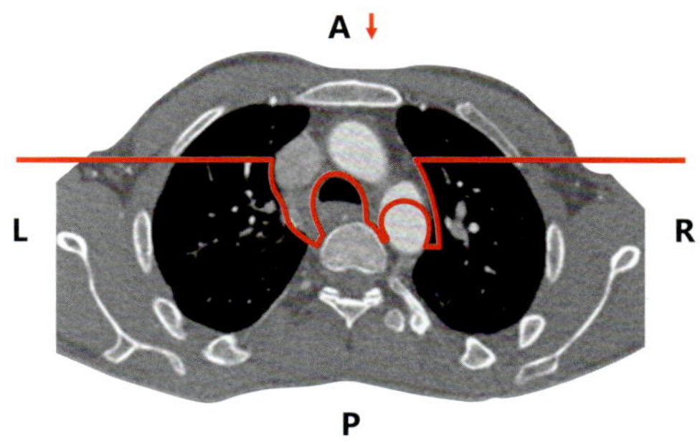

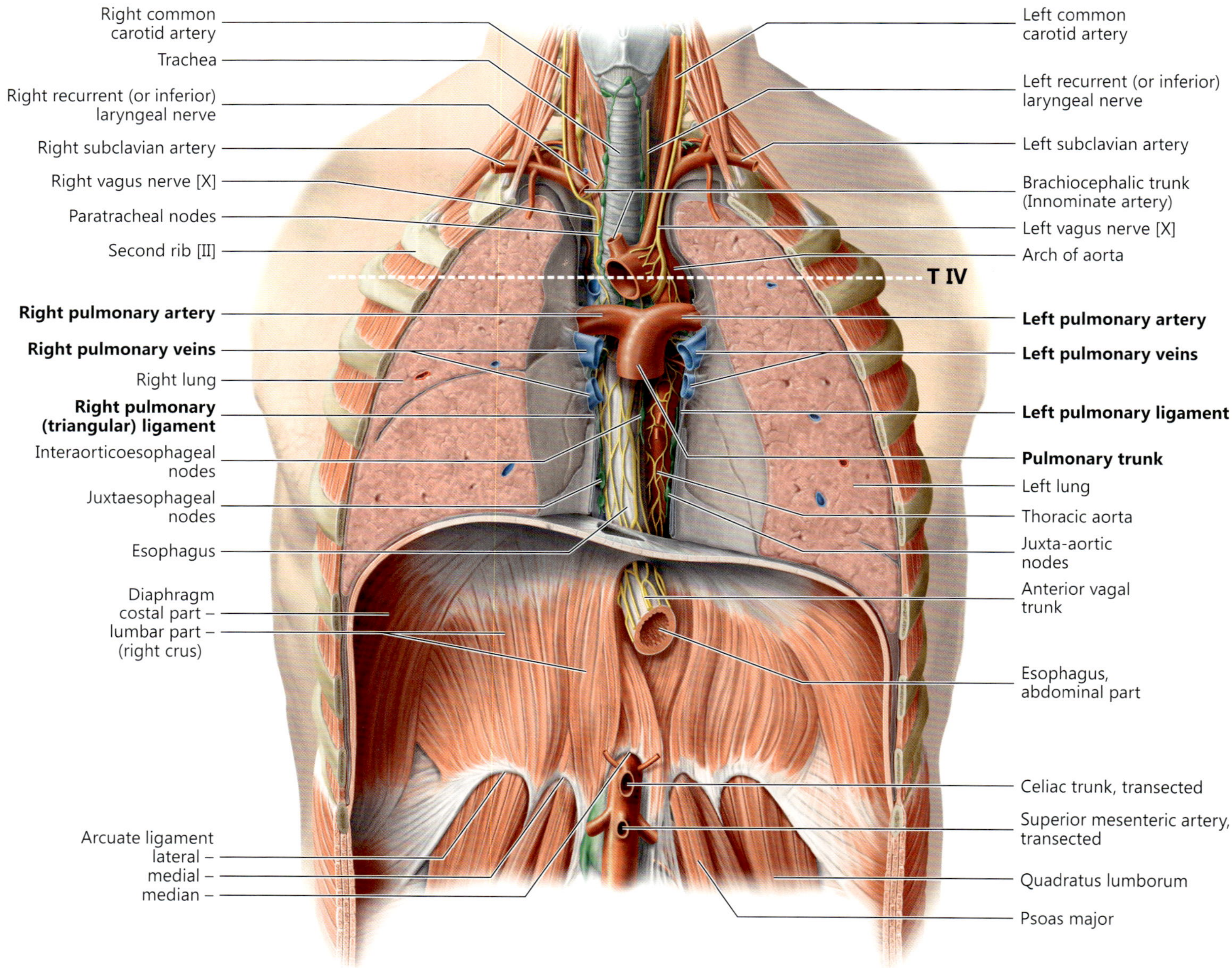

CONTENTS OF THE THORAX (in bold are the structures added with respect to the preceding table)
• Pulmonary arteries (right and left) • Pulmonary (triangular) ligaments (right and left) • Pulmonary trunk • Superior and inferior pulmonary veins (right and left)

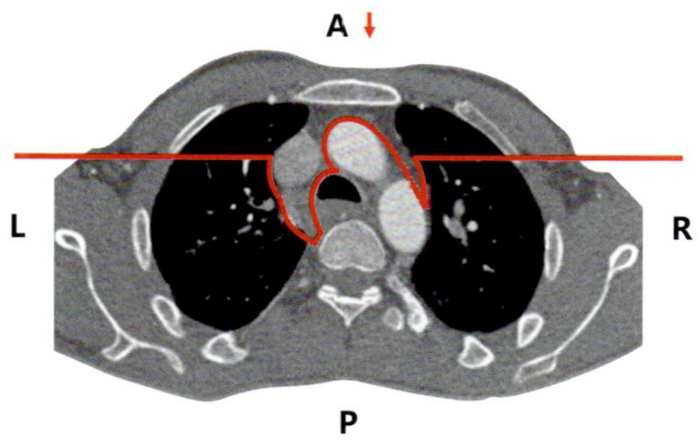

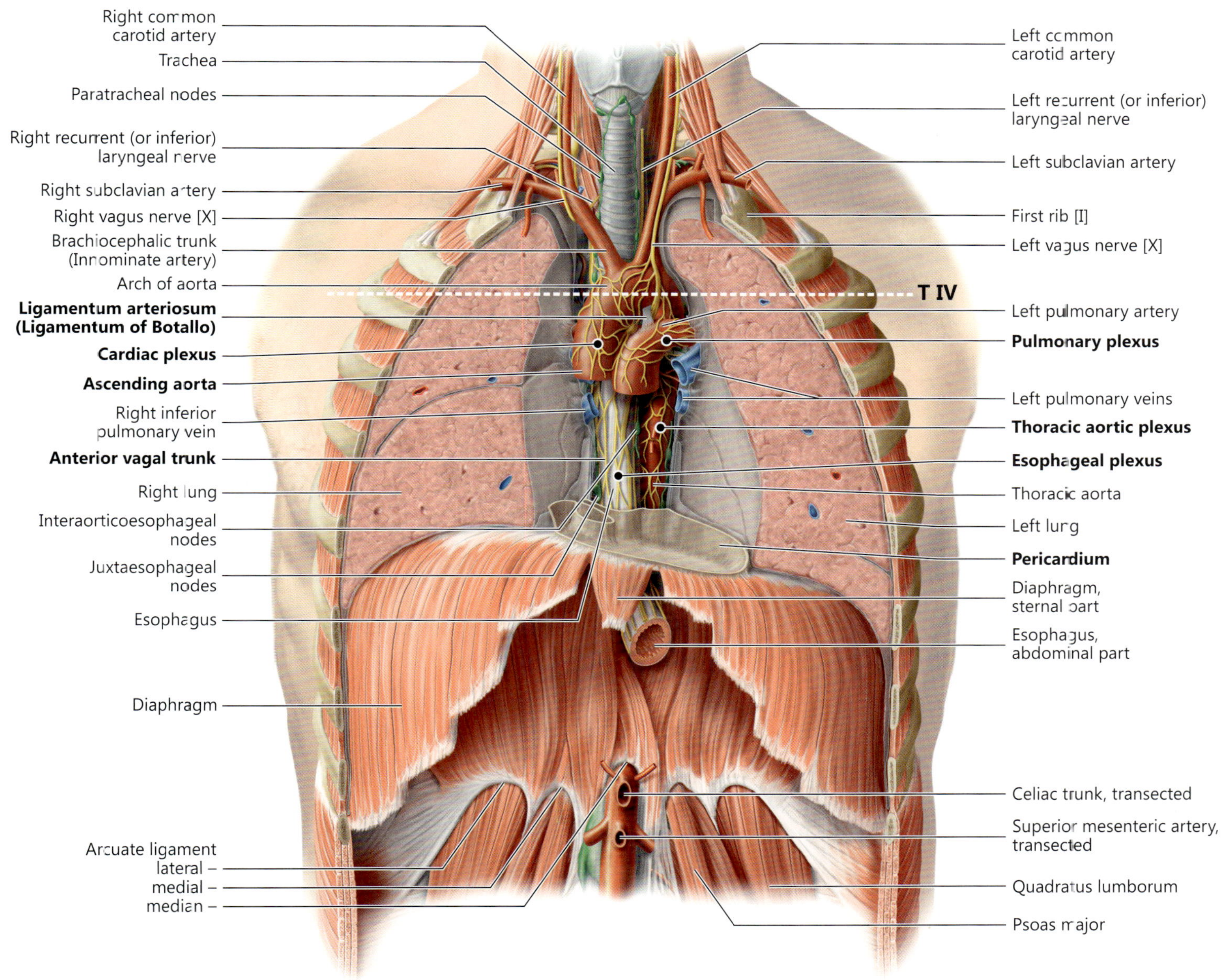

CONTENTS OF THE THORAX (in bold are the structures added with respect to the preceding table)
• Anterior vagal trunk • Ascending aorta • Cardiac plexus • Esophageal plexus • Ligamentum arteriosum (ligamentum of Botallo) • Pericardium (diaphragmatic surface) • Pulmonary plexus • Thoracic aortic plexus

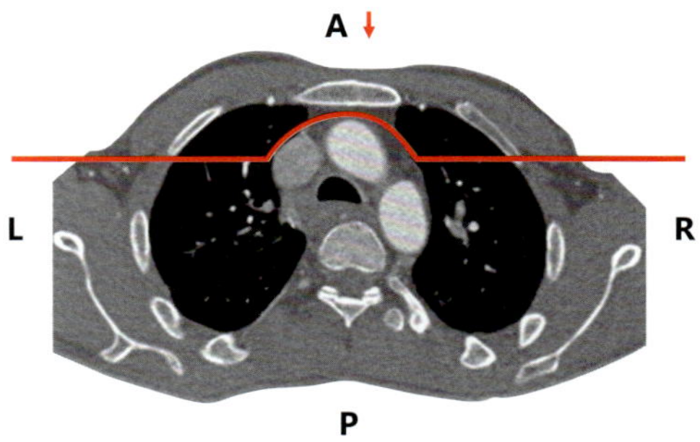

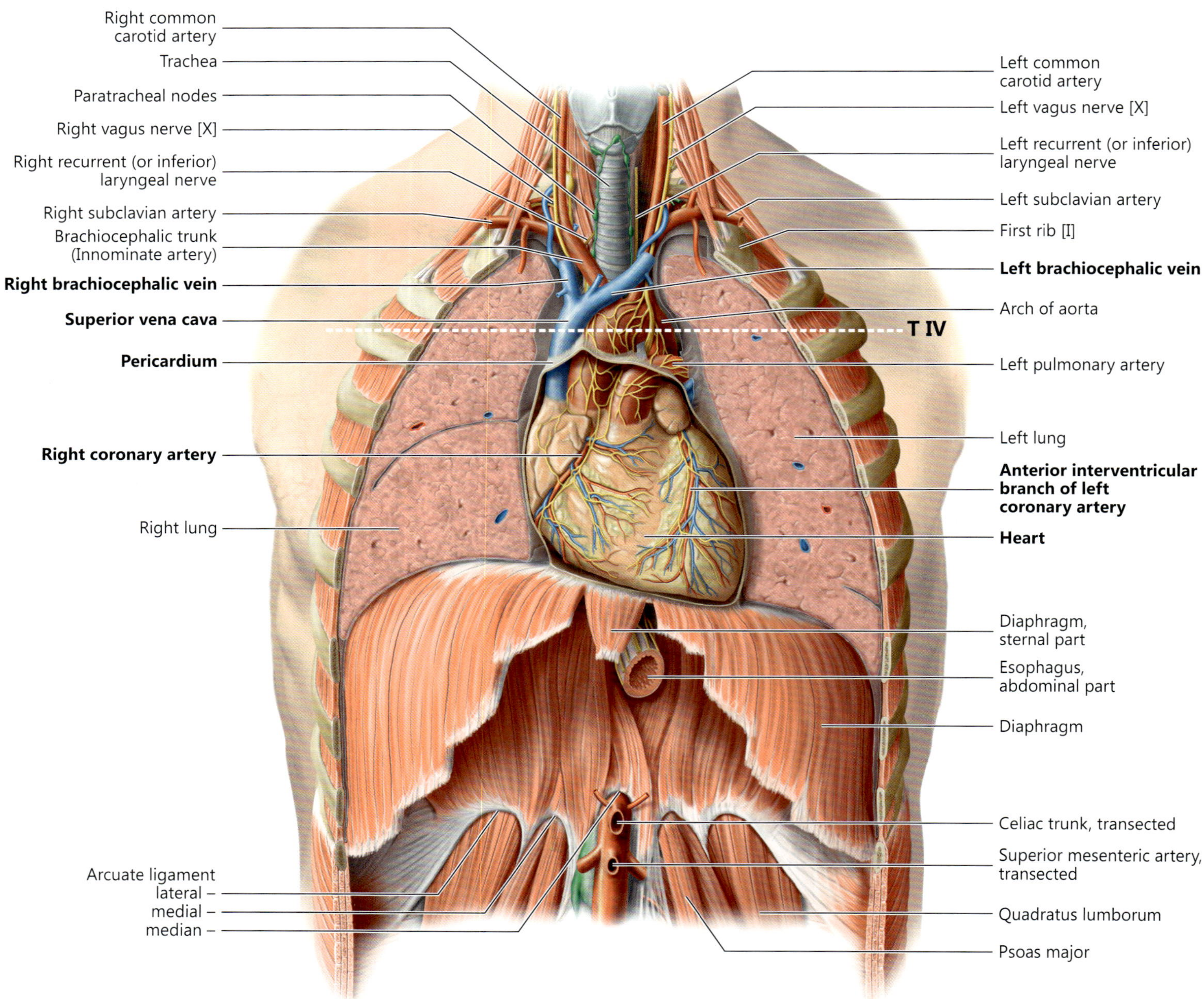

CONTENTS OF THE THORAX (in bold are the structures added with respect to the preceding table)
• Brachiocephalic veins (right and left) • Heart • Left coronary artery, anterior interventricular branch • Pericardium • Right coronary artery • Superior vena cava

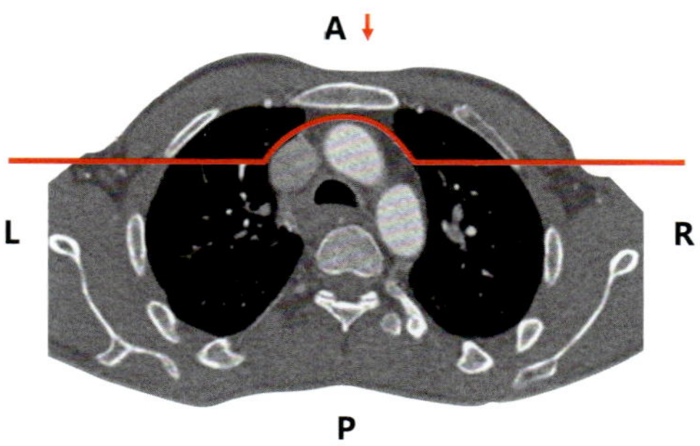

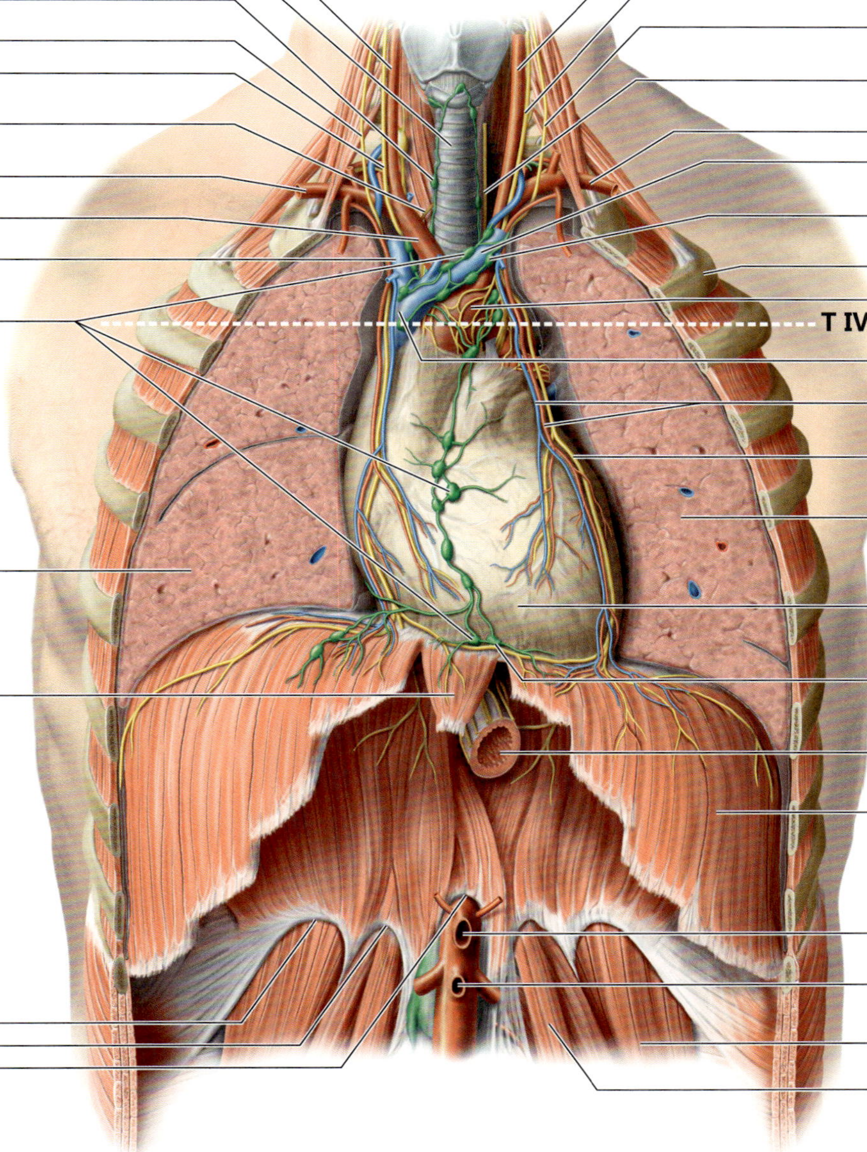

CONTENTS OF THE THORAX (in bold are the structures added with respect to the preceding table)
• Brachiocephalic nodes • Middle and anterior mediastinal nodes • Pericardiacophrenic arteries and veins (right and left) • Phrenic nerves (right and left) • Superior diaphragmatic nodes

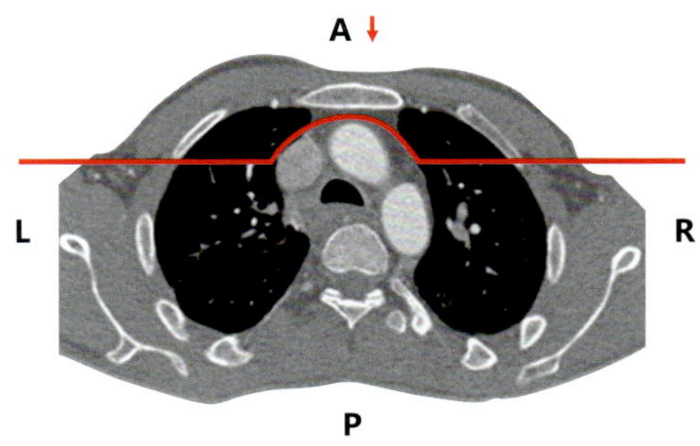

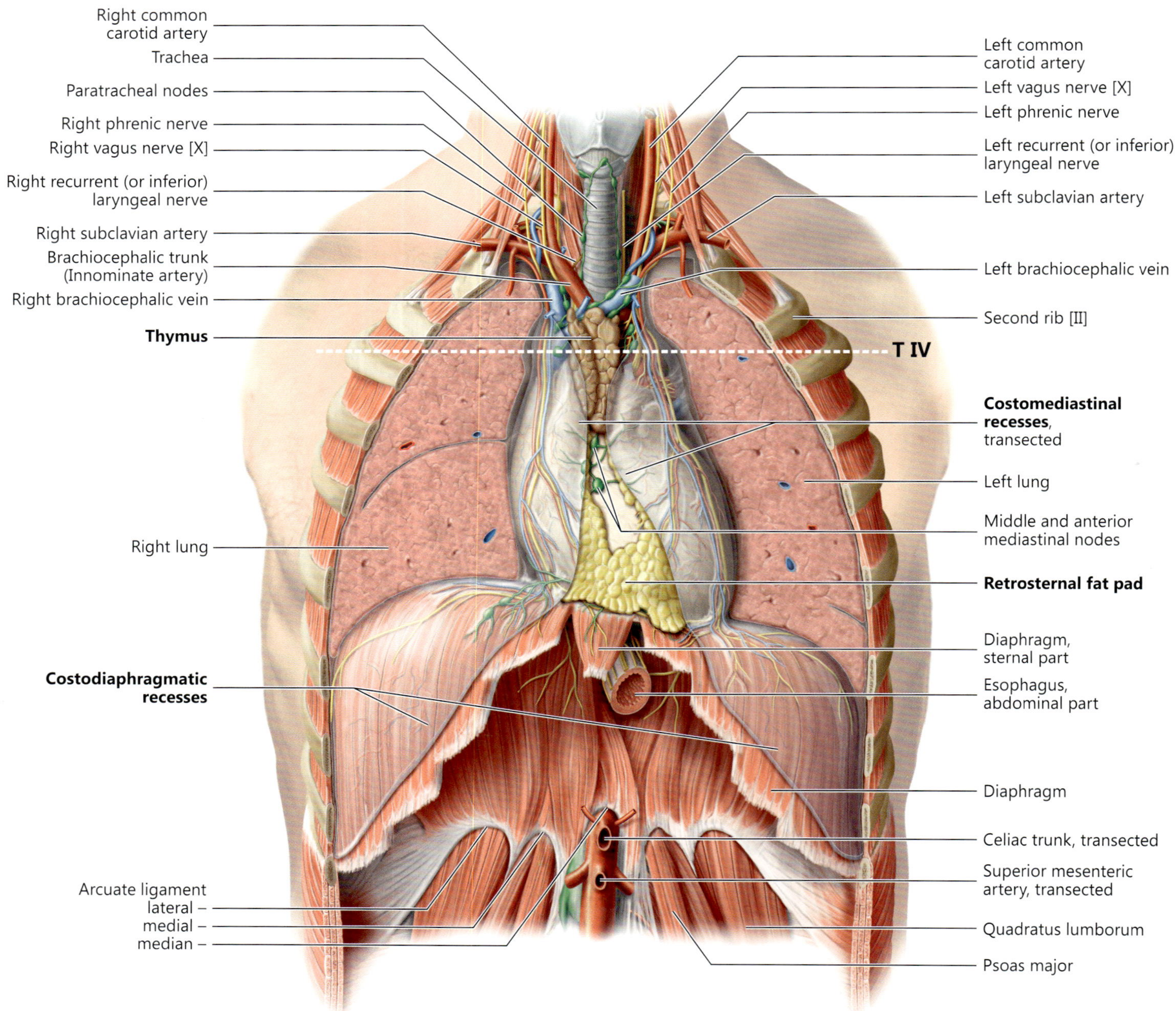

CONTENTS OF THE THORAX (in bold are the structures added with respect to the preceding table)
• Costodiaphragmatic recesses • Costomediastinal recesses • Retrosternal fat pad • Thymus or residual thymic tissue and fibroadipose tissue

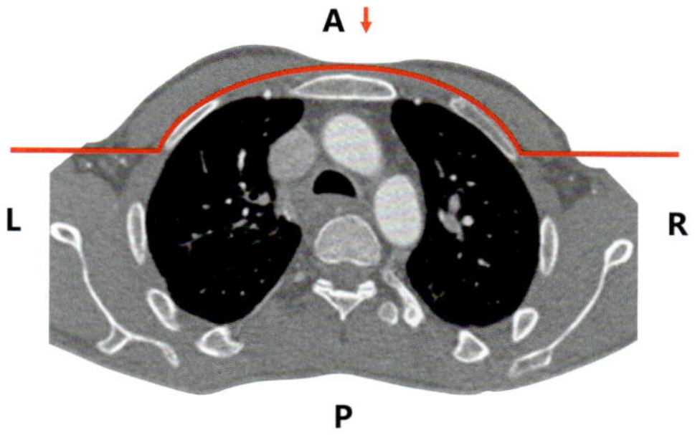

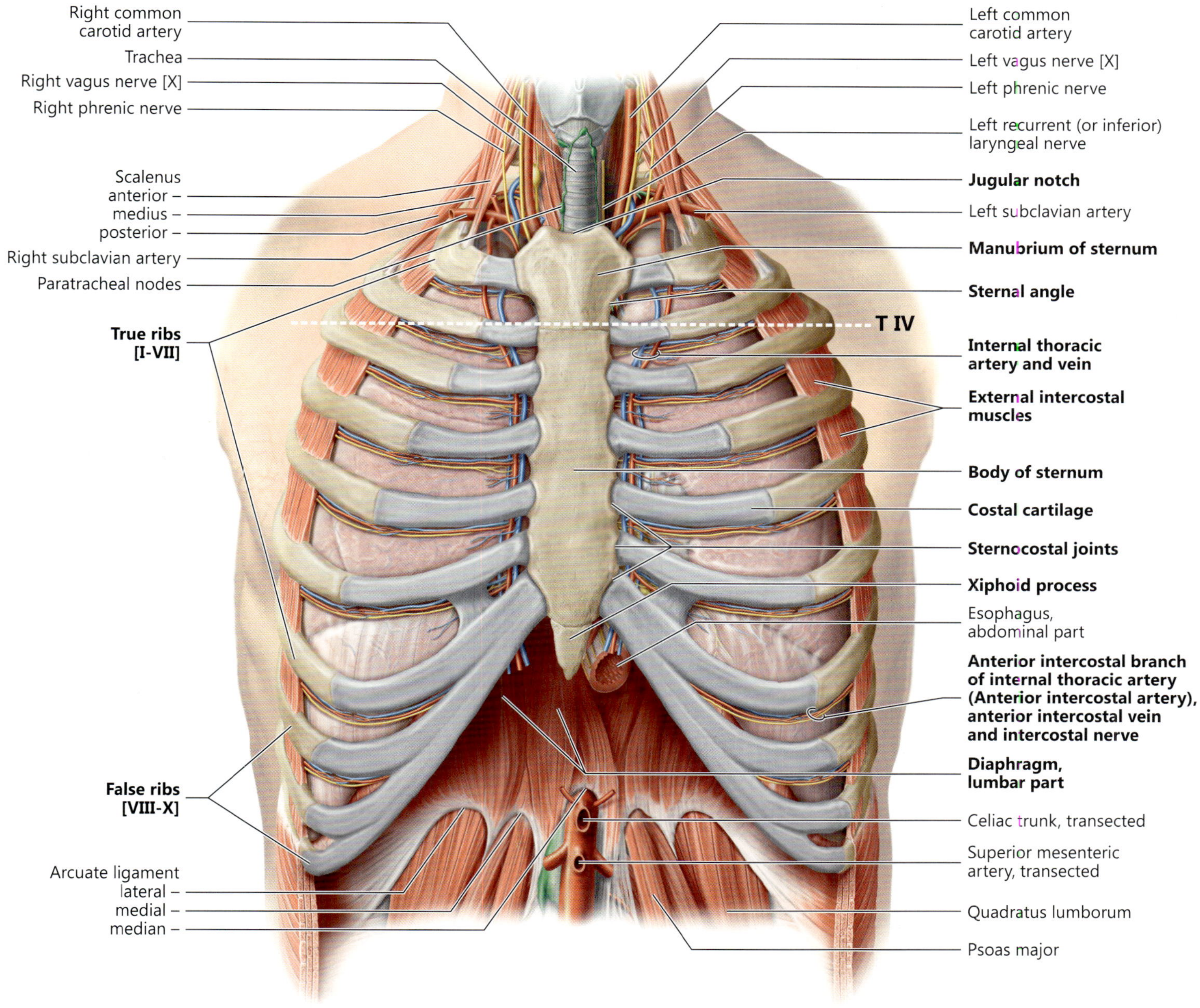

CONTENTS OF THE THORAX (in bold are the structures added with respect to the preceding table)	
• Anterior intercostal arteries • Anterior intercostal branches of internal thoracic artery • Anterior intercostal veins • Costal cartilage • Diaphragm (lumbar part) • External intercostal muscles	• Intercostal nerves • Internal thoracic arteries • Internal thoracic veins • Ribs • Sternocostal joints • Sternum

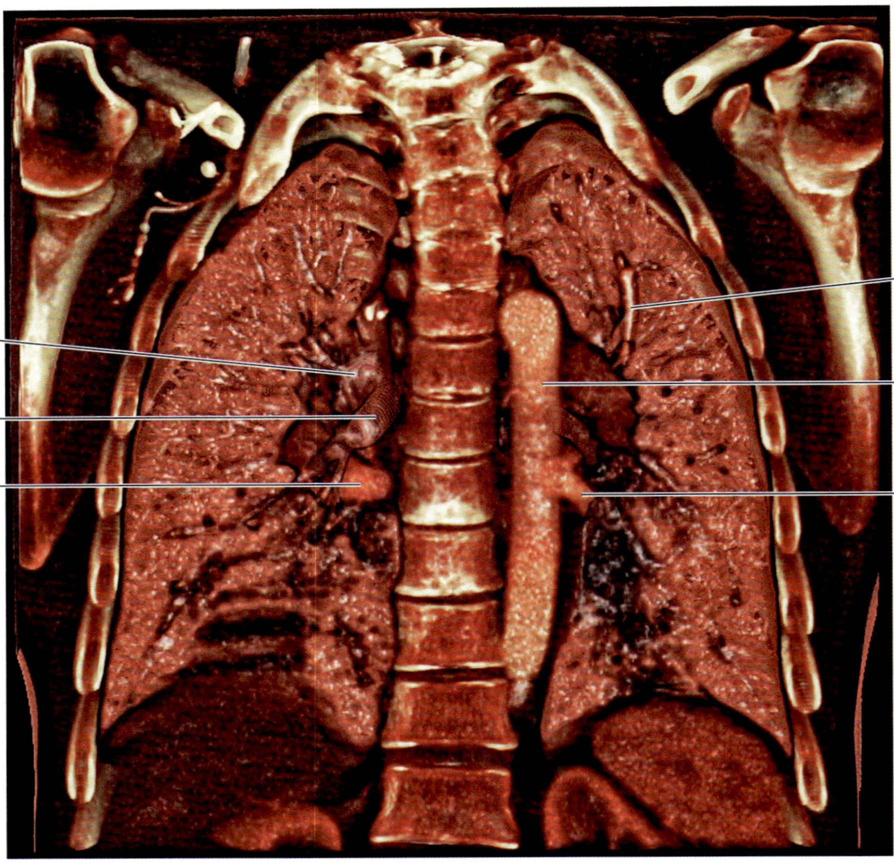

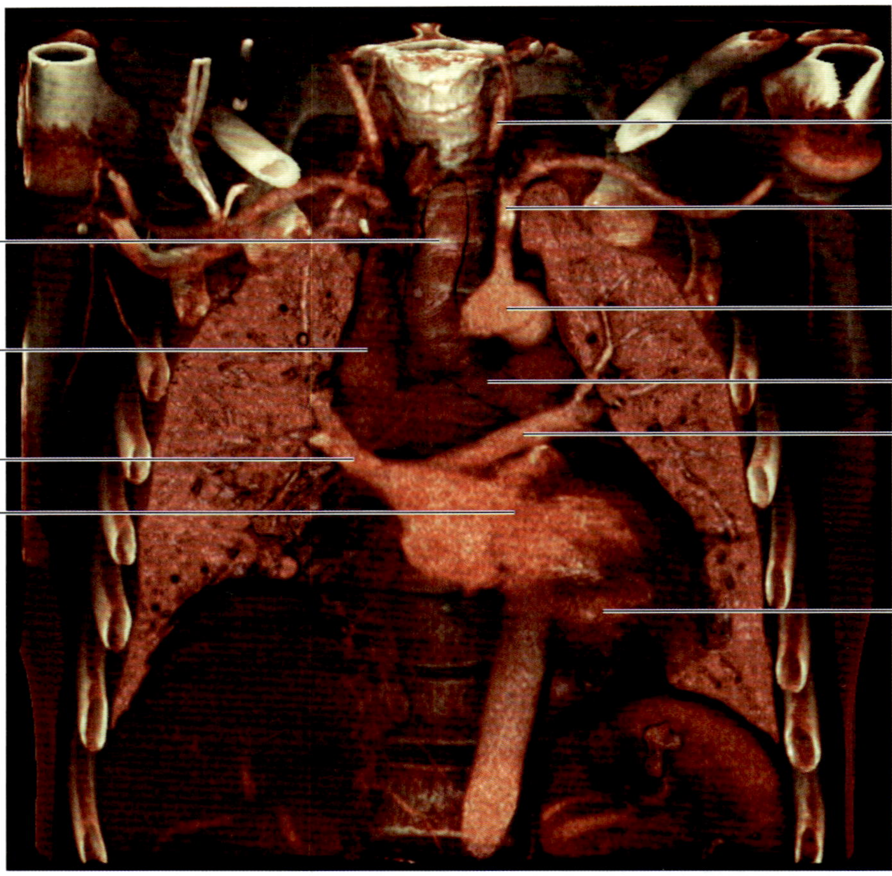

Series of four three-dimensional images of a living human body, obtained from frontal sections in dorso-ventral direction by CT scan using a 3D volume rendering technique. View from in front. The sequence starts with a section at the level of the vertebral bodies. To the left of the vertebral column, one can see the **descending aorta** (**thoracic aorta**) from its origin at T IV level to the point at which it moves in front of the column (transected). Superimposed on the aorta one can see a portion of the left **pulmonary veins**. In the successive image, one can see the left **atrium** and, above it, part of the **arch of the aorta** and its left branches.

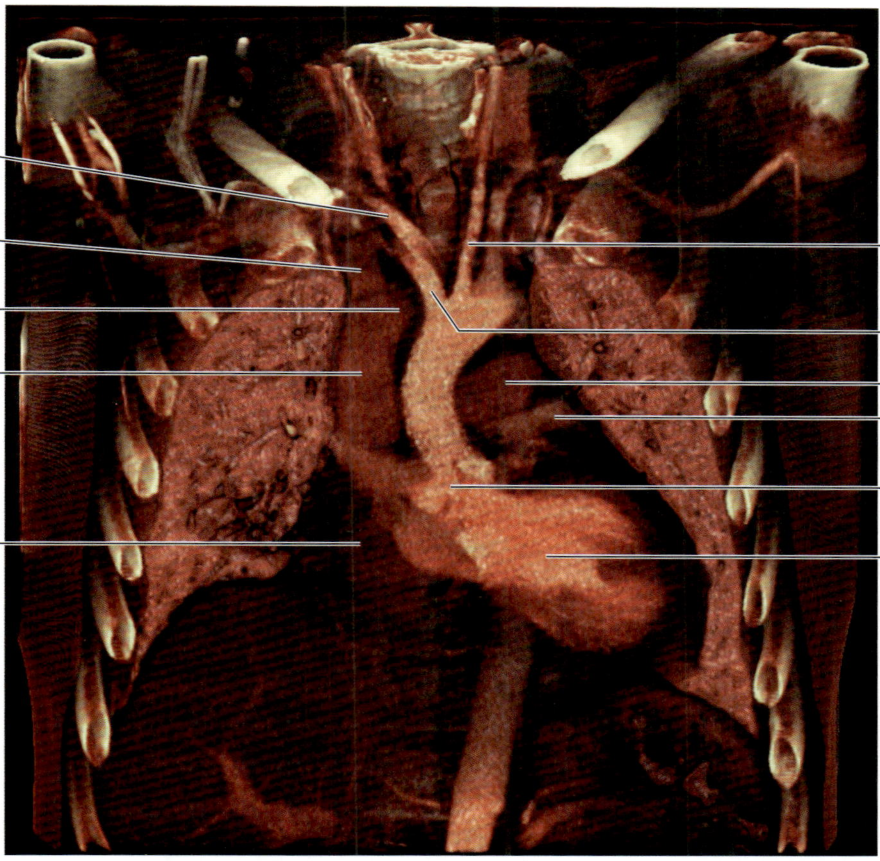

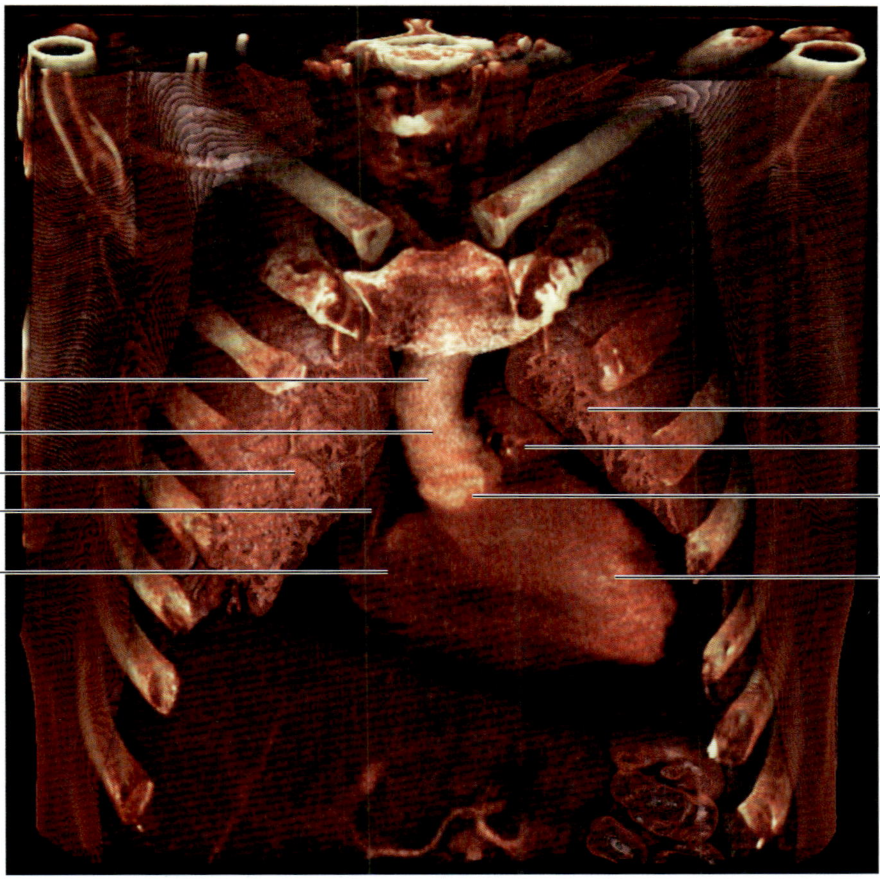

Following in sequence from the previous image, the section at the top shows the left **ventricle**, the **arch of the aorta** and its branches. The arch passes over the left pulmonary veins. To the right of the ascending aorta, one can see the superior vena cava. Note the aortic semilunar cusps.
In the section below, the lungs partially envelop the heart.

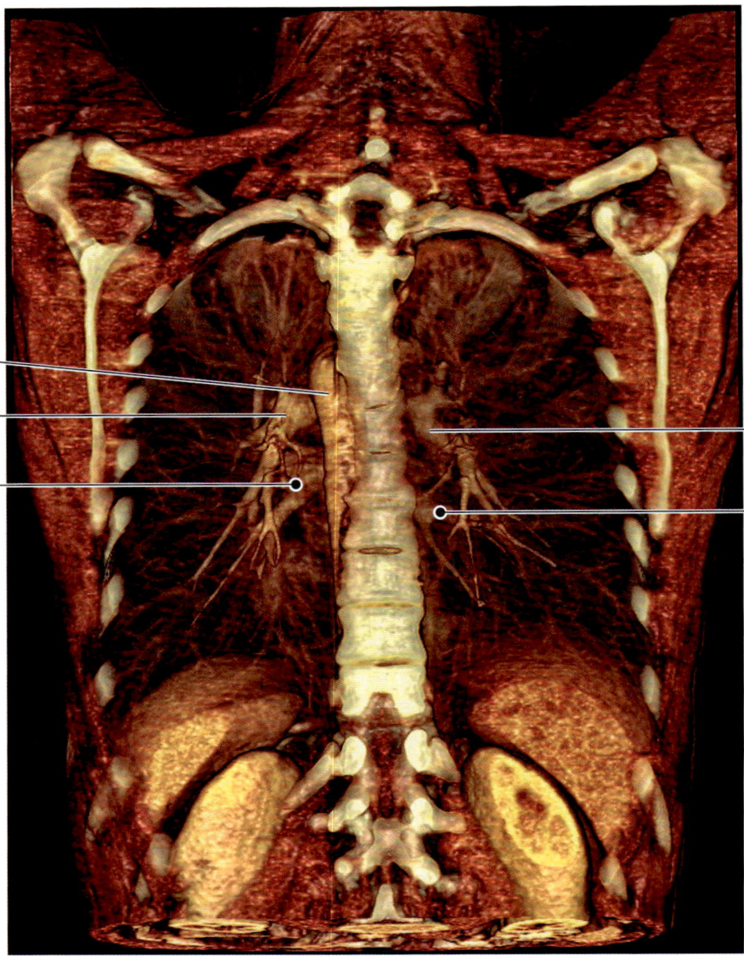

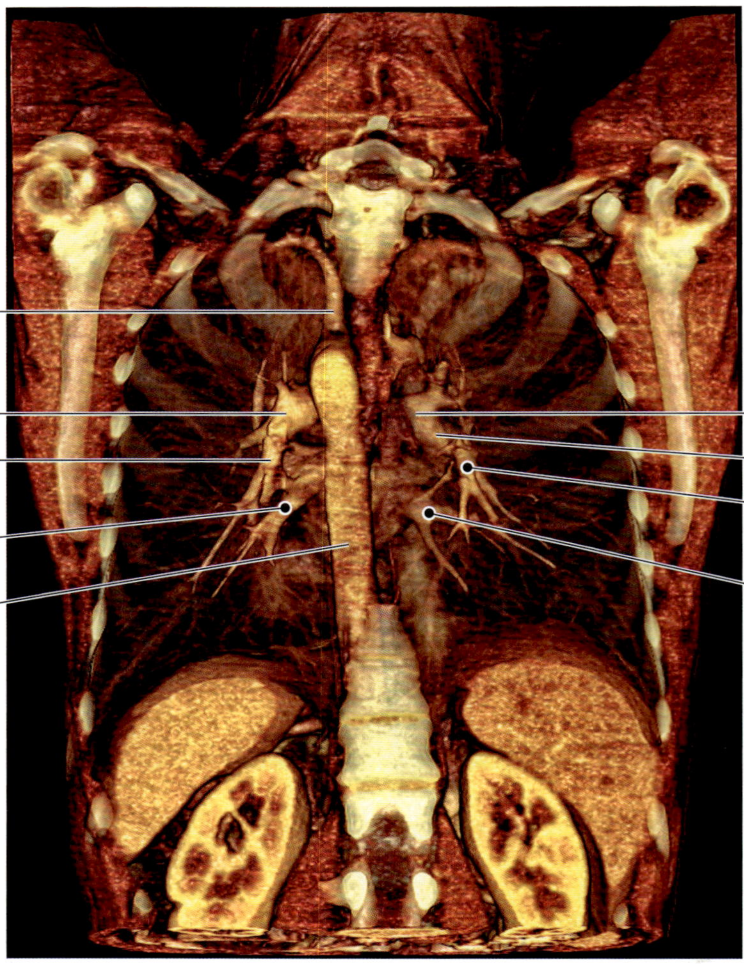

Series of two three-dimensional images of a living human body, obtained from frontal sections in a dorso-ventral direction by CT scan using a 3D volume rendering technique. View from behind. The sequence starts with a section at the level of the vertebral bodies. To the left of the vertebral column, one can see the **descending aorta** (**thoracic aorta**) and laterally, on both sides, the **pulmonary vessels**. The successive section passes in front of the vertebral column.

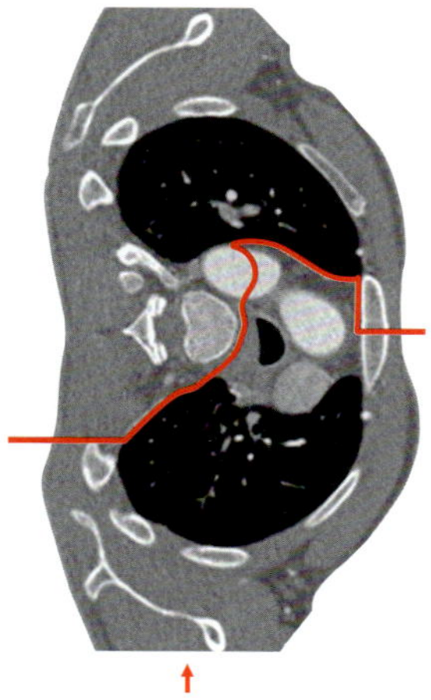

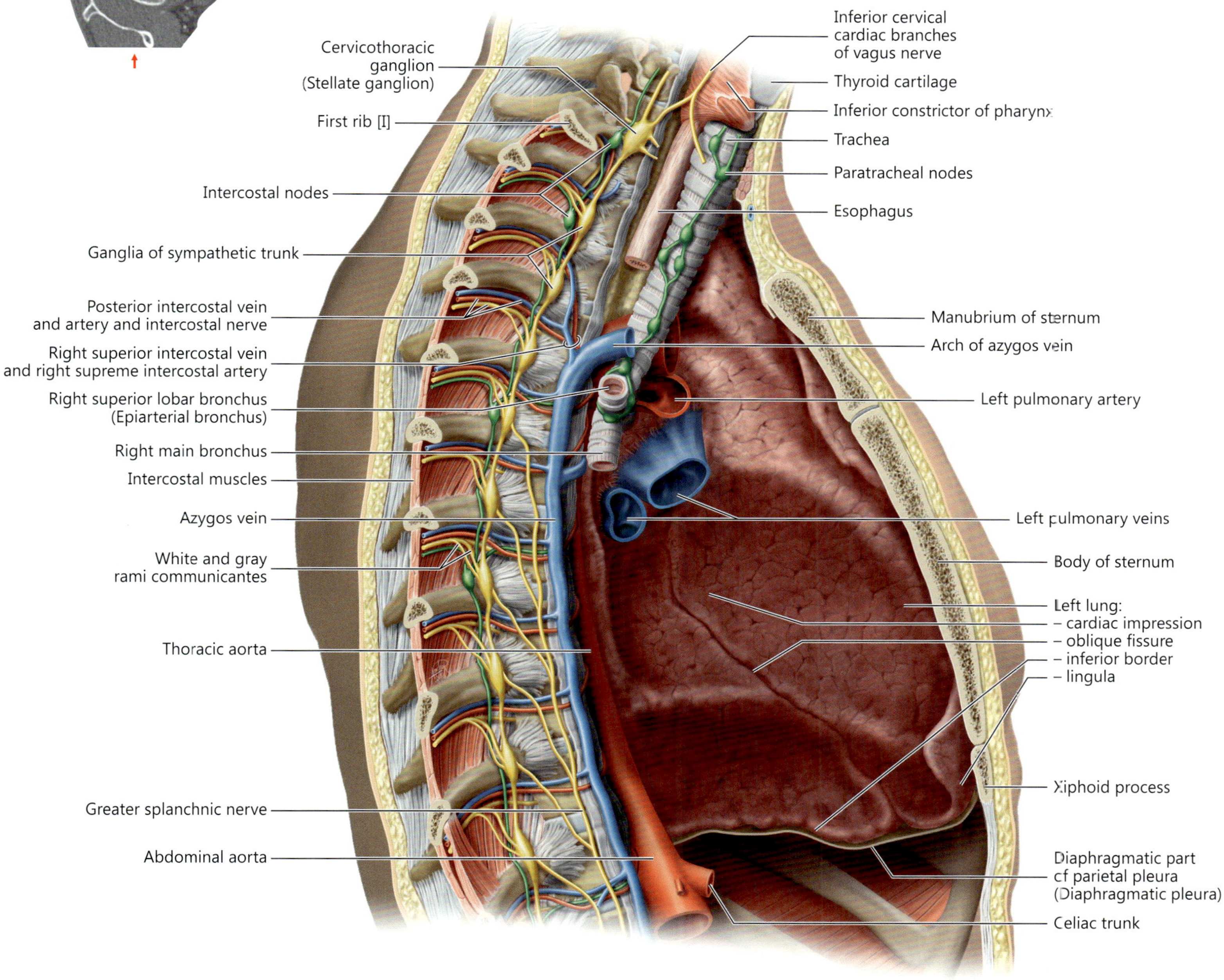

CONTENTS OF THE THORAX		
• Arch of azygos vein • Azygos vein • Esophagus • Ganglia of sympathetic trunk • Greater splanchnic nerve • Intercostal nodes • Left lung, cardiac impression • Left lung, inferior border	• Left lung, lingula • Left lung, oblique fissure • Left pulmonary artery • Left pulmonary veins • Paratracheal nodes • Posterior intercostal vein and artery and intercostal nerve • Right main bronchus	• Right superior intercostal vein • Right superior lobar bronchus (epiarterial bronchus) • Right supreme intercostal artery • Thoracic aorta • Trachea • White and gray rami communicantes

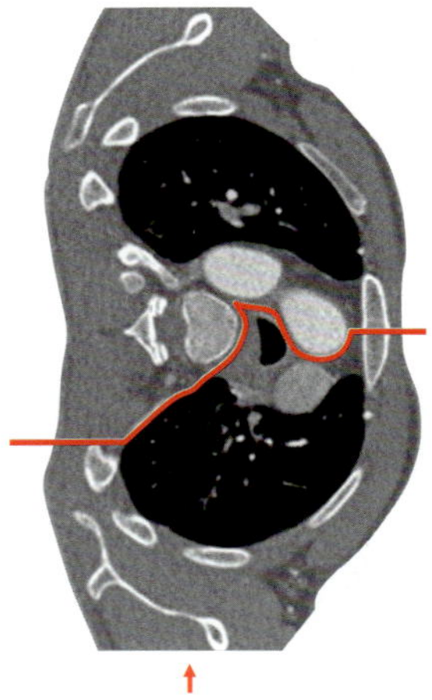

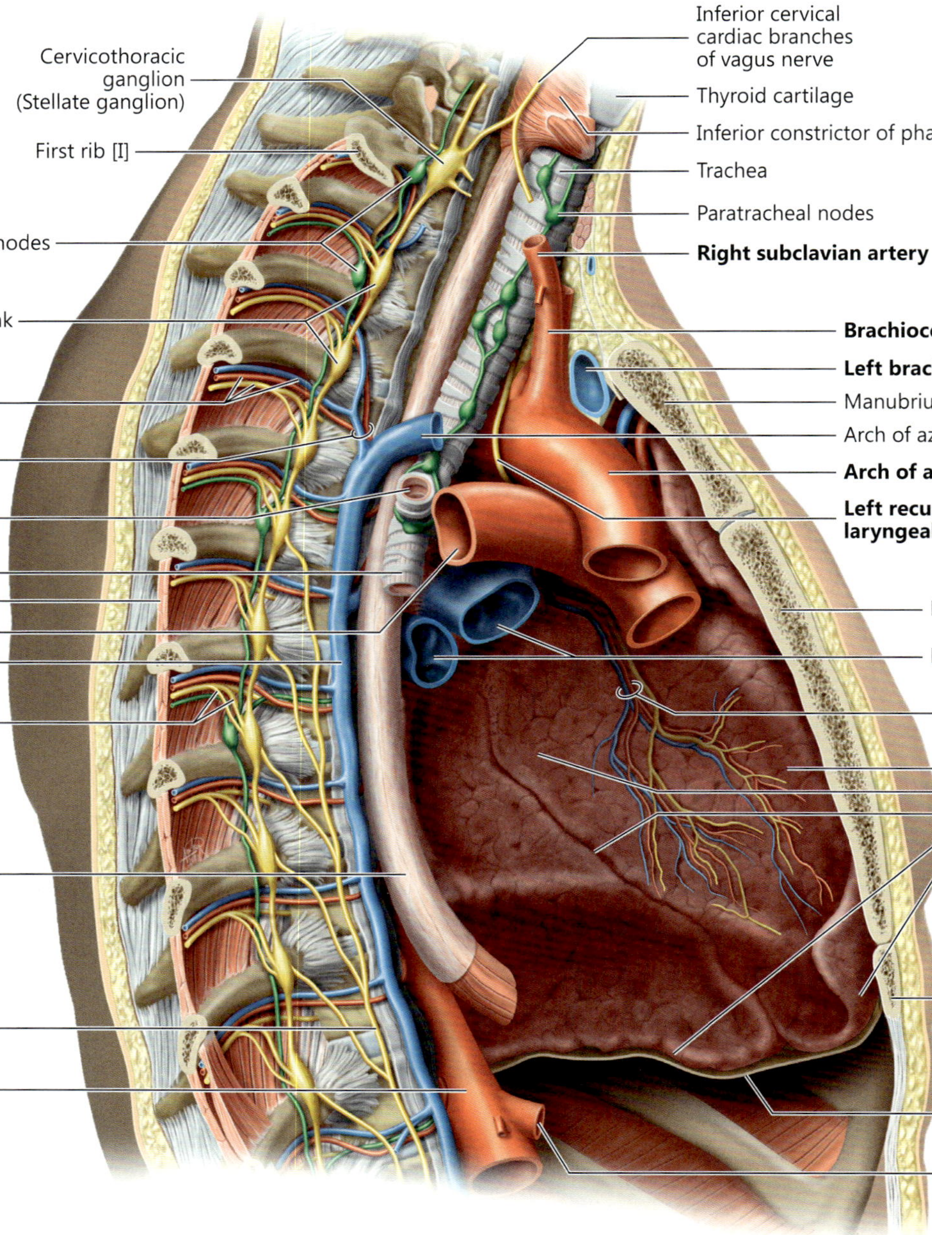

CONTENTS OF THE THORAX (in bold are the structures added with respect to the preceding table)
• Arch of aorta • Brachiocephalic trunk • Left brachiocephalic vein • Left pericardiacophrenic vessels • Left recurrent (or inferior) laryngeal nerve • Right pulmonary artery • Right subclavian artery

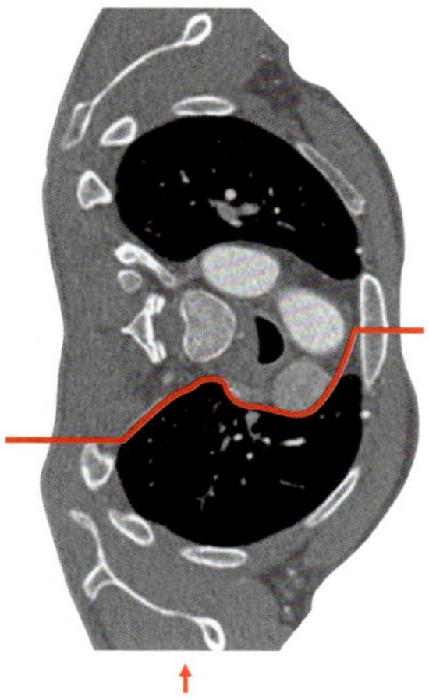

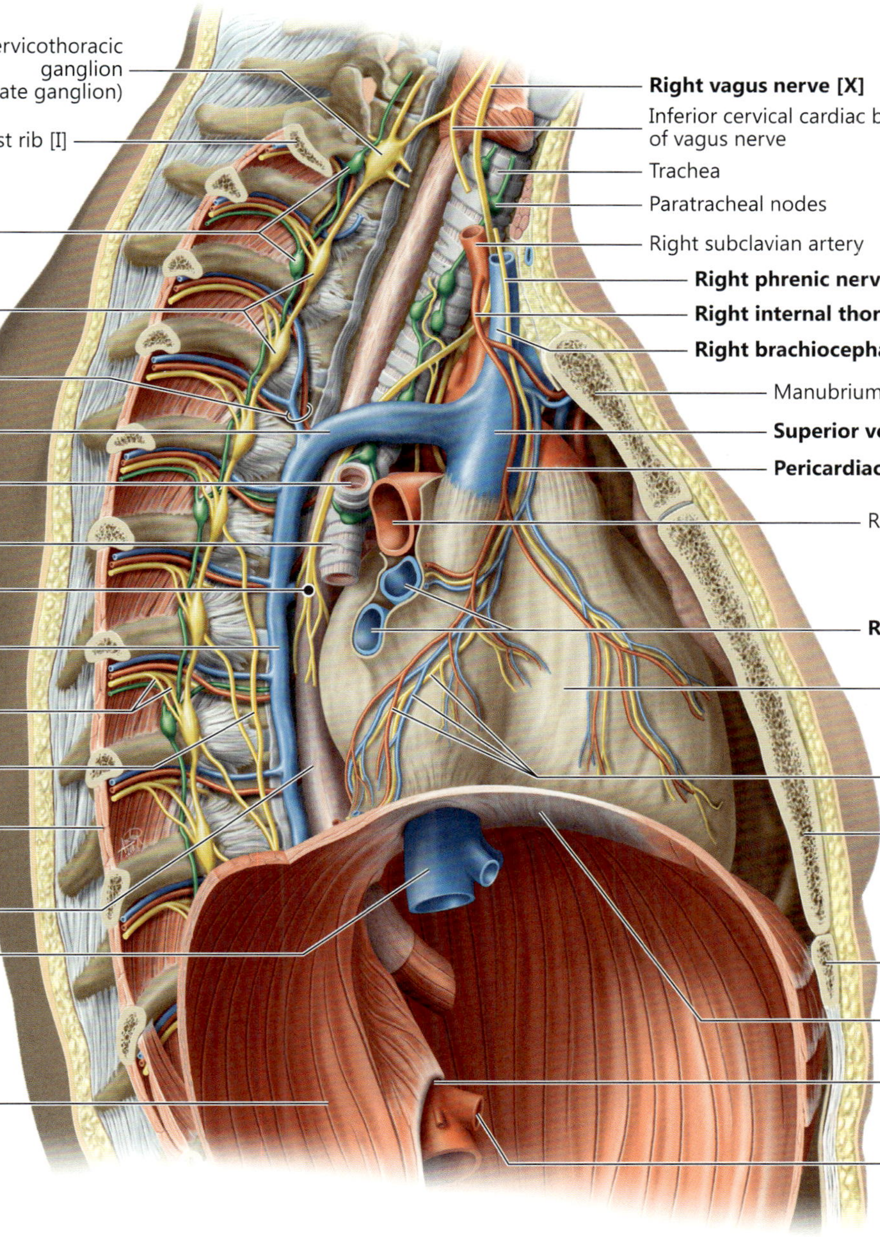

CONTENTS OF THE THORAX (in bold are the structures added with respect to the preceding table)	
• Aortic hiatus • Diaphragm • Esophageal plexus • Fibrous pericardium • Inferior vena cava • Pericardiacophrenic artery • Pericardiacophrenic veins	• Right brachiocephalic vein • Right internal thoracic artery • Right phrenic nerve • Right pulmonary veins • Right vagus nerve [X] • Superior vena cava

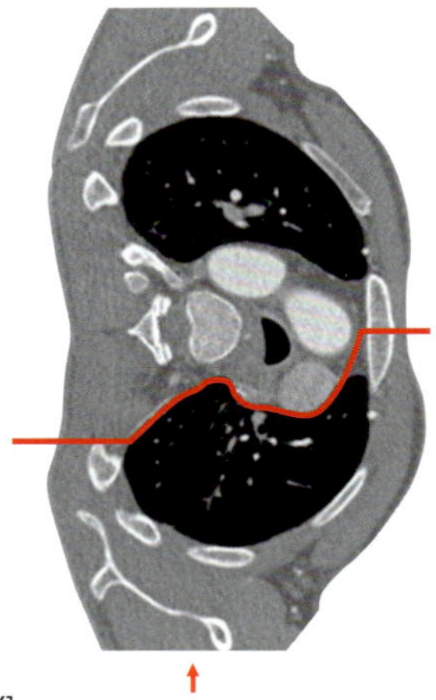

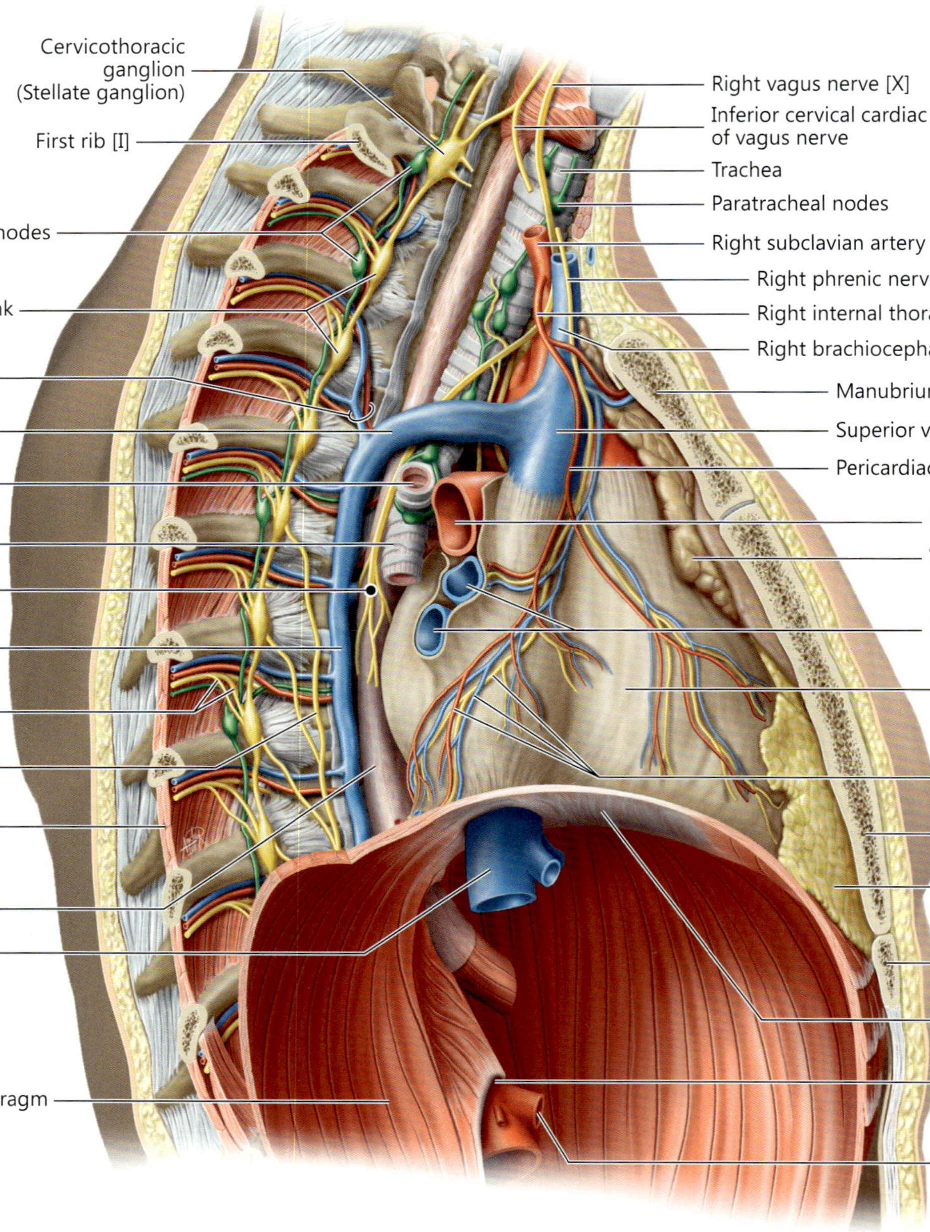

CONTENTS OF THE THORAX (in bold are the structures added with respect to the preceding table)
• Retrosternal fat pad • Thymus

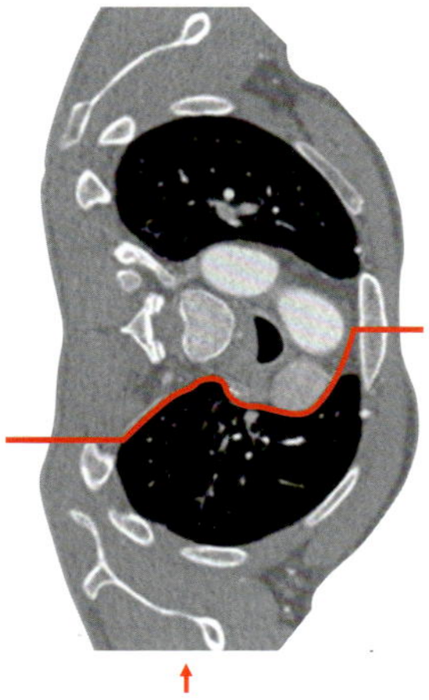

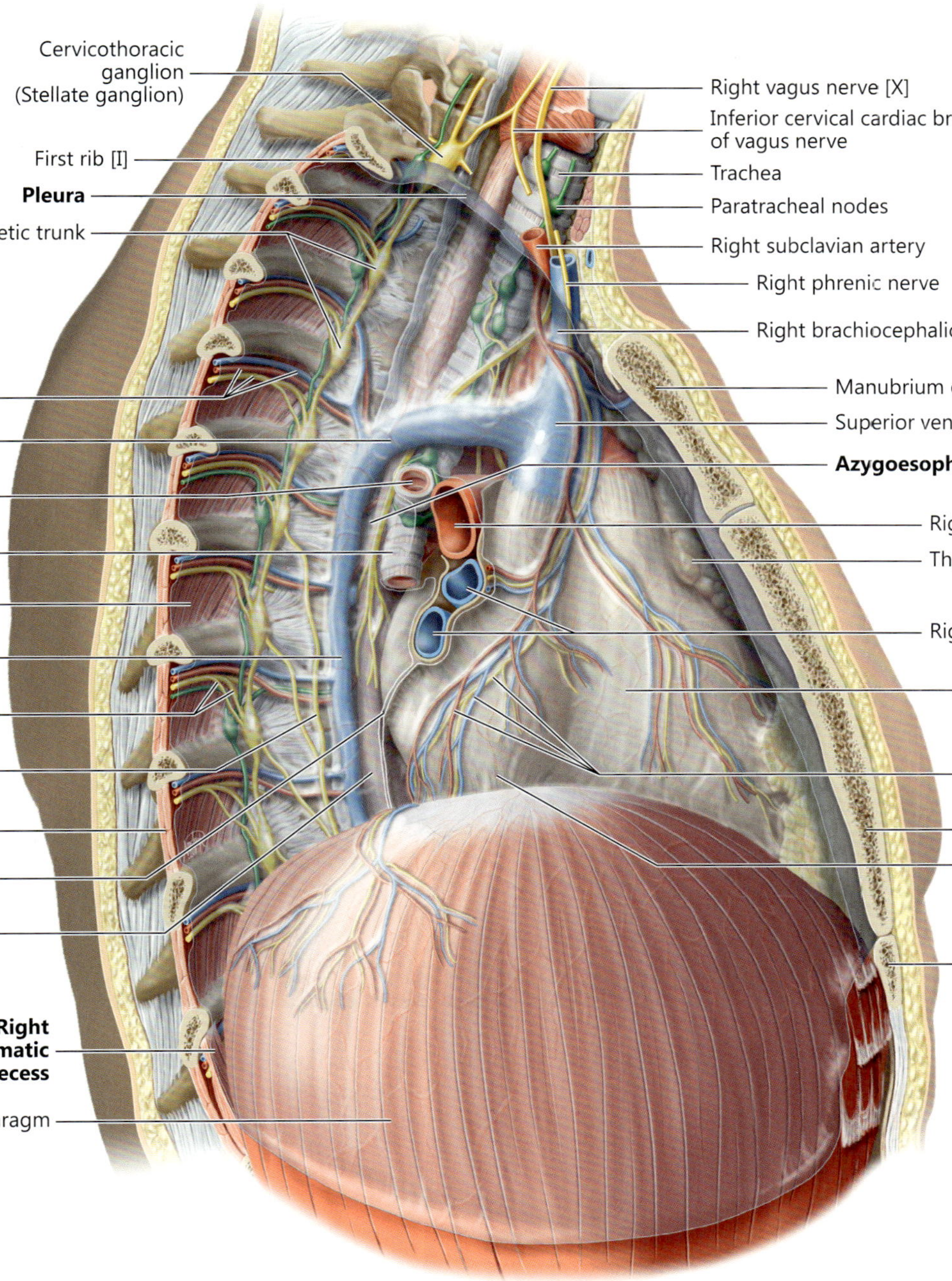

CONTENTS OF THE THORAX (in bold are the structures added with respect to the preceding table)

- Azygoesophageal recess
- Costal part of parietal pleura (costal pleura)
- Inferior vena cava
- Pleura
- Pulmonary (triangular) ligament
- Right costodiaphragmatic recess

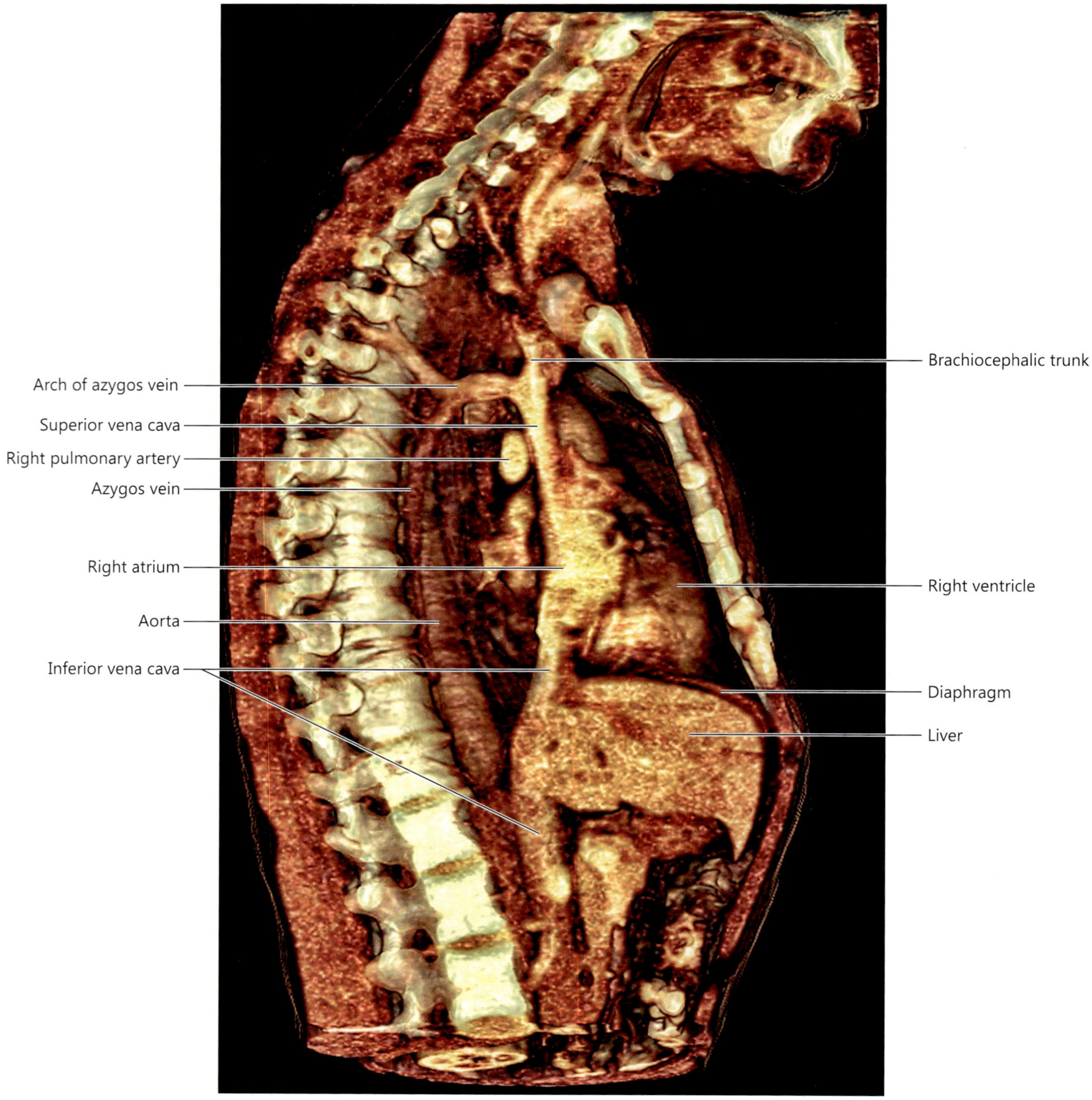

Three-dimensional image of a living human body, obtained by CT scan using a 3D volume rendering technique, in sagittal section. The **mediastinum** is observed from the right. The reconstruction enables one to see the right atrium and opening of the venae cavae. Clearly visible is the arch of the azygos vein on the root of the right lung.

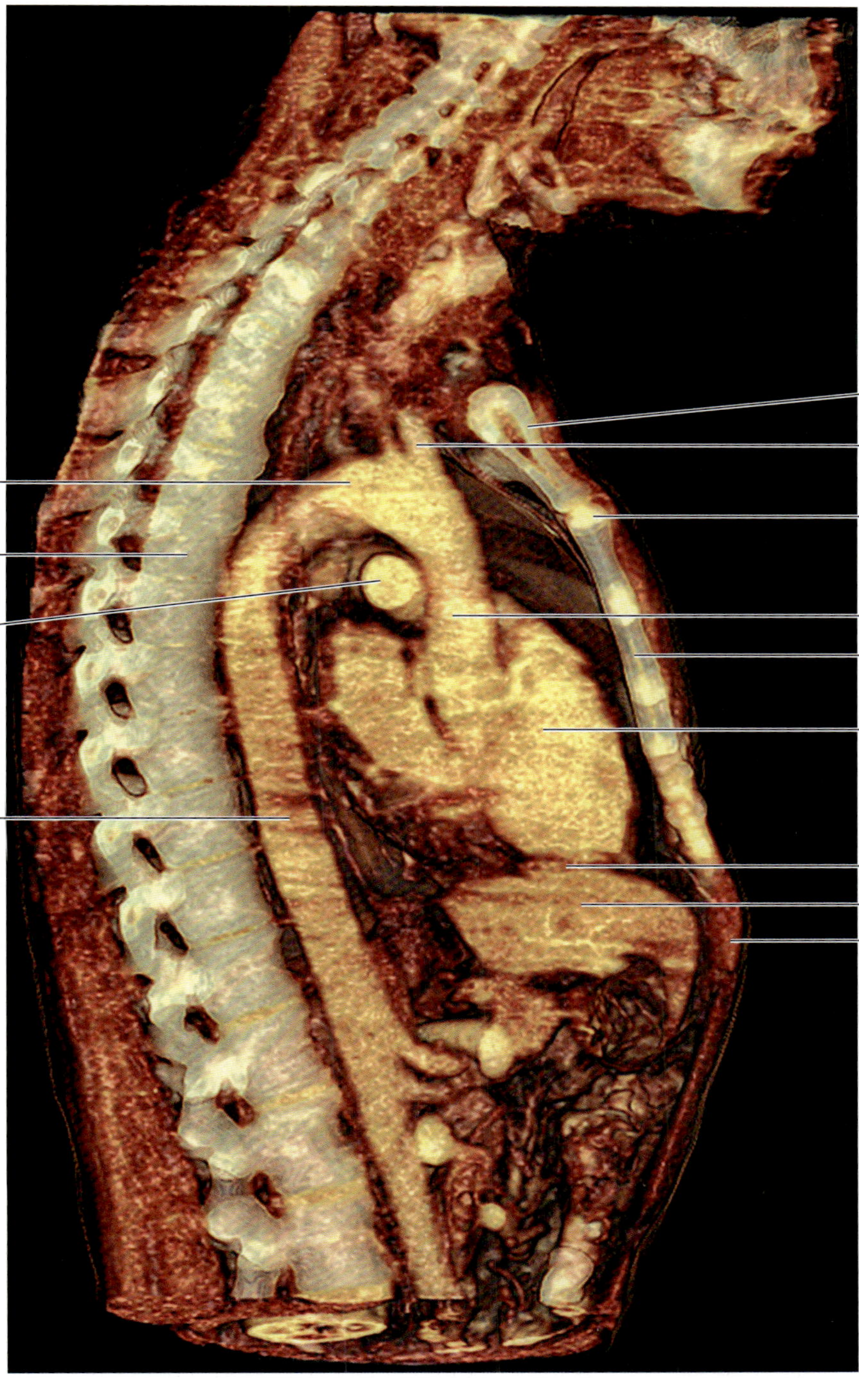

Three-dimensional image of a living human body, obtained by CT scan using a 3D volume, in sagittal section. The **mediastinum** is observed from the right. The reconstruction enables one to see the heart as well as the course of the ascending aorta, of the arch of the aorta, which passes over the root of the right lung, and of the thoracic aorta.

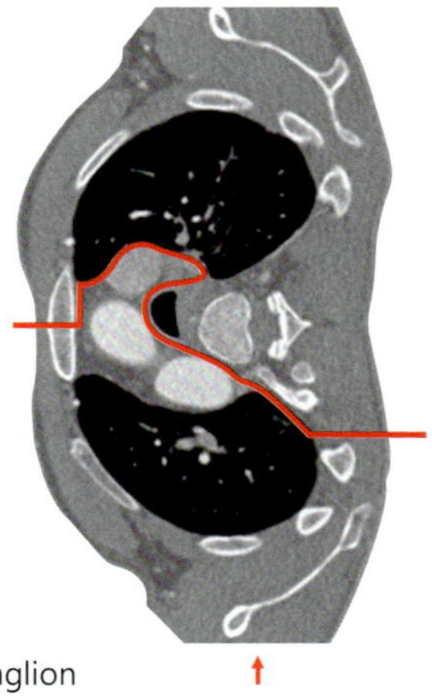

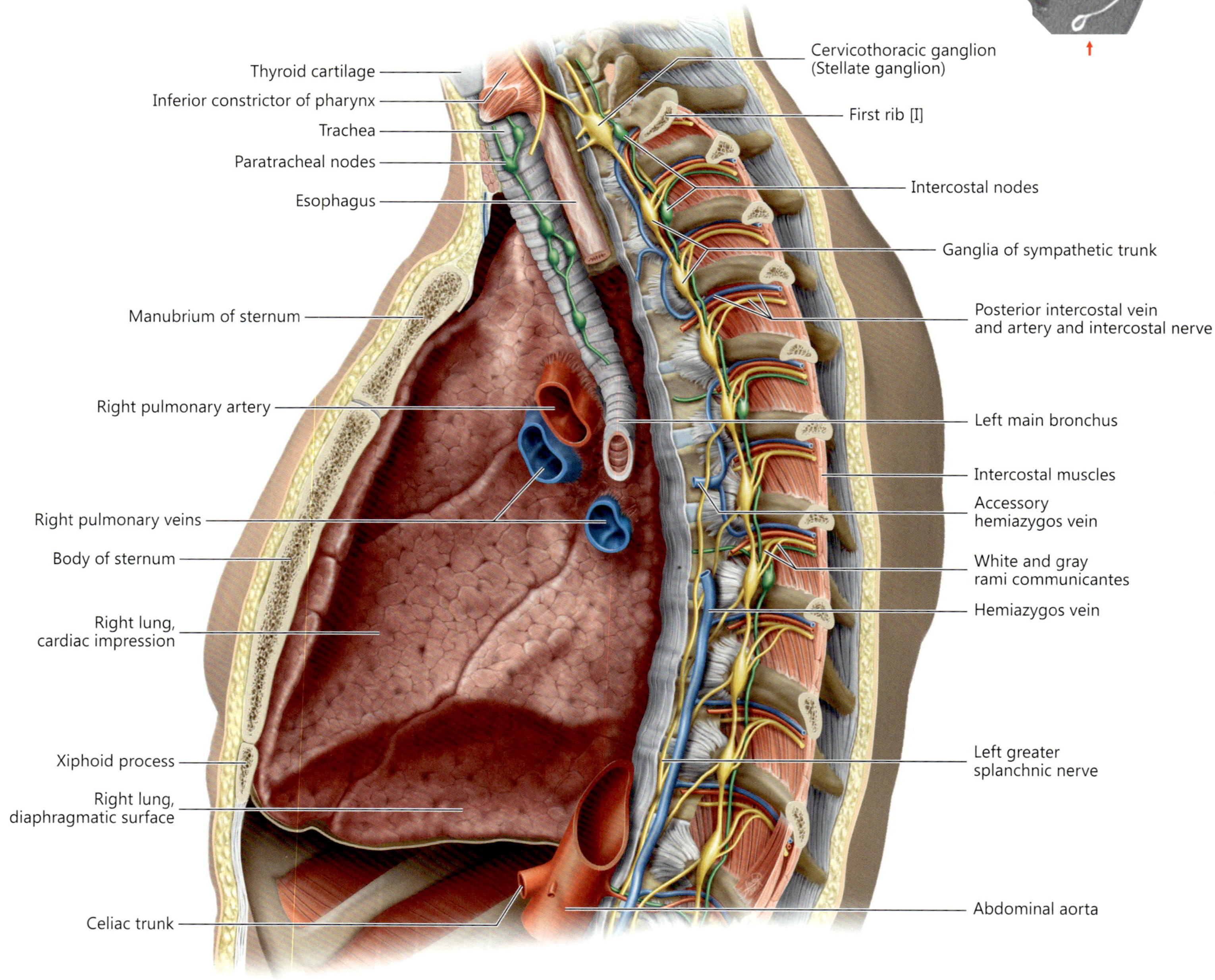

CONTENTS OF THE THORAX	
• Accessory hemiazygos vein • Esophagus • Ganglia of sympathetic trunk • Hemiazygos vein • Intercostal nodes • Left greater splanchnic nerve • Left main bronchus	• Paratracheal nodes • Posterior intercostal vein and artery and intercostal nerve • Right lung, cardiac impression • Right lung, diaphragmatic surface • Right pulmonary artery • Right pulmonary veins • Trachea

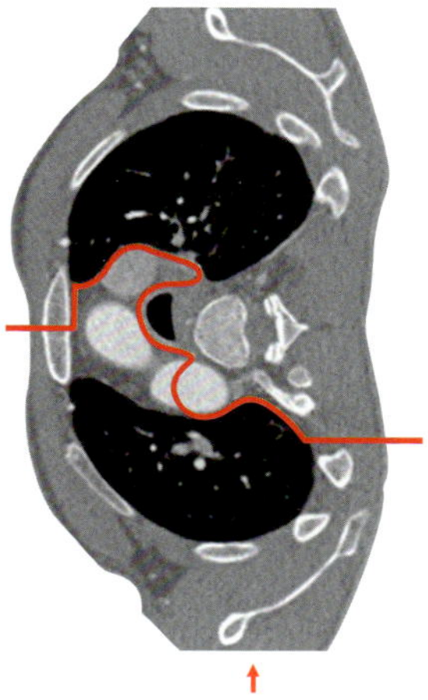

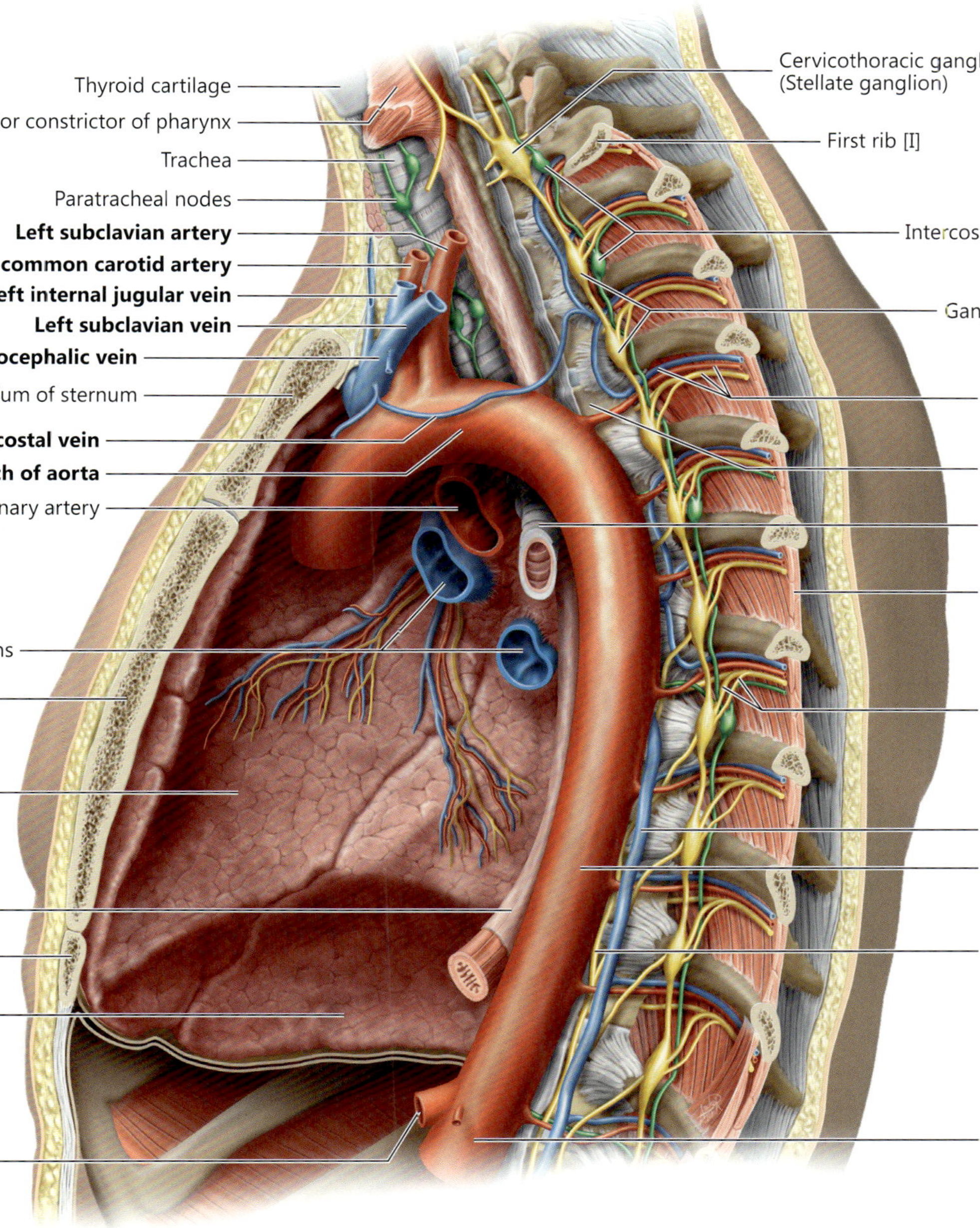

CONTENTS OF THE THORAX (in bold are the structures added with respect to the preceding table)
• Arch of aorta • Left brachiocephalic vein • Left common carotid artery • Left internal jugular vein • Left subclavian artery • Left subclavian vein • Left superior intercostal vein • Thoracic aorta

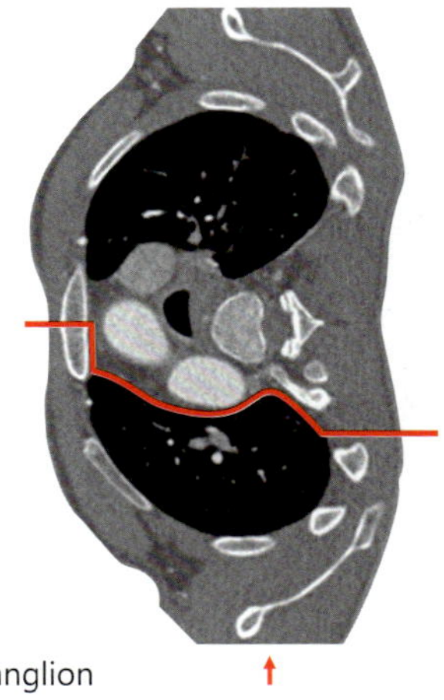

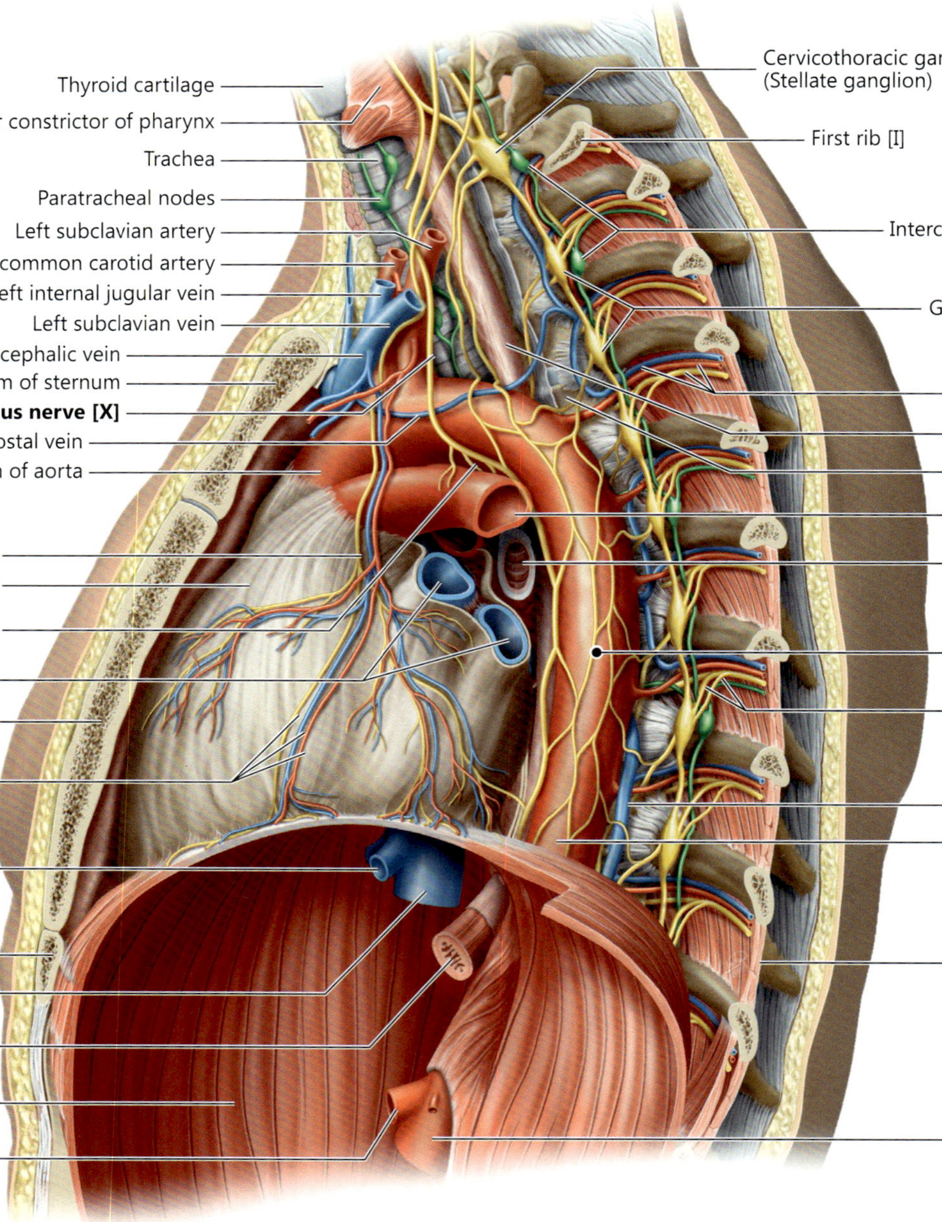

***CONTENTS OF THE THORAX** (in bold are the structures added with respect to the preceding table)*	
• Diaphragm • Fibrous pericardium • Inferior vena cava • Left phrenic nerve • Left pulmonary artery • Left pulmonary veins	• Left recurrent (or inferior) laryngeal nerve • Left vagus nerve [X] • Pericardiacophrenic artery • Pericardiacophrenic veins • Thoracic aortic plexus

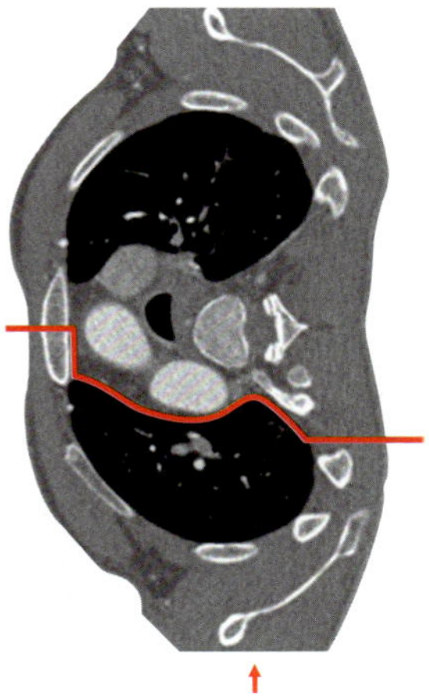

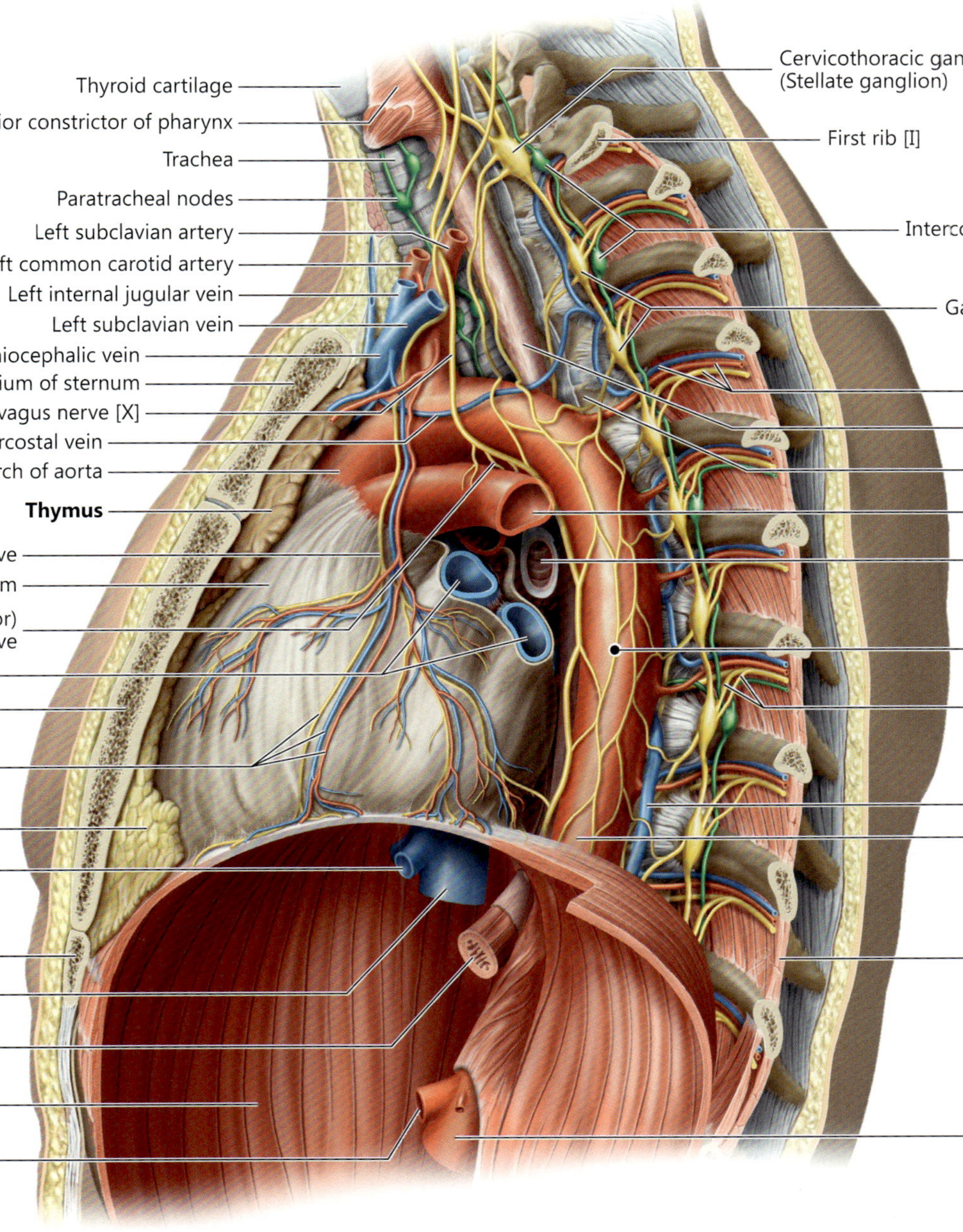

***CONTENTS OF THE THORAX** (in bold are the structures added with respect to the preceding table)*
• Retrosternal fat pad • Thymus

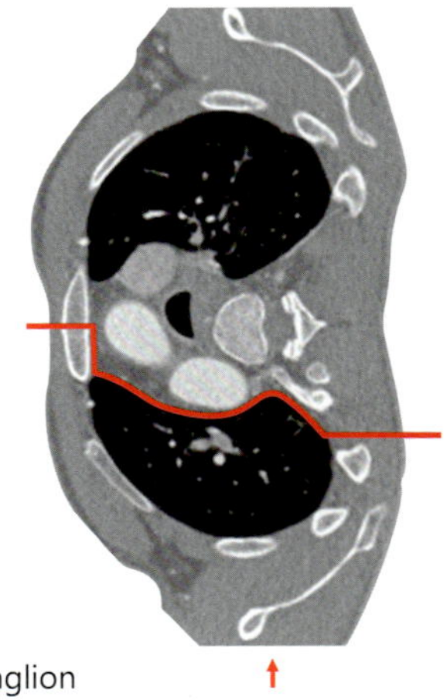

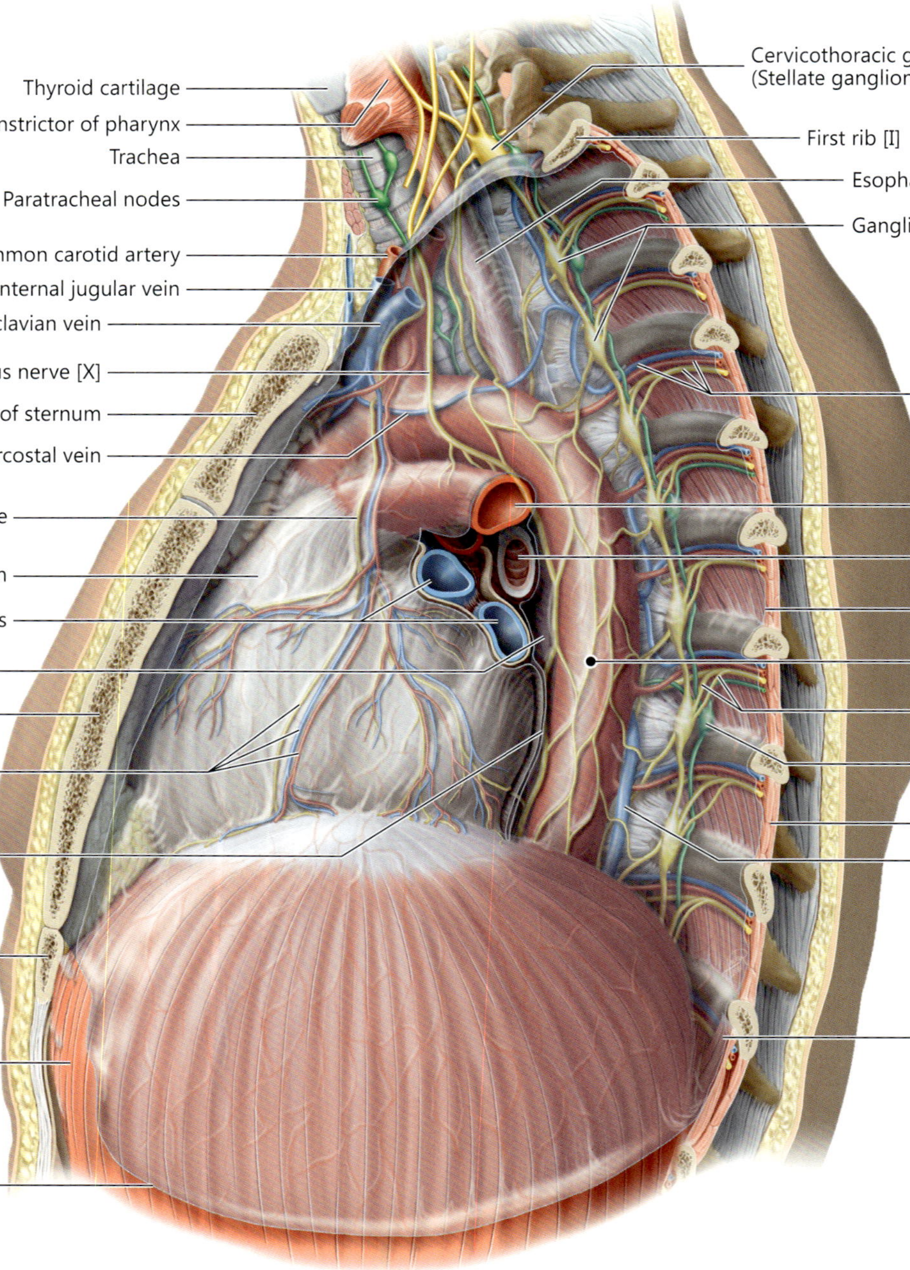

CONTENTS OF THE THORAX (in bold are the structures added with respect to the preceding table)
• Costal part of parietal pleura (costal pleura) • Diaphragmatic part of parietal pleura (diaphragmatic pleura) • Interaortoesophageal recess • Left costodiaphragmatic recess • Left pulmonary (triangular) ligament

The **esophagus** extends for approximately 25 cm, from C VI to T XI. It originates in the neck from the pharynx, at the level of the cricoid cartilage of the larynx (cervical part), and proceeds into the thorax, running in the posterior mediastinum and constituting the thoracic part, which divides in turn into an epibronchial part and a hypobronchial part. It terminates, after passing through the esophageal hiatus of the diaphragm, in the abdominal cavity at the level of the cardias (abdominal part).
The upper third of the esophagus presents also numerous striated muscle fibers, as a continuation of the inferior constrictor of the pharynx, participating in the swallowing process. In the inferior two-thirds, the muscular coat is composed exclusively of two layers of smooth muscle: an inner circular layer and an outer longitudinal layer (transected in the figure). The anterior surface of the esophagus adheres to the posterior surface of the trachea by means of bands of smooth muscle interspersed with loose connective tissue (tracheoesophageal muscle).
The esophagus does not have a sphincter at the level of the esophagogastric junction. This function is delegated to the diaphragmatic musculature at the level of the esophageal hiatus. At this site, some muscle layers of the diaphragm insert on the wall of the esophagus (phrenoesophageal muscle).

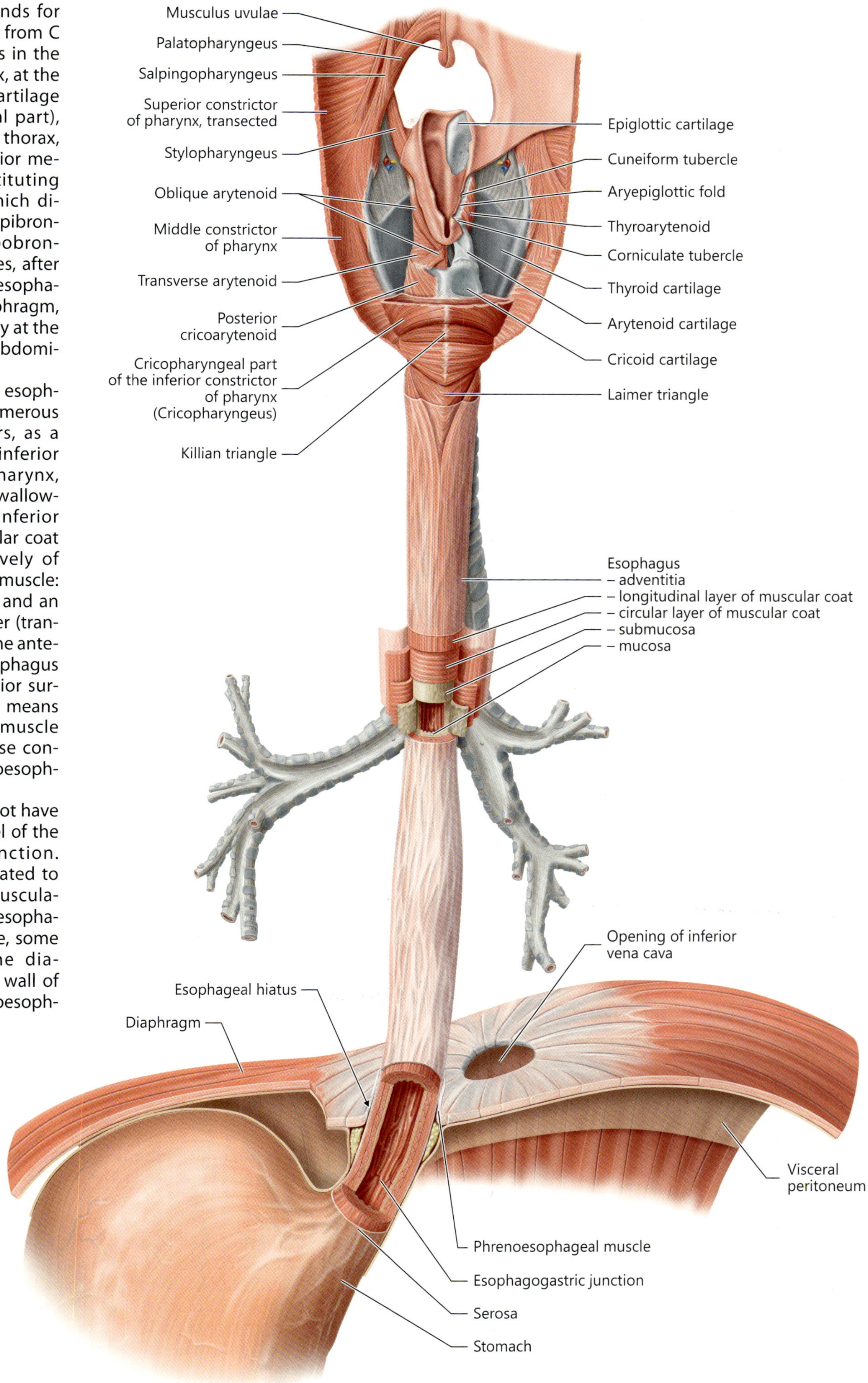

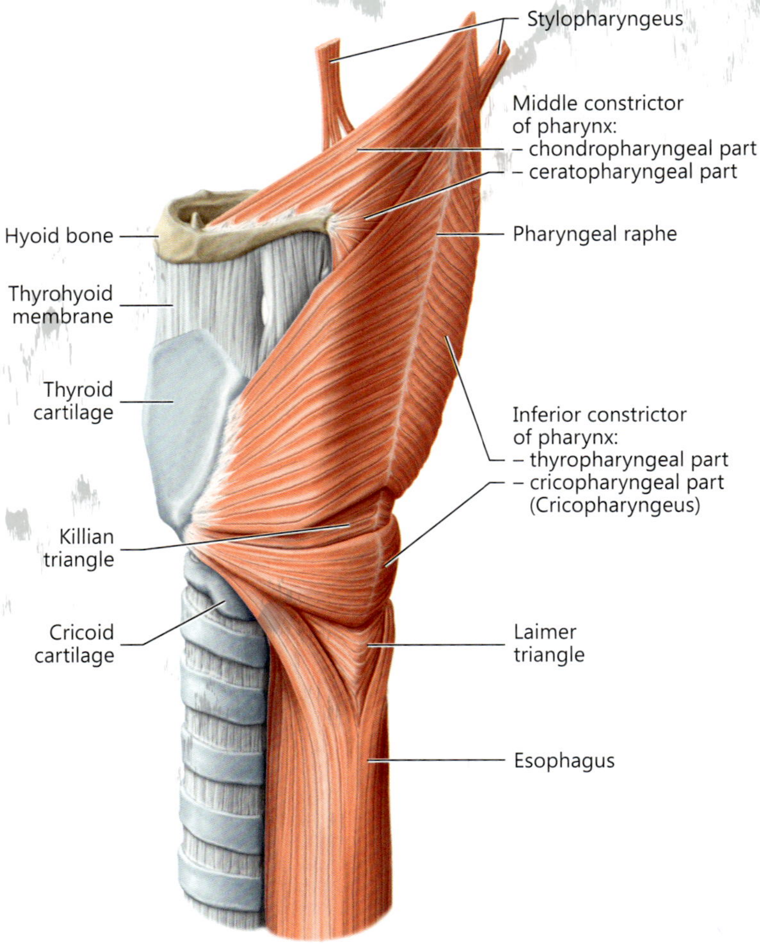

At the level of the pharyngoesophageal junction, there are two regions of minor wall resistance: the **Laimer triangle**, between the cricopharyngeal part of the inferior constrictor, above, and two layers of longitudinal muscle of the esophagus that insert on the sides of the cricoid cartilage of the larynx, below; and the **Killian triangle**, situated between the thyropharyngeal and cricopharyngeal parts of the inferior constrictor.

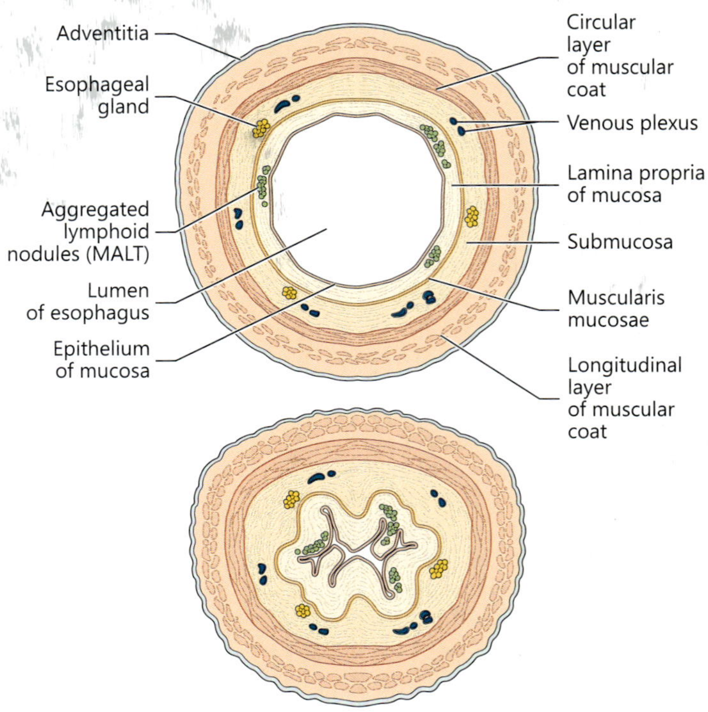

Schematic description of the **wall of the esophagus** in conditions of distension (above) and rest (below).

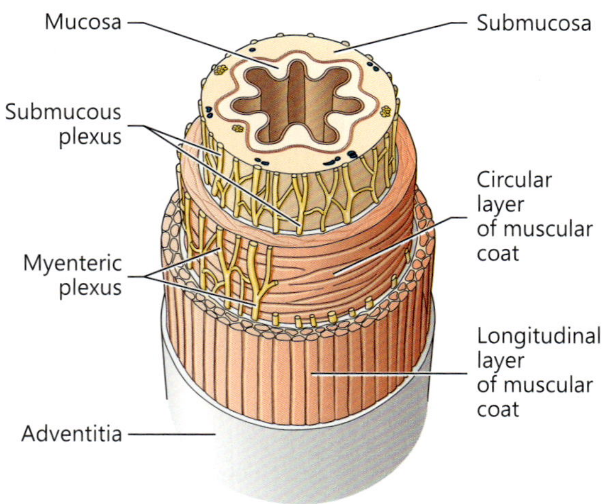

Two nerve plexuses regulate the motility of the gastrointestinal tract from the esophagus down to the rectum: the **submucous plexus** (**plexus of Meissner**), situated between the submucosa and the muscular coat, and the **myenteric plexus** (**plexus of Auerbach**) situated between the longitudinal and circular layers of muscular coat.

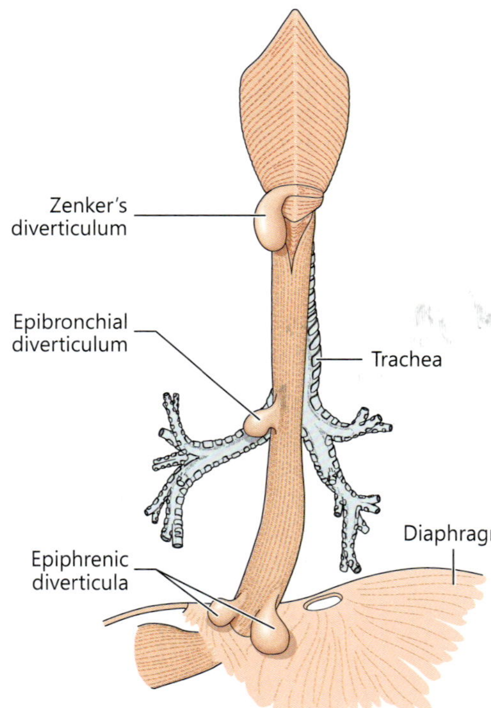

Prolapse of the submucosa and mucosa of the esophageal wall creates diverticula. The diverticula that form in the esophagus can be through pulsion or traction. To the first type belong Zenker's diverticula (pharyngoesophageal diverticula), at the level of the Killian or Laimer triangles, and epiphrenic diverticula, which form due to dyskinesia such as achalasia. Traction diverticula are due to the evaginations of the entire wall of the esophagus on account of adherences to surrounding tissues (e.g. lymph nodes). These adherences can be linked to inflammatory processes. The subsequent post-inflammatory sclerotic retraction exerts traction forming the diverticulum. A preferential site for these diverticula is at the epibronchial level.

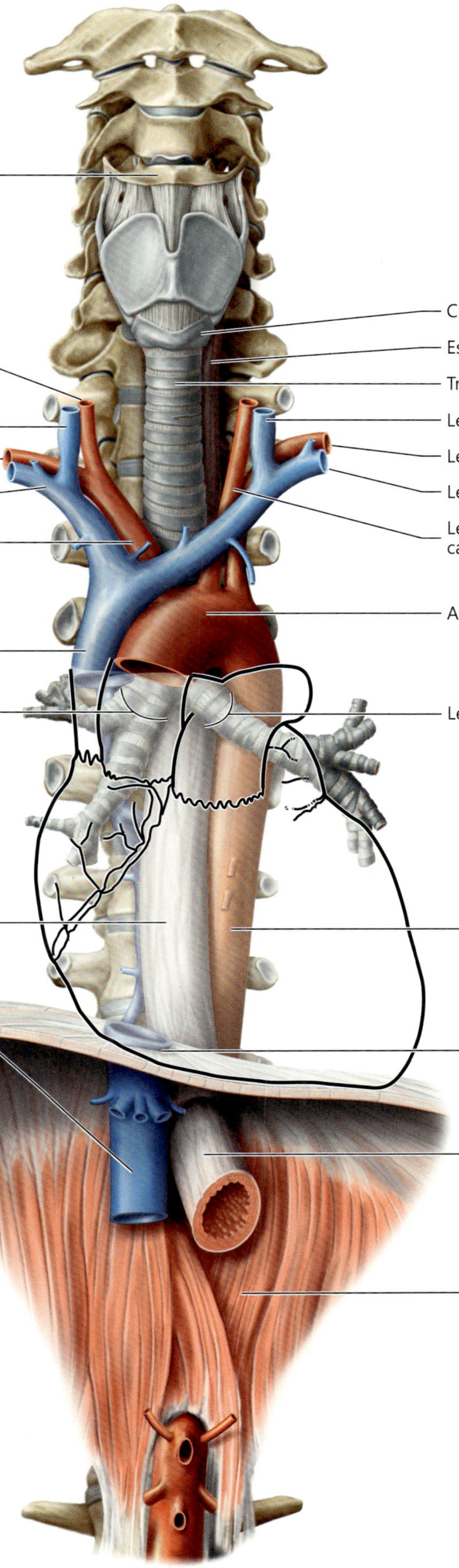

Relationship of the esophagus with contiguous structures, anterior view. At the origin of the esophagus, there is a point of constriction in proximity to the cricoid cartilage (the pharyngoesophageal constriction). In the epibronchial tract, the esophagus runs against the vertebral column in a slightly left paramedian position. The arch of the aorta, running on the left surface of the esophagus, induces a narrowing (aortic constriction) and a curvature on the frontal plane, with concavity to the right. Below the point where it crosses the arch of the aorta, the esophagus regains its paramedian position and is crossed anteriorly by the left bronchus, which induces another narrowing (bronchial constriction) adjacent to the preceding aortic constriction (clinically speaking, these are referred to as the thoracic or bronchoaortic constriction). A final esophageal constriction is found in the passage into the esophageal hiatus of the diaphragm.

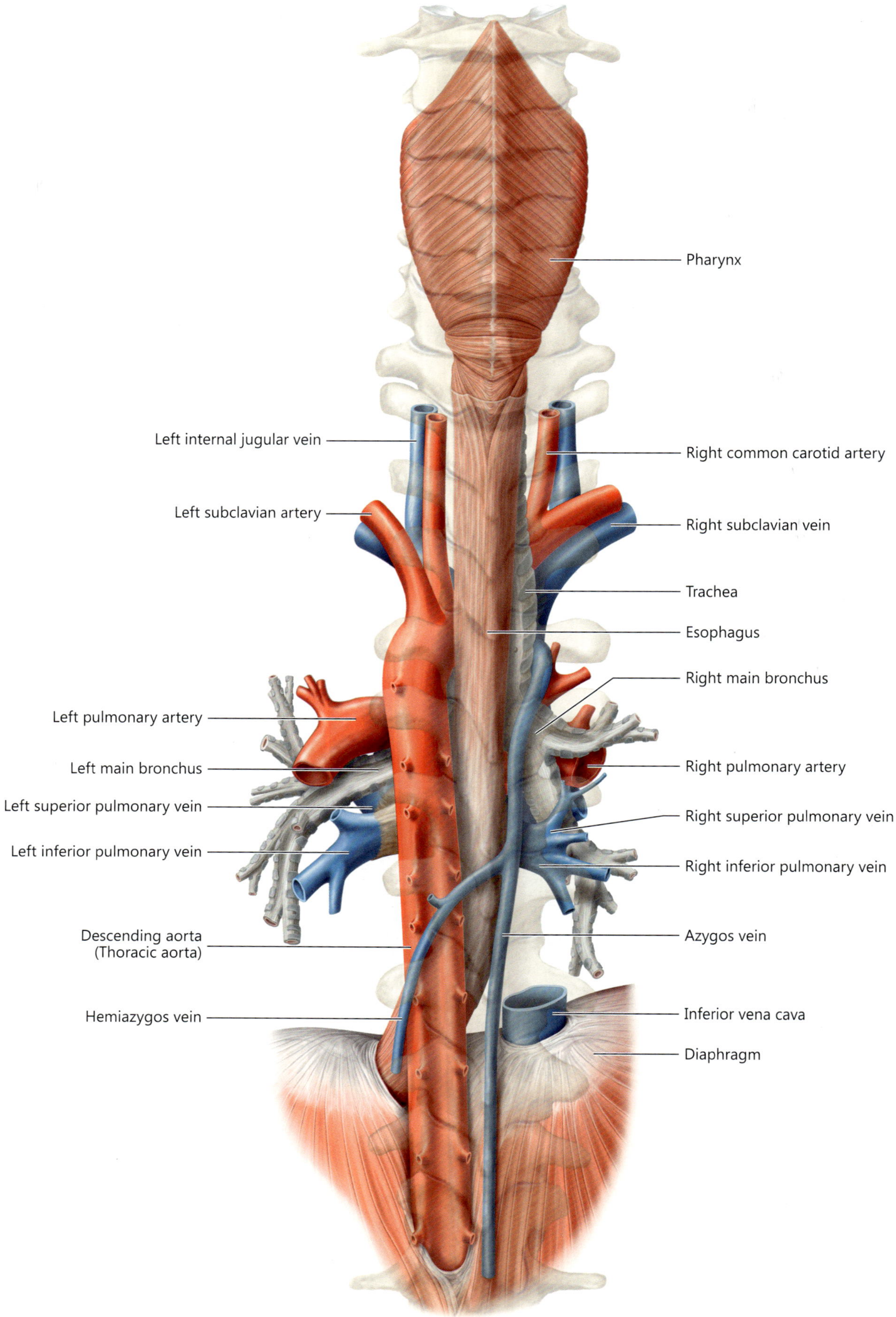

Relationships of the esophagus in the thorax, posterior view. Below the point where the esophagus is crossed by the arch of the aorta, the thoracic aorta shifts behind the esophagus and interposes itself between this and the opening of the hemiazygos vein into the azygos vein at the level of T VII-T VIII.

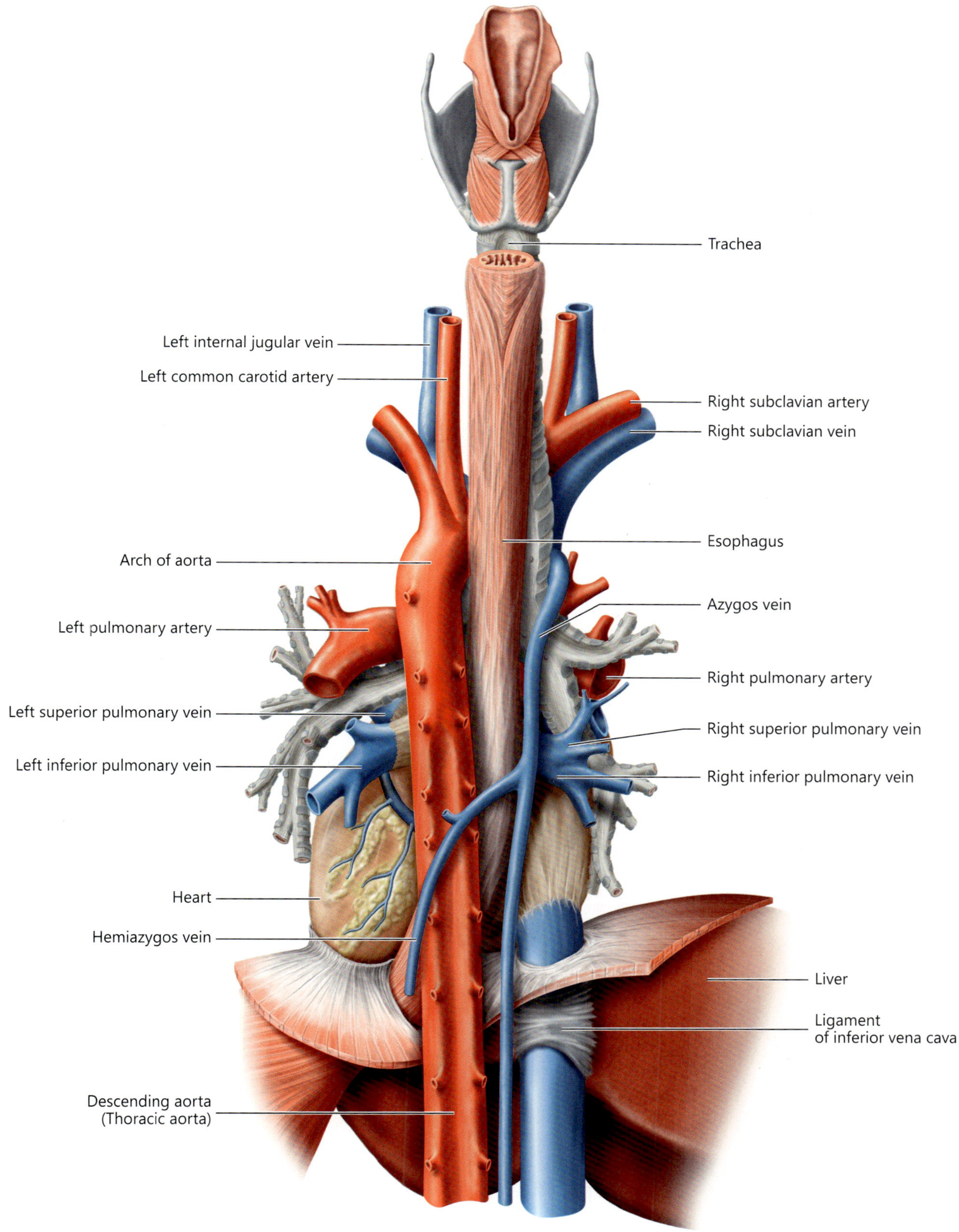

Relationship of the esophagus with the left atrium of the heart, posterior view. Resection of part of the diaphragm allows one to see the ligament of the inferior vena cava.

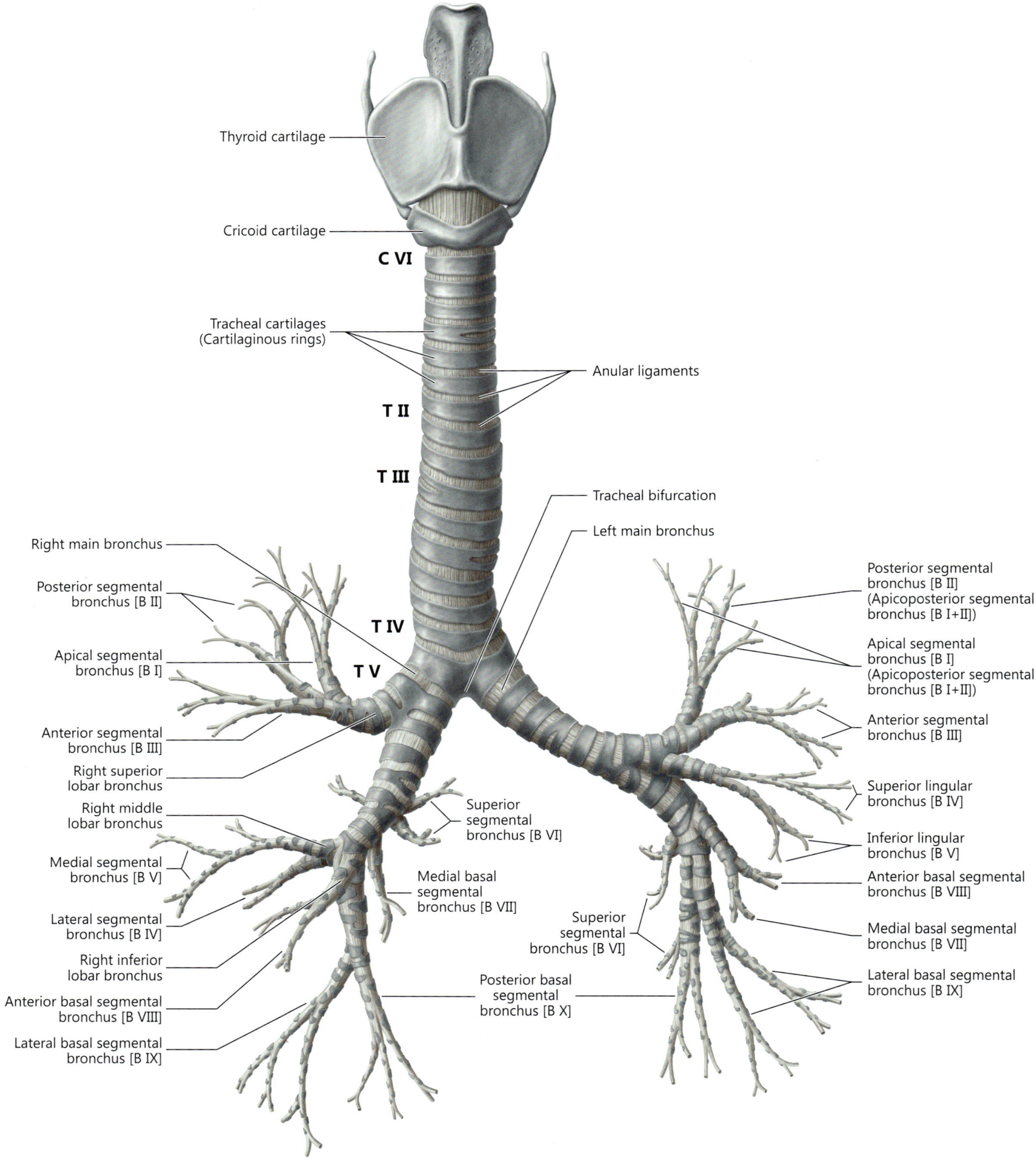

Anterior projection of the **trachea** and the bronchial subdivisions. The trachea extends from the cricoid cartilage of the larynx, at the level of C VI, to the superior border of T V. At this level it divides into the two main bronchi. The right main bronchus has a more vertical path than the left.

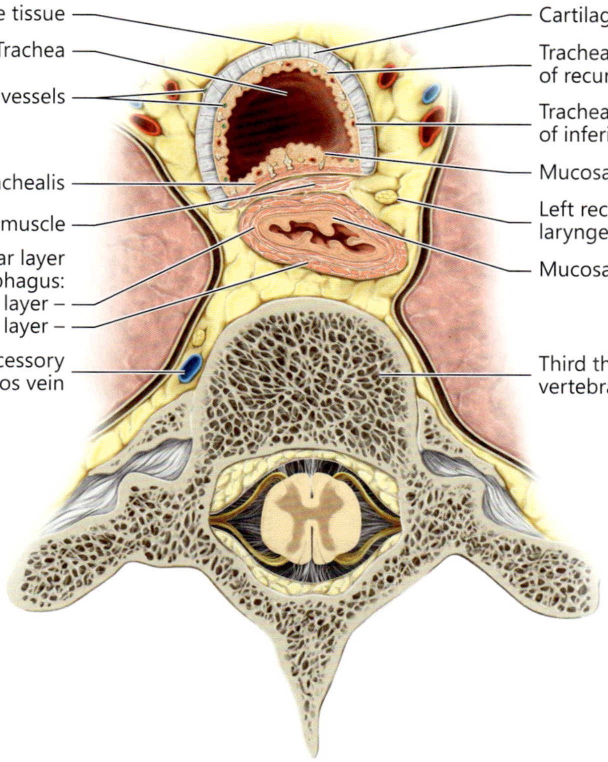

Transverse section of the posterior mediastinum at the level of T III, highlighting the **relationship between trachea and esophagus** (tracheoesophageal muscle).

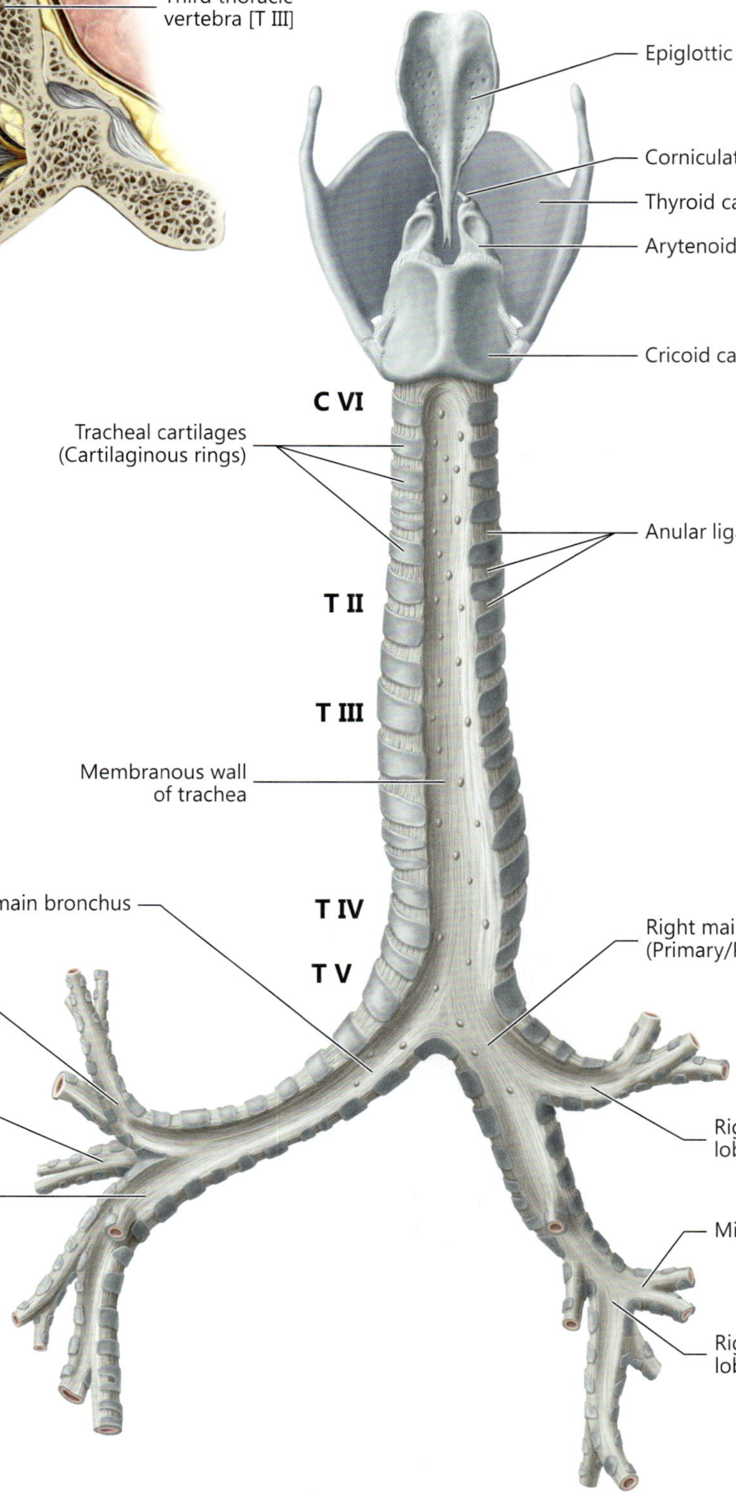

Posterior projection of the **trachea**. The tracheal rings are incomplete posteriorly and the continuity of the wall is ensured by the membranous part of the trachea.

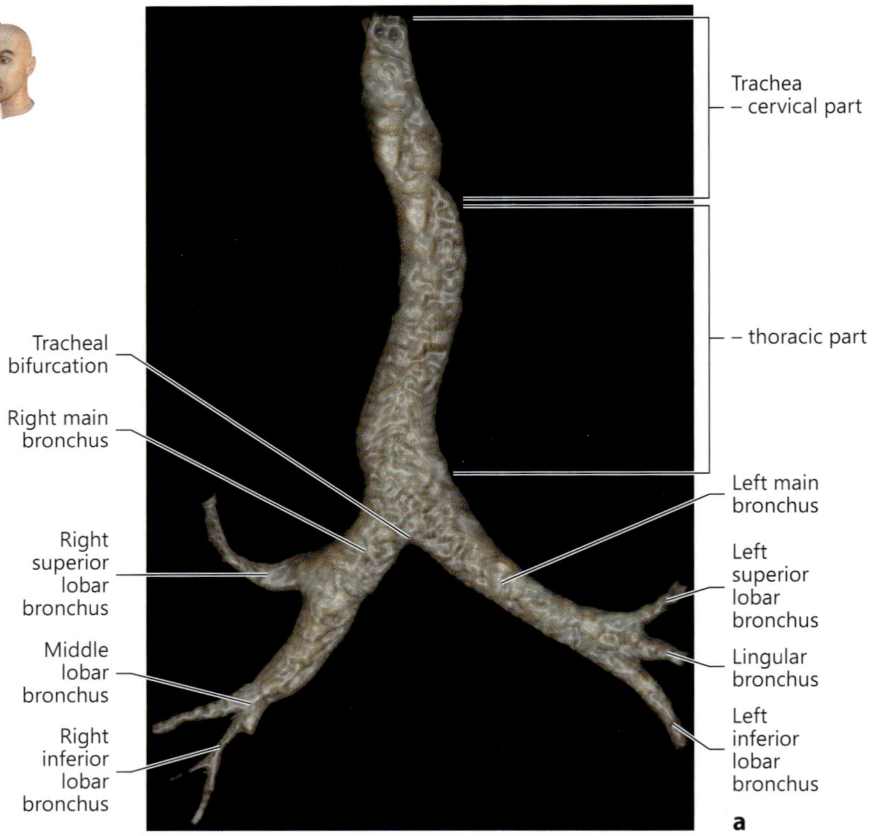

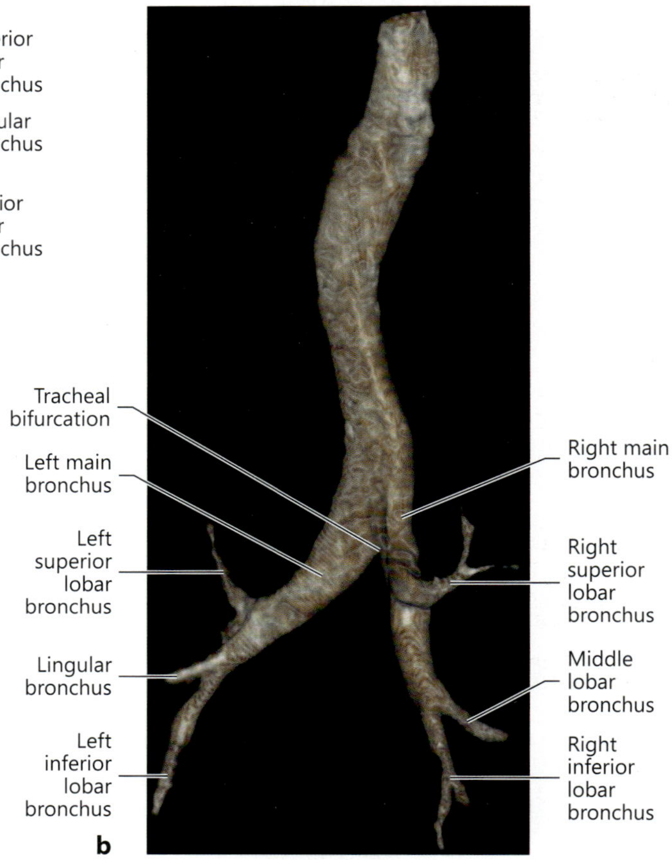

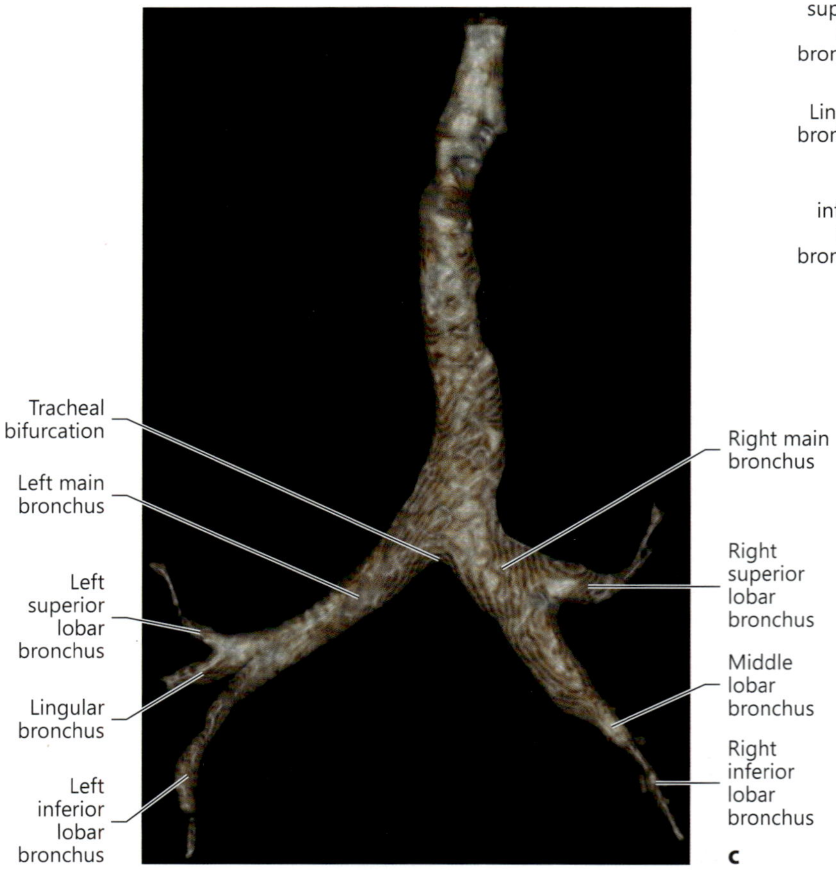

Three-dimensional image of a living human body, obtained by CT scan using a 3D volume rendering technique, showing the **trachea** and first bronchial branches. **a**, Anterior view. **b**, Right lateral view. **c**, Posterior view.

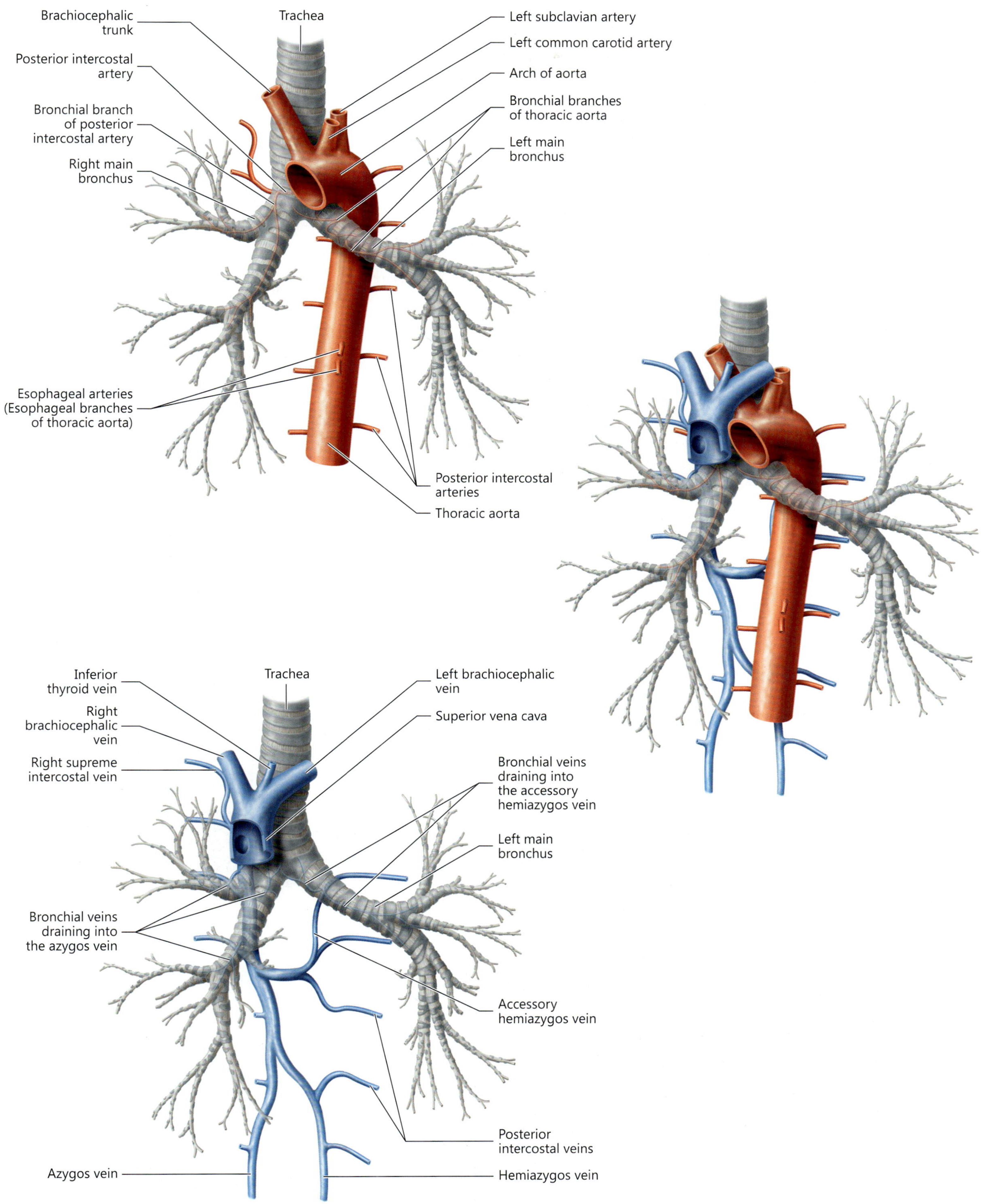

Frontal schematic view showing the relations between the principal **vessels of the large systemic circulation** and the **trachea**.

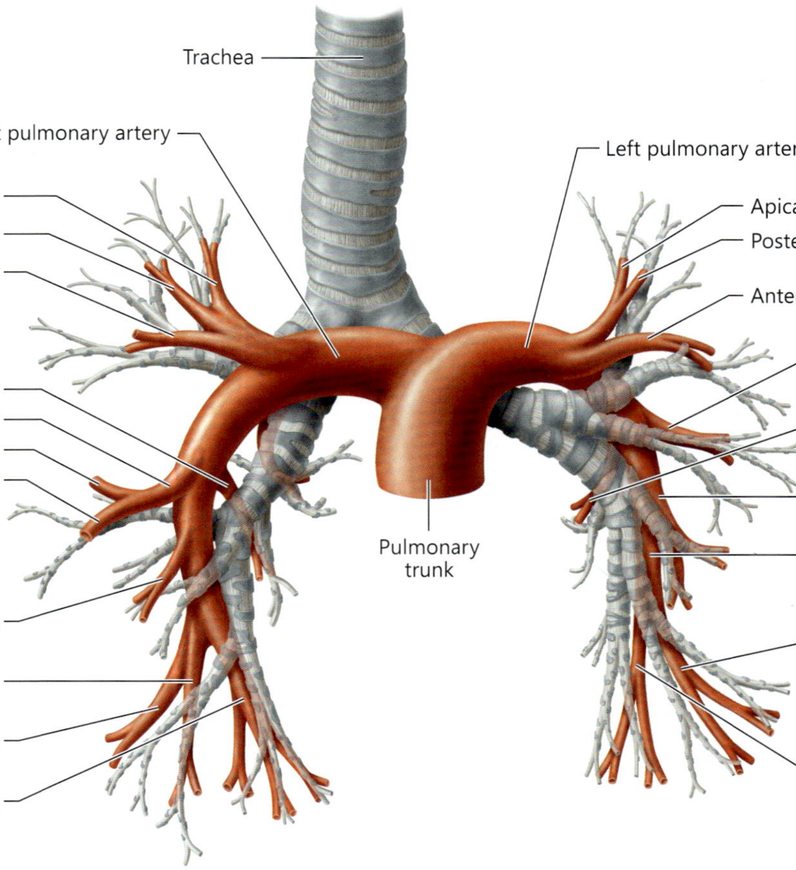

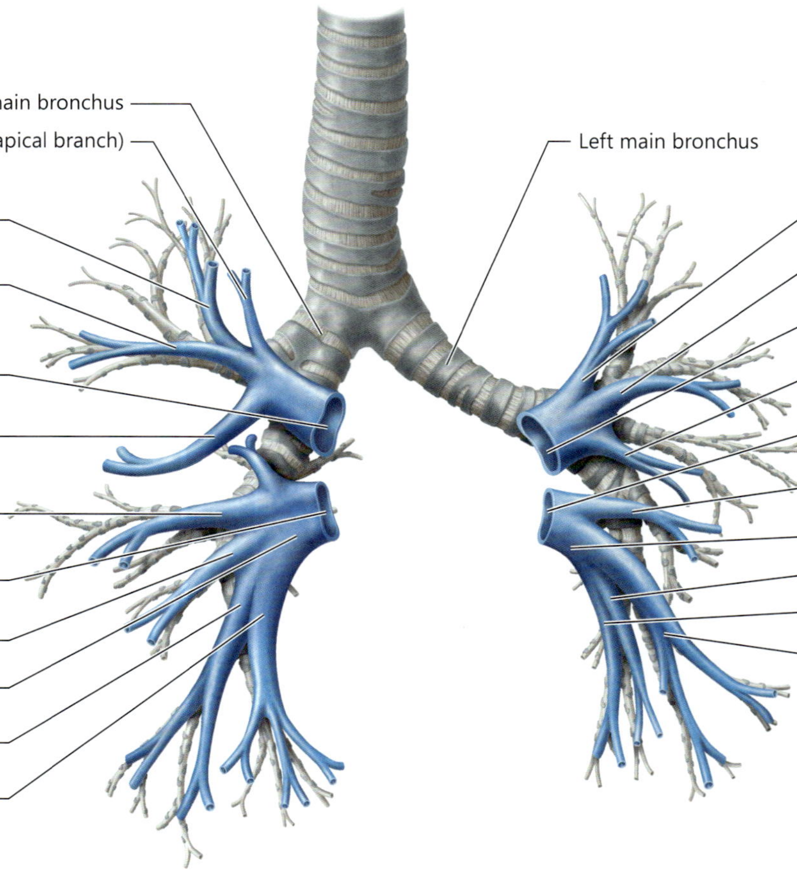

Frontal schematic view showing the relations between the principal vessels of the **pulmonary circulation** and the **trachea**. Above, pulmonary trunk and arteries. Below, pulmonary veins.

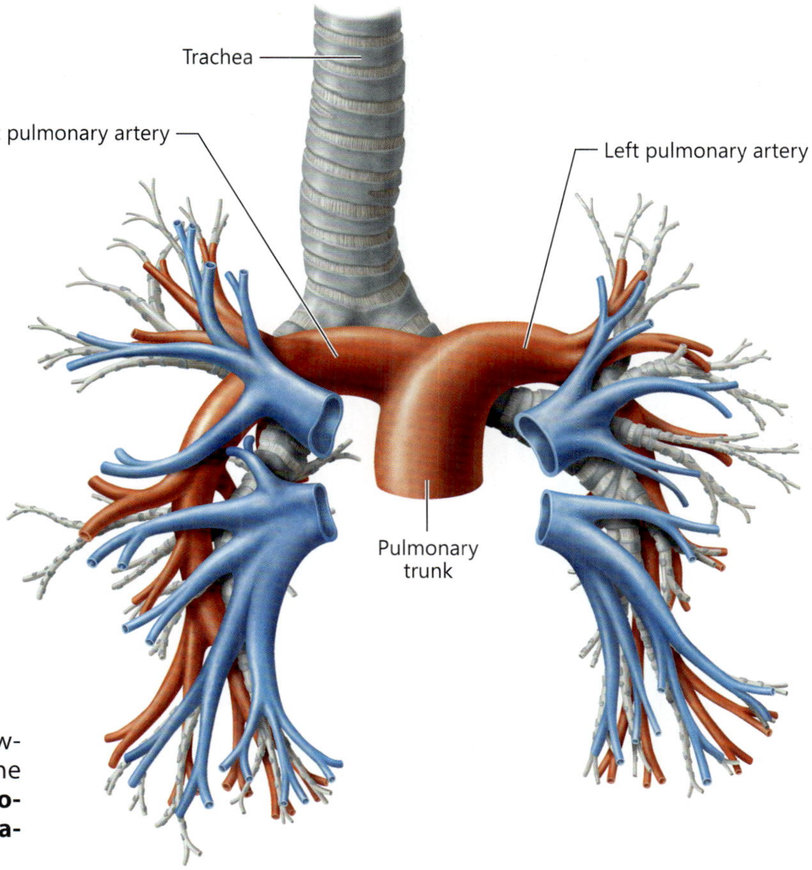

Frontal schematic view showing the relations between the principal vessels of the **pulmonary circulation** and the **trachea**.

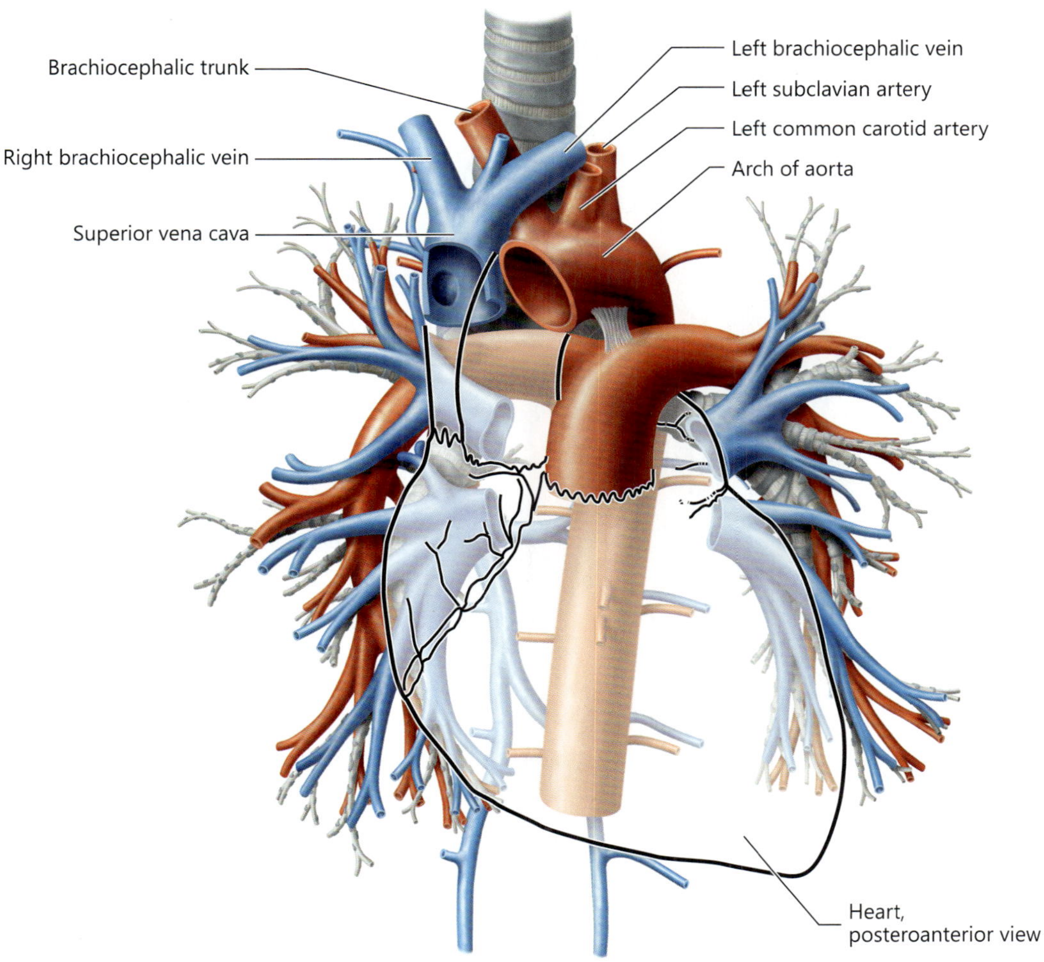

Frontal schematic view showing the relations between the principal vessels of the **large systemic circulation** and **pulmonary circulation**, and the **trachea**. The heart is drawn in transparency to highlight its relationship with the great vessels.

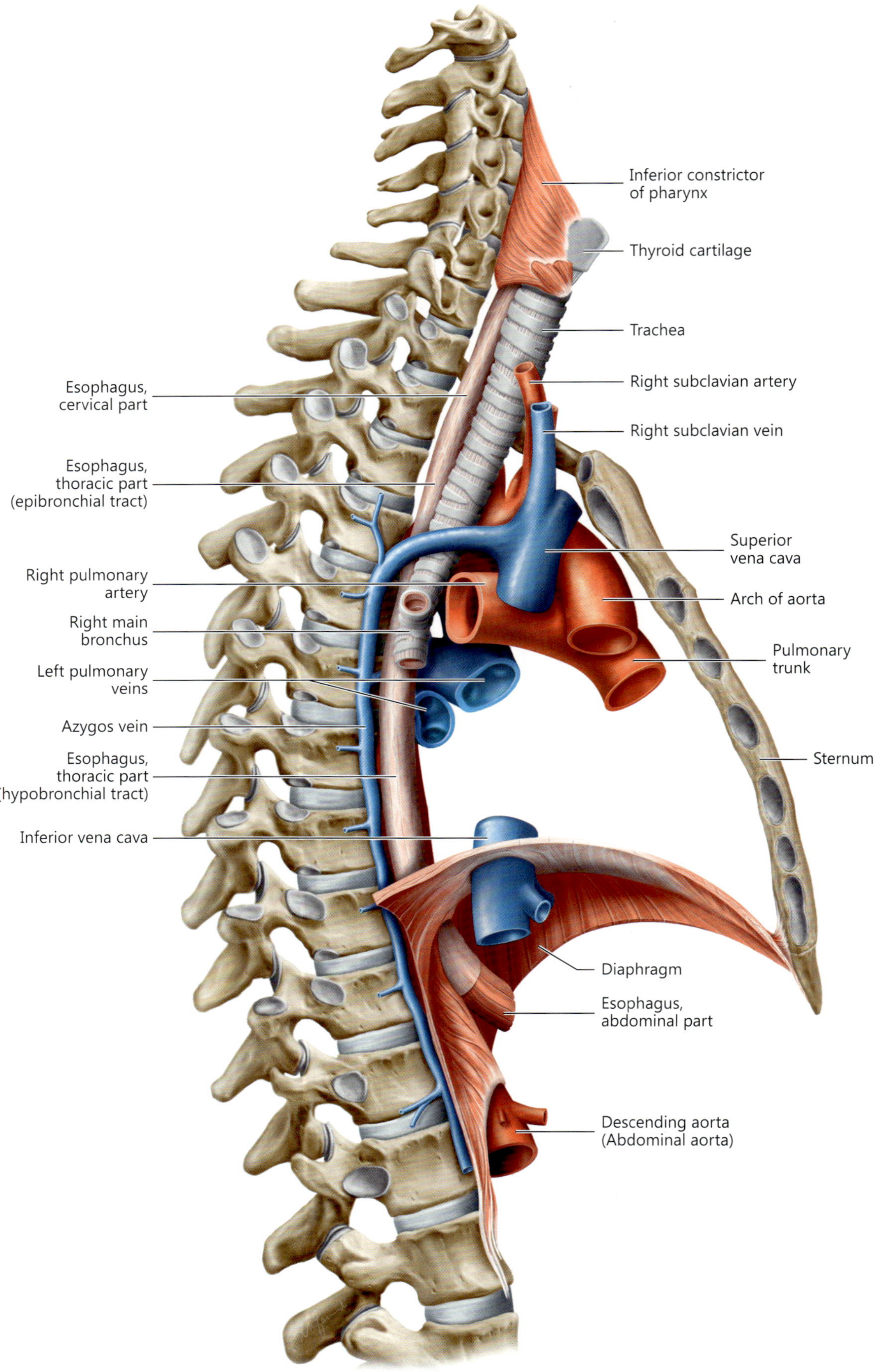

Schematic view showing the great vessels of the **large systemic circulation** and **pulmonary circulation** in relation to the **trachea** and **esophagus**. Right lateral projection.

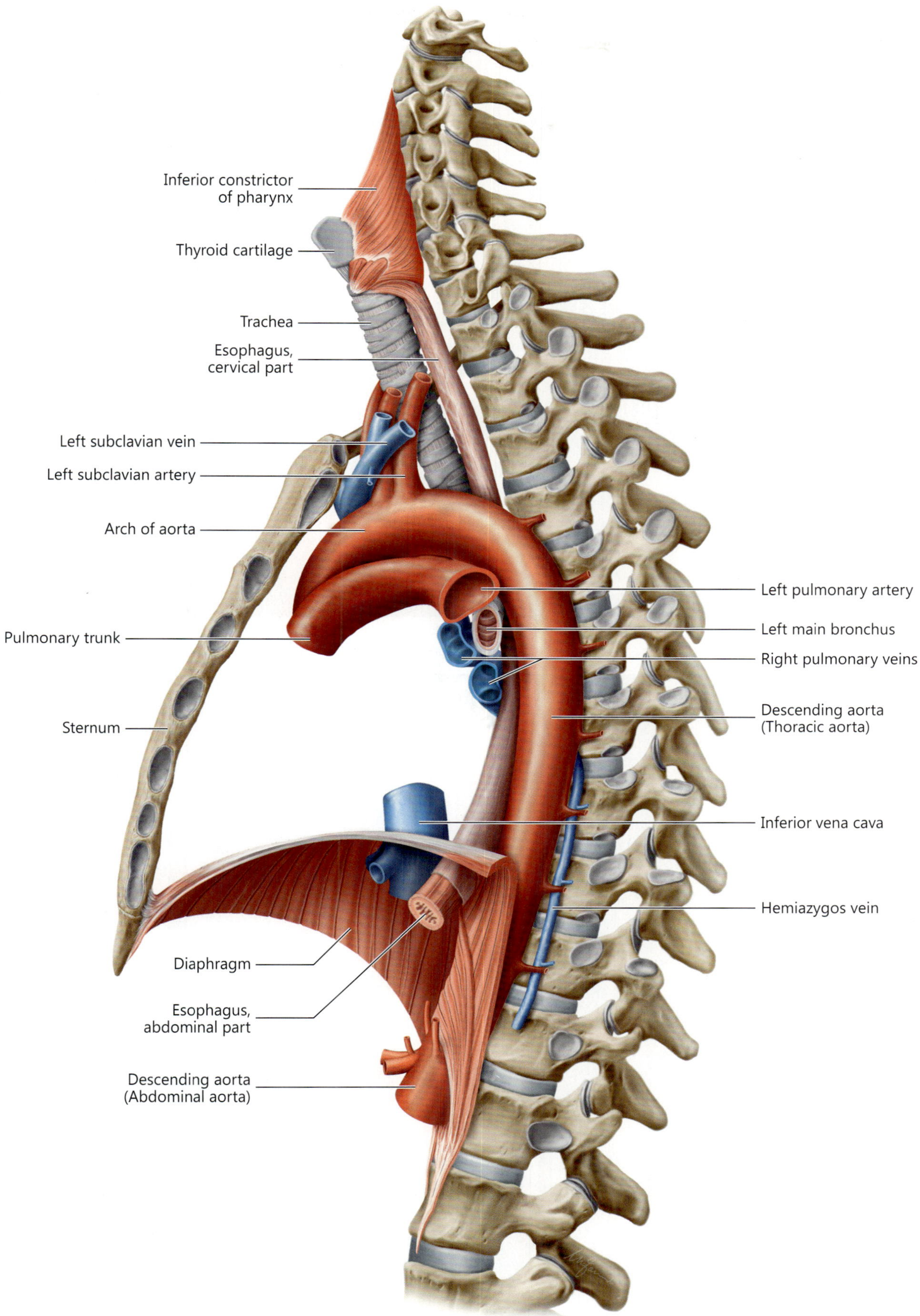

Schematic view showing the great vessels of the **large systemic circulation** and **pulmonary circulation** in relation to the **trachea** and **esophagus**. Left lateral projection.

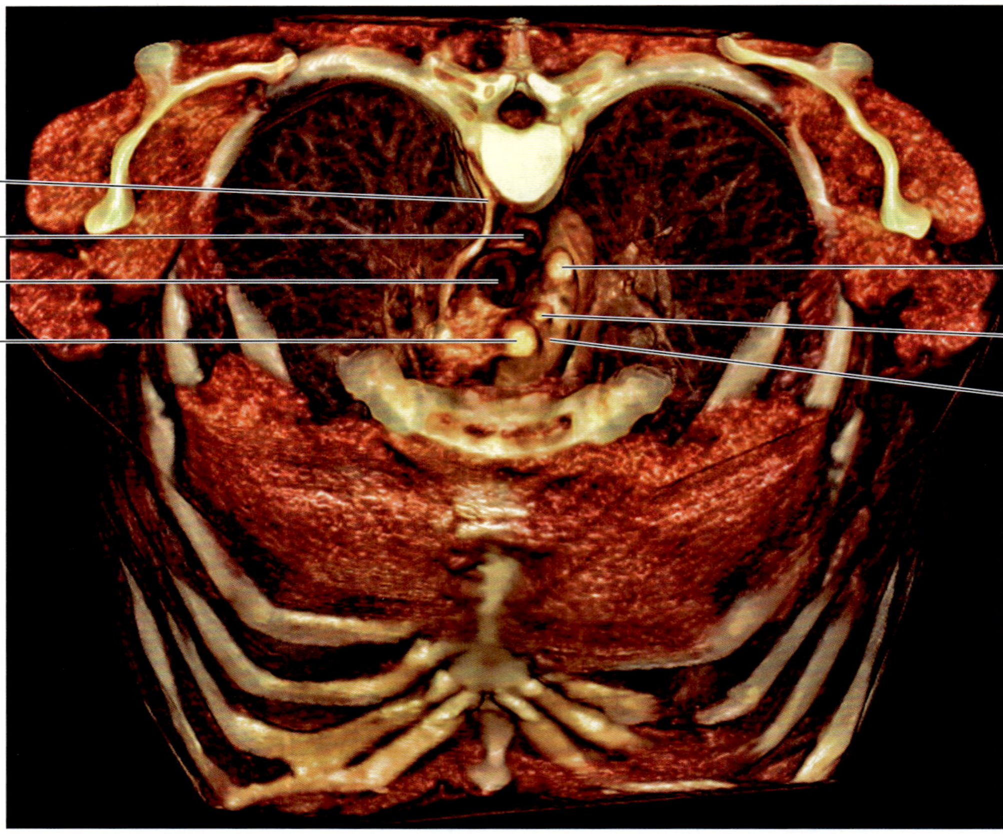

a

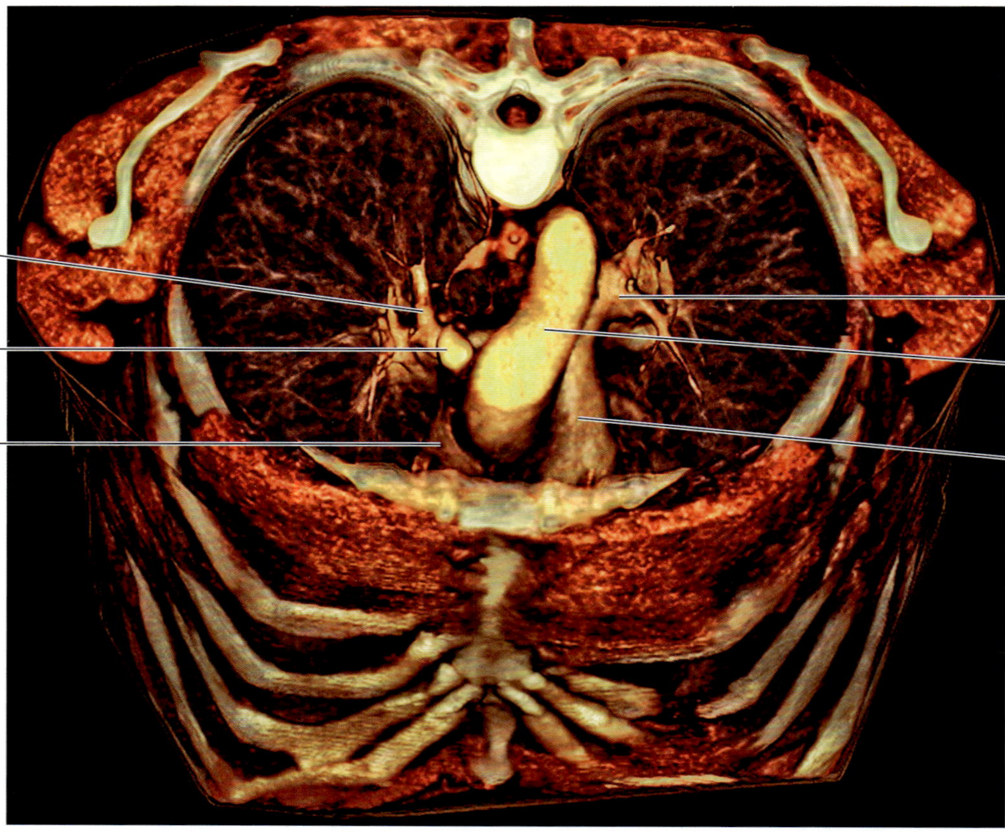

b

Series of four three-dimensional images of a living human body, obtained by CT scan using a 3D volume rendering technique, taken from transverse sections, in craniocaudal direction.

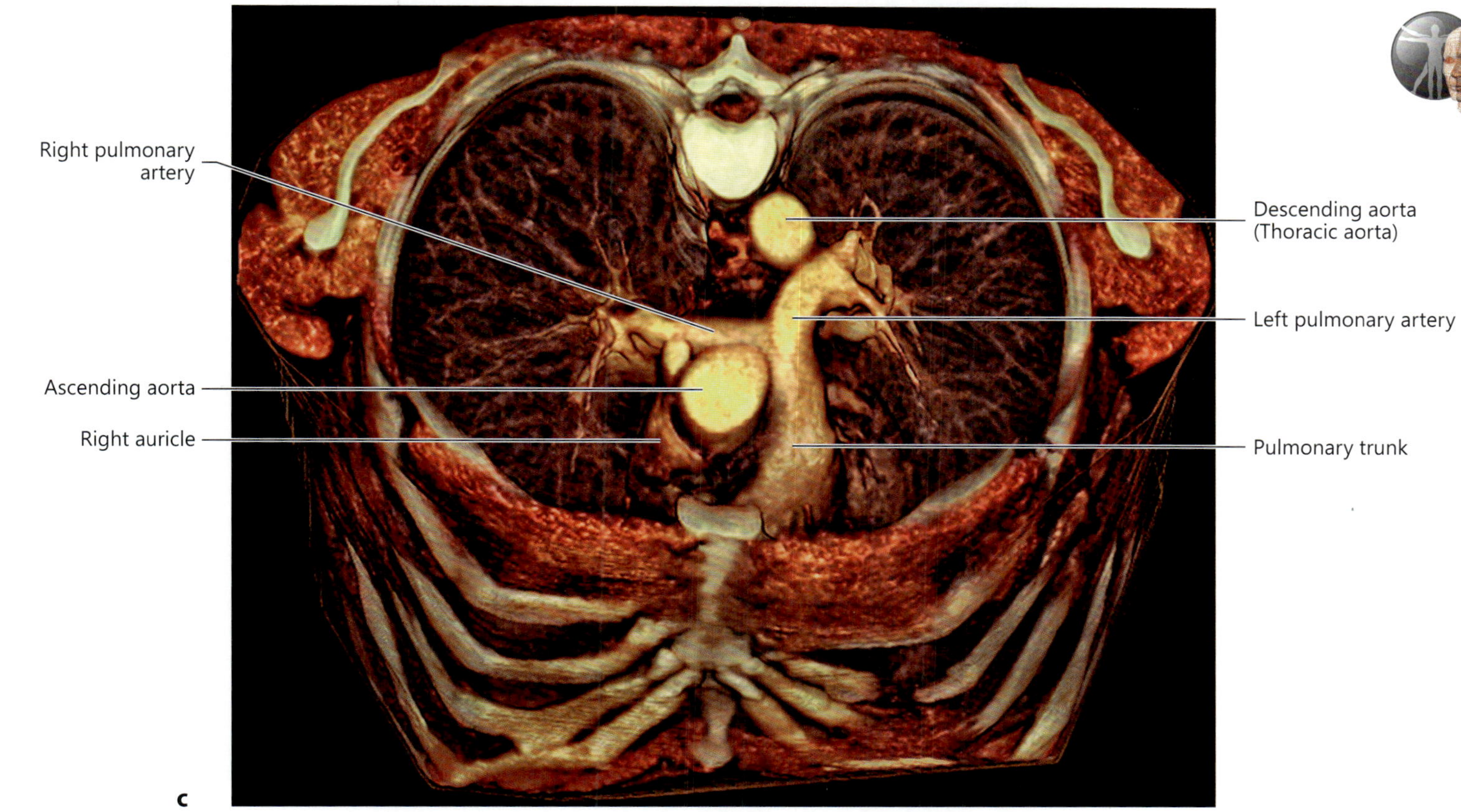

c

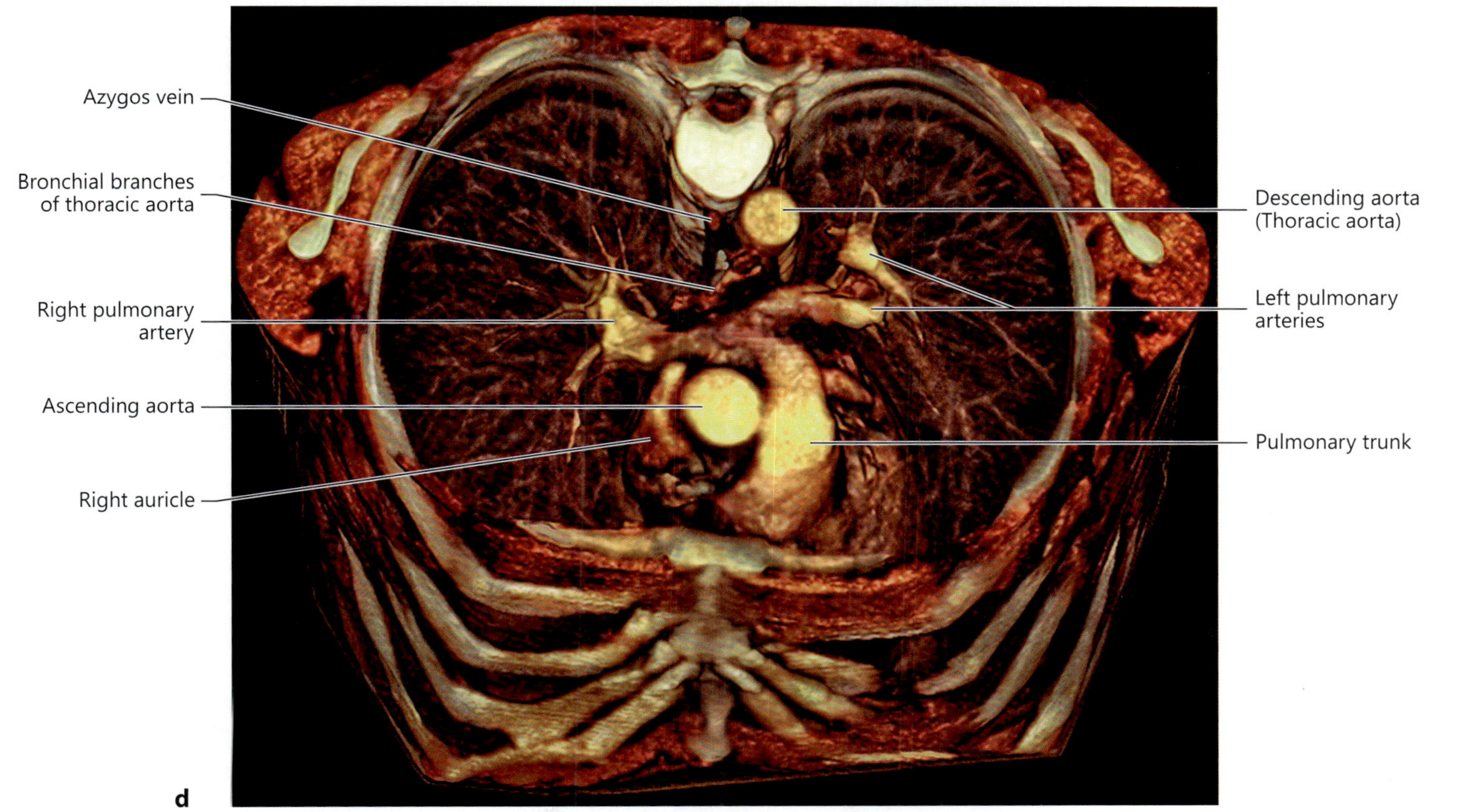

d

The sequence starts from the origin of the **collateral branches of the arch of the aorta** (**a**) and terminates at the origin of the **ascending aorta** from the aortic valve, in front, and the origin of the descending aorta (thoracic aorta), behind (**d**). The sequence shows the oblique orientation from front to back and from right to left of the arch (**b**), from its origin (right of the anterior median, or midsternal, line at the level of the 2nd sternocostal joint), to its termination on the left surface of T IV/T V (**c**). The arch of the aorta passes over the right pulmonary artery (**c**, **d**).

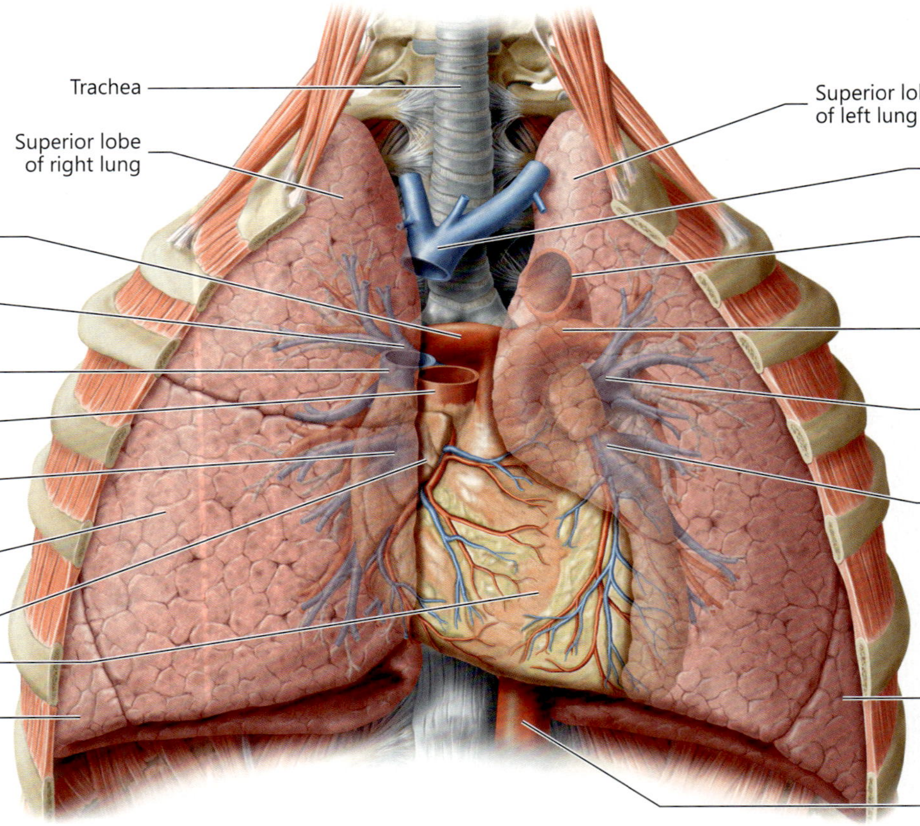

Position of the **heart in the middle mediastinum**. The lungs in transparency permit one to see the boundaries of the anterior surface of the heart (sternocostal surface). The apex is facing down and to the left. The heart, supported on the central tendon of the diaphragm, occupies most of the middle mediastinum, enveloped by the pericardial sac (not shown). The heart is situated between the 3rd and 6th costal cartilages, in front, and T V-T VIII, behind. Two-thirds of it extend to the left of the anterior median (midsternal) line, and one-third to its right. The anterior surface is partially covered by the anterior borders of the lungs and by the costomediastinal pleural recesses (not shown). The superior mediastinum contains the great vessels of the root of the heart.

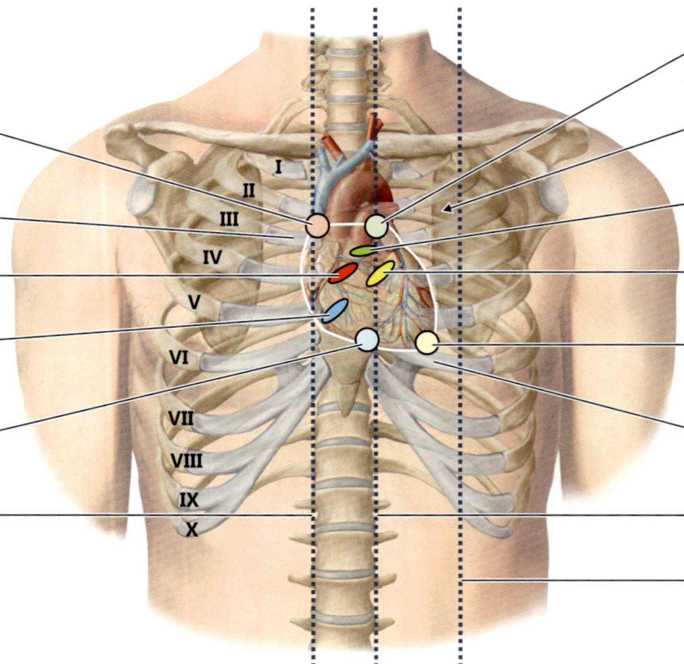

Boundaries of the projection of the heart on the thoracic cage (**surface projection of the heart**), with the projection of the heart valves and relative auscultation sites.

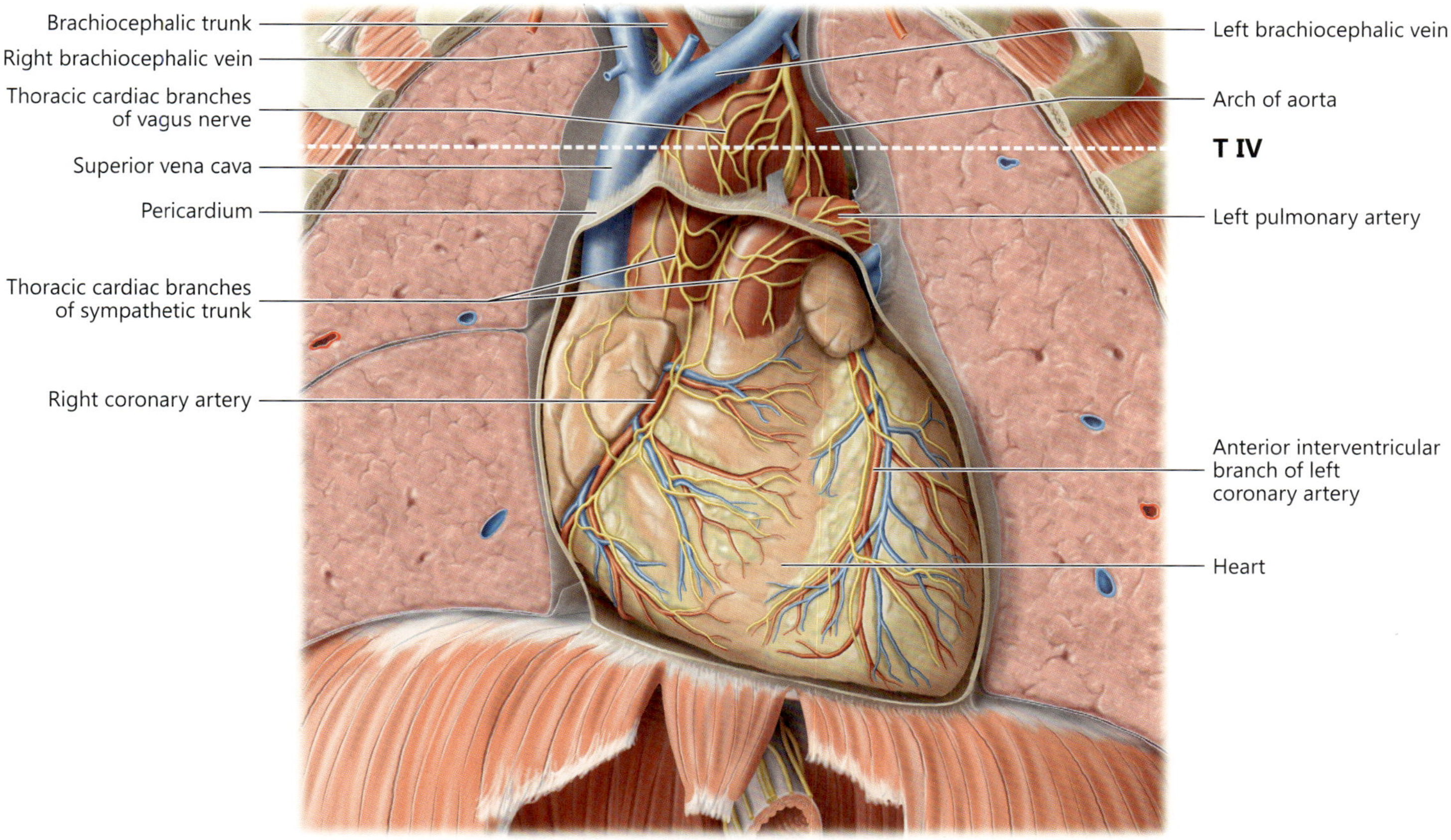

Position of the heart in the **middle mediastinum**. A close-up of the image on the page alongside.

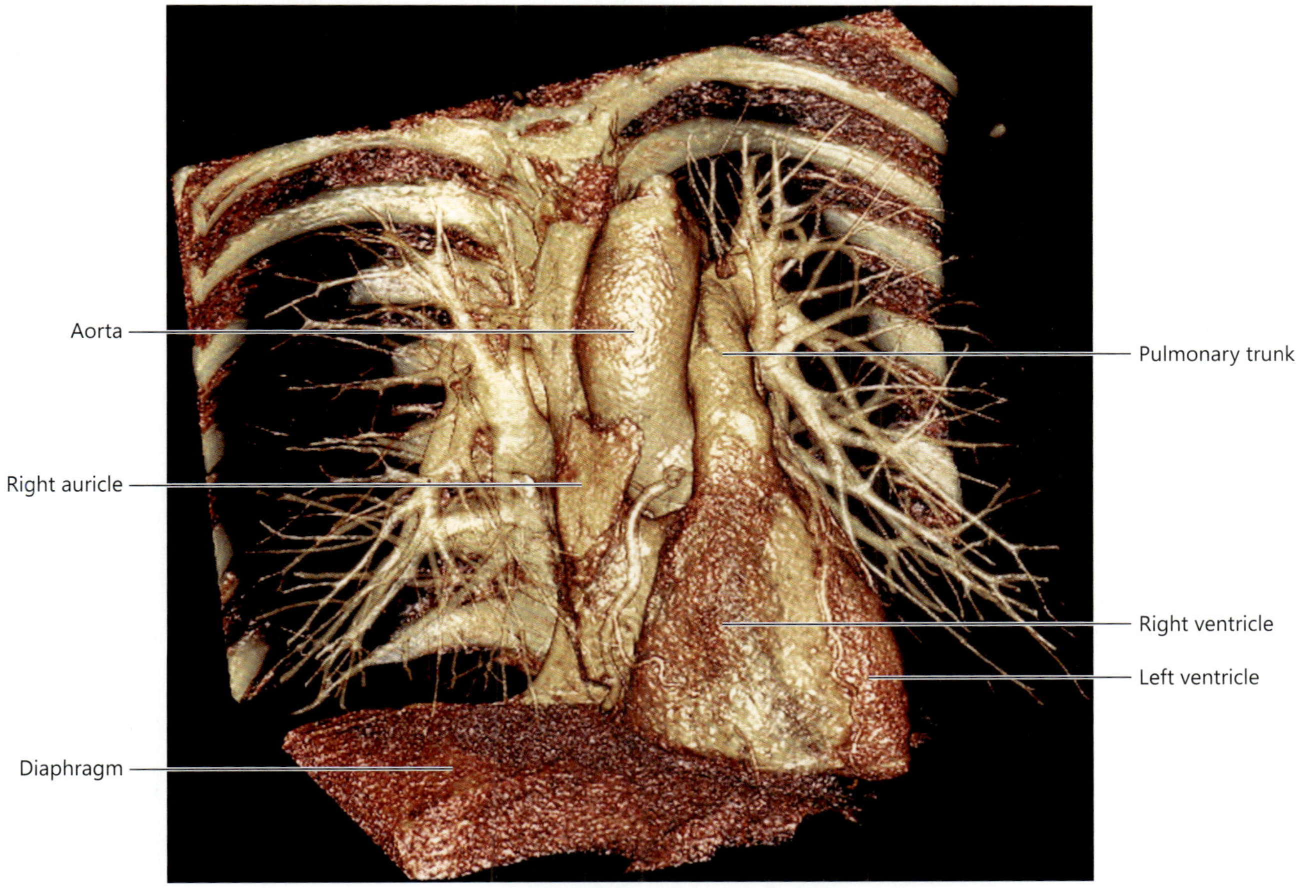

Three-dimensional image of a living human body, obtained by CT scan using a 3D volume rendering technique, showing the **heart**. One can see the ascending aorta, pulmonary trunk, bronchial tree and costal margin.

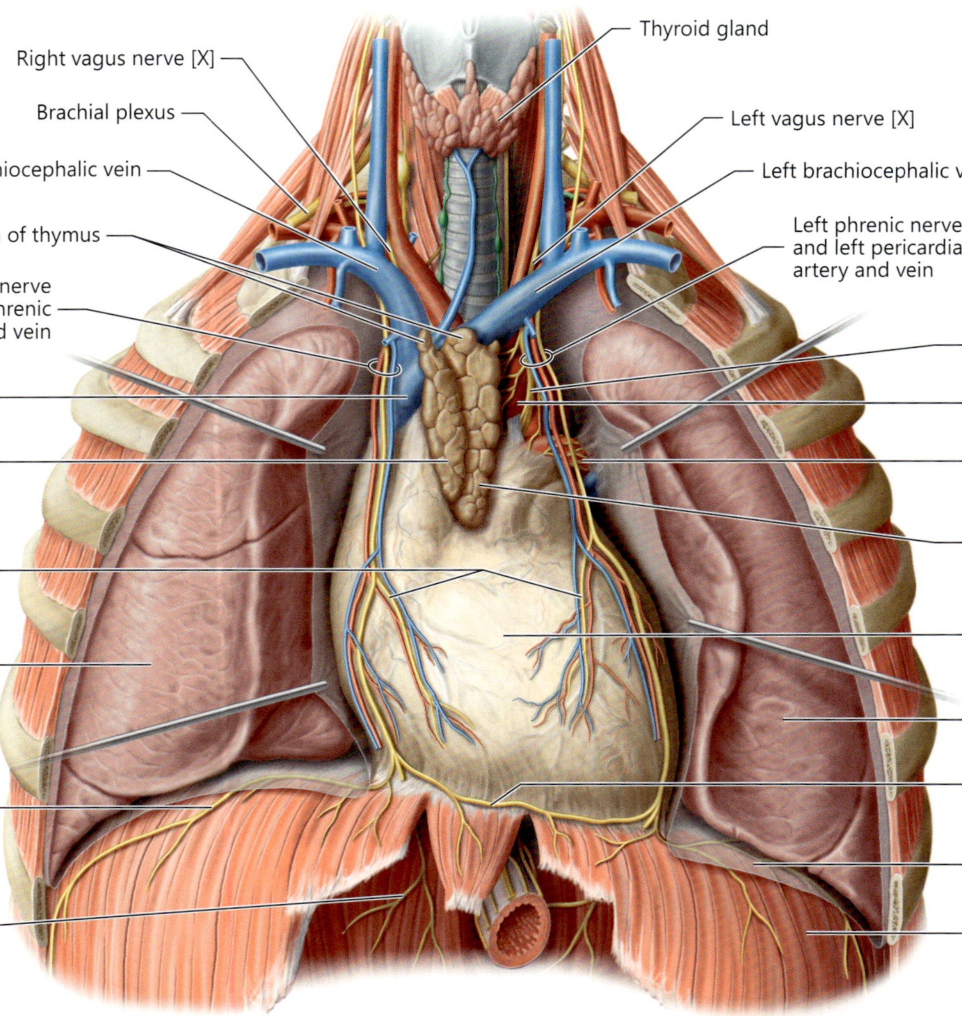

Position of the heart in the **middle mediastinum**. Pinning back the anterior borders of the lung and resecting the costomediastinal pleural recesses, one can see the pericardial sac and, between this and the mediastinal pleura on both sides, the pericardiacophrenic vessels and phrenic nerve. Situated in front are the middle and anterior mediastinal node chains. Before adult age, the point where the great vessels of the heart originate is in contact, in front, with the thymus. Successively, the thymus is replaced by fibroadipose tissue.

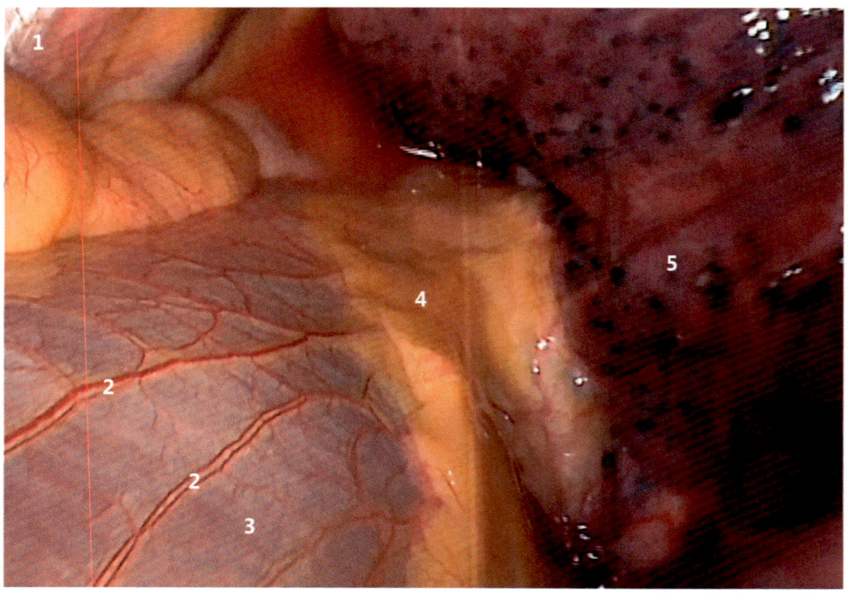

Image of the internal wall of the **thoracic cavity** obtained in the course of a diagnostic thoracoscopy. The endoscope is placed between the parietal and the visceral layers of the pleura. The view is from above. In the image, the lung has been moved aside and one can see the pericardium with the pericardial vessels.

1 Diaphragm
2 Pericardial branches of thoracic aorta
3 Pericardium
4 Pericardial fat
5 Lung

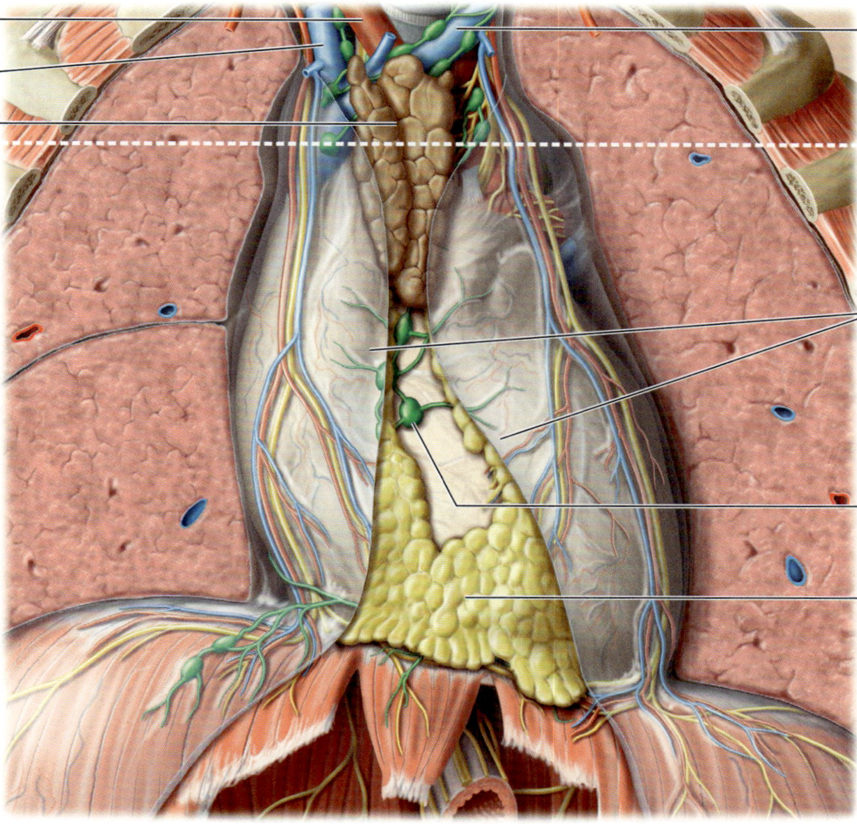

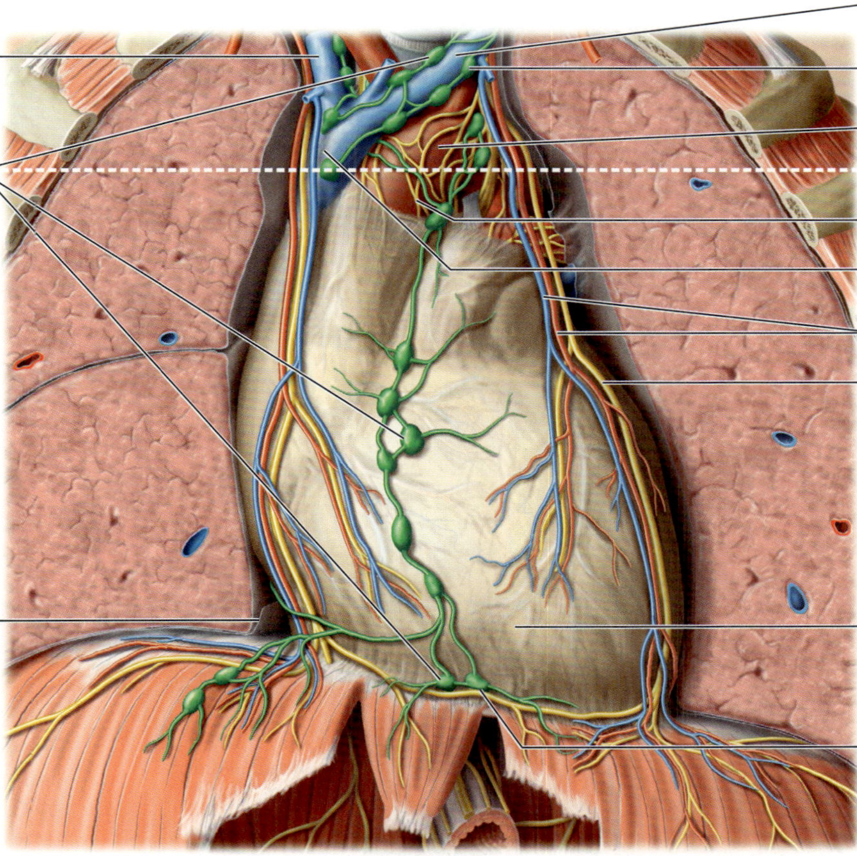

Representation of the **heart in situ**, in relation to the thymus gland and the mediastinal nodes.

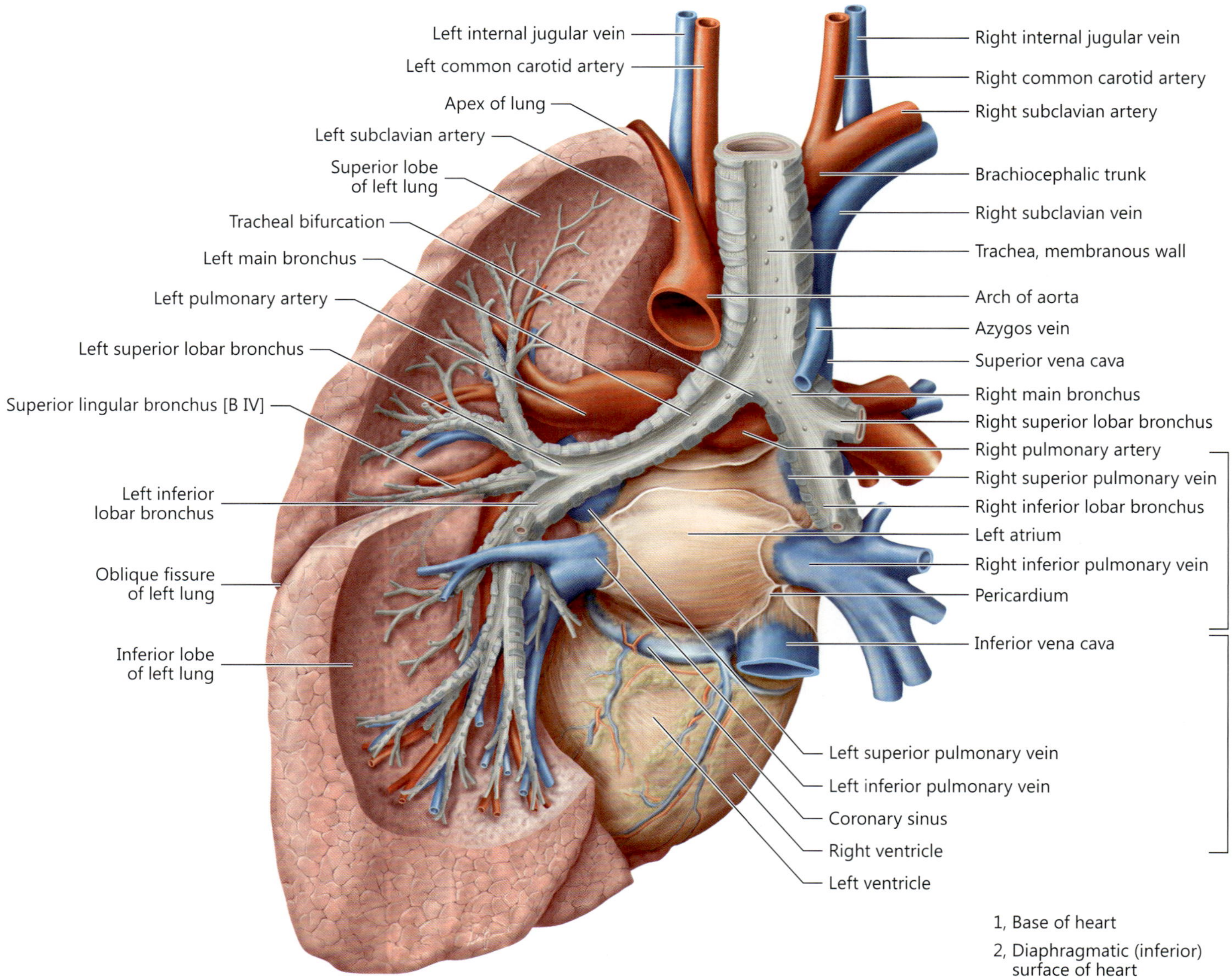

Position of the **heart in the middle mediastinum**, posterior view. The base of the heart faces upwards and to the right; the diaphragmatic or inferior surface, downwards and towards the back. The principal veins of the large and small circulation (the four pulmonary veins and the two venae cavae) arrive at the base of the heart. The vestibule (roof) of the left atrium represents the most posterior surface of the heart, in contact with the organs of the posterior mediastinum (esophagus and, further behind and to the left, the thoracic aorta, not shown).

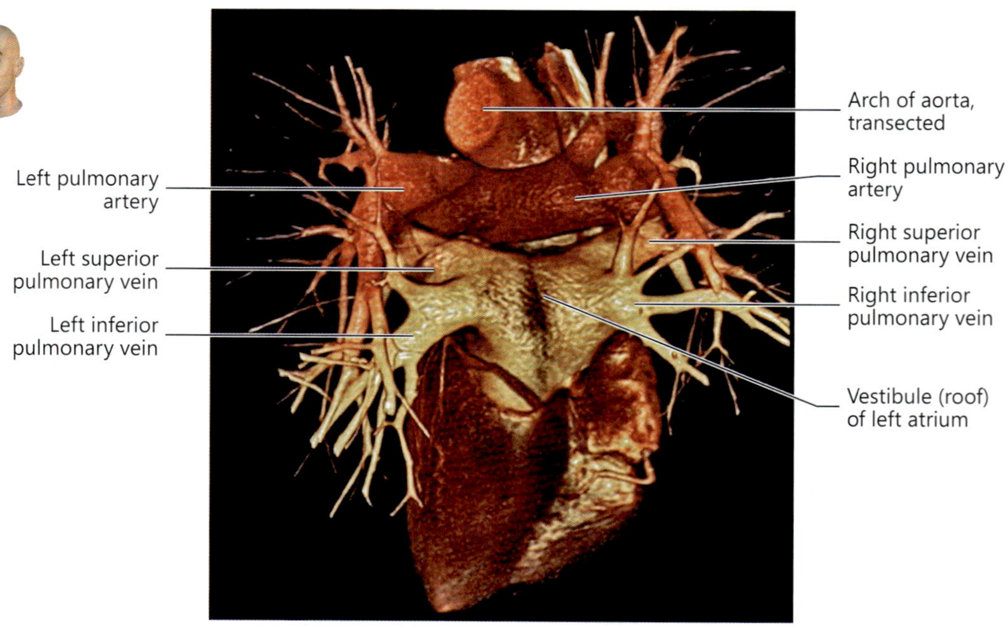

Three-dimensional image of a living human body, obtained by CT scan using a 3D volume rendering technique, taken from transverse sections, in craniocaudal direction, of the **heart**, posterior view. The projection is the same as that shown in the illustration above.

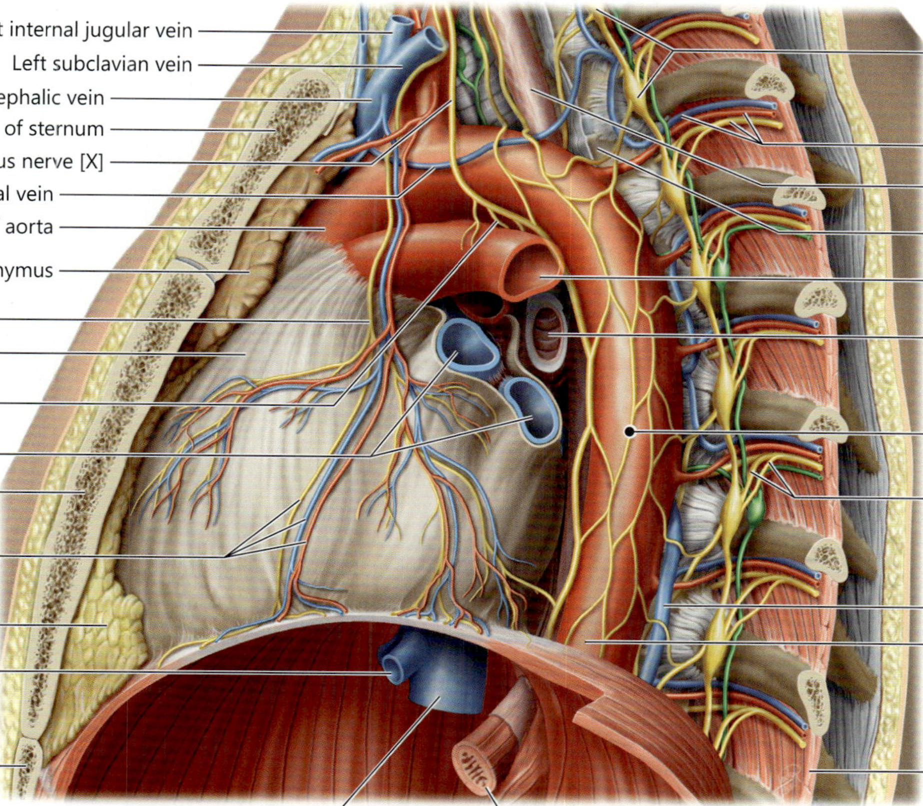

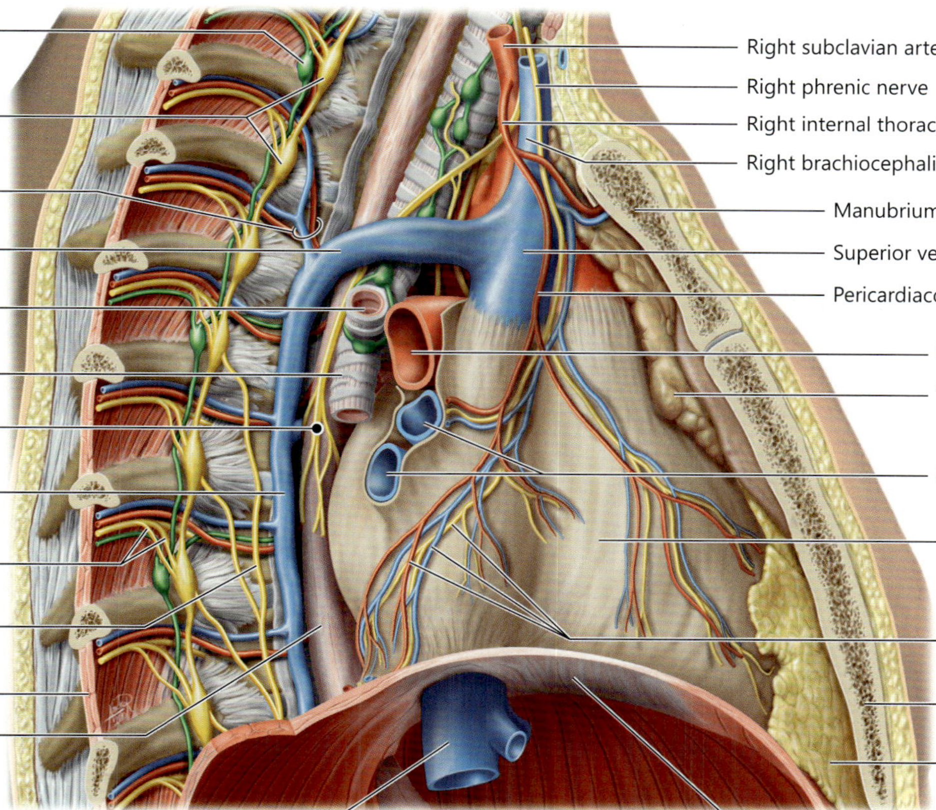

Left (above) and right (below) lateral view of the heart within the **pericardial sac**.

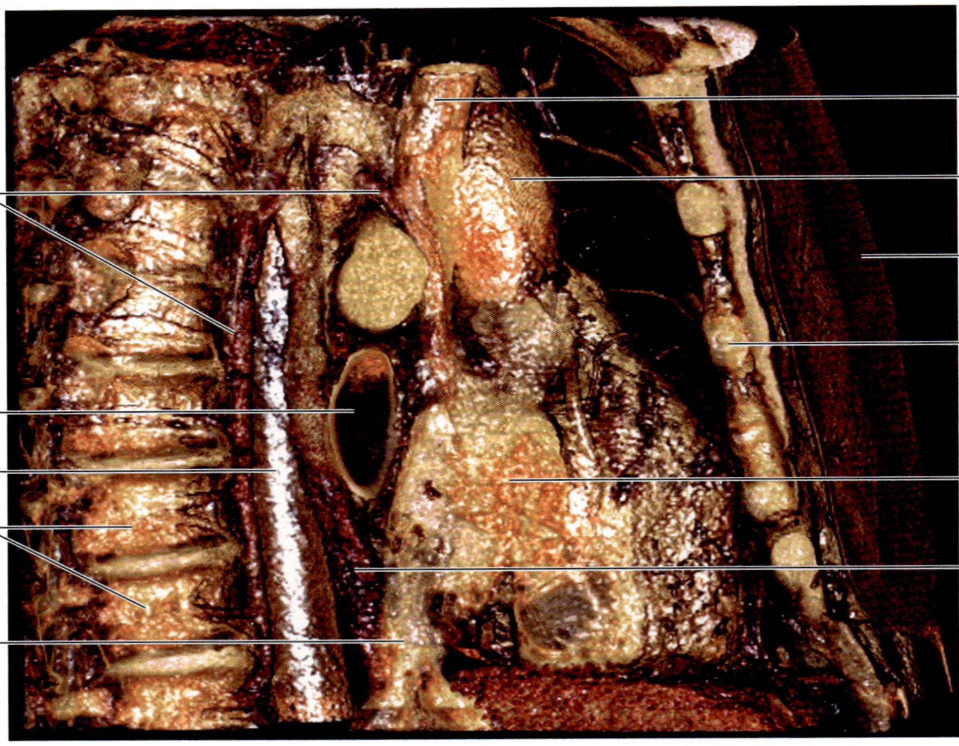

Three-dimensional image of a living human body, obtained by CT scan using a 3D volume rendering technique, taken from transverse sections, in craniocaudal direction, of the **mediastinum** observed from the right.

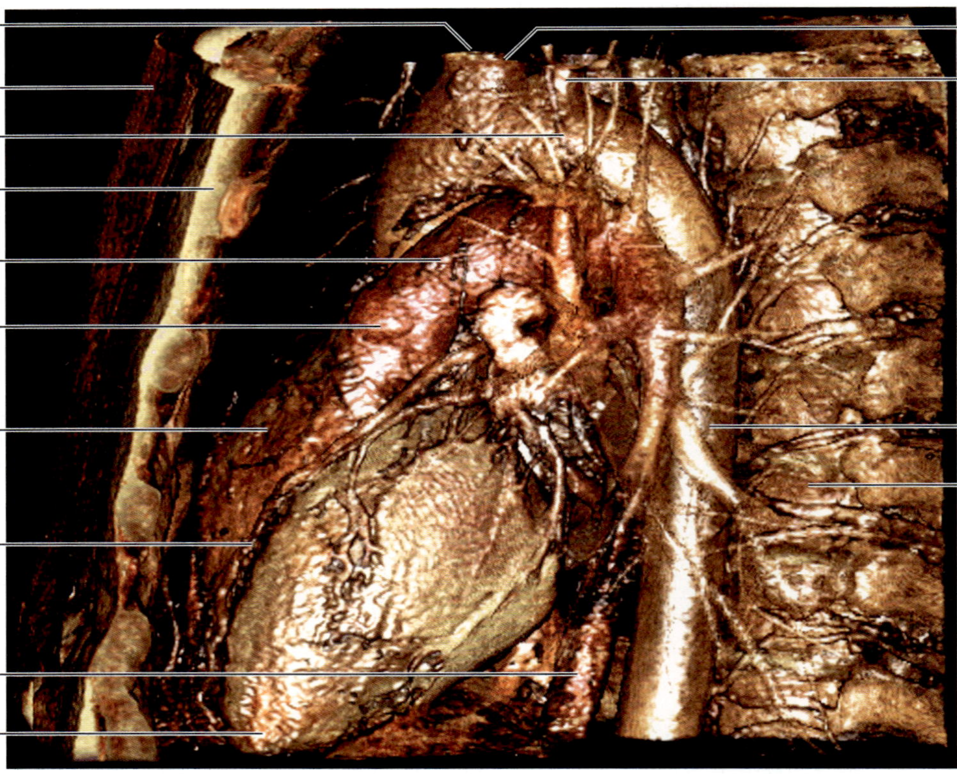

Three-dimensional image of a living human body, obtained by CT scan using a 3D volume rendering technique, taken from transverse sections, in craniocaudal direction, of the **mediastinum** observed from the left.

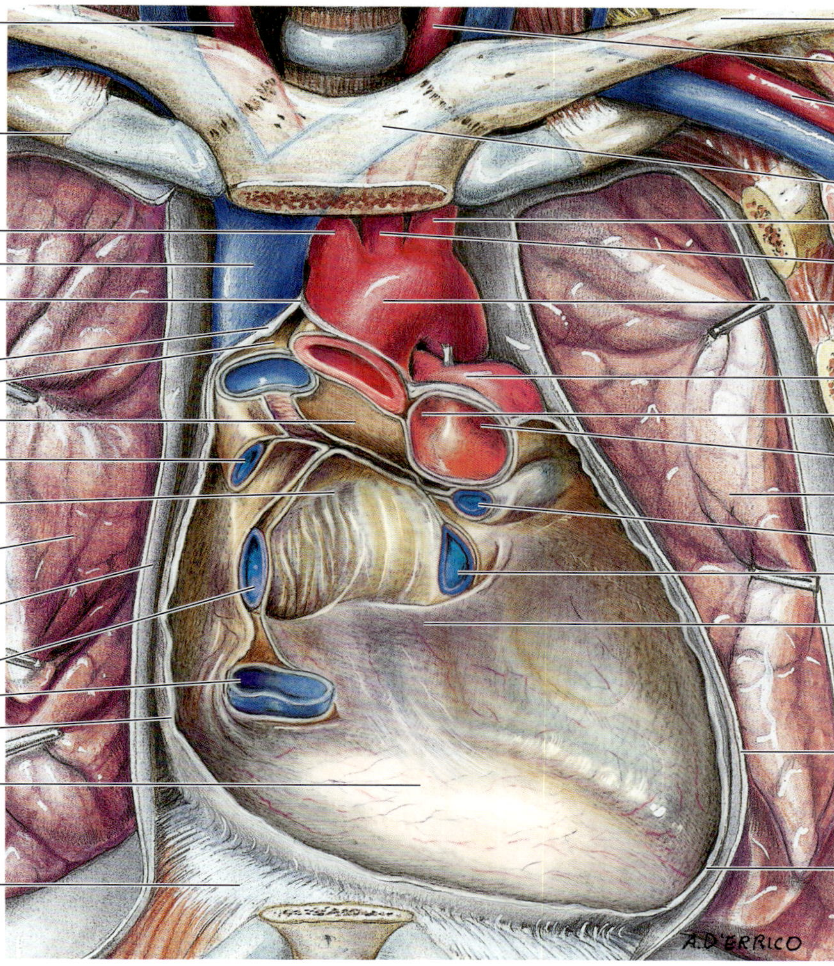

Pericardial sac after removal of the heart, anterior view. One can see the transverse pericardial sinus and oblique pericardial sinus.

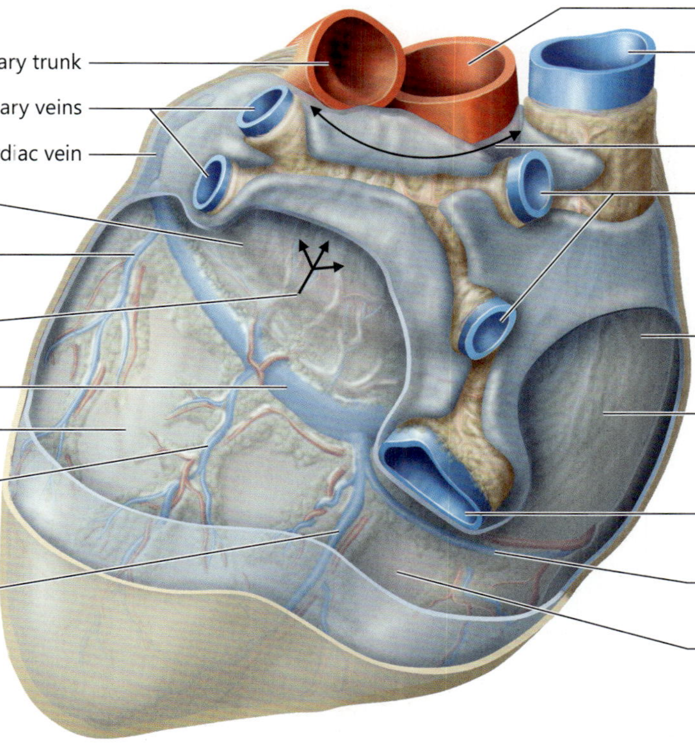

Reflection of the **visceral layer of serous pericardium** (epicardium) onto the parietal layer on the posterior wall of the heart. Arrows indicate the transverse pericardial sinus and oblique pericardial sinus.

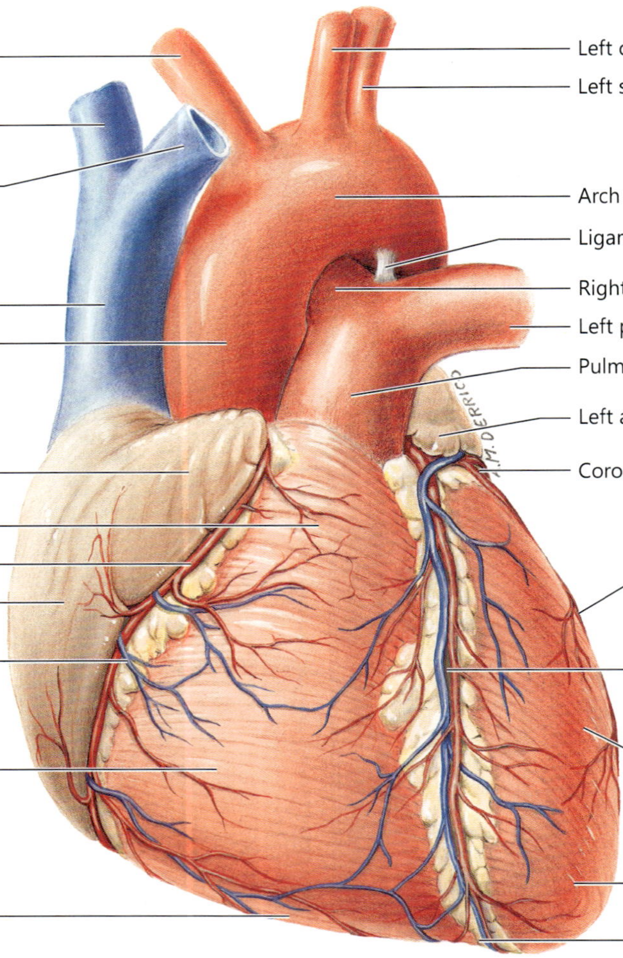

Anterior (**sternocostal**) **surface of the heart.**

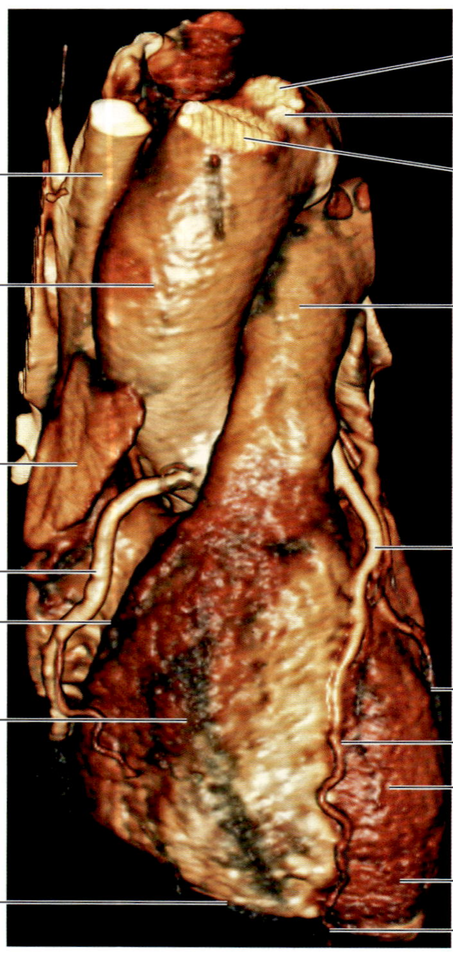

Three-dimensional image of a living human body, obtained by CT scan using a 3D volume rendering technique. **Anterior** (**sternocostal**) **surface** of the heart.

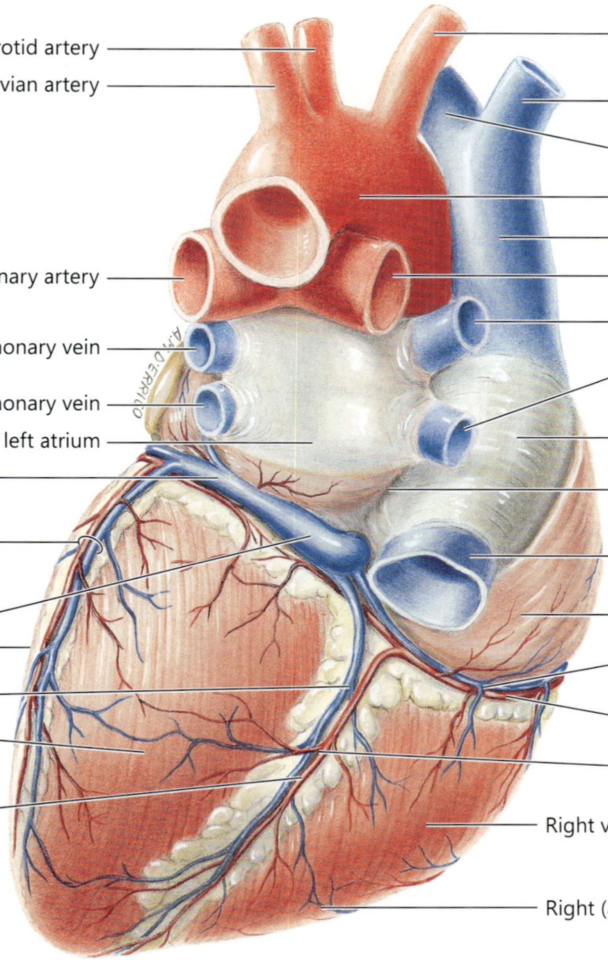

Diaphragmatic (inferior) surface of the heart.

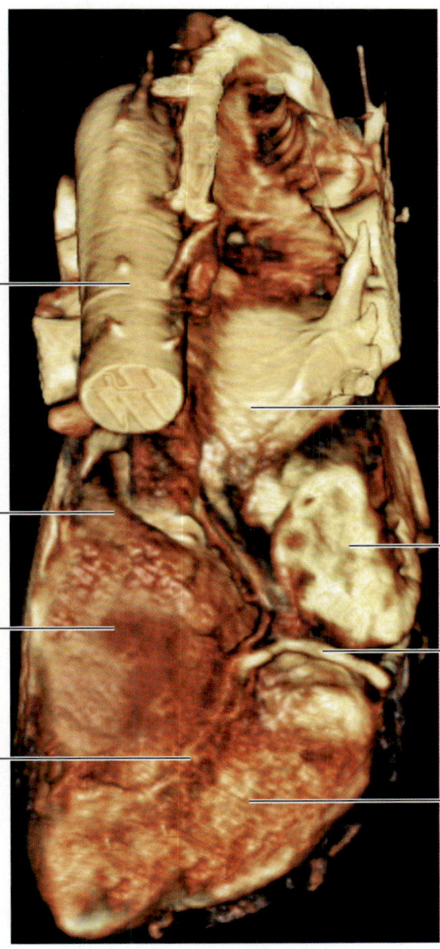

Three-dimensional image of a living human body, obtained by CT scan using a 3D volume rendering technique. **Diaphragmatic (inferior) surface** of the heart.

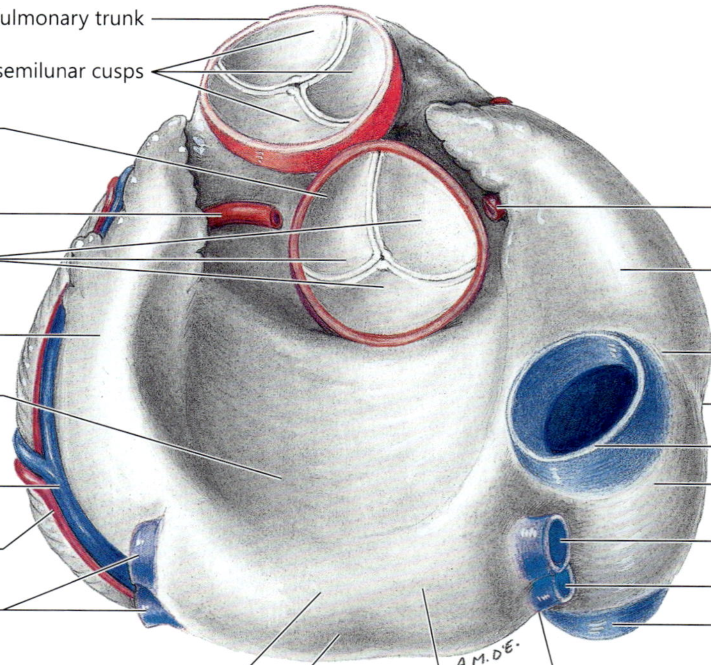

Base of the heart, viewed from above, after resection of the pulmonary trunk and aorta.

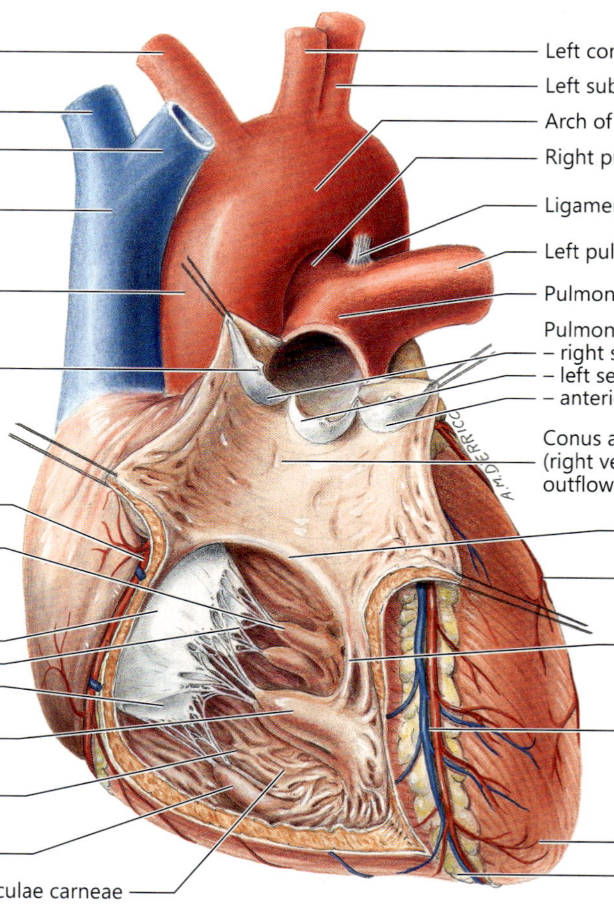

Cavity of the right ventricle, viewed from the front, after partial removal of the anterior wall. The conus arteriosus, open, highlights the valve.

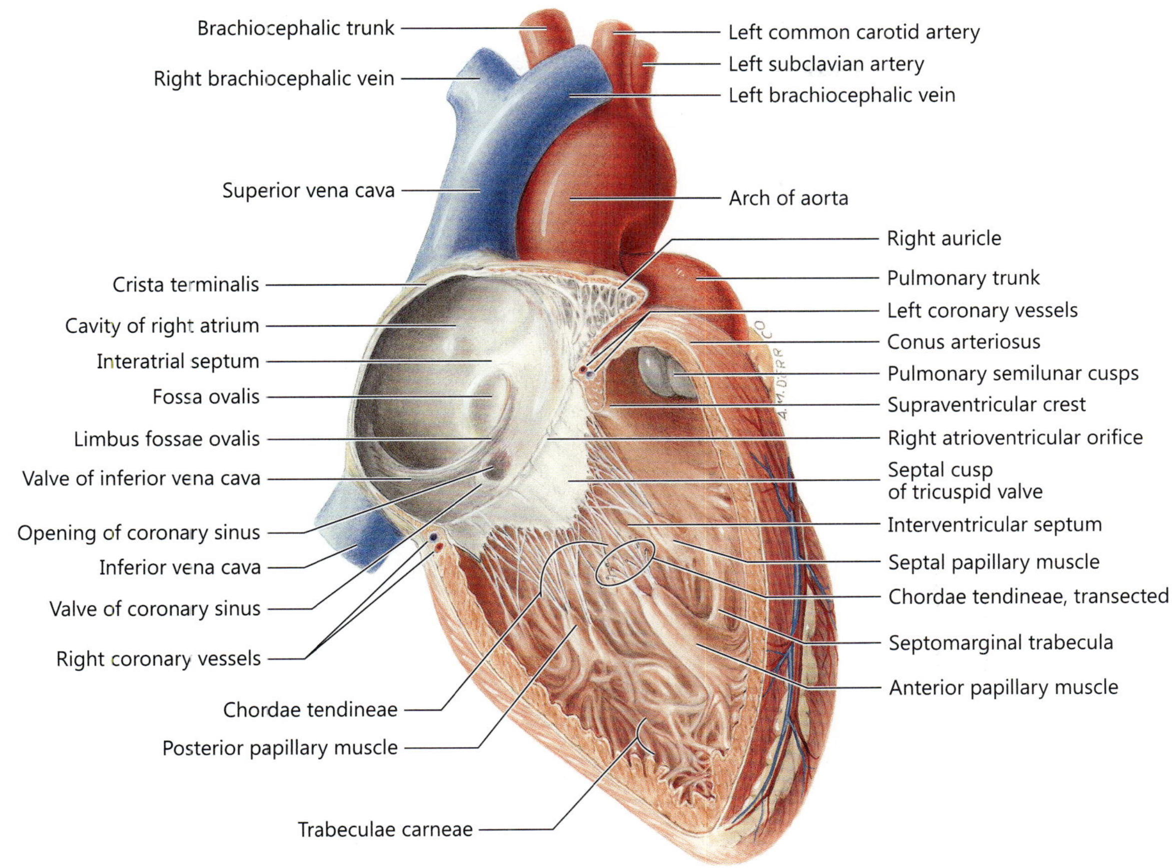

Cavities of the right ventricle and atrium, viewed from the right, after removal of the lateral wall of the sinus of the venae cavae, of the right auricle, and of the right (acute) margin of the heart.

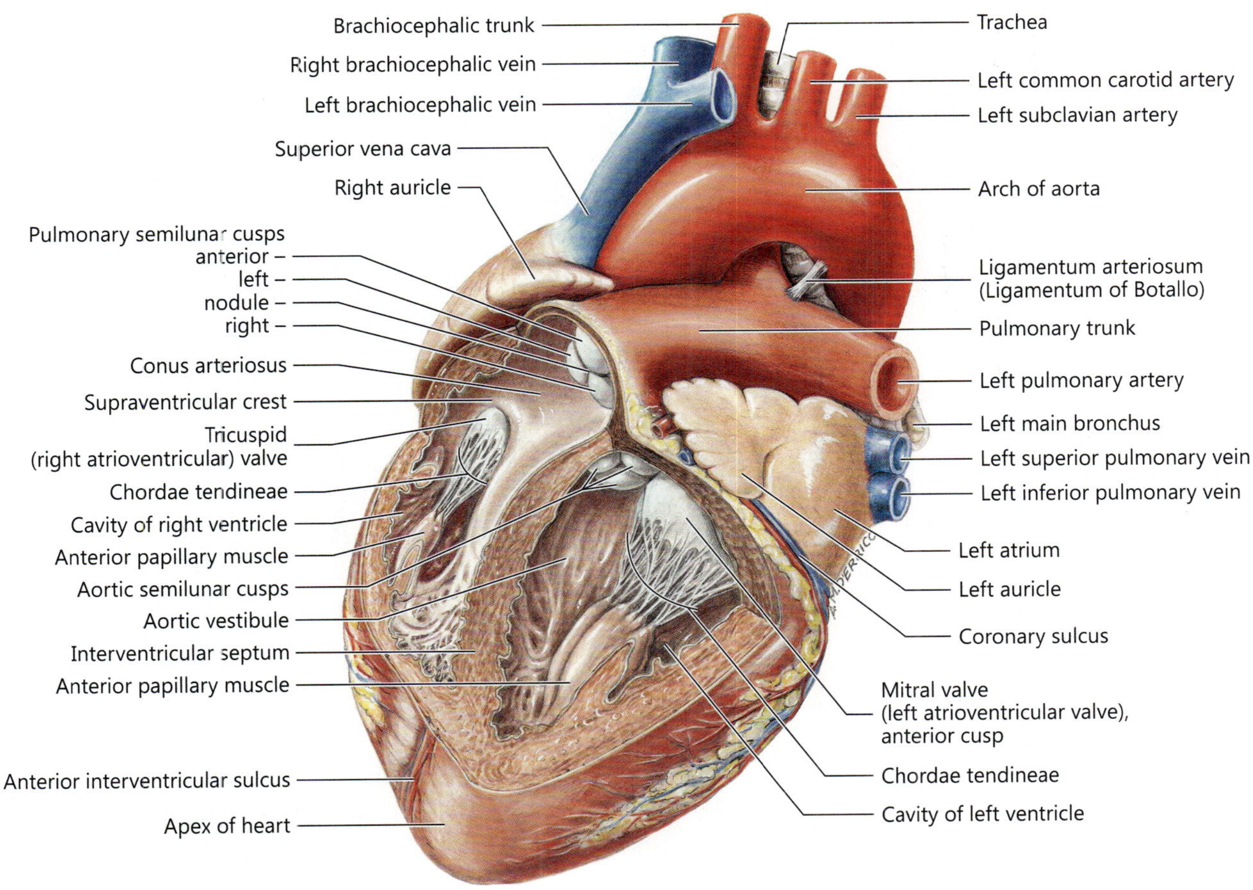

Cavities of the ventricles, viewed from the front and left, after partial removal of the anterior wall.

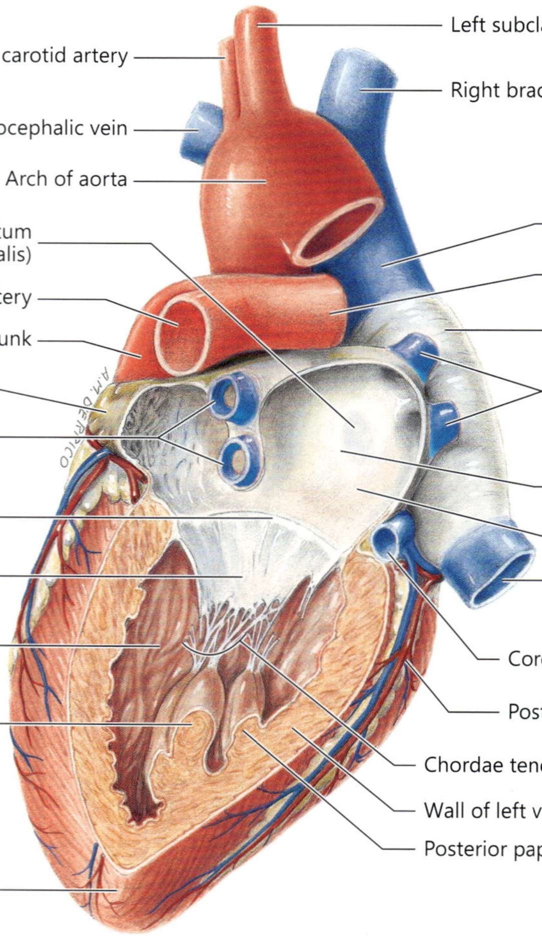

Left atrium and ventricle, viewed from the left, after removal of part of the auricle and of the left (obtuse) margin.

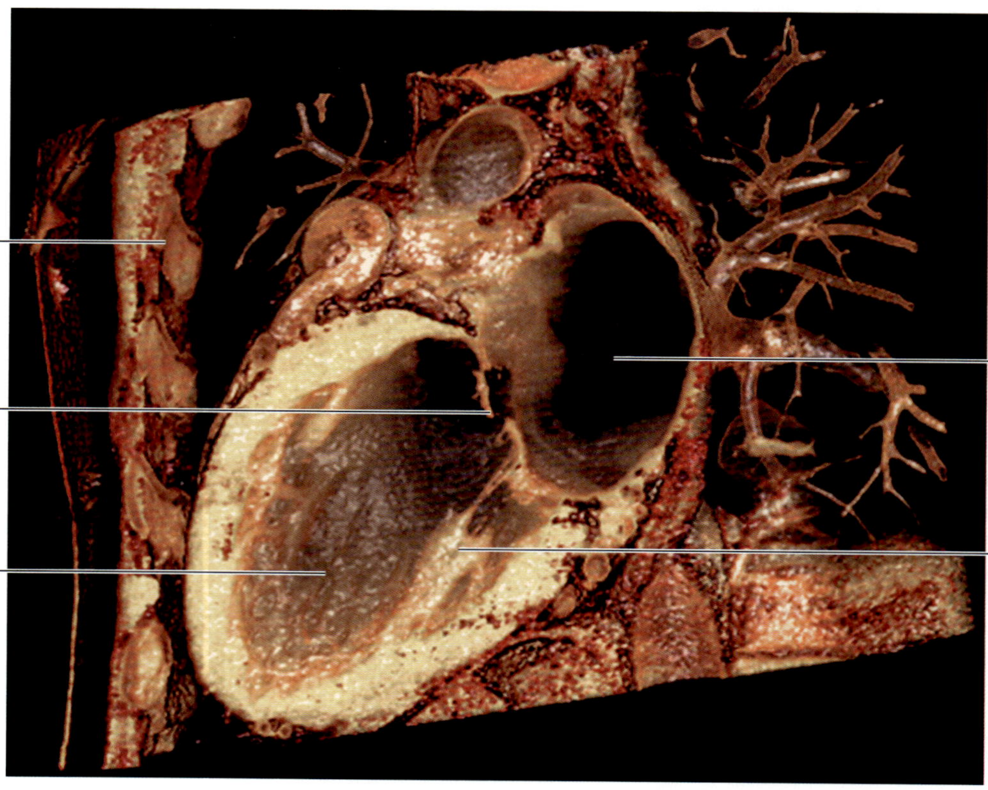

Three-dimensional image of a living human body, obtained by CT scan using a 3D volume rendering technique. **Cavities of the left atrium and ventricle**.

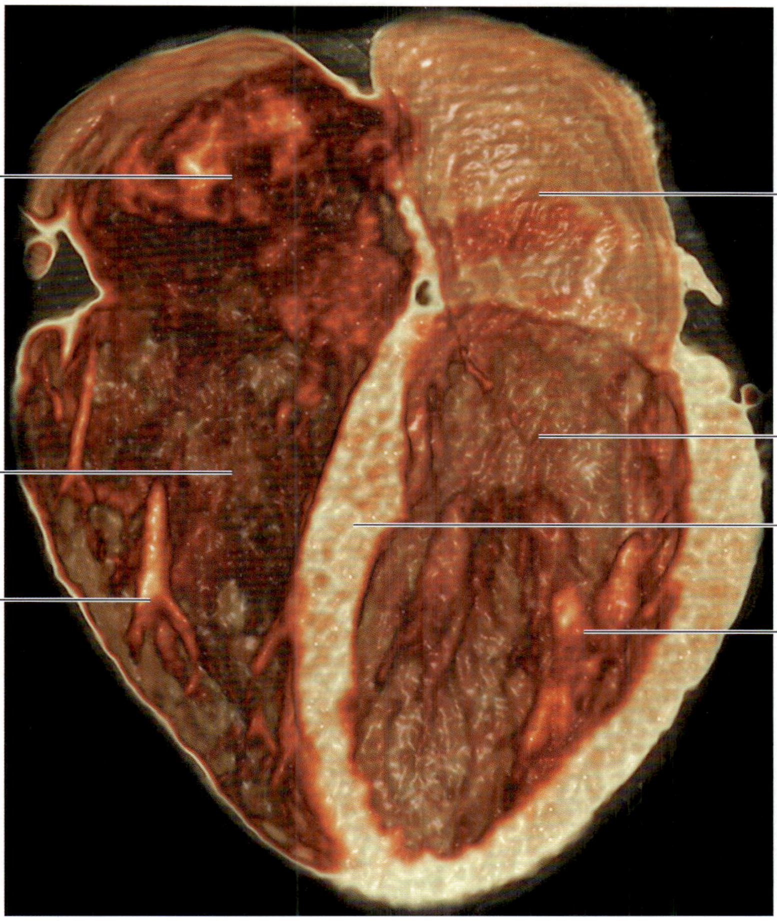

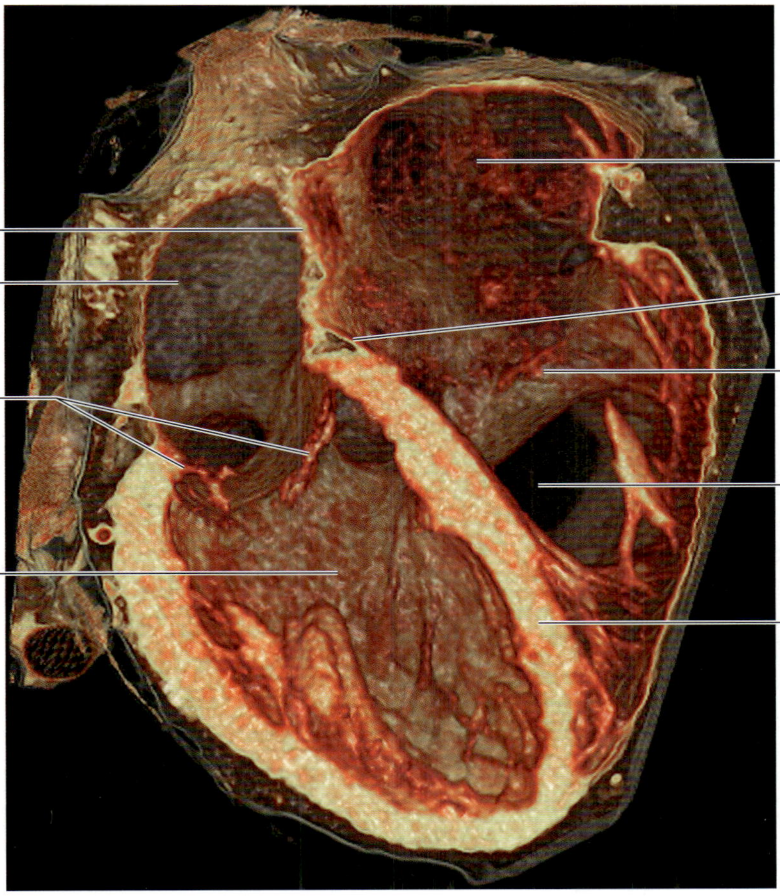

Three-dimensional image of a living human body, obtained by CT scan using a 3D volume rendering technique. **Atrial and ventricular cavities**. Above, view from the anterior (sternocostal) surface. Below, view from the diaphragmatic (inferior) surface.

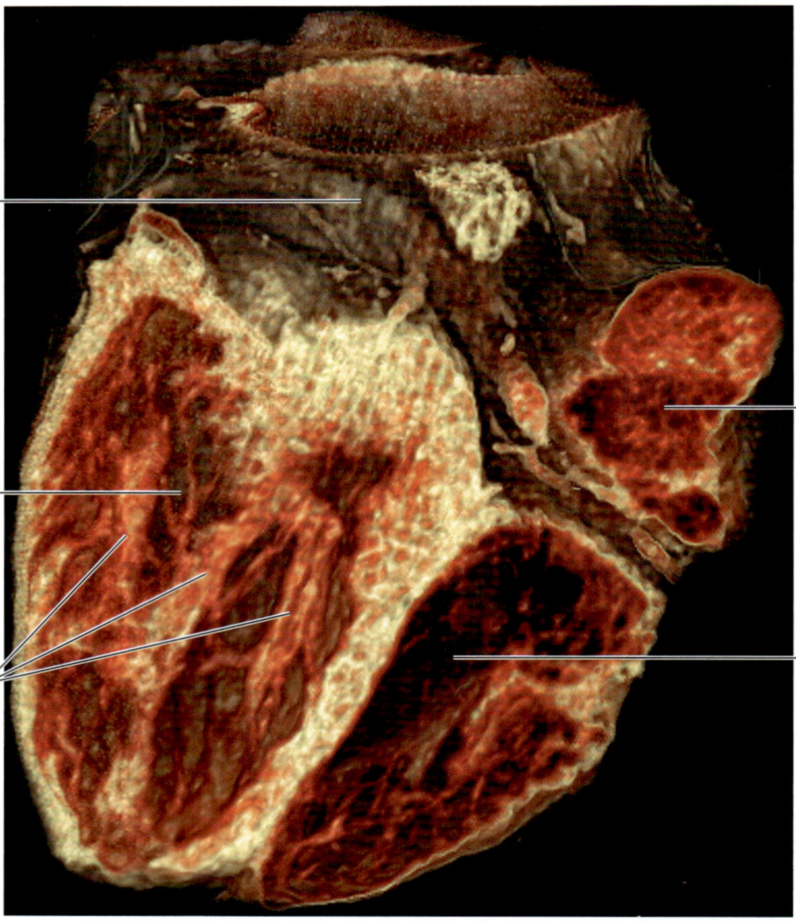

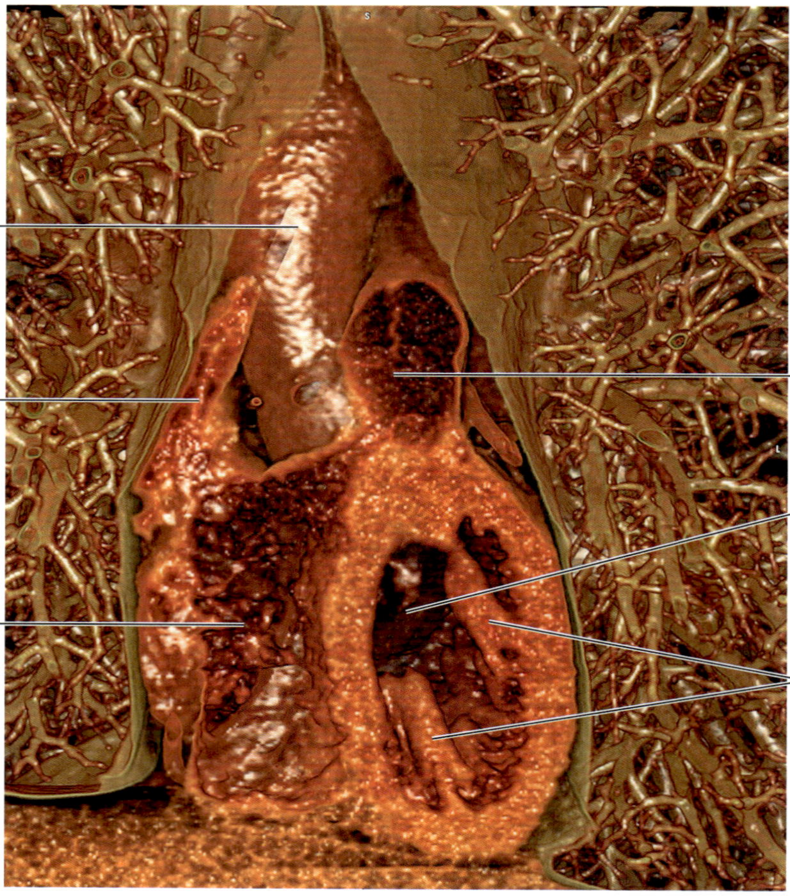

Three-dimensional image of a living human body, obtained by CT scan using a 3D volume rendering technique. **Atrial and ventricular cavities**. Above, view from the diaphragmatic (inferior) surface. Below, anterior view.

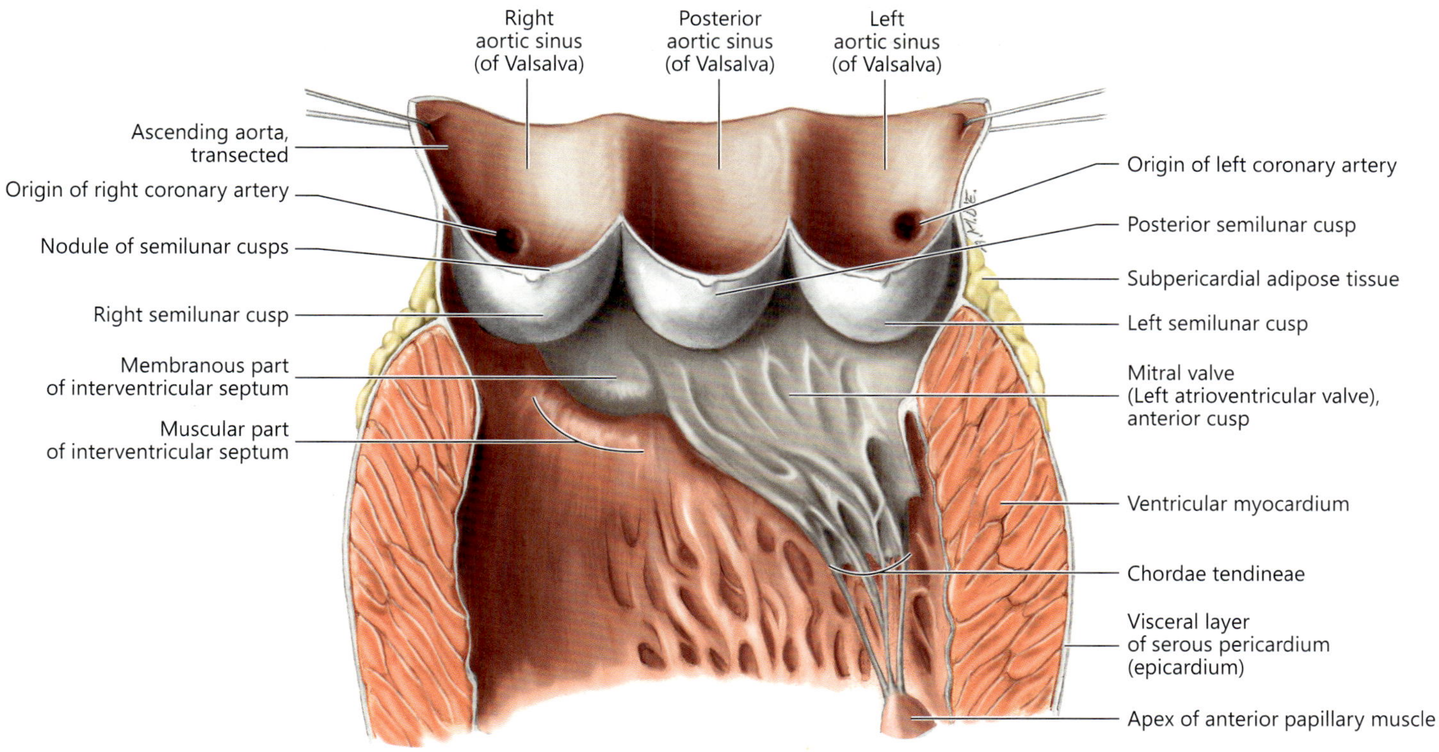

Section of the left ventricle, at the level of the aortic valve, corresponding to the areas inside the box in the figures on the following page. The aortic orifice has been resected and opened. One can see the **cusps of the aortic valve**. At the level of the right and left cusps, the orifices of the coronary arteries open. The membranous part of the interventricular septum extends in between the right and posterior cusps.

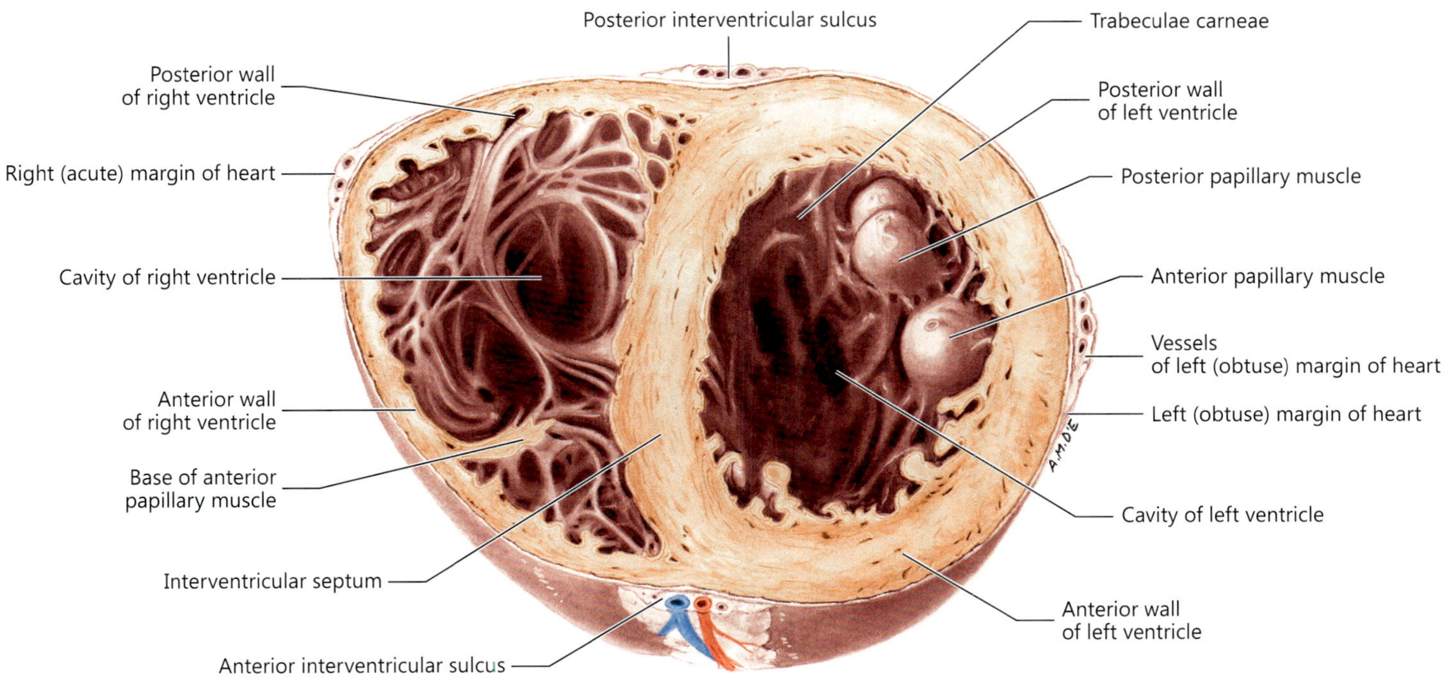

Transverse section of the ventricular part of the heart. Note the different thickness of the **ventricular walls**.

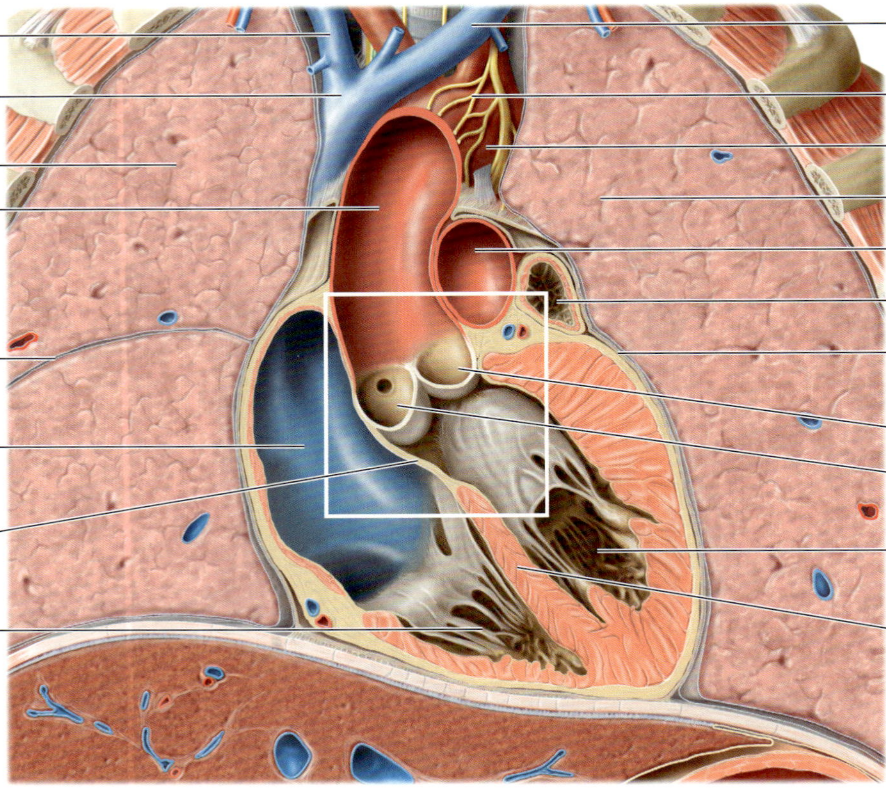

Heart in situ. Frontal section passing through the **aortic valve**. The region in the box corresponds to the area highlighted in the figure below.

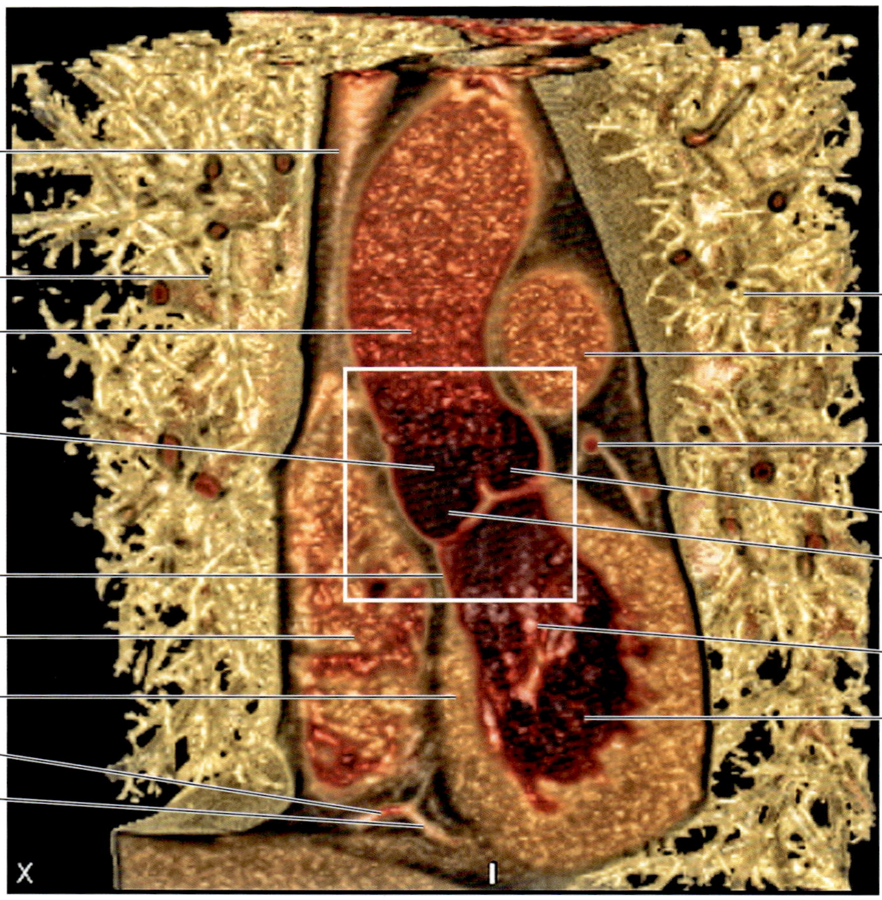

Three-dimensional image of a living human body, obtained by CT scan using a 3D volume rendering technique, showing the atrial and ventricular cavities, ascending aorta, pulmonary trunk and bronchial tree. Inside the box, the **aortic valve** is visible.

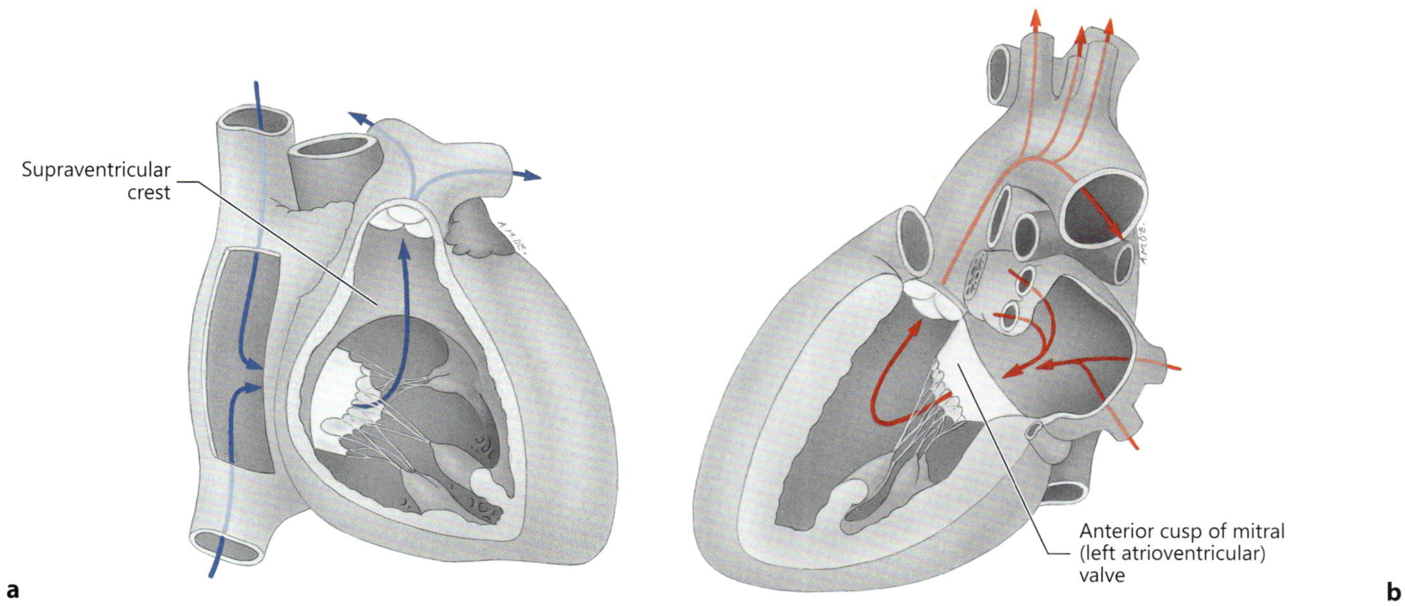

Schematic representation of the direction of **blood flow within the cardiac cavities**. The two pathways of flow into and out of the right ventricular cavity are separated by the supraventricular crest. In the left ventricular cavity, the two pathways are separated by the anterior cusp of the mitral valve.

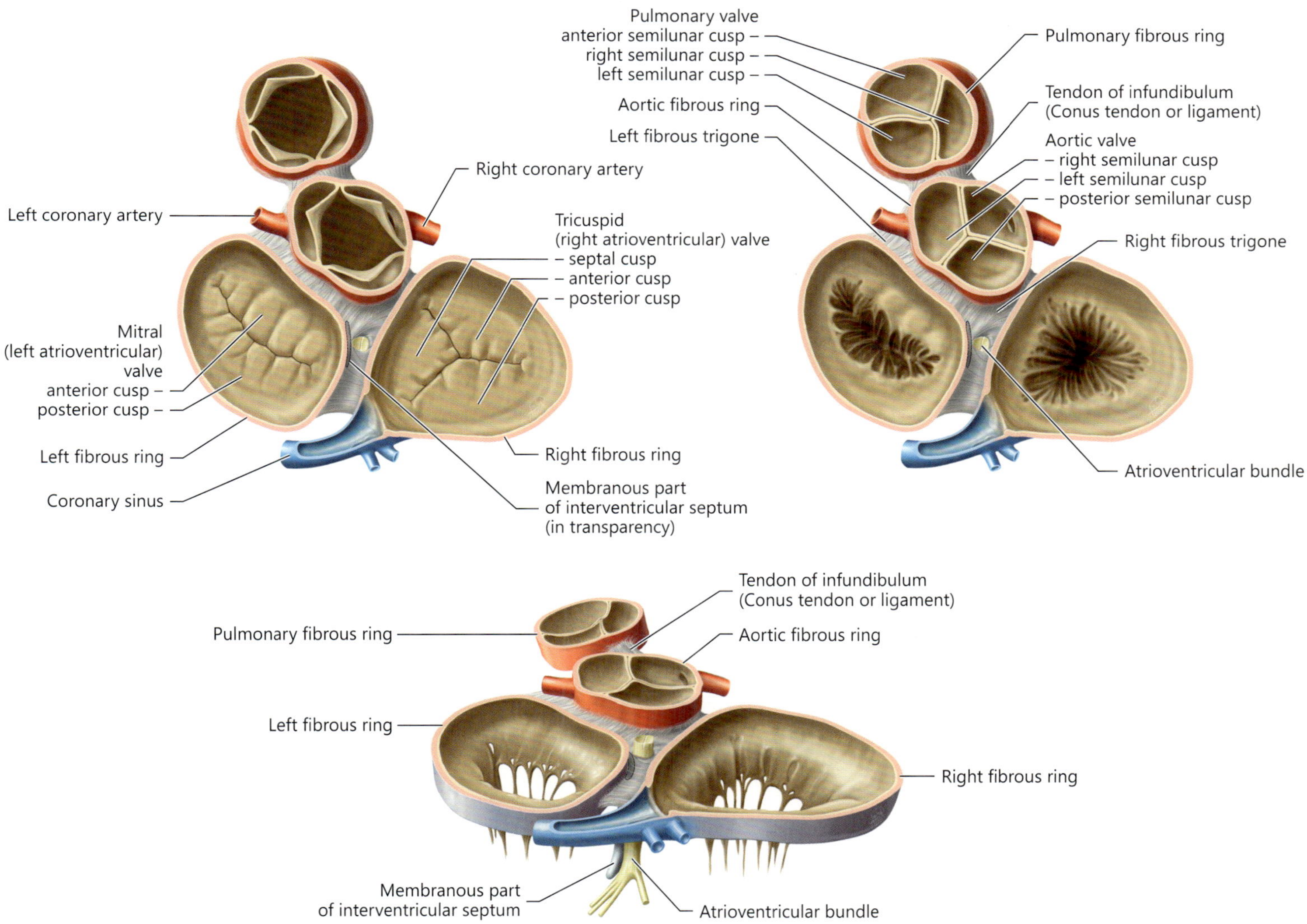

Fibrous skeleton of the heart. Above, the valves are shown in the phases of ventricular systole (on the left) and diastole (on the right). Below, projection viewed from behind. Note the oblique orientation of the pulmonary fibrous ring with respect to the fibrous rings of the atrioventricular and aortic valves. The atrioventricular bundle (of His) of the conduction system passes through the right fibrous trigone.

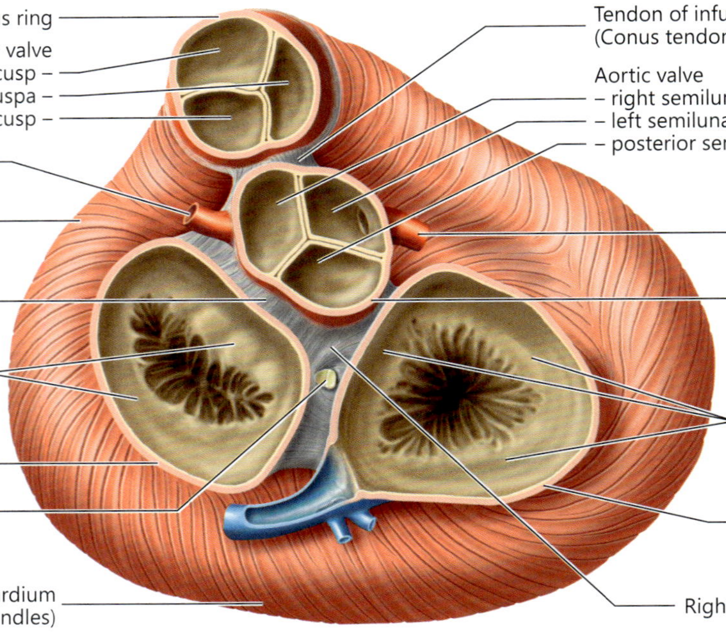

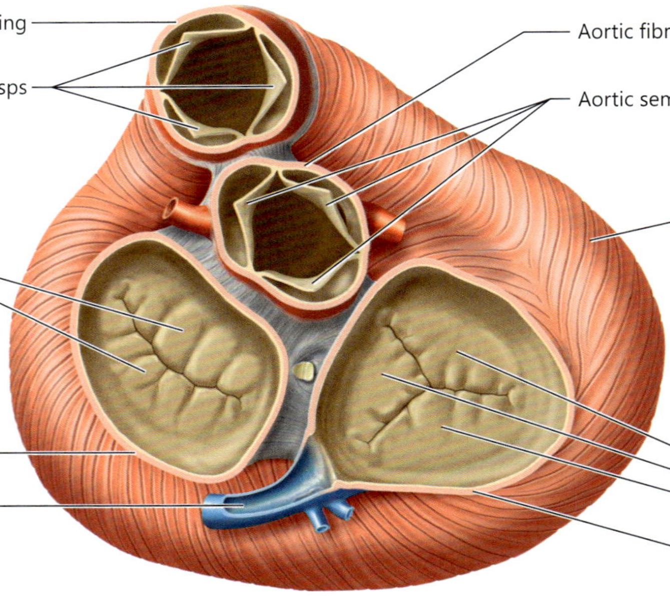

Base of the ventricles, viewed from above, after removal of the atria. Above, phase of ventricular diastole; below, phase of ventricular systole.

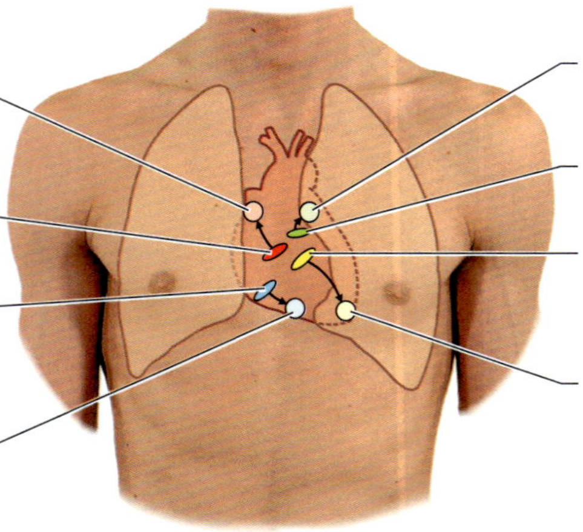

Projection onto the thoracic cage of the cardiac valves and relative **auscultation sites**.

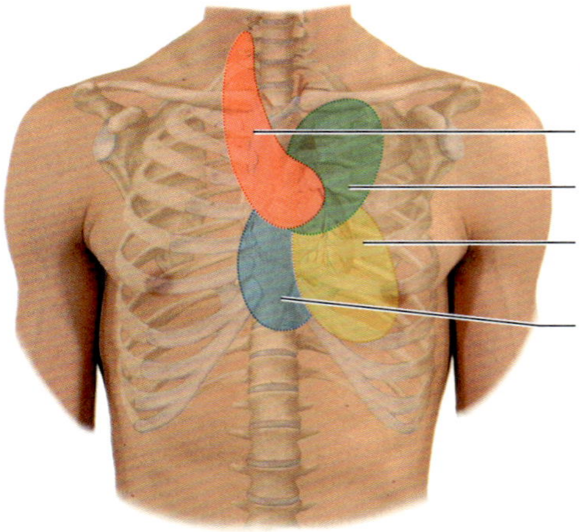

Sites of auscultation of pathological murmurs of the cardiac valves.

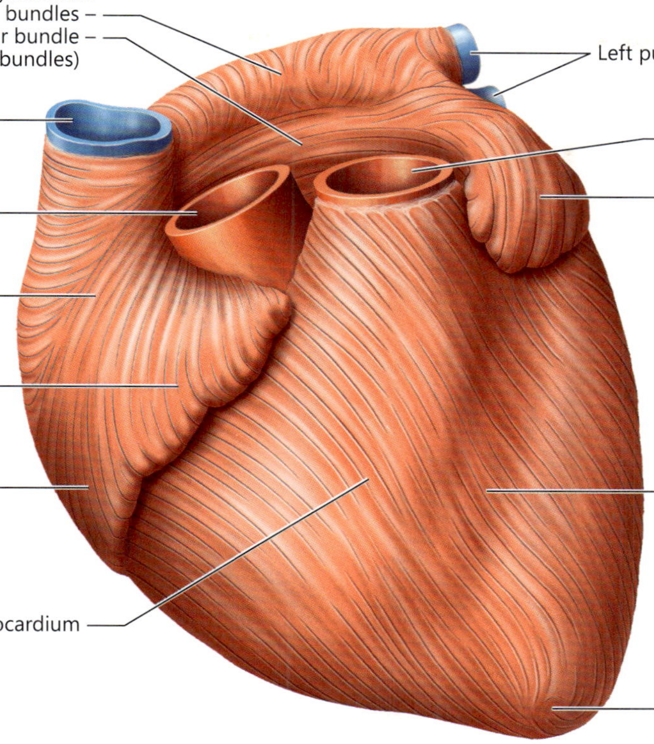

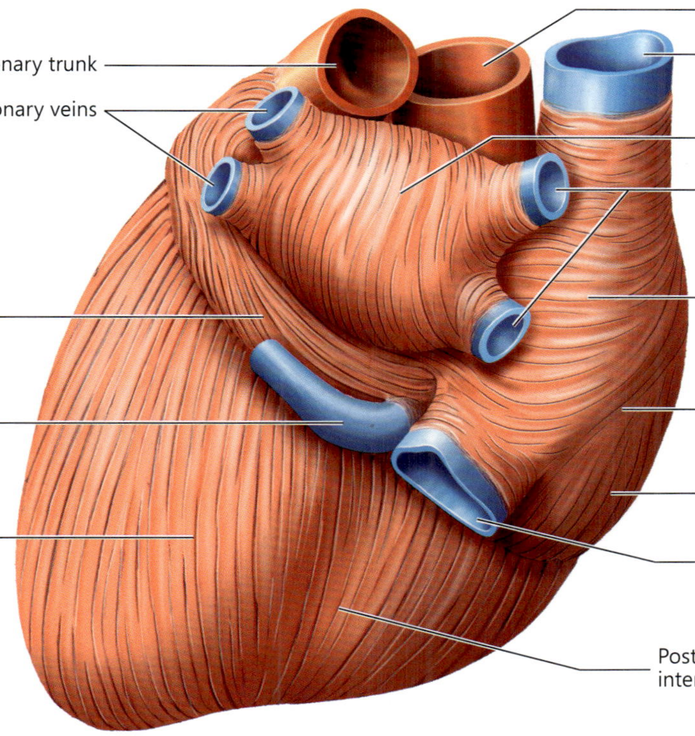

a, Disposition of the **atrial and ventricular muscles** on the anterior (sternocostal) surface of the heart; **b**, Disposition of the atrial and ventricular muscles on the diaphragmatic (inferior) surface and base of the heart.

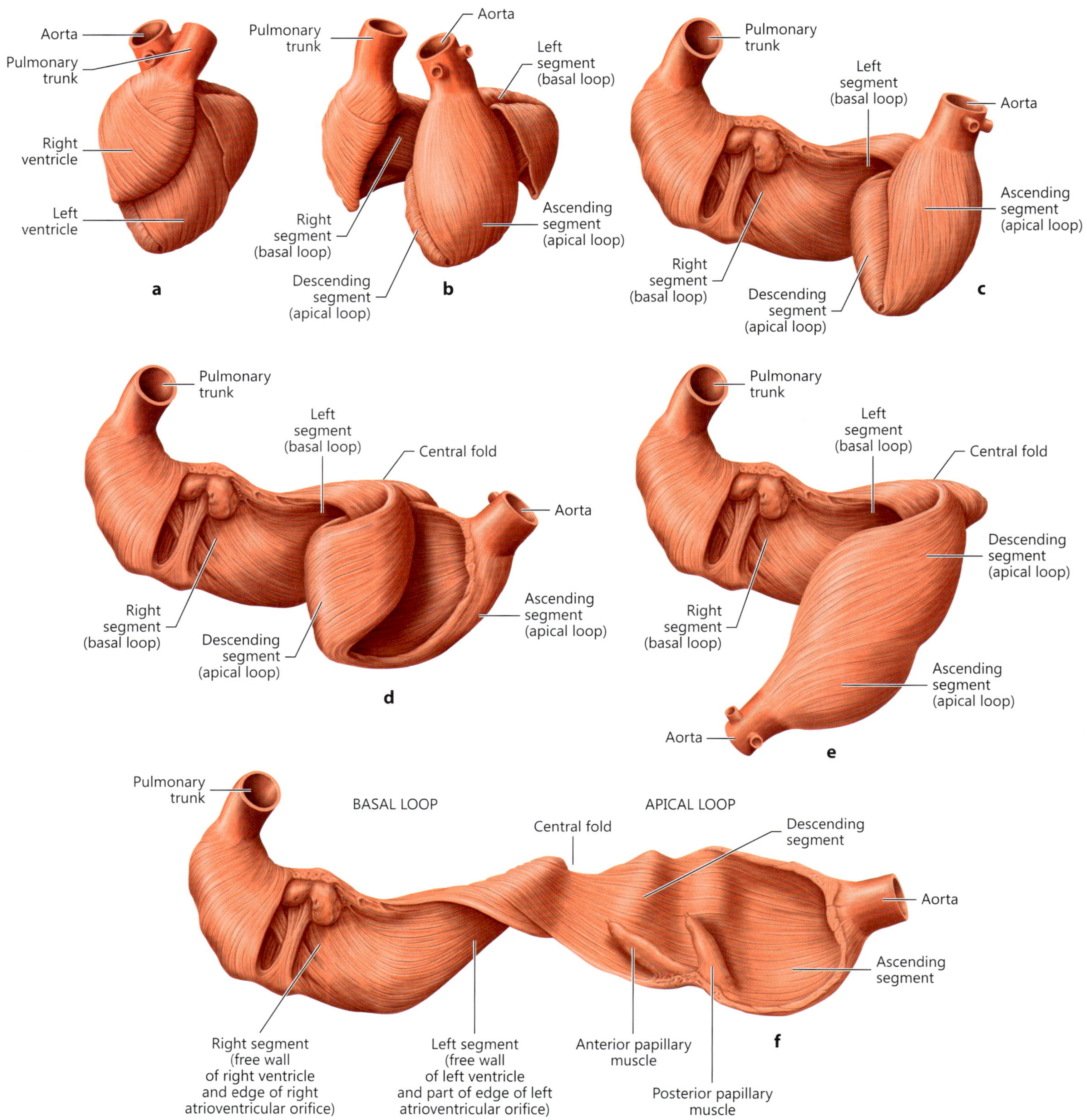

Schematic representation of the succession of unfolding phases of the heart according to Torrent-Guasp, so as to obtain the **ventricular myocardial band**, or helical ventricular myocardial band of Torrent-Guasp. In **a**, the initial arrangement of the myocardial fibers is represented in posterior view; in **b**, **c** and **d**, the successive phases of separation of the pulmonary trunk and right ventricular wall from the aorta are illustrated; note how it is possible to identify and separate, already in this phase, an ascending segment and a descending segment. With the successive pulling apart of the root of the pulmonary trunk from the aortic root, the two segments, ascending and descending, become separated from each other (**e**); in this phase one can identify, furthermore, a right segment and a left segment, corresponding to the basal portion of the heart, and separated from the ascending and descending segments by a zone known as the central fold, corresponding to the apex of the heart. With the complete unfolding of the myocardial structure, one obtains a continuous band of fibers extending from the pulmonary trunk to the aorta; in this band, it is possible to identify a zone corresponding to the basal portion of the heart, extending from the pulmonary trunk to the central fold and known as the basal loop, and an apical part of the heart, extending from the central fold to the aorta, which takes the name of apical loop. In the zone of the basal loop, the right and the left segments can be distinguished; the apical loop, instead, is composed of the descending and the ascending segments.

Schematic representation showing the direction of fibers of the **Torrent-Guasp band** in the initial position of the heart. One can see that this structure is composed of a single band of myocardial fibers that follow a winding helical path from the pulmonary trunk to the root of the aorta; one can easily distinguish the initial position of the descending and ascending segments separated by the central fold; these zones constitute the apical loop. Above the apical loop, one can detect the right and left segments that together constitute the basal loop.

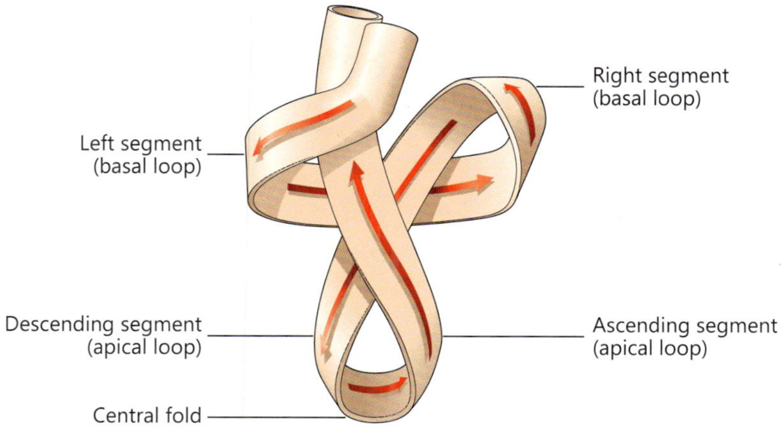

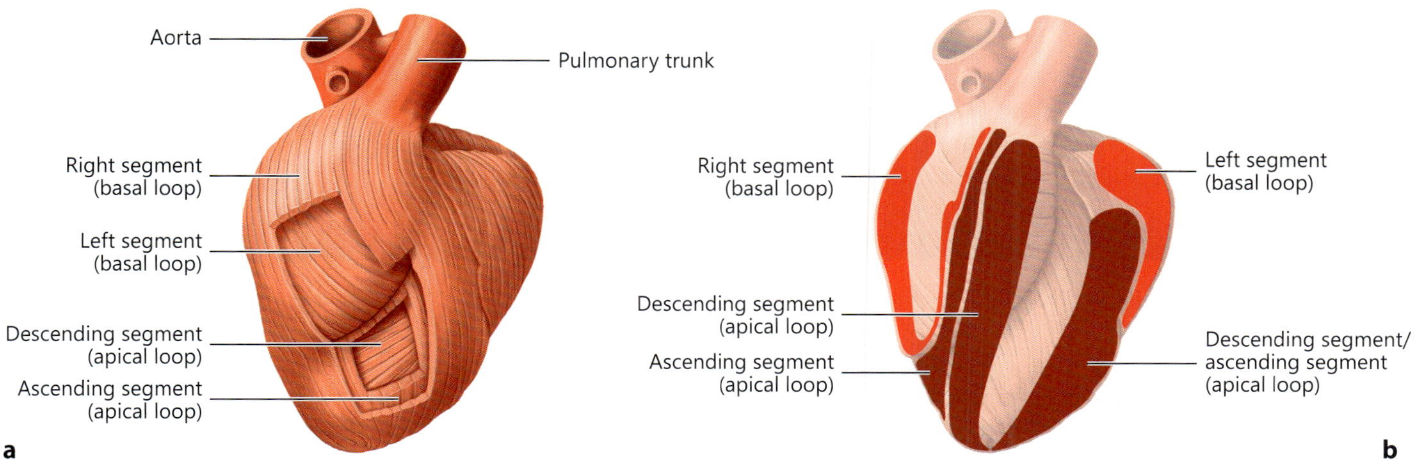

a, Schematic representation of the **Torrent-Guasp band** in which it is possible to distinguish the direction of the **myocardial band** both in correspondence to the apical loop, with the ascending and descending segments, as well as in correspondence to the basal loop, with the right and left segments. **b**, Schematic representation showing the position of the right and left segments with respect to the ascending and descending segments.

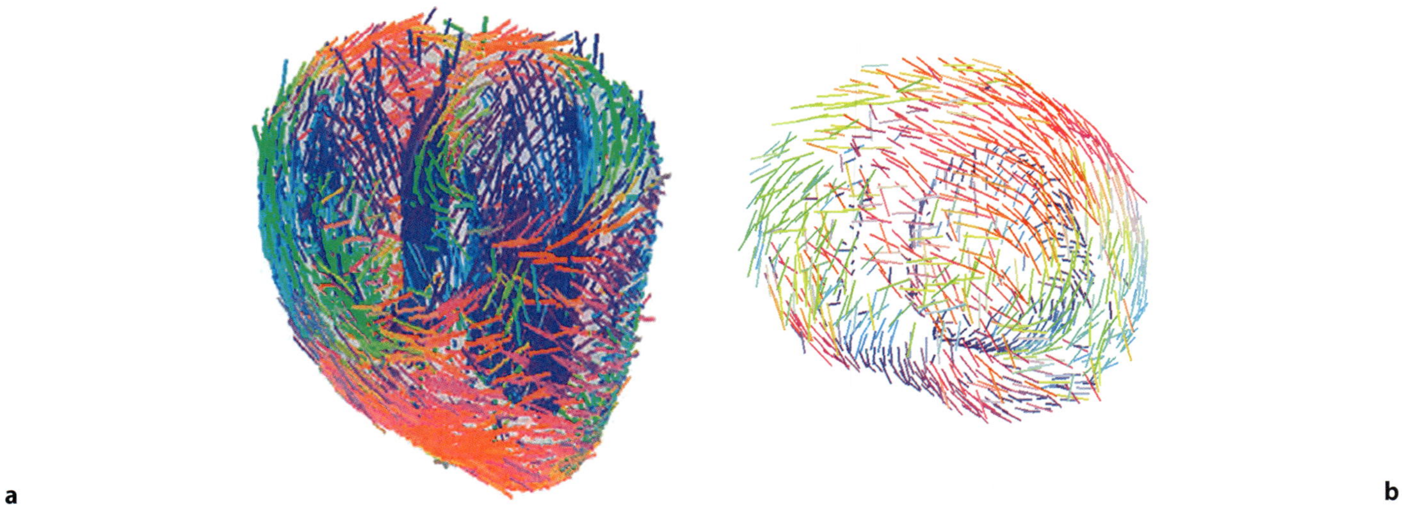

Organization of the **myocardial bands**, visualized by means of Diffusion Tensor Magnetic Resonance Imaging (DT-MRI). The fibers of the myocardium are organized into bundles. Bundles of myofibers of diverse orientation intersect generating a "mesh" (myocardial mesh). The myocardium is formed of three layers of fibers: subendocardial, intermediate and subepicardial. In the subendocardial layer, the fibers are oriented obliquely from the base towards the apex; in the intermediate layer they are oriented in a circular sense. The subepicardial layer is formed by the fibers of the subendocardial layer that re-emerge in correspondence to the apex and ascend in a longitudinal-oblique sense towards the base. The subepicardial fibers form an anticlockwise spiral, while the subendocardial ones form a clockwise spiral (**a**). In **b**, one can distinguish, looking down from above, the typical helical orientation that the myocardial fibers assume in correspondence to the apex of the heart.

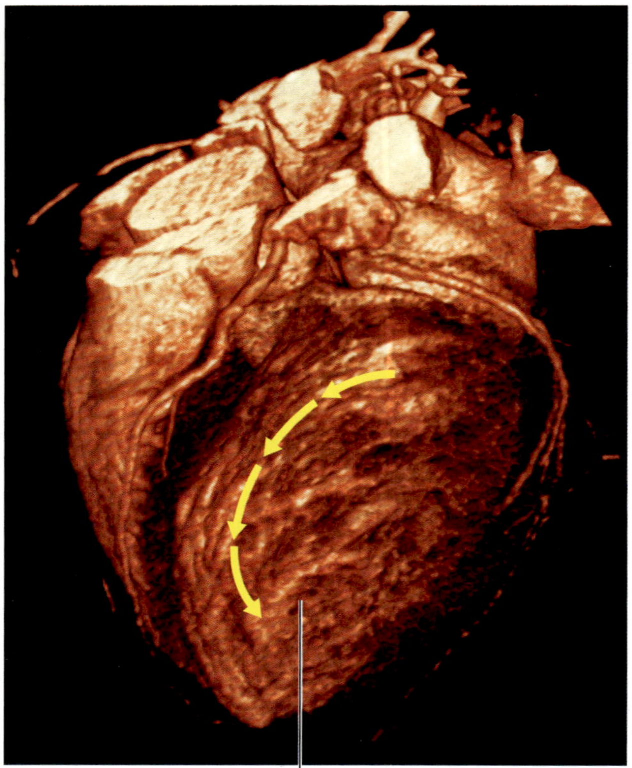

a

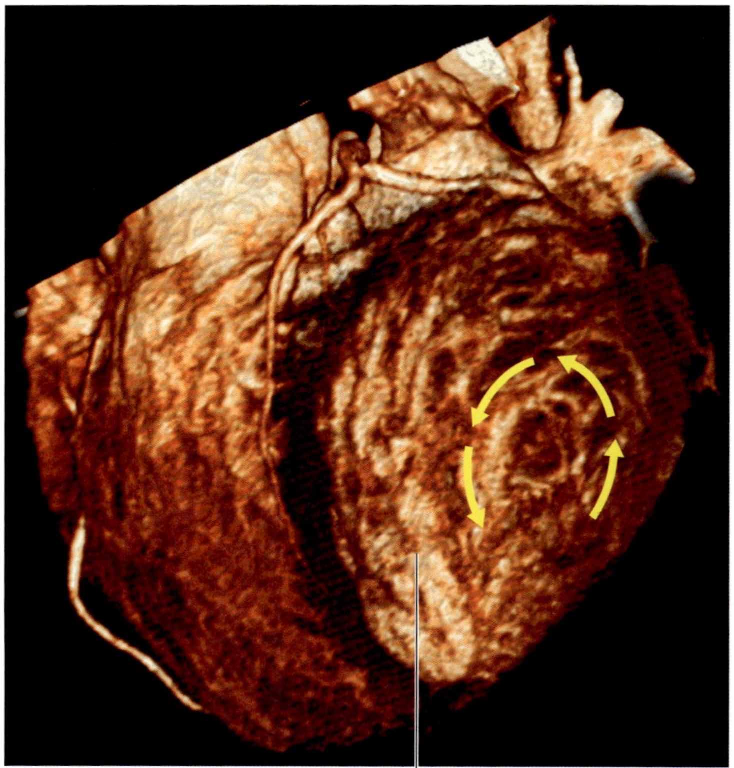

b

Three-dimensional image of the heart of a live subject, obtained using a 3D volume rendering technique. In **a**, one can see the oblique course of the **myocardial fibers of the left ventricle**. This course becomes helicoidal in correspondence to the ventricular cone (**b**). Thus is formed the helicoidal loop characteristic of the myocardial fibers of this zone.

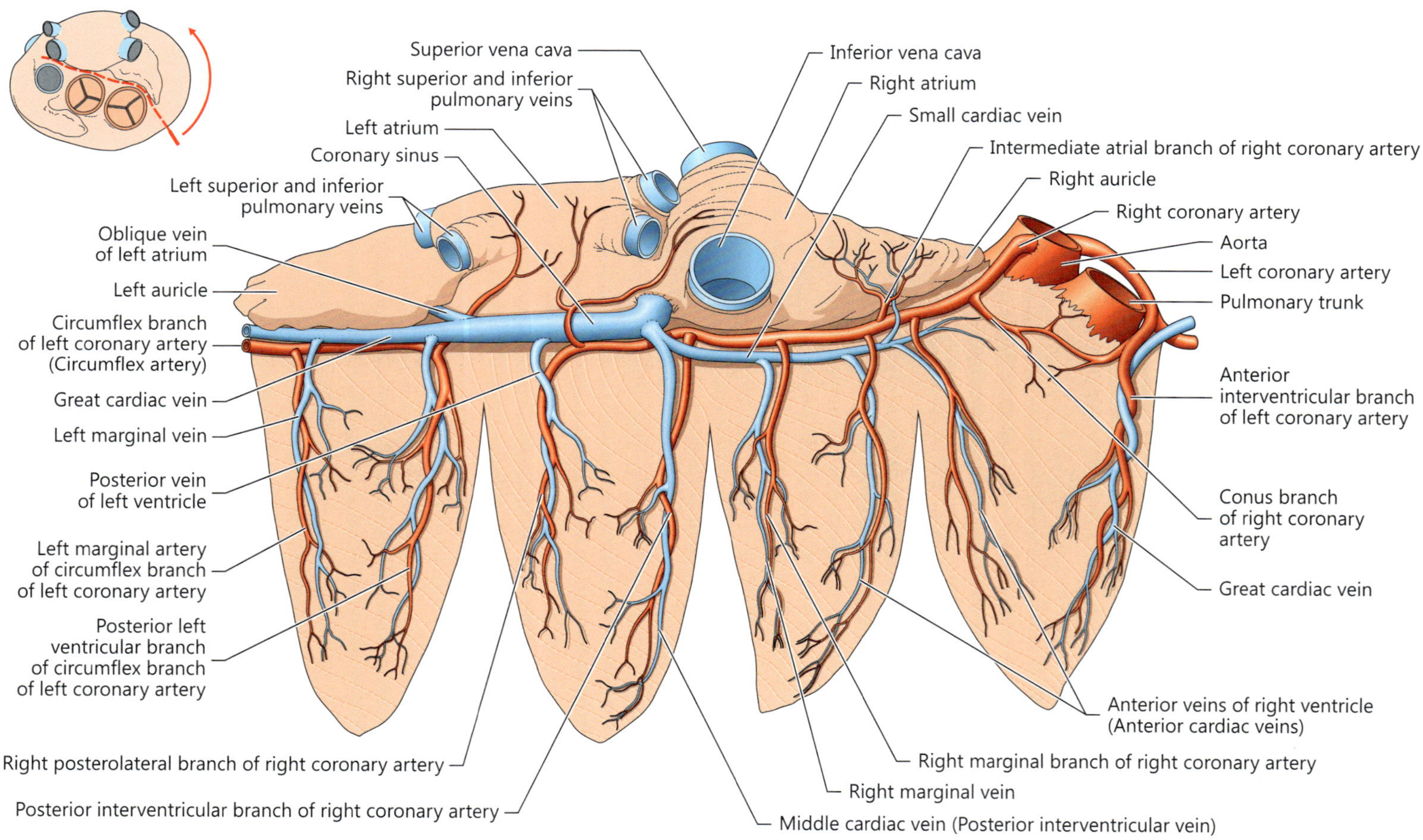

Aphylactic projection of the **cardiac vessels**. The reference relative to the cut surface is illustrated, viewed from above.

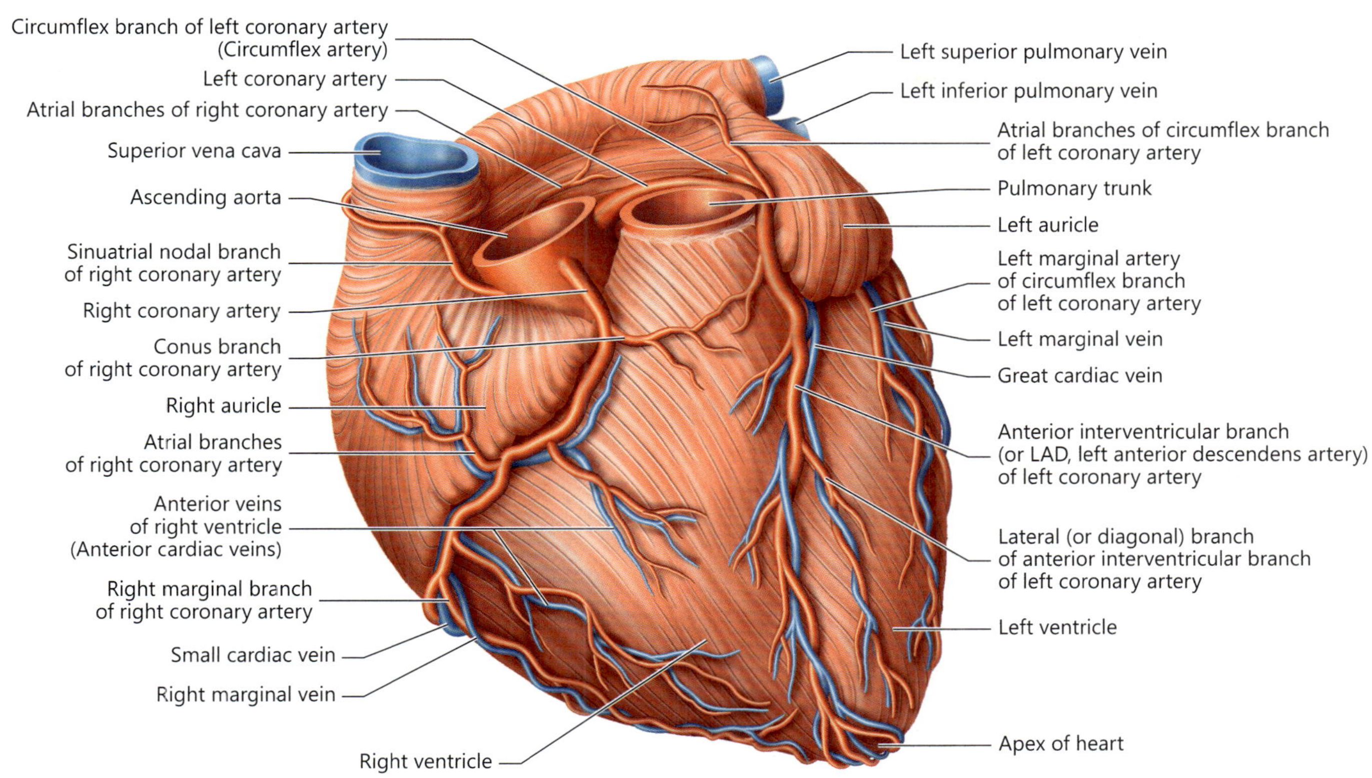

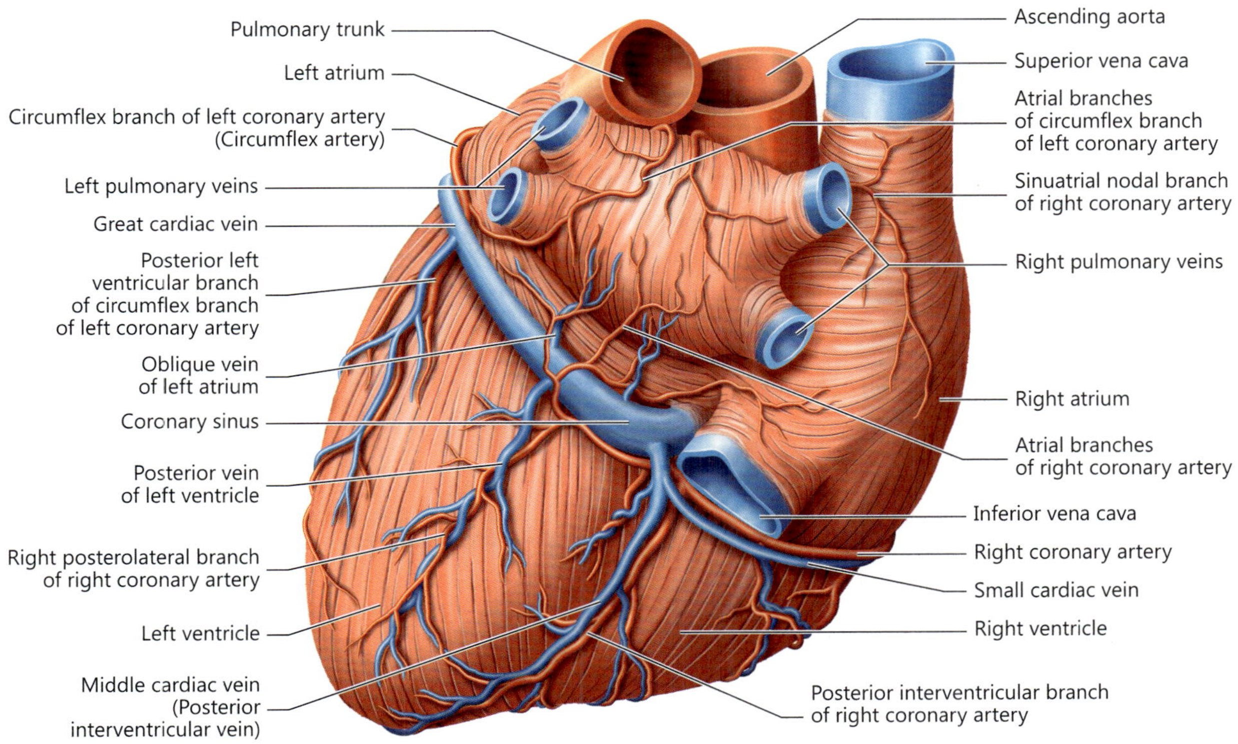

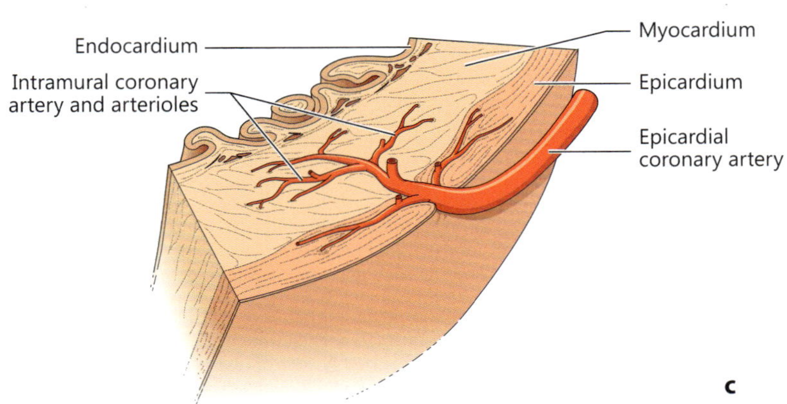

a, Distribution of the **arteries and veins of the heart** on the anterior (sternocostal) surface of the heart. **b**, Distribution of the arteries and veins of the heart on the diaphragmatic (inferior) surface and base of the heart. **c**, **Epicardial coronary artery**, with characteristics of a capacitance vessel opposing little resistance to the flow, and its ramification into intramural coronary artery and arterioles, with characteristics of resistance vessels that determine important variations of flow by dilating or contracting.

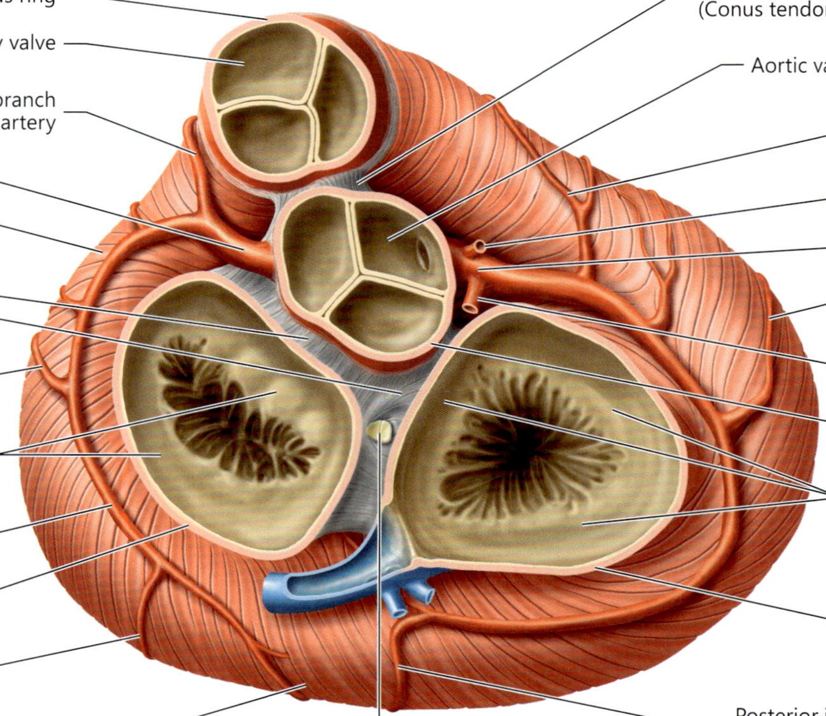

Course of the **coronary arteries** in the coronary sulcus visualized after removal of the atria.

ANATOMIC VARIANTS OF THE ARTERIES OF THE HEART
Conus branch (originates prevalently from the proximal tract of the right coronary artery) can also originate: – from the opening of the right coronary artery – from the aorta
Sinuatrial nodal branch (originates prevalently from the right coronary artery) can also originate: – from the circumflex branch of the left coronary artery – from the right coronary artery and from the circumflex branch of the left coronary artery – from the circumflex branch of the left coronary artery and from the pulmonary trunk – from the aorta
Length of the left coronary artery (prevalent length is 1-2 cm): – <1 cm – >2 cm
Intermediate branch of the left coronary artery (additional branch of the left coronary artery, which prevalently gives off only two branches: the circumflex branch and anterior interventricular branch)
Lateral (diagonal) branches arising from the anterior interventricular branch of the left coronary artery (prevalently two): – 1 – >2
Absence of interventricular septal branches arising from the anterior interventricular branch of the left coronary artery (prevalently present)
Marginal branches originating from the circumflex branch of the left coronary artery (prevalently two): – 1 – >2

ARTERIAL VASCULARIZATION OF THE CONDUCTION SYSTEM	
Sinuatrial node	Sinuatrial nodal branch of the right coronary artery (in approximately 65% of hearts) Sinuatrial nodal branch originating from the circumflex branch of the left coronary artery (in approximately 35% of hearts)
Atrioventricular node	Atrioventricular nodal branch or interventricular septal branch originating from the posterior interventricular branch of the right coronary artery (in approximately 80% of hearts) Atrioventricular nodal branch originating from the circumflex branch of the left coronary artery (in approximately 20% of hearts)
Atrioventricular bundle	Atrioventricular branches of the right coronary artery (in approximately 90% of hearts) Interventricular septal branches originating from the anterior interventricular branch of the left coronary artery (in approximately 10% of hearts)

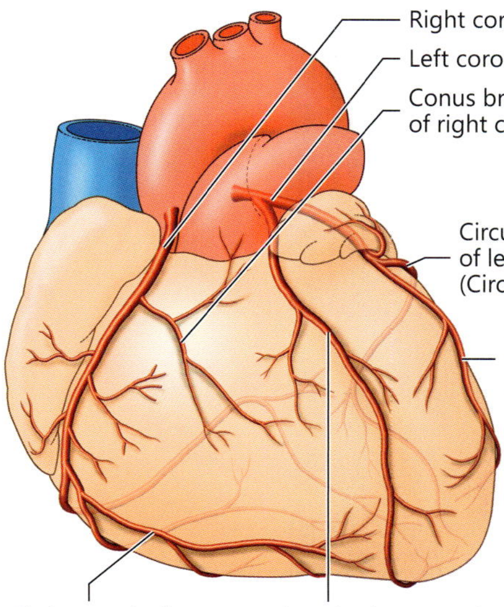

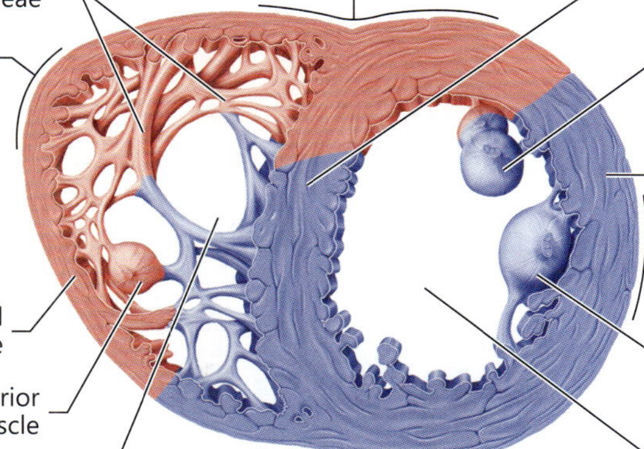

Common disposition: the right coronary artery supplies the posterior wall of the right ventricle and the posterior third of the interventricular septum.

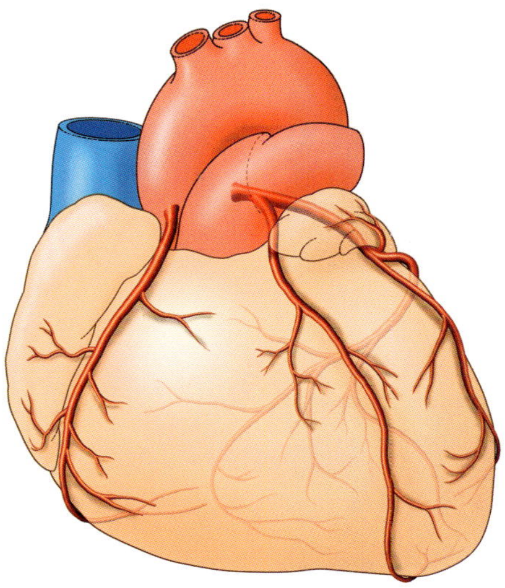

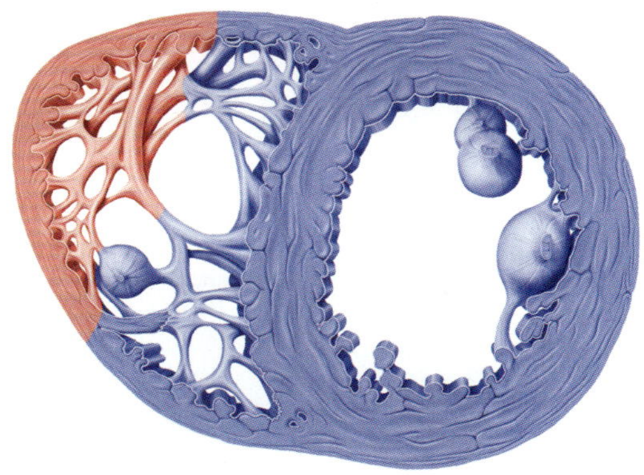

Left dominance: the left coronary artery supplies the posterior wall of the right and left ventricles and the interventricular septum.

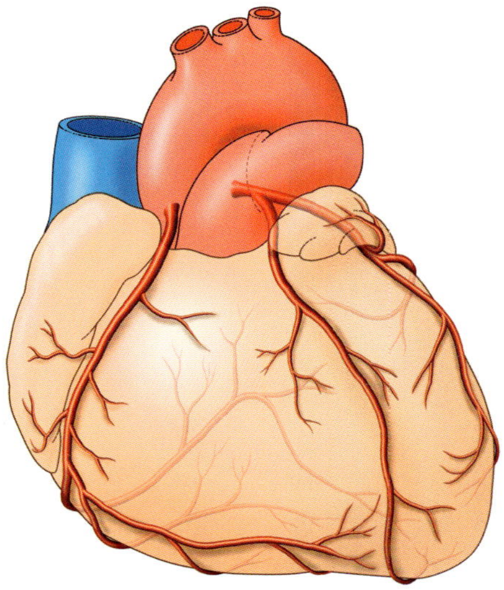

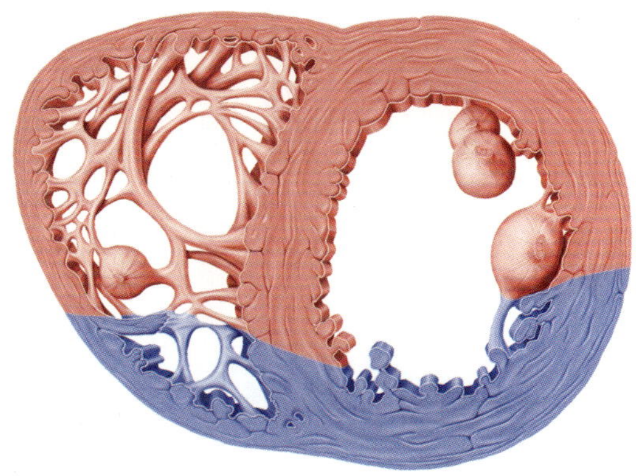

a

b Right dominance: the right coronary artery supplies the posterior wall of the right and left ventricles and the interventricular septum.

Principal variants in the **blood supply of the posterior wall of the heart and interventricular septum**: **a**, viewed from the front; **b**, transverse section of the ventricles showing the territory of supply of the right (in *red*) and left (in *blue*) coronary arteries.

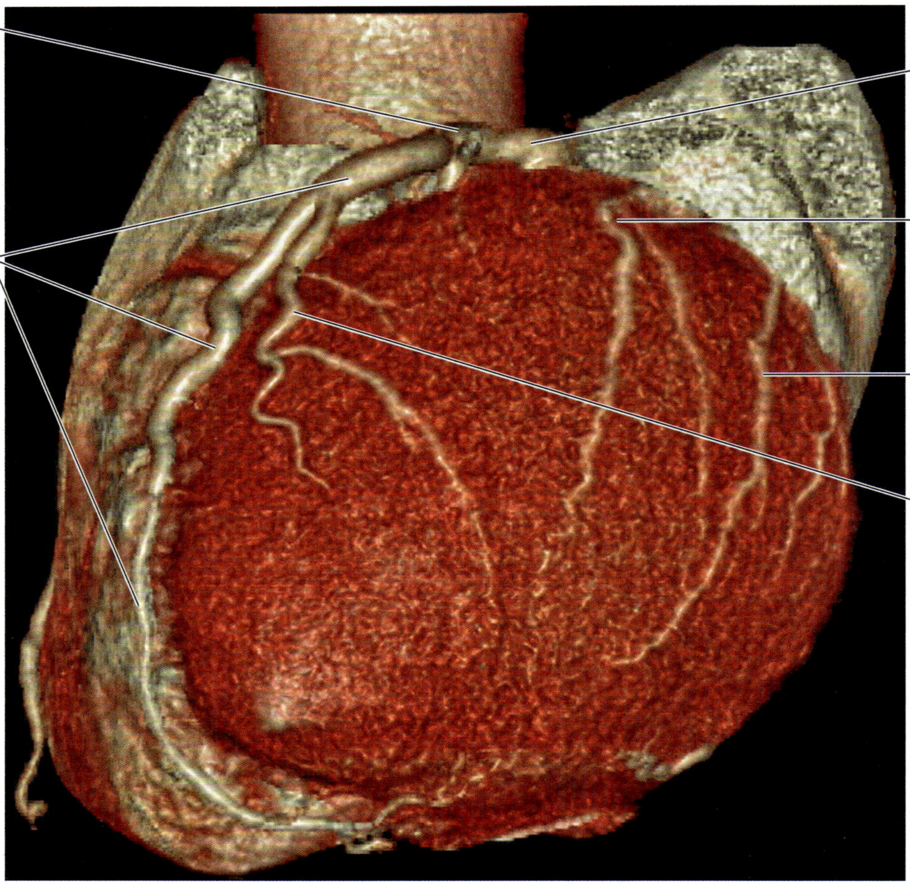

a

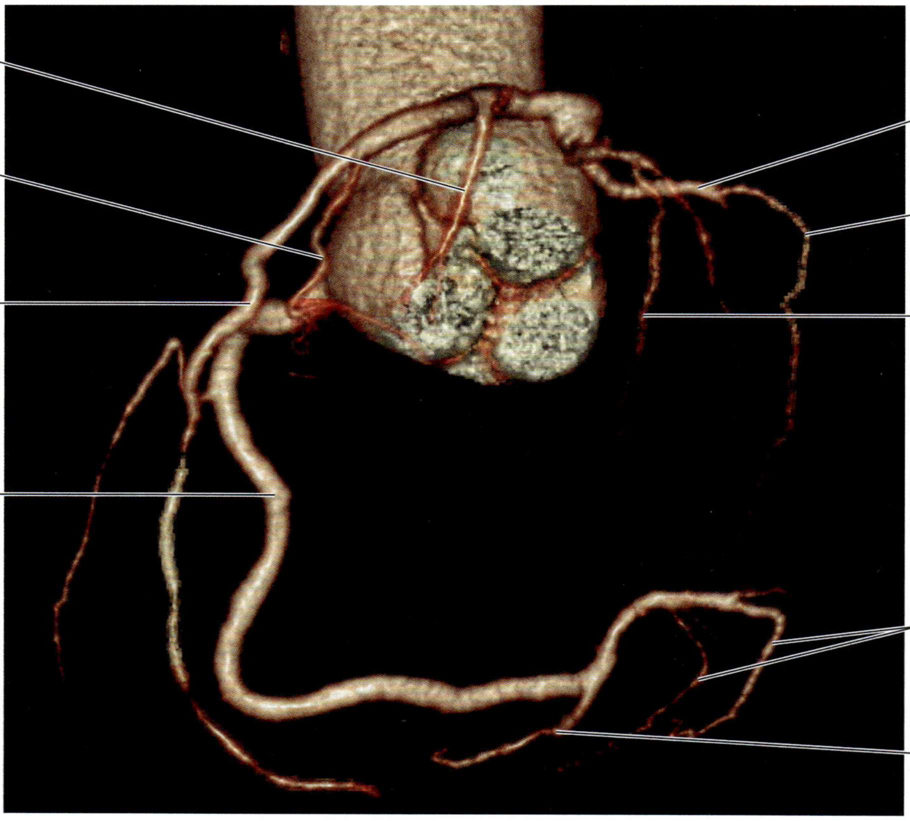

b

Three-dimensional images of a living human body, obtained by CT scan using a 3D volume rendering technique. **a**, The heart is viewed from the left (left or obtuse margin). The branches of the left coronary artery are highlighted. **b**, With the cardiac tissue extracted, also the right coronary artery and aortic sinuses are clearly displayed.

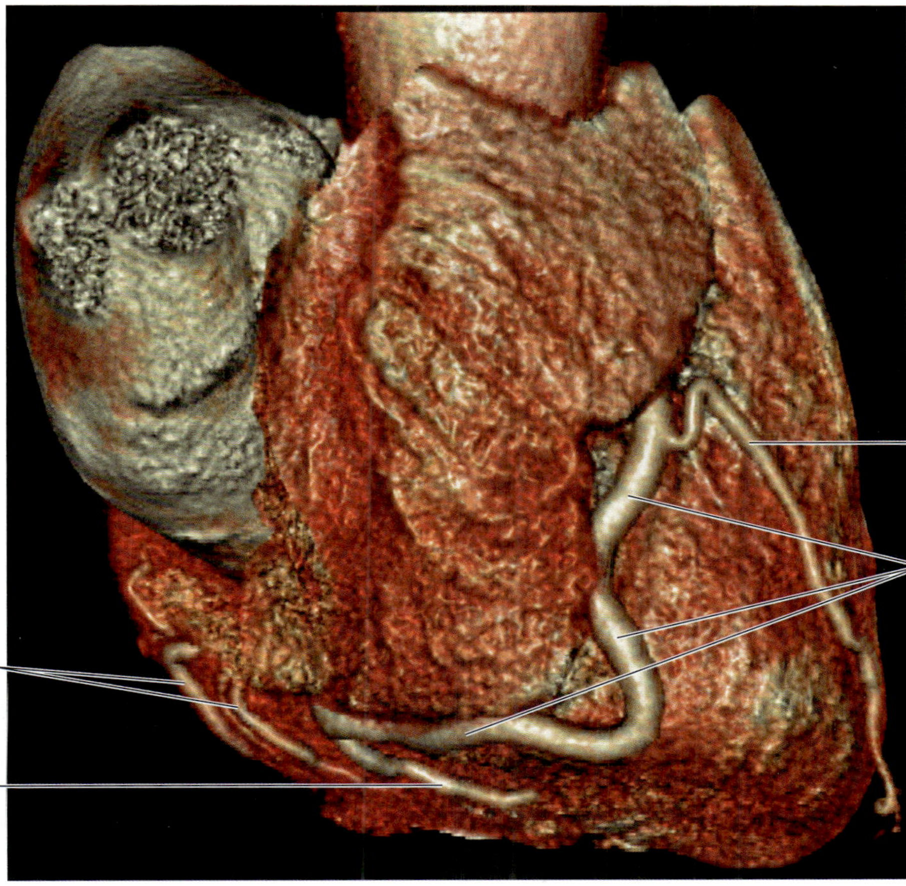

a

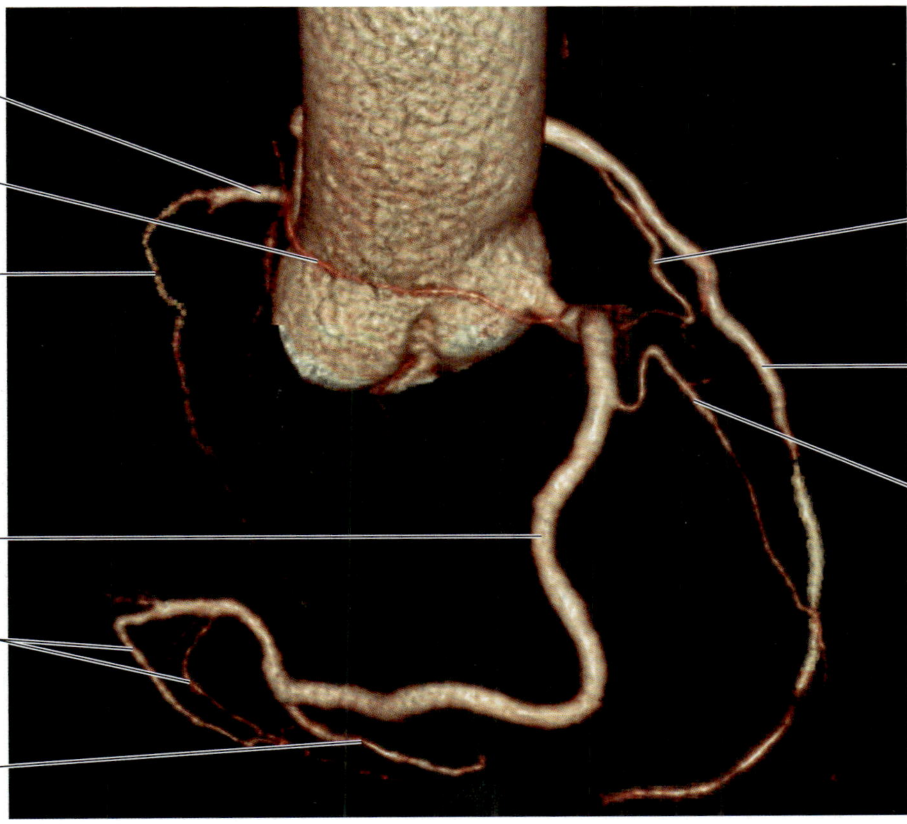

b

Three-dimensional images of a living human body, obtained by CT scan using a 3D volume rendering technique. **a**, The heart is viewed from the right (right or acute margin). The right coronary artery and some of its branches are displayed. **b**, With the cardiac tissue extracted, one can see the whole course of the right coronary artery and some branches of the left coronary artery.

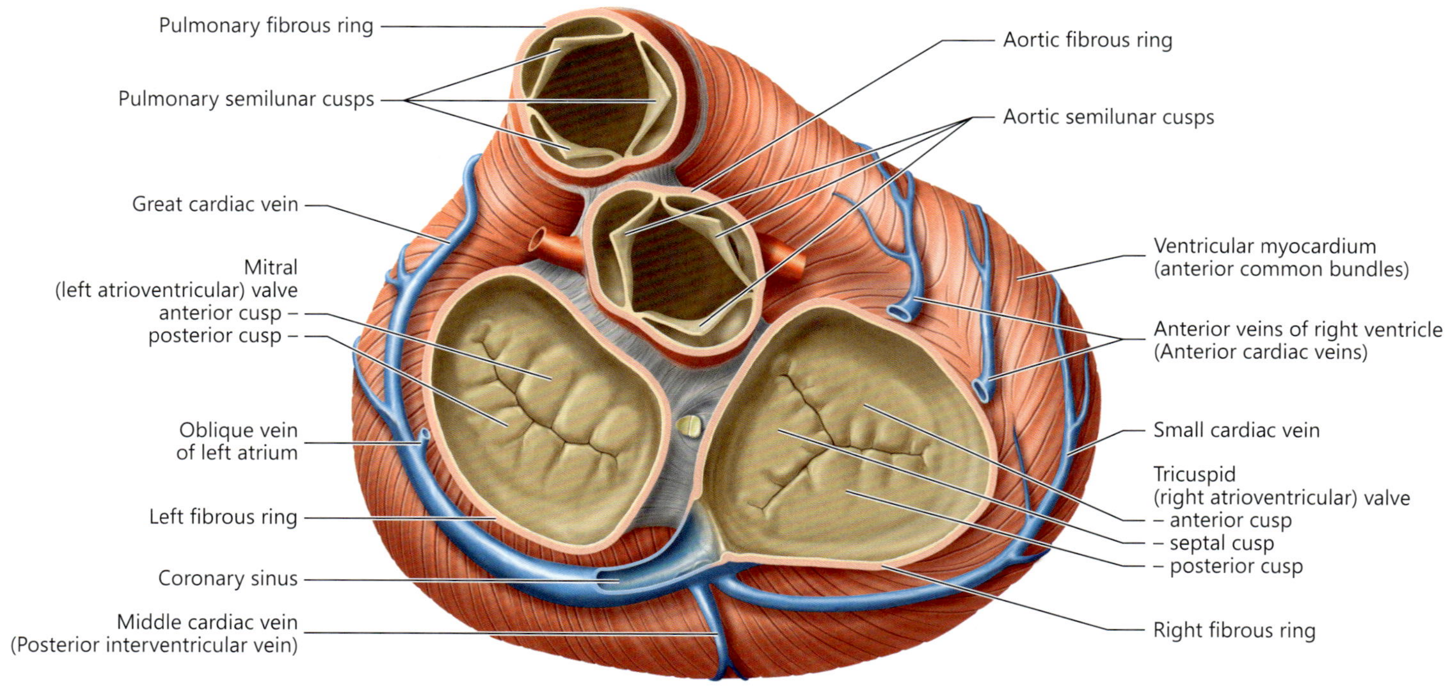

Course of the **cardiac veins** in the coronary sulcus visualized after removal of the atria.

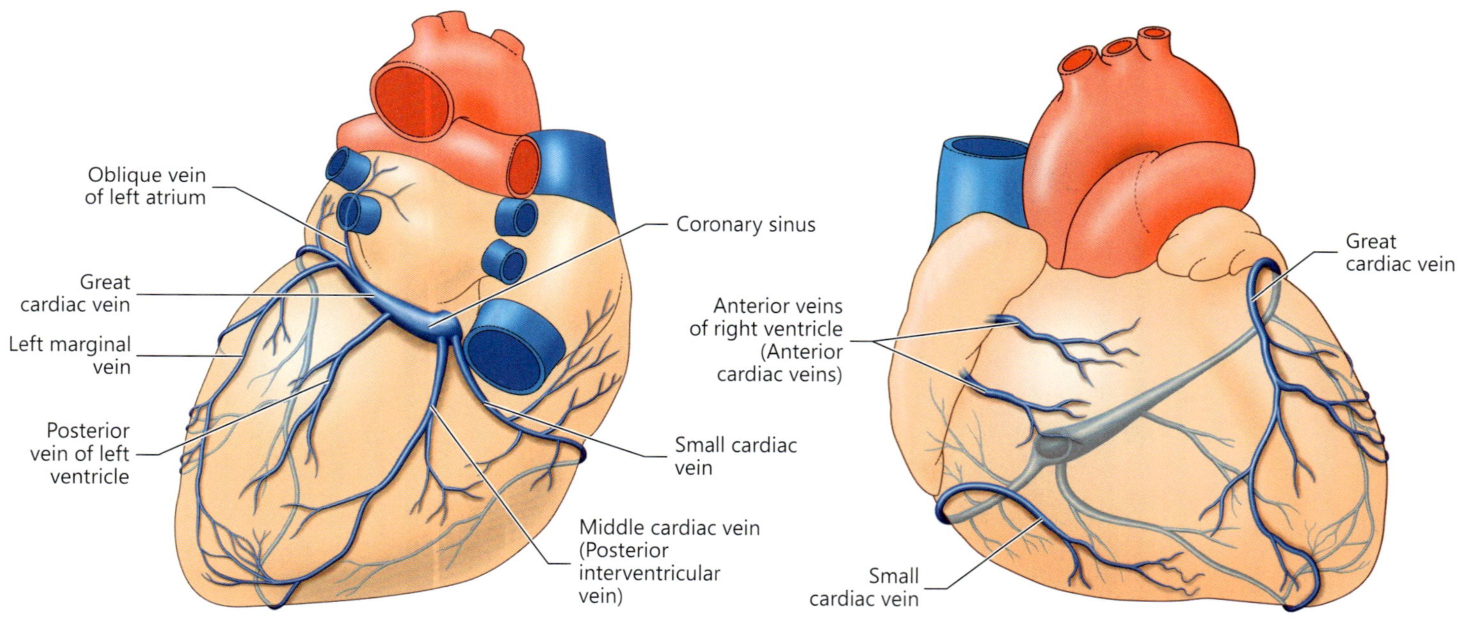

Territories of drainage of the principal **cardiac veins**. On the left, a posterior projection is shown; on the right, an anterior projection. With the myocardium drawn in transparency, one can see the distribution of the vessels over the whole surface of the heart.

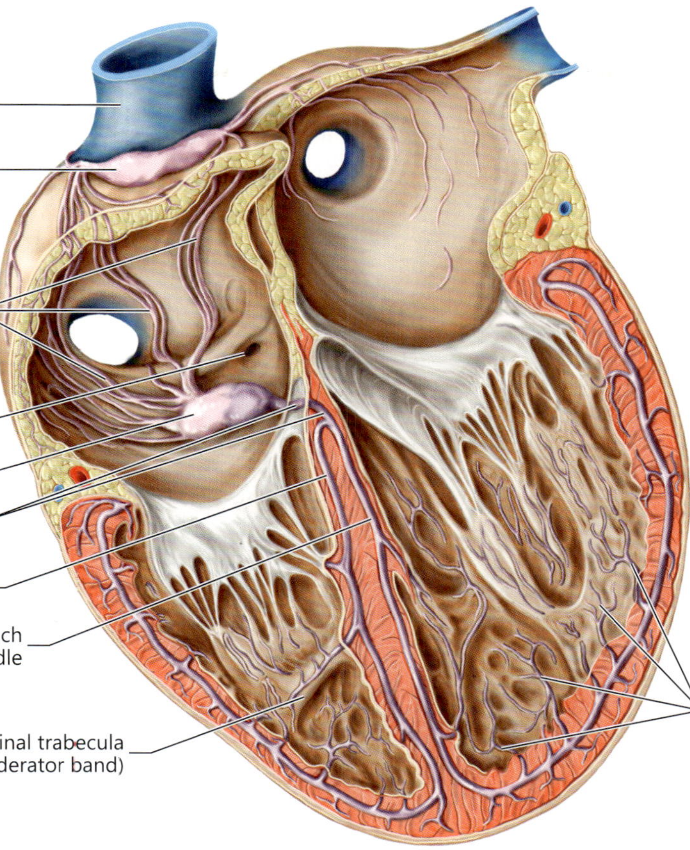

The **sinuatrial node** (SA node, pacemaker of the heart) is a structure formed of specific myocardium, with a semilunar shape, and approximately 10 mm in diameter. It has a more voluminous part situated between the opening of the superior vena cava and the crista terminalis and a thinner part extending downwards to the right. A series of myocardial bundles, which originate from the SA node and run on the anterior, posterior and lateral walls of the right atrium (internodal bundles), lead to the atrioventricular node (AV node). The AV node is situated in the right atrium, within Koch's triangle (see image below), in proximity to the right atrioventricular orifice. From the AV node departs the atrioventricular bundle (bundle of His) which, after crossing the fibrous skeleton of the heart, runs on the right side of the membranous part of the interventricular septum and separates into two branches (right and left). The right branch has a cylindrical form with a diameter of 1-2 mm. It runs on the right surface of the interventricular septum, first within the myocardium, and then in the subendocardial layer. It gives off a branch that enters the septomarginal trabecula (moderator band), which reaches the base of the anterior papillary muscle of the right ventricle. The left branch has the form of a flat tape. It first runs in the subendocardial layer to the left of the interventricular septum, and then divides into two or more branches that lead to the papillary muscles of the left ventricle.

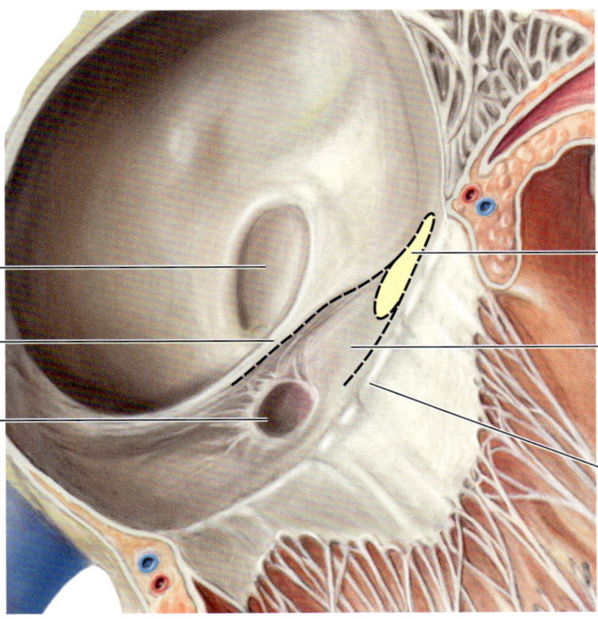

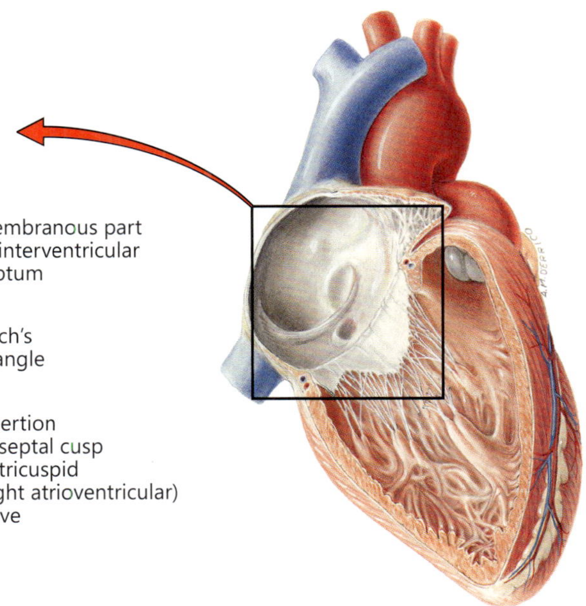

Koch's Triangle. Demarcates the area in which the atrioventricular node is situated.

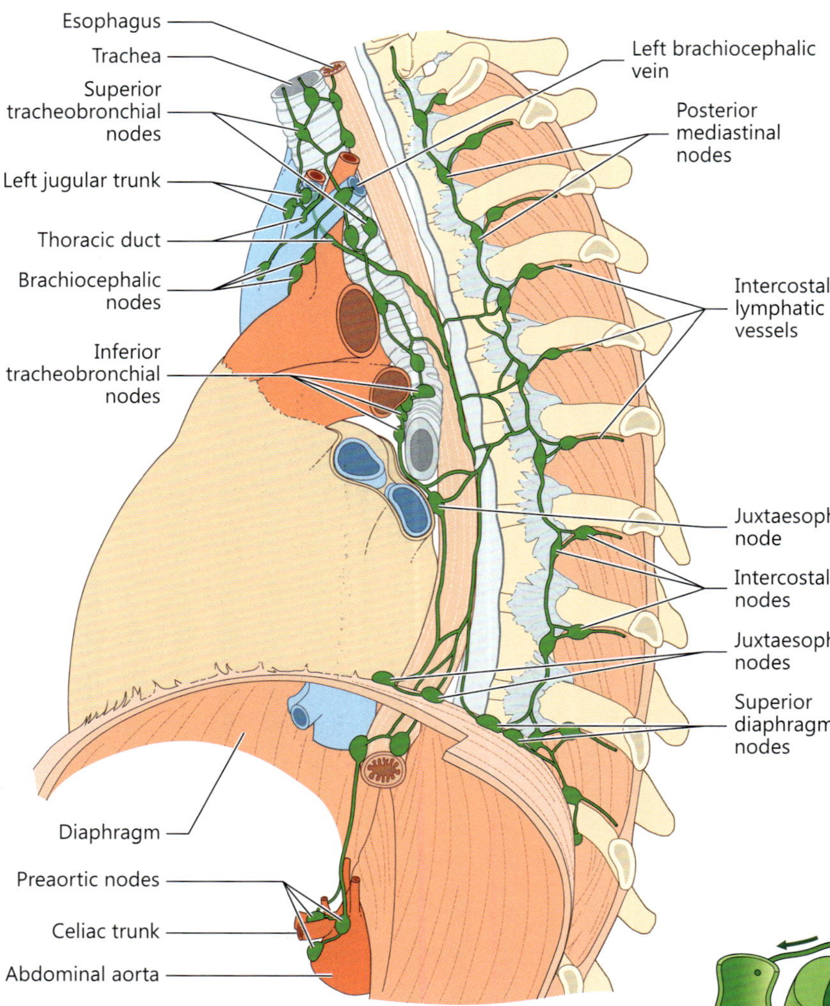

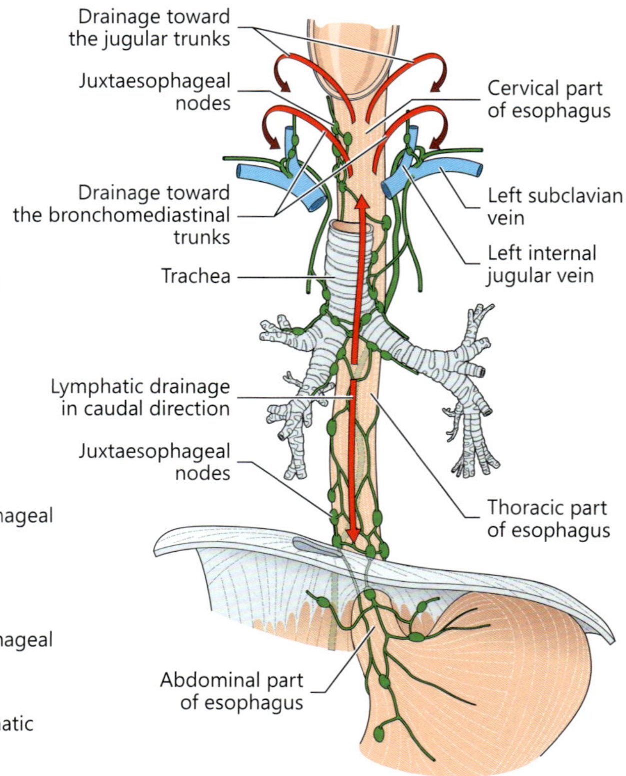

In the three images the **lymph nodes of the thorax** are represented in lateral and frontal projections and in a close-up detail of the transected esophageal wall. Arrows indicate the direction of **lymphatic flow**.

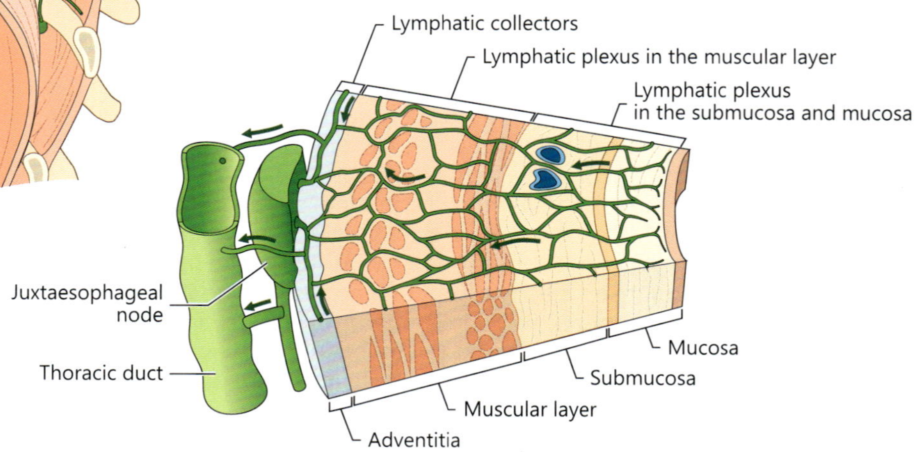

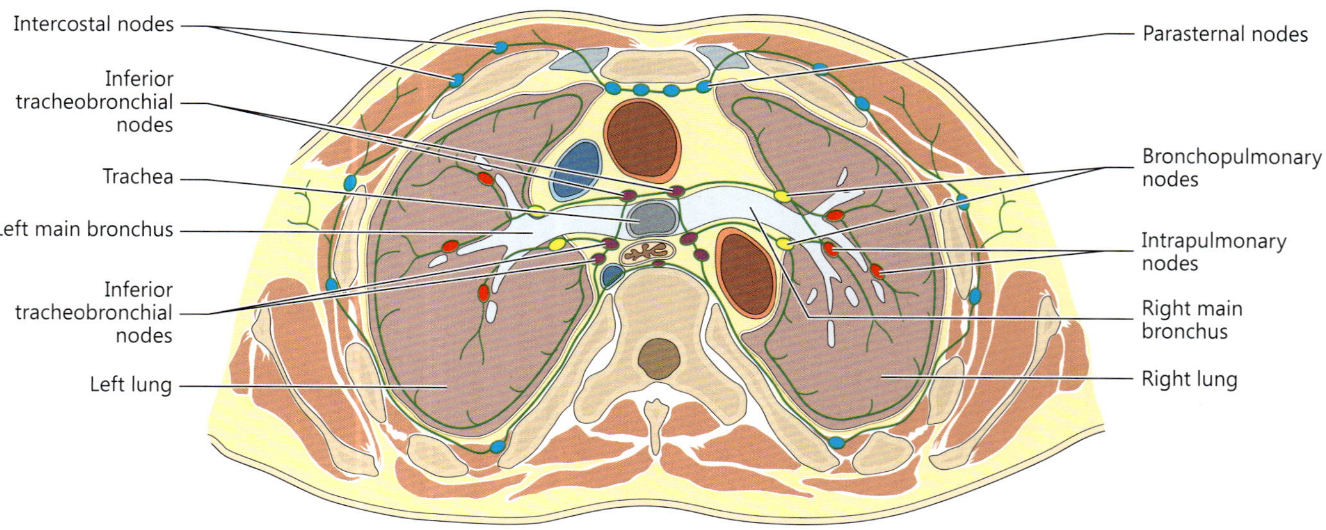

Principal **lymph nodes of the thorax** viewed in transverse section at the level of T IV.

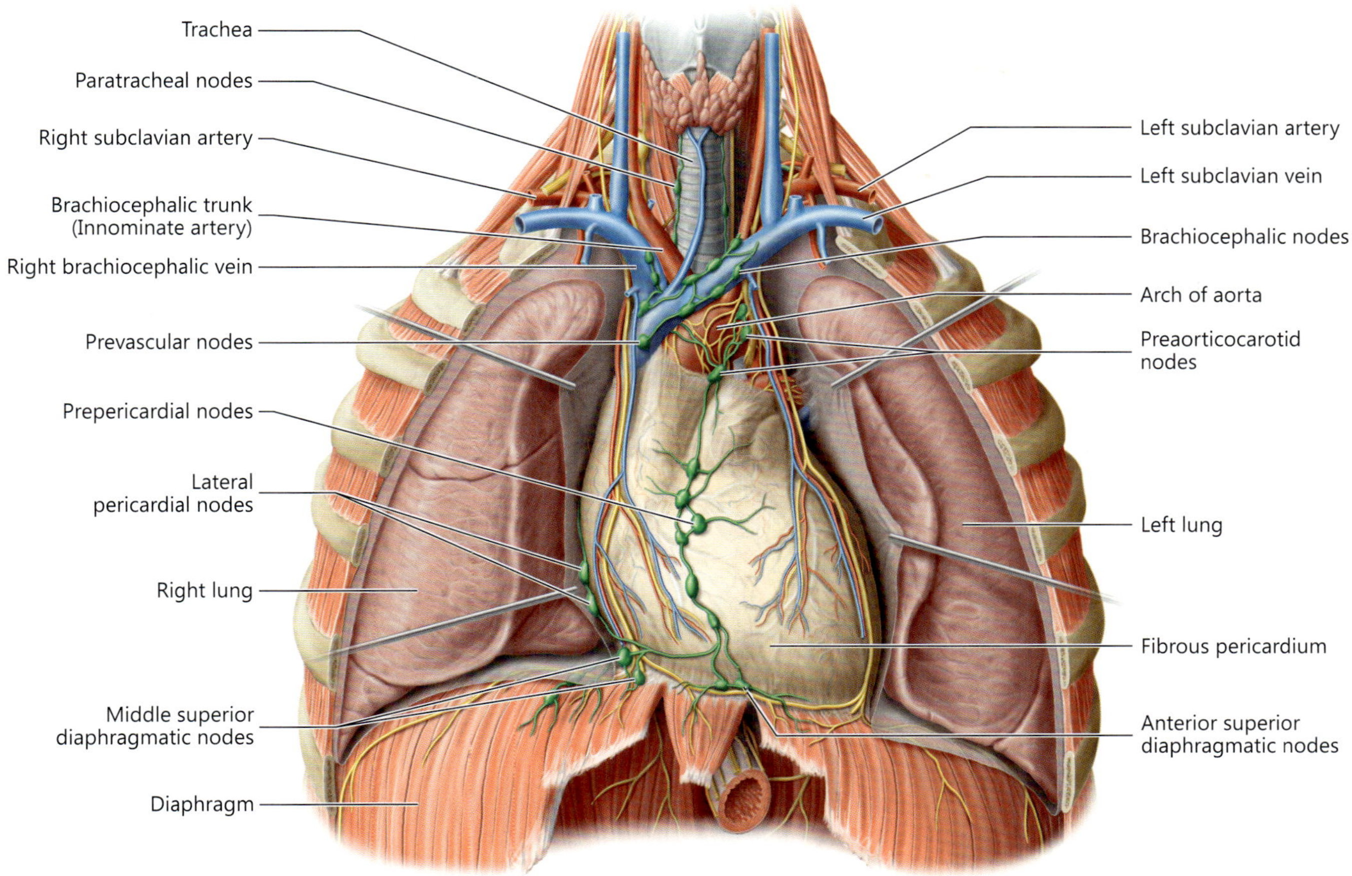

Middle and anterior mediastinal nodes.

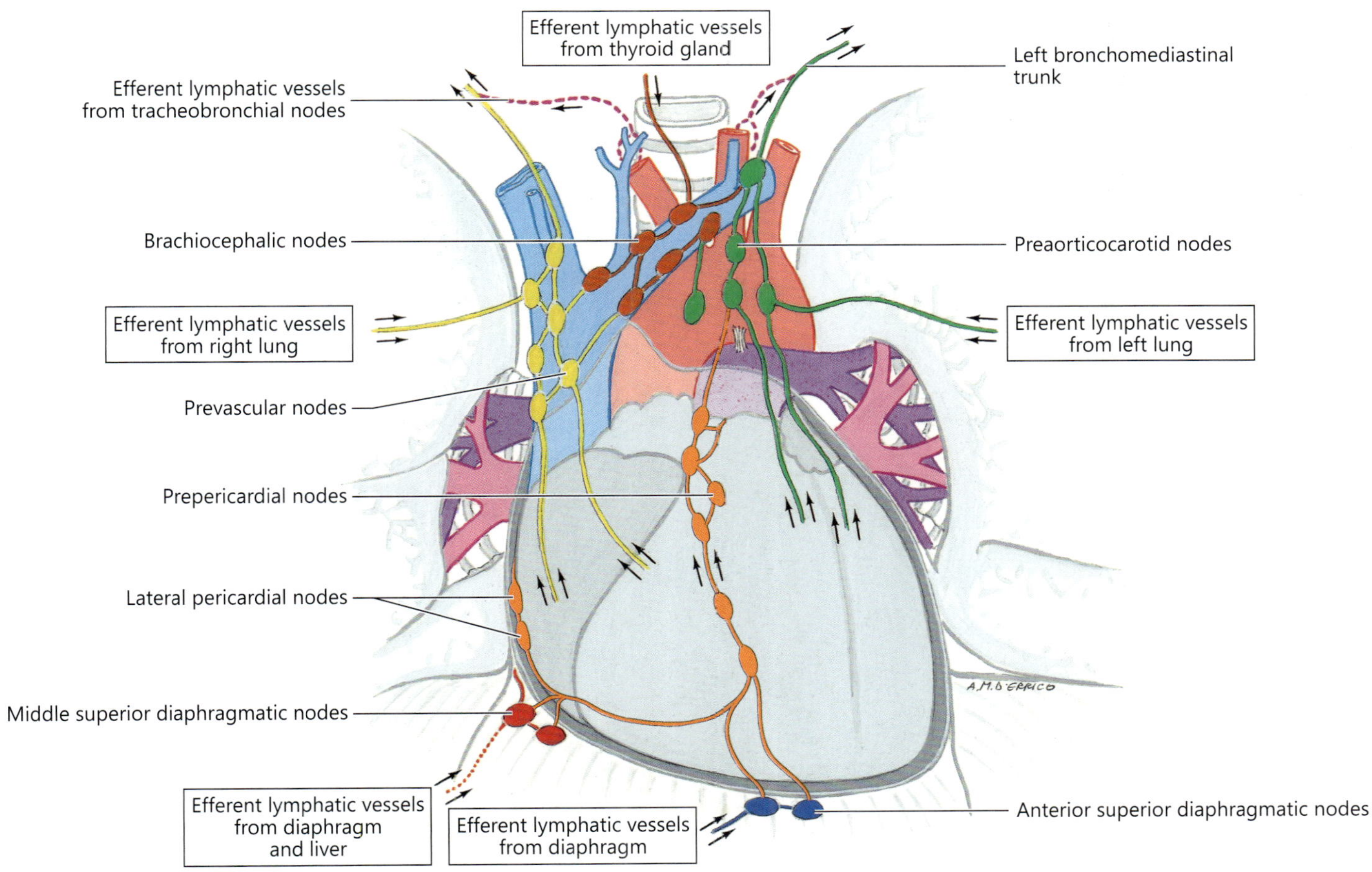

Superior and diaphragmatic mediastinal nodes. The principal pathways of drainage are indicated.

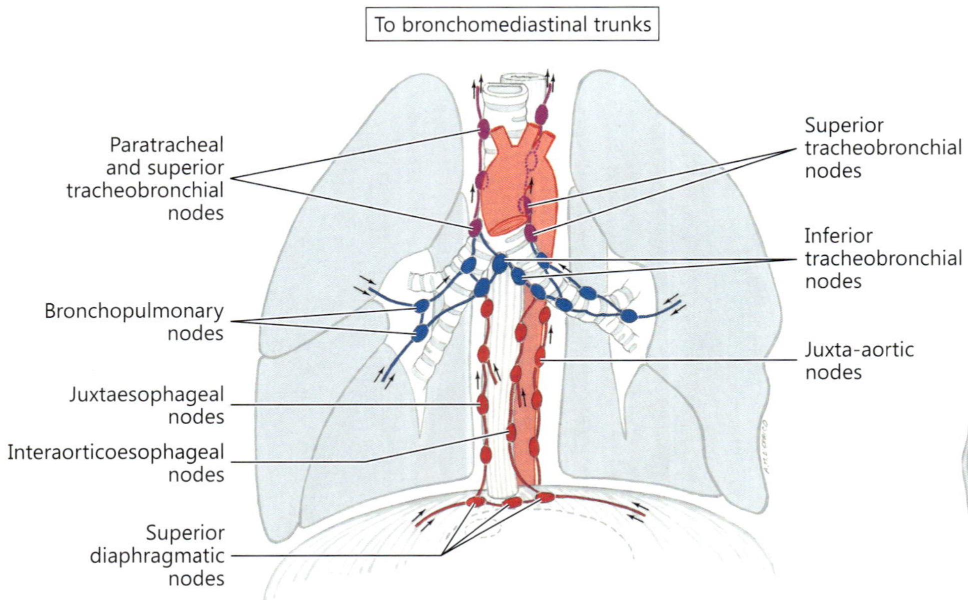

Schematic representation showing the **posterior mediastinal nodes** (in *red*), including the juxtaesophageal, interaorticoesophageal, juxta-aortic and superior diaphragmatic nodes, the **bronchopulmonary** and **inferior tracheobronchial nodes** (in *blue*), and the **superior tracheobronchial** and **paratracheal nodes** (in *purple*). The efferent collectors of the tracheobronchial nodes lead to the bronchomediastinal trunks. The arrows indicate the lymphatic drainage from the organs and the lymphatic flow towards the principal collecting trunks.

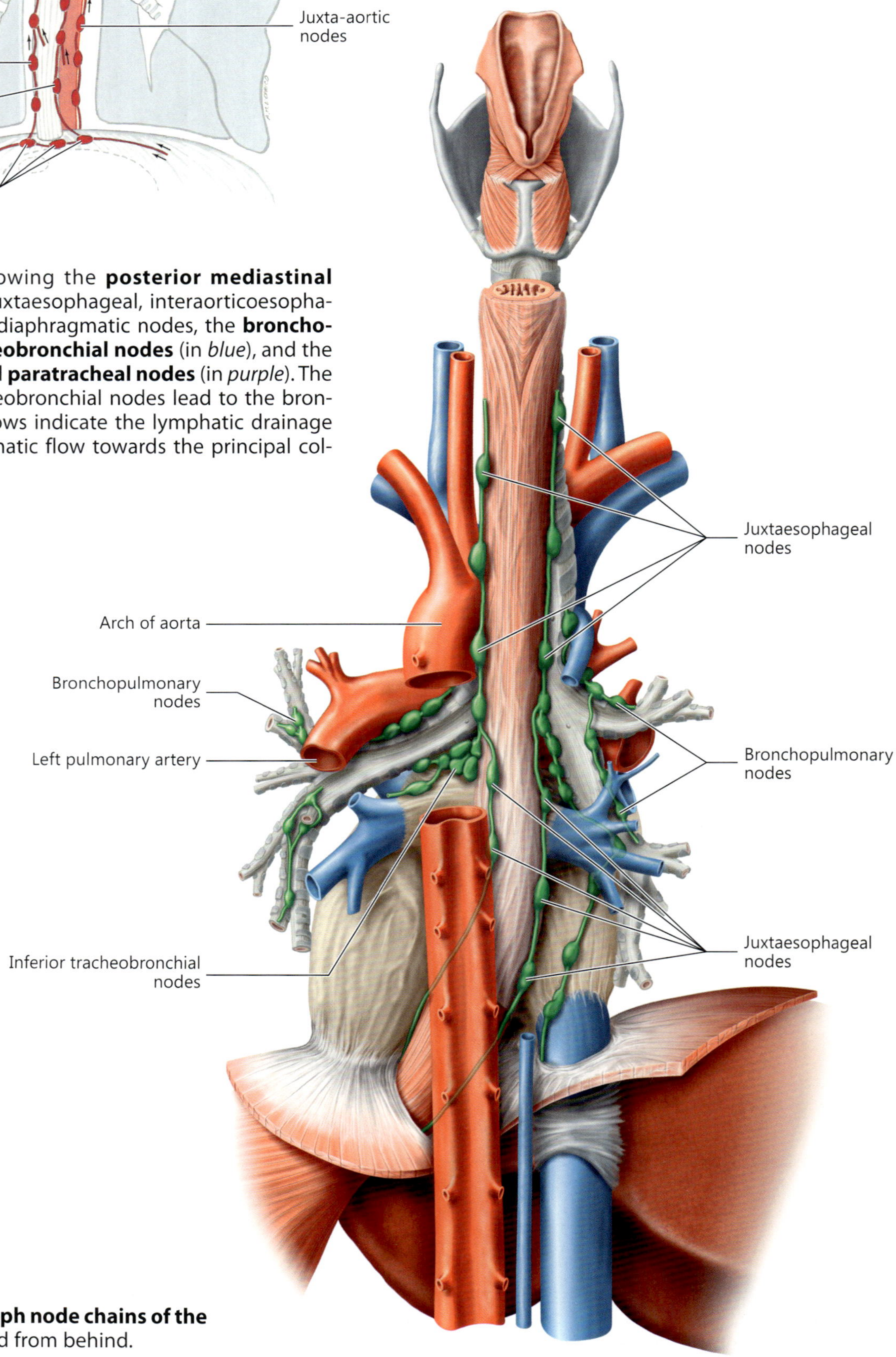

Distribution of the principal **lymph node chains of the posterior mediastinum**, viewed from behind.

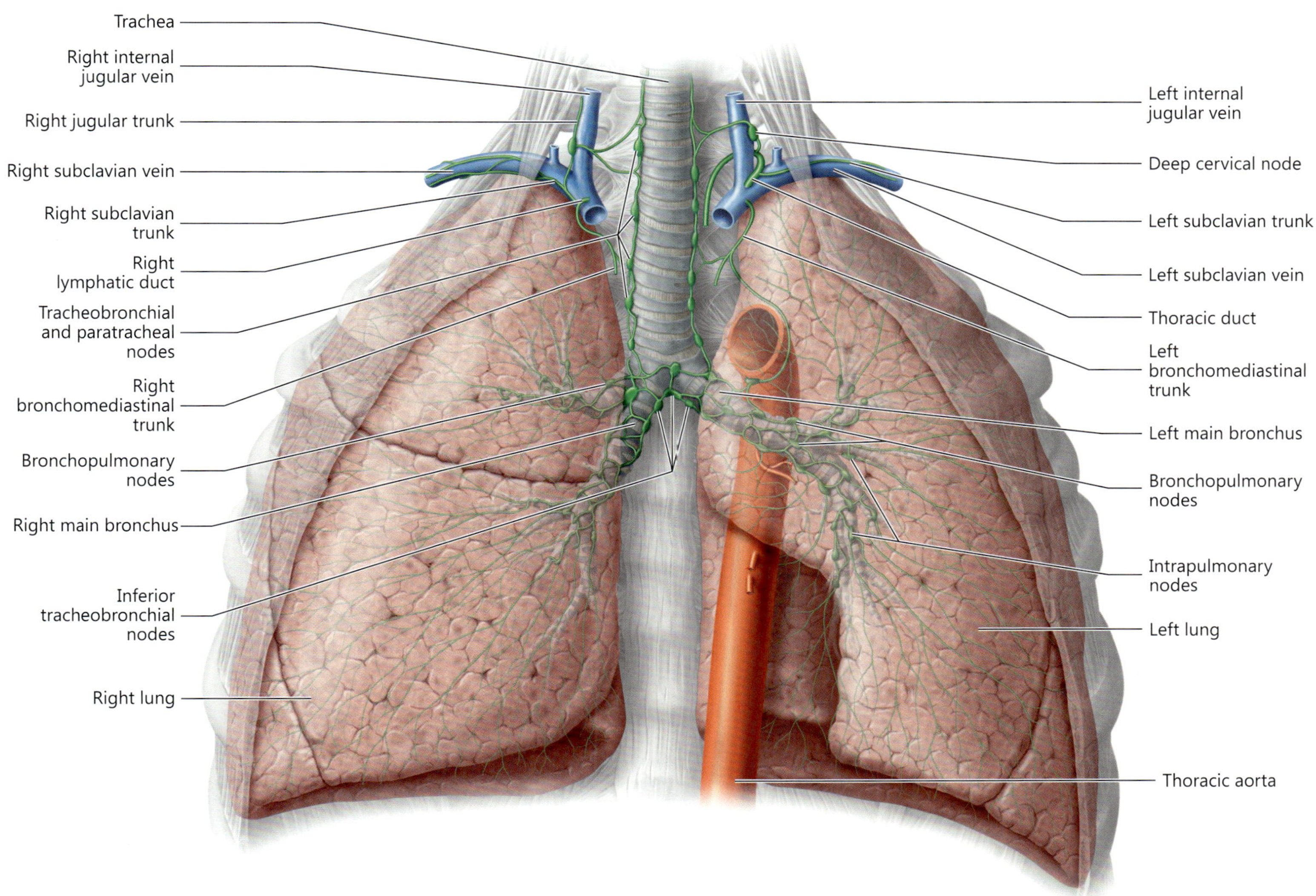

Lymphatic drainage of trachea, bronchi and lungs.

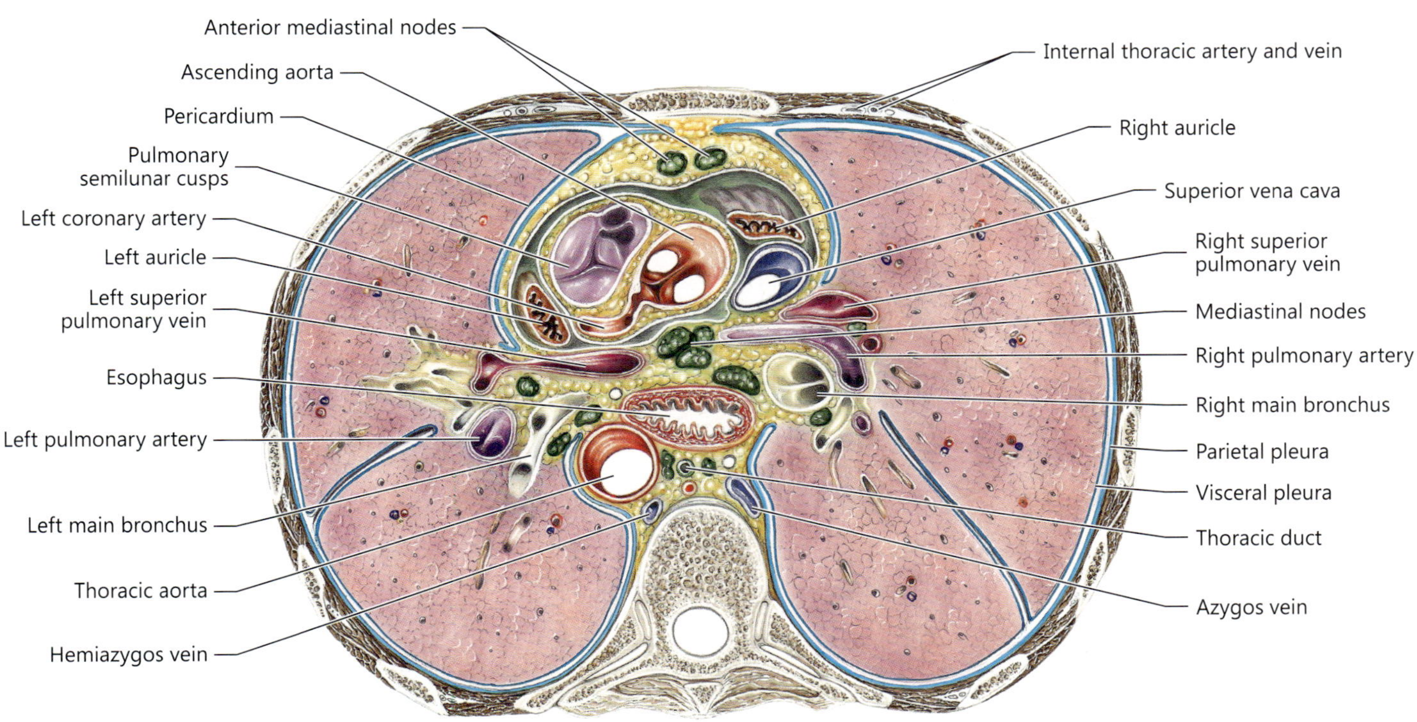

Transverse section passing through the fifth thoracic vertebra. Schematic representation of the **mediastinal nodes** in proximity to the tracheal bifurcation.

Schematic representation of the **innervation of the heart**. Parasympathetic and sympathetic fibers, both afferent and efferent, play a role in the innervation of the heart; they constitute the cardiac plexus. From the cardiac plexus originate fascicles of nerve fibers that travel to the heart, reaching the atrial and ventricular myocardium and the wall of the great vessels.

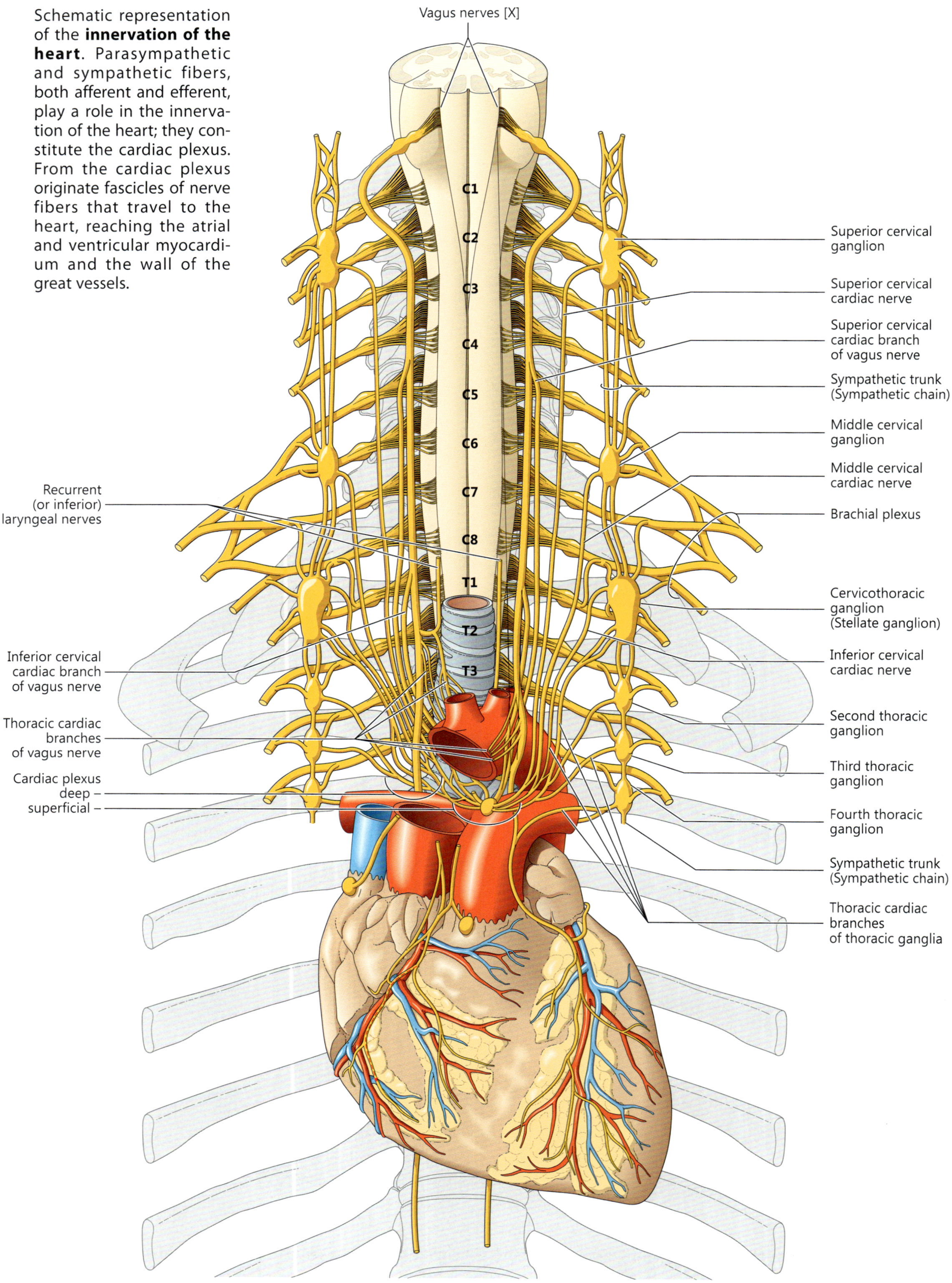

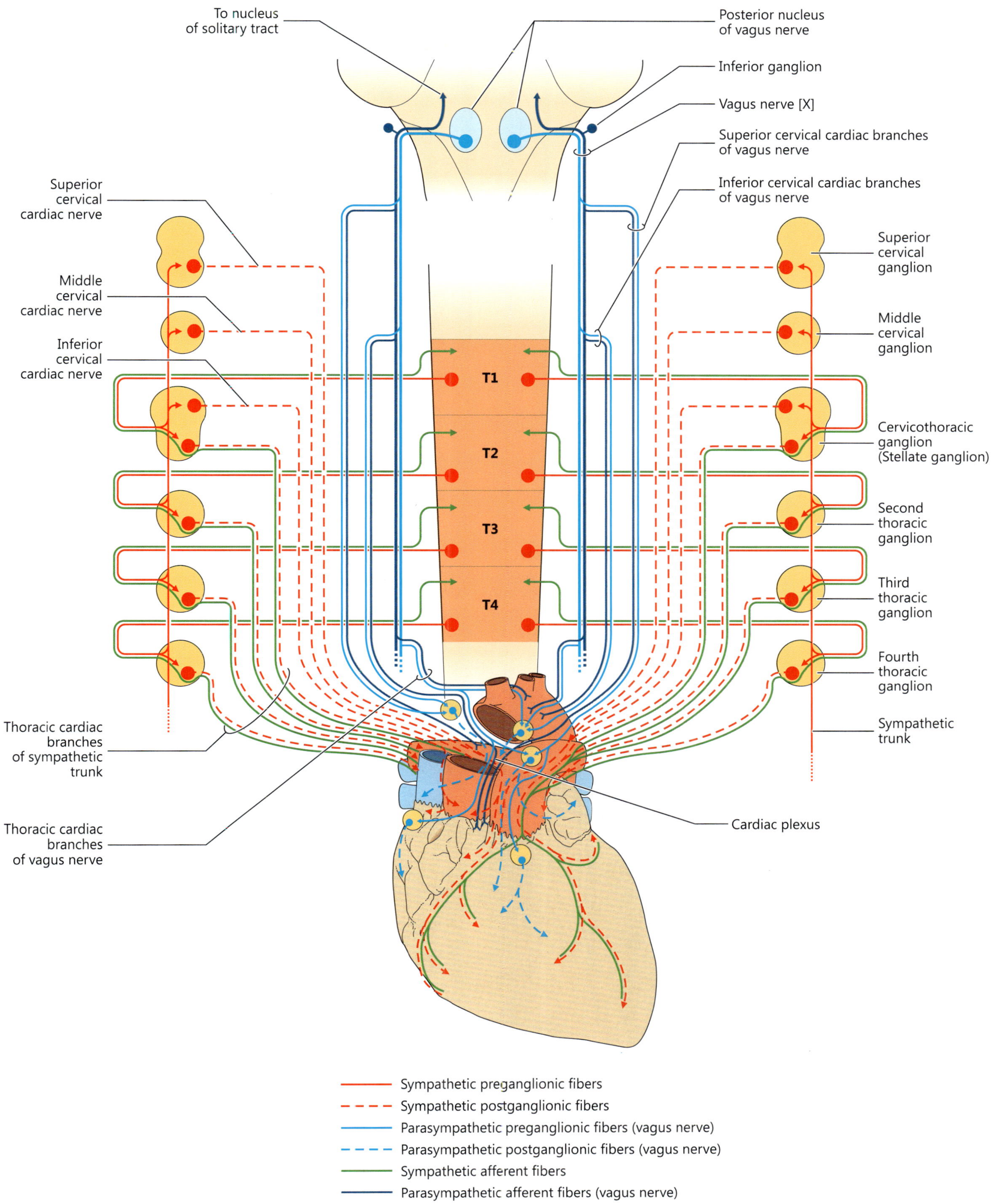

Schematic representation of the **innervation of the heart**. Detail of the composition of the nerve fibers that arrive at and depart from the cardiac plexus.

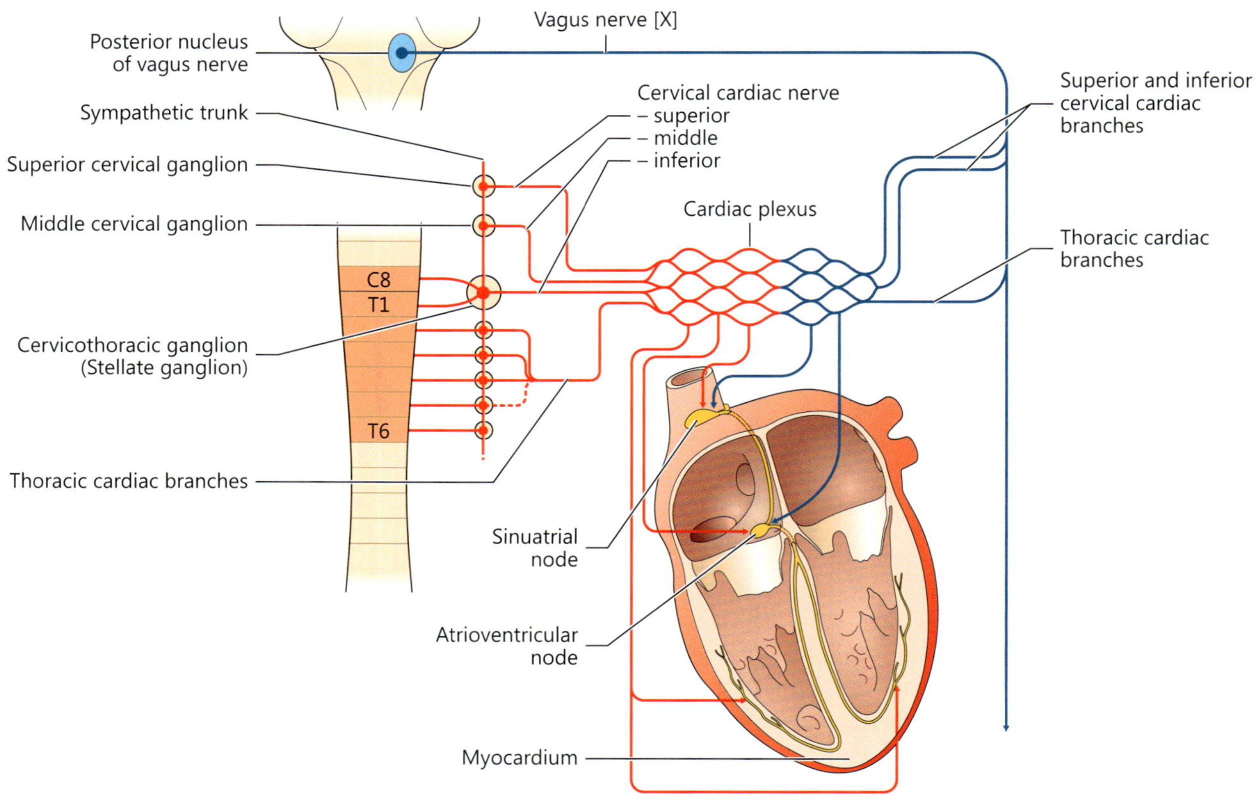

Schematic representation of the **innervation of the heart**. The innervation of the parasympathetic nervous system is carried out by cervical and thoracic collateral branches of the vagus nerve that travel to the cardiac plexus. Branches of the sympathetic nervous system coming from cervical and thoracic ganglia constitute the cardiac plexus. From the cardiac plexus depart branches of the parasympathetic nervous system for the sinuatrial node and atrioventricular node, and branches of the sympathetic system for the sinuatrial node, atrioventricular node, myocardium and coronary vessels. The cardiac conduction system, which generates cardiac impulses autonomously, is illustrated on page 357.

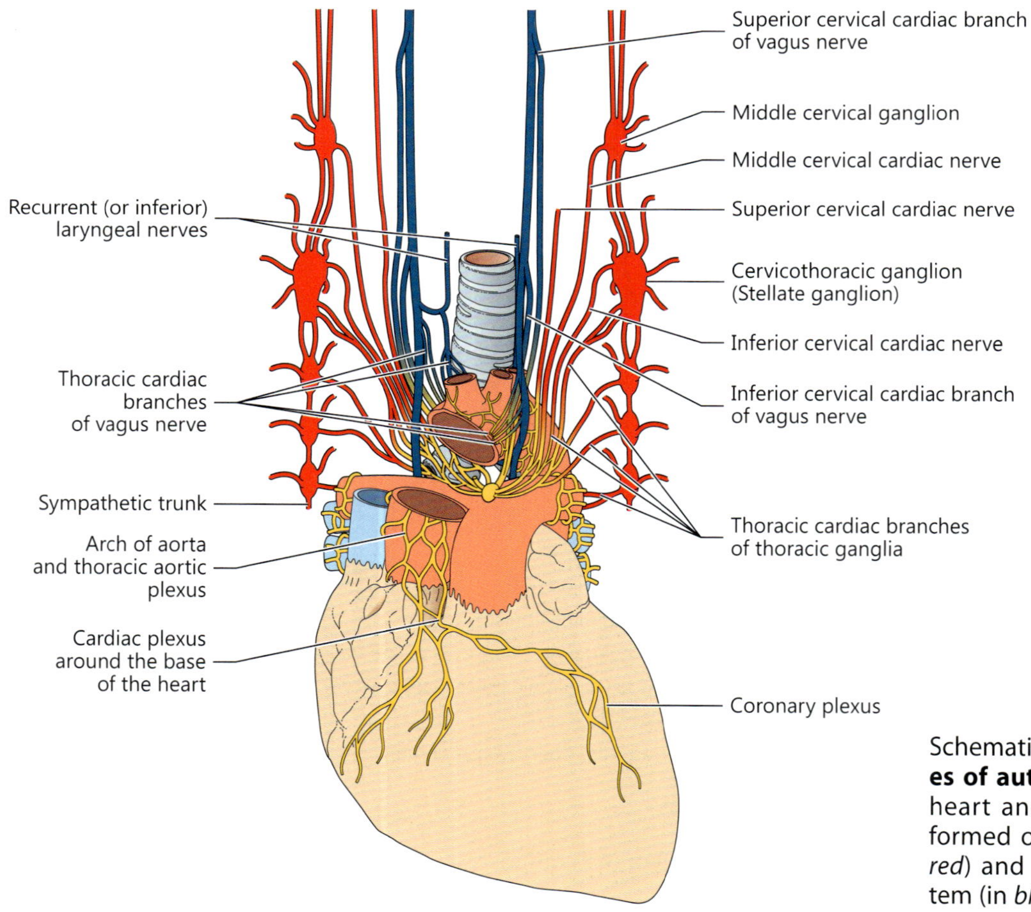

Schematic representation of the **plexuses of autonomic fibers** present on the heart and adjacent vessels, which are formed of fibers of the sympathetic (in *red*) and parasympathetic nervous system (in *blue*).

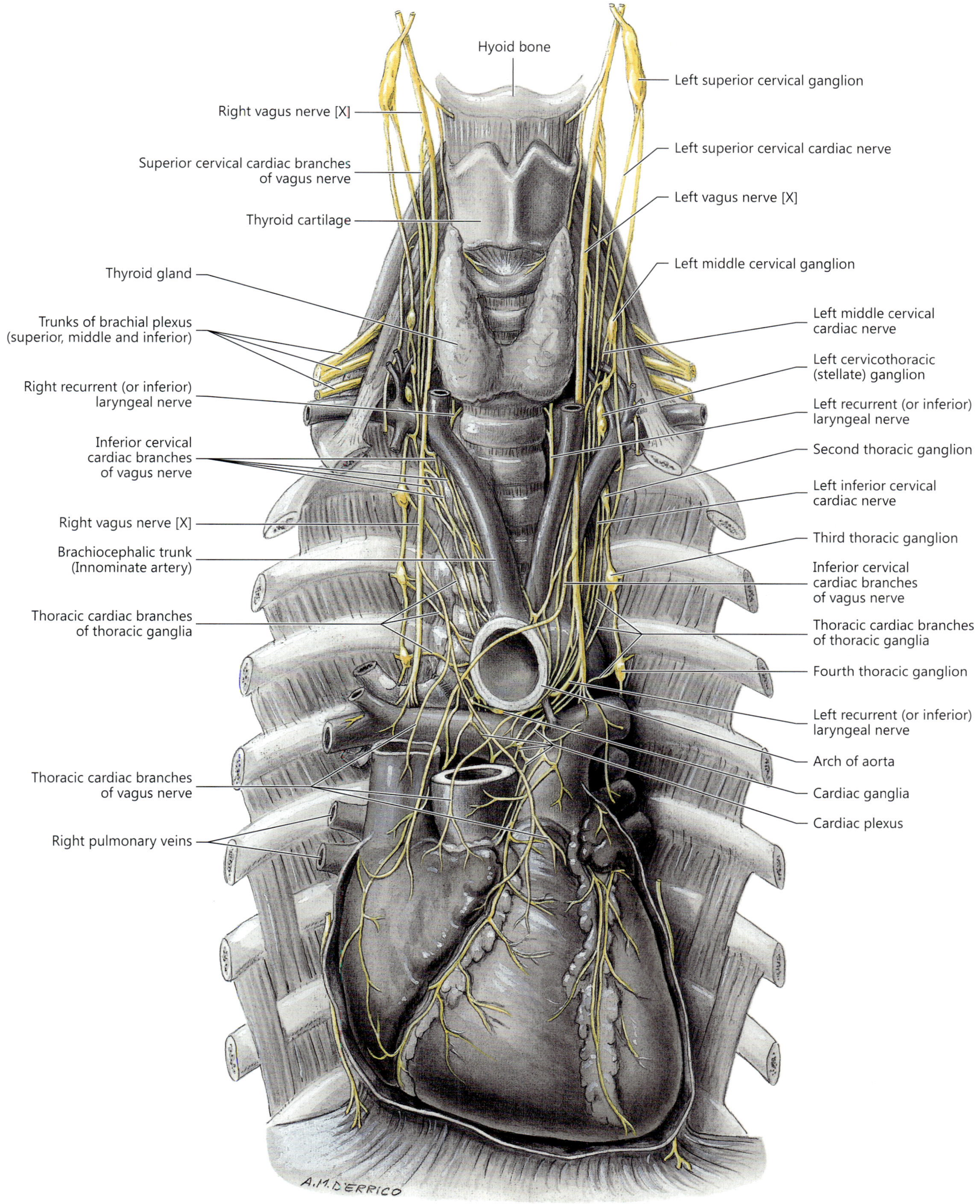

Schematic representation of the **innervation of the heart**. Through resection of the arch of the aorta, it is possible to see the exact location of the cardiac plexus, situated in correspondence to the base of the heart, between the arch of the aorta and the point of division of the pulmonary trunk. Removing the pericardium, one can see fascicles of nerve fibers accompanying the right and left coronary arteries, which travel to innervate the atrial and ventricular myocardium.

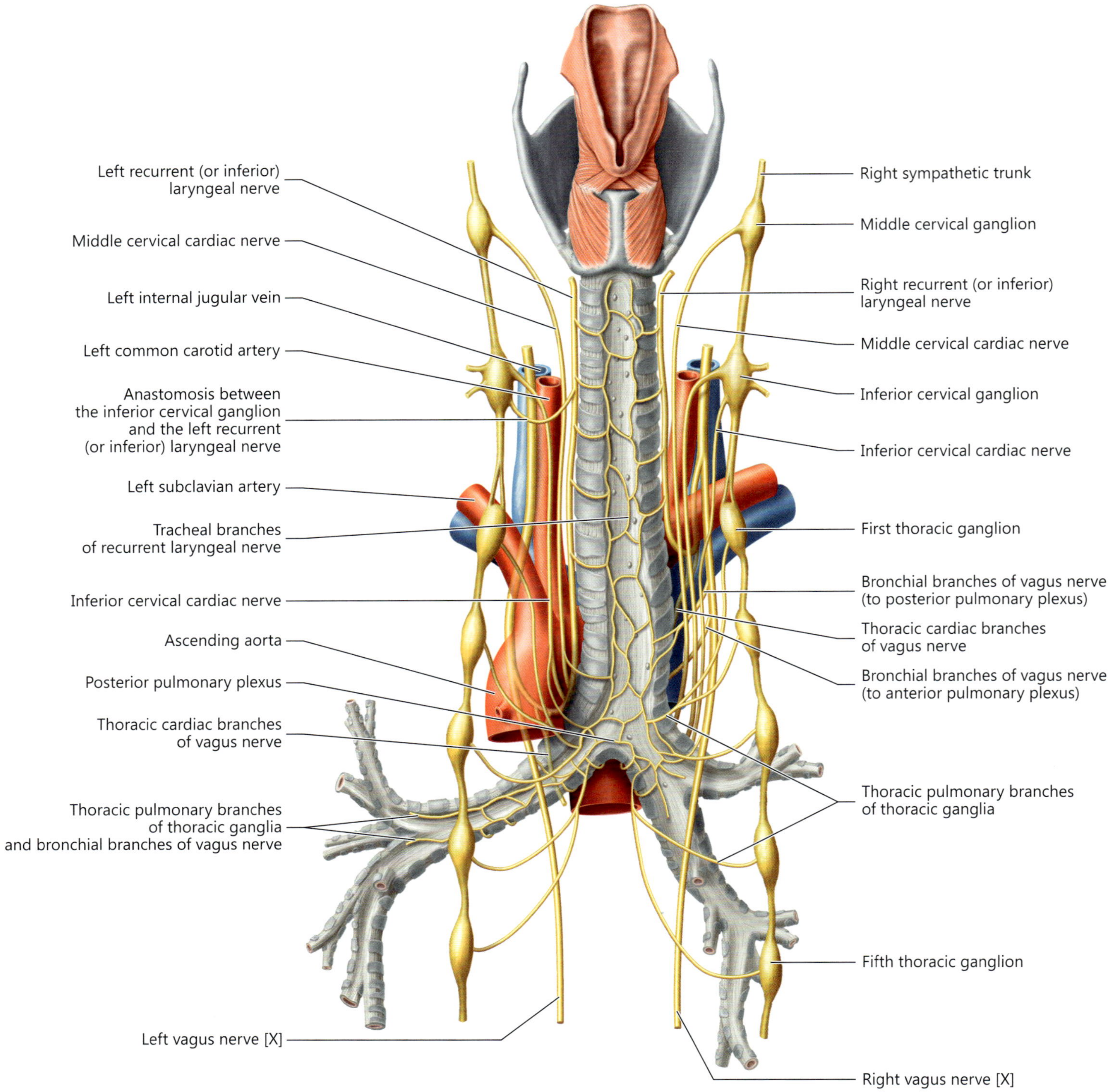

Representation of the **innervation of the heart**, posterior view. Note the course of the thoracic cardiac branches, coming from the vagus nerve, which join the cardiac plexus.

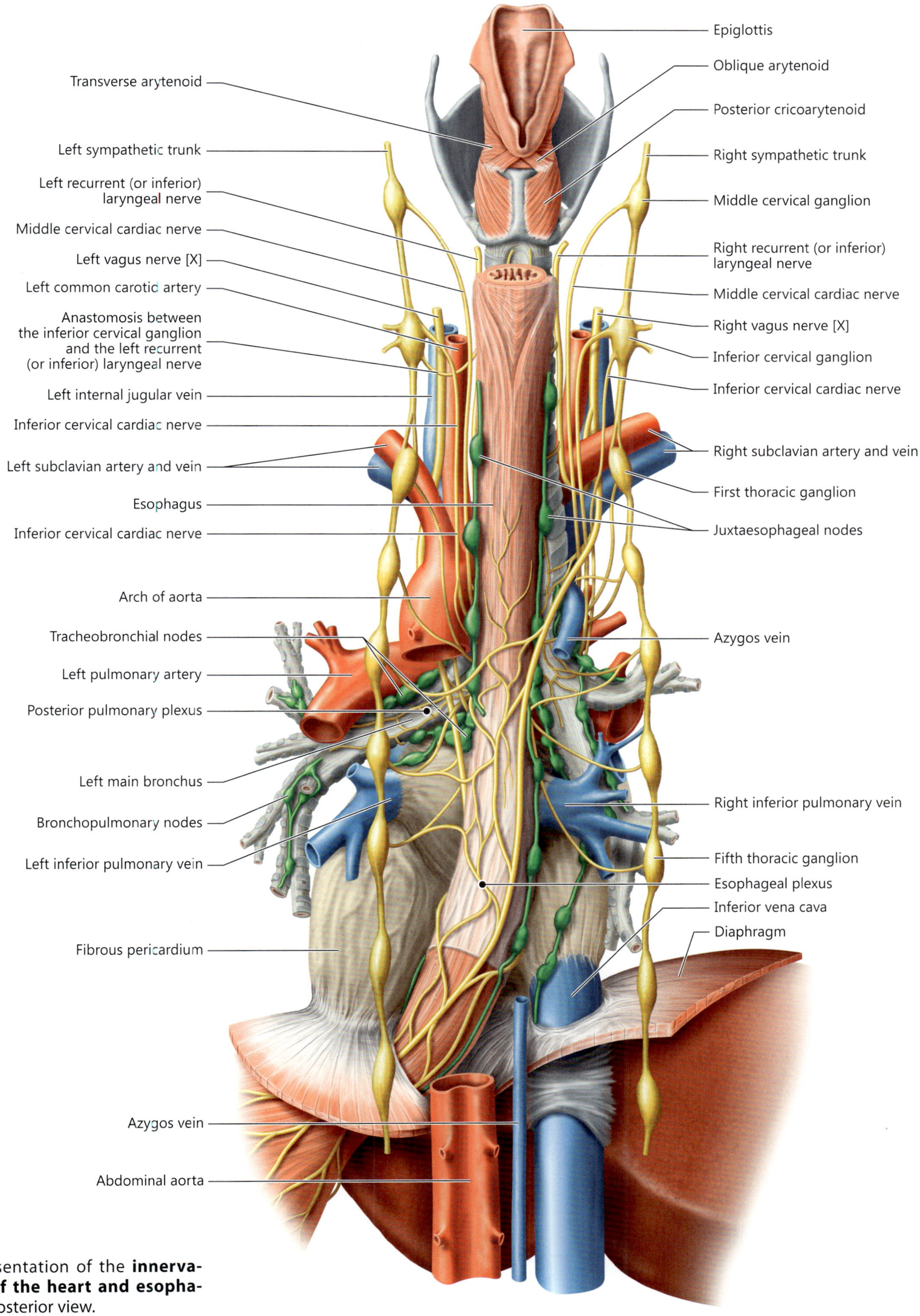

Representation of the **innervation of the heart and esophagus**, posterior view.

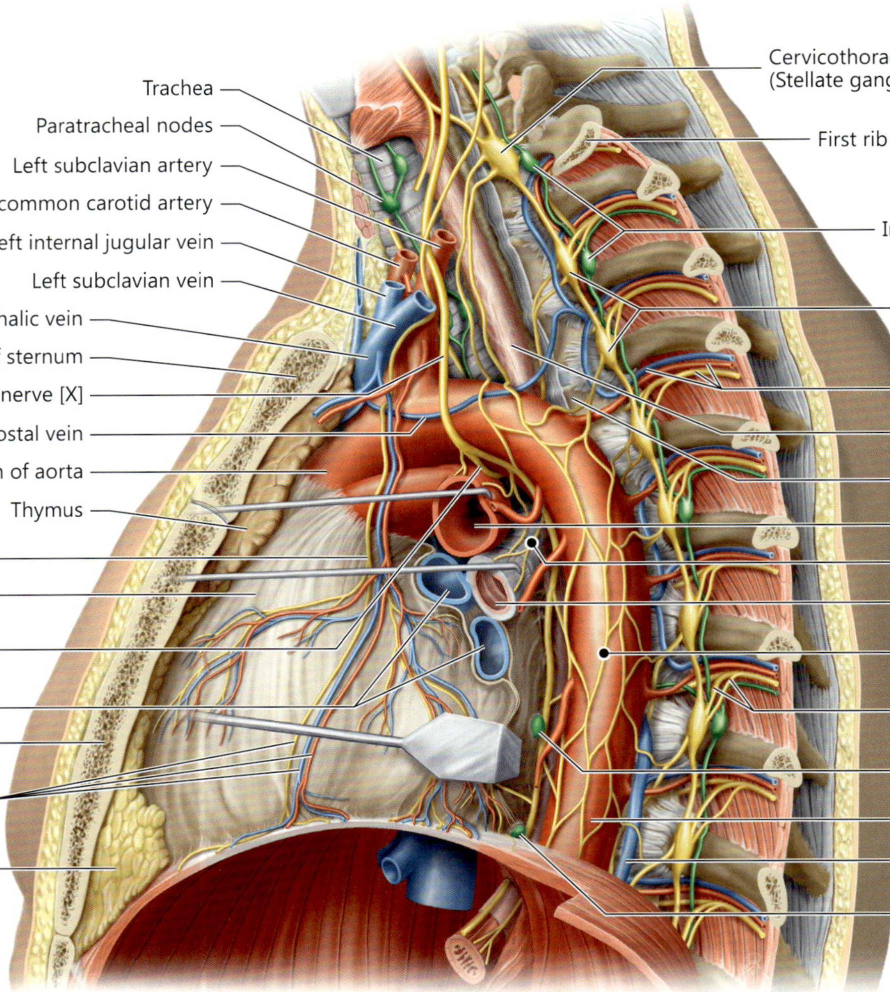

Schematic representation of the **innervation of the heart**, left lateral view. From the cervicothoracic (stellate) ganglion arise nerve fibers that descend towards the cardiac plexus. The pericardium is innervated by small branches coming from the right and left phrenic nerve.

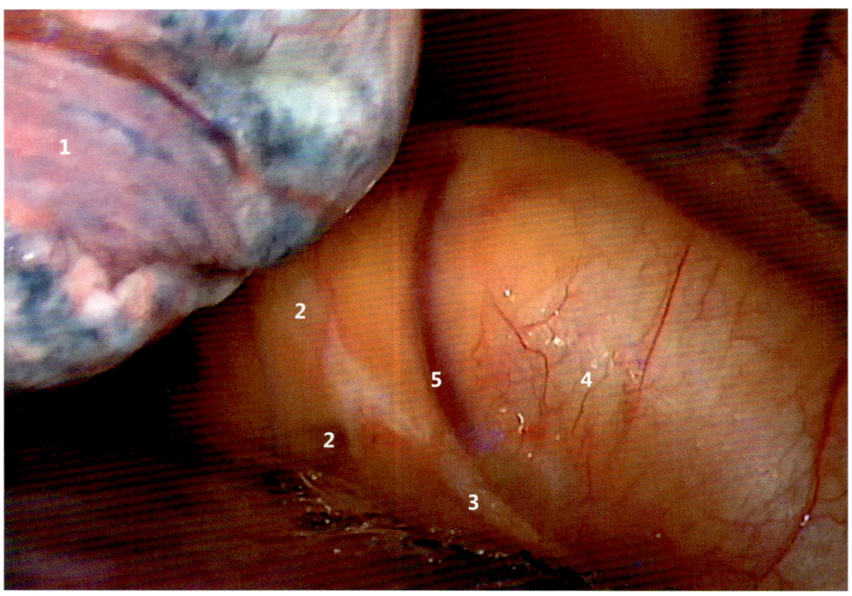

Image of the **arch of the aorta** obtained in the course of a thoracoscopy for diagnostic purposes. The endoscope is placed between the parietal and visceral layers of the pleura. View from above. One can observe the course of the vagus nerve on the lateral surface of the vessel.

1 Left lung
2 Vagus nerve [X]
3 Recurrent laryngeal nerve
4 Arch of aorta
5 Vasa vasorum

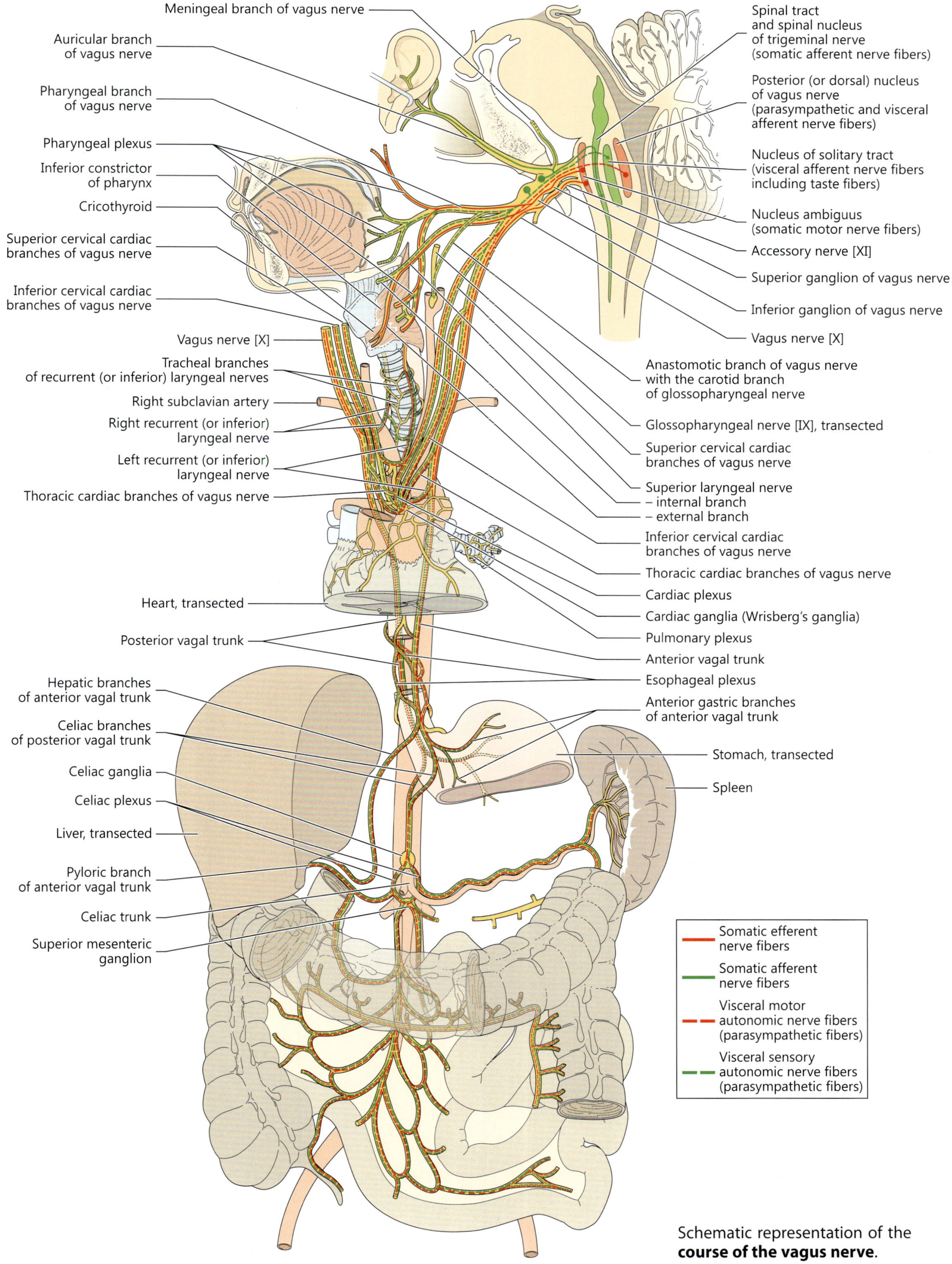

Schematic representation of the **course of the vagus nerve**.

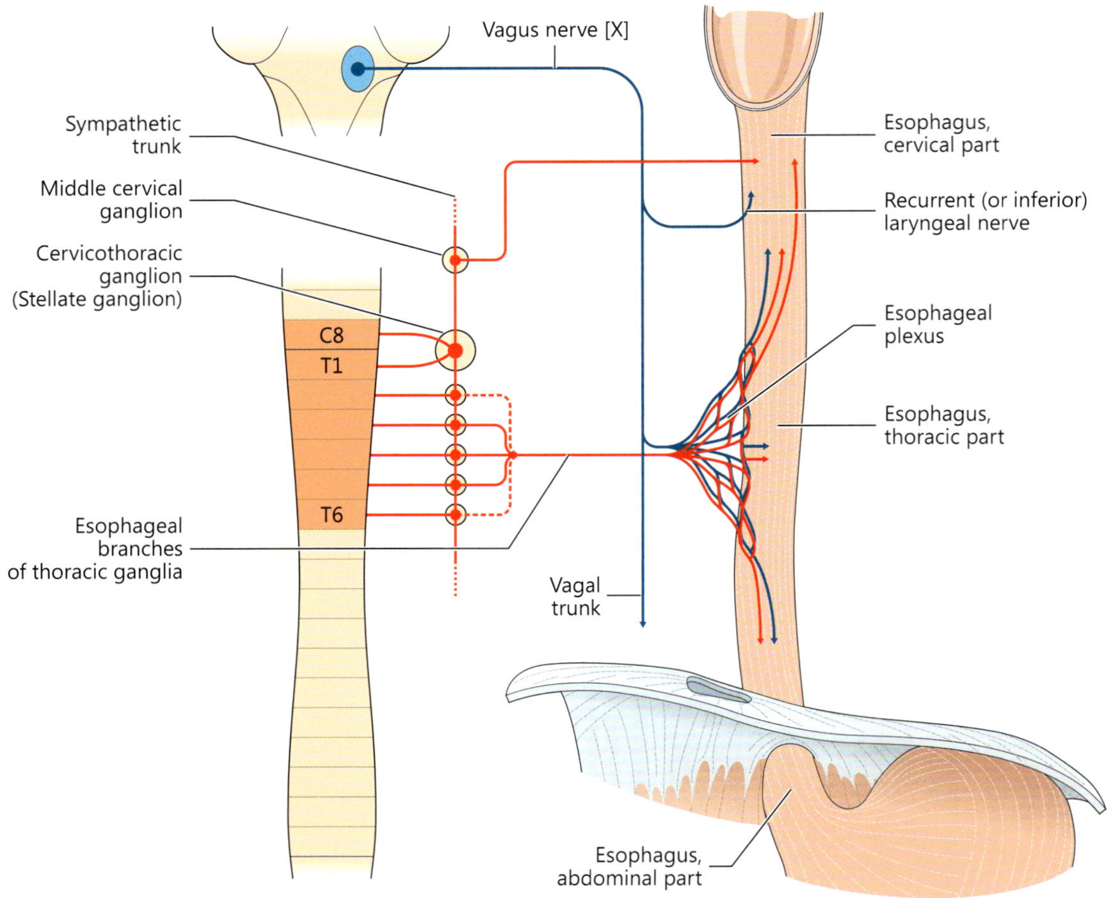

Schematic representation of the **innervation of the esophagus**. The parasympathetic nervous system innervation is carried out by esophageal branches of the laryngeal nerve coming from the vagus nerve. The fibers of the vagus nerve combine to form the esophageal plexus. From the thoracic ganglia of the sympathetic nervous system arise fibers that combine to form the esophageal plexus.

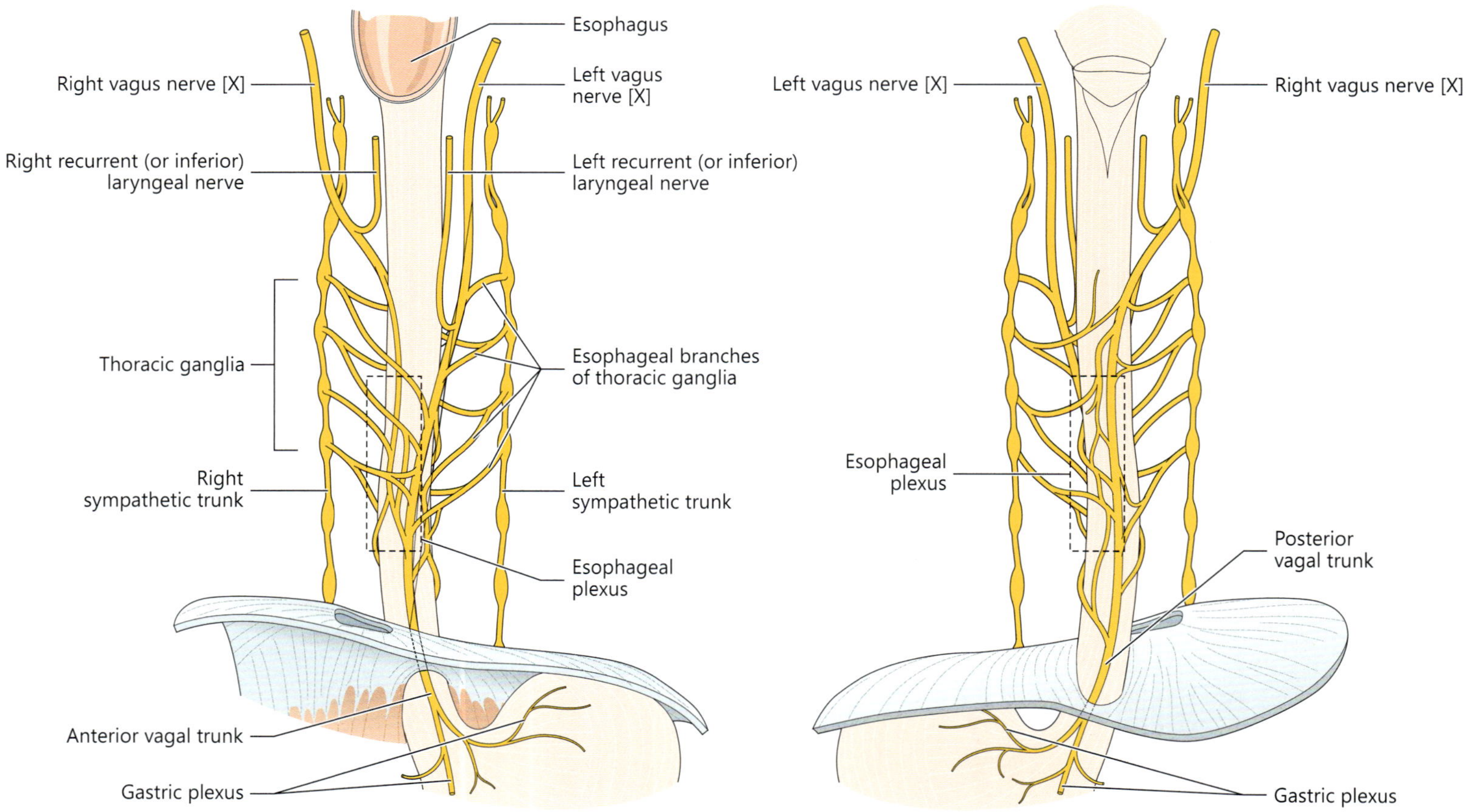

Schematic representation of the course of the (right and left) **vagus nerves** and their ramifications, which form the **plexuses of autonomic fibers on the esophagus**. Anterior and posterior views.

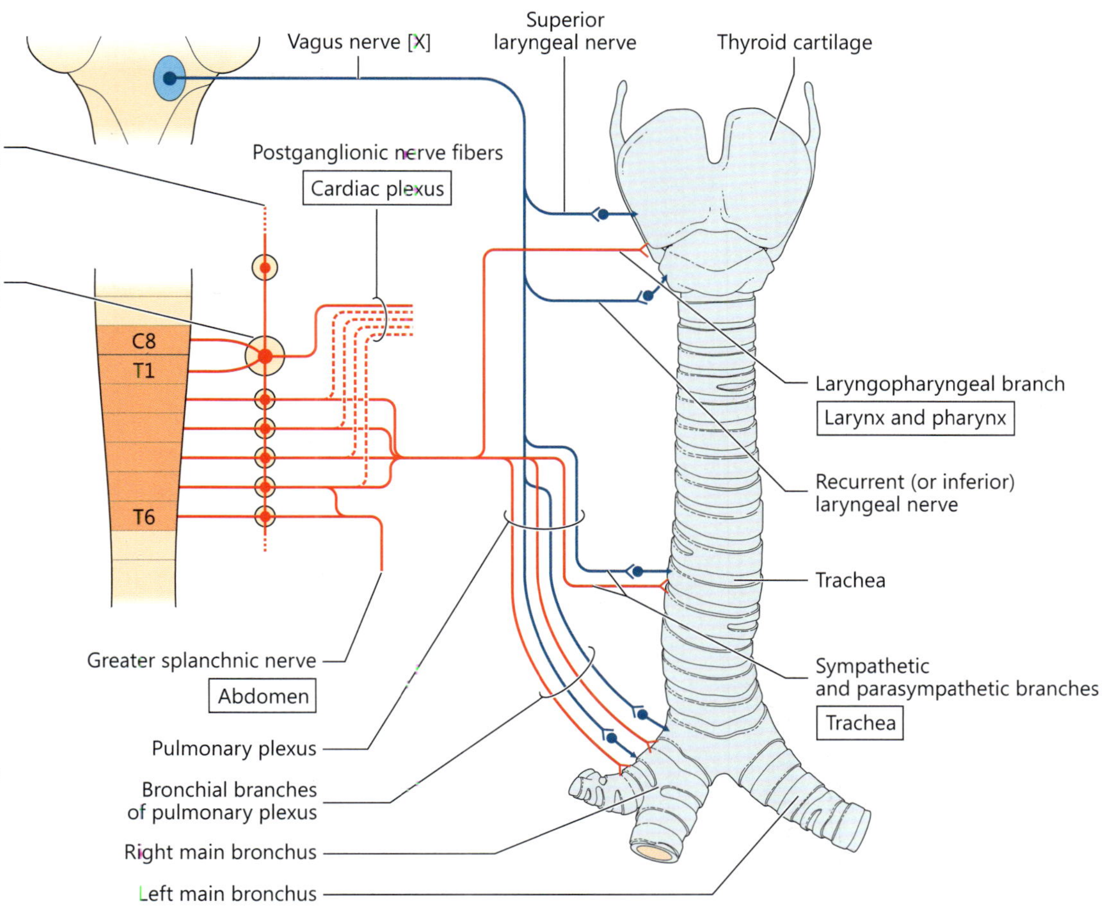

Schematic representation of the **innervation of the trachea and bronchial tree**. The parasympathetic nervous system innervation takes place through the laryngeal nerves, branches of the vagus nerve. The sympathetic nervous system innervation takes place through a few fibers towards the trachea and numerous postganglionic branches of the thoracic ganglia (thoracic pulmonary branches), which combine to form the pulmonary plexus.

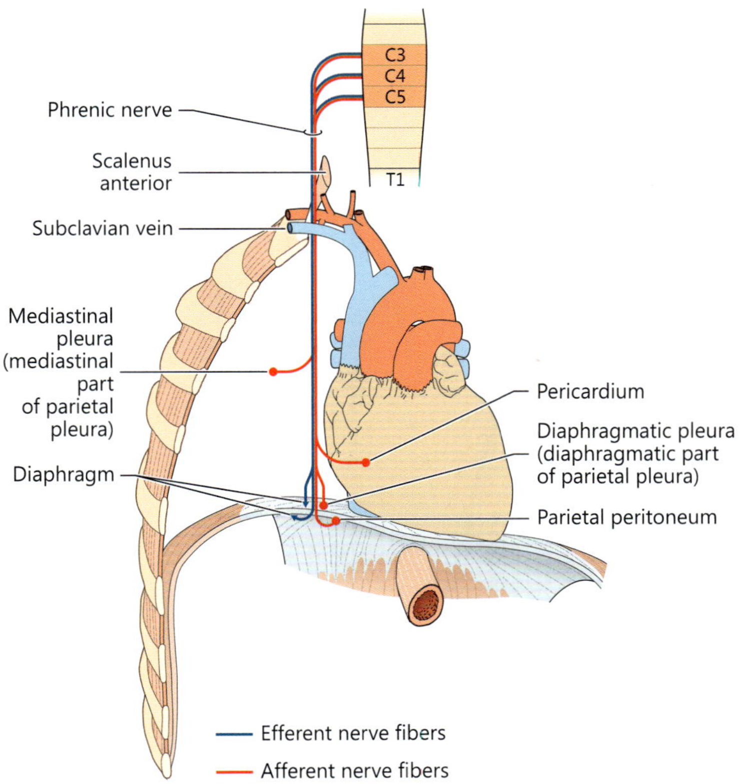

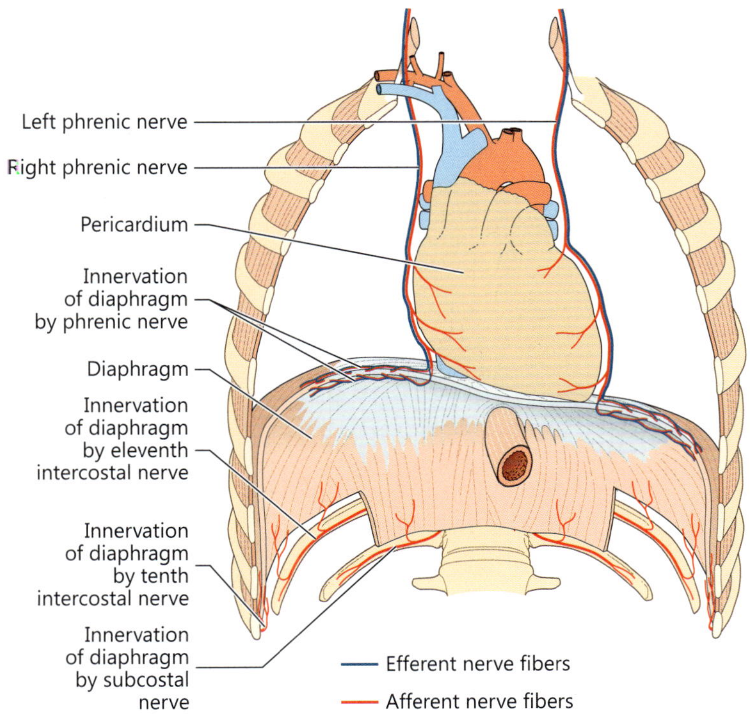

The **innervation of the diaphragm and pericardium** is carried out by the phrenic nerve (origin from C3-C5). The motor innervation of the diaphragm takes place also through the intercostal nerves X and XI and the subcostal nerve T12.

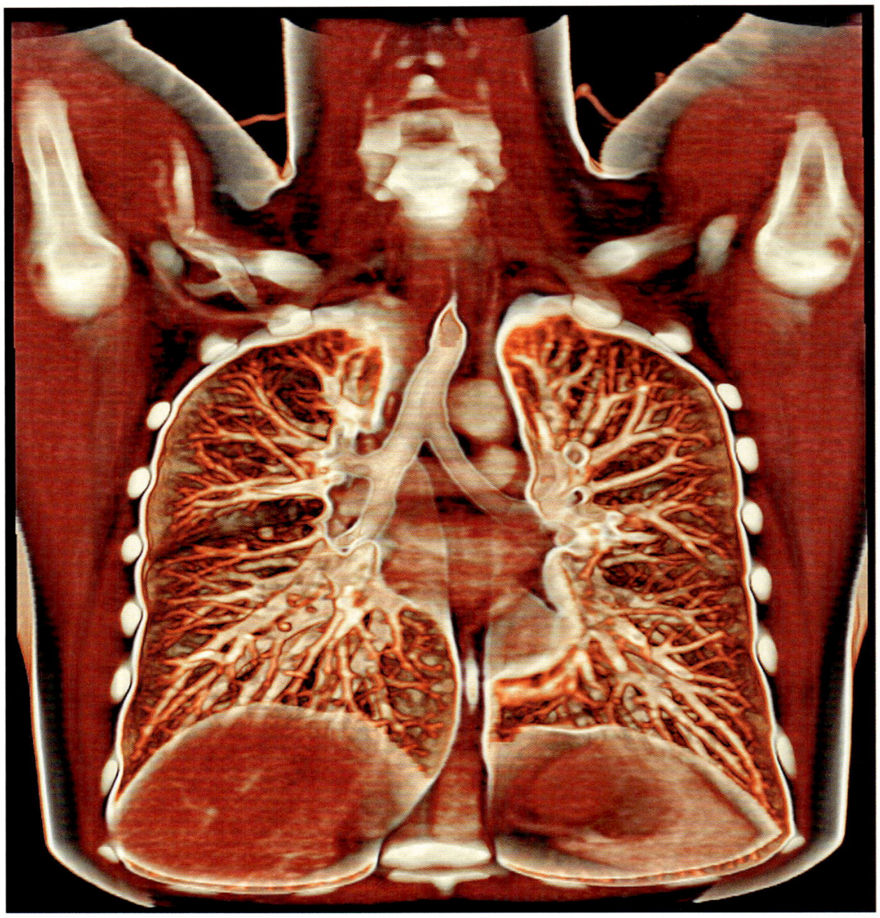

The lungs are contained within the **pleural compartments** inside the thoracic cavity (three-dimensional image of a living human body, obtained by CT scan using a 3D volume rendering technique).

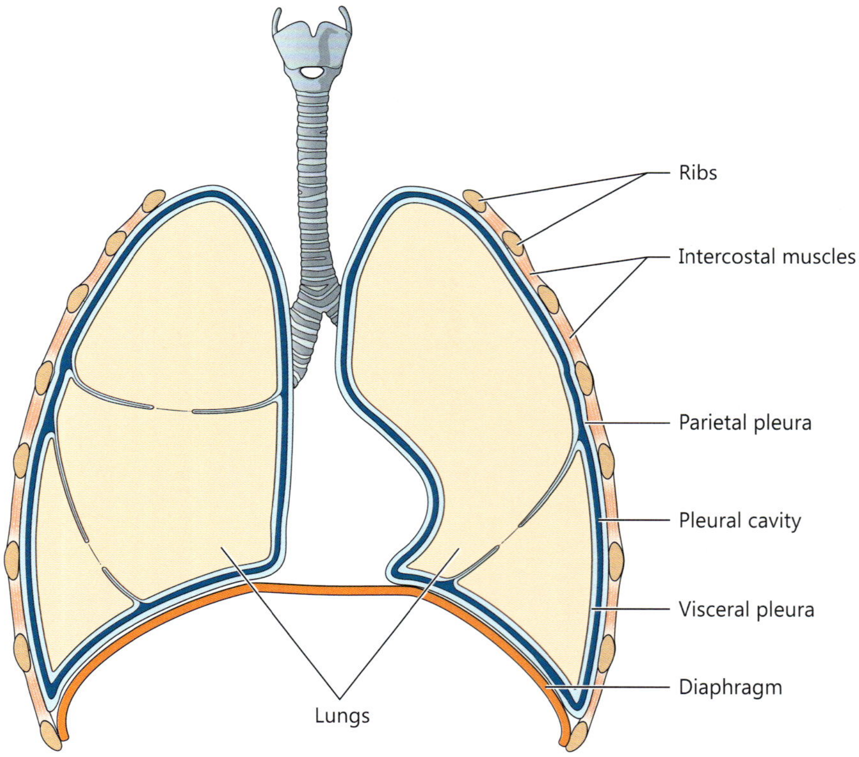

The **pleurae** consist of two layers: the parietal layer adheres closely to the endothoracic fascia, while the visceral layer adheres to the lung surface, extending into the pulmonary fissures. The two layers reflect one onto the other at the level of the hilum.

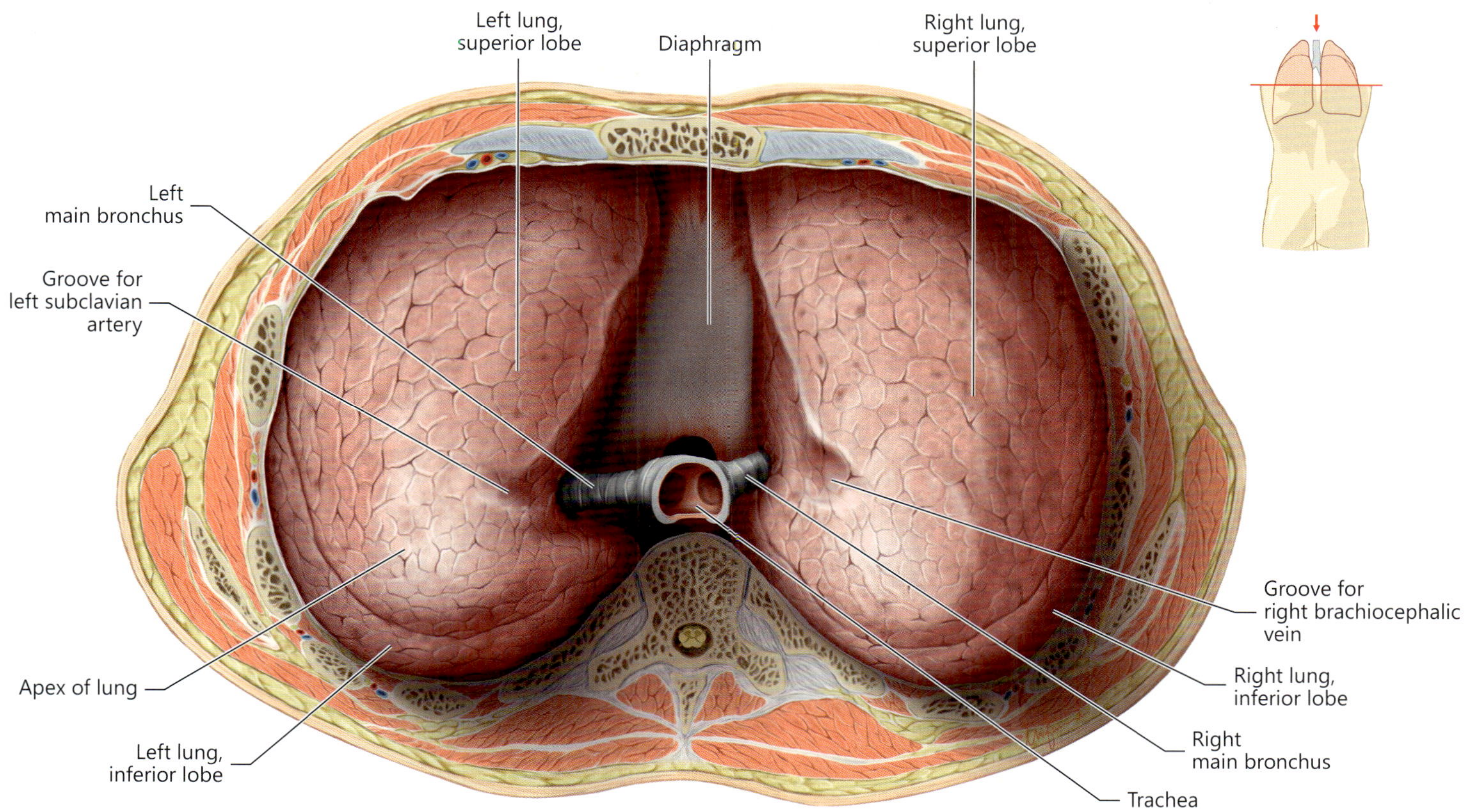

Lungs. View from above. The thoracic cage has been transected at the level of T V. The space of the mediastinum has been evacuated with the exception of the trachea.

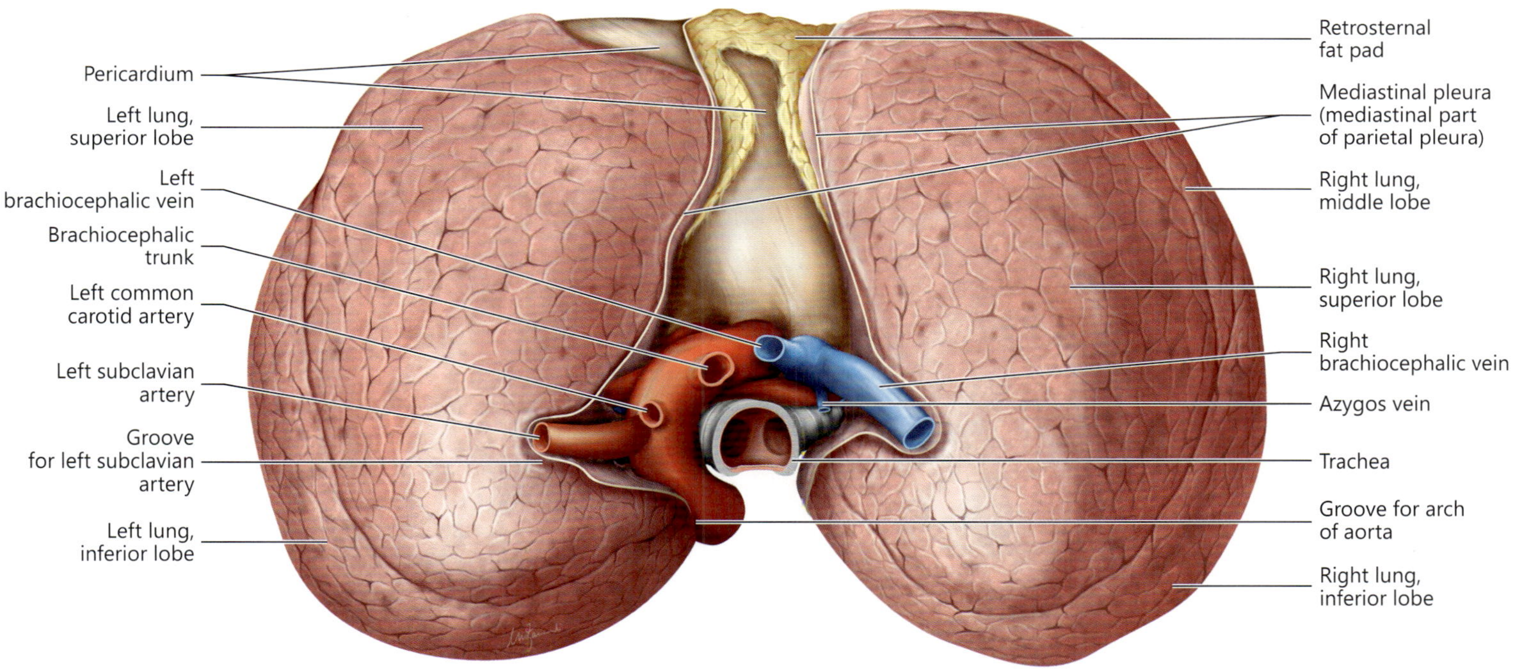

Lungs. View from above. The two mediastinal pleurae separate the pleural compartments from the mediastinum. On the medial surface of the lungs, the great vessels of the mediastinum and the heart leave deep impressions.

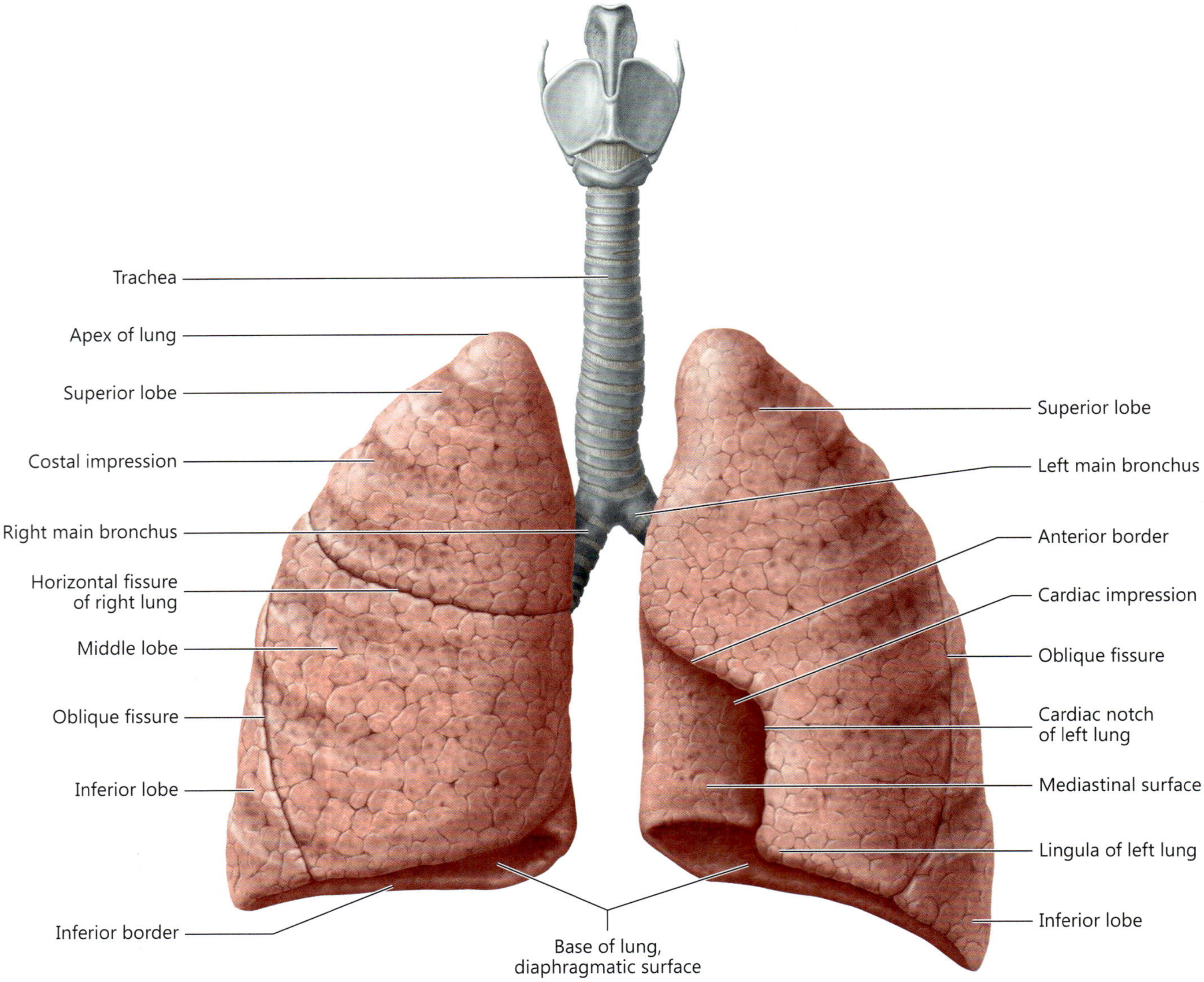

Lungs. Anterior view. One can see the pathway of the right pulmonary fissures and of the left pulmonary fissure. The left lung, smaller than the contralateral lung, presents only two lobes and contains the heart, which leaves the deep cardiac impression. The anterior border of the lung presents a notch (cardiac notch of the left lung), which renders it asymmetrical with respect to the contralateral one. The base of the lungs is concave towards the top and follows the line of the diaphragm. On the lateral surface of the lung, one can see the costal impressions created by the ribs.

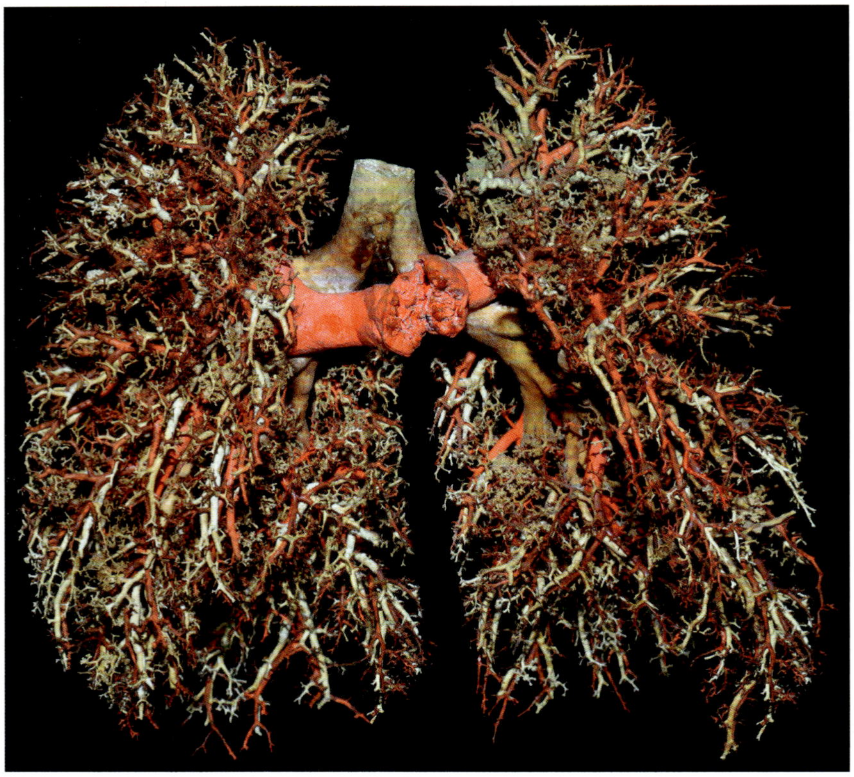

Tracheobronchial tree. Cast, anterior projection. In *yellow*, trachea and bronchial tree; in *red*, pulmonary trunk and its ramifications (courtesy of D. Sasse, Collezione anatomica, Basilea).

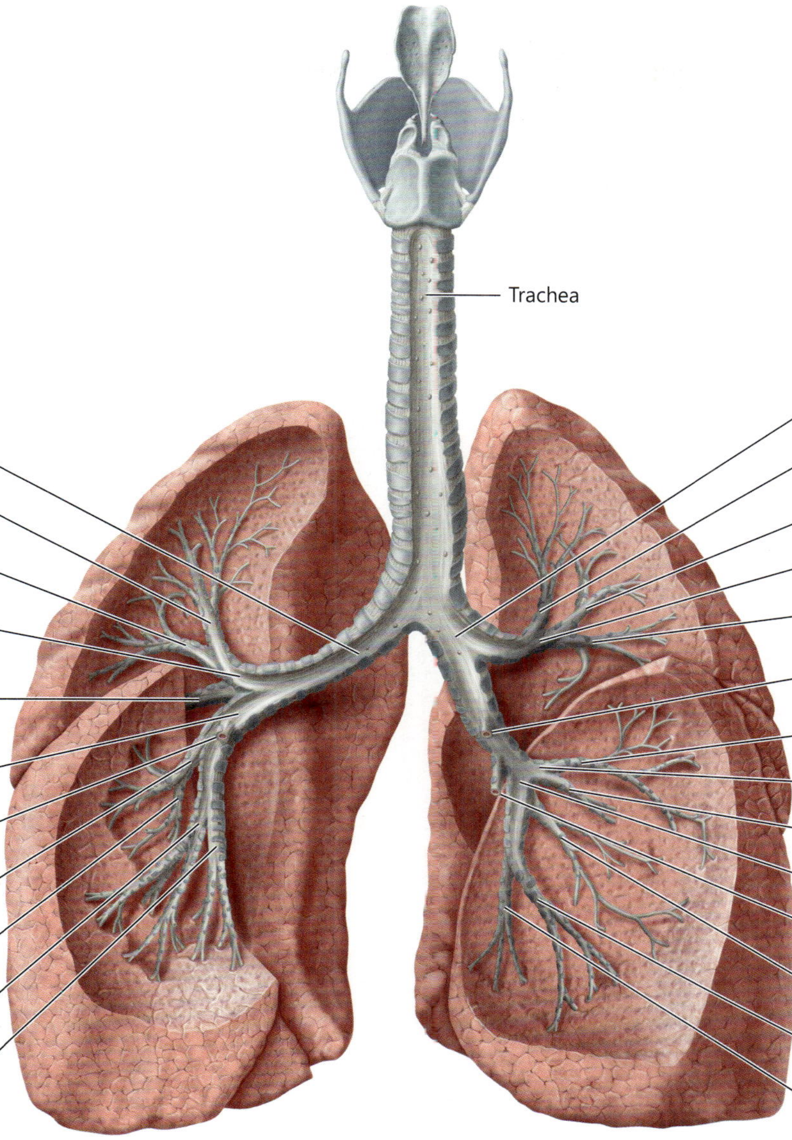

Tracheobronchial tree. Posterior view. Removal of part of the lung parenchyma makes it possible to see the ramifications of the bronchial tree.

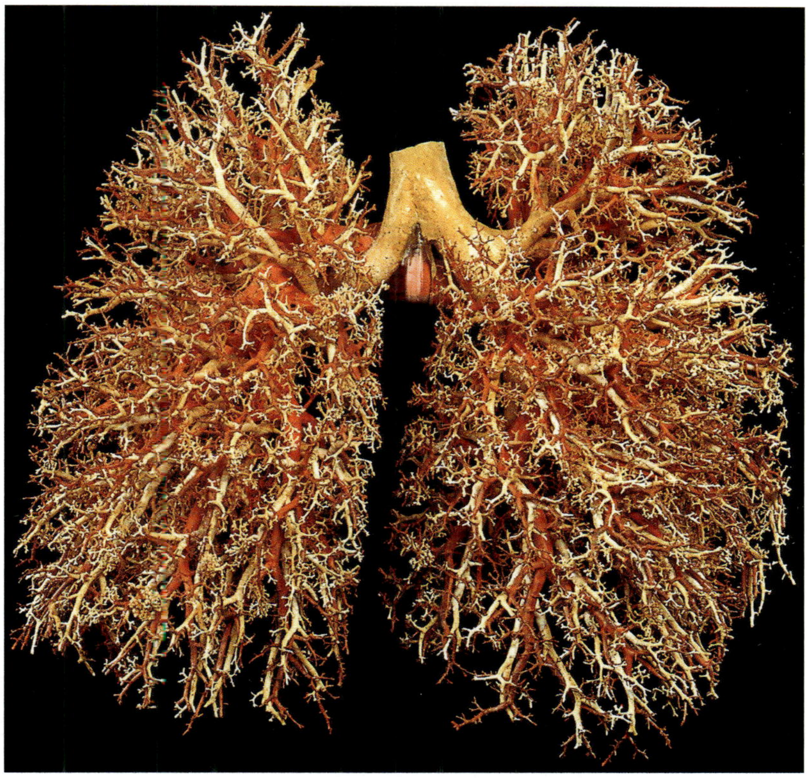

Tracheobronchial tree. Cast, posterior projection. In *yellow*, trachea and bronchial tree; in *red*, pulmonary trunk and its ramifications (courtesy of D. Sasse, Collezione anatomica, Basilea).

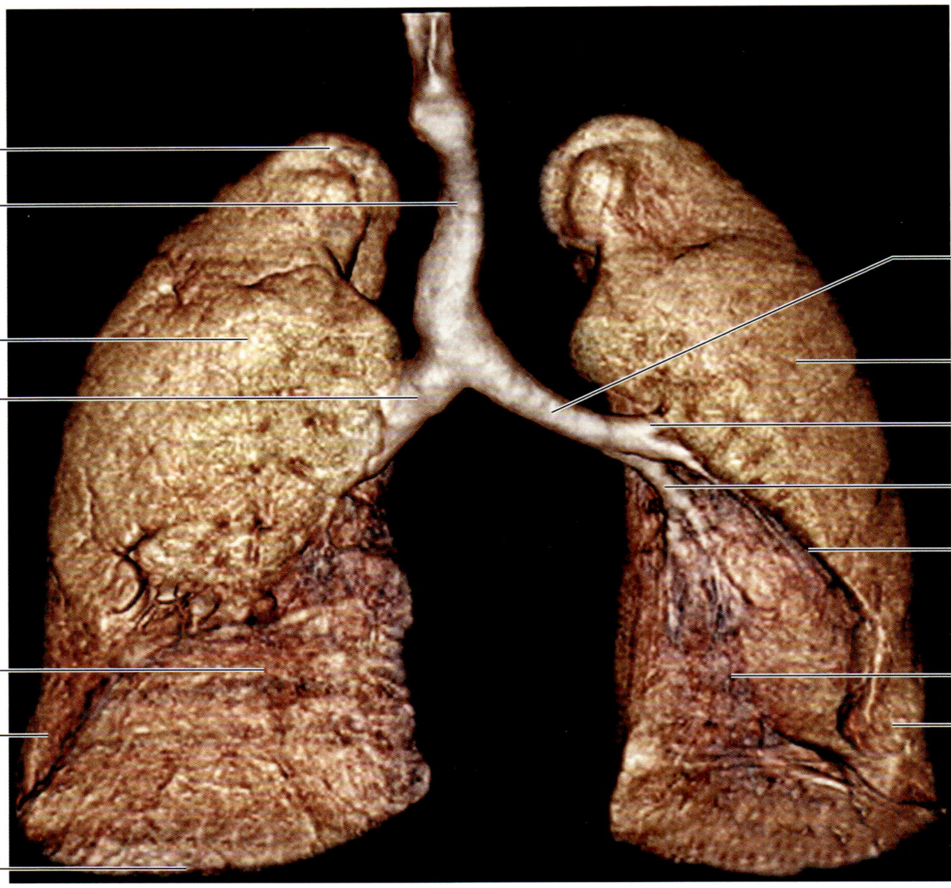

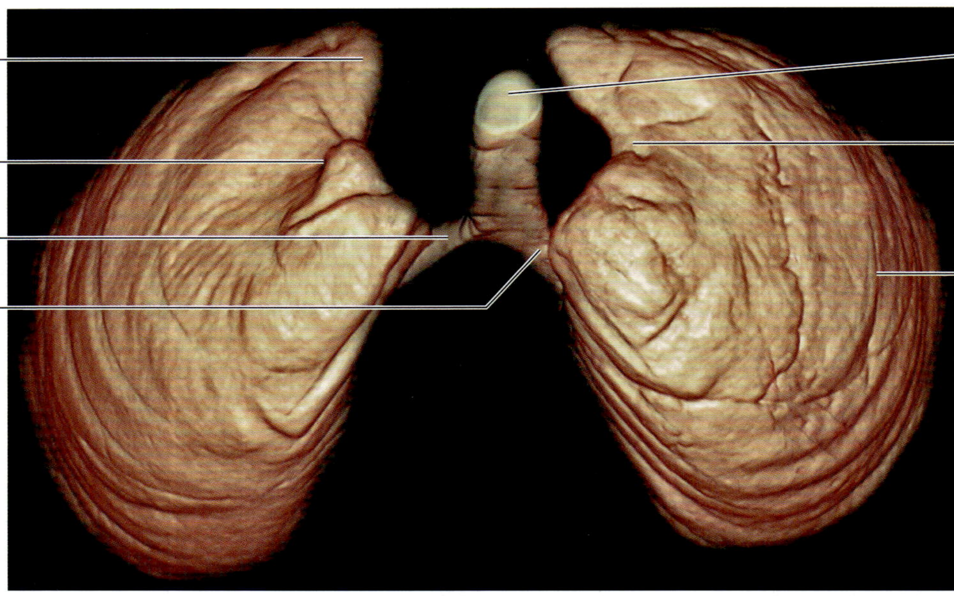

Three-dimensional images of a living human body, obtained by CT scan using a 3D volume rendering technique, showing the **lungs**. Above, anterior view; below, view from above.

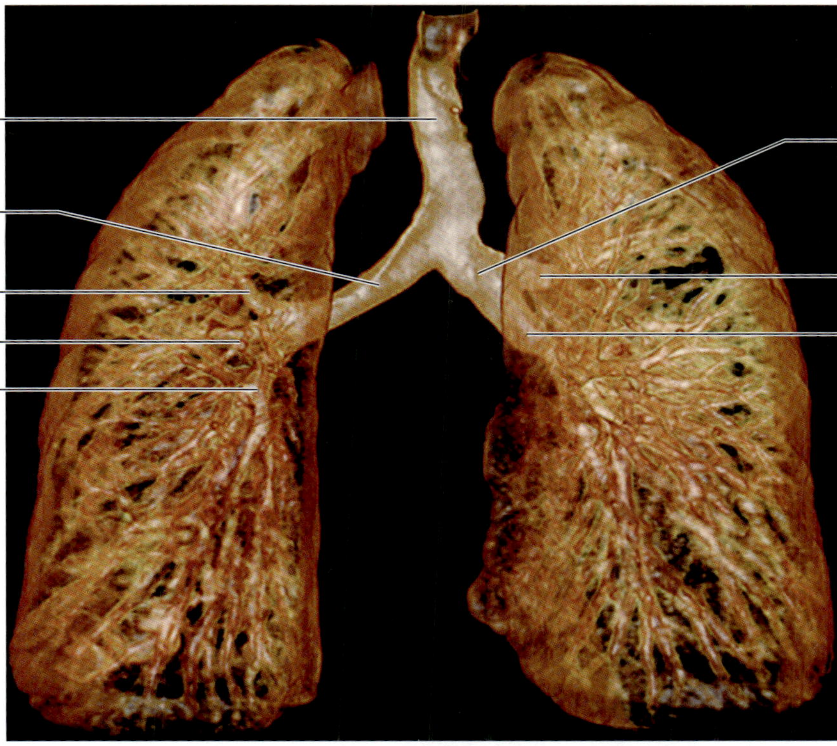

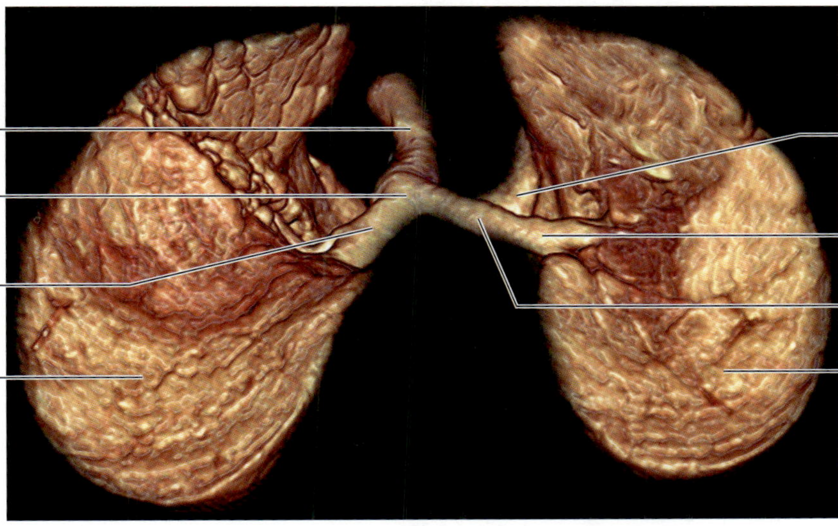

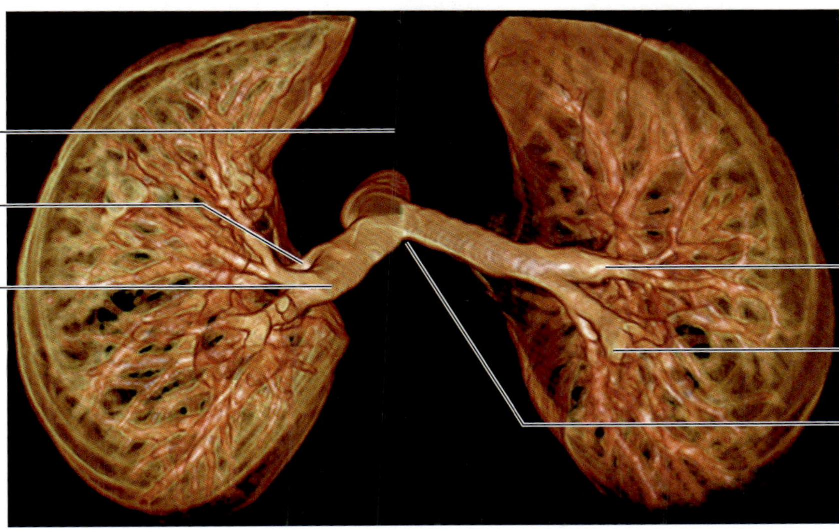

Three-dimensional images of a living human body, obtained by CT scan using a 3D volume rendering technique, showing the **lungs**. At the top, posterior view. The plane makes it possible to see the ramifications of the bronchial tree. In the center, view from the base of the lungs. At the bottom, same projection as the preceding but with removal of part of the lung parenchyma in order to display the bronchial ramifications.

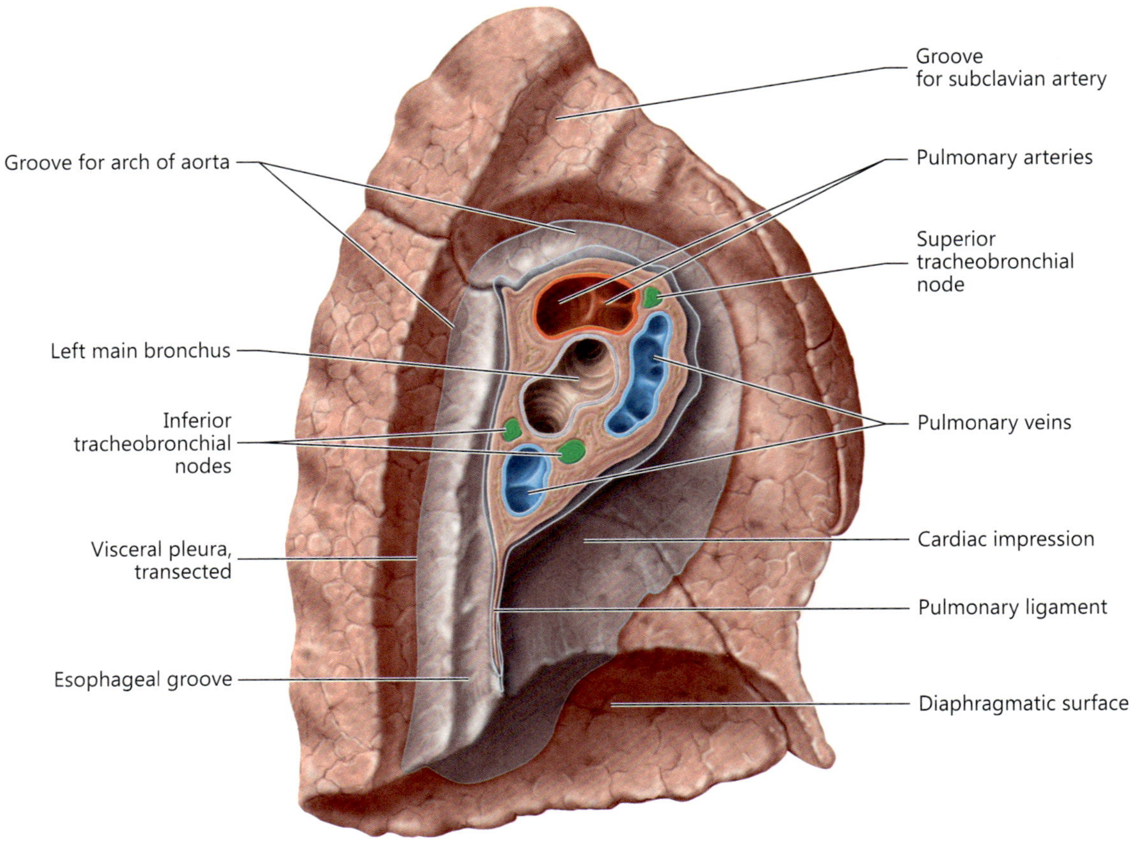

a

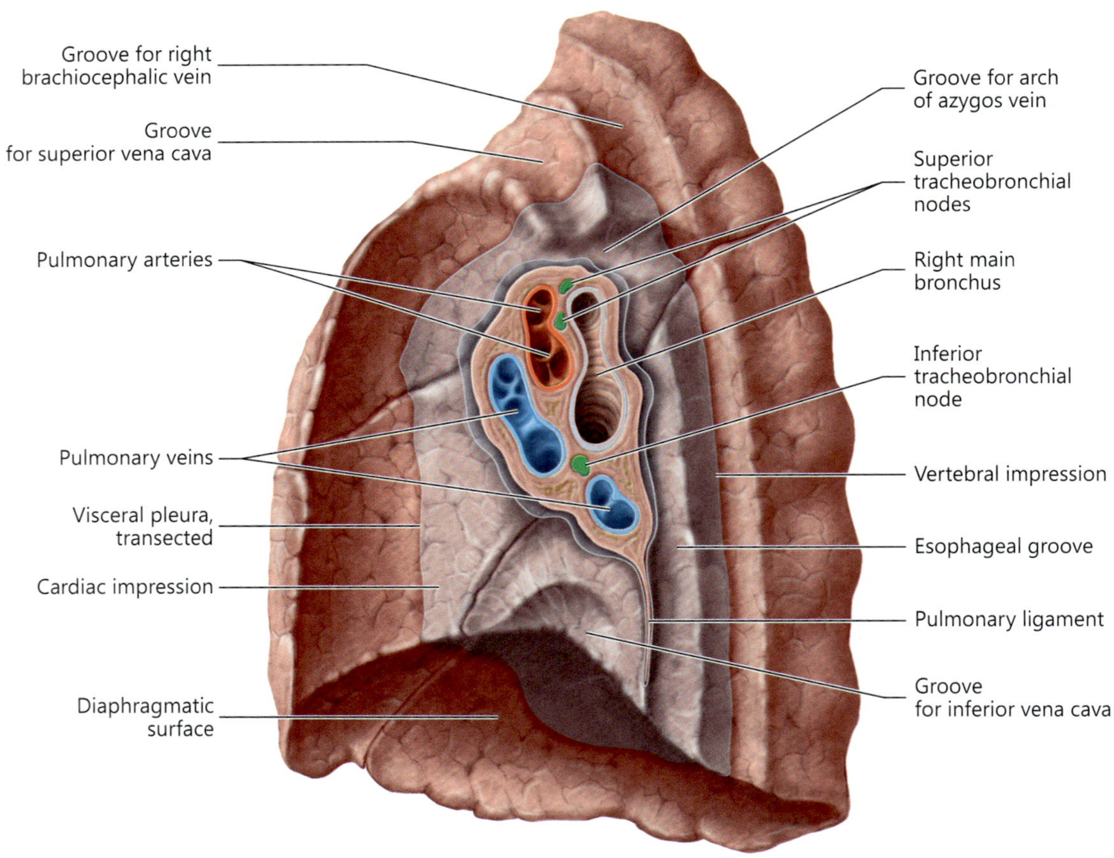

b

Lungs. a, Mediastinal surface of the left lung; **b**, mediastinal surface of the right lung. One can clearly see the deep impressions created by the structures of the mediastinum and by the thoracic vertebrae. One can also see the structures of the hilum and the position of the pulmonary arteries and veins relative to the bronchi. On the left, the pulmonary arteries are situated at a higher level than the bronchi, compared to the right hilum. The reflection of the parietal pleural layer onto the visceral one (not shown) at the level of the hilum contributes to the formation of the pulmonary (triangular) ligament.

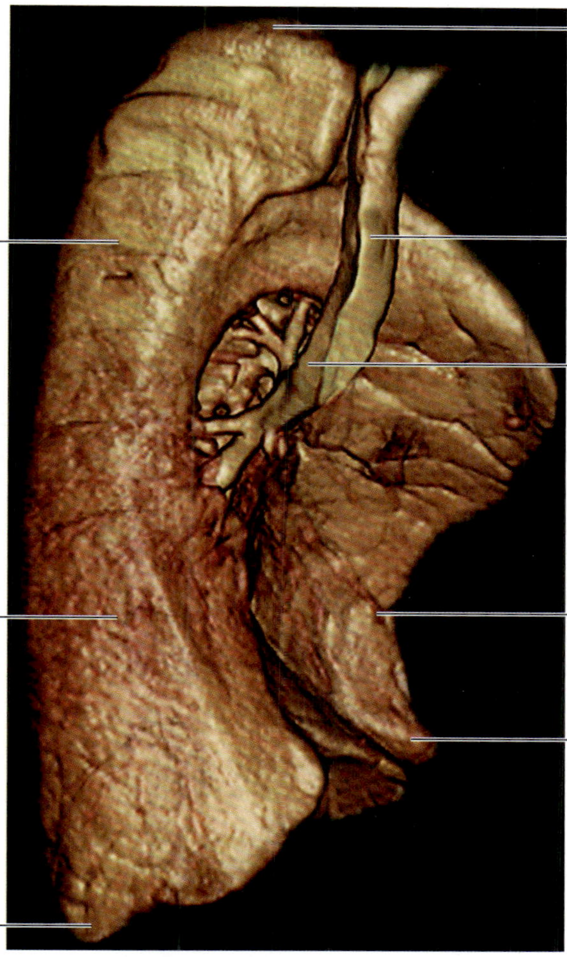

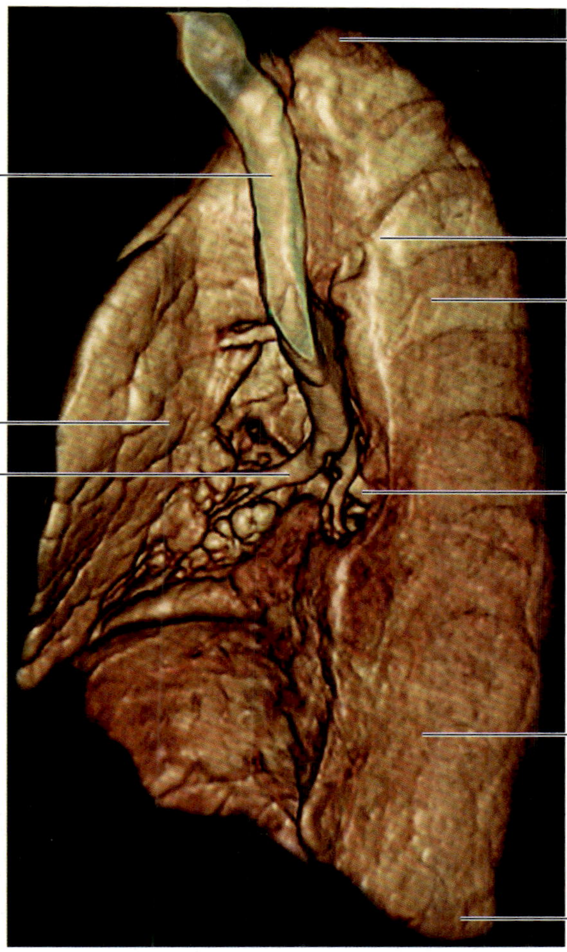

Three-dimensional images of a living human body, obtained by CT scan using a 3D volume rendering technique, showing the **lungs**. Above, medial view of the left lung. Below, medial view of the right lung.

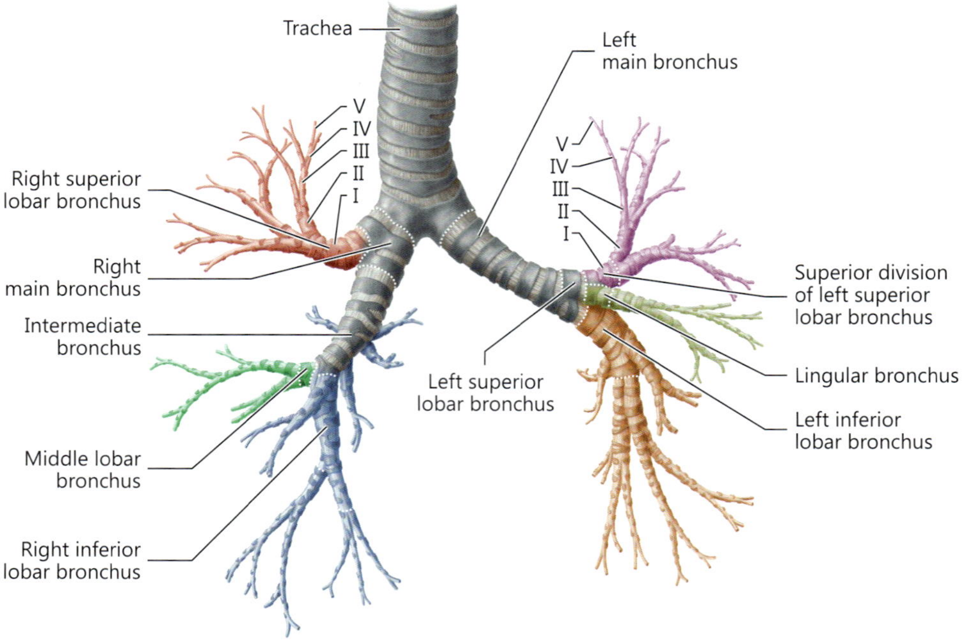

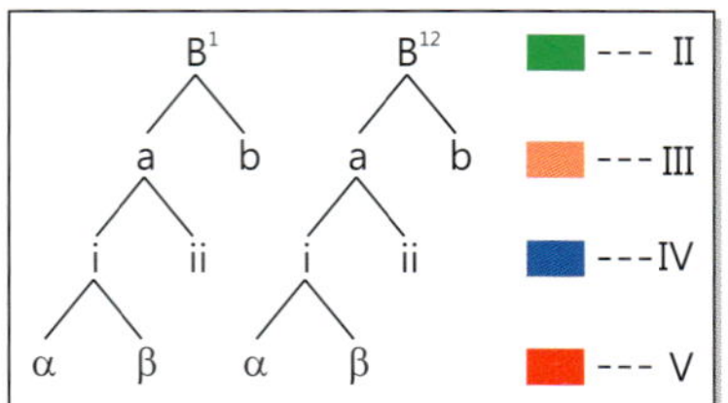

Ramification	*Order*
Lobar bronchus	I
(Division of lobar bronchus	I+II)
Segmental bronchus	II
Subsegmental bronchus	III
Subsubsegmental bronchus	IV
	V

Ikeda's classification with indication of the order of ramifications (I, II, III, IV and V) and the color code utilized in the figure below on the left.

Right lung

Superior lobe
- 1 Apical segmental bronchus [B I]
- 2 Posterior segmental bronchus [B II]
- 3 Anterior segmental bronchus [B III]

Middle lobe
- 4b Lateral segmental bronchus [B IV]
- 5b Medial segmental bronchus [B V]

Inferior lobe
- 6 Superior segmental bronchus [B VI]
- 7 Medial basal segmental bronchus [B VII]
- 8 Anterior basal segmental bronchus [B VIII]
- 9 Lateral basal segmental bronchus [B IX]
- 10 Posterior basal segmental bronchus [B X]

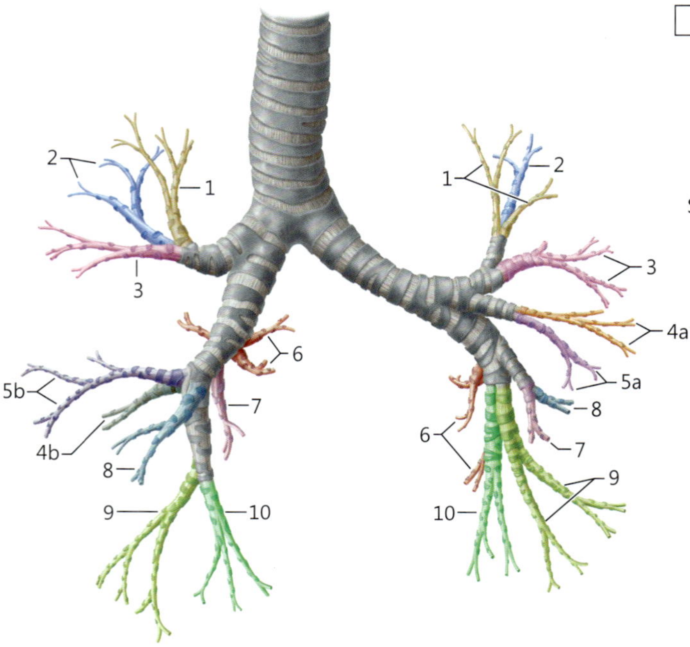

Left lung

Superior lobe
- 1 Apical segmental bronchus [B I] (Apicoposterior segmental bronchus [B I+II])
- 2 Posterior segmental bronchus [B II] (Apicoposterior segmental bronchus [B I+II])
- 3 Anterior segmental bronchus [B III]
- 4a Superior lingular bronchus [B IV]
- 5a Inferior lingular bronchus [B V]

Inferior lobe
- 6 Superior segmental bronchus [B VI]
- 7 Medial basal segmental bronchus [B VII]
- 8 Anterior basal segmental bronchus [B VIII]
- 9 Lateral basal segmental bronchus [B IX]
- 10 Posterior basal segmental bronchus [B X]

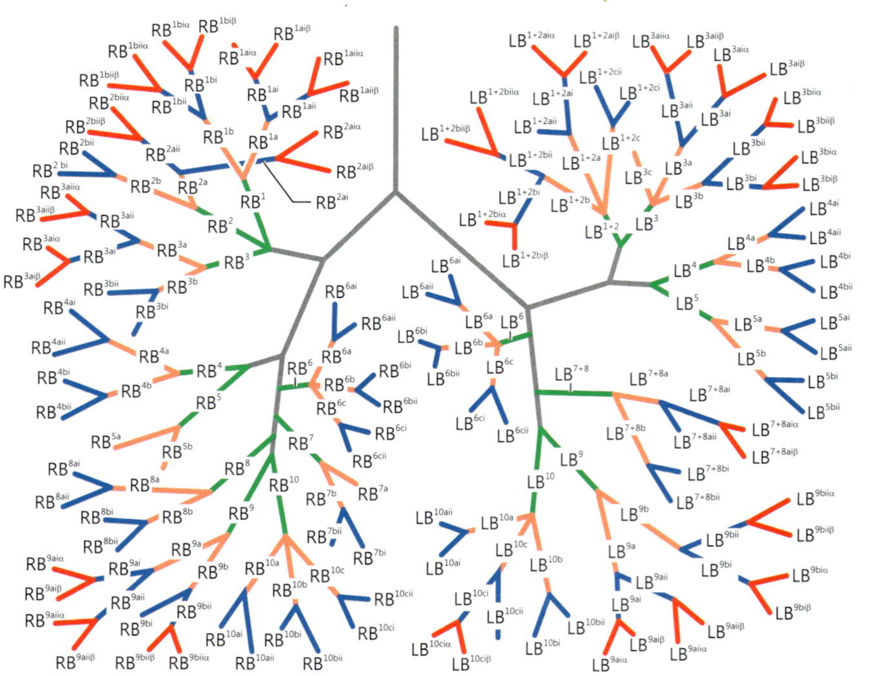

Nomenclature for describing the **lobar**, **segmental** and **subsegmental ramifications of the bronchi**. *Jackson-Huber* (figure at the top): the bronchi are named based on their spatial orientation (right/left, superior/inferior, anterior/posterior). *Boyden*: the classification reflects the subdivision of the bronchial tree (and of the accompanying arteries and veins) as it is observed during development. The segmental bronchi (order II) are indicated with the letter B followed by a number from 1 to 10 (1-3 superior lobe, 4-5 middle or lingular node, 6-10 inferior lobe). The introduction of flexible fiberoptic bronchoscopy by *Ikeda* has enabled a precise classification of the ramifications downstream of order II bronchi on the right (R) and left (L) side: the subsegmental ones (order III) are indicated by the letter a, b, c; the ramifications of order IV by i (anterior) and ii (posterior); and order V ramifications by α and β. The figure at the center shows the correspondence between the classifications of Jackson-Huber and Boyden regarding the segmental bronchi, while the figure on the side and the table above show the subsegmental classification of Ikeda.

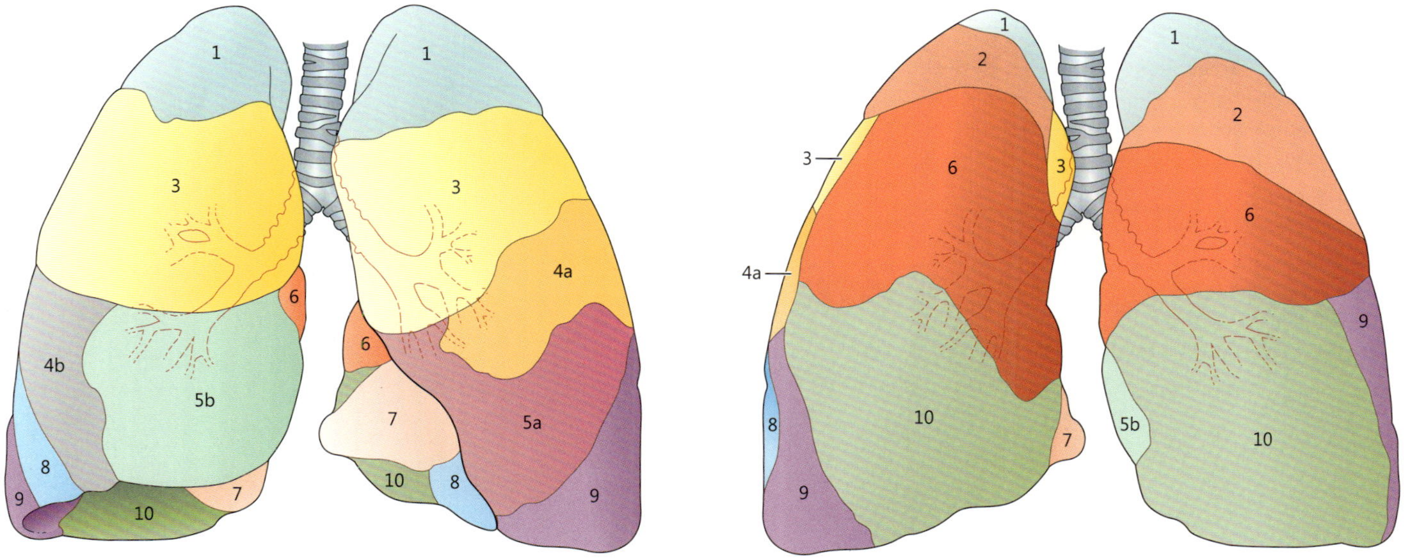

a **b**

Pulmonary zones (colors correspond to those indicated in the preceding figure). **a**, Anterior projection; **b**, posterior projection.

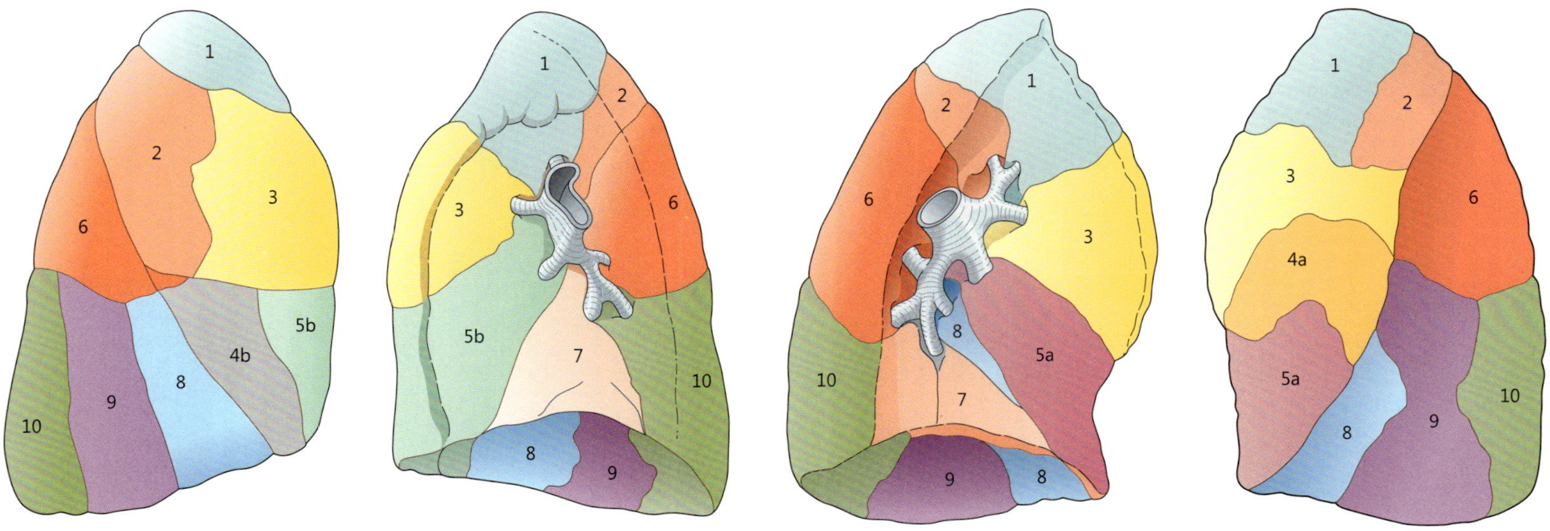

a **b**

Pulmonary zones. a, Lateral and mediastinal surface of the right lung; **b**, mediastinal and lateral surface of the left lung.

PULMONARY LOBES	*RIGHT BRONCHIAL TREE*			*LEFT BRONCHIAL TREE*		
	JACKSON-HUBER	*BOYDEN*	*IKEDA*	*JACKSON-HUBER*	*BOYDEN*	*IKEDA*
Superior lobe	Apical	B^1	RB^{1a}, RB^{1b} + ramifications (i, α)	Superior division		
	Posterior	B^2	RB^{2a}, RB^{2b} + ramifications (i, α)	apicoposterior	B^1, B^2	LB^{1+2a} + ramifications (i, α)
	Anterior	B^3	RB^{3a}, RB^{3b} + ramifications (i, α)	apicoanterior	B^3	LB^{3a}, LB^{3b} + ramifications (i, α)
				Lingular division		
				superior	B^4	LB^{4a}, LB^{4b} + ramifications (i, α)
				inferior	B^5	LB^{4a}, LB^{4b} + ramifications (i, α)
Middle lobe	Lateral	B^4	RB^{4a}, RB^{4b} + ramifications (i, α)			
	Medial	B^5	RB^{5a}, RB^{5b} + ramifications (i, α)			
Inferior lobe	Superior	B^6	RB^{6a}, RB^{6b} + ramifications (i, α)	Superior	B^6	LB^{6a}, LB^{6b} + ramifications (i, α)
	Medial basal	B^7	RB^{7a}, RB^{7b} + ramifications (i, α)	Anteromedial	B^7, B^8	LB^{7+8a} + ramifications (i, α)
	Anterior basal	B^8	RB^{8a}, RB^{8b} + ramifications (i, α)	Lateral basal	B^9	LB^{9a}, LB^{9b} + ramifications (i, α)
	Lateral basal	B^9	RB^{9a}, RB^{9b} + ramifications (i, α)	Posterior basal	B^{10}	LB^{10a}, LB^{10b} + ramifications (i, α)
	Posterior basal	B^{10}	RB^{10a}, RB^{10b} + ramifications (i, α)			

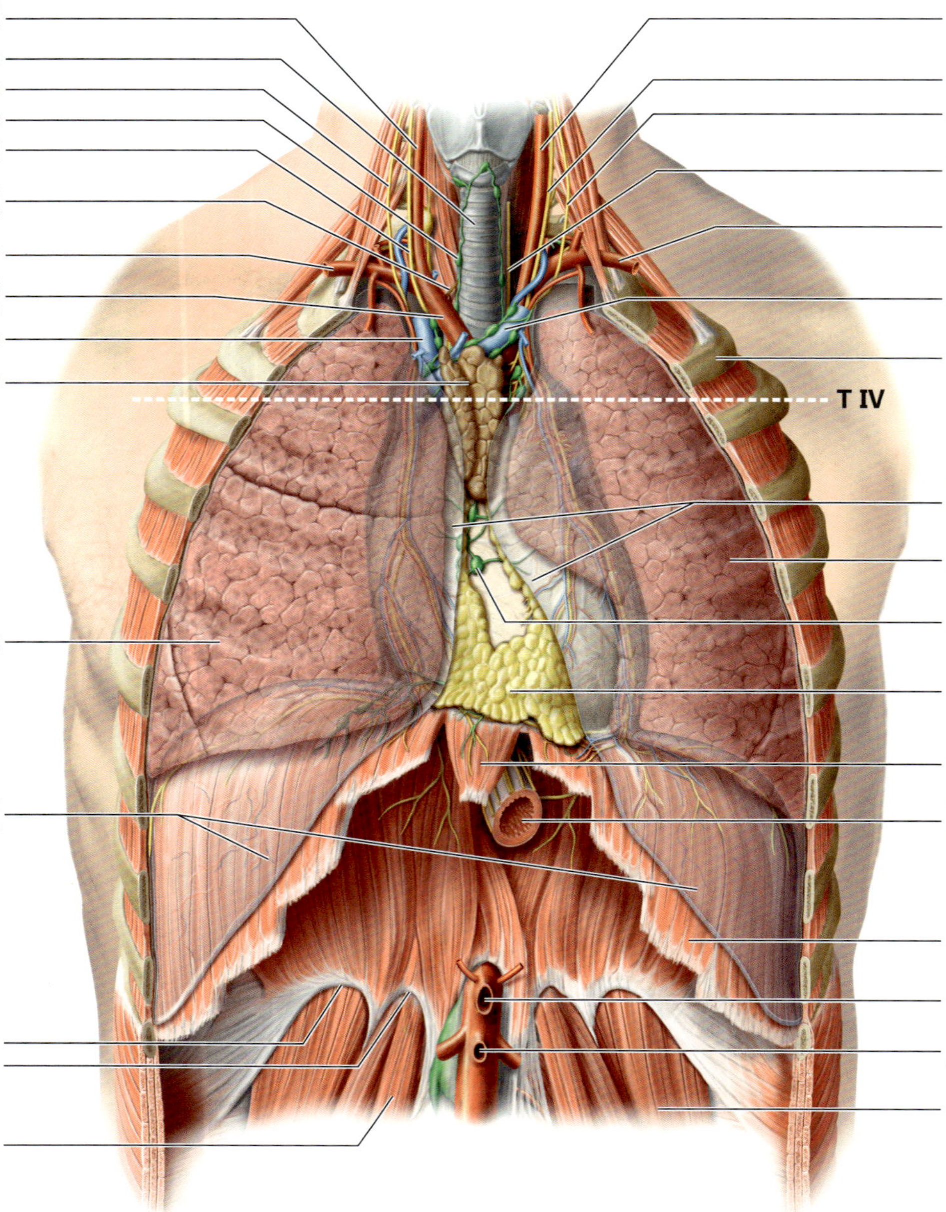

Anterior view of the **pleural lining of the lungs**, after resection of the anterior wall of the thoracic cage. In transparency, it is possible to see the course of the anterior and inferior borders of the lungs. The pleura exceeds the boundaries of the lung and extends downwards into the cavity formed between the diaphragm and the ribs (costodiaphragmatic recesses); forwards, it proceeds in front of the pericardium (costomediastinal recesses).

Pleurae

PLEURA	*DESCRIPTION*	*VASCULARIZATION, LYMPHATIC DRAINAGE, INNERVATION*
Visceral (pulmonary)	Membrane (serosa and subserosa) investing the surface of the lung penetrating into the depths of the interlobar fissures. At the level of the hilum of the lung it invests a short tract of the root of the lung, whereupon it continues in the mediastinal part of the parietal pleura	Bronchial arteries; pulmonary veins, bronchial veins. Bronchopulmonary nodes. Pulmonary plexus (sympathetic trunk)
Parietal		Intercostal arteries, internal thoracic artery, musculophrenic artery; superior vena cava. Intercostal, parasternal, superior diaphragmatic and celiac nodes. Intercostal, phrenic, vagus nerves and sympathetic trunk
– costal part (costal or costovertebral pleura)	Extending over the internal surface of the sternum, ribs, transversus thoracis, intercostal muscles, and lateral surface of the thoracic vertebrae, it continues medially in the mediastinal part, superiorly in the cervical pleura, and inferiorly in the diaphragmatic part	
– diaphragmatic part (diaphragmatic pleura)	Adhering to the diaphragm, it continues superiorly in the costal part and medially in the mediastinal part	
– mediastinal part (mediastinal pleura)	Extending between the sternum and vertebral column, it delimits the mediastinum, forming at the level of the root of the lung the pulmonary ligament	
– cervical pleura (dome of pleura)	Continuation upwards of the costal pleura to the point of covering the apex of lung	

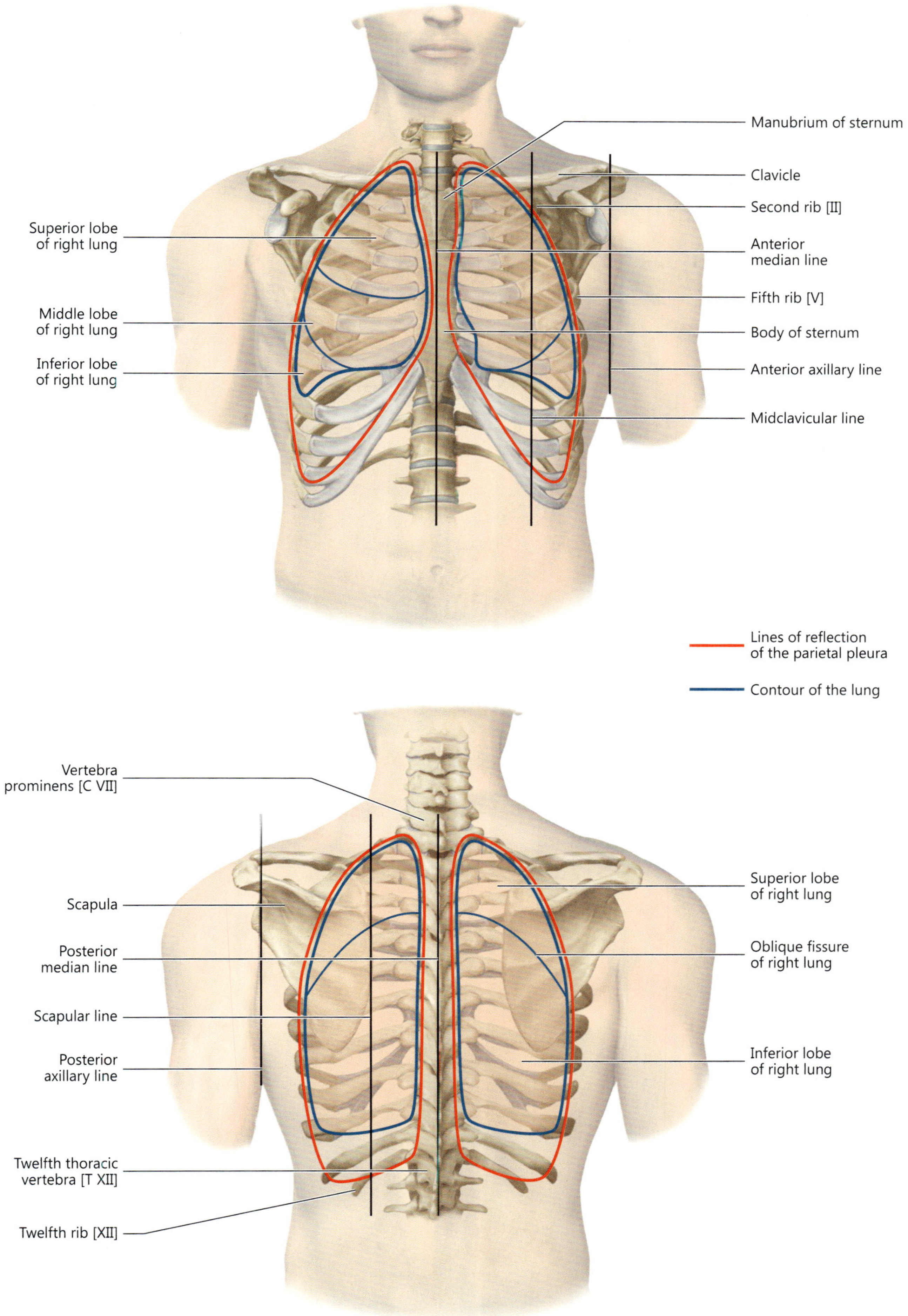

Pleural compartments. Projections onto the thoracic cage of the pulmonary (*blue*) and pleural (*red*) borders and fissures.

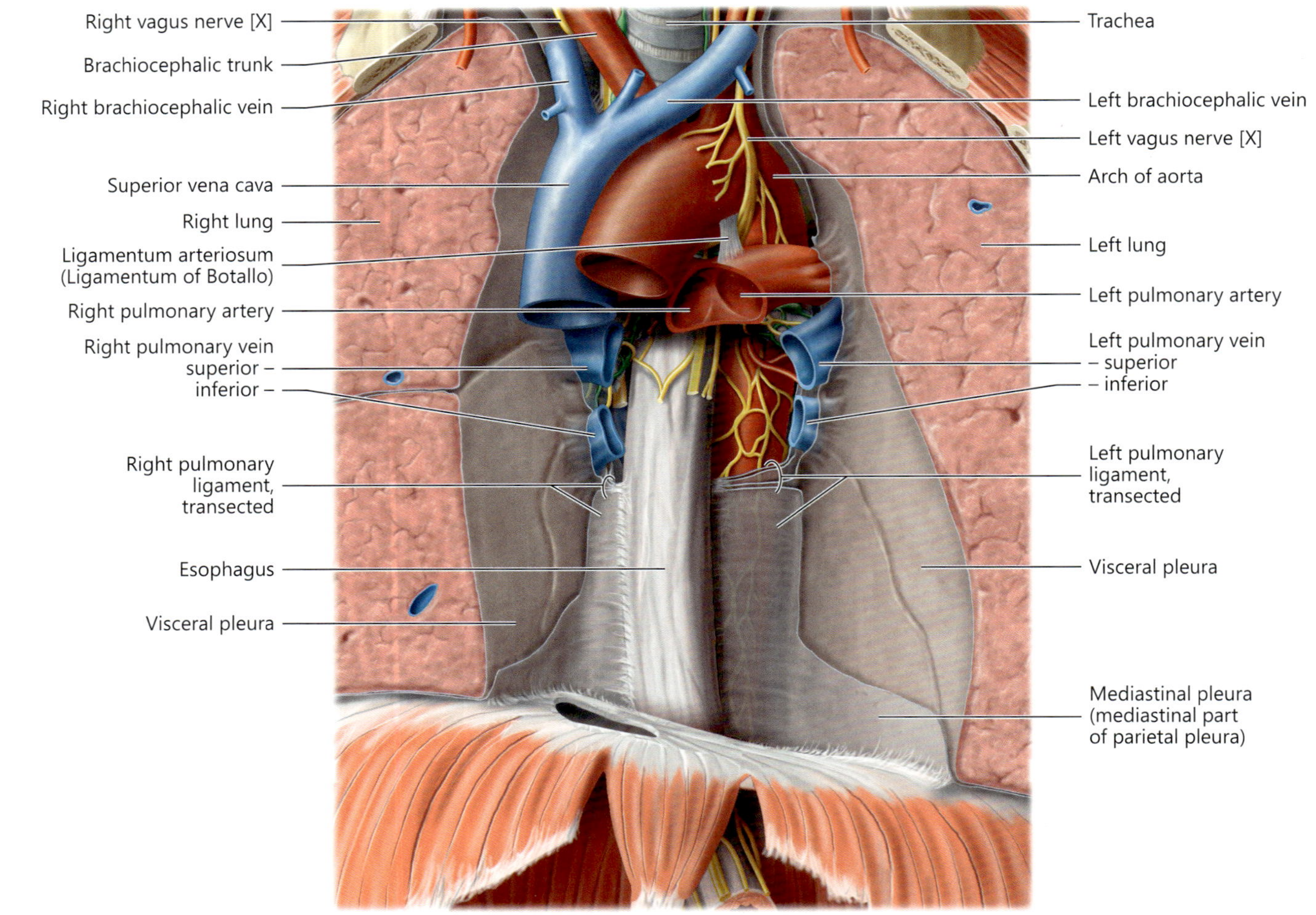

Frontal transection of the mediastinum. The great vessels of the pulmonary hilum and of the heart have been conserved. The **pulmonary ligament** is formed by the two reflections of mediastinal parietal pleura onto visceral pleura. The two reflections are located in front of and behind the hilum of the lung, respectively. Below the hilum, the two pleural reflections come together forming a ligament that extends medially to the lateral wall of the esophagus and below continues with the diaphragmatic pleura. By shifting the lung laterally, the pulmonary ligament is pulled, assuming a typical triangular shape.

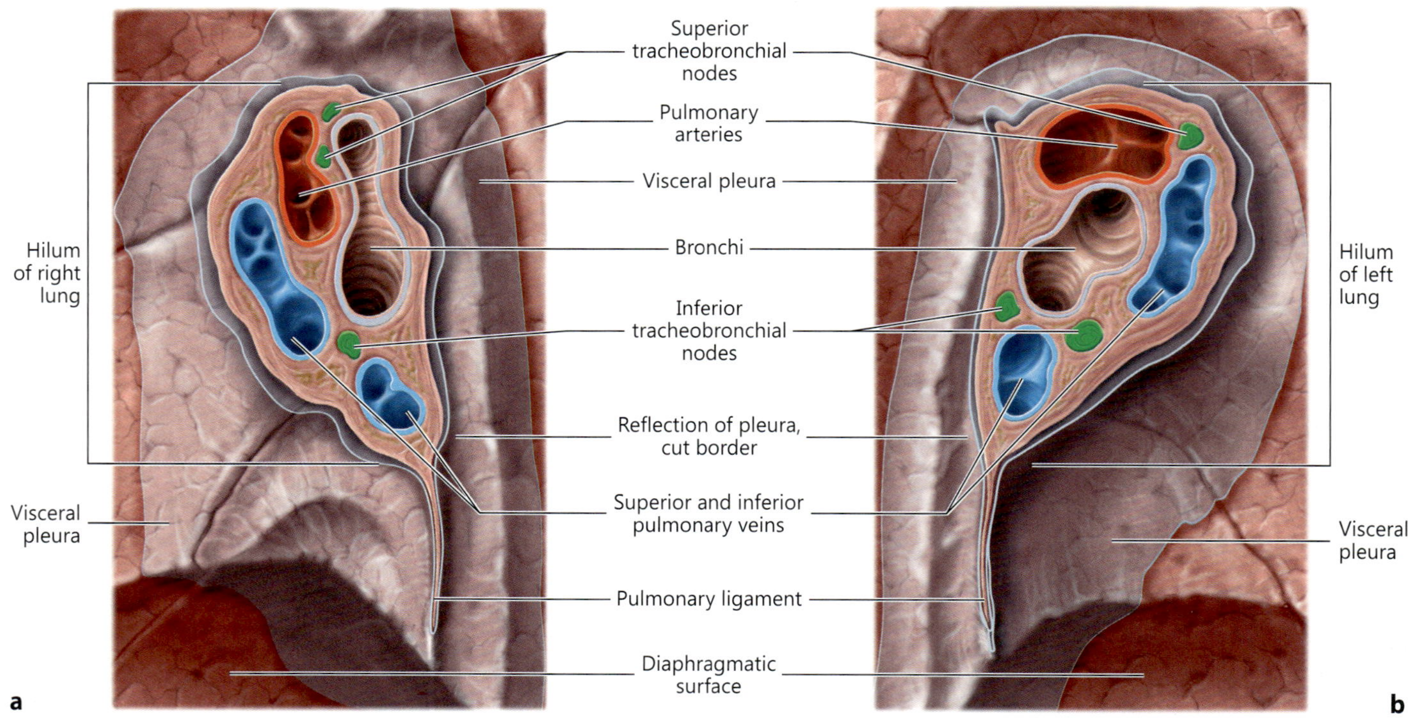

a, **Hilum of the right lung**; **b**, **hilum of the left lung**. In these images the **pulmonary ligament** (triangular ligament) appears in sagittal section.

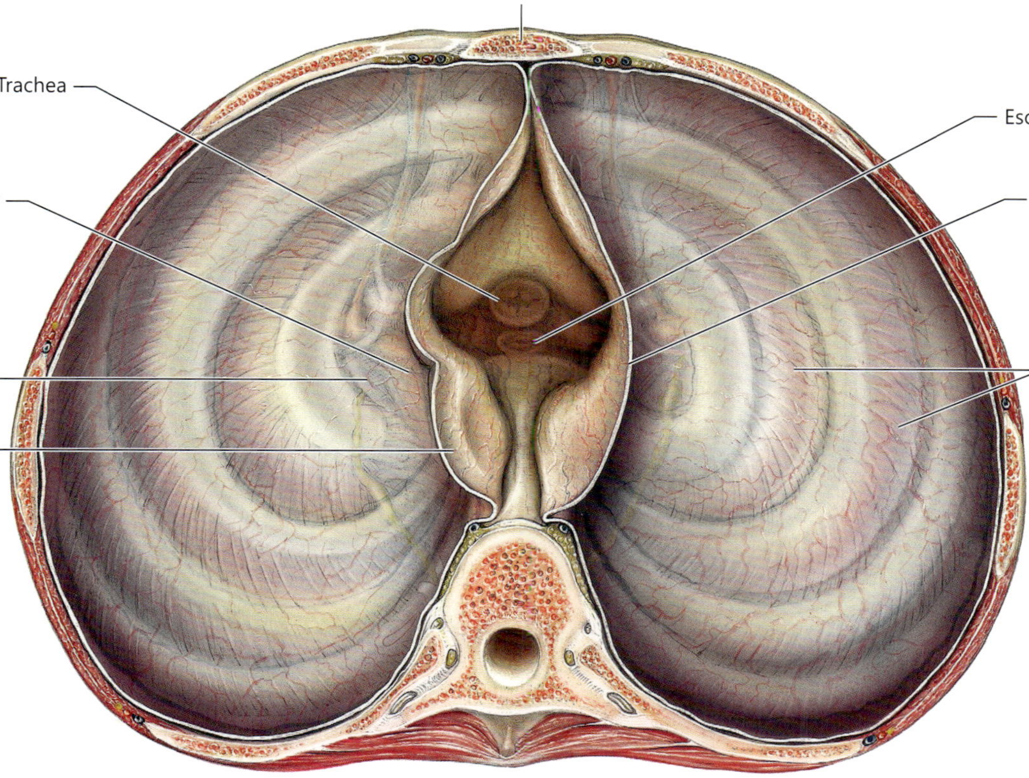

In this image one can see the **cervical pleurae** (domes of pleura) and the right and left **mediastinal pleurae**, viewed from below, after resection of the thorax, removal of the lungs and visceral pleural layers, and partial evacuation of the mediastinum.

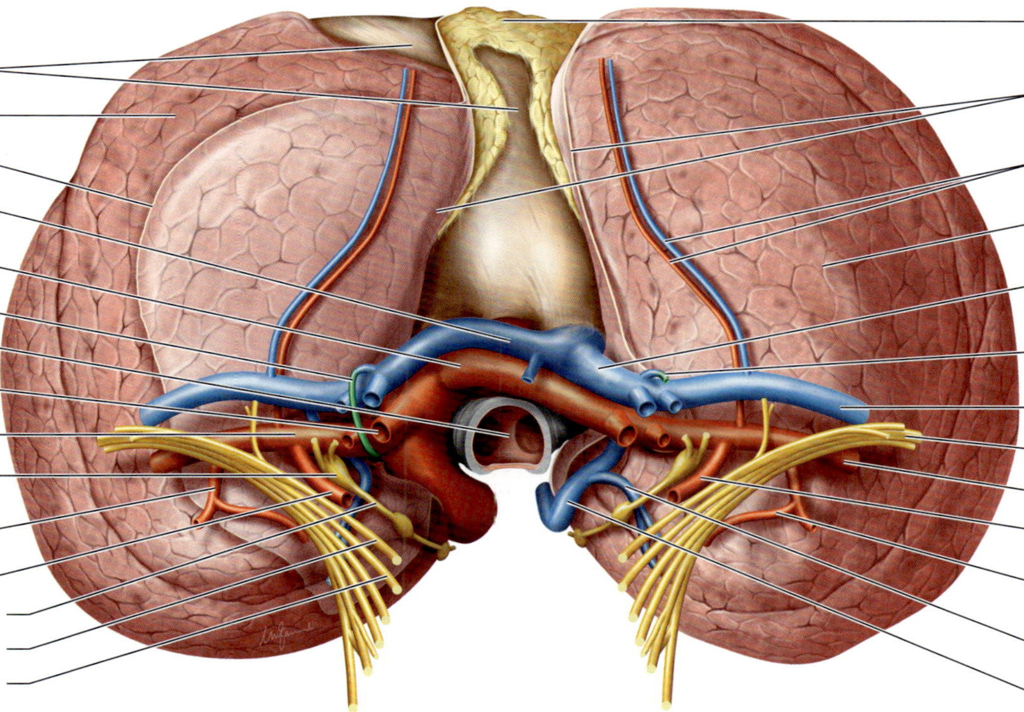

Relationship between the pulmonary apexes mediated by the **cervical pleurae**. In tumors of the apex of the lung (Pancoast tumor), an impairment of these structures is at the base of the typical symptomatologic features. In Bernard-Horner syndrome, impairment of the cervicothoracic (stellate) ganglion provokes palpebral ptosis, miosis, pseudoenophthalmos, and homolateral facial anhidrosis. In the Pancoast syndrome, these symptoms are associated with manifestations linked to the infiltration of the eighth cervical nerve and of the first and second thoracic nerve (pain in the shoulder and on the medial side of the upper limb). Impairment of the lymphatic ducts or of the subclavian veins can provoke lymphedema in the drainage sites or edema of the upper limb, respectively; impairment of the subclavian artery can manifest with cyanosis at the extremities of the upper limb.

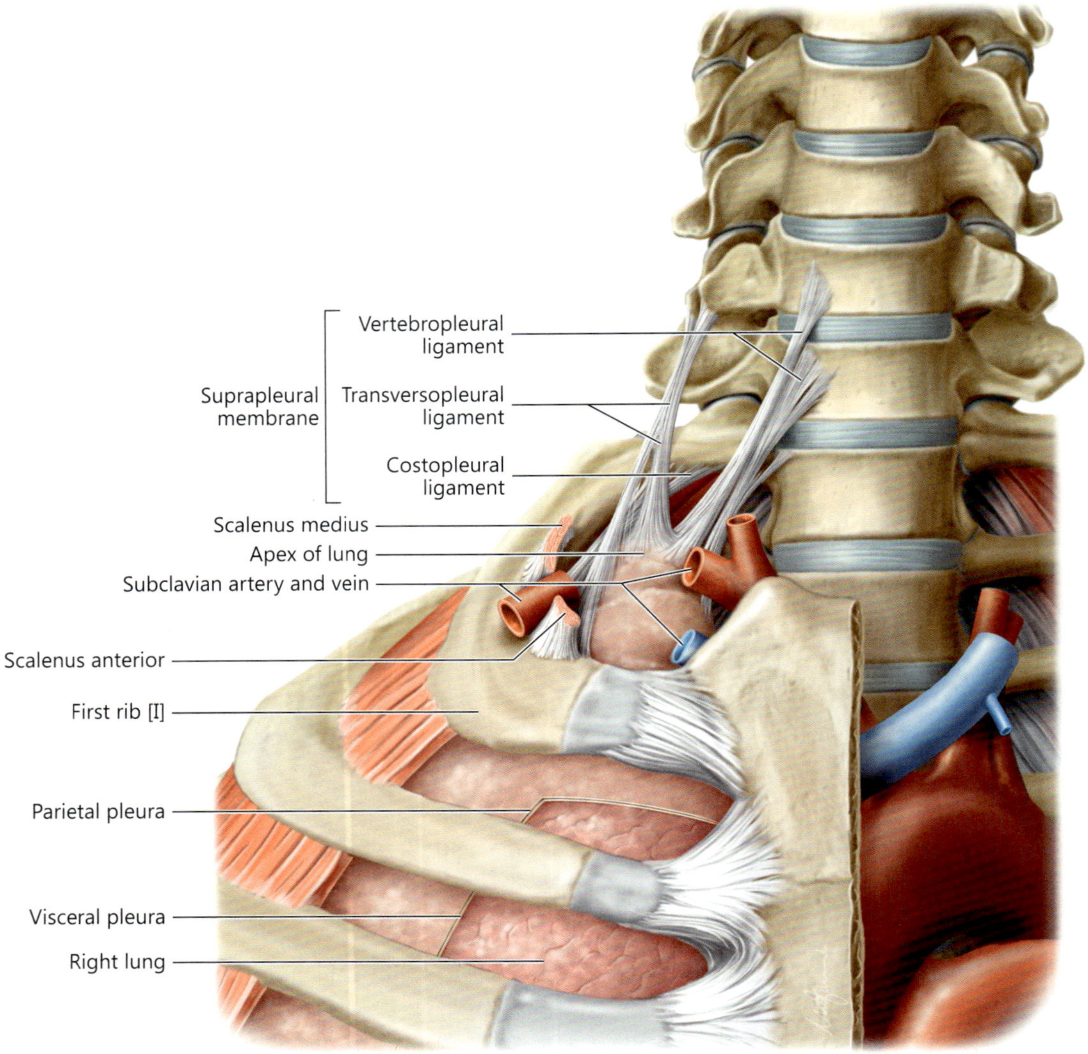

Schematic representation of the principal **means of attachment of the cervical pleura**.

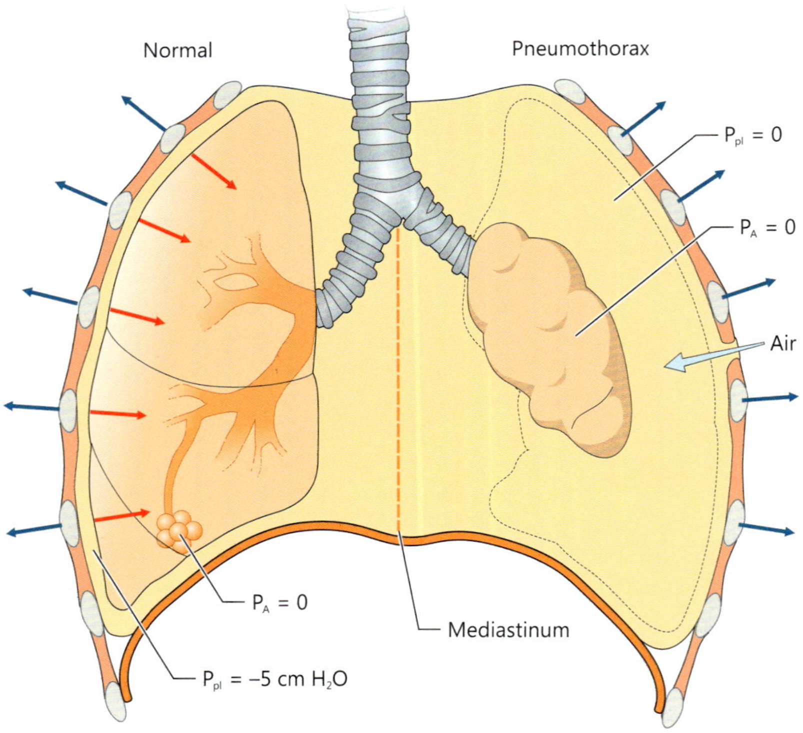

Rupture of the lung or of the thoracic wall, which places the pleural cavity in communication with atmospheric air, produces the condition known as **pneumothorax**, in which the transpulmonary pressure becomes zero and the lung's tendency to elastic recoil provokes its collapse. The barrier effect of the mediastinum impedes the collapse of the contralateral lung. In humans, the two pleural cavities (right and left) are not in communication with each other and, thus, in the case of pneumothorax, ventilation is nevertheless assured by the healthy lung.

The withdrawal of fragments of pulmonary tissue (**biopsy**) or the drainage of fluids or air from the pleural cavity is performed by the insertion of bioptic needles or needle-cannula through the intercostal spaces (**a1**). The procedure must take into account the position of the intercostal vascular and nervous structures. Needle biopsies involve inserting the needle on the superior border of the underlying rib. The procedure is often guided by a radiographic examination or CT scan.

Thoracentesis (**a2**) is a procedure used to drain fluids that have accumulated in the pleural recesses. The technique involves the insertion of a needle-cannula through an intercostal space. The patient is in a seated position and the upper limit of the pleural effusion is determined by percussion; local anesthesia is performed with a needle of reduced caliber and length in proximity to the zone of insertion of the thoracentesis needle. The insertion of the thoracentesis needle is performed 1 or 2 intercostal spaces below the upper limit of the effusion; the best site is between the posterior axillary line and a line placed 6-7 centimeters from the posterior median line. The thoracentesis needle is inserted at the level of the upper border of the inferior rib of the selected intercostal space. The needle, after being inserted a few centimeters aspirating repeatedly to verify if it has reached the effusion, is then positioned horizontally in the pleural cavity.

b, The positioning for **endopleural drainage**, indicated for the aspiration of fluids (gaseous in the case of pneumothorax, liquid in the case of effusions) from the pleural cavity, takes place through the introduction of a multi-fenestrated silicone catheter, with or without an introducer device, which penetrates through the skin and subcutaneous layer on the upper border of the underlying rib, in order to avoid lesions of the intercostal neurovascular bundle. When an introducer is used, as in the figure, it will penetrate through the intercostal muscles and proceed until it reaches the pleural cavity, whereupon it will implant the multi-fenestrated drainage.

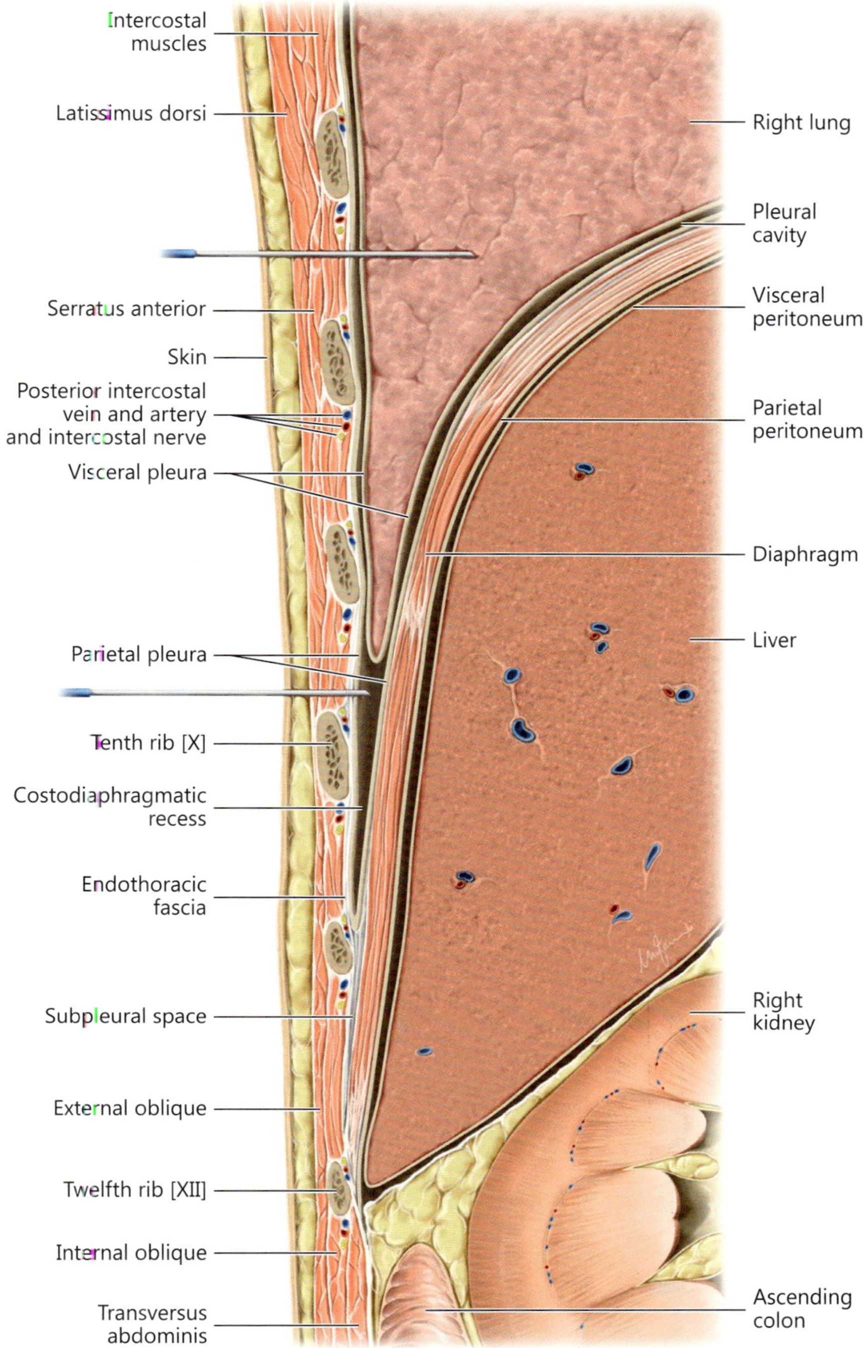

a

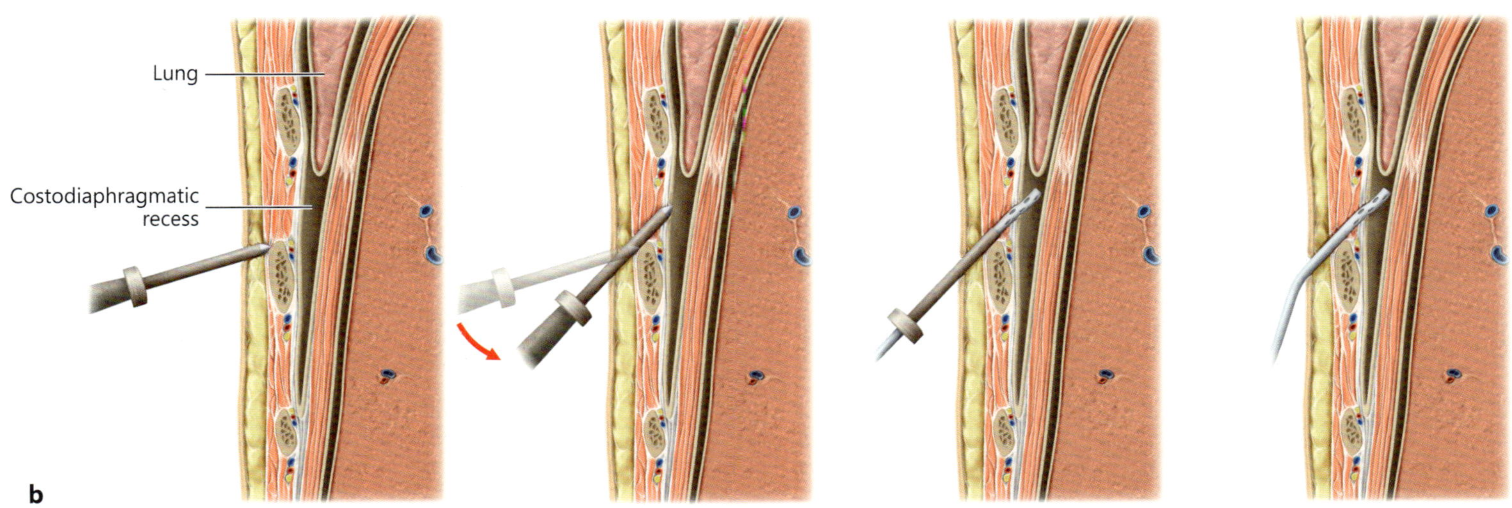

b

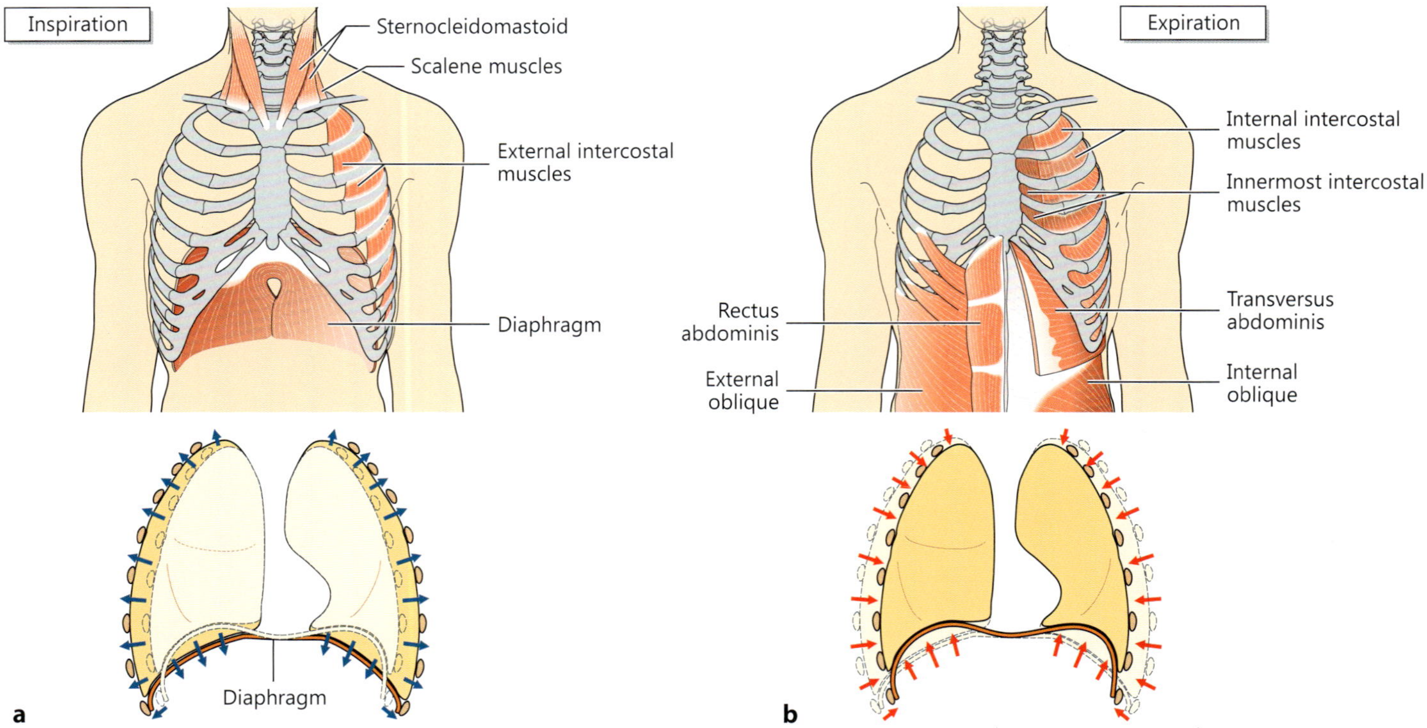

Flat muscle bundles run between two neighboring ribs, constituting the **intercostal musculature**. The external intercostal muscles of the posterior and lateral surface of the thoracic cage raise the ribs (**a**, inspiration); the internal intercostal muscles of the anterior surface lower the ribs (**b**, expiration).

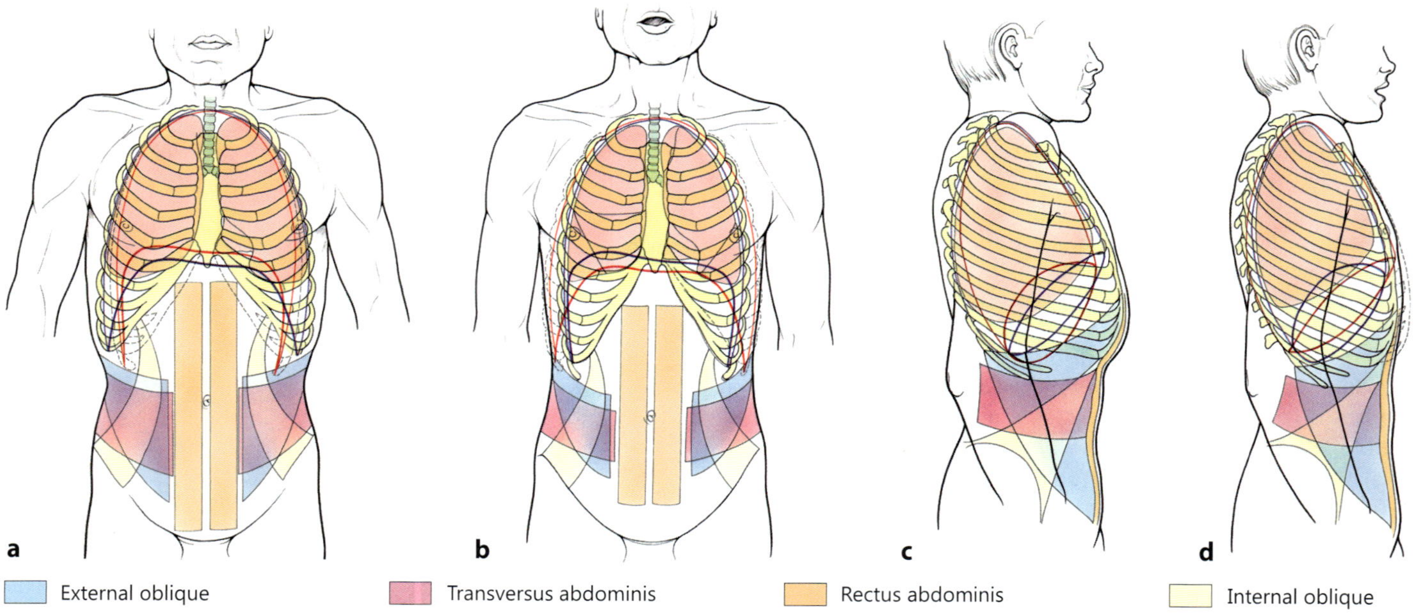

INSPIRATORY MUSCLES	*EXPIRATORY MUSCLES*
• Principal muscles: – diaphragm – external intercostal muscles – scalene muscles	• Principal muscles: – diaphragm – internal intercostal muscles – innermost intercostal muscles
• Secondary (accessory) muscles: – sternocleidomastoid – levatores costarum – serratus posterior superior – subclavius – pectoralis minor and pectoralis major – serratus anterior – latissimus dorsi – trapezius – levator scapulae – rhomboid major and rhomboid minor – muscles of back proper (erector spinae)	• Secondary muscles: – quadratus lumborum – transversus thoracis – subcostales – serratus posterior inferior – muscles of abdomen → A deficit of their muscle tone has major repercussions on respiratory function and, from this perspective, they can be considered as principal expiratory muscles. They act, however, as secondary expiratory muscles, pushing the contents of the abdominal cavity against the diaphragm, so reducing the thoracic cavity

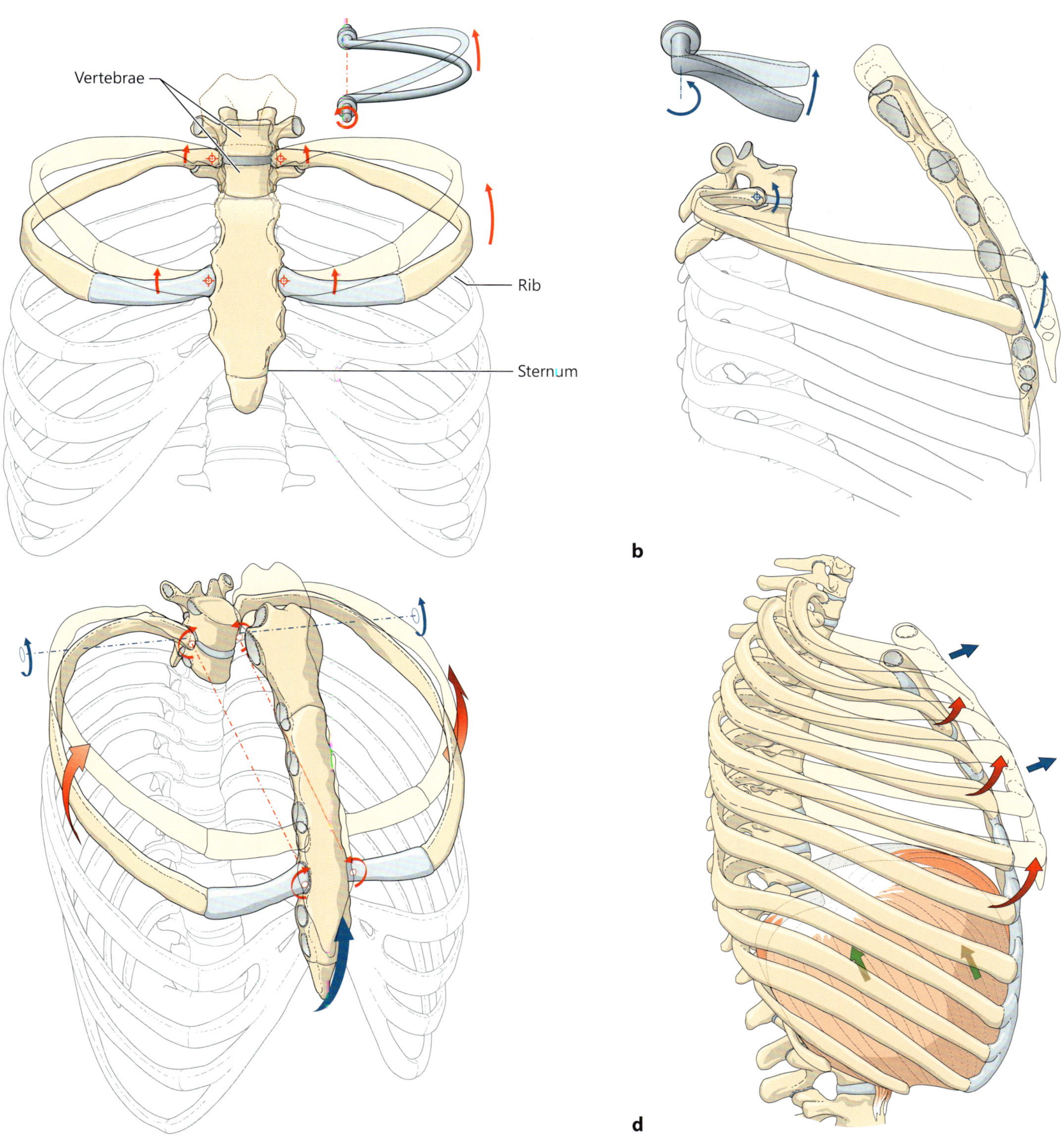

Respiration is based on a coordinated action of the diaphragm and the abdominal, thoracic and cervical muscles. The goal of the action of these muscles is to increase (inspiration) or reduce (expiration) the sagittal, vertical and transverse diameters of the thoracic cavity. In breathing at rest, their function is linked prevalently to the excursions of the diaphragm. In inspiration, the excursion downwards of the central tendon of the diaphragm is favored by the relaxation of the abdominal muscles. Once the maximum excursion consented by the residual abdominal pressure has been reached, a further contraction of the fibers of the diaphragm with respect to the blocked central tendon enables the ribs to be raised (**a**, **b**, **d** in the above figure). Elevation of the ribs enables the transverse (**a**, **c** in the above figure) and sagittal (**b**, **d** in the above figure) diameters of the thoracic cage to increase. The widening of the transverse diameter is the fruit of a movement analogous to lifting the handle of a bucket, which involves the costovertebral joints, from the 2nd to 12th, and the sternocostal joints, with the exclusion of the first. The widening of the sagittal diameter is, instead, linked to the raising of the sternum consequent upon the elevation of the ribs, with a movement that can be compared to that of a handle. The raising of the ribs is aided by the action of the external intercostal muscles (as illustrated in the figure above on the preceding page). The action is made more effective by the activity of the scalene muscles and sternocleidomastoid, which lift and hold the first rib. Expiration relies on the action of the internal intercostal muscles and the diaphragm. In the case of forced expiration, also the muscles of the abdomen (rectus abdominis, external oblique, internal oblique and transversus abdominis) enter into play. By contracting, they determine an increase of the intra-abdominal pressure, further raising the central tendon of the diaphragm, and they lower the ribs.

Schematic description of the final ramifications of the **bronchial tree**. The **pulmonary acinus** is formed of a group of alveoli which are intercommunicating through the alveolar pores. Furthermore, through the canals of Lambert a collateral ventilation is created, which interconnects all the pulmonary acini with the terminal bronchioles. Each acinus is aerated by a respiratory bronchiole, which, in turn, is one of the ramifications of a terminal bronchiole. The structure of the acini is completed by a dense satellite vascularization of the airways.

The **secondary pulmonary lobule** (today more simply called **pulmonary lobule**) is the smallest unit of the lung surrounded by connective tissue and has the following characteristics:

- it is distal to a small bronchiole
- 3-5 terminal bronchioles penetrate into it (for illustrative clarity, each drawing shows just one terminal bronchiole for each secondary pulmonary lobule)
- it includes a dozen pulmonary acini which are generated from these terminal bronchioles
- it has a polyhedral form with septa which are clearly visible in the peripheral (pleural) pulmonary lobules while they are thinner and partially discontinuous in the central pulmonary lobules (they become clearly visible also in the central lobules following fibrous diseases)
- it has a diameter of 1-2.5 cm (mean: 1.1-1.7 cm)
- in the interlobular septa run the smallest ramifications of the pulmonary veins and lymphatic vessels; other lymphatic vessels form in the peribronchovascular interstitium and then ascend the bronchial tree with the pulmonary arteries.

The **primary pulmonary lobule** is contained in the secondary pulmonary lobule and includes the alveolar ducts, the alveolar sacs and the alveoli distal to the last respiratory bronchiole (together with the vessels, nerves and associated connective tissue). It is estimated that each secondary pulmonary lobule is composed of 30-50 primary pulmonary lobules. The term "primary pulmonary lobule" is little used.

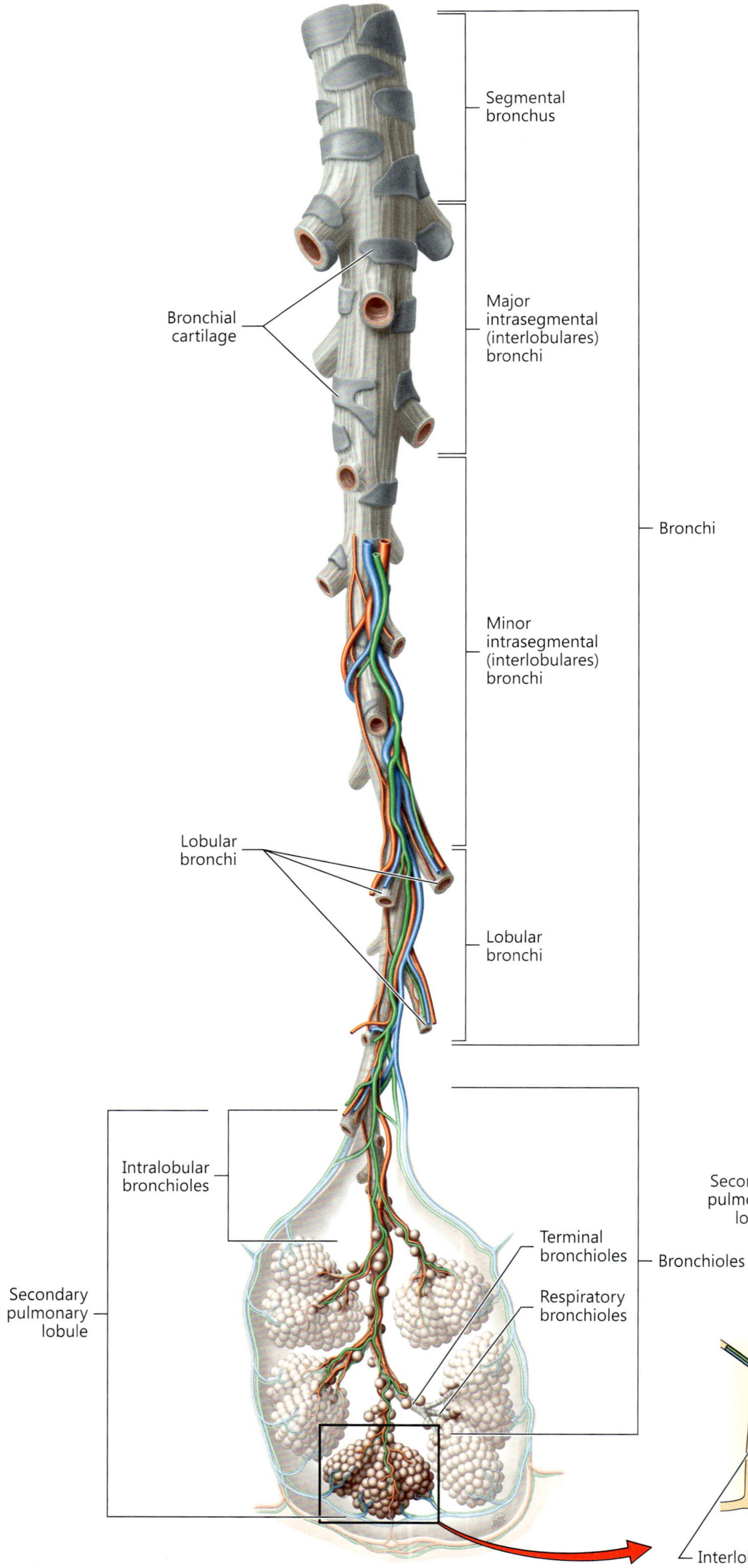

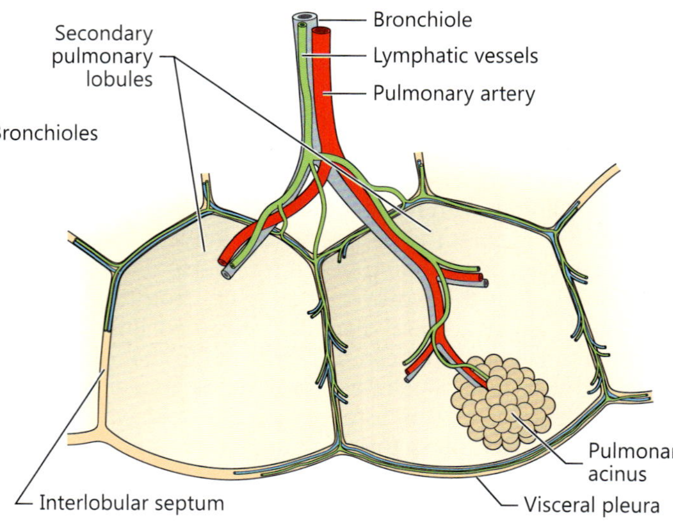

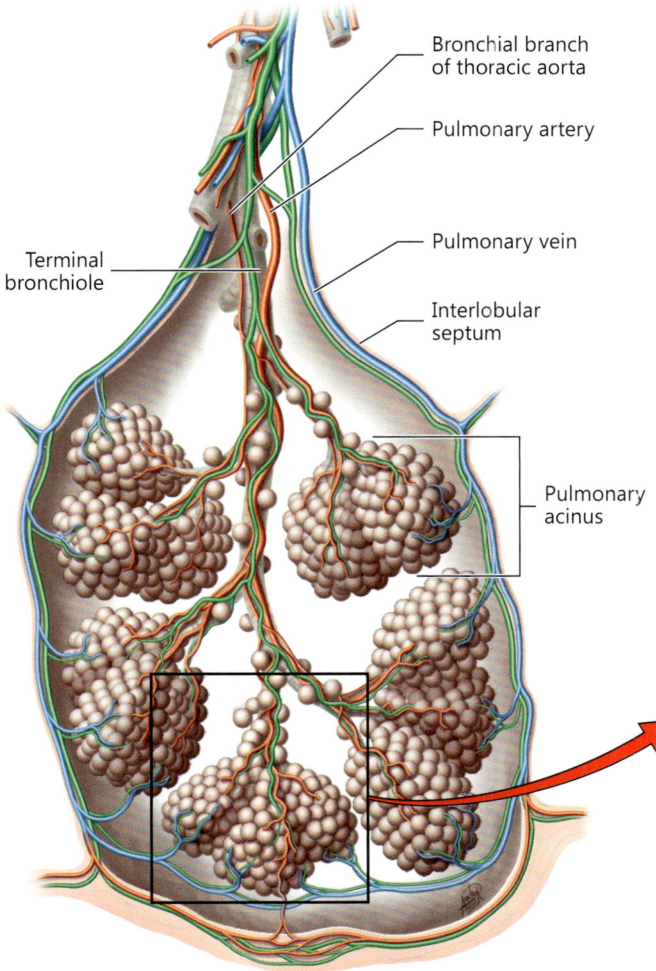

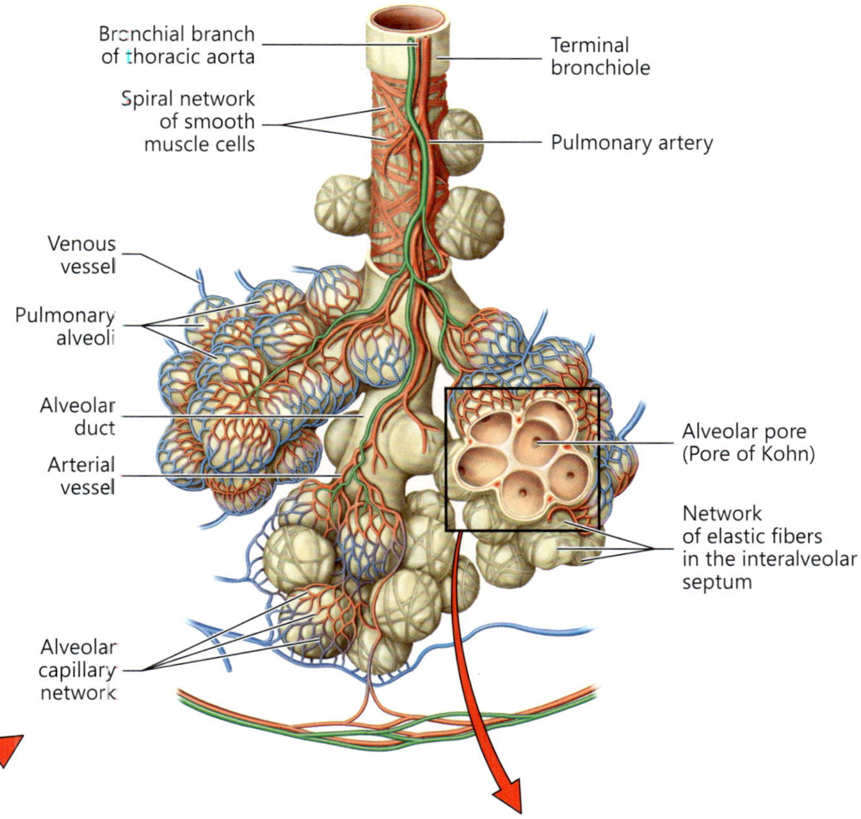

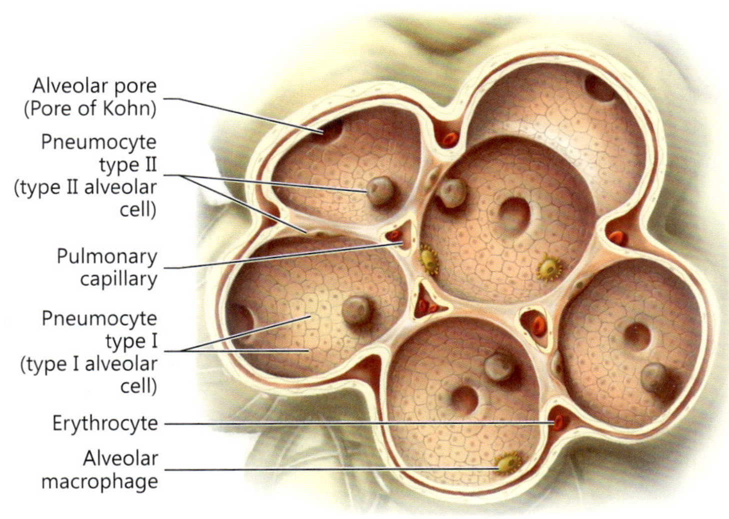

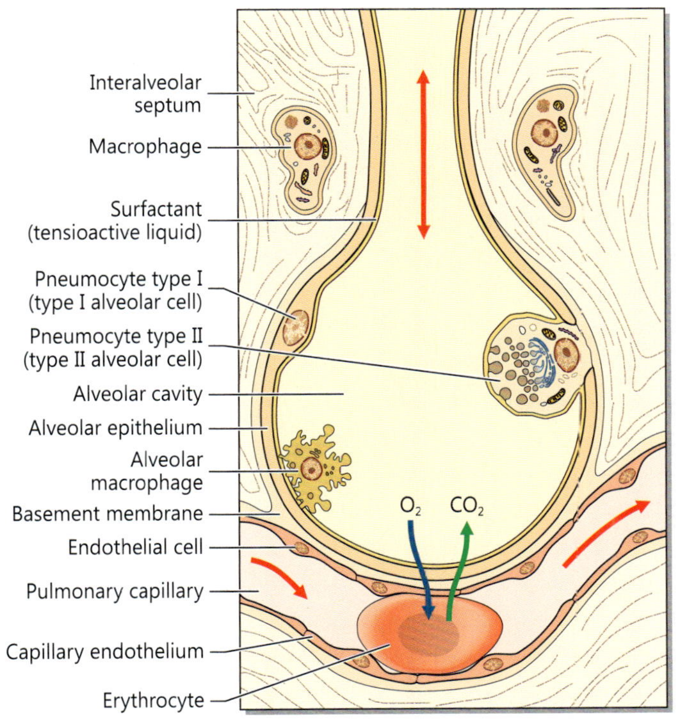

Schematic representation of a **pulmonary alveolus** and of the **blood-air barrier**.

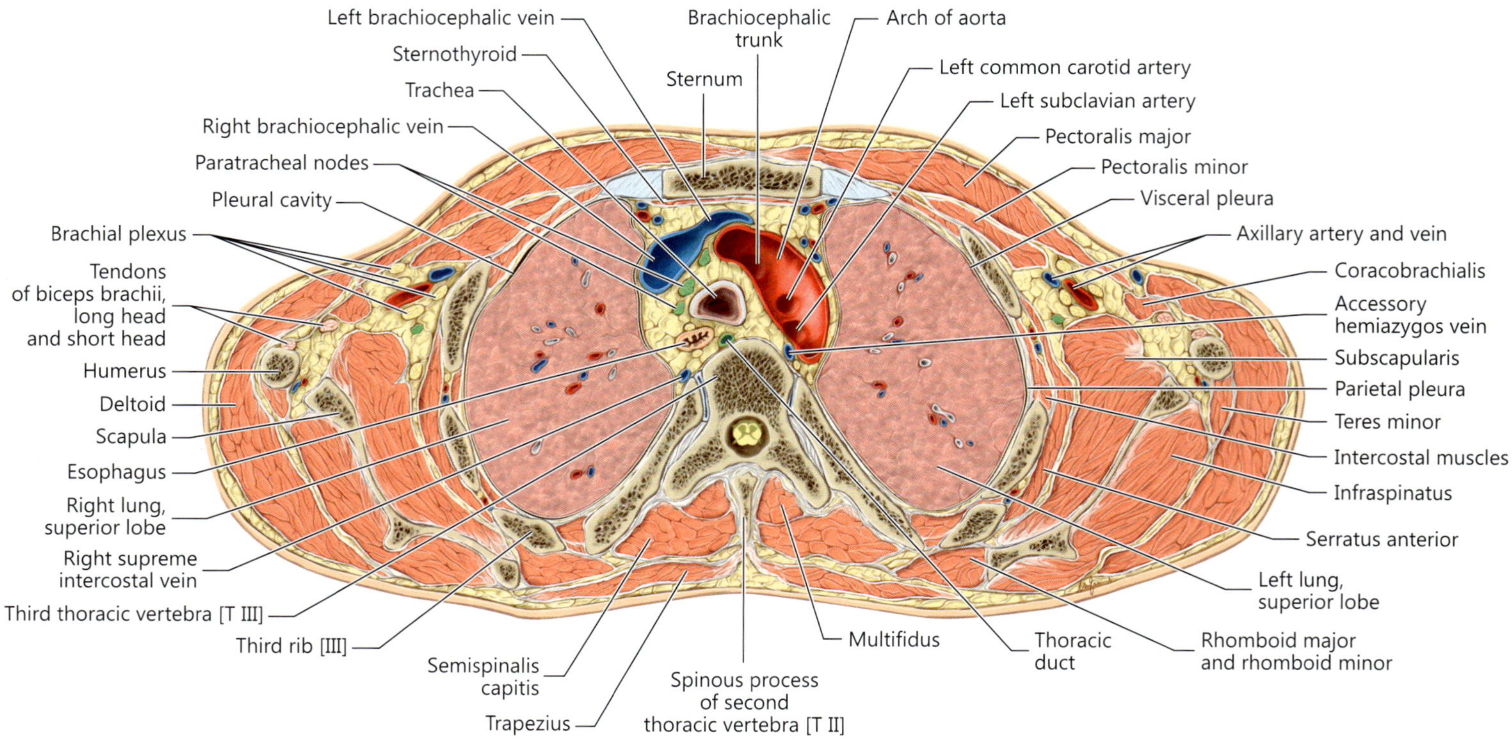

Thoracic cavity. Transverse section at the level of the arch of the aorta, viewed from below.

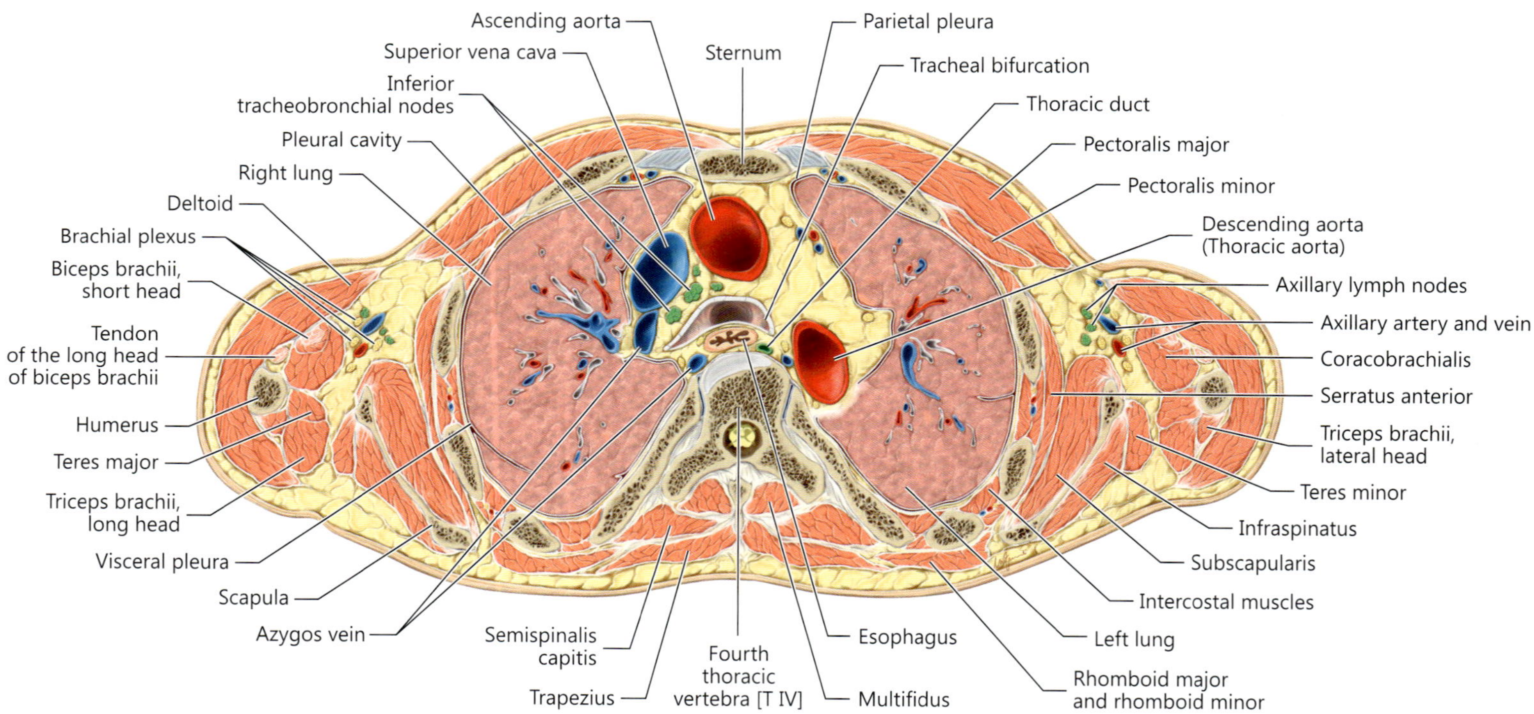

Thoracic cavity. Transverse section at the level of the fourth thoracic vertebra, viewed from below.

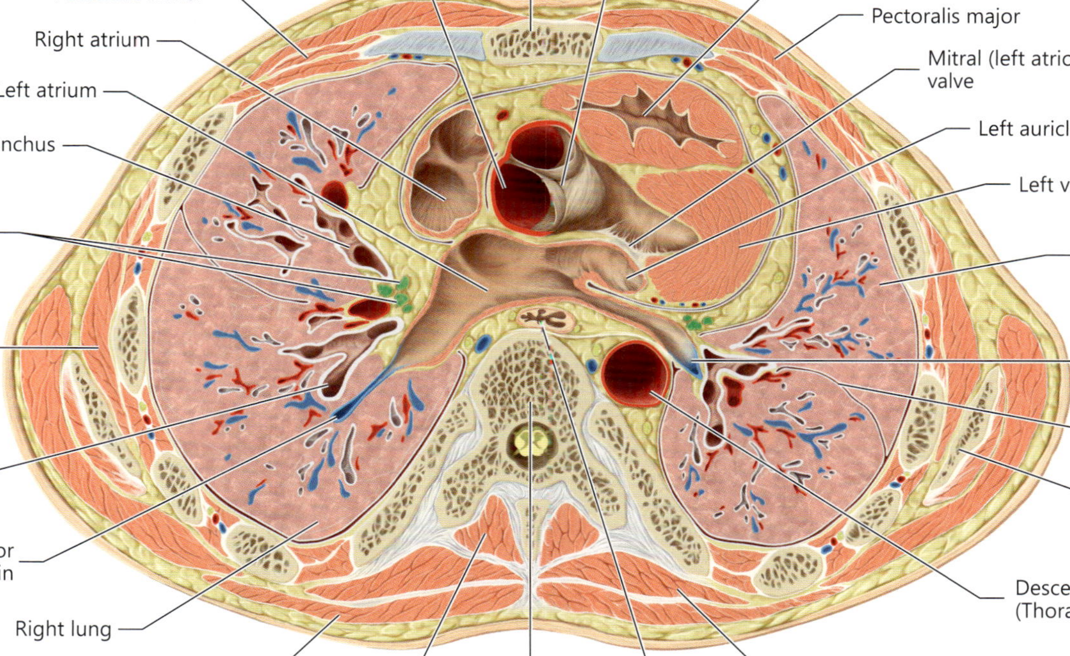

Thoracic cavity. Transverse section at the level of T VII, viewed from below.

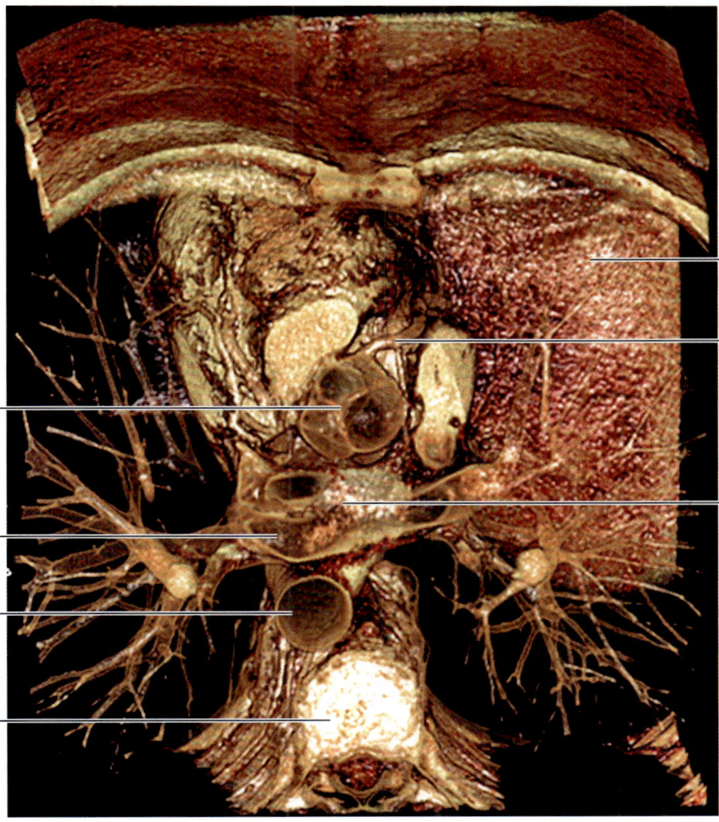

Three-dimensional image of a living human body, obtained by CT scan using a 3D volume rendering technique, showing the **organs of the thoracic cavity**, viewed from above, according to a plane passing through the seventh thoracic vertebra. The lungs and the diaphragm have been graphically removed in order to expose part of the diaphragmatic surface of the liver.

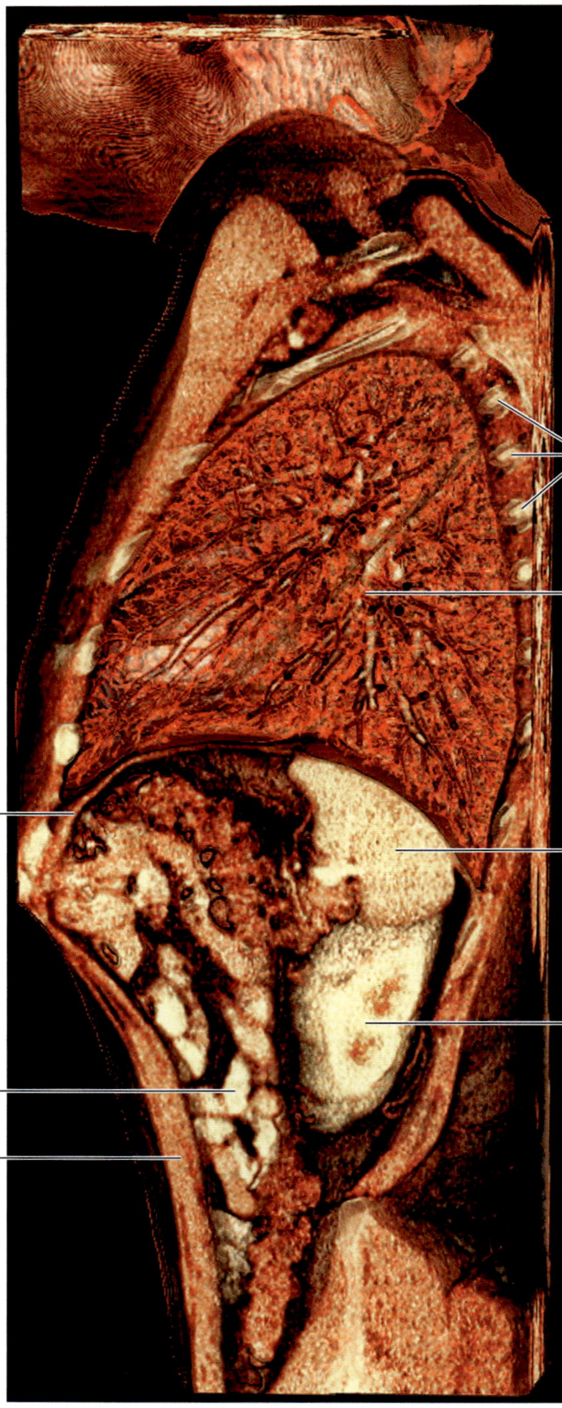

Three-dimensional image of a living human body, obtained using a 3D volume rendering technique. Left parasagittal section showing the position of the **organs of the abdominal and thoracic cavity**, in a plane passing through the spleen and the left kidney.

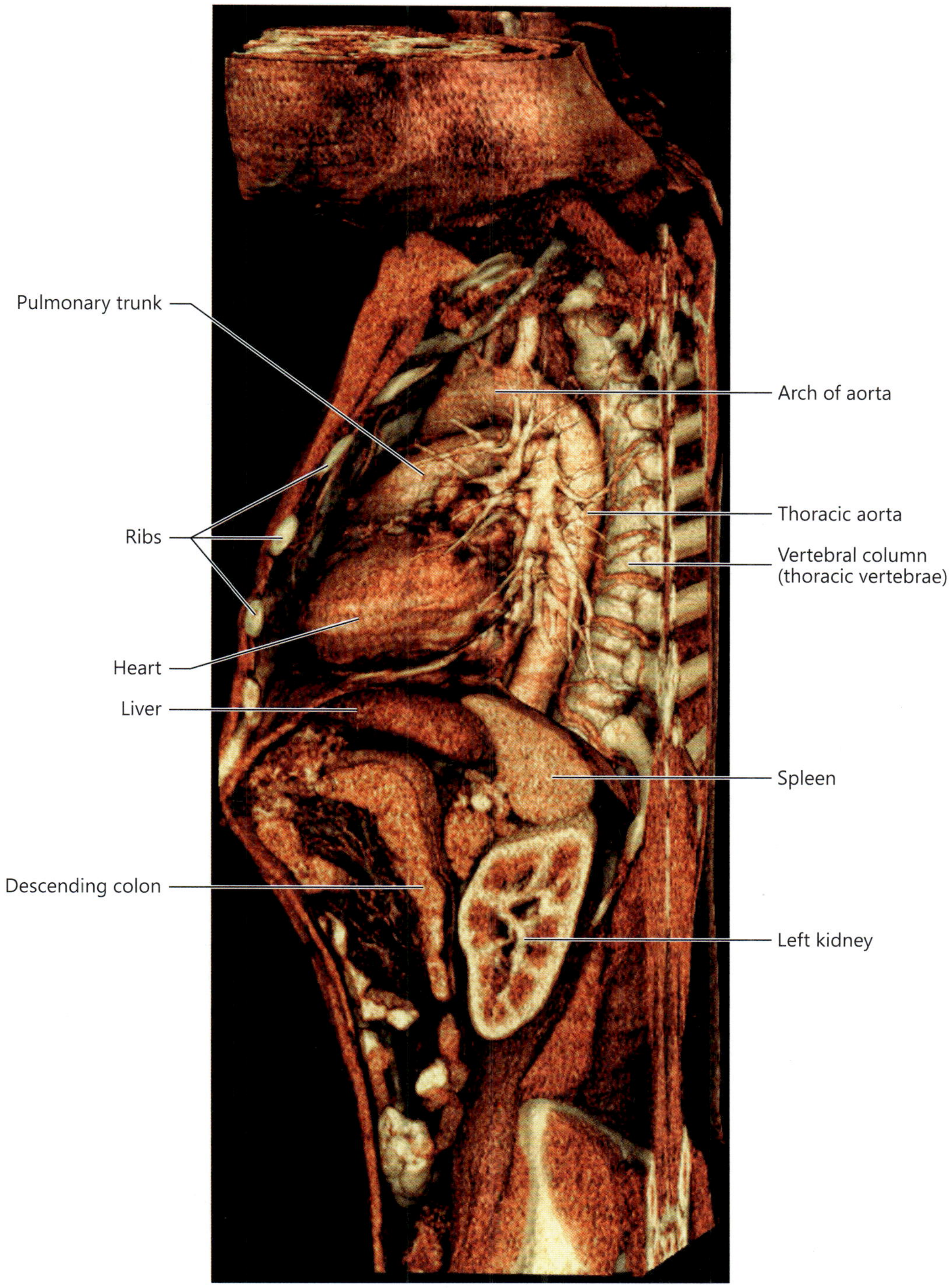

Three-dimensional image of a living human body, obtained using a 3D volume rendering technique. Left parasagittal section showing the position of the **organs of the abdominal and thoracic cavity**.

Index

This index provides that single entry words are organized under the relative structures (i.e. Artery(ies), Bone(s), Muscle(s), etc.). For example, the single entry word "Humerus" is under "Bone(s)" or the single entry word "Latissimus dorsi" is under "Muscle(s)". This concerns also structures like ducts, prominence, tracts, etc. Organs and regional structures (i.e. "Colon") have their main entries and subentries. Page numbers with "t" refer to tables.

C

N

U

V

VIRTUAL CAMPUS

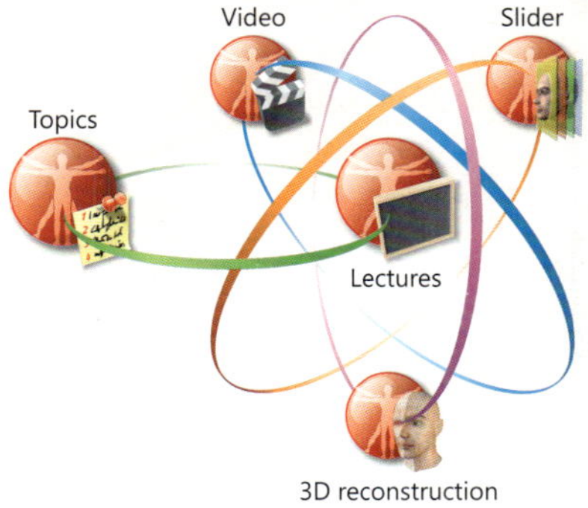

HUMAN ANATOMY *atlas*

FOUR STEPS TO ACCESS DIGITAL CONTENTS

1 CONNECT — Go to the website indicated on the label that includes the access code

2 REGISTER — Register at the Publisher Portal (first time only) to receive a username and password

3 ACCESS — Enter your username and password to access the restricted contents

4 TYPE THE CODE — On the web page dedicated to the work, type in the personal access code hidden under the protection of the label applied to this page

After the first access, digital contents will be available on the website entering your username and password.

The **online access** grants **personal license** limited to a **single user** for each code.
The **access is limited to the individual user** and is not a library or institutional access license.

Sharing of passwords and/or code is not allowed; any misuse of the personal code will invalidate and lock it. The access cannot be shared and will expire according as indicated in the license agreement subscribed at the first access. Further details may be provided upon acceptance of the license agreement. The use of the code is subject to acceptance of the conditions. Any violation or attempted violation of code protection will not be accepted.

To remove the protection
use a coin to scrape
with light pressure

EE5909003846
www.evirtualcampus.com/en

Hardware and software requirements:
PC with Windows, Macintosh or Linux,
latest generation of internet browser such as Internet Explorer (version 9 onwards), Firefox, Chrome etc.; Internet connection.

Technical Help Desk: by email at support@ediermes.com

Contributors
English Edition

Editors

Giuseppe Anastasi	University of Messina, Italy
Eugenio Gaudio	University of Roma (Sapienza), Italy
Carlo Tacchetti	University Vita-Salute San Raffaele, Milan, Italy

Editor English Edition

Estomih Mtui	Weill Cornell Medicine, New York, USA

Coauthors

Paolo Castano†	University Vita-Salute San Raffaele, Milan, Italy
Sergio Castorina	University of Catania, Italy
Ottavio Cremona	University Vita-Salute San Raffaele, Milan, Italy
Raffaele De Caro	University of Padova, Italy
Guido Macchiarelli	University of l'Aquila, Italy
Mario Rende	University of Perugia, Italy
Domenico Ribatti	University of Bari, Italy
Chiarella Sforza	University of Milan, Italy

Other contributors

Demetrio Milardi, University of Messina, Italy, *3D volume rendering images*

Web contents

Estomih Mtui, Weill Cornell Medicine, New York, USA
Angelo Favaloro, University of Messina, Italy
Antonio Esposito, University Vita-Salute San Raffaele, Milan, Italy
Maurizio Vertemati, Department of Biomedical and Clinical Sciences "L. Sacco" University of Milan, Italy

Giuseppe Anastasi
Eugenio Gaudio
Carlo Tacchetti

Estomih Mtui
Editor English Edition

HUMAN ANATOMY
Atlas

Volume 3

ABDOMEN
PELVIS AND PERINEUM
UPPER LIMB
LOWER LIMB
NERVOUS SYSTEM

edi·ermes

HUMAN ANATOMY - Atlas - Volume 3 - Giuseppe Anastasi, Eugenio Gaudio, Carlo Tacchetti (eds)
Estomih Mtui (ed English edition)

ISBN 978-88-7051-590-9

A book is the final product of a very complex series of operations that requires numerous tests on texts and images. It is almost impossible to publish a book with no errors. We will be grateful to those who find them and notify us. For enquiries or suggestions about this volume, please use the following address:

External relations - Edi.Ermes srl - Viale Enrico Forlanini, 65 - 20134 Milan (Italy)
Phone +39.02.70.21.121 - Fax +39.02.70.21.12.83

Drawings, Edi.Ermes Archive/Andrea Rossi Raccagni, Marco Fanuli, Raffaella Stilo, Maria Acquaviva, Anna Maria D'Errico, Ettore Orlandi

Printed in November 2017 by Faenza Printing Industries SpA
for Edi.Ermes - viale Enrico Forlanini, 65 - 20134 Milan, Italy
http://www.ediermes.com - Phone +39.02.70.21.121 - Fax +39.02.70.21.12.83

Foreword

The **Human Anatomy – Atlas** is in my opinion the finest Atlas of Human Anatomy that has ever been written. The main factors that make it so unique include the quality of both the anatomical and radiological images and the way they have been organized.

Artists and radiologists working in collaboration with Edi.Ermes in Milan have created a very high quality 3-D reconstructed images from CT scans and MRI's, used in the Atlas. This approach will give the student and the reader a very clear and accurate understanding of the three dimensional presentation of each structure covered in the Atlas.

All the images are accompanied by a very detailed information related to the anatomy and where appropriate clinical correlates have also been included. This approach will assist the student to relate each structure studied with a clinical condition.

In addition, the authorship of this Atlas has organized the images so that they present the flow of information needed by the student to build a clear image of the human anatomy.

An emphasis on cross sectional anatomy and lymphatic drainage has been made in this Atlas by including several illustrations and radiological images.

Each chapter of the Atlas covering both the anatomy and neuroanatomy contains an introduction, dissection movies and interactive slides used in building the anatomical regions with a stratigraphic approach.

It has been a great pleasure to collaborate with Edi.Ermes in preparing the English edition of this Atlas. I wish to thank in particular Raffaele Grandi the CEO of Edi.Ermes for allowing me to be part of the development team for this Atlas of Human Anatomy.

Estomih Mtui, MD
Editor English Edition

Professor of Clinical Anatomy in Radiology
Chief, Division of Anatomy
Department of Radiology
Weill Cornell Medicine
New York, USA

Contents Plan

Volume 1

Volume 2

Volume 3

Table of Contents

ABDOMEN

PELVIS AND PERINEUM

UPPER LIMB

LOWER LIMB

NERVOUS SYSTEM

ABDOMEN

Petros (Athens, 1928 - Milan, 2014), *Emiekeuh ethn amoohkh...*, mixed media on canvas-backed paper, cm 60 × 80, 1972

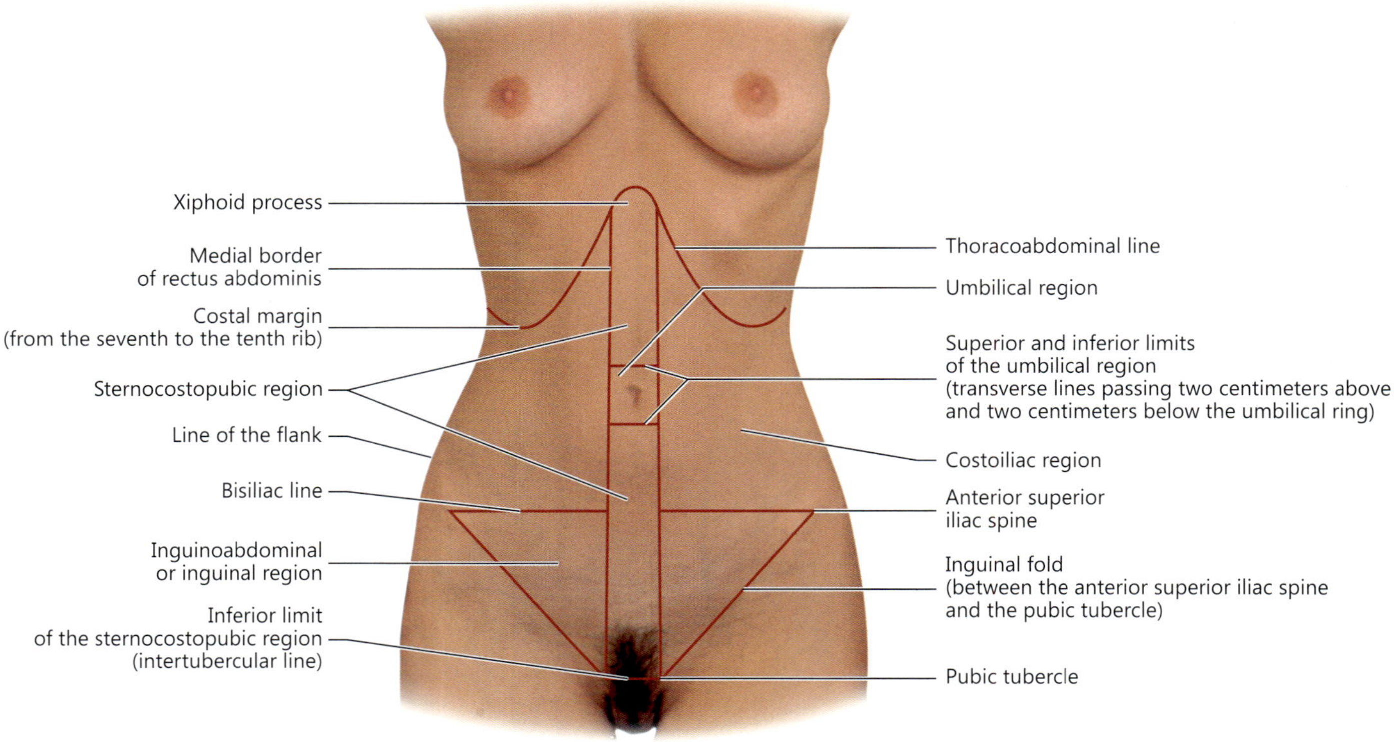

Projection of the abdominal cavity onto the anterior wall of the trunk. The thoracoabdominal line marks the upper boundary of the abdomen, while the two inguinal folds and the intertubercular line mark the lower boundary. The **anterolateral abdominal wall** comprises the sternocostopubic region and the left and right costoiliac regions; the former also includes the umbilical region, while the latter also comprise their respective inguinal or inguinoabdominal regions.

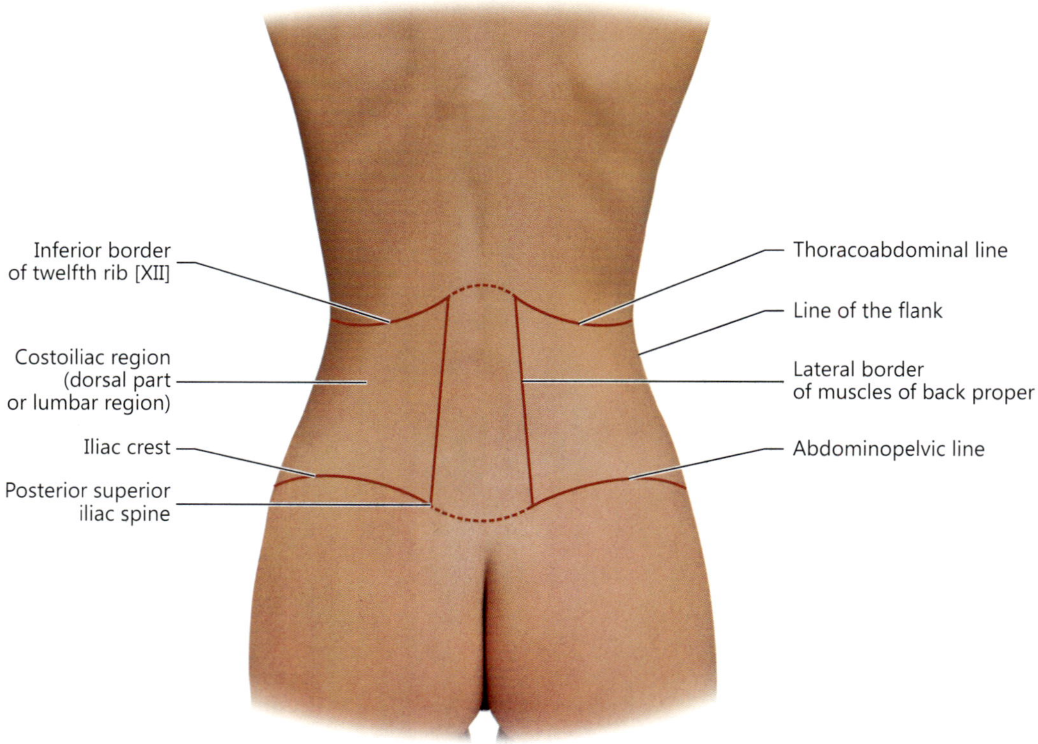

Projection of the abdominal cavity onto the posterior wall of the trunk. The thoracoabdominal line marks the upper boundary of the abdomen, while the abdominopelvic line marks the lower boundary. The **costoiliac region** covers most of the anterolateral abdominal wall, and extends posteriorly to include the lumbar region.

Boundaries

REGIONS OF THE ABDOMEN	BOUNDARIES OR LOCATION
Mesogastric region (including the umbilical region)	*Superior boundary*: subcostal line, in between the lines running from midpoint of clavicle to the pubic tubercle *Inferior boundary*: bisiliac line, in between the lines running from midpoint of clavicle to the pubic tubercle *Lateral boundaries*: lines running from midpoint of clavicle to the pubic tubercle, in between the subcostal and bisiliac lines Note: the umbilicus lies at the centre of the umbilical region, which belongs to the mesogastric region
Costoiliac region	*Superior boundary*: costal margin and inferior border of twelfth rib *Inferior boundary*: iliac crest and tract of the bisiliac line between the anterior superior iliac spine and the lateral border of rectus abdominis *Posterior boundary*: lateral border of muscles of back proper, in between the inferior border of twelfth rib and the iliac crest *Anterior boundary*: lateral border of rectus abdominis to its intersection with the bisiliac line
Epigastric region (epigastric fossa)	*Superior boundary*: xiphoid line, in between the lines running from midpoint of clavicle to the pubic tubercle *Inferior boundary*: subcostal line, in between the lines running from midpoint of clavicle to the pubic tubercle *Lateral boundaries*: lines running from midpoint of clavicle to the pubic tubercle, in between the xiphoid and subcostal lines
Right inguinal region or right groin or right inguinoabdominal region	*Superior boundary*: bisiliac line, in between the midclavicular line and the anterior superior iliac spine *Inferior bundary*: line between the anterior superior iliac spine and the pubic tubercle, corresponding to the inguinal fold *Medial boundary*: line running from midpoint of clavicle to the pubic tubercle, in between the bisiliac line and the pubic tubercle, or the lateral border of rectus abdominis Note: the inguinal region laterally overlaps with the iliac fossa region
Left inguinal region or left groin or left inguinoabdominal region	☞ Right inguinal region
Right hypochondrium	*Superior boundary*: xiphoid line, in between the line running from midpoint of clavicle to the pubic tubercle and the continuation of the anterior axillary line *Inferior boundary*: subcostal line, in between the line running from midpoint of clavicle to the pubic tubercle and the continuation of the anterior axillary line *Medial boundary*: line running from midpoint of clavicle to the pubic tubercle, in between the xiphoid and subcostal lines *Lateral boundary*: continuation of the anterior axillary line, in between the xiphoid and subcostal lines
Left hypochondrium	☞ Right hypochondrium
Right lateral abdominal region or right flank	*Superior boundary*: subcostal line, in between the line running from midpoint of clavicle to the pubic tubercle and the continuation of the anterior axillary line *Inferior boundary*: bisiliac line, in between the line running from midpoint of clavicle to the pubic tubercle and the anterior superior iliac spine *Medial boundary*: line running from midpoint of clavicle to the pubic tubercle, in between the subcostal and bisiliac lines *Lateral boundary*: continuation of the anterior axillary line, in between the subcostal line and the anterior superior iliac spine
Left lateral abdominal region or left flank	☞ Right lateral abdominal region
Lumbar region	*Superior boundary*: curved line that follows the inferior border of twelfth rib *Inferior boundary*: curved line that follows the iliac crest *Medial boundary*: vertical line tangential to the lateral border of muscles of back proper
– superior lumbar triangle (inconstant)	*Base*: twelfth rib and inferior border of serratus posterior inferior *Sides*: borders of quadratus lumborum (medially) and of internal oblique (laterally) *Floor*: transversalis fascia and aponeurosis of transversus abdominis *Roof*: external oblique and latissimus dorsi
– inferior lumbar triangle (inconstant)	*Base*: iliac crest (inferiorly) *Sides*: anterior border of latissimus dorsi (posteriorly) and posterior border of external oblique (anteriorly) *Floor*: thoracolumbar fascia and aponeurosis of internal oblique and transversus abdominis *Roof*: skin and investing layer
Umbilical region	*Superior boundary*: transverse line passing 2 cm above the umbilical ring *Inferior boundary*: transverse line passing 2 cm below the umbilical ring *Lateral boundaries*: medial borders of rectus abdominis Note: a region at the centre of the mesogastric region, or part of the sternocostopubic region
Pubic region	*Superior boundary*: bisiliac line, in between the lines running from midpoint of clavicle to the pubic tubercle *Inferior boundary*: intertubercular line *Lateral boundaries*: lines running from midpoint of clavicle to the pubic tubercle, in between the bisiliac and intertubercular lines Note: the above mentioned boundaries define the pubic region proper, while the pubic region (pelvis) is located below the intertubercular line
Sternocostopubic region	*Superior boundary*: xiphoid process and costal margin *Inferior boundary*: pubic symphysis, in between the pubic tubercles *Lateral boundaries*: medial borders of rectus abdominis
(Lumbar) vertebral region	A region of the back extending, at the abdomen, between the L I and L V spinous processes and the vertical lines that correspond to the lateral borders of muscles of back proper

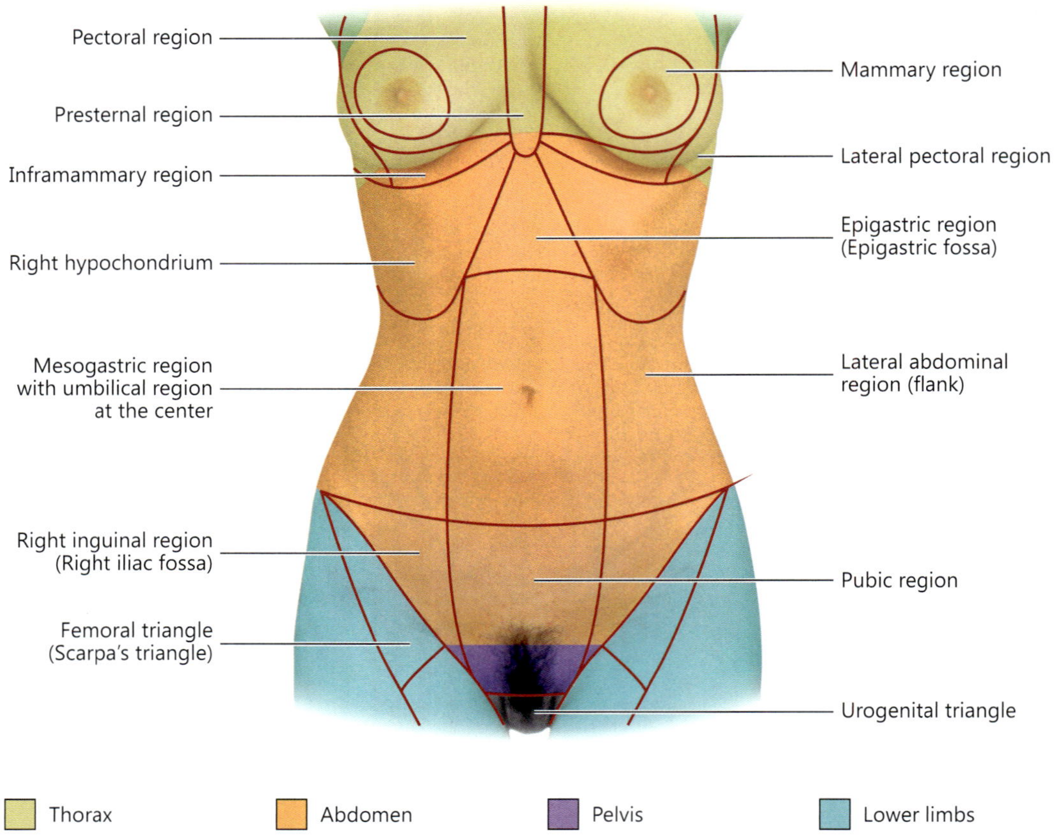

Regions of the trunk: **abdomen**. It is divided, from top to bottom, into epigastric, mesogastric and pubic regions, each comprising a median region and two lateral regions (right and left).

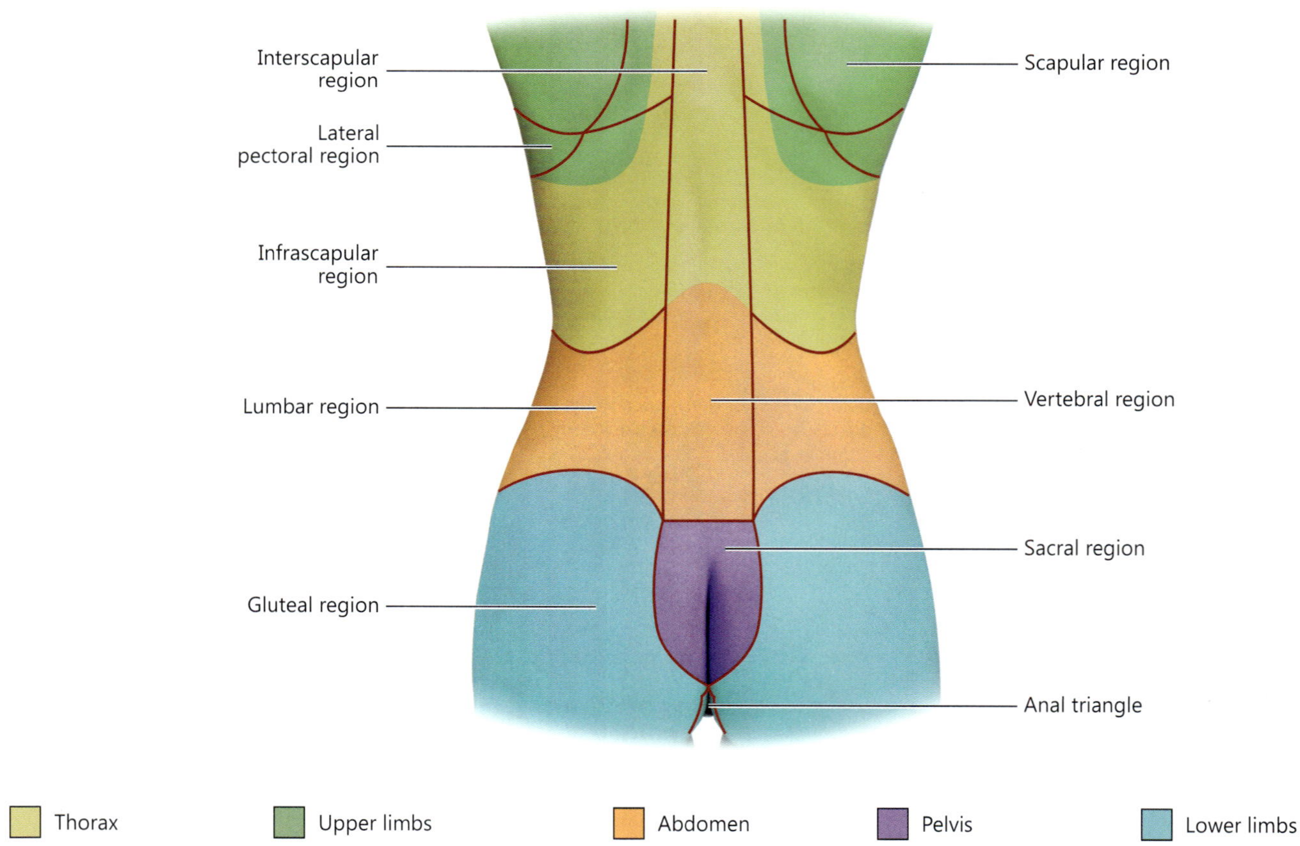

Regions of the back. Vertebral and lumbar regions of the abdomen.

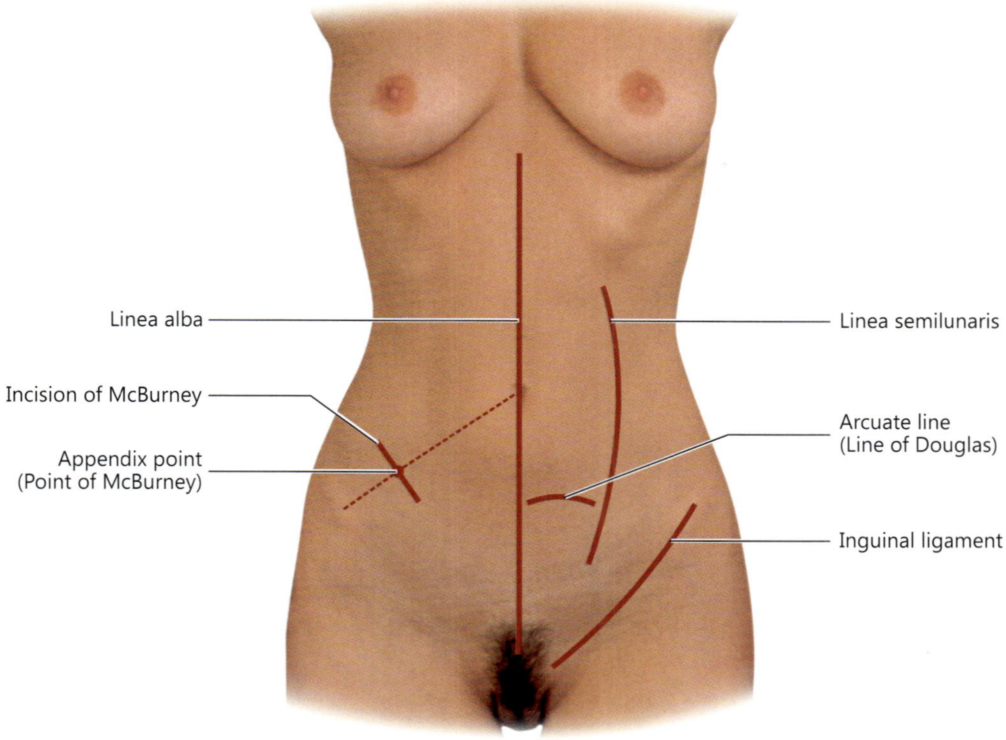

Lines and landmarks on the surface of the abdomen.

Lines and landmarks	
LINE OR LANDMARK	***LOCATION***
Appendix point (point of McBurney) (usually corresponding to the base of the appendix)	Between the lateral one-third and the medial two-thirds of a line joining the right anterior superior iliac spine and the umbilicus
Linea alba	A fibrous raphe extending from the xiphoid process of the sternum to the superior border of the pubic symphysis
Arcuate line (line of Douglas) (approximately 2-3 cm below the umbilicus)	A line marking the transition between the upper three-quarters (aponeurosis of internal oblique and transversus abdominis and transversalis fascia) and the lower one-quarter (transversalis fascia alone) of the posterior layer of rectus sheath
Linea semilunaris	A tendinous arch marking the lateral border of rectus abdominis
Inguinal ligament	A fibrous band running between the anterior superior iliac spine and the pubic tubercle
Incision of McBurney	An oblique incision along a perpendicular line to a line joining the right anterior superior iliac spine and the umbilicus, crossing the appendix point (point of McBurney), which provides access to the appendix

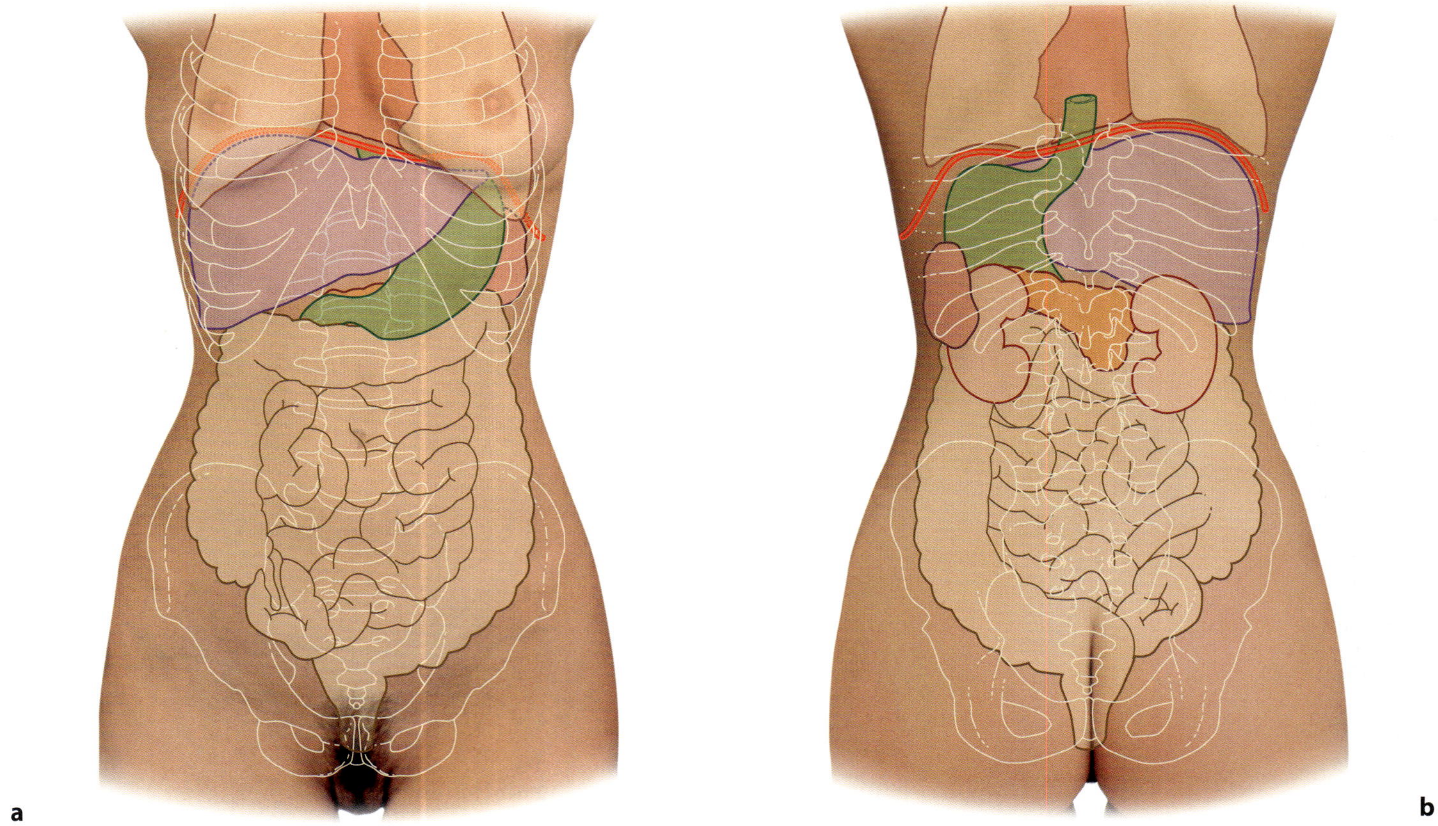

Projection of abdominal organs onto the anterior (**a**) and posterior (**b**) walls of the trunk. The surface projection depends on posture, age, body type, gender, nutritional status and phases of respiration.

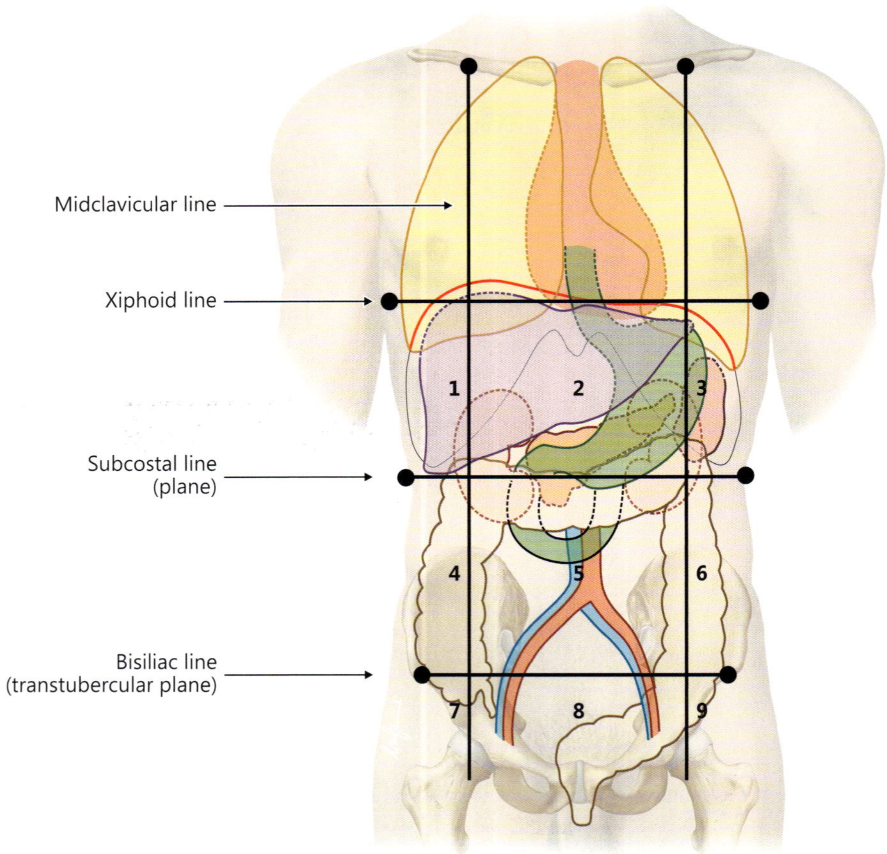

Subdivision of the **abdominal cavity** into quadrants. Boundaries and regions of the **anterior abdominal wall** as defined by the intersection of the midclavicular lines with the subcostal and transtubercular planes. The basisternal or xiphoid line represents the approximate level of the central tendon of the diaphragm. 1, Right hypochondrium; 2, epigastric region or epigastric fossa; 3, left hypochondrium; 4, right lateral abdominal region or right flank; 5, mesogastric region; 6, left lateral abdominal region or left flank; 7, right inguinal region or right inguinoabdominal region; 8, pubic region; 9, left inguinal region or left inguinoabdominal region.

Projection of the abdominal organs onto the regions of the anterior abdominal wall	
REGION	***ORGANS***
Epigastric region (epigastric fossa)	Stomach, liver (extending into the superior part of the region), pancreas
Right hypochondrium	Liver, right colic flexure (located deep in the region), right kidney (located deep in the region)
Left hypochondrium	Stomach, spleen (located in the deepest part of the region), liver (extending for a short lenght into the region), left colic flexure (located deep in the region), small intestine (first group of loops), left kidney (located deep in the region)
Mesogastric region	Small intestine (second and third part of the duodenum, deeply, and second group of loops), transverse colon (in the superior part of the region), hilar region of the right and left kidneys and corresponding renal pelvis
Right flank (right lateral abdominal region)	Ascending colon
Left flank (left lateral abdominal region)	Descending colon
Pubic region	Urinary bladder (distended), upper one-third of rectum (deeply)
Right groin (right inguinal or inguinoabdominal region) and right iliac fossa	Cecum and vermiform appendix (in the right iliac fossa), ascending colon (running for a short length into the right iliac fossa)
Left groin (left inguinal or inguinoabdominal region) and left iliac fossa	Small intestine (third group of loops in the left iliac fossa), descending colon (extending partly into the left iliac fossa), sigmoid colon (extending partly into the left iliac fossa)

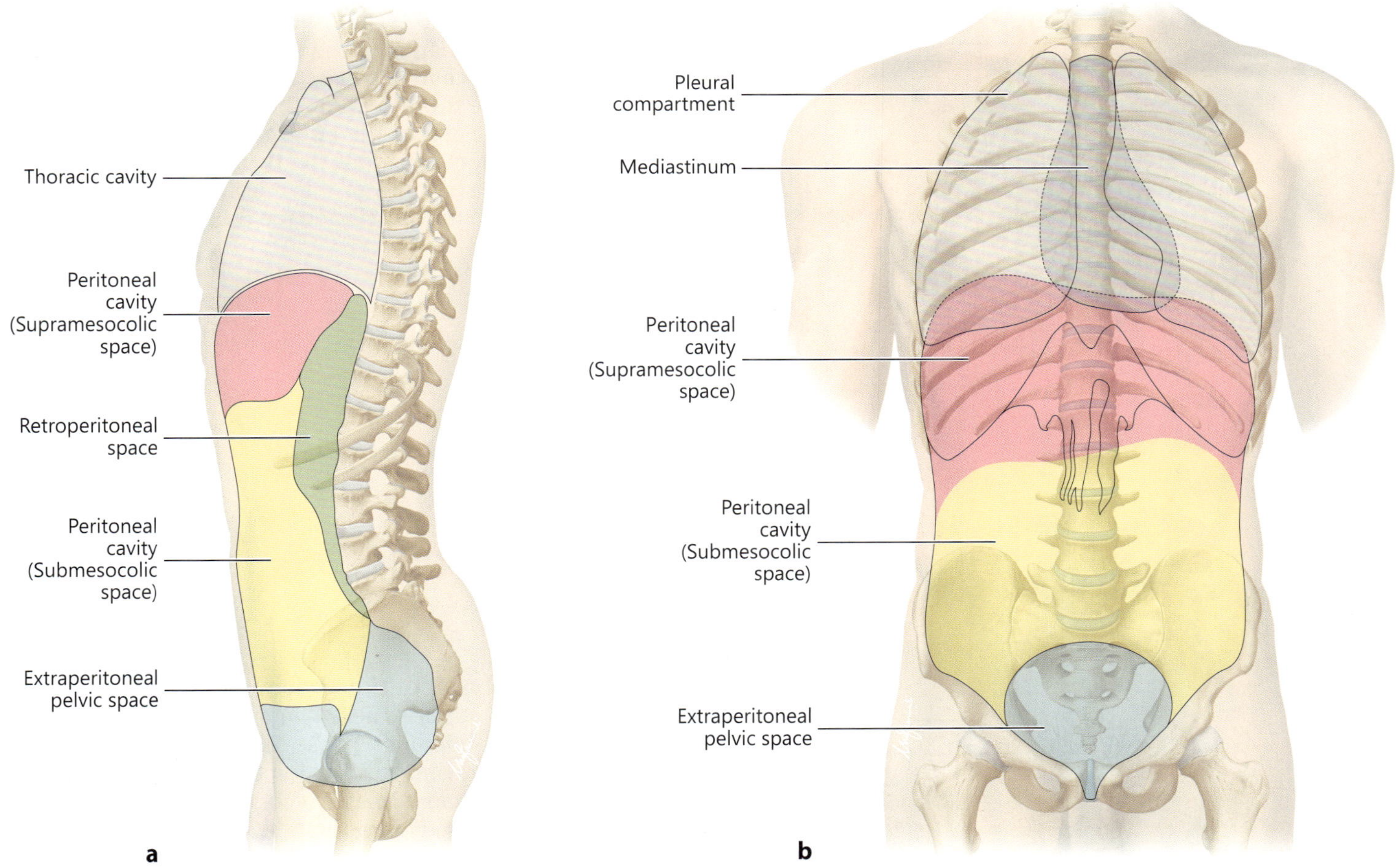

a, Lateral projection of the trunk, indicating the **cavities** and the **visceral spaces** of the **abdomen** (also defining the supramesocolic and submesocolic spaces of the peritoneal cavity) and the extraperitoneal space (retroperitoneal space, in *green*, and extraperitoneal or subperitoneal pelvic space, in *light blue*). **b**, Front projection.

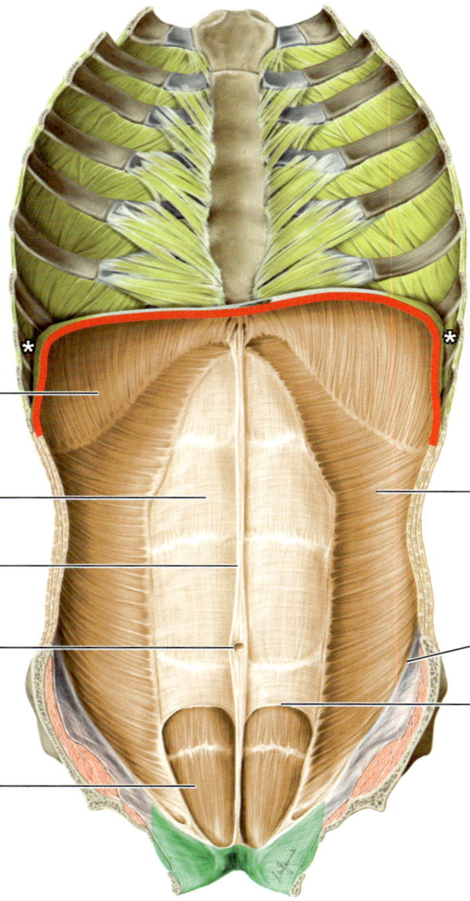

Anterior wall of the trunk, as seen from the inside. The boundaries of the surface projection of the **thoracic cavity** do not correspond to the boundaries of the **thoracic cage**. The latter also includes the superior part of the **abdominopelvic cavity**. The diaphragm separates the two cavities. The dome-shaped structure of this muscle creates recesses between the muscle and the thoracic cage (*).

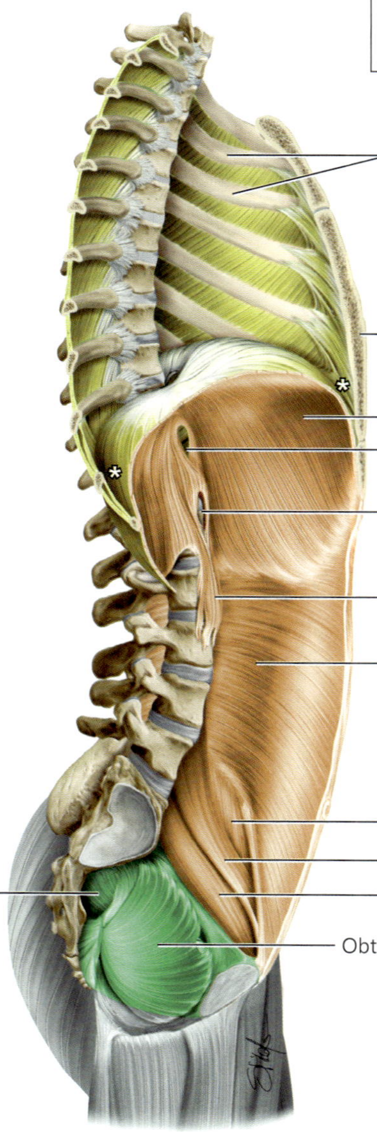

Thoracic cavity

Abdominal cavity

Pelvic cavity

Lateral wall of the abdominopelvic cavity, left internal surface. The left hemidiaphragm, the transversus abdominis, psoas minor, psoas major and iliacus are shown. The dome-shaped structure of the diaphragm creates recesses between the muscle and the thoracic cage (*). By removing the hip bone, on the right, the muscles of the pelvic wall can be observed: the **piriformis** forms part of the posterolateral wall of the lesser pelvis, and the **obturator internus** forms part of the anterolateral wall.

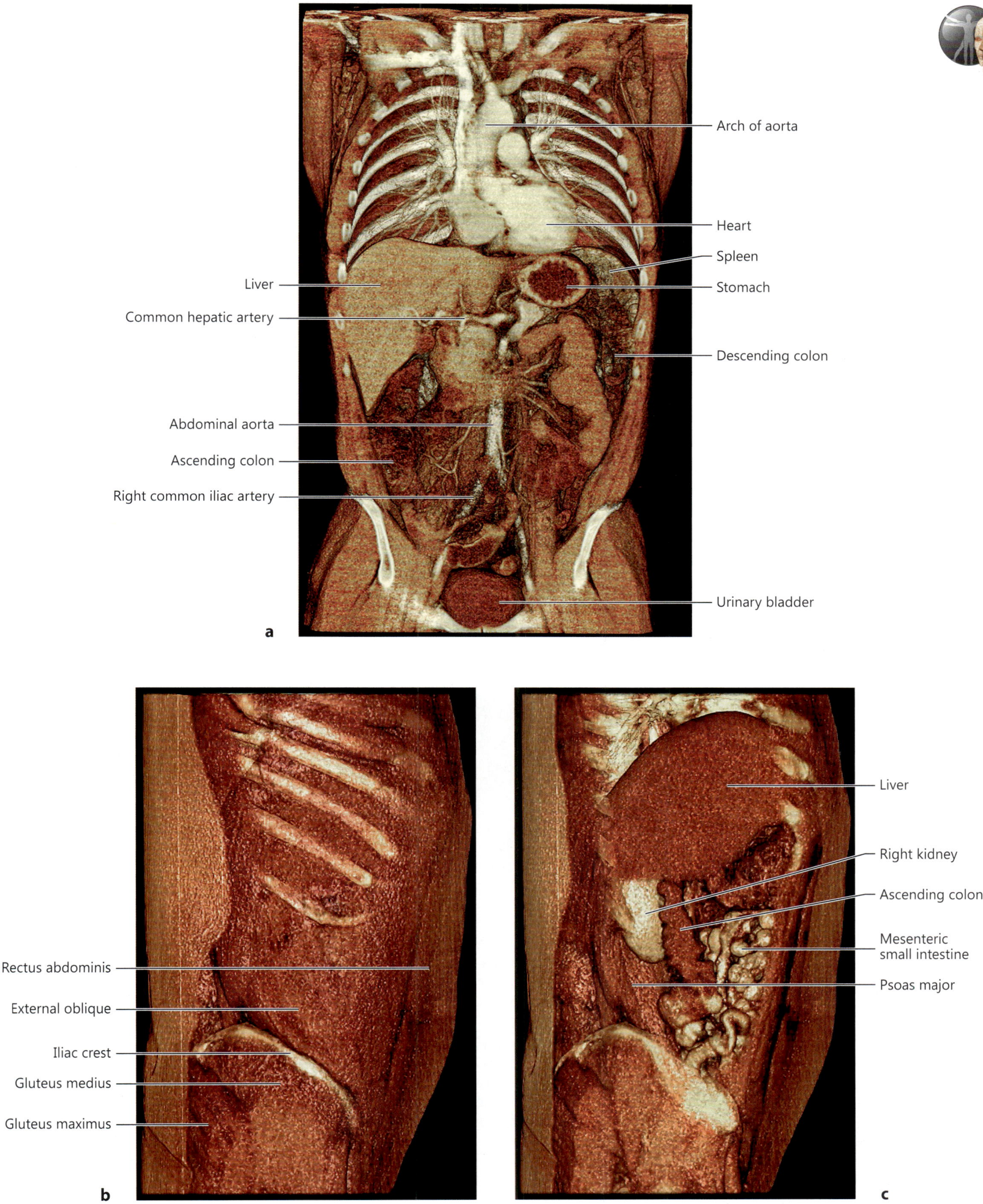

Three-dimensional images of a living human body, obtained by CT scan using a 3D volume rendering technique, showing the **content of the abdominal cavity** in a frontal section passing anterior to the urinary bladder (**a**). Image **b** shows a right lateral projection of the abdominal wall; image **c** shows a lateral projection of the content of the abdominal cavity after removal of the lateral abdominal muscles.

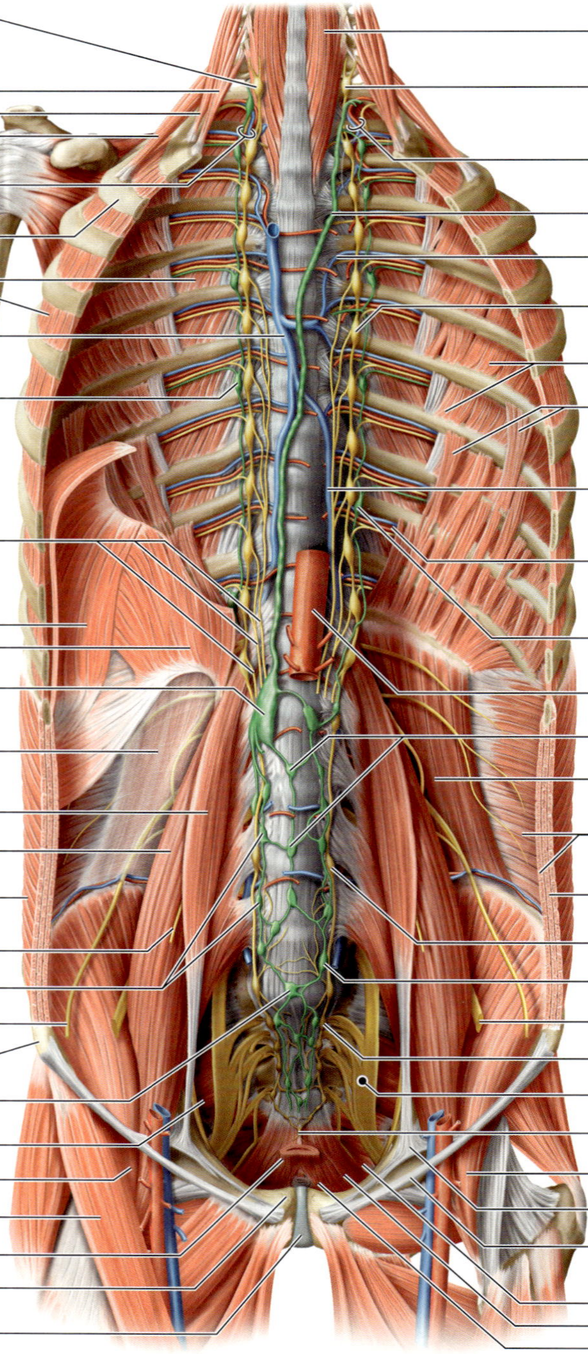

Muscles of the posterior part of the trunk. The psoas major and iliacus leave the pelvis through the muscular space, and insert on the lesser trochanter of the femur. The diaphragm has been transected on the left, so as to provide a view of the proximal insertion of the psoas muscles. On the right, the quadratus lumborum is invested by the anterior layer of thoracolumbar fascia. Along the lateral border of the muscle, this layer fuses with the middle layer (☞ Vol. 1, figure on page 403) of the same fascia, and with the deep lamina of posterior layer of thoracolumbar fascia (☞ Vol. 1, figure on page 404) to form the posterior insertion of the transversus abdominis and internal oblique, lateral to the muscles of back proper and above the iliac crest.
The figure also shows the sympathetic trunk, which extends all along the trunk on both the left and the right; the posterior intercostal arteries and veins with the intercostal nerves; the azygos and hemiazygos veins, which originate below the diaphragm and ascend in the posterior mediastinum; the thoracic duct and the cisterna chyli (the abdominal origin of the duct itself); and the sacral plexus, flat and triangular, whose base corresponds to the sacrum and whose apex marks the inferior edge of the greater sciatic foramen.

---

* The least splanchnic nerve is generally difficult to identify in dissections.

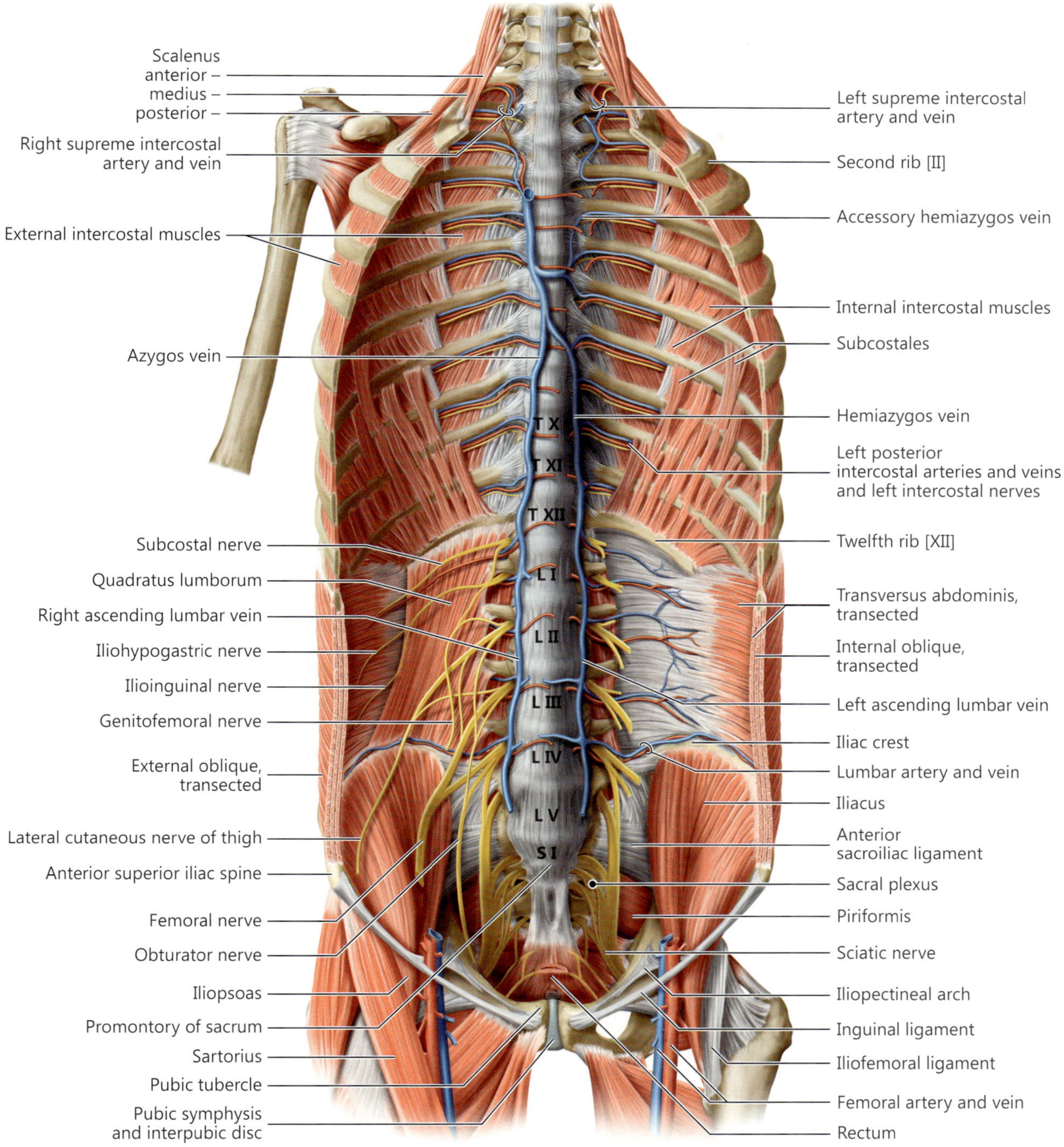

Muscles of the posterior part of the trunk. This diagram shows deeper layers than the previous figure. On the left, removal of the quadratus lumborum exposes the middle layer of thoracolumbar fascia. This layer fuses with the anterior layer (☞ Vol. 1, figure on page 402) of the same fascia, and with the deep lamina of posterior layer of thoracolumbar fascia (☞ Vol. 1, figure on page 404) to form the posterior insertion of the transversus abdominis and internal oblique, lateral to the muscles of back proper and above the iliac crest.
The figure also shows the posterior intercostal arteries and veins with the intercostal nerves; the azygos and hemiazygos veins, which originate below the diaphragm as a continuation of the right and left ascending lumbar veins, respectively, two anastomotic trunks joining the lumbar veins longitudinally; the anterior rami of lumbar nerves, forming the lumbar plexus, of which the iliohypogastric, ilioinguinal, genitofemoral nerves and the lateral cutaneous nerve of thigh, are collateral branches, while the obturator and femoral nerves are terminal branches; and the sacral plexus with the sciatic nerve.

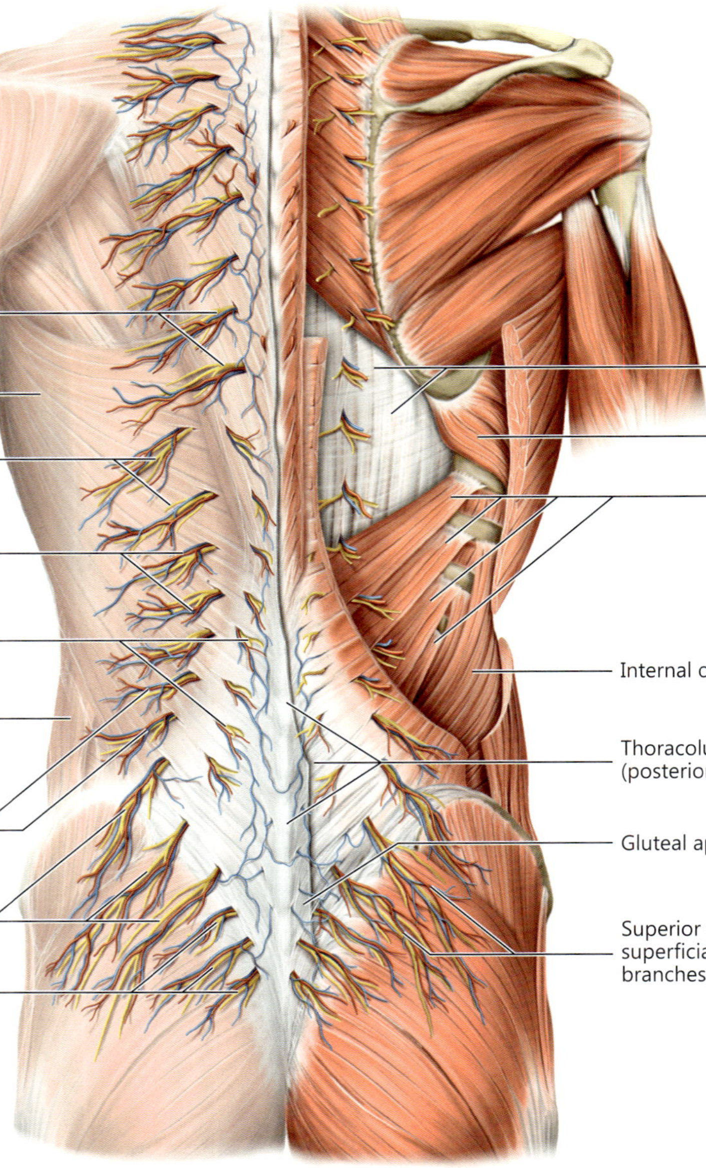

Stratigraphic sequence of the **muscles** and **fasciae of the posterior wall of the trunk**. On the right, the deltoid was removed, and the trapezius, latissimus dorsi and external oblique were partially transected to expose the underlying muscles and part of the thoracolumbar fascia (deep lamina of posterior layer). This lamina separates the deeper muscle layer from the superficial muscles (shown in the figure). The posterior layer of thoracolumbar fascia is thinner at the top, and it grows progressively thicker downwards. Laterally, it extends up to the costal angle. At the bottom, the deep lamina of posterior layer of thoracolumbar fascia runs deep underneath the serratus posterior inferior, reinforced by its insertions, and by those of the aponeurosis of the latissimus dorsi.
The medial and lateral branches of the posterior rami of spinal nerves are represented on the superficial muscles. The superior clunial nerves originate from the posterior rami of lumbar nerves, while the medial clunial nerves arise from the posterior rami of sacral nerves.

A complete stratigraphy of the muscles of trunk can be found in Volume 1, on page 404 onwards.

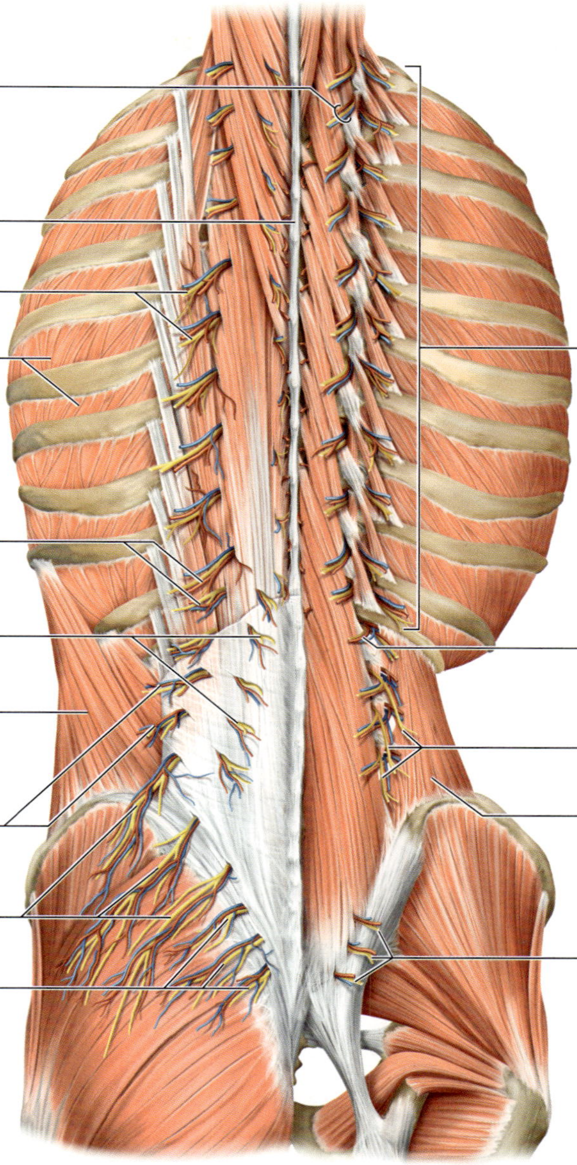

Stratigraphic sequence of the **muscles** and **fasciae of the posterior wall of the trunk**. On the left, the semispinalis capitis, splenius capitis, levator scapulae, rhomboid major and rhomboid minor were removed, and the trapezius, latissimus dorsi and serratus posterior inferior were transected to expose the muscles of the deep layer of the posterior wall of trunk. These muscles (longissimus thoracis, spinalis, iliocostalis thoracis and iliocostalis lumborum) are invested by the deep lamina of posterior layer of thoracolumbar fascia. On the right, the erector spinae was removed to expose the transversospinales.
The figure also shows the posterior rami of spinal nerves and the lateral and medial musculocutaneous branches, which originate from the dorsal branch of the posterior intercostal arteries.

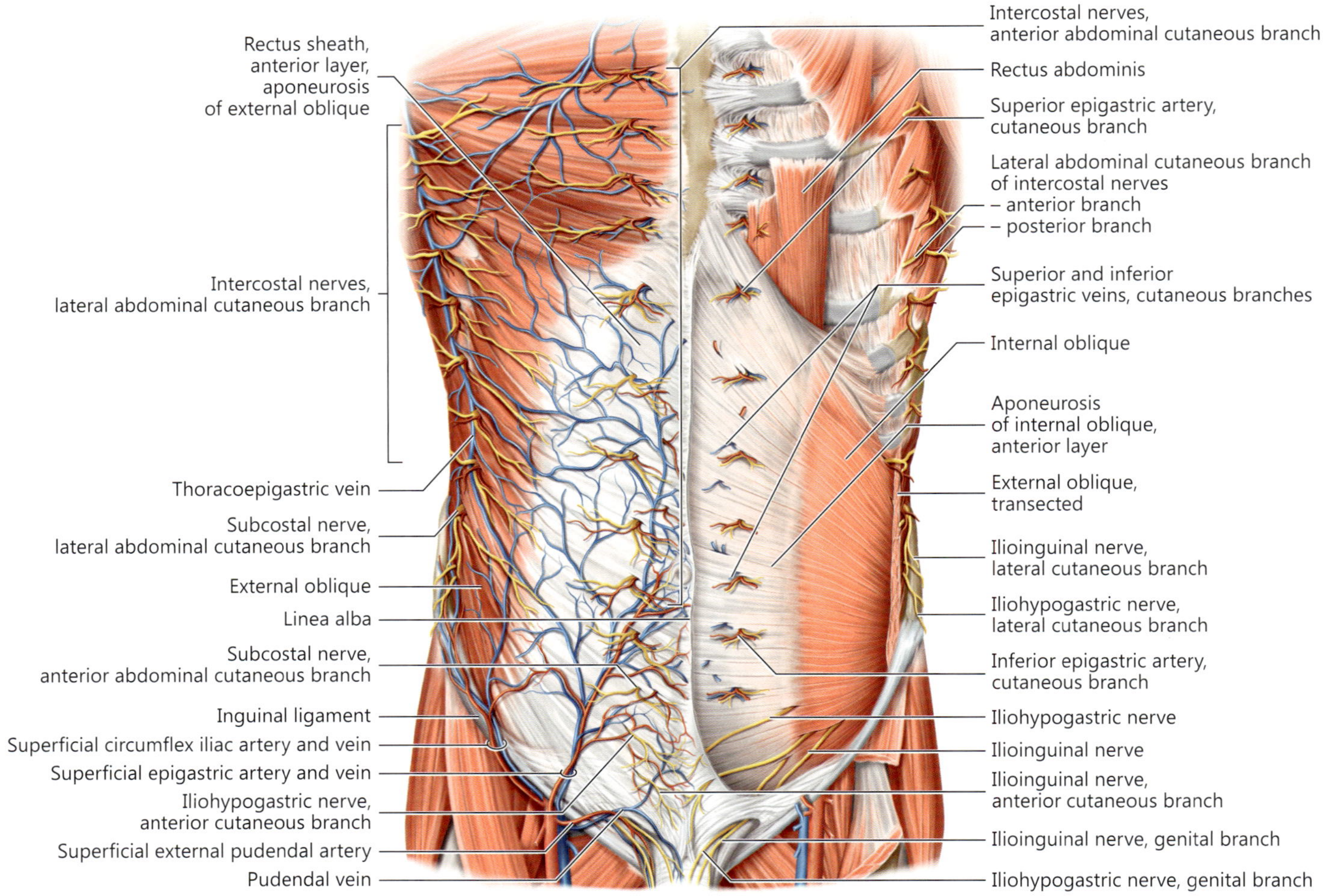
Rectus sheath, anterior layer, aponeurosis of external oblique
Intercostal nerves, lateral abdominal cutaneous branch
Thoracoepigastric vein
Subcostal nerve, lateral abdominal cutaneous branch
External oblique
Linea alba
Subcostal nerve, anterior abdominal cutaneous branch
Inguinal ligament
Superficial circumflex iliac artery and vein
Superficial epigastric artery and vein
Iliohypogastric nerve, anterior cutaneous branch
Superficial external pudendal artery
Pudendal vein
Intercostal nerves, anterior abdominal cutaneous branch
Rectus abdominis
Superior epigastric artery, cutaneous branch
Lateral abdominal cutaneous branch of intercostal nerves
– anterior branch
– posterior branch
Superior and inferior epigastric veins, cutaneous branches
Internal oblique
Aponeurosis of internal oblique, anterior layer
External oblique, transected
Ilioinguinal nerve, lateral cutaneous branch
Iliohypogastric nerve, lateral cutaneous branch
Inferior epigastric artery, cutaneous branch
Iliohypogastric nerve
Ilioinguinal nerve
Ilioinguinal nerve, anterior cutaneous branch
Ilioinguinal nerve, genital branch
Iliohypogastric nerve, genital branch

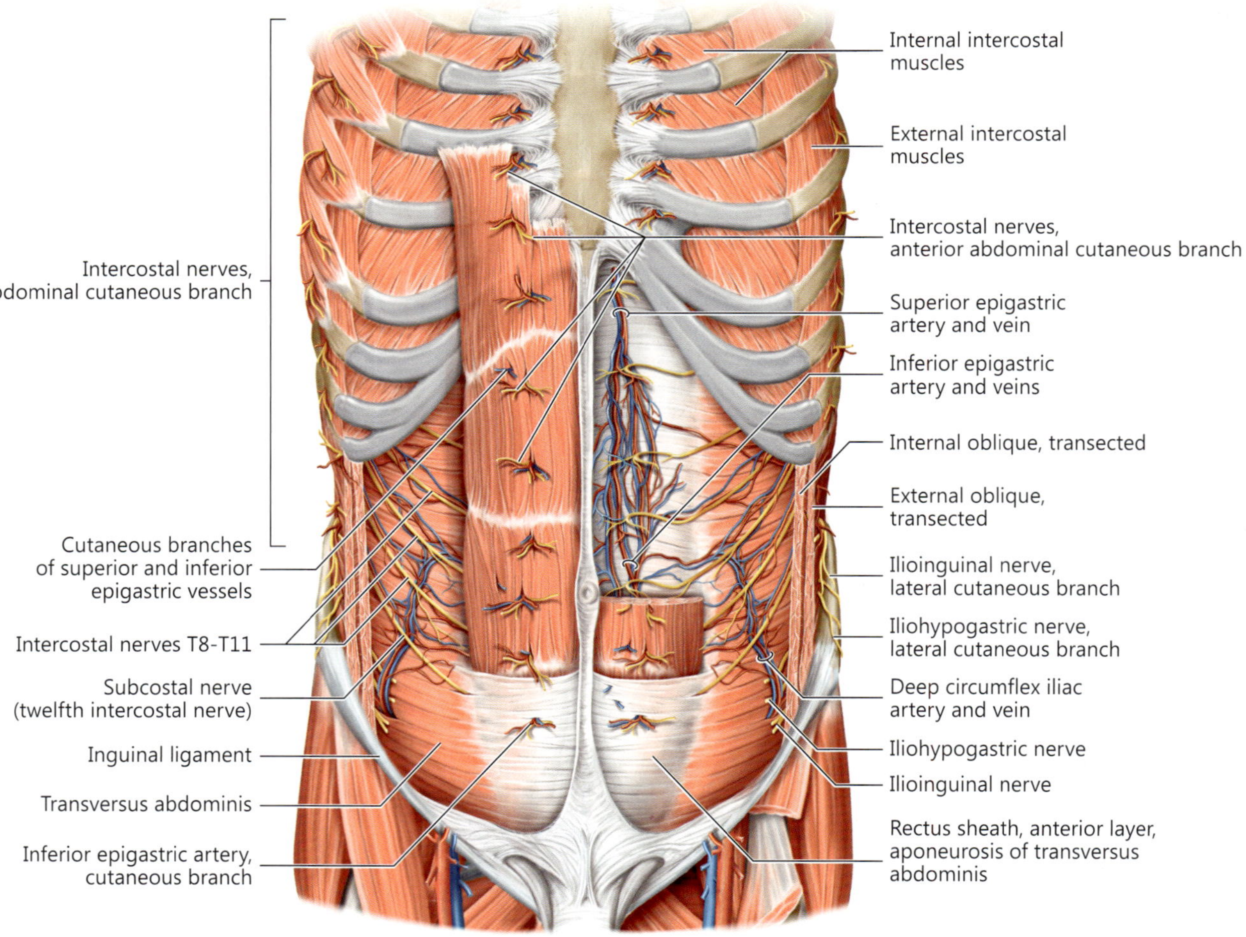
Intercostal nerves, lateral abdominal cutaneous branch
Cutaneous branches of superior and inferior epigastric vessels
Intercostal nerves T8-T11
Subcostal nerve (twelfth intercostal nerve)
Inguinal ligament
Transversus abdominis
Inferior epigastric artery, cutaneous branch
Internal intercostal muscles
External intercostal muscles
Intercostal nerves, anterior abdominal cutaneous branch
Superior epigastric artery and vein
Inferior epigastric artery and veins
Internal oblique, transected
External oblique, transected
Ilioinguinal nerve, lateral cutaneous branch
Iliohypogastric nerve, lateral cutaneous branch
Deep circumflex iliac artery and vein
Iliohypogastric nerve
Ilioinguinal nerve
Rectus sheath, anterior layer, aponeurosis of transversus abdominis

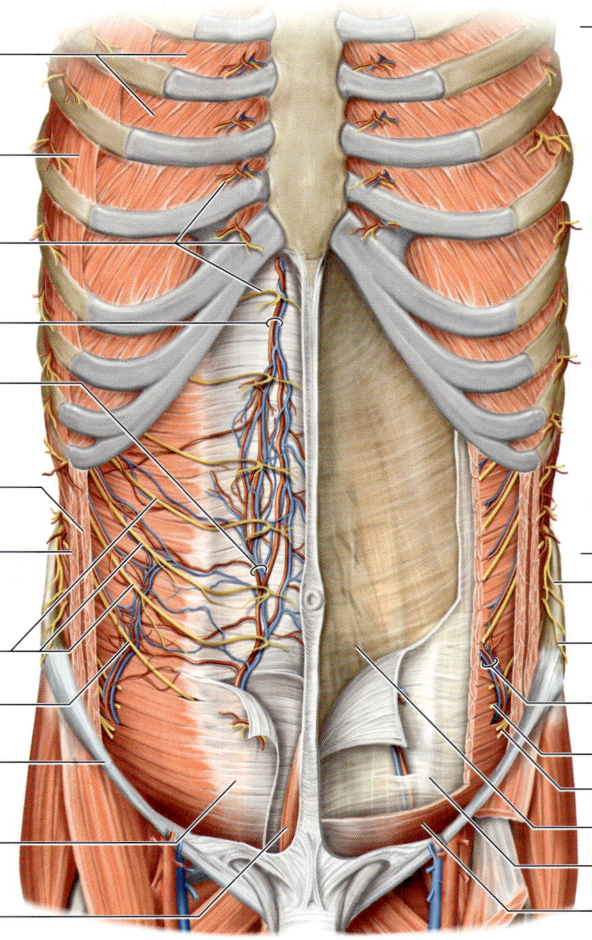

Stratigraphic sequence of the **muscles of the anterior wall of the trunk**. On the right, the rectus abdominis was removed, leaving the posterior layer of the rectus sheath in place. In the lower quarter below the arcuate line, this layer is made up of the transversalis fascia only. On the left, the posterior layer of rectus sheath was transected to expose the parietal peritoneum.
The last intercostal nerves take an increasingly downward oblique path and enter the abdominal wall, travelling between the transversus abdominis and the internal oblique up to the rectus abdominis, where they terminate as cutaneous branches on the anterior abdominal wall.

➤ Stratigraphic sequence of the **muscles of the anterior wall of the trunk**. The superficial layer is shown at the top right. On the left, the aponeurosis of external oblique was removed to expose the internal oblique and its aponeurosis. The aponeuroses of external oblique, internal oblique and transversus abdominis intermingle in the median line, forming the linea alba, which runs between the medial borders of the rectus abdominis.
The superficial epigastric artery arises from the femoral artery and ascends in front of the inguinal ligament, between the two layers of subcutaneous tissue, nearly as far as the umbilicus.

At the bottom right, as compared to the previous figure, the external oblique and its aponeurosis have been transected, along with the internal oblique, to expose the **rectus abdominis**, without the anterior layer of the rectus **sheath**. On the left, the upper three-quarters of the rectus abdominis were transected to expose the posterior layer of the rectus sheath.
The inferior epigastric artery arises from the external iliac artery, perforates the transversalis fascia, passes anterior to the arcuate line and ascends between the posterior layer of the rectus sheath and the posterior surface of the rectus abdominis, originating a number of branches that anastomose with those of the superior epigastric artery (terminal branch of the internal thoracic artery) at the umbilicus.

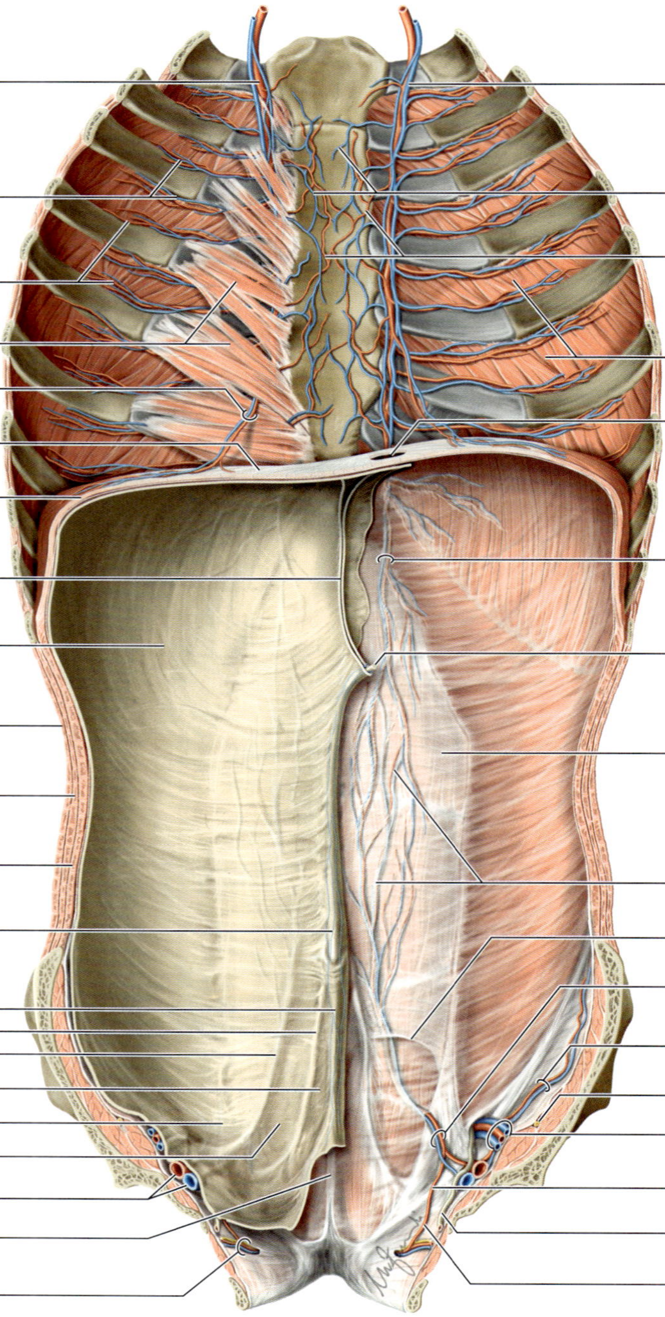

Muscles of the anterior abdominal wall in the male. Posterior view from inside the thoracic and abdominal cavities. Between the umbilical folds of peritoneum can be distinguished the **supravescical fossa**, located between the median and medial folds, the **medial inguinal fossa** (also referred to as inguinal triangle), between the medial and lateral folds, and the **lateral inguinal fossa**, lateral to the lateral umbilical fold.
The figure shows the internal thoracic artery, a branch of the subclavian artery, which gives the anterior and sternal intercostal branches, and splits into two terminal branches: the musculophrenic and superior epigastric arteries. The superior epigastric artery runs within the rectus sheath, and anastomoses with the inferior epigastric artery, a branch of the external iliac artery, at the umbilicus, hence establishing an anastomosis between the subclavian and external iliac arteries. The anastomosis between the pubic branches of the inferior epigastric and obturator arteries is also displayed.

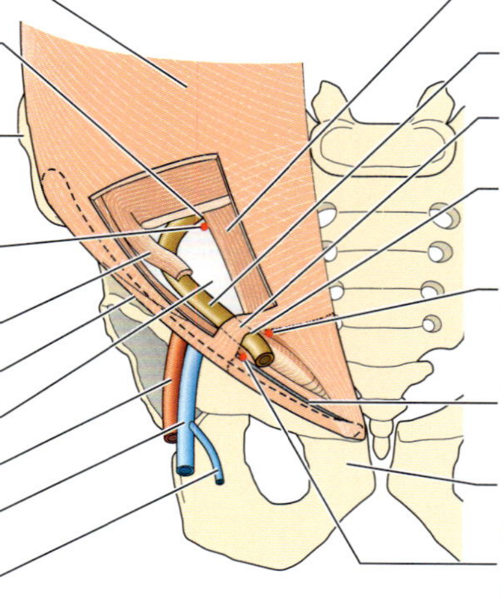

Anatomical pathways of formation of **inguinal and femoral hernias**.

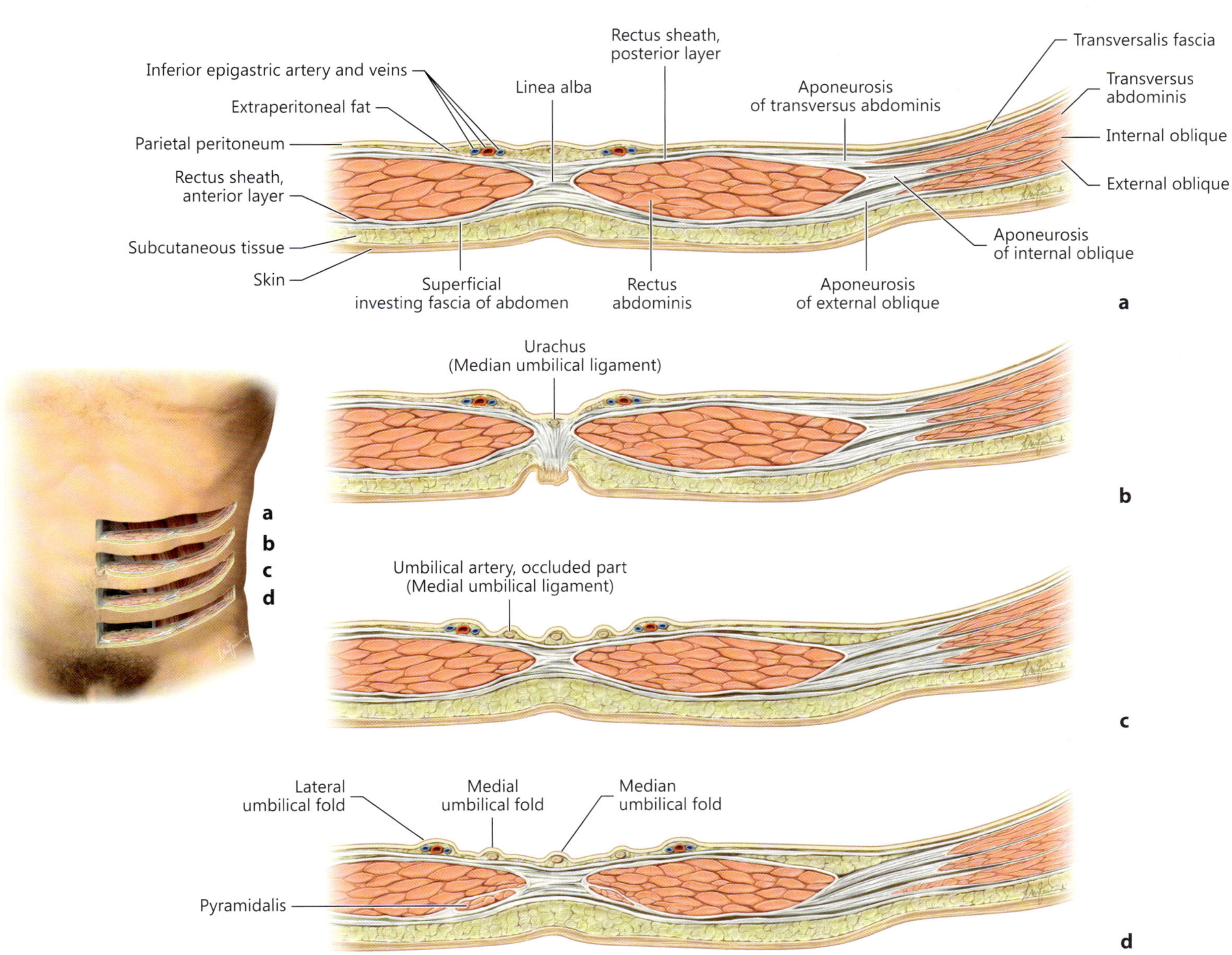

Transverse sections of the **anterior abdominal wall** taken above the umbilicus (**a**), at the umbilicus (**b**), below the arcuate line of rectus sheath (**c**) and near the pubis (**d**) (☞ Vol. 1, page 496).

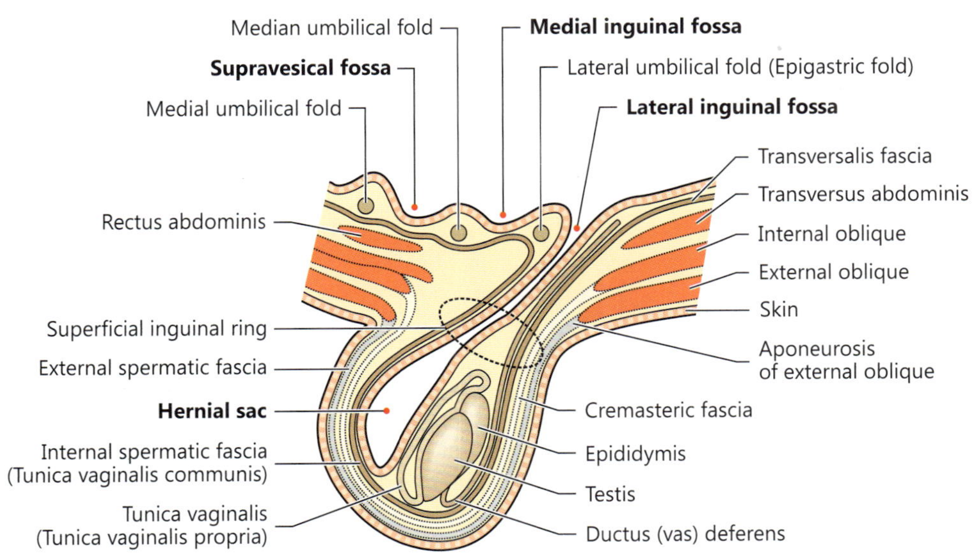

External oblique inguinal hernia. The direct hernia extrudes through the medial inguinal fossa.

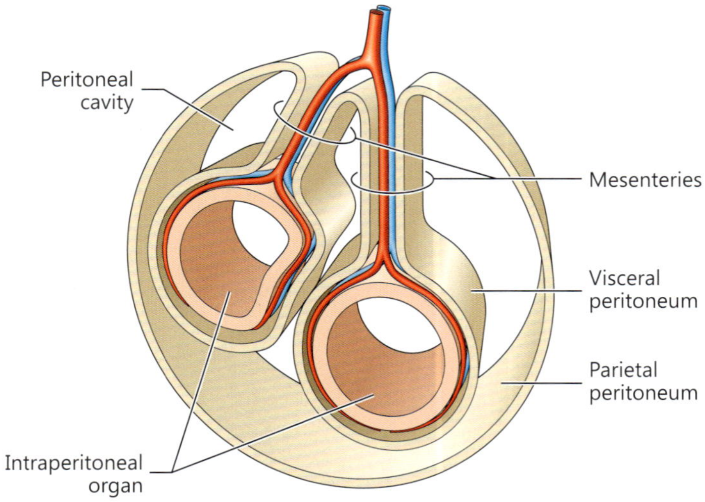

Mesenteries are a double peritoneal layer that arises from the peritoneum lining the abdominopelvic cavity (parietal peritoneum); the two layers are continuous to one another, and extend to the visceral organs, wrapping around them (visceral peritoneum). Between the layers are blood vessels and nerves that reach the organs, which are attached to the abdominal wall while retaining a certain degree of mobility.

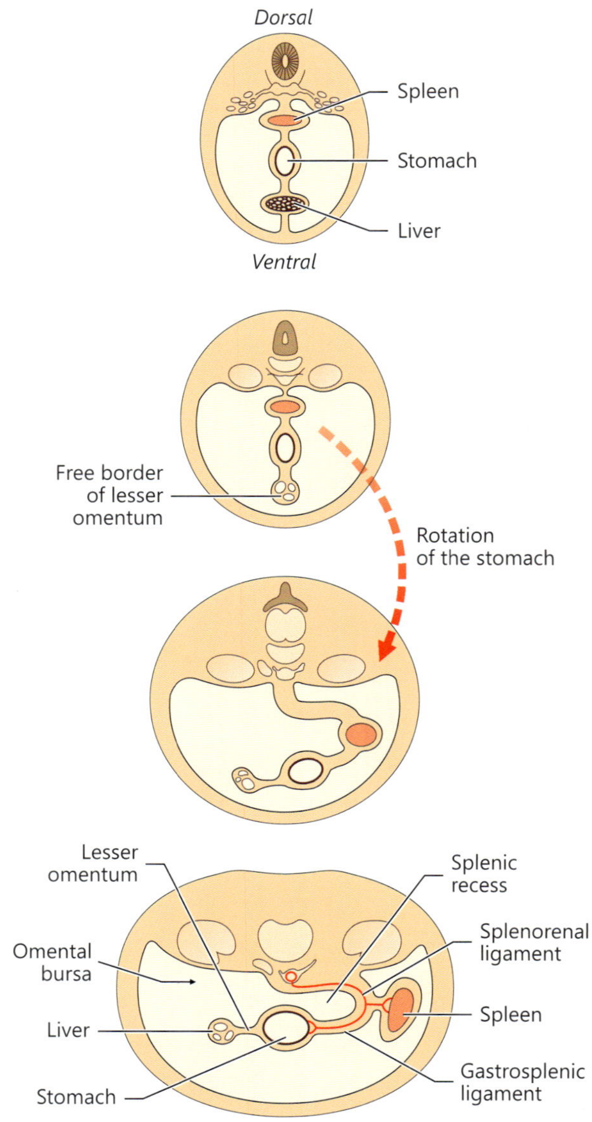

➤ Schematic diagrams along a transverse plane showing the **anlage** (embryonic rudiment development) **of spleen**, **liver** and **stomach** and the development of the **omental bursa**.

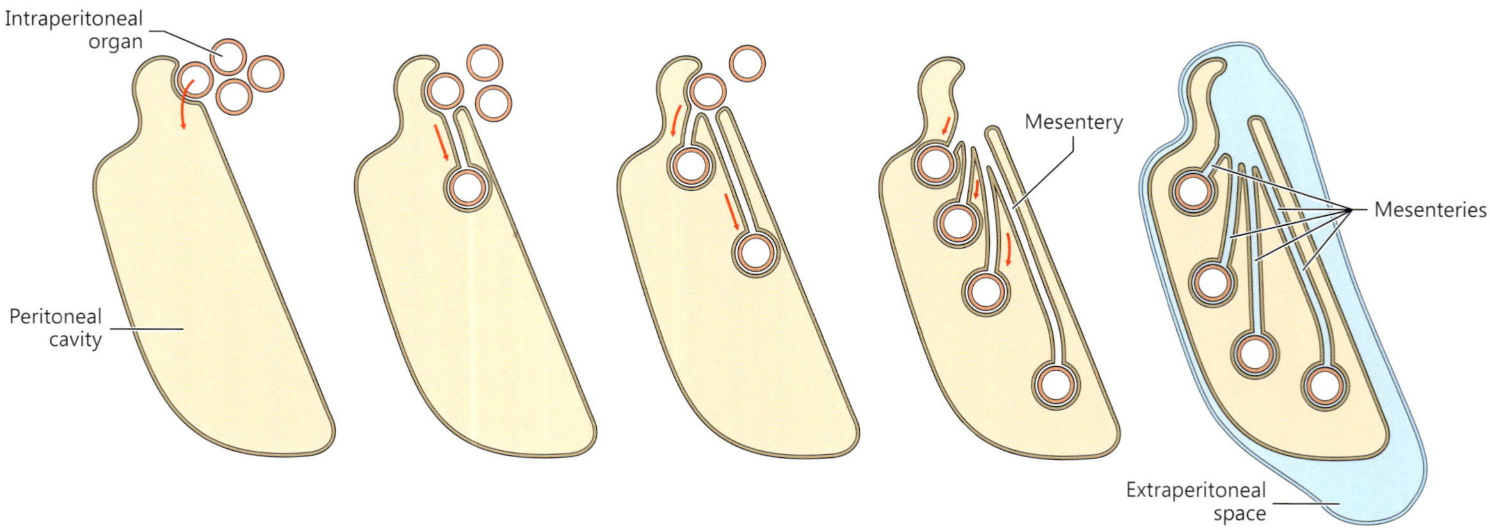

Organs completely invested by the peritoneum are suspended in the peritoneal cavity by mesenteries of variable length. The figure schematically illustrates the different phases of mesenteries development, as the parietal peritoneum is pushed towards the peritoneal cavity by a set of theoretical organs (depicted as circles in the diagram). Note that the close contact between the connective tissue (*light blue*) of the extraperitoneal space and the connective tissue between the two peritoneal layers forming the mesenteries is maintained throughout the process. Thus, the organs retain a direct relation to the extraperitoneal connective tissue, through which they receive blood and lymphatic vessels and nerves coming from the extraperitoneal space, and an indirect relation to the peritoneal cavity (*yellow*), mediated by the peritoneal investing layer.

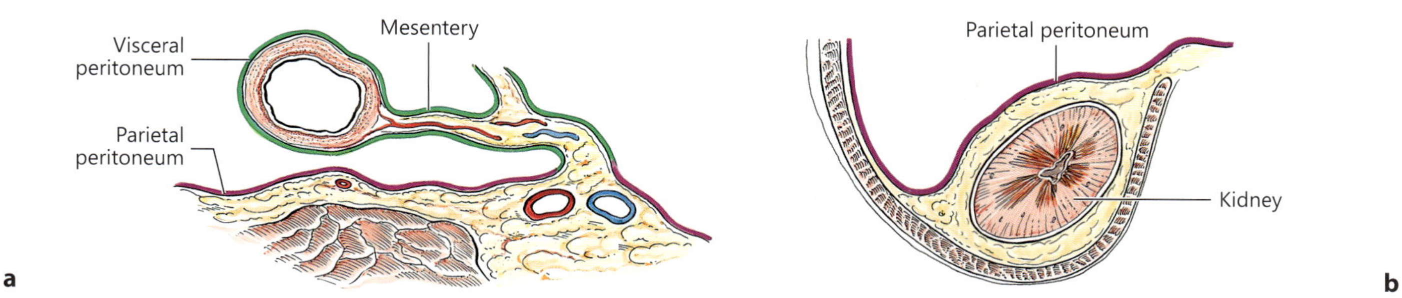

General location of the **organs in relation to the peritoneum**. Figure **a** shows the section of a hollow organ (intestine), which is almost entirely invested (approximately 90%) by the visceral peritoneum (*green*). When approaching a margin, the visceral peritoneum forms a serous lamina, composed of two layers (mesentery). Upon reaching the abdominal wall, the two layers turned, continuing in the parietal peritoneum (*red*). Figure **b** shows the section of a kidney, a retroperitoneal organ located deep to the parietal peritoneum (*red*), which invests exclusively the anterior surface of the organ (approximately 50% of the overall surface).

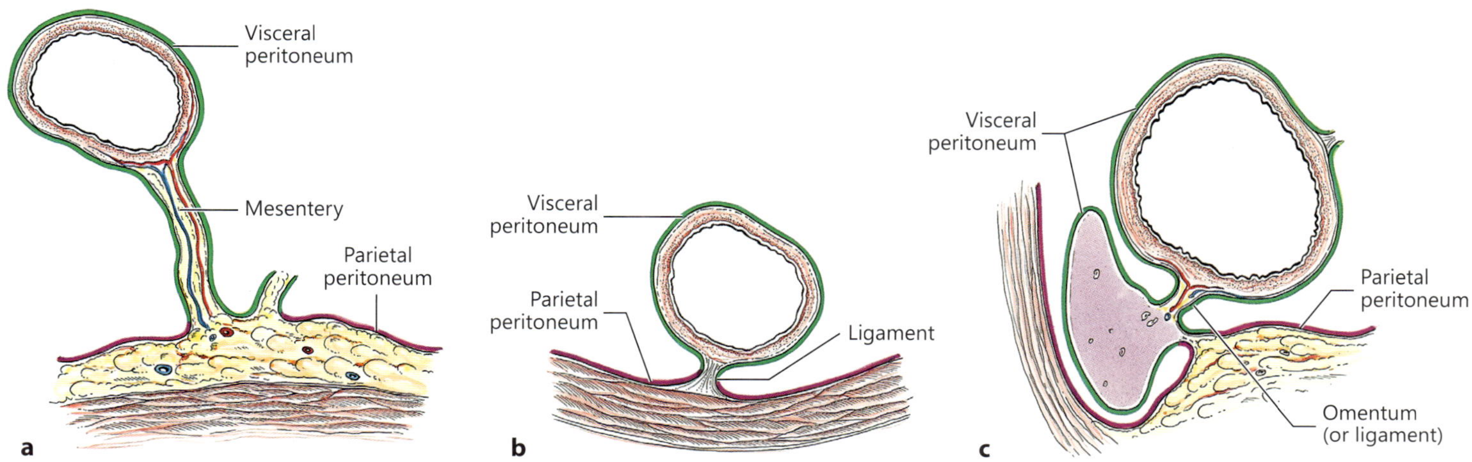

Mesenteries, **ligaments** and **omentum**. The mesentery (**a**) is a double layer of peritoneum that connects an organ to the abdominal wall, and often contains the vascular-nervous pedicle of the organ. The ligament (**b**) is a double layer of peritoneum that connects an organ to the abdominal wall; it is poor in vessels and, conversely, rich in fibrous connective tissue (providing support and attachment). The omentum (**c**) is a double layer of peritoneum that connects adjacent organs (in certain cases, the term epiploon can be replaced with ligament). Shown in *green* is the visceral peritoneum, shown in *red* is the parietal peritoneum.

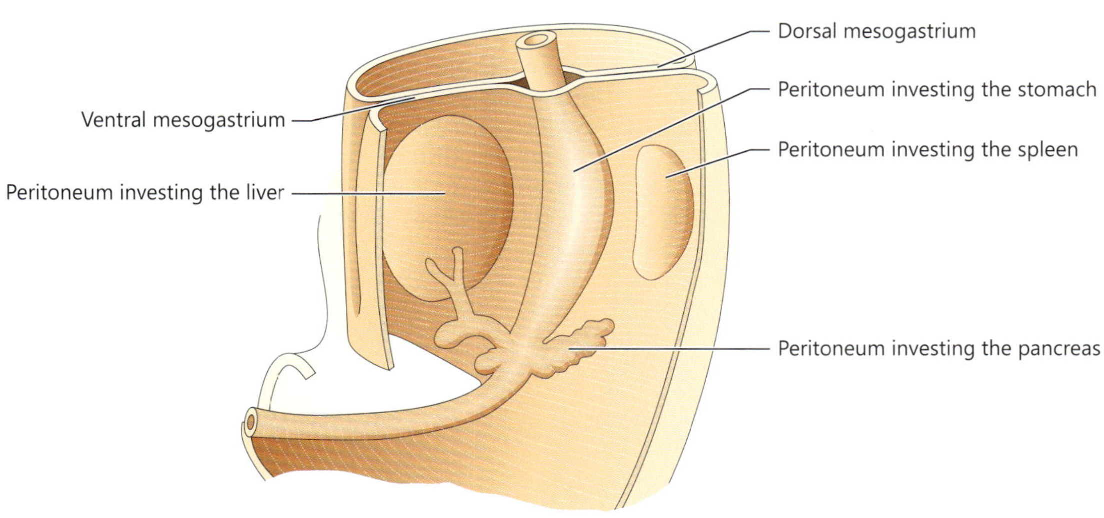

Organogenesis of the intestine and annex glands. Diagram displaying the supraumbilical segment of the peritoneal cavity and the anlagen of the organs it contains.

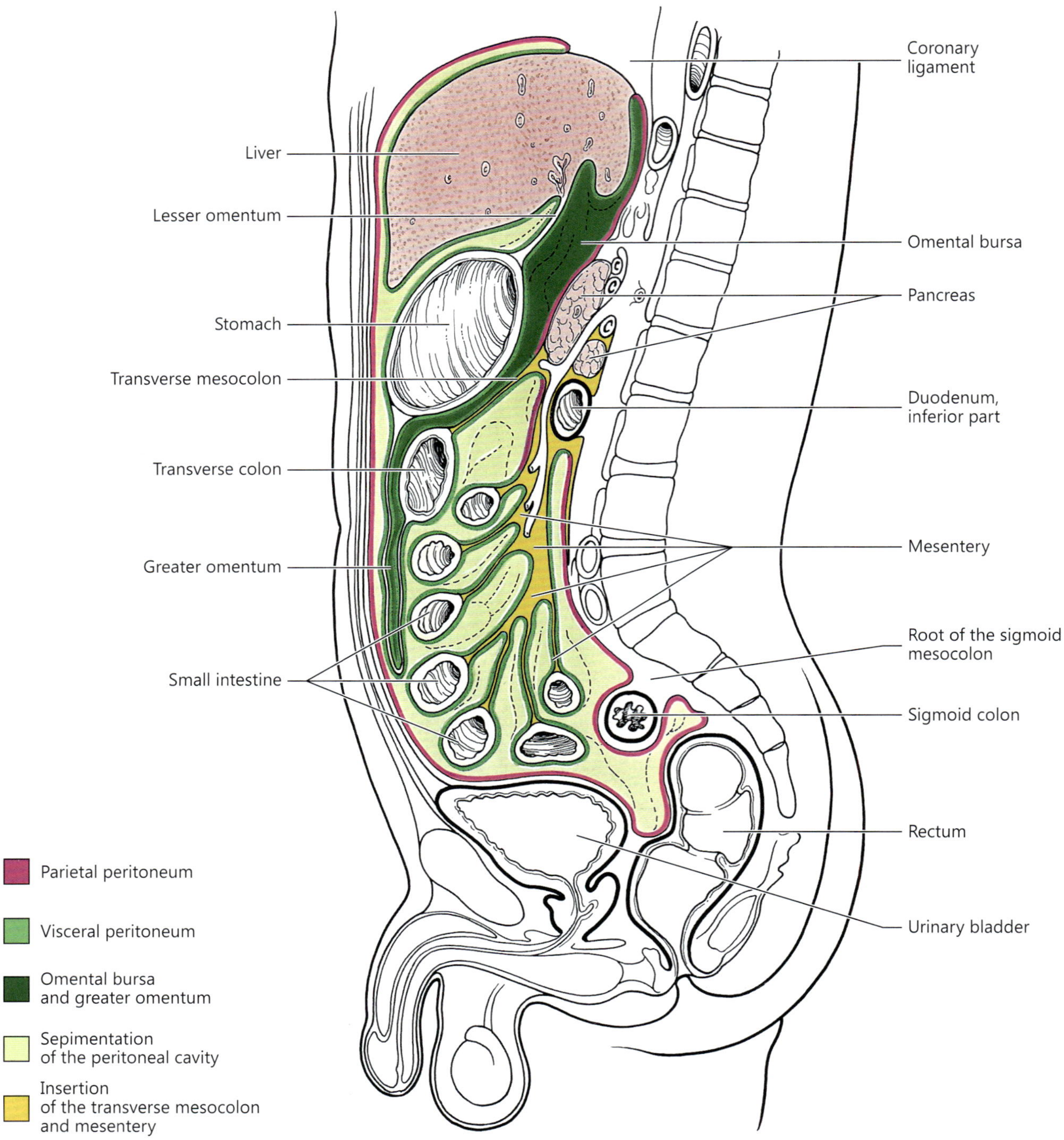

Sagittal section of the trunk showing the location of the **parietal** (*red*) **and visceral peritoneum** (*green*) in the abdominal and pelvic cavities. The *dark green* area comprises the omental bursa, or lesser sac, and the greater omentum. Outlined in *light green* is the sepimentation of the peritoneal cavity into pouches, recesses, excavations, fossae and diverticula by peritoneal compartments, ligaments, omenta or epiploon, and mesenteries. The *ochre* area displays the insertion of the transverse mesocolon and mesentery to the posterior abdominal wall. The transverse mesocolon divides the supramesocolic from the submesocolic spaces. The omental bursa extends posterior to the stomach and to the lesser omentum. The small intestine lies extensively in the submesocolic space.

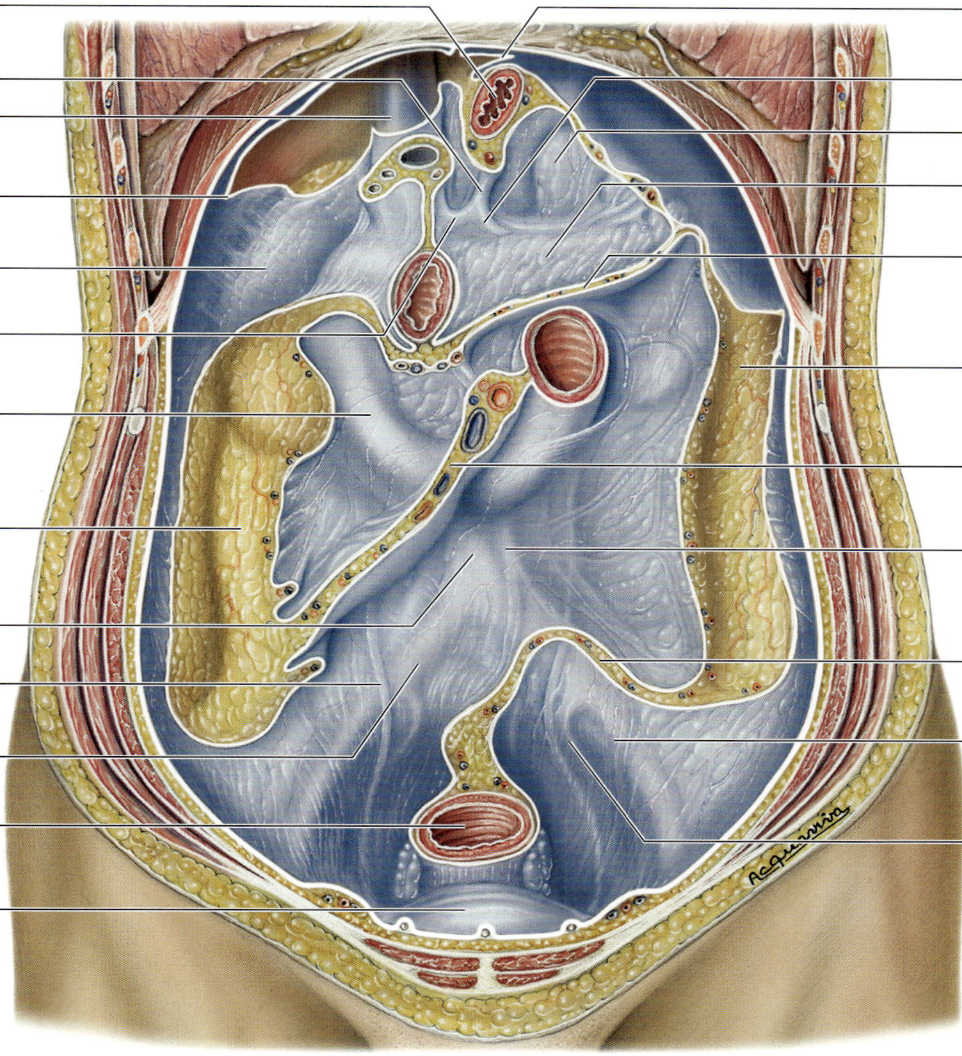

Frontal projection of the abdomen. All intraperitoneal organs and the ascending and descending colon have been removed from the abdominal cavity to expose the lines of reflection of the **parietal peritoneum** into the **visceral peritoneum**, in particular, the roots or lines of insertion into the posterior abdominal wall of the transverse mesocolon, the mesentery and the sigmoid mesocolon. Some retroperitoneal organs appear behind the parietal peritoneum, including the kidneys, the ureters, part of the duodenum, except the first 1 inch, and the pancreas, except the tail of pancreas. The transverse mesocolon separates the supramesocolic space of the peritoneal cavity from the submesocolic space. Its root runs obliquely upward, from right to left, between the lateral margins of the two kidneys. The line of insertion of the mesentery is directed obliquely from the duodenojejunal flexure, downward and to the right, to the superior part of the right sacroiliac joint and to the ileal orifice. The line of insertion of the sigmoid mesocolon has an inverted-V shape; the apex is located close to the left ureter, at the point where the ureter enters the pelvis, and close to the bifurcation of the left common iliac artery.

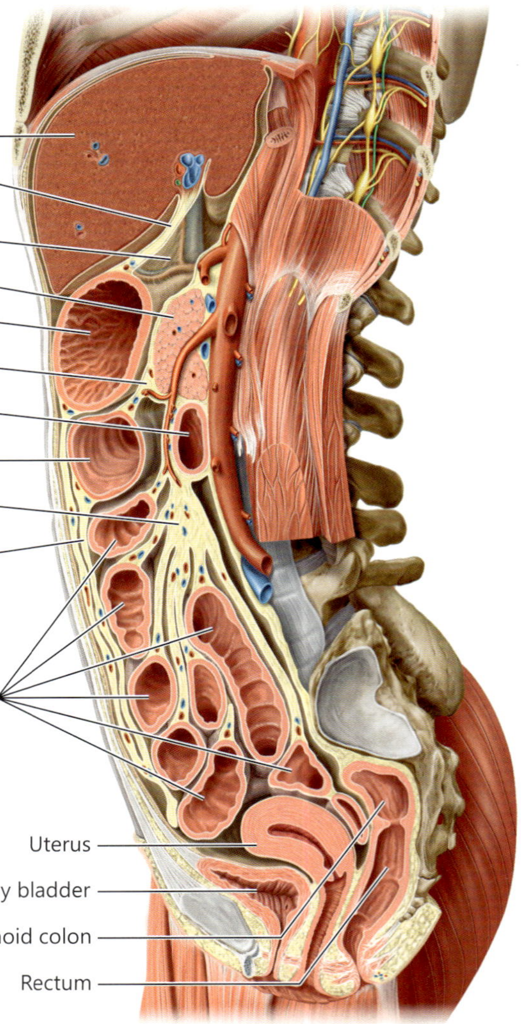

➤ Median sagittal section of the **abdomen** and **pelvis** in the female, left side view.

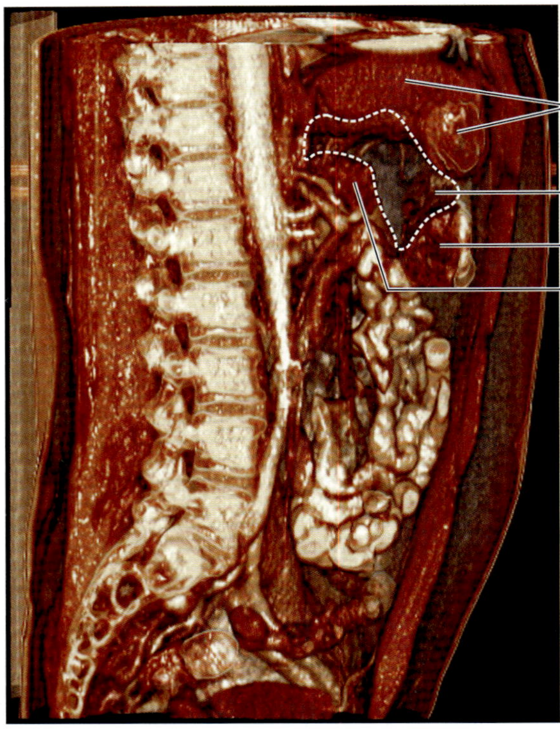

Three-dimensional image of a living human body, obtained by CT scan using a 3D volume rendering technique, showing a parasagittal section of the abdominal cavity. The image displays the **omental bursa** or lesser sac, which is a wide recess of the peritoneal cavity located behind the stomach.

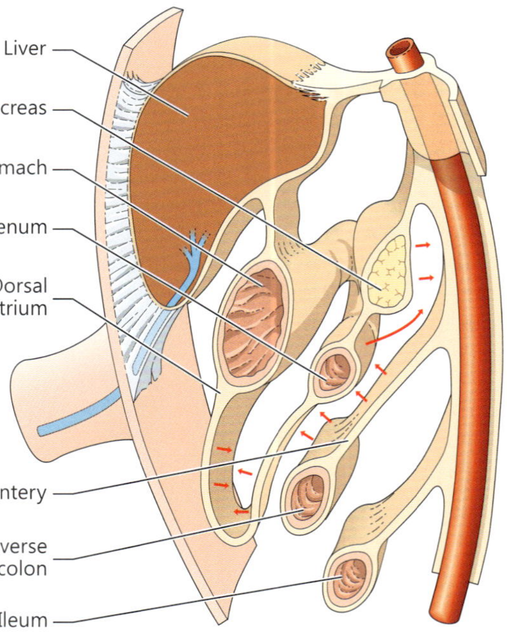

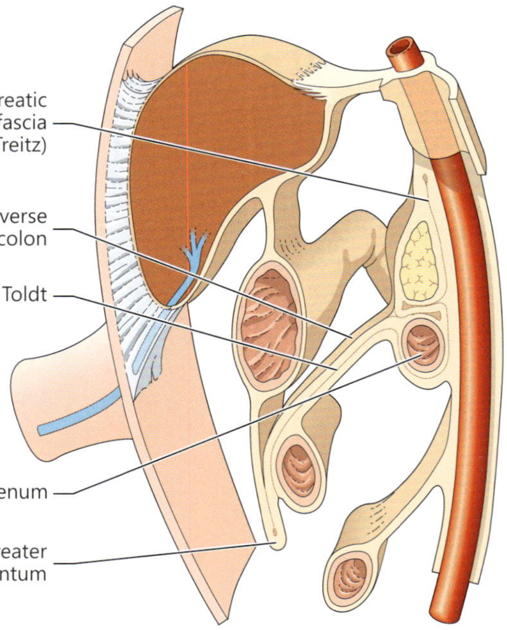

Greater omentum, **mesentery** and **transverse mesocolon**. Between the seventh and the eight week, an eversion of the dorsal mesogastrium at the greater curvature of stomach generates the greater omentum. As the stomach rotates along an anteroposterior axis passing through the midpoint of the lesser curvature, the greater omentum extends in front of and over the transverse colon, and over the superior layer of the transverse mesocolon, merging with it. The transverse mesocolon and the mesentery arise from the mesentery of the ansa umbilicalis. The fusion of the layers of the transverse mesocolon gives rise to a fibrous reinforced lamina, called fascia of Toldt. Posteriorly, the visceral peritoneum invests the pancreas, the second and the third part of the duodenum; it attaches to the posterior parietal peritoneum, determining the formation a fibrous lamina called the retropancreatic fascia (ligament of Treitz), which provides stable attachment for these two organs.

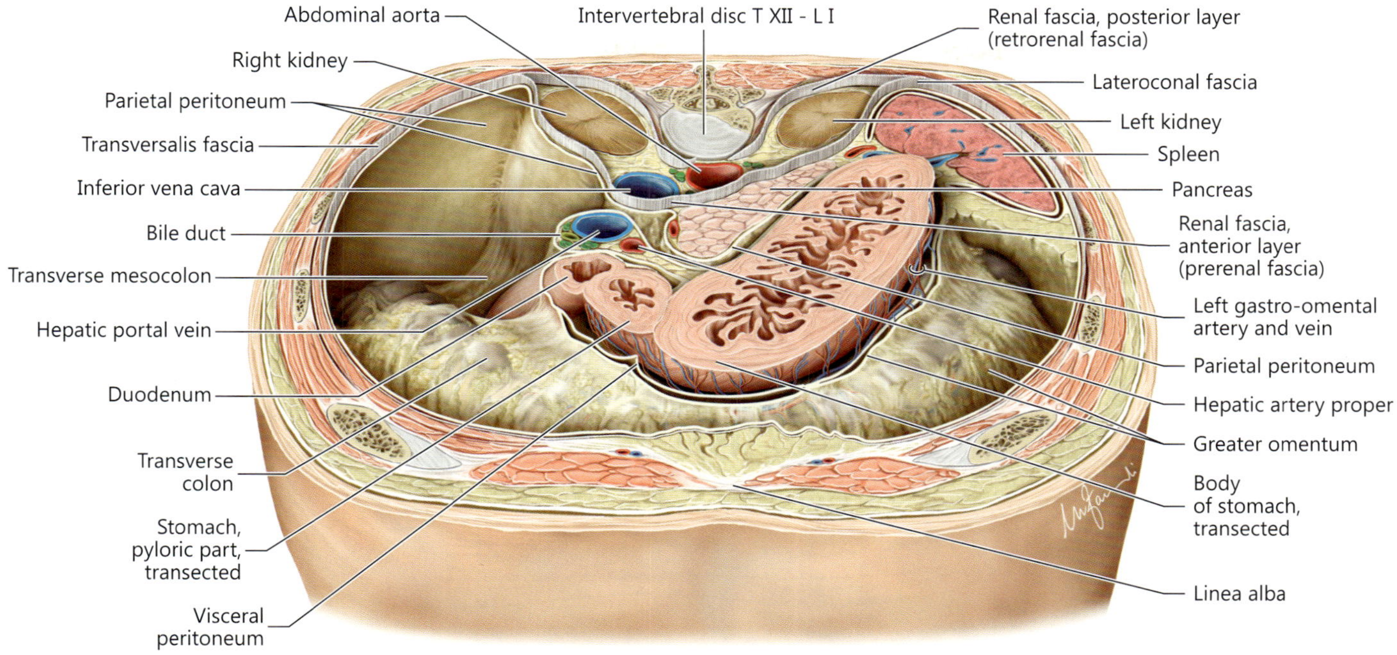

Transverse section of the trunk at the first lumbar vertebra, displaying the fascia lining the abdominal cavity and sepimenting the **retroperitoneal space**, forming the **renal space.** This space is encircled, anteriorly and posteriorly, by the prerenal and retrorenal fasciae, both of which insert on the lateroconal fascia. The laterocoronal fascia continues laterally as the transversalis fascia. The renal space contains the kidneys, the components of the renal hilum, and the suprarenal glands, with the supporting arteries and veins and is subdivided by the renal fascia into three major components, an anterior pararenal space, that is continuos across the midline, and a posterior pararenal space and a perirenal space, on each side, separated by the vertebral column and the psoas muscles. Thus, right and left renal spaces communicate with each other anteriorly to the vertebral column and posteriorly to the prerenal fascia. Superiorly, this space communicates with the posterior mediastinum (☞ Vol. 2, Mediastinum) and includes major vascular, nervous and lymphatic structures. The anterior and posterior renal fasciae fuse superiorly, closing the renal space, while remain separate inferiorly, leaving the renal space open. The connective tissue contained in the anterior pararenal space extends to the roots of the main mesenteries of the posterior parietal peritoneum, and contains nerves, blood and lymphatic vessels supplying the intraperitoneal organs.

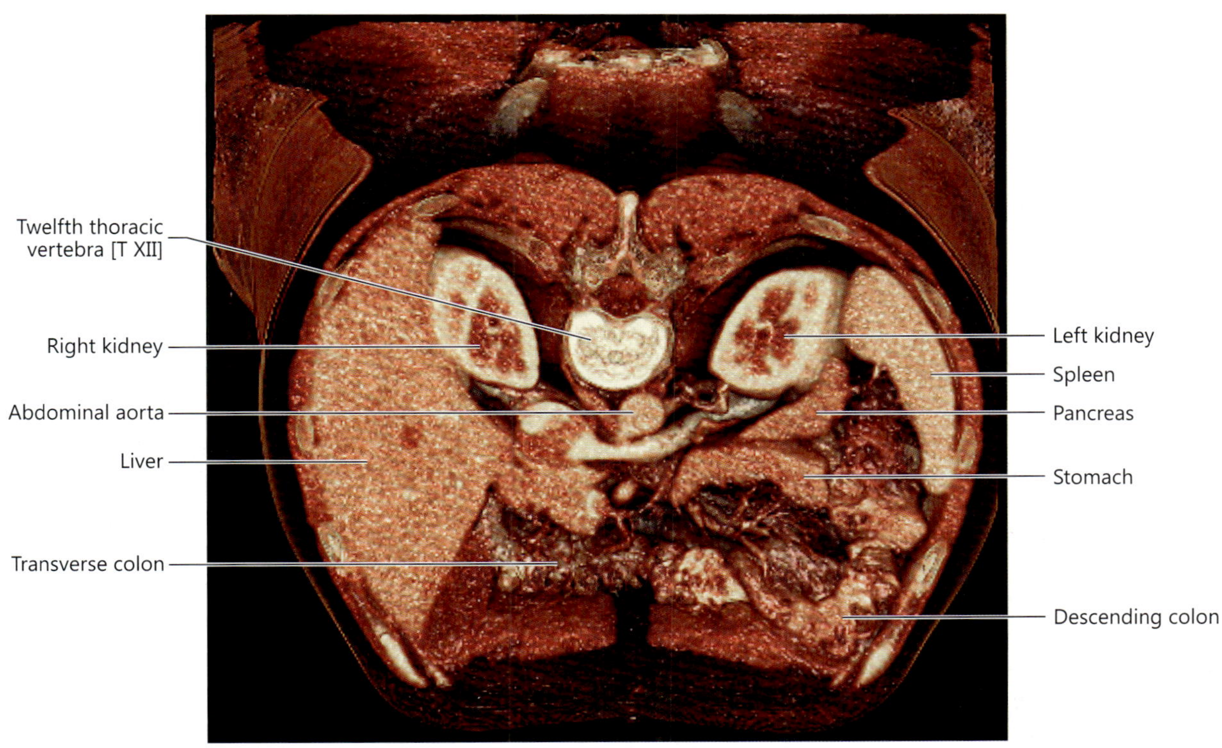

Three-dimensional image of a living human body, obtained by CT scan using a 3D volume rendering technique, showing the **abdominal organs** in a transverse section passing through the twelfth thoracic vertebra. As compared to the figure above, this section is rotated by 180 degrees relative to the radiological position; as such, the figure is seen from below, and the dorsal part is at the top.

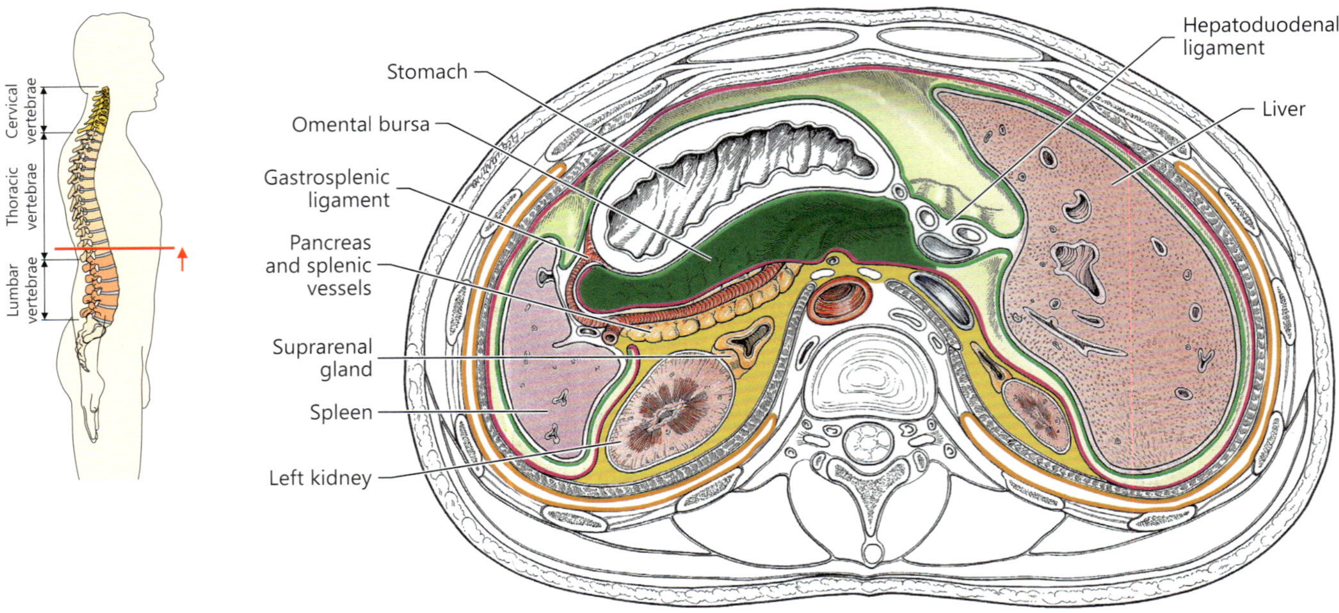

Transverse section of the trunk passing through the twelfth thoracic vertebra, showing the relations between the supramesocolic organs and the margins of the omental bursa. Anterior and lateral to the vertebral body, the retroperitoneal space and its organs can be observed. The visceral peritoneum is indicated in *green*, the parietal peritoneum in *red*, and the pleura of the costodiaphragmatic recess is indicated in *orange*.

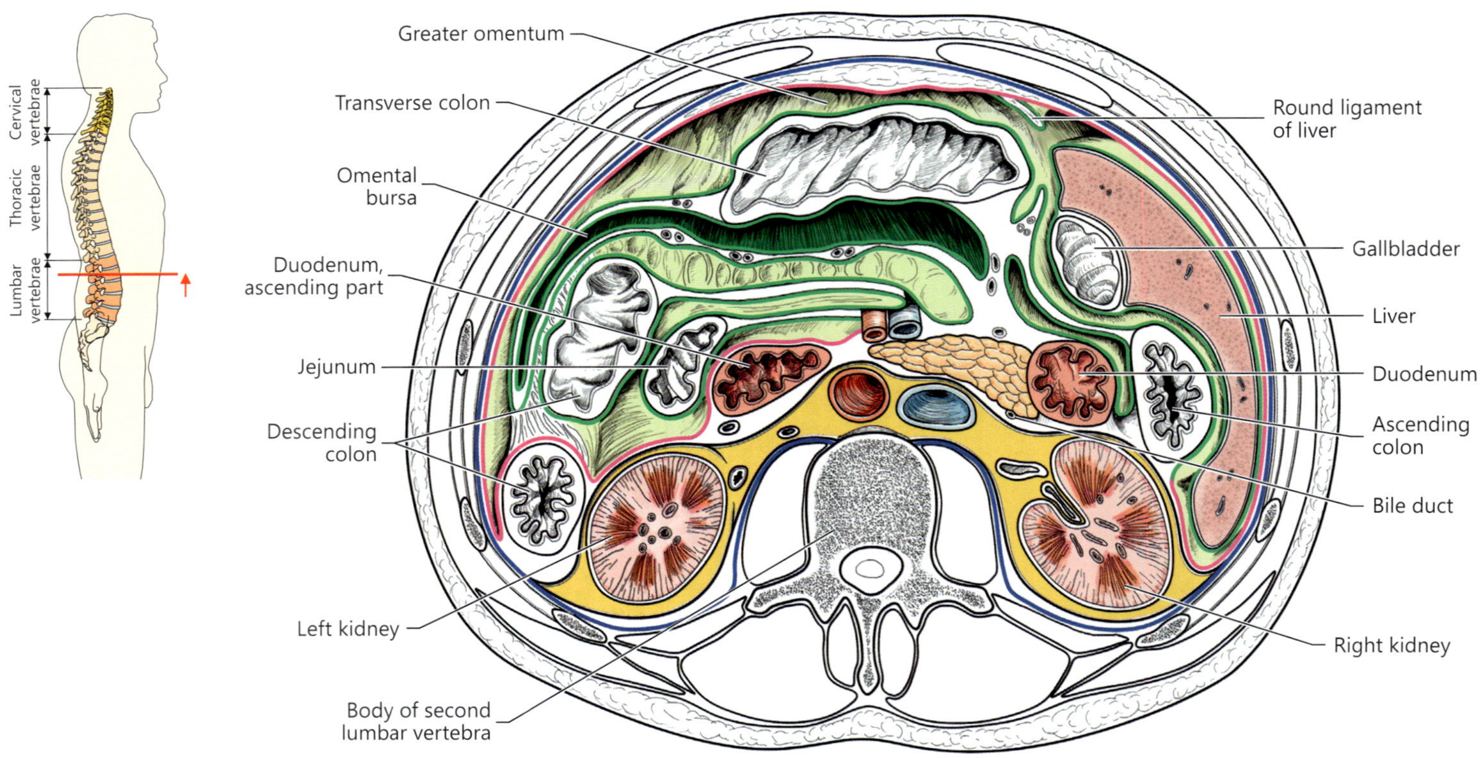

Transverse section of the trunk at the second lumbar vertebra. The visceral peritoneum is indicated in *green* and the parietal peritoneum in *red*. The renal fascia is indicated in *black*; the anterior (prerenal fascia) and posterior (retrorenal fascia) layers surround the renal space and continue laterally in the lateroconal fascia, which then fuses with the transversalis fascia, indicated in *blue*. The posterior layer attaches medially to the vertebral body, whereas the anterior layer merges with the contralateral layer, passing anteriorly to the kidney, the vertebral column and the great vessels, and is almost entirely invested by the posterior parietal peritoneum. When passing behind the mesocolon insertion the prerenal layer is reinforced by the fascia of Toldt, a connective lamina arising from the primitive mesocolon merging with the parietal peritoneum.

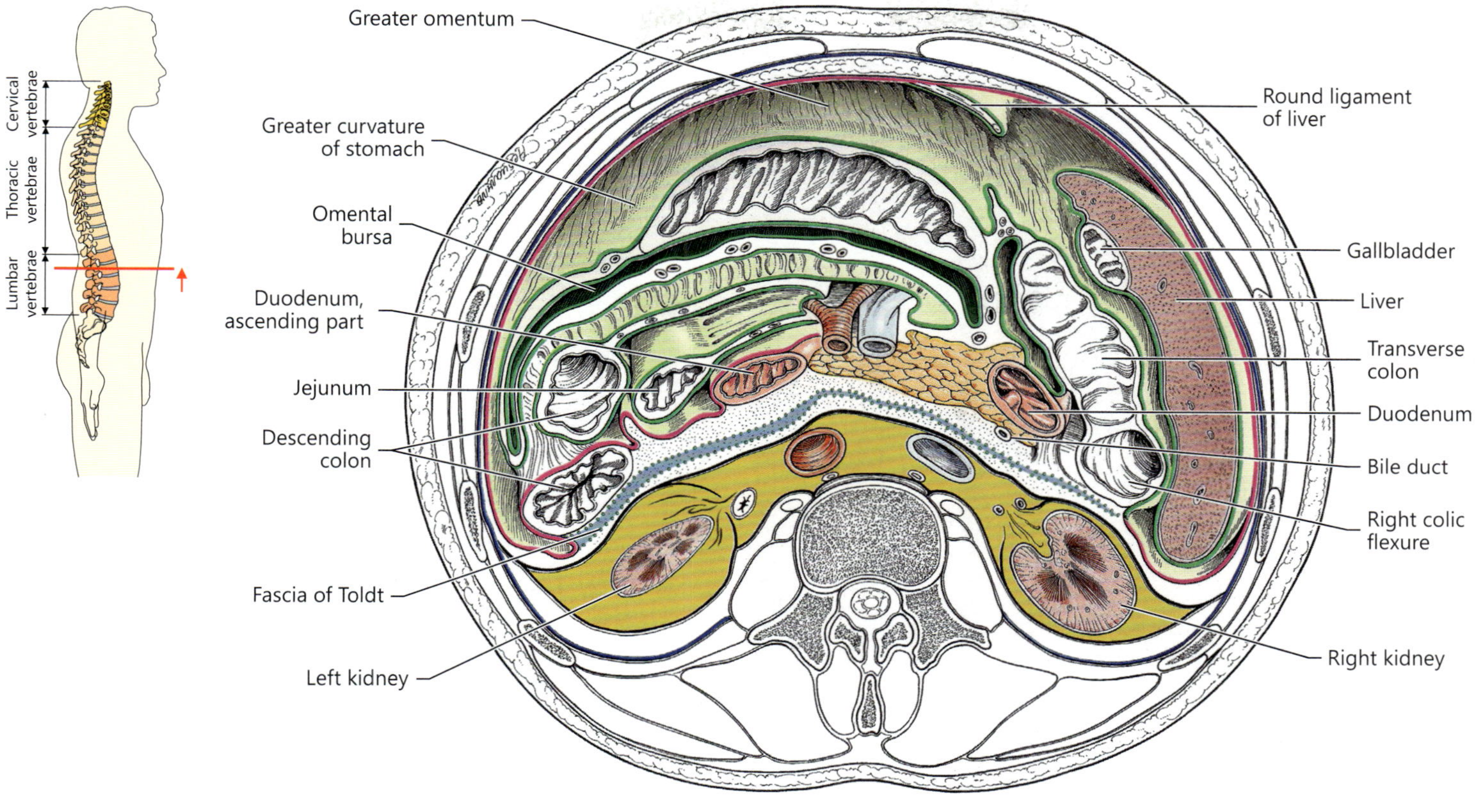

Transverse section of the trunk passing through the second lumbar vertebra. The section plane displays the relations between the organs in the zone of passage between the supramesocolic and the submesocolic planes. The dotted lines indicate attachment of primary peritoneal layers that originated the fascia of Toldt. The visceral peritoneum is indicated in *green* and the parietal peritoneum in *red*, the renal fascia is indicated in *black*, and the transversalis fascia in *blue*.

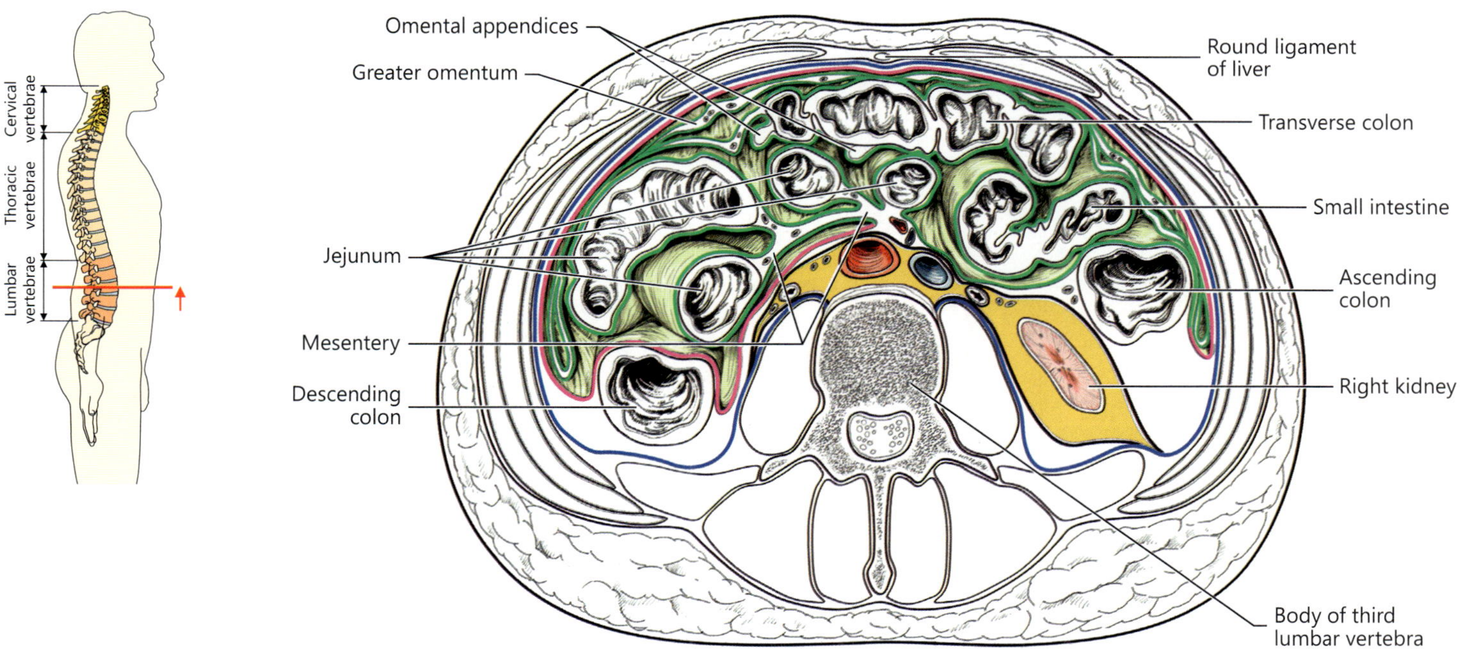

Transverse section of the trunk at the superior border of the third lumbar vertebra. The visceral peritoneum is indicated in *green* and the parietal peritoneum in *red*, the anterior (prerenal fascia) and posterior layers (retrorenal fascia) of the renal fascia are indicated in *black*. The *blue* outline defines the fascia investing the quadratus lumborum (anterior layer of the thoracolumbar fascia) and psoas (part of the iliac fascia belonging to the psoas), merging partly with the posterior layer of renal fascia. The *blue* outline also indicates the transversalis fascia; its anterior margin participates to the formation of the rectus sheath, and merges with the contralateral layer at the median line.

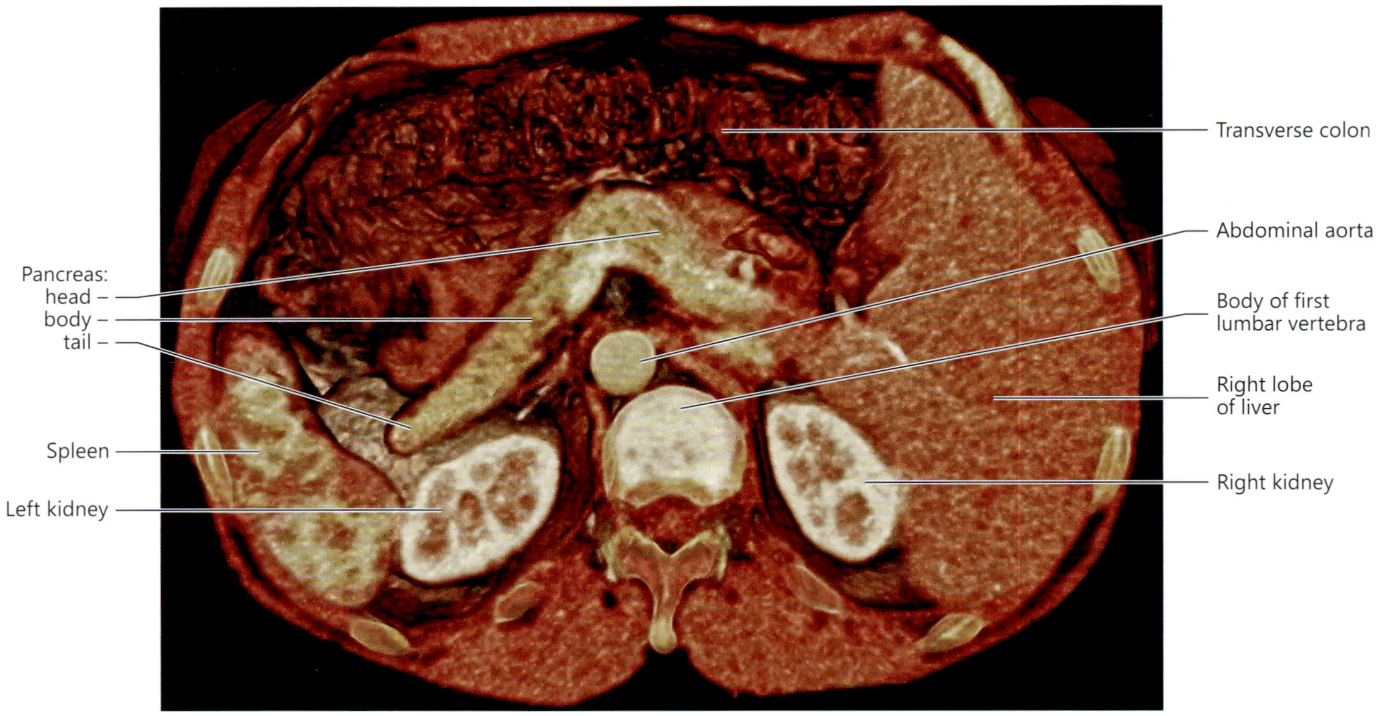

Three-dimensional image of a living human body, obtained by CT scan using a 3D volume rendering technique, showing the **abdominal organs** in a transverse section passing through the first lumbar vertebra.

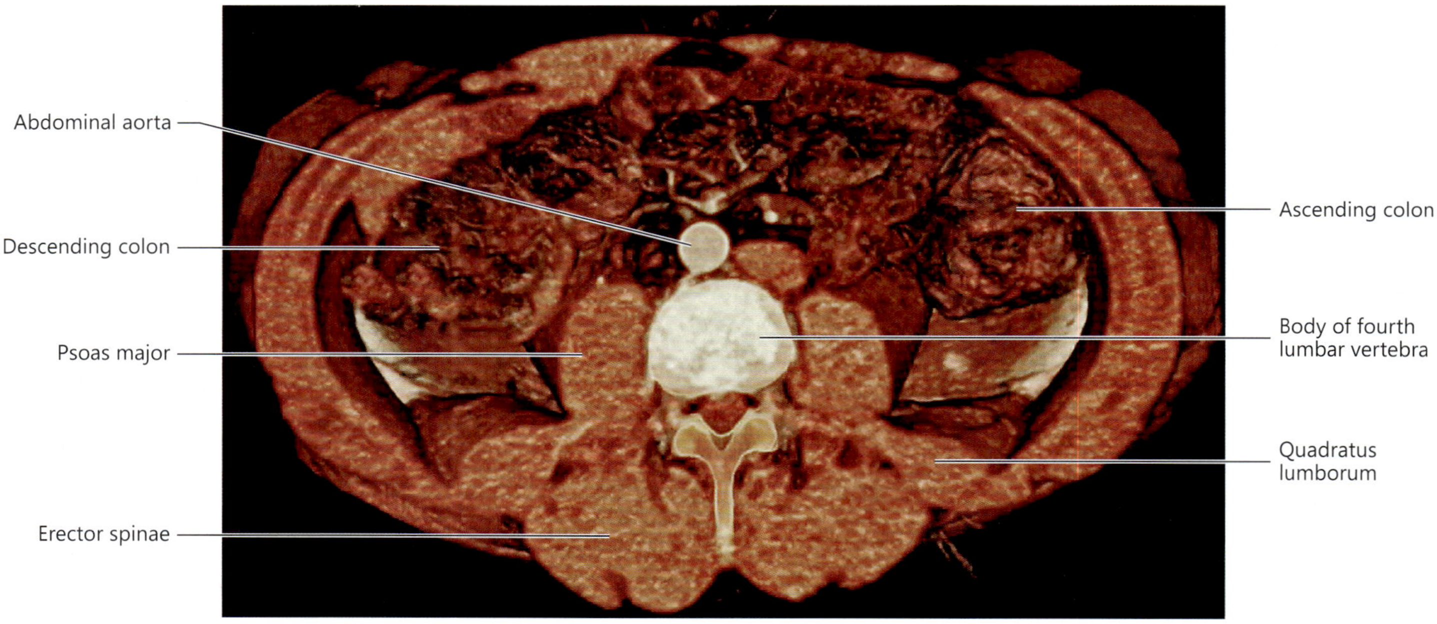

Three-dimensional image of a living human body, obtained by CT scan using a 3D volume rendering technique, showing the **abdominal organs** in a transverse section passing through the fourth lumbar vertebra.

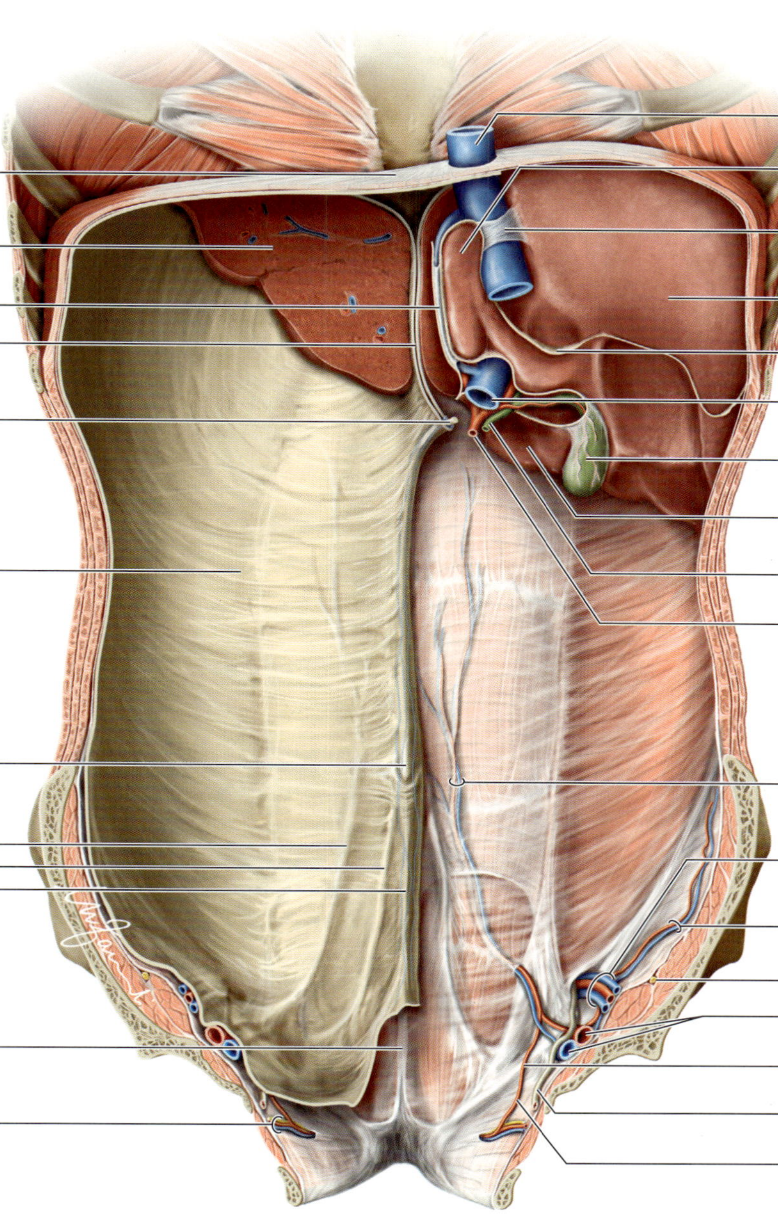

Posterior (inner) view of the **anterior abdominal wall** in the male. In the left half of the wall, the anterior parietal peritoneum forms the median, medial and lateral umbilical folds investing, respectively, the remnant of the urachus, the **medial umbilical ligament** and the inferior epigastric artery. In the right half of the wall, the anterior parietal peritoneum has been removed to expose the rectus sheath. The sheath is composed of two layers, anterior and posterior. The posterior layer (depicted in the figure) is formed, from back to front, by the transversalis fascia, the aponeurosis of transversus abdominis and the posterior splitting layer of the aponeurosis of the internal oblique. In the lower quarter, the posterior layer is made up of the transversalis fascia only. From the back to front, the anterior layer consists of the anterior splitting layer of the aponeurosis of the internal oblique, and of the aponeurosis of the external oblique. In the lower quarter, it further reinforced by the aponeurosis of the transversus abdominis.

The figure shows the visceral surface of liver, displaying: the **ligamentum venosum**, a remnant of the ductus venosus, which connects the left branch of the hepatic portal vein to the inferior vena cava in the fetus; the **round ligament of liver**, a remnant of the left umbilical vein, passing through the porta hepatis and the internal umbilical scar; the **falciform ligament**, composed of two layers of peritoneum, and running between the diaphragm and the supraumbilical part of the anterior abdominal wall, and the large sagittal groove on the diaphragmatic surface of liver, which separates the right and left lobes. The round ligament runs along the free border of the falciform ligament. The figure also displays the gallbladder, located in the homonymous fossa, the inferior vena cava, located in the groove for vena cava, the porta hepatis with the branches of division of the hepatic portal vein and the hepatic artery, and the bile duct that arises from the convergence of the common hepatic duct and the cystic duct.

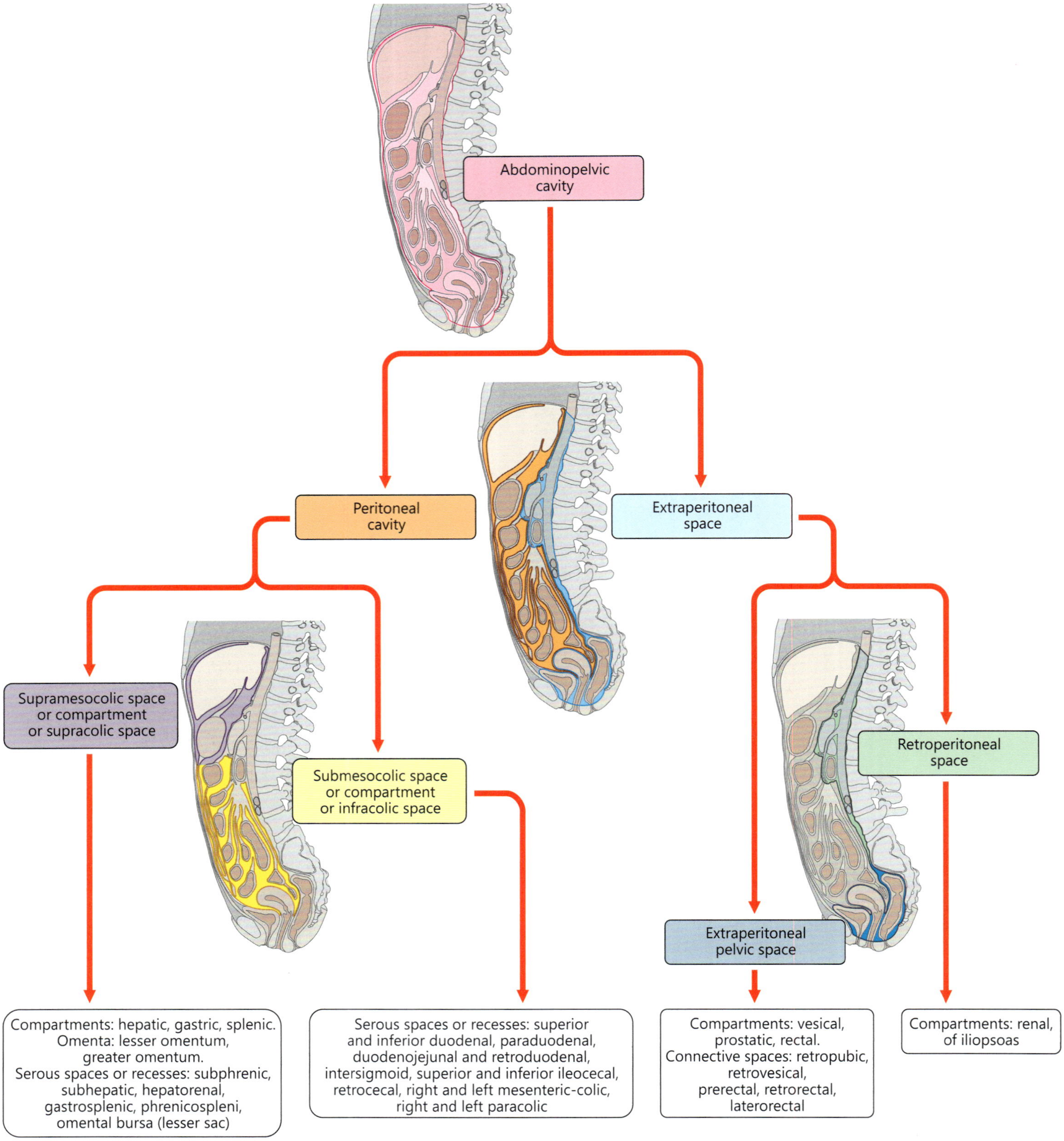

The **abdominopelvic cavity** is subdivided in the abdominal and the pelvic cavities, which are continuous with one another and separated by a virtual plane passing through the pelvic inlet. It contains a serous space, the **peritoneal cavity**, bounded by the parietal peritoneum and extending downwards to the pelvis, and two **connective spaces**: the **retroperitoneal space**, lying between the posterior parietal peritoneum and the abdominopelvic cavity wall, and the **extraperitoneal pelvic space**, lying between the inferior parietal peritoneum and the pelvic diaphragm. The superior or abdominal part of the peritoneal cavity is bounded by the diaphragm, superiorly, and by the conventional plane that separates it from the inferior or pelvic part, inferiorly. The transverse mesocolon and the transverse colon divide it into two spaces, or compartments, the **supramesocolic space** and the **submesocolic space**, communicating anteriorly through the space between the transverse colon and the internal surface of the anterior abdominal wall.

Content of the abdominal cavity

DENOMINATION OF THE SPACE	*CONTENTS*		
Supramesocolic space or compartment	*Organs*: – gallbladder – cystic duct – bile duct – hepatic ducts, right and left – common hepatic duct – duodenum, superior part – esophagus, abdominal part – liver – spleen – stomach *Ligaments*: – coronary – hepatocolic – hepatoduodenal – hepatoesophageal – hepatophrenic – hepatogastric – hepatorenal – falciform – gastrocolic – gastrophrenic	– gastrosplenic – phrenicocolic – phrenicosplenic – pancreaticosplenic – round of liver – right triangular – left triangular *Lymph nodes*: – hepatic – gastric, right/left – gastro-omental, right/left – pyloric – splenic *Arteries*: – cystic – hepatic – phrenic – gastric, left – gastric, right – posterior gastric – short gastric – gastro-omental, right	– gastro-omental, left – splenic *Veins*: – inferior vena cava – short gastric – gastric, right and left – gastro-omental, right – gastro-omental, left – hepatic portal vein – splenic *Nerves*: – hepatic plexus – gastric plexus – splenic plexus – hepatic branches of anterior vagal trunk – anterior gastric branches of anterior vagal trunk – posterior gastric branches of posterior vagal trunk – anterior vagal trunk – posterior vagal trunk
Submesocolic space or compartment	*Organs*: – appendix (vermiform appendix) – cecum – ascending colon – descending colon – sigmoid colon – transverse colon – jejunum – ileum *Ligaments and mesenteries*: – greater omentum – mesentery – mesoappendix – sigmoid mesocolon – transverse mesocolon *Lymph nodes*: – inferior mesenteric – – sigmoid	– superior mesenteric – – of appendix – – ileocolic – – juxtaintestinal mesenteric – – mesocolic – – – colic, right/middle/left – – – paracolic – – prececal – – retrocecal – – central superior mesenteric *Arteries*: – appendicular – colic, right – colic, middle – colic, left – jejunal – ileal – ileocolic	– inferior mesenteric – superior mesenteric – sigmoid *Veins*: – appendicular – colic, right – colic, middle – colic, left – jejunal – ileal – ileocolic – inferior mesenteric – superior mesenteric – sigmoid *Nerves*: – enteric plexus – inferior mesenteric plexus – superior mesenteric plexus
Retroperitoneal space	*Bones*: – bodies of lumbar vertebrae *Muscles*: – iliopsoas – quadratus lumborum – suspensory muscle of duodenum (Ligament of Treitz) *Organs*: – bile duct – thoracic duct – duodenum, descending, inferior and ascending parts – suprarenal gland – pancreas – kidney – ureter, abdominal part *Lymph nodes and trunks*: – cisterna chyli – thoracic duct – external iliac – internal iliac – right lumbar – – lateral caval – – precaval	– – postcaval – intermediate lumbar – left lumbar – – lateral aortic – – preaortic – – postaortic – intestinal trunk – lumbar trunk *Arteries*: – abdominal aorta – deep circumflex iliac – colic, middle – colic, left – inferior epigastric – inferior phrenic – iliolumbar – common iliac – external iliac – inferior mesenteric – superior mesenteric – lumbar – inferior pancreaticoduodenal – anterior and posterior superior pancreaticoduodenal	– ureteric branches of renal and testicular ♂ or ovarian ♀ arteries – renal – retroduodenal – median sacral – splenic – middle suprarenal – testicular ♂ or ovarian ♀ – celiac trunk *Veins*: – inferior vena cava – deep circumflex iliac – hepatic – inferior epigastric – inferior phrenic – iliolumbar – common iliac – external iliac – ascending lumbar – lumbar – inferior mesenteric – superior mesenteric

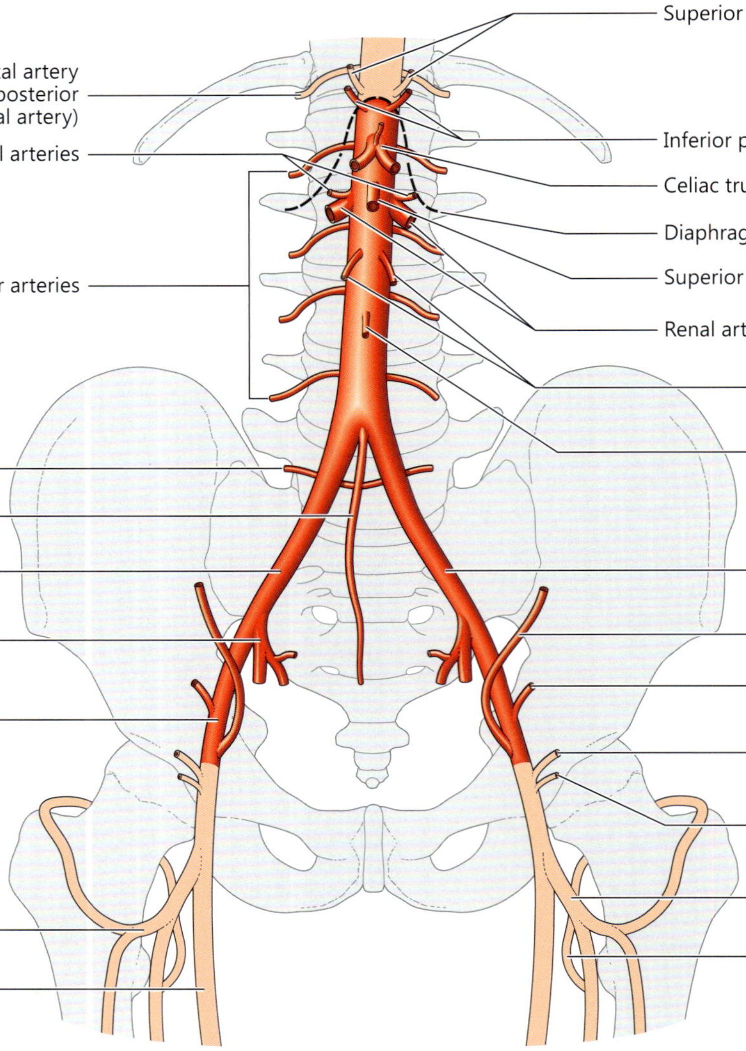

Schematic representation of the **abdominal aorta** and its branches. Unpaired branches of the abdominal aorta consist of the *celiac trunk*, originating anteriorly just below the aortic hiatus, the *superior mesenteric artery*, originating approximately 1 cm below the celiac trunk, and the *inferior mesenteric artery*, arising 3-4 cm above the abdominal aortic bifurcation into the common iliac arteries. Paired branches of the abdominal aorta consist of the *inferior phrenic arteries*, arising anteriorly just above the origin of the celiac trunk, the *middle suprarenal arteries*, originating laterally at the same level as the superior mesenteric artery, the *renal arteries*, arising laterally, 1 cm below the origin of the superior mesenteric artery, the *ovarian arteries*, in the female, or *testicular arteries*, in the male, branching anteriorly, below the renal arteries, the *lumbar arteries* (four on each side), originating posteriorly from the abdominal aorta, at the level of the first four lumbar vertebral bodies, and the *common iliac arteries*, originating at the sacroiliac joints level, and further branching into the external and internal iliac arteries. The abdominal aorta terminates at the fourth lumbar vertebra level, and continues as the median sacral artery.

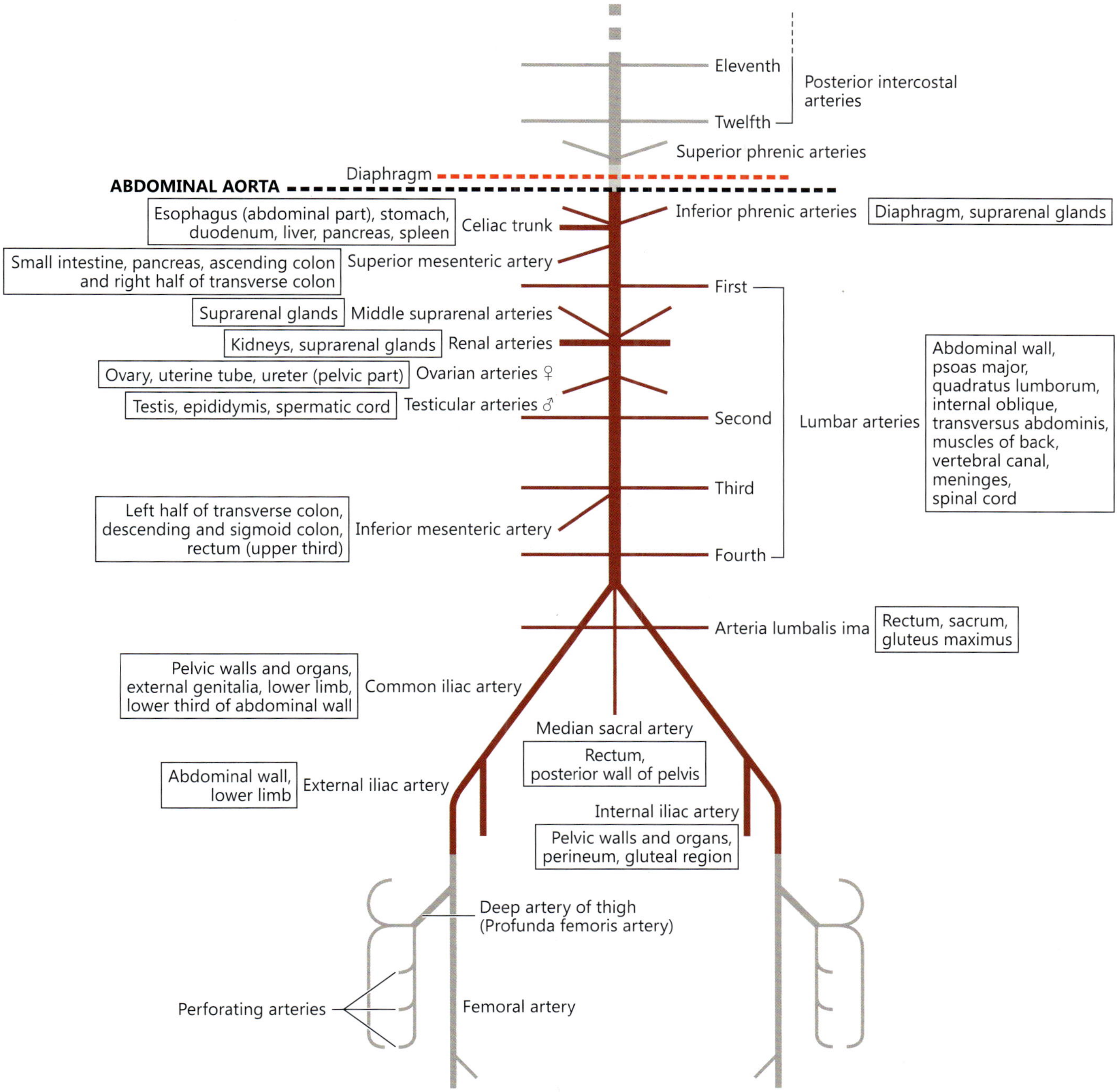

Schematic representation of the collateral and terminal branches of the **abdominal aorta**. Unpaired collateral branches are the celiac trunk, and the superior and inferior mesenteric arteries. Paired collateral branches are the inferior phrenic arteries, the middle suprarenal arteries, the ovarian or testicular arteries and the lumbar arteries. The median sacral artery is the terminal branch of the abdominal aorta. The lumbar arteries are parietal branches, as the inferior phrenic arteries, and continue the metameric series of posterior intercostal arteries on each side.

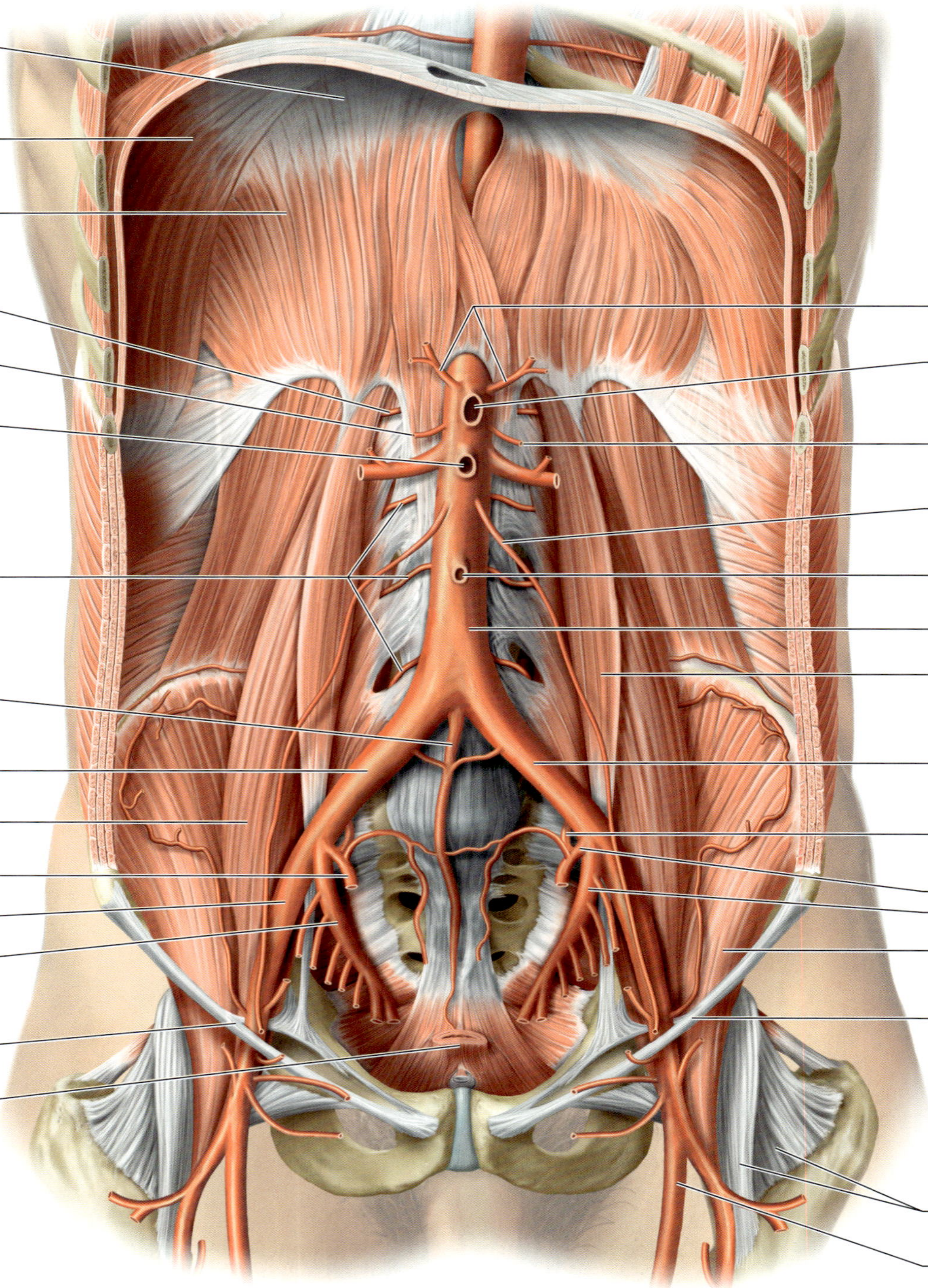

Course of the **abdominal aorta**. The abdominal aorta originates at the level of the aortic hiatus, corresponding to the inferior border of the twelfth thoracic vertebra and descends anterior to the vertebral column, slightly to the left of the median plane, to reach the fourth lumbar vertebra. Here, it bifurcates into the right and left common iliac arteries. The median sacral artery arises as an unpaired branch at the level of aortic bifurcation.

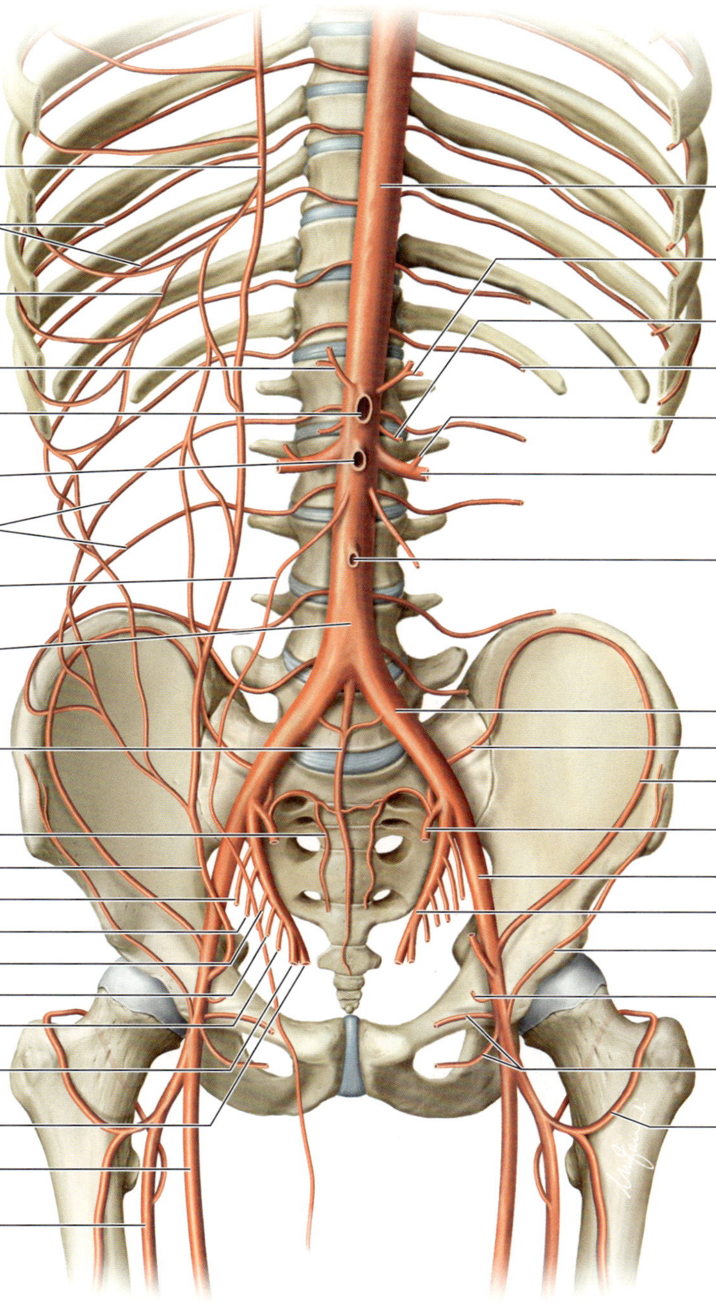

Parietal blood supply provided by the abdominal aorta. The figure shows the parietal branches of the abdominal aorta, namely the lumbar, inferior phrenic and medial sacral arteries. The internal and external iliac arteries also originate from the common iliac arteries. The obturator, internal pudendal and inferior gluteal arteries are the parietal branches of the internal iliac artery, originating from its anterior trunk or anterior division. The iliolumbar, lateral sacral and superior gluteal arteries arise from the posterior trunk, or posterior division, of the internal iliac artery. The inferior epigastric and deep circumflex iliac arteries originate from the external iliac artery, while the superficial epigastric, superficial circumflex iliac and superficial and deep external pudendal arteries arise from the femoral artery, into which the external iliac artery continues.

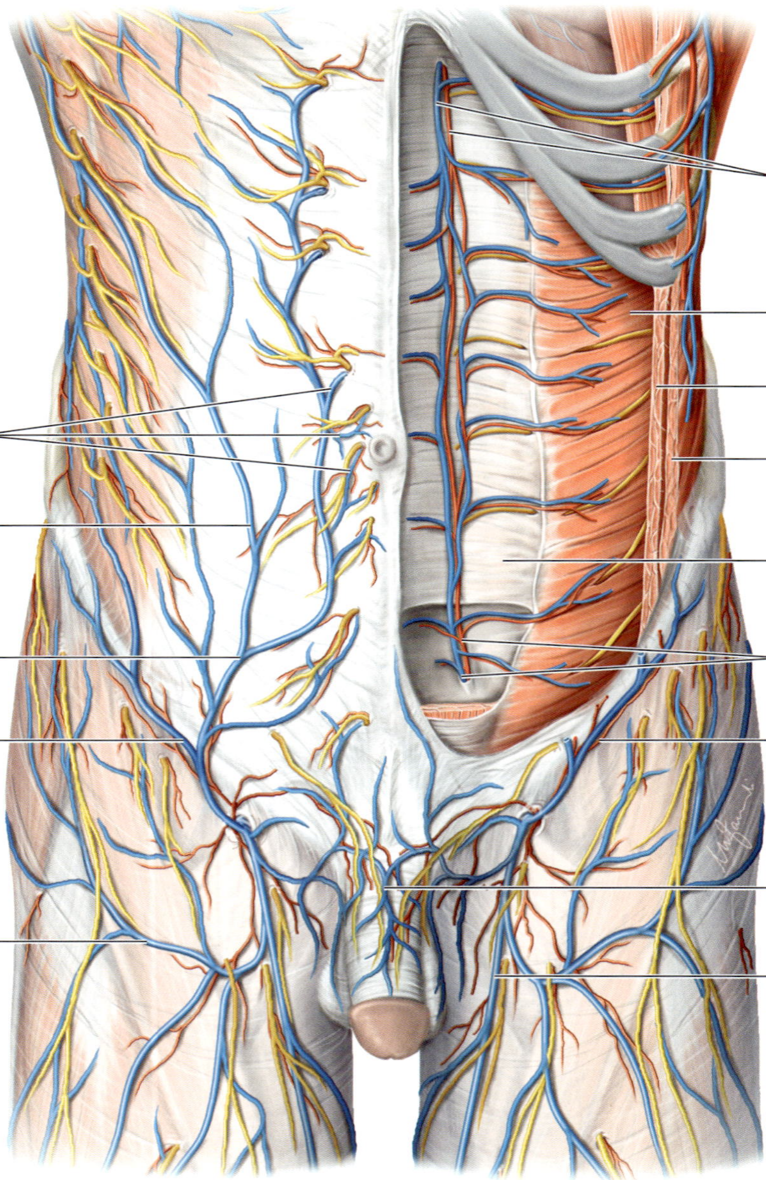

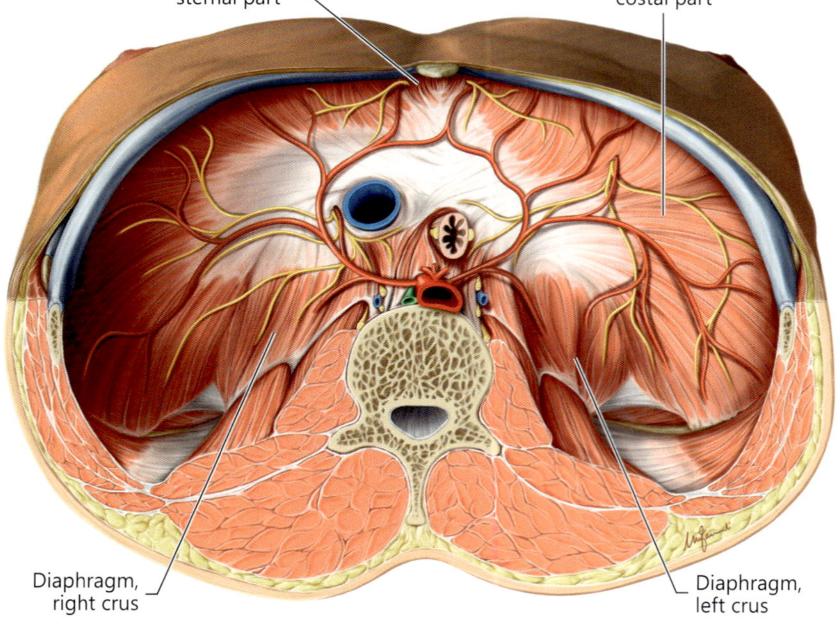

Inferior view of the **diaphragm** ☞ volume 2, page 252.

Parietal vascularization of the trunk: relations between the supradiaphragmatic and subdiaphragmatic vasculature. Some arterial anastomoses are established between the supra- and subdiaphragmatic regions of the trunk; in particular, anastomoses between the superior epigastric artery (a branch of the internal thoracic artery), pertaining to the subclavian artery, and the inferior epigastric artery (a branch of the external iliac artery). With reference to the venous vasculature, several **caval anastomoses** exist between the tributaries of the superior and of the inferior vena cava. Superficially, the lateral thoracic veins (tributaries of the axillary veins) anastomose with the superficial epigastric veins (tributaries of the femoral veins) via the thoracoepigastric veins. Deep within the rectus sheath, the superior epigastric veins (tributaries of the internal thoracic veins) anastomose with the inferior epigastric veins (tributaries of the external iliac veins). The anterior abdominal wall is supplied by the superior, inferior and superficial epigastric arteries, whose corresponding veins are satellite vessels, and by the ascending branch of the deep circumflex iliac artery.

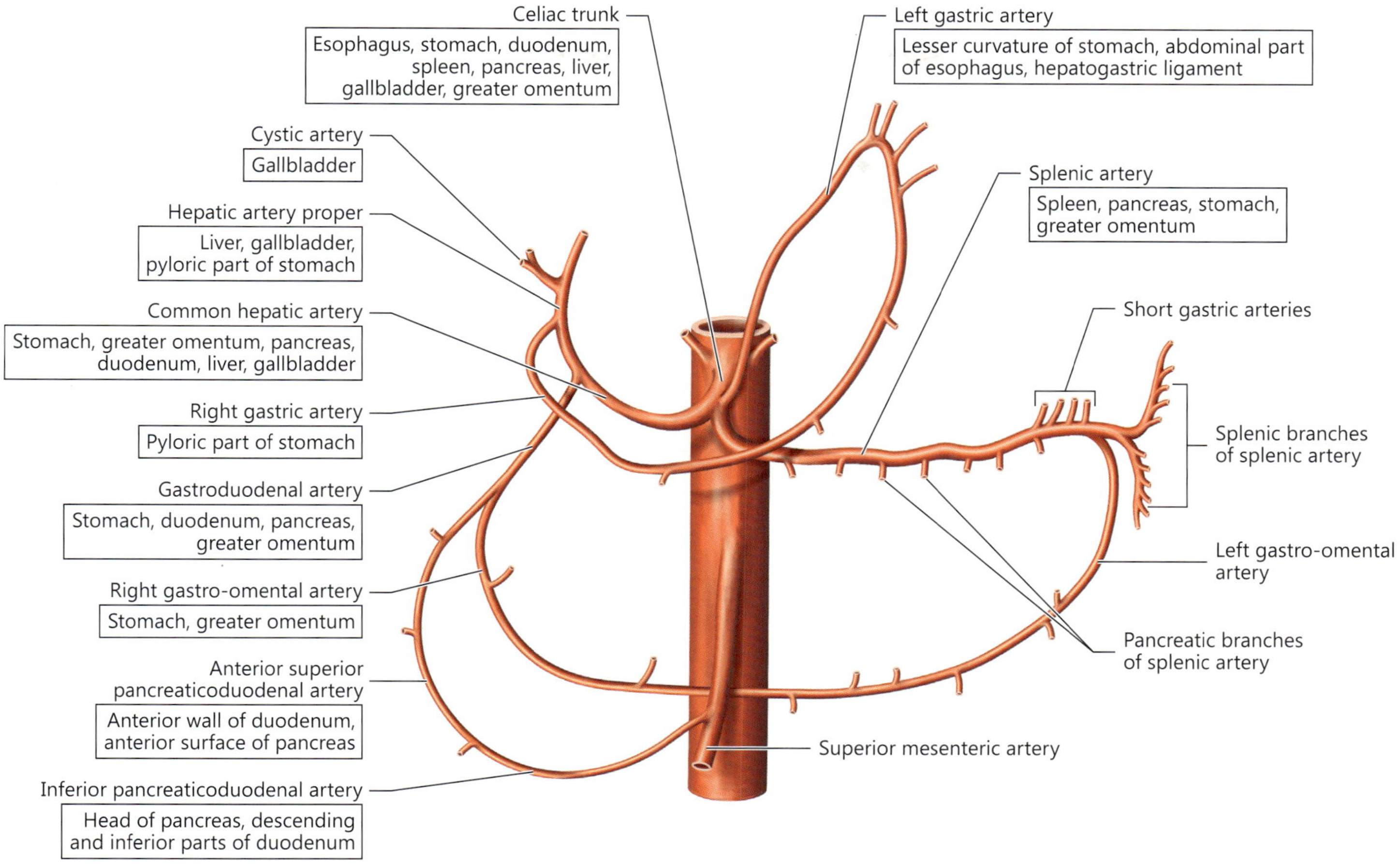

Schematic representation of the **celiac trunk** and its branches. Note the arterial arcades formed by the gastric arteries (right and left), and the gastro-omental arteries (right and left), and the anastomosis between the gastroduodenal artery and the superior mesenteric artery, via the pancreaticoduodenal vessels. A double anastomosis occurs when the pancreaticoduodenal vessels bifurcate to run anterior (anterior pancreaticoduodenal) and posterior (posterior pancreaticoduodenal) to the head of the pancreas.

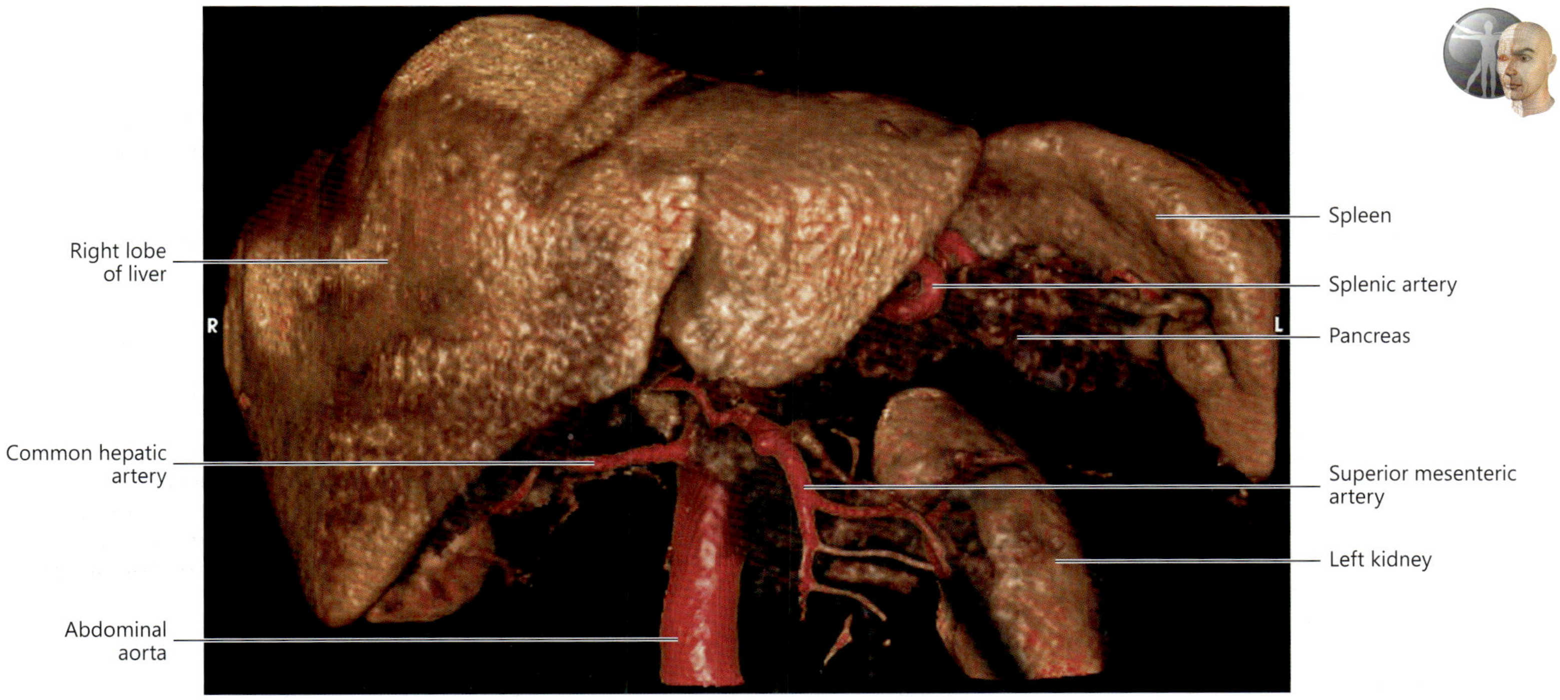

Three-dimensional image of a living human body, obtained by CT scan using a 3D volume rendering technique, showing some vessels coming from the **celiac trunk**; the image also displays the **superior mesenteric artery**.

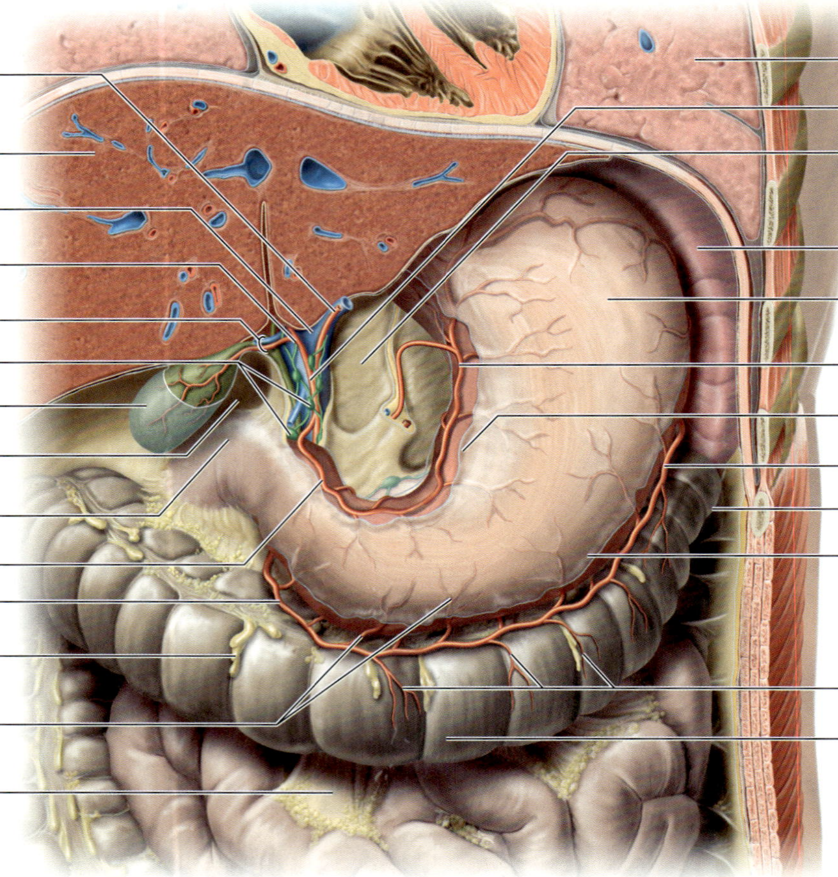

Branches of the celiac trunk. The right and left gastric arteries run one towards the other along the lesser curvature of the stomach, and anastomose by inosculation. The former is a branch of the hepatic artery proper (branch of the celiac trunk), and the latter is a direct branch of the celiac trunk. The right and left gastro-omental arteries anastomose by inosculation, forming an arcade along the greater curvature of the stomach. The former is a branch of the gastroduodenal artery (branch of the common hepatic artery); the latter is a branch of the splenic artery (branch of the celiac trunk).

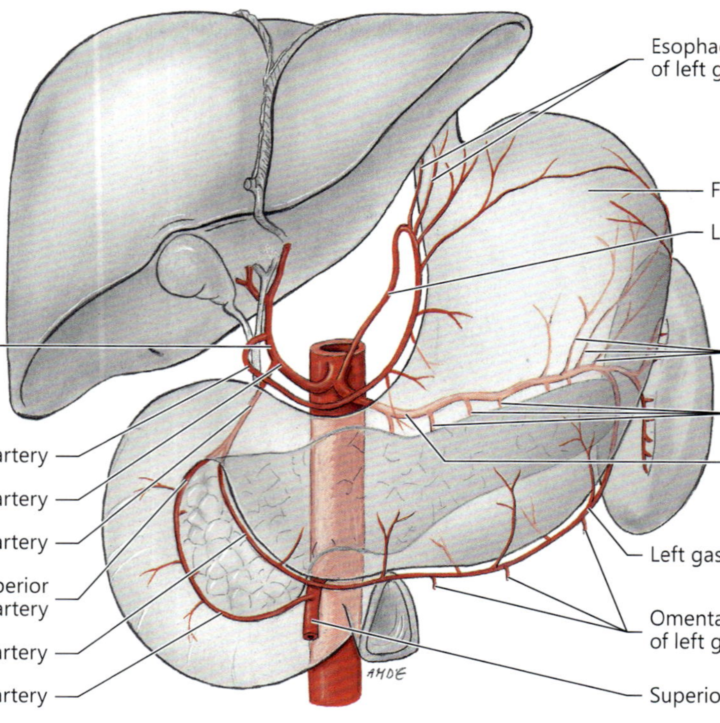

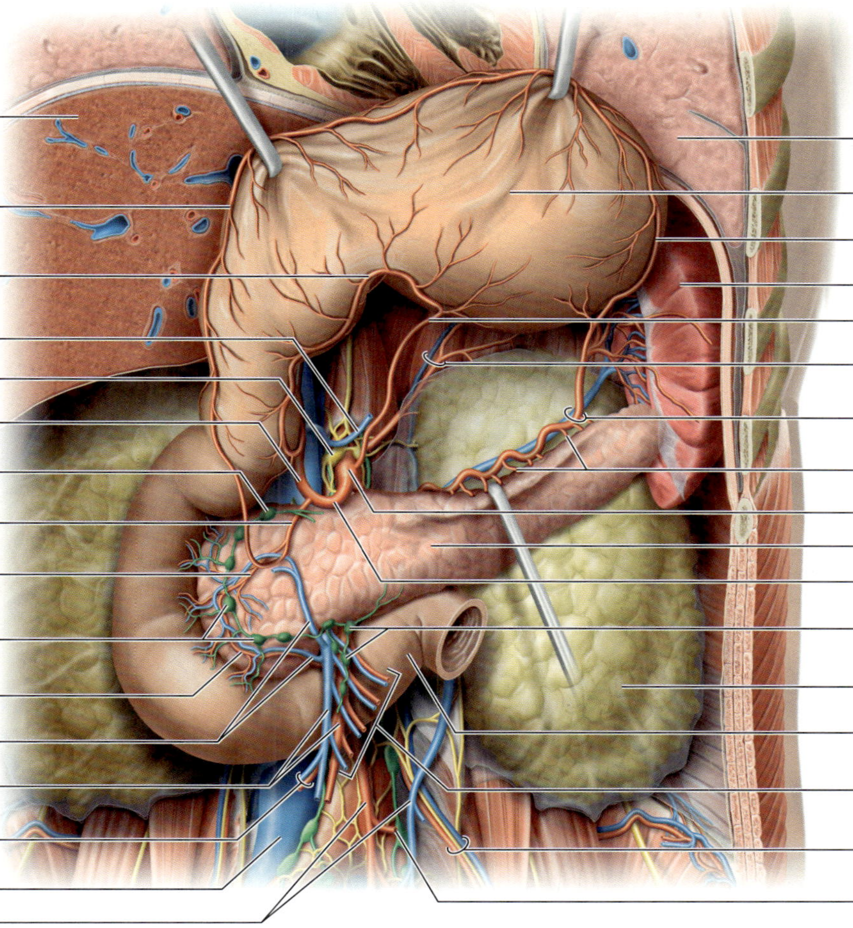

Preparation illustrating the **branches of the celiac trunk**. Measuring a little over 1 cm in length, the celiac trunk emerges from the anterior surface of the abdominal aorta at a ninety-degree angle, it runs forward almost horizontally and to the right, and divides into three branches: the left gastric artery, the common hepatic artery and the splenic artery.

◄
Schematic figure illustrating the arrangements of the **branches of the celiac trunk**, and the arterial arcades they form. The celiac trunk divides into three branches: the left gastric artery, the common hepatic artery and the splenic artery. The left gastric artery runs downward and to the right, along the lesser curvature of the stomach, to anastomose with the right gastric artery, a branch of hepatic artery proper. The gastroduodenal artery also arises from the common hepatic artery; in turn, it originates the right gastro-omental and the anterior and posterior superior pancreaticoduodenal arteries, the latter may originate either individually, or from a common trunk. The anterior superior pancreaticoduodenal artery descends within the groove between the duodenum and the head of the pancreas, and anastomoses with the anterior branch of the inferior pancreaticoduodenal artery, a branch of the superior mesenteric artery. The posterior superior pancreaticoduodenal artery descends to the right behind the head of the pancreas, and anastomoses with the posterior branch of the inferior pancreaticoduodenal artery. The splenic artery originates the short gastric arteries and the left gastro-omental artery, which anastomoses with the right gastro-omental artery after reaching the greater curvature of stomach.

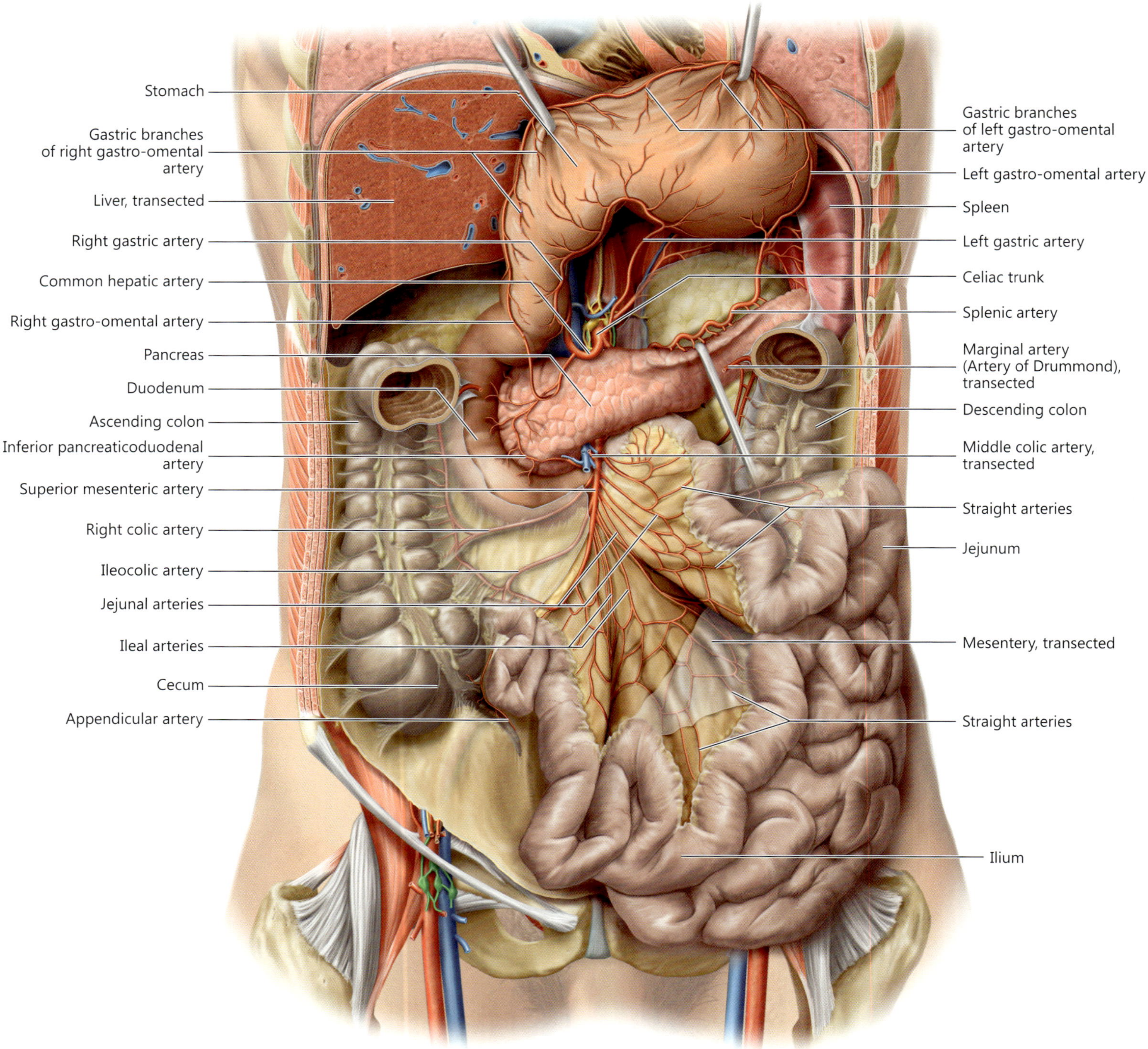

Preparation illustrating the **celiac trunk** and the **superior mesenteric artery**. The jejunal and ileal arteries, from 12 to 20 in number, arise from the superior mesenteric artery and travel between the two layers of the mesentery. Each artery bifurcates into a superior and an inferior branch, which connect to the corresponding division of the adjacent artery, hence forming multiple series of arcades. The straight arteries are vessels with a rectilinear course that arise from the terminal arcade, and reach both surfaces of each loop of the small intestine.

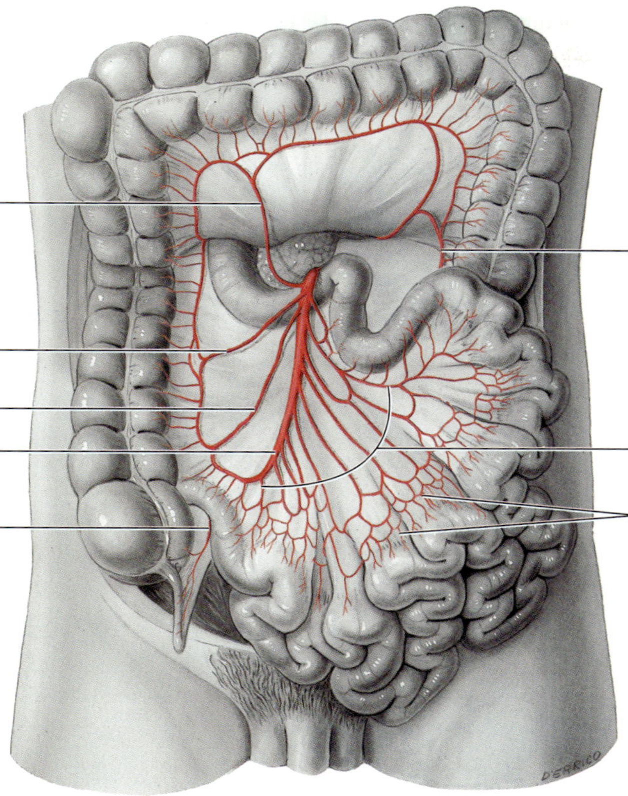

Preparation illustrating the course and relations of the **superior mesenteric artery**. The transverse colon has been elevated and the loops of small intestine have been moved to the lower right side to expose the artery and its branches.

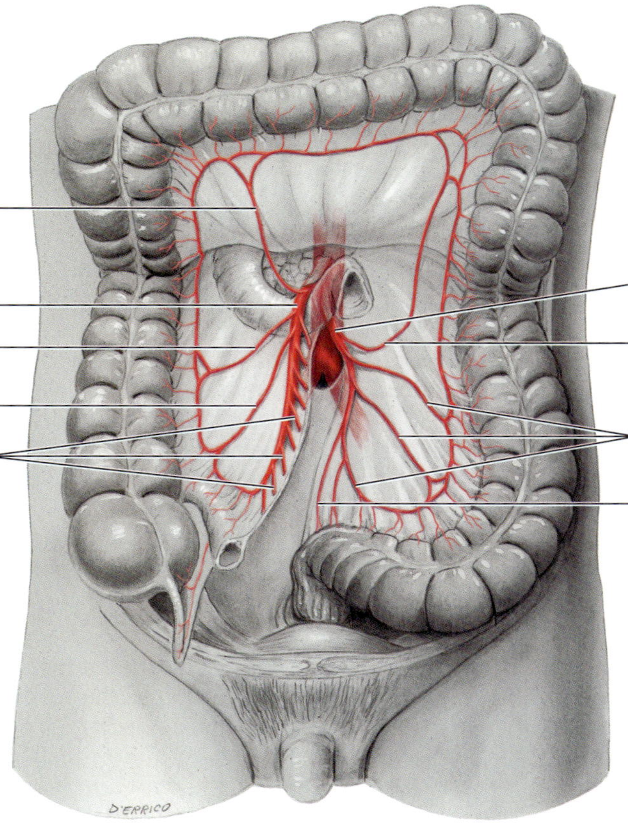

Preparation illustrating the **inferior mesenteric artery** and its branches. The entire mesenteric small intestine has been removed, and the transverse colon has been elevated to expose the entire arterial vasculature of the large intestine. Note a continuous, "marginal" arterial arcade of Drummond formed by the anastomoses between the arterial branches supplying the intestine.

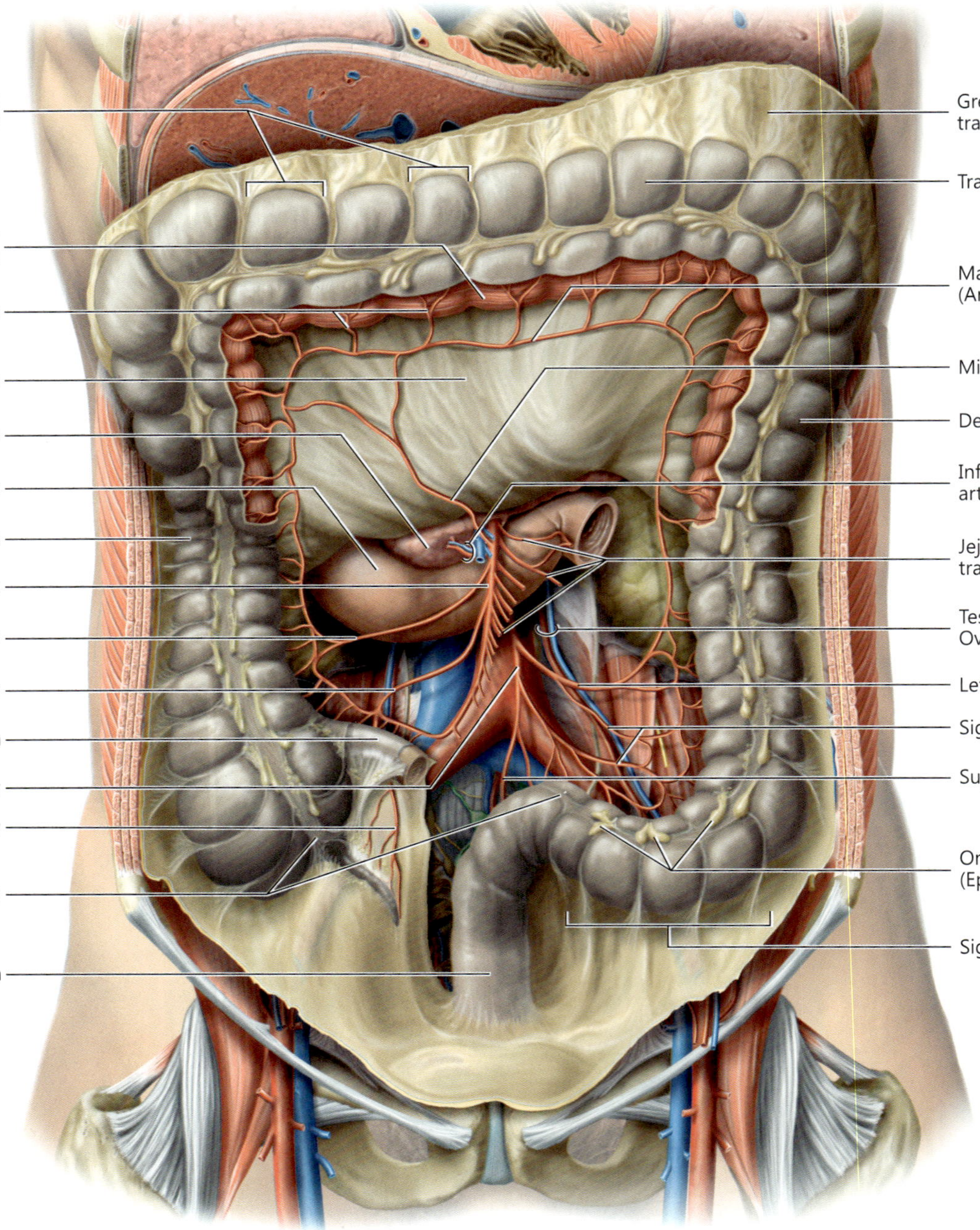

The small intestine has been removed and the transverse colon has been elevated to expose the branches of the **superior mesenteric artery**, comprising the inferior pancreaticoduodenal, jejunal, ileal, ileocolic, right colic and middle colic arteries. The marginal artery of the colon arises from the latter. The figure also displays the branches of the **inferior mesenteric artery**, comprising the left colic, sigmoid and superior rectal arteries. The left colic artery bifurcates: the ascending artery (or branch) anastomoses with the marginal artery of the colon, which, in turn, continues as the middle colic artery. The descending branch of left colic artery anastomoses with the sigmoid arteries, while the right branch of middle colic artery anastomoses with the right colic artery, which, in turn, anastomoses with the superior branch of the ileocolic artery. A major collateral circle (marginal arcade, artery of Drummond) is created between the superior and inferior mesenteric arteries, via the marginal artery of the colon, in the event of stenosis or occlusion of either vessel.

➤

Collateral branches of the **superior and inferior mesenteric arteries**. The superior mesenteric artery originates from the abdominal aorta, approximately 1 cm below the celiac trunk. It supplies blood to the small intestine (with the exception of the superior part of the duodenum), the head of the pancreas, the cecum, the ascending colon and the right half of the transverse colon, through the inferior pancreaticoduodenal artery, the jejunal and ileal (intestinal) arteries, the ileocolic artery, marginal artery of the colon, right colic and middle colic arteries. Occasionally, the right and middle colic arteries are replaced by one single artery, the right flexural artery; this artery can sometime be supernumerary to the right and middle colic arteries, arising from the superior mesenteric artery, in between the origins of the two colic arteries. The inferior mesenteric artery originates from the abdominal aorta, 3-4 cm above the aortic bifurcation. It supplies blood to the left half of transverse colon, the descending colon, the sigmoid colon and the rectum, through the left colic artery, the sigmoid arteries and the superior rectal artery. The ascending artery can arise directly from the inferior mesenteric artery. It is also referred to as intermesenteric artery, as it anastomoses with the marginal artery of the colon or the middle colic artery.

Mesenteric Arteries

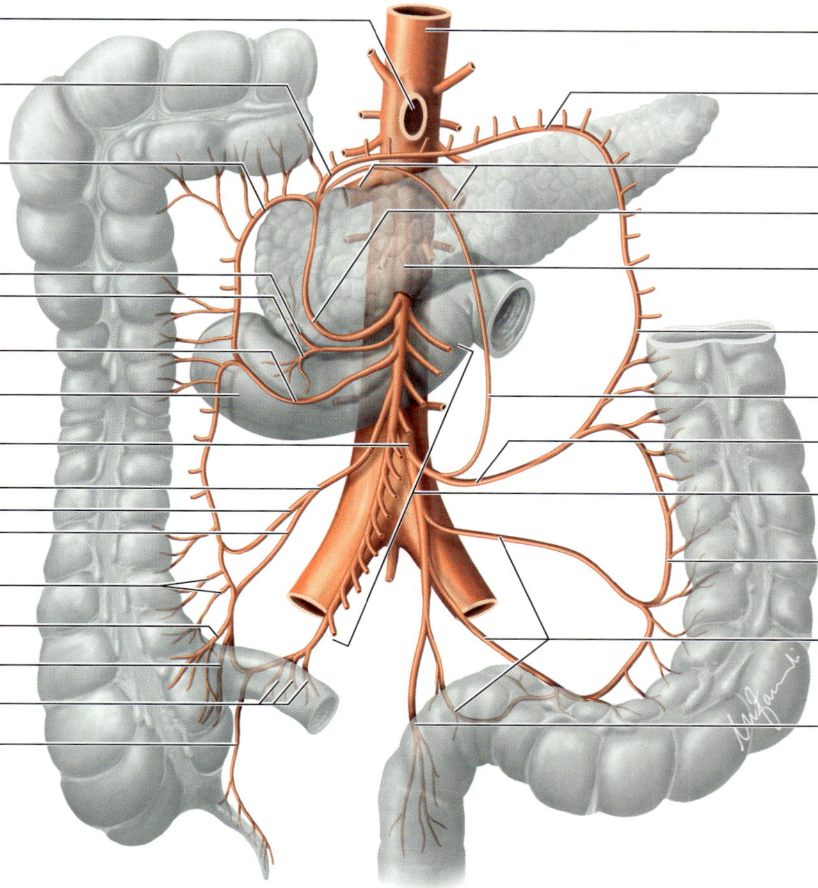

Abdominal aorta and corresponding unpaired autonomic branches (celiac trunk and mesenteric arteries); collateral branches of the superior and inferior mesenteric arteries: territories of supply

	ARTERIES	TERRITORIES OF SUPPLY
1	Abdominal aorta	Organs contained in the abdominal cavity, abdominal walls, spinal cord
2	Marginal artery (Artery of Drummond)	Colon (vessel running parallel to the wall of colon, consisting of the main trunks of the ileocolic, right colic, middle colic and left colic arteries, and of the arcades arising from these vessels)
3	Renal arteries	Kidneys, ureters, suprarenal glands
4	Middle colic artery	Transverse colon (right half)
5	Superior mesenteric artery	Small intestine (with the exception of the superior part of the duodenum), cecum, ascending and transverse (right half) colon
6	Ascending artery (ascending branch of left colic artery)	Transverse colon (left half)
7	Arc of Riolan	Anastomosis, not always present, between an ascending branch of the left colic artery and a descending branch arising from the left branch of the middle colic artery
8	Left colic artery	Transverse (left half) and descending colon
9	Jejunal arteries	Jejunum
10	Descending branch of left colic artery	Descending colon
11	Sigmoid arteries	Descending (inferior part) and sigmoid colon
12	Superior rectal artery	Rectum and internal anal sphincter
13	Appendicular artery	Appendix
14	Ileal arteries	Ileum
15	Posterior cecal artery	Cecum (posterior part)
16	Anterior cecal artery	Cecum (anterior part)
17	Straight arteries	Intestinal loops
18	Ileocolic artery	Ileum (distal part), cecum, appendix, ascending colon
19	Inferior mesenteric artery	Transverse (left half), descending, sigmoid colon and rectum (superior part)
20	Right colic artery	Ascending colon (superior part), right colic flexure, transverse colon (right half)
21	Inferior pancreaticoduodenal artery	Head of the pancreas, duodenum (descending and inferior parts)
22	Right branch of middle colic artery	Transverse colon (right half)
23	Left branch of middle colic artery	Transverse colon (left half)
24	Celiac trunk	Esophagus, stomach, duodenum, spleen, pancreas, liver, gallbladder, greater omentum

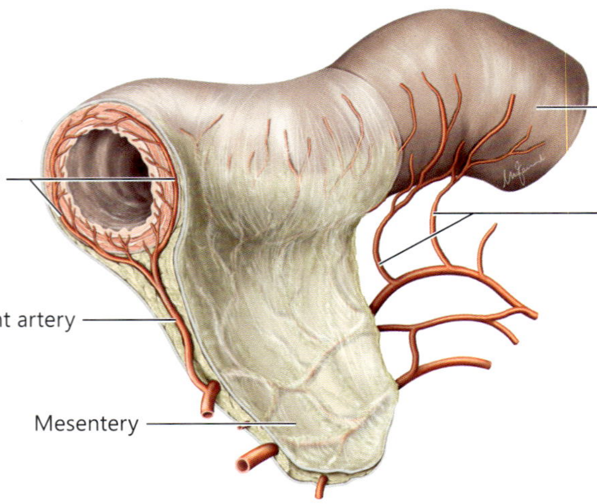

The **straight arteries** reach the concave margin of the small intestine and bifurcate, running along the opposite surfaces of the loop, then anastomose with their corresponding terminal branches on the convex margin.

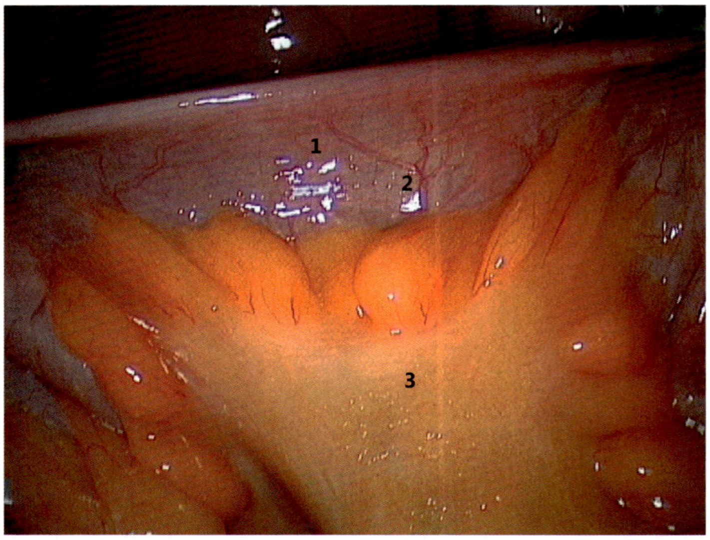

Image of a **jejunal loop** and its corresponding mesentery, taken during a diagnostic laparoscopy. The bifurcation of a straight artery can be observed.

1 Jejunal loop
2 Straight artery
3 Mesentery

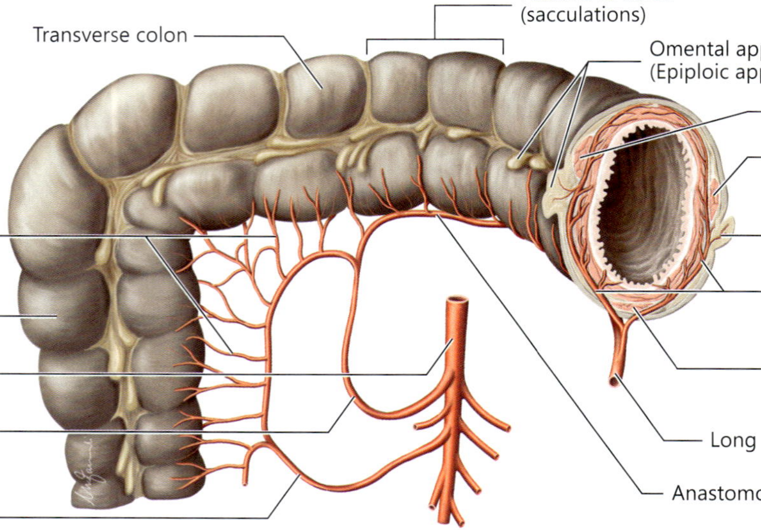

The **long straight arteries** attach below the free tenia of the colon, and then bifurcate into two branches embracing the wall of the colon.

➤

The **marginal artery of the colon** is an anastomotic arcade between the left colic artery, a branch of the inferior mesenteric artery, and the right colic artery, a branch of the superior mesenteric artery. The first jejunal artery can arise from the superior mesenteric artery via a common trunk with the inferior pancreaticoduodenal artery and a branch for the duodenojejunal flexure, which anastomoses with the other two branches.

Mesenteric Arteries

The **ileocolic artery**, via its ileal branch, anastomoses with the termination of the superior mesenteric artery (posterior view).

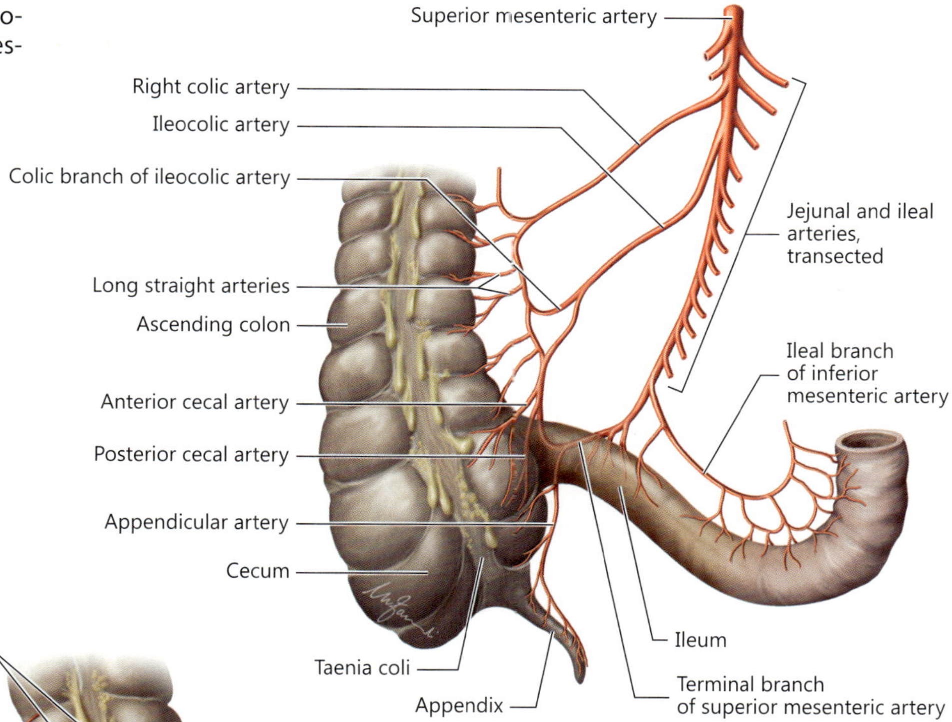

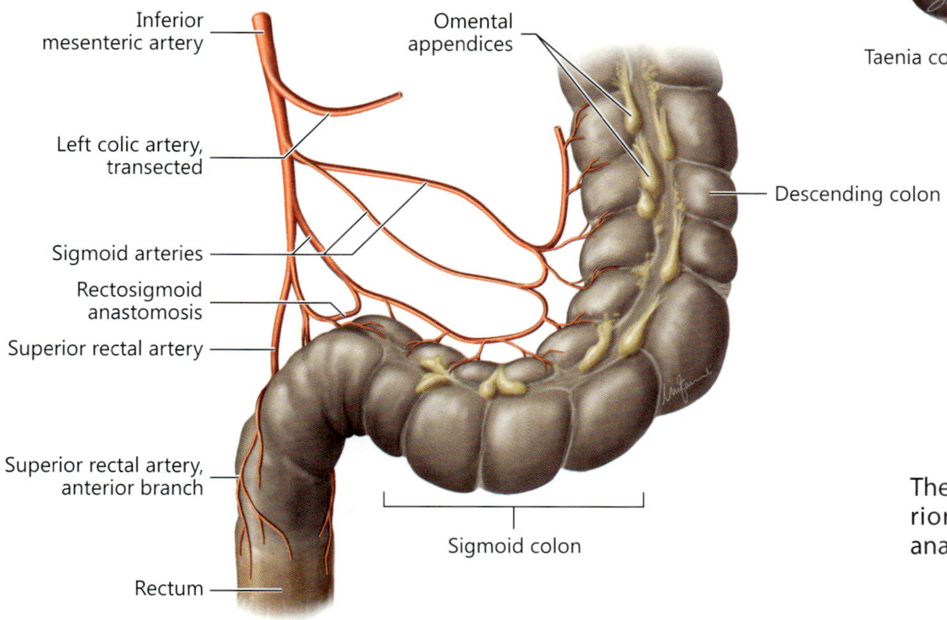

The last **sigmoid artery** connects to the superior rectal artery through the rectosigmoid anastomosis, which is not always present.

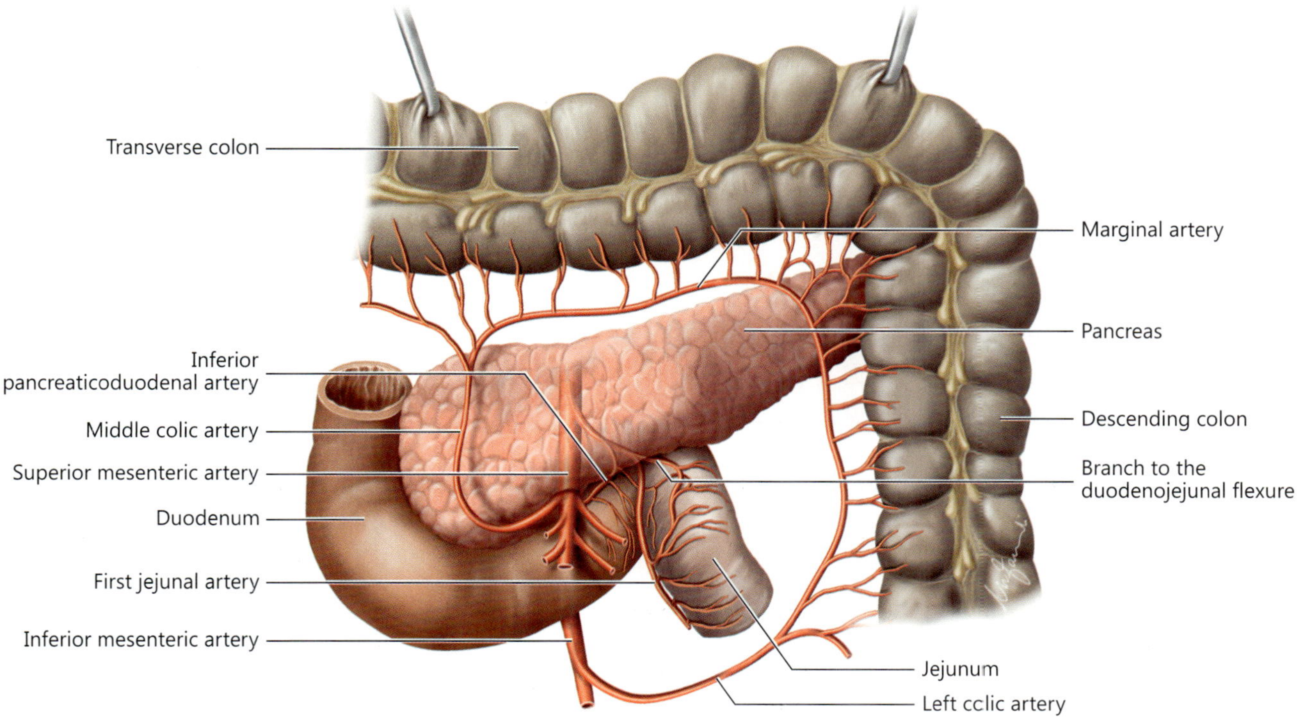

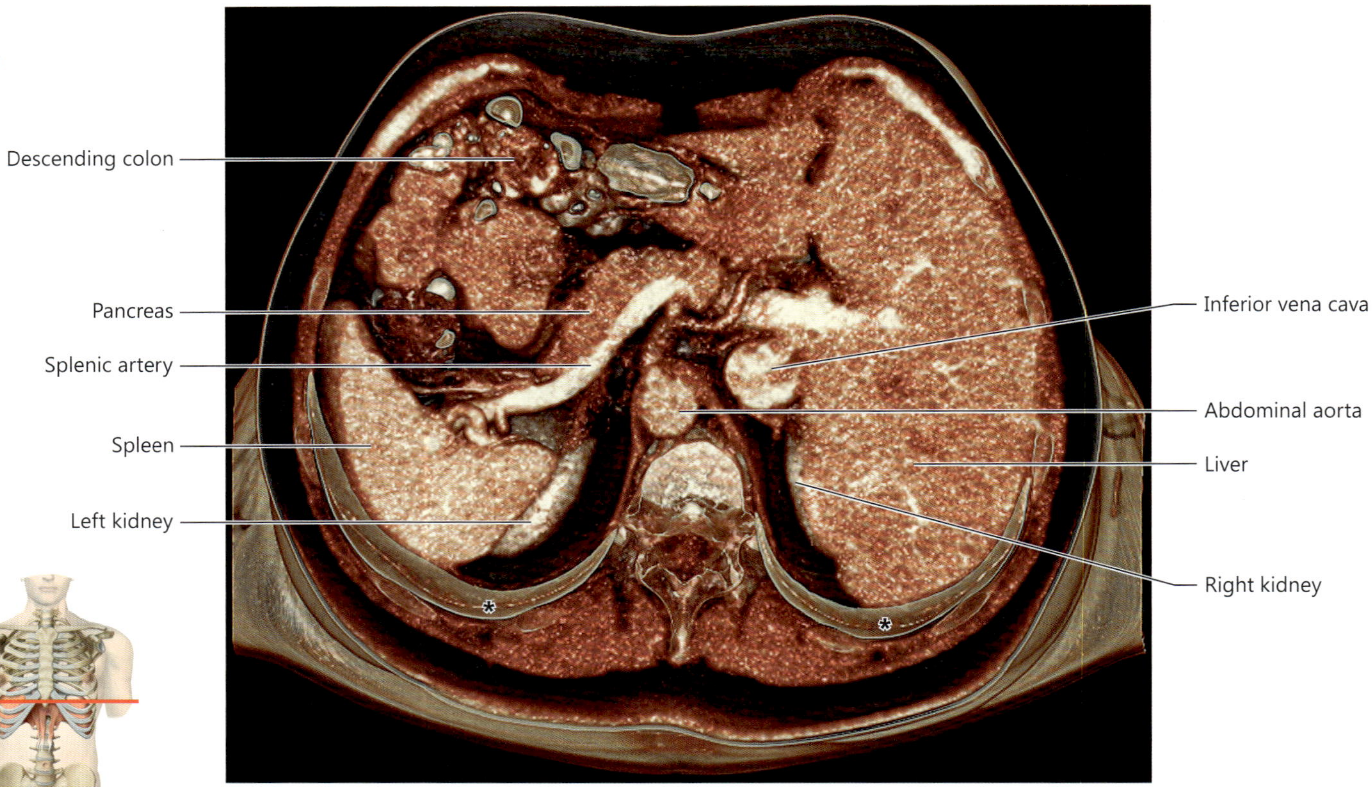

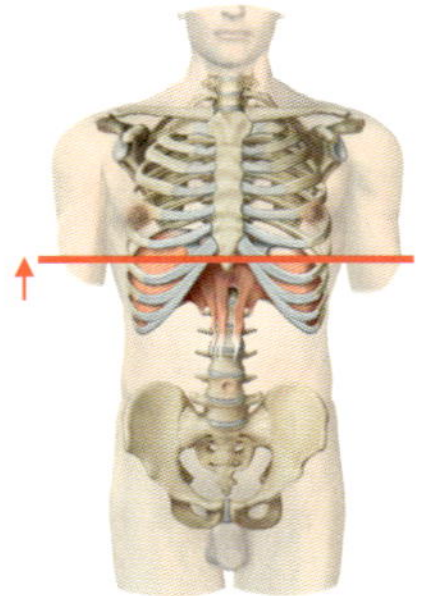

Three-dimensional image of a living human body, obtained by CT scan using a 3D volume rendering technique, showing the **abdominal organs** in a transverse section passing through the tenth thoracic vertebra, as seen from below. The scan parameters enable to highlight the left and right costodiaphragmatic pleural recesses (*).

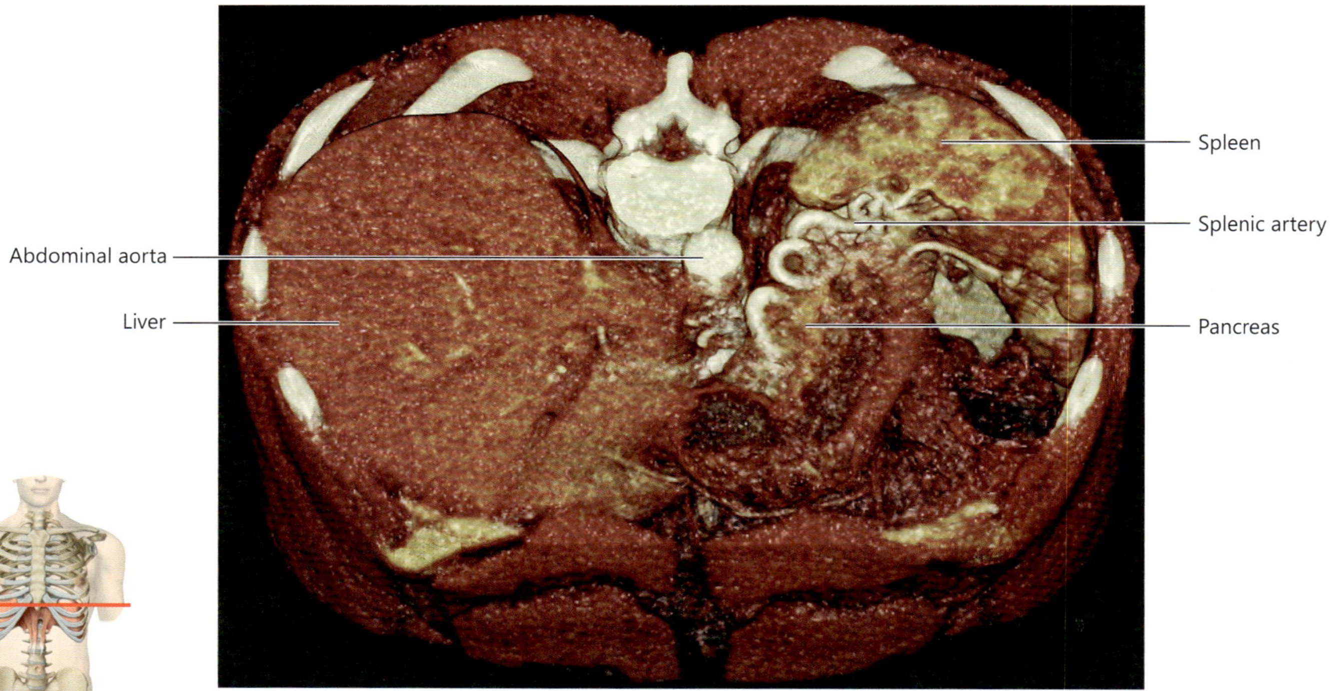

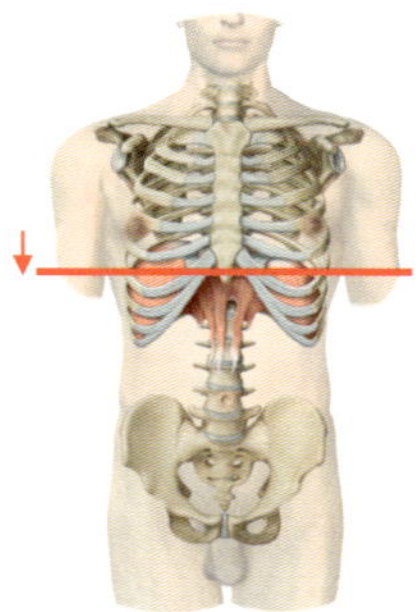

Three-dimensional image of a living human body, obtained by CT scan using a 3D volume rendering technique, showing the **abdominal organs** in a transverse section passing through the tenth thoracic vertebra, as seen from above.

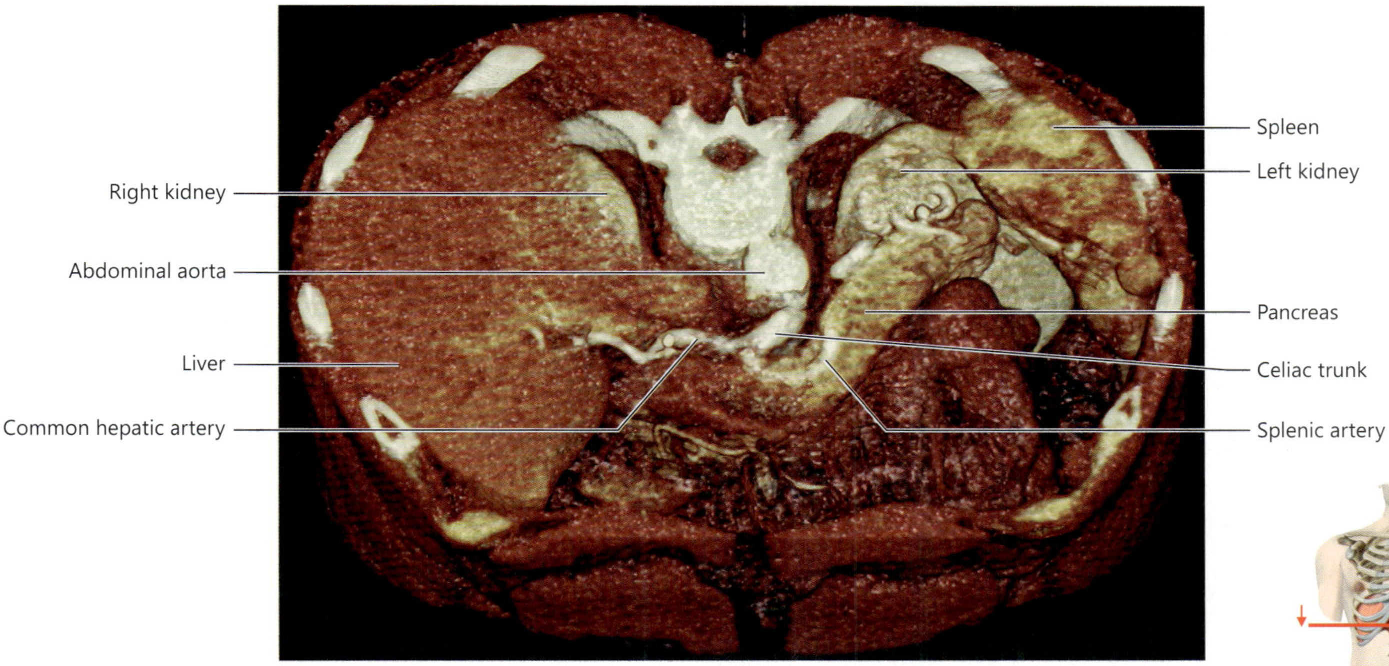

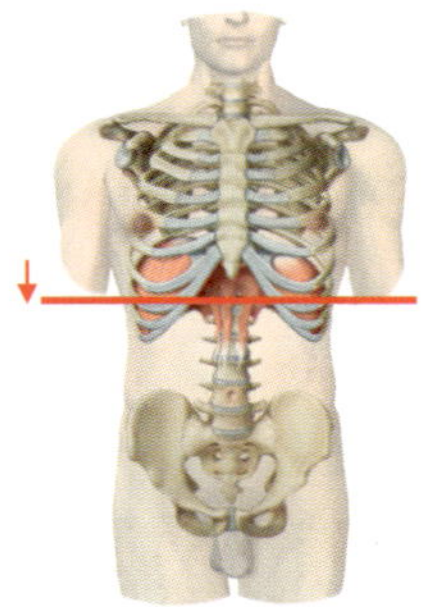

Three-dimensional image of a living human body, obtained by CT scan using a 3D volume rendering technique, showing the **abdominal organs** in a transverse section passing through the twelfth thoracic vertebra, as seen from above.

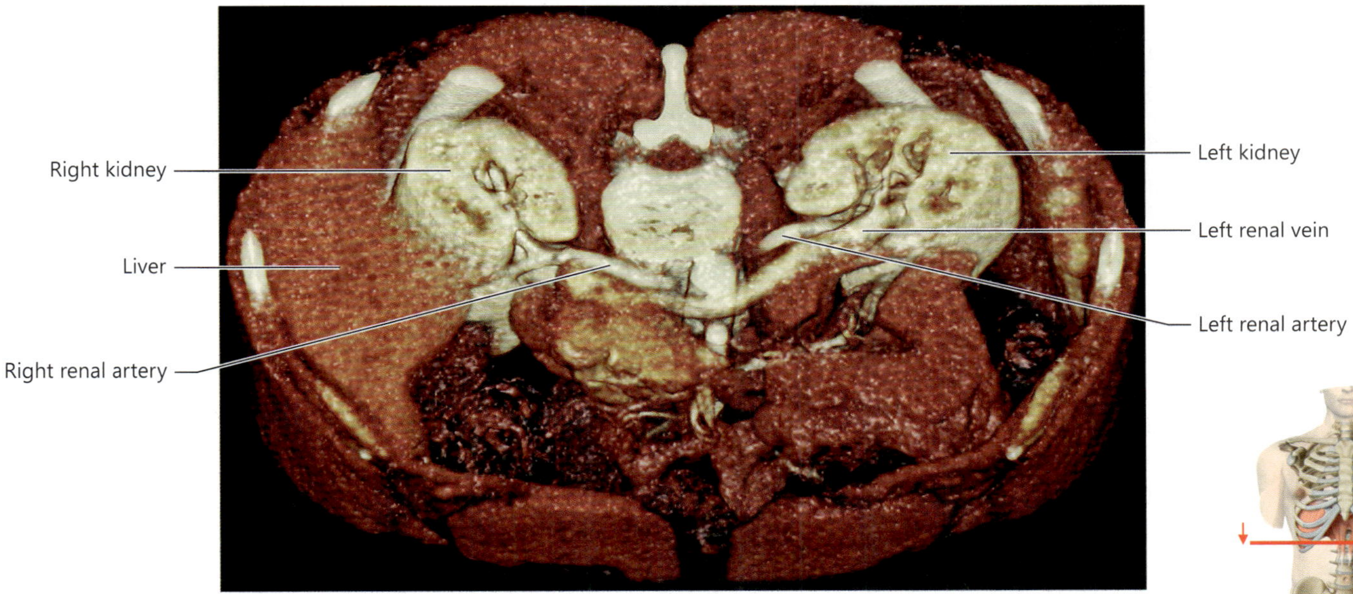

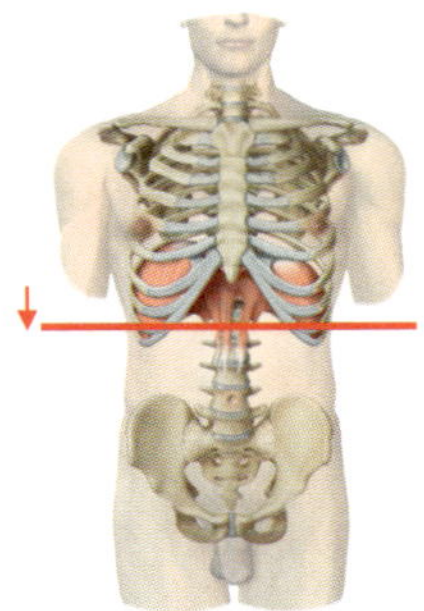

Three-dimensional image of a living human body, obtained by CT scan using a 3D volume rendering technique, showing the **abdominal organs** in a transverse section passing through the first lumbar vertebra, as seen from above.

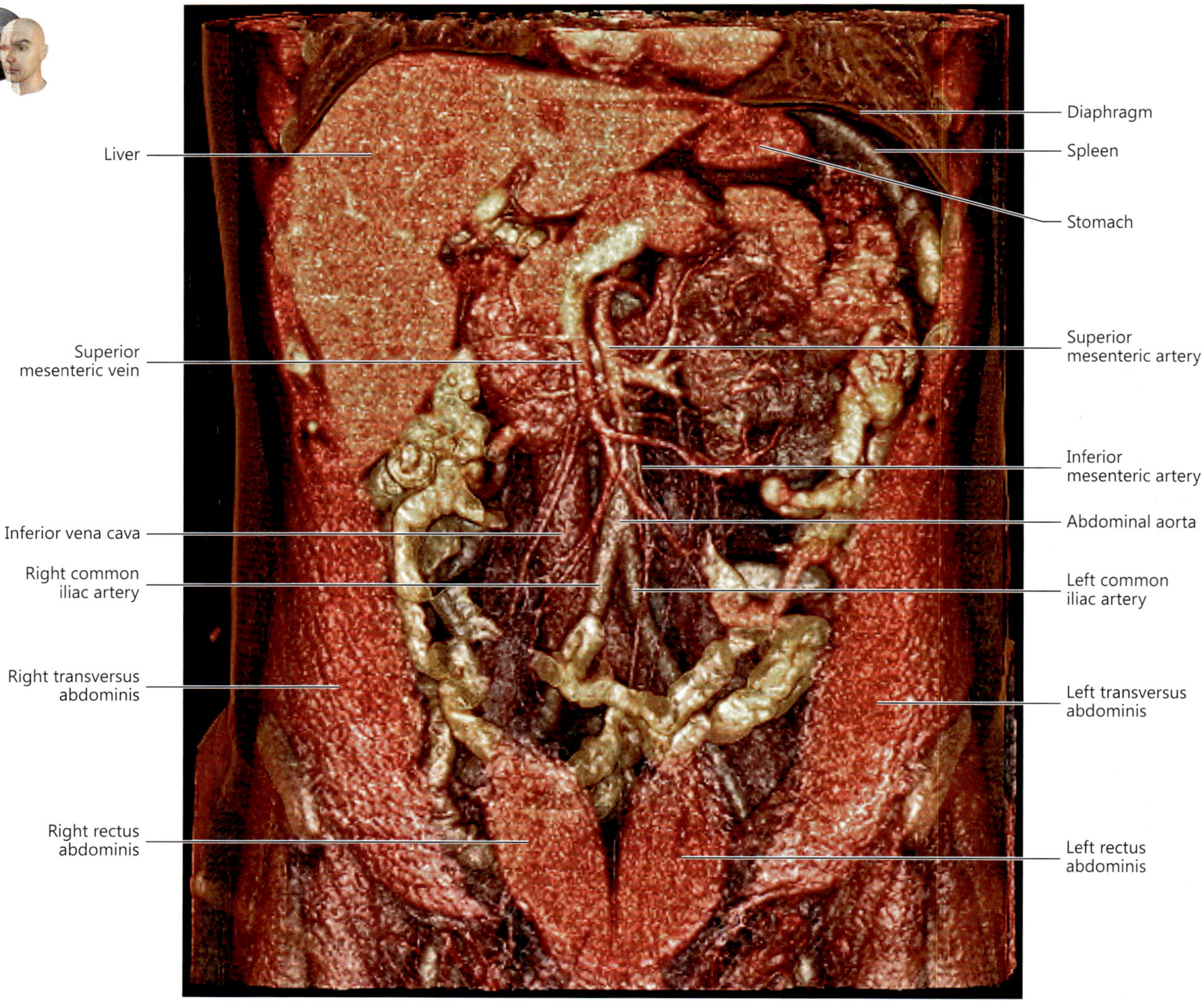

Three-dimensional image of a living human body, obtained by CT scan using a 3D volume rendering technique, showing the **mesenteric vessels** in frontal view.

Summary table of the abdominal arteries

ARTERY	ORIGIN	BRANCHES	TERRITORIES OF SUPPLY
abdominal aorta	Continuation of the thoracic aorta at the level of the aortic hiatus, usually at the level of the inferior border of the twelfth thoracic vertebra	*Visceral*: celiac trunk; superior mesenteric, inferior mesenteric, middle suprarenal, renal, testicular ♂, ovarian ♀ arteries *Parietal*: inferior phrenic, lumbar, median sacral arteries. It continues with the common iliac arteries at the level of the fourth lumbar vertebra	Organs contained in the abdominal cavity, abdominal walls, spinal cord
anterior cecal	Ileocolic		Anterior part of the cecum
anterior superior pancreaticoduodenal	Gastroduodenal	Pancreatic and duodenal	Anterior border of the duodenum and anterior surface of the pancreas
anterior, posterior, medial and lateral segmental arteries of liver	Right (anterior and posterior segmental arteries) and left (medial and lateral segmental arteries) branches of the hepatic artery proper		Segments of the liver
appendicular	Ileocolic		Appendix
arcuate	Interlobar of the kidney	Cortical radiate arteries	Parenchyma of the kidney
ascending	Inferior mesenteric		Left half of the transverse colon
celiac trunk (celiac artery)	Abdominal aorta, right below the aortic hiatus (at the level of the twelfth thoracic vertebra)	Left gastric, common hepatic, splenic arteries	Esophagus, stomach, duodenum, spleen, pancreas, liver, gallbladder, greater omentum
common hepatic	Celiac trunk	Gastroduodenal artery, hepatic artery proper, right gastric artery	Stomach, greater omentum, pancreas, duodenum, liver, gallbladder
common iliac (at the level of the fourth and fifth lumbar vertebrae)	Branches of divisions of the abdominal aorta	It terminates by branching into the external and internal iliac arteries	Pelvis and lower limb
cortical radiate (interlobular arteries)	Arcuate	Afferent and efferent glomerular arterioles	Renal glomerulus
cystic	Right branch of the hepatic artery proper	Superficial and deep	Gallbladder
dorsal pancreatic	Pancreatic branch of the splenic artery		Neck and body of the pancreas
gastroduodenal	Common hepatic	Supraduodenal (inconstant), posterior superior pancreaticoduodenal, retroduodenal, right gastro-omental, anterior superior pancreaticoduodenal arteries	Stomach, duodenum, pancreas, greater omentum
greater pancreatic	Pancreatic branch of the splenic artery		Tail of the pancreas
hepatic artery proper	Common hepatic	Right (or right hepatic artery), left (or left hepatic artery) and intermediate; right gastric artery (at times)	Liver, gallbladder, pyloric part of the stomach
ileocolic	Superior mesenteric	Anterior and posterior cecal arteries and appendicular artery; superior (anastomotic with the right colic artery) and inferior (anastomotic with the termination of the superior mesenteric artery), ileal and colic	Cecum, appendix, ascending colon, distal part of the ileum
inferior epigastric (in the tract running up to the umbilicus)	External iliac	Pubic, muscular and cutaneous; cremasteric artery ♂ or artery of round ligament of uterus ♀; anastomotic with the superior epigastric and subcostal arteries	Cremaster, spermatic cord, round ligament of uterus; peritoneum, muscles of abdomen, skin of the anterior abdominal wall
inferior mesenteric	Abdominal aorta	Ascending, left colic, sigmoid, superior rectal arteries	Transverse, descending and sigmoid colon and superior part of the rectum
inferior pancreatic	Pancreatic branch of the splenic artery or, at times, of the dorsal pancreatic artery		Body and tail of the pancreas, greater omentum

continued

Summary table of the abdominal arteries *(continued)*

ARTERY	*ORIGIN*	*BRANCHES*	*TERRITORIES OF SUPPLY*
inferior pancreaticoduodenal (duodenal)	Superior mesenteric or, at times, first jejunal	Anterior and posterior	Head of the pancreas, descending and inferior parts of the duodenum
inferior phrenic	Abdominal aorta	Superior suprarenal arteries	Diaphragm
inferior suprarenal	Renal		Suprarenal gland
interlobar of kidney	Segmental arteries of the kidney	Arcuate arteries	Kidney lobes
interlobular of liver	Right and left branches of the hepatic artery proper	Plexus outside each lobule of liver	Walls of the interlobular veins and of the interlobular bile ducts
intrarenal (interlobar, arcuate, perforating radiate arteries, vasa recta and capsular branches)	Renal		Kidney
jejunal and ileal (intestinal) (12-15 in number)	Superior mesenteric	Superior and inferior	Jejunum, ileum
left colic	Inferior mesenteric	Descending and ascending	Left half of the transverse colon and descending colon
left gastric	Celiac trunk	Esophageal	Lesser curvature of the stomach, abdominal part of the esophagus, hepatogastric ligament
left gastro-omental (left gastroepiploic)	Splenic	Gastric and omental	Stomach, greater omentum
lumbar (4 in number on each side)	Abdominal aorta	Dorsal and spinal; anastomotic with the more caudal posterior intercostal arteries, subcostal, iliolumbar, deep circumflex iliac and inferior epigastric arteries	Vertebrae, spinal cord and meninges, muscles and skin of the back
marginal (of Drummond)	Superior mesenteric		Intestinal wall
median sacral (at the level of the fourth and fifth lumbar vertebrae)	Abdominal aorta	Lateral sacral; arteriae lumbales imae	Sacrum, iliacus, rectum
middle colic	Superior mesenteric	Right and left	Transverse colon
middle suprarenal (middle capsular)	Abdominal aorta	Anastomotic with the superior suprarenal arteries, branches of the inferior phrenic artery, and with the inferior suprarenal artery, a branch of the renal artery	Suprarenal gland
of caudate lobe	Right and left branches of the hepatic artery proper		Caudate lobe of liver
ovarian ♀ (running up to the pelvic inlet)	Abdominal aorta	Ureteric ♀ and tubal ♀	Ovary, ureter, uterine tube, skin of the labium majus, inguinal region
posterior cecal	Ileocolic		Posterior part of the cecum
posterior gastric	Splenic		Posterior wall of the superior part of the body of the stomach, near the cardia and the fundus of the stomach
posterior superior pancreaticoduodenal	Gastroduodenal	Pancreatic and duodenal	Posterior border of the duodenum and posterior surface of the pancreas
prepancreatic	Pancreatic branch of the splenic artery		Neck and uncinate process of the head of the pancreas
renal	Abdominal aorta	Anterior, posterior, capsular and ureteric; inferior suprarenal and intrarenal arteries	Kidney, ureter, suprarenal gland

continued

Summary table of the abdominal arteries *(continued)*

ARTERY	ORIGIN	BRANCHES	TERRITORIES OF SUPPLY
retroduodenal	Gastroduodenal		Bile duct, superior and descending parts of the duodenum, head of the pancreas
right colic	Superior mesenteric or, at times, ileocolic	Descending and ascending	Ascending colon
right flexural	Superior mesenteric		Right colic flexure
right gastric	Common hepatic or, at times, hepatic artery proper		Pyloric part of the stomach
right gastro-omental (right gastroepiploic)	Gastroduodenal	Gastric and omental	Stomach, greater omentum
short gastric	Splenic		Fundus of the stomach
sigmoid (2 or 3 in number)	Inferior mesenteric	Anastomotic with the left colic artery and the superior rectal artery	Inferior part of the descending colon and sigmoid colon
splenic	Celiac trunk	Pancreatic and splenic; short gastric, left gastro-omental, posterior gastric arteries	Spleen, pancreas, stomach, greater omentum
subcostal (running between the transversus abdominis and internal oblique)	Thoracic aorta	Dorsal and spinal; anastomotic with the cutaneous branches of the inferior epigastric artery	Muscles of the thoracic wall and the abdominal wall; vertebrae, spinal cord and meninges, muscles and skin of the back
superficial epigastric (in the tract running up to the umbilicus)	Femoral	Anastomotic with the cutaneous branches of the inferior epigastric artery	Superficial inguinal nodes, abdominal integument
superior epigastric (running in the rectus sheath)	Internal thoracic	To the falciform ligament, anastomosing with the hepatic artery proper; anastomotic with the inferior epigastric artery	Skin, muscles and fasciae of the superior part of the abdominal wall; diaphragm, peritoneum, falciform ligament of the liver
superior mesenteric	Abdominal aorta	Inferior pancreaticoduodenal, jejunal, ileal, ileocolic, right colic, right flexural, middle colic, marginal arteries	Small intestine, proximal half of the colon
superior suprarenal (2 or 3 in number)	Inferior phrenic		Anterior surface of the suprarenal gland
superior, anterior superior, anterior inferior, inferior and posterior segmental arteries of the kidney	Anterior (superior, anterior superior, anterior inferior, inferior segmental arteries) and posterior branches (posterior segmental) of the renal artery		Parenchyma of the kidney
testicular ♂ (running up to the pelvic inlet)	Abdominal aorta	Ureteric ♂ and epididymal ♂	Testis, epididymis, perinephric fat, ureter, external iliac nodes, cremaster
to tail of pancreas	Pancreatic branch of the splenic artery		Tail of the pancreas

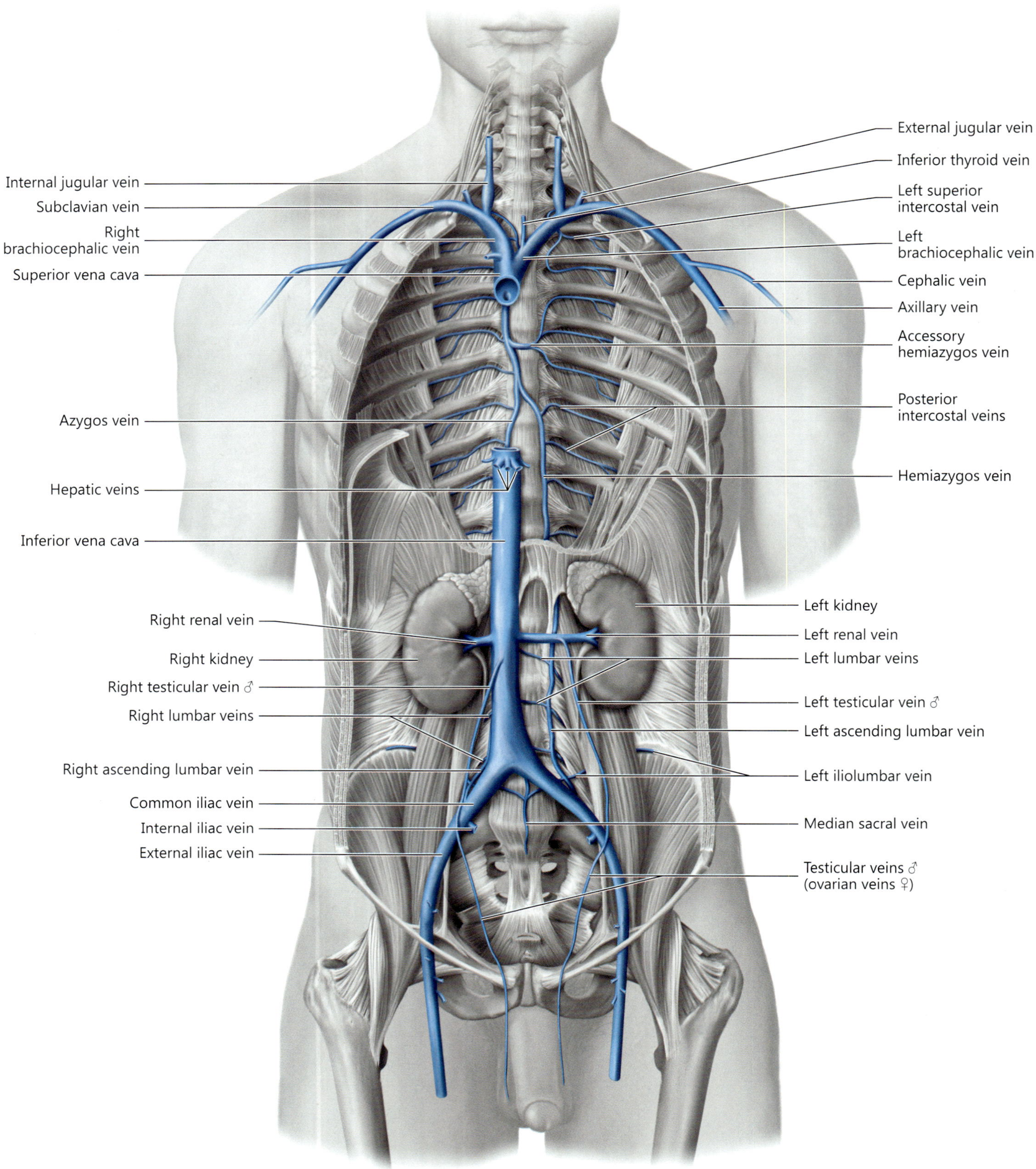

Schematic representation of the **superior** and **inferior venae cavae**, with their corresponding **collector veins**. The superior and inferior venae cavae provide visceral and parietal venous drainage for the supra- and subdiaphragmatic regions of the trunk, respectively. The azygos venous system (azygos, hemiazygos and accessory hemiazygos veins) constitutes a major anastomotic circle between the superior and the inferior vena cava.

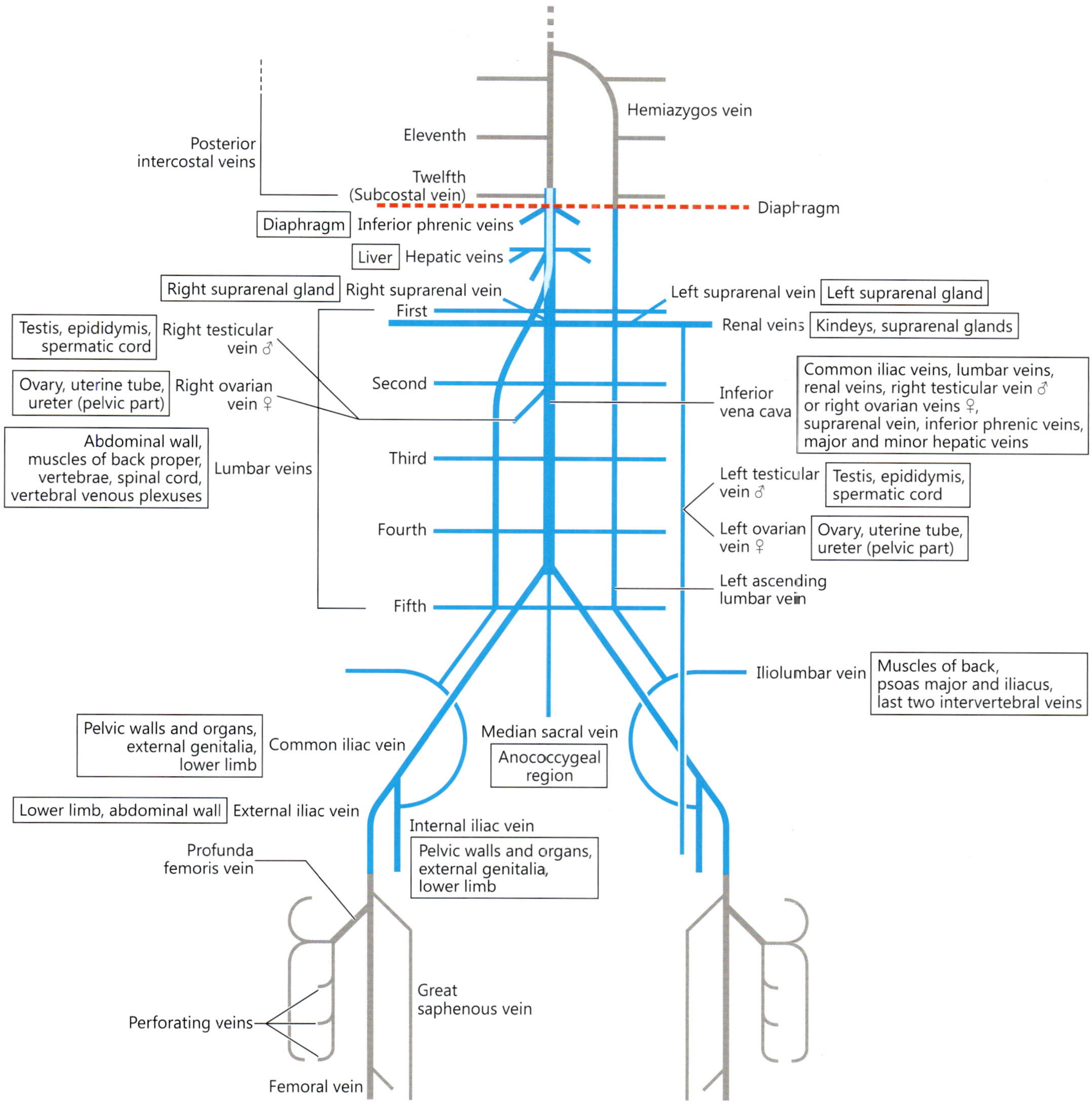

General diagram of the **inferior vena cava system**. The common iliac veins are its collectors of origin. Parietal tributaries of the inferior vena cava consist of the lumbar and inferior phrenic veins; visceral tributaries consist of the renal, right suprarenal, right testicular ♂, or right ovarian ♀, major and minor hepatic veins. The left suprarenal, left testicular ♂, or left ovarian ♀, veins open into the left renal vein.

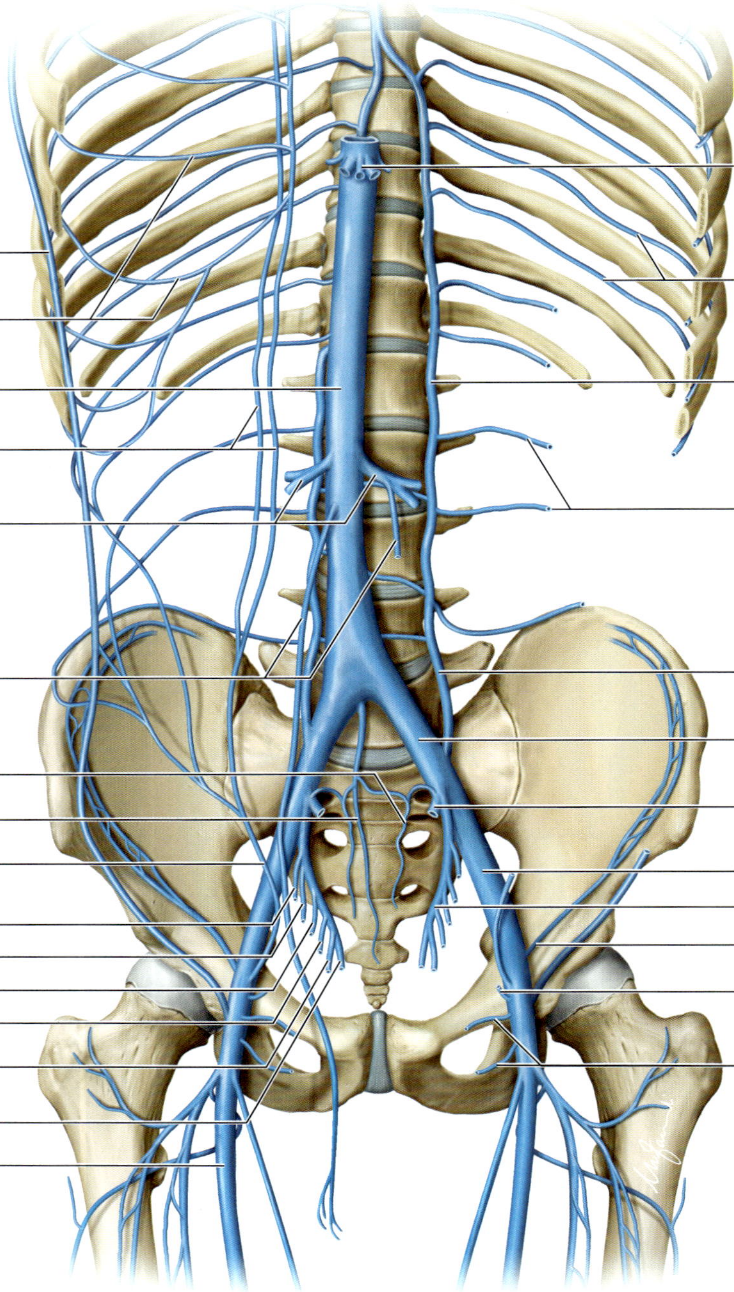

Schematic representation of the **superior** and **inferior venae cavae** and their major **tributaries**, and of the organization of the **deep parietal venous drainage of the trunk**. Anterior view.

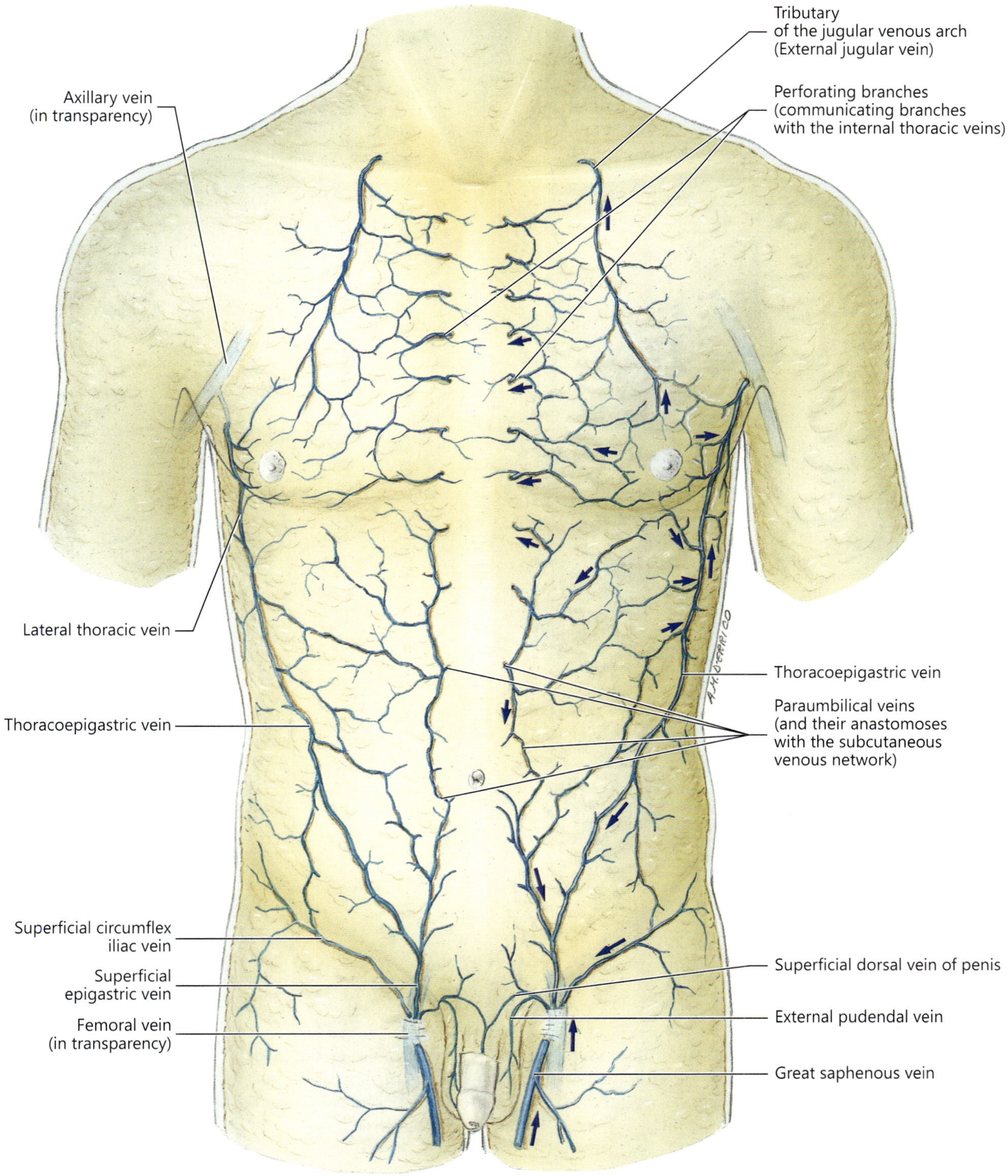

Cutaneous venous network of the anterolateral wall of the trunk. The arrows on the right half of the figure indicate the direction of the venous blood flow under normal conditions.

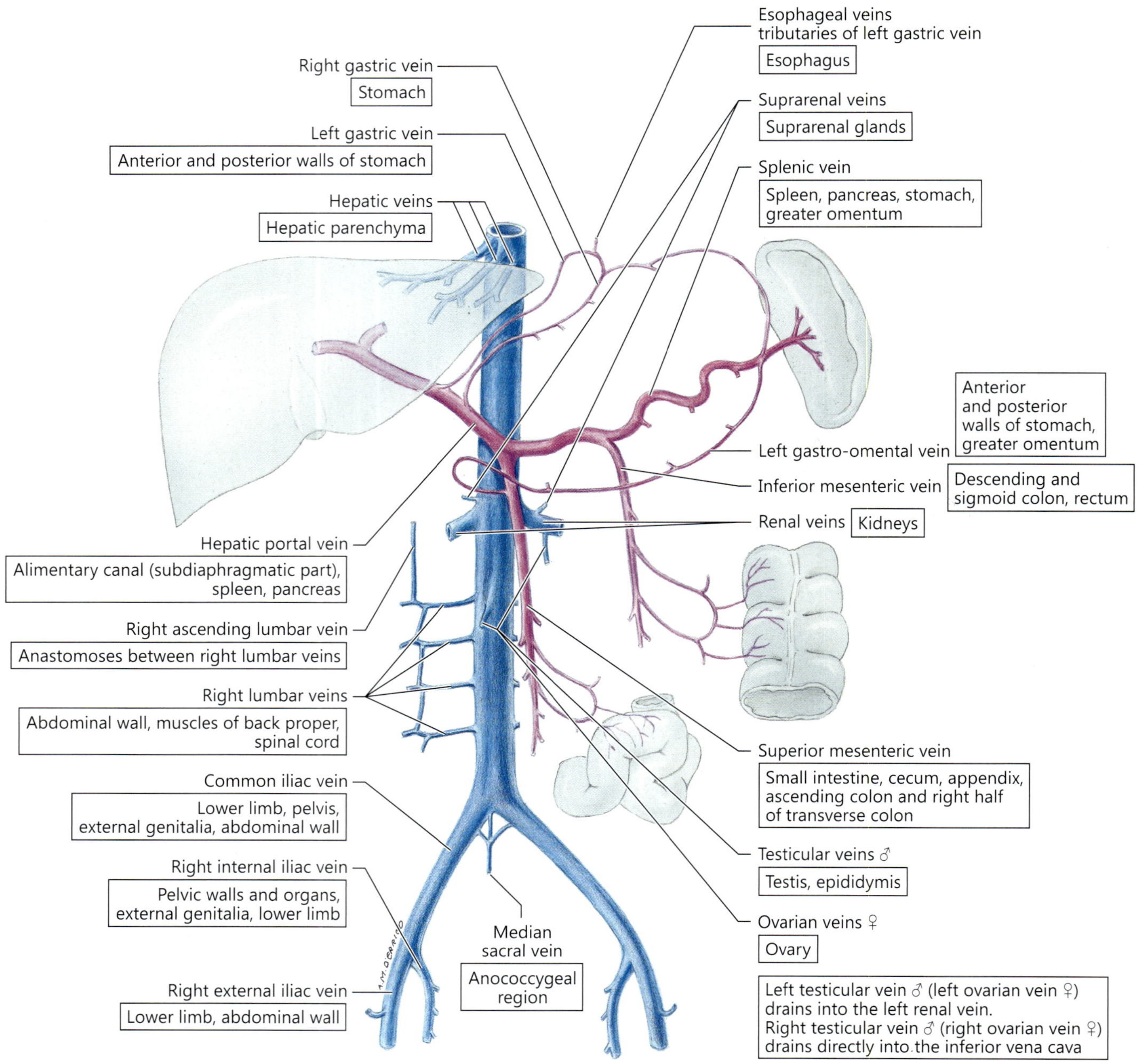

Schematic representation of the **hepatic portal vein** system and of the major **tributaries of the inferior vena cava**, with their corresponding drainage regions. The hepatic portal vein drains all the organs of the alimentary system within the peritoneal cavity as well as the spleen. After flowing through the hepatic circle, the portal vein blood drains into the hepatic veins and then into the inferior vena cava.

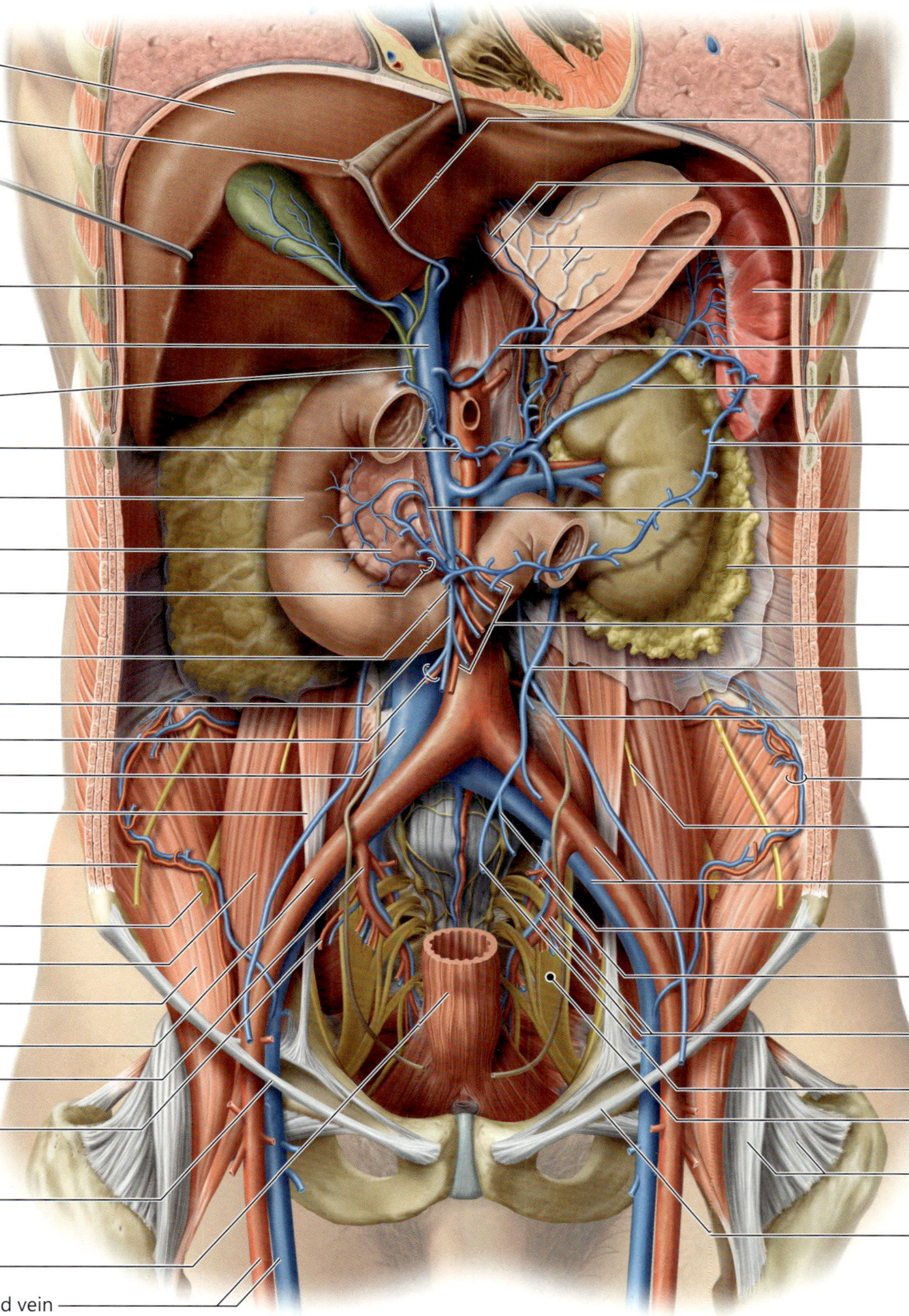

The **trunk of the hepatic portal vein** originates from the confluence of the superior mesenteric vein and a trunk formed by the convergence of the inferior mesenteric and splenic veins at the level of L I (transpyloric plane). It ascends obliquely up to the porta hepatis, where it divides in a T-shape into its right and left terminal branches. Tributary branches of the hepatic portal vein consist of the left and right gastric veins, the cystic vein and the umbilical vein (prenatally), which obliterates gradually after birth to become the round ligament of the liver. The left renal space was opened, transecting the prerenal layer and the perinephric fat. The figure displays the relations of the splenic vein with the kidney, interposing the removed layers (☞ page 98).

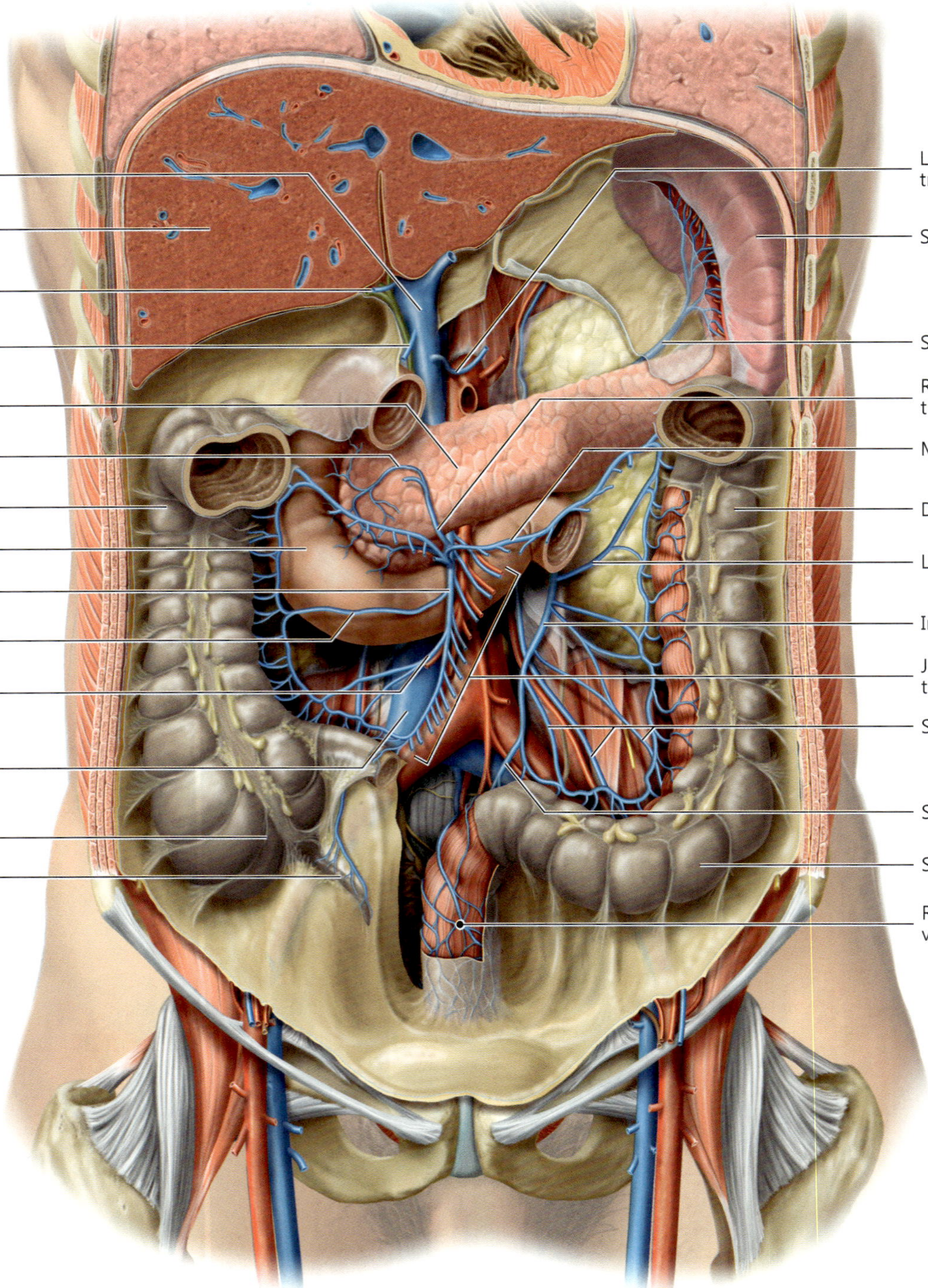

Anterior view of the **abdominal cavity**. The peritoneum and liver were transected, while the stomach, small intestine, mesentery, transverse colon and transverse mesocolon were removed. The figure displays the **trunks of origin of the hepatic portal vein**, formed by the confluence of the superior mesenteric and the splenic veins, into which the inferior mesenteric vein opens. Tributaries of the superior mesenteric vein include the jejunal and ileal veins, the ileocolic, right colic and middle colic veins, the pancreaticoduodenal and right gastro-omental veins. Tributaries of the inferior mesenteric vein include the left colic, sigmoid and superior rectal veins. The figure also shows the **tributaries of the hepatic portal vein**, notably the right and left gastric veins and the cystic vein.

Porto-caval Anastomoses

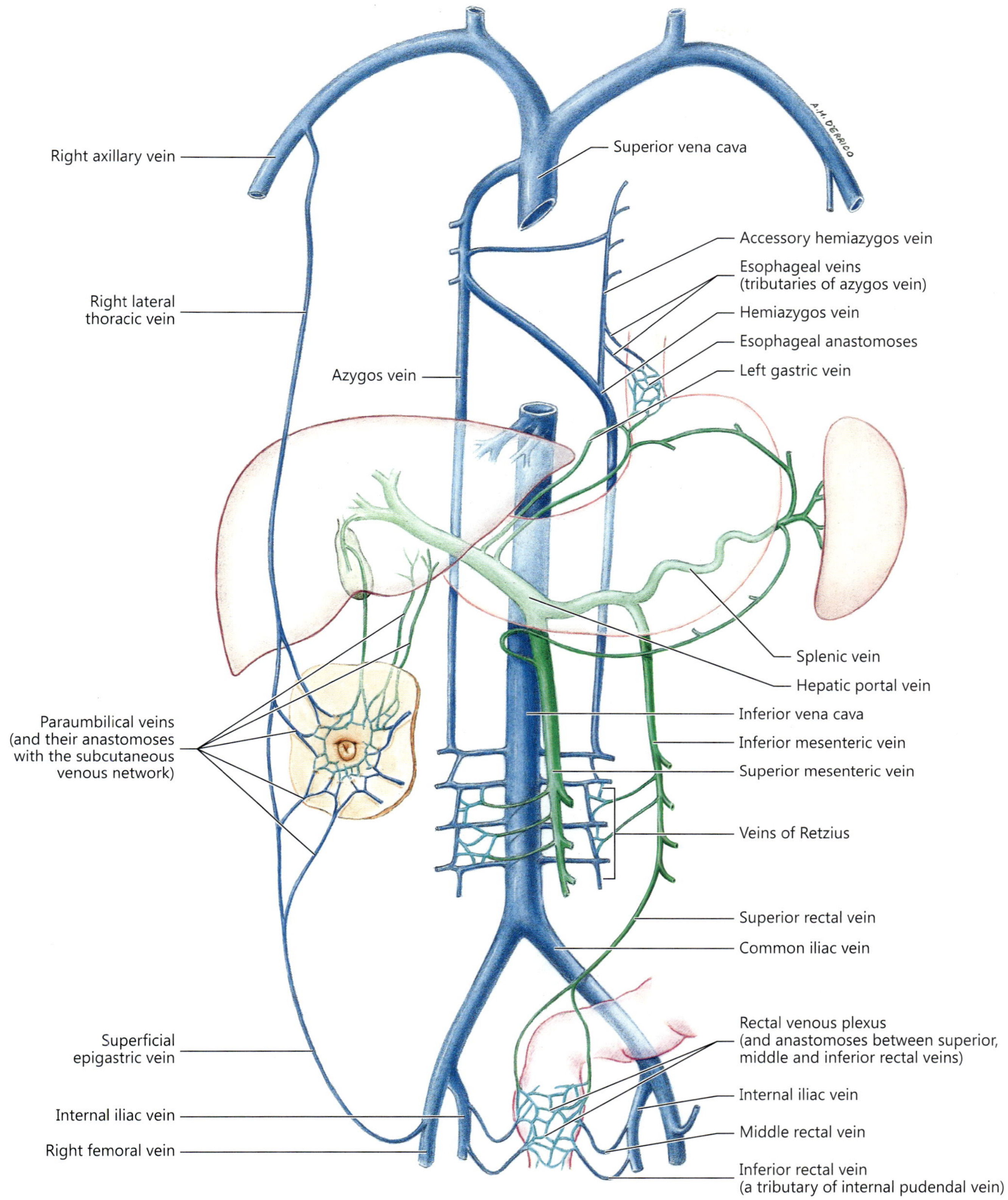

Schematic representation of the anastomoses between the superior and inferior vena cava systems, and between the hepatic portal system and the cava venous system. The **hepatic portal system** is indicated in *green*; the **cava venous system** is indicated in *blue*.

Summary table of the abdominal veins

VEIN	ORIGIN/LOCATION	DRAINAGE	CONFLUENCE
anterior and posterior spinal (in the lumbar segment of the vertebral column)	Longitudinal venous plexuses along the anterior and posterior surfaces of the spinal cord	Spinal cord	Anterior internal vertebral venous plexus
appendicular	Along the mesoappendix	Appendix	Ileocolic vein
arcuate	Base of renal pyramids	Cortical radiate veins and venulae rectae	Interlobar veins
ascending lumbar	Longitudinal anastomoses connecting the lumbar veins on each side, and which, overall, form a vertical trunk, anterior of the transverse processes of the lumbar vertebrae	Communicates with the iliolumbar vein; a major anastomosis between the inferior vena cava and superior vena cava (cava-caval anastomosis) through the lumbar veins	*Right side*: azygos vein *Left side*: hemiazygos vein
azygos (at its point of origin below the diaphragm)	Anterior surface of L I; continuation of the right ascending lumbar vein; enters the thoracic cavity running between the right and left crura of the diaphragm or through the aortic hiatus	Right ascending lumbar, right subcostal, posterior intercostal, hemiazygos, esophageal, mediastinal, pericardial and right bronchial veins	Superior vena cava
basivertebral (in the lumbar segment of the vertebral column)	Spongy bone of the vertebral bodies; emerge from the foramina on the posterior surfaces of the vertebral bodies	Vertebral bodies; communicating with the anterior external vertebral venous plexus	Anterior internal vertebral venous plexus
capsular	Fibrous capsule of the kidney	Fibrous capsule of the kidney	Renal veins
central	Center of the hepatic lobules where all the sinusoids pertaining to the hepatic lobule open into	Hepatic lobule	Sublobular veins
common iliac	By confluence of the external and internal iliac veins at the sacroiliac joint, from where it ascends up to L V	Lower limb, pelvis, external genitalia and abdominal wall; receives the median sacral, lateral sacral, iliolumbar and superior gluteal veins, which otherwise drain into the internal iliac vein	The two common iliac veins, right and left, come together at an acute angle at the L V level, forming the inferior vena cava, to the right of the median line
cortical radiate (interlobular veins)	From the center of the stellate veins	Renal cortex	Arcuate veins
cystic	Wall of the gallbladder; travels along with the cystic duct	Gallbladder	Right branch of the hepatic portal vein
esophageal (some of the esophageal veins that are tributaries of the azygos vein)	Abdominal part of the esophagus	Esophagus	Left gastric vein or azygos vein
hemiazygos (at its point of origin)	Left side of T XII; continuation of the left ascending lumbar vein; enters the thoracic cavity running between the right and left crura of the diaphragm or through the aortic hiatus	Left ascending lumbar vein, four of five caudal posterior intercostal veins, left subcostal, esophageal and mediastinal veins	Azygos vein
hepatic portal (approximately 8 cm long)	Posterior to the head of the pancreas, at the L I-L II level, by confluence of the superior mesenteric vein and a trunk formed by the convergence of the inferior mesenteric and splenic veins	Subdiaphragmatic part of the alimentary canal, spleen, pancreas and gallbladder; receives the cystic, umbilical (in the fetus), paraumbilical, posterior superior pancreaticoduodenal, left gastric and right gastric veins	Reaches the right edge of the porta hepatis, and divides in a T-shape into the right and left terminal branches
ileocolic	Satellite of the ileocolic artery	Terminal ileum, appendix, cecum, inferior part of the ascending colon; receives the appendicular vein	Superior mesenteric vein
iliolumbar	Iliacus and lumbar branches; runs posterior to the psoas major	Psoas major and iliacus; receives the last two intervertebral veins at the lumbar level and connects to the ascending lumbar vein	Common iliac vein or internal iliac vein

continued

Summary table of the abdominal veins *(continued)*

VEIN	ORIGIN/LOCATION	DRAINAGE	CONFLUENCE
inferior epigastric (deep epigastric)	Satellite of the inferior epigastric artery; descends laterally along the posterior surface of the rectus abdominis	Rectus abdominis and external oblique, subperitoneal tissue and skin at the level of the umbilicus; anastomoses with the superior epigastric and subcutaneous abdominal veins (cava-caval anastomosis) and with the paraumbilical veins (porto-caval anastomosis)	External iliac vein, approximately one centimeter above the inguinal ligament
inferior mesenteric	From the rectal venous plexus, on the rectum walls, as superior rectal vein; ascends posterior to the parietal peritoneum, up to the pancreas	Descending and sigmoid colon, rectum; receives the left colic, sigmoid and superior rectal veins	Splenic vein
inferior phrenic	Inferior surface of the diaphragm	Diaphragm	*Right side*: inferior vena cava *Left side*: left suprarenal vein (often a second vein opens into the left side of the inferior vena cava or into the left renal vein)
inferior vena cava (the largest vein in the body)	At the level of L V, from the convergence of the two common iliac veins; ascends vertically, right of the vertebral column; enters the thorax through the caval opening in the central tendon of the diaphragm	Lumbar, renal, inferior phrenic, major and minor hepatic, right suprarenal, right ovarian ♀, right testicular ♀ veins	Inferior and posterior part of the right atrium of the heart
interlobar of kidney	By confluence of the arcuate veins, between the renal pyramids	Kidney lobes	Renal veins
interlobular of liver	In the portal areas	Terminal branches of the hepatic portal vein	In the hepatic sinusoids
intervertebral (in the lumbar segment of the vertebral column)	Anterior and posterior internal vertebral venous plexuses; passes through the intervertebral foramen along with the roots and the first tract of the spinal nerve	Spinal cord; partly receives the anterior and posterior external vertebral venous plexuses as well	Lumbar veins
intrarenal (interlobar, arcuate, cortical radiate, stellate, venulae rectae)	Parenchyma of the kidney	Kidney	Renal veins
jejunal and ileal (intestinal) (usually 12-20 in number)	Jejunum and ileum; they run parallel to the superior mesenteric artery, between the layers of the mesentery	Walls of the jejunum and the ileum	Superior mesenteric vein
left colic	Satellite of the left colic artery	Left colic flexure and descending colon	Inferior mesenteric vein
left gastric (coronary vein)	Runs along the lesser curvature of the stomach, from right to left, up to the cardia, then turns backwards, to the right and downwards	Anterior and posterior walls of the stomach; receives a few esophageal veins at the level of the cardia	Hepatic portal vein
left gastro-omental (left gastroepiploic)	Runs along the greater curvature of the stomach, from right to left	Anterior and posterior walls of the stomach, greater omentum	Splenic vein
left ovarian ♀ (gonadal)	☞ right ovarian ♀	☞ right ovarian ♀	Left renal vein
left suprarenal	Hilum of the left suprarenal gland	Left suprarenal gland; receives one of the left inferior phrenic veins	Left renal vein
left testicular ♂ (spermatic or gonadal)	☞ right testicular ♂	☞ right testicular ♂	Left renal vein

continued

Summary table of the abdominal veins			*(continued)*
VEIN	***ORIGIN/LOCATION***	***DRAINAGE***	***CONFLUENCE***
lumbar (usually 4 or 5 in number on each side)	They arise at the intervertebral foramina of the lumbar column, by the convergence of different roots	External oblique, internal oblique, transversus abdominis and quadratus lumborum; muscles of back proper, vertebrae, spinal cord; receive the intervertebral veins at the level of the lumbar segment of the vertebral column	Posterior wall of the inferior vena cava
major hepatic (right, middle and left) **and minor hepatic**	Confluence of the interlobular veins	Hepatic parenchyma	*Superior group*: right, intermediate and left hepatic veins draining into the inferior vena cava, below the diaphragm (in 90% of cases, the intermediate and left hepatic veins form a common trunk) *Inferior group*: a few small veins draining into the inferior vena cava, and which emerge from the lower parts of the groove for the vena cava
middle colic	Satellite of the middle colic artery	Transverse colon	Superior mesenteric vein
pancreatic	Pancreas	Neck, body and tail of the pancreas	Splenic and superior mesenteric veins
pancreaticoduodenal	Head of the pancreas and proximal part of the duodenum	Pancreas and duodenum	Superior mesenteric vein
paraumbilical (accessory to the hepatic portal vein or of Sappey)	Four or five roots in the umbilical region (which can appear dilated in the case of portal hypertension, with a distinctive caput medusae appearance); run along the round ligament of liver	Skin around the umbilicus; anastomose with the thoracoepigastric veins, superficial epigastric vein and inferior epigastric vein (porto-caval anastomosis)	Hepatic portal vein, at the inferior border of the liver and the umbilical fissure
posterior superior pancreaticoduodenal	Near the duodenum on the posterior surface of the head of the pancreas	Posterior part of the upper half of the head of the pancreas	Hepatic portal vein
prepyloric (of Mayo)	Anterior to the pylorus	Pylorus	Right gastric vein
renal	Hilum of kidney	Kidney; receive the capsular, left suprarenal, left ovarian ♀, left testicular ♂, intrarenal veins and often, on the left, one of the inferior phrenic veins	Inferior vena cava (the right renal vein opens into the inferior vena cava slightly lower than the left renal vein)
right colic	Satellite of the right colic artery	Ascending colon	Superior mesenteric vein
right gastric (pyloric)	Runs along the lesser curvature of the stomach, at the pylorus, from left to right	Stomach; receives the prepyloric vein; communicates with the left gastric vein	Hepatic portal vein
right gastro-omental (right gastroepiploic)	Runs along the greater curvature of the stomach, from left to right	Greater omentum and parts of the stomach	Superior mesenteric vein
right ovarian ♀ (gonadal)	Venous plexus, in the broad ligament of uterus ♀	Ovary	Inferior vena cava
right suprarenal	Hilum of the right suprarenal gland	Right suprarenal gland	Inferior vena cava
right testicular ♂ (spermatic or gonadal)	Roots from the posterior border of the testis and from the head of the epididymis; they anastomose with each other and with posterior venous branches of the spermatic cord, originating the pampiniform plexus, whose anterior veins, after passing the inguinal canal, continue as one or two veins, running along the sides of the testicular artery ♂	Testis, epididymis	Inferior vena cava

continued

Summary table of the abdominal veins *(continued)*

VEIN	ORIGIN/LOCATION	DRAINAGE	CONFLUENCE
short gastric (4 or 5 in number)	Greater curvature of the stomach	Fundus of the stomach and left part of the greater curvature	Splenic vein
sigmoid	Satellites of the sigmoid arteries	Inferior part of the descending colon and sigmoid colon	Inferior mesenteric vein
splenic	Splenic hilum, by confluence of 5-6 splenic branches	Pancreatic, short gastric, left gastro-omental, inferior mesenteric veins	Hepatic portal vein
stellate	Renal cortex, near the fibrous capsule of kidney	Superficial part of the renal cortex	Interlobular veins of liver
subcostal	Satellite of the subcostal artery along the inferior border of the twelfth rib	Anterior abdominal wall	Azygos vein
subcutaneous abdominal	Anterior abdominal wall	Skin of the anterior abdominal wall; anastomose with the inferior epigastric vein (cava-caval anastomosis)	Superior epigastric veins
sublobular	Convergence of the central veins	Hepatic lobules	Hepatic veins
superficial circumflex iliac	Satellite of the superficial circumflex iliac artery	Inferolateral part of the abdominal wall and proximal part of the thigh; anastomoses with the superficial epigastric and thoracoepigastric veins (cava-caval anastomosis)	Great saphenous vein or femoral vein
superficial epigastric	Runs downwards and laterally from the umbilicus, towards the inguinal ligament	Inferior half of the abdominal wall; anastomoses with the thoracoepigastric (cava-caval anastomosis), superficial circumflex iliac and paraumbilical veins (porto-caval anastomosis)	Great saphenous vein
superior epigastric	Satellites of the superior epigastric artery; ascend posterior of the rectus abdominis towards the diaphragm	Anterosuperior part of the abdominal wall and part of the diaphragm; anastomoses with the inferior epigastric vein (cava-caval anastomosis)	Internal thoracic veins
superior mesenteric	In the right iliac fossa, from the roots that drain the terminal ileum, the cecum and the appendix; ascends between the two layers of the mesentery up to the pancreas	Small intestine, cecum, appendix, ascending colon and right half of the transverse colon; receives the jejunal, ileal, right gastro-omental, pancreatic, pancreaticoduodenal, ileocolic, right colic and middle colic veins	Hepatic portal vein
superior rectal	From the rectal venous plexus, located in the posterior and lateral walls of the rectum; after joining the left common iliac vessels, it continues as the inferior mesenteric vein	Rectum; anastomoses with the middle and inferior rectal veins (porto-caval anastomosis) through the rectal venous plexus	Inferior mesenteric vein
thoracoepigastric	Anterolateral wall of the trunk	Anastomose with the lateral thoracic, superficial epigastric and superficial circumflex iliac veins (cava-caval anastomosis) and with the paraumbilical veins (porto-caval anastomosis)	Axillary vein
umbilical (prenatally; it obliterates after birth to become the round ligament of liver)	Umbilical cord	Arterial blood from the placenta	Left branch of the hepatic portal vein and in the inferior vena cava, through the ductus venosus (Arantius' duct)

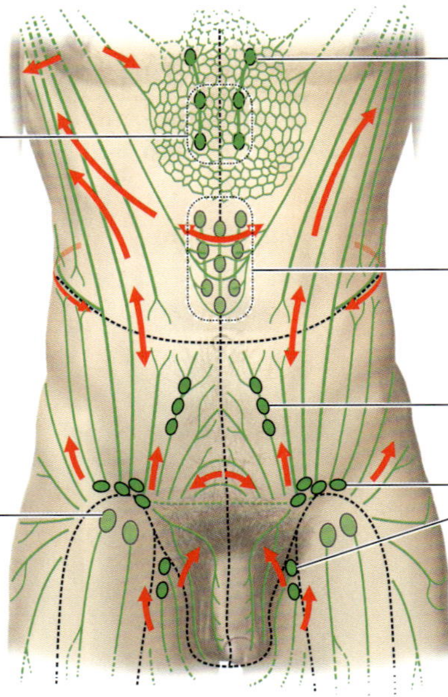

Projection of the main **lymphatic groups** and the flow pathways of lymphatic drainage on the body surface. Anterior view.

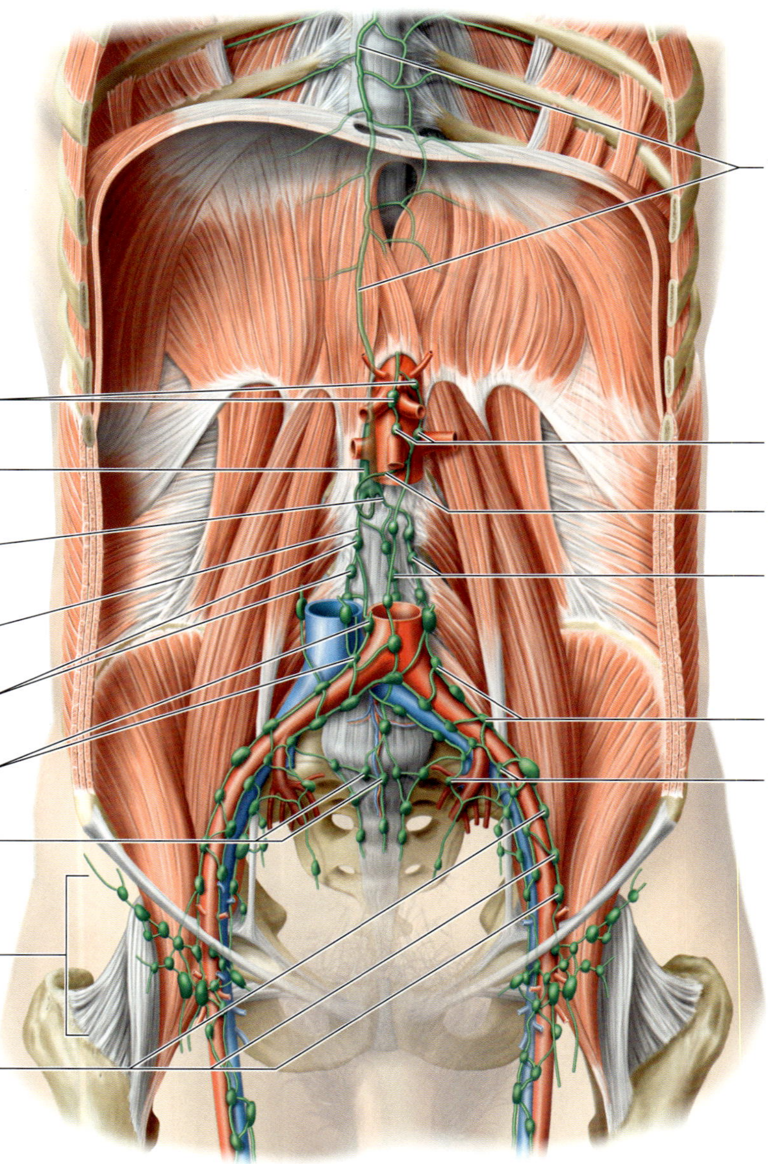

Lymph nodes of the extraperitoneal space in the abdomen and pelvis; anterior view after all organs have been removed. Note the lumbar and intestinal trunks opening into the cisterna chyli, where the thoracic duct originates. Most of the parietal drainage of the abdomen and pelvis depends on the retroperitoneal lymph nodes.

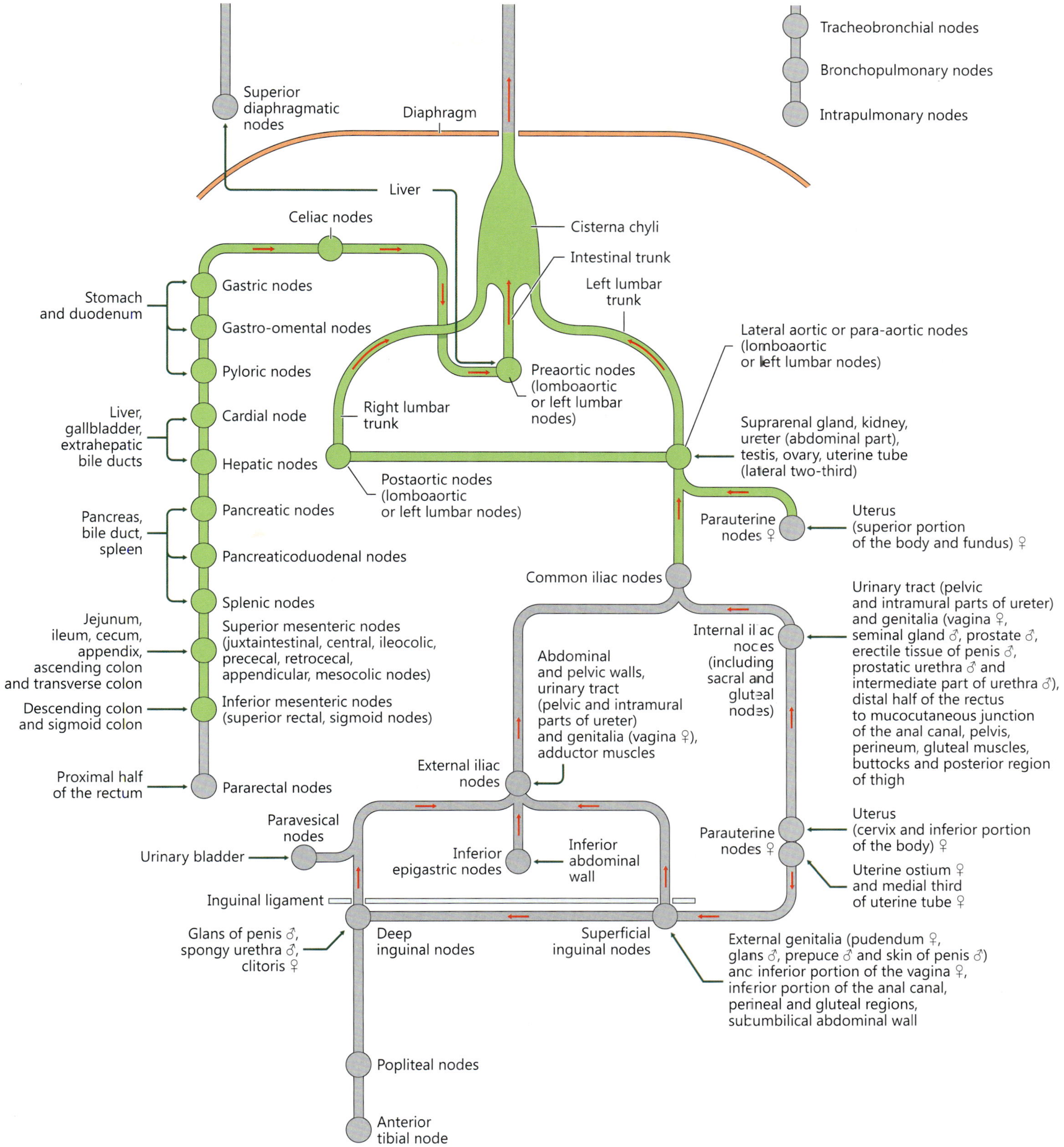

Diagram of the **visceral and parietal nodes of the abdomen**, with their corresponding territories of drainage and flow pathways.

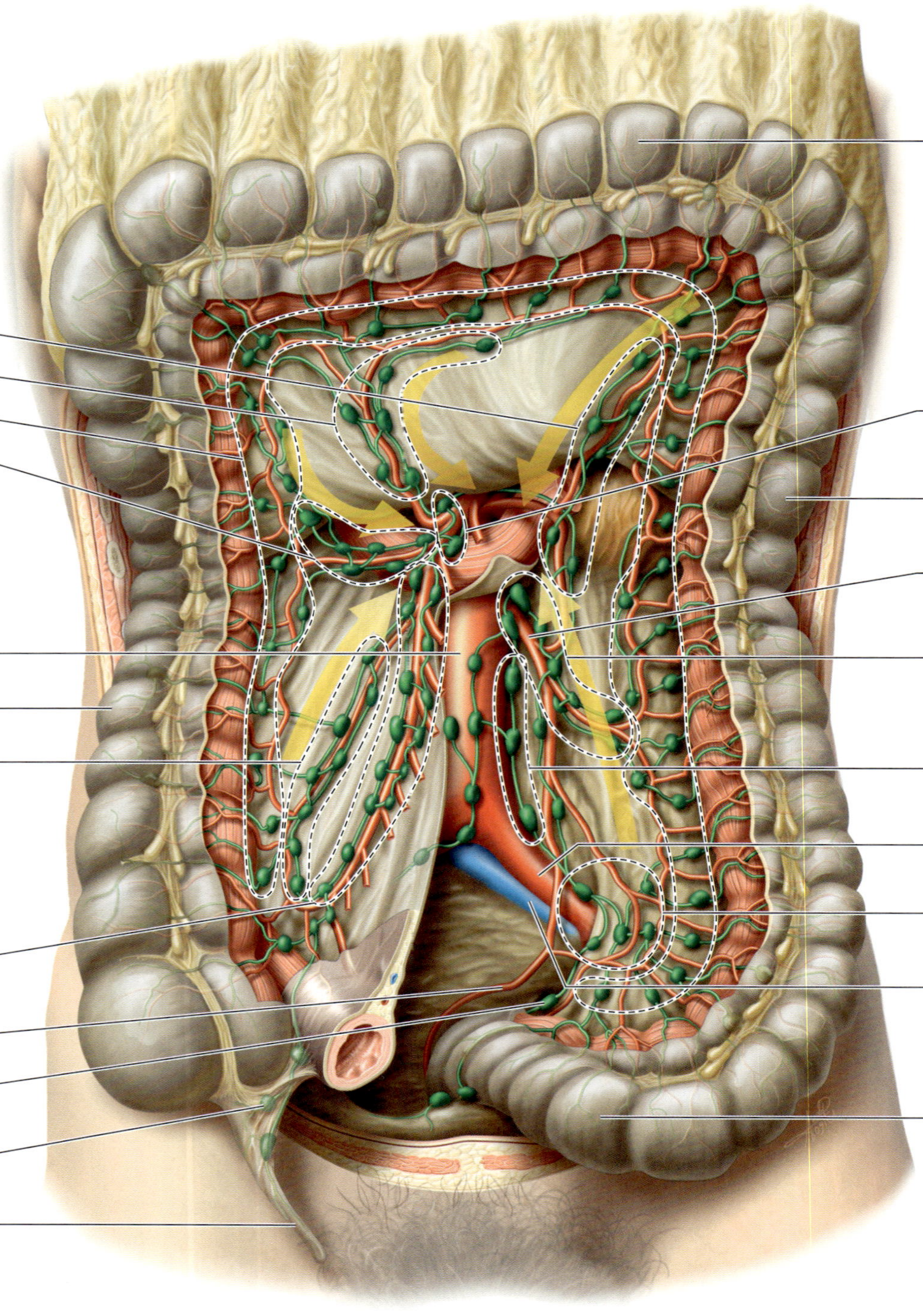

Main **superior and inferior mesenteric nodes**. The mesenteric nodes receive vessels from the abdominal part of the alimentary canal, distal to the duodenojejunal flexure, up to the superior half of the anal canal.

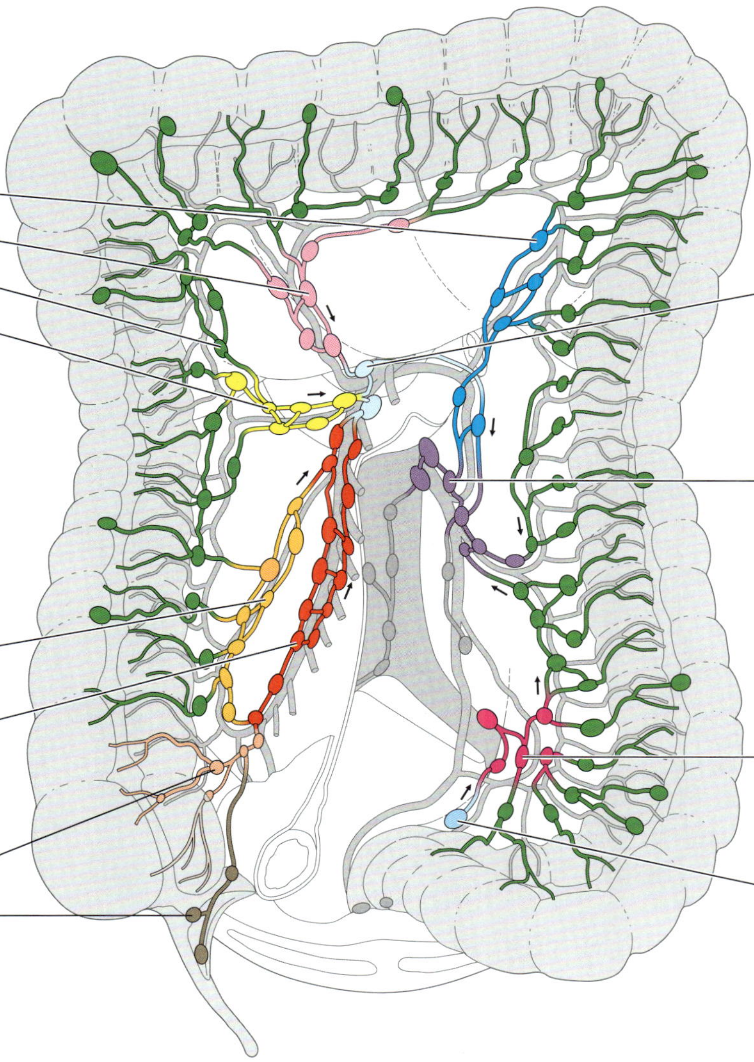

Lymphatic drainage of the colon. The lymph from the ascending and transverse colon is drained by the superior mesenteric nodes, while the lymph from the descending and sigmoid colon drains into the inferior mesenteric nodes.

Territories of drainage of the superior and inferior mesenteric nodes		
	LYMPH NODES	***TERRITORIES OF DRAINAGE***
1	Central superior mesenteric nodes	Jejunum, ileum, cecum, appendix, ascending and transverse colon
2	Inferior mesenteric nodes	Descending and sigmoid colon, proximal half of the rectum
3	Sigmoid nodes	Sigmoid colon
4	Superior rectal nodes	Proximal half of the rectum
5	Appendicular nodes	Appendix
6	Prececal nodes	Cecum
7	Inferior ileocolic nodes	Cecum and appendix
8	Superior ileocolic nodes	Ascending colon and cecum
9	Right colic nodes	Ascending colon
10	Paracolic nodes	Cecum and colon
11	Middle colic nodes	Right colic flexure and proximal two-third of the transverse colon
12	Left colic nodes	Descending colon
13	Mesocolic nodes	Large intestine

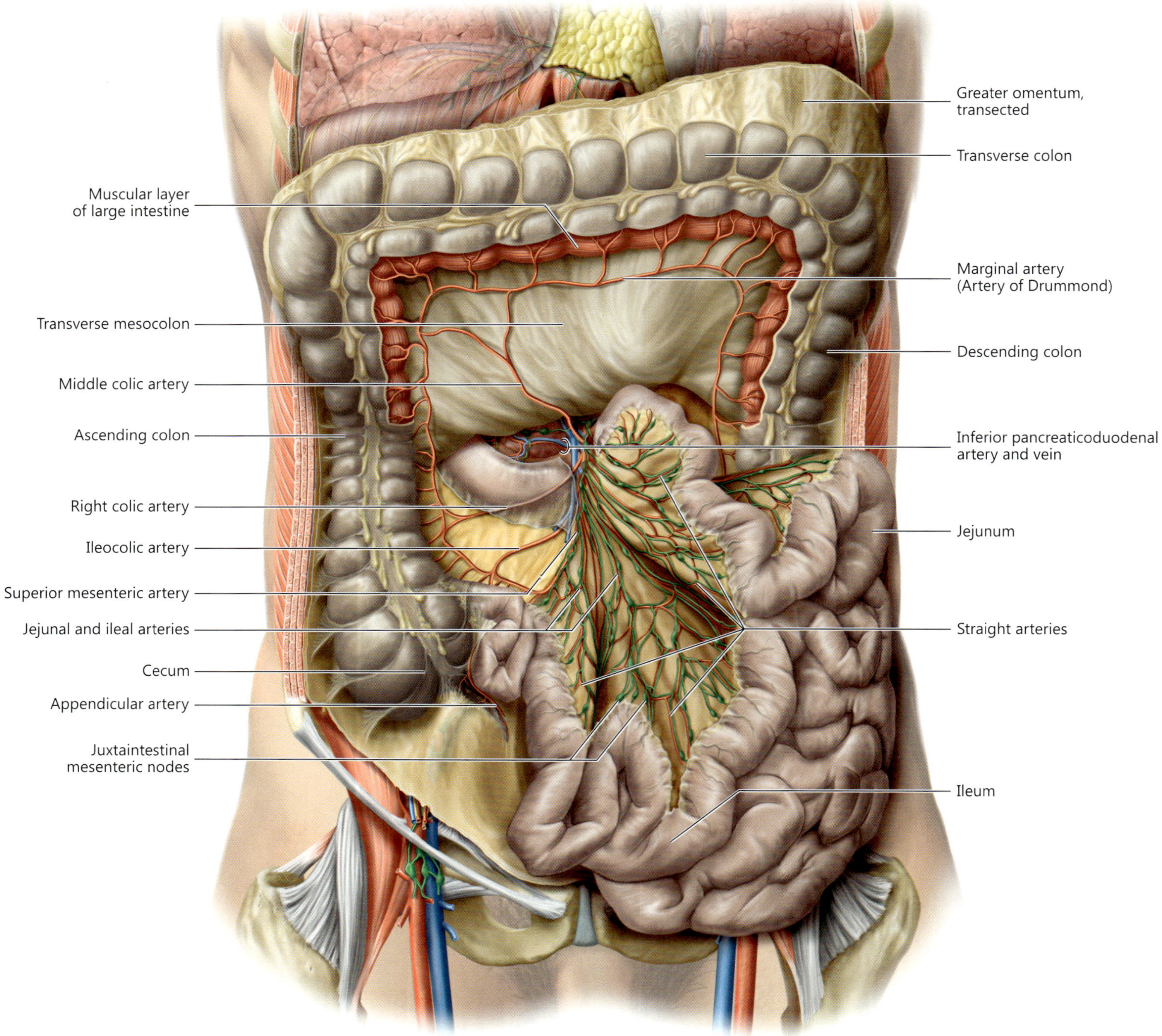

Distribution of **lymphatic vessels** and **lymph nodes** within the **mesentery**.

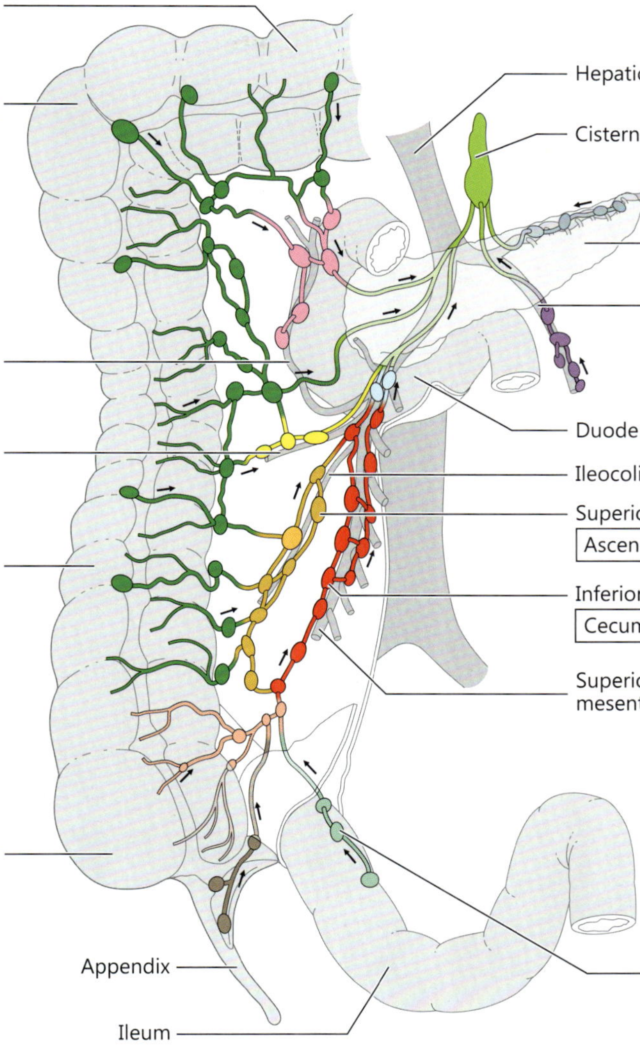

Lymph nodes along the **ileal branch of the ileocolic artery.** The lymphatic collectors of these nodes join those of the other superior ileocolic and mesenteric nodes and reach the intestinal trunk, which opens into the cisterna chyli.

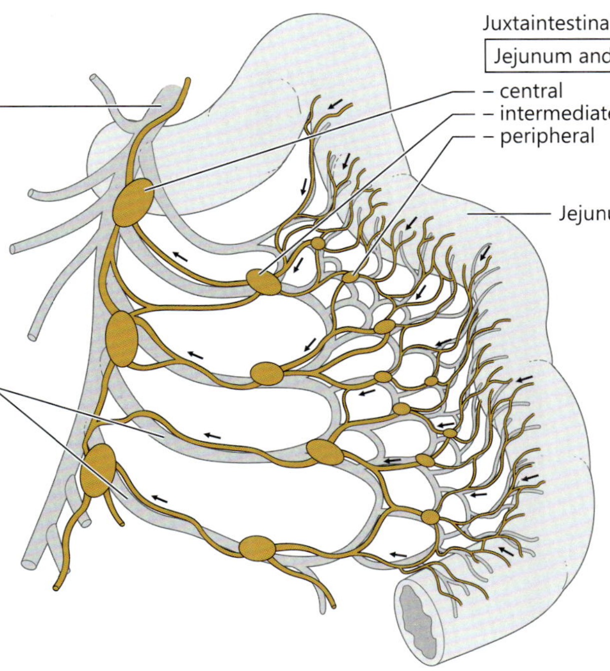

Lymphatic collectors and **juxtaintestinal mesenteric nodes** of a jejunal loop. The arrows indicate the lymph flow through three overlapping groups of juxtaintestinal mesenteric nodes, between the two layers of the mesentery.

Iliac, Inguinal and Aortic Lymph Nodes

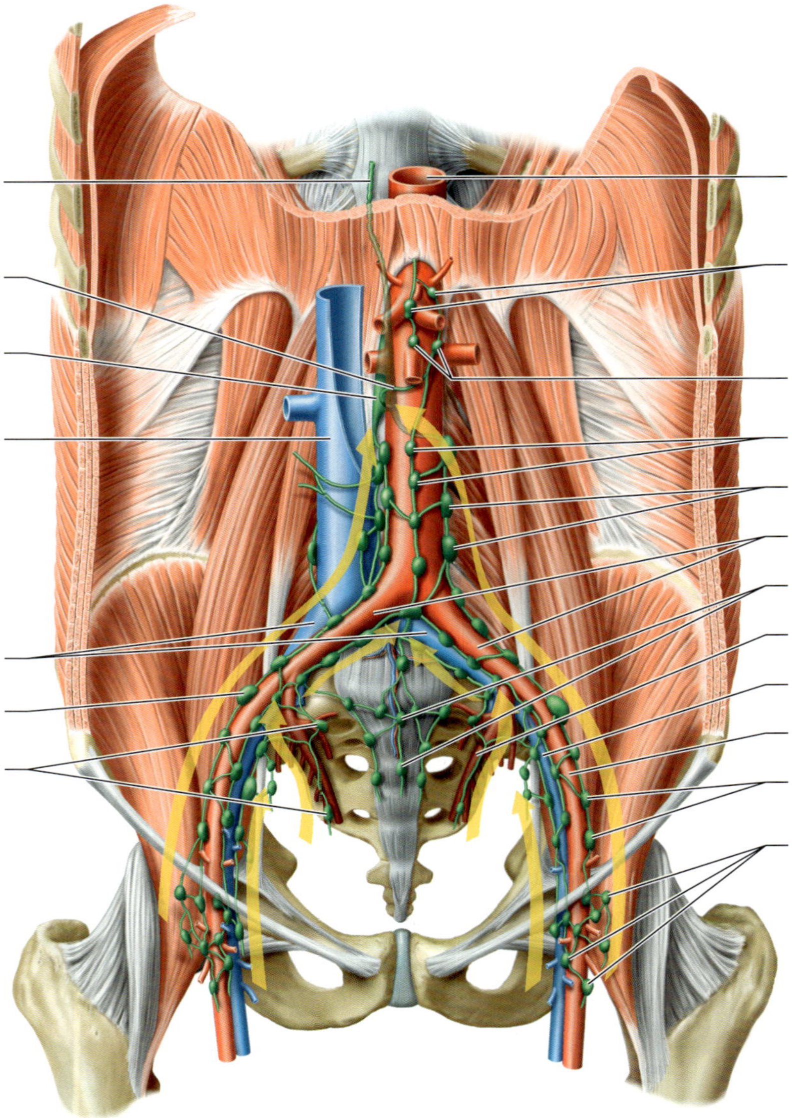

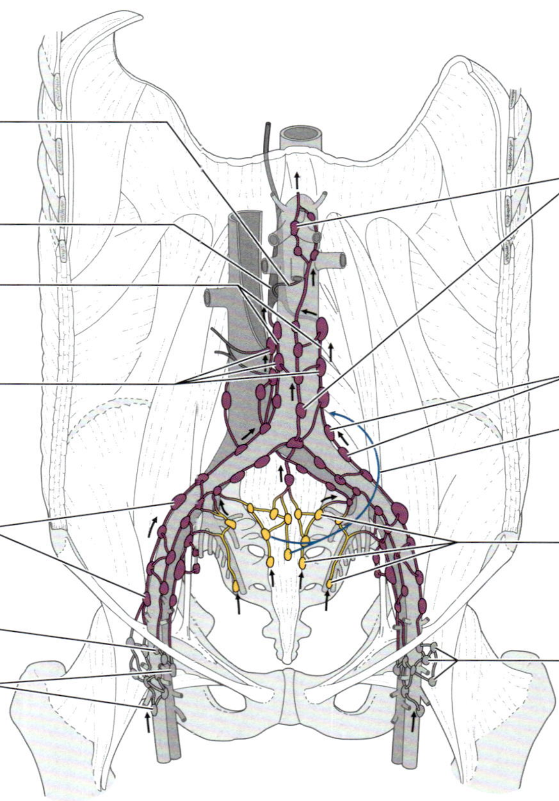

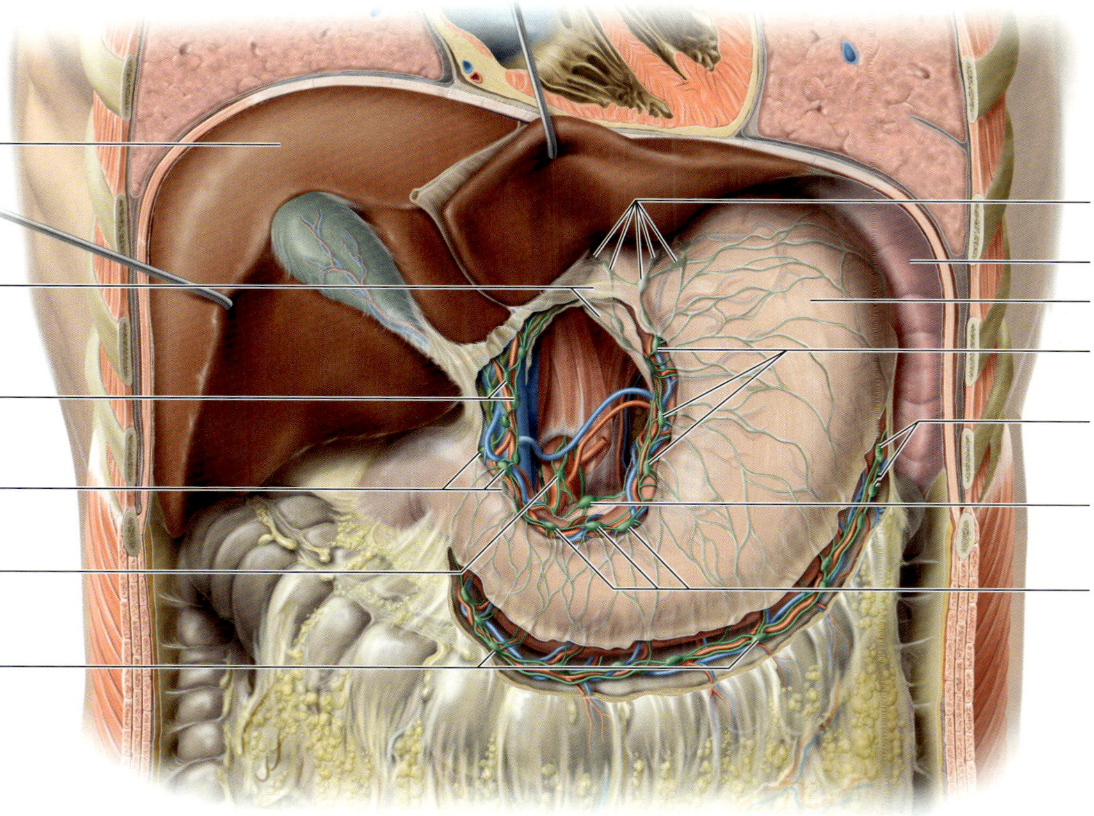

Lymphatic collectors and **lymph nodes of the gastric wall and the liver**. The hepatogastric ligament was removed.

Lymph nodes and territories of lymphatic drainage of the stomach. The region of the lesser curvature is the widest one and its collectors form the main lymphatic pathway, draining into the right and left gastric nodes, and into the nodes around the cardia. The collectors of the region of the greater curvature reach the right and left gastro-omental nodes and the pyloric nodes. The fundus region of the stomach drains into the splenic nodes. The gastric nodes, including those around the cardia, gastro-omental and pyloric nodes, are tributaries of the celiac nodes.

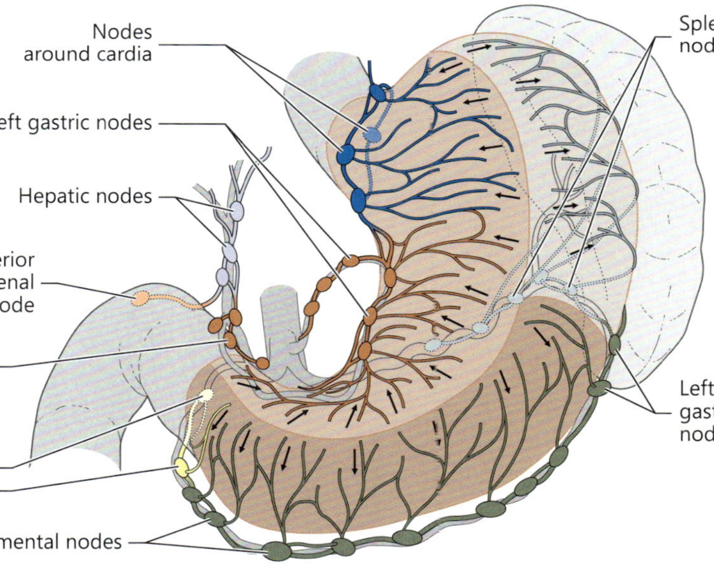

➤
At the top, **main lymphatic groups of the abdomen** (preaortic, celiac, superior and inferior mesenteric, lateral aortic nodes), the pelvis (sacral, common, external and internal iliac nodes) and the lower limb (superficial and deep inguinal nodes).

At the bottom, **lymphatic drainage pathways of the abdomen and the pelvis**. The intestinal trunk represents the efferent pathway of the celiac nodes, the terminal group of the preaortic nodes, and opens into the cisterna chyli, the abdominal confluence where the thoracic duct originates. The lumbar trunks, efferent vessels of the lateral aortic nodes, also open into the cisterna chyli, and receive the lymph from the common iliac nodes. The external iliac nodes, receiving lymph from the inguinal nodes, drain into the common iliac nodes.

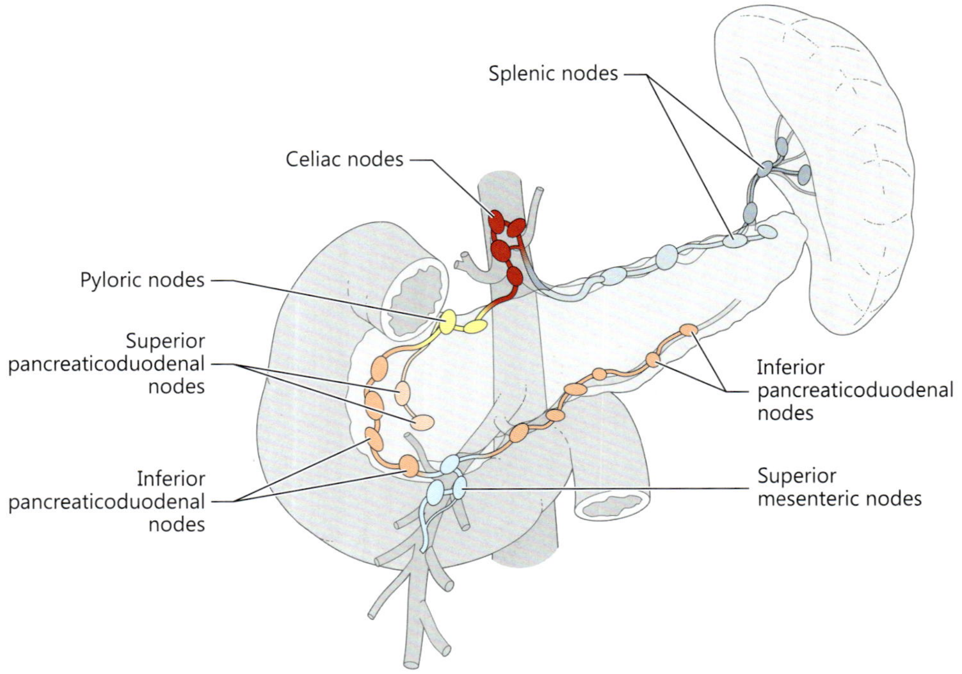

The **pancreaticoduodenal nodes** pass between the pancreas and the duodenum, above (superior nodes) and below (inferior nodes) the root of the transverse mesocolon, anterior and posterior to the head of the pancreas. The splenic nodes are situated along the superior border of the pancreas and along the splenic vessels. The efferent lymphatic vessels of these groups of nodes drain into the celiac and superior mesenteric nodes.

Summary table of the lymph nodes of the abdomen

LYMPH NODE	*GROUP*	*SITE*	*AFFERENTS*	*EFFERENTS*
		PARIETAL LYMPH NODES*		
Inferior diaphragmatic		Deep	Diaphragm and caudate lobe of the liver	Celiac, lateral aortic or superior mesenteric nodes
Inferior epigastric (three or four lymph nodes along the inferior epigastric artery)		Deep	Territory of distribution of the inferior epigastric vessels	External iliac nodes
Lumboaortic (left, intermediate and right lumbar nodes)		Deep		
– **lateral aortic** (or para-aortic nodes) including the lateral caval nodes (right lumbar nodes)	Lateral aortic (or para-aortic)		Suprarenal gland, kidney, abdominal part of the ureter, testis, ovary and lateral two-third of the uterine tube, fundus and superior portion of the body of uterus, posterior abdominal wall and, indirectly, through the iliac nodes, most of the pelvic organs, except the rectum, the subumbilical part of the abdominal wall (flat muscles of abdomen), pelvic wall, perineum and lower limb	Lumbar trunk
– **preaortic** including the precaval nodes (right lumbar nodes)	Preaortic		Subdiaphragmatic part of alimentary canal, up to the proximal half of the rectum (stomach, small intestine and large intestine), liver, gallbladder, bile ducts, pancreas and spleen	Celiac nodes
– **postaortic** including the postcaval nodes (right lumbar nodes)	Postaortic		The same as the lateral aortic nodes	The same as the lateral aortic nodes

continued

Summary table of the lymph nodes of the abdomen				*(continued)*
LYMPH NODE	*GROUP*	*SITE*	*AFFERENTS*	*EFFERENTS*
		VISCERAL LYMPH NODES		
Celiac	Preaortic	Deep	Gastric, hepatic, pancreaticosplenic nodes (abdominal part of the alimentary canal, proximal to the duodenojejunal flexure, liver, extrahepatic bile ducts, pancreas, spleen), superior and inferior mesenteric nodes (small intestine and large intestine up to the proximal half of the rectum)	Intestinal trunk
Hepatic – cystic – node of anterior border of omental foramen		Deep	Stomach, duodenum, liver, gallbladder, extrahepatic bile ducts, pancreas	Celiac nodes
Mesenteric – inferior – – superior rectal – – sigmoid – superior – – appendicular – – ileocolic – – juxtaintestinal – – mesocolic – – – right/middle/left colic – – – paracolic – – prececal – – retrocecal – – central superior	Preaortic	Deep	Abdominal part of the alimentary canal, distal to the duodenojejunal flexure, up to the proximal half of the rectum	Celiac nodes
Nodes around cardia (inconstant)		Deep	Cardia and abdominal part of the esophagus	Celiac nodes
Pancreatic (or pancreaticosplenic along with the splenic nodes) – inferior – superior		Deep	Pancreas, spleen, stomach	Celiac nodes
Pancreaticoduodenal – inferior – superior		Deep	Pancreas, duodenum	Celiac nodes
Pyloric – retropyloric (inconstant) – suprapyloric (inconstant) – subpyloric (inconstant)		Deep	Pylorus and initial part of the duodenum	Celiac nodes
Right/left gastric		Deep	Stomach, duodenum, abdominal part of the esophagus	Celiac nodes
Right/left gastro-omental		Deep	The same as the right/left gastric nodes	The same as the right/left gastric nodes
Splenic (or pancreaticosplenic along with the pancreatic nodes)		Deep	Inferior part of the bile duct, pancreas, spleen	Celiac nodes

* The lymph of the abdominal wall is drained by the deep collectors, satellites of the inferior epigastric and deep circumflex iliac arteries and tributaries of the external iliac nodes, located in the pelvis. The subumbilical part of the abdominal wall drains into the superficial inguinal nodes of the lower limb, which are tributaries of the external iliac nodes.

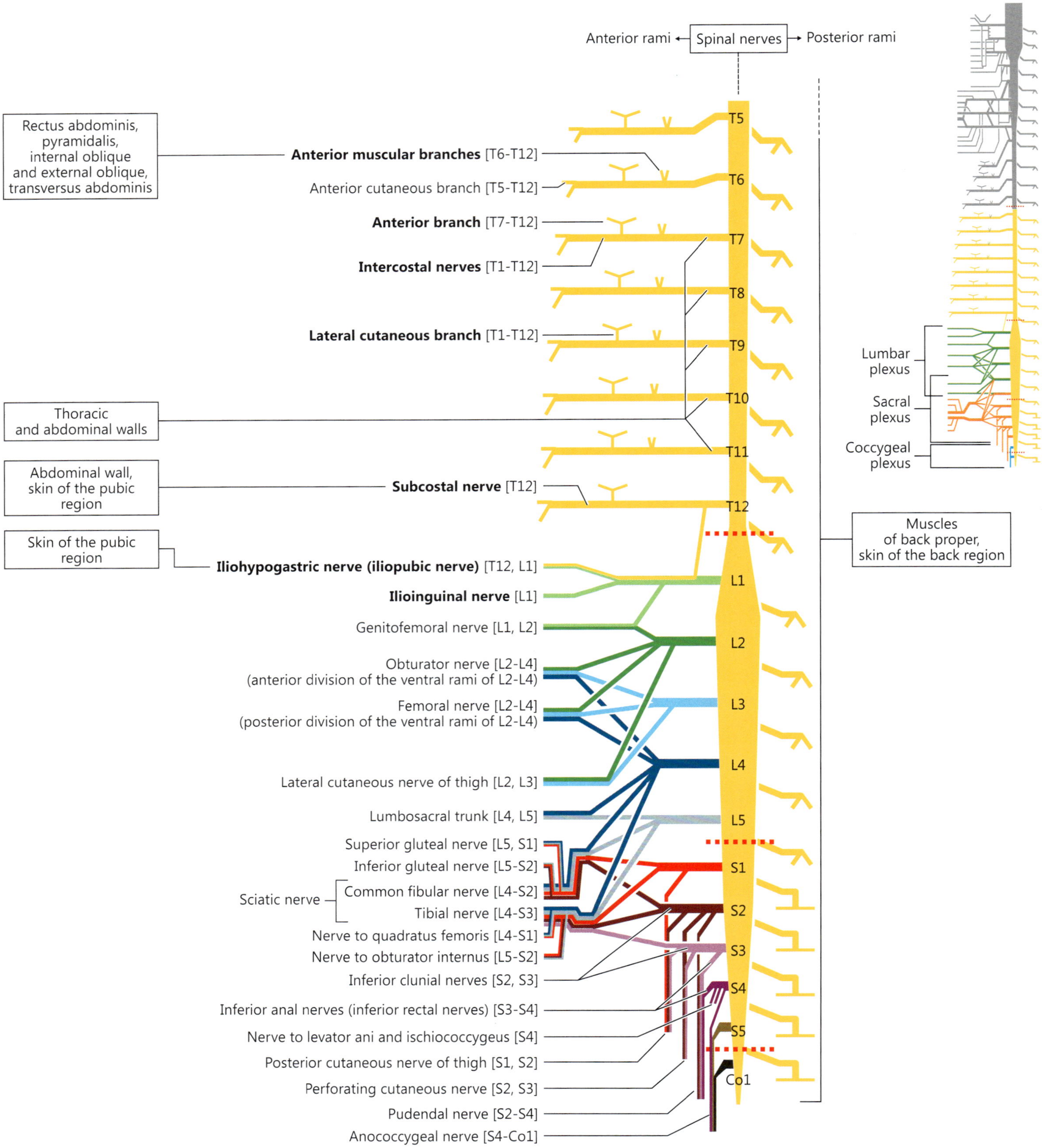

Schematic representation of the last **intercostal nerves** and of the **lumbar**, **sacral** and **coccygeal plexuses**. From top to bottom, on the left part of the figure, one can see the intercostal nerves from T5 to T12, which innervate the thoracoabdominal wall and the muscles of abdomen. Below is the lumbar plexus, formed by the anterior rami of L1-L4 and part of T12; anastomotic, collateral and terminal branches arise from this plexus. The sacral plexus, below the lumbar plexus, is formed by the anterior rami of L5, S1-S4, and part of L4. These roots merge into a common trunk, from where the sciatic nerve and several collateral branches arise. Below the sacral plexus is the coccygeal plexus, formed by the anterior rami of S5-Co1 and part of S4; it gives off visceral anterior rami and posterior rami, the main one being the anococcygeal nerve. The dorsal branches of the intercostal, lumbar, sacral and coccygeal nerves are depicted on the right part of the figure.

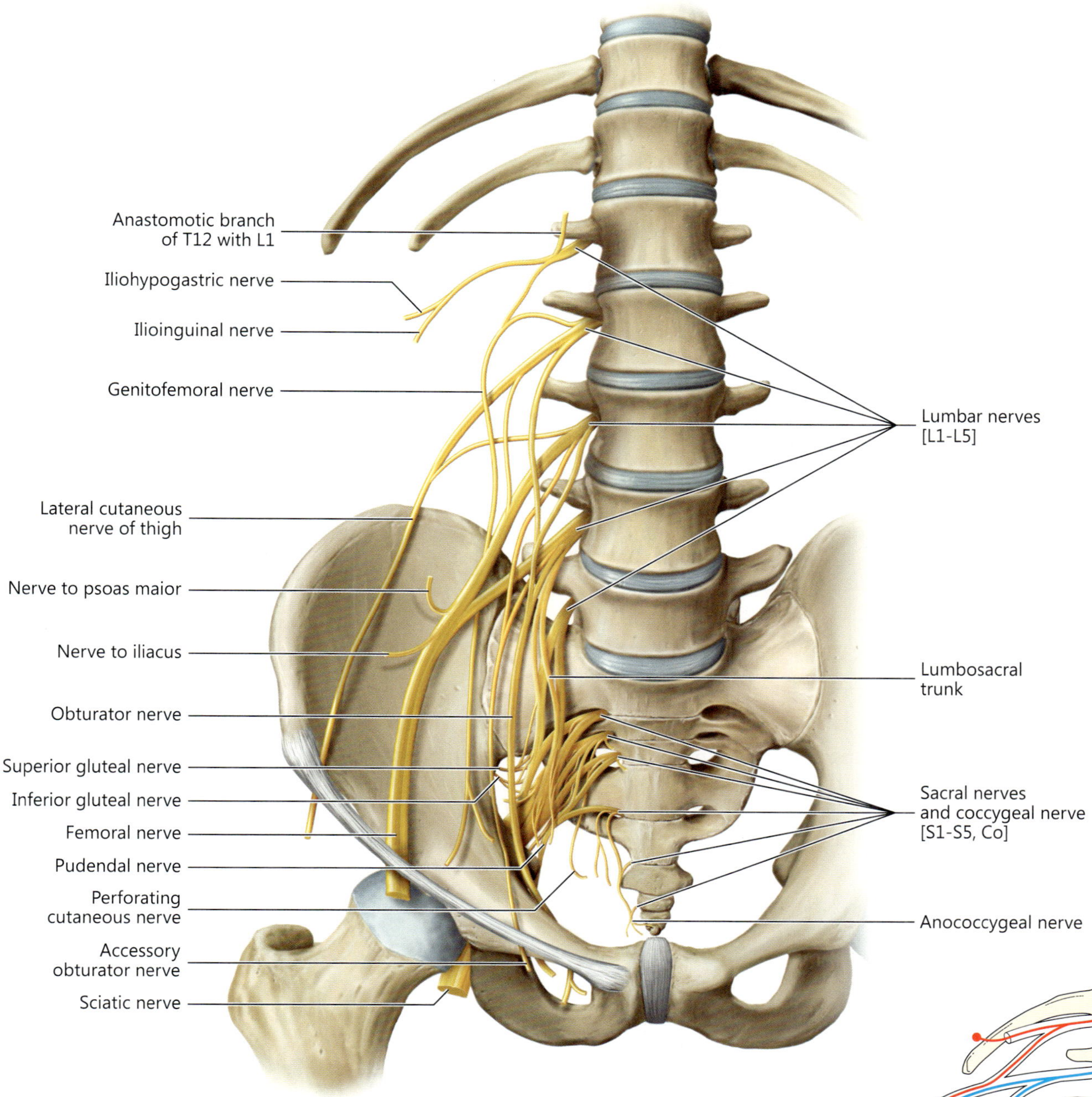

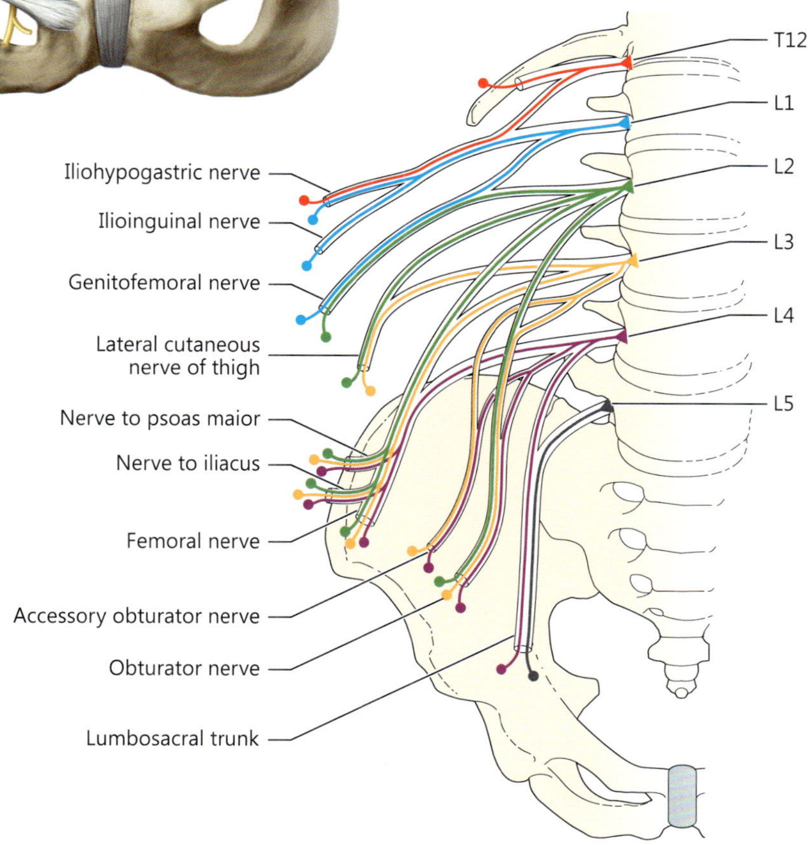

Lumbosacral plexus. The anterior rami of L1, L2, L3 and part of L4 form the lumbar plexus, with contributions from an anastomotic branch of T12. Each anterior ramus divides into three rami, two of which continue into the peripheral rami of the plexus, while the third one descends as an anastomotic branch to the anterior ramus of the underlying nerve. The anastomotic branch of L4, together with the anterior ramus of L5, forms the lumbosacral trunk, which, in turn, joins the anterior rami of S1, S2 and S3 to form the sacral plexus.

➤

Rami of the lumbar plexus. The anterior ramus of L1 originates the iliohypogastric and ilioinguinal nerves. The anterior ramus of L2 bifurcates into the genitofemoral nerve and lateral cutaneous nerve of thigh, and originates the superior roots of the femoral and obturator nerves. The anterior ramus of L3 gives off the middle roots of the femoral and obturator nerves. The anterior ramus of L4 originates the inferior roots of the femoral and obturator nerves.

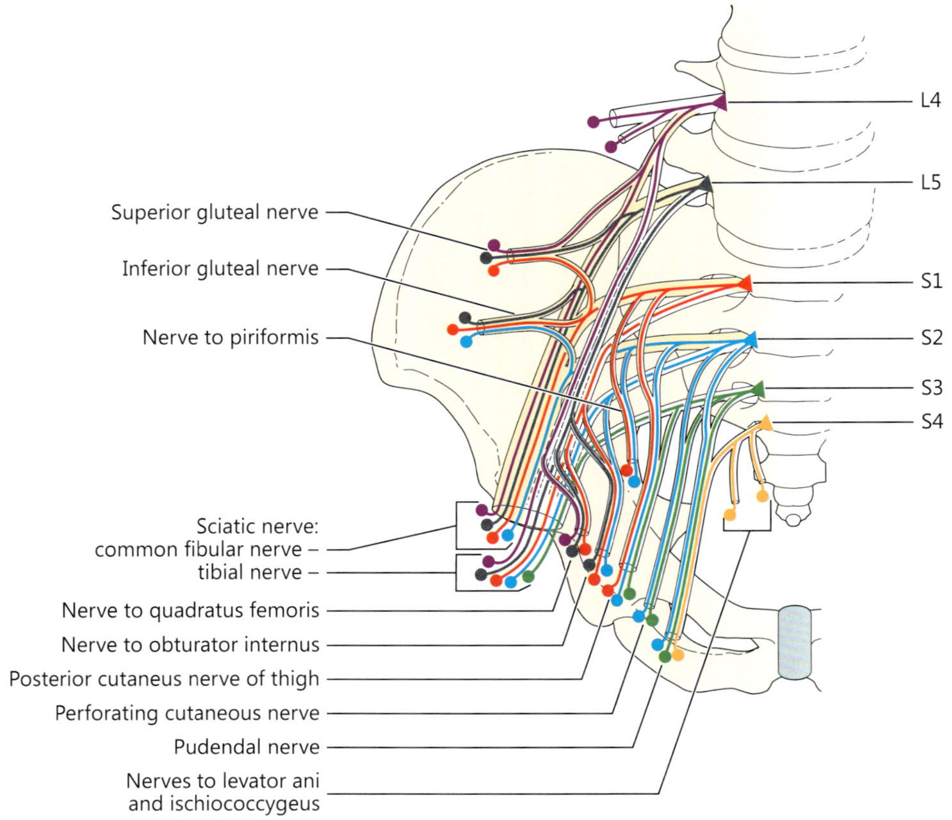

Anterior collateral branches of the sacral plexus: nerves to the obturator internus and quadratus femoris; posterior collateral branches: superior and inferior gluteal nerves, nerve to piriformis, posterior cutaneous nerve of thigh and perforating cutaneous nerve; terminal branches: sciatic nerve. The pudendal plexus (anterior ramus of S4 and anastomotic branches of S2 and S3 with S4) lies within the sacral plexus. It originates the pudendal nerve, which is a mixed nerve comprising sacral parasympathetic fibers.

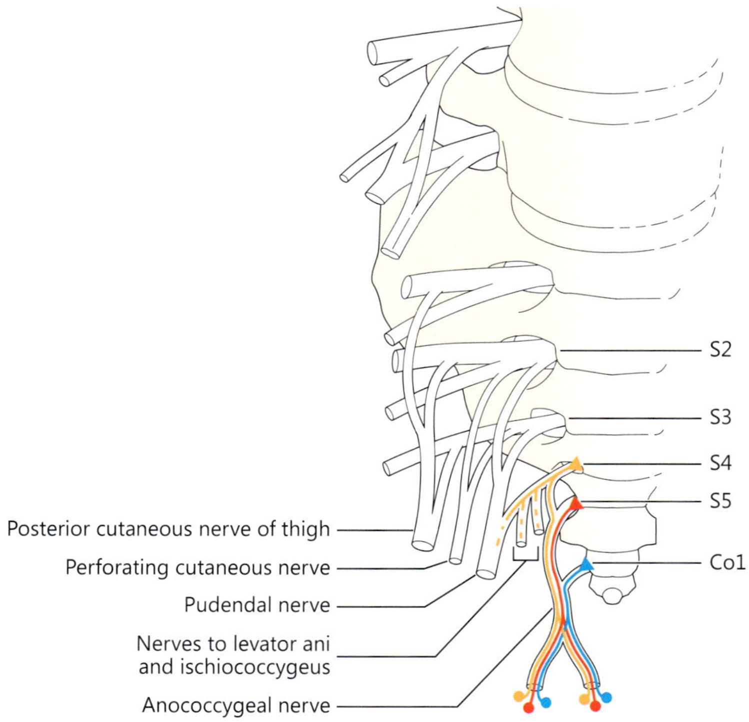

Rami of the coccygeal plexus. The anococcygeal nerve is the main posterior ramus, with fibers reaching the ischiococcygeus, and terminating as a cutaneous branch. The anterior rami of the plexus are composed of sacral parasympathetic fibers.

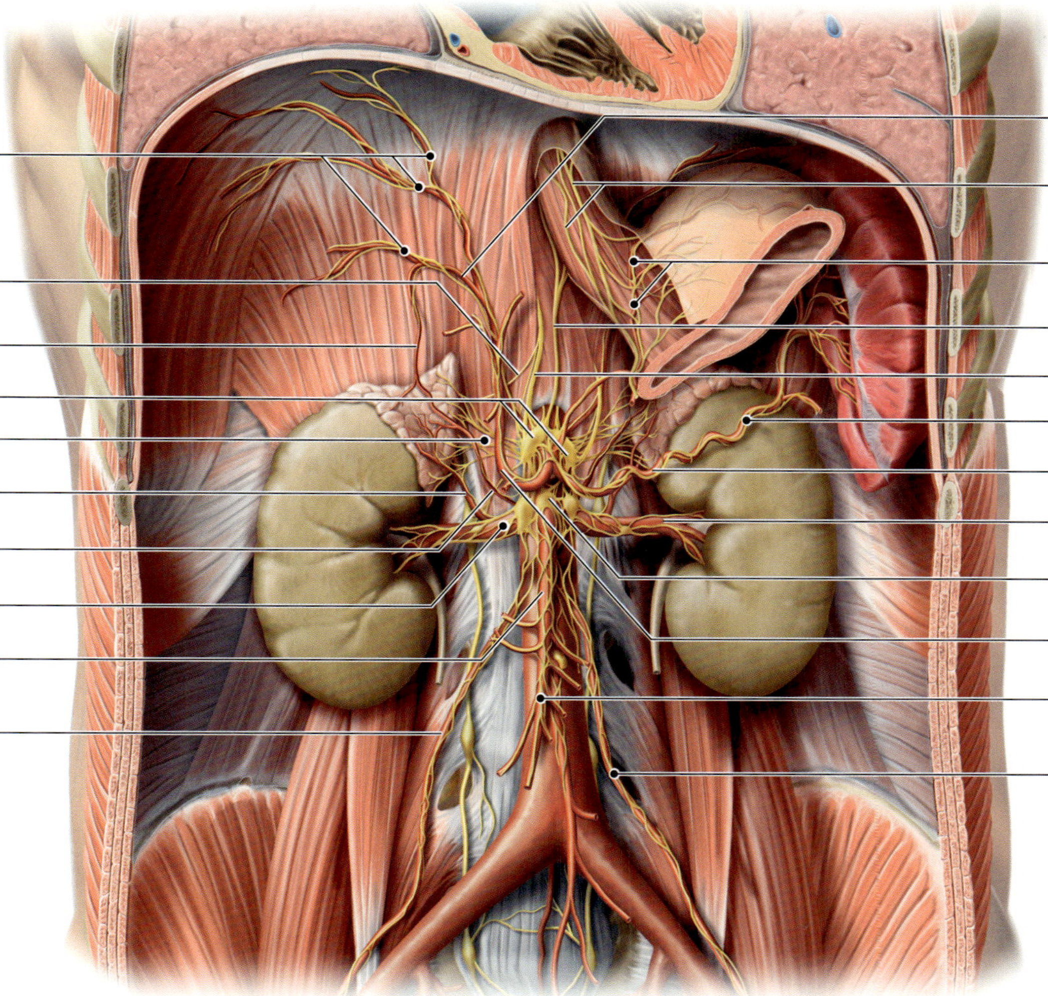

Celiac plexus. It extends in front of the aorta, between the celiac trunk and the superior mesenteric artery, and consists of the celiac ganglia and a network of fibers of diverse origin. The celiac plexus receives branches of the thoracic aortic plexus, and perivascular and visceral branches of the lumbar part of the sympathetic trunk, the greater and lesser splanchnic nerves, the posterior vagal trunk, as well as the terminal branches of the phrenic nerve. The efferents are represented by secondary plexuses along the collateral branches of the abdominal aorta.

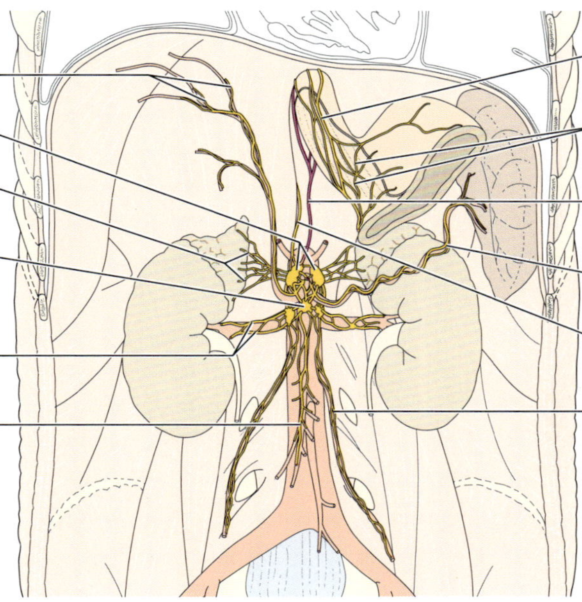

Diagram of the **celiac plexus**. The right celiac ganglion receives the greater splanchnic nerve, along with other sympathetic afferents, and the posterior vagal trunk, the termination of the right vagus nerve.

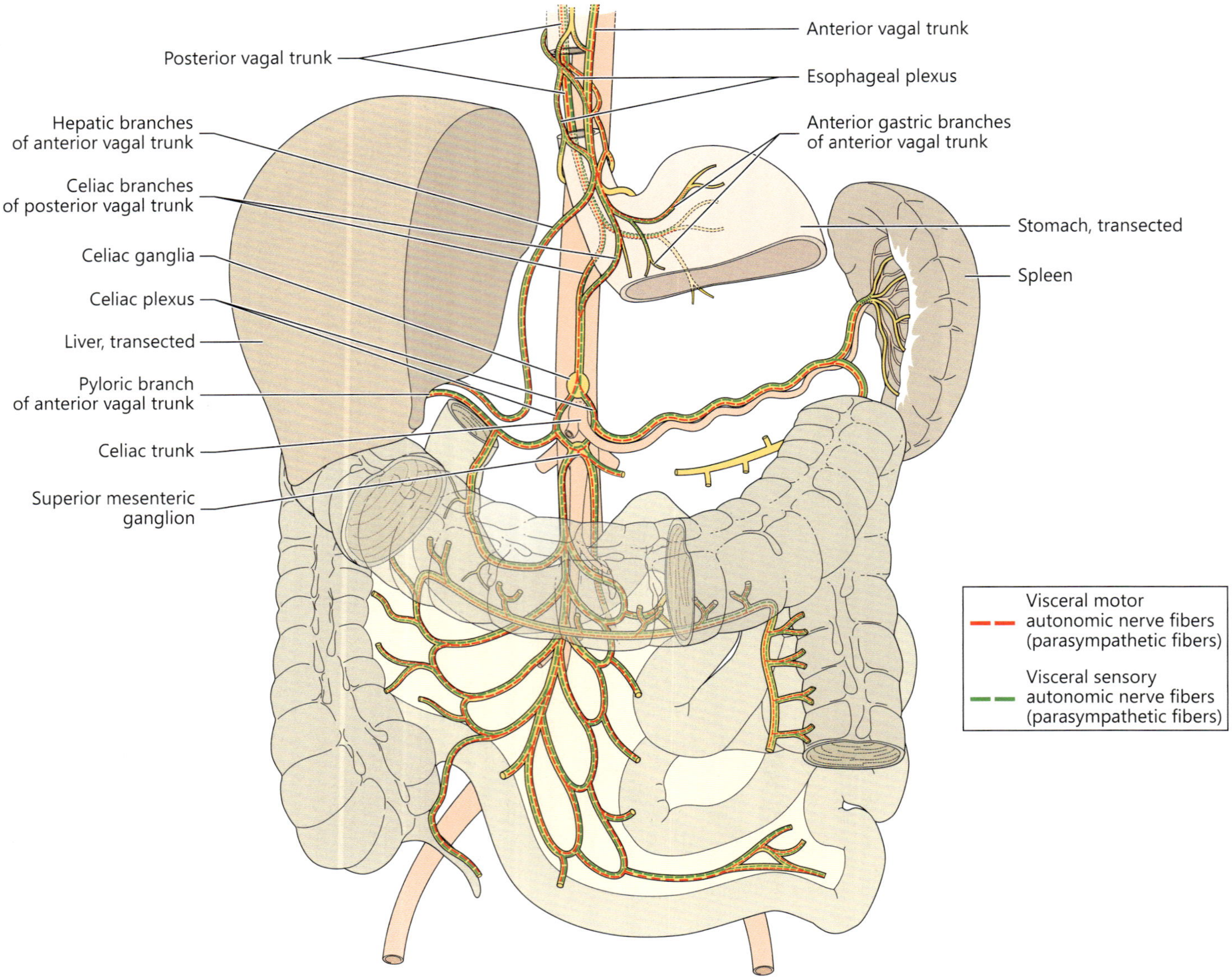

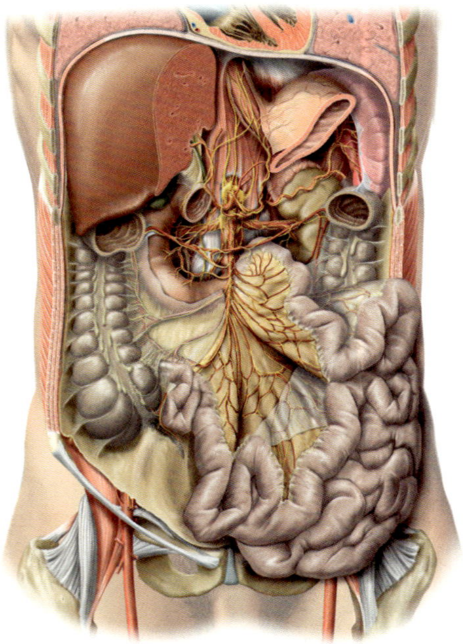

Schematic representation of the **course of the vagus nerve**. For the full course of the vagus nerve, see Volume 2, page 369.

For further details ☞ Submesocolic Space, Abdominal Plexuses, page 183.

Course and branches of the **iliohypogastric and ilioinguinal nerves**. The iliohypogastric (iliopubic) nerve provides the anterior and lateral cutaneous branches, a genital branch and muscular branches. From the ilioinguinal nerve originate a genital branch, which provides the anterior scrotal or labial nerves, as well as cutaneous and muscular branches.

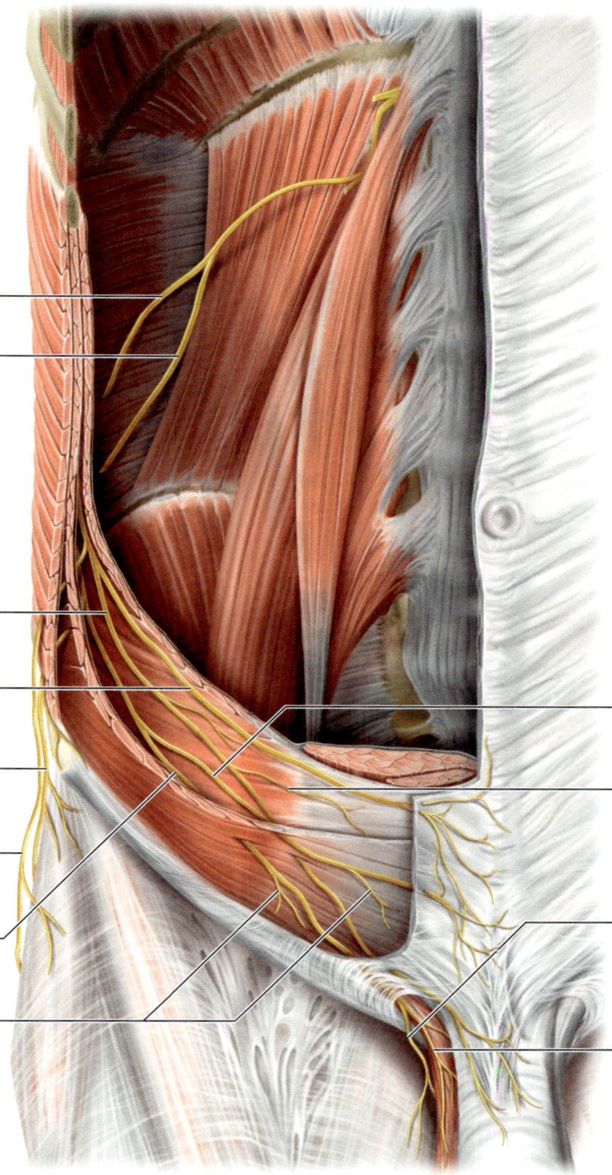

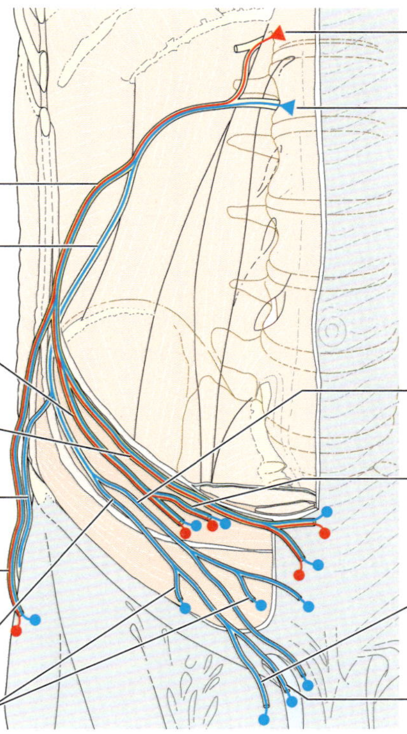

The **iliohypogastric nerve** (T12, L1) originates from the anterior ramus of the first lumbar nerve, and also receives contributions from the nervous fibers of the last intercostal nerve, by anastomosis between the anterior ramus of T12 and the anterior ramus of L1. The ilioinguinal nerve (L1) originates from the anterior ramus of the first lumbar nerve and runs below the hypogastric nerve, following the same course.

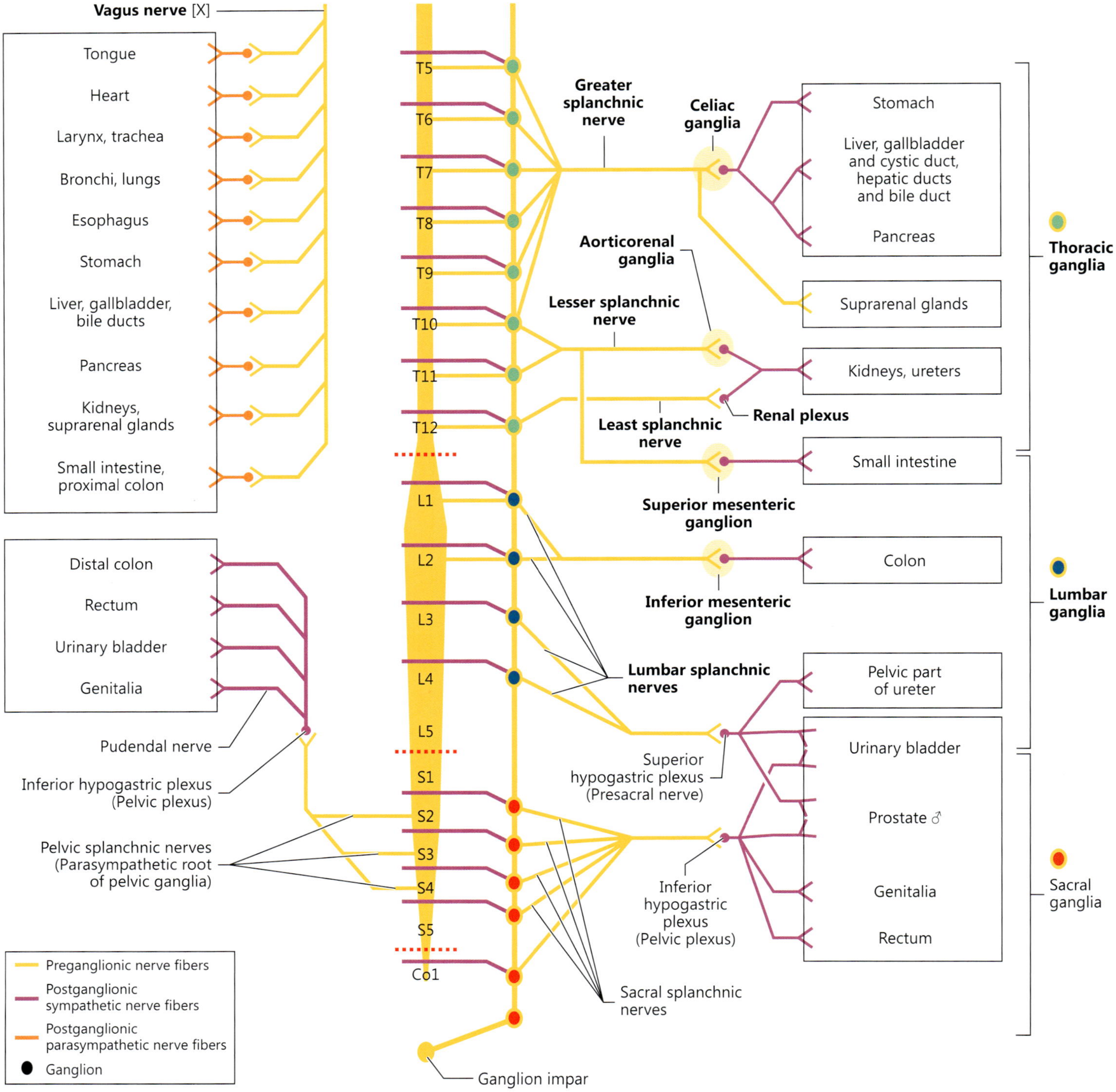

Schematic representation of the organization of the **sympathetic system** (on the right) and the **parasympathetic system** (on the left) (☞ Vol. 1, page 110). The autonomic division is the part of the central nervous system that controls the secretory functions of glands and wall motility of internal organs and vessels, thus regulating the vital functions of the organs and all body systems. The organs that are innervated by nerve fibers of the autonomic division are listed, in detail, in the boxes. The preganglionic nerve fibers are represented in *yellow*; in *purple*, the postganglionic fibers of the sympathetic nervous system, and in *orange* those of the parasympathetic nervous system. The parasympathetic (cranial and sacral) nervous system comprises the neuraxial visceral centers situated, respectively, in the brainstem and sacral segment of the spinal cord. Note, in particular, the elevated number of visceral efferent (motor) fibers of the parasympathetic part of the vagus nerve.

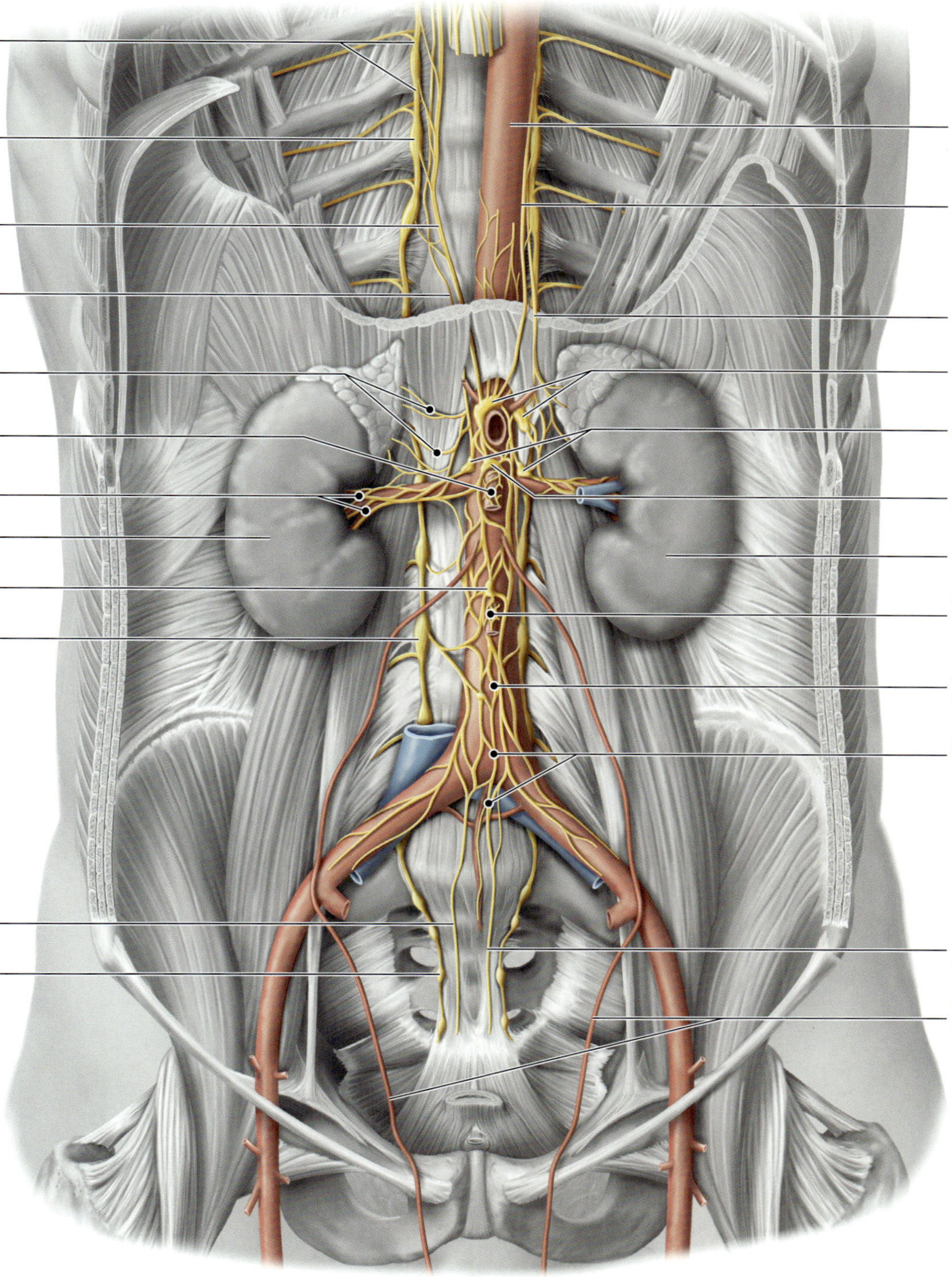

Coronal section of the trunk showing the arrangement of the **autonomic division**.

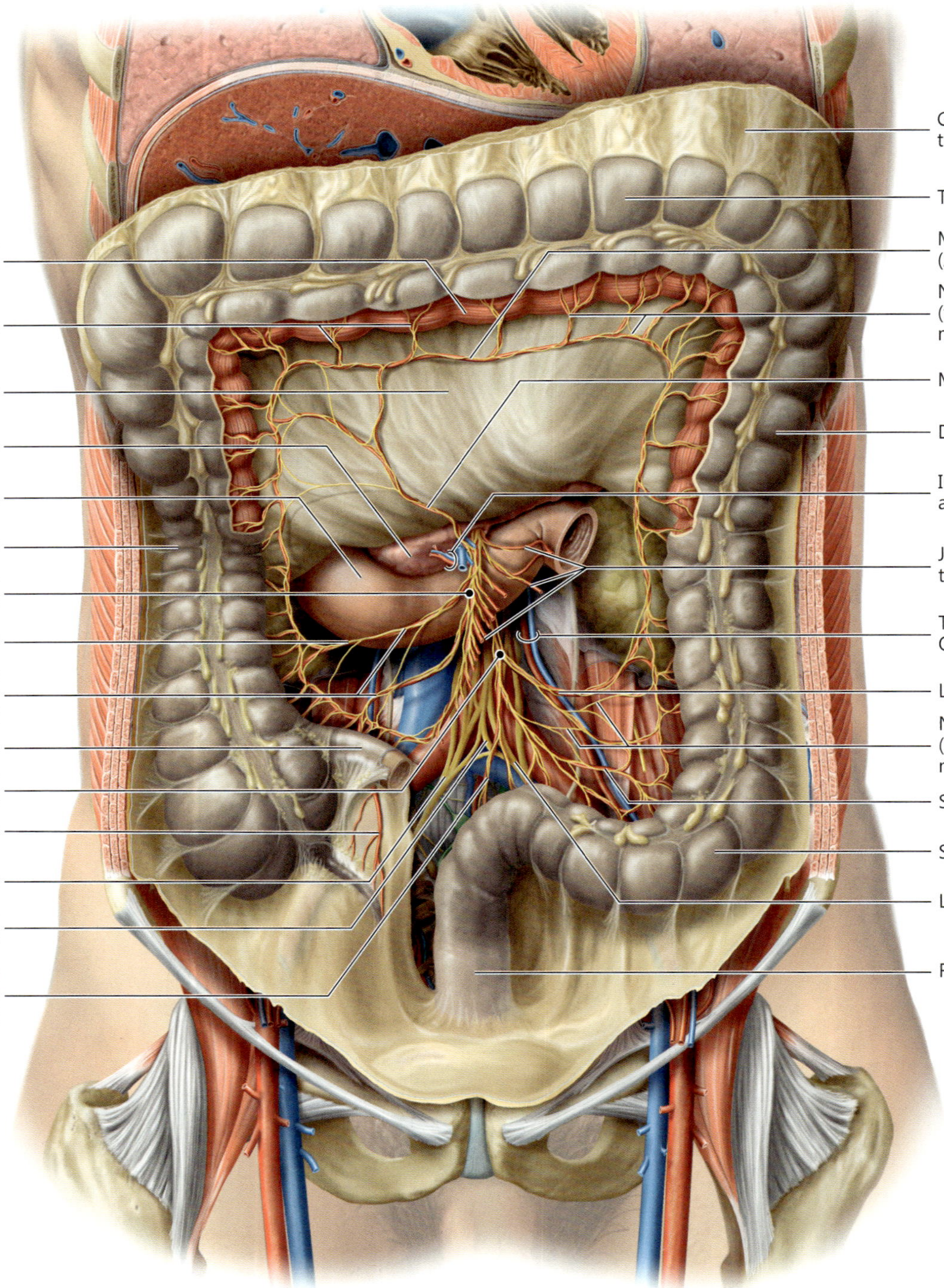

Superior and inferior mesenteric plexuses. The superior mesenteric plexus is composed of perivascular branches, displaced along the course of the superior mesenteric artery. These fibers are efferent from the celiac plexus. The inferior mesenteric plexus is composed of perivascular fibers displaced along the course of the inferior mesenteric artery. These fibers are efferent from the abdominal aortic plexus, which extends in front of the abdominal aorta, between the origin of the superior mesenteric artery and the aortic bifurcation.

Summary table of the abdominal nerves

NERVE	ORIGIN	RAMIFICATIONS	DISTRIBUTION
femoral [L2, L3, L4] (travels across the abdomen, through the psoas major, across the pelvis between the psoas major and iliacus up to the inguinal ligament, and across the thigh, where it reaches the femoral triangle running through the muscular space)	Lumbar plexus	Muscular and anterior cutaneous branches (nerves to psoas major, to iliacus, to pectineus, to quadriceps femoris, lateral and medial musculocutaneous branches) and saphenous nerve, which supplies the infrapatellar branch and the medial cutaneous nerve of leg	Psoas major, iliacus, iliopsoas, pectineus, adductor longus, sartorius, quadriceps femoris; skin of the anterior surface of the thigh, up to the knee, and of the superior part of the medial surface of the thigh, distal to the region that receives innervation from the ilioinguinal nerve; periosteum of the femur and of the patella, knee joint and patellar ligament; skin above the patella, of the medial part of the posterior surface of the thigh, of the medial surface of the leg, of the medial part of the posterior surface of the leg, above the medial malleolus, and of the medial part of the dorsum of foot
genitofemoral [L1, L2] (travels across the abdomen through the psoas major, across the pelvis below the psoas major fascia, across the perineum, with the genital branch running through the superficial inguinal ring, and across the thigh, with the femoral branch running through the vascular space)	Lumbar plexus	Genital and femoral branches	Skin of the scrotum, of the labium majus, of the superior part of the anterior region of the thigh, transversus abdominis, internal oblique and cremaster
greater splanchnic [T5-T9]	Four or five roots from the 5th (or 6th) to the 9th (or 10th) thoracic ganglion of the sympathetic trunk		Throughout the celiac plexus and the right celiac ganglion
hypogastric	Presacral plexus	A large, single nerve or a number of parallel fascicles connecting the presacral plexus with the inferior hypogastric plexuses	Through the inferior hypogastric plexuses
iliohypogastric [T12, L1] (iliopubic) (travels across the abdomen, anterior to the quadratus lumborum, then between the transversus abdominis and internal oblique)	Lumbar plexus	Anterior (genital branch of iliohypogastric nerve) and lateral cutaneous branches, muscular branches. The anterior cutaneous branch runs through internal oblique 2 cm medial to the anterior superior iliac spine and through the external oblique aponeurosis 3 cm above the superficial inguinal ring	Skin of the pubic region, of the external genitalia and of the gluteal region; rectus abdominis, external oblique and internal oblique, transversus abdominis, pyramidalis
ilioinguinal [L1] (follows the same course as the iliohypogastric nerve and across the perineum with the genital branch running through the superficial inguinal ring)	Lumbar plexus	Anterior scrotal ♂ or anterior labial ♀ nerves (genital branch), cutaneous and muscular branches	Anterior part of the scrotum and root of the penis in the male, labium majus and mons pubis in the female; skin of the pubic region, of the superior part of the medial surface of the thigh, of the buttock and above the anterior superior iliac spine; muscles of the abdominal wall
intercostal	Anterior rami of thoracic nerves [T1-T12]	Muscular, collateral, anterior and lateral pectoral cutaneous, medial and lateral mammary, anterior and lateral abdominal cutaneous branches; intercostobrachial nerves	Skin of the thoracoabdominal wall, muscles of the thorax and the abdomen
– **from the 6th to the 12th** – – **muscular branches**			Rectus abdominis, external oblique and internal oblique, transversus abdominis, pyramidalis
– **from the 7th to the 12th** – – **anterior abdominal cutaneous branches** (thoracoabdominal nerves)			Skin of the anterior wall of the abdomen
– **from the 8th to the 12th** – – **lateral abdominal cutaneous branches**			Skin of the lateral wall of the abdomen

continued

Summary table of the abdominal nerves *(continued)*

NERVE	ORIGIN	RAMIFICATIONS	DISTRIBUTION
lateral cutaneous nerve of thigh [L2, L3] (lateral femoral cutaneous) (travels across the abdomen through the psoas major, across the pelvis below the iliacus fascia, and across the thigh, after passing through the notch between the anterior superior and anterior inferior iliac spines)	Lumbar plexus	Gluteal and femoral branches	Skin of the buttock and of the lateral part of the anterior surface of the thigh
least splanchnic (lowest splanchnic)	Last thoracic ganglion of the sympathetic trunk		Through the renal plexus
lesser splanchnic	From the 9th and 10th thoracic ganglia of the sympathetic trunk		Celiac, renal and suprarenal plexuses
lumbar [L1-L5] (five pairs of spinal nerves)	Lumbar part of the spinal cord	Anterior ramus (lumbar plexus), posterior ramus (medial branch, lateral branch, posterior cutaneous branch, superior clunial nerves) and rami communicantes with the sympathetic trunk (white and gray rami of the 1st and 2nd lumbar nerve, gray rami of the 3rd and 4th lumbar nerve)	Through the anterior rami (lumbar plexus) to the intertransversarii laterales lumborum, psoas major, psoas minor and quadratus lumborum, to the skin of the pubic region, of the external genitalia, of the gluteal region (skin of the buttock) and of the anterior region of thigh (anterolateral and medial skin of the thigh), to the inferior part of the abdominal wall (rectus abdominis, external oblique and internal oblique, transversus abdominis, pyramidalis), to the cremaster, to the muscles and skin of the lower limb; through the posterior rami to the skin of the vertebral region and to the muscles of back proper, at the lumbar level
lumbar plexus	Anterior rami of the 1st, 2nd, 3rd and part of the 4th lumbar nerve and anastomotic branch from the 12th intercostal nerve	Iliohypogastric, ilioinguinal, genitofemoral, lateral cutaneous of thigh, obturator, accessory obturator, femoral nerves; lumbosacral trunk; anastomotic branches with the 12th thoracic nerve and with the 5th lumbar nerve	Intertransversarii laterales lumborum, psoas major, psoas minor, quadratus lumborum, skin of the pubic region, of the external genitalia, of the gluteal region (skin of the buttock) and of the anterior region of the thigh (anterolateral and medial skin of the thigh), inferior part of the abdominal wall (rectus abdominis, external oblique and internal oblique, transversus abdominis, pyramidalis), cremaster, muscles and skin of the lower limb
lumbar splanchnic [L1, L2] (two to four in number)	Lumbar ganglia of the sympathetic trunk		Through the celiac, abdominal aortic, inferior mesenteric and superior hypogastric plexuses
lumbosacral trunk [L4, L5]	Descending anastomotic branch of L5 for the anterior ramus of L5, and anterior ramus of L5	Forms the sacral plexus by joining the anterior rami of S1, S2 and S3	Through the sacral plexus
obturator [L2, L3, L4] (travels across the abdomen, where it descends through the psoas major, across the pelvis, running its lateral wall along with the obturator internus, and across the thigh, through the obturator canal)	Lumbar plexus	Anterior branch, supplies a cutaneous branch and muscular branches, and posterior branch, supplies an articular branch and muscular branches, and nerve to obturator externus	Obturator externus, adductor magnus, adductor longus, adductor brevis and adductor minimus, gracilis; hip joint and knee joint; skin of the inferior part of the medial surface of the thigh and of the internal surface of the knee
presacral (or **superior hypogastric plexus**)	Nerve plexus, at the level of the aortic bifurcation into the two common iliac arteries, consisting of the last two lumbar splanchnic nerves	Hypogastric nerve joining the perivascular and visceral peripheral branches of the sacral ganglia of the sympathetic trunk, and the pelvic splanchnic nerves of the sacral part of the parasympathetic system to form the inferior hypogastric plexuses (or pelvic plexuses), at the point of origin of the internal iliac arteries	Through the inferior hypogastric plexuses (or pelvic plexuses)

continued

Summary table of the abdominal nerves			*(continued)*
NERVE	***ORIGIN***	***RAMIFICATIONS***	***DISTRIBUTION***
subcostal [T12] (12th intercostal nerve)	Anterior ramus of the 12th thoracic nerve	Anastomotic branch with the lumbar plexus	Skin of the anterolateral abdominal wall and of the gluteal region, and muscles of the abdominal wall
sympathetic trunk – lumbar ganglia (4 or 5 in number)		Communicating branches with the lumbar nerves (white and gray rami of the 1st and 2nd lumbar nerves, gray rami of the 3rd and 4th lumbar nerves), perivascular and visceral peripheral branches (abdominal aortic plexus, phrenic ganglia, celiac plexus including the celiac ganglia, and hepatic, splenic, gastric, pancreatic and suprarenal plexuses, aorticorenal ganglia, superior mesenteric plexus including the superior mesenteric ganglion, renal plexus including the renal ganglia, ureteric plexus, ovarian ♀ or testicular ♂ plexus, inferior mesenteric plexus including the inferior mesenteric ganglion, superior rectal plexus, enteric plexus including the subserous, myenteric and submucous plexuses, iliac plexus and femoral plexus	Walls of the blood vessels and the abdominal organs, lamellar corpuscles (Pacinian corpuscles) in the peritoneal mesenteries, smooth muscle cells of the hepatic vessels, gallbladder, liver, spleen, gastric and intestinal glands, pyloric sphincter, sphincter of bile duct, sphincter of ampulla, sphincter of pancreatic duct, smooth muscle cells of the renal vessels
vagus [X]			
– **anterior vagal trunk**	Mainly from the left vagus nerve [X]	Anterior gastric branches, anterior nerve of lesser curvature, hepatic branches, pyloric branch	Through the gastric plexus (to most of the stomach, including the cardia and pylorus, to the superior and descending parts of the duodenum and to the head of the pancreas) and through the hepatic plexus
– **posterior vagal trunk**	Mainly from the right vagus nerve [X]	Posterior gastric branches, posterior nerve of lesser curvature, celiac branches, renal branches	Through the gastric plexus (to the posteroinferior wall of the stomach), the celiac plexus with related phrenic, hepatic, suprarenal, renal, ovarian ♀ or testicular ♂, splenic and superior mesenteric plexuses, and the right celiac ganglion (where the celiac branch terminates and continues the posterior vagal trunk)

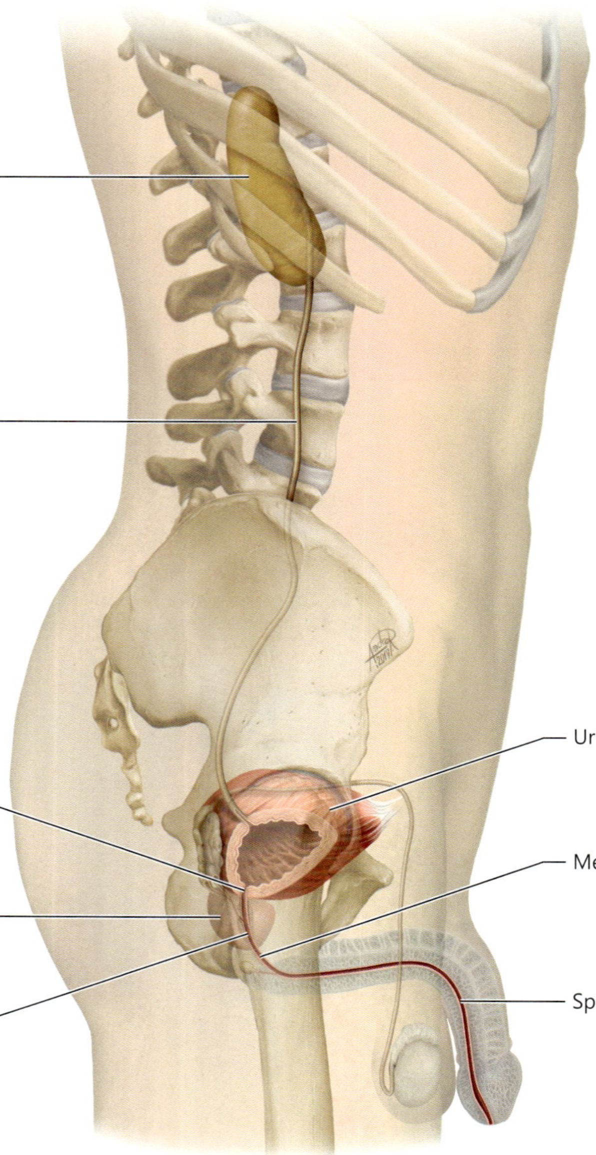

General organization of the **urinary system in the male**; right lateral projection. The kidneys are located posteriorly in the abdomen (lumbar regions). The urinary tract begins with the pyelocaliceal system stemming from the kidney, continues with the ureters, which run from the lumbar regions to the pelvis, to connect to the urinary bladder, and terminate with the urethra. The urethra originates from the urinary bladder in the pelvis; it terminates in the urogenital triangle (anterior perineum) in the female, whereas it continues along the penis to terminate in the glans in males.

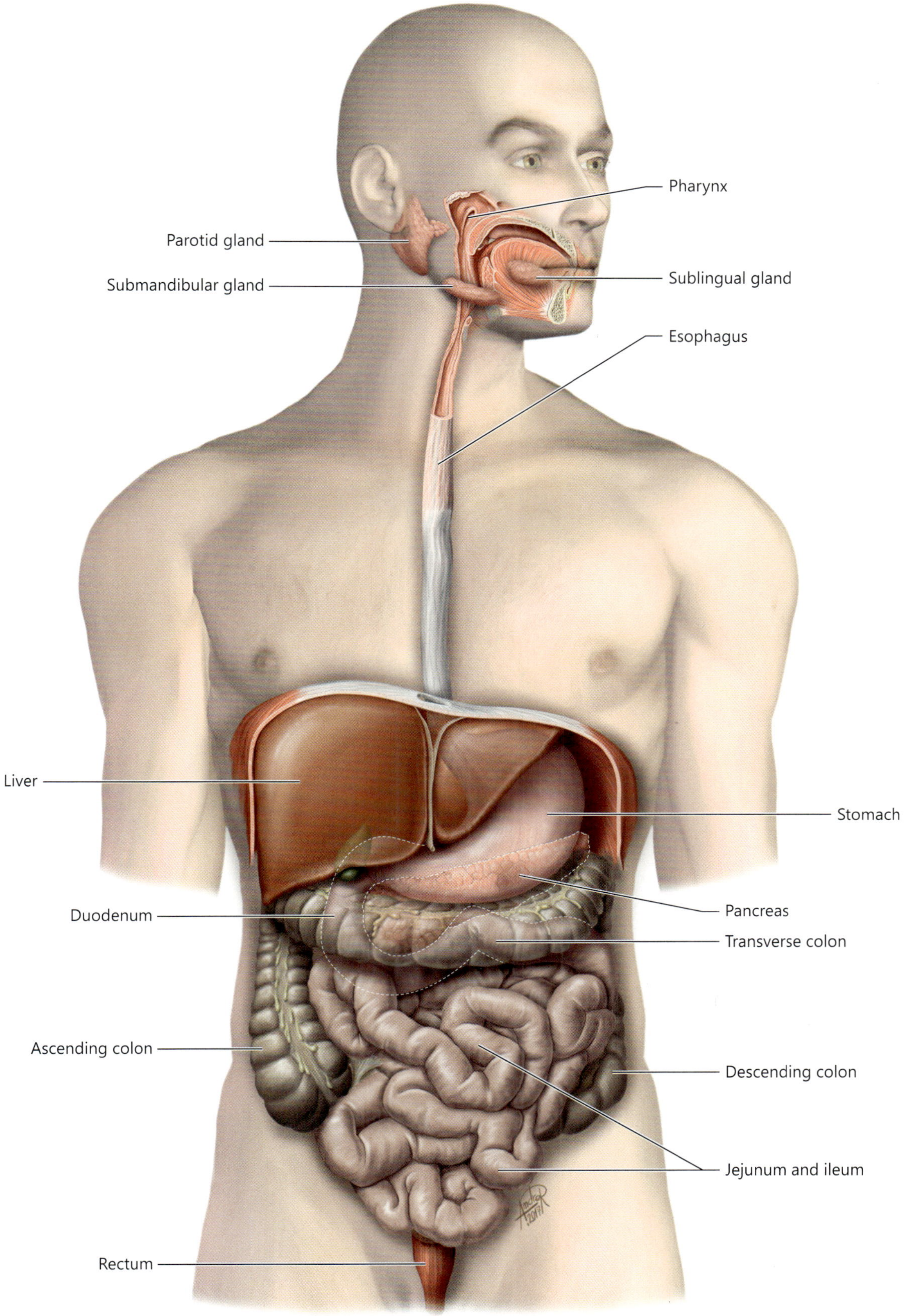

Organization of the **alimentary system**. The tube-shape characterizes long tracts of the alimentary canal, running from the head to the perineum, passing through the neck, the thorax, the abdomen and the pelvis and the perineum. Extraparietal glands are associated to the mouth (major salivary glands) and duodenum (liver and pancreas).

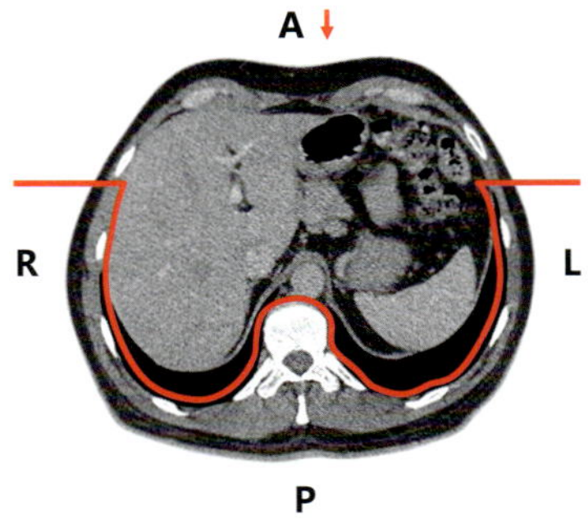

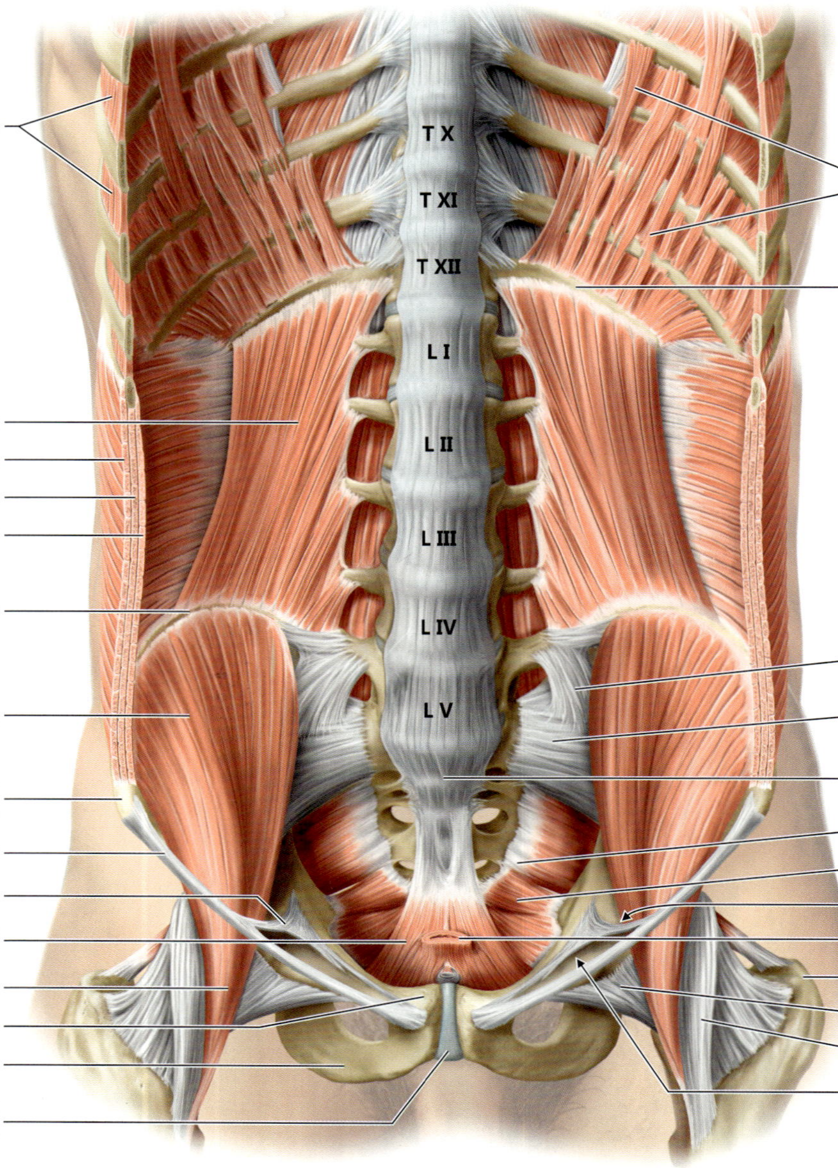

CONTENTS OF THE ABDOMEN		
• Anterior sacroiliac ligament • Anterior superior iliac spine • External oblique • Greater trochanter of femur • Iliac crest • Iliacus • Iliofemoral ligament • Iliolumbar ligament • Iliopectineal arch • Iliopsoas	• Inguinal ligament • Intercostal muscles • Internal oblique • Interpubic disc • Ischial tuberosity • Ischiococcygeus • Levator ani • Muscular space • Piriformis • Promontory of sacrum	• Pubic symphysis • Pubic tubercle • Pubofemoral ligament • Quadratus lumborum • Rectum • Subcostales • Transversus abdominis • Twelfth rib [XII] • Vascular space

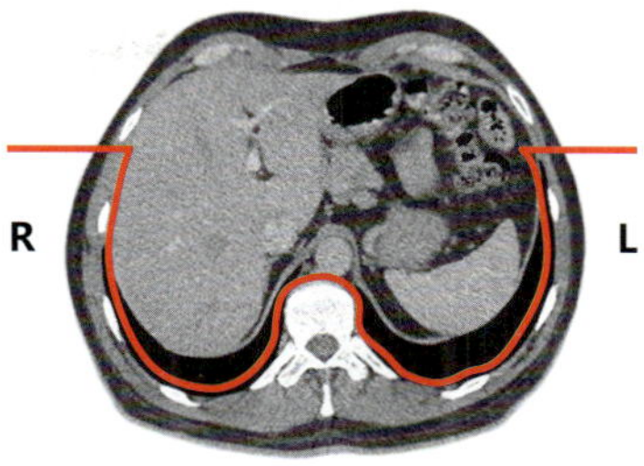

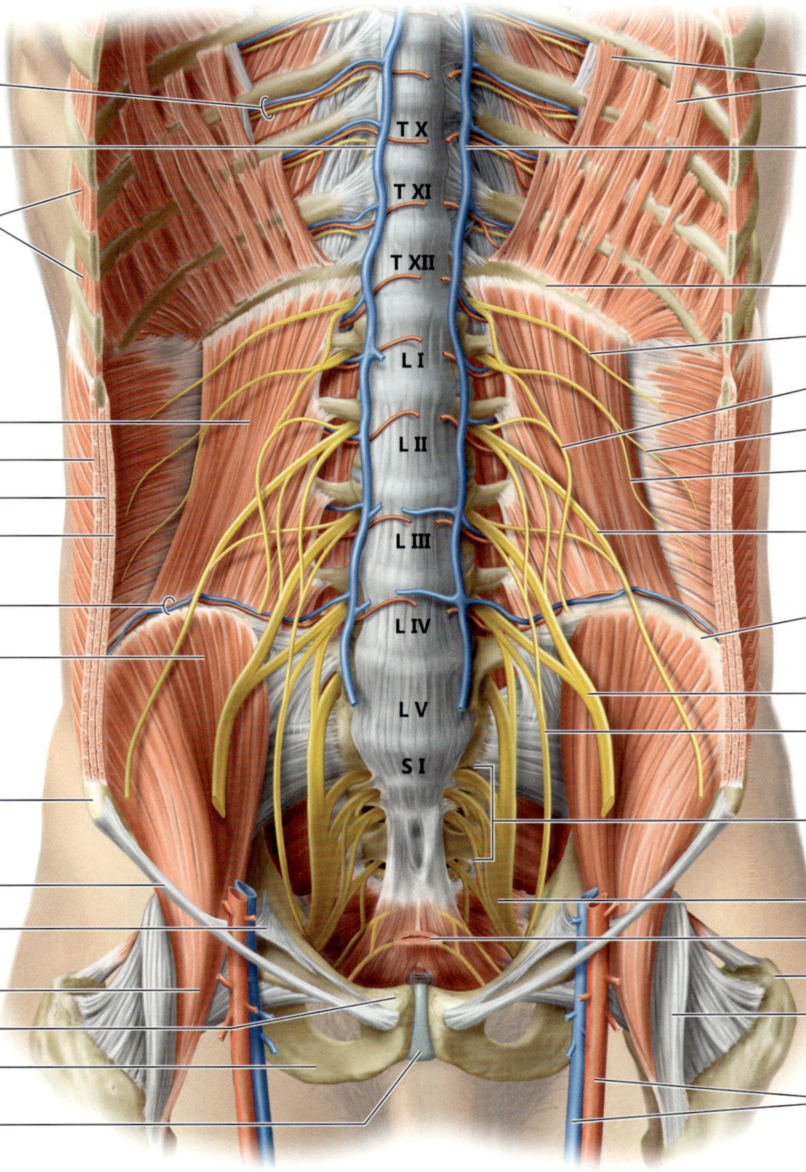

CONTENTS OF THE ABDOMEN (In bold are the structures added with respect to the preceding table)	
• Azygos vein • Femoral artery • Femoral nerve • Femoral vein • Genitofemoral nerve • Hemiazygos vein • Iliohypogastric nerve • Ilioinguinal nerve	• Lateral cutaneous nerve of thigh • Lumbar artery • Lumbar vein • Obturator nerve • Sacral plexus • Sciatic nerve • Subcostal nerve

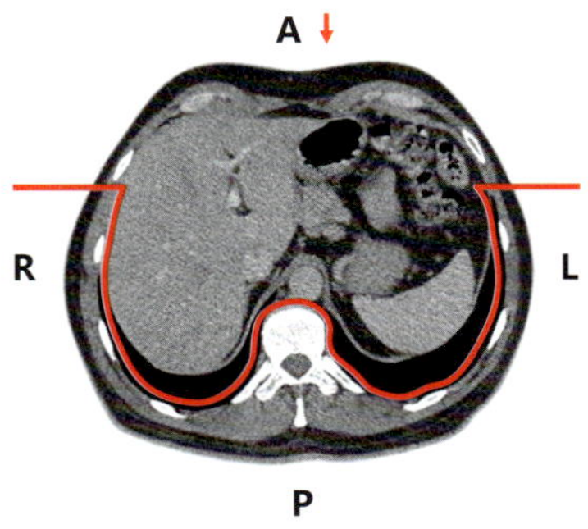

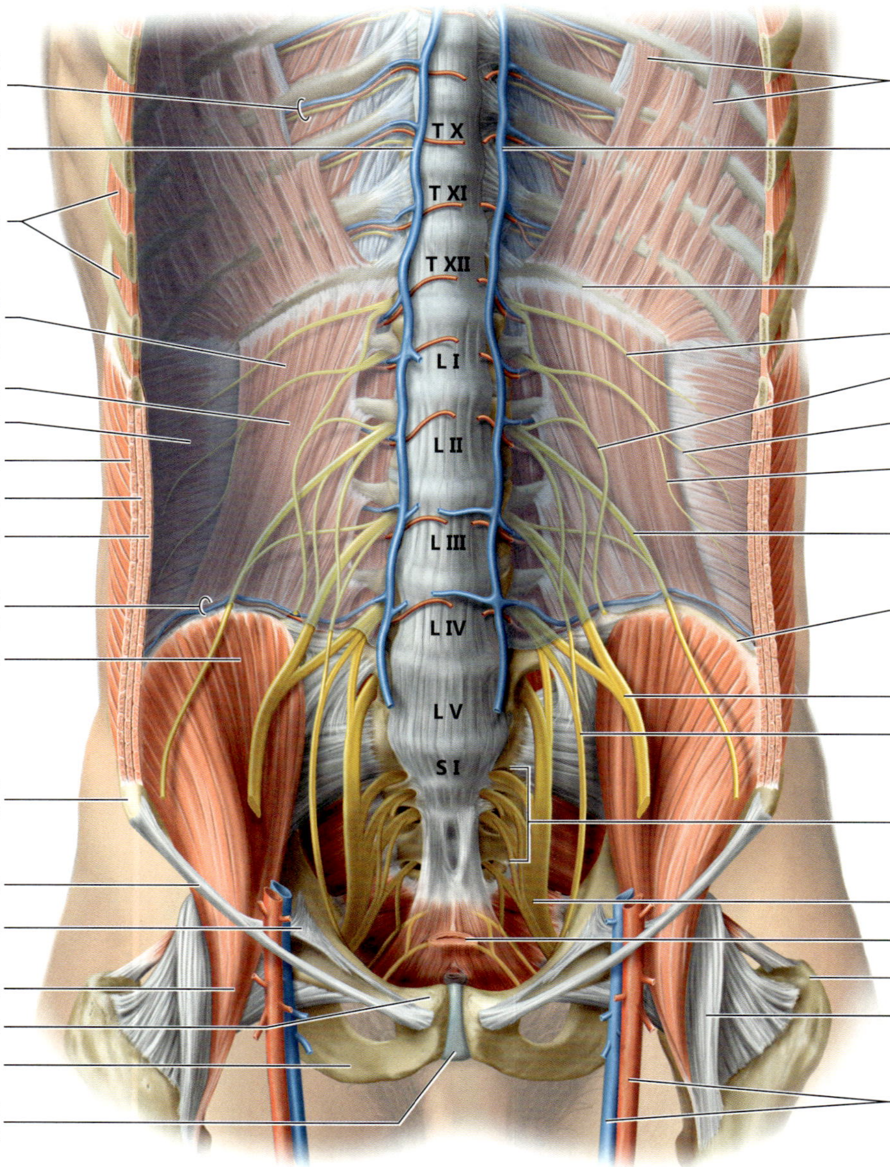

***CONTENTS OF THE ABDOMEN** (In bold are the structures added with respect to the preceding table)*
• Thoracolumbar fascia, anterior layer (Quadratus lumborum fascia) • Transversalis fascia, lateral part

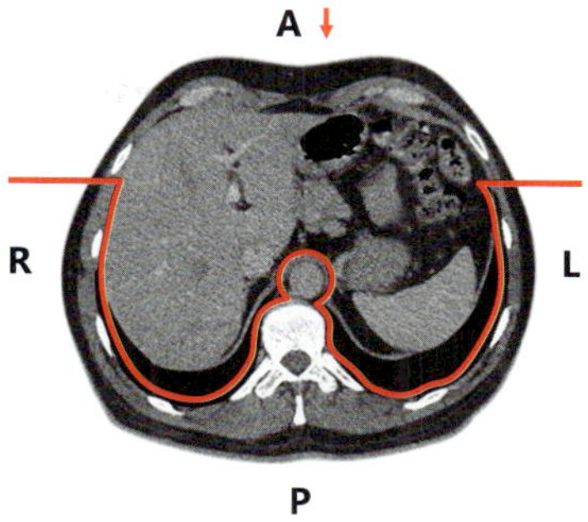

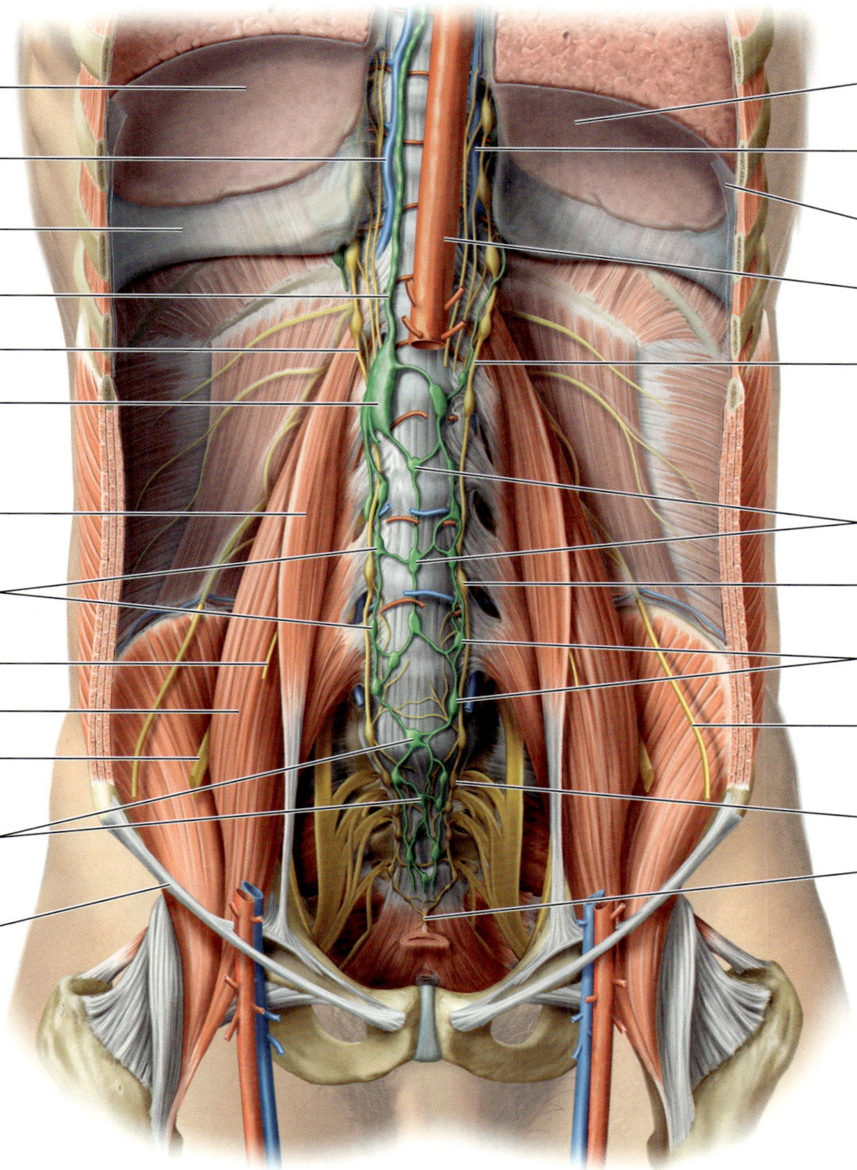

CONTENTS OF THE ABDOMEN (In bold are the structures added with respect to the preceding table)		
• Abdominal aorta • Cisterna chyli • Ganglion impar • Genitofemoral nerve • Lateral aortic nodes	• Left sympathetic trunk • Lumbar ganglia of sympathetic trunk • Para-aortic nodes • Postcaval nodes • Promontorial and sacral nodes	• Psoas major • Psoas minor • Right sympathetic trunk • Sacral ganglia of sympathetic trunk • Thoracic duct

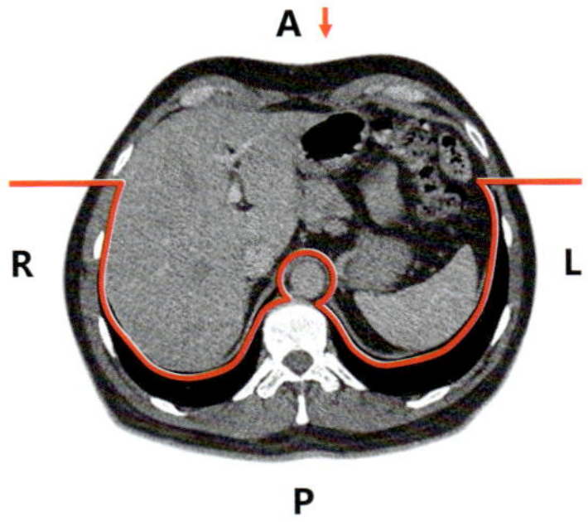

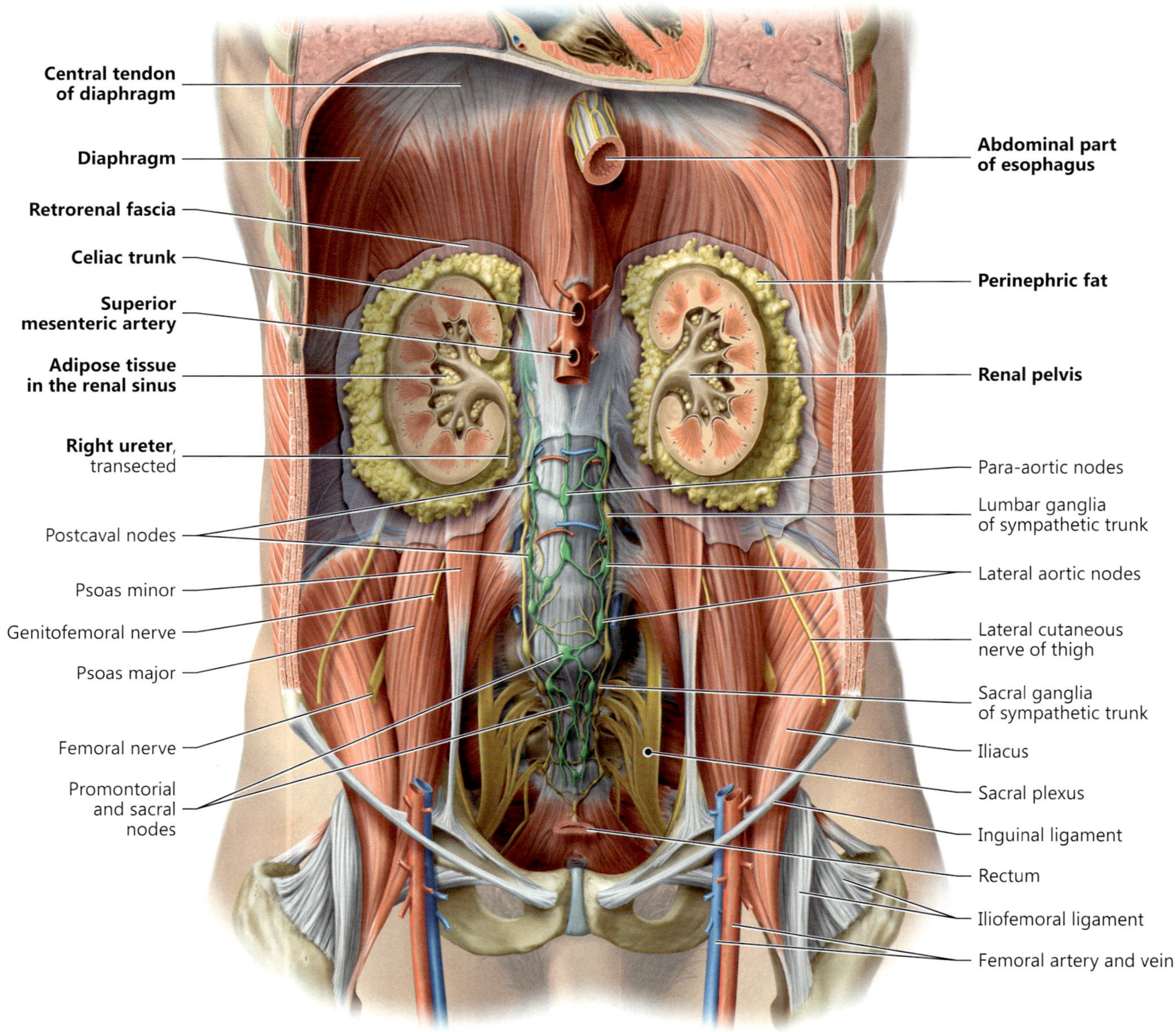

CONTENTS OF THE ABDOMEN (In bold are the structures added with respect to the preceding table)	
• Abdominal part of esophagus • Adipose tissue in the renal sinus • Celiac trunk • Central tendon of diaphragm • Diaphragm	• Perinephric fat • Renal pelvis • Retrorenal fascia • Right ureter • Superior mesenteric artery

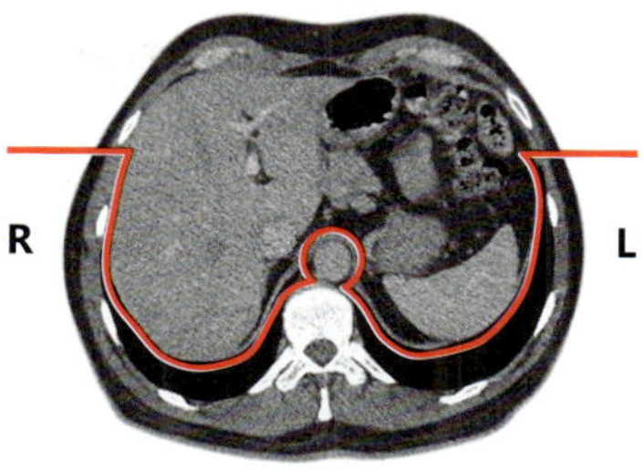

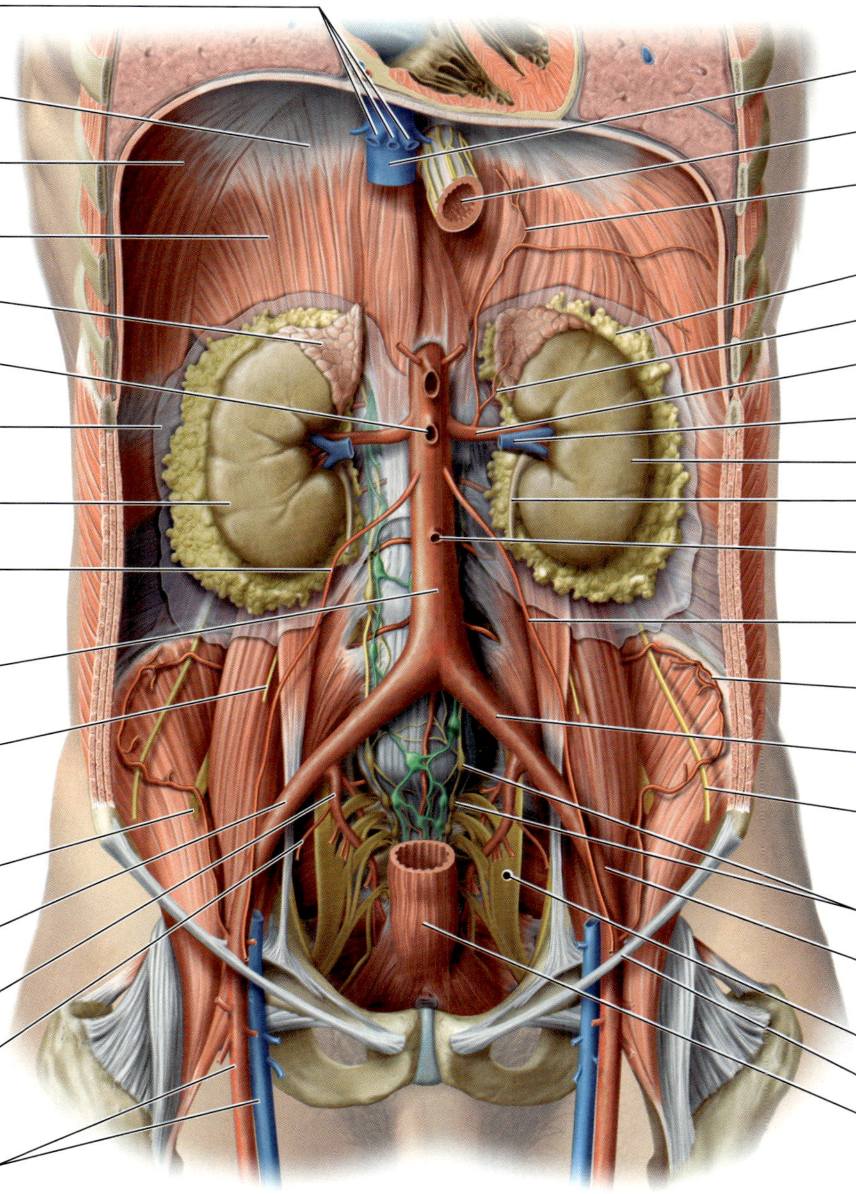

CONTENTS OF THE ABDOMEN (In bold are the structures added with respect to the preceding table)		
• Abdominal aorta • Deep circumflex iliac artery • Diaphragm, costal and lumbar parts • External iliac artery • Inferior mesenteric artery • Inferior phrenic artery	• Inferior suprarenal arteries • Inferior vena cava • Kidney • Left common iliac artery • Left renal artery • Left renal vein	• Major hepatic veins • Ovarian artery • Right internal iliac artery • Right suprarenal gland • Right umbilical artery • Testicular artery

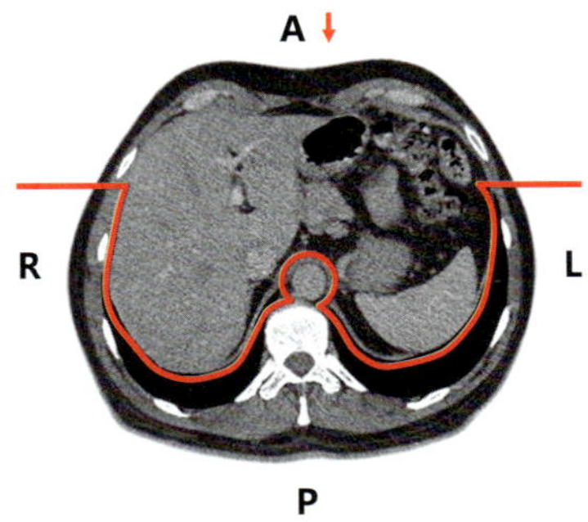

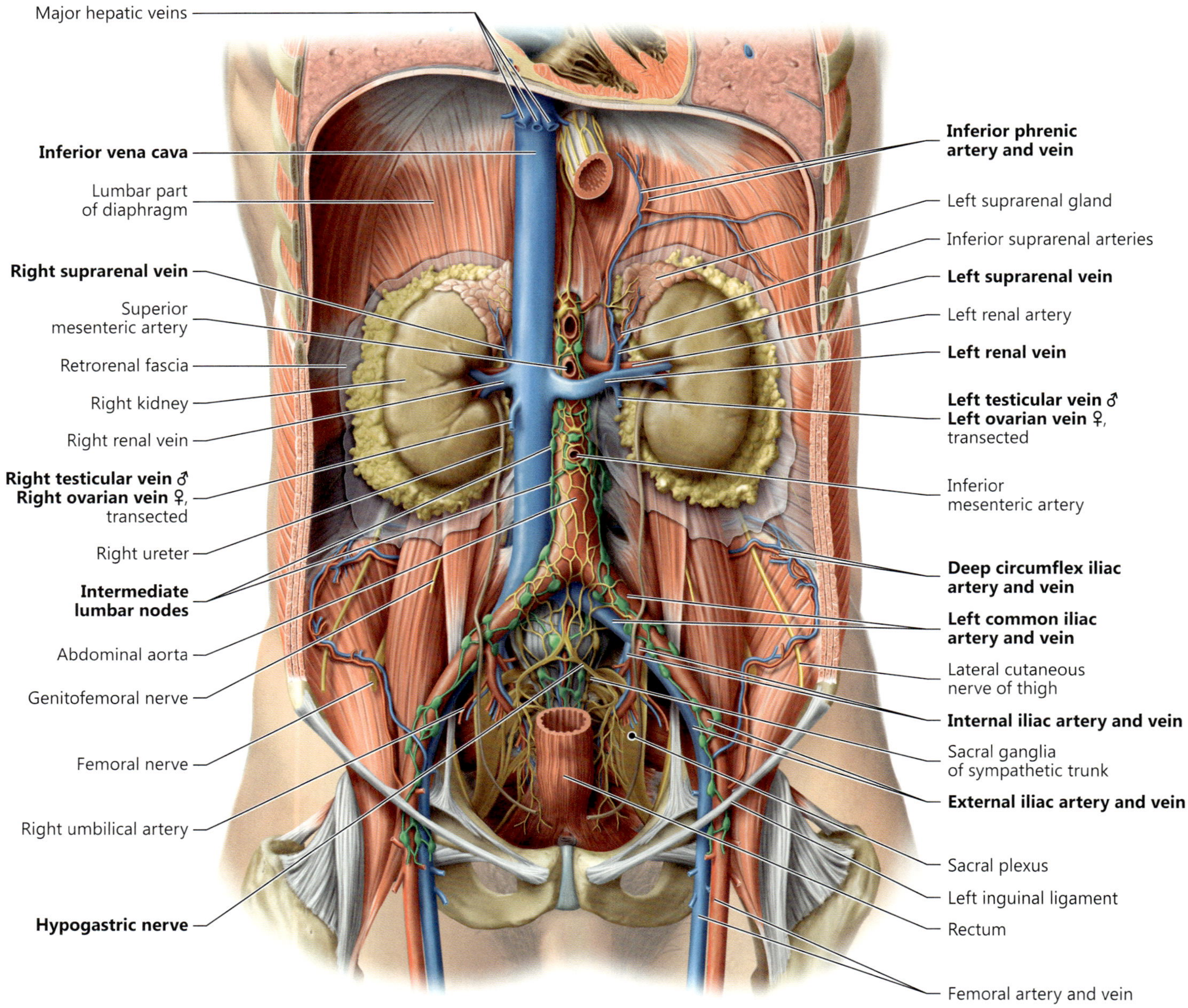

CONTENTS OF THE ABDOMEN (In bold are the structures added with respect to the preceding table)	
• Deep circumflex iliac artery • Deep circumflex iliac vein • External iliac artery • External iliac vein • Hypogastric nerve • Inferior phrenic artery • Inferior phrenic vein • Inferior vena cava • Intermediate lumbar nodes	• Internal iliac artery • Internal iliac vein • Left common iliac artery • Left common iliac vein • Left renal vein • Ovarian vein ♀ • Suprarenal vein • Testicular vein ♂

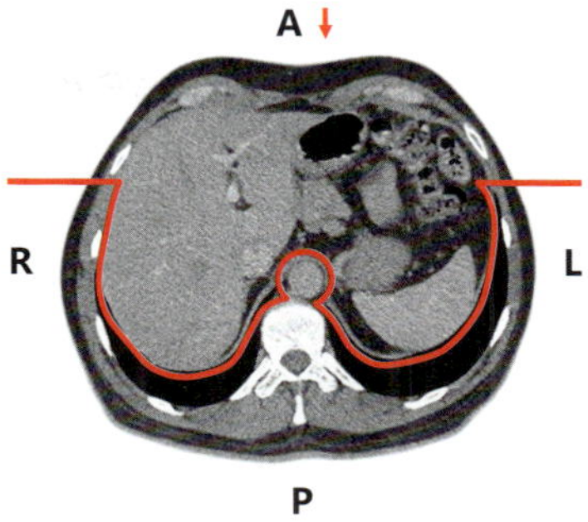

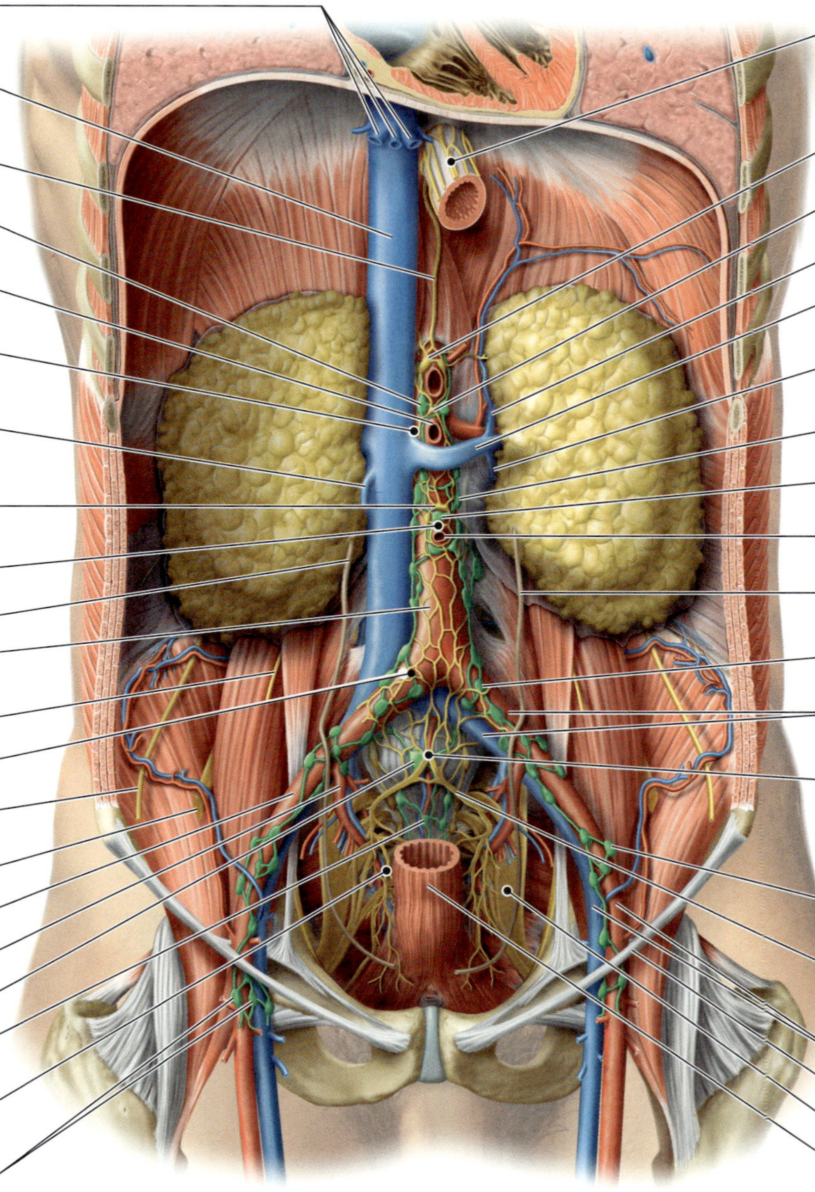

CONTENTS OF THE ABDOMEN (In bold are the structures added with respect to the preceding table)	
• Celiac branches of posterior vagal trunk • Celiac ganglia • Deep inguinal nodes • Esophageal plexus • Iliac plexus • Inferior mesenteric ganglion • Inferior mesenteric plexus	• Lateral cutaneous nerve of thigh • Left hypogastric nerve • Right inferior hypogastric plexus (Right pelvic plexus) • Superior hypogastric plexus (Presacral nerve) • Superior mesenteric ganglion • Superior mesenteric plexus

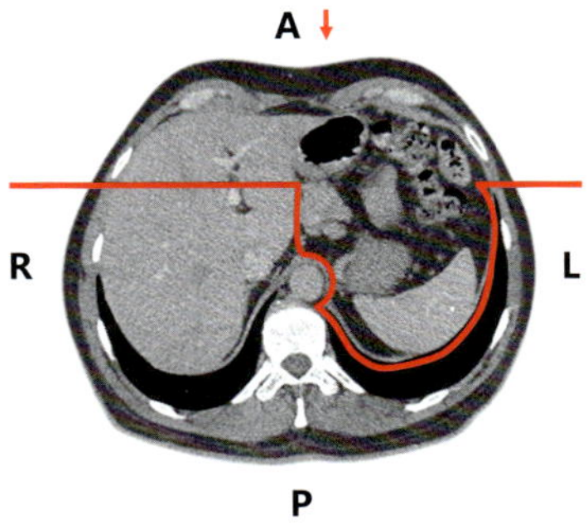

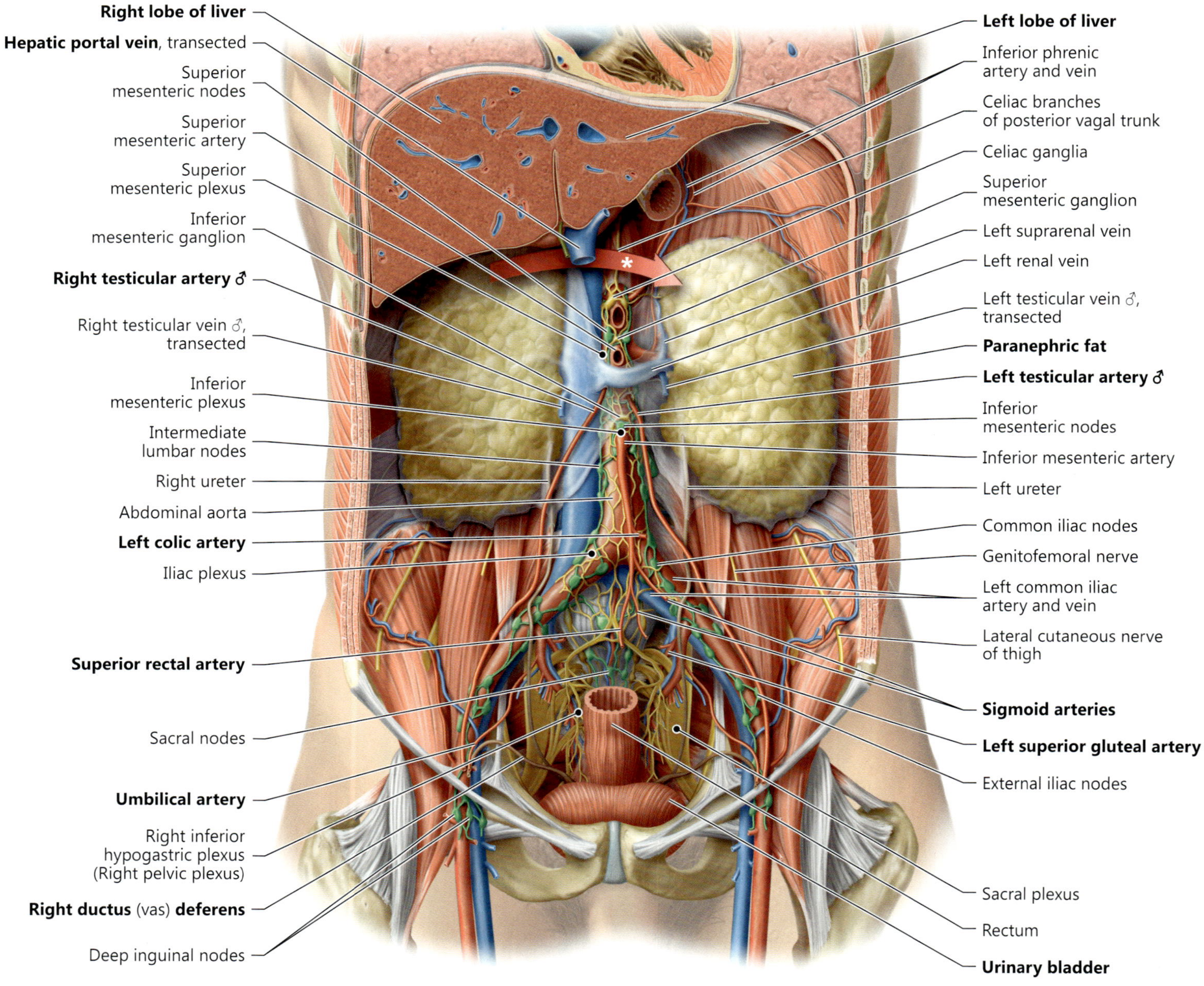

From this level on, the stratigraphy refers to the male, as reflected by the organization of the organs of the lesser pelvis, and by the course of the genital vessels (testicular artery in this figure). In females, the genital vessels head towards the lesser pelvis to reach the ovary. For a full description of the course of the ovarian vessels, ☞ Pelvis and Perineum, Contents of the Pelvis and Perineum, Female Frontal View, pages 259-262.

* The red arrow indicates the omental foramen (or epiploic foramen or foramen of Winslow), space between the terminal part of the hepatic portal vein, anteriorly, the inferior vena cava, posteriorly, the caudate lobe of liver, superiorly, and the first part of duodenum, inferiorly.

CONTENTS OF THE ABDOMEN (In bold are the structures added with respect to the preceding table)	
• Hepatic portal vein • Left colic artery • Left superior gluteal artery • Lobes of liver • Paranephric fat • Right ductus (vas) deferens	• Sigmoid arteries • Superior rectal artery • Testicular artery ♂ • Umbilical artery • Urinary bladder

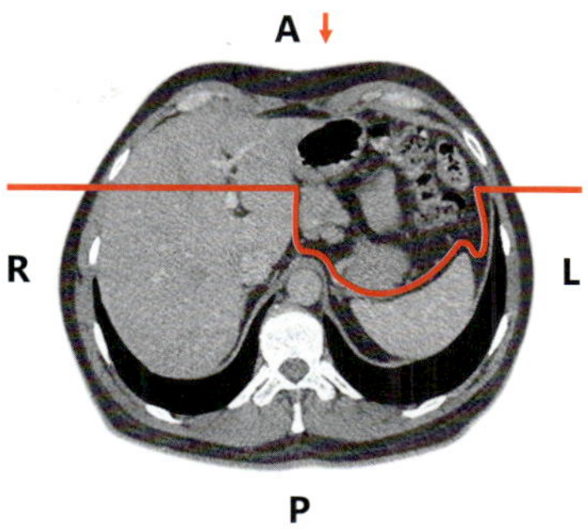

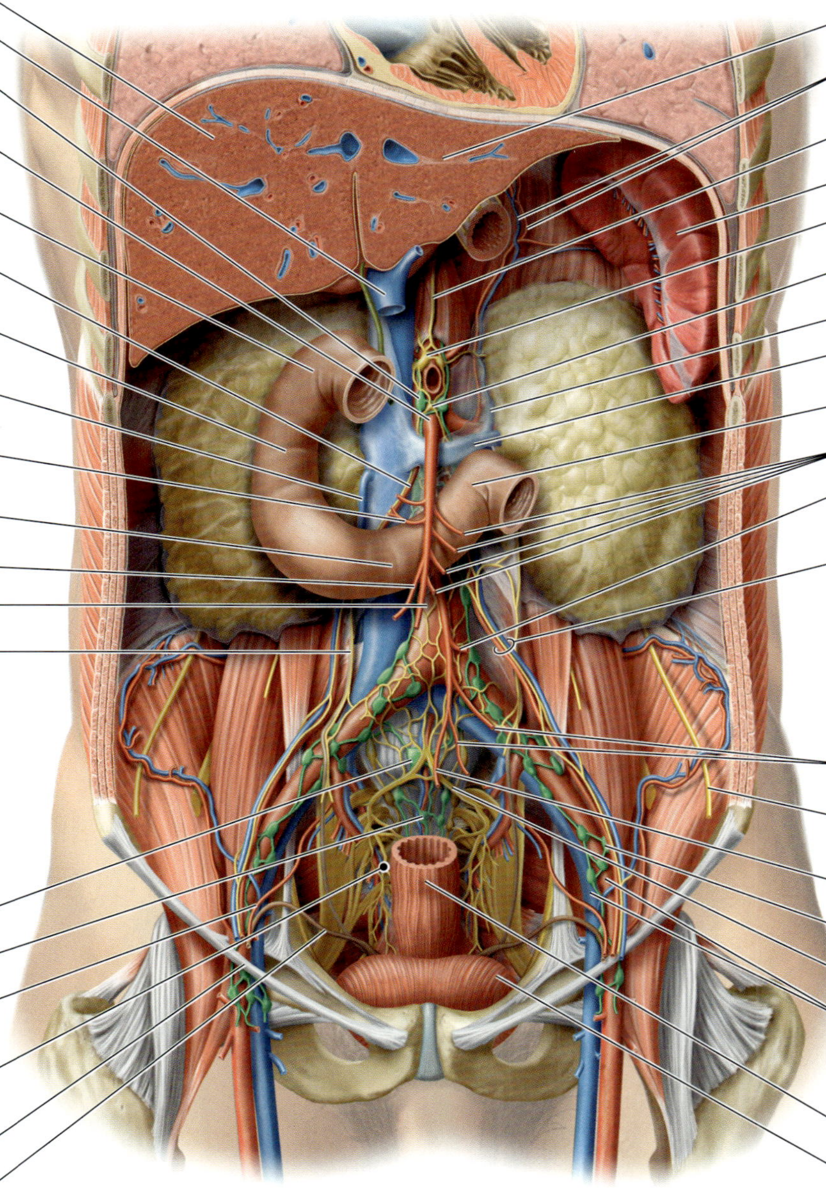

CONTENTS OF THE ABDOMEN (In bold are the structures added with respect to the preceding table)	
• Duodenum, ascending, descending, inferior, superior parts • Ileocolic artery • Inferior pancreaticoduodenal artery • Jejunal arteries • Left testicular artery ♂	• Left testicular plexus ♂ • Left testicular vein ♂ • Right ureter • Right ureteric plexus • Spleen

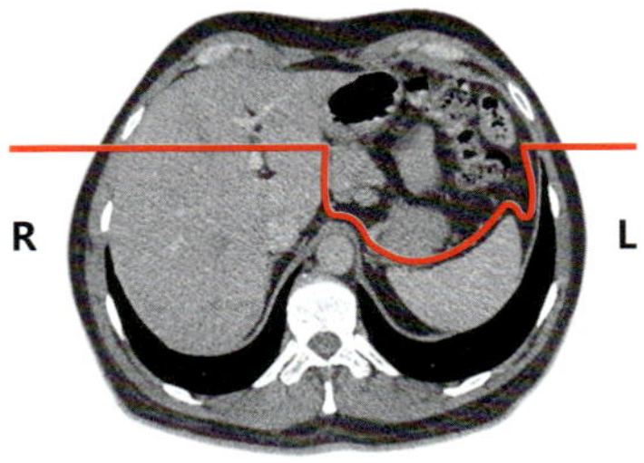

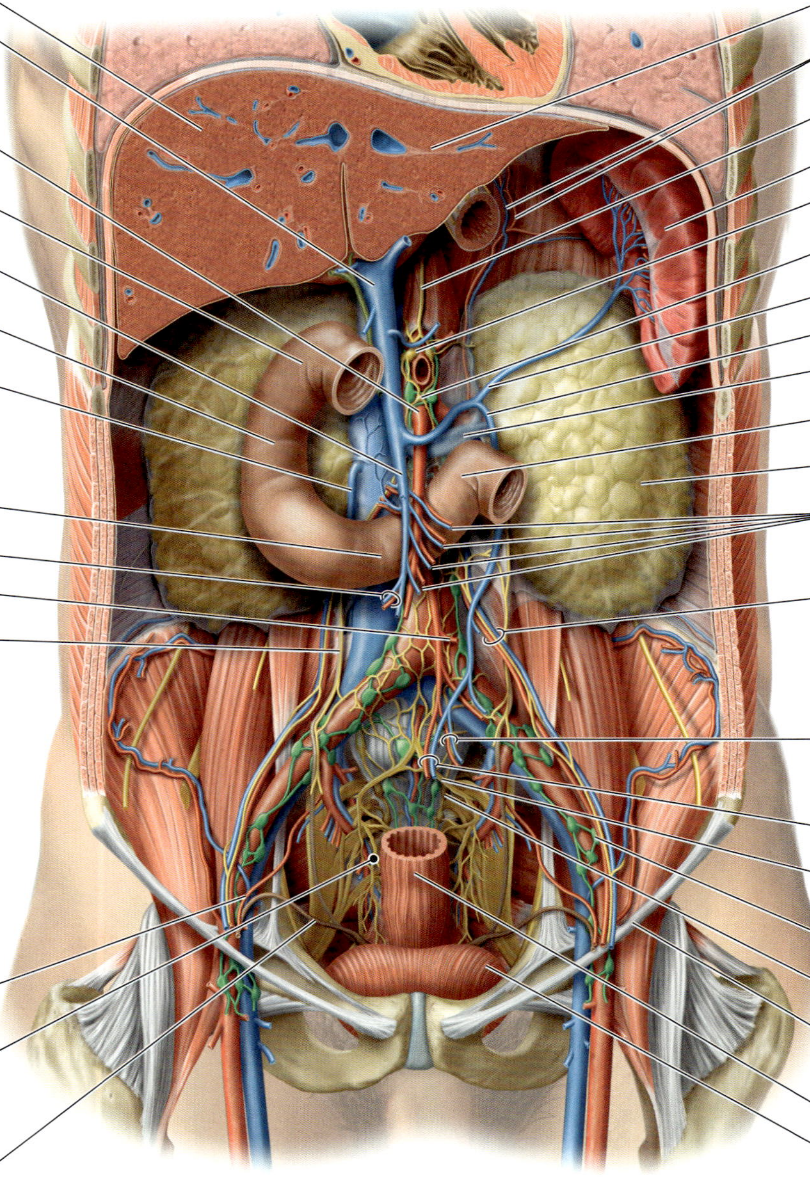

CONTENTS OF THE ABDOMEN (In bold are the structures added with respect to the preceding table)	
• Hepatic portal vein	• Sigmoid arteries
• Ileocolic artery	• Sigmoid veins
• Ileocolic vein	• Splenic vein
• Inferior mesenteric vein	• Superior mesenteric vein
• Jejunal arteries	• Superior rectal artery
• Jejunal veins	• Superior rectal vein
• Sacral ganglia of sympathetic trunk	

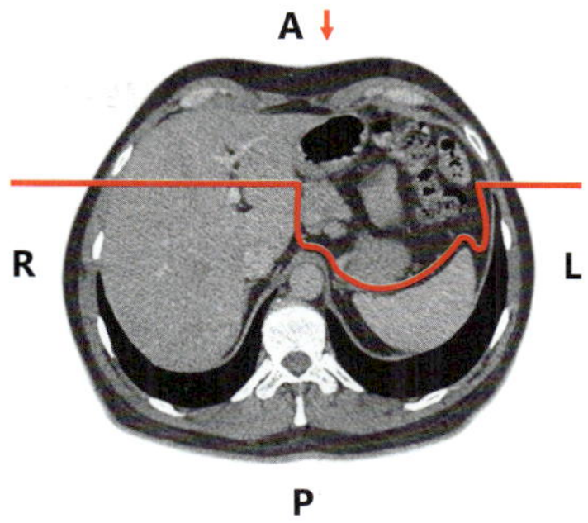

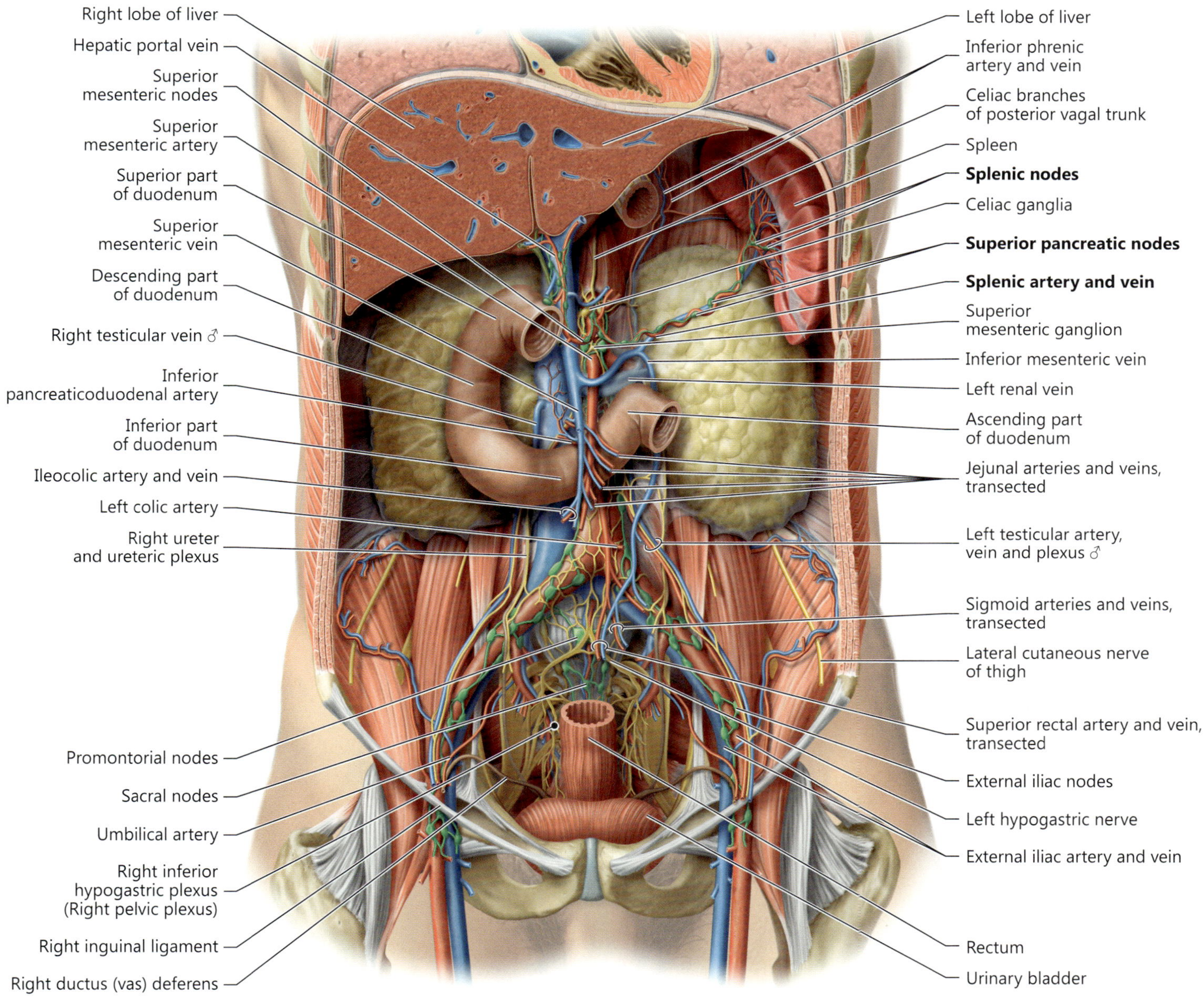

CONTENTS OF THE ABDOMEN (In bold are the structures added with respect to the preceding table)
• Splenic artery • Splenic nodes • Splenic vein • Superior pancreatic nodes

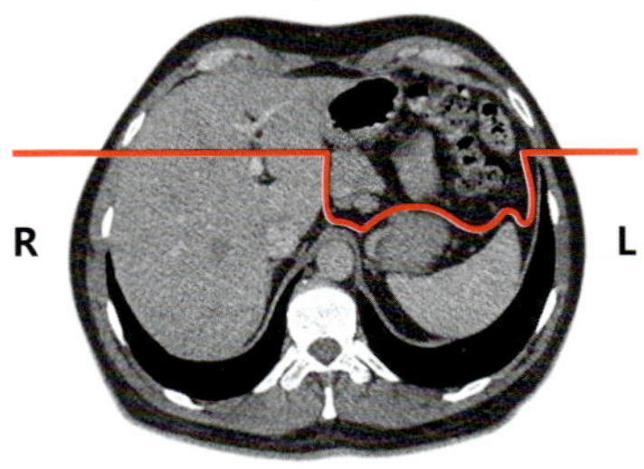

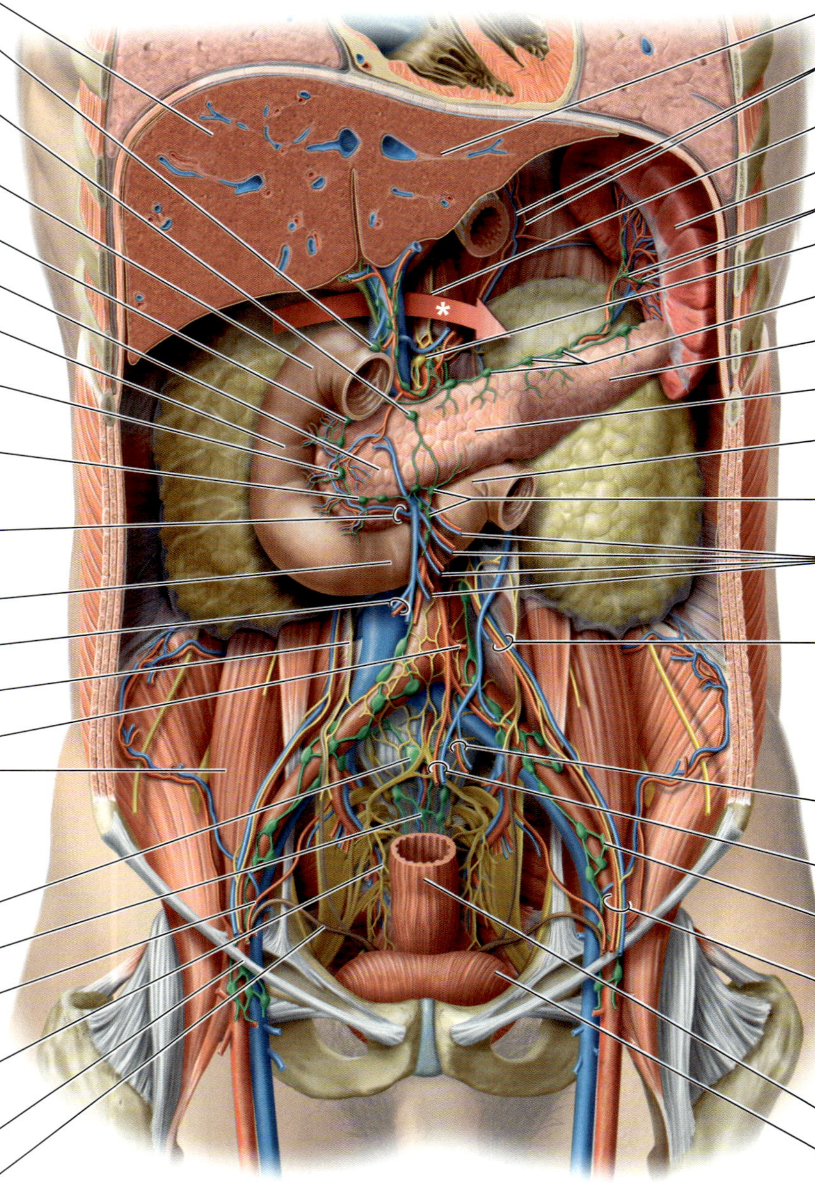

* The red arrow indicates the omental foramen (or epiploic foramen or foramen of Winslow), space between the terminal part of the hepatic portal vein, anteriorly, the inferior vena cava, posteriorly, the caudate lobe of liver, superiorly, and the first part of duodenum, inferiorly.

CONTENTS OF THE ABDOMEN (In bold are the structures added with respect to the preceding table)	
• Body of pancreas • Head of pancreas • Inferior pancreatic nodes • Inferior pancreaticoduodenal artery • Inferior pancreaticoduodenal nodes • Inferior pancreaticoduodenal vein	• Subpyloric nodes • Superior mesenteric nodes • Superior pancreaticoduodenal nodes • Suprapyloric node • Tail of pancreas

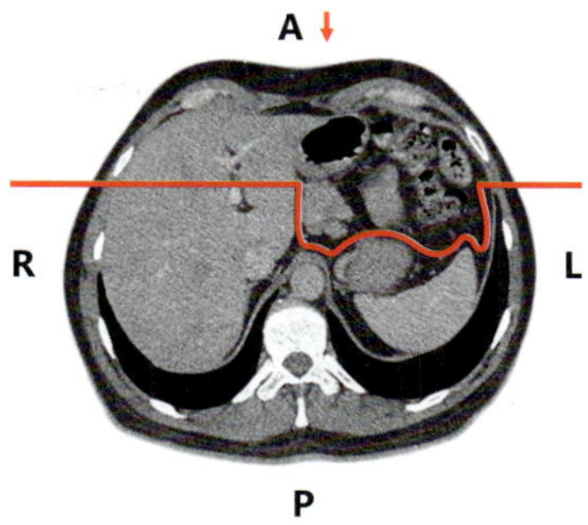

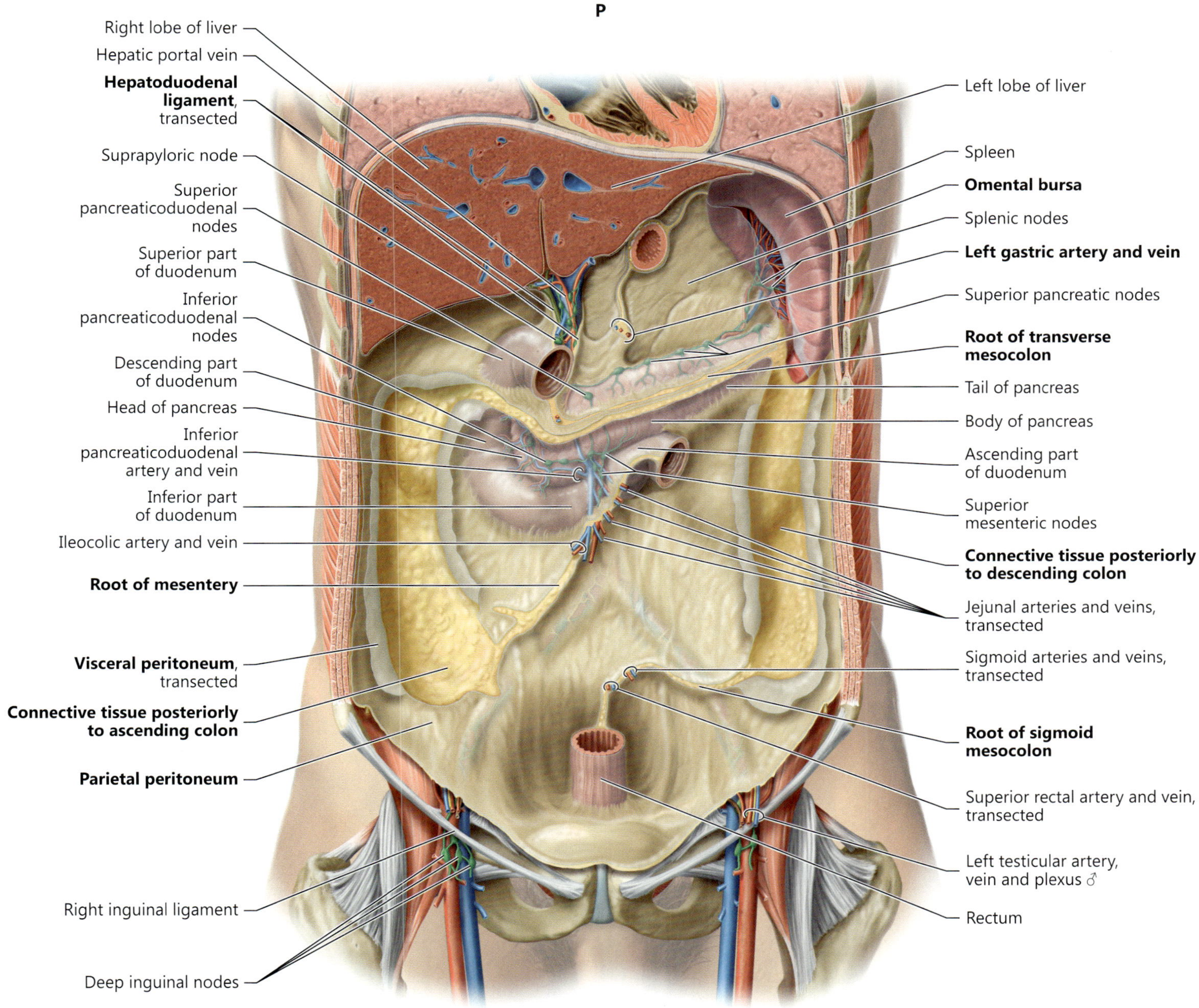

CONTENTS OF THE ABDOMEN (In bold are the structures added with respect to the preceding table)	
• Ascending colon, posterior connective tissue • Descending colon, posterior connective tissue • Hepatoduodenal ligament • Left gastric artery • Left gastric vein • Omental bursa	• Parietal peritoneum • Root of mesentery • Root of sigmoid mesocolon • Root of transverse mesocolon • Visceral peritoneum

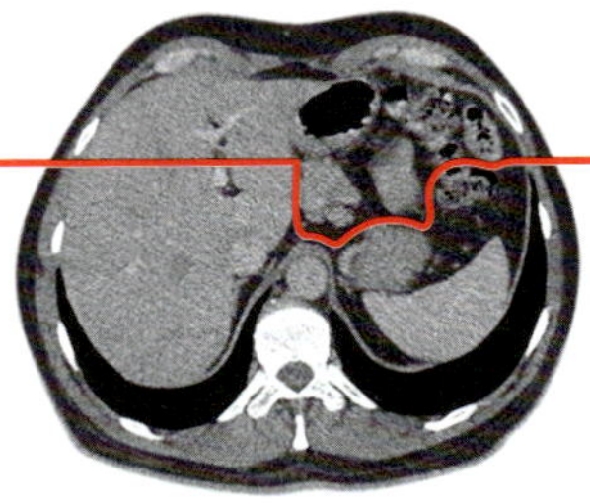

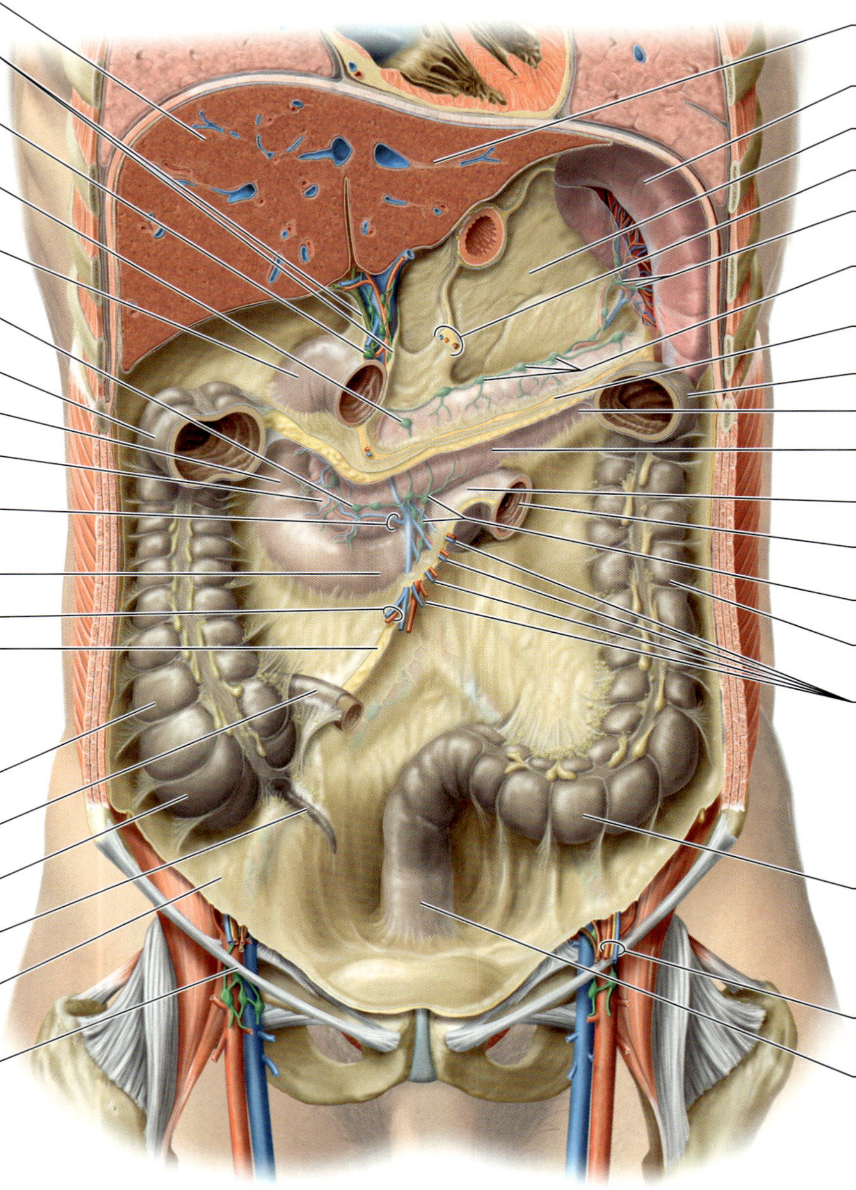

CONTENTS OF THE ABDOMEN (In bold are the structures added with respect to the preceding table)	
• Appendix • Ascending colon • Cecum • Descending colon • Duodenojejunal flexure	• Left colic flexure • Right colic flexure • Sigmoid colon • Terminal ileum

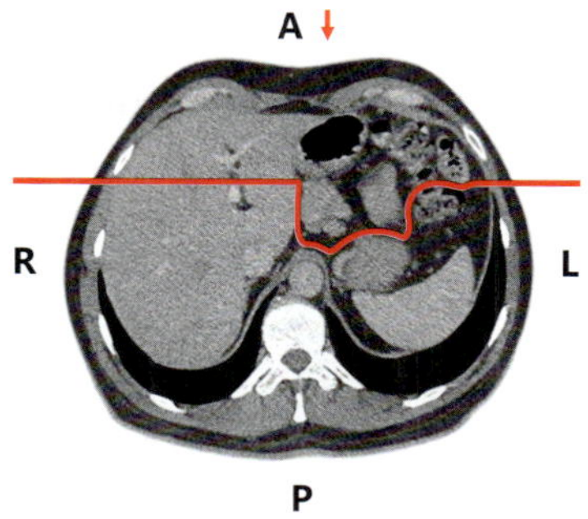

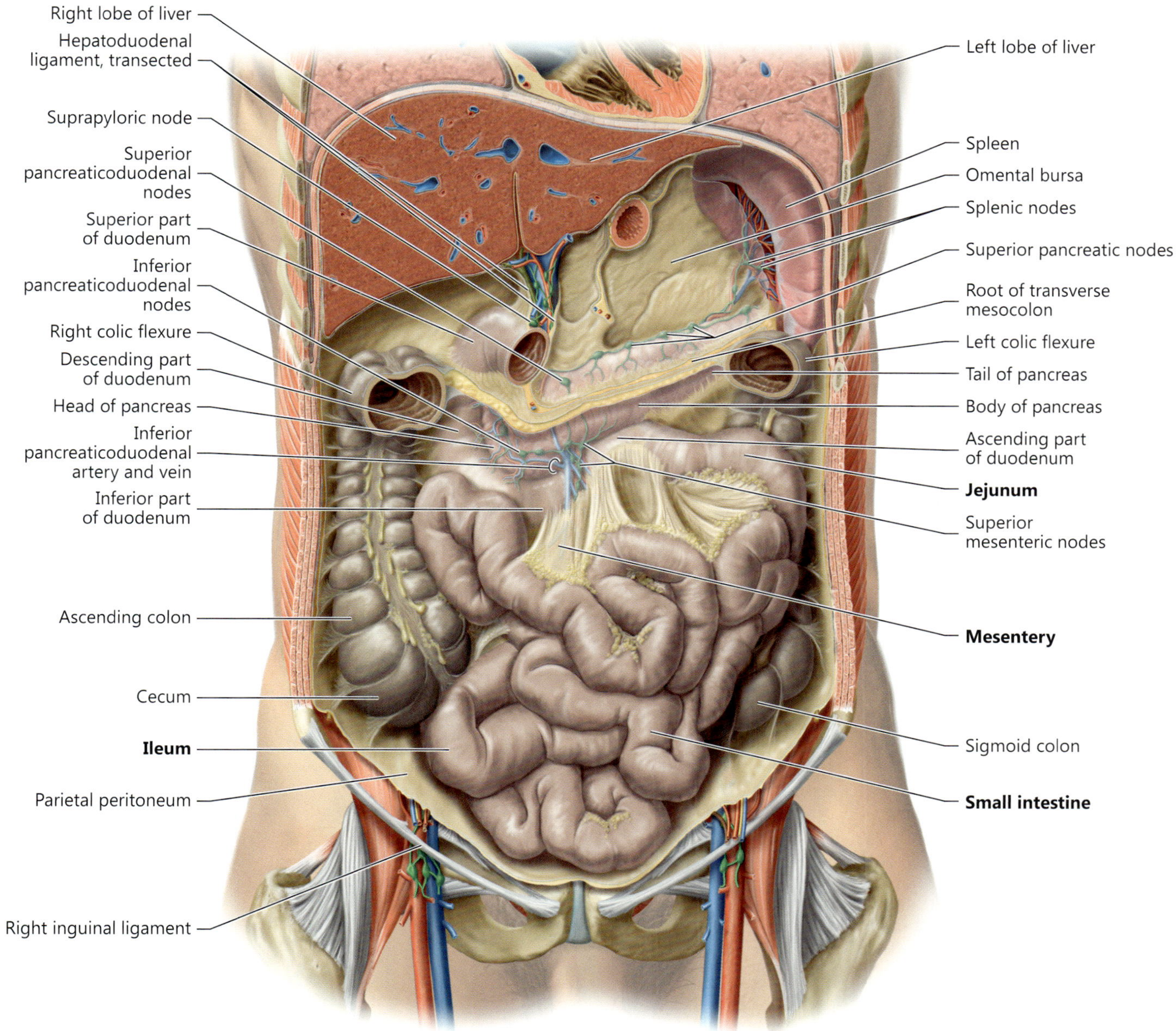

CONTENTS OF THE ABDOMEN (In bold are the structures added with respect to the preceding table)
• Ileum • Jejunum • Mesentery • Small intestine

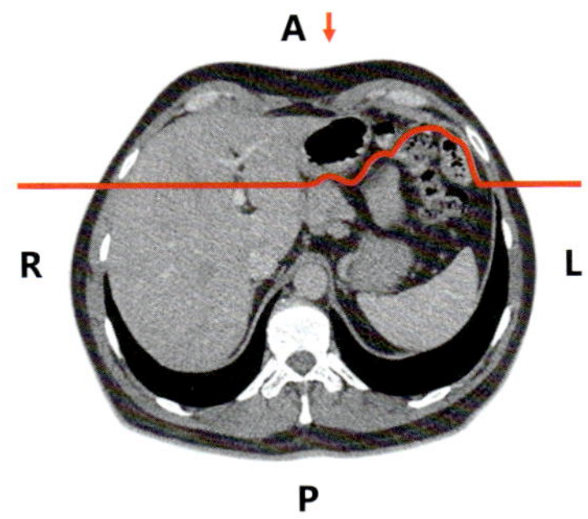

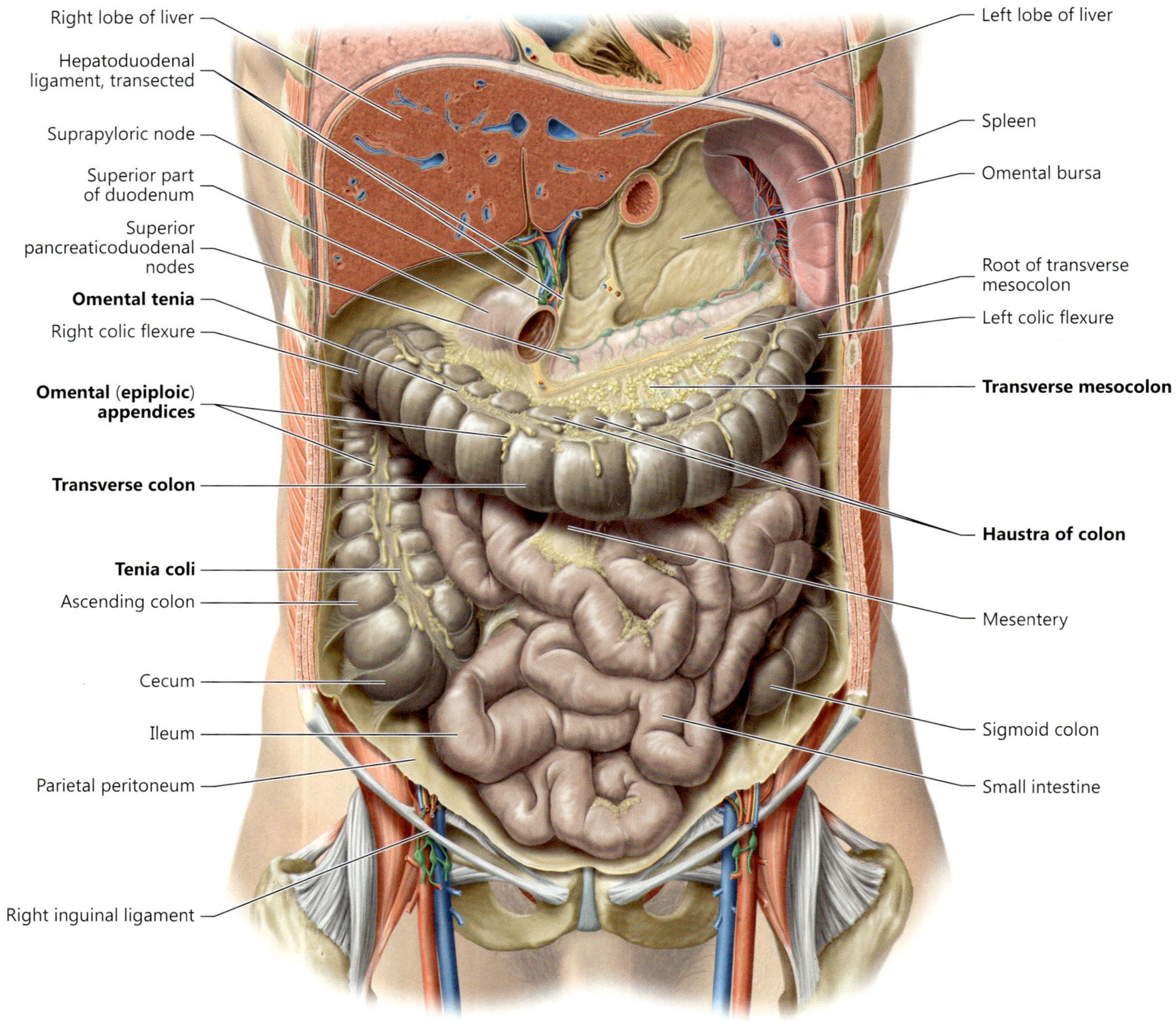

CONTENTS OF THE ABDOMEN (In bold are the structures added with respect to the preceding table)	
• Haustra of colon • Omental (epiploic) appendices • Omental tenia	• Tenia coli • Transverse colon • Transverse mesocolon

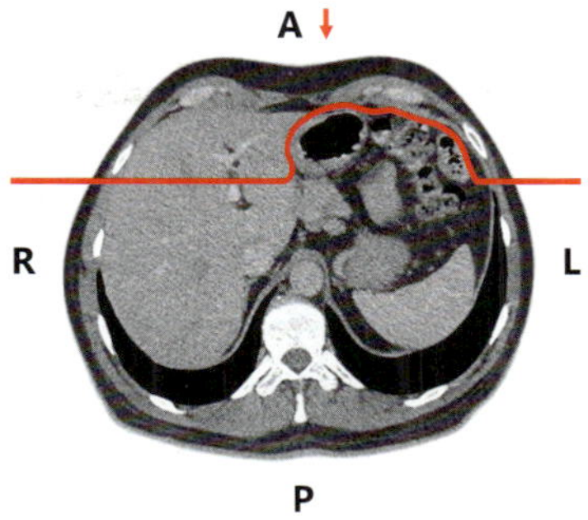

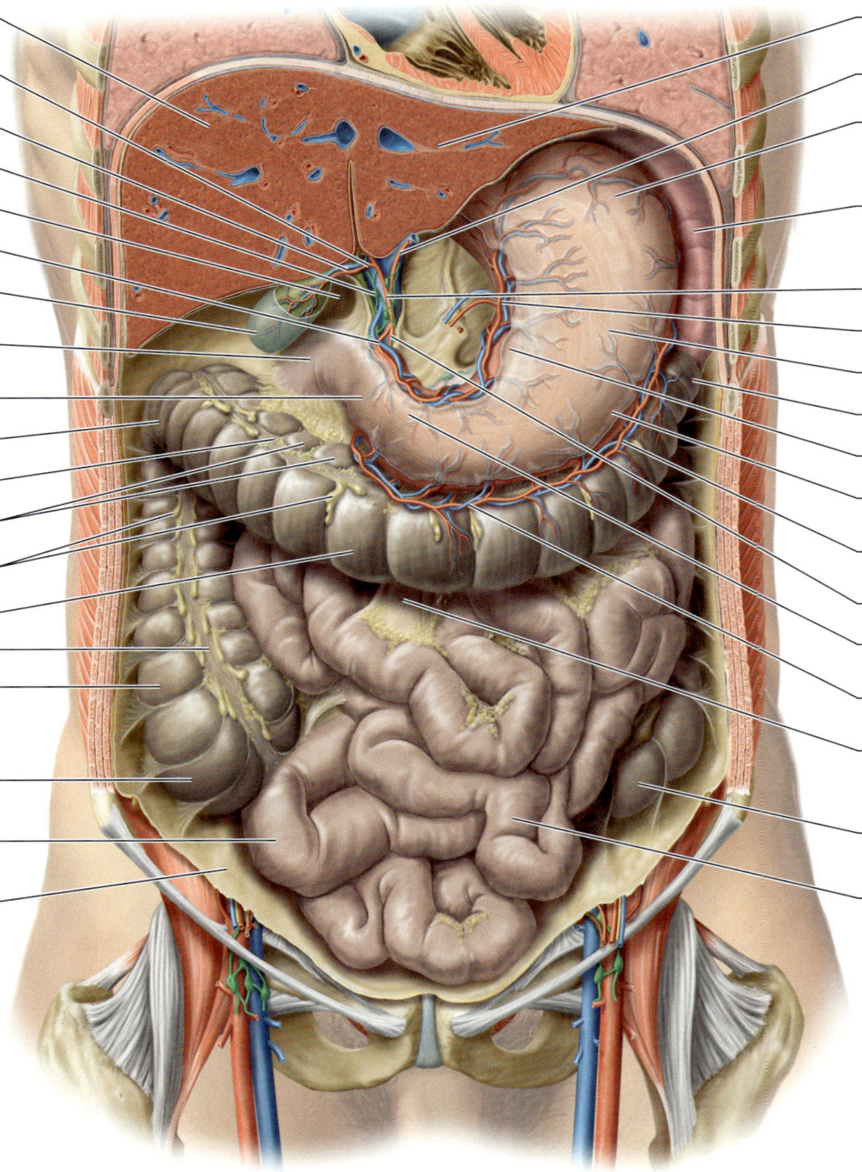

CONTENTS OF THE ABDOMEN (In bold are the structures added with respect to the preceding table)		
• Common hepatic artery • Cystic artery • Cystic vein • Gallbladder • Gastric artery • Gastric vein	• Gastro-omental artery • Gastro-omental vein • Hepatic artery proper • Hepatic artery proper, right and left branches • Hepatic nodes	• Omental foramen • Pyloric antrum • Pyloric canal • Stomach, body, fundus, lesser curvature and greater curvature

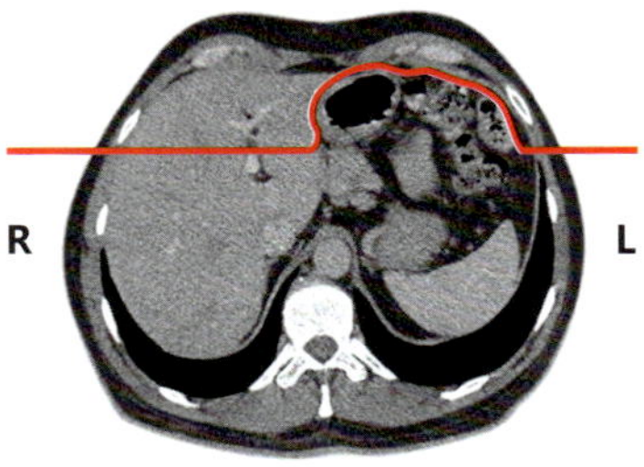

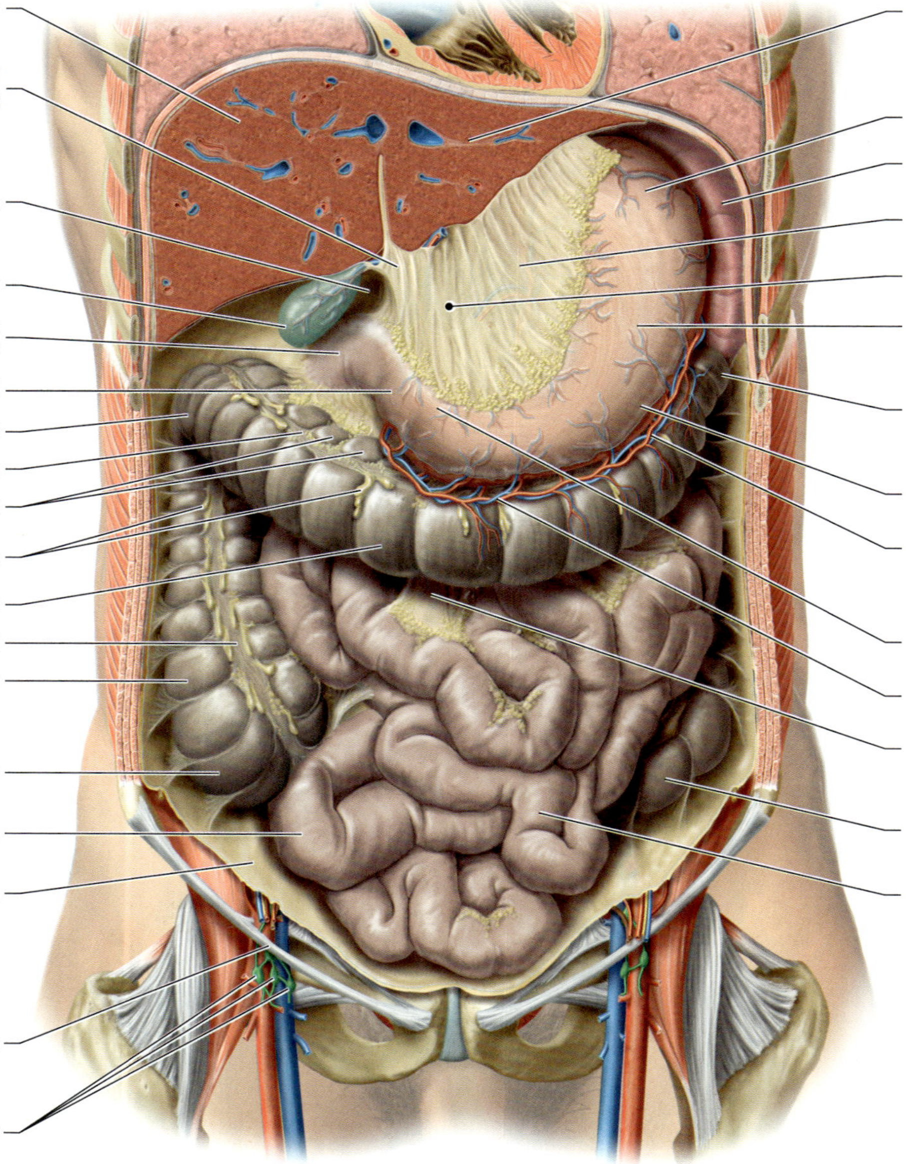

CONTENTS OF THE ABDOMEN (In bold are the structures added with respect to the preceding table)
• Hepatoduodenal ligament • Hepatogastric ligament • Lesser omentum

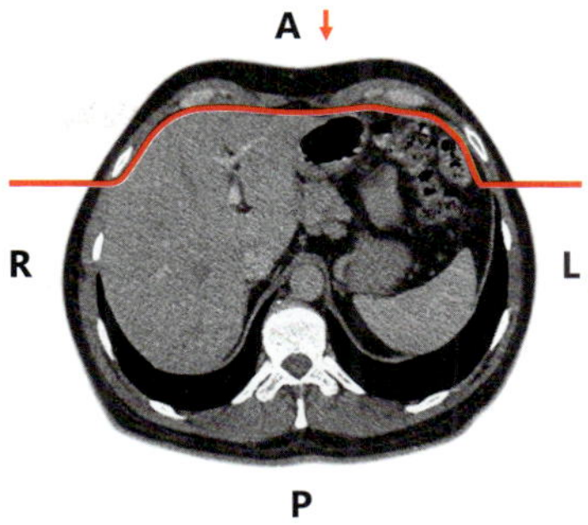

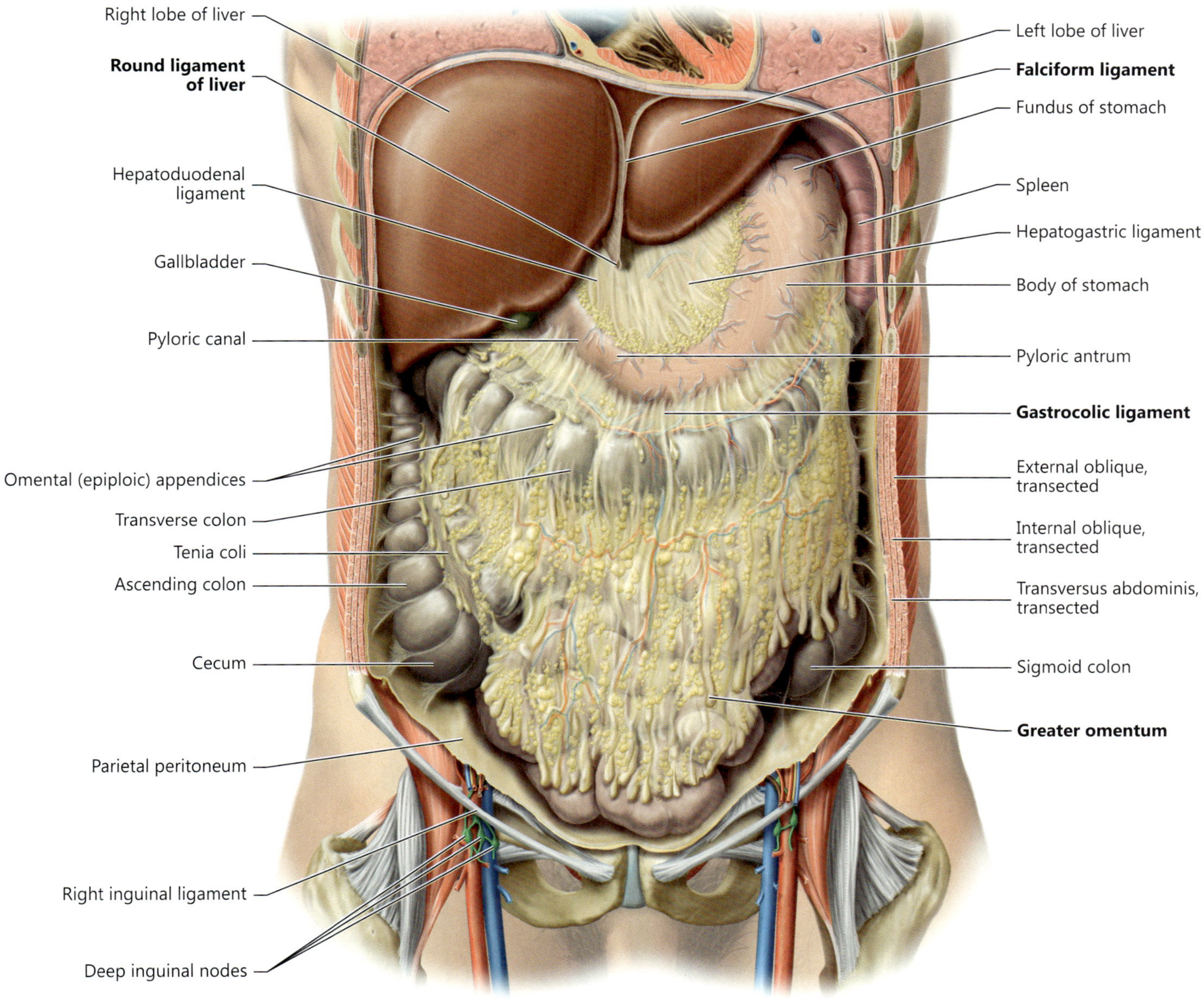

CONTENTS OF THE ABDOMEN (In bold are the structures added with respect to the preceding table)
• Falciform ligament • Gastrocolic ligament • Greater omentum • Round ligament of liver

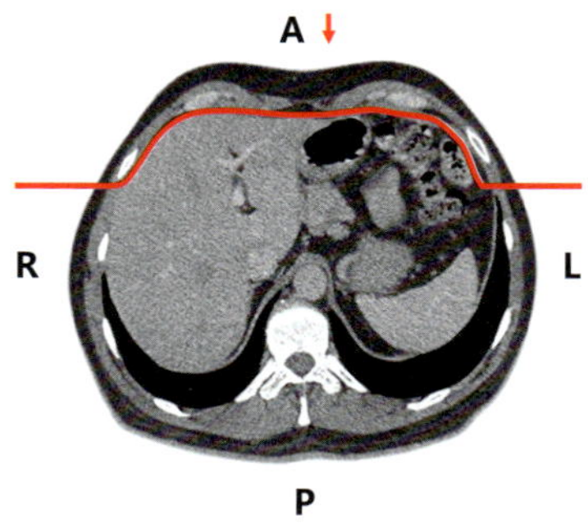

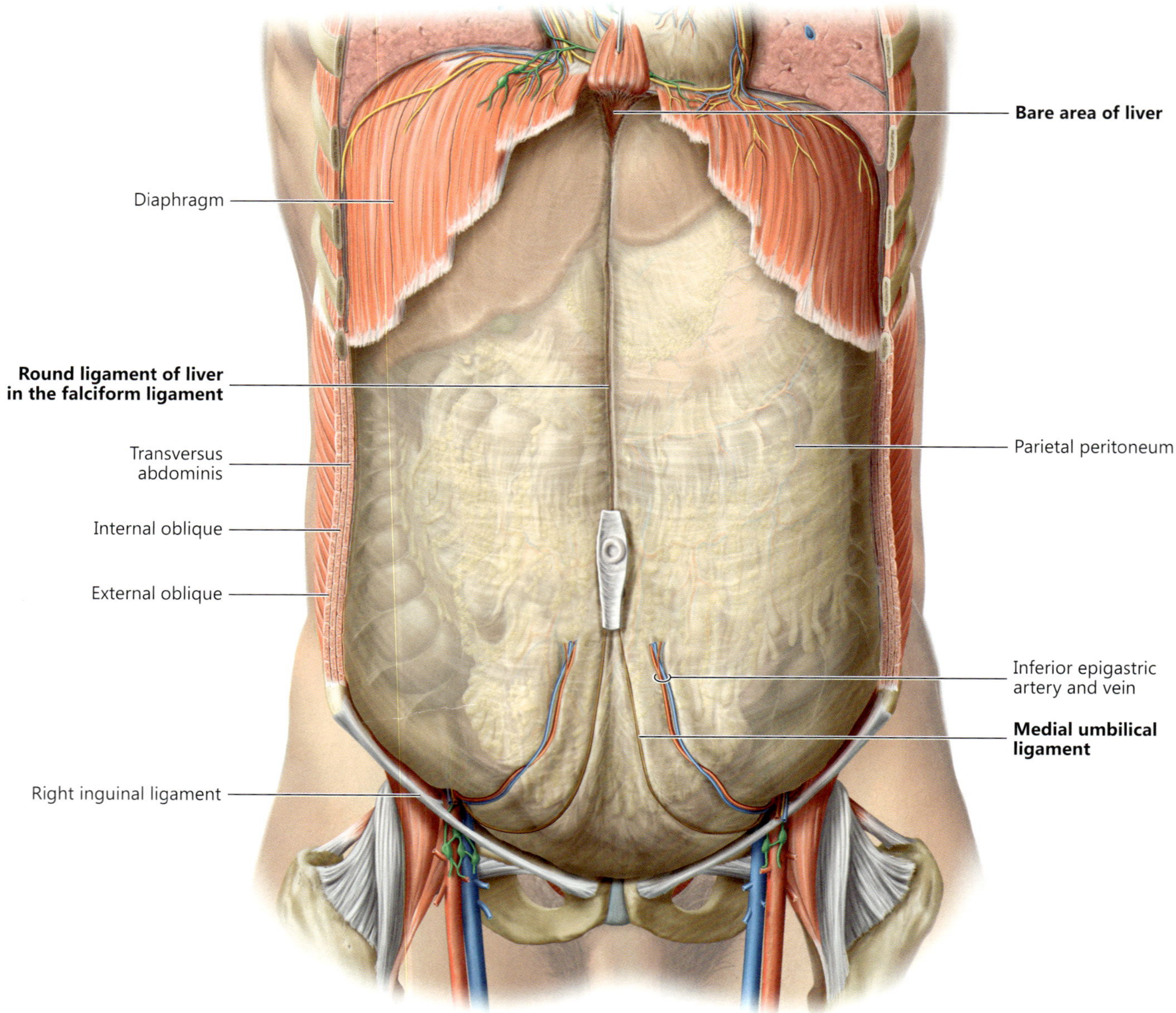

***CONTENTS OF THE ABDOMEN** (In bold are the structures added with respect to the preceding table)*
• Liver, bare area • Medial umbilical ligament • Round ligament of liver in the falciform ligament

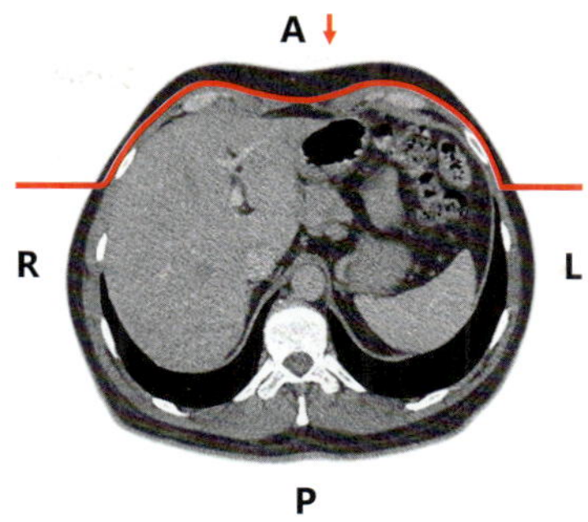

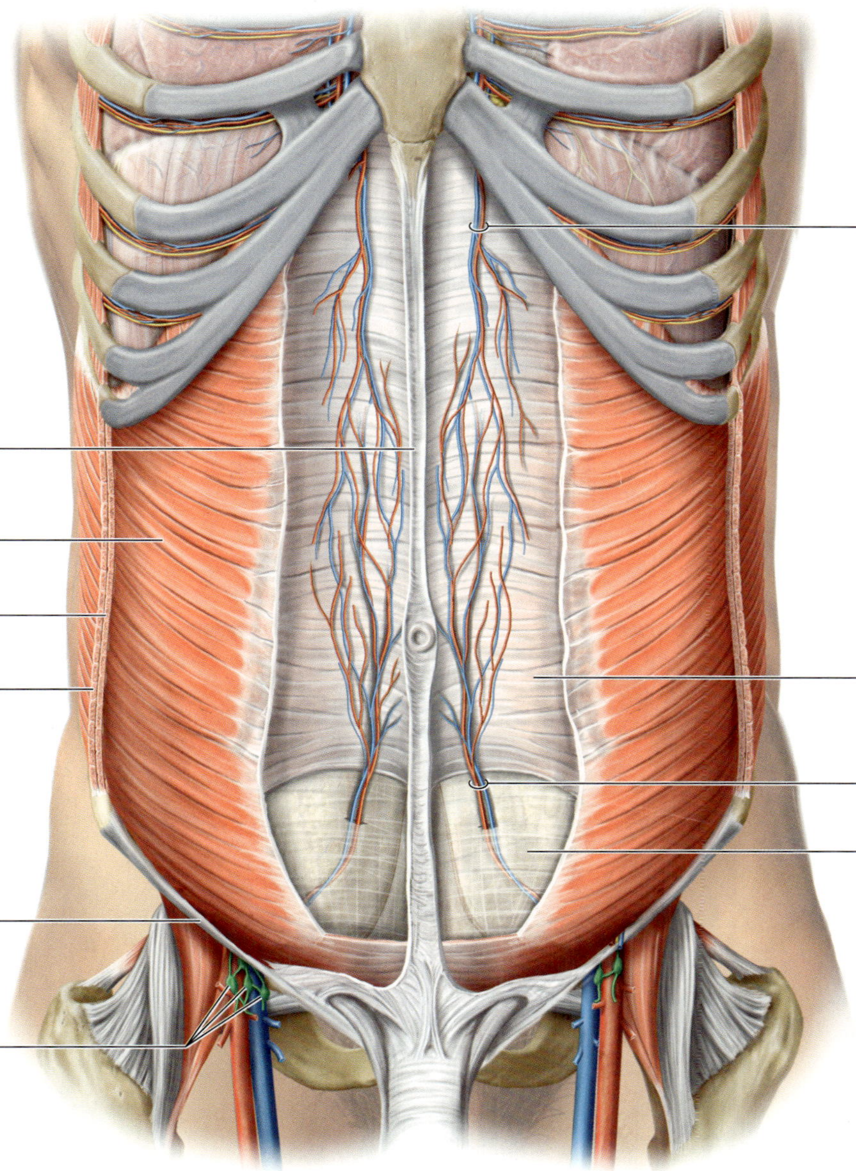

CONTENTS OF THE ABDOMEN (In bold are the structures added with respect to the preceding table)
• Linea alba • Rectus sheath, posterior layer, aponeurosis of transversus abdominis • Superior epigastric artery • Superior epigastric vein • Transversalis fascia

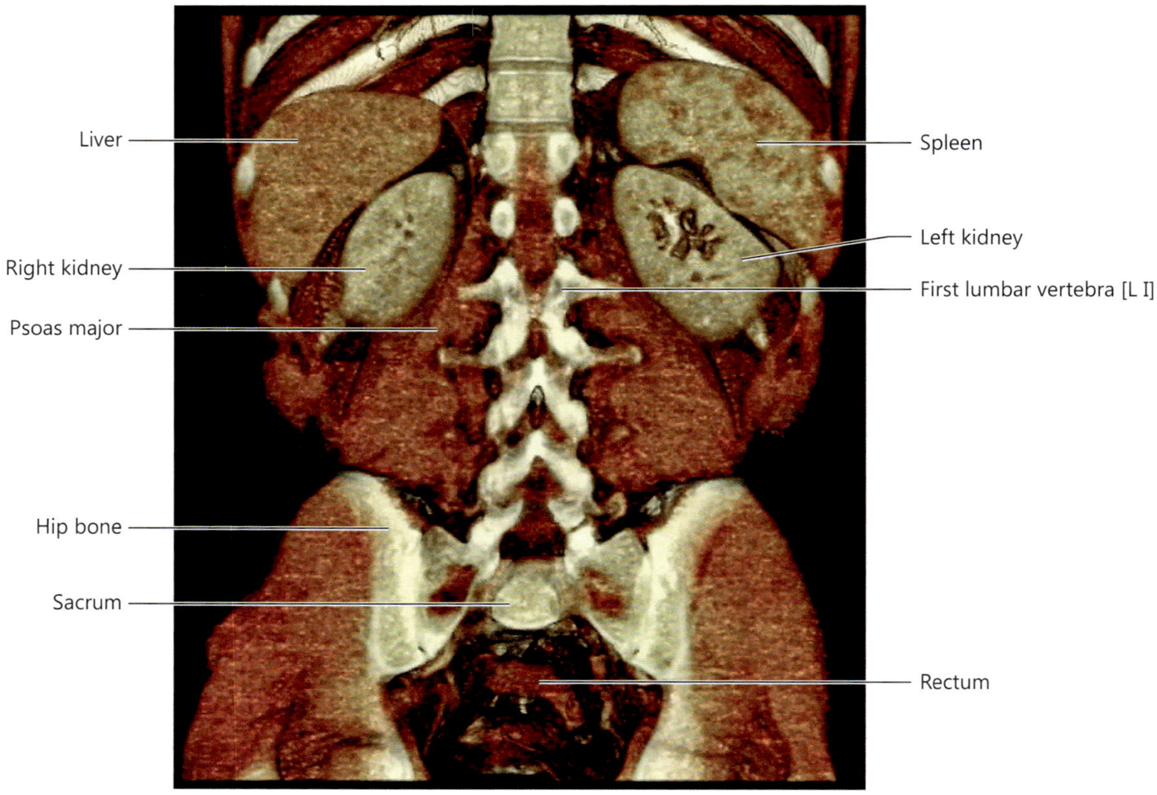

Three-dimensional image of a living human body, obtained by CT scan using a 3D volume rendering technique, showing **abdominal organs** displayed in a frontal section passing through the posterior superior iliac spine.

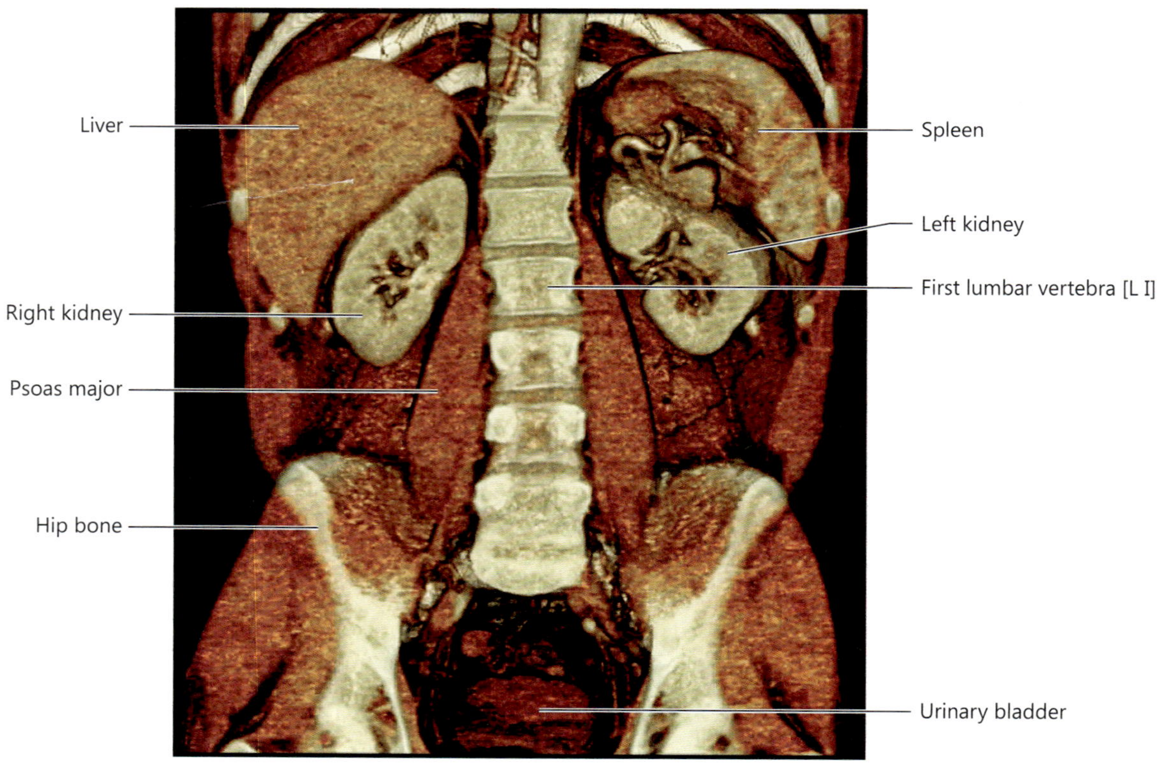

Three-dimensional image of a living human body, obtained by CT scan using a 3D volume rendering technique, showing **abdominal organs** displayed in a frontal section passing 2 cm in front of the posterior superior iliac spine.

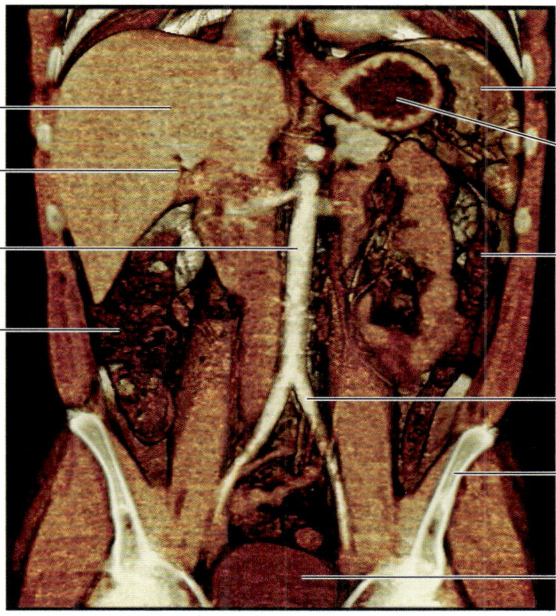

Three-dimensional image of a living human body, obtained by CT scan using a 3D volume rendering technique, showing **abdominal organs** displayed in a frontal section passing 1 cm behind the anterior superior iliac spine.

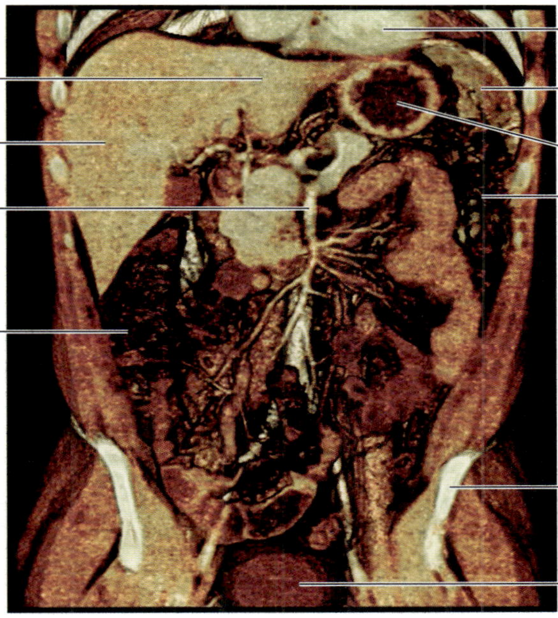

Three-dimensional image of a living human body, obtained by CT scan using a 3D volume rendering technique, showing **abdominal organs** displayed in a frontal section passing through the anterior superior iliac spine.

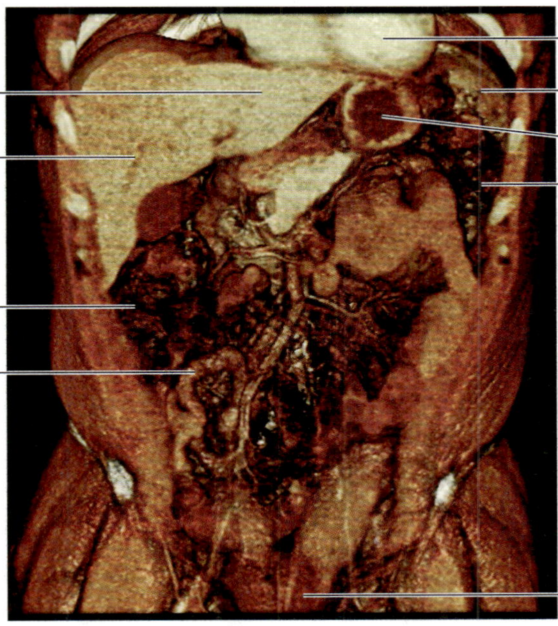

Three-dimensional image of a living human body, obtained by CT scan using a 3D volume rendering technique, showing **abdominal organs** displayed in a frontal section taken along a tangential plane to the posterior surface of the pubic symphysis.

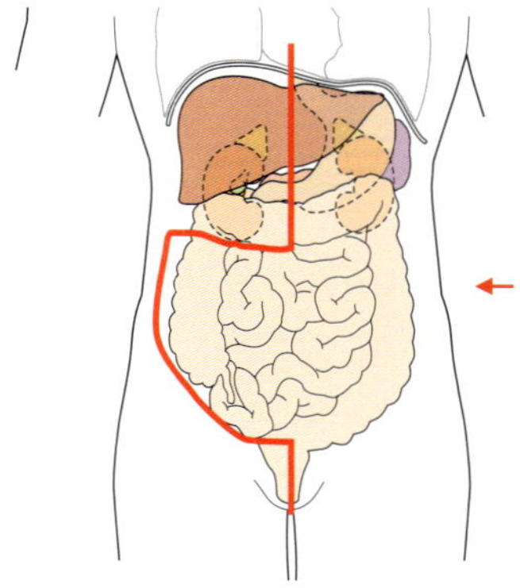

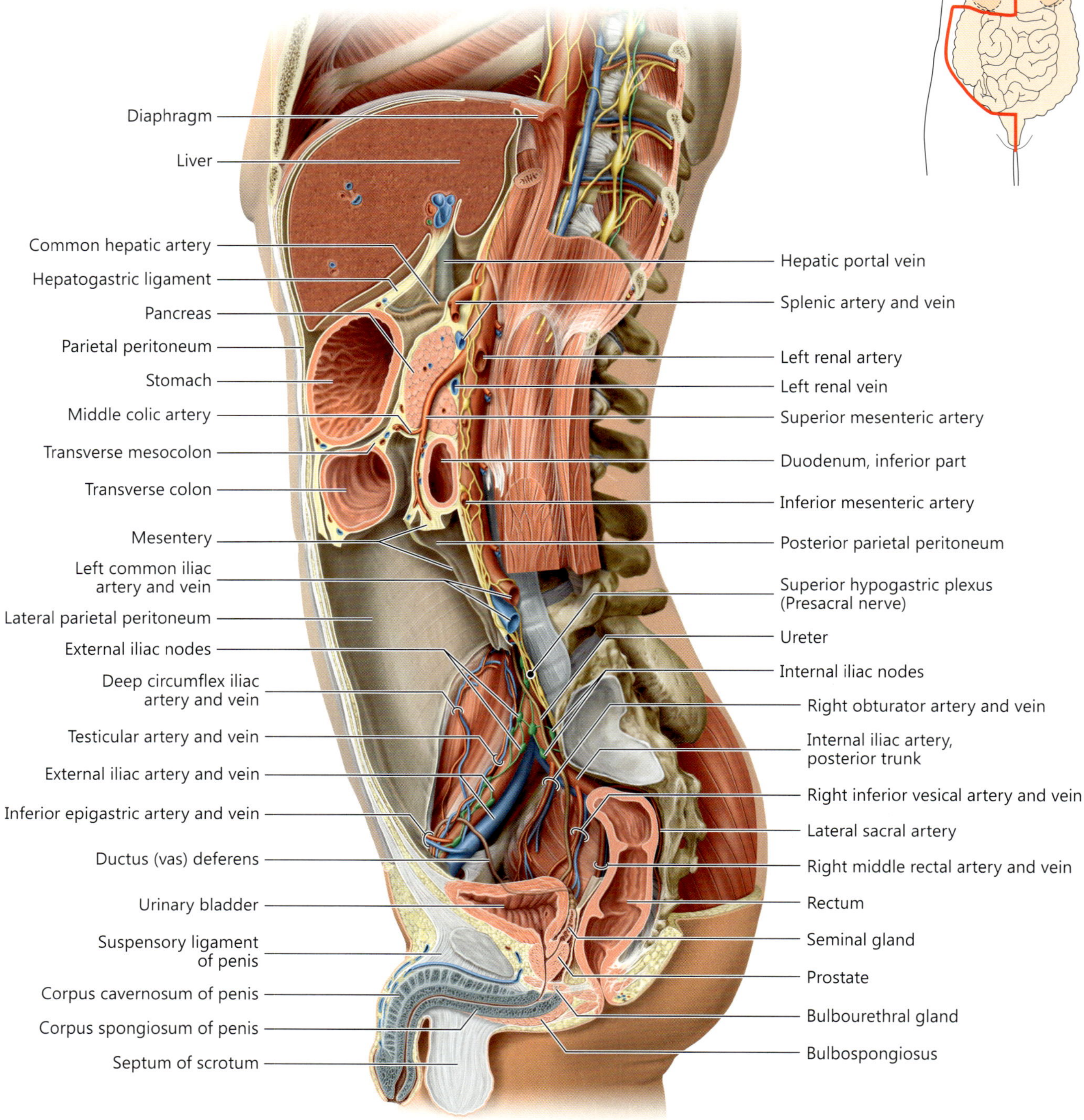

CONTENTS OF THE ABDOMEN

- Common hepatic artery
- Deep circumflex iliac artery
- Deep circumflex iliac vein
- Diaphragm
- Duodenum, inferior part
- External iliac artery
- External iliac nodes
- External iliac vein
- Hepatic portal vein
- Hepatogastric ligament
- Inferior epigastric artery
- Inferior epigastric vein
- Inferior mesenteric artery
- Internal iliac artery, posterior trunk
- Internal iliac nodes
- Lateral parietal peritoneum
- Lateral sacral artery
- Left common iliac artery
- Left common iliac vein
- Left renal artery
- Left renal vein
- Liver
- Mesentery
- Middle colic artery
- Ovarian artery
- Ovarian vein
- Pancreas
- Parietal peritoneum
- Posterior parietal peritoneum
- Right inferior vesical artery
- Right inferior vesical vein
- Right middle rectal artery
- Right middle rectal vein
- Right obturator artery
- Right obturator vein
- Splenic artery
- Splenic vein
- Stomach
- Superior hypogastric plexus (Presacral nerve)
- Superior mesenteric artery
- Transverse colon
- Transverse mesocolon
- Ureter

View from the Sagittal Plane in the Male

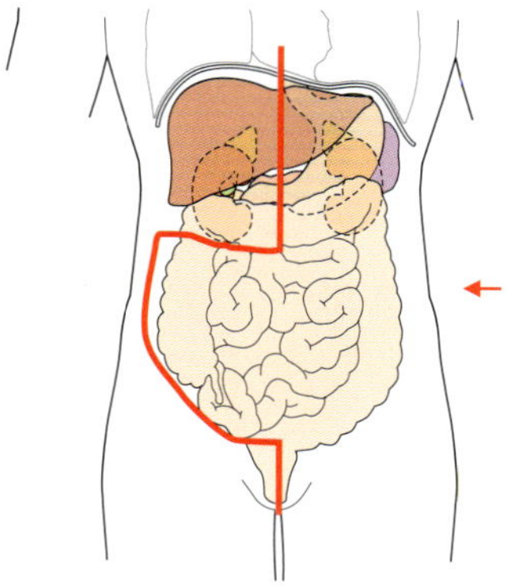

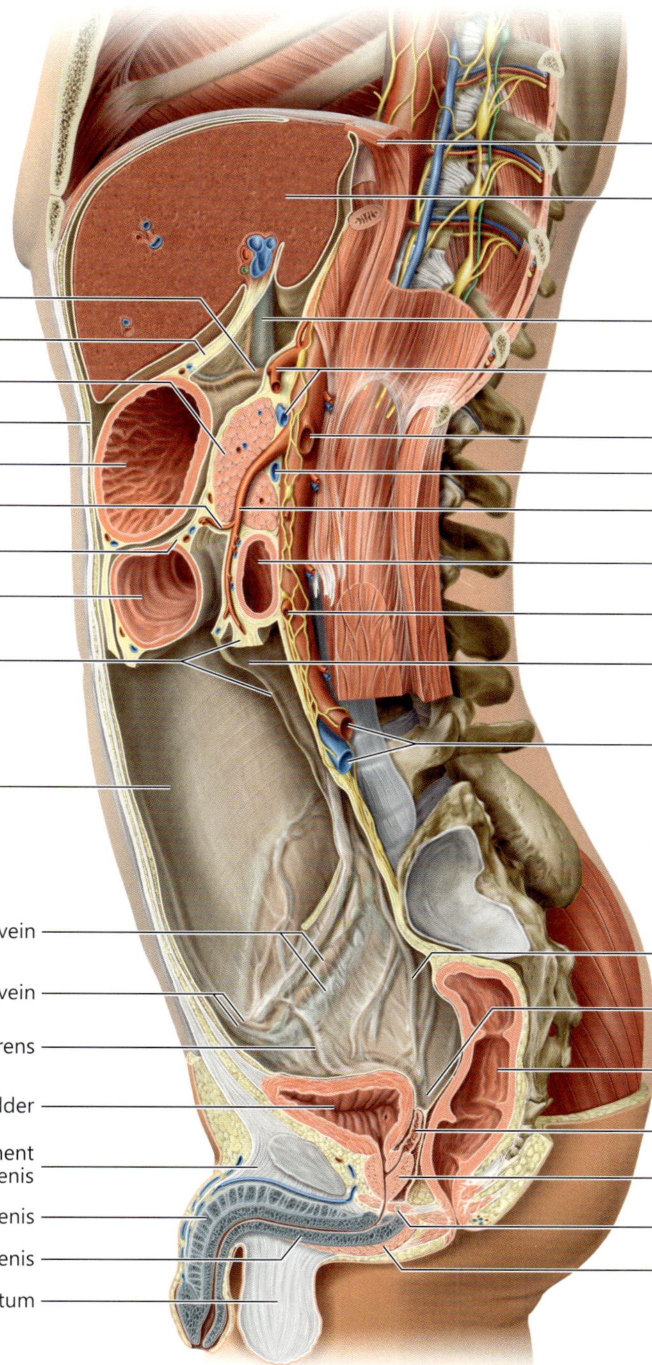

CONTENTS OF THE ABDOMEN (In bold are the structures added with respect to the preceding table)
• Rectovesical pouch

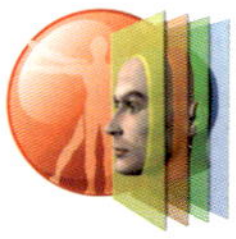

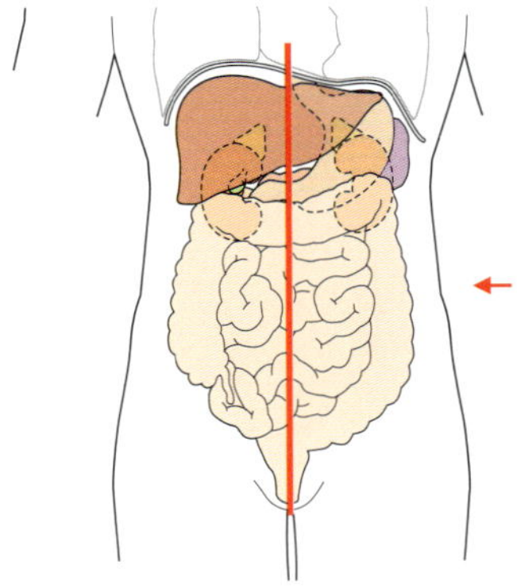

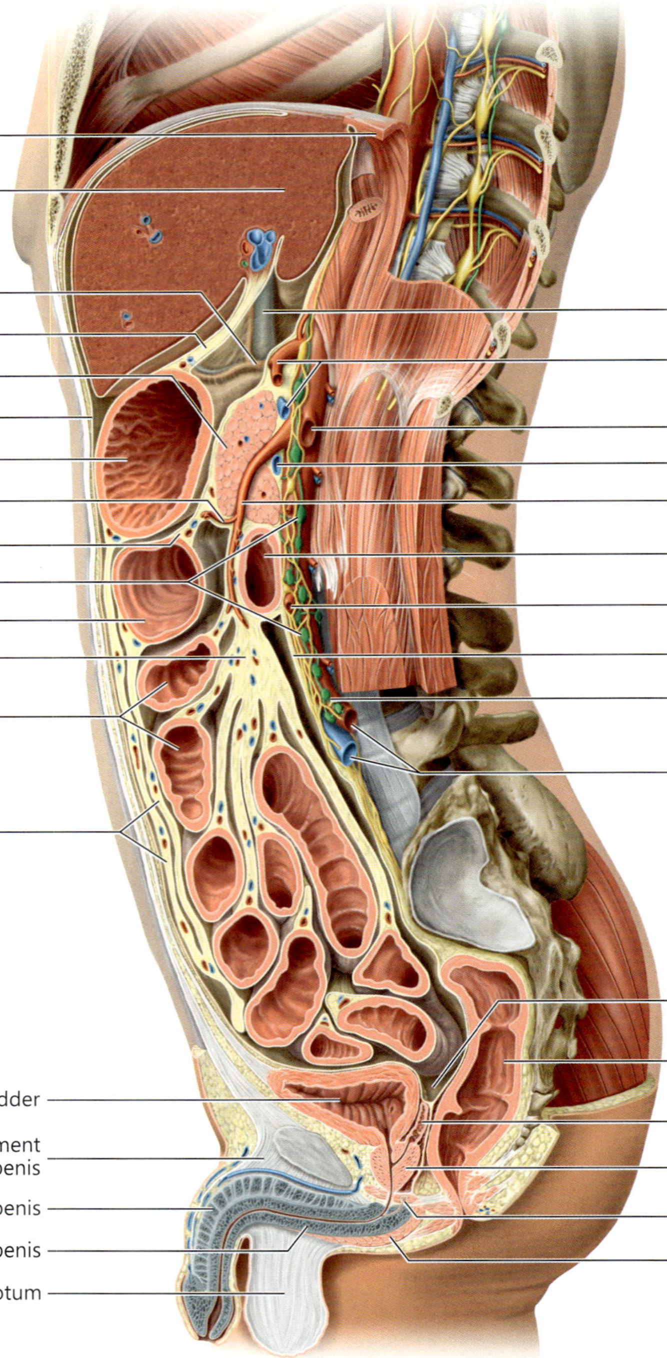

CONTENTS OF THE ABDOMEN (In bold are the structures added with respect to the preceding table)
• Common iliac nodes • Greater omentum • Lateral aortic nodes • Small intestine

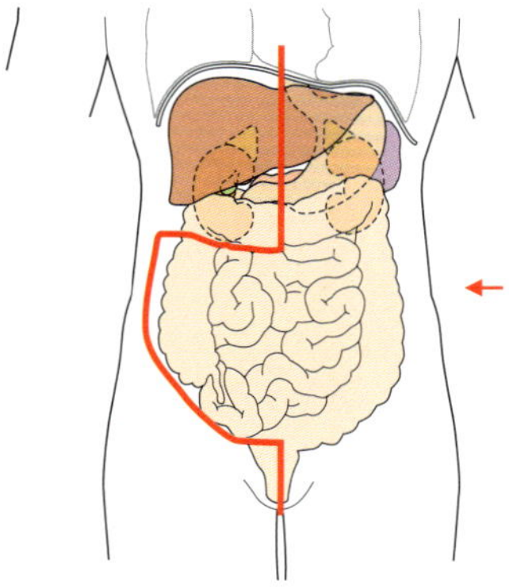

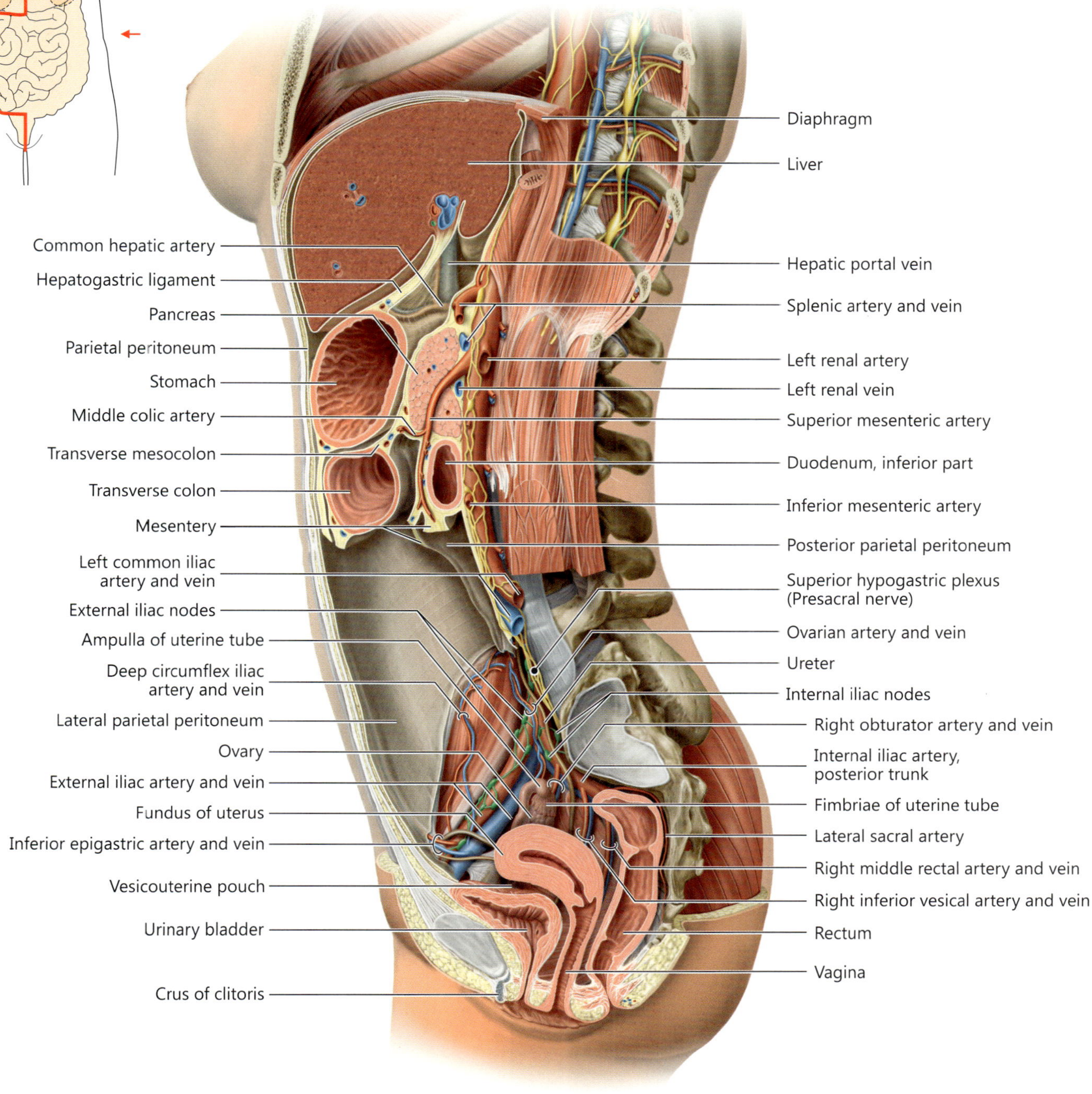

CONTENTS OF THE ABDOMEN

- Ampulla of uterine tube
- Common hepatic artery
- Crus of clitoris
- Deep circumflex iliac artery and vein
- Diaphragm
- Duodenum, inferior part
- External iliac artery and vein
- External iliac nodes
- Fimbriae of uterine tube
- Fundus of uterus
- Hepatic portal vein
- Hepatogastric ligament
- Inferior epigastric vein and artery
- Inferior mesenteric artery
- Internal iliac artery, posterior trunk
- Internal iliac nodes
- Lateral parietal peritoneum
- Lateral sacral artery
- Left common iliac artery and vein
- Left renal artery
- Left renal vein
- Liver
- Mesentery
- Middle colic artery
- Ovarian artery and vein
- Ovary
- Pancreas
- Parietal peritoneum
- Posterior parietal peritoneum
- Rectum
- Right inferior vesical artery and vein
- Right middle rectal artery and vein
- Right obturator artery and vein
- Splenic artery and vein
- Stomach
- Superior hypogastric plexus (Presacral nerve)
- Superior mesenteric artery
- Transverse colon
- Transverse mesocolon
- Ureter
- Urinary bladder
- Vagina
- Vesicouterine pouch

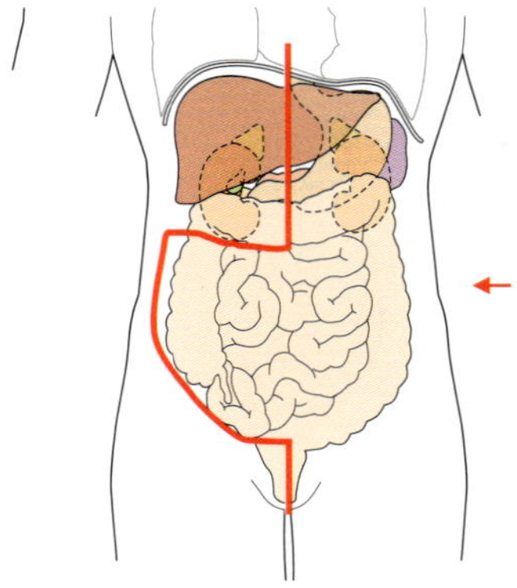

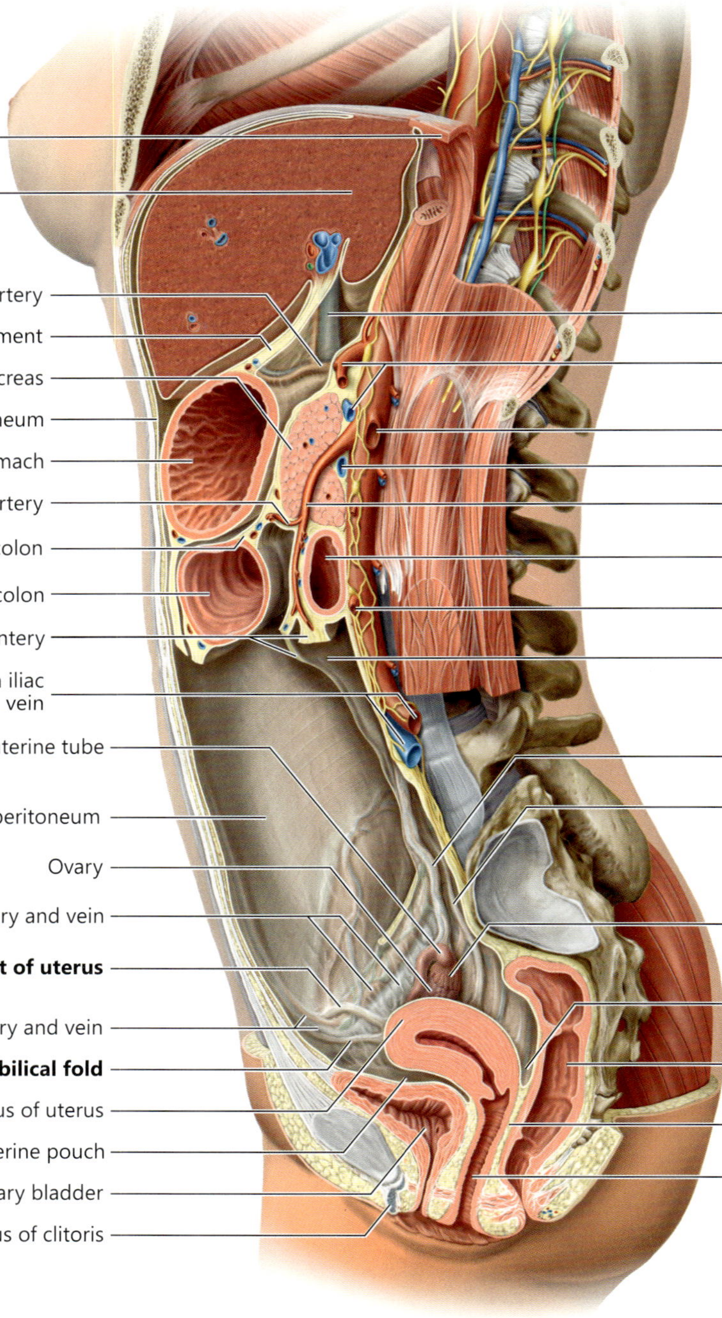

CONTENTS OF THE ABDOMEN (In bold are the structures added with respect to the preceding table)
• Medial umbilical fold • Round ligament of uterus • Suspensory ligament of ovary

View from the Sagittal Plane in the Female

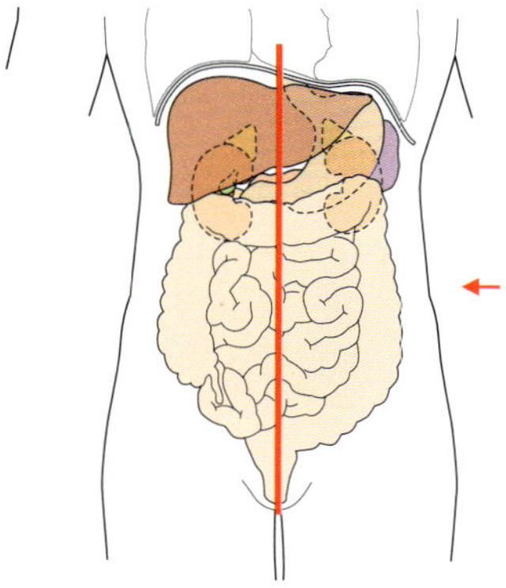

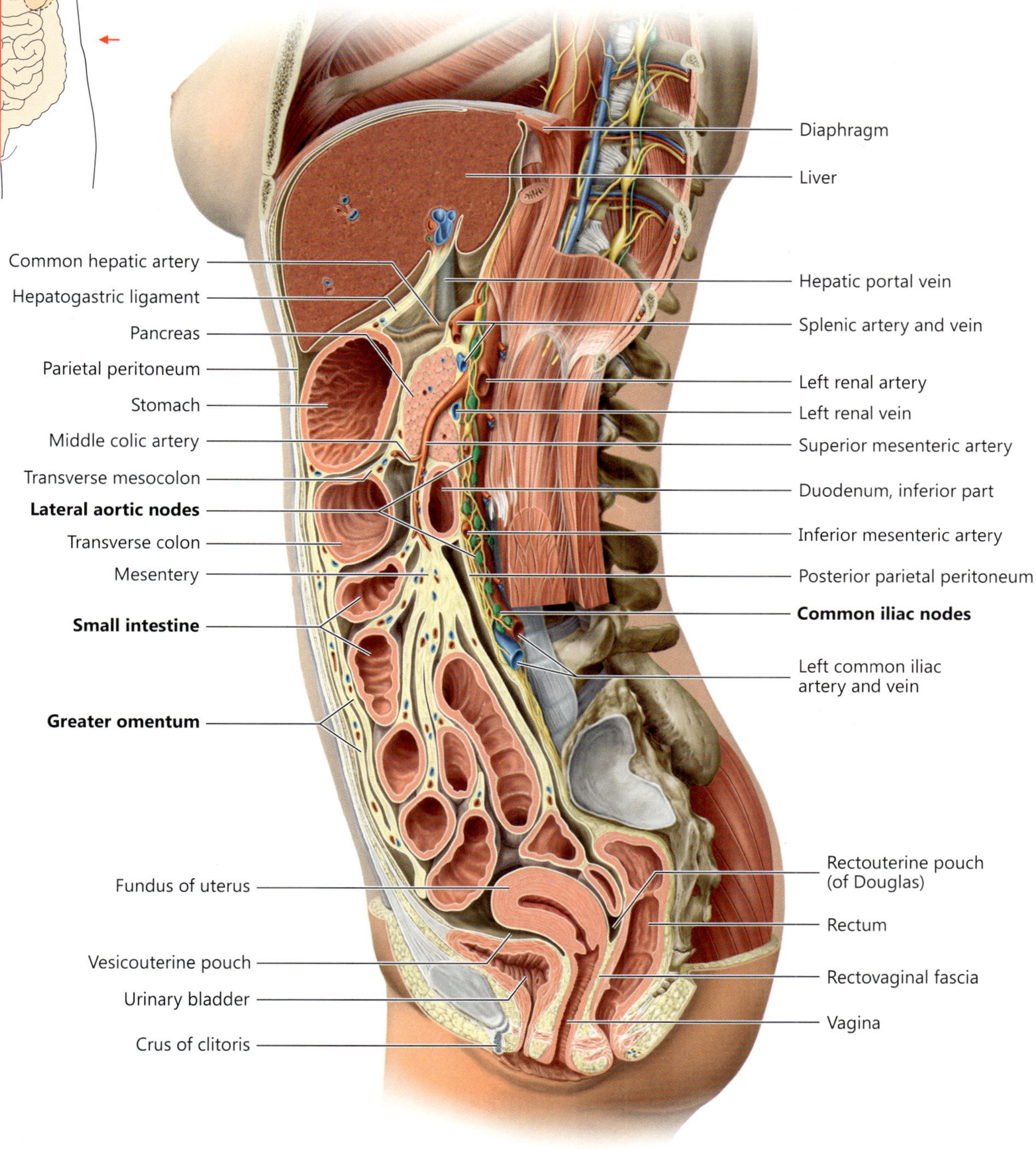

CONTENTS OF THE ABDOMEN (In bold are the structures added with respect to the preceding table)

- Common iliac nodes
- Greater omentum
- Lateral aortic nodes
- Small intestine

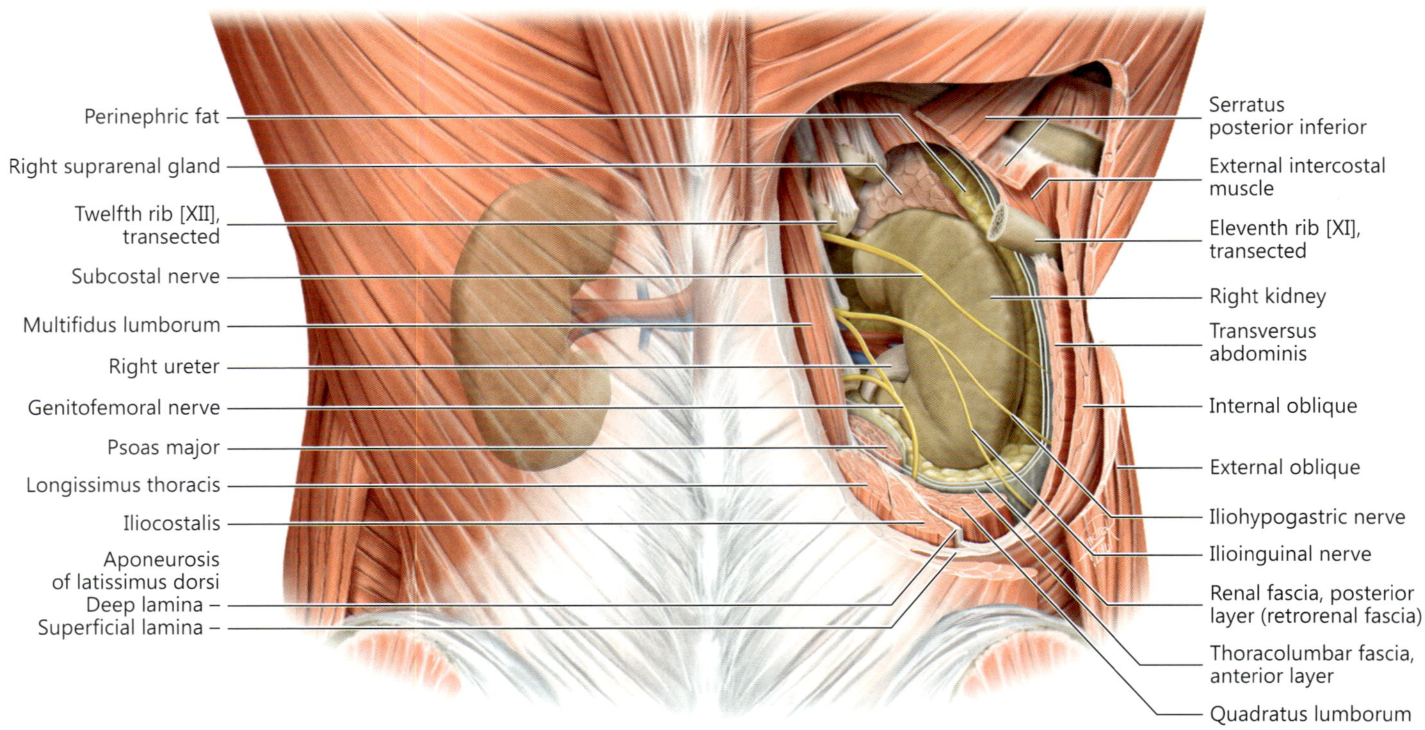

The right **renal space** was opened posteriorly. Below the twelfth rib, the posterior surface of the kidney rests against the quadratus lumborum, contacting the psoas major, medially, and the transversus abdominis, laterally. The figure displays the relations with the subcostal nerve, or 12th intercostal nerve, and with the iliohypogastric and ilioinguinal nerves, which descend obliquely, downwards and sideways, on the quadratus lumborum. Above the inferior border of twelfth rib, the kidney rests against the diaphragm.

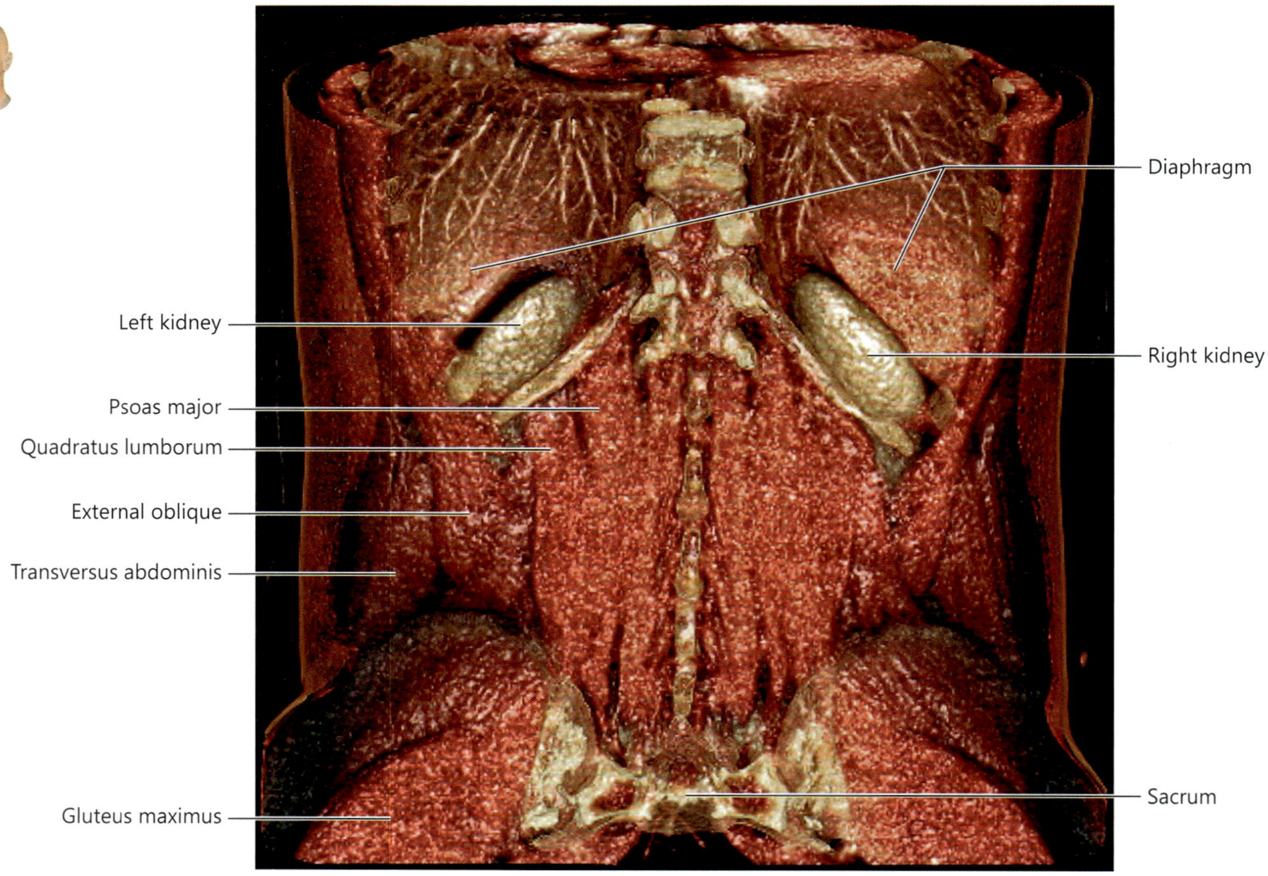

Three-dimensional image of a living human body, obtained by CT scan using a 3D volume rendering technique, showing the **posterior abdominal muscles** in posterior projection. The frontal section plane, tangential to the spinous processes of the eleventh and twelfth vertebrae, partly eliminates the ribs and the intercostal muscles, hence displaying the diaphragm and kidneys.

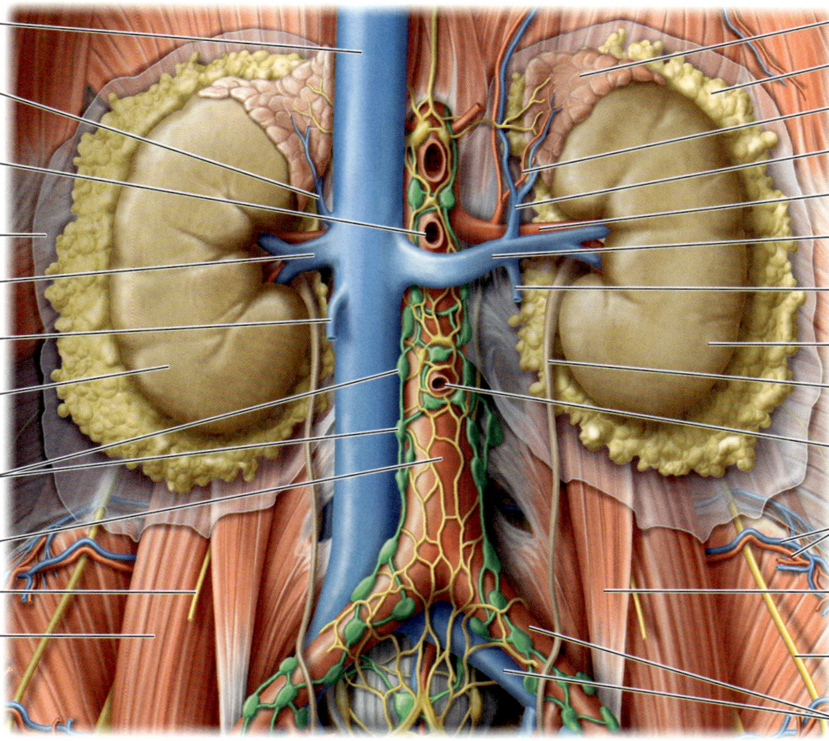

Course and relations of the **abdominal part of the ureter**. In the first part, the ureter runs behind the posterior parietal peritoneum (here removed along with the retroperitoneal fibroadipose tissue) corresponding to the lumbar region and to the iliac fossa. The intersection of the common iliac vessels on the left and the external iliac vessels on the right, causing the ureter to constrict at the marginal flexure, marks the transition from the abdominal to the pelvic part of the ureter.

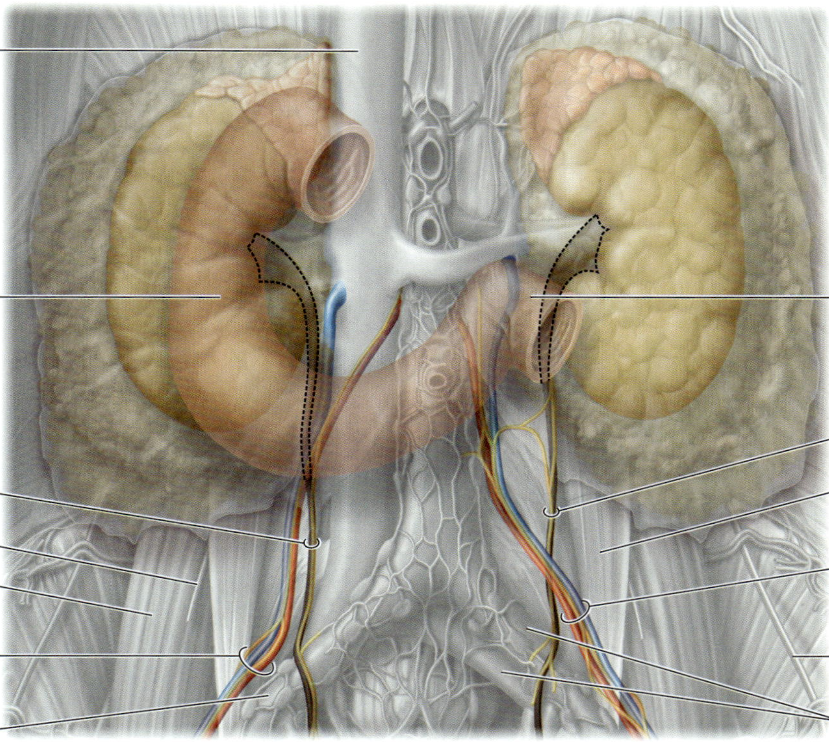

Course and main relations of the **ureter in the abdominal part** in the male. Posteriorly, it rests against the psoas major, crossed by the genitofemoral nerve and by the lateral cutaneous nerve of thigh. Anteriorly and superiorly, it is covered by the descending part of the duodenum (on the right) and by the duodenojejunal flexure (on the left). It is crossed by the genital vessels (testicular or ovarian artery and vein), laying laterally, further down, accompaging the ureter. Anteriorly, at the pelvic inlet, the ureter crosses the external iliac vessels on the right, and the common iliac vessels on the left. This describes the marginal flexure, and marks the transition from the abdominal to the pelvic part of the ureter. In the female, the ureters have a similar course; however, the ovarian vessels and nerves also run along the ureters from the marginal flexure to the pelvis.

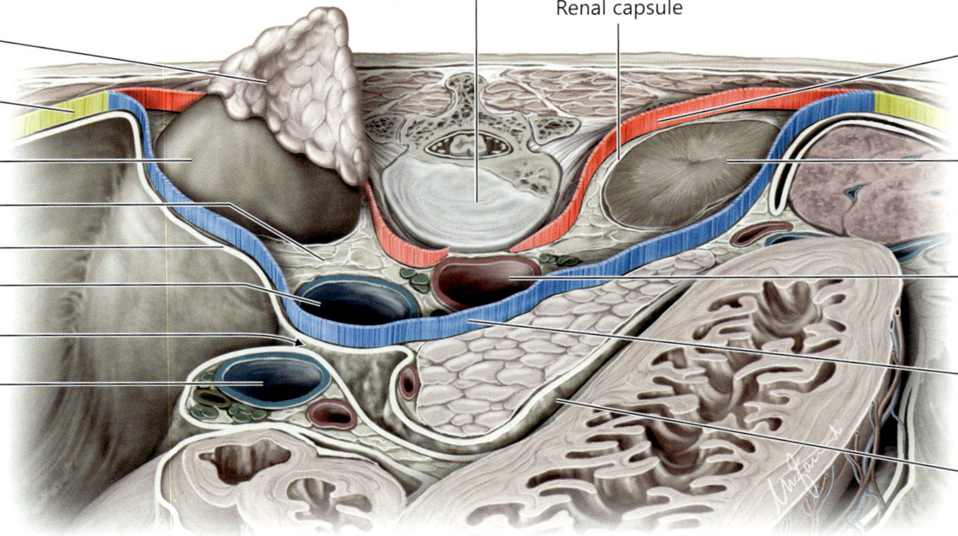

a

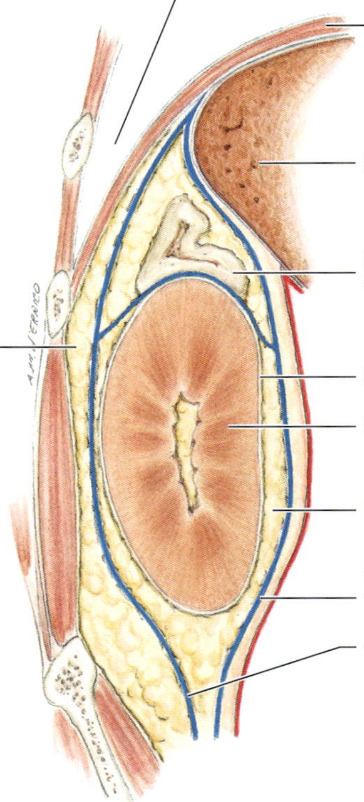

b

a, Transverse section passing through the **left kidney** as seen anteriorly. Layout of the organs located behind the posterior parietal peritoneum. The figure illustrates the course of the renal fascia. The anterior (prerenal, *blue*) and posterior (retrorenal, *red*) layers of the fascia bound the renal space, which contains the kidney, the renal pelvis, the renal vessels, the perinephric fat and the right suprarenal gland, with the suprarenal vessels. Laterally, the two layers of the fascia join together at the lateral border of the kidney which then fuser with the transversalis fascia (*yellow*). Medially, the retrorenal layer attaches to the vertebral body, while the prerenal layer runs anterior to the large prevertebral vessels, and continues fusing with its corresponding contralateral layer. **b**, Sagittal paramedian section passing through the right kidney. Layout of the organs located behind the posterior parietal peritoneum (*red*). The figure illustrates schematically the course of the renal fascia (*blue*). Superiorly, the two layers join together near the diaphragm, right behind the liver; at the bottom, the two layers remain separate and dissolve in the retroperitoneal connective tissue.

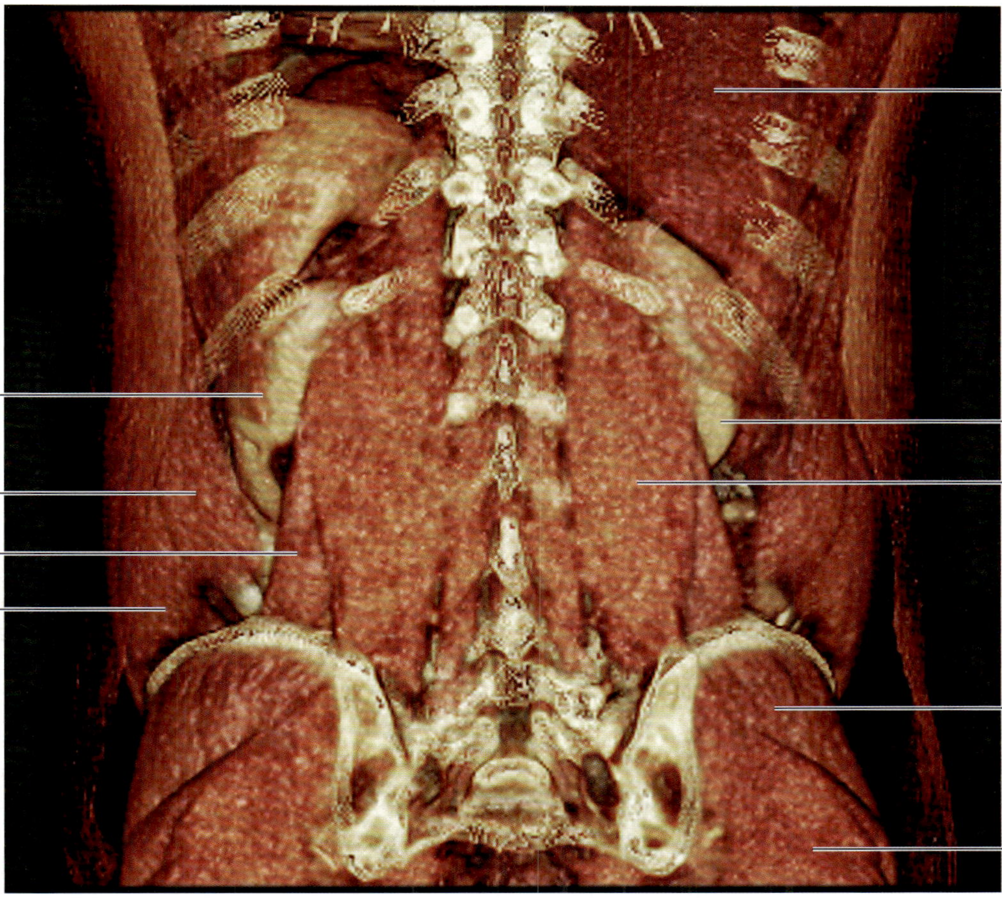

Three-dimensional image of a living human body, obtained by CT scan using a 3D volume rendering technique (posterior projection), showing the **posterior abdominal wall** and its relation to the kidneys.

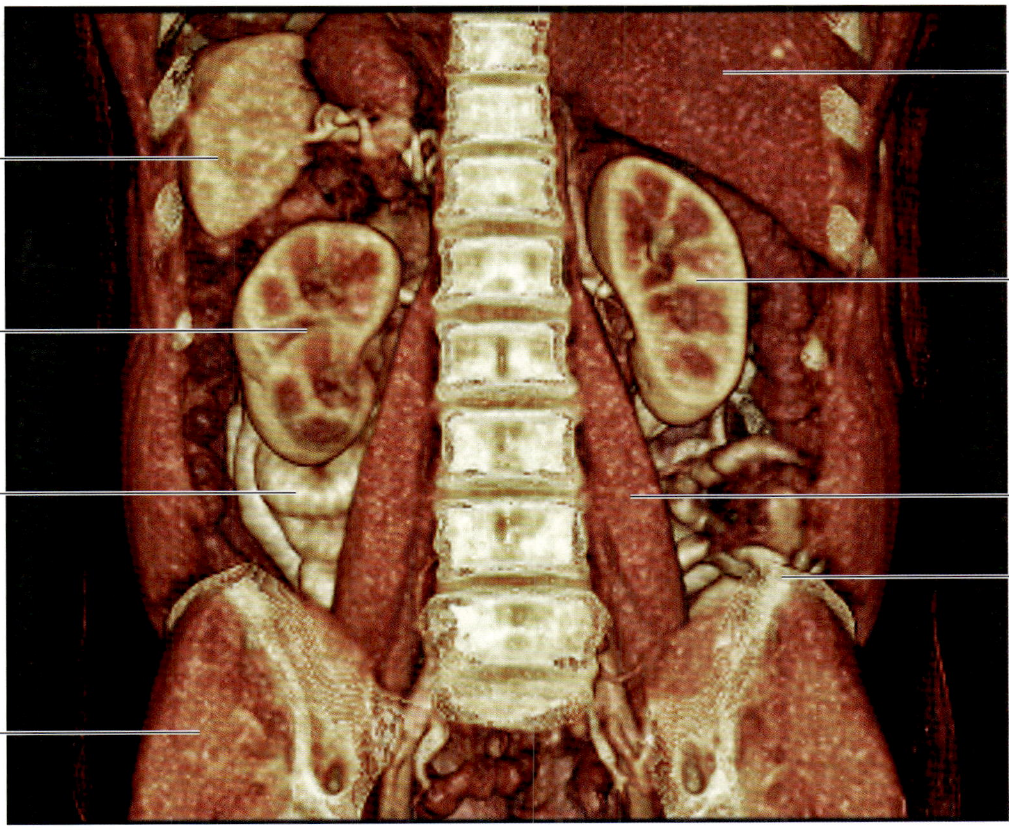

Three-dimensional image of a living human body, obtained by CT scan using a 3D volume rendering technique (posterior projection), showing some of the **abdominal organs**.

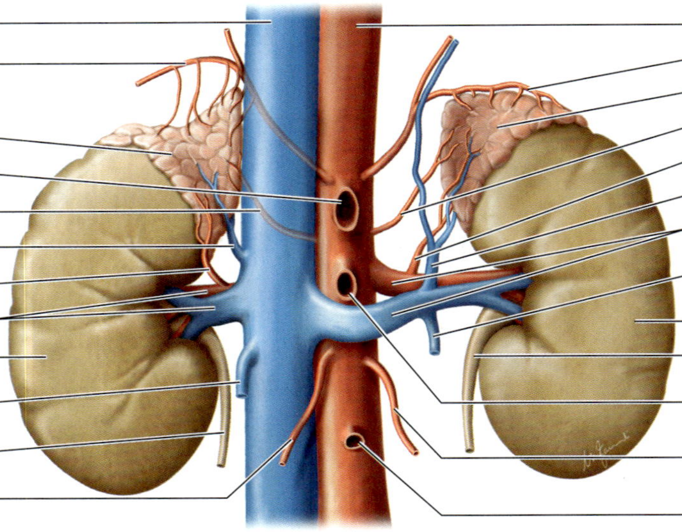

Relations and vascularization of the **suprarenal glands**. Two glands display differences in terms of shape and relations, as well as the different course of the supplying vessels, in particular: the right suprarenal vessel drains into the inferior vena cava, whereas the left one drains into the ipsilateral renal vein.

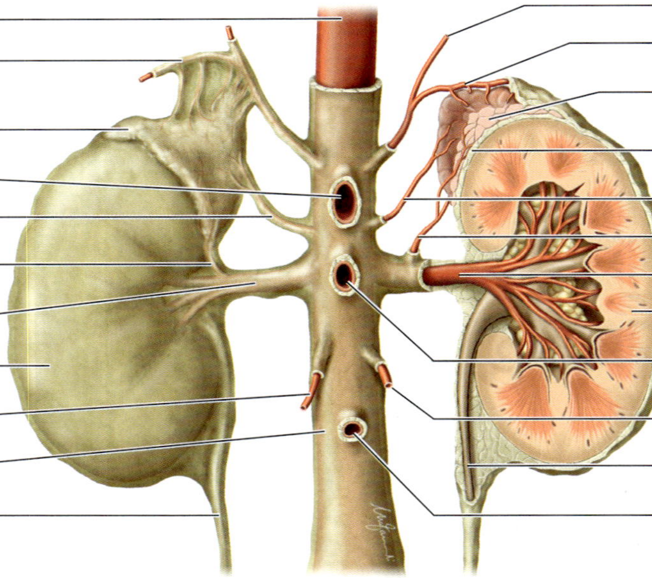

Vessel sheaths of the suprarenal gland, kidney and ureter.

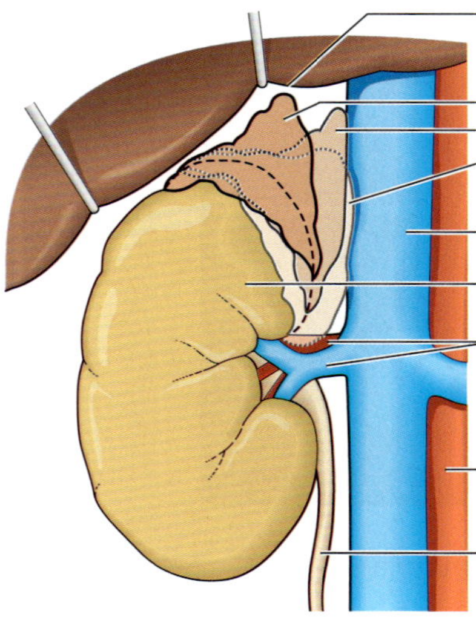

Positions of the suprarenal gland. A, high: the suprarenal gland covers the superior pole of the kidney. B, middle: the suprarenal gland lies in a superomedial position to the kidney. C, low: the suprarenal gland is located on the medial side of the kidney. C is the most frequently encountered position, whereas A is relatively rare, and corresponds to a fetal disposition persisiting in some adult subjects.

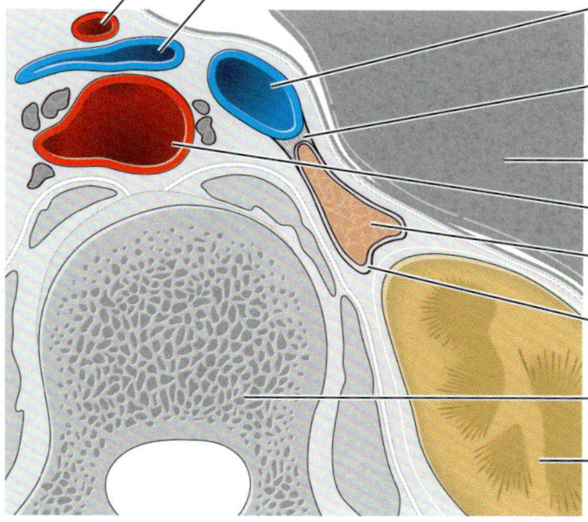

Transverse section at the second lumbar vertebra. There is a connective lamina between the inferior vena cava and the right suprarenal gland (**suprarenal-cava ligament**). It consists of a thickening of the sheath enclosing the inferior vena cava, which adheres to the vessel sheath of the suprarenal gland, a connective tissue surrounding the suprarenal blood vessels.

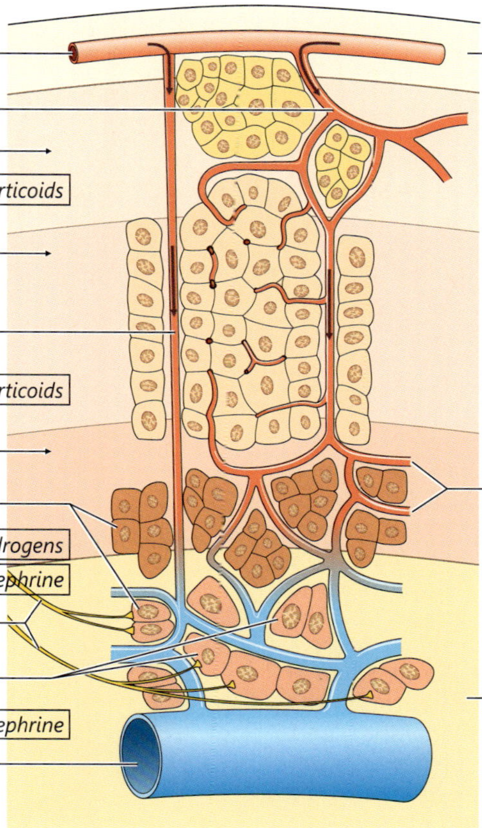

Diagram of the suprarenal circulation. Short and long arteries branch from the subcapsular plexus formed by the superior, middle and inferior suprarenal arteries. Short arteries enter the suprarenal cortex and branch into a network of cortical sinusoid capillaries, contacting the glandular cell cords. Venules arising from this network enter the medulla and form a network of medullary sinusoid capillaries. Long arteries run through the suprarenal cortex, without forming a capillary network, to reach the medulla and give rise to a network of medullary sinusoid capillaries. The deep suprarenal veins, arise from the medullary networks, join together into a central vein that emerges from the anterior surface of the gland, and drains into the inferior vena cava on the right, and into the left renal vein on the left. The cortisol released by the zona fasciculata of the cortex of the suprarenal gland, supplied by the short arteries, acts on the medullary cells, promoting the conversion of norepinephrine into epinephrine.

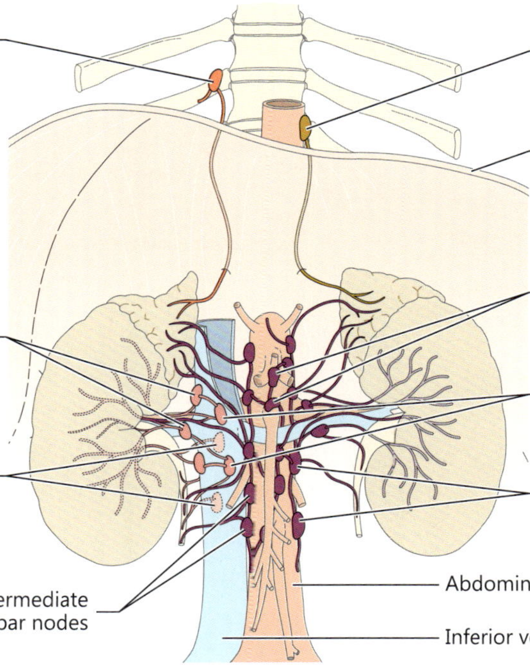

The **lymphatic vessels that drain the kidney and the suprarenal gland** are tributaries of the lumbar or lumboaortic nodes. The latter are divided into left lumbar nodes, which comprise preaortic, lateral aortic and postaortic nodes, intermediate lumbar nodes, between the aorta and the inferior vena cava, and right lumbar nodes, which include precaval, lateral caval and postcaval nodes.

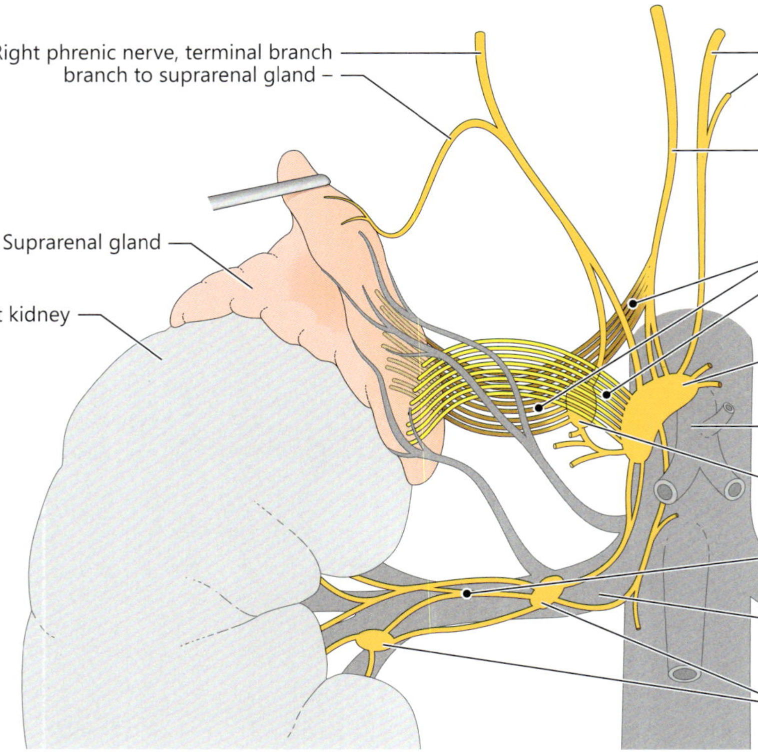

Innervation of the suprarenal gland. The nerves supplying the suprarenal gland arise from the suprarenal plexus, which, in turn, is an efferent from the celiac plexus and ganglia. The right celiac ganglion receives afferent sympathetic nerves, mainly the greater splanchnic nerve, as well as celiac branches from the posterior vagal trunk. Some terminal branches of the right phrenic nerve descend beyond the diaphragm, along the posterior abdominal wall, reaching the celiac plexus and the right suprarenal plexus.

The **superior suprarenal artery**, a branch of the inferior phrenic artery, enters the suprarenal gland at the apex, with its anterior and posterior branches. The middle suprarenal artery, a branch of the abdominal aorta, passes through the medial border of the gland, while the inferior suprarenal artery, a branch of the renal artery, enters the gland at the base.

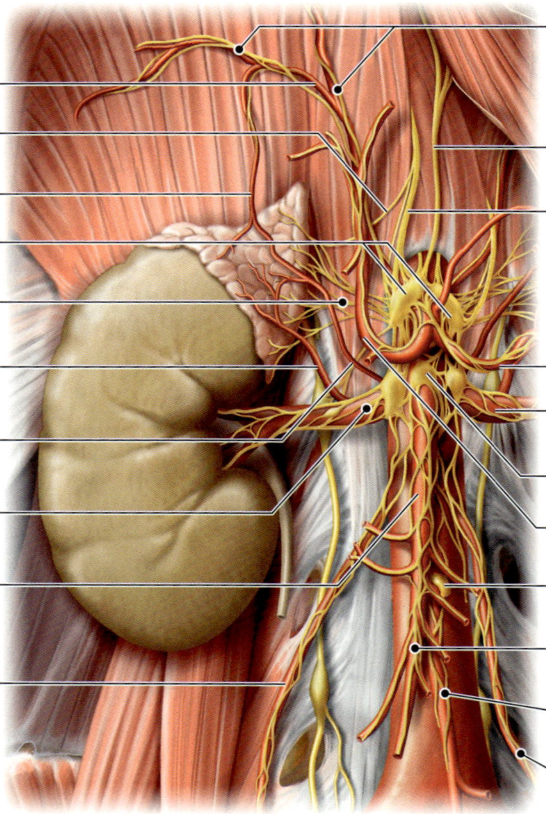

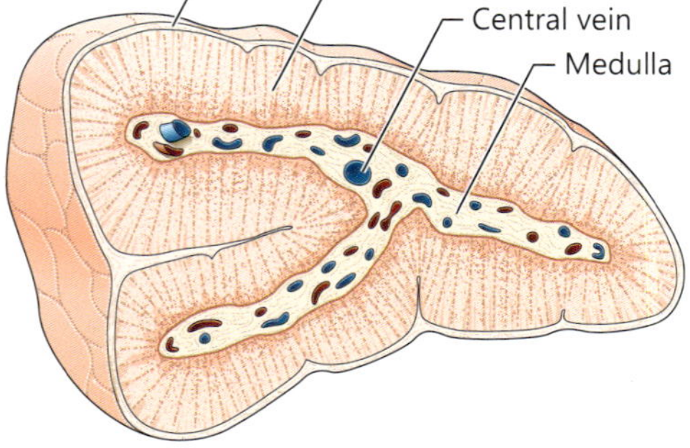

➤ Schematic structure of the **suprarenal gland**, seen in a section. The cortex is marked by *pink* dots and radial lines, while the medulla is indicated in *beige*, and features arterial and venous vessels.

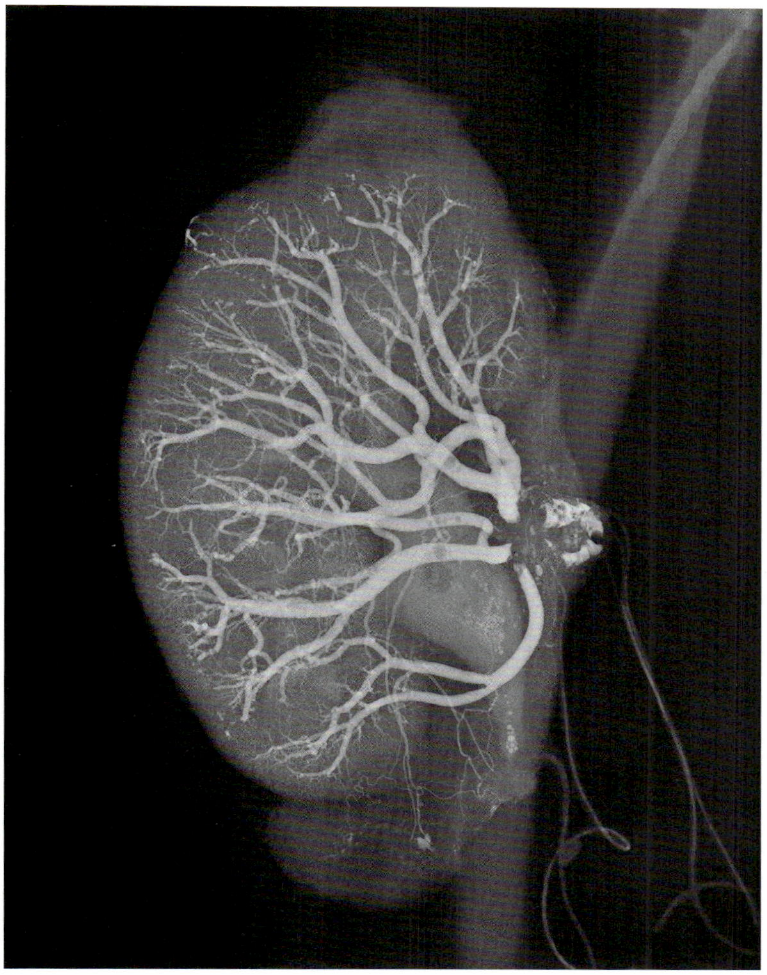

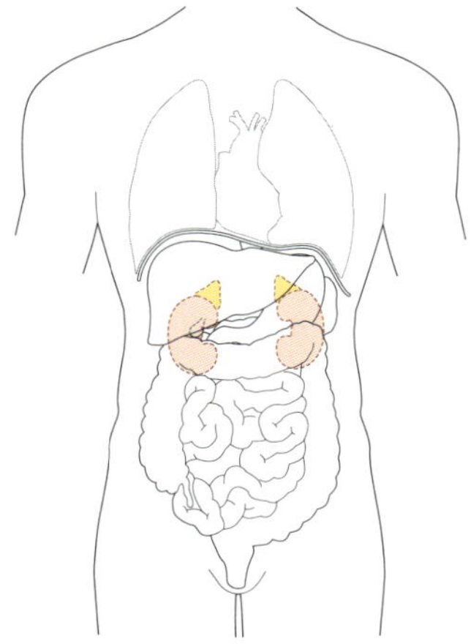

Renal arteriography. Arteries are highlighted radiographically by injecting the vessels with contrast medium. The interlobar, a´cuate and interlobular arteries can be clearly identified, based on their location, course and caliber. The interlobular arteries are quite thin and branch terminating with noumeorus vessels within the periphery of the cortex.

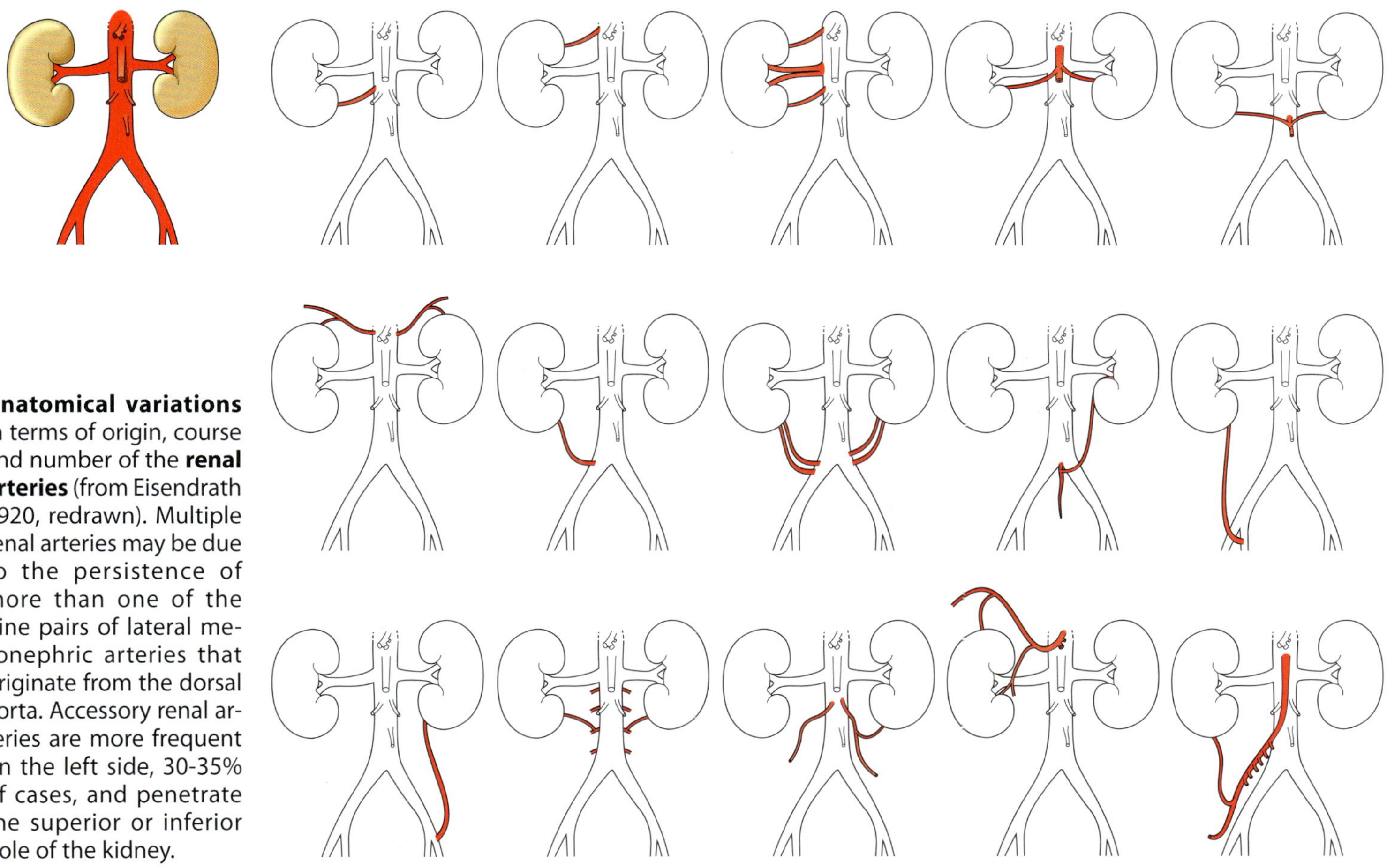

Anatomical variations in terms of origin, course and number of the **renal arteries** (from Eisendrath 1920, redrawn). Multiple renal arteries may be due to the persistence of more than one of the nine pairs of lateral mesonephric arteries that originate from the dorsal aorta. Accessory renal arteries are more frequent on the left side, 30-35% of cases, and penetrate the superior or inferior pole of the kidney.

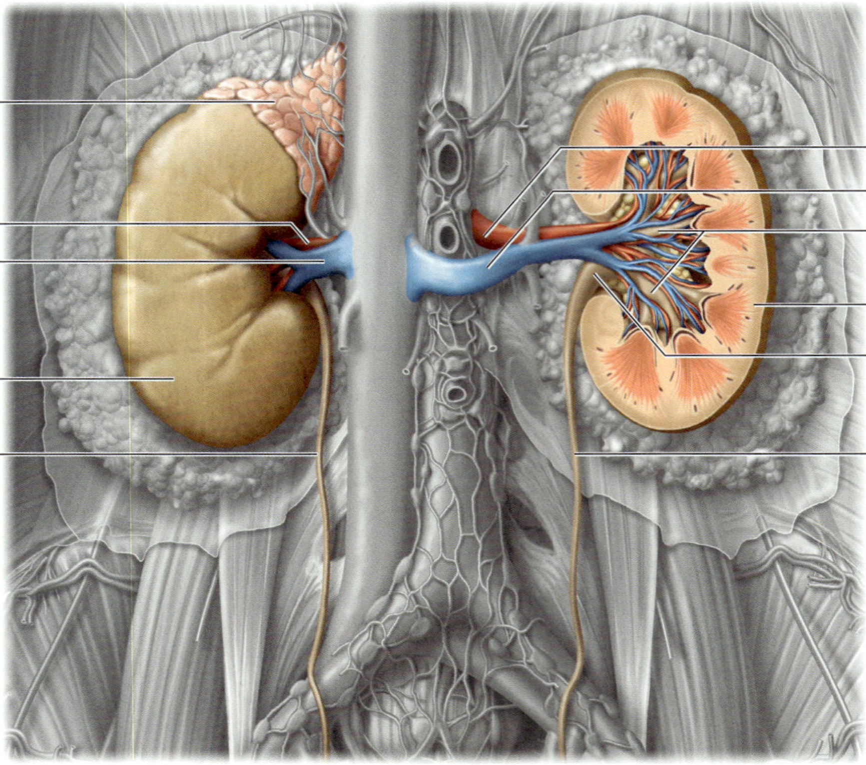

The **kidneys** are located in the posterior region of the abdomen, on either side of the vertebral column, in the **lumbar fossae**. On the right side, the perinephric fat surrounding the kidney within the renal space, bounded by the renal fascia, has been transected. On the left, a frontal section of the kidney shows the structures contained within the renal sinus.

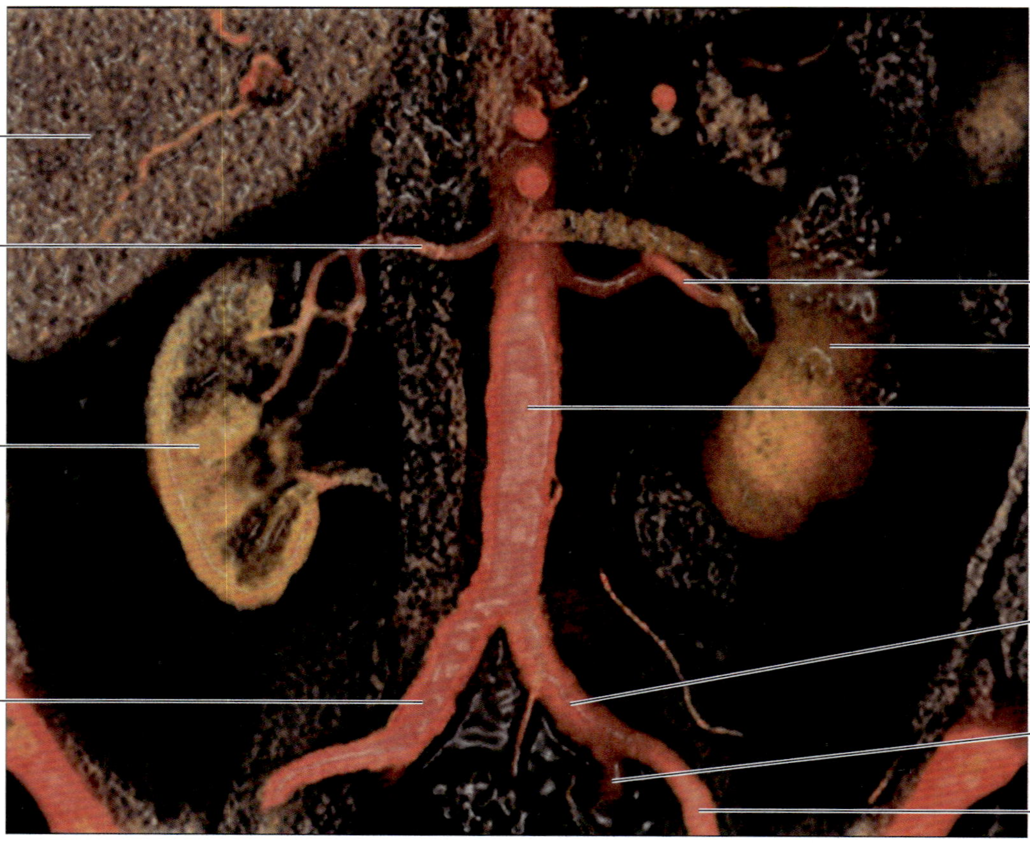

Three-dimensional image of a living human body, obtained by CT scan using a 3D volume rendering technique, showing **abdominal organs** in frontal section.

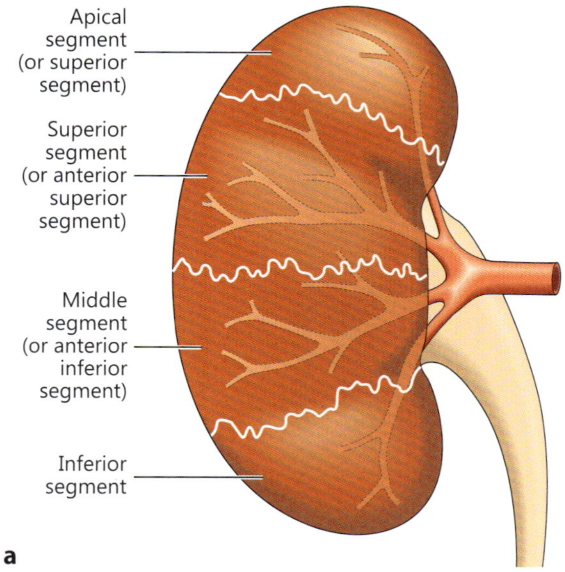

a

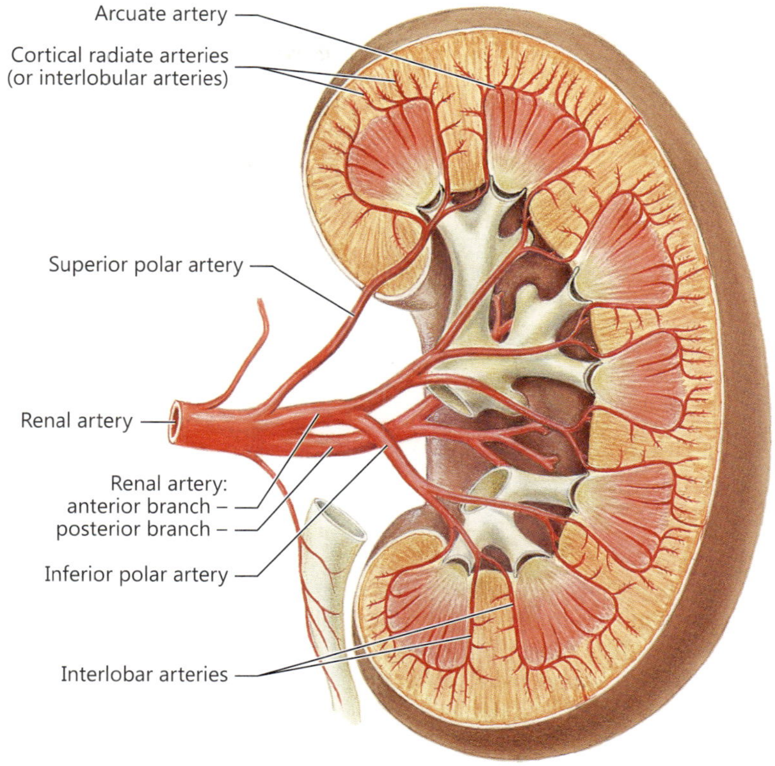

Frontal section of the **left kidney**, schematic representation of the **arterial vascularization** as seen anteriorly.

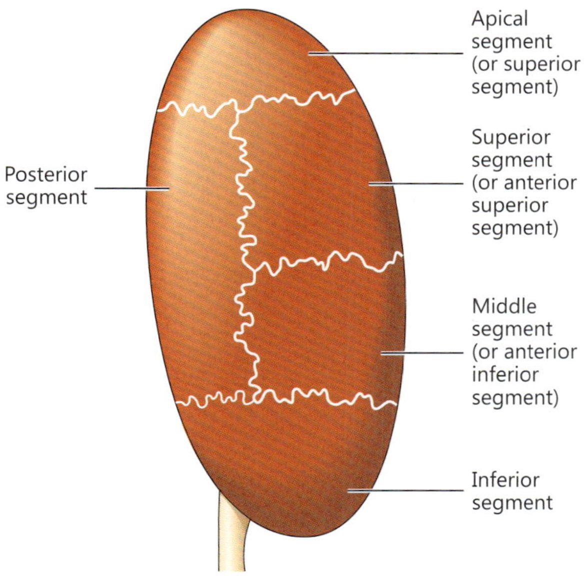

b

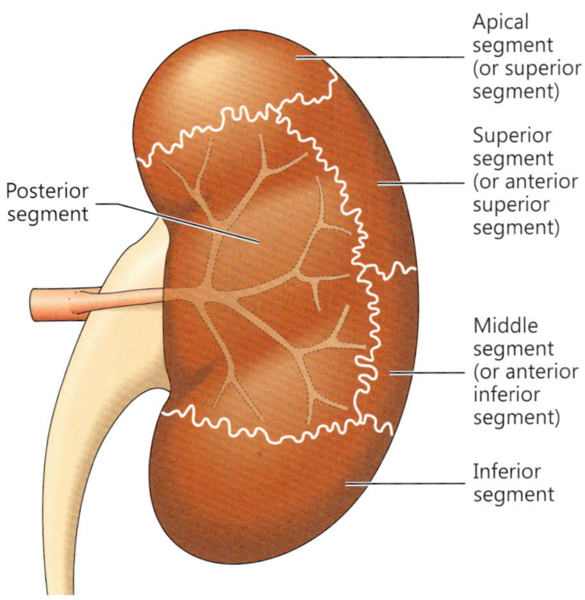

c

➤
Kidneys are organized in five **arterial vascular segments**, each of which is supplied by an individual segmental artery. The posterior (retropyelic) branch of division of the renal artery supplies the posterior segment of the kidney. The other four segmental arteries originate from the anterior (prepyelic) branch of division, and supply the apical, superior, middle and inferior segments, respectively, from top to bottom. **a**, Anterior view; **b**, lateral view; **c**, posterior view.

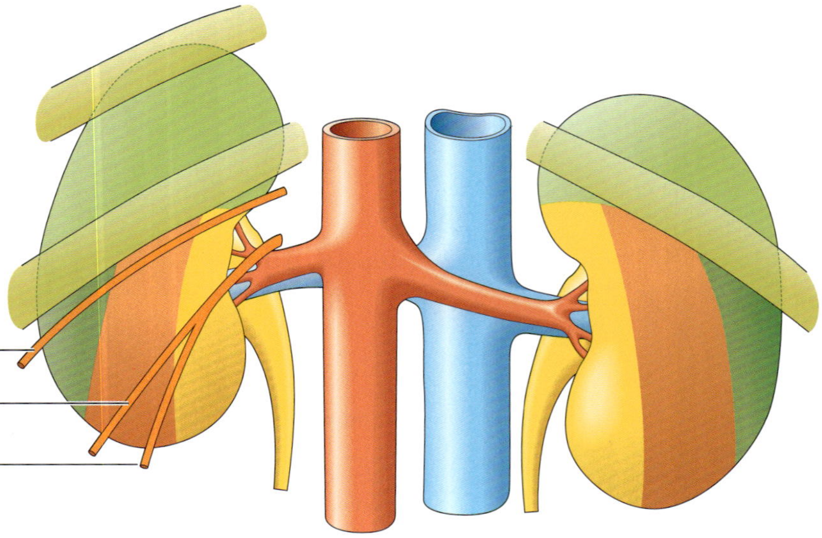

Schematic representation of the **relations of the posterior surfaces of the kidneys to the muscles and nerves**. The relation, from top to bottom, to the subcostal, iliohypogastric and ilioinguinal nerves is indicated on the left. The relations to the diaphragm (*light green*), transversus abdominis (*dark green*), quadratus lumborum (*ochre*), and psoas major (*yellow*), are shown. Left and right kidneys have similar posterior relations.

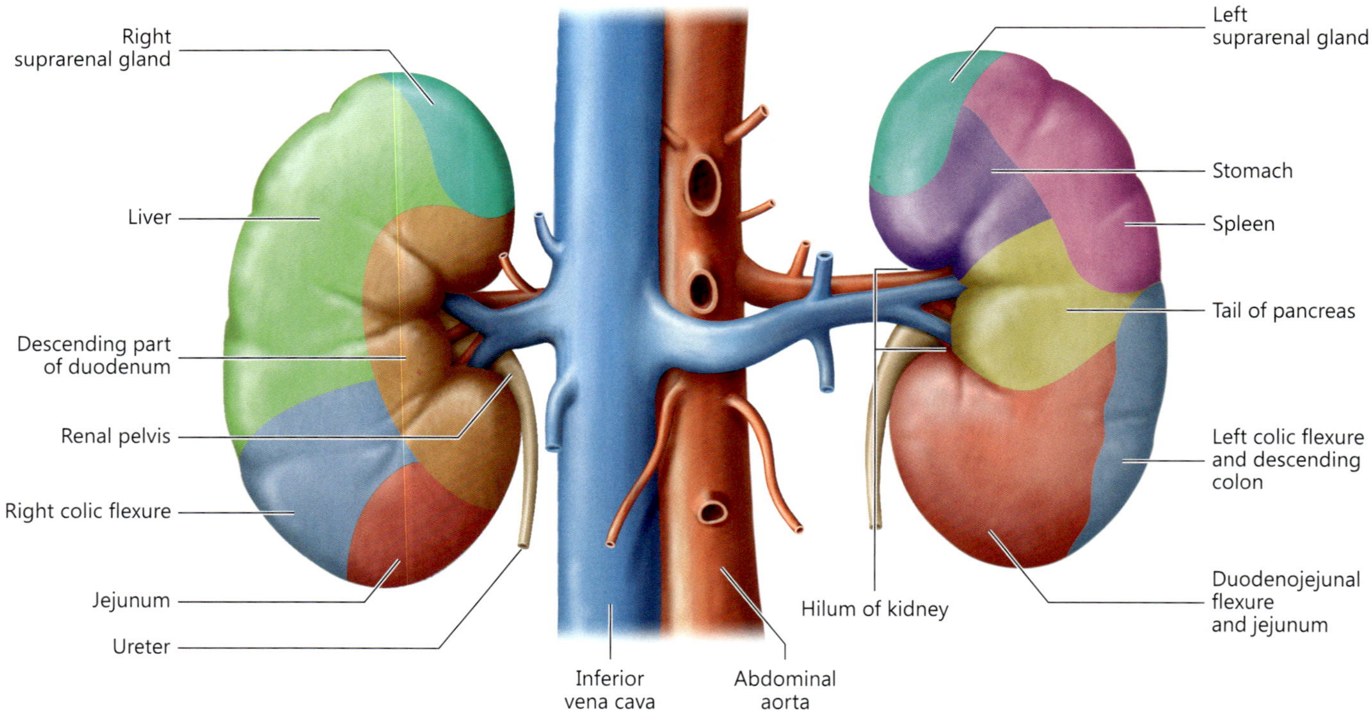

Schematic representation of the **relations** of the anterior surfaces **of the kidneys to the adjacent organs**. Areas that are related to different organs are highlighted. The anterior relations of the kidneys are different on each side. The relations with the liver, the jejunum, the spleen, the stomach and the duodenojejunal flexure have to take into account the peritoneum investing these organs.

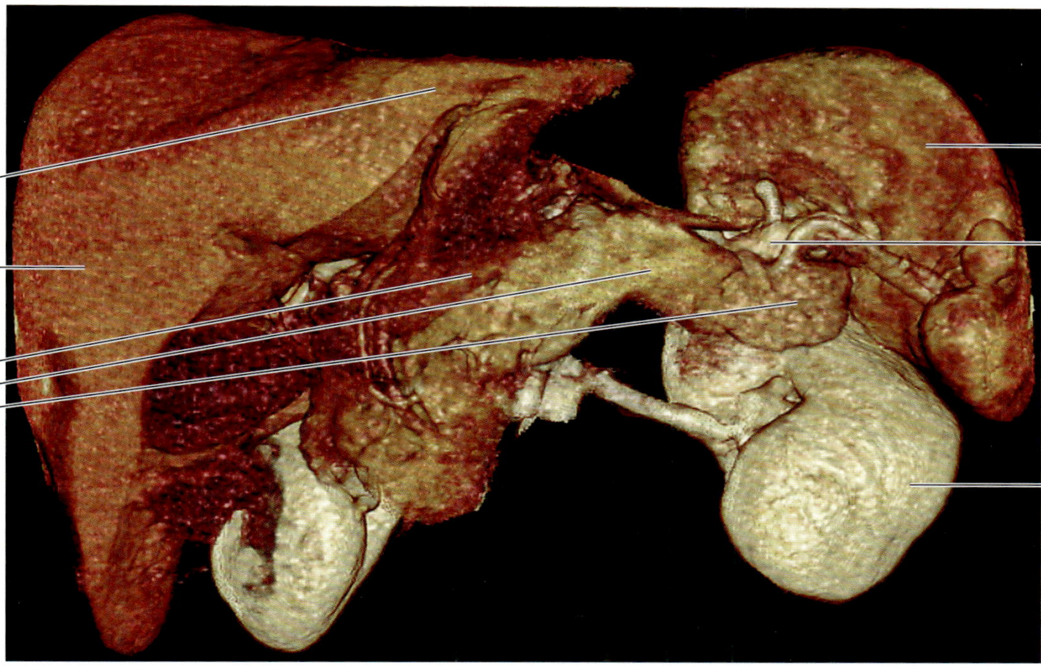

Three-dimensional image of a living human body, obtained by CT scan using a 3D volume rendering technique, showing the relations of the anterior surface of the **kidneys**.

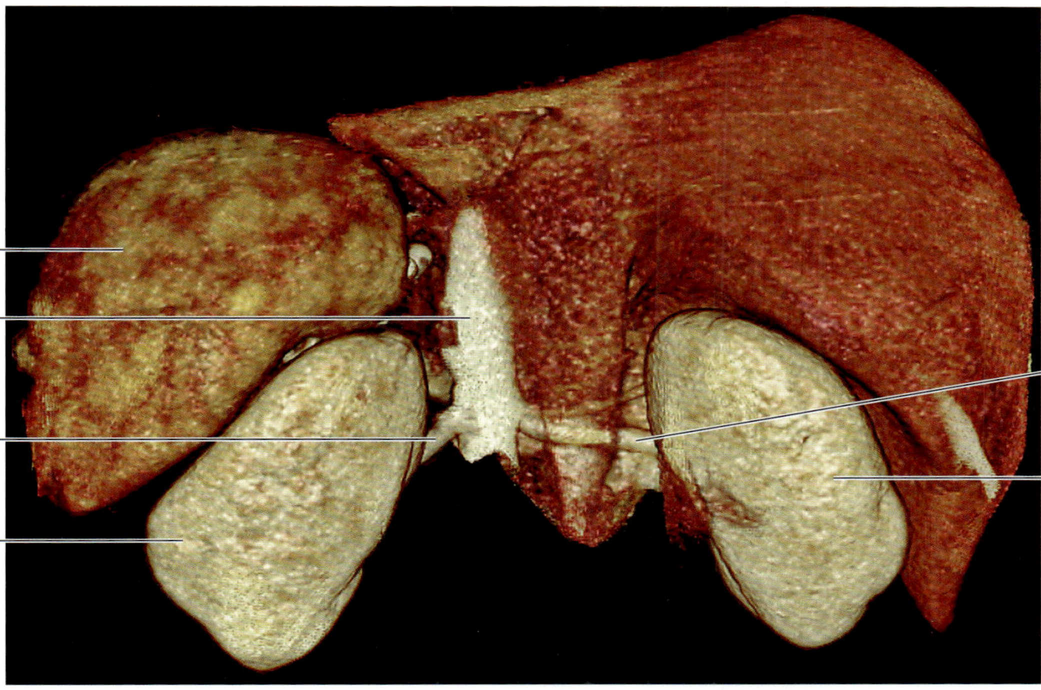

Three-dimensional image of a living human body, obtained by CT scan using a 3D volume rendering technique, showing the **kidneys** in posterosuperior projection.

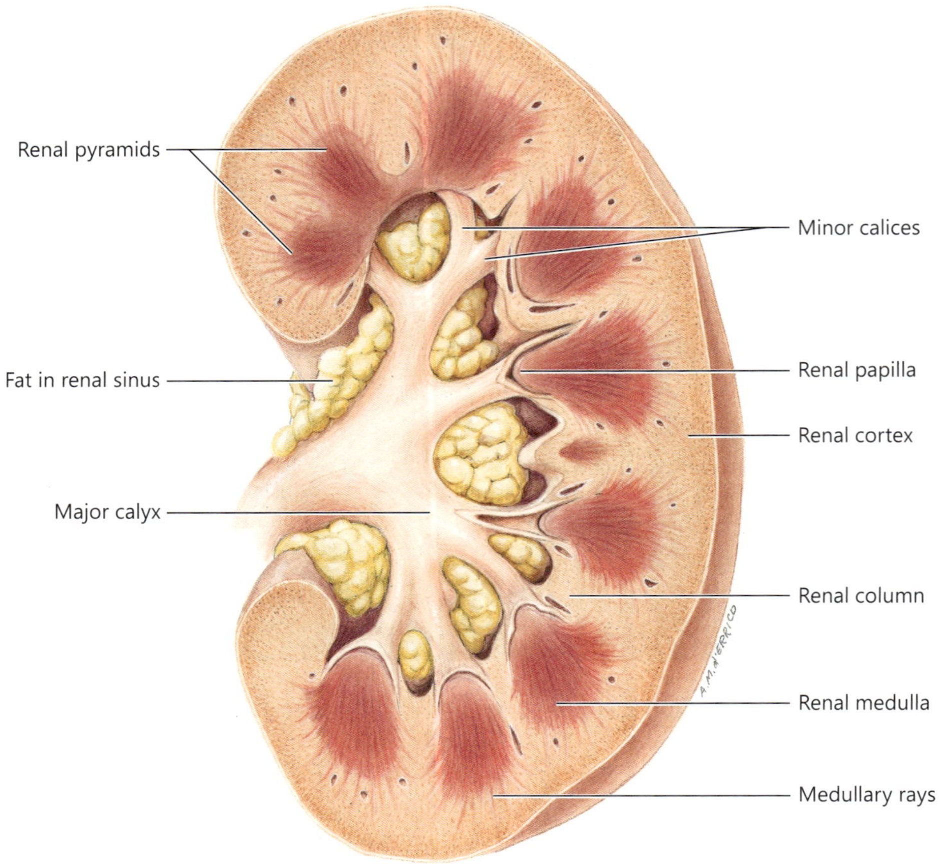

Frontal plane section of the **kidney** showing the **internal conformation**. The structures contained within the renal sinus are surrounded by the cellular adipose tissue that fills the renal sinus and is continuous with the perinephric fat through the hilum of the kidney.

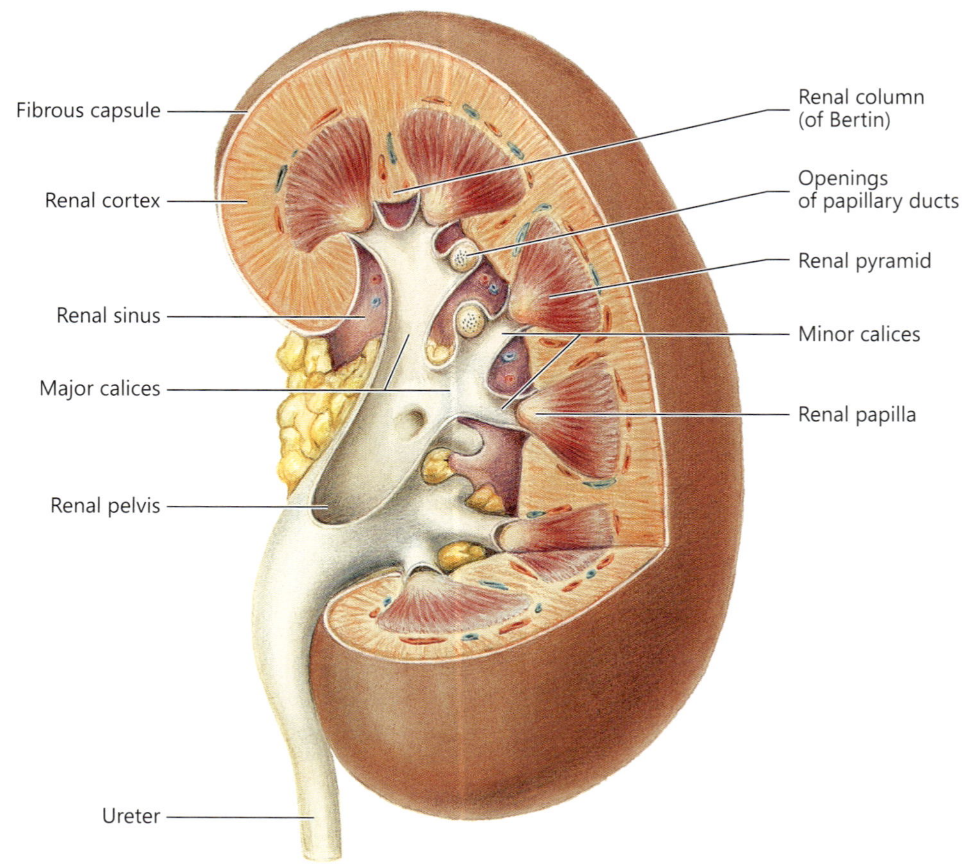

Internal conformation of the kidney in a three-dimensional representation. The kidney has been transected with two orthogonal cuts. The renal sinus wall surface is irregular due to the presence of more, the **renal papillae**, or less **renal columns** (of Bertin), evident prominences. Each papilla is surrounded by a minor calyx. The **minor calices** converge to form the **major calices**, which, in turn, open into the **renal pelvis**. The renal parenchyma displays an external layer of tissue, the renal cortex, and an inner tissue, the renal medulla, consisting of renal pyramids and columns.

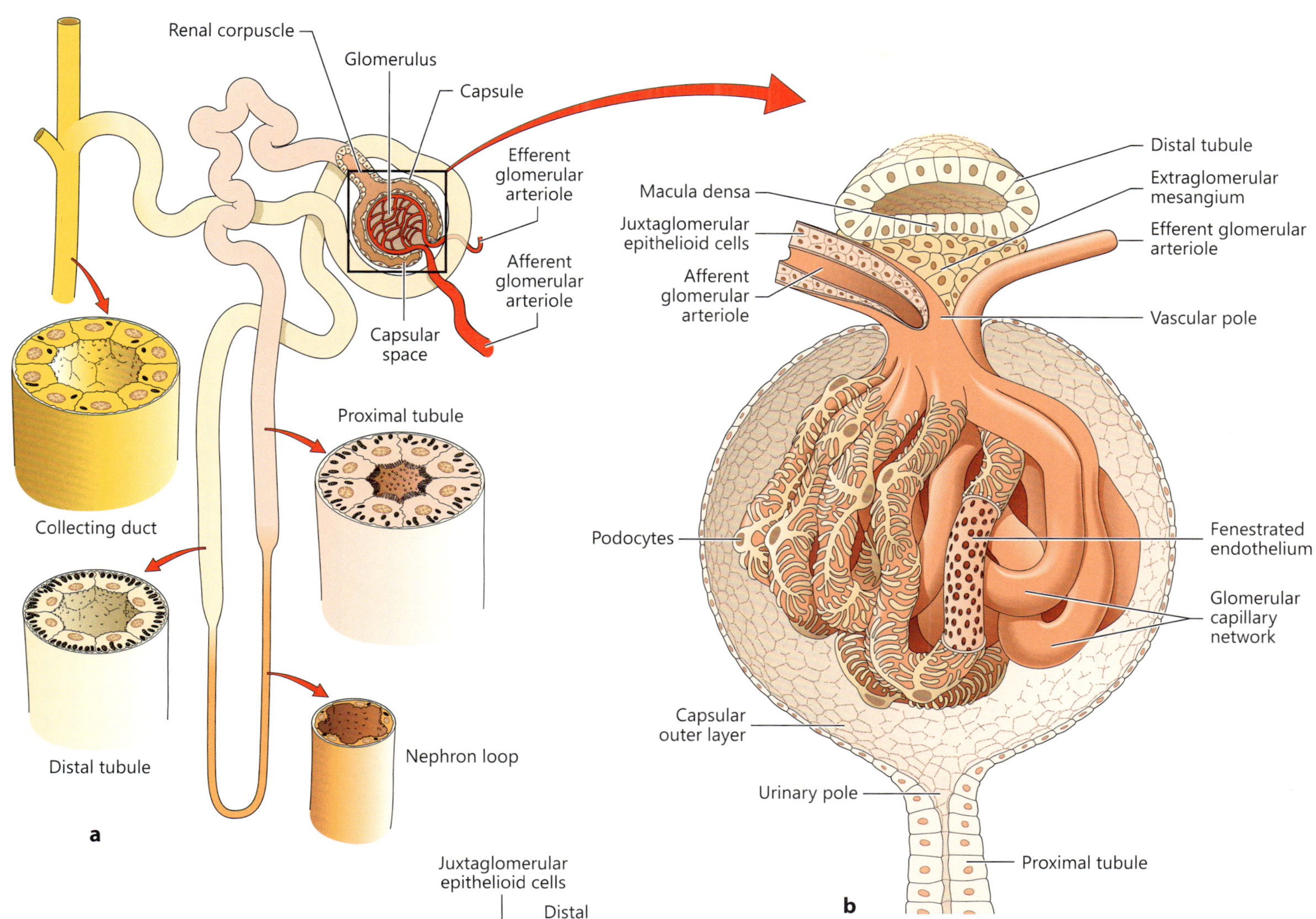

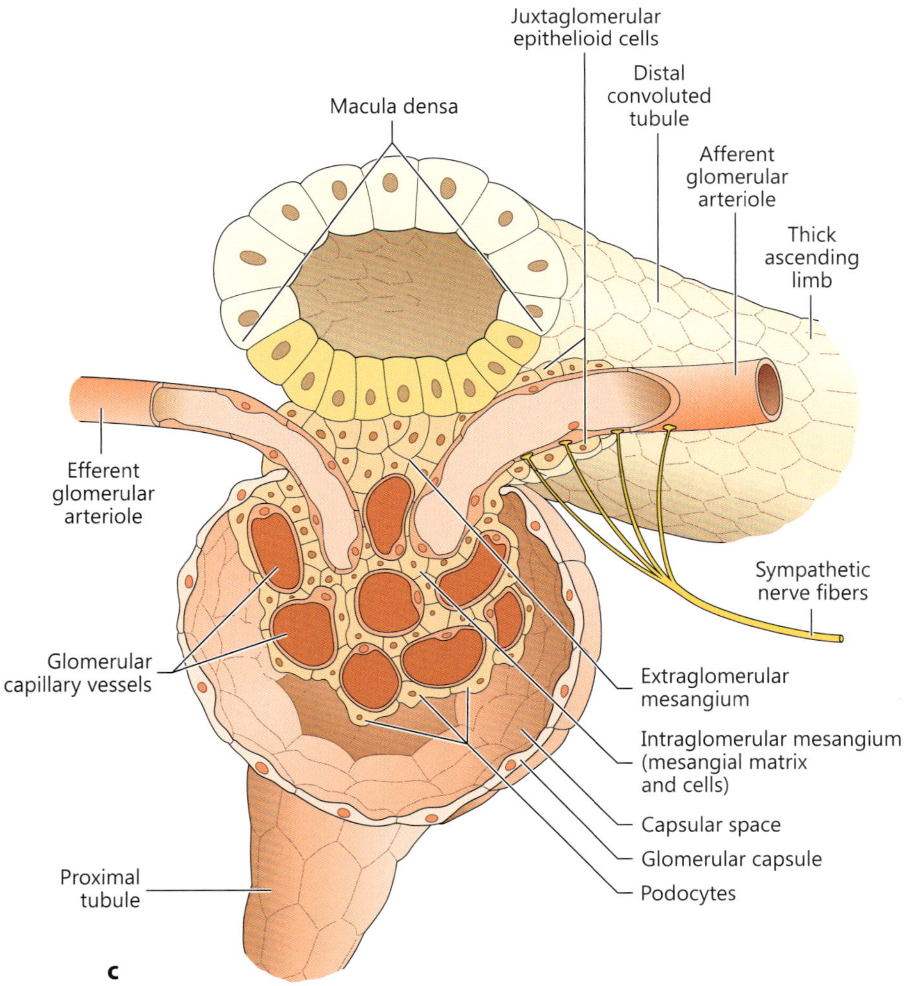

a, Schematic representation of the **nephron**. The functional unit of the kidney consist of a **renal corpuscle**, where plasma ultrafiltration occurs, and of a **renal tubule**, where the ultrafiltrate undergoes reabsorption to progressively form the urine.
b, Three-dimensional schematic reconstruction of the **renal corpuscle**. At the vascular pole, the parietal epithelial layer of the glomerular capsule invaginates to surround the glomerular capillaries and to form the visceral layer. The visceral layer consists of podocytes that wrap around the loops of glomerular capillaries, lined by fenestrated endothelium. At the vascular pole is present the juxtaglomerular apparatus. The capsular space continues into the lumen of the proximal tubule at the urinary pole.
c, Schematic diagram showing the organization of the **juxtaglomerular apparatus**. The macula densa is a part of the distal convoluted tubule whose cells lack the typical basal labyrinth. The extraglomerular mesangium is continuous with the intraglomerular mesangium. Renin-secreting juxtaglomerular cells are found in the walls of the afferent glomerular arteriole.

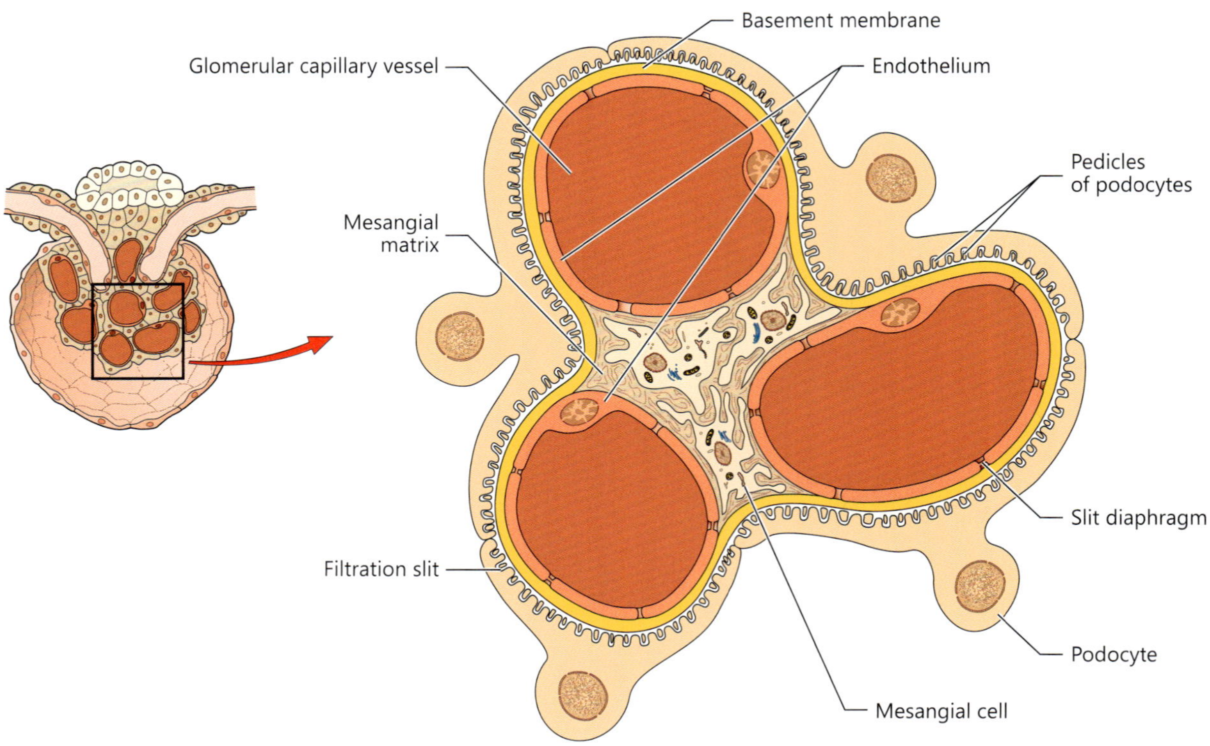

Schematic representation of a **renal lobule**. The spaces between the pedicels (foot processes) of the podocytes are called filtration slits. Some fenestrations of the endothelium are closed by diaphragms.

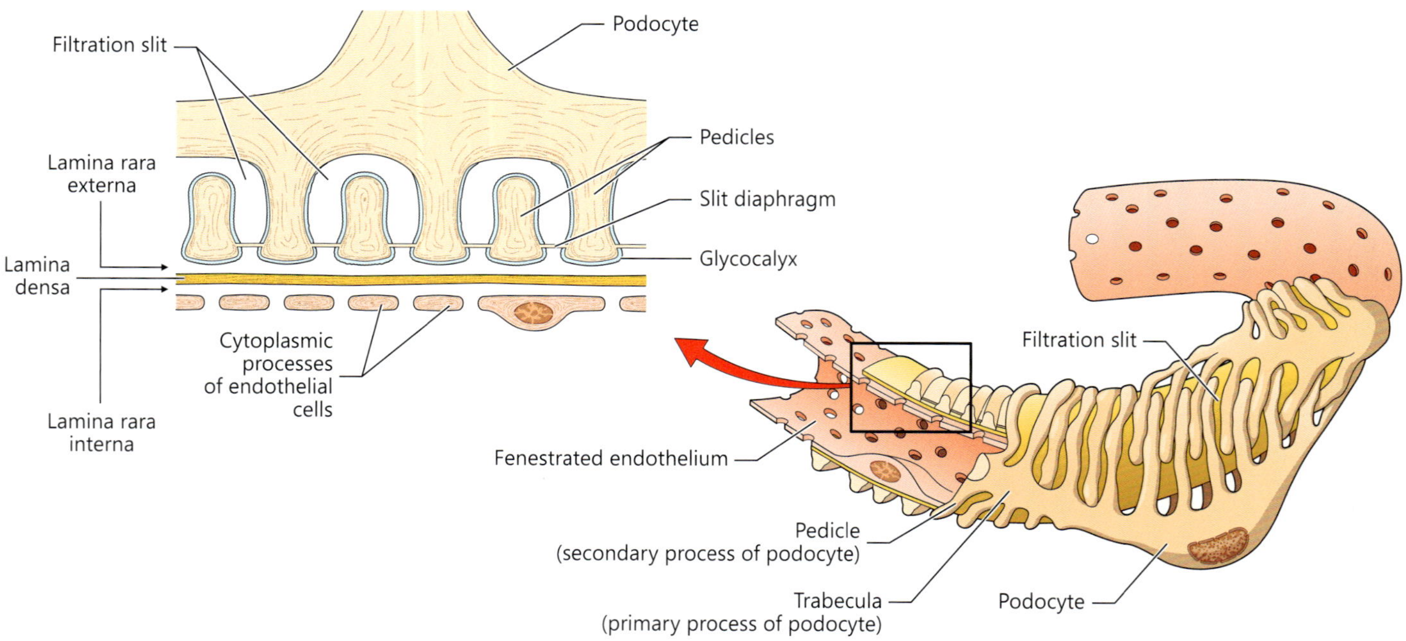

Relation between the **endothelium**, the **basement membrane** and the **pedicles of the podocytes**. These three structures constitute the kidney's ultrafiltration barrier. A schematic view of the glycocalyx lining the pedicles, the filtration slit diaphragms, the three layers of the basement membrane and the fenestrated endothelium is shown.

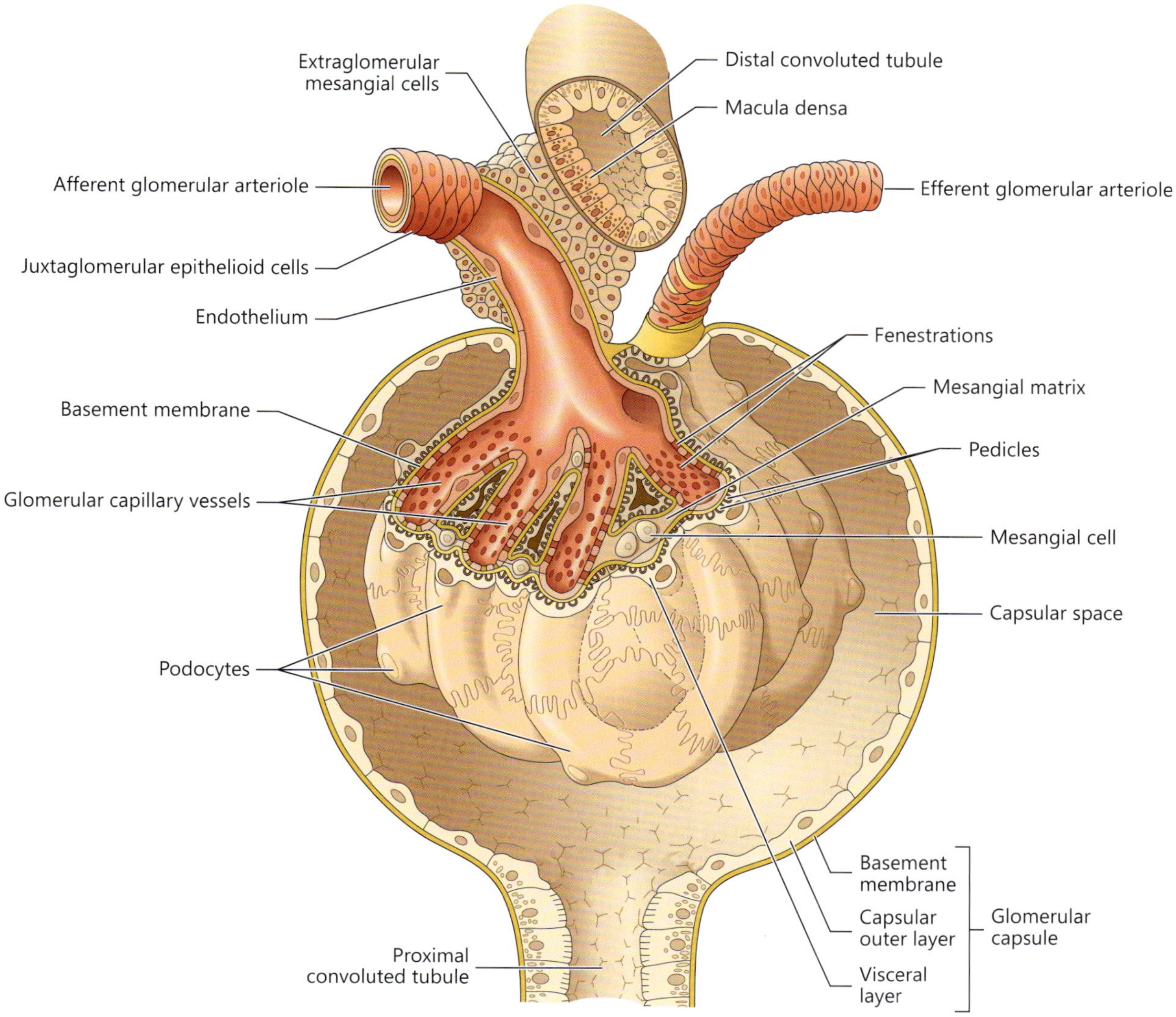

Schematic representation of **renal corpuscle** and **juxtaglomerular apparatus**.

The best way to understand the relation of the **glomerular capsule** to the **glomerulus** is to imagine that the proximal blind end (pouch) of the uriniferous tubule is invaginated to accommodate the glomerulus. The capsule, formed during the development phase, following an invagination of the renal tubule that accommodates the glomerular capillary vessels, is composed of two epithelial layers: the parietal and the visceral layer. The reflection point of the two layers (where the parietal becomes continuous with the visceral) occurs at the vascular pole, where the afferent glomerular arteriole originates the glomerulus, and the efferent glomerular arteriole forms from the glomerulus. In addition to a vascular pole, the renal corpuscle has a urinary pole. Here, the parietal epithelium of the capsule continuous with epithelium lining the proximal convoluted tubule. The parietal and the visceral layer of the corpuscle are separated by the capsular space where the preurine, an ultrafiltrate of the blood, is collected before draining into the lumen of the proximal convoluted tubule.

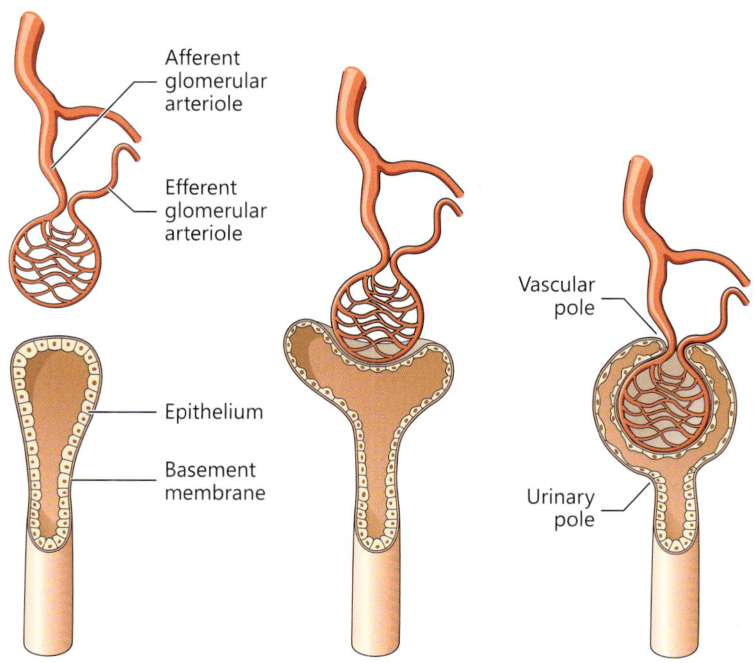

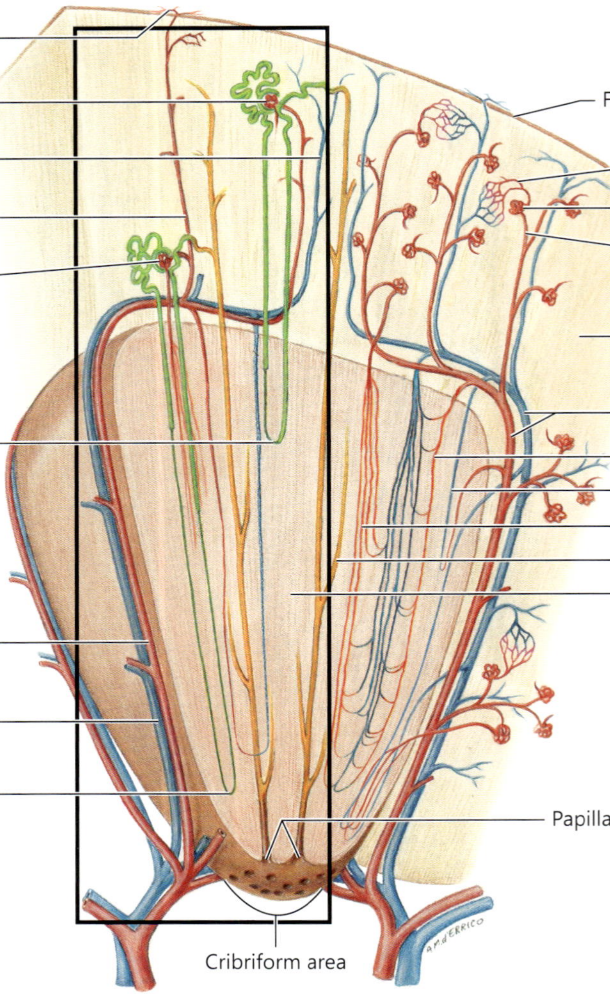

Schematic representation of the **blood supply of the renal parenchyma**. Note the typical distribution of the interlobar arteries, and the way the effluent venous circulation from the kidney is created.

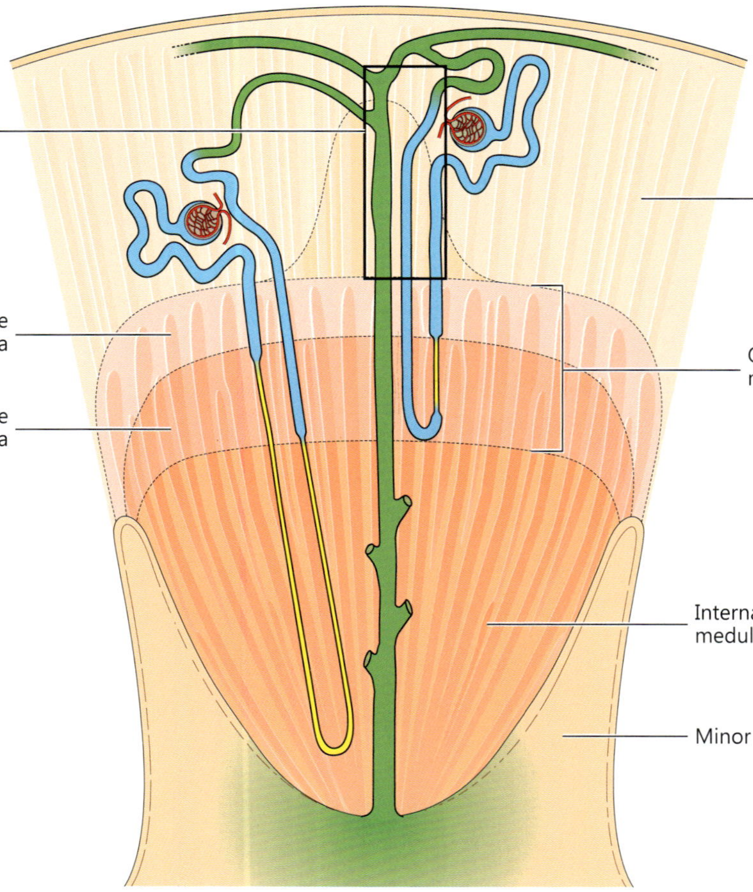

Schematic representation of the organization of a **renal pyramid** and the arrangement of the nephrons within the **renal parenchyma**: on the right, a cortical nephron; on the left, a juxtamedullary nephron. The medulla is divided into an inner and an outer zone; the latter is further divided into an inner and an outer stripe. The convoluted tubules are indicated in *blue*, the thin segments of the loops of Henle in *yellow*, while the connecting, collecting and papillary ducts are indicated in *green*.

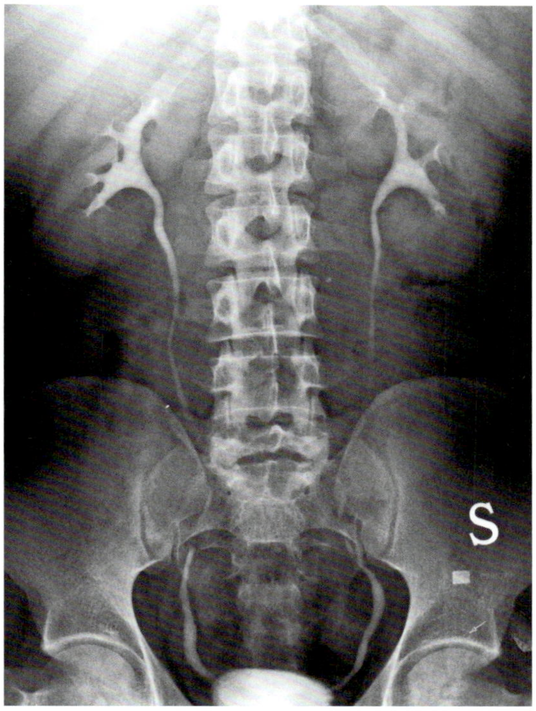

◄
Radiographic image of the **urinary tract**, with contrast medium injected intravenously (urography). The contrast medium excreted by the kidneys through the urine highlights the renal parenchyma and the urinary tract in a different way. From top to bottom, the contour of the kidneys, the major and minor calices, the renal pelves, the ureters and the vesical dome can be observed. In this subject, the renal pelves have a branched conformation. This urography clearly highlights the marginal flexure of the ureter, at the point of transition between the abdominal and the pelvic part of the duct.

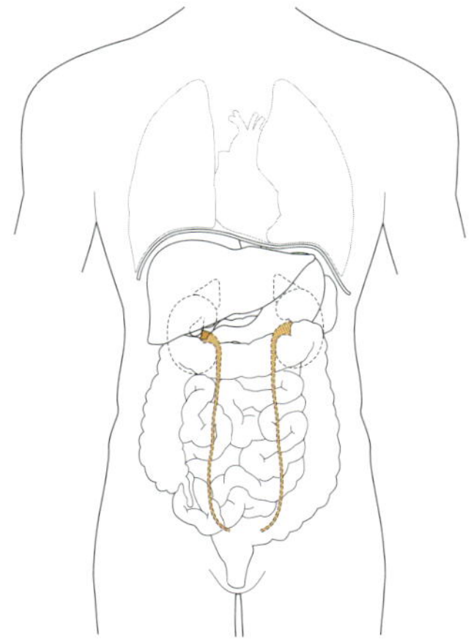

The **ureters** originate form the the renal pelvis, without a clear boundary, and drains into the urinary bladder, where it terminate through the urethral orifice, at the posterolateral angle of the trigone of the bladder. The abdominal part of the ureter has a slightly convex course, curving forward and medially; it crosses the iliac vessels and continues with the pelvic part at the pelvic inlet. The latter part describes a forward, upward and medially concave curve.

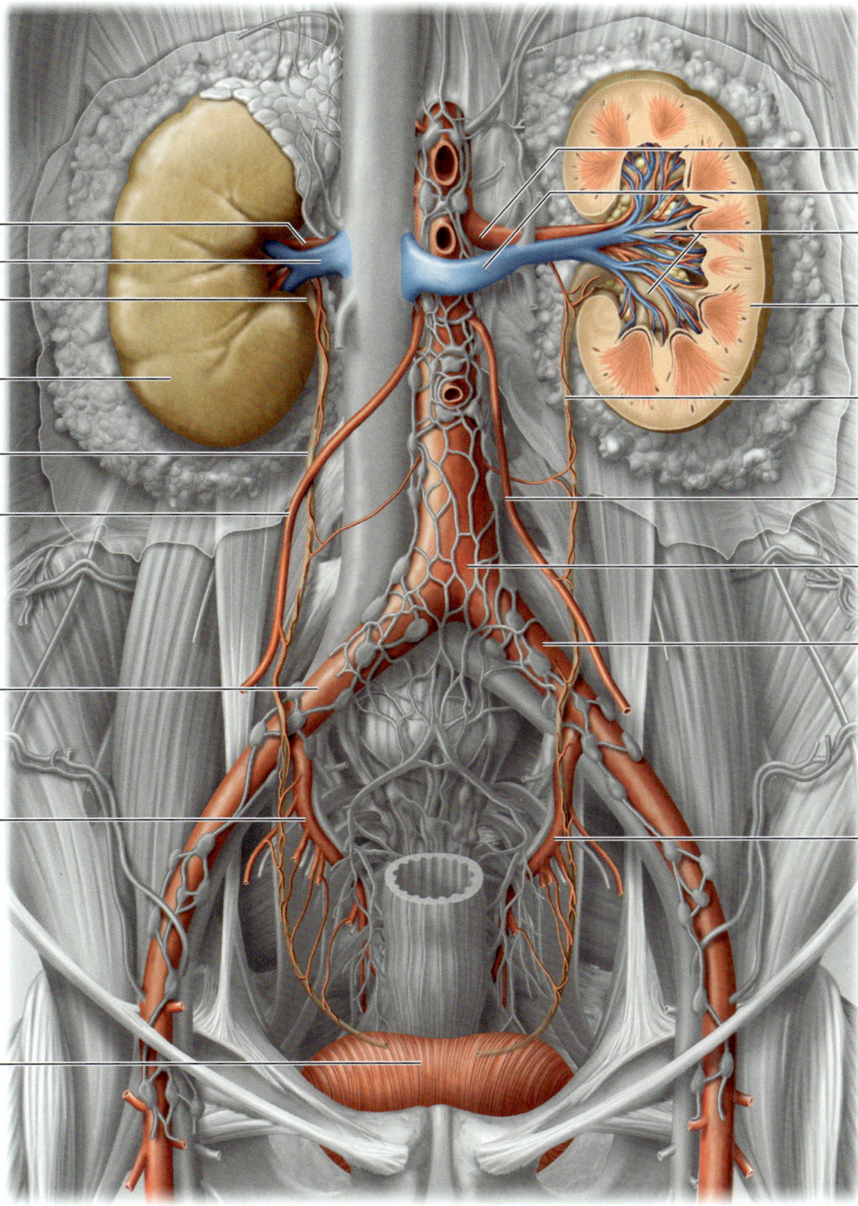

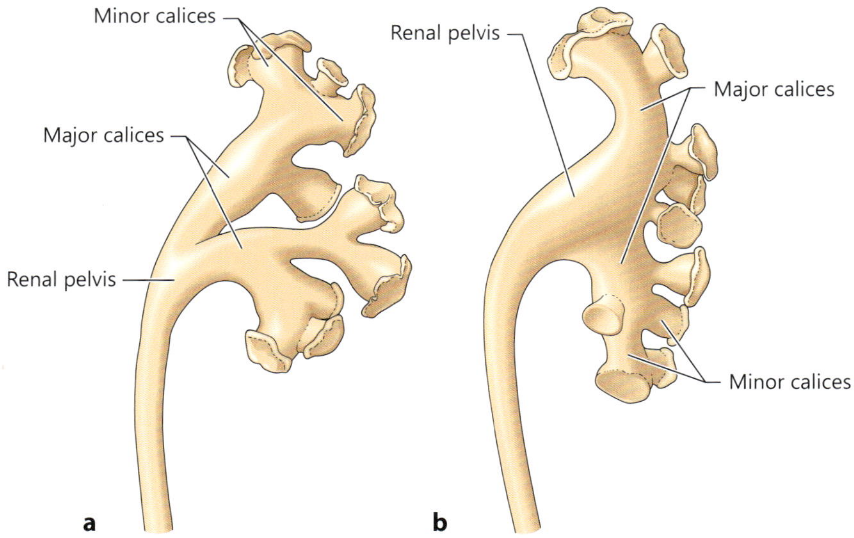

General conformation of the **pyelocalyceal complex**. **a**, Branched variant, marked by a considerable development of major and minor calices, and a narrow pelvis. **b**, Ampullary variant, marked by a broad pelvis and virtually absent the major calices.

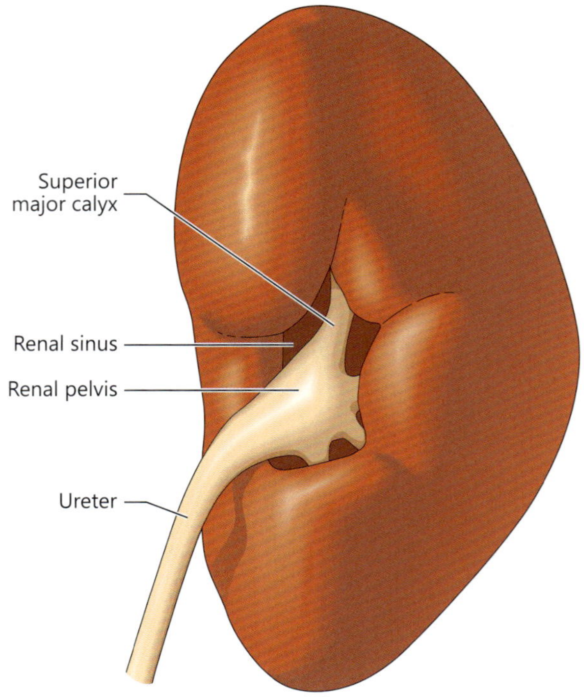

Conformation of the **pyelocalyceal complex** as seen from the medial border of the left kidney. Most of the renal sinus is occupied by the calices and the pelvis. The pelvis emerges from the renal sinus via the hilum, and then continues with the ureter.

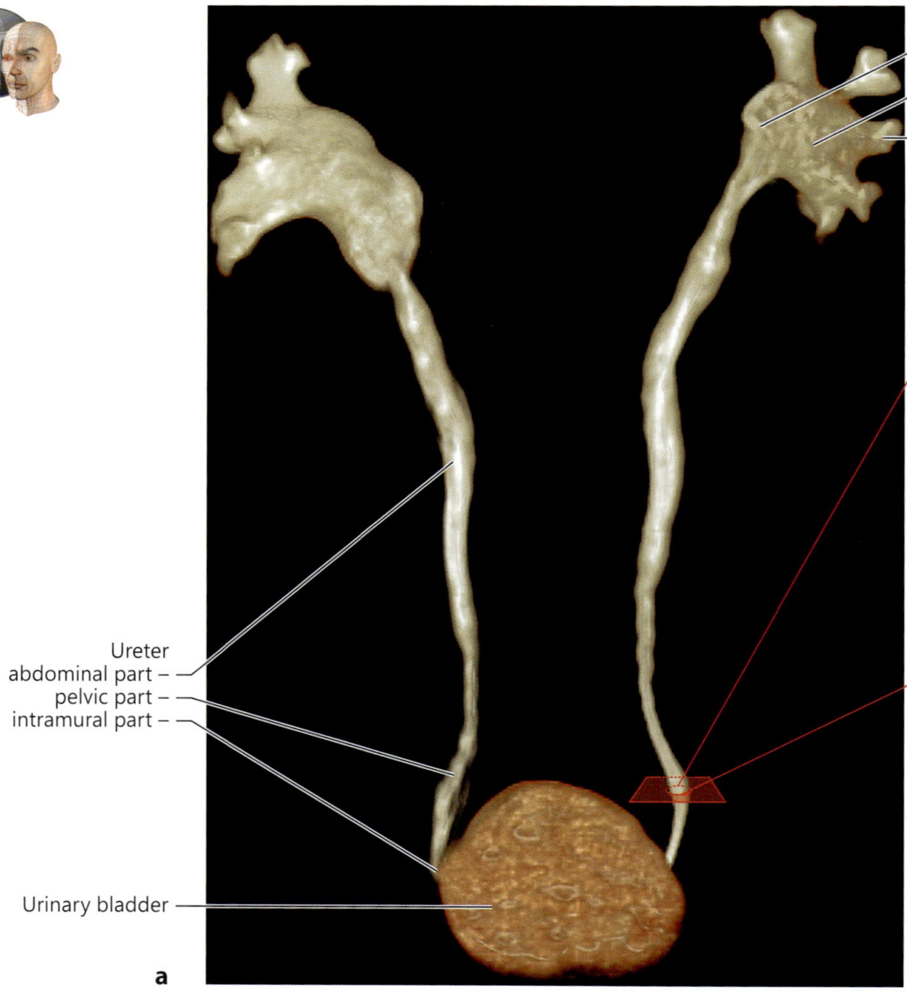

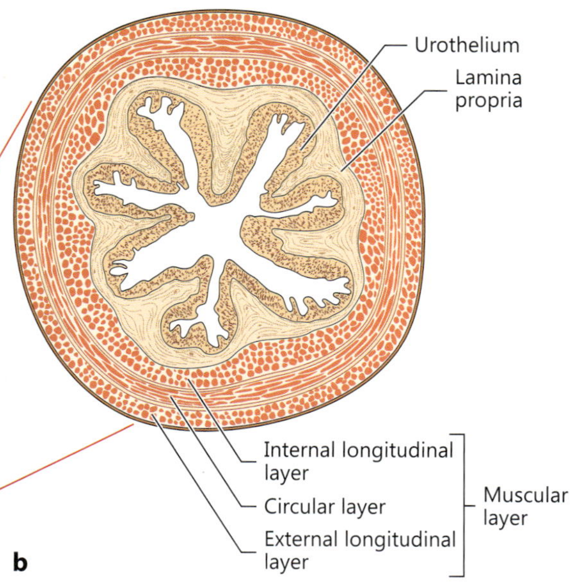

a, Three-dimensional image of a living human body, obtained by CT scan using a 3D volume rendering technique, showing the **organs of the urinary system**. **b**, Schematic representation of the ureter, showing the stellate configuration of the lumen and the arrangement of the layers of the muscular layer.

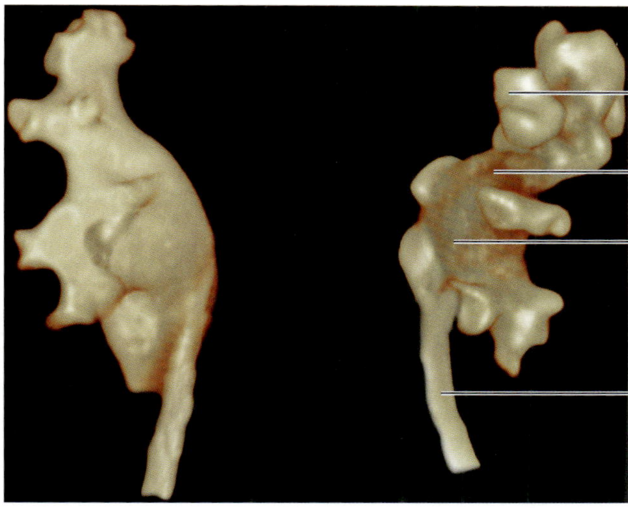

Three-dimensional image of a living human body, obtained by CT scan using a 3D volume rendering technique, showing the **renal pelvis** and the major and minor calices.

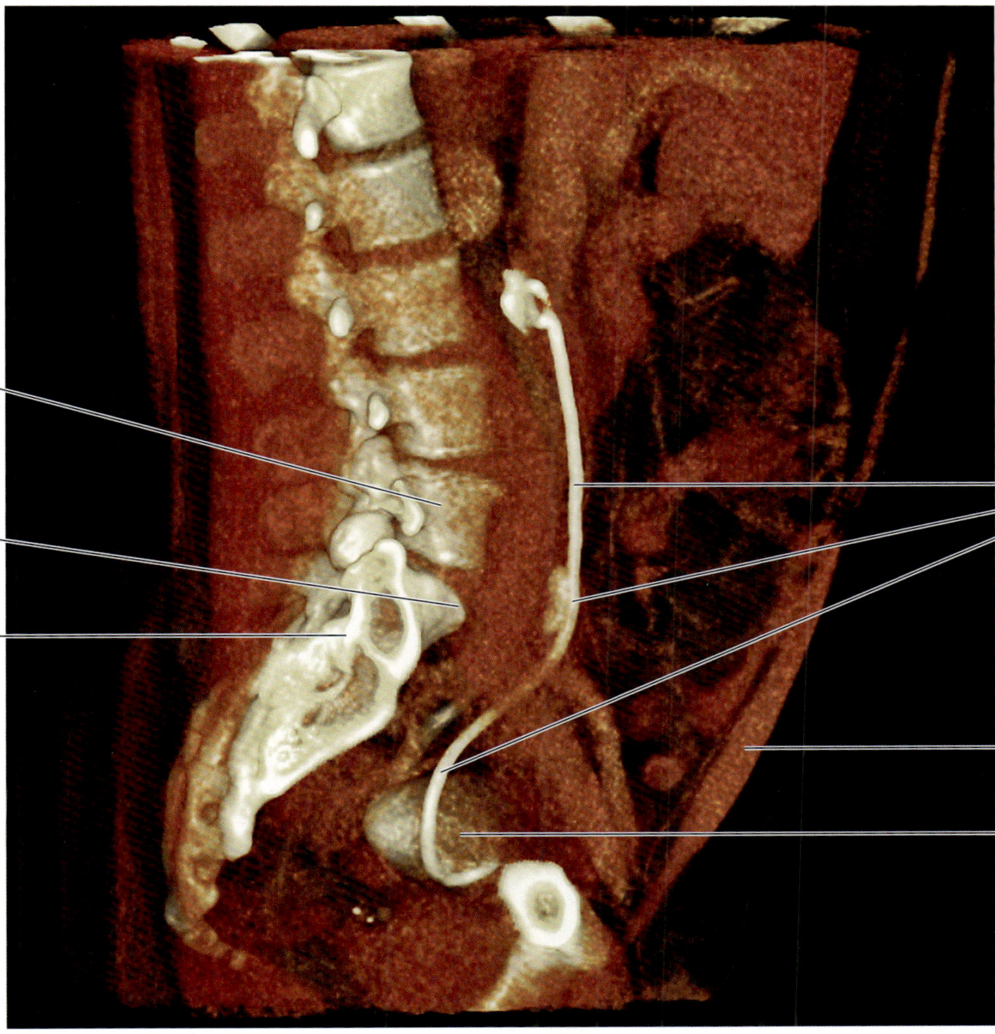

Three-dimensional image of a living human body, obtained by CT scan using a 3D volume rendering technique, showing the course of the **ureter** through the abdomen and the pelvis in lateral view.

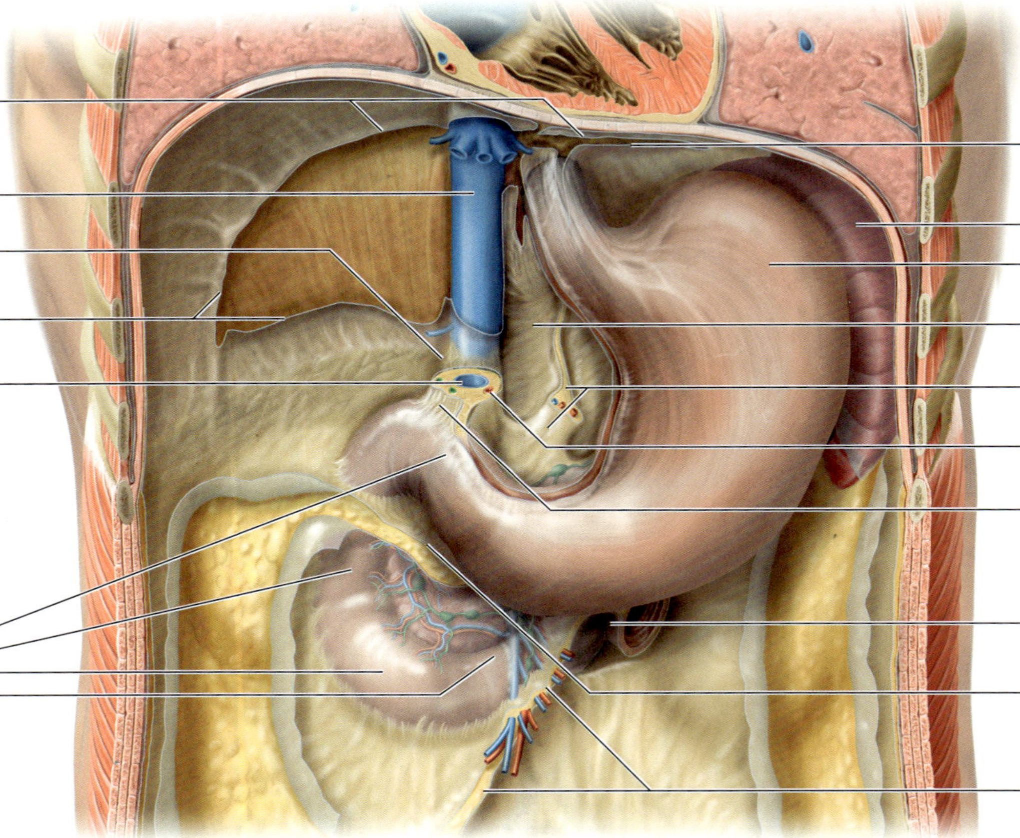

The **omental bursa**, or lesser sac, communicates with the greater peritoneal cavity through the omental foramen, located 5-6 centimeters right of the median line. The foramen is oriented in the sagittal plane and faces the right lateral abdominal wall. Anteriorly, the omental foramen is bounded by the right border of the hepatoduodenal ligament, in which run the components of the hepatic hilum, and posteriorly by the inferior vena cava.

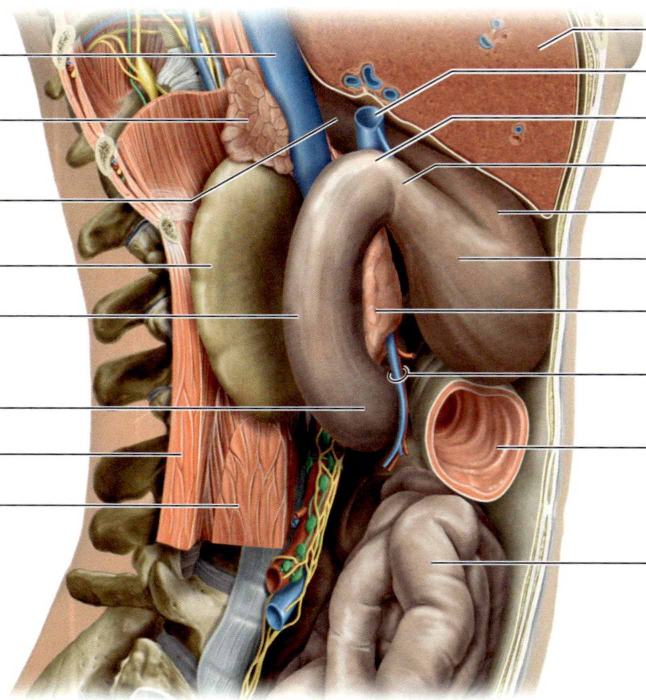

Right sagittal section of the **abdomen**, showing the incomplete ring shape of the duodenum, the head of the pancreas, lying within the concavity of the duodenum, and the pancreatic notch that accommodates the superior mesenteric vessels. The descending part of the duodenum, which is related posteriorly to the medial border of the right kidney, turns left at the inferior pole of the kidney, and continues with the inferior part.

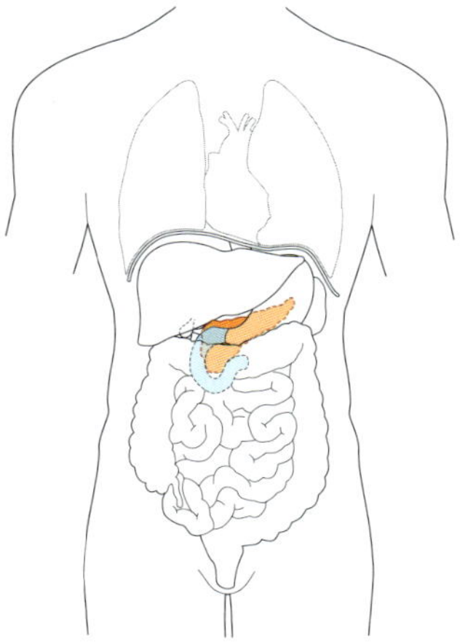

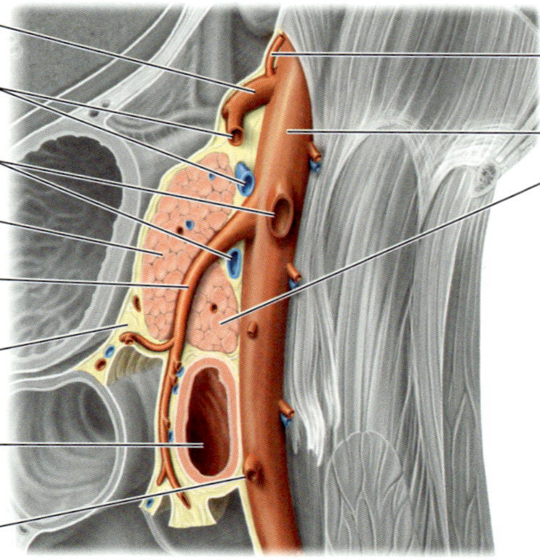

The parietal insertion of the **transverse mesocolon** divides the anterior surface of the head of the pancreas into a supramesocolic and a submesocolic part. The inferior medial end of the pancreas features the uncinate process, which is in relation to the superior mesenteric vein and artery, anteriorly.

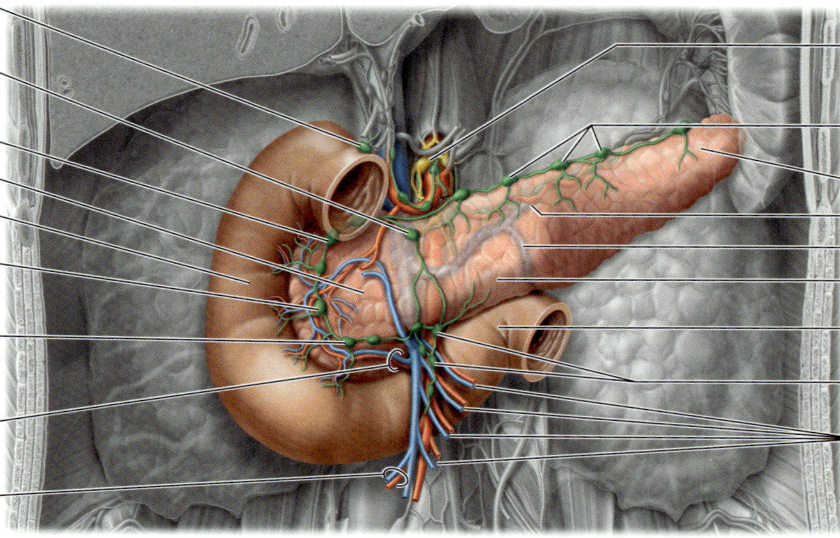

The **duodenum** is shaped as an incomplete ring, opened at the top and on the left. Its concavity surrounds the head of the pancreas; which, in turn, has a groove where the descending part of the duodenum is lodged. In this figure, the pancreas is displayed in transparency to show its posterior relations.

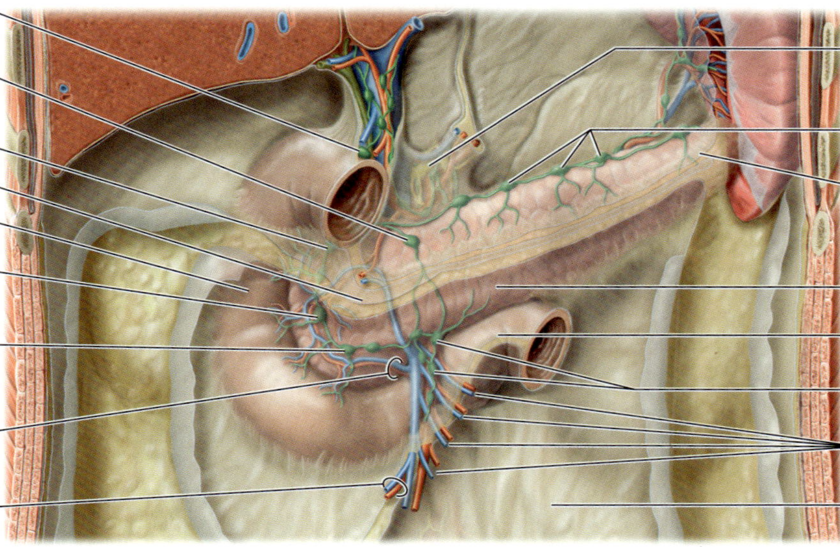

The **duodenum** is attached to the posterior abdominal wall, and is covered by the posterior parietal peritoneum on its anterior surface, while its superior part is entirely invested by the peritoneum. The pancreas is also attached to the posterior abdominal wall, and is invested anteriorly by the posterior parietal peritoneum.

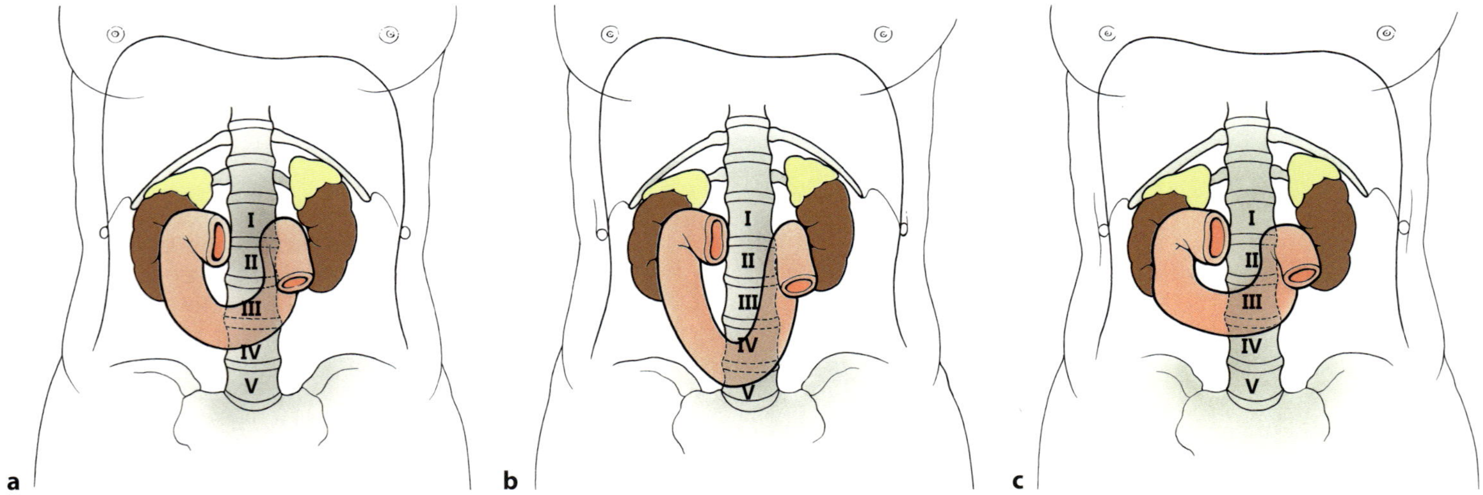

The three main morphological variants of the course of the **duodenum**. Figure **a** shows the typical morphology of this segment of the small intestine, where all parts can be easily distinguished. Figure **b** shows a V-shaped duodenum, where the inferior part is missing. Figure **c** shows a U-shaped duodenum, where the inferior part is especially pronounced. Of note, the inferior part of the V-shaped duodenum, extends as far as the fifth lumbar vertebra.

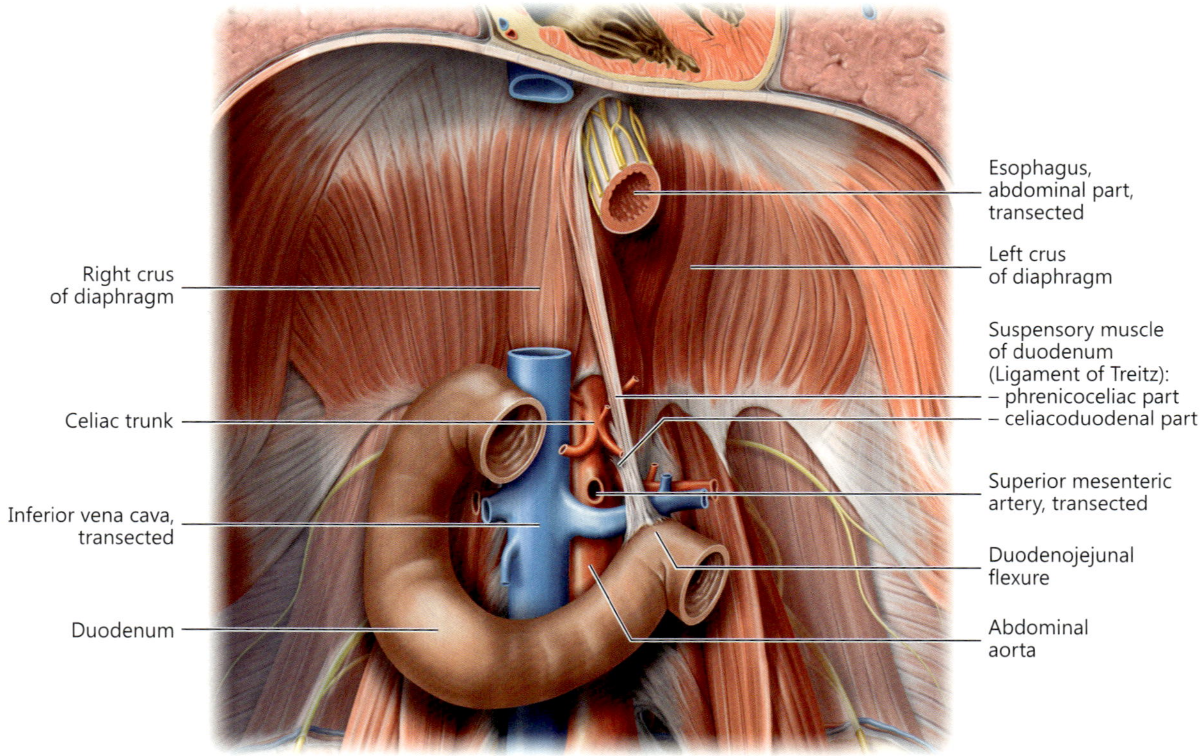

The **suspensory muscle of the duodenum** (or **ligament of Treitz**) is a muscle attaching to the duodenojejunal flexure and to the diaphragm. It is a landmark indicating the passage of the duodenum between the retroperitoneal and the peritoneal space (duodenojejunal flexure). The ligament is composed of a tendinous part ending with two muscular heads. The superior head consists of striated muscle, while the inferior head is composed of smooth muscle. The former originates largely from the right border of the esophageal hiatus, and, to a lesser extent, from the connective tissue surrounding the origin of the celiac trunk and of the superior mesenteric artery. The latter arises from the smooth fibers of the muscular layer of the duodenum, in the duodenojejunal region.

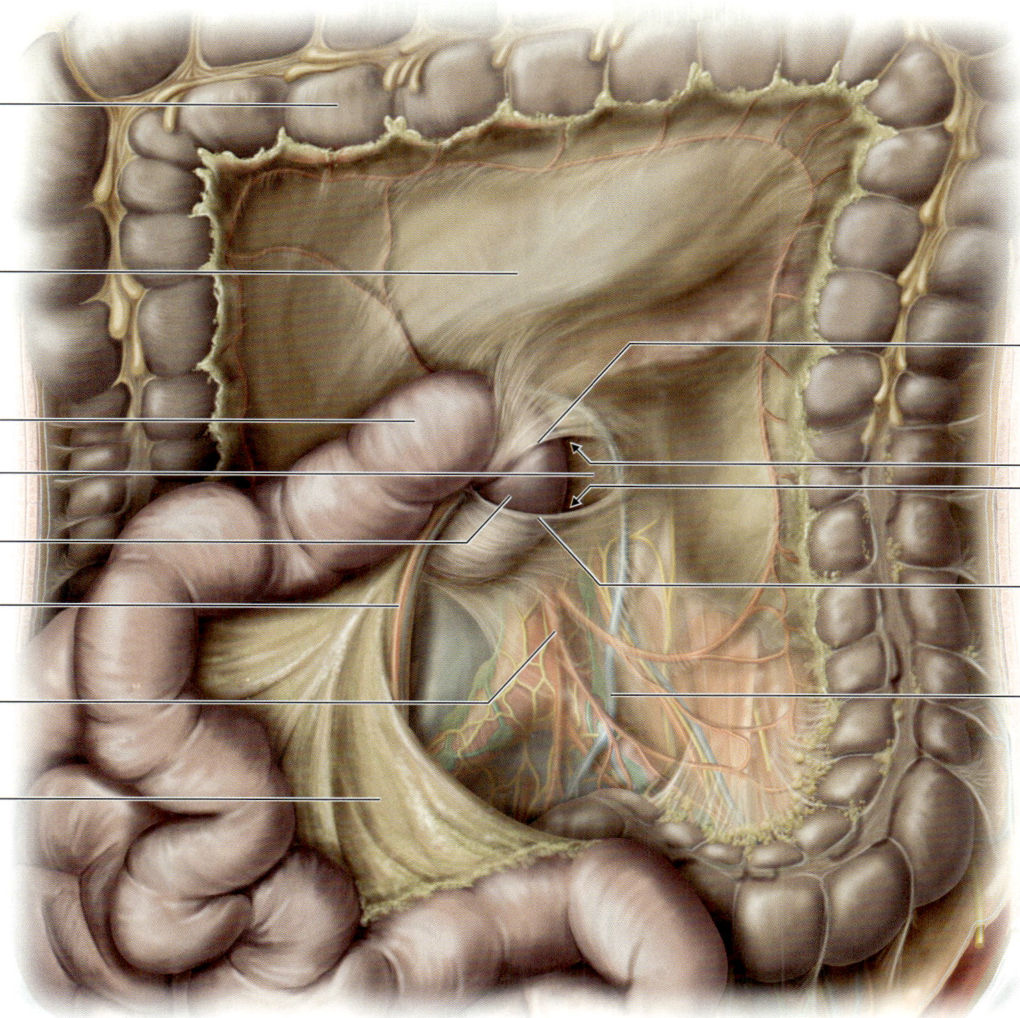

The **superior and inferior duodenal fossae** are serous spaces of the submesocolic compartment, located on the left side of the ascending part of the duodenum. The superior duodenal fossa lies anterior to the second lumbar vertebra and posterior to the superior duodenal fold, a triangular layer that extends between the left margin of the duodenojejunal flexure and the inferior mesenteric vein. The inferior duodenal fossa lies anterior to the third lumbar vertebra and posterior to the inferior duodenal fold, which accommodates the left colic artery.

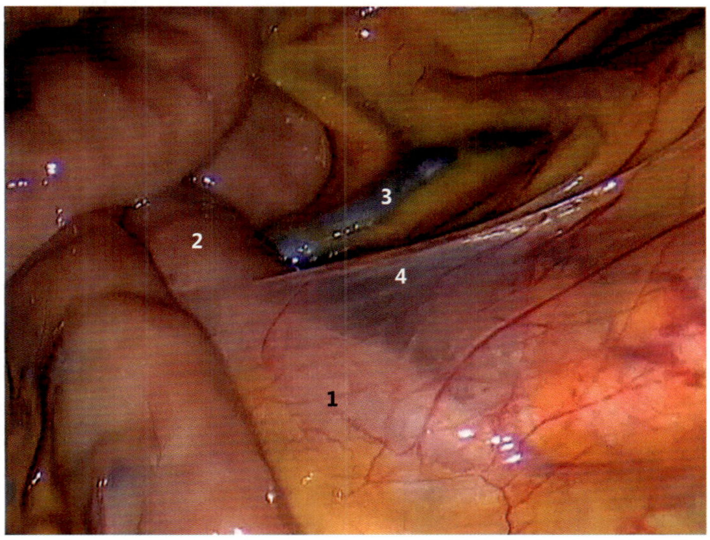

Image taken during a surgical laparoscopy showing the relations of the **ascending** (**fourth**) **part of the duodenum** to the mesenteric vessels and the jejunal loops.

1 Ascending part of the duodenum (fourth part)
2 First jejunal loop
3 Inferior mesenteric vein
4 Inferior duodenal fold

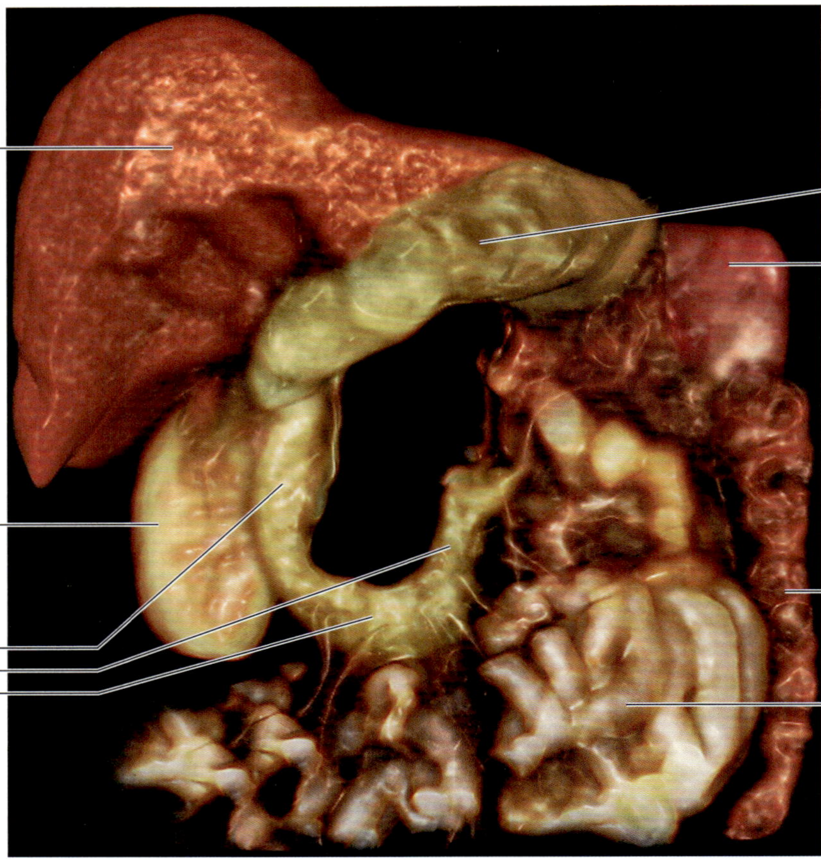

Three-dimensional image of a living human body, obtained by CT scan using a 3D volume rendering technique, showing some of the **organs contained in the abdominal cavity** in frontal projection.

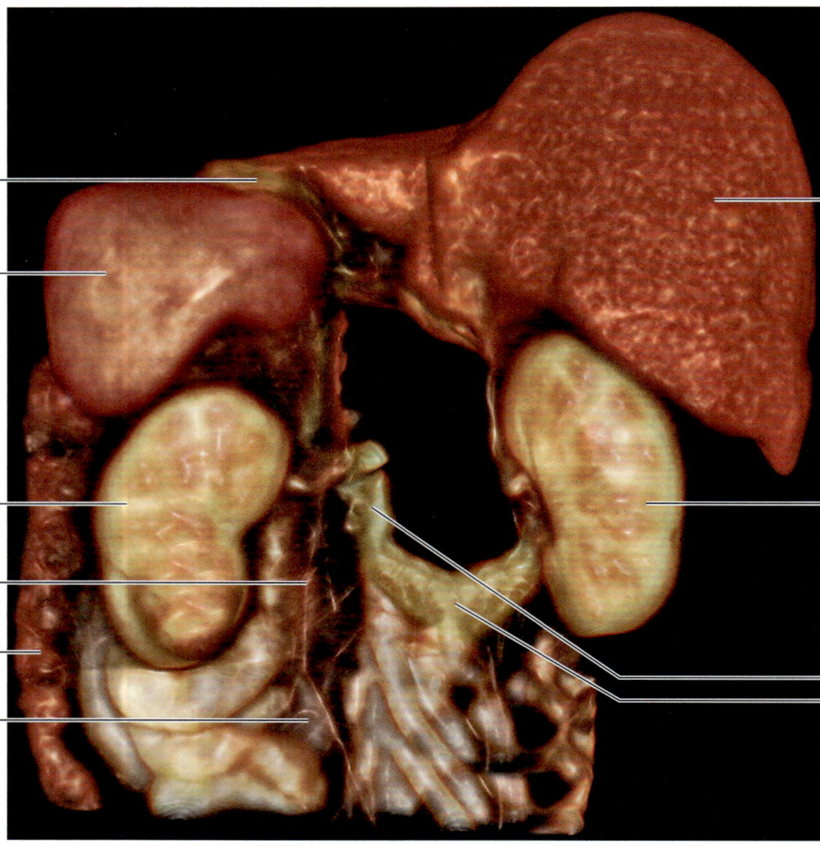

Three-dimensional image of a living human body, obtained by CT scan using a 3D volume rendering technique, showing **organs contained in the abdominal cavity** in posterior projection.

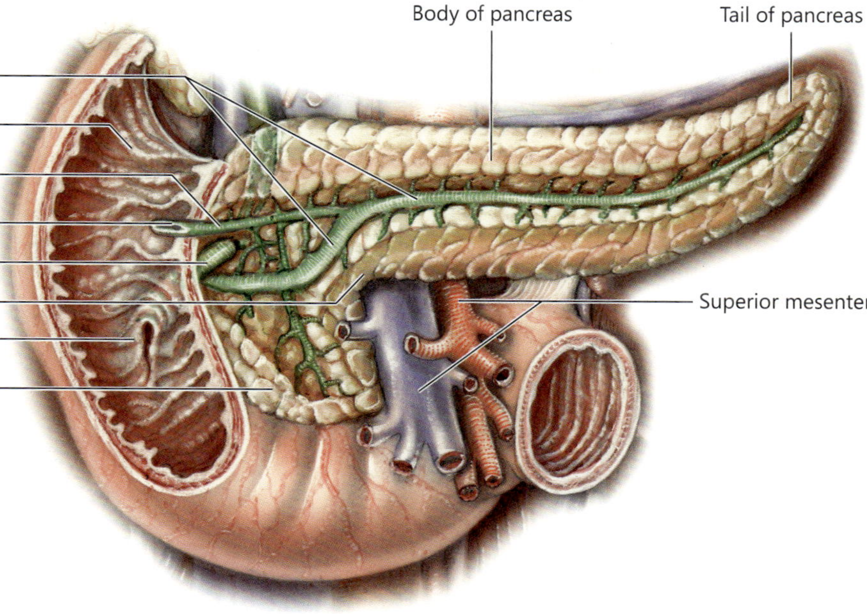

Section of the **duodenum** and **pancreas** showing the course of the pancreatic duct (of Wirsung) and of the accessory pancreatic duct (of Santorini).

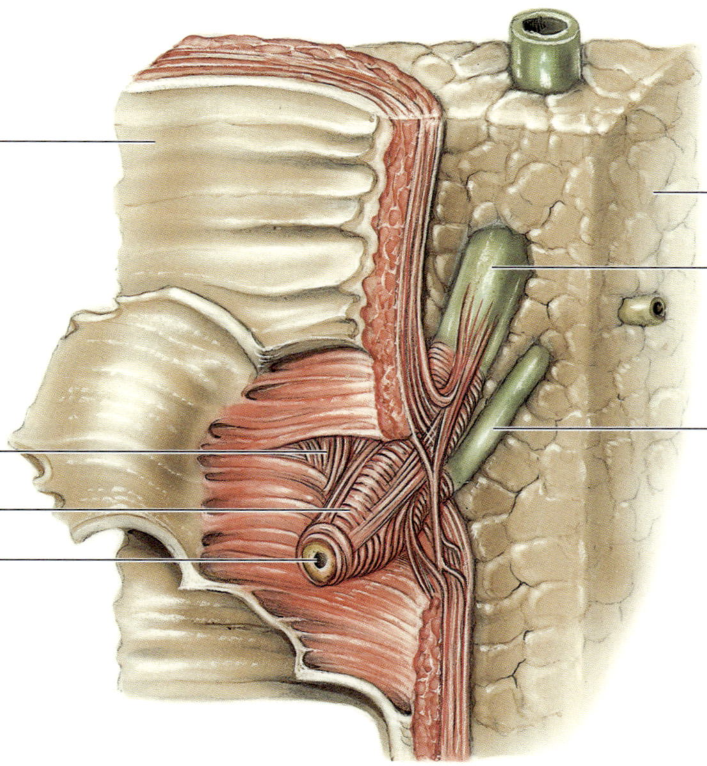

Reconstruction of the **sphincter of the ampulla** (or **sphincter of Oddi**) that surrounds the hepatopancreatic ampulla (or ampulla of Vater), displaying circular (sphincter of the ampulla) and longitudinal muscle fascicles. The bile and pancreatic ducts travel through an eyelet that is formed within the muscle layer of the duodenum. In the segment running closest to the ampulla, both ducts appear invested in an own circular muscle layer (sphincter of the bile duct and of the pancreatic duct).

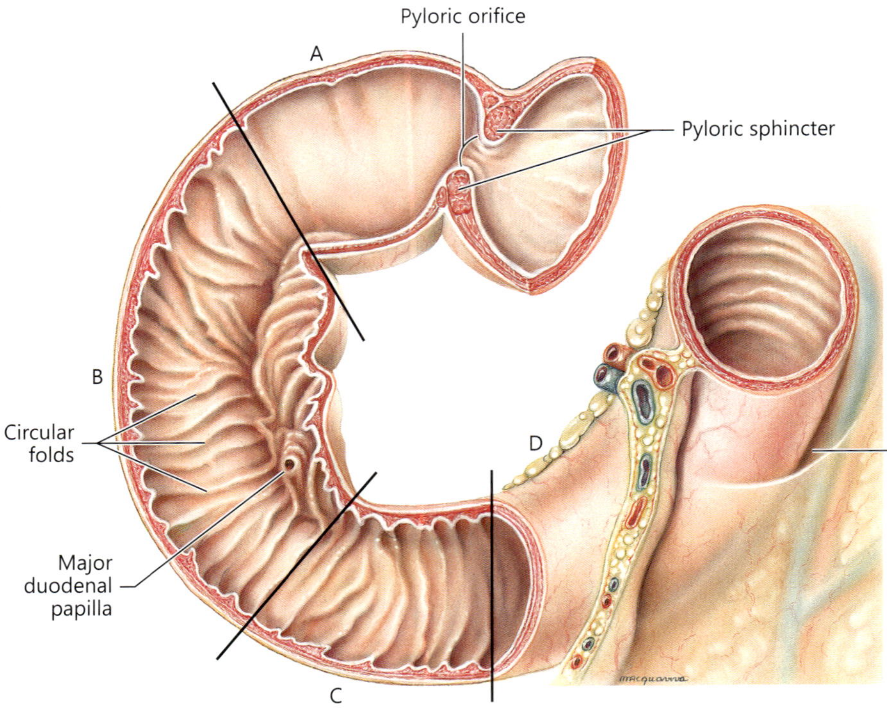

Internal configuration of the **duodenum** after removal of the anterior wall (the solid lines demarcate the four parts of the duodenum): A, superior or first part, comprising the ampulla; B, descending or second part; C, inferior or third part; D, ascending or fourth part. The first part of the duodenum is smooth, whereas circular folds appear at the beginning of the second part.

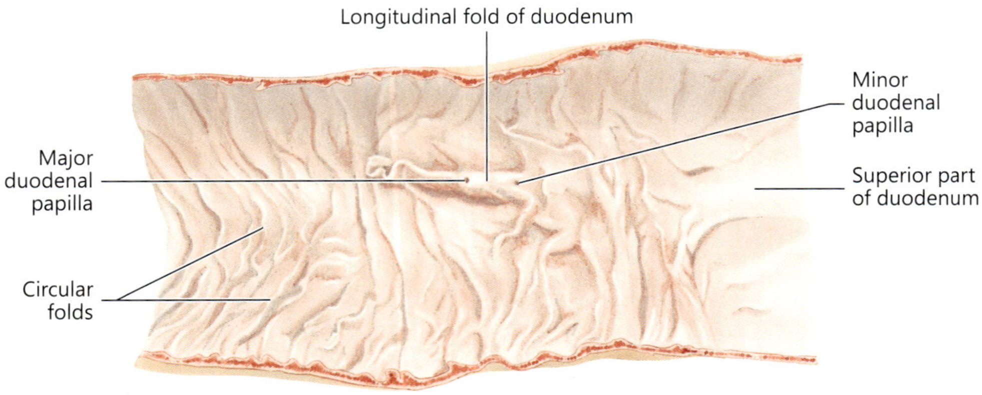

Internal configuration of the **duodenum** passing from the first to the second part. Note the transition from the smooth intestinal wall to the circular folds. The major and minor papillae are indicated in the second part of the duodenum.

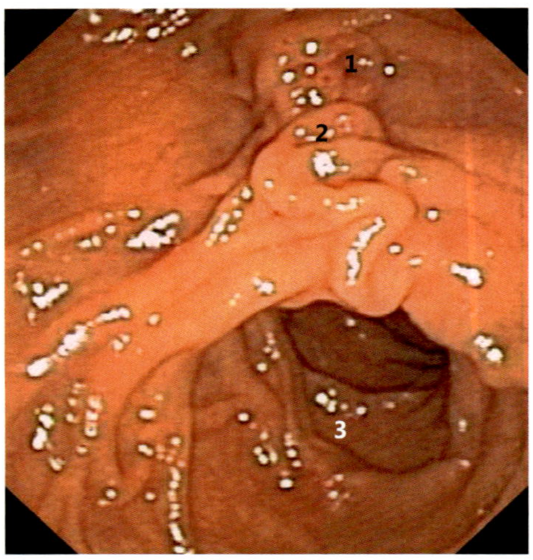

Image of the **duodenum** taken during a diagnostic duodenoscopy.

1 Duodenal papilla
2 Longitudinal fold
3 Lumen of the third part of the duodenum

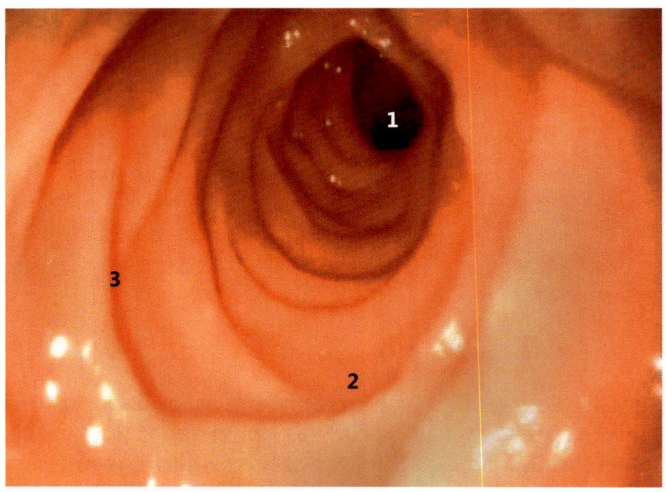

Image of the **duodenum** taken during a diagnostic duodenoscopy.

1 Duodenal lumen
2 Duodenal mucosa
3 Circular fold (Kerckring's valves)

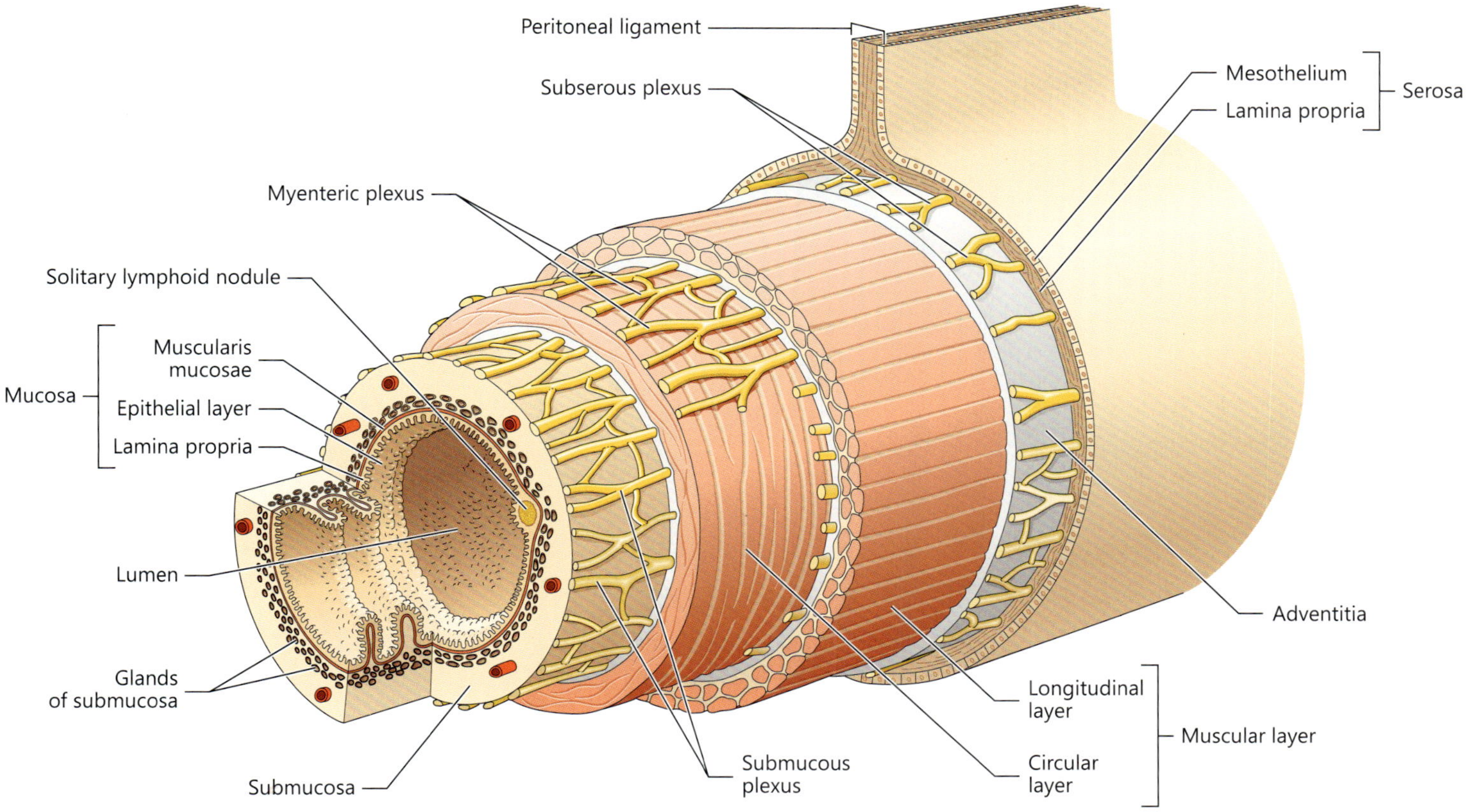

General organization of the intestinal canal. The muscular layer is composed of an inner circular and of an outer longitudinal layer of smooth muscle cells. The subserous, myenteric (or Auerbach's) and submucous (or Meissner's) plexuses constitute the enteric plexus of the autonomic part of the peripheral nervous system.

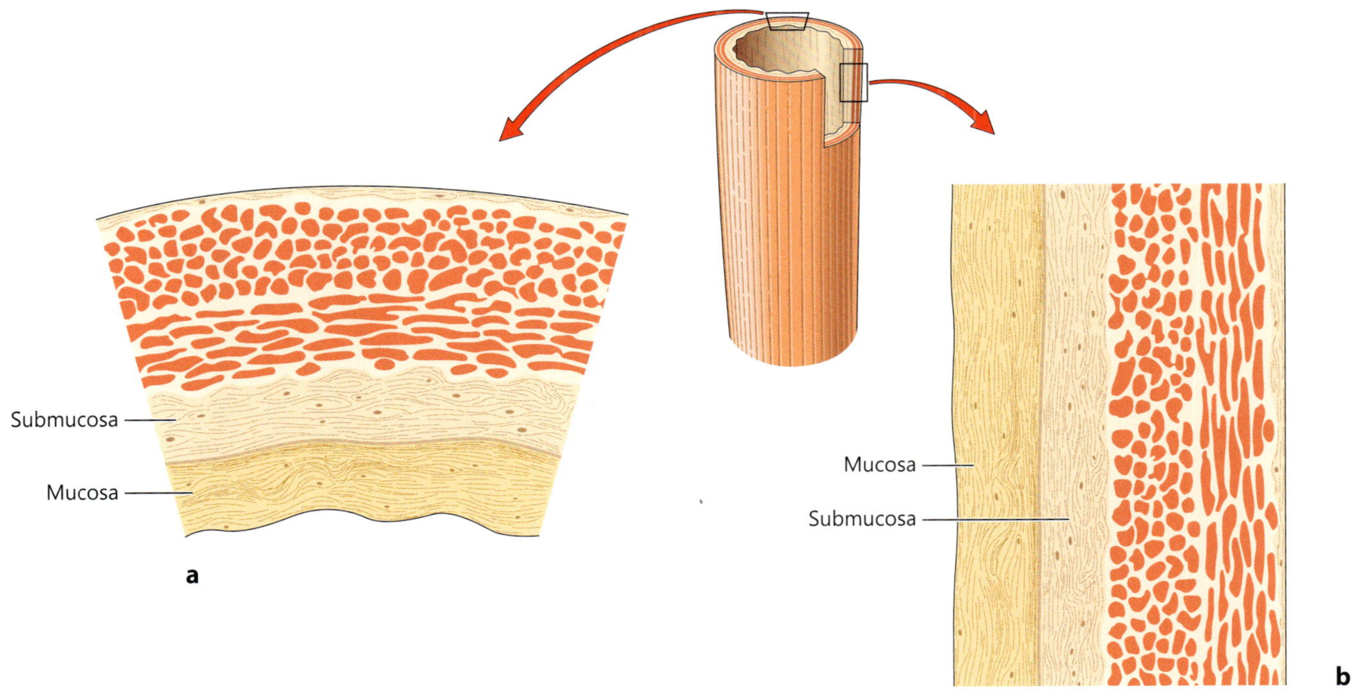

Different orientation of the outer longitudinal and inner circular layers of the muscular layer of an **organ of the alimentary system**, in a transverse (**a**) and longitudinal (**b**) section.

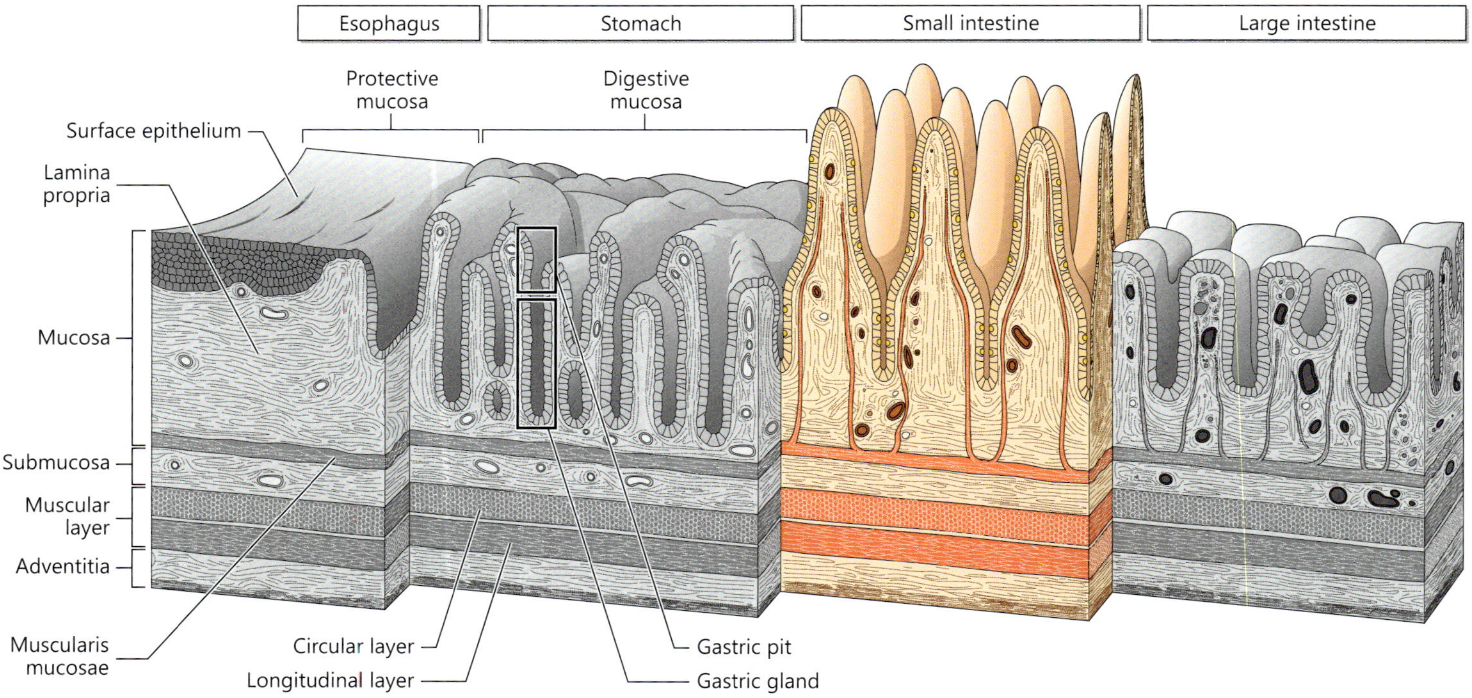

Summary of the **general organization of the hollow organs of the alimentary system** (esophagus, stomach, small intestine, large intestine). The portion of the small intestine is highlighted in color. The surface epithelium is simple columnar epithelium in the small intestine.

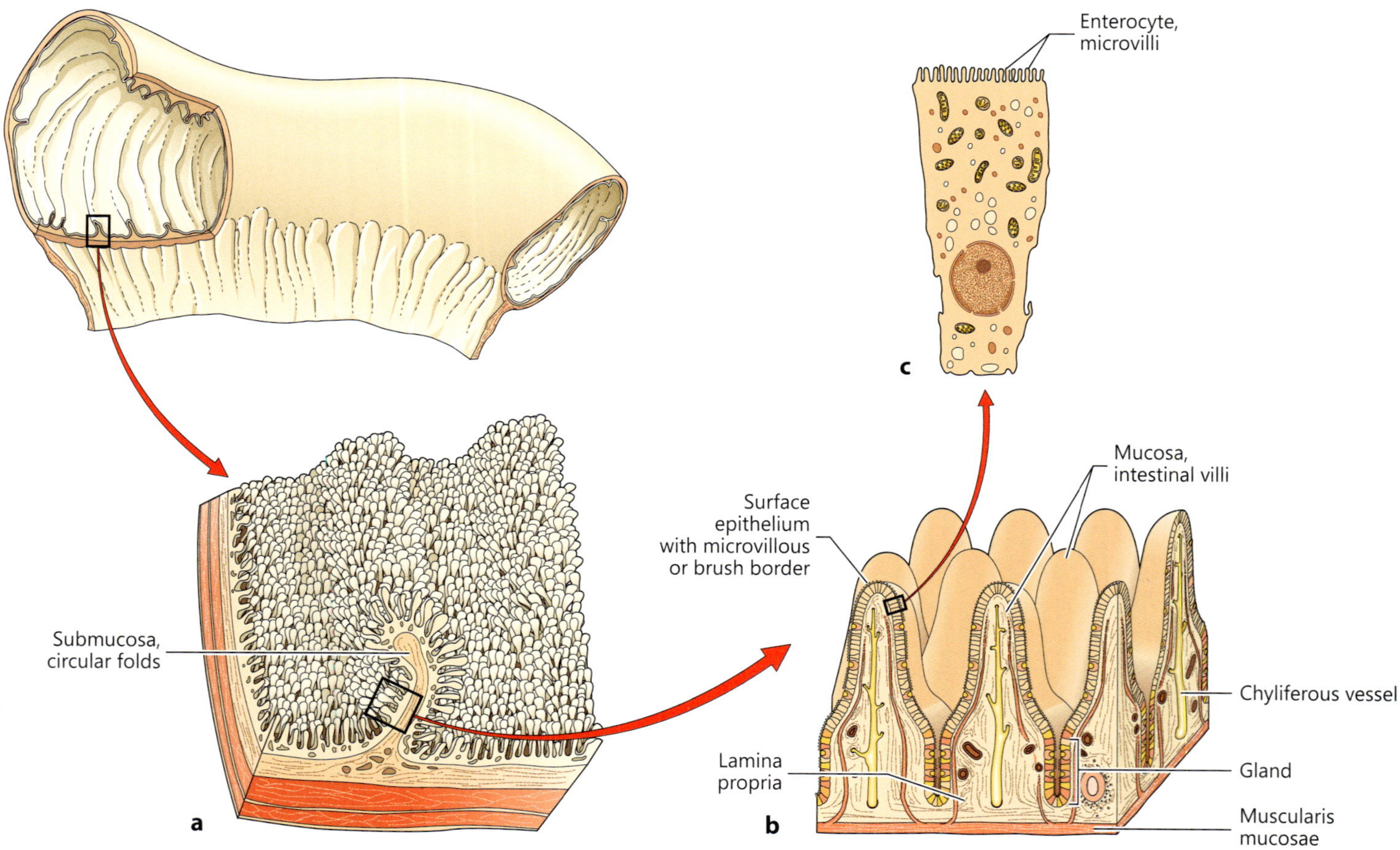

Representation of the three possible ways for increasing the **absorptive surface area of the small intestine**: elevations of the submucosa, the circular folds, (**a**); elevations of the mucosa, the intestinal villi (**b**) elevations of the apical cytoplasmic membrane of the enterocyte, the microvilli (**c**).

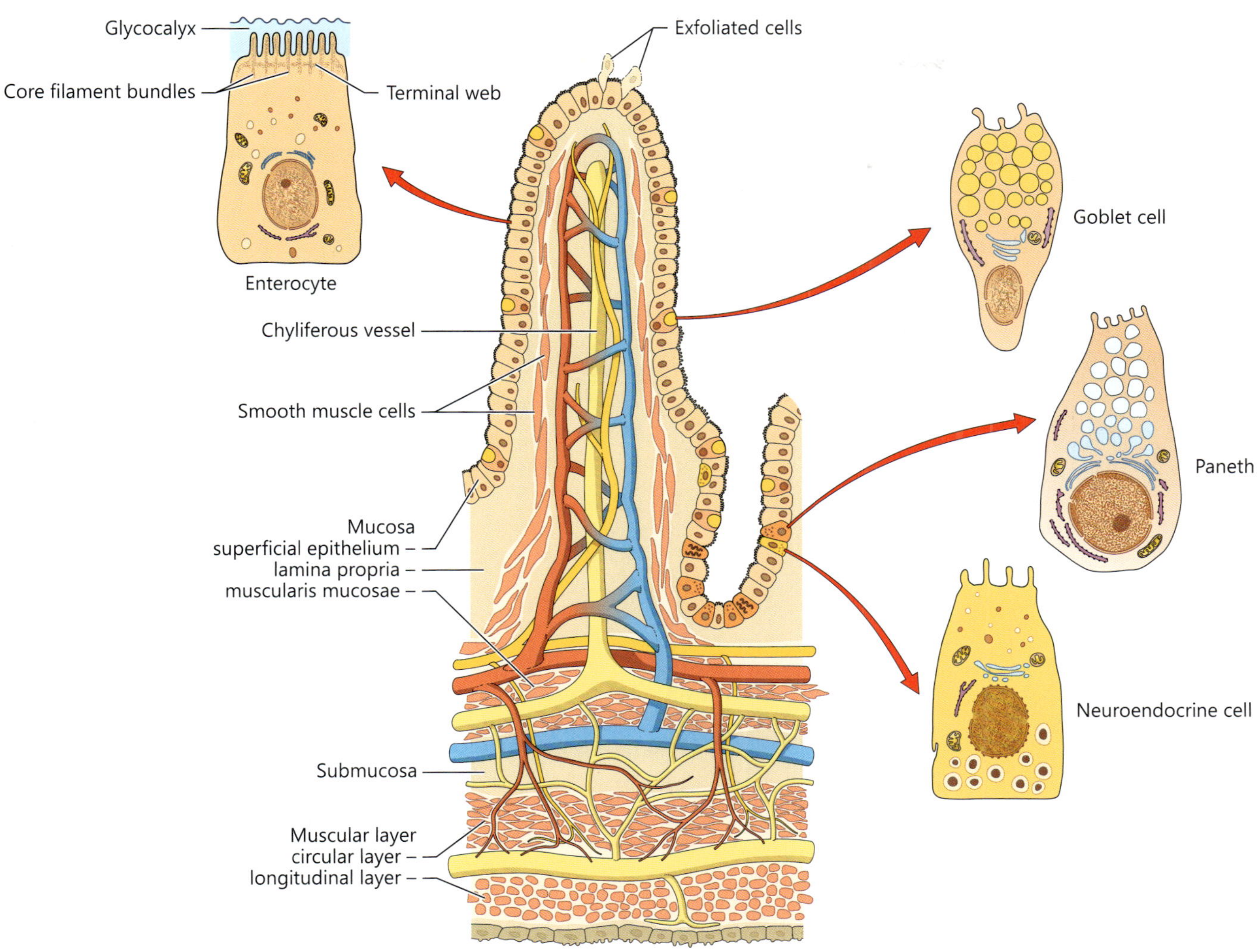

Diagram of the **arterial, venous and lymphatic organization of the intestinal villi**. The arteries terminate into a subepithelial capillary network drained by veins. The lymphatic circulation begins with the chyliferous vessels located centrally in the villi, draining into the lymphatic networks of the submucosa. The superficial, or investing, epithelium of the crypt-villus units has a striated border. The stromal axis of the villus contains an arteriole (*red*), a venule (*blue*), a lymphatic capillary (*yellow*) and smooth muscle cells. Note that some cells are undergoing mitosis at the bottom of the gland. The ultrastructural features of the main cell types of the unit are shown on the right.

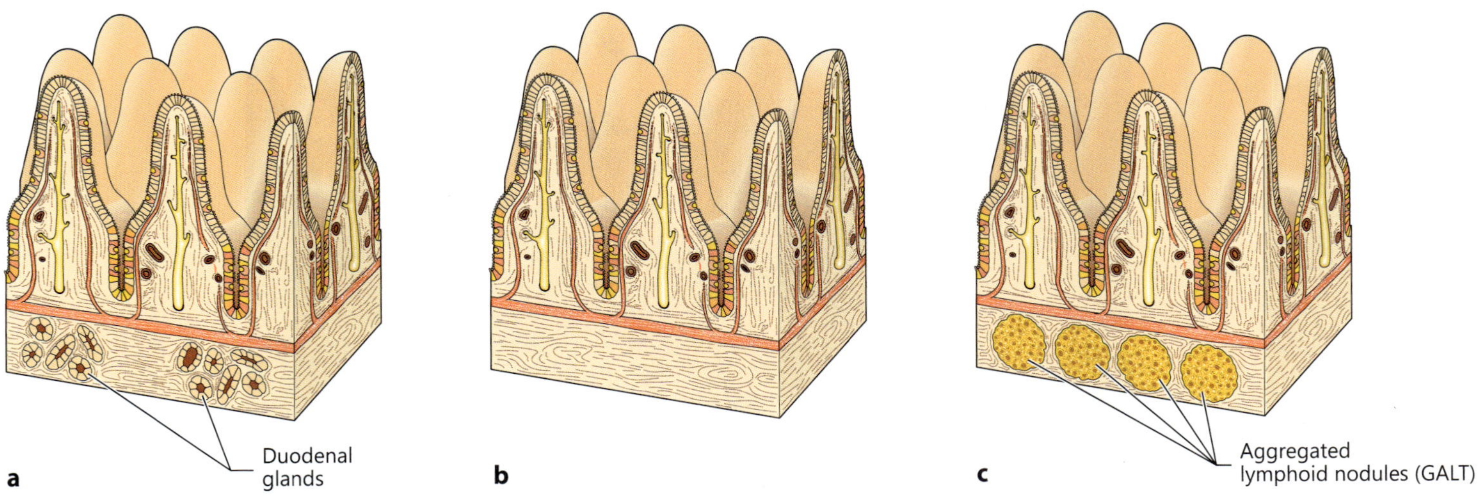

Morphologic criteria for the identification of the **duodenum** (**a**), **jejunum** (**b**) and **ileum** (**c**). Duodenum, compound tubular glands, the duodenal (or Brunner's) glands, can be found in the submucosa, **a**. Ileum, aggregated lymphoid nodules (an important part of the gut associated lymphoid tissue, GALT) can be found in the lamina propria of the mucosa, and extend to the submucosa (Peyer's patches, **c**). Jejunum lacks both (**b**).

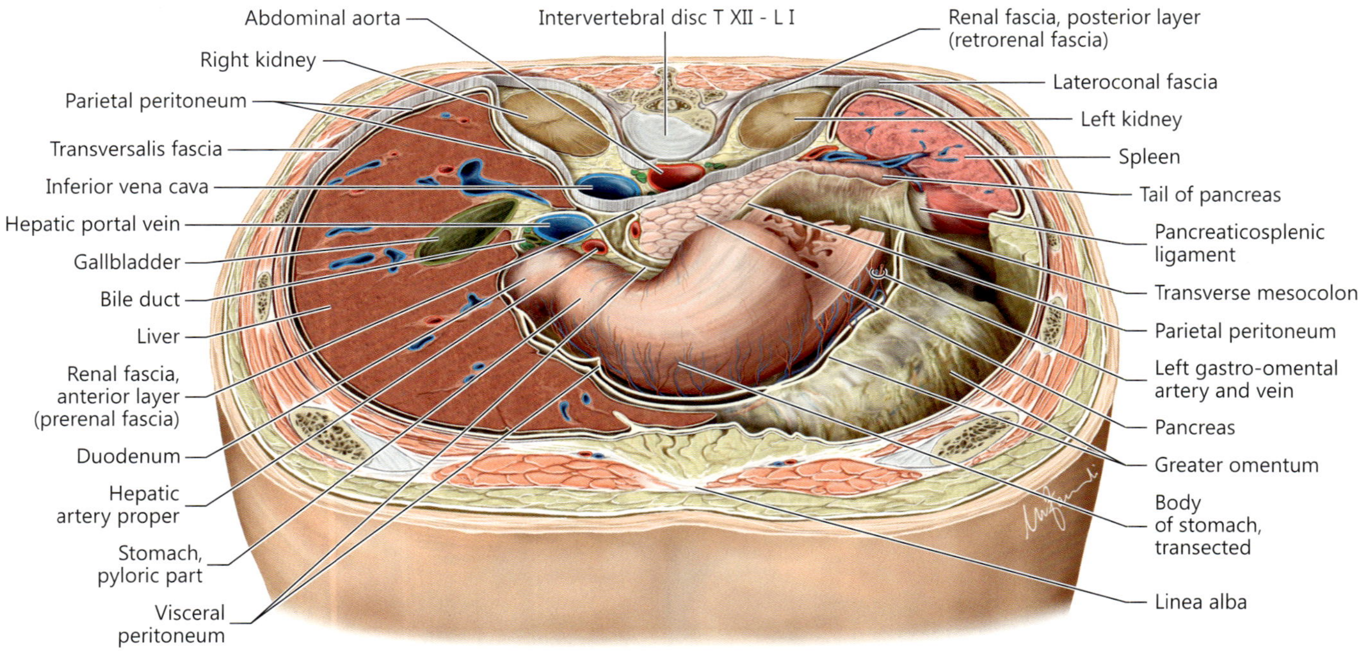

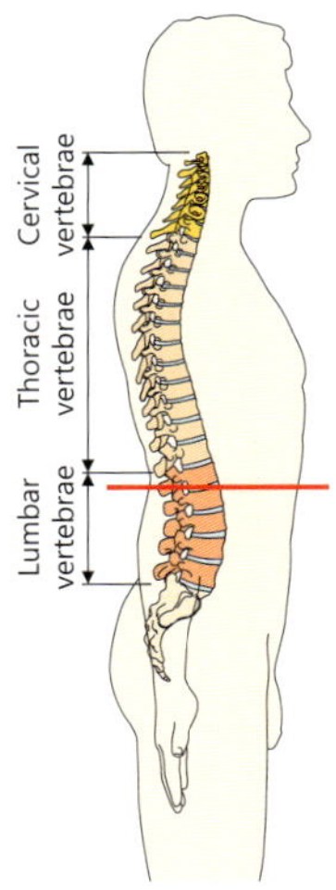

Transverse section of the trunk at the first lumbar vertebra. The fundus and part of the body of the stomach were removed to expose the relations to the spleen, the left kidney, the pancreas and the transverse mesocolon. The apex of the tail of the pancreas is connected to the splenic hilum by the **pancreaticosplenic ligament**.

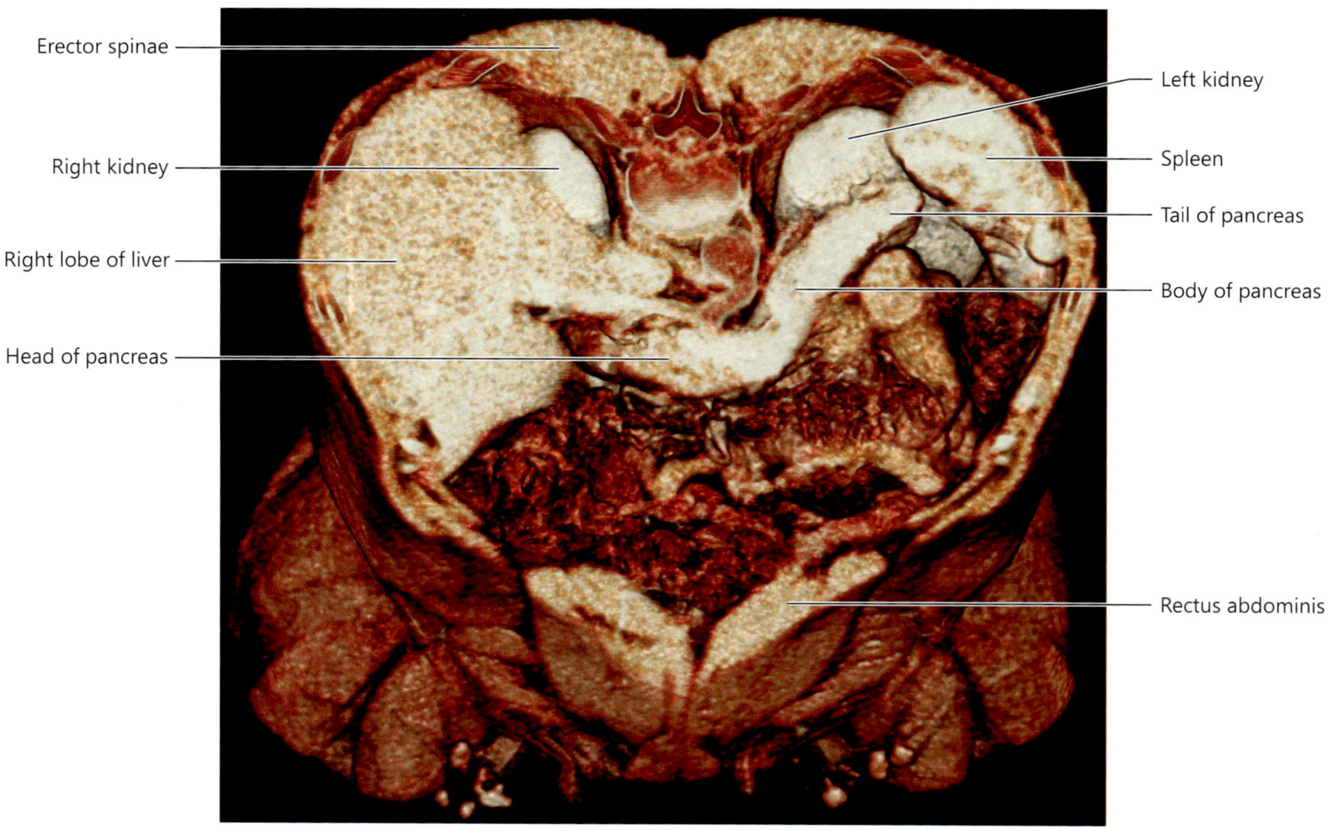

Three-dimensional image of a living human body, obtained by CT scan using a 3D volume rendering technique, showing the **abdominal organs** in a transverse section passing through the eleventh thoracic vertebra.

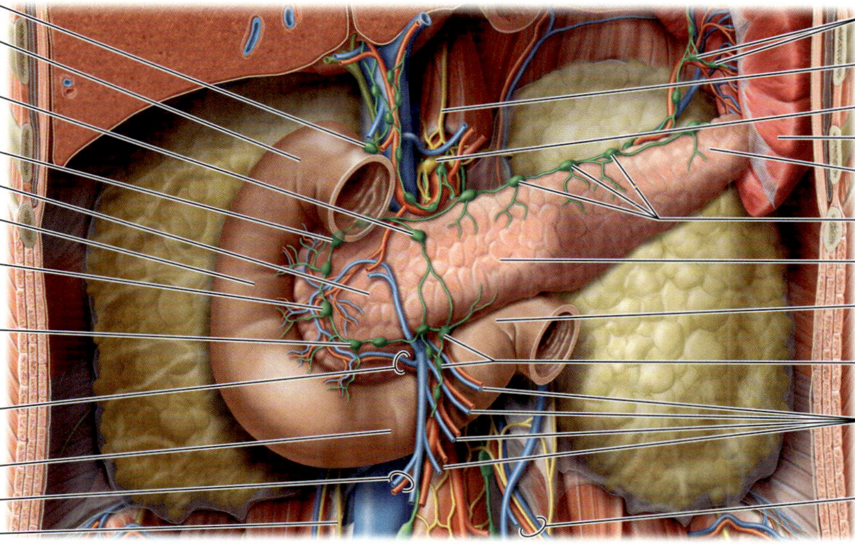

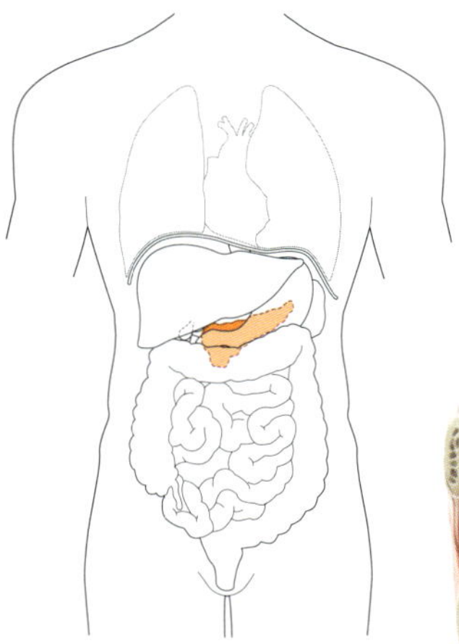

Anterior view of the tail, body, neck and head of the **pancreas**. The duodenal loop surrounds the head of the pancreas, hence providing attachment. The body of the pancreas, between the neck and the tail, is convex anteriorly; its superior border is angled upwards and to the left, and is related to the celiac trunk. The shape and size of the tail of the pancreas may vary considerably, from thin and flat to thick and rounded. The figure displays the superior pancreatic, superior pancreaticoduodenal, suprapyloric and subpyloric nodes, which drain into the celiac nodes, and the inferior pancreatic and inferior pancreaticoduodenal nodes, which drain into the superior mesenteric nodes.

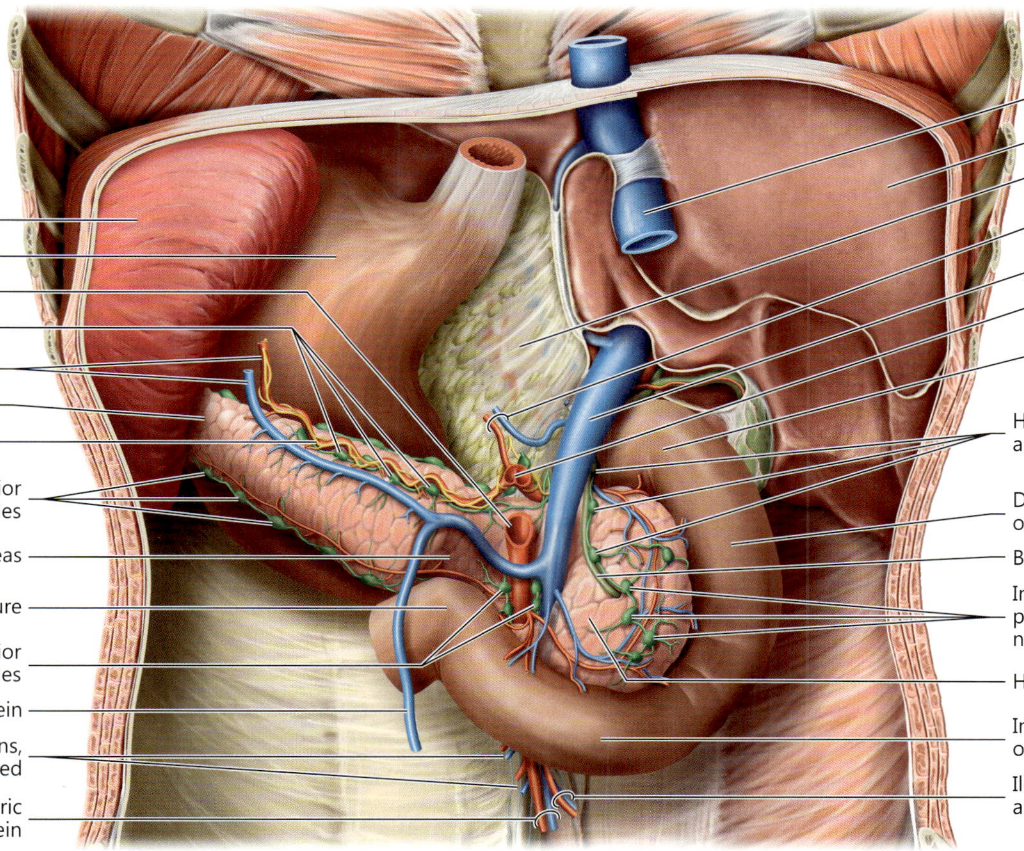

Posterior view of the tail, body, neck and head of the **pancreas**. A deep notch runs along the superior border of the body of the pancreas, accommodating the splenic vein and the splenic artery, a little above. The figure displays the splenic nodes, which drain into the celiac nodes, the hepatic nodes along the hepatic artery and the bile duct, which are also tributaries of the celiac nodes, and the inferior pancreatic and inferior pancreaticoduodenal nodes, which drain into the superior mesenteric nodes. The **lesser omentum** extends from the hilum of the visceral surface of the liver to the lesser curvature of the stomach and to the superior part of the duodenum. It can be divided into two parts: the hepatogastric ligament and the hepatoduodenal ligament; the latter has been transected in this figure to expose the structures of the porta hepatis that it contains.

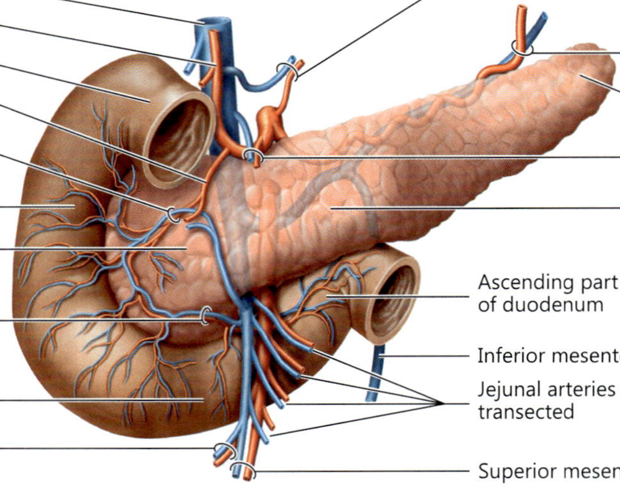

Anterior surfaces of the tail, body, neck and head of the **pancreas**, the latter lying within the concavity of the duodenum. The pancreas is depicted as semi-transparent to highlight its relations to the posterior vessels. The supramesocolic part of the head of the pancreas is crossed anteriorly by the gastroduodenal artery. The superior mesenteric artery and vein pass through the pancreatic notch.

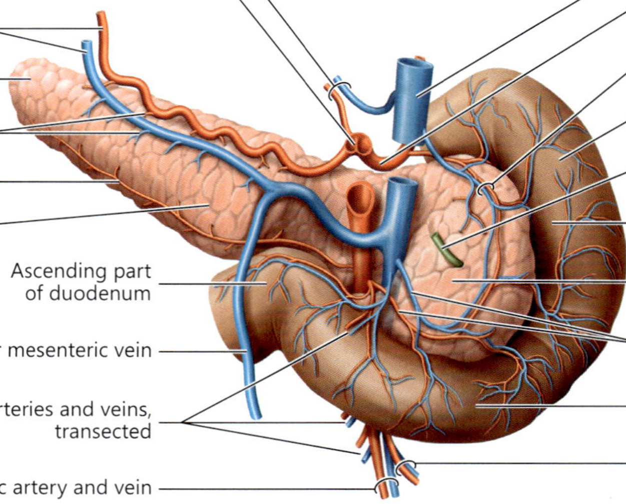

Posterior surfaces of the tail, body, neck and head of the **pancreas**, the latter lying within the concavity of the duodenum. The hepatic portal vein originates posterior to the neck of the pancreas, by confluence of the superior mesenteric vein and the trunk formed by the convergence of the inferior mesenteric and splenic veins.

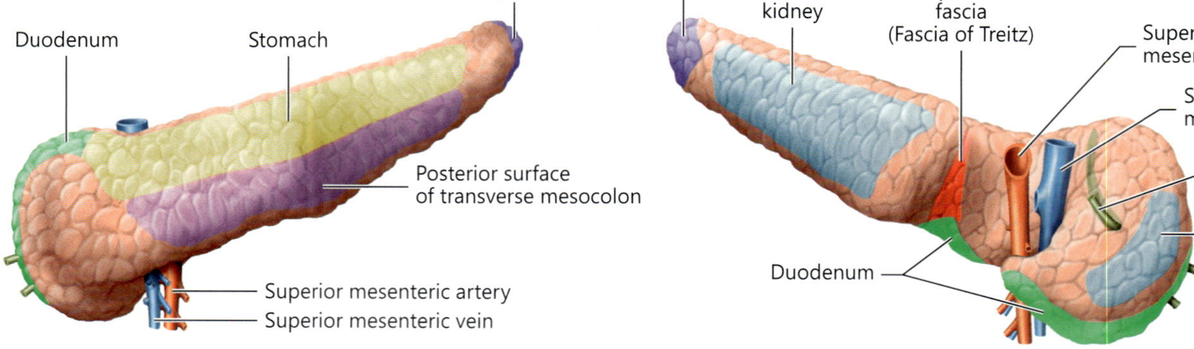

The anterior surface of the body of the **pancreas** (left) is related to the posterior surface of the stomach (area in *yellow*) through the omental bursa; the posterior surface of the head (right) is related to the bile duct; the posterior surface of the neck is related to the retropancreatic fascia, derived from the adhesion of the primitive mesoduodenum to the posterior parietal peritoneum. The posterior surface of the body of the pancreas is related to the anterior surface of the left kidney (area in *light blue*).

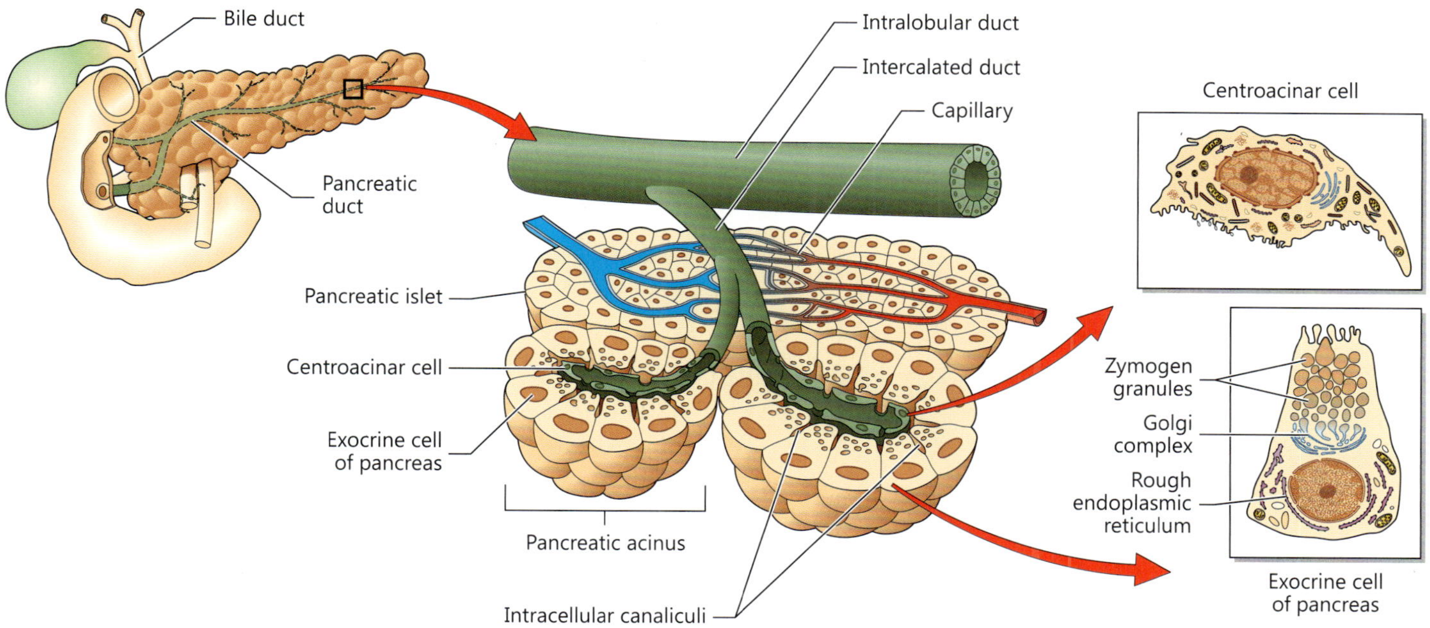

Diagram illustrating the organization of some **pancreatic acini** and their relations to the initial part of the excretory system. The intercalated duct extends up into the lumen of the acini to form the centroacinar cells.

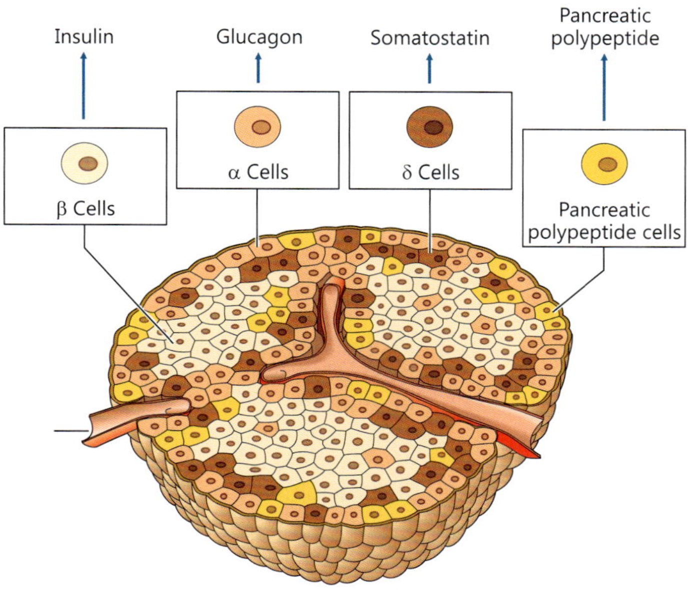

Diagram of a **pancreatic islet**. Pancreatic β cells are found mainly at the center, while α, δ and pancreatic polypeptide cells are located mainly in the periphery of the islet.

For the internal conformation of the pancreas, see also page 141 ☞ Duodenum, Structure.

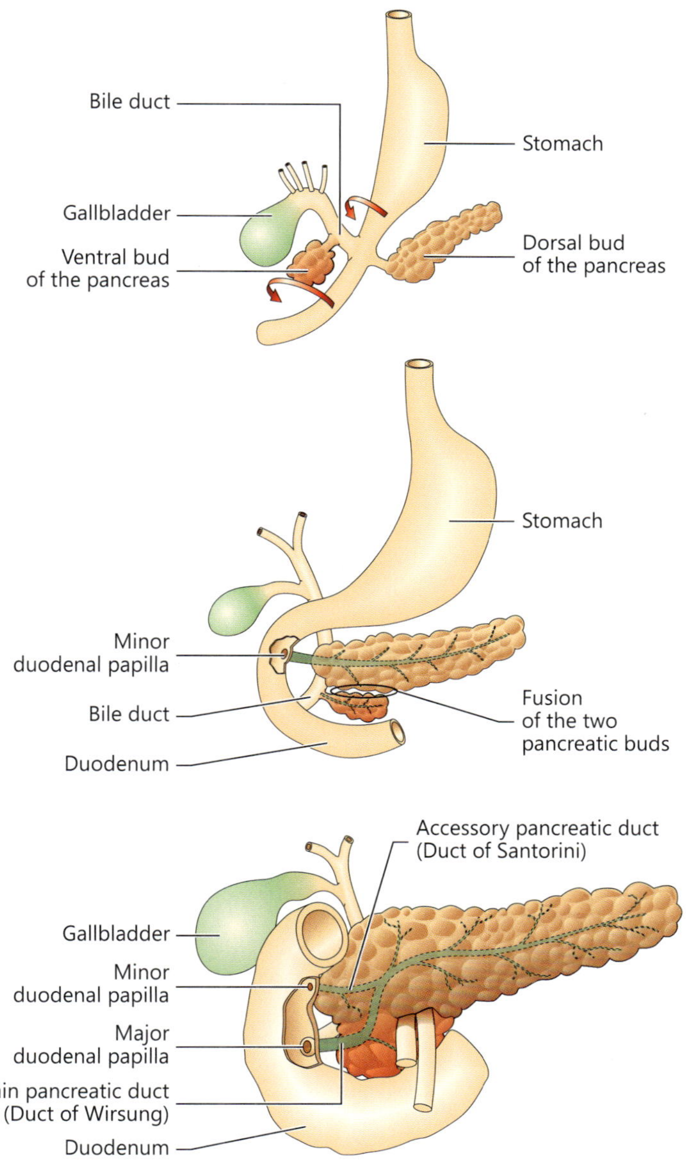

➤ **Development of the stomach**, **duodenum** and **pancreas**.

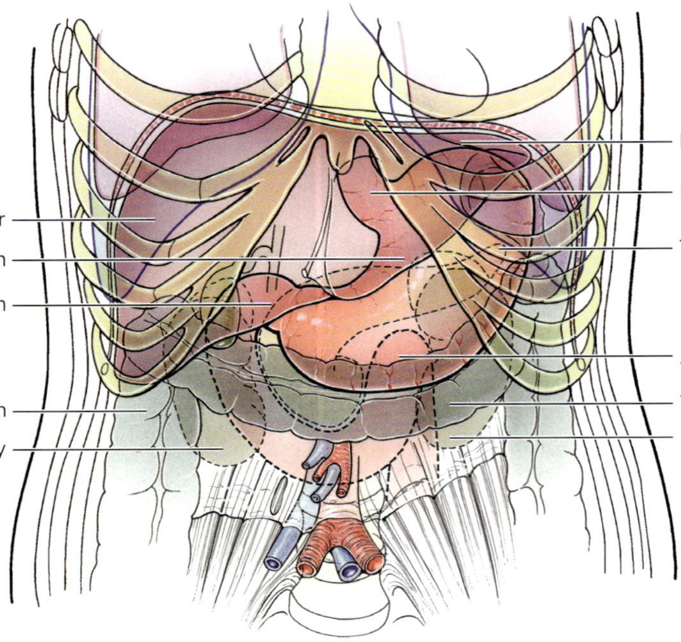

The transverse mesocolon, is a fold of peritoneum connecting the transverse colon to the posterior peritoneal wall, and divides the peritoneal cavity in two regions. The superior region is called the **supramesocolic space**, whereas the inferior region is called the **submesocolic space**. The supramesocolic space is bounded by the previsceral cavity anteriorly, the dorsolumbar wall posteriorly, the diaphragm superiorly and the colon inferiorly. The supramesocolic space contains the stomach, the first part and the superior half of the second part of the duodenum, the liver and the spleen. The figure highlights the relations of the anterior surface of the stomach to the thoracic cage and liver. The relations of the stomach to the pancreas, left kidney and transverse colon are shown in transparency.

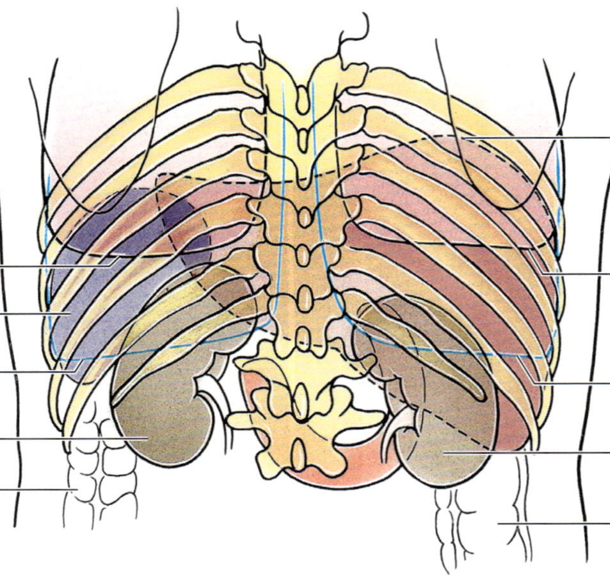

Projection area of the liver on the posterior wall of the trunk. It is highlighted the wide overlap of the area of the liver on the projection line of the bases of the lungs and of the costodiaphragmatic recesses.

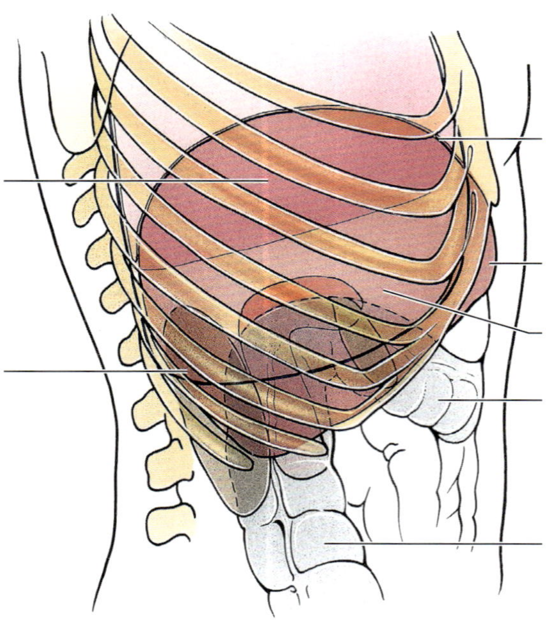

Projection area and **relations of the liver**; right lateral projection. Specifically, the relations to the base of the right lung and to the right costodiaphragmatic recess, as well as the relations to the right kidney and right colic flexure can be observed.

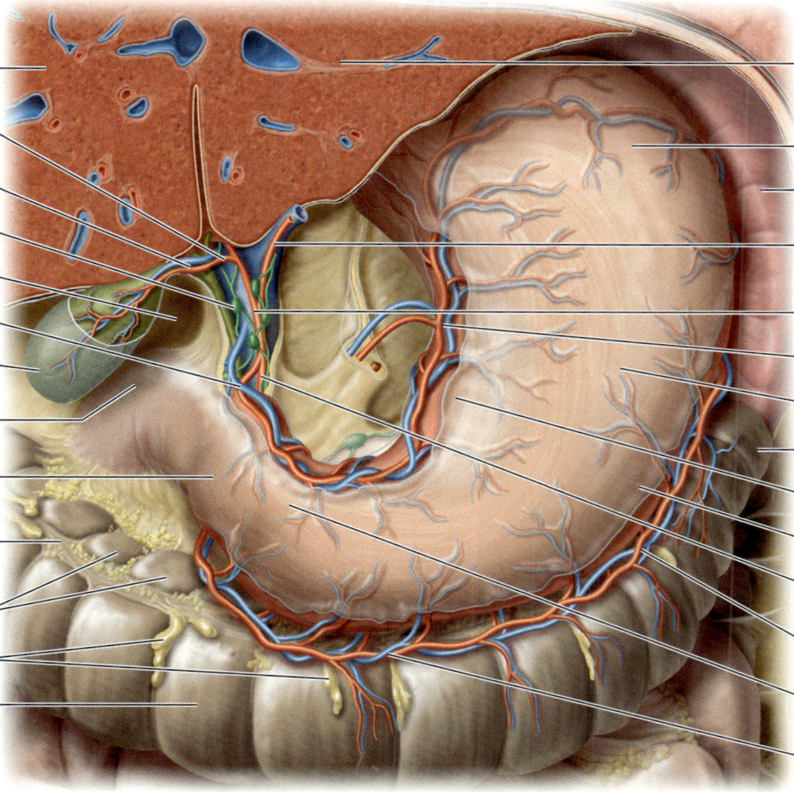

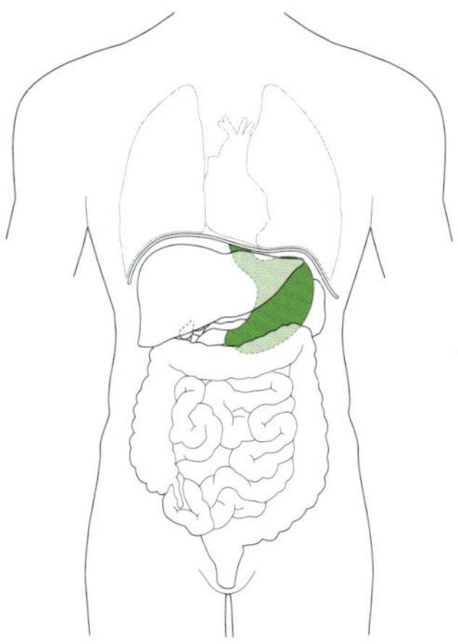

At the top, the lesser omentum was removed to expose the **structures of the porta hepatis**, run within the omentum pars tensa, corresponding to the hepatoduodenal ligament, and the left gastric vessels, contained within the pars flaccida, corresponding to the hepatogastric ligament. The figure also shows the omental (or epiploic) foramen, through which the omental bursa (or lesser sac) communicates with the greater peritoneal cavity.
At the bottom, the figure shows the anterior surface of the **lesser omentum**, which has a quadrangular shape. Three of the four margins are attached to the alimentary tract, the liver and the diaphragm, respectively; the fourth is free and corresponds to the concave right margin, which constitutes the anterior boundary of the omental foramen.

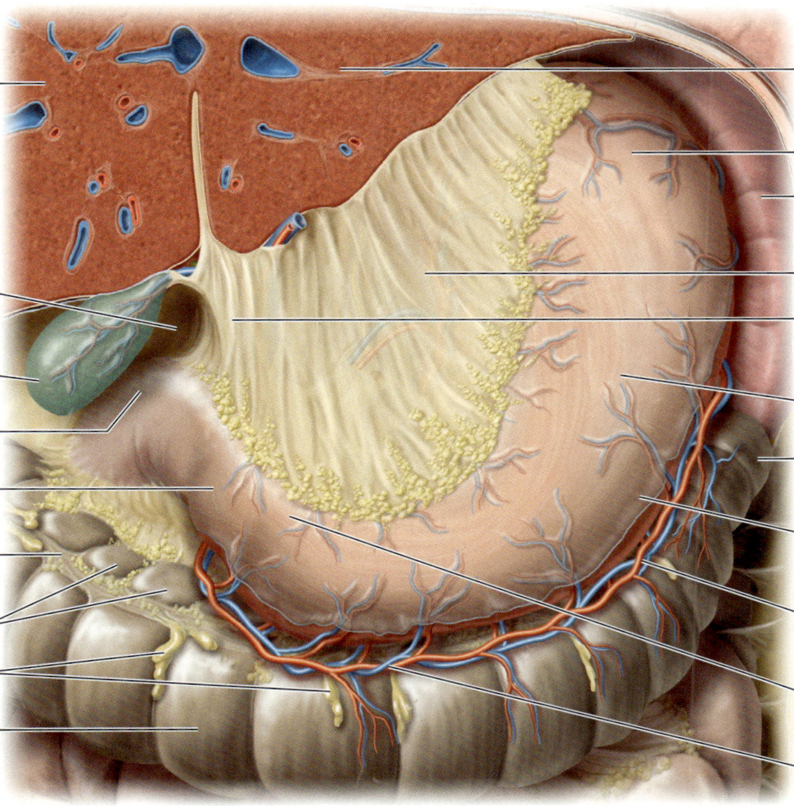

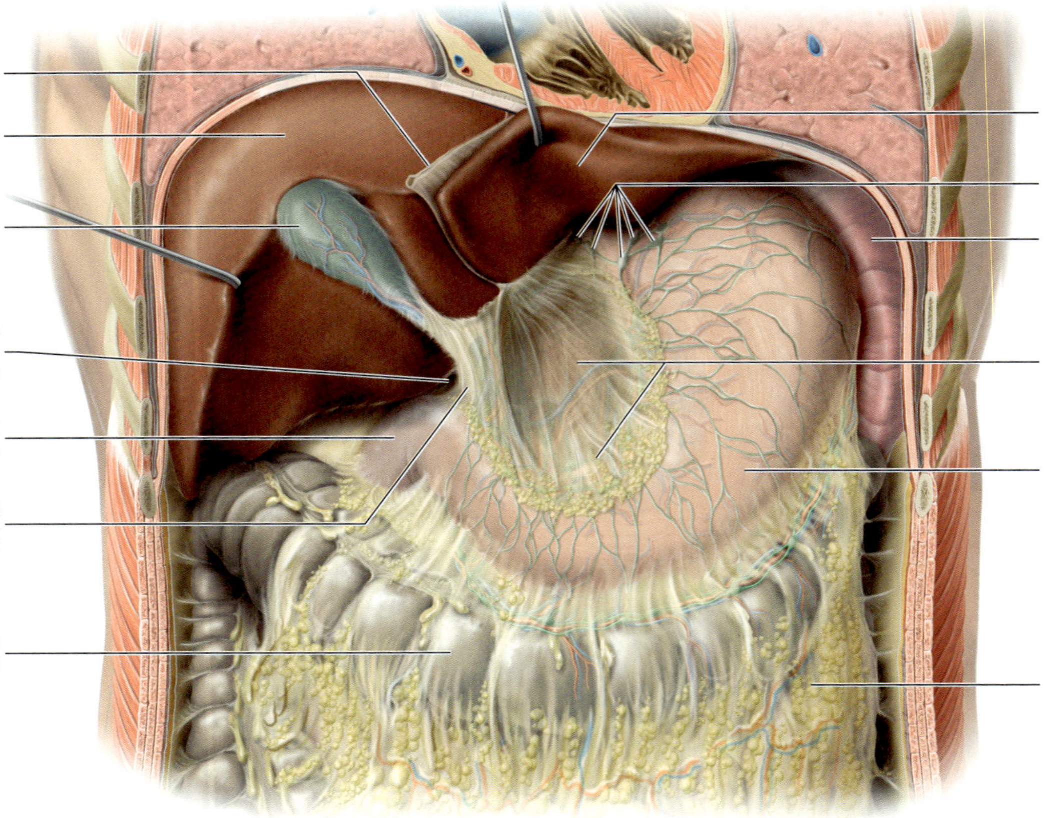

Anterior view of the **supramesocolic space**. The liver was turned upwards and to the right to expose the lesser omentum, which extends between the superior part of the duodenum, the lesser curvature of the stomach and the liver, forming the hepatoduodenal ligament, or pars tensa, and the hepatogastric ligament, or pars flaccida. The anterior surface of the lesser omentum corresponds to the layer of peritoneum that invests the anterior surface of the stomach, while the posterior surface corresponds to the layer of peritoneum that invests the posterior surface of the stomach and, hence, the anterior wall of the vestibule of the omental bursa, or lesser sac.

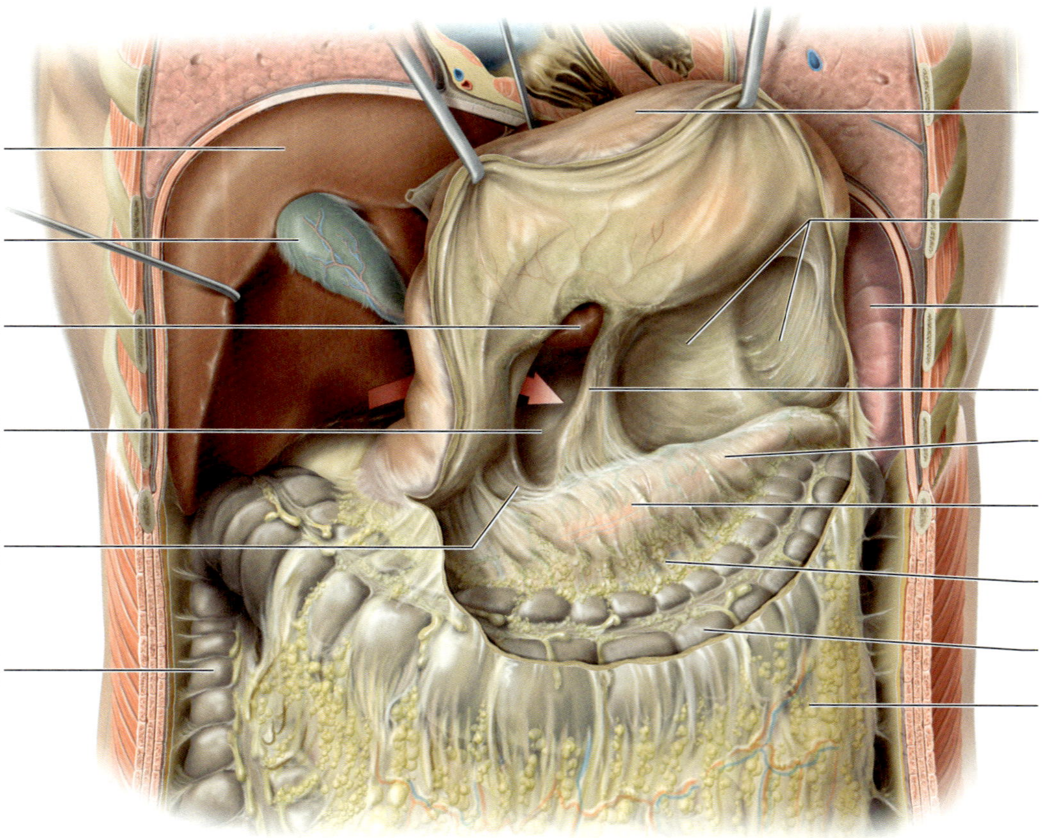

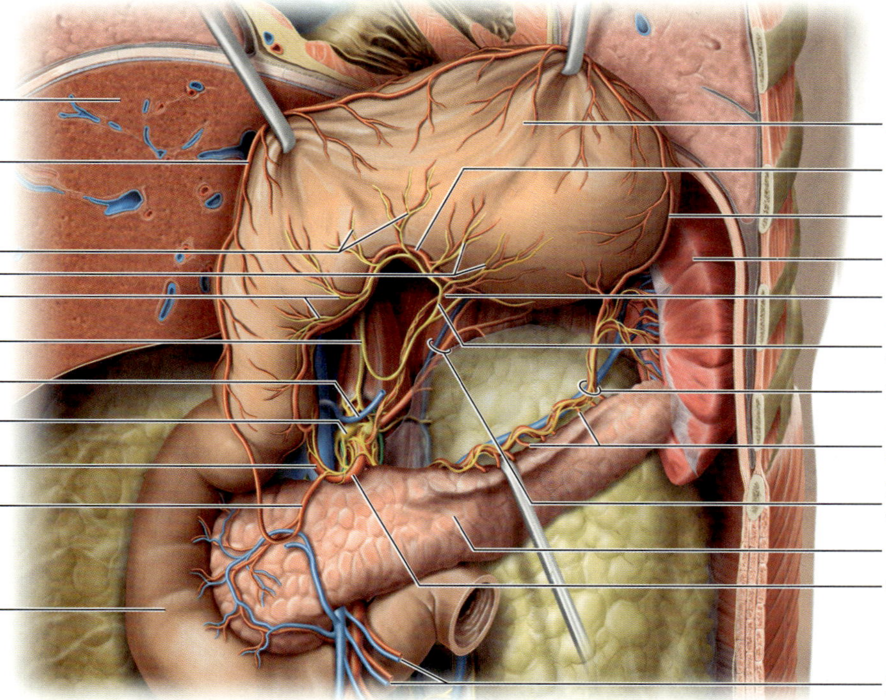

The **posterior vagal trunk** consists mainly of vagal fibers of the right vagus nerve. It gives origin to the posterior gastric branches, homologous to the anterior gastric branches of the anterior vagal trunk. The last and largest posterior gastric branch is termed the posterior nerve of the lesser curvature. The posterior vagal trunk also gives origin to a celiac branch; one of its poles reaches the right celiac ganglion, while the other one receives the right greater splanchnic nerve. This particular formation is known as ansa memorabilis, or Wrisberg's ansa.

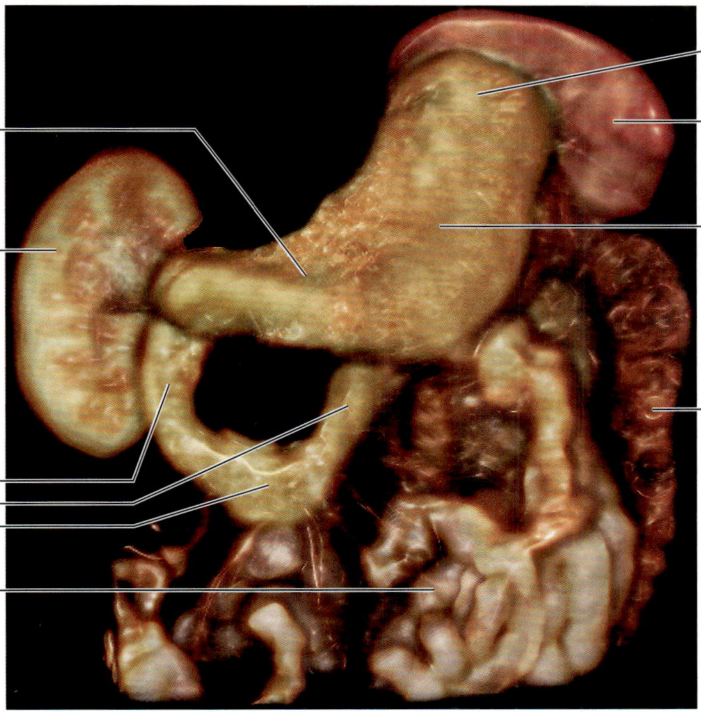

Three-dimensional image of a living human body, obtained by CT scan using a 3D volume rendering technique, showing some of the **organs contained in the abdominal cavity** in antero-superior projection.

➤

The **omental bursa**, or lesser sac, is a recess of the peritoneal cavity within the supramesocolic space, located posterior to the stomach. The vestibule of the omental bursa is situated behind the hepatogastric ligament, between the caudate lobe of the liver at the top, and the superior part of the duodenum at the bottom. The vestibule leads to a narrower portion, referred to as isthmus, which is bounded by the gastropancreatic fold; in turn, the isthmus leads to the omental bursa proper, whose anterior surface coincides with the posterior surface of the stomach and its corresponding peritoneum.

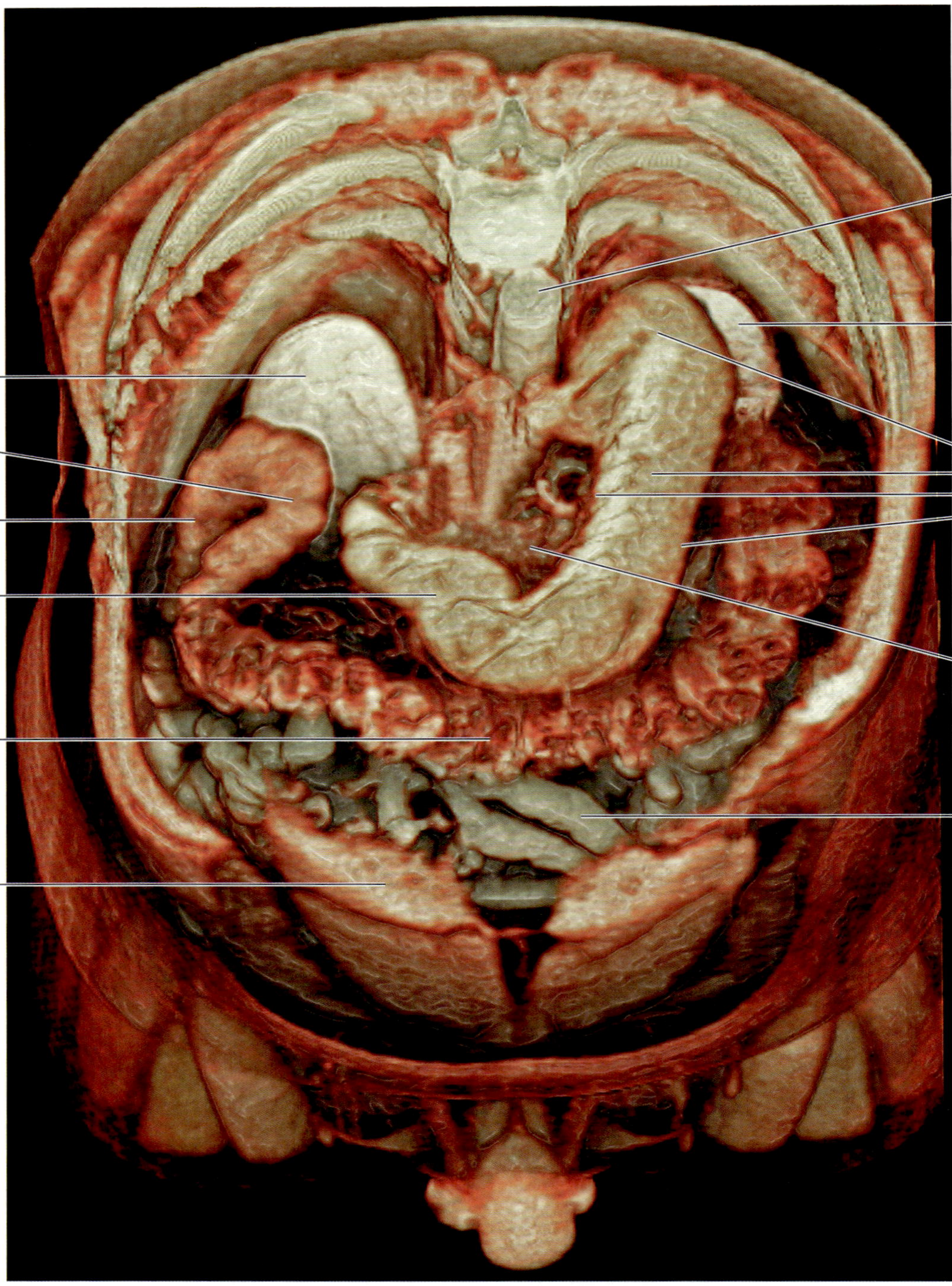

Three-dimensional image of a living human body, obtained by CT scan using a 3D volume rendering technique, showing the **abdominal organs** in a transverse section passing through the eleventh thoracic vertebra.

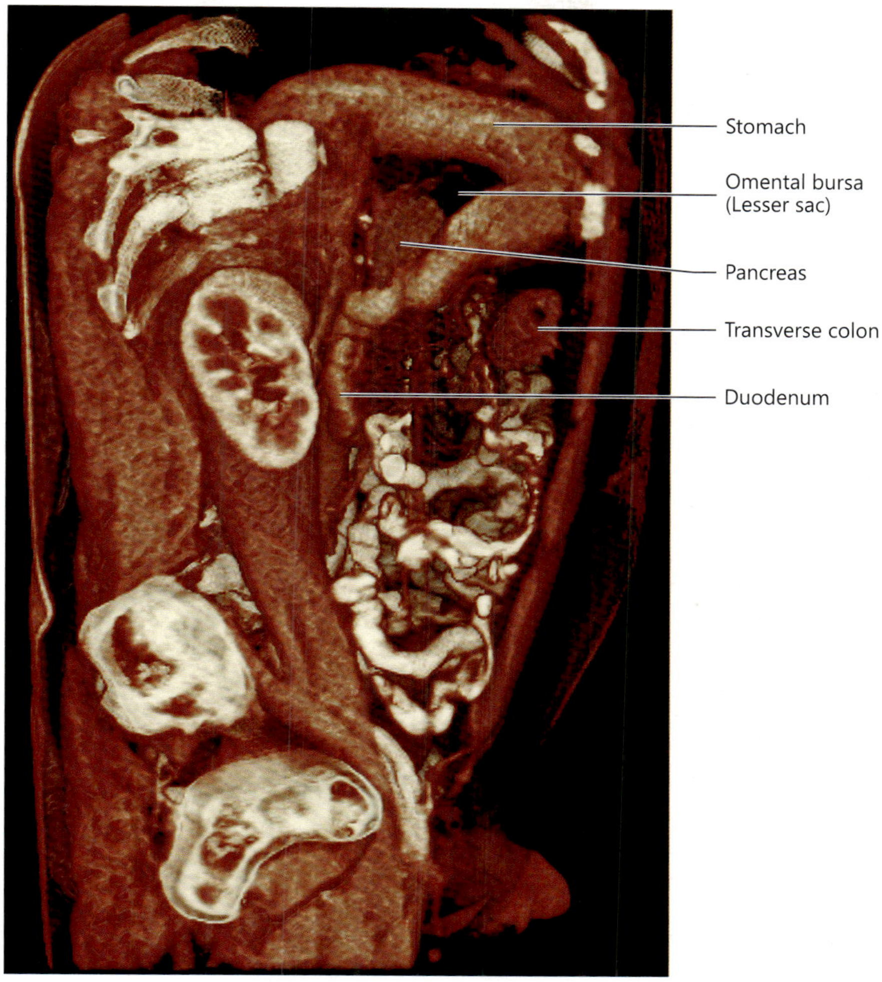

Three-dimensional image of a living human body, obtained by CT scan using a 3D volume rendering technique, showing the **organs contained in the abdominal cavity** in a sagittal section; right lateral view.

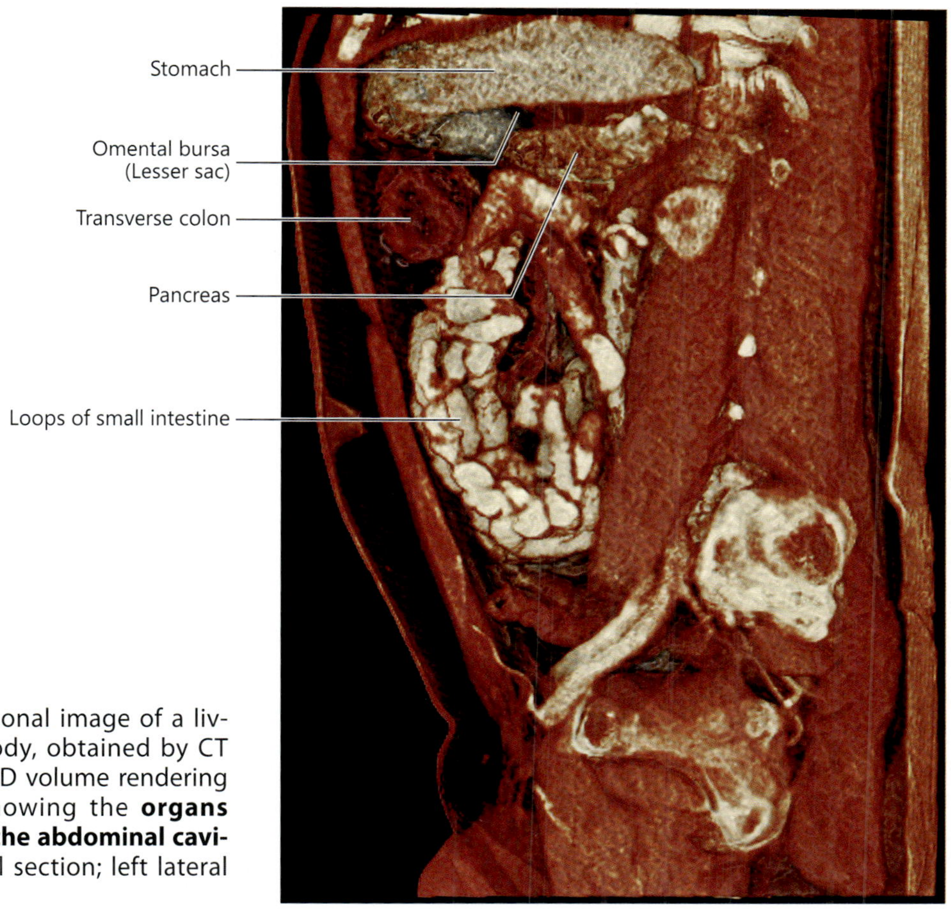

Three-dimensional image of a living human body, obtained by CT scan using a 3D volume rendering technique, showing the **organs contained in the abdominal cavity** in a sagittal section; left lateral view.

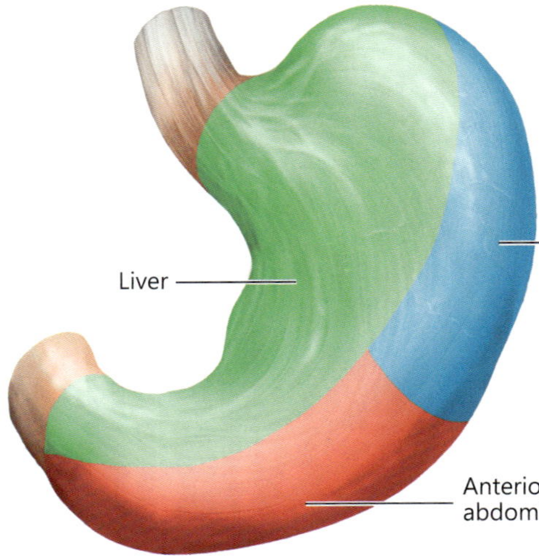

On the right, the superior part of the **anterior wall of the stomach** is covered by the left lobe of the liver, while on the left it is directly related to the diaphragm. The inferior part of the anterior wall of the stomach is covered for a short portion by the liver at the top, and contacts directly the anterior abdominal wall.

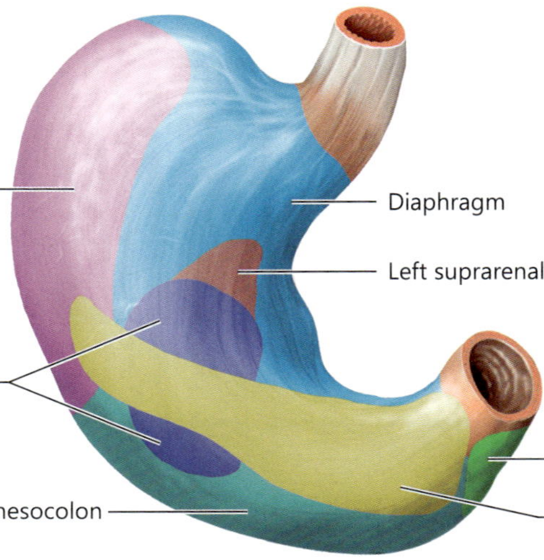

The **posterior wall of the stomach** is related to the organs of the supramesocolic space. From top to bottom, it is in contact with the diaphragm, the spleen, the left suprarenal gland, the left kidney, the pancreas and the transverse mesocolon.

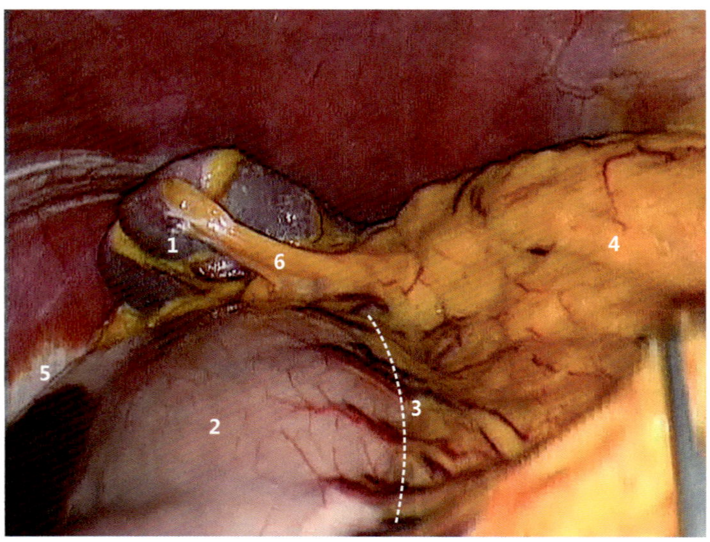

Image taken during a diagnostic laparoscopy, showing the **stomach** and the **spleen** in situ. The gastrosplenic ligament can be observed.

1 Spleen
2 Stomach
3 Greater curvature
4 Greater omentum
5 Central tendon of the diaphragm
6 Gastrosplenic ligament

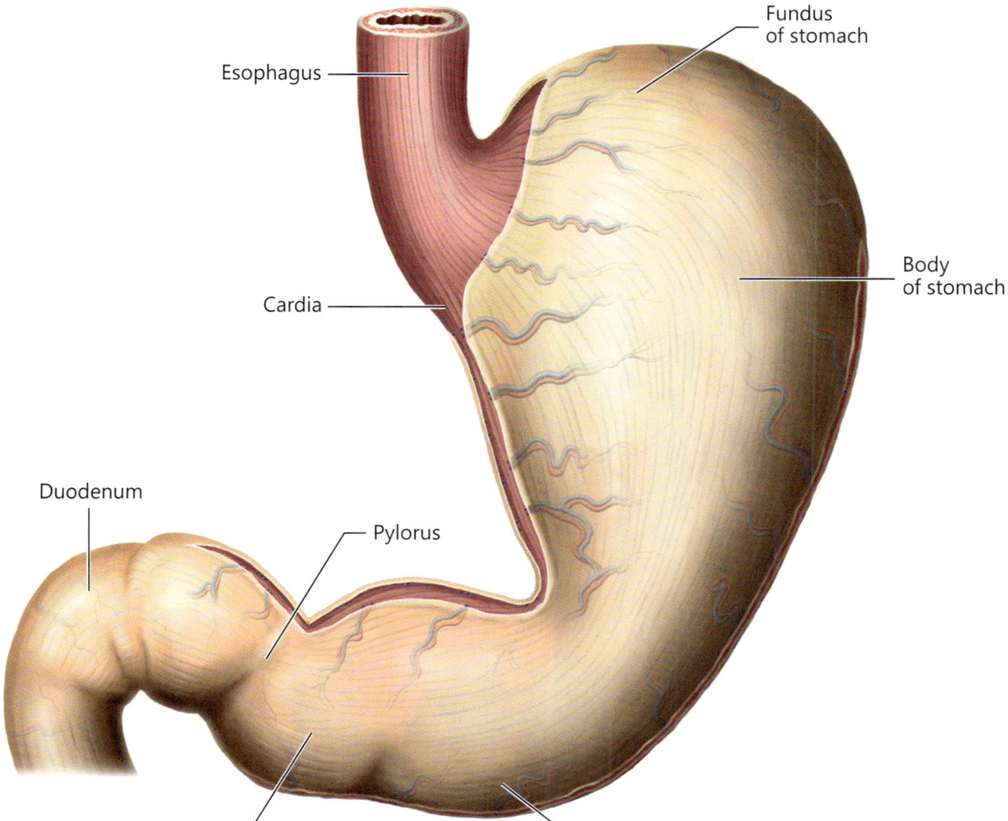

a

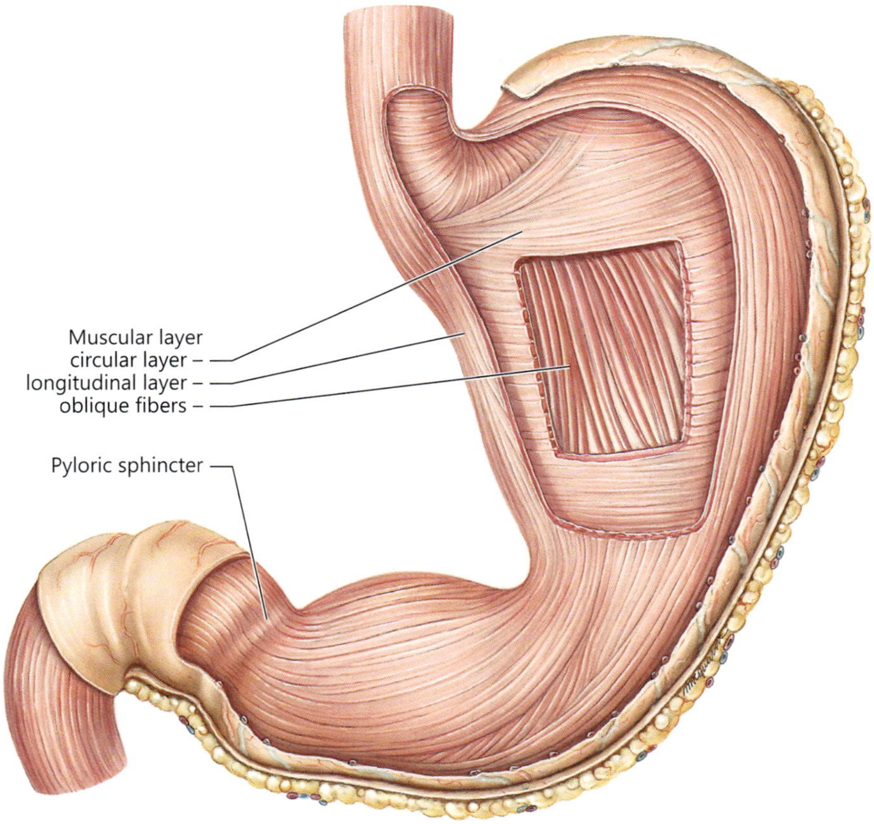

b

a, General conformation of the **stomach**. The organ, seen from an anterior projection, is invested by a serosa (visceral peritoneum). The stomach is continuous with the esophagus at the top, and with the duodenum at the bottom. **b**, Organization of the **muscular layer of the stomach**, comprising three layers of smooth muscle fibers: an inner oblique layer, an intermediate circular layer and an outer longitudinal layer. This arrangement of the muscle fibers enables the stomach to contract vigorously, reducing the size of the lumen and propelling the gastric contents towards the duodenum. The muscular layer causes the circular layer to thicken considerably at the pylorus, hence forming the pyloric sphincter.

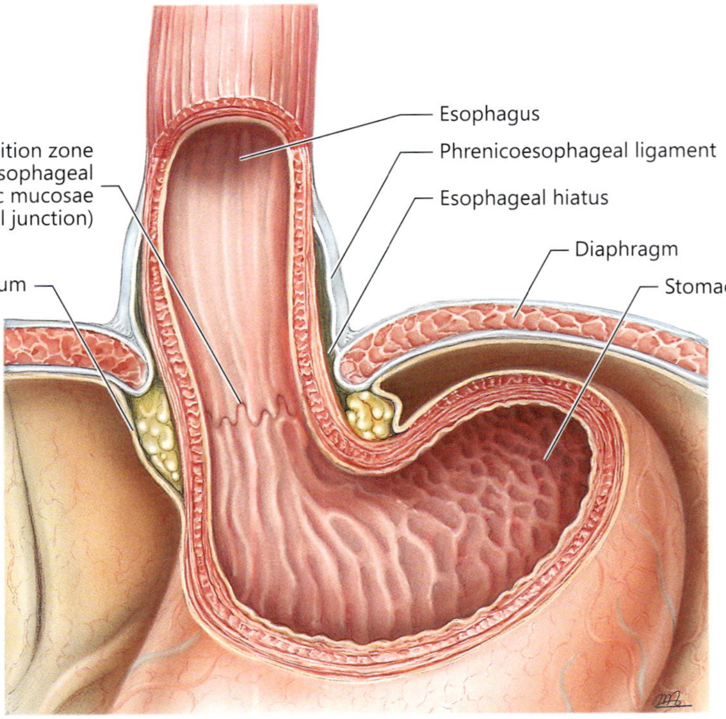

Frontal section of the **esophagus** and of the **stomach** at the **gastroesophageal junction**, showing the anatomical structures which contribute to keeping the abdominal part of the esophagus and the stomach within the abdominal cavity.

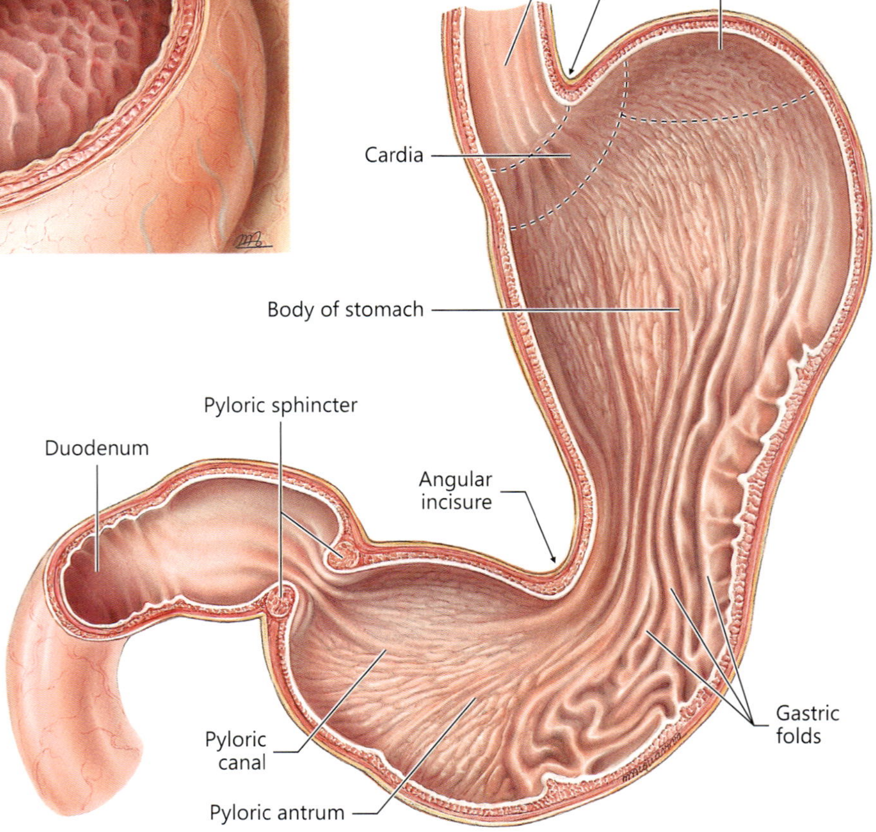

Internal configuration of the **stomach** in a frontal section of the organ. The mucosa and the deeper layers of the wall of the stomach evaginate, creating sets of folds that are oriented mainly craniocaudally. Gastric folds appear more even and prominent in the region that defines the short gastric way. The fundus of the stomach and the pyloric antrum appear relatively smooth.

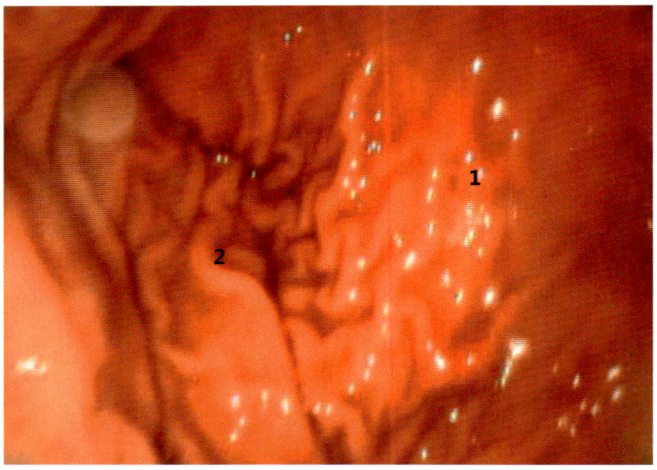

Image taken during a diagnostic endoscopy, showing the **inner wall of the stomach**.
1 Gastric mucosa
2 Gastric fold

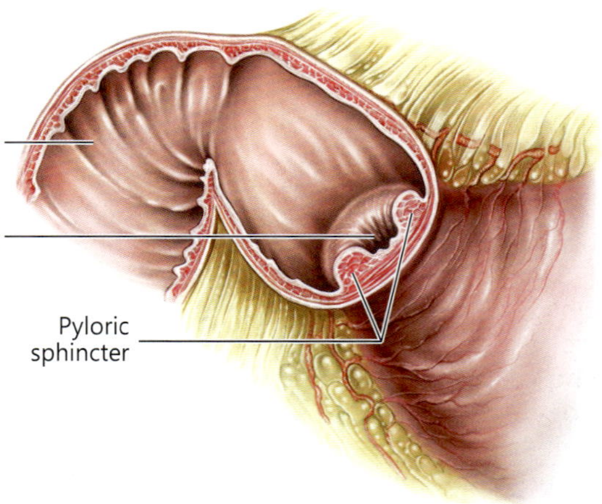

Internal configuration of the **pylorus** as seen from the duodenal side. Note the pylorus protruding into the duodenum, as the circular muscle layer thickens considerably (**pyloric sphincter**).

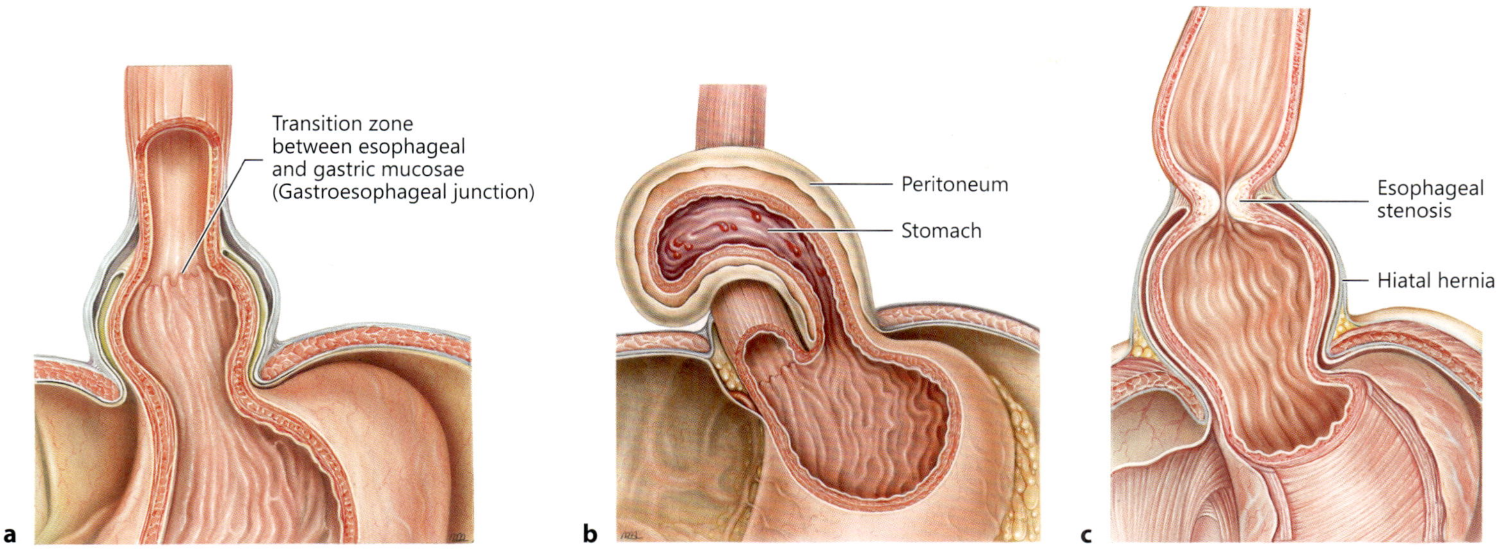

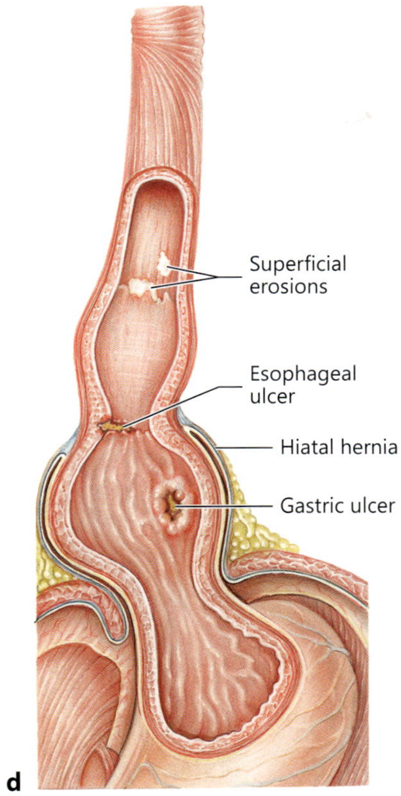

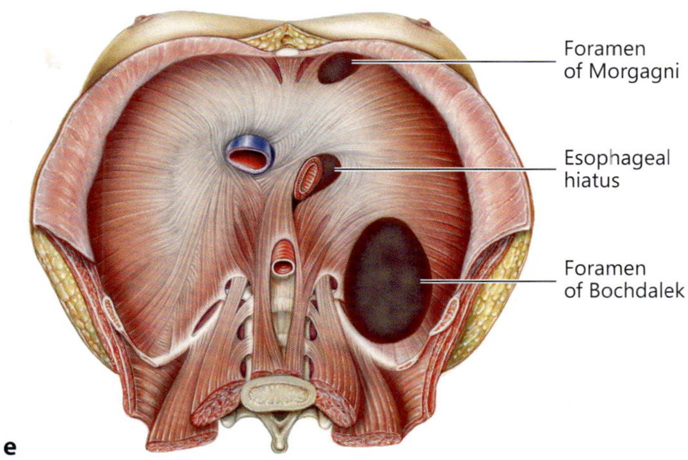

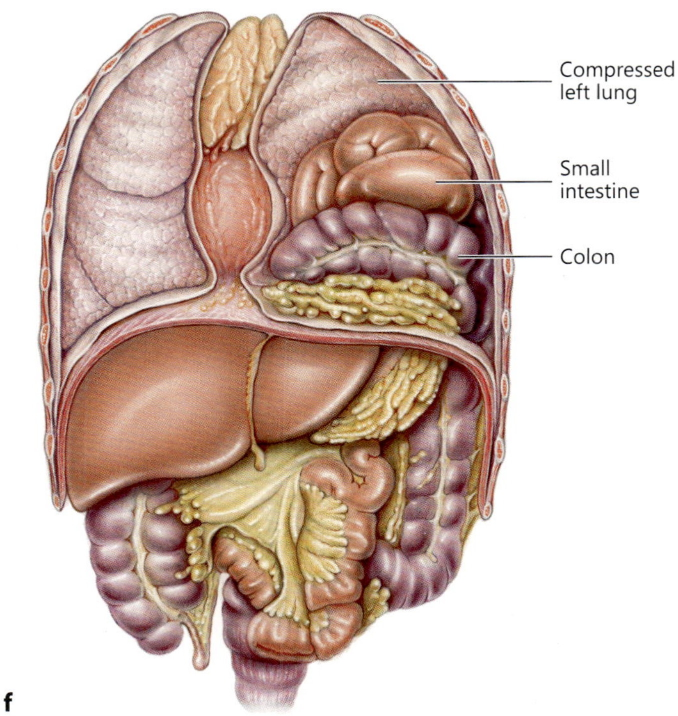

a, Sliding hiatal hernia. It occurs when the inferior part of the esophagus slides into the thoracic cavity, dragging the gastroesophageal junction and part of the stomach with it. **b**, Rolling or paraesophageal hiatal hernia. It occurs when a part of the stomach, not directly attached to the esophagus, migrates into the thoracic cavity through the esophageal hiatus, while the gastroesophageal junction remains in its normal position within the abdominal cavity. **c**, Esophageal stenosis associated with hiatal hernia. **d**, Hiatal hernia and different types of lesions caused by reflux esophagitis. **e-f**, Regions most frequently affected by diaphragmatic hernias (**e**) and consequences of a large hernia of the foramen of Bochdalek (**f**).

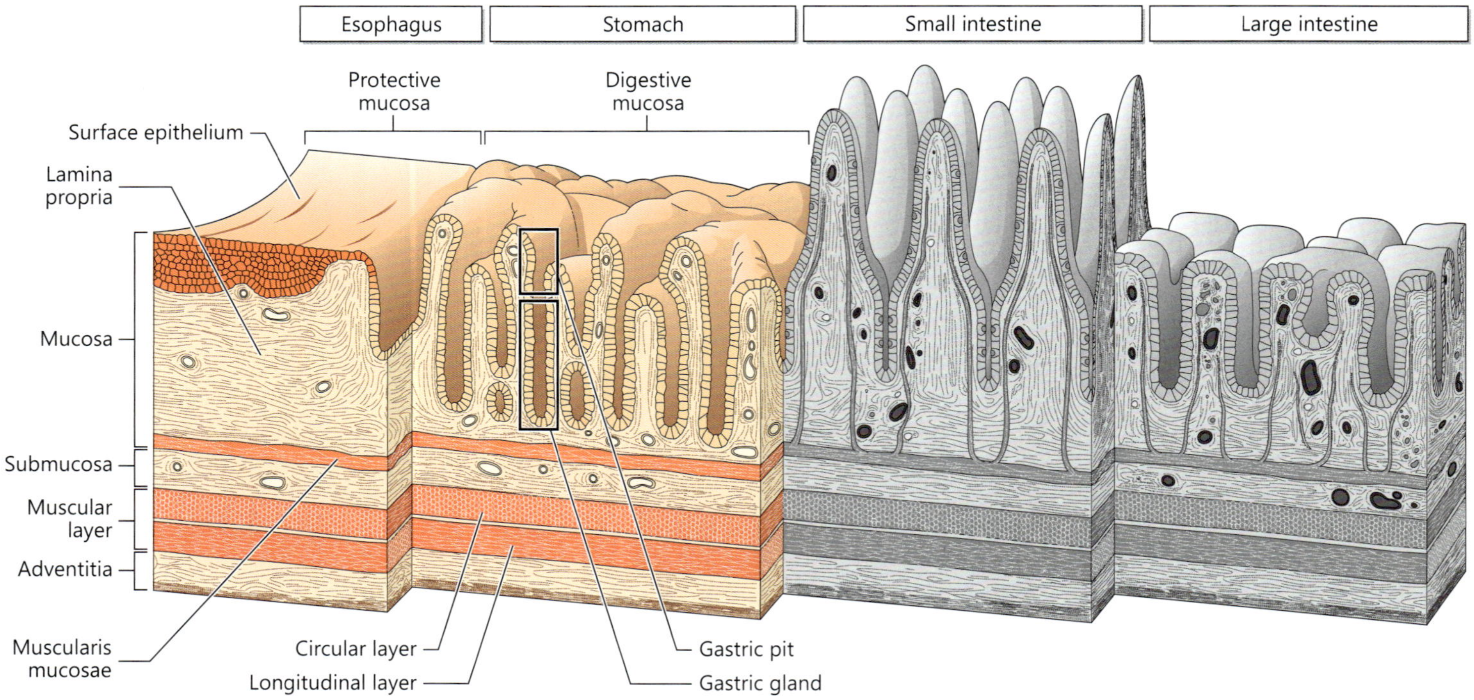

Summary of the **general organization of the hollow organs of the alimentary system** (esophagus, stomach, small intestine, large intestine). The portion of the esophagus and stomach is highlighted in color. The surface epithelium is stratified squamous epithelium in esophagus and single columnar epithelium in stomach.

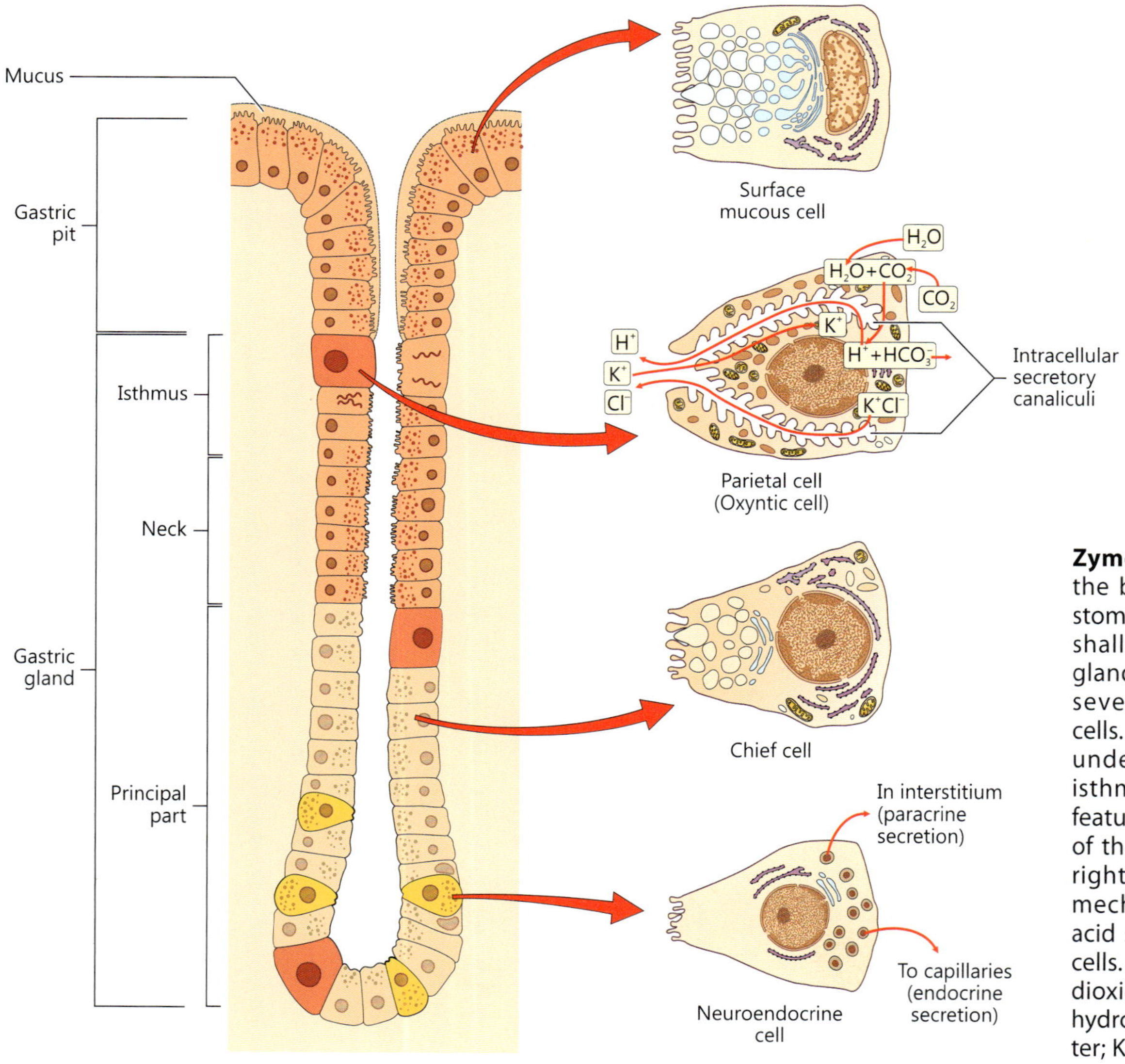

Zymogenic unit is found in the body and fundus of the stomach. It is characterized by shallow pits and elongated glands, the latter containing several chief (zymogenic) cells. Note that some cells are undergoing mitosis at the isthmus. The ultrastructural features of the main cell types of the unit are shown on the right. The figure shows the mechanism of hydrochloric acid secretion by the parietal cells. Cl^-, chlorine; CO_2, carbon dioxide; H^+, hydrogen; HCO_3^-, hydrogen carbonate; H_2O, water; K^+, potassium.

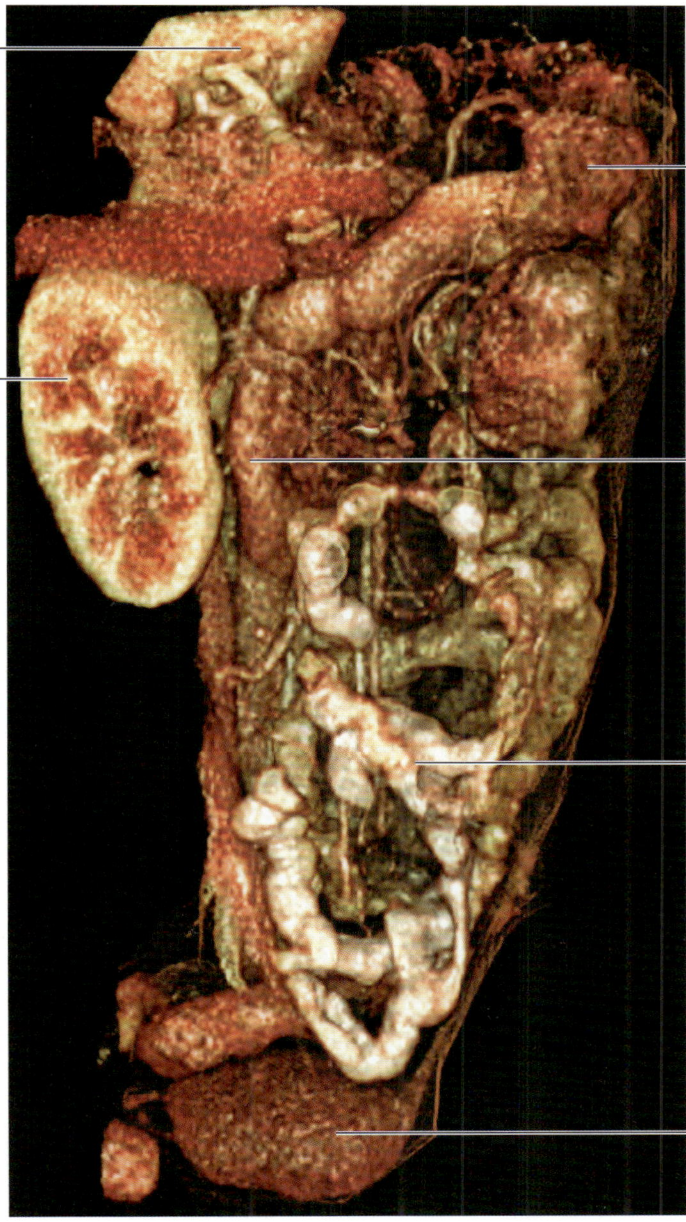

Three-dimensional image of a living human body, obtained by CT scan using a 3D volume rendering technique, showing the **organs contained in the abdominal cavity** in right lateral projection.

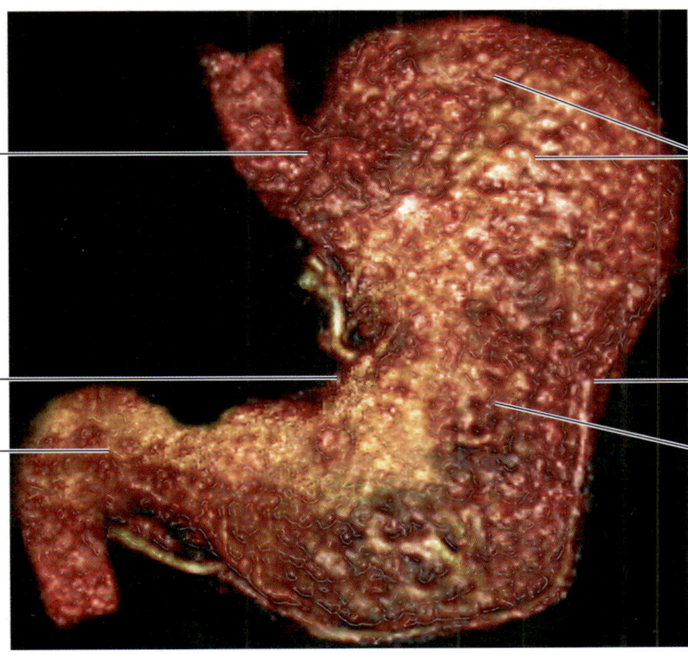

Three-dimensional image of a living human body, obtained by CT scan using a 3D volume rendering technique, showing the **stomach** in frontal view.

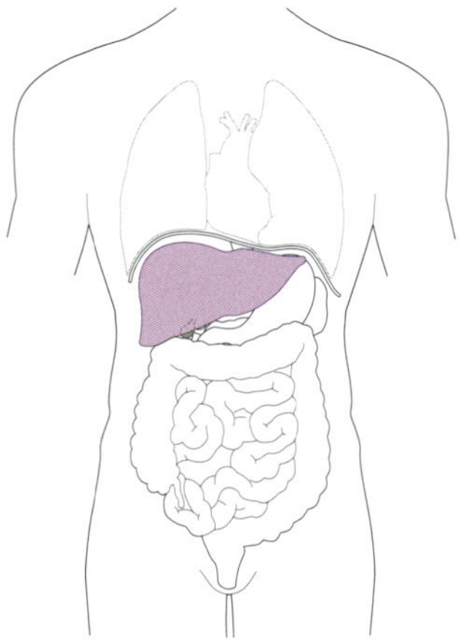

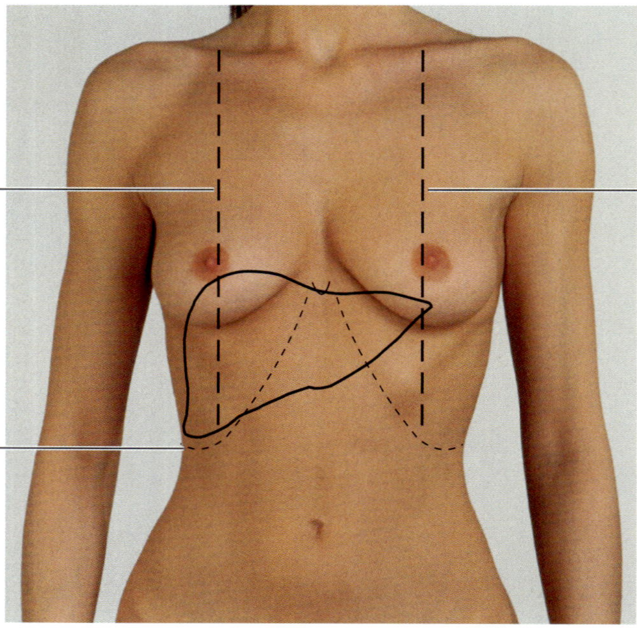

Size and shape of a normal **liver**, **projected onto the anterior wall of the trunk** (area of hepatic auscultation).

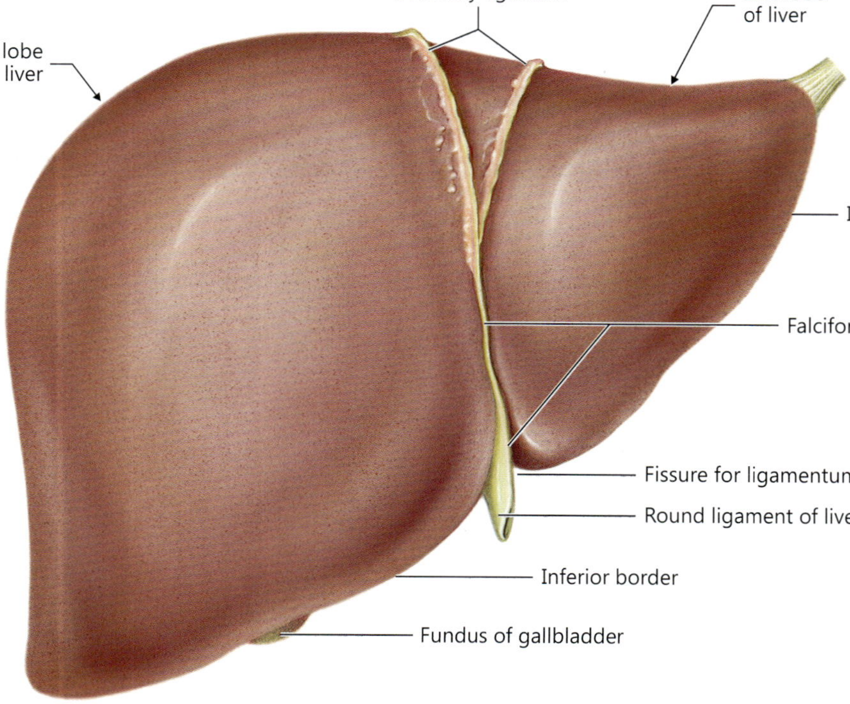

Diaphragmatic (or anterosuperior) surface of the **liver**. This surface is convex as it conforms to the concavity of the diaphragm. The fundus of the gallbladder can be seen protruding at the inferior border. The falciform ligament marks the boundary between the right and left lobes of the liver. It is composed of two layers, which enclose the round ligament at their anterior edge. Posteriorly, the two layers of the falciform ligament separate from each other to form the upper layer of the coronary ligament.

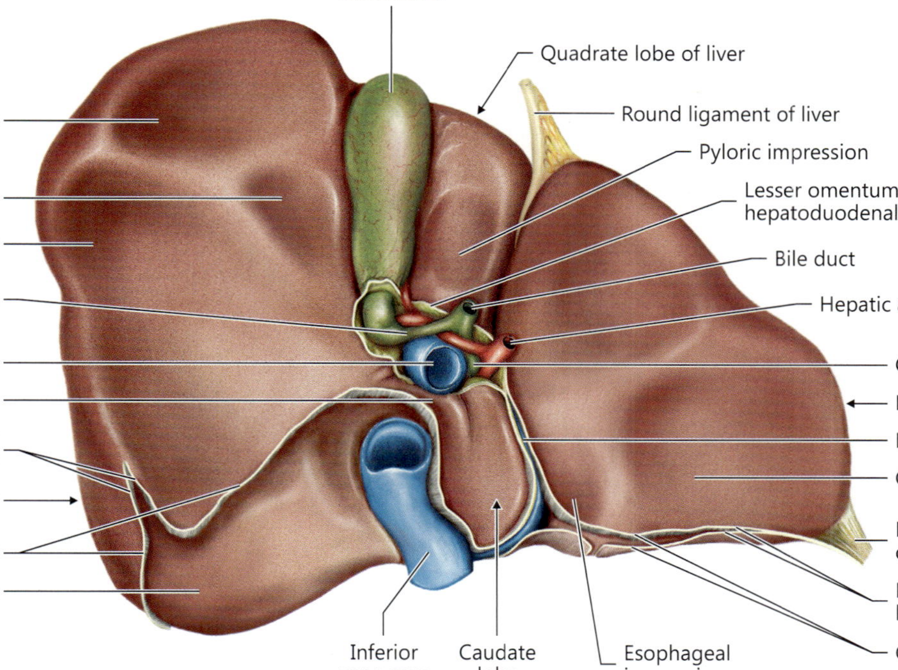

Visceral (posteroinferior) surface of the **liver**. Two sagittal grooves and a transverse groove, arranged in an H shape, divide the liver into four lobes (right, left, quadrate and caudate). The figure also displays the mutual relations between the main components of the porta hepatis (hepatic artery, hepatic portal vein and bile duct), which marks the transverse groove. The right sagittal groove comprises the gallbladder, anteriorly, and the inferior vena cava, posteriorly. The two segments of the left sagittal groove contain the round ligament and the ligamentum venosum.

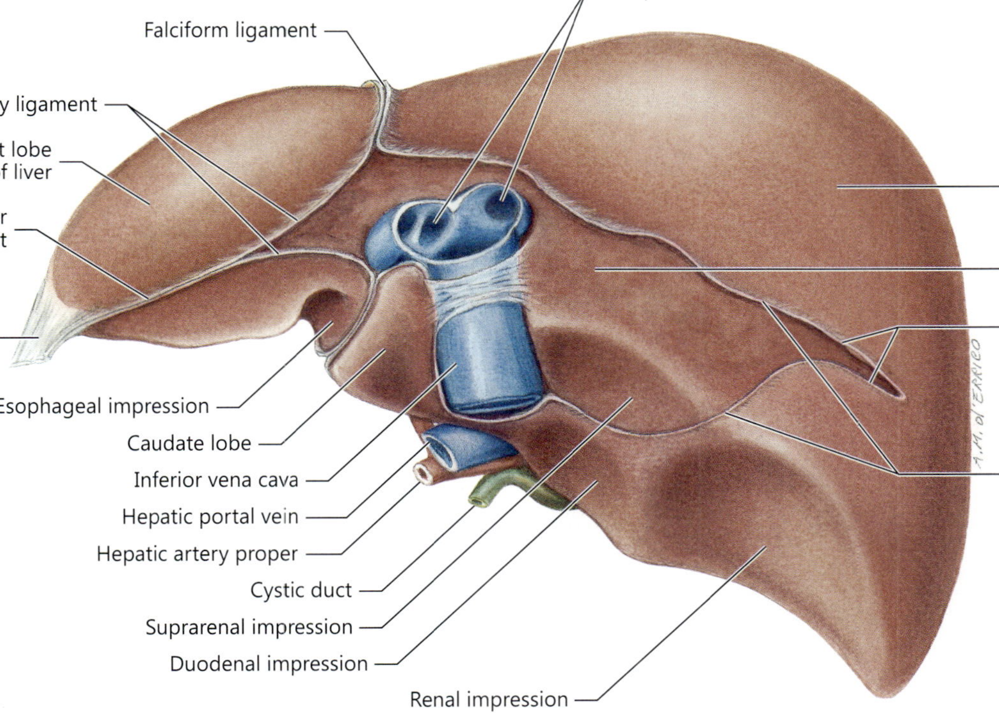

Obtuse posterior border of the **liver**. The visceral peritoneum reflects onto the diaphragmatic and posterior parietal peritoneum, leaving (on this surface) a large rhomboidal area devoid of serous covering (bare area). The inferior vena cava, which receives the hepatic veins, is found at the center of this area. At the lateral angles of the bare area, the superior visceral peritoneum adheres to the inferior layer, hence forming the left and right triangular ligaments.

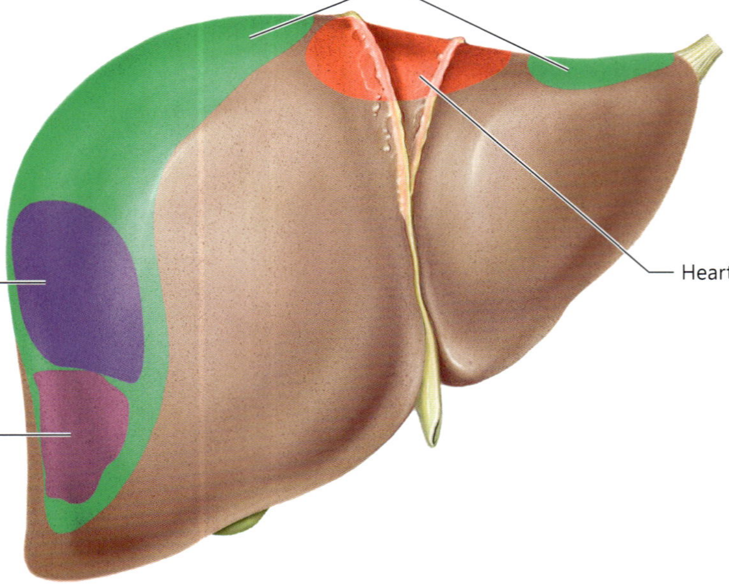

The **diaphragmatic surface of the liver** is related to the dome of the diaphragm, which separates it from the pleural cavity, the base of the right lung and the right costodiaphragmatic recess, on the right, and from the pericardium, the diaphragmatic surface and the apex of the heart (cardiac impression), on the left. The *green* surface indicates the contact area with the diaphragm, which separates the liver from the right lung and from the right costodiaphragmatic recess.

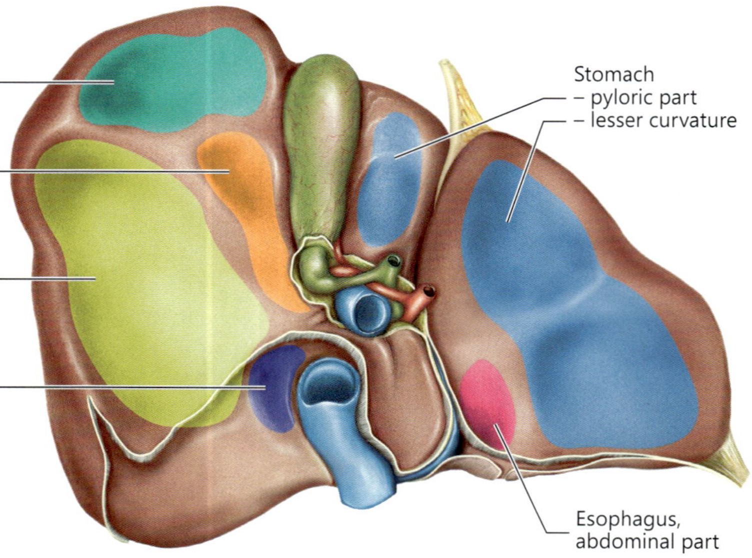

The **visceral surface of the right lobe of the liver** is marked by a number of impressions, formed by related underlying organs: anteriorly, the colic impression, formed by contact with the right colic flexure; posteriorly, the impressions of the right kidney and right suprarenal gland, as well as the duodenal impression, formed by contact with the supramesocolic tract of the descending part of the duodenum. The gastric impression is caused by the relation with the lesser curvature and part of the anterior surface of the stomach on the visceral surface of the left lobe of the liver, while the relation with the abdominal part of the esophagus causes the esophageal impression. The relation with the anterior surface of the pyloric part of the stomach causes the depression on the quadrate lobe surface.

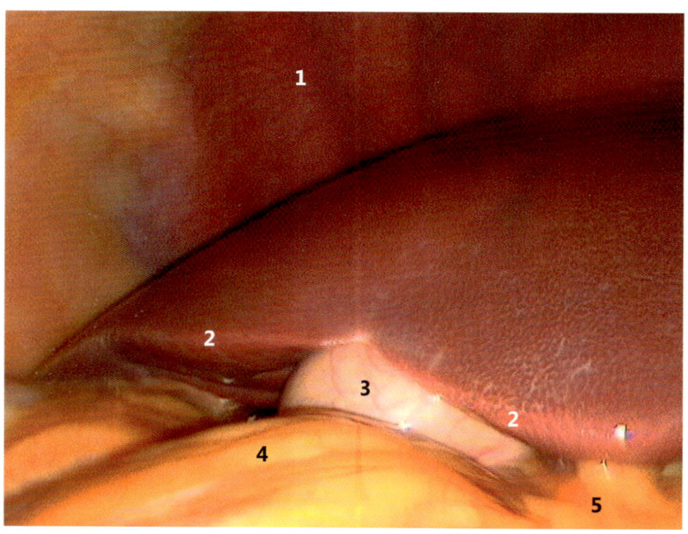

Image taken during a diagnostic laparoscopy, showing the **relations of the liver to the gallbladder**.

1 Diaphragm
2 Anterior border of the liver
3 Fundus of the gallbladder
4 Superior duodenal flexure
5 Lesser omentum

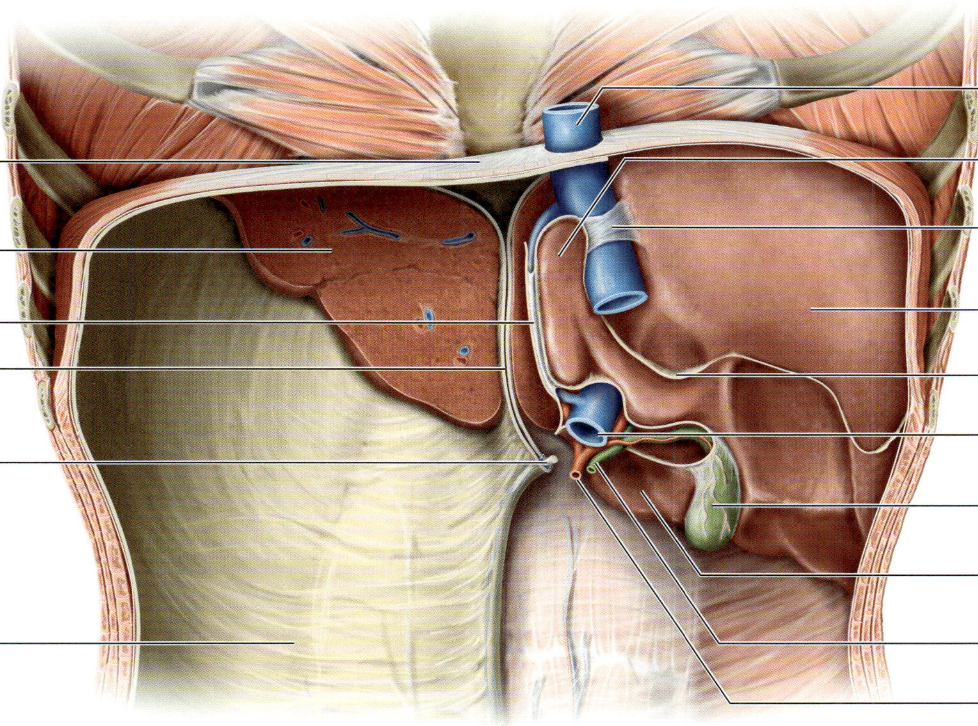

The figure shows the visceral (or posteroinferior) surface of the liver in posterior view, where the following can be observed: the **ligamentum venosum**, a remnant of the ductus venosus, the **round ligament of the liver**, a remnant of the left umbilical vein, the gallbladder, located in the homonymous fossa, the inferior vena cava, located in the groove for the vena cava, the porta hepatis with the branches of division of the hepatic portal vein and of the hepatic artery, and the bile duct. The figure also displays the **falciform ligament**, composed of two layers of peritoneum, and running between the diaphragm and the supraumbilical part of the anterior abdominal wall on one side, and the large sagittal groove of the diaphragmatic surface of the liver on the other side.

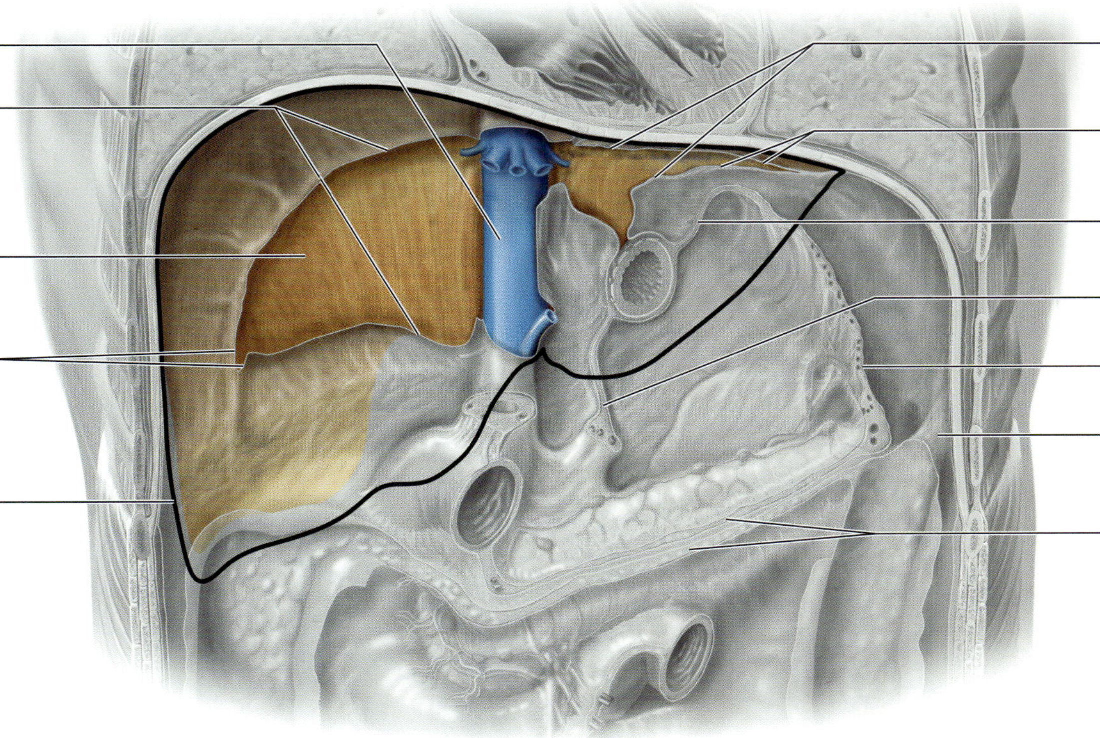

Two main **means of attachments** provide the **liver** with stable fixation to the diaphragm and to the posterior abdominal wall: the **inferior vena cava**, which adheres to the caval opening of the diaphragm and to the visceral surface of the liver, lodged in the groove for the vena cava, and the **connective tissue** between the inferior surface of the diaphragm and the bare area of the liver, devoid of peritoneal coverage, located at the posterior border of the liver, between the two layers of the coronary ligament.

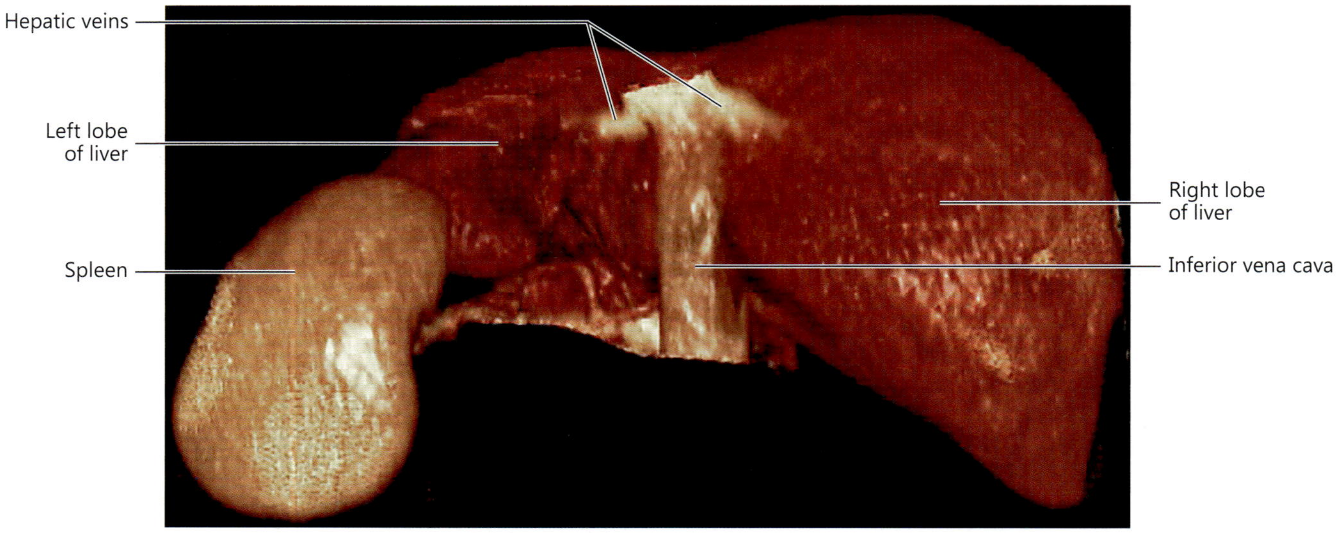

Three-dimensional image of a living human body, obtained by CT scan using a 3D volume rendering technique, showing the **supramesocolic organs** in posterior projection.

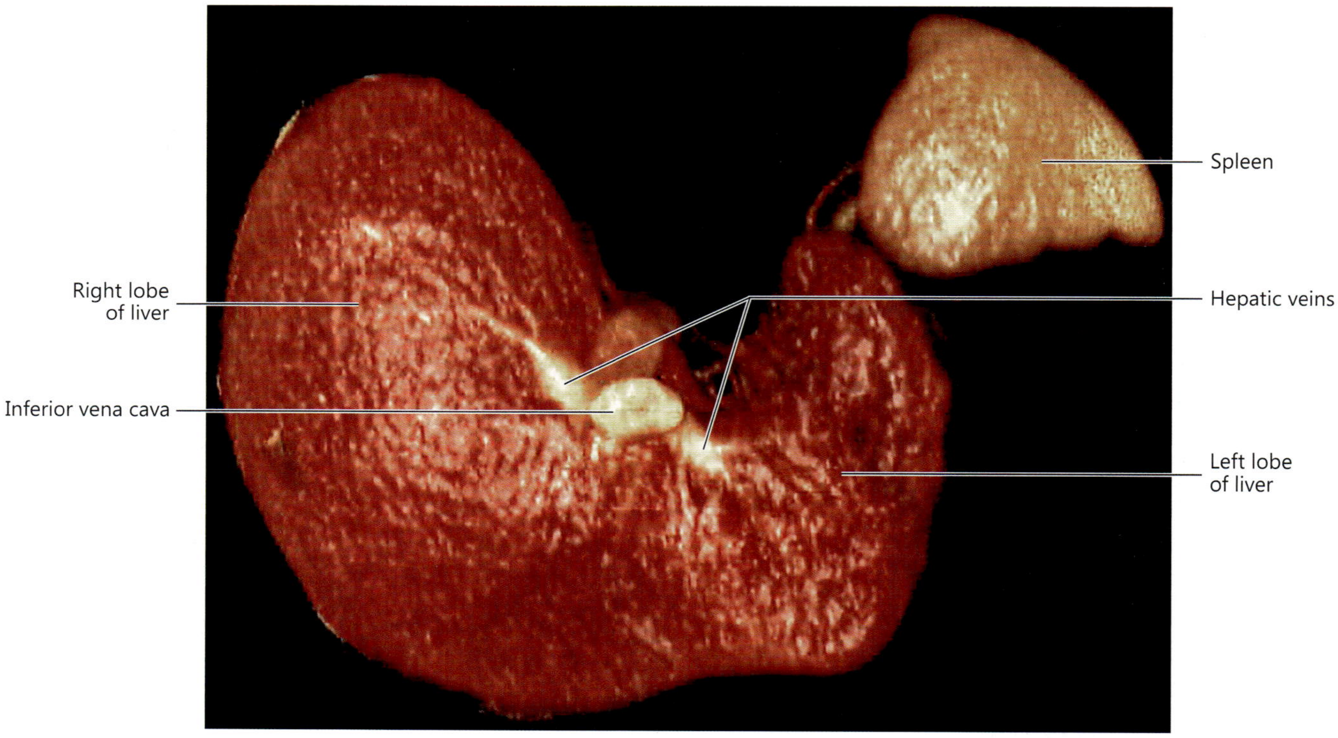

Three-dimensional image of a living human body, obtained by CT scan using a 3D volume rendering technique, showing the **supramesocolic organs** as seen from above.

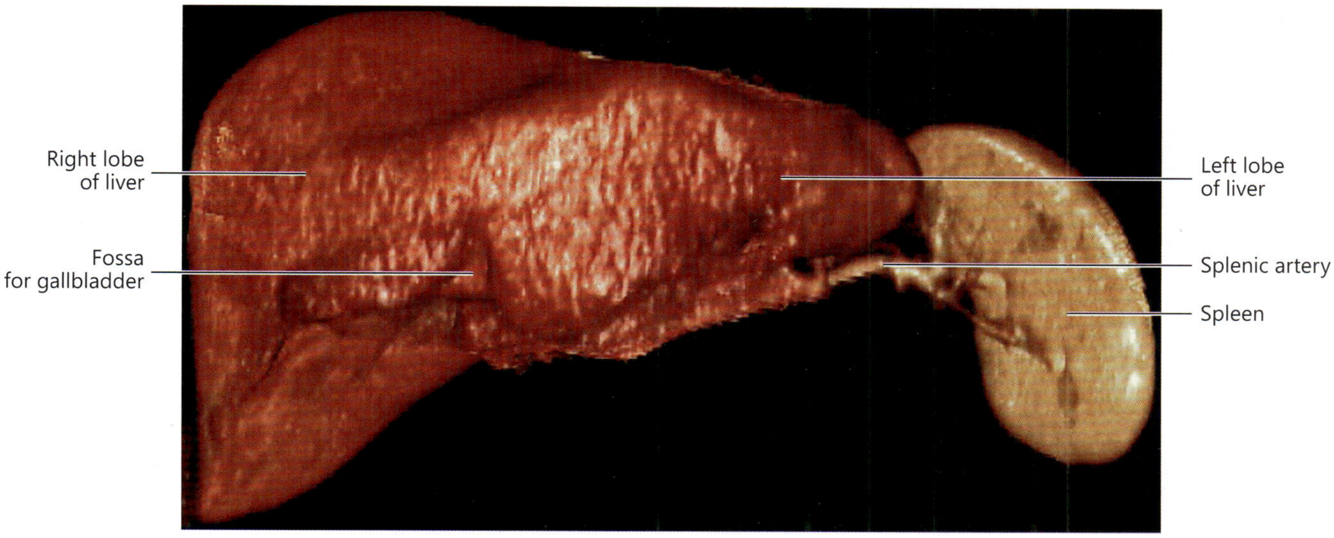

Three-dimensional image of a living human body, obtained by CT scan using a 3D volume rendering technique, showing the **liver** and the **spleen** in frontal projection.

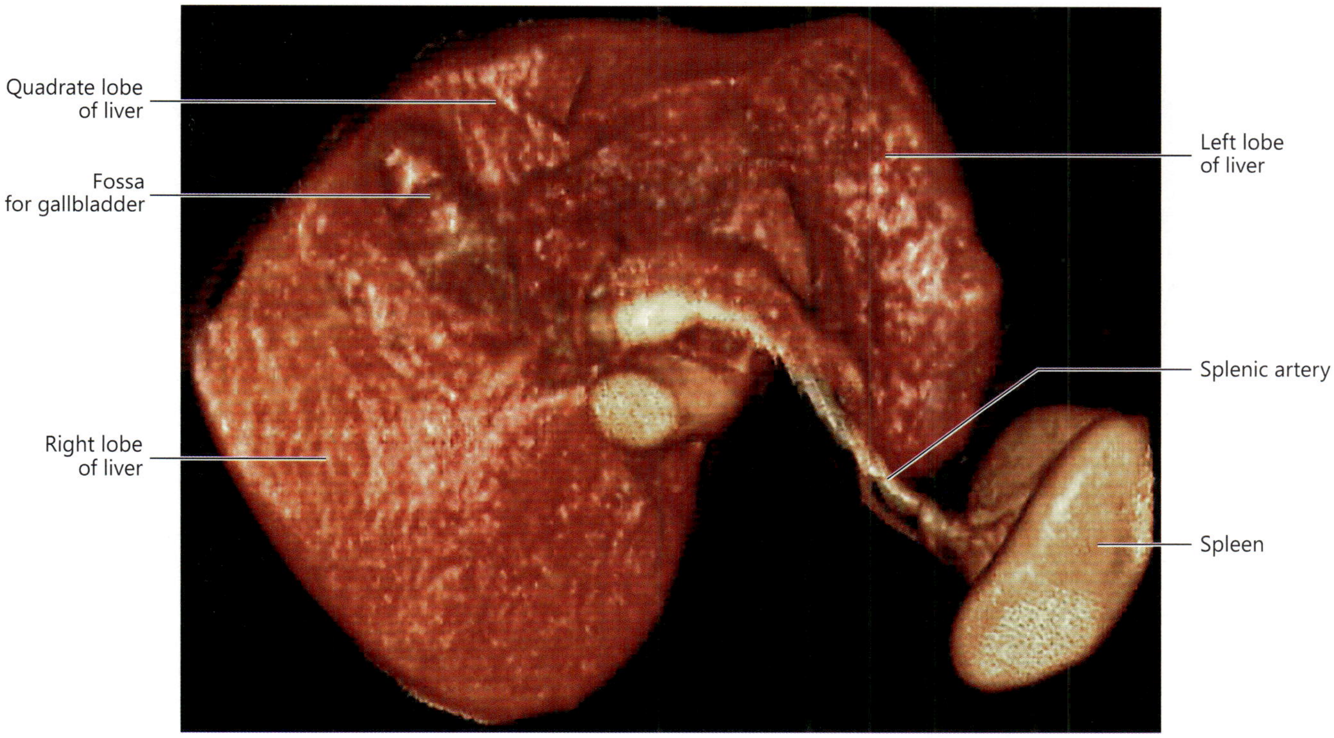

Three-dimensional image of a living human body, obtained by CT scan using a 3D volume rendering technique, showing the **liver** and the **spleen** as seen from below.

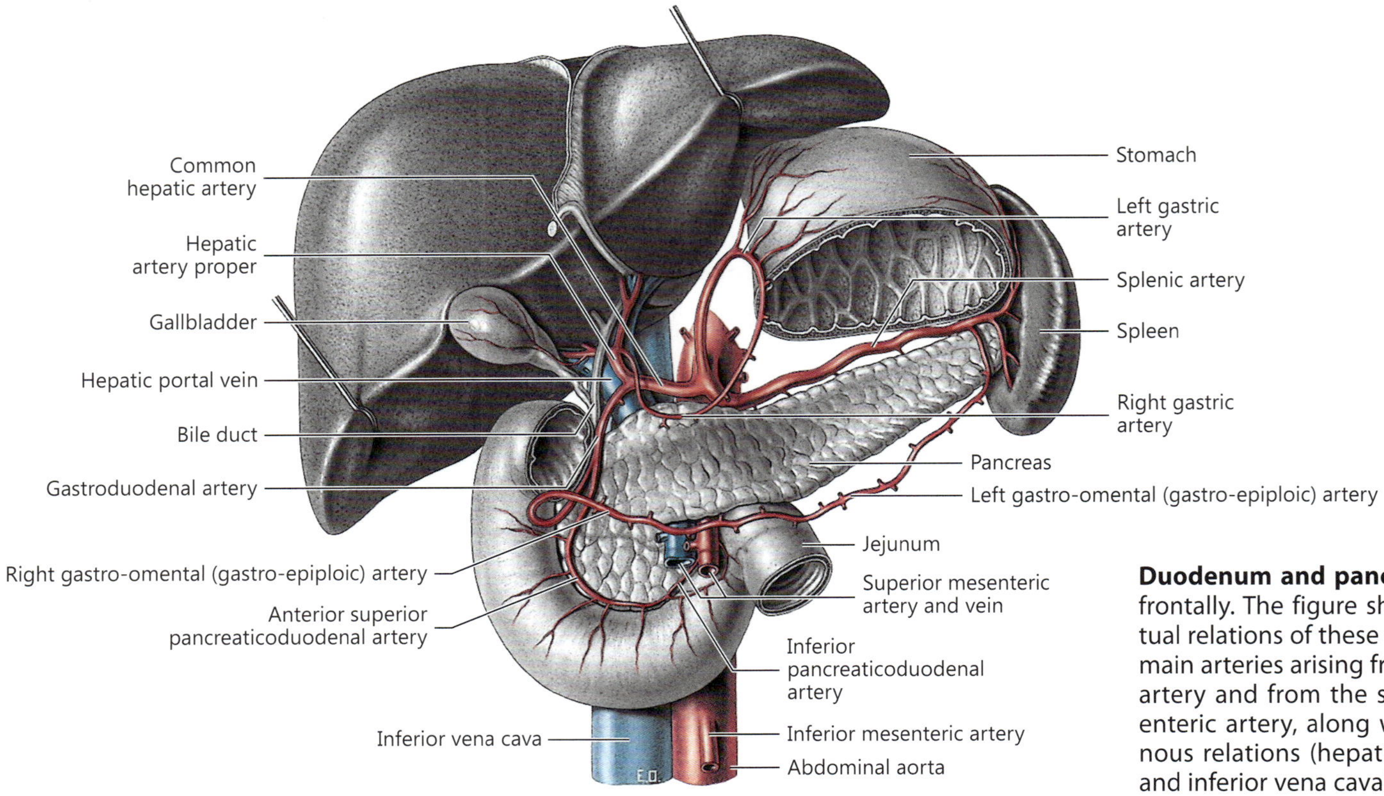

Duodenum and pancreas as seen frontally. The figure shows the mutual relations of these organs to the main arteries arising from the celiac artery and from the superior mesenteric artery, along with their venous relations (hepatic portal vein and inferior vena cava).

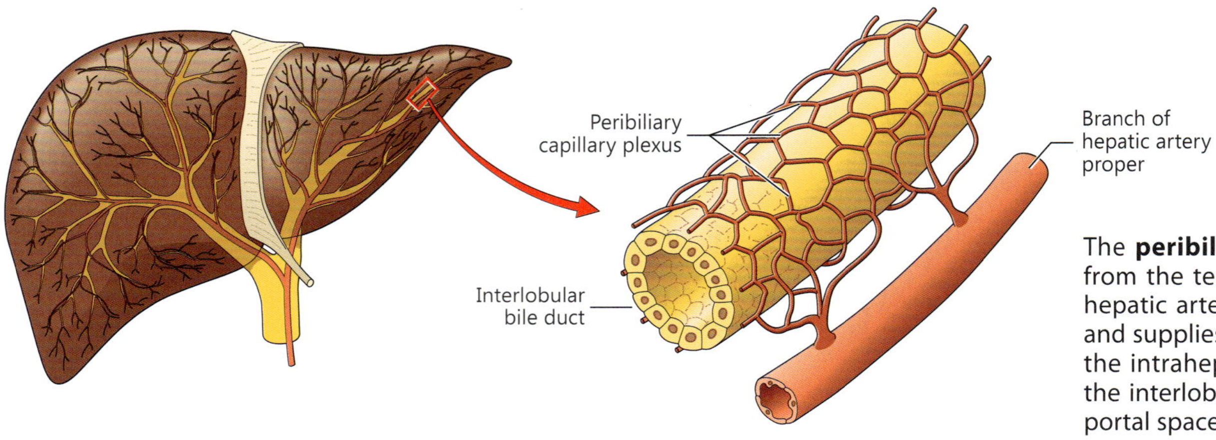

The **peribiliary plexus** originates from the terminal branches of the hepatic artery proper; it surrounds and supplies all the ramifications of the intrahepatic biliary tree, up to the interlobular ducts found in the portal spaces.

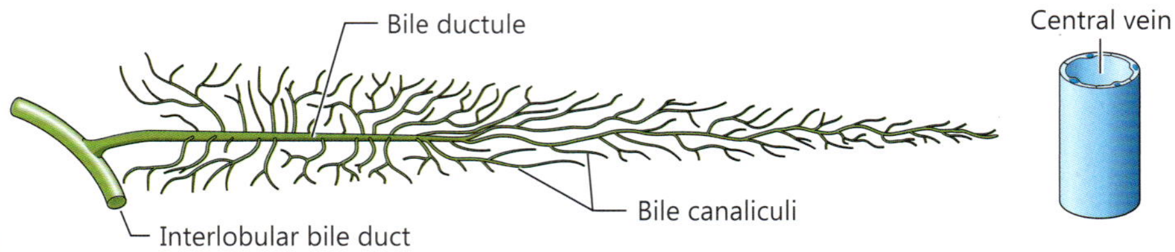

a

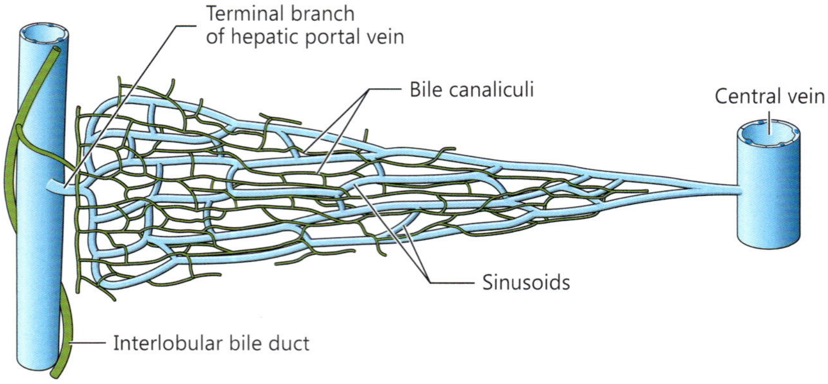

b

a, **Intralobular biliary system**, comprising the bile canaliculi, running between adjacent hepatocytes, which continue into the bile ductule which, in turn, opens into the interlobular bile duct. **b**, The sinusoids of the hepatic lobule originate from the terminal branch of the hepatic portal vein (the interlobular vein) and open into the central vein.

VASCULAR RAMIFICATIONS OF THE LIVER (→ ramification)

Hepatic artery proper
→ Right branch
  → Artery of the caudate lobe
    → Interlobular arteries
      → Interlobular capillaries
        → Peribiliary capillary plexus
          → Sinusoids
  → Anterior segmental artery (same as artery of the caudate lobe)
  → Posterior segmental artery (same as artery of the caudate lobe)
→ Left branch
  → Artery of the caudate lobe
    → Interlobular arteries
      → Interlobular capillaries
        → Peribiliary capillary plexus
          → Sinusoids
  → Medial segmental artery (same as artery of the caudate lobe)
  → Lateral segmental artery (same as artery of the caudate lobe)
→ Intermediate branch
  → Interlobular arteries
    → Interlobular capillaries
      → Peribiliary capillary plexus
        → Sinusoids

VASCULAR RAMIFICATIONS OF THE LIVER (→ ramification)

Hepatic portal vein
→ Right branch
  → Lateral branch
    → Anterior branches
      → Interlobular veins
        → Perilobular veins
          → Sinusoids
    → Posterior branches (same as anterior branches)
  → Paramedian branch
    → Anterior branches
      → Interlobular veins
        → Perilobular veins
          → Sinusoids
    → Posterior branches (same as anterior branches)
→ Left branch
  → Lateral branch
    → Interlobular veins
      → Perilobular veins
        → Sinusoids
  → Paramedian branch (same as lateral branch)
  → Caudate branch (same as lateral branch)

Sinusoids
→ Central veins
  → Sublobular veins
    → Right hepatic vein
      → Inferior vena cava
    → Intermediate hepatic vein (same as right hepatic vein)
    → Left hepatic vein (same as right hepatic vein)

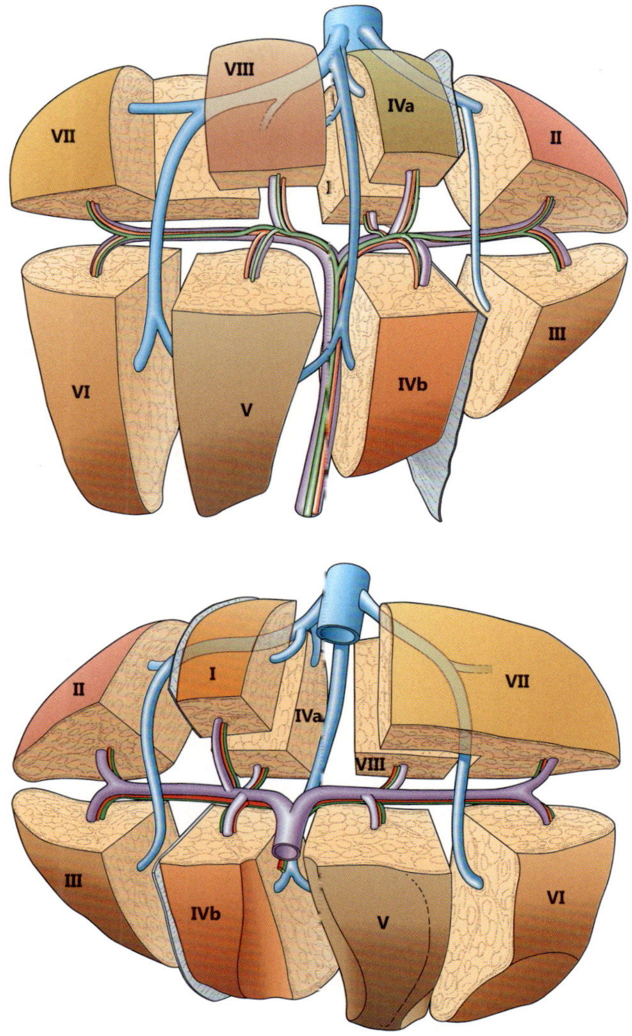

Hepatic Segments		*Irrigation*
I	dorsal - caudate lobe	Left branch of the hepatic portal vein
II	superolateral	
III	inferolateral	
IV a	medial superior	
IV b	medial inferior - quadrate lobe	
V	anteroinferior	Right branch of the hepatic portal vein
VI	posteroinferior	
VII	posterosuperior	
VIII	anterosuperior	

Distinction of the hepatic **segments** based on the branches of division of the hepatic portal vein. *Blue*, branches of hepatic veins; *violet*, branches of the hepatic portal vein; *red*, branches of the hepatic artery proper; *green*, branches of the hepatic duct. Blood supply to the segments is provided by an equal level of ramifications from the hepatic artery and the hepatic portal vein.

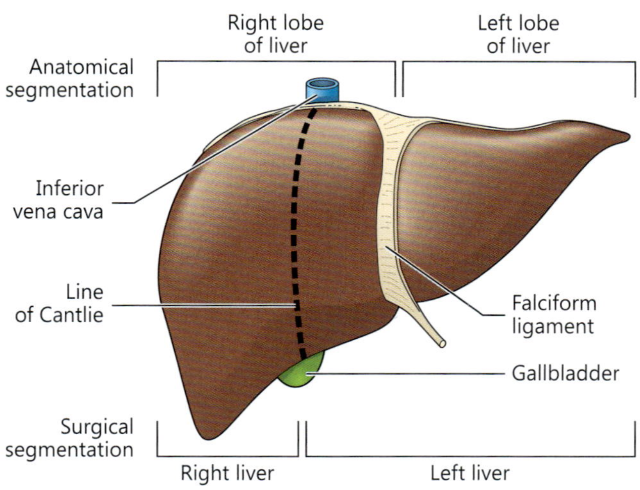

◄

The **line of Cantlie**, which is dashed onto the diaphragmatic surface of the liver, passes through the fossa for the gallbladder and the groove for the vena cava. It is an anatomical landmark for liver surgery, separating the right and left parts of the liver.

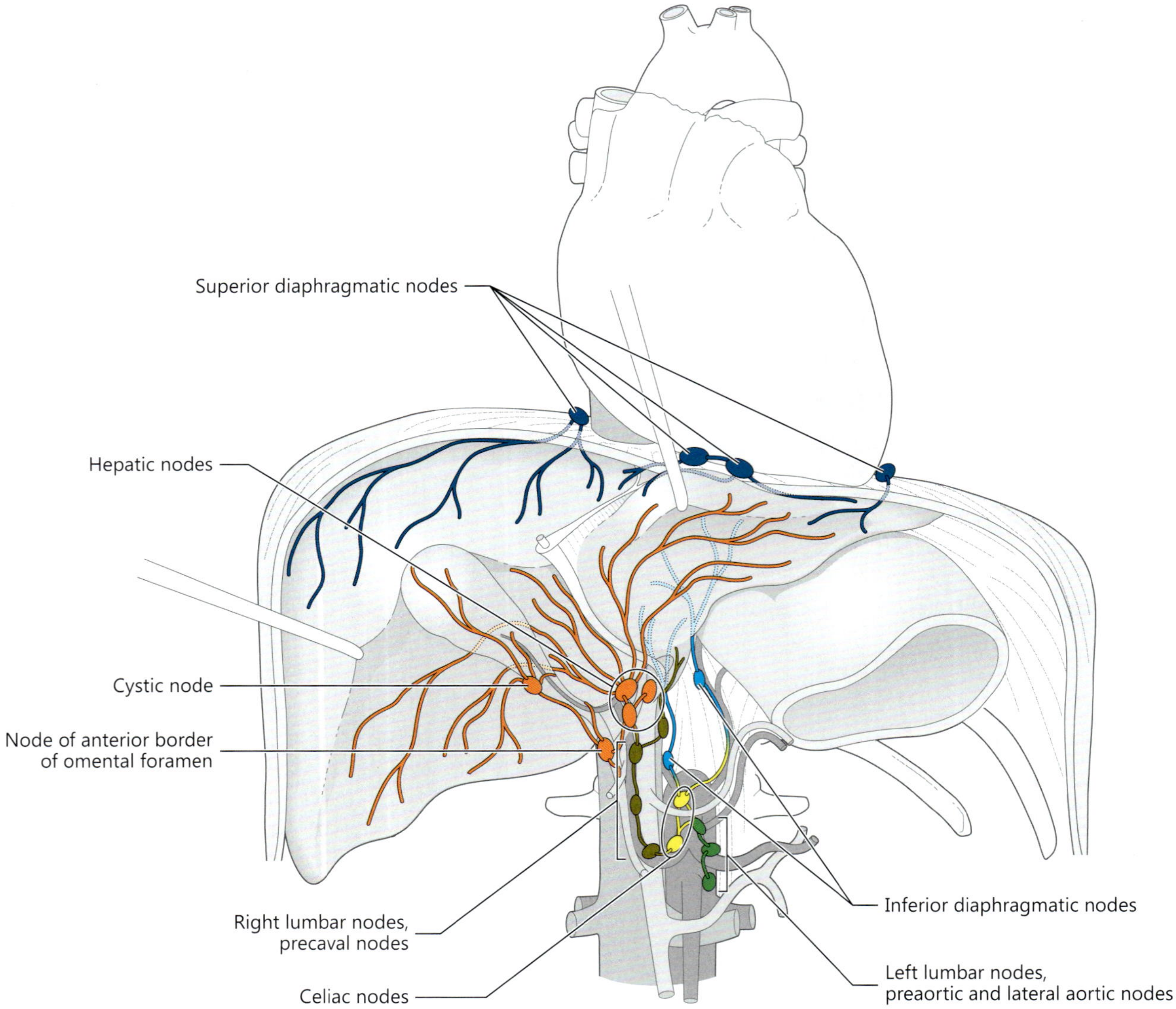

The **hepatic nodes** are located near the porta hepatis, in the lesser omentum, along the hepatic artery and the bile duct. They receive afferent lymphatic vessels from the liver, the stomach, the duodenum, the gallbladder and the pancreas, and drain into the celiac nodes. The superficial lymphatic vessels of the liver can drain into the hepatic and the left lumbar nodes; the lymphatic vessels coming from the posterior surface of the diaphragmatic surface and from the posterior part of the visceral surface drain into the superior diaphragmatic nodes, located around the opening of the inferior vena cava, behind the xiphoid process, and next to the termination of the phrenic nerve.

➤

Schematic representation of the **liver** and of its **parenchymal and vascular organization. a**, Diaphragmatic (or anterosuperior) surface. The falciform ligament marks the boundary between the liver right and left lobes. The hepatic artery and the hepatic portal vein enter the liver at the porta hepatis located in the visceral (or posteroinferior) surface. The inferior vena cava relates to the posterior surface or border of the liver, where it receives the hepatic veins. **b**, The hepatic lobules are polyhedral, appearing as polygonal in liver sections, and represent the structural units of the organ. Some animal species, but not humans, display by a thin boundary of connective tissue surrounding the lobules. The figure shows a three-dimensional reconstruction of these prism-shaped units. The (classical) hepatic lobule is traversed by the central vein, which receives the sinusoids running within the hepatic lobule. The connective tissue present between three or more adjacent lobules is termed portal area and contains the portal triad. **c**, The portal triad comprehend interlobular branches of the hepatic portal vein, of the hepatic artery and of the bile duct. **d**, Portal area bounded by the limiting hepatic plate, containing the triad. As a matter of simplification, the figure shows only two perisinusoidal spaces (or spaces of Disse) and one bile canaliculus, in a longitudinal section. The canal of Hering transports (*green* arrow) the bile from the bile canaliculus to the bile duct, and the periportal space (of Mall) collects the interstitial fluid from the perisinusoidal spaces and then carries it (*orange* arrow) into the lymphatic vessel. **e**, Main features of the cells of an hepatic cord. **f**, Structure of an hepatocyte. The figure also illustrates the relations between adjacent hepatocytes and with the hepatic sinusoids, in specific areas two adjacent hepatocytes form the boundaries of the bile canaliculi.

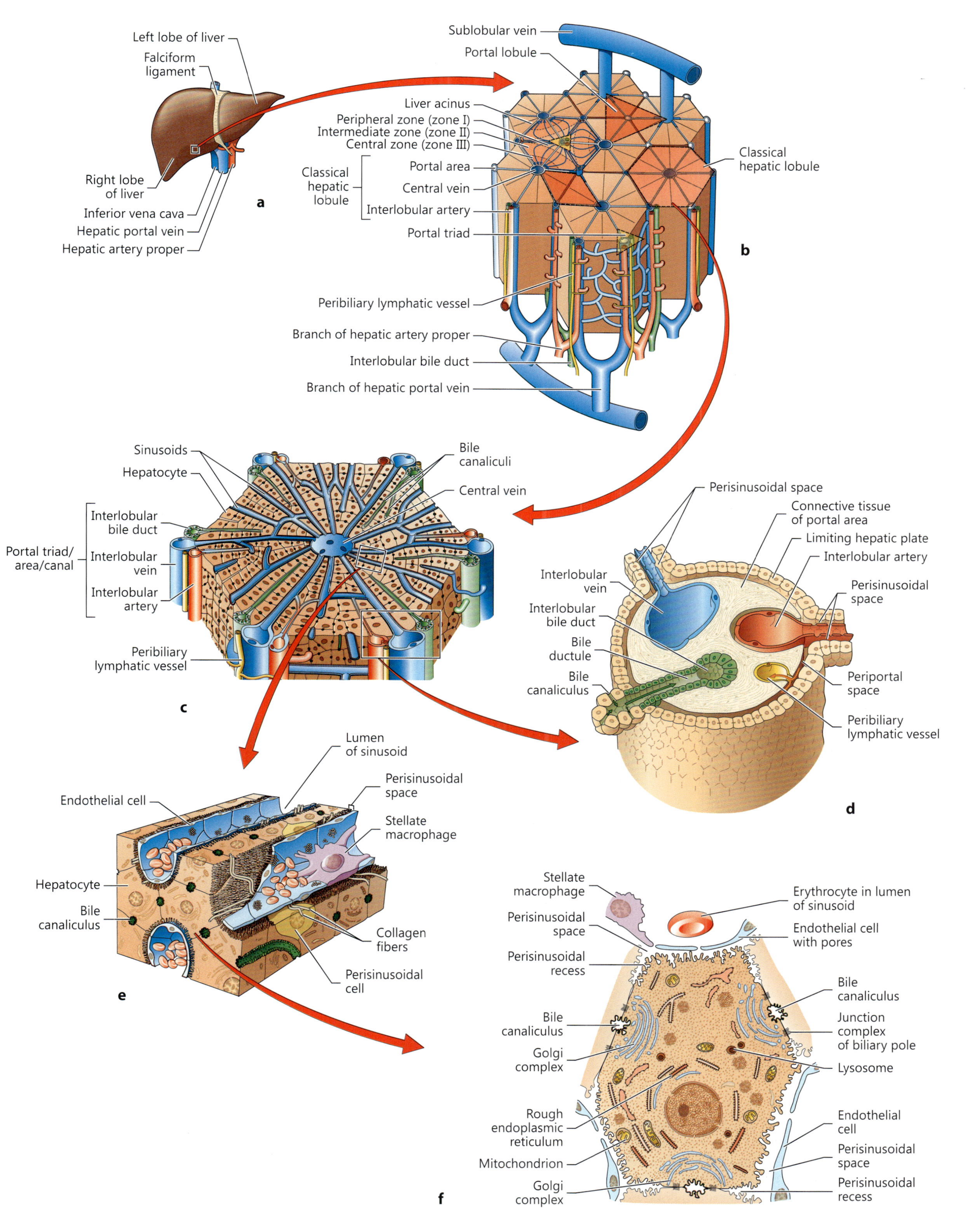

Left lobe of liver
Falciform ligament
Right lobe of liver
Inferior vena cava
Hepatic portal vein
Hepatic artery proper
a
Sublobular vein
Portal lobule
Liver acinus
Peripheral zone (zone I)
Intermediate zone (zone II)
Central zone (zone III)
Classical hepatic lobule
Portal area
Central vein
Interlobular artery
Portal triad
Classical hepatic lobule
b
Peribiliary lymphatic vessel
Branch of hepatic artery proper
Interlobular bile duct
Branch of hepatic portal vein
Sinusoids
Hepatocyte
Bile canaliculi
Central vein
Interlobular bile duct
Portal triad/ area/canal
Interlobular vein
Interlobular artery
Peribiliary lymphatic vessel
c
Perisinusoidal space
Connective tissue of portal area
Limiting hepatic plate
Interlobular artery
Interlobular vein
Interlobular bile duct
Perisinusoidal space
Bile ductule
Bile canaliculus
Periportal space
Peribiliary lymphatic vessel
d
Lumen of sinusoid
Endothelial cell
Perisinusoidal space
Stellate macrophage
Hepatocyte
Bile canaliculus
Collagen fibers
Perisinusoidal cell
e
Stellate macrophage
Erythrocyte in lumen of sinusoid
Perisinusoidal space
Endothelial cell with pores
Perisinusoidal recess
Bile canaliculus
Bile canaliculus
Junction complex of biliary pole
Golgi complex
Lysosome
Rough endoplasmic reticulum
Endothelial cell
Mitochondrion
Perisinusoidal space
Golgi complex
Perisinusoidal recess
f

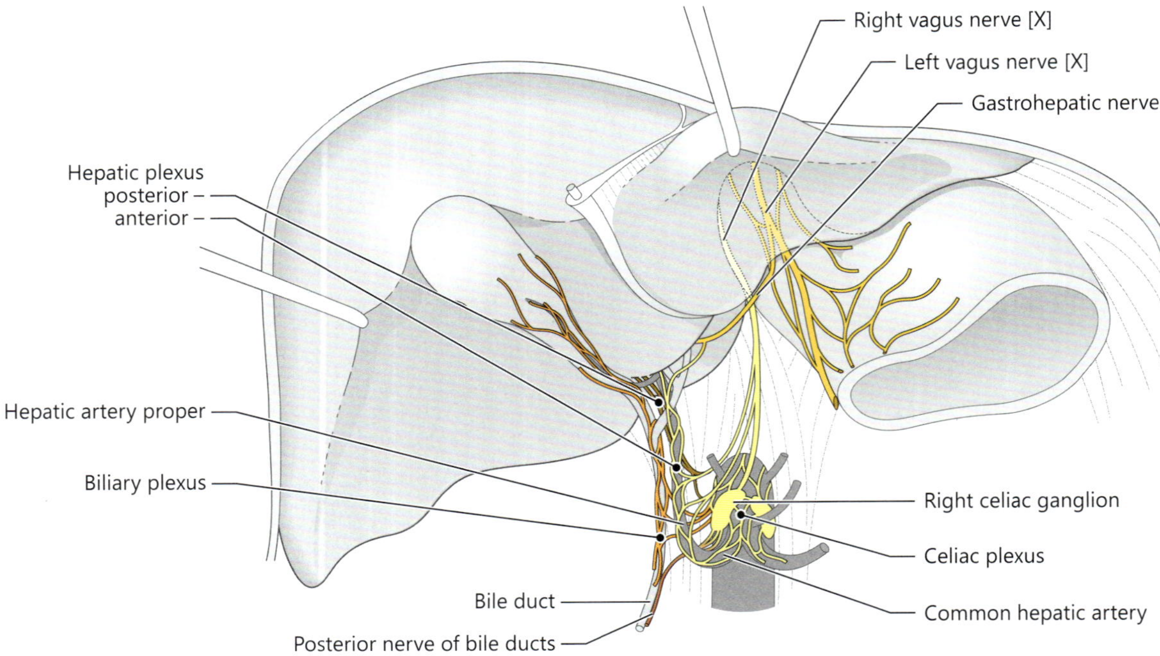

The **nerves of the liver** include sympathetic nerve fibers coming from the right celiac ganglion and of parasympathetic nerve fibers supplied by the right and left vagus nerves. The nerve branches intertwine forming the hepatic plexus, divided into an anterior and a posterior hepatic plexus. The anterior hepatic plexus receives the gastrohepatic nerve, arising from the left vagus nerve. Some nerve branches originating from the celiac plexus and from the right celiac ganglion reach the extrahepatic bile ducts, forming the biliary plexus, while others constitute the posterior nerve of the bile ducts.

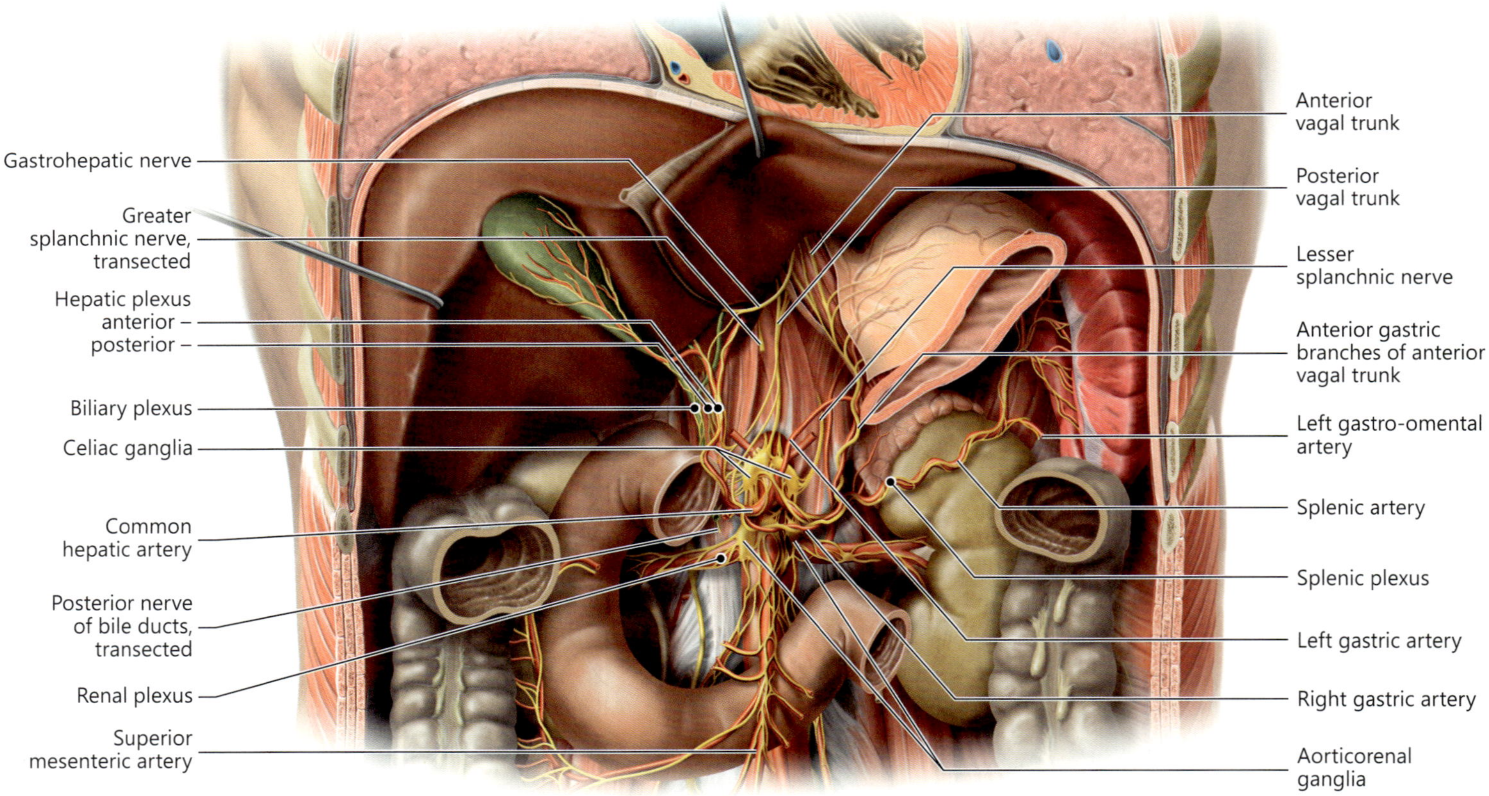

The **celiac plexus** extends in front of the aorta, between the celiac trunk and the superior mesenteric artery, and consists of the right and left celiac ganglia, and of a network of fibers of diverse origin. The celiac plexus receives the greater and lesser splanchnic nerves, as well as branches of the thoracic aortic plexus, branches of the lumbar part of the sympathetic trunk and terminal branches of the phrenic nerve, as well as the terminations of the right vagus nerve. The efferents are represented by secondary plexuses formed by perivascular branches that follow the course of the collateral arteries of the abdominal aorta.

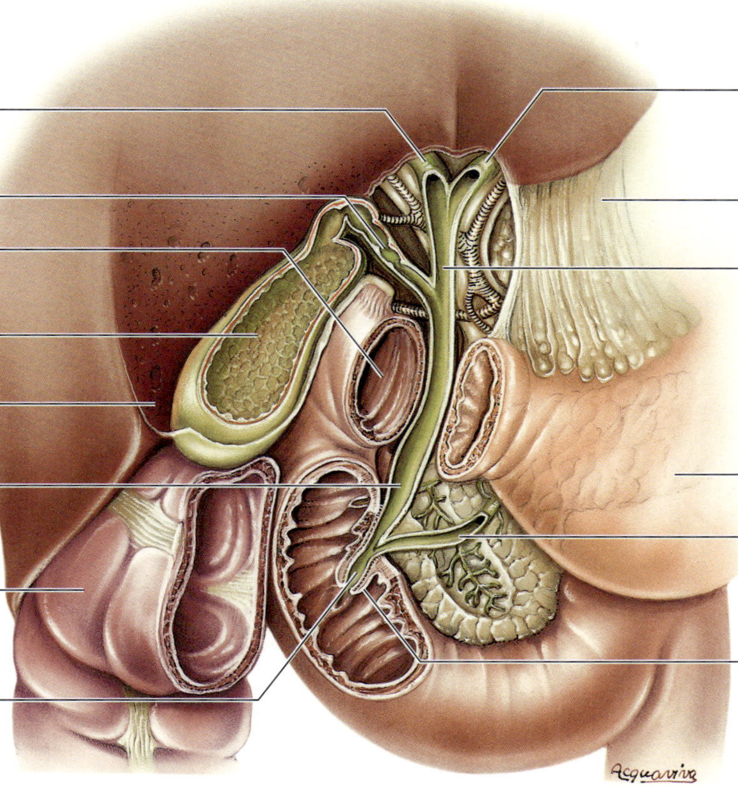

Course and relations of the **bile duct**, laid open in its entire length. The common bile duct runs behind the superior part of the duodenum (transected in this figure), contacts the pancreas and eventually passes through the wall of the descending part of the duodenum. The common bile duct and the main pancreatic duct converge and open into the major duodenal papilla, which was also transected.

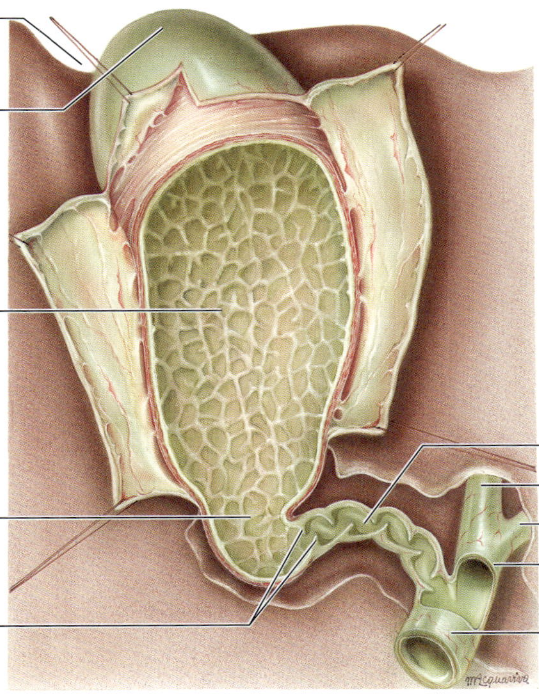

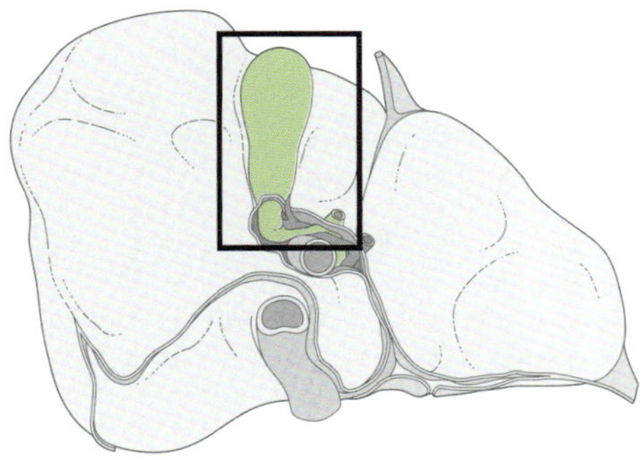

Macroscopic organization of the **extrahepatic bile ducts**.

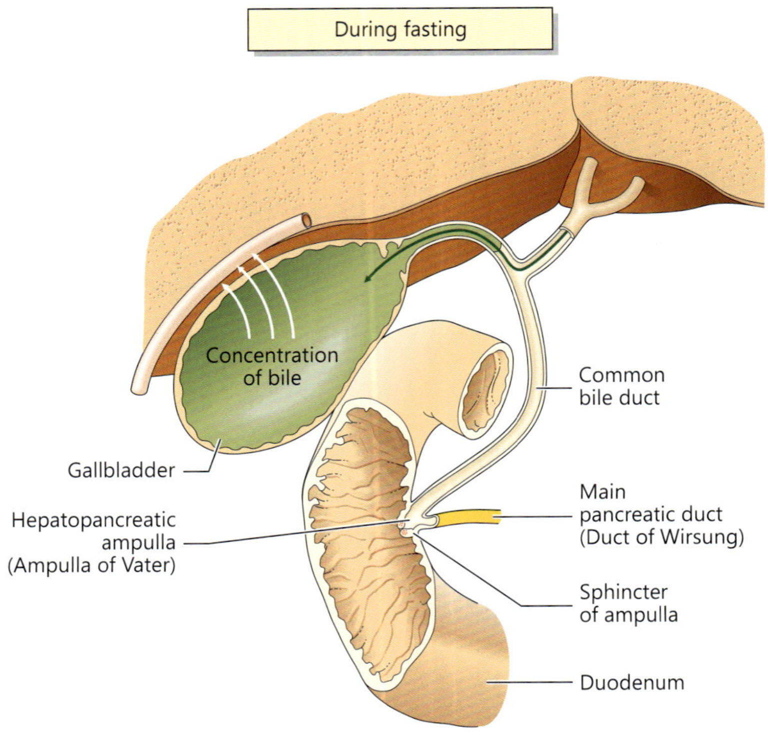

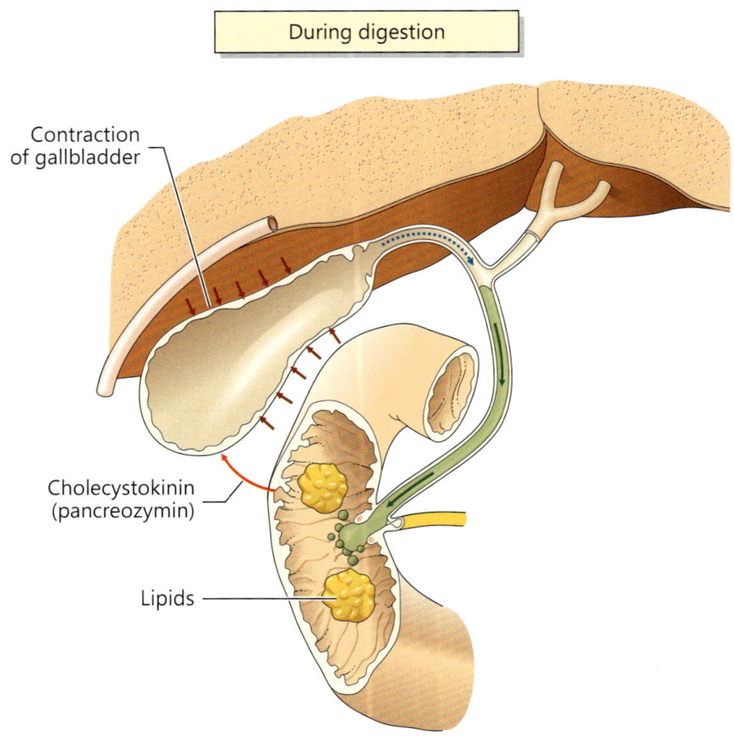

SEQUENCE OF THE BILE DUCTS (→ ramification)
Intrahepatic bile ducts
Bile canaliculi → Bile canals → Bile ductules → Interlobular bile ducts → Major intrahepatic bile ducts → Right hepatic duct → Left hepatic duct
Extrahepatic bile ducts
Right and left hepatic ducts → Common hepatic duct Common hepatic duct and cystic duct → Common bile duct Common bile duct and main pancreatic duct → Hepatopancreatic ampulla → Major duodenal papilla

Diagram of **bile** secretion by the hepatocytes during fasting and during digestion. In the former case, the bile is stored and concentrated in the gallbladder. During digestion, the secretion of cholecystokinin (pancreozymin) by the duodenal mucosa causes the muscles of the gallbladder to contract vigorously and the sphincter of ampulla or sphincter of Oddi to relax, leading to the bile release into the duodenum.

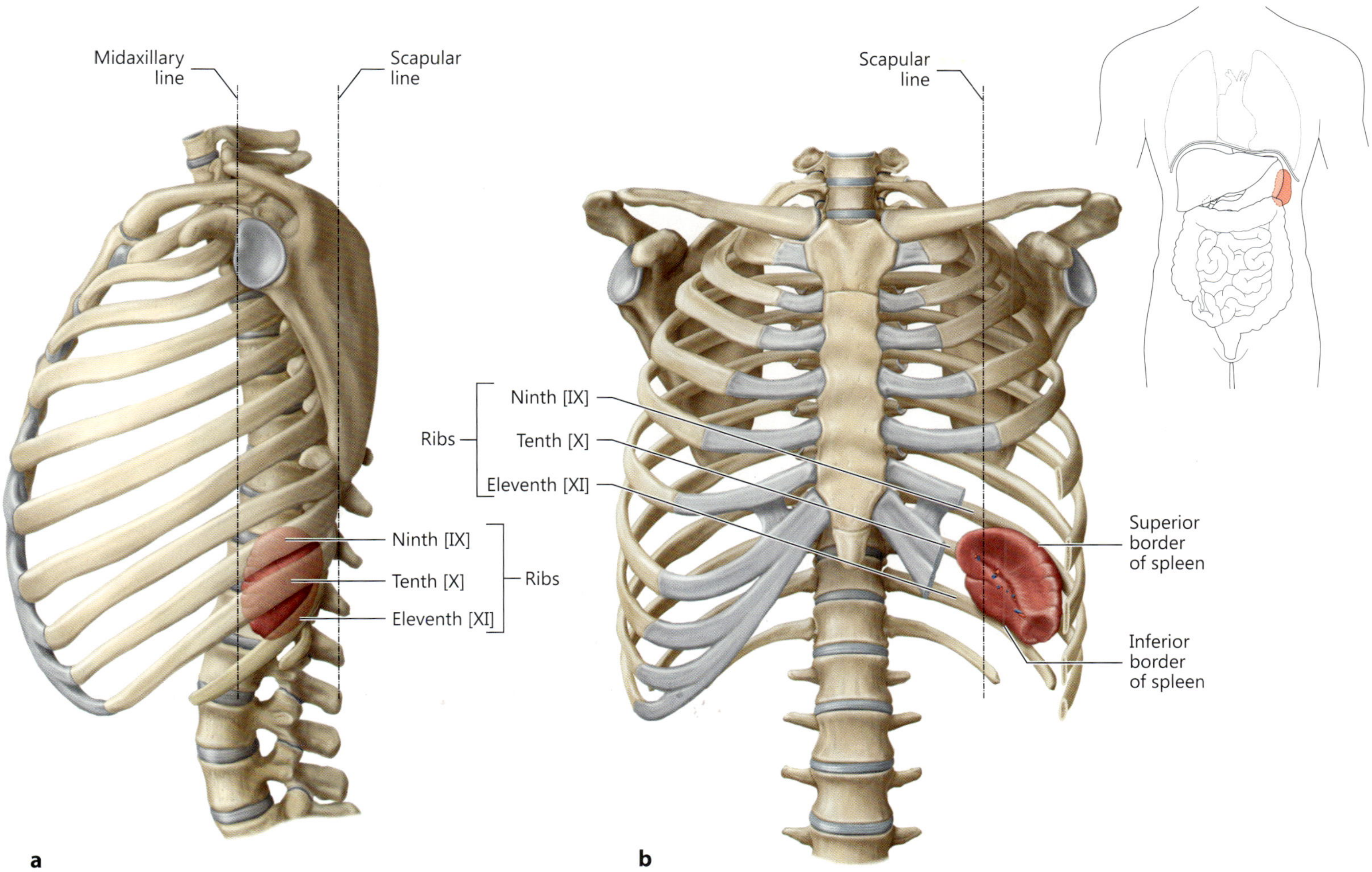

Lateral (a) and frontal (b) projections of the spleen onto the wall of the thoracic cage. The splenic area is located between the ninth [IX] and eleventh [XI] ribs, and between the midaxillary and scapular lines.

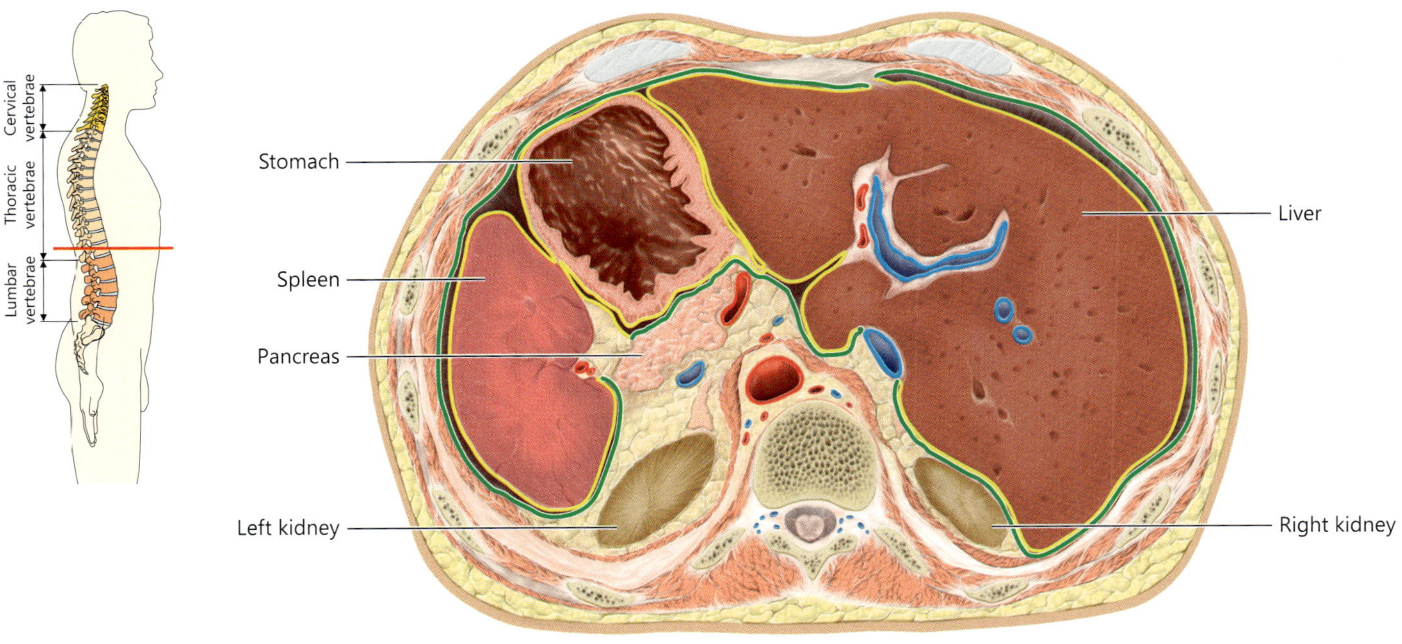

Transverse section of the abdomen taken at the **twelfth thoracic vertebra**, showing the relations of the lateral surface of the spleen to the diaphragm and the relations of the medial surface of the spleen to the kidney, pancreas and stomach. The peritoneum, indicated in *green* and *yellow*, appears as one single layer. To simplify the figure, the parts investing the abdominal cavity, or parietal peritoneum (*green* line), have been separated from those lining the abdominal organs (*yellow* line).

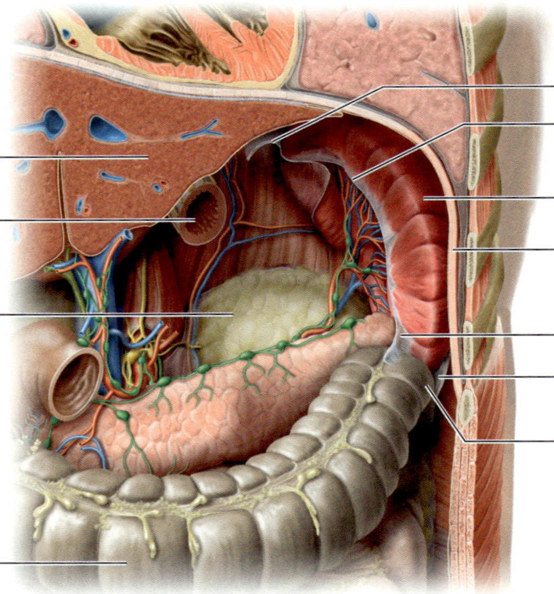
Phrenicosplenic ligament
Gastrosplenic ligament
Left lobe of liver
Spleen
Esophagus, transected
Diaphragm
Perinephric fat
Pancreaticosplenic ligament
Phrenicocolic ligament
Left colic flexure
Transverse colon

a

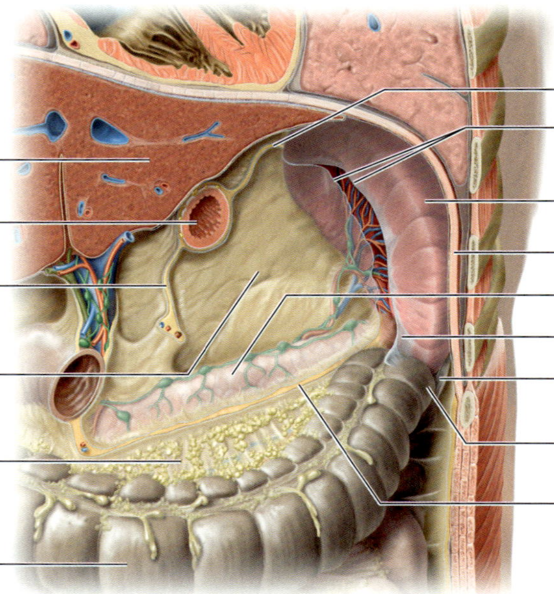
Phrenicosplenic ligament
Gastrosplenic ligament
Left lobe of liver
Spleen
Esophagus, transected
Diaphragm
Hepatogastric ligament
Pancreas
Pancreaticosplenic ligament
Omental bursa
Phrenicocolic ligament
Left colic flexure
Transverse mesocolon
Root of transverse mesocolon
Transverse colon

b

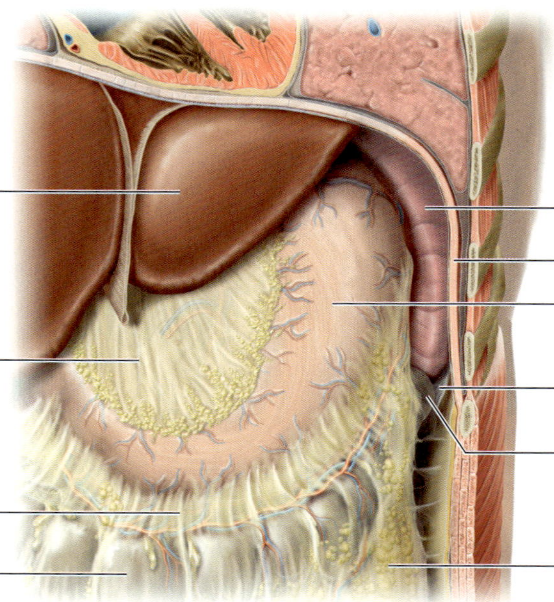
Left lobe of liver
Spleen
Diaphragm
Stomach
Lesser omentum
Phrenicocolic ligament
Left colic flexure
Gastrocolic ligament
Transverse colon
Greater omentum

c

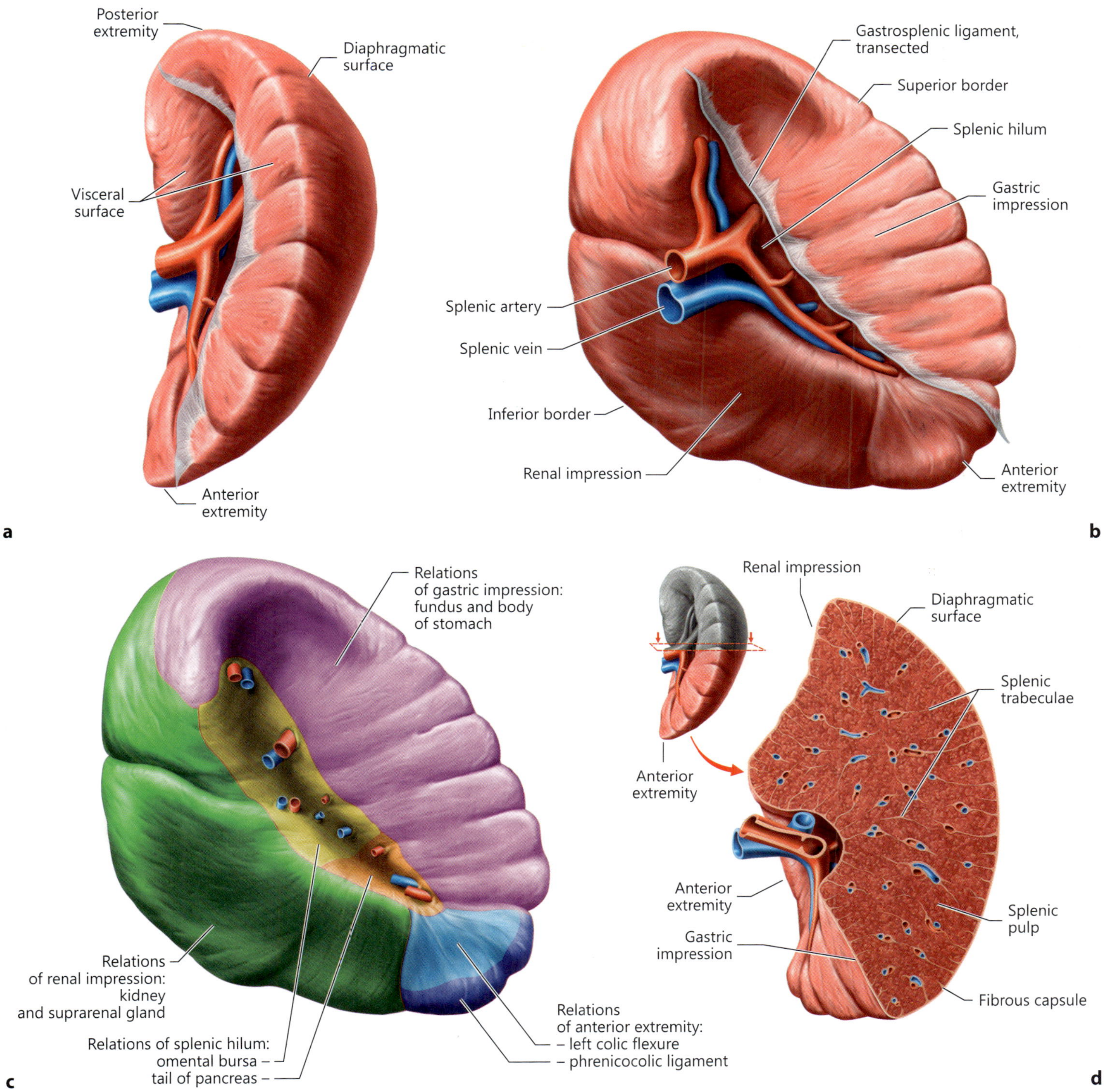

a, External configuration of the **spleen**, as seen frontally. The superior border is indented; the posterior extremity is rounded, while the anterior extremity is sharper. **b**, External configuration of the **spleen**, as seen medially. The visceral surface of the spleen contains the splenic hilum, which features the splenic tuberosity at its upper end, as well as the gastric and renal impressions. **c, Relations of the spleen**. **d**, Transverse section of the **spleen at the hilum**.

➤

Stratigraphic aspects of region of **omental bursa**. **a**, Posterior wall of retroperitoneal space. **b**, Posterior wall of omental bursa. **c**, Anterior wall of omental bursa. The omental bursa is located behind the stomach. Its left extension, or splenic recess, extends between the gastrosplenic ligament, anteriorly, and the pancreaticosplenic ligament, posteriorly, to reach the splenic hilum. The visceral peritoneum of the pancreaticosplenic ligament, which bounds the splenic space, originates the gastrosplenic ligament, between the vertical tract of the greater curvature of the stomach and the splenic hilum, and the pancreaticosplenic ligament, between the splenic hilum and the tail of the pancreas, and the phrenicosplenic ligament, between the diaphragm and the spleen. The posterior surface of the **lesser omentum** forms part of the anterior surface of omental bursa.

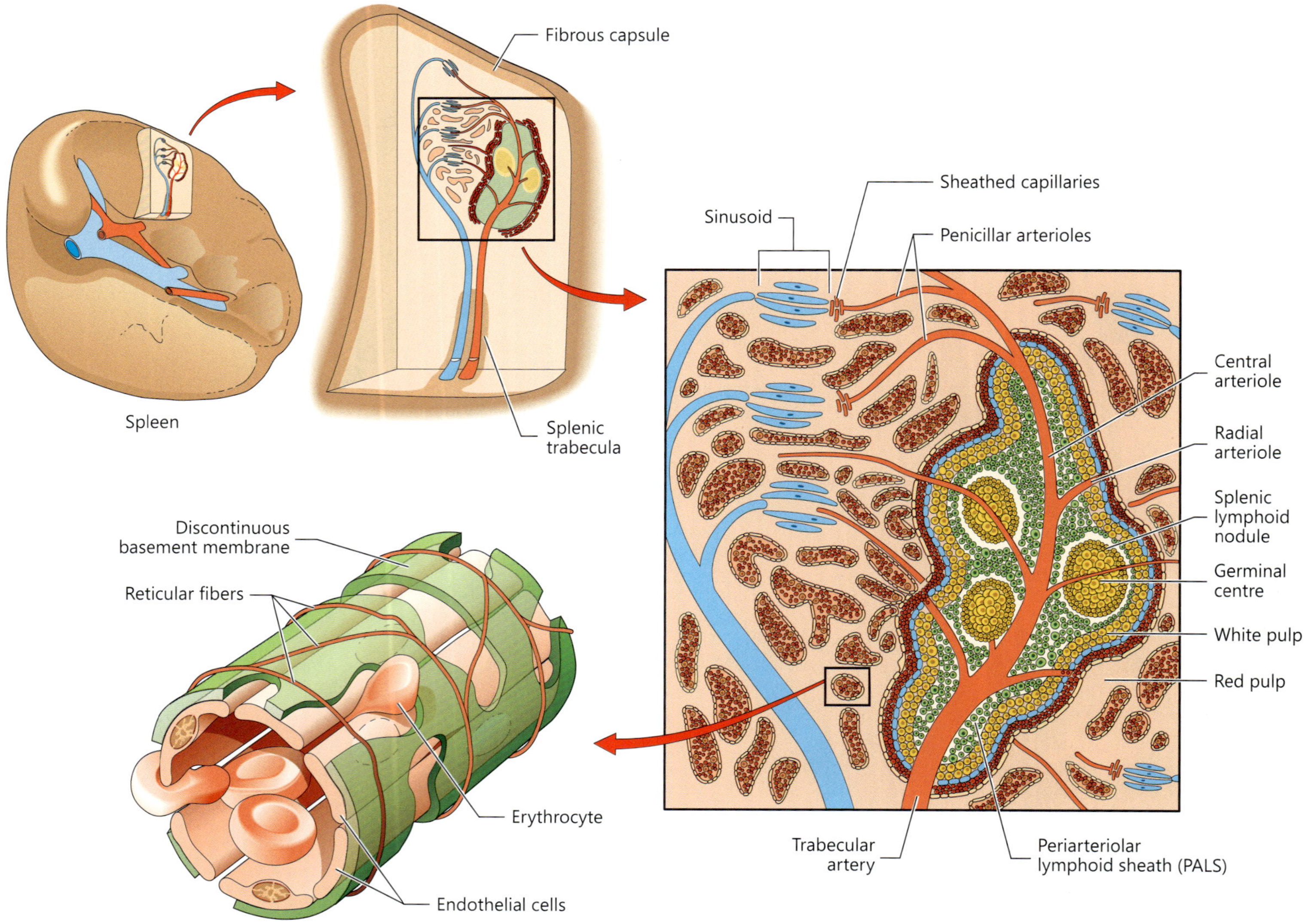

Diagram of the organization of the **marginal zone of the spleen** in humans. The collateral branches of the central artery, which travels axially across the splenic lymphoid nodules (surrounded by the PALS), reach a complex vascular system that surrounds the white pulp. This represents a transition zone between the white pulp and the cords of the red pulp. The structure of the venous sinuses can be compared to that of a barrel: the elongated endothelial cells can be considered as the staves, held together by the reticular cells, that can be equated with the hoops of the barrel. Erythrocytes pass through the venous sinus walls, squeezing through the intercellular spaces between the endothelial cells. The endothelial cells of the sinusoids rest against a discontinuous basement membrane, which allows cells to pass through.

Three-dimensional image of a living human body, obtained by CT scan using a 3D volume rendering technique, showing the **spleen** in inferior (**a**), frontal (**b**) and sagittal (**c**) projection.

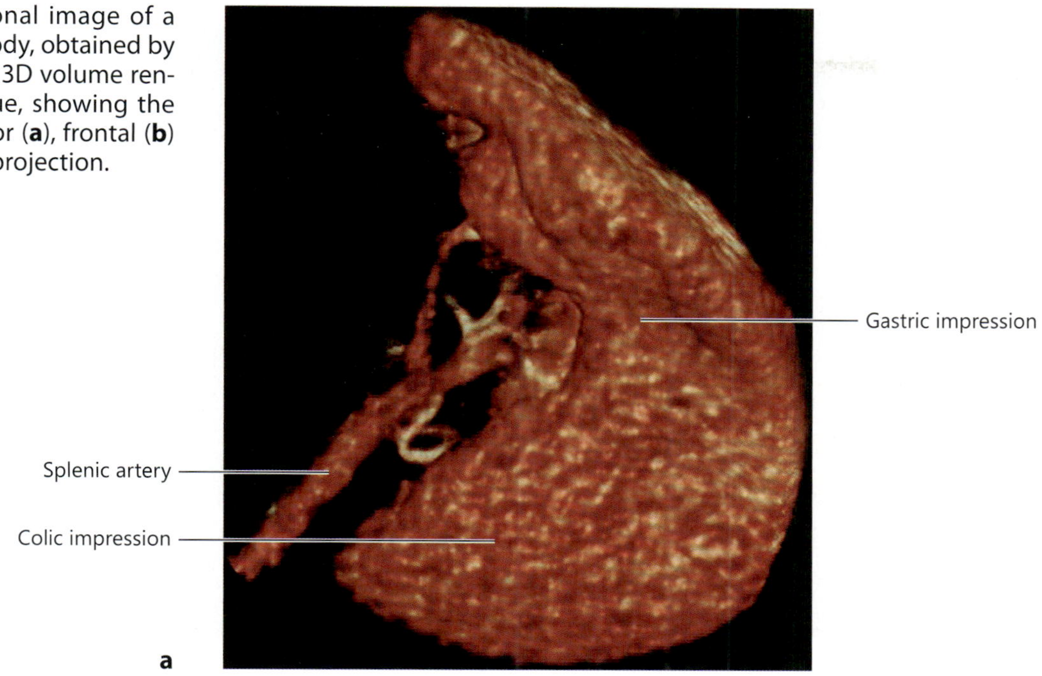

a

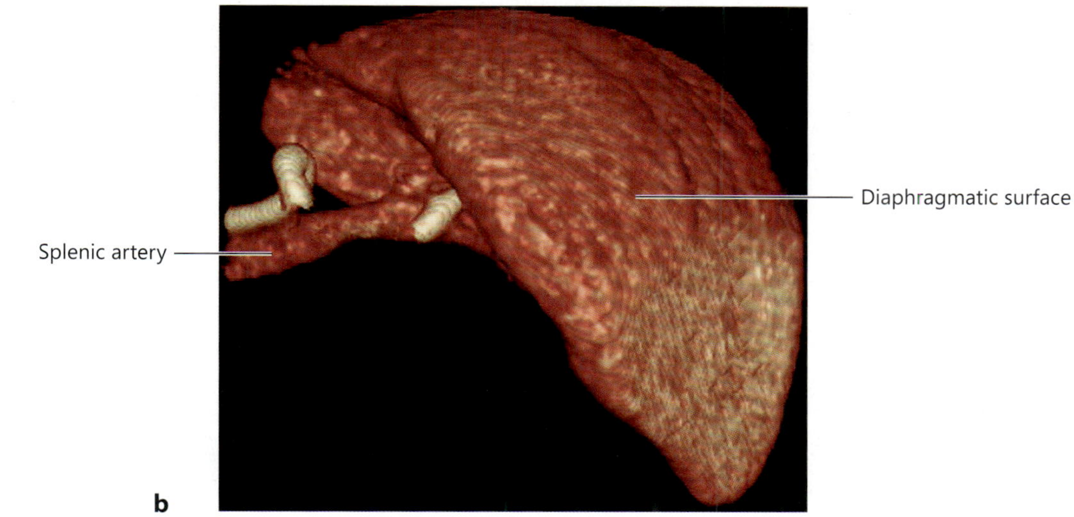

b

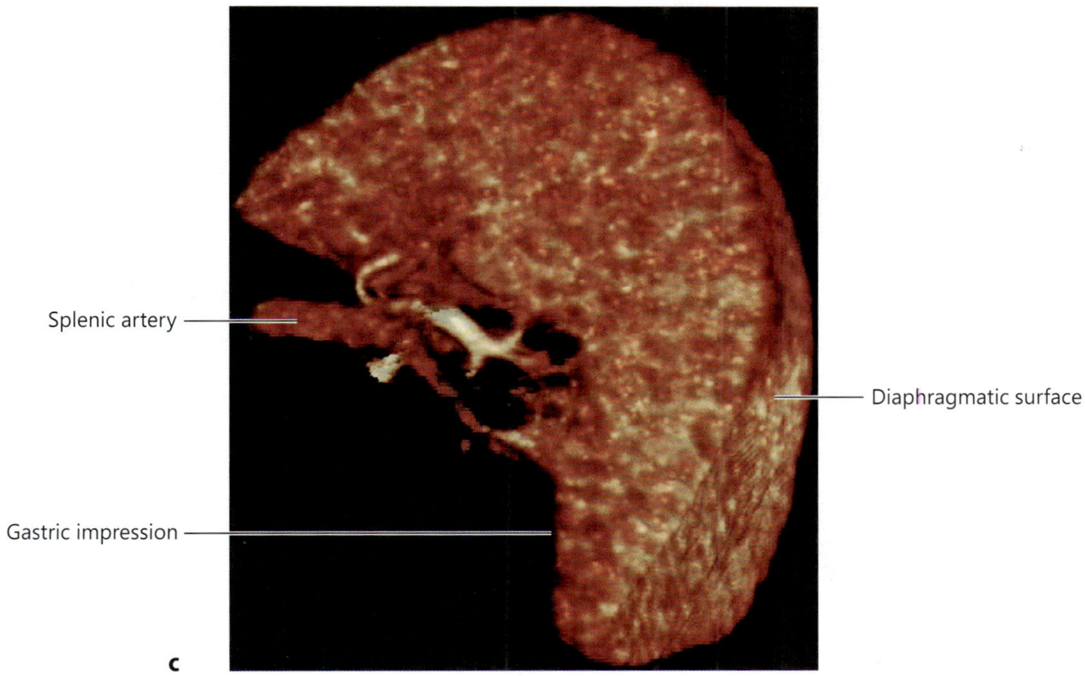

c

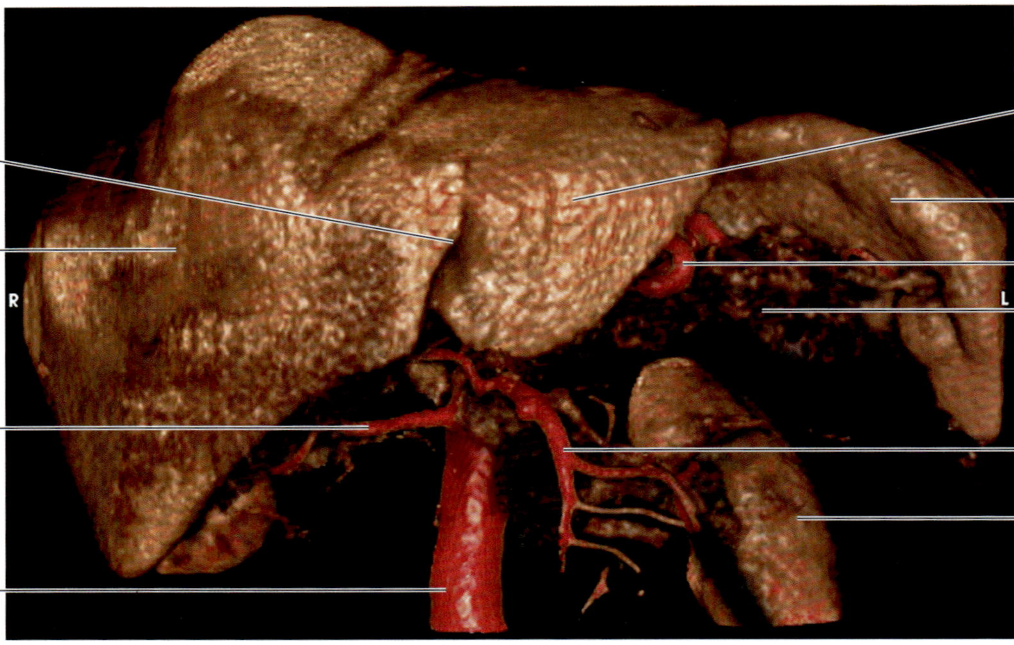

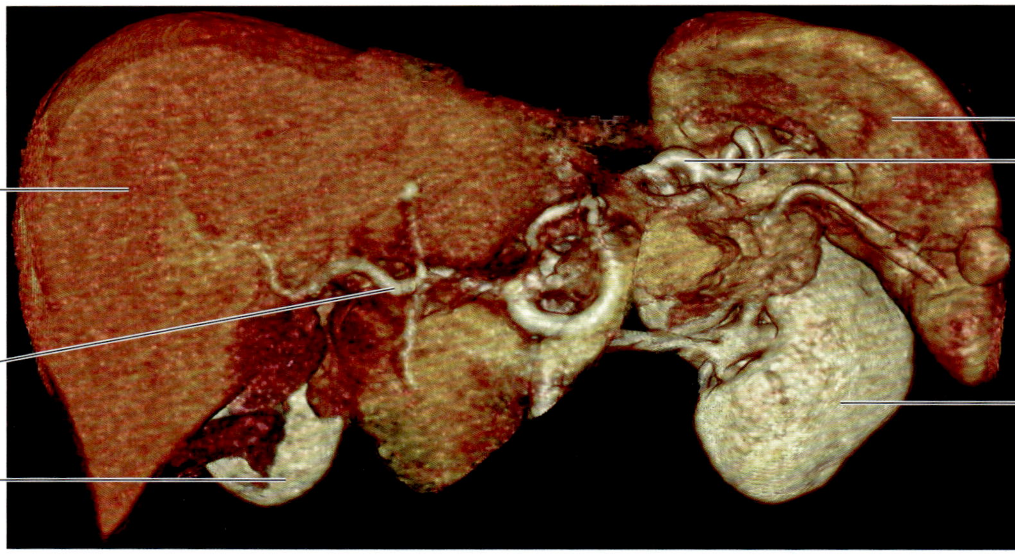

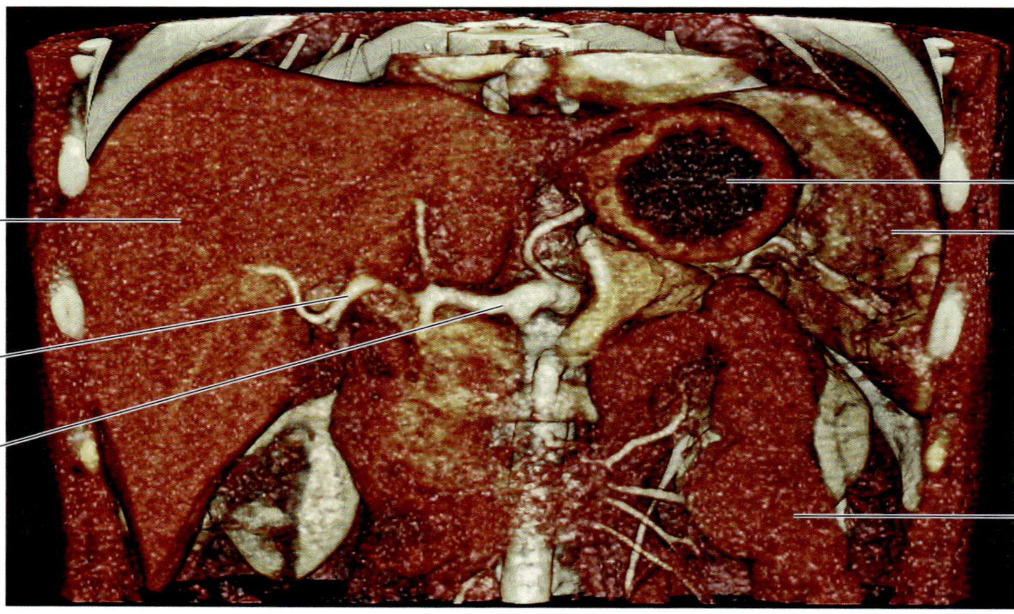

Three-dimensional image of a living human body, obtained by CT scan using a 3D volume rendering technique, showing the **abdominal organs** in frontal section.

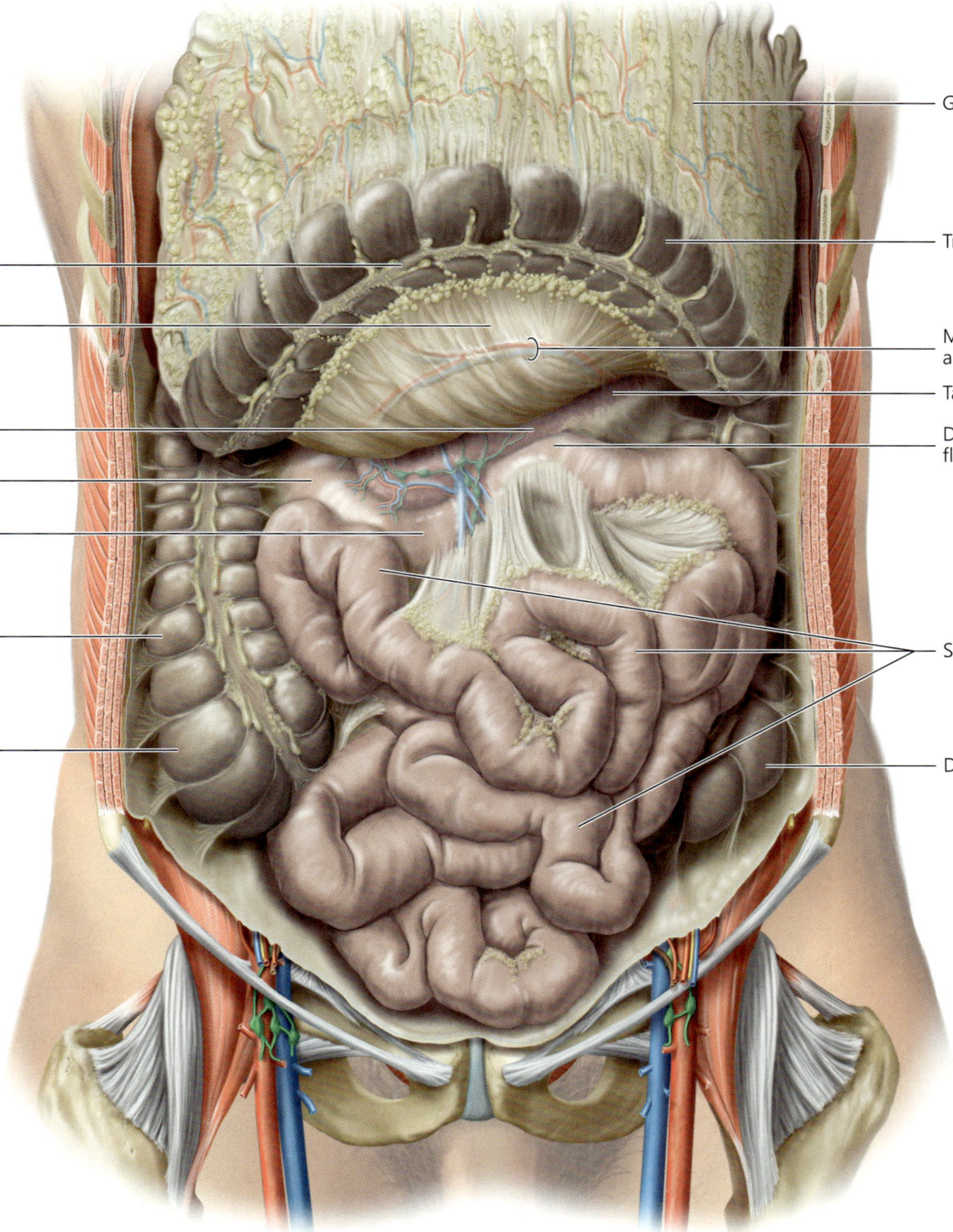

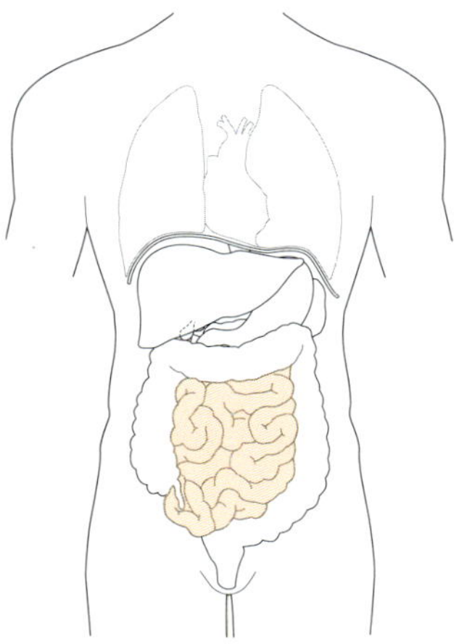

The submesocolic space is the part of the peritoneal cavity lying below the transverse mesocolon. The abdominal wall is transected and the greater omentum and transverse colon are elevated to show the arrangement of the **loops of the small intestine**. The sets of loops are packed regularly, and cover part of the ascending colon and the entire descending colon.

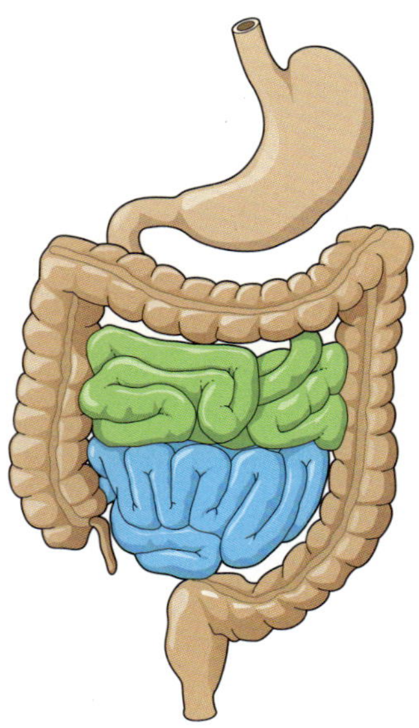

➤

The **mesenteric small intestine**, is also defined as movable portion of the small intestine because it lies within the free border of the mesentery. It is 6-7 meters in length and is arranged so as to form a number of convoluted loops that run transversally, vertically and obliquely, lying within the abdomen, below the transverse colon and the transverse mesocolon, and which, taken together, are termed inestinal "skein". It can be divided into two parts: the **jejunum** (*green*), corresponding to the proximal two-fifths, and the **ileum** (*light blue*), corresponding to the distal three-fifths; however, no real separation exists between these two segments. The mesenteric small intestine begins after the duodenum, at the duodenojejunal flexure, and terminates in the right iliac fossa, where it opens into the large intestine.

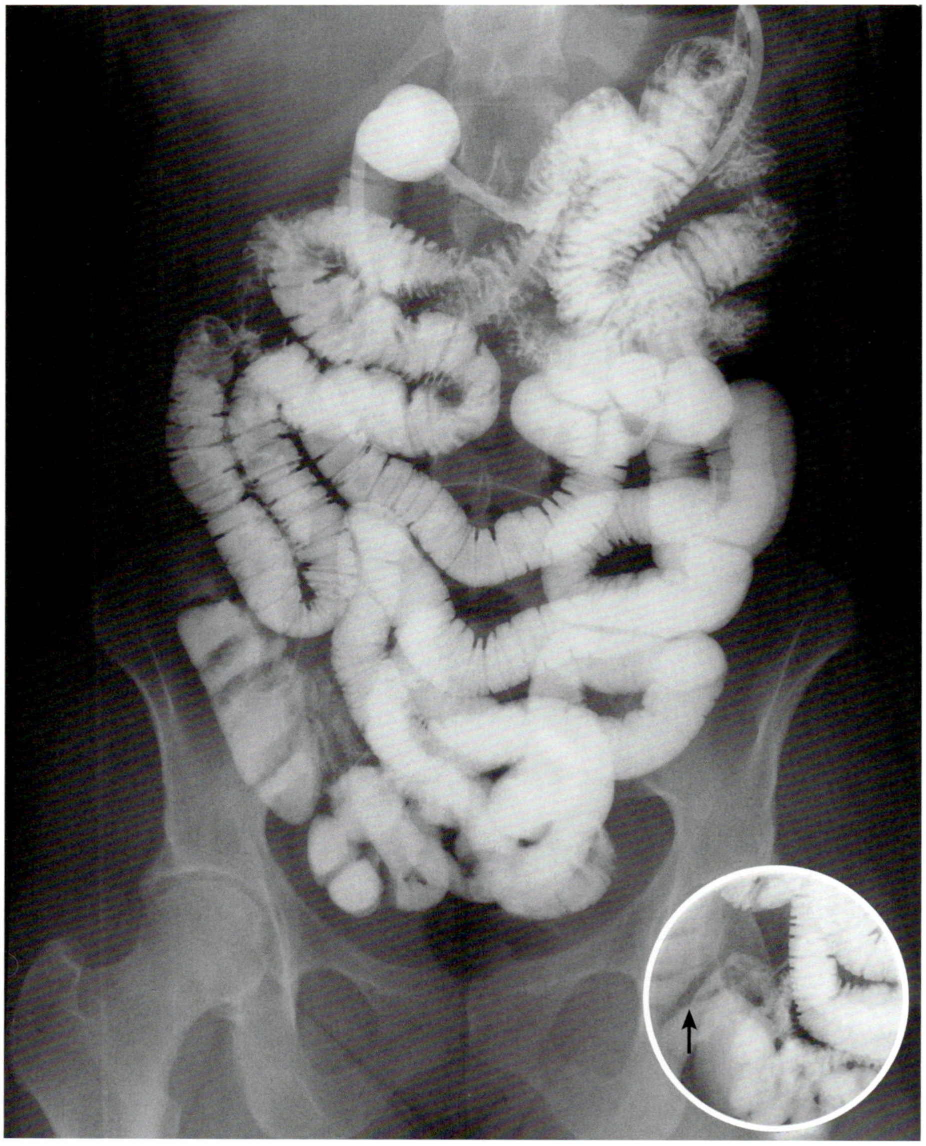

Radiographic image of the **small intestine**. The contrast medium clearly highlights the course of the intestinal loops. The notches denote the circular folds. The insert is a radiographic image of the last ileal loop, showing the two lips of the ileal orifice (arrow).

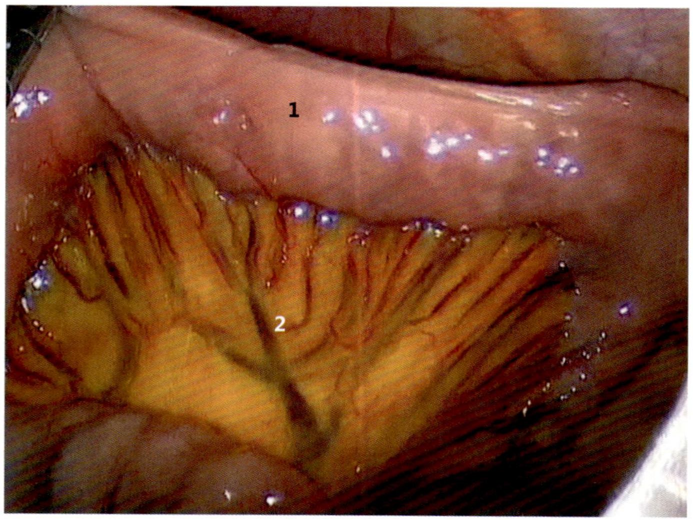

Image taken during a diagnostic laparoscopy, showing a **loop of the small intestine** with its corresponding **mesentery**, within which the terminal branches of the superior mesenteric vessels can be seen in transparency.

1 Loop of the small intestine
2 Mesentery and mesenteric vessels

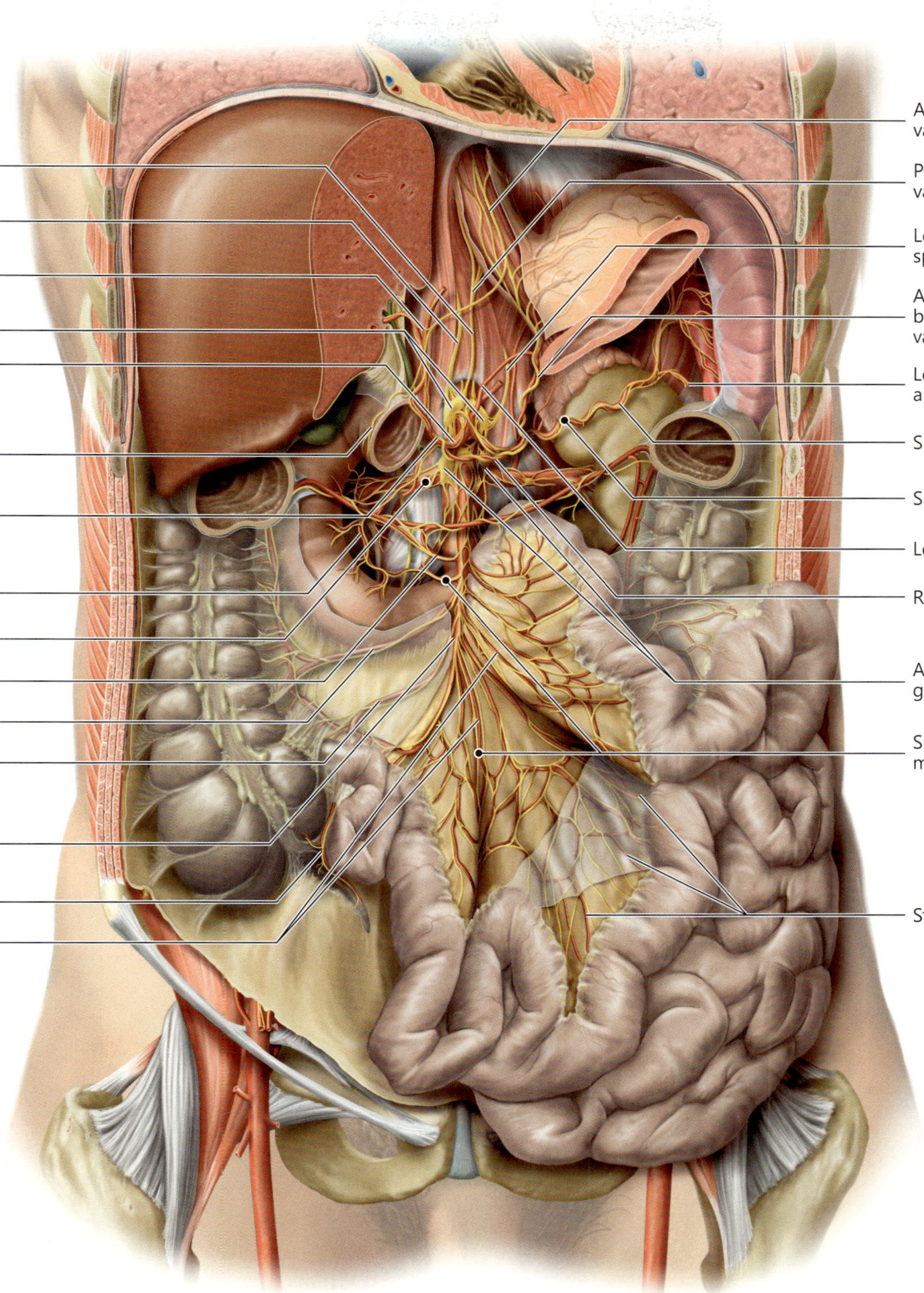

Visceral plexuses and ganglia of the abdomen. The figure shows the celiac ganglia, on either side of the celiac trunk, which receive a number of fibers of the celiac plexus, the greater and lesser splanchnic nerves and, in addition to the sympathetic afferents, the parasympathetic afferents of the celiac branch of the posterior trunk of the vagus nerve [X], which reaches the medial pole of the right celiac ganglion. The aorticorenal ganglia can be observed at the origin of the renal arteries. The figure also shows the renal, splenic and superior mesenteric plexuses, composed of the efferents of the celiac plexus around the collateral branches of the abdominal aorta. For the full course of the vagus nerve, see Volume 2, page 369 (☞ Vagus Nerve).

➤

Internal configuration of the **jejunum**. Note the regular sequence of the circular folds, which appear tightly packed.

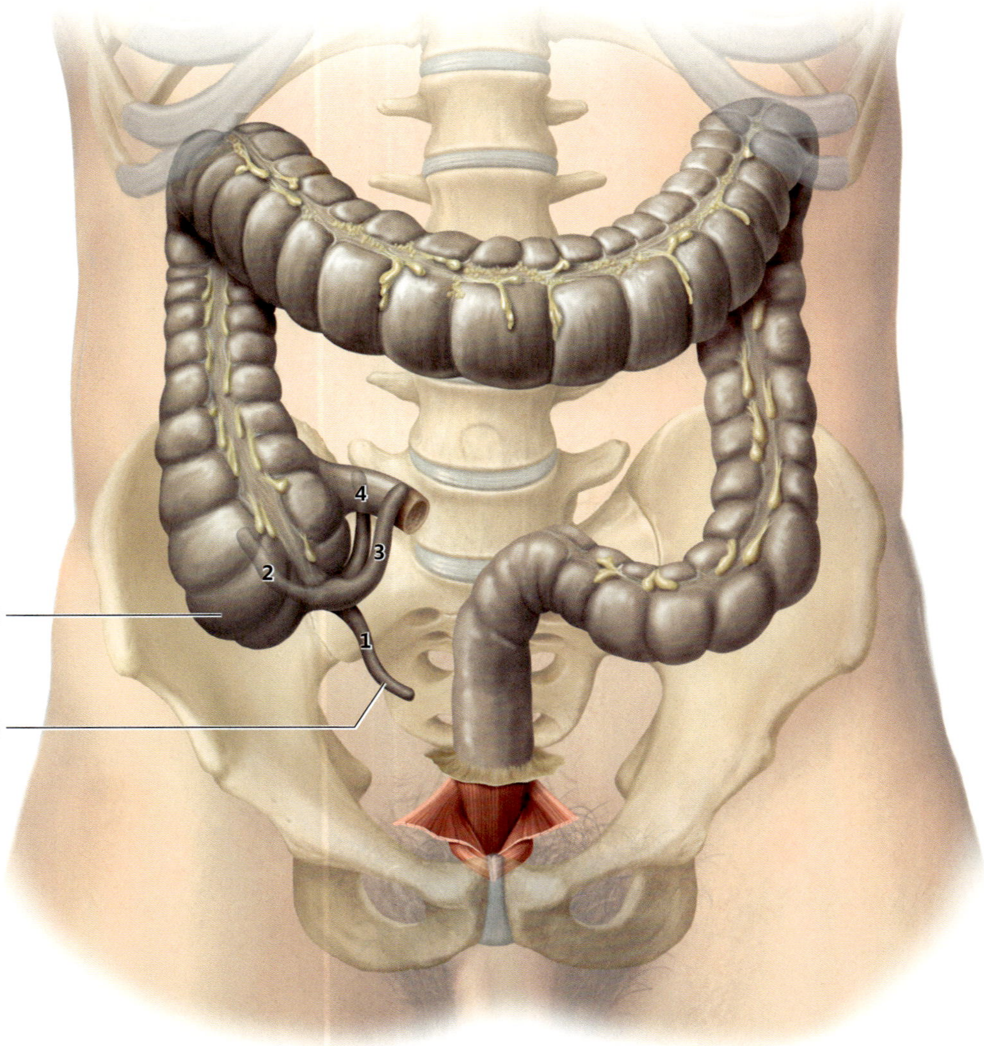

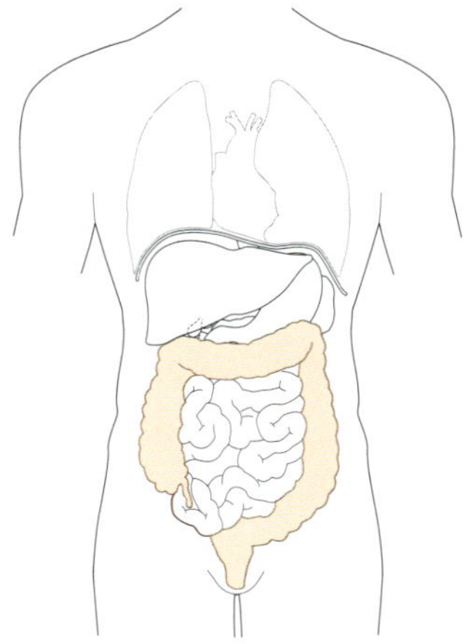

The **appendix** or **vermiform appendix** is located in the right inferior region of the abdomen, and appears as a continuation of the fundus of the cecum that protrudes into the abdominal cavity and might reach as far as the pelvis. The blind end of the appendix can move freely into different positions:

1 pelvic: the appendix is direct downwards, 20%
2 retrocecal/retrocolic: the appendix courses upwards, 15%
3 pre-ileal: the appendix is direct outwards, 2-5%
4 post-ileal: the appendix courses inwards, 2-5%

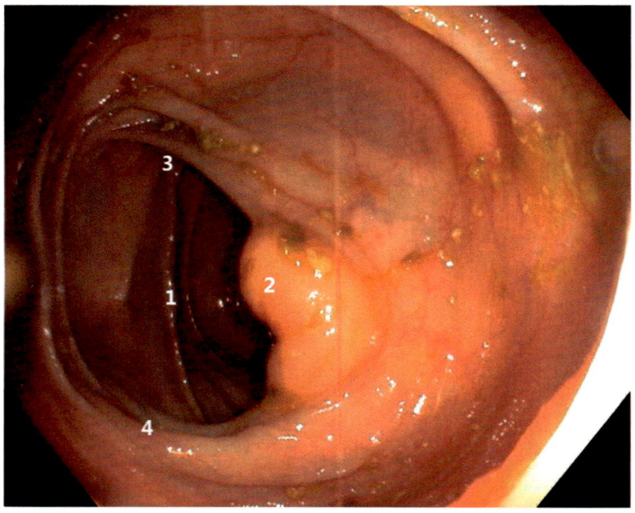

Image taken during a diagnostic colonoscopy, showing the **ileal orifice**.

1 Lumen of the ascending colon
2 Commissure of the ileal orifice
3 Ileocolic lip
4 Ileocecal lip

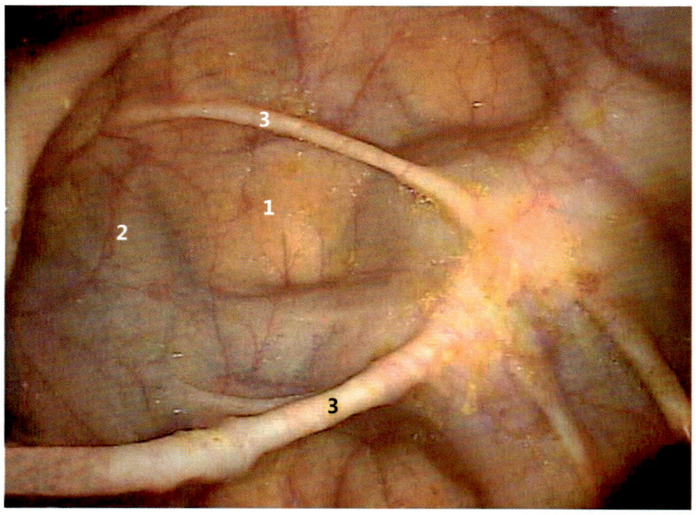

Image of the **hepatic flexure of the colon**, or colic flexure, taken during a diagnostic colonoscopy. Note that the mucosa is almost entirely smooth, as circular folds are not present in the large intestine.

1 Colic mucosa
2 Subcutaneous vascular network
3 Semilunar folds

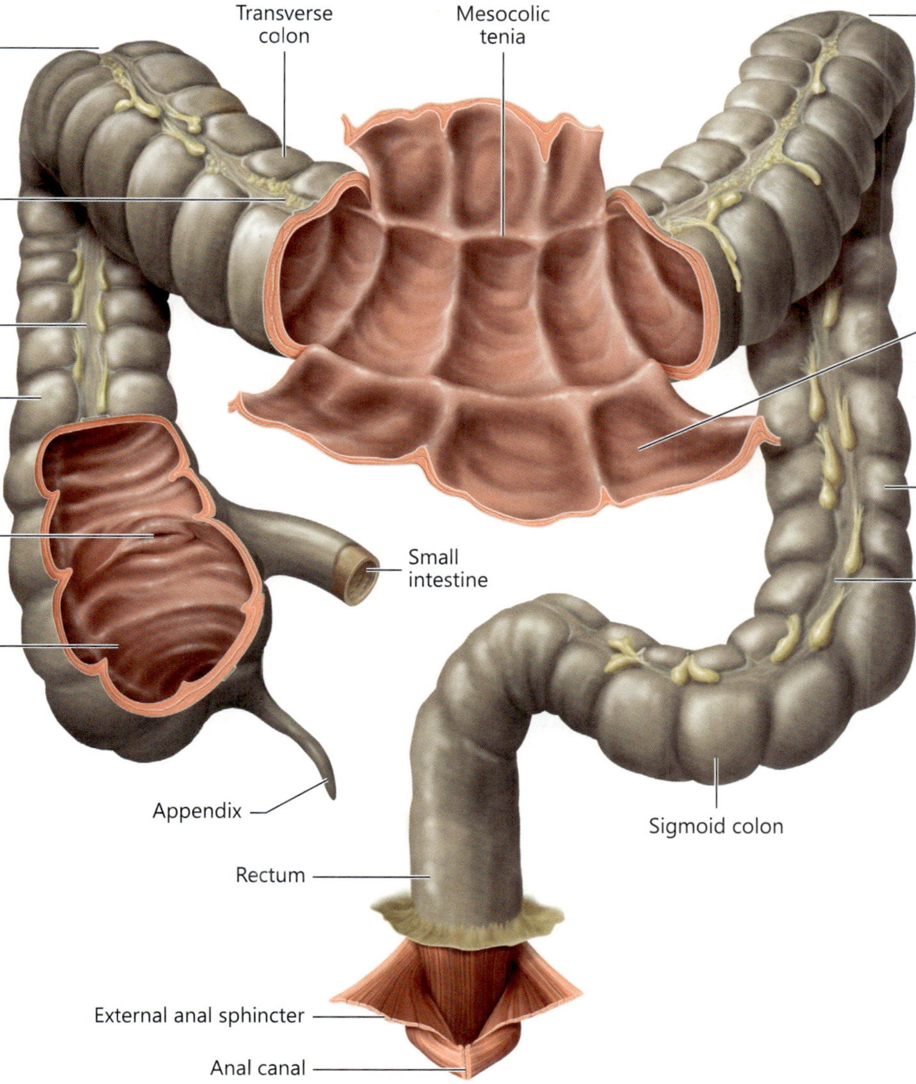

Internal and external conformation of the **large intestine**. The organ extends between the terminal loop of the ileum and the anal canal.

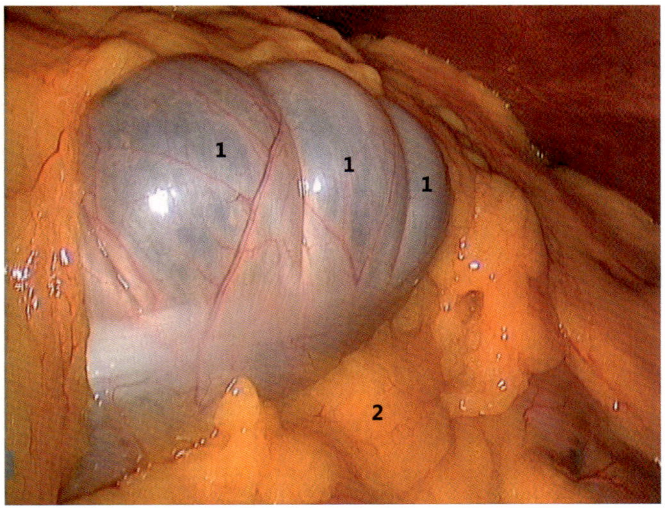

Image taken during a diagnostic laparoscopy, showing the **transverse colon** and the **greater omentum**.

1 Humps corresponding, on the internal surface, to the haustrae of the colon
2 Greater omentum

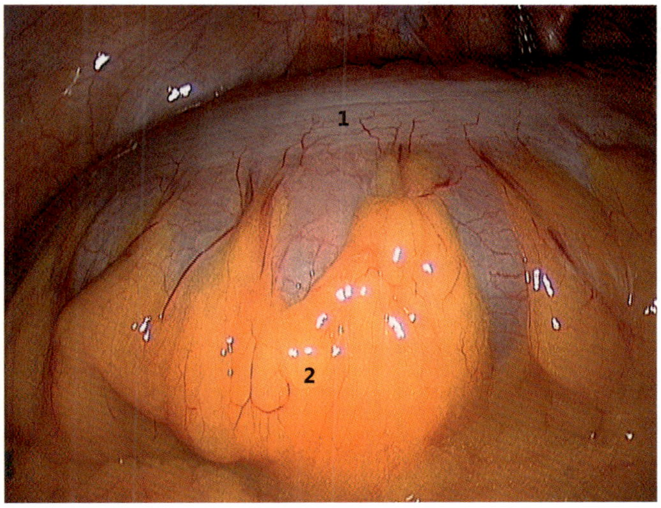

Image of the **descending colon** taken during a diagnostic laparoscopy, where the free tenia and the omental appendices can be observed.

1 Tenia coli
2 Omental appendices

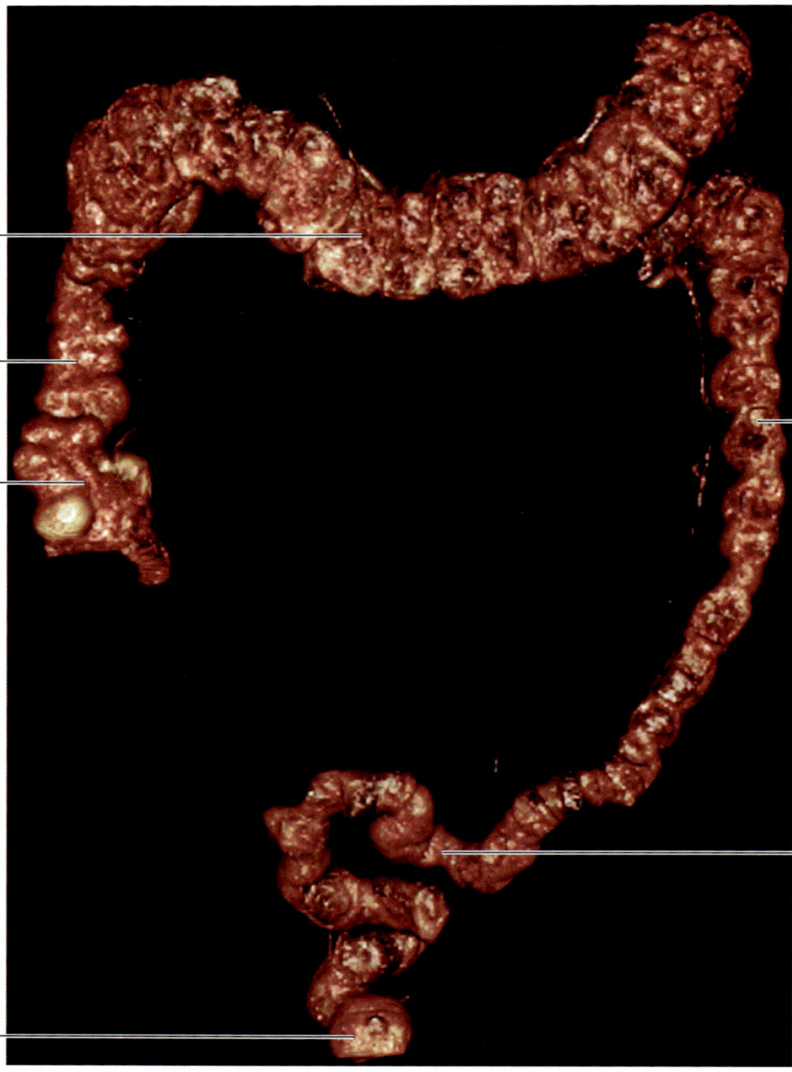

Three-dimensional image of a living human body, obtained by CT scan using a 3D volume rendering technique, showing the different parts of the **large intestine** in frontal projection.

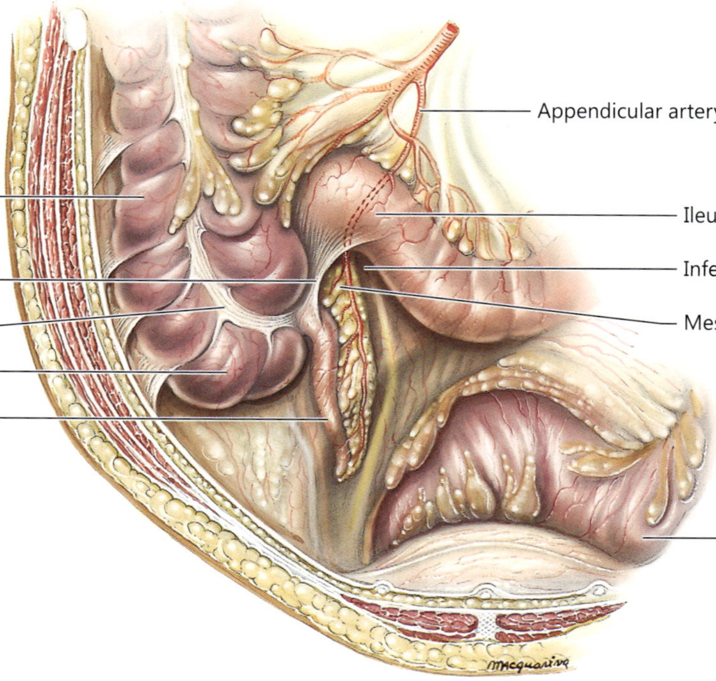

Relations of the **cecum** and of the **appendix** in the right iliac fossa. The figure also shows the main peritoneal structures and the arteries supplying the terminal segment of the ileum, the cecum and the appendix.

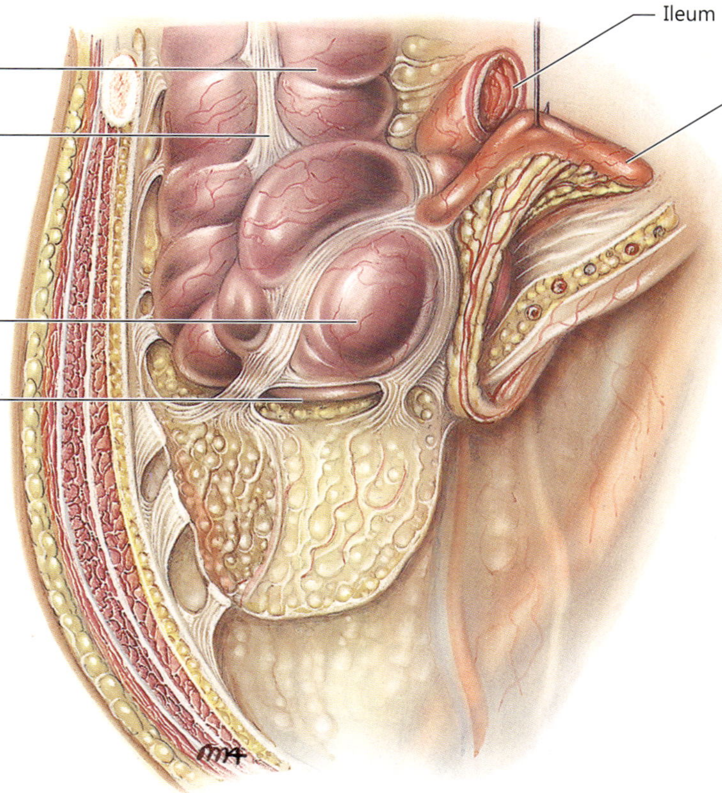

The **cecum** and the **appendix** have been elevated to expose how the parietal peritoneum reflects (from the posterior surface of the cecum) onto the visceral peritoneum. Retrocecal recesses can be found in this transitional zone.

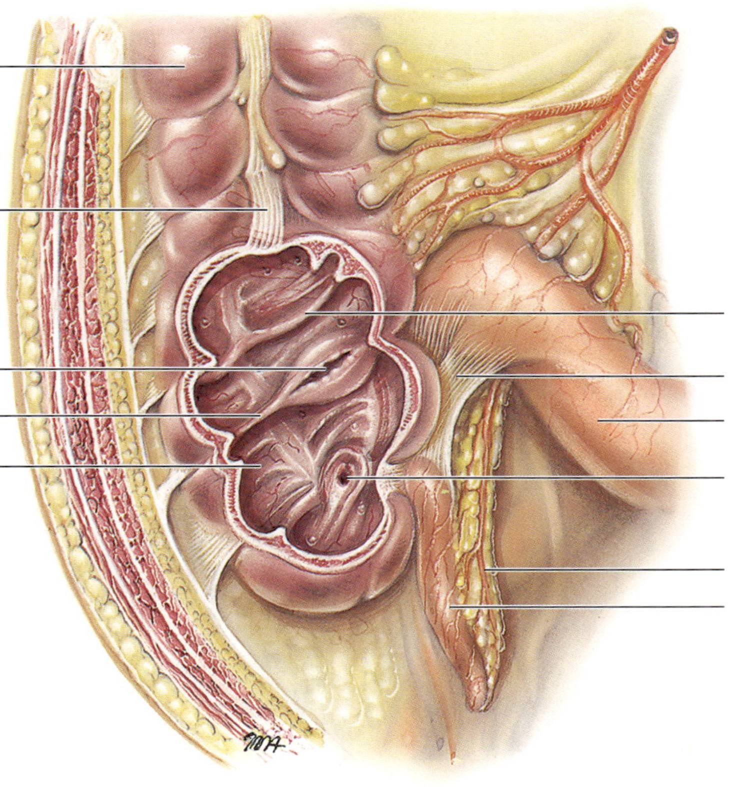

Internal configuration of the **cecum**, after removal of the anterior wall. At the border between the posterior and the medial wall, the ileal orifice can be observed superiorly, while the orifice of the vermiform appendix is found inferiorly.

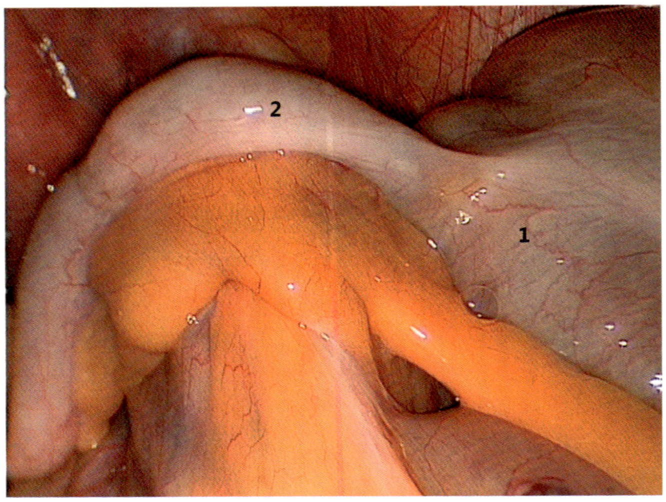

Image taken during a diagnostic laparoscopy, showing the **cecum** and the **appendix**.

1 Cecum
2 Appendix

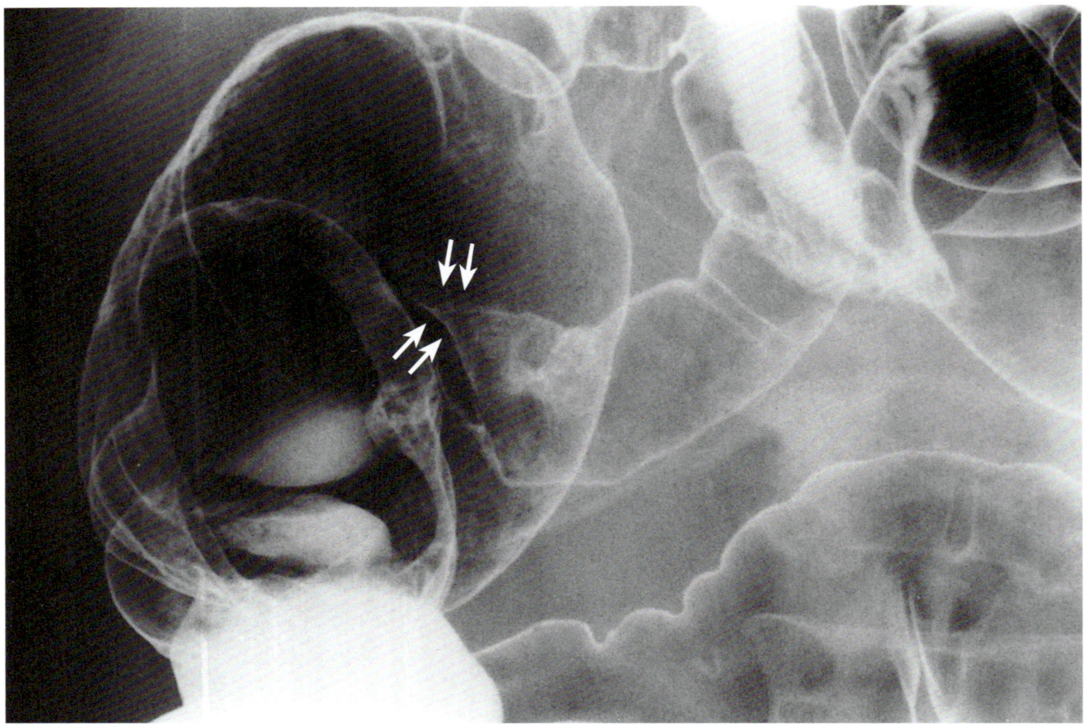

Double-contrast radiography showing the last **ileal loop** and the **cecum**. The arrows indicate the lips of the ileal orifice.

The **left colic flexure** bends at an acute angle, opening downward, forward and medially; it is located in the left hypochondrium, and is related to the anterior surface of the left kidney, posteriorly, and to the anterior extremity of the spleen, laterally. The figure also shows the posterior relations of the body of the pancreas to the terminal part of the inferior mesenteric vein, the superior pole of the left kidney and the left renal vein.

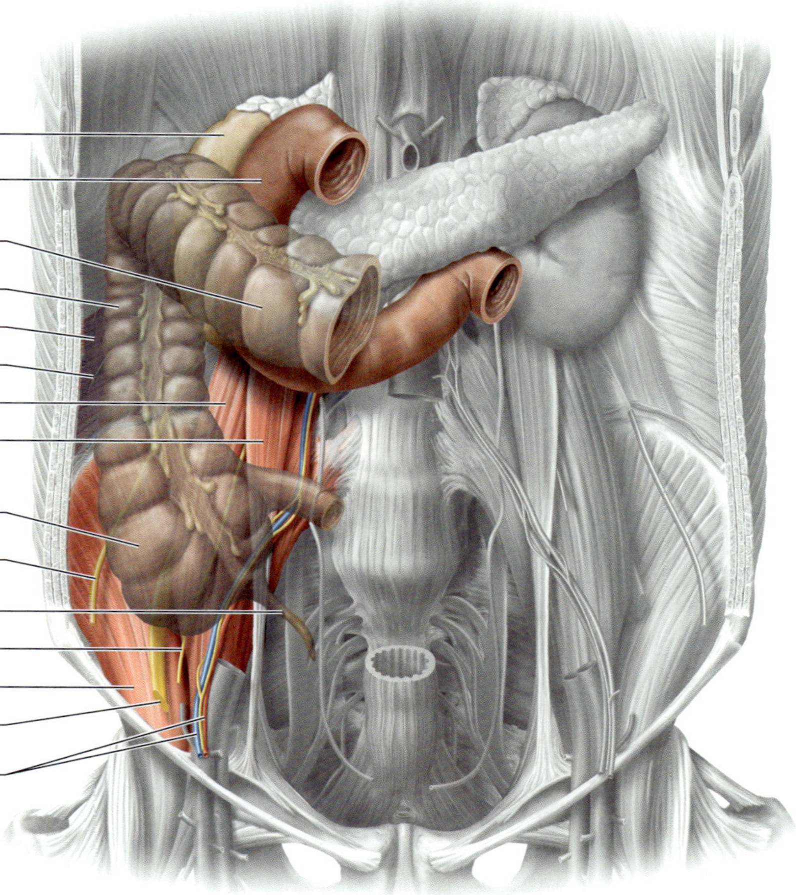

The **cecum** is located in the right iliac fossa. The organ is displayed in transparency to show its posterior relations. Its posterior and lateral surfaces are related to the iliacus, through the parietal peritoneum that invests the iliac fossa and the iliac fascia below the peritoneum. The medial surface is related to the psoas major and psoas minor, and receives the opening of the ileum in its superior part.

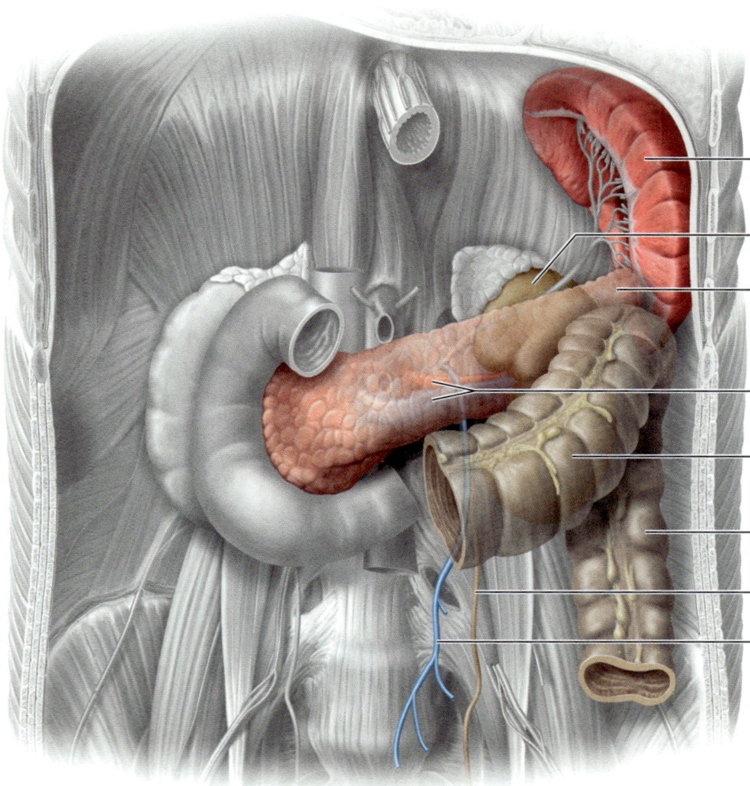

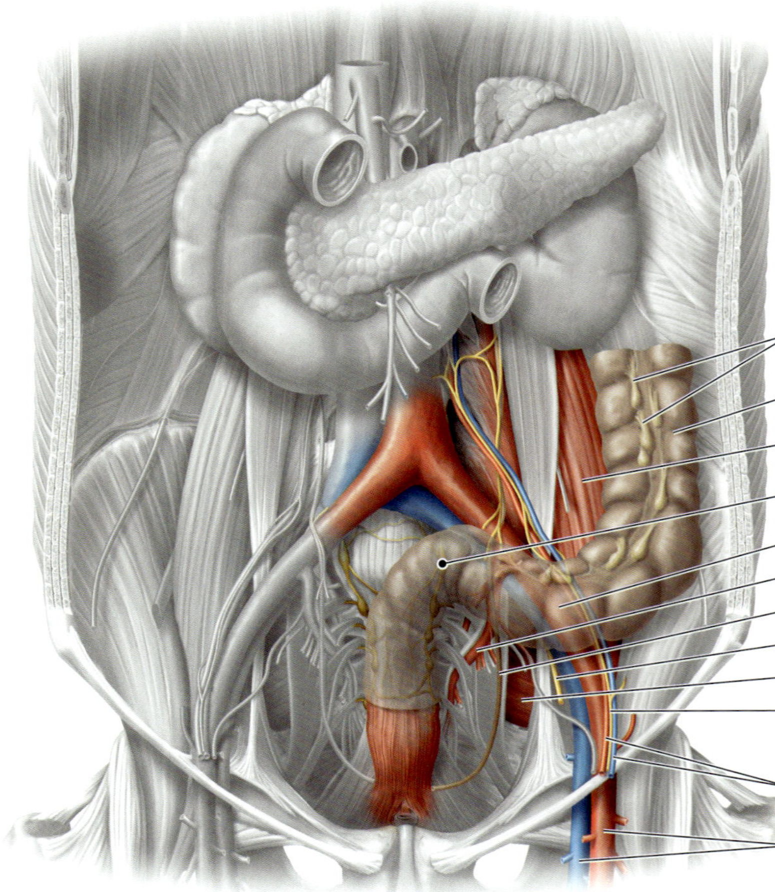

The **descending colon** travels downward, to reach the left iliac crest, where it continues as the sigmoid colon. The latter is divided, based on its course, into an iliac part, up to the medial border of the psoas major, and a pelvic part, which travels lateromedially against the posterior wall of the lesser pelvis, up to the third sacral vertebra, where it continues as the rectum.

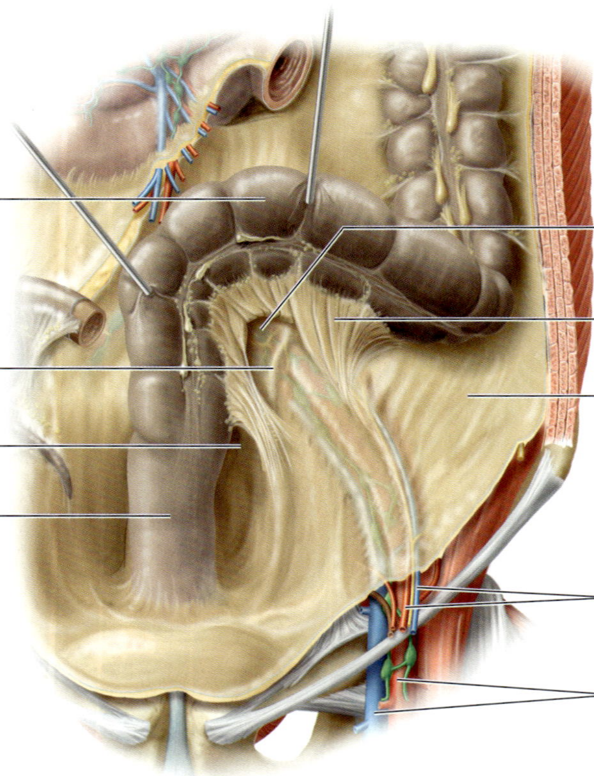

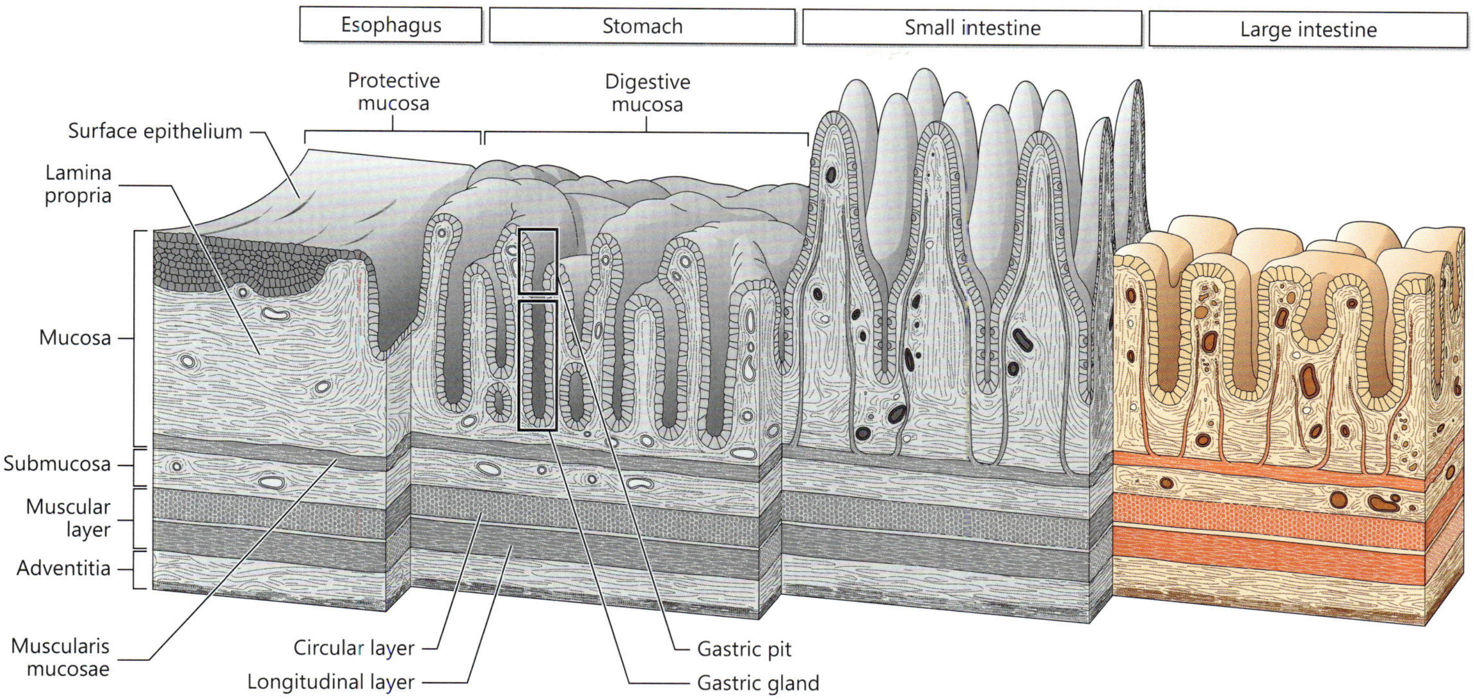

Summary of the **general organization of the hollow organs of the alimentary system** (esophagus, stomach, small intestine, large intestine). The portion of the large intestine is highlighted in color. The surface epithelium is simple columnar epithelium in the large intestine.

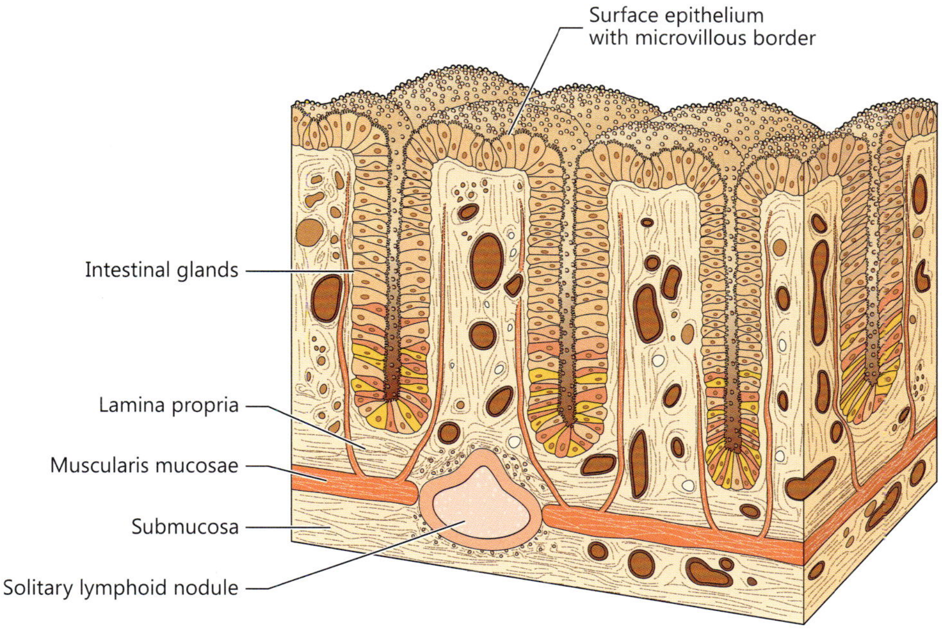

Diagram of the organization of the **colic mucosa**. The lining epithelium is discontinuous because of the glandular openings.

◄
The **sigmoid mesocolon** is rooted to the posterior abdominopelvic wall in an inverted-V shape, and is divided into a left and a right arm. The intersigmoid recess, located at the medial border of the psoas major, is bounded by the two segments of the parietal line of insertion of the sigmoid mesocolon, at the left common iliac artery level.

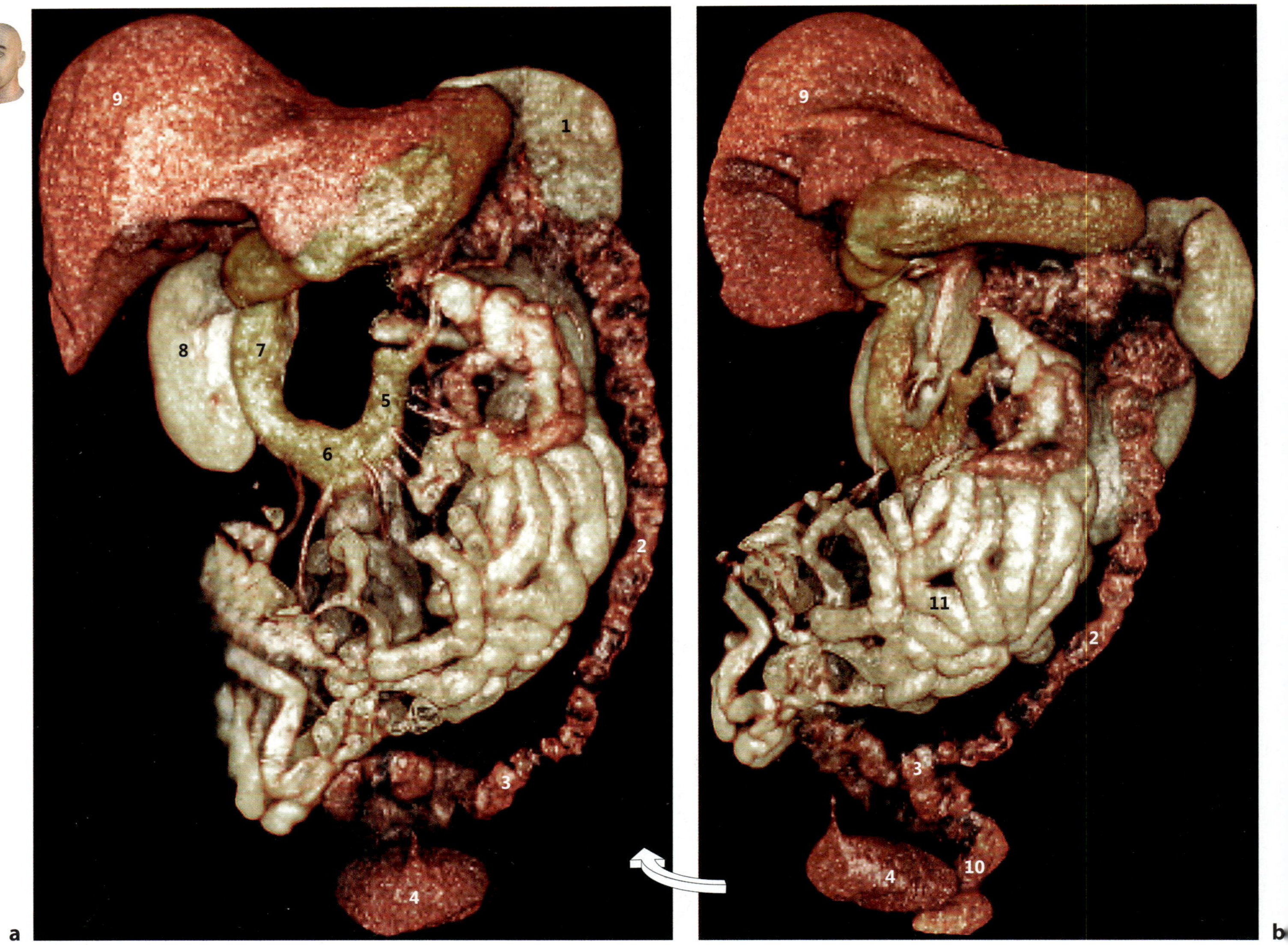

Three-dimensional image of a living human body, obtained by CT scan using a 3D volume rendering technique, showing the **organs contained in the abdominal cavity** in anterior (**a**), left lateral (**b**), right lateral (**c**) and posterior (**d**) projection.

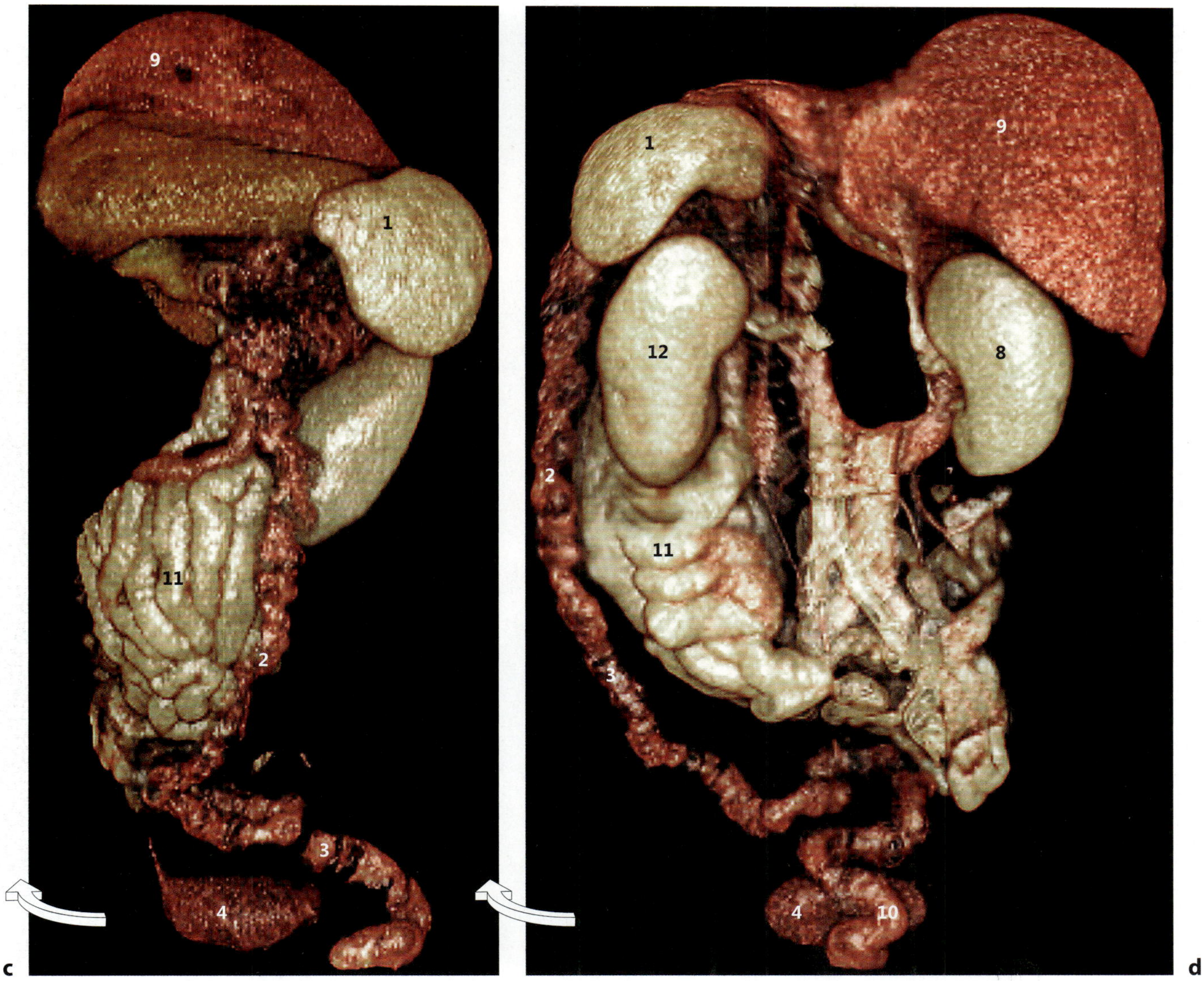

1 Spleen	7 Descending part of the duodenum
2 Descending colon	8 Right kidney
3 Sigmoid colon	9 Liver
4 Urinary bladder	10 Rectum
5 Ascending part of the duodenum	11 Mesenteric small intestine
6 Inferior part of the duodenum	12 Left kidney

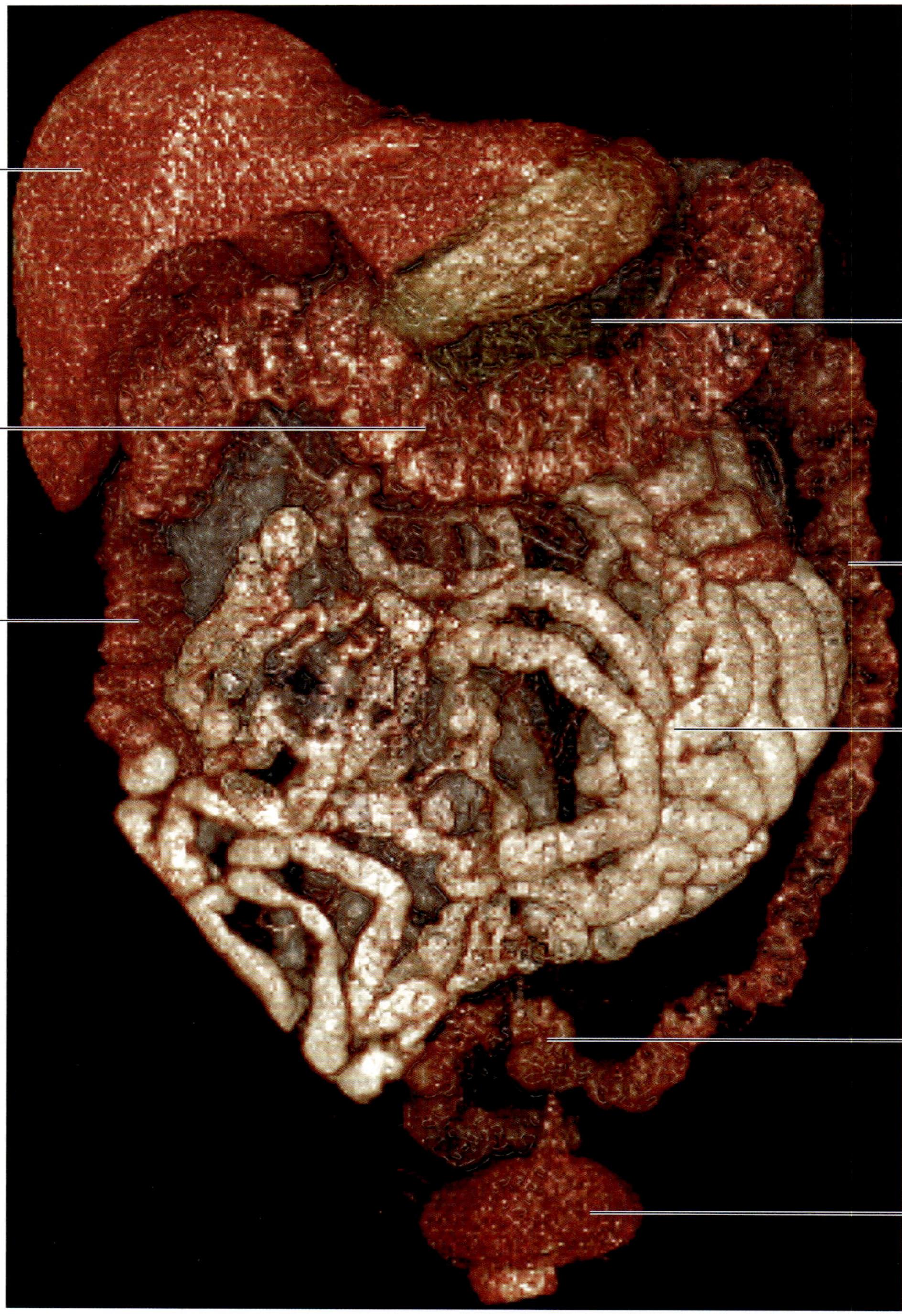

Three-dimensional image of a living human body, obtained by CT scan using a 3D volume rendering technique, showing the **organs contained in the abdominal cavity** in frontal view.

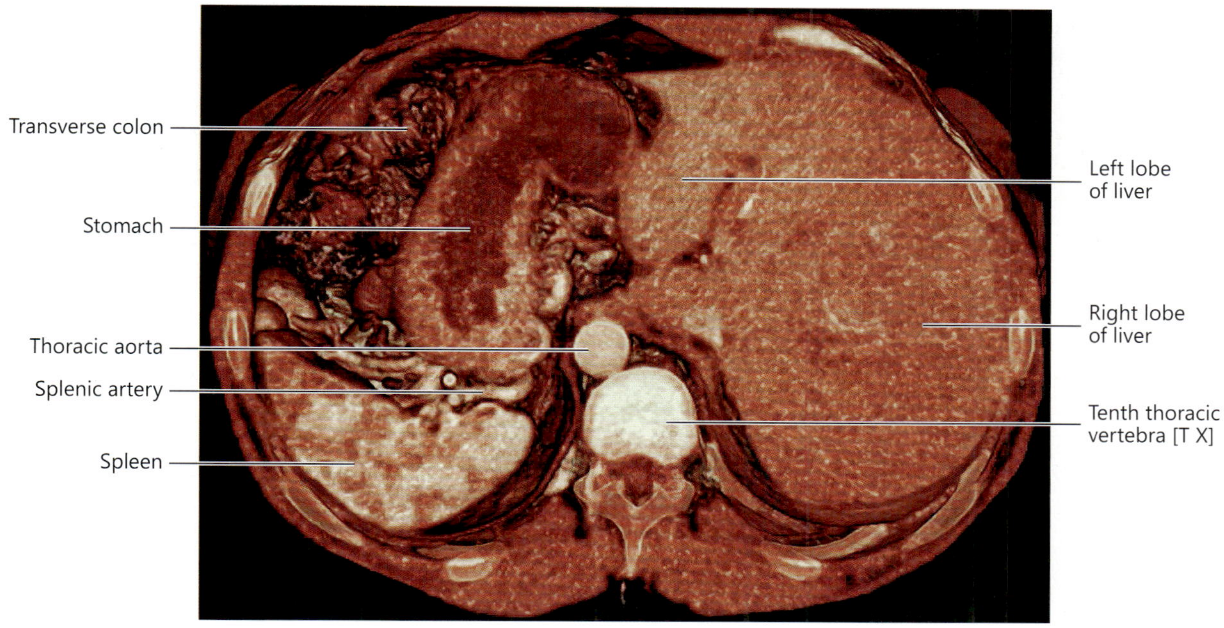

Three-dimensional image of a living human body, obtained by CT scan using a 3D volume rendering technique, showing the **organs of the supramesocolic space** in a transverse section passing through the tenth thoracic vertebra.

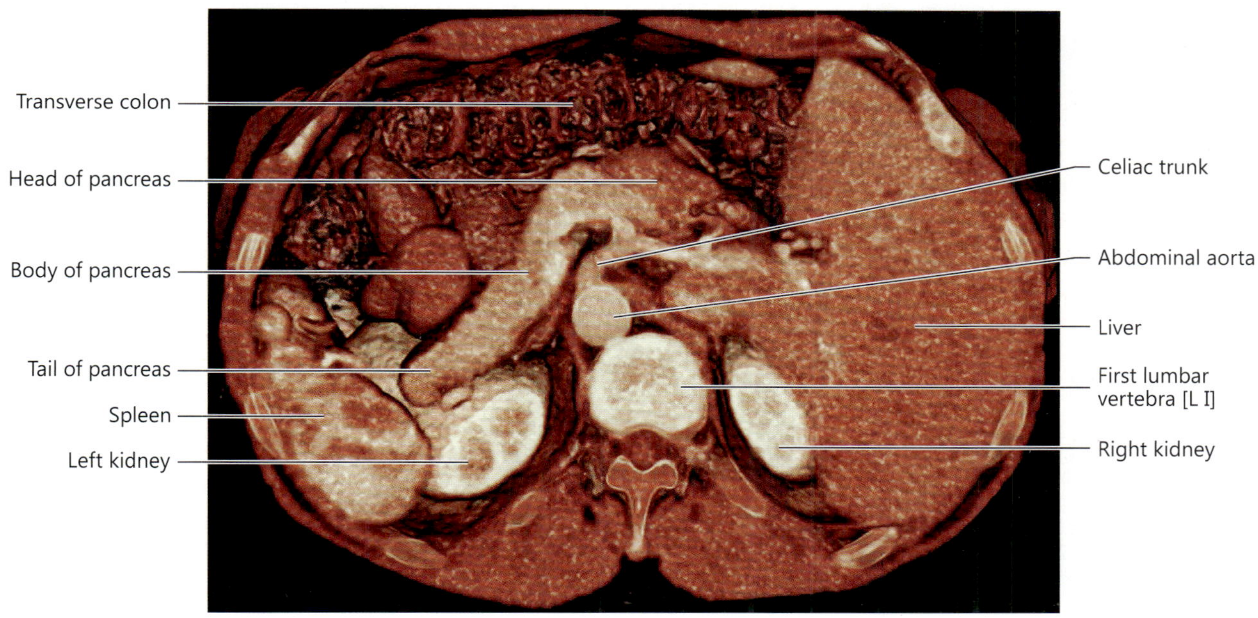

Three-dimensional image of a living human body, obtained by CT scan using a 3D volume rendering technique, showing some of the **abdominal organs** in a transverse section passing through the first lumbar vertebra.

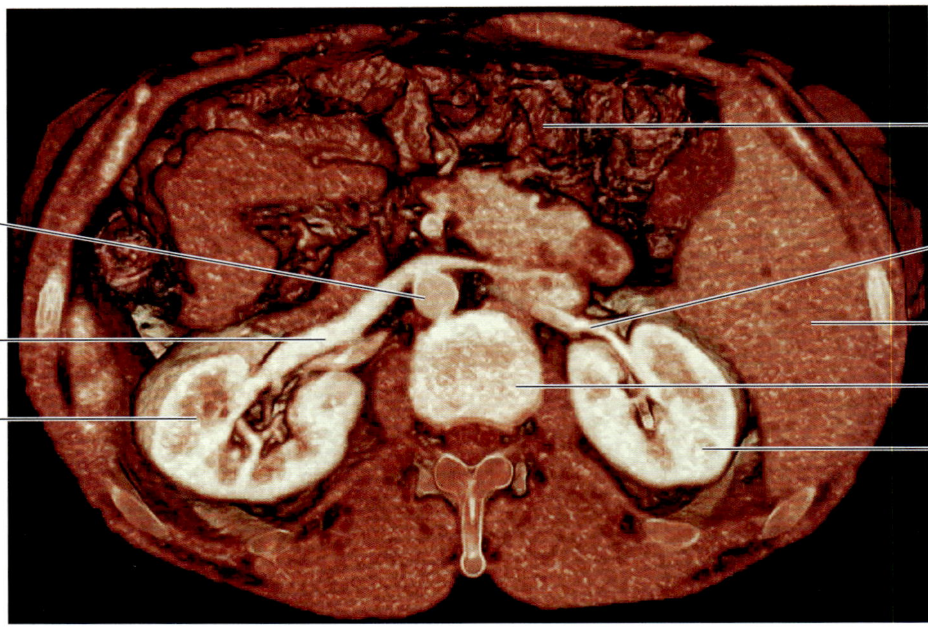

Three-dimensional image of a living human body, obtained by CT scan using a 3D volume rendering technique, showing some of the **abdominal organs** in a transverse section passing through the second lumbar vertebra.

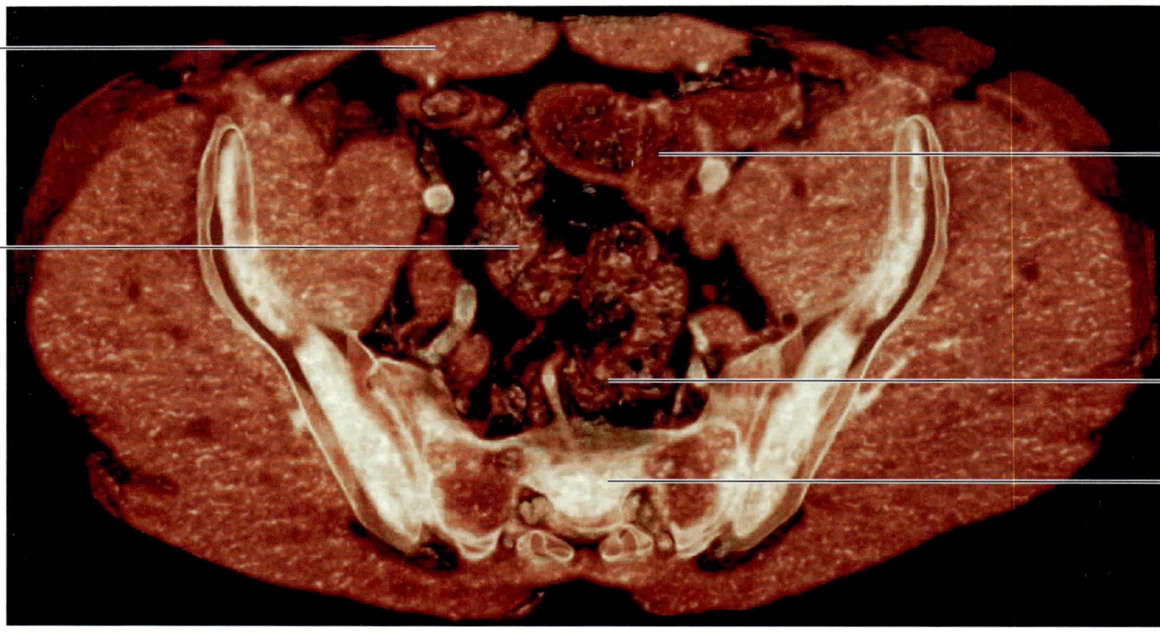

Three-dimensional image of a living human body, obtained by CT scan using a 3D volume rendering technique, showing some of the **organs contained in the pelvic cavity** in a transverse section passing between the first and the second sacral vertebrae.

PELVIS AND PERINEUM

Petros (Athens, 1928 - Milan, 2014), *P33*, oil on canvas, 50 × 70 cm, 1997

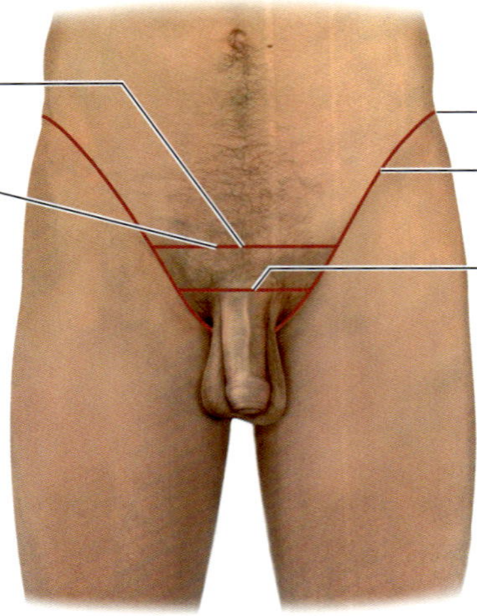

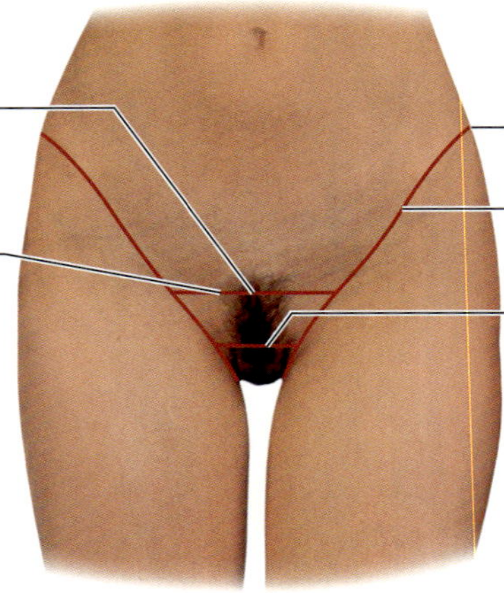

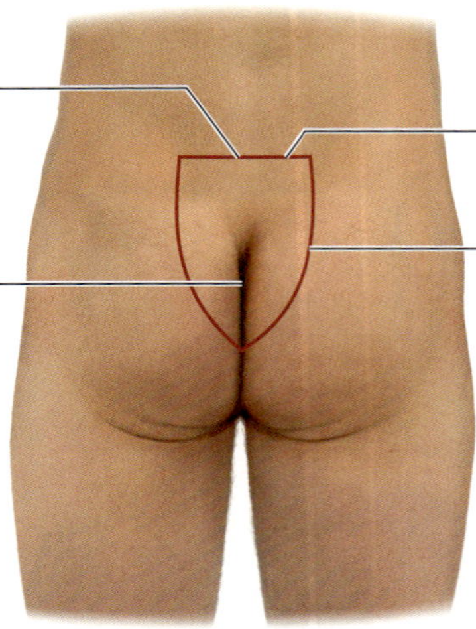

Projection of the boundaries of the pelvis on the body surface, in the male (**a** and **c**) and female (**b**). **a** and **b**, Anteriorly, the boundary between abdomen and pelvis corresponds to an horizontal line between the two pubic tubercles, separating the hypogastric region (abdomen) from the pubic region (pubis), while the boundary between pelvis and perineum corresponds to a plane passing through a line between the inferior pubic tubercles and the tip of the coccyx. The inguinal folds below the pubic tubercles define the superficial boundary between the pelvis and the lower limbs. **c**, Posteriorly, the boundary between abdomen and pelvis corresponds to a line passing through the lumbosacral joint, separating the lumbar segment of the vertebral column (abdomen) from the sacrococcygeal region (pelvis), while the boundary between pelvis and perineum is represented by the tip of the coccyx. The lateral margin of the sacrum marks posteriorly the boundary between pelvis and lower limb.

Regions of the pelvis and perineum

REGION	***BOUNDARIES OR LOCATION***
Pubic region (inferiorly to the hypogastric region or hypogastrium)	*Superior boundary*: horizontal line between the two pubic tubercles; *lateral boundaries*: inguinal folds; *inferior boundary*: pubic symphysis
Sacral region (or sacrococcygeal region)	*Superior boundary*: horizontal line passing through the lumbosacral joint; *lateral boundaries*: lateral margins of the sacrum; *inferior boundary*: tip of coccyx
Urogenital triangle (anterior perineum)	*Anterior angle*: subpubic angle; *sides of triangle*: ischiopubic branches; *base of triangle*: transverse line between the ischial tuberosities
Anal triangle (posterior perineum)	*Posterior angle*: tip of coccyx; *sides of triangle*: sacrotuberous ligaments; *base of triangle*: transverse line between the ischial tuberosities

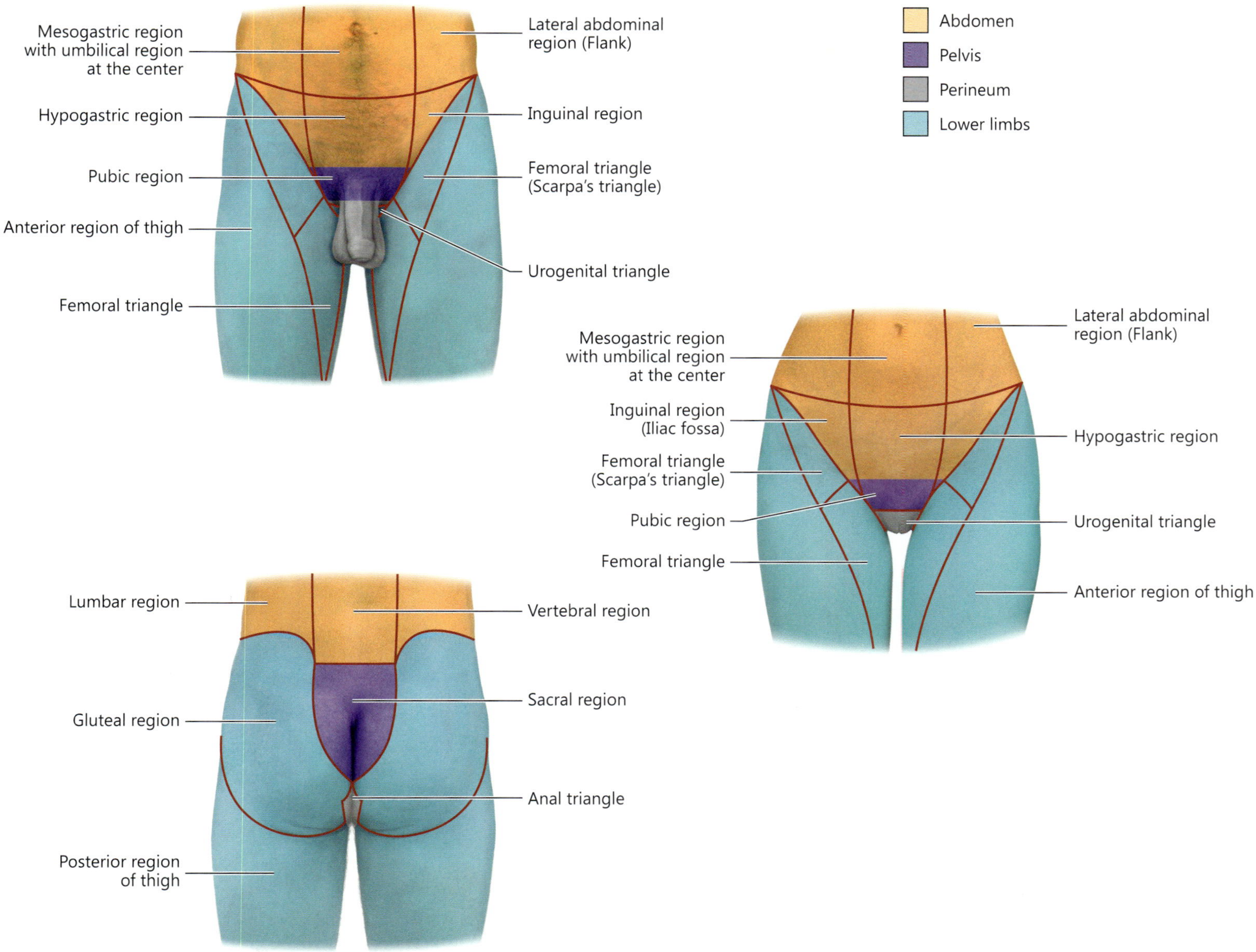

Regions of the pelvis and of the perineum in the male and in the female. In the pelvis, they are the pubic region, below the hypogastric region (or hypogastrium), and the sacral region. In the perineum, they are the urogenital and anal regions, or triangles (anterior perineum and posterior perineum). The external genitalia attach to the perineum. The pubic region anteriorly and the sacral region posteriorly delimit superficially the lesser pelvis, the most caudal space of the abdominopelvic cavity (☞ page 4).

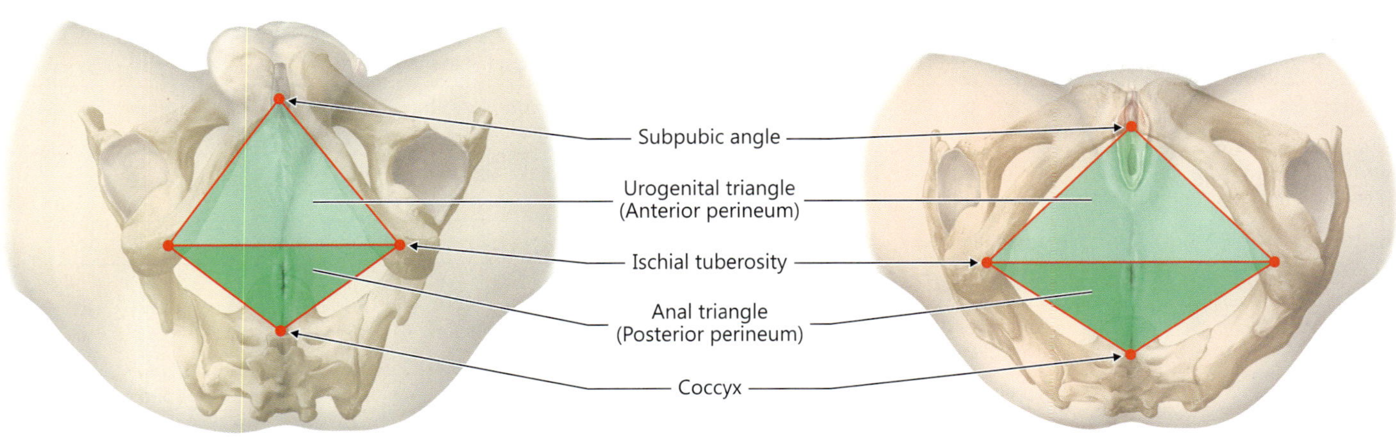

The **perineum** is composed of the soft tissue that closes the bottom of the pelvic cavity and can be subdivided into two triangles: the urogenital triangle, anteriorly, and the anal triangle, posteriorly. Attached to the posterior perineum is the anal canal (the terminal tract of the rectum); attached to the anterior perineum are the external genitalia.

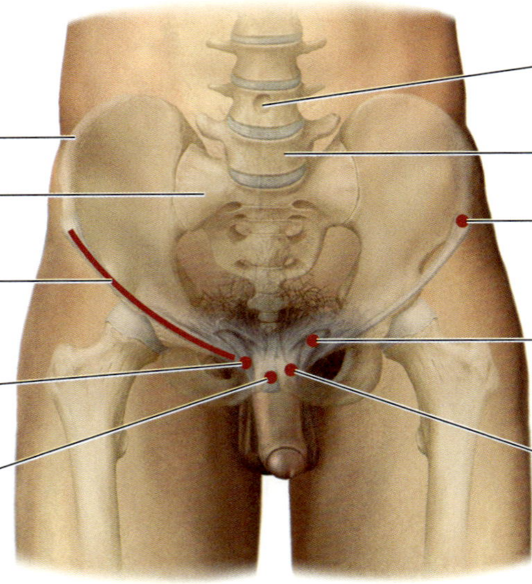

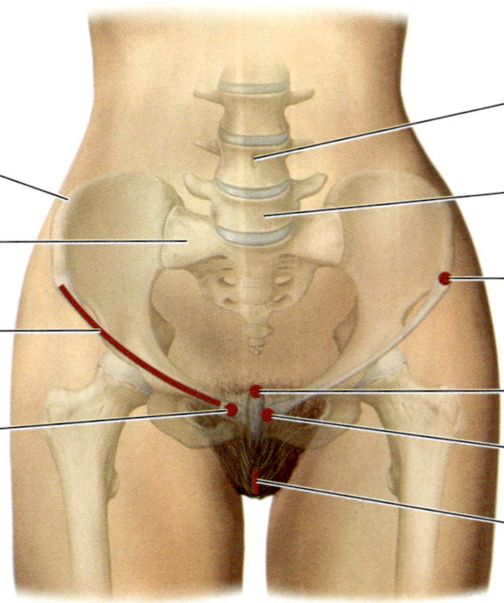

Landmarks of the pelvis	
STRUCTURE	***LOCATION***
Anterior superior iliac spine	Palpable projection through the integuments with which the iliac crest terminates in front
Fifth lumbar vertebra [L V]	The largest and most inferior of the lumbar vertebrae, articulated with the sacrum
Iliac crest	Superior margin of the ala of ilium corresponding to the superolateral margin of the greater pelvis
Inguinal ligament	Fibrous band which extends from the anterior superior iliac spine to the pubic tubercle
Mons pubis	Eminence formed of subcutaneous adipose tissue which continues above as the skin of the pubic region, below as the labia majora, and is bordered laterally by the inguinal folds
Posterior commissure of labia majora	Union, on the median line, of the posterior extremities of the labia majora
Pubic crest	Anterior margin of the body of the pubis between the pubic tubercle and pubic symphysis
Pubic symphysis	Joint binding anteriorly the two hip bones
Pubic tubercle	Eminence on the superior margin of the superior pubic ramus, laterally to the pubic crest, to which the inguinal ligament is attached
Sacrum	Formed by the fusion of the five sacral vertebrae, it constitutes together with the coccyx and hip bone the pelvic skeleton
Spermatic cord	Cord-like structure composed of the ductus deferens and accompanying vessels and nerves, which leads from the abdominal cavity into the scrotum through the inguinal canal
Umbilical region	Cicatricial formation located on the anterior surface of the abdomen, along the linea alba

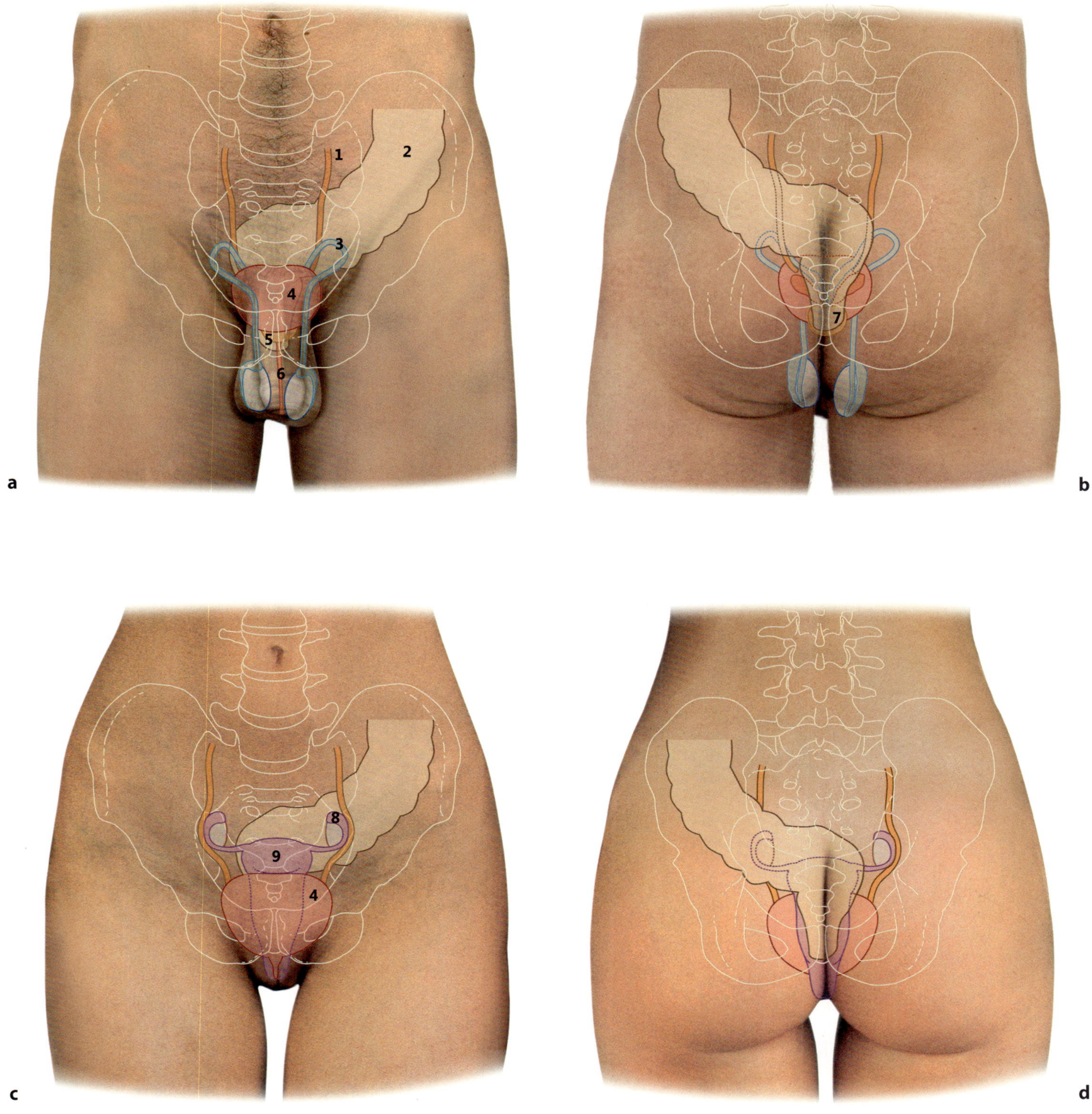

1 Ureter
2 Sigmoid colon
3 Ductus (vas) deferens
4 Urinary bladder
5 Prostate
6 Urethra
7 Rectum
8 Uterine tube
9 Uterus

Projection of the pelvic organs on the walls of the trunk: anteriorly (**a**) and posteriorly (**b**) in the male; anteriorly (**c**) and posteriorly (**d**) in the female.

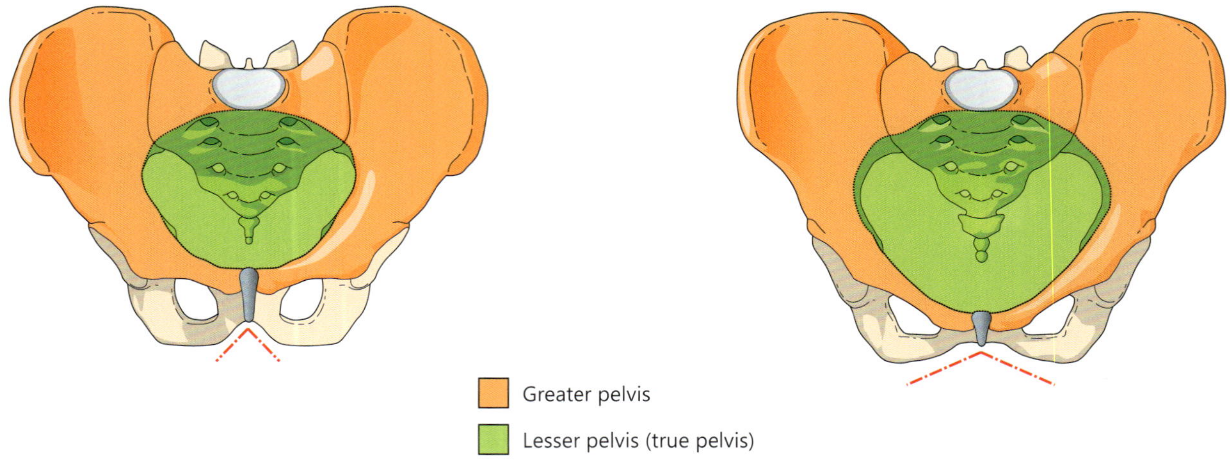

a b

The pelvis is subdivided into two parts: the **greater pelvis**, which belongs to the abdomen, and the **lesser pelvis**. The plane separating the two parts is inclined approximately 60° with respect to the horizontal plane and corresponds to the pelvic inlet. The superior boundary of the greater pelvis is represented by a plane passing tangential to the iliac crests of the two hip bones. The inferior boundary of the lesser pelvis is represented by the pelvic outlet.

The male pelvis (**a**) extends more vertically than the female one (**b**) which has larger transverse diameters and dimensions of the interior of the pelvic cavity. Furthermore in the female compared to the male pelvis, the plane of the pelvic inlet is more inclined by about 5 degrees with respect to the horizontal plane, the obturator foramen has a triangular shape (oval in the male), the alae of the iliac bones are more inclined laterally, the pubic symphysis has a smaller vertical diameter and the ischio-pubic rami are spaced further apart, about 110 vs. 70 degrees (namely pubic arch in the female and subpubic angle in the male).

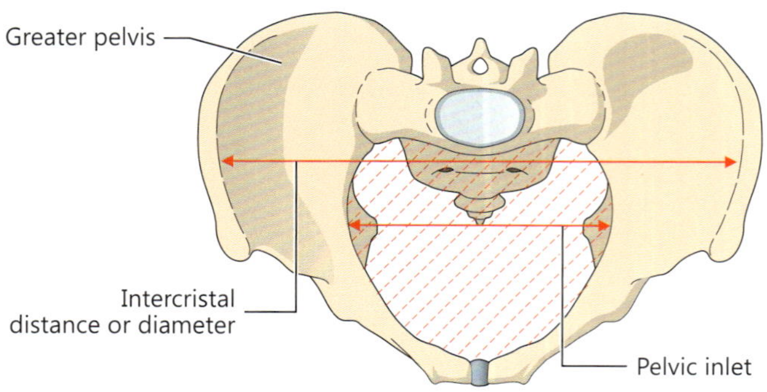

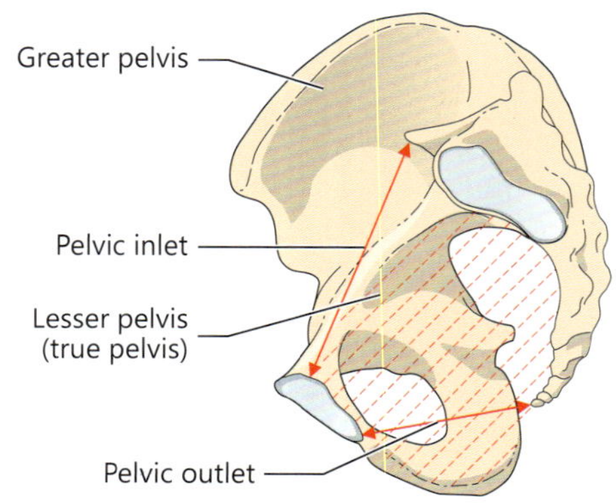

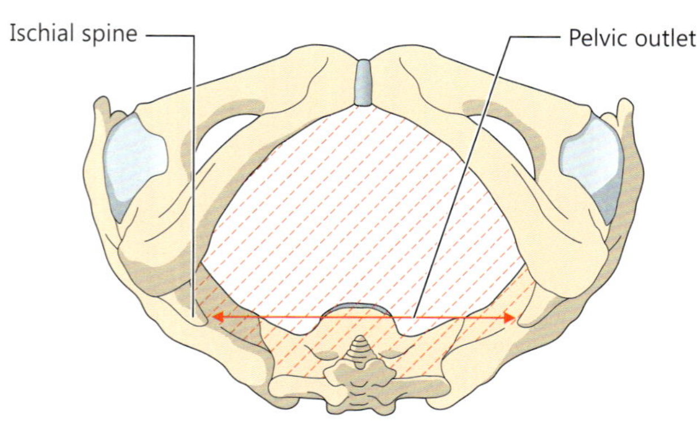

The **pelvic inlet** has an oval-shaped circumference formed by: the promontory of the sacrum, the anterior margin of the ala of the sacrum, the sacroiliac joint, the arcuate line of the ilium, the iliopubic ramus, the pecten pubis or pectineal line, the pubic tubercle, the pubic crest and the superior margin of the pubic symphysis. The **pelvic outlet** is delimited by the inferior margin of the pubic symphysis, the ischial tuberosity and the tip of the coccyx.

Schematic representation of the **pelvic organs** in a median sagittal section of the male pelvis, viewed from the left side.

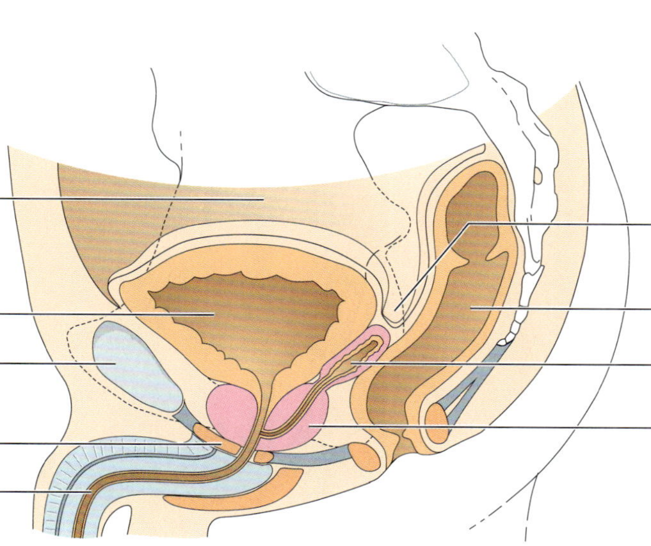

Schematic representation of the **pelvic organs** in a median sagittal section of the female pelvis, viewed from the left side.

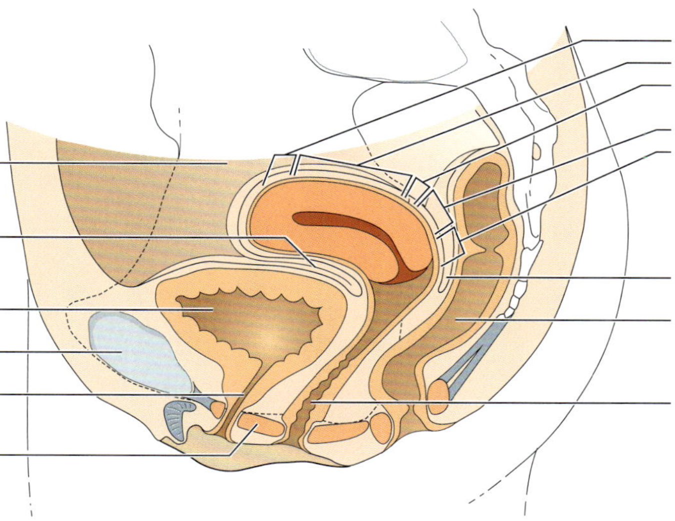

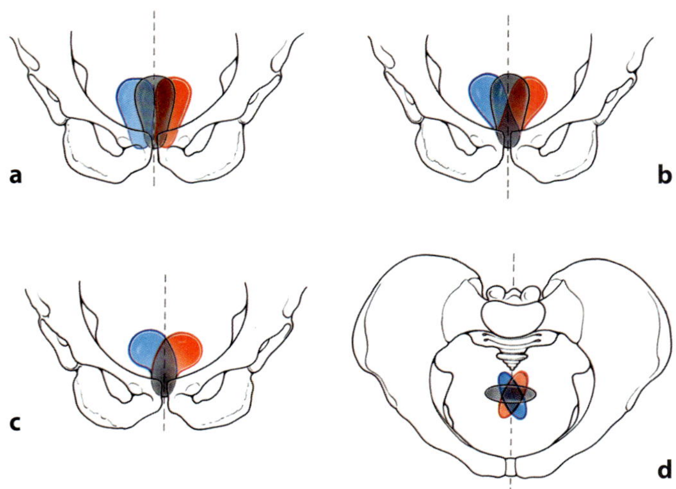

Positions of the uterus with respect to the frontal plane (**a**, **b**, **c**) and to its own axis (**d**). **a**, With respect to the normal (central) position, the uterus can sometimes be tilted in a right or left lateral position. **b**, Right and left lateroversion. In the lateroversion, an angle open laterally between the body of the utherus axis and the vagina axis is formed. **c**, Right and left lateral flexion. In the lateral flexion, an angle open laterally between the body and cervix of the uterus is formed. **d**, Displacement of the uterus, towards the right (dextroversion) or towards the left (levoversion).

Positions of the uterus in relation to the sagittal plane. **a**, With respect to the normal position, where the angle between the body of the uterus and vagina is normally open to the front (physiological anteversion), three different degrees of retroversion can be observed. In each of these, the major axis of the body of the utherus is modified with various degree of extension. Note that, in any case, the angle formed between the body and cervix axes of the uterus is unchanged. **b**, Retroflexion of the uterus. The angle between body and cervix, normally open to the front (physiological anteflexion), results instead open to the back. **c**, Severe anteflexion of the uterus due to an accentuated angle of physiological anteflexion.

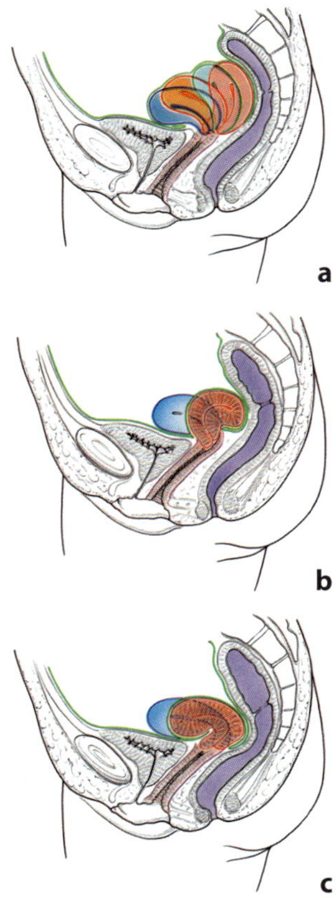

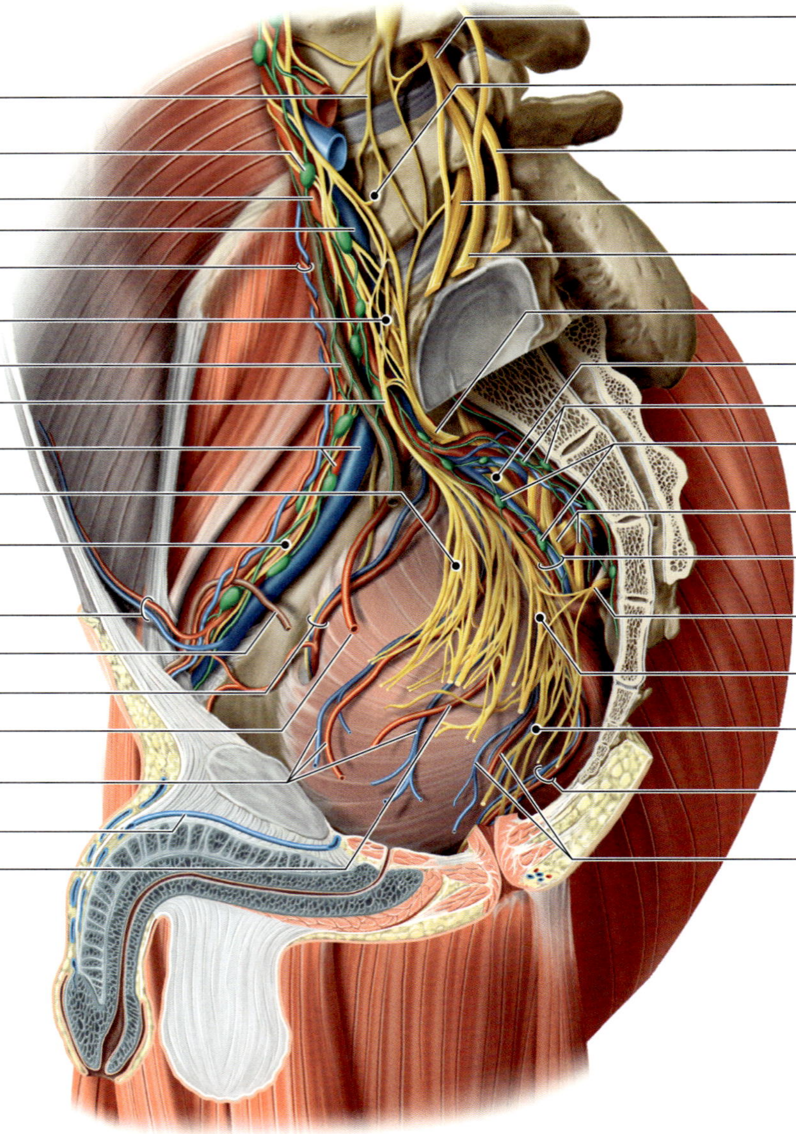

Left lateral view of a sagittal section of the **male pelvis** showing the vessels and nerves of the lateral wall of the extraperitoneal space. The inferior hypogastric plexus, or pelvic plexus, extends along the posterolateral wall of the pelvis, and is composed of sympathetic preganglionic and postganglionic fibers and of sacral parasympathetic preganglionic fibers; the pelvic splanchnic nerves, or nervi erigentes, constitute the parasympathetic root of the pelvic ganglia. The inferior hypogastric plexus also gives off numerous efferent fibers which, following the ramifications of the internal iliac artery, supply the pelvic organs.

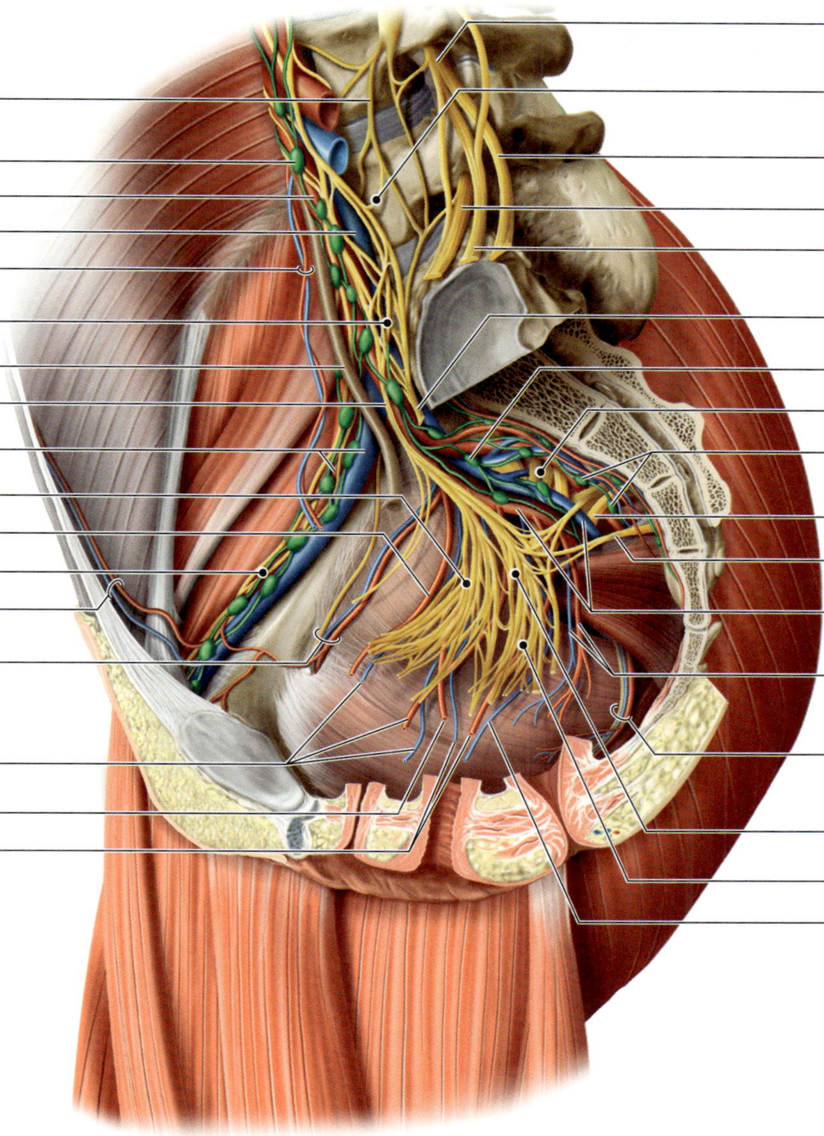

Left lateral view of a sagittal section of the **female pelvis** showing the vessels and nerves of the lateral wall of the extraperitoneal space. The inferior hypogastric plexus, or pelvic plexus, is composed of sympathetic preganglionic and postganglionic fibers. It also receives parasympathetic preganglionic fibers that run in the sacral nerves and sacral plexus constituting the pelvic splanchnic nerves. These then, following the visceral nerves and pudendal nerve, reach the pelvic ganglia satellite of the innervated organs.

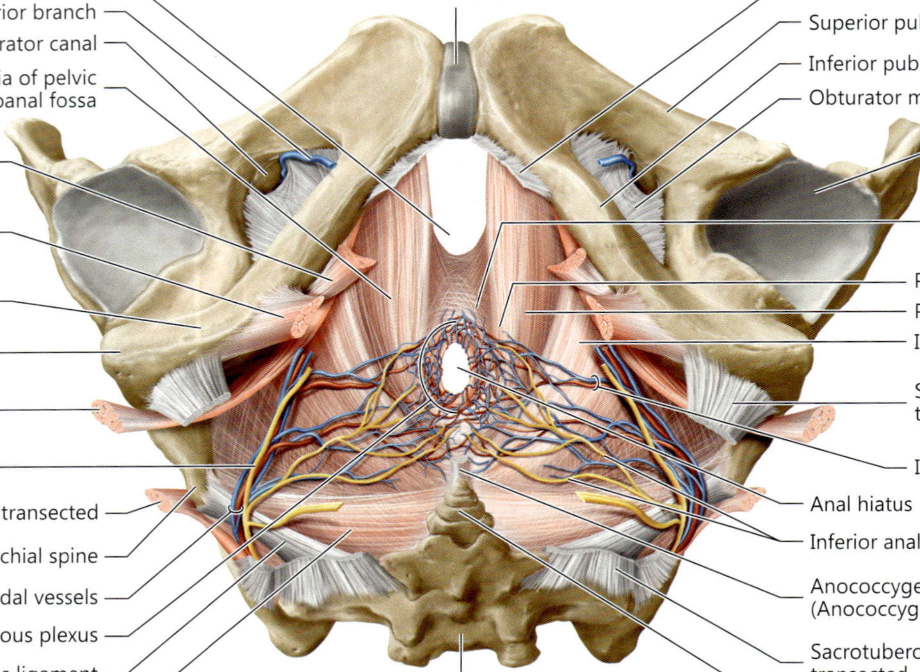

Vessels and nerves of the male posterior perineum (anal triangle) include the internal pudendal artery and vein and the pudendal nerve (in the ischioanal fossa), the inferior anal, or rectal, nerves, the inferior rectal artery and the inferior rectal veins, receiving the inferior rectal, or hemorrhoidal, plexus.

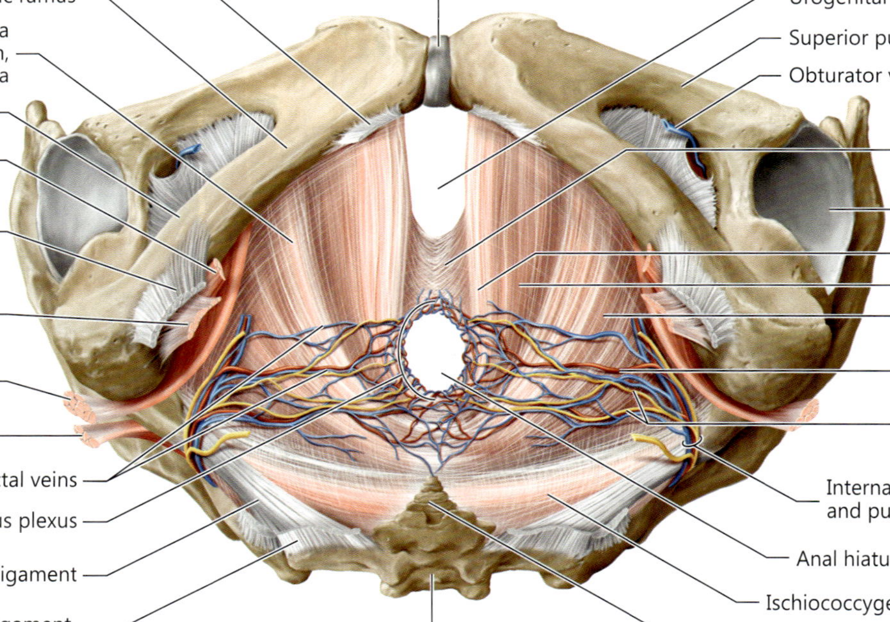

Vessels and nerves of the female posterior perineum (anal triangle) include the pudendal neurovascular bundle, composed of the internal pudendal artery and vein and the pudendal nerve, and its collateral branches in the ischioanal fossa.

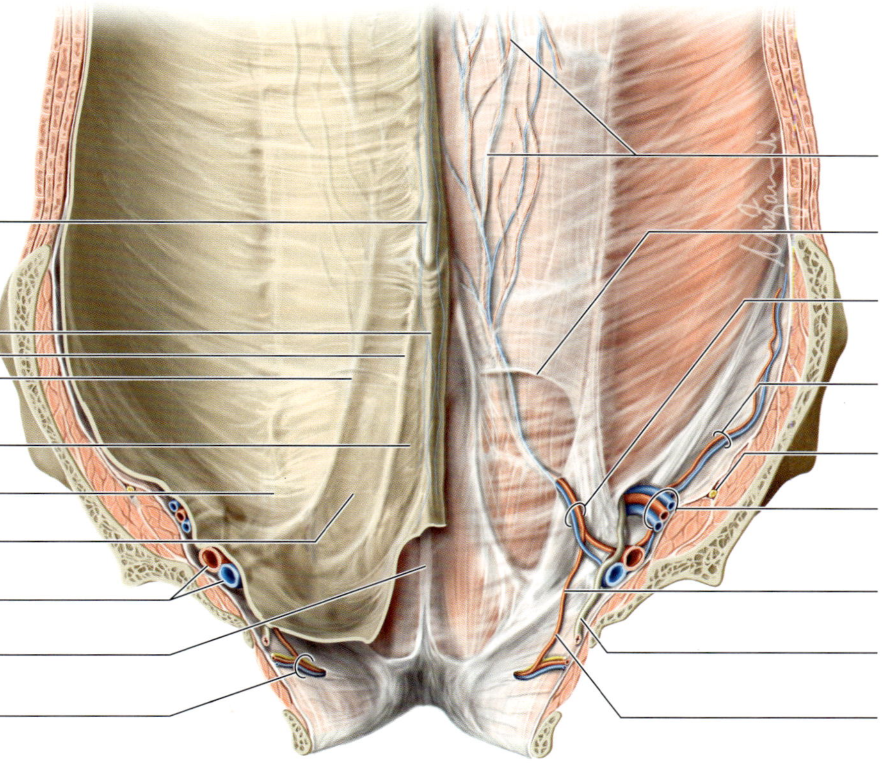

Muscles of the anterior wall of the abdomen. Posterior view from inside the abdominal cavity. The **supravesical fossa**, **medial inguinal fossa** (also called inguinal triangle) and **lateral inguinal fossa** form between the median and medial umbilical folds of the peritoneum, the medial and lateral folds, and laterally to the lateral fold, respectively, and represent potential sites for the formation of hernias. In the so-called direct inguinal hernias, intestinal ansae can protrude, pushing the peritoneum forward, through the medial inguinal fossa, corresponding in part to the posterior wall of the inguinal canal. Intestinal ansae can also protrude through the lateral inguinal fossa, corresponding to the deep inguinal ring, forming the external oblique inguinal hernias.

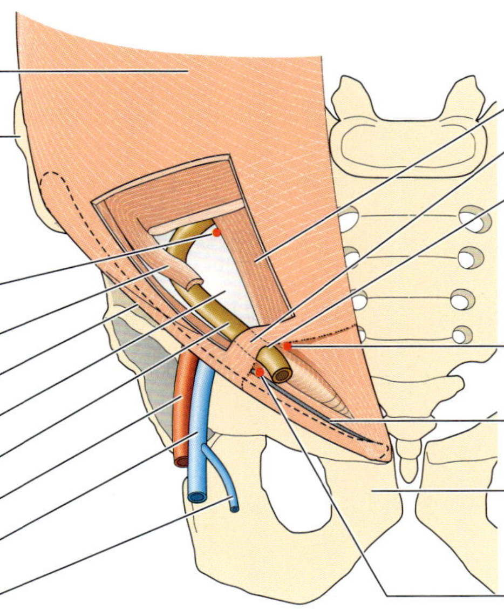

Direct, **external oblique**, and **crural** or **femoral inguinal hernias** in relation to the anatomical structures. The direct inguinal hernia exit through the median part of the posterior wall of the inguinal canal; the external oblique inguinal hernia exit through the deep inguinal ring; and the crural hernia exit through the femoral ring.

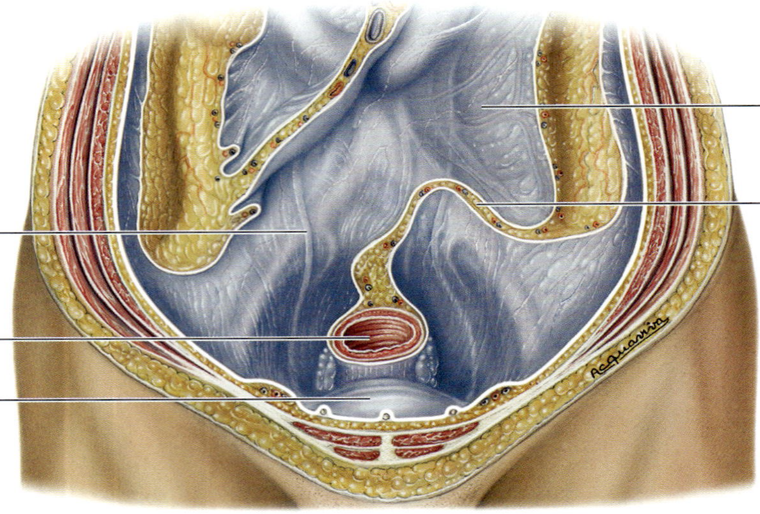

a

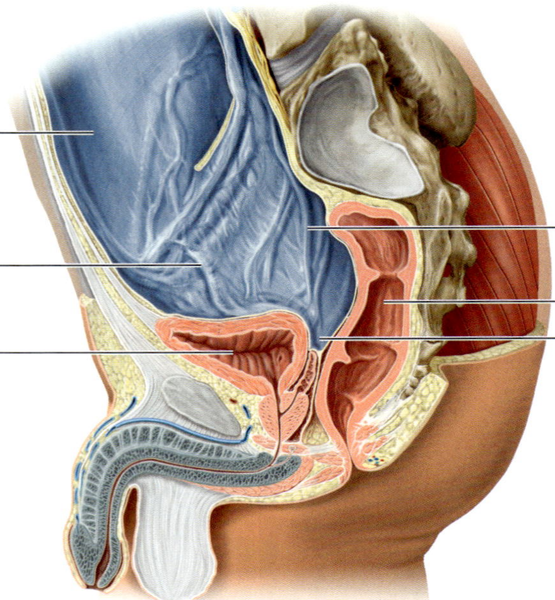

b

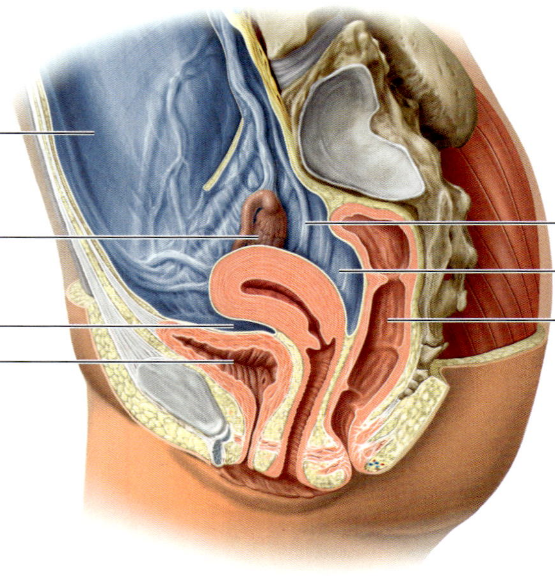

c

a, The inferior part of the **parietal peritoneum** invests the walls of the lesser pelvis and the organs contained within it for a variable extent. Weaving amongst them, the peritoneum creates serous spaces or cavities. **b**, Modalities of peritoneum investing of the male pelvic organs. **c**, Modalities of peritoneum investing of the female pelvic organs.

Contents of the pelvic space		
NAME OF THE SPACE	***CONTENTS***	
	MALE	FEMALE
Extraperitoneal pelvic space	*Muscles*: – deep and superficial trigonal – detrusor – obturator internus – pelvic diaphragm – – levator ani – – – iliococcygeus – – – pubococcygeus – – – – puboanalis – – – – puboperinealis – – – – puboprostaticus or levator prostatae – – – puborectalis – – ischiococcygeus – piriformis – pubovesicalis – rectovesicalis – vesicoprostaticus	*Muscles*: – deep and superficial trigonal – detrusor – obturator internus – pelvic diaphragm – – levator ani – – – iliococcygeus – – – pubococcygeus – – – – puboanalis – – – – puboperinealis – – – – pubovaginalis – – – puborectalis – – ischiococcygeus – piriformis – pubovesicalis – rectouterinus – rectovesicalis – vesicovaginalis
	Fascia and ligaments: – pelvic fascia – – extraperitoneal ligament – – parietal (endopelvic) – – – inferior of pelvic diaphragm – – – obturator – – – presacral – – – rectosacral – – – superior of pelvic diaphragm – – – – lateral ligament of bladder – – – – puboprostatic ligament or lateral puboprostatic – – – – pubovesical ligament or medial puboprostatic – – visceral – – – rectoprostatic fascia or rectovesical septum	*Fascia and ligaments*: – cardinal ligament – pelvic fascia – – extraperitoneal ligament – – parietal (endopelvic) – – – inferior of pelvic diaphragm – – – obturator – – – presacral – – – rectosacral – – – superior of pelvic diaphragm – – – – lateral ligament of bladder – – – – lateral pubovesical ligament – – – – medial pubovesical ligament – – visceral – – – rectovaginal fascia or septum – pubocervical ligament – round ligament of uterus – uterosacral or rectouterine ligament – vesicouterine ligament
	Organs: – ductus (vas) deferens – prostate – rectum – seminal glands – ureter, pelvic part – urinary bladder	*Organs*: – ovary – rectum – ureter, pelvic part – urinary bladder – uterine tube – uterus – vagina
	Lymph nodes: – common iliac – – medial, intermediate and lateral – – promontorial – – subaortic – external iliac – – interiliac – – medial, intermediate and lateral – – obturator – internal iliac – – sacral – – superior and inferior – pararectal – paravesical – – lateral vesical – – postvesical – – prevesical	*Lymph nodes*: – common iliac – – medial, intermediate and lateral – – promontorial – – subaortic – external iliac – – interiliac – – medial, intermediate and lateral – – obturator – internal iliac – – sacral – – superior and inferior – pararectal – parauterine – paravaginal – paravesical – – lateral vesical – – postvesical – – prevesical

continued

Contents of the pelvic space *(continued)*

NAME OF THE SPACE	*CONTENTS*	
	MALE	FEMALE
	Arteries: – external iliac – iliolumbar – inferior gluteal – inferior vesical – internal iliac – internal pudendal – lateral sacral – median sacral – middle rectal – obturator – superior gluteal – superior rectal – superior vesical – testicular – to ductus (vas) deferens – vesiculodeferential *Veins*: – external iliac – iliolumbar – inferior gluteal – internal iliac – internal pudendal – lateral sacral – median sacral – middle rectal – obturator – prostatic venous plexus – rectal venous plexus – sacral venous plexus – testicular – vesical – vesical venous plexus *Nerves, ganglia and plexuses*: – anococcygeal – coccygeal plexus – deferential plexus – hypogastric – inferior anal – inferior gluteal – inferior hypogastric plexus – inferior rectal plexus – middle rectal plexus – obturator – pelvic ganglia (of parasympathetic) – pelvic splanchnic – perforating cutaneous – posterior cutaneous of thigh – prostatic plexus – pudendal – sacral ganglia of sympathetic trunk – sacral plexus – sacral splanchnic – sciatic – superior anal – superior gluteal – superior hypogastric plexus – testicular plexus – to obturator internus – to piriformis – to quadratus femoris – vesical plexus	*Arteries*: – external iliac – iliolumbar – inferior gluteal – inferior vesical – internal iliac – internal pudendal – lateral sacral – median sacral – middle rectal – obturator – of round ligament of uterus – ovarian – superior gluteal – superior rectal – superior vesical – uterine – vaginal *Veins*: – external iliac – iliolumbar – inferior gluteal – internal iliac – internal pudendal – lateral sacral – median sacral – middle rectal – obturator – ovarian – rectal venous plexus – sacral venous plexus – uterine – uterine venous plexus – vaginal venous plexus – vesical – vesical venous plexus *Nerves, ganglia and plexuses*: – anococcygeal – coccygeal plexus – hypogastric – inferior anal – inferior gluteal – inferior hypogastric plexus – inferior rectal plexus – middle rectal plexus – obturator – ovarian plexus – pelvic ganglia (of parasympathetic) – pelvic splanchnic – perforating cutaneous – posterior cutaneous of thigh – pudendal – sacral ganglia of sympathetic trunk – sacral plexus – sacral splanchnic – sciatic – superior anal – superior gluteal – superior hypogastric plexus – to obturator internus – to piriformis – to quadratus femoris – uterusvaginal plexus – vaginal – vesical plexus

Contents of the perineum		
NAME OF THE SPACE	**CONTENTS**	
	MALE	FEMALE
Anterior perineum	*Muscles*: – bulbospongiosus – deep transverse perineal muscle – external urethral sphincter – ischiocavernosus – superficial transverse perineal muscle *Fascia, ligaments and compartments*: – deep perineal pouch (deep perineal space) – perineal body – perineal fascia – perineal membrane – perineal raphe – superficial perineal pouch (superficial perineal compartment or space) – transverse perineal ligament *Organs*: – bulb of penis and corpus spongiosum of penis – bulbourethral gland – intermediate part of urethra – root of penis (roots of corpora cavernosa) *Arteries*: – deep artery of penis – dorsal artery of penis – internal pudendal – of bulb of penis – perineal – urethral *Veins*: – deep dorsal vein of penis – deep veins of penis – internal pudendal – of bulb of penis – superficial dorsal veins of penis *Nerves*: – dorsal nerve of penis – perineal – pudendal	*Muscles*: – bulbospongiosus – compressor urethrae – deep transverse perineal muscle – external urethral sphincter – ischiocavernosus – sphincter urethrovaginalis – superficial transverse perineal muscle *Fascia, ligaments and compartments*: – deep perineal pouch (deep perineal space) – perineal body – perineal fascia – perineal membrane – perineal raphe – superficial perineal pouch (superficial perineal compartment or space) – transverse perineal ligament *Organs*: – bulb of vestibule – crus of clitoris (roots of corpora cavernosa) – external urethral orifice – external vaginal orifice – greater vestibular glands – urethra – vagina *Arteries*: – deep artery of clitoris – dorsal artery of clitoris – internal pudendal – of bulb of vestibule – perineal – urethral *Veins*: – internal pudendal – deep dorsal vein of clitoris – deep veins of clitoris – of bulb of vestibule – superficial dorsal veins of clitoris *Nerves*: – dorsal nerve of clitoris – perineal – pudendal
	MALE	FEMALE
Posterior perineum	*Muscles*: – external anal sphincter *Fascia, ligaments and compartments*: – anococcygeal body (anococcygeal ligament) – ischioanal fossa – perineal body – perineal raphe *Organs*: – anal canal *Arteries*: – inferior rectal – internal pudendal – perineal *Veins*: – inferior rectal – internal pudendal – rectal venous plexus *Nerves*: – inferior anal – perineal – pudendal	*Muscles*: – external anal sphincter *Fascia, ligaments and compartments*: – anococcygeal body (anococcygeal ligament) – ischioanal fossa – perineal body – perineal raphe *Organs*: – anal canal *Arteries*: – inferior rectal – internal pudendal – perineal *Veins*: – inferior rectal – internal pudendal – rectal venous plexus *Nerves*: – inferior anal – perineal – pudendal

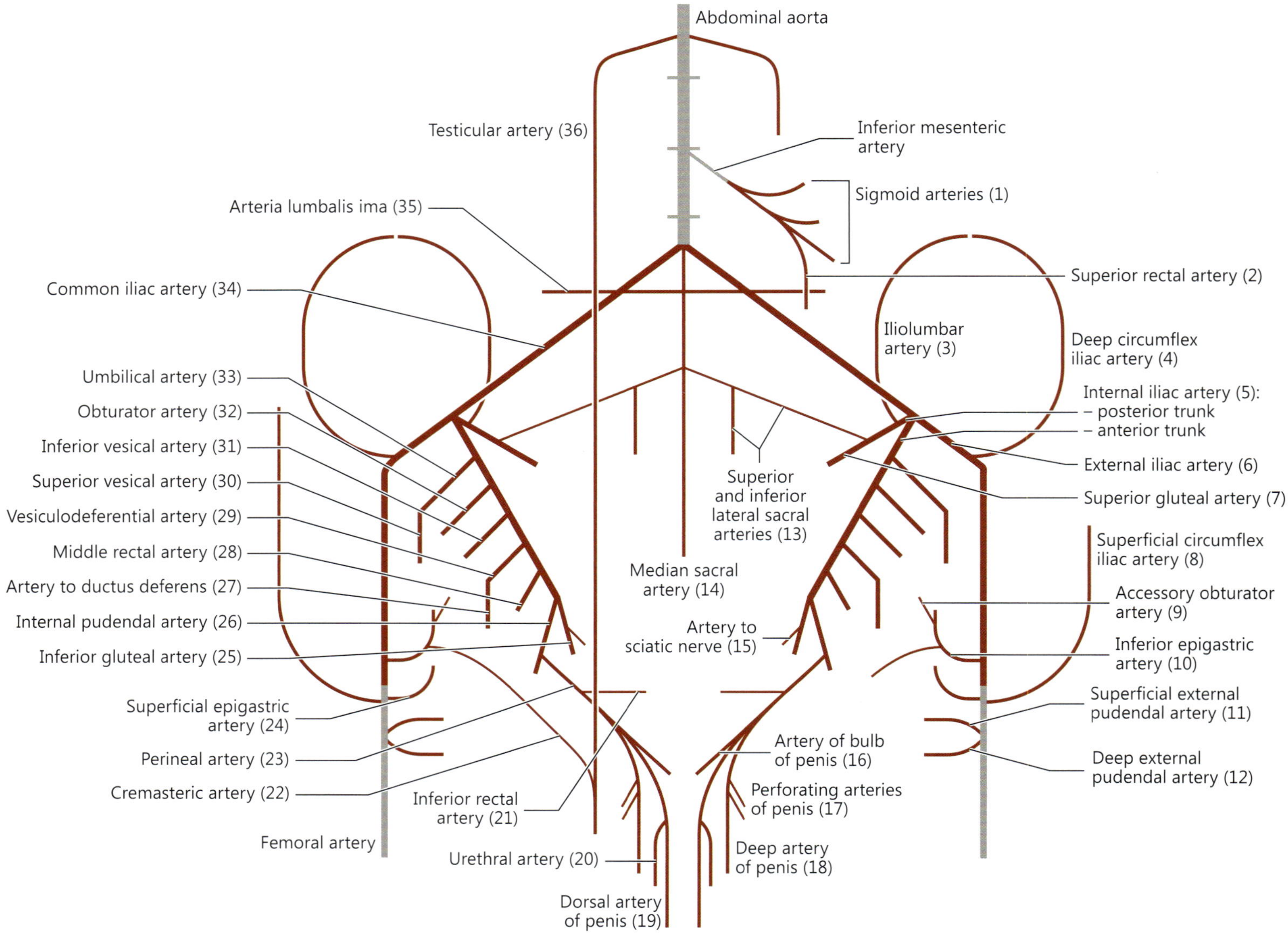

The organs and walls of the pelvis, the perineum and the gluteal region are supplied by the **internal iliac artery**, or hypogastric artery, a branch of the **common iliac artery** bifurcation. After its origin anterior to the sacroiliac joint, in proximity of the promontory of the sacrum, the **internal iliac artery** descends towards the superior margin of the greater sciatic foramen, where it divides into an anterior and a posterior trunk. The anterior trunck gives rise to visceral and parietal branches. The visceral branches are: the **superior vesical artery**, which stems from the part remaining patent of the umbilical artery, and the **inferior vesical**, **middle rectal** and **vesiculodeferential arteries**; the parietal branches are: the **obturator artery**, the **internal pudendal artery**, which supplies the perineum and external genitalia, and the **inferior gluteal artery**. The internal pudendal artery gives rise to the **inferior rectal** and **perineal arteries**, the **urethral artery**, the **artery of the bulb of the penis**, and the **dorsal**, **deep** and **perforating arteries of the penis**. The posterior trunk gives rise to parietal branches only: the **iliolumbar**, **lateral sacral** and **superior gluteal arteries**.

Arteries of the male pelvis and perineum: territories of supply		
	ARTERIES	***TERRITORIES OF SUPPLY***
1	Sigmoid arteries	Inferior part of the descending colon and sigmoid colon
2	Superior rectal artery	Rectum and internal anal sphincter
3	Iliolumbar artery	Psoas major, quadratus lumborum and iliacus, cauda equina
4	Deep circumflex iliac artery	Internal oblique and transversus abdominis, skin of the inferior part of the abdomen, sartorius and tensor fasciae latae
5	Internal iliac artery	Pelvic walls and organs, external genitalia, anal triangle, medial surface of the thigh
6	External iliac artery	Psoas major, inferior part of the abdominal wall, external genitalia, cremaster and ductus deferens
7	Superior gluteal artery	Piriformis and obturator internus, hip bone, gluteus maximus, gluteus medius and gluteus minimus, skin of the sacral region, hip joint
8	Superficial circumflex iliac artery	Skin of the groin, superficial inguinal nodes
9	Accessory obturator artery	Anastomoses between the pubic branch of inferior epigastric artery and the pubic branch of obturator artery
10	Inferior epigastric artery	Cremaster and spermatic cord, peritoneum, muscles of abdomen, skin of the anterior abdominal wall
11	Superficial external pudendal artery	The integument of the inferior part of the abdomen, the penis and the scrotum
12	Deep external pudendal artery	The integument of the scrotum and the perineum
13	Lateral sacral arteries	Sacral vertebrae and corresponding vertebral canal, skin and muscles of back proper at level of sacral segment of the vertebral column
14	Median sacral artery	Rectum, sacrum, coccyx
15	Artery to sciatic nerve	Sciatic nerve
16	Artery of bulb of penis	Bulb of penis, posterior part of the corpus spongiosum of penis and bulbourethral glands
17	Perforating arteries of penis	Glans penis and cavernous spaces of corpus spongiosum of penis
18	Deep artery of penis	Corpora cavernosa of penis
19	Dorsal artery of penis	Glans and prepuce of penis, skin and tunica albuginea of corpora cavernosa of penis
20	Urethral artery	Urethra, corpus spongiosum of penis
21	Inferior rectal artery	Anal skin, perineal part of the rectum and the external anal sphincter
22	Cremasteric artery	Cremaster and spermatic cord
23	Perineal artery	Perineum, external genitalia, bulbospongiosus and ischiocavernosus
24	Superficial epigastric artery	Superficial inguinal nodes and abdominal integument
25	Inferior gluteal artery	Skin of posterior surface of the thigh, levator ani, ischiococcygeus and piriformis, prostate and seminal glands, fundus of bladder, gluteus maximus, obturator internus, gemellus superior and gemellus inferior, quadratus femoris and posterior muscles of thigh, posterior region of thigh
26	Internal pudendal artery	Perineal muscles, anal canal, penis and scrotum
27	Artery to ductus deferens	Ductus deferens, ampulla of ductus deferens, prostate
28	Middle rectal artery	Muscular layer of lowest part of the rectum, prostate and seminal glands
29	Vesiculodeferential artery	Trigone and posterior part of the bladder, seminal glands
30	Superior vesical artery	Superior part of the bladder
31	Inferior vesical artery	Inferior part of the ureter and fundus of bladder, prostate and seminal glands
32	Obturator artery	Ilium, iliacus, pubis, bladder, obturator externus, pectineus, adductor muscles and gracilis, hip joint, muscles that attach to the ischial tuberosity
33	Umbilical artery	Ureter and urinary bladder, testes, seminal glands and ductus deferens
34	Common iliac artery	Pelvis and lower limb
35	Artery lumbalis ima	Sacrum, iliacus
36	Testicular artery	Ureter, epididymis, testis, cremaster

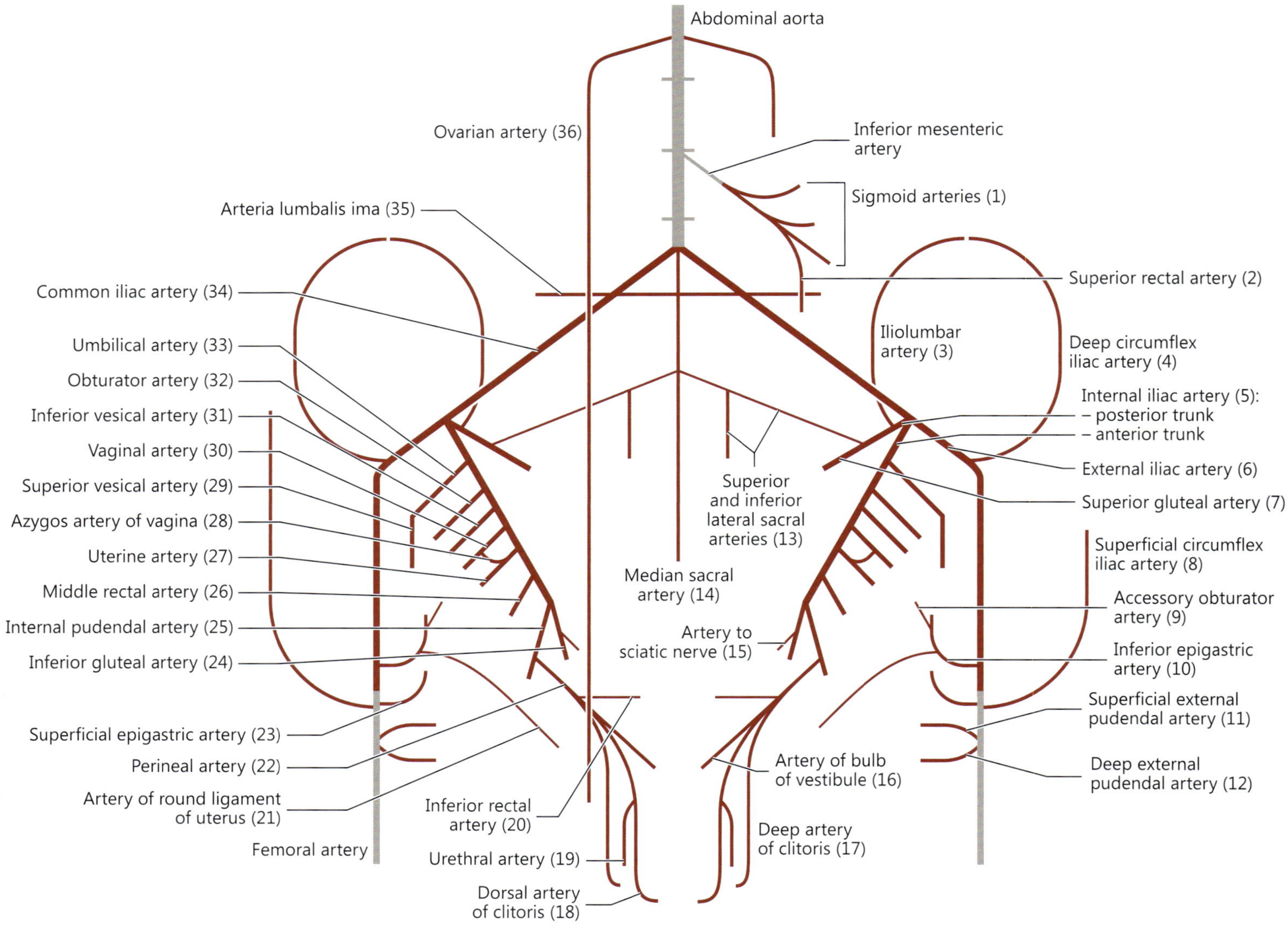

The organs and walls of the pelvis, the perineum and the gluteal region are supplied by the **internal iliac artery**, or hypogastric artery, a branch of the **common iliac artery** bifurcation. The **internal iliac artery**, divides into an anterior and a posterior trunk, at the level of the greater sciatic foramen. The anterior trunk gives rise to visceral and parietal branches. The visceral branches are: the **superior vesical artery**, which stems from the part remaining patent of the umbilical artery, and the **inferior vesical**, **middle rectal**, **uterine** and **vaginal arteries**; the parietal branches are: the **obturator artery**, the **internal pudendal artery**, which supplies the perineum and the external genitalia, and the **inferior gluteal artery**. The internal pudendal artery gives rise to the **inferior rectal** and **perineal arteries**, the **urethral artery** and the **arteries of the bulb of the vestibule**, the **dorsal** and **deep arteries of the clitoris**. The posterior trunk gives rise to parietal branches only: the **iliolumbar**, **lateral sacral** and **superior gluteal arteries**.

Arteries of the female pelvis and perineum: territories of supply		
	ARTERIES	TERRITORIES OF SUPPLY
1	Sigmoid arteries	Inferior part of the descending colon and sigmoid colon
2	Superior rectal artery	Rectum and internal anal sphincter
3	Iliolumbar artery	Psoas major, quadratus lumborum and iliacus, cauda equina
4	Deep circumflex iliac artery	Internal oblique and transversus abdominis, skin of the inferior part of the abdomen, sartorius and tensor fasciae latae
5	Internal iliac artery	Pelvic walls and organs, external genitalia, anal triangle, medial surface of the thigh
6	External iliac artery	Psoas major, inferior part of the abdominal wall, external genitalia, round ligament of uterus
7	Superior gluteal artery	Piriformis and obturator internus, hip bone, gluteus maximus, gluteus medius and gluteus minimus, skin of the sacral region, hip joint
8	Superficial circumflex iliac artery	Skin of the groin, superficial inguinal nodes
9	Accessory obturator artery	Anastomoses between the pubic branch of inferior epigastric artery and the pubic branch of obturator artery
10	Inferior epigastric artery	Round ligament of uterus, peritoneum, muscles of abdomen, skin of the anterior abdominal wall
11	Superficial external pudendal artery	Integument of the inferior part of the abdomen and labia majora
12	Deep external pudendal artery	Labia majora
13	Lateral sacral arteries	Sacral vertebrae and corresponding vertebral canal, skin and muscles of back proper at level of sacral segment of the vertebral column
14	Median sacral artery	Rectum, sacrum, coccyx
15	Artery to sciatic nerve	Sciatic nerve
16	Artery of bulb of vestibule	Bulb of vestibule and inferior part of the vagina
17	Deep artery of clitoris	Corpus cavernosum of clitoris
18	Dorsal artery of clitoris	Glans and prepuce of clitoris
19	Urethral artery	Urethra
20	Inferior rectal artery	Anal skin, perineal part of the rectum and the external anal sphincter
21	Artery of round ligament of uterus	Round ligament of uterus
22	Perineal artery	Perineum and external genitalia
23	Superficial epigastric artery	Superficial inguinal nodes and abdominal integument
24	Inferior gluteal artery	Skin of posterior surface of the thigh, levator ani, ischiococcygeus and piriformis, fundus of bladder, gluteus maximus, obturator internus, gemellus superior and gemellus inferior, quadratus femoris and posterior muscles of thigh, posterior region of thigh
25	Internal pudendal artery	Perineal muscles, anal canal, clitoris
26	Middle rectal artery	Muscular layer of lowest part of the rectum, vagina
27	Uterine artery	Uterine tube, round ligament of uterus, cervix, body of uterus, vagina, ovary
28	Azygos artery of vagina	Vagina
29	Superior vesical artery	Superior part of the bladder
30	Vaginal artery	Vagina, bulb of vestibule, fundus of bladder and adjacent parts of the rectum
31	Inferior vesical artery	Inferior part of the ureter and fundus of bladder, vagina and bulb of vestibule
32	Obturator artery	Ilium, iliacus, pubis, bladder, obturator externus, pectineus, adductor muscles and gracilis, hip joint, muscles that attach to the ischial tuberosity
33	Umbilical artery	Ureter and urinary bladder
34	Common iliac artery	Pelvis and lower limb
35	Artery lumbalis ima	Sacrum, iliacus
36	Ovarian artery	Ureter, ovary, uterus, round ligament of uterus

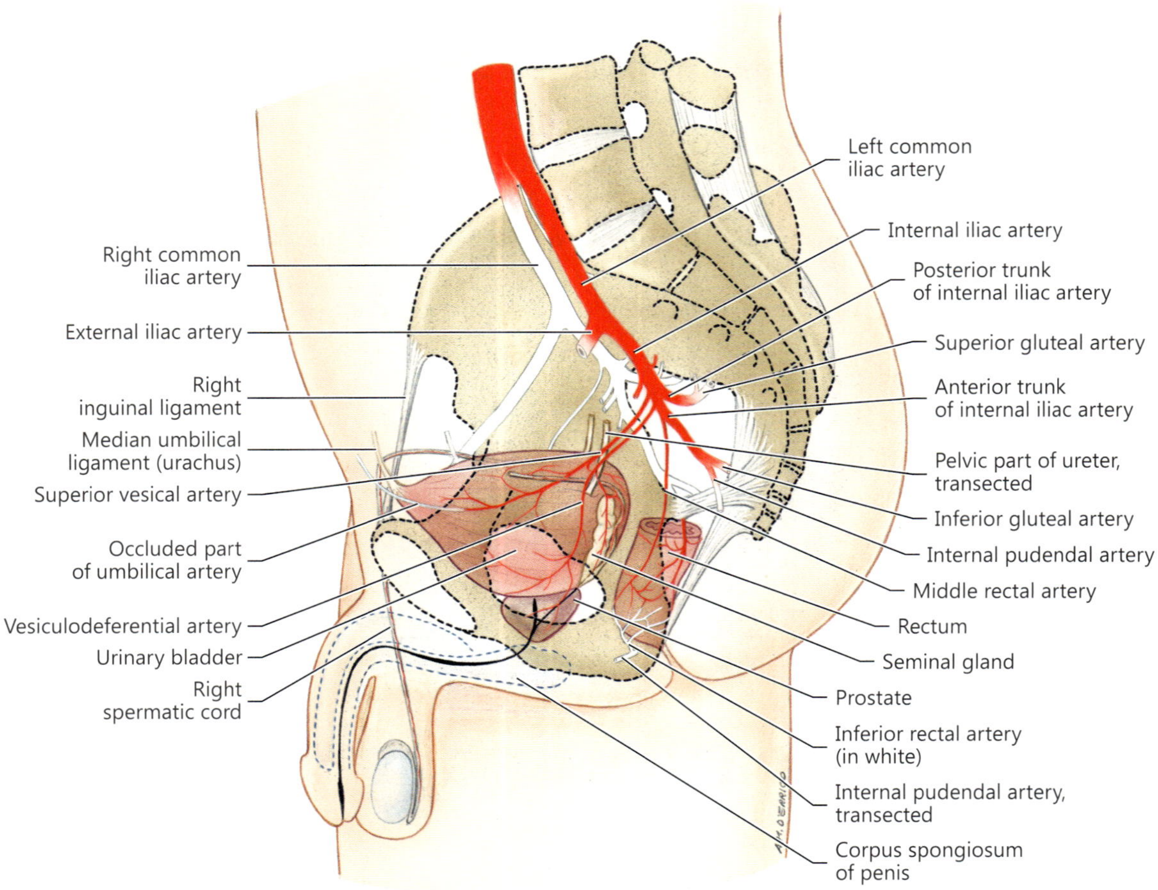

Scheme of the visceral branches of the anterior trunk of the left **internal iliac artery** and of the **pelvic organs**, in the male. The parietal branches of the anterior and the posterior trunks are transected at their origin and shaded in a muted white tone. In full white, the contralateral iliac vessels.

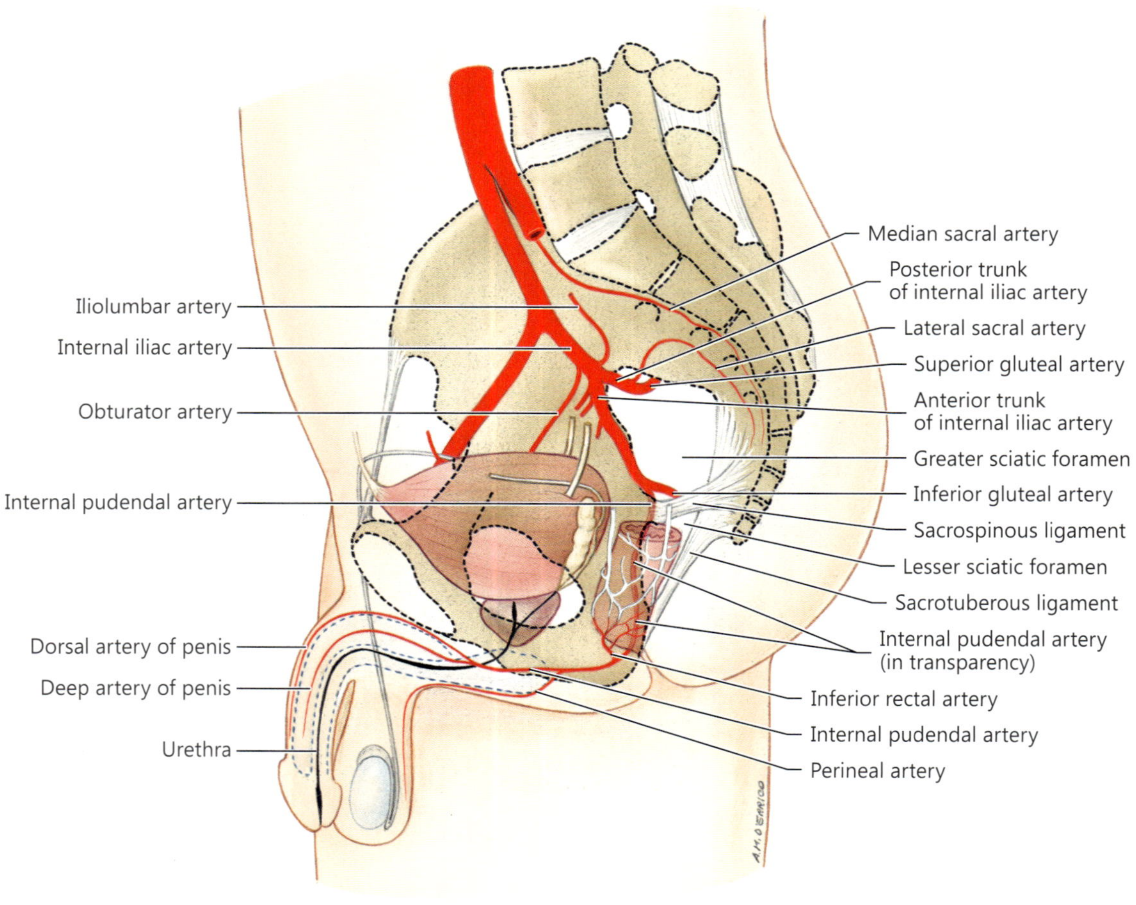

Semischematic illustration of the parietal branches of the anterior trunk, the branches of the posterior trunk of the right **internal iliac artery**, and the **pelvic organs**, in the male. The image in transparency shows the internal pudendal artery, even in the first part, while it emerges from the pelvic cavity through the greater sciatic foramen, enters the lesser sciatic foramen, close to the internal surface of the ischial tuberosity, and descend towards the perineal plane.

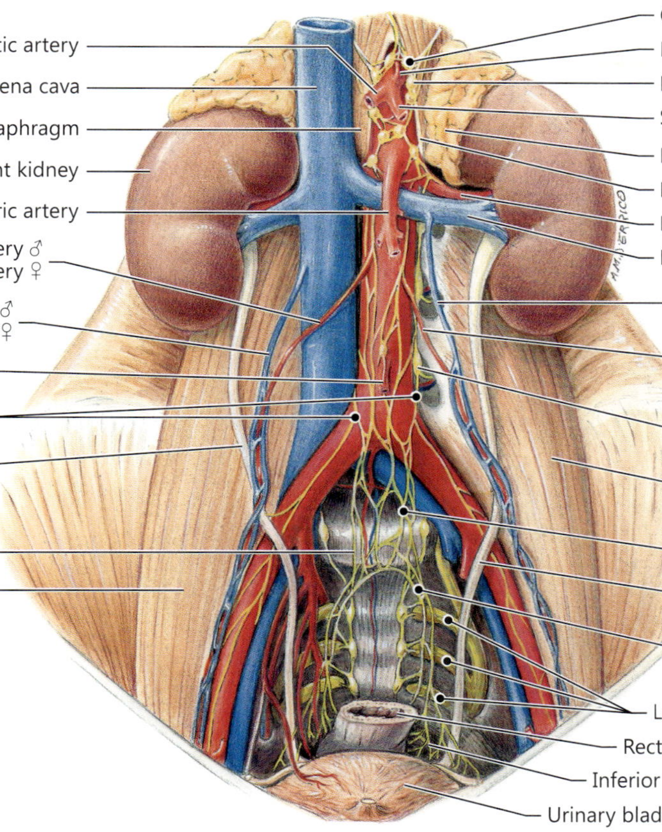

Course and relations of the **abdominal aorta**, **inferior vena cava** and **common iliac**, **internal iliac** and **external iliac arteries and veins**. The course and relations of the renal and genital arteries and veins are also displayed.

Visceral branches of the anterior trunk of the **internal iliac artery**, in the female. The uterus pulled up to the right side, to show the middle rectal, uterine, vaginal and superior vesical arteries. In the mesovarium, the terminal branches of the uterine artery join and run together with the terminal branches of the genital artery.

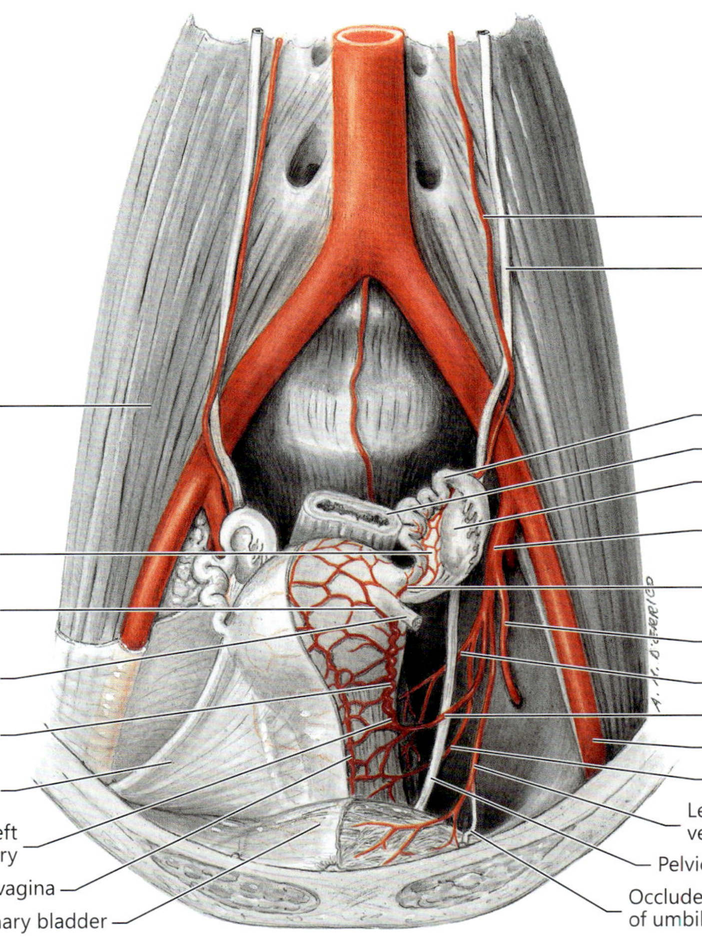

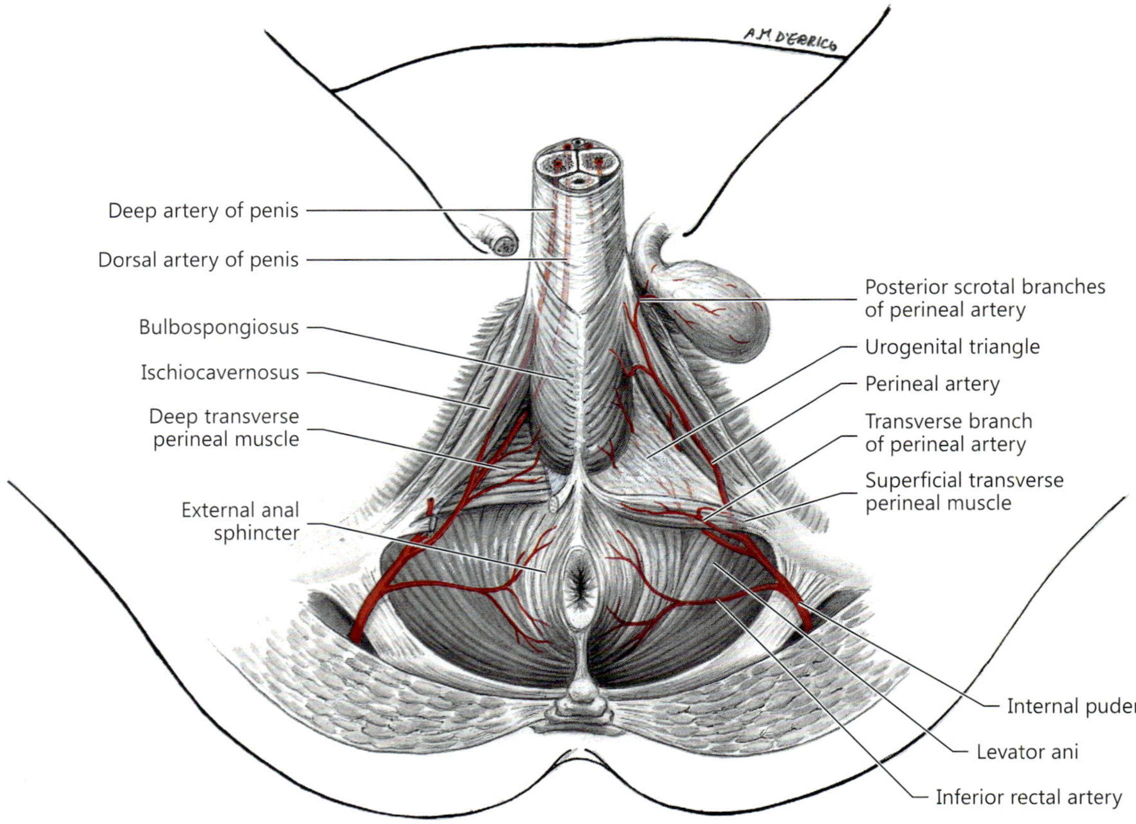

Course and distribution of the **internal pudendal artery** and its branches to the perineum and external genitalia, in the male. The corpora cavernosa of the penis and urethra have been transected to show the deep and dorsal arteries of the penis.

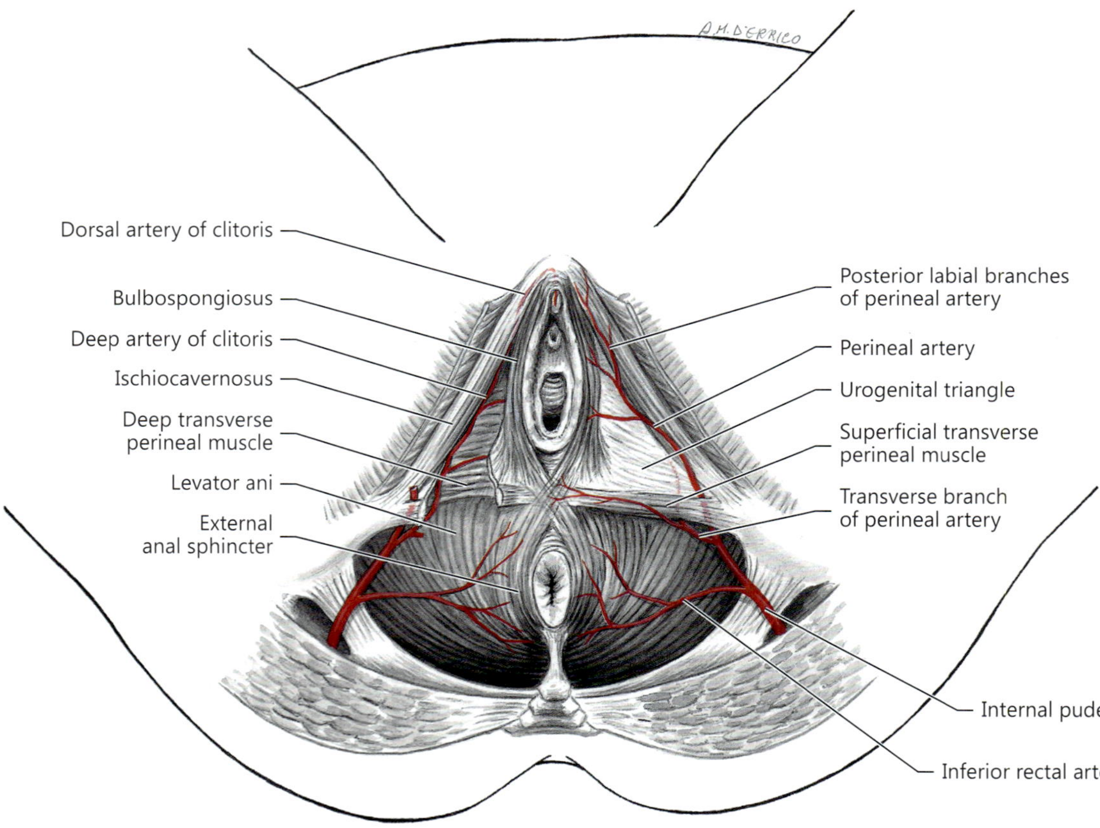

Course and distribution of the **internal pudendal artery** and its branches to the perineum and external genitalia, in the female.

Summary table of the arteries of the pelvis

ARTERY	ORIGIN	BRANCHES	TERRITORIES OF SUPPLY
accessory obturator (inconstant)	Pubic branch of the inferior epigastric artery	Anastomoses with the pubic branch of the obturator artery	Pubis
azygos artery of vagina ♀ (inconstant)	Vaginal branches of uterine artery which anastomose with branches of the vaginal artery		Vagina
common iliac (at level of sacroiliac joint)	Branches of division of the abdominal aorta	Terminates dividing into the external and internal iliac arteries	Pelvis and lower limb
cremasteric ♂ (external spermatic)	Inferior epigastric		Cremaster and spermatic cord
deep circumflex iliac (in its course behind the inguinal ligament)	External iliac	Ascending and anastomotic branches with the lumbar, inferior epigastric, superior gluteal, iliolumbar and superficial circumflex iliac arteries	Internal oblique and transversus abdominis, skin of the inferior part of the abdomen, sartorius and tensor fasciae latae
external iliac	Common iliac	Inferior epigastric and deep circumflex iliac	Psoas major, inferior part of the abdominal wall, external genitalia, cremaster and ductus deferens in the male, round ligament of uterus in the female
iliolumbar	Internal iliac	Lumbar, spinal and iliacus	Psoas major, quadratus lumborum and iliacus, cauda equina
inferior epigastric (deep epigastric) (in correspondence to its origin, immediately prior to the inguinal ligament)	External iliac	Pubic, muscular and cutaneous branches; cremasteric artery ♂ or artery of round ligament of uterus ♀; anastomotic branches with the superior epigastric and subcostal arteries	Cremaster and spermatic cord in the male; round ligament of uterus in the female; peritoneum, muscles of abdomen, skin of the anterior abdominal wall
inferior gluteal (to sciatic nerve)	Internal iliac	Muscular, visceral and coccygeal branches; artery to sciatic nerve	Skin of posterior surface of the thigh, levator ani, ischiococcygeus and piriformis, prostate and seminal glands in the male, fundus of bladder, gluteus maximus, obturator internus, gemellus superior and gemellus inferior, quadratus femoris and posterior muscles of thigh, posterior region of thigh
inferior rectal (inferior hemorrhoidal)	Internal pudendal	Anastomotic branches with the superior rectal artery, middle rectal artery and inferior rectal artery contralateral, and with perineal artery	Anal skin, perineal part of the rectum and external anal sphincter
inferior vesical (vesico-prostatic ♂ or vesico-vaginal ♀)	Internal iliac or, at times, from the vesiculodeferential artery ♂	Prostatic ♂	Inferior part of the ureter and fundus of bladder; prostate and seminal glands in the male; vagina and bulb of vestibule in the female
internal iliac (hypogastric)	Common iliac	*Anterior trunk*: obturator, superior gluteal, inferior gluteal, umbilical, inferior vesical, uterine ♀, vaginal ♀, middle rectal, internal pudendal arteries *Posterior trunk*: iliolumbar, lateral sacral arteries	Pelvic walls and organs, external genitalia, anal triangle, medial surface of the thigh
internal pudendal (from its origin to the inferior part of margin of the greater sciatic foramen)	Internal iliac	Posterior labial ♀ and posterior scrotal ♂; inferior rectal, perineal, urethral arteries, artery of bulb of vestibule ♀, artery of bulb of penis ♂, dorsal artery of clitoris ♀, dorsal artery of penis ♂, deep artery of clitoris ♀, deep artery of penis ♂, perforating arteries of penis ♂	Perineum, external genitalia
lateral sacral (generally 2 in number, a superior and an inferior one)	Internal iliac	Spinal	Sacral vertebrae and corresponding vertebral canal, skin and muscles of back proper at level of sacral segment of the vertebral column

continued

Summary table of the arteries of the pelvis *(continued)*

ARTERY	*ORIGIN*	*BRANCHES*	*TERRITORIES OF SUPPLY*
lumbales imae	Median sacral		Sacrum, iliacus
median sacral (middle sacral or coccygeal)	Posterior surface of the abdominal aorta just above its bifurcation	Arteriae lumbales imae; lateral sacral arteries and anastomotic branches with the lumbar branch of iliolumbar artery and with the lateral sacral arteries	Rectum, sacrum, coccyx
middle rectal (middle hemorrhoidal)	Internal iliac	Prostatic ♂, vaginal ♀ and anastomotic with the superior and inferior rectal arteries	Muscular layer of lowest part of the rectum; prostate and seminal glands in the male; vagina in the female
obturator	Internal iliac	Pubic, acetabular, anterior and posterior	Ilium, iliacus, pubis, bladder, obturator externus, pectineus, adductor muscles and gracilis, hip joint, muscles that insert on the ischial tuberosity
of round ligament of uterus ♀	Inferior epigastric		Round ligament of uterus
ovarian ♀ (gonadal)	Anterior surface of the abdominal aorta a little below the origin of the renal artery	Ureteric ♀, tubal ♀ and anastomotic with the uterine artery ♀	Ureter, ovary, uterus, round ligament of uterus
sigmoid (left colic arteries) (2 or 3 in number)	Inferior mesenteric	Anastomotic branches with the left colic artery and the superior rectal artery	Inferior part of the descending colon and sigmoid colon
superficial circumflex iliac (in correspondence to its origin, immediataly below the origin of the superficial epigastric artery)	Femoral	Anastomotic branch with the deep circumflex iliac artery	Skin of the groin, superficial inguinal nodes
superficial epigastric (in correspondence to its origin, approximately 1 cm below the inguinal ligament)	Femoral	Anastomotic branches with the cutaneous branches of the inferior epigastric artery	Superficial inguinal nodes and abdominal integument
superior gluteal	Internal iliac	Superficiale and superior and inferior deep branches	Piriformis and obturator internus, hip bone, gluteus maximus, gluteus medius and gluteus minimus, skin of the sacral region, hip joint
superior rectal (superior hemorrhoidal)	Terminal branch of the inferior mesenteric artery	Anastomotic branches with the middle and inferior rectal arteries	Rectum and internal anal sphincter
superior vesical	Umbilical		Superior part of the bladder
testicular ♂ (gonadal)	Anterior surface of the abdominal aorta a little below origin of the renal artery	Ureteric ♂, epididymal ♂ and anastomotic with the artery to ductus deferens ♂	Ureter, epididymis, testis, cremaster
to ductus deferens ♂ (deferential)	Umbilical artery or, at times, from one of the superior vesical arteries or vesiculodeferential artery	Anastomotic branches with the testicular artery	Ductus deferens, ampulla of ductus deferens, prostate
to sciatic nerve	Inferior gluteal		Sciatic nerve
umbilical	Internal iliac	Ureteric; vesiculodeferential ♂, to ductus deferens ♂ and superior vesical arteries; continues as cord of umbilical artery	Ureter and urinary bladder; testes, seminal glands and ductus deferens in the male
uterine ♀	Internal iliac	Helicine ♀, vaginal ♀, ovarian ♀, tubal ♀	Uterine tube, round ligament of uterus, cervix, body of uterus, vagina, ovary

continued

Summary table of the arteries of the pelvis			*(continued)*
ARTERY	***ORIGIN***	***BRANCHES***	***TERRITORIES OF SUPPLY***
vaginal ♀	Internal iliac		Vagina, bulb of vestibule, fundus of bladder and adjacent parts of the rectum
vesiculodeferential ♂	Umbilical		Trigone and posterior part of the bladder, seminal glands

Summary table of the arteries of the perineum			
ARTERY	***ORIGIN***	***BRANCHES***	***TERRITORIES OF SUPPLY***
deep artery of clitoris ♀	Internal pudendal		Corpus cavernosum of clitoris
deep artery of penis ♂ (of the corpus cavernosum penis)	Internal pudendal		Corpora cavernosa of penis
deep external pudendal	Femoral	Anterior labial ♀, anterior scrotal ♂, inguinal; anastomotic with the branches of the perineal artery	Integument of the scrotum and perineum in the male, labia majora in the female
dorsal artery of clitoris ♀	Internal pudendal		Glans and prepuce of clitoris
dorsal artery of penis ♂	Internal pudendal		Glans and prepuce of penis, skin and tunica albuginea of corpora cavernosa of penis
inferior rectal (inferior hemorrhoidal)	Internal pudendal	Anastomotic branches with the superior rectal artery, middle rectal artery, inferior rectal artery contralateral and perineal artery	Anal skin, perineal part of the rectum and the external anal sphincter
internal pudendal	Internal iliac	Inferior rectal, perineal, urethral, of bulb of vestibule ♀, of bulb of penis ♂ arteries, dorsal artery of clitoris ♀, dorsal artery of penis ♂, deep artery of clitoris ♀, deep artery of penis ♂, perforating arteries of penis ♂	Perineal muscles, anal canal, penis and scrotum in the male; clitoris in the female
of bulb of penis ♂	Internal pudendal		Bulb of penis, posterior part of the corpus spongiosum of penis, bulbourethral glands
of bulb of vestibule ♀	Internal pudendal		Bulb of vestibule and inferior part of the vagina
perforating arteries of penis ♂	Internal pudendal or from the dorsal artery of penis		Glans and cavernous spaces of corpora cavernosa of penis
perineal (superficial perineal)	Internal pudendal	Posterior labial ♀, posterior scrotal ♂	Perineum, external genitalia, bulbospongiosus and ischiocavernosus
superficial external pudendal	Femoral	Anastomotic branches with the branches of the internal pudendal artery	Integument of inferior part of the abdomen; integument of the penis and scrotum in the male, labia majora in the female
testicular ♂ (gonadal)	Anterior surface of the abdominal aorta a little below origin of the renal artery	Ureteric ♂, epididymal branches ♂ and anastomotic with the artery to ductus deferens ♂	Ureter, epididymis, testis, cremaster
umbilical	Internal iliac	Ureteric; vesiculodeferential artery ♂, artery to ductus deferens ♂ and superior vesical arteries; continues on as cord of the umbilical artery	Ureter and urinary bladder; testes, seminal glands and ductus deferens in the male
urethral	Internal pudendal		Urethra; corpus spongiosum of penis in the male
vaginal ♀	Internal pudendal		Vagina, bulb of vestibule, fundus of bladder and adjacent parts of the rectum

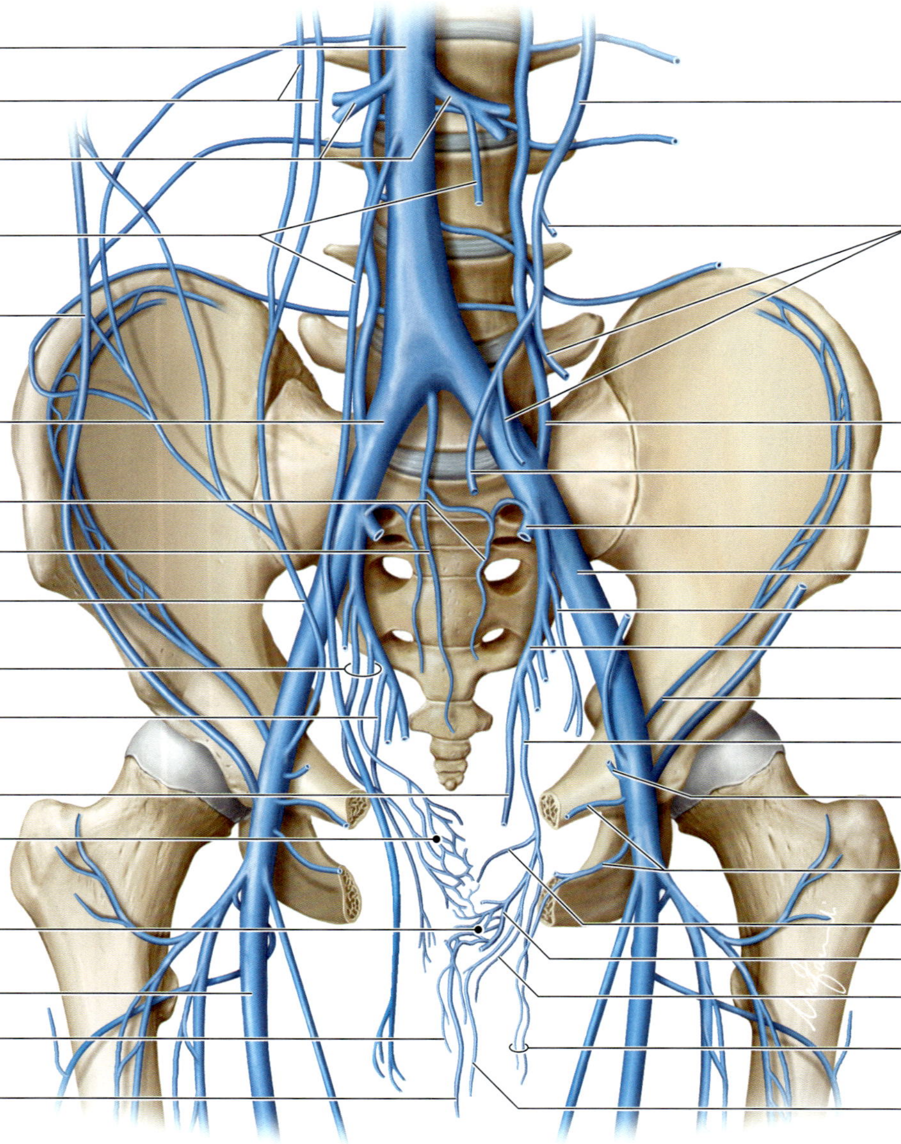

Veins in the male pelvis and perineum. The vesical and prostatic venous plexuses are altogether named vesico-prostatic venous plexus that is located laterally to the prostate and the inferior part of the bladder and extends in front of the bladder and prostate, behind the pubic symphysis. Here it constitutes the pudendal plexus, receiving the deep dorsal vein of the penis. The vesico-prostatic venous plexus by means of the vesical veins, and the pudendal plexus by means of the internal pudendal, drain into the internal iliac vein.

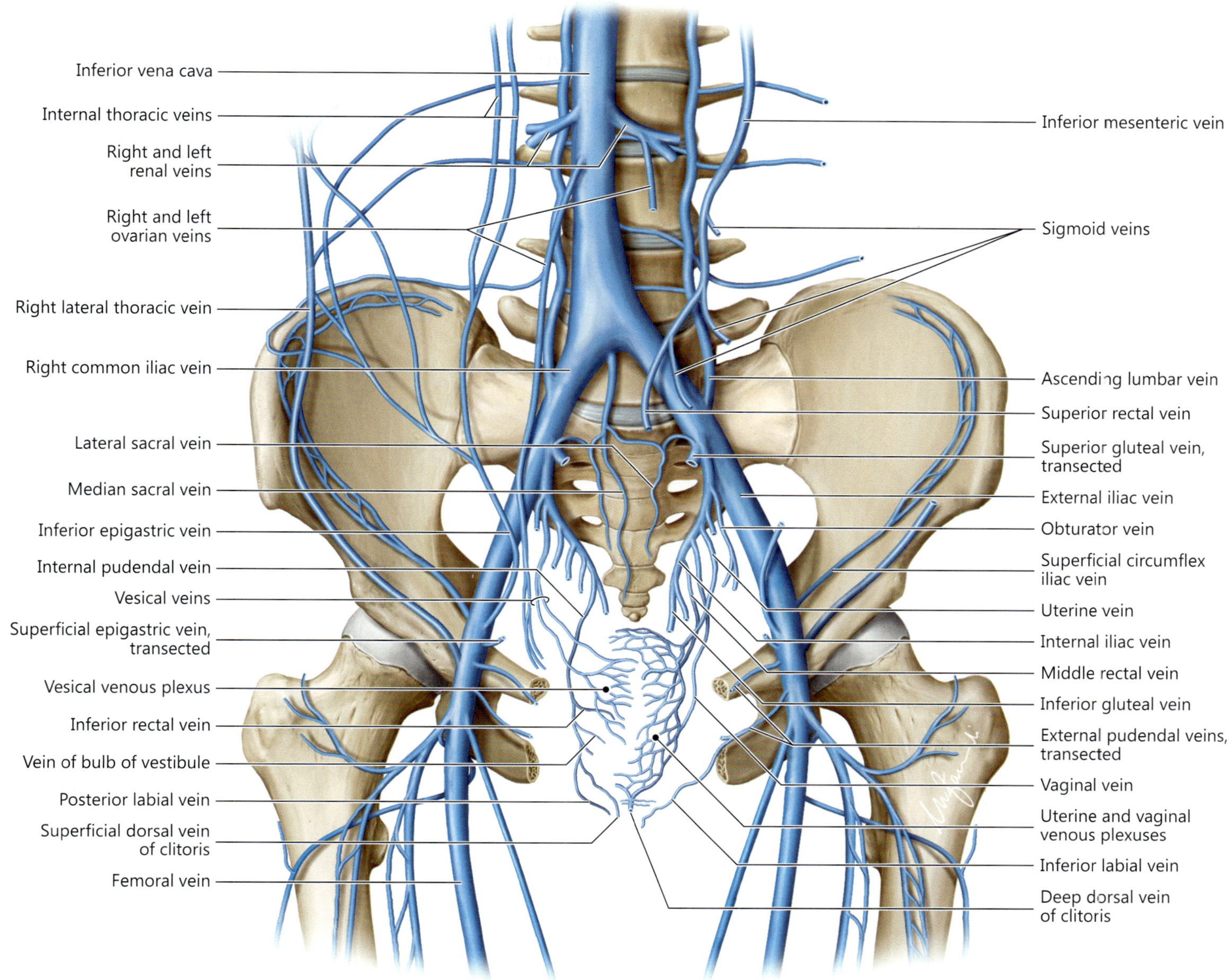

Veins in the female pelvis and perineum. The vesical, uterine and vaginal venous plexuses are also known as vesico-vaginal and utero-vaginal venous plexuses. The vesico-vaginal plexus drains into the internal iliac vein, by means of the vesical and vaginal veins. The utero-vaginal venous plexus, located at the sides of the uterus, in the thickness of the broad ligaments, extends downwards to the level of the superior part of the vagina, and drains, through the uterine veins, into the internal iliac vein.

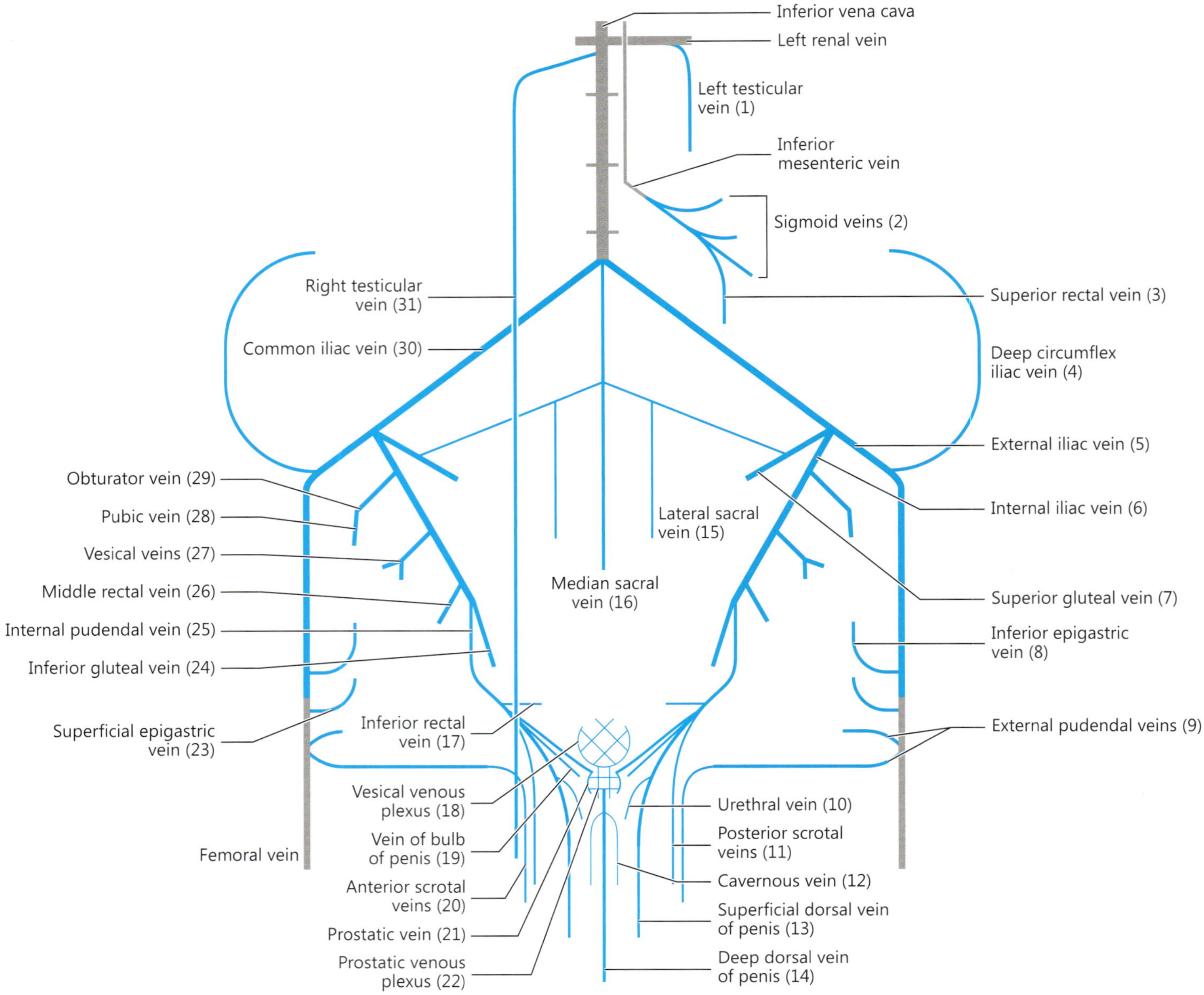

The **internal iliac vein**, or **hypogastric vein**, in the male drains blood from the pelvis and its content, the external genitalia, and part of the lower limb. It originates at the level of the superior border of the greater sciatic foramen and ascends anteriorly along the sacroiliac joint to the upper extremity, where joins the **external iliac vein** to form the **common iliac vein**. The internal iliac vein receives visceral branches from the venous plexuses surrounding the pelvic organs, i.e. the rectal, vesical and prostatic plexuses, the **internal pudendal vein**, receiving the **inferior rectal veins** and **deep veins of the penis**, the **middle rectal vein** and the **vesical veins**. Affluent parietal branches of the internal iliac vein are the **obturator**, **superior** and **inferior gluteal**, and **lateral sacral veins** and the **iliolumbar vein**. The latter can also drain into the common iliac vein.

Veins in the male pelvis and perineum: territories of drainage		
	VEINS	***TERRITORIES OF DRAINAGE***
1	Left testicular vein	Testis and epididymis
2	Sigmoid veins	Inferior part of the descending colon and sigmoid colon
3	Superior rectal vein	Rectum; anastomoses with the middle and inferior rectal veins (porto-caval anastomosis) through the rectal venous plexus
4	Deep circumflex iliac vein	Satellite veins of the deep circumflex iliac artery and small tributary branches of the inferior epigastric vein
5	External iliac vein	Lower limb and inferior abdominal wall; receives the deep circumflex iliac and inferior epigastric veins
6	Internal iliac vein	Pelvic organs, external genitalia, quadratus femoris and adductor muscles, hip joint, gluteal region and posterior region of thigh; receives the superior and inferior gluteal veins, obturator veins, median and lateral sacral veins, vesical veins, deep dorsal vein of penis through the vesical and prostatic plexuses (pudendal venous plexus), middle rectal veins, internal pudendal vein
7	Superior gluteal vein	Gluteal region; anastomosis with the inferior gluteal veins
8	Inferior epigastric vein	Rectus abdominis and external oblique muscles, subperitoneal tissue and skin at level of the umbilical region; anastomoses with the superior epigastric and subcutaneous abdominal veins (cava-caval anastomosis) and with the paraumbilical veins (porto-caval anastomosis)
9	External pudendal veins	Pubic region and external genitalia; they receive the superficial dorsal vein of penis and anterior scrotal veins; anastomoses with the subcutaneous abdominal veins (cava-caval anastomosis)
10	Urethral vein	Urethra and corpus spongiosum of penis
11	Posterior scrotal veins	Posterior part of the scrotum
12	Cavernous vein	Cavernous spaces of corpora cavernosa
13	Superficial dorsal vein of penis	Skin and subcutaneous tissue of penis
14	Deep dorsal vein of penis	Body of penis and glans penis; receives the cavernous veins and deep veins of penis
15	Lateral sacral vein	Skin and muscles of back proper at the sacral level; receives the intervertebral veins; anastomoses between the lateral sacral veins and with the median sacral vein (sacral venous plexus)
16	Median sacral vein	Anococcygeal region; anastomoses with the lateral sacral veins (sacral venous plexus)
17	Inferior rectal vein	Rectum; anastomoses with the superior (porto-caval anastomosis) and middle rectal veins through the rectal venous plexus
18	Vesical venous plexus	Urinary bladder and prostate; anastomoses with the prostatic venous plexus
19	Vein of bulb of penis	Expansive posterior part of the corpus spongiosum of penis
20	Anterior scrotal veins	Anterior part of the scrotum
21	Prostatic vein	Prostatic venous plexus
22	Prostatic venous plexus	Prostate; anastomoses with the vesical and rectal venous plexuses
23	Superficial epigastric vein	Inferior half of the abdominal wall; anastomoses with the thoracoepigastric (cava-caval anastomosis), superficial circumflex iliac and paraumbilical veins (porto-caval anastomosis)
24	Inferior gluteal vein	Posterior region of thigh and buttocks; anastomoses with the superior gluteal, medial circumflex femoral and perforating veins
25	Internal pudendal vein	Perineum (urogenital and anal triangles) and external genitalia; receives the posterior scrotal veins, veins of bulb of penis and inferior rectal veins
26	Middle rectal vein	Rectum, urinary bladder, prostate and seminal glands; anastomoses with the superior (porto-caval anastomosis) and inferior rectal veins through the rectal venous plexus
27	Vesical veins	Vesical venous plexus; anastomoses with the prostatic venous plexus
28	Pubic vein	Communication between the obturator vein and external iliac vein
29	Obturator vein	External genitalia, quadratus femoris and adductor muscles, hip joint; anastomoses with the medial circumflex femoral veins
30	Common iliac vein	Lower limb, pelvis, external genitalia and abdominal wall; receives the median sacral, lateral sacral, iliolumbar and superior gluteal veins, otherwise draining into the internal iliac vein
31	Right testicular vein	Testis and epididymis

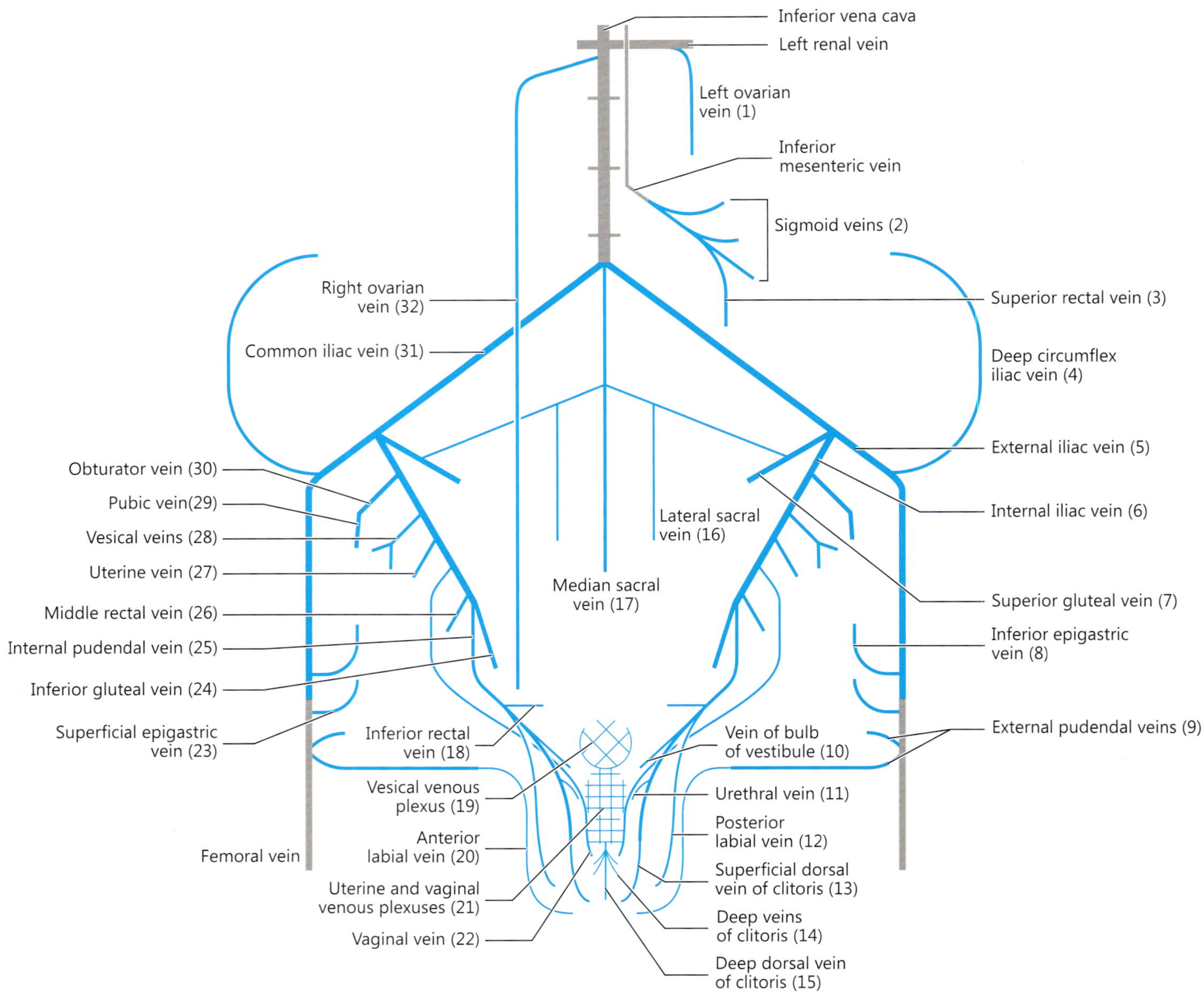

In addition to the **vesical and rectal venous plexuses**, other plexuses surround the female pelvic organs. In particular these are the **uterine and vaginal venous plexuses**, draining into the **internal iliac vein**, or **hypogastric vein**, by means of the **uterine veins**. The internal iliac vein drains blood from the pelvis and its content, the external genitalia, and part of the lower limb. It receives affluent visceral branches, i.e. the **internal pudendal vein**, receiving the **inferior rectal veins** and the **deep veins of the clitoris**, the **middle rectal vein**, receiving the rectal venous plexus together with the **inferior** and **superior rectal veins**, and the **vesical veins**, receiving the vesical venous plexus. Affluent parietal branches of the internal iliac vein are the **obturator**, **superior** and **inferior gluteal**, and **lateral sacral veins** and the **iliolumbar vein**. The latter can also drain into the common iliac vein.

Veins in the female pelvis and perineum: territories of drainage		
	VEINS	**TERRITORIES OF DRAINAGE**
1	Left ovarian vein	Ovary
2	Sigmoid veins	Inferior part of the descending colon and sigmoid colon
3	Superior rectal vein	Rectum; anastomoses with the middle and inferior rectal veins (porto-caval anastomosis) through the rectal venous plexus
4	Deep circumflex iliac vein	Satellite veins of the deep circumflex iliac artery and small tributary branches of the inferior epigastric vein
5	External iliac vein	Lower limb and inferior abdominal wall; receives the deep circumflex iliac and inferior epigastric veins
6	Internal iliac vein	Pelvic organs, external genitalia, quadratus femoris and adductor muscles, hip joint, gluteal region and posterior region of thigh; receives the superior and inferior gluteal veins, obturator veins, median and lateral sacral veins, vesical veins, deep dorsal vein of clitoris through the vesical and vaginal venous plexuses (pudendal venous plexus), uterine, middle rectal, internal pudendal veins
7	Superior gluteal vein	Gluteal region; anastomoses with the inferior gluteal veins
8	Inferior epigastric vein	Rectus abdominis and external oblique, subperitoneal tissue and skin at level of the umbilical region; anastomoses with the superior epigastric and subcutaneous abdominal veins (cava-caval anastomosis) and with the paraumbilical veins (porto-caval anastomosis)
9	External pudendal veins	Pubic region and external genitalia; they receive the superficial dorsal vein of clitoris and anterior labial veins; anastomoses with the subcutaneous abdominal veins (cava-caval anastomosis)
10	Vein of bulb of vestibule	Bulb of vestibule
11	Urethral vein	Urethra
12	Posterior labial vein	Posterior part of the labia majora, labia minora, vestibule of the vagina
13	Superficial dorsal vein of clitoris	Skin and subcutaneous tissue of the clitoris
14	Deep veins of clitoris	Crus of clitoris
15	Deep dorsal vein of clitoris	Body and glans of clitoris; receives the deep veins of clitoris
16	Lateral sacral vein	Skin and muscles of back proper at the level sacral; receives the intervertebral veins; anastomoses between the lateral sacral veins and with the median sacral vein (sacral venous plexus)
17	Median sacral vein	Anococcygeal region; anastomoses with the lateral sacral veins (sacral venous plexus)
18	Inferior rectal vein	Rectum; anastomoses with the superior (porto-caval anastomosis) and middle rectal veins through the rectal venous plexus
19	Vesical venous plexus	Urinary bladder; anastomoses with the uterine and vaginal venous plexuses
20	Anterior labial vein	Anterior part of the labia majora, mons pubis
21	Uterine and vaginal venous plexuses	Uterus and vagina; anastomoses with the vesical venous plexus
22	Vaginal vein	Vaginal venous plexus; anastomoses with the vesical, uterine and rectal venous plexuses
23	Superficial epigastric vein	Inferior half of the abdominal wall; anastomoses with the thoracoepigastric (cava-caval anastomosis), superficial circumflex iliac and paraumbilical veins (porto-caval anastomosis)
24	Inferior gluteal vein	Posterior region of thigh and buttocks; anastomoses with the superior gluteal, medial circumflex femoral and perforating veins
25	Internal pudendal vein	Perineum (urogenital and anal triangles) and external genitalia; receives the posterior labial veins, vein of bulb of vestibule and inferior rectal veins
26	Middle rectal vein	Rectum, urinary bladder and vagina; anastomoses with the superior (porto-caval anastomosis) and inferior rectal veins through the rectal venous plexus
27	Uterine vein	Uterine venous plexus; anastomoses with the vaginal and rectal venous plexuses
28	Vesical veins	Vesical venous plexus; anastomoses with the vaginal venous plexus
29	Pubic vein	Communication between the obturator vein and external iliac vein
30	Obturator vein	External genitalia, quadratus femoris and adductor muscles, hip joint; anastomoses with the medial circumflex femoral veins
31	Common iliac vein	Lower limb, pelvis, external genitalia and abdominal wall; receives the median sacral, lateral sacral, iliolumbar and superior gluteal veins (which, otherwise, drain into the internal iliac vein)
32	Right ovarian vein	Ovary

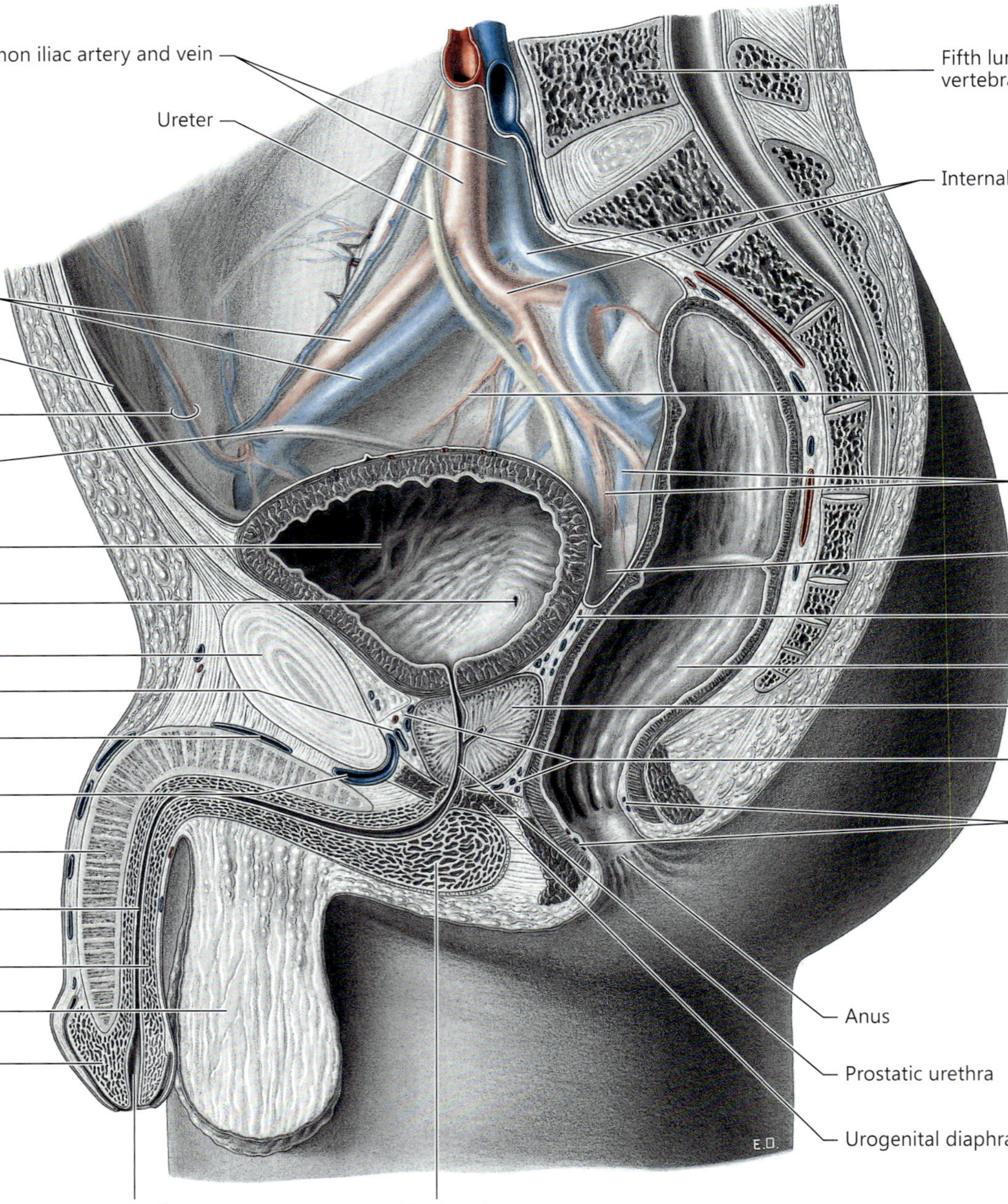

Median sagittal section of the male pelvis, right side, showing the course and relations of the **common iliac**, **internal iliac** and **external iliac veins** and **arteries**.

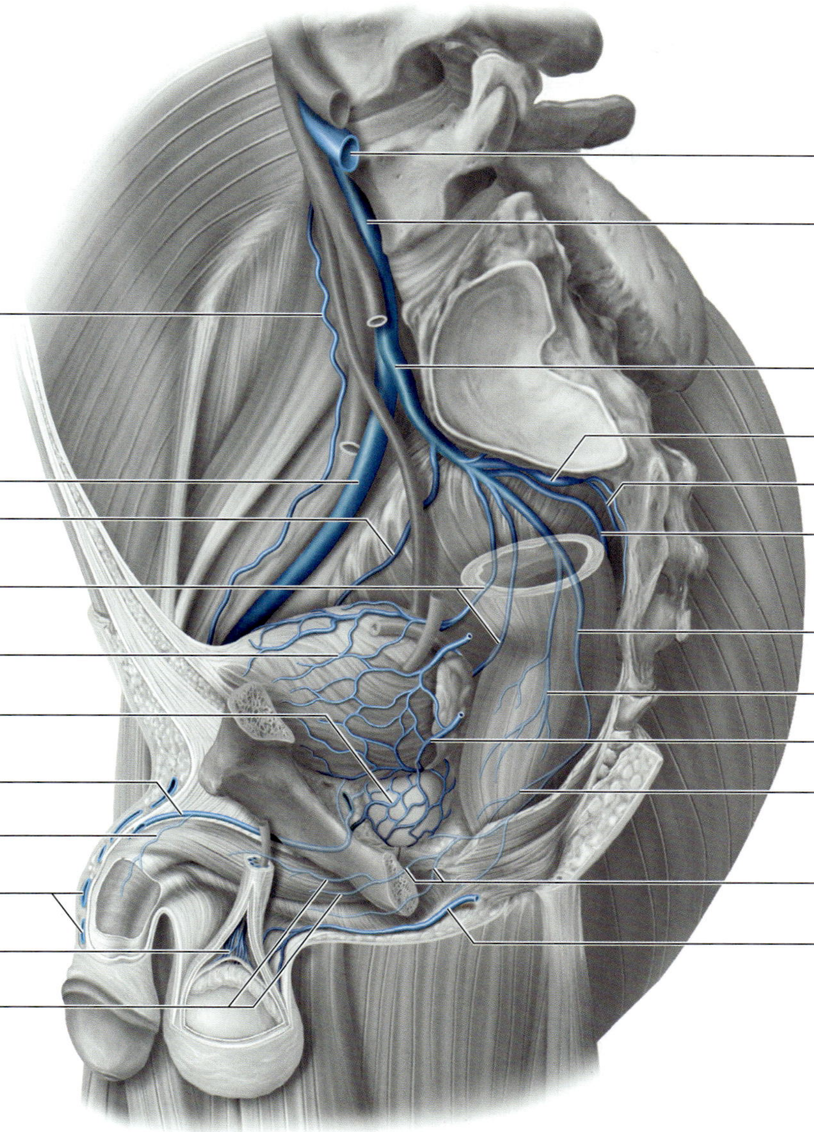

Veins in the male **pelvis and perineum**. The deep veins of the penis, veins of the bulb of the penis, deep dorsal vein of the penis and the posterior scrotal veins join to form the internal pudendal vein, a satellite of the homonymous artery. The internal pudendal vein drains into the internal iliac vein, and receives the inferior rectal veins and is in communication with the prostatic venous plexus. The vesical venous plexus, anastomosed with the prostatic venous plexus, is drained by means of the vesical veins, present in a variable number, which in turn drain into the internal iliac vein.

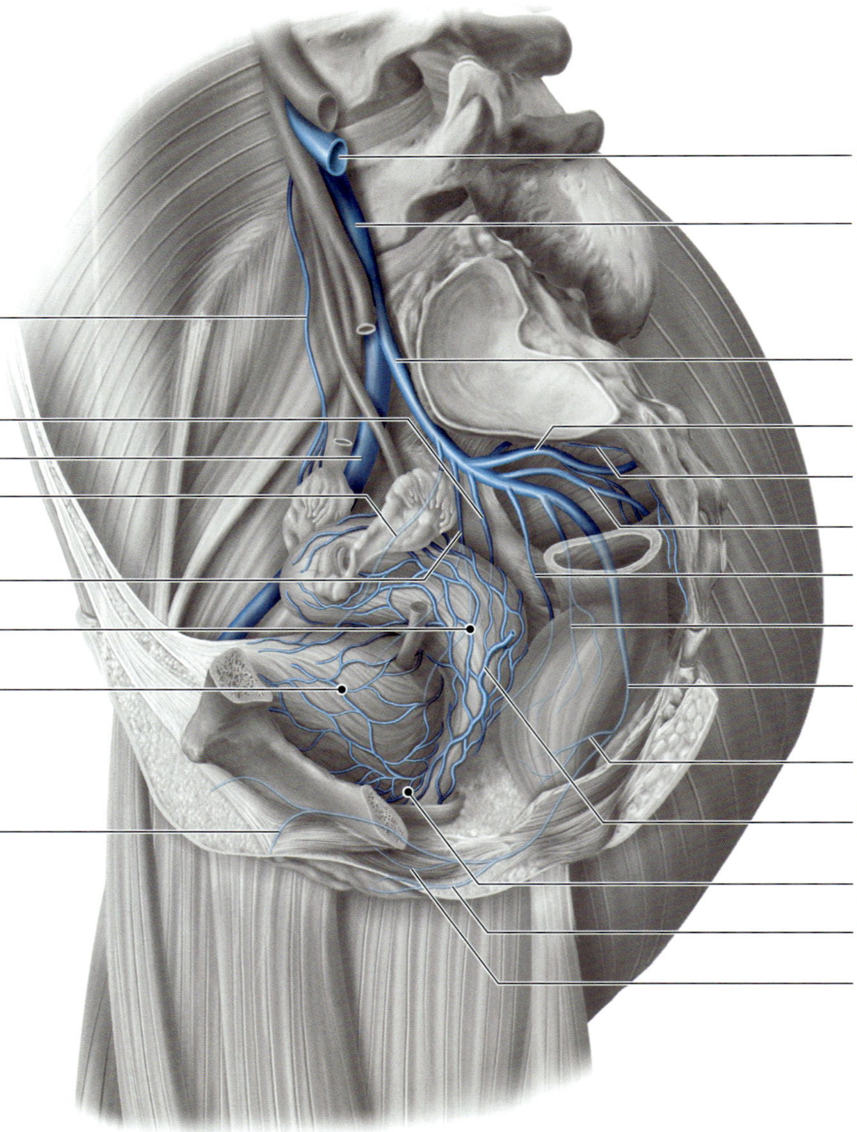

Veins in the female **pelvis and perineum**. The uterine and vaginal venous plexuses communicate anteriorly with the vesical venous plexus and, posteriorly, with the rectal or hemorrhoidal venous plexus. The plexuses (uterine, vaginal, vesical and middle rectal or hemorrhoidal) drain into the internal iliac vein. The rectal venous plexus is also drained by the superior rectal vein, a root of the inferior mesenteric vein, and the inferior rectal vein, a tributary of the internal pudendal vein.

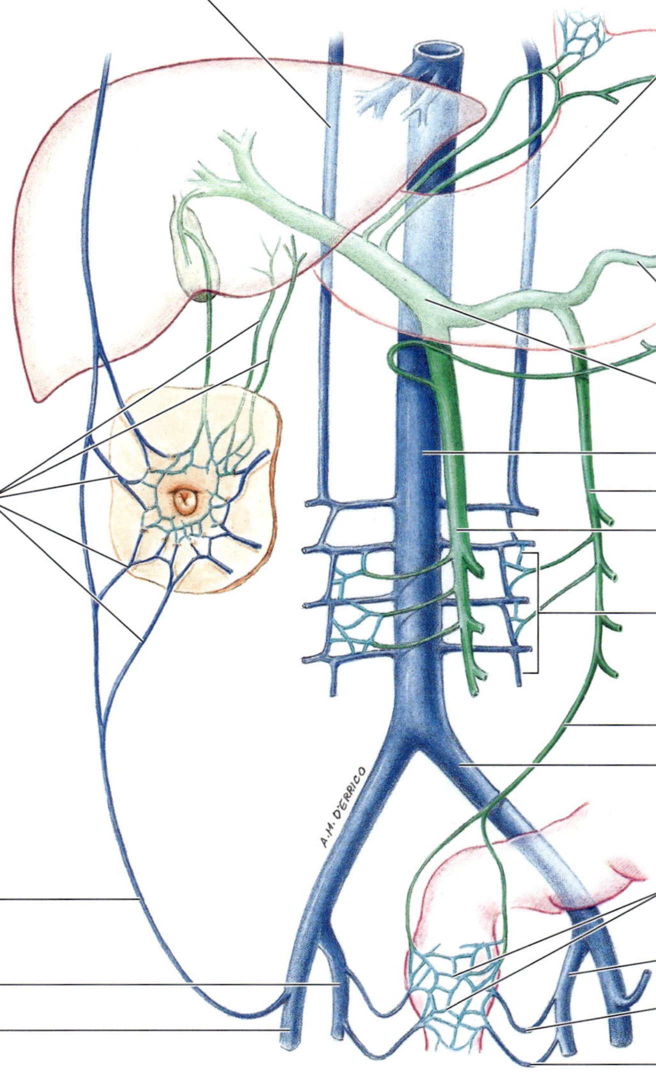

Schematic representation of the anastomoses between the hepatic portal vein system and the cava venous system. In *green*, the **system of the hepatic portal vein**, in *blue*, the **cava venous system**. In the pelvis, the porto-caval anastomoses form within the rectal, or hemorrhoidal, venous plexus draining into the superior rectal or hemorrhoidal vein, a root of the inferior mesenteric vein, the middle rectal or hemorrhoidal vein, a tributary of the internal iliac vein, and the inferior rectal or hemorrhoidal vein, a tributary of the internal pudendal vein.

Summary table of the veins of the pelvis

VEIN	ORIGIN/LOCATION	DRAINAGE	CONFLUENCE
anterior and posterior spinal (in the sacral segment of the vertebral column)	Longitudinal venous plexuses along the anterior and posterior surfaces of the spinal cord	Spinal cord	Anterior internal vertebral venous plexus
basivertebral (in the sacral segment of the vertebral column)	Spongy bone of the vertebral bodies; they emerge from foramina located on posterior surface of the vertebral bodies	Vertebral bodies; in communication with the anterior external vertebral venous plexus	Anterior internal vertebral venous plexus
common iliac	Confluence of the external and internal iliac veins at level of the sacroiliac joint, from where it ascends to the level of L V	Lower limb, pelvis, external genitalia and abdominal wall; receives the median sacral, lateral sacral, iliolumbar and superior gluteal veins (which, otherwise, drain into the internal iliac vein)	At the level of L V the two right and left common iliac veins unite at an acute angle forming the inferior vein cava to the right of the median line
deep circumflex iliac vein	Common trunk formed by satellite veins of the deep circumflex iliac artery	Satellite veins of the deep circumflex iliac artery and small tributary branches of the inferior epigastric vein	External iliac vein, approximately 2 cm above the inguinal ligament
external iliac	Continuation of the femoral vein below the inguinal ligament; enters the pelvis through the vascular space and joins with the internal iliac vein forming the common iliac vein at level of the sacroiliac joint	Lower limb and inferior abdominal wall; receives the deep circumflex iliac and inferior epigastric veins, a tributary of which is the pubic vein or accessory obturator vein	Common iliac vein
external pudendal	Single branches which go from the pubic region and external genitalia to the femoral triangle	Pubic region and external genitalia; they receive the superficial dorsal veins of clitoris ♀ or of penis ♂, anterior labial ♀ or anterior scrotal veins ♂; anastomoses with the subcutaneous abdominal veins (cava-caval anastomosis)	Great saphenous vein or femoral vein
inferior epigastric (deep epigastric)	Satellite of the inferior epigastric artery; descends laterally on the posterior surface of the rectus abdominis	Rectus abdominis and external oblique, subperitoneal tissue and skin at level of the umbilical region; anastomoses with the superior epigastric and subcutaneous abdominal veins (cava-caval anastomosis) and with the paraumbilical veins (porto-caval anastomosis)	External iliac vein, approximately 1 cm above the inguinal ligament
inferior gluteal	Satellites of the inferior gluteal artery, they enter the pelvis through the inferior part of the greater sciatic foramen	Posterior region of thigh and buttocks; anastomoses with the superior gluteal, medial circumflex femoral and perforating veins	Internal iliac vein
inferior rectal (inferior hemorrhoidal)	From the inferior part of the rectal venous plexus, situated in correspondence to the posterior and lateral walls of the rectum	Rectum; anastomoses with the superior (porto-caval anastomosis) and middle rectal veins through the rectal venous plexus	Internal pudendal vein
internal iliac (hypogastric)	At level of superior border of the greater sciatic foramen; it ascends along the sacroiliac joint and at the upper extremity joins with the external iliac vein to form the common iliac vein	Pelvic organs, external genitalia, quadratus femoris and adductor muscles, hip joint, gluteal region and posterior region of thigh; receives the superior and inferior gluteal veins, obturator veins, median and lateral sacral veins, vesical veins, deep dorsal vein of clitoris ♀ or of penis ♂, through the vesical and prostatic venous ♂ or vaginal ♀ (pudendal venous plexus), uterine ♀, middle rectal, internal pudendal	Common iliac vein
internal pudendal	Originates inferiorly to the pubic symphysis from a branch of the deep dorsal vein of penis ♂ or of clitoris ♀; runs in the lateral wall of the ischioanal fossa; enters the pelvis passing through the inferior part of the greater sciatic foramen	Perineum (urogenital and anal triangles) and external genitalia; receives the posterior labial ♀, or posterior scrotal veins ♂, vein of bulb of vestibule ♀ or of penis ♂, inferior rectal veins	Internal iliac vein
intervertebral (in the sacral segment of the vertebral column)	Anterior and posterior internal vertebral venous plexuses; passes through the intervertebral foramen accompanied by the roots and initial part of the spinal nerve	Spinal cord; in part, also the anterior and posterior external vertebral venous plexuses drain into it	Lateral sacral veins

continued

Summary table of the veins of the pelvis			*(continued)*
VEIN	*ORIGIN/LOCATION*	*DRAINAGE*	*CONFLUENCE*
lateral sacral	Margins of sacrum	Skin and muscles of back proper at sacral level; receives the intervertebral veins; anastomoses between the lateral sacral veins and with the median sacral vein (sacral venous plexus)	Common iliac vein or internal iliac vein
left ovarian ♀ (gonadal)	☞ right ovarian ♀	☞ right ovarian ♀	Left renal vein
left testicular ♂ (internal spermatic or gonadal)	☞ right testicular ♂	☞ right testicular ♂	Left renal vein
median sacral	Pelvic surface of sacrum	Anococcygeal region; anastomoses with the lateral sacral veins (sacral venous plexus)	Common iliac vein or internal iliac vein
middle rectal (middle hemorrhoidal)	From the rectal venous plexus, situated in correspondence to the posterior and lateral walls of the rectum	Rectum; urinary bladder, prostate and seminal glands in the male, vagina in the female; anastomoses with the superior (porto-caval anastomosis) and inferior rectal veins through the rectal venous plexus	Internal iliac vein
obturator	Satellites of the obturator artery, they enter the pelvis through the superior part of the obturator foramen	External genitalia, quadratus femoris and adductor muscles, hip joint; anastomoses with the medial circumflex femoral veins	Internal iliac vein
prostatic ♂	At sides of the prostate	Prostatic venous plexus ♂; anastomoses with the vesical and rectal venous plexuses	Internal iliac vein
pubic (accessory obturator vein or pubic branch of inferior epigastric vein)	Originates from the obturator vein in the obturator canal, ascends along the pubic branch of inferior epigastric artery	Communication between the obturator vein and external iliac vein	External iliac vein
right ovarian ♀ (gonadal)	Venous plexus in the broad ligament of uterus ♀	Ovary	Inferior vein cava
right testicular ♂ (internal spermatic or gonadal)	Roots from the posterior border of testis and head of epididymis which anastomose with one another and with posterior venous branches of the spermatic cord, giving rise to the pampiniform plexus, whose anterior veins, after passing the inguinal canal, proceed on in one or two veins at the sides of the testicular artery ♂	Testis, epididymis	Inferior vein cava
sigmoid	Satellites of the sigmoid arteries	Inferior part of the descending colon and sigmoid colon	Inferior mesenteric vein
superficial epigastric	Runs from the umbilical region downwards and laterally towards the inguinal ligament	Inferior half of the abdominal wall; anastomoses with the thoracoepigastric (cava-caval anastomosis), superficial circumflex iliac and paraumbilical veins (porto-caval anastomosis)	Great saphenous vein
superior gluteal	Satellites of the superior gluteal artery, they enter the pelvis through the greater sciatic foramen	Gluteal region; anastomoses with the inferior gluteal veins	Internal iliac vein
superior rectal (superior hemorrhoidal)	From the rectal venous plexus, situated in correspondence to the posterior and lateral walls of the rectum; ascends, crossing the left common iliac vessels, and continues on as the inferior mesenteric vein	Rectum; anastomoses with the middle and inferior rectal veins (porto-caval anastomosis) through the rectal venous plexus	Inferior mesenteric vein
urethral	Urethra	Urethra, corpus spongiosum of penis in the male	Internal pudendal vein through the prostatic venous plexus or the vaginal venous plexus
uterine ♀	Root of the broad ligament of uterus ♀	Uterine venous plexus ♀; anastomoses with the vaginal ♀ and rectal venous plexuses	Internal iliac vein

continued

Summary table of the veins of the pelvis			(continued)
VEIN	ORIGIN/LOCATION	DRAINAGE	CONFLUENCE
vaginal	Around the vagina	Vaginal venous plexus ♀; anastomoses with the vesical, uterine ♀ and rectal venous plexuses	Internal iliac vein
vesical	Fundus of bladder	Vesical venous plexus; anastomoses with the prostatic ♂ and vaginal venous plexuses ♀	Internal iliac vein

Summary table of the veins of the perineum			
VEIN	ORIGIN/LOCATION	DRAINAGE	CONFLUENCE
anterior labial ♀	Pudendum	Anterior part of the labia majora, mons pubis	External pudendal veins
anterior scrotal ♂	Scrotum	Anterior part of scrotum	External pudendal veins
cavernous ♂	Corpora cavernosa of penis	Cavernous spaces of corpora cavernosa	Deep dorsal vein of penis
deep dorsal vein of clitoris ♀ (dorsal of clitoris)	Median unpaired vein on dorsal surface of the clitoris, situated deeply to the fascia of clitoris	Body and glans of clitoris; receives the deep veins of clitoris	Vesical and vaginal ♀ venous plexuses (pudendal venous plexus) draining into the internal iliac vein
deep dorsal vein of penis ♂ (dorsal of penis)	Median unpaired vein on dorsum of the penis, situated deeply to the fascia of penis	Body of penis and glans penis; receives the cavernous veins and deep veins of penis	Vesical and prostatic ♂ venous plexuses (pudendal venous plexus) draining into the internal iliac vein
deep veins of clitoris ♀	Roots of the corpora cavernosa of clitoris, deeply to the fascia of clitoris	Crus of clitoris	Through the deep dorsal vein of clitoris ♀ into the vesical and vaginal ♀ venous plexuses (pudendal venous plexus) which drain into the internal iliac vein
deep veins of penis ♂	Roots of the corpus spongiosum and corpora cavernosa of penis, deeply to the fascia of penis	Root of penis	Through the deep dorsal vein of penis ♂ into the vesical and prostatic ♂ venous plexuses (pudendal venous plexus) which drain into the internal iliac vein
external pudendal	Single branches that go from the pubic region and external genitalia to the femoral triangle	Pubic region and external genitalia; they receive the superficial dorsal veins of clitoris ♀ or of penis ♂, anterior labial ♀ and anterior scrotal ♂ veins; anastomoses with the subcutaneous abdominal veins (cava-caval anastomosis)	Great saphenous vein or femoral vein
internal pudendal	Originates inferiorly to the pubic symphysis from a branch of the deep dorsal vein of penis ♂ or of clitoris ♀; runs in the lateral wall of the ischioanal fossa; enters the pelvis passing through the inferior part of the greater sciatic foramen	Perineum (urogenital and anal triangles) and external genitalia; receives the posterior labial ♀, posterior scrotal ♂, of bulb of vestibule ♀ or of penis ♂, inferior rectal veins	Internal iliac vein
of bulb of penis ♂	Bulb of penis	Expansive posterior part of the corpus spongiosum of penis	Internal pudendal vein
of bulb of vestibule ♀	Bulb of vestibule	Bulb of vestibule	Internal pudendal vein
posterior labial ♀	Pudendum	Posterior part of the labia majora, labia minora, vestibule of the vagina	Internal pudendal vein
posterior scrotal ♂	Scrotum	Posterior part of scrotum	Internal pudendal vein
superficial dorsal veins of clitoris ♀	Pair of veins, on both sides, located on dorsal surface of the clitoris, superficially to the fascia of clitoris	Skin and subcutaneous tissue of clitoris	External pudendal veins
superficial dorsal veins of penis ♂	Pair of veins, on both sides, located on dorsum of the penis, superficially to the fascia of penis	Skin and subcutaneous tissue of penis	External pudendal veins

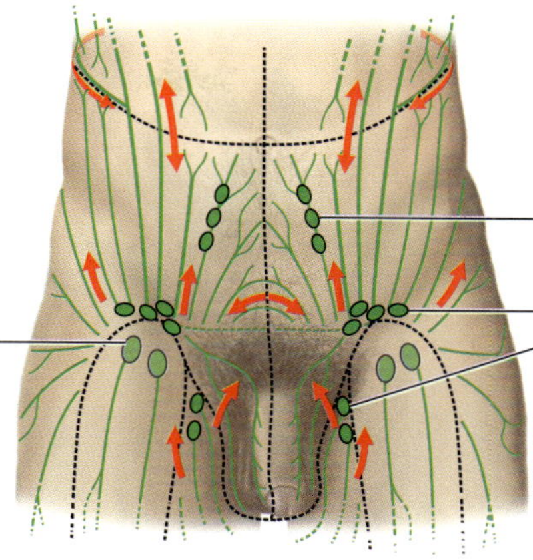

Projection of the **inguinal** and **common iliac nodes** on the body surface. The inguinal nodes are tributaries of the external iliac nodes, which in turn are tributaries of the common iliac nodes, draining lymph into the lateral aortic nodes.

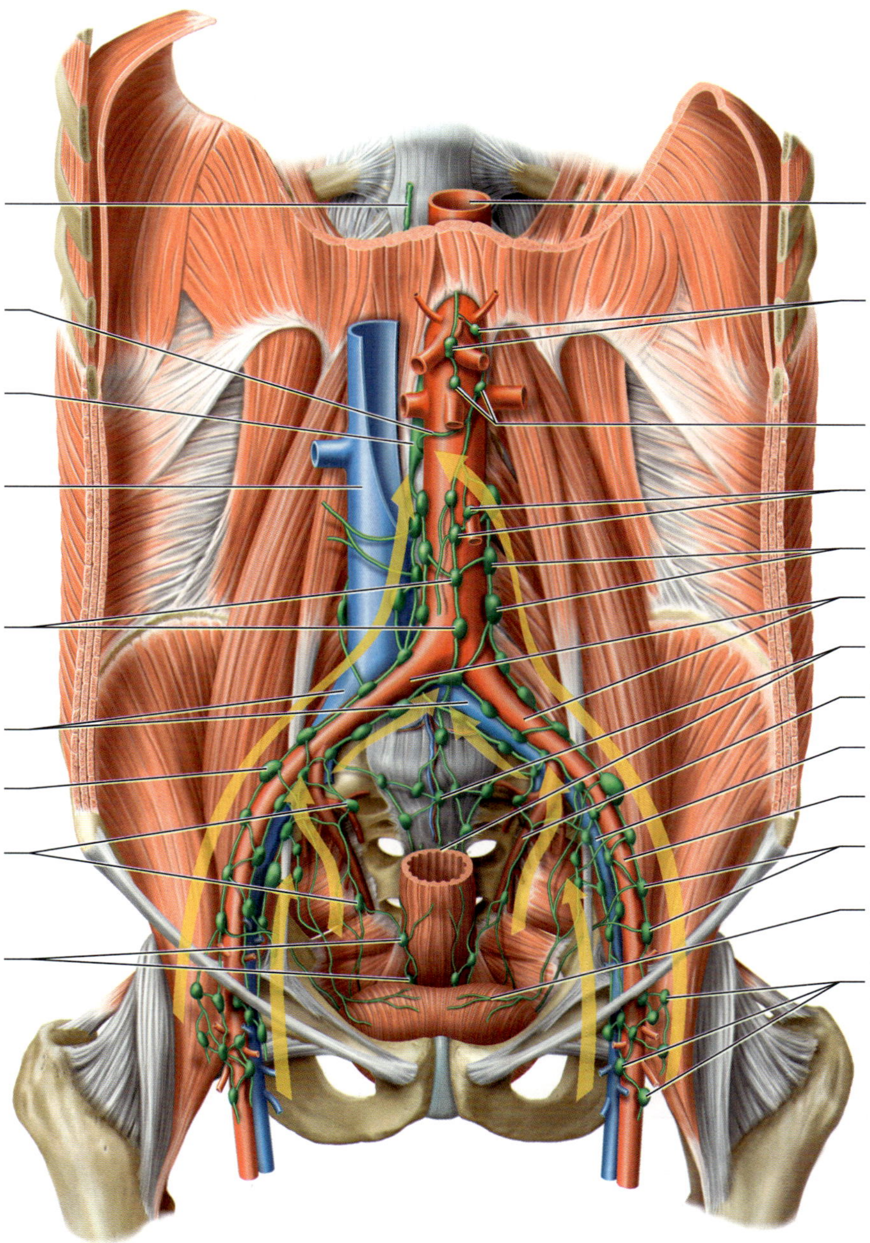

Parietal lymph nodes of the pelvis: common iliac, external iliac and internal iliac nodes. One can also see the inguinal lymph nodes, receiving the lymphatic collectors of the lower limb and from the walls of the abdomen, gluteal region, external genitalia and perineum.

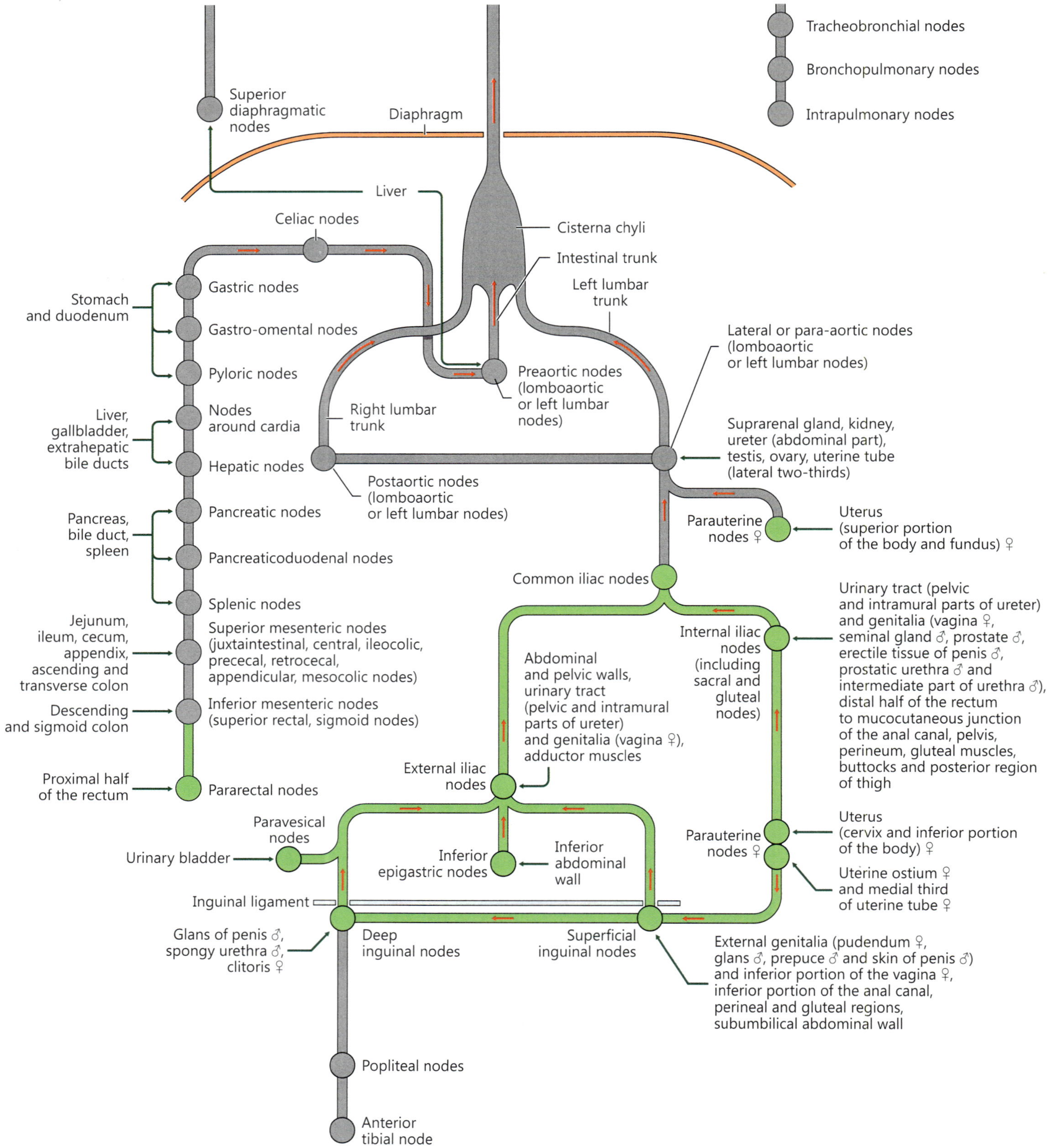

Schematic representation of the **lymphatic drainage of the pelvis and perineum**. The common iliac nodes receive the lymphatic collectors from the internal iliac nodes (draining the pelvic organs, perineum, gluteal region and posterior femoral region) and from the external iliac nodes (draining the abdominal and pelvic walls, ureter, vagina and adductor muscles of the thigh). The external iliac nodes also receive lymphatic collectors from the inguinal nodes (draining the subumbilical abdominal wall, perineum, external genitalia, gluteal region and lower limb).

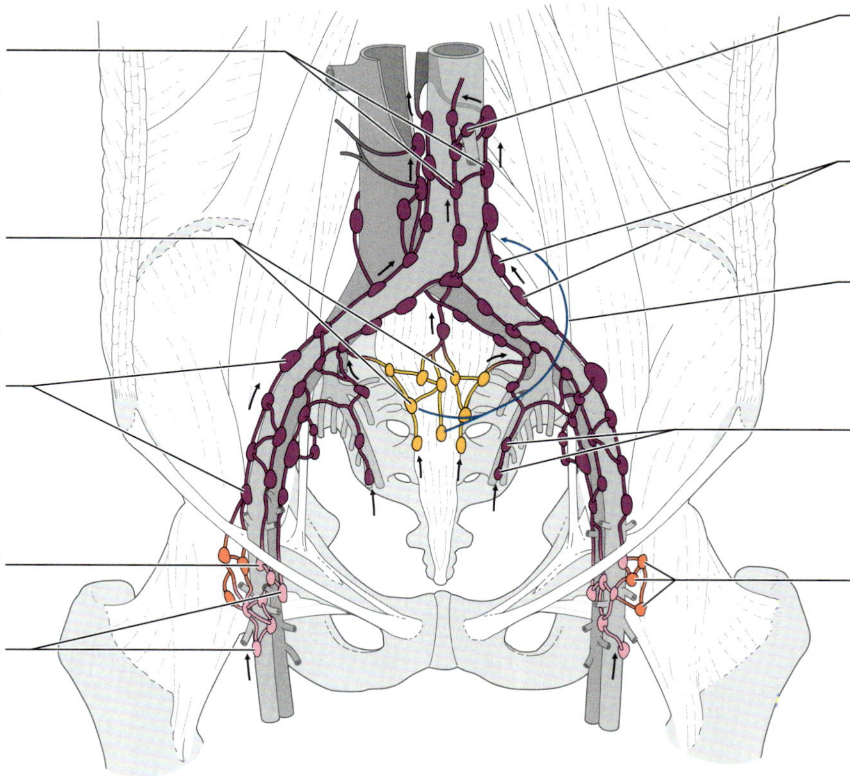

Parietal lymph nodes of the pelvis. The external iliac nodes receive collectors from the inguinal nodes, the abdominal and pelvic walls, genitalia, urinary tract and the adductor muscles. They send efferent collectors to the common iliac nodes, which also receive collectors from the internal iliac nodes. The internal iliac nodes, in turn, receive collectors from the urinary tract and genitalia and from the perineum, rectum, pelvic cavity and posterior muscles of the thigh and buttocks. The common iliac nodes also receive collectors from the pelvic organs and drain into the lateral aortic nodes.

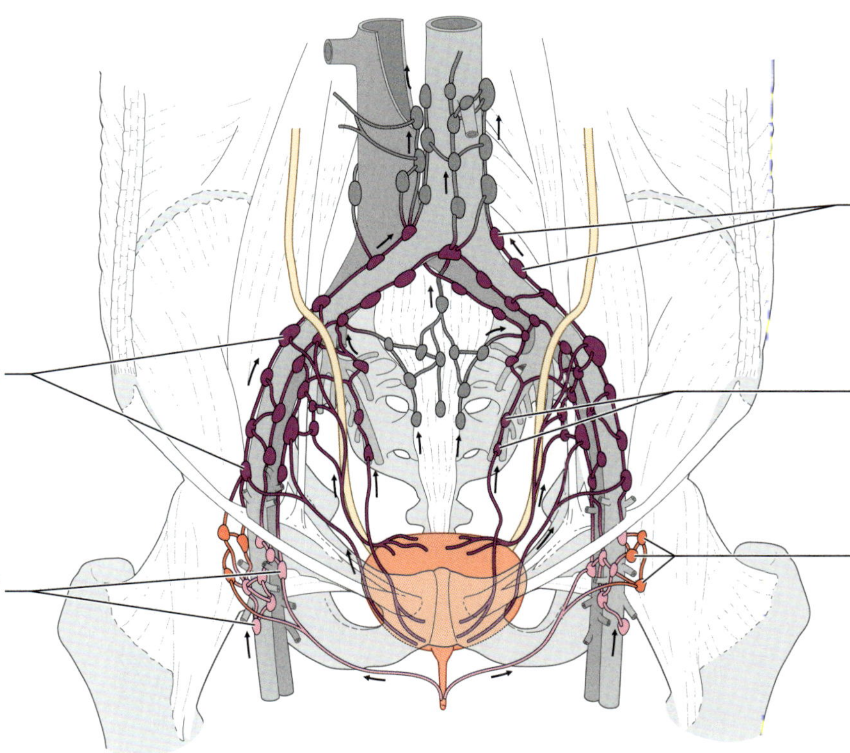

The **lymphatic drainage of the bladder** is carried out mostly by the external iliac nodes and partly by the internal iliac nodes which receive the lymphatic vessels of the female urethra and of the preprostatic, prostatic and intermediate parts of the male urethra, while the spongy urethra drains prevalentely into the deep inguinal nodes.

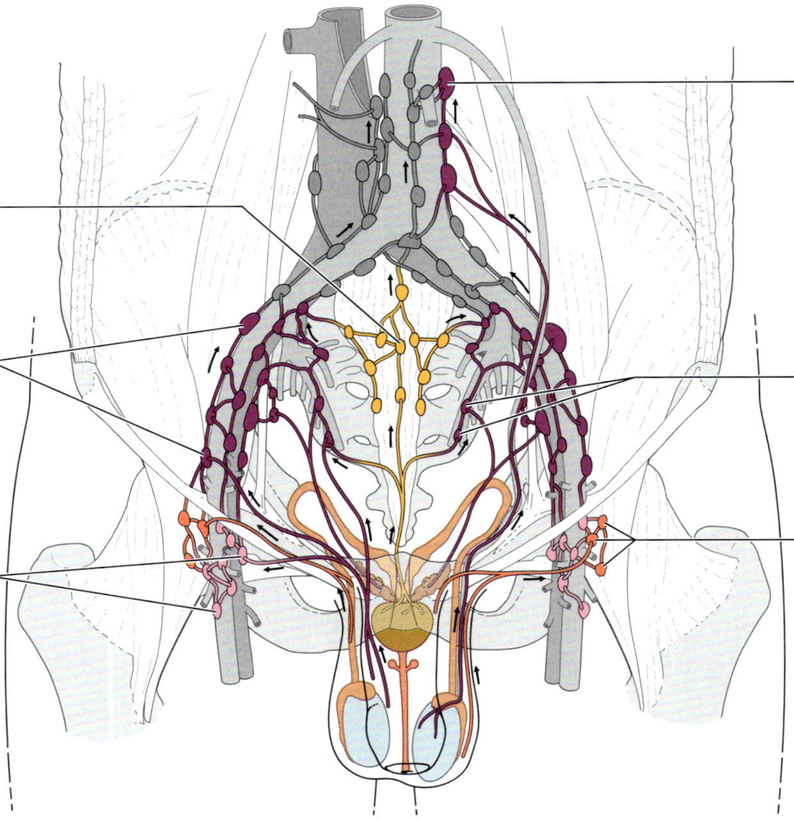

Lymphatic drainage of the male genitalia. The testis drains into the lateral aortic nodes through lymphatic collectors running in the spermatic cord. The lymphatic collectors coming from the seminal glands, prostate, erectile tissue of the penis and urethra drain towards the internal iliac nodes. The scrotum drains into the superficial inguinal nodes. Lymph from the glans penis drains into the deep inguinal nodes.

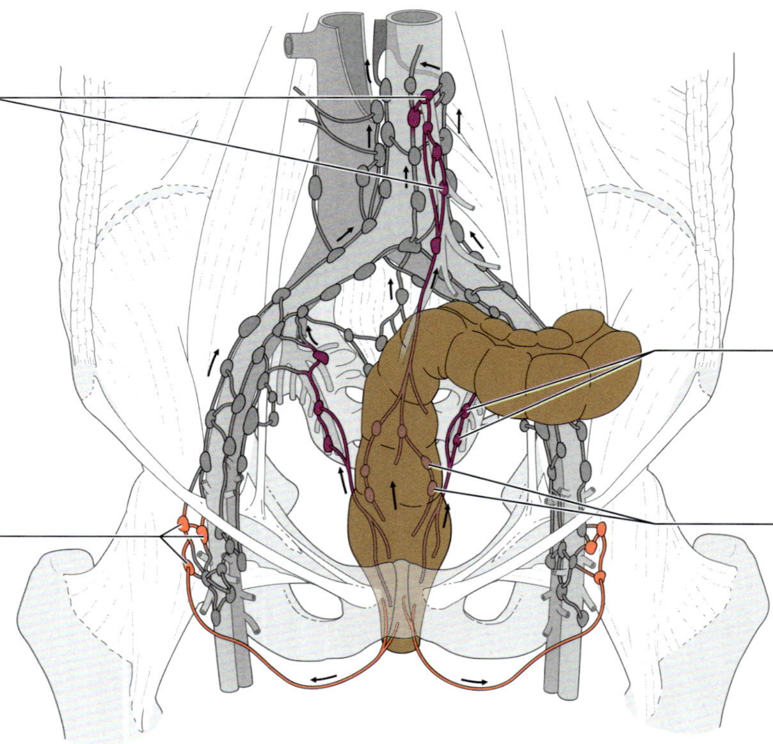

Lymphatic drainage of the rectum. Lymph of the proximal half of the rectum is collected by the pararectal nodes, which drain into the inferior mesenteric nodes. The lymphatic vessels coming from the distal half of the rectum and from the proximal part of the anal canal up to the anal transition zone are tributaries of the internal iliac nodes. The part of the anal canal distal to the anal transition zone drains into the superomedial group of the superficial inguinal nodes.

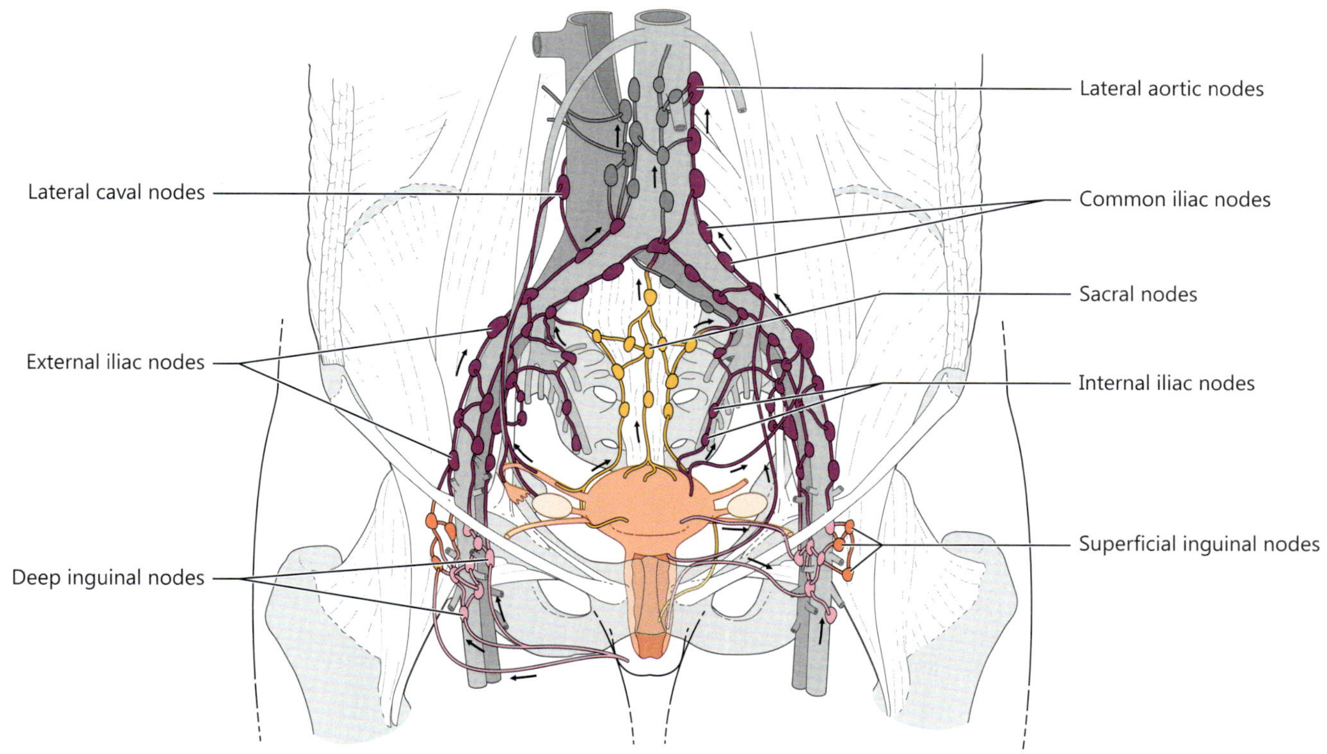

Lymphatic drainage of the female genitalia. The ovary, lateral two-thirds of the uterine tube, and base and superior part of the body of the uterus drain into the lateral aortic nodes. The lymphatic collectors from the parauterine nodes of the inferior part of the body of the uterus and the cervix, and those from the paravaginal nodes of the superior part of the vagina, drain towards the external and internal iliac nodes. The medial third of the uterine tube, the part of uterus adjacent to the uterine ostium of the tube, the inferior part of the vagina and the vulva all drain into the superficial inguinal nodes. The female urethra drains into the internal iliac nodes and the glans clitoris into the deep inguinal nodes.

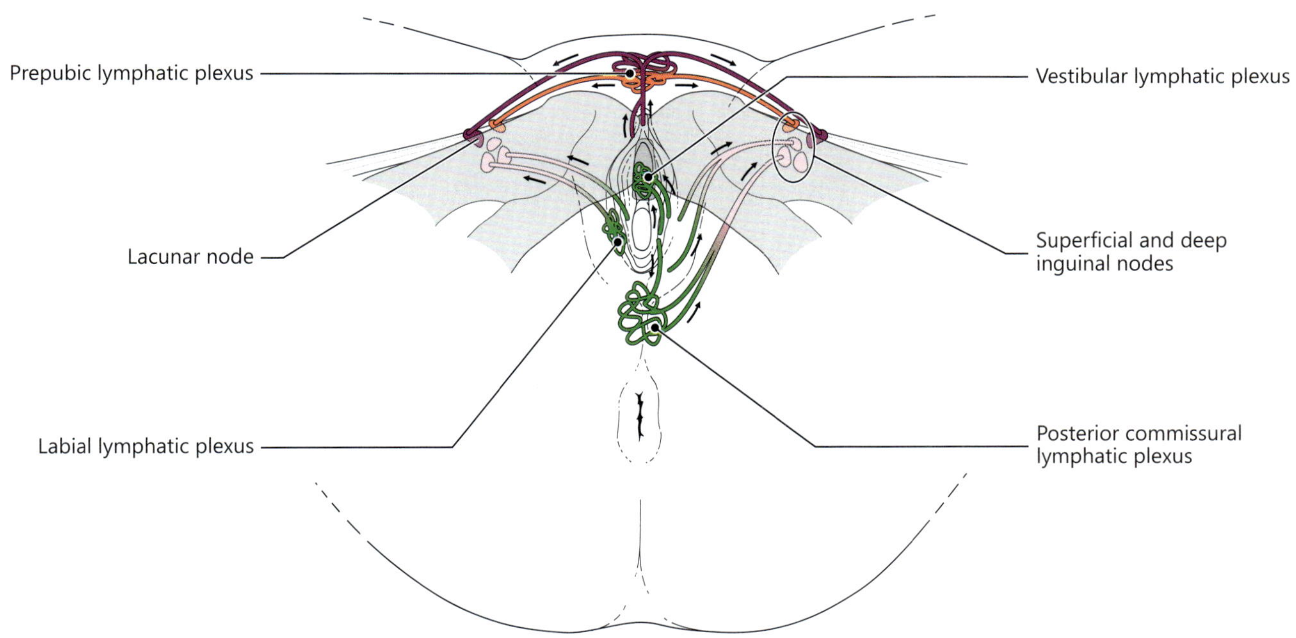

Lymphatic drainage of the vulva runs towards the superficial inguinal nodes through lymphatic plexuses located in the labia majora. The vestibule of the vagina drains into the pubic or presymphyseal lymphatic plexus, efferent to the lacunar node, if present, belonging to the external iliac nodes, and into the deep inguinal nodes.

Summary table of the lymph nodes of the pelvis

LYMPH NODE	*GROUP*	*SITE*	*AFFERENTS*	*EFFERENTS*
			PARIETAL LYMPH NODES	
Common iliac – **promontorial** – **intermediate** – **lateral** – **medial** – **subaortic**	Chains interposed in the common iliac plexus	Deep	Middle and pelvic parts of ureter, external and internal iliac nodes	Lateral aortic nodes
External iliac – **interiliac** – **intermediate** – **intermediate lacunar** (inconstant) – **lateral lacunar** (inconstant) – **medial lacunar** (inconstant) – **lateral** – **medial** – **obturator**	Chains interposed in the external iliac plexus	Deep	Pelvic part of ureter, pelvic and abdominal walls (deep parietal drainage), efferent collectors of the inguinal lymph nodes, urinary bladder, cervix and inferior region of the body of uterus, vagina, adductor muscles of the thigh	Common iliac nodes
Internal iliac – **gluteal** – – **inferior** – – **superior** – **sacral**	Chains interposed in the internal iliac plexus	Deep	Lower half of the rectum and part of the anal canal proximally to anal transition zone, prostatic urethra, intermediate part of urethra and spongy urethra in the male, female urethra, cervix and inferior region of the body of uterus, vagina, prostate and seminal glands, ductus deferens, corpora cavernosa of penis, perineum, gluteal muscles and muscles of the posterior region of thigh	Common iliac nodes
			VISCERAL LYMPH NODES	
Pararectal (**anorectal**)		Deep	Upper half of the rectum	Inferior mesenteric nodes through the sigmoid nodes
Parauterine ♀		Deep	Uterus	External and internal iliac nodes, lateral aortic and superficial inguinal nodes
Paravaginal ♀		Deep	Vagina	External and internal iliac nodes and superficial inguinal nodes
Paravesical – **prevesical** – **postvesical** – **lateral vesical**		Deep	Urinary bladder	External iliac nodes
			LYMPH NODES OF LOWER LIMB WITH TERRITORY OF PELVIC DRAINAGE	
Inguinal		Superficial and deep		
– **deep** – – **distal** – – **intermediate** (inconstant) – – **proximal** (inconstant) (**node of Cloquet**)		Deep	Deep efferent collectors of the lower limb, external genitalia (glans penis or of clitoris), some efferent collectors of the superficial inguinal nodes	External iliac nodes
– **superficial**		Superficial		
– – **inferior** (**inferomedial** and **inferolateral**) – – **superolateral** – – **superomedial**	Vertical chain Oblique superior chain Oblique superior chain		Superficial efferent collectors of the lower limb except those of posterior part of lateral surface, tributaries of the popliteal nodes whose deep collectors drain towards the deep inguinal nodes, part of the anal canal distally to the anal transition zone, anterolateral and posterior abdominal walls (superficial parietal drainage), gluteal region, external genitalia (glans, prepuce and skin of penis, inferior part of the vagina, pudendum), perineum and perianal region, parts of the subumbilical abdominal wall adjacent to the inguinal ligament, medial third of the uterine tube and parts of the uterus in the zone of the uterine ostium	External iliac nodes and to a minor extent the deep inguinal nodes

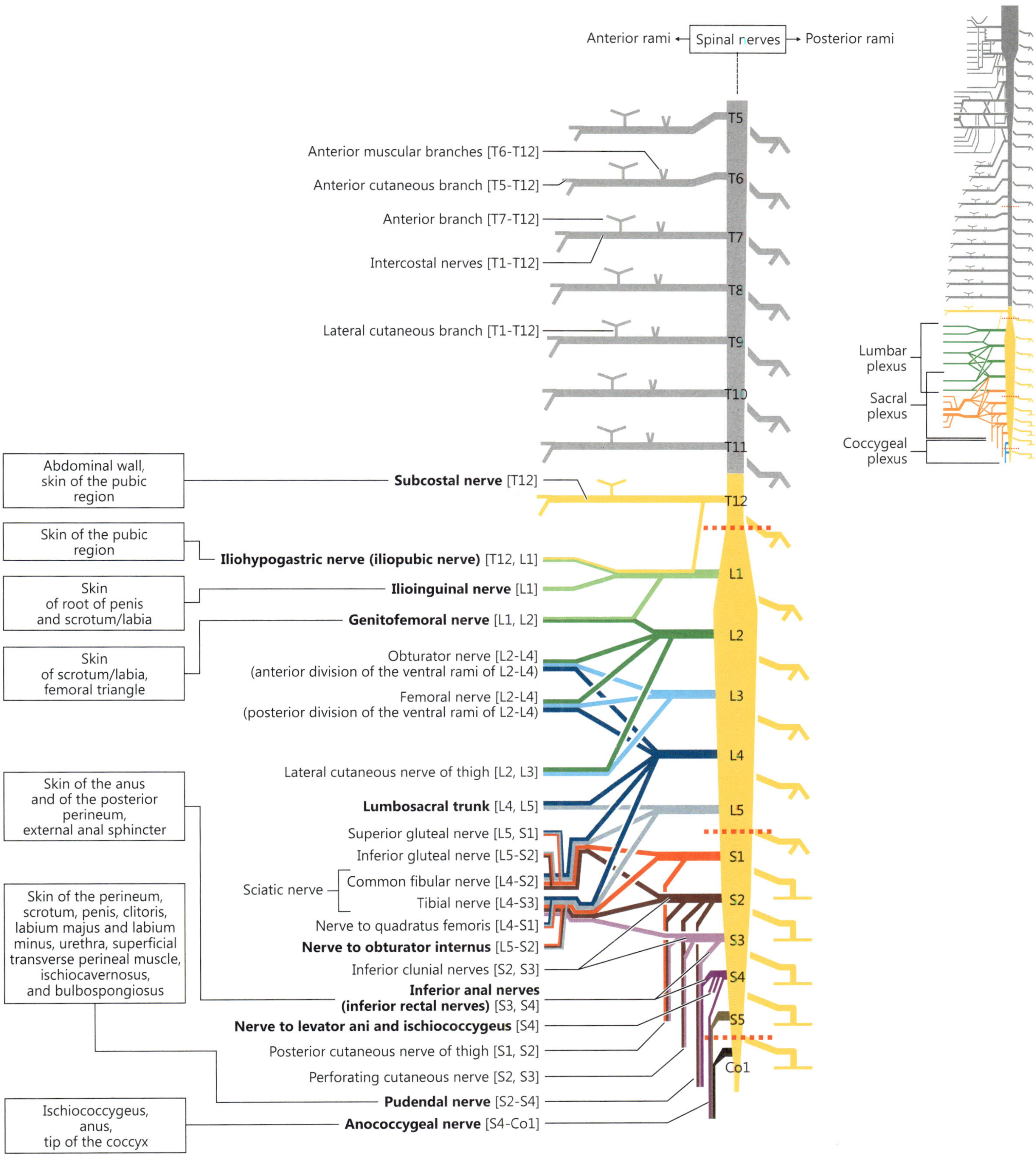

Schematic representation of the **lumbar**, **sacral** and **coccygeal plexuses**. The lumbar plexus is composed of the anterior rami of L1-L4 with a contribution from T12; from this plexus arise anastomotic, collateral and terminal branches. Following the lumbar plexus is the sacral plexus composed of the anterior rami of L5, S1-S4, with a contribution from L4. These roots join together in a common trunk originating the sciatic nerve and diverse collateral branches. Below the sacral plexus is the coccygeal plexus, formed by the anterior rami of S5-Co1 with a contribution from S4; it provides anterior, visceral, and posterior branches, the main posterior branch being the anococcygeal nerve. In the right part of the picture are represented the dorsal branches of the intercostal, lumbar, sacral and coccygeal nerves.

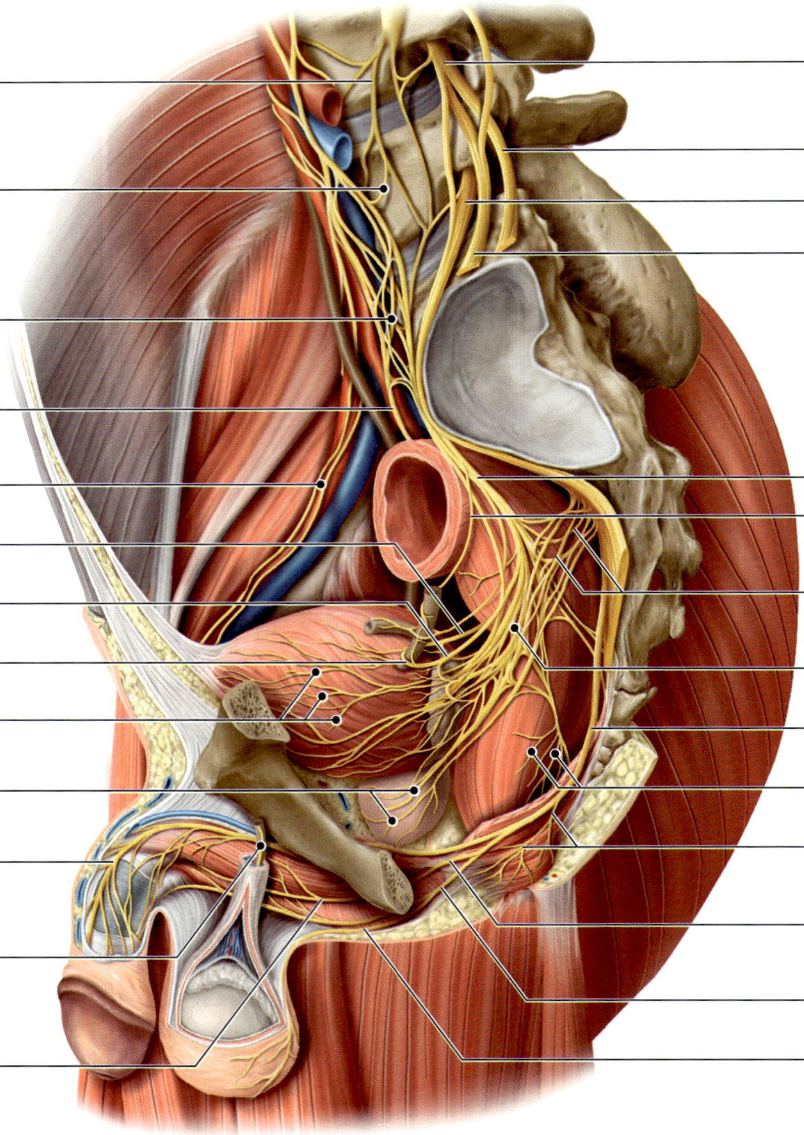

Superior and inferior hypogastric plexuses. The superior hypogastric plexus originates from the abdominal aortic plexus, occupies the space subjacent to the aortic bifurcation and is formed, in its initial portion, by the presacral nerve, which descends in the pelvis assuming a plexiform aspect, and divides into two branches, the hypogastric nerves, whose paths diverge to encompass the rectum and then reach the inferior hypogastric plexus, or pelvic plexus. The superior hypogastric plexus is anastomosed with the sympathetic trunk, while the inferior one, anastomosing with the sacral plexus, receives, through the 2nd, 3rd and 4th sacral nerve, preganglionic nerve fibers of the sacral parasympathetic trunk.

➤

Innervation of the male pelvis. Left lateral view of the right hemipelvis after the rectum has been turned down and the bladder turned up. The nerves to the seminal glands and prostate come from the inferior hypogastric, or pelvic, plexus and form the prostatic plexus. Also the nerves to the ductus deferens, which form the deferential plexus, come from the inferior hypogastric plexus.

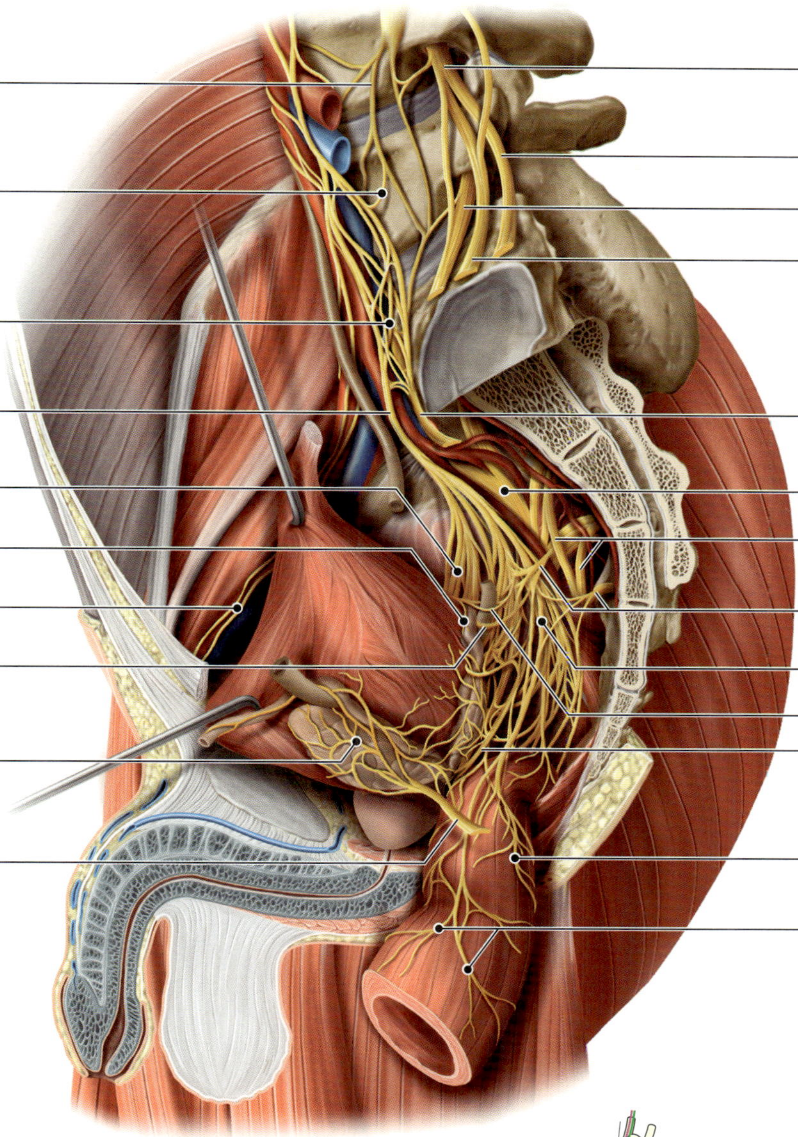

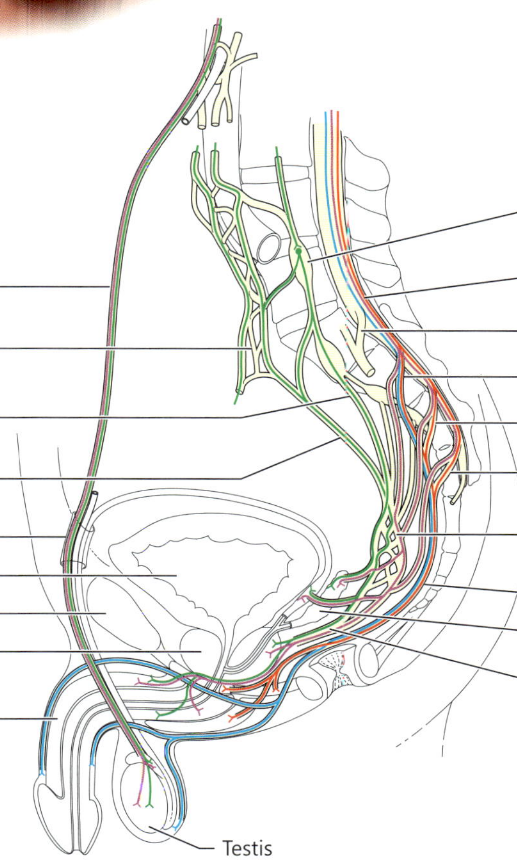

➤ **Sympathetic** (*green*) and **parasympathetic** (*red*) **innervation of the male pelvis and perineum.** The pudendal nerve is composed of somatic motor and sensory (*blue*) fibers, and also carries parasympathetic fibers of the pelvic splanchnic nerves.

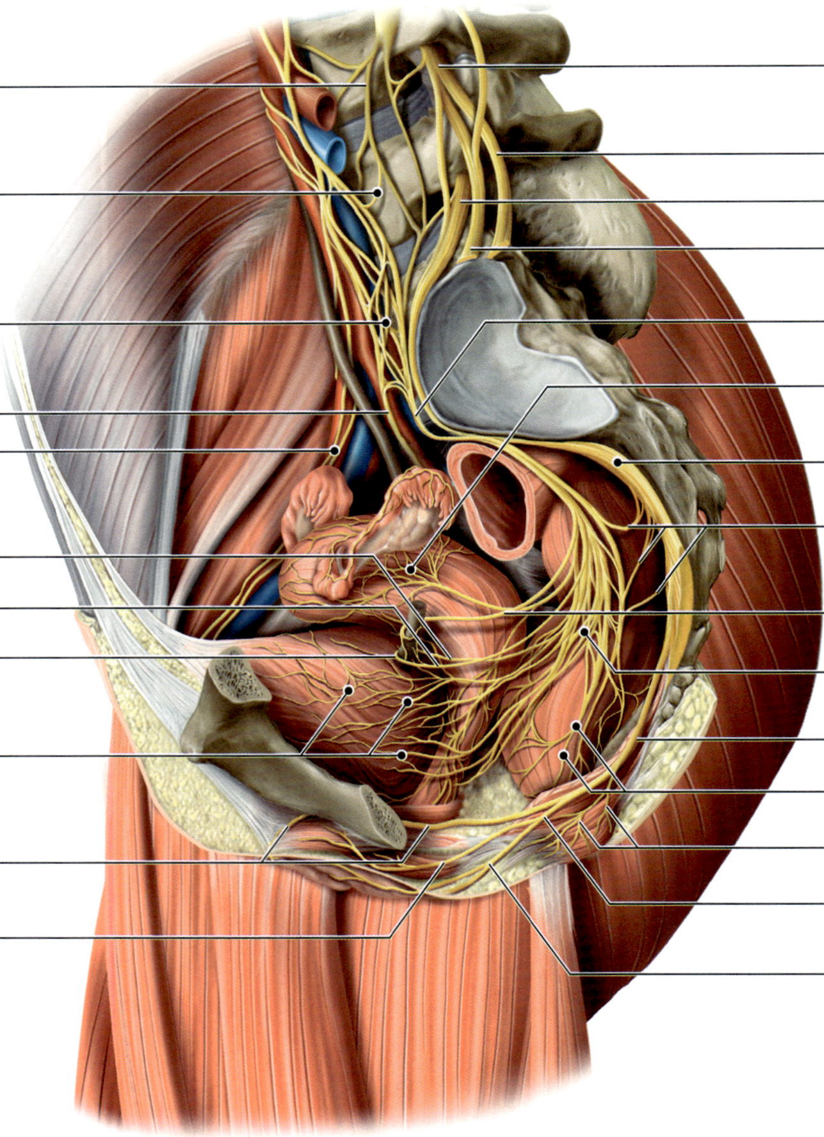

The **inferior hypogastric plexus**, or pelvic plexus, is the most extensive plexus of the sympathetic trunk, and extends from the front to the back of the pelvis, running along the posterolateral walls, encompassing the back and sides of the rectum (forming the middle and inferior rectal plexuses). Extending forwards it surrounds the lateral walls of the bladder (forming the vesical plexus) and, in the female, reaches the side of the cervix and the vagina (forming the uterusvaginal plexus). The pudendal nerve, a mixed nerve containing mostly fibers of the sacral parasympathetic system, departs from the anterior surface of the sacral plexus and exits the pelvis through the greater sciatic foramen. It enters the ischiorectal fossa through the lesser sciatic foramen and, after giving off the inferior anal or rectal nerves, divides into its terminal branches, the perineal nerve and the dorsal nerve of the clitoris (in the female) or of the penis (in the male).

➤

Innervation of the female pelvis. Left lateral view of the right hemipelvis after the uterus and rectum have been turned downwards. The right inferior hypogastric, or pelvic, plexus extends along the lateral wall of the rectum through the middle and inferior rectal plexuses. One can see the pelvic splanchnic nerves, constituting the parasympathetic root of the pelvic ganglia, which go from the anterior rami of the 2nd, 3rd and 4th sacral nerve to the inferior hypogastric plexus. From here they distribute to the pelvic organs. Some of the fibers ascend to the superior hypogastric and to the abdominal aortic plexuses, reaching through the inferior mesenteric plexus the distal third of the transverse colon and the descending colon.

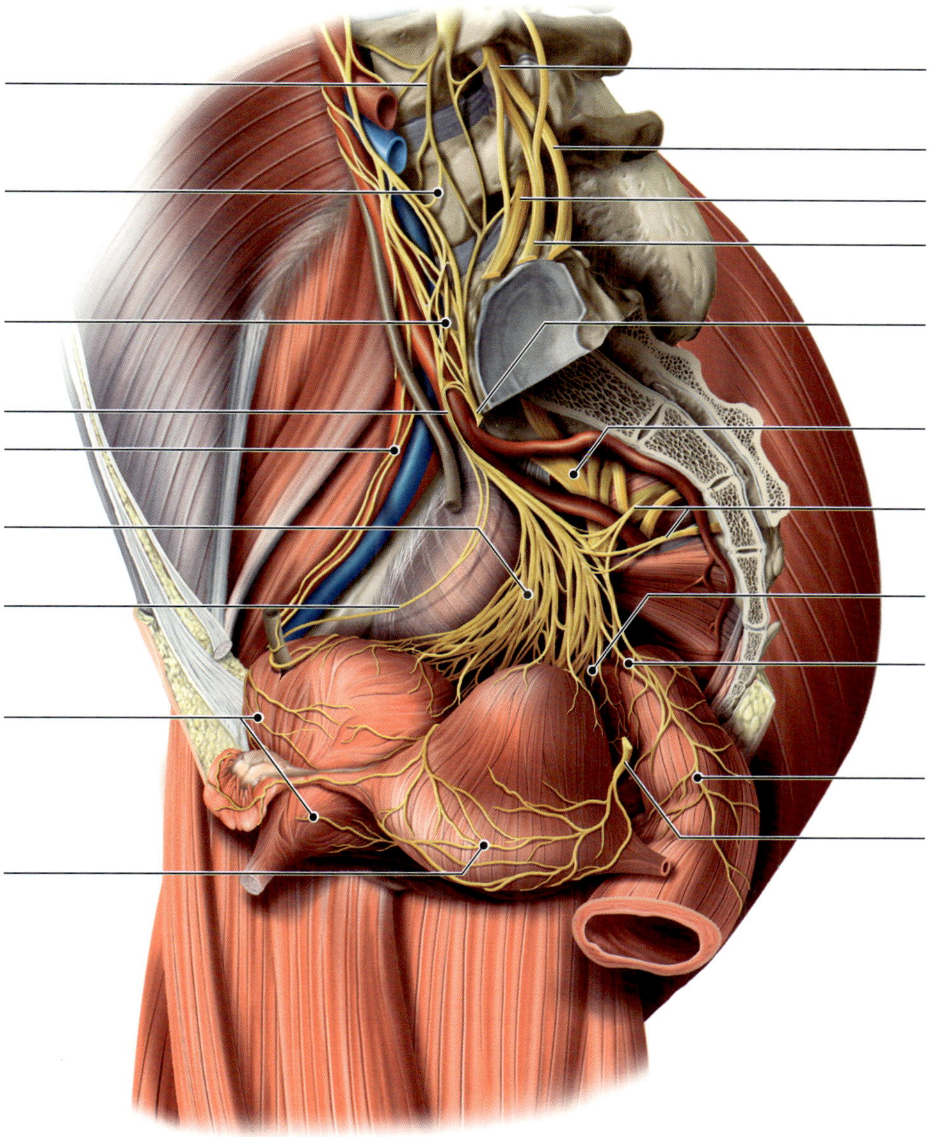

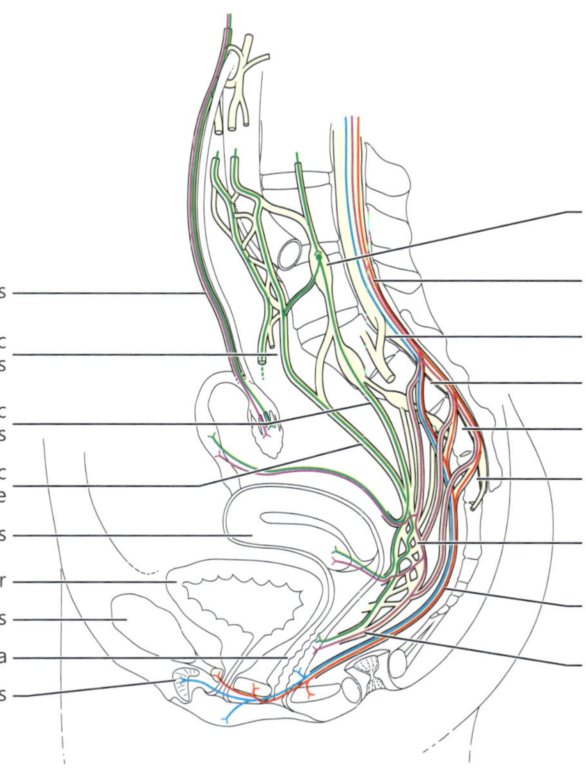

➤ **Sympathetic** (*green*) and **parasympathetic** (*red*) **innervation of the female pelvis and perineum.** The pudendal nerve is composed of somatic motor and sensory (*blue*) fibers, and also carries parasympathetic fibers of the pelvic splanchnic nerves.

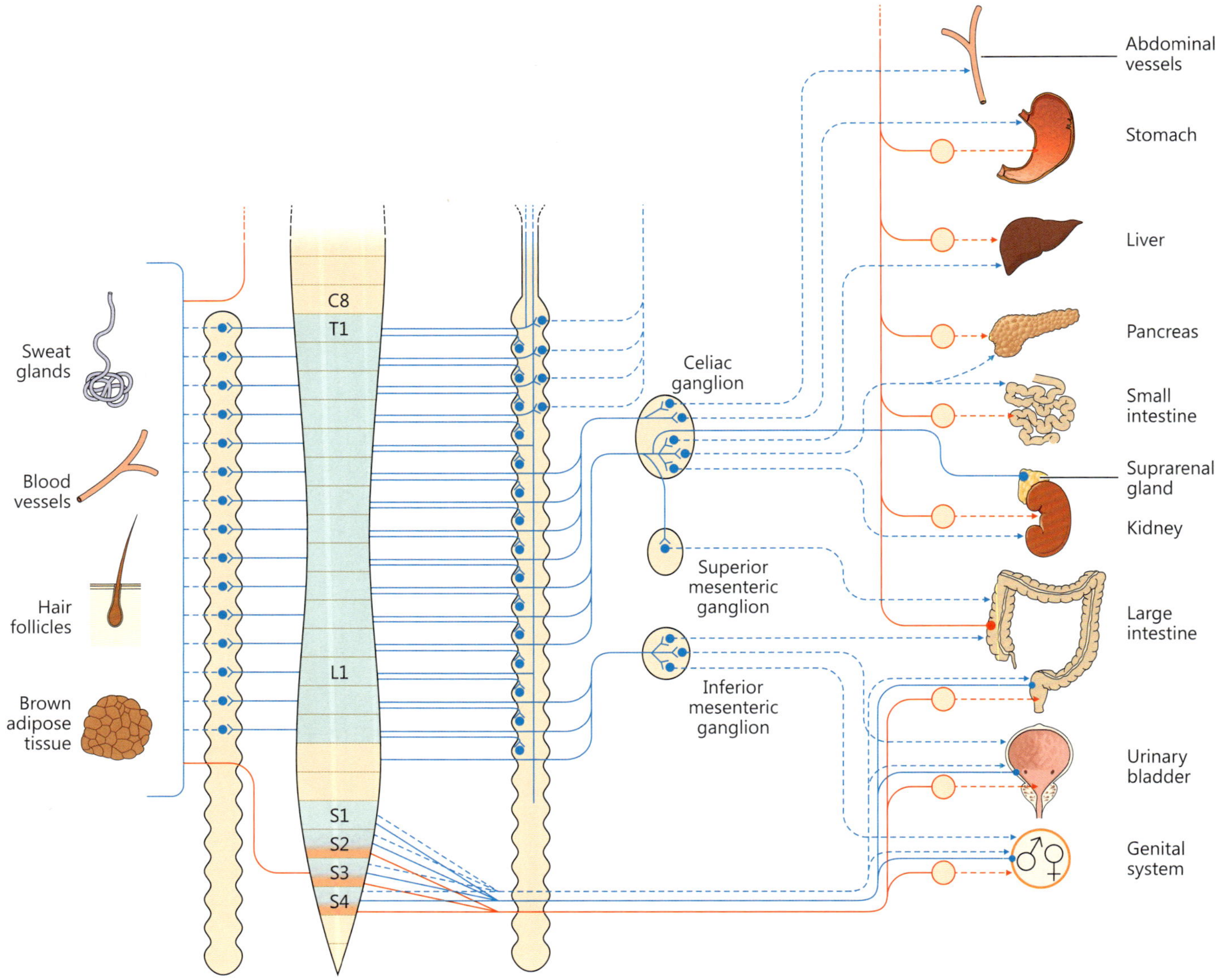

General organization and territories of innervation of the sympathetic (*blue*) and **parasympathetic** (*red*) **systems**. The **sympathetic** innervation of the distal third of the transverse colon, left colic flexure, descending colon, pelvic organs and genitalia occurs through the lumbar and sacral splanchnic nerves. The **lumbar splanchnic nerves** originate from the lumbar segment of the sympathetic trunk. They contain preganglionic visceral effector sympathetic fibers, arising from the intermediolateral nucleus of the lateral horn of the gray matter of the spinal cord, in the segments from T12 to L2 or L3, and visceral sensory fibers. The preganglionic fibers reach the abdominal aortic plexus and successively the inferior mesenteric plexus, and interrupt their course in the inferior mesenteric ganglion. From the ganglion originate the postganglionic fibers to the final territories of innervation. The **sacral splanchnic nerves** originate from the sacral segment of the sympathetic trunk, and carry mainly postganglionic visceral effector sympathetic fibers (from the ganglia of the sacral segment of the sympathetic trunk), and preganglionic ones (from the segments S1 to S4 of the spinal cord), and visceral sensory fibers. The postganglionic and preganglionic fibers of the sacral splanchnic nerves reach the inferior hypogastric, or pelvic, plexus through which they are distributed out; the preganglionic fibers interrupt their course in the visceral ganglia situated near or within the wall of the innervated organs which originate the postganglionic fibers.
The **parasympathetic** innervation of the distal part of the descending colon, iliopelvic colon, pelvic organs and genitalia occurs through the **pelvic splanchnic nerves**, or nerves erigentes, whose visceral effector parasympathetic fibers originate from the sacral parasympathetic nuclei of the intermediate column of the spinal cord in the segments from S2 to S4, run in the sacral nerves from the 2nd to the 5th and reach, through the sacral plexus and the pudendal nerve, the inferior hypogastric or pelvic plexus. From here they are distributed, together with the pudendal nerve, interrupting their course in the dislocated pelvic ganglia near the organs to be innervated. The visceral sensory fibers of the lumbar, sacral and pelvic splanchnic nerves have their real origin in the cells of the spinal ganglia attached to the posterior roots of the spinal nerves.

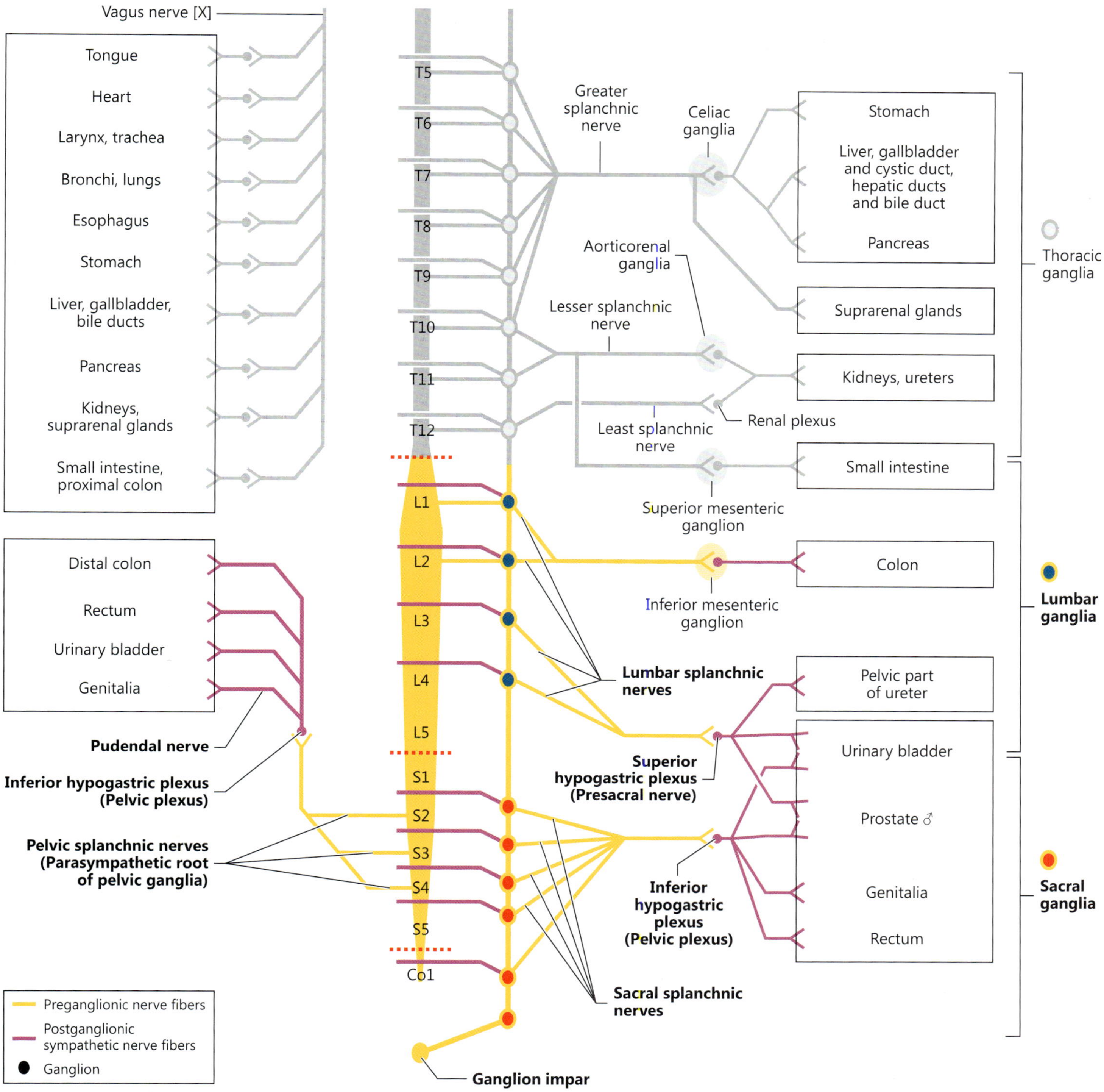

Schematic representation of the organization of the **sympathetic system** (on the right) and the **parasympathetic system** (on the left) (☞ Vol. 1, page 110). The autonomic division is the part of the central nervous system that controls the secretory functions of glands and wall motility of internal organs and vessels, thus regulating the vital functions of the organs and all body systems. The organs that are innervated by nerve fibers of the autonomic division are listed, in detail, in the boxes. The preganglionic nerve fibers are represented in *yellow*; in *purple*, the postganglionic fibers of the sympathetic nervous system, and in *orange* those of the parasympathetic nervous system. The parasympathetic (cranial and sacral) nervous system comprises the neuraxial visceral centers situated, respectively, in the brainstem and sacral segment of the spinal cord. Note, in particular, the elevated number of visceral efferent (motor) fibers of the parasympathetic part of the vagus nerve.

Summary table of the nerves of the pelvis

NERVE	ORIGIN	RAMIFICATIONS	DISTRIBUTION
coccygeal [Co]	Coccygeal part of the spinal cord	Filaments	Ischiococcygeus and skin of the anococcygeal region through the coccygeal plexus
coccygeal plexus	Anterior branches of the 4th and 5th sacral nerve and of the coccygeal nerve	Visceral anterior branches (sacral parasympathetic fibers) and somatic posterior branches; gray rami communicantes with the sympathetic trunk; anococcygeal nerve	Pelvic organs through the inferior hypogastric plexus; ischiococcygeus and skin of the anococcygeal region
deferential plexus ♂	Inferior hypogastric plexus	Filaments	Ductus deferens, seminal glands
hypogastric	Superior hypogastric plexus	Single large nerve or some parallel bundles linking the superior hypogastric plexus to the inferior hypogastric plexuses	Through the inferior hypogastric plexuses
inferior hypogastric plexus or pelvic plexus (paired)	Sympathetic fibers from the superior hypogastric plexus and sacral ganglia of the sympathetic trunk, and sacral parasympathetic fibers from the pudendal plexus	Middle and inferior rectal, vesical, deferential ♂, prostatic ♂, uterusvaginal ♀ plexuses	Pelvic organs
inferior rectal plexus (inferior hemorrhoidal plexus)	Inferior hypogastric plexus	Superior anal nerves	Rectum and anal canal superiorly to the pectinate line
lumbosacral trunk	Descending anastomotic branch of L4 to the anterior ramus of L5 and anterior ramus of L5	Formation of sacral plexus from union with the anterior branches of S1, S2 and S3	Through the sacral plexus
middle rectal plexus (middle hemorrhoidal plexus)	Inferior hypogastric plexus	Filaments	Rectum
pelvic splanchnic (nervi erigentes)	Anterior branches of 2nd, 3rd and 4th sacral nerve (parasympathetic root of pelvic ganglia)	Filaments	Pelvic organs and genitalia through the inferior hypogastric plexus; descending colon and distal third of the transverse colon through the superior hypogastric plexus
prostatic plexus ♂	Inferior hypogastric plexus	Filaments	Prostate, seminal glands
pudendal plexus	Anterior branches of 2nd, 3rd and 4th sacral nerve	Visceral (sacral parasympathetic fibers) and somatic branches; pudendal nerve; anastomotic branches with the sacral plexus	Urinary bladder, rectum and vagina through the visceral branches; levator ani, ischiococcygeus and external anal sphincter, anal skin, skin of posterior perineum and of inferomedial surface of buttock through the somatic branches; male and female genitalia, rectal ampulla, muscles and skin of perineum, anal skin and skin of scrotum through the pudendal nerve
sacral [S1-S5] (five pairs of spinal nerves)	Sacral part of the spinal cord	Anterior and posterior branches	Muscles and skin of the perineum and of the lower limb and hip joint through the sacral plexus; skin and muscles of the sacral region, buttocks and gluteus maximus
sacral plexus	Union of the lumbosacral trunk [L4-L5] with the anterior branches of the 1st, 2nd and 3rd sacral nerve	Anterior and posterior branches: nerve to obturator internus, nerve to piriformis, nerve to quadratus femoris, branches to the gemellus superior and gemellus inferior, superior and inferior gluteal nerves, posterior cutaneous nerve of thigh and perforating cutaneous nerve; terminal branch: sciatic nerve; gray rami communicantes with the sympathetic trunk; anastomotic branches with the pudendal and femoral plexuses	Muscles and skin of the lower limb, hip joint

continued

Summary table of the nerves of the pelvis *(continued)*

NERVE	ORIGIN	RAMIFICATIONS	DISTRIBUTION
sacral splanchnic	Sacral ganglia of sympathetic trunk	Filaments	Pelvic organs and vessels through the inferior hypogastric plexus
superior anal	Inferior rectal plexus	Filaments	Rectum and anal canal superiorly to the pectinate line
superior rectal plexus (superior hemorrhoidal plexus)	Inferior mesenteric plexus	Filaments	Rectum
sympathetic trunk – **sacral ganglia** (four in number) and **ganglion impar**		Perivascular and visceral branches constituting the inferior hypogastric plexus; sacral splanchnic nerves; gray rami communicantes with the sacral nerves	Pelvic organs and vessels
to obturator internus [L5, S1], **to piriformis** [L5, S1, S2], **to quadratus femoris** [L4, L5, S1], **superior** [L4, L5, S1] **and inferior** [L5, S1, S2] **gluteal**, **posterior cutaneous nerve of thigh** [S1, S2, S3] and **perforating cutaneous** [S2, S3], **sciatic** [L4, L5, S1, S2, S3] (to their origin in the pelvis)	☞ Sacral plexus		
uterusvaginal plexus ♀	Inferior hypogastric plexus	Filaments	Uterus, vagina, clitoris
vaginal ♀	Uterusvaginal plexus ♀	Filaments	Vagina
vesical plexus	Inferior hypogastric plexus	Filaments	Urinary bladder

Summary table of the nerves of the perineum

NERVE	ORIGIN	RAMIFICATIONS	DISTRIBUTION
anococcygeal	Coccygeal plexus	Muscular branches and cutaneous branch	Ischiococcygeus and skin of the anococcygeal region
anterior labial ♀	Ilioinguinal nerve	Filaments	Anterior part of the labia majora
anterior scrotal ♂	Ilioinguinal nerve	Filaments	Anterior part of the scrotum and root of penis
cavernous nerves of clitoris ♀	Inferior hypogastric plexus	Filaments	Corpora cavernosa of clitoris
cavernous nerves of penis ♂	Inferior hypogastric plexus	Filaments	Corpora cavernosa and corpus spongiosum of penis
dorsal nerve of clitoris ♀	Perineal nerves	Filaments	Clitoris and urethra
dorsal nerve of penis ♂	Perineal nerves	Filaments	Penis and urethra
external spermatic (genital branch of genitofemoral nerve)	Genitofemoral nerve	Filaments	Skin of the scrotum, labia majora
genital branch of iliohypogastric nerve	Iliohypogastric nerve	Filaments	Skin of the external genitalia
inferior anal (inferior rectal)	Pudendal nerve	Cutaneous and muscular branches	Skin of the anal triangle and of the posterior perineum and external anal sphincter
perineal	Pudendal nerve	Inferior anal nerves, dorsal nerve of clitoris ♀ or dorsal nerve of penis ♂; superficial branch: posterior labial ♀ or posterior scrotal nerves ♂; deep branch from which originate muscular branches	Skin of the anterior perineum, scrotum, corpora cavernosa and skin of penis, labia majora, clitoris and part of the labia minora, superficial transverse perineal muscle, ischiocavernosus and bulbospongiosus, bulb of penis, bulb of vestibule, mucosa of urethra, skin of the anal triangle and of posterior perineum, and external anal sphincter
perineal branches of posterior cutaneous nerve of thigh	Posterior cutaneous nerve of thigh	Filaments	Skin of perineum, posterior part of the scrotum, posterior part of the labia majora
posterior labial ♀	Perineal nerves	Filaments	Posterior part of the labia majora, vestibule of the vagina
posterior scrotal ♂	Perineal nerves	Filaments	Posterior part of the scrotum
pudendal [S2, S3, S4]	Pudendal plexus	Inferior anal and perineal nerves	Male and female genitalia, rectal ampulla, muscles and skin of perineum, anal skin and skin of the scrotum

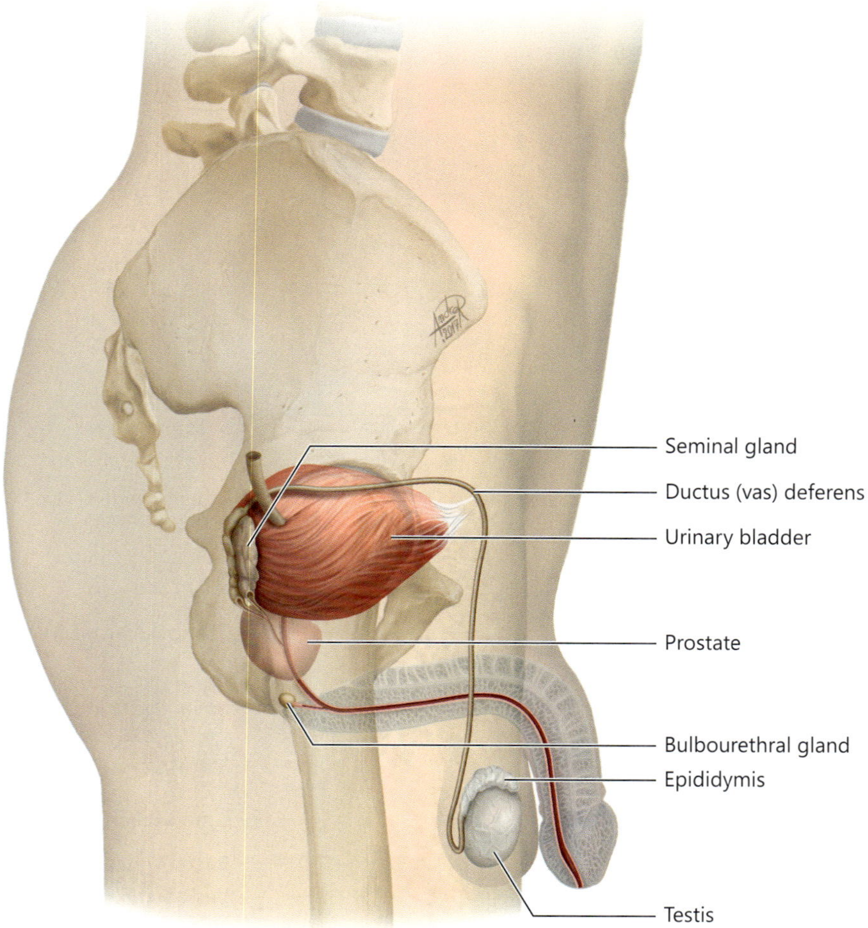

General organization of the **male genital system**. The various organs of the apparatus, shown in right lateral projection, are located principally in the pelvis and anterior perineum. However, the ductus deferens (vas deferens), after crossing the inguinal canal, have a brief excursion in the abdomen, before descending in the pelvis. Assocciated to the sperm tract are several glands: the seminal glands, that release their secretion into the urethra by mean of the ejaculatory ducts, the prostate, and the bulbourethral glands, directly releasing their secretion into the urethra. The image also displays the bladder, its close relations with the prostate, the seminal glands and the ductus deferens.

General organization of the **female genital system**. The gonads (ovaries) and most of the female genital tract are situated in the lesser pelvis. The inferior part of the vagina and the external genitalia occupy the anterior perineal region. Indicated in *gray* are the round ligaments of the uterus which, curving at the level of the inlet of the inguinal canal, run through it terminating in the thickness of the connective tissue of the labium majus.

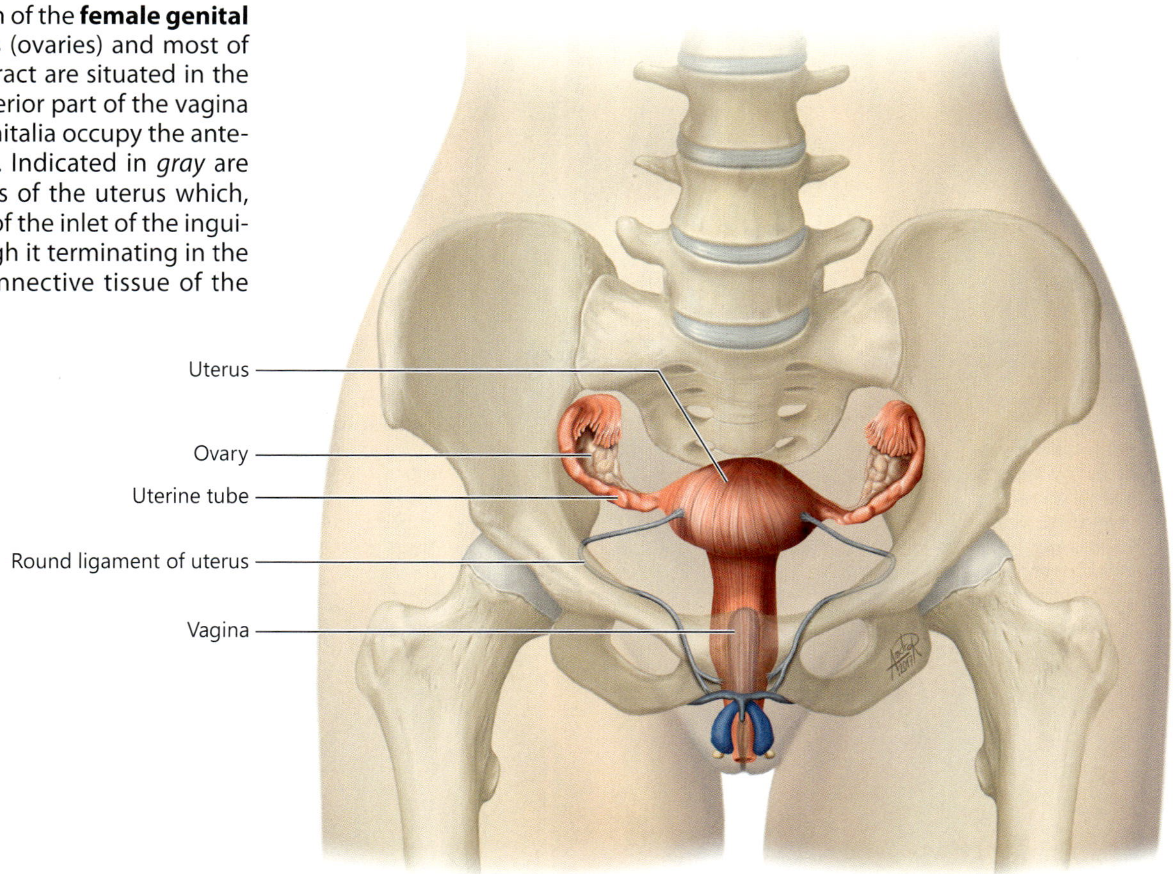

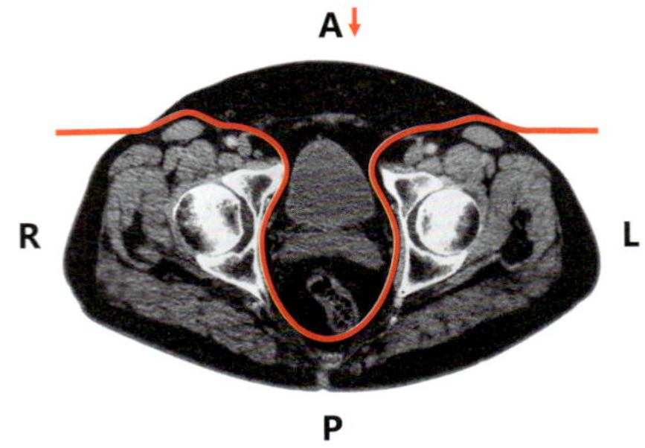

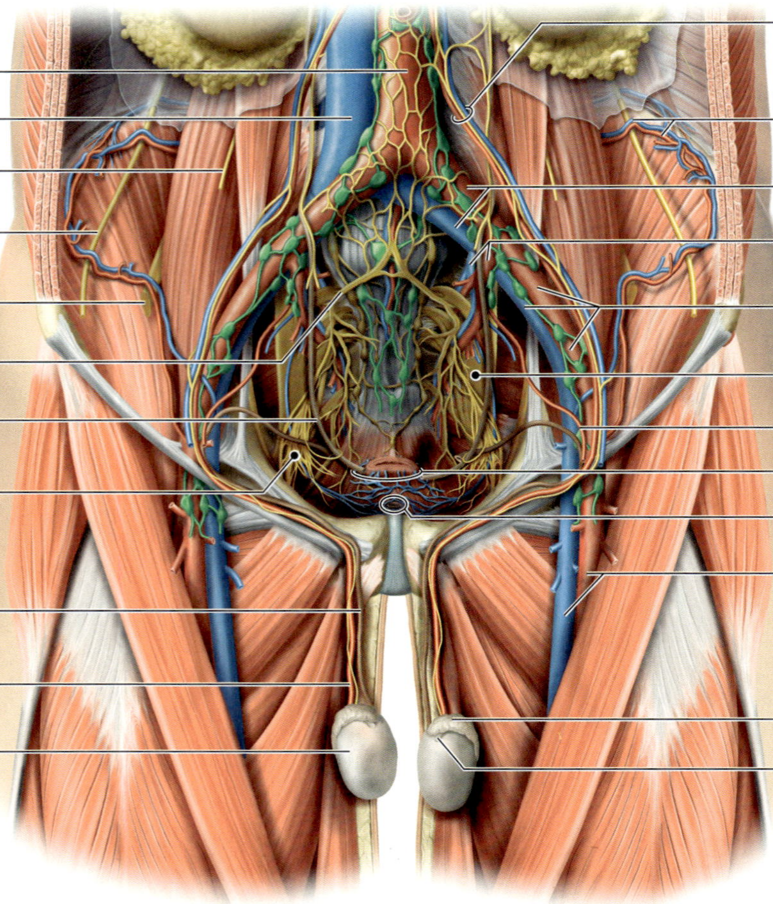

CONTENTS OF THE PELVIS AND PERINEUM (in bold are the structures added with respect to the table on page 92, Contents of the Abdomen)	
• Ductus deferens (Vas deferens) • Epididymis • Hypogastric nerve • Inferior hypogastric plexus (Pelvic plexus) • Internal iliac artery • Internal iliac vein • Pelvic ureter • Prostatic venous plexus	• Rectal venous plexus • Testicular artery • Testicular plexus • Testicular vein • Testis • Umbilical artery • Vesical venous plexus

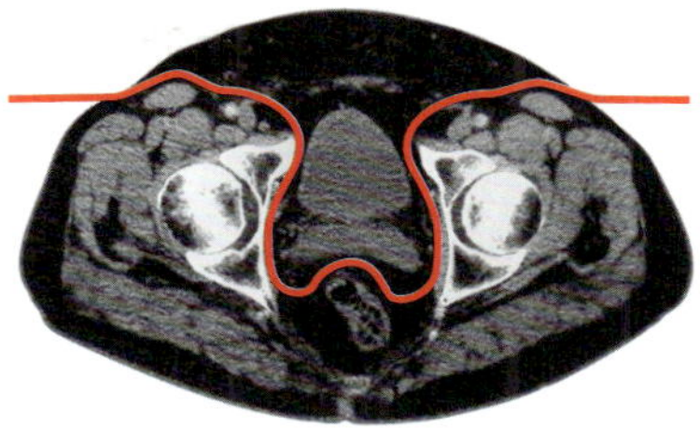

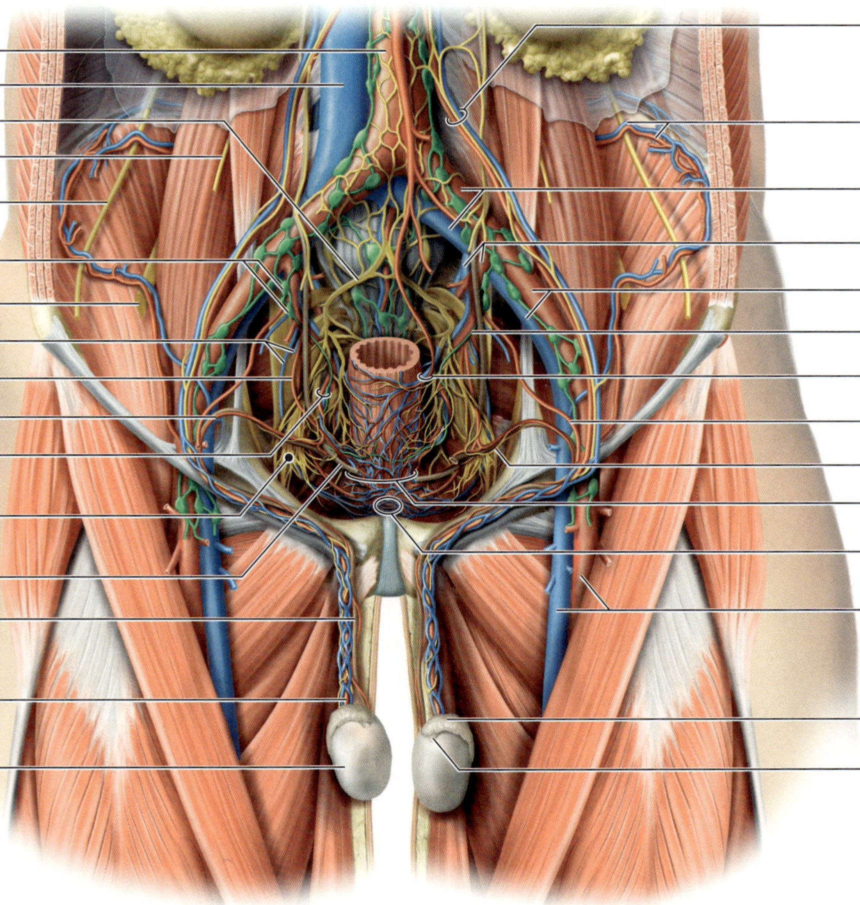

CONTENTS OF THE PELVIS AND PERINEUM (in bold are the structures added with respect to the preceding table)	
• Artery to ductus (or vas) deferens • Deferential plexus • Inferior vesical artery • Internal iliac nodes • Middle rectal artery	• Middle rectal vein • Rectal lymphatic vessels • Superior rectal artery • Superior rectal vein • Vesical veins

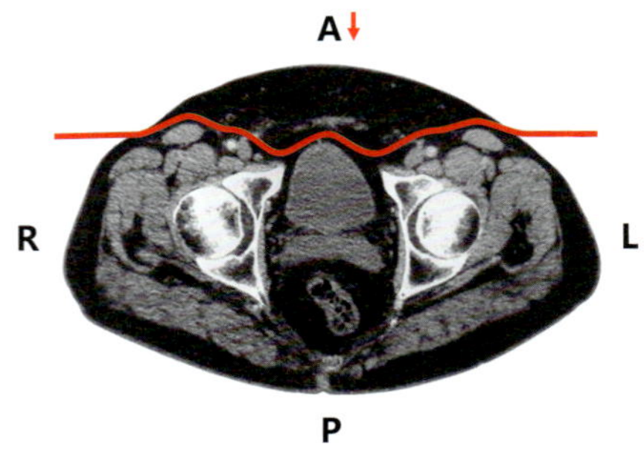

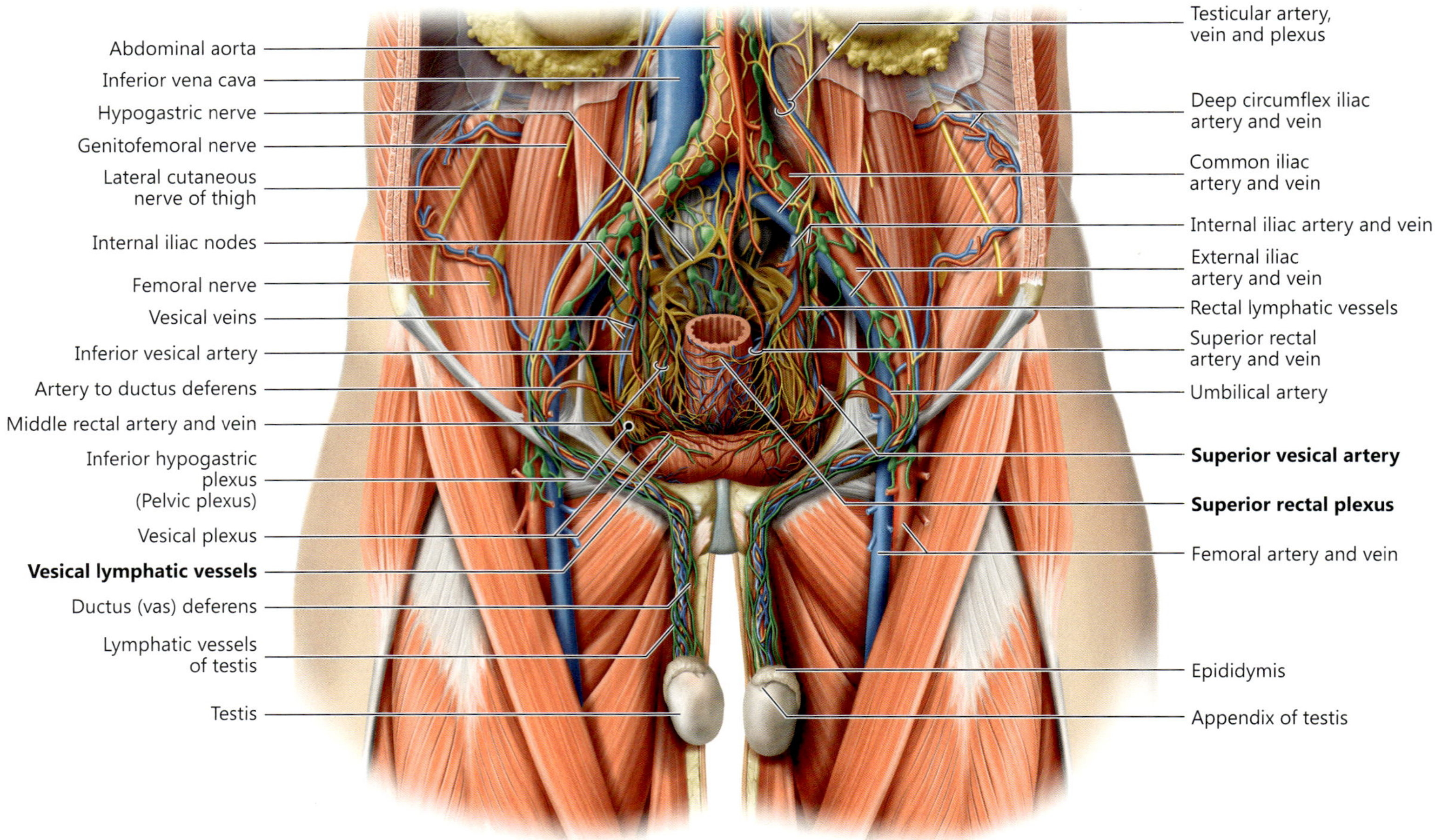

CONTENTS OF THE PELVIS AND PERINEUM (in bold are the structures added with respect to the preceding table)
• Superior rectal plexus • Superior vesical artery • Vesical lymphatic vessels

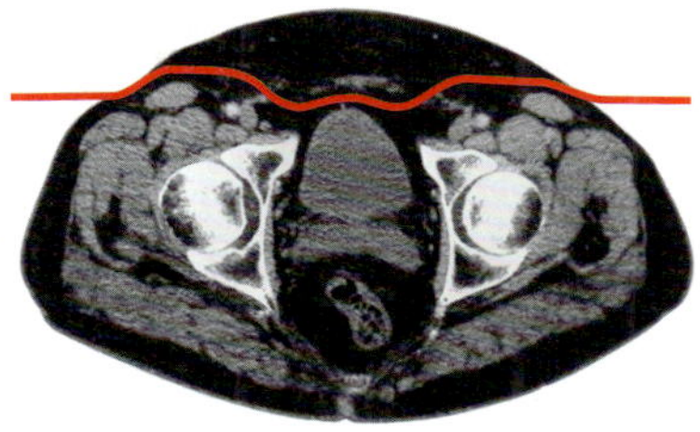

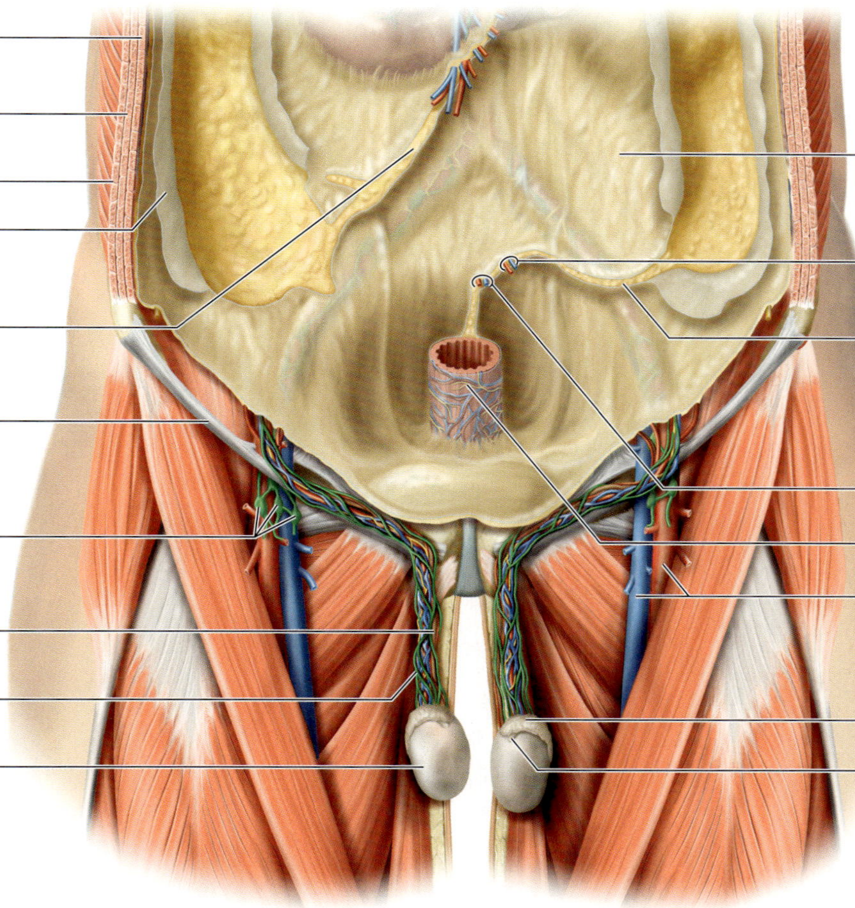

CONTENTS OF THE PELVIS AND PERINEUM (in bold are the structures added with respect to the preceding table)	
• Deep inguinal nodes • Inguinal ligament • Parietal peritoneum • Root of mesentery	• Root of the sigmoid mesocolon • Sigmoid arteries • Sigmoid veins • Visceral peritoneum

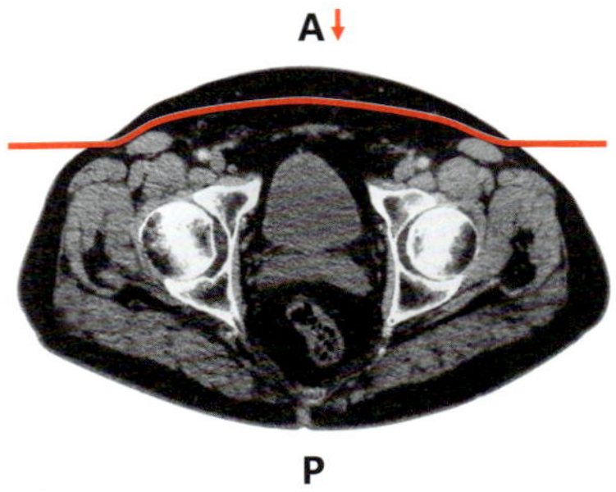

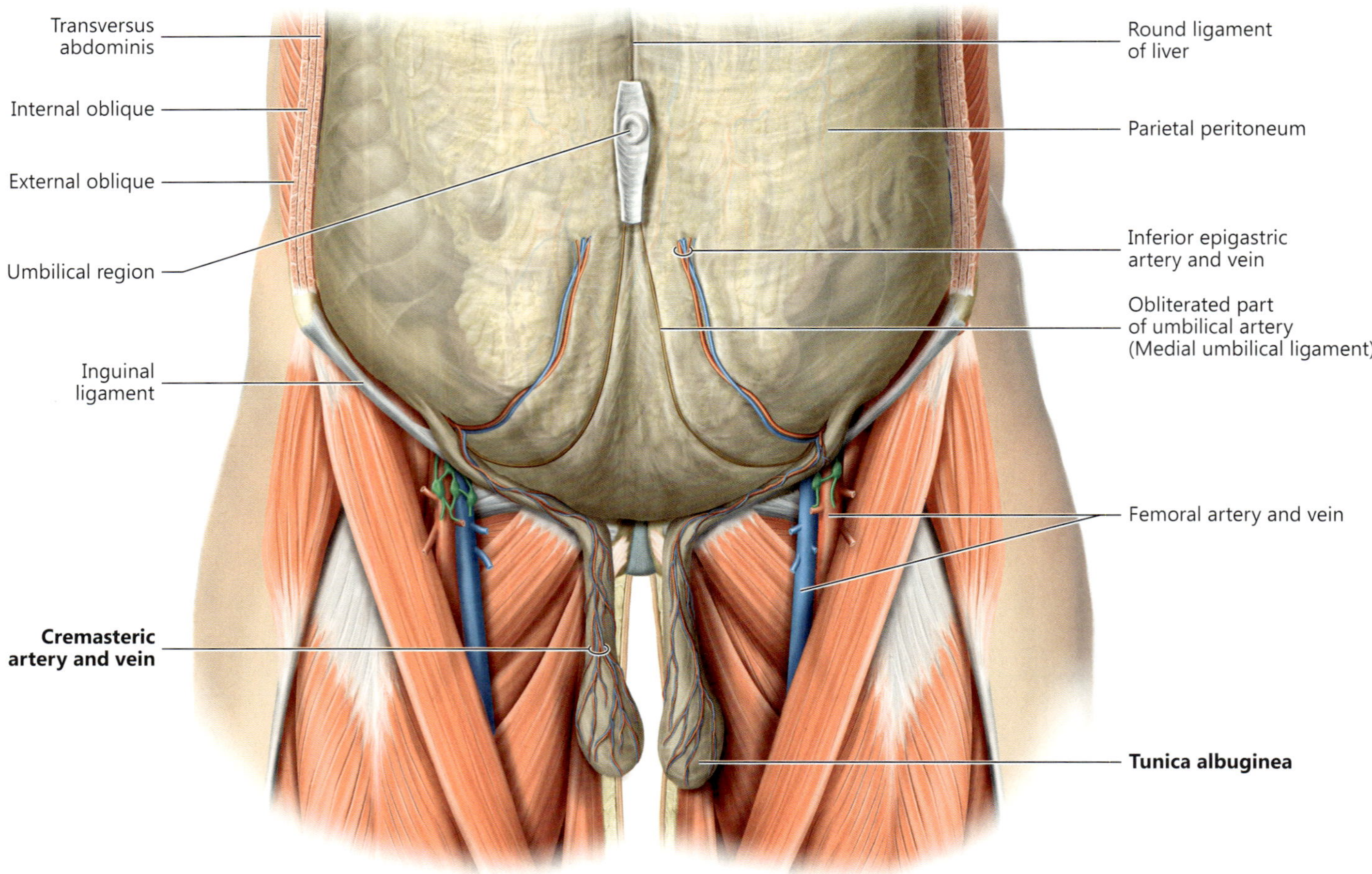

CONTENTS OF THE PELVIS AND PERINEUM (in bold are the structures added with respect to the preceding table)
• Cremasteric artery • Cremasteric vein • Tunica albuginea

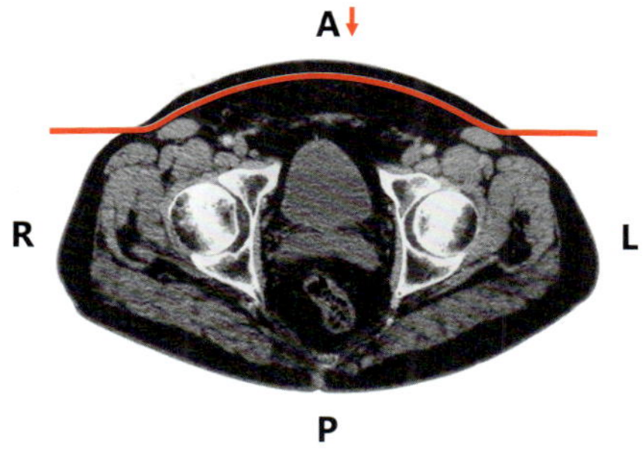

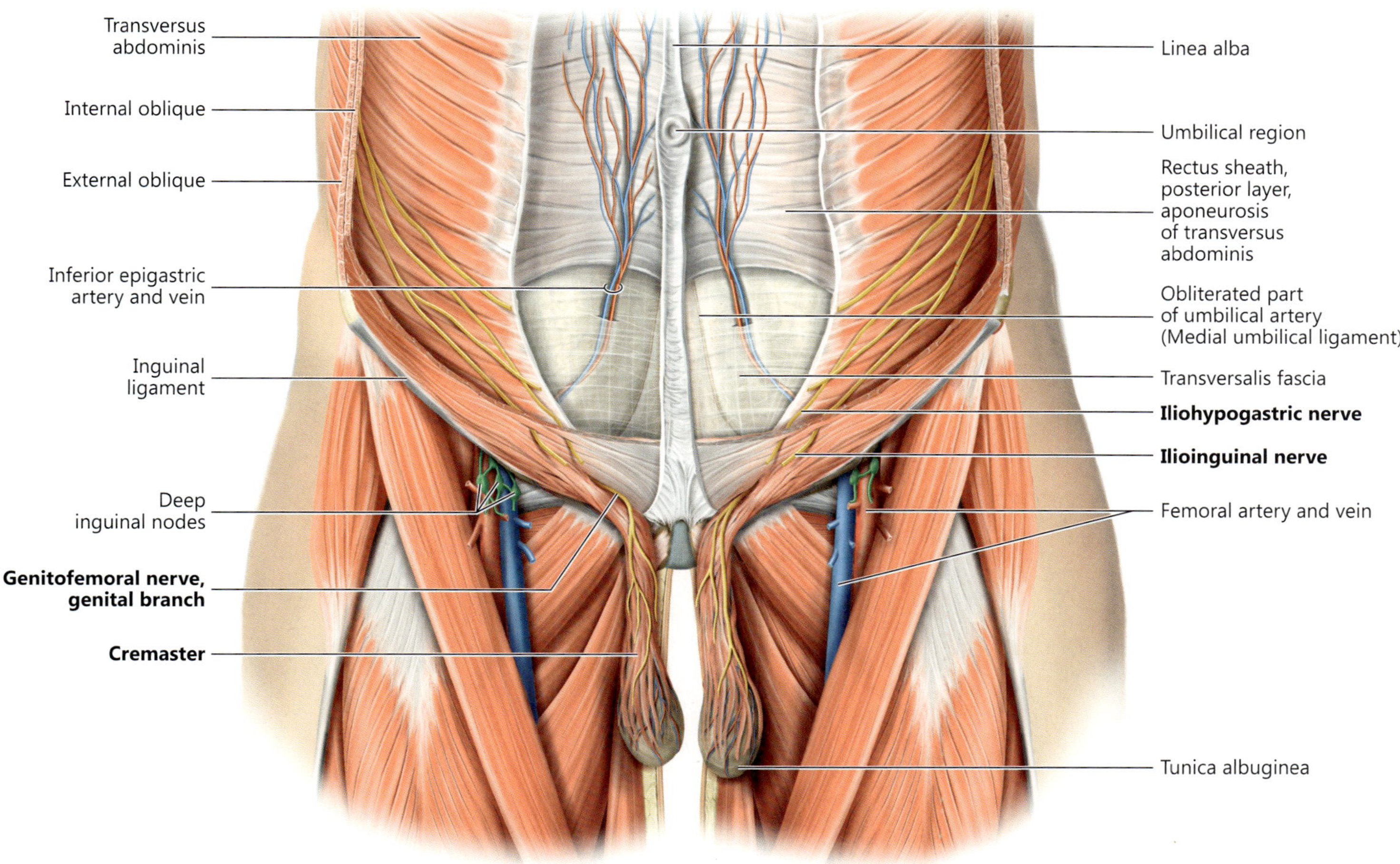

CONTENTS OF THE PELVIS AND PERINEUM (in bold are the structures added with respect to the preceding table)	
• Cremaster • Genitofemoral nerve, genital branch	• Iliohypogastric nerve • Ilioinguinal nerve

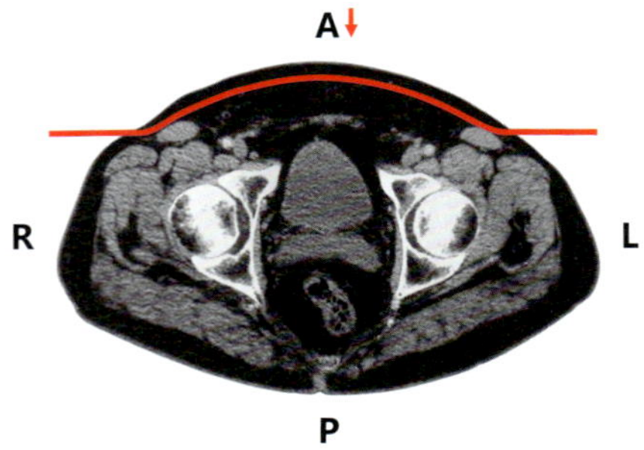

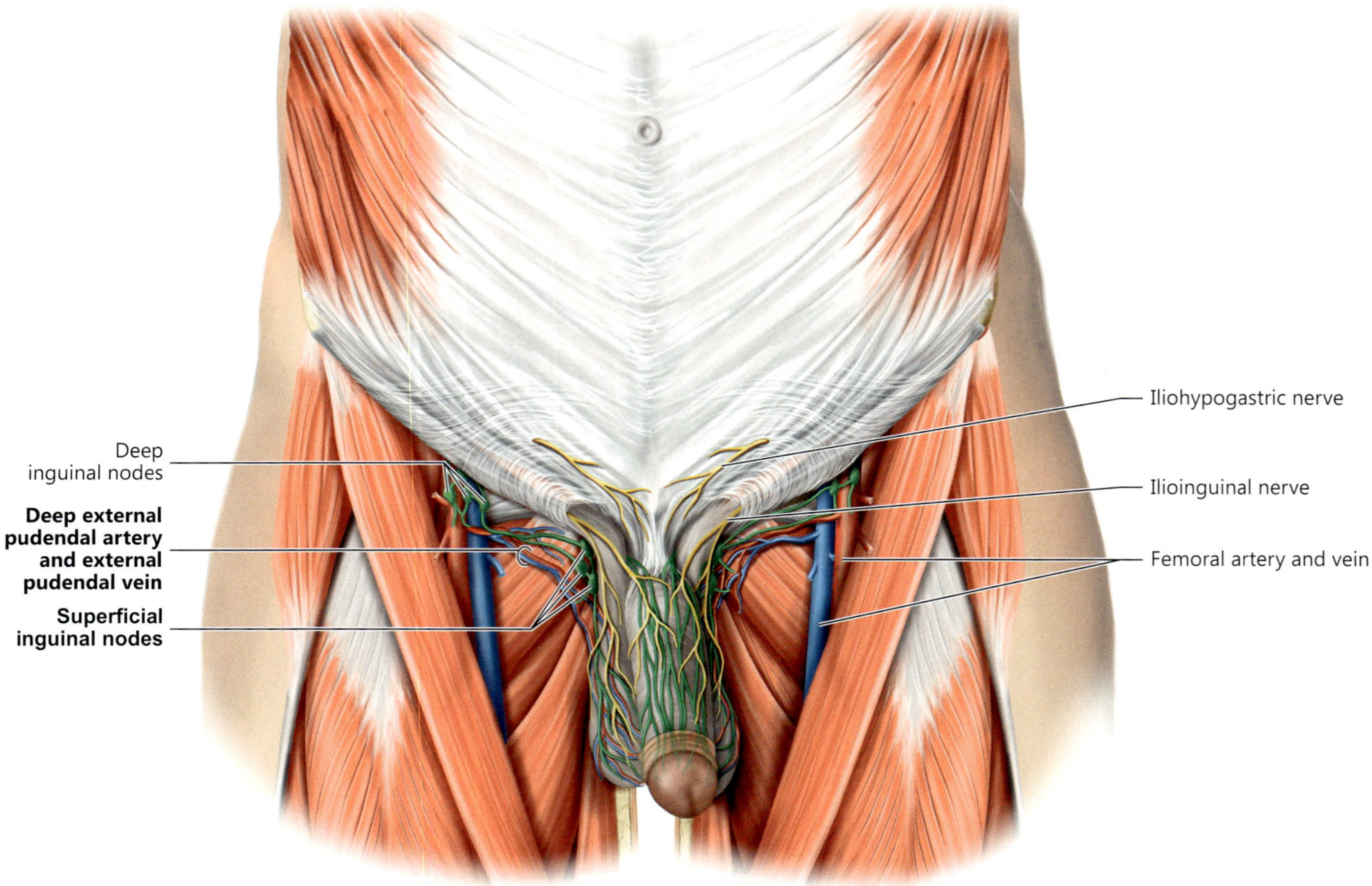

CONTENTS OF THE PELVIS AND PERINEUM (in bold are the structures added with respect to the preceding table)	
• Deep external pudendal artery • External pudendal vein	• Superficial inguinal nodes

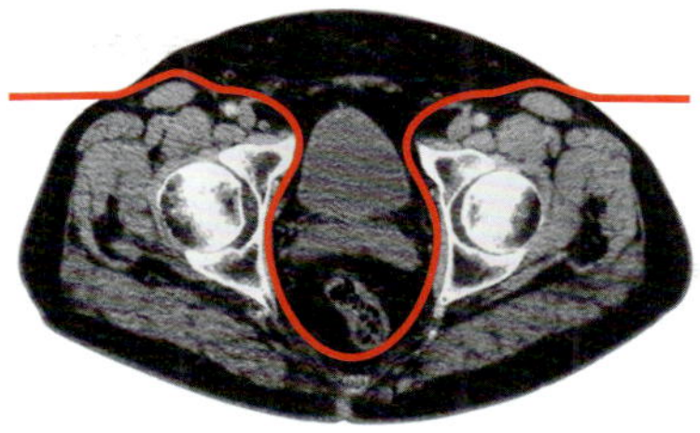

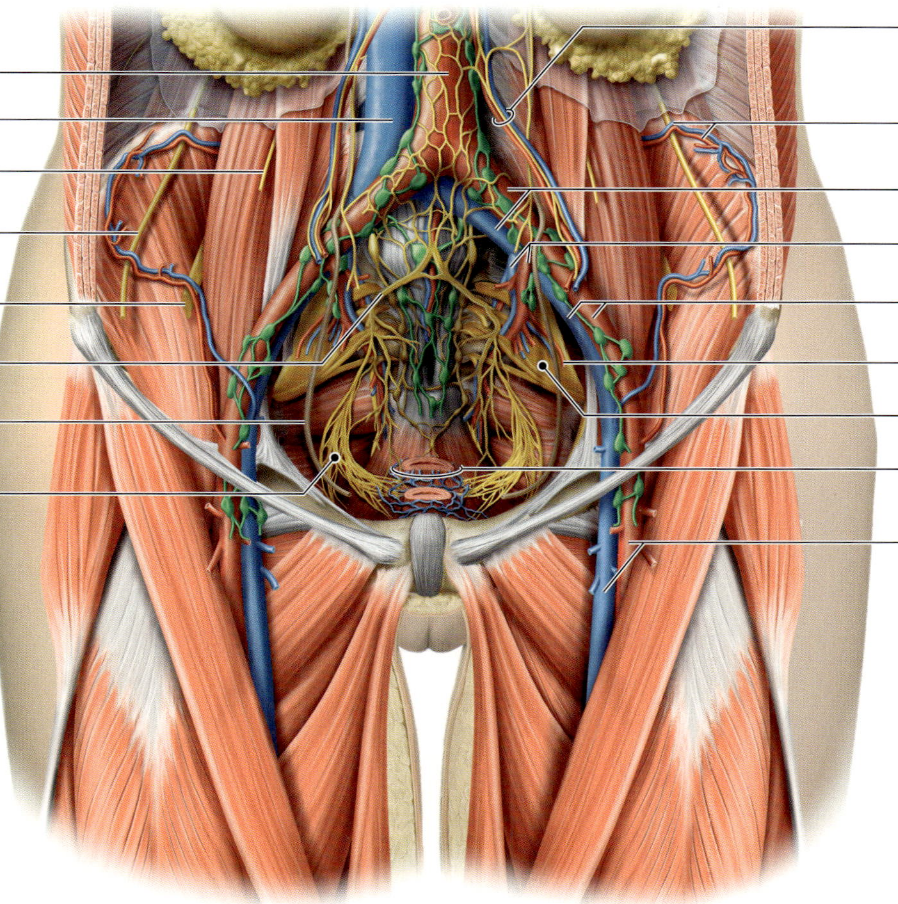

***CONTENTS OF THE PELVIS AND PERINEUM** (in bold are the structures added with respect to the table on page 92, Contents of the Abdomen)*	
• Hypogastric nerve • Inferior hypogastric plexus (Pelvic plexus) • Internal iliac artery • Internal iliac vein • Ovarian artery	• Ovarian plexus • Ovarian vein • Pelvic ureter • Rectal venous plexus • Umbilical artery

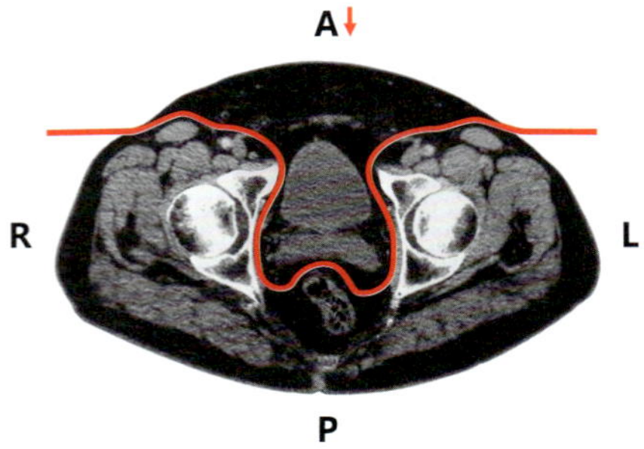

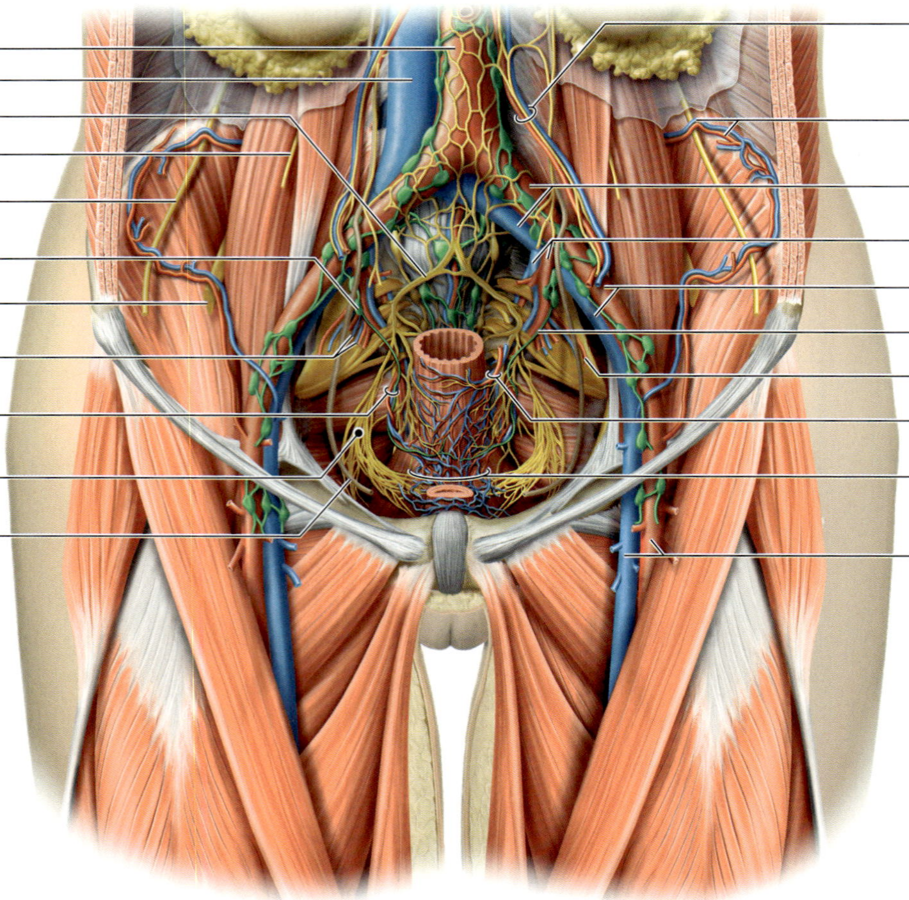

CONTENTS OF THE PELVIS AND PERINEUM (in bold are the structures added with respect to the preceding table)	
• Internal iliac nodes • Middle rectal artery • Middle rectal vein • Rectal lymphatic vessels	• Superior rectal artery • Superior rectal vein • Vesical veins

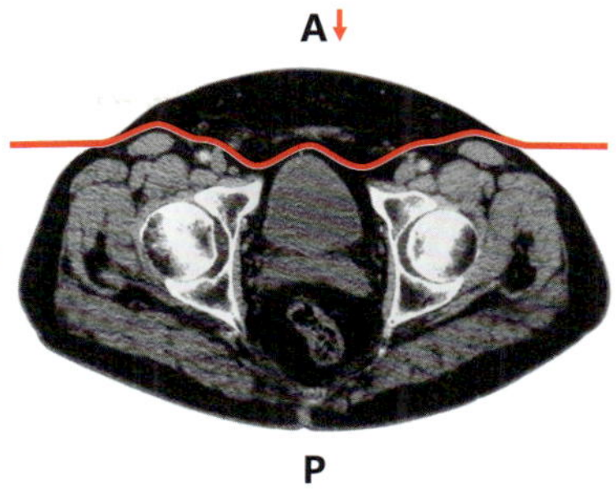

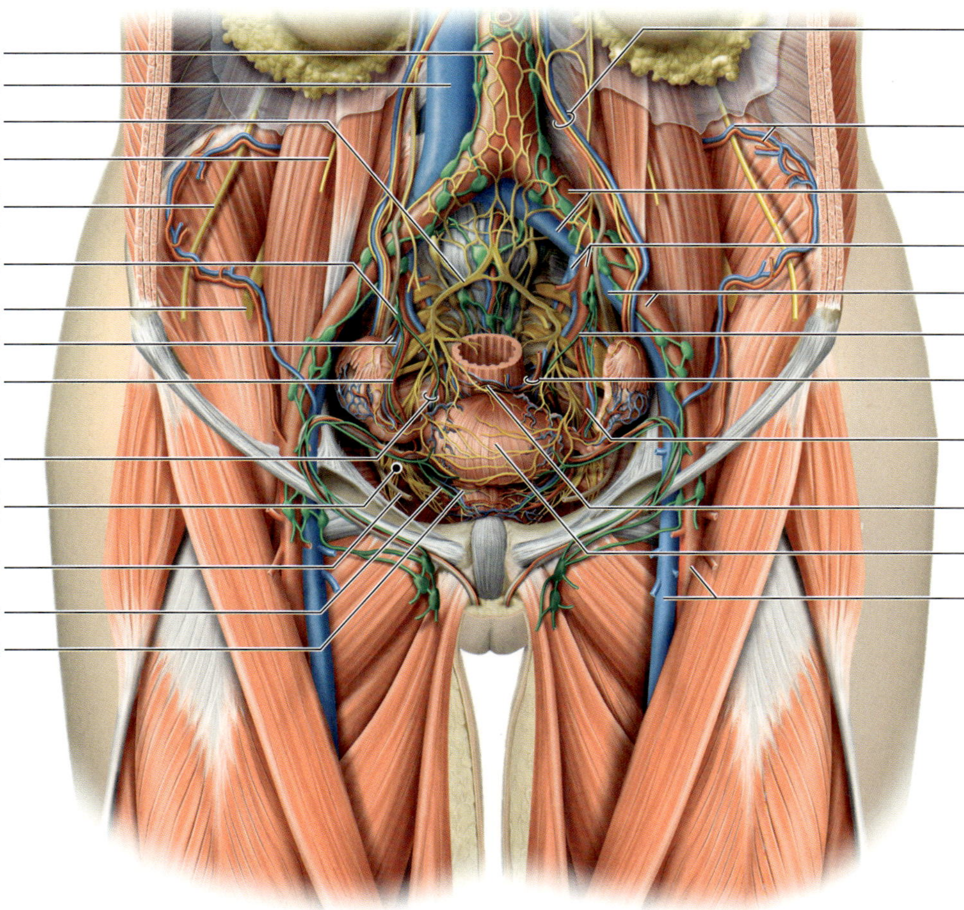

CONTENTS OF THE PELVIS AND PERINEUM (in bold are the structures added with respect to the preceding table)	
• Inferior vesical artery • Superior rectal plexus • Superior vesical artery	• Uterus • Vesical lymphatic vessels • Vesical plexus

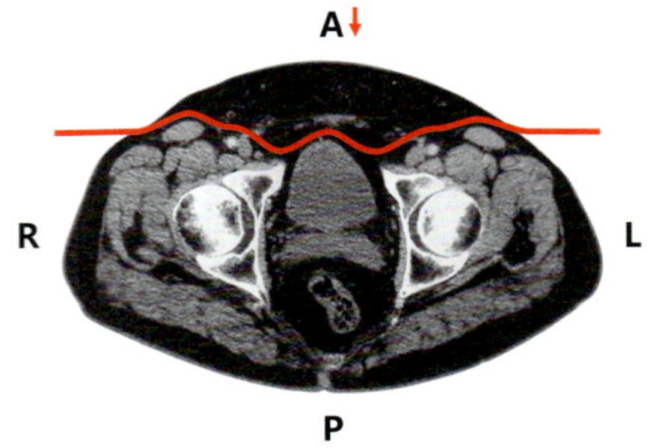

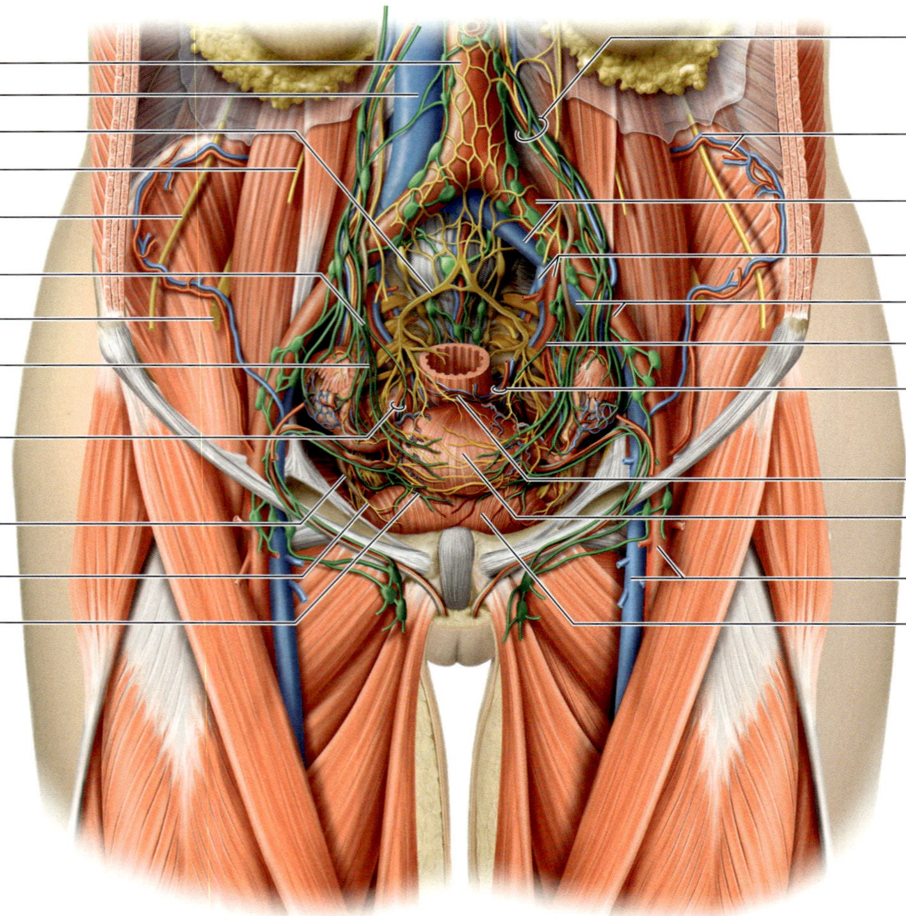

CONTENTS OF THE PELVIS AND PERINEUM (in bold are the structures added with respect to the preceding table)	
• Uterine lymphatic vessels	• Urinary bladder

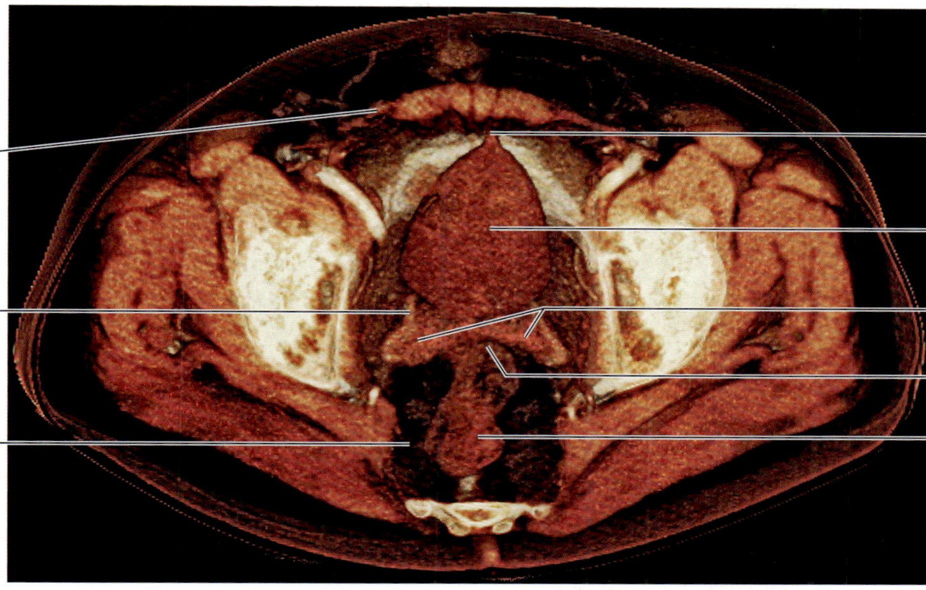

a

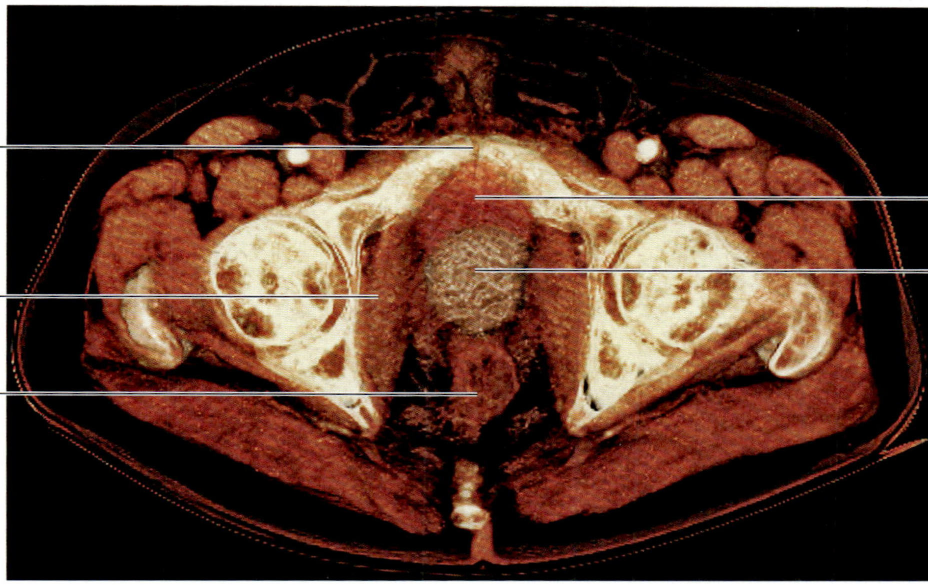

b

Three-dimensional images of a living human body, obtained by CT scan using a 3D volume rendering technique, of transverse sections of the trunk, passing above the bladder (**a**), and through the prostate (**b**), showing the contents of the **male pelvis**.

a 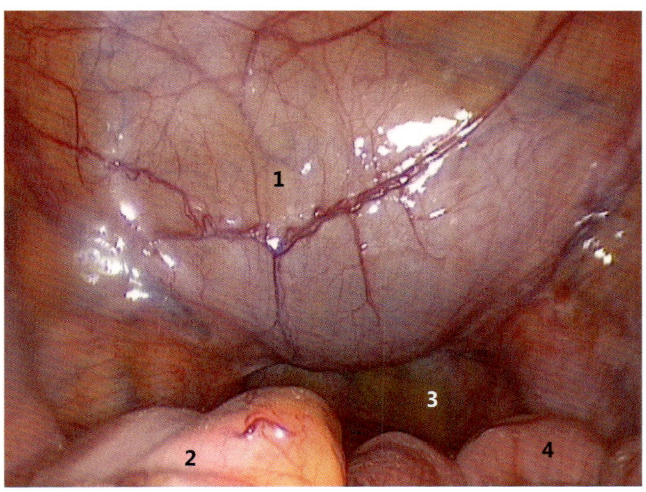

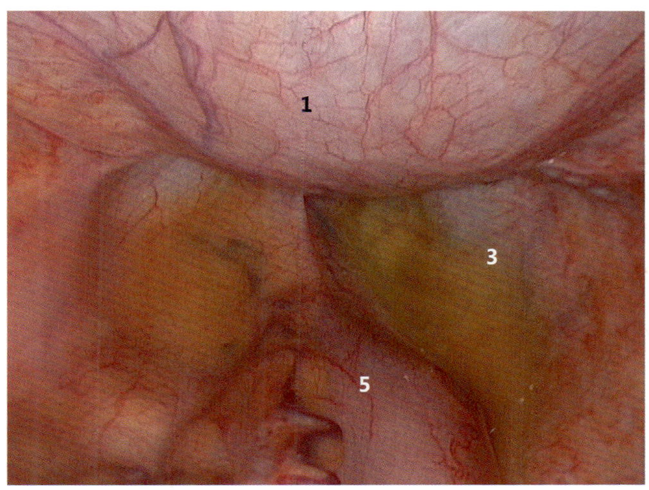 b

Images obtained in the course of a diagnostic laparoscopy of the male pelvis showing the **bladder** and **sigmoid colon** (**a**) and, on a lower plane, the **rectovesical pouch** (**b**).

1, Urinary bladder
2, Sigmoid colon
3, Rectovesical pouch
4, Ileal loop
5, Rectum

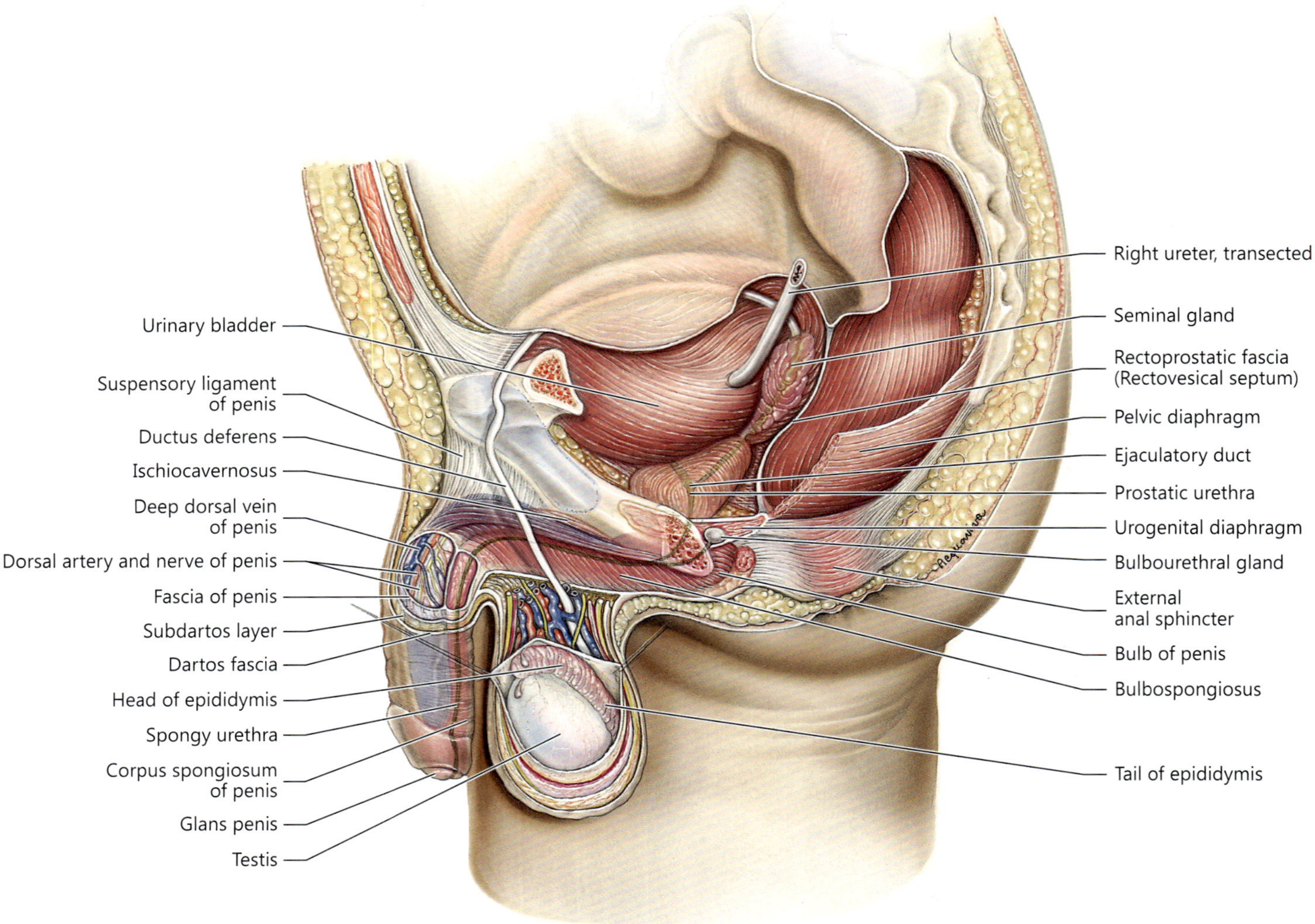

Conformation and relations of the bladder in the **male pelvis** of an adult subject. The bladder in conditions of vacuity has a flattened ovoid shape and does not exceed the superior margin of the pubic symphysis.

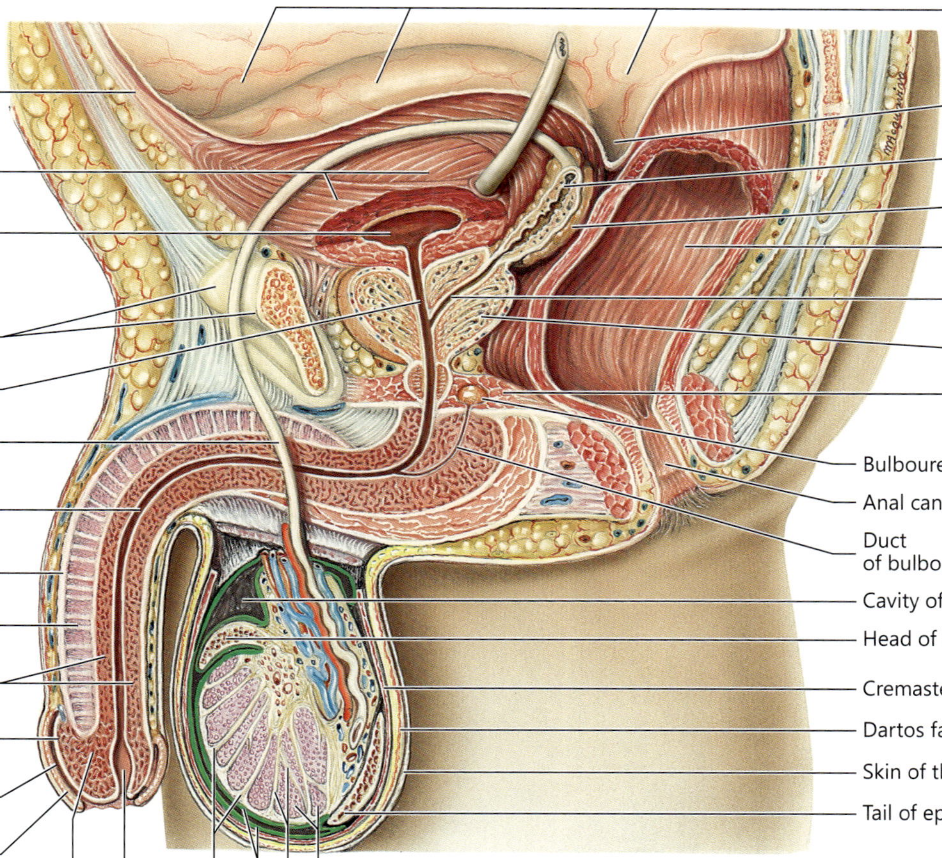

Schematic representation of the **pelvis** and **perineum of the adult male** showing the position and relations of the bladder, rectum and various organs of the male genital apparatus.

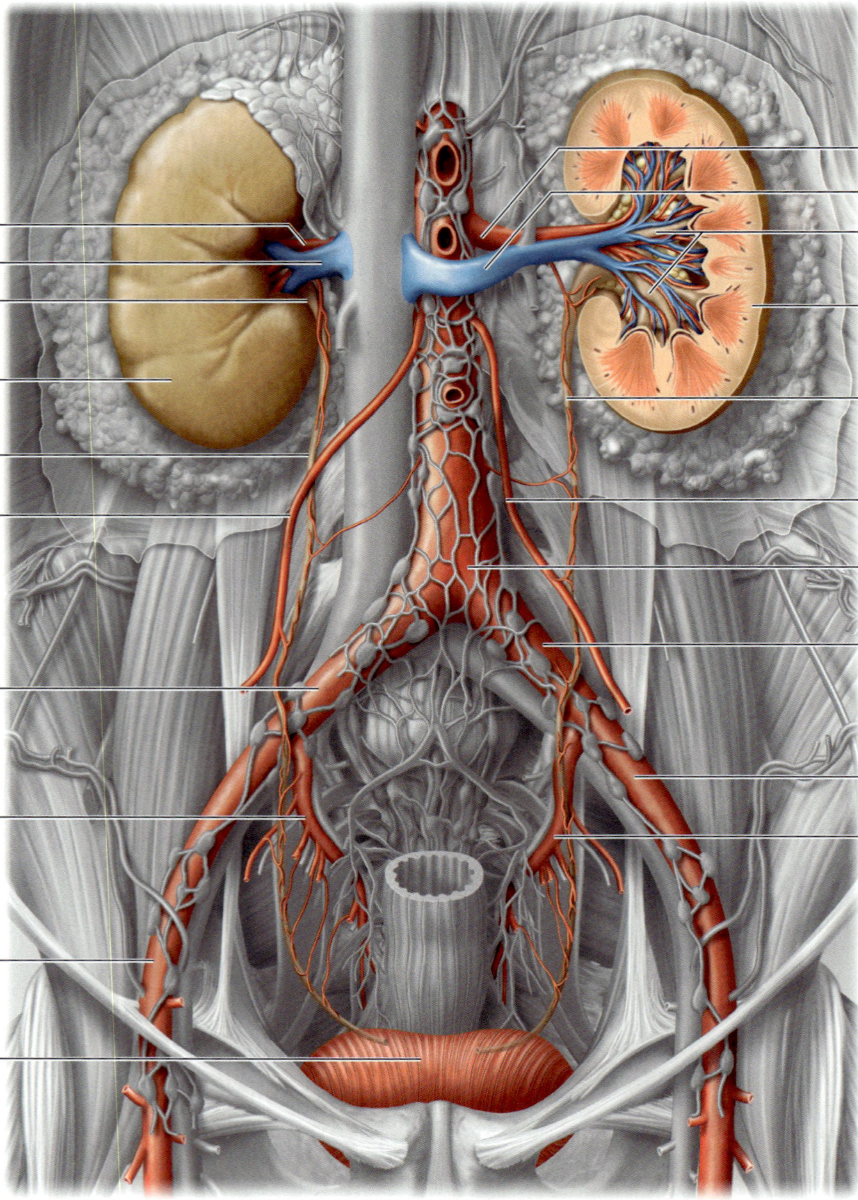

The **ureters** continue from the renal pelvis wall and terminate connecting to the urinary bladder through the ureteric orifice, situated in the posterolateral angle of the trigone of the bladder. The abdominal part of the ureter describes a mild convex curve forwards and medially; at the pelvic inlet, circles the iliac vessels, and continues in the pelvic part of the ureter, forming a concave curve oriented forwards, above and medially.

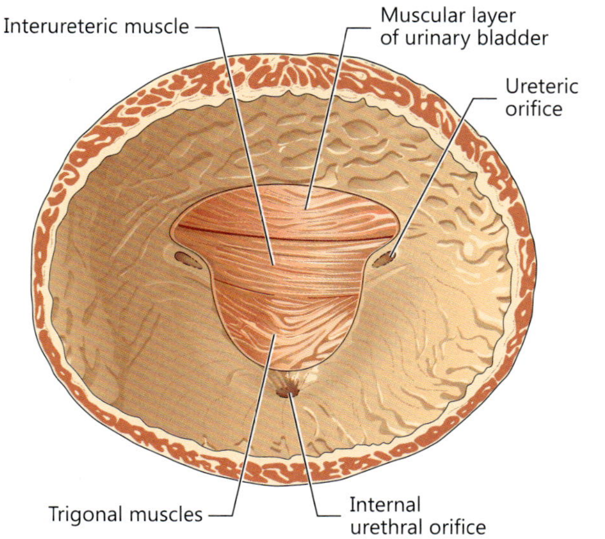

Internal configuration of the **bladder**, after removal of the mucosa in the region of the trigone; note the organization of the innermost trigonal muscles which converge towards the ureteric orifices and internal urethral orifice.

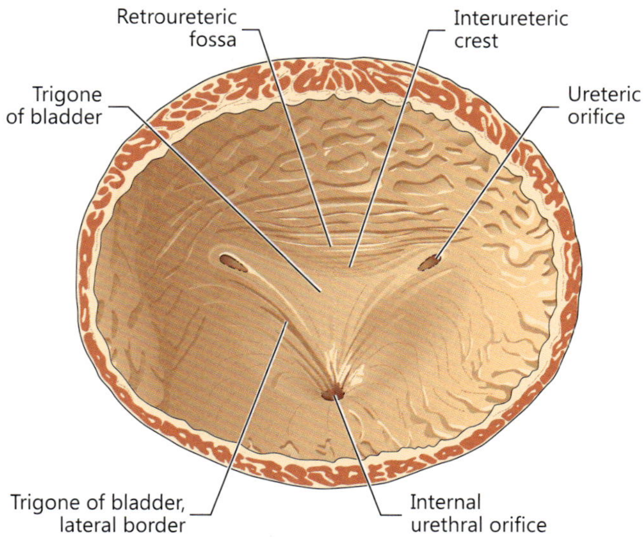

Internal configuration of the **bladder**. The region of the trigone is delimited by the two ureteric orifices and the internal urethral orifice. The innermost bundles of the muscle layer raise the vesical mucosa and form rugae.

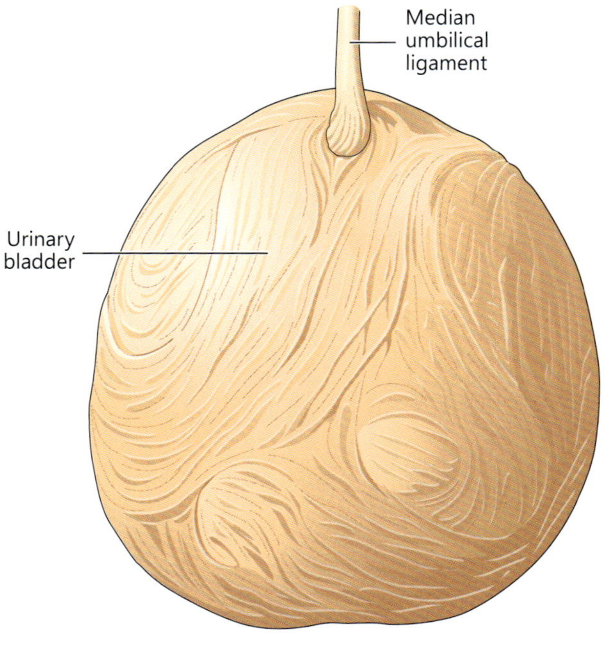

External conformation of the **bladder** viewed in anterior projection, after removal of the visceral peritoneum. The surface of the organ appears sculpted in a plexiform arrangement by robust muscular bundles.

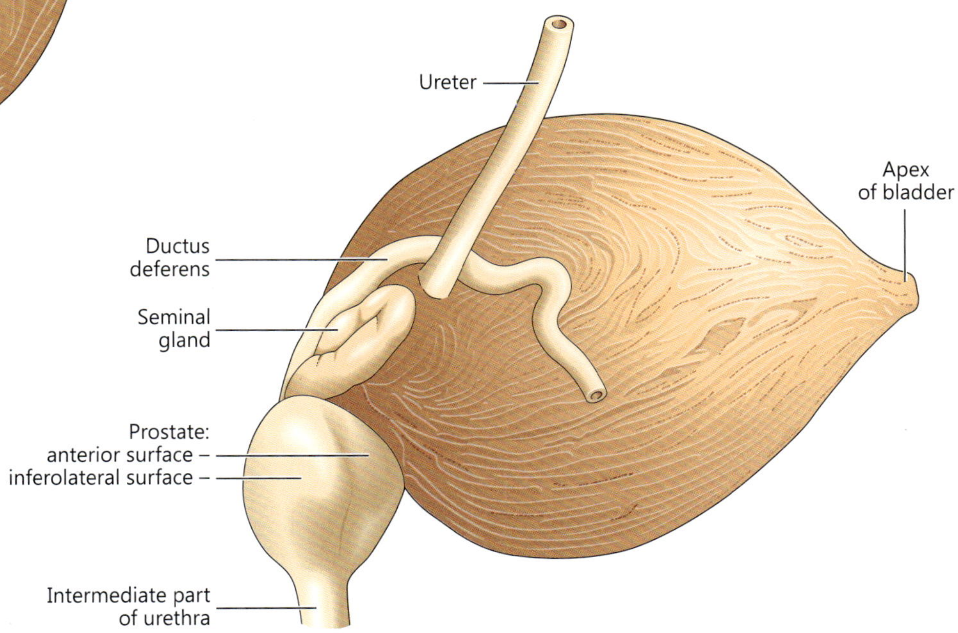

Relations of the right pelvic **ureter** with the **bladder**, and homolateral, seminal gland and ductus deferens. Note the interesection between the ureter and ductus deferens. The bladder is as it would appear in an anatomical preparation after removal of the visceral peritoneum.

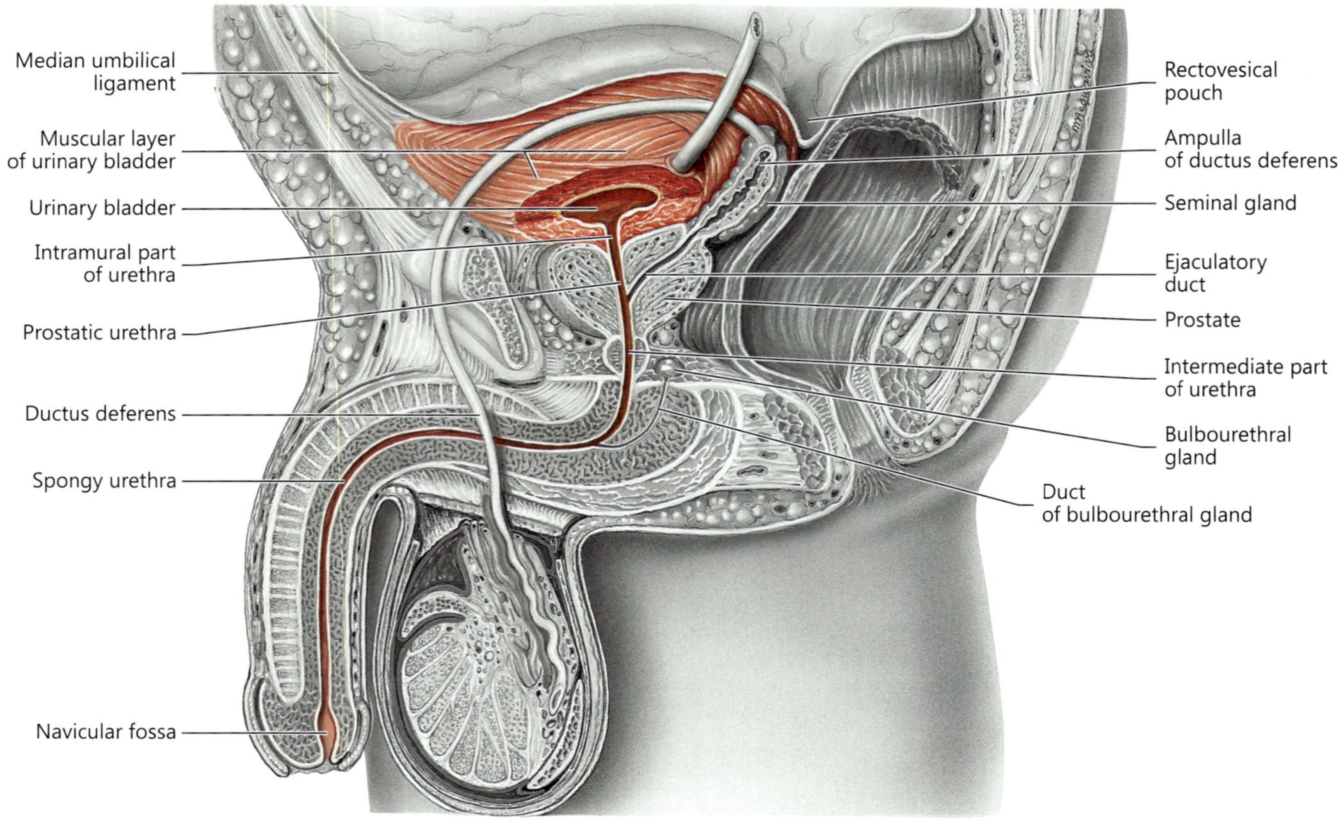

The **male urethra** is divided into a preprostatic initial portion, transporting exclusively urine, a second prostatic urethra part passing through the prostate, a third intermediate part, the membranous urethra, crossing the whole thickness of the urogenital triangle, and the spongy urethra, enveloped by the corpus spongiosum.

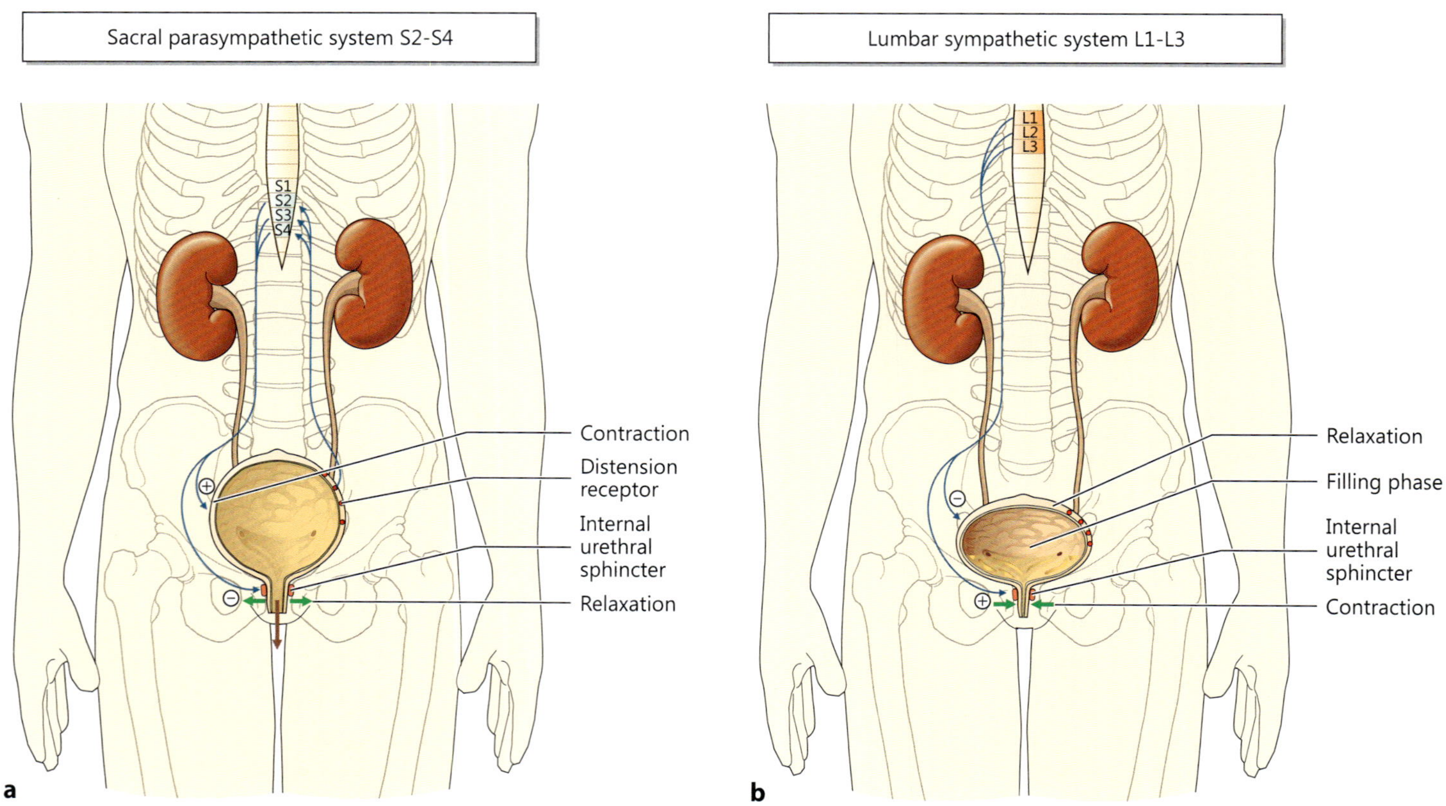

Innervation of the bladder. In the full bladder (**a**) receptors sensitive to distension determine the relaxation of the smooth sphincter and bladder emptying, through the parasympathetic pathways involving the sacral spinal cord. In the empty bladder (**b**) sympathetic tone prevails: the bladder remains relaxed and the smooth sphincter is contracted.

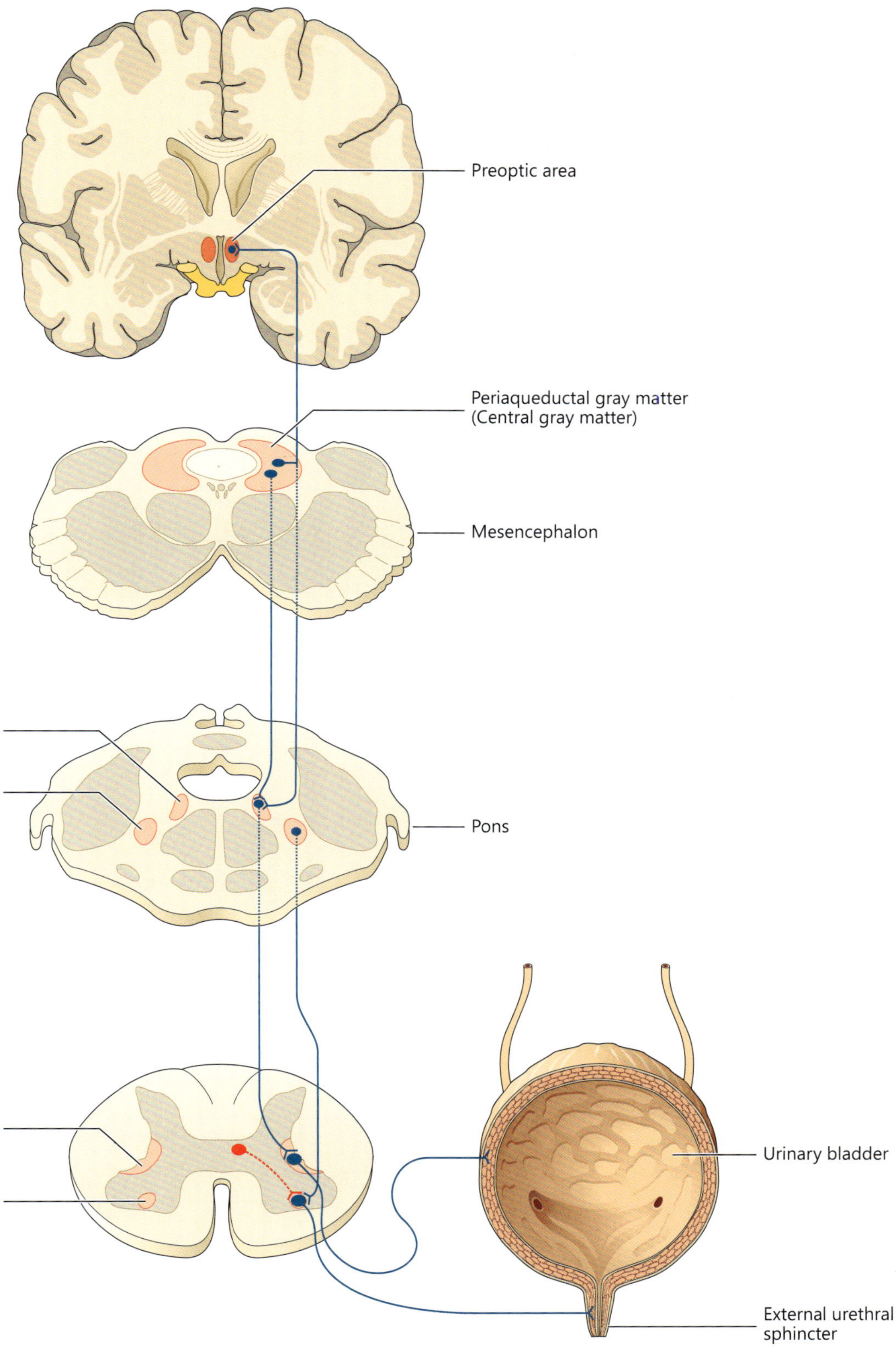

Brainstem and spinal cord pathways and centers involved in the control of continence. The lateral pontine tegmentum contains two separate regions involved in the control of urination (centers of urination): the M-region of the pons, located in the dorsomedial part of the lateral pontine tegmentum, and the L-region, located in the ventrolateral pontine tegmentum. Onuf's nucleus is situated in lamina X of the anterior horn of the gray matter of the spinal cord between the neuromeres S1 and S3 and innervates the external urethral sphincter. The preganglionic axons of the visceral effector parasympathetic fibers which determine the contraction of the detrusor of the bladder and the relaxation of the internal urethral sphincter arise from the sacral visceral effector column in lamina VII of the anterior horn of the gray matter of the spinal cord between the neuromeres S2 and S4 and reach the pelvic ganglia near or in the wall of the bladder.

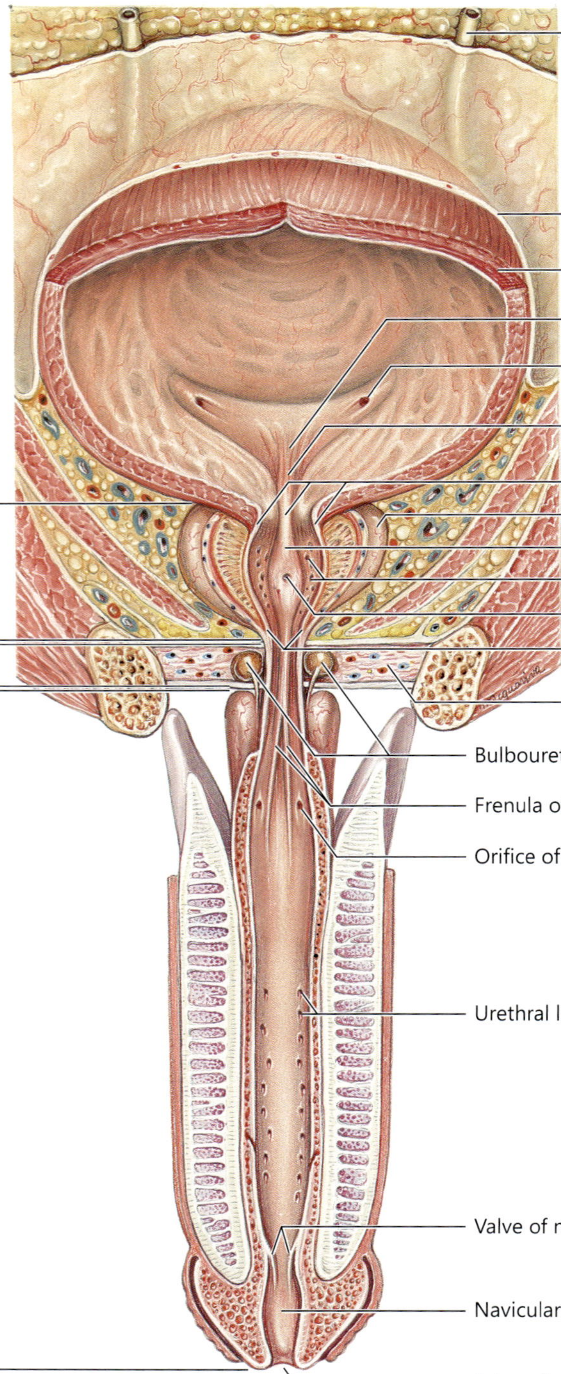

Schematic representation of the whole complex of the **bladder** and **male urethra**; the anterior wall of the urethra has been removed to show the internal configuration of the various parts. Note the succession of narrow vs. dilated segments.

➤

a, Anterior view of the **prostate**. The base has a triangular shape, with one posterior side and two anterior sides converging in the front. The anterior surface, almost vertical, is uniformly convex. The inferolateral surfaces are rounded. The posterior surface is oriented obliquely downward and forward.
b, Transverse section of the **prostate** at the level of the ejaculatory ducts. The division into right and left lobes of the prostate is due to the presence of a shallow median longitudinal sulcus on the posterior surface of the prostate, utilized as a landmark in the palpation of the organ through rectal exploration. The superomedial and anteromedial lobules, which originate from the middle lobe in the course of development, become hypertrophic in the elderly, creating a valve-like occlusion of the internal urethral orifice.
c, Median sagittal section of the male pelvis. The **prostate** is situated in the lesser pelvis, behind the pubic symphysis and in front of the rectal ampulla. The prostatic section of the urethra runs through the prostate, between the anterior and the middle third of the organ. The ejaculatory ducts open into the prostatic urethra, after crossing the posterior part of the prostate, directed forward and inferiorly. At the top, the base of the prostate is contiguous with the neck of the bladder, while the apex, at the bottom, surrounds the passage between the prostatic urethra and intermediate part of the urethra.

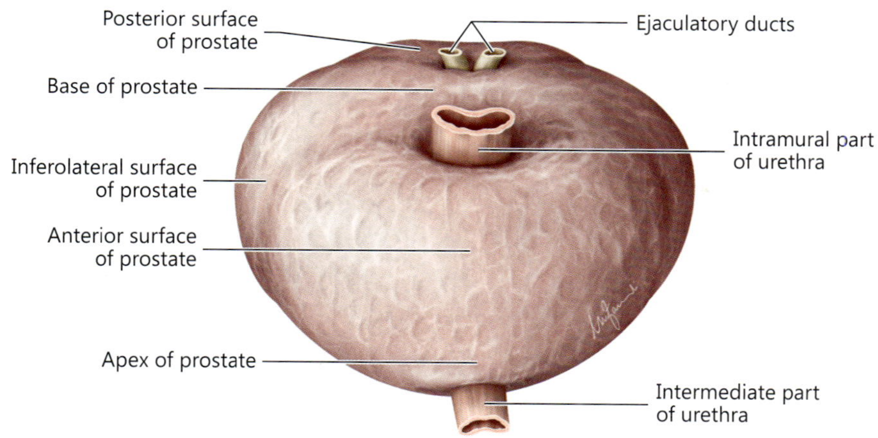
Posterior surface of prostate
Ejaculatory ducts
Base of prostate
Intramural part of urethra
Inferolateral surface of prostate
Anterior surface of prostate
Apex of prostate
Intermediate part of urethra

a

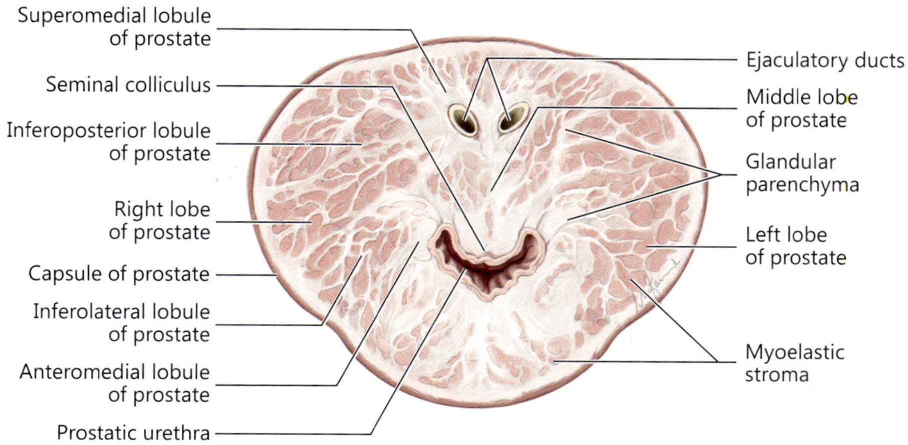
Superomedial lobule of prostate
Ejaculatory ducts
Seminal colliculus
Middle lobe of prostate
Inferoposterior lobule of prostate
Glandular parenchyma
Right lobe of prostate
Left lobe of prostate
Capsule of prostate
Inferolateral lobule of prostate
Myoelastic stroma
Anteromedial lobule of prostate
Prostatic urethra

b

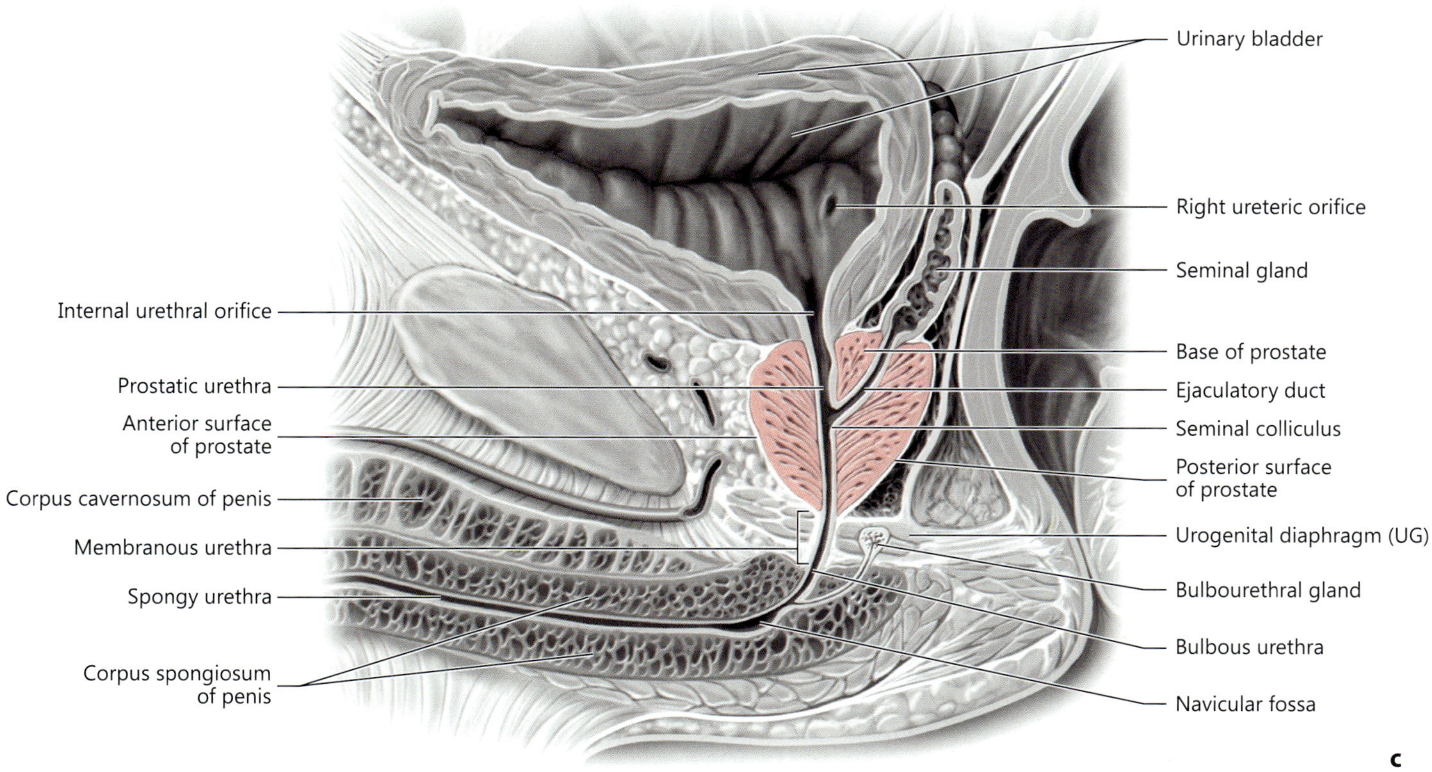
Urinary bladder
Right ureteric orifice
Seminal gland
Internal urethral orifice
Base of prostate
Prostatic urethra
Ejaculatory duct
Seminal colliculus
Anterior surface of prostate
Posterior surface of prostate
Corpus cavernosum of penis
Urogenital diaphragm (UG)
Membranous urethra
Bulbourethral gland
Spongy urethra
Bulbous urethra
Corpus spongiosum of penis
Navicular fossa

c

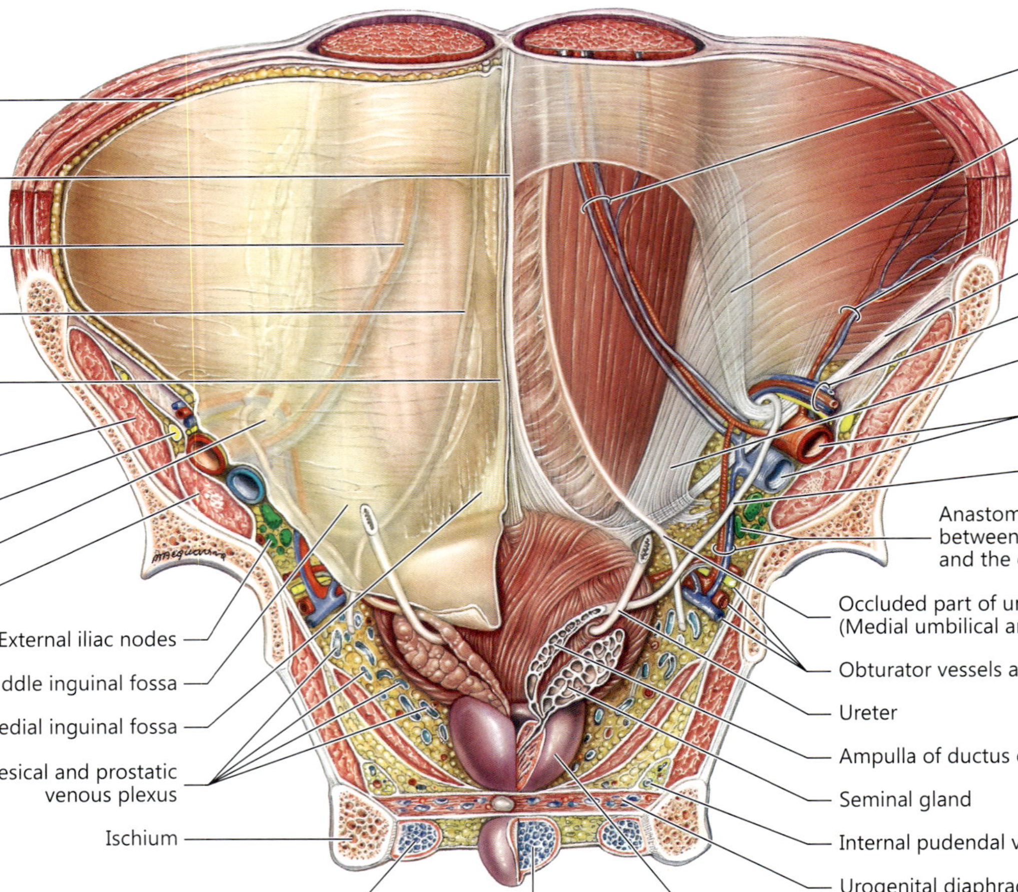

Posterior relations of the bladder in the male and disposition of the vesical and prevesical peritoneum. The preparation is viewed from behind after demolition of the posterior pelvic wall and removal of all the retrovesical visceral and vascular formations. On the right side, the anterior parietal peritoneum has been removed.

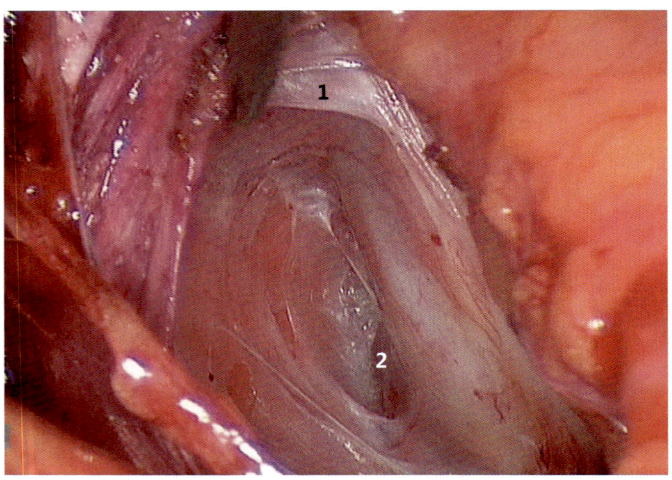

Image obtained in the course of a laparoscopy of a male pelvis showing the **prostate** on the right and, on the left in correspondence to the instrument, the **pelvic fascia** separated from the muscle fibers of the pelvic floor. The pelvic fascia is whitish in color.

1, Pelvic fascia
2, Prostate

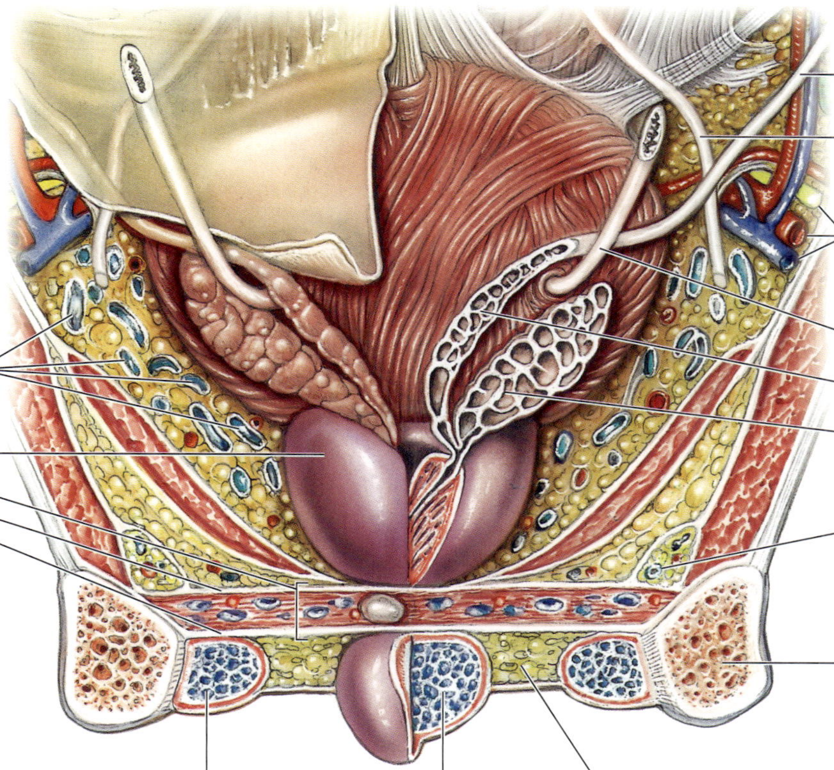

The first section, or **pelvic tract**, **of the ureter**, in the male, runs attached to the lateral wall of the pelvis, crosses the obturator vessels and nerve laterally, and reaches the pelvic floor; it then proceeds forward and medially, crossed anteriorly by the ductus deferens in correspondence to the lateral margin of the bladder; finally, in its last section, it wedges between the vesical wall and the anterior surface of the seminal gland. The right half-side of the image displays a frontal transection of the ampulla of the ductus deferens, seminal gland and prostate, to show their internal conformation.

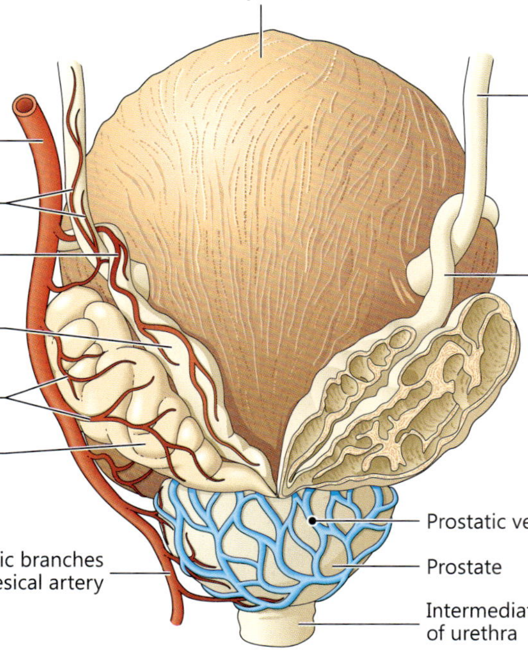

Relations of the last section of the **ductus deferentes**, **ampullae of the ductus deferentes** and **seminal glands** with the ureters, bladder and prostate. The preparation is viewed from behind; the ampulla of the ductus deferens and the right seminal gland have been opened to show the irregularity of the lumen. The inferior vesical artery (also known as the vesicoprostatic artery in the male), is a branch of the internal iliac artery or of the vesiculodeferential artery, and provides the ureteric and prostatic branches and branches to the seminal gland and, sometimes, an artery of the ductus deferens, which can also originate from the vesiculodeferential artery or from one of the superior vesical arteries.

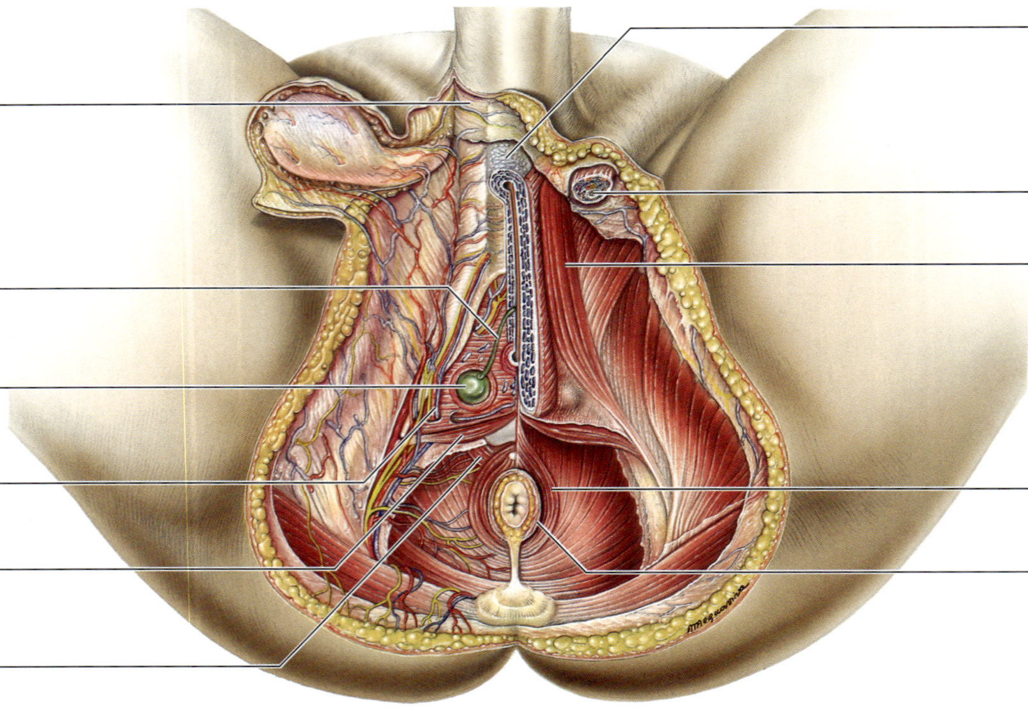

Preparation of the male anterior perineum illustrating the position and the **relations of the bulbourethral glands and their excretory ducts.** On the right, the floor of the superficial perineal pouch has been demolished and the bulbourethral glands can be seen in relation to the roof of the pouch itself.

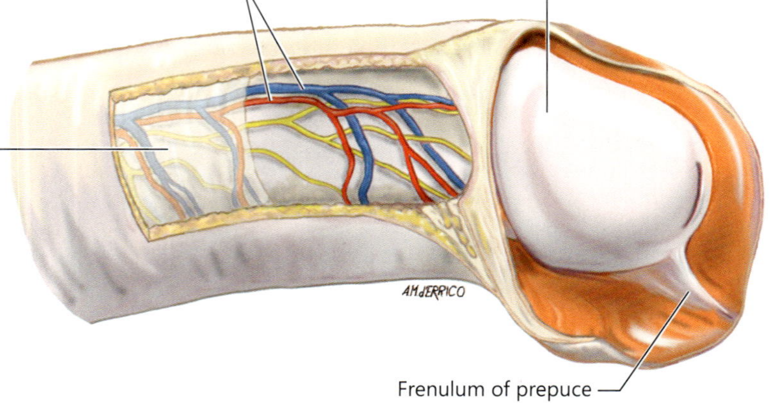

Relations of the prepuce, partly demolished and spread out, with **the gland**. Inferiorly, the two formations are connected by the frenulum of the prepuce.

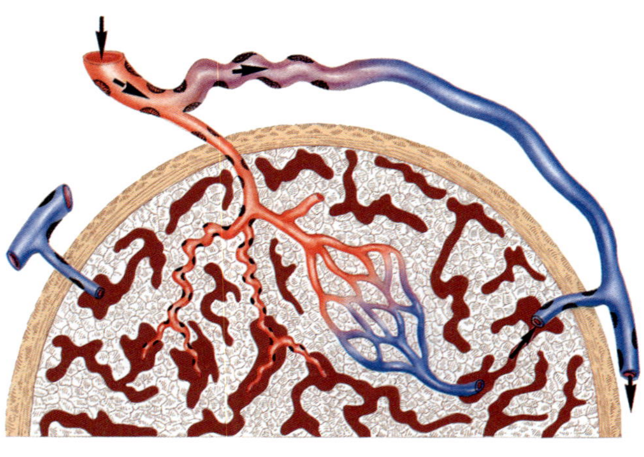

Vascular organization of the penis. Most of the arterial blood reaches the nutritive capillary network between the walls of the cavernous spaces.

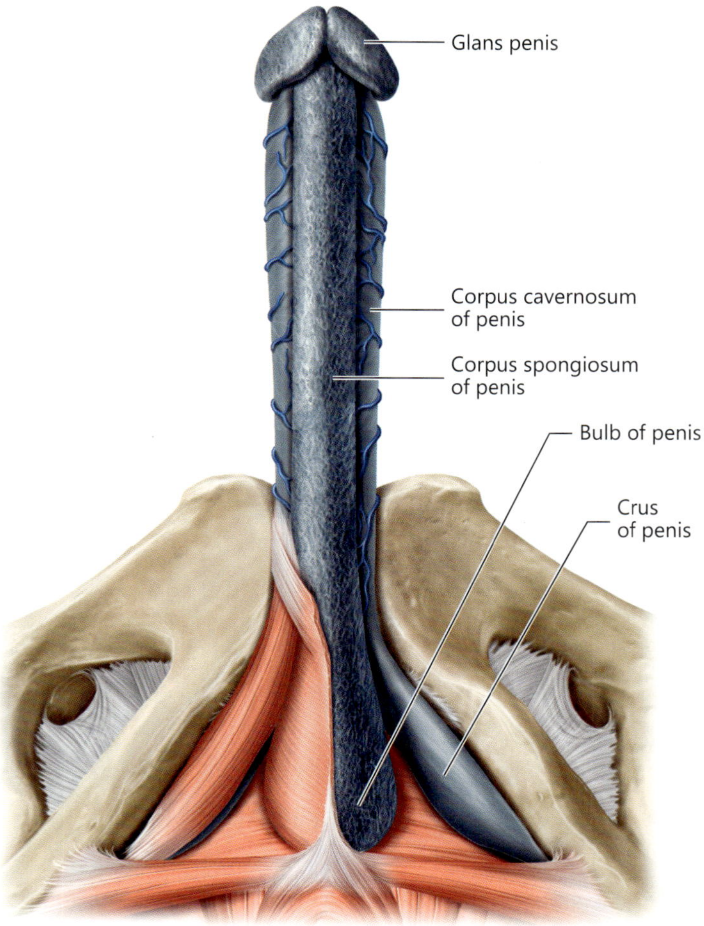

Organization of the **corpora cavernosa** and of the **corpus spongiosum of the penis** (in *blue*) viewed from the ventral side. The corpus spongiosum proceeds, without apparent interruption, into the glans penis. Note the relation of the roots of the corpora cavernosa of the penis with the ischiopubic rami of the hip bone.

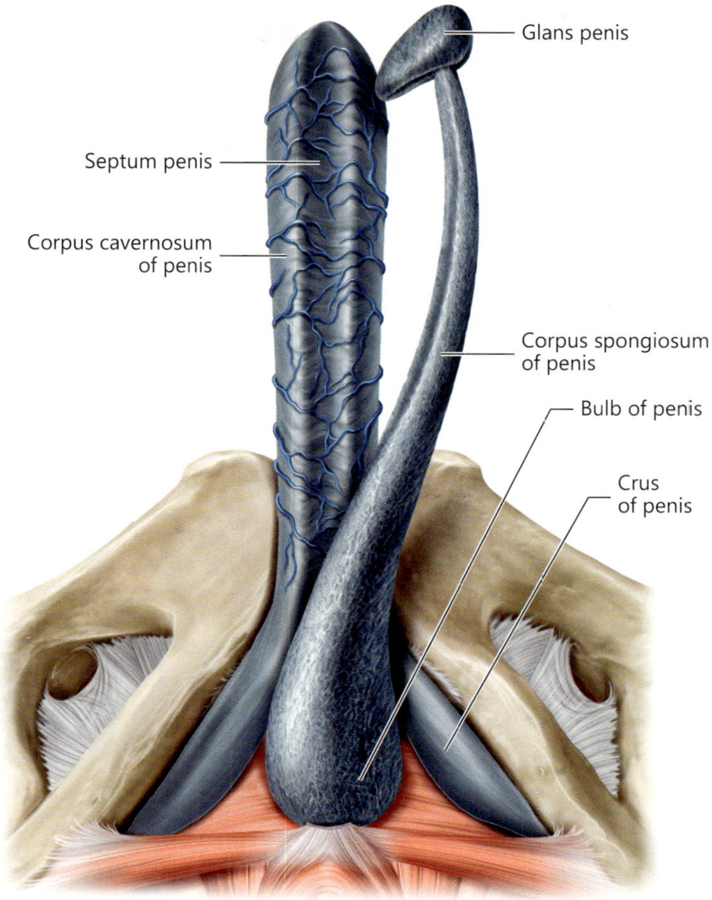

Corpora cavernosa and **corpus spongiosum of penis** partially separated. The corpora cavernosa of the penis are viewed from the ventral or inferior surface; connected to the corpus spongiosum of the penis, viewed from the side, is the glans penis. The two corpora cavernosa of the penis are juxtaposed delimiting ventrally a sulcus (septum penis) which contains the corpus spongiosum of the penis.

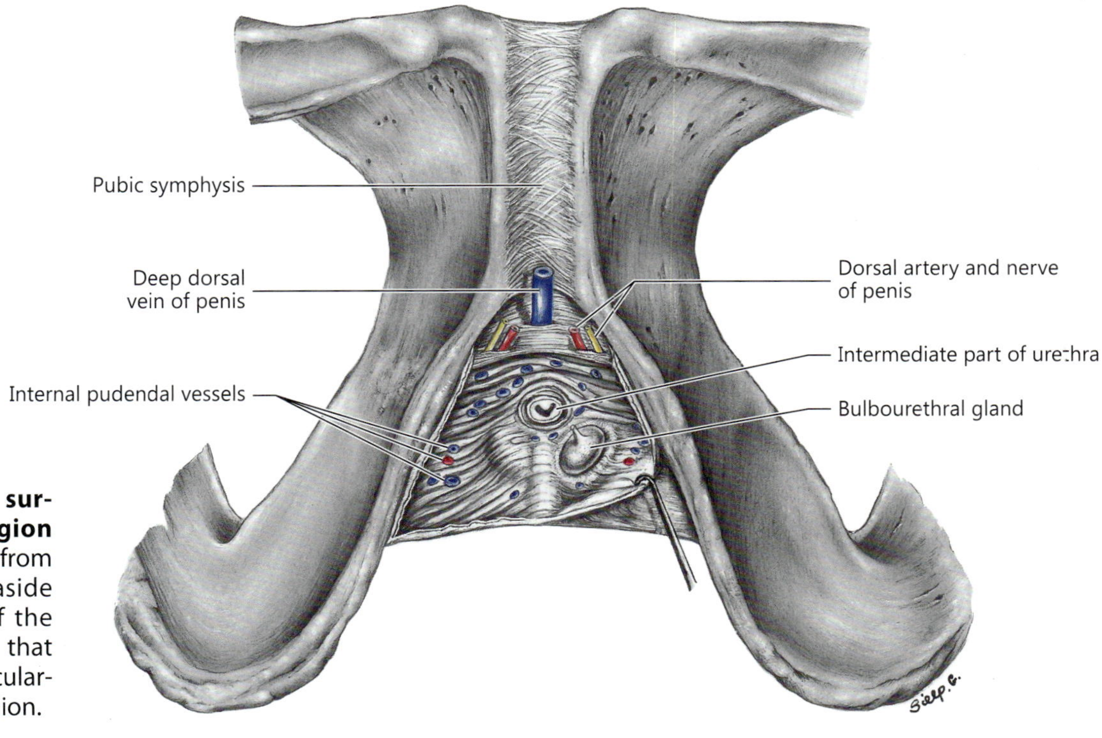

Preparation of the **inferior surface of the urogenital region** (urogenital triangle) viewed from the front. The hook pulls aside part of the inferior layer of the middle perineal fascia. Note that the urethra passes perpendicularly through the urogenital region.

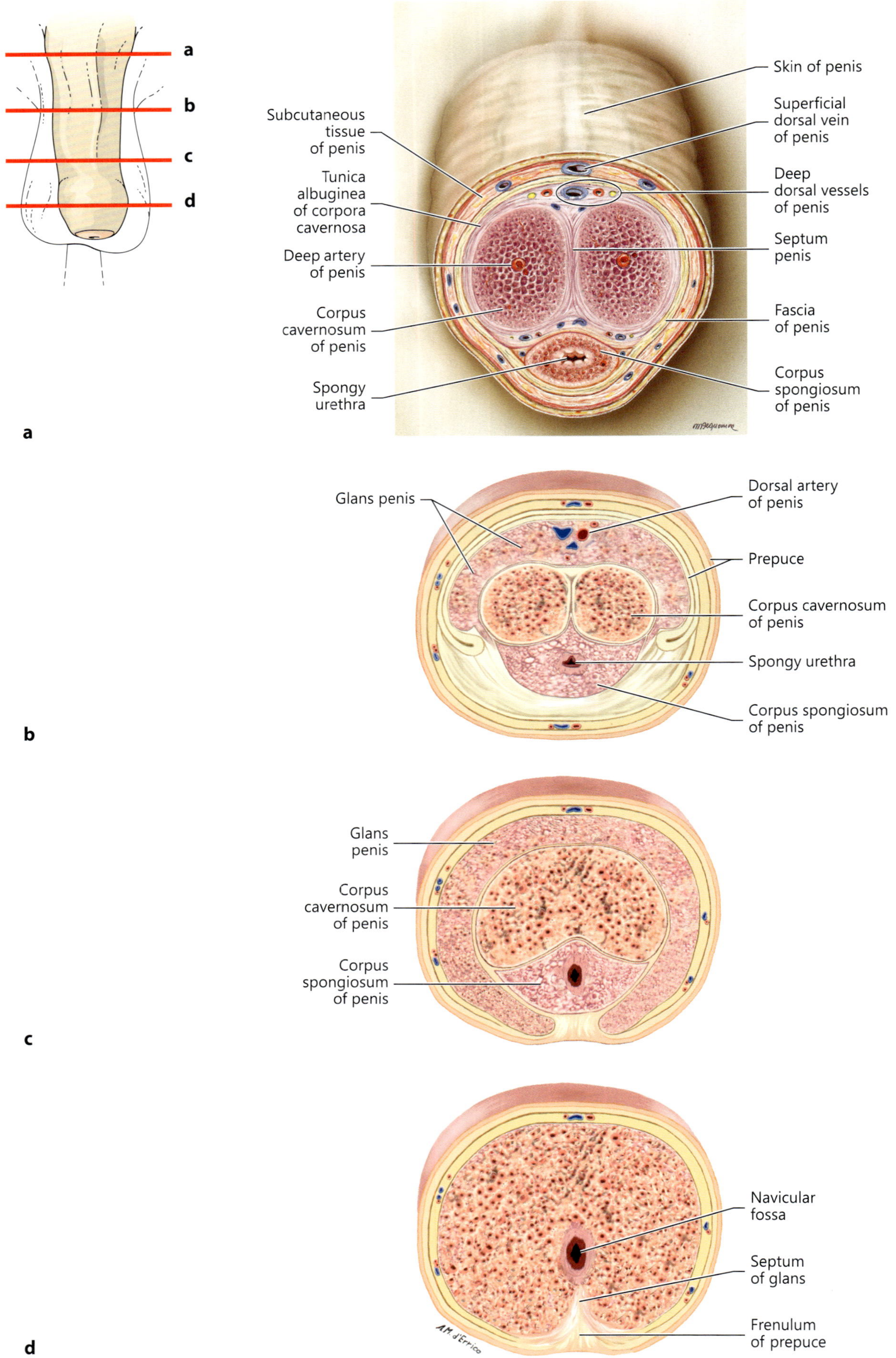

Transverse sections of the penis passing through the body (**a**), the neck (**b**), the corona (**c**) and the body of the glans penis (**d**). At the different levels one can see the organization and reciprocal relations of the diverse cavernous formations. Note also the different position of the urethra in the organ at the different levels.

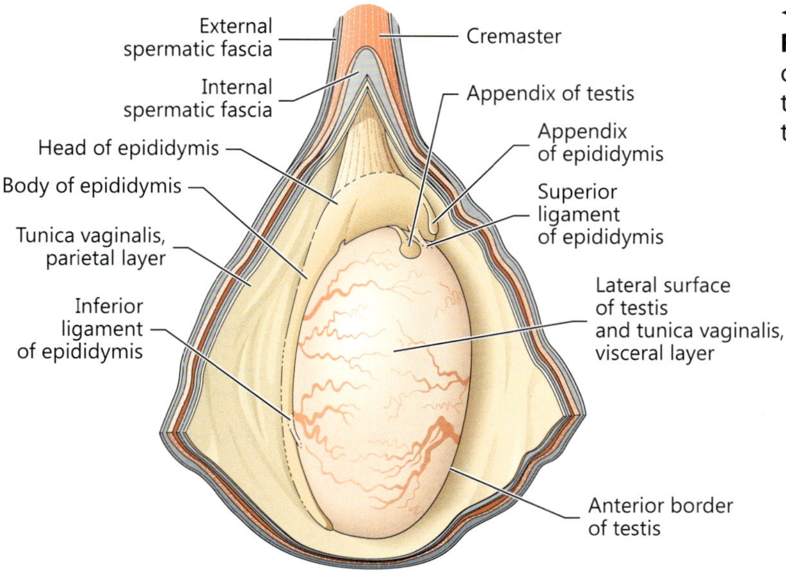

➤ **Right testis** viewed in anterior projection after opening of the tunica vaginalis. Above the testis is the head of the epididymis, while, below, it is anchored to the wall of the scrotal sac.

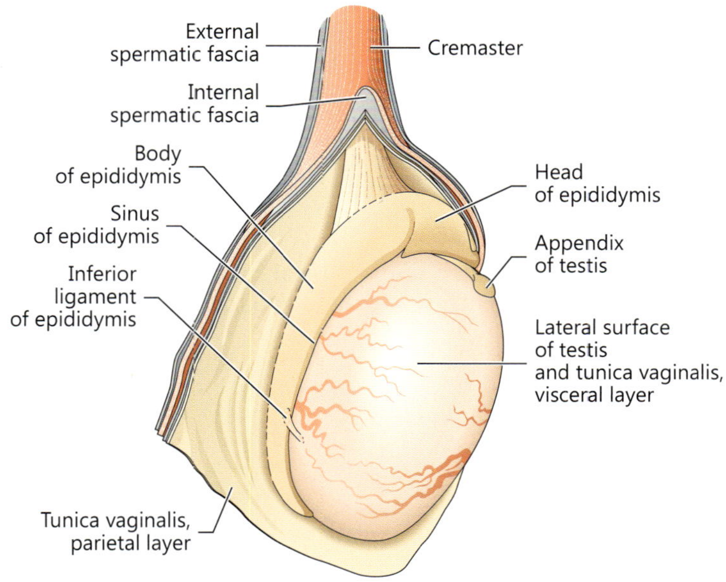

➤ **Right testis** viewed in lateral projection showing the extensive relation established between the posterior margin of the testis and the various parts of the epididymis.

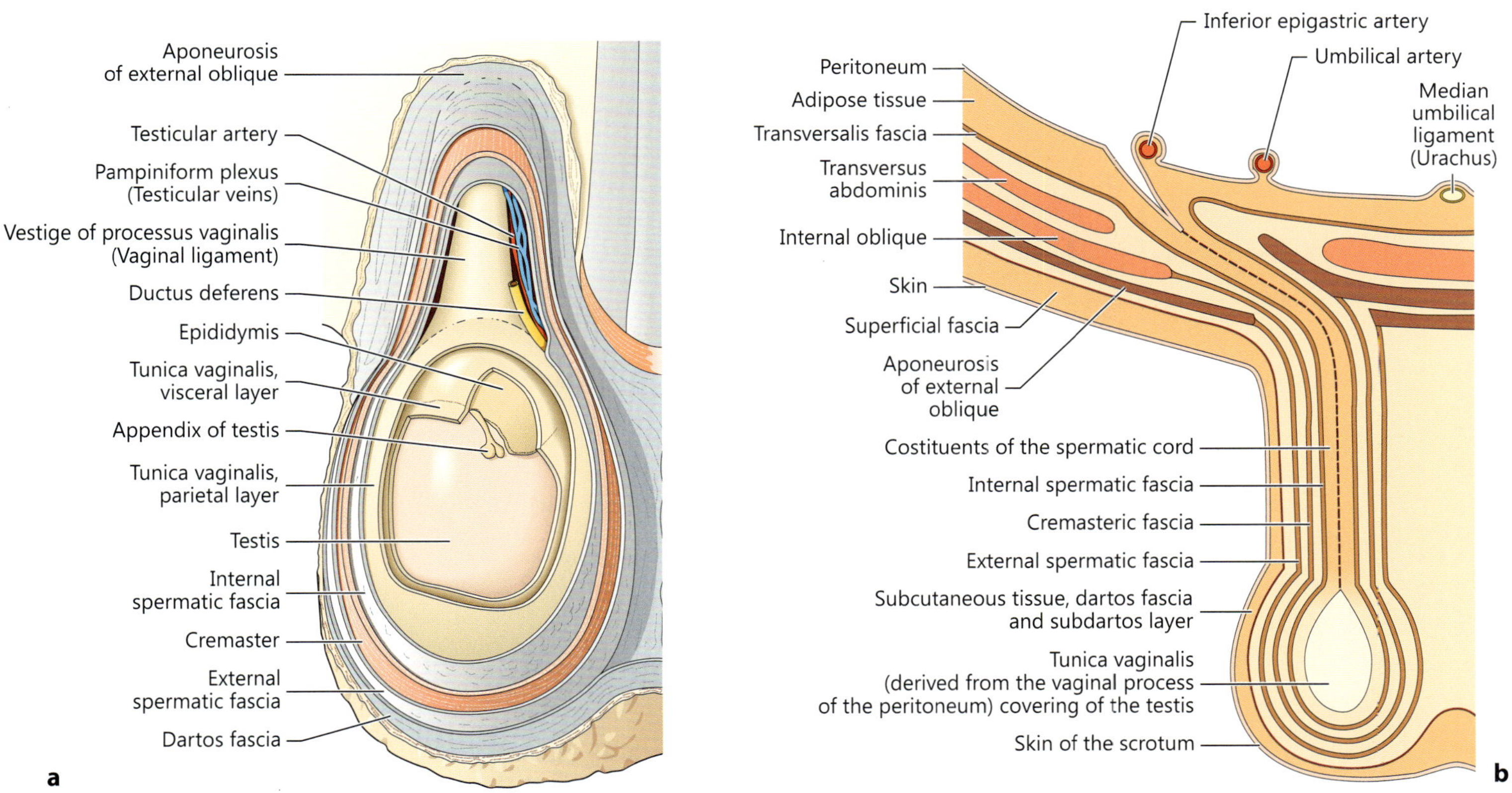

a, **Right testis** viewed in anterior projection after opening of its investing tunicae. **b**, Schematic representation of the **testis** and of its **tunicae**. Some tunicae of the testis are in continuity with the layers covering the spermatic cord. In this transverse section, the scrotum has been raised in a horizontal position.

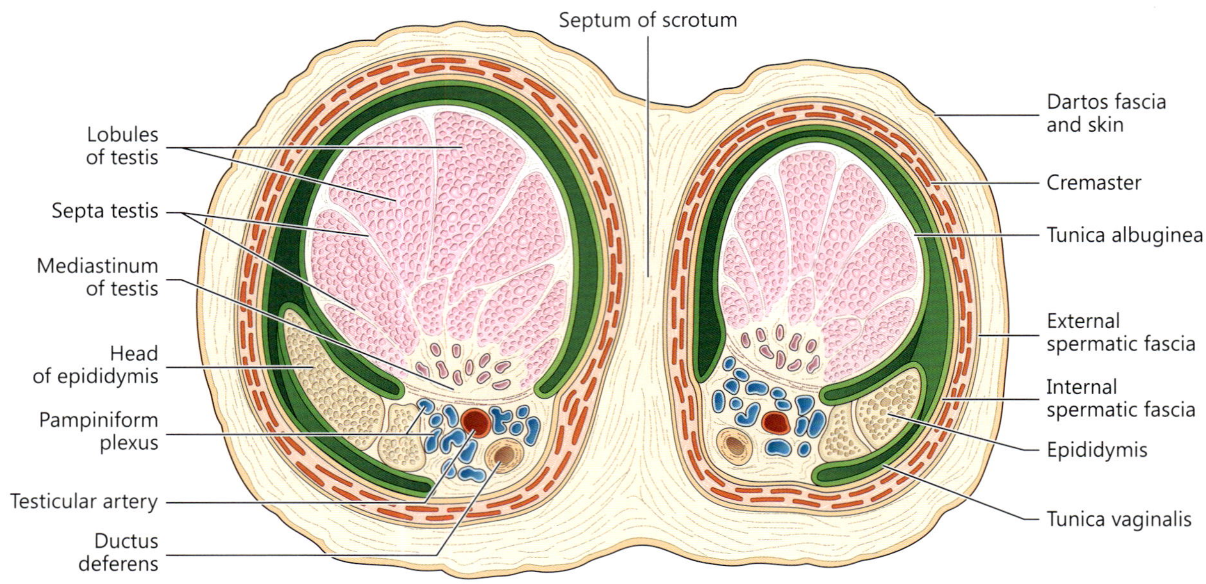

Transverse section of the **scrotal sac** and its contents displaying the lobular structure of the testis, the relations of the gonad with the epididymis and the ductus deferens, and the various tunicae investing the testis and the wall of the scrotal sac.

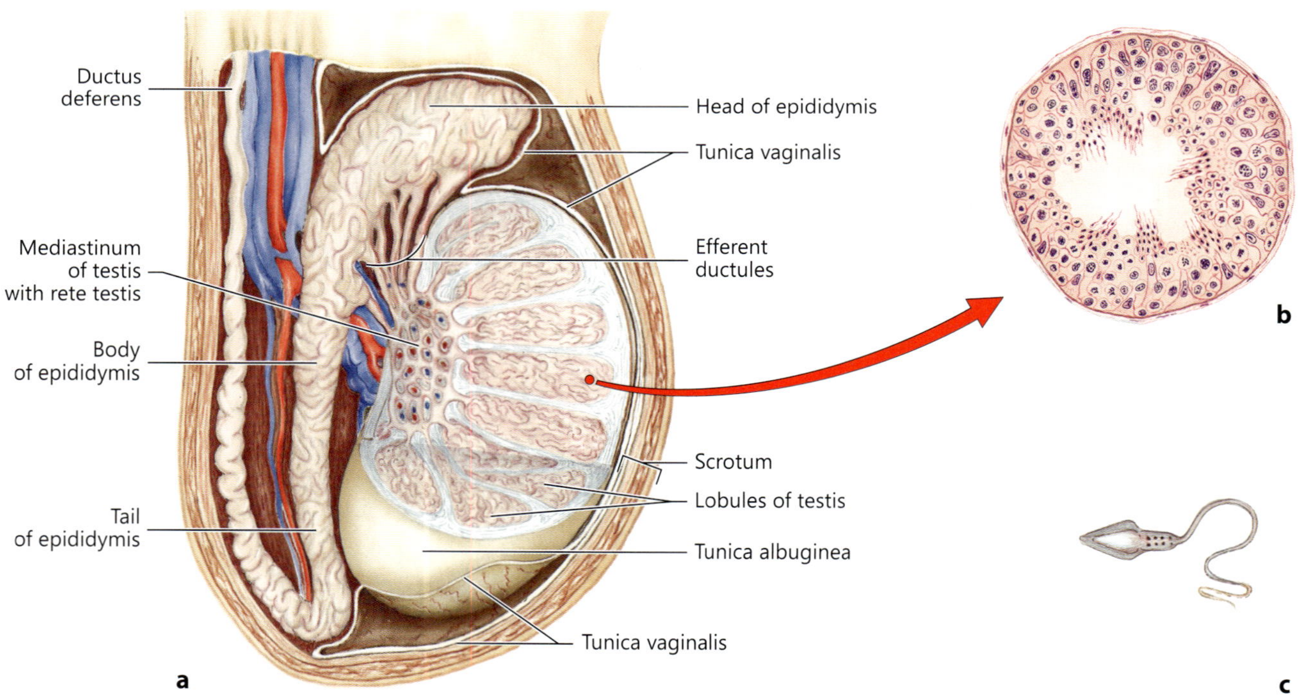

a, Overall structure of the **testis** and of the initial segment of the sperm canal. The testis (shown in sagittal section) is composed of the seminiferous tubules in which the differentiation of the gametes takes place; the epididymis forms the initial part of the extratesticular sperm canal. **b**, Internal organization of the testis, **transverse section of a seminiferous tubule**. The seminiferous tubules, with their convoluted pattern, are contained in compartments which converge towards the posterior margin of the testis. Inside the seminiferous tubules are the gametes, whose maturation proceeds from the periphery of the tubule (basal compartment) towards the central lumen (adluminal compartment), where the differentiated ones collect (**spermatozoa**, **c**).

Scheme illustrating the **organization of the testis and initial intratesticular and extratesticular parts of the spermatic pathways**. Note the continuity established between the distinct sections: convoluted seminiferous tubules, straight seminiferous tubules, rete testis (situated in the mediastinum of the testis or Highmore body) and efferent ductules (at the superior pole of the mediastinum), which perforate the tunica albuginea to leave the testis and enter into the head of the epididymis. The lobules of the testis contain numerous convoluted seminiferous tubules. One convoluted seminiferous tubule has been elongated to show that both its extremities converge to the mediastinum of the testis. The tubule then proceeds on into the straight seminiferous tubule (where spermatogenesis no longer occurs) and, from there, into the rete testis.

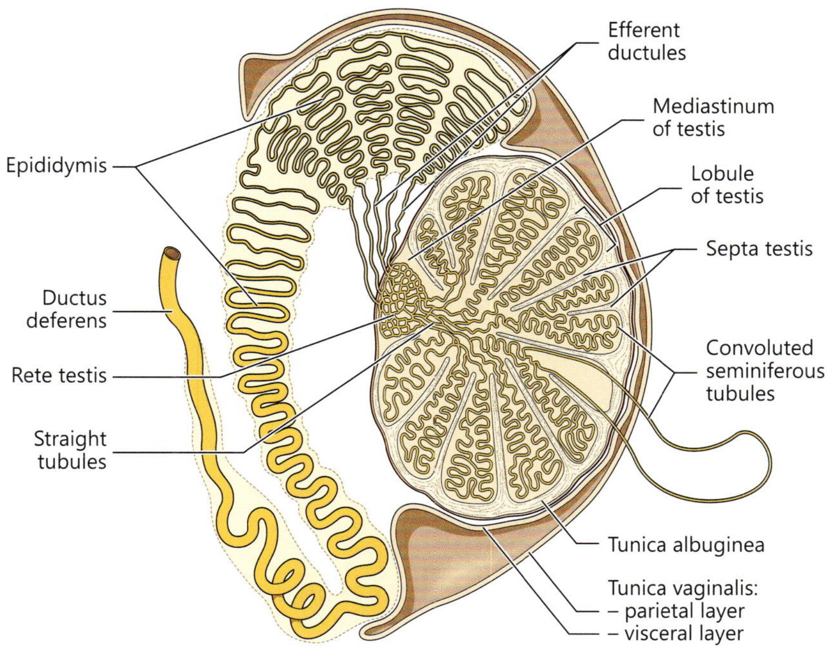

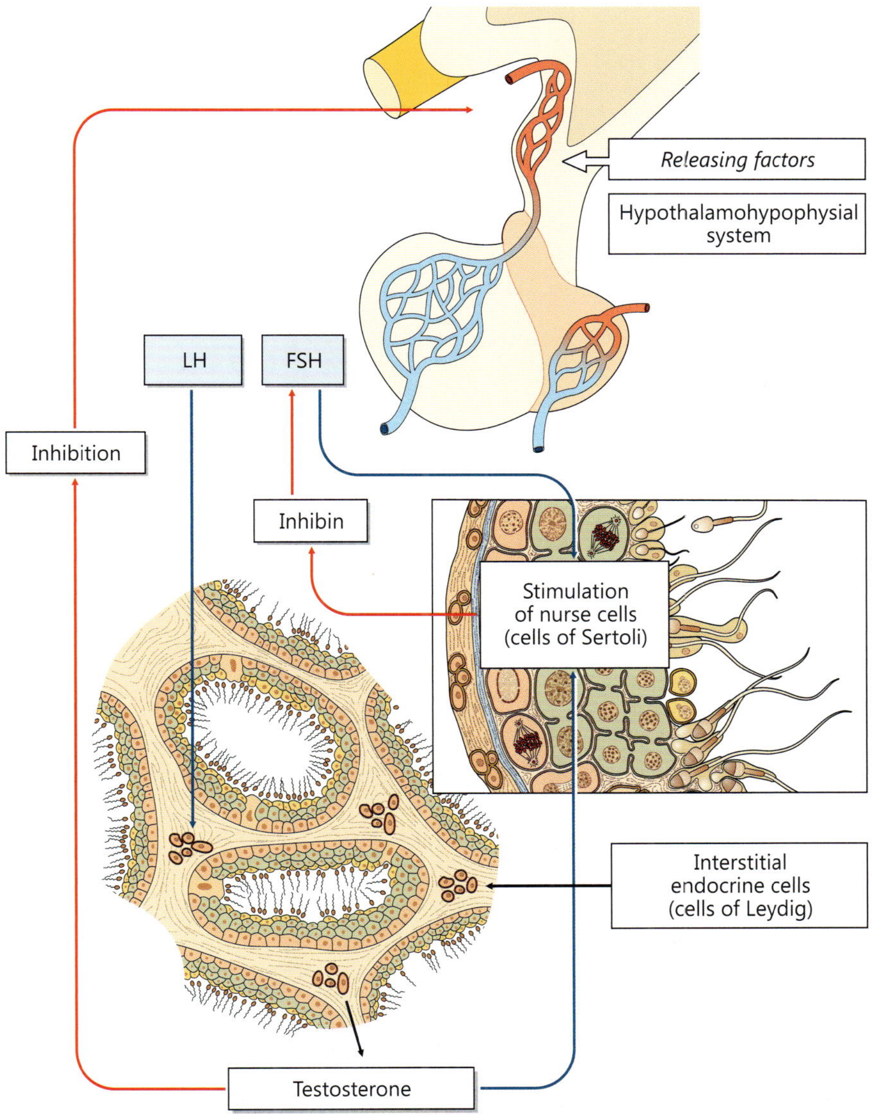

Schematic representation of the **regulation of spermatogenesis** and endocrine activity of the testis. The *blue* arrows indicate stimulation and the *red* arrows indicate inhibition. FSH, follicle stimulating hormone; LH, luteinizing hormone.

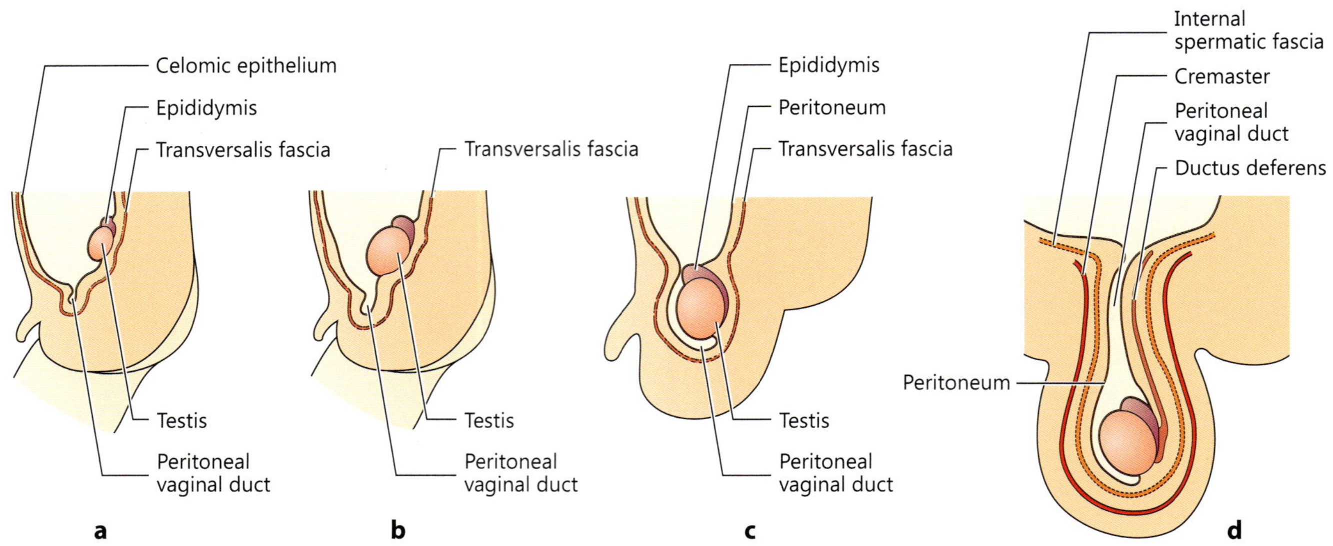

Four **successive stages of the descent of the testis** from the pelvic cavity (**a**, **b**) into the scrotal sac (**c**, **d**). In the last stage (**d**) the peritoneal vaginal duct is still indicated as patent, but it is destined to close itself.

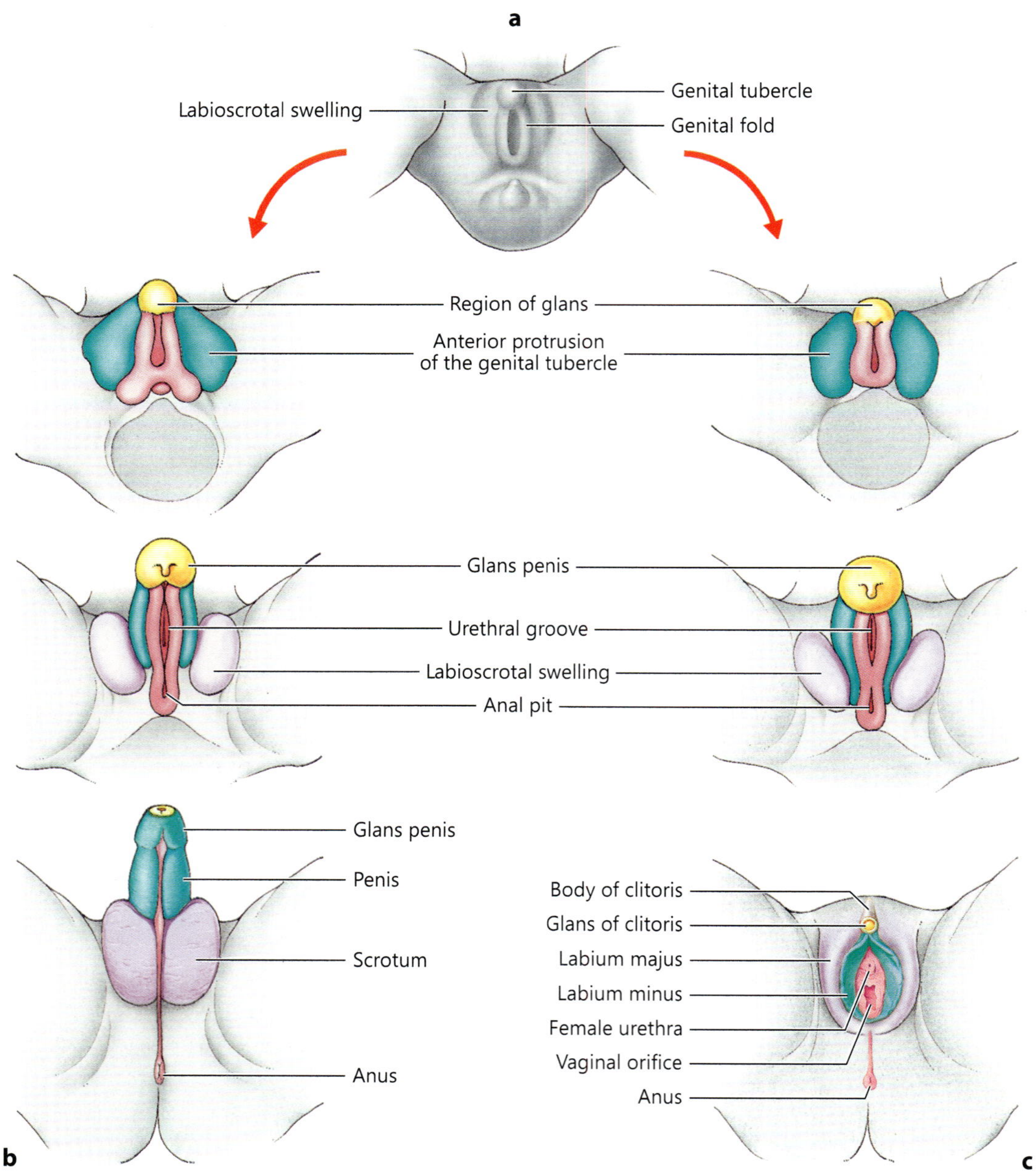

Development of the external genitalia. a, The external genitalia, similarly to the internal genitalia, remain undifferentiated until approximately the seventh week of intrauterine life. Before that date the embryos of the two sexes present a genital tubercle which has, ventrally, a urethral groove. The groove is in turn delimited laterally by genital (or urethral) folds and around these one can see a labioscrotal swelling. **b**, In the male, the urethral groove extends longitudinally to then close completely, while the genital folds unite to form the penis. The labioscrotal swellings of the two sides unite to form the scrotum. **c**, In the female, the urethral groove elongates and remains open, while the genital folds form the labia minora. The structure of the vulva and of the labia majora is eventually completed through the transformation of the labioscrotal swellings.

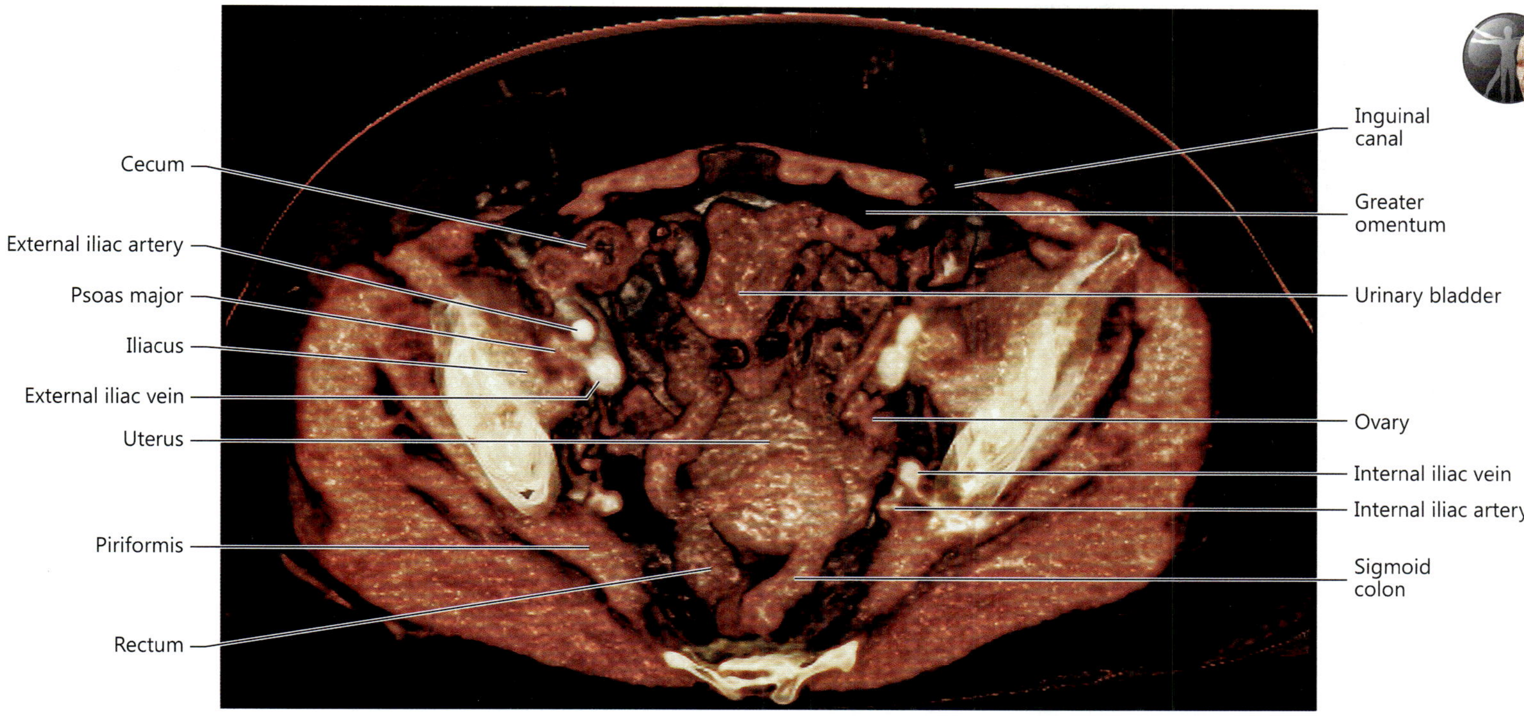

Three-dimensional image of a living human body, obtained by CT scan using a 3D volume rendering technique, of a transverse section of the trunk passing above the uterus, showing the contents of the **female pelvis**.

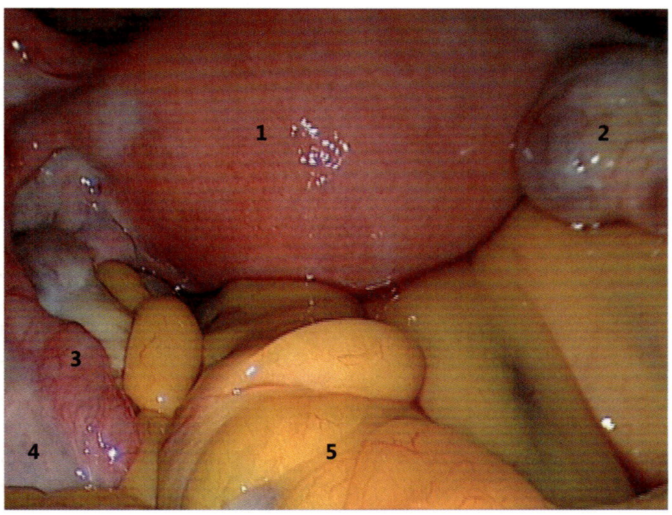

Image obtained in the course of a diagnostic laparoscopy of the female pelvis showing some **organs of the pelvis**.

1 Uterus
2 Right ovary
3 Left uterine tube
4 Left ovary
5 Sigmoid colon

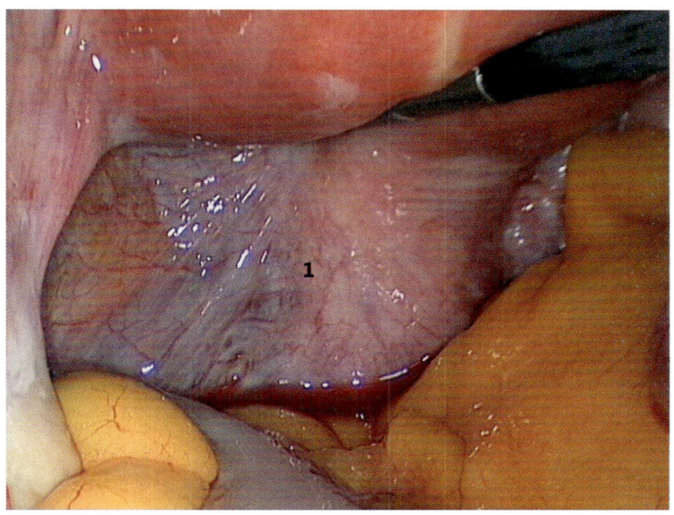

Image obtained in the course of a diagnostic laparoscopy of the female pelvis showing the **rectouterine pouch** (pouch of Douglas), after the uterus has been shifted forward.

1 Rectouterine pouch (Pouch of Douglas)

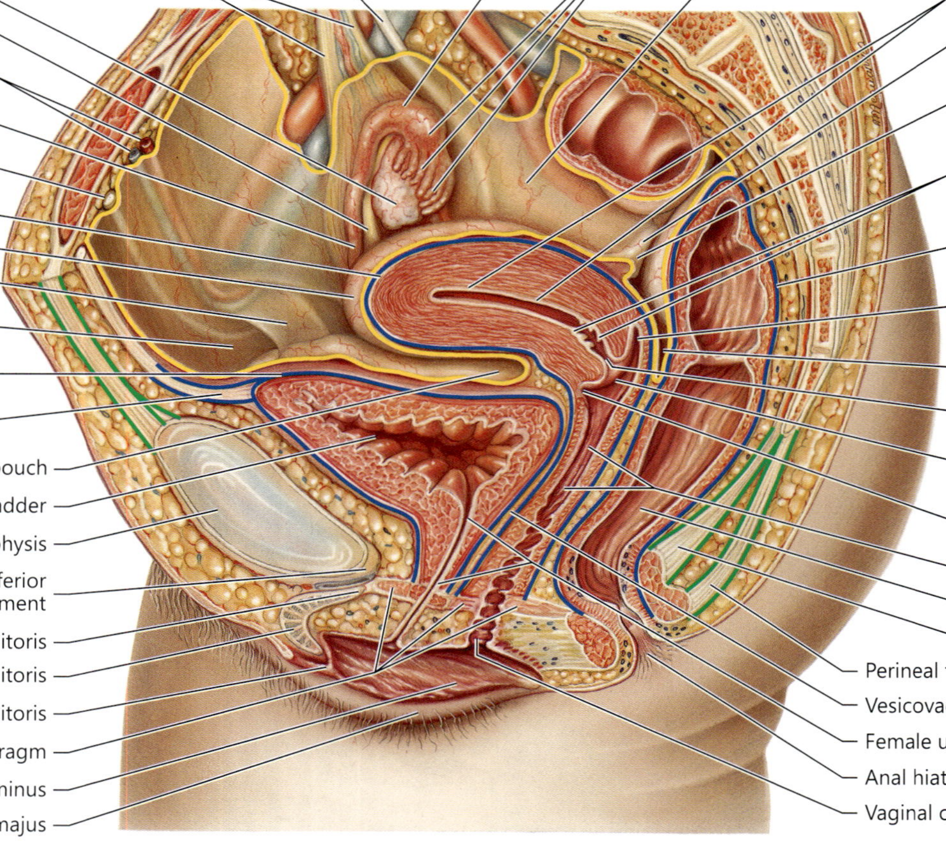

Median sagittal section of the **female pelvis**. Highlighted are: the peritoneum (in *yellow*), transected to highlight the rectouterine and vesicouterine pouches; the main connective tissue formations of the subperitoneal space, surrounding the bladder, uterus and rectum (in *blue*); the fasciae of the anterior and perineal abdominal muscles (in *green*).

➤

Evaluation of urethral hypermobility (**Q-tip test**). In gynecological position, after injecting a sterile gel into the urethra, a lubricated stick is inserted into the external urethral orifice and advanced along the urethra until it passes through the internal urethral orifice and reaches the vesicourethral junction. At rest (**a**), the angle with respect to the horizontal plane of insertion of the stick is between 10° and 15° in nulliparous women. On the request to "push" or make a cough (**b**), the angle between the stick and the horizontal plane should not vary more than 30°. If the amplitude is greater (index of a more mobile urethra), this points, in the case of urinary incontinence, to stress incontinence (i.e. due to exertion).

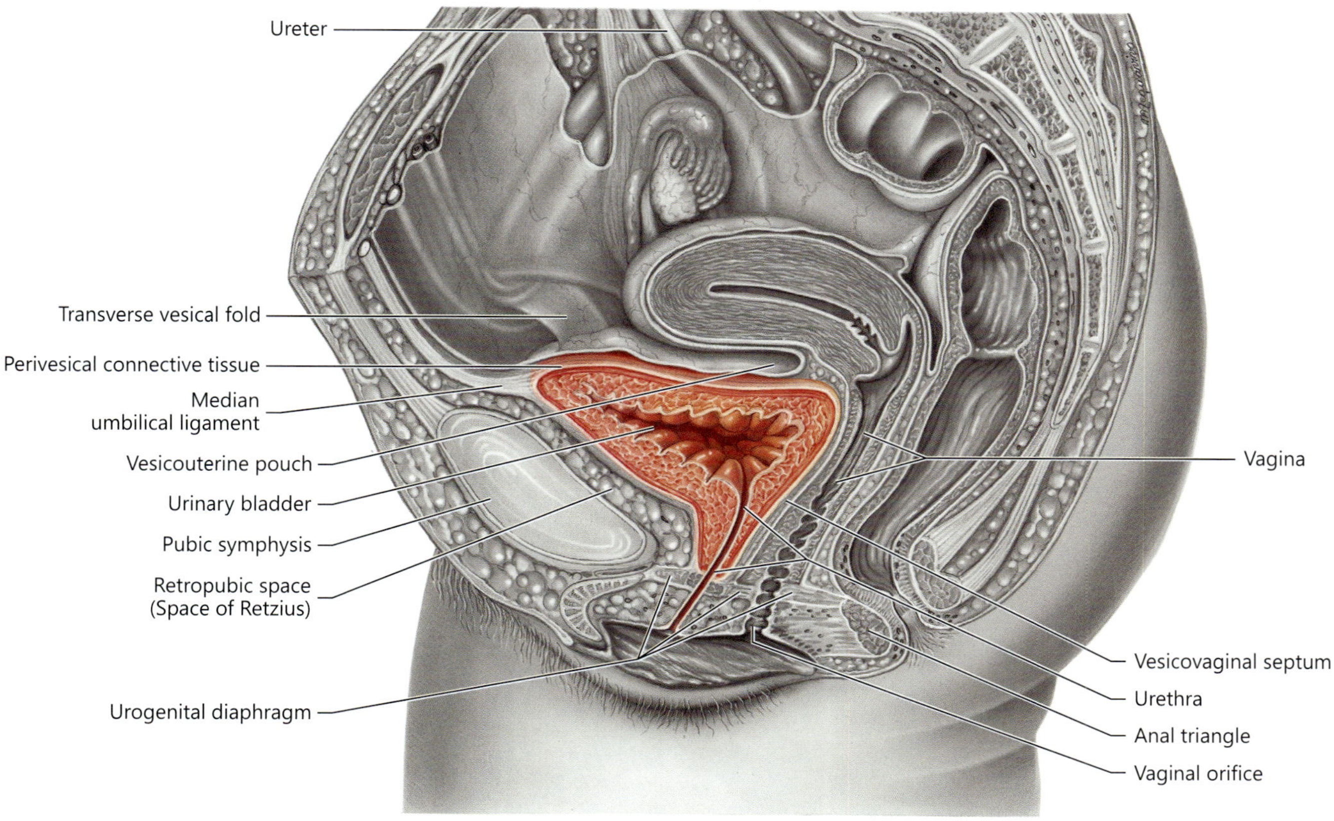

The **female urethra** originates in correspondence to the anterior apex of the trigone of the bladder, runs obliquely down and forwards crossing the pelvic diaphragm, which fixes it to the pelvic floor. Successively, it passes through the urogenital region (urogenital triangle) and terminates in correspondence to the roof of the vaginal vestibule. The urogenital diaphragm is a musculofascial formation comprising the deep transverse perineal muscle and external urethral sphincter, lying between the superior fascia of the urogenital diaphragm and the perineal membrane (inferior fascia of the urogenital diaphragm).
For the innervation of the bladder ☞ Male Pelvis, Urinary Bladder, pages 268-269.
For the fasciae and muscles of the region ☞ Volume 1, Musculoskeletal System, Pelvis and Perineum, Muscles of the Perineum, pages 505-509.

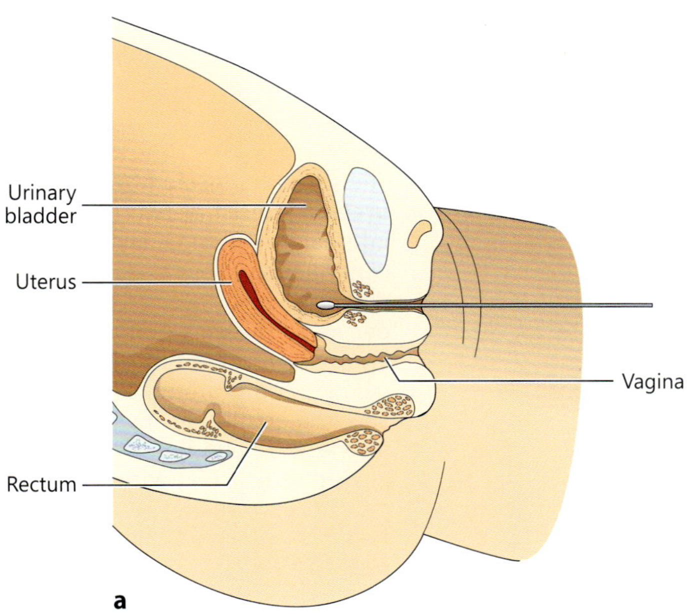

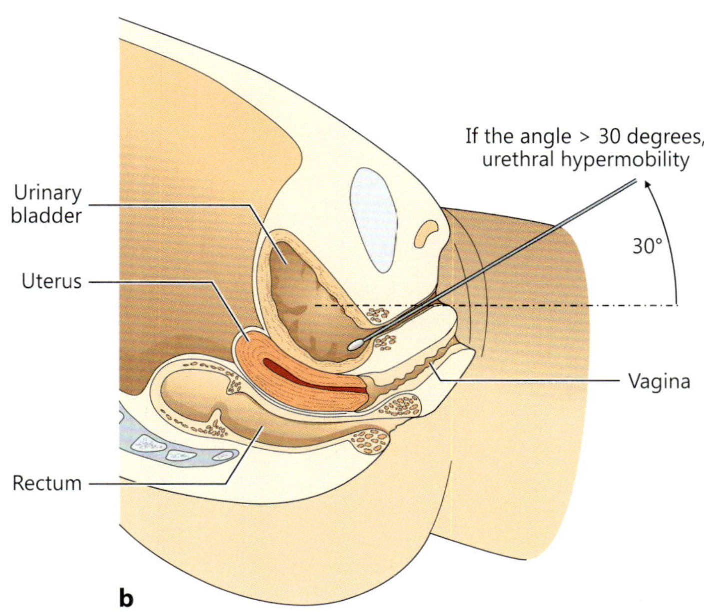

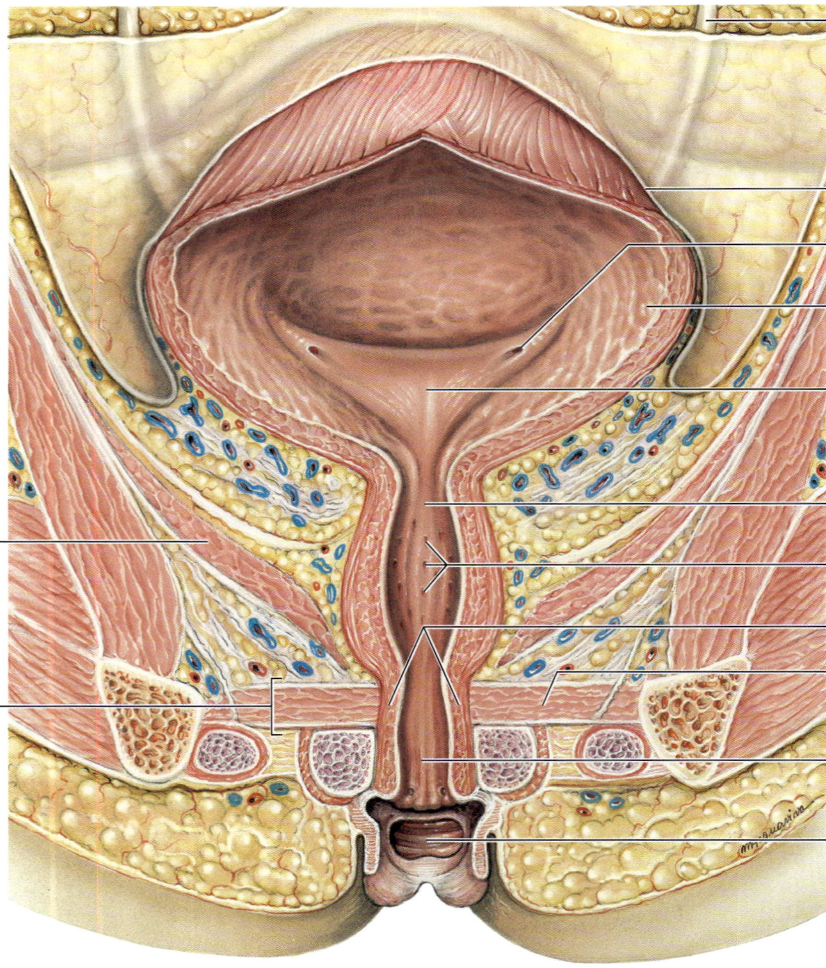

Schematic representation of the **female bladder** and **urethra** as they appear in a frontal section passing through the urethra. The female urethra, crosses the pelvic floor, and opens after a short course into the anterior portion of the vestibule.

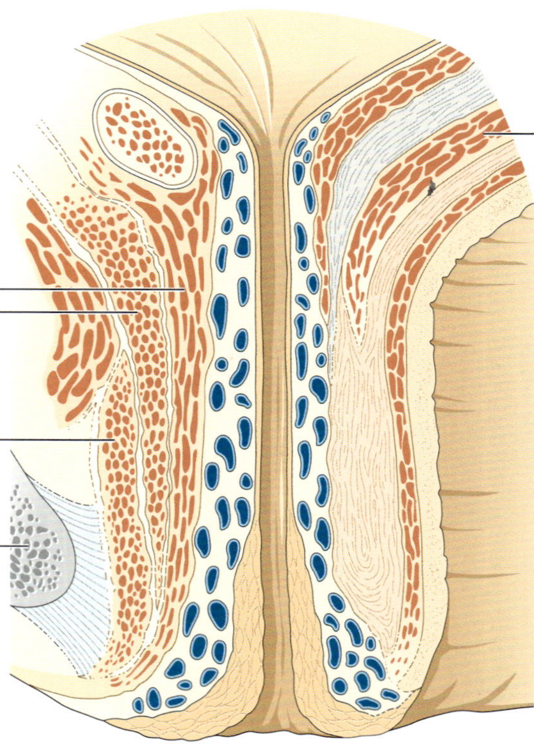

The **smooth muscle fibers of the urethra** consist of two layers: an internal longitudinal layer, which continues on above in the longitudinal muscular layer of the bladder, and a thicker external circular layer which constitutes the internal or smooth urethral sphincter. The striated muscle fibers are represented by bundles of fibers arranged in a circular pattern which form the external or striated urethral sphincter, also known as the rhabdosphincter or rhabdomyosphincter.

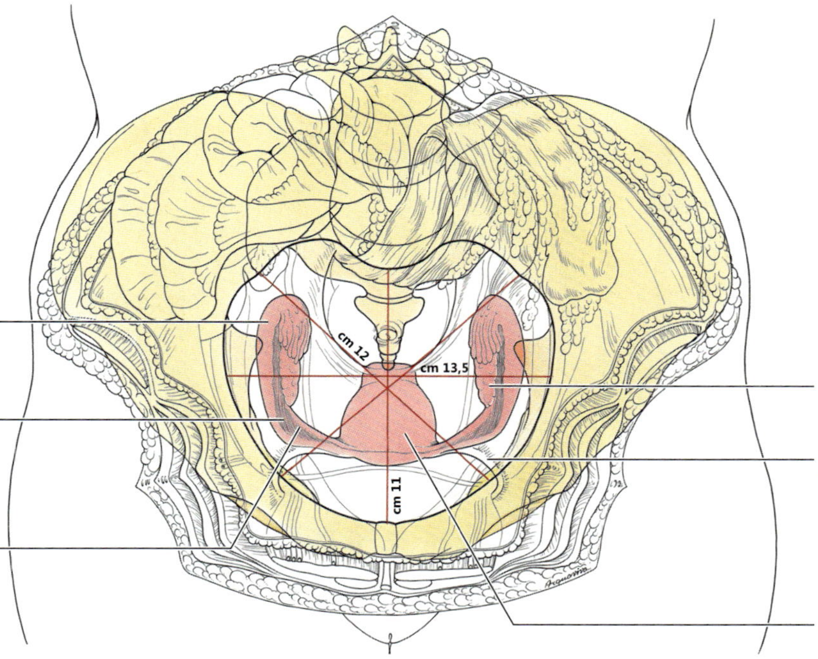

a

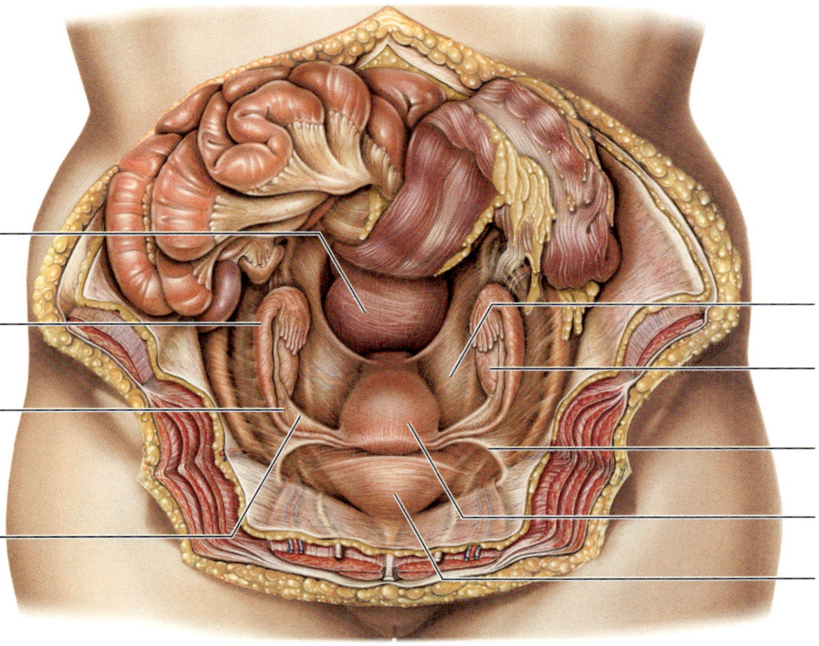

b

The **pelvic organs of the female genital system**, viewed from above. In **a**, scheme of the anatomical preparation illustrated in **b**. Some of the skeletal relations, the limits of the pelvic inlet and its main diameters are indicated. The preparation in **b** illustrates the relations of the ovaries and genital tract (tube and uterus); note also the principal means of attachment of the ovaries.

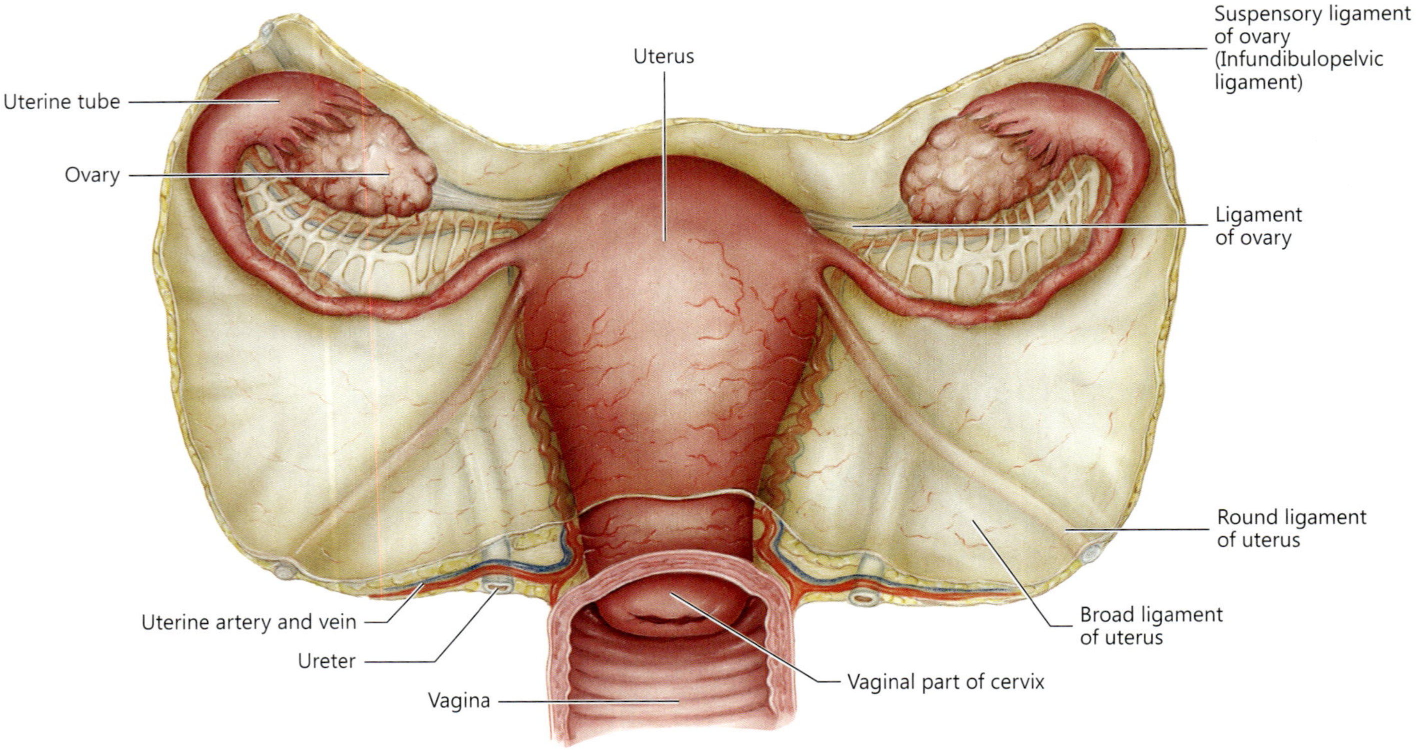

Ovaries, **tube** and **uterus**. Note the form of the ovary and its surface, which appears irregular in fertile age due to the presence of mature follicles, corpora lutea and connective tissue scars.

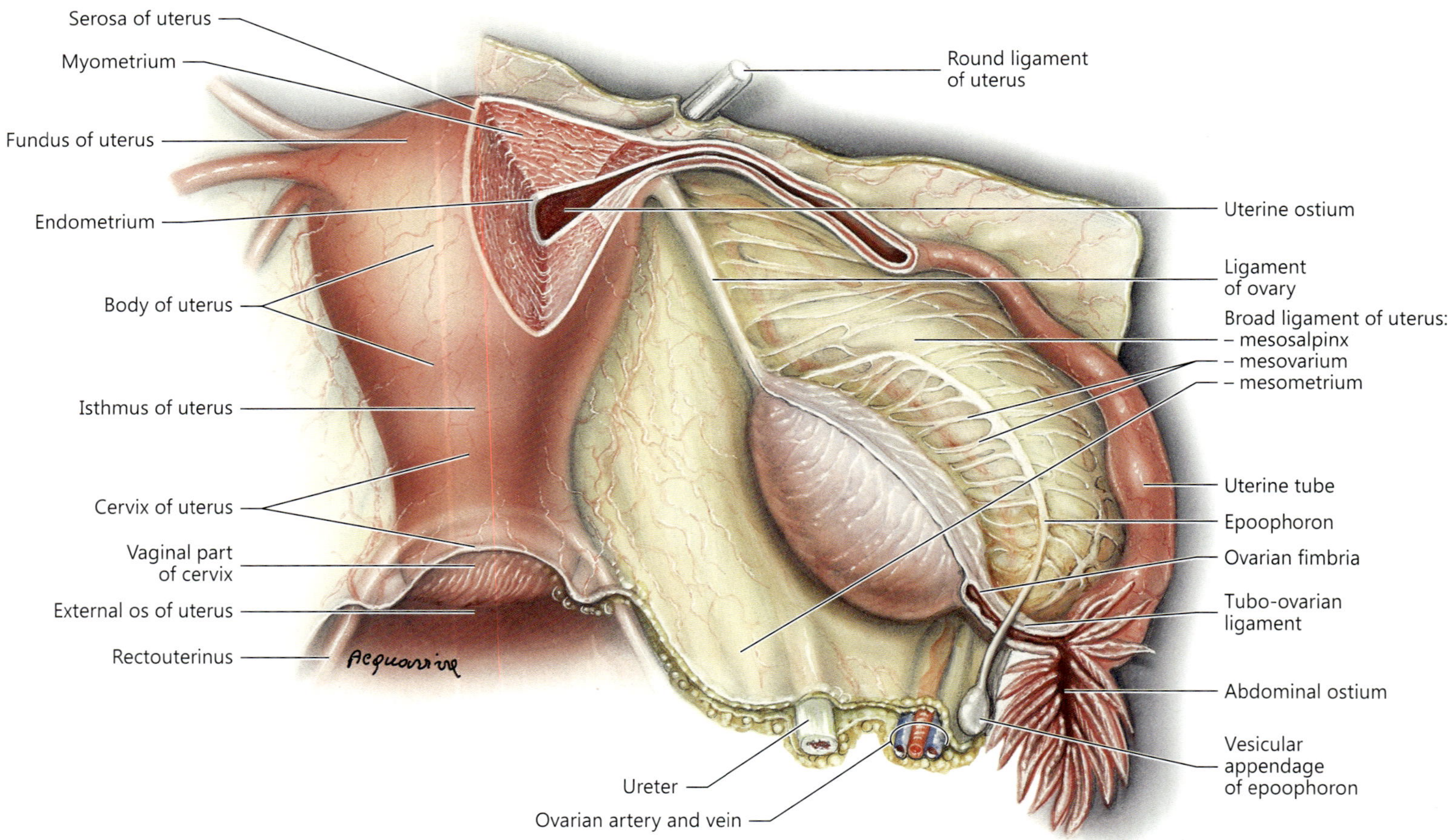

General conformation of the uterus and of its appendages in a young subject. The preparation is viewed in posterior projection. The uterine wall has been demolished in correspondence to the right superolateral angle to show the intramural portion of the tube and its entry into the uterine cavity.

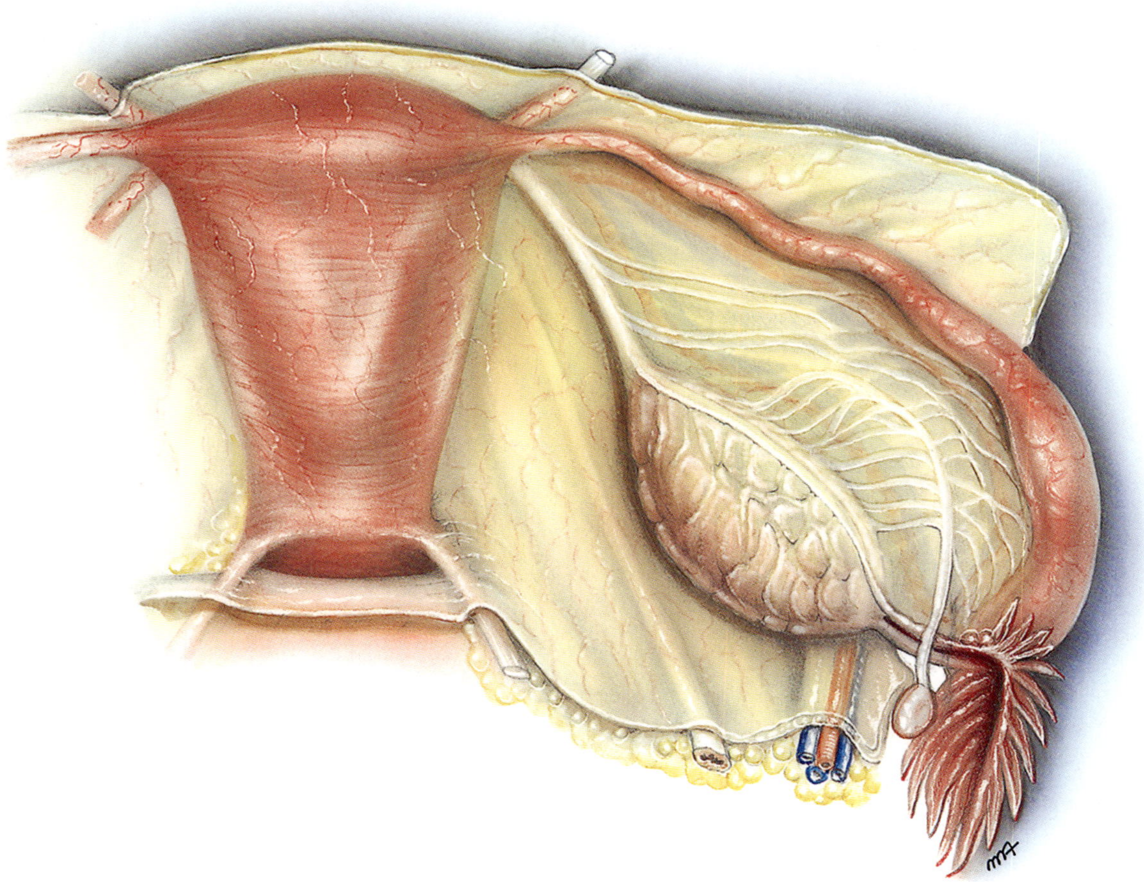

a

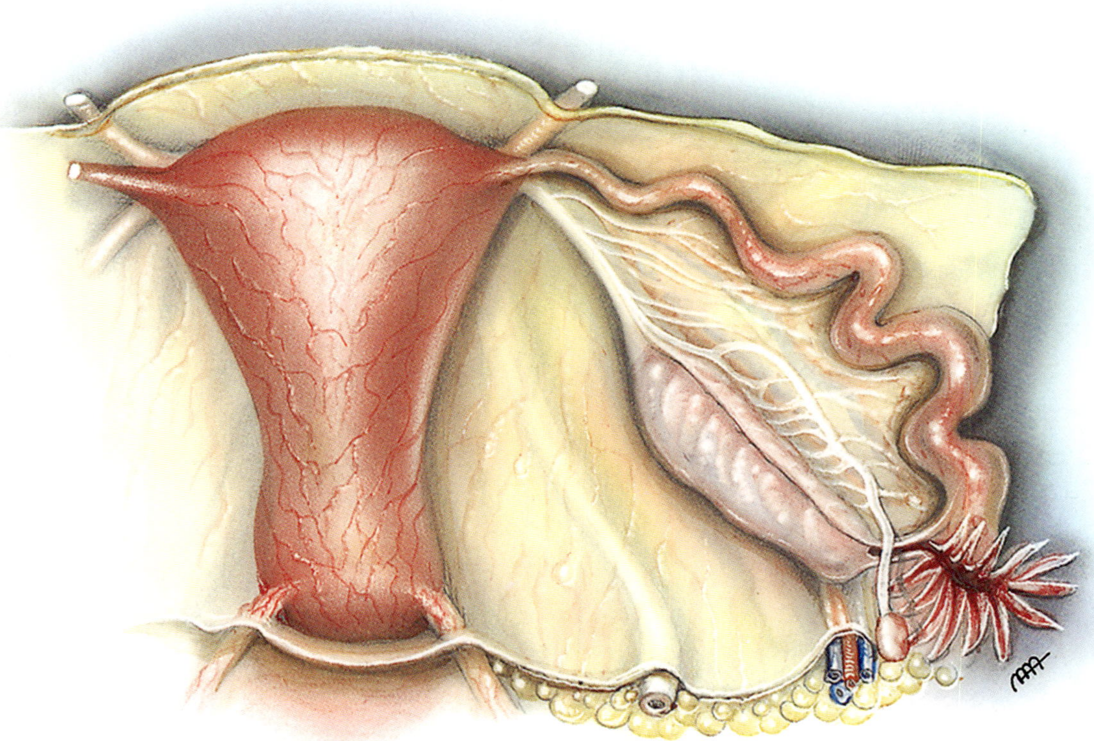

b

General conformation of the uterus and its appendages in an elderly subject (**a**) and in a child (**b**). The preparations are viewed in posterior projection.

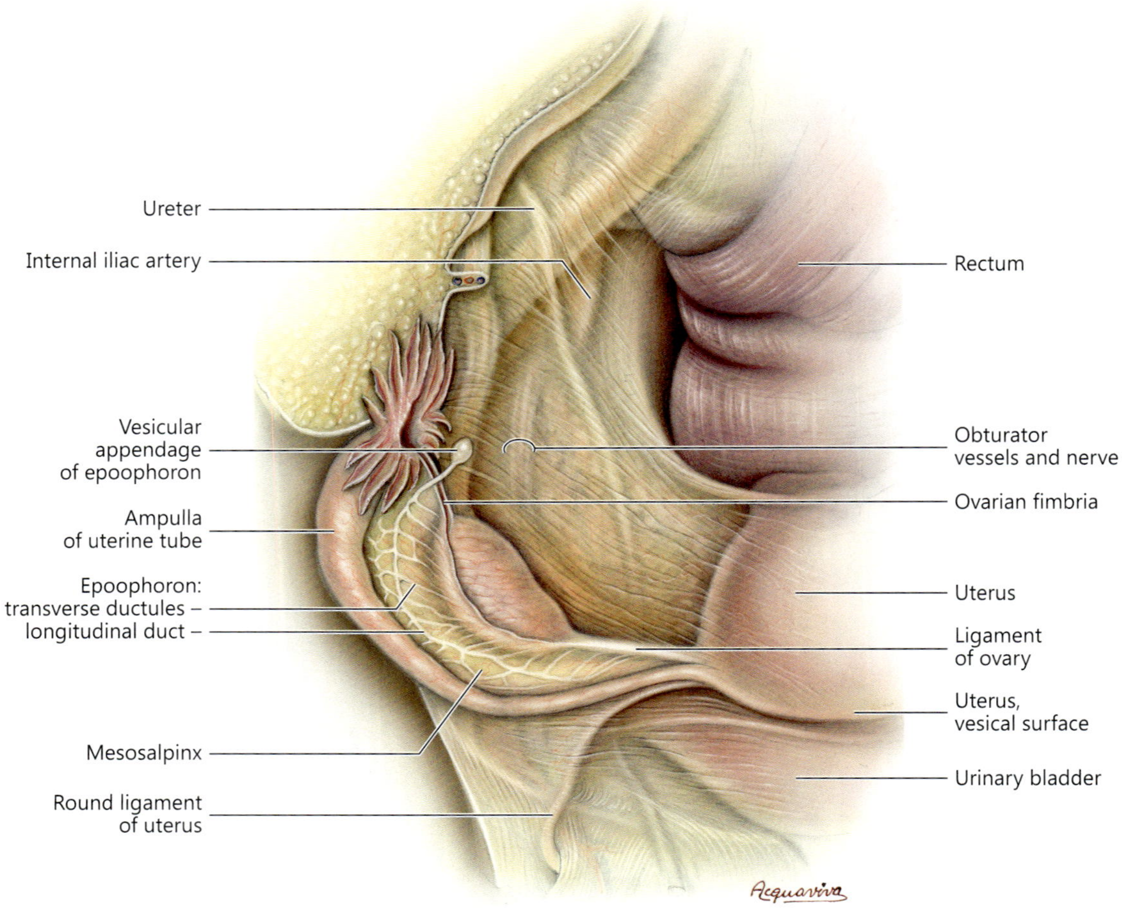
Ureter
Internal iliac artery
Vesicular appendage of epoophoron
Ampulla of uterine tube
Epoophoron: transverse ductules – longitudinal duct –
Mesosalpinx
Round ligament of uterus
Rectum
Obturator vessels and nerve
Ovarian fimbria
Uterus
Ligament of ovary
Uterus, vesical surface
Urinary bladder
Acquaviva

a

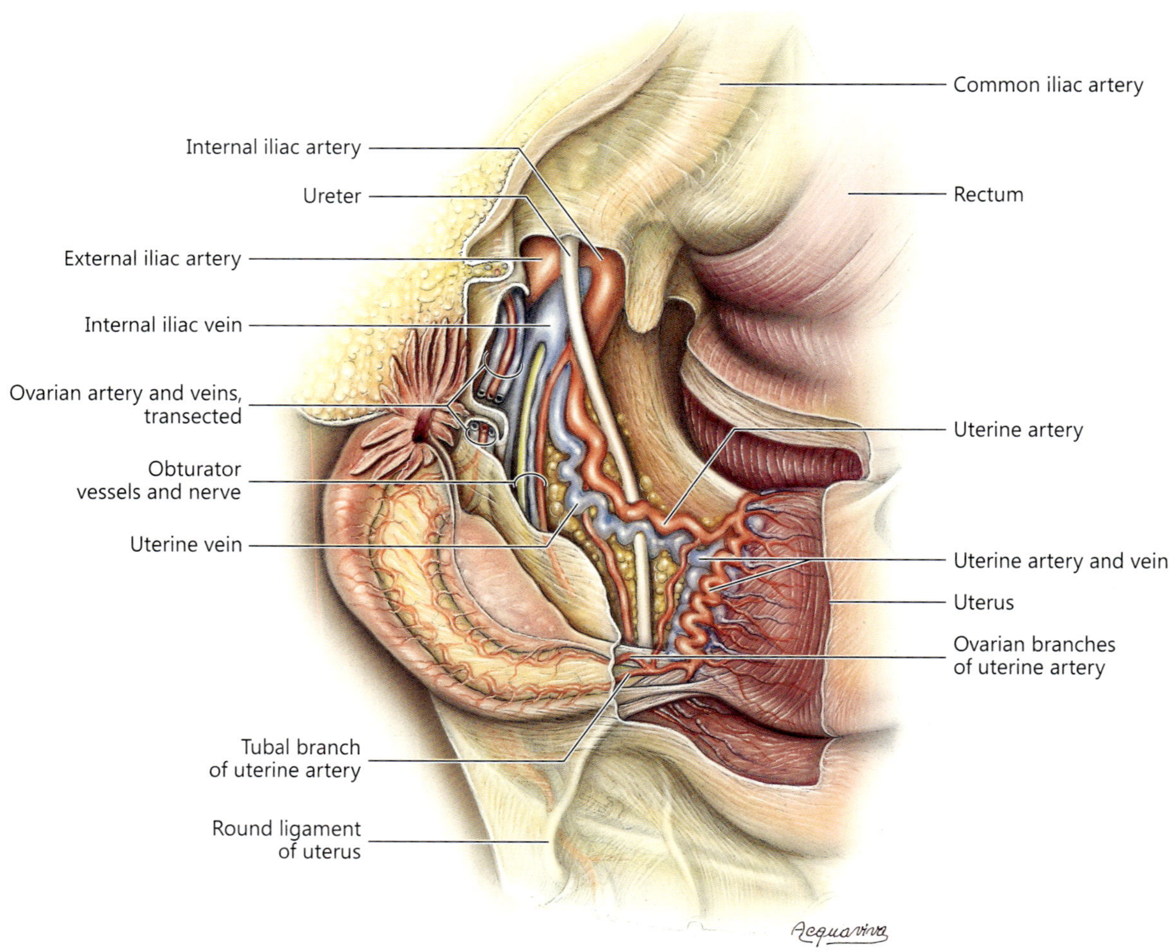
Common iliac artery
Internal iliac artery
Ureter
External iliac artery
Internal iliac vein
Ovarian artery and veins, transected
Obturator vessels and nerve
Uterine vein
Tubal branch of uterine artery
Round ligament of uterus
Rectum
Uterine artery
Uterine artery and vein
Uterus
Ovarian branches of uterine artery
Acquaviva

b

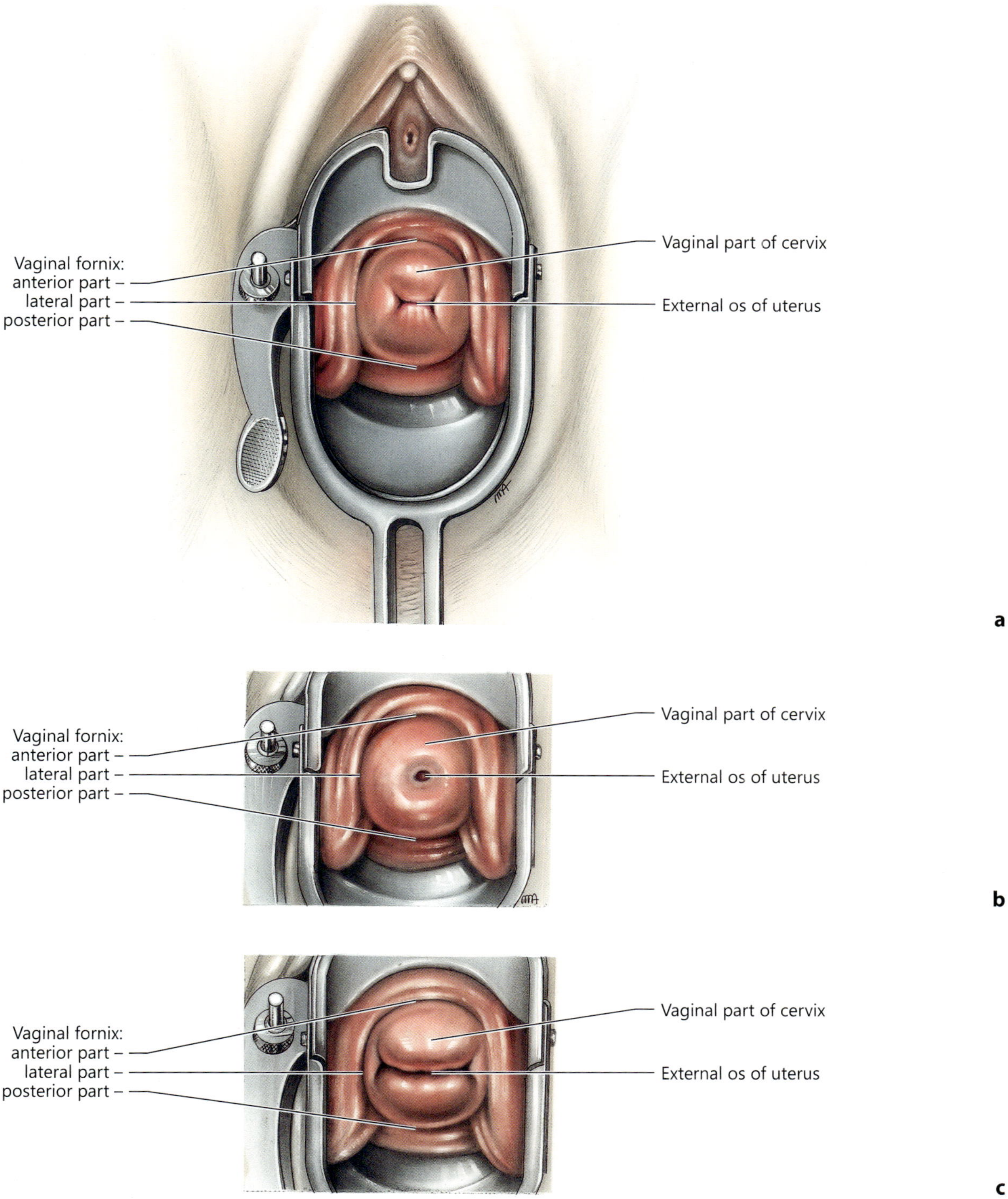

Characteristics of the **vaginal part of the cervix** and **external os of the uterus** viewed after dilatation of the vagina. In **a**, the image presented is that of a primiparous subject, in **b**, a nullipara, and in **c**, a multipara. In this last case, the external os of the uterus appears as a wide fissure bordered by an anterior and a posterior lip.

➤

a, Principal means of attachment and relations of the **ovary**, viewed from above. The thickness of the mesosalpinx contains some embryonic remnants: the epoophoron, containing longitudinal duct receiving 10-15 transverse ductules, and representing the residue of the caudal part of the mesonephric duct; the vesicular appendage of the epoophoron, or hydatid of Morgagni, a remnant of the cranial part of the mesonephric duct, can be either attached to the ovarian fimbria or connected to the broad ligament of the uterus through the pedicle, which is fixed to the free margin of the mesosalpinx, in which the longitudinal duct of the epoophoron continues. The peritoneum has been maintained intact, covering the lateral wall of the pelvis and extending over all the subperitoneal formations. **b**, The parietal pelvic peritoneum and much of the visceral one have been removed. This makes it possible to see the relations that the **ureter** and subperitoneal neurovascular formations have with the **female gonad** and the genital tract.

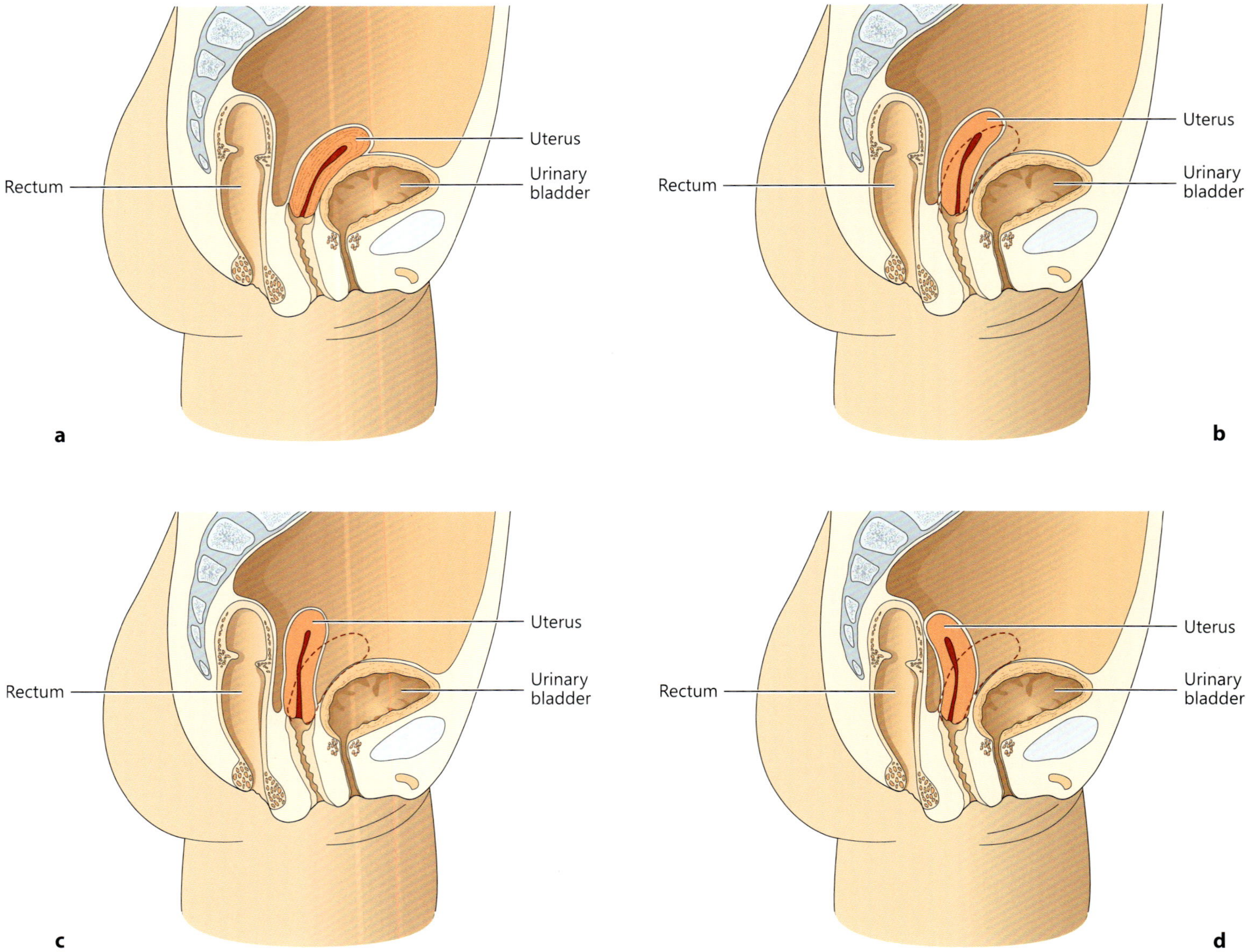

Schematic representation of some of the possible variations of the **position of the uterus**. **a**, Normal position (anteversion and anteflexion). Retroversion, a shift of the axis of the uterus backwards with respect to the axis of the vagina, to a modest degree (**b**), and to a more marked degree (**c**). **d**, Retroflexion, a flexion of the body of the uterus backwards with respect to the axis of the cervix.

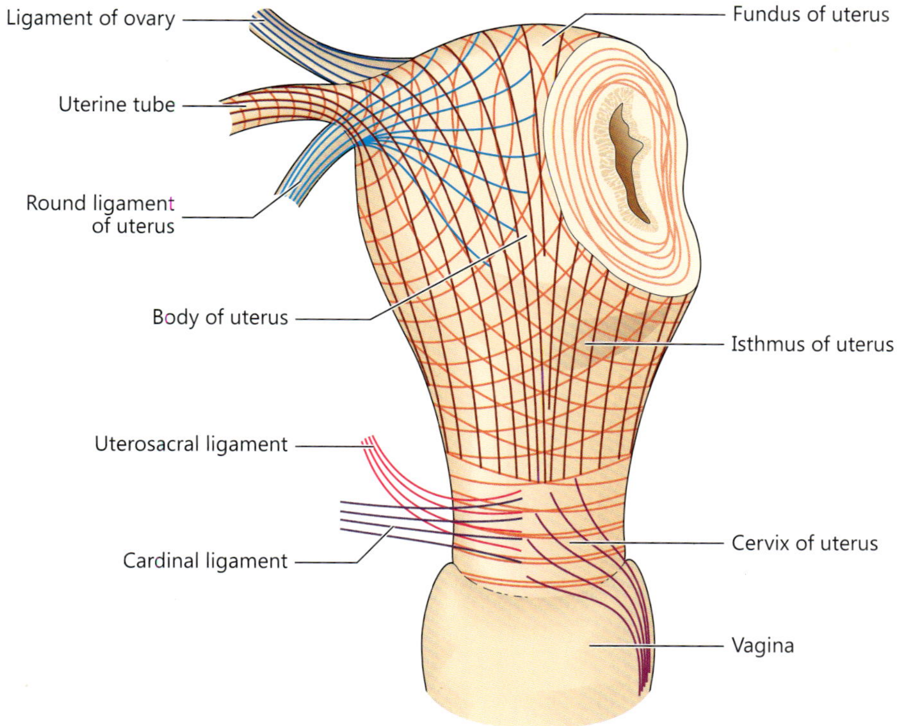

Schematic representation of the double-spiral pattern of the bundles of smooth muscle cells of the middle, or vascular, layer of the **myometrium**. The two spirals pass through the various planes without interruption (in a configuration of eight), one directed to the right, the other to the left. They are very steeply inclined in the body and fundus of the uterus, where they are in continuity with the smooth muscle cells of the uterine tube. In correspondence to the cervix they have, instead, an almost horizontal orientation. In the external layer of the myometrium, there are bundles of smooth muscle cells oriented in both a circular and a longitudinal direction, which are in continuation with the smooth muscle cells contained in the uterine ligaments.

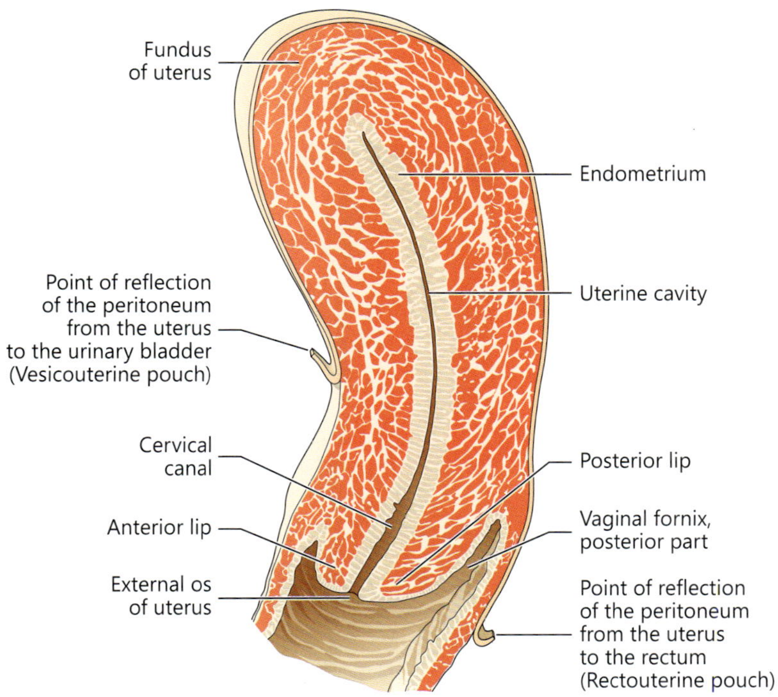

Sagittal section of the uterus. The anterior and posterior walls of the organ are very close to each other, so much so that, on the anteroposterior plane, the lumen appears virtual.

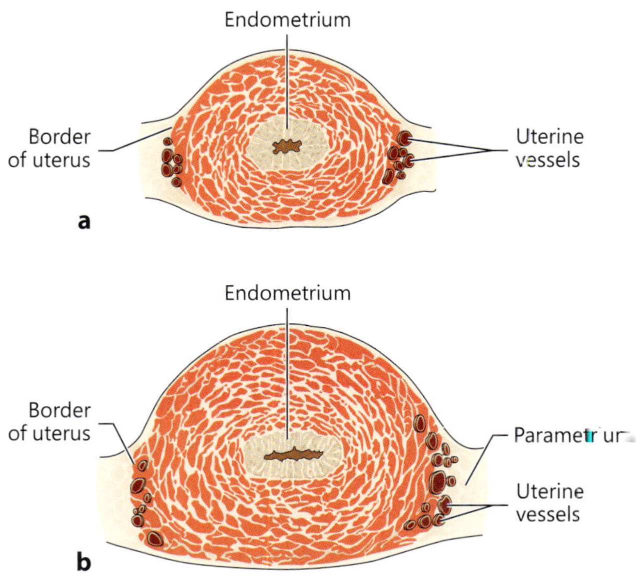

Transverse sections of the uterus conducted in correspondence to the cervix (**a**) and body (**b**) of the uterus. While the cervical canal appears more or less circular in transection, the lumen of the body appears like a horizontal fissure.

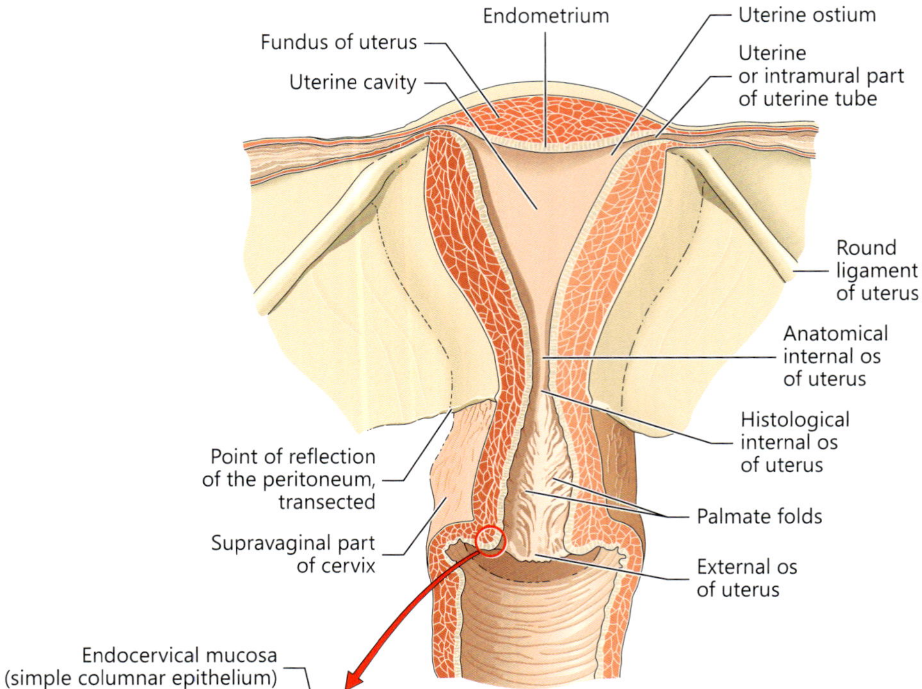

Uterus of a nullipara. The uterine cavity is viewed after demolition of the anterior wall of the organ. In correspondence to the body and isthmus, the lumen of the uterus has a smooth appearance. In the posterior wall of the cervical canal, the system of palmate folds determines the 'tree of life' formation. In contrast to the uterus of the neonate (figure below), the region of the isthmus, at the limits between the body and cervix, is very distinct.
Site of the squamocolumnar junction of the uterine cervix in normal conditions (**a**), and in the course of erosion (ectropion) (**b**). In the zone of cervical erosion, simple columnar epithelium is transformed into stratified squamous epithelium (squamous metaplasia) (**c**), on which the onset of cervical carcinoma can occur.

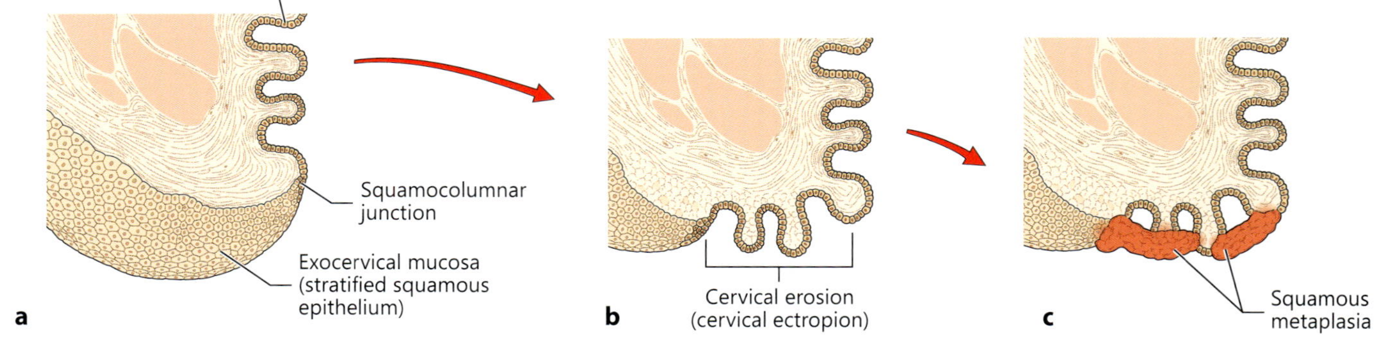

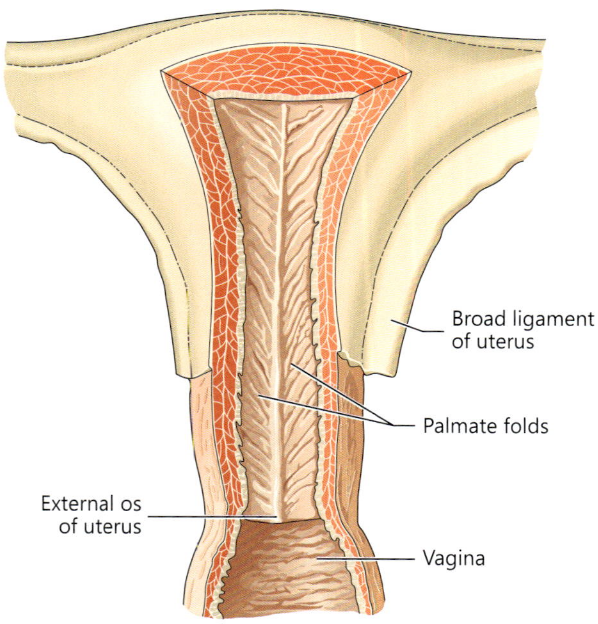

Uterus of the neonate. With the anterior wall demolished, one can note the characteristics of the internal configuration of the uterine cavity. The system of palmate folds seen in the cervical canal proceeds, although in a less accentuated manner, through the isthmus and into the body of the uterus.

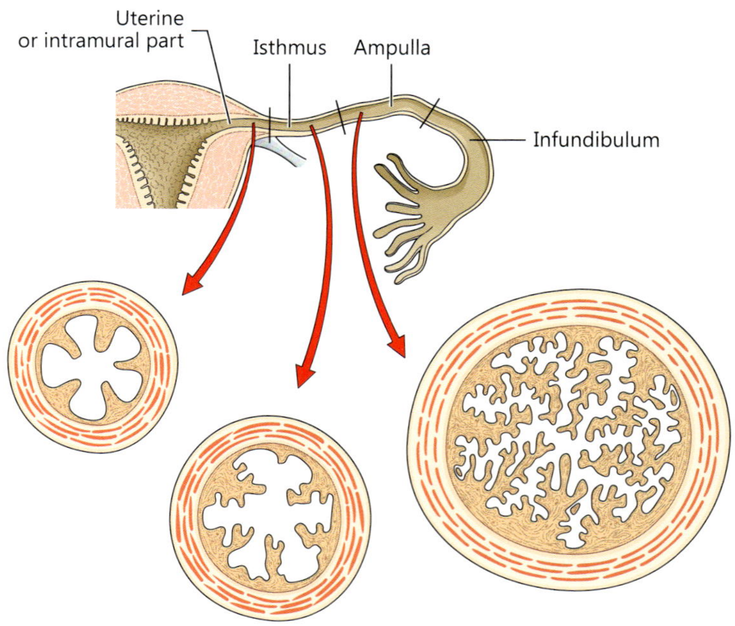

Different aspects of the **morphology of the uterine tube**, in particular of the endometrium, according to the segments examined. Note the extensive development and irregularity of the folds of the mucosa in the ampullar segment which give it a labyrinthine aspect.

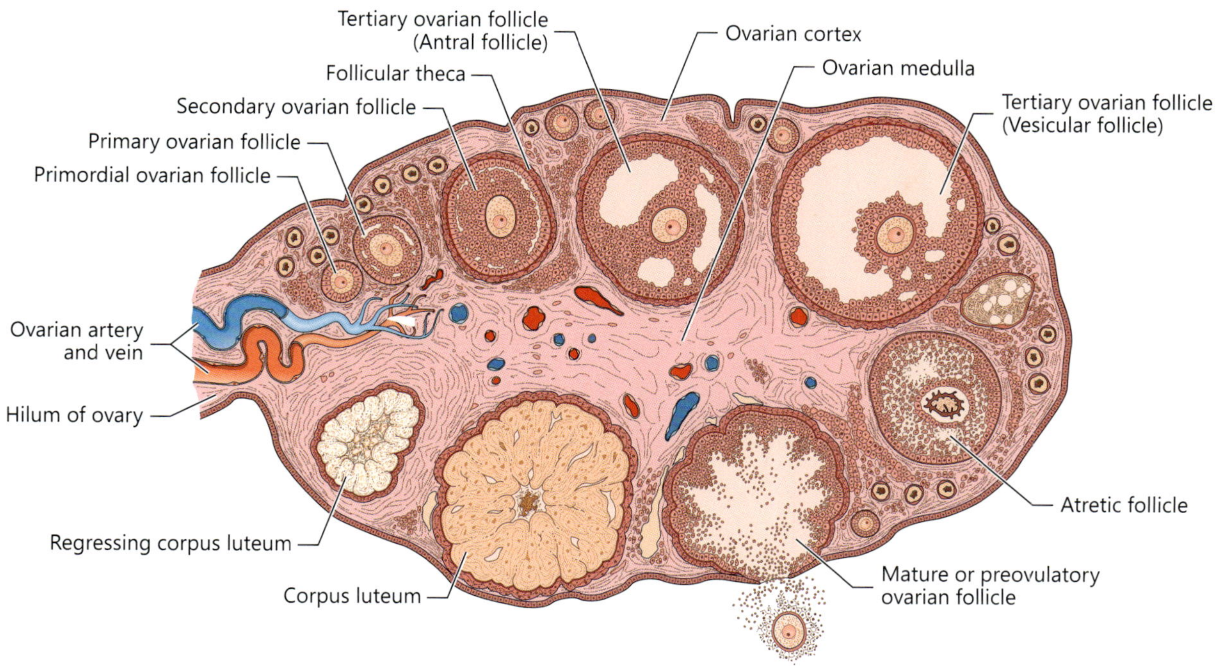

Transverse section of the **ovary** illustrating its structure and cycle. At birth, the ovarian cortex contains a superficial zone of primordial ovarian follicles. Many of these degenerate during the prepubertal period or during the fertile period, after beginning their cyclic maturation, by atresia (atretic follicle) in the various phases of maturation from primordial ovarian follicle to primary, secondary and tertiary ovarian follicle. After puberty, groups of primordial ovarian follicles become activated in each cycle, and their development takes more than one cycle. Usually only one follicle becomes dominant in each cycle, reaches maturity (mature or preovulatory ovarian follicle) and releases the oocyte during ovulation. The two ovaries ovulate generally in alternate cycles. After ovulation, the remains of the mature ovarian follicle become a corpus luteum. If the oocyte is not fertilized, the corpus luteum degenerates becoming atretic (a corpus albicans which remains as a fibrous scar of the ovary).

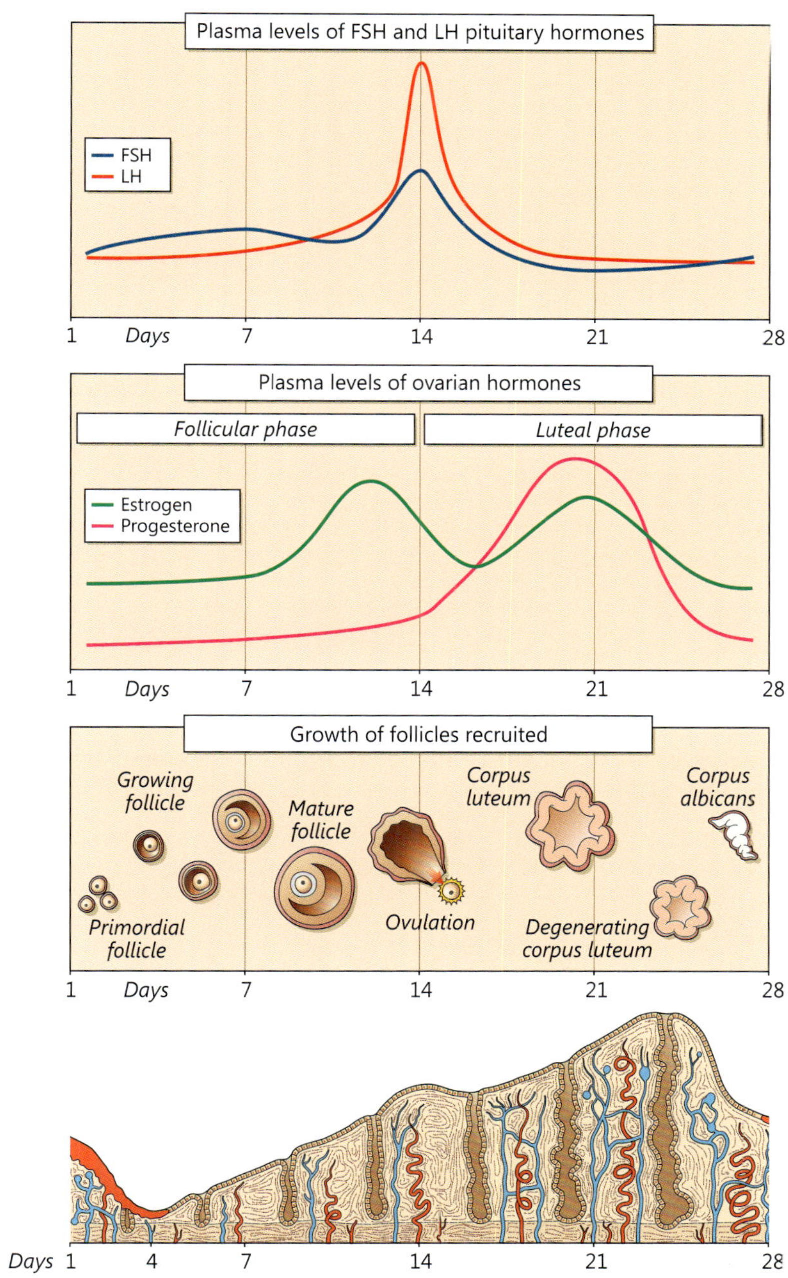

➤ The hormones produced by the ovary act on the endometrium and regulate the development and activity of its glands. Thus, in each **ovarian cycle**, the endometrium modifies its own structure to assume the characteristics most suitable for the possible implantation of the embryo. The scheme illustrates the modifications of the endometrium during the diverse phases of an ovarian cycle.

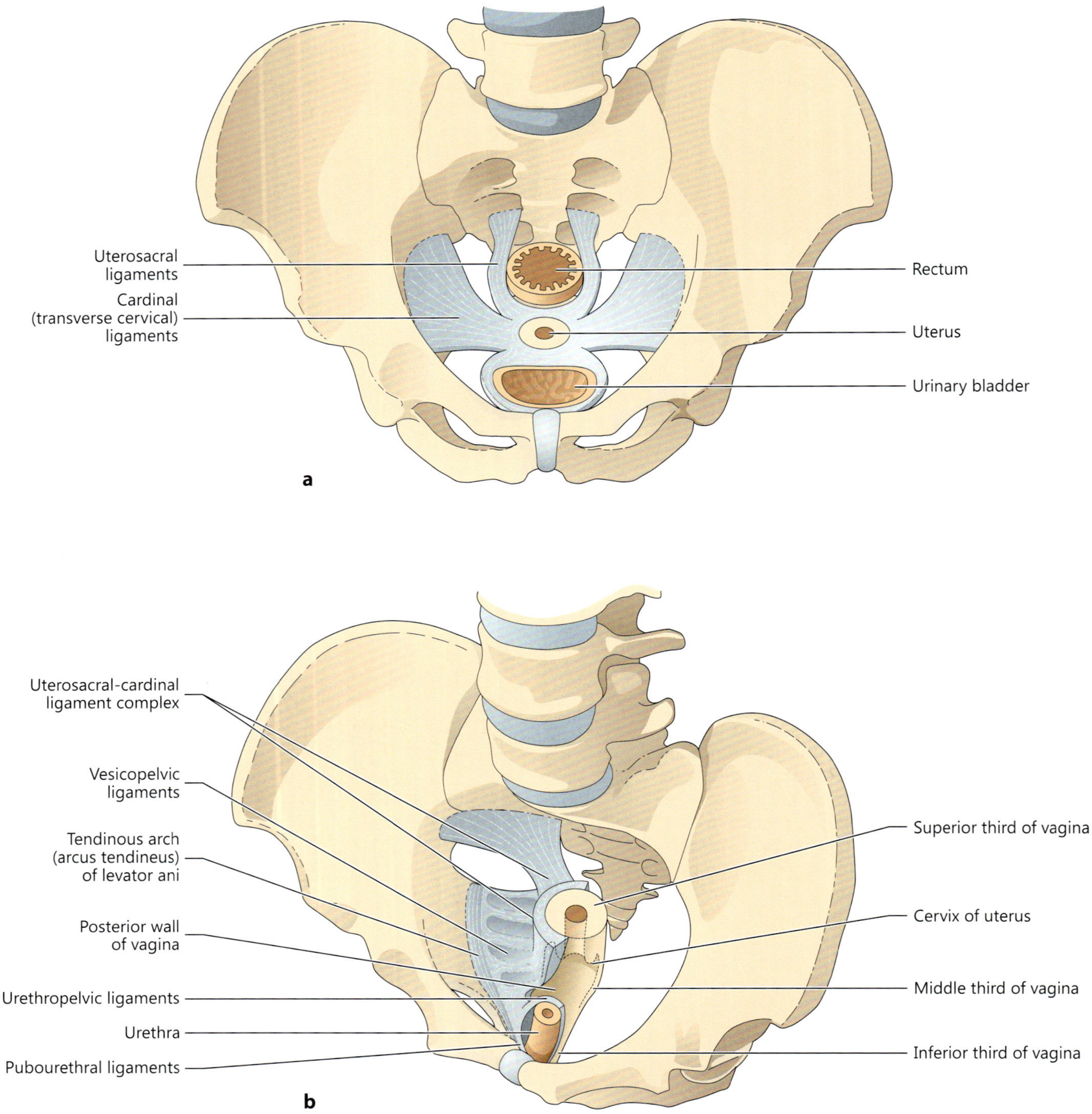

The **suspensory system of the pelvic organs** is composed of a lateroposterior system of attachment (cardinal and uterosacral ligaments) (**a**) and an anterior system of attachment (pubourethral, vesicopelvic and urethropelvic ligaments) (**b**).

a-c, Constrictor muscles of the rectum (puborectalis) and **of the vagina** (pubovaginalis) and **means of attachment depending on the pelvic fascia. d**, The **striated muscle connected to the urethra and vagina** constitutes a continuous mass of muscle tissue, the urogenital sphincter, but one one can nevertheless distinguish two parts: one surrounding the urethra (the external urethral sphincter), and the other surrounding the urethra and vagina (sphincter urethrovaginalis). In addition to these two parts are the fibers attached to the ischiopubic rami, which constitute the compressor urethrae (sphincter). From the base of the bladder, within the pelvis, the urogenital sphincter extends into the perineum, around the vagina, passing through the urogenital hiatus of the pelvic diaphragm.

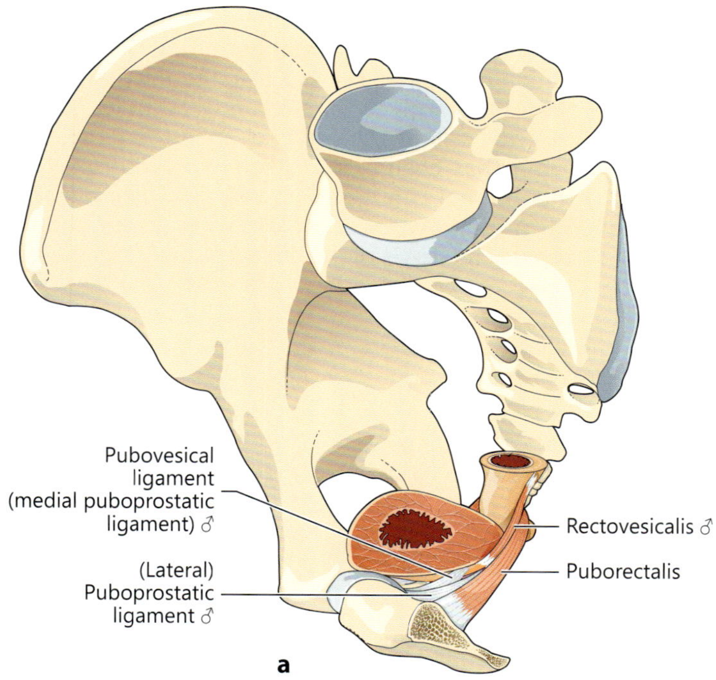

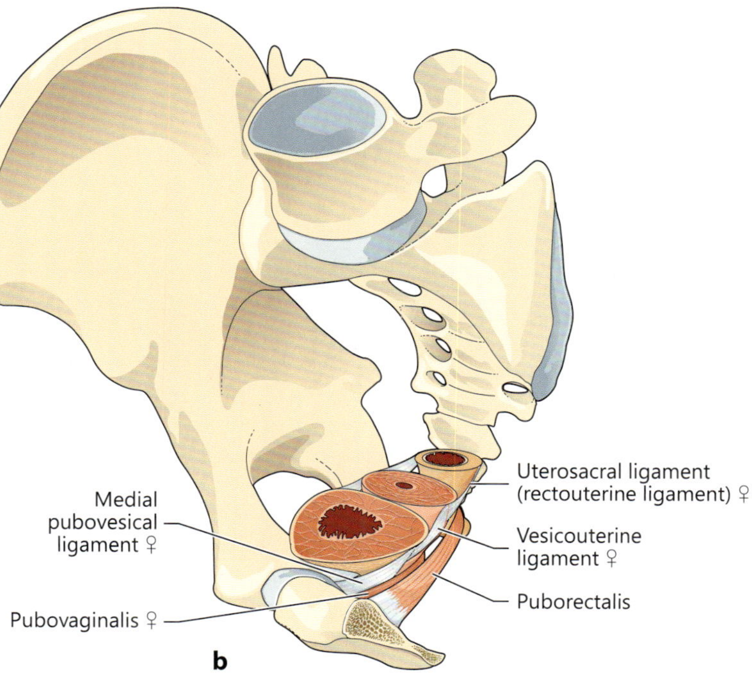

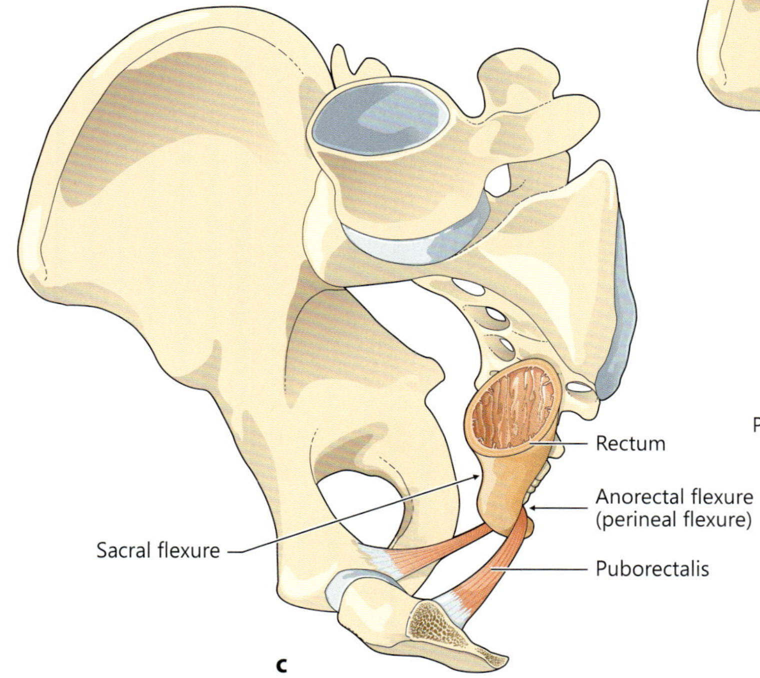

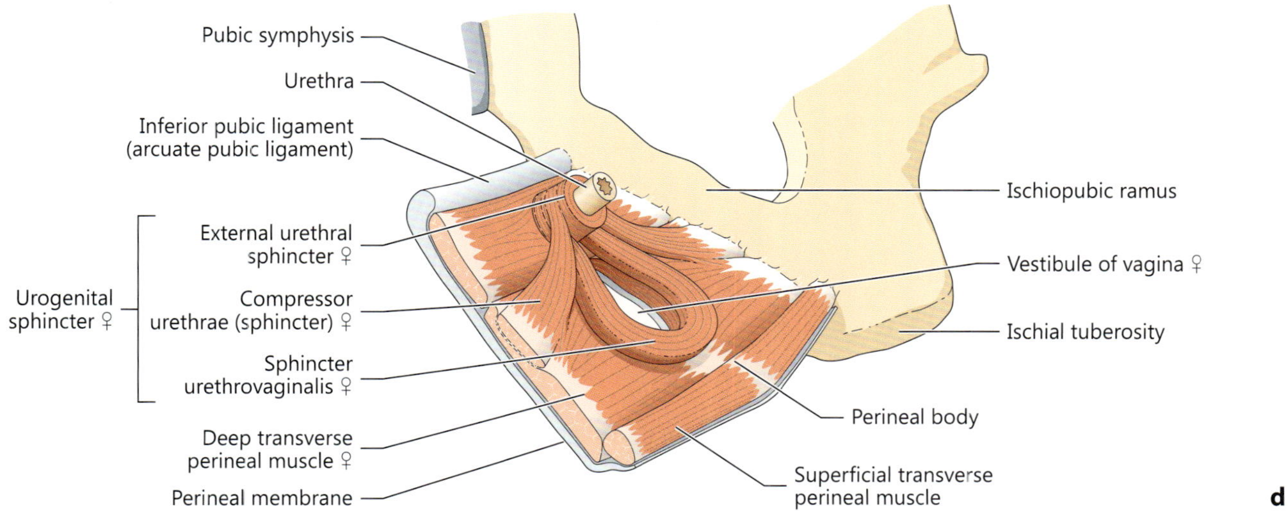

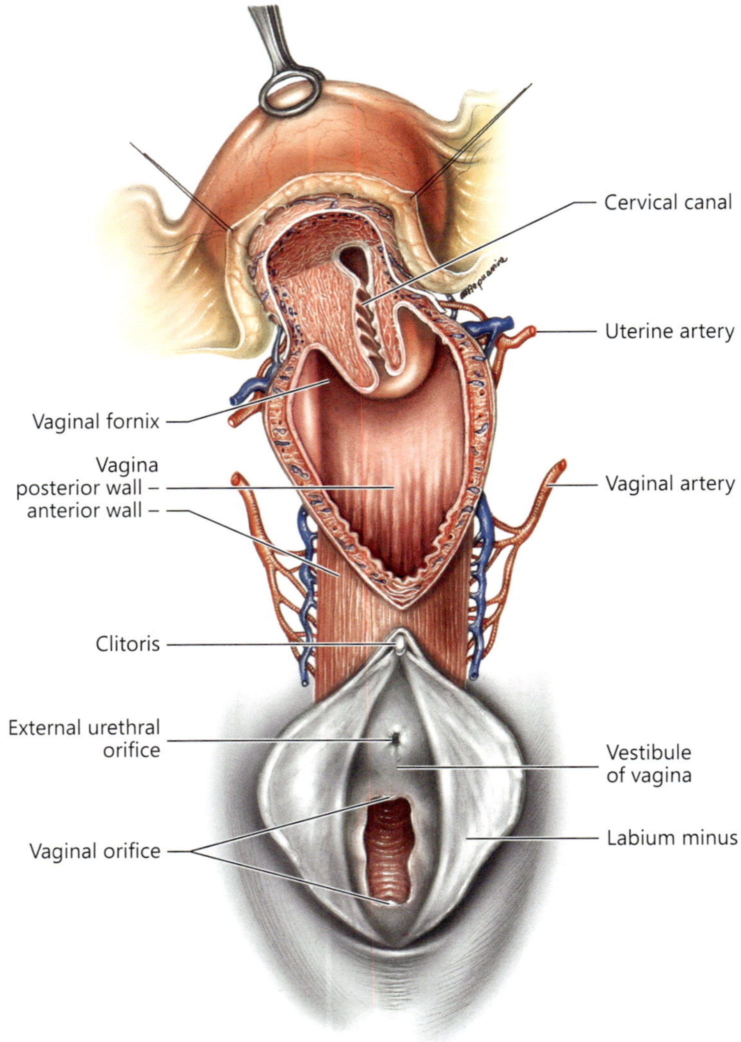

Preparation of the last segment of the **female genital tract** and **external genitalia**. The illustration highlights, in particular, the relations of the vagina with the uterus and vestibule. The anterior wall of the cervix of the utherus and vagina has been amply demolished to display the internal configuration of the two organs. The principal vascular pedicles which supply the vagina and cervix of the utherus are also visible.

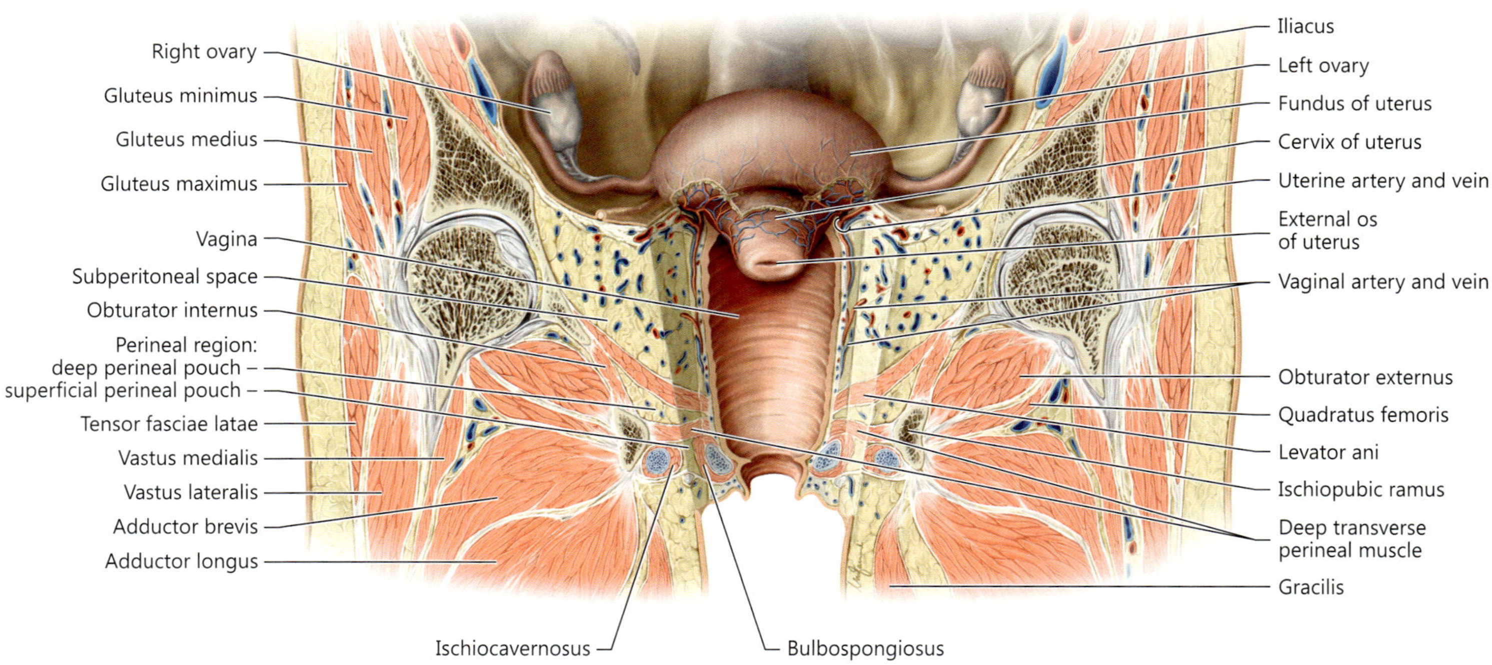

Female pelvis and perineum. Anterior view. Frontal section of the lesser pelvis passing through the plane of the urethra and urinary bladder. In the female, the bladder appears more elongated transversally and flattened in an anteroposterior sense. The female urethra crosses the pelvic diaphragm and after a short course in the anterior perineum opens into the vestibule.

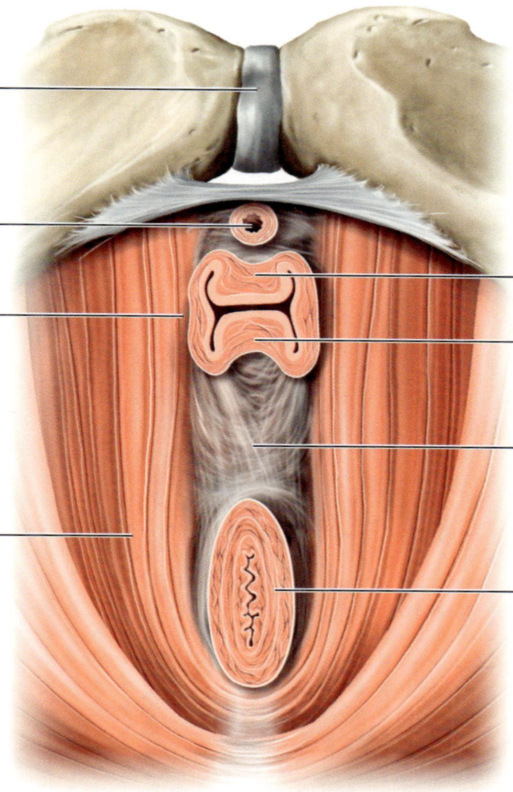

Transverse section of the **vagina**, **urethra** and **rectum** at the level of the perineum. The body of the vagina presents, in the state of vacuity, an anterior and a posterior surface. At the level of the perineum, the vaginal and urethral walls are almost fused, the dense connective tissue between them being often only about 1 cm.

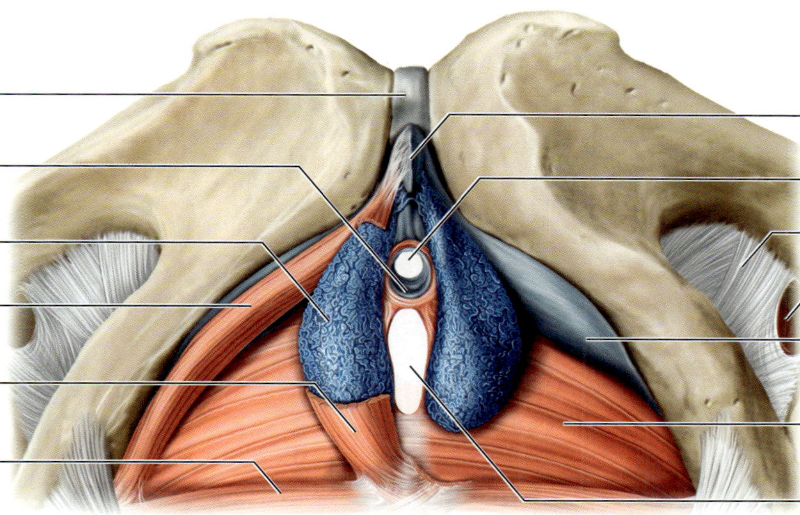

The **erectile formations attached to the female external genitalia** are represented in *blue*. One can see the skeletal relations of the clitoris, vestibular bulbs, and of the spongy tissue formed around the terminal part of the urethra. The crura of the clitoris, enveloped by the ischiocavernosus muscles, extend along the ischiopubic rami. The inferior fascia of urogenital diaphragm is not shown.

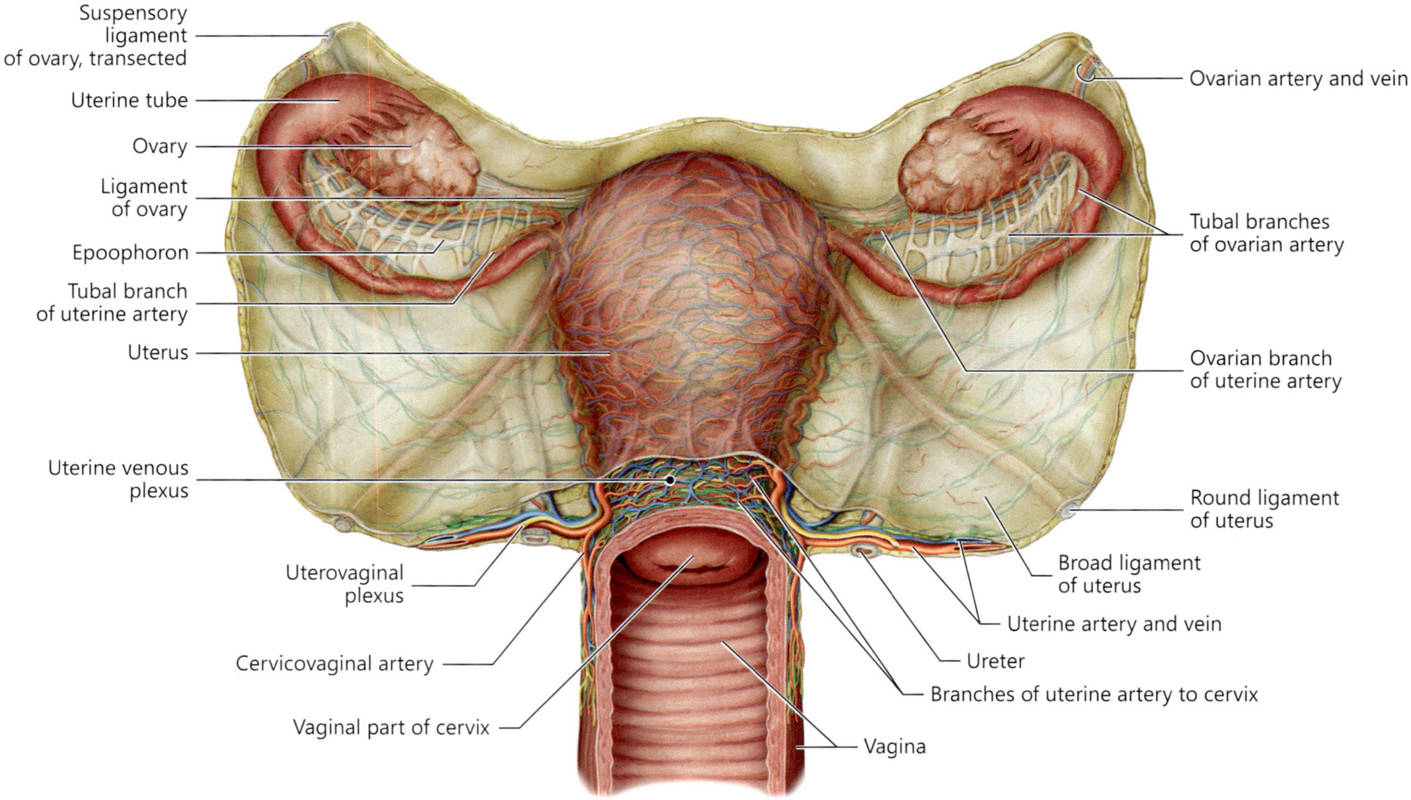

The branches of the **uterine artery** which supply the body of the uterus anastomose abundantly at the level of the middle, or vascular, layer of the myometrium, and from this arterial network arise the branches to the endometrium. At the base of the broad ligament, in the direction of the lateral wall of the pelvis, the uterine vessels form part of the cardinal or transverse cervical ligament.

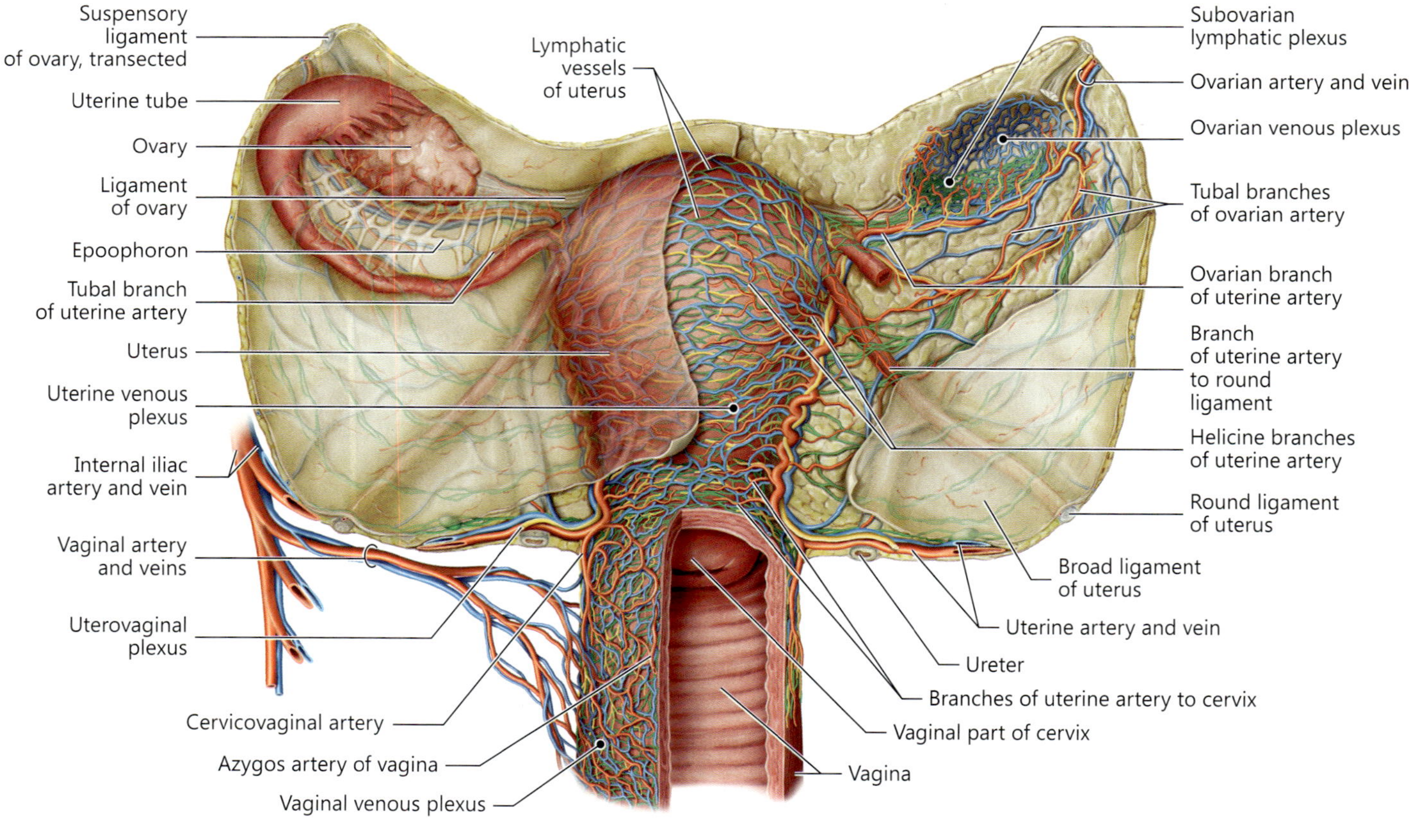

The **uterine artery**, a branch of the internal iliac artery, gives off vaginal branches, amongst which the azygos artery of the vagina, helicine branches, a tubal branch which anastomoses with the tubal branches of the ovarian artery, and an ovarian branch which terminates joining to flow full canal with the ovarian artery.

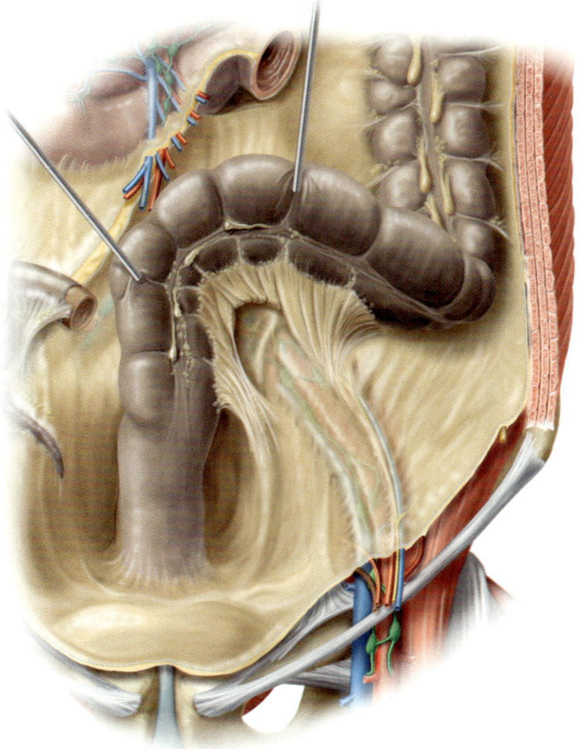

a

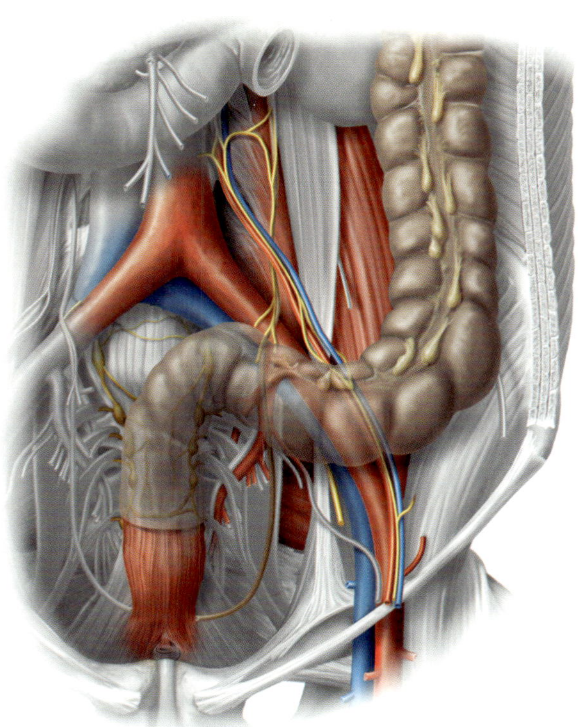

b

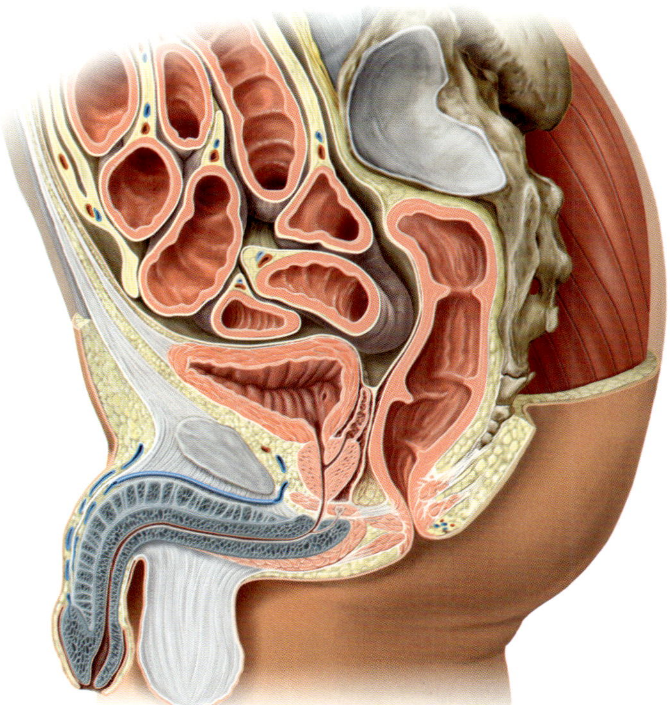

c

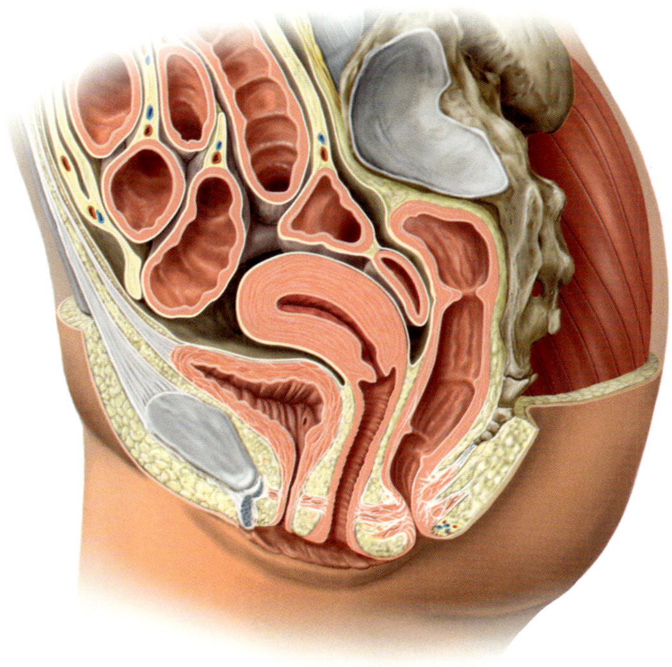

d

Principal relations of the **rectum**. The peritoneum invests partially the anterior surface of the rectum and the lateral walls, but the posterior surface is without any investment. The superior third of the rectum relates anteriorly with the intestinal ansae and corresponds to the rectovesical pouch in the male and rectouterine pouch in the female. The subperitoneal part of the pelvic rectum is in relation, through the rectoprostatic fascia, with the bladder, prostate, ductus deferentes and seminal glands in the male, and, through the rectovaginal fascia, with the vagina in the female (**a** and **b**, ☞ page 190; **c**, ☞ page 112; **d**, ☞ page 115).

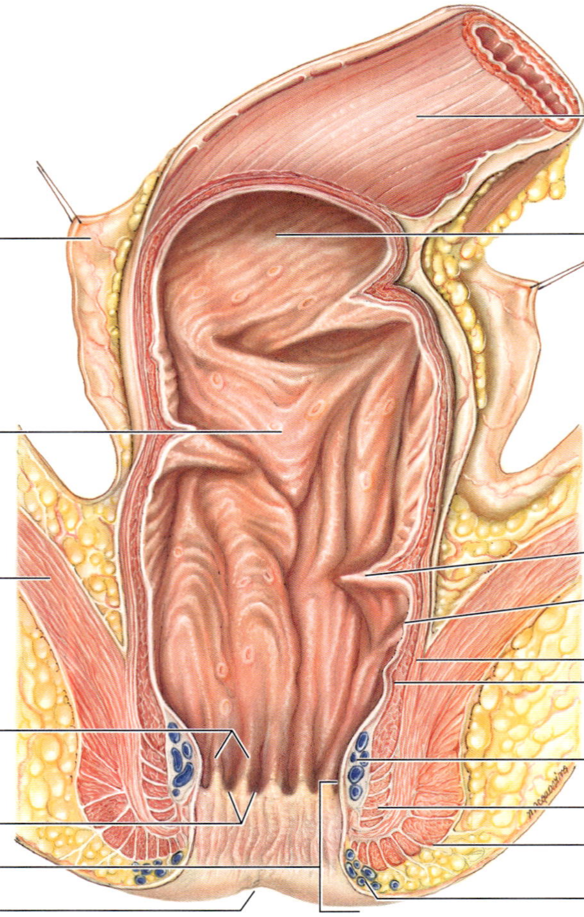

Frontal section of the **rectum**. The internal configuration of the rectal ampulla and anal canal is illustrated, as well as the sphincter muscular and venous vascular formations.

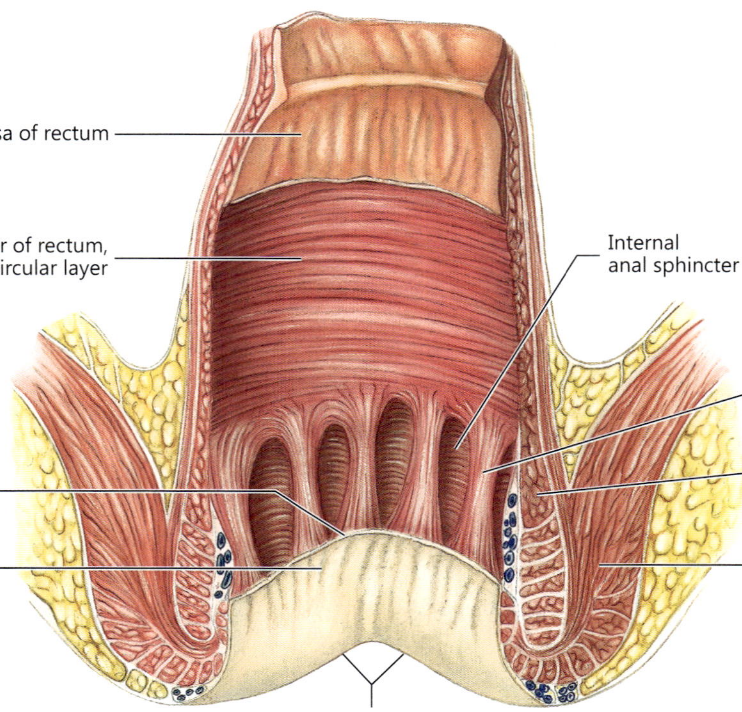

With the mucosa of the **rectum** removed, one can see the muscle formations which make the superimposing mucosa lining irregular. In the anal columns, the muscular bundles have a longitudinal orientation.

Mucosa of the rectum and of the anal and perianal skin		
PART OF RECTUM AND PERIANAL SKIN	**TYPE OF INVESTING EPITHELIUM**	**GLANDS**
Rectal ampulla (pelvic part of rectum)	Simple columnar	Intestinal
Anal canal (perineal part of rectum)		
– anal transition zone (anal columns)	Transition from simple columnar to nonkeratinized stratified squamous (squamocolumnar junction)	Anal
– below the pectinate line (anal pecten)	Nonkeratinized stratified squamous	
– anocutaneous line	Transition from nonkeratinized stratified squamous to keratinized stratified squamous	
– anal skin	Keratinized stratified squamous	Sebaceous
Perianal skin	Keratinized stratified squamous	Sebaceous and apocrine sweat (circumanal)

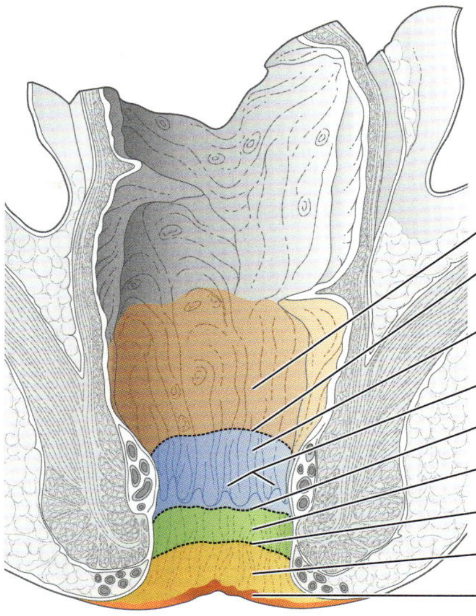

The transition zone of the **anal canal**, of variable height and position, includes islands of squamous epithelium interspersed in the mucosa lining, which is composed of columnar epithelium. It is distinct from the pectinate line (or dentate, or mucocutaneous line), formed by the anal valves and sinuses, which represents the junction between the superior and inferior parts of the anal canal.

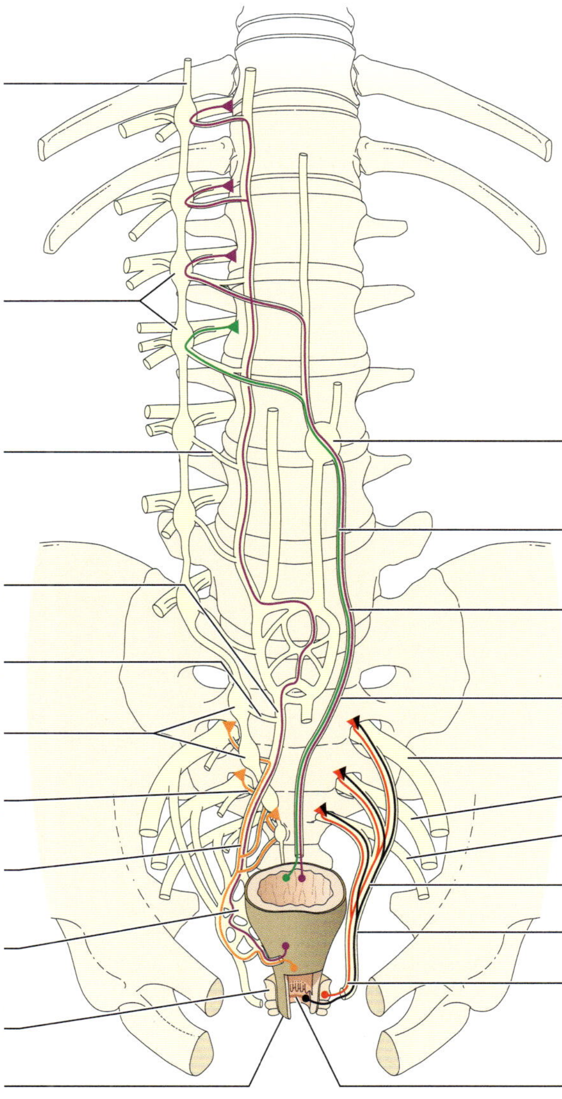

Innervation of the rectum. The rectal ampulla is innervated by the lumbar (sympathetic) splanchnic nerves, while the anal canal is innervated by the sacral (sympathetic) and pelvic (parasympathetic) splanchnic nerves. The pudendal nerve innervates the external anal sphincter and the anocutaneous zone of the anal canal through the inferior anal nerves. The visceral sensory information is collected by afferent fibers through the lumbar splanchnic nerves.

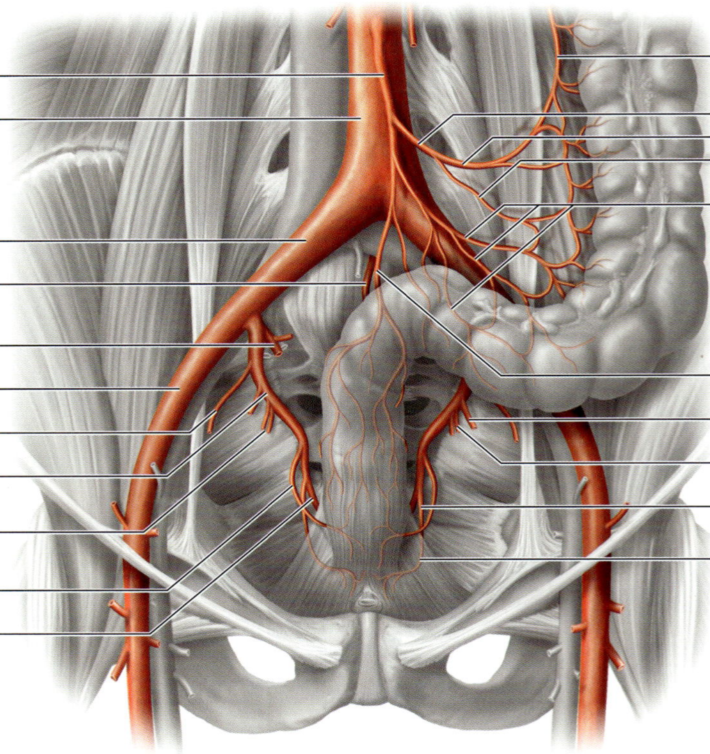

Arteries of the rectum. The rectum is supplied by the superior rectal artery, a branch of the inferior mesenteric artery, the middle rectal artery, a branch of the internal iliac artery, and the inferior rectal artery, a branch of the internal pudendal artery, which in turn is a branch of the internal iliac artery. As well, it receives minor branches of the median sacral artery.

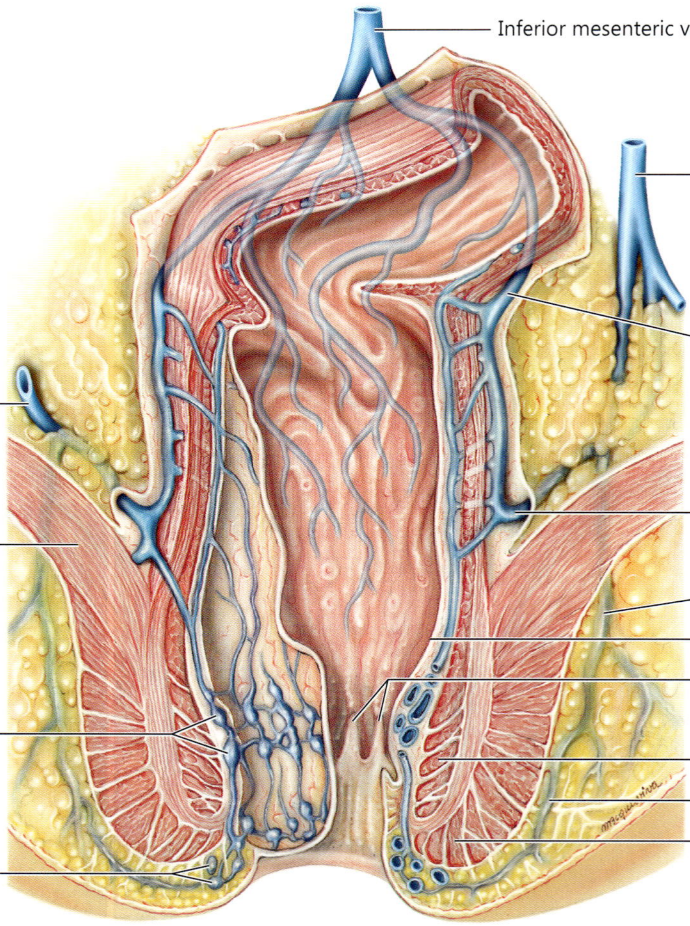

Venous vascularization of the rectum. The illustration shows the distribution of the veins of the hemorrhoidal plexuses.

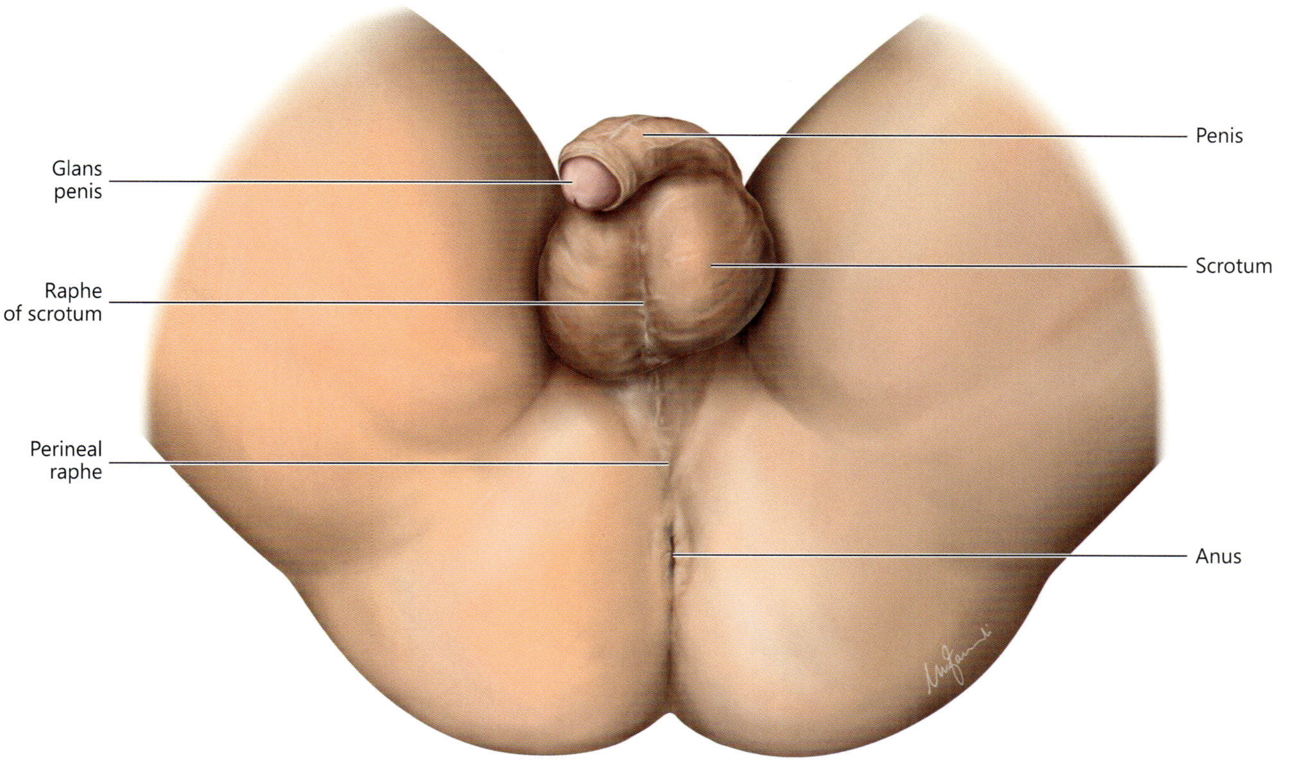

Male external genitalia. In the flaccid penis, the distal extremity, corresponding to the glans penis, is covered by the prepuce, a cutaneous fold, from which it remains separated through the preputial cavity. The scrotum is characterized by an inferior extremity that is broader, free and generally more elongated in the left half, which sustains the testes, and a superior extremity that is narrower and adheres to the pubic region. On the anterior surface of the scrotum lies the body of the penis in the state of flaccidity.

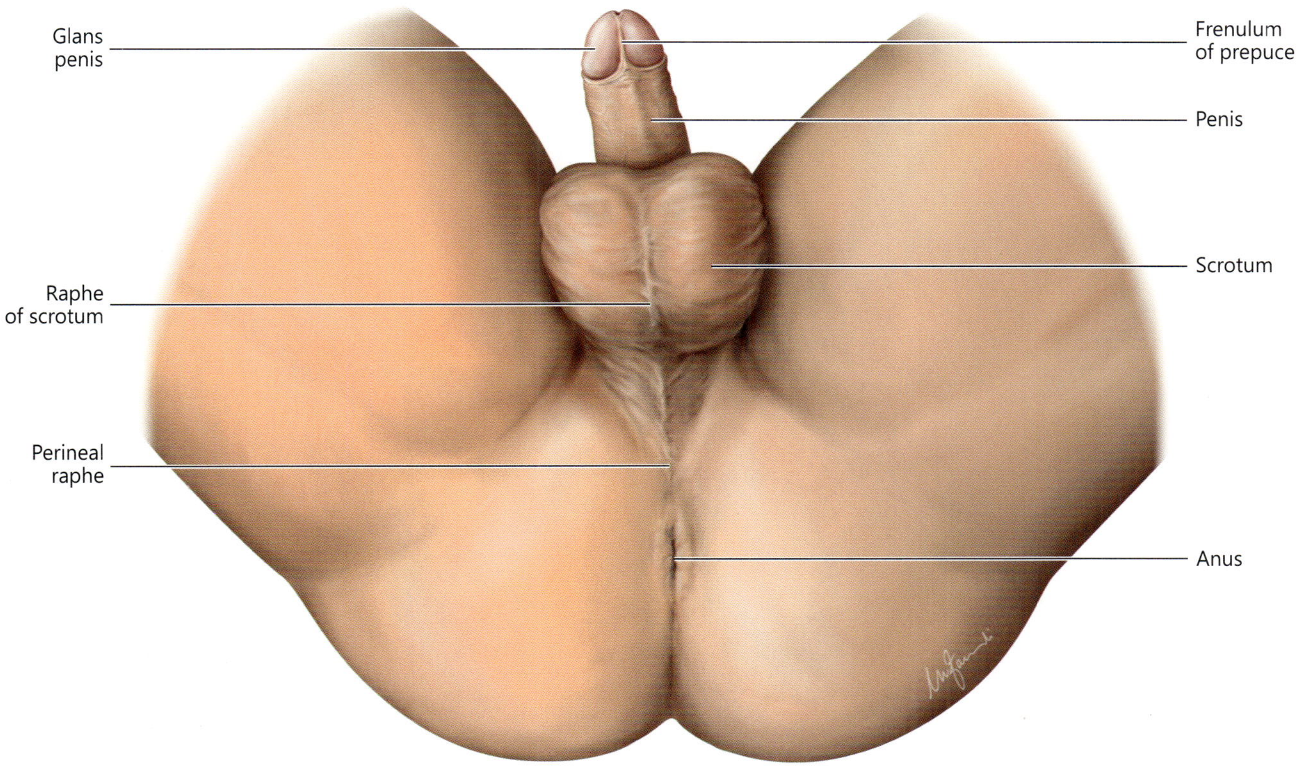

Male external genitalia. In the state of erection, the body of the penis rises from the scrotum and moves towards to the anterior wall of the abdomen, aligning itself with the root of the penis. The corona of the glans becomes visible, in particular without the prepuce, separated from the neck of the glans by the balanopreputial sulcus. The raphe of the scrotum, which continues on the inferior surface of the penis (raphe of the penis) and on the perineum (perineal raphe), represents the line of fusion of the genital and scrotal folds.

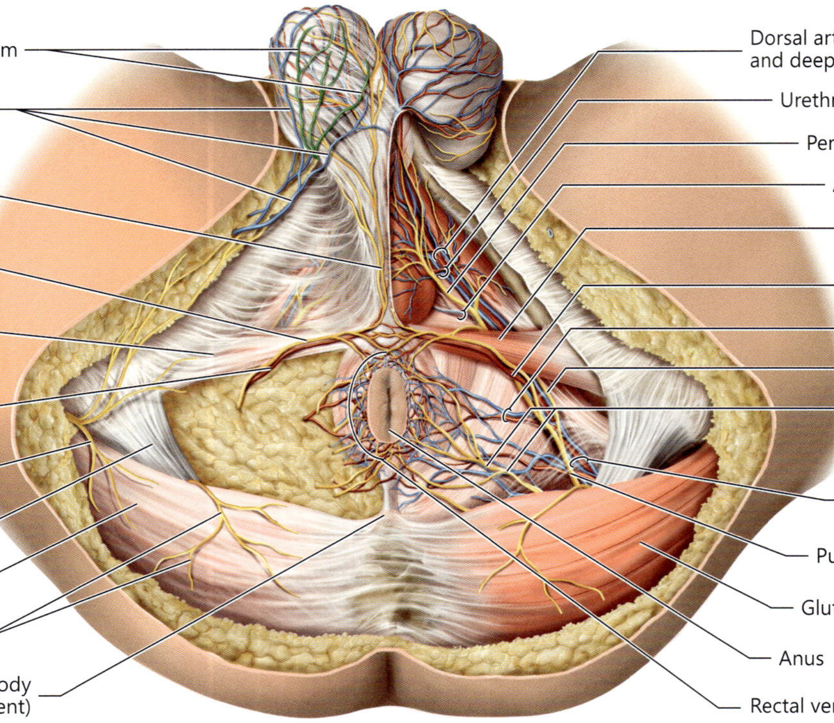

Male perineum. On the right, with the skin and subcutaneous tissue removed, one can see, anteriorly, the perineal fascia and, posteriorly, the adipose tissue of the ischioanal or ischiorectal fossa. On the left, with the perineal fascia transected and the fat body of the ischioanal fossa removed, one can see, anteriorly, the ischiocavernosus, which envelops the crus of the penis, the bulbospongiosus, which envelops the bulb of the penis, the superficial transverse perineal muscle and the vessels and nerves running between these formations, and, posteriorly, the vessels and nerves of the ischioanal fossa.

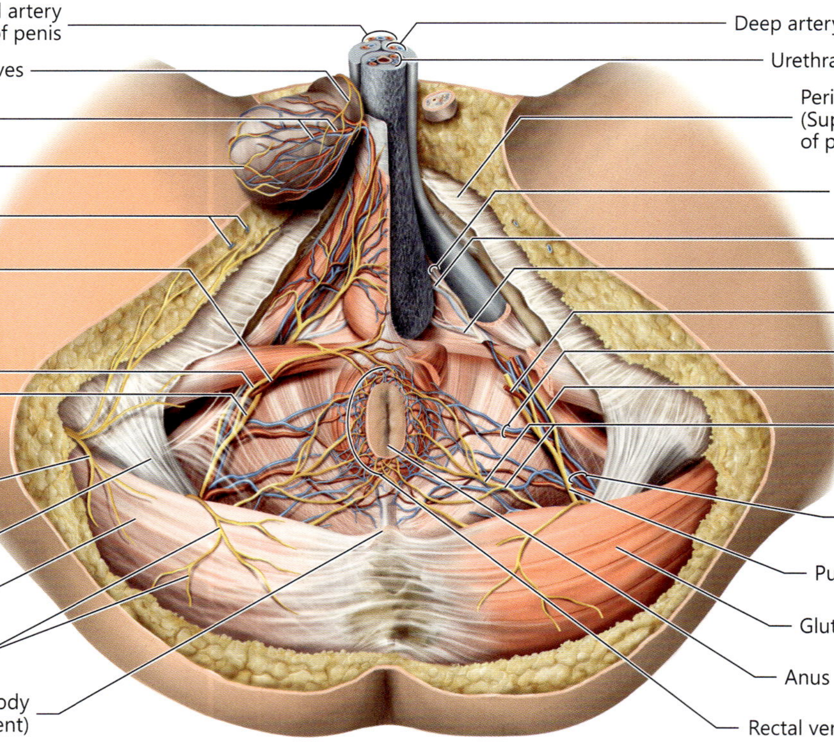

Male perineum. On the right, with the perineal fascia transected and the fat body of the ischioanal fossa removed, one can see, anteriorly, the ischiocavernosus, the bulbospongiosus, the superficial transverse perineal muscle and the vessels and nerves running between these formations, and, posteriorly, the vessels and nerves of the ischioanal fossa, i.e. the perineal, internal pudendal and inferior rectal arteries, the veins which accompany them and the pudendal nerve which divides into the perineal nerves. On the left, with the ischiocavernosus, bulbospongiosus and superficial transverse perineal muscles transected, the contents of the superficial perineal pouch are visible, in particular the bulb and crus of the penis.

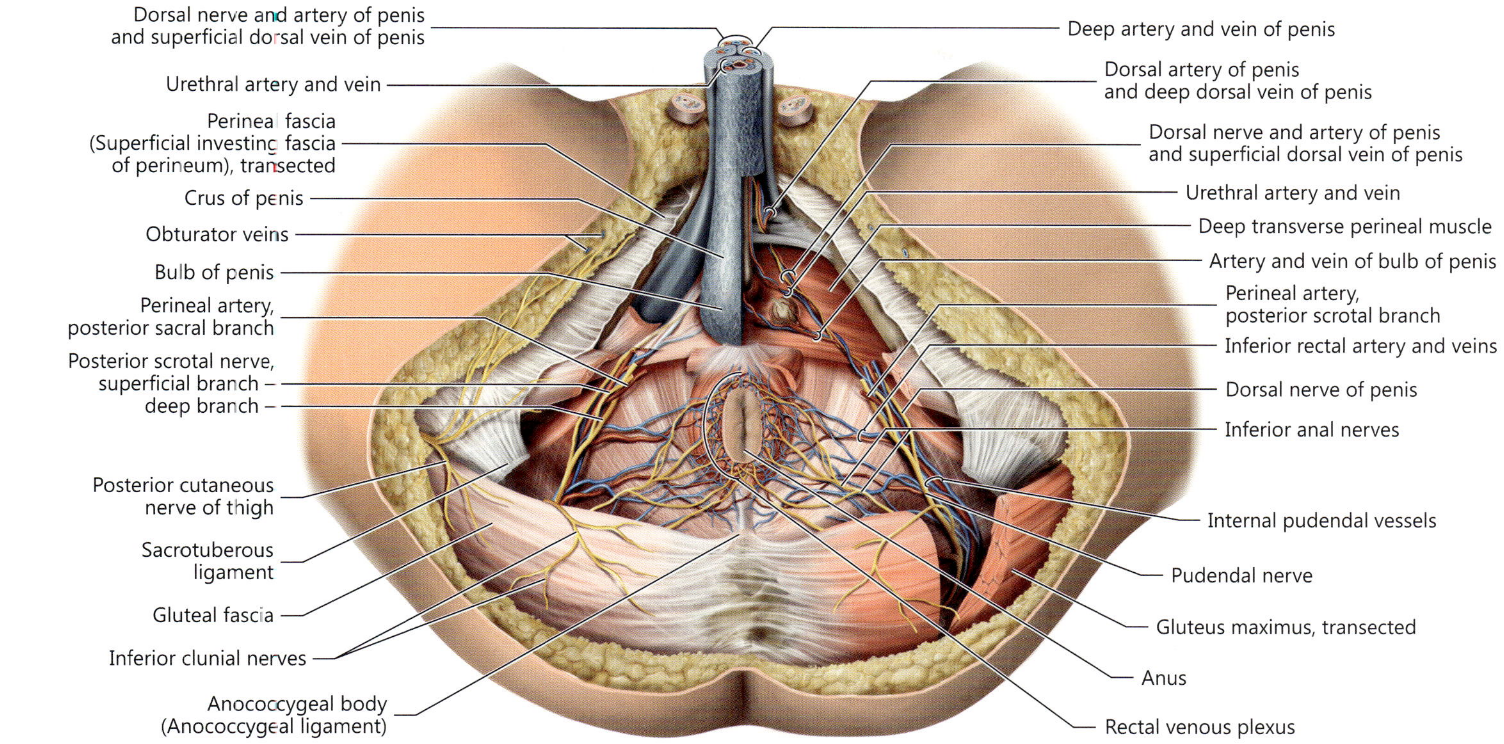

Male perineum. On the right, after transection of the perineal fascia and the ischiocavernosus, bulbospongiosus and superficial transverse perineal muscles, and removal of the fat body of the ischioanal fossa, one can see, anteriorly, the bulb and crus of the penis and, posteriorly, the vessels and nerves of the ischioanal fossa. On the left, after transecting the bulb and crus of the penis, and the inferior fascia of the perineal membrane (inferior fascia of the urogenital trigone), the deep transverse perineal muscle is visible and one can see the bulbourethral gland, contained in the thickness of the inferior fascia of the perineal membrane.

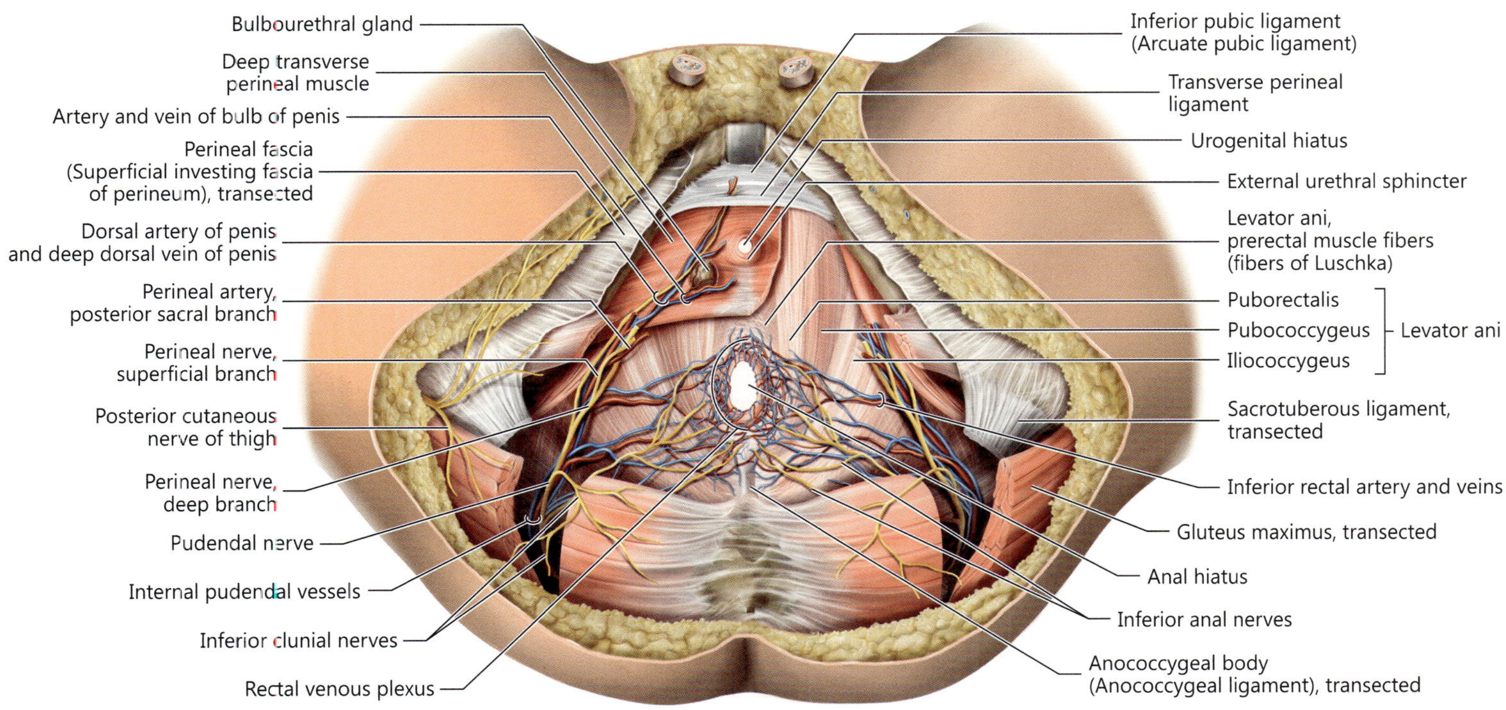

Male perineum. Anteriorly, with the deep transverse perineal muscle transected on the left, one can see the levator ani invested by the inferior fascia of the pelvic diaphragm. Posteriorly, on the right and left, the ischioanal fossae can be seen, situated between the inferior fascia of the pelvic diaphragm and the lateral walls of the lesser pelvis. These walls continue further on posteriorly, in a space situated between the levator ani and the gluteus maximus.

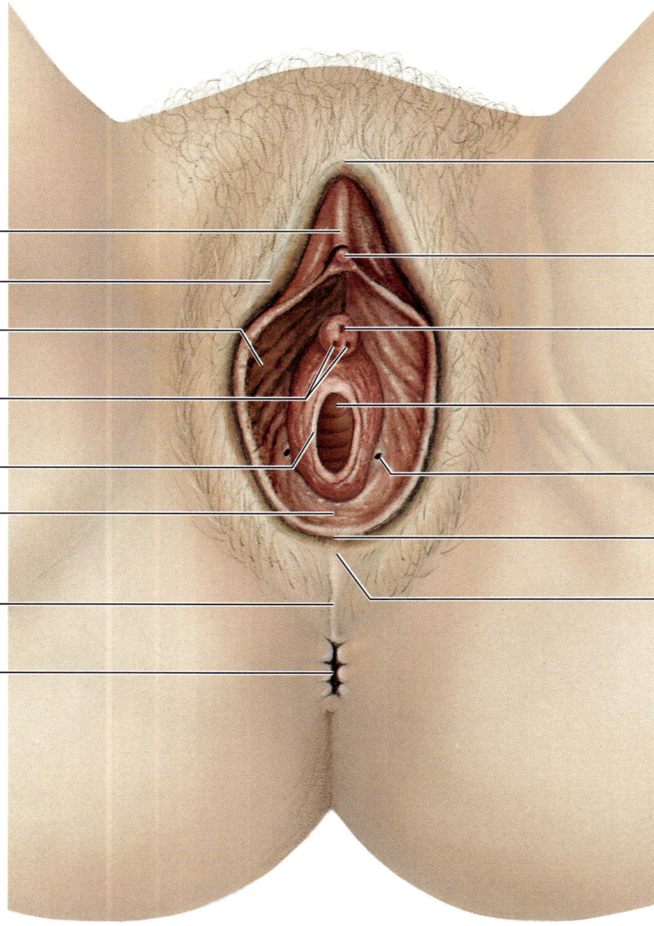

External genitalia of a virgin. The hymen, of a circular shape, delimits the vaginal orifice which opens into the vestibule.

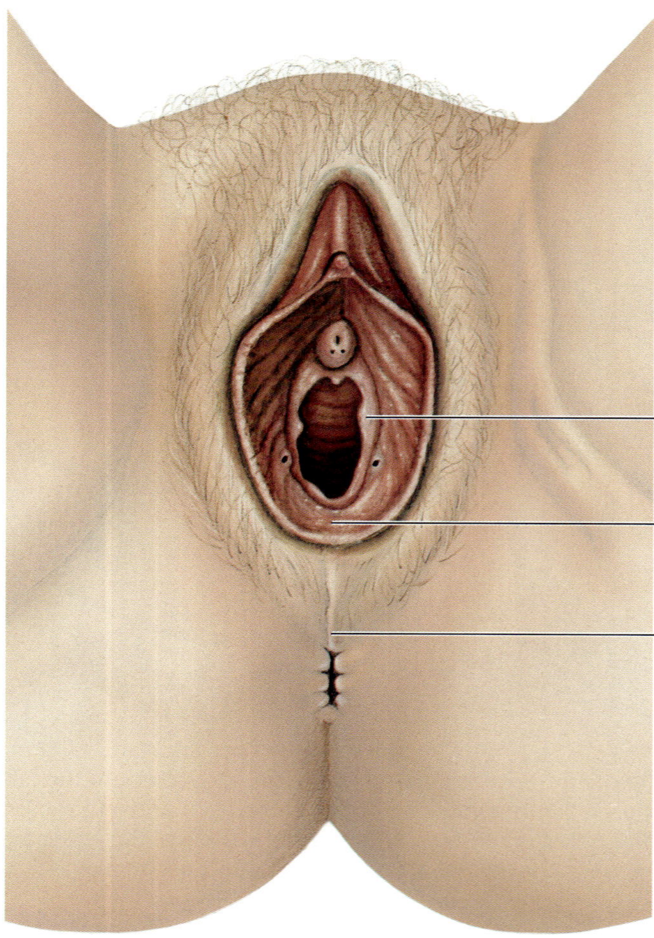

External genitalia of a multipara. Remains of the hymen can be seen, in the form of carunculae hymenales. Note also, with respect to the figure above, the greater extension of the labia minora.

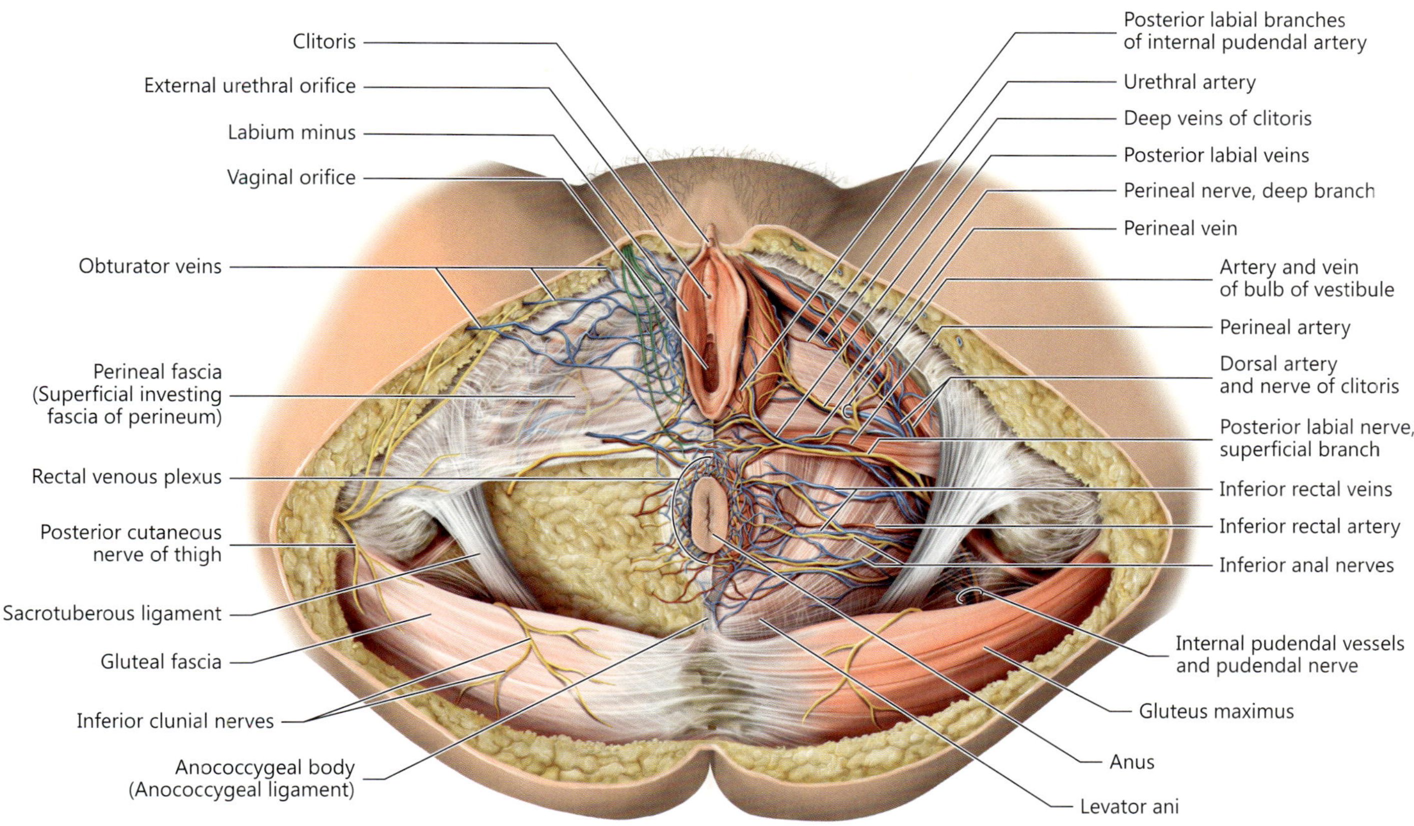

Vessels and nerves of the female perineum. On the right, removal of the skin and subcutaneous tissue. On the left, removal of the perineal fascia and of the adipose tissue of the ischioanal fossa.

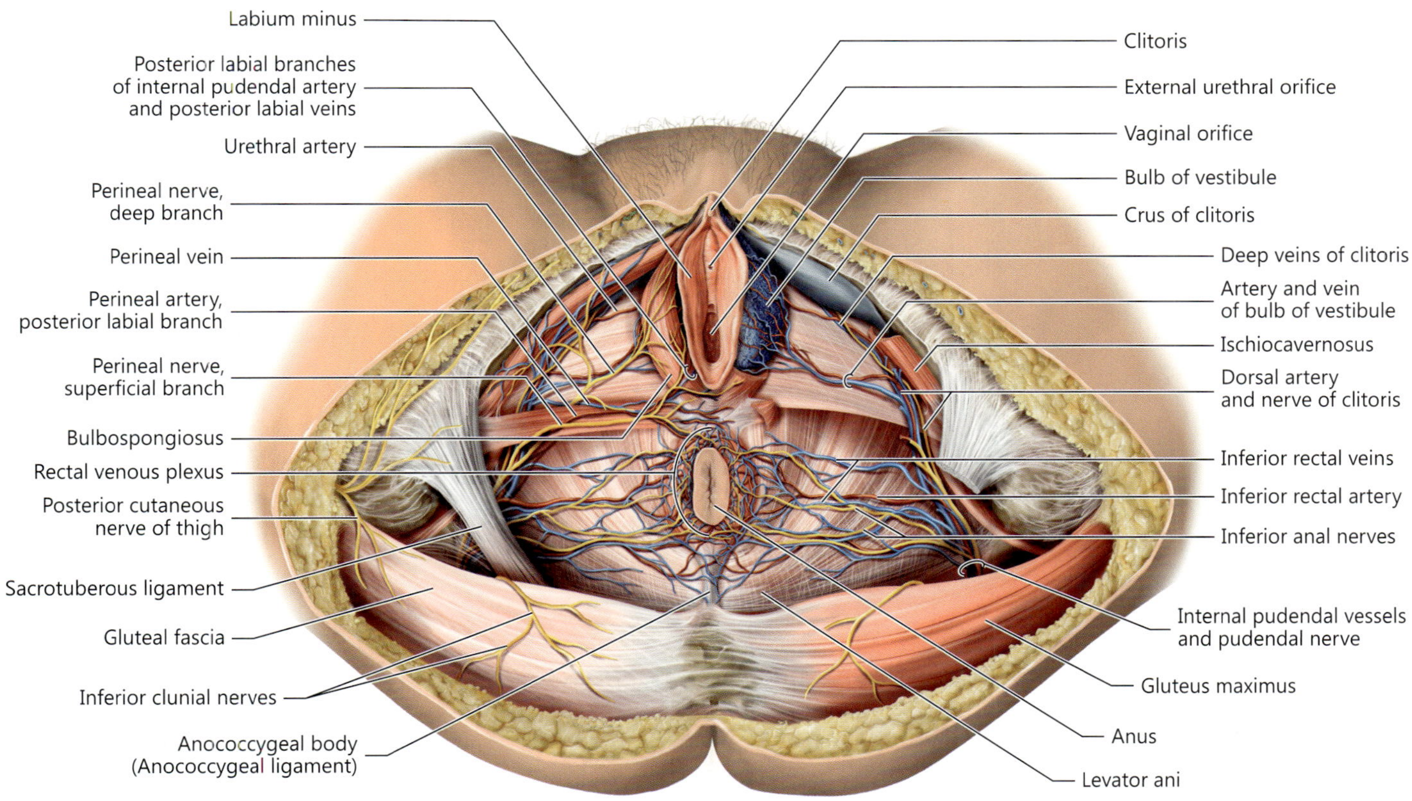

Vessels and nerves of the female perineum. On the right, removal of the skin, subcutaneous tissue, perineal fascia and adipose tissue of the ischioanal fossa. On the left, removal of the ischiocavernosus, bulbospongiosus and superficial transverse perineal muscles, and transection of the sacrotuberous ligament and pelvic fascia.

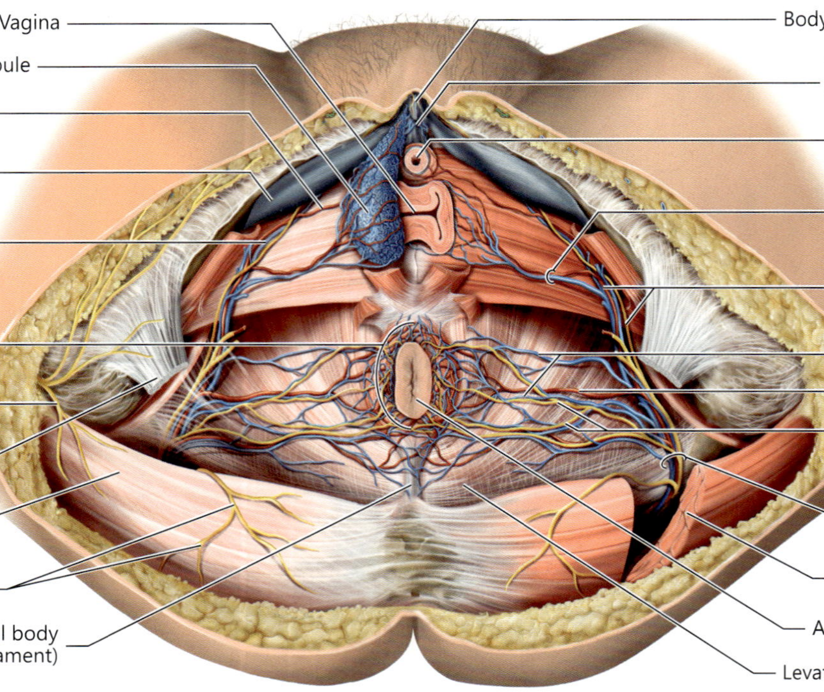

Vessels and nerves of the female perineum. On the right, removal of the skin, subcutaneous tissue, perineal fascia, adipose tissue of the ischioanal fossa, ischiocavernosus, bulbospongiosus and superficial transverse perineal muscles, the sacrotuberous ligament, pelvic fascia, and labium minus. On the left, removal of the bulb of the vestibule and of the gluteal fascia, and transection of the gluteus maximus.

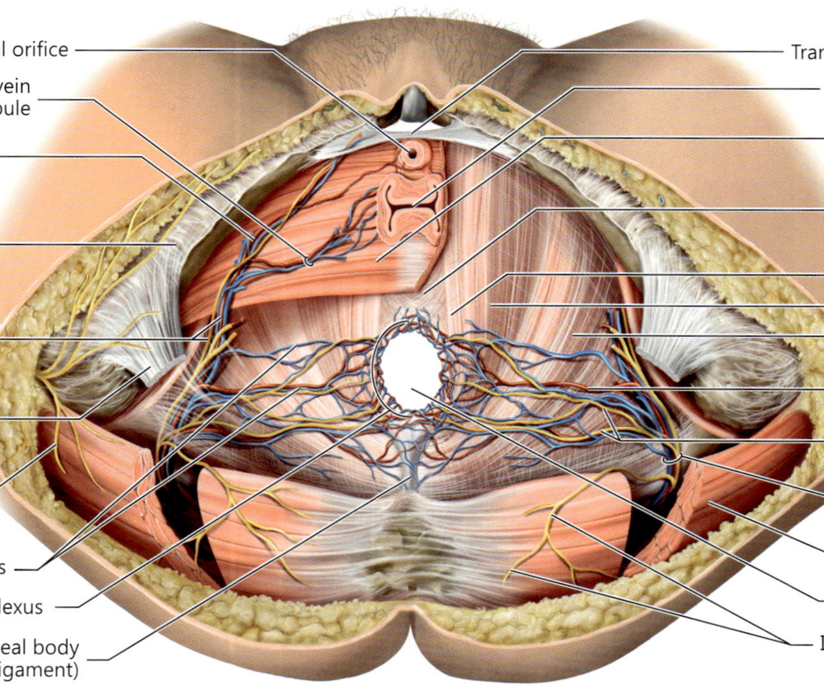

Female perineum. Anteriorly, on the right, after transection of the skin, subcutaneous tissue, perineal fascia, ischiocavernosus, bulbospongiosus and superficial transverse perineal muscles, corpus cavernosum of the clitoris, bulb of the vestibule, and of the inferior layer of the perineal membrane, one can see the deep transverse perineal muscle with the vessels and nerves of the anterior perineum, branches of the internal pudendal vessels and pudendal nerve. On the median line, the urethra and vagina can be seen and, on the left, with the deep transverse perineal muscle and the superior layer of the perineal membrane transected, one can see the levator ani invested by the inferior fascia of the pelvic diaphragm.

UPPER LIMB

Petros (Athens, 1928 - Milan, 2014), *Bestiaire GS 56*, mixed media, cm 50 × 70, 1990

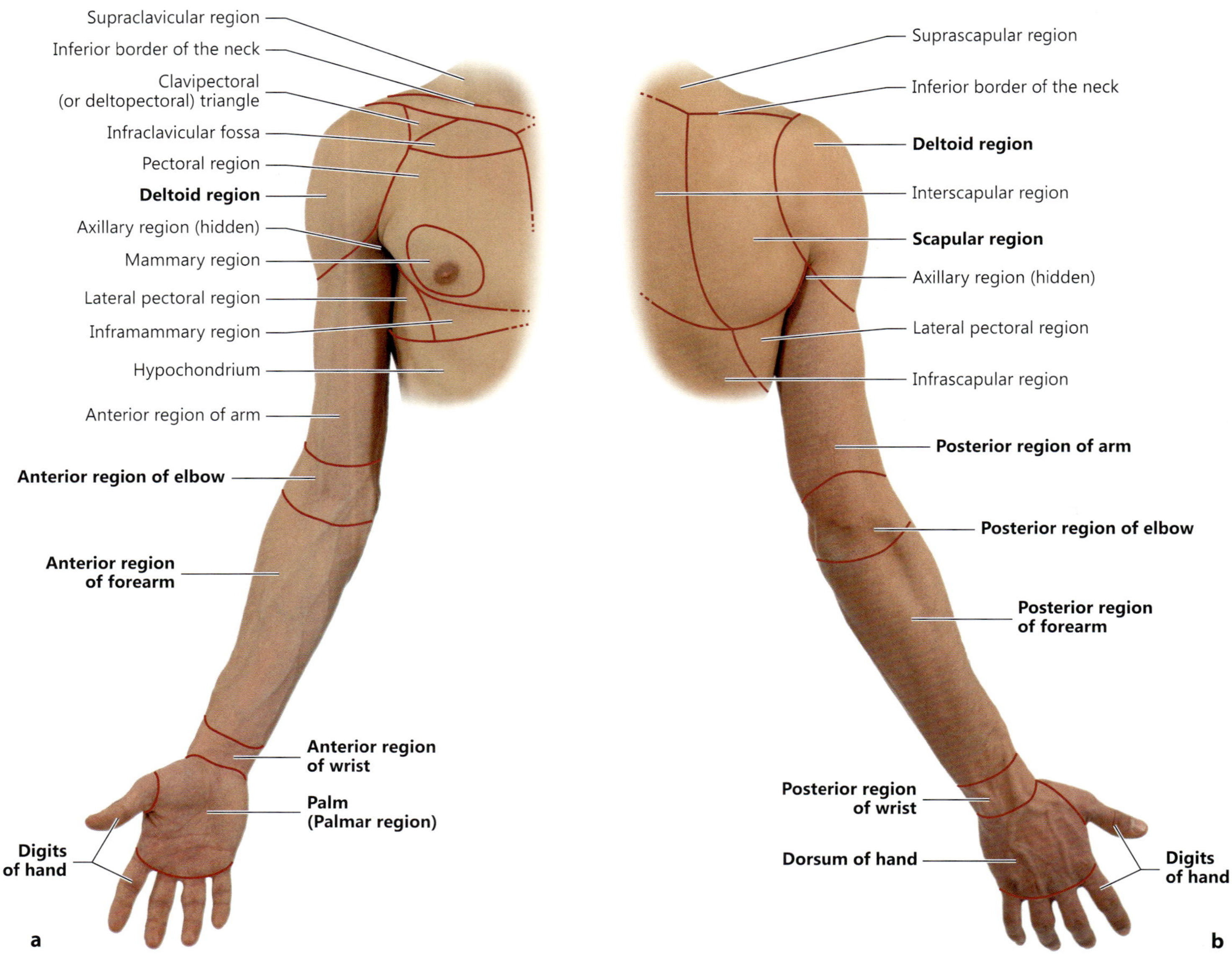

Landmark lines and points determining the anterior (**a**) and posterior (**b**) topographic regions **of the upper limb** (indicated in bold type).
For the **landmark points** ☞ Limbs, Topographic Landmarks, vol. 1 page 117.

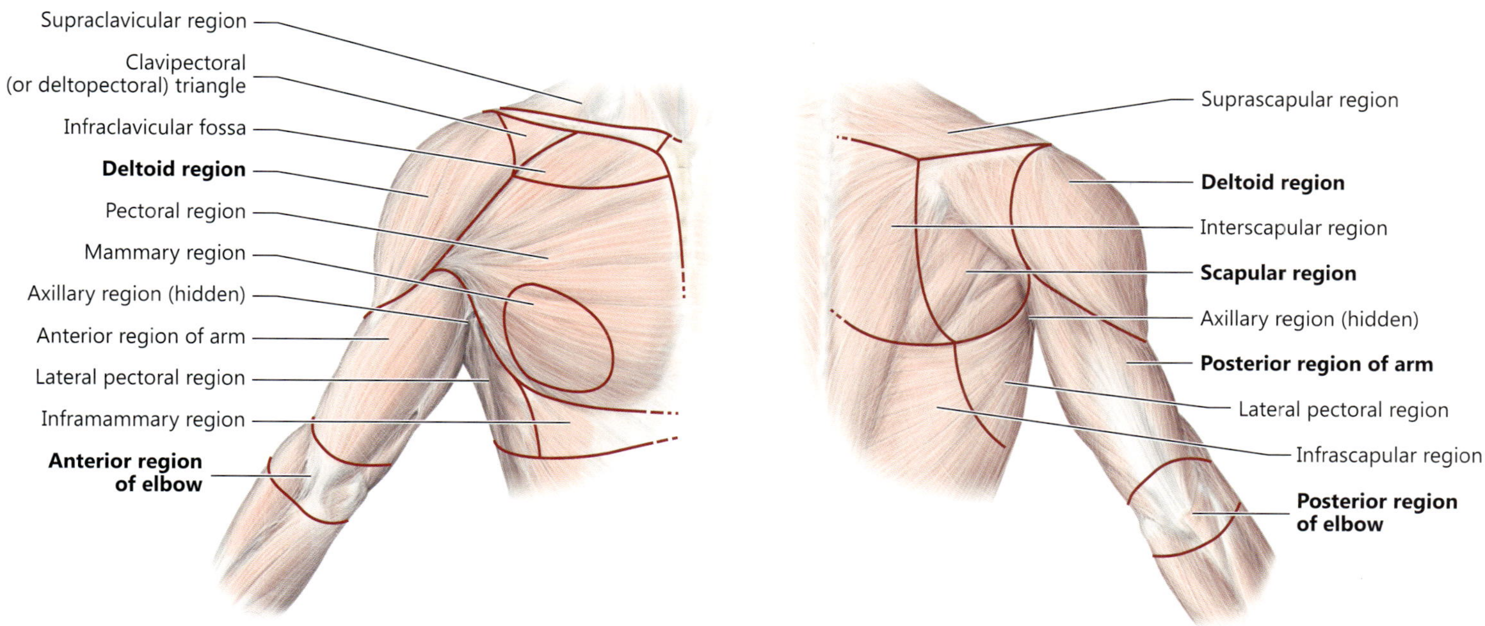

The proximal third of the upper limb and the shoulder muscles and the projection of the corresponding **superficial topographic regions** are shown after removal of the cutaneous and subcutaneous layers.

REGIONS OF THE UPPER LIMB	*BOUNDARIES OR LOCATION*
Anterior	
– of the arm (anterior brachial)	Anterior margin of the deltoid and inferior margin of the pectoralis major (superiorly); horizontal plane passing approximately 3 cm above the medial epicondyle and the lateral epicondyle (inferiorly)
– of the wrist (anterior carpal)	Horizontal plane passing above the head of the ulna (superiorly); horizontal plane passing under the pisiform (inferiorly)
– of the elbow (anterior cubital, with cubital fossa)	Horizontal plane passing approximately 3 cm above the medial epicondyle and the lateral epicondyle (superiorly); horizontal plane passing approximately 3 cm under the medial epicondyle and the lateral epicondyle (inferiorly)
– of the forearm (anterior antebrachial)	Horizontal plane passing approximately 3 cm inferior to the medial epicondyle and the lateral epicondyle (superiorly); horizontal plane passing above the head of the ulna (inferiorly)
Axillary (of the axilla base)	Inferior margin of the pectoralis major (anteriorly); inferior margins of the teres major and the latissimus dorsi (posteriorly). The medial and lateral boundaries are shown by lines that join the anterior and posterior boundaries on the lateral wall of the thorax and on the medial surface of the arm
Deltoid	External third of the clavicle, acromion, spine of the scapula (superiorly); anterior margin of the deltoid (anteriorly); posterior margin of the deltoid (posteriorly); deltoid tuberosity of the humerus on which converge the anterior and posterior margins of the deltoid (inferiorly)
Dorsum (dorsal region) of the hand	Horizontal plane passing under the pisiform and the tubercle of the scaphoid (superiorly); line, proximal concave, passing through the digitopalmar folds (inferiorly)
Palm (palmar region)	☞ Dorsum
Posterior	
– of the arm (posterior brachial)	☞ Anterior of the arm
– of the wrist (posterior carpal)	☞ Anterior of the wrist
– of the elbow (posterior cubital)	☞ Anterior of the elbow
– of the forearm (posterior antibrachial)	☞ Anterior of the forearm
Scapular	Dorsal region between the superior, medial and lateral borders of the scapula, highlighted by the muscles which are inserted into it: trapezius, latissimus dorsi, levator scapulae, rhomboid major, rhomboid minor, deltoid, supraspinatus, infraspinatus, teres major, teres minor, subscapularis

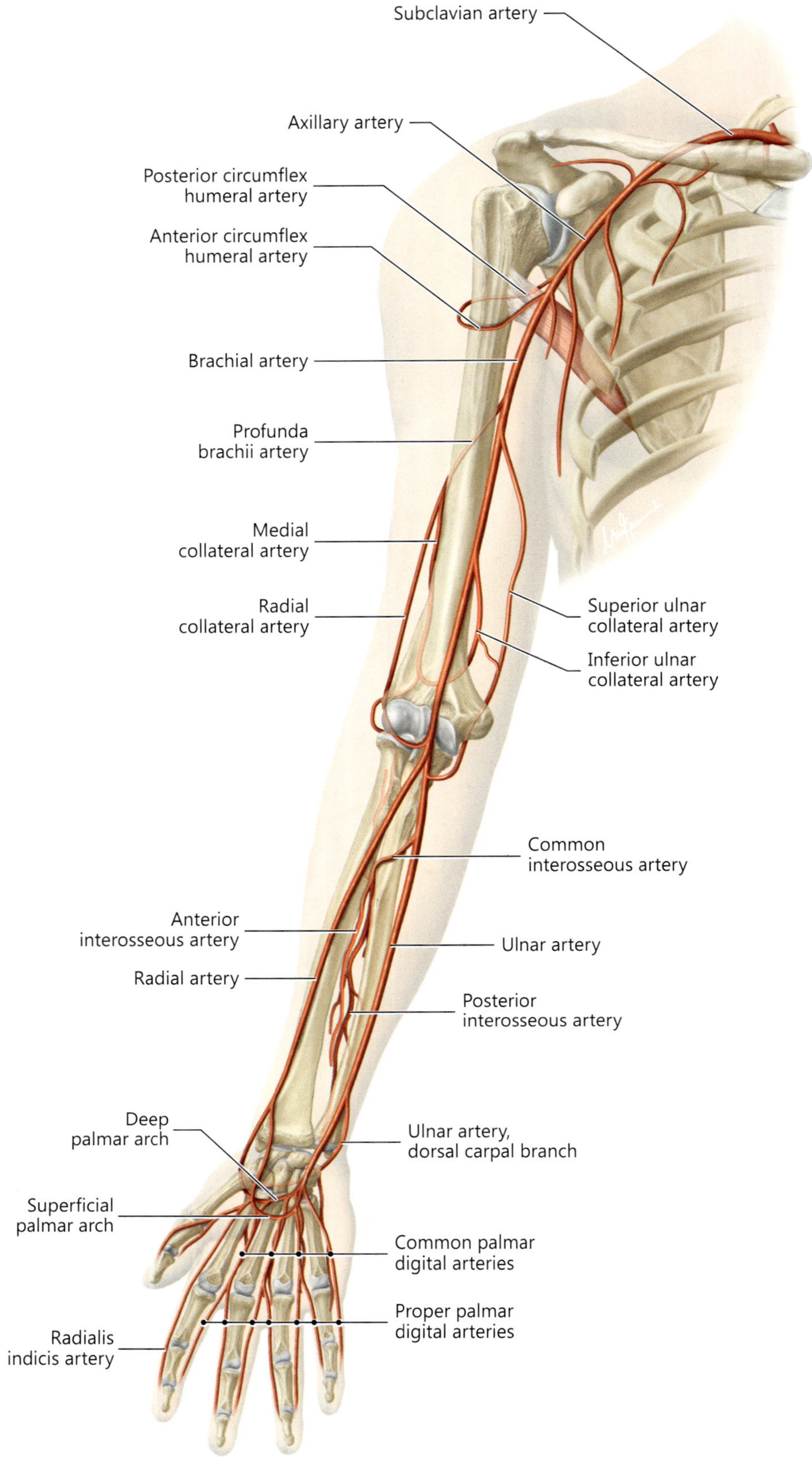

Schematic representation of the principal right **upper limb arteries**, anterior view.

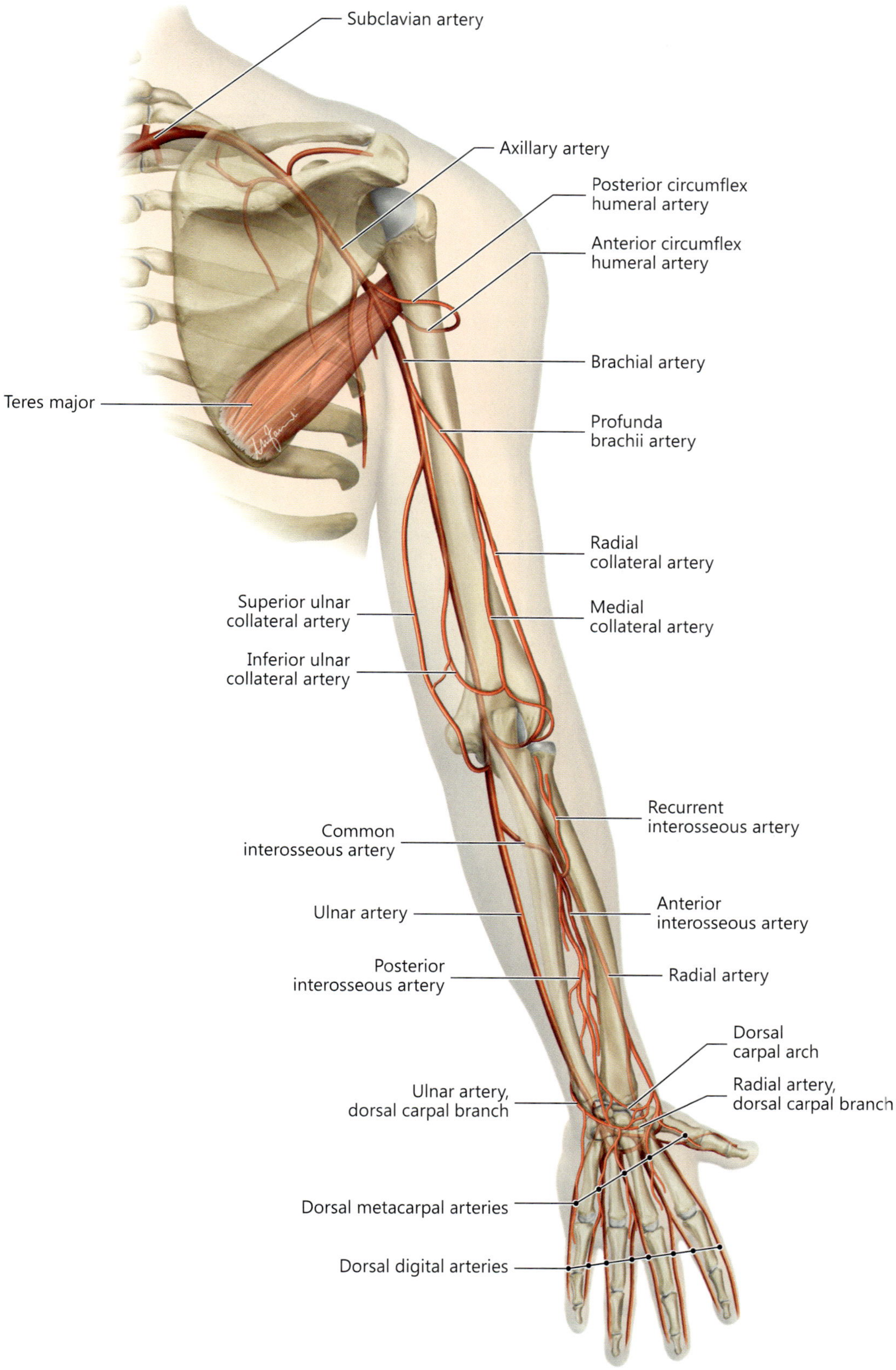

Schematic representation of the principal right **upper limb arteries**, posterior view.

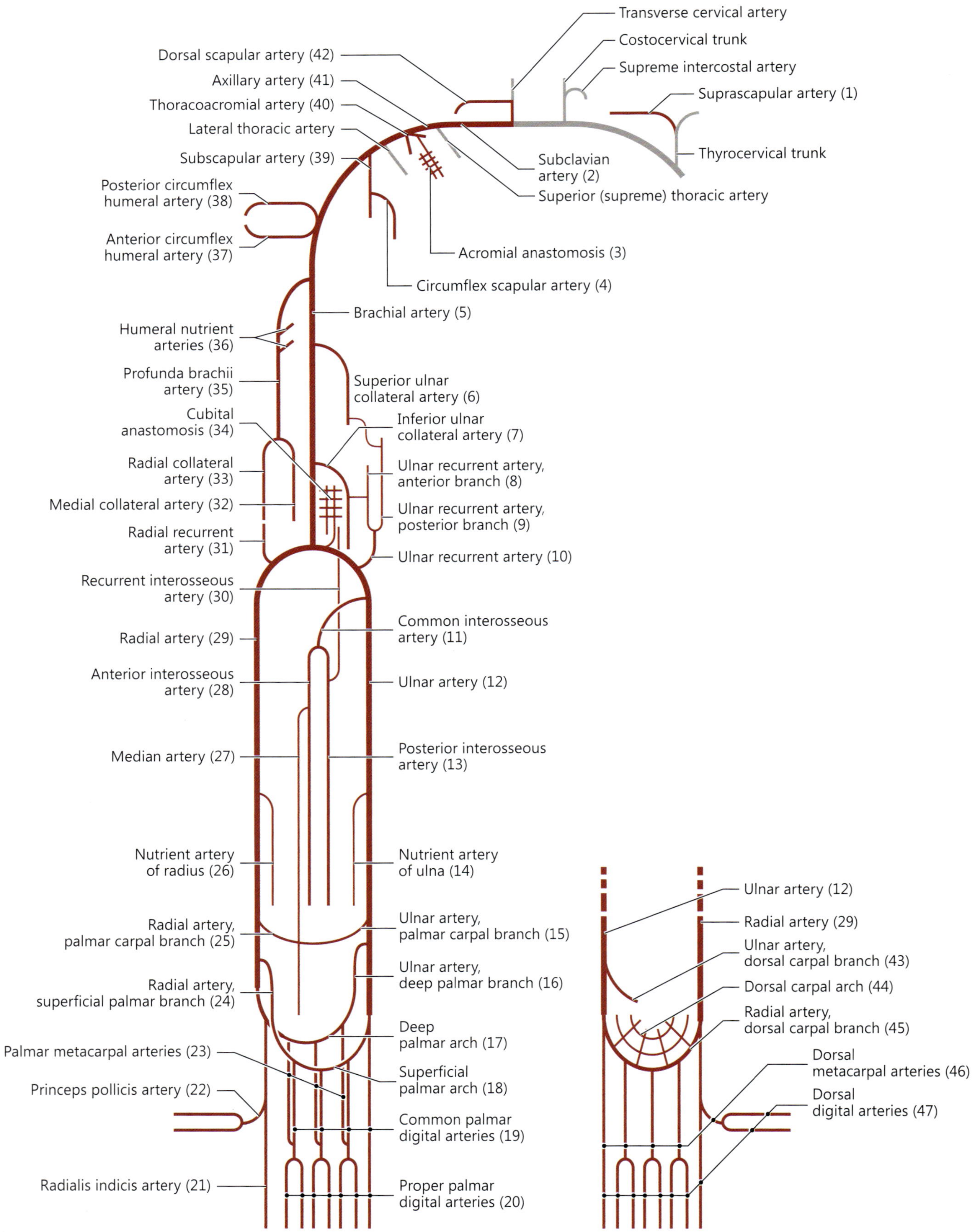

Arterial tree of the upper limb.

Arteries of the upper limb: territories of supply		
	ARTERIES	*TERRITORIES OF SUPPLY*
1	Suprascapular artery	Skin of the thorax, clavicle and scapula, skin above the acromion, acromioclavicular and glenohumeral joints, supraspinatus and infraspinatus
2	Subclavian artery	Head, neck, superior part of the thorax
3	Acromial anastomosis	Acromion, deltoid
4	Circumflex scapular artery	Long head of the triceps brachii, teres minor, teres major, deltoid and subscapularis, glenohumeral joint
5	Brachial artery	Muscles of the upper limb, elbow joint
6	Superior ulnar collateral artery	Triceps brachii, elbow joint
7	Inferior ulnar collateral artery	Triceps brachii and pronator teres, elbow joint
8	Ulnar recurrent artery, anterior branch	Brachialis and pronator teres
9	Ulnar recurrent artery, posterior branch	Elbow joint, adjacent muscles and skin
10	Ulnar recurrent artery	Brachialis and pronator teres, elbow joint and adjacent skin
11	Common interosseous artery	Forearm, radius and ulna
12	Ulnar artery	Forearm, wrist and hand
13	Posterior interosseous artery	Deep posterior part of the forearm
14	Nutrient artery of ulna	Ulna
15	Ulnar artery, palmar carpal branch	Palm of the hand
16	Ulnar artery, deep palmar branch	Palm of the hand
17	Deep palmar arch	Palmar interossei and dorsal interossei, and bases of the metacarpals
18	Superficial palmar arch	Palm of the hand
19	Common palmar digital arteries	Digits of the hand
20	Proper palmar digital arteries	Lateral surfaces of each finger, nail matrix
21	Radialis indicis artery	Medial surface of the index finger
22	Princeps pollicis artery	Lateral and medial surfaces of the thumb, dorsal interossei
23	Palmar metacarpal arteries	Interossei and lumbricals, metacarpals
24	Radial artery, superficial palmar branch	Palm of the hand
25	Radial artery, palmar carpal branch	Palm of the hand
26	Nutrient artery of the radius	Radius
27	Median artery	Accompanies and supplies the median nerve up to the palm of the hand
28	Anterior interosseous artery	Deep anterior part of the forearm, radius and ulna
29	Radial artery	Muscles of the forearm and of the hand, radius, skin of the dorsum of the hand and of the palmar surface of the thumb, lateral surface of the index finger, carpal joints
30	Recurrent interosseous artery	Posterior surface of the elbow joint
31	Radial recurrent artery	Supinator, brachioradialis and brachialis, elbow joint
32	Medial collateral artery	Triceps brachii and anconeus, elbow joint
33	Radial collateral artery	Triceps brachii, brachioradialis and brachialis, elbow joint
34	Cubital anastomosis	Elbow
35	Profunda brachii artery	Deltoid and muscles of the arm, including the triceps brachii, humerus and elbow joint
36	Humeral nutrient arteries	Humerus
37	Anterior circumflex humeral artery	Long head of the biceps brachii and muscles of the arm, deltoid and coracobrachialis, tendon of the pectoralis major, head of the humerus and glenohumeral joint
38	Posterior circumflex humeral artery	Deltoid, teres minor and triceps brachii, glenohumeral joint and neck of the humerus
39	Subscapular artery	Scapular region and glenohumeral joint
40	Thoracoacromial artery	Pectoralis major and pectoralis minor, thorax, deltoid and subclavius, sternoclavicular joint and acromion
41	Axillary artery	Pectoralis major and pectoralis minor, muscles of the shoulder and the arm, acromion, sternoclavicular and glenohumeral joints, breast
42	Dorsal scapular artery	Levator scapulae, trapezius, rhomboid major and rhomboid minor
43	Ulnar artery, dorsal carpal branch	Dorsum of the hand
44	Dorsal carpal arch	Dorsum of the hand
45	Radial artery, dorsal carpal branch	Dorsum of the hand
46	Dorsal metacarpal arteries	Dorsal surface of the fingers
47	Dorsal digital arteries	Dorsal surface of the fingers

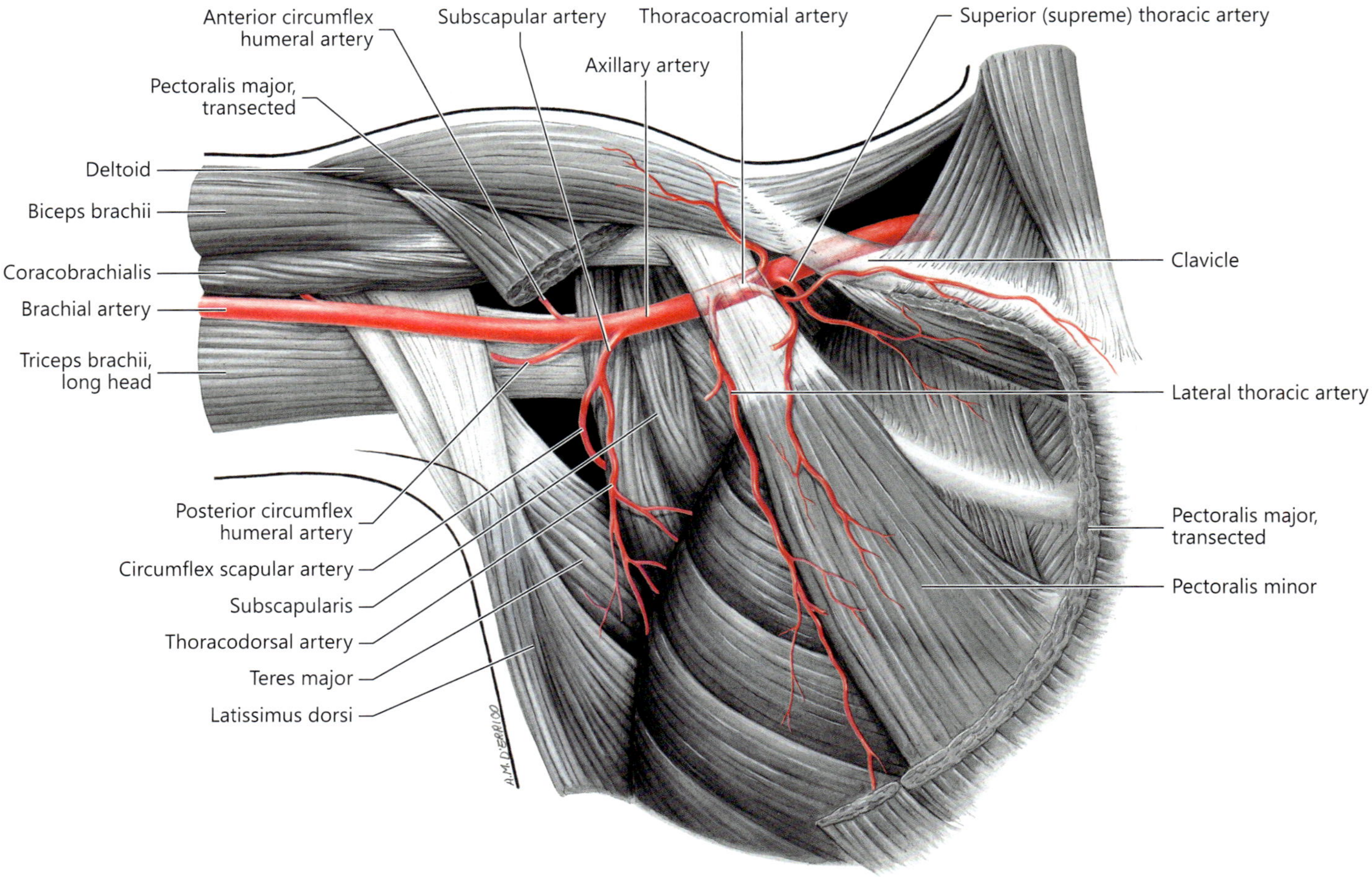

Right **axillary artery** and its branches; the pectoralis major has been transected and removed to show the relations of the artery with the pectoralis minor.

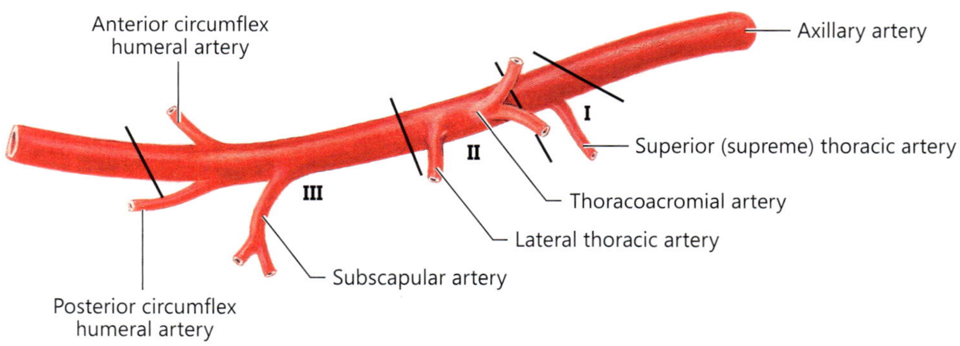

Schematic representation of the **axillary artery** and its three parts with their respective collateral branches. The three black strokes indicate the limits of the parts I, II and III (proximal, intermediate and distal tracts).

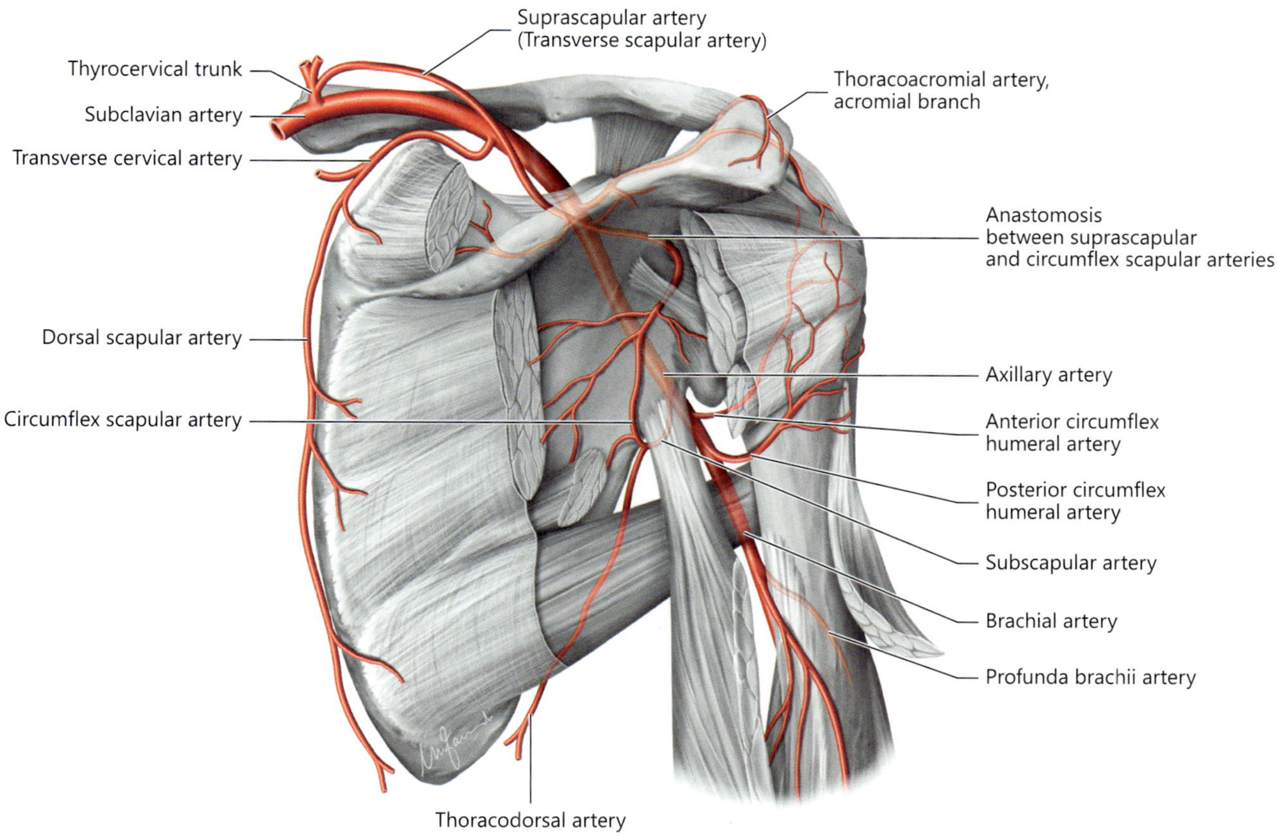

Acromial anastomosis. In evidence, the anastomoses between the dorsal scapular artery (branch of the transverse cervical artery or of the thyrocervical trunk) and the circumflex scapular artery (branch of the subscapular artery), and between the suprascapular artery or transverse scapular artery (branch of the thyrocervical trunk) and the acromial branch of the thoracoacromial artery (branch of the axillary artery). Right side, posterior view.

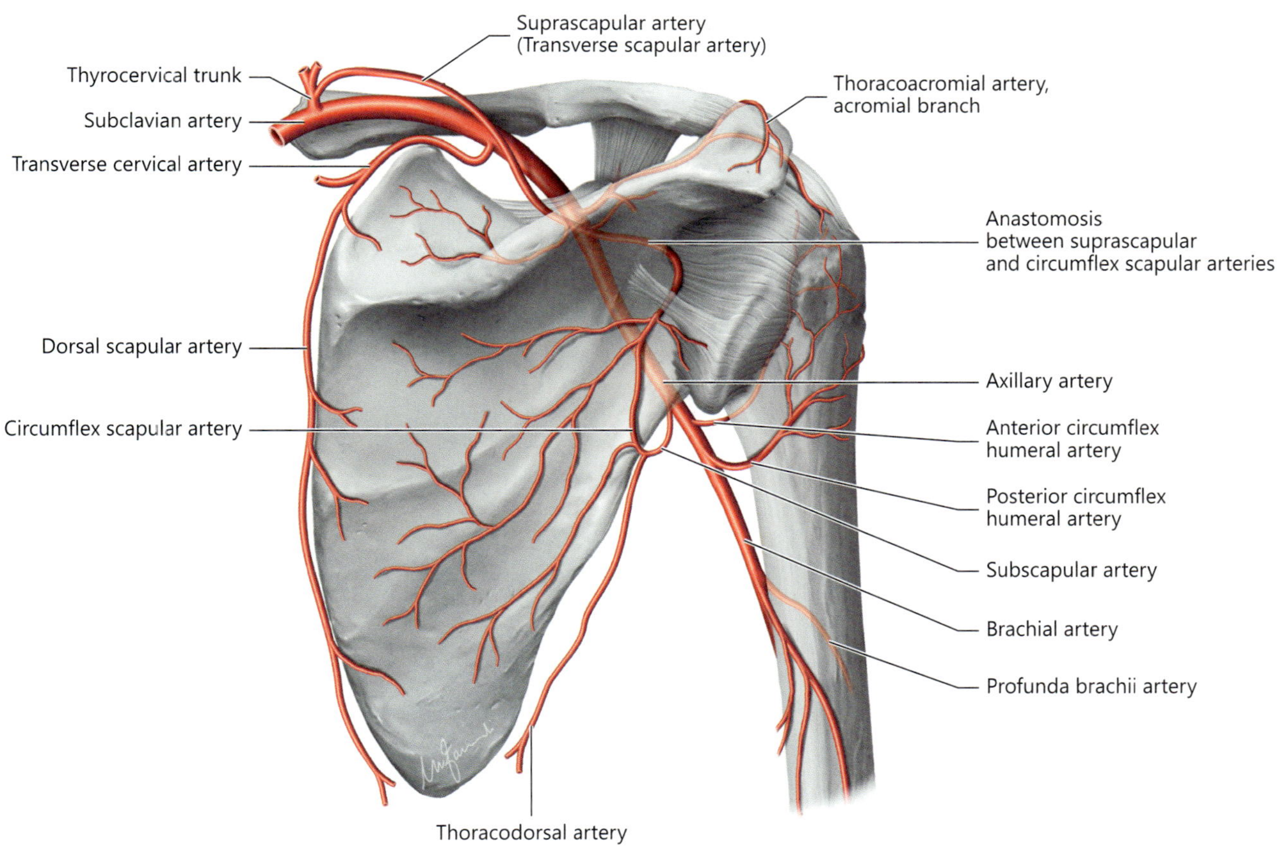

Acromial anastomosis. In evidence, the anastomoses between the transverse cervical artery and the circumflex scapular artery, and between the suprascapular artery (or transverse scapular artery) and the thoracoacromial artery. Right side, posterior view.

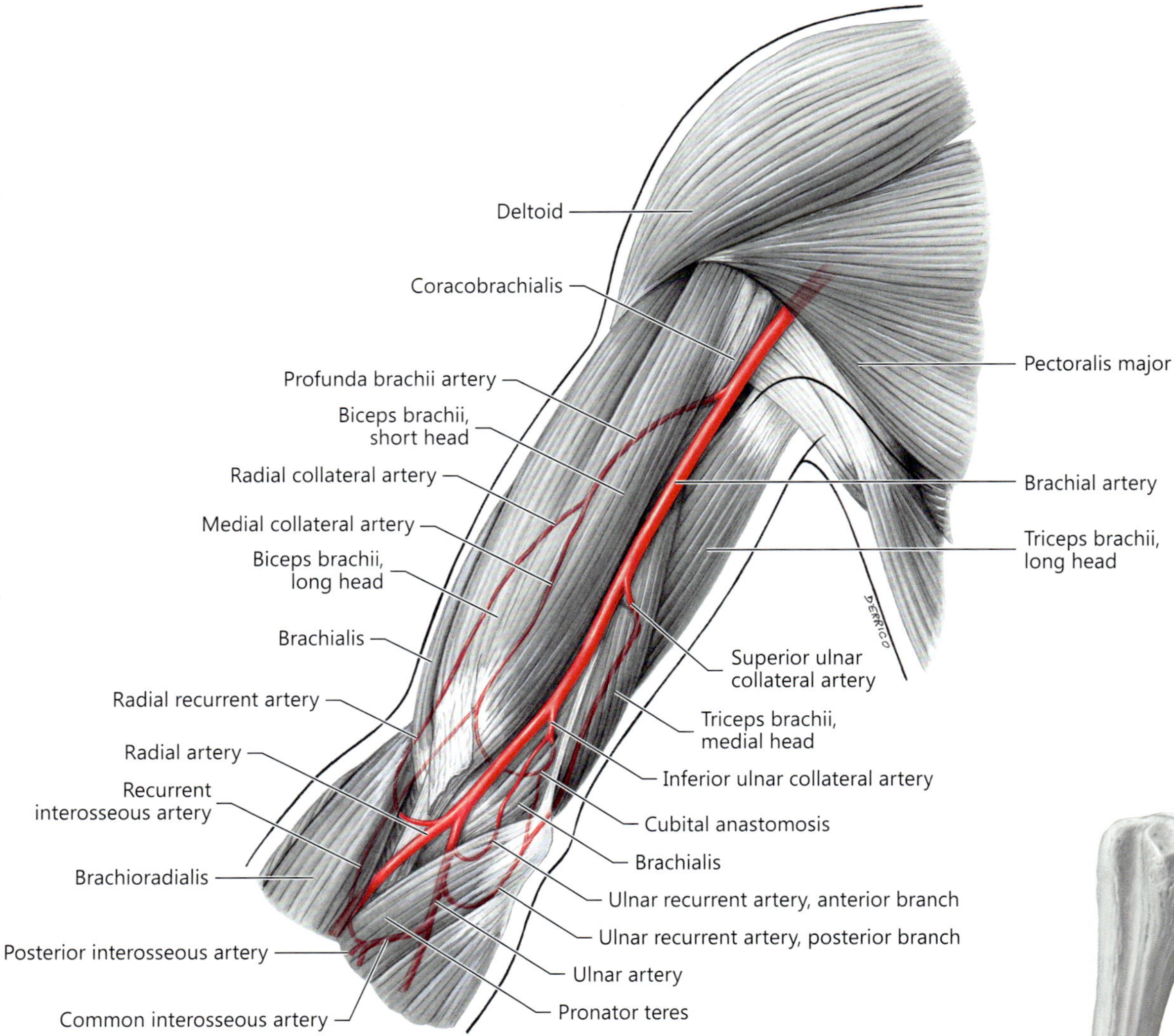

Right **brachial artery** and its branches. In evidence, the arteries that contribute to forming the cubital anastomosis.

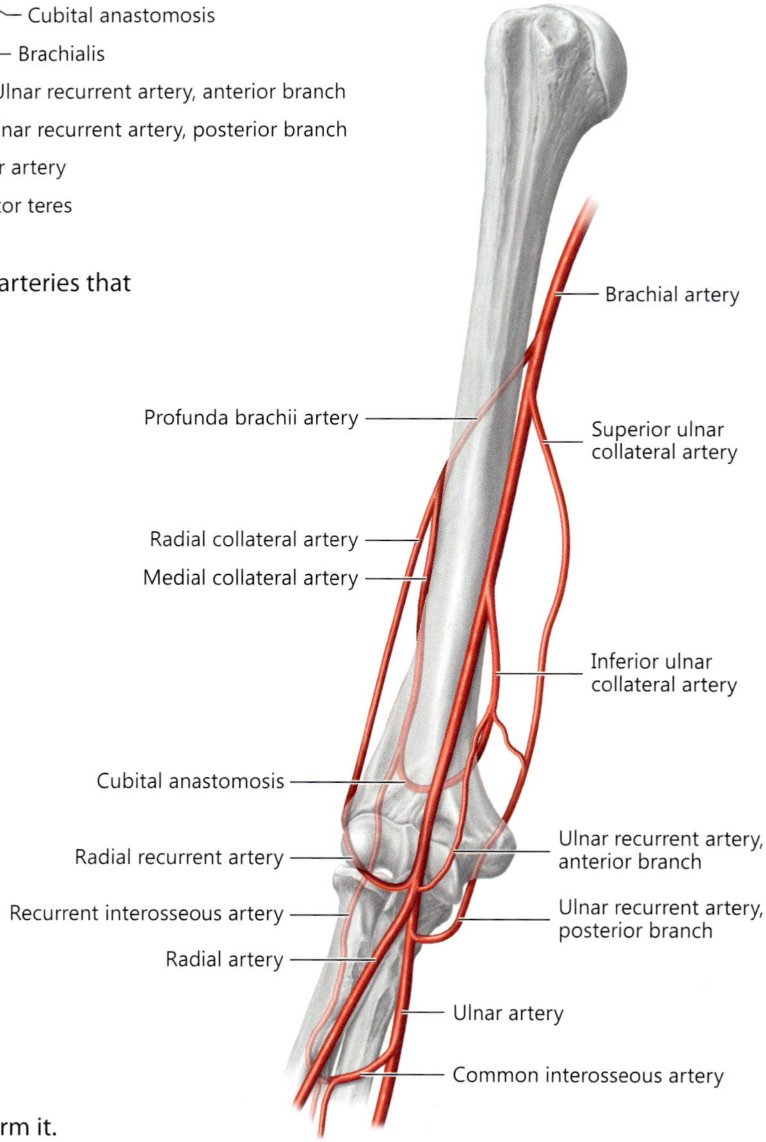

The right **cubital anastomosis** and the branches that form it.

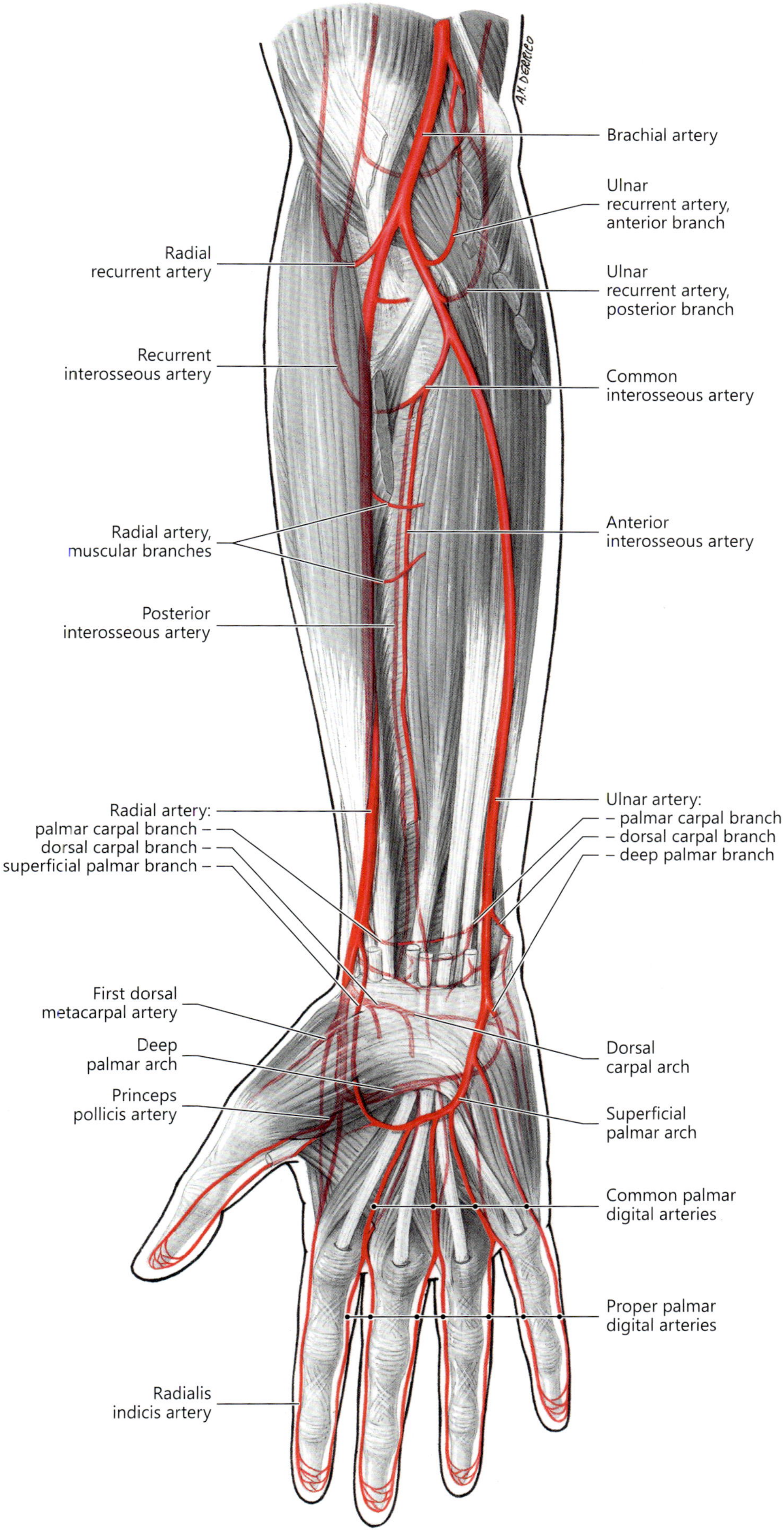

Radial and ulnar arteries and their collateral and terminal branches. In the foreground, the superficial palmar arch; the deep palmar arch and the dorsal carpal arch are transparent. Right limb, anterior view.

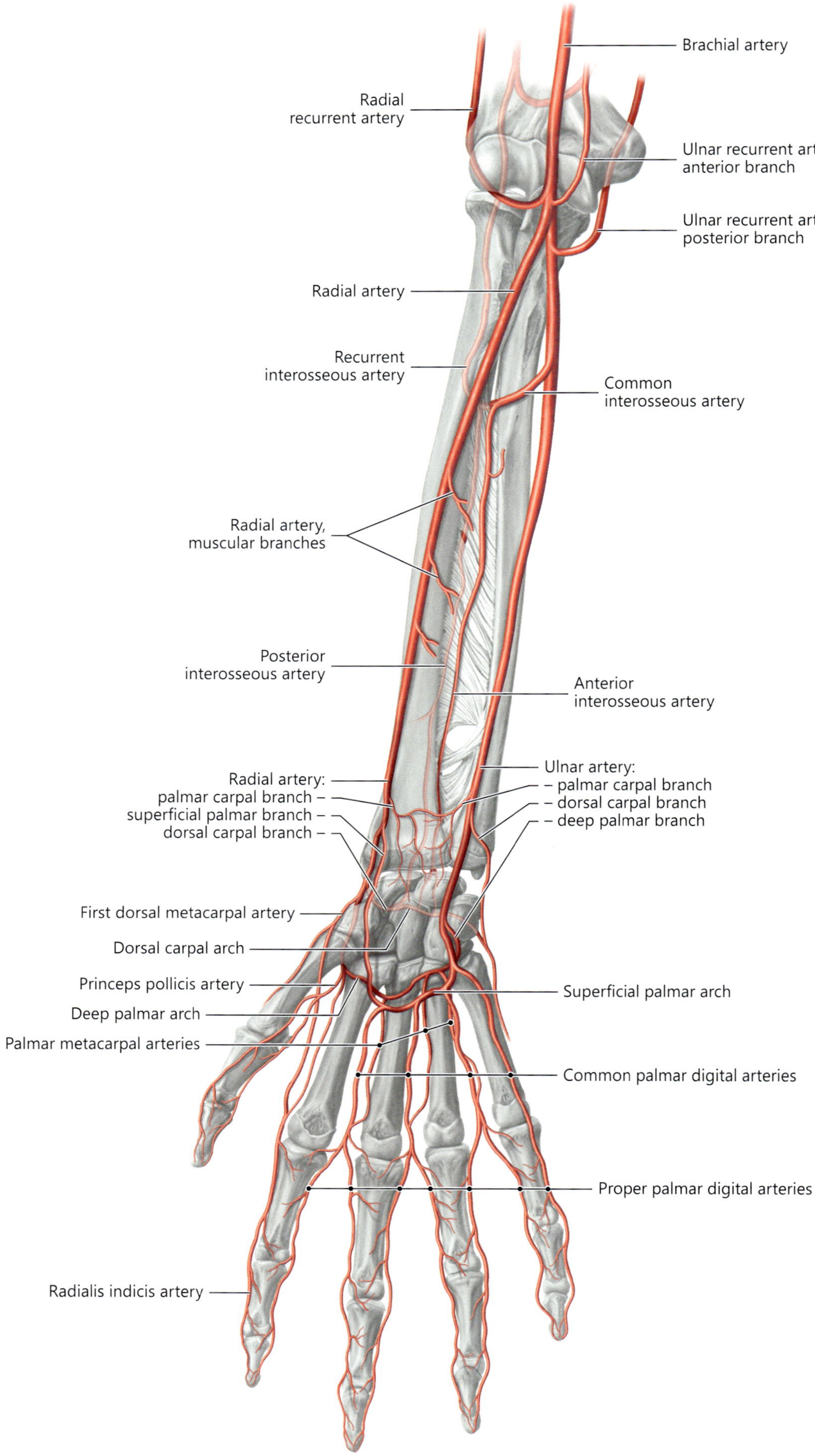

Representation, projected onto the skeleton, of the **radial and ulnar arteries**, their collateral branches, the palmar arches and the dorsal carpal arch, from which the most part of the arterial branches of the hand originates. Right limb, anterior view.

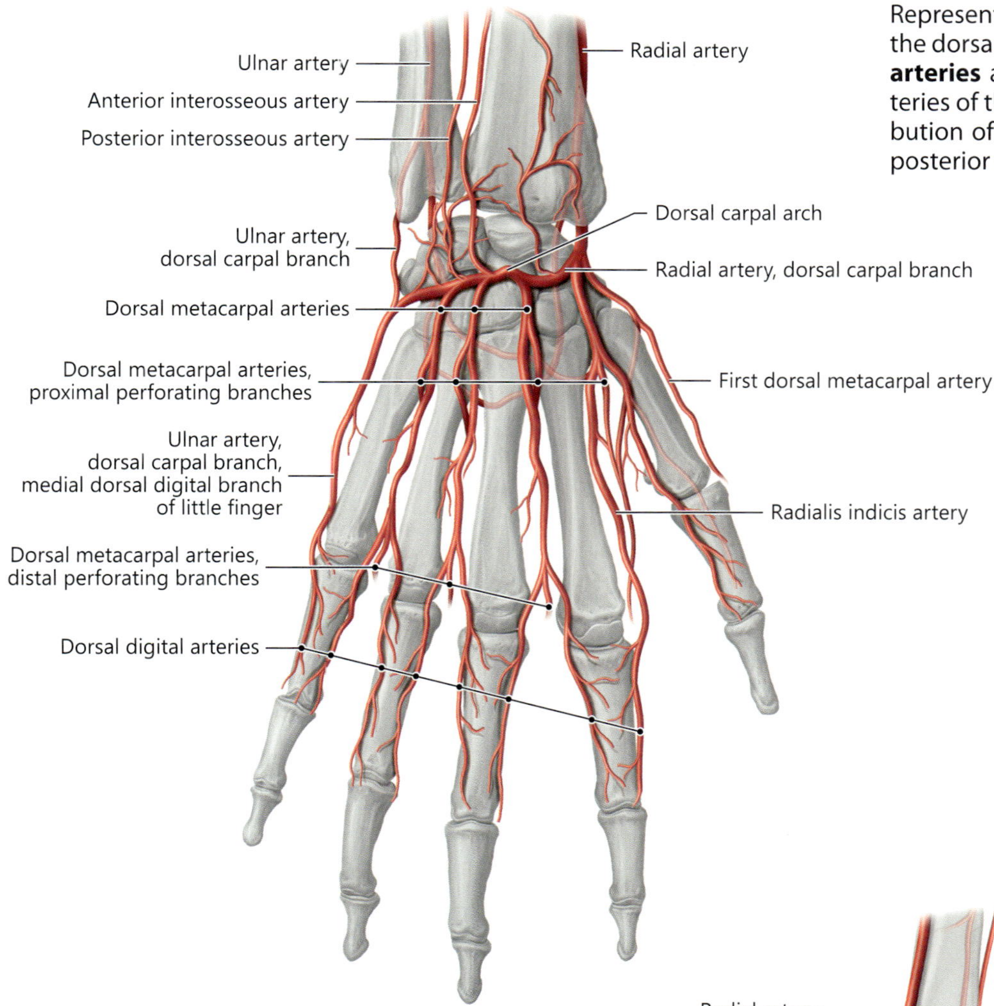

Representation, projected onto the skeleton, of the dorsal carpal arch, of the **dorsal metacarpal arteries** and of their anastomoses with the arteries of the palmar region. One notes the distribution of the dorsal digital arteries. Right limb, posterior view.

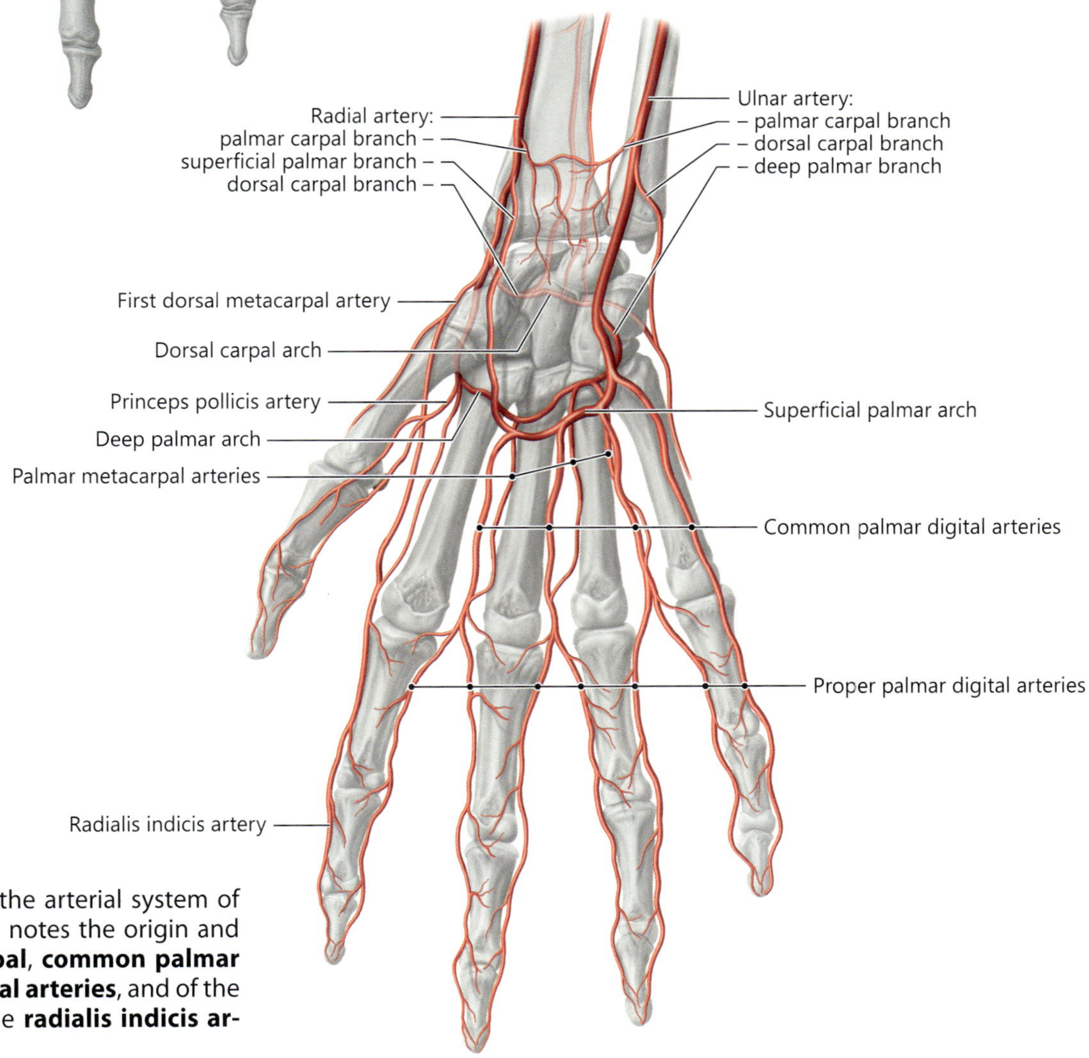

Projection onto the skeleton of the arterial system of the palm of the right hand. One notes the origin and course of the **palmar metacarpal**, **common palmar digital** and **proper palmar digital arteries**, and of the **princeps pollicis artery** and the **radialis indicis artery**. Right limb, anterior view.

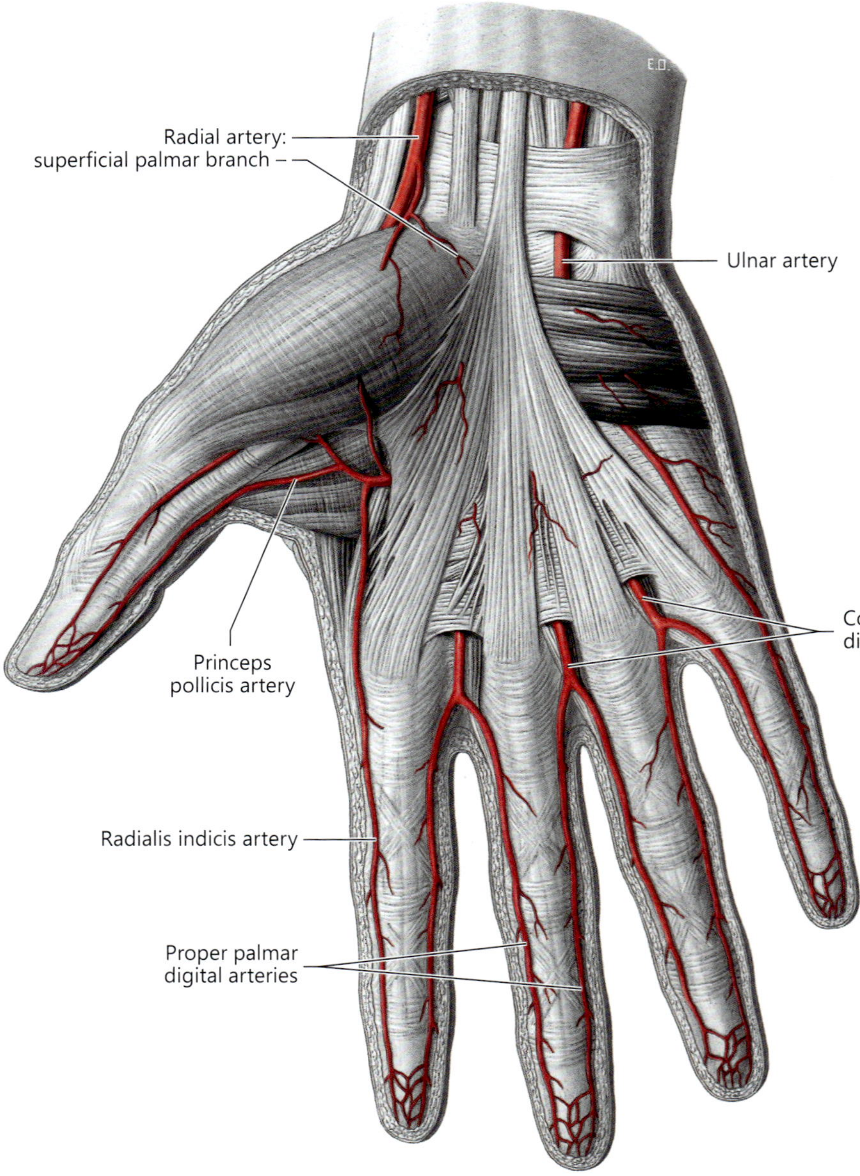

Representation of the **palmar digital arteries**, the **princeps pollicis artery** and the **radialis indicis artery**. The palmar aponeurosis is maintained in place. Right limb, anterior view.

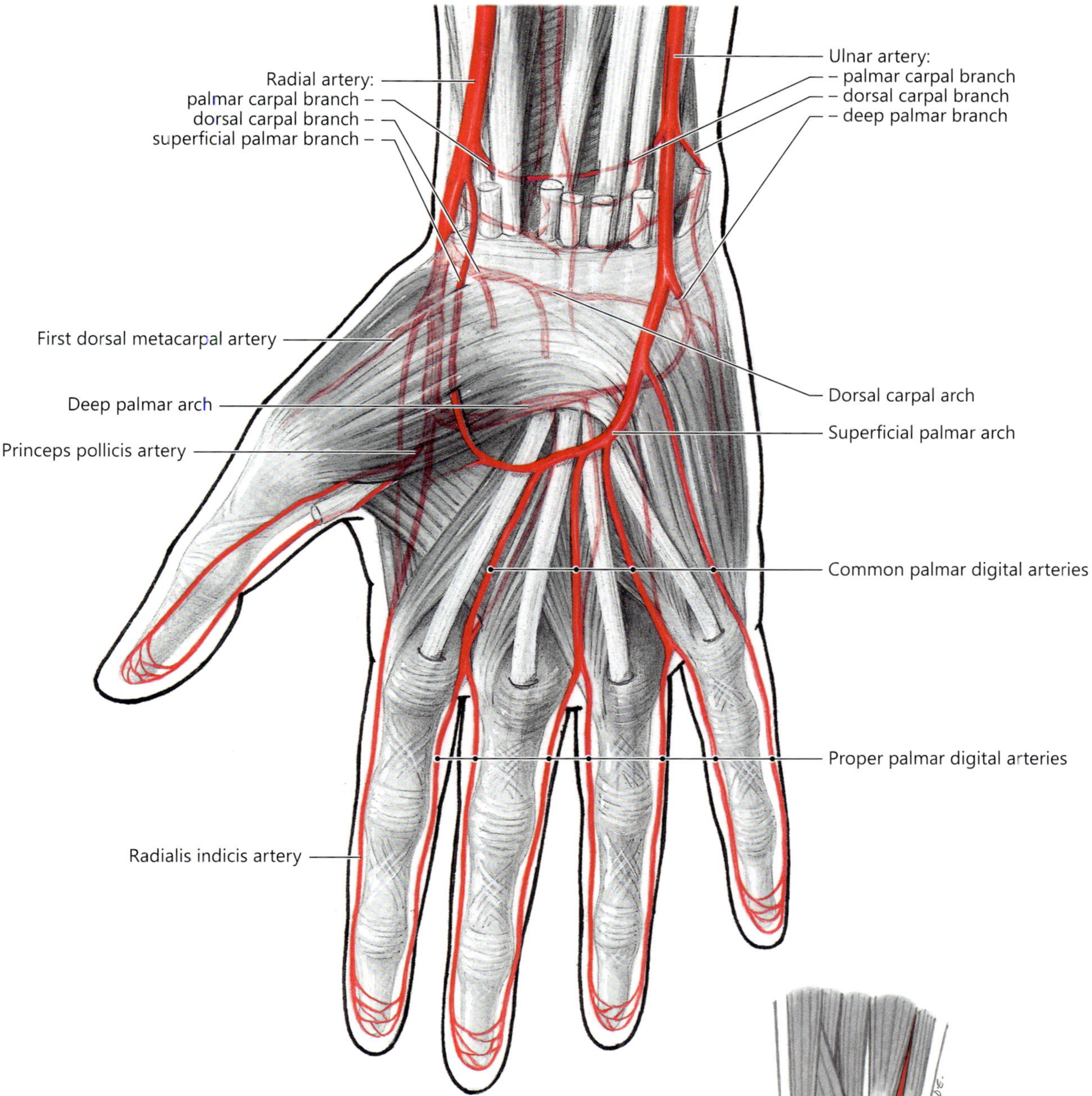

Arterial system of the palm of the right **hand**.

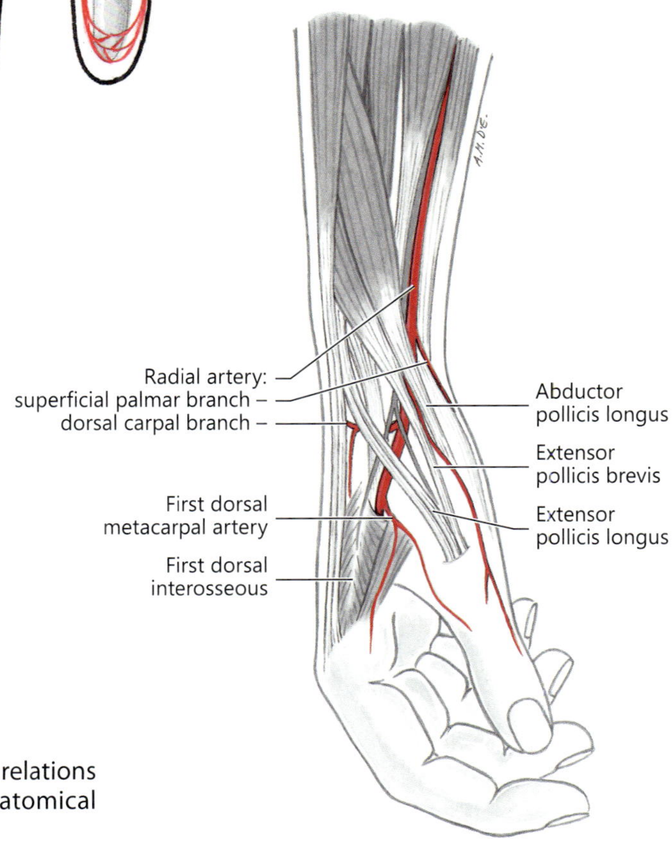

Course and branches of the **radial artery in the wrist**. The relations of the radial artery with the tendons that delimit the anatomical snuffbox are shown.

Summary table of the arteries of the upper limb

ARTERY	*ORIGIN*	*BRANCHES*	*TERRITORIES OF SUPPLY*
anterior branch of ulnar recurrent artery	Ulnar, immediately distal at the elbow joint		Brachialis and pronator teres
anterior circumflex humeral	Axillary		Long head of the biceps brachii and muscles of the arm, deltoid and coracobrachialis, tendon of the pectoralis major, head of the humerus and glenohumeral joint
anterior interosseous (volar interosseous)	Common interosseous	Median artery	Deep anterior part of the forearm, radius and ulna
axillary	Continuation of the subclavian artery, in correspondence with the outer border of the first rib	*Proximal tract*: superior thoracic artery *Intermediate tract*: thoracoacromial and lateral thoracic arteries *Distal tract*: subscapular, anterior circumflex humeral and posterior circumflex humeral arteries; subscapular branches	Pectoralis major and pectoralis minor, muscles of the shoulder and the arm, acromion, sternoclavicular and glenohumeral joints, breast
brachial (humeral)	Continuation of the axillary artery, in correspondence with the tendon of the teres major	Profunda brachii, superior and inferior ulnar collateral, radial and ulnar arteries	Muscles of the upper limb, elbow joint
circumflex scapular	Subscapular		Long head of the triceps brachii, teres minor, teres major, deltoid and subscapularis, glenohumeral joint
common interosseous	Ulnar, immediately distal at the radial tuberosity	Anterior interosseous and posterior interosseous arteries	Forearm, radius and ulna
common palmar digital (volar digital)	Superficial palmar arch	Proper palmar digital arteries	Digits of the hand
deep palmar arch	Radial	Palmar metacarpal arteries and perforating branches; anastomosis with the deep palmar branch of the ulnar artery	Palmar interossei and dorsal interossei, and bases of the metacarpals
dorsal carpal arch	Dorsal carpal branch of the radial artery		Dorsum of the hand
dorsal digital	Dorsal metacarpal arteries and the first dorsal metacarpal artery		Dorsal surface of the fingers
dorsal metacarpal (including the first dorsal metacarpal)	Dorsal carpal branch of the radial artery, dorsal carpal arch or radial artery	Dorsal digital arteries and proximal and distal perforating branches	Dorsal surface of the fingers
dorsal scapular (deep branch)	Thyrocervical trunk, transverse cervical artery or subclavian artery		Levator scapulae, trapezius, rhomboid major and rhomboid minor
humeral nutrient	Profunda brachii artery, at approximately halfway along the arm		Humerus
inferior ulnar collateral (anastomotica magna)	Brachial, at approximately 5 cm from the elbow		Triceps brachii and pronator teres, elbow joint
medial collateral	Profunda brachii artery		Triceps brachii and anconeus, elbow joint
median artery	Anterior interosseous		Accompanies and supplies the median nerve up to the palm of the hand
nutrient artery of the radius	Radial		Radius
nutrient artery of ulna	Ulnar		Ulna
palmar metacarpal (palmar interosseous)	Deep palmar arch		Interossei and lumbricals, metacarpals
posterior branch of ulnar recurrent artery	Ulnar, more distal with respect to the origin of the anterior branch		Elbow joint and adjacent muscles and skin

continued

Summary table of the arteries of the upper limb			(continued)
ARTERY	***ORIGIN***	***BRANCHES***	***TERRITORIES OF SUPPLY***
posterior circumflex humeral	Axillary, in correspondence with the distal border of the subscapularis		Deltoid, teres minor and triceps brachii, glenohumeral joint and neck of the humerus
posterior interosseous (dorsal interosseous)	Common interosseous	Recurrent interosseous artery and perforating branch	Deep posterior part of the forearm
princeps pollicis	Radial		Lateral and medial surfaces of the thumb, dorsal interossei
profunda brachii (deep artery of arm)	Brachial	Humeral nutrient, medial collateral and radial collateral arteries; deltoid branch	Deltoid and muscles of the arm, including the triceps brachii, humerus and elbow joint
proper palmar digital (collateral digital)	Common palmar digital		Lateral surfaces of each finger, nail matrix
radial	Brachial	Radial recurrent artery, nutrient artery of radius, princeps pollicis and radialis indicis arteries; muscular, palmar carpal, superficial palmar and dorsal carpal branches; deep palmar arch	Muscles of the forearm and of the hand, radius, skin of the dorsum of the hand and of the palmar surface of the thumb, lateral surface of the index finger, carpal joints
radial collateral	Profunda brachii artery		Triceps brachii, brachioradialis and brachialis, elbow joint
radial recurrent	Radial, immediately distal at the elbow		Supinator, brachioradialis and brachialis, elbow joint
radialis indicis	Radial		Medial surface of the index finger
recurrent interosseous	Posterior interosseous		Posterior surface of the elbow joint
subscapular	Axillary	Circumflex scapular and thoracodorsal arteries	Scapular region and glenohumeral joint
superficial palmar arch	Ulnar	Common palmar digital arteries	Palm of the hand
superior ulnar collateral (inferior profunda)	Brachial, in the distal half of the arm		Triceps brachii, elbow joint
suprascapular (transverse scapular)	Thyrocervical trunk	Acromial branch	Skin of the thorax, clavicle and scapula, skin above the acromion, acromioclavicular and glenohumeral joints, supraspinatus and infraspinatus
thoracoacromial	Axillary	Acromial, clavicular, deltoid and pectoral branches; acromial anastomosis	Pectoralis major and pectoralis minor, thorax, deltoid and subclavius, sternoclavicular joint and acromion
ulnar	Brachial, immediately distal at the elbow	Ulnar recurrent artery, nutrient artery of ulna and common interosseous artery; dorsal carpal, palmar carpal and deep palmar branches; articular network of the elbow; superficial palmar arch	Forearm, wrist and hand
ulnar recurrent	Ulnar	Anterior (anterior branch of ulnar recurrent artery) and posterior (posterior branch of ulnar recurrent artery) branches	Brachialis and pronator teres, elbow joint and adjacent skin

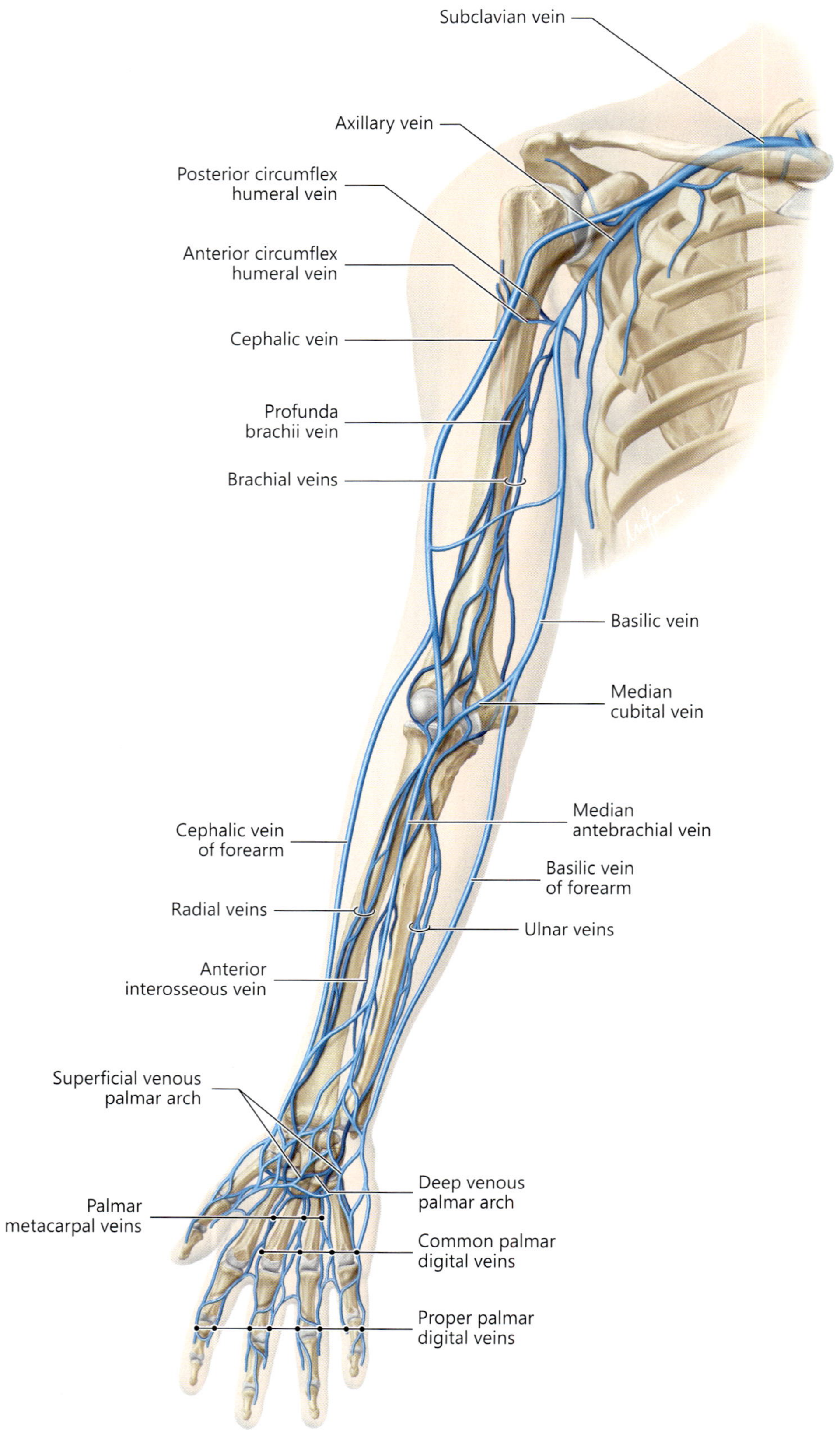

Schematic representation of the principal **superficial and deep veins of the** right **upper limb**, anterior view.

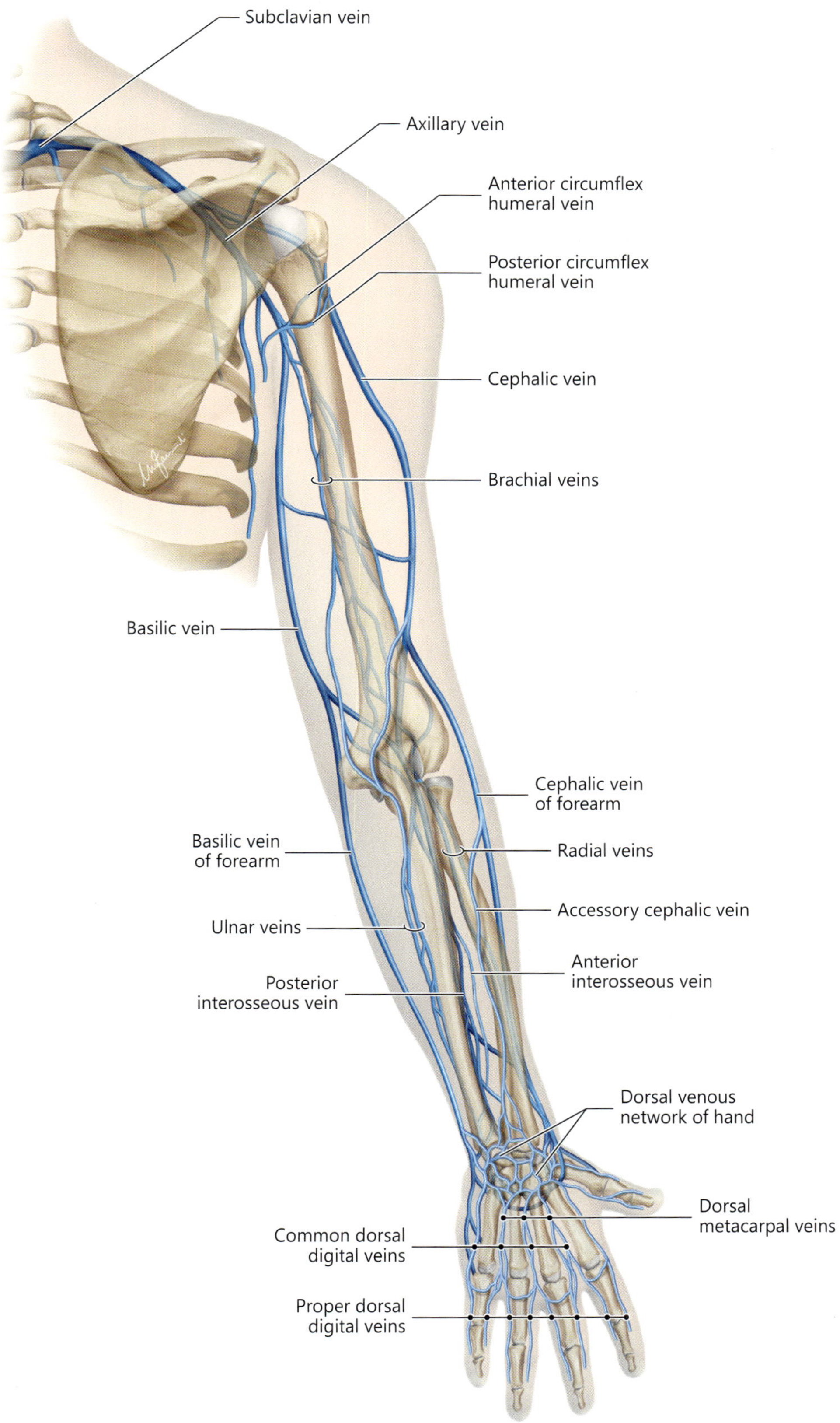

Schematic representation of the principal **superficial and deep veins of the** right **upper limb**, posterior view.

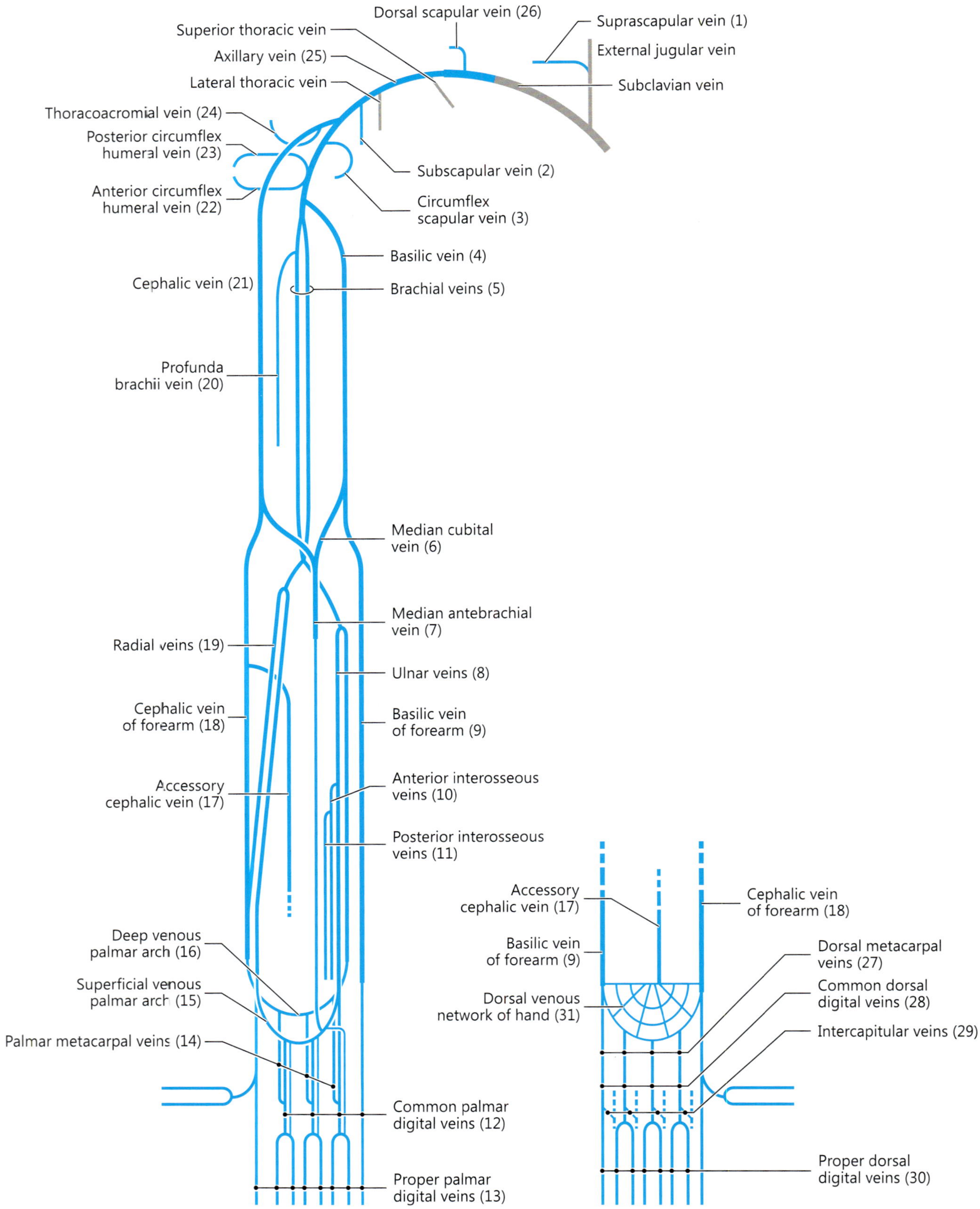

Venous tree of the upper limb.

Veins of the upper limb: territories of drainage		
	VEINS	**TERRITORIES OF DRAINAGE**
1	Suprascapular vein	Glenohumeral joint, scapula, clavicle, and contiguous muscles and skin
2	Subscapular vein	Inferior part of the scapular region
3	Circumflex scapular vein	Venous branches from the infraspinous fossa
4	Basilic vein	Cutaneous venous branches of the medial surface of the arm, and median cubital vein
5	Brachial veins	Ulnar and radial veins
6	Median cubital vein	Cephalic vein of the forearm
7	Median antebrachial vein	Superficial venous palmar arch
8	Ulnar veins	Deep venous palmar arch, superficial venous branches in correspondence with the wrist and anterior interosseous veins
9	Basilic vein of forearm	Dorsal venous network of the hand and cutaneous venous branches from the dorsal and volar surfaces of the forearm
10	Anterior interosseous veins	Anterior part of the forearm and posterior interosseous veins
11	Posterior interosseous veins	Posterior part of the forearm
12	Common palmar digital veins	Digits of the hand
13	Proper palmar digital veins	Digits of the hand
14	Palmar metacarpal veins	Palmar metacarpal region
15	Superficial venous palmar arch	Proper and common palmar digital veins
16	Deep venous palmar arch	Palmar metacarpal veins
17	Accessory cephalic vein	Dorsal venous network of the hand
18	Cephalic vein of forearm	Dorsal venous network of the hand and cutaneous venous branches from the volar and dorsal surfaces of the forearm
19	Radial veins	Dorsal metacarpal veins
20	Profunda brachii vein	Deep regions of the arm
21	Cephalic vein	Venous branches from the lateral parts of the arm and shoulder, thoracoacromial and accessory cephalic veins
22	Anterior circumflex humeral vein	Venous branches from the muscles and from the skin of the anterior surface of the arm
23	Posterior circumflex humeral vein	Venous branches from the muscles adjacent to the glenohumeral joint
24	Thoracoacromial vein	Acromion and coracoid process of the scapula, sternoclavicular joint, venous branches from the deltoid, subclavius, pectoralis major and pectoralis minor
25	Axillary vein	Basilic, brachial, cephalic and profunda brachii veins
26	Dorsal scapular vein	Posterior surface of the scapula
27	Dorsal metacarpal veins	Dorsal metacarpal region
28	Common dorsal digital veins	Digits of the hand
29	Intercapitular veins	Connect the dorsal digital and proper palmar veins
30	Proper dorsal digital veins	Digits of the hand
31	Dorsal venous network of hand	Digital and dorsal metacarpal veins

Representation of the **superficial veins of the** right **upper limb**. In the image on the left, one notes the course of the cephalic vein in the right arm and in the deltopectoral sulcus, where it ascends up to its opening into the axillary vein.

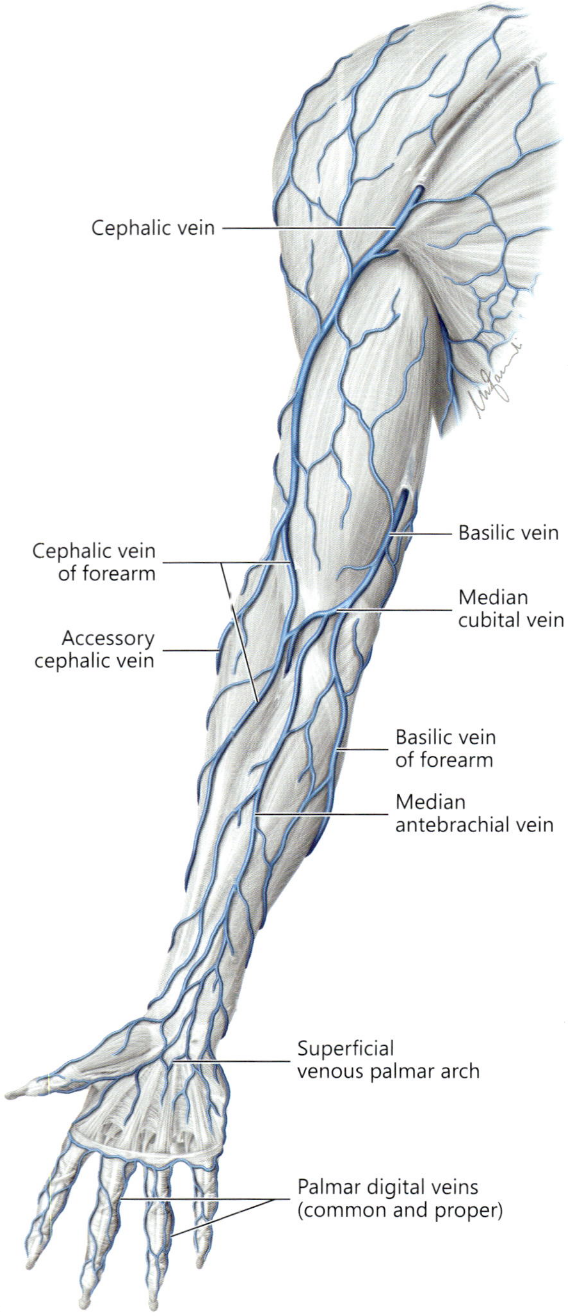

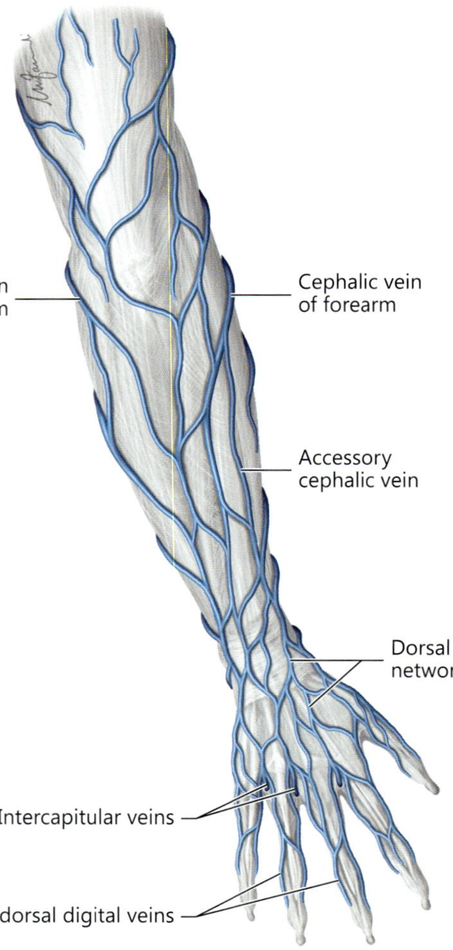

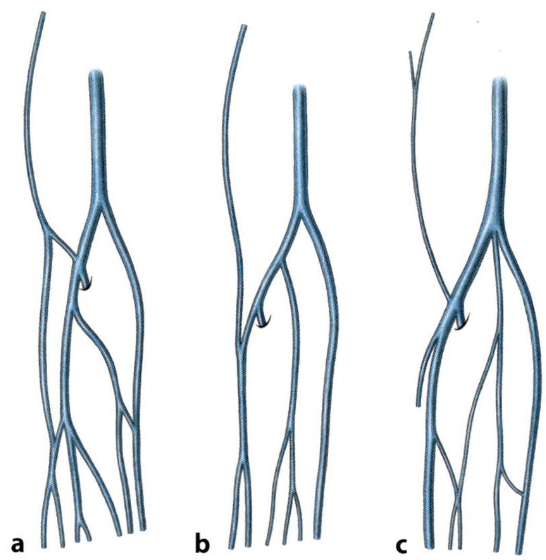

Principal types of distribution of the **superficial veins of the upper limb in the anterior region of the elbow**. In **b** the most frequent distribution is represented, **a** and **c** show less common systems.

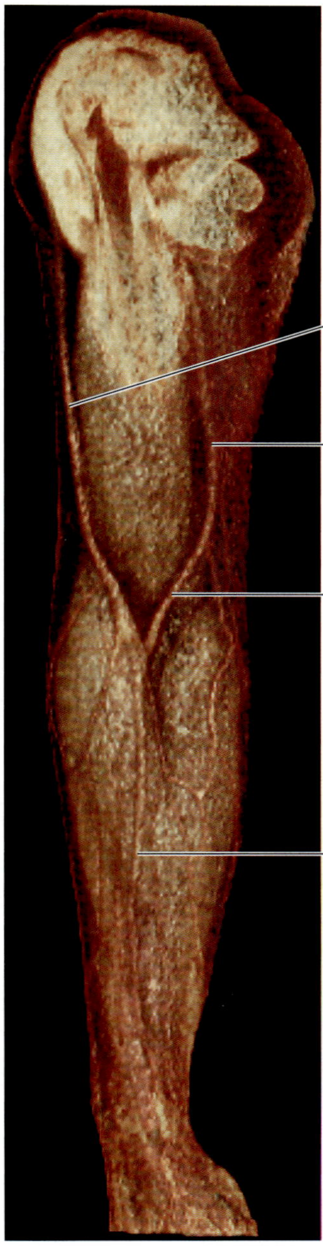

a

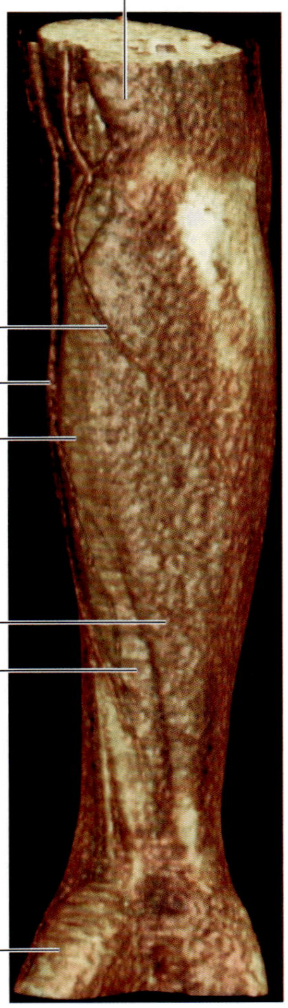

b

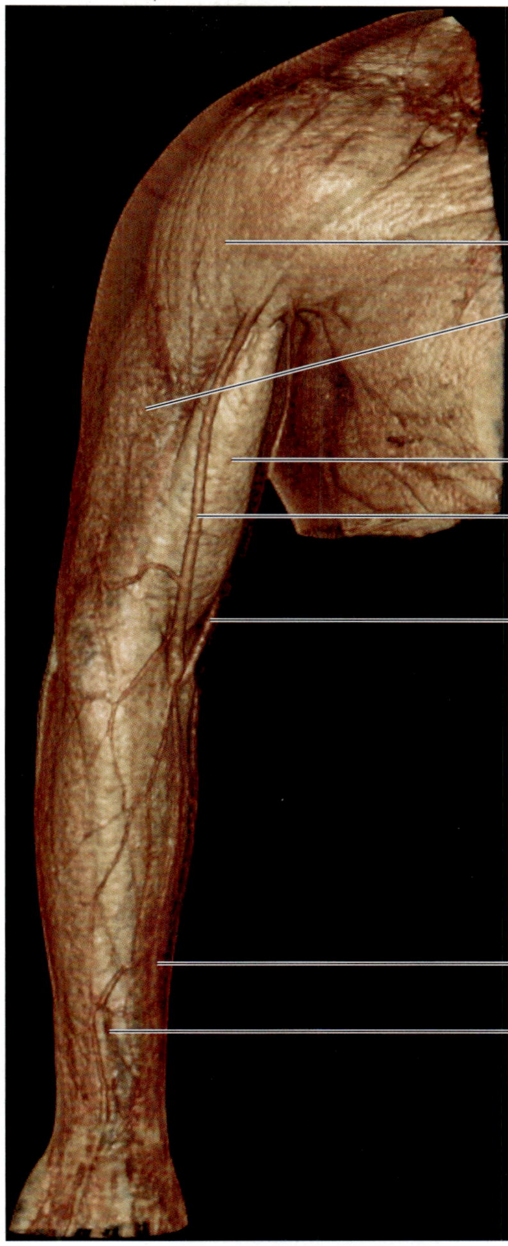

Three-dimensional image of a living human body, obtained by CT scan using a 3D volume rendering technique, showing the **muscles** and the **venous circulation of the** right **arm**; in pronation.

➤

Three-dimensional images of a living human body, obtained by CT scan using a 3D volume rendering technique, showing the **muscles** and the **venous circulation of the** left **arm** (**a**) and of the right **forearm** (**b**); in supination.

Three-dimensional image of a living human body, obtained by CT scan using a 3D volume rendering technique, showing the **muscles** and the **venous circulation of the** right **forearm and hand**.

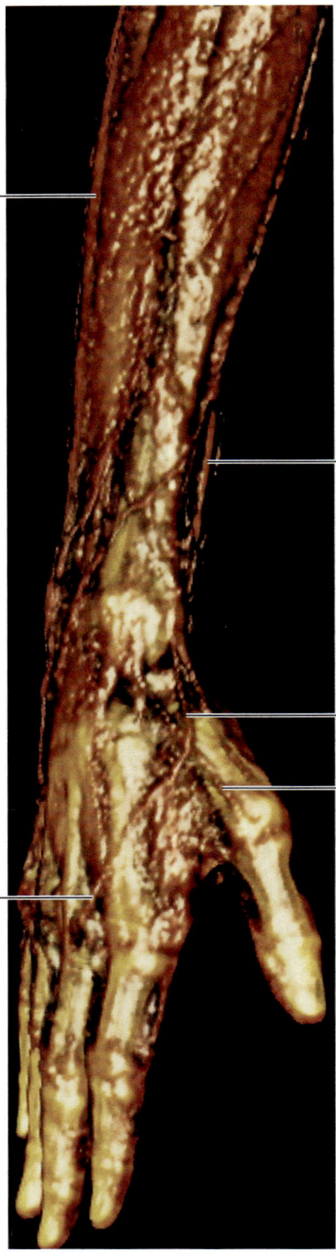

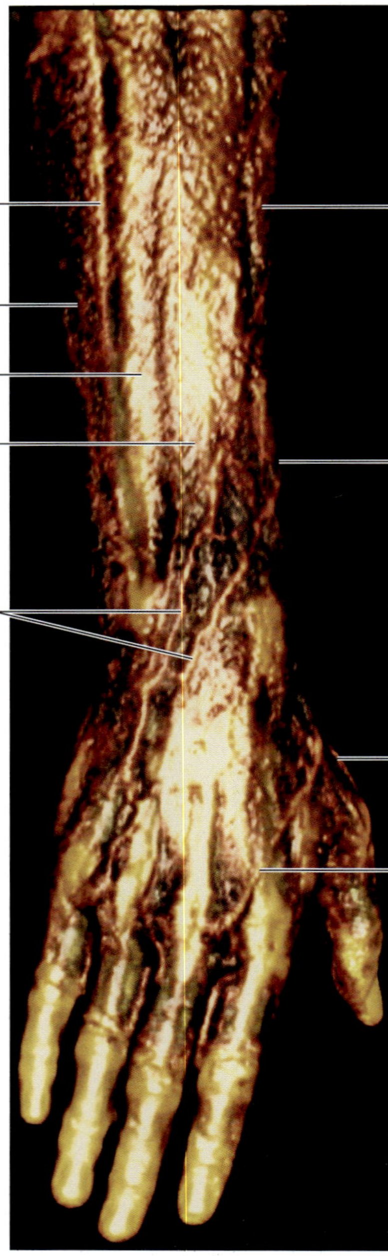

Three-dimensional image of a living human body, obtained by TC scan using a 3D volume rendering technique, showing the **muscles** and the **venous circulation of the** right **forearm**; in pronation.

Summary table of the veins of the upper limb			
VEIN	***ORIGIN/LOCATION***	***DRAINAGE***	***CONFLUENCE***
accessory cephalic	Radial side of the forearm	Dorsal venous network of the hand	Cephalic vein, at the level of the elbow
anterior circumflex humeral	Satellite of the anterior circumflex humeral artery	Venous branches from the muscles and from the skin of the anterior surface of the arm	Axillary vein
anterior interosseous	Satellites of the anterior interosseous artery	Anterior part of the forearm; connect with the ulnar and radial veins and receive the posterior interosseous veins	Ulnar veins
axillary	From the inferior border of the teres major to the external border of the first rib	Basilic and brachial veins, from whose junction originate the axillary vein, cephalic vein and profunda brachii vein	Subclavian vein, in correspondence with the external border of the first rib
basilic	Continuation of the basilic vein of the forearm, along the medial border of the biceps brachii	Median cubital vein, cutaneous venous branches from the medial surface of the arm; anastomosis with the cephalic vein and the brachial veins	Connects to the brachial veins to form the axillary vein
basilic vein of forearm	From the dorsal surface of the forearm, on the ulnar side, to the volar surface of the forearm in its proximal third	Dorsal venous network of the hand and cutaneous venous branches from the dorsal and volar surfaces of the forearm	Continues in the basilic vein
brachial	Positioned on each side of the brachial artery	Radial and ulnar veins	Connects to the basilic vein to form the axillary vein
cephalic vein	Ascends along the lateral border of the biceps brachii and then in the deltopectoral sulcus; gives rise to the median cubital vein in the anterior region of the elbow	Venous branches from the lateral parts of the arm and shoulder, thoracoacromial and accessory cephalic veins	Axillary vein, just behind the clavicle
cephalic vein of the forearm	Ascends on the radial side of the volar surface of the forearm	Dorsal venous network of the hand and cutaneous venous branches of the volar and dorsal surfaces of the forearm	Continues in the cephalic vein
circumflex scapular	Satellite of the circumflex scapular artery	Venous branches from the infraspinous fossa	Connects to the thoracodorsal vein, forming the subscapular vein
deep venous palmar arch	Palm of the hand	Palmar metacarpal veins	Dorsal venous network of the hand and superficial veins of the volar surface of the forearm
dorsal metacarpal	The last four fingers	Dorsal metacarpal region	Dorsal venous network of the hand
dorsal scapular	Satellite of the dorsal scapular artery	Posterior surface of the scapula	Subclavian vein
dorsal venous network of hand	Principal return system of the veins of the hand, positioned superficially on the dorsum of the hand	Digital and dorsal metacarpal veins	Cephalic (radial side) and basilic (ulnar side) veins; there is often a third drainage pathway in the accessory cephalic vein
intercapitular	Positioned at the level of the heads of the metacarpals	Connect the dorsal digital and proper palmar veins	Dorsal digital and proper palmar veins
median antebrachial (median vein of forearm)	From the base of the thumb to halfway of the volar surface of the forearm	Superficial venous palmar arch	Basilic vein, cephalic vein or median cubital vein
median cubital	Runs obliquely through the anterior region of the elbow	Cephalic vein of the forearm	Basilic vein
palmar metacarpal	Satellites of the palmar metacarpal arteries	Palmar metacarpal region	Deep venous palmar arch
posterior circumflex humeral	Satellite of the posterior circumflex humeral artery	Venous branches from the muscles adjacent to the glenohumeral joint; anastomosis with the anterior circumflex humeral, thoracoacromial and suprascapular veins	Circumflex scapular vein

continued

Summary table of the veins of the upper limb *(continued)*

VEIN	ORIGIN/LOCATION	DRAINAGE	CONFLUENCE
posterior interosseous	Satellites of the posterior interosseous artery	Posterior part of the forearm	Anterior interosseous veins
profunda brachii	Satellite of the profunda brachii artery	Deep regions of the arm	Axillary vein
proper and common dorsal digital	Origin from the distal extremities of the fingers	Digits of the hand	Form the dorsal digital network of the hand
proper and common palmar digital	Origin from the distal extremities of the fingers	Digits of the hand	Form the superficial venous palmar arch
radial	Satellites of the radial artery	Dorsal metacarpal veins	Brachial veins, at the level of the elbow
subclavian	Extension of the axillary vein, from the middle point of the clavicle to the sternoclavicular joint	Superior thoracic region, upper limb and some structures of the head and of the neck	Connects to the internal jugular vein to form the brachiocephalic vein
subscapular	Satellite of the subscapular artery	Inferior part of the scapular region	Axillary vein
superficial venous palmar arch	Palm of the hand	Proper and common palmar digital veins	Dorsal venous network of the hand and superficial veins of the volar surface of the forearm
suprascapular	From the posterior surface of the scapula, passing through the suprascapular notch, and then running parallel to the clavicle	Glenohumeral joint, scapula, clavicle and contiguous muscles and skin	External jugular vein or subclavian vein
thoracoacromial (acromiothoracic)	Satellite of the thoracoacromial artery	Acromion and coracoid process of the scapula, sternoclavicular joint, venous branches from the deltoid, subclavius, pectoralis major and pectoralis minor	Cephalic vein
ulnar	Satellites of the ulnar artery	Deep venous palmar arch, superficial venous branches in correspondence with the wrist and anterior interosseous veins	Brachial veins

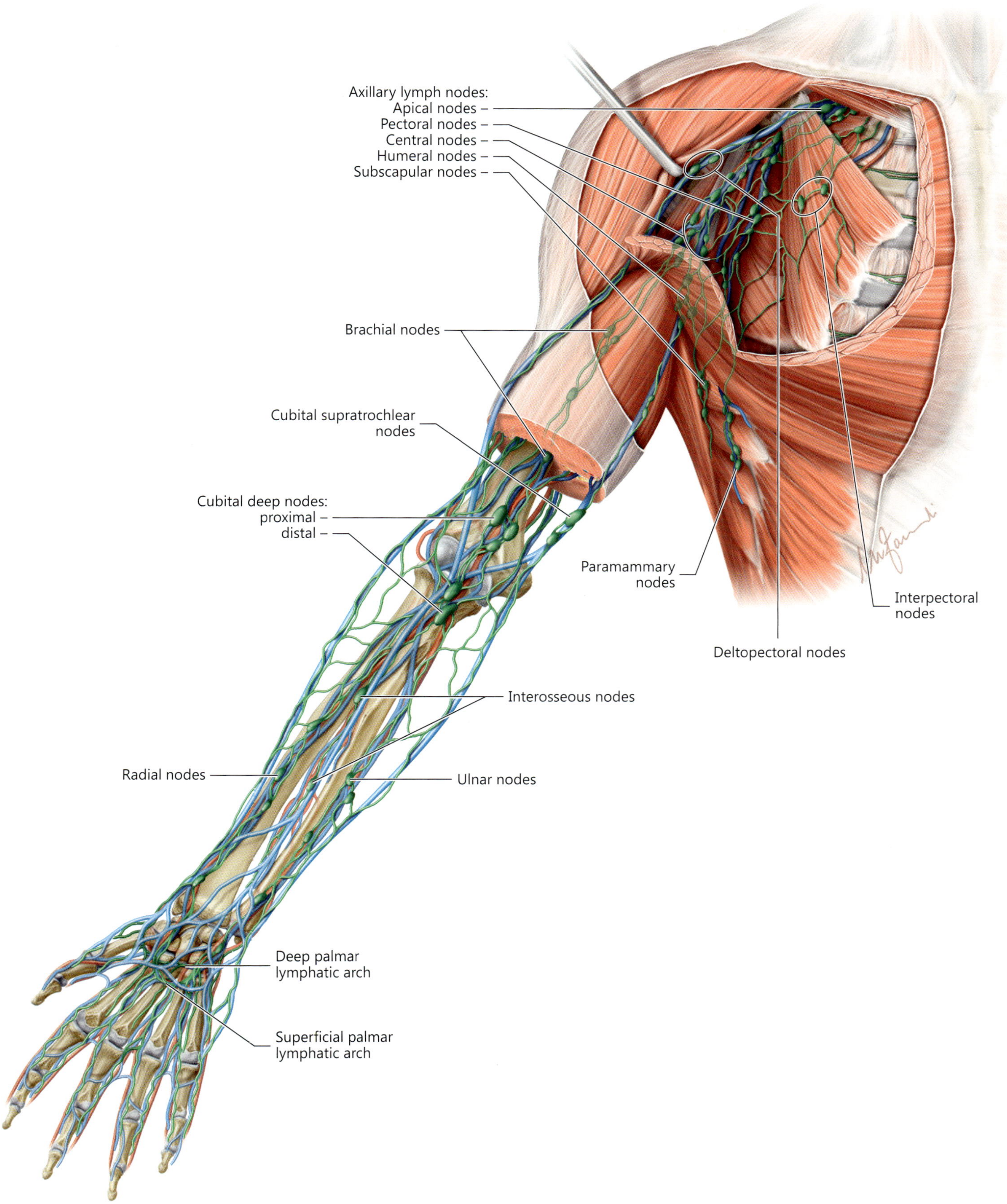

The lymphatic drainage of the upper limb converges to the **axillary lymph nodes**, which also drain an extensive homolateral superficial region of the trunk, and more than 75% of the lymph coming from the breast. Other lymph nodes of the upper limb, fewer in number and variable, are distributed along the course of the lymphatic collectors that, from the hand, forearm and arm, reach the axillary lymph nodes.

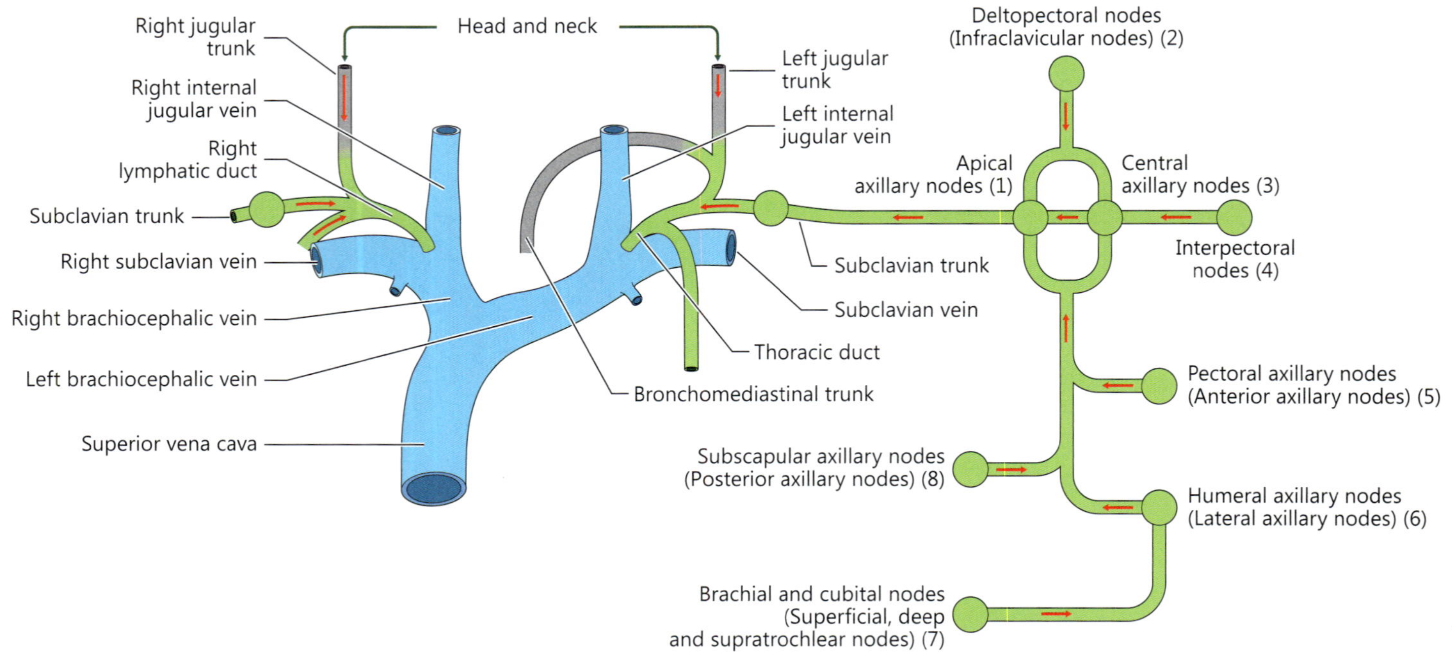

Schematic representation of the principal **lymphatic groups** and **lymphatic ducts**. In evidence, the lymph nodes, the ducts and their territories of drainage of the upper limb (correlate with the numbers in the table on the side).

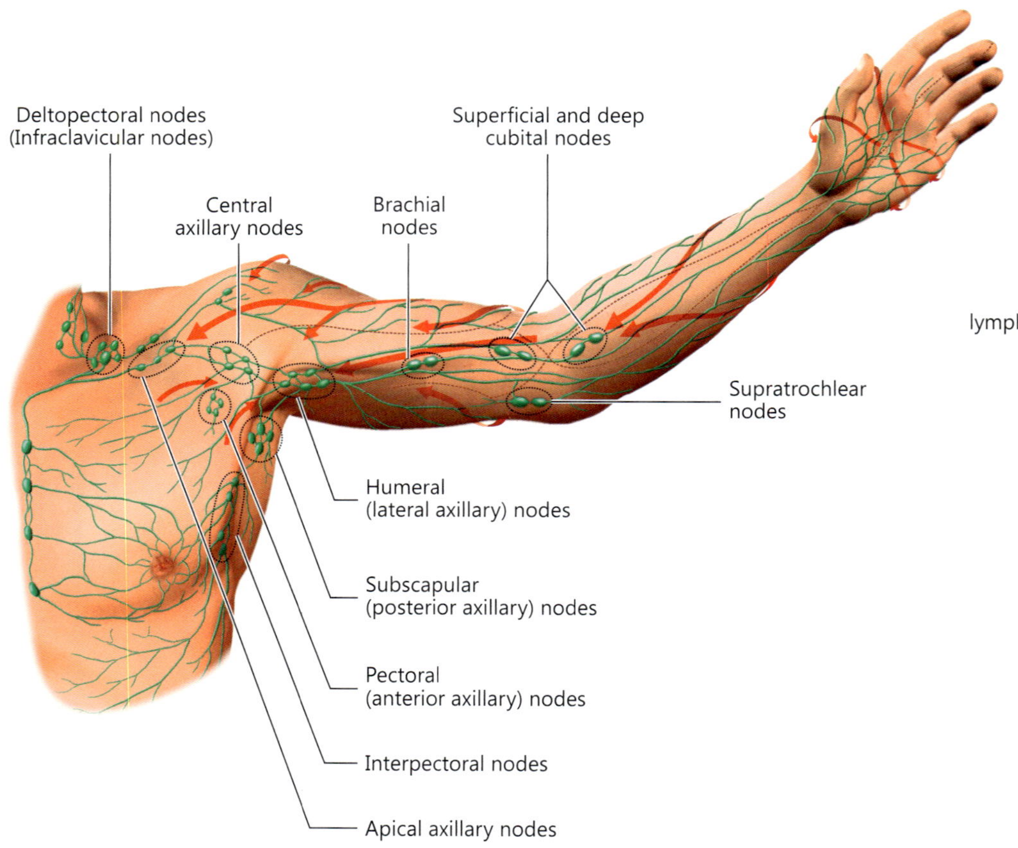

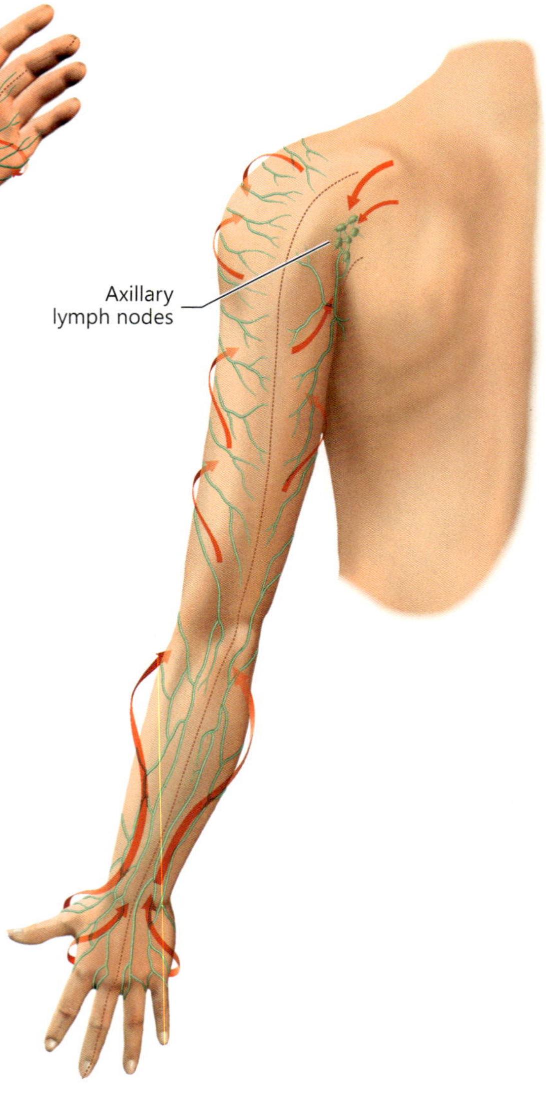

The **lymph nodes of the upper limb** are prevalently collected in the axillary lymphatic group. Fewer additional nodes, variable in numbers and small in size, mostly isolated, are distributed along the course of the collectors running from the hand, the forearm and the arm to the axillary lymph nodes. The paramammary lymph nodes, or inferior pectoral lymph nodes, are found along the lateral margin of the mammary gland.

Lymph nodes of the upper limb: territories of drainage

	LYMPH NODES	TERRITORIES OF DRAINAGE
1	Apical axillary nodes	From the central axillary, deltopectoral and interpectoral lymph nodes and, to a lesser degree, from the superior peripheral mammary region and from the lateral surface of the proximal part of the arm
2	Deltopectoral nodes (infraclavicular nodes)	From the upper limb through the lymphatic vessels that accompany the cephalic vein
3	Central axillary nodes	From the humeral, pectoral and subscapular axillary nodes and from the interpectoral nodes
4	Interpectoral nodes	From the anterior wall of the thorax and from the breast through the deep lymphatic vessels
5	Pectoral axillary nodes (anterior axillary nodes)	From the most part of the breast and from the anterior walls of the thorax and of the breast, up to the umbilicus
6	Humeral axillary nodes (lateral axillary nodes)	From the whole upper limb, except the lymphatic vessels that accompany the cephalic vein
7	Brachial and cubital nodes (superficial, deep and supratrochlear nodes)	From the hand and from the forearm
8	Subscapular axillary nodes (posterior axillary nodes)	From the posteroinferior part of the neck, from the scapular region and from the posterior walls of the thorax, and from the abdomen up to the iliac crest

Summary table of the lymph nodes of the upper limb

LYMPH NODES	GROUP	SITE	AFFERENTS	EFFERENTS
Anterior and posterior interosseous (sometimes present)		Deep	Collectors from the hand and from the forearm	Brachial collectors
Axillary		Deep		
– **apical**	Apical (infraclavicular)		Efferent collectors from the other axillary groups	Subclavian trunk
– **central**	Central		Efferent collectors from the other axillary groups	Apical axillary nodes
– **humeral** (lateral axillary	Lateral (brachial)		Superficial and deep collectors of the upper limb	Central and apical axillary nodes
– **pectoral** (anterior axillary)	Anterior (pectoral or thoracic)		Collectors from the teguments and from the muscles of the anterolateral thoracic wall, from the central and lateral mammary regions, from the skin and from the muscles of the supraumbilical region of the abdominal wall	Central and apical axillary nodes
– **subscapular nodes** (posterior axillary)	Posterior (subscapular)		Collectors from the skin and from the superficial muscles of the posterior walls of the thorax and the abdomen, from the scapular region and from the posteroinferior part of the neck	Central and apical axillary nodes
Brachial		Deep	Brachial collectors	Axillary lymph nodes
Cubital		Superficial and deep	Collectors from the hand and from the forearm	Brachial collectors
– **deep**		Deep		
– **supratrochlear** (epitrochlear)		Superficial		
– **superficial**		Superficial		
Deltopectoral (infraclavicular)		Deep	Collectors of the upper limb that accompany the cephalic vein	Axillary lymph nodes
Interpectoral (retropectoral)		Deep	Anterior thoracic wall and breast through the deep collectors	Axillary lymph nodes
Of the palm of the hand (sometimes present)		Superficial	Collectors from the hand	Brachial collectors
Radial (sometimes present)		Deep	Collectors from the hand and from the forearm	Brachial collectors

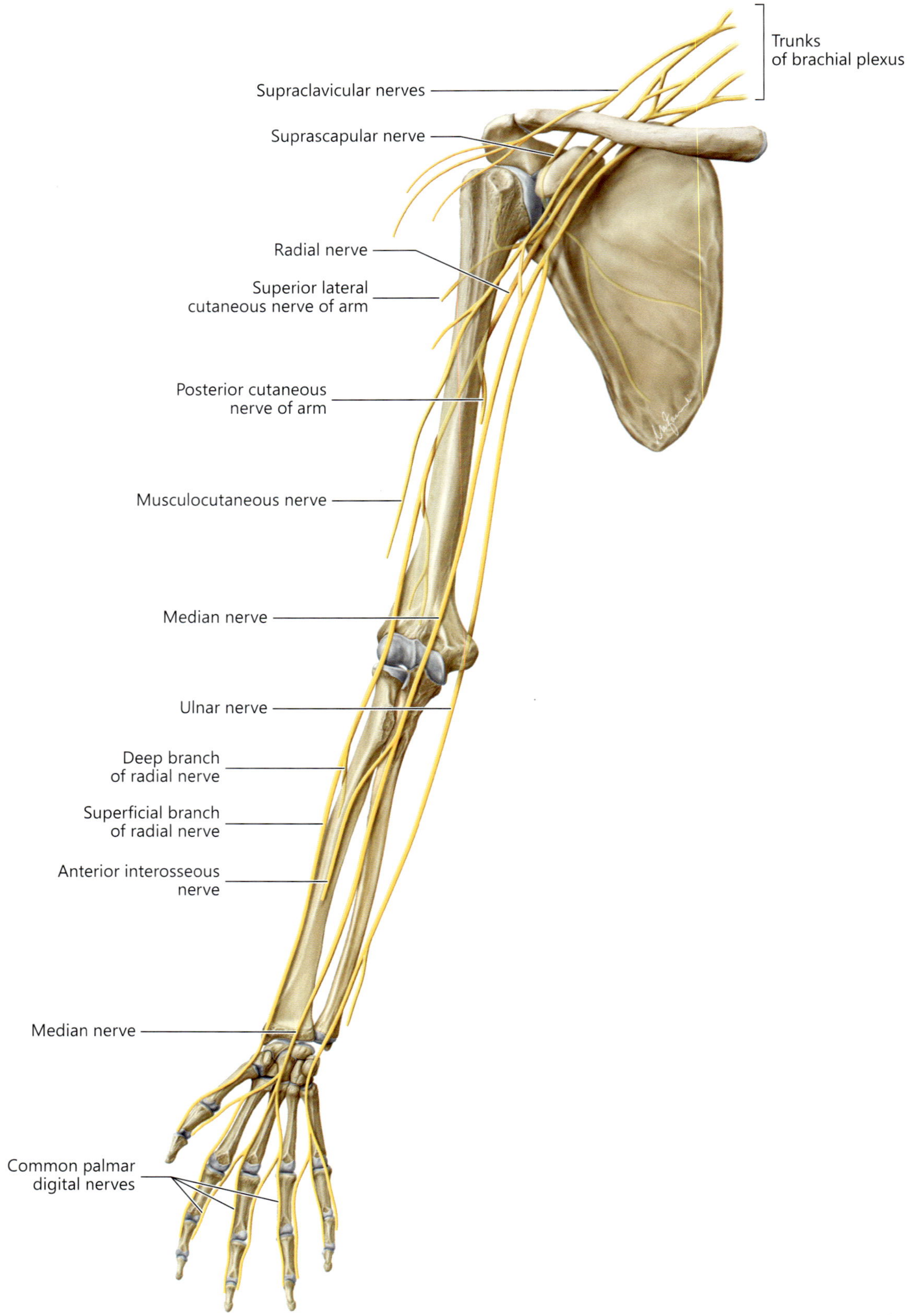

Schema of the **spinal nerves** of the right **upper limb**. Anterior view. The brachial plexus is formed by the anterior rami of the fifth, sixth, seventh and eighth cervical nerves and the first thoracic nerve. Even with frequent individual variations, the brachial plexus is characterized by a series of anastomoses and subdivisions of its nerves that form three trunks (primary), superior, medial and inferior, branching the cords (or secondary trunks), posterior, lateral and medial, from which originate the terminal branches, represented by the musculocutaneous, median and ulnar nerves, and medial cutaneous nerve of the forearm and medial cutaneous nerve of the arm.

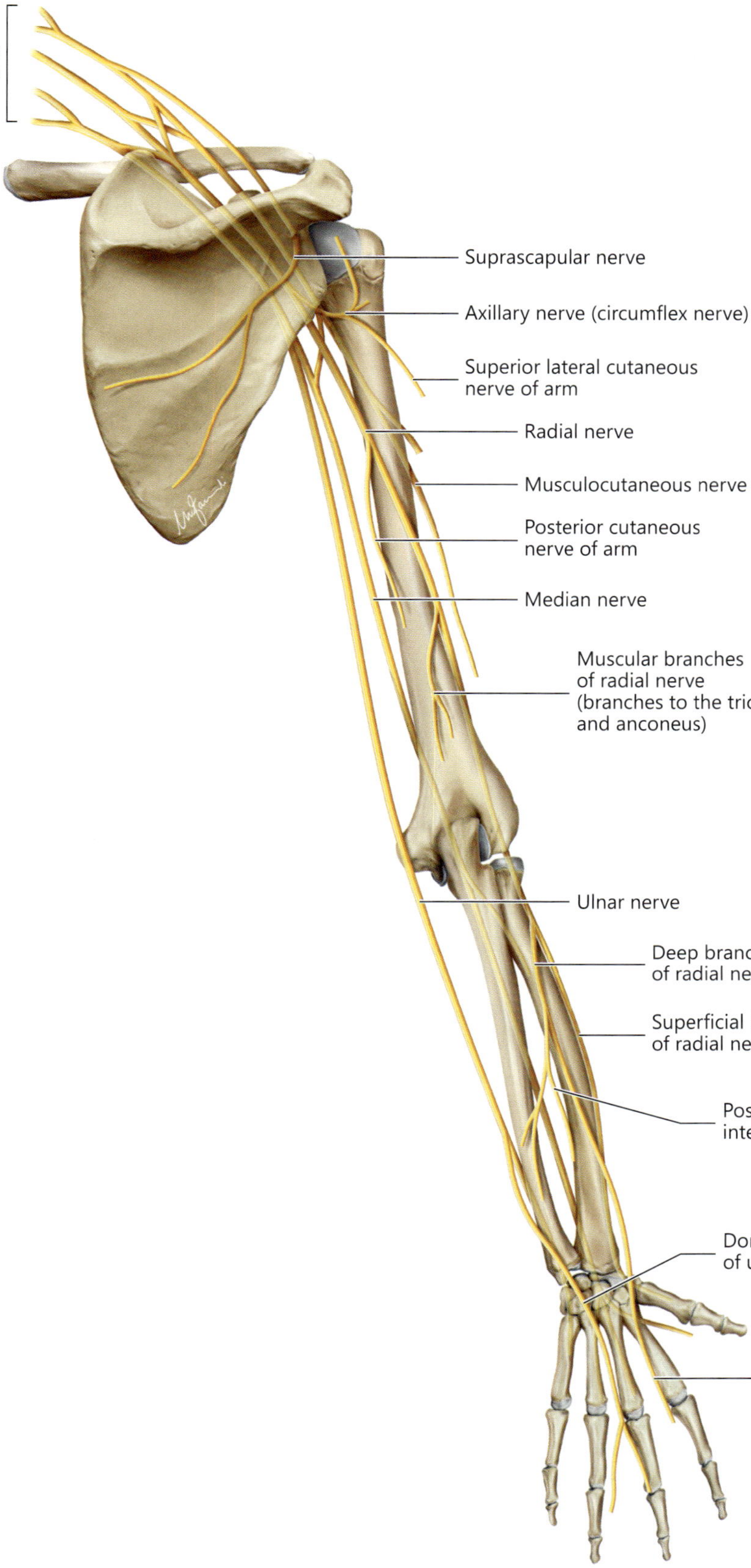

Schema of the **spinal nerves** of the right **upper limb**. Posterior view. The trunks (primary) of the brachial plexus are directed laterally towards the axilla and, just above or behind the clavicle, each of them divides into an anterior and a posterior branch. The posterior branches of the three primary trunks become a single trunk, called posterior cord; the anterior divisional branches of the superior and medial trunks fuse to form the lateral cord, while the anterior divisional branch of the inferior trunk remains independent and takes the name of medial cord.

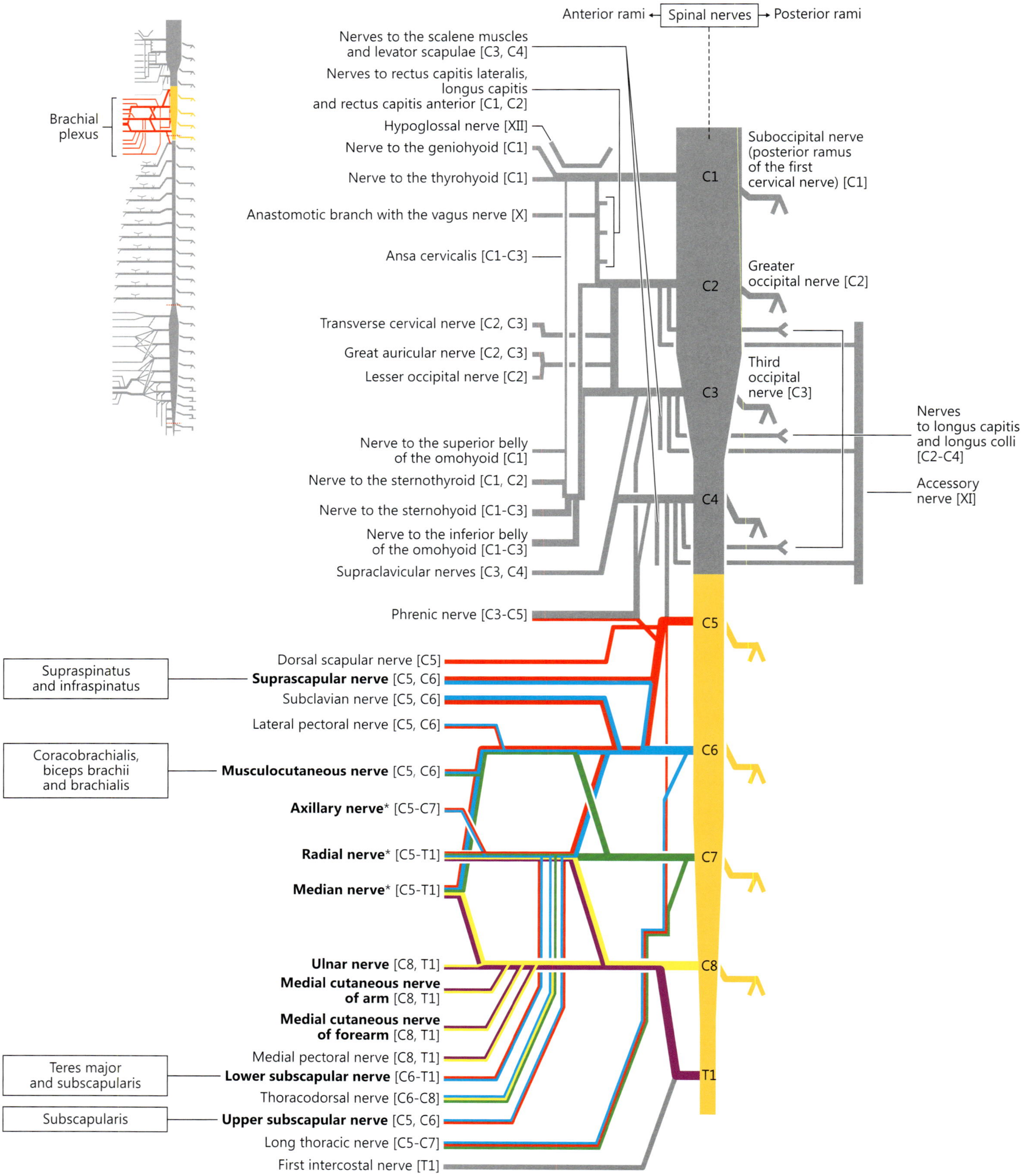

Schema of the superior, medial and inferior trunks of the **brachial plexus** and their anterior and posterior divisions, from which originate the lateral, medial and posterior cords. The terminal branches of the plexus are the radial, axillary, musculocutaneous, median and ulnar nerves, and medial cutaneous nerve of the arm and medial cutaneous nerve of the forearm; the collateral branches are the dorsal scapular, long thoracic, subscapular, thoracodorsal, subclavian, medial pectoral, lateral pectoral and suprascapular nerves.

* For the distribution territories ☞ table on pages 346-347.

Representation of the **brachial plexus** in relation to the skeleton of the vertebral column, the thorax and the upper limb.

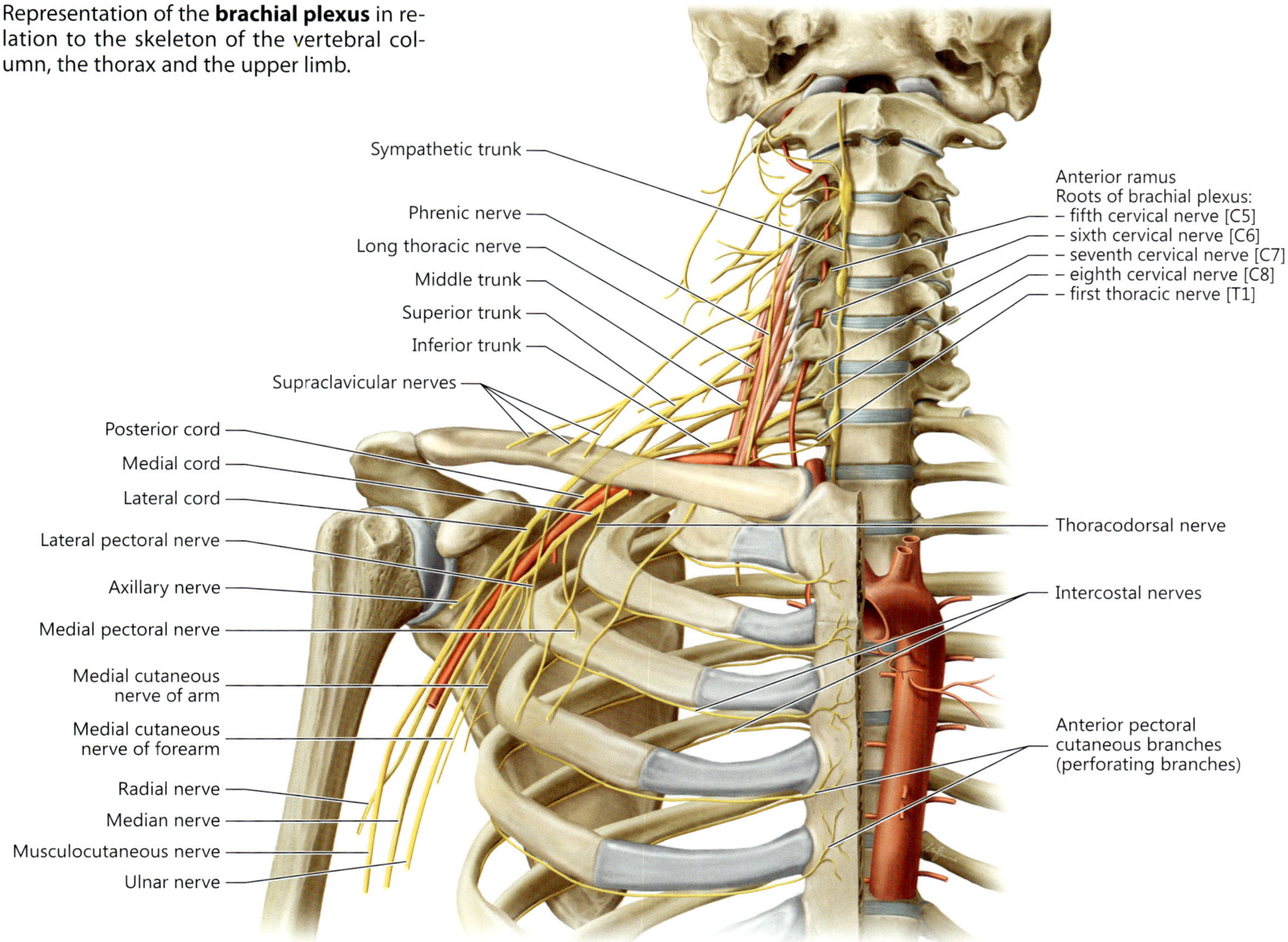

Schematical representation of the structure of the **brachial plexus**, with the formation of the trunks (superior, medial and inferior) and the cords. Note the position of the anterior (ant. b), posterior (post. b) and terminal (term b.) branches. The phrenic nerve [C3, C4, C5] belongs to the cervical plexus; the first intercostal nerve to the thoracic nerves (anterior ramus of the first thoracic nerve).

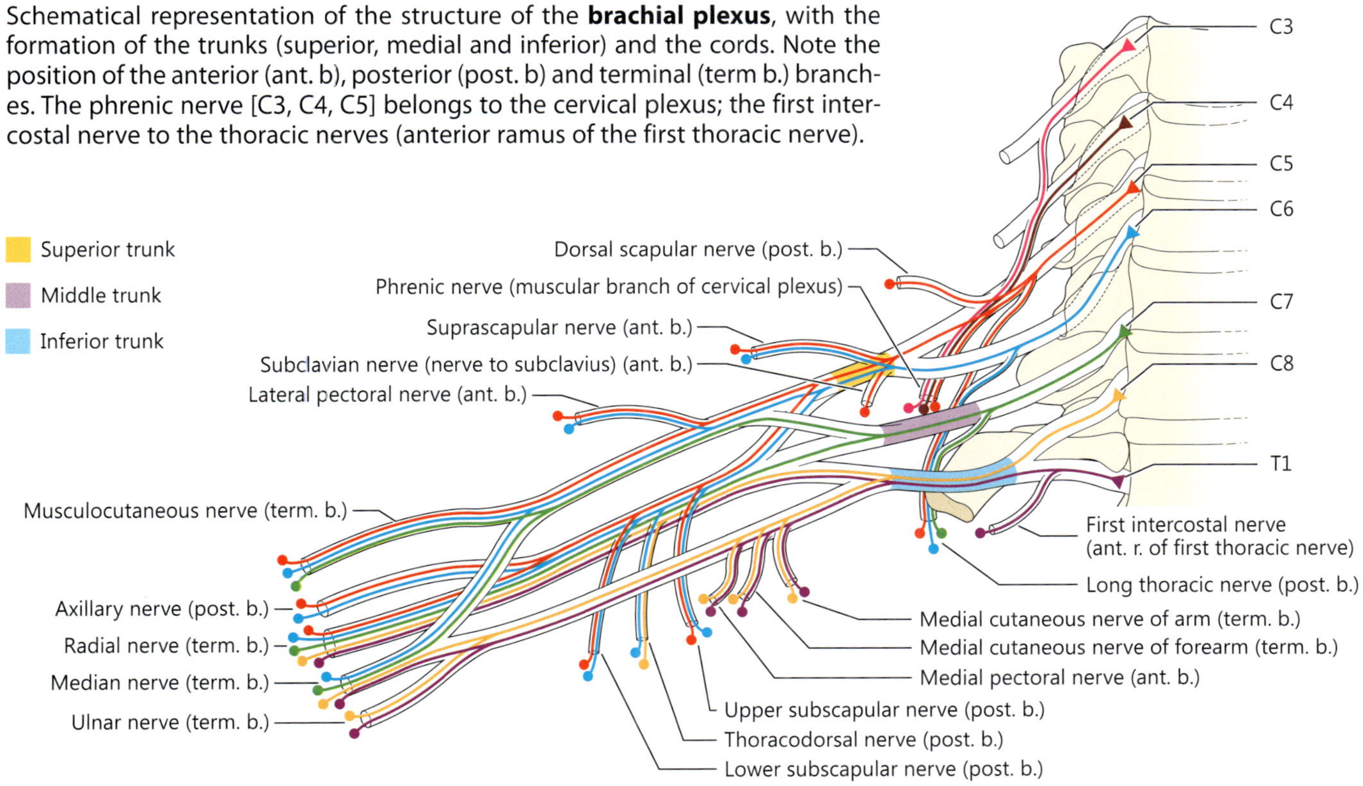

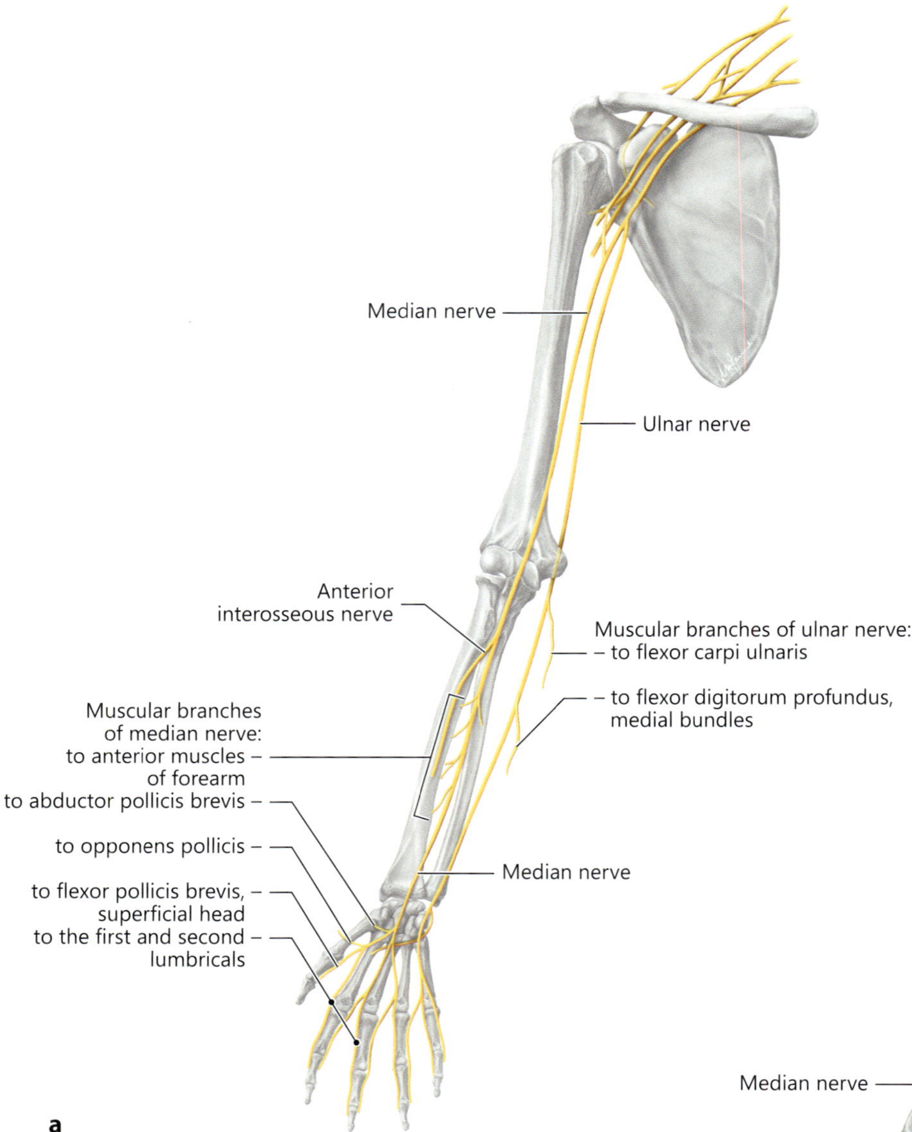

a

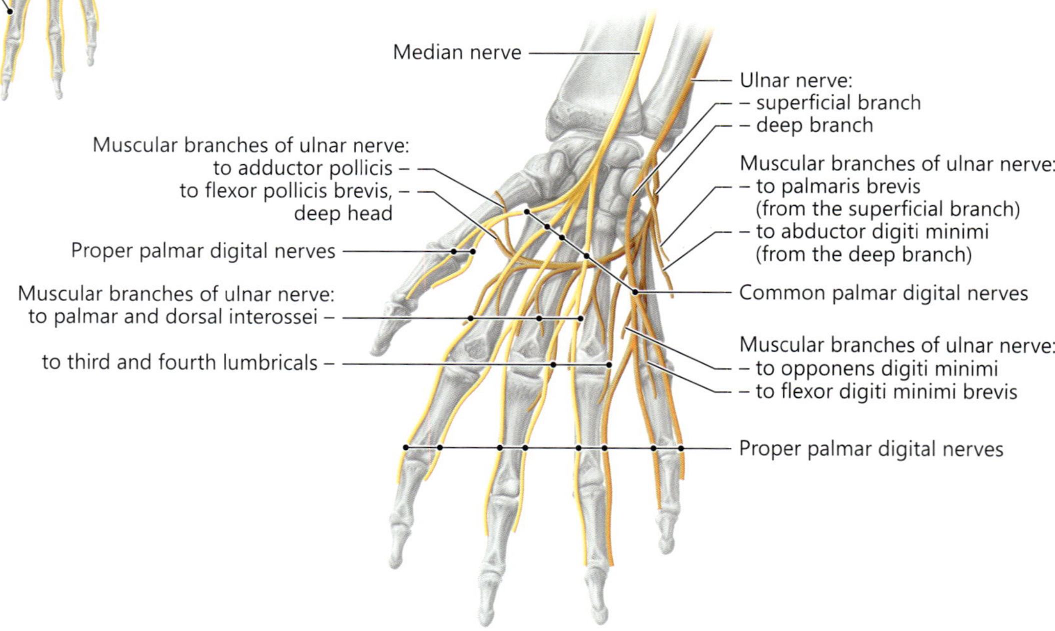

b

a, Schema of the course and distribution of the **ulnar and median nerves**. In the palmar region, the median nerve gives rise to muscular and cutaneous branches in the lateral half of the hand. Moreover, it supplies the greater part of the anterior muscles of the forearm; a small part of these being innervated by the ulnar nerve.

b, Schema of the distribution of the terminal branches of the **ulnar nerve** and of the cutaneous branches of the **median nerve** of the hand. The cutaneous branches of the ulnar nerve terminate at the fourth and fifth finger, the muscular ones are distributed in a region expanding laterally. The cutaneous branches of the median nerve are distributed to the thumb, the index finger and the middle finger and the lateral part of the ring finger.

➤

a, Schema of the course and distribution of the **musculocutaneous nerve**. At the level of the anterior region of elbow, the nerve changes its name in lateral cutaneous nerve of the forearm and divides into two branches, anterior and posterior. The nerve supplies muscular branches only in the arm.

b, Schema of the course and distribution of the **medial cutaneous nerve of the arm** and **medial cutaneous nerve of the forearm**. These nerves are exclusively sensitive and supply the medial half of the arm and the forearm, expanding to both the dorsal and ventral surfaces of the limb.

c, Schema illustrating the course, distribution and branching of the **radial nerve**. The nerve supplies muscular branches, and the cutaneous branches for the dorsal surface of the arm and the forearm.

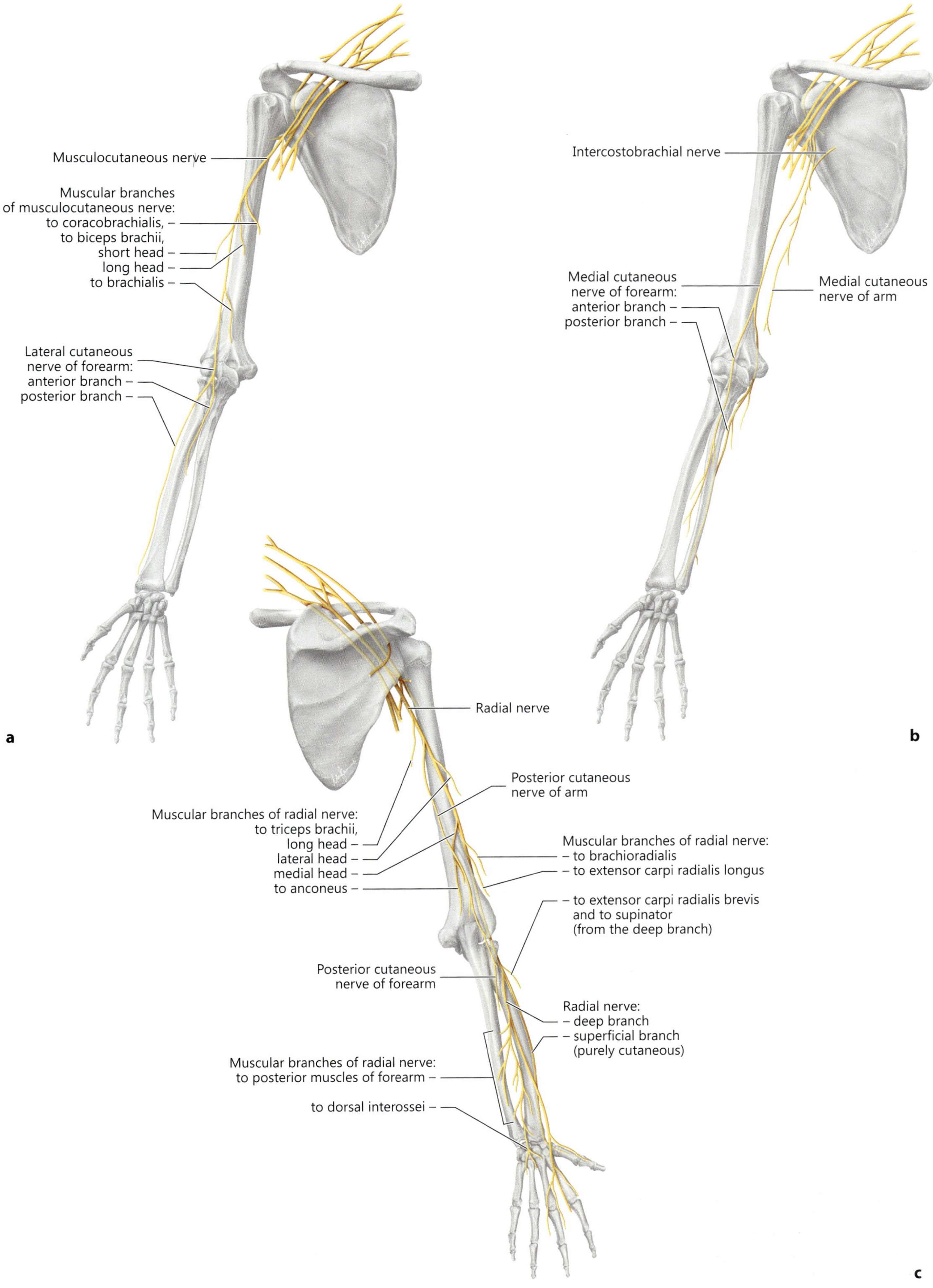
Musculocutaneous nerve
Muscular branches of musculocutaneous nerve:
to coracobrachialis,
to biceps brachii,
short head
long head
to brachialis
Lateral cutaneous nerve of forearm:
anterior branch
posterior branch
a
Intercostobrachial nerve
Medial cutaneous nerve of forearm:
anterior branch
posterior branch
Medial cutaneous nerve of arm
b
Radial nerve
Posterior cutaneous nerve of arm
Muscular branches of radial nerve:
to triceps brachii,
long head
lateral head
medial head
to anconeus
Muscular branches of radial nerve:
– to brachioradialis
– to extensor carpi radialis longus
– to extensor carpi radialis brevis and to supinator (from the deep branch)
Posterior cutaneous nerve of forearm
Radial nerve:
– deep branch
– superficial branch (purely cutaneous)
Muscular branches of radial nerve:
to posterior muscles of forearm
to dorsal interossei
c

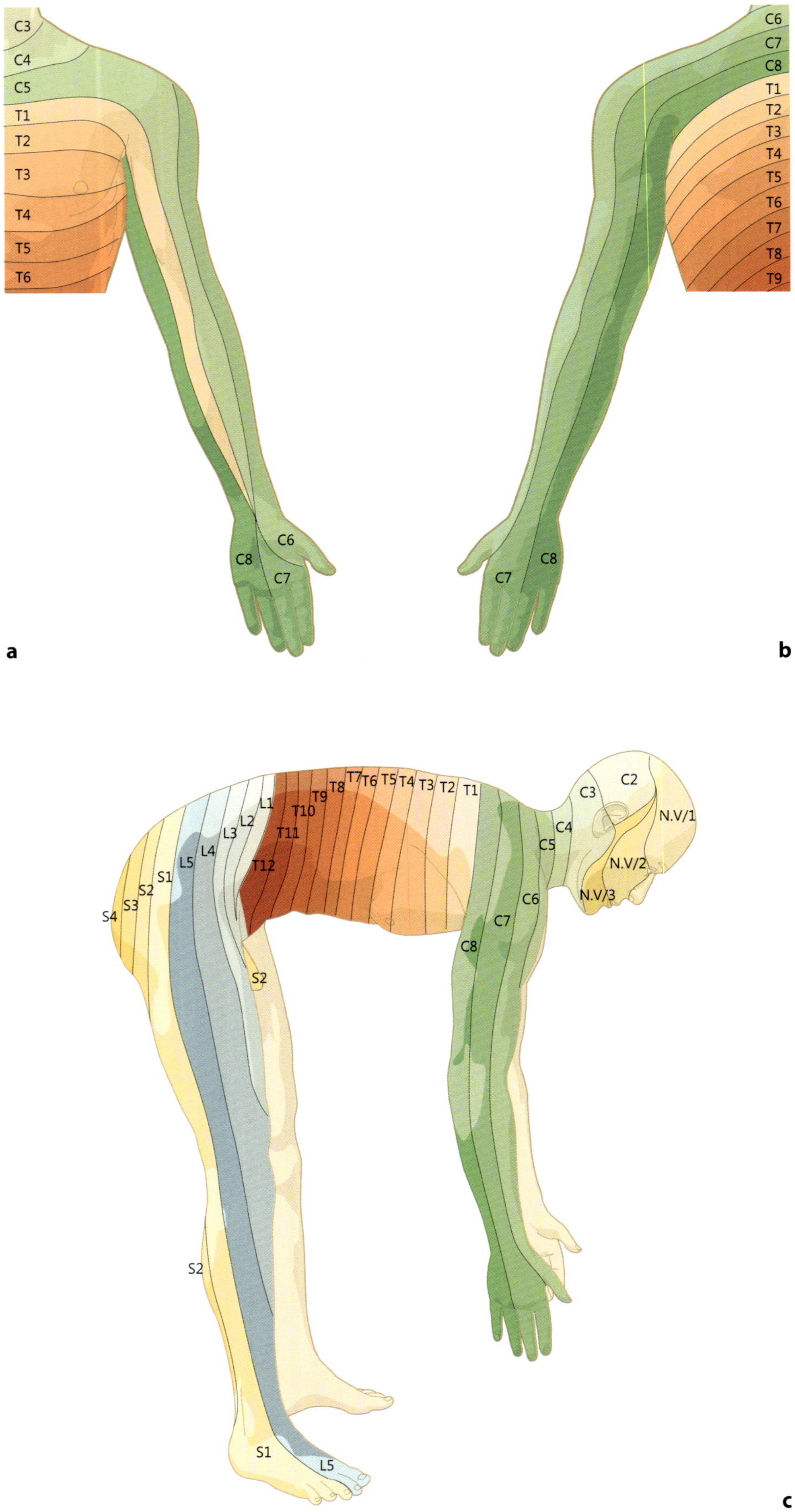

Dermatomes of the upper limb; anterior view (**a**), posterior view (**b**) and lateral view (**c**).

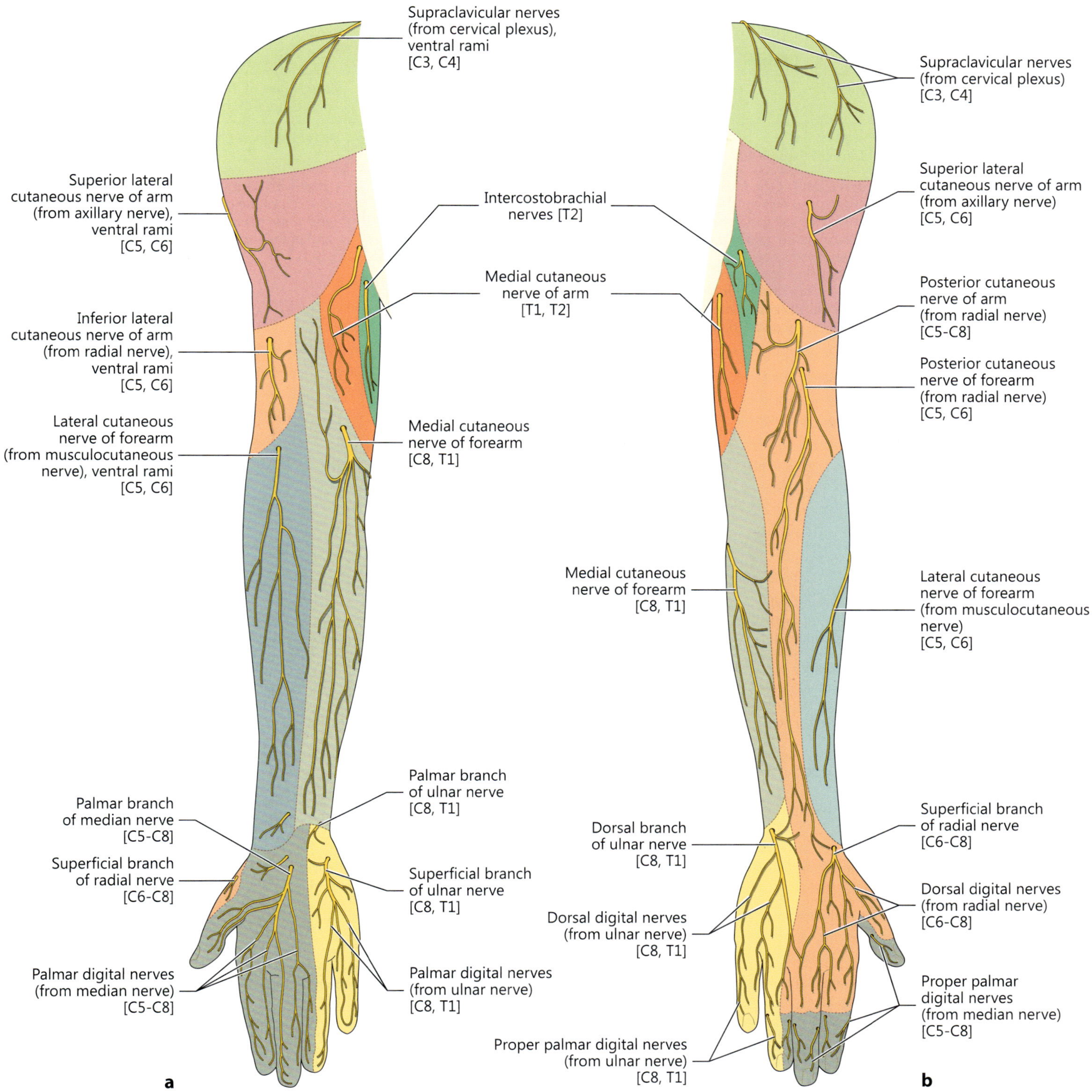

Territories of **cutaneous nerves** of the **upper limb**. **a**, Ventral surface; **b**, dorsal surface. The free part of the limb is supplied by the medial, lateral (superior and inferior parts of the arm) and posterior cutaneous nerves of the arm and the forearm. The sensitive innervation of the hand depends on the radial, median and ulnar nerves. For the dermatomes of the upper limb, ☞ previous page.

Summary table of the nerves of the upper limb

NERVE	ORIGIN	RAMIFICATIONS	DISTRIBUTION
anterior interosseous [C7, C8]	Median	Filaments	Greater part of the anterior muscles of the forearm
axillary [C5, C6, C7]	Posterior cord of the brachial plexus	Muscular branches and superior lateral cutaneous nerve of the arm	Deltoid and teres minor, and surrounding skin
brachial plexus	Anterior rami of the fifth, sixth, seventh and eighth cervical nerves and of the first thoracic nerve (taking part in the constitution of the plexus, also a small part of fibers from the anterior rami of the fourth cervical nerve and of the second thoracic nerve)	Superior, middle and inferior trunks and its anterior and posterior divisions, from which originate the lateral (from the union of the anterior divisions of the superior and middle trunks), medial (from the anterior division of the inferior trunk) and posterior (from the union of the posterior divisions of the three trunks) cords. Terminal branches of the brachial plexus: from the posterior cord originate the radial and axillary nerves, from the lateral cord the musculocutaneous nerve and the lateral root of the median nerve, from the medial cord the medial root of the median nerve, and the ulnar nerve, the medial cutaneous nerve of the arm and the medial cutaneous nerve of the forearm. Collateral branches of the brachial plexus: dorsal scapular, long thoracic, subscapular, thoracodorsal, subclavian, medial pectoral, lateral pectoral, suprascapular, muscular. Anastomotic branches: with the cervical plexus, the sympathetic trunk and the anterior ramus of the second thoracic nerve	Upper limb (terminal branches), muscles of the pectoral girdle and some muscles of dorsum and of the thorax (collateral branches)
common palmar digital [C5, C6, C7, C8, T1]	Median and superficial branch of the ulnar nerve	Proper palmar digital nerves	Skin of the palmar and lateral surfaces of the fingers; lumbricals
dorsal digital [C5, C6, C7, C8, T1]	Dorsal branch of the ulnar nerve and superficial branch of the radial nerve	Filaments	Skin of the dorsal surface of the fingers
dorsal scapular [C5]	Roots of the superior trunk of the brachial plexus	Filaments	Levator scapulae, rhomboid major, rhomboid minor and serratus posterior superior
inferior lateral cutaneous nerve of arm [C5, C6]	Radial	Filaments	Skin of the lateral surface of the inferior part of the arm
intercostobrachial [T2, T3]	Second and, frequently, third intercostal nerve	Filaments	Skin of the medial surface and of the posterior surface of the arm; axilla
intermediate supraclavicular (middle supraclavicular) [C2, C3]	Cervical plexus	Filaments	Skin surrounding the pectoralis major and deltoid
lateral cutaneous nerve of forearm [C6, C7]	Musculocutaneous	Anterior and posterior branches	Skin on the radial side of the forearm
lateral pectoral [C5, C6]	Lateral cord of the brachial plexus	Filaments	Skin of the superoanterior part of the thorax, below the clavicle, pectoralis major and pectoralis minor
lateral supraclavicular (posterior supraclavicular) [C2, C3]	Cervical plexus	Filaments	Skin of the superior and posterior parts of the shoulder
long thoracic [C5, C6, C7]	Roots of the superior and middle trunks of the brachial plexus	Filaments	Serratus anterior
medial cutaneous nerve of arm [C8, T1]	Medial cord of the brachial plexus	Filaments	Skin of the medial surface of the arm
medial cutaneous nerve of forearm [C8, T1]	Medial cord of the brachial plexus	Anterior and posterior branches	Skin surrounding the biceps brachii and on the ulnar side of the forearm
medial pectoral [C8, T1]	Medial cord of the brachial plexus	Filaments	Pectoralis minor and pectoralis major

continued

Summary table of the nerves of the upper limb *(continued)*

NERVE	ORIGIN	RAMIFICATIONS	DISTRIBUTION
median [C5, C6, C7, C8, T1]	From the medial and lateral cords of the brachial plexus via two roots, medial and lateral	Anterior interosseous nerve and common palmar digital nerves (providing the proper palmar digital nerves), and muscular branches (including the recurrent branch), palmar branches and communicating branches with the ulnar nerve	Anterior muscles of the forearm (muscles attaching to the medial epicondyle of the humerus), muscles of the thenar eminence (abductor pollicis brevis, opponens pollicis and flexor pollicis brevis) and the first and second lumbrical; skin of the anterior region of the wrist, of the palm of the hand on the radial side, of the palmar surface of the first three fingers and of half of the fourth (the plane passing through the axis of the ring finger constitutes the boundary with the medial region to which the ulnar nerve is distributed)
musculocutaneous [C5, C6]	Lateral cord of the brachial plexus	Muscular branches and lateral cutaneous nerve of the forearm	Coracobrachialis, brachialis and biceps brachii; skin of the lateral part of the forearm
posterior cutaneous nerve of arm [C5, C6, C7, C8]	Radial	Filaments	Skin of the posterior surface of the arm
posterior cutaneous nerve of forearm [C5, C6, C7, C8]	Radial	Filaments	Skin of the posterior surface of the lower half of the arm, and of the posterior surface of the forearm
posterior interosseous [C7, C8]	The deep branch of the radial nerve become the posterior interosseous nerve after piercing the supinator muscle	Filaments Deep branching of the radial nerve	Wrist joint and carpal joints; extensor carpi radialis brevis, supinator Extensor digitorum, extensor digiti minimi, extensor carpi ulnaris, abductor pollicis longus, extensor pollicis brevis, extensor pollicis longus and extensor indicis
proper palmar digital [C5, C6, C7, C8, T1]	Common palmar digital	Filaments	Skin of the palmar and dorsal surfaces of the distal phalanges of the fingers
radial [C5, C6, C7, C8, T1]	Posterior cord of the brachial plexus	Posterior cutaneous nerve of the arm, inferior lateral cutaneous nerve of the arm and posterior cutaneous nerve of the forearm and deep (from which originate the posterior interosseous nerve), superficial (from which originate the communicating branch with the ulnar nerve and the dorsal digital nerves) and muscular branches	Posterior muscles of the arm (triceps brachii and anconeus), lateral (brachioradialis, extensor carpi radialis longus and extensor carpi radialis brevis) and posterior muscles of the forearm (extensor digitorum, extensor digiti minimi, extensor carpi ulnaris, supinator, abductor pollicis longus, extensor pollicis brevis, extensor pollicis longus, extensor indicis), and dorsal interossei; skin of a part of the posterior surface of the arm, of the posterior median region of the forearm and of the wrist, of the lateral half of the dorsum of the hand and of the dorsal surface of the first two fingers and part of the third (the plane passing through the axis of the middle finger constitutes the boundary with the medial region to which the ulnar nerve is distributed), with the exception of the dorsal surface of the medial and distal phalanges of the index finger and middle finger (to which the median nerve is distributed)
subclavian [C6]	Superior trunk of the brachial plexus	Filaments	Subclavius
subscapular [C5, C6, C7, C8, T1] **upper** **lower**	Posterior cord of the brachial plexus	Filaments	 Subscapularis Subscapularis and teres major
superior lateral cutaneous nerve of arm [C5, C6]	Axillary	Filaments	Skin of the lateral surface of the superior part of the arm
suprascapular [C5, C6]	Superior trunk of the brachial plexus or from one of its roots	Filaments	Supraspinatus and infraspinatus
thoracodorsal [C6, C7, C8]	Posterior cord of the brachial plexus	Filaments	Latissimus dorsi
ulnar (cubital) [C8, T1]	Medial cord of the brachial plexus	Muscular, dorsal (from which originate the dorsal digital nerves), palmar, superficial (from which originate the common palmar nerves; and from these, the proper palmar nerves) and deep branches	Flexor carpi ulnaris and flexor digitorum profundus, adductor pollicis, flexor pollicis brevis, palmaris brevis, abductor, opponens digiti minimi and flexor digiti minimi brevis, palmar interossei and dorsal interossei, third and fourth lumbrical; medial skin of the carpus (dorsal and palmar surfaces) and skin of the hand (the plane passing through the axis of the ring finger constitutes the boundary with the lateral region to which the median nerve is distributed, while the plane passing through the axis of the middle finger constitutes, on the dorsal surface, the boundary with the lateral region to which the radial nerve is distributed)

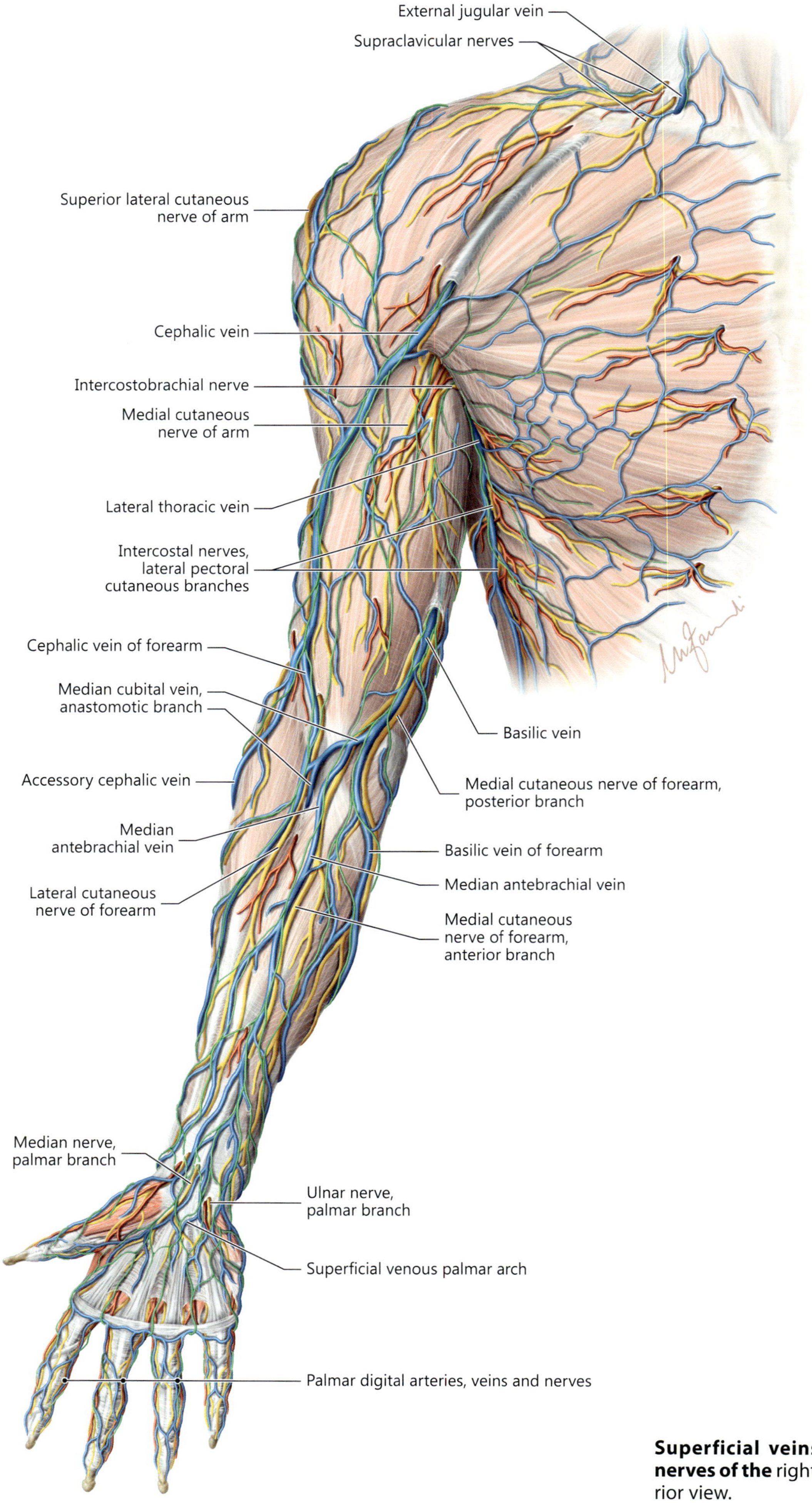

Superficial veins and cutaneous nerves of the right **upper limb**, anterior view.

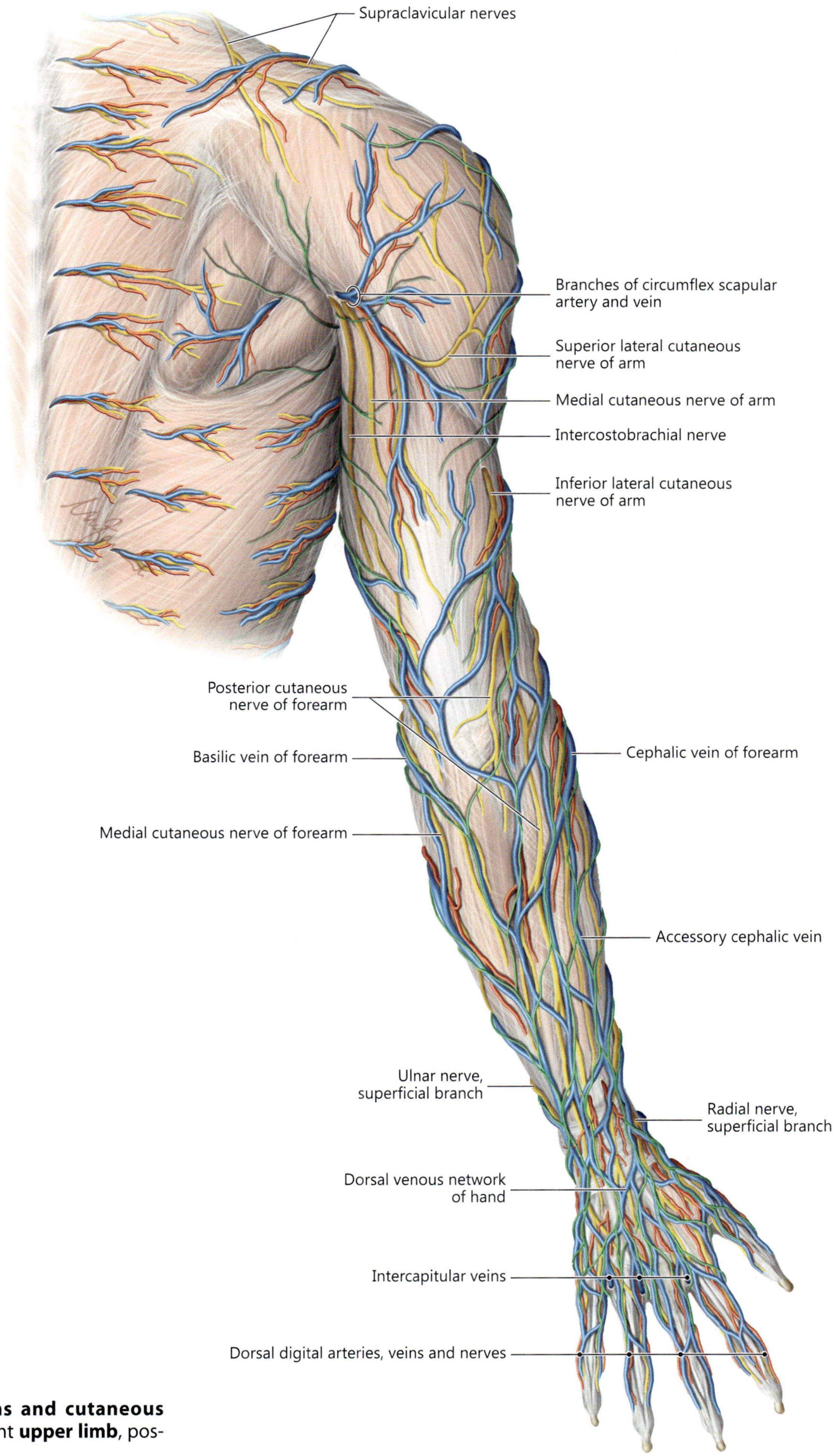

Superficial veins and cutaneous nerves of the right **upper limb**, posterior view.

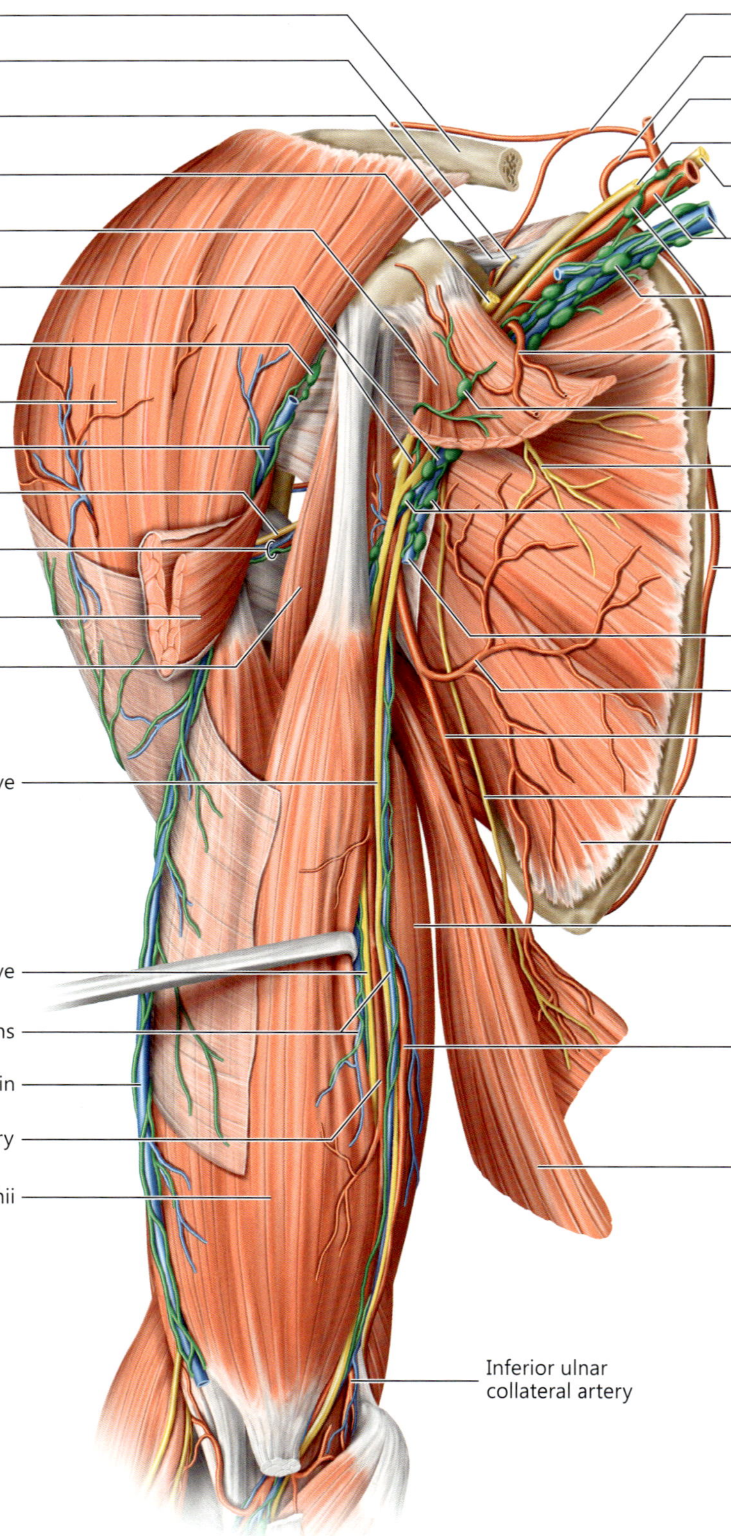

Right arm, anterior view. The **ulnar nerve** has its origin in the axilla from the medial cord of the brachial plexus and penetrates the anterior compartment of the arm, together with the **brachial artery** and the **median nerve** (positioned laterally). While descending towards the elbow, it carries posteriorly and reaches the posterior compartment of the arm. The **thoracoacromial artery** originates from the anterior surface of the second part of the axillary artery and, underneath the pectoralis major, divides into the pectoral, acromial, deltoid and clavicular branches. The **subscapular artery** is the most conspicuous branch of the axillary artery, detaching itself at the level of the inferior margin of the subscapularis, and terminating on the lateral border of the scapula, where it supplies the thoracoacromial and circumflex scapular arteries. The **cephalic vein** runs in the arm, along the anterolateral surface of the biceps brachii and, superiorly, in the sulcus between the deltoid and pectoralis major, crossing the clavipectoral or deltopectoral triangle, where it flows into the axillary vein.

* In this image, the transverse cervical artery arises from the thyrocervical trunk and not directly from the subclavian artery.

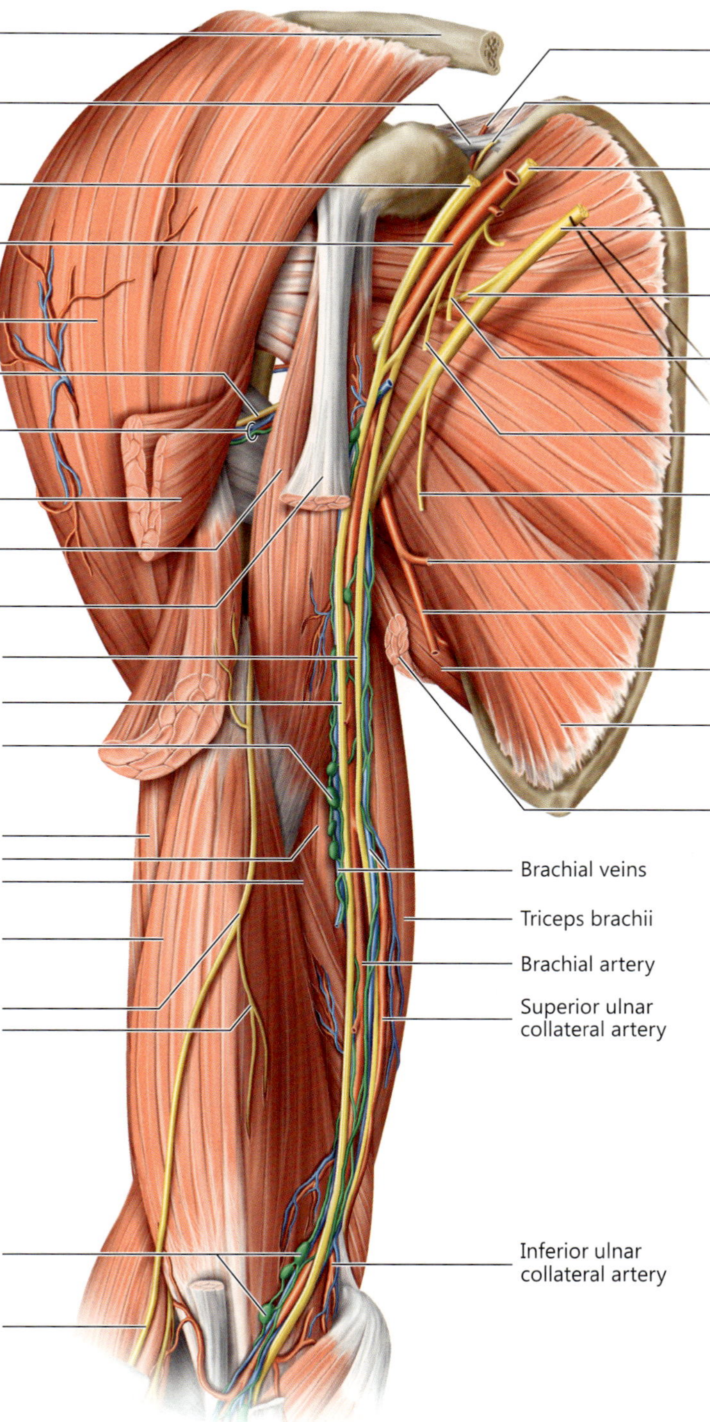

Right arm, anterior view. By removing the short head of the biceps brachii one can observe the course of the **musculocutaneous or perforating nerve**, which reaches the anterior compartment of the arm penetrating the full thickness of the coracobrachialis, and of the **median nerve**, including its origin formed by the V-shaped convergence of the medial and lateral cords of the brachial plexus. The **brachial artery**, continuation of the axillary artery, runs along the medial side of the anterior surface of the arm, in the sulcus between the coracobrachialis and biceps brachii, and descends, with an almost rectilinear course, shifting itself towards the median line. At the limits of the anterior region of the elbow, it divides into its terminal branches, represented by the radial and ulnar arteries. The **brachial veins**, satellites of the homonymous artery, flow in the axillary vein at the level of the inferior margin of the pectoralis major. The **median brachial vein** normally receives the basilic vein.

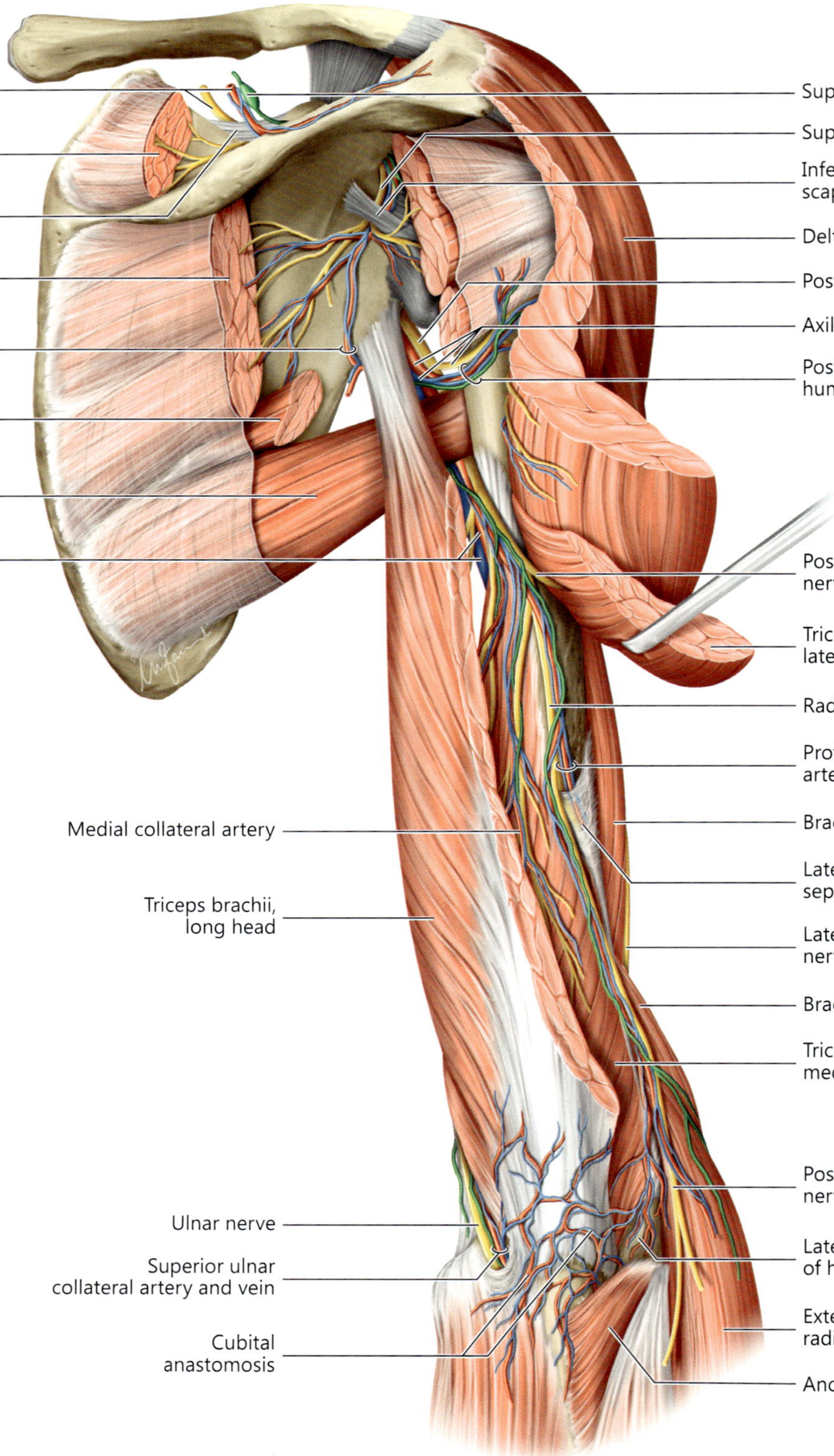

Right arm, posterior view. The removal of the lateral head of the triceps brachii enables the observation of the course of the **radial nerve**, which has its origin in the axilla as a continuation of the posterior cord of the brachial plexus, and which arrives at the posterior compartment of the arm, passing between the long head and the medial head of the triceps brachii. The cubital anastomosis is constituted by the **superior and inferior ulnar collateral arteries**, branches of the brachial artery, and by the **anterior and posterior radial and ulnar recurrent arteries**, branches of the radial and ulnar arteries, respectively. It is found in the dorsal region of the elbow, against the osseous plane and the tendon of the triceps brachii.

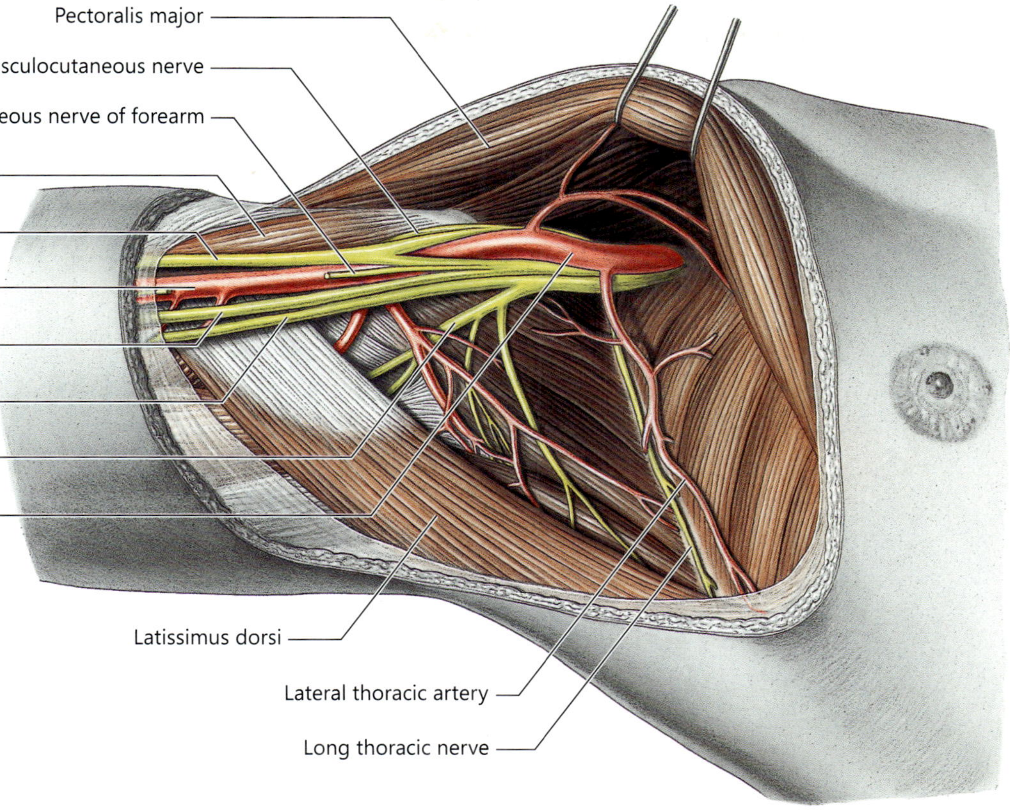

Relations of the cords (or secondary trunks) and the terminal branches of the **brachial plexus** with the **axillary and brachial arteries**. The vascular and nervous structures occupy a deep position in the axilla.

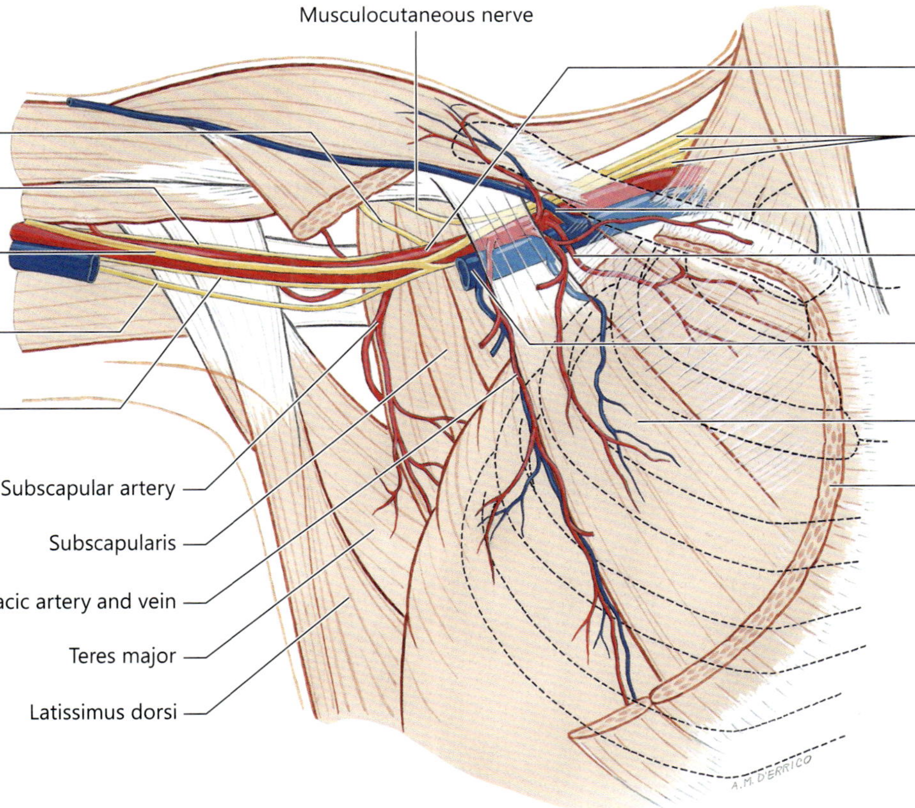

Schematic representation of the relations of the **axillary artery** and the **axillary vein** with the branches of the **brachial plexus**. The axillary vein has been partially eliminated; the pectoralis major has been transected and removed, in order to highlight the vascular and nervous structures of the axilla. Right side, anterior view.

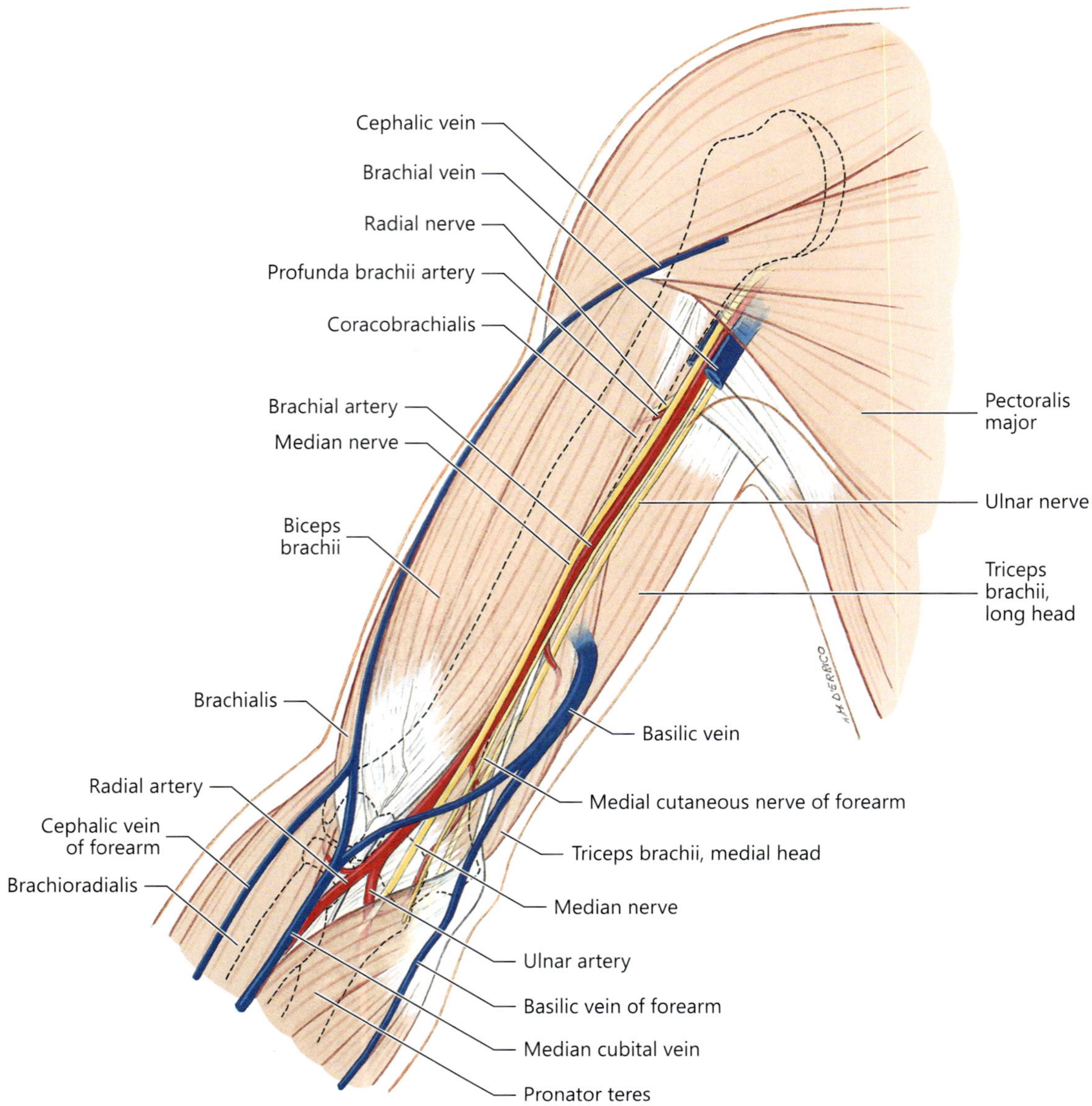

Schematic representation of the course and the relations of the right **brachial artery**. The skeletal segments, outlined with a dotted line, give useful references on the arrangement of the artery.

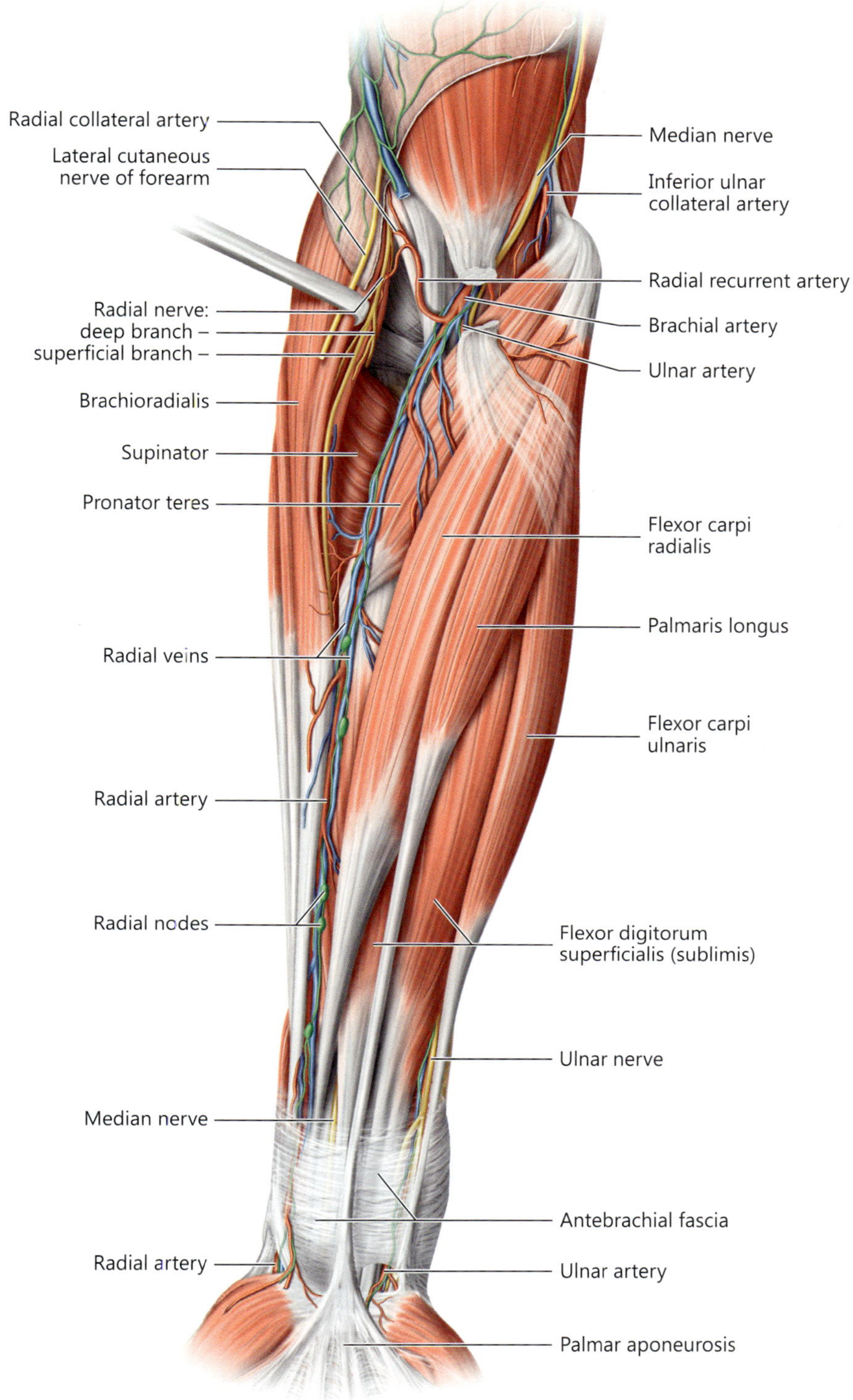

Right forearm, anterior view. The **radial artery** is covered by the belly of the brachioradialis, and becomes superficial when the muscle changes into a tendon, running directly beneath the antebrachial fascia, laying on the osseous surface of the styloid process and the distal part of the radius, where its pulse can be palpated.

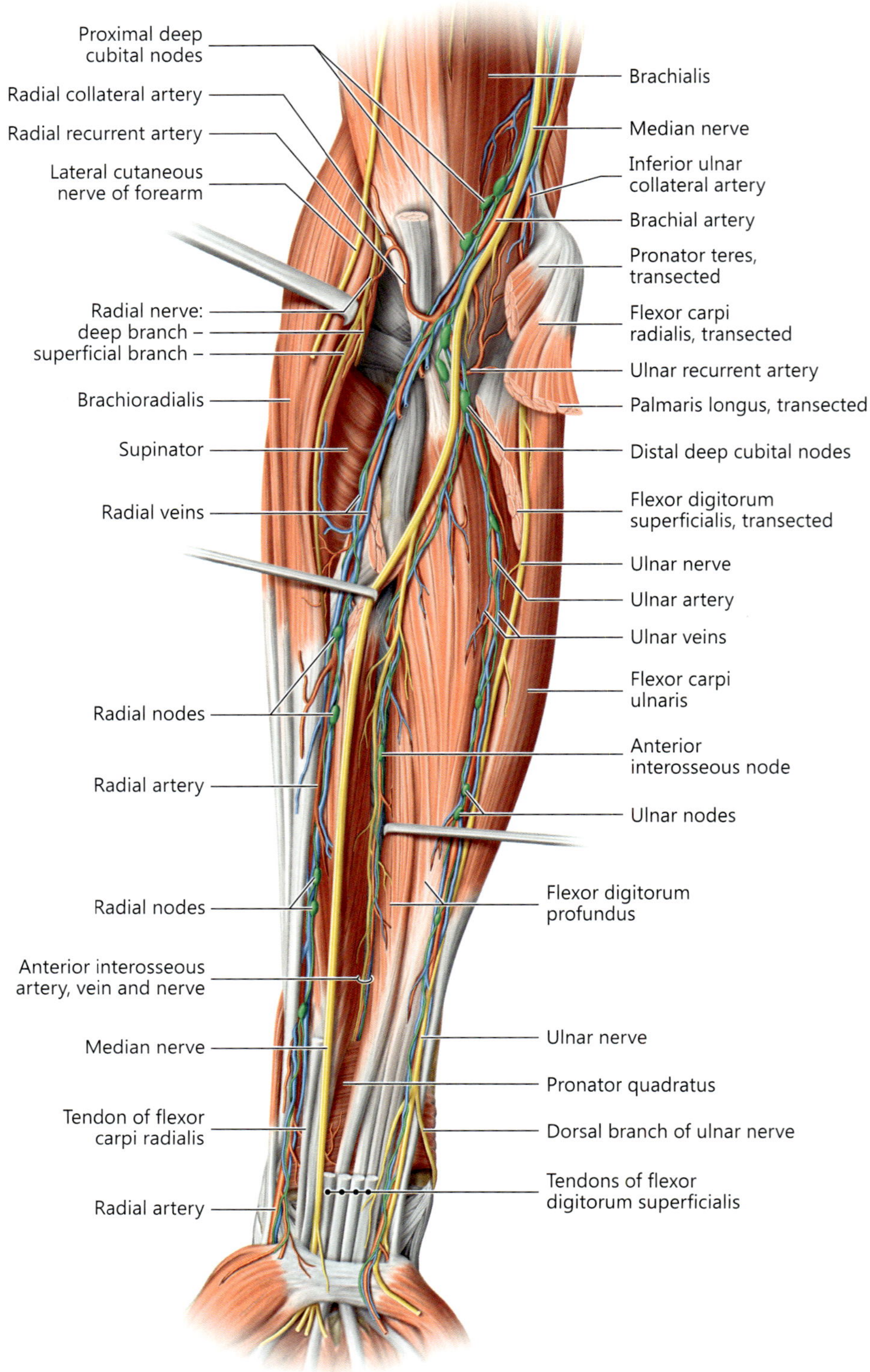

Right forearm, anterior view. The removal of the pronator teres, flexor carpi radialis, palmaris longus and flexor digitorum superficialis, enables the observation of the **median and ulnar nerves** course in the forearm. The **anterior interosseous nerve**, branch of the median nerve, descends deeply on the interosseous membrane, running underneath the pronator quadratus.

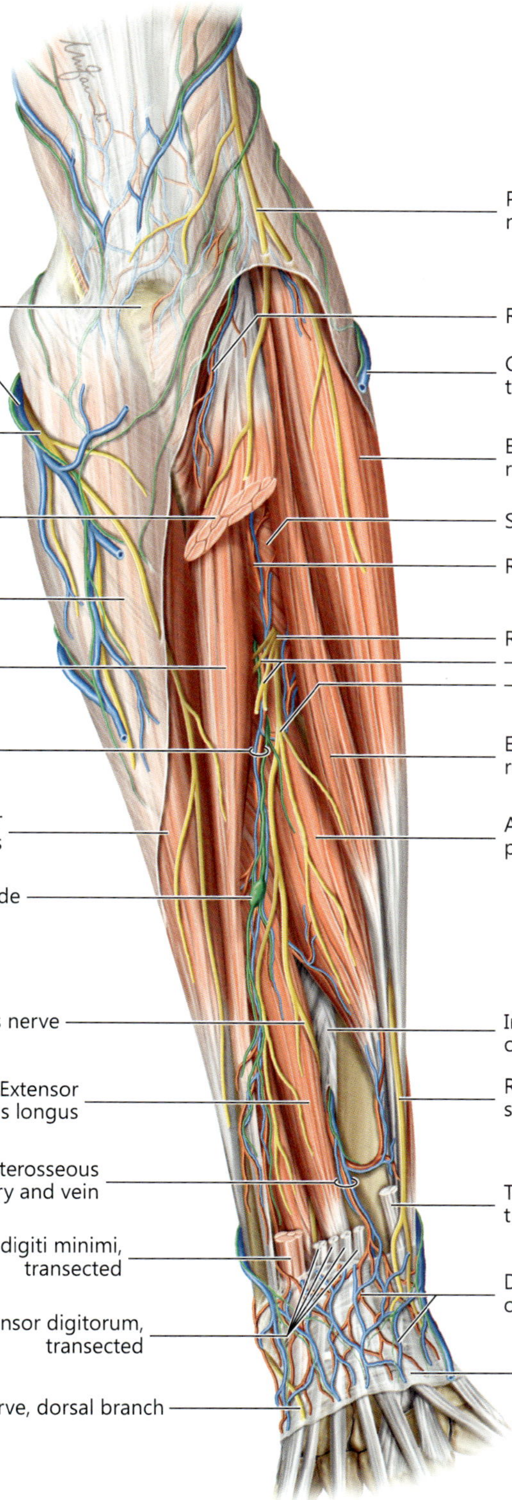

Right forearm, posterior view. The extensor digitorum and the extensor digiti minimi have been removed. Together with the extensor carpi ulnaris, they form the superficial layer of the posterior muscles of the forearm. One can observe the deep branch of the **radial nerve**, that runs from the anterior region of the elbow to the posterior region of the forearm, penetrating the full thickness of the supinator, and changing its name to **posterior interosseous nerve** in the distal third of the forearm.

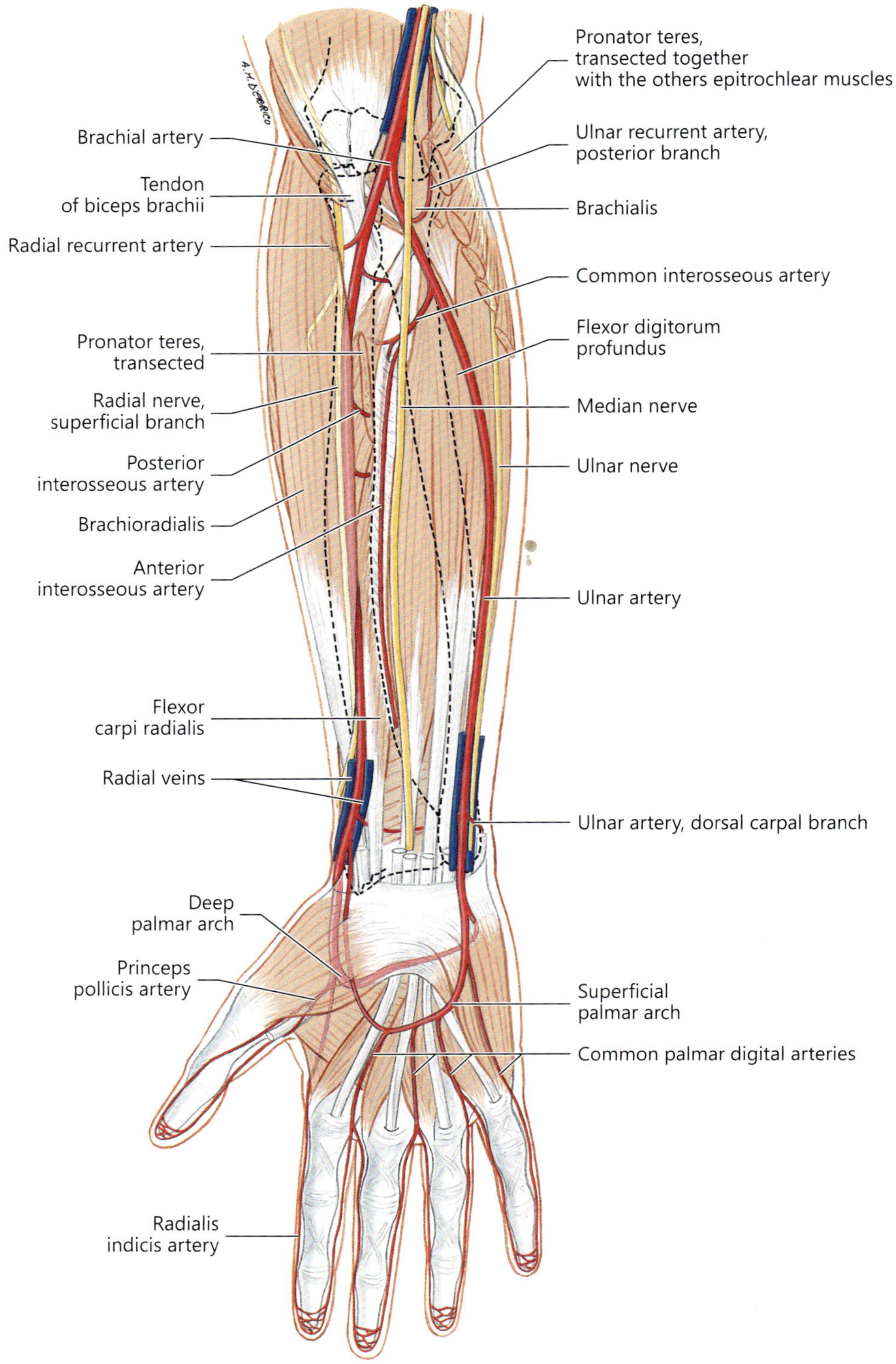

Schematic representation of the course and the relations of the **radial and ulnar arteries of the forearm** and of their terminal branches in the palmar region of the hand.

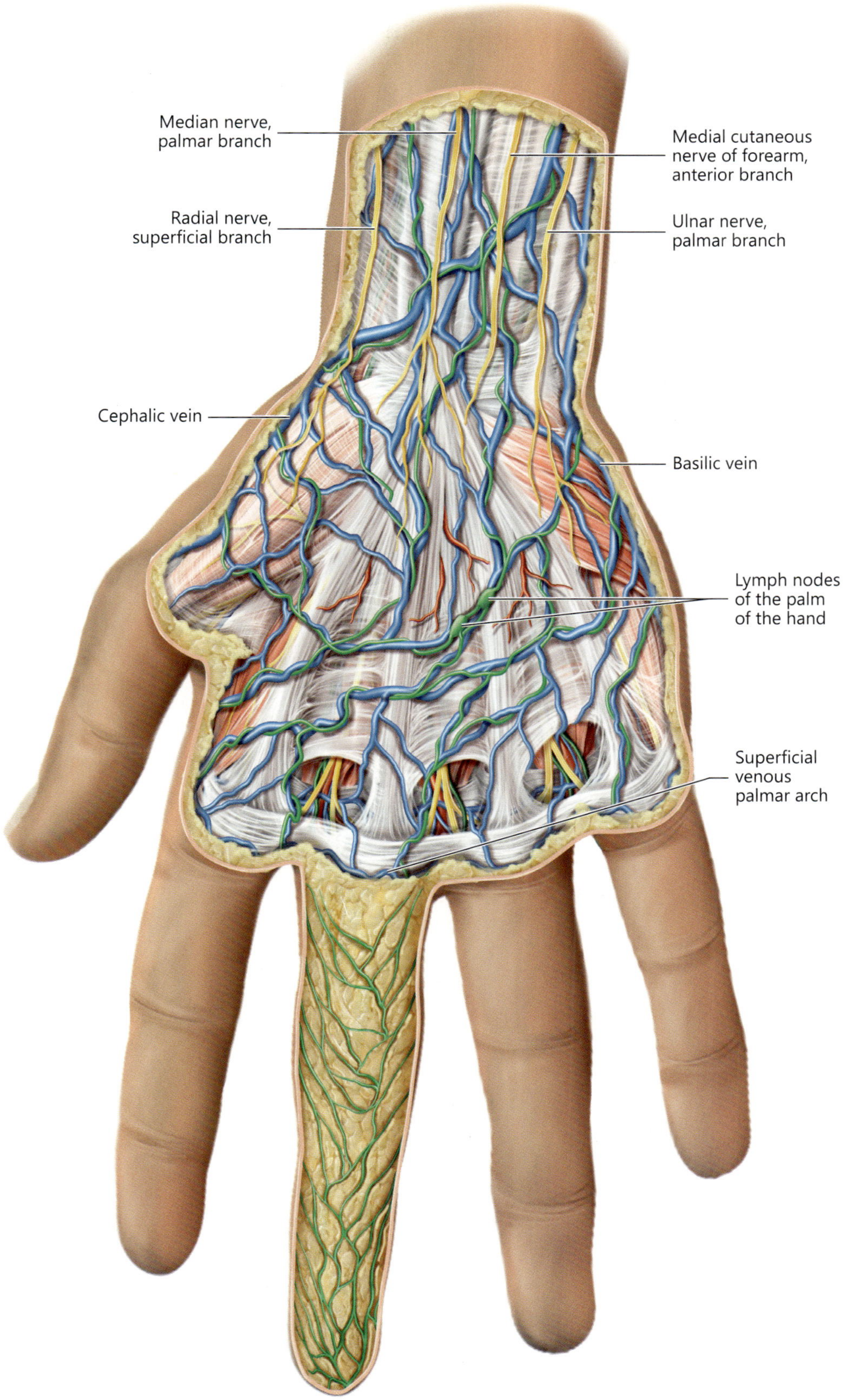

Hand, frontal view (palmar). One can observe the **cephalic vein**, originating from the dorsal venous network of the hand in the radial side, runs upward reaching the volar surface of the forearm, the **basilic vein**, originating from the dorsal venous network of the hand in the ulnar side, and the **superficial venous palmar arch**.

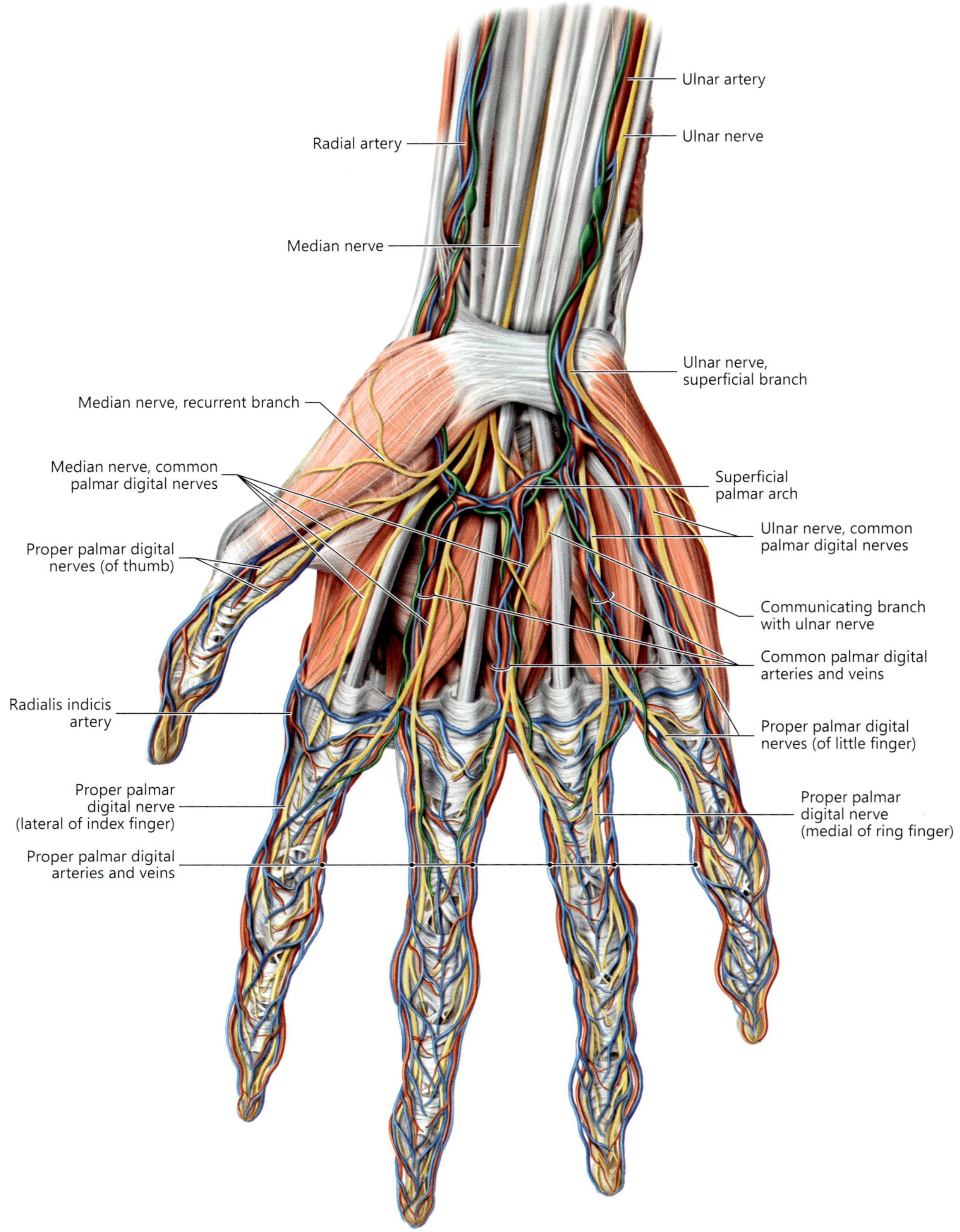

Hand, frontal view (palmar). The **median nerve**, runs in the carpal tunnel, between the tendinous sheath of the flexor pollicis longus and the tendinous sheath of the flexor digitorum profundus, and ends in its terminal branches in the palm of the hand: the **recurrent branch of the median nerve** (or nerve of thenar eminence) and the **common and proper palmar digital nerves** (of the thumb, of the index finger, of the middle finger and of the ring finger). From the superficial palmar branch of the ulnar nerve originate the **common palmar nerves** (one for the little finger, the other for the little finger and the ring finger), from which originate the **proper palmar digital nerves** (of the little finger and the ring finger).

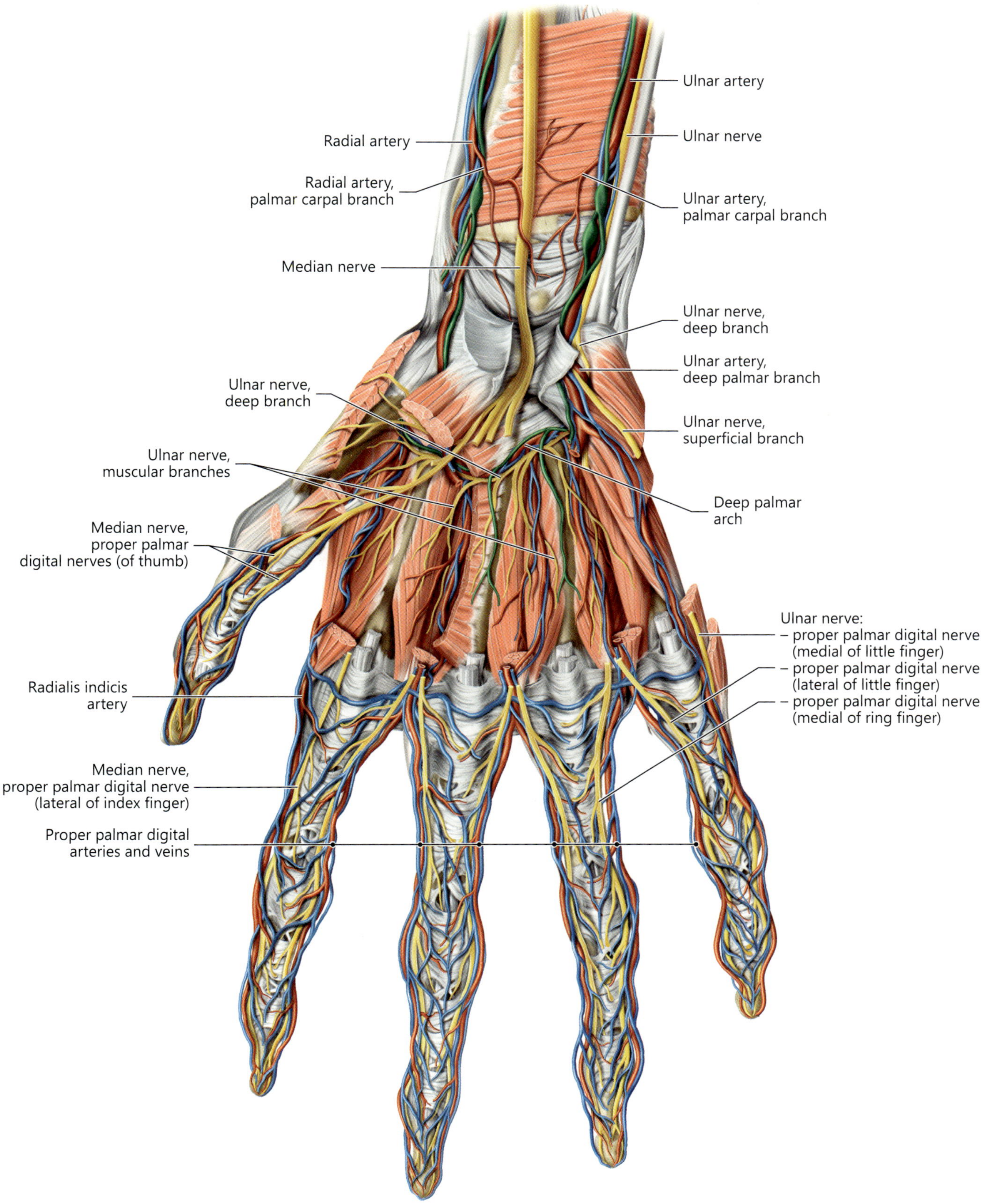

Hand, frontal view (palmar). The deep branch of the **ulnar nerve**, a prevalently motor nerve, runs deeper into the palm of the hand, together with the deep palmar branch of the **ulnar artery**, following the three central metacarpals, and forming an arch, concave laterally and superiorly, where numerous muscular branches originate.

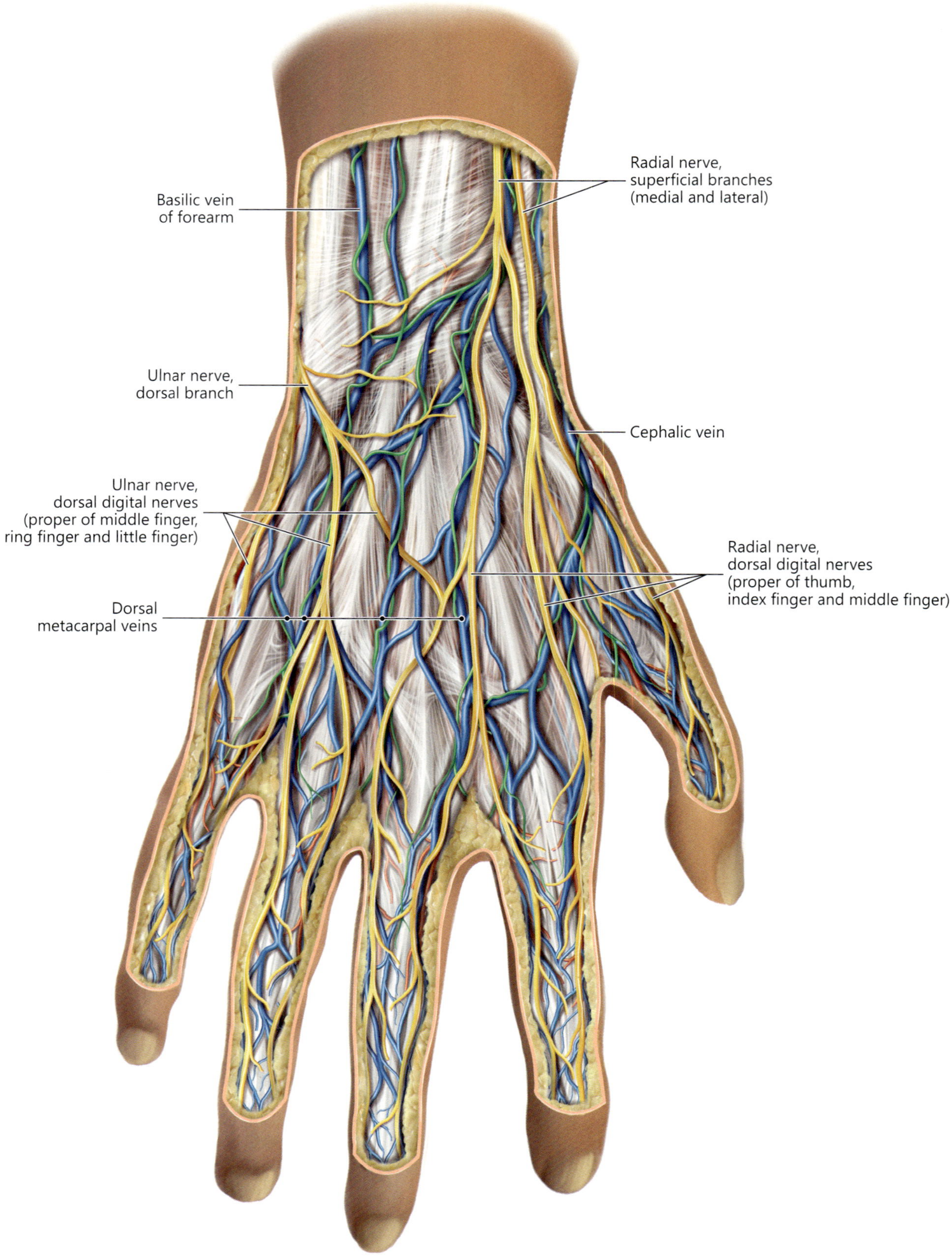

Hand, posterior view (dorsal). From the superficial branch of the **radial nerve**, here seen as a medial and a lateral branch, originate the **dorsal digital nerves** that are distributed to the skin of the dorsal surface of the first three fingers, excluding the distal phalanges of the index finger and the middle finger, to which the median nerve is distributed. From the dorsal branch of the **ulnar nerve** (or dorsal nerve of the hand) originate the **dorsal digital nerves** that are distributed to the skin of the dorsal surface of the little finger and, for the proximal phalanges only, to the skin of the ring and middle fingers (for the latter, only to the medial surface).

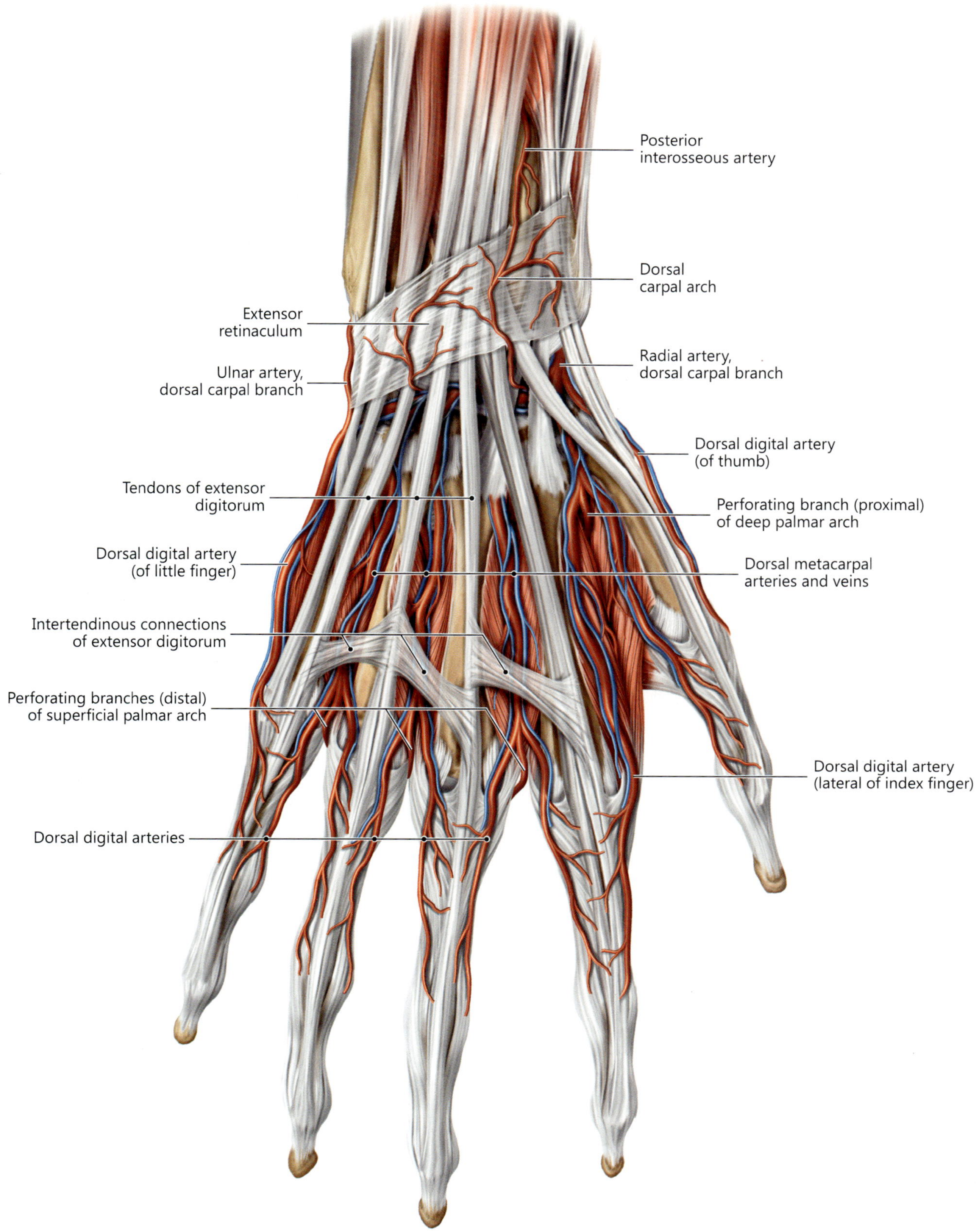

Hand, posterior view (dorsal). **Dorsal metacarpal arteries** and their anastomoses with the **palmar metacarpal arteries** through the perforating branches.

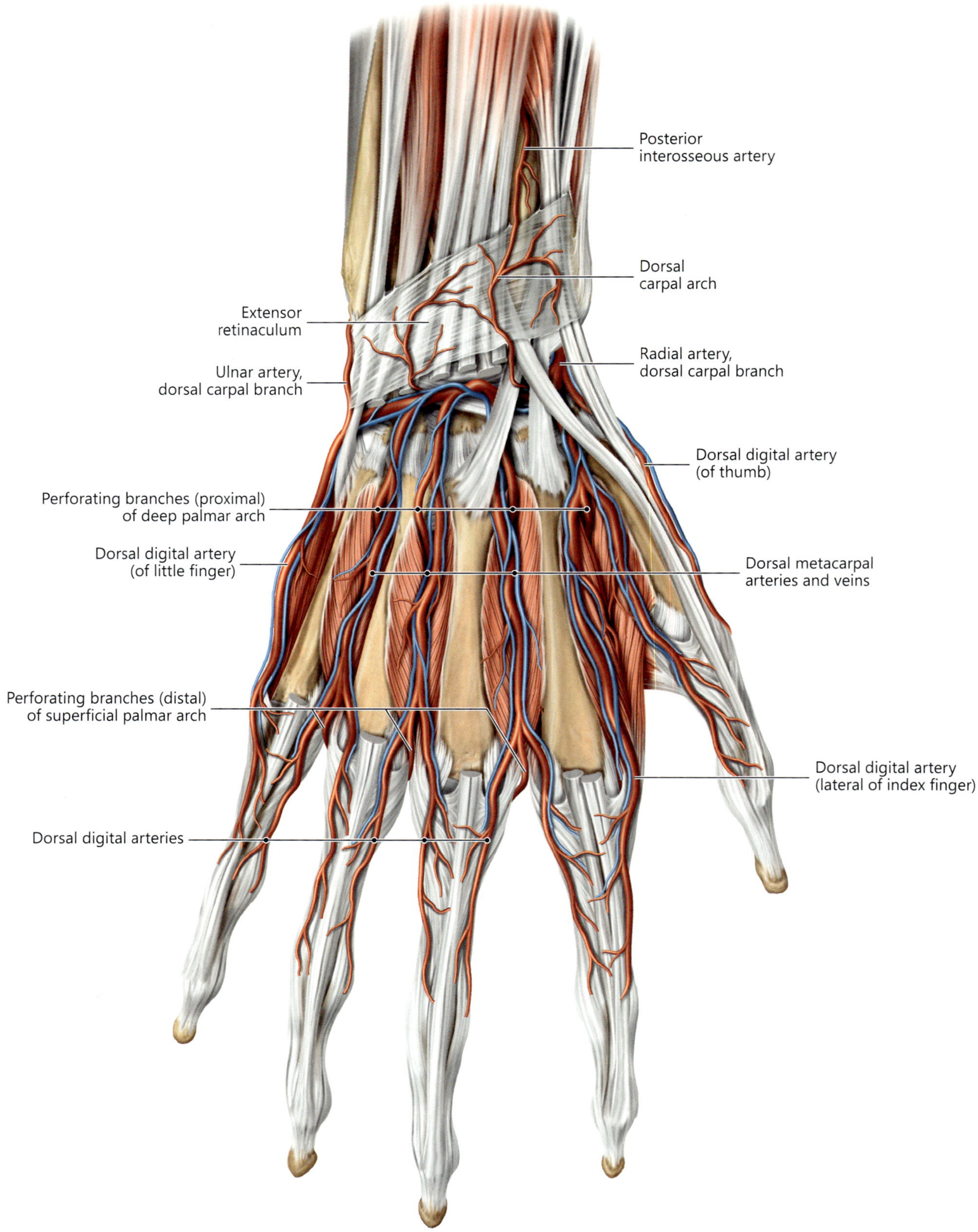

Hand, posterior view (dorsal). The **dorsal carpal arch** derives from the anastomosis of the dorsal carpal branch of the **radial artery** with the dorsal carpal branch of the **ulnar artery**. From the dorsal carpal arch originate the **dorsal metacarpal arteries** which connect, through perforating branches, with the deep **palmar arch** and the superficial one, both immediately after their origin and before their bifurcation in the **dorsal digital arteries**, respectively.

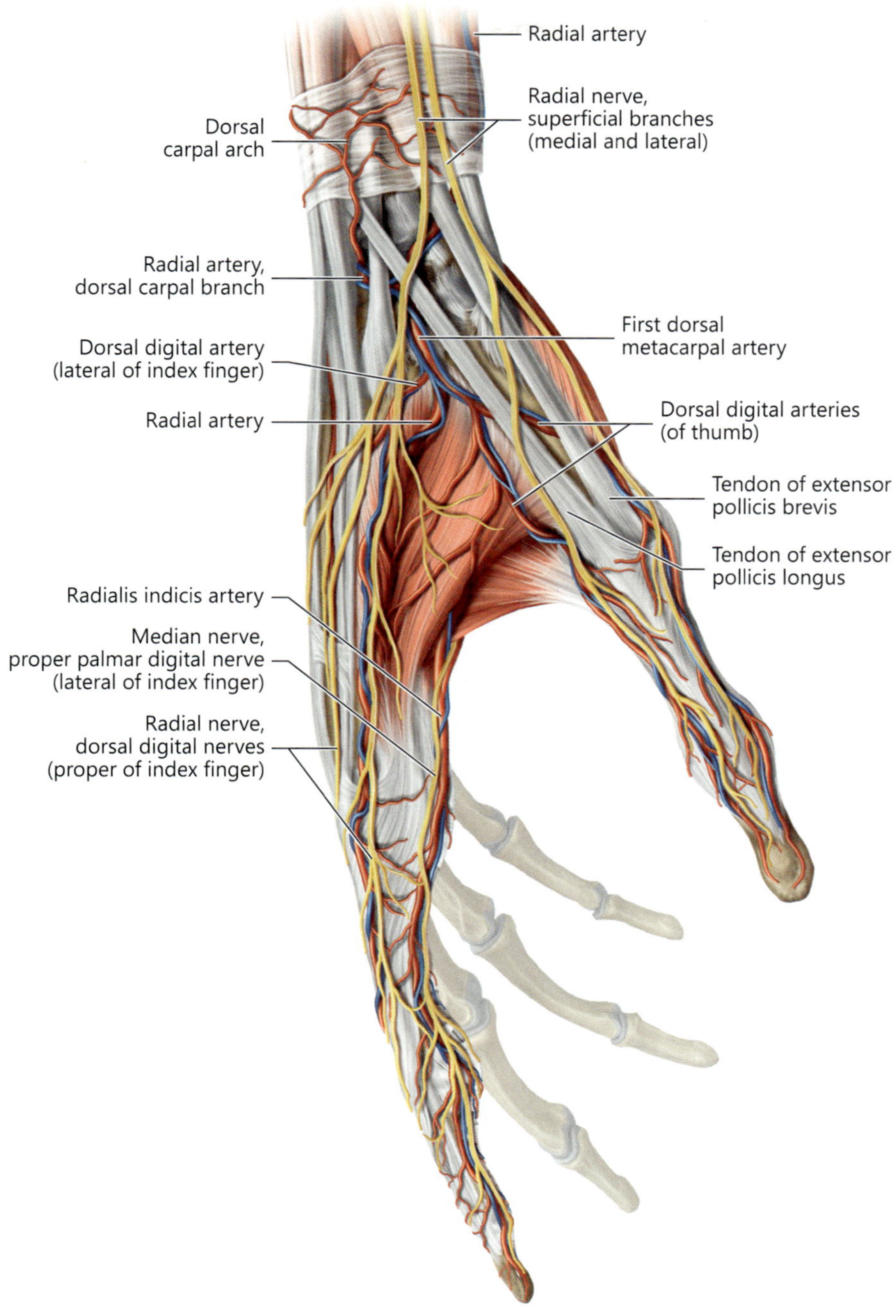

Hand, lateral view. Course and branches of the **radial artery** in the carpus. Here are shown the relations of the radial artery with the tendons of the extensor pollicis longus and extensor pollicis brevis, which delimit the anatomic snuffbox.

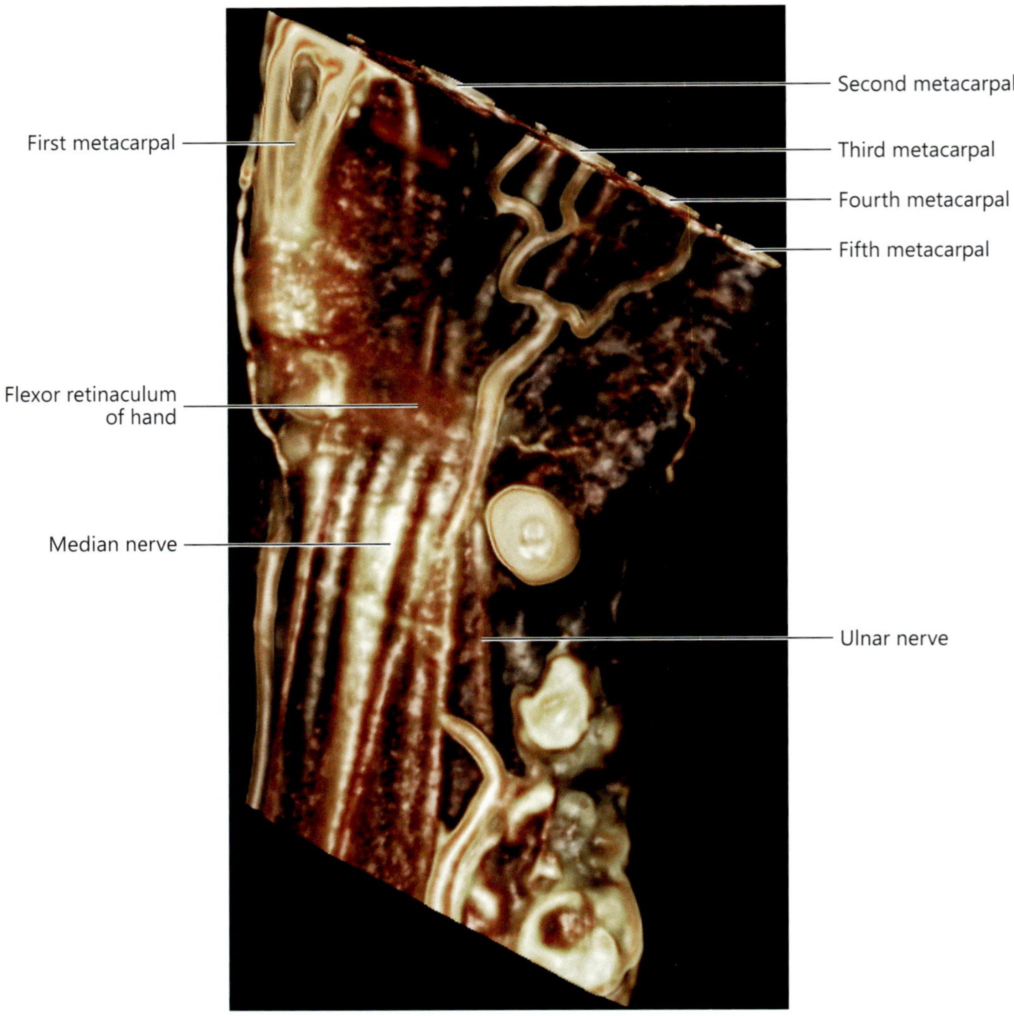

Three-dimensional image of a living human body, obtained by CT scan using a 3D volume rendering technique, showing the **volar surface** of the left **hand**. The hand has been transected at the level of the distal third of the metacarpals, and rotated slightly outwards.

LOWER LIMB

Petros (Athens, 1928 - Milan, 2014), *La ripetizione divina*, oil on canvas, cm 140 × 190, 1991

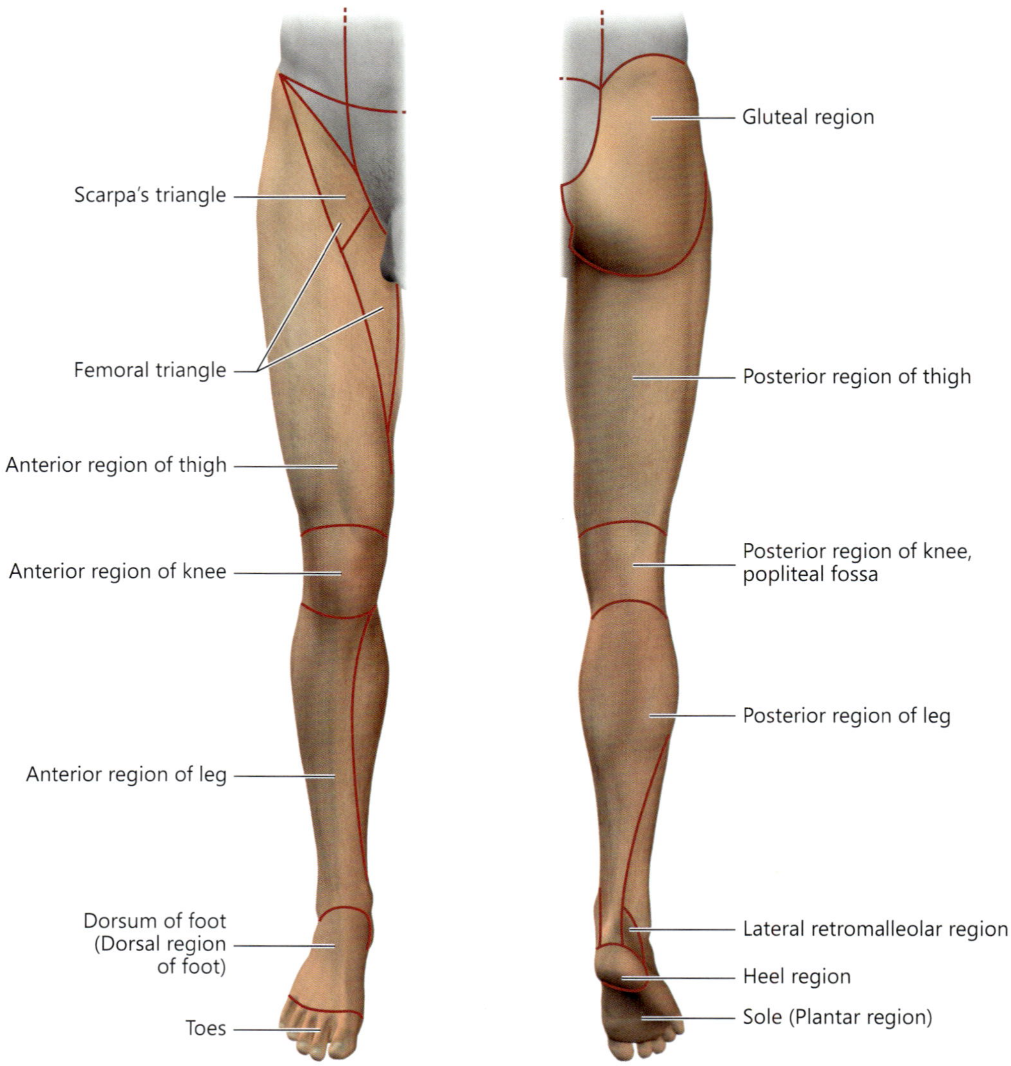

Landmark lines and points determining the anterior (**a**) and posterior (**b**) boundaries of the **lower limb** with the abdomen and the pelvis. **a**, Anteriorly, the boundary between abdomen and lower limb (between inguinal region and anterior femoral region) is made evident by a line drawn from the anterior superior iliac spine to the pubic tubercle and is marked superficially by the inguinal fold. **b**, Posteriorly, the boundary between abdomen and lower limb is marked by the iliac crest that, starting from the posterior superior iliac spine, divides the lateral abdominal region (flank) and the costoiliac region from the gluteal region, while the boundary between pelvis and lower limb is marked by the lateral margin of the sacrum. The figure displays also the surface topographic regions of the lower limb.
For the landmark points ☞ General Anatomy, Limbs, Topographic Landmarks, vol. 1 page 118.

➤

The **lower limb regions** are delimited by lines marked in correspondence to bone prominences (iliac spines, sciatic tuberosity, greater trochanter, medial and lateral epicondyles, patella margin, medial and lateral malleoli) or muscles visible through the cutaneous layers (i.e. the sartorius and adductor longus, which constitute the sides of the Scarpa's triangle).

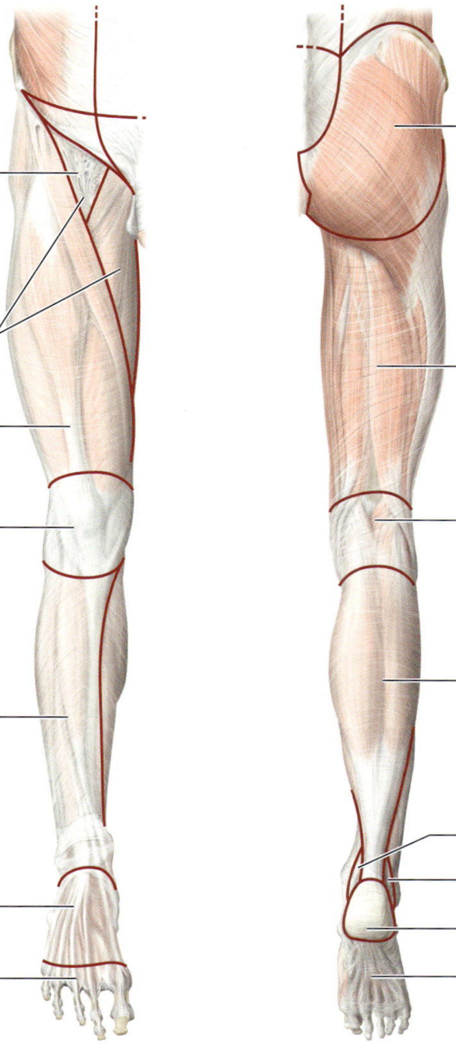

REGIONS OF THE LOWER LIMB	*BOUNDARIES OR LOCALIZATION*
Anterior	
– of knee	Horizontal line drawn approximately 3 cm above the patella (superiorly); horizontal plane passing through the tuberosity of the tibia (inferiorly)
– of thigh (anterior femoral)	Iliac crest (superiorly); sartorius (medially); horizontal line drawn approximately 3 cm above the patella (inferiorly)
– – femoral triangle	Medial margin of the sartorius (laterally); inguinal ligament (superiorly); lateral margin of the gracilis (medially)
– – – Scarpa's triangle	Inguinal ligament (superiorly); medial margin of the sartorius (laterally); lateral margin of the adductor longus (medially)
– of leg (anterior crural)	Horizontal plane passing through the tuberosity of the tibia (superiorly); line joining the lateral and medial malleoli, passing approximately 2 cm below the line of the ankle joint (inferiorly); anterior margin of the tibia (medially)
Of calcaneus (calcaneal)	Corresponds to the calcaneal tuberosity
Dorsum (dorsal region) of foot	Line joining the lateral and medial malleoli passing approximately 2 cm below the line space of the ankle joint (superoposteriorly); posterior concave line passing through the commissure of the toes (inferoanteriorly)
Gluteal	Iliac crest (superiorly); gluteal fold (inferiorly); intergluteal cleft and curve line, medially concave, between the intergluteal cleft and the iliac crest (medially)
Sole (plantar region)	Calcaneal tuberosity (posteriorly); anterior convex line passing through the roots of the toes or digitoplantar sulcus (anteriorly)
Posterior	
– of knee (with popliteal fossa)	Horizontal line drawn approximately 3 cm above the medial and lateral condyles of the femur (superiorly); horizontal plane passing through the tuberosity of the tibia (inferiorly). The popliteal fossa is a diamond shape depression observed when the knee is flexed, formed by two triangles united at the base: the upper one, or femoral, is limited medially by the semitendinous and semimembranous and laterally by the two heads of the biceps femoris; the lower one, or tibial, is delimited by the two lateral and medial heads of the gastrocnemius
– of thigh (posterior femoral)	Gluteal fold (superiorly); horizontal line drawn approximately 3 cm above the medial and lateral condyles of the femur (inferiorly)
– of the leg (posterior crural)	☞ Anterior of the leg
Lateral retromalleolar	Between the lateral malleolus and the calcaneal tendon

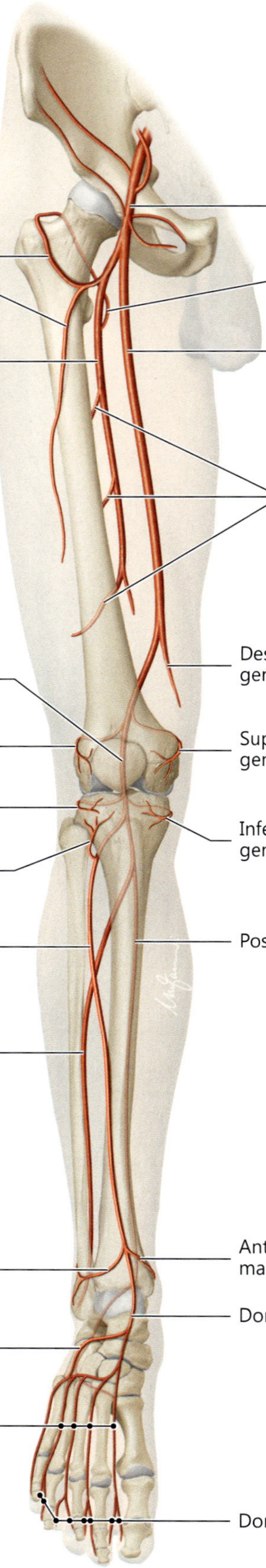

Representation of the **main arteries of the** right **lower limb**. Anterior view. The arterial arches of the foot are represented schematically.

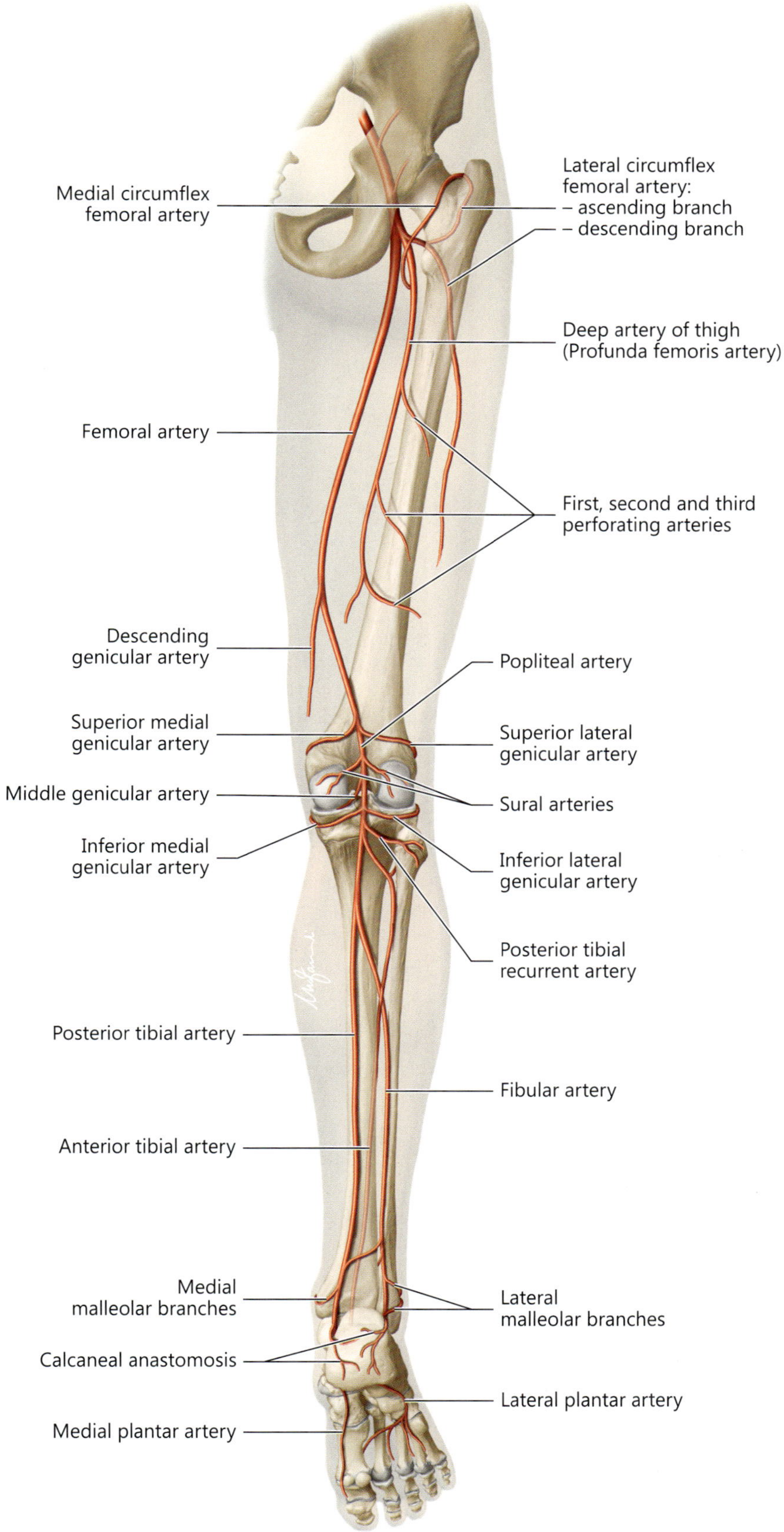

Representation of the **main arteries of the** right **lower limb**. Posterior view. The arterial arches of the foot are represented schematically.

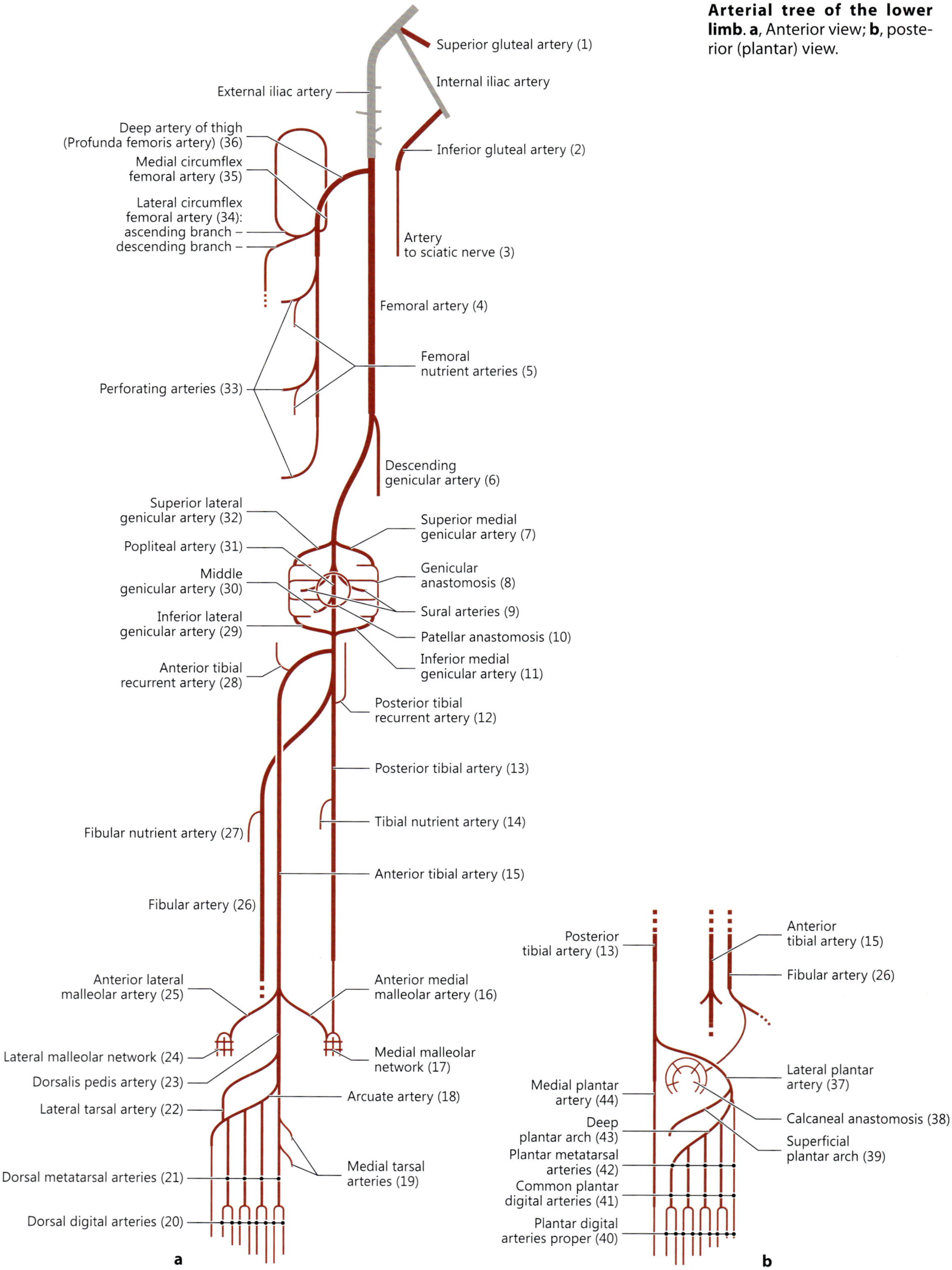

Arterial tree of the lower limb. a, Anterior view; **b**, posterior (plantar) view.

Arteries of the lower limb: territories of supply		
	ARTERIES	***TERRITORIES OF SUPPLY***
1	Superior gluteal artery	Piriformis, obturator internus, gluteus maximus, gluteus medius, gluteus minimus; skin of the sacral region; hip bone and hip joint
2	Inferior gluteal artery	Buttock and posterior region of the thigh
3	Artery to sciatic nerve	Associated to the sciatic nerve, penetrates and descends inside the nerve to the inferior part of the thigh
4	Femoral artery	Skin of the anterior abdominal wall, groin and perineum; thigh muscles, external genitalia, inguinal lymph nodes
5	Femoral nutrient arteries	Femur
6	Descending genicular artery	Skin of the anterior surface of the knee
7	Superior medial genicular artery	Anterior muscles of the thigh
8	Genicular anastomosis	Articular surfaces of femur and tibia, capsule and synovial membrane of the knee joint, patella and subcutaneous tissue around the patella
9	Sural arteries	Gastrocnemius, soleus, plantaris and overlying skin
10	Patellar anastomosis	Patella and subcutaneous tissue around the patella
11	Inferior medial genicular artery	Capsule of the knee joint and popliteus
12	Posterior tibial recurrent artery	Knee joint and tibiofibular joint; popliteus
13	Posterior tibial artery	Tibialis posterior, flexor digitorum longus, flexor hallucis longus and medial part of the soleus, skin of the medial surface of the leg
14	Tibial nutrient artery	Tibia
15	Anterior tibial artery	Tibialis anterior, extensor hallucis longus, and extensor digitorum longus
16	Anterior medial malleolar artery	Medial malleolus, ankle joint and overlying skin
17	Medial malleolar network	Ankle joint
18	Arcuate artery	Dorsal region of the foot
19	Medial tarsal artery	Medial surface of the foot
20	Dorsal digital arteries	Dorsal surface of the toes
21	Dorsal metatarsal arteries	Dorsal region of the foot
22	Lateral tarsal artery	Tarsal bone, overlying skin and extensor digitorum brevis
23	Dorsalis pedis artery	Tarsal bone and extensor hallucis longus and extensor digitorum longus
24	Lateral malleolar network	Ankle joint
25	Anterior lateral malleolar artery	Lateral malleolus, lateral surface of the calcaneus, inferior tibiofibular joint and ankle joint and overlying skin
26	Fibular artery	Soleus and other muscles of the deep layers of the leg, lateral and posterior surfaces of the ankle, calcaneus
27	Fibular nutrient artery	Fibula
28	Anterior tibial recurrent artery	Anterior and lateral regions of the knee joint, tibialis anterior, patella
29	Inferior lateral genicular artery	Capsule of the knee joint and popliteus
30	Middle genicular artery	Ligaments and synovial membrane of the knee joint
31	Popliteal artery	Thigh muscles in the knee region, femur, patella and tibia
32	Superior lateral genicular artery	Anterior thigh muscles
33	Perforating arteries	Adductor brevis, adductor magnus, biceps femoris and gluteus maximus and muscles of the posterior region of the thigh
34	Lateral circumflex femoral artery	Greater trochanter of the femur and quadriceps femoris
35	Medial circumflex femoral artery	Hip joint and posterior thigh muscles
36	Deep artery of thigh (Profunda femoris artery)	Vastus medialis, pectineus, adductor longus, adductor brevis and adductor magnus
37	Lateral plantar artery	Quadratus plantae, abductor hallucis, flexor digitorum brevis and flexor digiti minimi brevis, skin of the lateral margin of the sole of the foot
38	Calcaneal anastomosis	Calcaneal tendon, calcaneus
39	Superficial plantar arch	Plantar region of the foot
40	Plantar digital arteries proper	Adjacent toe edges
41	Common plantar digital arteries	Toes
42	Plantar metatarsal arteries	Plantar region of the foot
43	Deep plantar arch	Plantar region of the foot, interossei muscles, toes
44	Medial plantar artery	Muscles, bones and joints of the medial part of the plantar region of the foot, skin of the medial margin of the foot

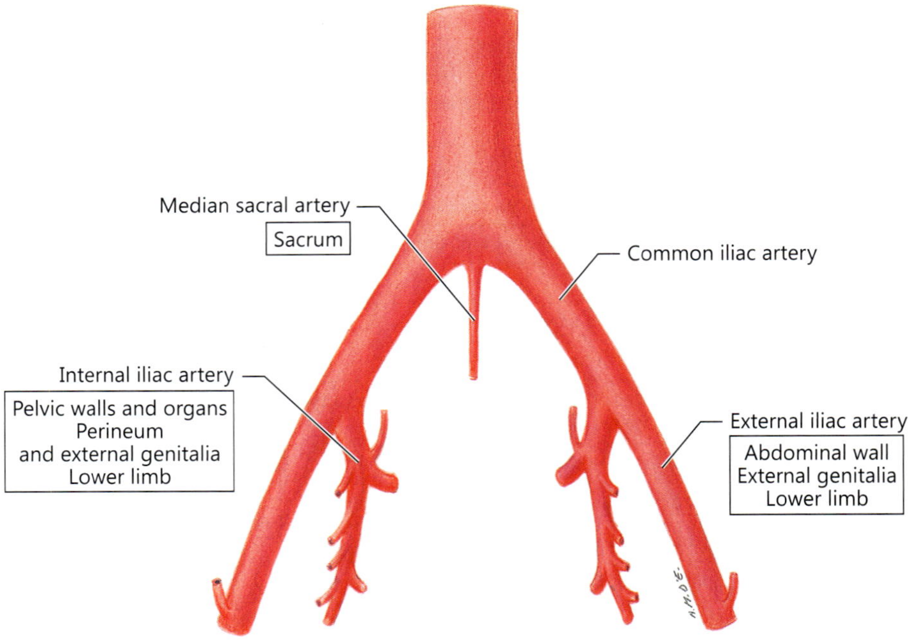

Schematic representation of the large collateral arteries of the terminal tract of the abdominal aorta. The common iliac artery with the two branch divisions, **internal iliac artery** and **external iliac artery**, provides the blood supply to the pelvic and lower limb regions. The territories of supply, respectively, are indicated next to the arteries.

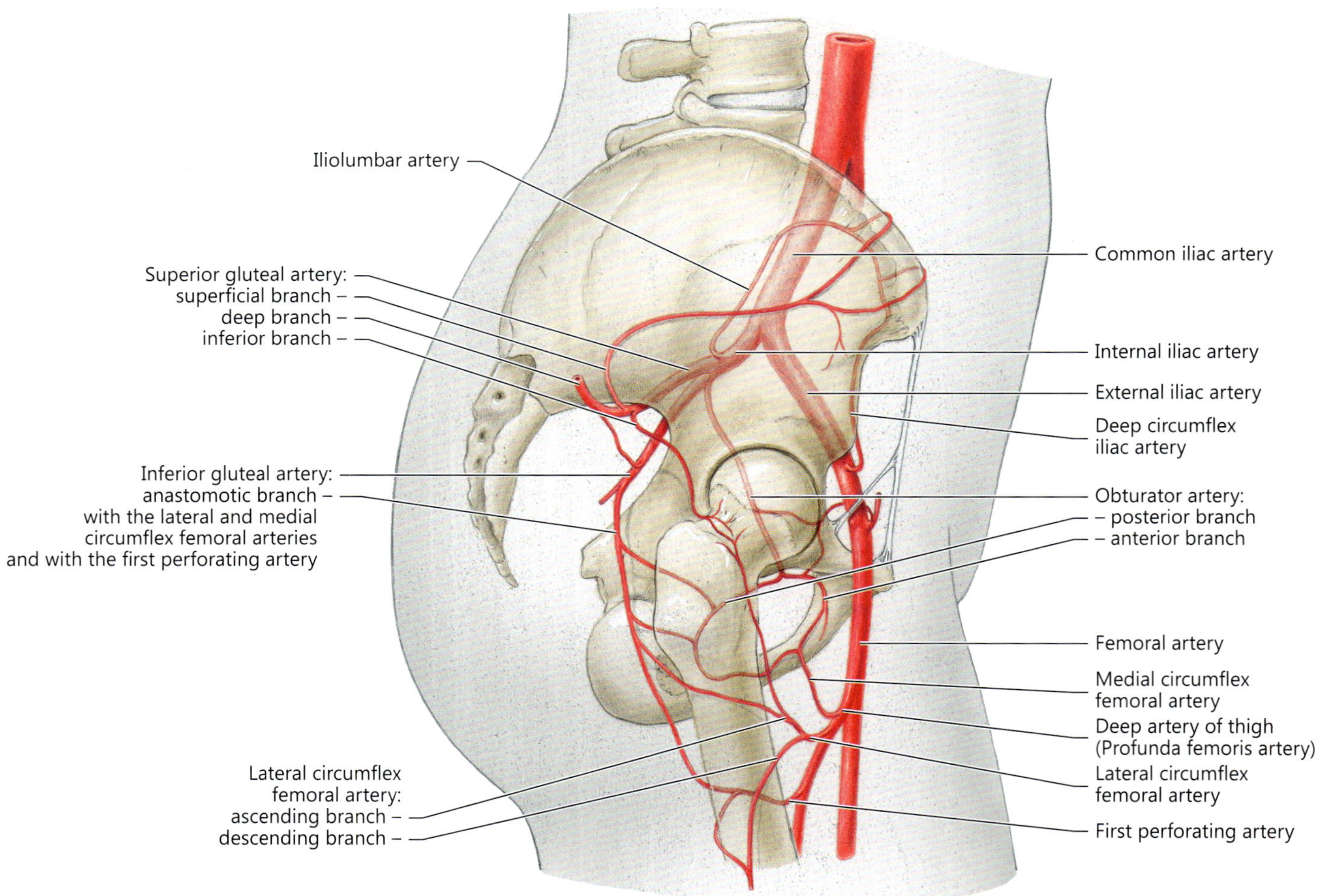

Anastomotic circle of the hip (right side). Several anastomotic branches connect the internal iliac, external iliac and femoral arteries to each other.

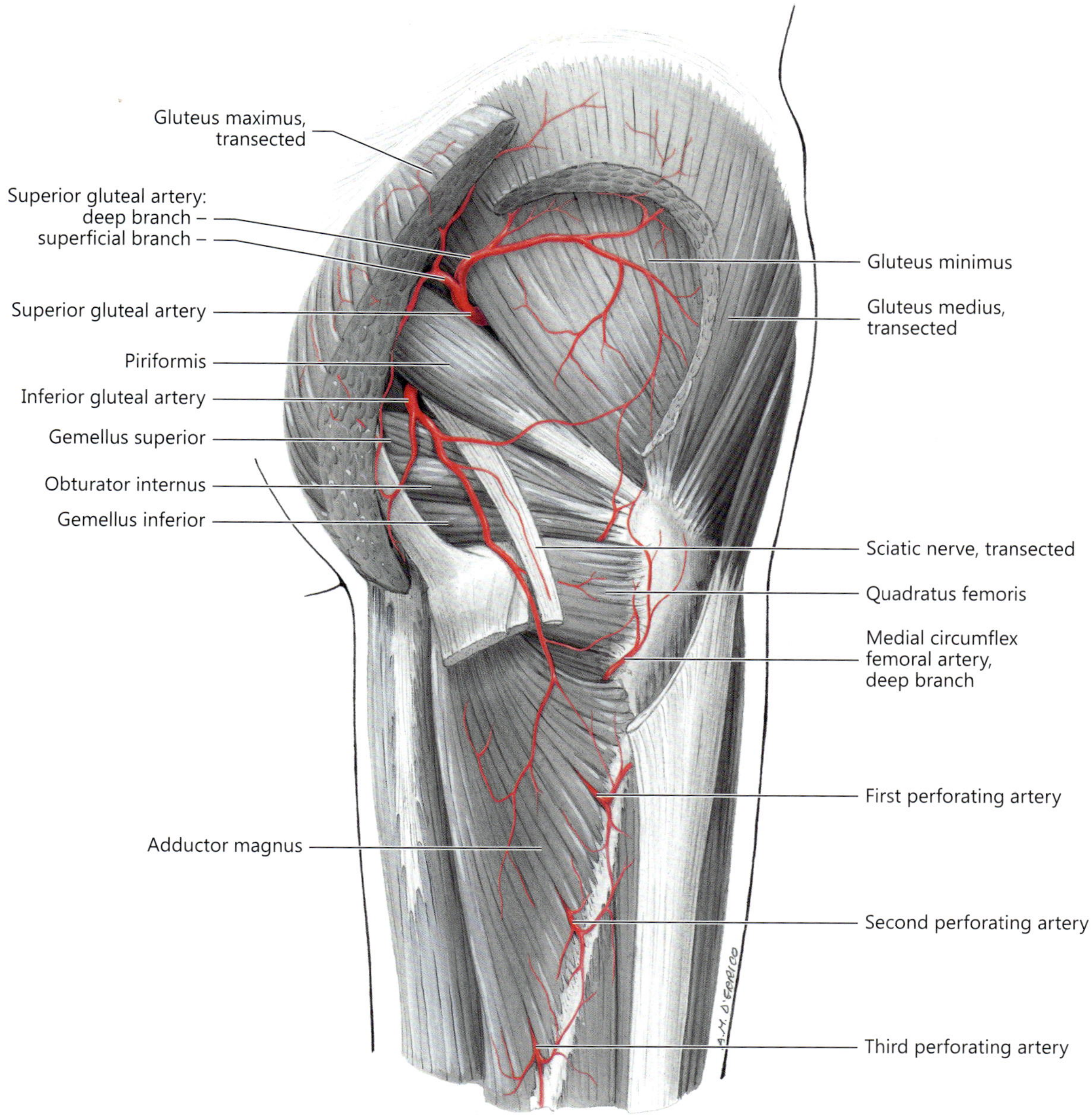

Course and relationships of the **gluteal arteries**: right side. Note the anastomoses between the branches of the inferior gluteal artery and those of the first perforating artery (of the deep artery of the thigh) and medial circumflex femoral artery that connect the internal iliac artery to the femoral artery.

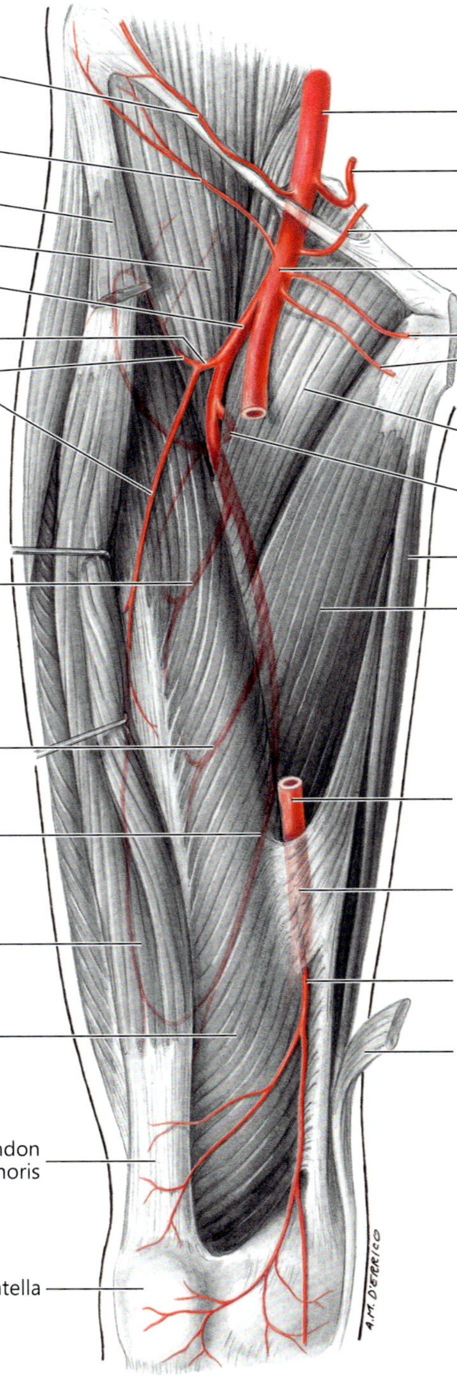

Arteries of the anterior region of the right thigh. The **femoral artery** tract, comprised between the origin of the deep artery of the thigh and the superior foramen of the adductor canal, has been removed in order to show the course and the branches of the deep artery of the thigh.

Right **genicular anastomosis**. The numerous anastomoses between the branches of the femoral, popliteal, anterior tibial and posterior tibial arteries, offer a good alternative of collateral circuits.

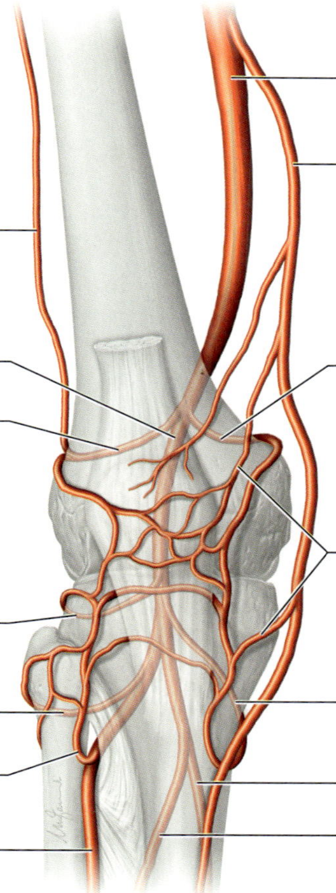

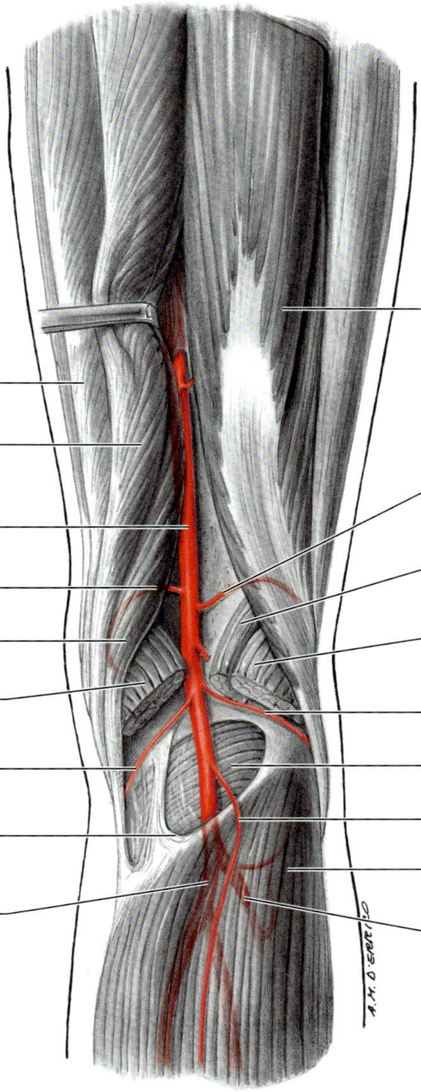

Course and branches of the right **popliteal artery**. The gastrocnemius has been transected and removed in order to show the tendinous arch of the soleus below to which the popliteal artery passes.

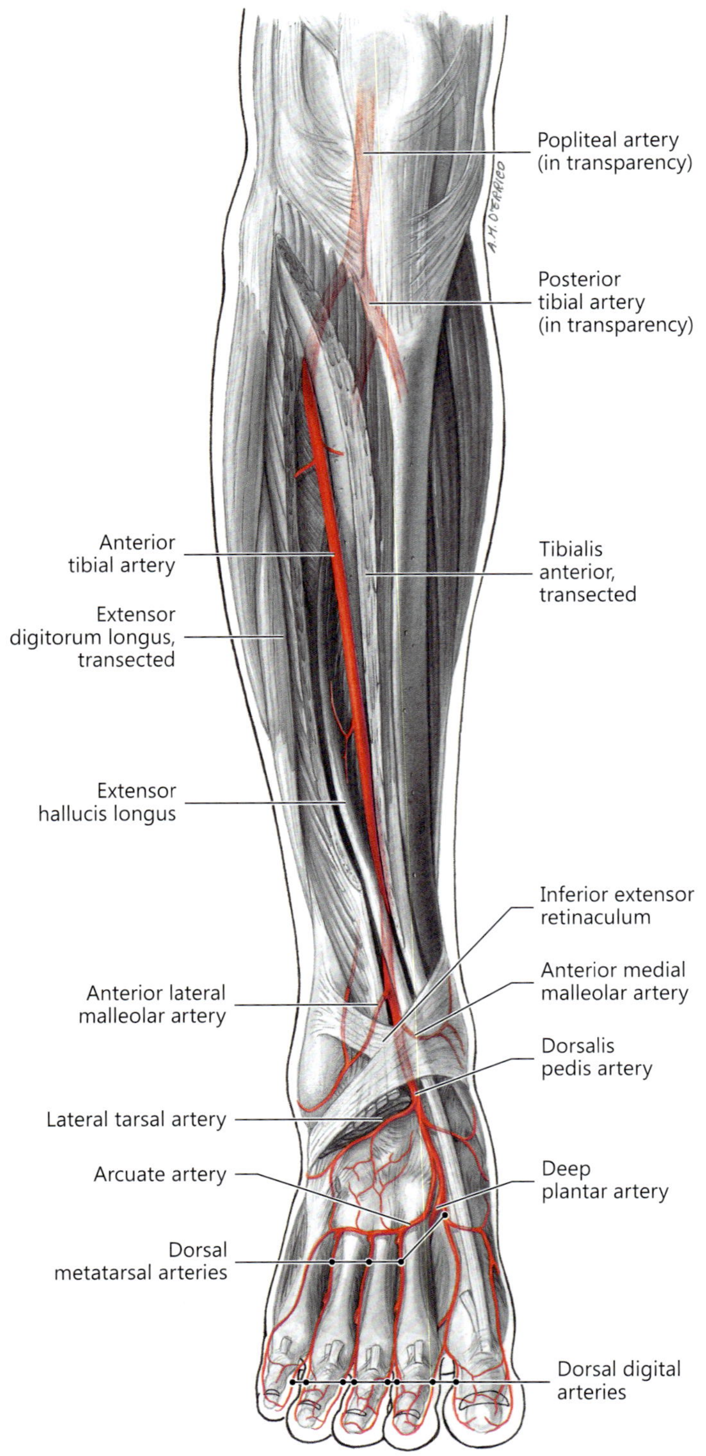

Course and branches of the right **anterior tibial** and **dorsalis pedis arteries**. The **arcuate artery** is in evidence.

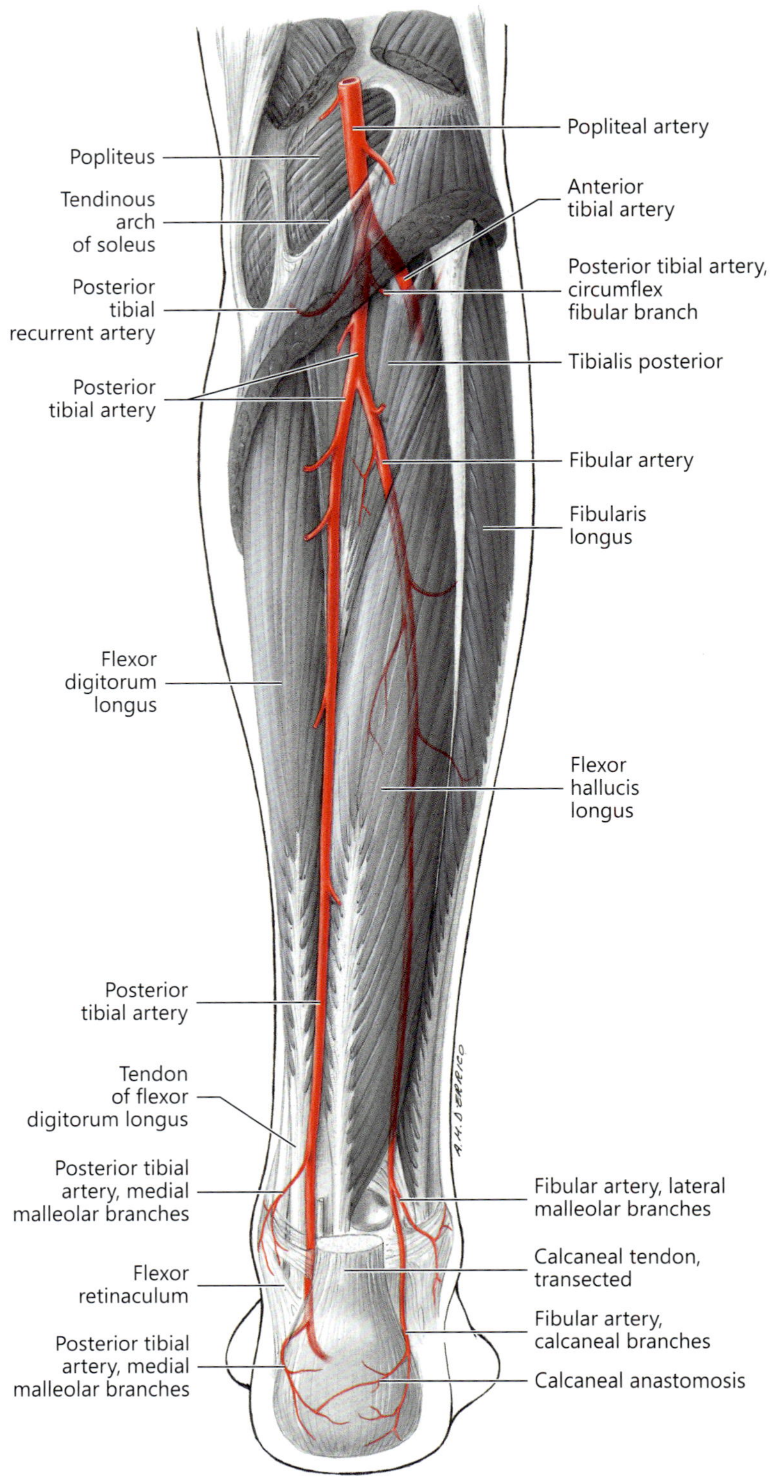

Course and branches of the right **posterior tibial artery**. The gastrocnemius and soleus have been transected and removed.

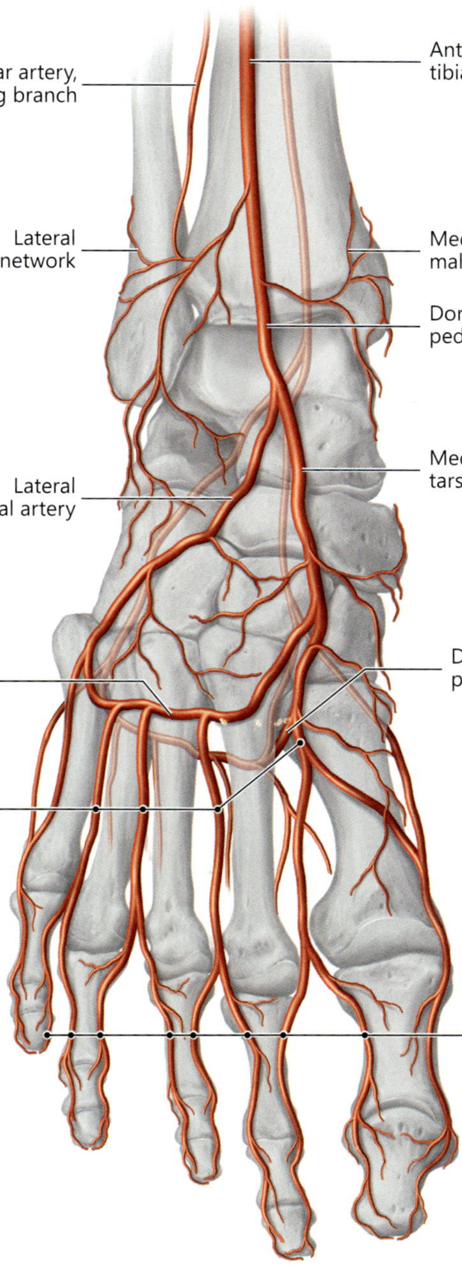

Arteries of the dorsum of the foot. From the convex margin of the arcuate artery, collateral branch of the dorsalis pedis artery, originate the dorsal metatarsal arteries that, in turn, bifurcate giving two dorsal digital arteries to each of them. The first dorsal metatarsal artery originates from the dorsalis pedis artery and supplies the great toe and the medial part of the second toe.

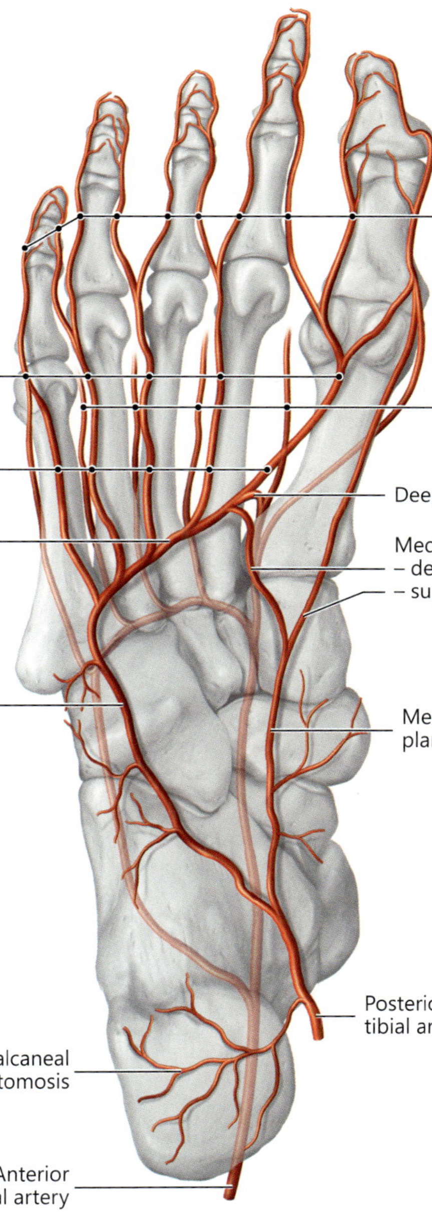

Arteries of the sole of the foot. The deep plantar arch develops from the base of the fifth metatarsal to the first interosseous metatarsal space, forming a distolateral convexity, deeply positioned and in contact with the osseous plane.

Summary table of the arteries of the lower limb

ARTERY	ORIGIN	BRANCHES	TERRITORIES OF SUPPLY
anterior lateral malleolar	Anterior tibial	Participates to the formation of the lateral malleolar network	Lateral malleolus, lateral surface of the calcaneus, inferior tibiofibular joint and ankle joint and overlying skin
anterior medial malleolar	Anterior tibial	Participates to the formation of the medial malleolar network	Medial malleolus, ankle joint and overlying skin
anterior tibial	Popliteal, under the tendinous arch of the soleus	Anterior tibial recurrent, posterior tibial recurrent, anterior lateral malleolar, anterior medial malleolar and dorsalis pedis arteries; lateral malleolar network	Tibialis anterior, extensor hallucis longus, and extensor digitorum longus
anterior tibial recurrent	Anterior tibial		Anterior and lateral regions of the knee joint, tibialis anterior, patella
arcuate (of the foot or metatarsal)	Dorsalis pedis		Dorsal region of the foot
artery to sciatic nerve	Inferior gluteal		Associated to the sciatic nerve, penetrates and descends inside the nerve to the inferior part of the thigh
calcaneal anastomosis	Fibular		Calcaneal tendon, calcaneus
common plantar digital	Plantar metatarsal	Plantar digital arteries proper	Toes
deep artery of thigh (profunda femoris)	Femoral	Medial circumflex femoral (from which the superficial, deep, acetabular, ascending and descending branches originate), lateral circumflex femoral (from which the ascending, descending and transverse branches originate) and perforating (from which the femoral nutrient arteries originate) arteries	Vastus medialis, pectineus, adductor longus, adductor brevis and adductor magnus
deep plantar	Dorsalis pedis	Anastomosis with the lateral plantar artery, participating to the formation of the deep plantar arch	Inferior and adjacent surfaces of the great toe and second toe
deep plantar arch (plantar arterial arch)	Lateral plantar	Plantar metatarsal arteries (from which the perforating branches and common plantar digital arteries originate, from which the plantar digital arteries proper originate)	Plantar region of the foot, interossei muscles, toes
descending genicular (highest genicular)	Femoral	Saphenous and articular branches; participates to the formation of the genicular anastomosis	Skin of the anterior surface of the knee
dorsal digital	Dorsalis pedis		Dorsal surface of the toes
dorsal metatarsal	Dorsalis pedis		Dorsal region of the foot
dorsalis pedis (dorsal artery of foot)	Continuation of the anterior tibial artery, at the level of the ankle joint	Lateral tarsal, medial tarsal, arcuate, dorsal metatarsal, dorsal digital and deep plantar arteries; anastomosis with lateral plantar artery	Tarsal bone and extensor hallucis longus and extensor digitorum longus
femoral	Continuation of the external iliac artery, distally to the inguinal ligament	Superficial epigastric, superficial circumflex iliac, superficial external pudendal and deep external pudendal (from which the inguinal, anterior labial and anterior scrotal branches originate), descending genicular (from which the saphenous and articular branches originate) arteries and deep artery of thigh	Skin of the anterior abdominal wall, groin and perineum; thigh muscles, external genitalia, inguinal lymph nodes
femoral nutrient	Perforating		Femur
fibular (peroneal)	Posterior tibial	Fibular nutrient artery and perforating, communicating and lateral malleolar branches, from which the calcaneal branches originate; calcaneal anastomosis	Soleus and other muscles of the deep layers of the leg, lateral and posterior surfaces of the ankle, calcaneus

continued

Summary table of the arteries of the lower limb			*(continued)*
ARTERY	*ORIGIN*	*BRANCHES*	*TERRITORIES OF SUPPLY*
fibular nutrient	Fibular		Fibula
genicular anastomosis	Popliteal		Articular surfaces of femur and tibia, capsule and synovial membrane of the knee joint, patella and subcutaneous tissue around the patella
inferior gluteal (sciatic)	From the anterior tract of the internal iliac artery	Artery to sciatic nerve; anastomosis with the lateral circumflex femoral, medial circumflex femoral and first perforating arteries	Levator ani, ischiococcygeus, piriformis, gluteus maximus, obturator internus, gemellus superior, gemellus inferior, quadratus femoris and adductor magnus, superior part of posterior muscles of the thigh; hip joint; buttock and posterior region of the thigh; bladder, prostate, seminal gland
inferior lateral genicular	Popliteal	Participates to the formation of the genicular anastomosis	Capsule of the knee joint and popliteus
inferior medial genicular	Popliteal	Participates to the formation of the genicular anastomosis	Capsule of the knee joint and popliteus
lateral circumflex femoral (lateral circumflex of the thigh or lateral femoral)	Deep artery of thigh	Ascending, descending and transverse branches	Greater trochanter of the femur and quadriceps femoris
lateral malleolar network	Anterior tibial		Ankle joint
lateral plantar	Posterior tibial	Plantar metatarsal arteries (from which perforating branches and common plantar digital arteries originate, from which, in turn, the plantar digital arteries proper originate); deep plantar and superficial plantar arches; anastomosis with the deep plantar artery	Quadratus plantae, abductor hallucis, flexor digitorum brevis and flexor digiti minimi brevis, skin of the lateral margin of the sole of the foot
lateral tarsal (tarsal)	Dorsalis pedis	Participates to the formation of the dorsal network of the foot	Tarsal bone, overlying skin and extensor digitorum brevis
medial circumflex femoral (medial circumflex of the thigh or medial femoral)	Deep artery of thigh	Superficial, deep, acetabular, ascending and descending branches	Hip joint and posterior thigh muscles
medial malleolar network	Posterior tibial		Ankle joint
medial plantar (internal plantar)	Posterior tibial	Deep and superficial branches	Muscles, bones and joints of the medial part of the plantar region of the foot, skin of the medial margin of the foot
medial tarsal	Dorsalis pedis		Medial surface of the foot
middle genicular (azygos articular)	Popliteal		Ligaments and synovial membrane of the knee joint
patellar anastomosis	Popliteal		Patella and subcutaneous tissue around the patella
perforating (first, second and third perforating artery)	Deep artery of thigh	Femoral nutrient arteries	Adductor brevis, adductor magnus, biceps femoris and gluteus maximus and muscles of the posterior region of the thigh
plantar digital arteries proper	Common plantar digital		Adjacent toe edges
plantar metatarsal	Lateral plantar	Common plantar digital arteries (from which the plantar digital arteries proper originate) and perforating branches	Plantar region of the foot
popliteal	Continuation of the femoral artery, at the level of the adductor hiatus	Superior lateral genicular, superior medial genicular, middle genicular, sural, inferior lateral genicular, inferior medial genicular, anterior tibial and posterior tibial arteries; genicular anastomosis and patellar anastomosis	Thigh muscles in the knee region, femur, patella and tibia

continued

Summary table of the arteries of the lower limb *(continued)*

ARTERY	*ORIGIN*	*BRANCHES*	*TERRITORIES OF SUPPLY*
posterior tibial	Popliteal	Tibial nutrient, medial plantar, lateral plantar arteries and fibular arteries; circumflex fibular, medial malleolar and calcaneal branches; medial malleolar network	Tibialis posterior, flexor digitorum longus, flexor hallucis longus and medial part of the soleus, skin of the medial surface of the leg
posterior tibial recurrent	Anterior tibial		Knee joint and tibiofibular joint; popliteus
superficial plantar arch	Lateral plantar		Plantar region of the foot
superior gluteal	Direct continuation of the posterior division of the internal iliac artery	Superficial and deep branches (from which the superior and inferior branches originate)	Piriformis, obturator internus, gluteus maximus, gluteus medius, gluteus minimus; skin of the sacral region; hip bone and hip joint
superior lateral genicular	Popliteal	Participates to the formation of the genicular anastomosis	Anterior thigh muscles
superior medial genicular	Popliteal	Participates to the formation of the genicular anastomosis	Anterior thigh muscles
sural	Popliteal, at the level of the knee joint		Gastrocnemius, soleus, plantaris and overlying skin
tibial nutrient (the most voluminous of all the nutrient arteries)	Posterior tibial		Tibia

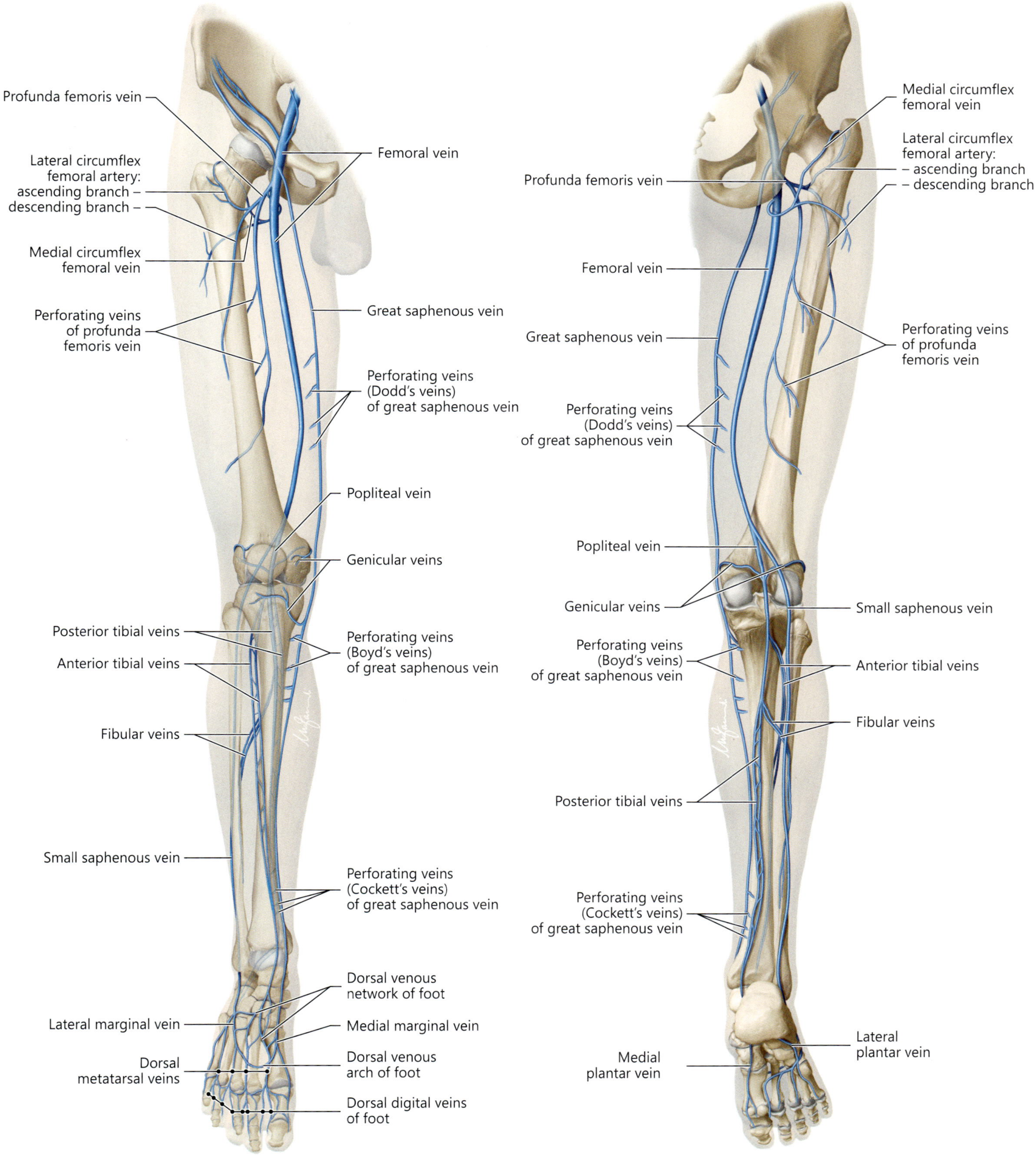

Representation of the right **superficial veins of the lower limb**. Of note is the origin of the great saphenous vein from the medial marginal vein, and of the small saphenous vein from the lateral marginal vein. The opening of the small saphenous vein in the popliteal vein is visible for transparency.

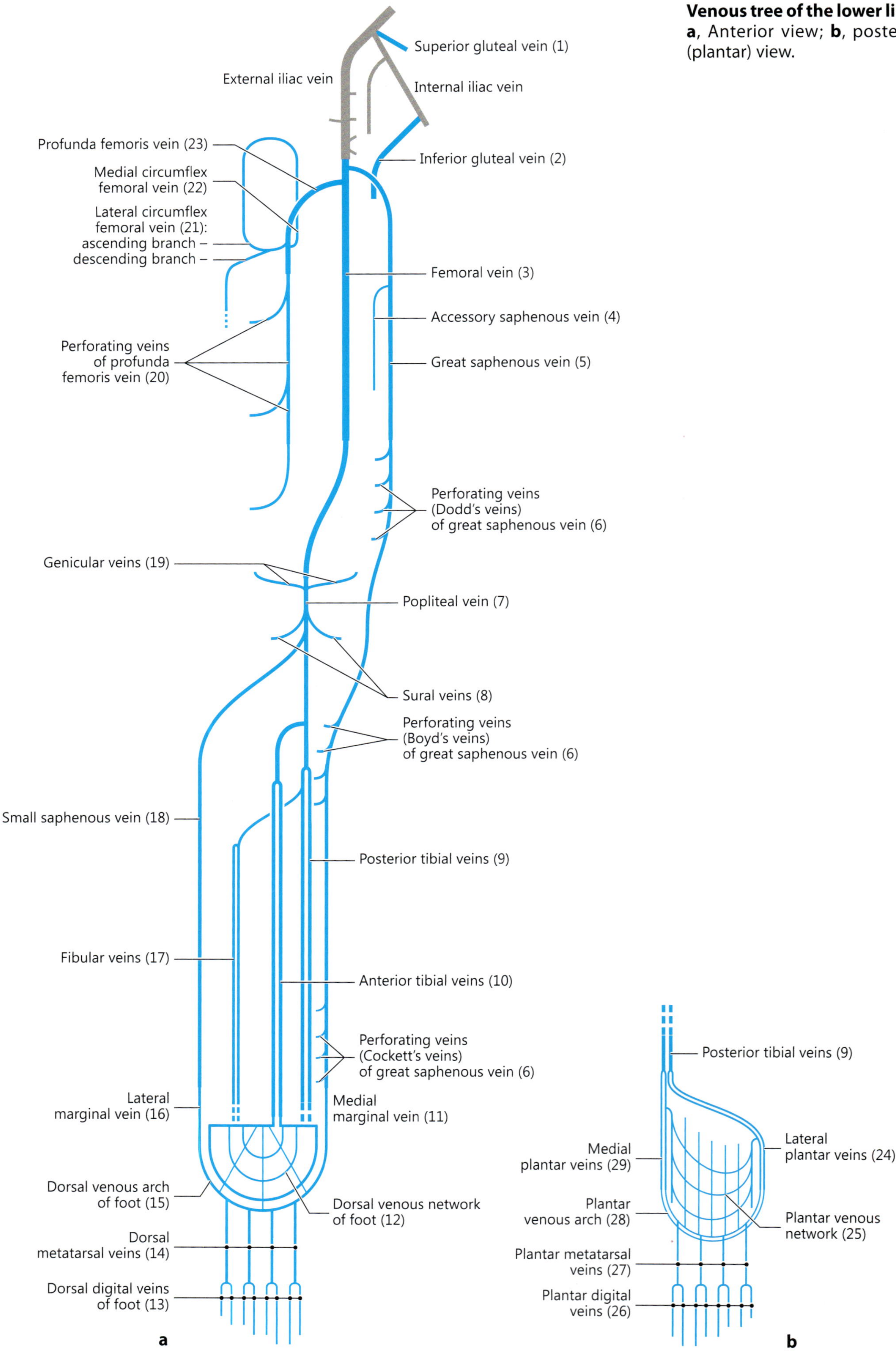

Venous tree of the lower limb. **a**, Anterior view; **b**, posterior (plantar) view.

	Veins of the lower limb: territories of drainage	
	VEINS	**TERRITORIES OF DRAINAGE**
1	Superior gluteal vein	Gluteal region
2	Inferior gluteal vein	Posterosuperior region of the thigh and buttock
3	Femoral vein	Superficial and deep veins of the foot, leg and greater part of the thigh
4	Accessory saphenous vein	Posteromedial surface of the thigh
5	Great saphenous vein	Dorsal venous network of the foot, medial part of the plantar venous network, anteromedial region of the leg, posteromedial surface of the thigh
6	Perforating veins of great saphenous vein (Dodd's veins, Boyd's veins, Cockett's veins)	Territory of drainage of the great saphenous vein
7	Popliteal vein	Knee joint and gastrocnemius
8	Sural veins	Soleus and gastrocnemius
9	Posterior tibial veins	Plantar region of the foot and muscles of the posterior region of the leg
10	Anterior tibial veins	Dorsal region of the foot
11	Medial marginal vein	Dorsal venous arch of the foot
12	Dorsal venous network of foot	Dorsal region of th foot, which receives the superficial veins
13	Dorsal digital veins of foot	Intercapitular veins (anastomotic branches of the plantar digital veins)
14	Dorsal metatarsal veins	Metatarsals and adjacent muscles
15	Dorsal venous arch of foot	Dorsal venous network of the foot and plantar venous arch
16	Lateral marginal vein	Dorsal venous arch of the foot
17	Fibular veins	Calcaneus, muscles of the leg, tibiofibular syndesmosis
18	Small saphenous vein	Dorsal venous network of the foot, lateral part of the plantar venous network, posterior region of the leg, posterior superficial region of the thigh
19	Genicular veins	Deep genicular venous plexus
20	Perforating veins of profunda femoris vein	Posterior and deep medial muscles of the thigh
21	Lateral circumflex femoral vein	Muscles of the thigh, particularly the quadriceps femoris
22	Medial circumflex femoral vein	Posterior muscles of the thigh and hip joint
23	Profunda femoris vein	Anteromedial region of the thigh
24	Lateral plantar veins	Plantar venous arch
25	Plantar venous network	Plantar metatarsal veins
26	Plantar digital veins	Venous plexus on the plantar surface of the foot
27	Plantar metatarsal veins	Metatarsals and adjacent muscles
28	Plantar venous arch	Plantar metatarsal veins and plantar venous network
29	Medial plantar veins	Plantar venous arch

In the venous circulation, the blood flows from the superficial veins towards the deep veins. The main superficial veins flow directly into the deep veins (for example, the great saphenous vein into the femoral vein). However, numerous perforating veins connect the two venous planes, enabling **blood flow** from the superficial veins towards the deep veins. The direction of the flow is regulated by the presence of valves. The venous return to the heart from the subdiaphragmatic regions is favored by the compression exerted by the muscle bellies on the deep veins running in the intermuscular connective septa (venous muscular pump) and by the aspiration exerted on the thoracic venous vessels during inspiration.

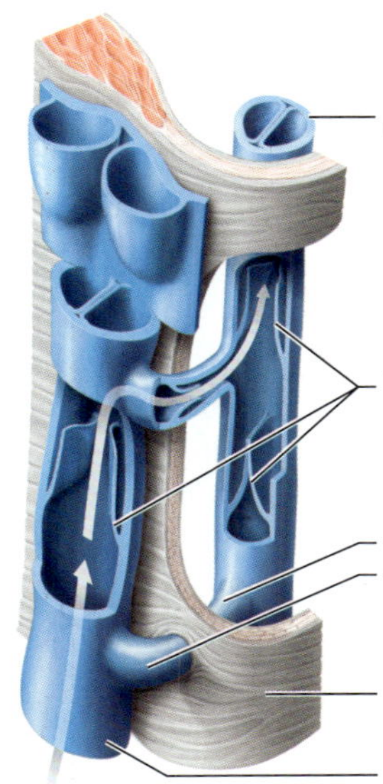

Dorsum of the foot. The **dorsal venous network of the foot**, on the surface, discharges in the **venous arch of the foot** which, through the **medial and lateral marginal veins**, flows, respectively, in the **great saphenous and small saphenous veins.** The **deep veins of the dorsum of the foot** start at the level of the dorsal intermetatarsal spaces as dorsal metatarsal veins and give rise to the **anterior tibial veins**.

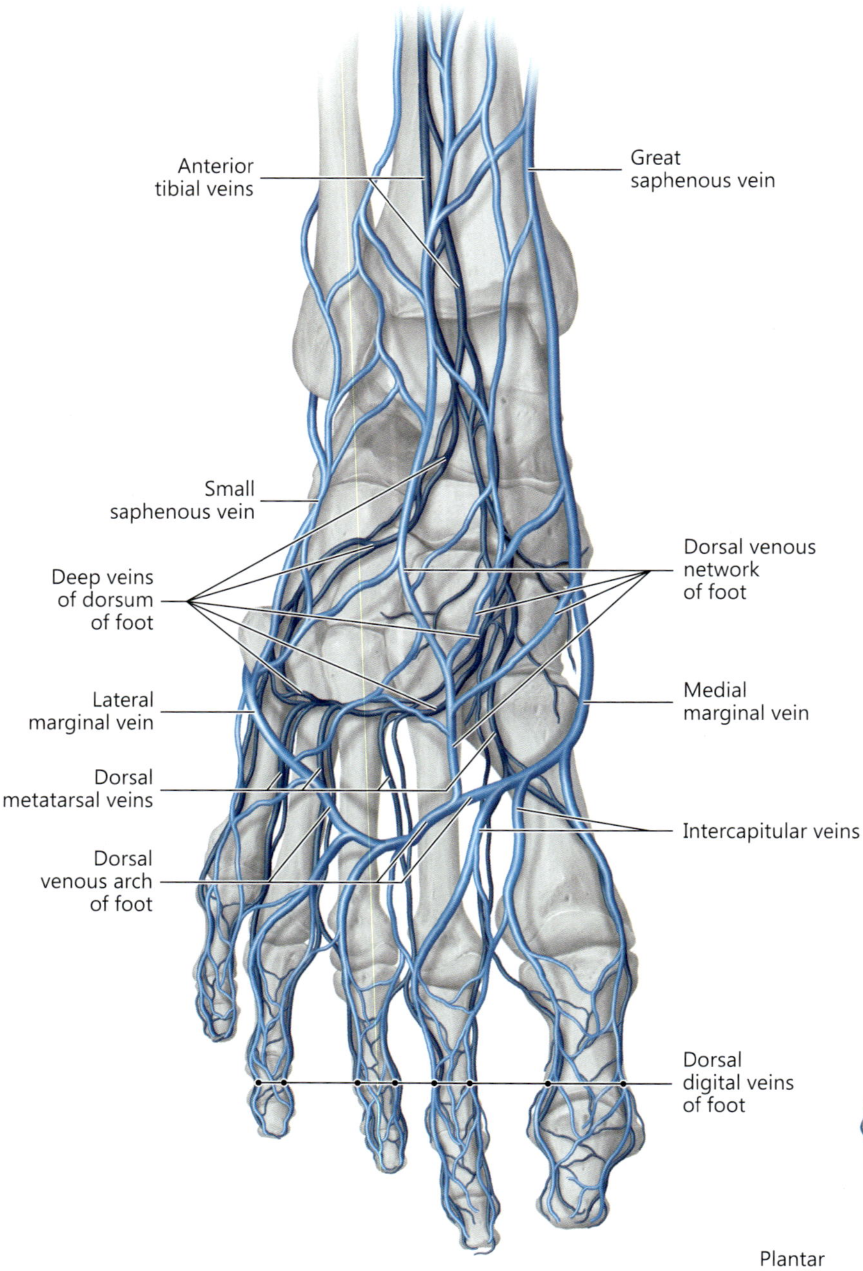

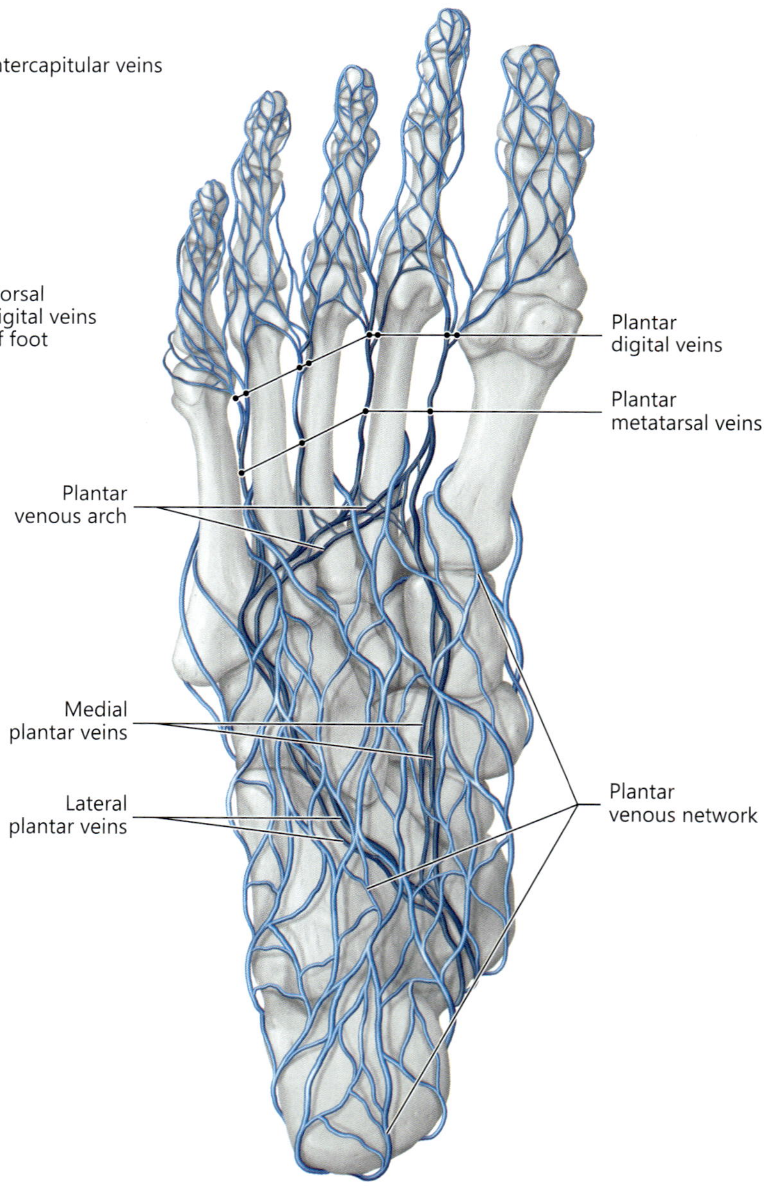

Veins of the sole of the foot. The four **plantar metatarsal veins**, originating from the confluence of the **plantar digital veins**, run in the respective intermetatarsal spaces and open in the **plantar venous arch**. The plantar venous arch opens in the two **lateral plantar veins**, the confluence of which with the **medial plantar veins** forms the posterior tibial veins. The **plantar venous network** extends in the subcutaneous tissue of the sole of the foot and opens in the plantar venous arch and in the medial and lateral marginal veins.

Right foot, medial view. **Deep veins of the dorsum and sole of the foot.** The dorsal metatarsal veins originate the **anterior tibial veins**. The plantar metatarsal veins drain in the plantar venous network on the surface and originate, through the plantar venous arch and the medial and lateral plantar veins, the **posterior tibial veins.** The deep veins of the leg accompany one another to the homonymous arteries, two for each artery.

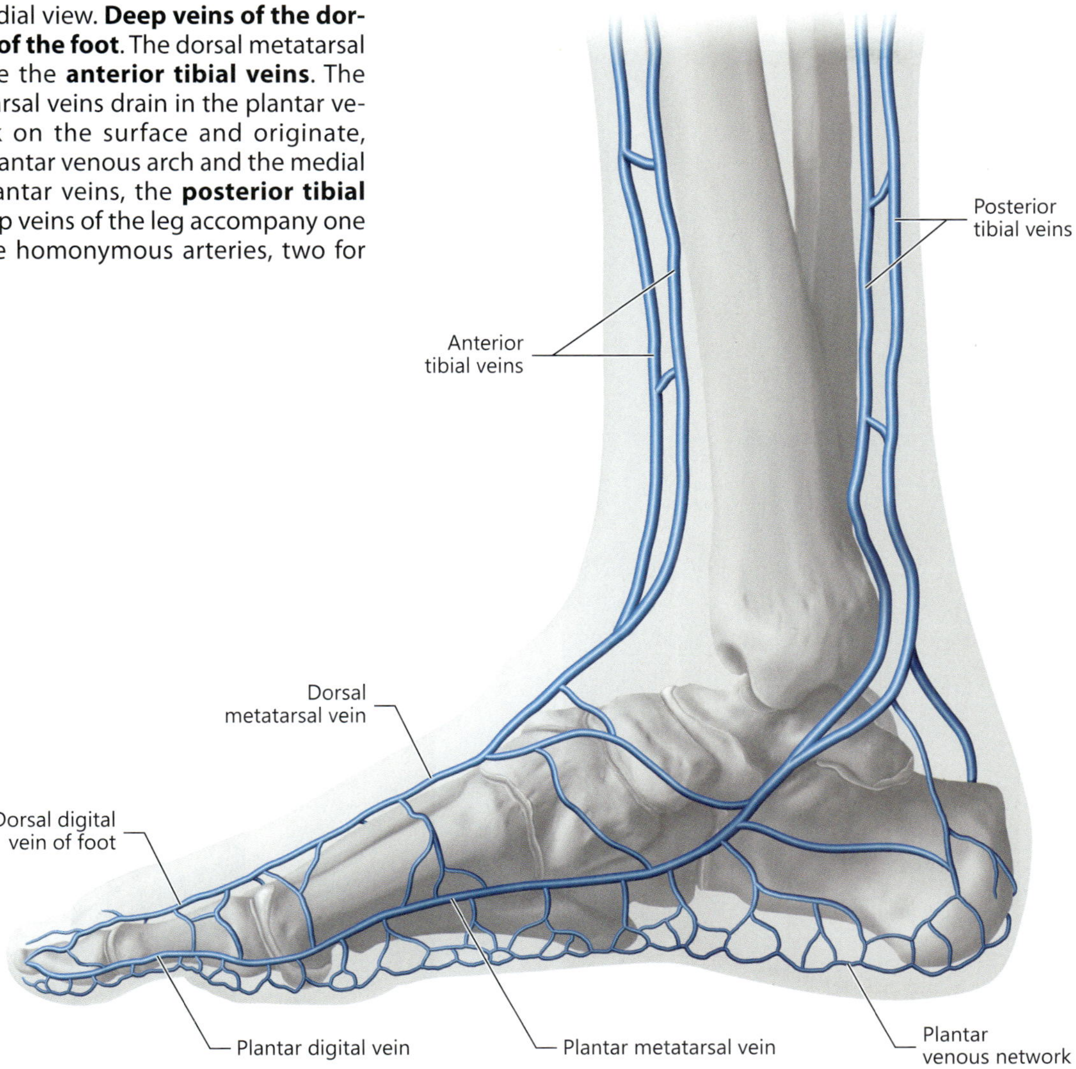

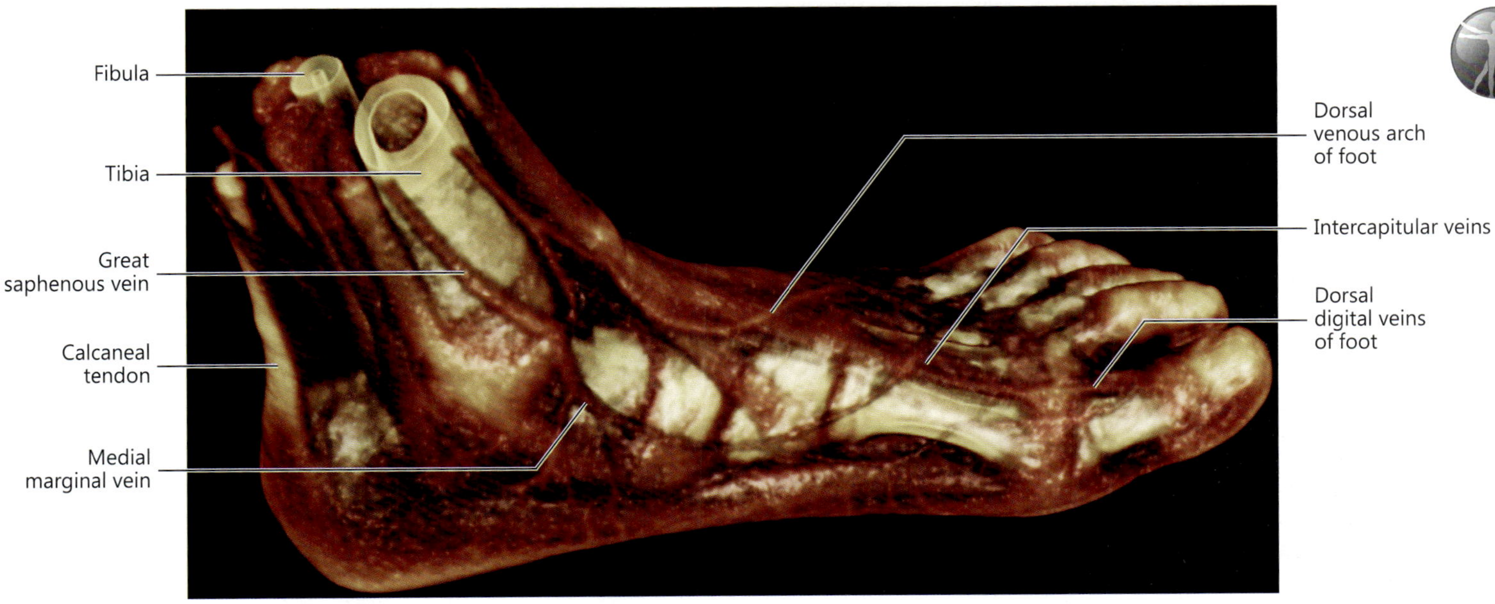

Three-dimensional image of a living human body, obtained by CT scan using a 3D volume rendering technique, showing the **superficial venous circulation of the** left **foot**; medial view.

Three-dimensional image of a living human body, obtained by CT scan using a 3D volume rendering technique, showing the muscles and the **venous circulation of the thigh and leg**; medial view.

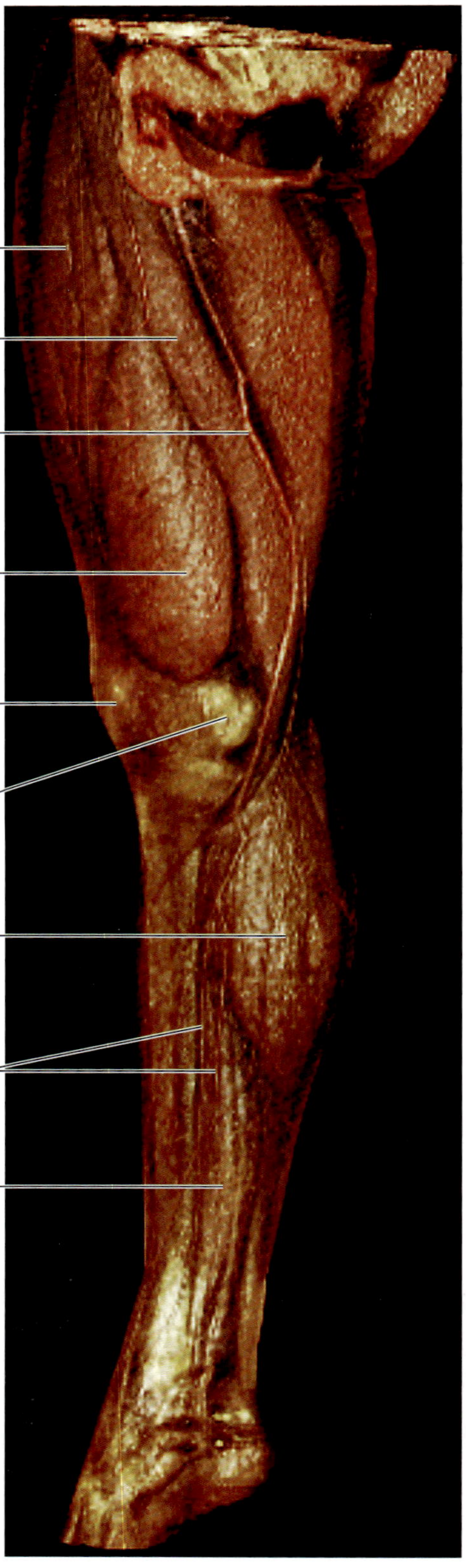

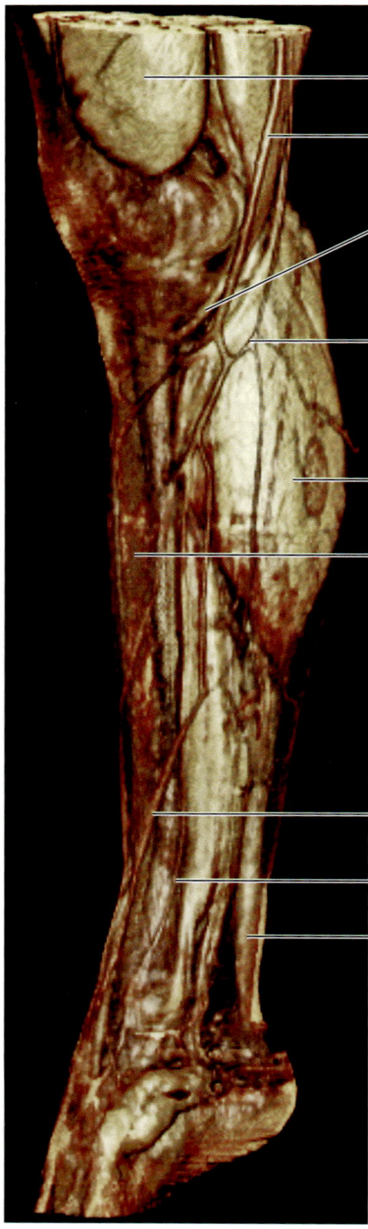

Three-dimensional image of a living human body, obtained by CT scan using a 3D volume rendering technique, showing the **venous circulation of the leg**; medial view.

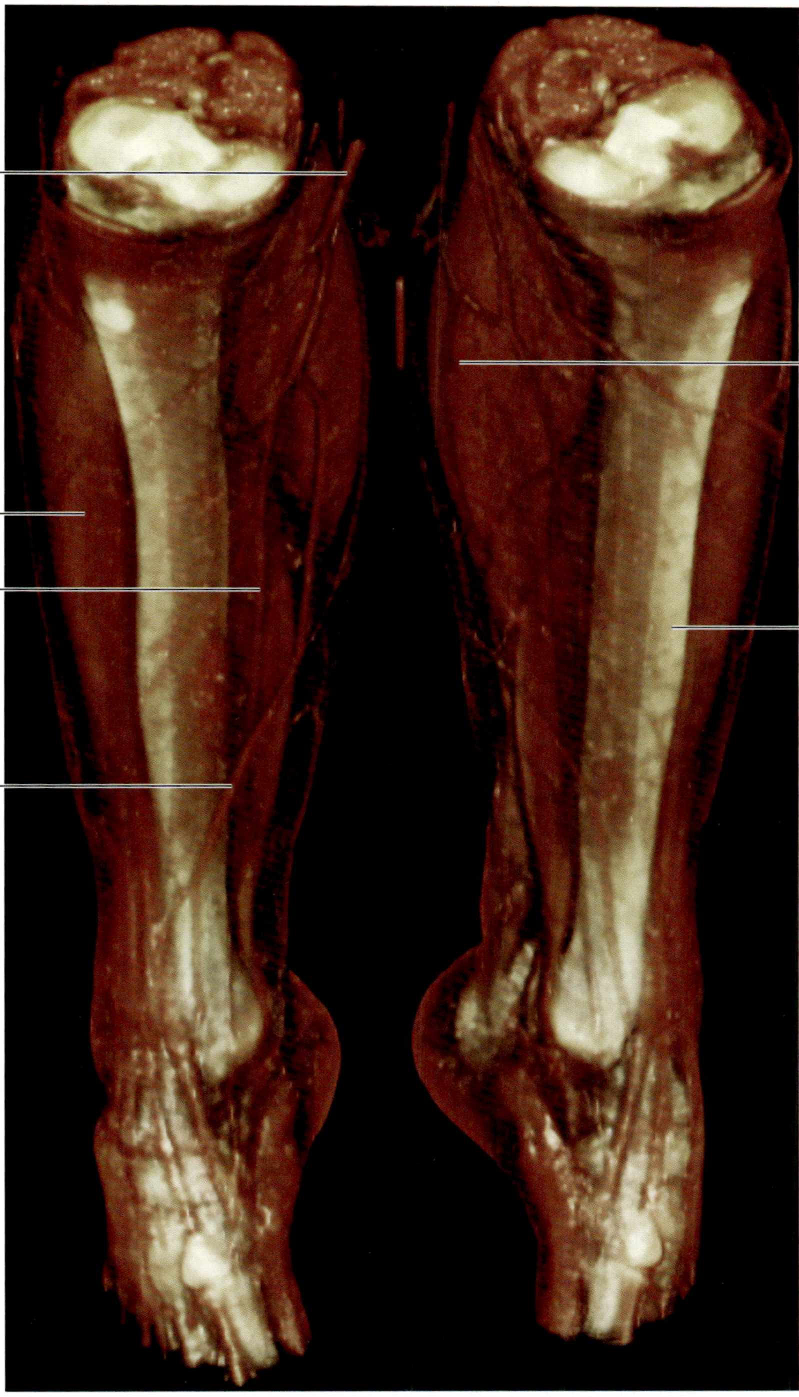

Three-dimensional image of a living human body, obtained by CT scan using a 3D volume rendering technique, showing the **venous circulation of the leg**; anterosuperior view.

Summary table of the veins of the lower limb

VEIN	ORIGIN/LOCATION	DRAINAGE	CONFLUENCE
accessory saphenous	Arises at the level of the posteromedial surface of the thigh and flows through the great saphenous vein at a variable height	Posteromedial surface of the thigh	Great saphenous vein
anterior tibial	Originate from the dorsal metatarsal veins and ascend along the anterior region of the leg, passing through the interosseous membrane to reach the posterior region of the leg	Dorsal region of the foot, which receives the deep veins	Popliteal vein
dorsal digital veins of foot	Along the margins of the toes	Intercapitular veins (anastomotic branches of the plantar digital veins)	Dorsal metatarsal veins
dorsal metatarsal (common digital)	Dorsal intermetatarsal spaces	Metatarsals and adjacent muscles; receive the dorsal digital veins	Anterior tibial veins
dorsal venous arch of foot	At the level of the metatarsals	Dorsal venous network of the foot and plantar venous arch	Medial and lateral marginal veins
dorsal venous network of foot	Dorsum of the foot	Dorsal region of the foot, which receives the superficial veins	Dorsal venous arch of foot
femoral	Continuation of the popliteal vein, starting at the level of the termination of the adductor canal and ending behind the inguinal ligament	Superficial and deep venous circulation of the foot, leg and greater part of the thigh; receives the profunda femoris and great saphenous veins; the superficial circumflex iliac, superficial epigastric and external pudenda veins drain directly or, indirectly, by way of the great saphenous vein, into the femoral vein; even the superficial dorsal veins of the penis ♂ and the clitoris ♀ drain into the femoral or great saphenous veins	External iliac vein, at the level of the inguinal ligament
femoropopliteal	Superficial vein, not always present, that descends and joins the small saphenous vein before it flows into the popliteal vein	Posterior superficial region of the thigh	Small saphenous vein
fibular (peroneal)	From the lateral surface of the calcaneus up to below the knee, runing through the posterior region of the leg	Calcaneus, muscles of the leg, tibiofibular syndesmosis	Posterior tibial veins
genicular	Knee	Deep genicular venous plexus	Popliteal vein
great saphenous (long saphenous) (the longest vein in the body)	Continuation of the medial marginal vein of the foot, 2-3 cm in front of the medial malleolus, ascends to the femoral triangle, opening into the femoral vein 3-4 cm before the inguinal ligament	Dorsal venous network of foot, medial part of the plantar venous network, anteromedial region of the leg, posteromedial surface of the thigh; receives the medial marginal and accessory saphenous veins; the superficial circumflex iliac, superficial epigastric and external pudenda veins drain into the great saphenous vein or, directly, into the femoral vein; also the superficial dorsal veins of the penis ♂ and clitoris ♀ drain into the great saphenous vein or into the femoral vein; anastomosis with the small saphenous vein	Femoral vein
inferior gluteal (sciatic)	From the proximal part of the posterior region of the thigh, they enter the pelvis by way of the great sciatic foramen	Posterosuperior region of the thigh and buttock; anastomosis with the medial circumflex femoral veins and with the perforating veins, tributaries of the profunda femoris vein	Internal iliac vein
intercapitular	Pass through the interdigital spaces	Anastomosis of the plantar digital veins with the dorsal digital veins	Connect the plantar venous arch with the veins of the dorsum of the foot
lateral circumflex femoral	Approximately halfway, laterally, of the superior part of the femur	Muscles of the thigh, particularly the quadriceps femoris	Profunda femoris vein

continued

Summary table of the veins of the lower limb			*(continued)*
VEIN	***ORIGIN/LOCATION***	***DRAINAGE***	***CONFLUENCE***
lateral marginal	Lateral margin of the foot, in the dorsal surface	Dorsal venous arch of the foot	Small saphenous vein
lateral plantar	Satellites of the lateral plantar artery	Plantar venous arch	Posterior tibial veins
medial circumflex femoral	Approximately halfway, medially, of the superior part of the femur	Posterior muscles of the thigh and hip joint	Profunda femoris vein
medial marginal	Medial margin of the foot, in the dorsal surface	Dorsal venous arch of the foot	Great saphenous vein
medial plantar	Satellites of the medial plantar artery	Plantar venous arch	Posterior tibial veins
obturator	From the proximal part of the adductor region of the thigh to the pelvis, by way of the obturator foramen	Joint and muscles of the hip	Internal iliac vein
perforating of great saphenous vein	Along the course of the great saphenous vein	Territory of drainage of the great saphenous vein	Deep veins of the lower limb
– **Boyd's veins**	Medial part of the upper third of the leg		Posterior tibial veins
– **Cockett's veins**	Medial part of the lower third of the leg		Posterior tibial veins
– **Dodd's veins**	Adductor canal		Femoral vein
perforating of profunda femoris vein	Run through the adductor magnus to reach the posterior region of the thigh	Posterior and deep medial muscles of the thigh	Profunda femoris vein
plantar digital	Plantar surface of the toes	Venous plexus on the plantar surface of the foot	Plantar metatarsal veins
plantar metatarsal	Plantar intermetatarsal spaces	Metatarsals and adjacent muscles; receive the plantar digital veins	Plantar venous network and plantar venous arch
plantar venous arch	Toe roots	Plantar metatarsal veins and plantar venous network	Dorsal venous arch of foot and medial and lateral plantar veins
plantar venous network	Sole of the foot, proximal to the plantar venous arch	Plantar metatarsal veins	Plantar venous arch, medial and lateral marginal veins and superficial veins of the leg
popliteal	From the junction of the anterior and posterior tibial veins	Knee joint and gastrocnemius; receives the sural and genicular veins	Femoral vein, in correspondence with the adductor hiatus
posterior tibial	Satellites of the posterior tibial artery, originate from the junction of the medial and lateral plantar veins	Plantar region of the foot, which receives the deep veins, and the muscles of the posterior region of the leg; receive the fibular veins and communicating branches with the superficial veins of the leg	Popliteal vein
profunda femoris (deep vein of thigh)	Satellite of the deep artery of the thigh (profunda femoris artery)	Anteromedial region of the thigh; receives the perforating veins (connection between the popliteal and inferior gluteal veins) and the medial and lateral circumflex femoral veins (connection between the profunda femoris, inferior gluteal and obturator veins)	Femoral vein
small saphenous (short saphenous)	Continuation of the lateral marginal vein, from the ankle to the popliteal fossa, in the posterior region of the leg	Dorsal venous network of the foot, lateral part of the plantar venous network, posterior region of the leg, posterior superficial region of the thigh; anastomosis with the great saphenous vein and the deep venous circulation	Popliteal vein
superior gluteal (gluteal)	Enter the pelvis by way of the great sciatic foramen, in the suprapiriform canal	Gluteal region; anastomosis with the inferior gluteal veins	Internal iliac vein
sural	Satellites of the sural arteries in the calf	Soleus and gatrocnemius	Popliteal vein

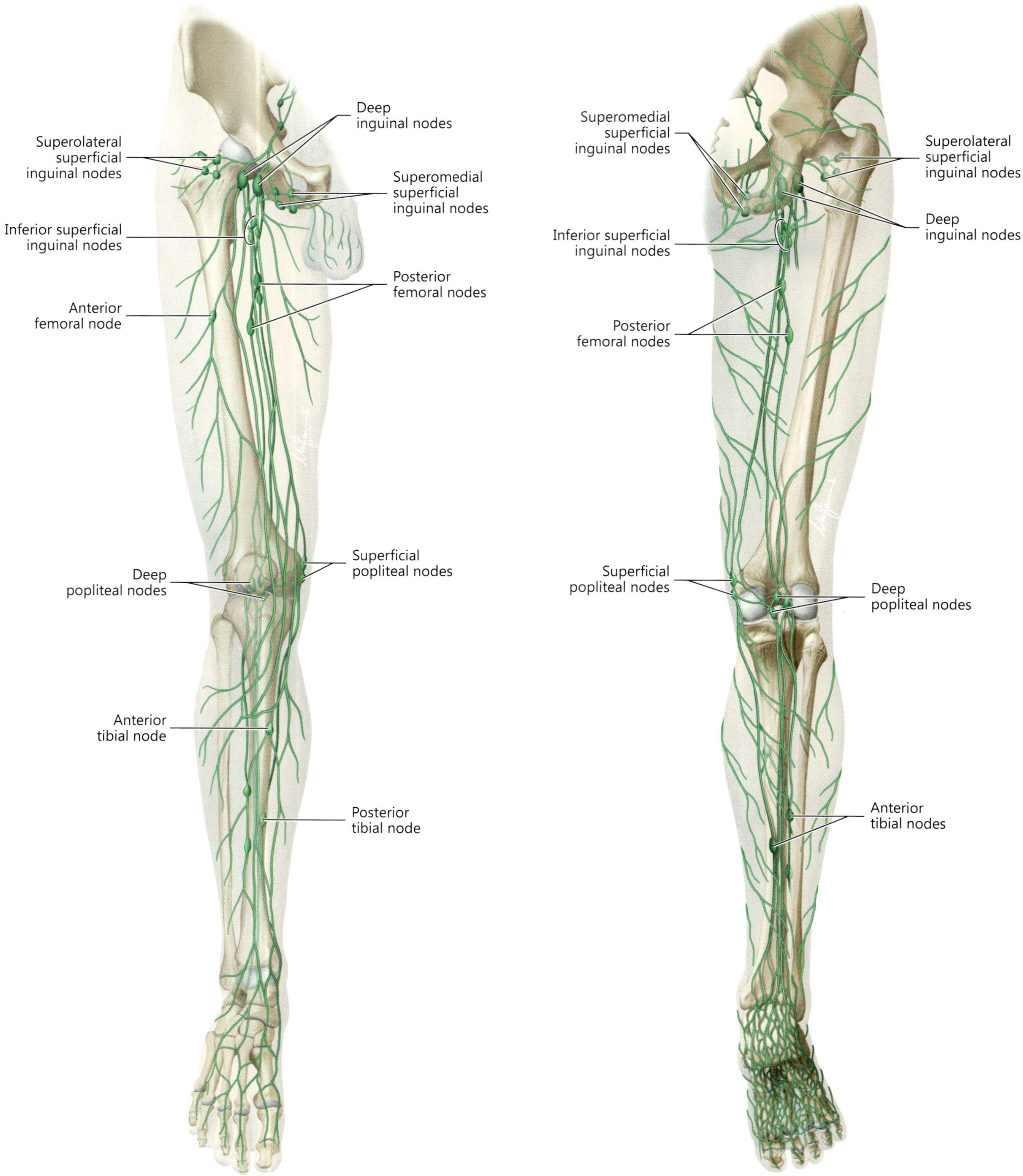

Lymph nodes of the lower limb. The lymphatic ducts of the lower limb converge in the inguinal lymphatic group, in which superficial inguinal lymph nodes are present in two chains: superior oblique, that follows the inguinal ligament on the surface (superomedial and superolateral nodes) and a vertical one, aligned with the great saphenous vein (inferior nodes), and deep inguinal nodes. These are found in the femoral triangle, medially to the femoral vein, and drain, like the superficial inguinal nodes, into the external iliac nodes.

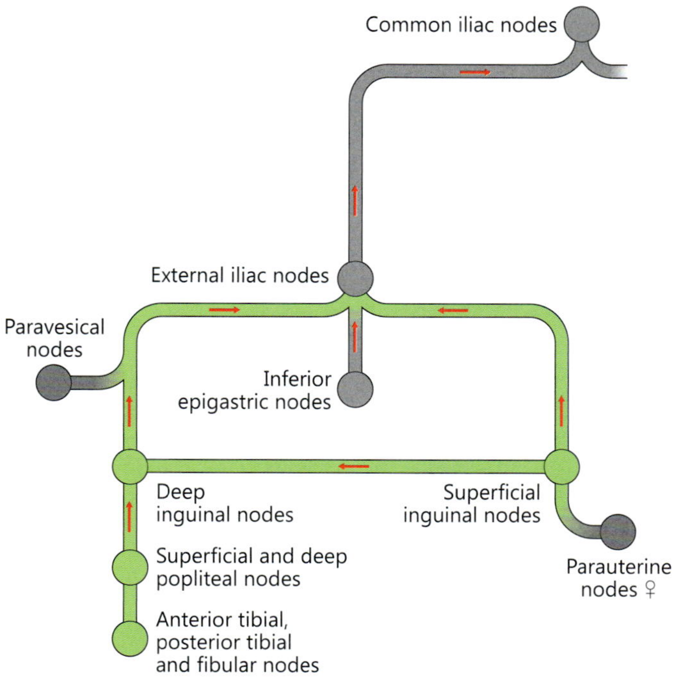

Schema of the **lymph nodes of the lower limb** with their territories of drainage and flow ducts.

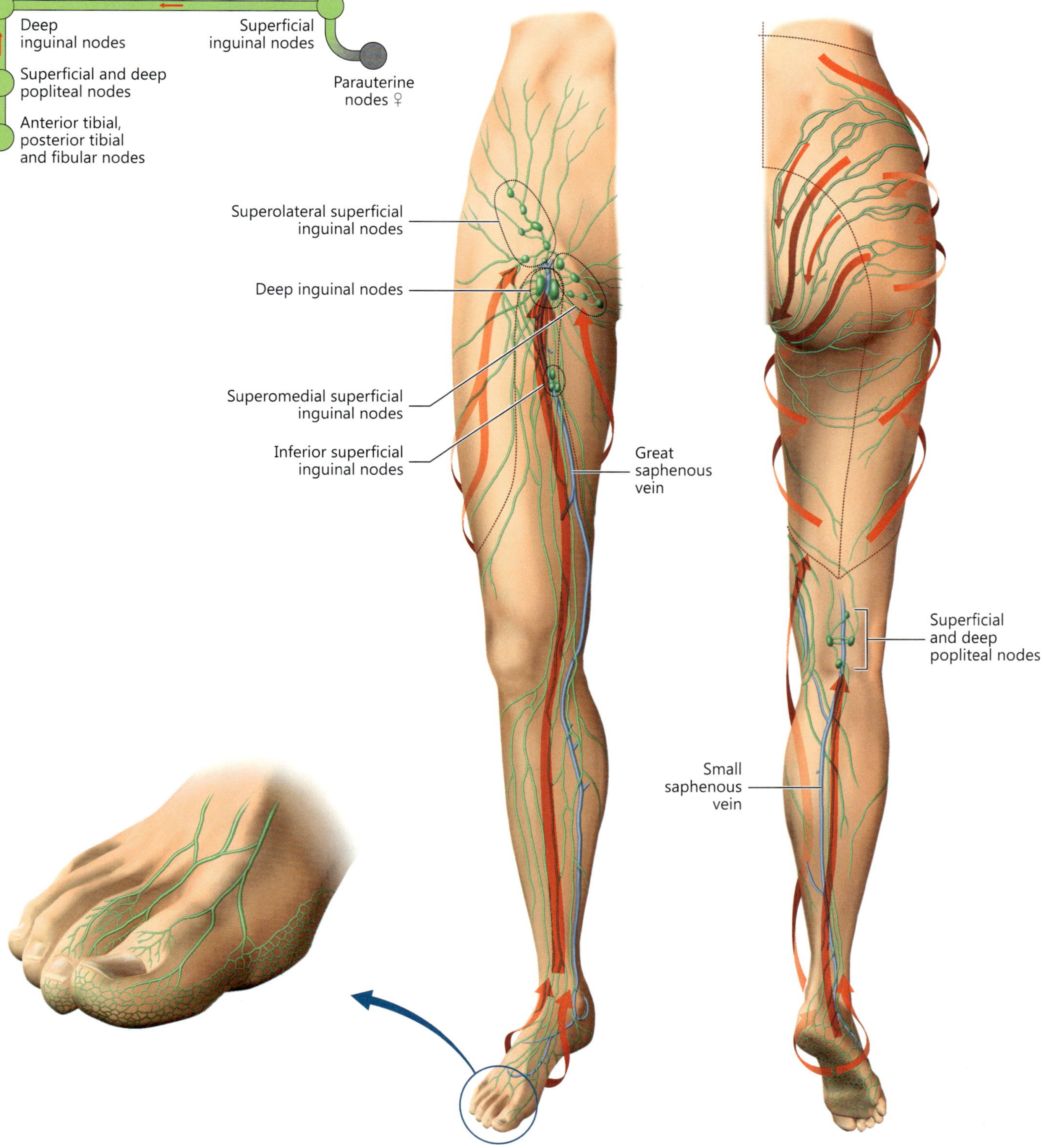

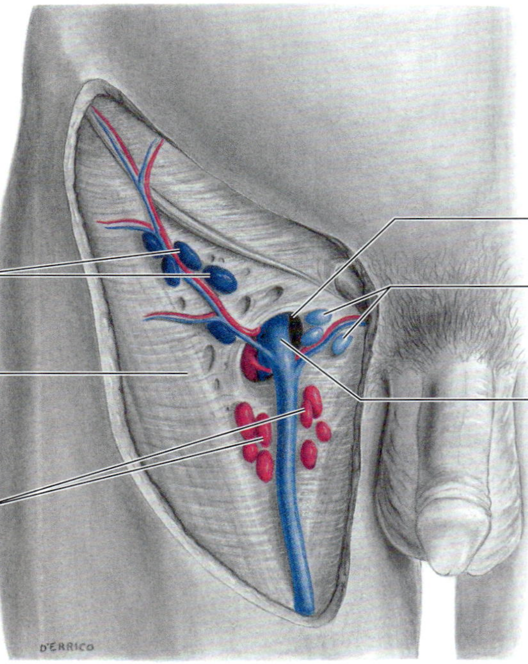

Semischematic representation of the **superficial inguinal nodes**. In *blue*, the superomedial and superolateral node groups; in *red*, the inferior node group.

Summary table of the lymph nodes of the lower limb

LYMPH NODE	*GROUP*	*SITE*	*AFFERENTS*	*EFFERENTS*
Femoral – anterior – posterior		Deep	Deep collectors of the thigh	Deep inguinal nodes
Fibular (peroneal)		Deep, along the course of the satellite limphatic collectors of the fibular (peroneal) artery	Satellite collectors of the fibular (peroneal) vessels	Popliteal nodes
Inguinal		Superficial and deep		
– deep		Deep, in the femoral triangle, medially to the femoral vein	Deep collectors of the lower limb (satellites of the femoral vessels), abdominal walls, superficial gluteal region, external genitalia (glans penis penis and glans of clitoris), perineum, some efferent collectors of the superficial inguinal nodes	External iliac nodes
– – distal		Below the cribriform fascia		
– – intermediate		Below the cribriform fascia		
– – proximal (superior or of Cloquet)		Below the cribriform fascia, on the external margin of the lacunar ligament		
– superficial		Superficial	Superficial collectors of the lower limb (satellites of the great saphenous vein), inferior part of the anal canal (distally to the mucocutaneous junction), superficial gluteal region, external genitalia, inferior part of the vagina, perineum, perianal region, subumbilical abdominal wall, round ligament of the uterus	External iliac nodes and, to a small extent, deep inguinal nodes
– – inferior (inferomedial and inferolateral)	Vertical chain	Arranged along the last tract of the great saphenous vein		
– – superolateral	Superior oblique chain	Below the inguinal ligament, laterally		
– – superomedial	Superior oblique chain	Below the inguinal ligament, medially		
Popliteal – deep – superficial		Superficial and deep Deep, in the popliteal fossa Superficial, in the popliteal fossa	Deep (efferent collectors of the tibial lymph nodes) and superficial (satellite of the small saphenous vein) collectors of the leg	Deep inguinal nodes
Tibial		Deep	Satellite collectors of the tibial vessels	Popliteal nodes
– anterior		Along the course of the satellite lymphatic collectors of the anterior tibial artery	Satellite collectors of the anterior tibial vessels	
– posterior		Along the course of the satellite lymphatic collectors of the posterior tibial artery	Satellite collectors of the posterior tibial vessels	

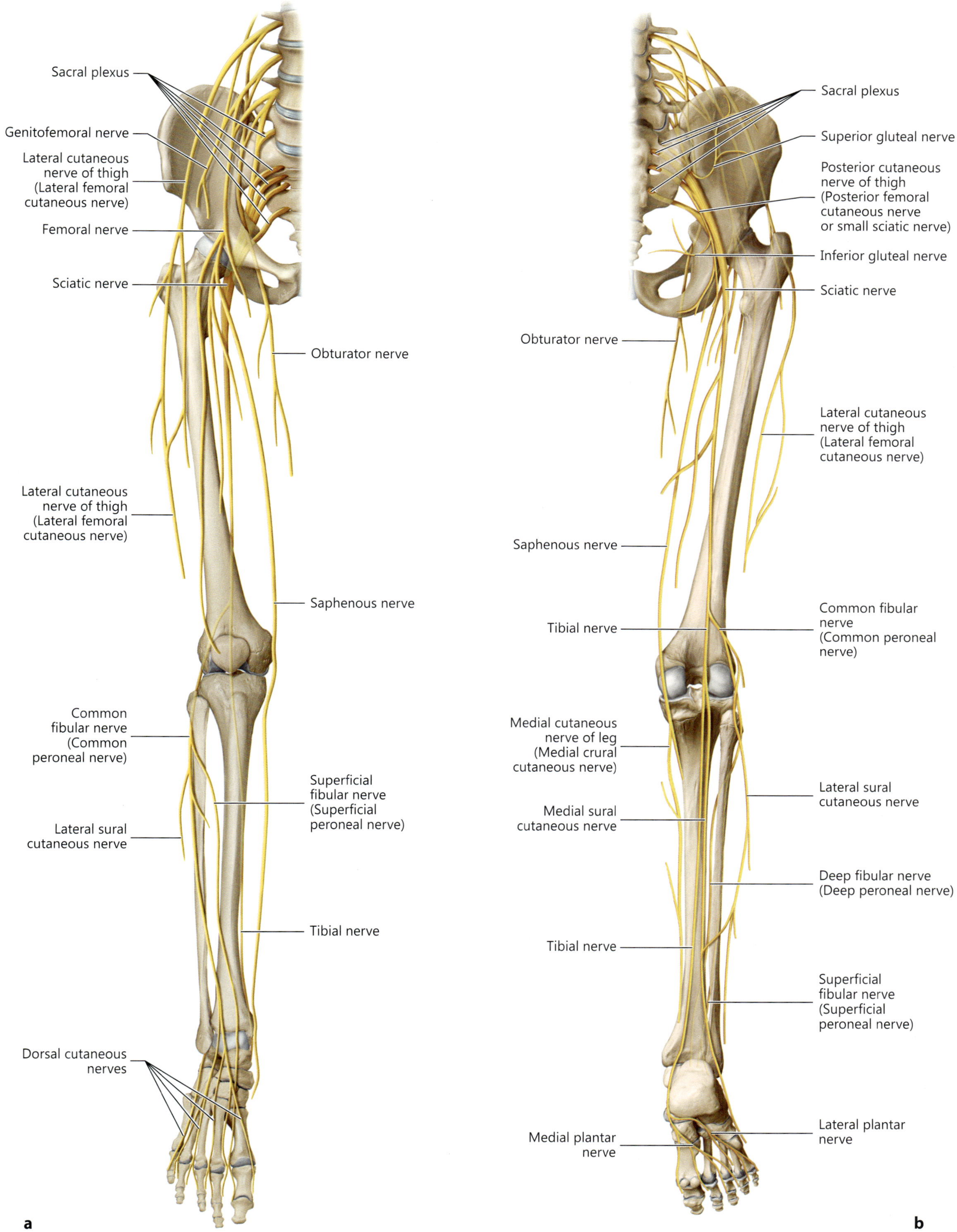

Schema of the **spinal nerves** of the **lower limb**. **a**, Anterior view; **b**, posterior view.

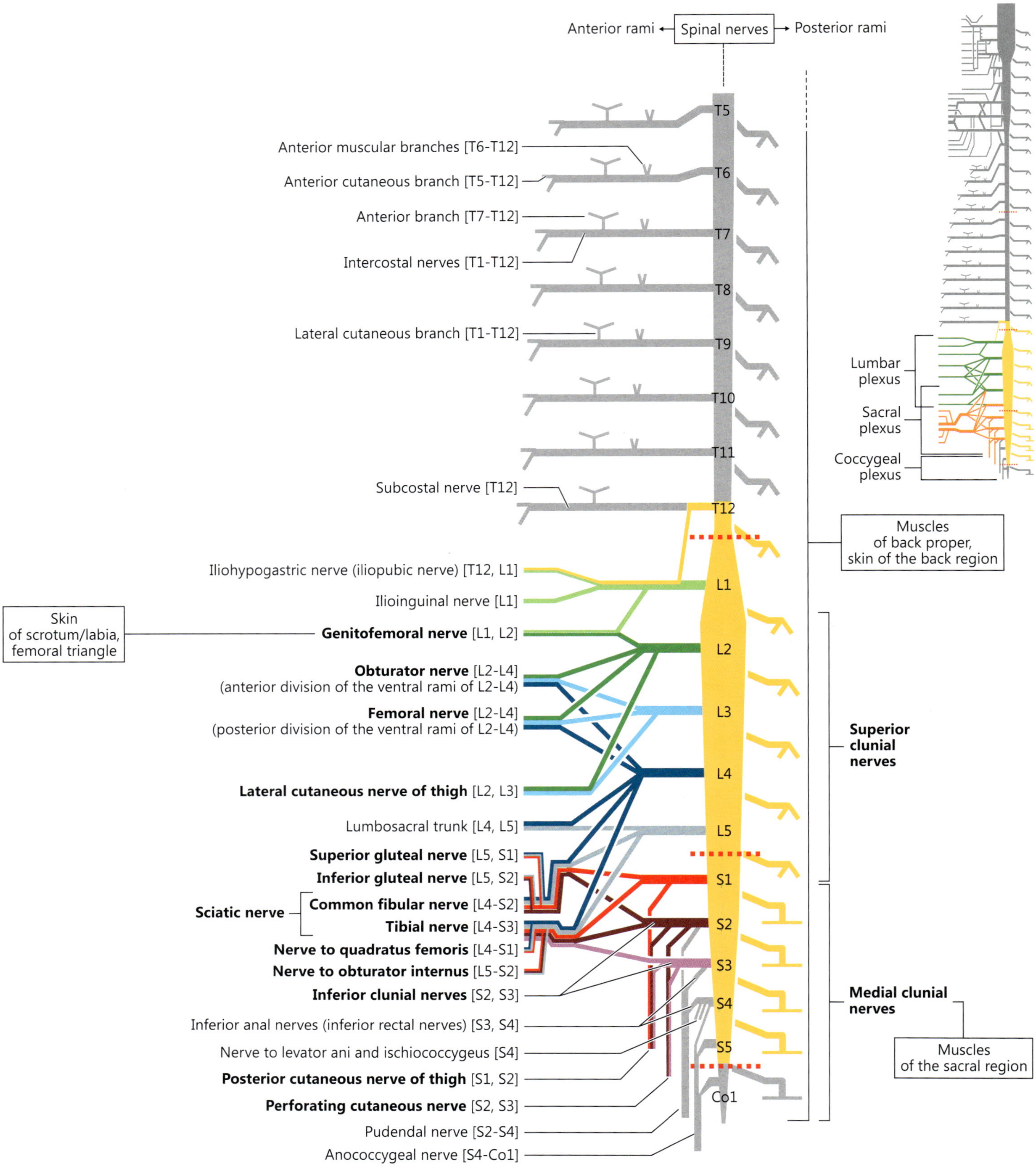

Schema of the lumbar plexus, formed by the anterior rami of the 1st, 2nd and 3rd and, in part, the 4th lumbar nerve, **and sacral plexus**, formed by the union of the lumbosacral trunk (anterior ramus of the 5th lumbar nerve that receives an anastomosis from the 4th lumbar nerve) with the anterior rami of the 1st, 2nd and 3rd sacral nerve. The long collateral branches of the lumbar plexus consist of the iliohypogastric, ilioinguinal, genitofemoral nerves and lateral cutaneous nerve of the thigh, the short ones from branches of the psoas major and psoas minor, quadratus lumborum and intertransversarii laterales lumborum; while the terminal branches are made up of obturator and femoral nerves. Collateral branches of the sacral plexus separate from the anterior surface of the plexus (for the gemellus superior and gemellus inferior, obturator internus and quadratus femoris) and from the posterior surface (superior and inferior gluteal nerves, posterior cutaneous nerve of the thigh, perforating cutaneous nerve and nerve to piriformis). The terminal branch of the sacral plexus is constituted by the sciatic nerve.

Rami of the lumbar plexus. The L1 anterior ramus gives rise to the iliohypogastric and ilioinguinal nerves. The L2 anterior ramus divides into the genitofemoral nerve and lateral cutaneous nerve of the thigh and gives rise to the superior roots of the femoral and obturator nerves. The L3 anterior ramus provides the medial roots of the femoral and obturator nerves. The L4 anterior ramus gives rise to the inferior roots of the femoral and obturator nerves.

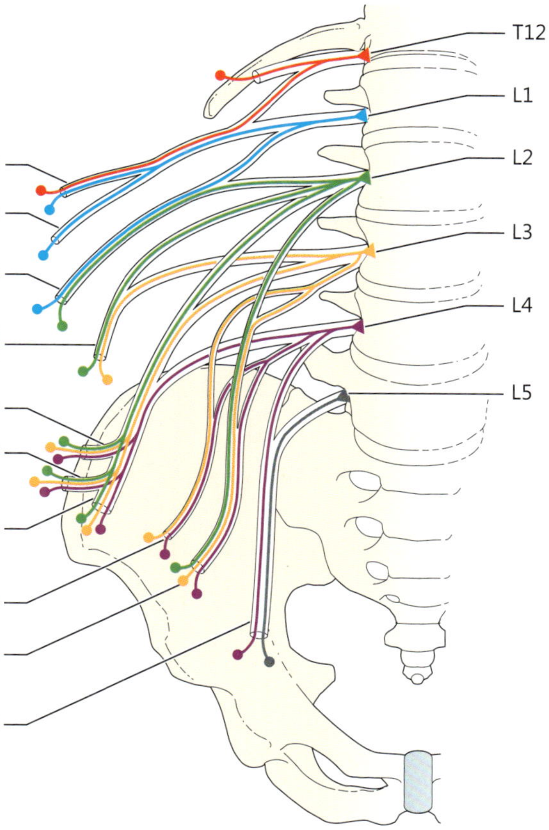

Anterior collateral branches of the sacral plexus: nerves to the obturator internus and quadratus femoris; posterior collateral: superior and inferior gluteal nerves, to the piriformis, posterior cutaneous nerve of the thigh and perforating cutaneous nerve; terminal: sciatic nerve. In the sacral plexus region is distinguished the pudendal plexus (S4 anterior ramus and the anastomotics of S2 and S3 with S4) from which arise the pudendal nerve, a mixed nerve with sacral parasympathetic fibers.

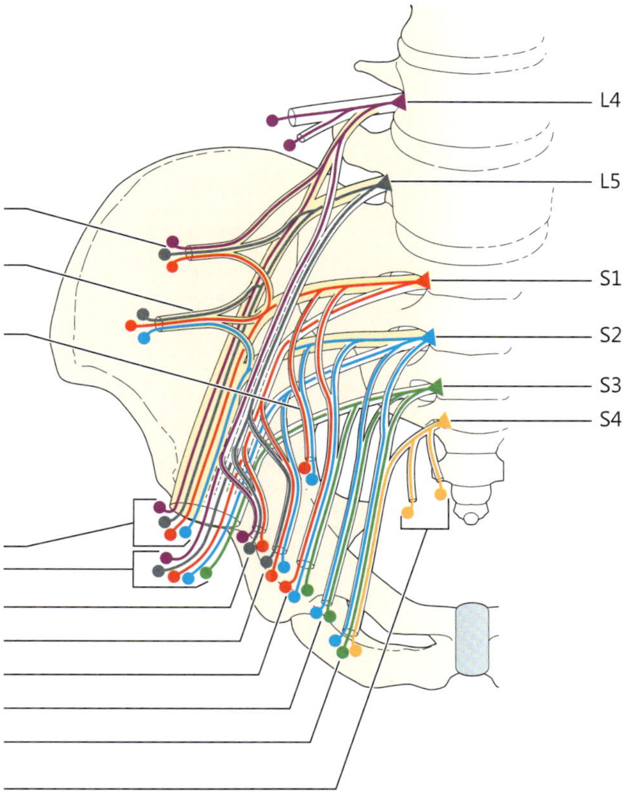

Nerves of the Thigh and Leg

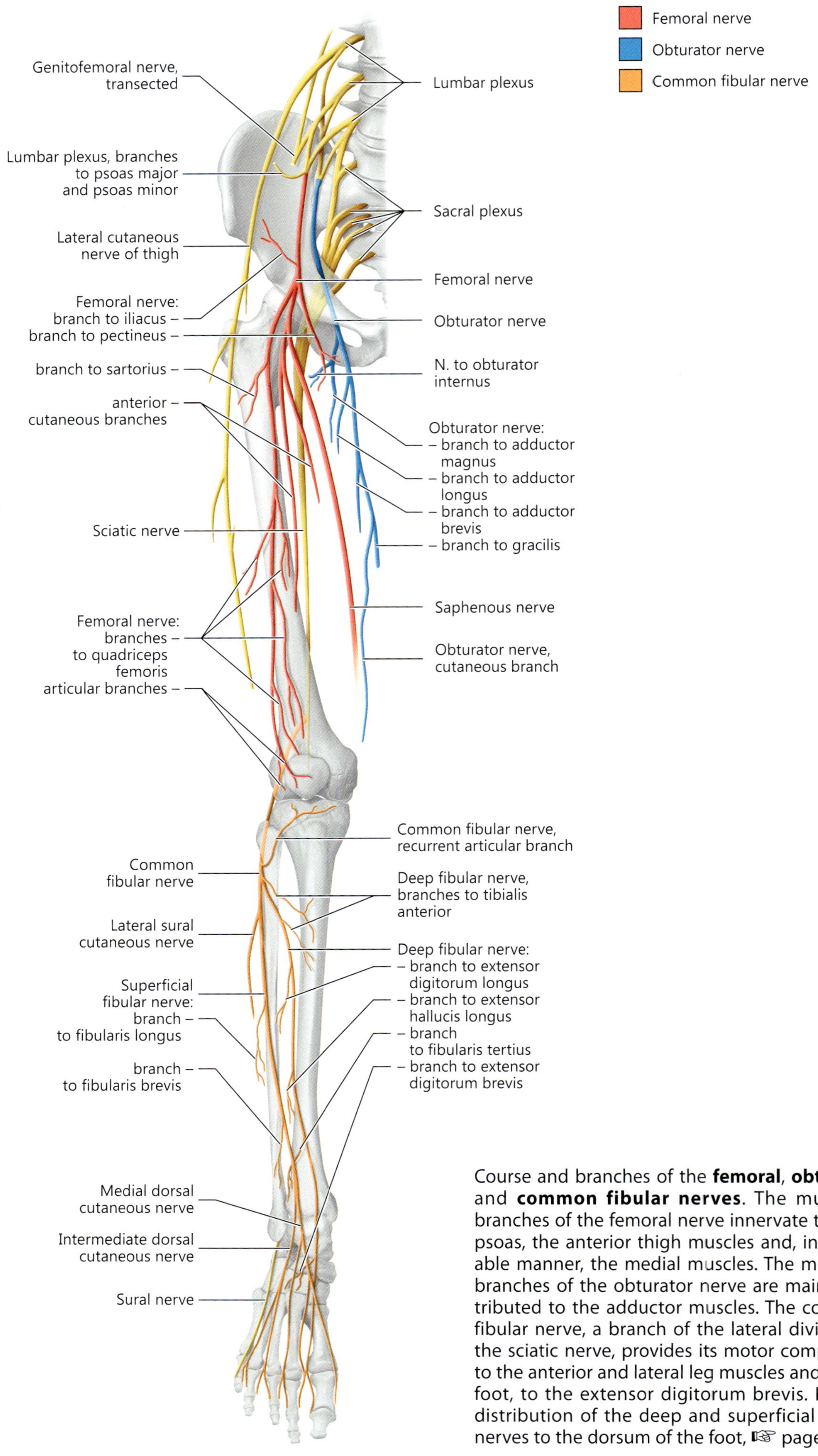

Course and branches of the **femoral**, **obturator** and **common fibular nerves**. The muscular branches of the femoral nerve innervate the iliopsoas, the anterior thigh muscles and, in a variable manner, the medial muscles. The muscular branches of the obturator nerve are mainly distributed to the adductor muscles. The common fibular nerve, a branch of the lateral division of the sciatic nerve, provides its motor component to the anterior and lateral leg muscles and, in the foot, to the extensor digitorum brevis. For the distribution of the deep and superficial fibular nerves to the dorsum of the foot, ☞ page 400.

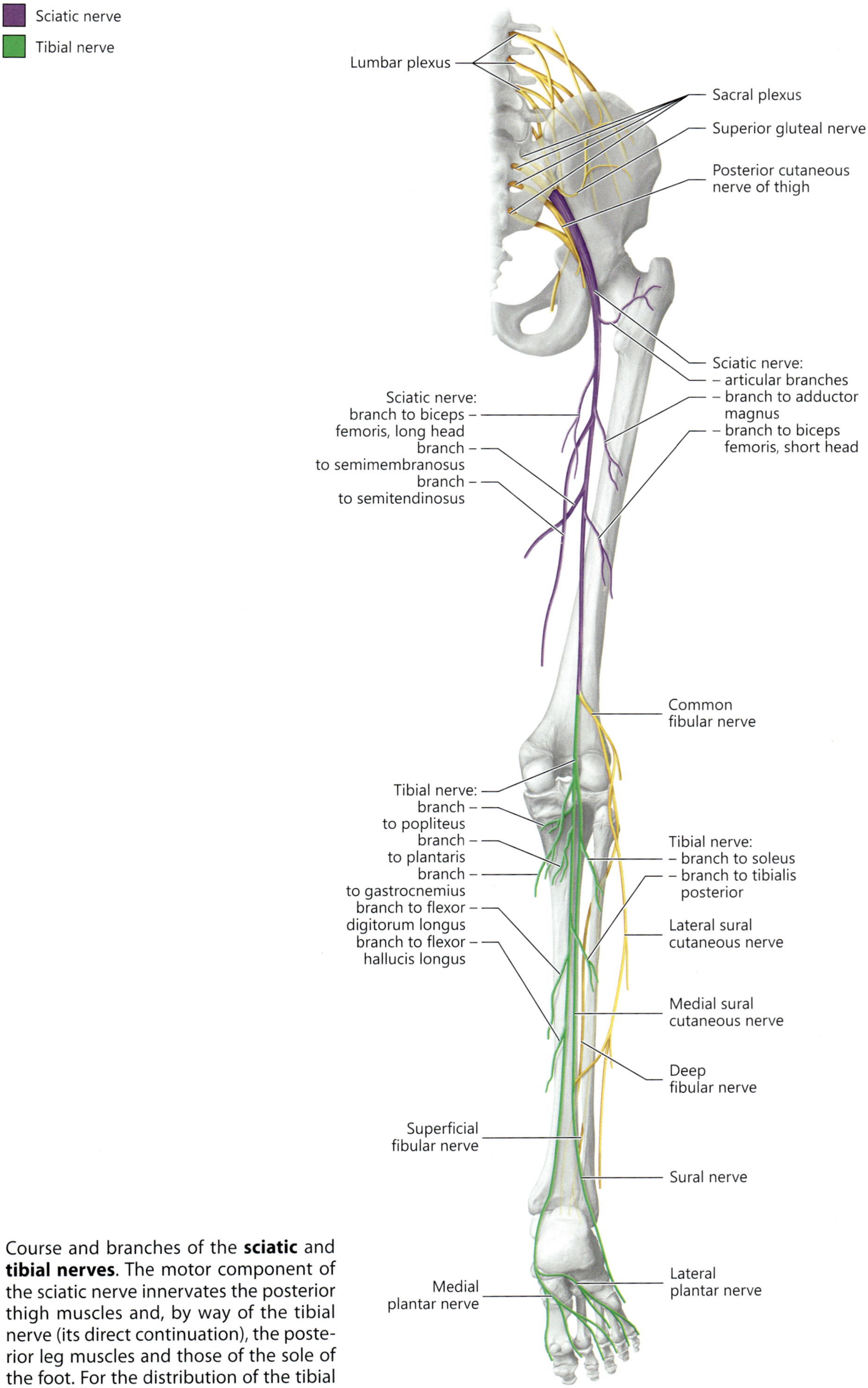

Course and branches of the **sciatic** and **tibial nerves**. The motor component of the sciatic nerve innervates the posterior thigh muscles and, by way of the tibial nerve (its direct continuation), the posterior leg muscles and those of the sole of the foot. For the distribution of the tibial nerve to the sole of the foot, ☞ the successive page.

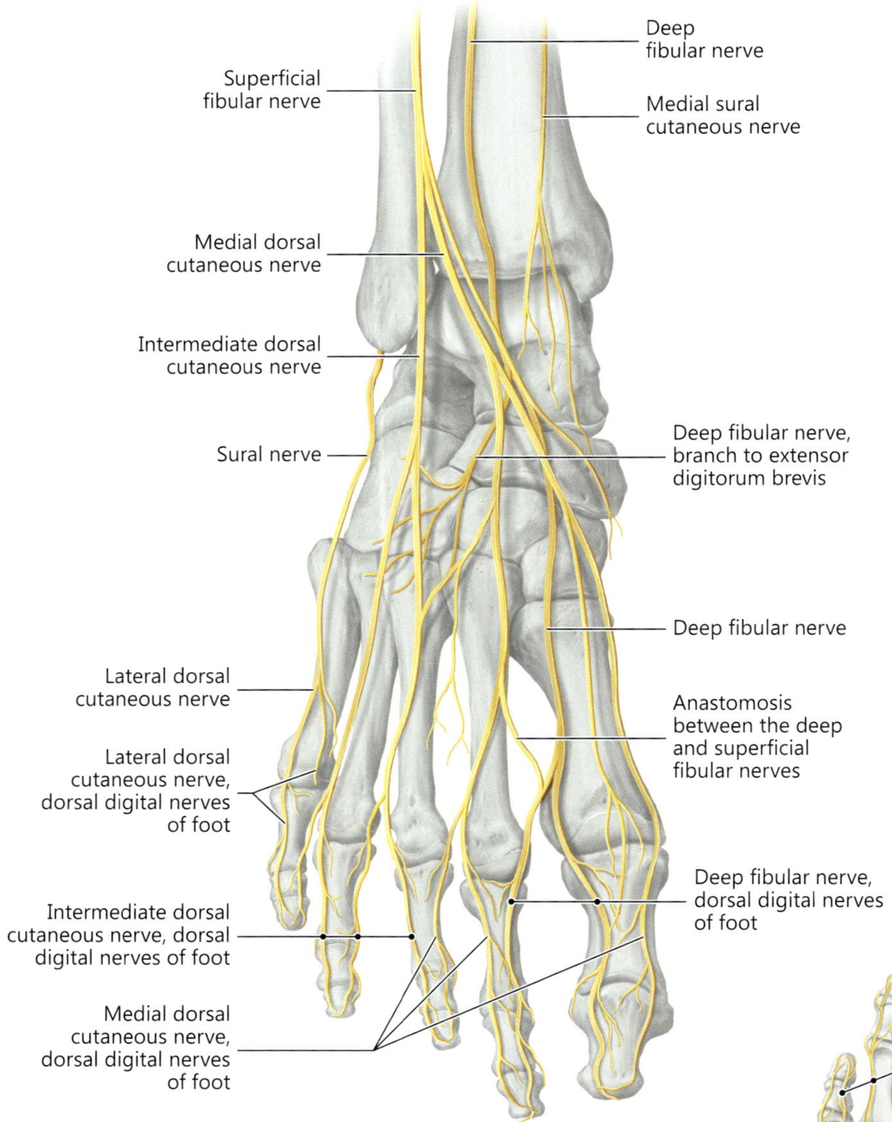

Nerves of the dorsum of the foot. The cutaneous innervation is supplied by the deep fibular nerve, by the medial dorsal and intermediate dorsal cutaneous nerves, the branches of the superficial fibular nerve, and by the lateral dorsal cutaneous nerve, a branch of the sural nerve, which derives from the anastomosis between the medial sural cutaneous nerve, a branch of the tibial nerve, and the lateral sural cutaneous nerve, a branch of the common fibular nerve. For the distribution of the deep and superficial fibular nerves to the anterior region of the leg, ☞ page 398.

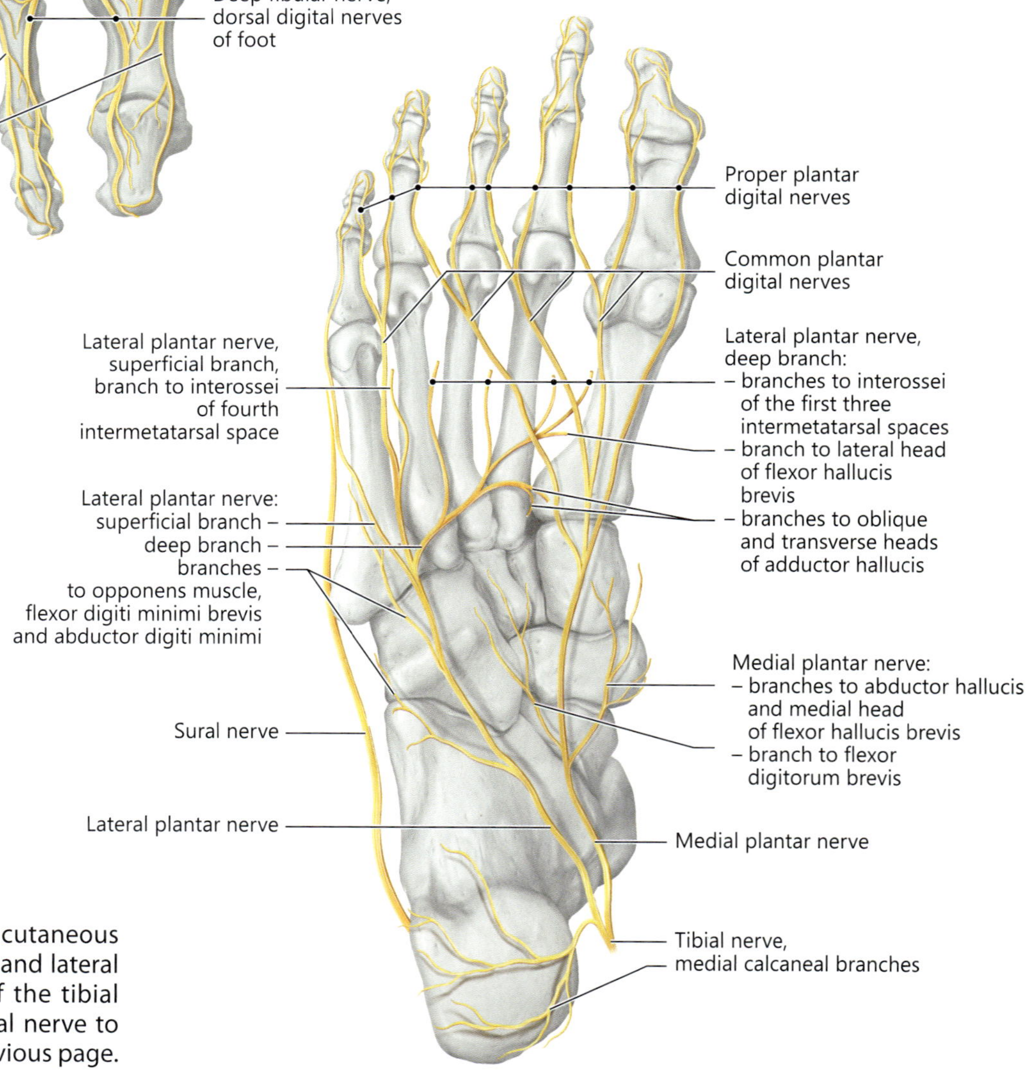

Nerves of the sole of the foot. The cutaneous innervation is supplied by the medial and lateral plantar nerves, terminal branches of the tibial nerve. For the distribution of the tibial nerve to the posterior region of the leg, ☞ previous page.

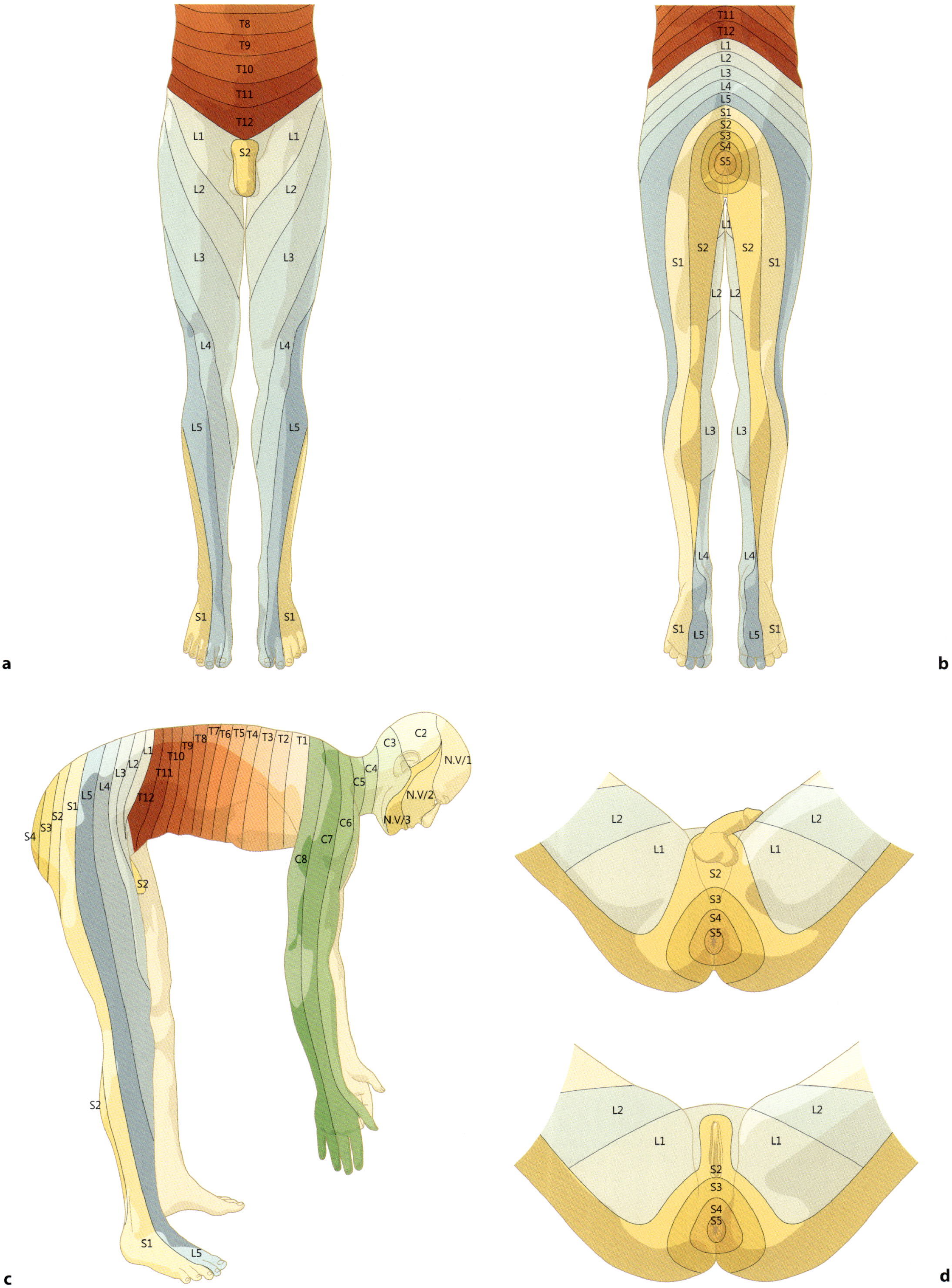

Dermatomes of the lower limb; anterior view (**a**), posterior view (**b**), lateral view (**c**) and peripherally to the perineal region (**d**).

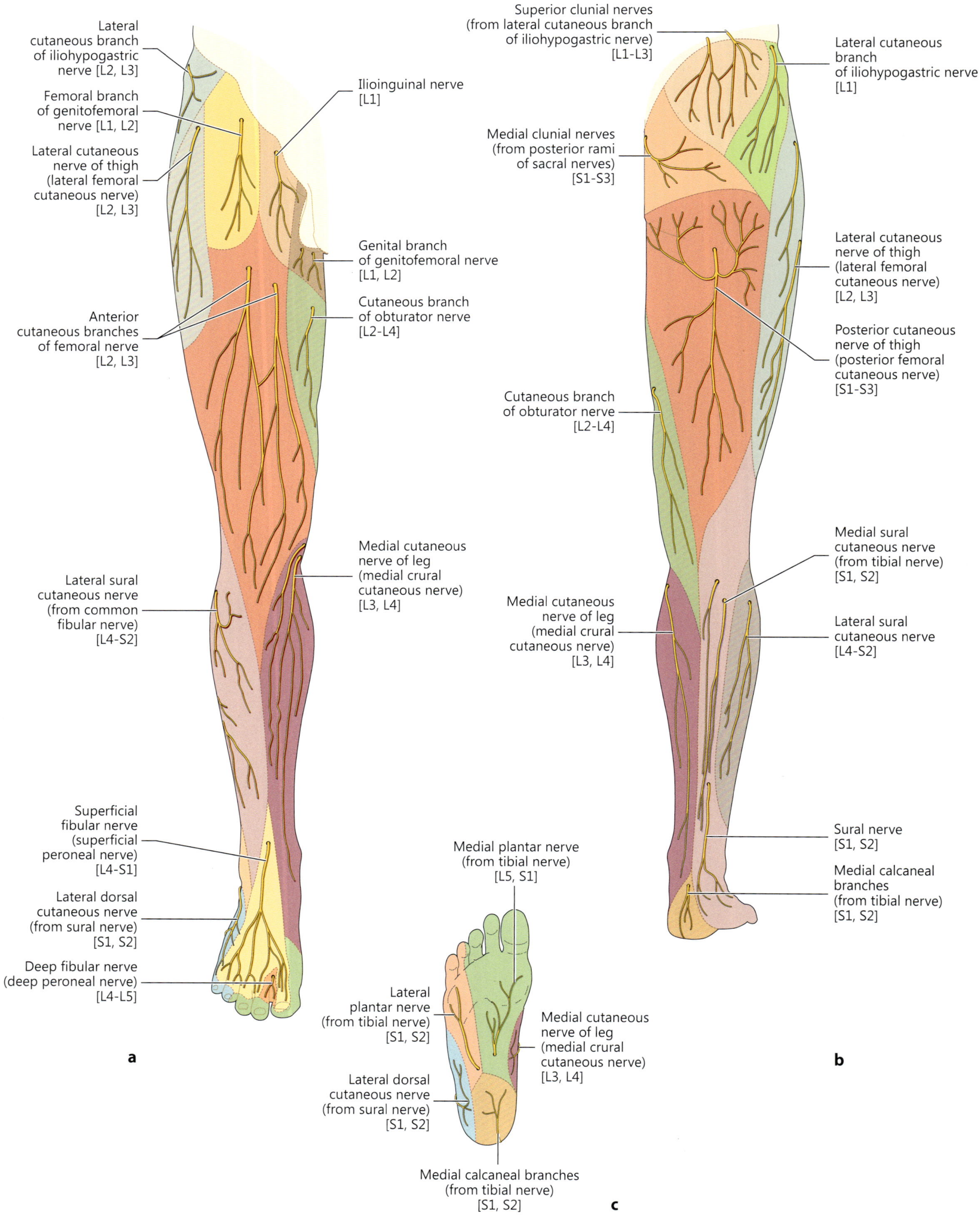

Territories of the **cutaneous nerves** of the **lower limb**. **a**, Anterior surface; **b**, posterior surface; **c**, plantar surface. For dermatomes, ☞ previous page.

Summary table of the nerves of the lower limb

NERVE	ORIGIN	RAMIFICATIONS	DISTRIBUTION
accessory obturator [L3, L4]	Lumbar plexus, anterior rami of the third and fourth lumbar nerves	Filaments	Pectineus and hip joint
common fibular (common peroneal) [L4, L5, S1]	Lateral branch of the sciatic nerve division	Lateral sural cutaneous, superficial fibular (from which the muscular branches and the medial dorsal cutaneous and intermediate dorsal cutaneous nerves, with the dorsal digital nerves of foot, originate) and the deep fibular nerves (from which the muscular branches and the dorsal digital nerves of foot originate); sural communicating branch	Anterior and lateral muscles of the leg and extensor digitorum brevis; skin of the lateral half of the leg, of the neck and of the dorsum of the foot (excluding the margins); dorsal surface of the toes up to the distal phalanx, with the exception of the lateral part of the dorsal surface of the little toe; knee joint, tibiofibular joint and tibiofibular syndesmosis, ankle and foot joints
common plantar digital [L5, S1, S2]	From the medial plantar nerve and from the superficial branch of the lateral plantar nerve	Proper plantar digital nerves	Skin of the plantar surface of the toes
deep fibular (anterior tibial) [L4, L5, S1]	Medial branch of common fibular nerve division	Dorsal digital nerves of the foot and muscular branches	Tibialis anterior, extensor digitorum longus, fibularis tertius, extensor digitorum brevis and extensor hallucis longus; great toe dorsal skin, and that of the second toe
dorsal digital nerves of foot [L4, L5, S1]	From the deep fibular nerve and lateral, medial and intermediate dorsal cutaneous nerves	Filaments	Skin of the dorsum of the toes (the deep fibular nerve is distributed to the great toe and the second toe, the intermediate dorsal cutaneous nerve innervates the third toe and part of the fourth toe)
femoral [L1, L2, L3, L4]	Lumbar plexus, anterior rami of the second, third and fourth lumbar nerves (also contains fibers of the anterior ramus of the first lumbar nerve)	Saphenous nerve, from which originate the medial cutaneous nerve of the leg (or medial crural cutaneous nerve) and the infrapatellar branch and muscular (for the sartorius and the quadriceps femoris) and anterior cutaneous branches (of which, the occasionally distinctive medial and intermediate cutaneous nerves of the femur); as terminal branches of the femoral nerve, they can sometimes be distinctive in an anterior trunk, which gives rise to the medial (for the pectineus and adductor longus and for the skin of the superomedial surface of the thigh) and lateral (for the sartorius muscle and for the skin of the anterior surface of the thigh) muscolocutaneous nerves, and a posterior trunk which gives rise to the nerve to the quadriceps femoris and the saphenous nerve	Psoas major, iliacus and iliospoas, quadriceps femoris, pectineus, sartorius and adductor longus; skin of the anteromedial surface of the thigh, the knee, the leg and the dorsum of the foot; hip and knee joints; femoral artery
genitofemoral [L1, L2]	Lumbar plexus, anterior ramus of the second lumbar nerve	Genital and femoral branches	Skin of the anterosuperior region of the thigh (innervated by the femoral branch); transverse, internal oblique and cremaster; skin of the scrotum and the labia majora
iliohypogastric (iliopubic) [T12, L1]	Lumbar plexus, anterior ramus of the first lumbar nerve	Anterior and lateral cutaneous, genital and muscular branches	Skin of the buttock (innervated by the lateral cutaneous branch), of the external genitalia and of the pubic region; external oblique, internal oblique, transversus abdominis, rectus abdominis and pyramidalis
ilioinguinal [L1]	Lumbar plexus, anterior ramus of the first lumbar nerve	Genital, cutaneous, muscular, anterior scrotal ♂ and anterior labial ♀ branches	Skin of the buttock and medial surface of the thigh; anterior part of the scrotum or labia majora; skin of the pubic region and muscles of the abdominal wall
inferior clunial [S1, S2, S3]	Posterior cutaneous nerve of the thigh	Filaments	Skin of the inferior and lateral parts of the buttock
inferior gluteal [L4, L5, S1]	Sacral plexus, anterior roots of the fourth and fifth lumbar nerves and of the first sacral nerve	Filaments	Gluteus maximus
intermediate dorsal cutaneous [L4, L5, S1]	Terminal branch of the superficial fibular nerve	Dorsal digital nerves	Skin of the dorsum of the foot, the third toe and part of the fourth toe

continued

Summary table of the nerves of the lower limb *(continued)*

NERVE	ORIGIN	RAMIFICATIONS	DISTRIBUTION
interosseous nerve of the leg (crural interosseous) [L4, L5, S1, S2]	Tibial	Filaments	Interosseous membrane of the leg and tibiofibular syndesmosis
lateral cutaneous nerve of the thigh (lateral femoral cutaneous) [L2, L3]	Lumbar plexus, anterior rami of the second and third lumbar nerves	Gluteal and femoral branches	Skin of the buttock and the anterolateral surface of the thigh
lateral dorsal cutaneous [S1, S2]	Sural	Dorsal digital nerves	Skin of the dorsolateral part of the foot, the dorsal surface of the little toe and the external half of the fourth toe
lateral plantar (external plantar) [S1, S2]	Branch of tibial nerve division	Superficial (from which the common and proper plantar digital nerves originate) and deep branches	Quadratus plantae, abductor digiti minimi, flexor digiti minimi brevis, interossei of the first three intermetatarsal spaces, third and fourth lumbrical, oblique and transverse heads of the adductor hallucis and lateral head of the flexor hallucis brevis
lateral sural cutaneous [L4, L5, S1]	Common fibular	Anastomosis with the medial sural cutaneous nerve, from which the sural nerve originates	Skin of the posterolateral surface of the lower third of the leg
lumbar plexus	Anterior rami of the first, second, third lumbar nerves and part of the fourth lumbar nerve (an anastomotic branch of the last intercostal nerve also contributes to its formation)	Iliohypogastric, ilioinguinal and first anastomotic loop nerves (from the anterior ramus of the first lumbar nerve); lateral cutaneous nerve of thigh, genitofemoral and second anastomotic loop nerves, which separates in the superior root of the obturator nerve and in the superior root of the femoral nerve (from the anterior ramus of the second lumbar nerve); medial root of the obturator nerve, medial root of the femoral nerve and third anastomotic loop (from the anterior ramus of the third lumbar nerve); inferior root of the femoral nerve, inferior root of the obturator nerve and the fourth anastomotic loop (from the anterior branch of the fourth lumbar nerve); the anterior ramus of the fifth lumbar nerve connects with the fourth anastomotic loop forming the lumbosacral trunk; anastomotic branches with the twelfth thoracic nerve and with the fifth lumbar nerve and white and grey rami connecting with the sympathetic trunk	Abdomen, pelvis, perineum and lower limb
lumbosacral plexus	☞ lumbar plexus and sacral plexus		
medial clunial [S1, S2, S3]	Posterior rami of the first, second and third sacral nerves (with possible variability also of the fourth and fifth)	Filaments	Skin of the medial part of the buttocks
medial dorsal cutaneous [L4, L5, S1]	Terminal branch of the superficial fibular nerve	Dorsal digital nerves	Skin of the dorsum of the great toe, the second toe and part of the third toe
medial plantar (internal plantar) [L4, L5]	Branch of tibial nerve division	Common plantar digital nerves (from which the proper plantar digital nerves originate)	Abductor hallucis, medial head of the flexor hallucis brevis, flexor digitorum brevis and the first and second lumbrical; articular and cutaneous branches of the medial part of the sole of the foot
medial sural cutaneous nerve [L5, S1, S2]	Tibial	Anastomosis with the lateral sural cutaneous nerve, from which the sural nerve originates	Skin of the posteromedial surface of the lower third of the leg
obturator [L2, L3, L4]	Lumbar plexus, anterior rami of the second, third and fourth lumbar nerves	Anterior and posterior rami, cutaneous, articular and muscular branches	Adductor muscles (adductor longus, adductor brevis and adductor magnus), gracilis and obturator externus; skin of the medial surface of the thigh; hip and knee joints
perforating cutaneous [S2, S3]	Sacral plexus, anterior rami of the second and third sacral nerves	Filaments	Skin overlying the gluteus maximus

continue

Summary table of the nerves of the lower limb *(continued)*

NERVE	ORIGIN	RAMIFICATIONS	DISTRIBUTION
posterior cutaneous nerve of thigh (posterior femoral cutaneous) [S1, S2, S3]	Sacral plexus, anterior rami of the first, second and third sacral nerves	Inferior clunial nerves and perineal branches	Gluteal region, perineum, and posterior surface of the thigh and leg
proper plantar digital [L5, S1, S2]	Common plantar digital	Filaments	Skin of the plantar surface of the toes and dorsal extremities (the medial plantar nerve innervates the first three toes and part of the fourth toe, and the dorsal extremity of the first three toes; the lateral plantar nerve innervates the little toe, part of the fourth toe and the dorsal extremity of the last two toes); flexor digiti minimi brevis and opponens digiti minimi and interossei of the fourth intermetatarsal space
sacral plexus	Connection of the lumbosacral trunk with the anterior rami of the first, second and third sacral nerves	Anterior (nerves to the obturator internus and to the quadratus femoris) and posterior (superior and inferior gluteal nerves, to piriformis muscle and posterior cutaneous nerve of thigh) collateral branches and terminal branch represented by the sciatic nerve; the anterior ramus of the third sacral nerve which is connected to the anastomotic arch with the anterior ramus of the fourth sacral nerve forming the pudendal plexus, to whose formation also contributes a small part of the second sacral nerve and whose terminal branch is the pudendal nerve	Perineum and lower limb
saphenous (internal saphenous) [L2, L3, L4]	Terminal branch of the femoral nerve	Medial cutaneous nerve of the leg (or medial crural cutaneous nerve) and infrapatellar branch	Skin of the medial surface of the leg and foot; knee joint
sciatic nerve [L4, L5, S1, S2, S3]	Sacral plexus, anterior rami of the fourth and fifth lumbar nerves and of the first, second and third sacral nerves	Tibial and common fibular nerves; muscular branches	Muscles of the posterior region of the thigh, of the leg and of the foot; skin of the leg and foot, except a medial strip in which the saphenous nerve is distributed; hip and knee joints
superficial fibular (musculocutaneous) [L4, L5, S1]	Lateral branch of common fibular nerve division	Medial dorsal cutaneous and intermediate dorsal cutaneous nerves (from which the dorsal digital nerves of foot originate) and muscular branches	Fibularis longus and fibularis brevis; skin of the inferior and lateral parts of the leg
superior clunial [L1, L2, L3]	Posterior roots of the first, second and third lumbar nerves	Filaments	Skin of the superior part of the buttocks
superior gluteal [L4, L5, S1]	Sacral plexus, anterior roots of the fourth and fifth lumbar nerves and of the first sacral nerve	Filaments	Gluteus medius and gluteus minimus and tensor fasciae latae
sural (short saphenous) [S1, S2]	Derives from the anastomosis between the medial sural cutaneous nerve and the lateral sural cutaneous nerve	Lateral dorsal cutaneous nerve and lateral calcaneal branches	Skin of the posterolateral surface of the lower third of the leg
tibial nerve (internal popliteal) [L4, L5, S1, S2]	Direct continuation of the sciatic nerve	Interosseous nerve of leg, medial sural cutaneous nerve, sural (from which the lateral dorsal cutaneous nerve and the lateral calcaneal branches originate), medial plantar (from which the common plantar digital nerves with the proper plantar digital nerves originate) and lateral plantar (from which a superficial branch, with common and proper plantar digital nerves, and a deep branch originate); muscular and medial calcaneal branches	Posterior muscles of the leg and of the sole of the foot; skin of the distal tract of the posterolateral region of the leg, of the calcaneus, of the sole and of the lateral margin of the foot; plantar surface of the toes, dorsal surface of the distal phalanges and the lateral part of the dorsal surface of the little toe; knee, ankle and foot joints
to obturator internus [L5, S1, S2]	Sacral plexus, anterior ramus	Filaments	Obturator internus and gemellus superior
to piriformis [S1, S2]	Sacral plexus, anterior roots of the first and second sacral nerves	Filaments	Piriformis
to quadratus femoris [L4, L5, S1]	Sacral plexus, anterior rami of the fourth and fifth lumbar nerves and of the first sacral nerve	Filaments	Quadratus femoris and gemellus inferior; hip joint

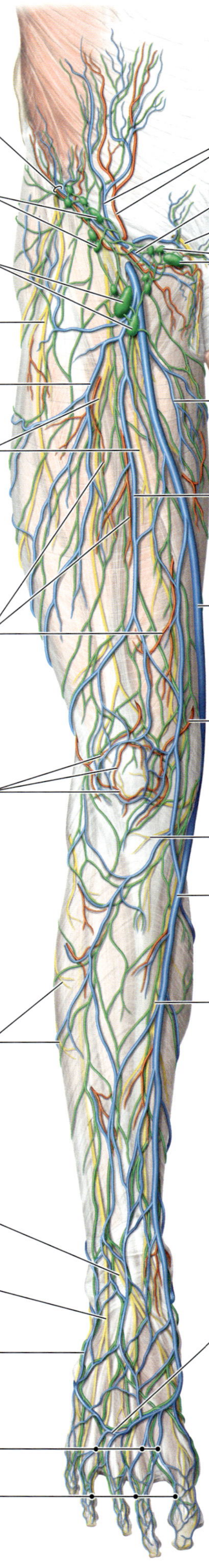

Right lower limb, anterior view. The **great saphenous vein** is subcutaneous throughout its course, supported first by the deep fascia of the leg and then by the fascia lata. Arriving at Scarpa's triangle, it runs deeper, perforating the cribriform fascia (saphenous opening) and then opens into the femoral vein. The **small saphenous vein** is only subcutaneous in its first tract, while when higher, at a variable height, it perforates the deep fascia of the leg becoming subfascial, and flows into the popliteal vein.

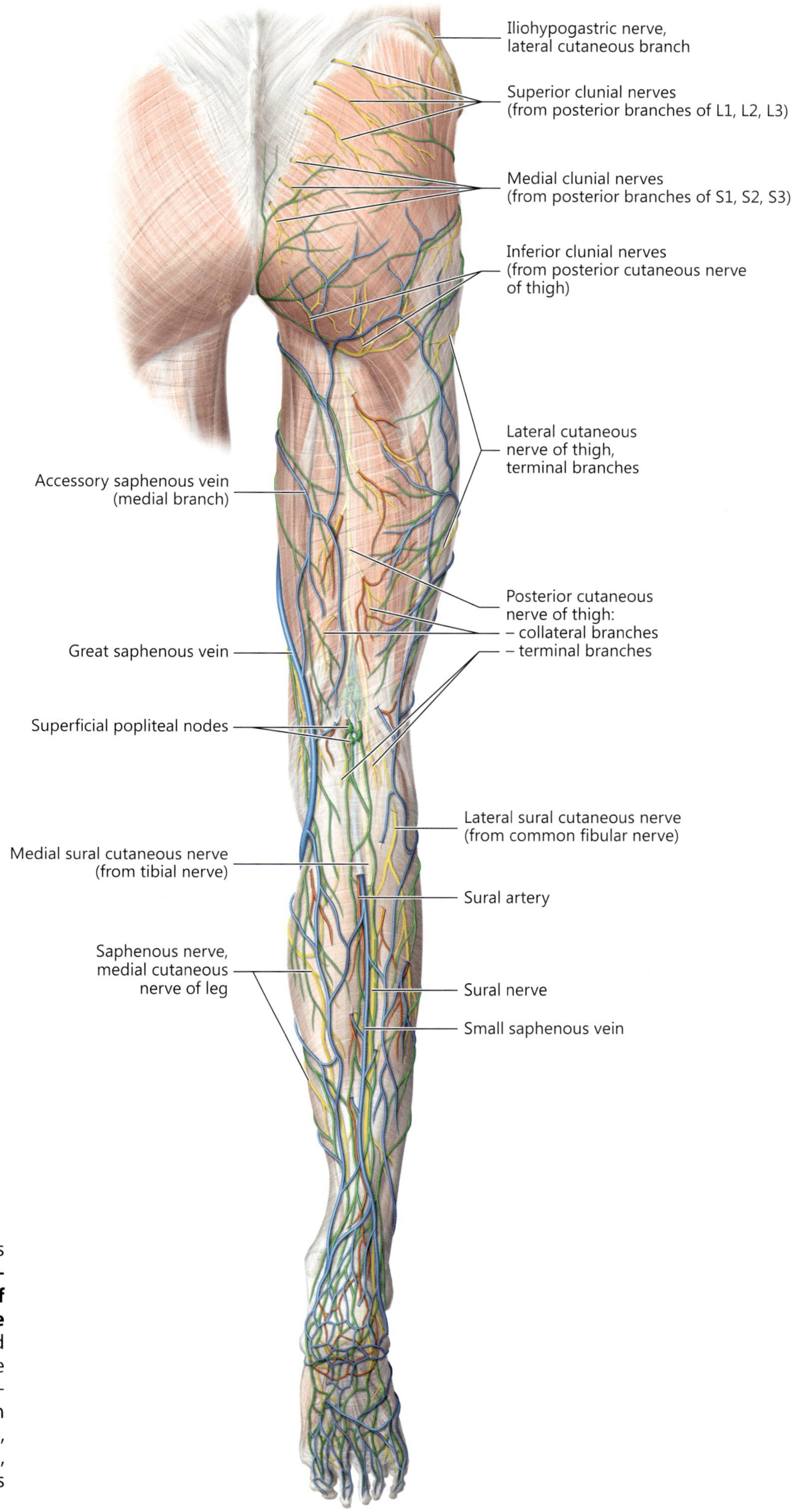

Right lower limb, posterior view. It is represented by the **cutaneous innervation** of the **posterior region of thigh and posterior region of the leg**. The **great saphenous vein** and **small saphenous vein** can also be observed. The first rises along the medial surface of the leg, describing an anterior concave curve; the second, rising along the calcaneal tendon, first on its lateral margin, then on its posterior surface.

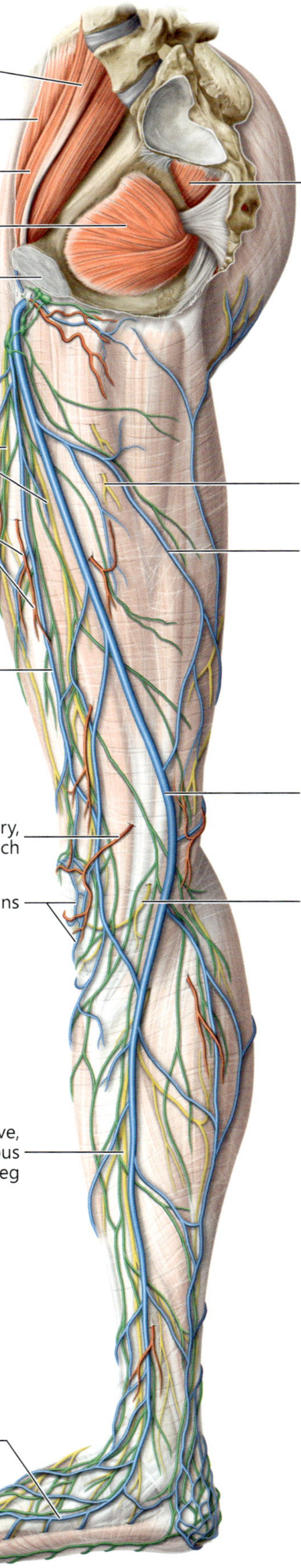

Right lower limb, medial view. The **great saphenous vein**, originating as a continuation of the medial marginal vein of the foot, 2-3 cm in front of the medial malleolus, runs along the medial surface of the leg, describing an anterior concave curve behind the medial condyle of the femur. In the thigh, it follows the sartorius up to the saphenous opening, or oval fossa, of the femoral triangle, where is runs deeper, perforates the cribriform fascia and flows into the femoral vein, at a distance of 3-4 cm from the inguinal ligament.

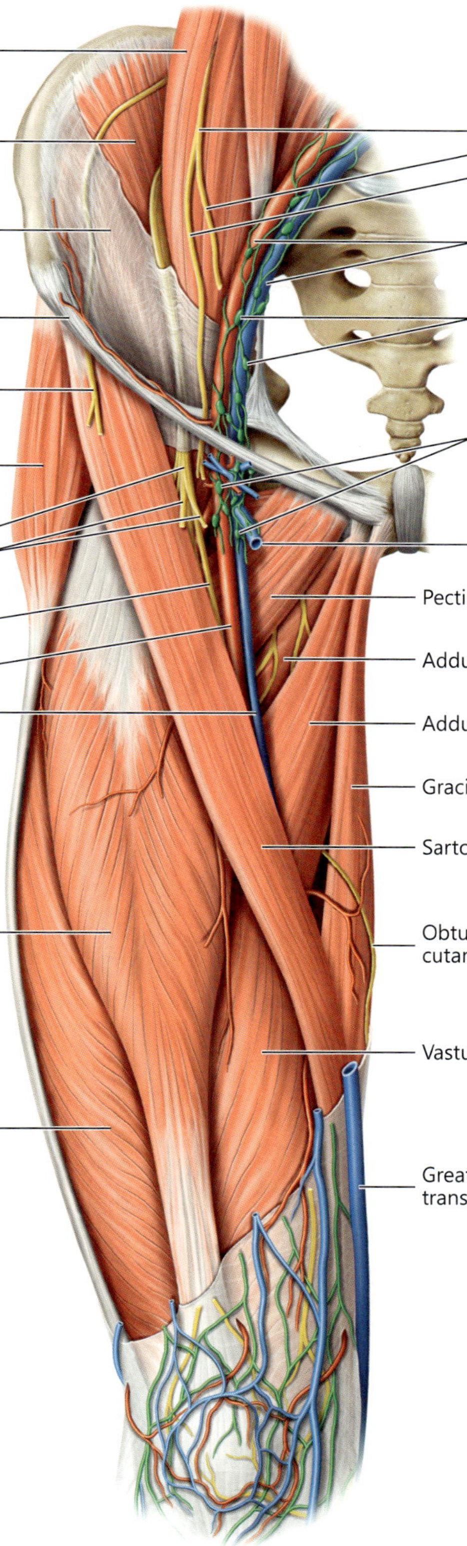

Right thigh, anterior region of thigh. The **femoral artery**, a continuation of the external iliac artery, originates from behind the inguinal ligament, halfway between the anterior superior iliac spine and the pubic symphysis, occupying the lateral angle of the vascular space. It follows the **femoral nerve** laterally which, in correspondence with the inguinal ligament, continues with the iliopsoas in the muscular space. The **femoral vein**, which in its course follows the femoral artery, is found medially to the artery in the femoral triangle and in the vascular space and terminates behind the inguinal ligament, continuing in the external iliac vein.

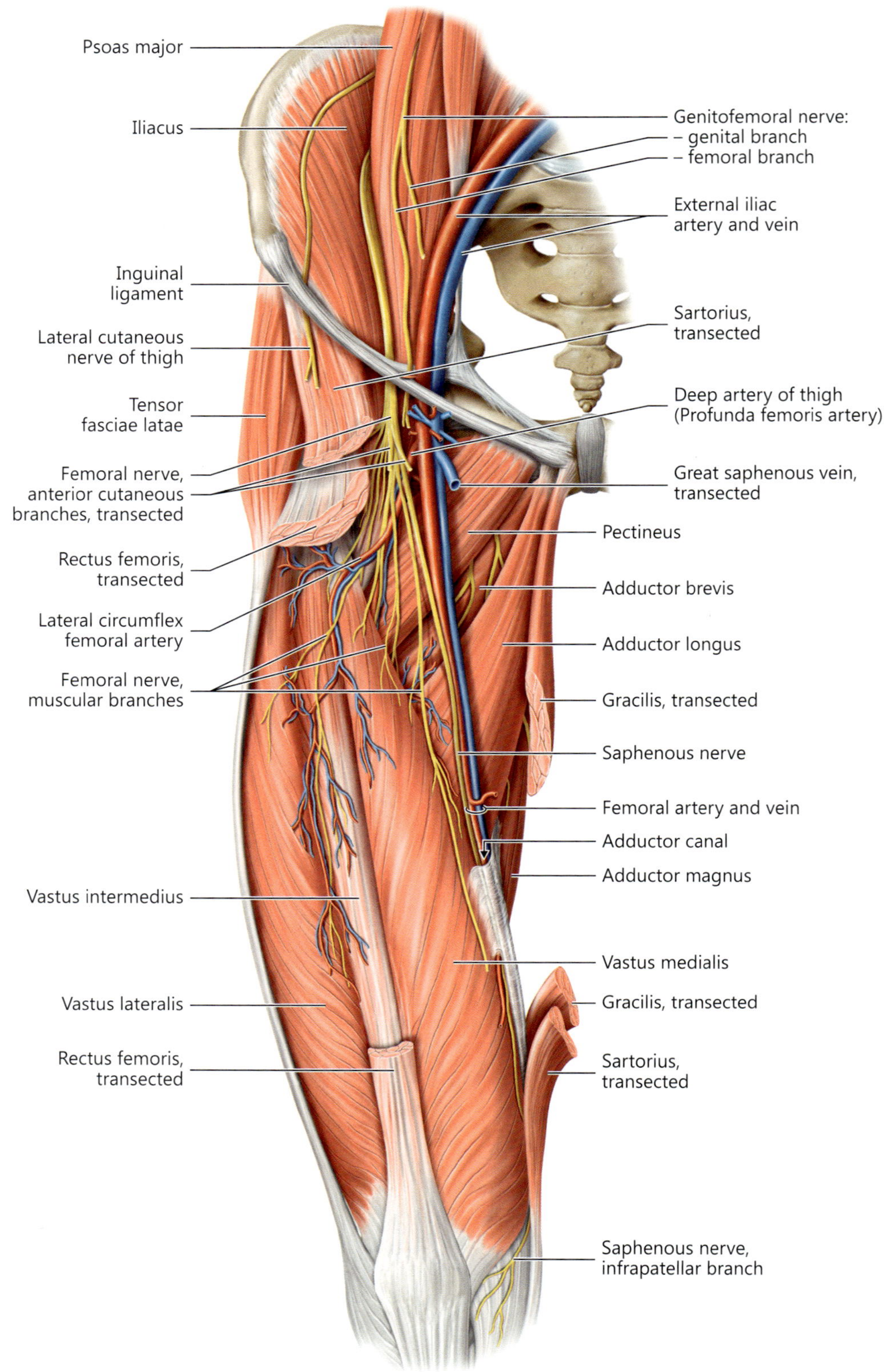

Right thigh, anterior region of the thigh. The **femoral artery** is observed after removal of the sartorius, rectus femoris and gracilis. The artery runs along with the **saphenous nerve** and the **femoral vein**, descends into the adductor canal, located between the medial margin of the vastus medialis and the lateral margin of the adductor magnus, covered by a membrane stretched between these two muscles. Superficially to the canal, it runs the sartorius; at a deep level, forming the floor, is the adductor longus. In the femoral triangle, the femoral nerve divides into its anterior cutaneous terminal branches, muscular branches, that innervate the sartorius and femoral quadriceps, and saphenous. The opening of the great saphenous vein into the femoral vein is also observed. This takes place at a distance of approximately 3-4 cm from the inguinal ligament.

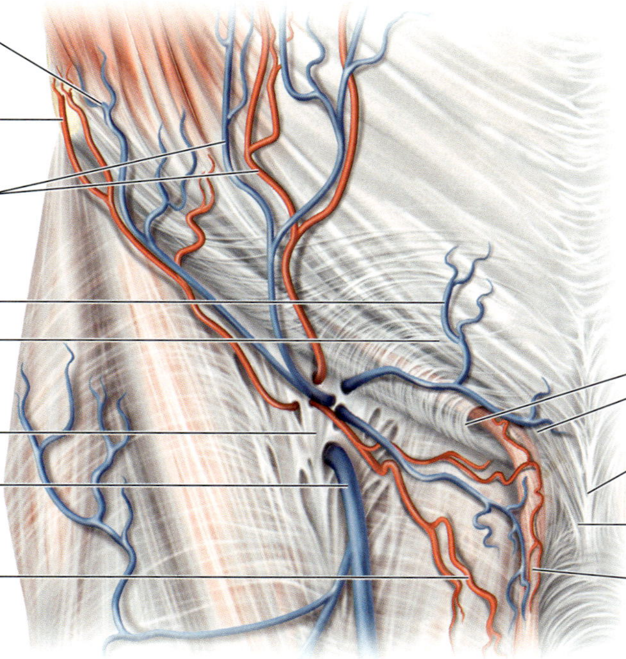

Inguinal region and inferior part of the anterior abdominal wall of the male. The superficial **inguinal ring** is occupied by the spermatic cord and the ilioinguinal nerve (in the female it is occupied by the round ligament of the uterus and the ilioinguinal nerve). In the scrotal region, the spermatic cord is also covered externally by the cremasteric fascia. The fibers of the aponeurosis of the external oblique, which originate from the anterior superior iliac spine, run towards the pubic tubercle, forming the lateral crus of the ring; those reaching the pubic symphysis constitute the medial crus. The suspensory and fundiform ligaments of the penis (equivalent to the suspensory and fundiform ligaments of the clitoris of the female) constitute the suspensory system of the penis (suspensory system of the clitoris in the female). The first originates from the pubic symphysis, the second from the linea alba; both of which merge with the fascia of the penis.

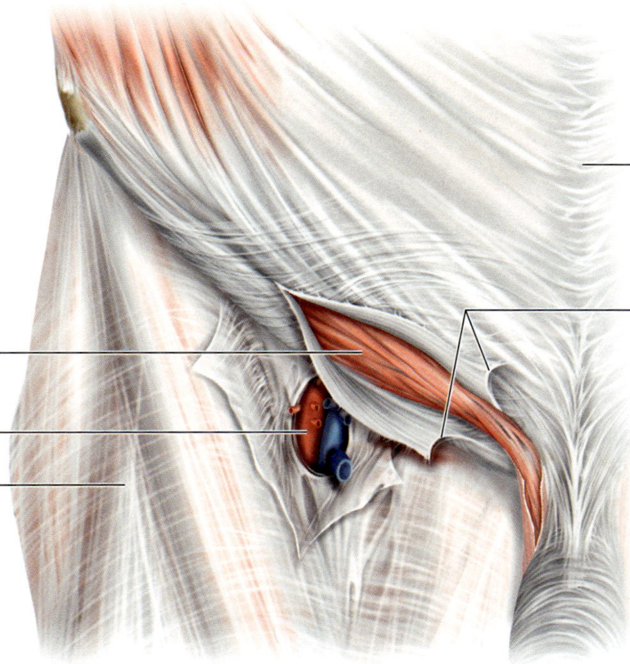

Inguinal region and inferior part of the anterior abdominal wall of the male. The anterior wall of the **inguinal canal** has been transected. The wall is formed by the aponeurosis of the external oblique, which adheres to the anterior margin of the inguinal ligament, making evident the cremaster that covers the inguinal and scrotal parts of the spermatic cord (in the female the cremaster is formed by a few fibers that descend on the round ligament of the uterus). The cremasteric fascia (or external spermatic) covers the scrotal part of the spermatic cord and separates from the aponeurosis of the external oblique at the level of the external inguinal ring.

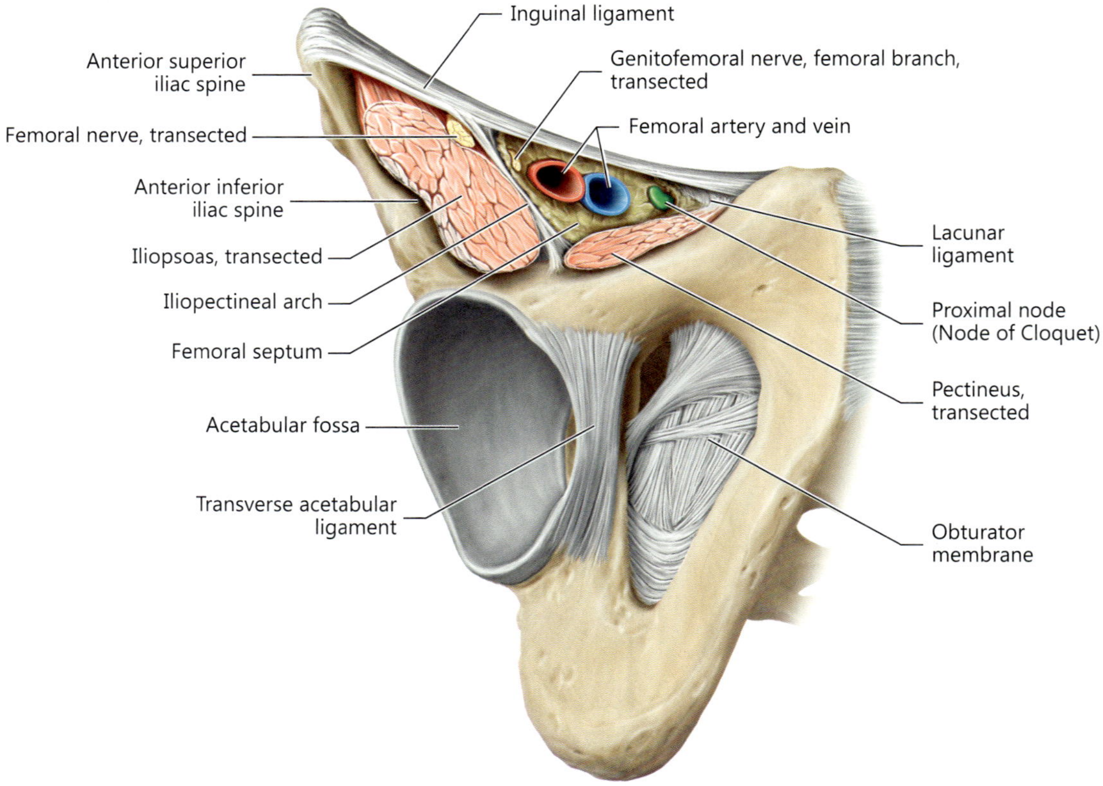

The **lacunar ligament**, which medially delimits the vascular space, originates from the inguinal ligament in correspondence with its medial part, and inserts itself into the pectineal line. The **iliopectineal arch**, which separates the **vascular space** (medial) from the **muscular space** (lateral), is an extension of the iliac fascia (anterior investing layer of the iliopsoas), that reaches and adheres to the posterior margin of the inguinal ligament, continues therefore on the inferior surface, and arrives anteriorly and attaches at the iliopectineal eminence, between the ischium and pubis. Anteroinferior view of the right hip.

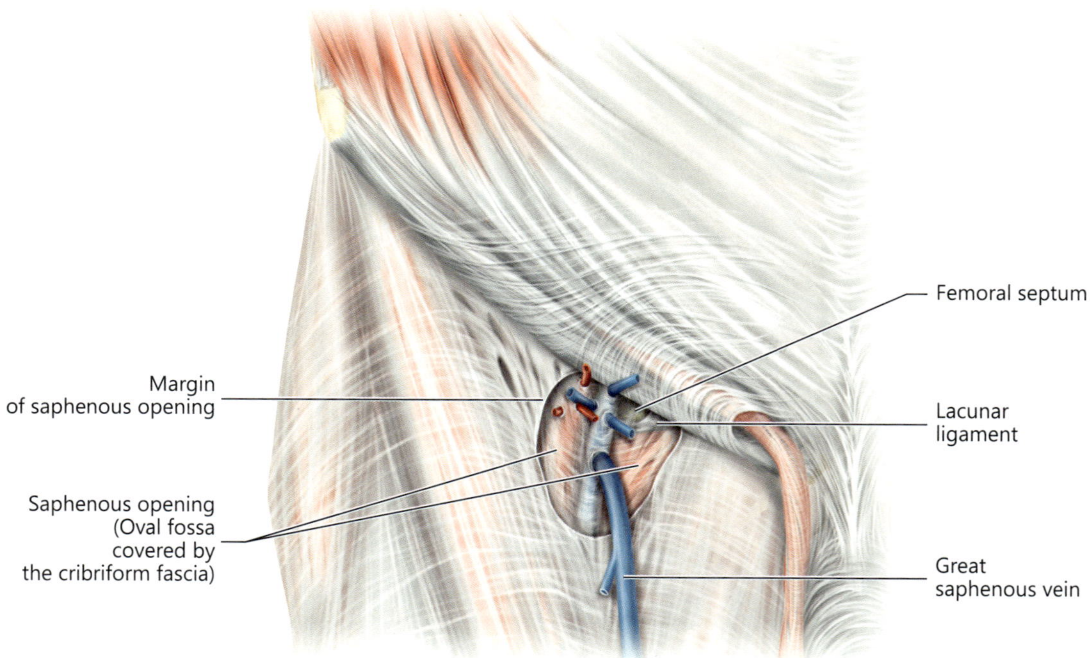

Saphenous opening. Also called oval fossa, it is covered by the **cribriform fascia**, which is perforated by the great saphenous vein and by other small vessels that cross the opening.

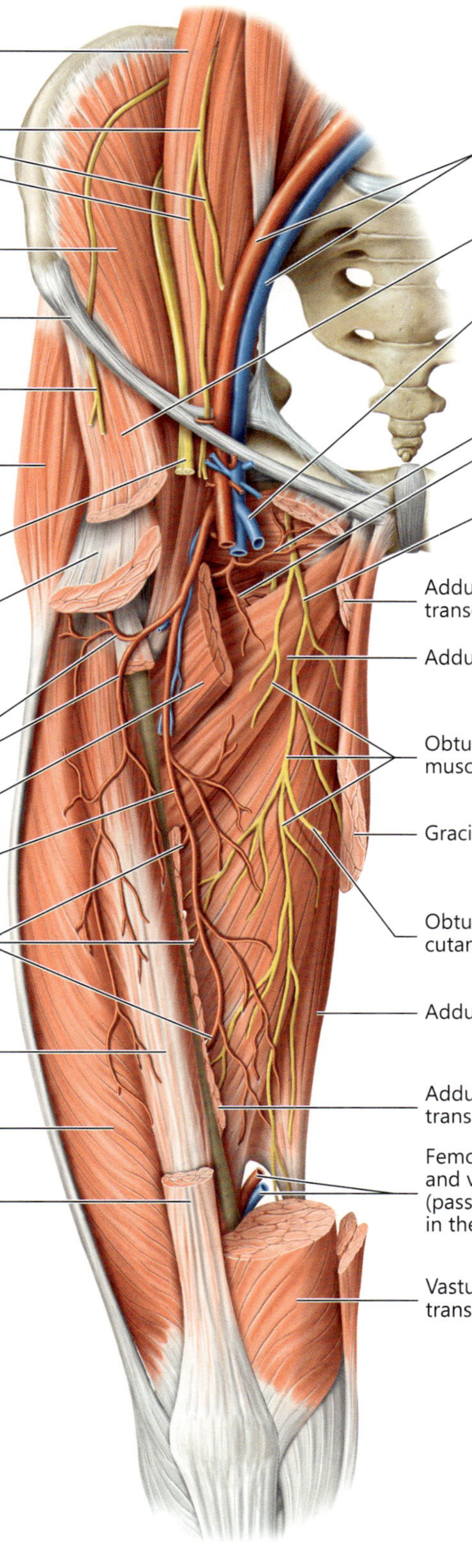

Right thigh, anterior view. Transecting the sartorius, pectineus, rectus femoris, gracilis, adductor longus and vastus medialis it is possible to observe the **deep artery of thigh** (or profunda femoris artery) that, originating from the posterolateral surface of the femoral artery, descends between the vastus medialis and pectineus, then between the adductor longus and adductor brevis, and finally between the adductor longus and adductor magnus. Here it supplies the **perforating arteries** (usually three) that reach the posterior region of the thigh, crossing the adductor muscle insertions. It is also displayed the **obturator nerve** which divides into its anterior and posterior terminal branches which, in turn, supply the muscular branches and, respectively, a cutaneous branch and an articular branch.

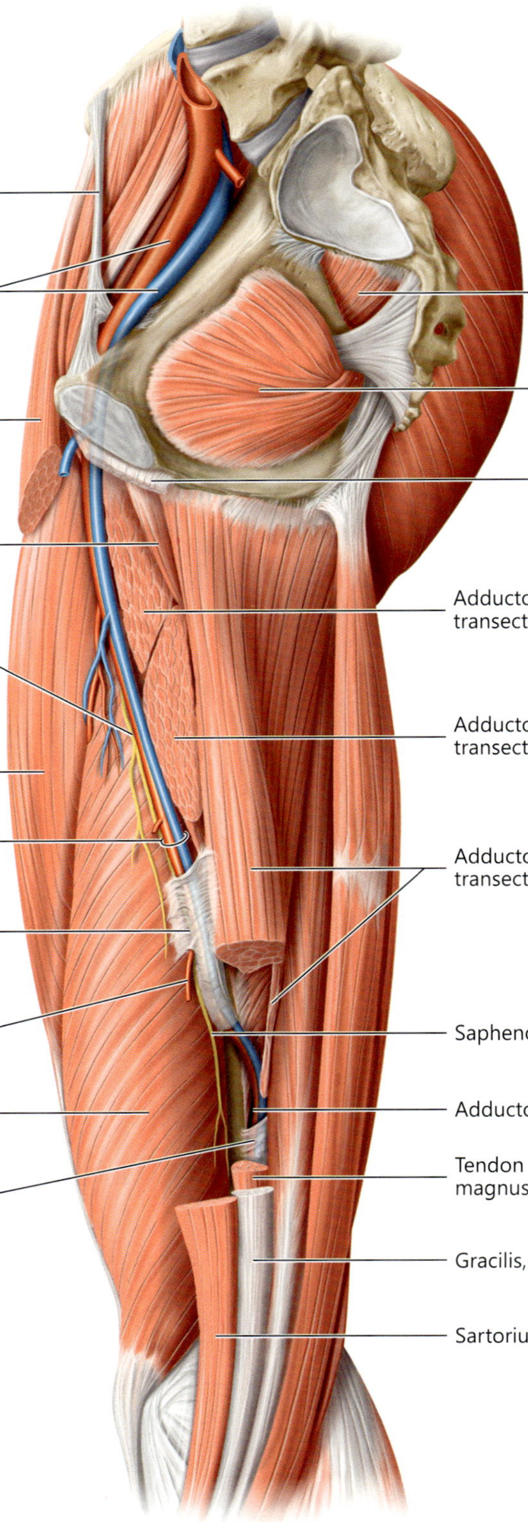

Right thigh, medial view after removing the sartorius, gracilis, adductor longus, adductor brevis and, in part, the adductor magnus. The **femoral artery,** along with the **saphenous nerve** and the **femoral vein**, descends into the adductor canal, a groove formed by the vastus medialis and adductor longus, covered anteriorly by a fibrous layer and by the sartorius. By way of the adductor hiatus, an opening dug into the tendinous insertion of the adductor magnus, the femoral vessels reach the posterior part of the knee.

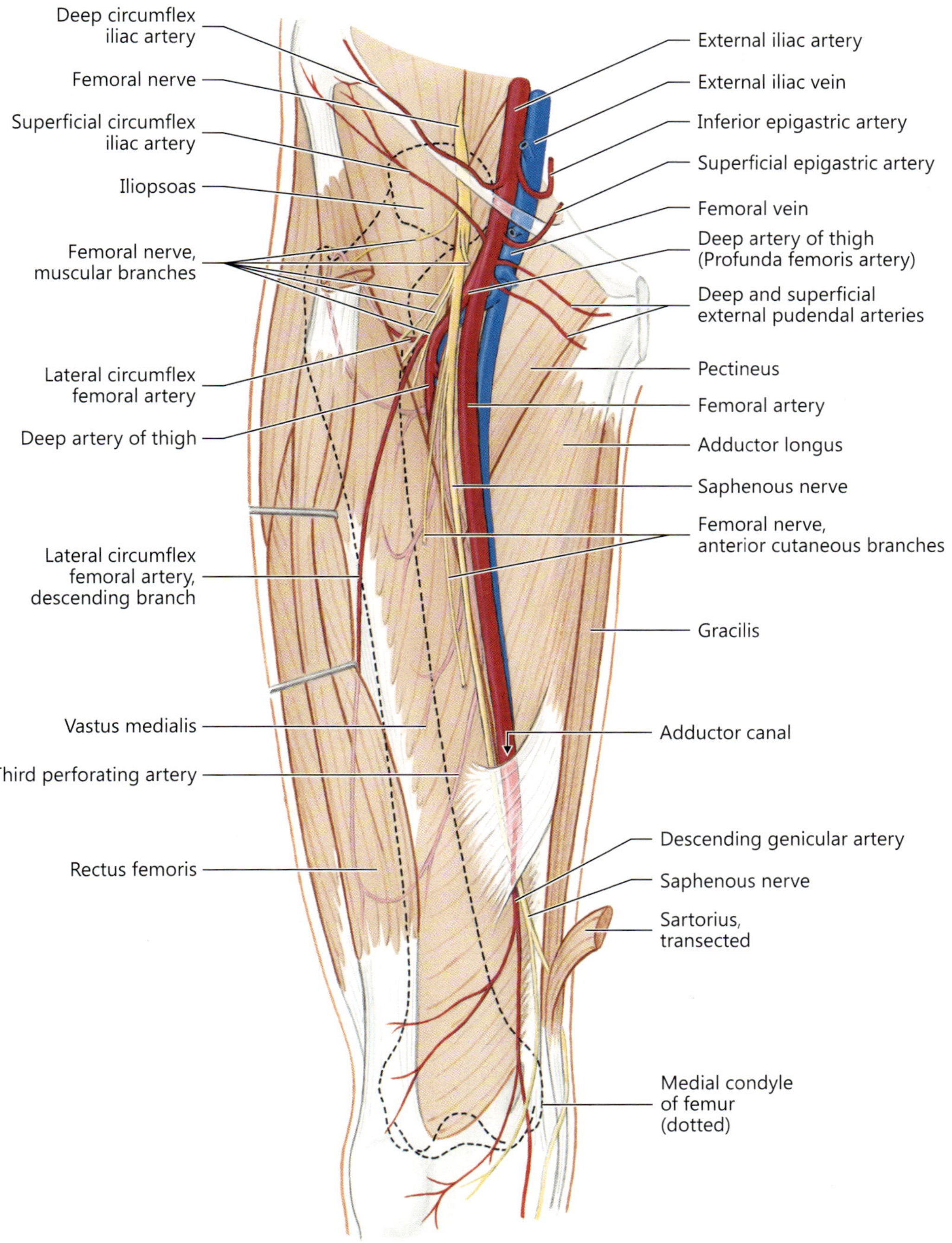

Schematic representation of the course and the relationships of the **femoral artery** and **vein** in the right anterior region of the thigh. The sartorius has been transected and removed.

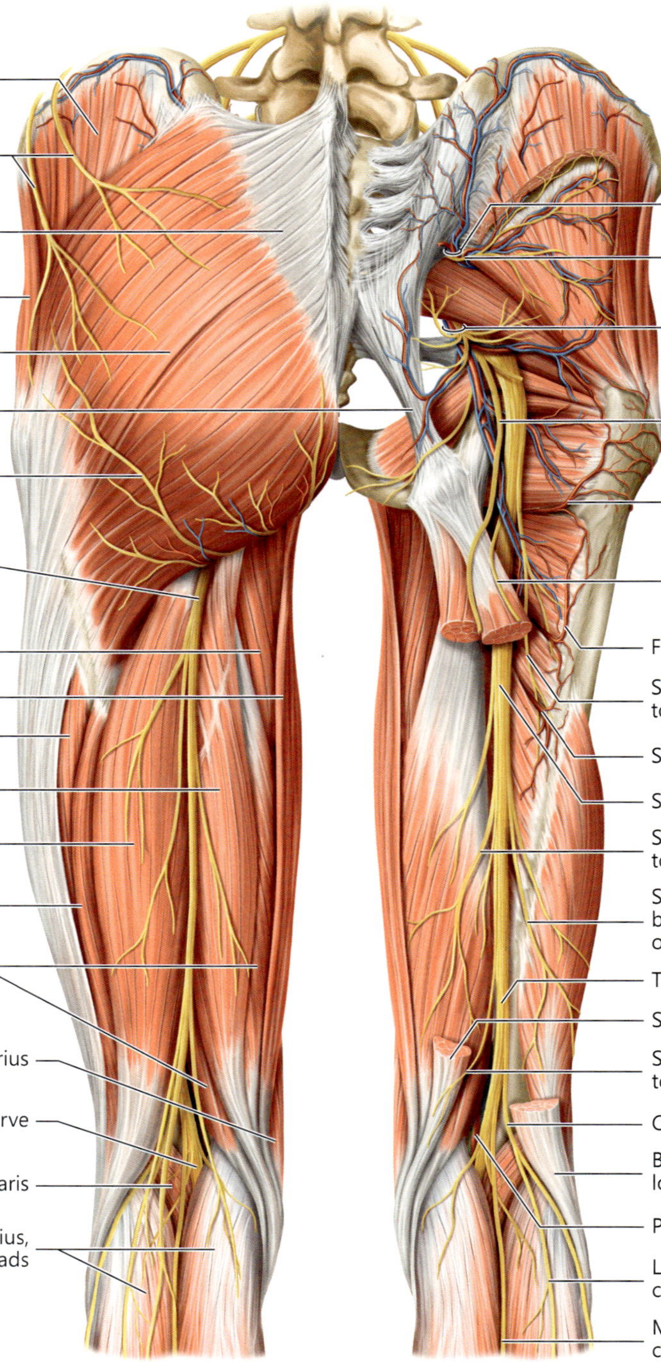

Gluteal region and posterior region of the thigh. On the left, the course of the **posterior cutaneous nerve of thigh** which, passing the gluteal fold, arrives posteriorly in the thigh, in a subfascial position between the biceps femoris and semitendinosus. Once in the popliteal fossa, it divides into its terminal branches, which continue on the superior surface of the calf. On the right, having removed the gluteal muscles and transected the semitendinosus and the long head of the biceps femoris, it appears the course of the **sciatic nerve** which, at the superior angle of the popliteal fossa, becomes superficial, dividing into the **tibial** and **common fibular nerves**.

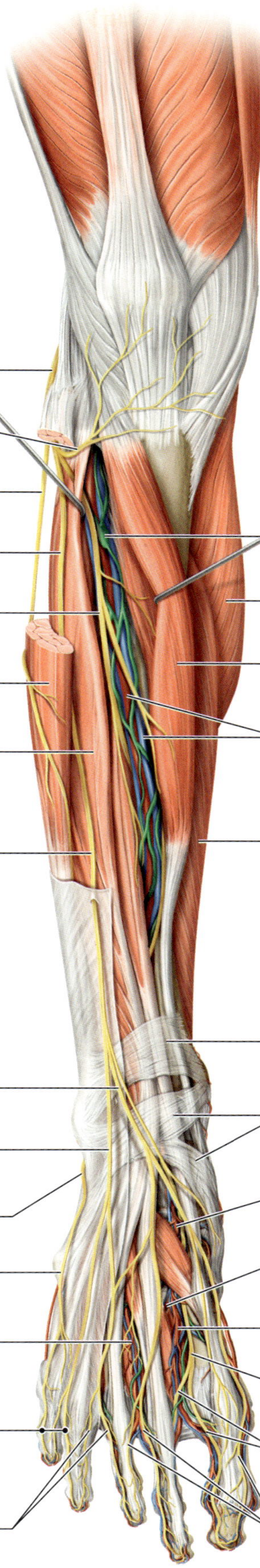

Leg and dorsum of the foot. Course of the **superficial fibular nerve**, the lateral division of the **common fibular nerve**. It descends vertically, passing between the origins of the fibularis longus, and then between the latter and the extensor digitorum longus. At the level of the lower third of the leg, the superficial fibular nerve perforates the deep fascia of the leg, becoming superficial and dividing into the **medial dorsal and intermediate dorsal cutaneous nerves**.

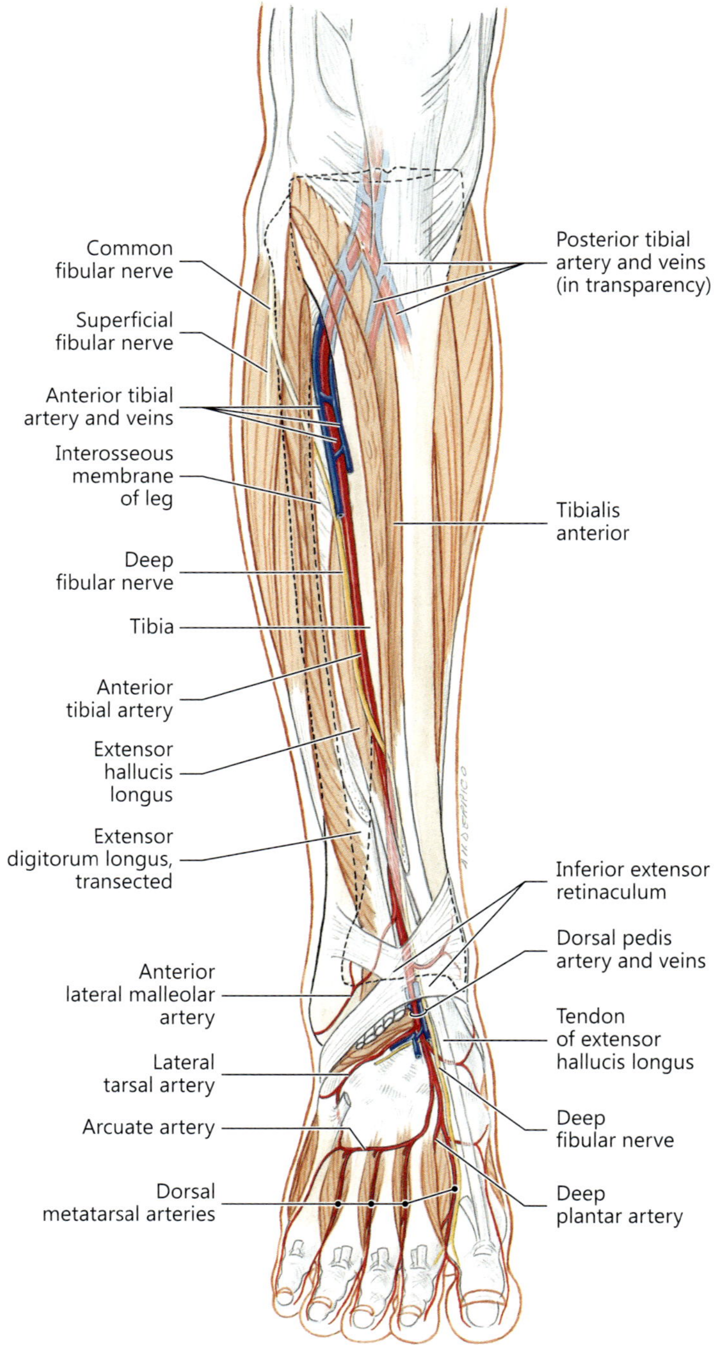

Schematic representation of the relationships of the **anterior tibial** and **dorsalis pedis arteries**; right side. The dotted line highlights the profile of the skeletal segments.

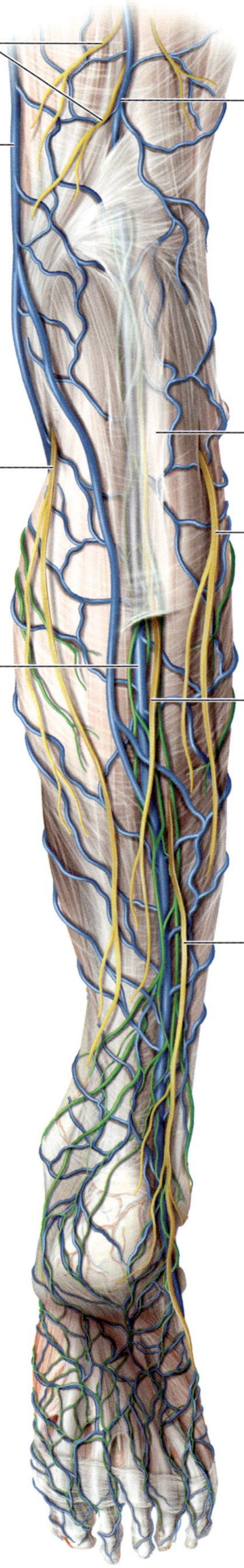

➤ Leg and sole of the foot. The **medial sural cutaneous nerve** originates from the tibial nerve in the popliteal fossa and running downwards next to the small saphenous vein, crosses the deep fascia of the leg. Once subcutaneous, it anastomoses with the **lateral sural cutaneous nerve**, forming the **sural nerve**. It is also displayed the **saphenous nerve**, terminal branch of the femoral nerve.

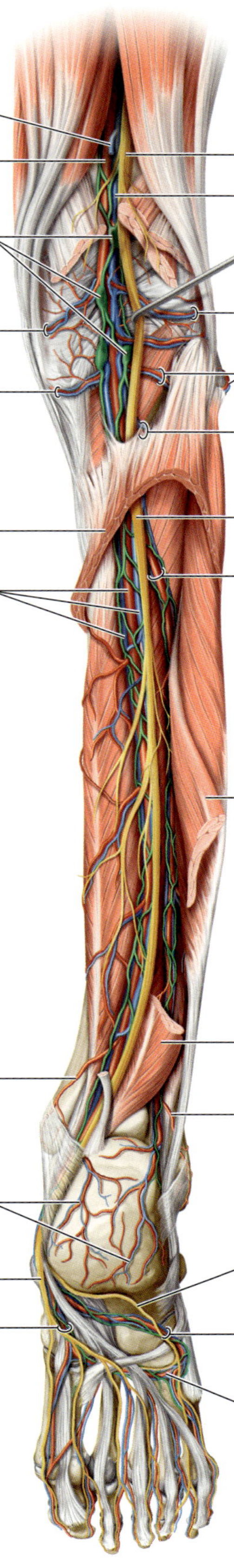

➤ Popliteal fossa and posterior region of the leg. After removal of the gastrocnemius, soleus and plantaris and transection of the flexor hallucis longus, it appears the **tibial nerve**, a direct extension of the sciatic nerve, which runs towards the medial malleolus and descends into the calcaneus, dividing into the **plantar nerves**.

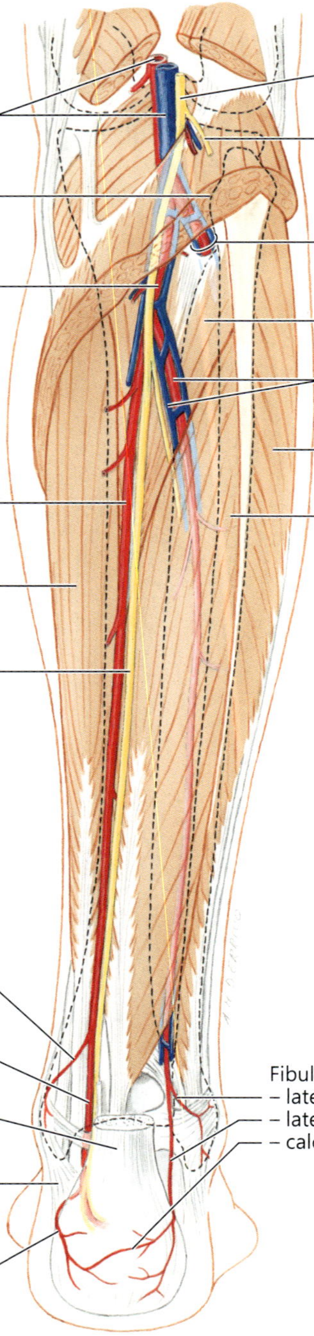

Schematic representation of the relationships of the **posterior tibial artery** and the **fibular artery** in the right leg. The triceps surae has been transected and removed.

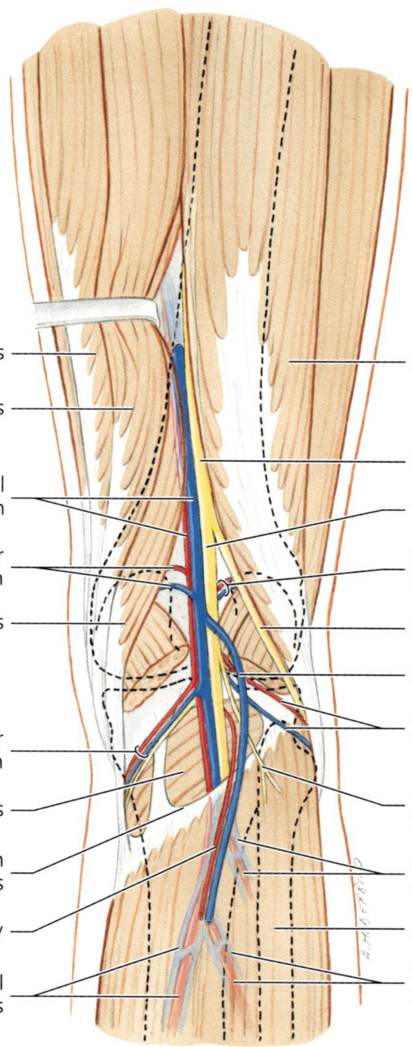

Schematic representation of the relationships of the **popliteal artery** and its branches. The dotted line highlights the profile of the skeletal segments.

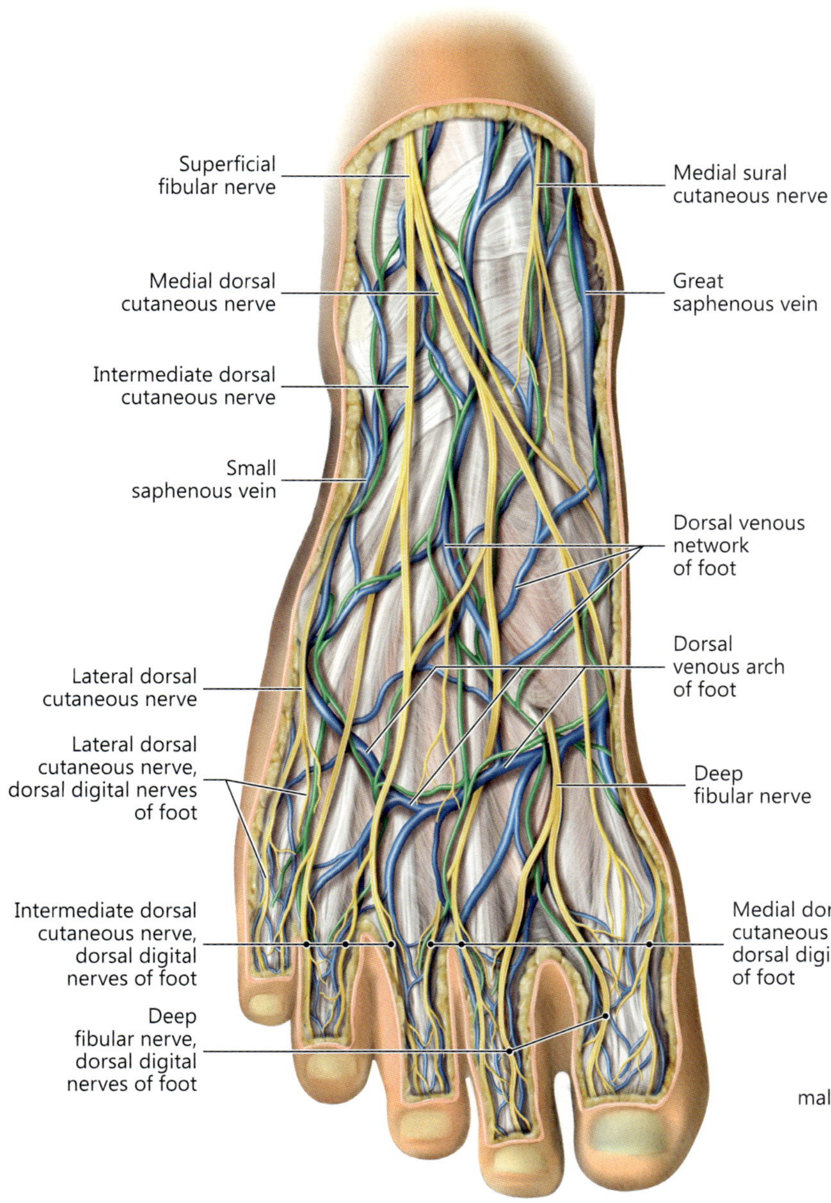

Dorsum of the foot. **Dorsal venous network of the foot** and divisions of the superficial fibular nerve, represented by the **medial dorsal and intermediate dorsal cutaneous nerves**. The **lateral dorsal cutaneous nerve** is the continuation on the dorsum of the foot of the sural nerve, originating from the anastomosis between the medial and lateral sural cutaneous nerves.

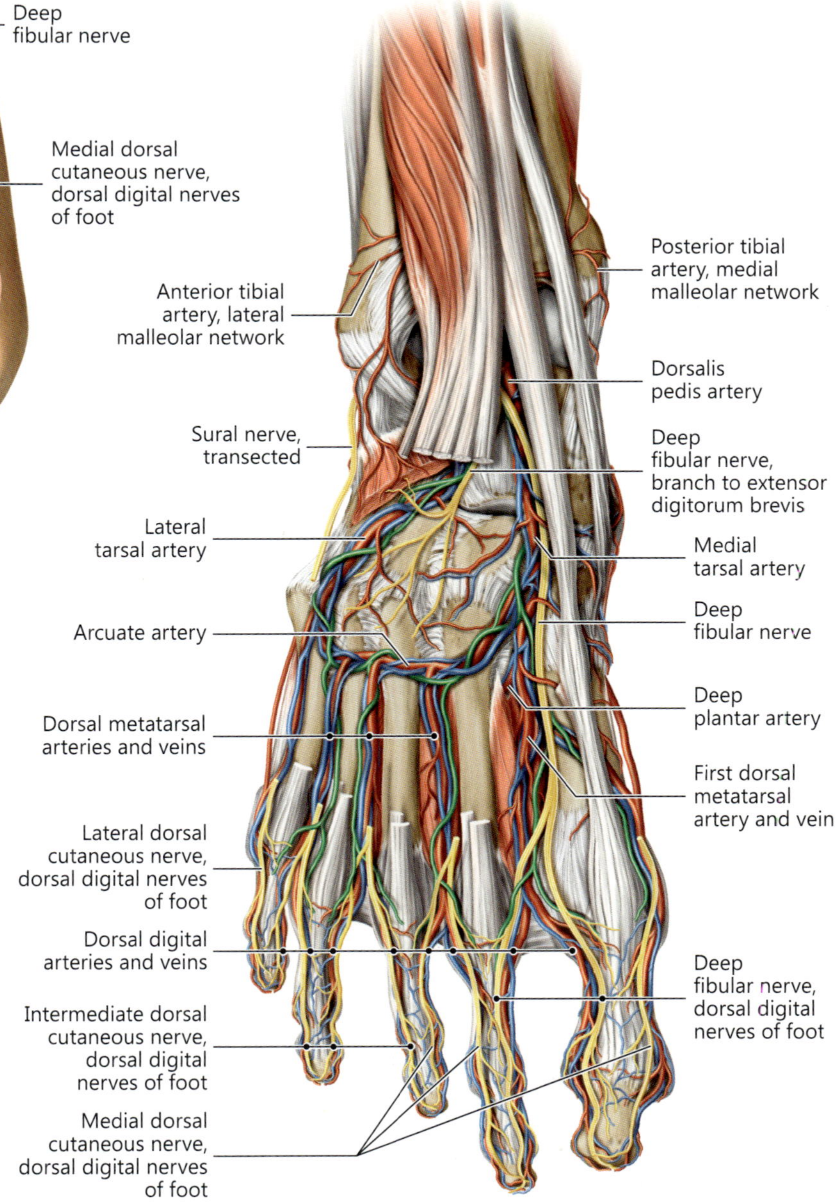

Vessels and deep nerves of the dorsum of the foot. The **arcuate artery** originates from the dorsalis pedis artery before its terminal branches, represented by the **first dorsal metatarsal artery** and by the deep plantar artery. The **dorsal metatarsal arteries**, second, third and fourth, originate from the arcuate artery. The **deep fibular nerve** crosses the dorsum of the foot, running along the lateral margin of the tendon of the extensor hallucis longus.

Sole of the foot. The sole of the foot is completely covered by a **network of superficial lymphatic vessels**, most of which reach the dorsum of the foot passing through the medial and lateral margins. These lymphatic collectors run next to the great saphenous vein, for the most part, and empty into the superficial inguinal nodes. Some of them follow the course of the small saphenous vein and drain into the popliteal nodes.

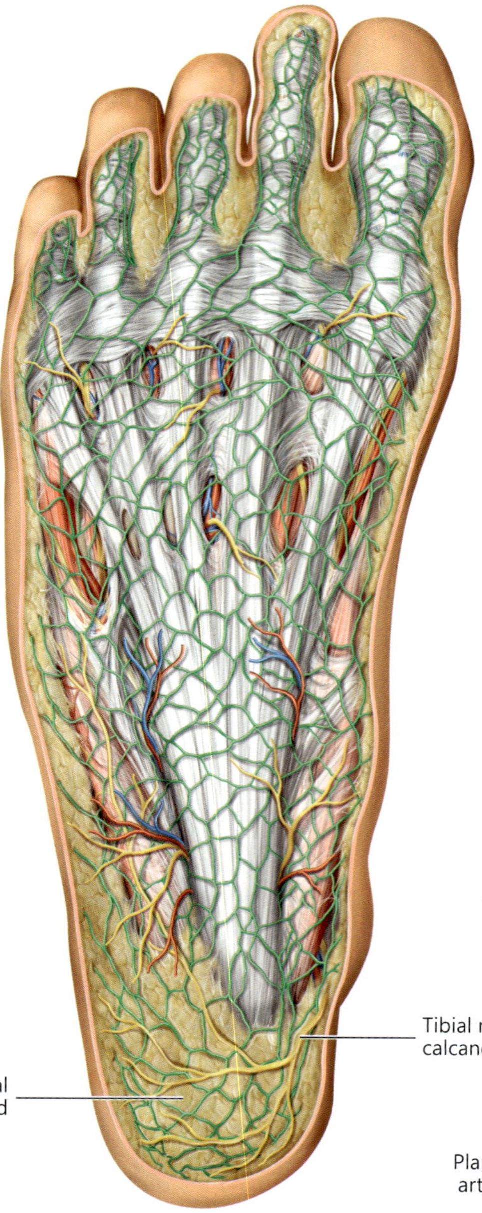

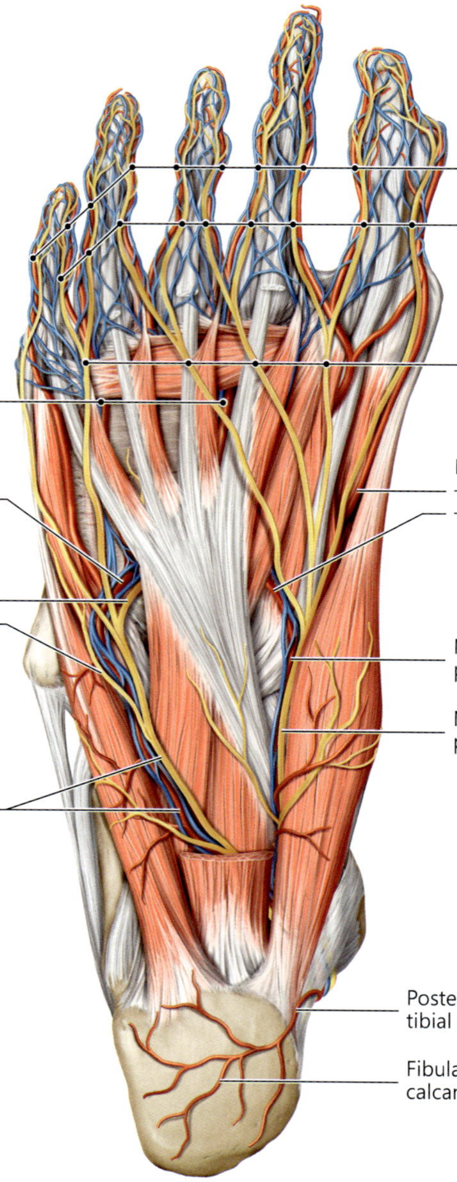

Nerves of the sole of the foot. The **medial plantar nerve** runs along the medial part of the sole of the foot, together with the medial plantar vessels that cross its path superficially, and then run along satellites in a medial position. At the base of the metatarsals it divides into a proper plantar digital nerve, which runs along the medial margin of the great toe, and into three common plantar digital nerves, from which the proper plantar digital nerves originate. The **lateral plantar nerve** divides into a superficial branch, which gives rise to a common plantar digital nerve with two proper plantar digital nerves, and a deep branch which gives rise to muscular branches.

Sole of the foot. On a deep plane are shown the plantar metatarsal arteries and veins, and the deep branch of the lateral plantar nerve, which runs medially and describes a posterior concave ansa, following the deep plantar arch.

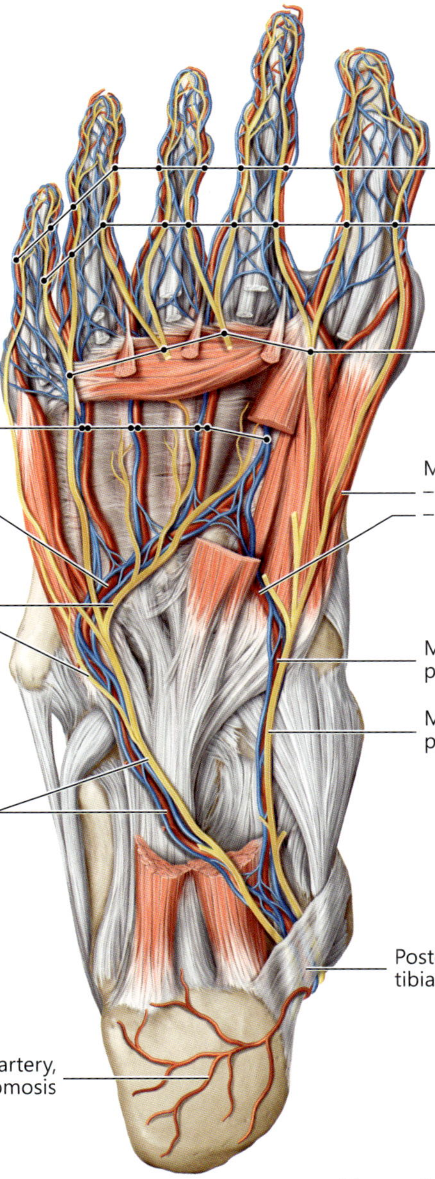

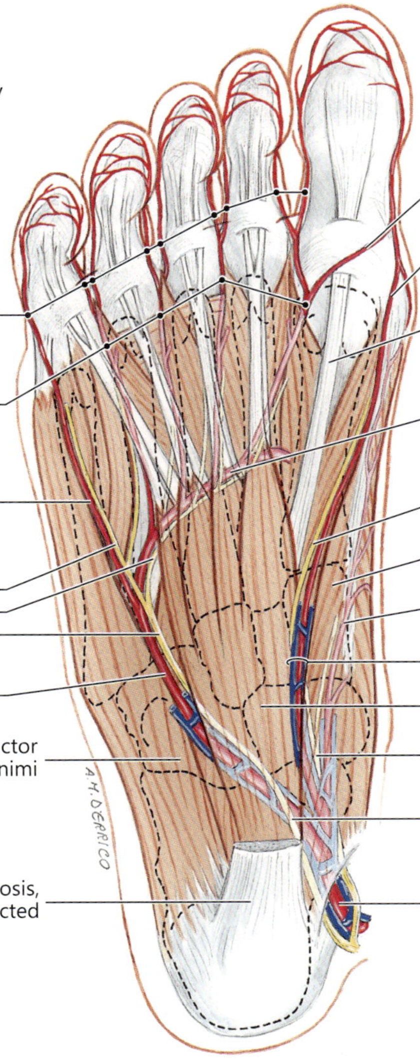

Schematic representation of the course and relationships of the **lateral** and **medial plantar arteries** of the right foot. The deep plantar arch is in transparency.

NERVOUS SYSTEM

Petros (Athens, 1928 - Milan, 2014), *Biologikos dialogos*, oil on canvas, cm 32 × 40, 1993

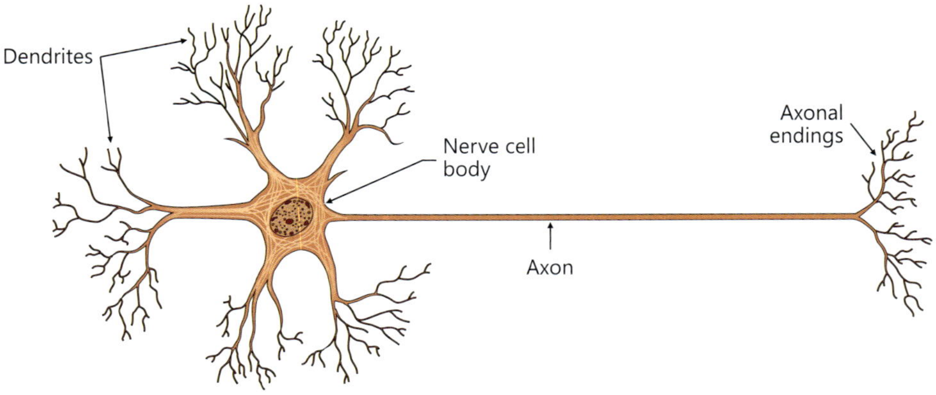

Graphic representation of a **neuron**.

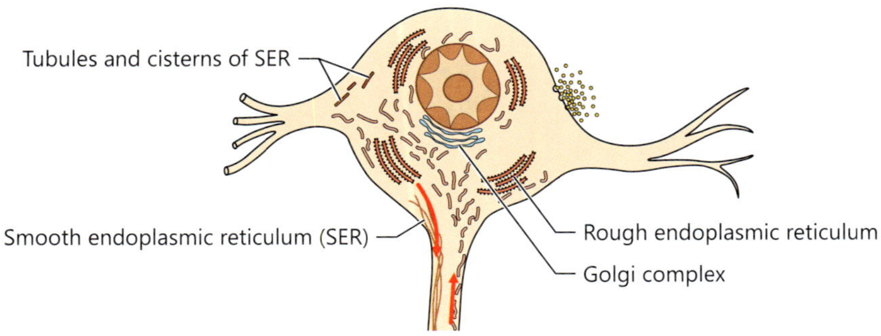

The **soma** contains the nucleus and the cytoplasmic organelles.

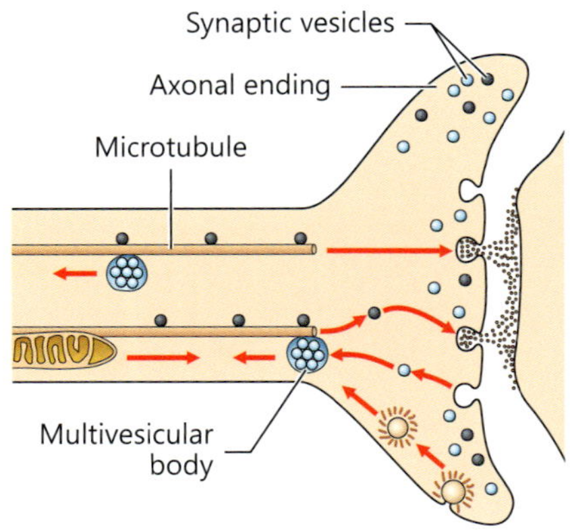

In the axon, there is an **axonal flow of chemical neurotransmitters** contained in the synaptic vesicles and transported towards the presynaptic membrane along the microtubules.

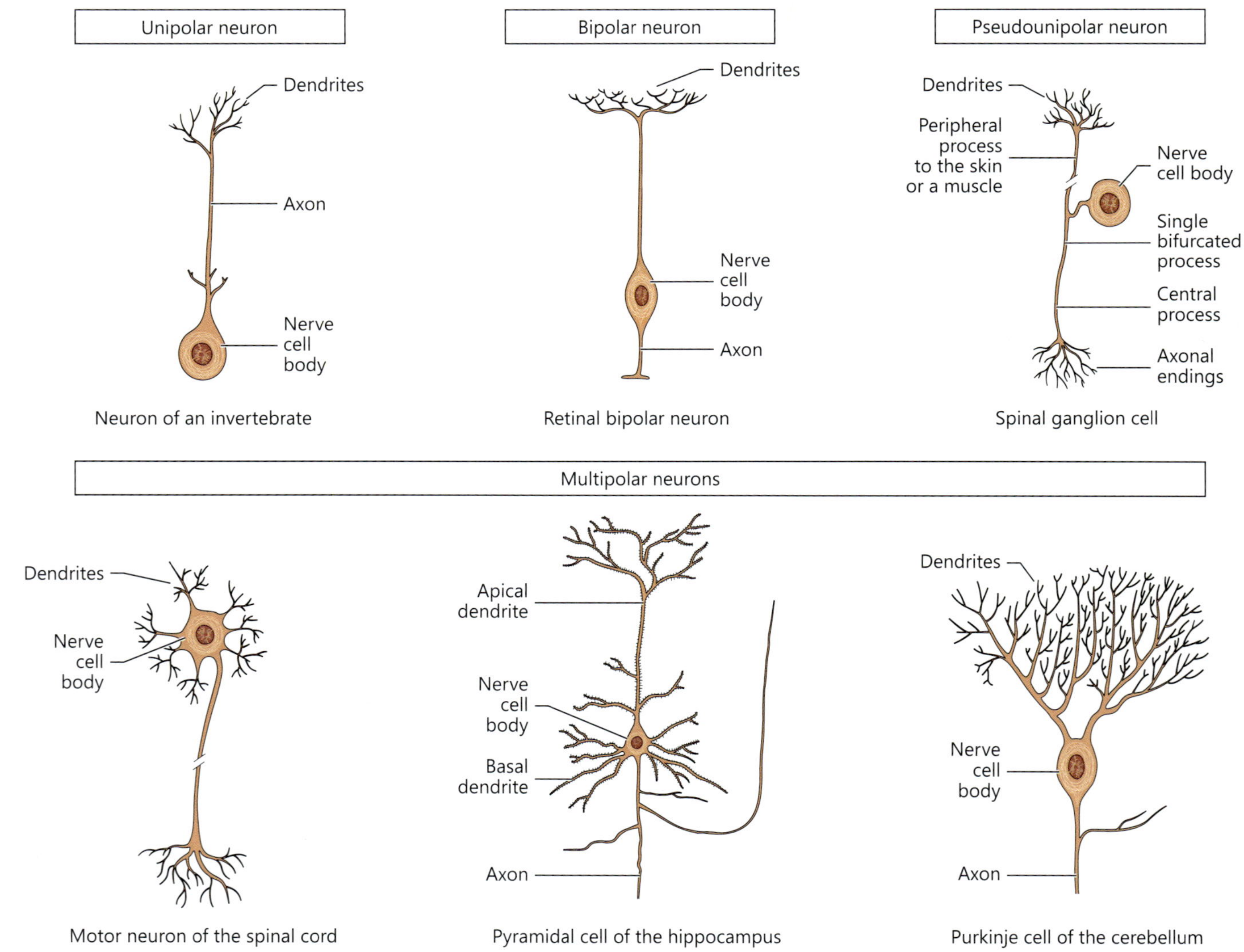

Classification of the **nerve cells** based on the number of processes originating from the cell body.

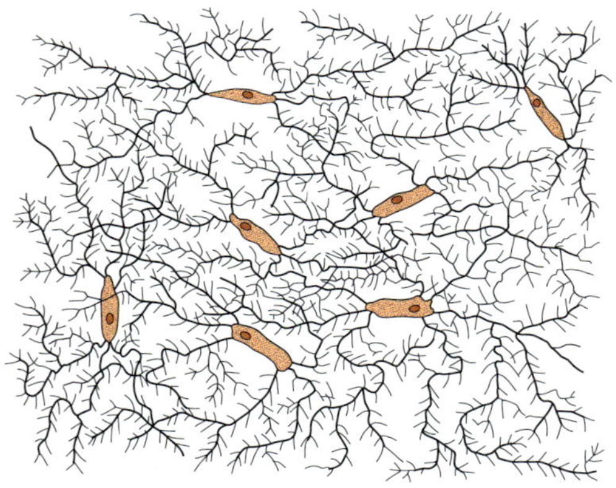

Organization of the **microglia**. The small spindle-shaped cells display cytoplasmatic processes densely interdigitated with those stemming from the adjacent elements.

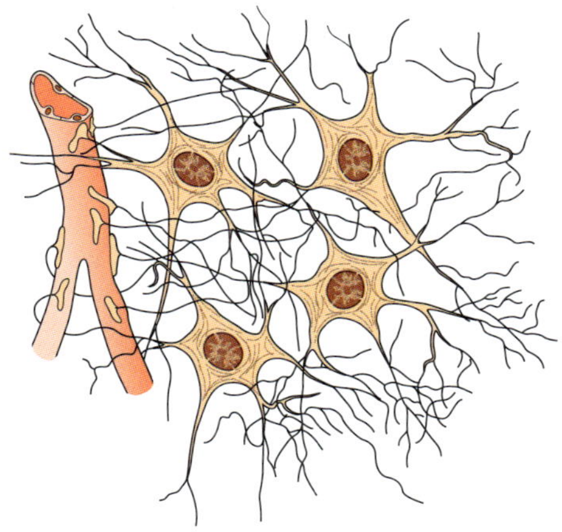

Astrocytes in relation with a thin vase; glial cell processes are thin and long.

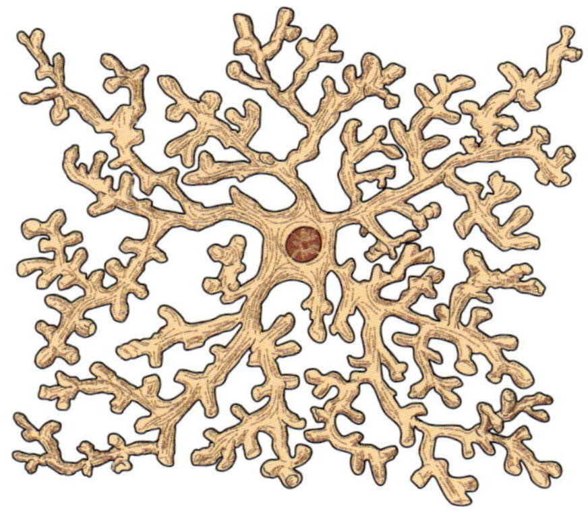

The **protoplasmatic astrocyte** displays many abundantly branched processes.

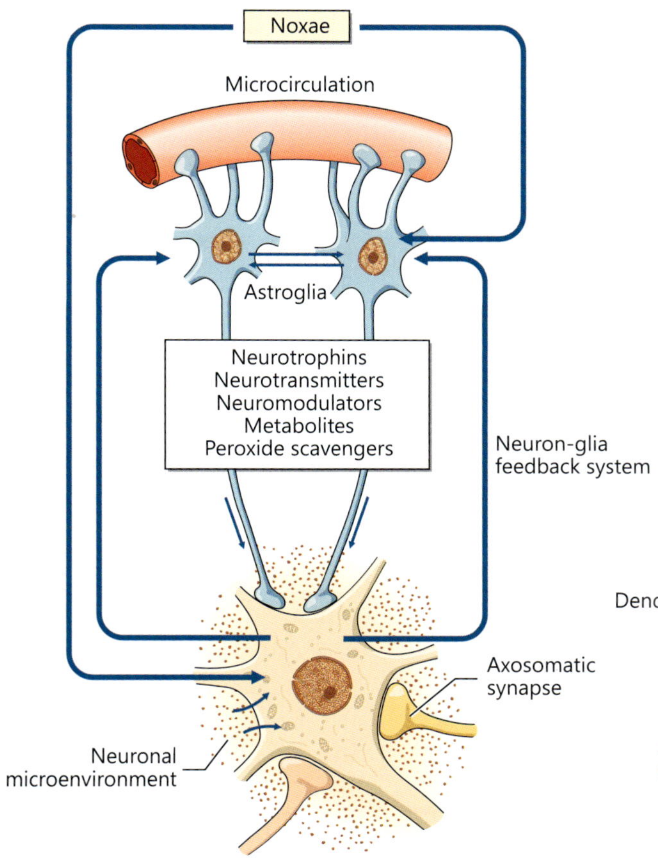

Diagram illustrating the **functional connections between glia, neurons and vascular components**.

An **oligodendrocyte** or a Schwann cell can wrap more than one axon at the same time, without forming a myelin sheath, for this reason these **fibers** will be defined **nonmyelinated**.

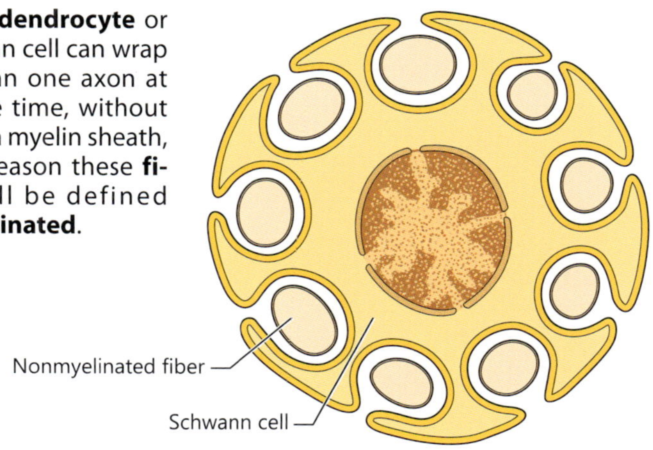

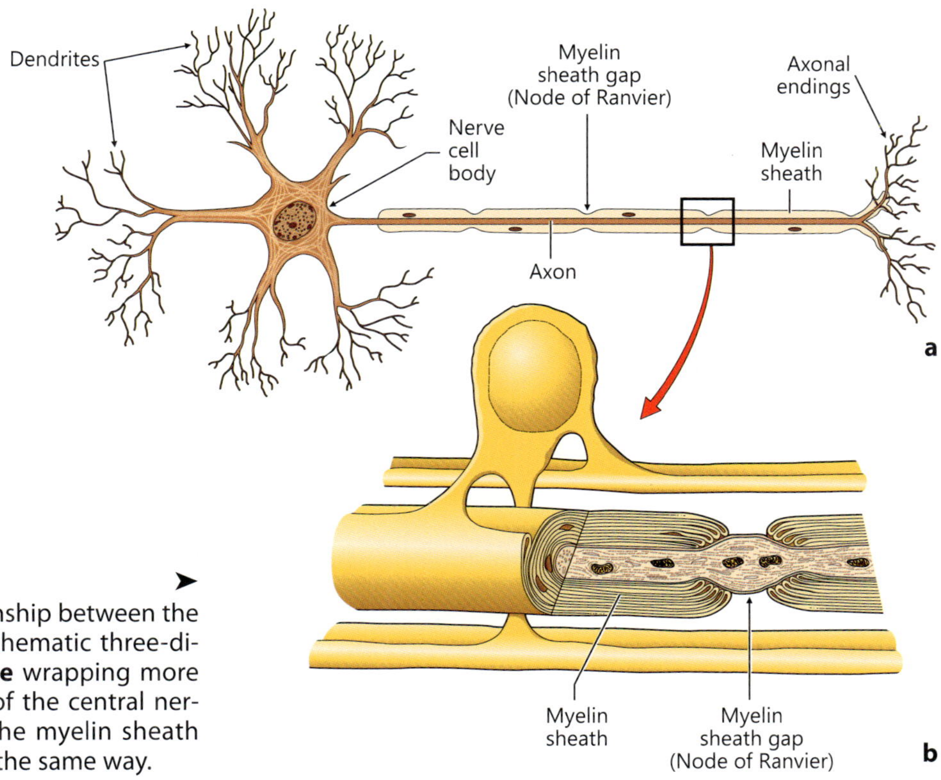

a, Schematic diagram illustrating the close relationship between the **myelin sheath** and the **axon of a neuron**. **b**, Schematic three-dimensional representation of an **oligodendrocyte** wrapping more axons, constituting the myelinated nerve fibers of the central nervous system. In the cranial and spinal nerves, the myelin sheath formed by the Schwann cells invests the axons in the same way.

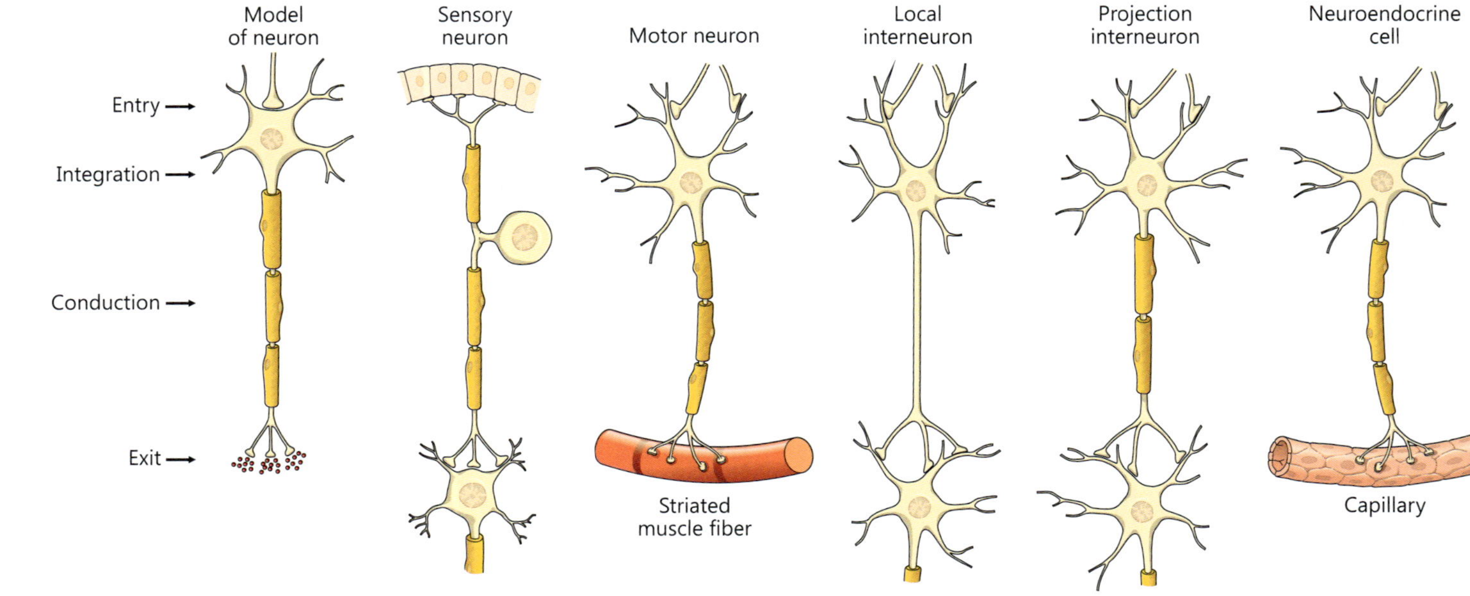

The **neurons** share four functional elements related to the occurrence of electrical events: an input element or receptive element, a trigger element transforming the receptor potential into action potential, a conduction element represented by the axon and an output (or secretory) element.

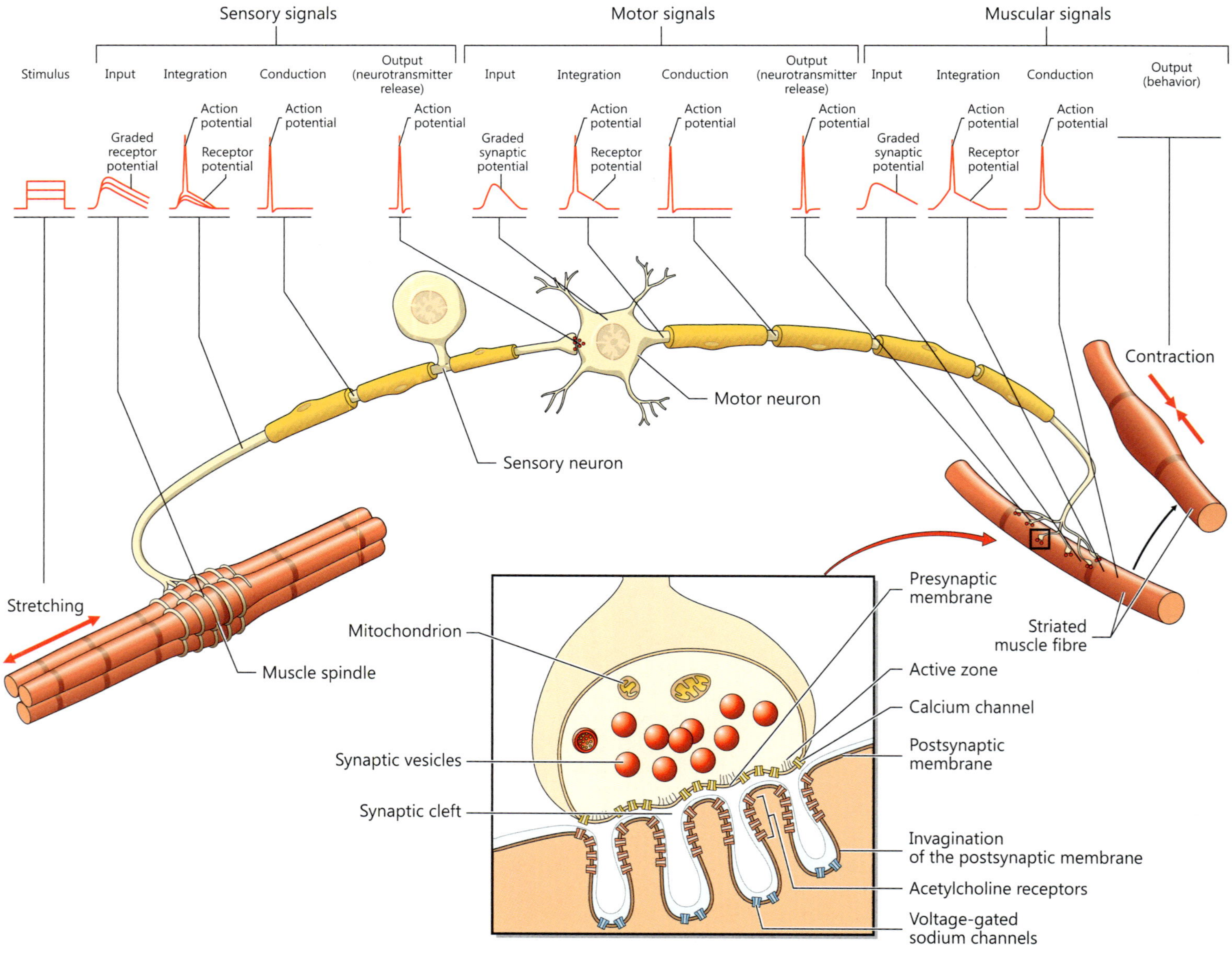

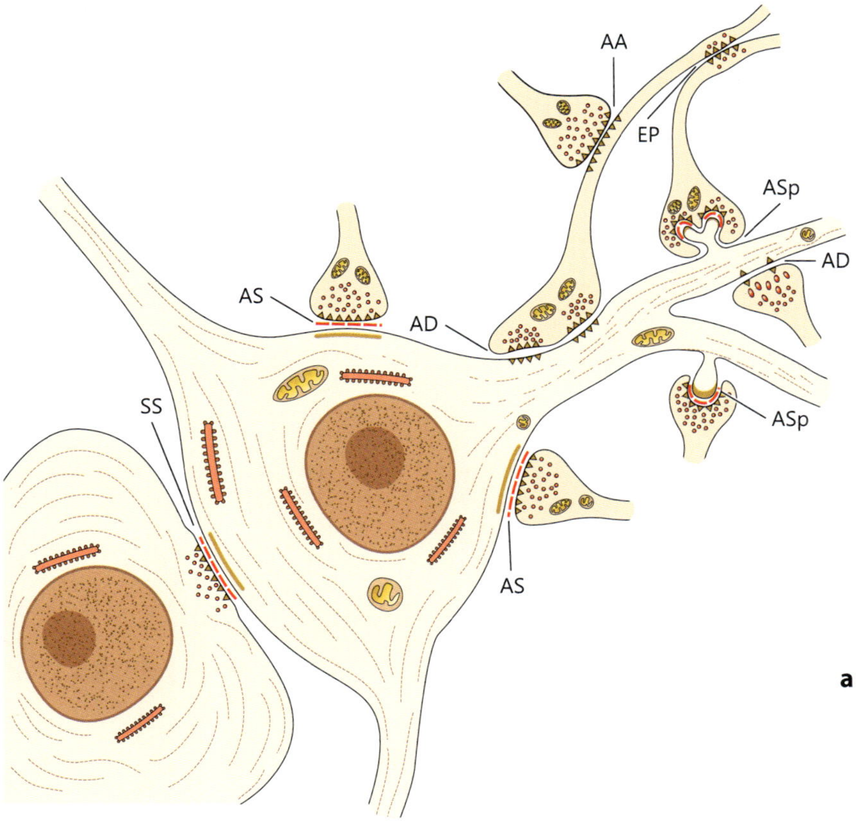

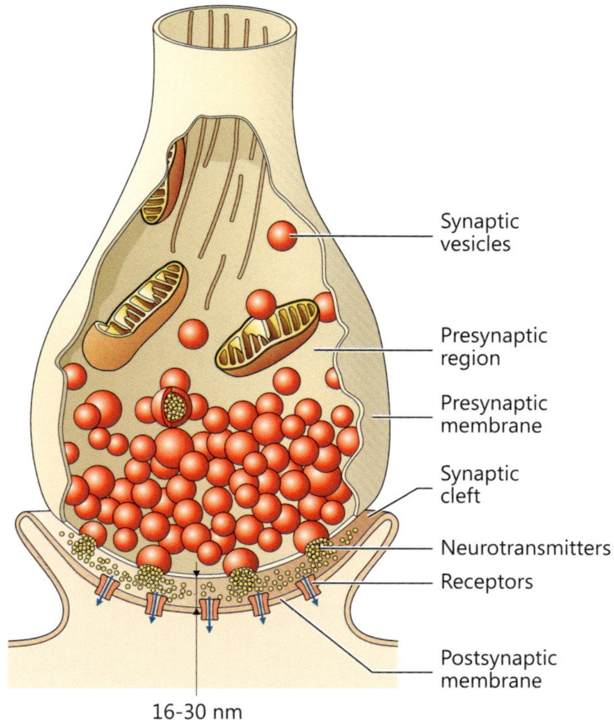

a

Schematic diagram showing different **types of synapses** occurring among neurons.

Synapses:
AA, axoaxonic, synapses between two axons
AD, axodendritic, synapses between one axon and one dendrite
AS, axosomatic, synapses between one axon and the soma of a neuron
ASp, axospinous, synapses between one axon and a dendritic spine
EP, synapses en-passant
SS, somato-somatic, synapses between two soma

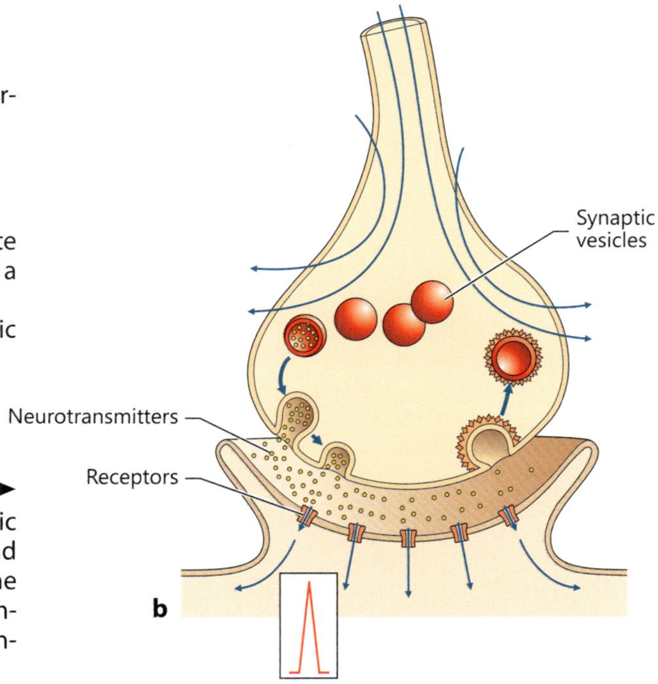

b

In a **chemical synapse**, the depolarization of the presynaptic terminal determines the exocytosis of the synaptic vesicles and the release of the neurotransmitter in the synaptic cleft (**a**). The neurotransmitter binds to receptors on the postsynaptic membrane causing an excitatory or inhibitory postsynaptic potential (**b**).

Sequence of electrical events occurring in different parts of the neuron
The neurotransmitter interaction with the receptors present on the postsynaptic membrane promotes a synaptic potential with the same characteristics of a receptor potential, i.e. is a graduated potential. This signal is transmitted passively toward the trigger area preceeding the beginning of the myelin sheath. At this level, if the receptor potential, or synaptic potential, is sufficiently high, it triggers an action potential that spreads actively along the axon until it reaches the synaptic endings. At this level, the action potential is named secretion potential, as it induces an output signal followed by the release of chemical neurotransmitters in the synaptic cleft. At this site, the neurotransmitters bind to receptors located on the outer surface of the interacting neuron or of effector elements (e.g. striated or smooth muscle, or glandular epithelium).
In the box, **axon terminal of a neuromuscular junction**. The input of calcium ions following the opening of the calcium channels in the presynaptic membrane causes the release of acetylcholine from the synaptic vesicles into the synaptic cleft. The binding of acetylcholine to the receptors present on the postsynaptic membrane determines the opening of the sodium channels of the postsynaptic membrane.

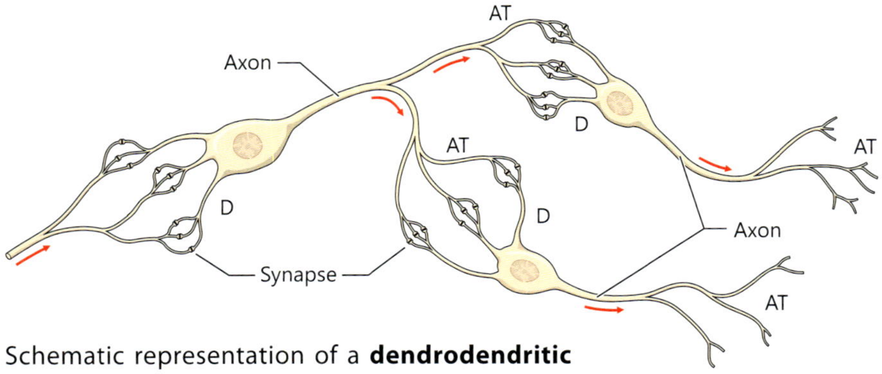

Schematic representation of a **dendrodendritic synapse**.
D, dendrite
AT, axon terminal

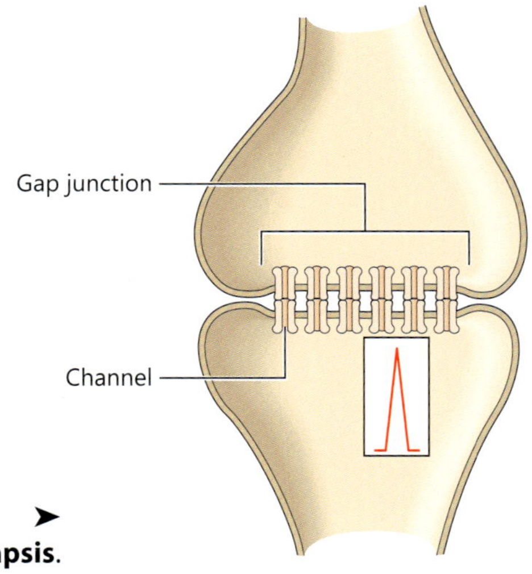

Schematic representation of gap junctions forming an **electric synapsis**.

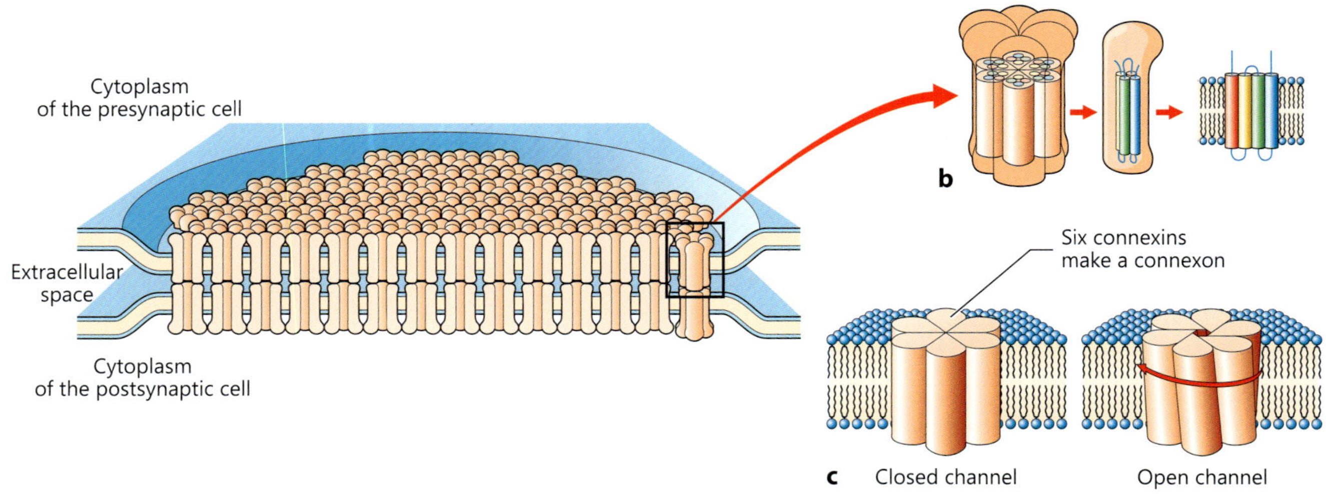

Three-dimensional model of an electric junction. **a**, In electrical synapses, the pre- and postsynaptic cells are coupled through electrical junctions. These are formed by hemichannels, called connexons, in the membrane of contiguos cells in contact, connecting their cytoplasm and allowing the flow of ions from cell to cell. **b**, Each hemichannel consists of six connexins, each of which is costituted by four hydrophobic transmembrane segments. **c**, The connexins form a central pore which opens upon rotation clockwise (approximately 0.9 nm) of the subunits.

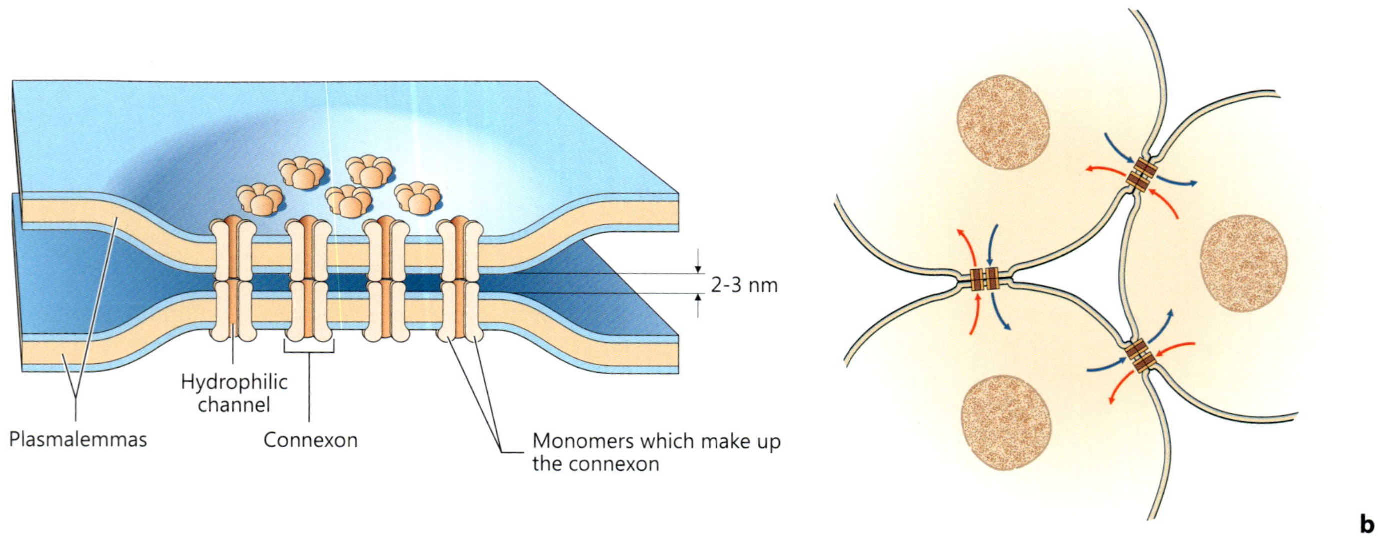

a, **Three-dimensional model of electric synapsis**. The two membranes are very close, but without fusion between the outer layers. **b**, The passage of ions through the connexons is bidirectional.

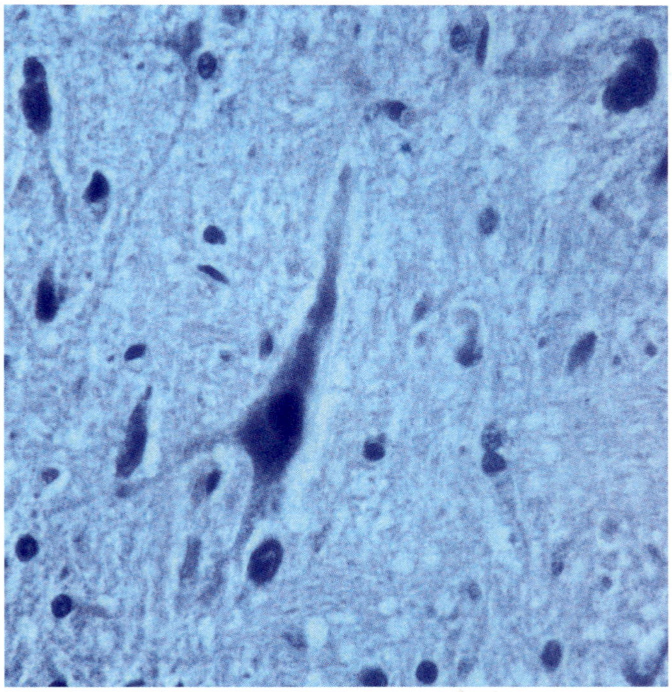

a

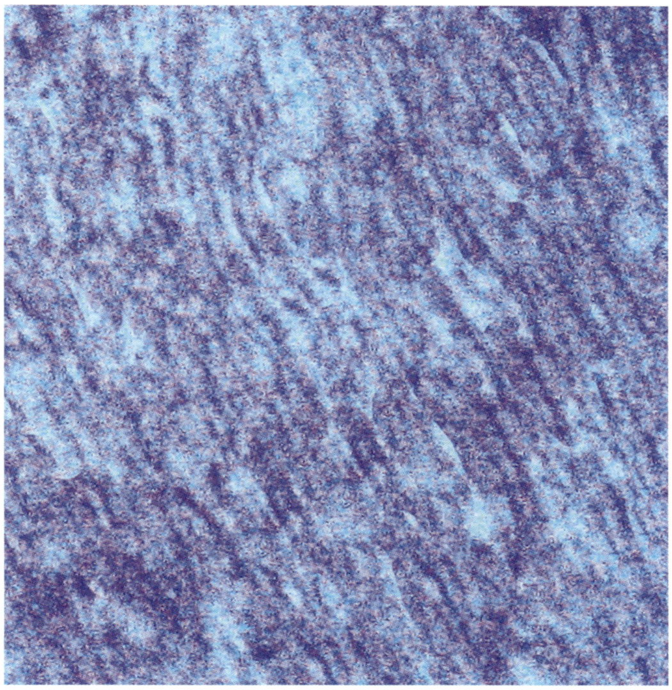

b

Gray matter consists mainly of cell bodies (**a**); **white matter** consists of myelinated and nonmyelinated axons (**b**).

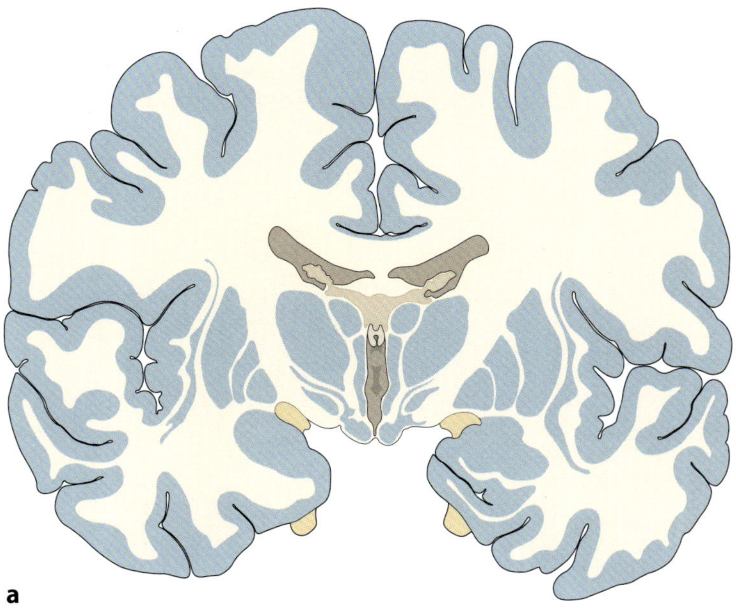

a

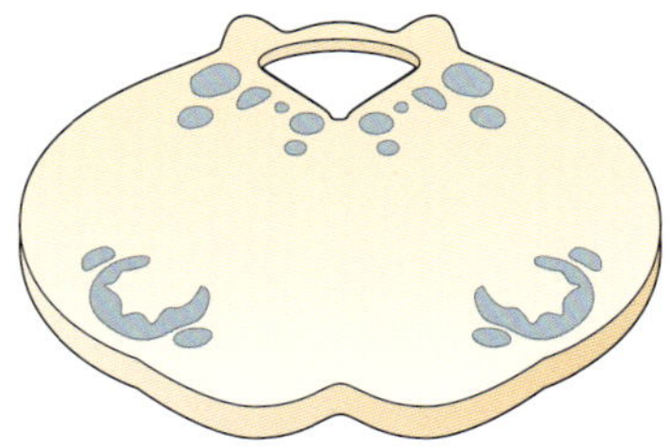

b

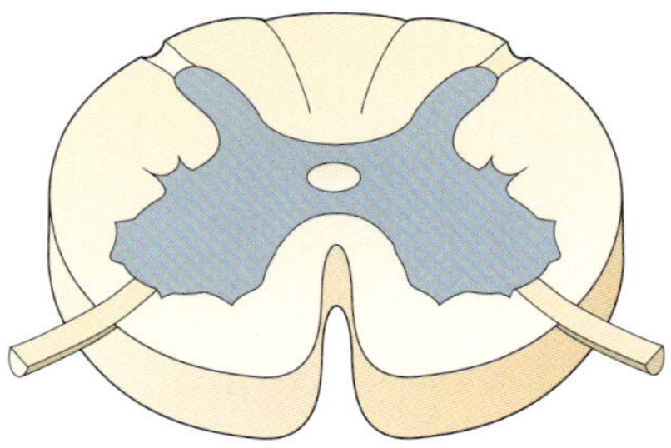

c

Distribution of the **gray matter** and **white matter** in different parts of the neuraxis: **a**, in the telencephalon; **b**, in the myelencephalon (medulla oblongata or bulb); **c**, in the spinal cord. In the telencephalon, the gray matter is located on the surface, where it forms the cortex, and, deeply, in the context of the white matter, constitutes the basal nuclei; in the myelencephalon, the gray matter is distributed in numerous nuclei interlayered to bundles of white matter; in the spinal cord, the gray matter organizes in a compact form around the central channel, whereas the white matter constitutes a peripheral cover.

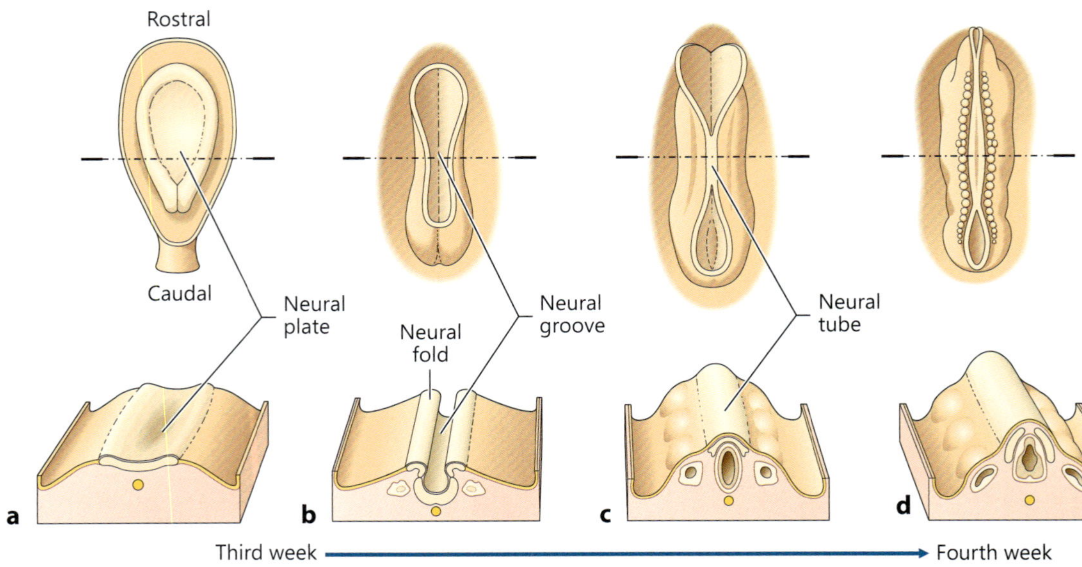

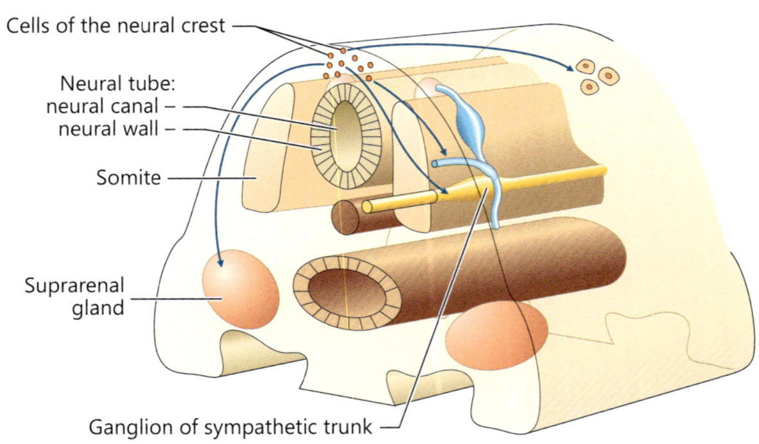

Graphic representation of the very first stages of the **organogenesis**. The structures of the nervous system form from the neural plate, or by thickening of the ectoderm located at the level of the amniotic sac floor (**a**). In the third week of embryonic life, two neural folds form from the plate (**b**), and eventually join to form the neural tube and the neural canal (**c** and **d**).

◄
Differentiation of the **central** (from the neural tube) and the **peripheral nervous system** (from the neural crest).

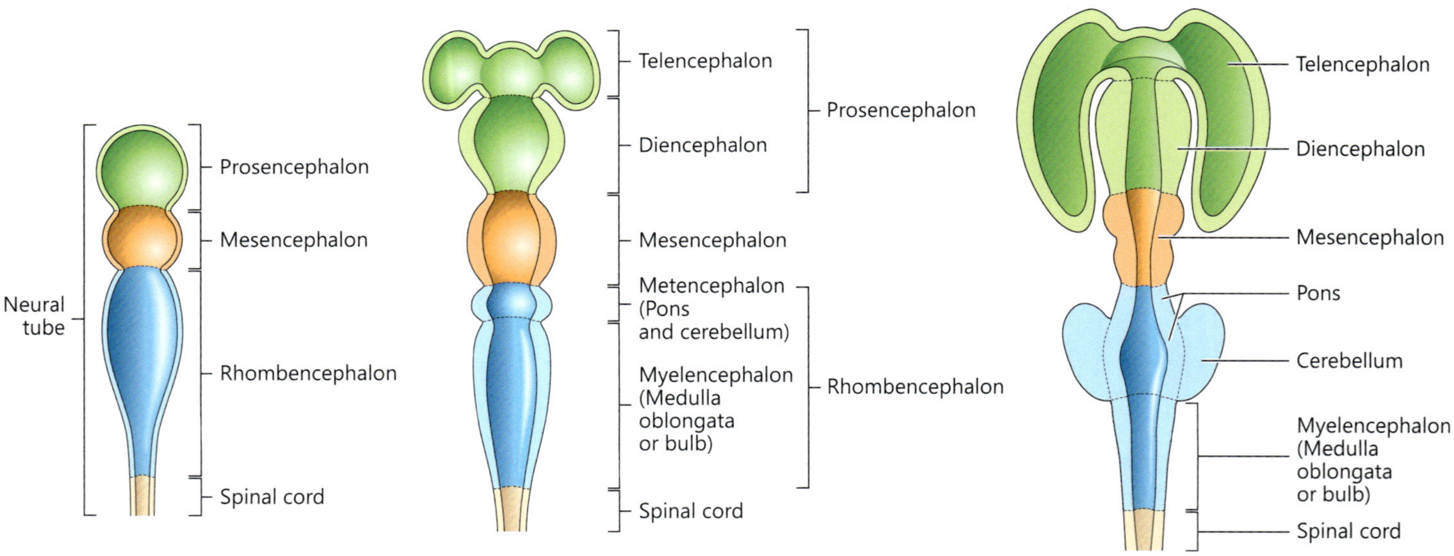

Evolution of the cranial part of the **neural tube**: transition from a three-vesicle structure to a five-vesicle structure.

►
Starting timescale of the **development of the central nervous system portions**. The three primary encephalic vesicles, represented by prosencephalon, mesencephalon and rhombencephalon, originate from the neural tube after the closure of the neural plate space. From the prosencephalic vesicle originate the vesicle of the telencephalon and that of the diencephalon and from the rhombencephalic vesicle originate the vesicle of the metencephalon, from which the pons and the cerebellum derive from, and that of the myelencephalon from which the medulla oblongata originates. The mesencephalon vesicle does not undergo further divisions.

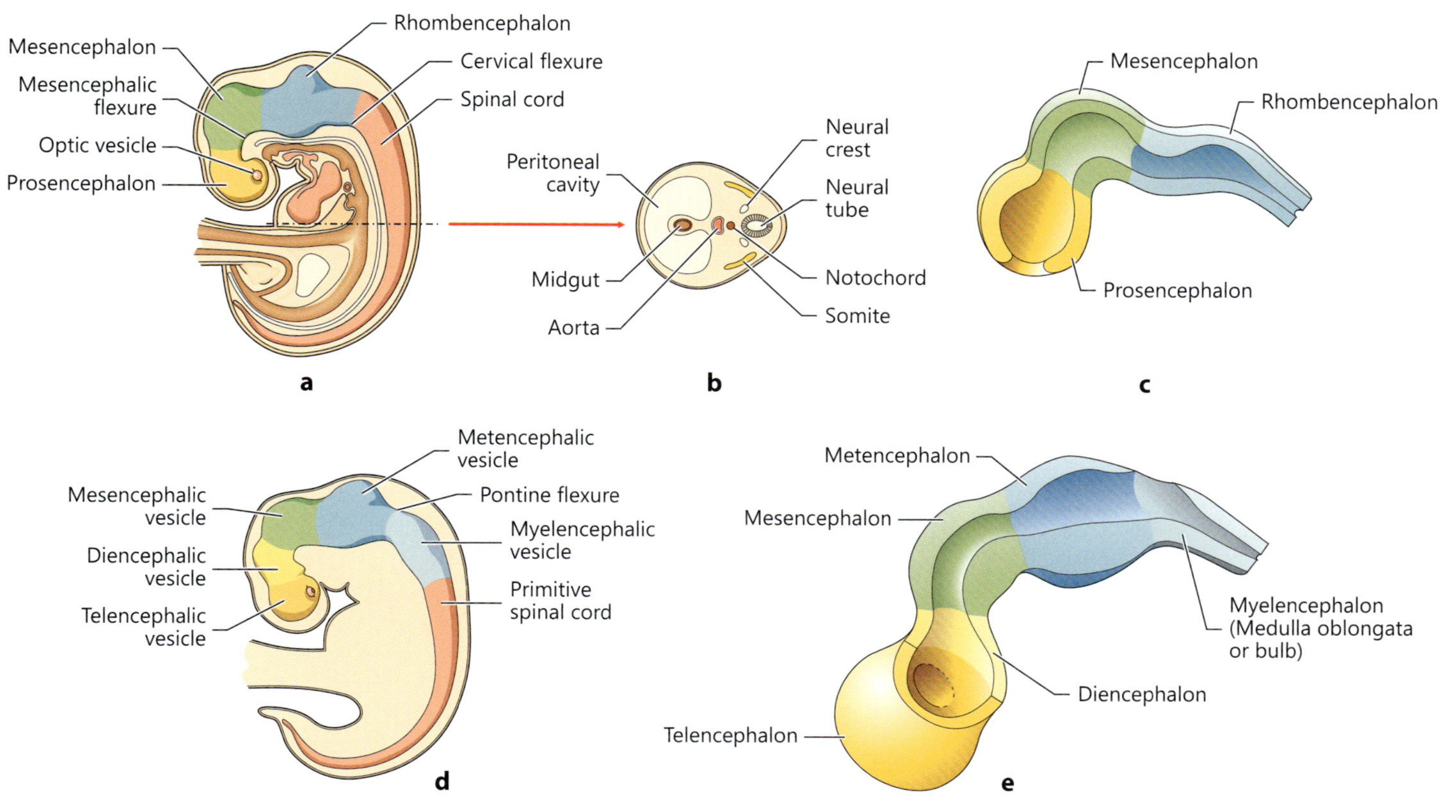

In **a** schematic representation of a 28-day embryo that displays the three primary encephalic vesicles: **prosencephalon** or anterior encephalon, **mesencephalon** and **rhombencephalon**; **b**, transverse section showing the relationship between the **neural tube** and the **notochord**; **c**, three-dimensional reconstruction of the encephalic vesicles alone. **d**, Central nervous system of a 5-week embryo; **e**, three-dimensional reconstruction of the encephalic vesicles isolated.

	Embryonic period - weeks (days)								*Fetal period - weeks (days)*							*Birth*
	2 (7-13)	*3 (14-21)*	*4 (22-28)*	*5 (29-35)*	*6 (36-42)*	*7 (43-49)*	*8 (50-56)*	*9 (57-63)*	*10 (64-70)*	*12 (78-84)*	*14 (92-98)*	*16 (106-112)*	*20 (134-140)*	*28 (192-196)*	*36 (246-252)*	*38*
Ectodermal germ layer																
Notochord	•															
Neural tube			•													
Prosencephalon				•												
Telencephalon					•											
Paleocortex							•									
Corpus striatum						•										
Neocortex							•									
Cerebral hemispheres					•[1]						•[2]			•[3]		
Diencephalon					•											
Epithalamus, thalamus, hypothalamus, infundibulum						•										
Mesencephalon				•												
Tectum, tegmentum					•											
Cerebral peduncles						•										
Rhombencephalon				•												
Metencephalon					•											
Cerebellum, pons						•										
Myelencephalon					•											
Medulla oblongata (Bulb)						•										
Spinal cord				•												

•[1], beginning of expansion; •[2], beginning of subdivision into lobes and sulci; •[3], beginning of the formation of gyri

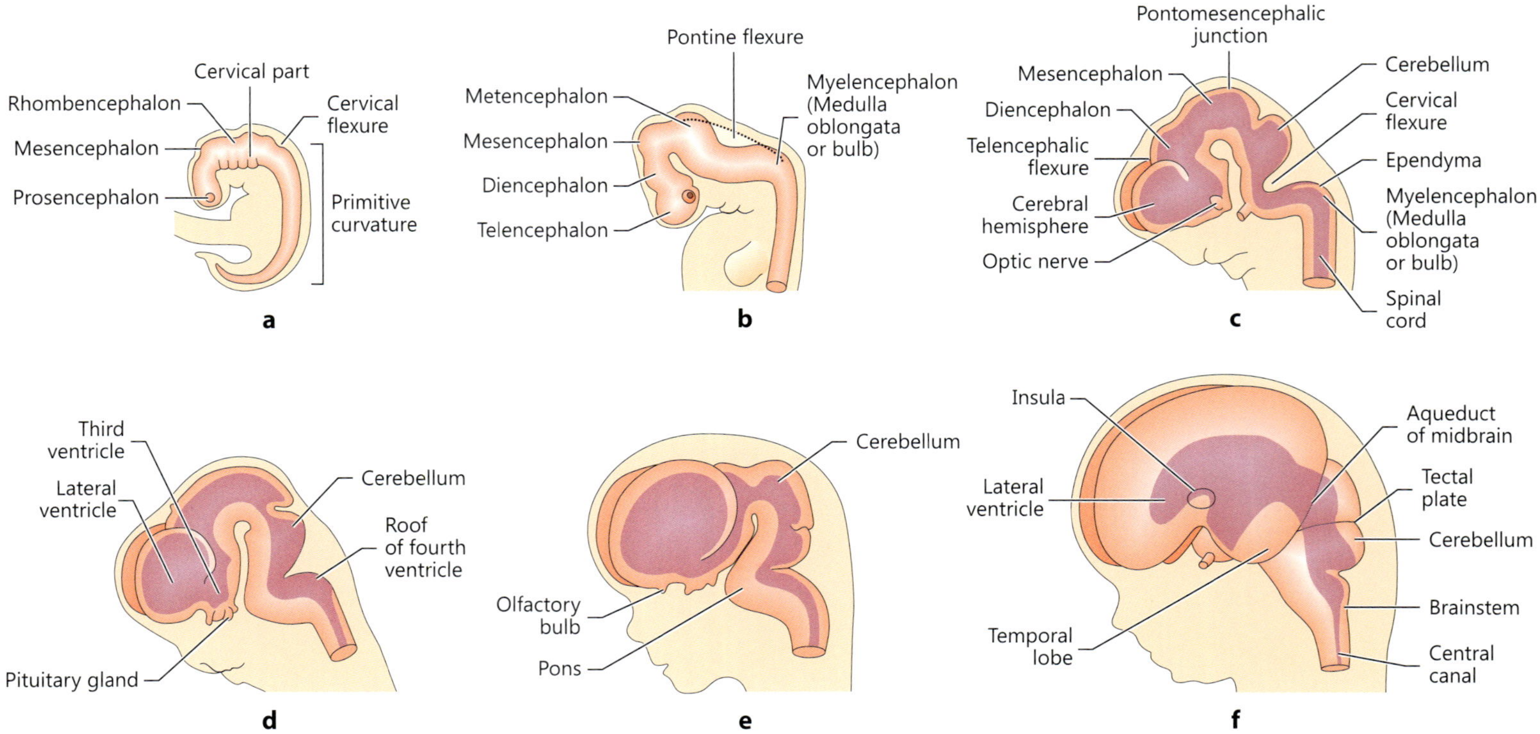

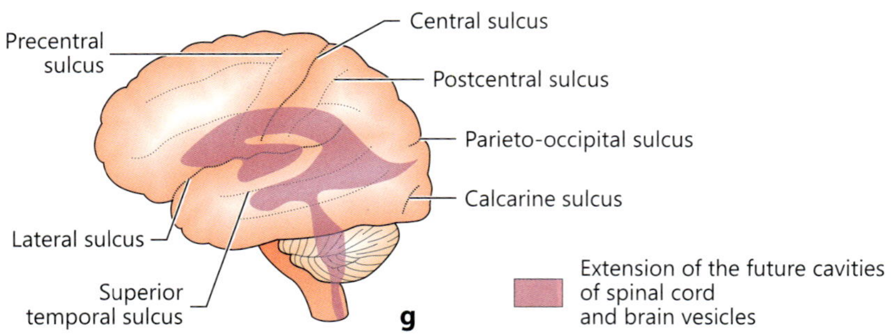

Brain development at the twenty-fourth (**a**), thirty-fifth (**b**, fortieth (**c**), fiftieth (**d**), sixtieth (**e**), seventy-second day (**f**), and at the end (**g**). **g**, Surface configuration of the telencephalon at birth. A lissencephalic surface configuration is common in the early stage of development. At birth the surface of the **telencephalon** becomes irregular due to the appearance of sulci and fissures that identify the gyri of the telencephalon. In transparency, ventricular cavities originating from the primitive neural tube are observed.

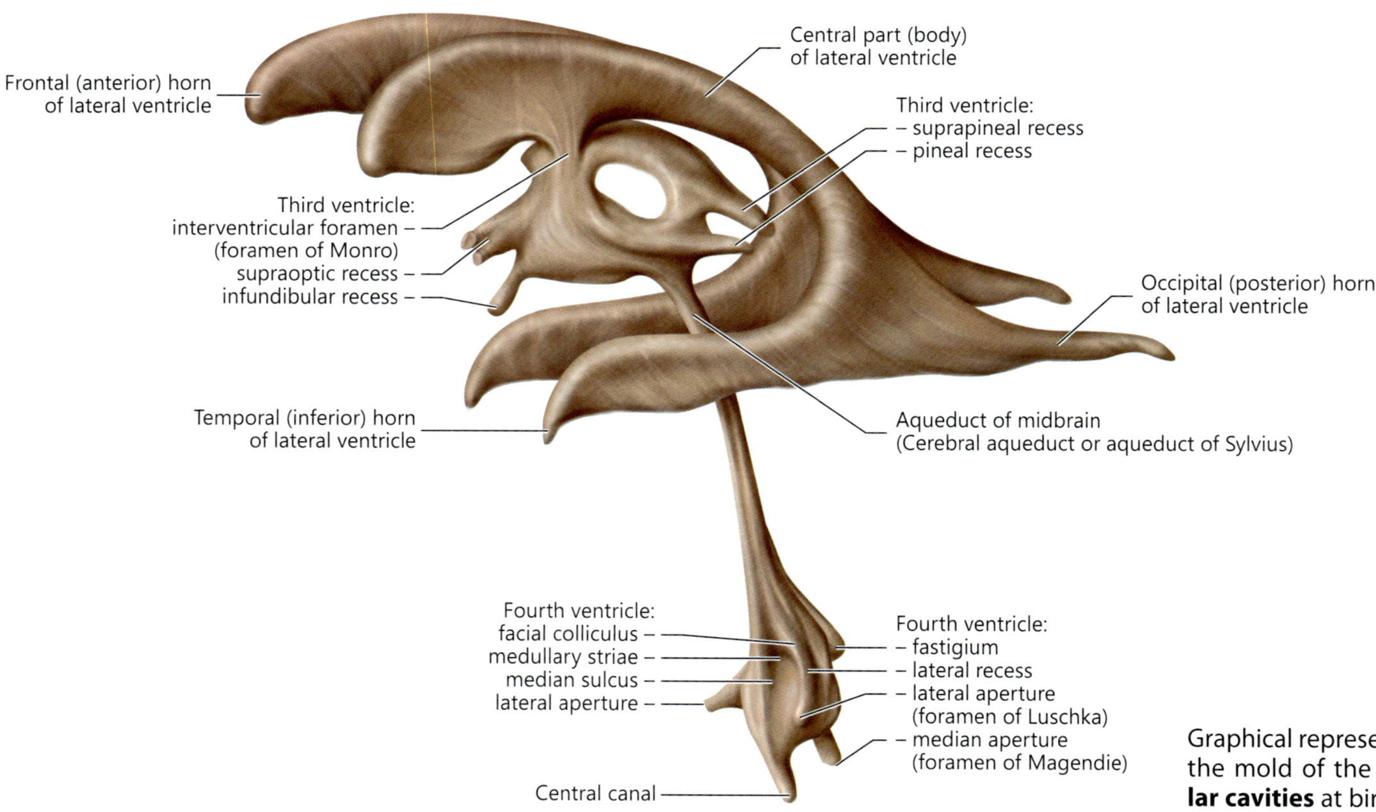

Graphical representation of the mold of the **ventricular cavities** at birth.

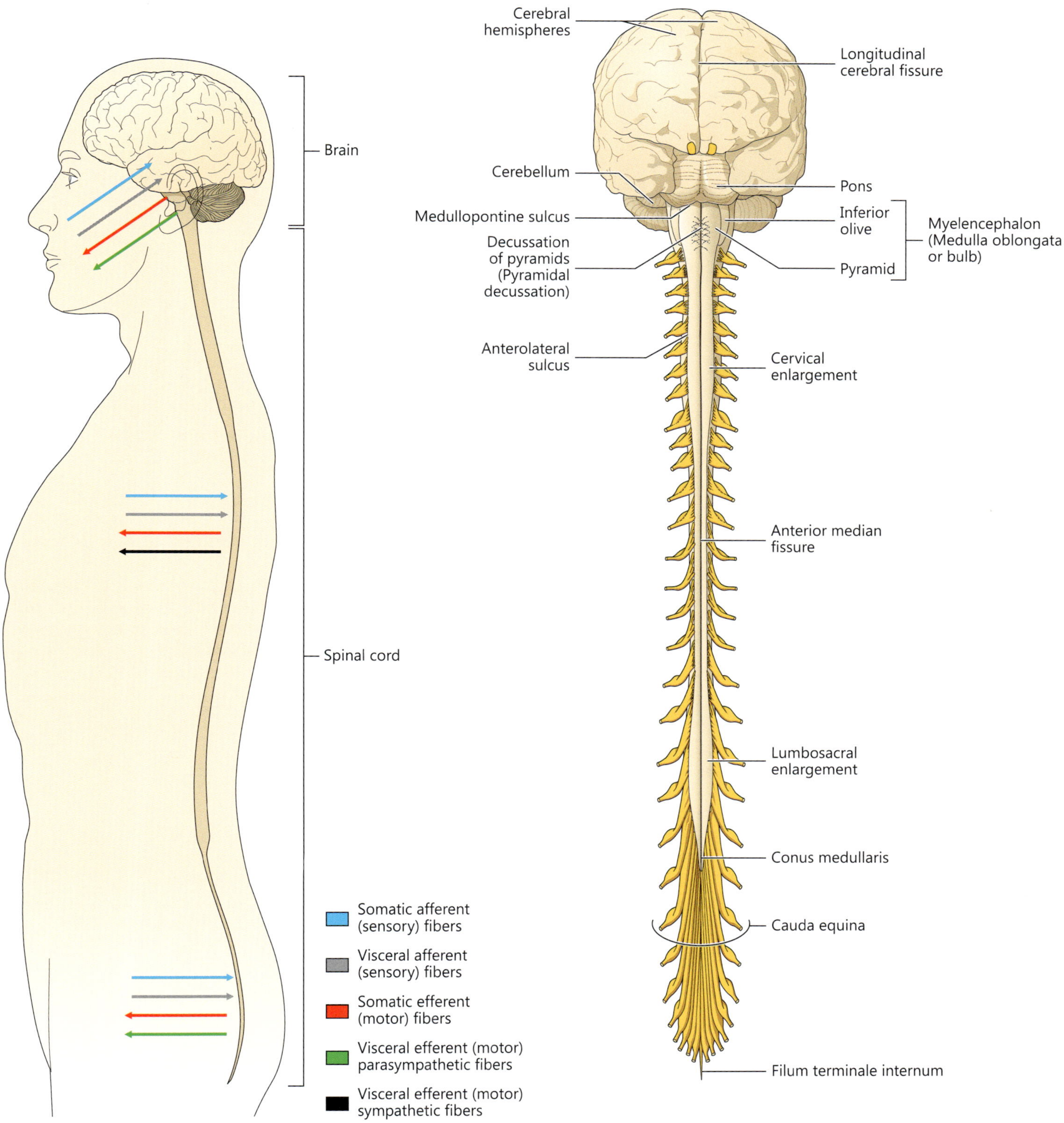

Schematic representation of the organization of the **neuraxis**. The neuraxis is formed by the **brain** and the **spinal cord**; the latter is contained in the vertebral canal between the superior border of the atlas and the 2nd lumbar vertebra.

Thirty-three pairs of spinal nerves emerge from the **spinal cord**. Out of the three pairs of coccygeal nerves, only the first one runs towards the periphery. The remaining two do not go beyond the dural sac and are included in the filum terminale. Each spinal nerve is formed by two roots that converge before emerging from the vertebral canal in a nerve which is defined mixed nerve and consists of four components:
- viscerosensory
- visceroeffector
- somatosensory
- somatomotor.

DIVISION OF THE SPINAL NERVES	
• 8 cervical [C1-C8]	Mixed
• 12 thoracic [T1-T12]	Mixed
• 5 lumbar [L1-L5]	Mixed
• 5 sacral [S1-S5]	Mixed
• 1 coccygeal [Co]	Mixed

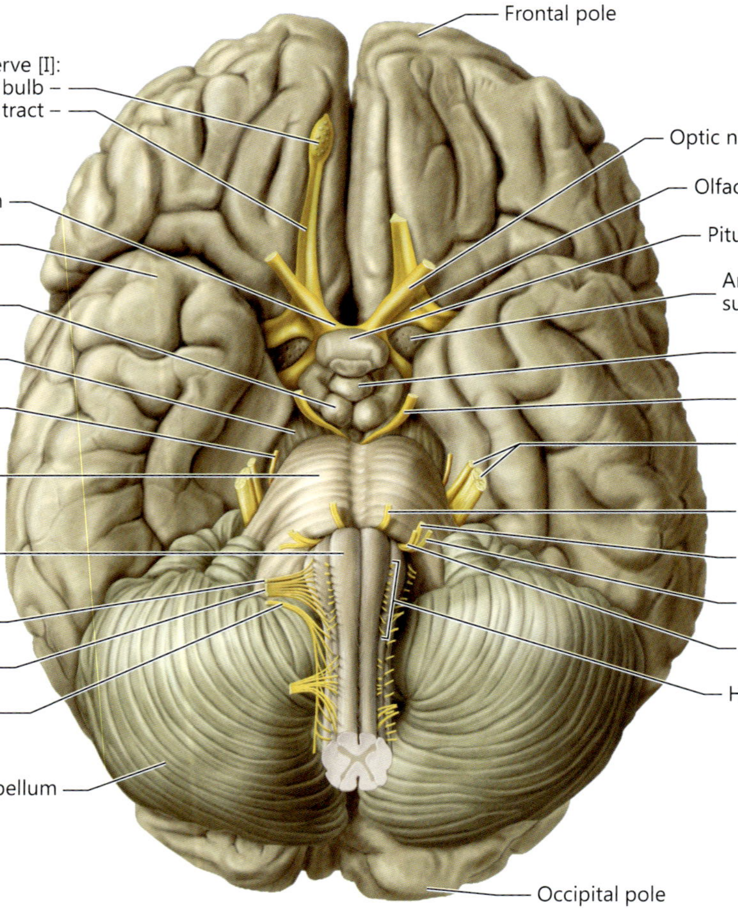

External configuration of the **brain** seen in the ventral surface. Most of the cranial nerves emerge from the ventral surface of the brainstem.

	CRANIAL NERVES	NUMBER OF COMPONENTS	COMPONENT	EMERGENCE [III-XII] OR COURSE [I AND II]	EXIT FROM THE CRANIUM [III-XII] OR ENTRY IN THE CRANIUM [I AND II]
I	Olfactory nerve	1	Somatosensory specific	From the cribriform plate of the ethmoid to the olfactory bulb	Foramina of the cribriform plate of the ethmoid
II	Optic nerve	1	Somatosensory specific	From the optic disc to the optic chiasm	Optic canal
III	Oculomotor nerve	2	Somatomotor Visceroeffector	Interpeduncular fossa	Superior orbital fissure
IV	Trochlear nerve	1	Somatomotor	Dorsal surface of the mesencephalon	Superior orbital fissure
V	Trigeminal nerve	2	Somatomotor Somatosensory	Ventrolateral surface of the pons	Ophthalmic nerve: superior orbital fissure Maxillary nerve: foramen rotundum Mandibular nerve: foramen ovale
VI	Abducens nerve	1	Somatomotor	Medullopontine sulcus	Superior orbital fissure
VII	Facial nerve	4	Somatomotor Somatosensory Visceroeffector Viscerosensory	Supraolivary fossa of the bulb	Internal acoustic meatus
VIII	Vestibulocochlear nerve	1	Somatosensory specific	Retro-olivary fossa of the bulb	Internal acoustic meatus
IX	Glossopharyngeal nerve	4	Somatomotor Somatosensory Visceroeffector Viscerosensory	Posterolateral sulcus of the bulb	Jugular foramen
X	Vagus nerve	4	Somatomotor Somatosensory Visceroeffector Viscerosensory	Posterolateral sulcus of the bulb	Jugular foramen
XI	Accessory nerve	1	Somatomotor	Posterolateral sulcus of the bulb and anterolateral sulcus of the spinal cord	Jugular foramen
XII	Hypoglossal nerve	1	Somatomotor	Anterolateral sulcus of the bulb	Hypoglossal canal

Twelve pairs of **cranial nerves** emerge from the brain. They differ from the spinal nerves because, except for the V pair or trigeminal, originate from one root only, and, with the exception of the VII, IX and X nerves, neither of them can be defined mixed nerve, as they include only one or two of the four components observed in the spinal nerves. Similar to the spinal nerves, the trigeminal nerve [V] has two roots, one motor and one sensory, connected to a ganglion.

External surface of the cranial base. The **foramina** are shown with the emergence of the corresponding vessels and nerves.

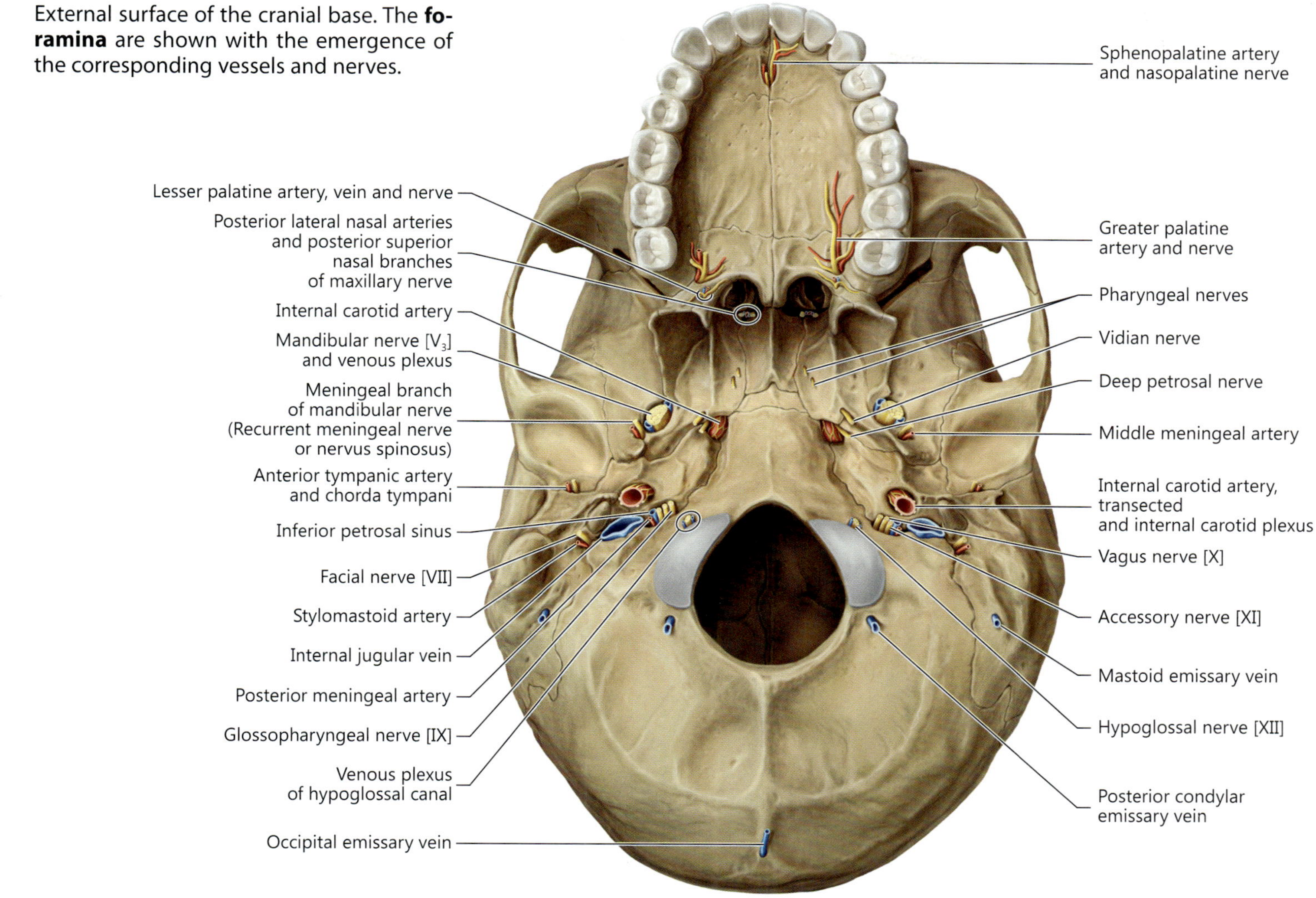

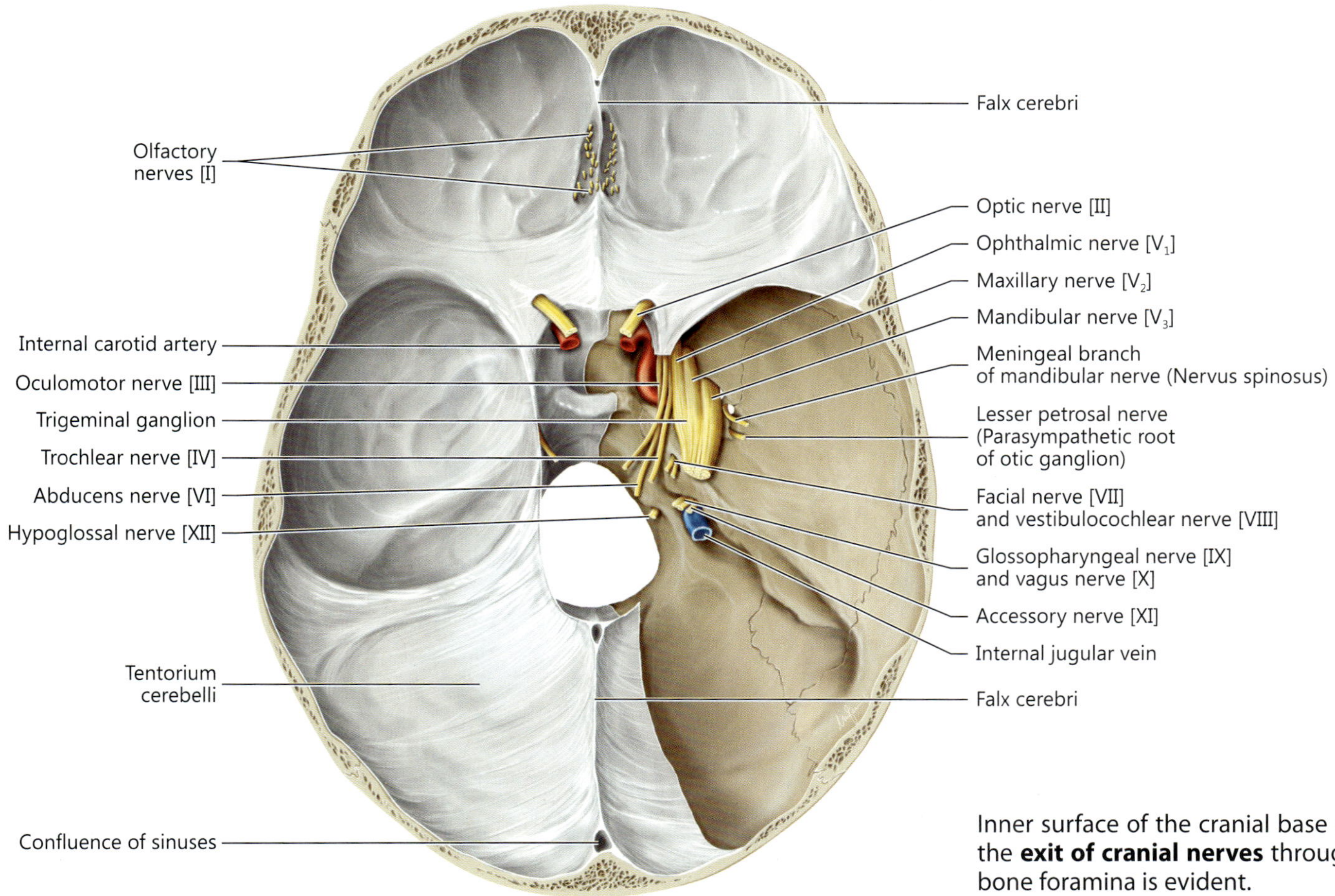

Inner surface of the cranial base where the **exit of cranial nerves** through the bone foramina is evident.

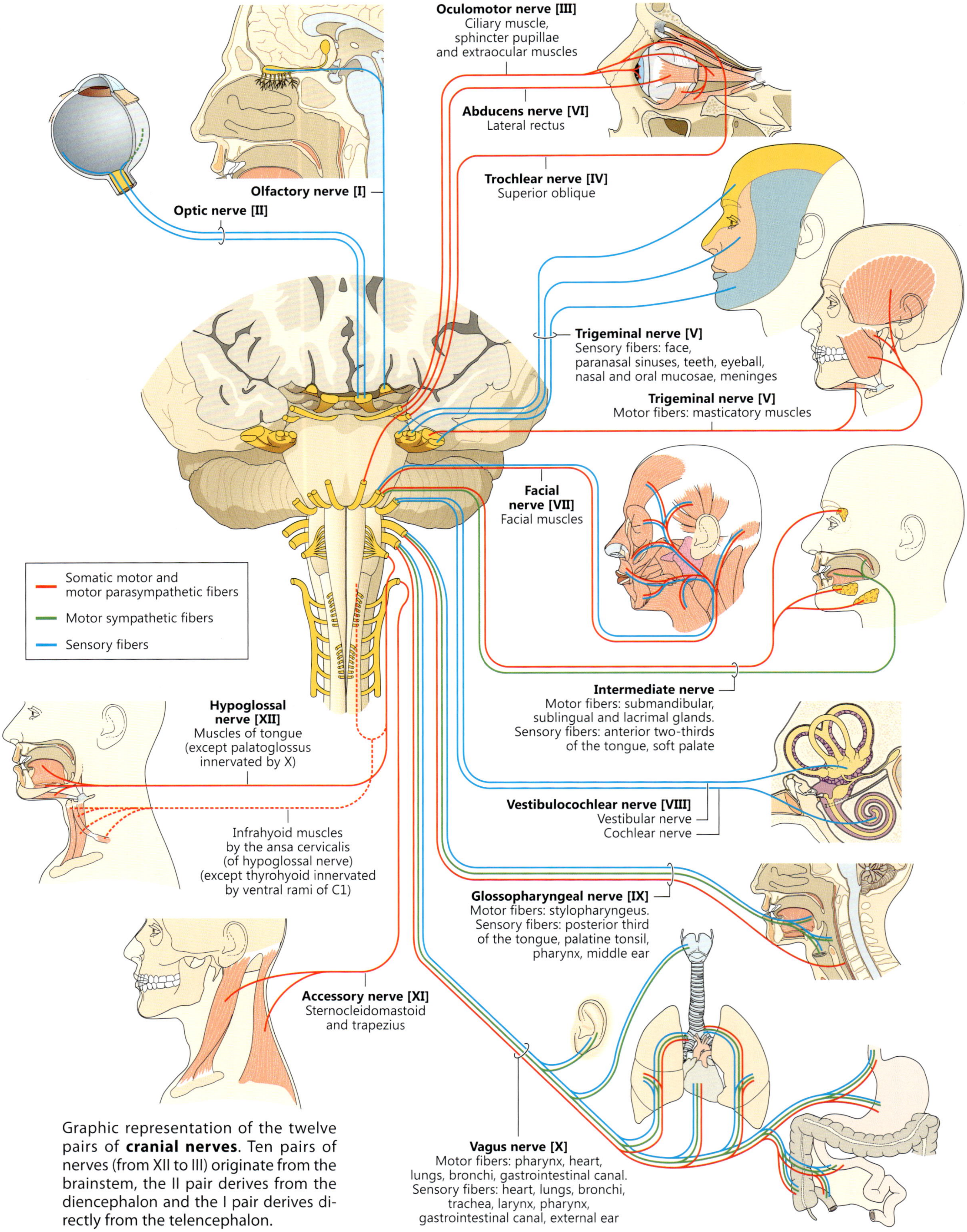

Graphic representation of the twelve pairs of **cranial nerves**. Ten pairs of nerves (from XII to III) originate from the brainstem, the II pair derives from the diencephalon and the I pair derives directly from the telencephalon.

Classification, origin, course and distribution of the cranial nerves					
NERVE		**NERVE COMPOSITION**	**REAL ORIGIN OF THE NERVE FIBERS**	**INDICATIONS ON THE NERVE COURSE AND ON ITS COMPONENTS**	**PERIPHERAL DISTRIBUTION**
I pair	Olfactory	SSS	Olfactory epithelium	Filaments of the olfactory nerve → cribriform plate → olfactory bulb	Olfactory part of the mucosa of the nasal cavity
II pair	Optic	SSS	Ganglionic layer of the retina	Optic nerve → optic canal → optic chiasm → optic tract	Retina
III pair	Oculomotor	SM	Nucleus of the oculomotor nerve (First column)	Cavernous sinus lateral wall → superior orbital fissure → orbital cavity	Superior rectus, inferior rectus, medial rectus, inferior oblique, levator palpebrae superioris
		VM	Accessory nuclei of the oculomotor nerve (Edinger-Westphal nucleus)	Oculomotor nerve → branch to ciliary ganglion → ciliary ganglion → short ciliary nerves	Sphincter (constrictor) pupillae and ciliary muscle
IV pair	Trochlear	SM	Nucleus of the trochlear nerve (First column)	Cavernous sinus lateral wall → superior orbital fissure	Superior oblique
V pair	Trigeminal	SM	Motor nucleus of the trigeminal nerve (Second column)	Motor root of the trigeminal nerve → mandibular nerve → foramen oval → infratemporal fossa	Masticatory muscles, anterior belly of digastric, milohyoid, tensor tympani, tensor veli palatini
		GSS	Sensory nuclei of the trigeminal nerve (spinal nucleus, principal or chief pontine nucleus, mesencephalic nucleus)	Ophthalmic, maxillary and mandibular nerves → trigeminal ganglion → sensory root of the trigeminal nerve	Skin of the face, mucosa of the mouth and nose, teeth, conjunctiva
		GSS	Mesencephalic nucleus of the trigeminal nerve	Mandibular and maxillary nerves → sensory root of the trigeminal nerve	Masticatory muscles (neuromuscular spindles) and dental alveoli
VI pair	Abducens	SM	Nucleus of the abducens nerve (First column)	Cavernous sinus → superior orbital fissure → orbital cavity	Lateral rectus
VII pair*	Facial	SM	Nucleus of the facial nerve (Second column)	Internal acoustic meatus → facial canal → stylomastoid foramen → parotid space → face	Muscles of facial expression, stapedius, posterior belly of digastric
		VM	Lacrimal nucleus	Facial nerve → greater petrosal nerve → pterygopalatine ganglion → zygomatic nerve, greater and lesser palatine nerves, and posterior superior medial and lateral nasal branches of the maxillary nerve	Lacrimal gland, glands of the mucosa of the nasal cavity and palate
	Intermediate	VM	Superior salivatory nucleus	Intermediate nerve → chorda tympani → lingual nerve → submandibular and sublingual ganglia	Submandibular and sublingual glands
		VS	Nucleus of the solitary tract	Lingual nerve → chorda tympani → geniculate ganglion → intermediate nerve	Papillae of the tongue (taste) anterior to the terminal sulcus of the tongue (anterior two third of the tongue)
		GSS	Spinal nucleus of the trigeminal nerve	Posterior auricular nerve → facial nerve	Skin of the auricular and mastoid region

SM, somatic motor; VM, visceral motor; SSS, specific somatosensory; GSS, general somatosensory; VS, visceral sensory
* VII pair, facial nerve (facial nerve proper and intermediate nerve)

continued

Classification, origin, course and distribution of the cranial nerves *(continued)*

NERVE		NERVE COMPOSITION	REAL ORIGIN OF THE NERVE FIBERS	INDICATIONS ON THE NERVE COURSE AND ON ITS COMPONENTS	PERIPHERAL DISTRIBUTION
VIII pair	Vestibular	SSS	Vestibular nuclei (external white wing)	Utricular, utriculoampullary, saccular and ampullary nerves → vestibular ganglion → vestibulocochlear nerve → internal acustic meatus	Ampullary crest and macula of the utricle and macula of the saccule
	Cochlear	SSS	Anterior and posterior coclear nuclei	Peripheral extensions of the neurons of the cochlear ganglion → cochlear ganglion → central extensions of the neurons of the cochlear ganglion in the modiolus of the cochlea → vestibulocochlear nerve → internal acoustic meatus	Spiral organ
IX pair	Glossopharyngeal	SM	Nucleus ambiguus	Glossopharyngeal nerve → jugular foramen → pharyngeal branches of the glossopharyngeal nerve	Superior constrictor and stylopharyngeus
		VM	Inferior salivatory nucleus	Glossopharyngeal nerve → tympanic nerve → lesser petrosal nerve → optic ganglion → auriculotemporal nerve	Parotid and minor salivary glands
		GSS	Spinal nucleus of the trigeminal nerve	Tympanic plexus → tympanic nerve → glossopharyngeal nerve → superior ganglion of the glossopharyngeal nerve → jugular foramen	Mucosa of the tympanic cavity and the pharyngotympanic tube and the skin of the auricle
		VS	Nucleus of the solitary tract	Lingual, pharyngeal and tonsillary branches of the glossopharyngeal nerve → tympanic nerve; carotid branch of the glossopharyngeal nerve → glossopharyngeal nerve → inferior ganglion of the glossopharyngeal nerve → jugular foramen	Vallate papillae and mucosa of the third posterior of the tongue and palate (taste); carotid body and carotid sinus
X pair	Vagus nerve	SM	Nucleus ambiguus (Second column)	Jugular foramen → external branch of the superior laryngeal nerve and the recurrent laryngeal nerve	Pharyngeal and laryngeal muscles
		VM	Posterior nucleus of the vagus nerve	Course of the vagus nerve in the neck, thorax and abdomen	Organs of the neck, thorax and abdomen; collateral branches of the aorta
		GSS	Nucleus of the solitary tract	Auricular branch of the vagus nerve → superior ganglion of the vagus nerve → vagus nerve	Auricle
		VS	Spinal nucleus of the trigeminal nerve	All collateral branches of the vagus nerve → inferior ganglion of the vagus nerve → vagus nerve	Organs of the neck, thorax and abdomen; aortic sinuses and para-aortic bodies; collateral branches of the aorta
XI pair	Accessory	SM	Nucleus of the accessory nerve	Spinal root of the accessory nerve → foramen magnum → jugular foramen → accessory nerve	Sternocleidomastoid and trapezius
XII pair	Hypoglossal	SM	Nucleus of the hypoglossal nerve (First column)	Hypoglossal nerve → hypoglossal canal → lingual branches of the hypoglossal nerve	Muscles of the tongue and some suprahyoid muscles

SM, somatic motor; VM, visceral motor; SSS, specific somatosensory; GSS, general somatosensory; VS, visceral sensory

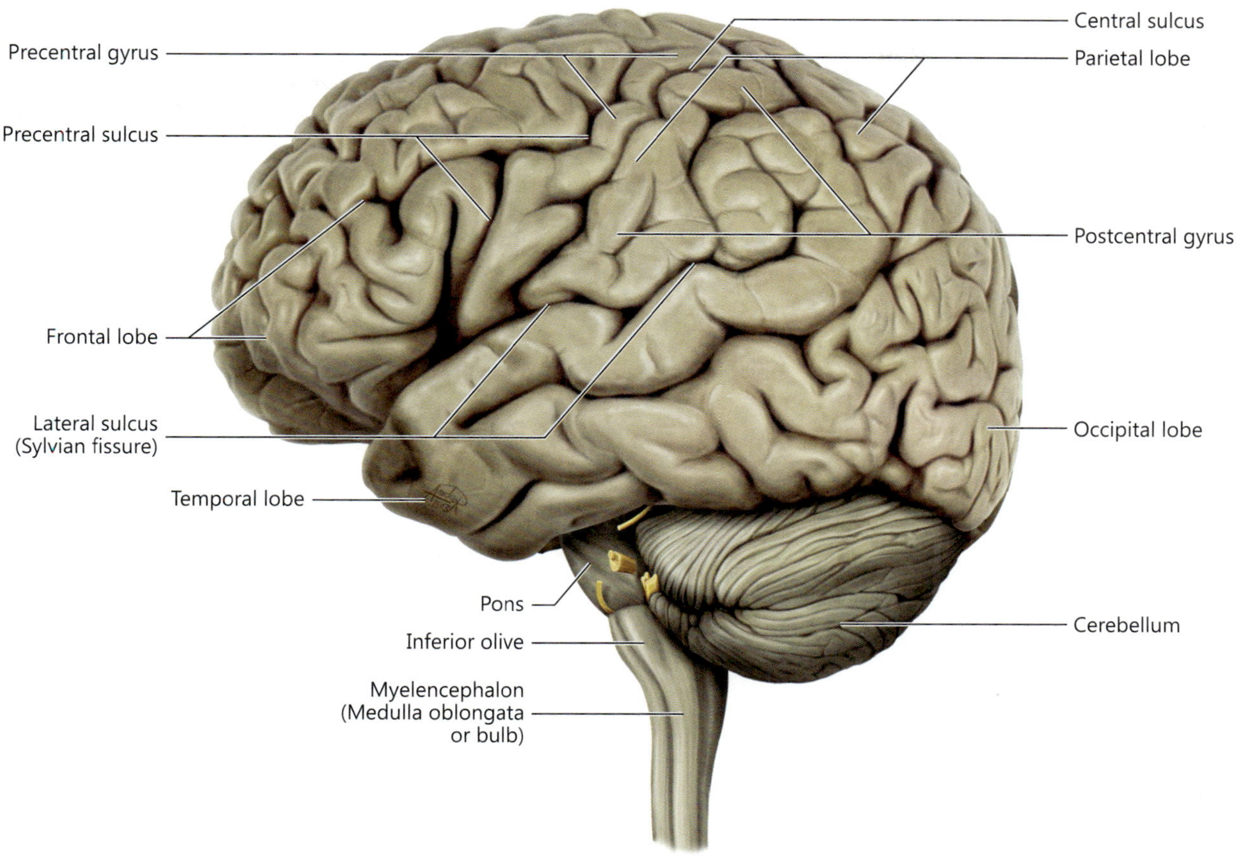

External configuration of the **brain** seen in lateral projection.

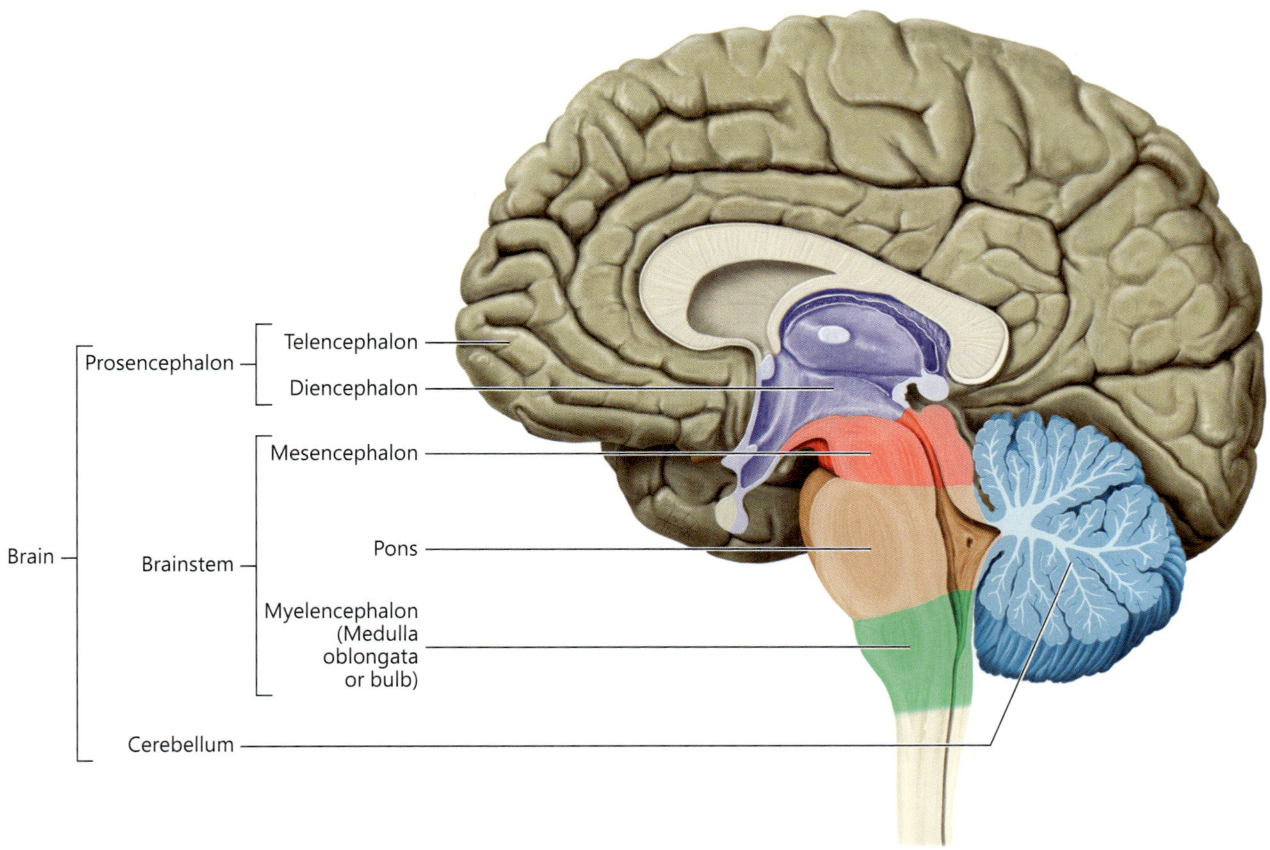

External configuration of the **brain** seen in medial projection after sagittal section of the brainstem, cerebellum and diencephalon.

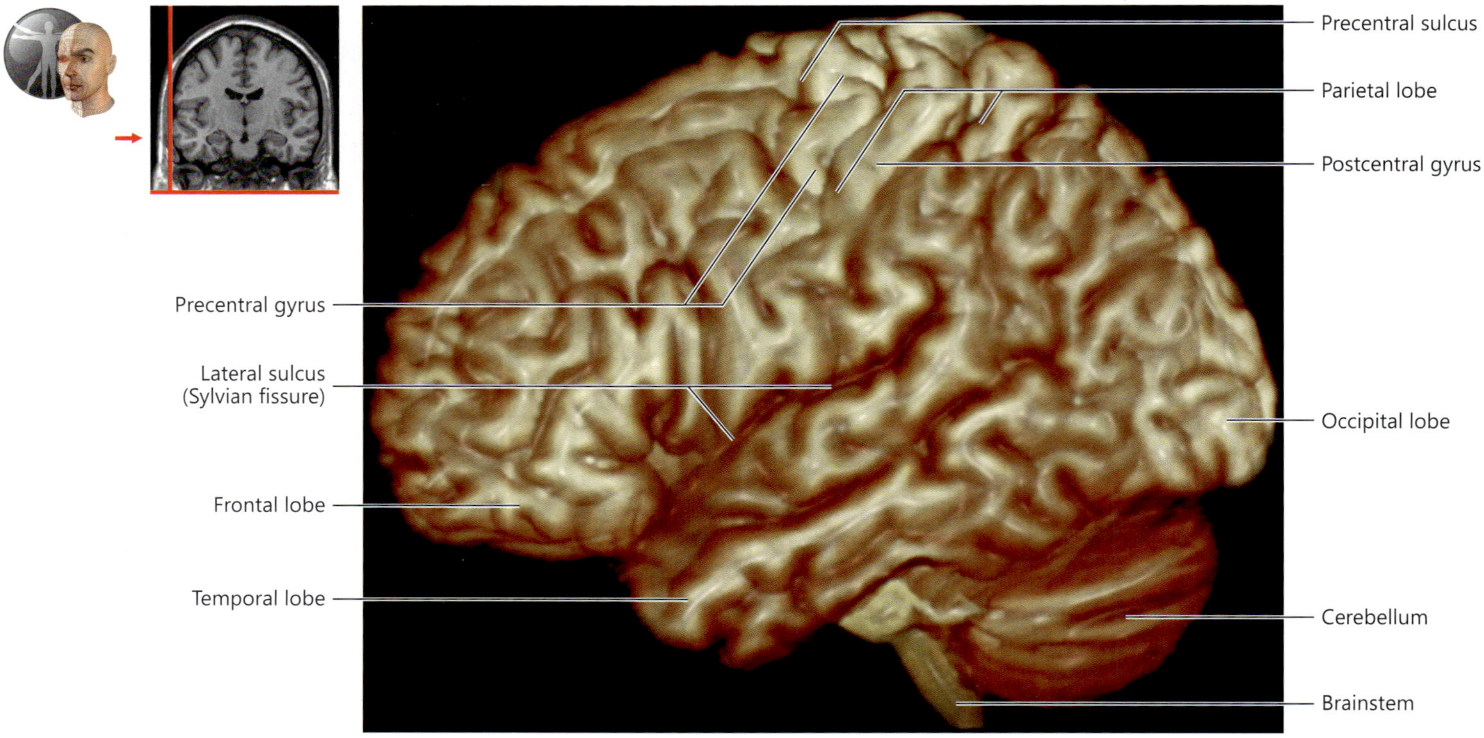

Three-dimensional image of a living human body, obtained by NMR using a 3D volume rendering technique, showing the external conformation of the **brain** in lateral projection.

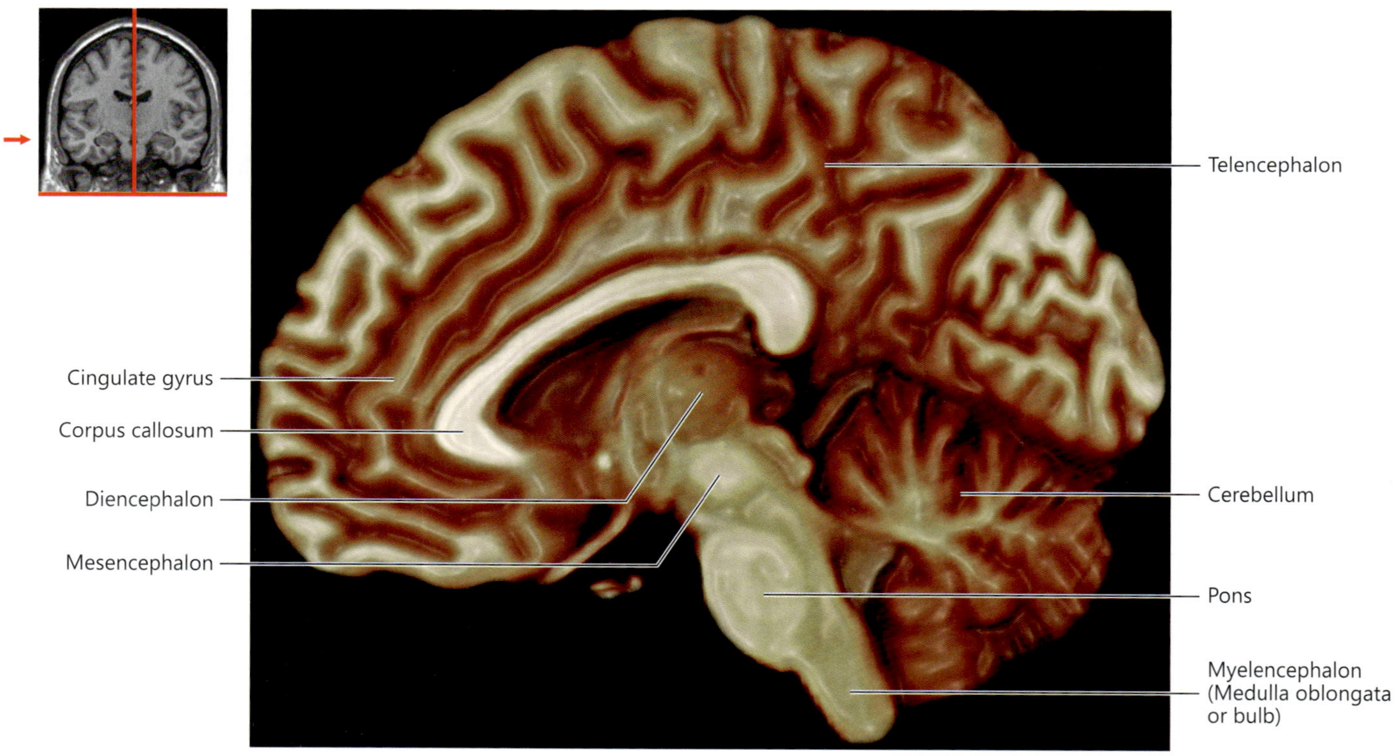

Three-dimensional image of a living human body, obtained by NMR using a 3D volume rendering technique, showing the **brain** in sagittal section.

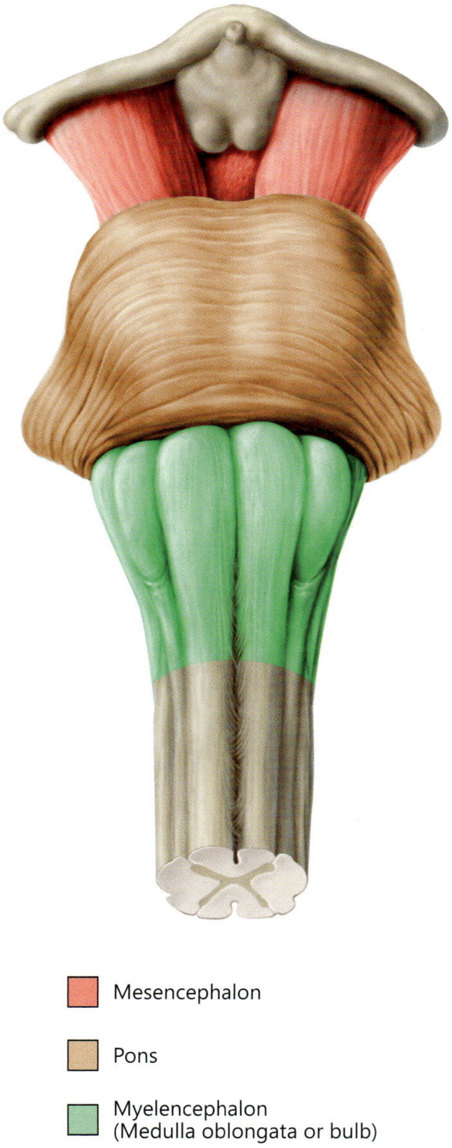

External configuration of the different parts of the **brainstem**.

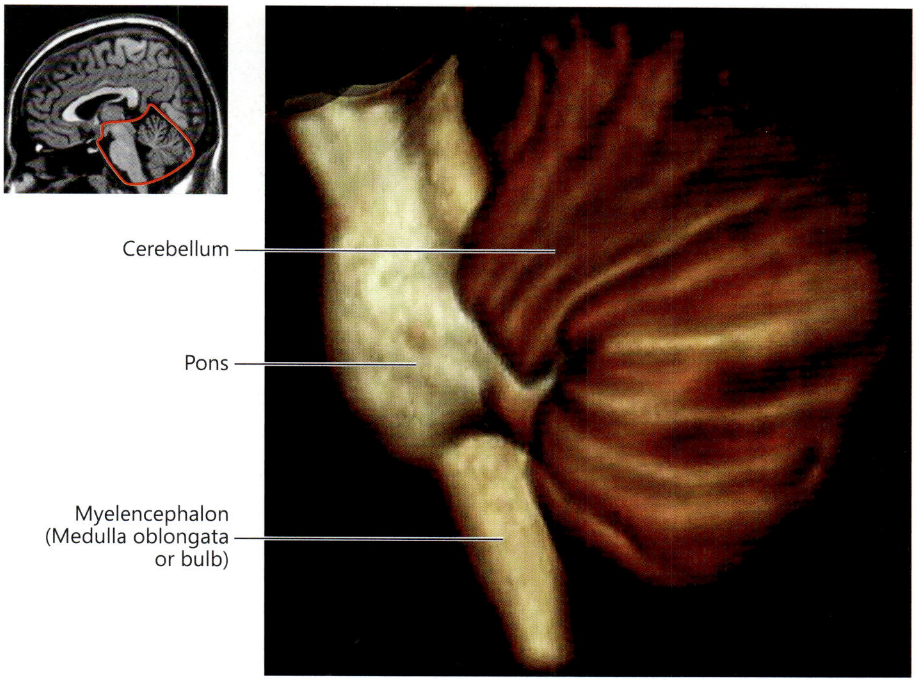

Three-dimensional image of a living human body, obtained by NMR using a 3D volume rendering technique, showing the **cerebellum** and its relationships with the brainstem in left lateral projection.

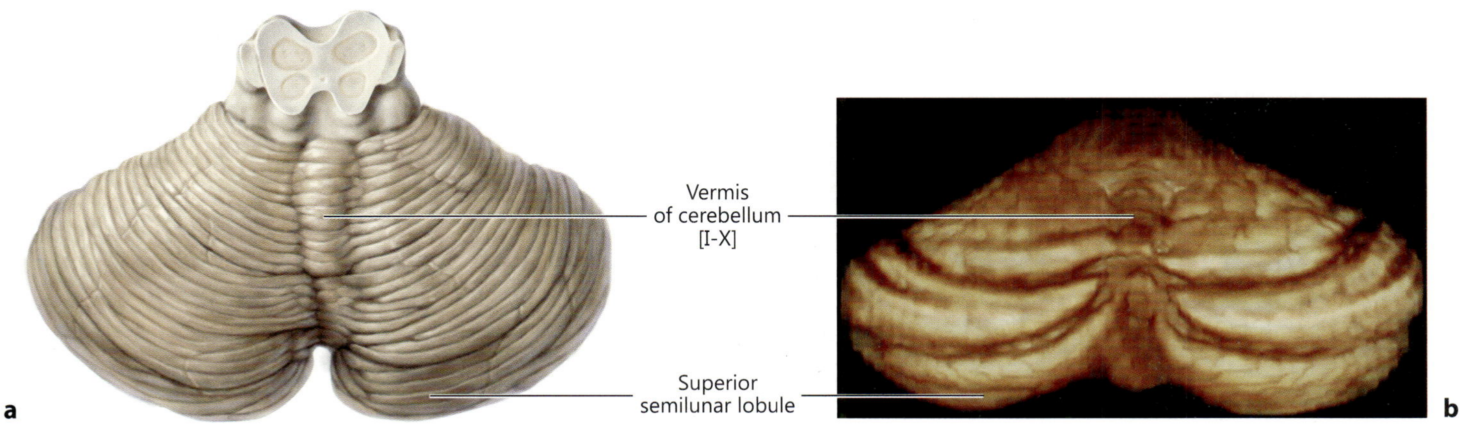

a, External configuration of the superior surface of the **cerebellum**. **b**, Three-dimensional image of a living human body, obtained by NMR using a 3D volume rendering technique, showing the superior surface of the **cerebellum**. The cerebellar surface is marked by deep fissures that delimit distinct territories.

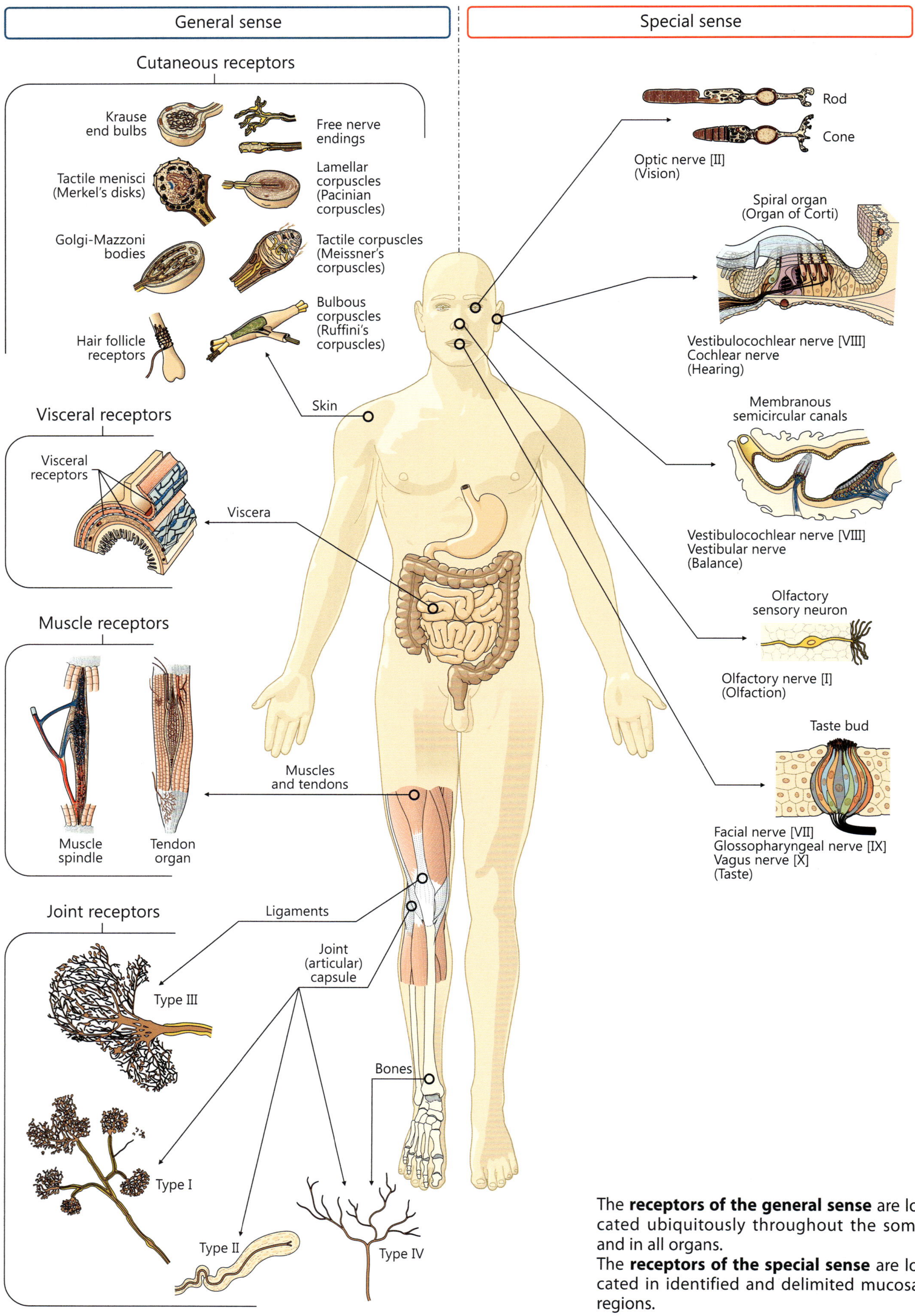

The **receptors of the general sense** are located ubiquitously throughout the soma and in all organs.
The **receptors of the special sense** are located in identified and delimited mucosal regions.

Receptors mediating the general sense			
RECEPTOR	**LOCATION**	**FUNCTION**	**ADAPTATION**
Bulbous corpuscles (Ruffini's corpuscles)	Epidermis	Skin deformation	Slow
Free nerve endings	Skin	Temperature, mechanical stimuli (touch, pressure stretching) or pain (nociception)	Rapid Intermediate Slow
Golgi-Mazzoni bodies	Subcutaneous tissue of the fingers	Movements and vibrations of the fingers	Rapid
Hair follicle receptors	Hairs	Hairs movements	Rapid
Krause end bulbs	Dermis	Touch	Slow
Lamellar corpuscles (Pacinian corpuscles)	Subcutaneous tissue, interosseous membrane, viscera	Vibrations	Rapid
Muscle spindles	Muscles	Stretching and speed of muscle lengthening	Slow Rapid
Tactile corpuscles (Meissner's corpuscles)	Mostly hairless skin	Rubbing, trembling	Rapid
Tactile menisci (Merkel's disks)	Epidermis, hair follicles	Pressure, objects characteristics	Slow
Tendon organs	Tendons	Muscle contraction	Slow

Receptors mediating the special sense			
RECEPTOR	**LOCATION**	**FUNCTION**	**ADAPTATION**
Cone	Retina	Night time vision	Rapid
Membranous semicircular canals	Vestibular labyrinth	Balance	Rapid
Olfactory sensory neuron	Olfactory epithelium	Olfaction	Rapid
Rod	Retina	Chromatic perception (day time vision)	Slow
Spiral organ (organ of Corti)	Cochlea (internal ear)	Hearing	Rapid
Taste bud	Papillae of the tongue	Taste	Rapid

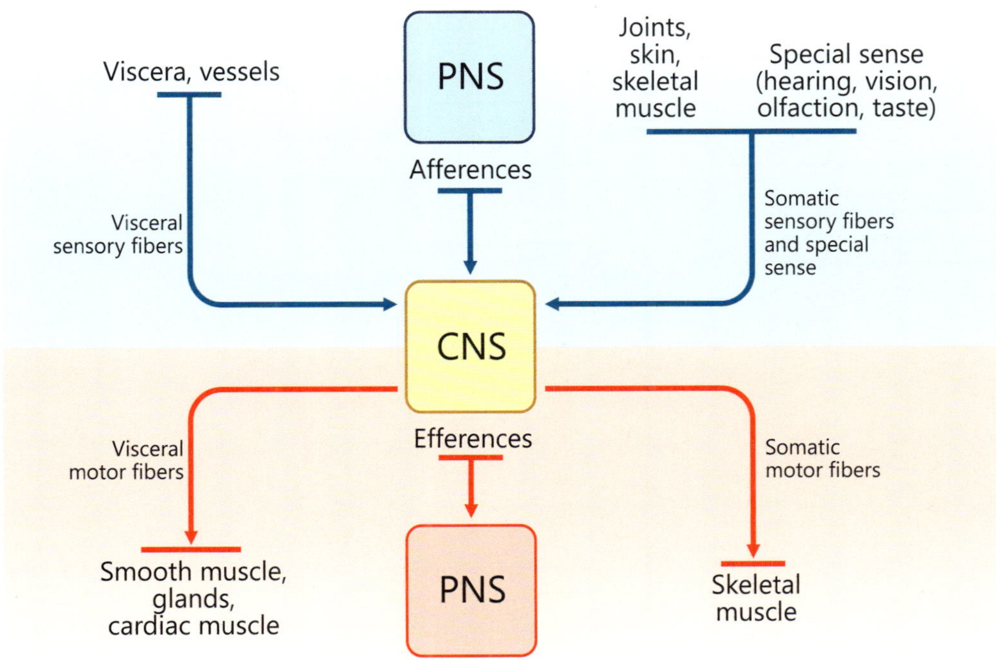

Schema of the **nervous system organization**. The central nervous system receives sensory information (general sense afferences) from the viscera (visceral sensory fibers), the soma (somatic sensory fibers) and special sense information from the sense organs, through the peripheral nervous system. These informations are integrated to develop cognitive elements and, accordingly, promote appropriate motor responses. These responses are addressed through the peripheral nervous system to the smooth muscles of the viscera and glands via visceral motor fibers, and to the skeletal muscle through somatic motor fibers.

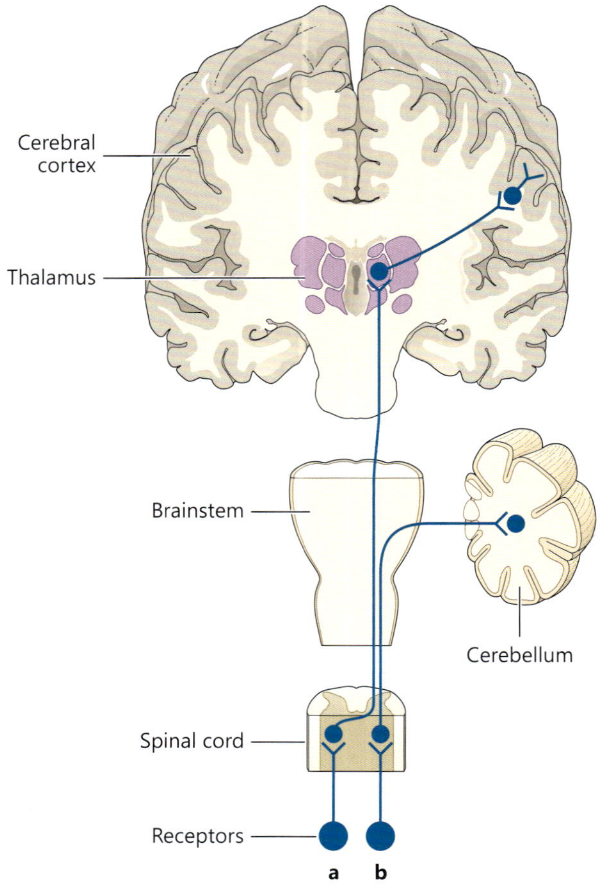

Sensory pathways are constituted by a sequence of three types of neurons placed in series that originate from the receptor and reach the cerebral cortex, where the stimulus reaches the level of consciousness. The first order neuron is pseudounipolar, it originates from the receptor, or it is itself part of the receptor, and reaches the laminae of the spinal cord or the brainstem nuclei, where it forms synapses with the second order neuron, or projection interneuron. The axon of the second order neuron crosses the midline (decussation) and goes up to the thalamus, where it performs the second synapse with the projection thalamocortical neuron. The latter neuron reaches the cerebral cortex in correspondence of the so-called primary or unimodal areas; from here a series of neurons originate, these are not always precisely quantifiable, defined with the virtual term of "fourth order neuron", which transfer the signal to higher-order areas with cognitive functions (**a**).
There are also pathways that originate from proprioceptors and are addressed to the cerebellar cortex, which transfer key informations for the modulation of the motor gesture and posture. These pathways are constituted by two types of neurons placed in series, one of which travels from the receptor to the spinal cord or the brainstem nuclei, where it generates synapses with the second order neuron that reaches the cerebellar cortex through the cerebellar peduncles (**b**).
This tri- or bineuronal pattern is followed by all the spinal and cranial nerves, except for certain nerves responsible for special sense, such as the olfactory, optic and cochlear nerve.

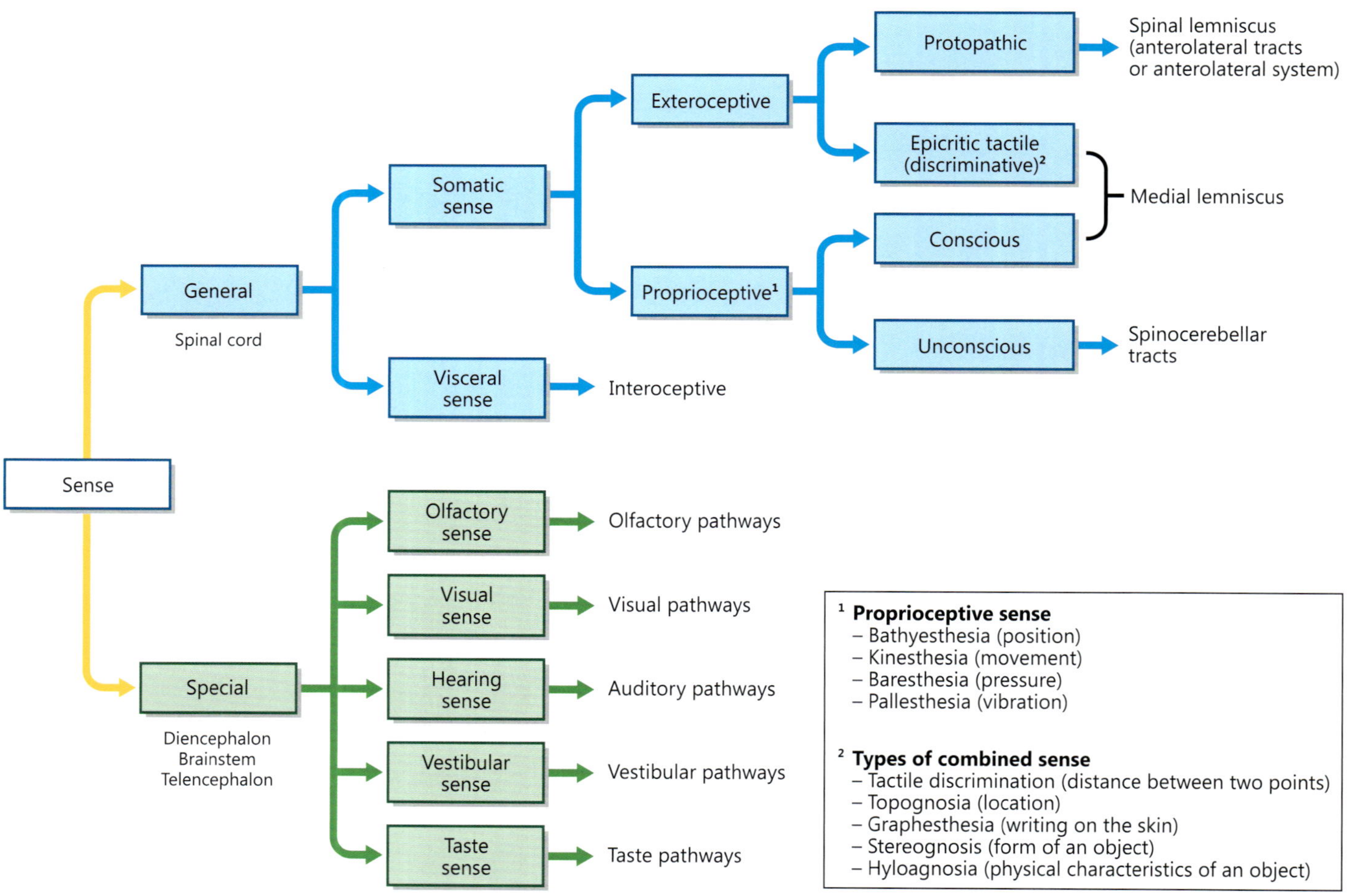

[1] **Proprioceptive sense**
- Bathyesthesia (position)
- Kinesthesia (movement)
- Baresthesia (pressure)
- Pallesthesia (vibration)

[2] **Types of combined sense**
- Tactile discrimination (distance between two points)
- Topognosia (location)
- Graphesthesia (writing on the skin)
- Stereognosis (form of an object)
- Hyloagnosia (physical characteristics of an object)

Schematic summary of the **general and special sense pathways**.

The motor gesture is governed by two pathways, the **direct pathway** and the **indirect pathway**, which connect three levels or hierarchies of the central nervous system: the cerebral cortex (**cortical hierarchy**), the specific nuclei of the brainstem (**brainstem hierarchy**) and the laminae VIII and IX of the spinal cord (**spinal hierarchy**). The direct pathway is represented by the corticospinal tracts, or pyramidal tracts, consisting of the upper motor neuron, or pyramidal neuron, originating from motor areas; the pathway descends through the lateral part of the posterior limb of the internal capsule, descends along the cerebral crus and the pons and reaches the bulbar pyramids; at the level of the spinomedullary junction, approximately 90% of the axons crosses the median line and passes into the contralateral funiculus, forming the lateral corticospinal tract (LCST); approximately 10% of the axons remains ipsilateral and cross at the level of the neuromere corresponding to the emergence of the spinal nerve, generating the direct pyramidal tracts. Both the direct and the crossed pyramidal tracts form synapses with α and γ motor neurons (lower motor neurons) located in the lamina IX of the spinal cord (the synapsis is almost always mediated by short interneurons lodged in the lamina VIII). Each α motor neuron innervates a number of muscle fibers forming the **motor unit**.
The indirect pathway provides an upper motor neuron that transfers the motor plane to the specific nuclei of the brainstem; these nuclei receive cerebellar afferents that modulate their firing rate; from these nuclei originate the extrapyramidal pathways terminating in synapses with the motor neuron synapsis formed by the upper and lower motor neurons of the pyramidal tracts. By acting as facilitators or inhibitors on the motor synapsis, the purpose of the extrapyramidal tracts is to change the number of the motor units recruited for the motor scheme to be realized.

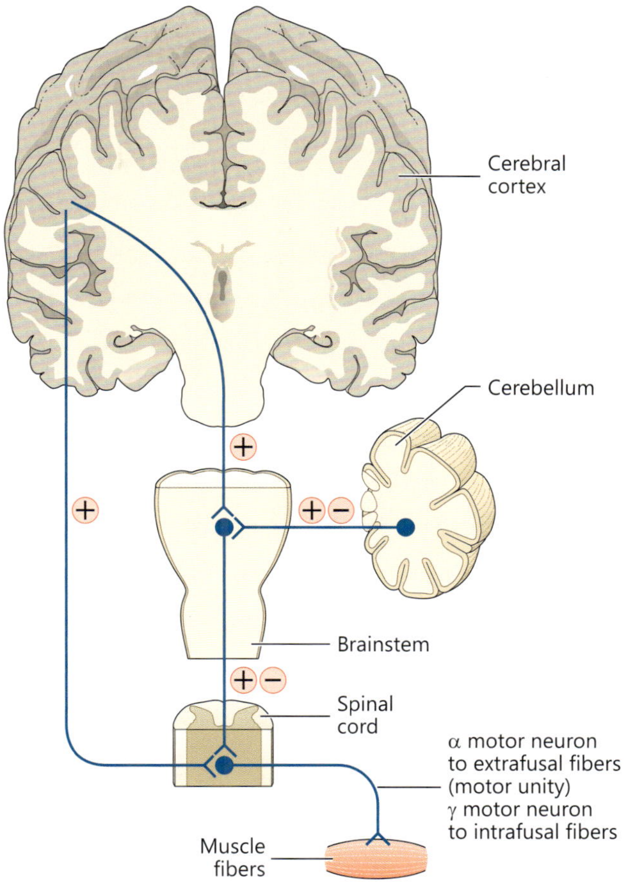

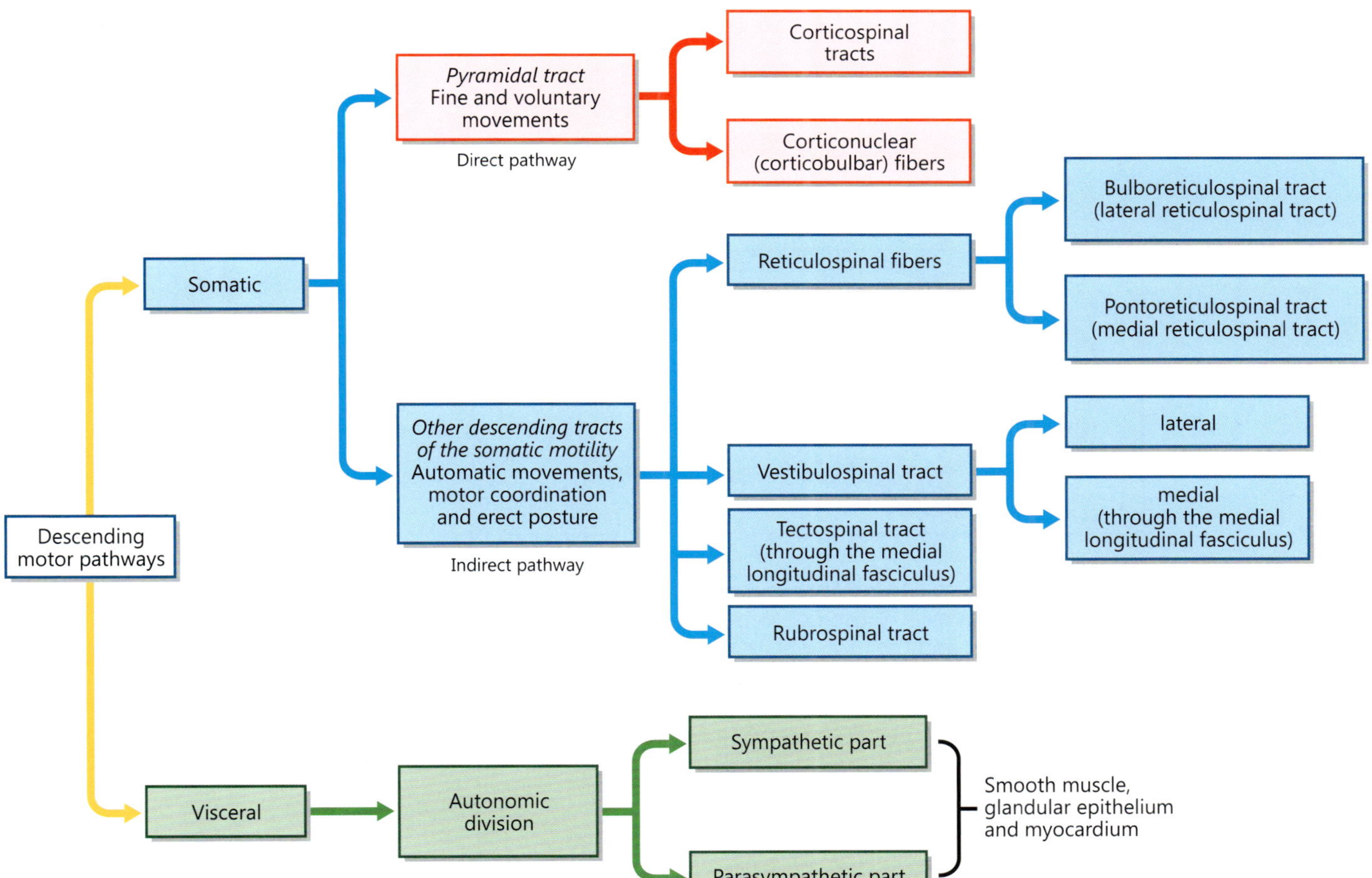

Schematic summary of the **motor pathways**.

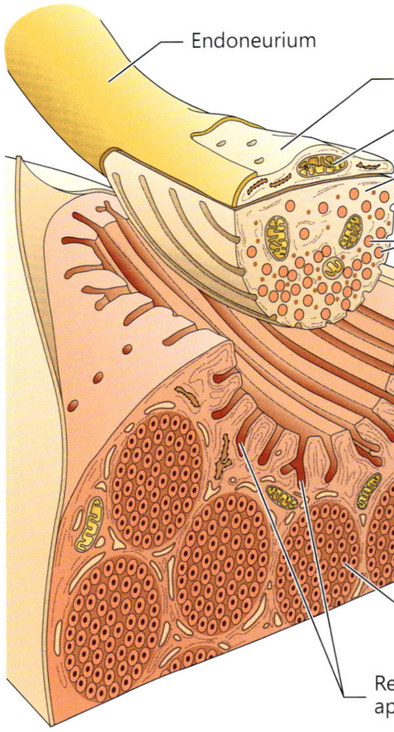

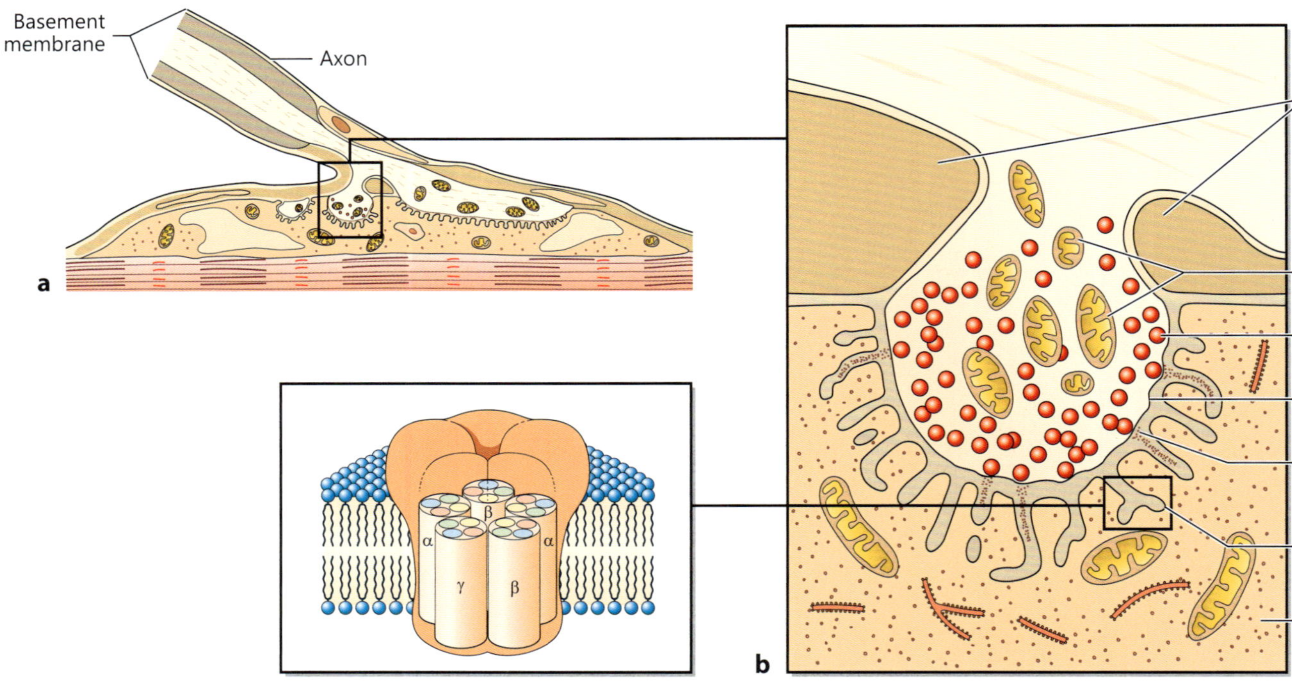

Schematic representation of a **synapsis** or **neuromuscular junction**. The motor fiber ends numerous expansions, called synaptic boutons, covered with Schwann cells. These boutons interact with specialized regions of the muscle fibers, called motor plates. Between the synaptic boutons and the motor plate there is a space of about 25 nm, the synaptic cleft. Inside the presynaptic boutons are numerous mitochondrions and synaptic vesicles containing acetylcholine, a ligand for specific receptors located in the junctional folds of the postsynaptic membrane, i.e. on the muscle fiber.
The lower box on the left shows the structure of the acetylcholine receptors. The acetylcholine nicotinic receptor is a membrane glycoprotein composed of five subunits: two α subunits and three subunits, β, γ, and δ, which contribute to form a central pore.

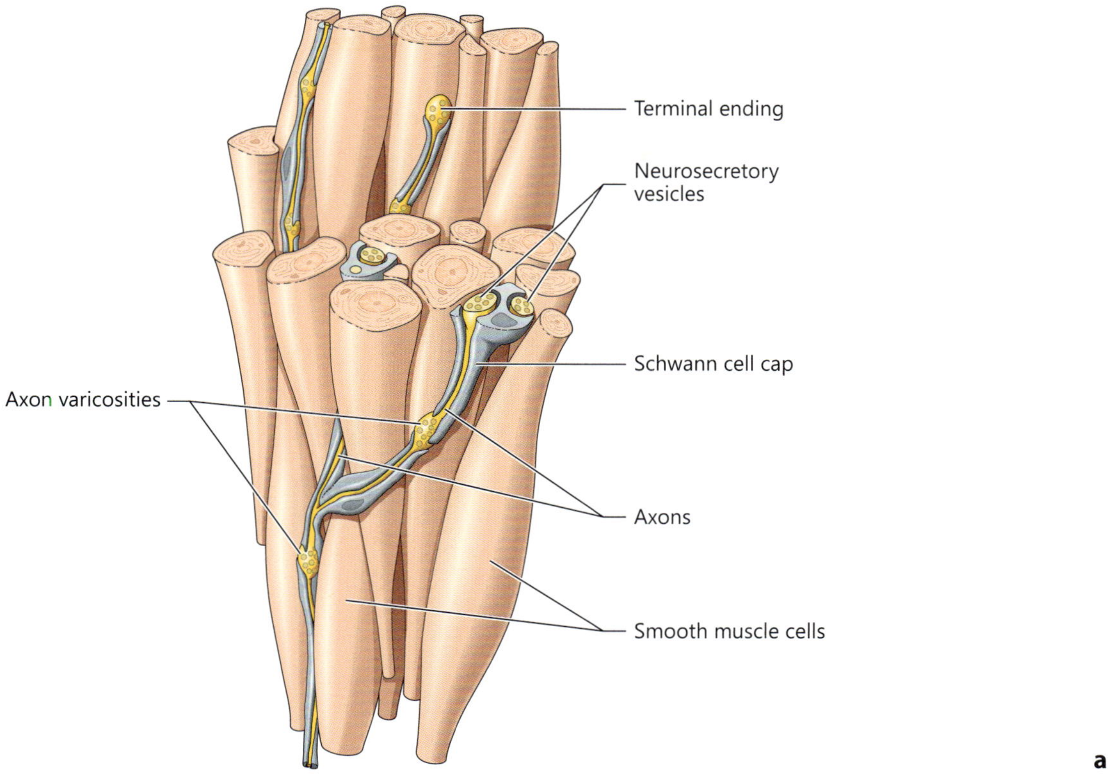

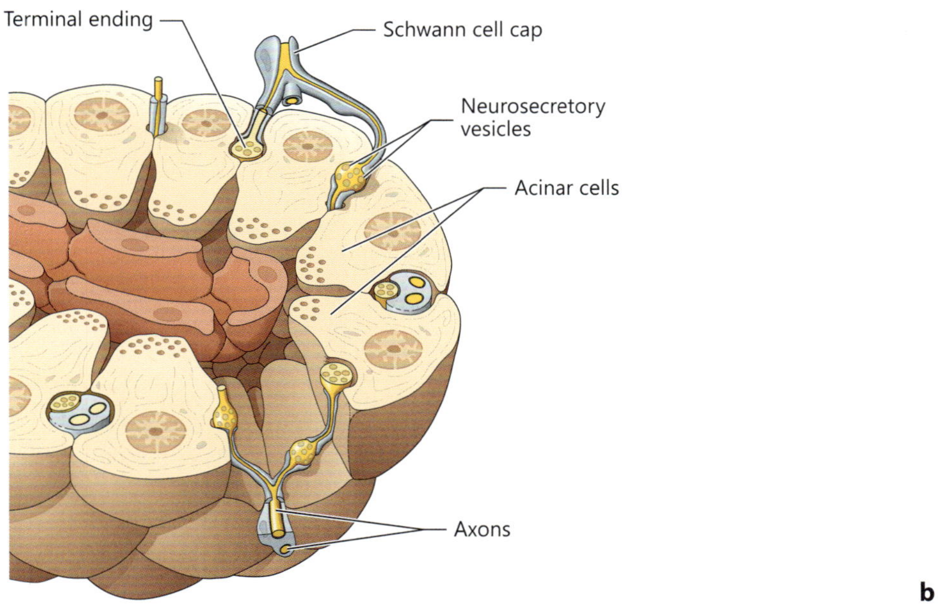

Schematic representation illustrating the innervation of the **smooth muscle cells** (**a**) and **glandular epithelia** (**b**).

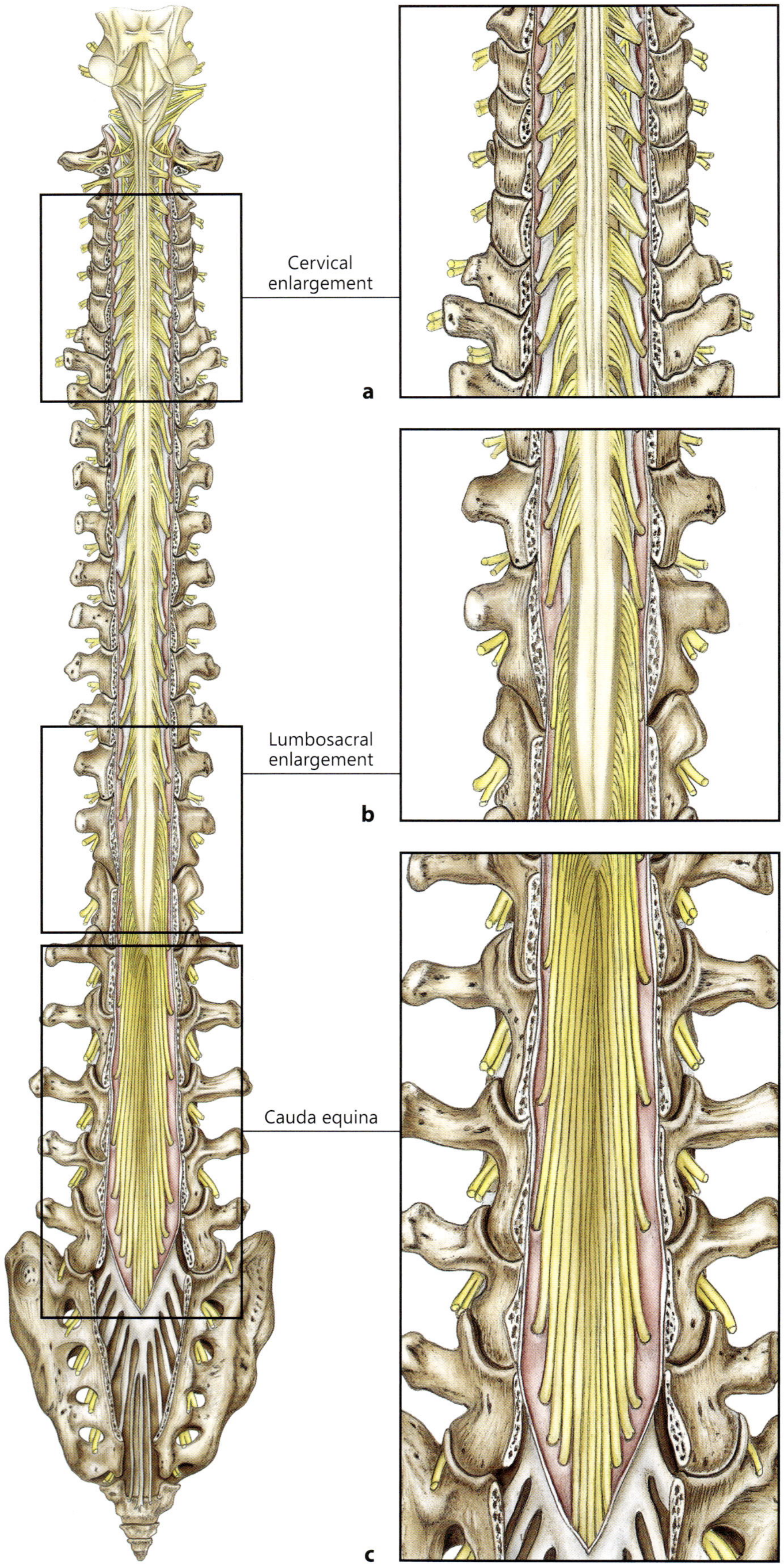

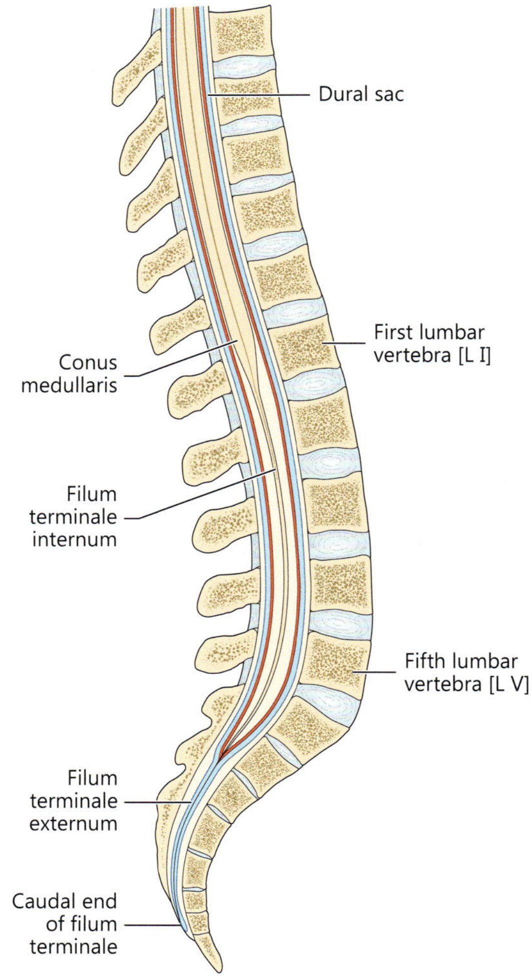

Median sagittal section of the terminal tract of the **spinal cord**. In the adult, the conus medullaris ends between the 1st and 2nd lumbar vertebra. The caudal end of the filum terminale binds to the dorsal surface of the coccygeal vertebrae. The dural sac closes, around the filum terminale, at the level of the second sacral segment. The dura mater is represented in *blue*, the external layer of the arachnoid mater in *red*. The arachnoid mater ends at 2nd sacral vertebra. Below the conus medullaris, the roots of the last lumbar nerves, of the sacral nerves and of the coccygeal nerve gather around the filum terminale forming the cauda equina.

The **vertebral canal** is observed in posterior projection after laminectomy and removal of the spinous processes of the vertebrae. Inside the canal the spinal cord is wrapped by the meninges. In correspondence of the two regions deputed to the innervation of the upper and lower limbs, the spinal cord presents two fusiform enlargements, the cervical (**a**) and the lumbosacral enlargements (**b**), respectively. At its caudal ends, the spinal cord shrinks and forms the conus medullaris, the cauda equina and then the filum terminale (**c**), generate from its apex.

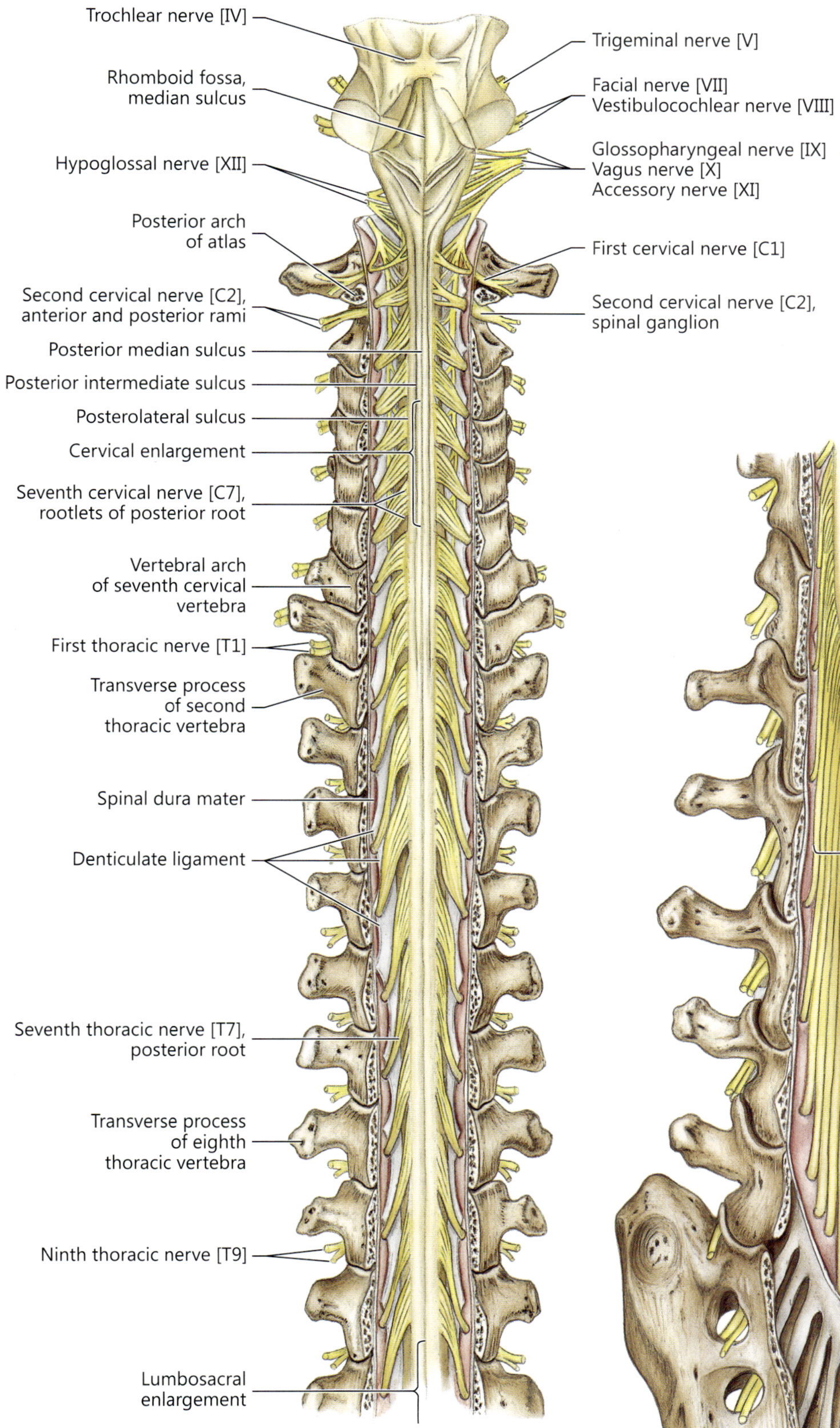

➤ **Spinal cord** and **spinal dura mater** in the vertebral canal. The canal has been opened posteriorly.

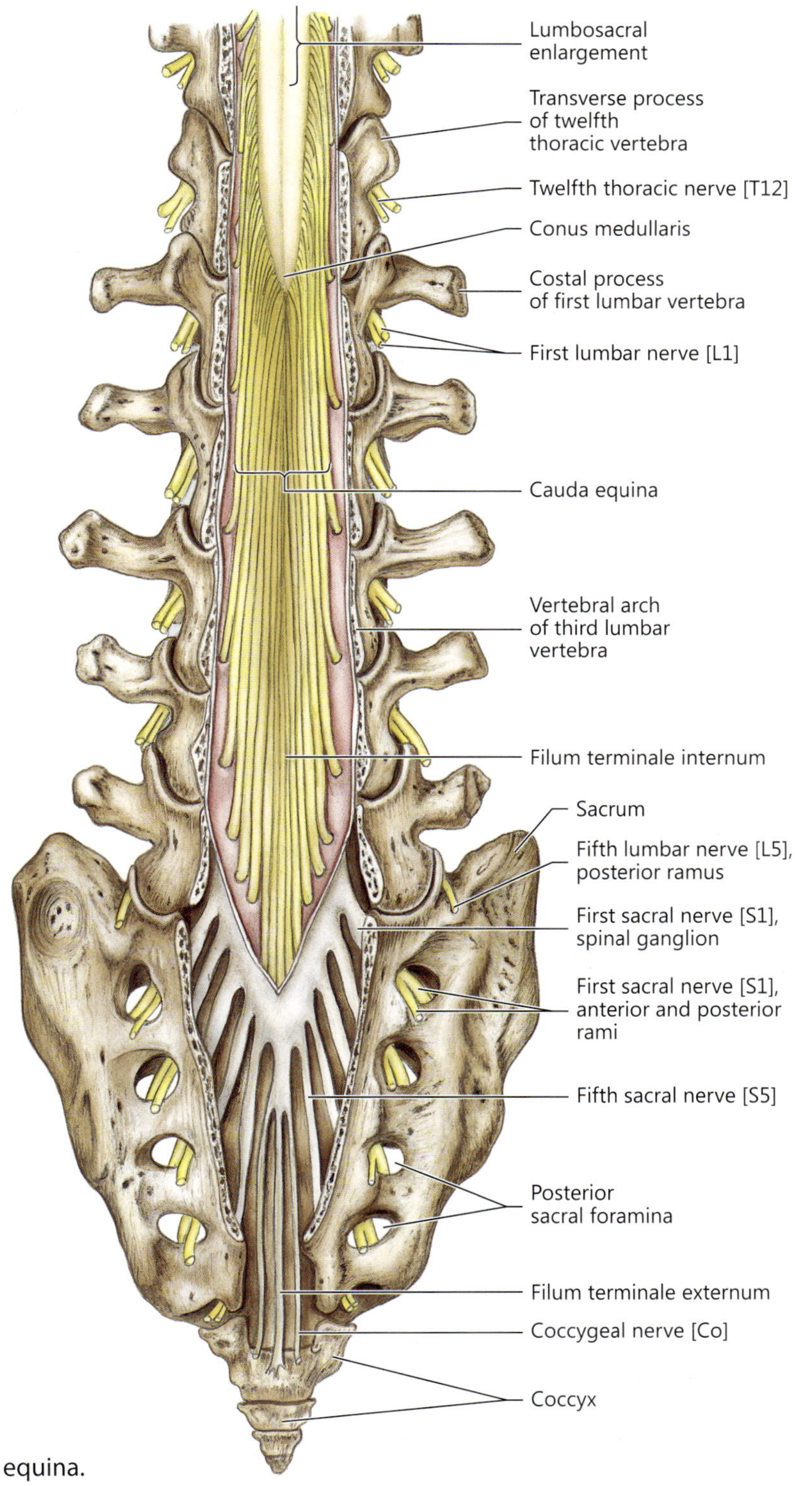

Cauda equina.

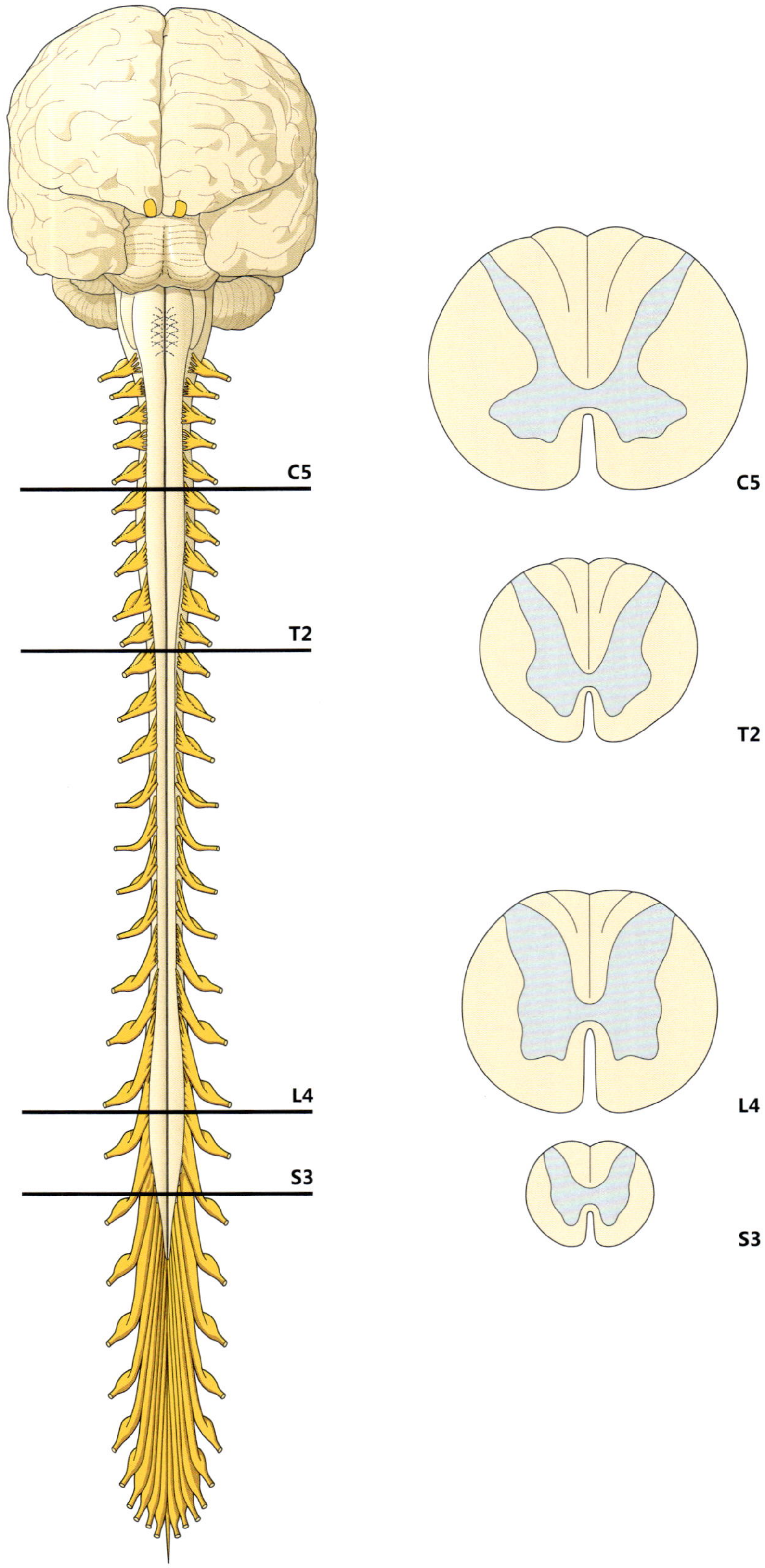

On the left, the **external configuration of the spinal cord** in anterior projection, after removal of the meninges; the cervical and lumbosacral enlargements are highlighted. On the right, the **conformation** and the **structure of the white and the gray matters**, in transverse sections of different tracts of the spinal cord. The gray matter increases in size at the level of the cervical and lumbosacral enlargements.

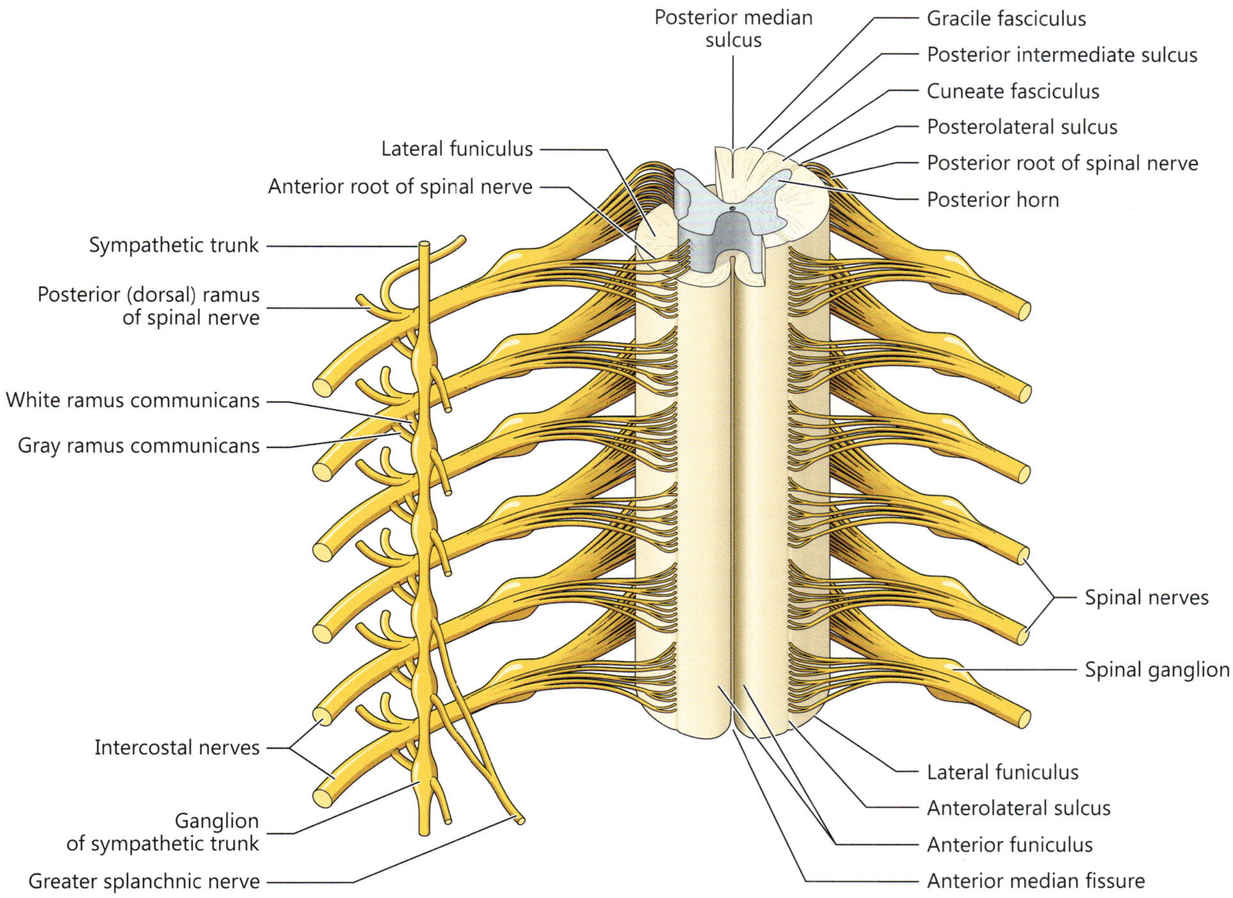

A segment of the **spinal cord** in anterior projection showing the anterior median fissure, the anterolateral sulcus and the anterior roots of the spinal nerves emerging from it.

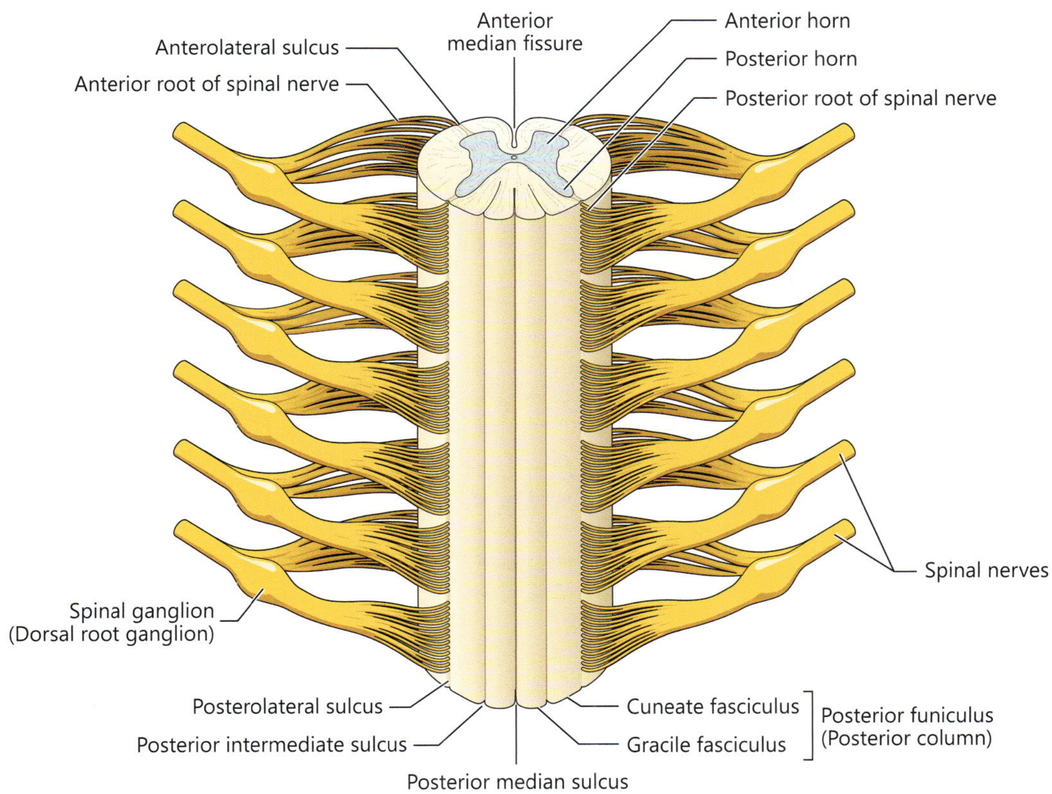

A segment of the **spinal cord** in posterior projection showing the posterior median, the posterior intermediate and the posterolateral sulci.

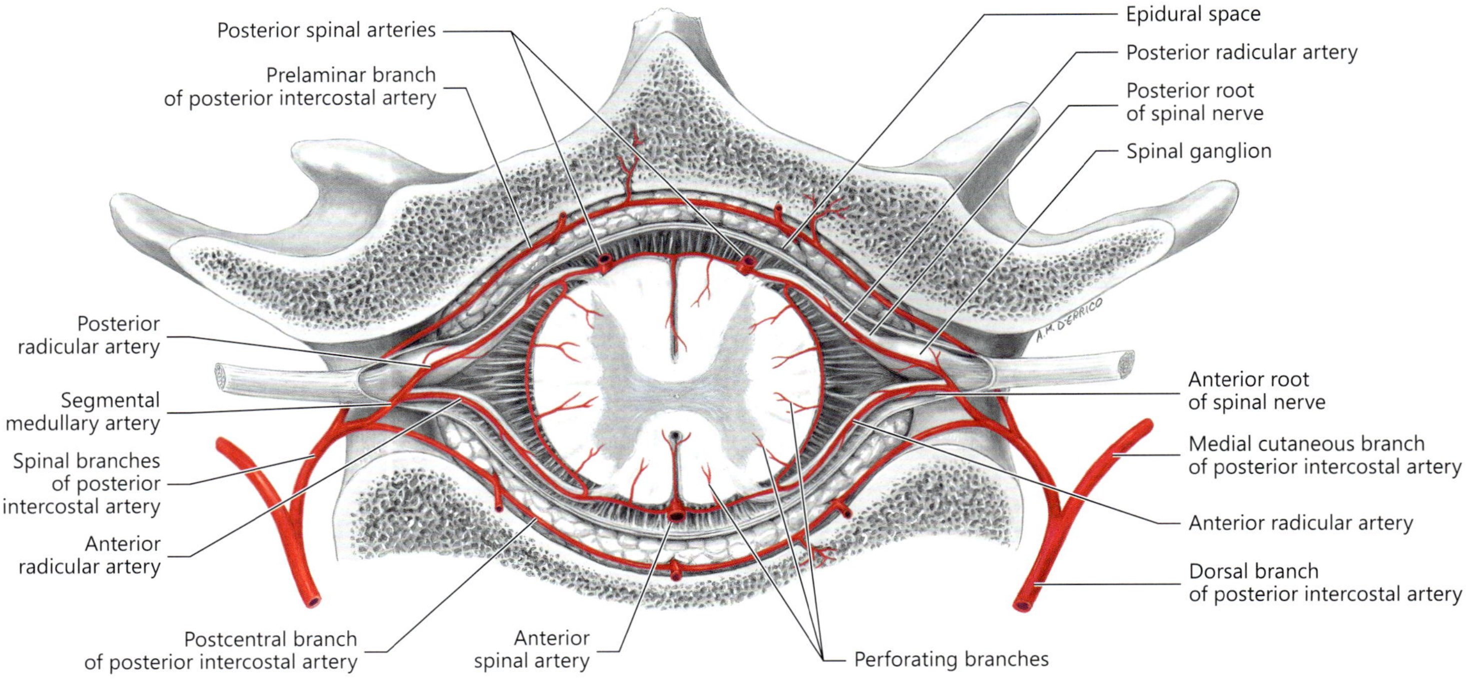

Representation of the **spinal cord arterial circulation** in a transverse section of the dorsal vertebral column. Spinal segmental arteries reach the spinal cord through the intervertebral foramina. At the level of the dorsal vertebral column these arteries are the spinal branches of the posterior intercostal arteries, or lateral spinal arteries. In the extramedullary plexus, the anterior and posterior spinal arteries are also evident. Each lateral spinal artery is connected with those above and below by, anterior and posterior, longitudinal anastomoses, which extend from the inside of the neurocranium to the cauda equina. From the anastomotic tracts originates the perimedullary plexus that releases the nutrient arteries of the spinal cord, the perforating branches.

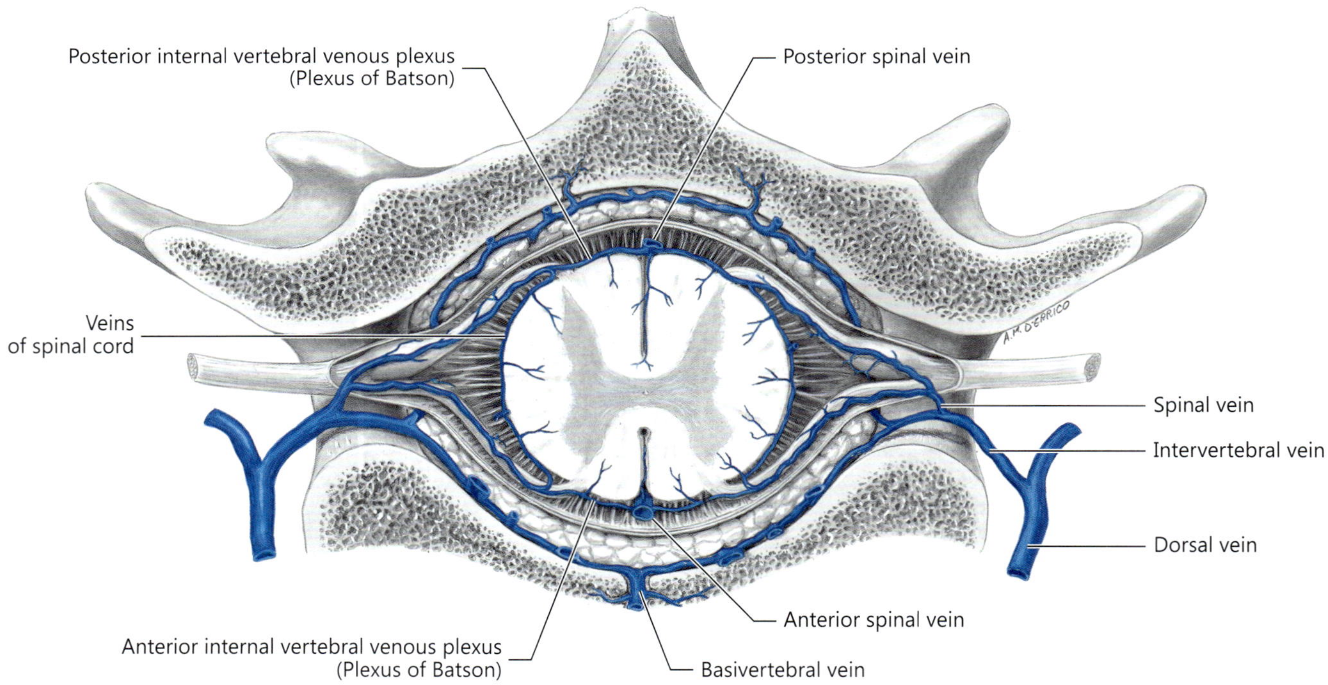

Representation of the **internal vertebral venous plexuses** in a transverse section of the dorsal vertebral column, located in the epidural space and in the spinal dura mater. They extend longitudinally from the foramen magnum to the apex of the sacrum.

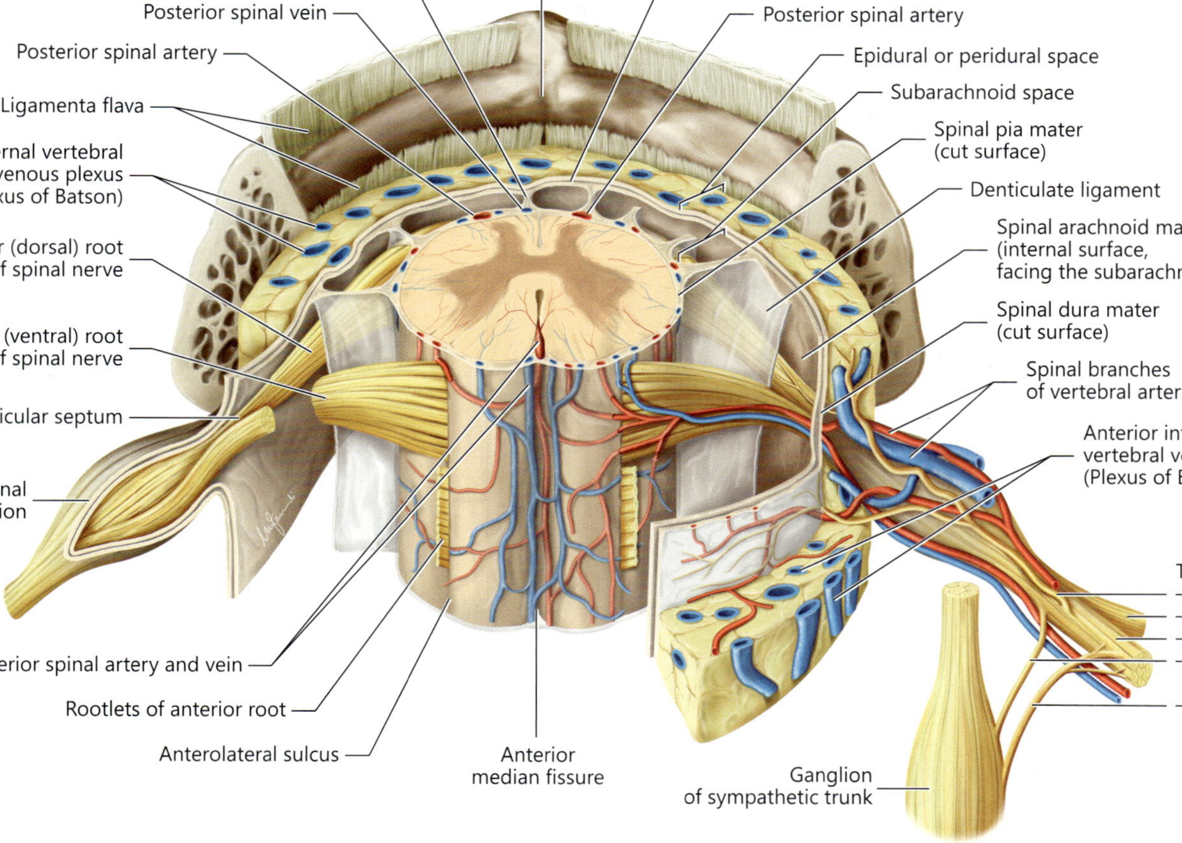

Three-dimensional representation of a **transverse section of the spinal cord** passing through the first cervical vertebra. Note the relationships between the spinal cord and the meninges and spinal vessels.

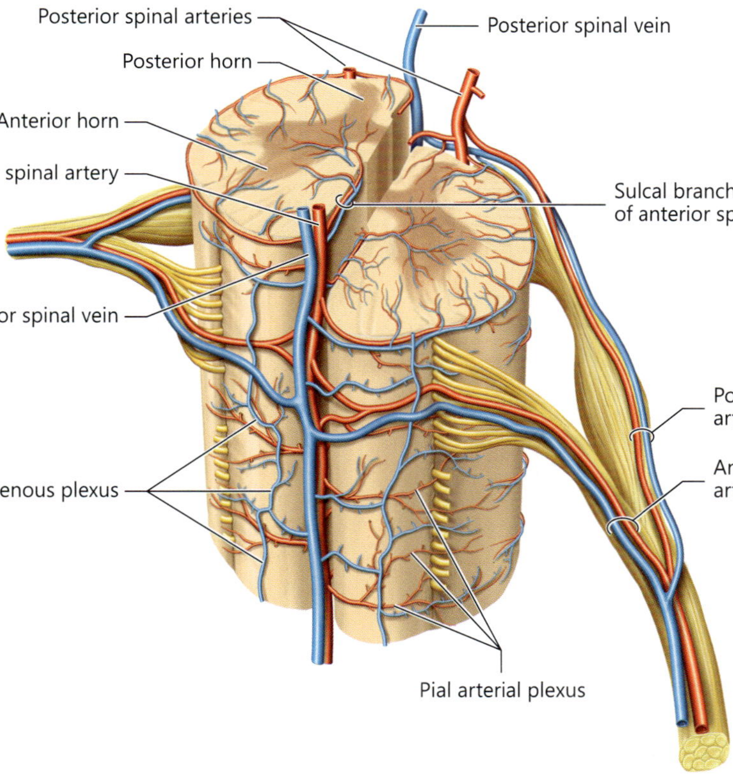

Vascular organization of a **neuromere of the spinal cord** in transverse section. The anterior and posterior spinal arteries supply branches that penetrate into the spinal cord, where the gray matter is more richly vascularized than the white matter.

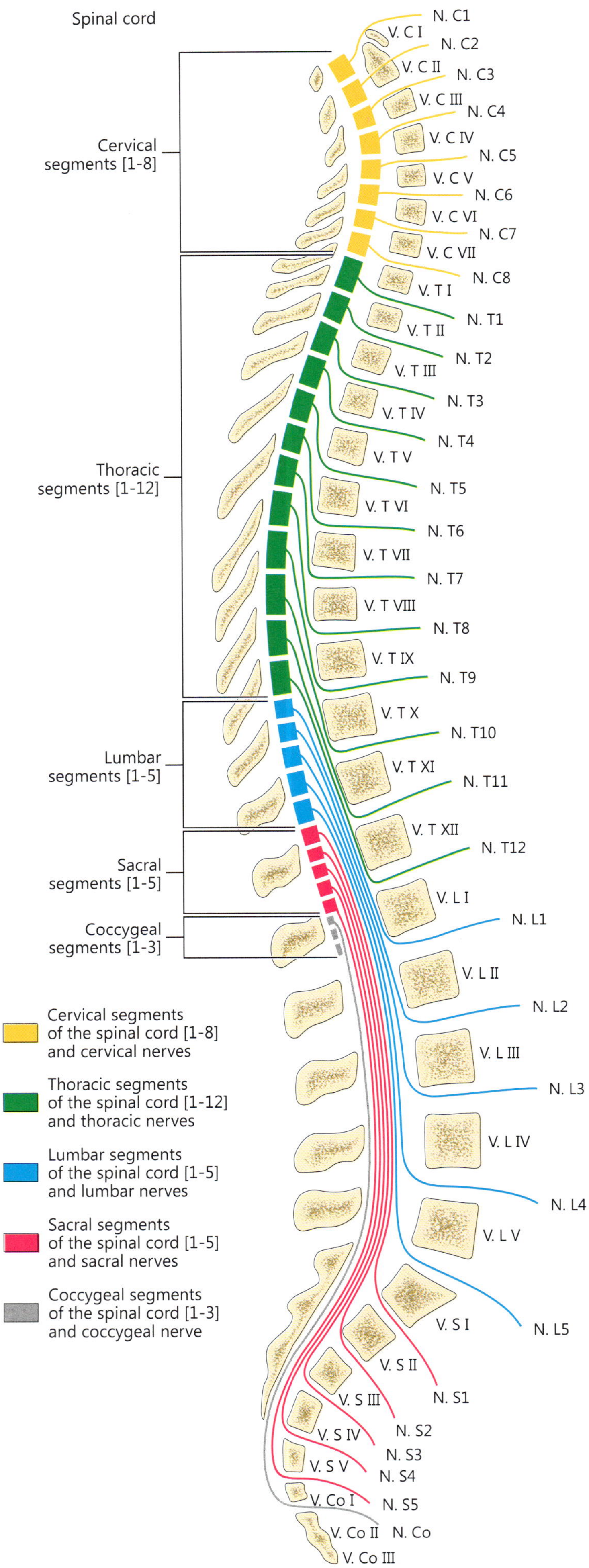

Main dermatomes at neurological examination

NEUROMERE	CORRESPONDING SENSORY AREA
C2	Posterior nuchal area
C4	Shoulder
C6	Thumb
C7	Middle finger
C8	Little finger
T4-T5	Nipple
T10	Umbilicus
L1	Groin
L2-L3	Thigh
S1	Foot
S4-S5	Genitals

Myomeres of major clinical relevance

NEUROMERE	CORRESPONDING SENSORY AREA
C5	Deltoid
C6	Biceps brachii
C7	Triceps brachii
T1	Muscles of the hypothenar eminence
L4	Quadriceps femoris
L5	Extensor hallucis longus
S1	Gastrocnemius
S3-S4	External anal sphincter

Schematic representation illustrating the relationships of the **spinal cord** and **spinal nerves** with the vertebral column. The spinal cord and the vertebral column are divided into cervical, thoracic, lumbar, sacral and coccygeal tracts; each tract comprises different segments, or neuromeres, equal in number to the vertebrae forming the corresponding tract of the vertebral column (with the exception of the cervical column vertebral which presents 8 neuromers instead of 7). Each neuromere gives rise to an anterior root, or motor root, and a posterior root, or sensory root, on each side; the muscular territory innervated by each pair of motor roots is defined myomere and the cutaneous territory innervated by each pair of sensory roots is called dermatome.
The tables show the most significant cutaneous and muscle areas innervated by dermatomes and myomeres, respectively.

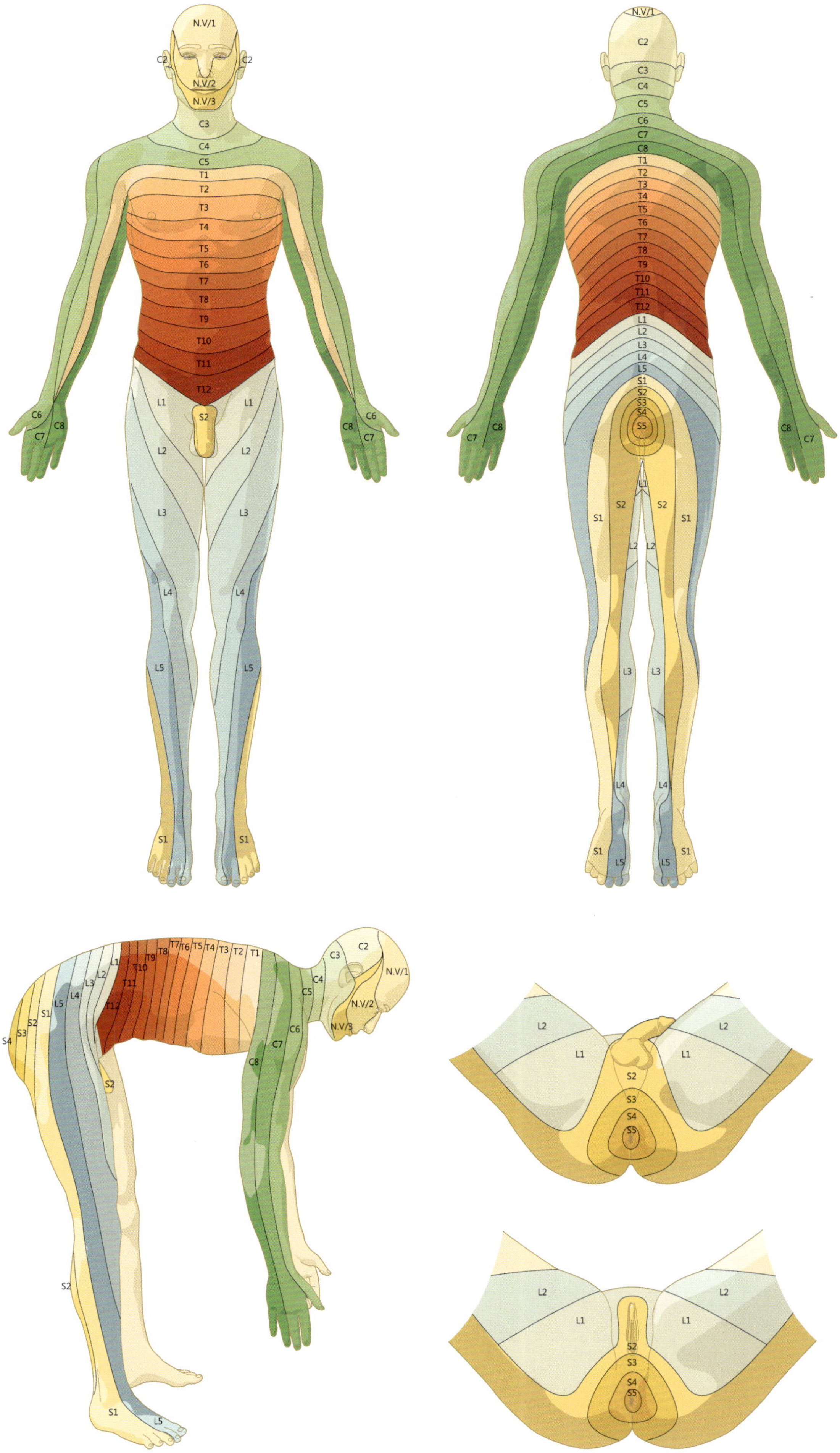

Schematic representation of the **dermatomes of the head**, **trunk** and **limbs**.

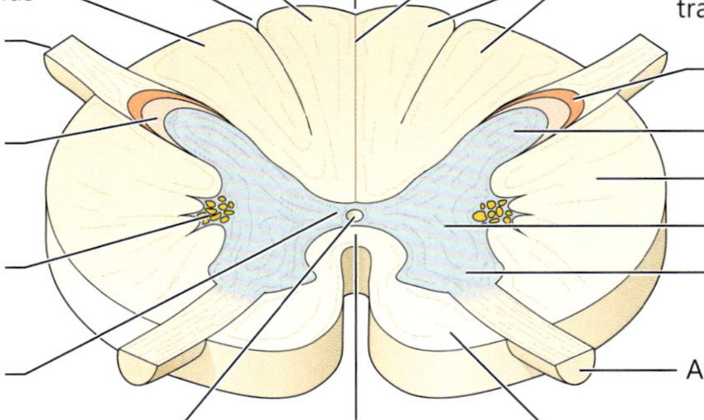

Schematic representation of the internal conformation of the **spinal cord** in transverse section showing the distribution of the gray matter and white matter and the spinal nerves roots.

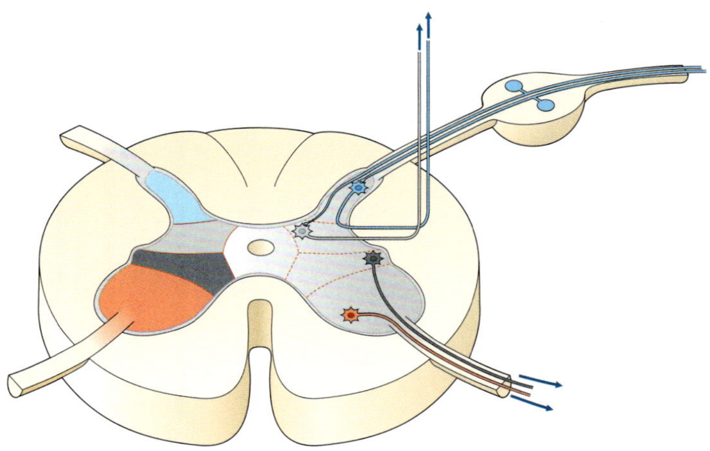

Schematic representation of the internal conformation of the **spinal cord** in transverse section. Right side, types of neurons found in the gray matter of the spinal cord, that is somatomotor (*red*), visceroeffector (*dark gray*), viscerosensory (*light gray*) and somatosensory (*blue*) cells. Left side, somatomotor (*red*), visceroeffector (*dark gray*), viscerosensory (*light gray*) and somatosensory (*blue*) territories.

Laminar organization of the gray matter of the spinal cord		
SPINAL LAMINAE	***LOCATION***	***CHARACTERISTICS***
I	Apex of the posterior horn (marginal nucleus or spongy zone)	Marginal neurons with dendritic ramification which extends parallel to the entire apex of the posterior horn
II	Head of the posterior horn (gelatinous substance of Rolando)	Neurons of small dimensions on which nociceptive fibers end and from which intersegmental and projective axons start Synaptic glomeruli
III	Neck of the posterior horn (nucleus proprius)	Neurons of different dimensions Numerous projective elements
IV	Neck of the posterior horn (nucleus proprius)	Neurons of different dimensions Numerous projective elements
V	Neck of the posterior horn	Neurons of different dimensions Numerous projective elements
VI	Base of the posterior horn	Particularly noticeable at the level of cervical and lumbosacral enlargements Small and medium size neurons
VII	Lateral horn (intermediate column or zone) and anterior horn	Neurons of different dimensions Numerous premotor elements (Renshaw cells)
VIII	Anterior horn	Neurons of different dimensions Numerous commissural elements
IX	Anterior horn	Divided into several columns consisting of motor neurons
X	Anterior and posterior gray commissures	Few neurons of small size

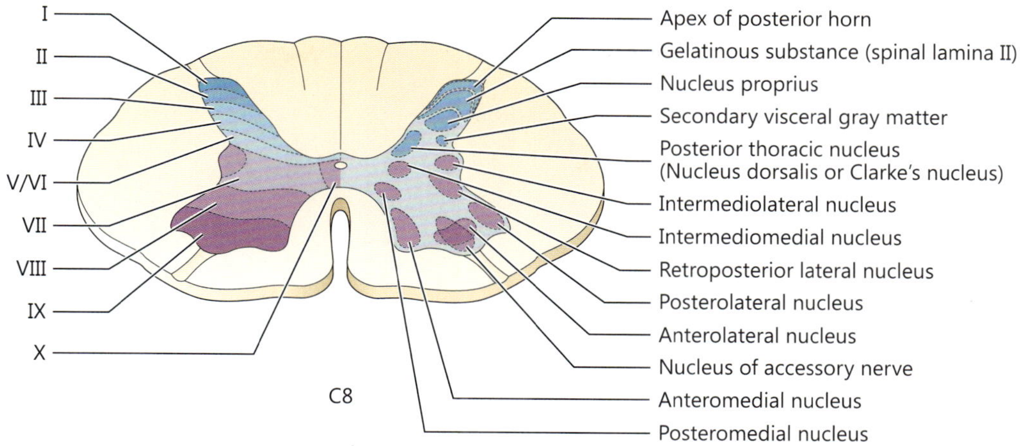

a

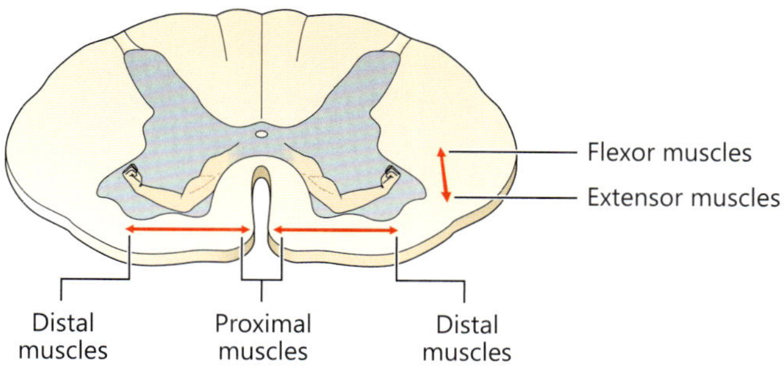

b

Cervical portion of the **spinal cord**, subdivision of the nuclear groups in laminae and the gray matter in cell groups (**a**). **b**, Diagram illustrating the somatotopic organization of motor neurons in the cervical enlargement. Note that the motor neurons to the extensor muscles are placed ventrally to those to the flexor muscles.

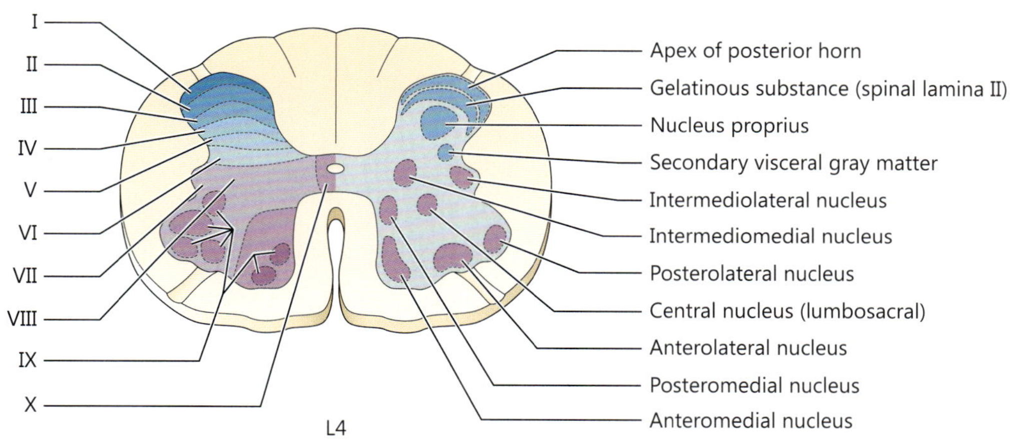

a

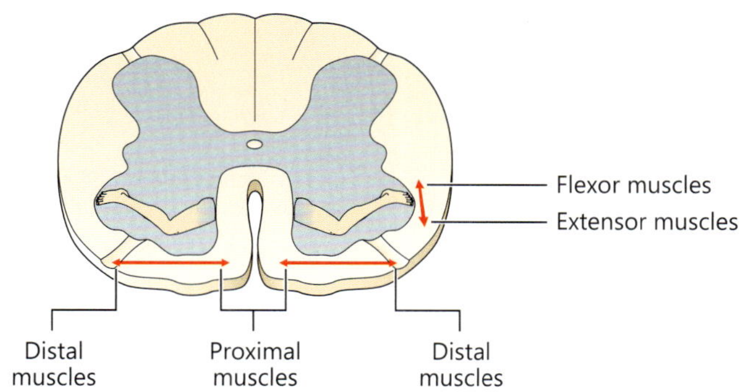

b

Lumbosacral portion of the **spinal cord**, subdivision of the nuclear groups in laminae and the gray matter in cell groups (**a**). **b**, Diagram illustrating the somatotopic organization of motor neurons in the lumbosacral enlargement. Note that the motor neurons to the extensor muscles are placed ventrally to those to the flexor muscles.

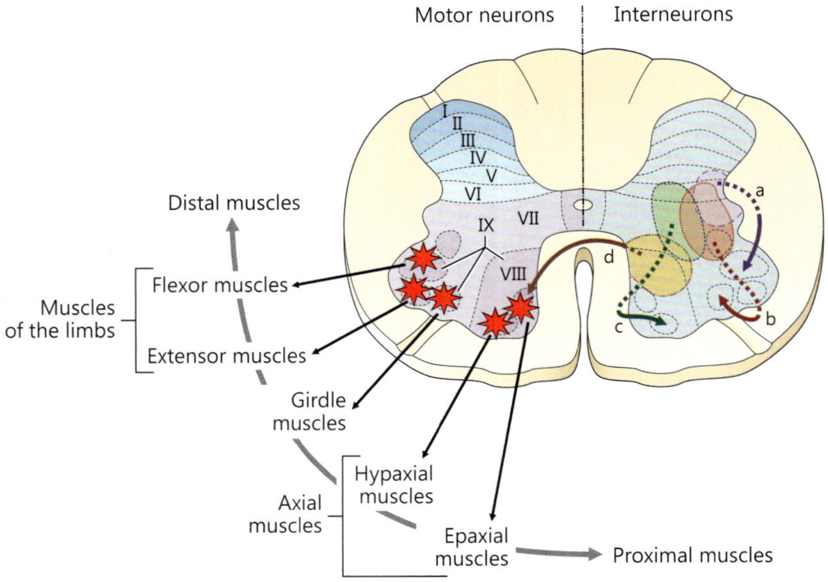

Diagram illustrating the higher complexity of the organization of **motor columns** at the level of the cervical and lumbosacral enlargements, where the motor neurons for the girdle muscles and axial muscles are found; in other words, the motor neurons for the distal muscles (muscles of the limbs) and proximal muscles (axial muscles) (left side of the image). In the right side of the image are represented the projections of interneurons of laminae VII and VIII, on the motor neurons of lamina IX.

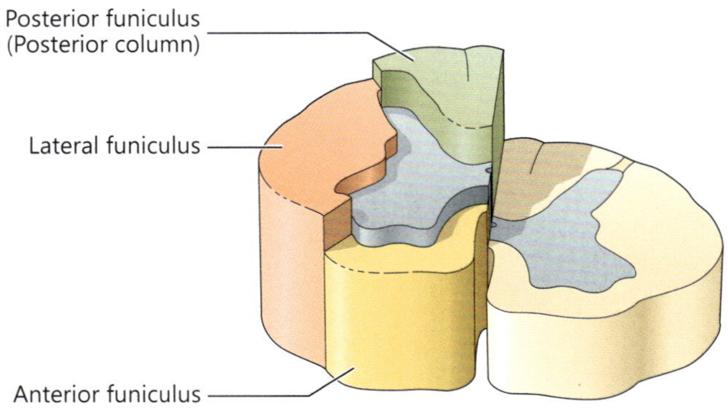

Three-dimensional visualization of the distribution of the **white matter** at the level of the spinal cord.

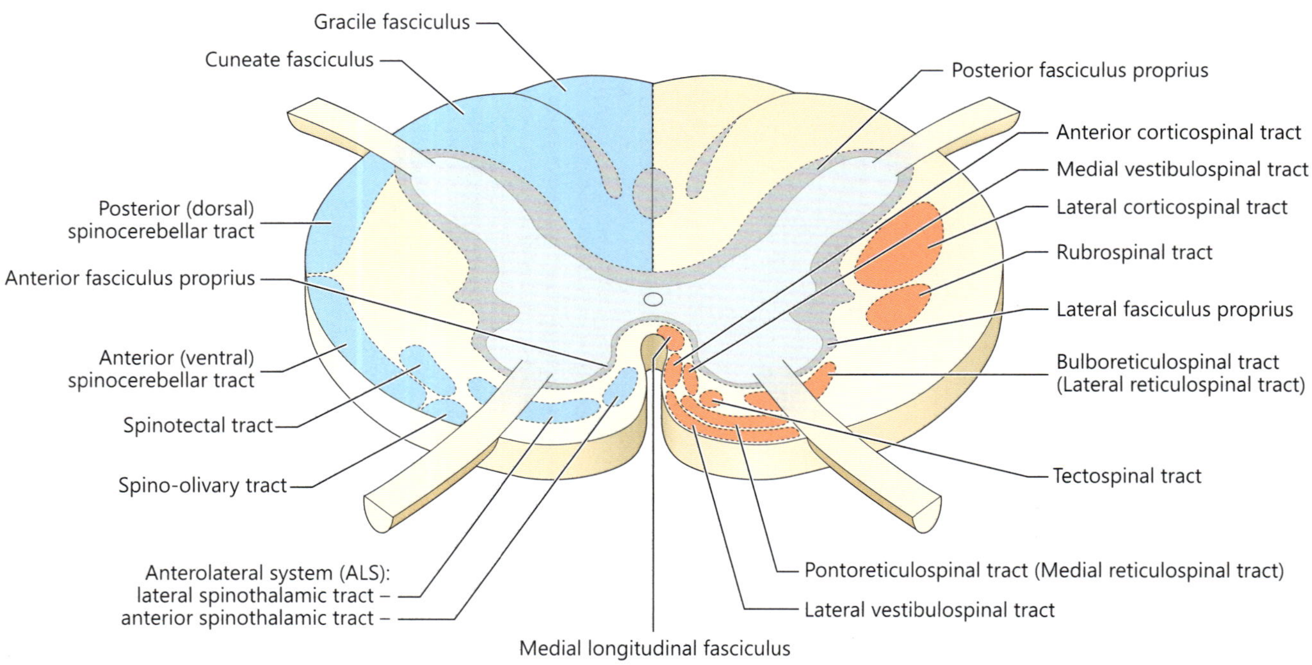

Distribution of the main **descending** (*red*), **ascending** (*blue*) and **association** (*gray*) **projection tracts** into the funiculi of the white matter of the spinal cord, seen in transverse section. The descending pathways are located mainly in the anterolateral funiculus whereas the ascending pathways in the lateral and posterior funiculi.

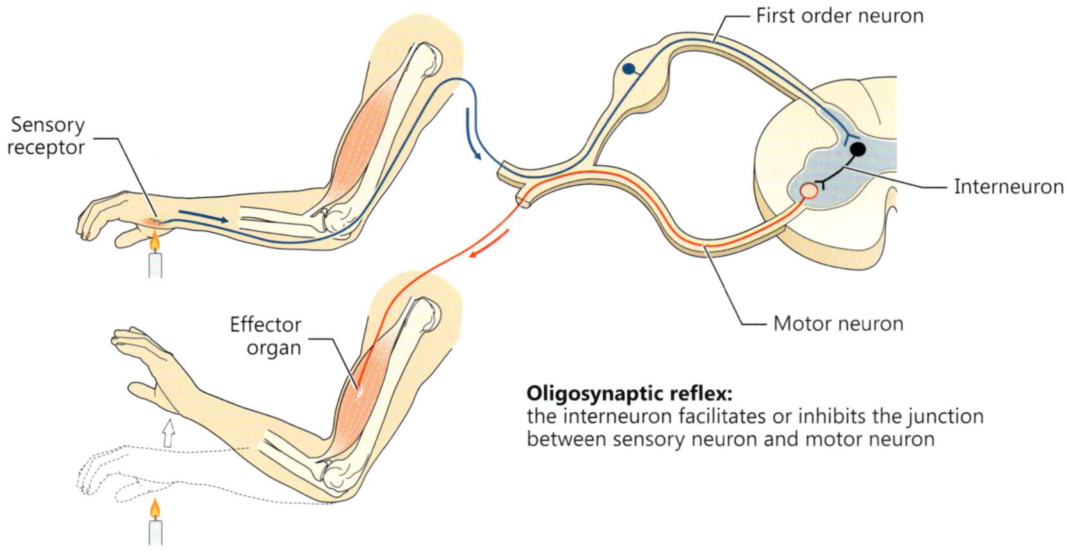

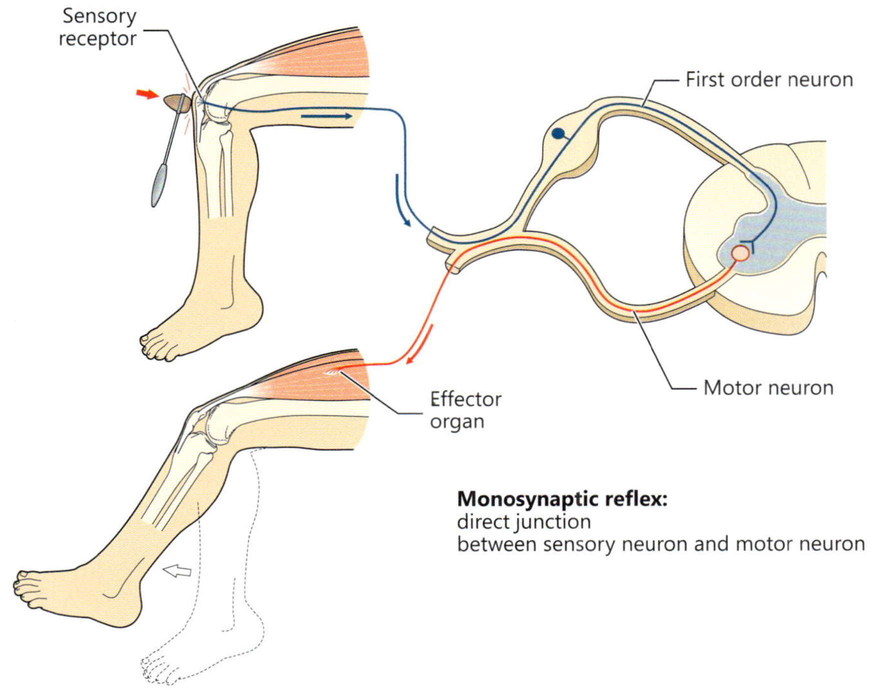

Schematic representation of the **proprioceptive reflex arc**. Above it is shown an oligosynaptic reflex arc; below, a monosynaptic reflex arc.

Antero Lateral System (ALS) - Lateral Compartment

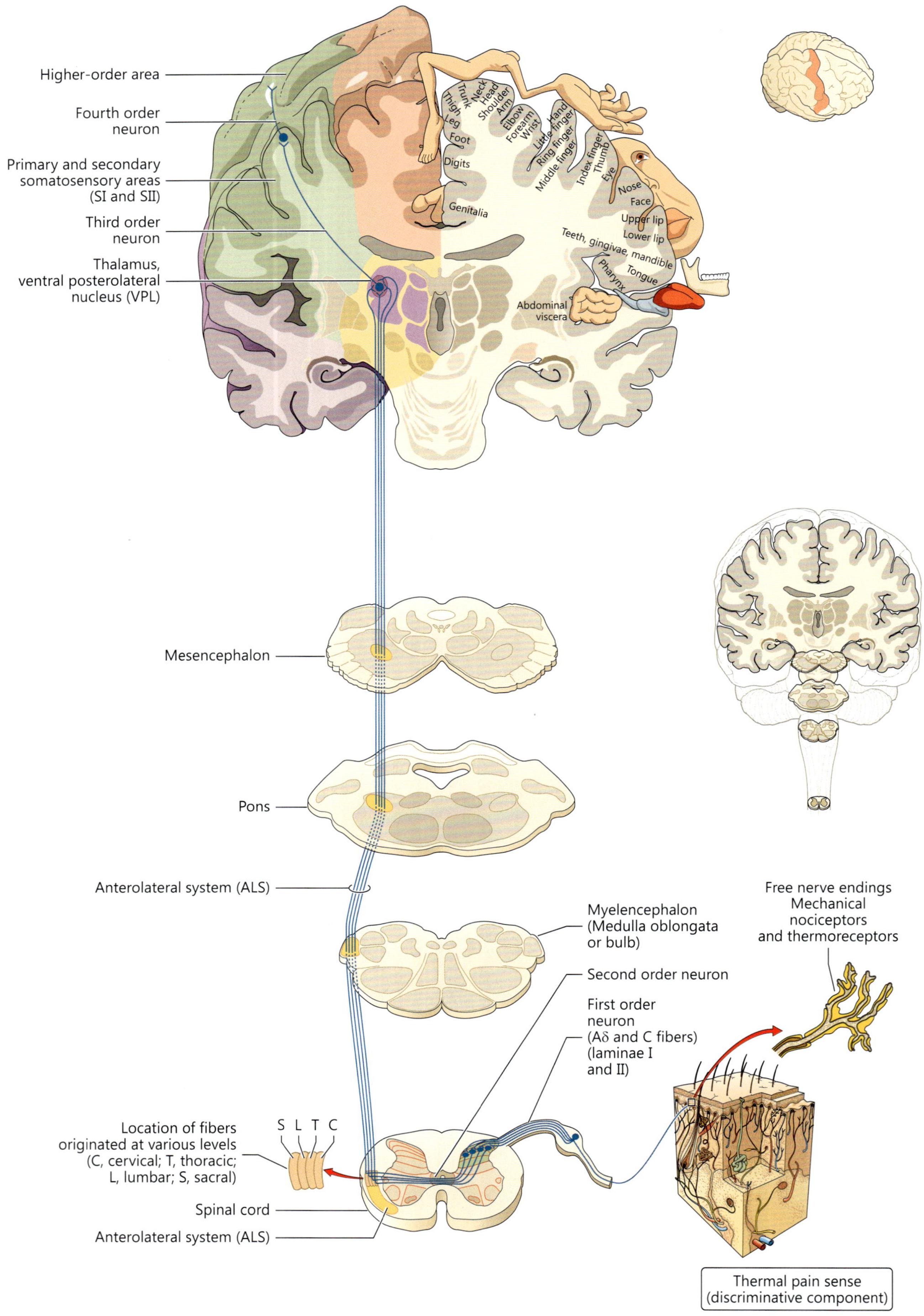

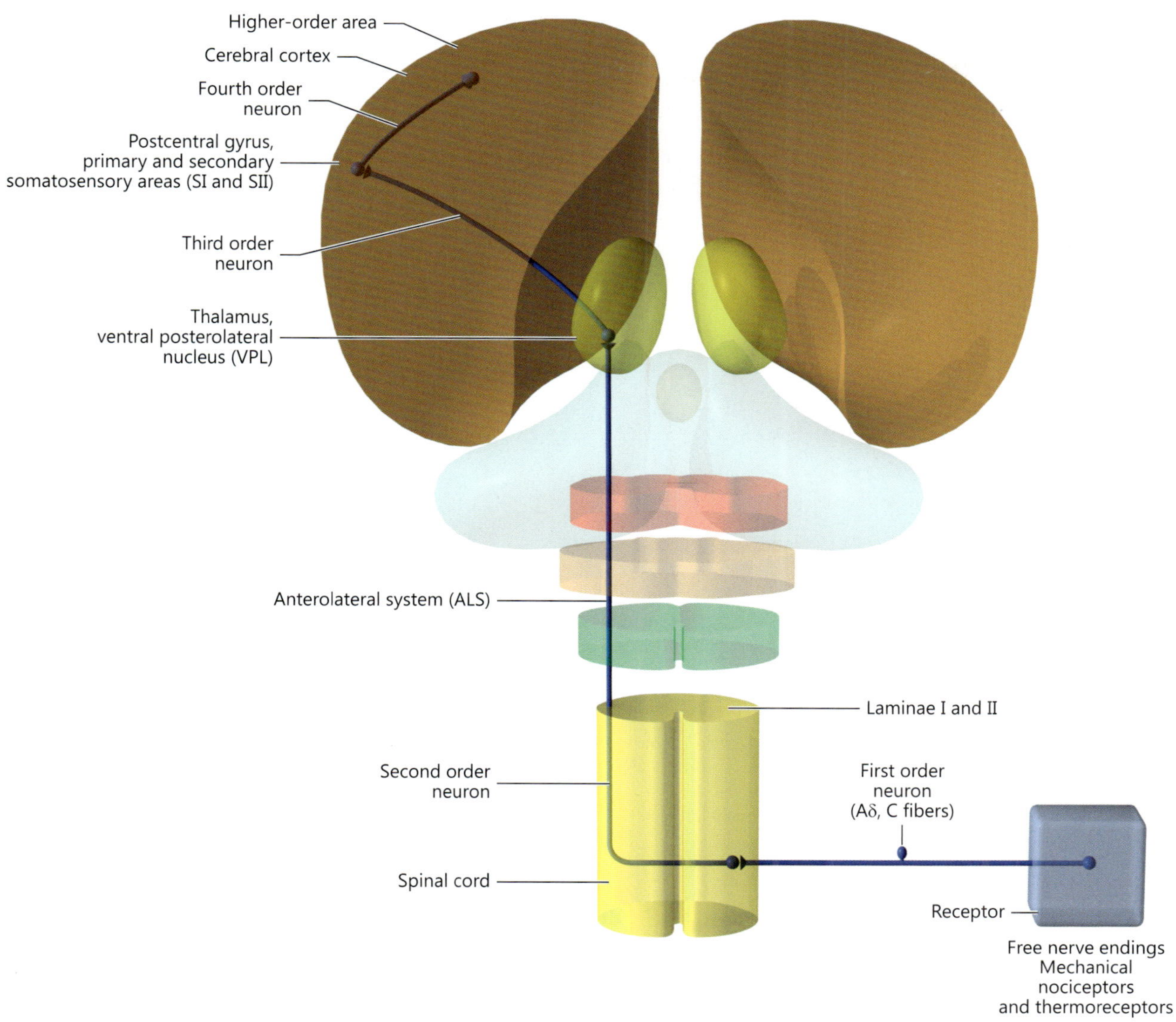

Anterolateral system - Lateral spinothalamic tract (lateral compartment) or spinal lemniscus	
Sense	Nociceptive (dolorific and thermal), discriminative component
First order neuron (pseudounipolar neuron)	From the mechanical, thermal and polymodal nociceptors (free nerve endings) to the laminae I and II of the spinal cord
Second order neuron (deutoneuron)	From the laminae I and II of the spinal cord it crosses the midline, it runs in the lateral funiculus and goes up along the brainstem (anterolateral system) until the ventral posterolateral nucleus (VPL) of the thalamus
Third order neuron (thalamocortical projection neuron)	From the VPL nucleus to the primary somatosensory area (SI)
Fourth order neuron	From the primary and secondary somatosensory areas (SI and SII) to the higher-order areas

The **lateral spinothalamic tract** occupies the lateral compartment of the anterolateral system and it is responsible for pain reception; this tract is composed of two compartments, the lateral compartment (or neospinothalamic), and the medial compartment (or paleospinothalamic). The lateral compartment, so named because it pertains to the ventral posterolateral nucleus of the thalamus, is responsible for the discriminative perception of pain, in terms of quality and location. Signal remodulation systems, present at the level of the laminae I and II (☞ following pages), can inhibit the retransmission of pain.

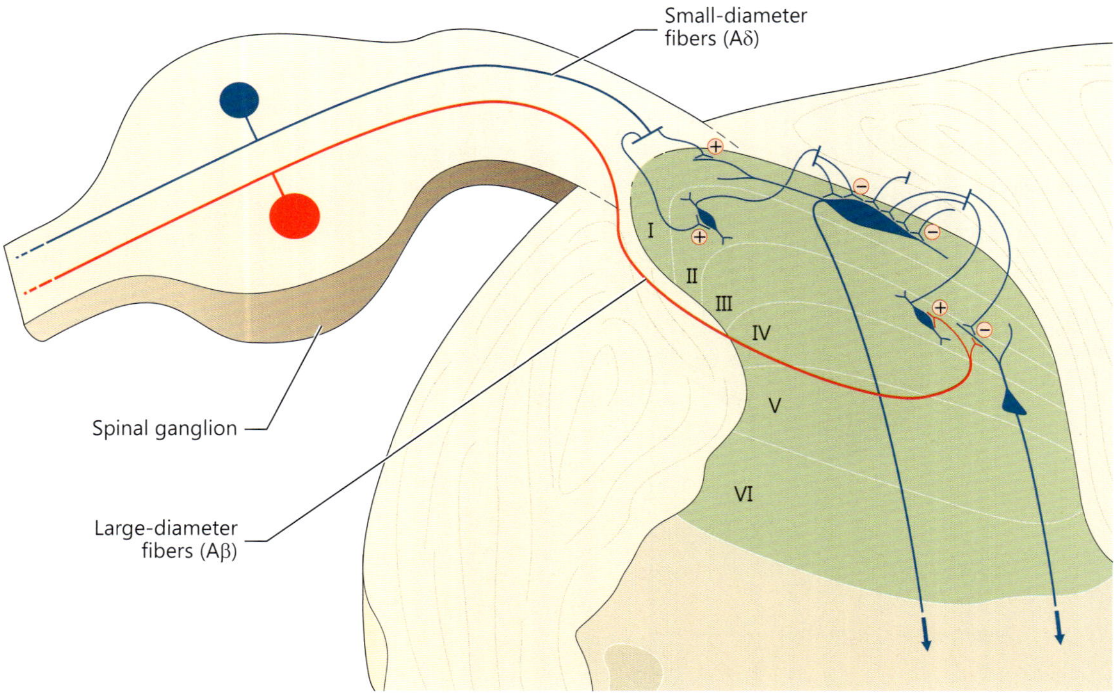

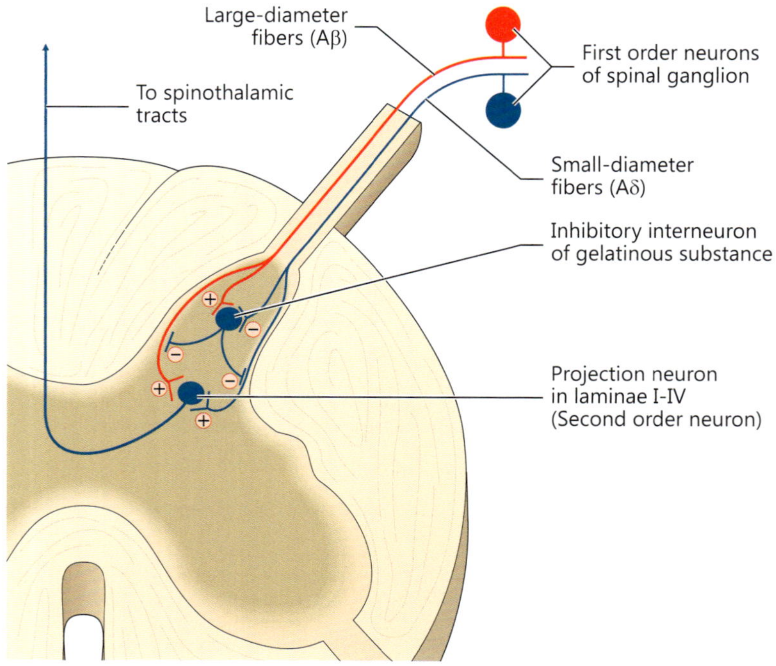

The activation of the large diameter and low activation threshold Aβ fibers, which collect tactile, not nociceptive sense, close at the level of the spinal cord the **"gate" controlling pain transmission**. This transmission is constituted by circuits established between interneurons of the gelatinous substance (of Rolando) and the projection neurons of laminae I, II, III and IV of the gray matter of the posterior horn of the spinal cord. The activation of the inhibiting interneurons of the gelatinous substance, inhibits the transmission of impulses conducted from the small diameter and high activation threshold Aδ fibers, which collect thermal, vibrational and nociceptive (acute pain) sense and blocks pain transmission through the ascending pathways of the spinal cord.

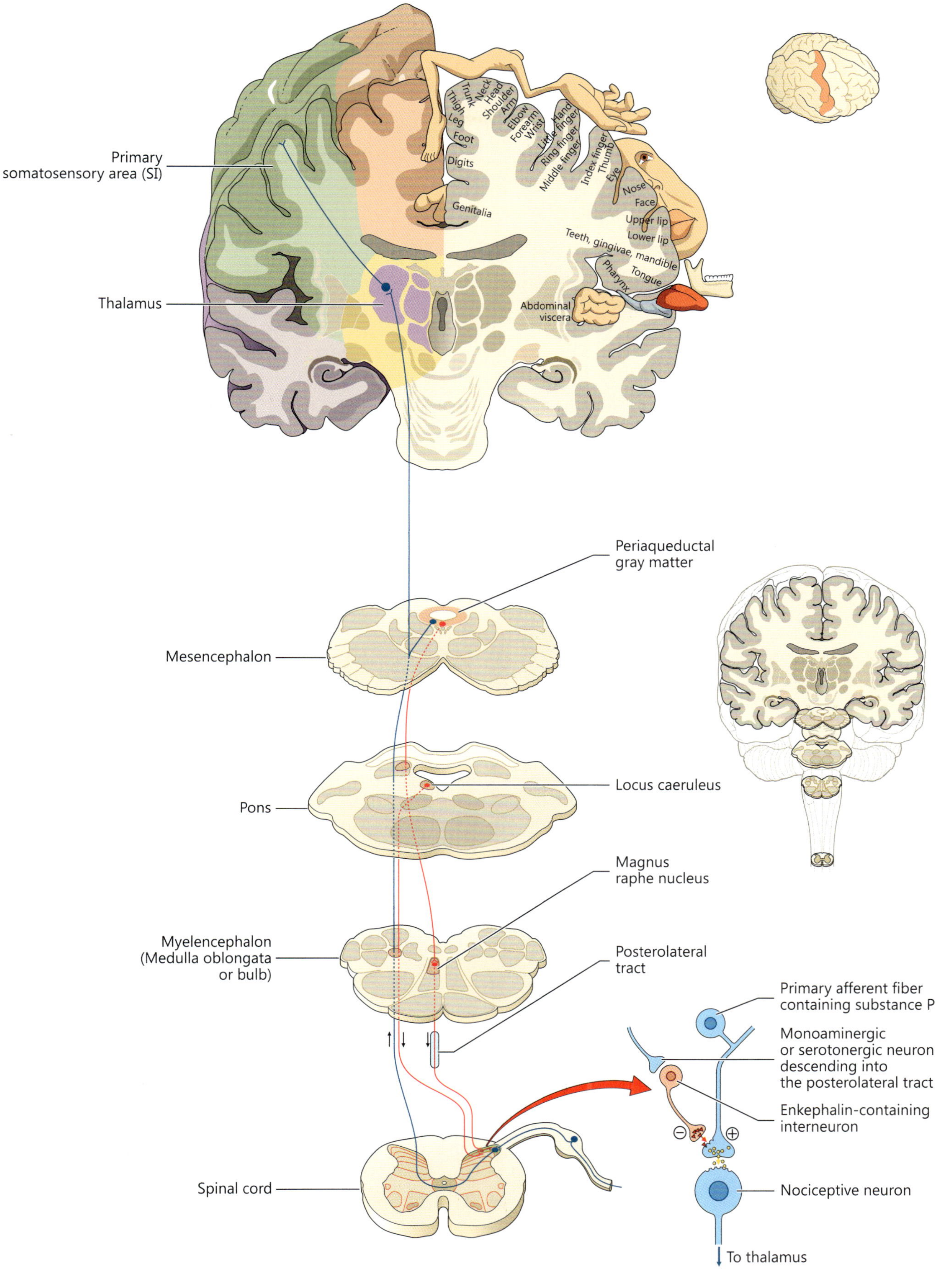

Diagram illustrating the role of the **monoaminergic and serotonergic neurons of the brainstem** in the modulation of nociceptive neurons in the posterior horn of the gray matter of the spinal cord, through the interneurons of the gelatinous substance (of Rolando). By releasing enkephalins, the interneurons interrupt the excitatory transmission of ganglionic cells by acting with an inhibitory presynaptic control on primary afferent fibers, blocking the release of substance P on nociceptive neurons in response to painful stimuli.

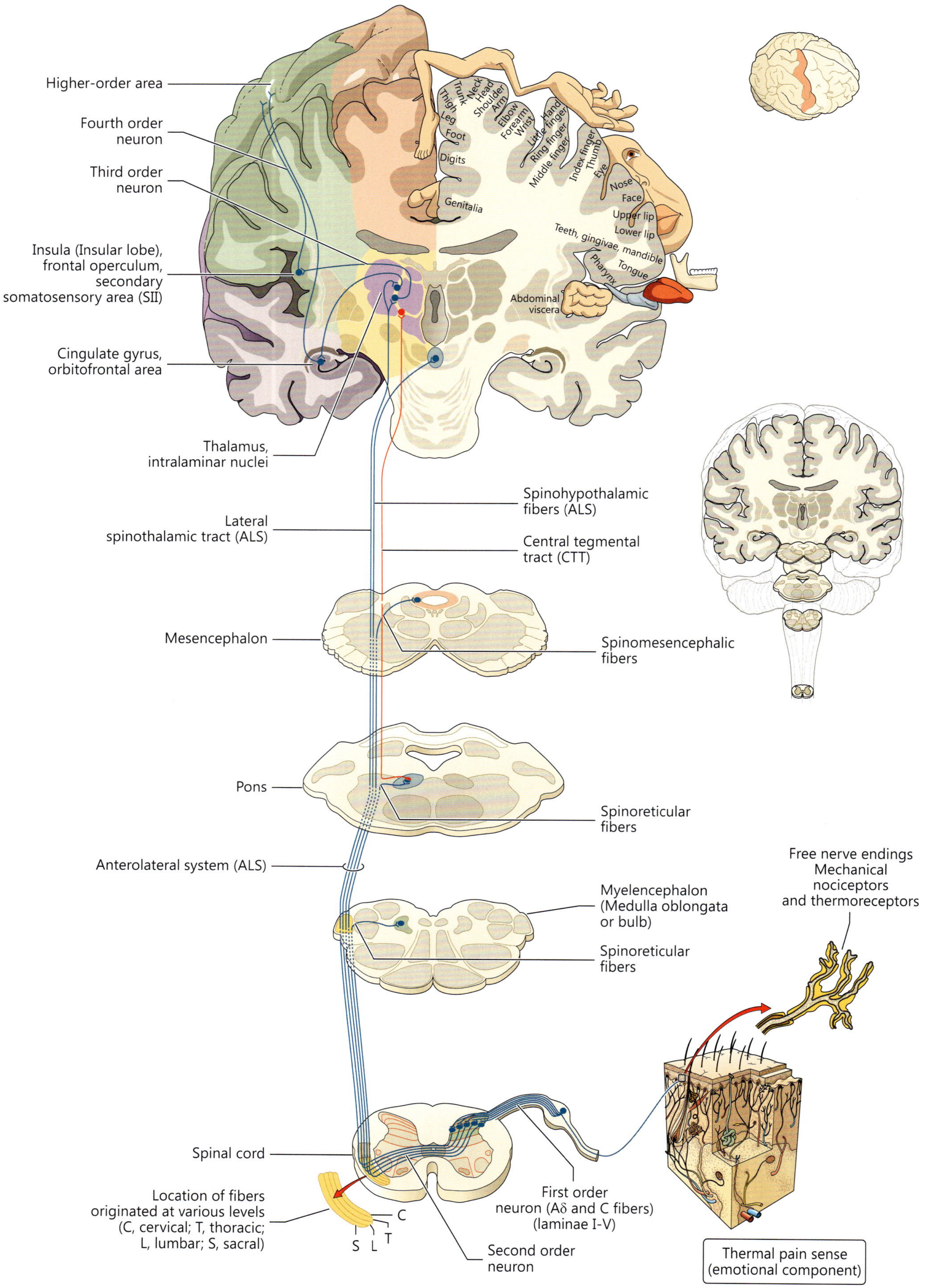
Higher-order area
Fourth order neuron
Third order neuron
Insula (Insular lobe), frontal operculum, secondary somatosensory area (SII)
Cingulate gyrus, orbitofrontal area
Thalamus, intralaminar nuclei
Toes
Genitalia
Digits
Foot
Leg
Thigh
Trunk
Neck
Head
Shoulder
Arm
Elbow
Forearm
Wrist
Hand
Little finger
Ring finger
Middle finger
Index finger
Thumb
Eye
Nose
Face
Upper lip
Lower lip
Teeth, gingivae, mandible
Tongue
Pharynx
Abdominal viscera
Spinohypothalamic fibers (ALS)
Lateral spinothalamic tract (ALS)
Central tegmental tract (CTT)
Mesencephalon
Spinomesencephalic fibers
Pons
Spinoreticular fibers
Anterolateral system (ALS)
Myelencephalon (Medulla oblongata or bulb)
Spinoreticular fibers
Free nerve endings Mechanical nociceptors and thermoreceptors
Spinal cord
Location of fibers originated at various levels (C, cervical; T, thoracic; L, lumbar; S, sacral)
C
T
L
S
First order neuron (Aδ and C fibers) (laminae I-V)
Second order neuron
Thermal pain sense (emotional component)

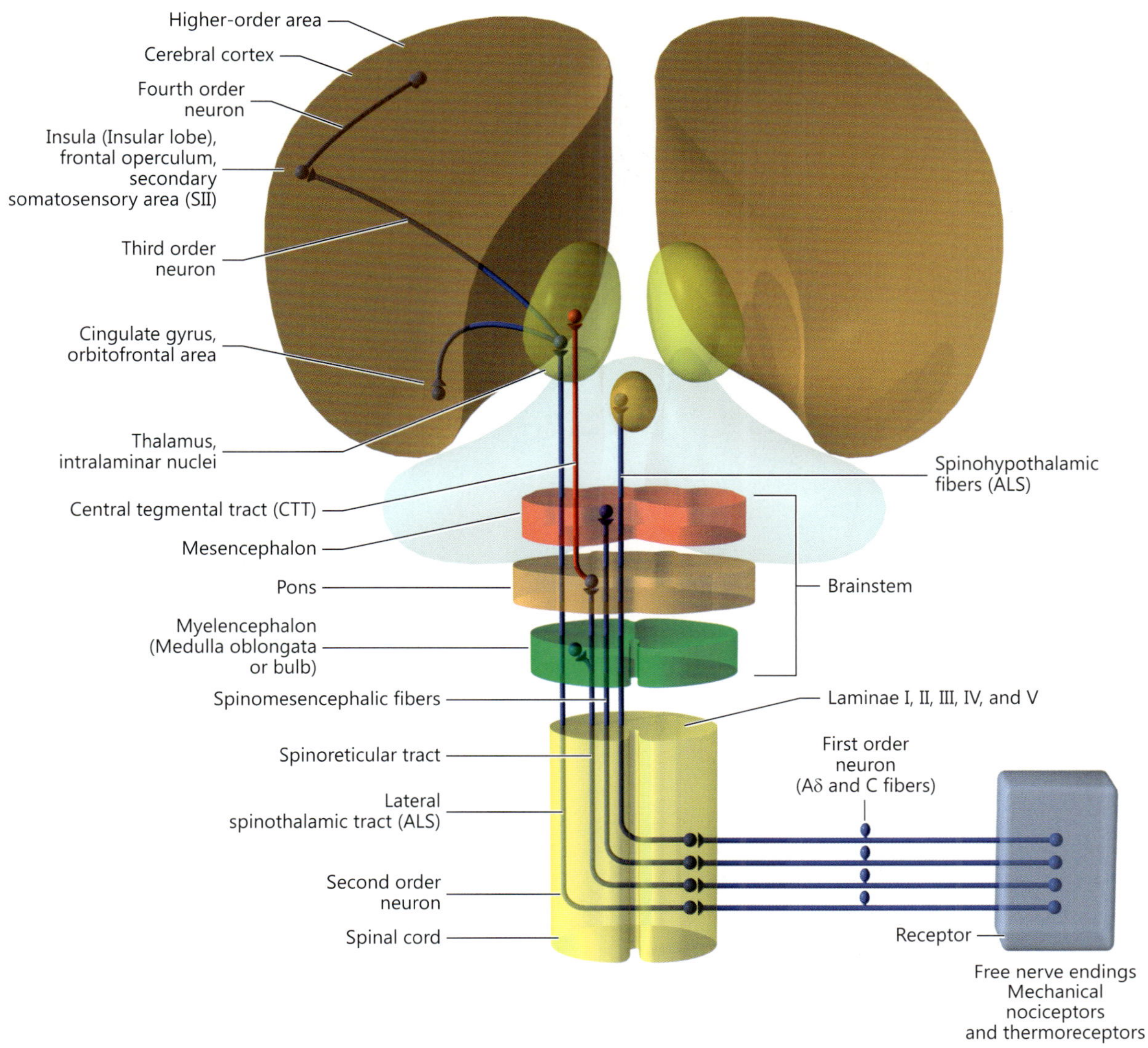

Anterolateral system - Lateral spinothalamic tract (medial compartment) or spinal lemniscus	
Sense	Nociceptive (dolorific and thermal), affective-motivational component
First order neuron (pseudounipolar neuron)	From the mechanical, thermal and polymodal receptors (free nerve endings) to the laminae I and V of the spinal cord
Second order neuron (deutoneuron)	From the laminae I and V of the spinal cord it crosses the midline, it runs in the lateral funiculus and goes up along the brainstem (anterolateral system, ALS) until: – intralaminar nuclei and median nuclei of the thalamus (lateral spinothalamic tract) – hypothalamus (spinohypothalamic fibers) – mesencephalon: central gray matter and superior colliculus (spinomesencephalic fibers) – reticular nuclei of the myelencephalon and pons (spinoreticular fibers)
Third order neuron (thalamocortical projection neuron)	From the intralaminar nuclei and from the median nuclei of the thalamus to the insular and orbitofrontal cortex
Fourth order neuron	From the insula (insular lobe), from the secondary somatosensory area (SII), from the orbitofrontal area and the cingulate gyrus to the higher-order areas

The **medial compartment of the lateral spinothalamic tract**, so named because it reaches the intralaminar nuclei and the median nuclei of the thalamus, is responsible for the nociceptive (dolorific and thermal), affective-motivational components. The medial component is formed by the lateral spinothalamic tract that reaches the thalamus, as spinal lemniscus, by spinohypothalamic fibers, spinomesencephalic fibers (central gray matter and superior colliculi) and spinoreticular fibers, directed to the nuclei of the bulbopontine reticular formation. The component of the spinomesencephalic fibers that reach the central gray matter triggers the mechanism of pain control.

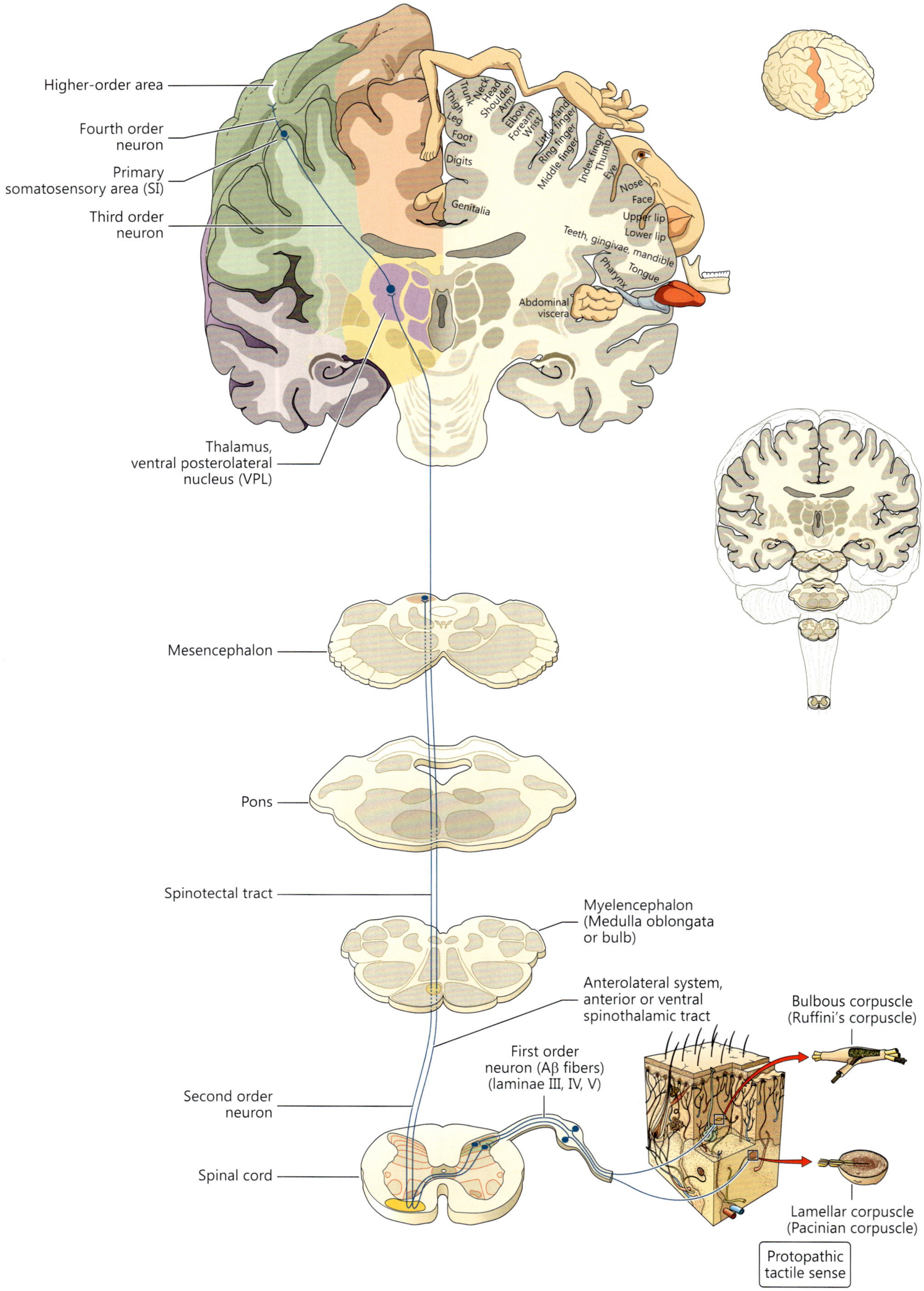
Higher-order area
Fourth order neuron
Primary somatosensory area (SI)
Third order neuron
Thalamus, ventral posterolateral nucleus (VPL)
Neck
Head
Trunk
Shoulder
Arm
Thigh
Leg
Elbow
Forearm
Wrist
Hand
Little finger
Ring finger
Middle finger
Foot
Digits
Index finger
Thumb
Eye
Nose
Face
Genitalia
Upper lip
Lower lip
Teeth, gingivae, mandible
Tongue
Pharynx
Abdominal viscera
Mesencephalon
Pons
Spinotectal tract
Myelencephalon (Medulla oblongata or bulb)
Anterolateral system, anterior or ventral spinothalamic tract
Bulbous corpuscle (Ruffini's corpuscle)
First order neuron (Aβ fibers) (laminae III, IV, V)
Second order neuron
Spinal cord
Lamellar corpuscle (Pacinian corpuscle)
Protopathic tactile sense

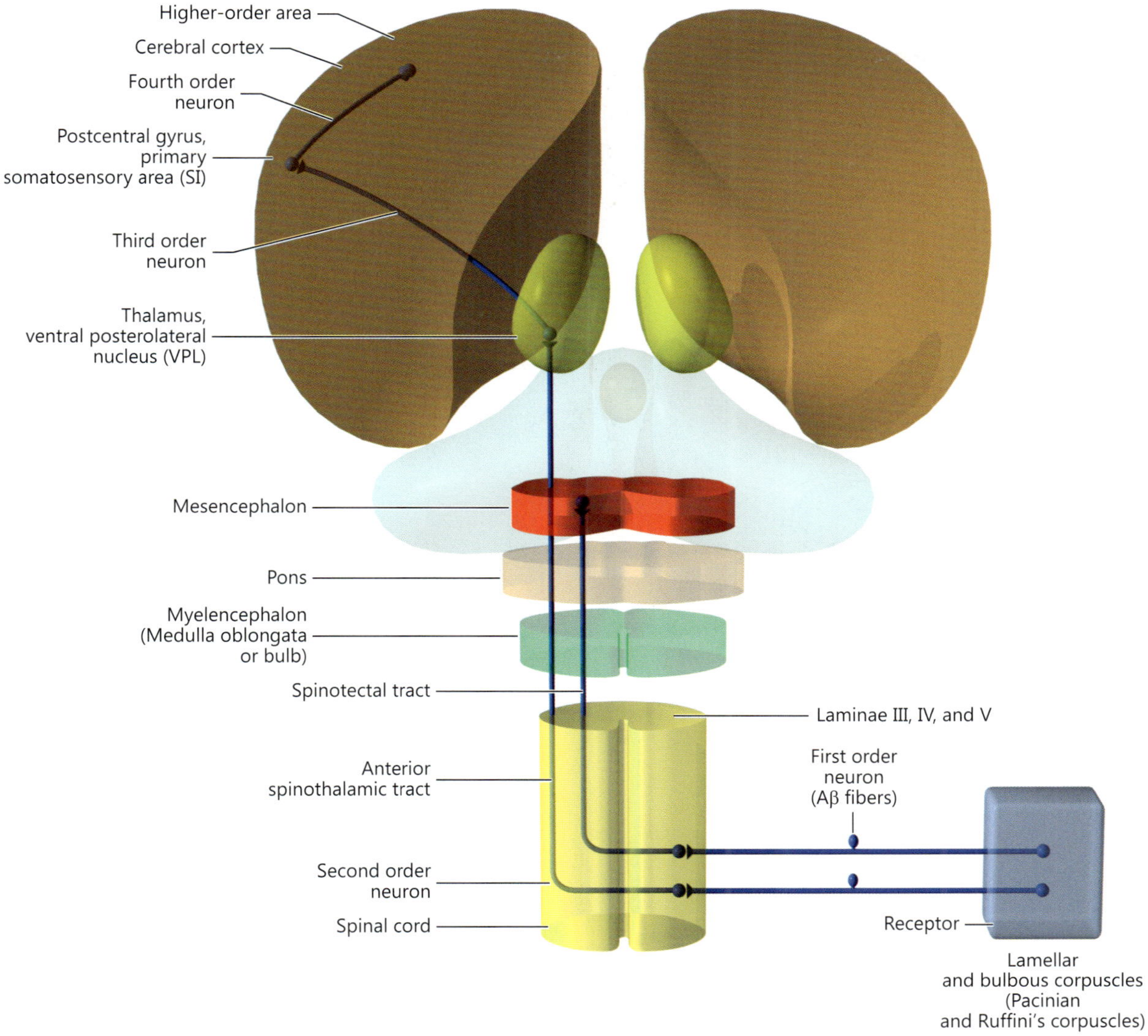

Anterolateral system - Ventral spinothalamic and spinotectal tracts or spinal lemniscus	
In the anterior compartment it is also included the spino-olivary tract that proceeds from muscular proprioceptors and reaches the inferior olive (☞ Inferior Olive, page 520)	
Sense	Protopathic tactile (deep or crude tactile)
First order neuron (pseudounipolar neuron)	From cutaneous mechanoreceptors, lamellar and bulbous corpuscles (Pacinian and Ruffini's corpuscles), mainly to the laminae from III to V of the spinal cord
Second order neuron (deutoneuron)	From the laminae III and V of the spinal cord it crosses the midline, it runs in the anterior funiculus (anterior spinothalamic tract) and goes up along the brainstem (spinal lemniscus) until the ventral posterolateral nucleus (VPL) of the thalamus
Third order neuron (thalamocortical projection neuron)	From the VPL nucleus of the thalamus to the primary somatosensory area (SI)
Fourth order neuron	From the primary somatosensory area (SI) to the higher-order areas

The **anterior compartment of the anterolateral system** is occupied by the ventral spinothalamic tract, which travels together with the spinotectal tract (spinotectal tract that runs in the lateral funiculus) destined to the superior colliculus. Both carry information about the crude touch and in particular the spinotectal tract plays an important role in the multi-sensory integrative function performed by the superior colliculus.

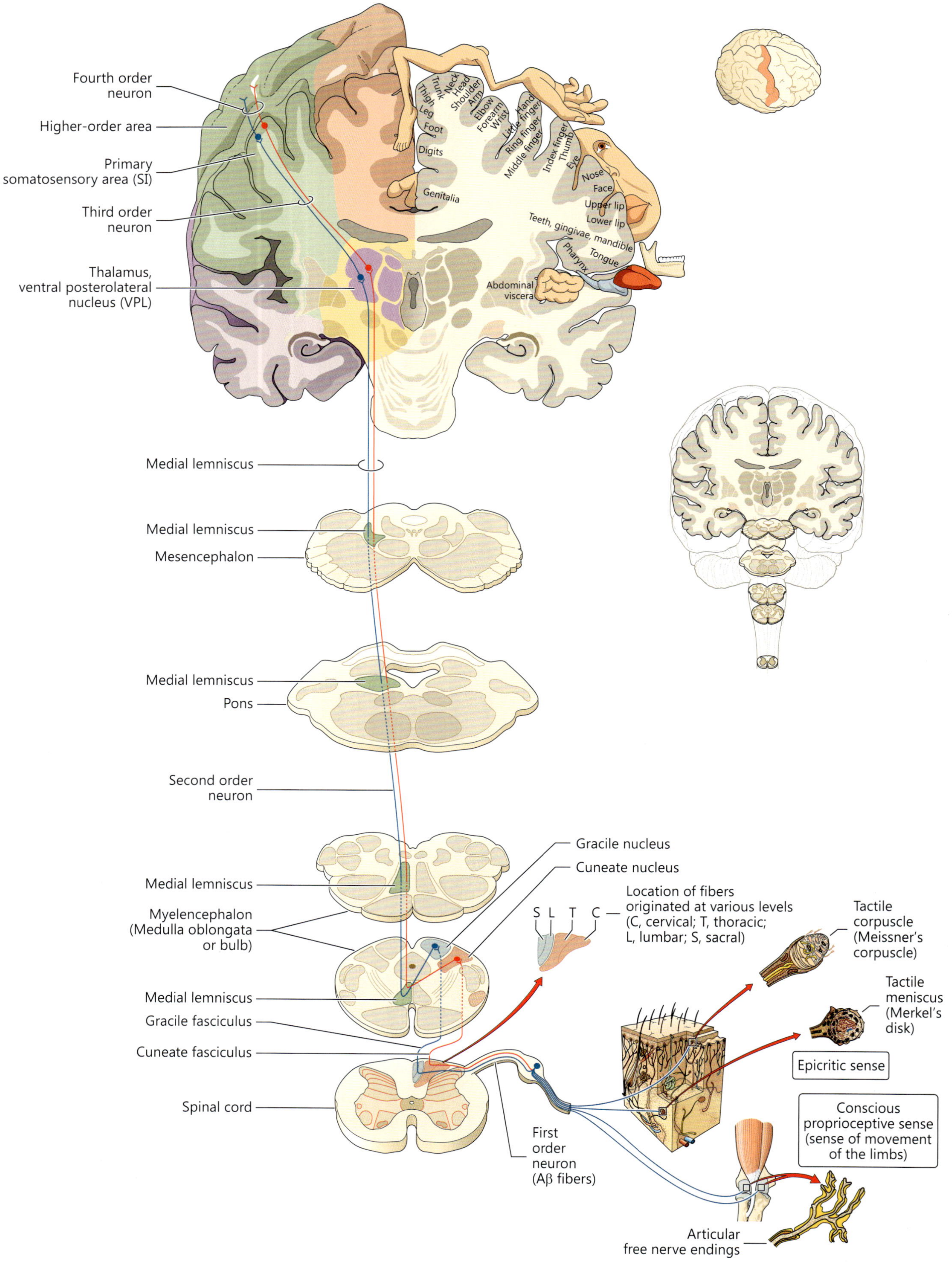
Fourth order neuron
Higher-order area
Primary somatosensory area (SI)
Third order neuron
Thalamus, ventral posterolateral nucleus (VPL)
Neck
Head
Trunk
Shoulder
Thigh
Arm
Leg
Elbow
Forearm
Wrist
Hand
Little finger
Foot
Ring finger
Middle finger
Index finger
Thumb
Digits
Eye
Nose
Face
Genitalia
Upper lip
Lower lip
Teeth, gingivae, mandible
Pharynx
Tongue
Abdominal viscera
Medial lemniscus
Medial lemniscus
Mesencephalon
Medial lemniscus
Pons
Second order neuron
Gracile nucleus
Cuneate nucleus
Medial lemniscus
Location of fibers originated at various levels (C, cervical; T, thoracic; L, lumbar; S, sacral)
S L T C
Tactile corpuscle (Meissner's corpuscle)
Myelencephalon (Medulla oblongata or bulb)
Tactile meniscus (Merkel's disk)
Medial lemniscus
Gracile fasciculus
Cuneate fasciculus
Epicritic sense
Spinal cord
Conscious proprioceptive sense (sense of movement of the limbs)
First order neuron (Aβ fibers)
Articular free nerve endings

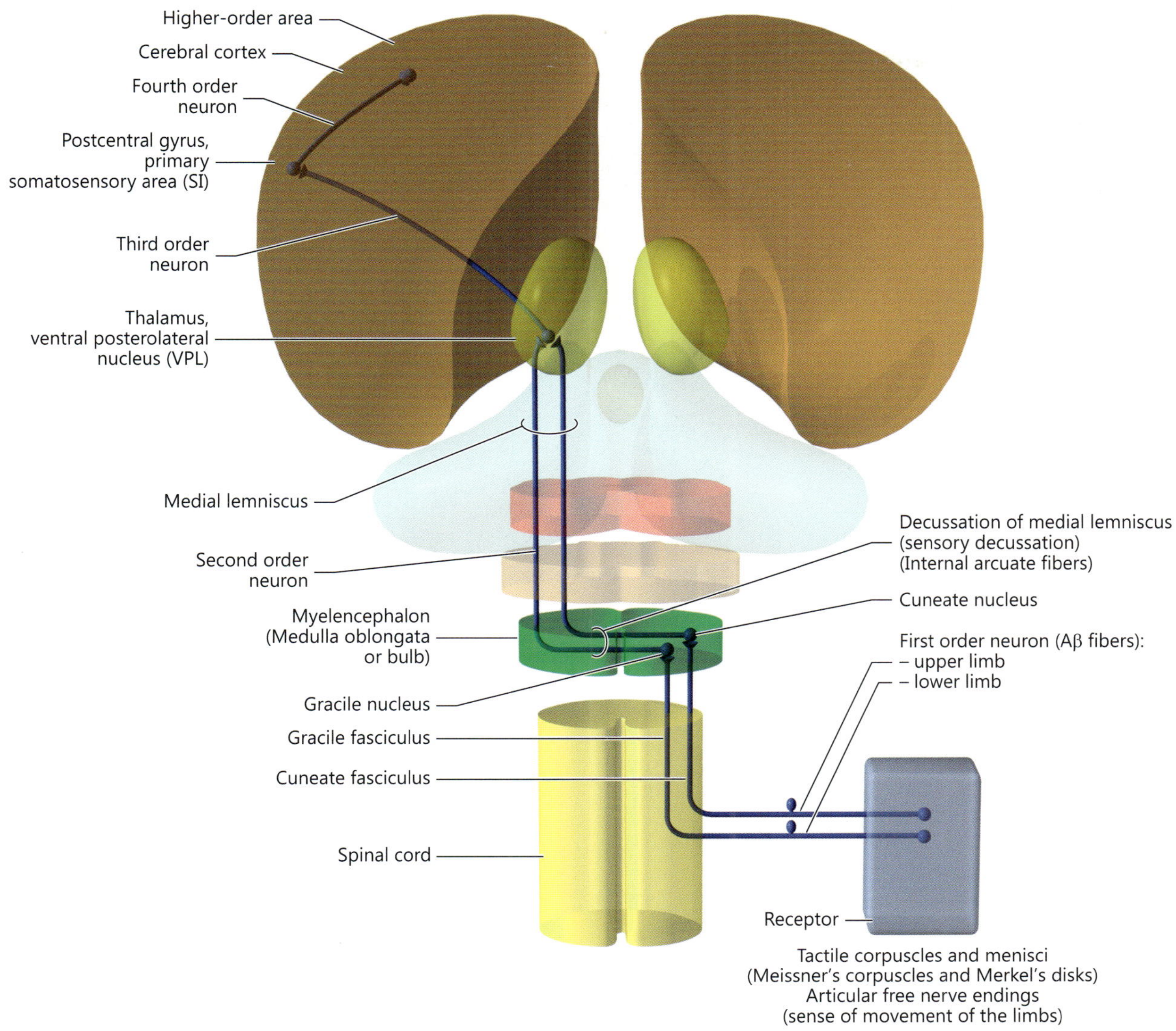

Gracile and cuneate fasciculi (system of the dorsal columns) - Medial lemniscus	
Sense	Epicritical tactile and conscious proprioceptive (motion sensitivity of the limbs or kinesthetic sensitivity)
First order neuron (pseudounipolar neuron)	From the Meissner's corpuscles, the Merkel's disks and the free articular nerve endings directly to the posterior funiculus of the spinal cord from where they go up homolaterally (gracile and cuneate fasciculi) till the gracile and cuneate nuclei of the bulb
Second order neuron (deutoneuron)	From the gracile and cuneate nuclei of the bulb, the deutoneurons cross the midline and go up contralaterally, forming the medial lemniscus (ML), until the ventral posterolateral nucleus (VPL) of the thalamus (ML is somatotipically organized)
Third order neuron (thalamocortical projection neuron)	From the VPL nucleus of the thalamus to the primary somatosensory area (SI)
Fourth order neuron	From the primary somatosensory area (SI) to the higher-order areas

The **gracile and cuneate fasciculi** form part of the so-called **system of the dorsal columns - medial lemniscus**, responsible for conducting the epicritical tactile sense and the conscious proprioceptive sense (motion sensitivity of the limbs or kinesthetic sensitivity). The characteristic of this system is the homolateral distribution of the centripetal extension of the ganglionic cells, that arrange themselves homolaterally in the posterior funiculus (dorsal columns) of the white matter of the spinal cord (gracile and cuneate fasciculi), and run upward without forming synapses in the spinal cord. The synapsis with the deutoneuron occurs in the bulb at the level of the gracilis and cuneate nuclei: for this reason, these cells are defined bulbopetal neurons and not spinopetal neurons as those forming synapses in the posterior horn of the spinal cord. After crossing the median line, the axons of the deutoneuron form the tract called medial lemniscus that reaches the ventral posterolateral nucleus of the thalamus. As the axons progressively associate with the tract proceeding caudocranially, they arrange themselves starting from the midline gradually occupying more and more lateral regions. Moreover, the fibers of tactile sense are located more deeply compared to those of kinesthetic sense which arrange themselves more superficially.

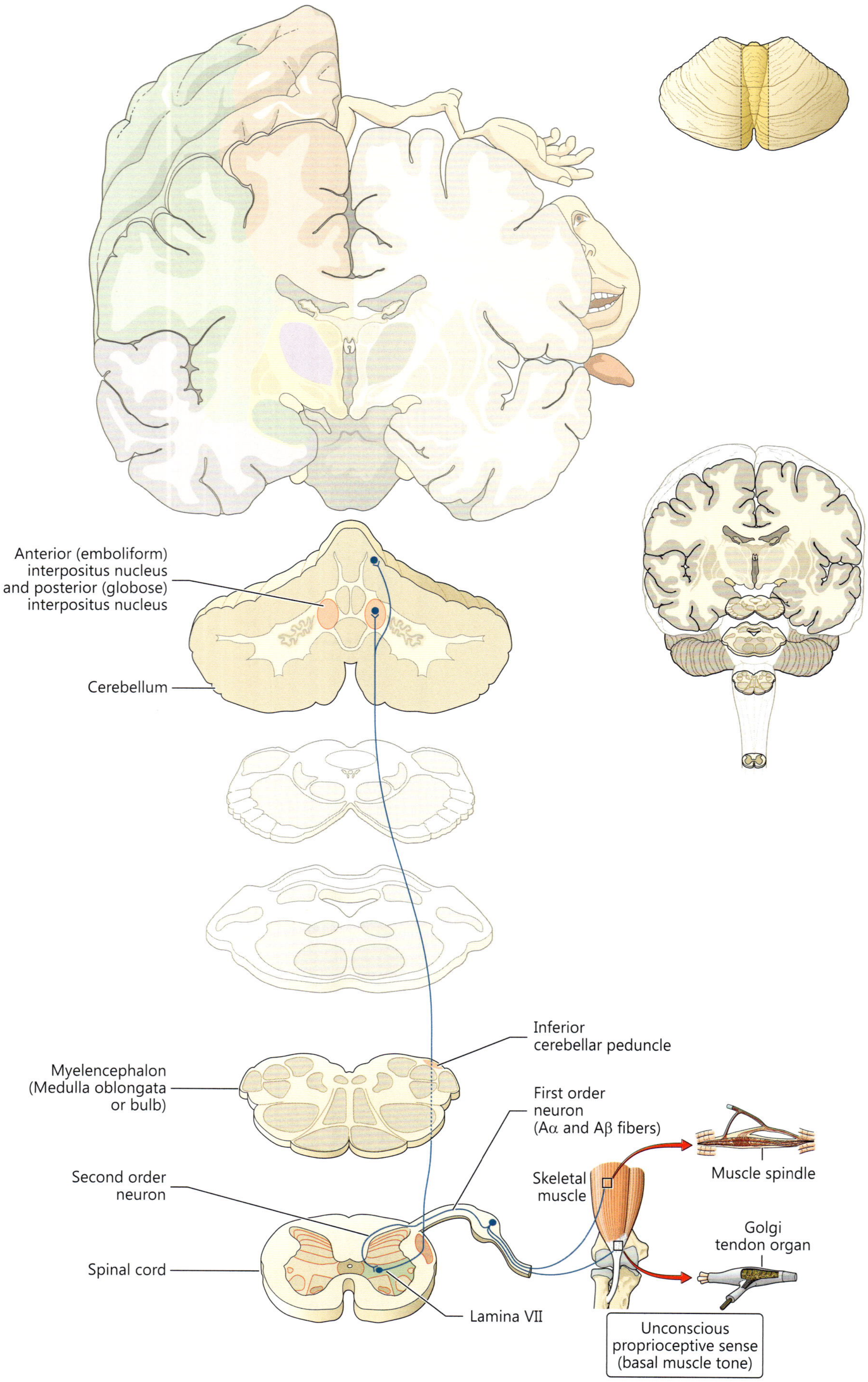

Anterior (emboliform) interpositus nucleus and posterior (globose) interpositus nucleus
Cerebellum
Inferior cerebellar peduncle
Myelencephalon (Medulla oblongata or bulb)
First order neuron (Aα and Aβ fibers)
Muscle spindle
Skeletal muscle
Second order neuron
Golgi tendon organ
Spinal cord
Lamina VII
Unconscious proprioceptive sense (basal muscle tone)

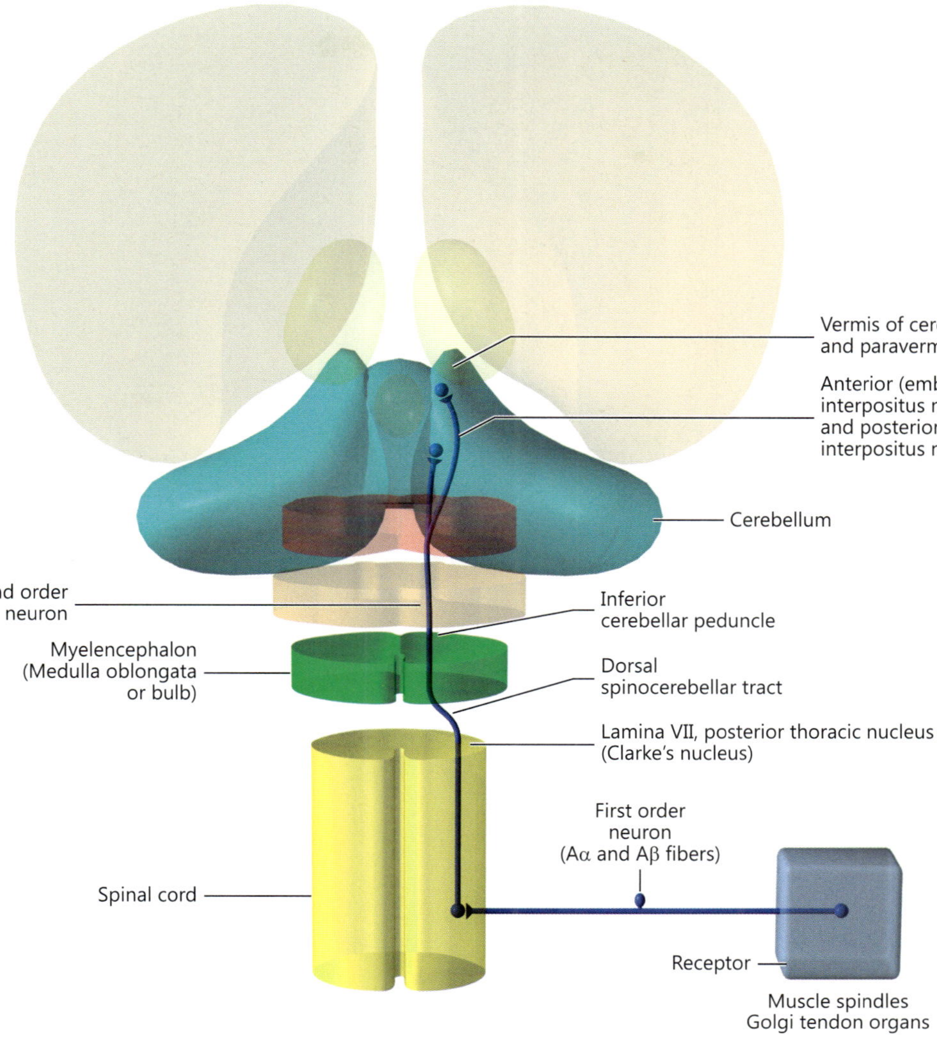

Posterior (dorsal) spinocerebellar tract (or direct or Flechsig's tract)	
Sense	Unconscious proprioceptive (basal muscle tone)
First order neuron (pseudounipolar neuron)	From the neuromuscular spindles and tendon organs (Golgi) to the lamina VII (posterior thoracic nucleus or Clarke's nucleus from T1 to L2)*
Second order neuron (deutoneuron)	From the Clarke's nucleus the deutoneurons send the axon ipsilaterally in the lateral funiculus, from where it runs upward (posterior or dorsal spinocerebellar tract), through the bulb and the inferior cerebellar peduncle, directly to the cerebellar cortex (vermis and paravermian regions)

* This tract carries information from the lower limb and trunk. The same information related to the upper limb reach the cuneate nucleus of the bulb through the cuneate fasciculus and from the cuneate nucleus they reach the same regions of the cerebellar cortex forming the cuneocerebellar fibers.

The **dorsal** or direct **spinocerebellar tract** is part of the spinocerebellar system and is appointed to conduct the unconscious proprioception sense (basal muscle tone or sense of limbs positioning). This pathway is bineuronal as the tract runs upward from the spinal cord to the bulb and from here through the inferior cerebellar peduncle it reaches the cerebellar cortex. The fibers of this tract receive information sometimes also from few muscles or from a single muscle that function as agonists during movement.

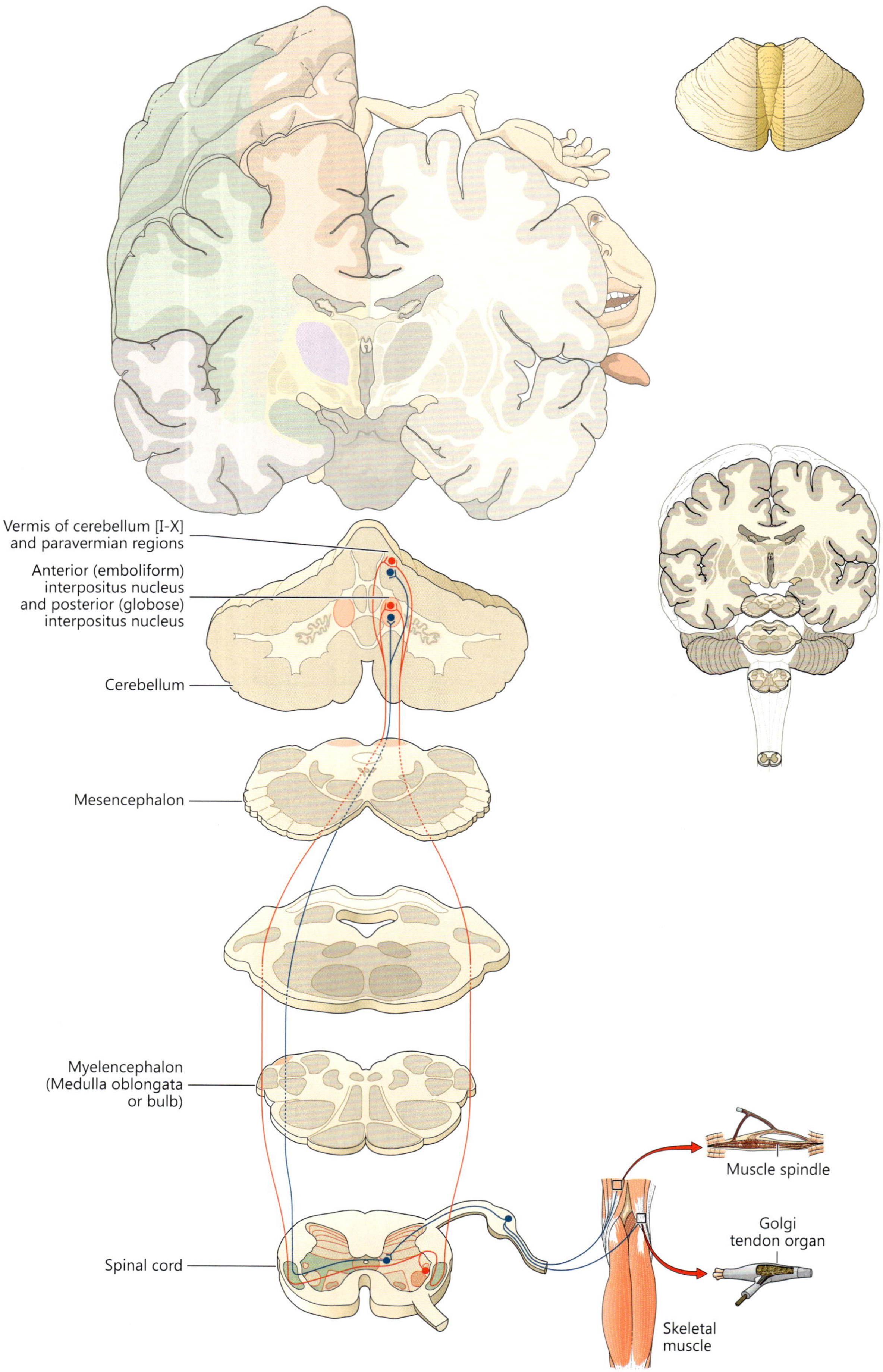
Vermis of cerebellum [I-X] and paravermian regions
Anterior (emboliform) interpositus nucleus and posterior (globose) interpositus nucleus
Cerebellum
Mesencephalon
Myelencephalon (Medulla oblongata or bulb)
Spinal cord
Muscle spindle
Golgi tendon organ
Skeletal muscle

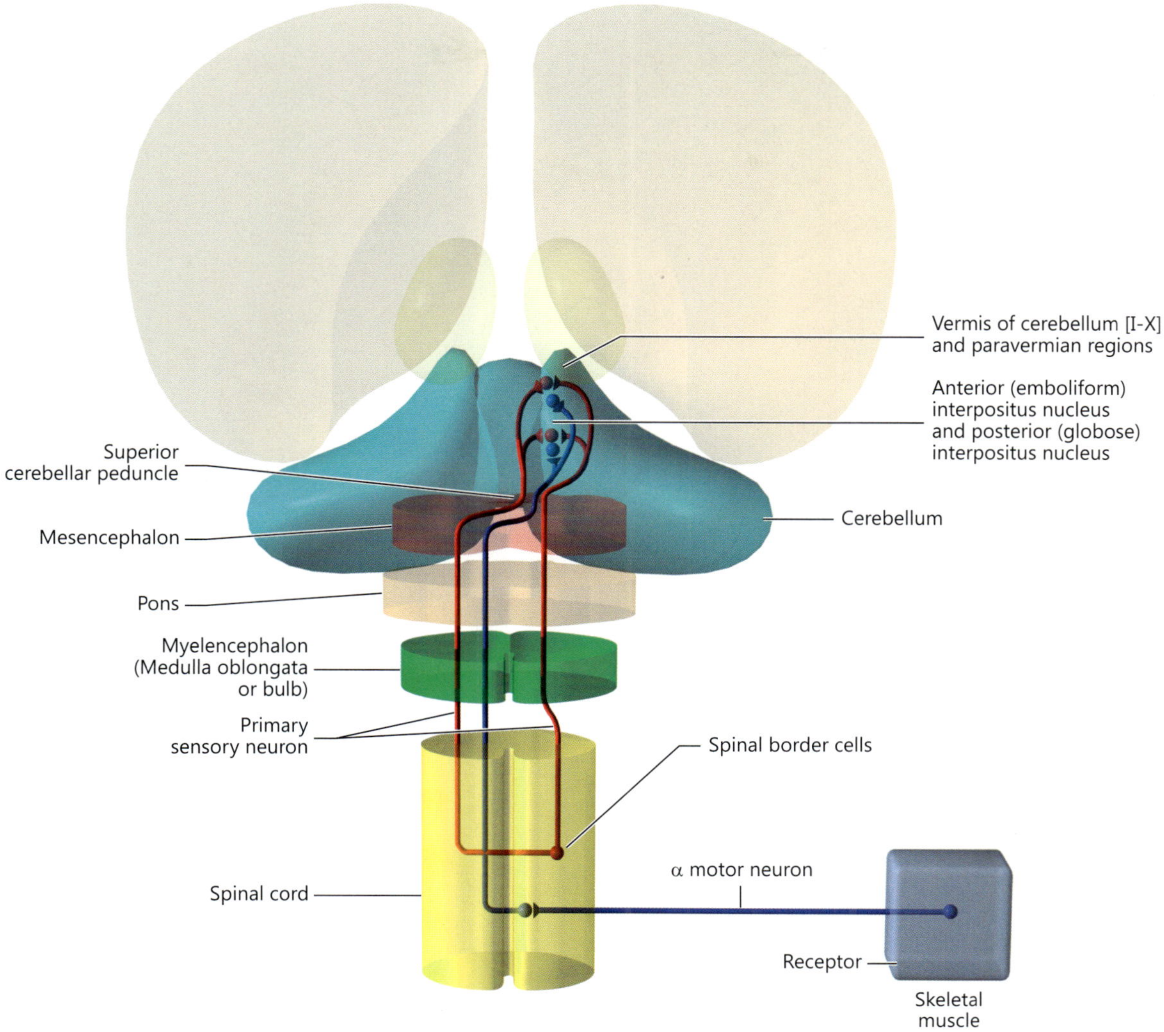

Anterior (or ventral) spinocerebellar tract (or crossed or Gowers's tract)	
Sense	Unconscious proprioception (sense of static position)
First order neuron (pseudounipolar neuron)	From the neuromuscular spindles and tendon organs (Golgi) to the lamina VII (intermediomedial nucleus of Bechterew)
Second order neuron (deutoneuron)	From the intermediomedial nucleus of Bechterew the axons of the deutoneurons cross the median line, reach the contralateral lateral funiculus, where they run upward (anterior or crossed spinocerebellar tract), through the bulb and pons and, once they reach the mesencephalon, they cross the median line for the second time; from here through the superior cerebellar peduncle, they arrive at the cerebellar cortex (vermis and paravermian regions). It has long been known that there is a considerable amount of fibers that originate from cells localized mainly along the lateral margin of the anterior horn (called spinal border cells by Cooper and Sherrington) and which is part of the anterior spinocerebellar tract of both sides

The **ventral** or crossed **spinocerebellar tract** is part of the spinocerebellar system and is appointed to conduct the unconscious proprioception sense (basal muscle tone or sense of limbs positioning). This pathway is bineuronal as the tract runs upward from the spinal cord through the bulb and pons and, after a second crossing in the mesencephalon (decussation of the superior cerebellar peduncles or Wernekink's peduncles), it reaches the cerebellar cortex (*blue* line) through the superior cerebellar peduncle. This tract receives information from large receptive fields (even an entire limb) and seems implicated in postural information. Some authors believe that there is a large body of fibers that originates, through interneurons, directly from the α motor neurons of the spinal cord (*red* line) and informs the cerebellum on the number and type of active motor units.

Pyramidal System - Direct and Crossed Pyramidal Tracts

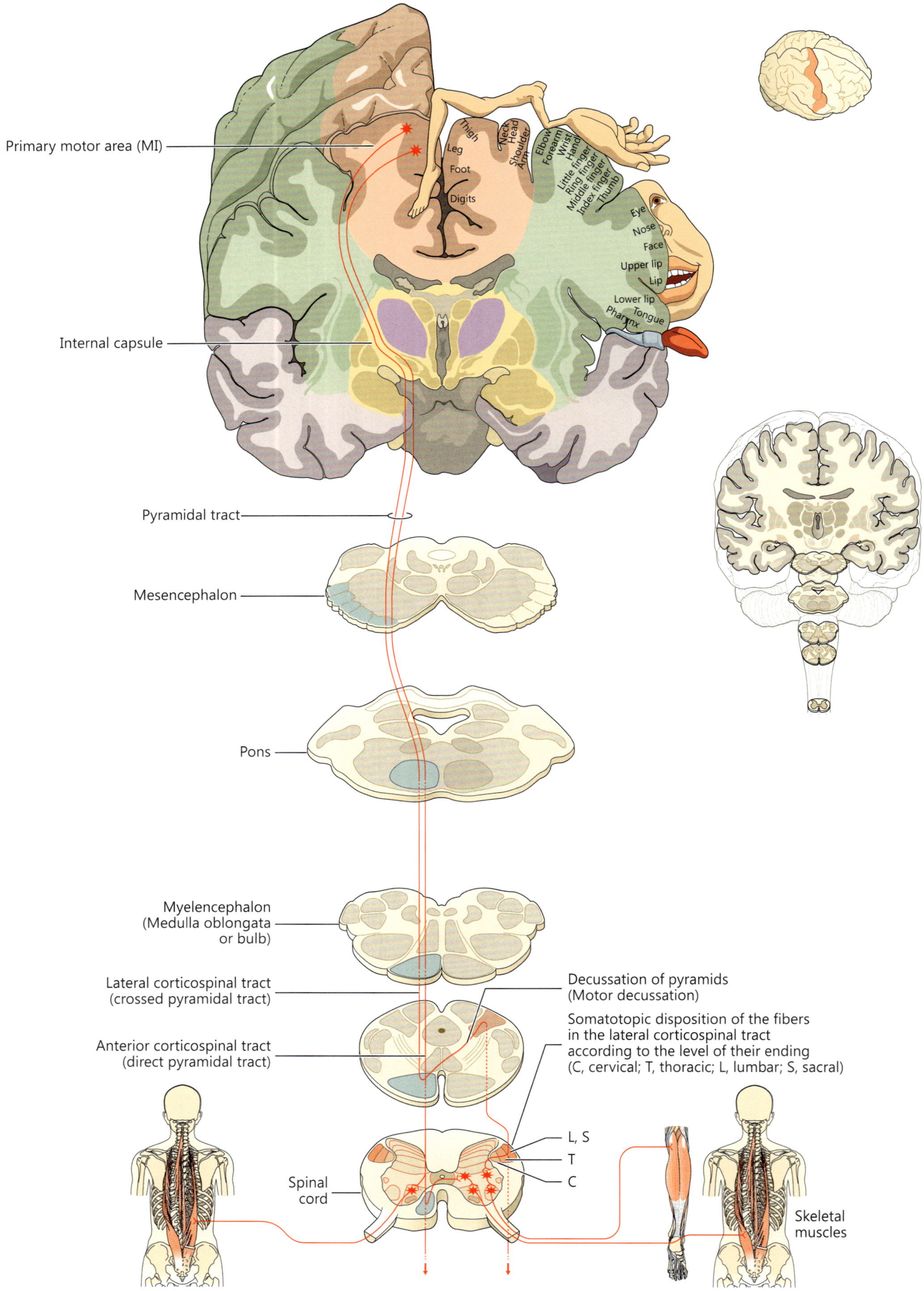

Pyramidal System - Direct and Crossed Pyramidal Tracts

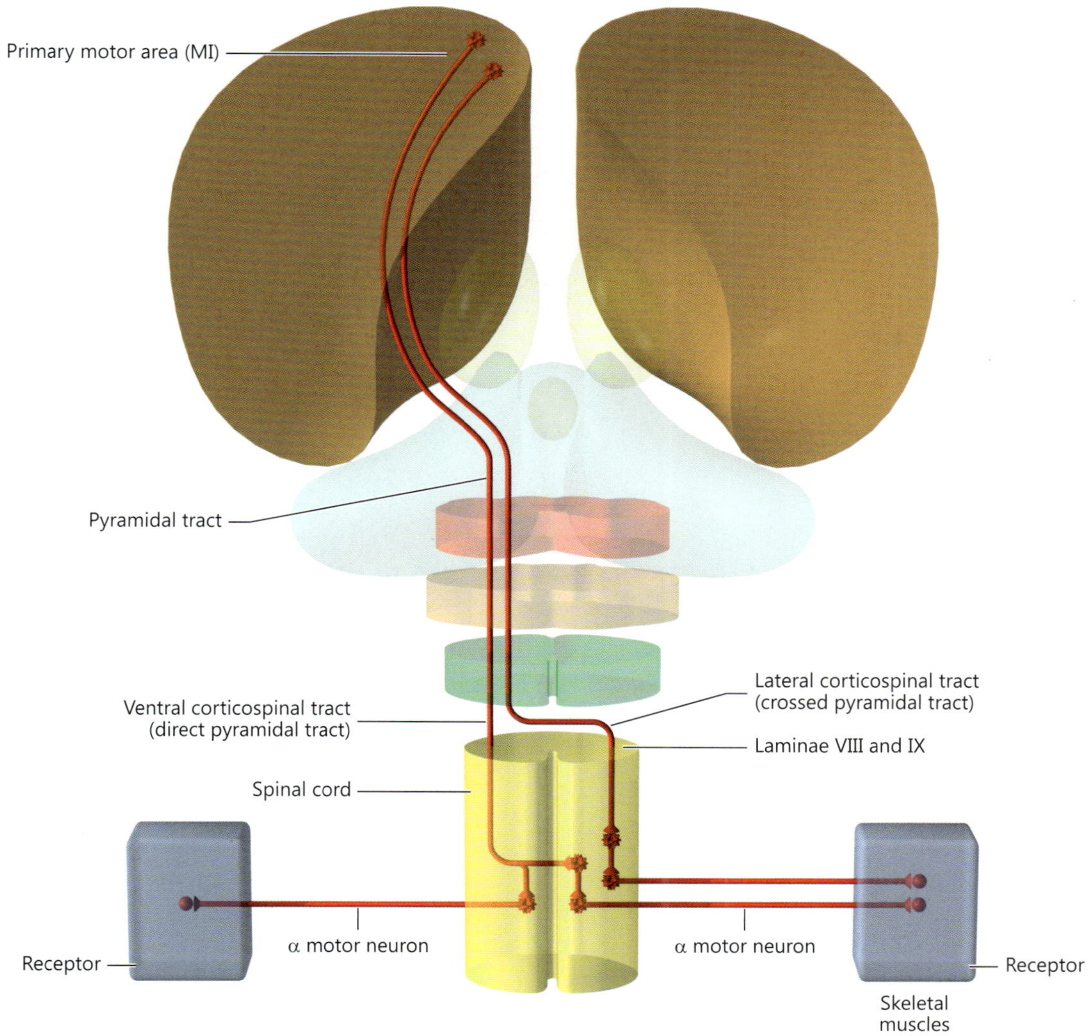

Pyramidal system - Direct and crossed pyramidal tracts	
Upper motor neuron	From the giant pyramidal neurons of the III and V layers of the motor and premotor areas, from somatosensory area, from parietal association area (5, 7 areas) to the α and γ motor neurons of lamina IX of the spinal cord (also through short interneurons present in the lamina VIII)
Lower motor neuron	*Direct pyramidal tracts*: from the α and γ motor neurons of the lamina IX of the anterior homo- and contralateral horn (medial region) of gray matter of the spinal cord to the proximal muscles *Crossed pyramidal tracts*: from the α and γ motor neurons of the lamina IX of the anterior contralateral horn (ventrolateral region) of gray matter of the spinal cord to the distal muscles

The **pyramidal tracts** consist of the motor neuron, or pyramidal neuron, which: originates from the motor and premotor areas, from somatosensory area, from parietal association area (5, 7 areas); crosses the posterior limb of the internal capsule, descends along the cerebral crus and the pons reaching the bulbar pyramids. At the level of the narrow part of the bulb approximately 90% of the axons cross the median line and passes into the lateral funiculus of the opposite side, forming the *crossed pyramidal tracts*, while approximately 10% stays ipsilateral constituting the *lateral corticospinal tracts* (LCST). Both these tracts form synapses with α and γ motor neurons localized in the lamina IX of the spinal cord (the synapsis is almost always mediated by short interneurons located in the lamina VIII).
The axons of the crossed pyramidal tracts form synapses with the motor units of the spinal cord, placed laterally, which innervate the contralateral distal muscles; the axons of the direct pyramidal tracts form synapses with the homolateral and contralateral motor units placed medially, innervating the proximal muscles of both sides.

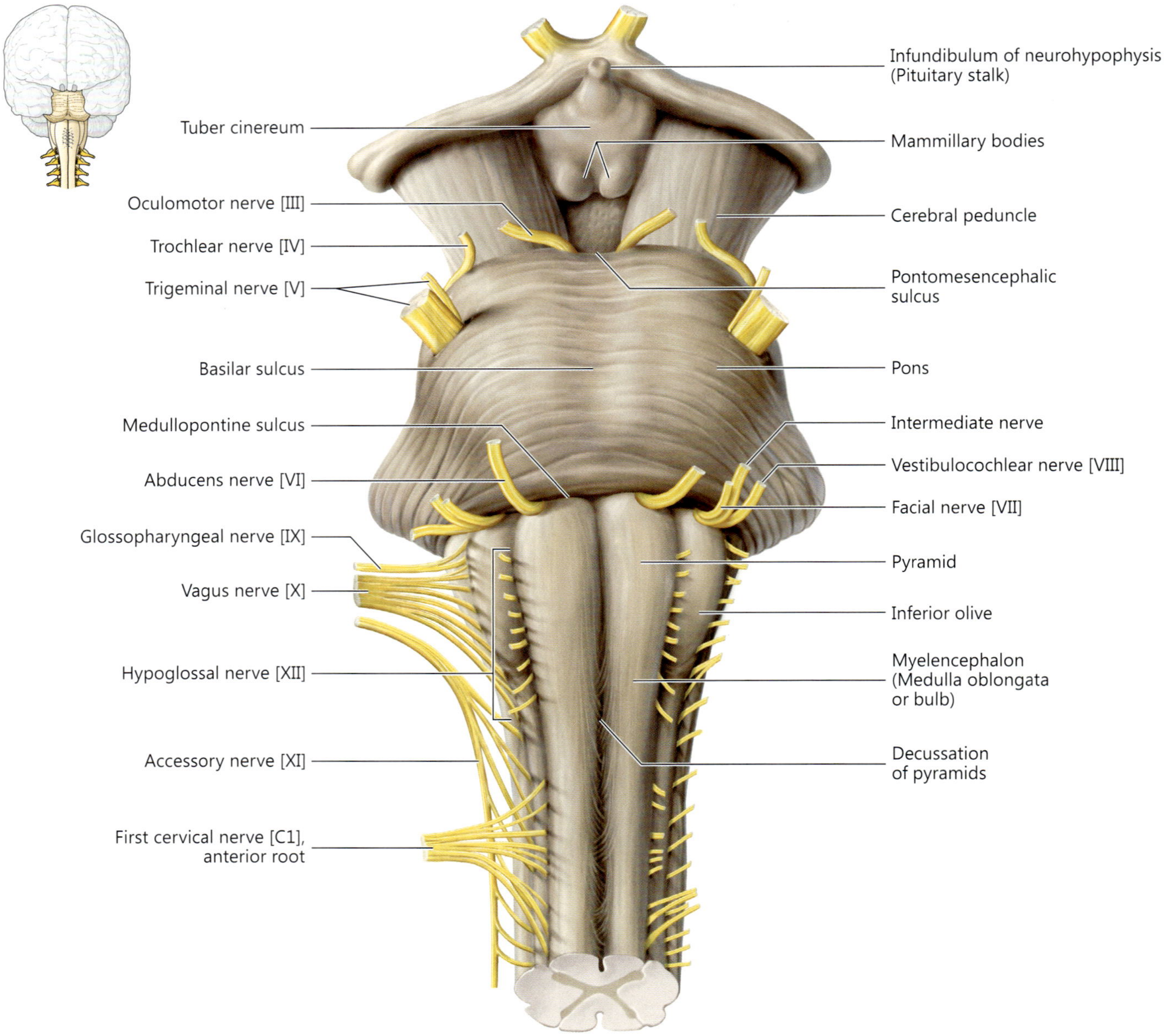

Schematic representation of the ventral surface of the **brainstem**. Are visible the apparent emergences of the cranial nerves of which 9 out of 12 originate from the ventrolateral surface of the brainstem.

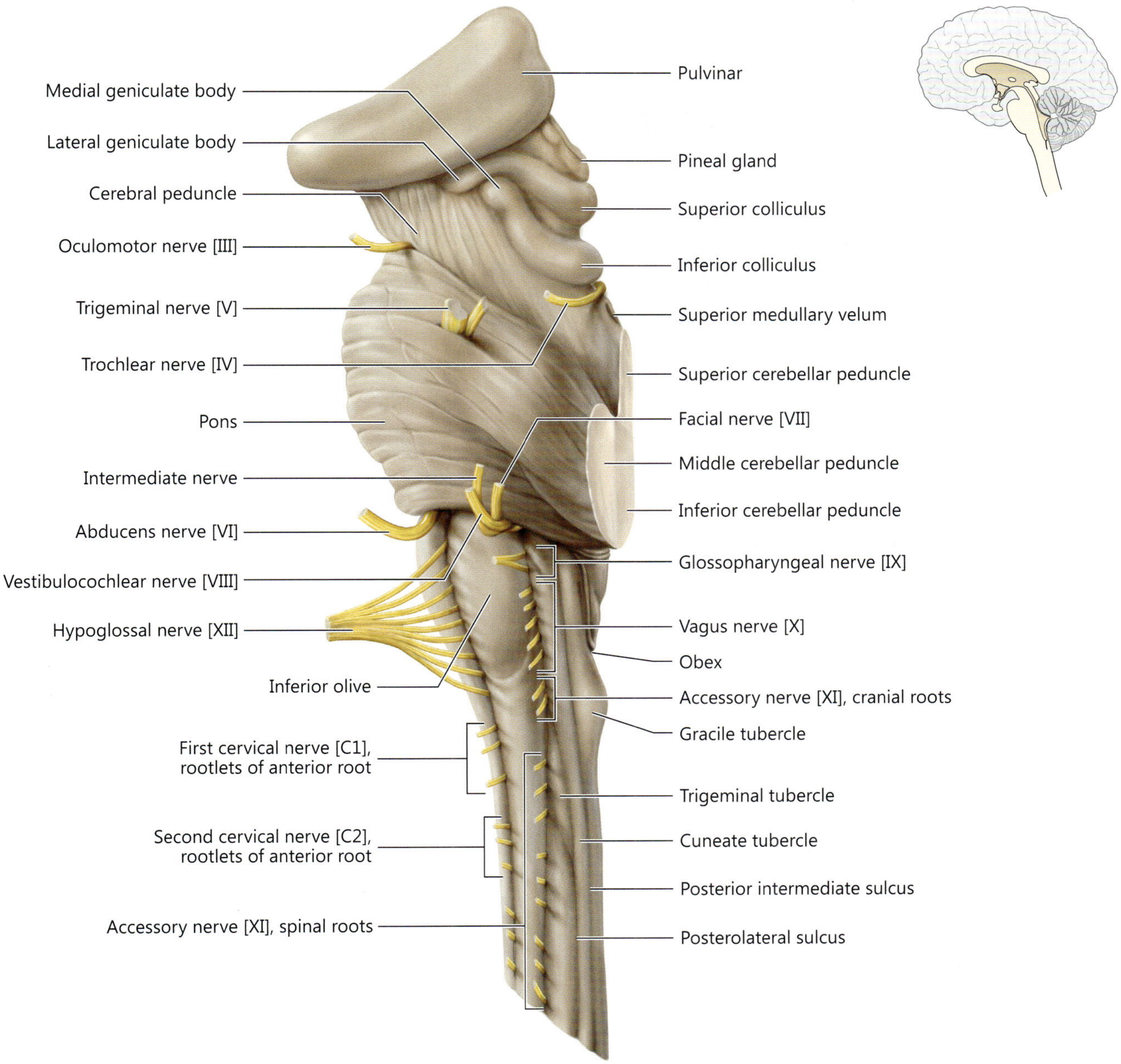

Schematic representation of the external configuration of the **brainstem** in posterolateral projection.

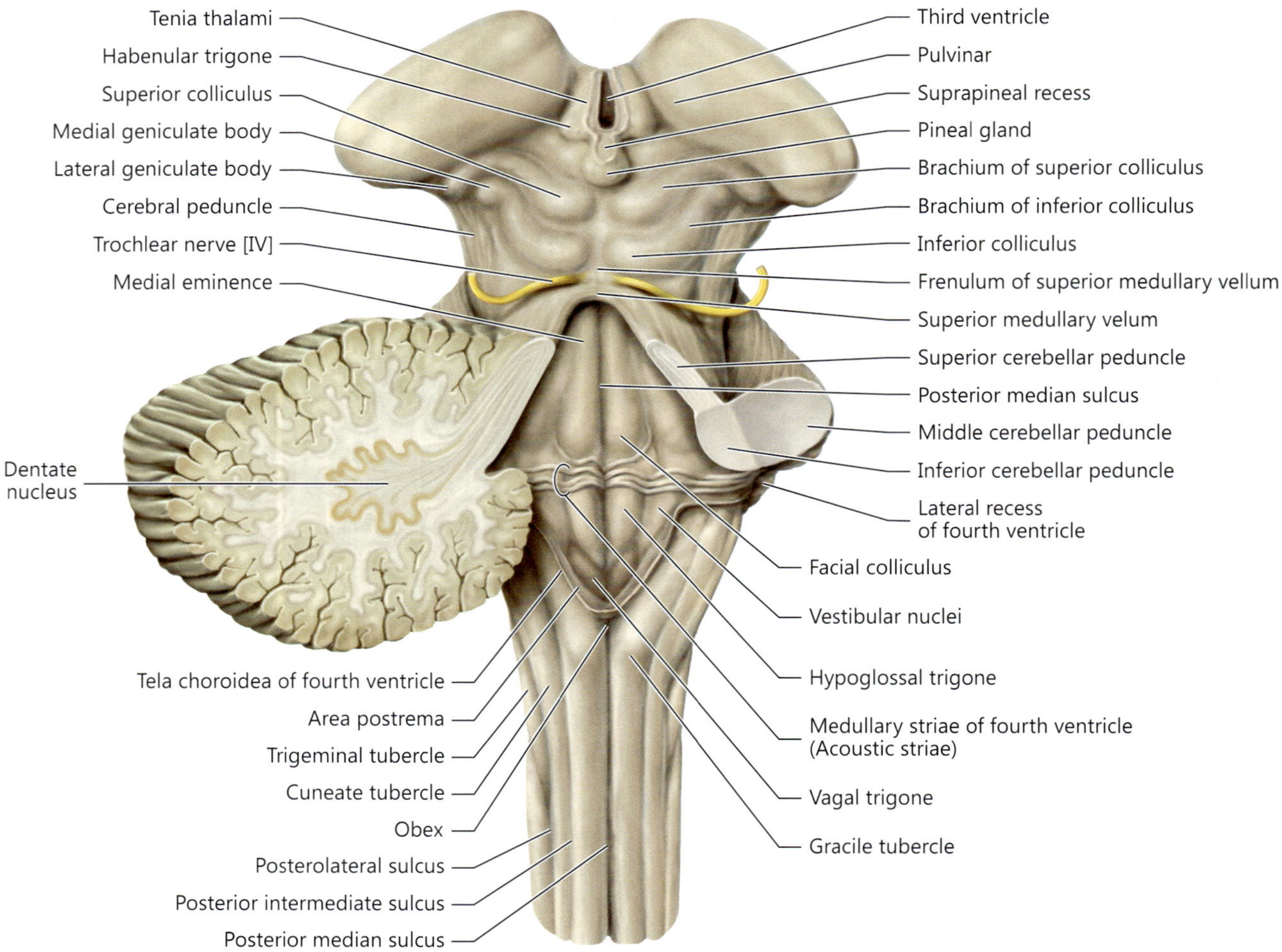

External configuration of the **brainstem**, seen posteriorly after resection of the right cerebellar peduncles and partial removal of the cerebellum, to highlight the rhomboid fossa; the fossa forms the floor of the fourth ventricle and is formed by a lower bulbar triangle and a superior pontine triangle. Only one nerve, the trochlear [IV], emerges from the dorsal surface of the brainstem.

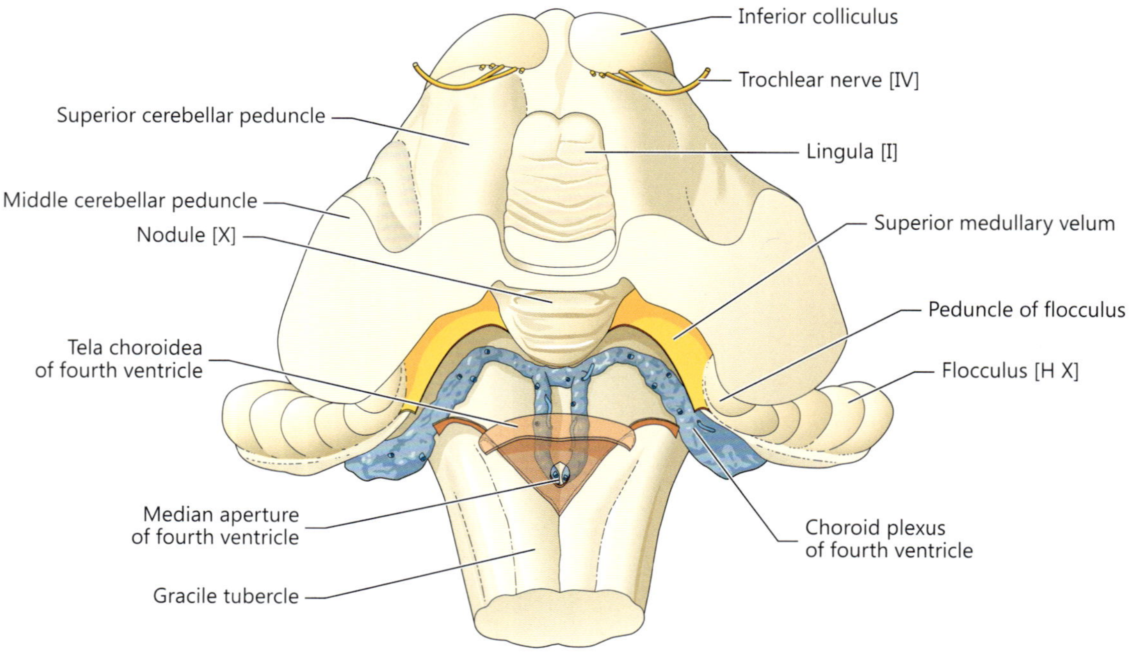

Conformation of the **roof of the fourth ventricle**, prepared by extensive demolition of the cerebellum. The superior medullary velum is covered by the lingual [I]. The tela choroidea of the fourth ventricle is indicated with *red*; the choroid plexuses annexed to the roof of the ventricle itself are indicated with *blue*.

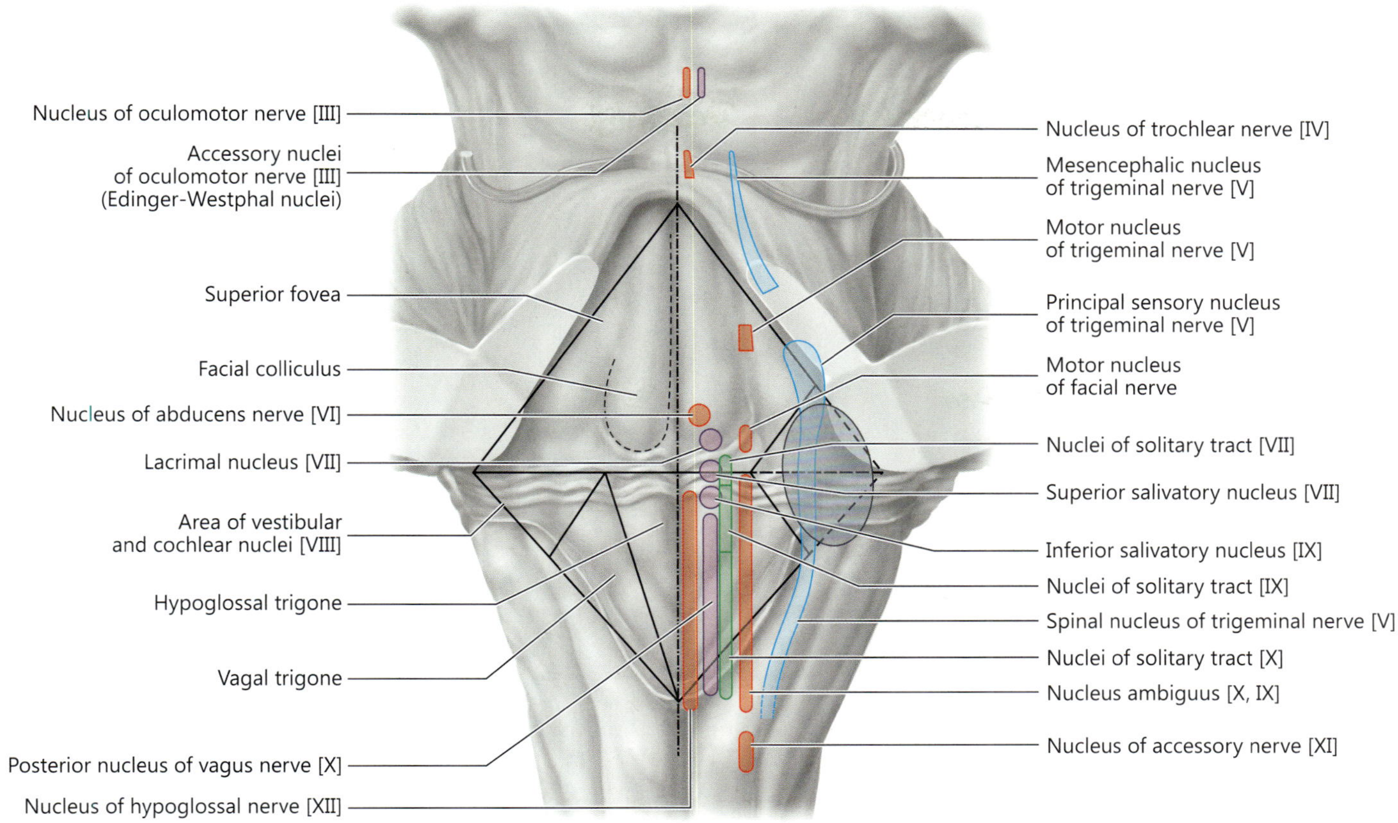

Schematic representation of the external configuration of the **rhomboid fossa**, seen posteriorly after resection of cerebral peduncles and removal of the cerebellum. The rhomboid fossa is divided into two triangles, superior, or pontine, and inferior, or bulbar. Here are represented the **nuclei of the cranial nerves** placed ventrally on the floor of the fourth ventricle. *Light blue*, the general sensory somatic nuclei; *green*, the general and special visceral sensory nuclei; *red* the somatic motor nuclei and *purple* the general and special visceral motor nuclei. In the circumscribed oval area and corresponding to the lateral area of the bulbar and pontine triangles are located the special somatic sensory nuclei.

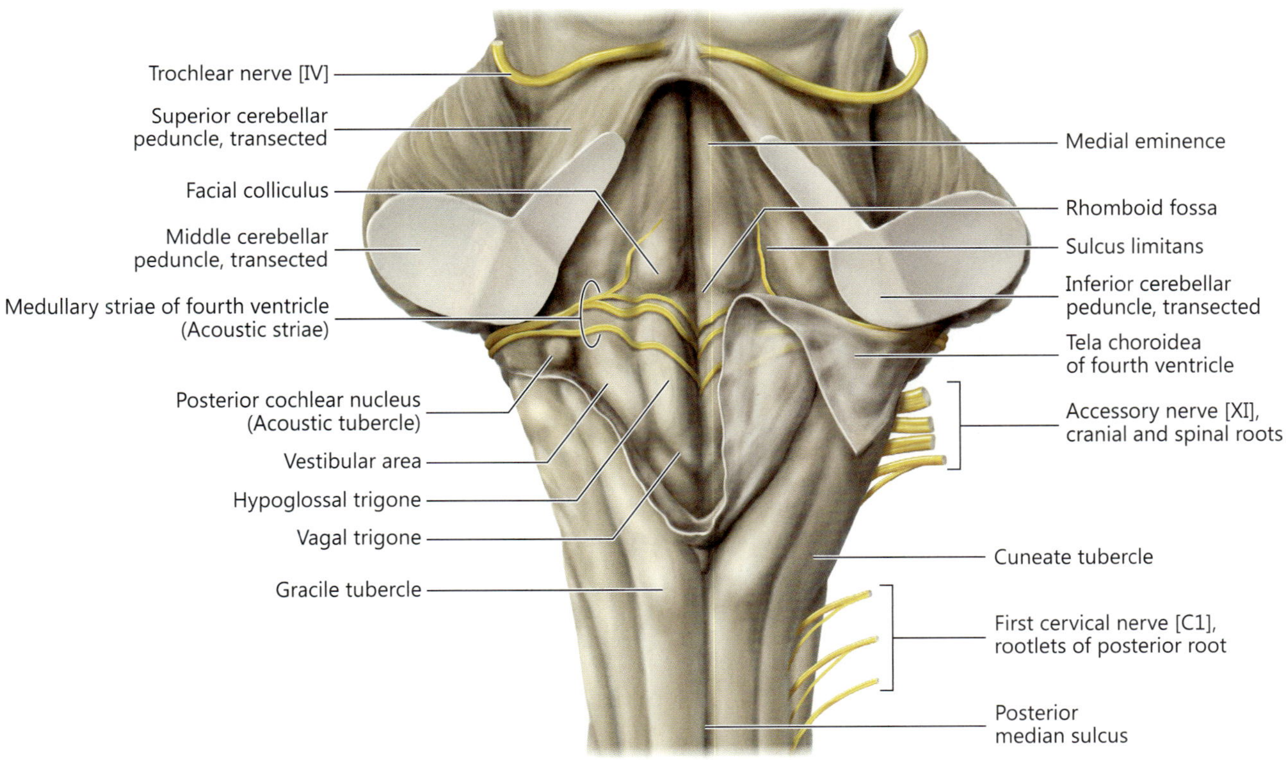

Schematic representation of the external configuration of the **brainstem** in posterior projection showing, in details, the superior, or pontine, triangle and the inferior, or bulbar, triangle of the fourth ventricle.

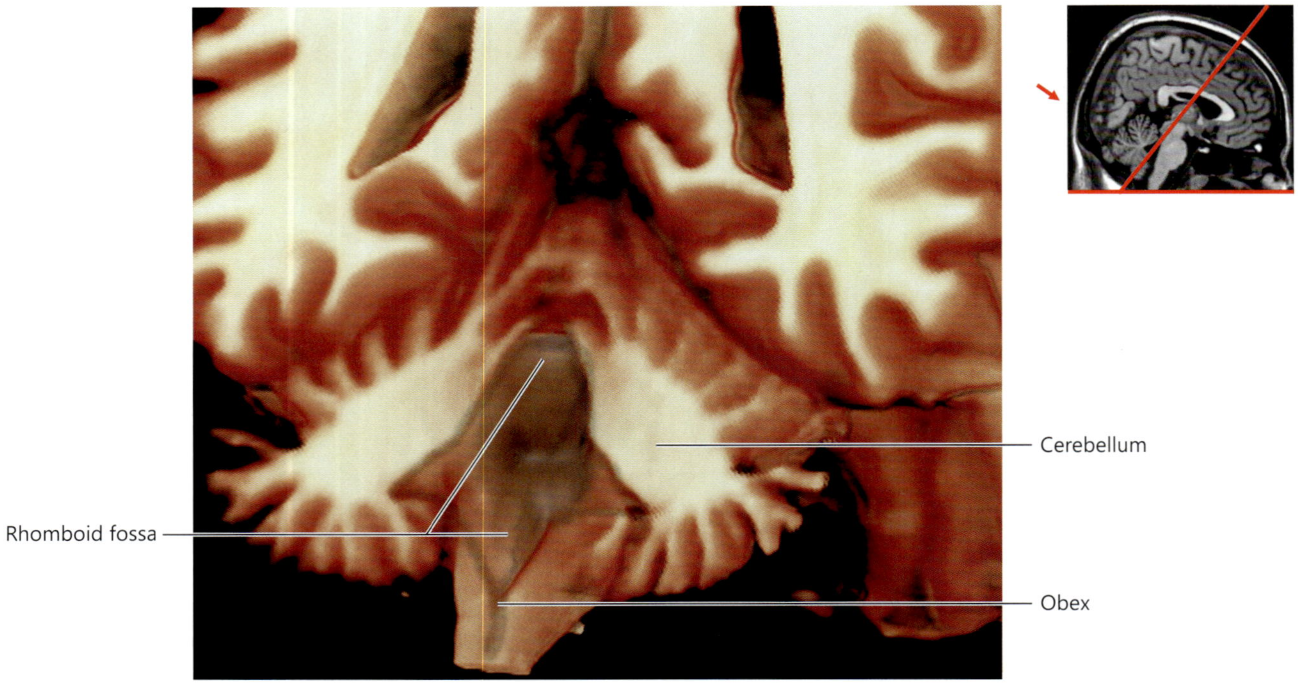

Three-dimensional image of a living human body, obtained by NMR using a 3D volume rendering technique, showing the **fourth ventricle** in a posterior view.

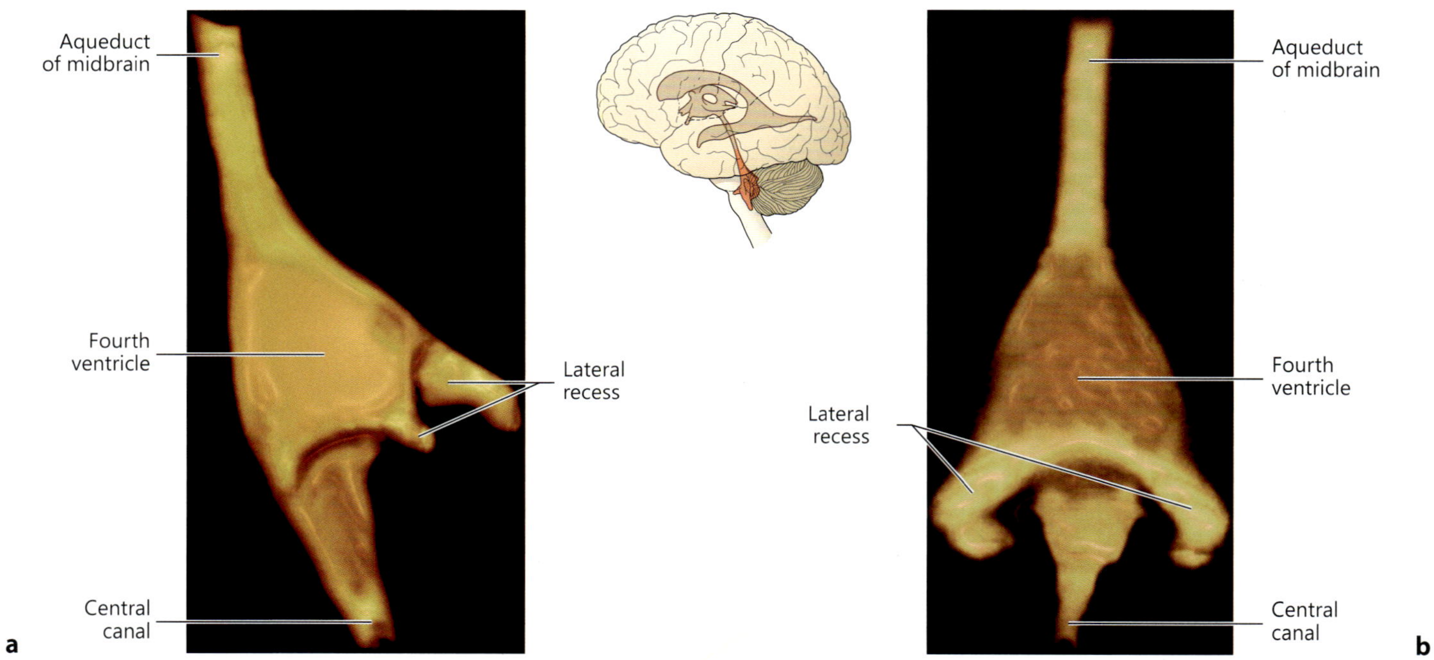

Three-dimensional image of a living human body, obtained by NMR using a 3D volume rendering technique, so as to present the mold of the **fourth ventricle** in different positions; posterolateral (**a**) and posterior (**b**) view.

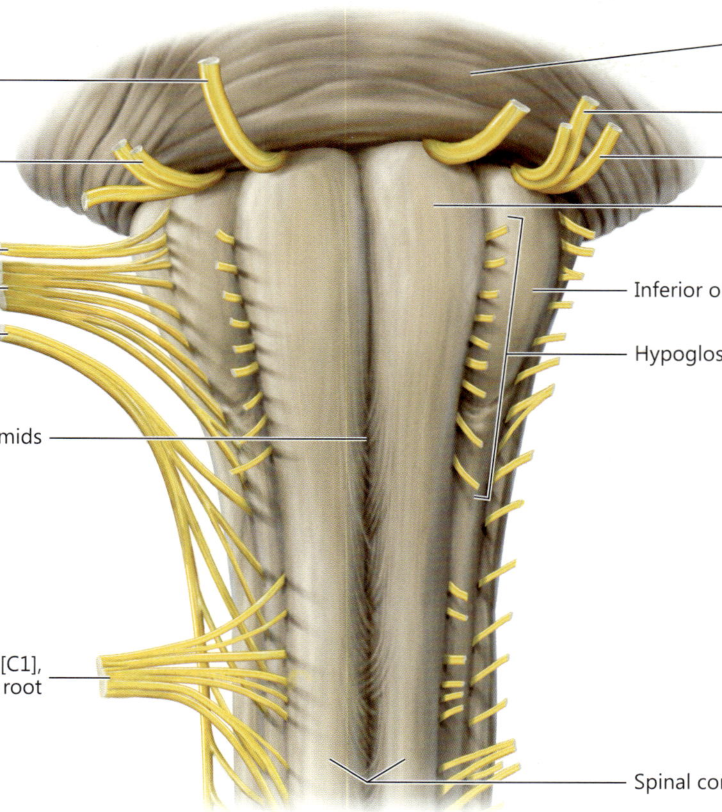

Myelencephalon, ventral surface. The radicular fibers of the hypoglossal nerve [XII] emerge along the anterolateral sulcus of the bulb; due to their position, these fibers correspond to the anterior roots of the spinal nerves. The accessory [XI], vagus [X] and glossopharyngeal [IX] nerves emerge along the posterolateral sulcus, proceeding from bottom to top, in line with the posterior roots of the spinal nerves. Higher up, where the posterolateral sulcus ends in the medullopontine sulcus, emerge the vestibulocochlear nerve [VIII], immediately medially the intermediate nerve and facial nerve [VII], and above the pyramid nerve the abducens nerve [VI].

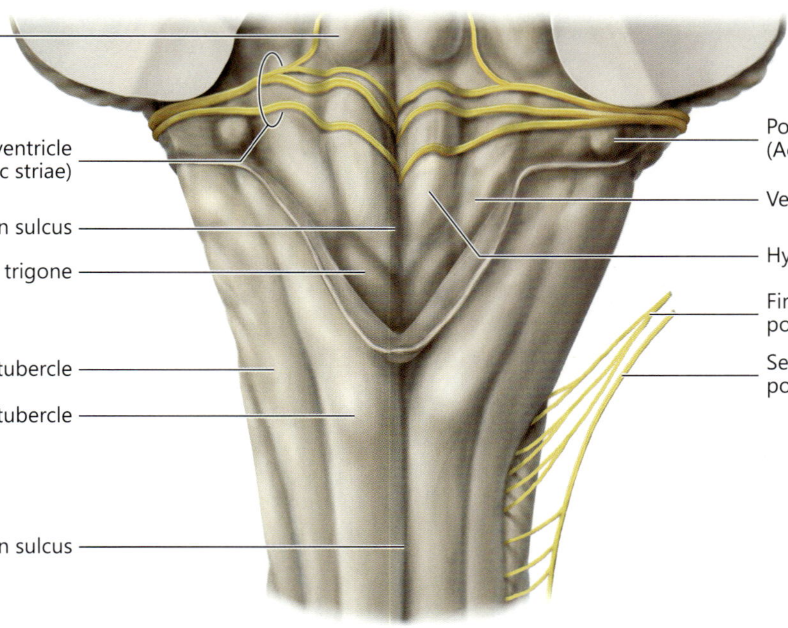

Myelencephalon, dorsal surface. Removing the cerebellar peduncles and the cerebellum highlights the rhomboid fossa; the fossa forms the floor of the fourth ventricle and is formed by a lower bulbar triangle and a superior pontine triangle. From the posterior cochlear nucleus, or acoustic tubercle, at the border between the bulb and pons, originate transverse fibers (medullary striae or acoustic striae) that reach the lateral and medial nuclei of the trapezoid body (or posterior nucleus of the trapezoid body) homolateral and contralateral.

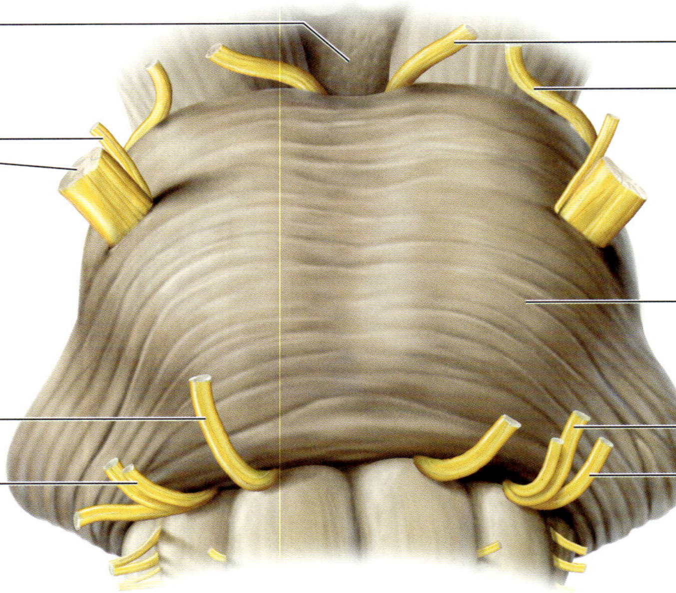

Pons, ventral surface. The trigeminal nerve [V] emerges from the pons with its great sensory root and superomedially to this, the small motor root. From the lateral margin of the interpeduncular fossa of the mesencephalon emerges the oculomotor nerve [III]. The trochlear nerve [IV] is the only cranial nerve emerging from the posterior surface of the brainstem, at the border between the tectal plate and the superior medullary velum.

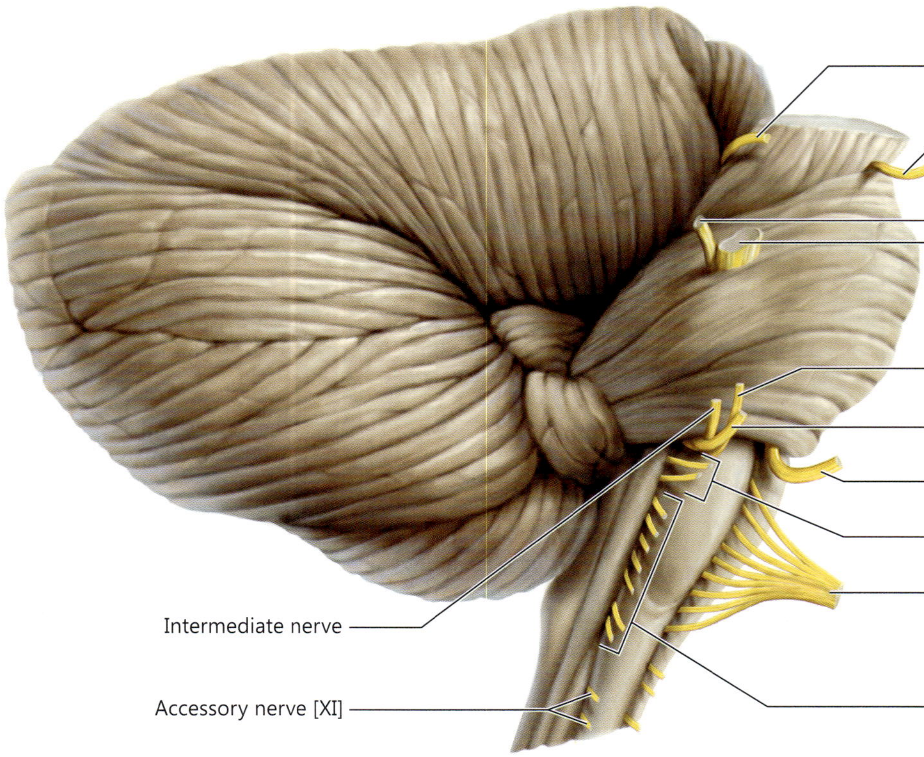

Myelencephalon and pons, lateral surface. Where the posterolateral sulcus of the bulb terminates in the medullopontine sulcus, there is the emergence of the vestibulocochlear nerve [VIII] and, immediately medially, those of the intermediate and facial [VII] nerves. The abducens nerve [VI] emerges in the medullopontine sulcus just above the pyramids. Along the anterolateral sulcus of the bulb appear the radicular fibers of the hypoglossal nerve [XII], whereas the accessory [XI], vagus [X] and glossopharyngeal [IX] nerves emerge along the posterolateral sulcus.

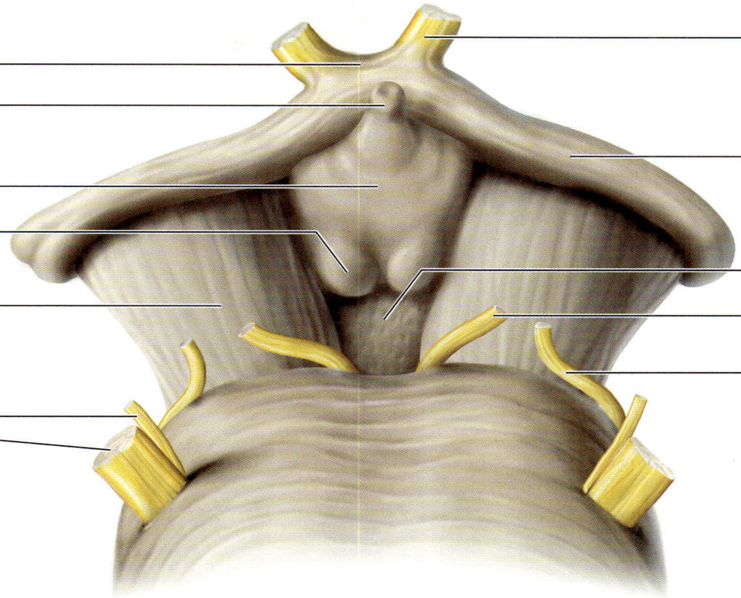

Mesencephalon, ventral surface. The cerebellar peduncles and the interpeduncular fossa are visible. Above, it is also visible the inferior surface of the diencephalon with a transected infundibulum of the neurohypophysis.

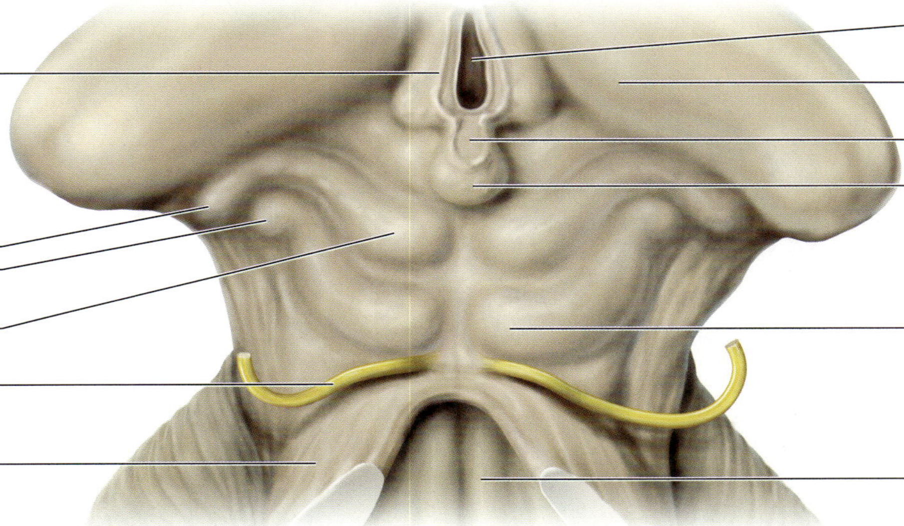

Mesencephalon, dorsal surface. Are mainly evident the colliculi, the pineal gland, and the emergence of the trochlear nerve.

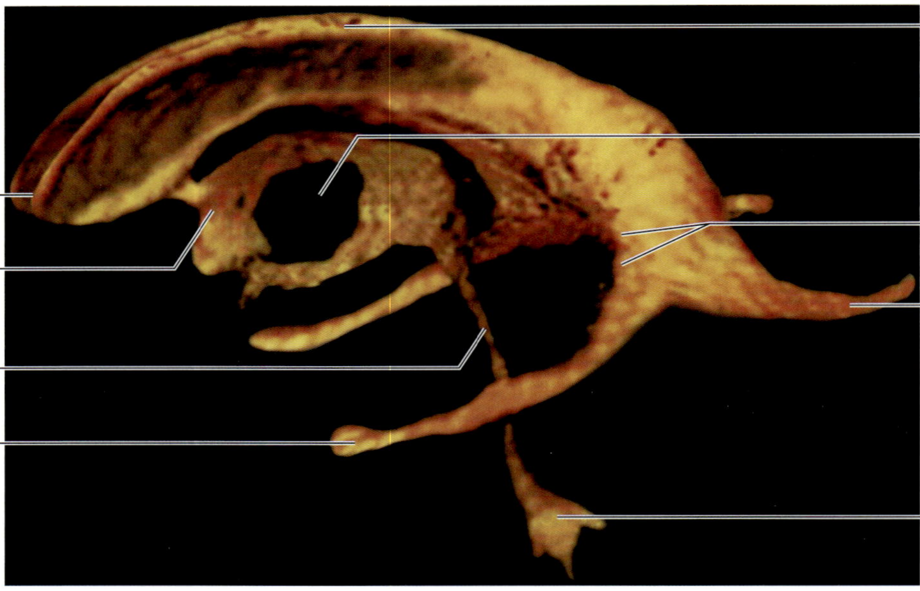

Three-dimensional image of a living human body, obtained by NMR using a 3D volume rendering technique, so as to present the mold of the **ventricular cavities** in lateral projection.

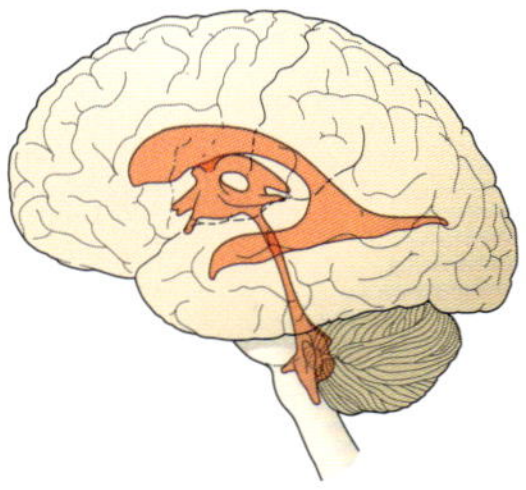

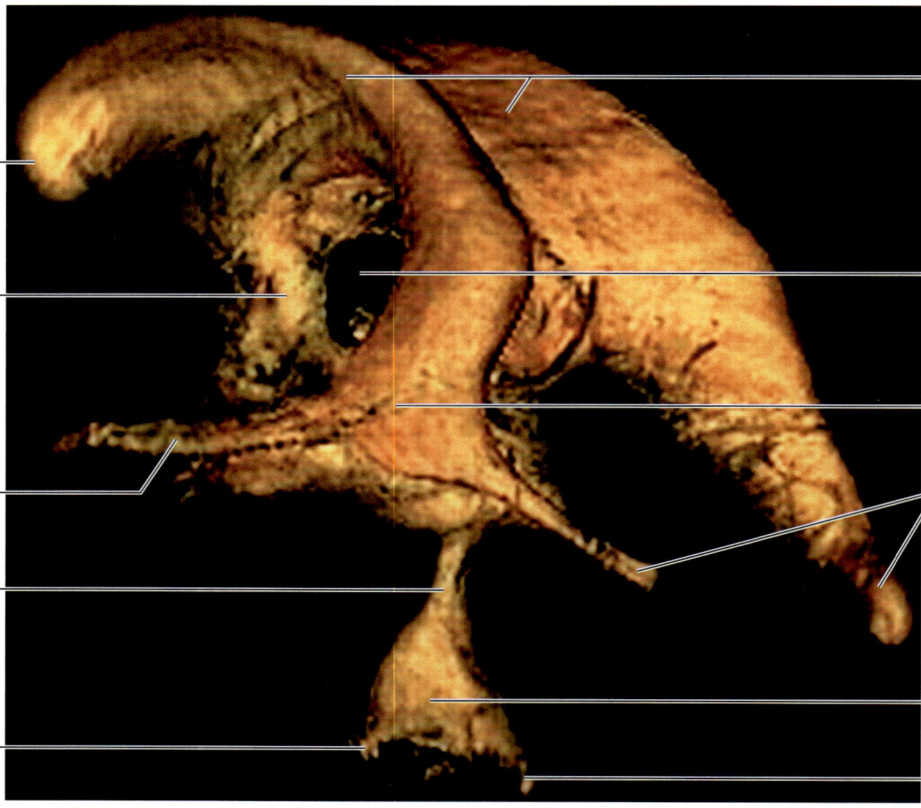

Three-dimensional image of a living human body, obtained by NMR using a 3D volume rendering technique, so as to present the mold of the **ventricular cavities** in posterolateral projection.

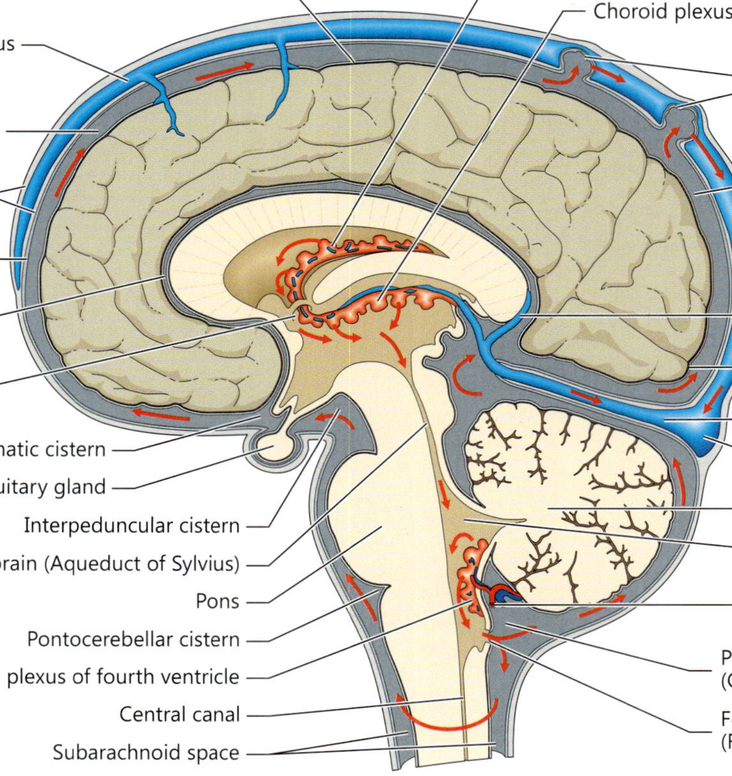

a, The meninges are the dura mater, the arachnoid mater and the pia mater. Between the arachnoid mater and the pia mater lies the subarachnoid space containing the **cerebrospinal fluid**, which mainly forms through the choroid plexuses and circulates inside the ventricles and, through the median (foramen of Magendie) and lateral (foramen of Luschka) openings of the fourth ventricle, it reaches the subarachnoid space, where it is reabsorbed through the arachnoid villi and granulations. **b**, The **lumbar puncture** or **spinal tap**, through which a sample of cerebrospinal fluid is collected for diagnostic purposes, is performed by inserting a needle in the subarachnoid space below the termination of the spinal cord (between L III and L IV in adult and L IV and L V newborn).

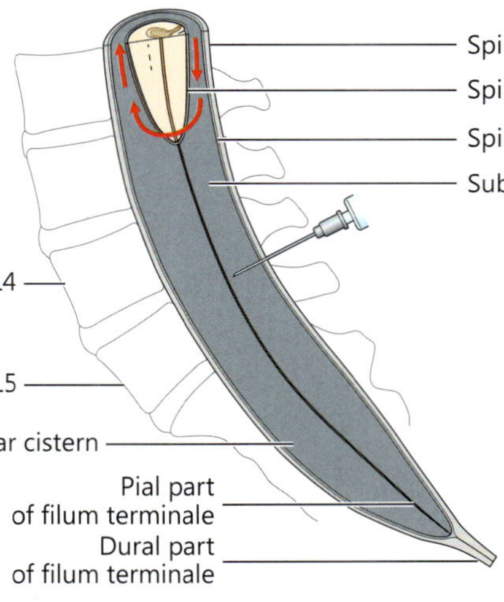

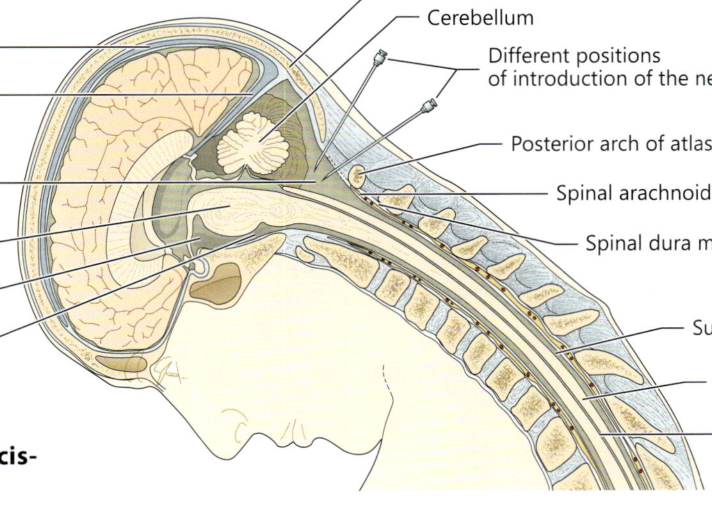

Puncture the **cerebellomedullary cistern** (suboccipital puncture).

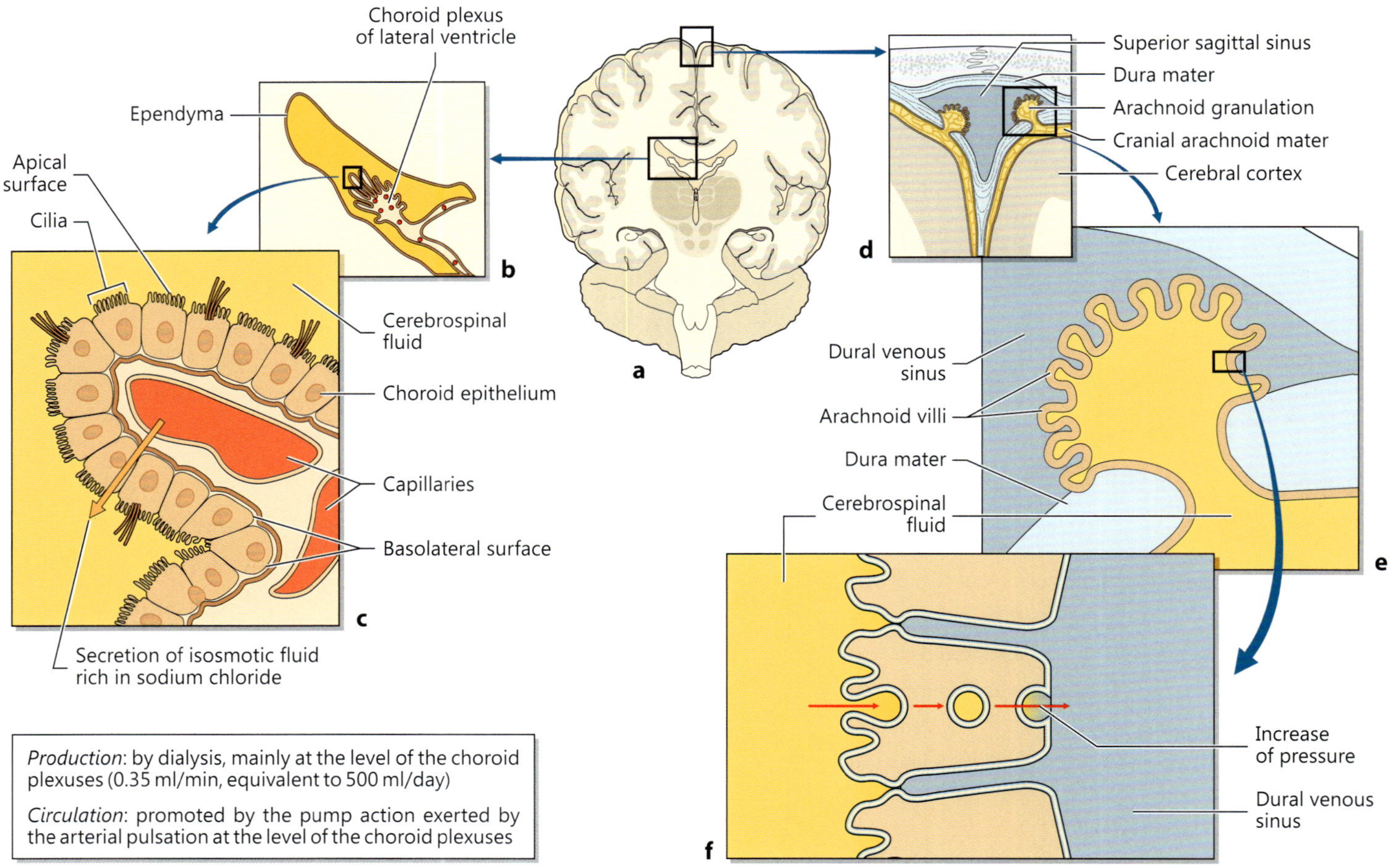

Production: by dialysis, mainly at the level of the choroid plexuses (0.35 ml/min, equivalent to 500 ml/day)

Circulation: promoted by the pump action exerted by the arterial pulsation at the level of the choroid plexuses

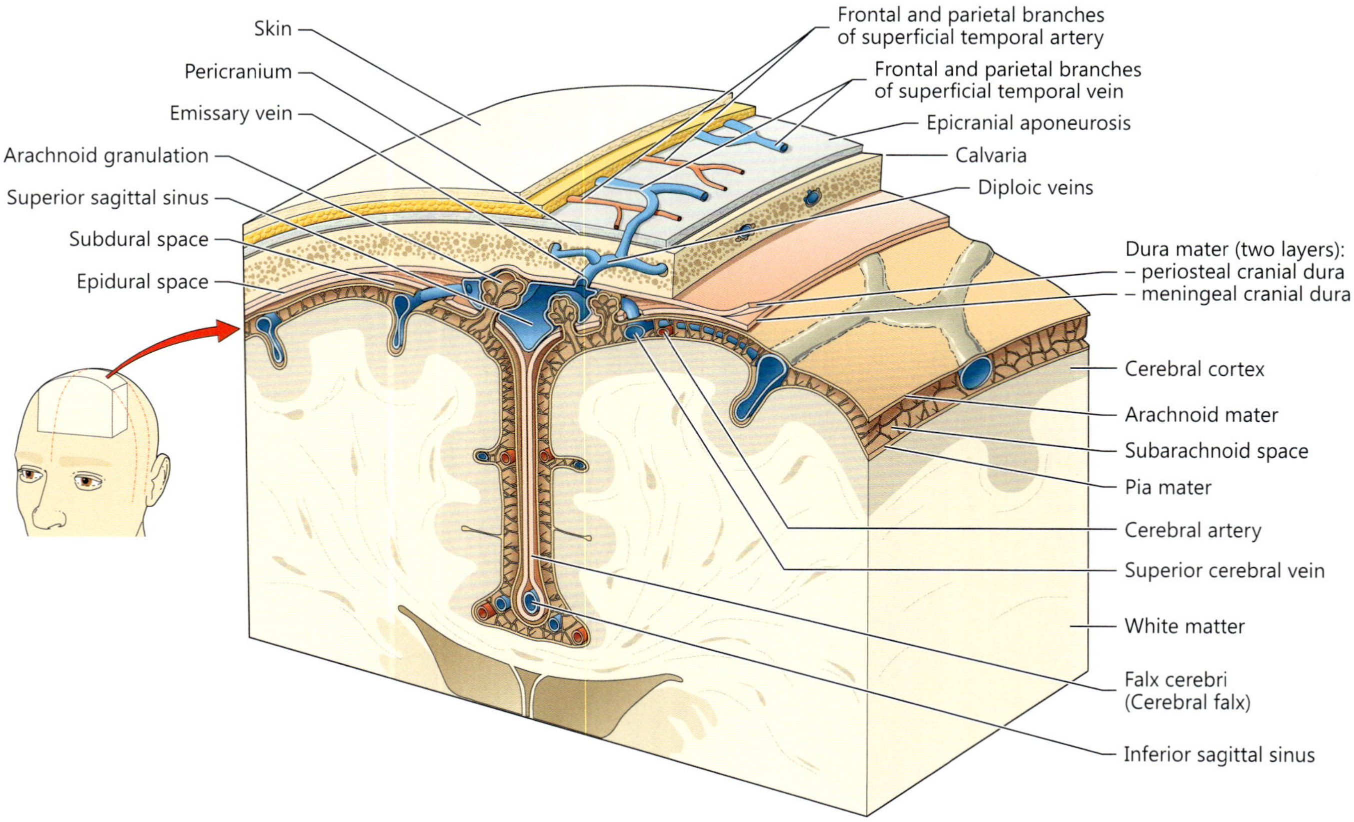

Disposition of the **superior sagittal sinus** in a frontal section of the cranium. Note the triangular shape of the sinus which opens out, to the left, into a cerebral vein and, to the right, it forms a blood lake where the diploic veins run into.

◄ Production and reabsorption of the **cerebrospinal fluid**. In **a** it is shown a coronal section of the cerebral hemispheres. The boxed area in the center is enlarged in **b** to show the choroid plexus of the lateral ventricle, where the cerebrospinal fluid is produced. **c**, Shows the organization of the choroidal epithelial cells; their basolateral surface faces the underlying capillaries. The mechanism of production by dialysis of the cerebrospinal fluid is schematically outlined. The boxed area above is enlarged in **d** to show the arachnoid granulations, site of reabsorption of the cerebrospinal fluid from the subarachnoid space to the superior sagittal sinus. In **e** are evident the arachnoid villi and in **f** is schematically illustrated the reabsorption mechanism.

Lateral ventricles
(choroid plexuses → production)

↓ Interventricular foramen (Foramen of Monro)

Third ventricle

↓ Aqueduct of midbrain

Fourth ventricle (choroid plexuses → production) ⇄ Central canal of spinal cord

↓ Median and lateral apertures (or foramen of Magendie and foramen of Luschka)

Subarachnoid space
(arachnoid granulations and villi, dural sinuses → reabsorption)

Dynamics of the **cerebrospinal fluid**.

General characteristics of the **meninges**.

- **Encephalic meninges**
  - Extradural (or epidural) space
  - **Cranial dura mater**
    - Trigeminal cave
    - Falces
      - Falx cerebri
      - Falx cerebelli
    - Dural venous sinuses
      - Cavernous sinus (anterior and posterior intercavernous)
      - Marginal sinus
      - Occipital sinus
      - Inferior petrosal sinus
      - Superior petrosal sinus
      - Petrosquamous sinus
      - Straight sinus
      - Inferior sagittal sinus
      - Superior sagittal sinus
      - Sphenoparietal sinus
      - Sigmoid sinus
      - Transverse sinus
    - Tentoria
      - Diaphragma sellae
      - Tentorium cerebelli
  - Subdural space
  - **Cranial arachnoid mater**
    - Arachnoid granulations
    - Arachnoid trabeculae
  - Subarachnoid space (Cerebrospinal fluid)
    - Subarachnoid cisterns
      - Cisterna ambiens
      - Posterior cerebellomedullary cistern (Cisterna magna)
      - Lateral cerebellomedullary cistern
      - Chiasmatic cistern
      - Cistern of lateral cerebral fossa
      - Cistern of lamina terminalis
      - Interpeduncular cistern
      - Pericallosal cistern
      - Pontocerebellar cistern
      - Quadrigeminal cistern (Cistern of great cerebral vein)
  - **Cranial pia mater**
    - Choroid enlargement, telae and plexuses
      - Choroidal enlargement
      - Choroid plexus of lateral ventricle
      - Choroid plexus and tela choroidea of fourth ventricle
      - Choroid plexus and tela choroidea of third ventricle
- **Spinal meninges**
  - Epidural space
  - **Spinal dura mater**
  - **Spinal arachnoid mater**
  - Subarachnoid space (Cerebrospinal fluid)
    - Lumbar cistern
  - **Spinal pia mater**
  - (Spinal dura mater and spinal pia mater:)
    - Filum terminale externum (dural part of filum terminale)
    - Filum terminale internum (pial part of filum terminale)
    - Denticulate ligament
    - Intermediate cervical septum

The dura mater is also known as pachymeninx, while the term leptomeninx indicates the arachnoid mater and pia mater; accordingly, the subarachnoid space is also known as leptomeningeal space.

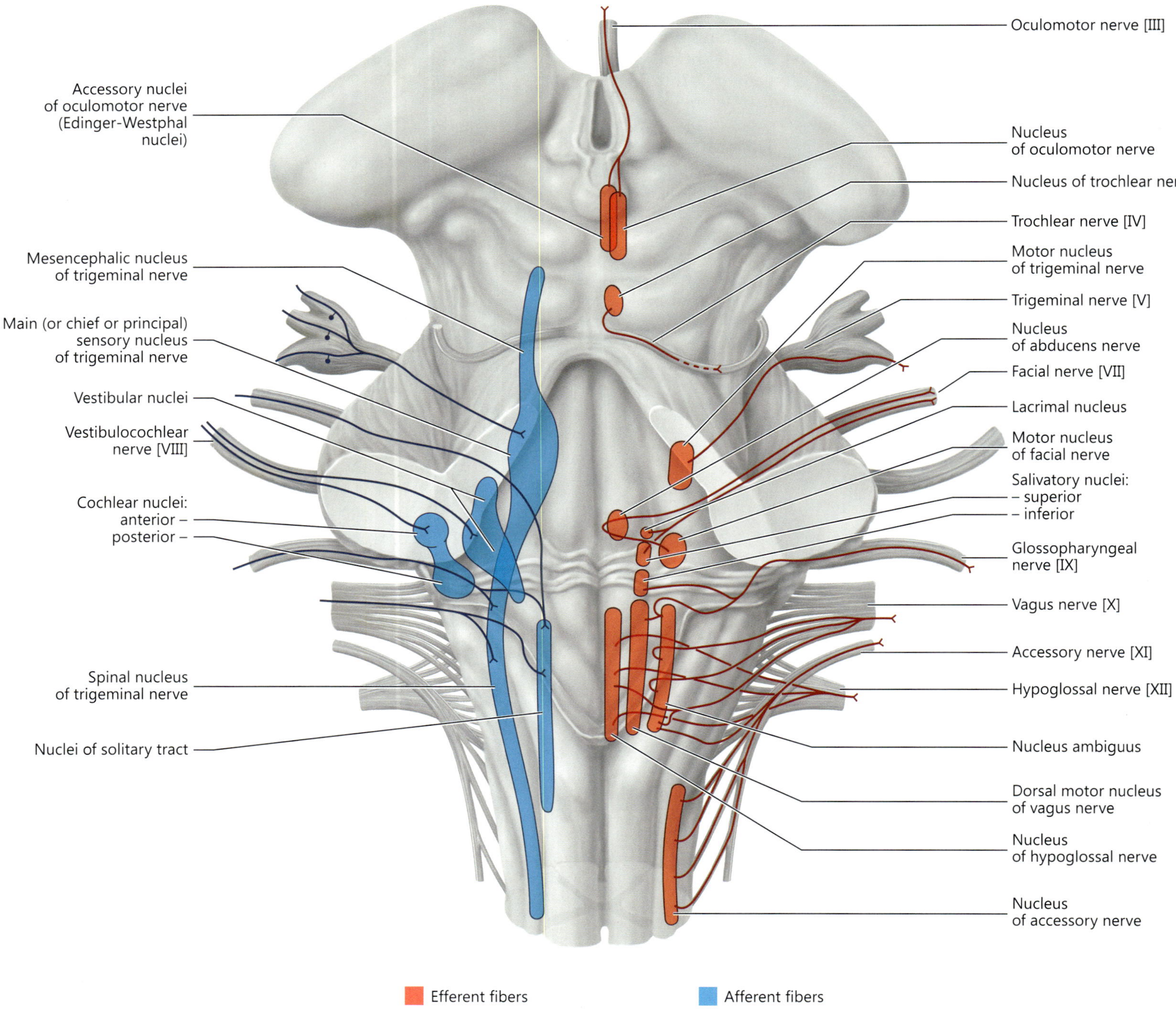

Schematic representation of the **brainstem** in posterior projection showing the cranial nerves nuclei in transparency (*blue*, the somatic and visceral sensory nuclei; *red*, the somatic motor and visceral effector nuclei).

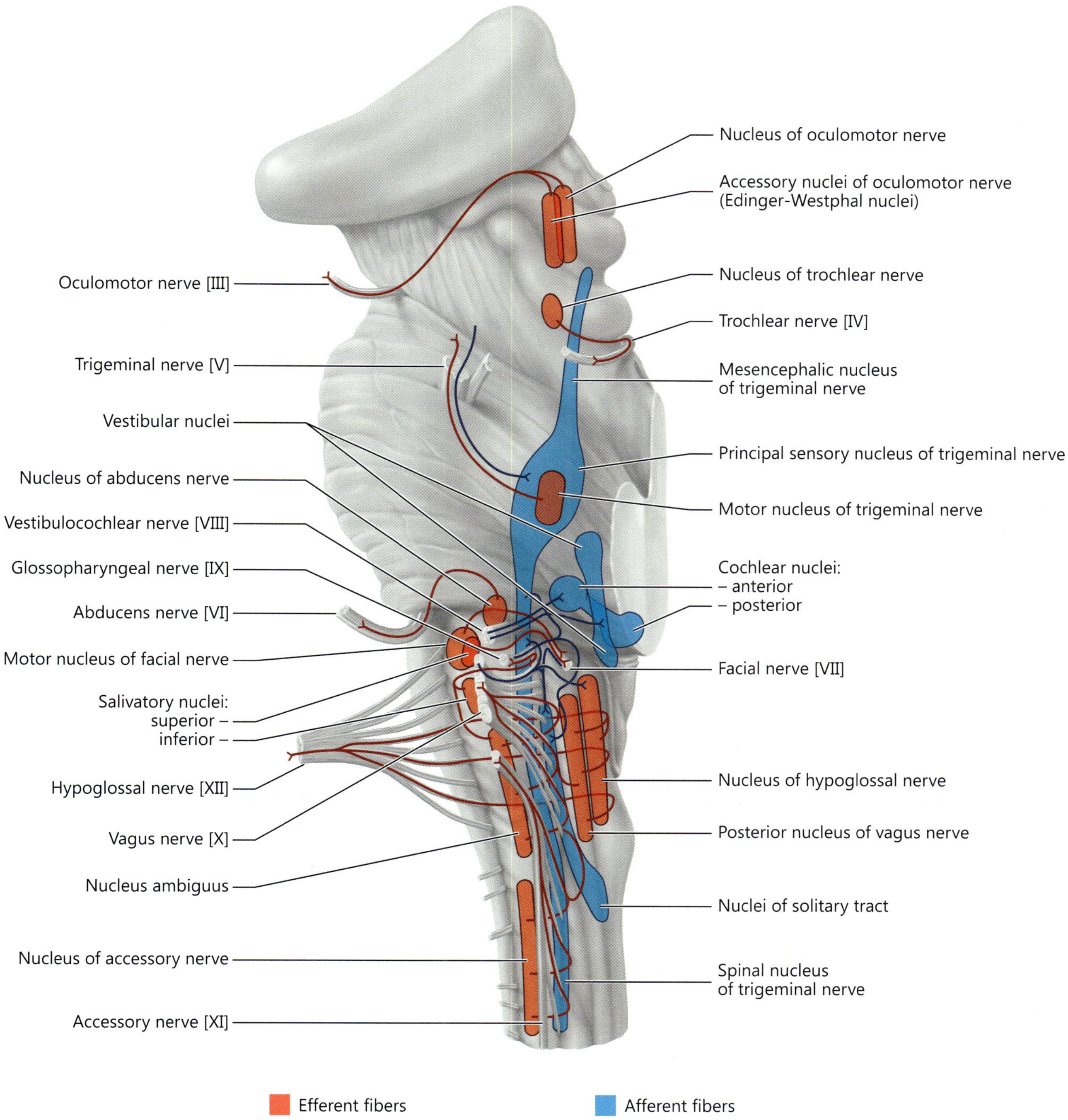

Schematic representation of the **brainstem** seen from the left side, showing the cranial nerves nuclei in transparency (*blue*, the somatic and visceral sensory nuclei; *red*, the somatic motor and visceral effector nuclei).

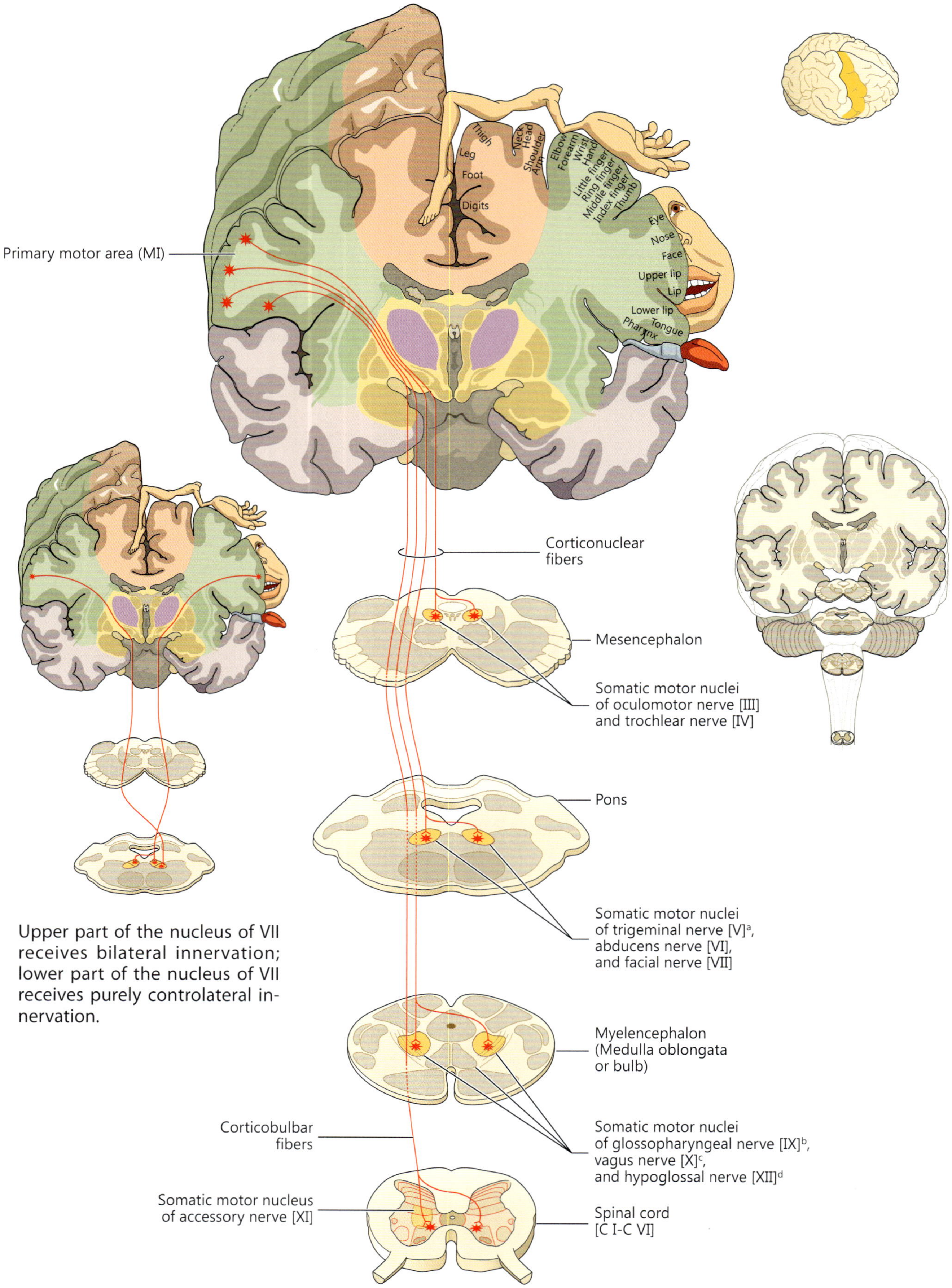

Upper part of the nucleus of VII receives bilateral innervation; lower part of the nucleus of VII receives purely controlateral innervation.

[a] The nucleus of V receives bilateral innervation.
[b] The nucleus of IX receives bilateral innervation.
[c] The nucleus of X receives bilateral innervation.
[d] The nucleus of XII receives more crossed than uncrossed innervation.

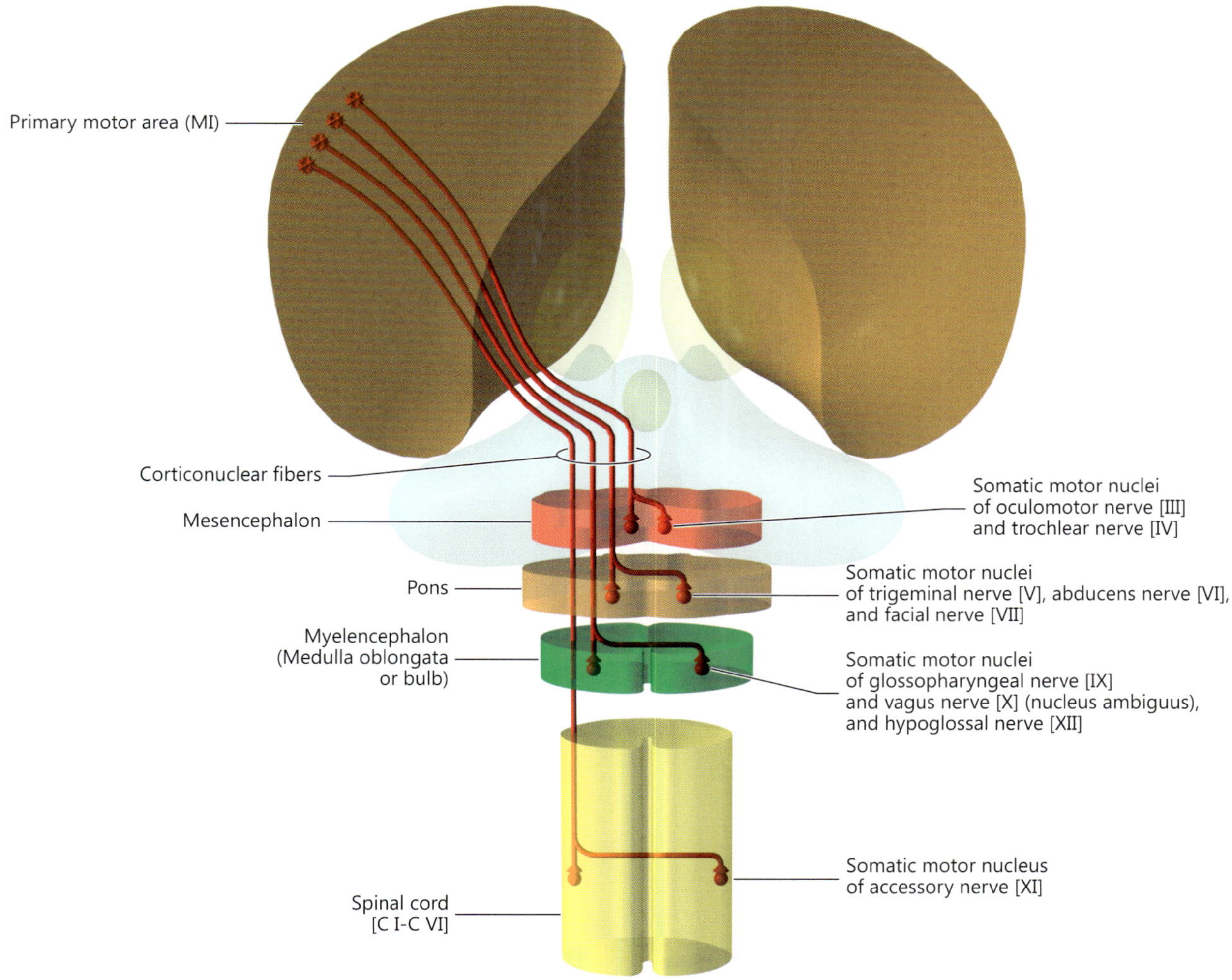

Corticonuclear tract	
Upper motor neuron	From the pyramidal neurons of the III and V layers of the motor and premotor areas of the cerebral cortex, bilaterally to the somatic motor neurons of the cranial nerves
Lower motor neuron	From the α and γ motoneurons of the cranial nerves to the extrinsic muscles of the eyeball, face, masticatory, tongue, pharynx and larynx and the sternocleidomastoid and trapezius

The **corticonuclear tracts** comprise the superior motor neuron, or pyramidal neuron, which originates from the motor and premotor areas (region of the face, tongue, pharynx and larynx) of the cerebral cortex, crosses the posterior limb of the internal capsule and descends along the brainstem, forming branches in succession that cross the median line and bilaterally take synapses (such as the direct pyramidal tract) with the inferior motor neuron (α and γ motor neurons) placed in the somatic motor nuclei of the cranial nerves; in this respect it is worth remembering that α and γ motor neurons in the spinal root of the accessory nerve are located in the ventromedial part of the lamina IX of the spinal cord from C1 to C6.

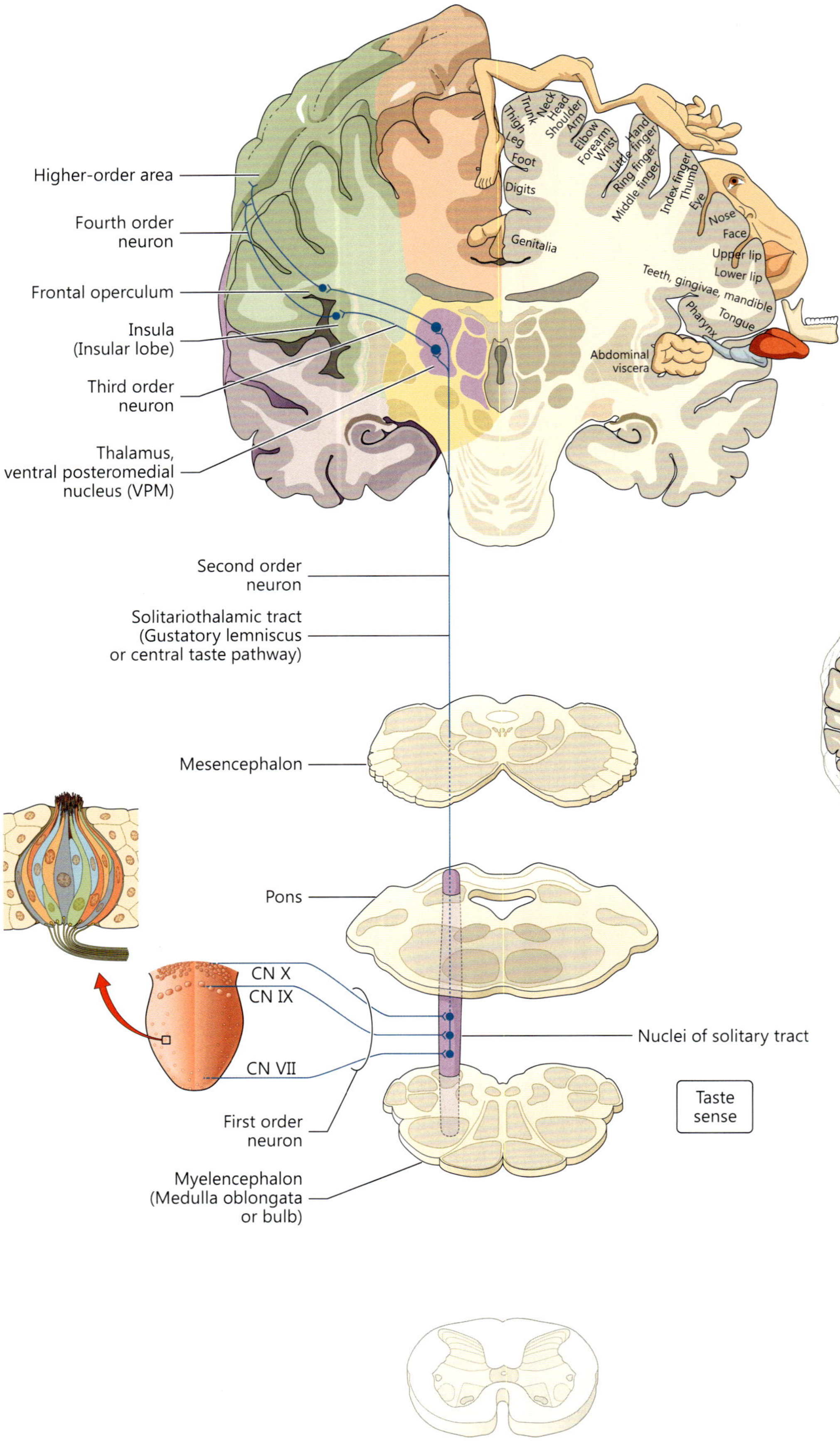
Higher-order area
Fourth order neuron
Frontal operculum
Insula (Insular lobe)
Third order neuron
Thalamus, ventral posteromedial nucleus (VPM)
Thigh
Trunk
Neck
Head
Shoulder
Arm
Elbow
Forearm
Wrist
Hand
Little finger
Ring finger
Middle finger
Index finger
Thumb
Eye
Nose
Face
Upper lip
Lower lip
Teeth, gingivae, mandible
Tongue
Pharynx
Leg
Foot
Digits
Genitalia
Abdominal viscera
Second order neuron
Solitariothalamic tract (Gustatory lemniscus or central taste pathway)
Mesencephalon
Pons
CN X
CN IX
CN VII
Nuclei of solitary tract
First order neuron
Myelencephalon (Medulla oblongata or bulb)
Taste sense

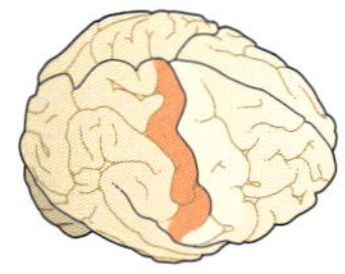

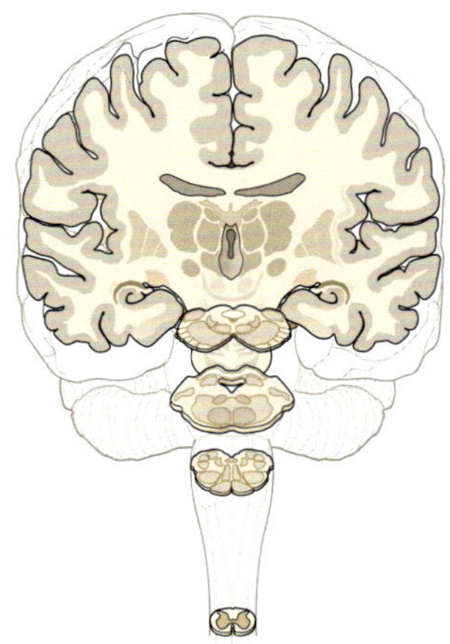

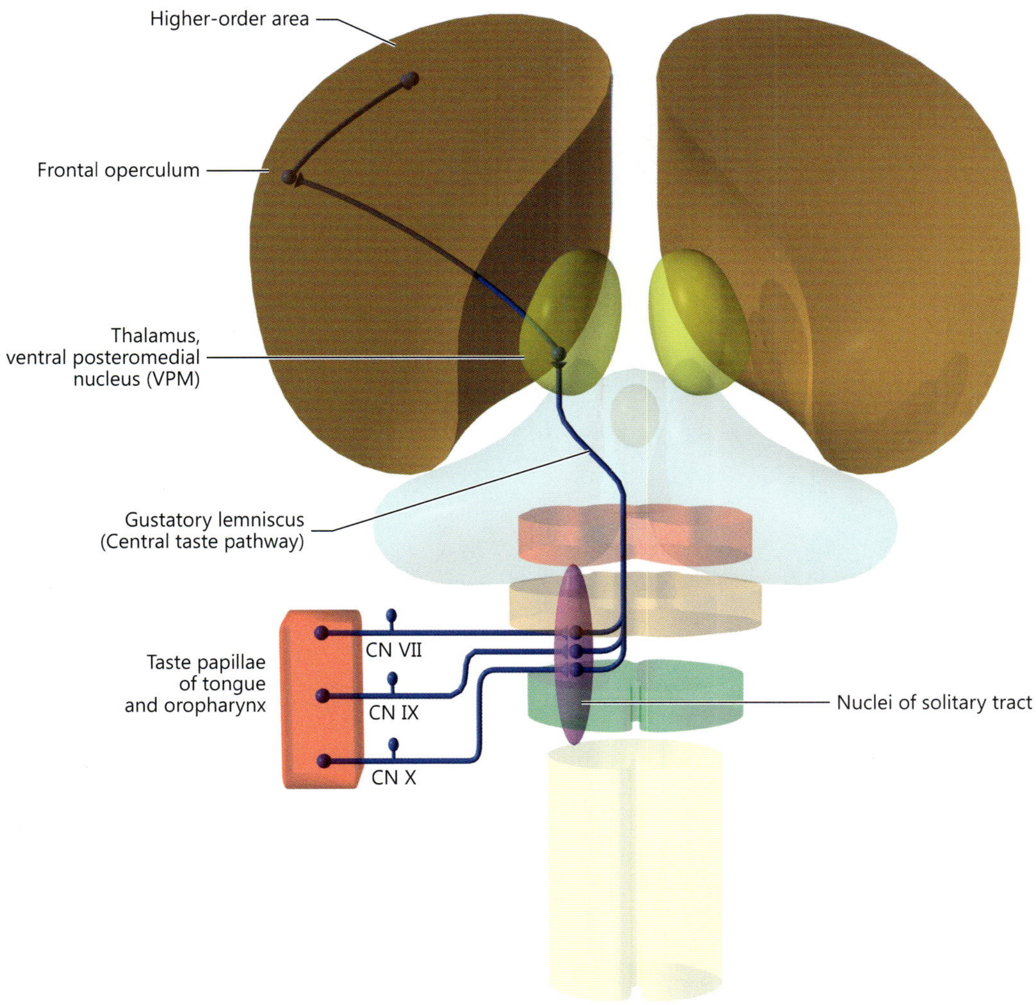

Solitariothalamic tract - Gustatory lemniscus	
Sense	Taste
First order neuron (pseudounipolar neuron)	From the sensory cells of the taste buds to the rostral portion of the nuclei of the solitary tract (visceral sensory nucleus common to cranial nerves VII, IX and X)
Second order neuron (deutoneuron)	From the nuclei of the solitary tract to the ventral posteromedial nucleus (VPM) of the thalamus (there are also projective bundles that from the nuclei of the solitary tract are directed to the hypothalamus and the amygdaloid body)
Third order neuron (thalamocortical projection neuron)	From the VPM nucleus of the thalamus to both the primary somatosensory area (frontal operculum zone) and to the insula
Fourth order neuron	From the frontal operculum and insula to the higher-order areas

The **solitariothalamic tract** (or gustatory lemniscus or central taste pathway) originates from the sensory cells of the taste buds (in all taste buds there are sensory cells for each type of taste – salty, sour, sweet, bitter and umami), placed in the papillae of the tongue and pharynx. In detail, stimuli coming from the taste buds:

- located in the anterior third of the tongue are collected by the sensitive fibers of the intermediate nerve, component of the facial nerve [VII];
- located in the posterior third of the tongue (terminal sulcus of the tongue and root of the tongue) are collected by the sensitive fibers of the glossopharyngeal nerve [IX];
- located in the oropharyngeal region and the region of the dorsal surface of the epiglottis are collected by the sensitive fibers of the vagus nerve [X].

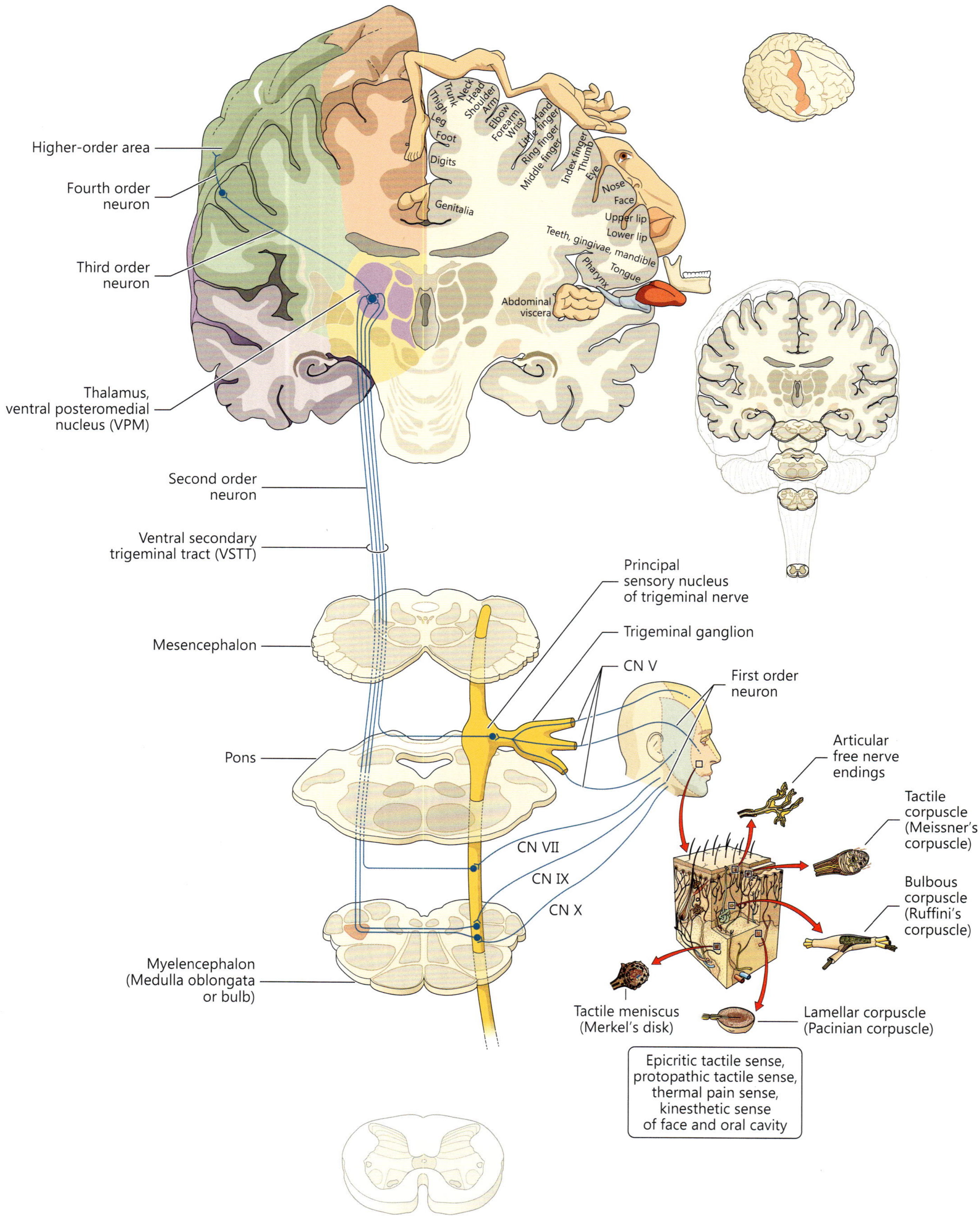
Higher-order area
Fourth order neuron
Third order neuron
Thalamus, ventral posteromedial nucleus (VPM)
Second order neuron
Ventral secondary trigeminal tract (VSTT)
Mesencephalon
Pons
Myelencephalon (Medulla oblongata or bulb)
Thigh
Trunk
Neck
Head
Shoulder
Arm
Elbow
Forearm
Wrist
Hand
Little finger
Ring finger
Middle finger
Index finger
Thumb
Eye
Nose
Face
Upper lip
Lower lip
Teeth, gingivae, mandible
Tongue
Pharynx
Abdominal viscera
Leg
Foot
Digits
Genitalia
Principal sensory nucleus of trigeminal nerve
Trigeminal ganglion
CN V
First order neuron
Articular free nerve endings
Tactile corpuscle (Meissner's corpuscle)
Bulbous corpuscle (Ruffini's corpuscle)
CN VII
CN IX
CN X
Tactile meniscus (Merkel's disk)
Lamellar corpuscle (Pacinian corpuscle)
Epicritic tactile sense, protopathic tactile sense, thermal pain sense, kinesthetic sense of face and oral cavity

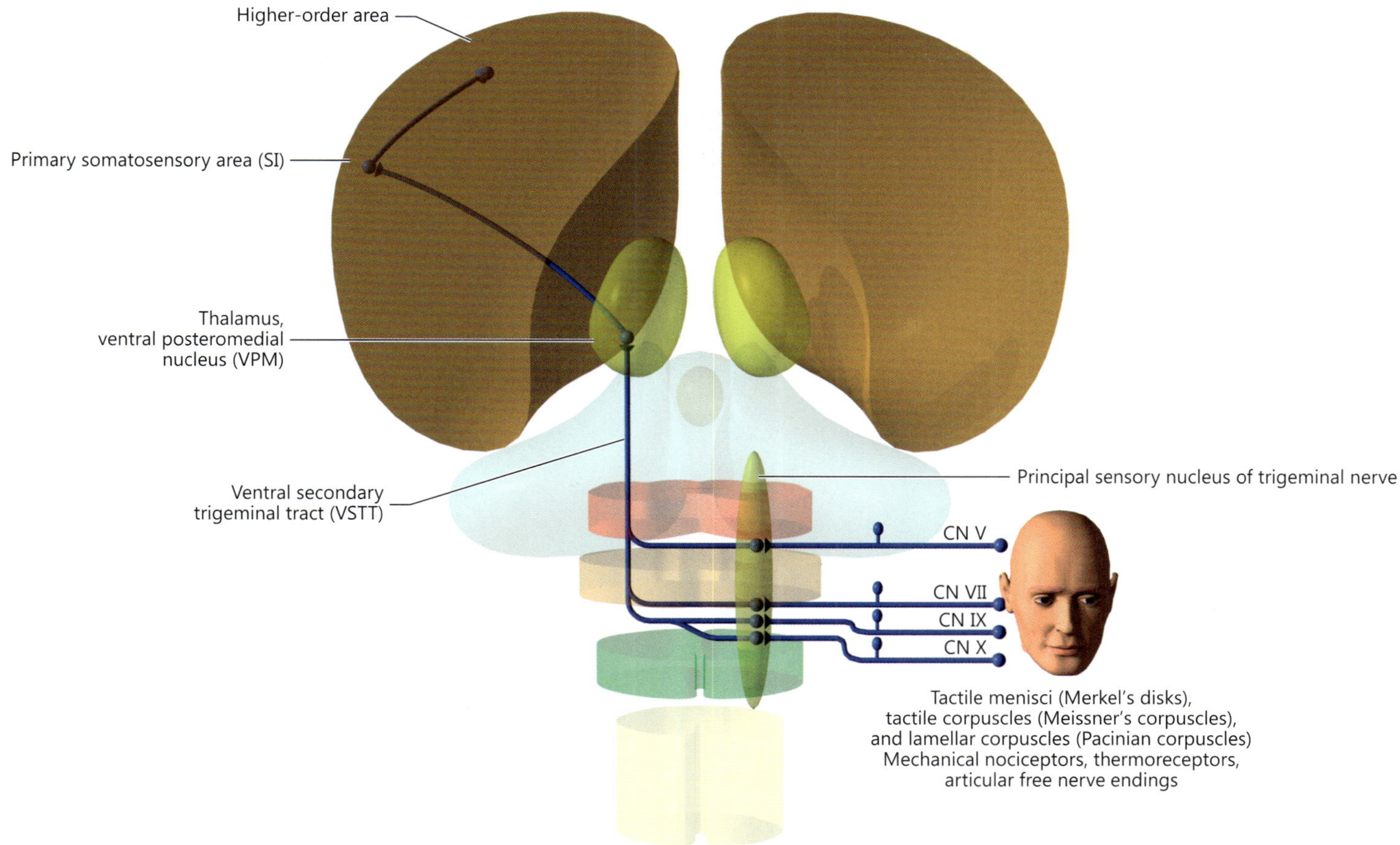

Trigeminothalamic tract - Trigeminal lemniscus (Quintothalamic tract)	
Sense	Tactile, thermal pain and conscious proprioceptive
First order neuron (pseudounipolar neuron)	From the mechanoreceptors, nociceptors and muscular articular proprioceptors to the sensory nuclei of the trigeminal nerve (spinal, principal and mesencephalic nuclei)
Second order neuron (deutoneuron)	From the sensory nuclei of the trigeminal nerve it crosses the median line, runs in the contralateral funiculus and goes along the brainstem up to the ventral posteromedial nucleus (VPM) of the thalamus
Third order neuron (thalamocortical projection neuron)	From the VPM nucleus of the thalamus to the primary somatosensory area
Fourth order neuron	From the primary somatosensory area to the higher-order areas

The **trigeminothalamic tract** (or trigeminal lemniscus or quintothalamic) delivers tactile sense, thermal pain sense and conscious proprioceptive (kinesthetic) sense of the face; after synapse in the principal nucleus of the trigeminal nerve, the bundles of axons cross the midline and run upward along the medial lemniscus, until the ventral posteromedial nucleus (VPM) of the thalamus. It is important to remember that the principal sensory nucleus of the trigeminal nerve is reached by somatosensory fibers also of the VII, IX and X pair of cranial nerves.

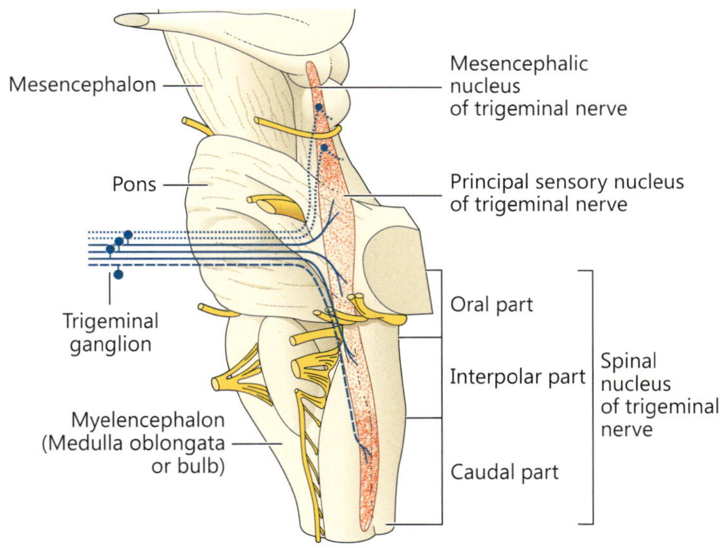

The **somatosensory nucleus** of the trigeminal nerve nucleus is constituted by the principal nucleus, localized in the pons, by a descending extension or spinal nucleus, localized in the bulb (it continues with the laminae I and II of the spinal cord), in turn craniocaudally divided in oral nucleus, interpolar and caudal nucleus, and by an ascending extension or mesencephalic nucleus. The spinal nucleus harbors the synapses for crude touch sensitivity (protopathic and nociceptive), in the principal pontine nucleus are located synapses for tactile discriminative (epicritic) sense and in the mesencephalic nucleus are present synapses for conscious proprioceptive (kinesthetic) sense, for unconscious proprioceptive sense and for sense of face and oral cavity.

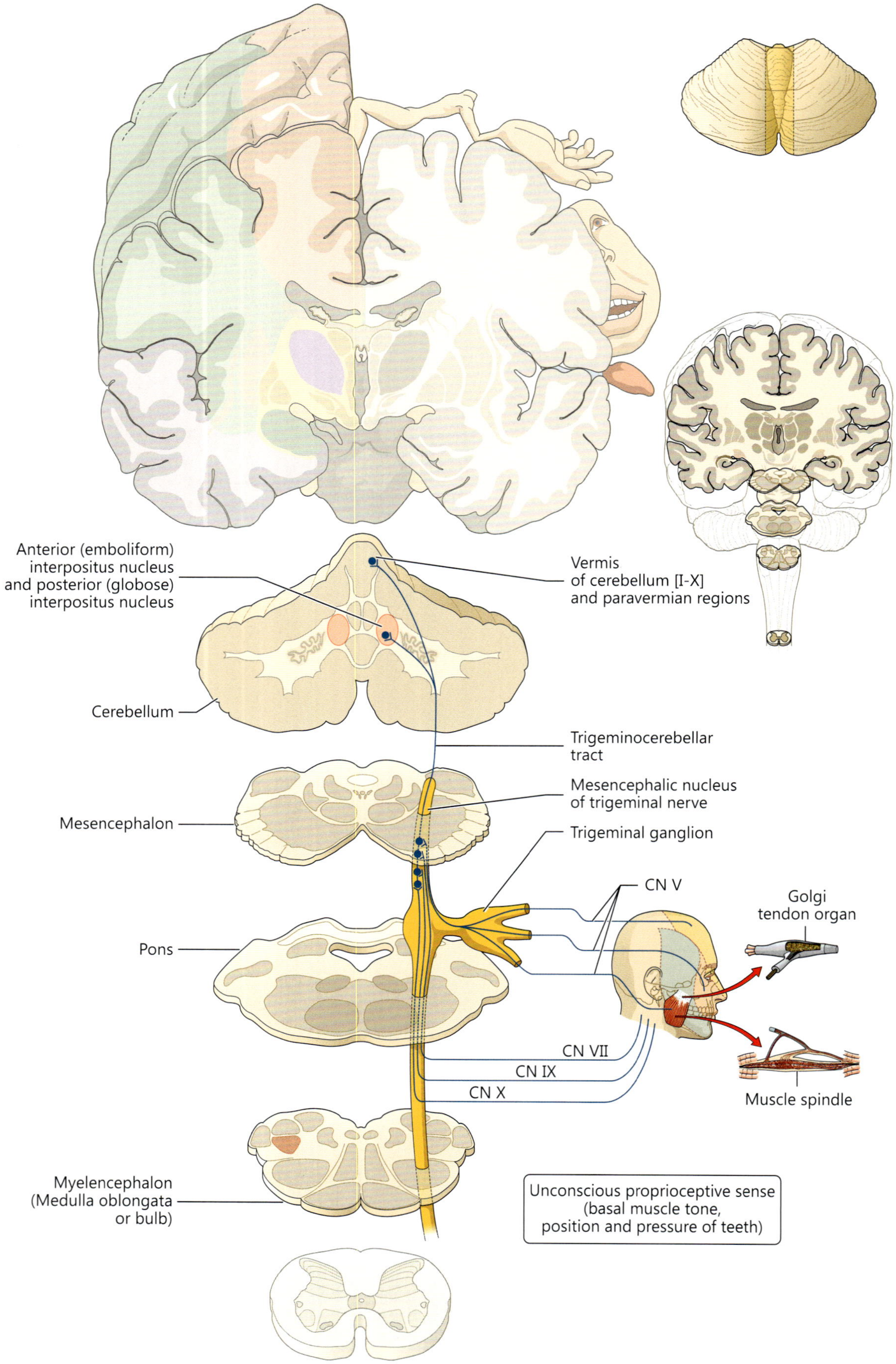
Anterior (emboliform) interpositus nucleus and posterior (globose) interpositus nucleus
Vermis of cerebellum [I-X] and paravermian regions
Cerebellum
Trigeminocerebellar tract
Mesencephalic nucleus of trigeminal nerve
Mesencephalon
Trigeminal ganglion
CN V
Golgi tendon organ
Pons
CN VII
CN IX
CN X
Muscle spindle
Myelencephalon (Medulla oblongata or bulb)
Unconscious proprioceptive sense (basal muscle tone, position and pressure of teeth)

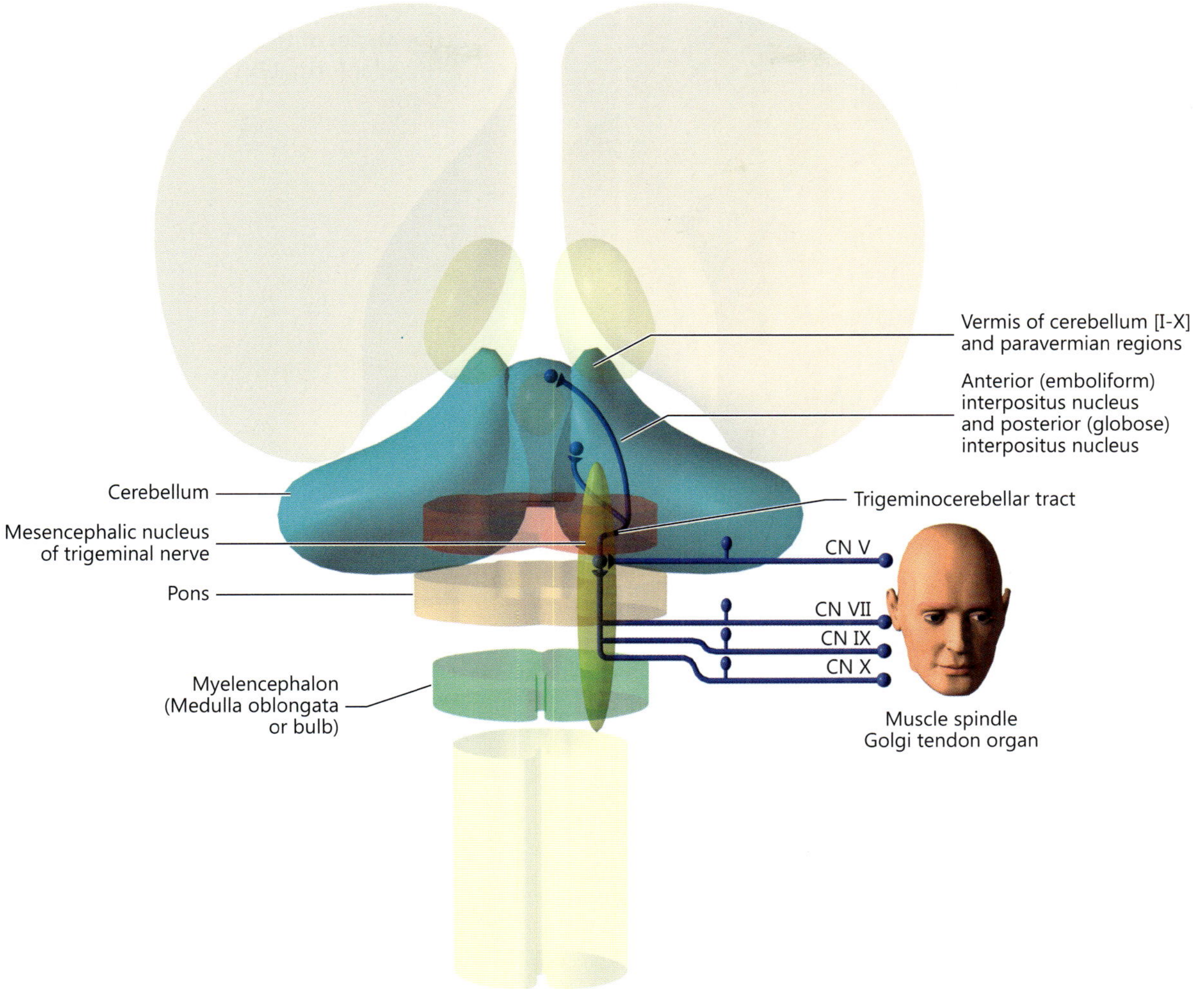

Trigeminocerebellar tract	
Sense	Unconscious proprioceptive
First order neuron (pseudounipolar neuron)	From the neuromuscular spindles, the tendon organs (Golgi) and from the receptors of the desmodentium (or periodontal fiber) to the mesencephalic nucleus of the trigeminal nerve
Second order neuron (deutoneuron)	From the mesencephalic nucleus of the trigeminal nerve to the cerebellar cortex of the vermis [I-X] and the paravermian regions (spinocerebellum-paleocerebellum)

The **trigeminocerebellar tract** carries the unconscious proprioceptive sense that originates from the masticatory muscles, muscles of the face and extraocular muscles and from the receptors of the desmodentium (or periodontal fiber)* and after synapse in the mesencephalic nucleus of the trigeminal nerve, through the superior cerebellar peduncle, reaches the cerebellar cortex of the vermis [I-X] and paravermian regions.

* The free nerve endings representing the receptors of the desmodentium and muscles and tendons proprioceptors have the most part of cell body, not in the trigeminal ganglion (or Gasser's), but directly in the mesencephalic nucleus of the trigeminal nerve.

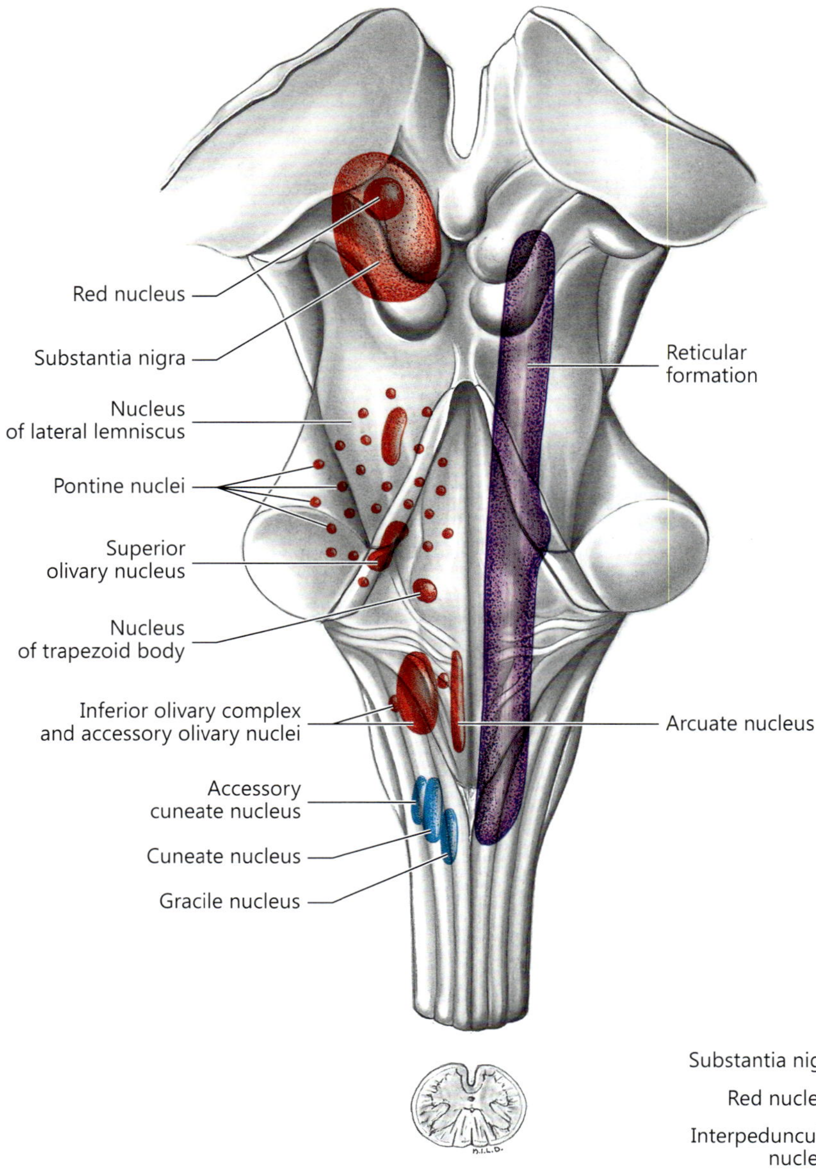

Schematic three-dimensional reconstruction showing the **nuclei of the brainstem**. Note the extension of the reticular formation in the three segments of the brainstem.

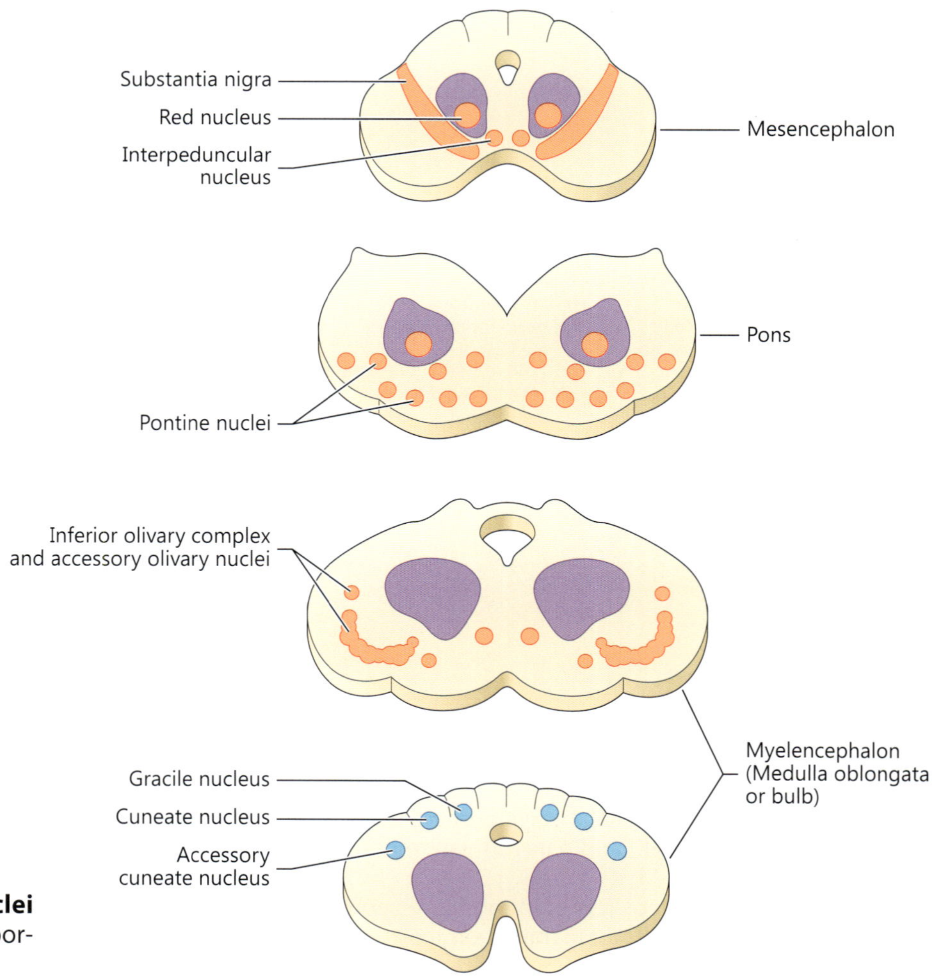

Schematic representation showing the **nuclei of the brainstem** at the level of the three portions.

Schematic representation of the brainstem in dorsal projection; on the left side of the image it is possible to see the subdivision of the **nuclei** within the reticular formation (*orange* area), in the three segments of the brainstem.

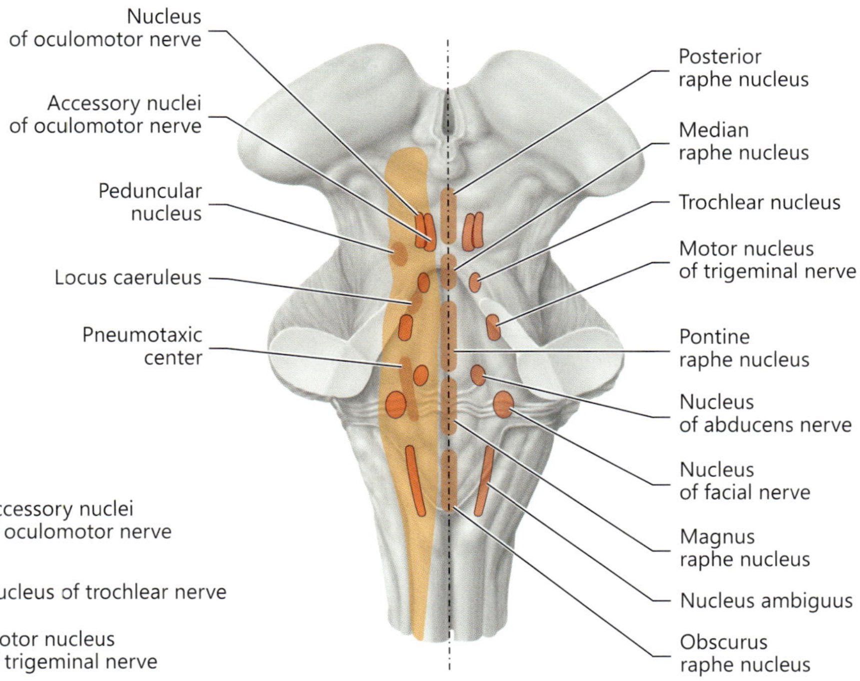

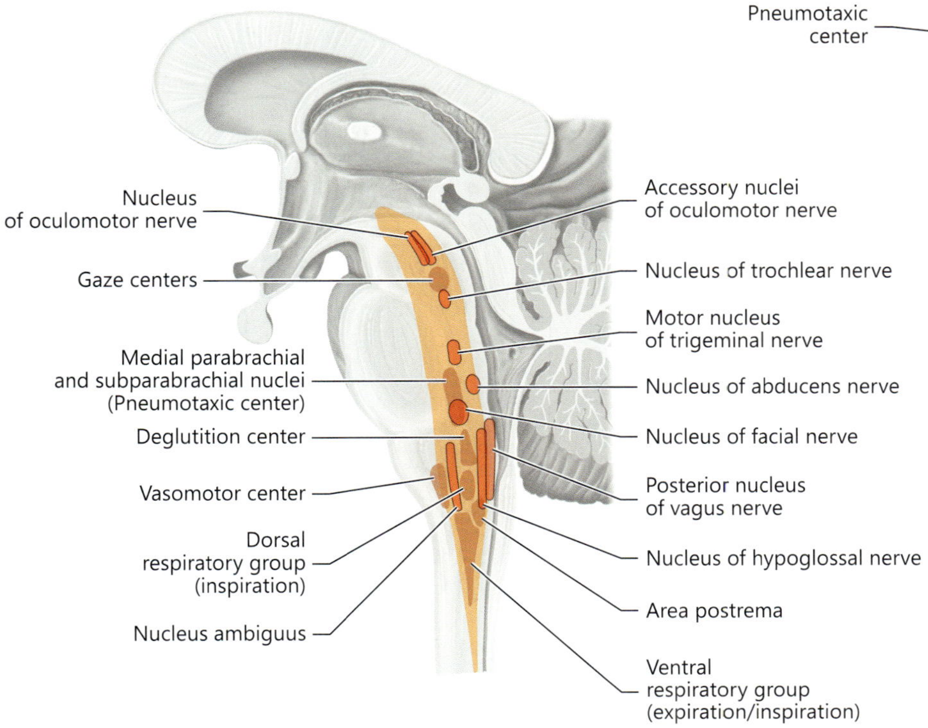

Schematic representation of the brainstem in sagittal projection showing the structure and relationships of the reticular formation (*orange* area). The **reticular formation** is an area of the brainstem, located between the nuclei of the cranial nerves and the nuclei in the brainstem, where important centers of the vegetative pathway are located.

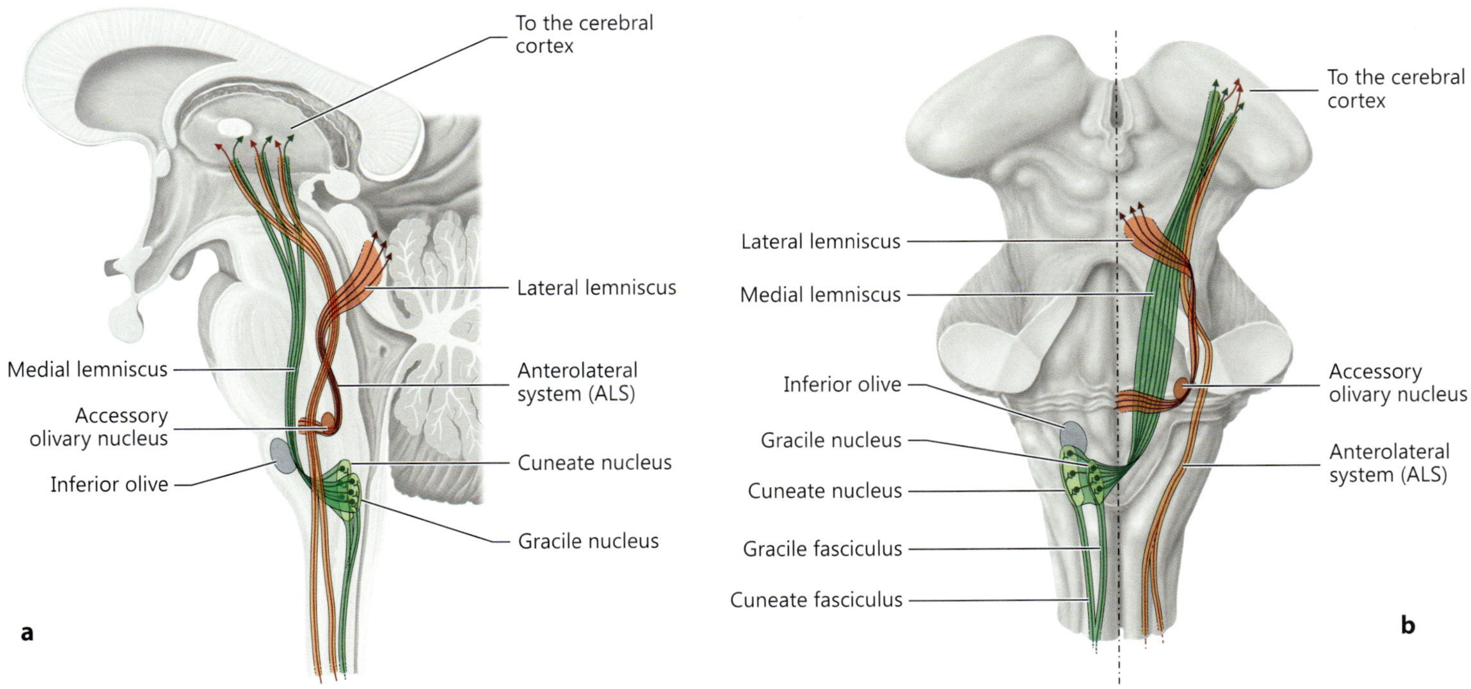

Schematic representation showing the course of the **ascending fascicles** through the brainstem, in left sagittal section (**a**) and posterior view (**b**).

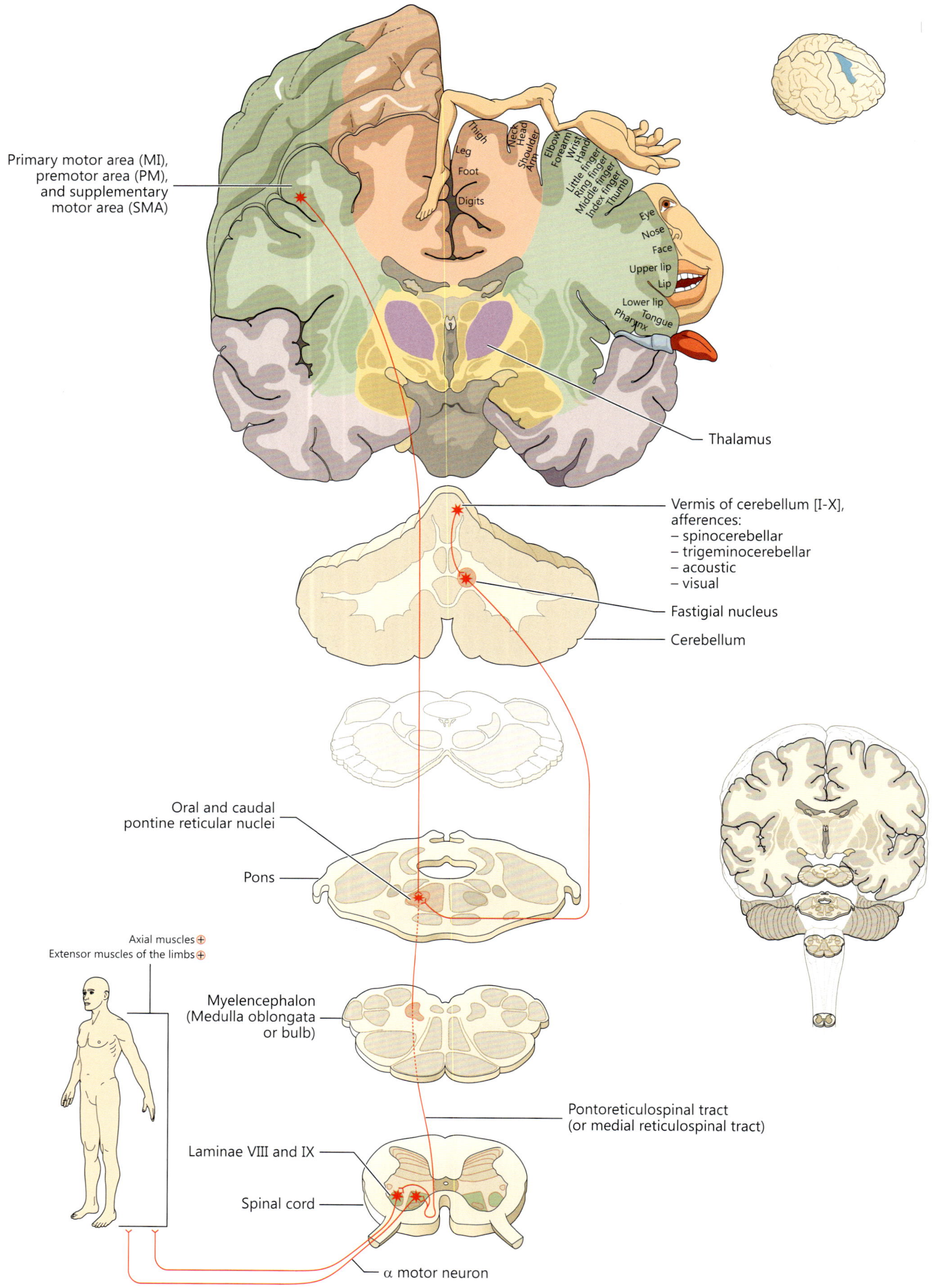
Primary motor area (MI), premotor area (PM), and supplementary motor area (SMA)
Thigh
Leg
Foot
Digits
Neck
Head
Shoulder
Arm
Elbow
Forearm
Wrist
Hand
Little finger
Ring finger
Middle finger
Index finger
Thumb
Eye
Nose
Face
Upper lip
Lip
Lower lip
Tongue
Pharynx
Thalamus
Vermis of cerebellum [I-X], afferences:
– spinocerebellar
– trigeminocerebellar
– acoustic
– visual
Fastigial nucleus
Cerebellum
Oral and caudal pontine reticular nuclei
Pons
Axial muscles
Extensor muscles of the limbs
Myelencephalon (Medulla oblongata or bulb)
Pontoreticulospinal tract (or medial reticulospinal tract)
Laminae VIII and IX
Spinal cord
α motor neuron

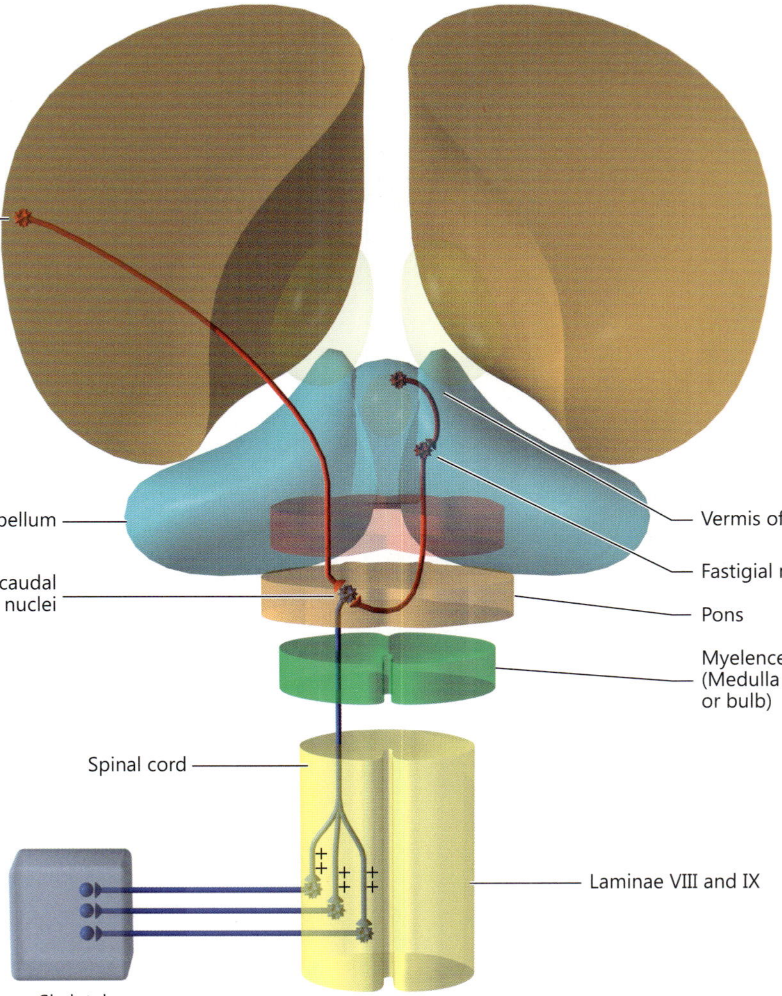

Pontoreticulospinal tract (or medial reticulospinal tract)		
Upper motor neuron	From the primar motor (MI) and premotor (PM) areas and supplementary motor area (SMA) to the rostral and caudal pontine reticular nuclei	
Oral and caudal pontine reticular nuclei	From the oral and caudal pontine reticular nuclei to the laminae VIII and IX of all the spinal cord	*Afferences* – from the fastigial nucleus (it receives both from the archicerebellum and the vermis of cerebellum) – from the spinal cord (spinoreticular tract)
Lower motor neuron	The α and γ motor neurons of the spinal cord innervating the axial muscles and the extensor muscles of the limbs are stimulated	

The **pontoreticulospinal tract** (or medial reticulospinal tract) originates from the oral and caudal pontine reticular nuclei and ends in the laminae VII and IX of the entire spinal cord and plays a facilitation function on axial muscles and the extensor muscles of the limbs. These nuclei receive afferences from the same areas of the cerebral cortex where the pyramidal tract originates, from the fastigial nucleus (nucleus of the archicerebellum to which the cortex of the vermis of cerebellum projects) and the spinal cord.

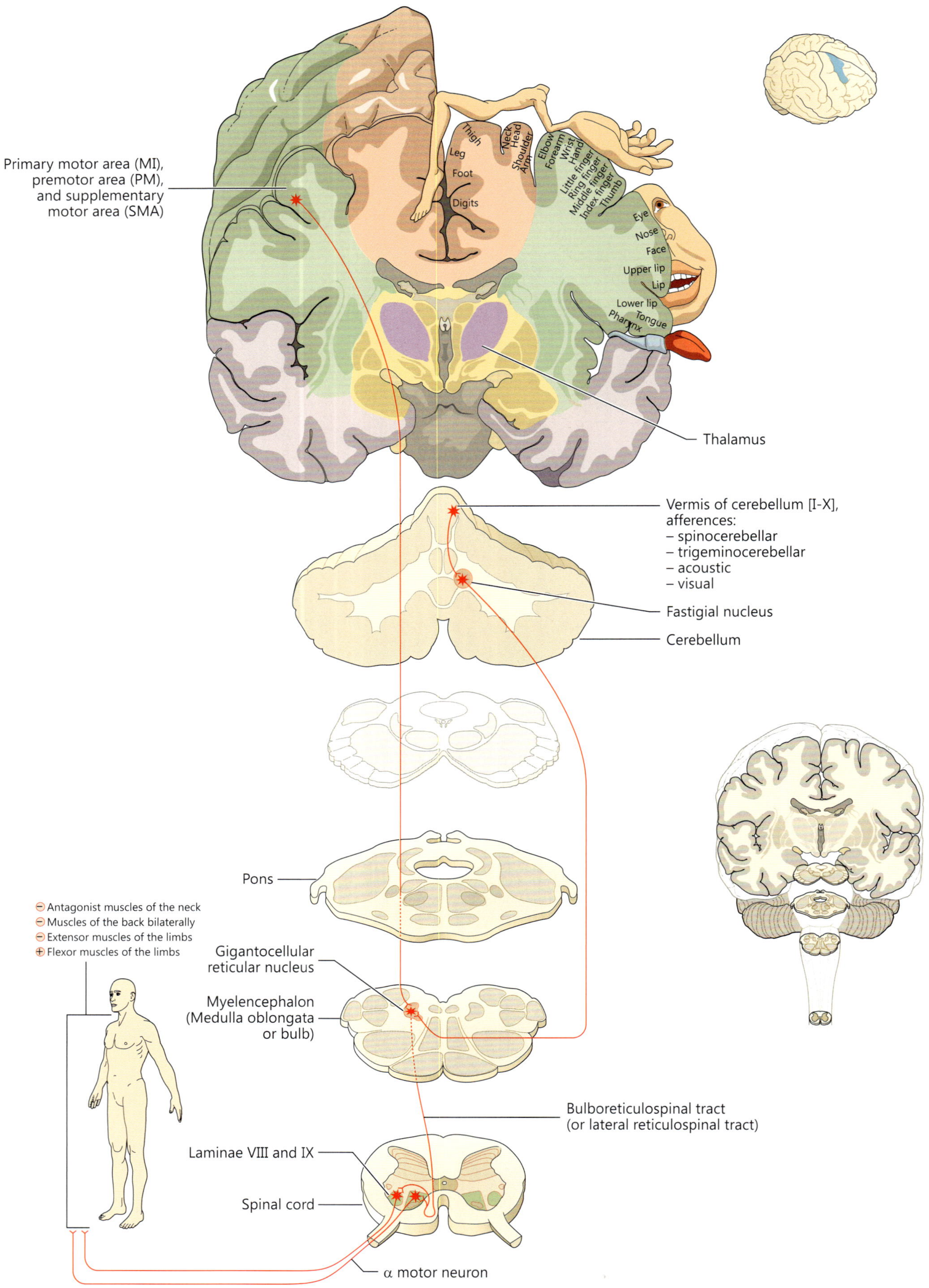
Primary motor area (MI), premotor area (PM), and supplementary motor area (SMA)
Thigh
Leg
Foot
Digits
Neck
Head
Shoulder
Arm
Elbow
Forearm
Wrist
Hand
Little finger
Ring finger
Middle finger
Index finger
Thumb
Eye
Nose
Face
Upper lip
Lip
Lower lip
Tongue
Pharynx
Thalamus
Vermis of cerebellum [I-X], afferences:
– spinocerebellar
– trigeminocerebellar
– acoustic
– visual
Fastigial nucleus
Cerebellum
Pons
⊖ Antagonist muscles of the neck
⊖ Muscles of the back bilaterally
⊖ Extensor muscles of the limbs
⊕ Flexor muscles of the limbs
Gigantocellular reticular nucleus
Myelencephalon (Medulla oblongata or bulb)
Bulboreticulospinal tract (or lateral reticulospinal tract)
Laminae VIII and IX
Spinal cord
α motor neuron

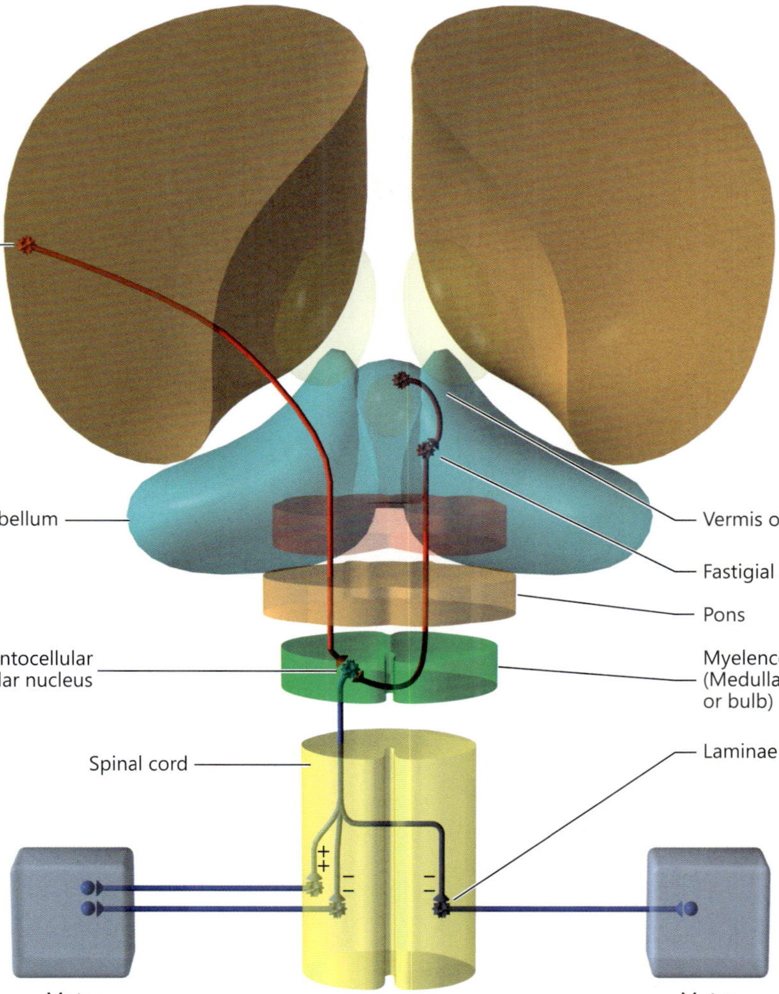

Lateral bulboreticulospinal tract (or lateral reticulospinal tract)		
The **reticulospinal tract** (pontoreticulospinal and lateral bulboreticulospinal tracts) coordinates posture and movement integrating the vestibular signals, the signals of muscle proprioceptors and those proceeding from other sensory afferences upon commands by the cerebral cortex		
Upper motor neuron	From the primary motor (MI) and premotor (PM) areas and supplementary motor area (SMA) to the gigantocellular reticular (bulbar) nucleus	
Gigantocellular reticular nucleus	From the gigantocellular reticular nucleus, that receives fibers from the lateral reticular (bulbar) nucleus, to the laminae VIII and IX of all the spinal cord	*Afferences* – from the fastigial nucleus (it receives both from the archicerebellum and the vermis of cerebellum) – from the spinal cord (spinoreticular tract)
Lower motor neuron	The α and γ motor neurons of the spinal cord innervating the flexor muscles of the limbs are facilitated. The α and γ motor neurons of the spinal cord innervating the extensor muscles of the limbs and bilaterally the muscles of the neck and back are inhibited	

The **lateral bulboreticulospinal tract** originates from the gigantocellular reticular nucleus, that also receives fibers from the lateral reticular nucleus, and ends in the laminae VIII and IX of all the spinal cord. This tract has a bilateral inhibitory function on the muscles of the neck and back; it also has an inhibitory function on the extensor muscles and facilitates the flexor muscles of the limbs. These nuclei receive afferences from the same areas of the cerebral cortex where the pyramidal tract originates, from the fastigial nucleus (nucleus of the archicerebellum to which the cortex of the vermis of the cerebellum projects) and the spinal cord.

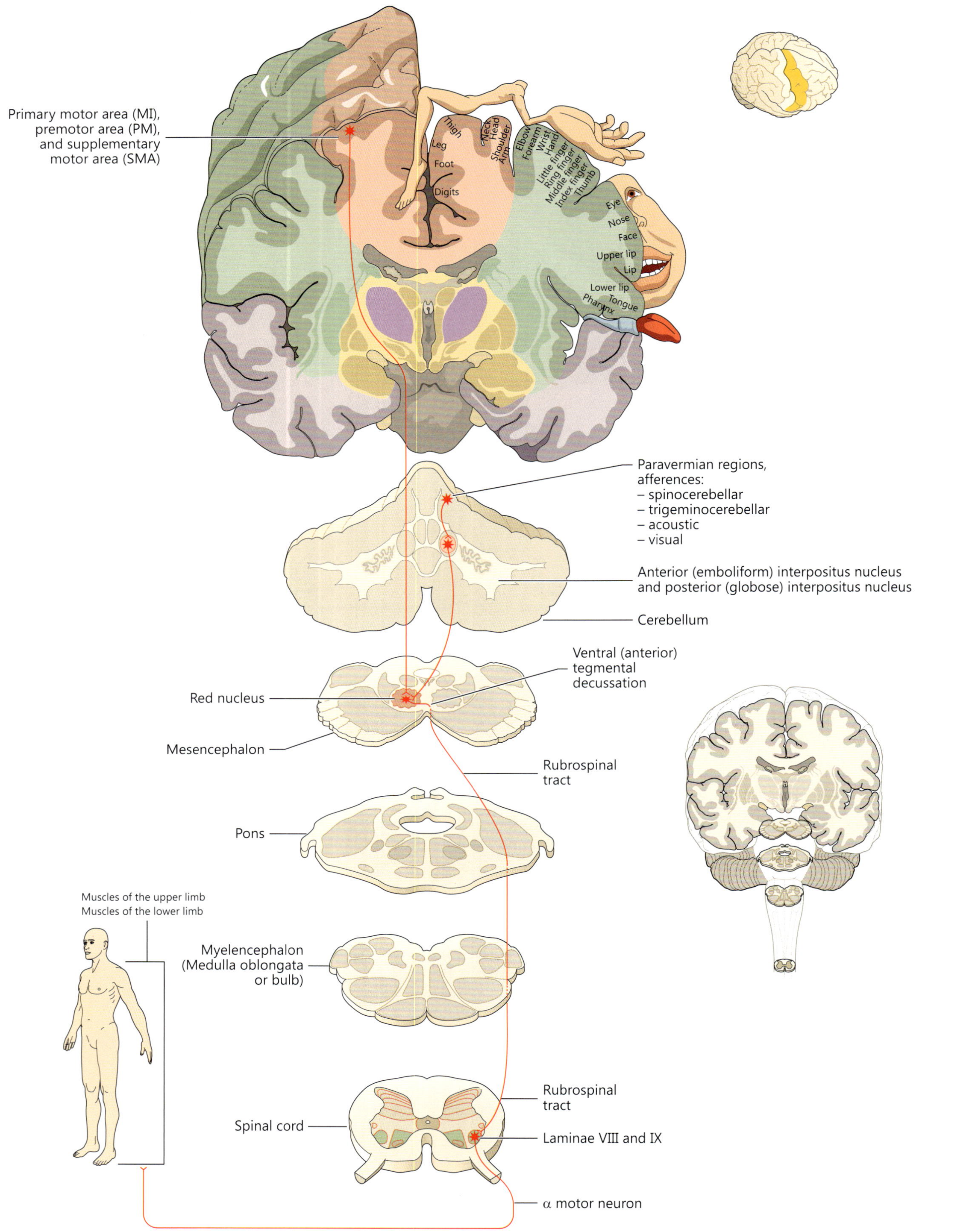
Primary motor area (MI), premotor area (PM), and supplementary motor area (SMA)
Thigh
Leg
Foot
Digits
Neck
Head
Shoulder
Arm
Elbow
Forearm
Wrist
Hand
Little finger
Ring finger
Middle finger
Index finger
Thumb
Eye
Nose
Face
Upper lip
Lip
Lower lip
Tongue
Pharynx
Paravermian regions, afferences:
– spinocerebellar
– trigeminocerebellar
– acoustic
– visual
Anterior (emboliform) interpositus nucleus and posterior (globose) interpositus nucleus
Cerebellum
Ventral (anterior) tegmental decussation
Red nucleus
Mesencephalon
Rubrospinal tract
Pons
Muscles of the upper limb
Muscles of the lower limb
Myelencephalon (Medulla oblongata or bulb)
Rubrospinal tract
Spinal cord
Laminae VIII and IX
α motor neuron

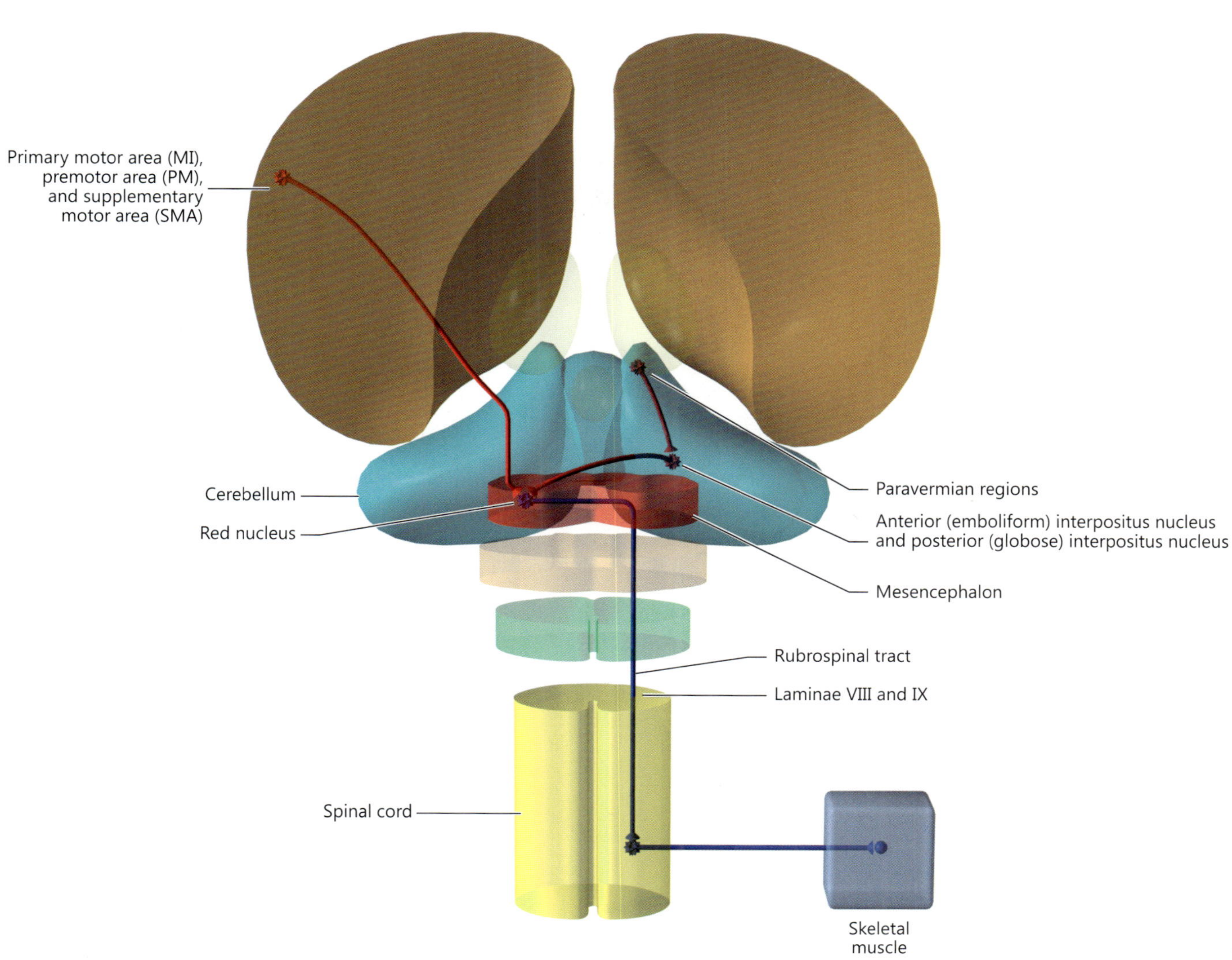

Rubrospinal tract		
Upper motor neuron	From the primary motor (MI), premotor (PM) and supplementary motor (SMA) areas to the magnocellular part of the red nucleus	
Red nucleus	From the magnocellular part of the red nucleus to the laminae VIII and IX of all the spinal cord	*Afferences* – from the anterior (emboliform) interpositus nucleus and posterior (globose) interpositus nucleus (nuclei of paleocerebellum receiving afferences from the paravermian regions of the cerebellar cortex; these nuclei are also simply referred to as interpositus nucleus)
Lower motor neuron	The α and γ motor neurons of the spinal cord innervating the flexor muscles of the limbs are facilitated	

The **rubrospinal tract** originates from the magnocellular part of the red nucleus, in the context of which there is a somatotopic structure, it immediately decussates across the midline, descending in the opposite antimere, and ending in the laminae VIII and IX of the entire spinal cord. This tract has a facilitation function on the flexor muscles of the limbs. The red nucleus receives afferences from the same motor and premotor areas from which the pyramidal tract originates and from the globose and emboliform nuclei (nuclei of the paleocerebellum that receive afferences from the paravermian regions of the cerebellar cortex).

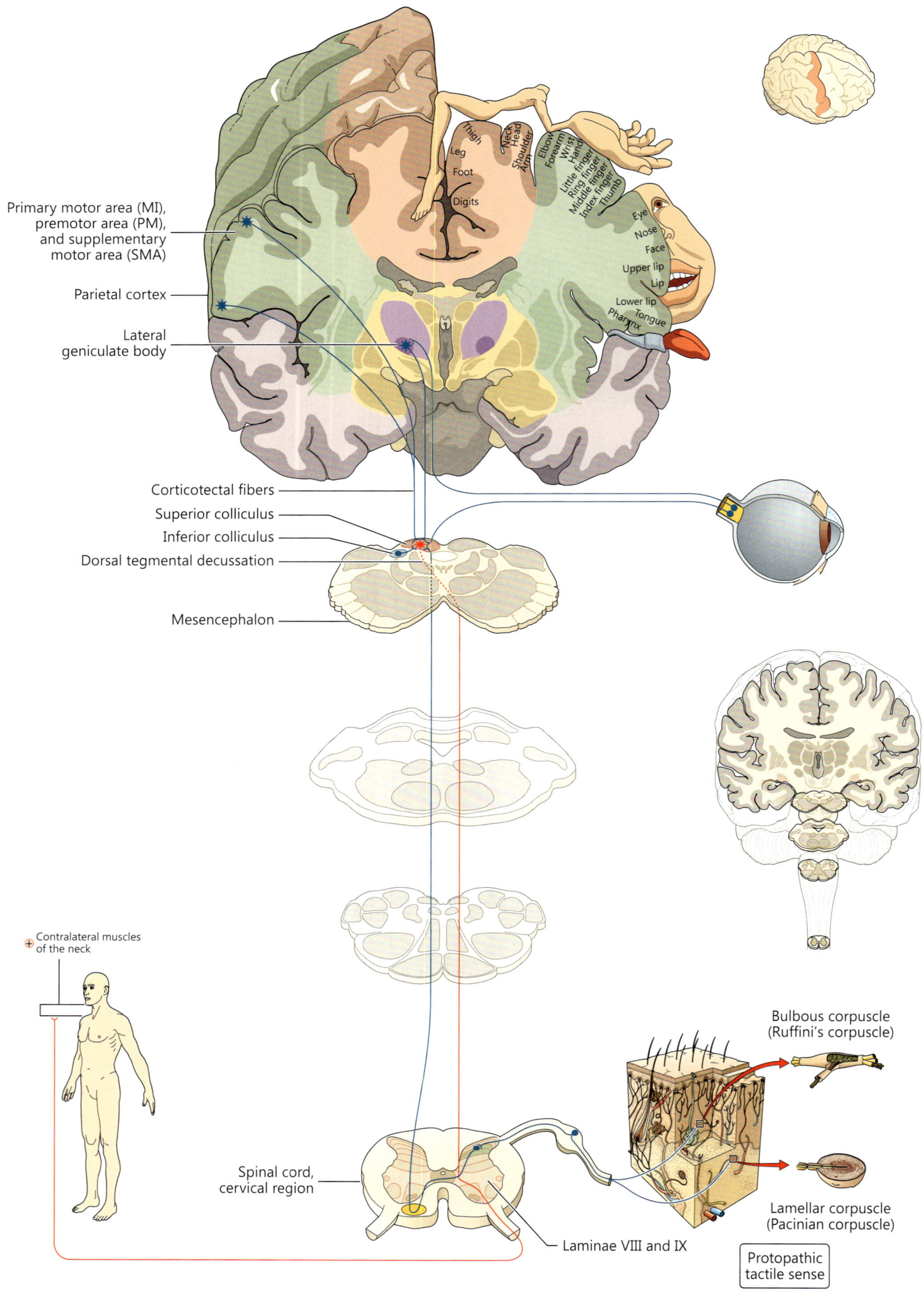
Primary motor area (MI), premotor area (PM), and supplementary motor area (SMA)
Parietal cortex
Lateral geniculate body
Corticotectal fibers
Superior colliculus
Inferior colliculus
Dorsal tegmental decussation
Mesencephalon
Contralateral muscles of the neck
Spinal cord, cervical region
Laminae VIII and IX
Bulbous corpuscle (Ruffini's corpuscle)
Lamellar corpuscle (Pacinian corpuscle)
Protopathic tactile sense
Thigh
Leg
Foot
Digits
Neck
Head
Shoulder
Arm
Elbow
Forearm
Wrist
Hand
Little finger
Ring finger
Middle finger
Index finger
Thumb
Eye
Nose
Face
Upper lip
Lip
Lower lip
Tongue
Pharynx

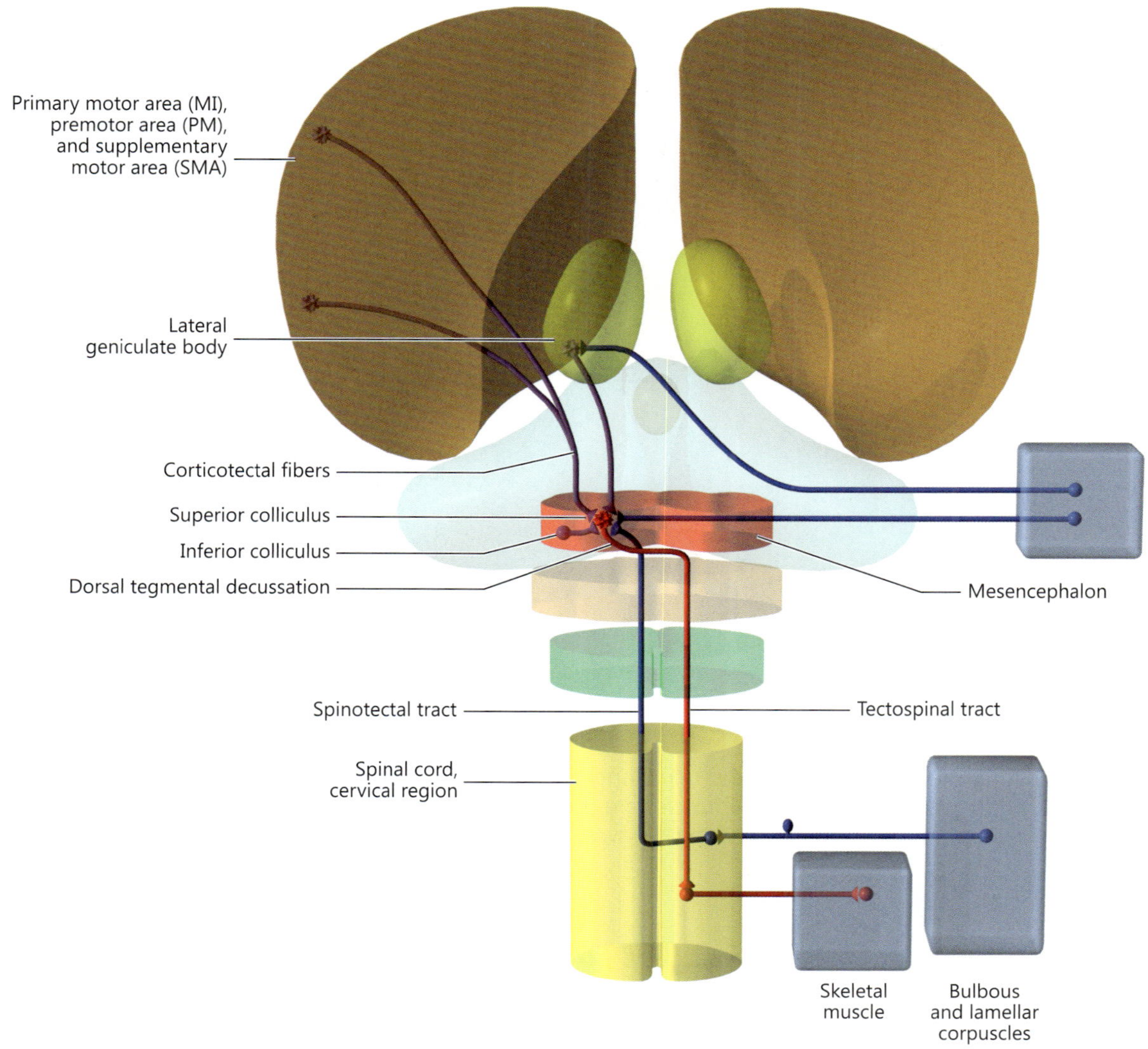

Tectospinal tract		
The **tectospinal tract**, the only pathway among the extrapyramidal ones which does not receive afferences from the cerebellum, it is important for coordinating the head and eyes and, therefore, also for maintaining posture in relation to the visuo-spatial environment.		
Upper motor neuron	From the primary motor (MI), premotor (PM) and supplementary motor (SMA) areas to the superior colliculus	
Superior colliculus	From the superior colliculus mainly to the laminae VIII and IX of the spinal cord bilaterally	*Afferences* – motor visual areas of the cerebral cortex – lateral geniculate body – inferior colliculus – spinal cord (spinotectal tract)
Lower motor neuron	The α and γ motor neurons of the spinal cord innervating the contralateral and homolateral muscles of the neck are facilitated. The α and γ motor neurons of the spinal cord innervating the homolateral muscles of the neck are also inhibited	

The **tectospinal tract** originates from the superior colliculus and ends mainly in laminae VIII and IX of the cervical spinal cord bilaterally. This tract has a facilitation function on the contralateral muscles of the neck and both facilitation and inhibition function on the homolateral muscles of the neck. The superior colliculi receive afferences from the same areas of the cerebral cortex from which the pyramidal tract originates, from the motor visual areas of the cerebral cortex, the lateral geniculate body, the inferior colliculus and the spinal cord through the spinotectal tract.

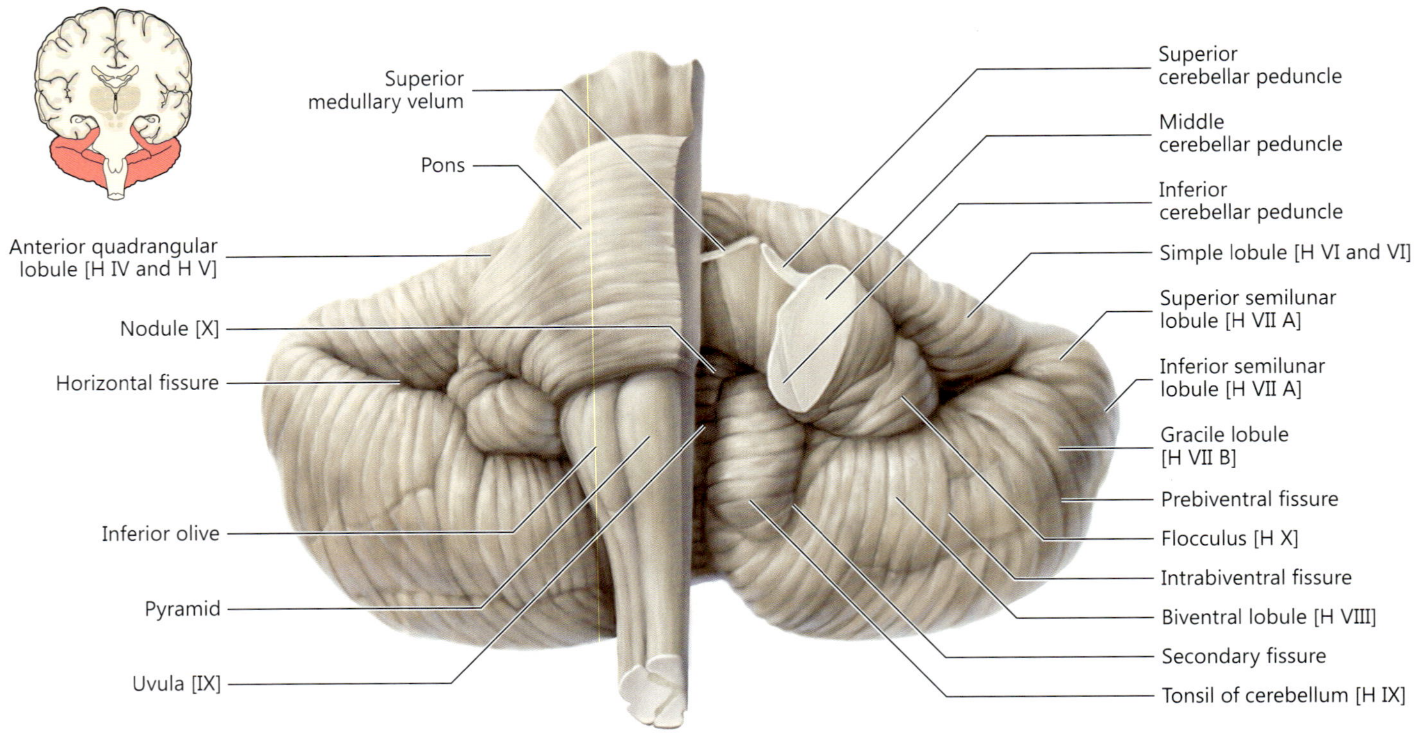

Brainstem and **cerebellum**, anterior view. To the left, the uvula and nodule, the lobules of the vermis of the cerebellum, the tonsil of the cerebellum and sections of the cerebellar peduncles are shown after removing bulb and pons. The superior medullary velum is a surface layer of white matter present between the two superior cerebellar peduncles.

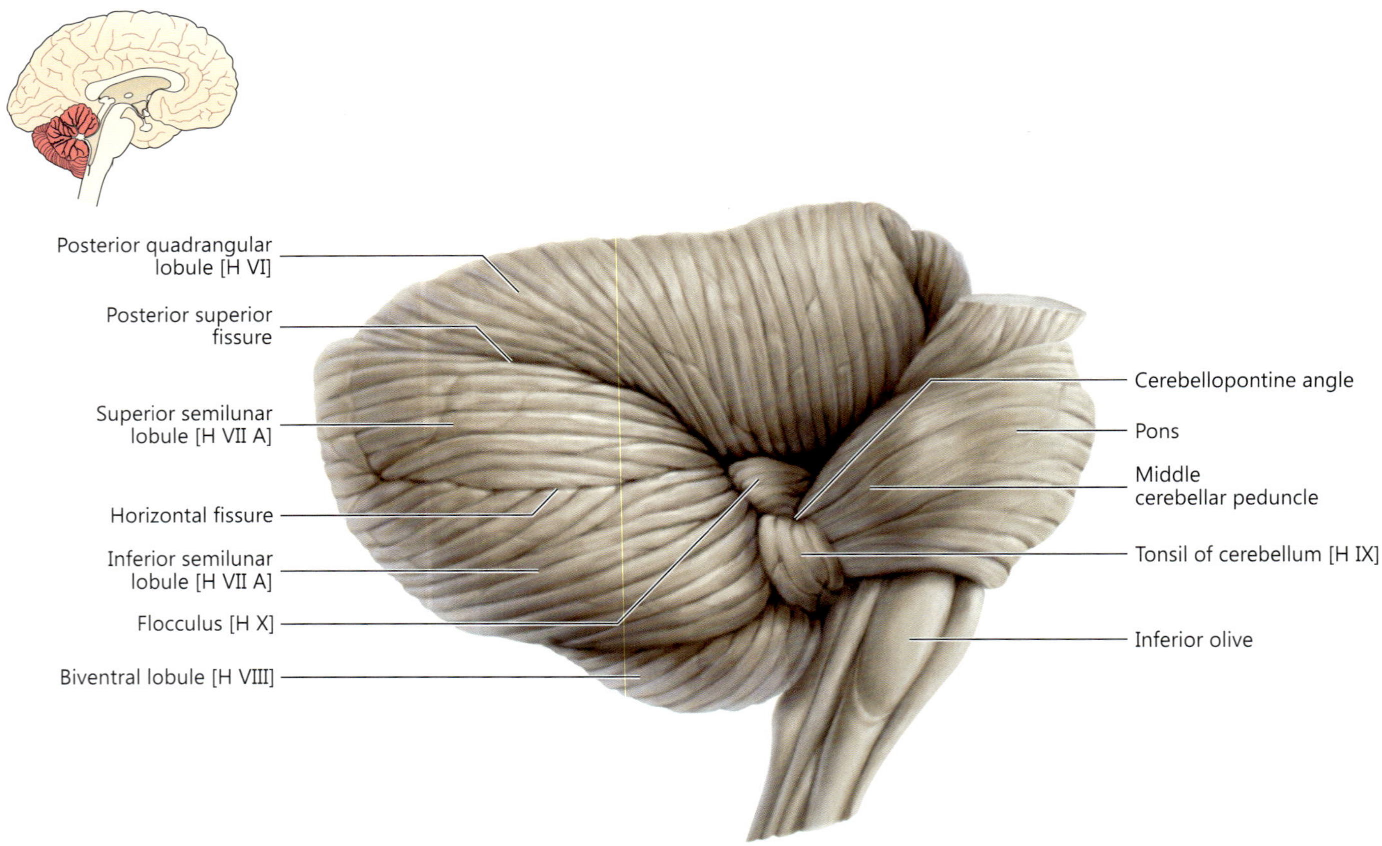

Cerebellum, lateral view. The cerebellopontine angle is a triangular area bordered medially and dorsally by the pons, positioned anteriorly, and the cerebellum positioned posteriorly, in correspondance of the angle formed by the ventral surface of the cerebellum at the level of the flocculus.

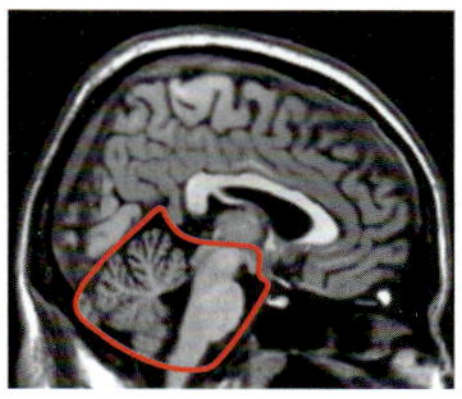

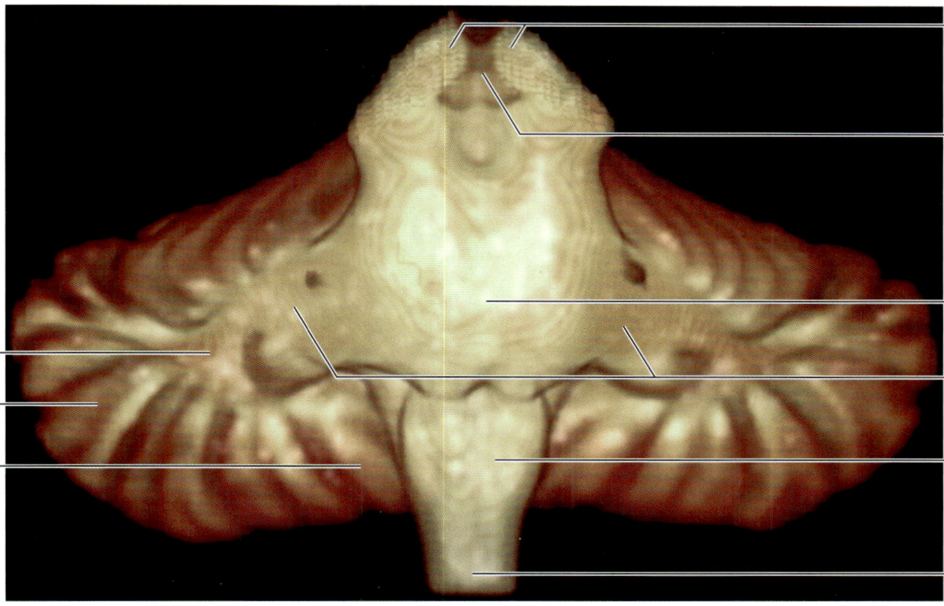

Three-dimensional image of a living human body, obtained by NMR using a 3D volume rendering technique, showing the **cerebellum** and its relationships with the brainstem in anterior projection.

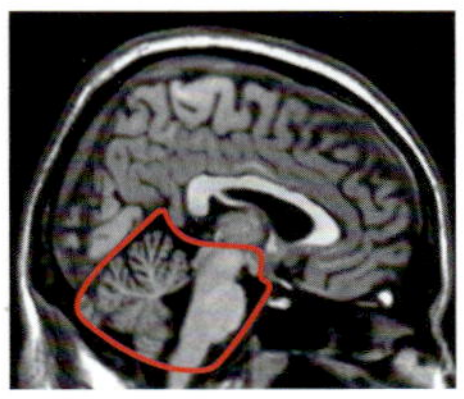

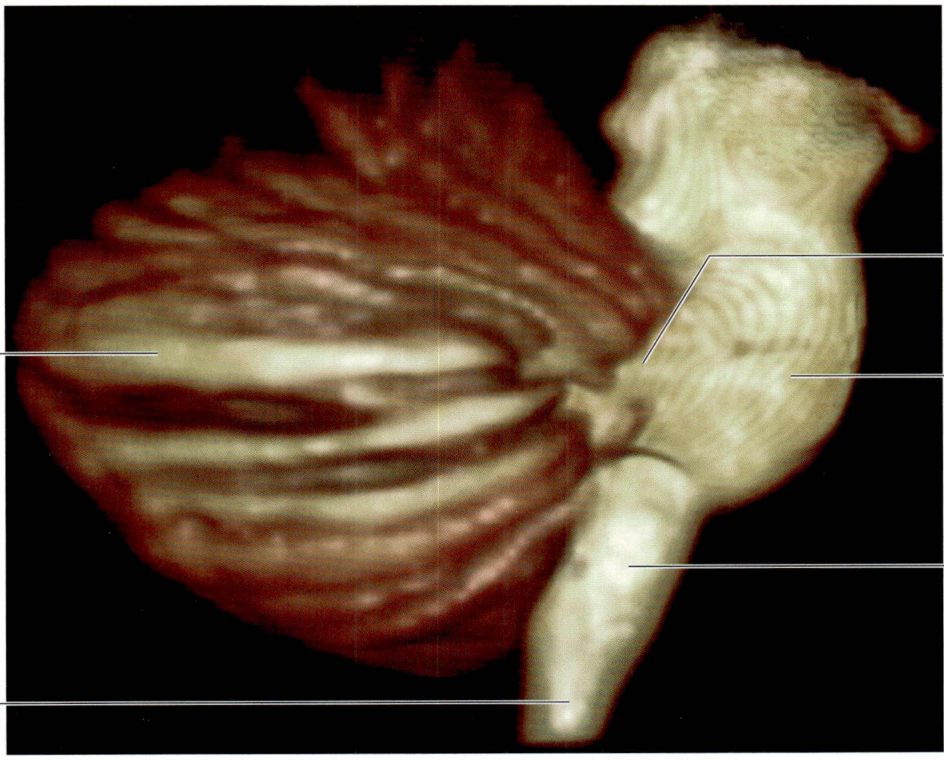

Three-dimensional image of a living human body, obtained by NMR using a 3D volume rendering technique, showing the **cerebellum** and its relationships with the brainstem in lateral projection.

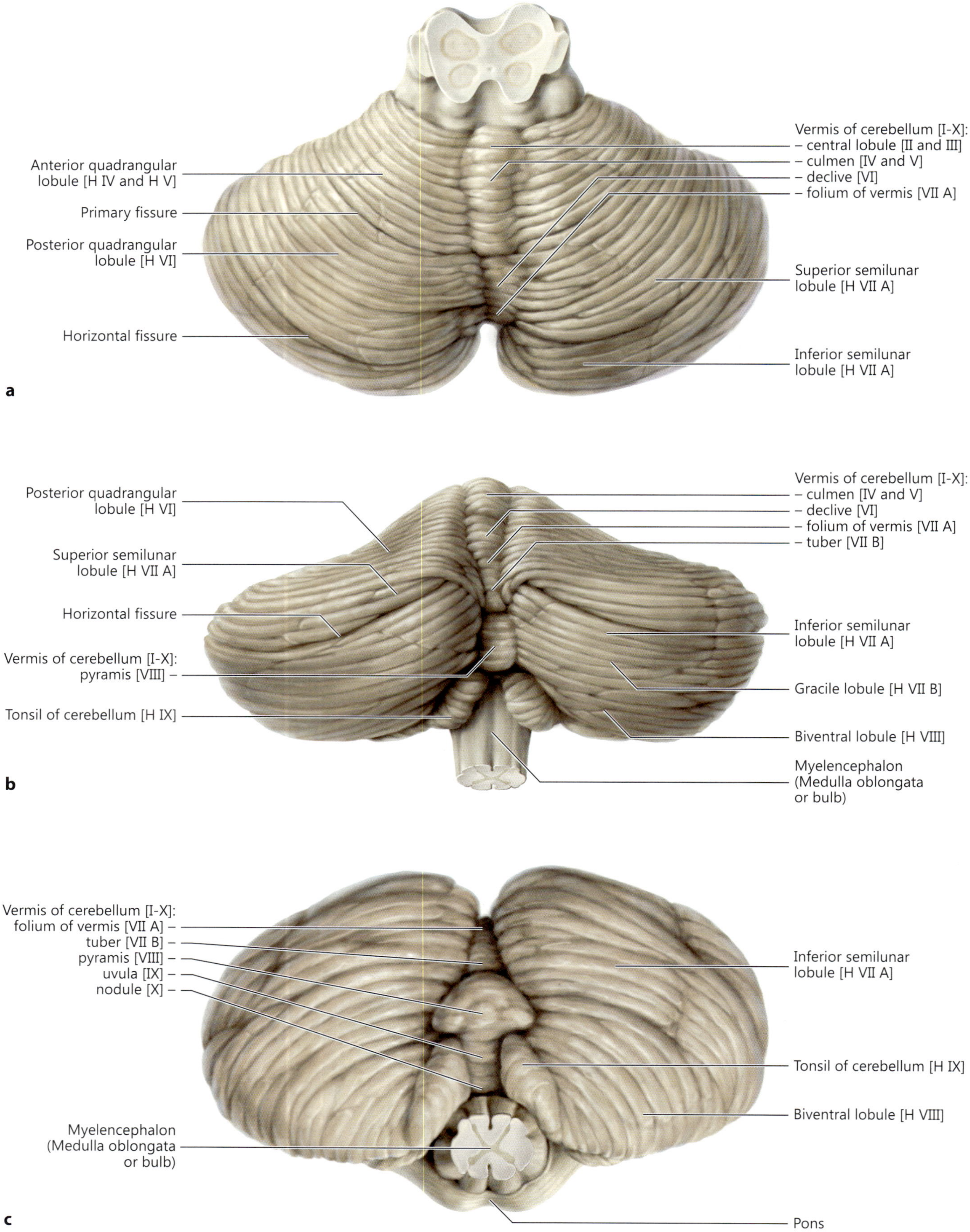

Schematic representation showing the **cerebellum** seen from above (**a**), in posterior projection (**b**) and seen from below (**c**). The vermis territory is bounded by two sulci and the cerebellar surface is covered by deep fissures that separate distinct territories.

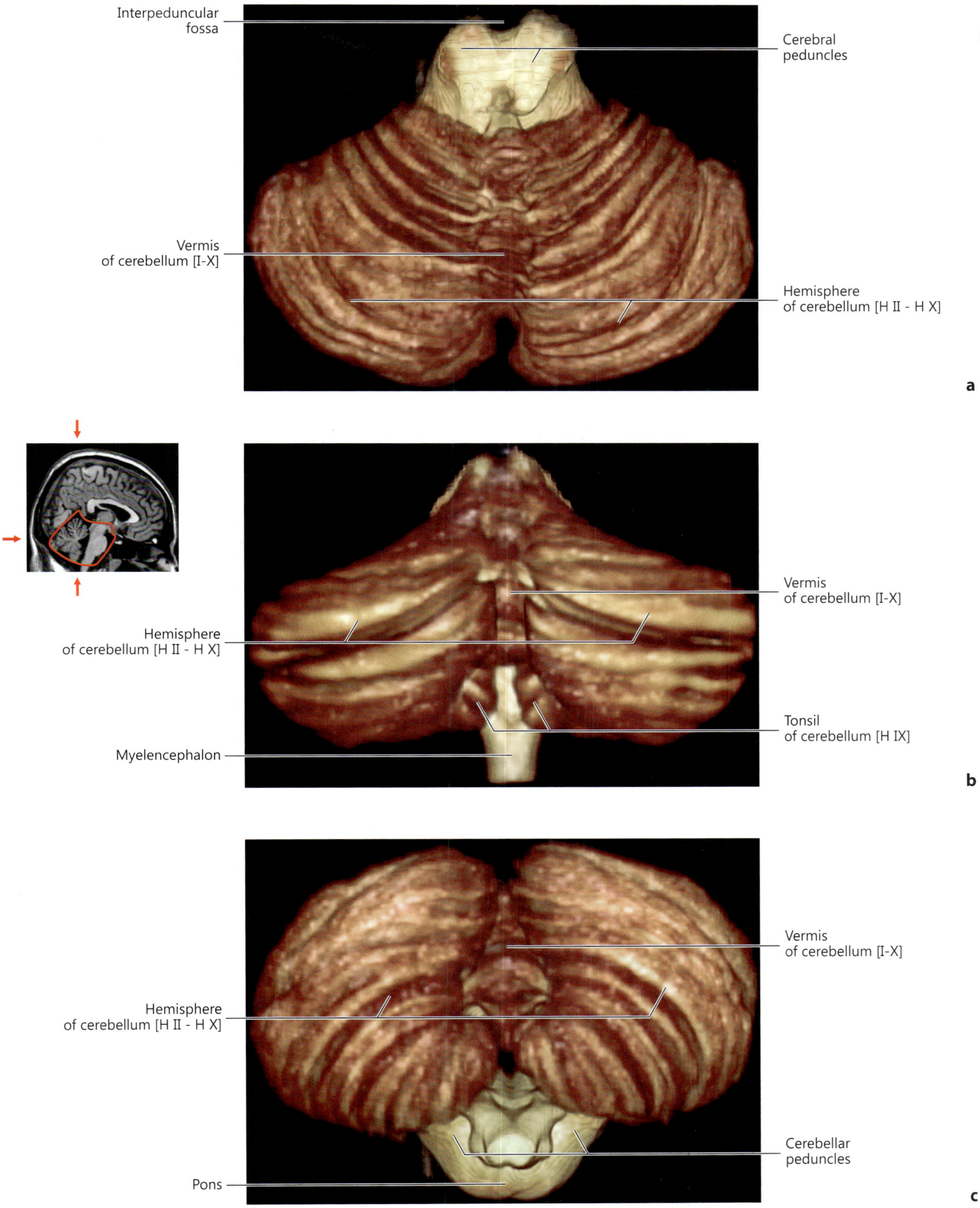

Three-dimensional image of a living human body, obtained by NMR using a 3D volume rendering technique, showing the **cerebellum** and its relationships with the brainstem in superior (**a**), posterior (**b**) and inferior (**c**) projection.

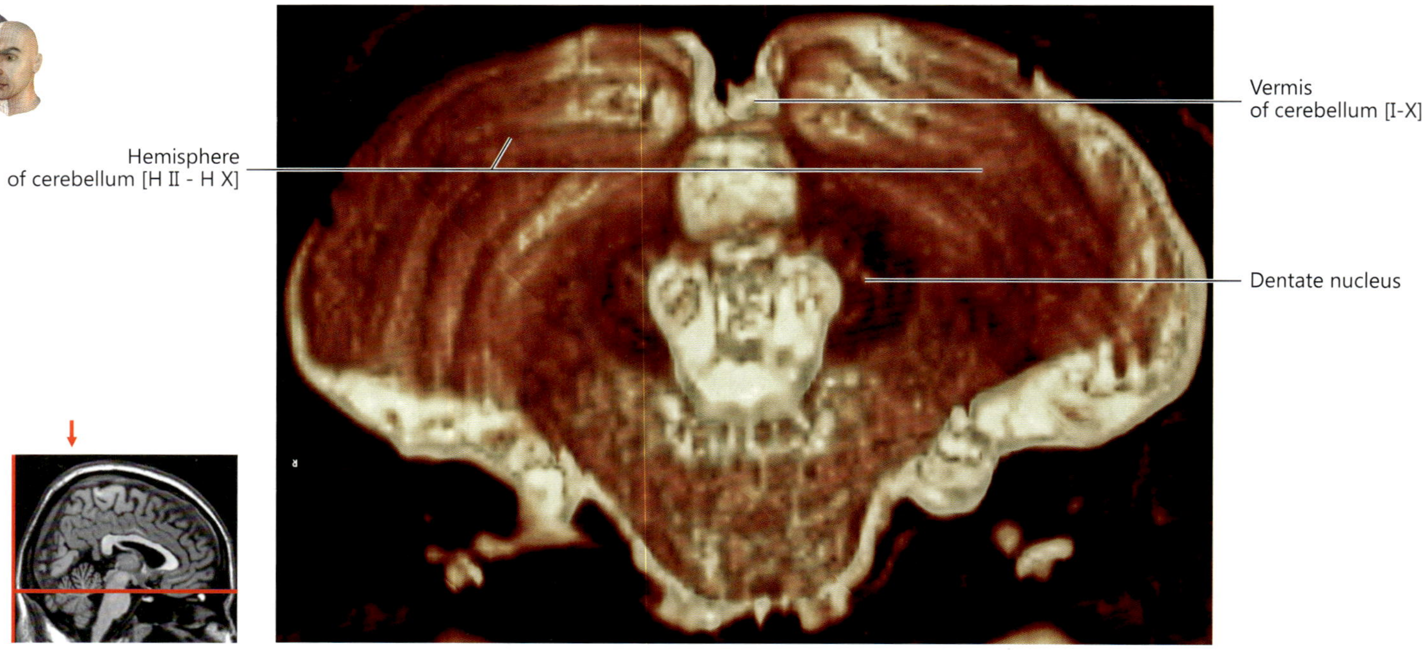

Three-dimensional image of a living human body, obtained by NMR using a 3D volume rendering technique, showing a section of the **cerebellum** passing through the superior cerebellar peduncles.

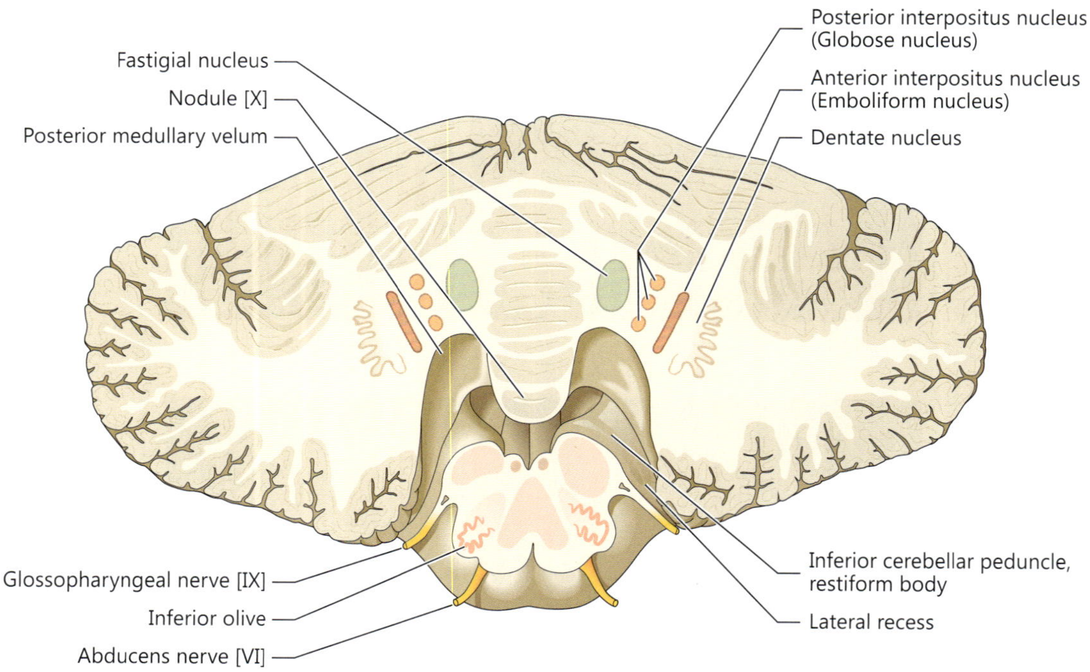

Sections of the **cerebellum** and brainstem, at the level of the lateral recess of the fourth ventricle, display their internal conformation; the cutting plane crosses the dentate nucleus, the most voluminous among the intrinsic nuclei of the cerebellum. The surface gray coat of the cerebellar cortex lies on top of the white matter of the cerebellum.

➤

Schematic representation showing how the **cerebellum** is divided into lobes (**a**, median sagittal section; **b**, superior surface; **c**, inferior surface). The cerebellum is functionally divided into three portions: the vermis [I-X], located centrally, the intermediate zones and the lateral areas of the hemispheres [H II-H X]. The primary fissure divides the anterior lobe from the posterior lobe; moreover, it is shown how the posterolateral fissure divides the flocculonodular lobe from the anterior and posterior lobes.

Lobes and lobules of the cerebellum delimited by the fissures on its surface		
LOBES	***LOBULES***	
	VERMIS [I-X]	***CEREBELLAR HEMISPHERES [H II-H X]***
Anterior lobe	Lingula [I]	
	Precentral or postlingual fissure	
	Central lobule [II and III] – anterior or ventral part [II] – posterior or dorsal part [III]	Wing of the central lobule [H II and III H] – inferior or ventral part [H II] – superior or dorsal part [H III]
	Preculminate or postcentral fissure	
	Culmen [IV and V] – anterior or ventral part [IV]	
	Intraculminate fissure	
	– posterior or dorsal part [V]	Anterior quadrangular lobule [H IV and H V] – anterior or ventral part [H IV] – posterior or dorsal part [H V]
	Primary fissure; Preclival fissure	
Posterior lobe	Declive [VI] Simple lobule [VI]	Simple lobule [H VI] Posterior quadrangular lobule [H VI]
	Posterior superior fissure or postclival fissure	
	Folium of vermis [VII A]	Semilunar lobules or ansiform lobule [H VII A] – superior semilunar lobule or first crus of ansiform lobule [H VII A]
	Horizontal or intercrural fissure	
		– inferior semilunar lobule or second crus of ansiform lobule [H VII A]
	Lunogracile or ansoparamedian fissure	
	Tuber [VII B]	Gracile or paramedian lobule [H VII B]
	Prebiventral or prepyramidal fissure	
	Pyramis [VIII]	Biventral lobule [H VIII] – lateral part of the biventral lobule or pars copularis of the paramedian lobule [H VIII A]
	Intrabiventral or anterior inferior fissure	
		– medial part of the biventral lobule or dorsal parafloccularis [H VIII B]
	Secondary or postpyramidal fissure	
	Uvula [IX]	Tonsil of cerebellum or ventral paraflocculus [H IX]
	Posterolateral fissure	
Flocculonodular lobe	Nodule [X] Peduncle of flocculus	Flocculus [H X]

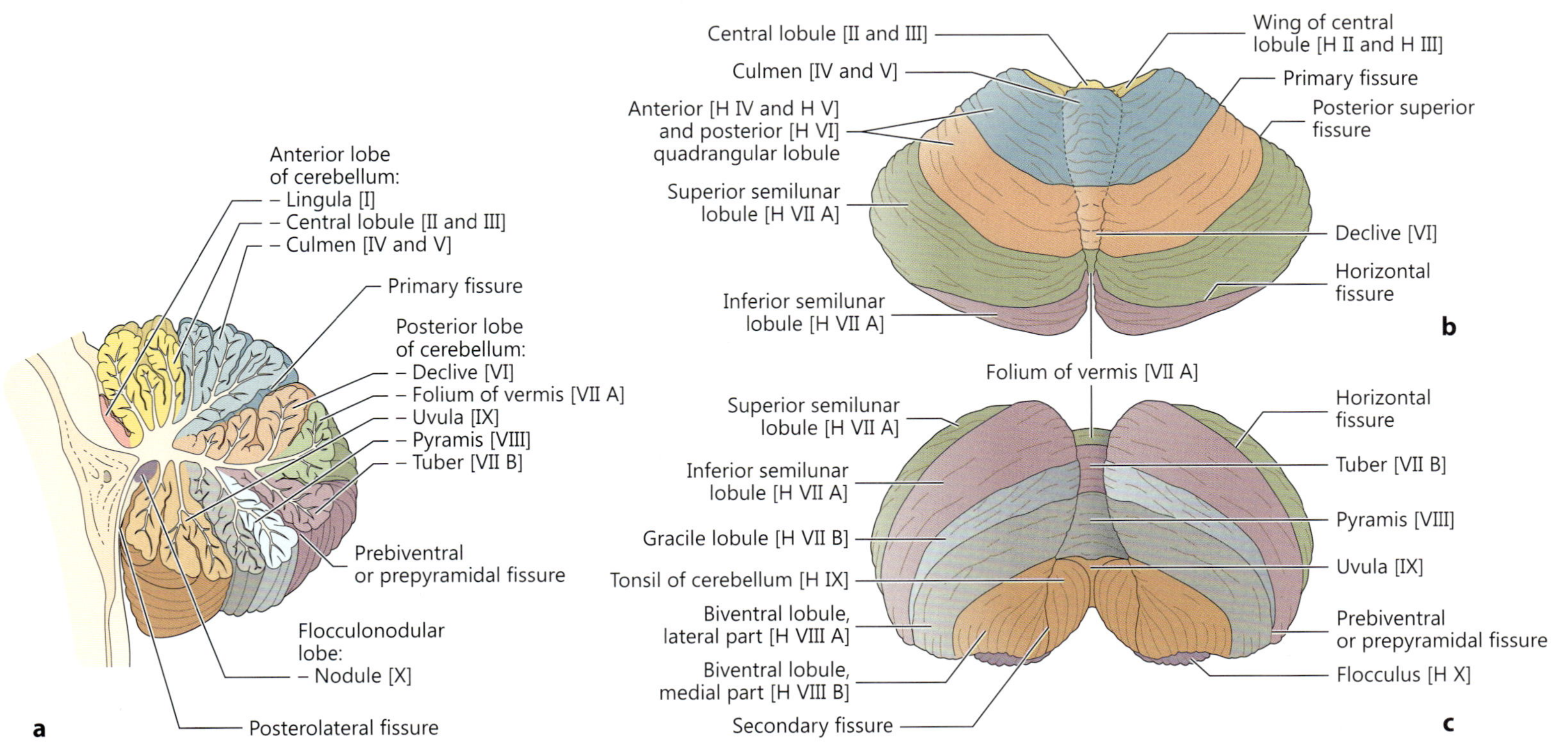

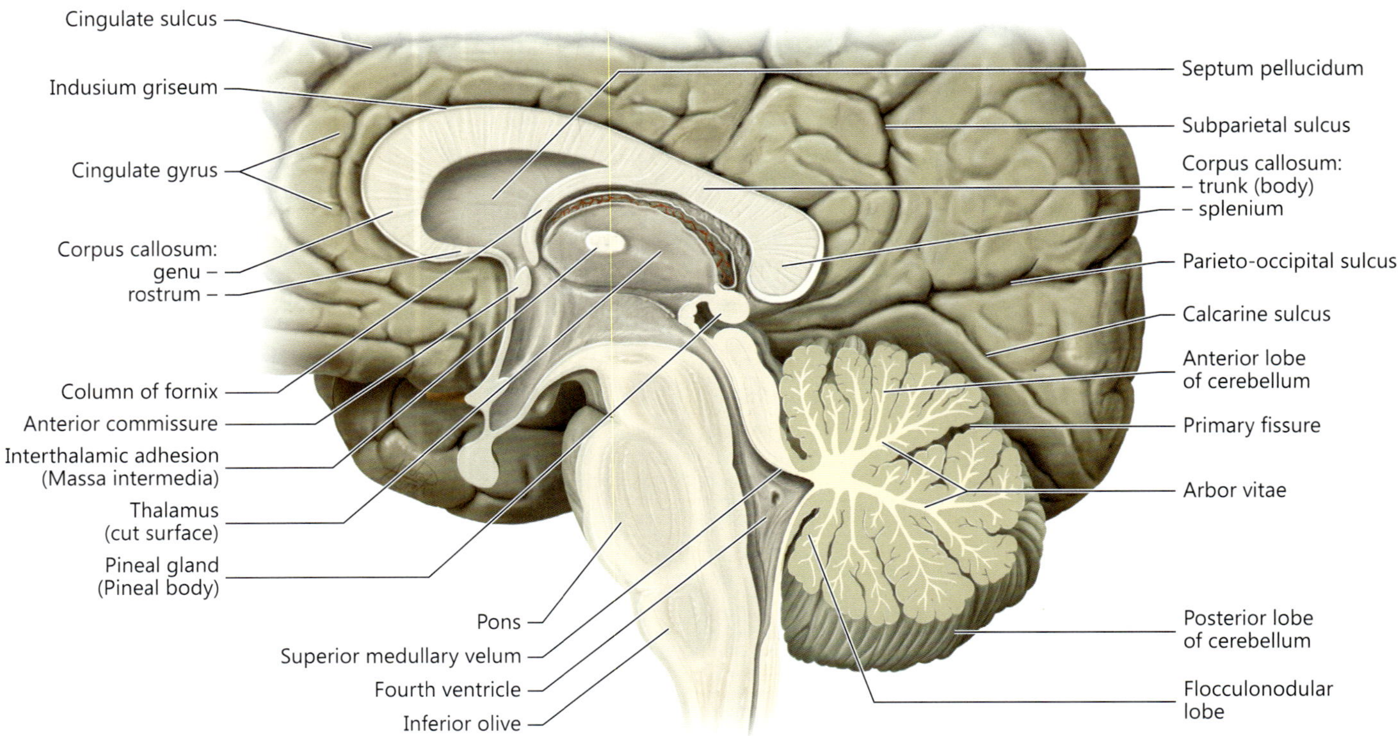

Median sagittal section of the **brain**, left lateral view. The cingulate gyrus, which forms part of the limbic lobe, the corpus callosum, the septum pellucidum and the fornix are particularly evident. The medial surface of the thalamus connects to the medial surface of the contralateral thalamus by a bridge of gray matter, the interthalamic adhesion.

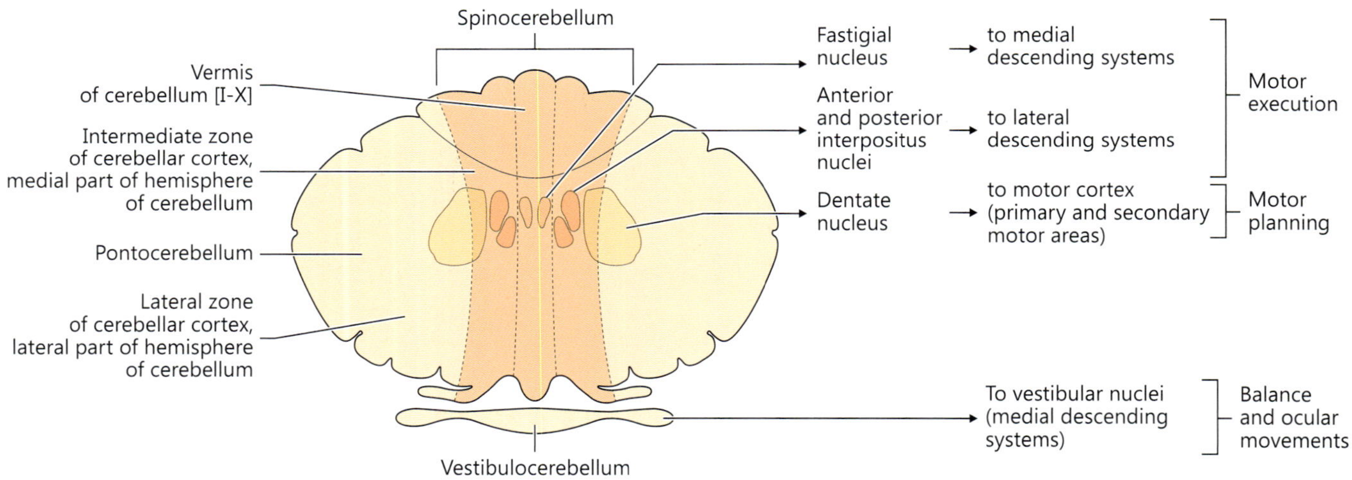

a

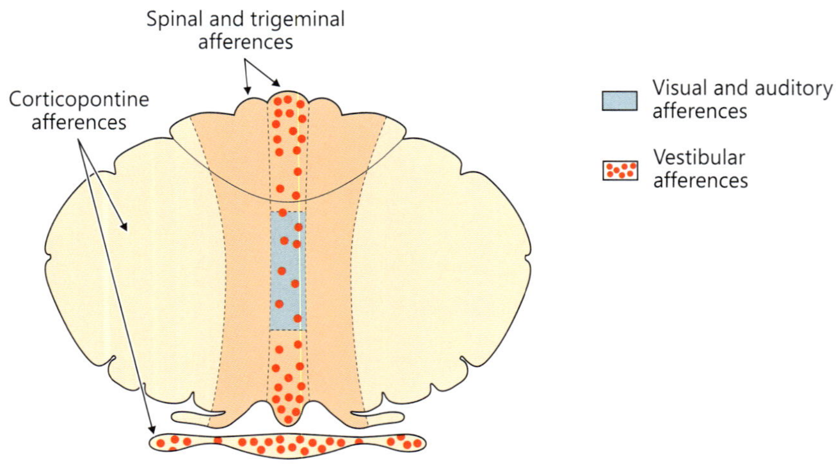

b

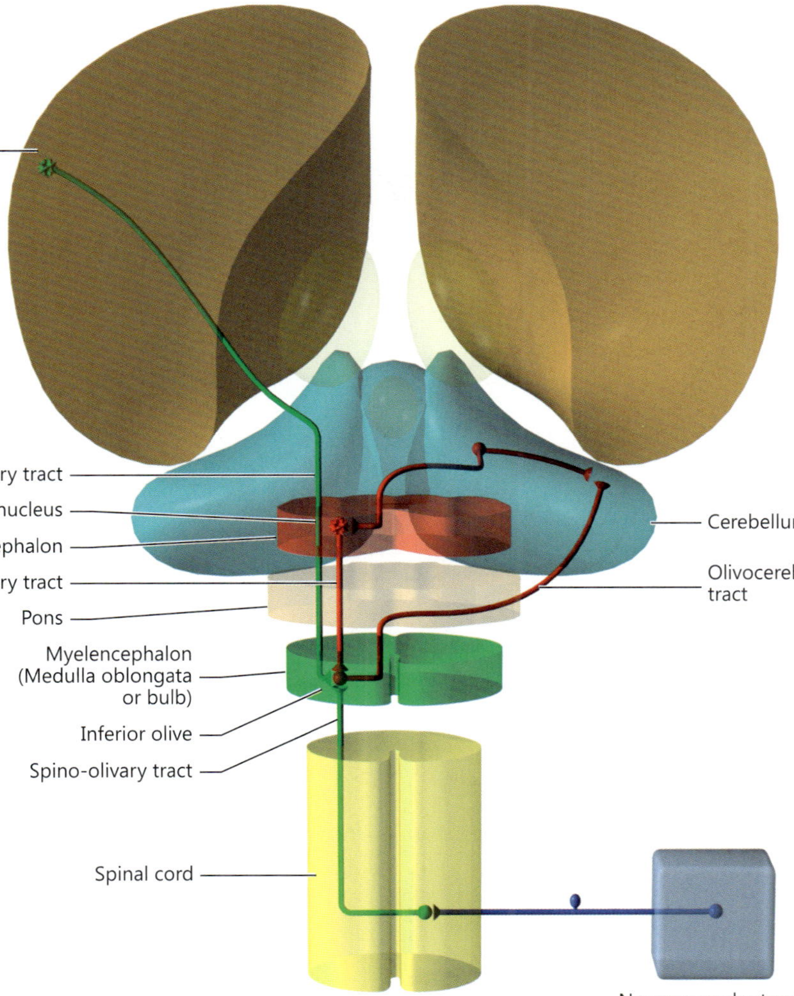

Inferior olive (acquisition of motor parameters) (*green* in the graphs)	
Afferences	From the primary motor (MI) and premotor (PM) areas of the cerebral cortex and from the laminae VI and VII of the spinal cord to the inferior olive
Efferences	From the inferior olive to the entire cerebellar cortex (archi-, paleo- and neocerebellum)

Inferior olive (memorization of motor parameters) (*red* in the graphs)*	
Neuron I	From the inferior olive to the cerebellar cortex
Neuron II	From the nuclei of the cerebellum and, in particular, from the dentate nucleus to the parvocellular part of the red nucleus (neorubro)
Neuron III	From the neorubro to the inferior olive

* The order in roman numbers of the three neurons is only virtual because it is a closed circuit that is repeated until the parameters change and new parameters are acquired.

The cerebellum is involved in the construction of the motor gesture acting on the inferior motor neuron through the **vestibulocerebellum circuits** (☞ Vestibular system, lateral and medial vestibulospinal tracts) and through the **paleocerebellum circuits** (☞ Brainstem, reticolospinal tracts - medial and lateral - and rubrospinal); it also acts on the superior motor neuron through the neocerebellum circuit (*blue*) (☞ Neocerebellum circuit, next page). In particular, in the cooperation between the cerebral cortex and the cerebellar cortex, the inferior olive plays a key role, both in learning the parameters that characterize the specific motor action (*green*) and in memorizing the same parameters (*red*).

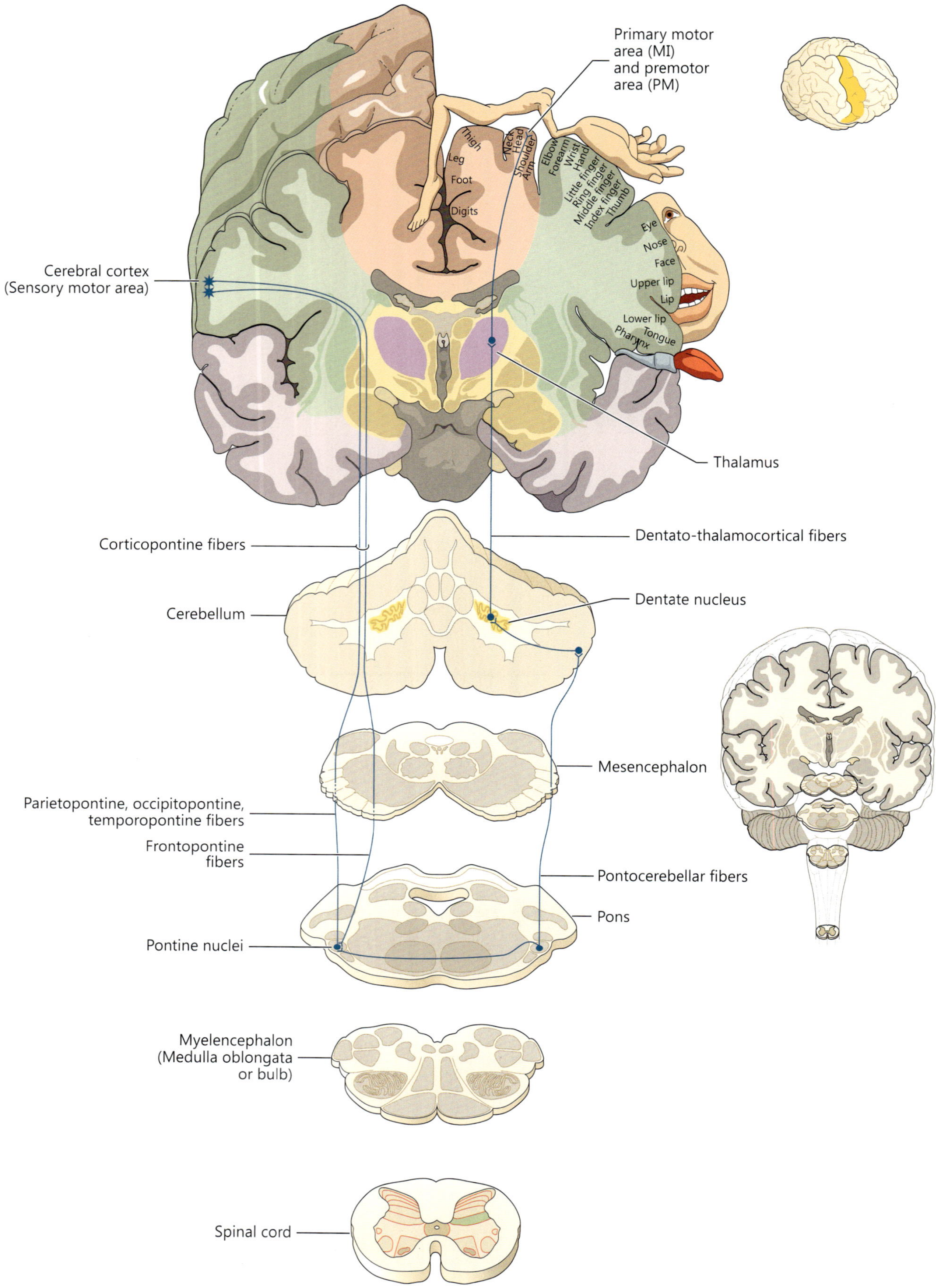
Primary motor area (MI) and premotor area (PM)
Thigh
Leg
Foot
Digits
Neck
Head
Shoulder
Arm
Elbow
Forearm
Wrist
Hand
Little finger
Ring finger
Middle finger
Index finger
Thumb
Eye
Nose
Face
Upper lip
Lip
Lower lip
Tongue
Pharynx
Cerebral cortex (Sensory motor area)
Thalamus
Corticopontine fibers
Dentato-thalamocortical fibers
Cerebellum
Dentate nucleus
Mesencephalon
Parietopontine, occipitopontine, temporopontine fibers
Frontopontine fibers
Pontocerebellar fibers
Pons
Pontine nuclei
Myelencephalon (Medulla oblongata or bulb)
Spinal cord

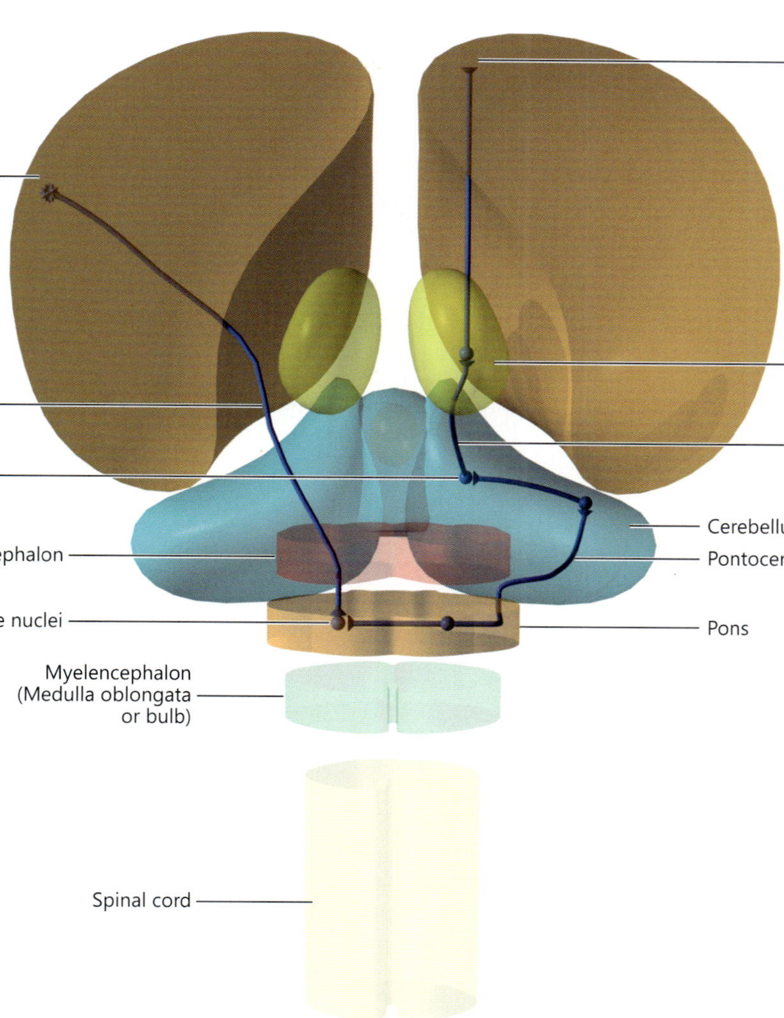

Neocerebellum circuit*	
First order neuron (I)	From the sensory motor areas of the cerebral cortex to the pontine nuclei
Second order neuron (II)	From the pontine nuclei to the neocerebellum cortex
Third order neuron (III)	From the neocerebellum cortex to the dentate nucleus of the cerebellum in the opposite side
Fourth order neuron (IV)	From the dentate nucleus of the cerebellum to the ventral lateral (VL) nuclei of the thalamus
Fifth order neuron (V)	From the VL nuclei of the thalamus to the primary motor (MI) and premotor (PM) areas

* It is called virtually, first-order neuron (I), only because it is the first order neuron to be mentioned in a circuit that can be described as "closed" or better said "reverberant".

The **neocerebellum** receives afferences from the sensory motor cerebral cortex through the pontine nuclei and reprojects to the motor cortex through the dentate nucleus and the thalamus. In other words, the neocerebellum is the only subcortical structure acting on the first motor neuron changing motor gesture also during its execution.

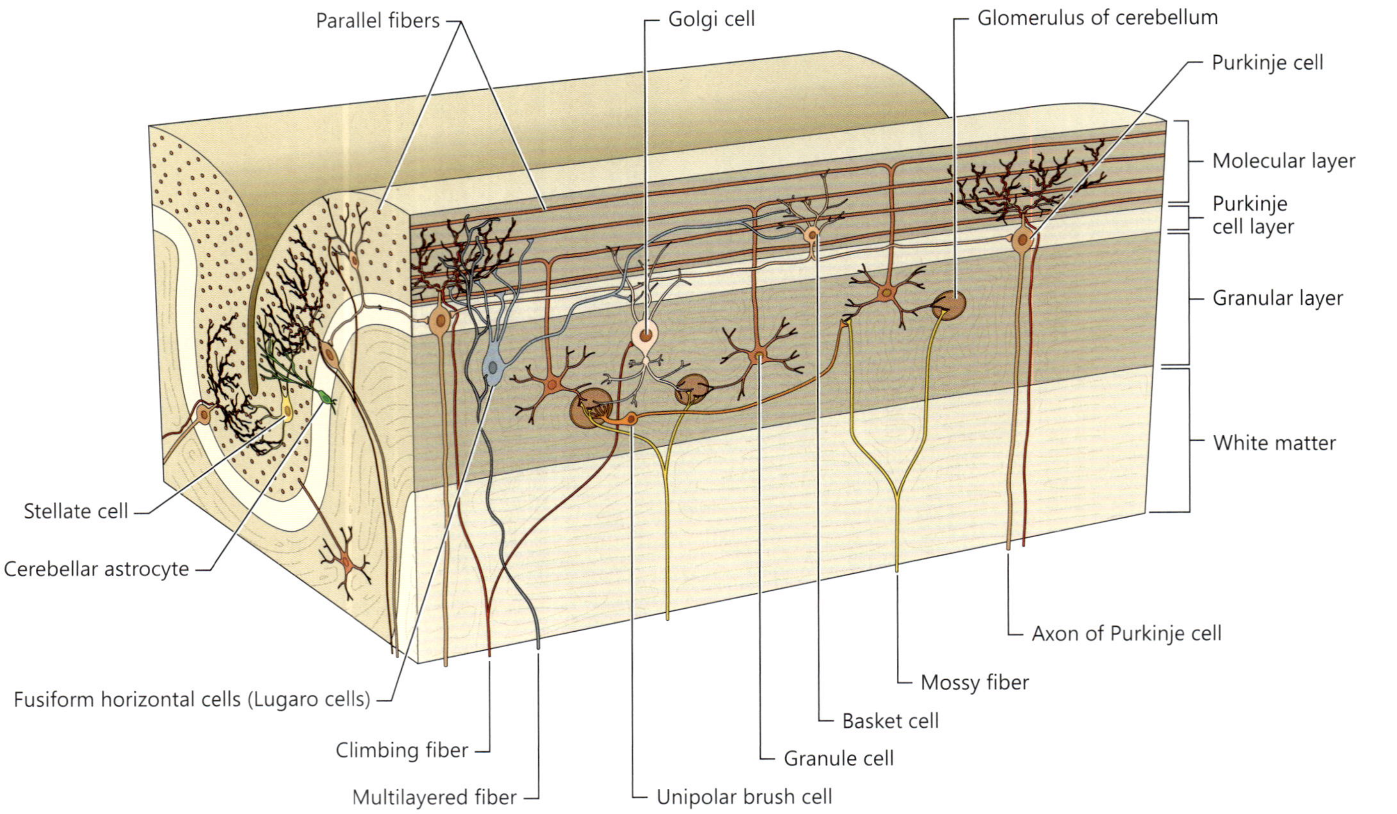

Schematic representation showing the structure of the **cerebellar cortex** in which three layers are clearly distinguishable (molecular layer, Purkinje cell layer and granular layer) and seven types of neurons (Purkinje, granule, basket, stellate, Golgi, Lugaro and unipolar brush cells).

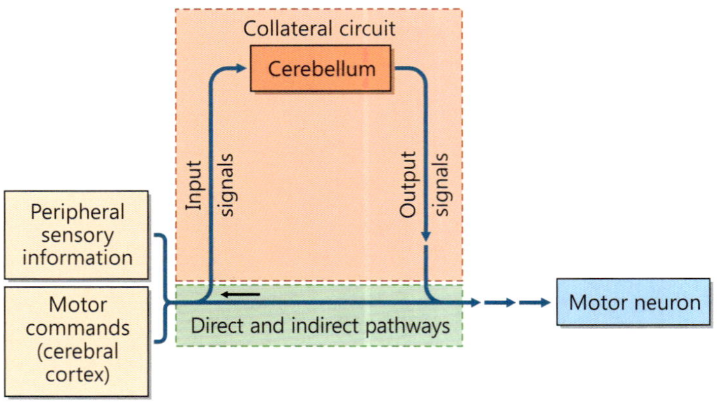

Functional role of the cerebellum within the nervous system. The motor commands (motor program) reach the spinal **motor centers** both directly (direct pathway) and through intermediate synaptic stations on the brainstem (indirect pathway), on which the cerebellum operates by changing the discharge frequency. Although the cerebellum, by receiving both peripheral sensory afferences and the motor programs from the cerebral cortex, may appear as a redundant collateral circuit not essential to the execution of movements, its activity is crucial. On both incoming motor and sensory signals the cerebellum (arrow "input signals") operates integration and elaboration processes and sends back the resulting signals to both spinal motor neurons and to the cerebral cortex. This activity is essential to finalize movements, maintain posture and optimize the performance.

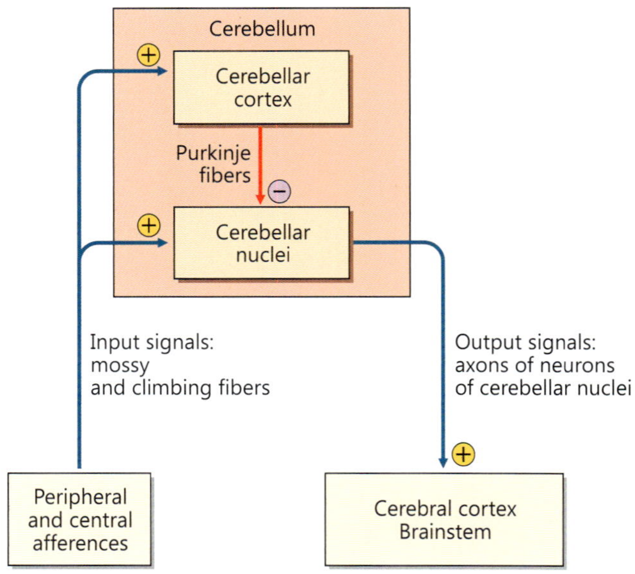

General organization of information flow in the cerebellum. The input signals reach the cerebellum through two types of nerve fibers: **mossy fibers** and **climbing fibers**. Both innervate the nuclei and the cerebellar cortex. The cortex projects to the cerebellar nuclei, which, in most of the cases, are the only output pathway from the cerebellum ("output signals").

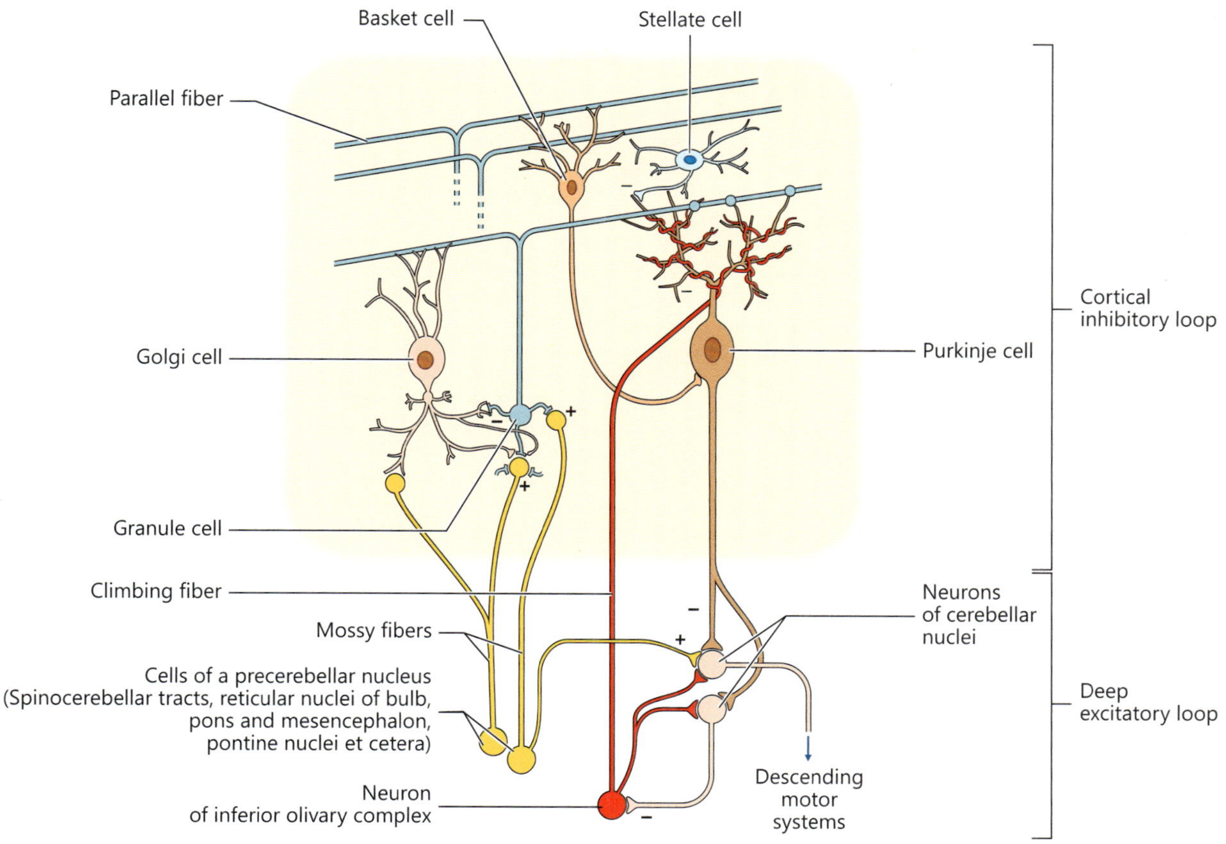

Functional module of the cerebellar cortex. The cerebellar cortex can be represented by a sequence of parallel modules (also parallel to the axis of the cortical lamellae), each consisting of two integrated loops: a deep loop, or of the cerebellar nuclei, carrying an excitatory output to the descending motor systems (indirect pathway), and a cortical loop, or of the cerebellar cortex, carrying an inhibitory output from the Purkinje cells, to the cerebellar nuclei. The afferences conducted by both mossy fibers and climbing fibers act as activators on both cerebellar nuclei, through collateral fibers, and the cerebellar cortex; in the latter, the integration by the inhibitory interneurons signals result in an inhibitory output signal carried by the axons of the Purkinje cells to the cerebellar nuclei. Therefore, the axons of the nuclei neurons are projected to the descending motor systems conveying a signal that represents the result of the comparison between peripheral facilitation and inhibition of the cerebellar cortex.

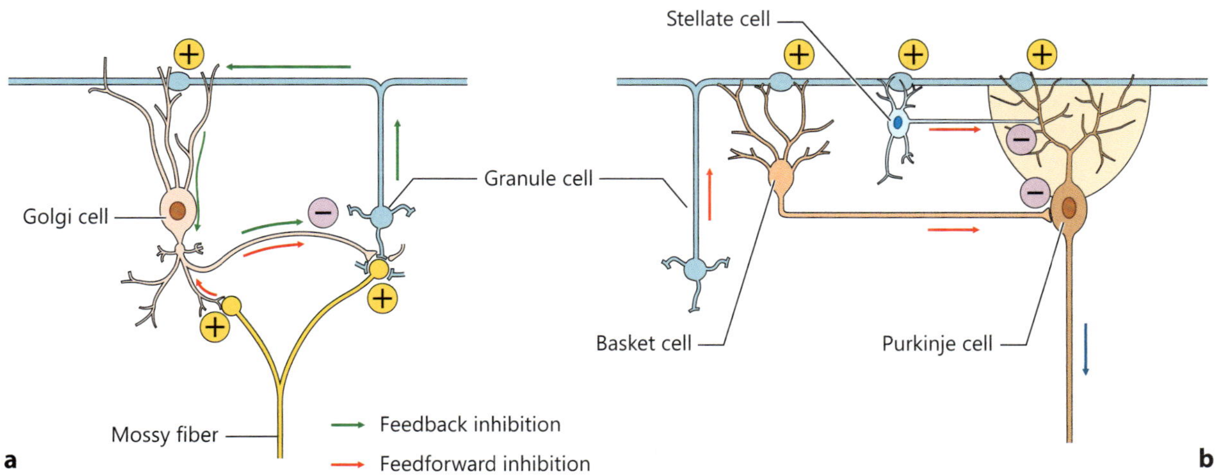

Role of the **inhibitory interneurons** in the cortical circuits of the cerebellum. **a**, The Golgi cell receives synapses from the mossy fibers, directely, and from the granules. The inhibition on the granules activity occurs through two pathways: mossy fiber-Golgi cell-granule (feedforward inhibition), and mossy fiber-granule-Golgi cell-granule (feedback inhibition). **b**, The interneurons of the molecular layer, the stellate cells and the basket cells receive synapses from the parallel fibers and inhibit the Purkinje cells. Therefore, the result is a feedforward inhibitory circuit (*red* arrows). The inhibitory interneurons limit spreading and excessive signals over time (Golgi cells) and spreading to other adjacent modules (stellate cells and basket cells).

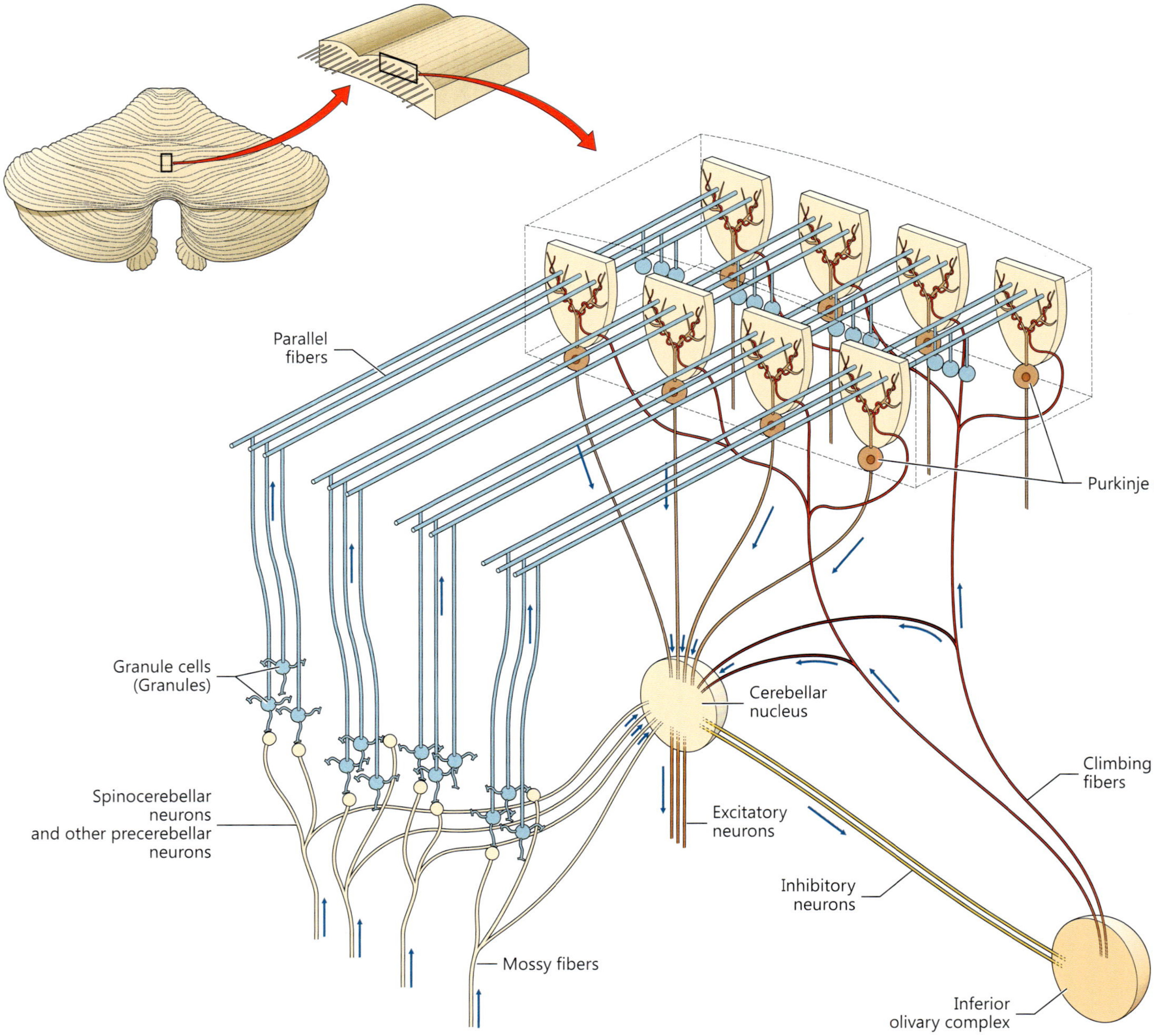

Three-dimensional organization of the cerebellar cortex. Each module consists of numerous granules, stimulated by mossy fibers, activating the overlying Purkinje cells through a layer of parallel fibers running transversally along the cerebellar lamellae. This organization pattern allows more modules to be connected to each other over a distance of several millimeters. The climbing fibers activate a small group of Purkinje cells (six-seven) whose dendritic tree is arranged transversely to the parallel fibers. The inhibitory cells of the afferent and efferent pathways of the cerebellar cortex control system (Golgi cells and basket cells) are not represented.

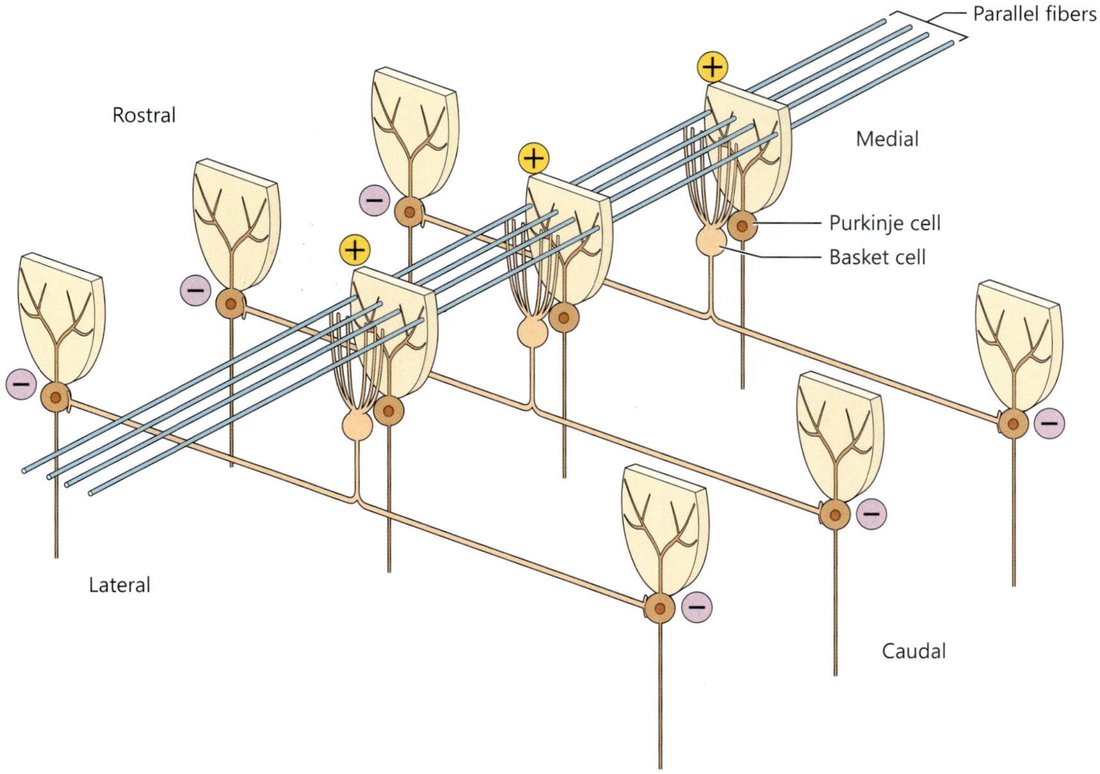

The axons of the basket cells run on a plane transversal to that of the cerebellar lamellae parallel fibers. Consequently, the basket cells activated by a bundle of parallel fibers, do not inhibit the Purkinje cells located along the same bundle, but those located in more rostral or more caudal contiguous rows. Therefore, this is a **lateral inhibition circuit** and operates by increasing the contrast between the row of Purkinje cells excited by the parallel fibers (+) compared to the most rostral and caudal rows, not excited by that bundle and inhibited by basket cells (–).

Structure of the cerebellar cortex

LAYER	*NEURONAL POPULATION*	*FIBERS*
Molecular (300 μm)	Stellate cells Basket cells Fusiform cells	Dendritic arborization Parallel fibers –
Purkinje cell (50-70 μm)	Purkinje cells	Afferent and efferent fibers of different origin
Granular (from 100 to 400-500 μm)	Granule cells (either small granules or granules) Golgi cells (or big granules) Interneurons (Golgi type II cells)	Preterminal and terminal ramifications of mossy and climbing fibers Axons of the Purkinje cells –

Functional characteristics of the neurons of the cerebellar cortex

NEURON	*ACTION*	*FUNCTION*
Purkinje cells	Inhibitory	Corticofugal neurons (they project to the cerebellar nuclei)
Granule cells	Excitatory	Input port of afferences to the cerebellar cortex: the axons move vertically towards the molecular layer where they T bifurcate, forming parallel fibers; the dendrites of granule cells, along with the axons of the Golgi cells and the terminations of the mossy fibers, form the cerebellar glomeruli (they activate the Purkinje cells through the parallel fibers)
Basket cells	Inhibitory	Control system of the efferent pathways from the cerebellar cortex (they inhibit the Purkinje cells)
Golgi cells	Inhibitory	Control system of the afferent pathway to the cerebellar cortex (presynaptic inhibition at glomerular level)
Stelate cells	Inhibitory	Control system of the efferent pathways from the cerebellar cortex (they inhibit the Purkinje cells)

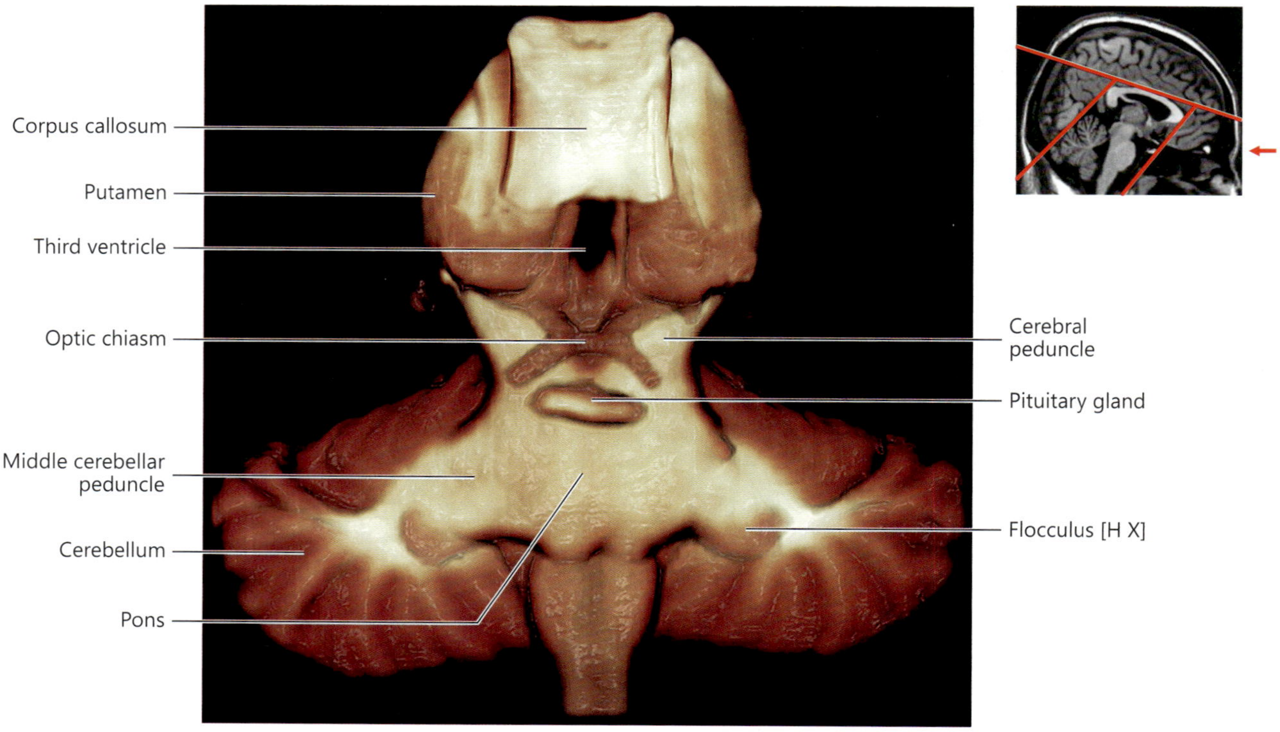

a

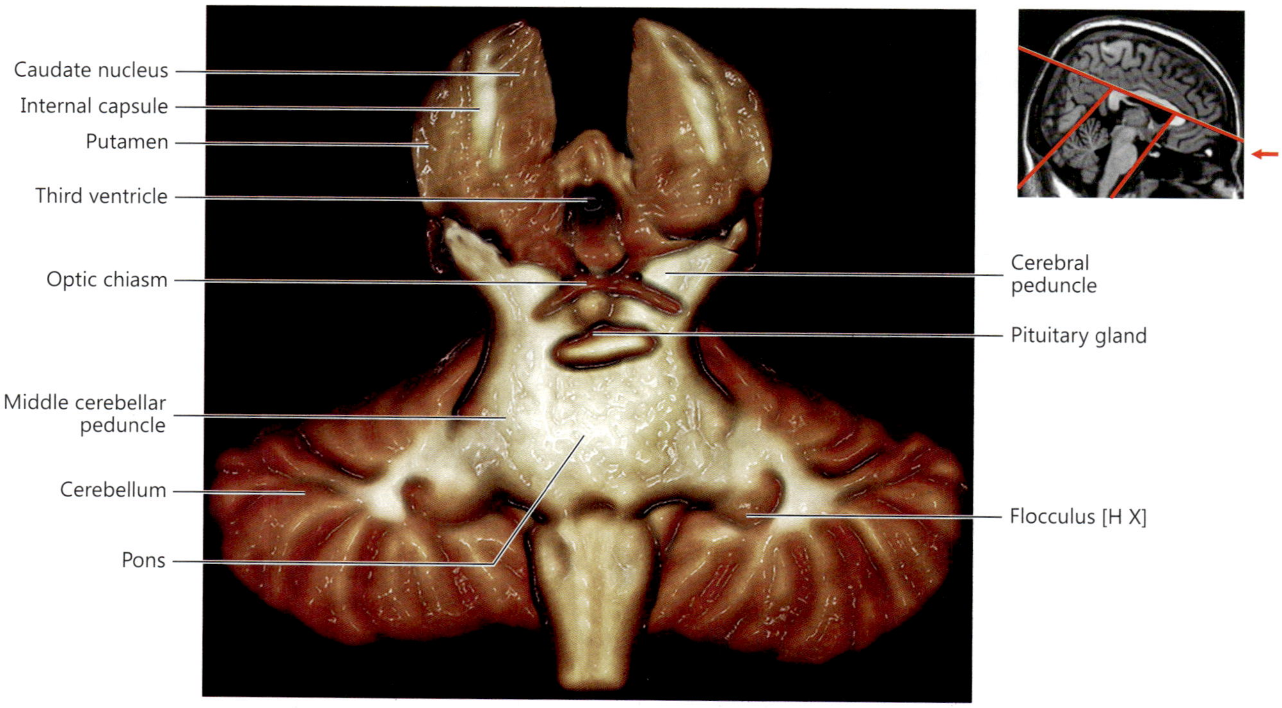

b

Three-dimensional image of a living human body, obtained by NMR using a 3D volume rendering technique, showing the **subcortical brain structures** in anterior projection: in **a** the corpus callosum is evident; in **b** the corpus callosum has been removed.

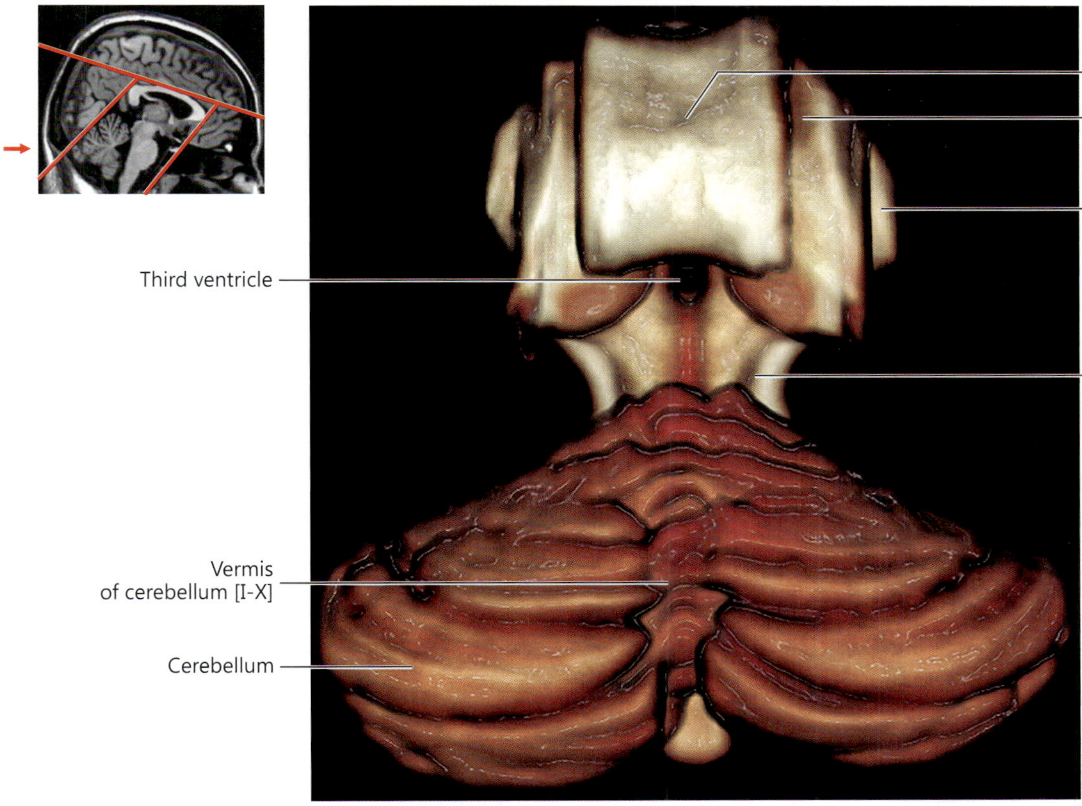

a

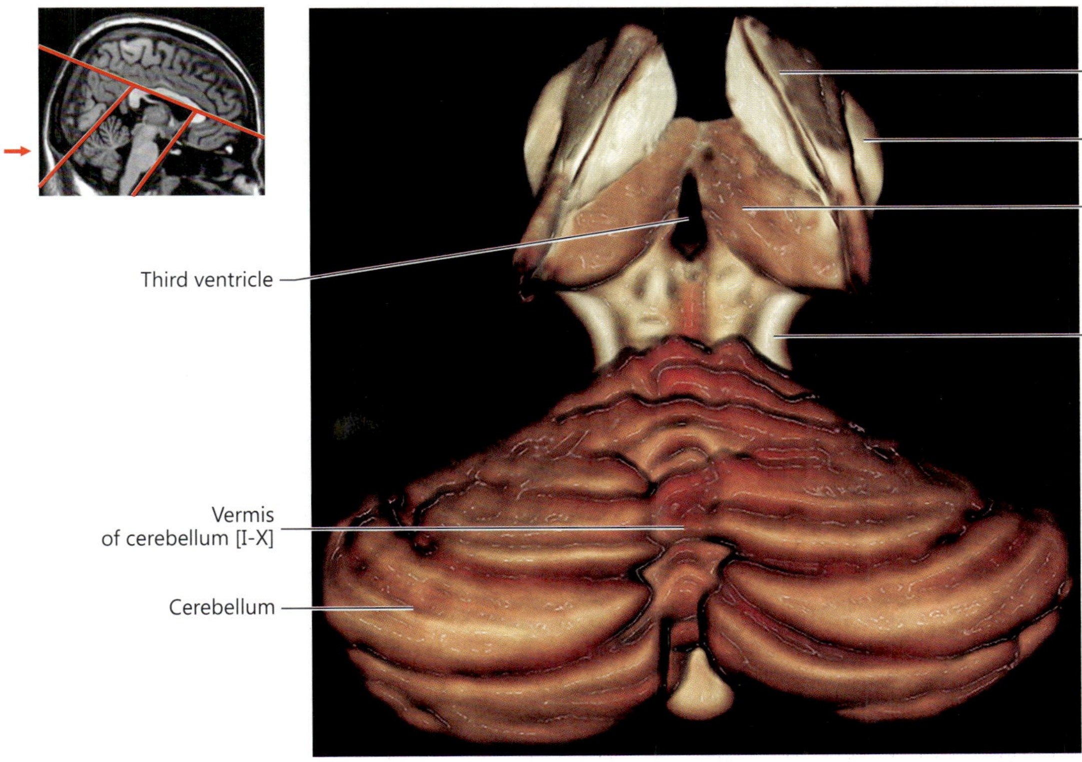

b

Three-dimensional image of a living human body, obtained by NMR using a 3D volume rendering technique, showing the **subcortical brain structures** in posterior projection: in **a** the corpus callosum is evident; in **b** the corpus callosum has been removed.

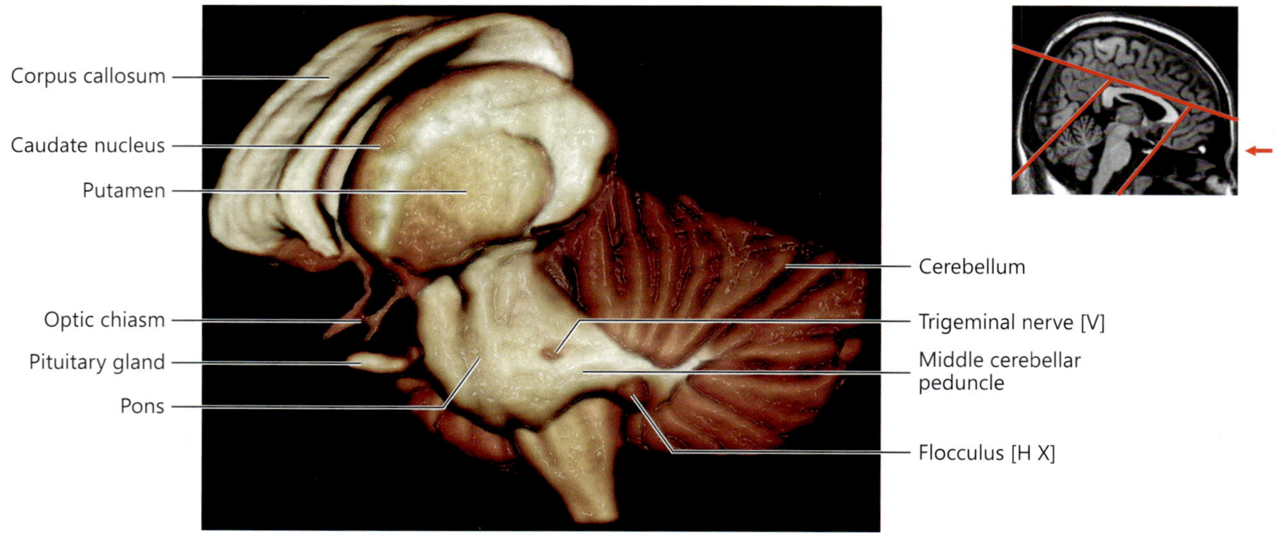

a

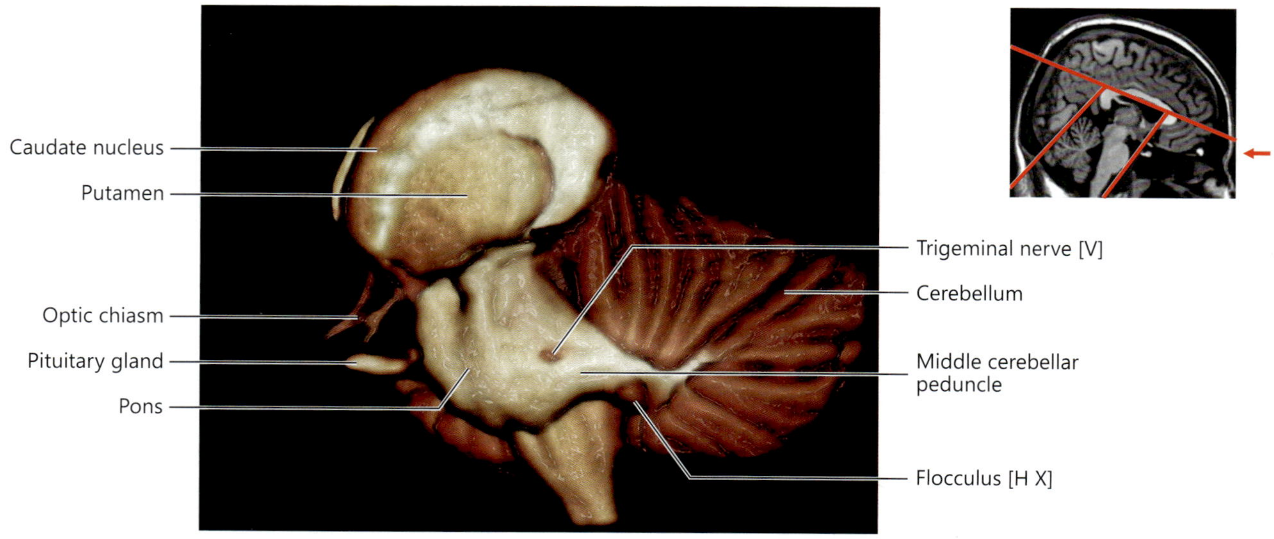

b

Three-dimensional image of a living human body, obtained by NMR using a 3D volume rendering technique, showing the **subcortical brain structures** in anterolateral projection: in **a** the corpus callosum is evident; in **b** the corpus callosum has been removed.

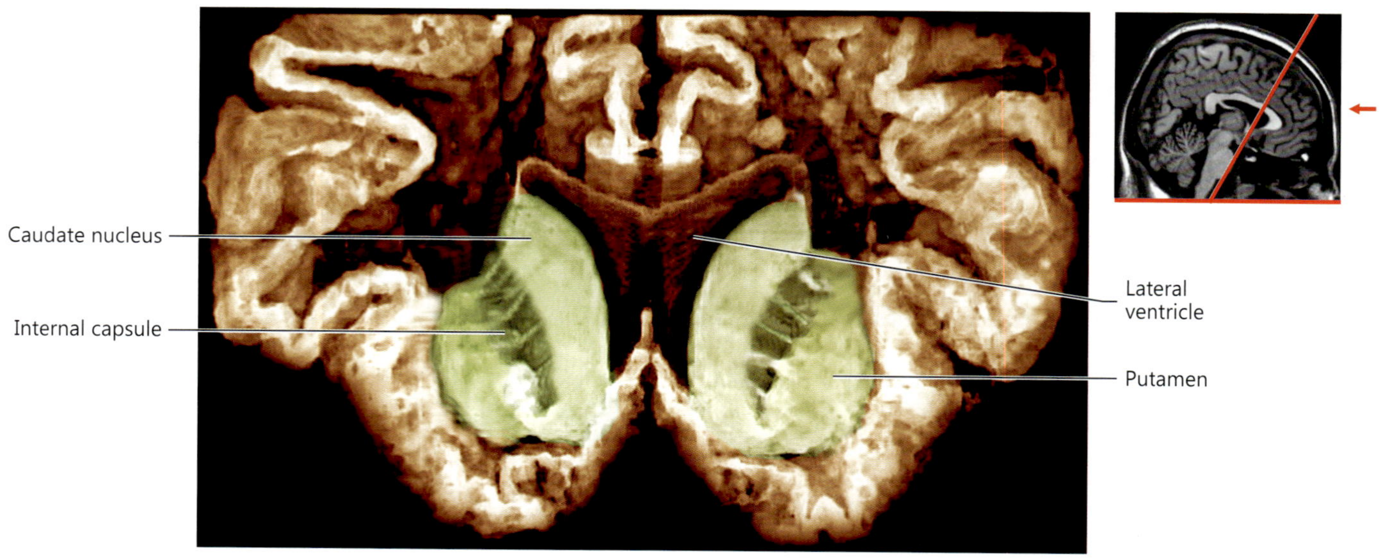

Three-dimensional image of a living human body, obtained by NMR using a 3D volume rendering technique, upon eliminating the white matter, which shows the **caudate nucleus** and the **putamen** in an anterosuperior section passing through the head of the caudate nucleus.

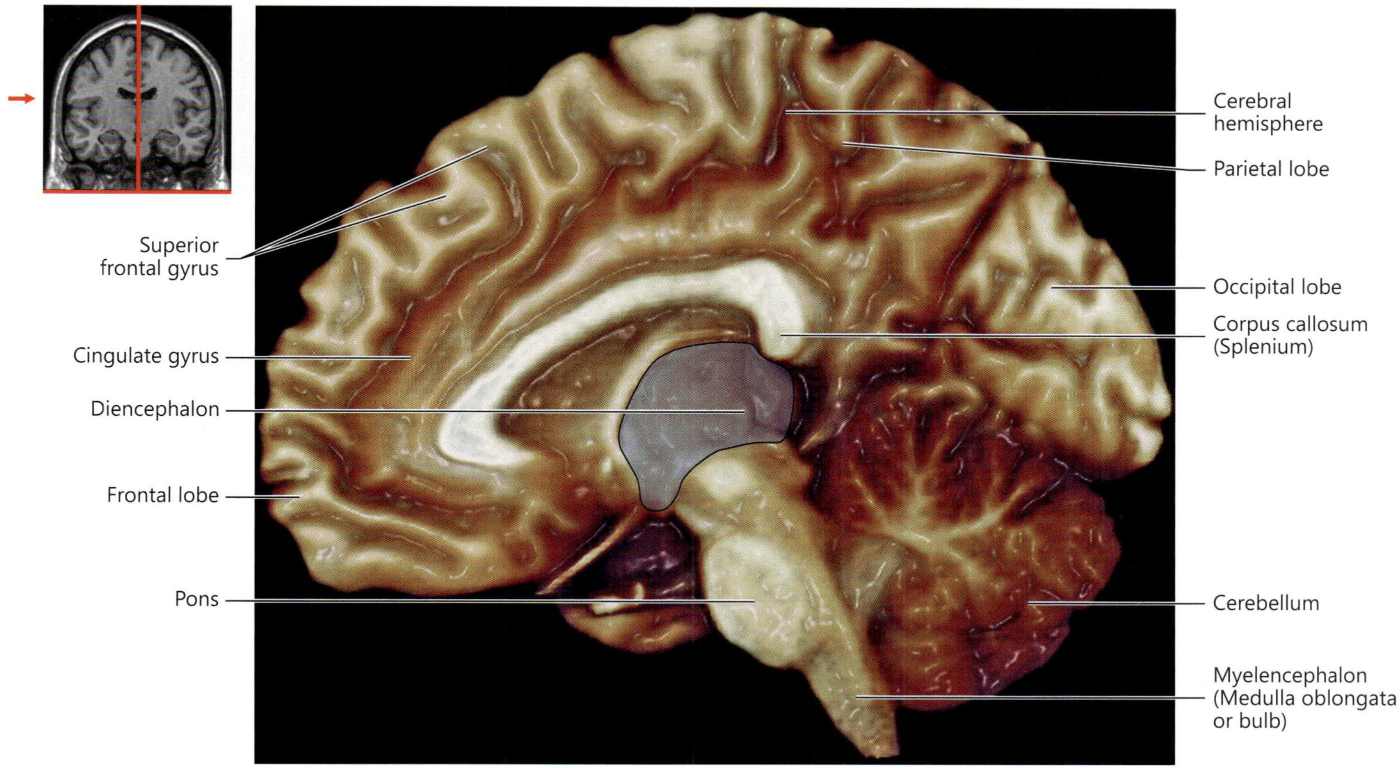

Three-dimensional image of a living human body, obtained by NMR using a 3D volume rendering technique, showing the medial position of the **diencephalon** following the section of the corpus callosum and brainstem.

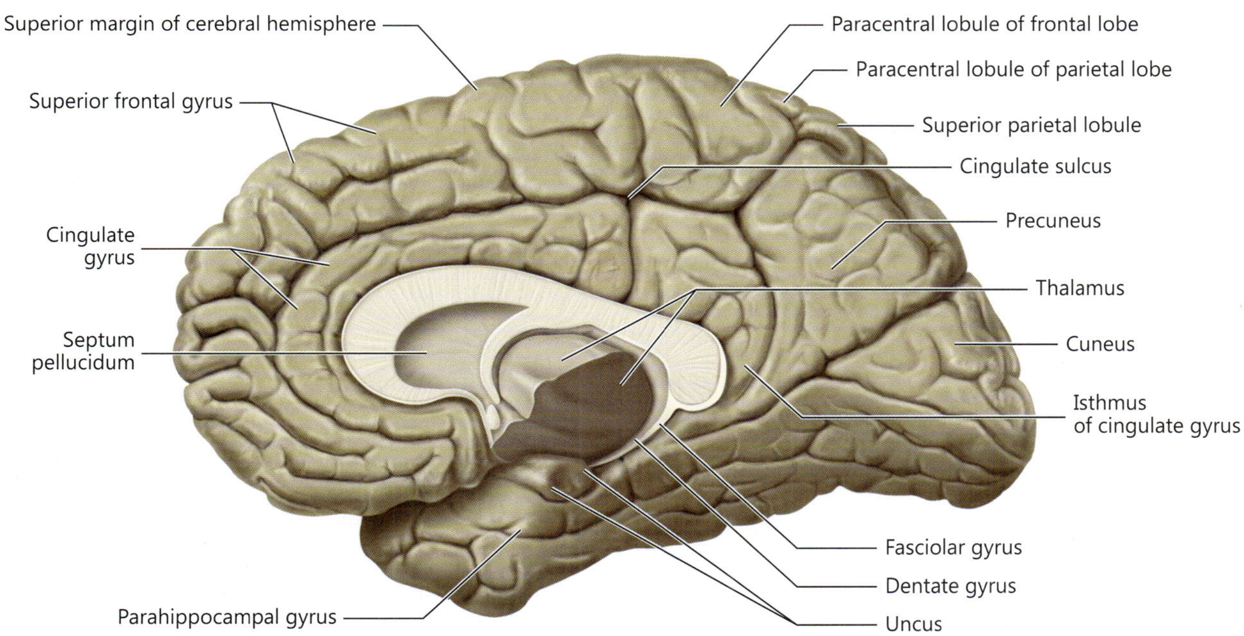

Right cerebral hemisphere, median sagittal section of the brain, medial projection. The corpus callosum and the diencephalon have been transected. Note the centroencephalic position of the thalamus.

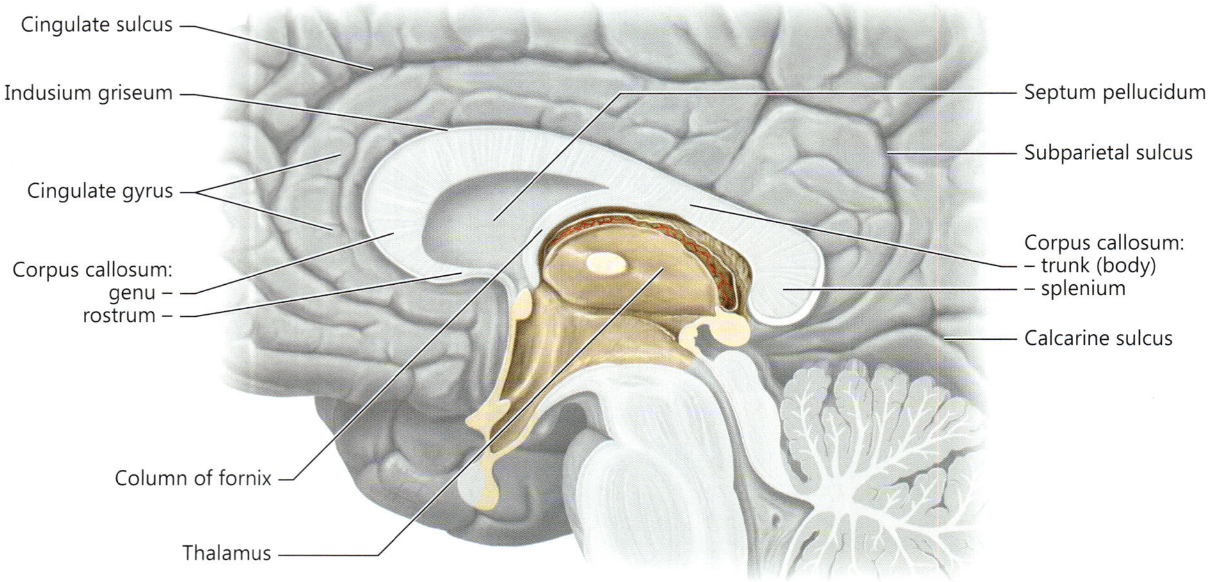

Schematic representation showing the position and relationships of the **diencephalon** (colored).

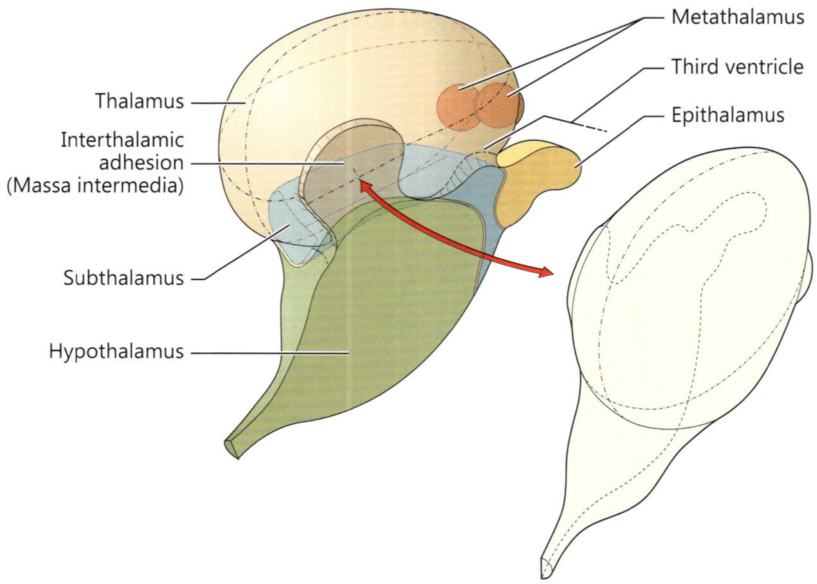

Craniocaudal subdivision of the diencephalon
Epithalamus
Metathalamus (Geniculate nuclei of the thalamus)
Thalamus
Subthalamus
Hypothalamus

Schematic subdivision of the **diencephalon**.

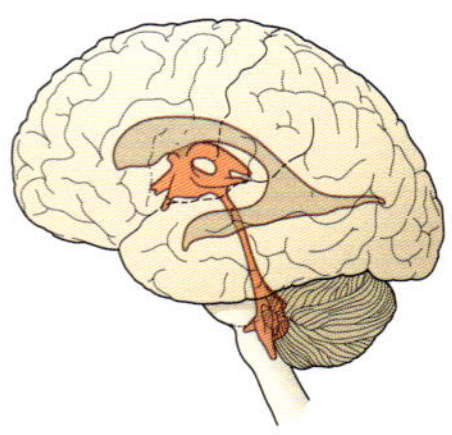

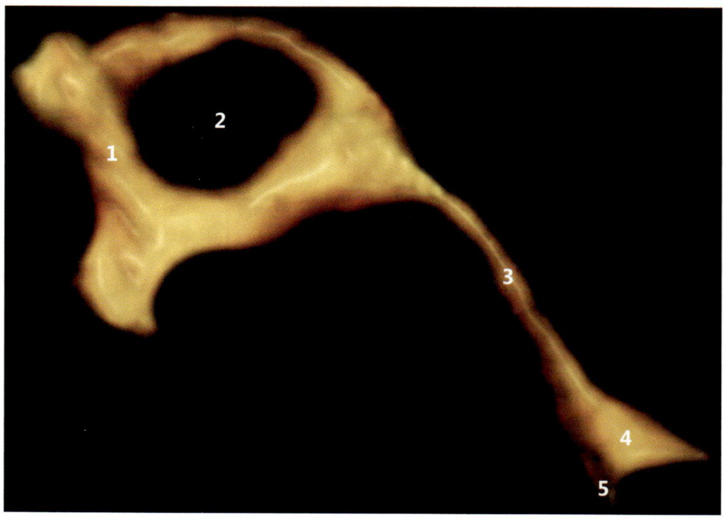

Three-dimensional image of a living human body, obtained by NMR using a 3D volume rendering technique, showing the mold in lateral projection of the **third ventricle**, **fourth ventricle** and the **aqueduct of the midbrain**.

1, Third ventricle
2, Interthalamic adhesion
3, Aqueduct of the midbrain
4, Fourth ventricle
5, Central canal

Anatomical subdivision and functional characterization of the diencephalon regions

REGIONS	LOCATION	ANATOMICAL FORMATIONS	NUCLEI	FIBERS	FUNCTION
Subthalamus	Ventral part of the diencephalon, placed laterally to the hypothalamus and in direct continuity with the tegmentum of midbrain		Subthalamic nucleus (or Luys' body) Zona incerta Nuclei of perizonal fields (tegmental or prerubral) or nuclei of medial, dorsal and ventral fields [H, H1 and H2] Nucleus of ansa lenticularis	Subthalamic fasciculus Lenticular fasciculus (field H2 of Forel) Ansa lenticularis Dentatothalamic fibers Rubrothalamic fibers Thalamic fasciculus (field H1 of Forel)	Control of somatic motility within the group of corticobasal circuits of the extrapyramidal system
Hypothalamus	Below the hypothalamic sulcus, around the cavity of the third ventricle, which forms the inferior part, lateral walls and floor	Preoptic area (part of the telencephalon assigned to the hypothalamus from a functional point of view) Optic chiasm with (optic tracts) and perichiasm region Lamina terminalis Tuber cinereum and infundibulum of the neurohypophysis Mammillary bodies	Medial, lateral and median preoptic and periventricular nuclei Supraoptic nucleus Anterior hypothalamic nucleus Paraventricular nucleus of the hypothalamus Lateral hypothalamic area Lateral tuberal nuclei Ventromedial nucleus of the hypothalamus Dorsomedial nucleus Perifornical nucleus Arcuate (or infundibular) nucleus Posterior nucleus of the hypothalamus (or posterior hypothalamic) Supramammillary, dorsal and ventral premammillary and lateral and medial mammillary nuclei	Medial forebrain bundle Hippocampothalamic tract Corticohypothalamic fibers Fibers of the stria terminalis (or amygdalohypothalamic tract) Thalamohypothalamic tract Mammillary peduncle tract Retinohypothalamic tract Principal mammillary tract Mammillothalamic fasciculus Mammillotegmental fasciculus Periventricular fibers Hypothalamohypophysial tract Supraopticohypophysial tract Paraventriculohypophysial tract	Control of centers of the brainstem and spinal cord related to visceral effector activities Secretion of neurohormones (vasopressin and oxytocin) and regulation, through the production of releasing hormone (RH), of the activity of the adenohypophysis. Other important functions such as the control of body temperature, food and water intake and sexual activity
Epithalamus	Corresponds to the posterior part of the third ventricle roof	Habenular trigone Habenular commissure Stria medullaris of the thalamus Pineal glands Posterior commissure of the epithalamus Subcommissural organ	Medial and lateral habenular nuclei Nucleus of the posterior commissure	Stria medullaris of the thalamus Fibers from the hippocampus Fibers from the olfactory tubercle Fibers from the anterior perforated substance Fibers from the septal nuclei Fibers from the preoptic area Fibers from the anterior nuclei of the thalamus Tectohabenular tract Habenulointerpeduncular tract (or fasciculus retroflexus of Meynert) Posterior commissure of the epithalamus Fibers from the pretectal nuclei Fibers from the nucleus of the posterior commissure Commissural fibers from the superior colliculi Fibers from the medial and lateral habenular nuclei Fibers from interstitial nucleus (or of Cajal) and from the nucleus of posterior commissure (or of Darkschewitsch)	Intermediate station of a visceral descending effector system connected to the limbic system

continued

Anatomical subdivision and functional characterization of the diencephalon regions *(continued)*

REGIONS	LOCATION	ANATOMICAL FORMATIONS	NUCLEI	FIBERS	FUNCTION
Thalamus	Above the hypothalamic sulcus and forms the superior part of the lateral walls of the third ventricle		Anterior nuclei of the thalamus – anteroventral – anterodorsal – anteromedial Medial nuclei of the thalamus – mediodorsal (o medial dorsal) – medioventral (o medial ventral) Dorsal nuclei of the thalamus – lateral dorsal (o dorsolateral) – posterior lateral – pulvinar Ventral nuclei of the thalamus – ventrobasal – – posterolateral ventral – – posteromedial ventral – ventral medial – – ventral medial basal – – ventral medial principal – – submedial – ventral posterior inferior – ventral lateral – – anterior ventrolateral – – posterior ventrolateral – ventral anterior – ventral intermediate – ventral posterolateral – ventral posterior internal – ventral posterior parvocellular Intralaminar nuclei of the thalamus – centromedian – parafascicular – paracentral – central lateral – central medial Posterior nuclei of the thalamus – nucleus limitans – posterior – suprageniculate Reticular nucleus of the thalamus Median nuclei of thalamus – paratenial – paraventricular nuclei of the thalamus – reuniens – rhomboid nucleus	Perithalamic white matter – corona radiata – – anterior radiation of the thalamus (anterior or frontal peduncle) – – central radiation of the thalamus (superior or parietal peduncle) – – posterior radiation of the thalamus (posterior or occipital peduncle) – – inferior radiation of the thalamus (inferior or temporal peduncle) – tenia thalami – stria medullaris of the thalamus Intrathalamic white matter – medial or internal (Y-shaped) medullary lamina – lateral or external medullary lamina	Association center connected to the cerebral cortex, to which it is connected through direct tracts in both directions Integrating center interposed on all the tracts of general sense and those special (apart from the olfactory), which come directly or indirectly (through the reticular formation)
Metathalamus	Ventrally to the thalamus and in close relationship with the mesencephalon	Medial geniculate body Lateral geniculate body	Dorsal, ventral and medial magnicellular nuclei of the medial geniculate body Dorsal and ventral (or pregeniculate) nuclei of the lateral geniculate body	Dorsal nucleus of the medial geniculate body – fibers from and to the nuclei of the trapezoid body, superior olivary, of the lateral lemniscus and the inferior colliculus (through the brachium of the inferior colliculus) – acoustic radiation (geniculotemporal fibers) Ventral nucleus of medial geniculate body – fibers to the pulvinar and the ventral and dorsal nuclei of the thalamus – fibers to the homonymous contralateral nucleus (through the ventral supraoptic commissure or of Gudden) – fibers from the suprageniculate nucleus – fibers from the posterior nuclei of the thalamus Ventral nucleus of lateral geniculate body – fibers from the optic tract Dorsal nucleus of the lateral geniculate body – fibers from the optic tract – optic radiation (geniculocalcarin fibers)	Modulator center of special sense impulses (acoustic for medial geniculate bodies, visual for lateral geniculate bodies), in the last tract of the projective pathway, through connections in both directions with the cerebral cortex

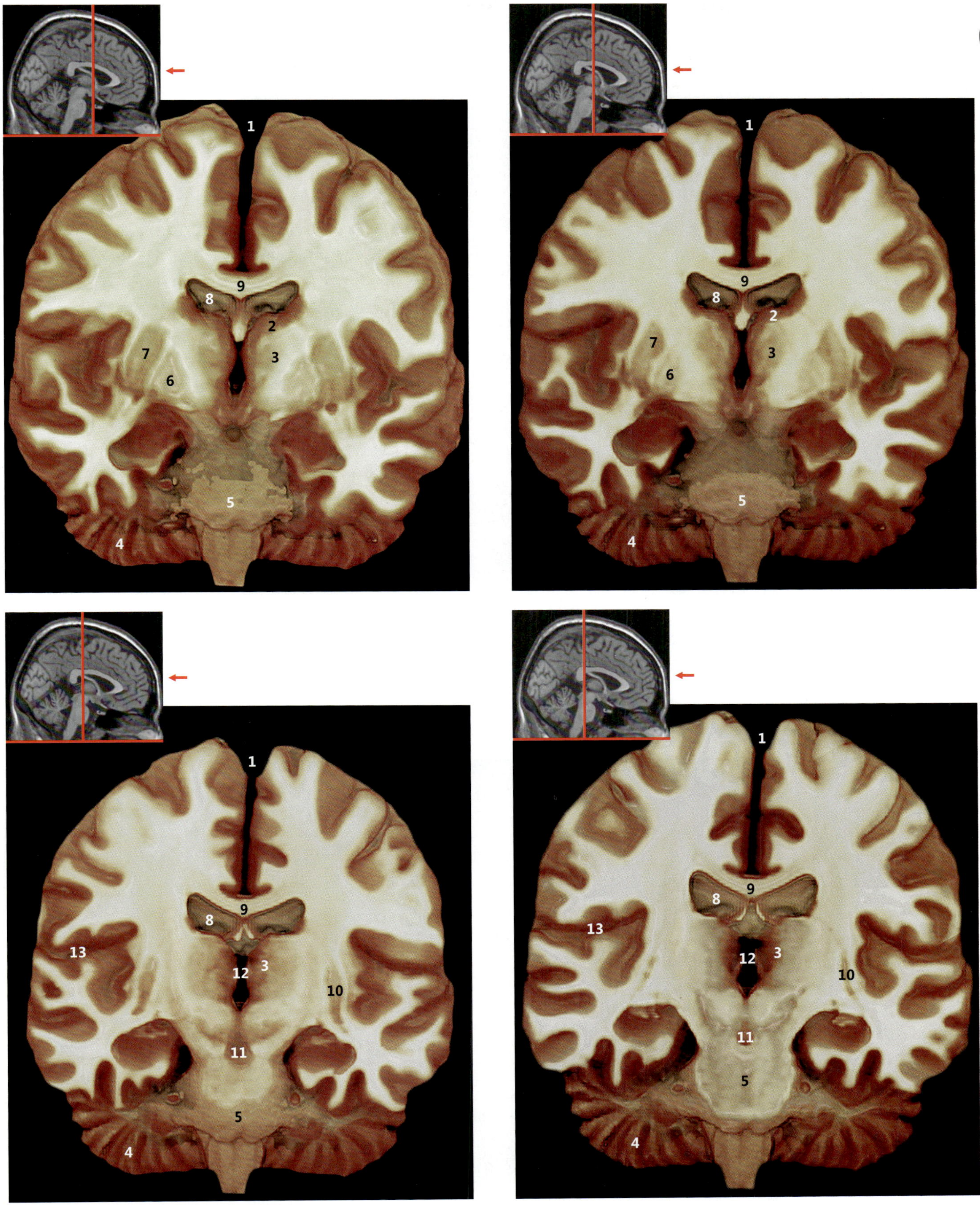

Three-dimensional images of a living human body, obtained by NMR using a 3D volume rendering technique, of coronal sections of the **brain** in frontal-occipital direction showing the relationship of the diencephalon with the basal nuclei and the formations of the brainstem.

1, Longitudinal cerebral fissure
2, Caudate nucleus
3, Thalamus
4, Cerebellum
5, Pons
6, Pallidum
7, Putamen
8, Laterale ventricle
9, Corpus callosum
10, Claustrum
11, Aqueduct of the midbrain
12, Third ventricle
13, Lateral sulcus

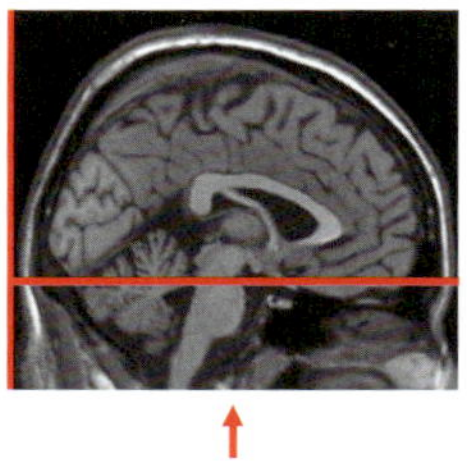

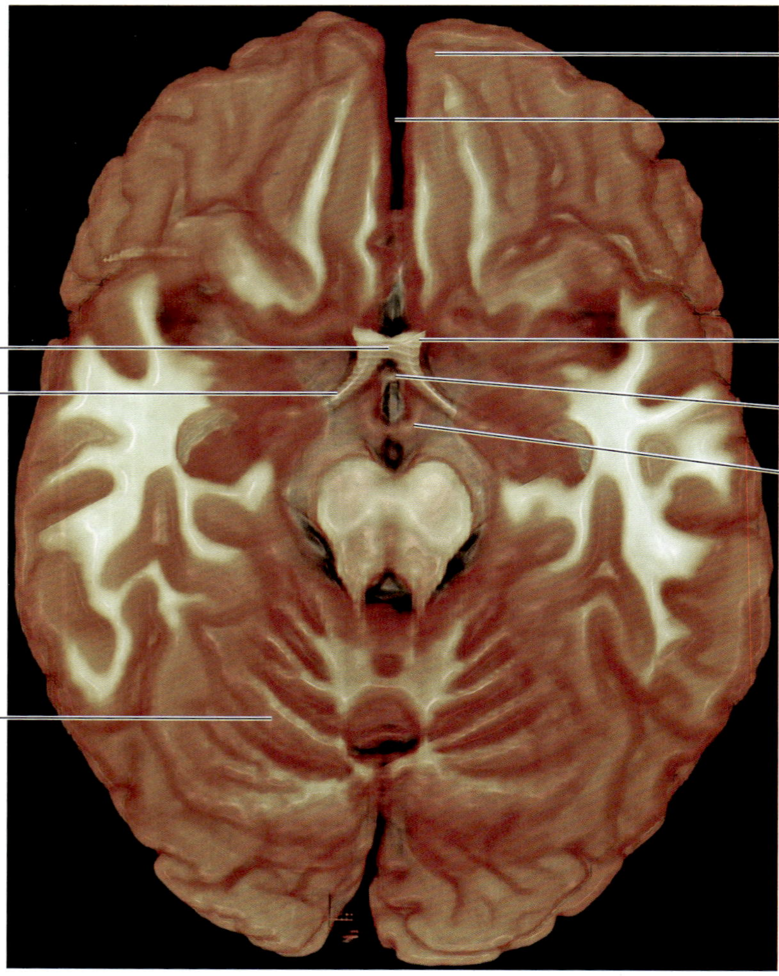

Three-dimensional image of a living human body, obtained by NMR using a 3D volume rendering technique, showing the inferior surface of the **diencephalon**.

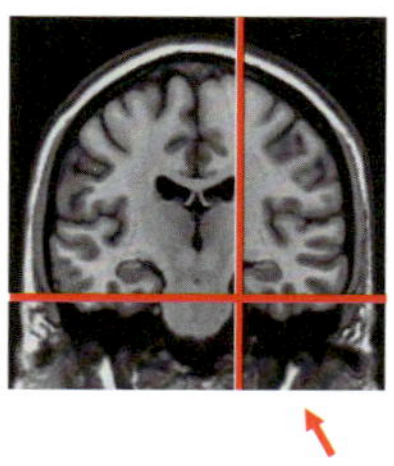

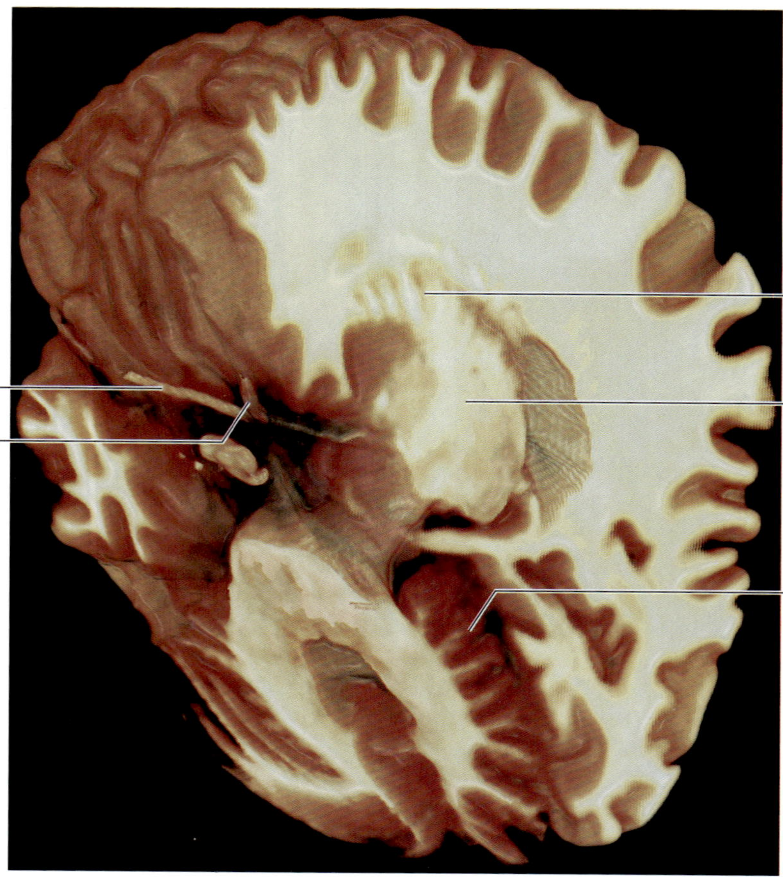

Three-dimensional image of a living human body, obtained by NMR using a 3D volume rendering technique, showing the inferior surface of the **diencephalon** in anteroinferior projection.

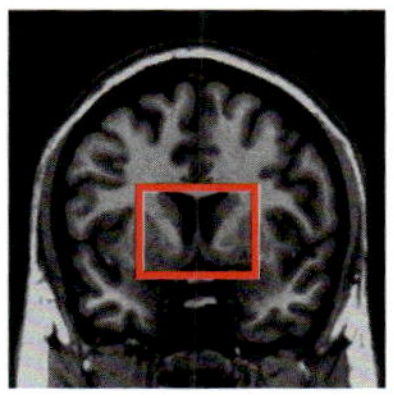

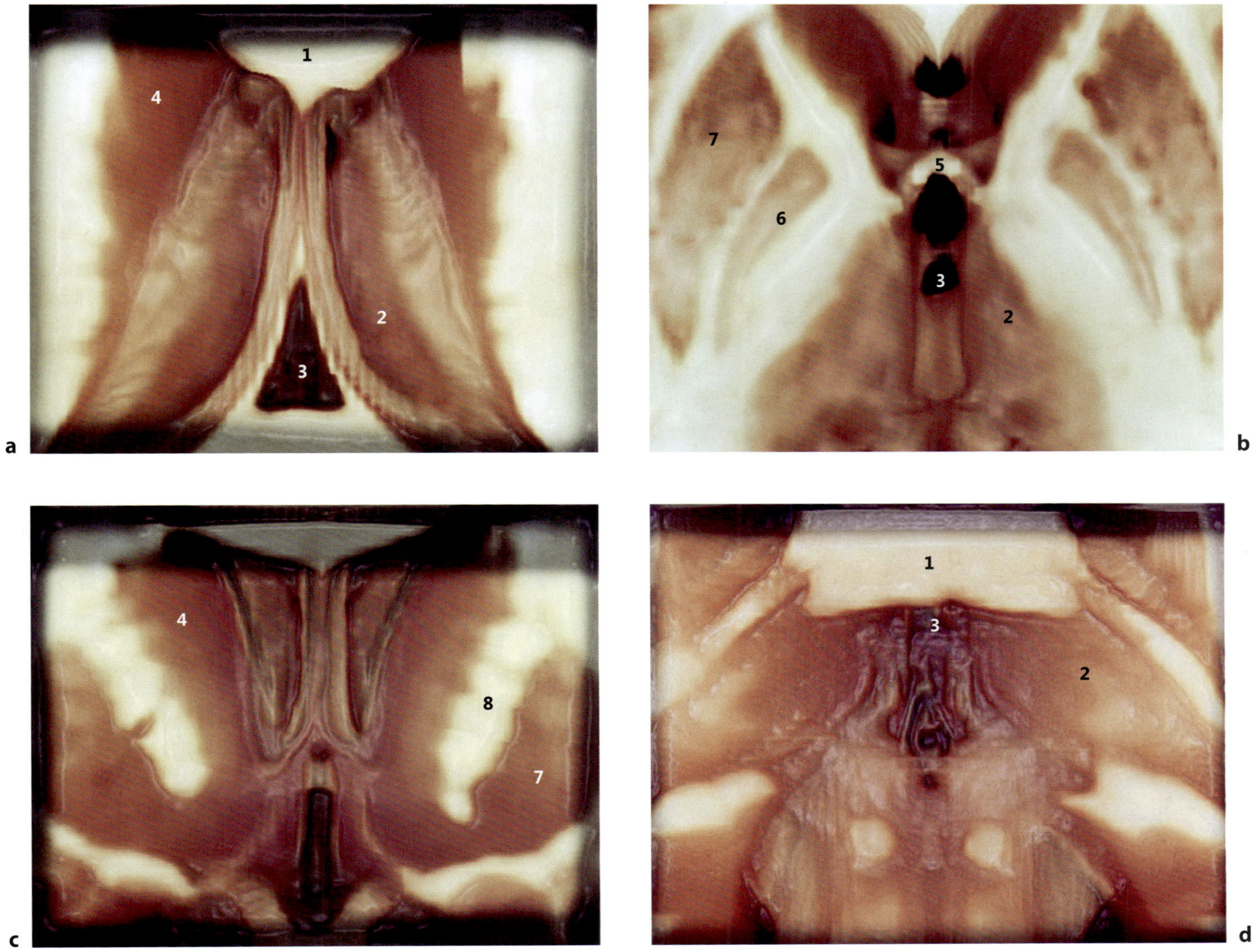

Three-dimensional images of a living human body, obtained by NMR using a 3D volume rendering technique, showing the superior (**a**), inferior (**b**), anterior (**c**) and posterior (**d**) surfaces of the **diencephalon**.
The inferior surface should be considered vertical in that it was obtained through a section plane running above the infundibulum of the third ventricle and was carried out to highlight the inferior portion of the internal capsule and its relationships with the diencephalon.

1, Corpus callosum
2, Thalamus
3, Third ventricle
4, Caudate nucleus
5, Optic chiasm
6, Pallidum
7, Putamen
8, Internal capsule

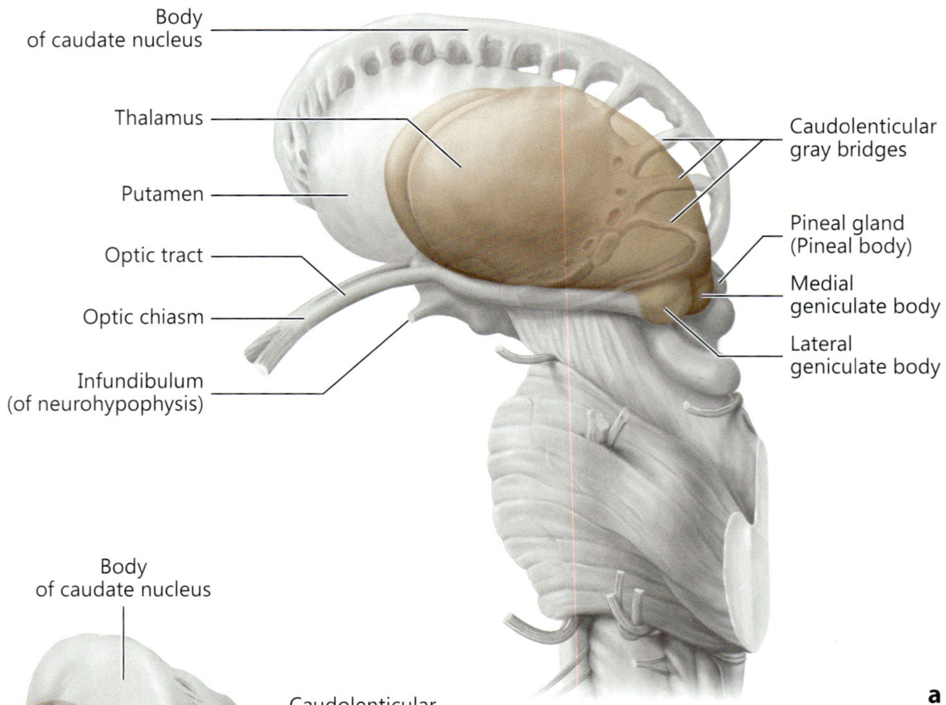

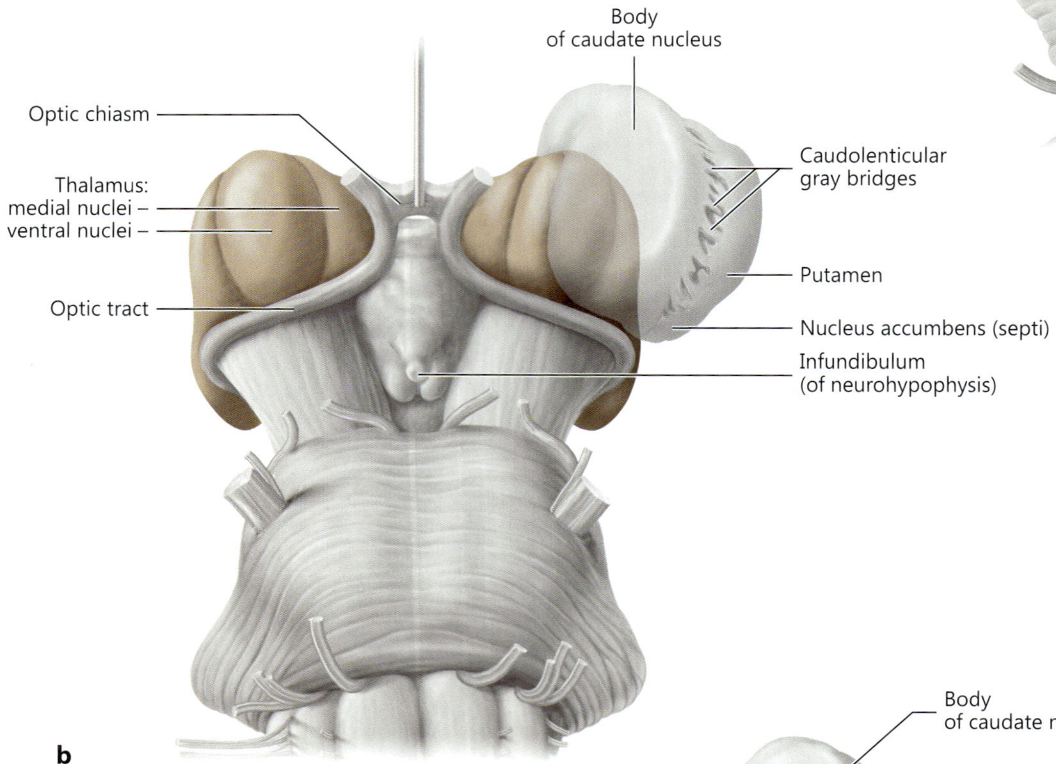

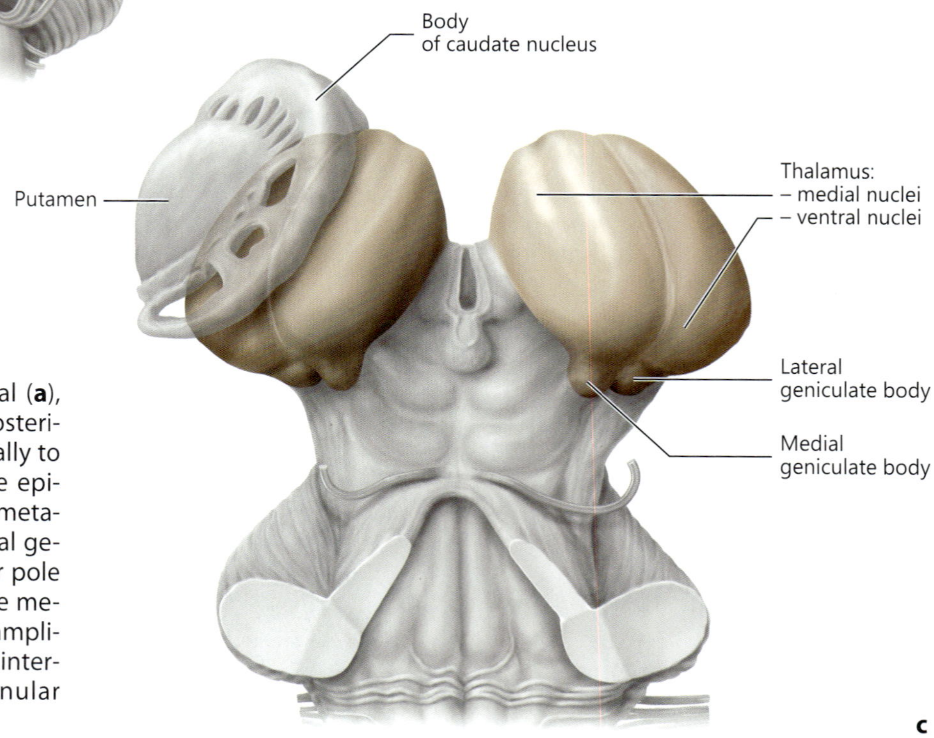

Thalamus and caudate nucleus in lateral (**a**), anterior (**b**) and posterior (**c**) projection. Posteriorly and medially to the thalami and rostrally to the tectal plate are the formations of the epithalamus, including the pineal gland. The metathalamus, formed by the medial and lateral geniculate bodies, lies beneath the posterior pole of the thalamus. The distance between the medial surfaces of the due thalami has been amplified to allow a more separate view of the interthalamic structures, such as the habenular trigones and the pineal gland.

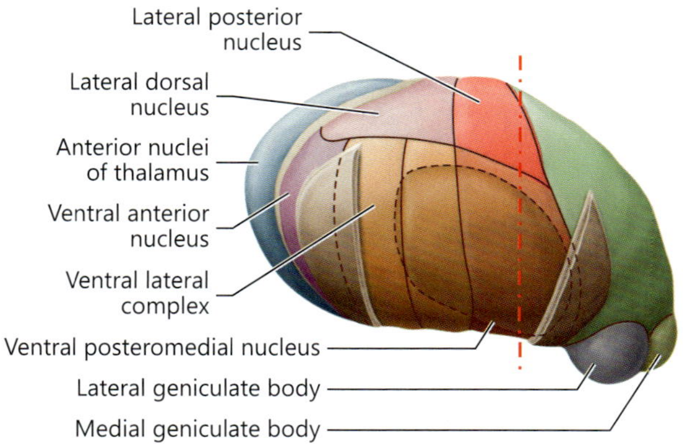

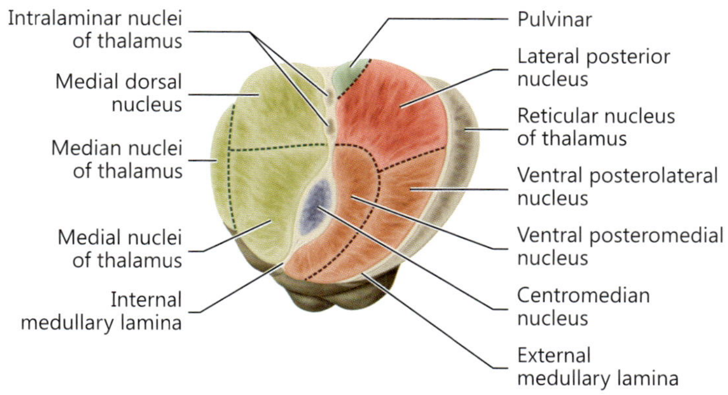

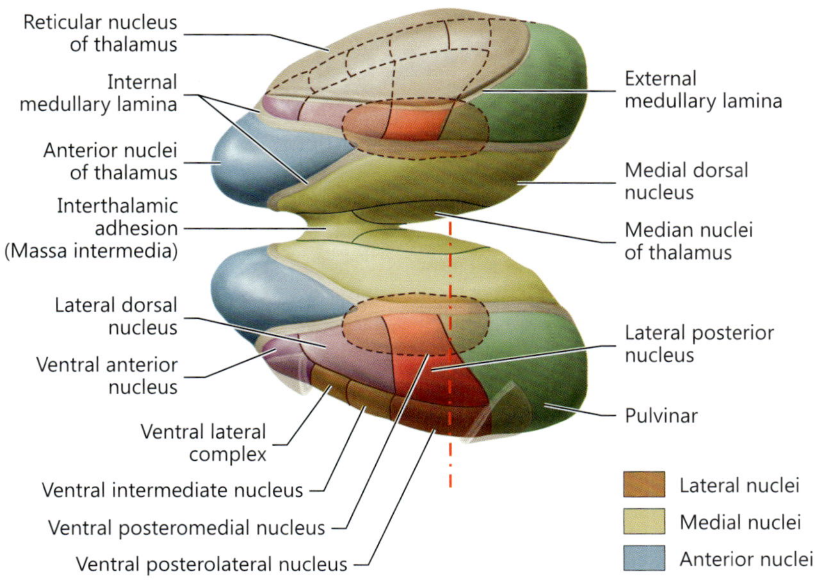

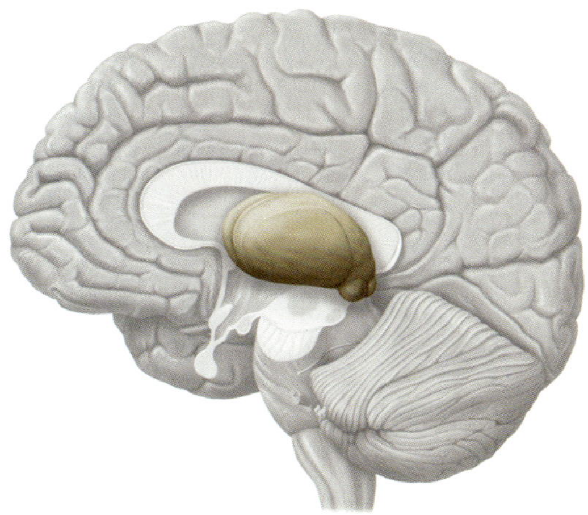

The **medial medullary lamina** separates the medial nuclei of the thalamus from a lateral nuclear group, divided in turn into the dorsal and ventral nuclei, and, in correspondence of the anterior pole, with a Y division, it circumscribes the anterior nuclei. In its thickness are comprised the intralaminar nuclei, whereas the median nuclei of thalamus are located along the surface of the third ventricle and in the interthalamic adhesion.

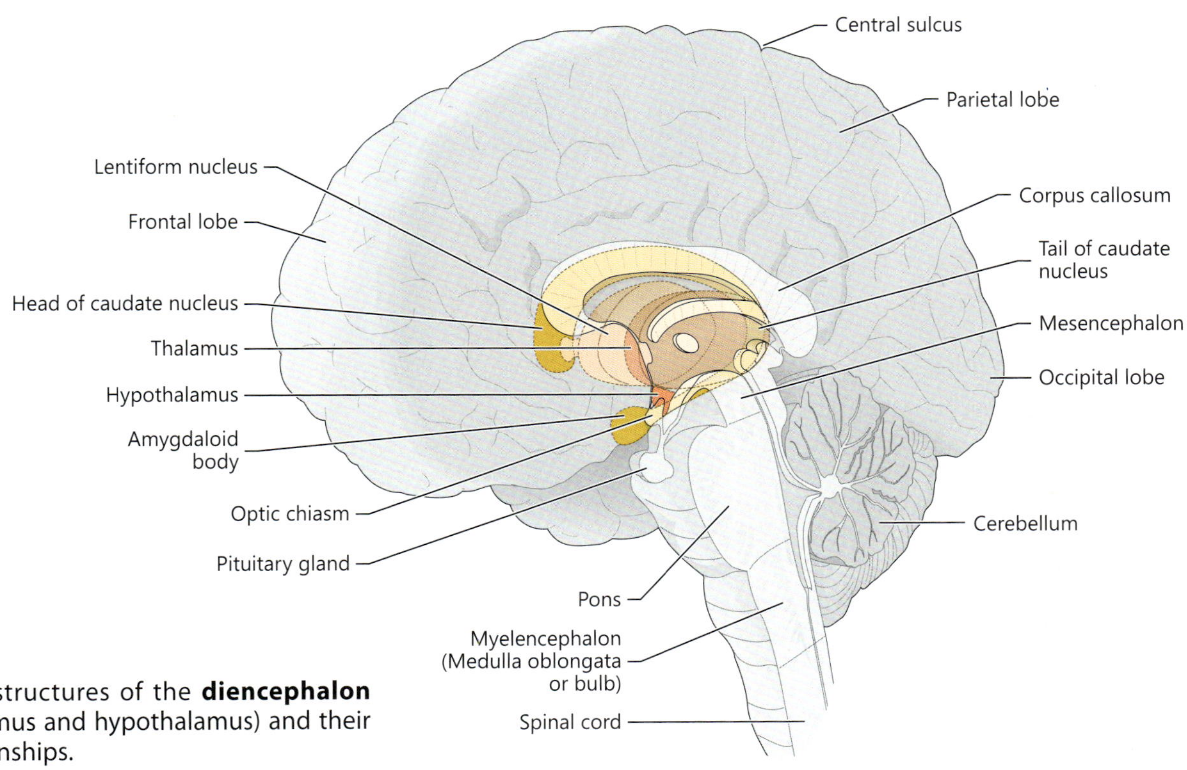

Main structures of the **diencephalon** (thalamus and hypothalamus) and their relationships.

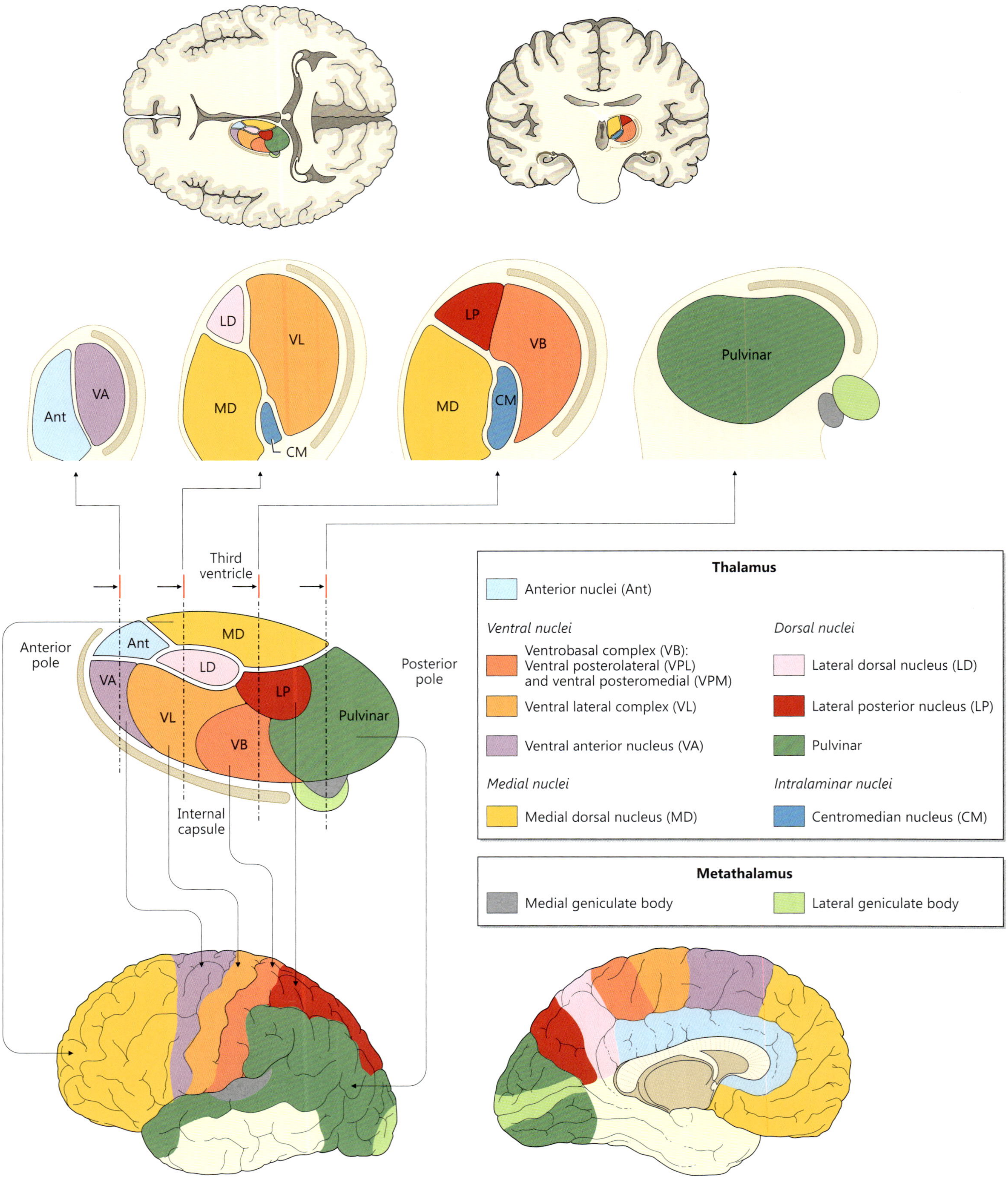

Schematic representation of the main **cortical projections of the nuclei of the thalamus** (the lateral, with their dorsal and ventral nuclei, and medial, with their dorsal, ventral and medial magnocellular nuclei, geniculate bodies of the metathalamus can be defined as geniculate nuclei of the thalamus) on the lateral and medial surfaces of the left telencephalic hemisphere, as indicated by color matching.

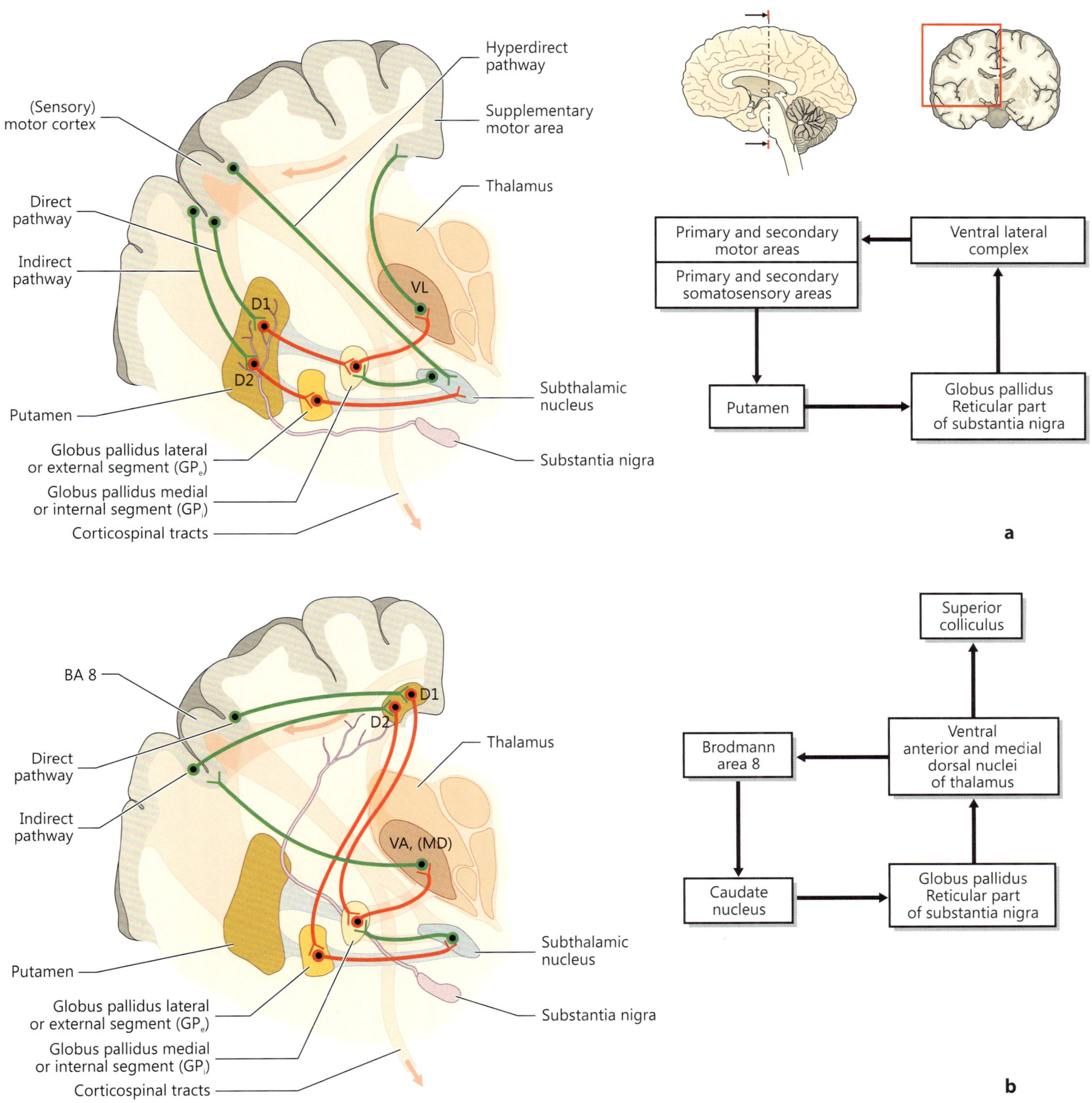

Organization of basal nuclei in the **motor** (**a**) **and oculomotor** (**b**) **circuits or loops**. The motor loop is organized in three pathways: direct, indirect and hiperdirect. The direct pathway (facilitatory) starts from the sensory motor cortex (primary and secondary motor areas, MI and MII, and primary and secondary somatosensory areas, SI and SII) and stimulates a specific nucleus of the neostriatum: the putamen. From here, inhibitory axons (GABAergic) leave to reach the globus pallidus medial or internal segment (GP$_i$). The lack of GP$_i$ activation allows the ventral lateral nuclei (VL) of the thalamus to stimulate the MII cortex, in particular the supplementary motor area, facilitating movement. The indirect pathway (inhibitory) starts always from the sensory motor cortex and stimulates the putamen, which inhibitory neurons, interspersed along this pathway, project to the globus pallidus lateral or external segment (GP$_e$); GP$_e$ inhibition allows the subthalamic nucleus to stimulate the GP$_i$. This signal will inhibit the VL nuclei, reducing their stimulating action on the MII cortex and therefore the motor gesture. The hyperdirect pathway (inhibitory) provides a direct cortical stimulation on the subthalamic nucleus, which will activate the GP$_i$, resulting in the thalamus inhibition and block of the motor gesture. The projections of the substantia nigra act on D1 and D2, type 1 and 2 receptors for dopamine, respectively. Glutamatergic projections (excitatory) are indicated in *green*, whereas the GABAergic projections (inhibitory) are colored in *red*. The oculomotor loop is divided into two pathways that follow the organization of the direct and indirect pathways just described. Only the areas and nuclei vary, they are represented by the area 8 of Brodmann (BA 8) and by the ocular motor fields for the cortex, by the caudate nucleus for the neostriatum and the ventral anterior and medial dorsal nuclei (VA and MD) for the thalamus, intercalated on this pathway.

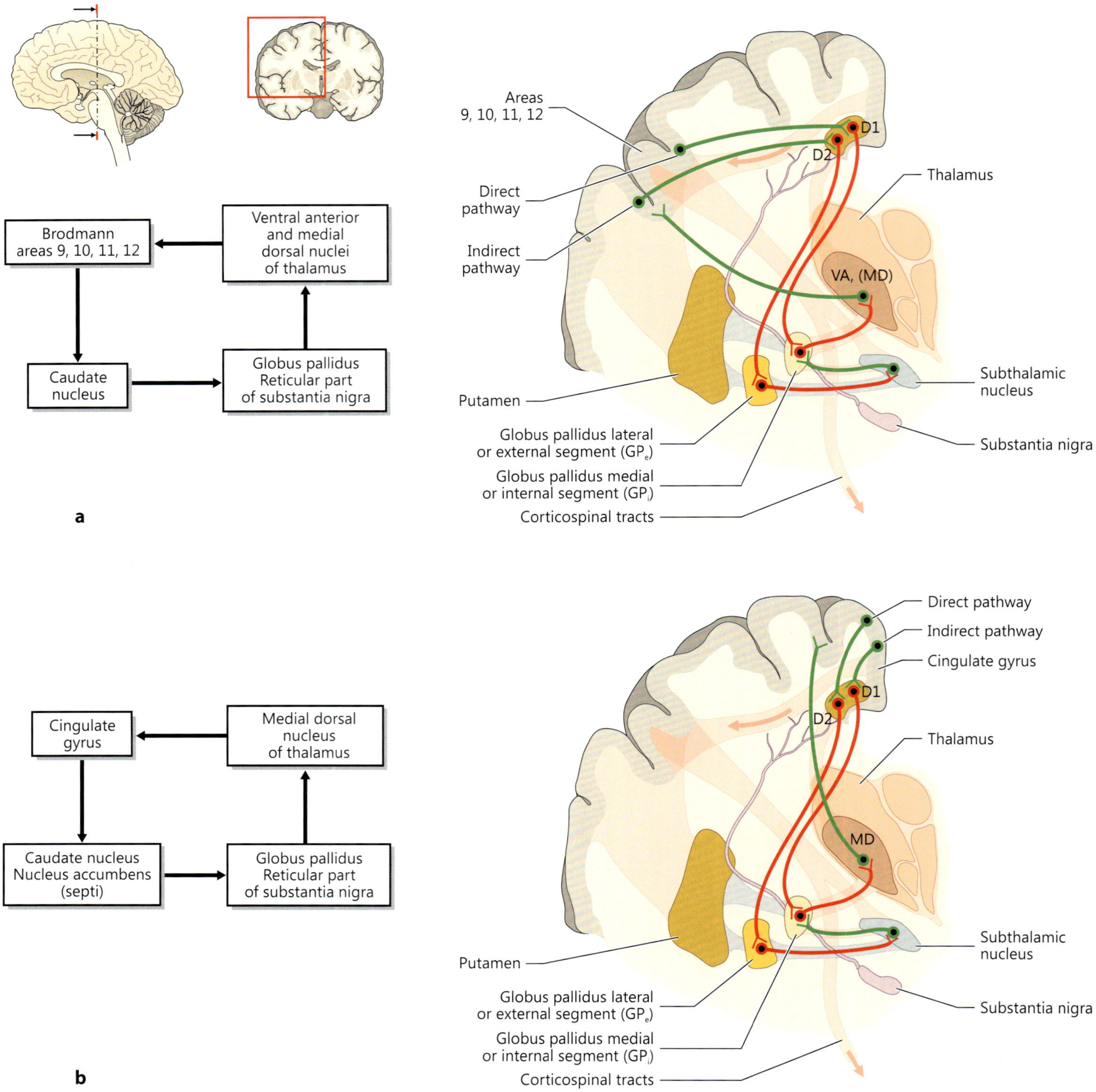

Organization of basal nuclei in the **cognitive** (**a**) **and limbic** (**b**) **circuits or loops**. The cognitive loop is organized in two pathways, similarly to the oculomotor loop illustrated and described in the previous page. Differently to the oculomotor loop, the areas of the cortex involved in the cognitive loop are the areas of Brodmann from 9 to 12 in the prefrontal cortex, whereas the nuclei involved, e.g. the caudate nucleus for the neostriatum and the ventral anterior (VA) and medial dorsal (MD) nuclei to the thalamus, intercalated on this pathway, remain unchanged. The limbic loop is organized in a direct and an indirect pathway, similarly to motor loop illustrated and described in the previous page. Differently to the motor loop, the area of the cortex involved is the cingulate gyrus, and the nuclei are the nuclei of neostriatum, represented by the caudate nucleus and the nucleus accumbens (septi), and the nuclei of the thalamus, represented by the MD nucleus, intercalated on this tract.

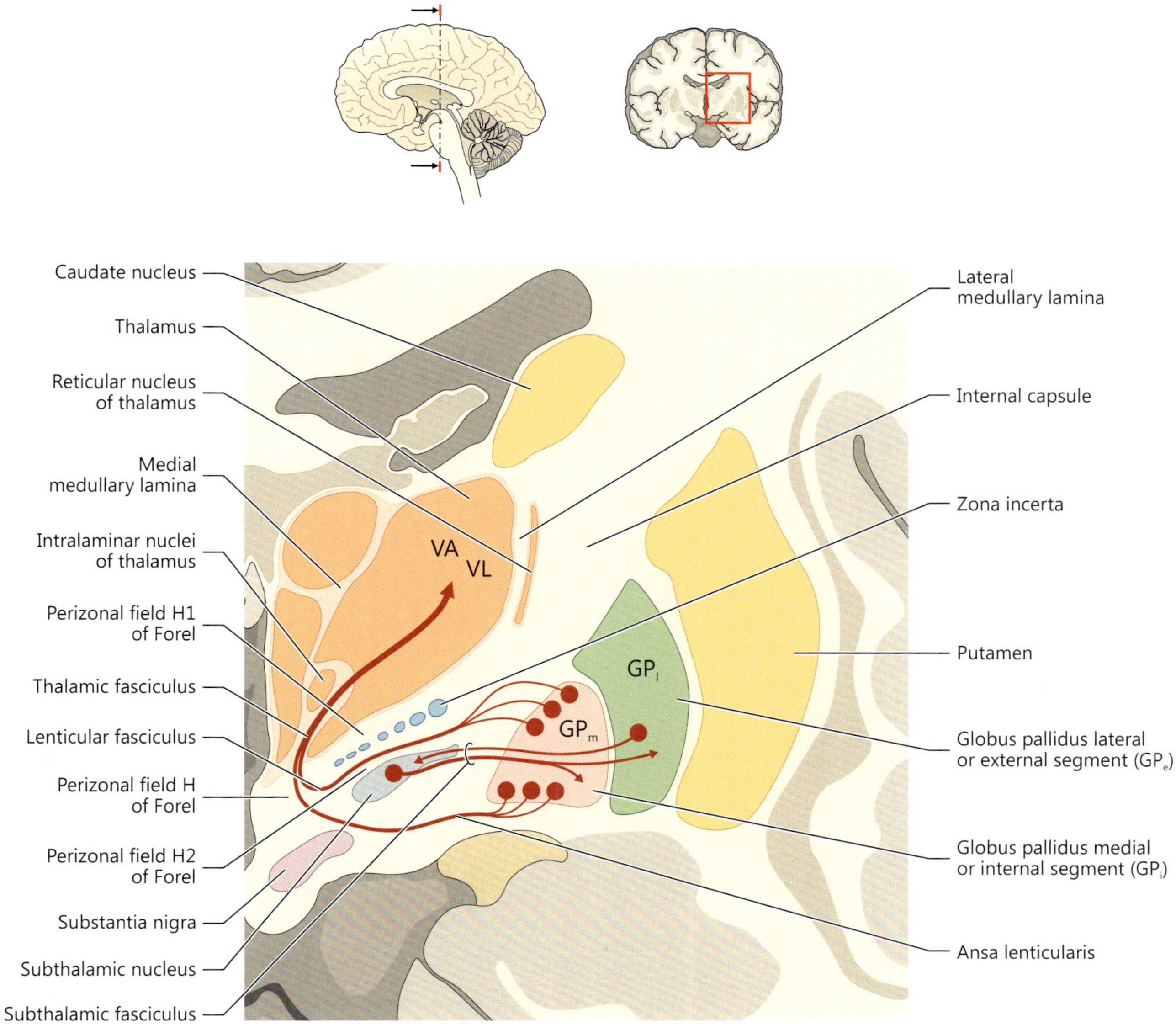

Relationships between the **thalamus** and the **basal nuclei**. Schematic representation of a front section passing through the basal nuclei, the thalamus and subthalamus. The basal nuclei are represented by the caudate nucleus, the putamen and globus pallidus (nuclei of the telencephalon); joint formations with the circuits of the basal nuclei are also the subthalamic nucleus (located in the subthalamus) and the substantia nigra (located in the mesencephalon, between the cerebral crus and the tegmentum).

Classification, afferences, efferences, function and propagation of the cortical projections of the main nuclei of the thalamus and metathalamus (nuclei of the geniculate bodies)

NUCLEI	*AFFERENCES*	*EFFERENCES*	*FUNCTION*	*PROPAGATION OF CORTICAL PROJECTION*
		∗ ANTERIOR NUCLEAR GROUP (FIRST ORDER NUCLEI) ∗		
○ **Anterior nucleus/i** ○	Mammillary bodies, entorhinal cortex	Cingulate gyrus → hippocampus → fornix → mammillary bodies	Emotional behaviour; memory circuits of recent facts	++
– dorsal portion (or anterodorsal nucleus)				
– medial portion (or anteromedial nucleus)	Brainstem			
– ventral portion (or anteroventral nucleus)				

continued

Classification, afferences, efferences, function and propagation of the cortical projections of the main nuclei of the thalamus and metathalamus (nuclei of the geniculate bodies) *(continued)*

NUCLEI	*AFFERENCES*	*EFFERENCES*	*FUNCTION*	*PROPAGATION OF CORTICAL PROJECTION*
		∗ MEDIAL NUCLEAR GROUP OR MEDIAL NUCLEI (HIGHER ORDER NUCLEI) ∗		
○ **Medial dorsal nucleus (or dorsomedial)** ○			Modulation of behavioral activities and of affective processes. Plays a role also in the working-memory. Lesions of the medial dorsal nucleus produce transient amnesia, confusion, reduction of the anxiety, tension and aggression	++
□ – medial portion □	Brainstem, ventral tegmental area, reticular part of the substantia nigra, anterior dorsal cingulate gyrus, globus pallidus			
– – magnicellular (medial) part	Entorhinal cortex, piriform cortex, amygdaloid body	Reticular formation of the thalamus, orbitofrontal cortex, ventromedial prefrontal cortex		
– – parvocellular (lateral) part	Orbitofrontal cortex	Reticular formation of the thalamus, dorsolateral prefrontal cortex, frontopolar part of the frontal cortex		
□ – caudodorsal portion □				
– – caudodorsal nucleus	Brainstem, ventral tegmental area, reticolar part of the substantia nigra, anterior dorsal cingulate gyrus, globus pallidus			
□ – lateral portion □				
– – lateral nucleus	Brainstem, ventral tegmental area, reticular part of the substantia nigra	Reticular formation of the thalamus, globus pallidus, anterior dorsal cingulate cortex, ventrolateral prefrontal cortex, dorsolateral prefrontal cortex, frontal eye fields		
○ **Medioventral nucleus** ○	Hypothalamus, other thalamic nucleii	Prefrontal cortex, olfactory cortex		
		∗ LATERAL NUCLEAR GROUP ∗		
○ **Ventral nuclei** ○				
– ventral anterior nucleus (first order nucleus)	Globus pallidus, reticular part of the substantia nigra	Premotor cortex, supplementary motor cortex, prefrontal cortex	Key role in selecting the start and stop of the motor program to run	+++
– ventral lateral nucleus/i (and ventral intermediate) (first order nucleus)	Globus pallidus, cerebellar nuclei, reticular part of the substantia nigra	Primary motor cortex, premotor cortex, supplementary motor cortex	Modulation of motor gesture during its execution and postural adjustments	++
– ventrobasal complex				
– – ventral posterolateral nucleus (VPL) (first order nucleus)	Medial lemniscus, anterolateral system	Primary and secondary somatosensory cortex	Transmission of somatic sensitivity tracts of the trunk and limbs. The periphery is represented somatotopically	+
– – ventral posteromedial nucleus (VPM) (first order nucleus)	Trigeminal lemniscus (trigeminothalamic tract), gustatory lemniscus	Primary and secondary somatosensory cortex, gustatory cortex	Transmission of somatic sensitivity tracts of the head and mouth and of the gustatory pathways The periphery is represented somatotopically	+
○ **Nuclei of geniculate bodies** ○				
– nuclei of the lateral geniculate body (first order nuclei)	Retina	Primary visual cortex, superior colliculus	Transmission of optic tracts	+

continued

Classification, afferences, efferences, function and propagation of the cortical projections of the main nuclei of the thalamus and metathalamus (nuclei of the geniculate bodies) *(continued)*

NUCLEI	*AFFERENCES*	*EFFERENCES*	*FUNCTION*	*PROPAGATION OF CORTICAL PROJECTION*
– nuclei of the medial geniculate body (first order nuclei)	Inferior colliculus, TRAPEZOID BODY, SUPERIOR OLIVARY NUCLEUS, NUCLEI OF LATERAL LEMNISCUS	Primary auditory cortex	Transmission of acoustic tracts	+
○ **Dorsal nuclei (or lateral)** ○			Involvement in the visual responses. Regulation of perceptive (in particular, pain modulation), cognitive and anamnestic functions	
– dorsal (lateral) nucleus	Pretectal area, superior colliculus	Retrosplenial cingulate gyrus, parahippocampal cortex, presubiculum		+
– lateral (posterior) nucleus	Parietal cortex, superior colliculus	Retrosplenial cingulate gyrus, parahippocampal cortex,		++
○ **Pulvinar (pulvinar nuclei) (higher order nuclei)** ○			Involvement in cognitive, attention and memory processes, generated by visual stimuli	+
□ – lateral portion (or lateral pulvinar nucleus) □				
– – dorsal part	Superior colliculus	Inferior parietal cortex, dorsolateral prefrontal cortex		
– – ventral part	Superior colliculus	Visual cortex		
□ – inferior portion (or inferior pulvinar nucleus) □				
– – centrolateral part	Medial and lateral geniculate bodies, superior colliculus	Corteccia visiva		
– – centromedial part	Superior colliculus	Middle temporal cortex (superior portion)		
– – medial part	Retina, superior colliculus	Middle temporal cortex		
– – posterior part	Superior medial temporal cortex, superior colliculus	Middle temporal cortex (superior portion)		
□ – medial portion (or medial pulvinar nucleus) □				
– – medial part	Amygdaloid body, superior colliculus	Superior temporal cortex, inferior temporal cortex, orbitofrontal cortex, dorsolateral prefrontal cortex		
– – lateral part	Superior colliculus	Superior temporal cortex, inferior temporal cortex		
∗ **INTRALAMINAR NUCLEI (HIGHER-ORDER NUCLEI)** ∗			Cortical activation; integration of sensory and motor impulses; control of the level of vigilance and its motor behavior	
○ **Anterior nuclear group** ○	Pontomesencephalic reticular formation, cerebellar nuclei, spinal cord (medial compartment of the lateral spinothalamic tract)			+++
– central medial nucleus	Granular prefrontal cortex, medial limbic cortex	Parietal cortex, (medial and basal portions), frontal cortex (medial and basal portions)		
– paracentral nucleus	Granular prefrontal cortex, medial limbic cortex	Parietal cortex (lateral portion), frontal cortex (lateral portion)		
– central lateral nucleus	Motor cortex, parietal cortex	Parietal cortex (lateral portion), frontal cortex (lateral portion)		
○ **Posterior nuclear group** ○				+++
– centromedian nucleus	Globus pallidus medial (or internal) segment	Motor cortex, striatum (or neostriatum)		
– parafascicular nucleus	Premotor cortex	Cortex adjacent to the rhinal sulcus, cingulate gyrus, striatum (or neostriatum)		

continued

Classification, afferences, efferences, function and propagation of the cortical projections of the main nuclei of the thalamus and metathalamus (nuclei of the geniculate bodies) *(continued)*				
NUCLEI	***AFFERENCES***	***EFFERENCES***	***FUNCTION***	***PROPAGATION OF CORTICAL PROJECTION***
∗ MEDIAN NUCLEI OF THALAMUS (HIGHER-ORDER NUCLEI) ∗	Hypothalamus		They are critical for the maintenance of the sleep-wake cycle and for limbic functions	++
Paratenial nucleus	Prefrontal cortex, reticular nucleus of the thalamus	Anterior cingulate cortex, prelimbic cortex, insula, piriform cortex, entorhinal cortex, nucleus accumbens, corpus striatum, subiculum, claustrum, amygdaloid complex (or amygdaloid body)		
Paraventricular nucleus	Ventral tegmental area, posterior (or dorsal) and median (or superior central) nuclei of raphe nuclei, medullopontine reticular formation, subiculum	Prefrontal cortex, insula, nucleus accumbens, olfactory tubercle, subiculum, claustrum, amygdaloid body, hypothalamus, lateral septal nucleus (or lateral septum)		
Rhomboid nucleus	Ventral tegmental area, central gray matter (or periaqueductal gray) peduncular nucleus (or pedunculopontine), reticular part of the substantia nigra, parahippocampal cortex, parabrachial nuclei (or parabrachial complex)	Primary and secondary somatosensory cortex, prefrontal cortex, posterior parietal cortex, occipital cortex, temporal cortex, dorsal striatum, ventral striatum, olfactory tubercle, amygdaloid body, insula, hippocampus, claustrum		
Nucleus reuniens	Ventral tegmental area, posterior (or dorsal) and median (or superior central) raphe nuclei, entorhinal cortex, subiculum, claustrum, ventral and dorsal nuclei of the lateral geniculate body, parabrachial nuclei (or parabrachial complex)	Prefrontal cortex, anterior cingulate cortex, entorhinal cortex, insula, retrosplenial cortex, perirhinal cortex, hippocampus		
∗ RETICULAR FORMATION OF THE THALAMUS (FIRST ORDER NUCLEUS) ∗				
Reticular nucleus of thalamus	Collaterals of the thalamocortical fibers, cerebral cortex, pontomesencephalic ascending reticular formation, amygdaloid body, GLOBUS PALLIDUS, SUBSTANTIA NIGRA, LOCUS CAERULEUS, PARABRACHIAL NUCLEI, HYPOTHALAMUS	Intralaminar nuclei, median nuclei of thalamus and retransmission nuclei of the thalamus	It modulates the thalamocortical efferences; given the connections with the amygdaloid body, it is called the "emotional gatekeeper"	

Note: ∗ ... ∗, between asterisks are indicated the nuclear groups of the thalamus; ○ ... ○, between circles are indicated the nuclei of each nuclear group; □ ... □, between squares are indicated the subdivision portions of a nucleus; with underlined character are indicated the double-turn connections and with underlined UPPERCASE character the quantitatively reduced afferences. The diffusivity to the cerebral cortex is indicated by a progressive number of signs +: one single + indicates specific projection nuclei, three + signs refer to nonspecific projection nuclei, two + stands for an intermediate condition between the two. The higher degree nuclei are involved in superior cognitive processes and mainly connect more areas of the cerebral cortex, whereas the first degree nuclei retransmit signals from the subcortical structures.

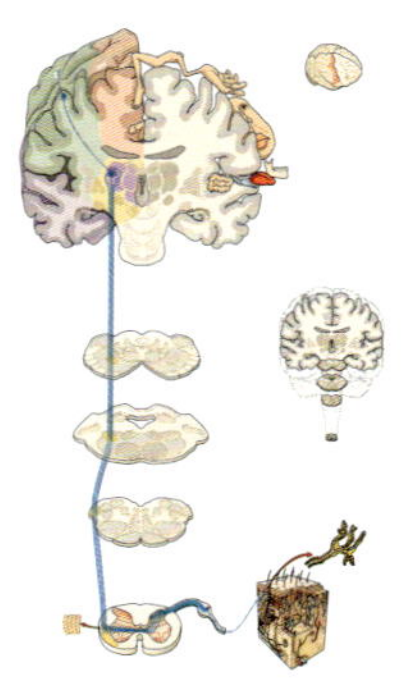

Antero lateral system (ALS) - Lateral and medial compartments (☞ pages 462 and 466).

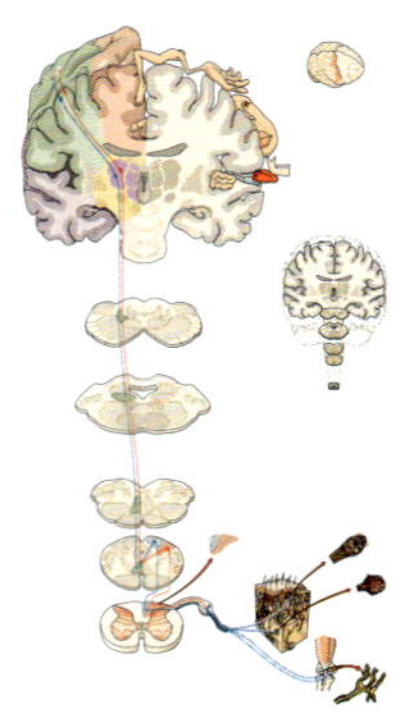

Dorsal column - Medial lemniscal pathway (DC-ML) (☞ page 470).

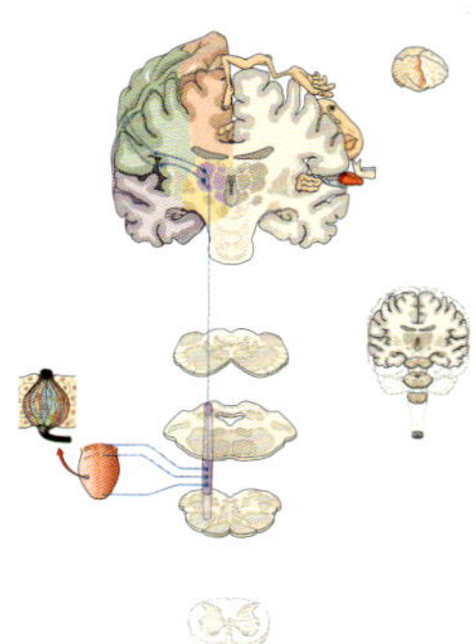

Solitariothalamic tract - Gustatory lemniscus (☞ page 494).

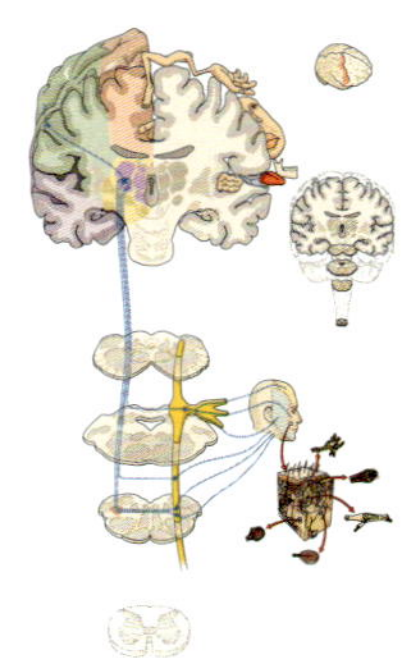

Ventral secondary trigeminal tract (VSTT) - Trigeminal lemniscus (☞ page 496).

Connexions of the thalamus

BUNDLES	LOCATION	ORIGIN	PROJECTION	FUNCTION
Ansa lenticularis	It bends, following a lateromedial course, around the ventral margin of the internal capsule	Medial and lateral parts, of globus pallidus medial segment Putamen	It stops, in part, in the nucleus of the ansa lenticularis and continues dorsally, with the H2 field of Forel, to form the H field of Forel Pallidohypothalamic tract (from the H field of Forel to the ventromedial nucleus of the hypothalamus, in relationship with the columns of the fornix)	Circuit between cortical areas involved in planning and executing movements, striatum, globus pallidus, motor thalamus (ventral anterior and ventral lateral nuclei), motor cortex
Dentatothalamic and rubrothalamic fibers	Along the dorsal surface of the zona incerta of the subthalamus	Dentate nucleus and red nucleus	Thalamus (for the most part to the ventral nuclei and in small part to the intralaminar nuclei)	Modulation of motor commands of the pyramidal tracts with motion programming
Gustatory lemniscus	Ipsilaterally ascending fibers from the nucleus of the solitary tract through the medial lemniscus (in many species)	Gustatory pathway of the nuclei of the solitary tract (of the vagus, glossopharyngeal and intermediate nerves)	From the ventral posteromedial nucleus of the thalamus to the gustatory cortex (anterior part of the insula and adjacent parainsular region, frontal operculum and inferior frontal gyrus) Fibers to the hypothalamus	Central gustatory tract
Habenulointerpeduncular tract	From the habenular trigone to the basal region of the mesencephalon	Lateral and medial habenular nuclei	Interpeduncular nucleus and from this, through the reticular formation to the spinal cord	Integration of olfactory stimuli, of somatic and visceral senses
Lenticular fasciculus	It crosses the internal capsule opening its way between the subthalamic nucleus and the zona incerta	Medial part of medial globus pallidus medial segment	It continues in the H2 field of Forel, thus reaching the fibers of the ansa lenticularis, to form the H field of Forel; it stops, in part, at the nuclei of the zona incerta and the medial perizonal field [H]	Circuit between cortical areas involved in planning and executing movements, striatum, globus pallidus, motor thalamus (ventral anterior nucleus and ventral lateral nuclei), motor cortex
Medial lemniscus	The axons of the neurons of the gracile and cuneate nuclei cross the median line moving contralaterally, through the internal arcuate fibers, forming the medial lemniscus which goes up in the brainstem changing orientation in the pons and mesencephalon	Gracile and cuneate nuclei	Ventral posterolateral nucleus (VPL) of the thalamus	It delivers the discriminative aspects of sensitivity (epicritic sense), in particular the sense of position and movement of the body parts, the tactile recognition of the shape and other characteristics of objects and of the variations of position of the tactile stimuli
Spinal lemniscus (anterior and lateral spinothalamic tracts and spinotectal tract)	Bulb, posterior part of the pons and tegmentum of the midbrain	Funicular neurons (in the laminae IV, VI and VII) of all neuromers of the spinal cord	Ventral posterolateral nucleus of the thalamus	It delivers the protophatic sense (tactile non discriminative, thermal and dolorific)
Subthalamic fasciculus	It crosses the internal capsule	Medial and lateral parts of the medial globus pallidus medial segment	Subthalamic nucleus Subthalamopallidal fibers to the medial part of the medial globus pallidus medial segment	Corticobasal circuit of the extrapiramidal system involved in controlling somatic motility
Thalamic fasciculus	Along the dorsal surface of the zona incerta of the subthalamus, with rostrolateral direction	It is formed by the confluence of the H field of Forel fibers with dentatothalamic fibers and rubrothalamic fibers	Pallidothalamic fibers to the ventral anterior and ventral lateral nuclei of the thalamus Collateral to the zona incerta of the subthalamus	Circuit between cortical areas involved in planning and executing movements, striatum, globus pallidus, motor thalamus (ventral anterior and ventral lateral nuclei), motor cortex
Trigeminal lemniscus – from the spinal nucleus of trigeminal nerve – from the principal sensory nucleus of the trigeminal nerve	 Associated to the spinothalamic tracts Associated to the medial lemniscus	 Spinal nucleus of the trigeminal nerve Principal sensory nucleus of the trigeminal nerve	 Crossed at the ventral posteromedial nucleus of the thalamus Crossed at the ventral posteromedial nucleus of the thalamus	It delivers the tactile, thermific and dolorific sense of the skin of the face, of the mucosae of the oral and nasal cavities, of the eye and the proprioceptive sense of the facial and masticatory muscles

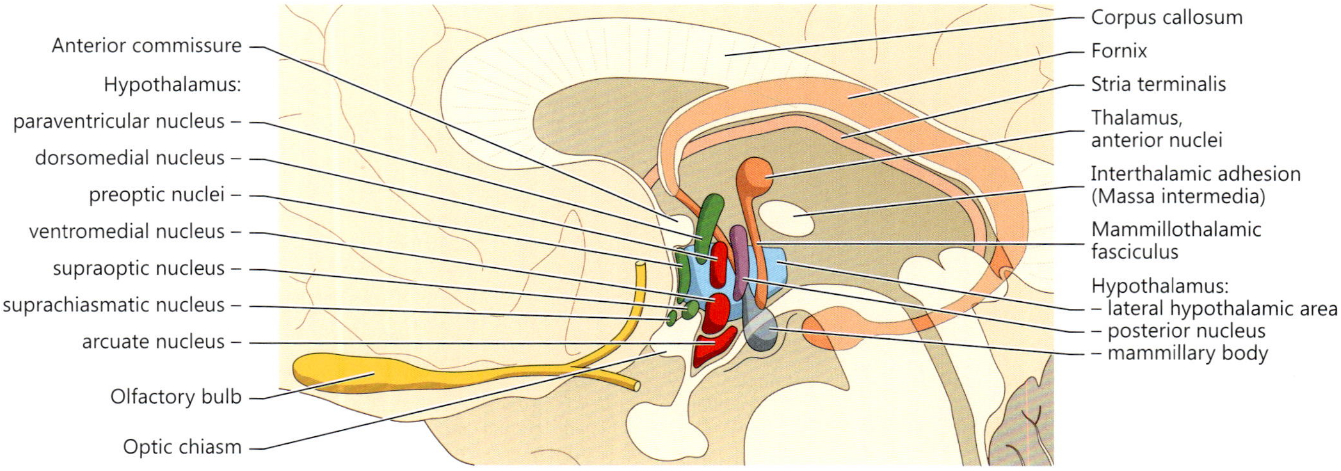

Lateral wall of the **third ventricle** with median diencephalic formations.

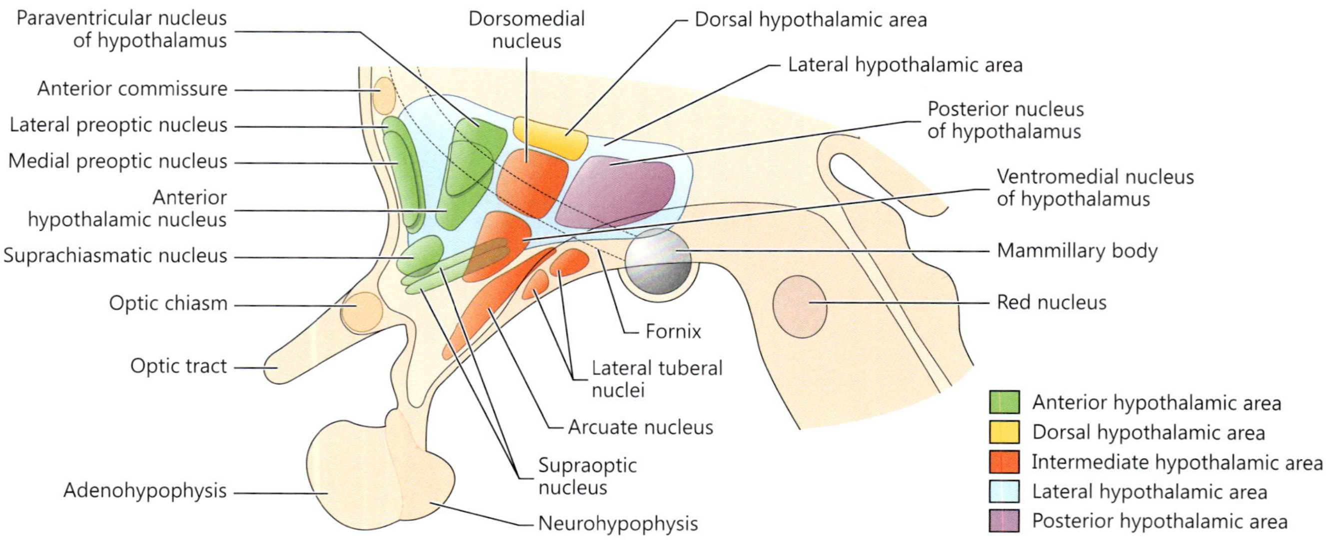

Schematic representation of the **hypothalamus** in sagittal section showing the hypothalamic nuclei.

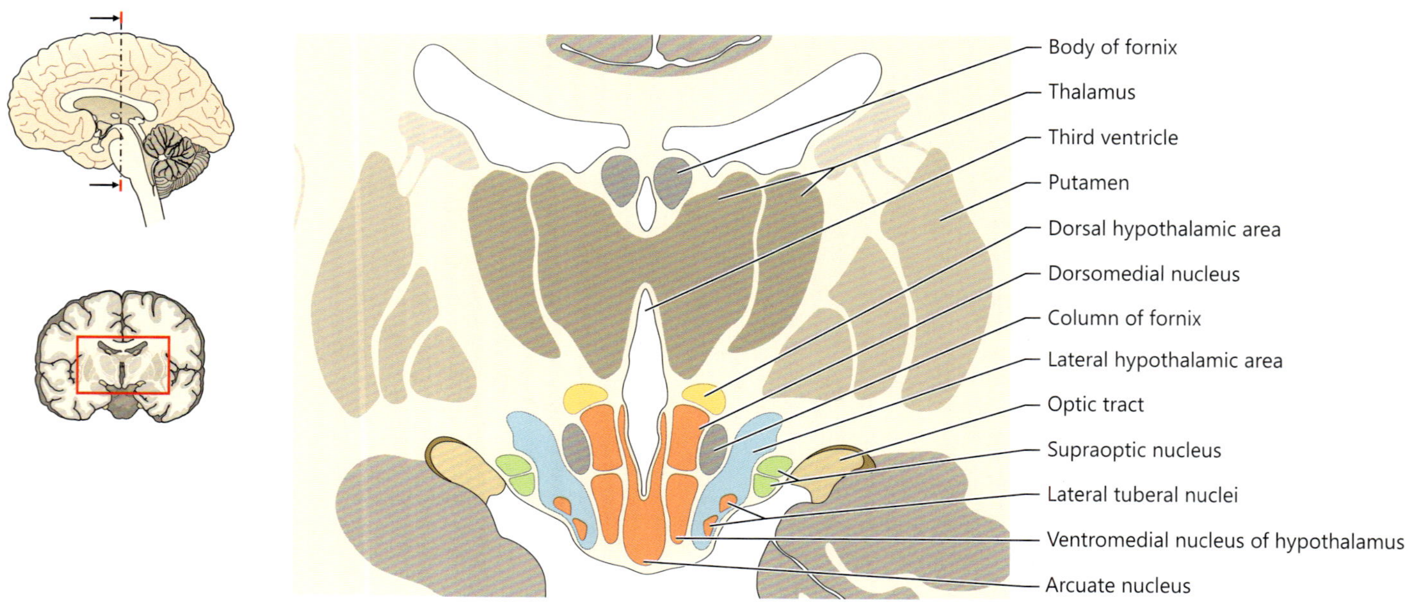

Schematic representation of the **hypothalamus** in frontal section showing the hypothalamic nuclei.

Subdivision of the hypothalamus in regions and zones with the corresponding nuclei			
AREAS OR REGIONS	***LATERAL ZONE (occupied for the most part by a widespread neuronal matrix)***	***MEDIAL ZONE (relatively rich in cells)***	***PERIVENTRICULAR ZONE (with few parallel layers of small cells alternated with thin amyelinic fibers)***
Anterior hypothalamic area (or anterior hypothalamic region), including the preoptic and supraoptic or suprachiasmatic areas	Lateral preoptic nucleus Suprachiasmatic nucleus Supraoptic nucleus	Interstitial nuclei of anterior hypothalamus Anterior hypothalamic nucleus Medial preoptic nucleus Median preoptic nucleus	Anterior periventricular nucleus Paraventricular nucleus of the the hypothalamus Periventricular preoptic nucleus
Dorsal hypothalamic area (or dorsal hypothalamic region), including the retrochiasmatic area or region	Endopeduncular nucleus	Dorsomedial nucleus	Nucleus of the ansa lenticularis
Intermediate hypothalamic area (or intermediate hypothalamic region), including the infundibular area	Lateral tuberal nuclei	Dorsomedial nucleus Ventromedial nucleus of the hypothalamus Dorsal nucleus of the hypothalamus Arcuate (or infundibular) nucleus	Periventricular nucleus Posterior periventricular nucleus
Lateral hypothalamic area	Lateral tuberal nuclei Perifornical nucleus	Tuberomammillary nucleus	
Posterior hypothalamic area (or posterior hypothalamic region)		Medial (or principal) nucleus of mammillary body Lateral nucleus of mammillary body Dorsal premammillary nucleus Ventral premammillary nucleus Supramammillary nucleus	Posterior nucleus of the hypothalamus

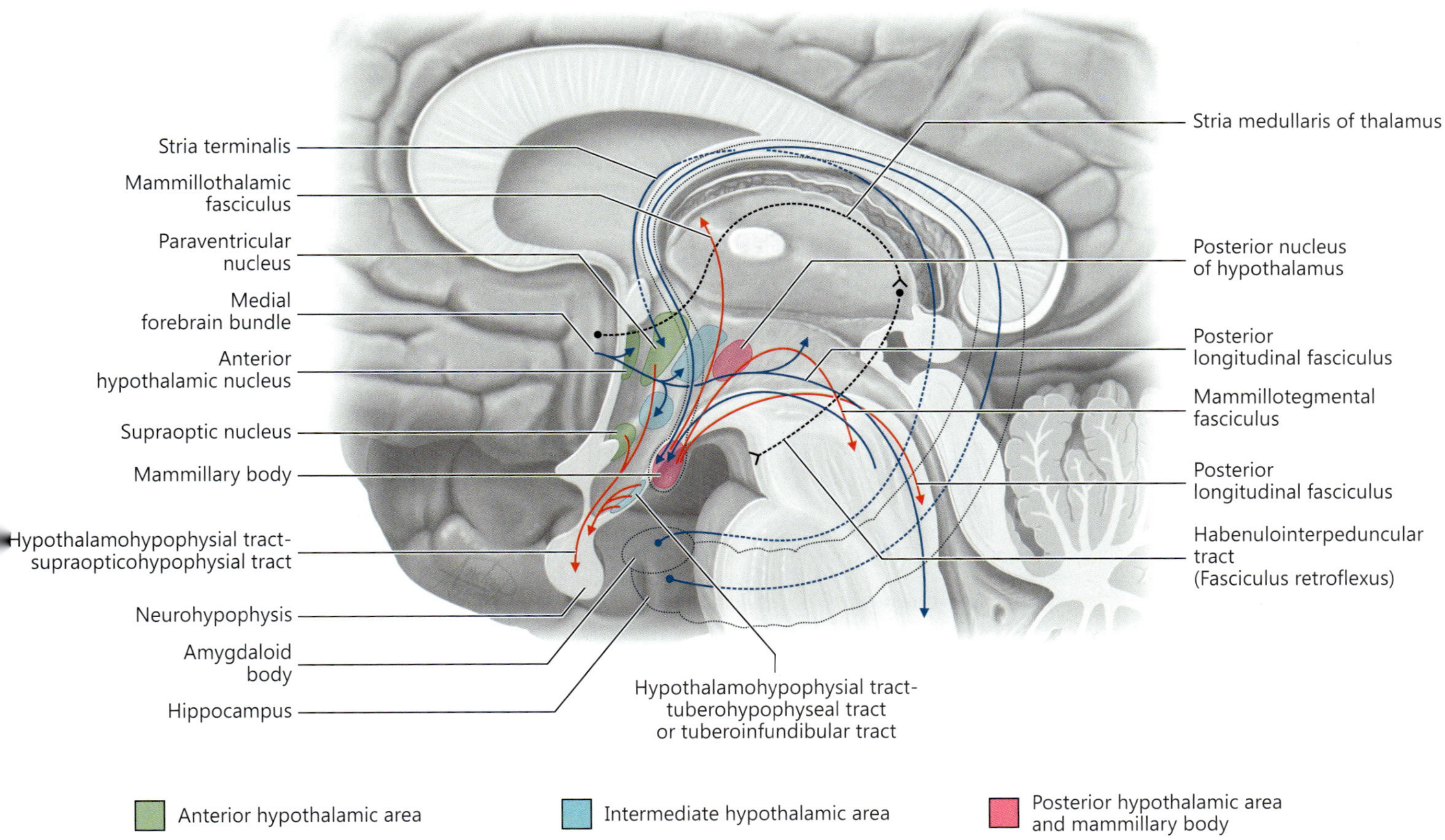

Schematic representation showing part of the brain in sagittal section with the main afferences (*blue* lines) and efferences (*red* lines) of the **hypothalamus**.

Afferences and efferences of the main nuclei of the hypothalamus

NUCLEI	MAIN AFFERENCES	MAIN EFFERENCES	FUNCTION
Anterior hypothalamic nucleus	Amygdalothalamic tract (through the stria medullaris of the thalamus)	Areas of the medial zone of the hypothalamus, medial preoptic area and ventromedial nucleus of hypothalamus Projections also to the amygdaloid body, to the median nuclei of the thalamus, to the subthalamus and to the mesencephalon	Together with the preoptic region, regulation of body temperature, sweating and respiratory activity Production of releasing hormones
Arcuate (or infundibular) nucleus	From the subfornical and vascular organs of lamina terminalis and the posterior and median raphe nuclei	To the medial preoptic, paraventricular of the hypothalamus and perifornical nuclei	Production of GHRH and other releasing hormones Secretion of dopamine in the hypophyseal portal system
Dorsomedial nucleus	From the suprachiasmatic nucleus and from almost all the other main nuclei and areas of the hypothalamus	To the preoptic area, to the locus caeruleus and the perifornical nucleus	Reception of leptine levels Stimulation of gastrointestinal functions Secretion of atrial natriuretic peptide
Lateral nucleus of mammillary body	Mammillary peduncle tract (from the nuclei of the tegmentum of the midbrain)	Mammillothalamic fasciculus (anterodorsal nucleus of the thalamus)	Control of the activity of the autonomic nervous system and of emotional instinctive activities
Lateral tuberal nuclei	Medial forebrain bundle Corticohypothalamic fibers	Periventricular and tuberohypophysial fibers	Production of releasing hormones
Medial nucleus of mammillary body	Hippocampohypothalamic tract	Mammillothalamic fasciculus (to the anteroventral and anteromedial nuclei of the thalamus) Mammillotegmental fasciculus (to the nuclei of the tegmentum of the midbrain)	Control of the activity of the autonomic nervous system and of emotional instinctive activities
Paraventricular nucleus of the the hypothalamus	Retinohypothalamic fibers through the optic tract	To the supraopticohypophysial tract for the neurohypophysis	Oxytocin secretion
Perifornical nucleus	Medial forebrain bundle and precommissural fibers from the fornix Corticohypothalamic and thalamohypothalamic fibers	Intrahypothalamic and to the locus caeruleus	Hunger and satiety centers Orexin secretion Increase in blood pressure Regulation of the sleep-wake rhythm
Posterior nucleus of the hypothalamus	Medial forebrain bundle	Periventricular fibers and posterior longitudinal fasciculus	Thermoregulation (shiver center) and control of the activity of the autonomic nervous system Production of releasing hormones
Preoptic area (medial, median, lateral and periventricular preoptic nuclei)	Intrahypothalamic	Intrahypothalamic	Termoregulation
Suprachiasmatic nucleus	Retinohypothalamic and geniculohypothalamic fibers	Intrahypothalamic (preoptic area, subparaventricular area, area between the arcuate and the ventromedial nuclei of the hypothalamus)	Control of circadian rhythms
Supraoptic nucleus	Medial forebrain bundle (from the anterior olfactory centers of telencephalon) Corticohypothalamic fibers Retinohypothalamic fibers through the optic tract Amygdalothalamic tract (through the stria medullaris of the thalamus)	To the thalamus (mediodorsal nucleus), to the tectal plate, to the posterior longitudinal fasciculus (or of Schütz) to the supraopticohypophysial tract for the neurohypophysis	Vasopressin secretion
Ventromedial nucleus of the the hypothalamus	From the amygdaloid body through the fibers of the stria terminalis	Areas of the medial zone of the the hypothalamus	Thirst and satiety center

Connections of the hypothalamus

FIBERS	LOCATION	ORIGIN	PROJECTION	FUNCTION
		AFFERENT PATHWAYS		
Amygdalohypothalamic tract through the stria terminalis and the ventral amygdalofugal pathway	Partly in the stria terminalis and partly ventrally to the lenticular nucleus	Ventral half of the amygdaloid body	Preoptic area Anterior nucleus of the hypothalamus Supraoptic nucleus Septal area Perifornical nucleus Amygdaloid body	Connections between amygdaloid body and the hypothalamus
Corticohypothalamic tract	Partly in the medial forebrain bundle	Frontal lobe	Preoptic area e lateral zone of the hypothalamus	Cortical influences on the hypothalamus
Hippocampohypothalamic tract	Fornix	Hippocampus	Medial nucleus of mammillary body Precommissural fibers to the septal nuclei, to the preoptic and lateral hypothalamic areas	Connections with the ventral hippocampus
Medial forebrain bundle (septo-hypothalamic olfactohypothalamic and corticohypothalamic fibers)	Complex of fibers running longitudinally in the hypothalamus	Periamygdaloid region and septal nuclei	Lateral zone of the the hypothalamus	Connection between anterior olfactory centers of the telencephalon and lateral zone of hypothalamus
Thalamohypothalamic tract	Inferior thalamic radiation	Mediodorsal nucleus of the thalamus	Rostral half of the amygdaloid body Preoptic area Perifornical nucleus	It carries regulatory impulses of cortical origin to the hypothalamus
		EFFERENT PATHWAYS		
Hypothalamohypophysial tract (supraopticohypophysial tract) and tuberoinfundibular pathway	Infundibulum of the neurohypophysis	Supraoptic nucleus Paraventricular nucleus Lateral tuberal nuclei	Neurohypophysis Median eminence	Transport of vasopressin, oxytocin and of releasing hormones
Mammillary peduncle tract	The ascending fibers from the dorsal tegmentum of the mesencephalon terminate mainly in the mammillary bodies, whereas those proceeding from the ventral tegmentum reach the more rostral levels of the hypothalamus and the limbic structures through the medial fasciculus of the telencephalon	Nuclei of the tegmentum of the midbrain	Lateral nucleus of mammillary body	Conduction to the hypothalamus of impulses of the general visceral and special gustatory sense and of the general somatic and special acoustic sense
Periventricular fibers	Ventrally with respect to the aqueduct of the midbrain for the direct fibers to the brainstem	Supraoptic nucleus Lateral tuberal nuclei Posterior hypothalamic nucleus	Mediodorsal nucleus of the thalamus Tectal plate Periaqueductal (central) gray matter and reticular formation of the mesencephalon and pons through the posterior longitudinal fasciculus (or of Schültz)	Control, through the reticular formation, of the activity of somatomotor and visceroeffector neurons of the brainstem and spinal cord
Principal mammillary tract				
– mammillothalamic fasciculus	Dorsal component of the principal mamillary fasciculus that wraps the mammillary nuclei	Medial and lateral mammillary body nuclei	Direct pathway to the anteromedial and anteroventral nuclei of the thalamus from the medial nucleus of mammillary body Direct and crossed pathways to the anterodorsal nucleus of the thalamus from the lateral mammillary nucleus	Projection of limbic impulses to the thalamus
– mammillotegmental fasciculus	Dorsally with respect to the mammillary nuclei	Medial nucleus of mammillary body	Nuclei of the tegmentum of the mesencephalon	Influence on the somatic and visceral motor centers of the telencephalon
Retinohypothalamic fibers	Optic nerves, optic chiasm and optic tracts	Retina	Preoptic area Supraoptic nucleus Paraventricular nucleus of the hypothalamus	Regulation of hypothalamic functions through visual stimuli

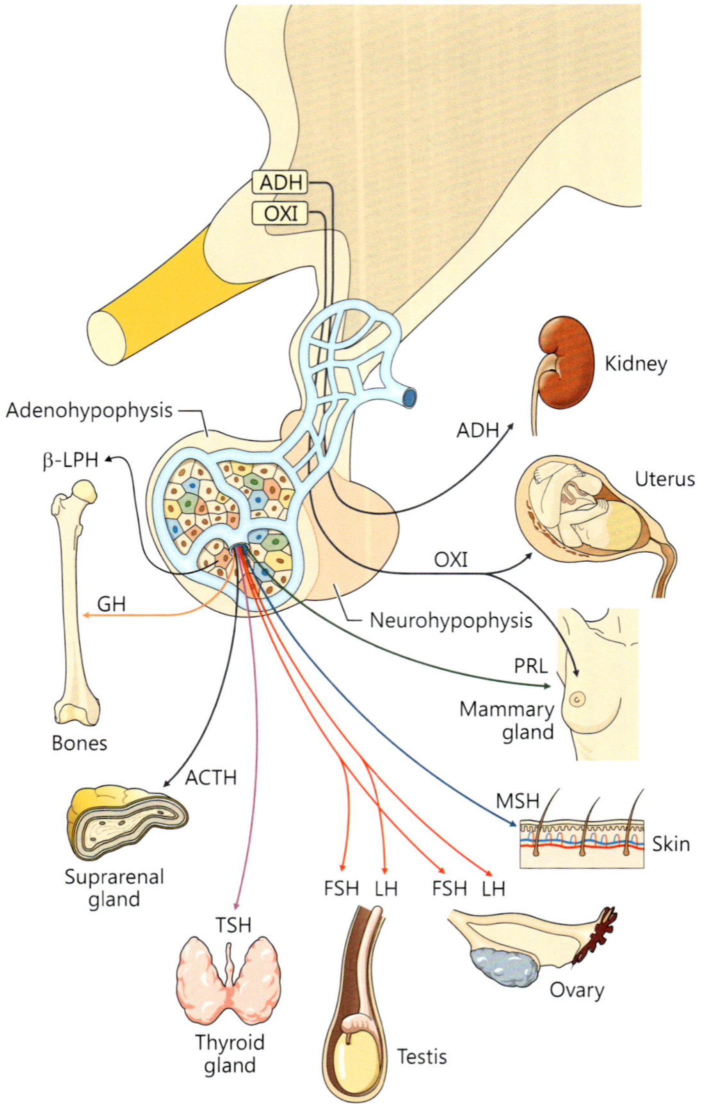

➤ Schematic representation showing the endocrine correlations among the **hypothalamus**, the **pituitary gland** (or hypophysis) and the **target organs**.
ACTH, adrenocorticotropic hormone
ADH, antidiuretic hormone
β-LPH, β-lipotropin
FSH, follicle-stimulating hormone
GH, growth hormone
LH, luteinizing hormone
MSH, melanocyte-stimulating hormone
OXT, oxytocin
PRL, prolactin
TSH, thyroid-stimulating hormone

Below, schematic representation showing the conformation of the **hypothalamic-hypophysial portal system**. The adenohypophysis is supplied by the superior hypophysial artery, originated from the cerebral arterial circle (circle of Willis), it penetrates in the median eminence and in the superior part of the infundibulum of the neurohypophysis, where it forms the primary capillary plexus of the hypothalamic-hypophysial portal system; from this point the long and short portal veins of the hypophysis originate and, through the infundibulum, enter the adenohypophysis, onto the secondary capillary plexus formed by the sinusoids of the glandular stroma. From this plexus originate the hypophysial veins tributaries of the neighboring sinuses of the dura mater. The neurohypophysis receives, instead, the inferior hypophysial artery, branch of the internal carotid artery, forming a capillary plexus inside, from which the tributary hypophysial veins of the surrounding sinuses of the dura mater originate.

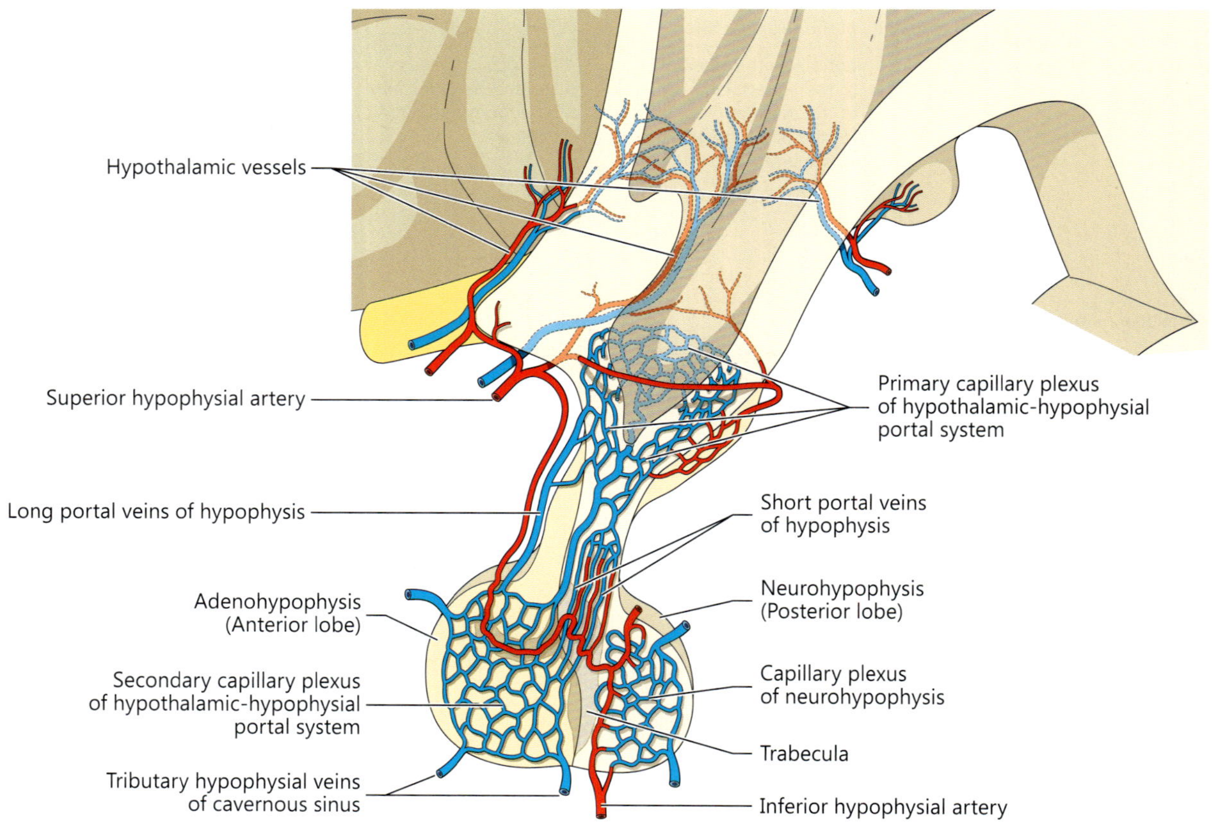

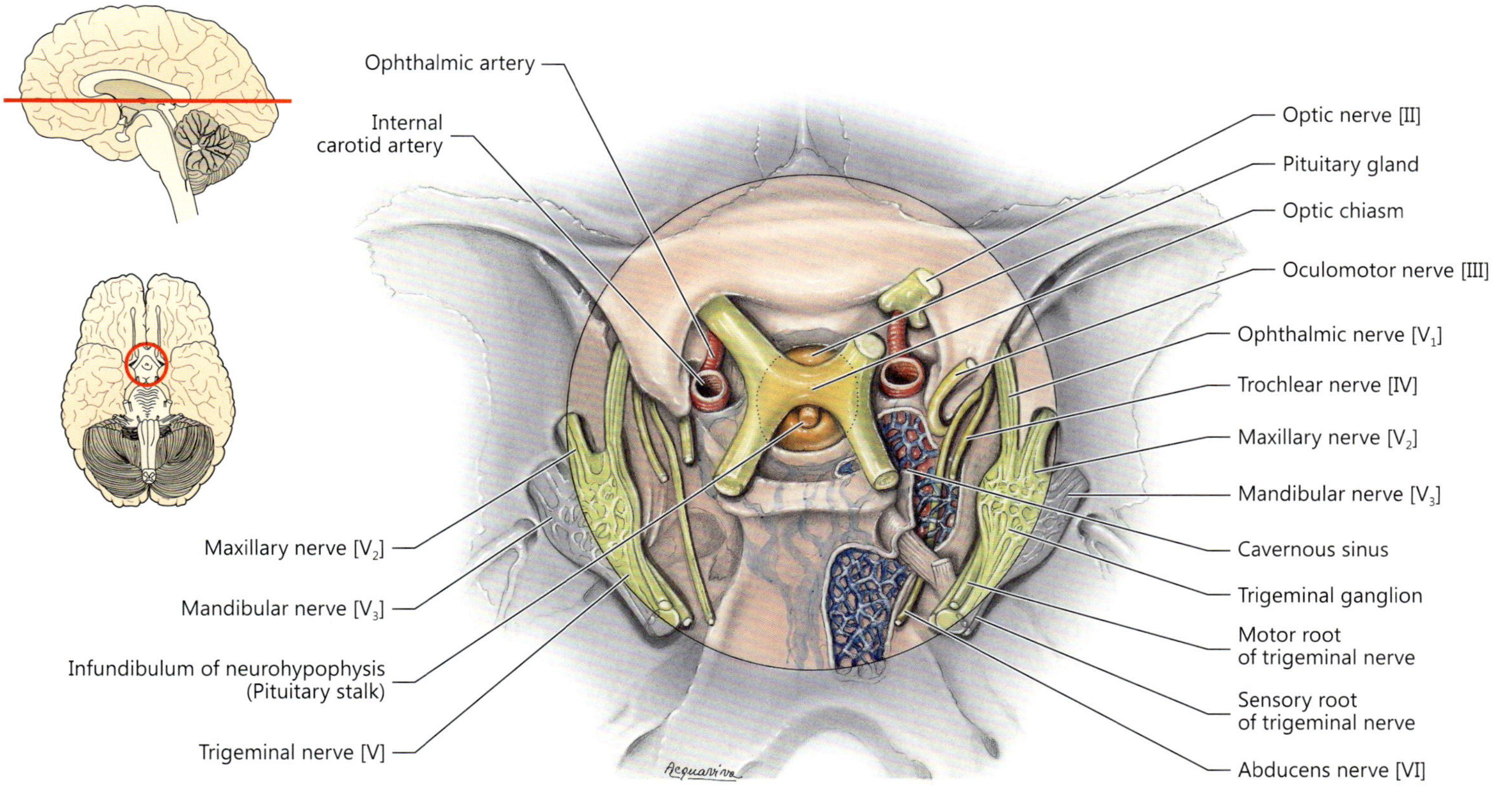

Relationships of the **pituitary gland** (or hypophysis) seen from above. This projection demonstrates the close relationships established between the pituitary gland and the optic chiasm and the lateral relationships that the gland has with the cavernous sinus and the vascular nervous formations it hosts.

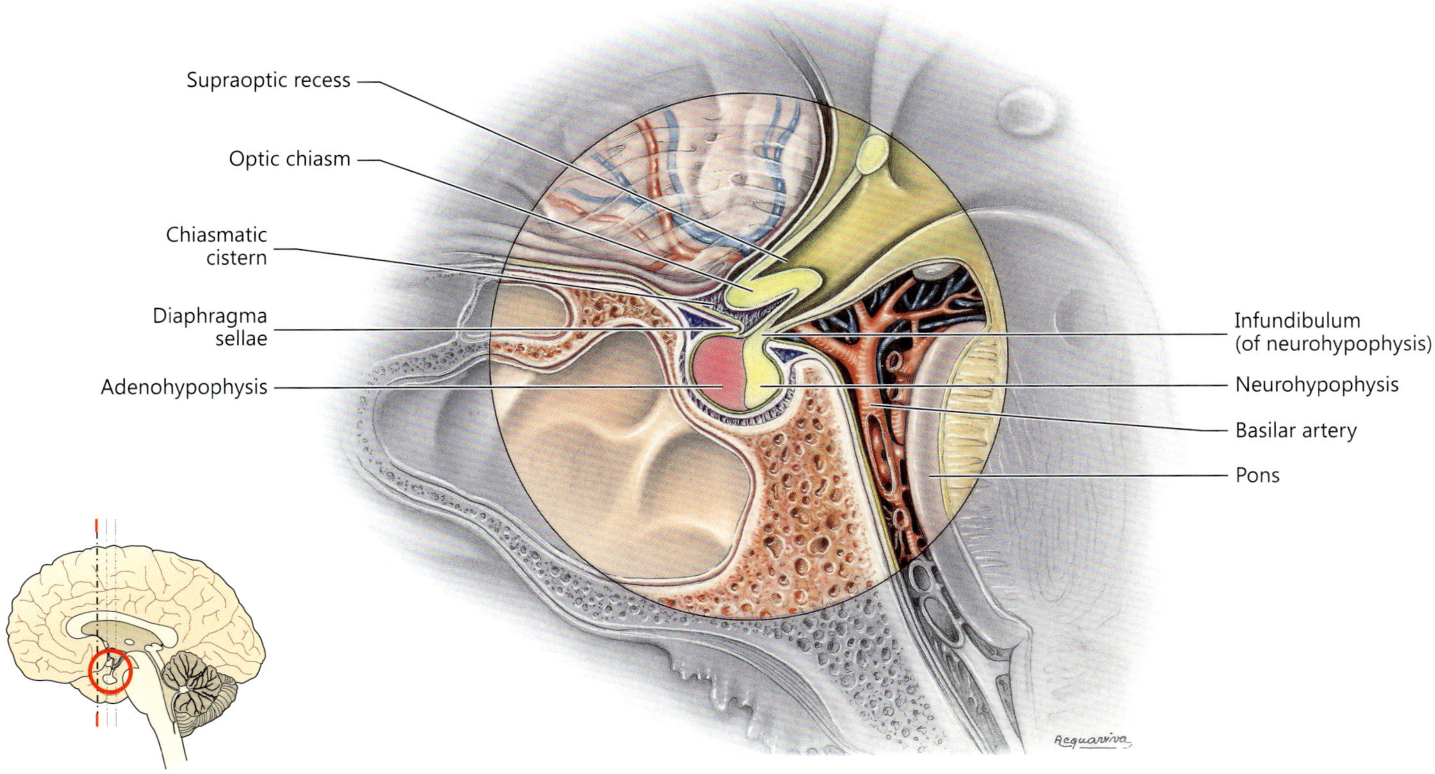

Median sagittal section showing the main relationships of the **pituitary gland** (or hypophysis). The main dural (diaphragma sellae) and subarachnoid formations (chiasmatic cistern) in relation with the gland are as well shown.

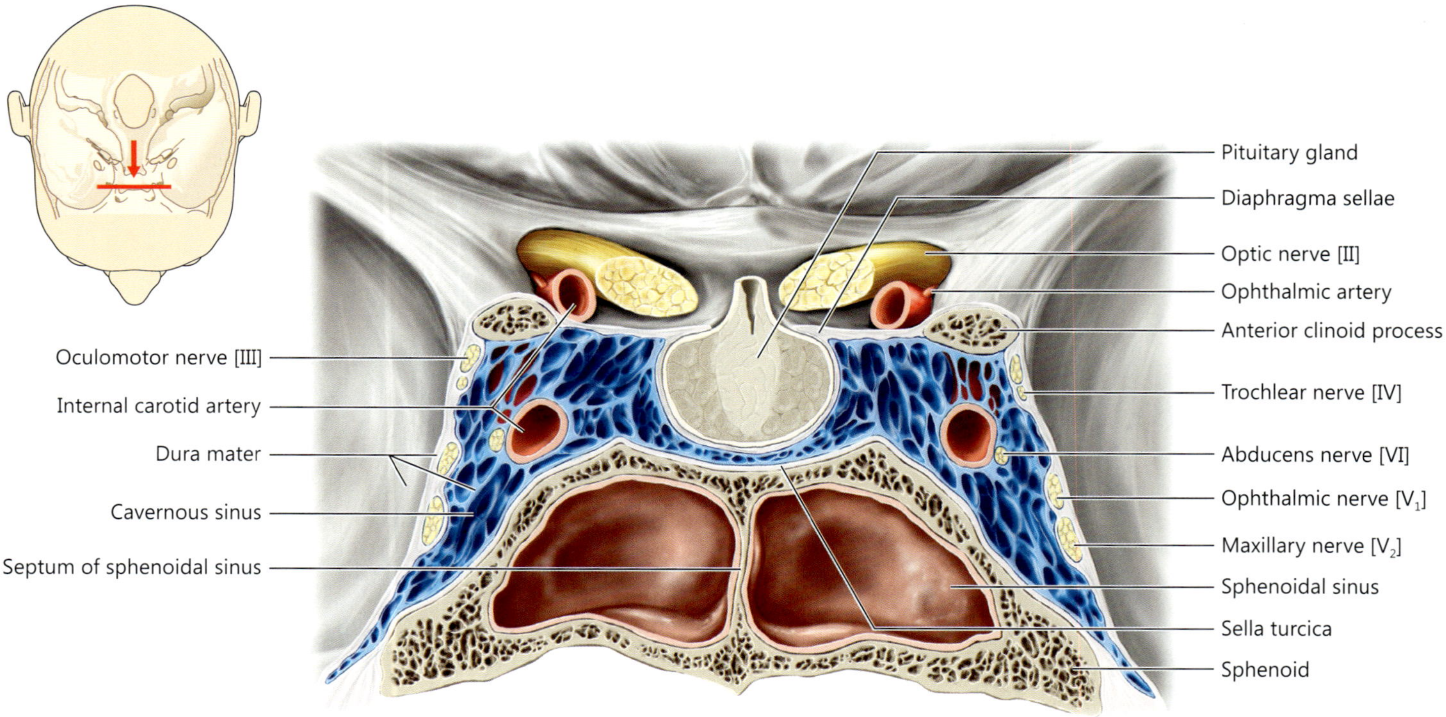

Schematic representation showing a frontal section of the cranium passing in front of the **pituitary gland** (or hypophysis). Note the relationships of the gland with the cavernous and the sphenoid sinuses.

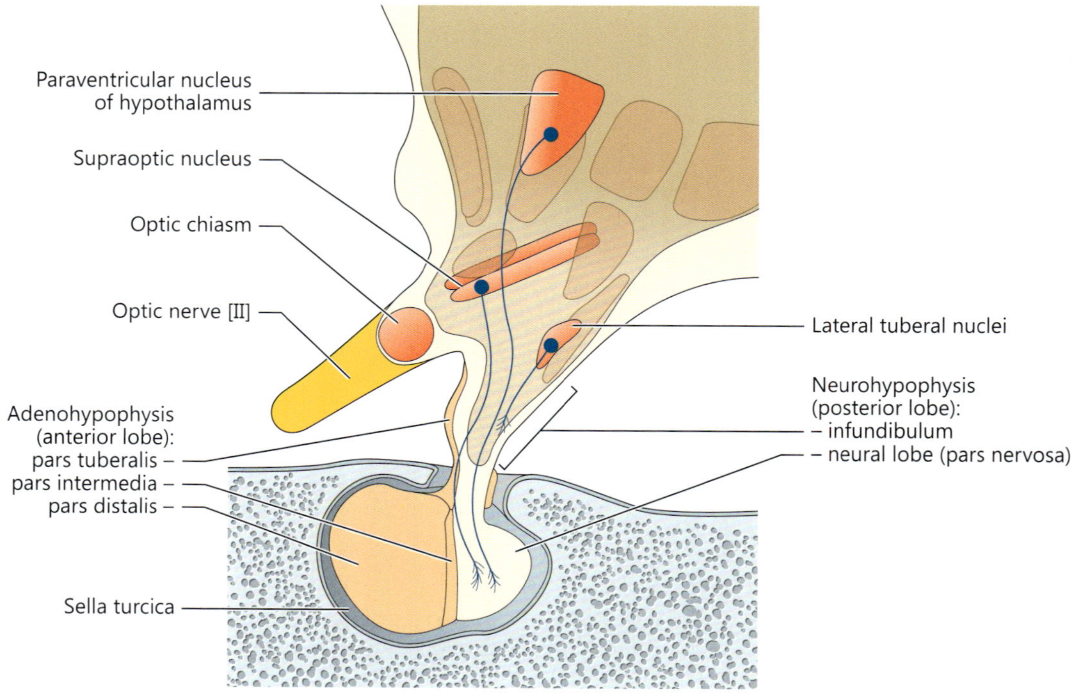

Diagram illustrating the nervous relationships of the **hypothalamus** with the **neurohypophysis**, aside from the constituent parts of the **pituitary gland** (or hypophysis).

Diagram of the afferences and efferences of the **subthalamic nucleus** (or Luys' body). The main input for the subthalamic nucleus comes from the globus pallidus lateral segment. The activity of the globus pallidus medial segment is affected by the subthalamic nucleus, whose output is directed also to the globus pallidus lateral segment and the substantia nigra.

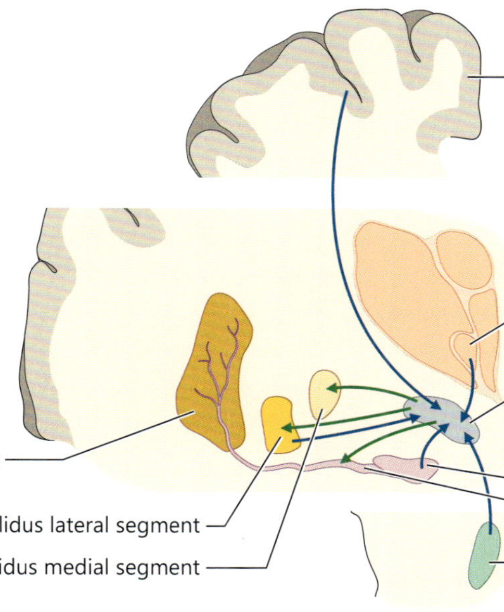

Nuclei of the subthalamus

NUCLEI OR ZONE	NEURONAL POPULATION	LOCATION	AFFERENCES	EFFERENCES	FUNCTION
Subthalamic nucleus (or Luys' body)	Few neurons in the medial part, numerous neurons in the lateral part	Medial face of the peduncular portion of the internal capsule	Subthalamic fasciculus from the globus pallidus lateral segment Fibers from the frontal cortex Fibers from the compact part of the substantia nigra Fibers from the centromedian and parafascicular nuclei of the thalamus Fibers from the posterior raphe nucleus	Fibers to the globus pallidus medial and lateral segments Fibers to the reticular part of the substantia nigra	Center of the extrapyramidal system that controls the start of locomotion and determines the progression forward
Zona incerta	Not well delimited group of cells that continue laterally with the reticular nucleus of the thalamus	Between the thalamus and the subthalamic nucleus	Fibers from the precentral cortex Collaterals of the pallidothalamic tract (in the ansa lenticularis)	To the superior colliculus, to the tegmentum of the pons and mesencephalon, to the thalamus and the spinal cord	Intermediate stations between the globus pallidus medial segment and the reticular formation of the mesencephalon
Nuclei of the perizonal (or tegmental) fields and nucleus of the ansa lenticularis	Small clusters of neurons dispersed among the fibers of the ansa lenticularis	Spread within the ansa lenticularis	From the putamen and the globus pallidus medial segment	Fibers to the brainstem (until the inferior olivary complex) through the central tegmental tract	

Connections of the subthalamus

FASCICULUS, ANSA OR FIBERS	CONNECTION
Subthalamic fasciculus (through the internal capsule)	From the globus pallidus lateral segment to the subthalamic nucleus From the subthalamic nucleus to the globus pallidus medial and lateral segments
Lenticular fasciculus (through the internal capsule)	From the globus pallidus medial segment to the zona incerta and to the nuclei of the perizonal fields (or of Forel) It continues in the (perizonal) H2 field of Forel, reaching the fibers of the ansa lenticularis, to form the (perizonal) H field (or prerubro) of Forel
Ansa lenticularis	From the putamen and the globus pallidus medial segment to the nucleus of the ansa lenticularis, proceeding partly, together with the (perizonal) H2 field of Forel in the (perizonal) H field (or prerubro) of Forel
Dentatothalamic and rubrothalamic fibers	They proceed, together with the (perizonal) H field (or prerubro) of Forel, in the thalamic fasciculus or (perizonal) H1 field of Forel Through the thalamic fasciculus, to the ventral lateral complex and intralaminar nuclei of the thalamus
Thalamic fasciculus	Confluence of the fibers of the (perizonal) H field (or prerubro) of Forel with the dentatothalamic and rubrothalamic fibers From the globus pallidus medial segment and the putamen to the ventral anterior nuclei and ventral lateral complex of the thalamus

The connections of the subthalamus with the basal nuclei (globus pallidus medial segment and putamen) identify this structure as a territory forming part of the group of the corticobasal circuits of the extrapyramidal system, involved in the control of somatic motility.
For the connections of the subthalamus ☞ Thalamus-Basal Nuclei Relationships in page 541.

Schematic representation of the external configuration of the brainstem in posterolateral projection showing the position of the **pineal gland** (or pineal body) and its relationships with the posterior pole of the thalamus with the mesencephalon.

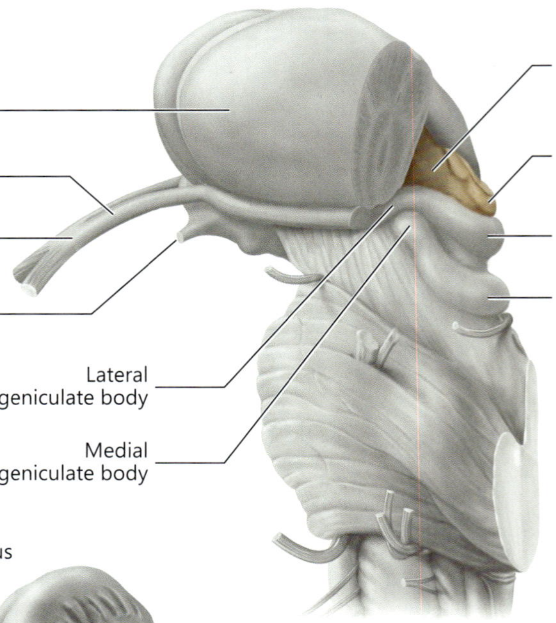

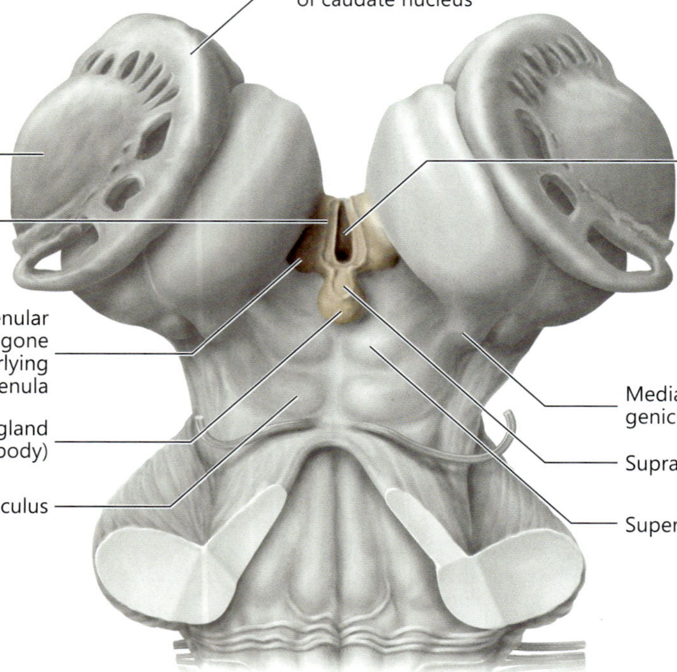

Schematic representation of the external configuration of the mesencephalon in the posterior projection showing the location of the **pineal gland** (or pineal body), formation of the epithalamus (including also the habenular trigone and the underlying abenula) which occupies a small area located posterosuperiorly on the roof of the third ventricle.

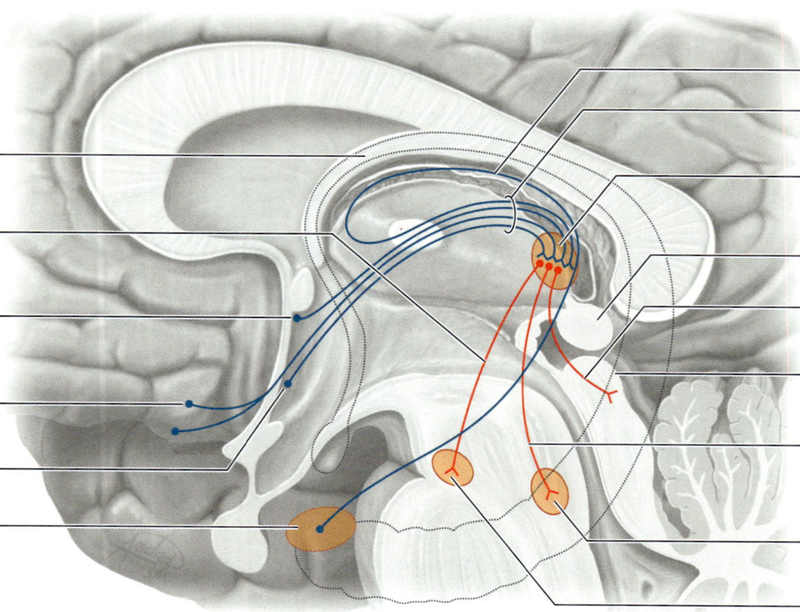

Nuclei of the epithalamus				
NUCLEI	**LOCATION**	**AFFERENCES**	**EFFERENCES**	**FUNCTION**
Habenular nuclei – medial – lateral	Roof of the third ventricle in the habenular trigone	Stria medullaris of the thalamus: – fibers from the amygdaloid body (through the stria terminalis) – fibers from the hippocampus (through the fornix) – fibers from the olfactory tubercle, from the anterior perforated substance and from the septal nuclei – fibers from the preoptic area – fibers from the anterior nuclei of the thalamus – tectohabenular tract Stria terminalis: fibers from the amygdaloid body	To the interpeduncular nucleus through the habenulointerpeduncular tract or fasciculus retroflexus (or of Meynert), to the superior colliculus through the habenulotectal tract, to the posterior tegmental nucleus through the habenulotegmental tract	Intermediate station on a visceral descending effector system operating mainly as a response to stimula of limbic origin

Connections of the epithalamus				
BUNDLES	**LOCATION**	**ORIGIN**	**PROJECTION**	**FUNCTION**
Habenular commissure	Anterior (or superior) labrum of the pineal recess	Medial and lateral habenular nuclei of one side	Contralateral medial and lateral habenular nuclei	Associative white matter
Habenulointerpeduncular tract or fasciculus retroflexus (or of Meynert)	It surrounds, in a craniocaudal direction, the posterior part of the medial nuclei of the thalamus and goes through the red nucleus	Medial and lateral habenular nuclei	Interpeduncular nucleus and from this one, through the reticular formation, to the spinal cord	Regulation of visceral activities
Habenulotectal tract	In the mesencephalon	Habenula	Superior colliculus	Olfactory impulses to the superior colliculus
Habenulotegmental tract	Tegmentum of the midbrain	Habenula	Posterior tegmental nucleus and, through the posterior longitudinal fasciculus, to the salivatory nuclei and the motor nuclei of the masticatory muscles	Olfactory stimuli which cause saliva secretion
Posterior commissure of the epithalamus	Inferior labrum of the pineal recess, right above the superior colliculi	Fibers from the pretectal nuclei Fibers from the nucleus of the posterior commissure Commissural fibers among the superior colliculi Fibers from the interstitial nucleus (or of Cajal) and from the nucleus of the posterior commissure (or of Darkschewitsch) Fibers from the medial and lateral habenular nuclei	Fibers to the accessory nuclei of the oculomotor nerve	Conduction of stimuli for pupillary reflexes to light and accommodation
Stria medullaris of the thalamus	Superficial part of the medial surface of the thalamus, until the medial margin of the habenular trigone	Fibers from the hippocampus Fibers from the olfactory tubercle, from the anterior perforated substance and the septal nuclei Fibers from the preoptic area Fibers from the anterior nuclei of the thalamus Tectohabenular tract	Homo and contralateral medial and lateral habenular nuclei (through the habenular commissure)	Associative bundle among gray matter formations of the limbic system
Stria terminalis	Along the lateral margin of the ventricular surface of the thalamus	Amygdaloid body	Fibers from the stria terminalis to the habenula	Limbic system

◄
Schematic representation of the brain in sagittal section showing the afferences (*blue*) and the efferences (*red*) of the **habenula**, which also forms an intermediate synaptic station for the transmission of afferent olfactory impulses.

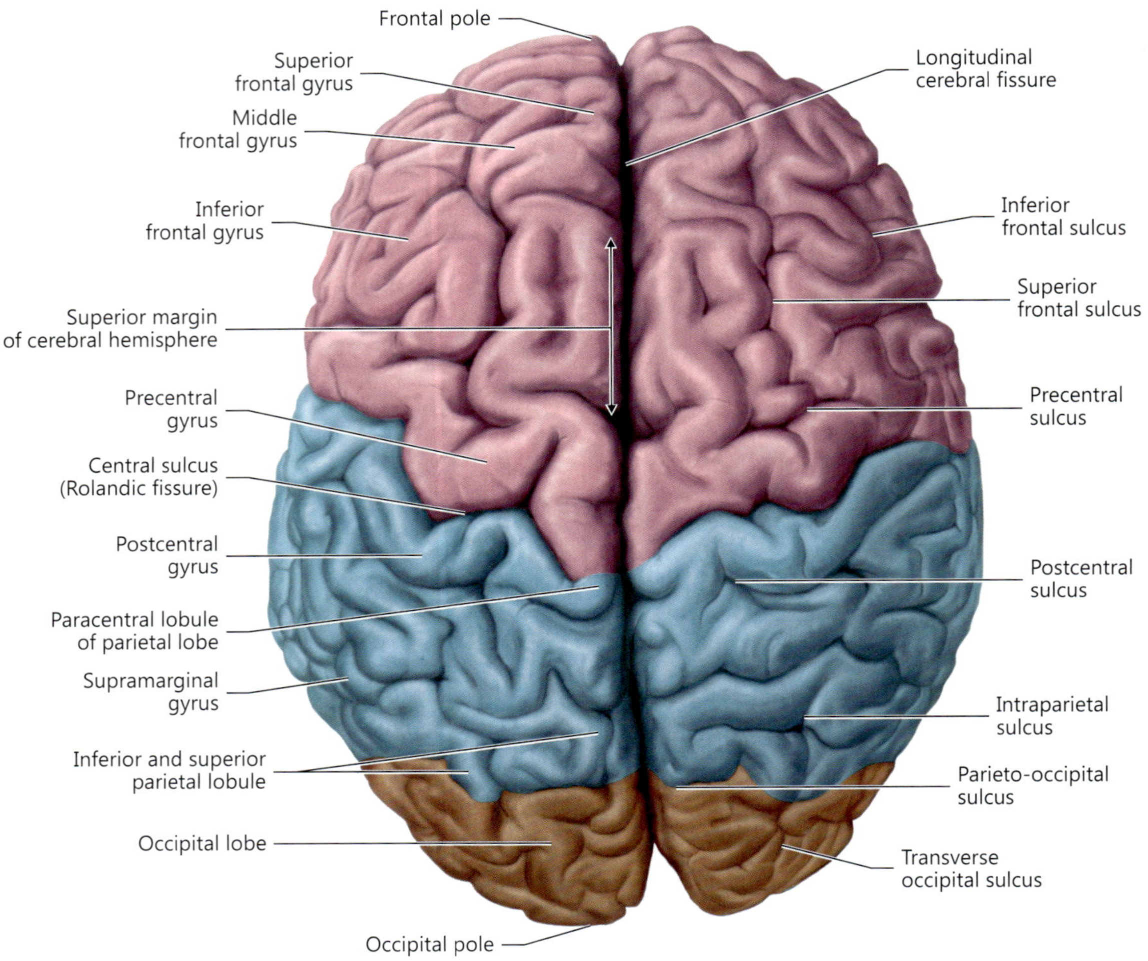

External configuration of the **telencephalon** seen from above. The two cerebral hemispheres are separated by the deep sagittal or interhemispheric sulcus represented by the longitudinal cerebral fissure. The surface becomes irregular due to the presence of sulci and gyri.

- Frontal lobe
- Parietal lobe
- Temporal lobe
- Occipital lobe
- Limbic lobe

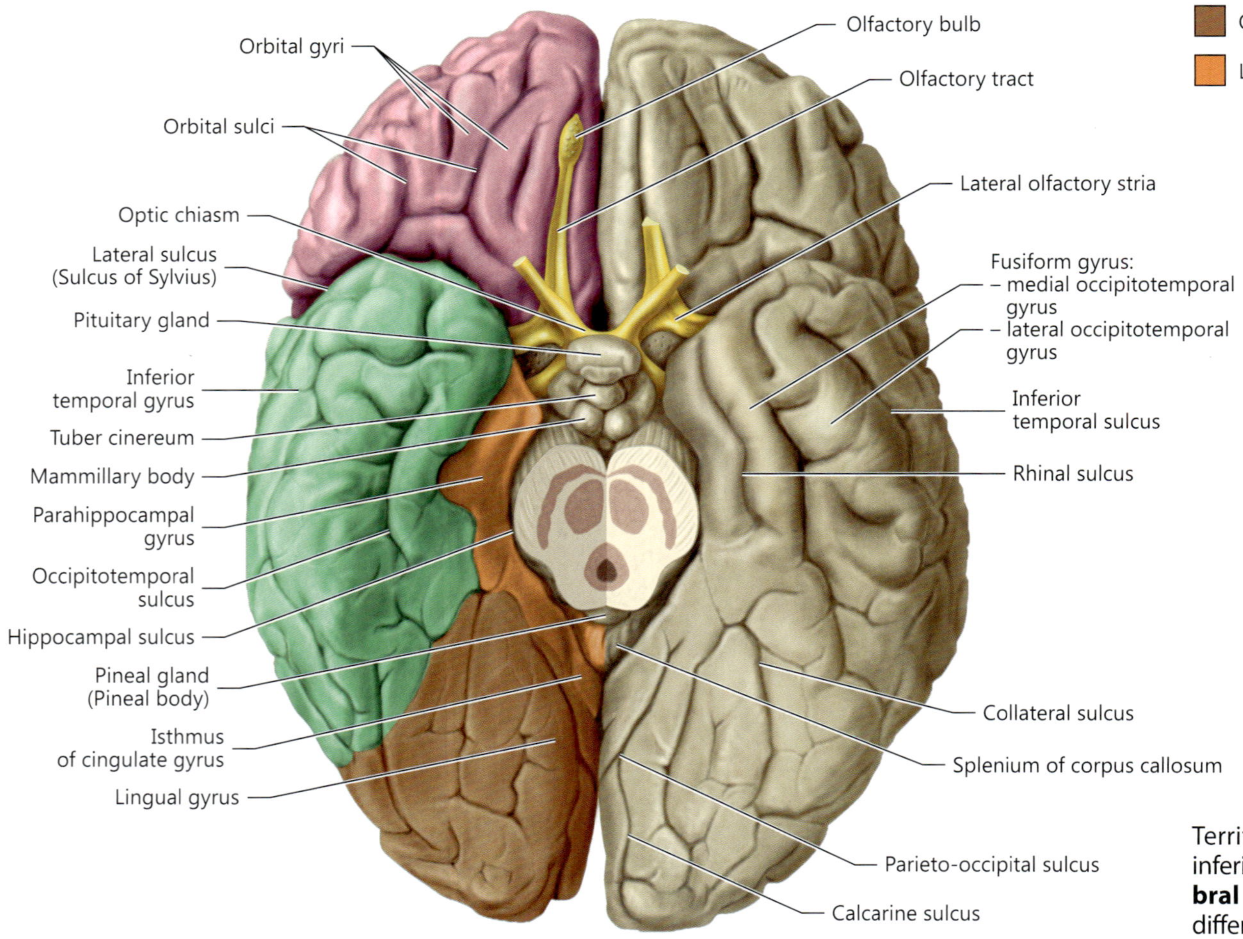

Territorial division of the inferior surface of the **cerebral hemispheres** in the different lobes and gyri.

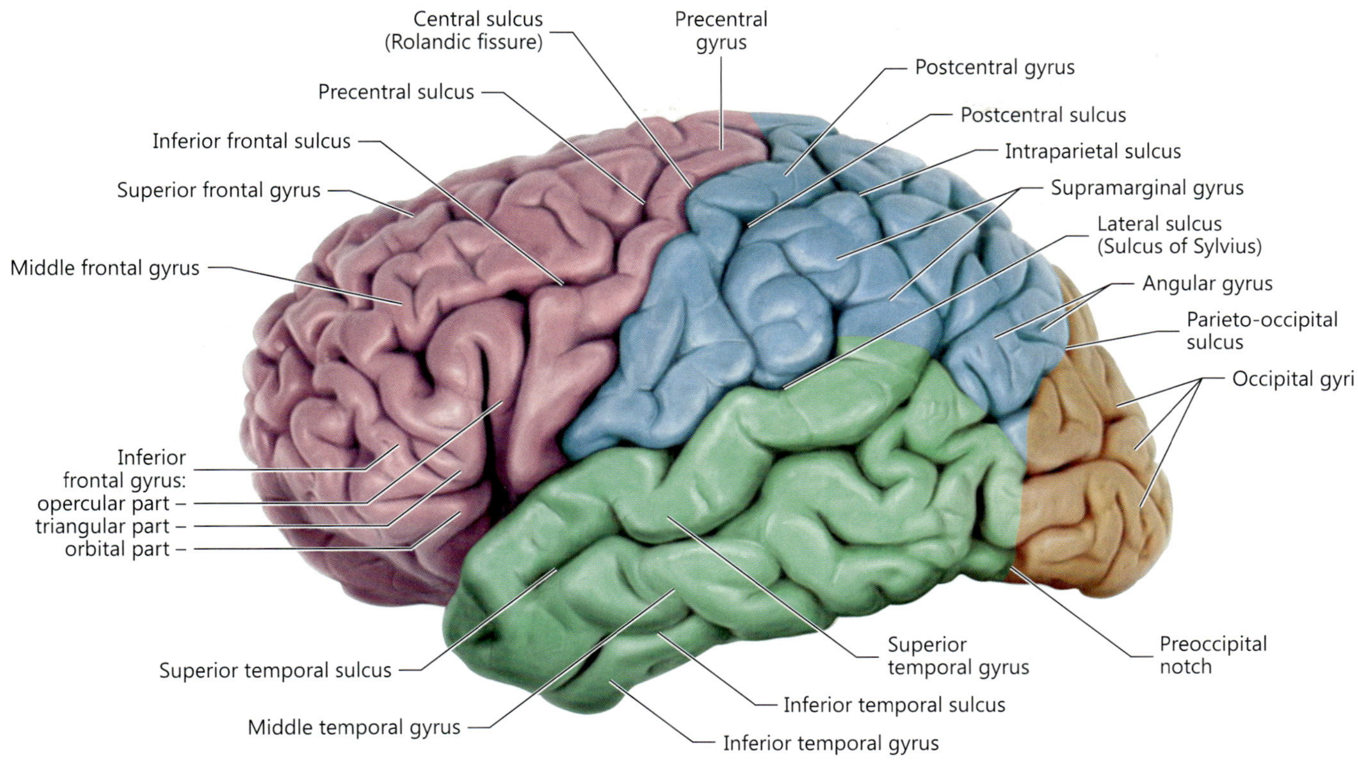

Territorial division of the superolateral surface of the left **cerebral hemisphere** in the different lobes and gyri.

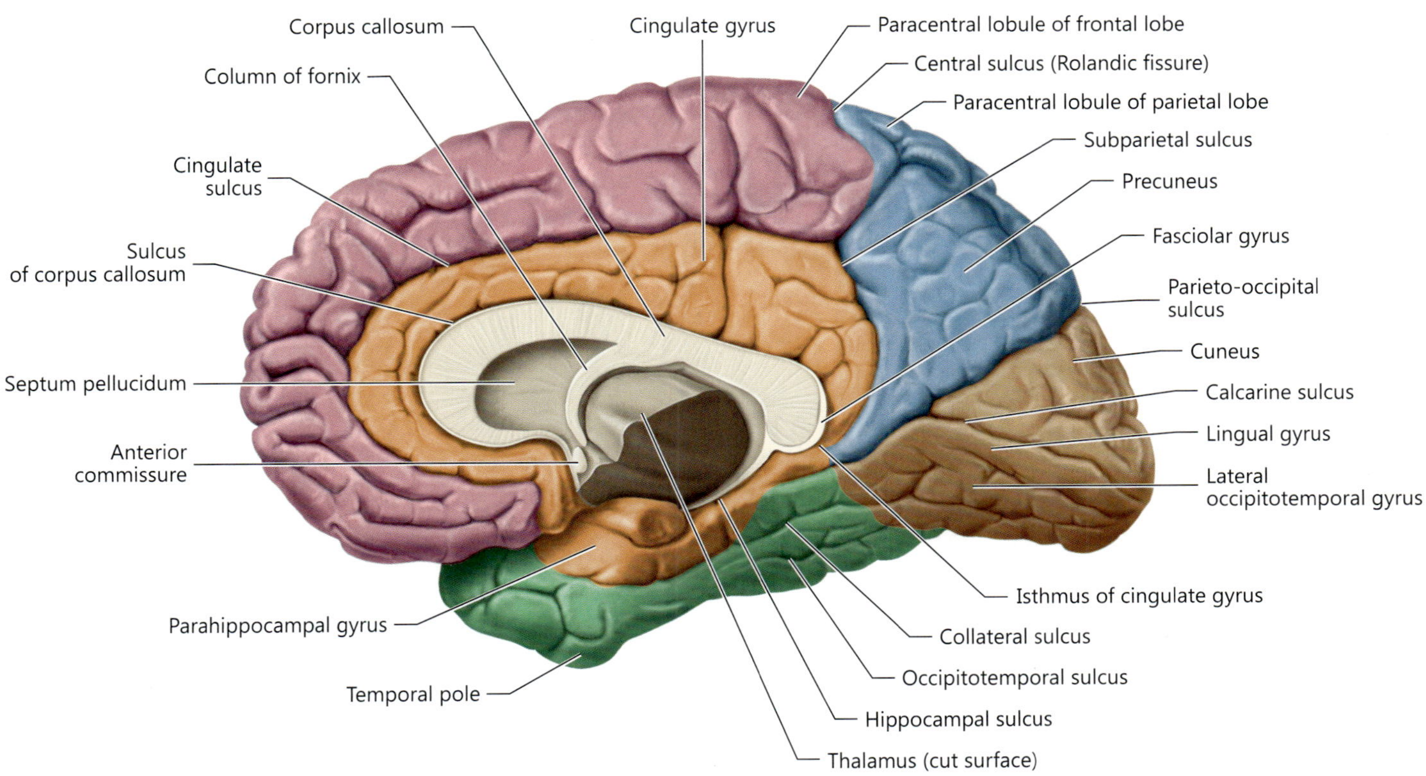

Territorial division of the medial surface of the right **cerebral hemisphere** in the different lobes and gyri.

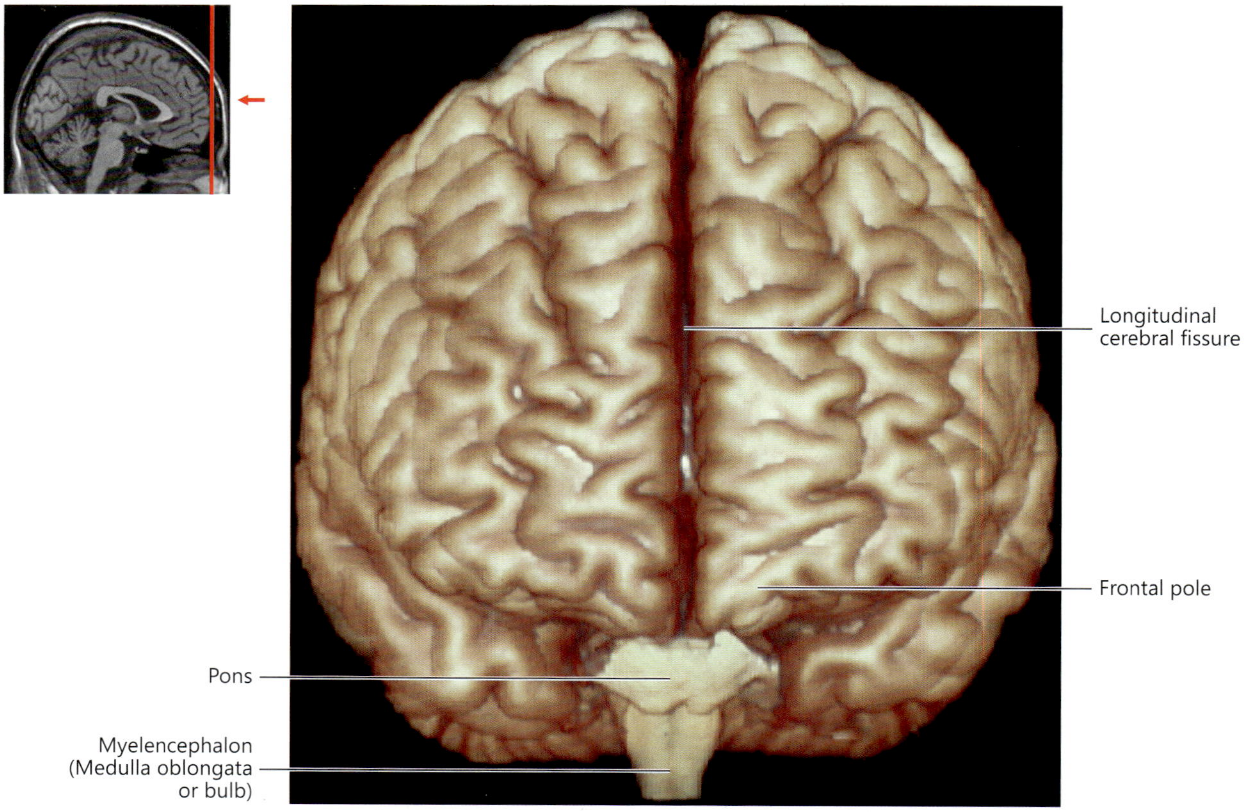

Three-dimensional image of a living human body, obtained by NMR using a 3D volume rendering technique, showing the **telencephalon** in anterior projection.

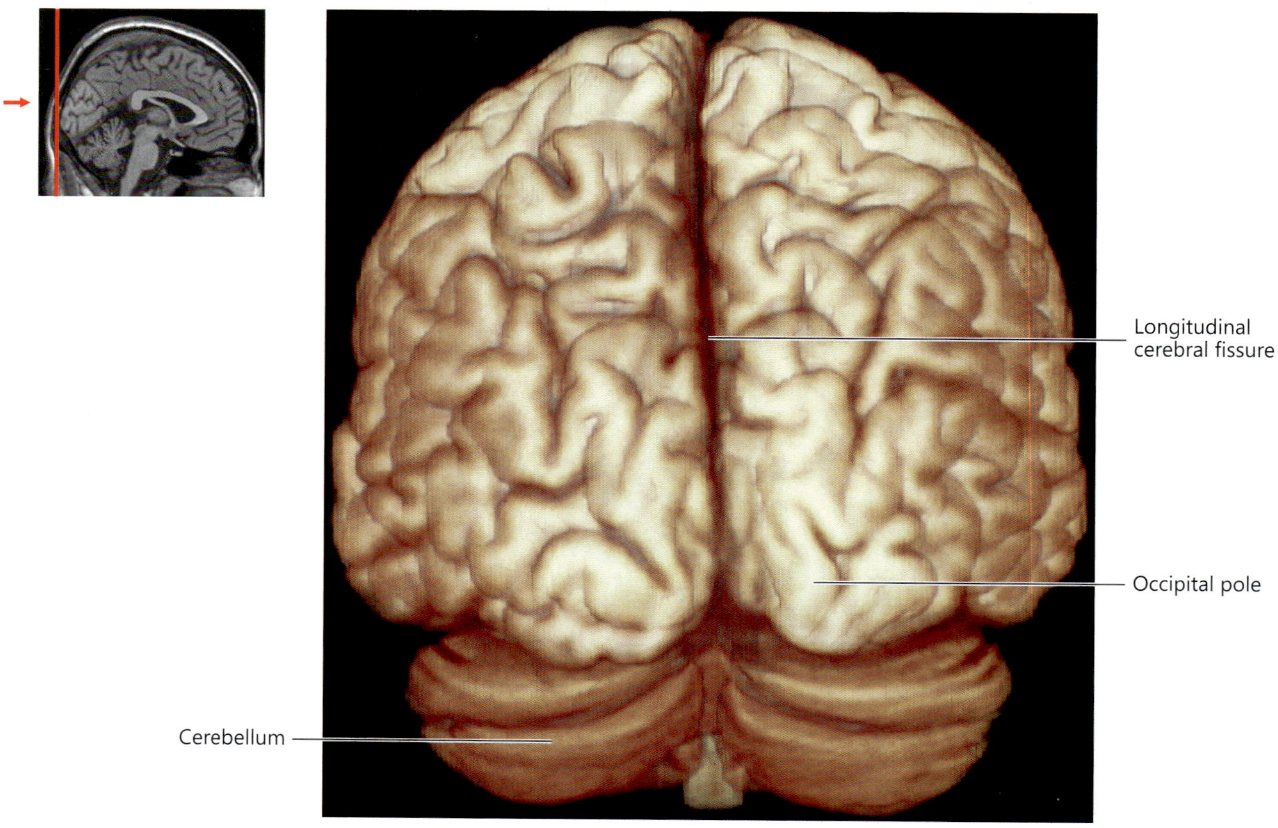

Three-dimensional image of a living human body, obtained by NMR using a 3D volume rendering technique, showing the **telencephalon** in posterior projection.

Three-dimensional image of a living human body, obtained by NMR using a 3D volume rendering technique, showing the external configuration of the **telencephalon** seen from above.

1, Frontal pole
2, Longitudinal cerebral fissure
3, Superior frontal gyrus
4, Inferior frontal sulcus
5, Superior frontal sulcus
6, Precentral sulcus
7, Central sulcus
8, Postcentral sulcus
9, Intraparietal sulcus
10, Parieto-occipital sulcus
11, Occipital pole
12, Transverse occipital sulcus
13, Occipital gyri
14, Paracentral lobule of the parietal lobe
15, Postcentral gyrus
16, Central sulcus (of Rolando)
17, Precentral gyrus
18, Superior margin of the cerebral hemisphere
19, Inferior frontal gyrus
20, Middle frontal gyrus

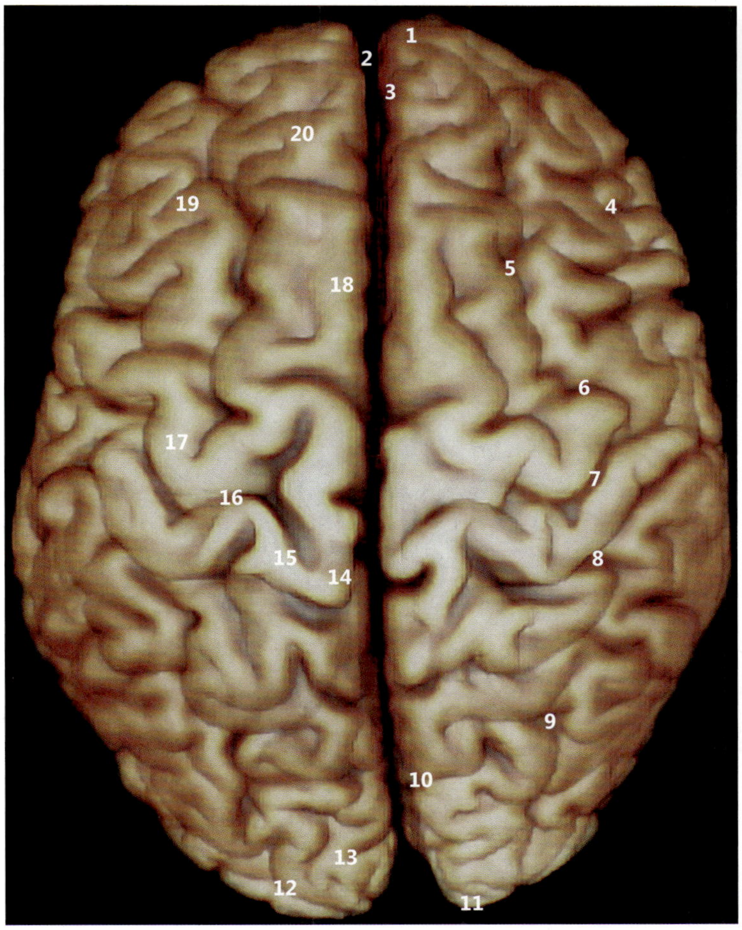

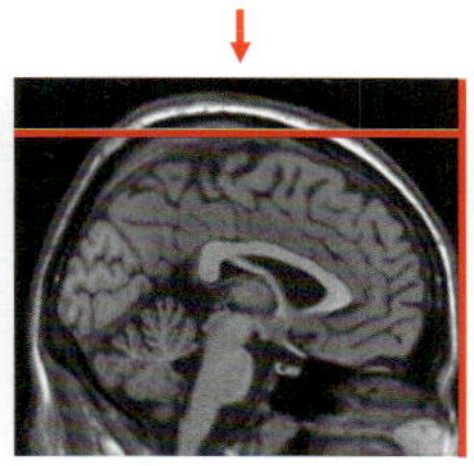

Three-dimensional image of a living human body, obtained by NMR using a 3D volume rendering technique, showing the external configuration of the **telencephalon** seen from below.

1, Frontal pole
2, Orbital gyri
3, Longitudinal cerebral fissure
4, Olfactory sulcus
5, Straight gyrus
6, Temporal pole
7, Optic nerve [II]
8, Pituitary gland
9, Uncus
10, Mammillary body
11, Pons
12, Inferior olive
13, Spinal cord
14, Vermis of cerebellum [I-X]
15, Occipital pole
16, Hemisphere of cerebellum [H II - H X]
17, Inferior temporal gyrus
18, Lateral sulcus (of Sylvius)
19, Inferior frontal gyrus
20, Superior frontal gyrus

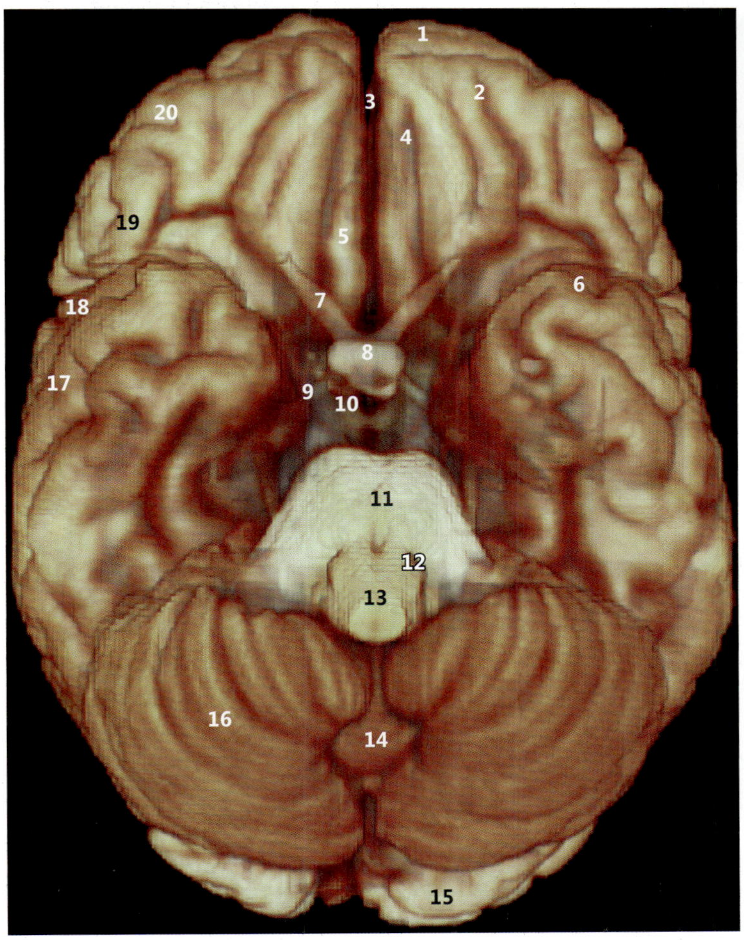

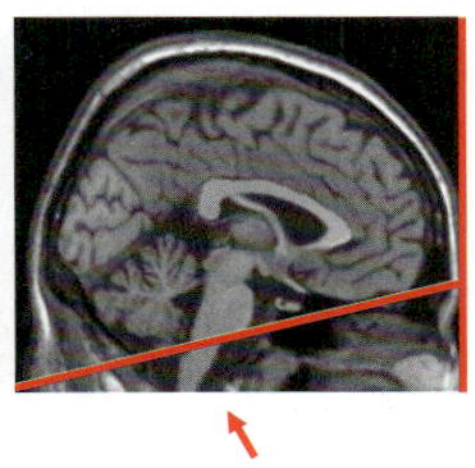

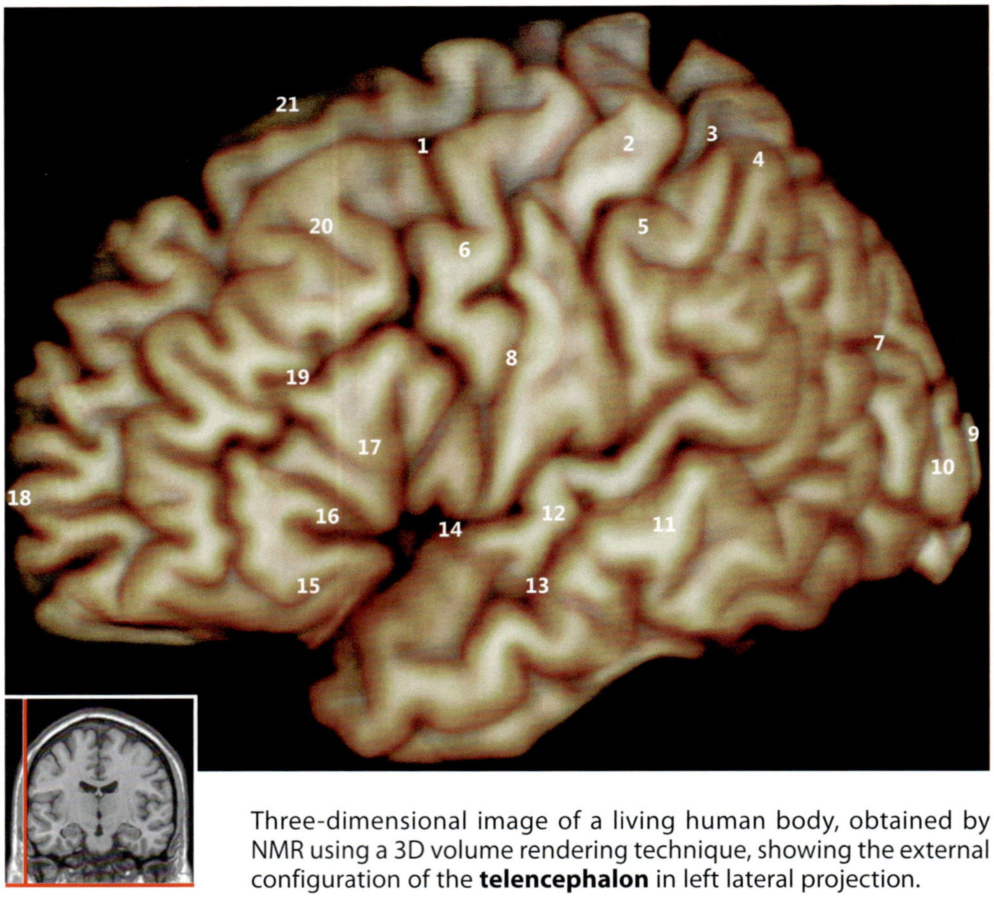

Three-dimensional image of a living human body, obtained by NMR using a 3D volume rendering technique, showing the external configuration of the **telencephalon** in left lateral projection.

1, Precentral sulcus
2, Postcentral gyrus
3, Intraparietal sulcus
4, Angular gyrus
5, Supramarginal gyrus
6, Precentral gyrus
7, Parieto-occipital sulcus
8, Central sulcus (of Rolando)
9, Occipital pole
10, Occipital gyri
11, Middle temporal gyrus
12, Superior temporal gyrus
13, Superior temporal sulcus
14, Lateral sulcus (of Sylvius)
15, Inferior frontal gyrus orbital part
16, Inferior frontal gyrus triangular part
17, Inferior frontal gyrus opercular part
18, Frontal pole
19, Inferior frontal sulcus
20, Middle frontal gyrus
21, Superior margin of the cerebral hemisphere

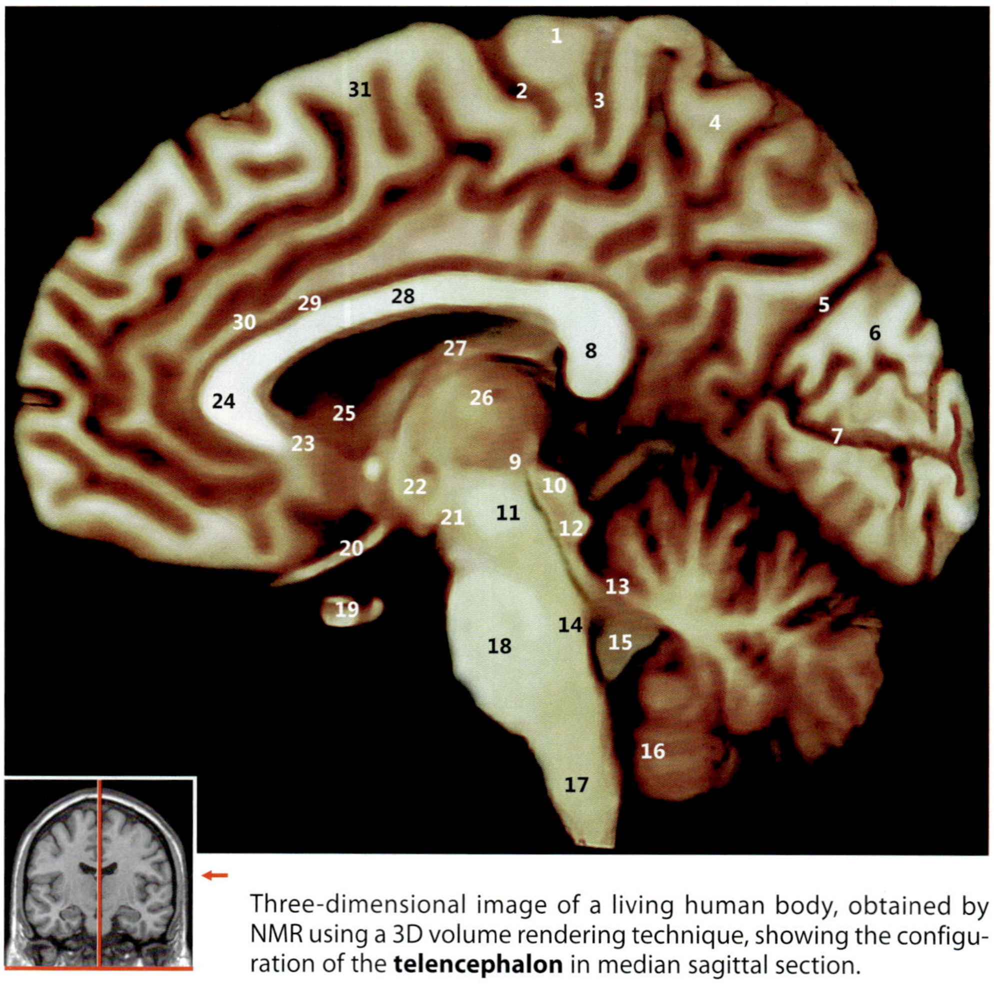

Three-dimensional image of a living human body, obtained by NMR using a 3D volume rendering technique, showing the configuration of the **telencephalon** in median sagittal section.

1, Anterior paracentral gyrus
2, Paracentral sulcus
3, Central sulcus (of Rolando)
4, Precuneus
5, Parieto-occipital sulcus
6, Cuneus
7, Calcarine sulcus
8, Splenium of corpus callosum
9, Posterior commissure
10, Superior colliculus
11, Tegmentum of midbrain
12, Inferior colliculus
13, Superior medullary velum
14, Tegmentum of pons
15, Fourth ventricle
16, Tonsil of cerebellum [H IX]
17, Myelencephalon (medulla oblongata or bulb)
18, Pons
19, Pituitary gland
20, Optic chiasm
21, Interpeduncular fossa
22, Hypothalamus
23, Rostrum
24, Genu of corpus callosum
25, Septum pellucidum
26, Thalamus
27, Fornix
28, Trunk of corpus callosum
29, Sulcus of corpus callosum
30, Cingulate gyrus
31, Superior frontal gyrus

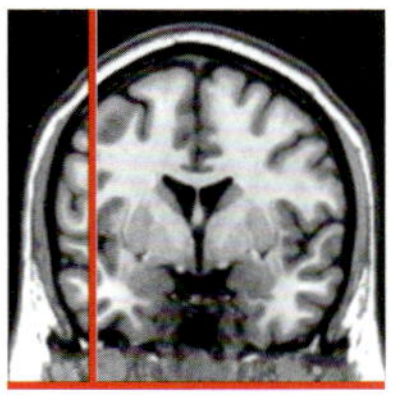

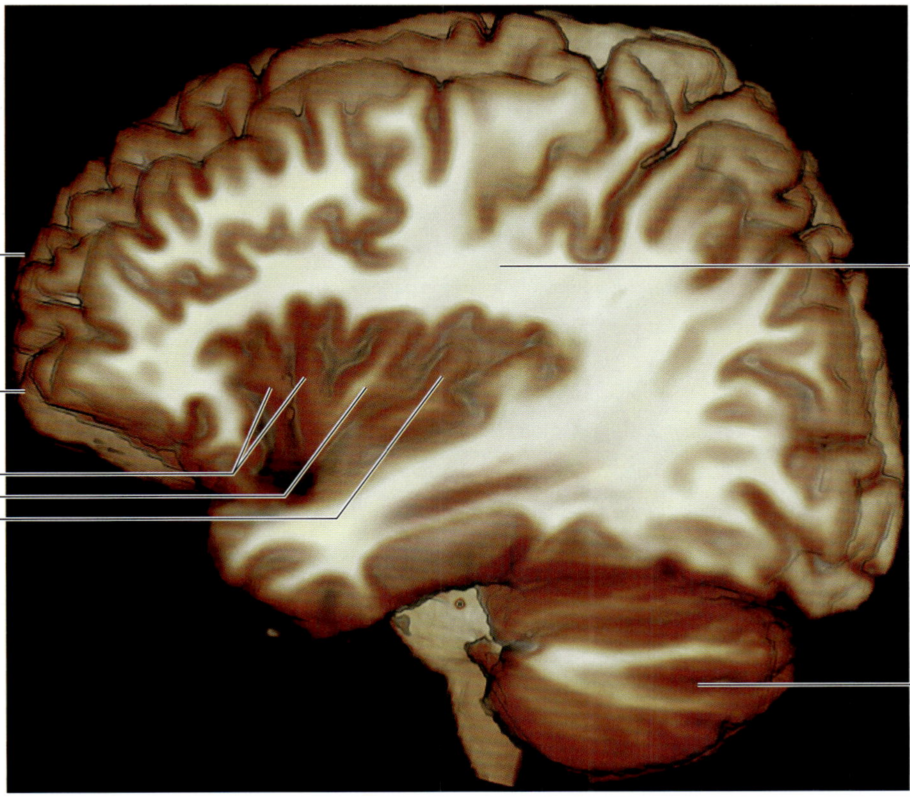

Three-dimensional image of a living human body, obtained by NMR using a 3D volume rendering technique, showing the **insula** or **insular lobe** in left parasagittal section. The precentral gyrus, not always present, is comprised between the short posterior gyrus and the central sulcus of the insula.

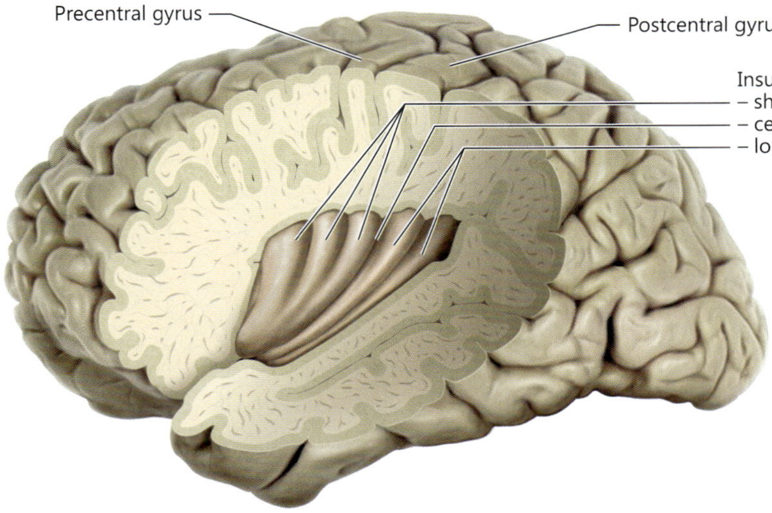

The entire **insular lobe** is uncovered on the lateral surface of the right cerebral hemisphere after removing all the opercula.

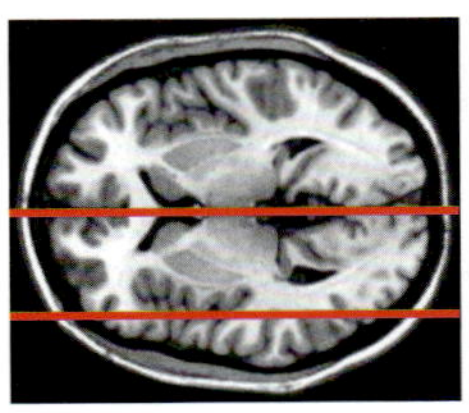

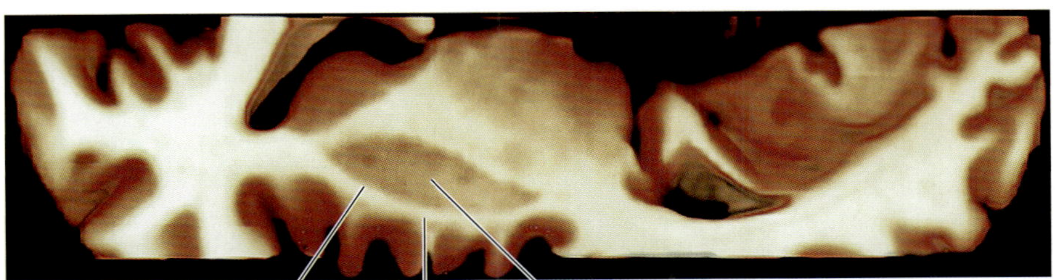

Three-dimensional image of a living human body, obtained by NMR using a 3D volume rendering technique, showing the relationship of the **insula** with the **claustrum** and **putamen**, separated by the external capsule. The claustrum lies between the external capsule and the insula.

Superficial subdivision of the telencephalon		
CEREBRAL HEMISPHERES	The telencephalon is formed by two cerebral hemispheres, separated by the longitudinal cerebral fissure (or interhemispheric fissure) hosting the falx cerebri. The two hemispheres present three faces each (medial, lateral, inferior) and three rounded margins (superior, inferomedial, inferolateral)	
PRIMARY SULCI	Sulci of the surface of the brain that divide the cerebral cortex and, deepening down, delimit lobes and gyri	
	SUPEROLATERAL FACE OF THE CEREBRAL HEMISPHERE	
Lateral sulcus (or of Sylvius)	It separates the temporal lobe from the frontal and parietal lobes	
Central sulcus (or of Rolando)	It separates the frontal lobe from the parietal lobe	
Parieto-occipital sulcus (or external perpendicular)	It separates the occipital lobe from the parietal and temporal lobes	
	MEDIAL FACE OF THE CEREBRAL HEMISPHERE	
Cingulate sulcus	It separates the limbic lobe from the frontal and parietal lobes and the latter among them	
Calcarine sulcus	It does not outline any lobe; in the area of the occipital lobe it separates the cuneus from the lingual gyrus	
Parieto-occipital sulcus (or internal perpendicular)	It separates the occipital lobe from the parietal lobe	
	INFERIOR FACE OF THE CEREBRAL HEMISPHERE	
Cingulate sulcus	It separates the limbic lobe from the temporal lobe	
Parieto-occipital sulcus	It separates the occipital lobe from the parietal lobe	
LOBES	***GYRI***	***SECUNDARY SULCI***
Frontal lobe		
Superolateral face of the cerebral hemisphere	Precentral gyrus (or ascending frontal) Superior frontal gyrus Middle frontal gyrus Inferior frontal gyrus – orbital part – triangular part – opercular part (or area of Broca) Frontal operculum	Superior frontal sulcus Inferior frontal sulcus Precentral sulcus Paracentral sulcus Olfactory sulcus Paraolfactory sulci Orbital sulci
Medial and inferior faces of the cerebral hemisphere	Medial frontal gyrus Anterior paracentral gyrus of the paracentral lobule Orbital gyri (medial, intermediate e lateral) Straight gyrus Medial olfactory gyrus Lateral olfactory gyrus Subcallosal area including the paraterminal gyrus Paraolfactory area including the paraolfactory gyri	
Parietal lobe		
Superolateral face of the cerebral hemisphere	Postcentral gyrus Superior parietal lobule Inferior parietal lobule including the parietal operculum Supramarginal gyrus Angular gyrus (or preoccipital)	Intraparietal sulcus Postcentral sulcus Subparietal sulcus Marginal branch (or marginal sulcus)
Medial and inferior faces of the cerebral hemisphere	Posterior paracentral gyrus of the paracentral lobule Precuneus (or quadrilateral lobe)	
Temporal lobe		
Superolateral face of the cerebral hemisphere	Superior temporal gyrus (or acoustic area) including the temporal operculum Anterior and posterior transverse temporal gyri Middle temporal gyrus Inferior temporal gyrus	Transverse temporal sulcus Superior temporal sulcus Inferior temporal sulcus Collateral sulcus Occipitotemporal sulcus
Medial and inferior faces of the cerebral hemisphere	Inferior temporal gyrus Medial occipitotemporal gyrus (lingual gyrus or lobule) Lateral occipitotemporal gyrus (fusiform gyrus or lobule)	
Occipital lobe		
Superolateral face of the cerebral hemisphere	Superior occipital gyrus Middle occipital gyrus (or lateral or descending) Inferior occipital gyrus	Semilunar sulcus Transverse occipital sulcus Occipitotemporal sulcus
Medial and inferior faces of the cerebral hemisphere	Cuneus Lingual gyrus Medial occipitotemporal gyrus (lingual gyrus or lobule) Lateral occipitotemporal gyrus (fusiform gyrus or lobule)	

continued

Superficial subdivision of the telencephalon		*(continued)*
Insula or **insular lobe** (superolateral face of the cerebral hemisphere)	Insular gyri – Short gyri of insula (anterior group of the insular gyri) – – first gyrus (anterior) – – second gyrus (middle) – – third gyrus (posterior) – Long gyri of insula (posterior group of the insular) – – long gyrus of insula (or fourth gyrus) – – fifth gyrus Limen insulae (or gyrus ambiens)	Central sulcus of insula Central sulcus of insula
Limbic lobe (medial and inferior faces of the cerebral hemisphere) The limbic lobe is also assigned the hippocampus, including the Ammon's horn (or hippocampus properly speaking) (archicortex or archipallium)	Parahippocampal gyrus (or gyrus of hippocampus) – Uncus Fimbria of hippocampus Cingulate gyrus (or gyrus of corpus callosum) – Isthmus of the cingulate gyrus Fasciolar gyrus Dentate gyrus	Sulcus of corpus callosum Hippocampal sulcus Fimbriodentate sulcus Collateral sulcus Rhinal sulcus

Data of brain topography	
TELENCEPHALON	***CRANIUM***
Superior margin of the cerebral hemisphere	15 mm from the median line, between the glabella and the inion
Lateral sulcus (or sulcus of Sylvius)	Shortly behind the squamous suture and parallel to it, on a line between the nasion and the lambda, ending where the line between the nasion and the lambda meet with the superior temporal line
Central sulcus (or fissure of Rolando)	Parallel to the coronal suture and oblique, with respect to the median line, below and forward, according to an angle of approximately 70°
Parieto-occipital sulcus	15 mm from the median line, between the lambda and the asterion
Inferolateral margin of the cerebral hemisphere	From the glabella to the pterion, passing 1 cm above the zygomatic process of the frontal bone, and from there to the asterion, passing through the zygomatic arch, ending at the inion

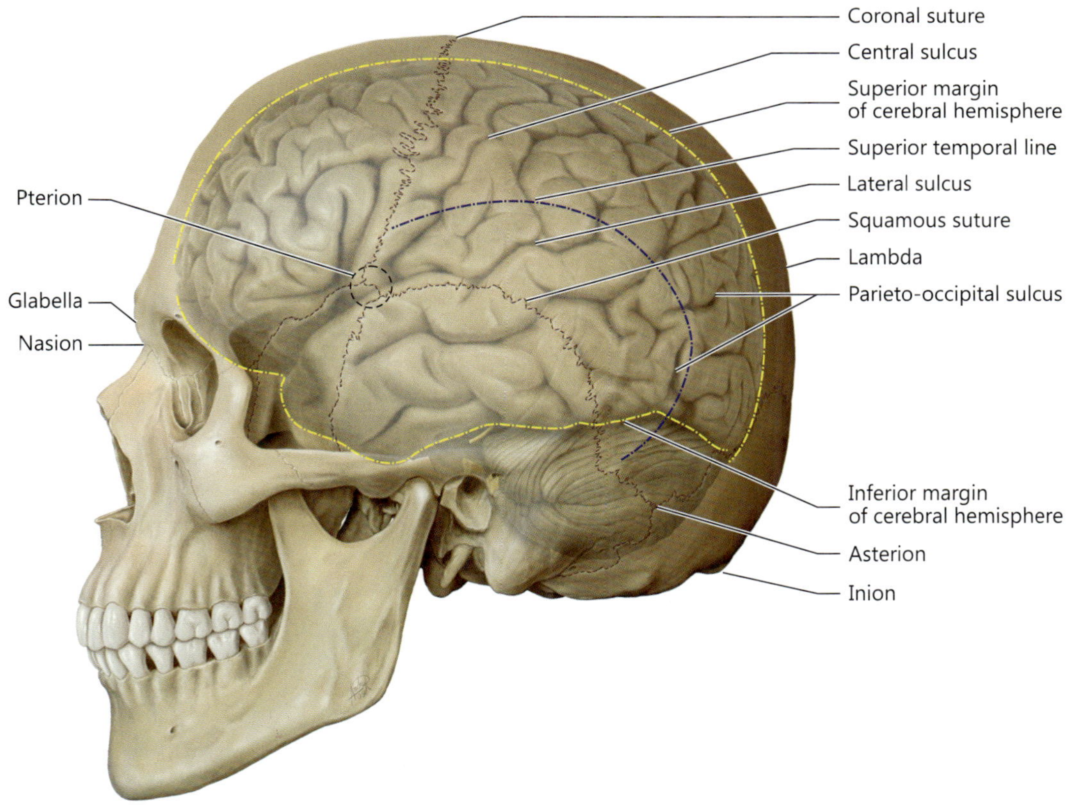

Lines, **sutures** and **craniometric landmarks** to project the central, lateral and parieto-occipital sulci of the superolateral face of the cerebral hemisphere and its superior and inferolateral margins on the surface of the skull.
For the **craniometric points** ☞ vol.1 Head, Topography and Craniometry.

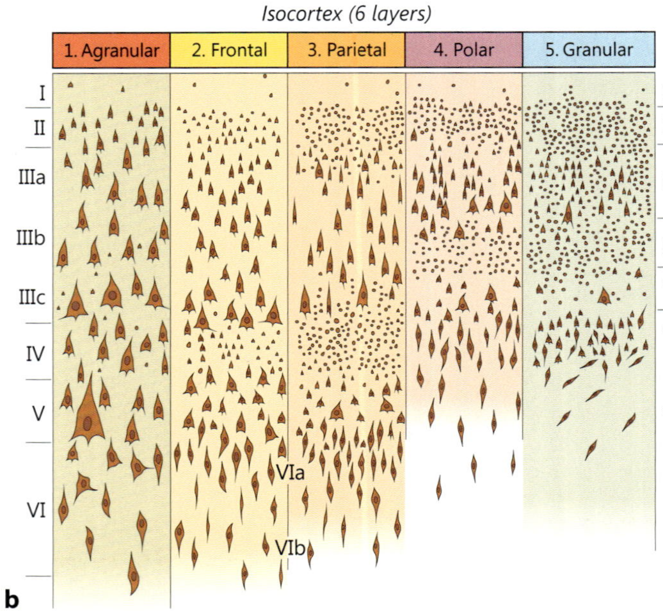

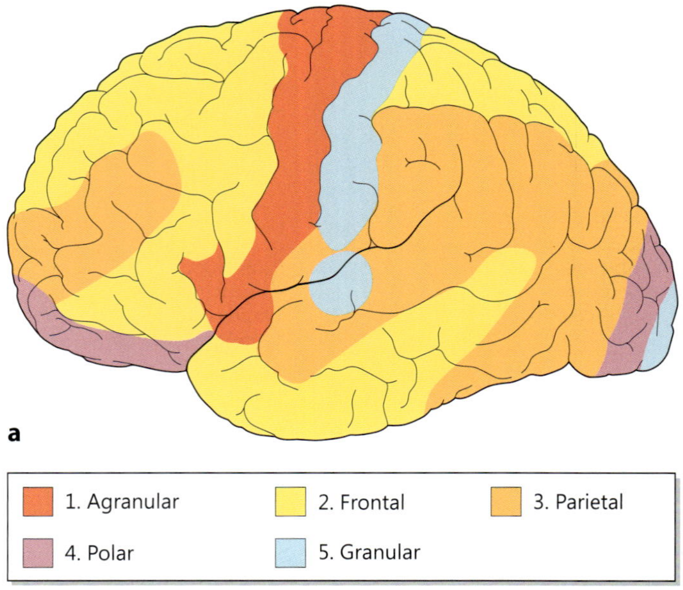

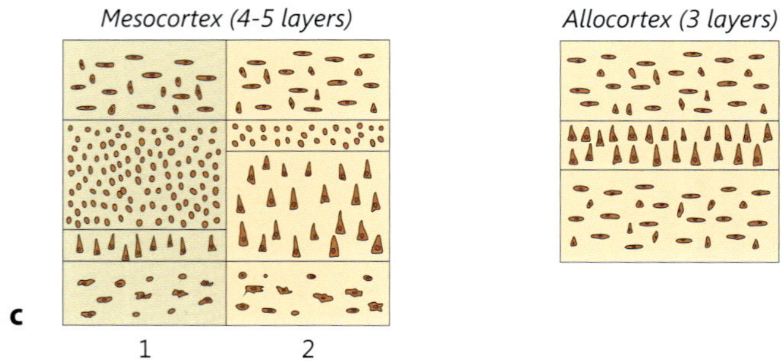

a, Distribution of the different cytoarchitectures of the **cerebral cortex**. **b** and **c**, Diagram illustrating the cytoarchitectures of the cerebral cortex divided into *isocortex*, *mesocortex* and *allocortex*. The *isocortex* (**b**) displays an agranular structure, in which the pyramidal cells are prevalent, and a granular structure, characterized by the prevalence of granules. In the *allocortex* (*mesocortex*) (**c**), the granules are prevalent in the dentate gyrus (1) whereas the pyramidal cells are more represented in the ventral hippocampus (2).

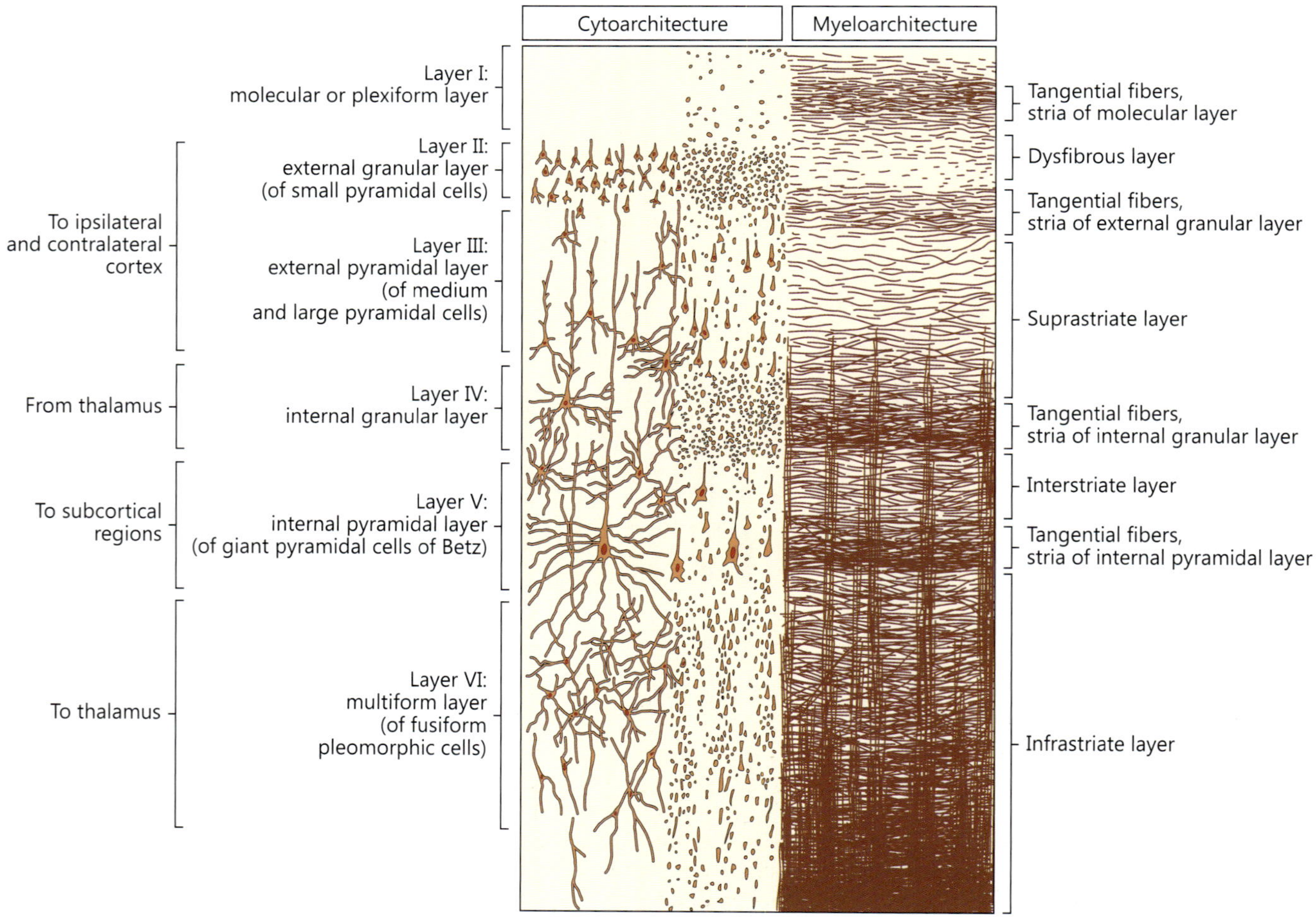

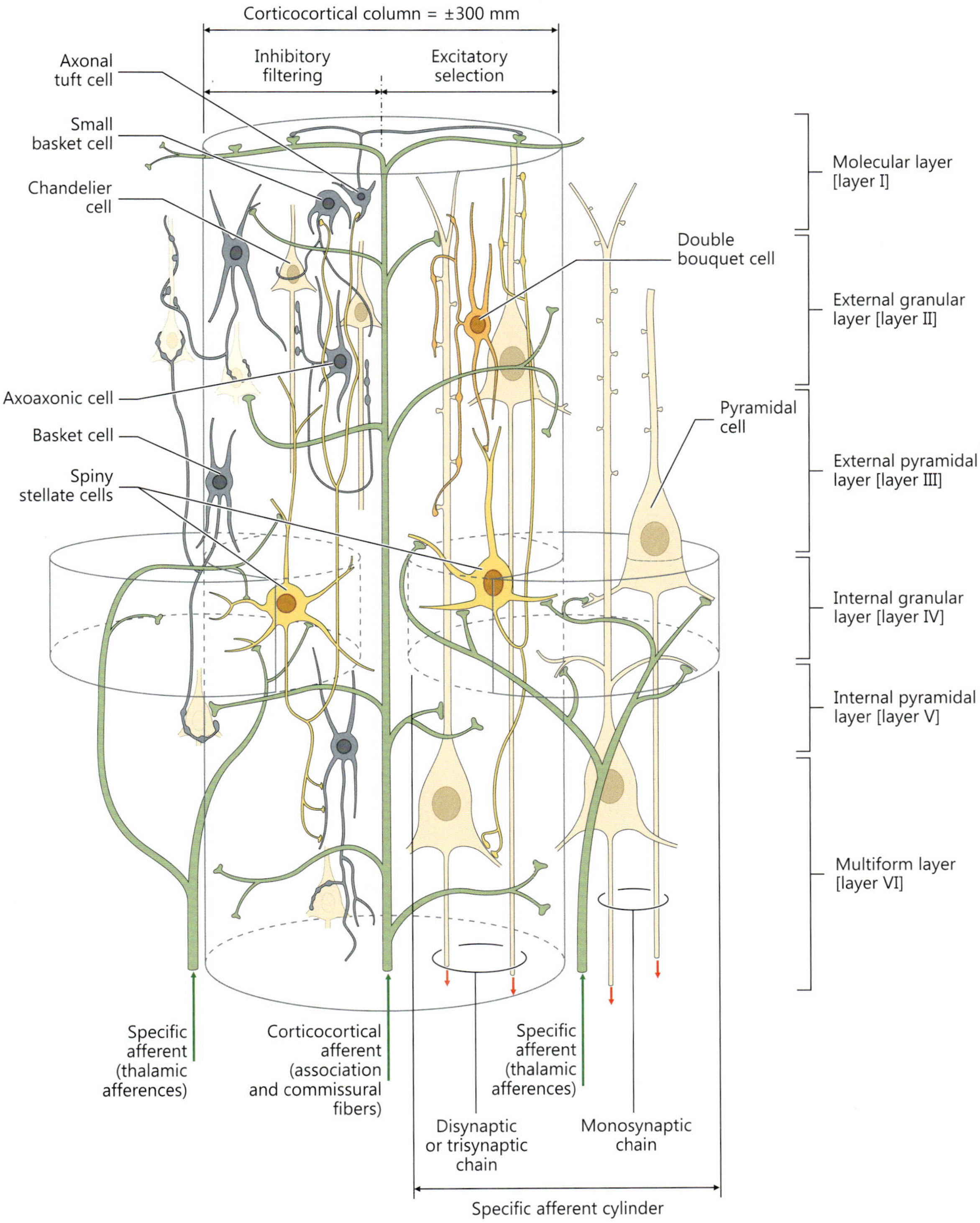

Cytoarchitecture of the cortical columns. Note the pyramidal cells and the different types of interneurons (stellate or granule neurons) that take part in the formation of a column. Inhibitory cells are colored black. Note the distribution of specific (thalamocortical) inputs to the interneurons of the layer IV and of the corticocortical inputs to the apical dendrites.

◄
Left side of the diagram. **Cytoarchitecture of the cerebral cortex**; an agranular field (on the left) and a granular field (on the right) are shown. Right side of the diagram. Course of the main myelinated fiber systems of the cortex and overall myeloarchitecture. The thalamic neurons project to the telencephalic cortex sending their axons to make synapses mainly with the dendrites of the neurons contained in the layer IV, and with the dendrites, present in the layer IV, branching from cell bodies of neurons present in other layers. The pyramidal neurons of the layers II, III, V and VI project to other cortical areas and subcortical structures: most of layers II and III pyramidal neurons project to ipsilateral and contralateral cortical areas (through the corpus callosum); the pyramidal neurons of the layer V project to the striatum, the brainstem and spinal cord; the pyramidal neurons of the layer VI project to the thalamus. The layer I contains mostly dendrites of neurons located in the deeper layers.

Lateral view

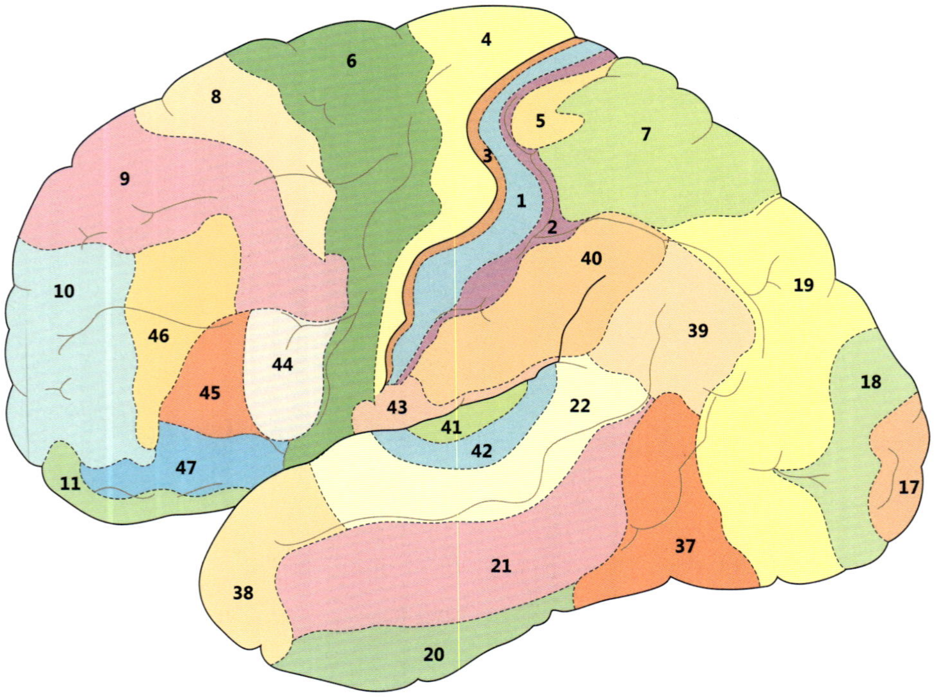

Medial view

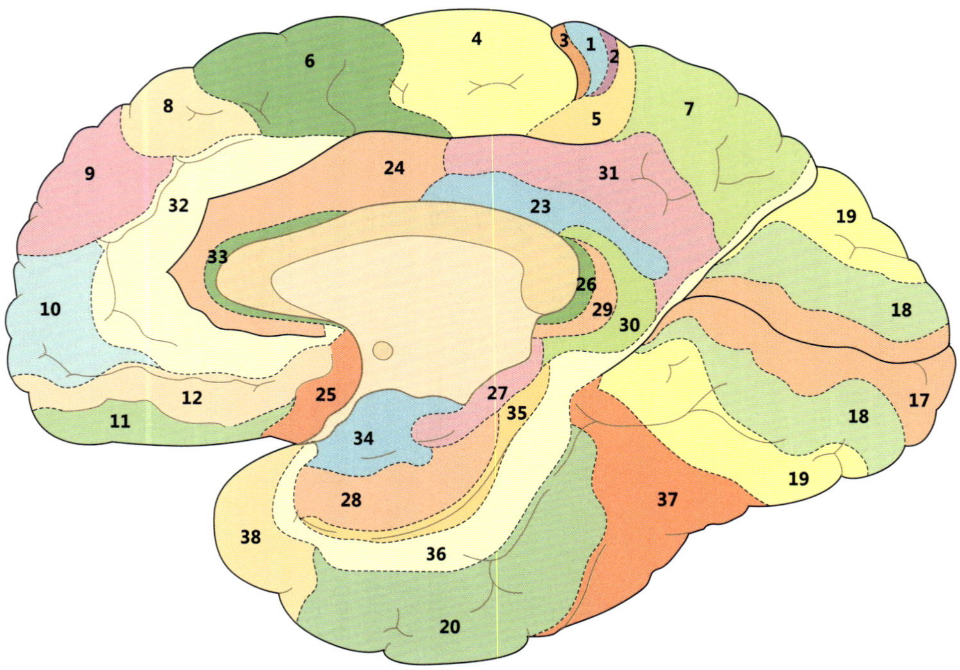

Cytoarchitectonic map of the human cortex according to Brodmann. The different architectonic areas are represented with colors and numbers.

Brodmann areas				
AREA	**NAME**	**TYPE**	**FUNCTION**	**LOCATION**
3, 1 e 2 (sequence so indicated by convention)	Primary somatosensory cortex (postcentral gyrus)	Primary somatosensory area (SI)	Area that converts impulses of the general sense of the contralateral half of the body into conscious feelings	Parietal lobe
4	Primary motor cortex (postcentral gyrus)	Primary motor area (MI)	Area in which the voluntary motor impulses for all muscles of the contralateral half of the body are generated, except for the extraocular muscles	Frontal lobe
5	Posterior parietal association area	Association area	Area that allows the integration of tactile and position informations with visual informations	Superior parietal gyrus
6	Premotor cortex and supplementary motor cortex (secondary motor cortex and supplementary motor)	Premotor (PM) and supplementary motor (SMA) areas	Area from which originate the extrapyramidal tracts responsible for the control and regulation of voluntary movements and for the postural tone and semi-voluntary and automatic movements associated to voluntary movements	Frontal lobe
7	Posterior parietal association area	Association area	Area that allows the integration of tactile and position informations with visual informations	Superior parietal gyrus
8	Dorsolateral prefrontal cortex (frontal eye fields included)	Prefrontal area	Motor area from which originate voluntary motor impulses that govern the eyes movements (area of the voluntary oculocephalogyric motricity)	Frontal lobe
9	Dorsolateral prefrontal cortex	Prefrontal area	Intellective or association area with eminently psychic function	Frontal lobe
10	Anterior prefrontal cortex (more rostral part of the superior and middle frontal gyri)	Frontopolar area	Intellective or association area with eminently psychic function	Frontal lobe
11	Orbitary gyri and straight gyrus and part of the rostral zone of the superior frontal gyrus	Orbitofrontal area	Intellective or association area with eminently psychic function	Frontal lobe
12	Between the superior frontal gyrus and the inferior rostral sulcus (previously considered part of area 11)	Orbitofrontal area	Association area for the emotional responses influencing the visceral functions	Limbic system
13	Part of the insula	Insular cortex	Association area for the emotional responses influencing the visceral functions	Limbic system
14	Part of the insula and ventromedial prefrontal cortex (only in subhuman primates)	Prefrontal area	Association area for the emotional responses influencing the visceral functions	Limbic system
15	Part of the insula closest to the temporal lobe and anterior part of the temporal lobe (only in subhuman primates)	Anterior region of the temporal lobe	Association area for the emotional responses influencing the visceral functions	Limbic system
16	Insula	Insular cortex	Association area for the emotional responses influencing the visceral functions	Limbic system
17	Primary visual cortex	Primary visual area (VI)	Area in which luminous stimuli are transformed into conscious visual sensations	Occipital lobe
18	Secondary visual cortex	Secondary visual area (VII)	Area site of visual memory (responsible also for the conjugated movements of the eyes in following a moving object and for gaze fixation)	Occipital lobe
19	Association visual cortex	Tertiary visual area (VIII)	Area site of visual memory (responsible also for the conjugated movements of the eyes in following a moving object and for gaze fixation)	Occipital lobe
20	Inferior temporal gyrus	Association area connected with the pulvinar	Association area of polysensorial integration (reception center of the vestibular tracts)	Temporal lobe

continued

Brodmann areas				(continued)
AREA	**NAME**	**TYPE**	**FUNCTION**	**LOCATION**
21	Middle temporal gyrus	Vestibular projection area	Association area of polysensorial integration	Temporal lobe
22	Superior temporal gyrus (the caudal part of area 22 is considered site of the area of Wernicke)	Area of Wernicke	Comprehension of spoken language and localization of mnemonic memories of sounds	Temporal lobe
23	Posterior cingulate cortex, ventral part; divided in 23a, adjacent to the sulcus of the corpus callosum and closer to the area 30, and 23b	Cingulate gyrus and subcallosal area	Individual self protection and preservation of the species	Limbic system
24	Anterior cingulate cortex, ventral part	Cingulate gyrus and subcallosal area	Individual self protection and preservation of the species	Limbic system
25	Cortex or area subgenualis (involved in depression)	Cingulate gyrus and subcallosal area	Individual self protection and preservation of the species	Limbic system
26	Ectosplenial area	Cingulate gyrus and subcallosal area	Individual self protection and preservation of the species	Limbic system
27	Piriform cortex	Cingulate gyrus and subcallosal area	Individual self protection and preservation of the species	Limbic system
28	Entorhinal cortex, posterior part	Cingulate gyrus and subcallosal area	Individual self protection and preservation of the species	Limbic system
29	Cingulate cortex, retrosplenial part	Cingulate gyrus and subcallosal area	Individual self protection and preservation of the species	Limbic system
30	Part of the cingulate cortex	Cingulate gyrus and subcallosal area	Individual self protection and preservation of the species	Limbic system
31	Posterior cingulate cortex dorsal part	Cingulate gyrus and subcallosal area	Individual self protection and preservation of the species	Limbic system
32	Anterior cingulate cortex, dorsal part	Cingulate gyrus and subcallosal area	Individual self protection and preservation of the species	Limbic system
33	Part of the anterior cingulate cortex	Cingulate gyrus and subcallosal area	Individual self protection and preservation of the species	Limbic system
34	Anterior entorhinal cortex	Parahippocampal gyrus	Individual self protection and preservation of the species	Limbic system
35	Perirhinal cortex	Parahippocampal gyrus	Individual self protection and preservation of the species	Limbic system
36	Parahippocampal cortex	Parahippocampal gyrus	Individual self protection and preservation of the species	Limbic system
37	Lateral and medial occipitotemporal gyri	Fusiform gyrus	Association area of polysensorial integration	Temporal lobe
38	More rostral part of the superior and middle temporal gyri	Temporopolar area	Association area of polysensorial integration	Temporal lobe
39	Angular gyrus (area considered by someone as part of the area of Wernicke)	Inferior parietal area	Area connected with the pulvinar in which the true recognition of perceptions or gnosic process with the integration of vision, hearing and general sense takes place	Parietal lobe
40	Supramarginal gyrus (area considered by someone as part of the area of Wernicke)	Inferior parietal area	Area connected with the pulvinar in which the true recognition of perceptions or gnosic process with the integration of vision, hearing and general sense takes place	Parietal lobe
41	Primary cortex and of auditory association	Primary acoustic area (AI)	Area in which acoustic impulses are transformed into conscious auditory sensations	Temporal lobe
42	Primary cortex and of auditory association	Association acoustic area	Area in which acoustic impulses are transformed into conscious auditory sensations	Temporal lobe

continued

Brodmann areas				(continued)
AREA	**NAME**	**TYPE**	**FUNCTION**	**LOCATION**
43	Subcentral area (between the insula and the precentral and postcentral gyri)	Area of Broca and primary gustatory area (GI)	Taste perception area	Insula and frontal lobe
44	Opercular part of the inferior frontal gyrus and part of the area of Broca	Area of Broca	Area governing the motor strategies of language	Frontal lobe
45	Triangular part of the inferior frontal gyrus and part of the area of Broca	Area of Broca	Intellective or association area with eminently psychic function	Frontal lobe
46	Dorsolateral prefrontal cortex	Prefrontal area	Intellective or association area with eminently psychic function	Limbic system
47	Inferior frontal gyrus	Orbitofrontal area	Association area for the emotional responses influencing the visceral functions	Limbic system
48	Small part of the medial surface of the temporal lobe	Retrosubicular area	Area described by Brodmann, but non included in the maps of cortical areas	Limbic system
49	Parasubiculum (only in some rodents)	Transition area between the presubiculum and the rhinal sulcus	Area of integration of spatial orientation according to the direction of the head	Limbic system
52	Parainsular area	Dorsal surface of the temporal lobe in the lateral sulcus	There are no known functions associated to this area	Temporal lobe

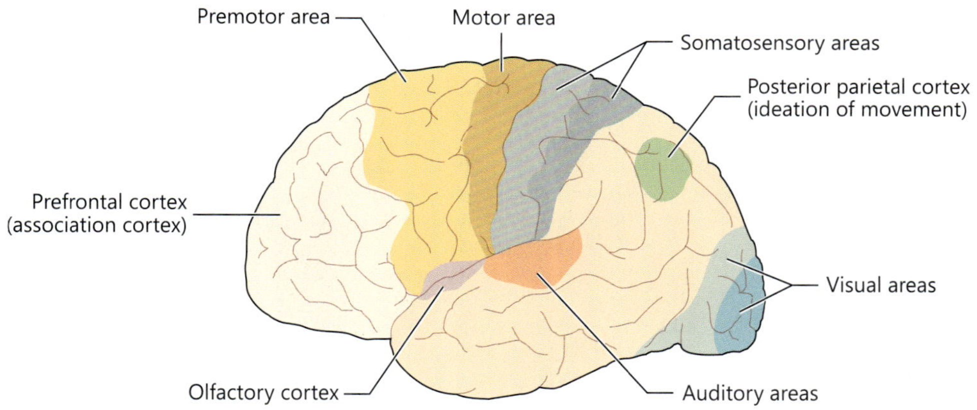

Anatomical and functional subdivision of the cerebral hemisphere, lateral view. The specific functional location is limited to a small part of the cerebral cortex, much of which has roles and functions not exactly definable and, in any case, modifiable.

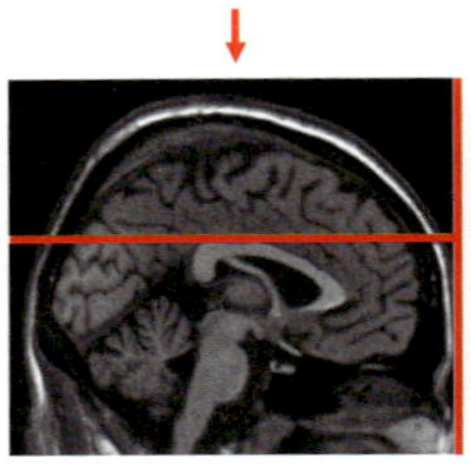

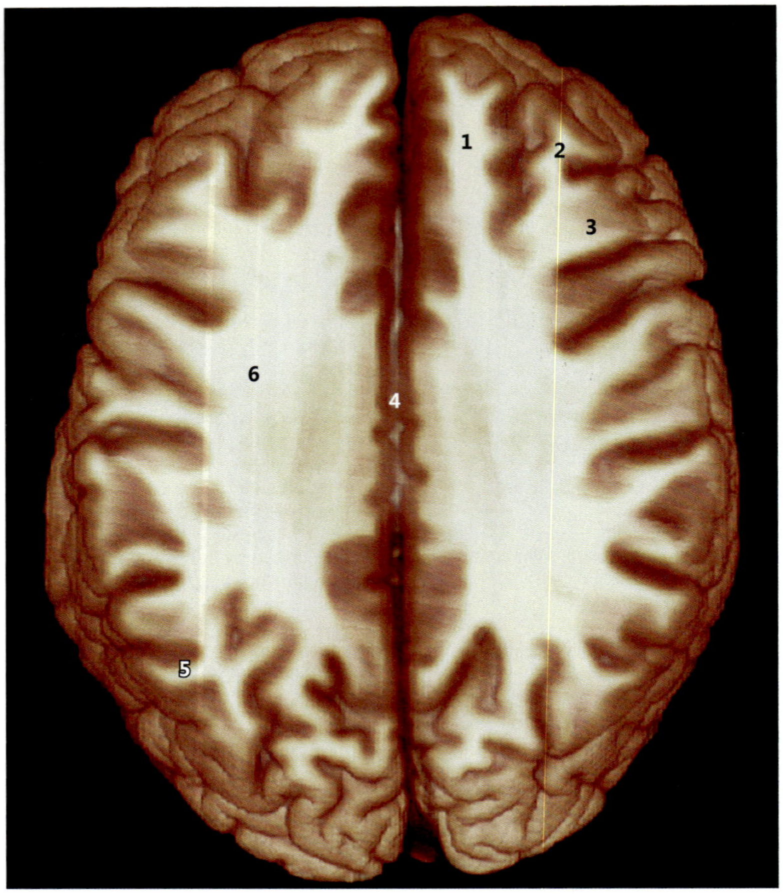

Three-dimensional image of a living human body, obtained by NMR using a 3D volume rendering technique, showing a transverse section of the **telencephalon** passing above the corpus callosum.

1, Superior frontal gyrus
2, Middle frontal gyrus
3, Inferior frontal gyrus
4, Falx cerebri
5, Angular gyrus
6, Central white matter of the cerebral hemisphere (centrum semiovale)

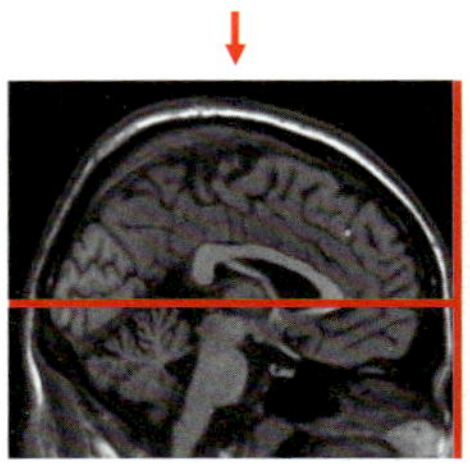

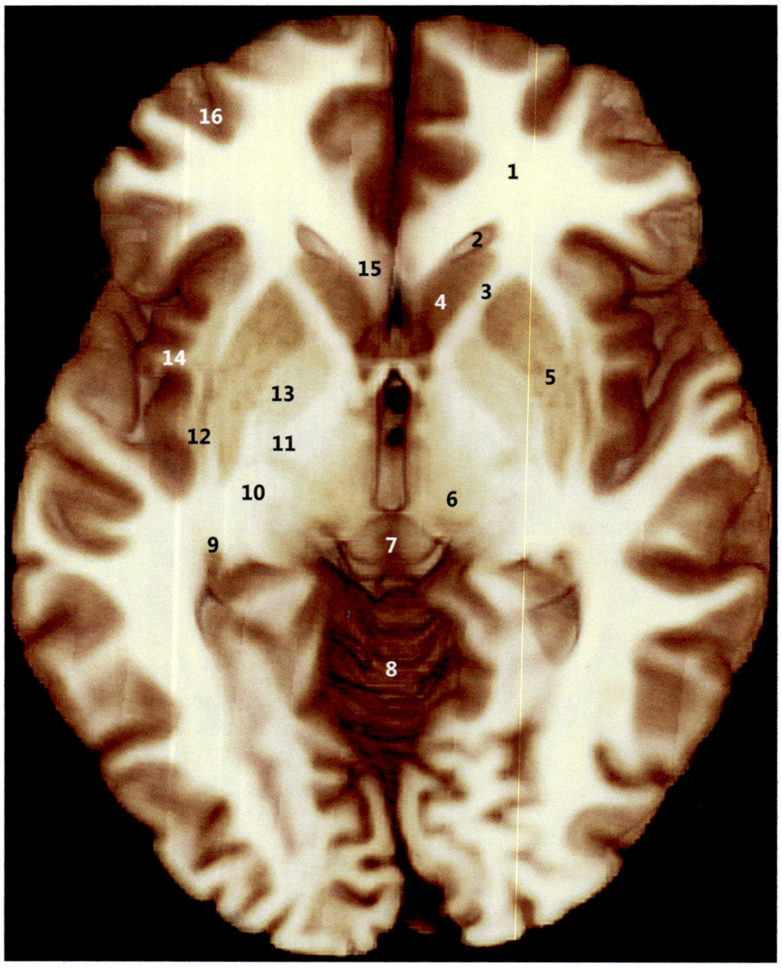

Three-dimensional image of a living human body, obtained by NMR using a 3D volume rendering technique, showing a transverse section of the **telencephalon** passing through the head of the caudate nucleus.

1, Minor forceps
2, Frontal horn of lateral ventricle
3, Anterior limb of internal capsule
4, Head of caudate nucleus
5, Putamen
6, Thalamus
7, Tectal plate
8, Vermis of cerebellum [I-X]
9, Tail of caudate nucleus
10, Posterior limb of internal capsule
11, Globus pallidus medial segment
12, Claustrum
13, Globus pallidus lateral segment
14, Insula
15, Genu of corpus callosum
16, Superior frontal gyrus

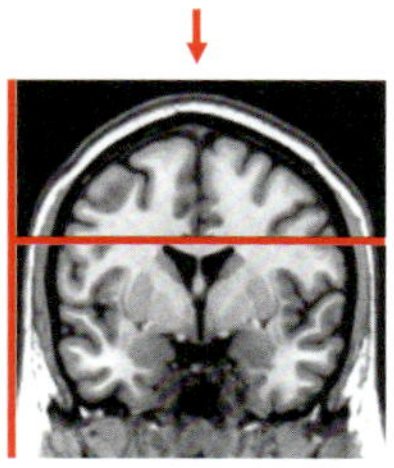

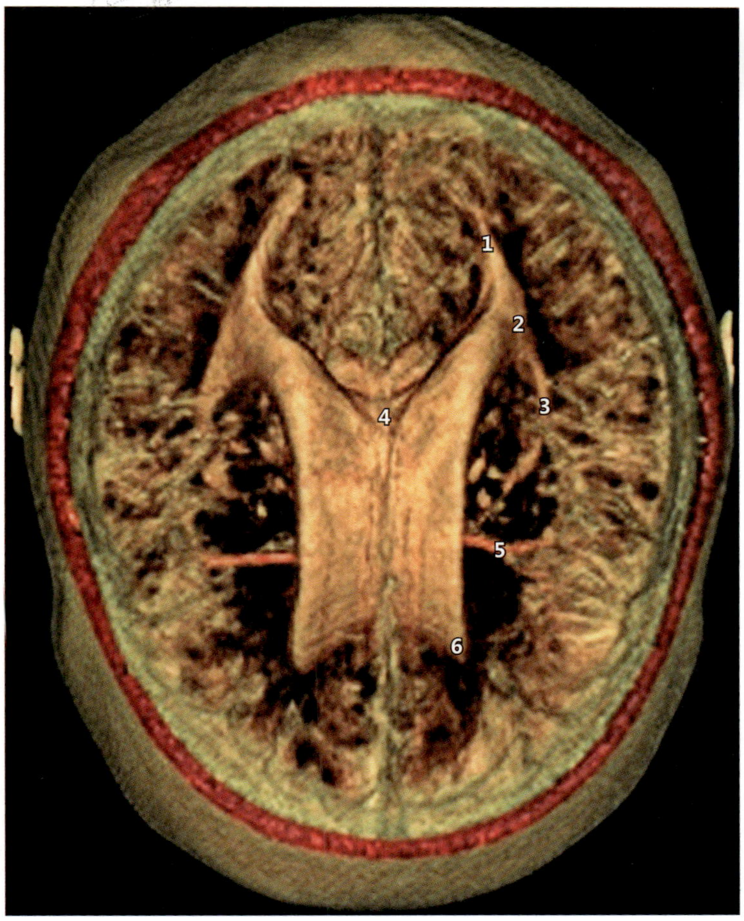

Three-dimensional image of a living human body, obtained by NMR using a 3D volume rendering technique, showing a transverse section of the cranium and brain. The **ventricular cavities** are highlighted by applying a specific rendering parameter.

1, Occipital (posterior) horn
2, Collateral trigone
3, Temporal horn
4, Third ventricle
5, Anterior communicating artery
6, Frontal (anterior) horn

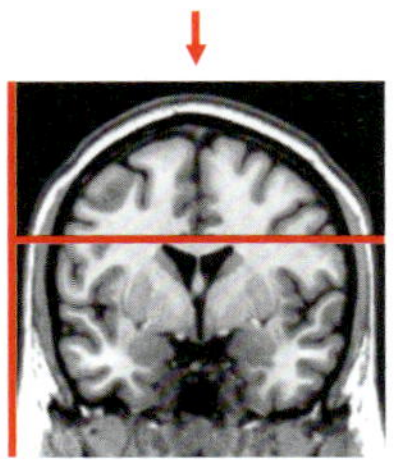

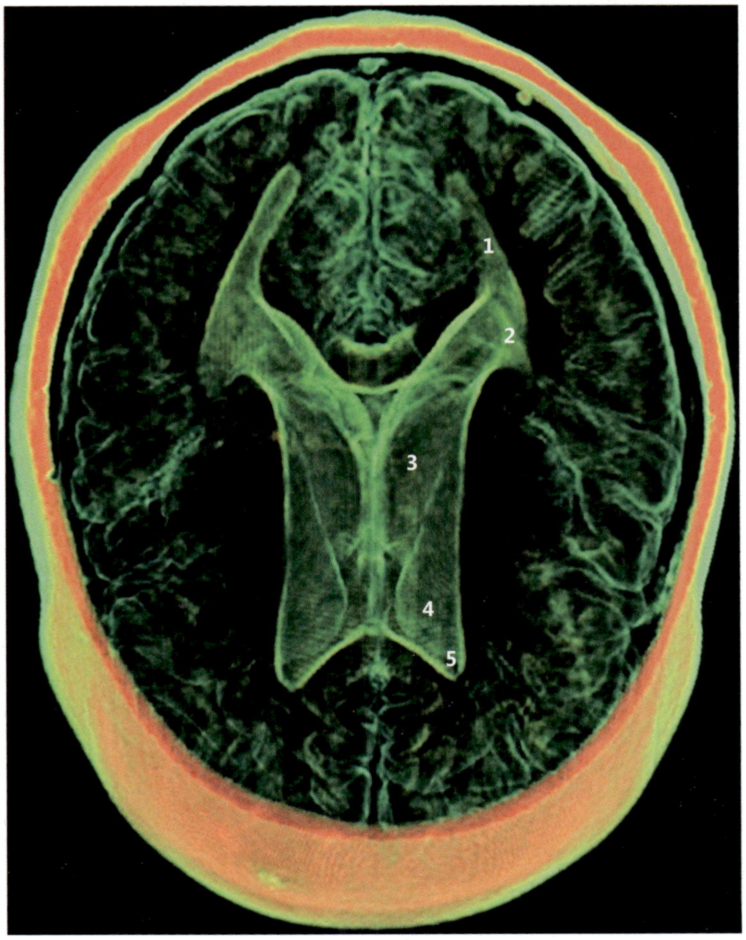

Three-dimensional image of a living human body, obtained by NMR using a 3D volume rendering technique, showing a transverse section of the cranium and brain. The **ventricular cavities** are highlighted by applying a transparent rendering parameter, able to identify of the underlying structures as well.

1, Occipital (posterior) horn
2, Collateral trigone
3, Impression of thalamus
4, Impression of head of caudate nucleus
5, Frontal (anterior) horn

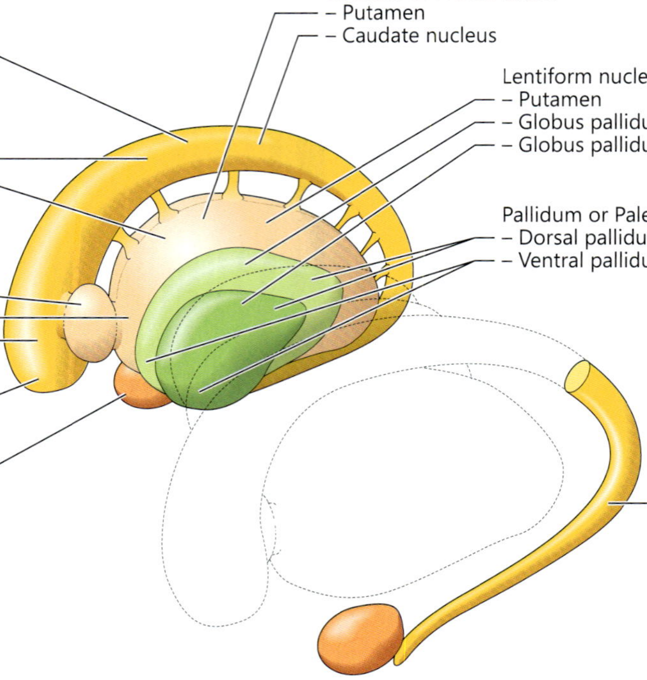

The **basal nuclei**, located in the basal telencephalon, form the corpus striatum, subdivided, phylogenetically into the striatum, or neostratium (caudate nucleus and putamen), and the pallidum, or paleostriatum (globus pallidus lateral segment and globus pallidus medial segment). The striatum and the pallidum consit of a dorsal and a ventral part. The ventral striatum and pallidum also belong to the amygdaloid body or complex, and include some of its structures. Altogether, putamen and globus pallidus are named lentiform nucleus, because of the form they assume. The anatomo-functional circuit of the basal nuclei involves also the subthalamic nucleus, subthalamic formation and substantia nigra situated in the cerebral peduncle of the mesencephalon (☞ page 580).

Principal nuclei of the telencephalon		
Amygdaloid body or complex Septal area (dorsal, lateral and medial septal nuclei, septofimbrial nucleus, triangular nucleus of septum) Claustrum Corticomedial nuclei (anterior amygdaloid area, medial, central and cortical amygdaloid nuclei, nucleus of lateral olfactory tract) Basolateral nuclei (lateral and medial basal amygdaloid nuclei, interstitial and lateral amygdaloid nuclei) Ventral pallidum Basal substance (basal nucleus, bed nucleus of stria terminalis and sublenticular extended amygdala) Innominate substance Ventral striatum (including the nucleus accumbens)		
Basal nuclei (or basal ganglia)	*Subdivision*	*Organization*
	Caudate nucleus Lentiform nucleus (or lenticular nucleus) Putamen Globus pallidus lateral (or external) segment Globus pallidus medial (or internal) segment	Corpus striatum Striatum or Neostriatum (divided by the internal capsule into the caudate nucleus and putamen) Dorsal striatum (dorsolateral parts of the putamen and caudate nucleus) Ventral striatum (ventromedial parts of the putamen and caudate nucleus, components of the amygdaloid body or complex and includes the nucleus accumbens) Pallidum or Paleostriatum Dorsal pallidum (globus pallidus dorsal to the anterior commissure) Ventral pallidum (globus pallidus ventral to the anterior commissure, component of the amygdaloid body or complex)

Principal afferences and efferences of the telencephalon nuclei

NUCLEUS, CORPUS, SUBSTANCE OR AREA	AFFERENCES	EFFERENCES	FUNCTION
Nucleus accumbens (nucleus of the ventral striatum)	Corticostriatal fibers (from the cingulate gyrus and the temporal cortex) and amygdalostriatal fibers	Amygdaloid body	Limbic system and motivational circuits
Amygdaloid body or complex	Olfactory bulb, frontal and temporal cortex, limbic lobe and hippocampus, intralaminar nuclei, reticular formation of the pons and the mesencephalon	Frontal and prefrontal cortex, cingulate gyrus, fibers of the stria terminalis for the hypothalamus, amygdalofugal pathway and amygdalostriatal fibers (olfactory striae, fibers for the caudate nucleus, for the nucleus accumbens and for the thalamus), commissural fibers for the contralateral amygdaloid body across the anterior commissure	Olfactory pathways, limbic system, emotional circuit and control of autonomic nervous system
– corticomedial nuclei (anterior amygdaloid area, medial, central and cortical amygdaloid nuclei, nucleus of lateral olfactory tract) and basolateral nuclei (lateral and medial basal amygdaloid nuclei, interstitial and lateral amygdaloid nuclei)	Fibers from the olfactory bulb through the lateral olfactory stria (to the corticomedial nuclei) and from the amygdalopiriform transition area (to the basolateral nuclei) and from the reticular formation of the pons and mesencephalon	Fibers to the anterior and intermediate hypothalamic areas, through the stria terminalis and lateral hypothalamic area, fibers at the innominate substance, to the septal are and to the olfactory tubercle, fibers to the medial dorsal nucleus of the thalamus, fibers to the cingulate gyrus and the superior, middle and inferior temporal,gyri, fibers to the insula, to the claustrum and the rostral part of the putamen	Centers of the limbic system relating to the visceral activities and to emotional behaviour
– basal substance and septal area	Other formations of the amygdaloid body, frontal and temporal cortex, insula, hypothalamus, reticular formation of the pons and of the mesencephalon	Diffuse projections to the cerebral cortex and to the basal nuclei.	Limbic system, memory and cognitive processes, state of alertness and wakefulness
– innominate substance	Pallidum	Other formations of the amygdaloid body, pallidum, hypothalamus	Memory and cognitive processes
Claustrum	Parietal and temporal cortex (neocortex), associative fibers of the telencephalon	Frontal cortex (neocortex), thalamus	Assimilation of the information underlying the conscious perception
Basal nuclei (or basal ganglia)			
– caudate nucleus	Direct and cruciform corticostriatal fibers (from the frontal and parietal cortex), from the amygdaloid body, thalamostriatal fibers (from the intralaminar nuclei of the thalamus, the medial dorsal nucleus and the median nuclei of the thalamus), nigrostriatal dopaminergic fibers (form the compact part of the substantia nigra) and subthalamostriatal fibers (from the subthalamic nucleus)	Striatopallidal (to the globus pallidus lateral segment), striatonigral (to the substantia nigra), striatosubthalamic (to the subthalamic nucleus) and striatothalamic fibers	Subcortical center of the extrapyramidal system
– putamen	Direct and cruciform corticostriatal fibers (from the frontal and parietal cortex), thalamostriatal fibers (from the centromedian nucleus), nigrostriatal dopaminergic fibers (from the compact part of the substantia nigra) and subthalamostriatal fibers (from the subthalamic nucleus)	Striatopallidal (to the globus pallidus lateral and medial segments), striatonigral (to the substantia nigra), striatosubthalamic (to the subthalamic nucleus) and striatothalamic fibers	Subcortical center of the extrapyramidal system
– pallidum	Striatopallidal fibers (from the caudate nucleus and the putamen), subthalamopallidal fibers (to the globus pallidus medial segment), nigropallidal fibers (from the substantia nigra), thalamopallidal fibers (from the medial dorsal nucleus and from the intralaminar nuclei of thalamus)	Lenticular fasciculus and ansa lenticularis (pallidothalamic fibers to the ventral anterior, ventral lateral, medial dorsal, centromedian and ventral posteriomedial muclei of the thalamus), thalamic fasciculus (fibers to the hypothalamus), pallidorubral fibers (to the red nucleus), pallidosubthalamic fibers (to the subthalamic nucleus), pallidotegmental fibers (from the globus pallidus medial segment), pallidonigral fibers (to the substantia nigra), pallidoreticular fibers (to the reticular formation of the mesencephalon), to the zona incerta, to the superior colliculi, to the medial and lateral habenular nuclei	Subcortical center of the extrapyramidal system

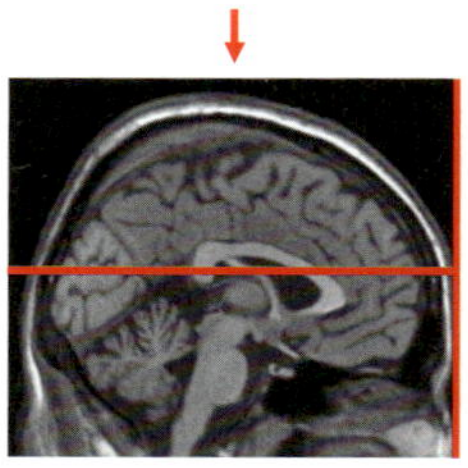

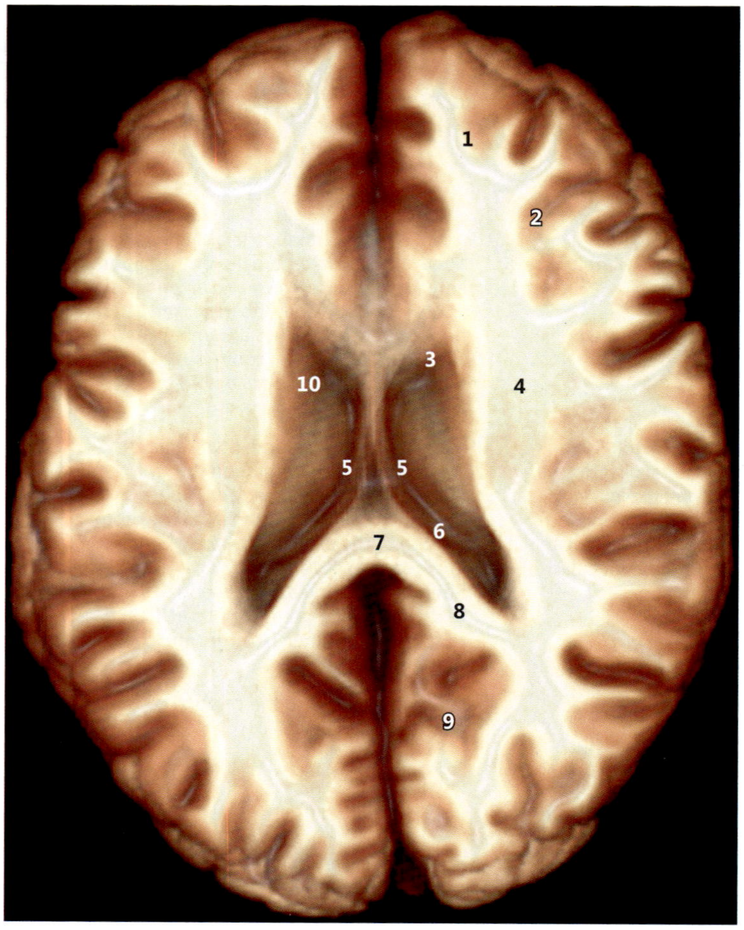

Three-dimensional image of a living human body, obtained by NMR using a 3D volume rendering technique, showing a transverse section of the **telencephalon** passing above the head of the caudate nucleus.

1, Superior frontal gyrus
2, Middle frontal gyrus
3, Lateral ventricle
4, Corona radiata
5, Thalamus
6, Choroid plexus of lateral ventricle
7, Splenium of corpus callosum
8, Major forceps
9, Parieto-occipital sulcus
10, Head of caudate nucleus

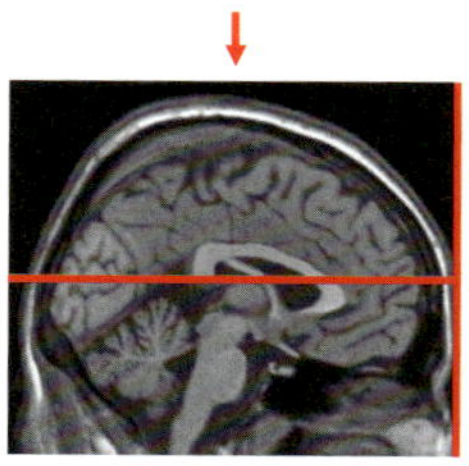

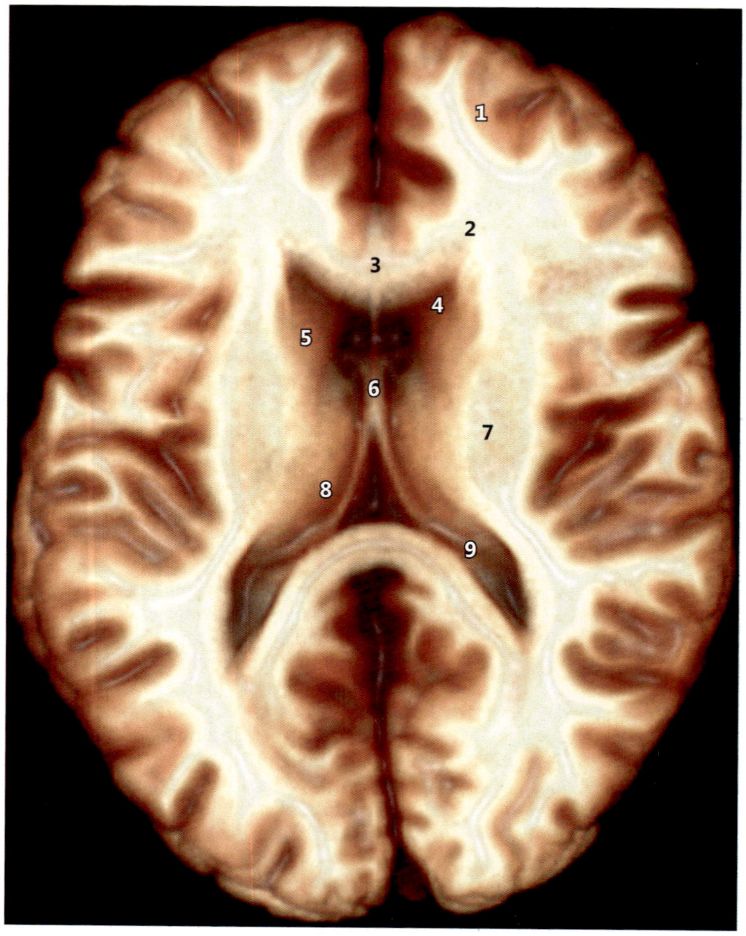

Three-dimensional image of a living human body, obtained by NMR using a 3D volume rendering technique, showing a transverse section of the **telencephalon** passing through the genu of the corpus callosum.

1, Superior frontal gyrus
2, Minor forceps
3, Genu of corpus callosum
4, Frontal (anterior) horn of lateral ventricle
5, Head of caudate nucleus
6, Column of fornix
7, Putamen
8, Thalamus
9, Occipital (posterior) horn of lateral ventricle

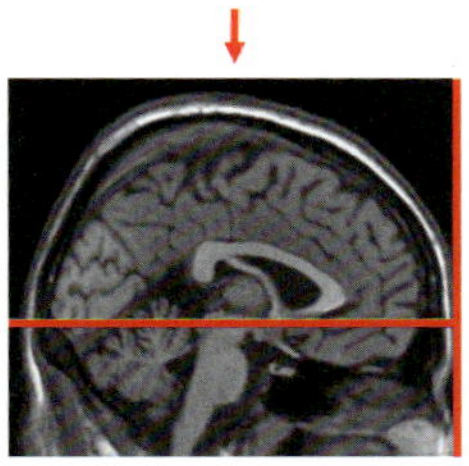

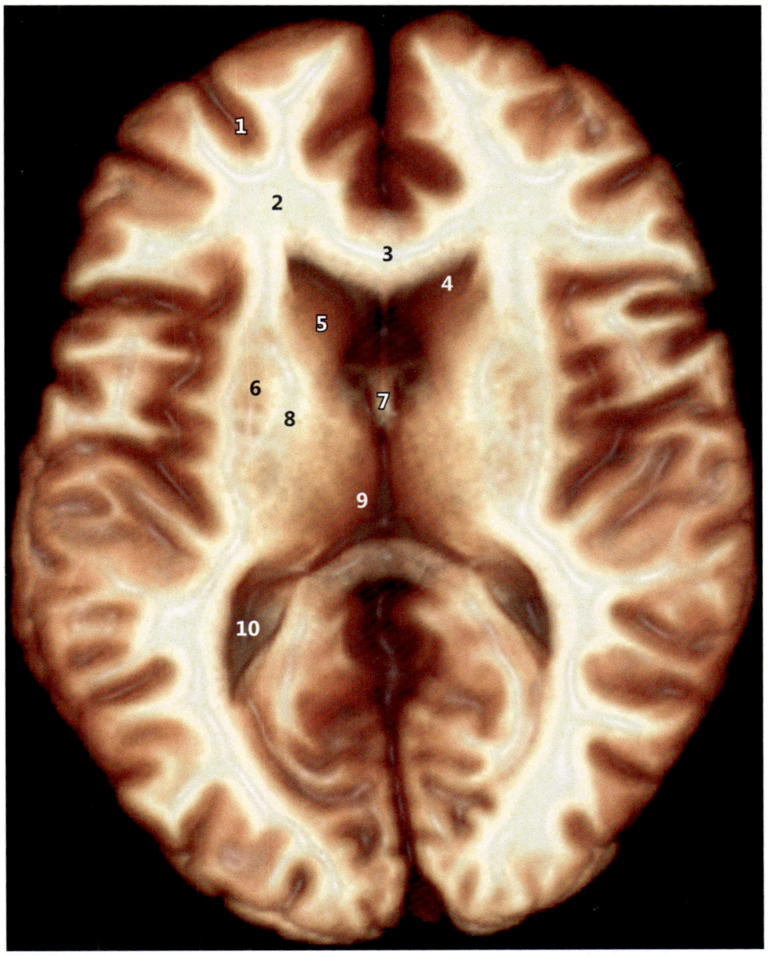

1, Superior frontal gyrus
2, Minor forceps
3, Genu of corpus callosum
4, Frontal (anterior) horn of lateral ventricle
5, Head of caudate nucleus
6, Putamen
7, Column of fornix
8, Pallidum
9, Thalamus
10, Occipital (posterior) horn of lateral ventricle

Three-dimensional image of a living human body, obtained by NMR using a 3D volume rendering technique, showing a transverse section of the **telencephalon** passing 2 mm below the genu of the corpus callosum.

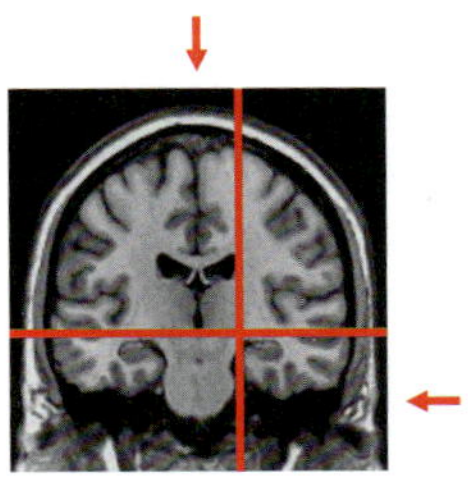

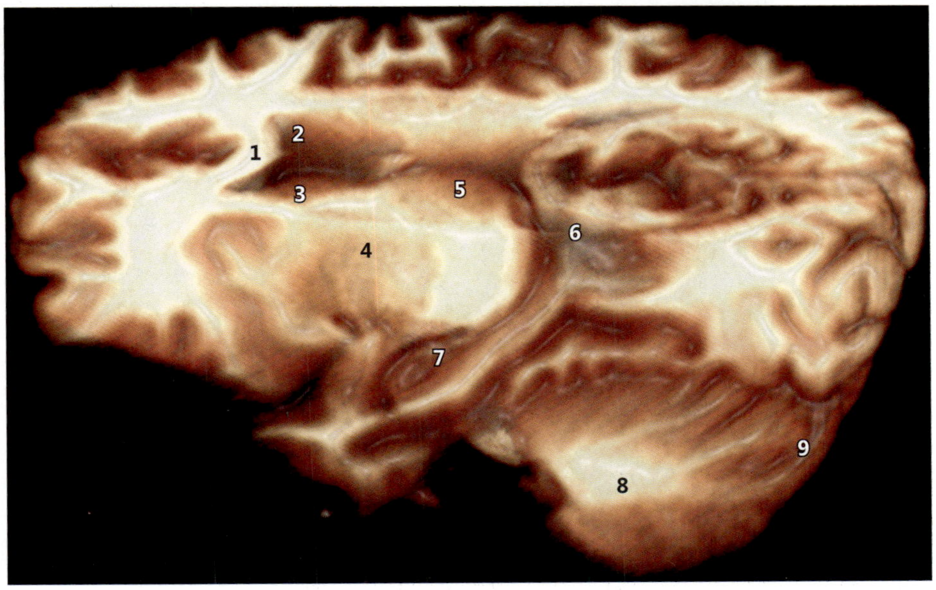

1, Genu of corpus callosum
2, Frontal (anterior) horn of lateral ventricle
3, Head of caudate nucleus
4, Putamen
5, Thalamus
6, Occipital (posterior) horn of lateral ventricle
7, Temporal horn of lateral ventricle
8, White matter of cerebellum
9, Cerebellum

Three-dimensional image of a living human body, obtained by NMR using a 3D volume rendering technique, showing a transverse section of the **telencephalon** passing above the head of the caudate nucleus, viewed from above and laterally; a parasagittal section has also been included to show the collateral trigone and the temporal horn of the lateral ventricle.

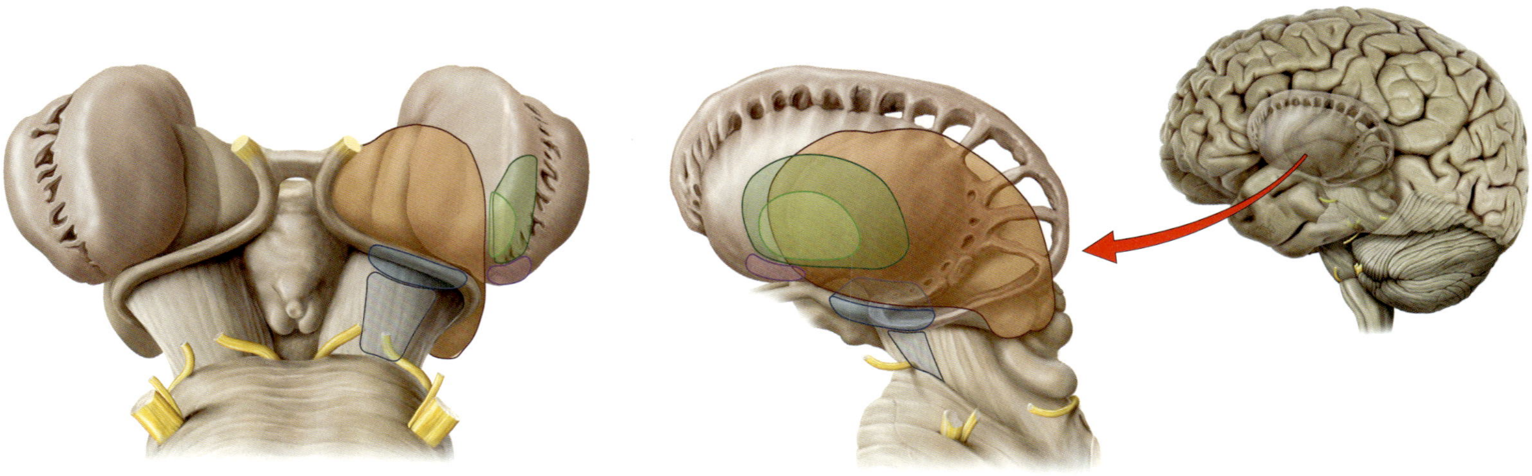

- Thalamus
- Globus pallidus lateral segment
- Globus pallidus medial segment
- Ventral pallidum
- Subthalamic nucleus
- Substantia nigra

Anterior and lateral projections of the **basal nuclei**, formations of gray matter in the basal telencephalon. The anatomo-functional circuit of the basal nuclei also includes the subthalamic nucleus, the subthalamic formation, and the substantia nigra located in the cerebral peduncle of the mesencephalon. Altogether the basal nuclei form the corpus striatum, and comprise the caudate nucleus and the lentiform nucleus, the latter formed by putamen, globus pallidus lateral segment and globus pallidus medial segment. The caudate nucleus and the putamen, separated by the internal capsule, form the striatum, or neostriatum, divided in ventral and dorsal striatum, while the pallidum (globus pallidus lateral segment and globus pallidus medial segment) forms the paleostriatum, and is divided in ventral pallidum and dorsal pallidum (☞ table and figure page 574). There is an anatomical continuity between the caudate nucleus and the putamen due to the numerous connecting gray matter bridges. The lateral face of the thalamus is in close relation to the internal capsule, lateral to which is the globus pallidus, whereas the caudate nucleus is situated above and lateral to the superior face of the thalamus and medial to the internal capsule.

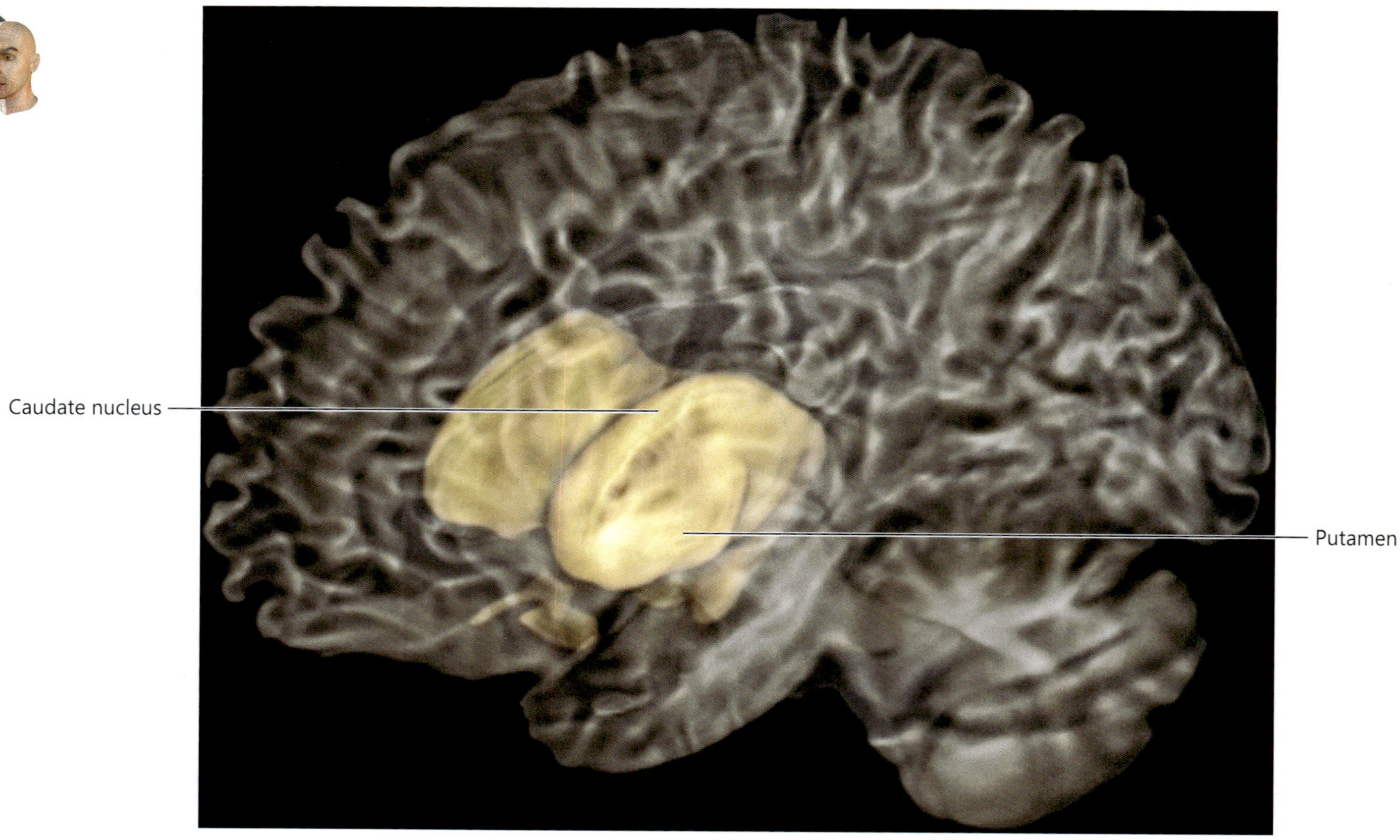

Three-dimensional image of a living human body, obtained by NMR using a 3D volume rendering technique, in which the cerebral hemispheres can be observed in transparency, seen from above, anteriorly and laterally, to show the **basal nuclei** (displayed through an opacity parameter).

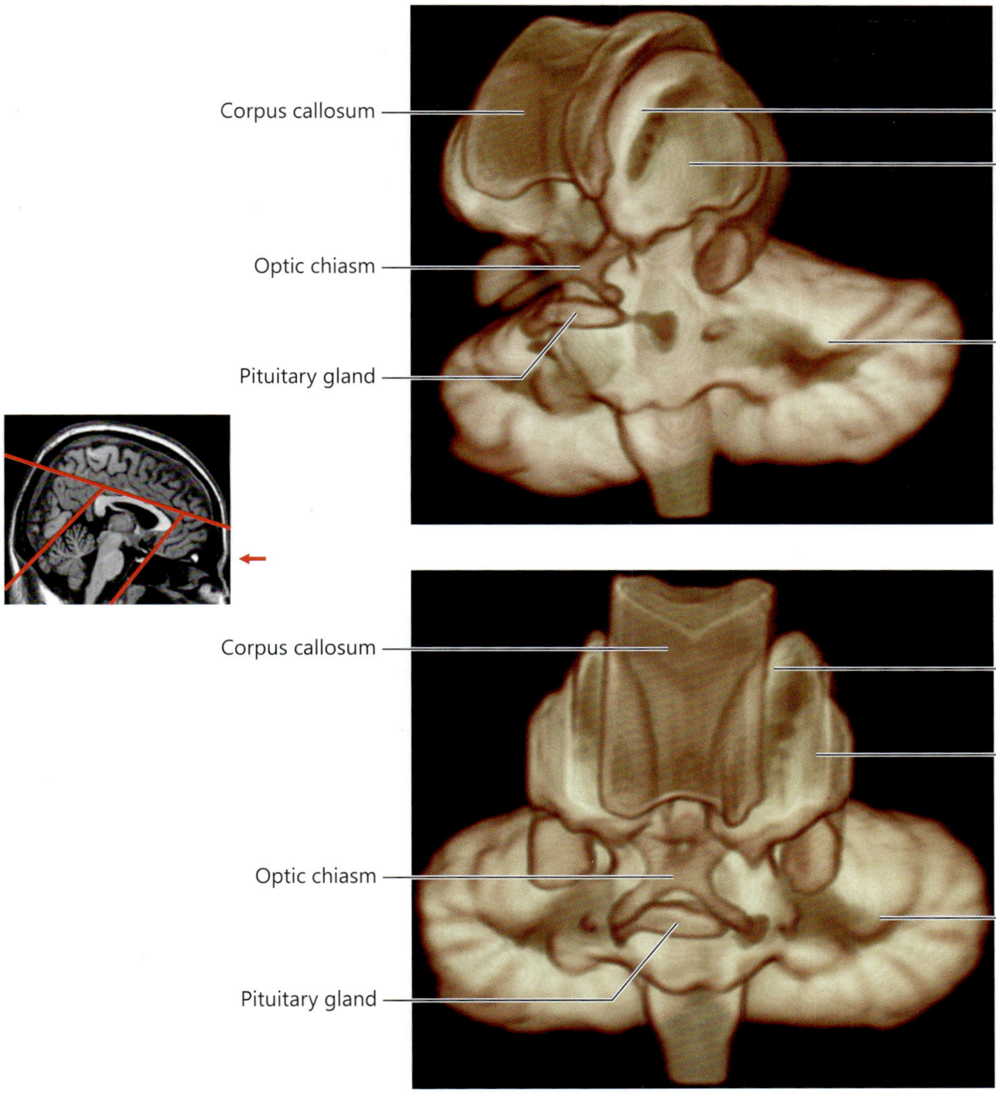

Three-dimensional image of a living human body, obtained by NMR using a 3D volume rendering technique, showing the **basal nuclei** in anterolateral projection (**a**) and in anterior projection (**b**).

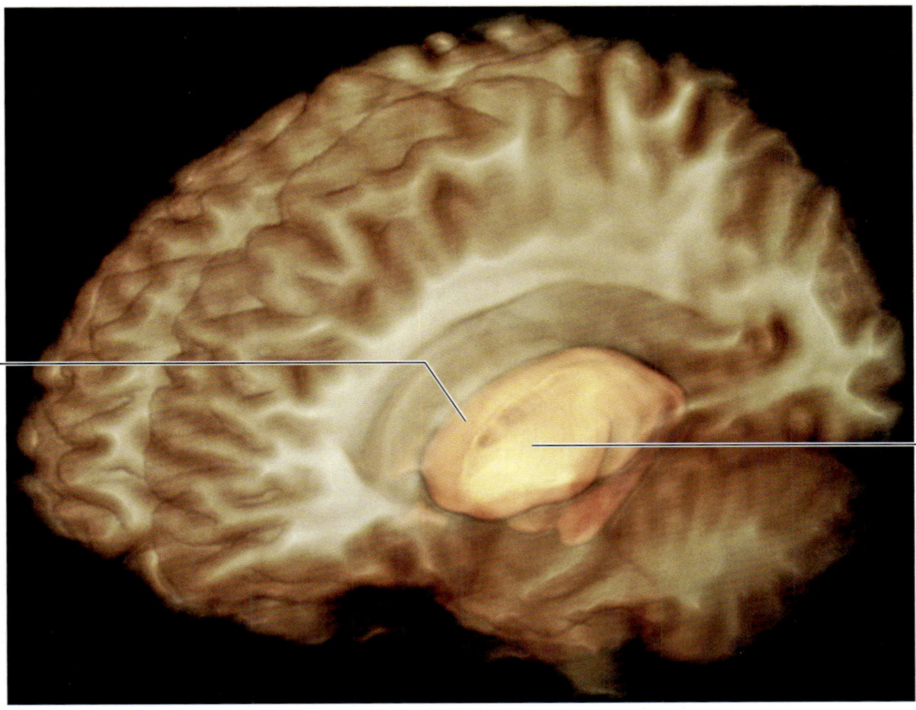

Three-dimensional image of a living human body, obtained by NMR using a 3D volume rendering technique, after removal of the left cerebral hemisphere, showing the **basal nuclei** in anterolateral projection.

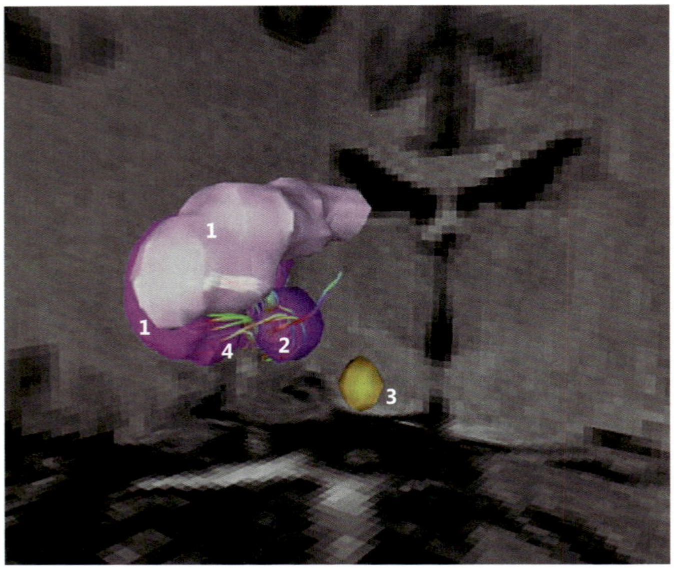

a

b

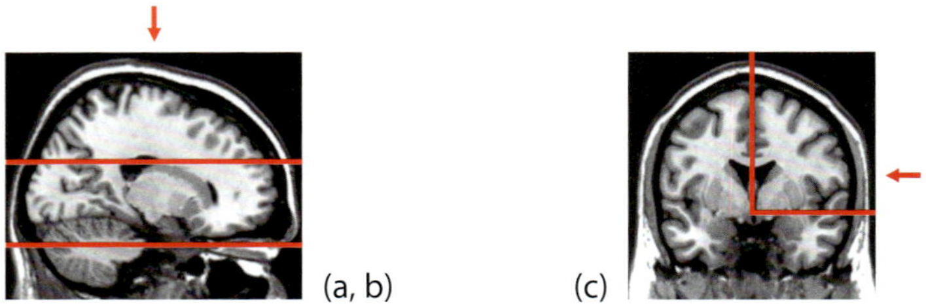

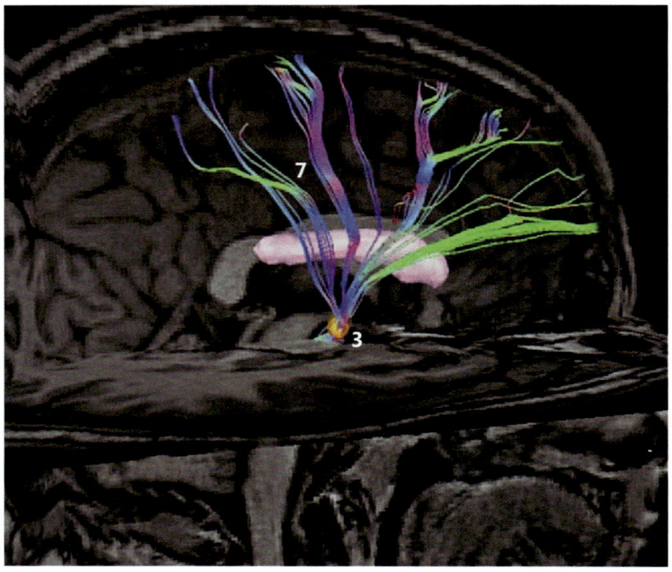

c

Three-dimensional image of a tractography obtained by diffusion-weighted MRI sequences showing the **direct pathway** (**a**), the **indirect pathway** (**b**) and the **hyperdirect pathway** (**c**).

1, Neostriatum
2, Globus pallidus medial or internal segment (GP_i)
3, Subthalamic nucleus (Corpus of Luys)
4, Neostriatopallidal fibers
5, Globus pallidus lateral or external segment (GP_e)
6, Pallidosubthalamic fibers
7, Corticosubthalamic fibers

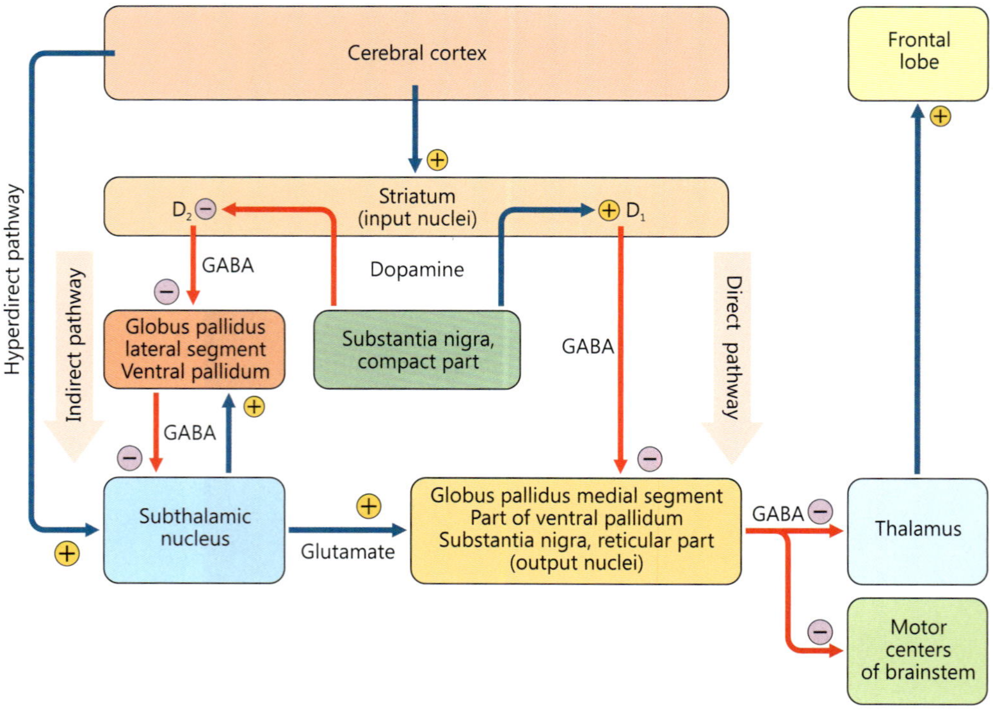

Simplified diagram of the **direct**, **indirect** and **hyperdirect pathways in the basal nuclei**. The activity of the two principal nuclei projecting to the thalamocortical pathway (reticular part of the substantia nigra and internal segment of the globus pallidus) is regulated by the dopamine released in the striatum by the dopaminergic neurons localized in the compact part of the substantia nigra. This regulation takes place through a direct GABAergic pathway, with inhibitory function, and through an indirect pathway. The output from the indirect pathway consists of excitatory glutamatergic fibers from the subthalamic nucleus. The dopamine has an excitatory role on the striatum D1 receptors, and inhibitory on the D2 receptors. *Blue* lines indicate excitatory pathways and *red* lines inhibitory pathways. For the classification of the basal nuclei ☞ Principal nuclei of the telencephalon, page 574. For the principal afferences and efferences of the basal nuclei ☞ Nuclei of the Telencephalon, pages 574-575.

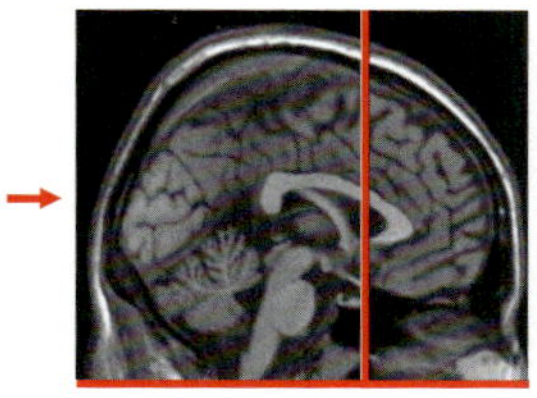

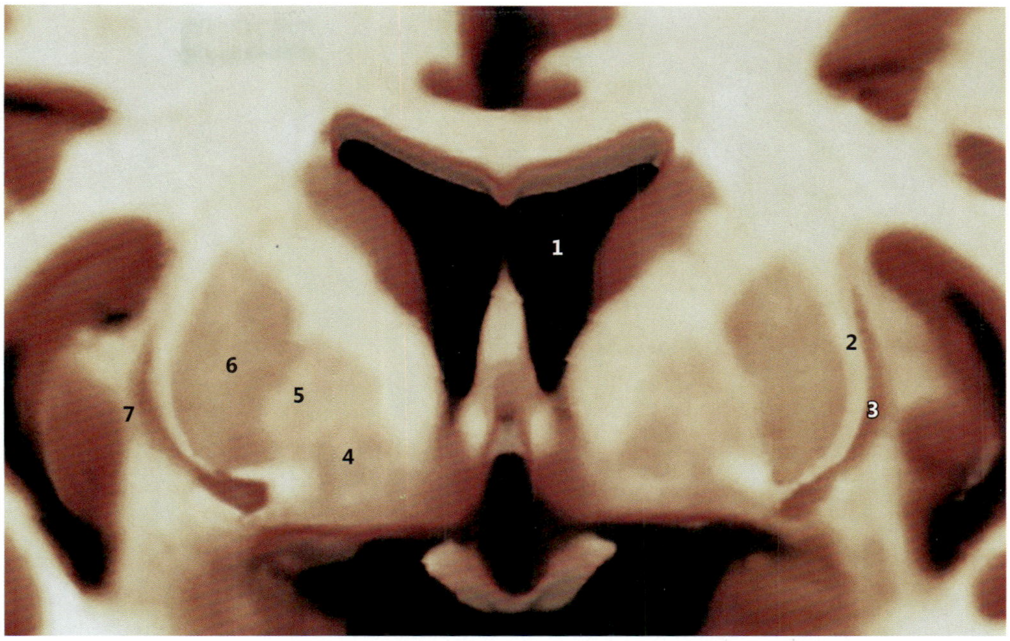

Three-dimensional image of a living human body, obtained by NMR using a 3D volume rendering technique, of a coronal section passing through the rostrum of the corpus callosum showing the **claustrum**.

1, Lateral ventricle
2, External capsule
3, Claustrum
4, Globus pallidus medial segment
5, Globus pallidus lateral segment
6, Putamen
7, Extreme capsule

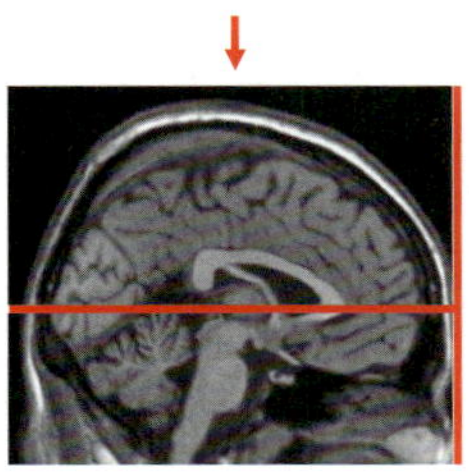

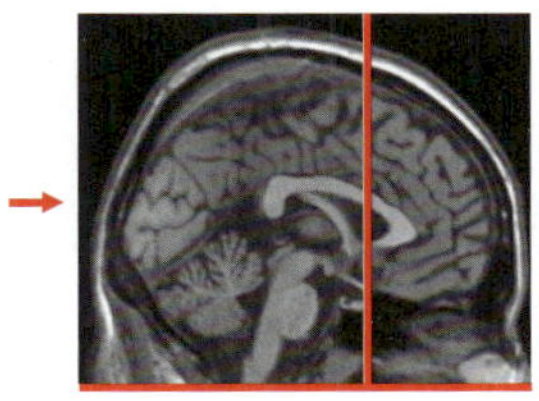

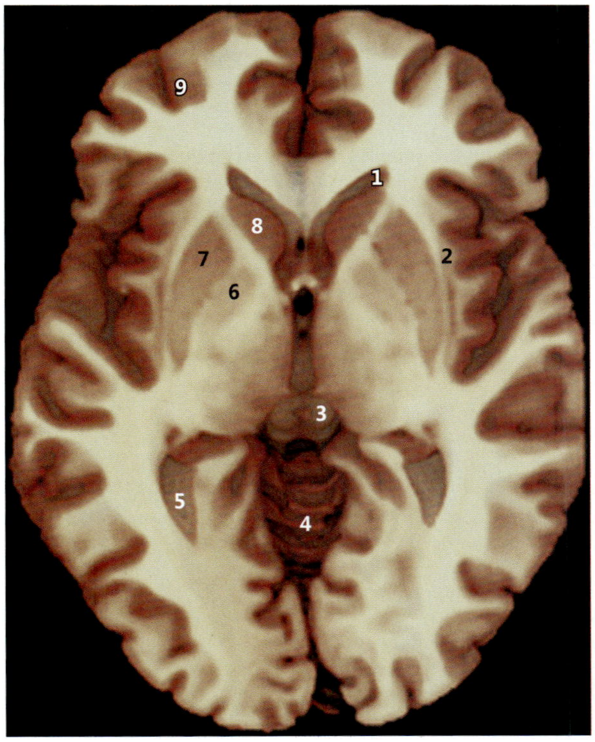

a

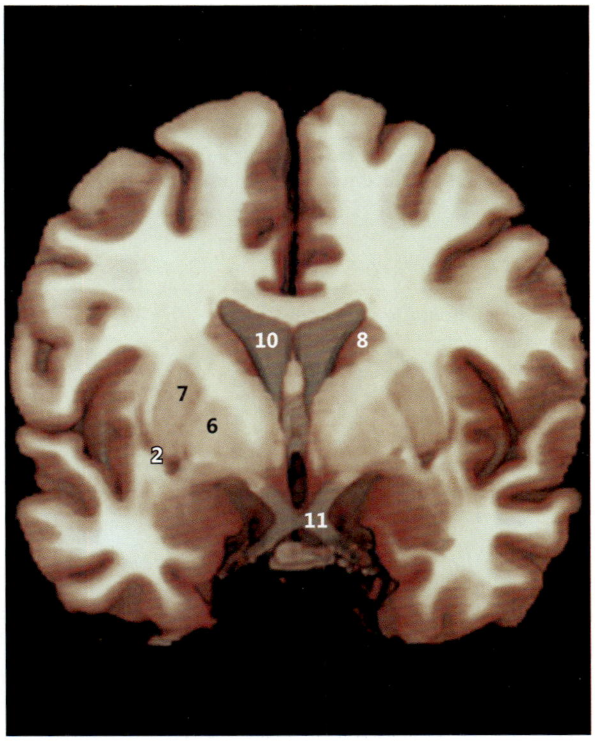

b

Three-dimensional image of a living human body, obtained by NMR using a 3D volume rendering technique, of a transverse section passing through the genu of the corpus callosum (**a**) and of a coronal section passing through the rostrum of the corpus callosum (**b**), showing the **claustrum**.

1, Frontal horn of lateral ventricle
2, Claustrum
3, Tectal plate (Quadrigeminal plate)
4, Vermis of cerebellum [I-X]
5, Occipital (posterior) horn of lateral ventricle
6, Globus pallidus, lateral or esternal segment (GP_e)
7, Putamen
8, Head of claudate nucleus
9, Superior frontal gyrus
10, Lateral ventricle
11, Optic chiasm

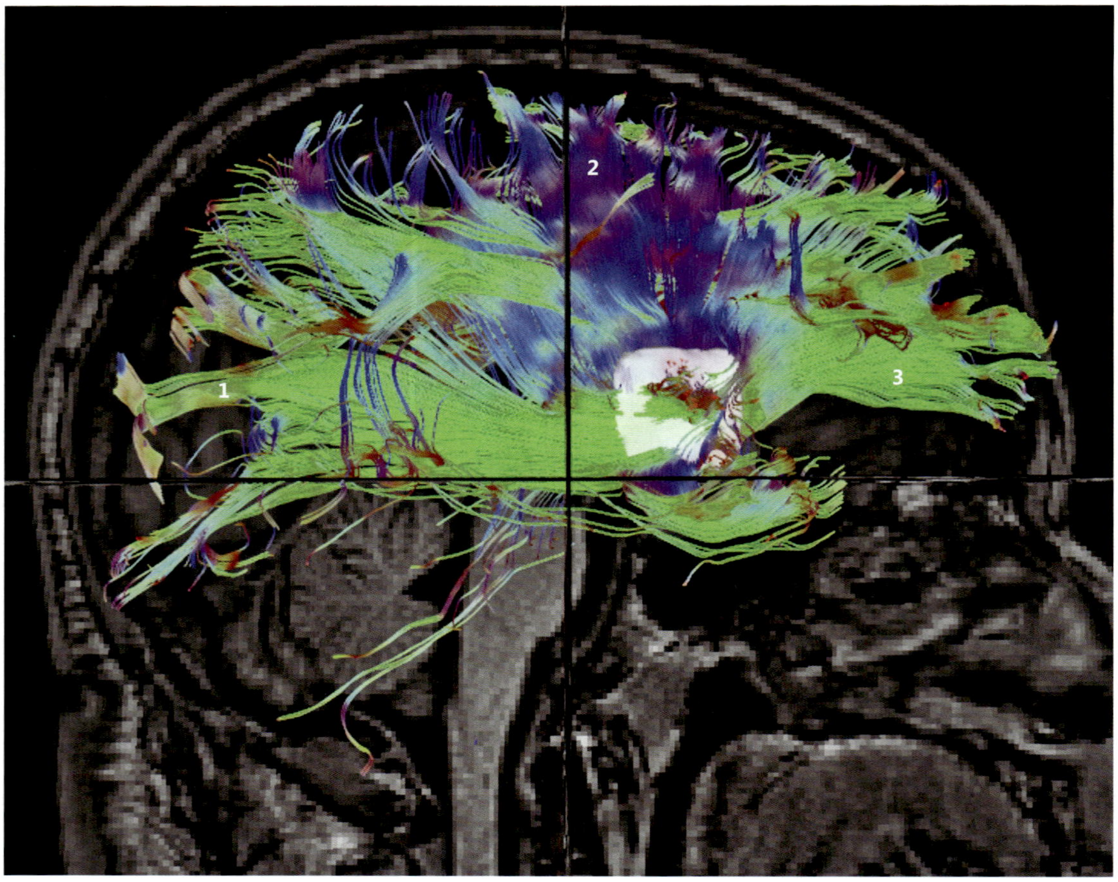

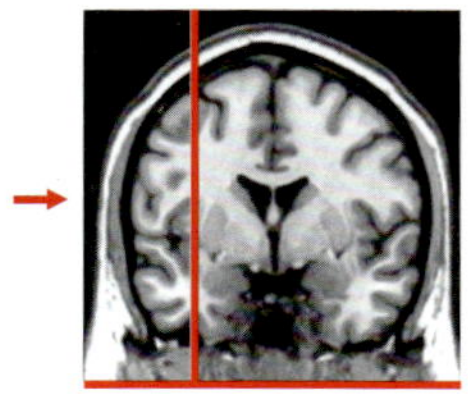

1, Connections with the occipital lobe
2, Connections with the parietal lobe
3, Connections with the frontal lobe

Three-dimensional image of a living human body, obtained by diffusion-weighted MRI using a tractography technique, showing the main color-coded **connections of the claustrum** following the positioning of the fibers: *green* fibers following the anteroposterior direction, *blue*, the superoinferior direction and, *red*, the laterolateral direction. In *white*, the VOI (Volume of Interest) corresponding to the claustrum can be identified in transparency.

Table showing the functional correspondence of the connections between claustrum and Brodmann areas, the basal nuclei and through the corpus callosum (☞ also Principal afferences and efferences of the telencephalon nuclei, page 575)

BRODMANN AREAS, BASAL NUCLEI, CORPUS CALLOSUM	*FUNCTIONAL CORRESPONDENCE*
Areas 11 and 12 (obitofrontal areas)	Executive function, cognitive control
Areas 8, 9 and 10 (prefrontal and frontopolar areas)	Executive function, cognitive control
Area 6 (supplementary motor area)	Visual and somatosensory guide of movements
Area 4 (primary motor area)	Control of voluntary movements
Areas 3, 1, 2 (primary somatosensory area)	Basic somatosensory functions
Areas 5 and 7 (primary somatosensory areas)	Visual-motor coordination and processing of proprioceptive and tactile information (supplied by the mechanoreceptors and proprioceptors)
Areas 41 and 42 (primary auditory area and association area)	Processing of auditory information
Areas 17, 18 and 19 (primary, secondary and tertiary visual areas)	Processing of visual information
Areas 22 and 39 (area of Wernicke)	Comprehension of speech
Areas 43, 44 and 45 (area of Broca)	Production of speech
Area 34 (parahippocampal gyrus)	Memory and learning
Area 31 (cingulate gyrus)	Recognition and processing of emotions
Caudate nucleus, putamen and globus pallidus	Motor control
Corpus callosum	Interhemispheric connections

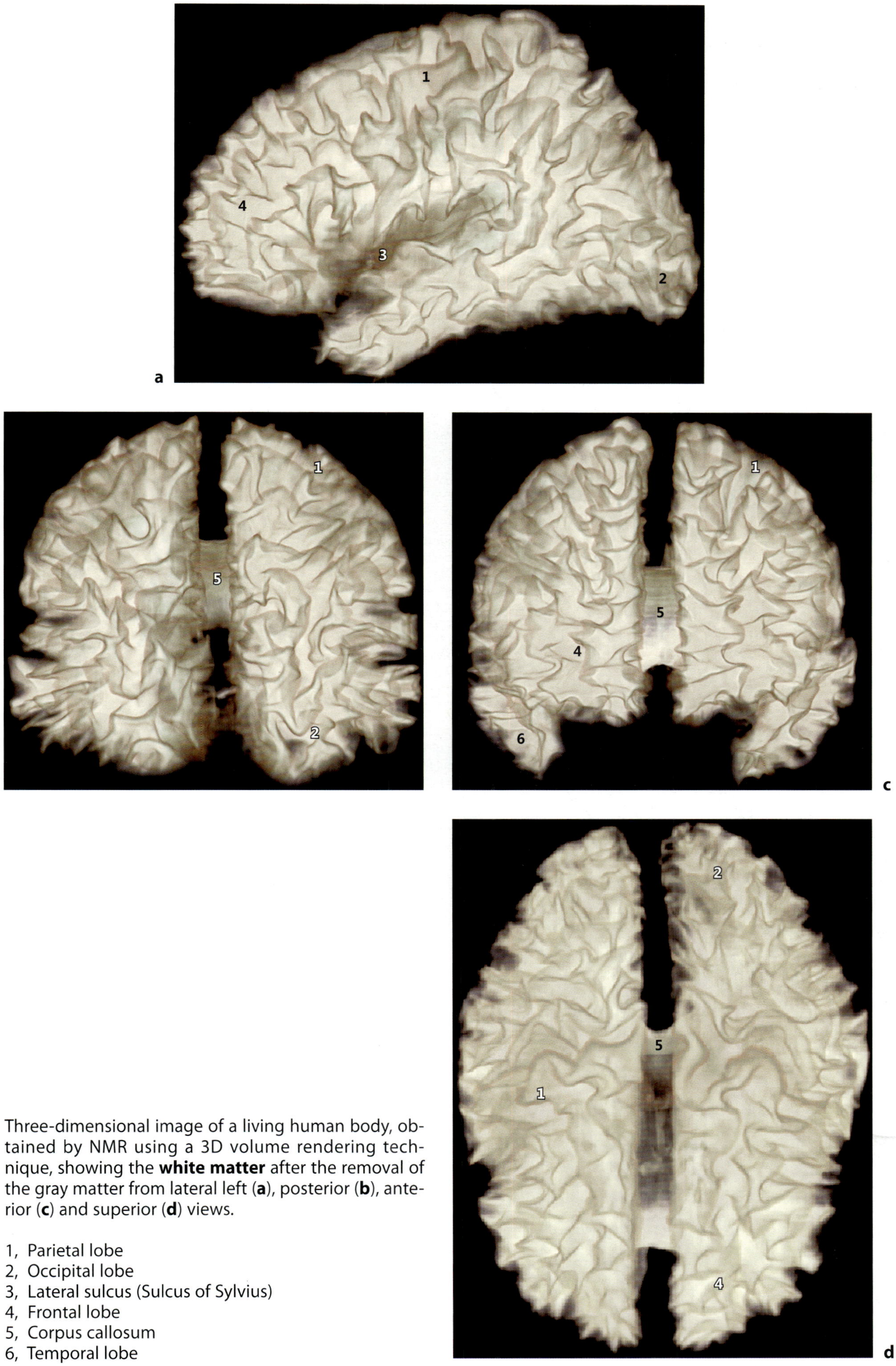

Three-dimensional image of a living human body, obtained by NMR using a 3D volume rendering technique, showing the **white matter** after the removal of the gray matter from lateral left (**a**), posterior (**b**), anterior (**c**) and superior (**d**) views.

1, Parietal lobe
2, Occipital lobe
3, Lateral sulcus (Sulcus of Sylvius)
4, Frontal lobe
5, Corpus callosum
6, Temporal lobe

Organization of the white matter of the telencephalon		
RIGHT CEREBRAL HEMISPHERE	***INTERHEMISPHERIC FORMATIONS***	***LEFT CEREBRAL HEMISPHERE***
Association fibers of interhemispheric telencephalon and in the extreme capsule and projection fibers in the corona radiata, and in the external and internal capsules	Commissural fibers of the telencephalon (corpus callosum fibers, hippocampal commissure or commissure of the fornix, anterior commissure)	Association fibers of interhemispheric telencephalon and in the extreme capsule and projection fibers in the corona radiata, and in the external and internal capsules

Composition of the centrum semiovale			
Association fibers of the telencephalon	Interhemispheric association fibers (horizontal anterior-posterior course in the arcuate fibers of the telencephalon)	Short association fibers (run close to the cortex, associating adjacent gyri) Long association fibers (joining in bundles connecting cortical territories mainly from different lobes)	Uncinate fasciculus (from frontal lobe to temporal lobe) Cingulum (forms the white matter of the limbic lobe) Superior or arcuate and inferior longitudinal fasciculi (from the frontal lobe to the occipital and temporal lobes) Superior or subcallosal and inferior occipitofrontal fasciculi (from frontal lobe to the temporal and occipital lobes) Vertical occipital fasciculi (lateral and caudal fibers from the superior to the inferior margin of the occipital lobe) and horizontal occipital fasciculi (cuneus and lingual fibers)
Projection fibers (in the corona radiata, in the internal capsule and the external capsule)	Corticofugal fibers	Corticothalamic fibers Mesencephalic corticonuclear fibers Pontine corticonuclear fibers Bulbar corticonuclear fibers Corticoreticular fibers Corticorubral fibers Corticotectal fibers Corticospinal fibers Frontopontine fibers Parietopontine fibers Thalamoparietal fibers Temporopontine fibers Occipitopontine fibers Occipitotectal fibers Corticotegmental fibers	
	Corticopetal fibers	Optic radiation (geniculocalcarine fibers) Acoustic radiation (geniculotemporal fibers) Thalamic radiations (anterior, central and posterior)	

Organization of the interhemispheric formations containing the commissural fibers of the telencephalon (homotopic fibers between areas corresponding to the two hemispheres and heterotopic fibers between different areas) and projection fibers			
Anterior commissure			
– anterior part (or olfactory part)	Connects the anterior olfactory nucleus of one side to the contralateral anterior olfactory nucleus and the olfactory bulb, controlling, by reflex pathway, the activity of the olfactory bulb through a comparable system to that of the gamma circuit of the spinal cord		
– posterior part	Connects the middle and inferior temporal gyri of both sides		
Corpus callosum			
– rostrum	Foremost part		
– genus	Curved anterior part at the bottom that straightens out in the rostrum		
– trunk	Free part (median, visibly opening up the interhemispheric fissure) and lateral parts (enter into the centrum semiovale creating the radiations of the corpus callosum, containing commissural fibers taking both horizontal and transverse courses	Transverse striae, medial (or of Lancisi) and lateral longitudinal striae (or teniae tectae) and indusium griseum	Formations of the dorsal hippocampus on the superior face of the trunk of the corpus callosum (the inferior face relates to the septum pellucidum anteriorly and to the fornix posteriorly)
– splenium	Posterior part	Fasciolar gyrus or fasciola cinerea (thin ribbon on both sides of the splenium)	Formation of the dorsal hippocampus

continued

Organization of the interhemispheric formations containing the commissural fibers of the telencephalon (homotopic fibers between areas corresponding to the two hemispheres and heterotopic fibers between different areas) and projection fibers *(continued)*		
Fornix (commissural organ of the limbic system)		
– body (central part)		Fimbria-fornix system (connections of the hippocampus with the septal nuclei, or septohippocampal pathway, the mammillary bodies and the thalamus)
– columns (funiculi of white matter where the body continues anteriorly)	Precommissural fibers	From the hippocampus to the septal area, to the orbital cortex and to the cingulate gyrus
	Postcommissural fibers	From the subiculum of the hippocampus to the preoptic area, to the mammillary bodies and to the anterior nuclei of the thalamus
– crus (funiculi of white matter where the body continues posteriorly, that continue in the alveus and in the fimbria of the hippocampus up to the uncus of the parahippocampal gyrus)	Commissure of the fornix (hippocampal commissure or psalterium) between the crura of the fornix	Temporoammonic pathway: fibers from the entorhinal cortex (area 28 of Brodmann) of one side to the Ammon's horn of the contralateral hippocampus
Septum pellucidum (formation of white matter) and **septal nuclei** (formations of gray matter)	Fibers (peduncle of septum pellucidum connecting to the tenia thalami and the thalamus) to the anterior perforated substance and to the medial and lateral habenular nuclei through the stria medullaris of the thalamus.	

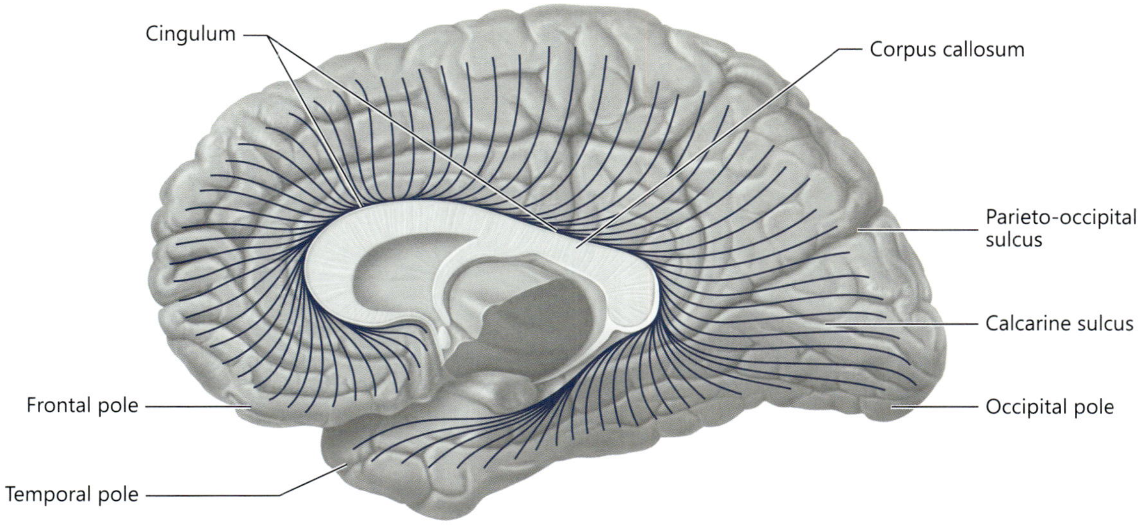

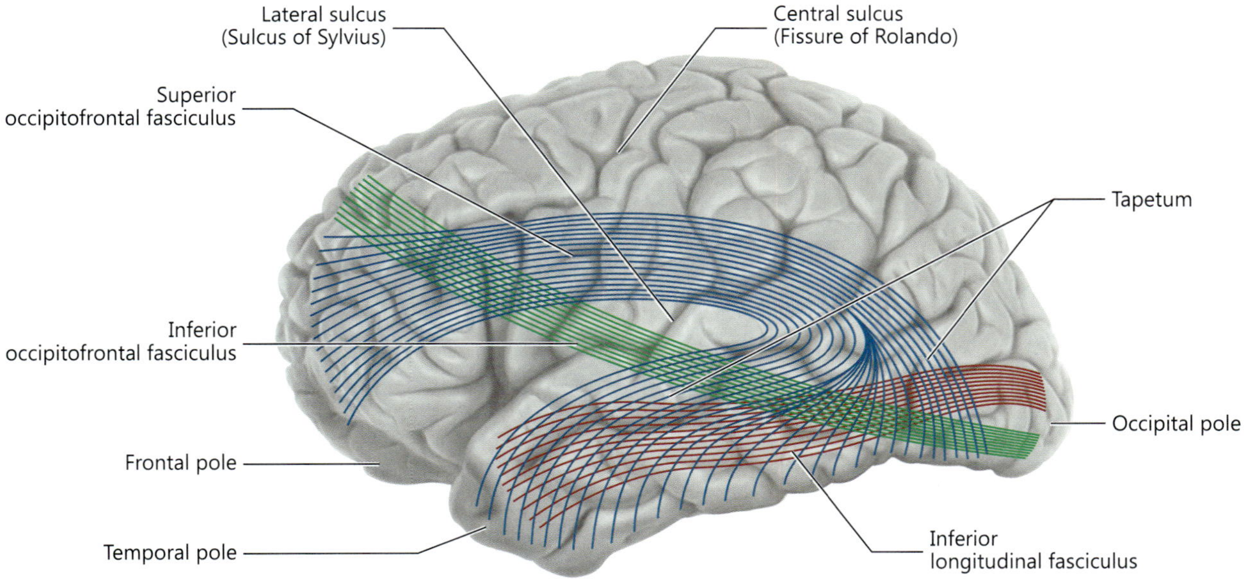

Schematic representation illustrating the organization of the **association fibers of the telencephalon**.

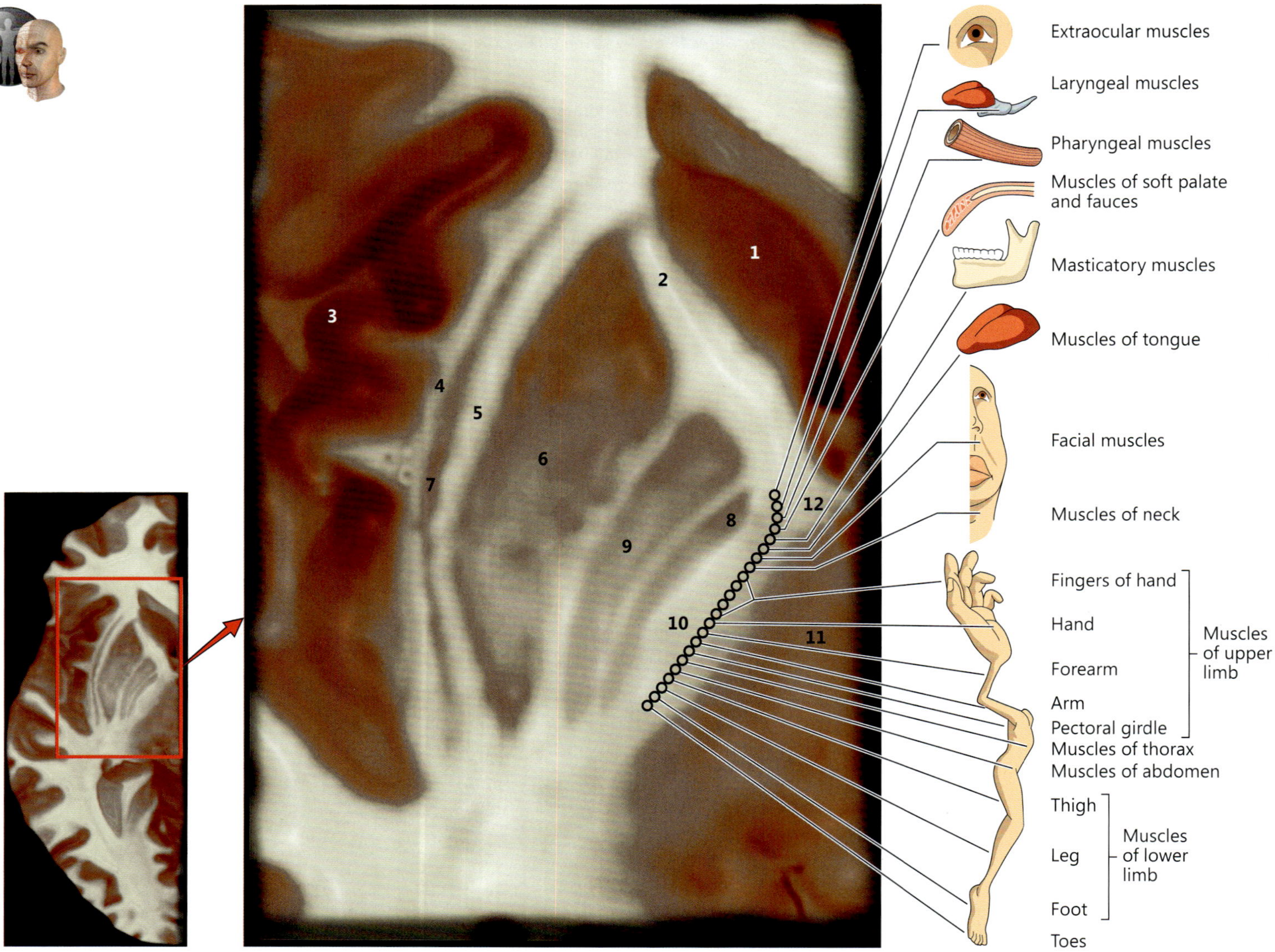

Schematic diagram showing the anatomic organization of the **system of the capsules**, in particular the internal capsule. Note the corticospinal tract passing through the genu and the anterior three-quarters of the posterior limb of the internal capsule with a somatotopic structure.

1, Head of caudate nucleus
2, Anterior limb of internal capsule
3, Insula
4, Extreme capsule
5, External capsule
6, Putamen
7, Claustrum
8, Globus pallidus medial segment
9, Globus pallidus lateral segment
10, Posterior limb of internal capsule
11, Thalamus
12, Genu of internal capsule

General structure of the capsules

CAPSULE	*LOCATION*	*FIBER*	*ORIGIN*	*PROJECTION*
External capsule	Between the claustrum and the convex lateral surface of the putamen	Corticotegmental fibers	Cerebral cortex (frontal, temporal, anterior parietal and insular)	Tegmentum of midbrain (passing beneath the lenticular nucleus and sending fibers to the putamen)
Extreme capsule	Between the insular cortex and the claustrum	It especially contains association fibers between the cortex of the parietal, temporal and insular lobes		
Internal capsule (anterior limb, genu, posterior limb)	Between the thalamus and the caudate nucleus, medially, and the lentiform nucleus, laterally	Corticopetal fibers (mainly fibers of the thalamic radiation) Corticofugal fibers (in the ventral part of the internal capsule)	Thalamus (for the most part) Cerebral cortex (except the thalamoparietal fibers)	Cerebral cortex Brainstem and spinal cord

Constitution of the internal capsule				
PARTS	***LOCATION***	***FIBERS***	***ORIGIN***	***PROJECTION***
Anterior limb	Between the head of the caudate nucleus and the lentiform nucleus, laterally	Anterior thalamic radiation	Anterior and medial dorsal nuclei of thalamus Prefrontal cortex and cingulate gyrus	Prefrontal cortex and cingulate gyrus Anterior and medial dorsal nuclei of thalamus
		Frontopontine fibers	Prefrontal cortex	Pontine nuclei
Posterior limb	Anteriorly related to the thalamus and laterally to the lentiform nucleus, behind which it extends in a caudal direction	Corticospinal fibers	Motor area	Spinal cord
		Corticorubral fibers	Premotor area	Red nucleus (ipsilaterally)
		Central thalamic radiation	Ventral anterior, ventral lateral, ventral posterolateral and ventral posteromedial nuclei of the thalamus	Brodmann areas 4 and 6
		Corticoreticular fibers	Precentral cortex (mainly)	Reticular formation of the mesencephalon and of the pons and from those to the cerebellum, to the nuclei of the cranial nerves and to the spinal cord
		Corticothalamic fibers	Cerebral cortex	Thalamus
		Thalamoparietal fiber	Thalamus	Parietal cortex
		Parietopontine fibers	Parietal cortex	Pontine nuclei
– sublenticular part		Corticothalamic fibers	Cerebral cortex	Thalamus
		Acoustic radiation (geniculotemporal fibers)	Medial geniculate bodies	Acoustic areas of the superior temporal gyrus
		Temporopontine fibers	Temporal cortex	Pontine nuclei
		Corticotectal fibers	Cerebral cortex	Tectal plate
– retrolenticular part		Optic radiation (genicolocalcarine fibers)	Lateral geniculate bodies	Visual areas of the occipital lobe
		Occipitopontine fibers	Occipital cortex	Pontine nuclei
		Occipitotectal fibers	Occipital cortex	Tectal plate
		Optic radiation (genicolocalcarine fibers)	Lateral geniculate bodies	Visual areas of the occipital lobe
		Posterior thalamic radiation	Pulvinar	Parietal-temporal-occipital association areas
Genu	Between the anterior and posterior limbs of the internal capsule	Corticonuclear fibers:		
		– to the oculomotor nuclei of the cranial nerves	Brodmann area 8 (center of the convergence)	Nuclei of the oculomotor (bilaterally), trochlear (homolaterally) and abducens (contralaterally) nerves
		– to the somatomotor nuclei of the other cranial nerves	Brodmann area 4 in the prefrontal cortex	Somatomotor nuclei of the trigeminal, facial, glossopharyngeal, vagus and accessory nerves (bilaterally) and of the hypoglossal nerve (contralaterally)

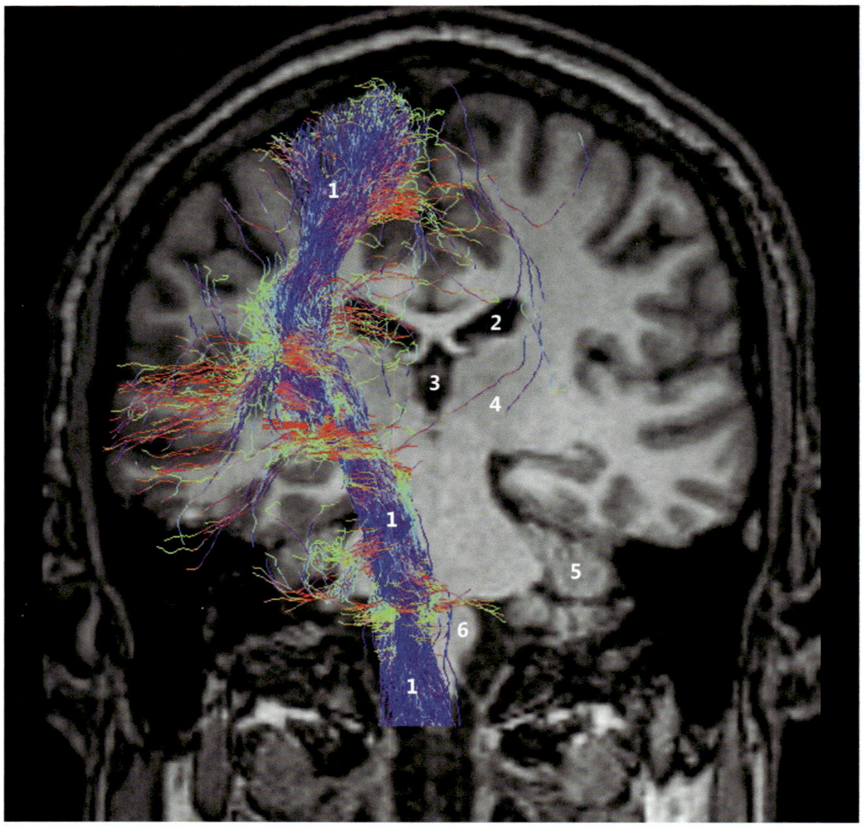

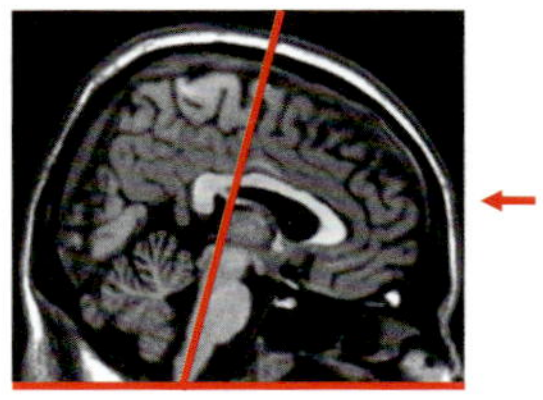

1, Pyramidal tract
2, Lateral ventricle
3, Third ventricle
4, Internal capsule
5, Cerebellum
6, Myelencephalon (Medulla oblongata or bulb)

Three-dimensional image of a living human body, obtained by diffusion tensor nuclear magnetic resonance (DTNMR) using a tractography technique. The **projections of white matter** to the corticospinal tract are highlighted with colours tracking the spatial orientation of the fibers. In *green* fibers following the anteroposterior direction, in *blue* the superoinferior direction, and in *red* the laterolateral direction.

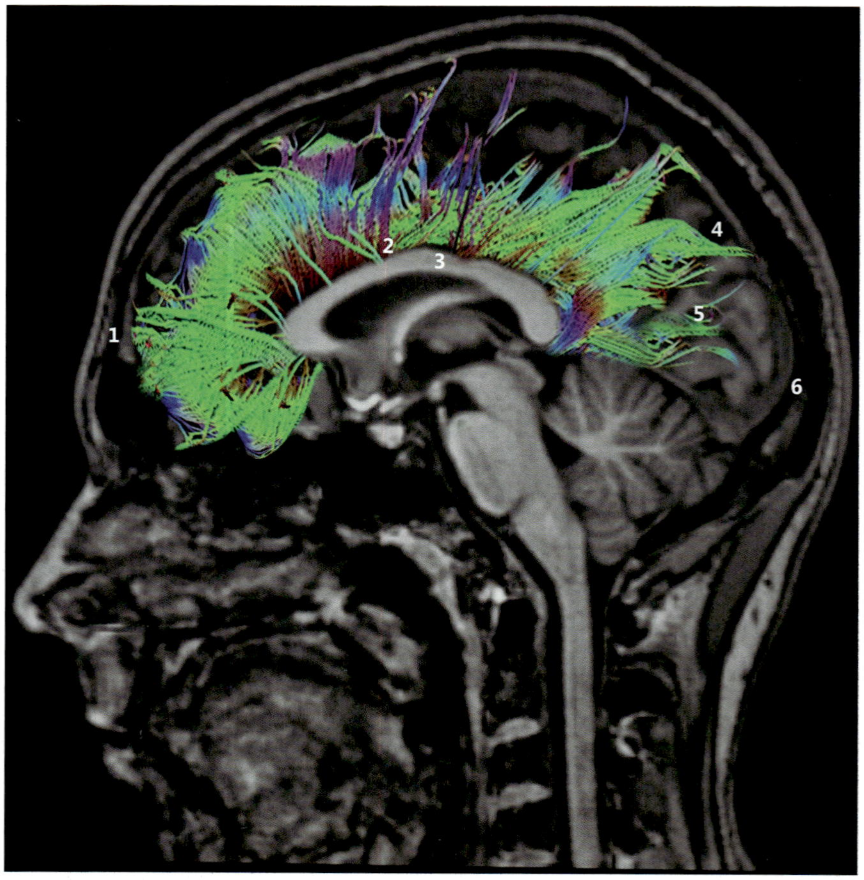

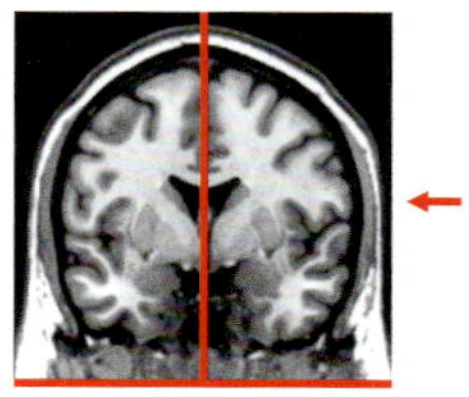

1, Frontal pole
2, Cingulum
3, Corpus callosum
4, Parieto-occipital sulcus
5, Calcarine sulcus
6, Occipital pole

Tractography image obtained by DTNMR scan of a living human body, showing the **association bundles of the cingulum**.

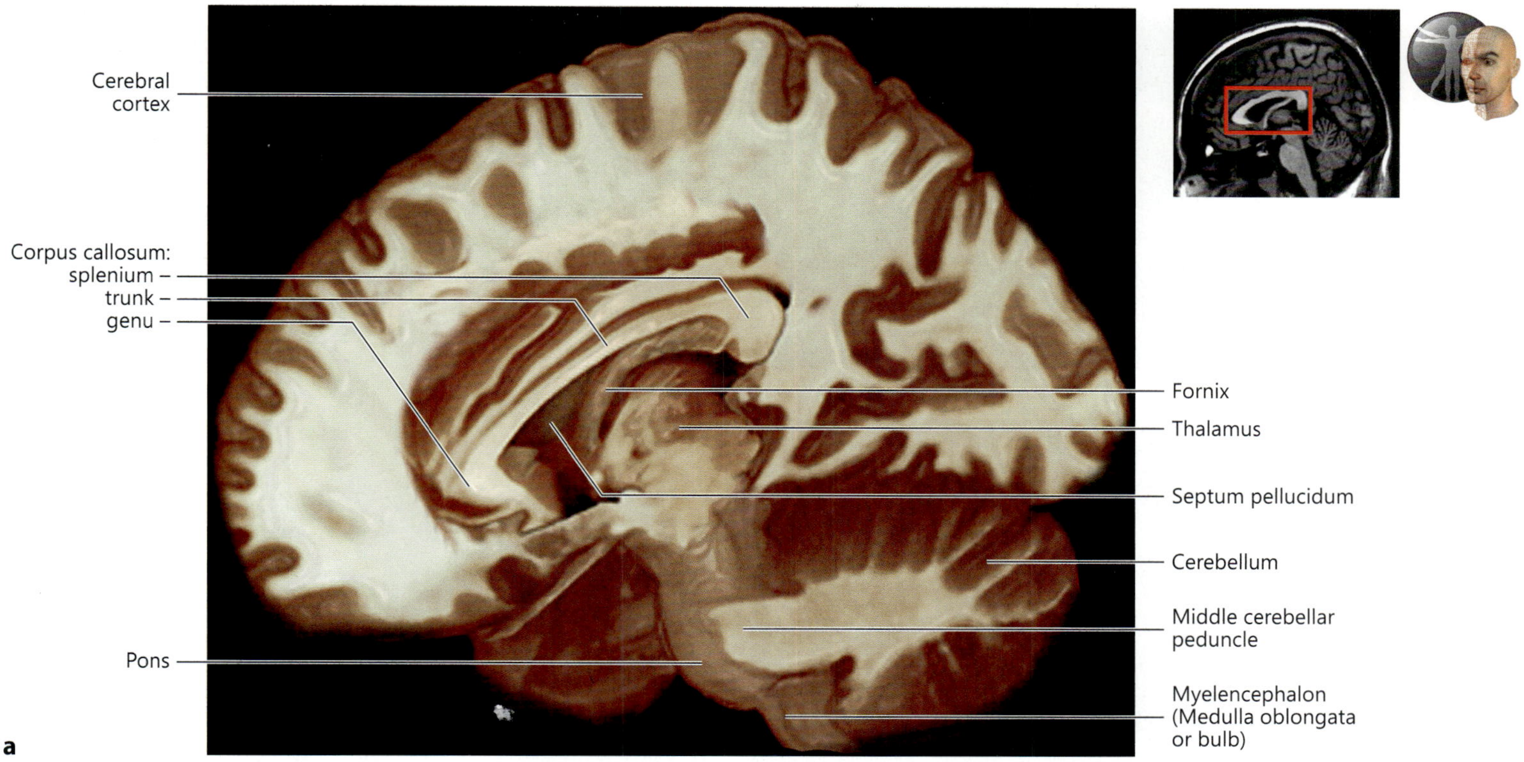

a

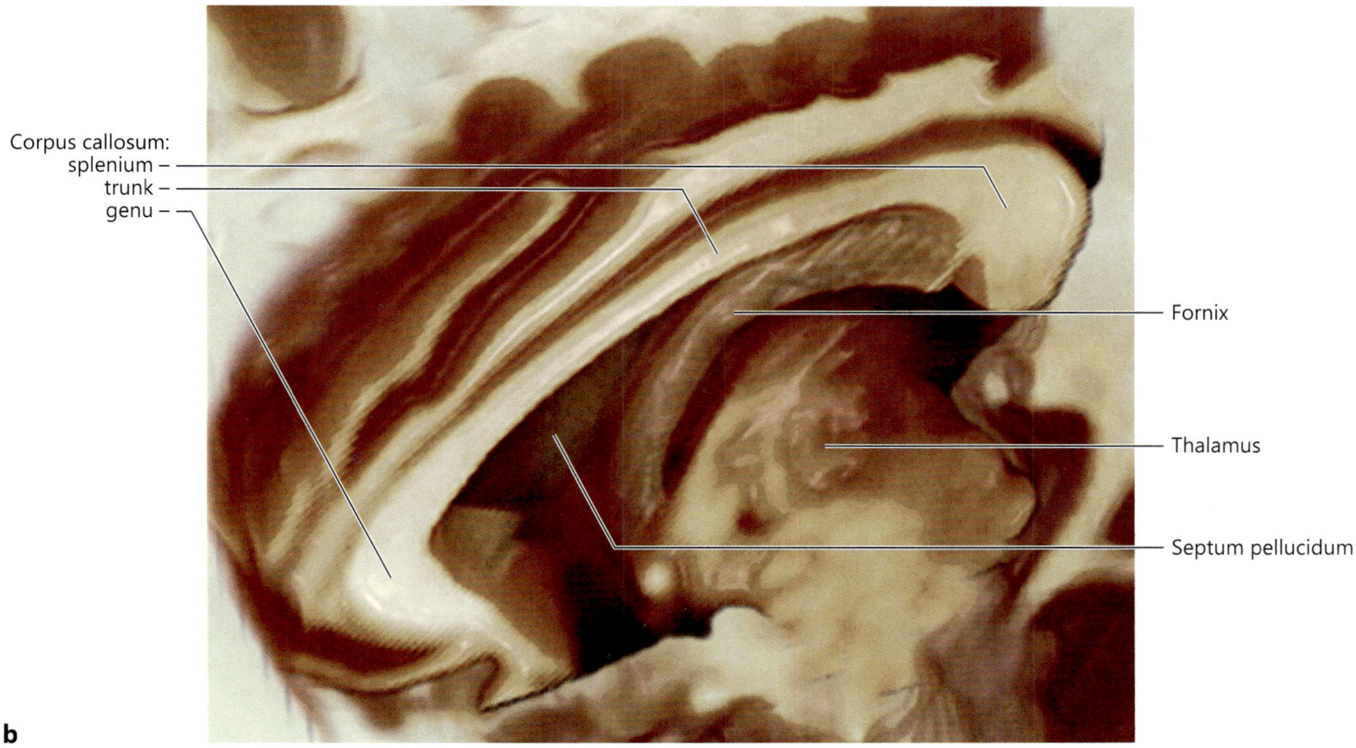

b

Three-dimensional images of a living human body, obtained by NMR using a 3D volume rendering technique, showing the **corpus callosum** in a median sagittal section of the brain (**a**) and in detail (**b**).

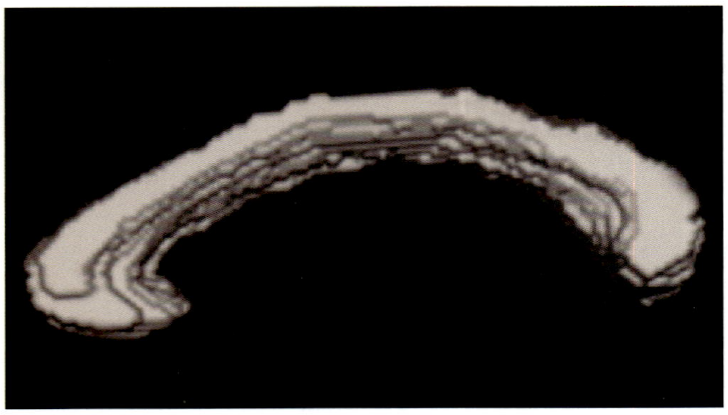

a

b

Three-dimensional images of a living human body, obtained by NMR using a 3D volume rendering technique, showing the segmentation of the **corpus callosum** (**a**) and chromatically subdivided into seven segments according to Witelson (Witelson SF., Brain, 112: 799-835, 1989) (**b**).

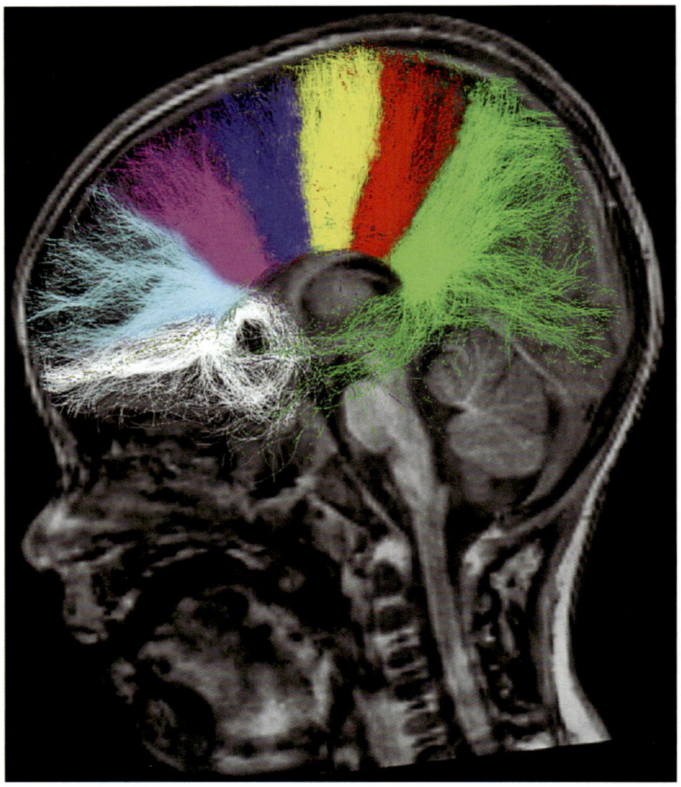

Three-dimensional image of a living human body, obtained by DTNMR using a tractography technique, showing the afferences and efferences of the **corpus callosum** chromatically highlighted according to the seven segments subdivision by Witelson (Witelson SF., Brain, 112: 799-835, 1989).

Constitution of the corpus callosum

PARTS	*RADIATION OF CORPUS CALLOSUM*	*ORIGIN*	*TERMINAL*	*FUNCTION*
Genu and rostrum	Minor forceps	Large pyramidal cells of the cortex of the anterior and orbital parts of the frontal lobe on one side	Cortex of anterior and orbital parts of the contralateral frontal lobe	Important, even if not yet well specified, in the functioning of the brain. It's lesions determine mental disturbances, incoordination of ideas, alterations of character and specifically alterations in the execution of movements (motor apraxia)
Trunk	Tapetum	Large pyramidal cells of one side of the cortex, at the level of:	Contralateral cortex, at the level of:	
	– external superior fibers	– posterior parts of the frontal and temporal lobes and the entire parietal lobe	– posterior parts of the frontal and temporal lobes and the entire parietal lobe	
	– internal superior fibers	– medial frontal gyrus, paracentral lobule of the parietal lobe, superior parietal lobule, cingulate gyrus and superior marginal parts of the precentral and postcentral gyri	– medial frontal gyrus, paracentral lobule of the parietal lobe, superior parietal lobule, cingulate gyrus and superior marginal parts of the precentral and postcentral gyri	
	– middle fibers	– middle frontal gyrus, middle parts of the precentral and postcentral gyri and inferior parietal lobule	– middle frontal gyrus, middle parts of the precentral and postcentral gyri and inferior parietal lobule	
	– inferior fibers	– parietal operculum and, in part, temporal lobe	– parietal operculum and, in part, temporal lobe	
Splenium	Major forceps	Large pyramidal cells of the middle temporal gyrus, of the cingulate gyrus and of the occipital and temporal cortex of one side	Middle temporal gyrus, cingulate gyrus and contralateral occipital and temporal cortex	

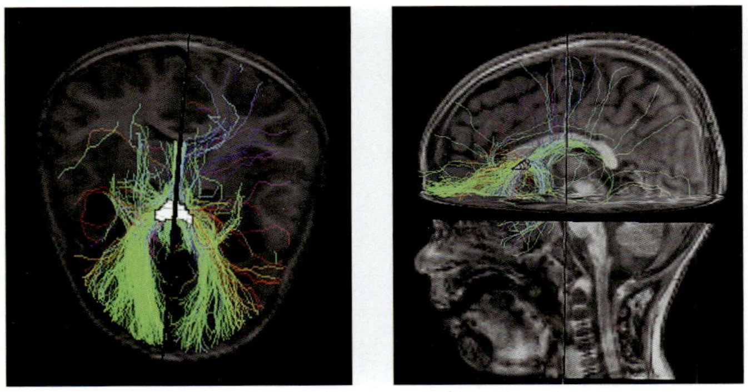

a Rostrum b

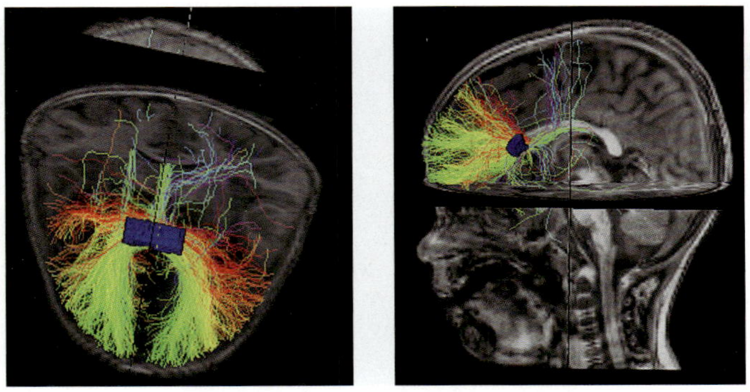

a Genu b

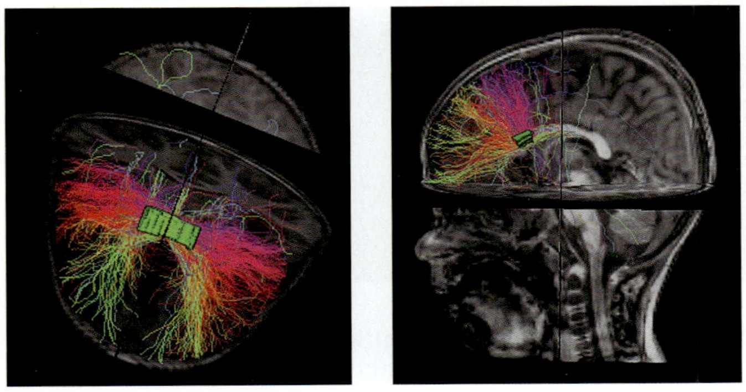

a Rostral trunk b

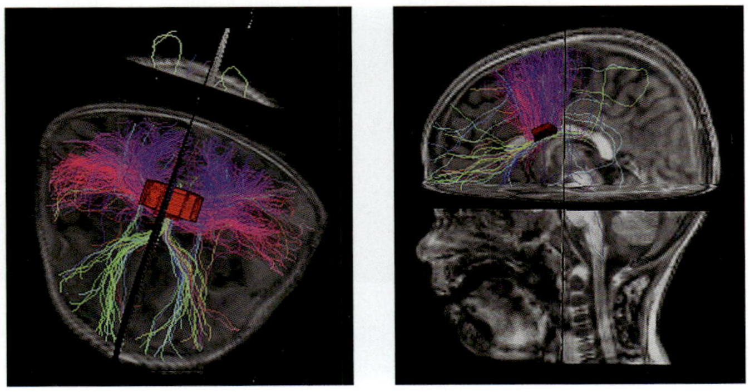

a Anterior half of trunk b

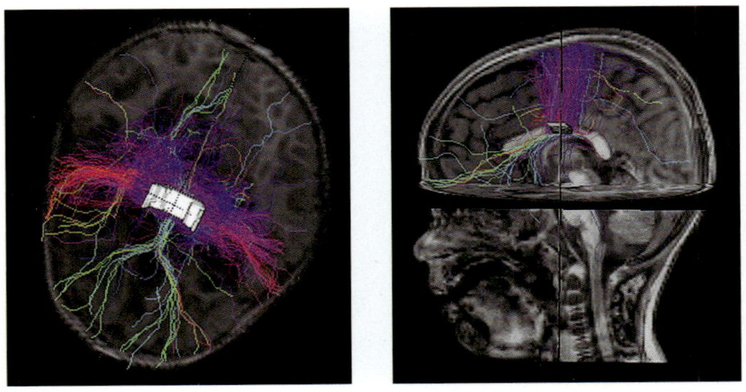

a Posterior half of trunk b

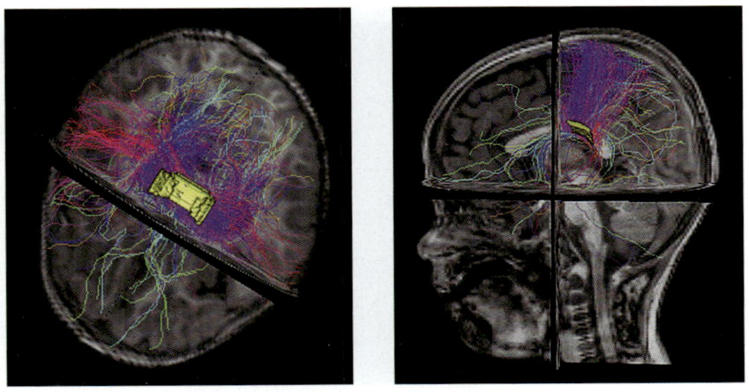

a Isthmus b

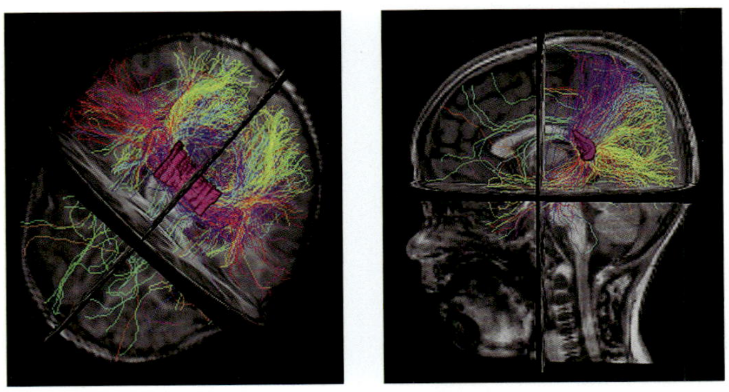

a Splenium b

The **transcallosal connections** of the seven segments of the corpus callosum (Witelson SF., Brain, 1989), are highlighted using a tractography technique; as seen from above (**a**) and medially (**b**).

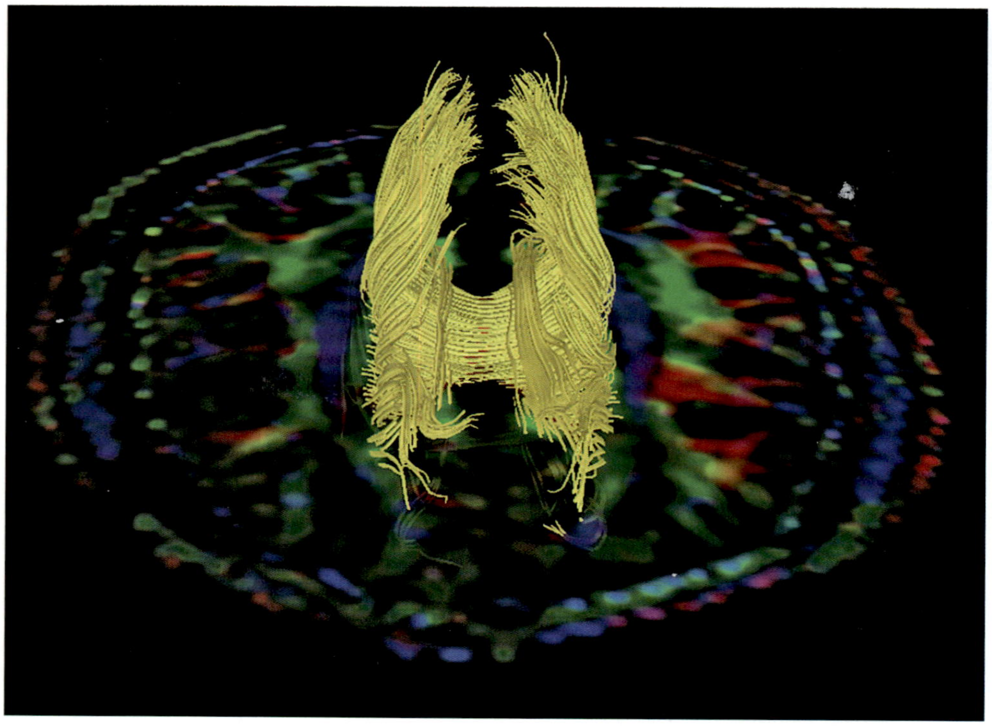

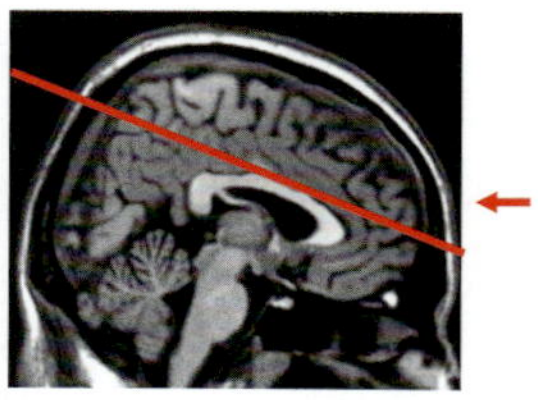

Three-dimensional image of a living human body, obtained by DTNMR using a tractography technique. **Commissural fibers of the corpus callosum** in anterosuperior projection; the transverse section shows the callosal tracts emergence, and the anisotropic map.

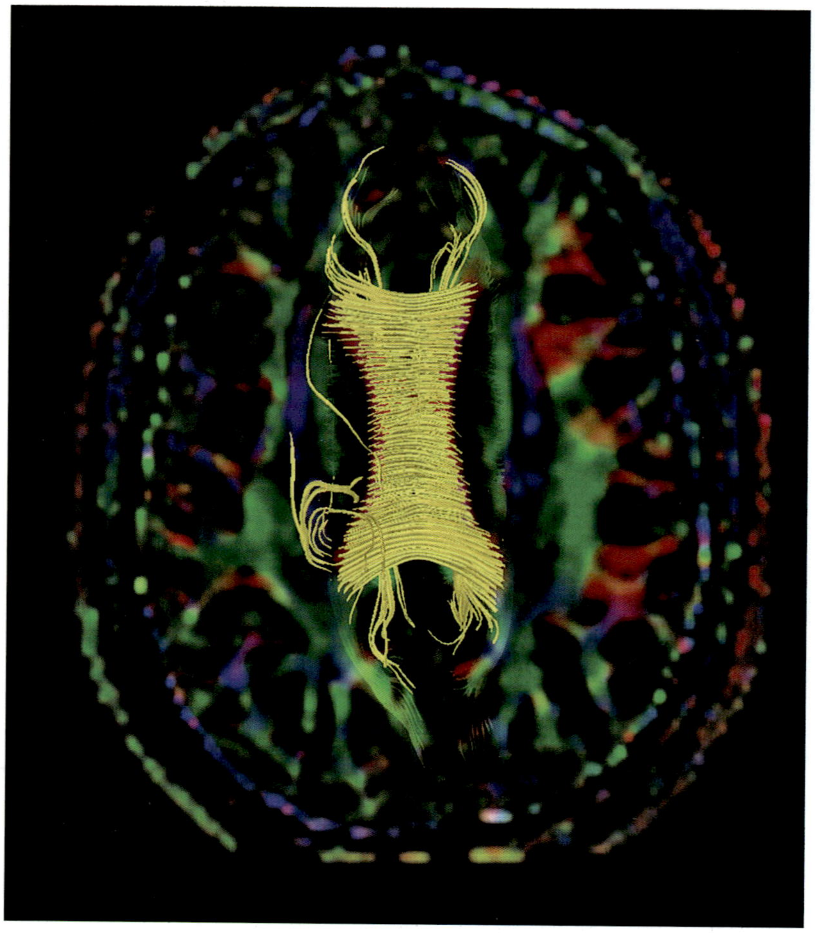

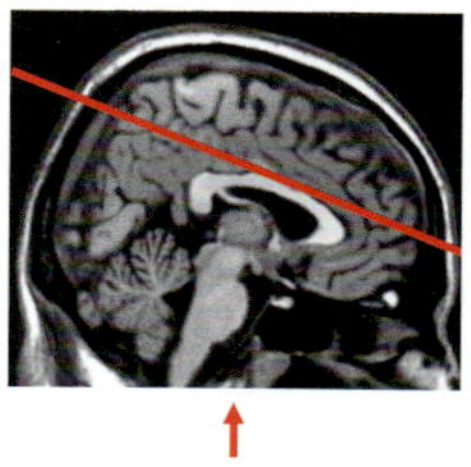

Three-dimensional image of a living humam body, obtained by DTNMR using a tractography technique. **Commissural fibers of the corpus callosum** seen from below; the transverse section shows the callosal tracts emergence, and the anisotropic map.

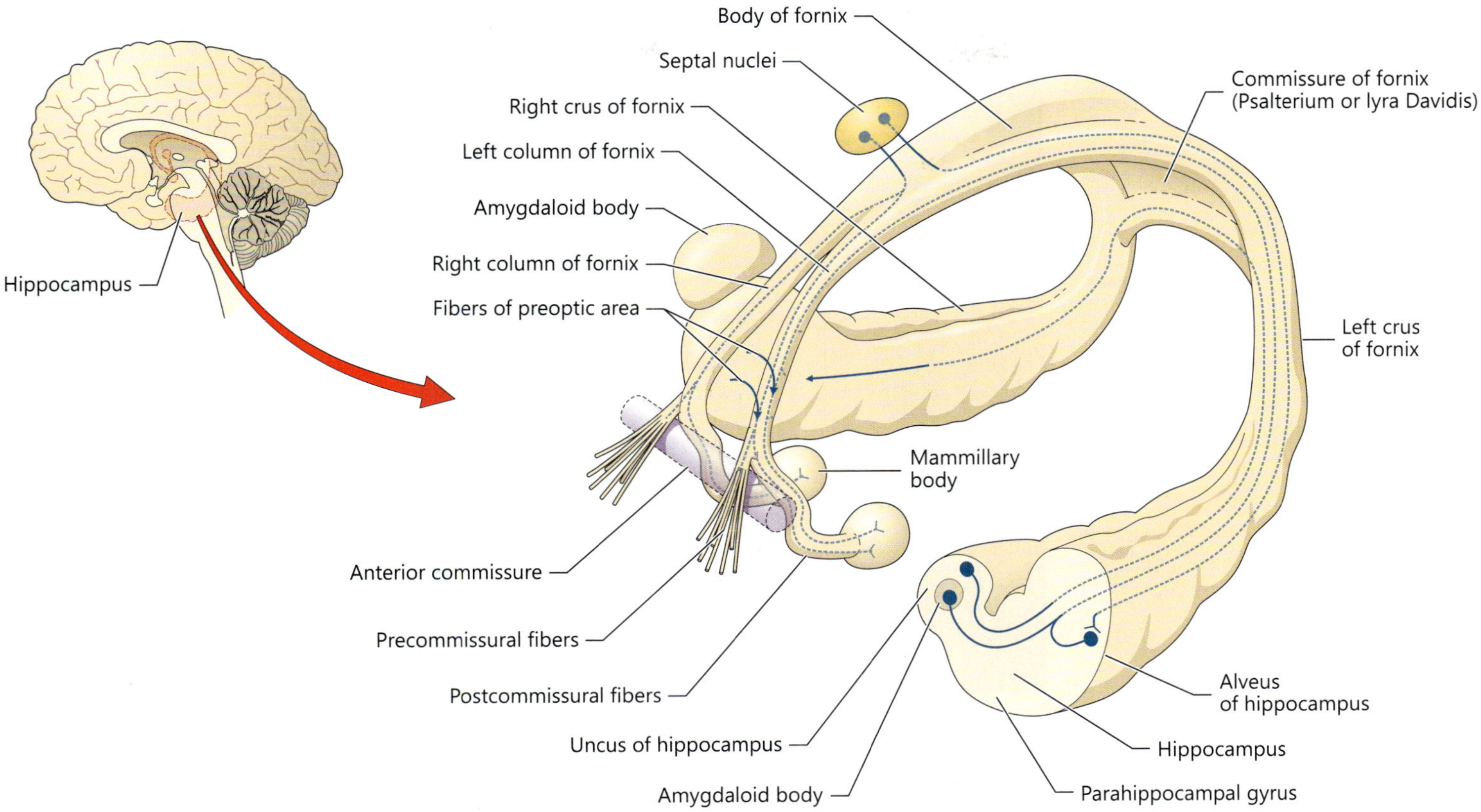

Scheme of the principal fiber systems running in the **fornix**. The commissural and projection fibers are represented. These fibers connect the hippocampus, respectively, with the septal nuclei (precommissural fornix) and with the preoptic area and mammillary bodies (postcommissural fornix).

Structure of the fornix

FIBERS	*ORIGIN*	*COURSE*	*TERMINAL*
Transverse (commissural)	Pyramidal neurons of the cortex of the hippocampus, of one side	Pass contralaterally through the psalterium (between the crura of the fornix), forming a commissural system between the two hippocampi, called the commissure of the fornix (or of the hippocampus or of the Ammon's horn)	Cortex of the contralateral hippocampus
Longitudinal (ipsilateral)	Pyramidal neurons of the cortex of the hippocampus and, in part, of the parahippocampal gyrus	Surround the thalamus	Homolateral mammillary body
Olfactory fasciculus of the hippocampus (described by Zuckerkandl)	Cortex of the hippocampus	Detaches itself from the columns of the fornix and passes under the septum pellucidum, and in front of the anterior commissure, arriving in the region of the anterior perforated substance, and continues in the diagonal band	Parahippocampal gyrus

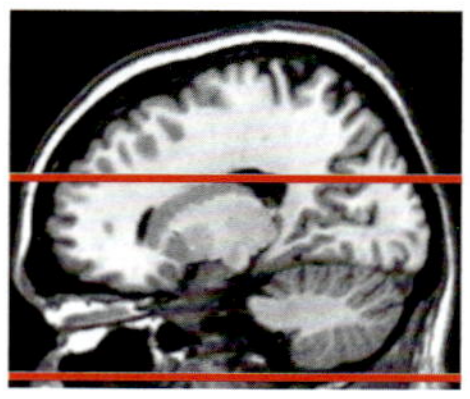

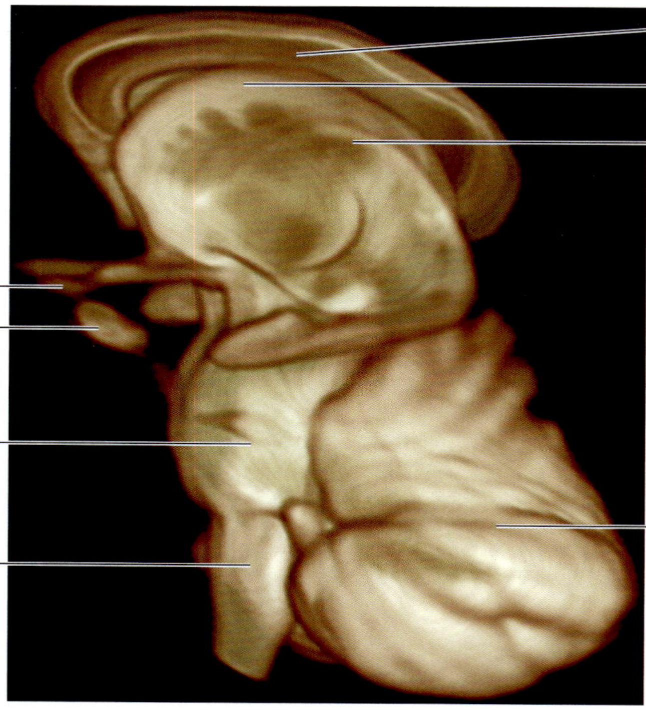

a

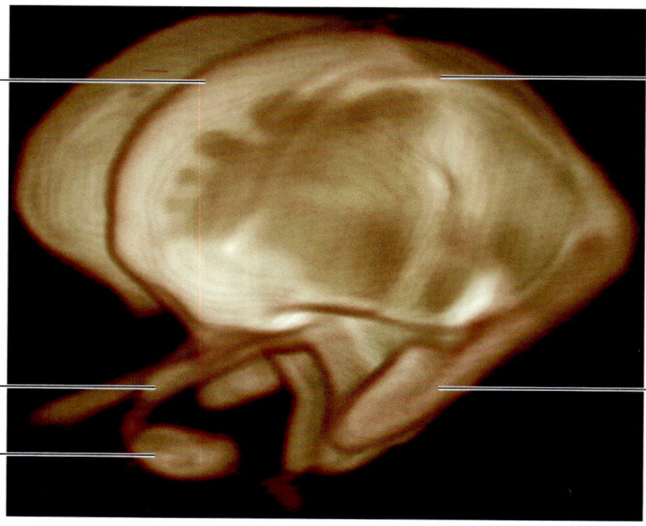

b

Three-dimensional images of a living human body, obtained by NMR using a 3D volume rendering technique, showing the **fornix** after applying the intermediate transparency and opacity parameters to the diencephalon (**a**); detail of the previous image (**b**).

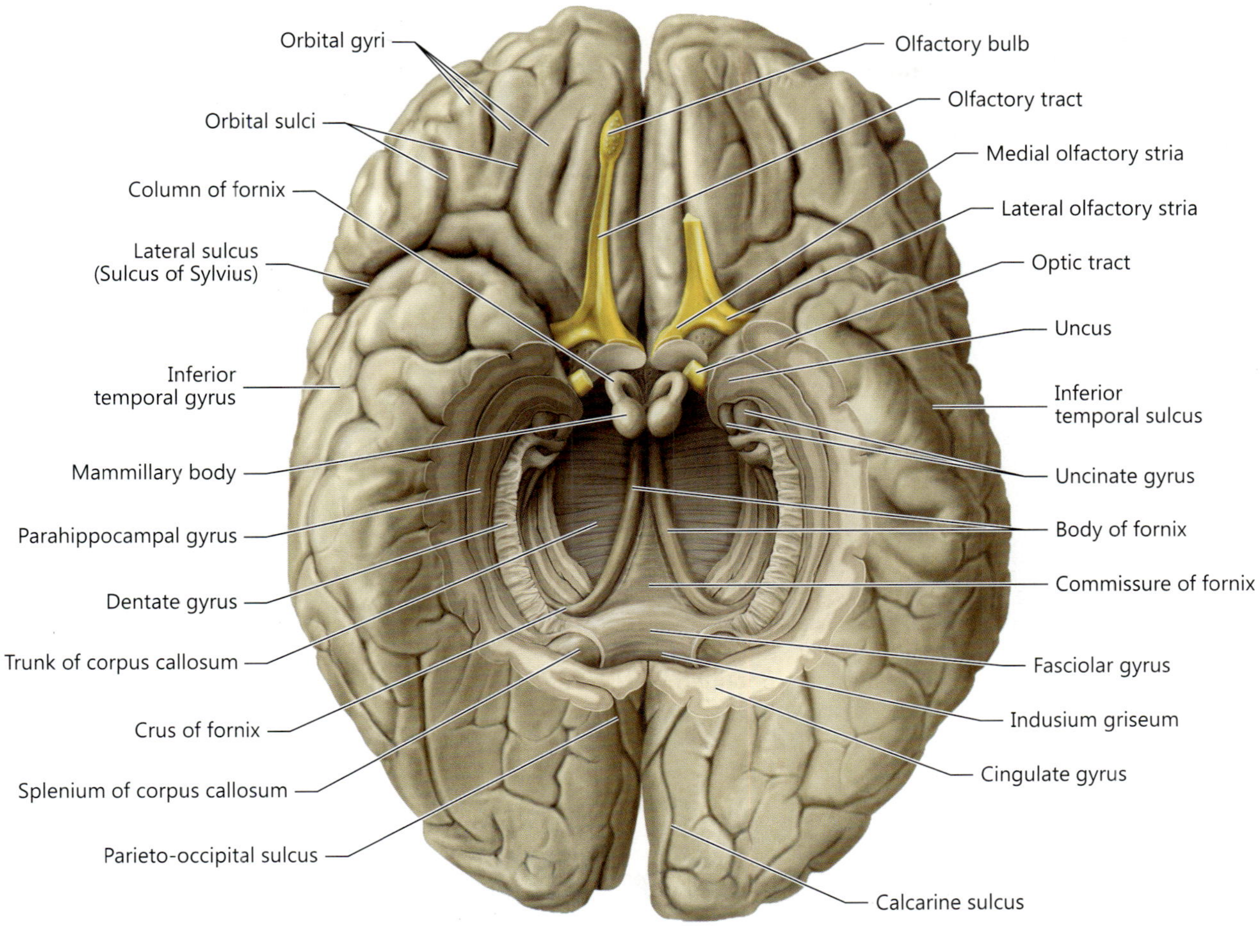

Brain, seen from below, partial removal of the inferior surface highlights the **hippocampus**, the **fornix** and the high structures of the limbic system.

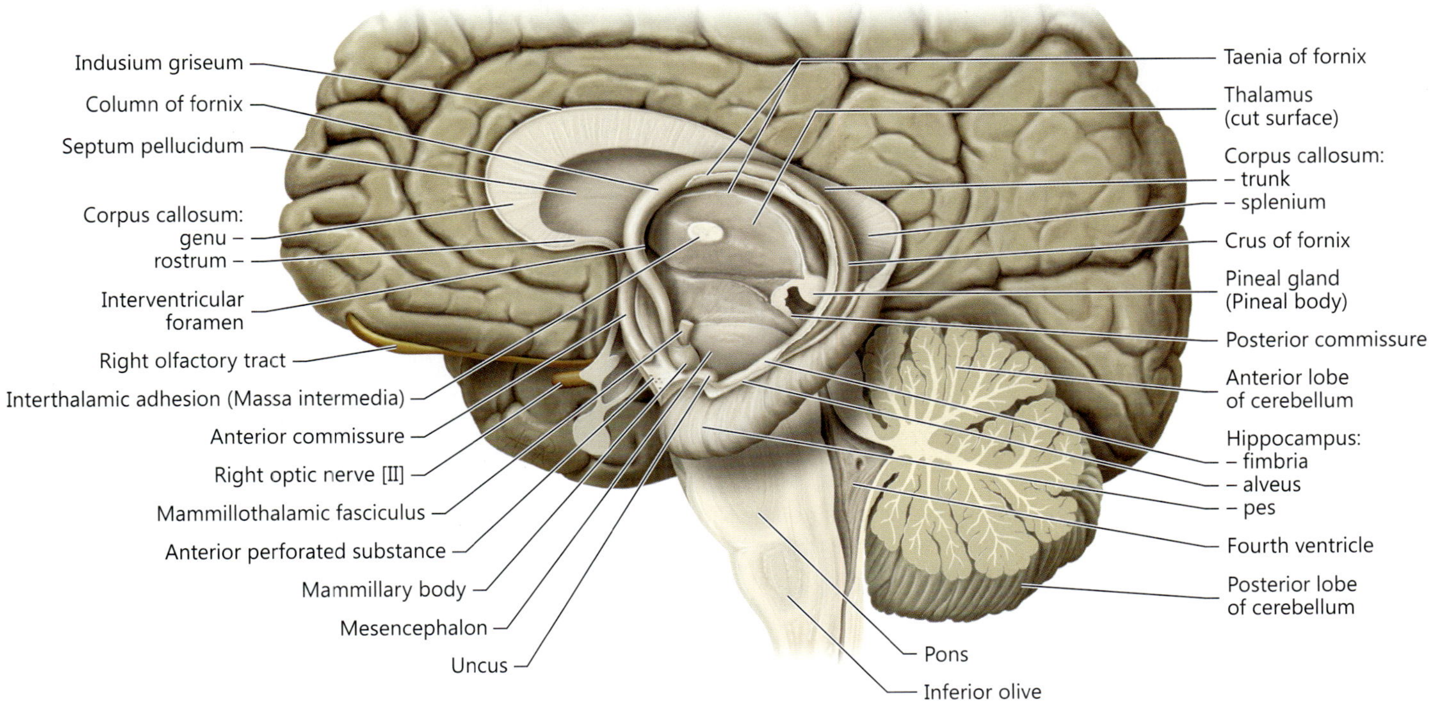

Medial face of the right hemisphere showing the **structures of the limbic system.**

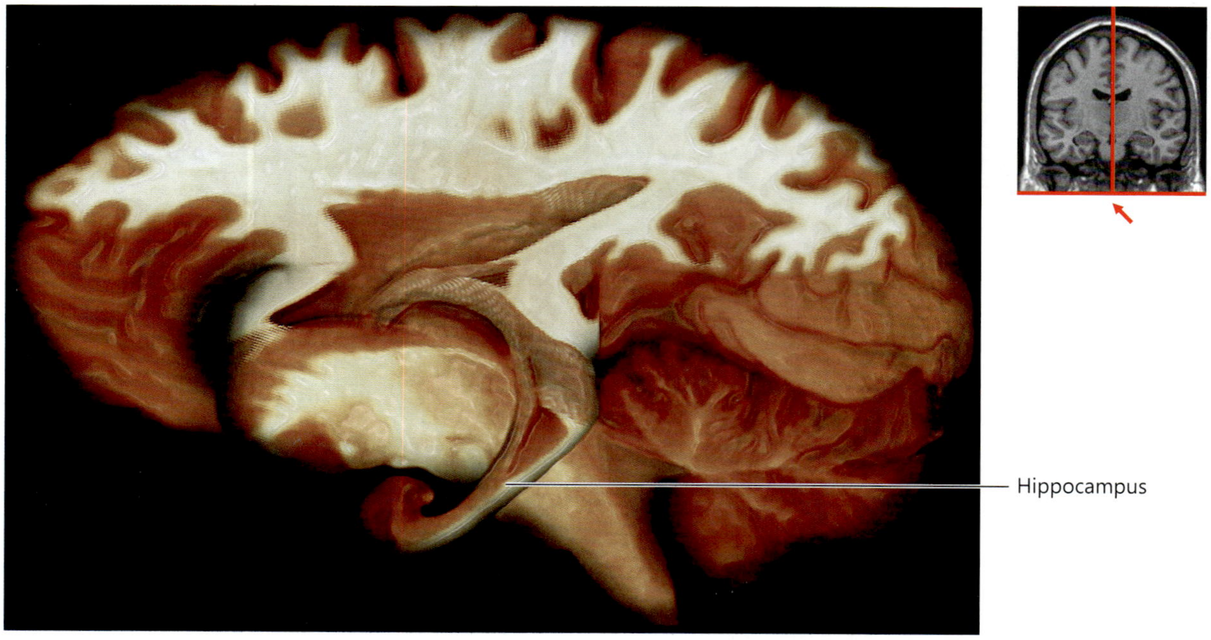

Three-dimensional image of a living human body, obtained by NMR using a 3D volume rendering technique. Transverse and sagittal section of the brain highlights the **hippocampus**.

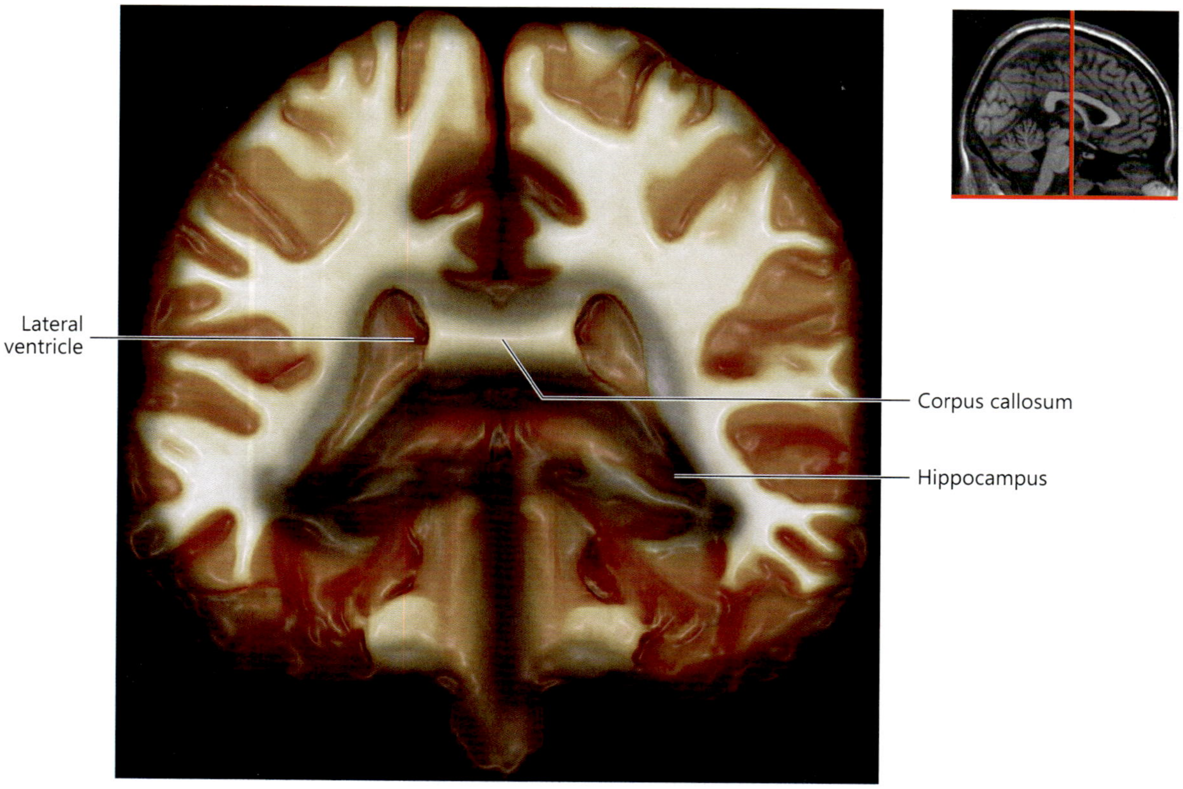

Three-dimensional image of a living human body, obtained by NMR using a 3D volume rendering technique. A coronal section has been carried out to expose the corpus callosum and the temporal horns of the lateral ventricles; in the latter, the **hippocampus** is evident in the posteroinferior wall.

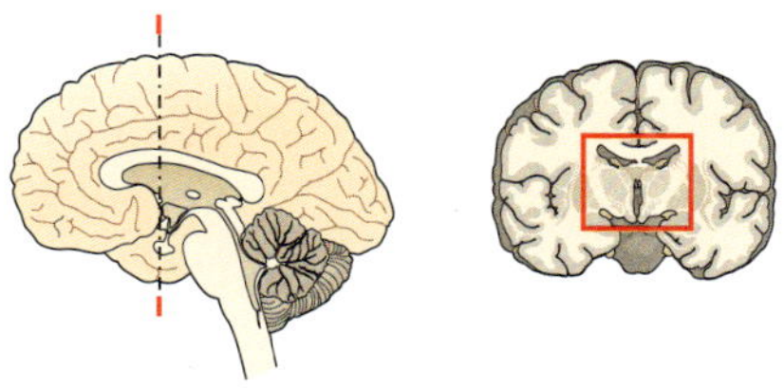

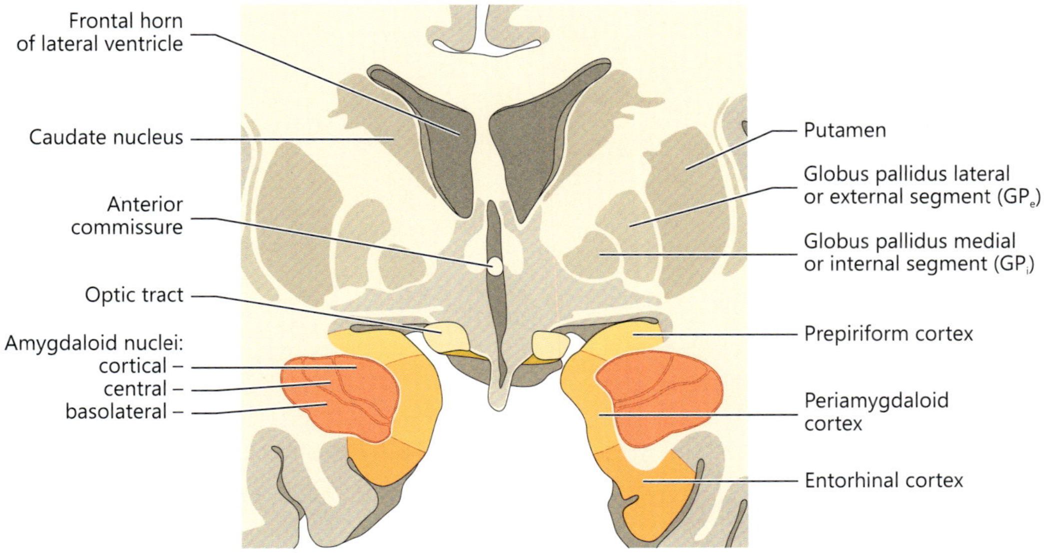

Amygdaloid body or **complex**. The amygdaloid body is seen as a voluminous aggregate of subcortical nuclei maintaining a strict relationship with the cortical structures.

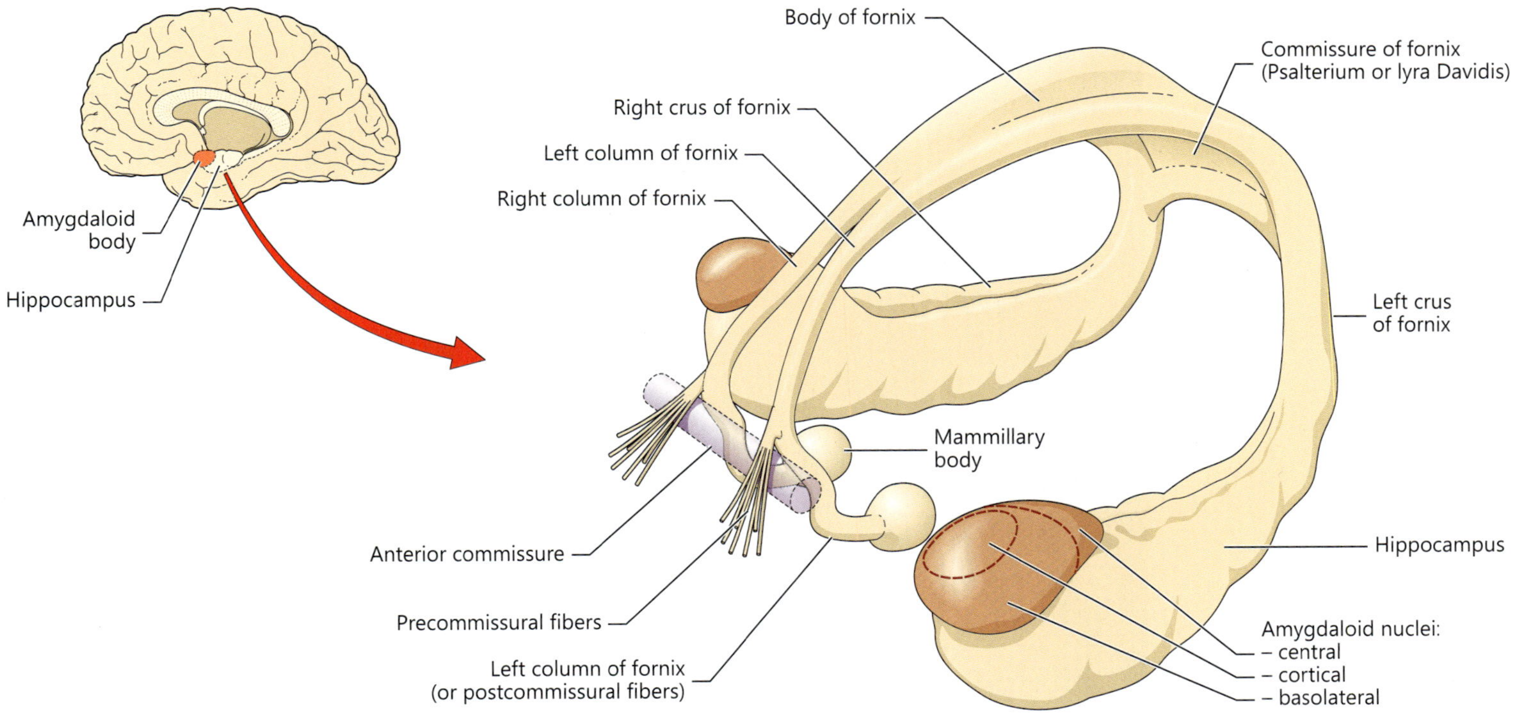

The **amygdaloid body** is formed by a group of nuclei found in the limbic lobe under the uncus of the parahippocampal gyrus. The cortical nucleus is situated near the cortex of the uncus and receives olfactory information from the lateral olfactory stria and, indirectly, from the primary olfactory cortex. The basolateral nucleus, continues in part with the cortex of the parahippocampal gyrus, and receives olfactory information directly from the primary olfactory cortex. The central nucleus projects to the lateral hypothalamic area, to the periaqueductal gray matter of the mesencephalon and to the bed nucleus of stria terminalis.

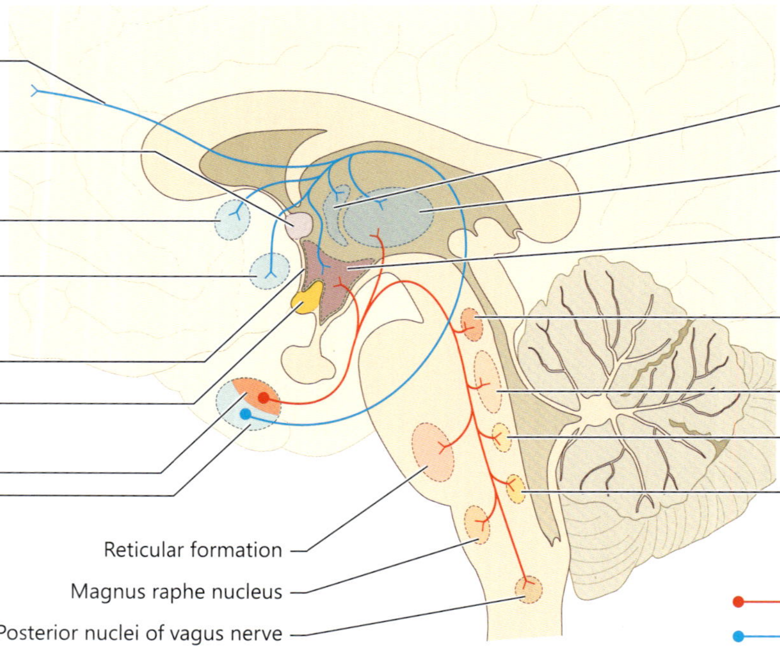

Schematic representation of the **amygdalofugal pathway** and of the **stria terminalis**.

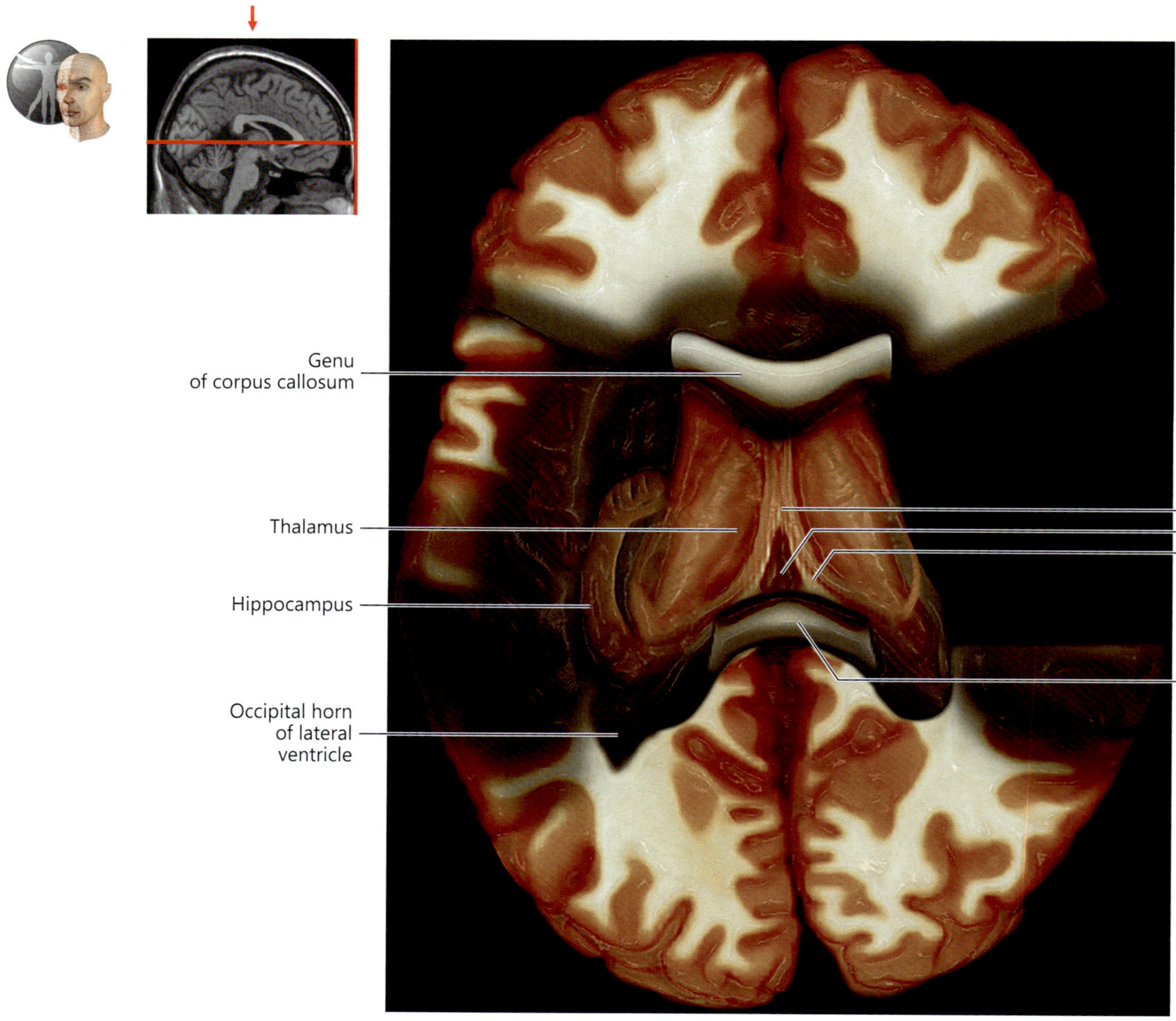

Three-dimensional image of a living human body, obtained by NMR using a 3D volume rendering technique. Parts of the brain have been removed to show **corpus callosum**, **thalamus** and **hippocampus**.

Limbic system				
COMPONENTS				***FUNCTION***
Telencephalic formations	Gray	Limbic lobe	Cingulate gyrus Fasciolar gyrus Parahippocampal gyrus Dentate gyrus Indusium griseum	Integration of the visceral, olfactory and somatic stimuli; fundamental in determining some behavioral characteristics (homeostatic responses). Memorization processes (short-term memory)
		Hippocampus	Hippocampus proper, or Ammon's horn Region I or CA1 Region II or CA2 Region III or CA3 Region IV or CA4 Digitations Pes Parasubiculum Presubiculum Subiculum	
		Amygdaloid body or complex	Amygdaloid nuclei and areas, septal nuclei, basal substance, claustrum, nucleus accumbens, nucleus of lateral olfactory tract, periamygdaloid cortex, anterior olfactory nucleus, diagonal band, innominate substance, anterior perforated substance, olfactory islets, olfactory bulb, olfactory trigone, olfactory tubercle	
		Basal nuclei	Ventral striatum, ventral pallidum	
	White	Fornix Anterior commissure Alveus of the hippocampus Fimbria of the hippocampus Ansa peduncularis Fasciculus peduncularis Olfactory peduncle, olfactory tract and olfactory striae		
Diencephalic formations		Hypothalamus Epithalamus Thalamus	Anterior nuclei	

"Dorsal hippocampus" is defined as the highest part of the hippocampus, which corresponds to its dorsal tip, also defined as "septal pole". The term "hippocampal formation" refers to Ammon's horn and the dentate gyrus as a whole.

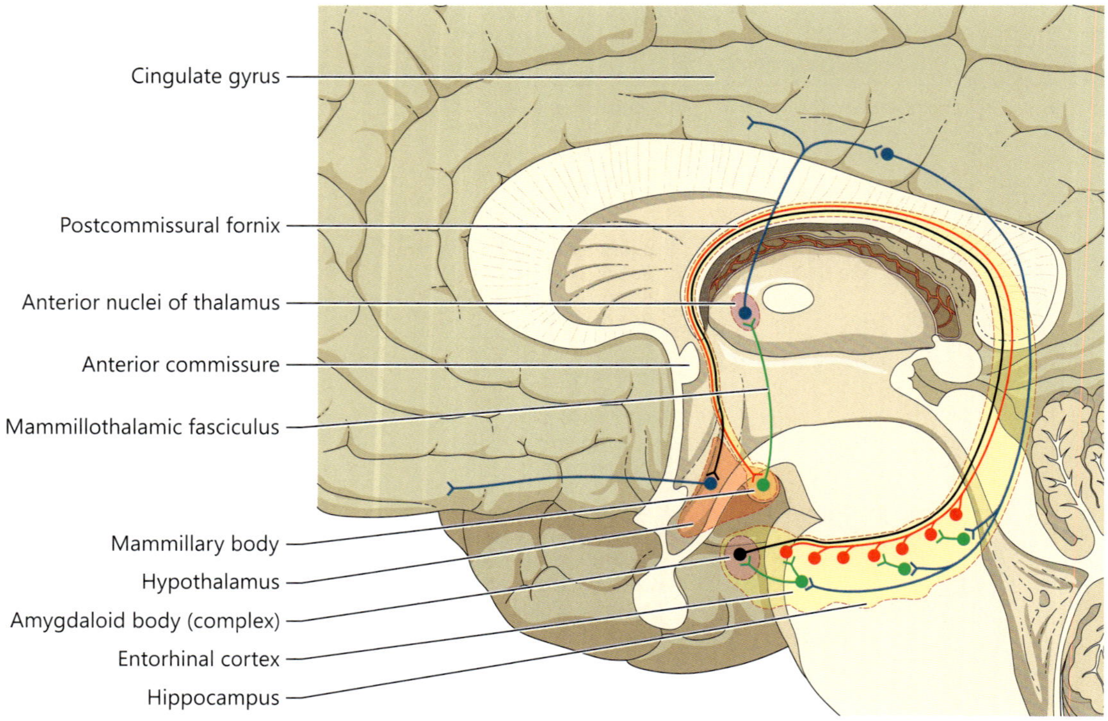

a

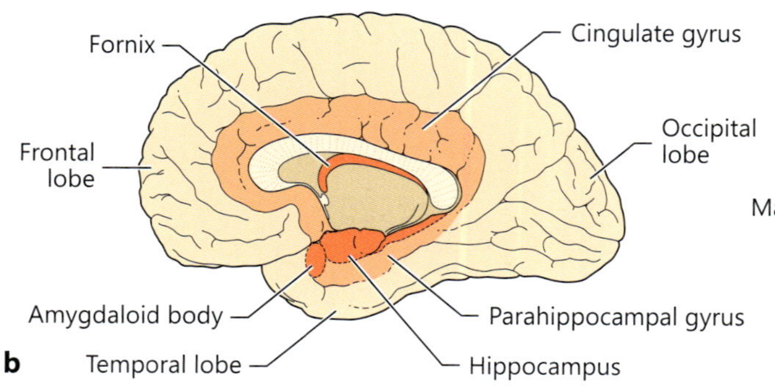

b

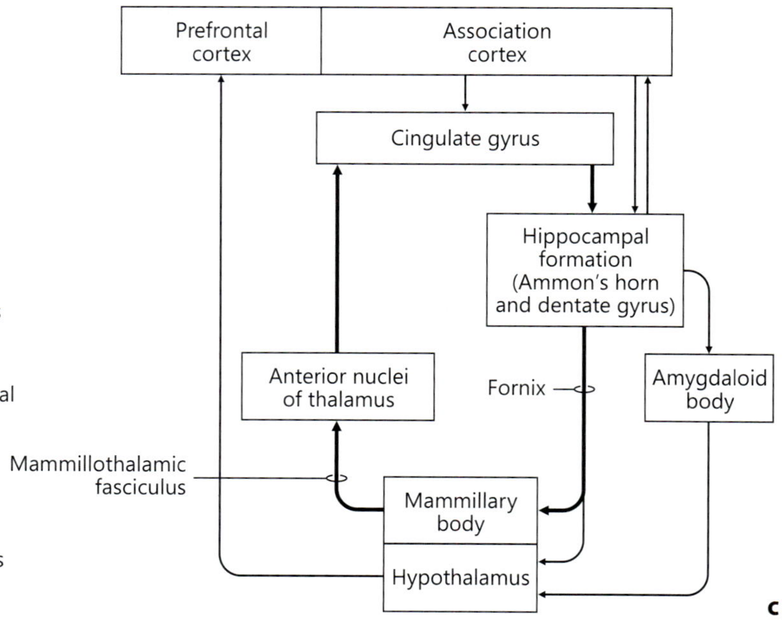

c

a, The **Papez circuit** is a "closed" circuit that connects the diencephalon to the cerebral cortex of the cingulate gyrus, through a continuous flow of information. The information, originating from the hippocampus, for example, reaches the mammillary bodies of the hypothalamus through the fornix and, from there, is transmitted to the anterior nuclei of the thalamus (mammillothalamic fasciculus or bundle of Vicq d'Azyr); from the thalamic nuclei, the information is transmitted to the cingulate gyrus, and eventually returns to the hippocampus. Mc Lean, based on the description of Papez, introduced the term "limbic system", and completed the Papez circuit by adding further synaptic stations (1952).

The sensorial information that comes from the association cortex enters the Papez circuit, through both the cingulate gyrus and the two-way connections with the hippocampus; from there, it is also retransmitted to the amygdaloid body, from here projected to the hypothalamus and then to the prefrontal cortex. In conclusion, the Papez circuit, redesigned by Mc Lean, is a circuit through which the vegetative life and that of relationships coexist and are brought to the level of conscience, acting as a motivation for the choice of motor control.

b, The **limbic system** comprises the various structures (darker areas) of the medial part of the brain that form a ring around the brainstem and the corpus callosum. **c**, There exists a close correlation between the limbic system structures and the Papez circuit (thick black lines). Many of the limbic system structures are involved in the declarative memory. The thin black lines indicate the Mc Lean integrations to the Papez circuit.

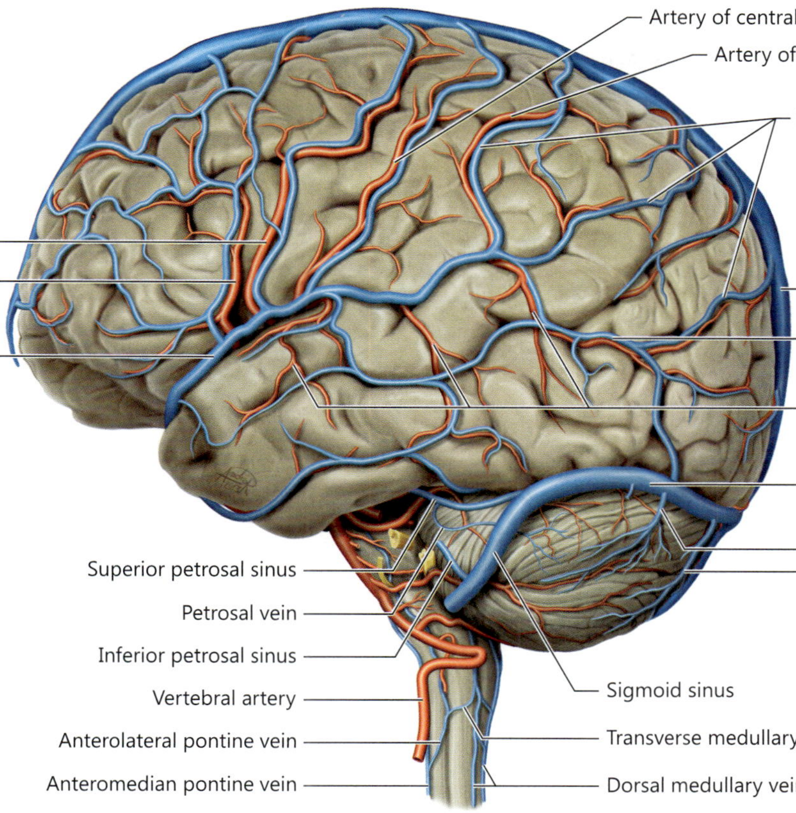

Vascularization of the superolateral face of the cerebral hemisphere. The superior and inferior terminal branches (cortical branches) of the middle cerebral artery and the superficial cerebral veins, in particular, the superior cerebral veins and the superficial middle cerebral vein, are shown.

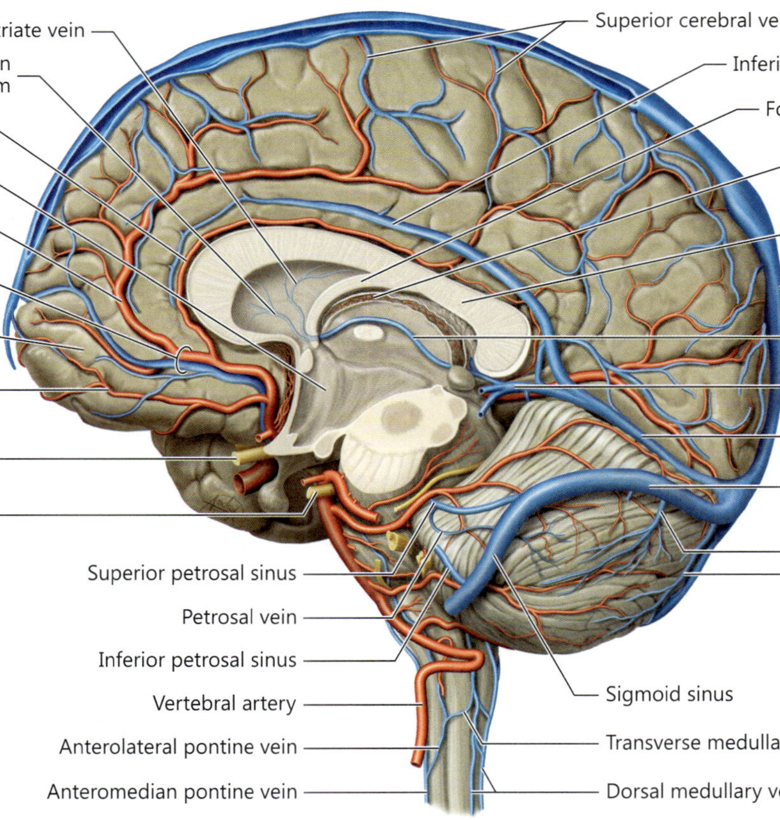

Vascularization of the medial and inferior faces of the cerebral hemisphere. The branches of the postcommunicating part (segment A2) of the anterior cerebral artery, in particular, the medial frontobasal, callosomarginal and pericallosal arteries, are shown. The right and left internal cerebral veins flow into the great cerebral vein, or of Galen, which opens at the anterior extremity of the straight sinus.

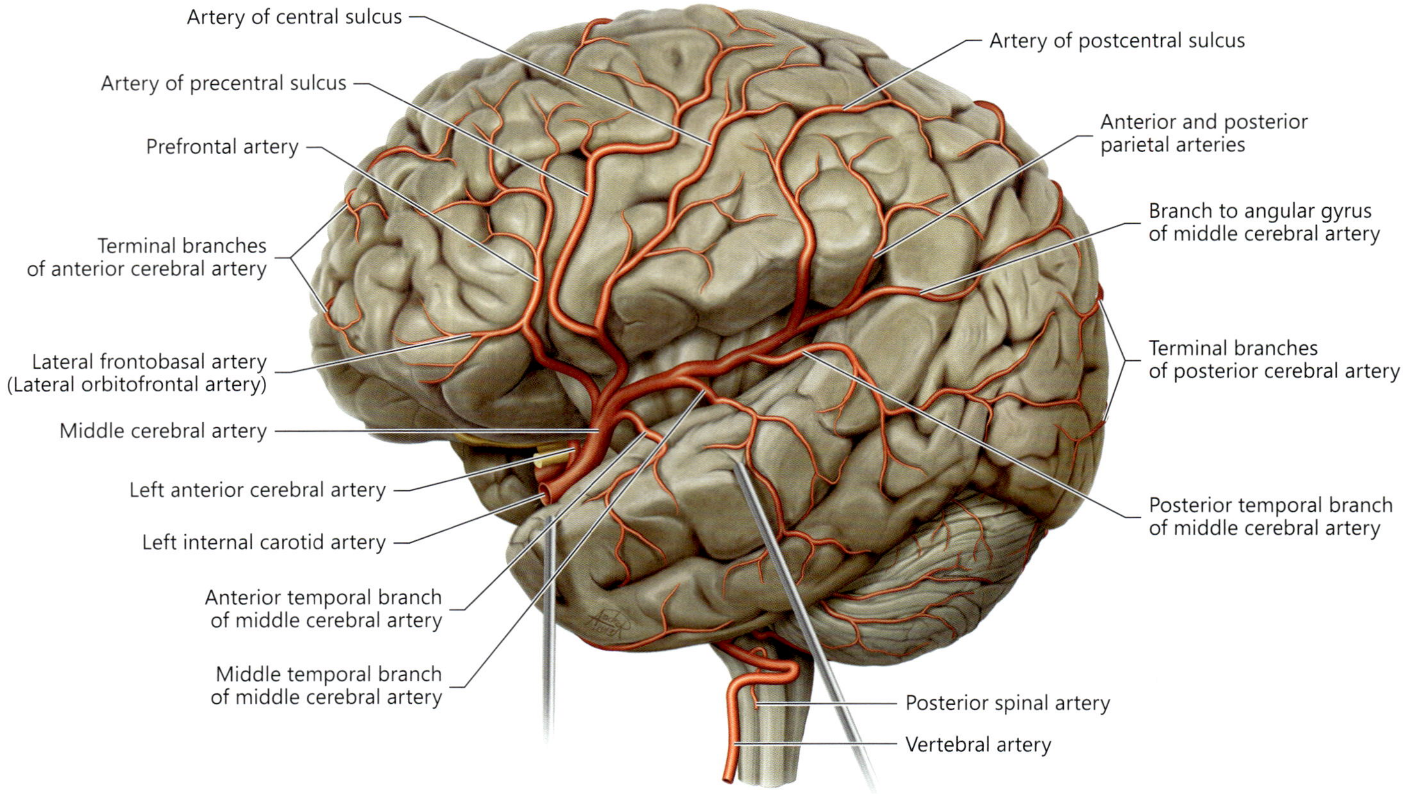

Middle cerebral artery. Lateral view of the left cerebral hemisphere. The labra of the lateral sulcus, on the superolateral face of the hemisphere, have been opened to show the insular part of the middle cerebral artery.

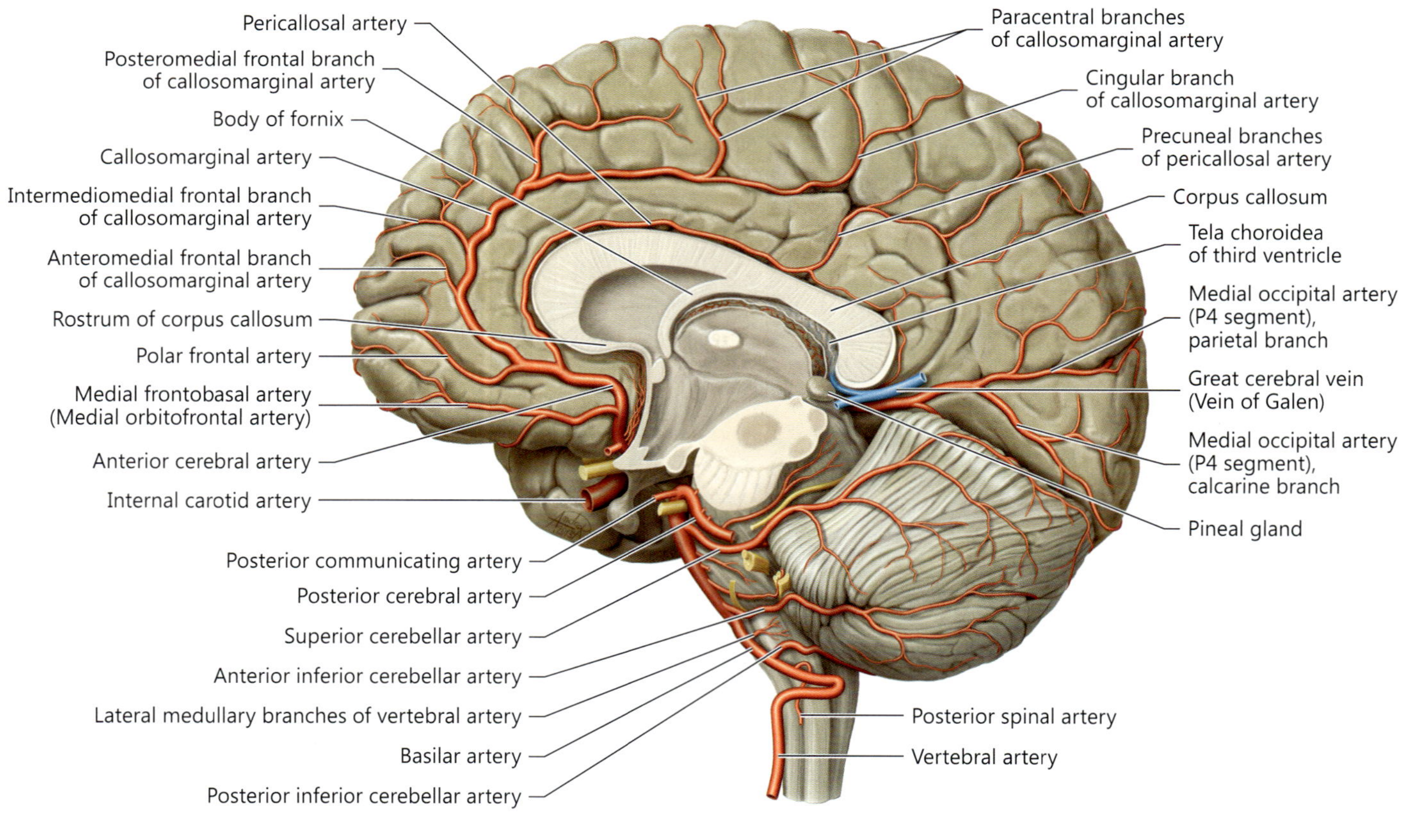

Anterior and **posterior cerebral arteries**. Medial view of the right cerebral hemisphere.

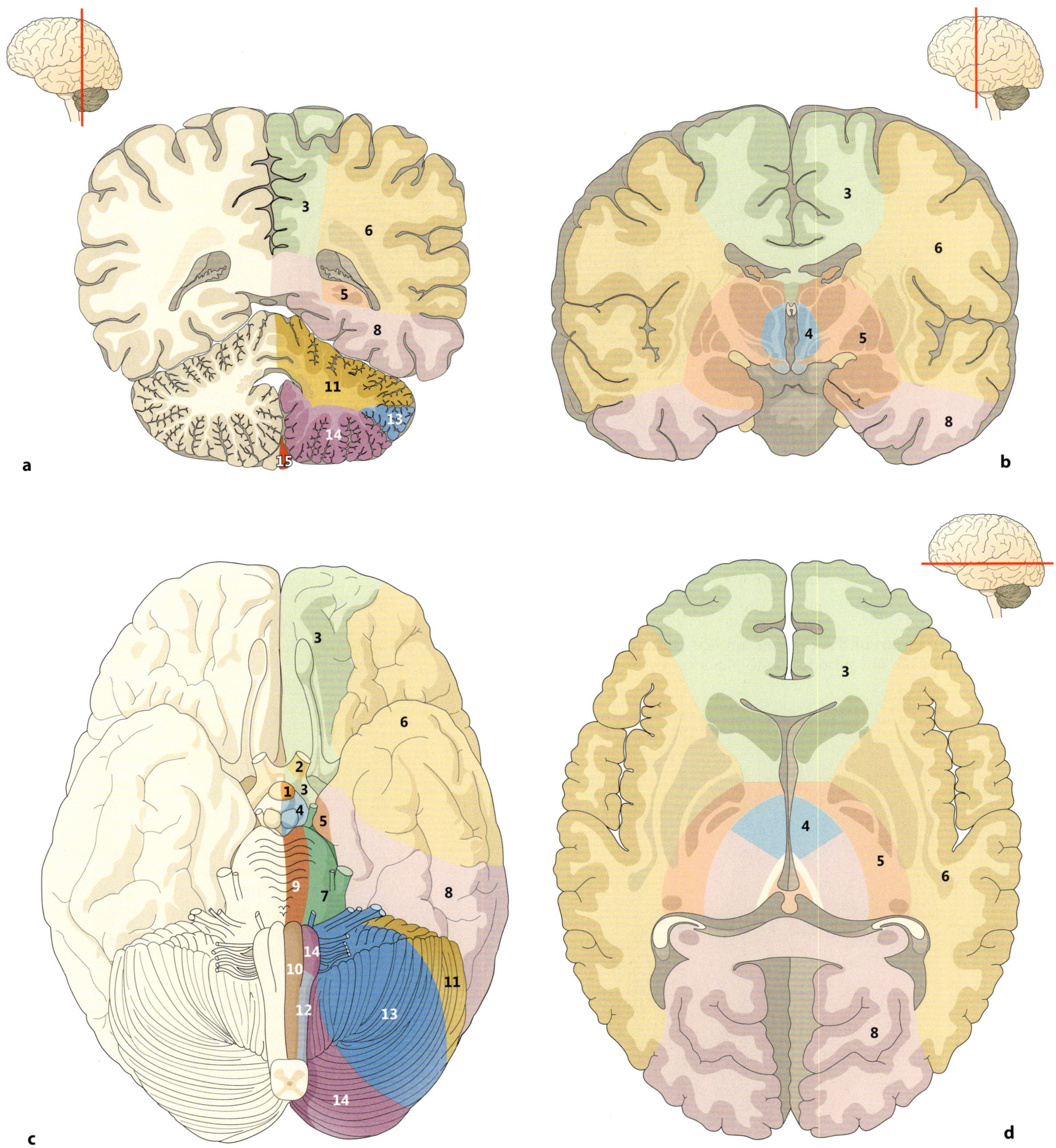

Grahical representation of the **cerebral and cerebellar territories of supply**. Frontal sections passing through the posterior extremity of the hippocampus (**a**) and through the mammillothalamic fasciculus (**b**), inferior projection of the cerebral hemispheres, of the brainstem and of the cerebellum (**c**) and horizontal section passing through the middle part of the diencephalon (**d**).
For the **circle of Willis** ☞ Head and Neck, Arterial Vascularization, volume 2, page 32.

1, Internal carotid artery
2, Ophthalmic artery
3, Anterior cerebral artery
4, Posterior communicating artery
5, Anterior choroidal artery
6, Middle cerebral artery
7, Pontine arteries, lateral branches
8, Posterior cerebral artery
9, Pontine arteries, medial branches
10, Anterior spinal artery
11, Superior cerebellar artery
12, Vertebral artery
13, Anterior inferior cerebellar artery
14, Posterior inferior cerebellar artery
15, Posterior spinal artery

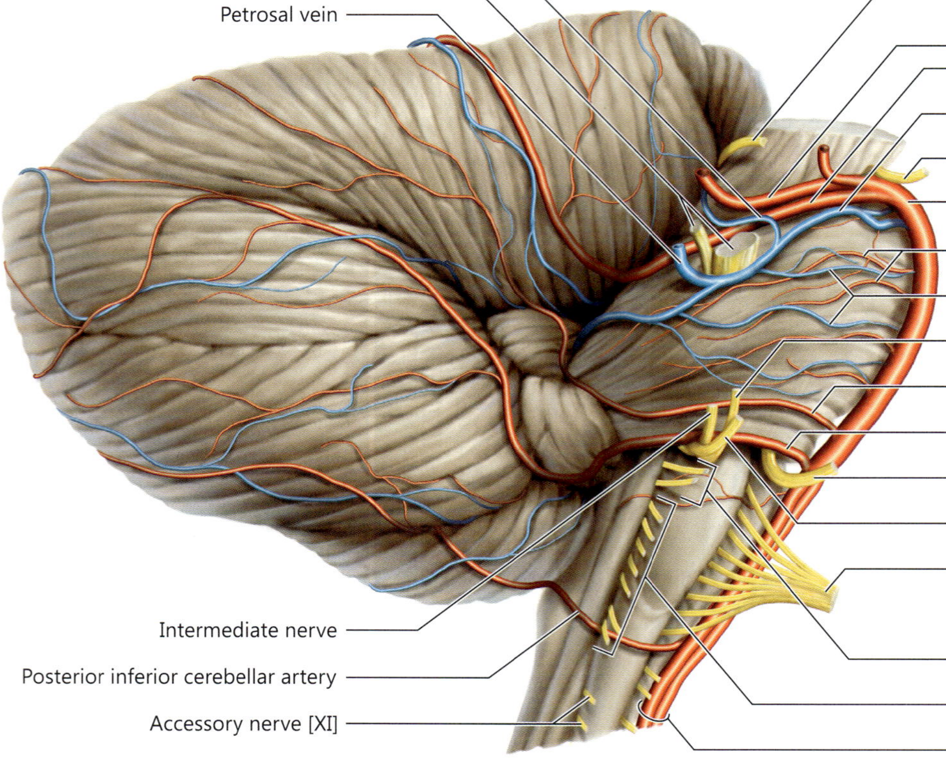

Vascularization of the cerebellum. The posterior inferior cerebellar artery, the vertebral artery branch, and the anterior inferior cerebellar and superior cerebellar arteries, branches of the basilar artery, are shown.

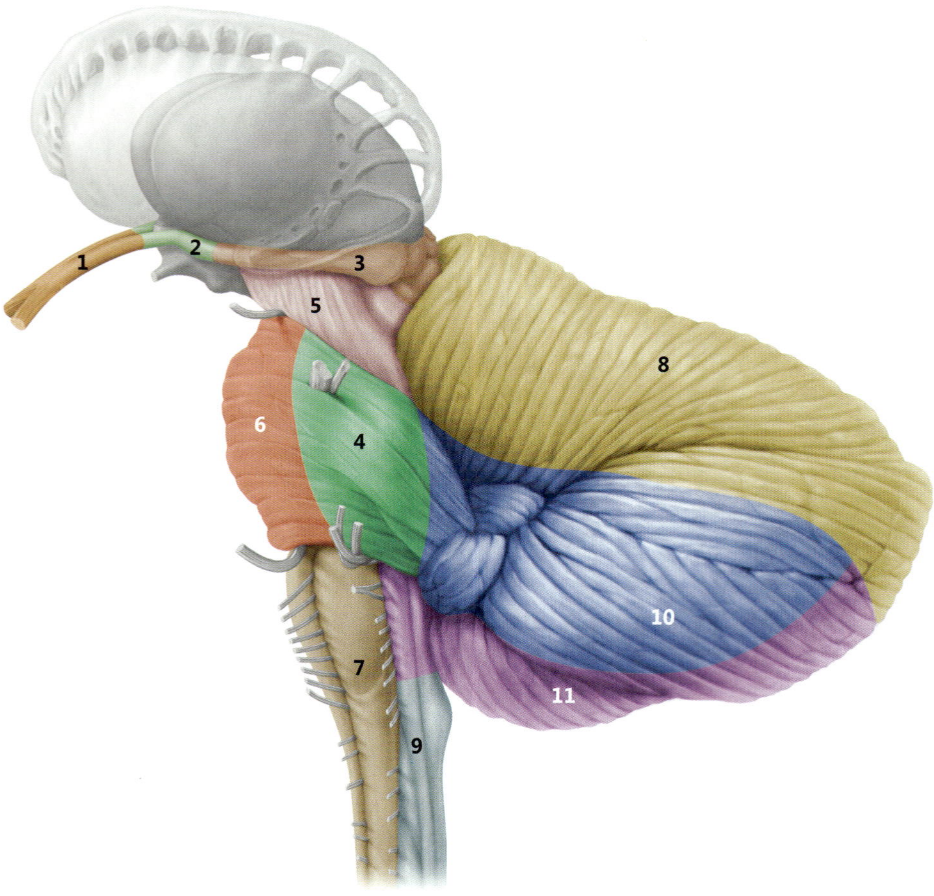

Territories of **vascularization of the myelencephalon**, **the pons** and **the cerebellum**, seen in left lateral projection. The distribution territories of the vertebrobasilar arterial system and the anterior spinal artery are represented with different colors. One also observes the distribution territories in the mesencephalon and some structures of the diencephalon of the internal carotid, anterior and posterior cerebral, anterior choroidal arteries, and the posterior choroidal branches of the posterior cerebral artery.

1, Internal carotid artery
2, Anterior cerebral artery
3, Anterior choroidal artery
4, Pontine arteries, lateral branches
5, Posterior cerebral artery
6, Pontine arteries, medial branches
7, Anterior spinal artery
8, Superior cerebellar artery
9, Vertebral artery
10, Anterior inferior cerebellar artery
11, Posterior inferior cerebellar artery

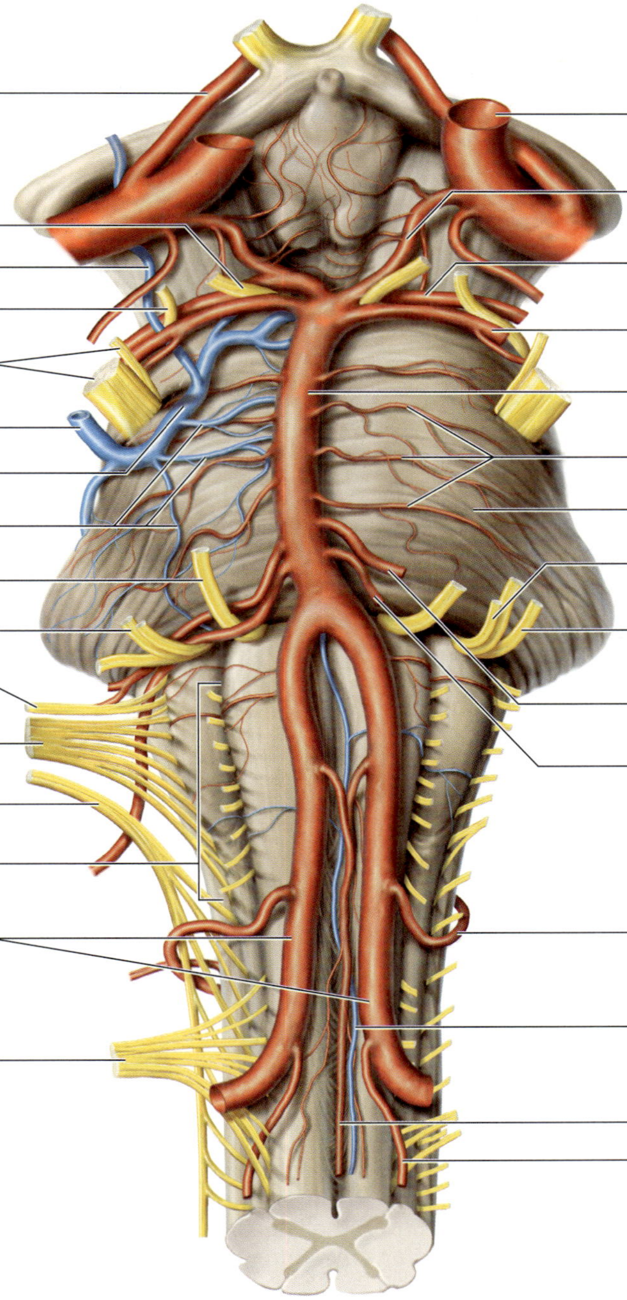

The **basilar artery** extends from the medullopontine sulcus to the pontomesencephalic sulcus, terminating at the level of the interpeduncular fossa, where it divides into the two **posterior cerebral arteries**. Its collateral branches are the anterior inferior cerebellar artery, the labyrinthine artery, the pontine arteries and the superior cerebellar artery.

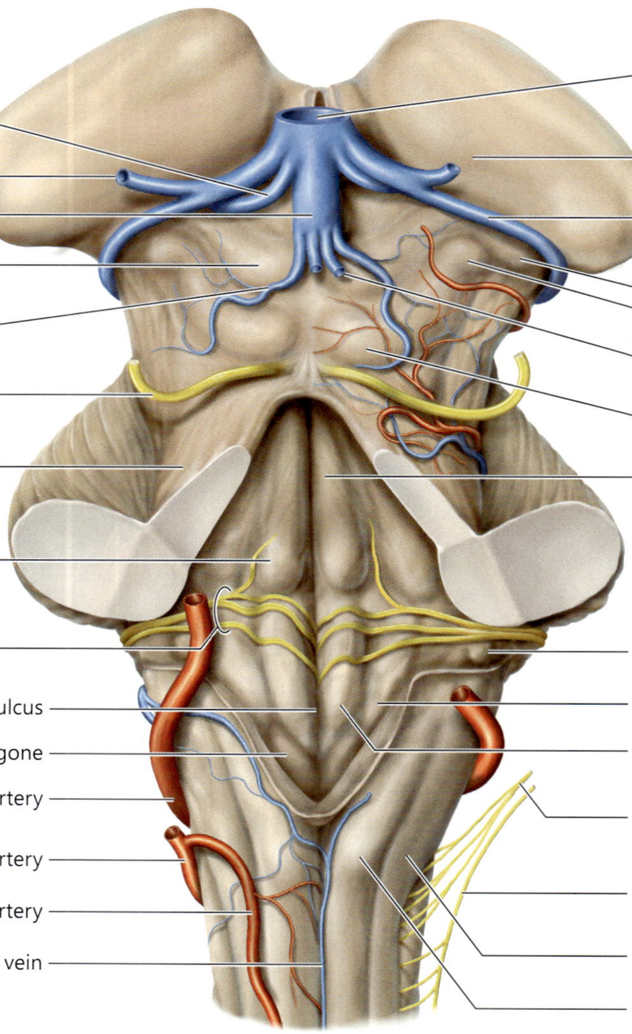

Brainstem and fourth ventricle, posterior view after the removal of the cerebellum. The **deep cerebral veins**, coming from the basal nuclei, from the lateral ventricles and from the centrum semiovale join in two voluminous right and left trunks; right, the **internal cerebral veins**, having received collateral branches, flow into a single (odd and median) trunk, the **great cerebral vein**, or of Galen, which opens at the anterior extremity of the straight sinus. The **basal vein**, coming from the base of the brain, and originating from the anterior cerebral vein, runs from the anterior perforated substance of the amygdaloid body to the great cerebral vein. The **lateral mesencephalic vein** runs in the lateral sulcus of the mesencephalon and drains in the basal vein. The **precentral cerebellar vein** is a single median trunk that originates in the precentral fissure of the cerebellum and runs into the great cerebral vein.

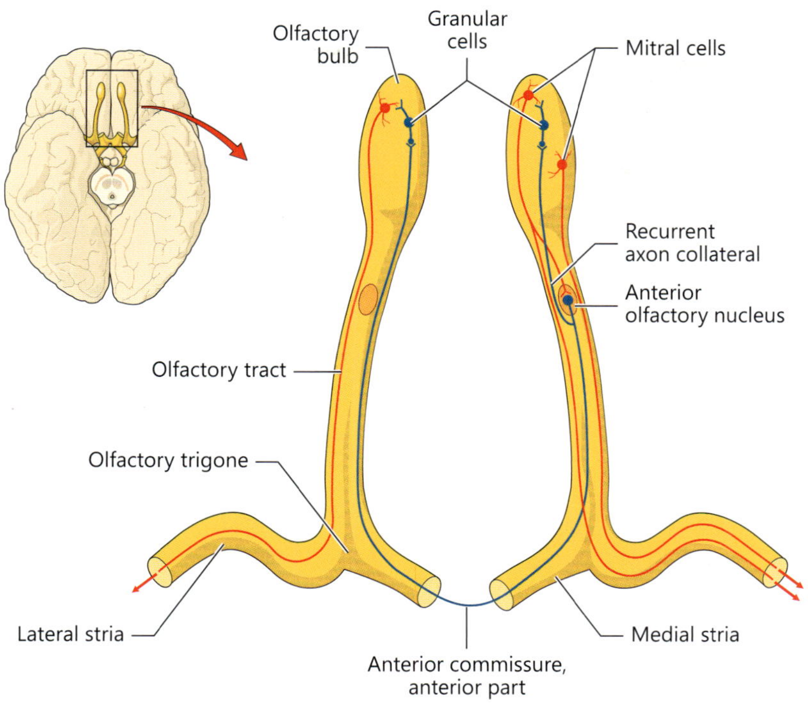

Schematic representation that illustrates the connections of the **olfactory bulbs** through the olfactory tract and the olfactory striae, medial longitudinal (or of Lancisi) stria and lateral longitudinal stria (or stria tecta).

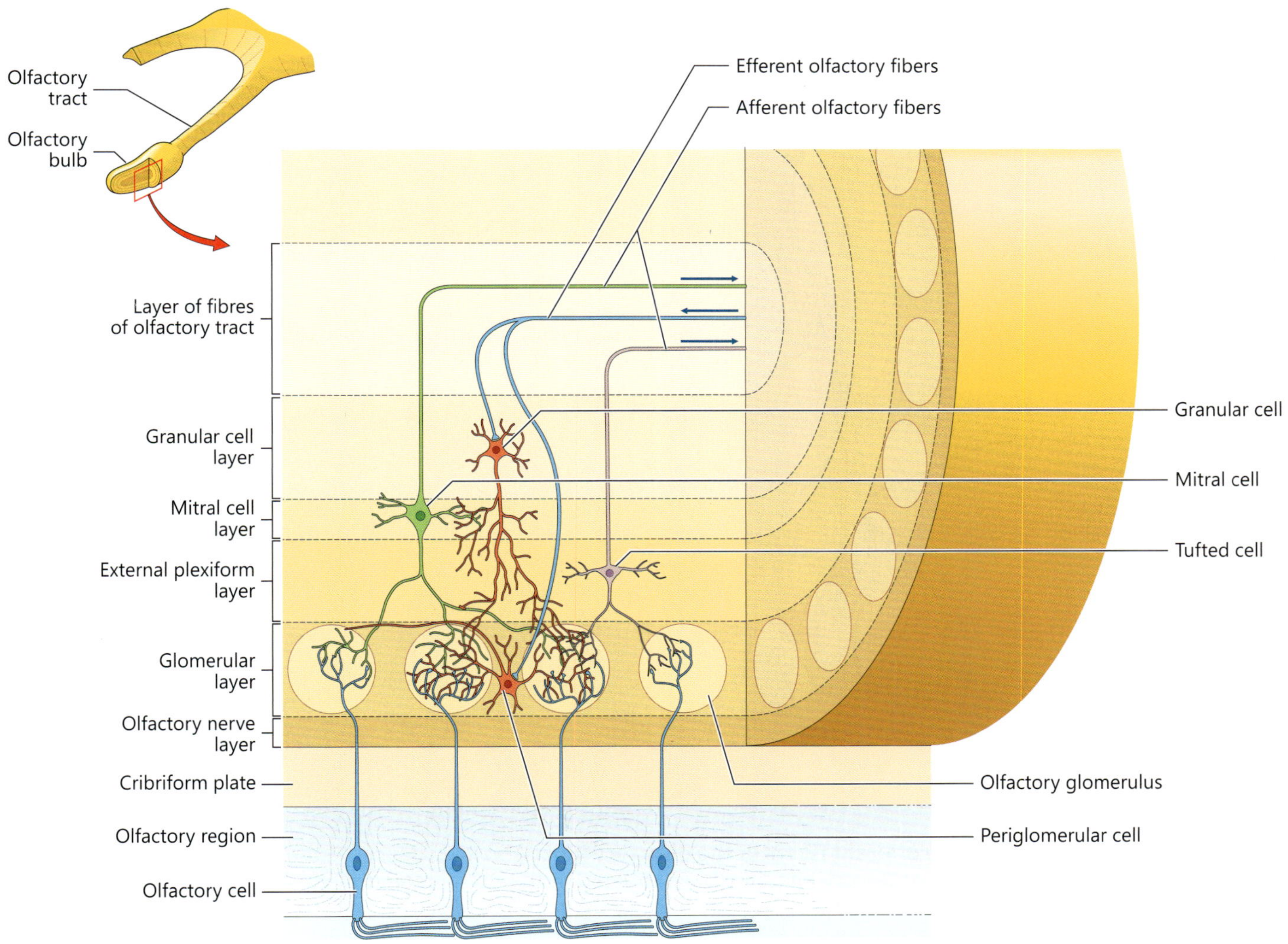

Schematic representation that illustrates the cytoarchitecture and the synaptic connections of the **olfactory bulb**. The olfactory cells in the area of the olfactory nasal mucosa, are shown.

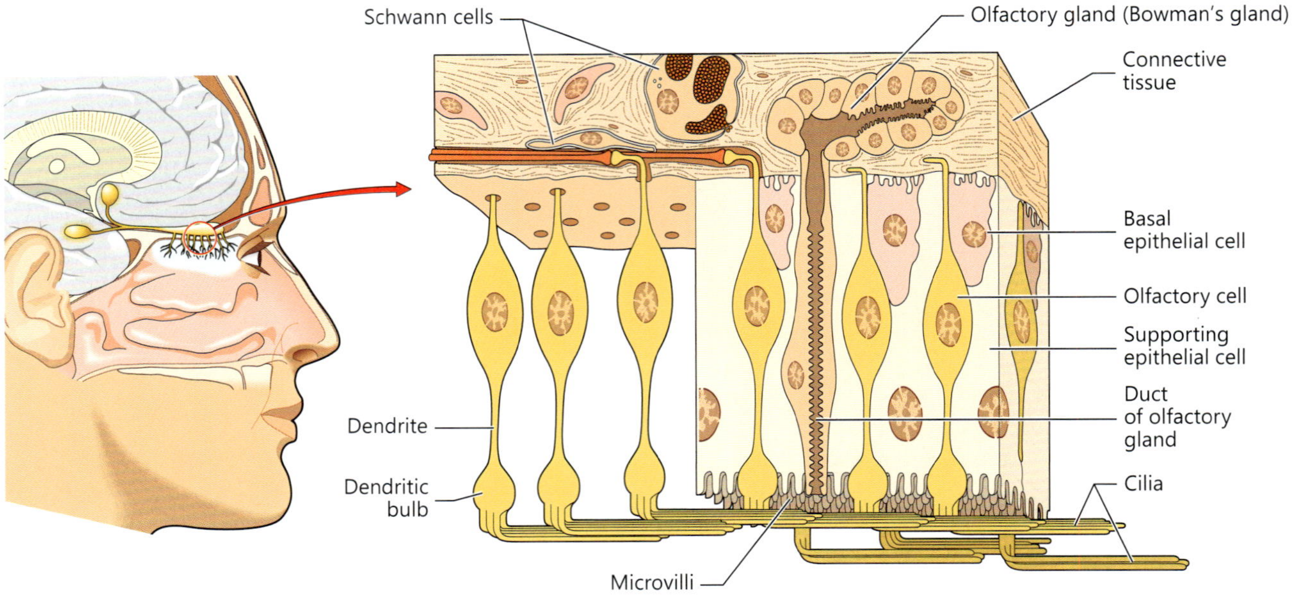

Organization of the **olfactory mucosa**. This covers the vault of the nasal cavities, the top part of the nasal septum and the superior face of the superior concha. The mucosa displays a lining epithelium, formed by olfactory cells, supporting cells and basal cells, and a lamina propria. The olfactory cell is characterized by two extensions, superficial and deep: the first, reaches the surface of the epithelium where it expands to form the dendritic bulb or olfactory vesicle, from which irradiate ten to twenty cilia; the second, is the olfactory axon, which crosses the lamina propria and the cribriform plate of the ethmoid, and penetrates into the olfactory bulb. The basal cell wraps around the olfactory cell axon, forming an apparatus similar to that formed by the Schwann cell around the axon. The olfactory glands (Bowman's glands) are found in the lamina propria. These glands are simple tubulooalveolar with seromucous secretion. The secretion of these glands, poured onto the surface of the olfactory mucosa, forms a solvent for odorous substances.

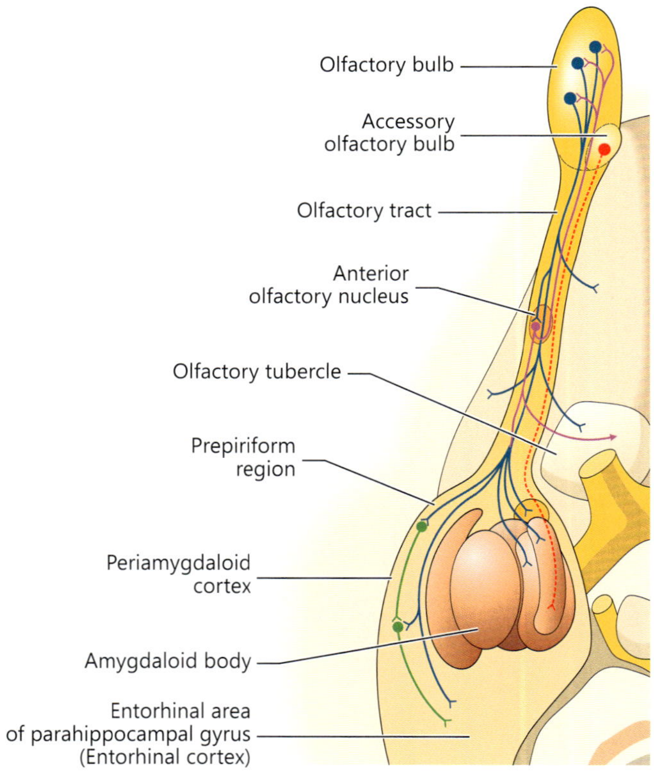

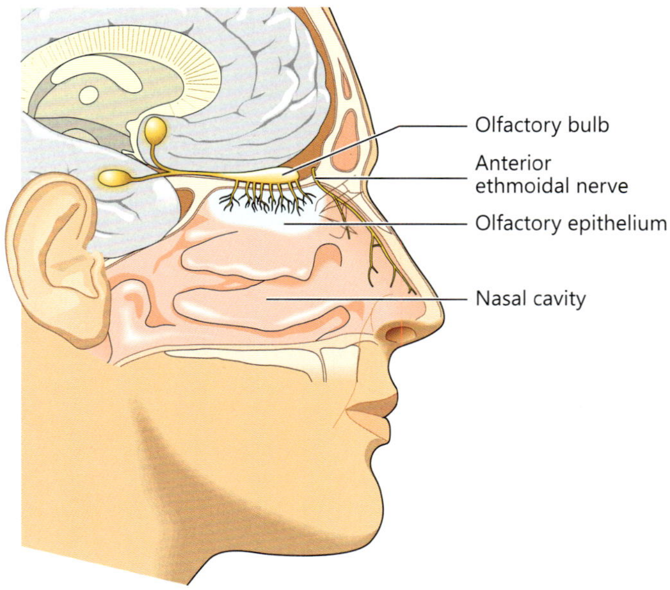

Course and terminal centers of the **efferent fibers of the olfactory bulb**. Also in evidence are the projections from the piriform cortex to the entorhinal area. The accessory olfactory bulb, situated in the dorsal region of the principal olfactory bulb, gives rise to an independent pathway, parallel to that originating from the principal olfactory bulb. Axons coming from the vomeronasal organ, whose sensorial epithelium regresses in human beings during fetal development, join the accessory olfactory bulb pathways. Moreover, there is no proof, in the adult, of a connection between the vomeronasal organ and the brain. In addition, some studies have reported a function similar to that of the vomeronasal organ, to a small formation found in the nasal septum of some individuals, however its presence in human beings still remains controversial.

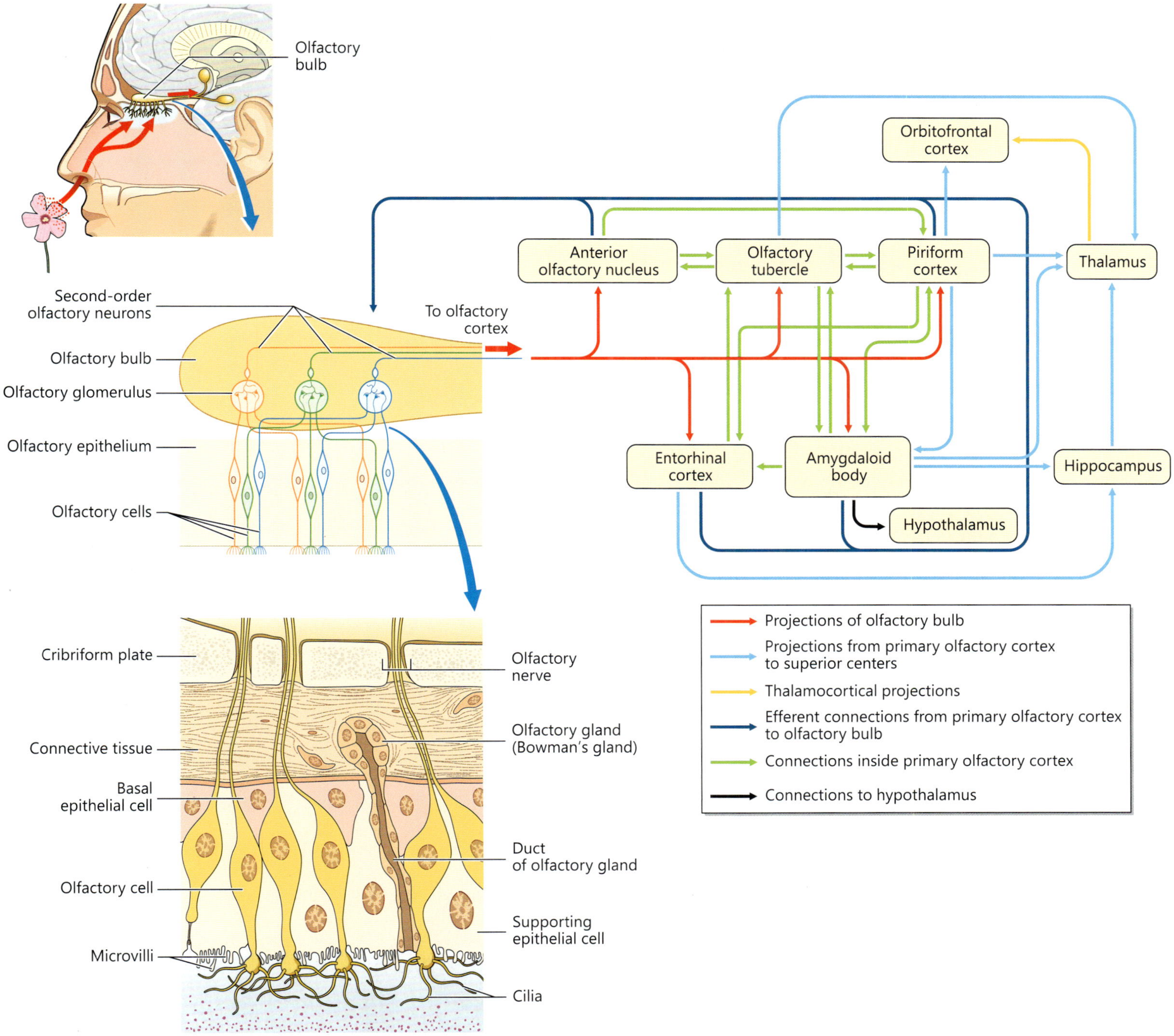

Schematic organization of the **olfactory system**. Binding of odorous molecules to specific protein receptors leads to signals transmitted from the axons of the mitral cells and the tufted cells of the olfactory bulb, through the olfactory tract, to the olfactory cortex, formed by: the anterior olfactory nucleus, olfactory tubercle, piriform cortex, entorhinal cortex and amygdaloid body. The anterior olfactory nucleus projects to the piriform cortex, both ipsi- and contralateral, but also to the olfactory bulb. The olfactory tubercle is the only region of the olfactory cortex that does not have efferent connections with the olfactory bulb, and mainly projects to the medial dorsal nucleus of the thalamus. The piriform cortex, which receives the greatest number of olfactory bulb projections, has direct connections with the orbitofrontal cortex, but also with the medial dorsal nucleus of the thalamus. The entorhinal cortex mainly projects to the hippocampus, but also to the olfactory bulb, while the amygdaloid body projects both to the hypothalamus and, again, to the olfactory bulb. All the regions of the primary olfactory cortex are linked by a complex network of intracortical connections.

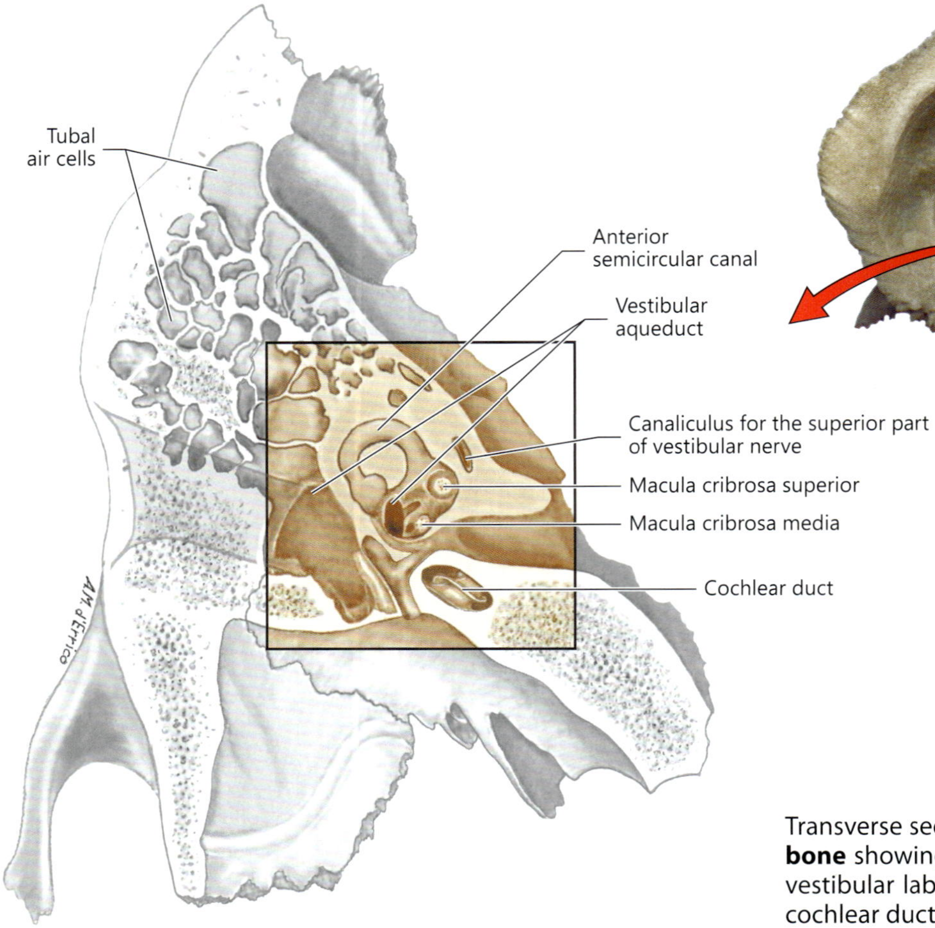

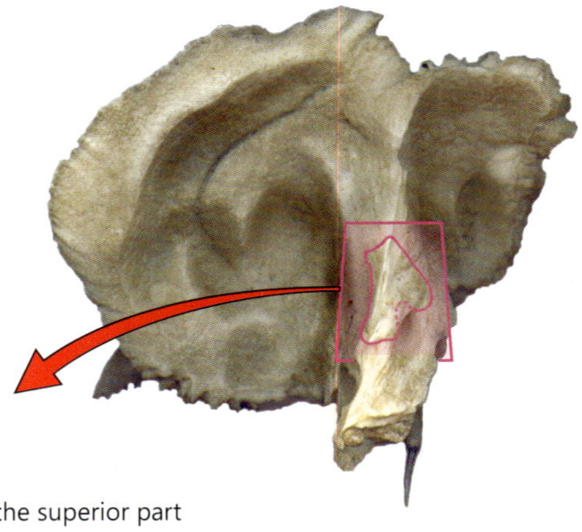

Transverse section of the **pyramid of temporal bone** showing some of the components of the vestibular labyrinth and, partially, a turn of the cochlear duct.

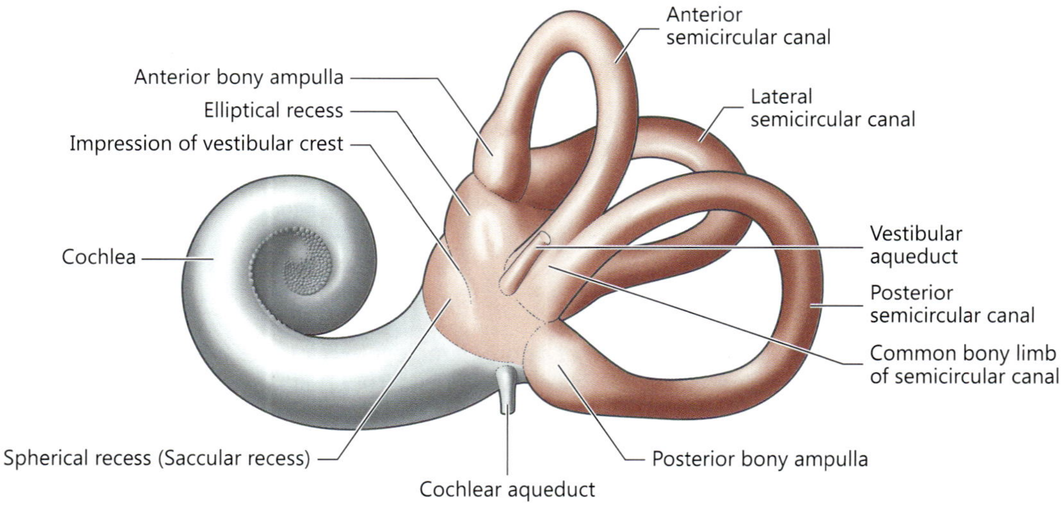

Schematic image showing the cast of the **bony labyrinth** in medial projection: in *gray*, the cochlear labyrinth; in *pink*, the vestibular labyrinth.

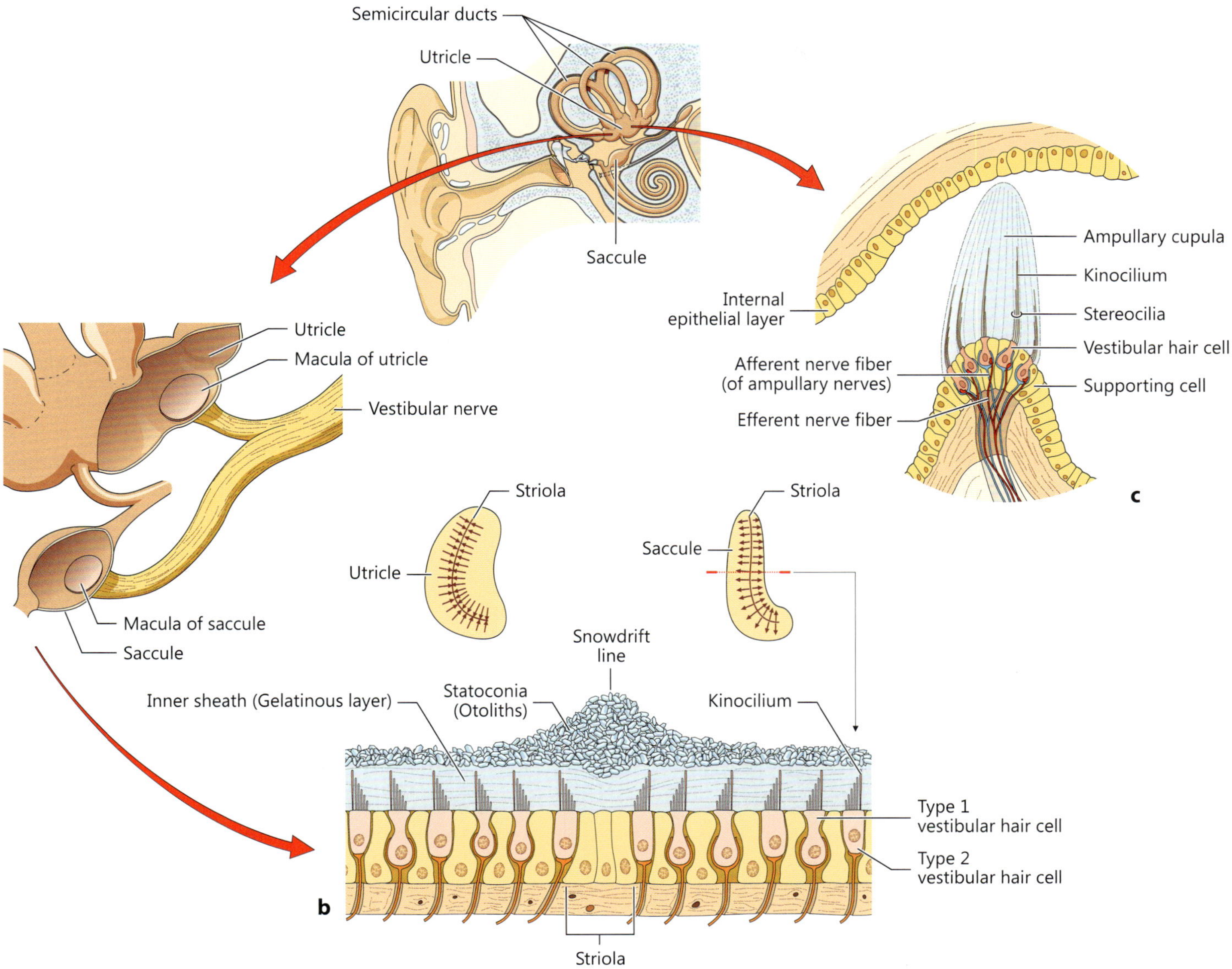

a, Schematic representation of the position of the **maculae of the utricle** and **the saccule**. **b**, Schematic representation of the **structural organization of the maculae of the utricle and the saccule**. In the macula of the utricle the hair cells have the kinocilium facing the striola, while in the macula of the saccule the kinocilium is facing in the opposite to the striola. The sensory cells have kinocilium and stereocilia stemming from the lamina or apical cuticular plate; above these structures is a gelatinous cupola embedding the otoliths. **c**, Schematic representation of the structure of the **ampullary crest** of the semicircular ducts. The ampullary crest is formed by supporting and sensory cells. The cilia of the sensory cells (stereocilia and kinocilia) are immersed in a gelatinous cupola containing no otoliths. The epithelium and the overlying gelatinous mass form the ampullary cupula, whose axis is transversal with respect to the major axis of the semicircular duct.

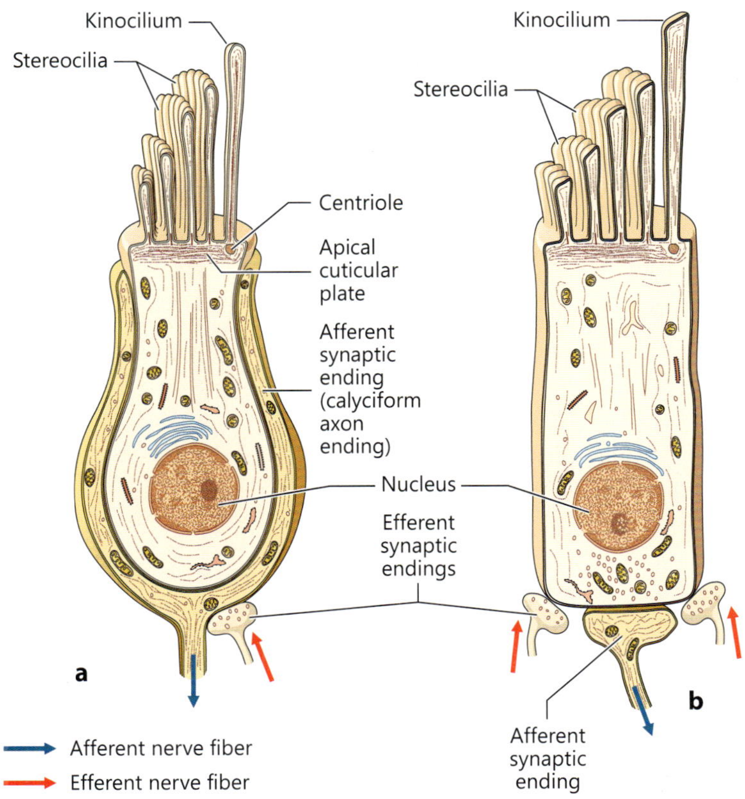

➤ Schematic representation illustrating the ultrastructure of the **sensory elements of the maculae of the utricle, saccule and ampullary crests. a**, Type 1 vestibular hair cell; **b**, type 2 vestibular hair cell. The two cell types display different cytoneural junctions.

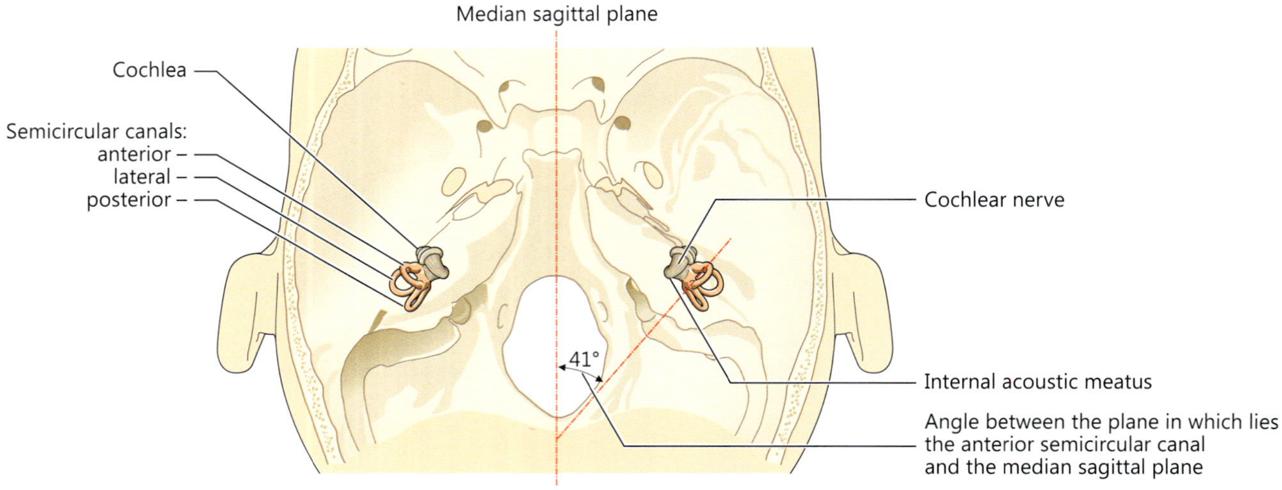

Schematic representation illustrating the symmetrical arrangement of the **semicircular canals** of the two sides. The lateral semicircular canals of the two sides, being oriented on the same horizontal plane, form a functional couple. The anterior semicircular canals of one side is instead parallel to the posterior canal of the opposite side; thus forming a different functional couple.

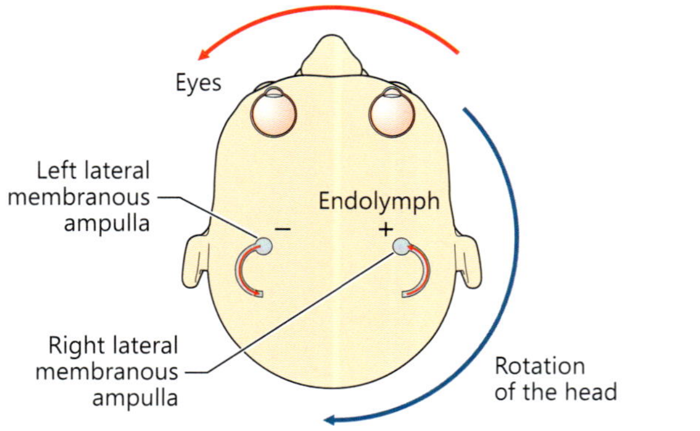

a

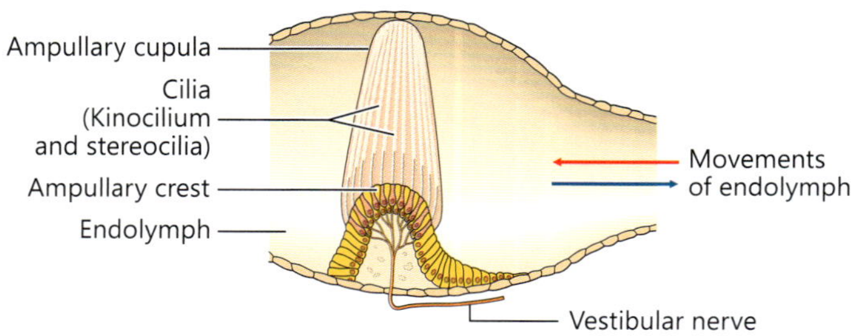

b

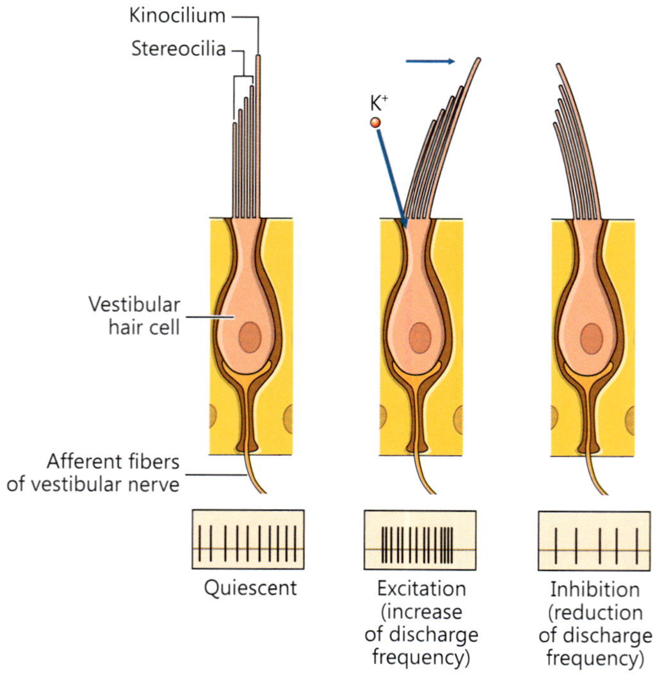

c

a, **Angular acceleration of the head** and **spatial arrangement of the semicircular canals**. The activation of the membranous semicircular canals causes slow movements of the eyes in the direction opposite to the shifting of the head, allowing the maintaining of constant vision during the rotation of the head. In fact, in the presence of angular accelerating events, the endolymph in the membranous labyrinth of the semicircular canals (or semicircular ducts) moves in the opposite direction to the movement of the head. The movement of the **endolymph** causes the deflection of the ampullary crests (**b**). The movement can take place in the direction of the membranous ampullae (current ampullipeta, *red* arrow), or opposite to it (current ampullifuga, *blue* arrow), depending on the direction of the angular acceleration. **c**, The **deflection of the ampullary crests** in an ampullipeta direction, determines the excitation, i.e. depolarization, of the ciliate receptor elements in the ampullary crests, whereas in an ampullifuga direction, it determines inhibition, i.e. hyperpolarization. These phenomena are due to the bending of the stereocilia and kinocilium of the receptor cells. The receptor cells are depolarized when the movement of the endolymph causes the stereocilia to bend towards the kinocilium, while the opposite movement causes hyperpolarization. Below, is shown the discharge frequency in the afferent fiber of the vestibular nerve, when at rest, in excitation and in inhibition. Therefore, if the head rotates towards the right, as represented in **a**, the endolymph rotates towards the left causing the activation of the nervous fibers of the ampullary crests on the right and the inhibition of those on the left.

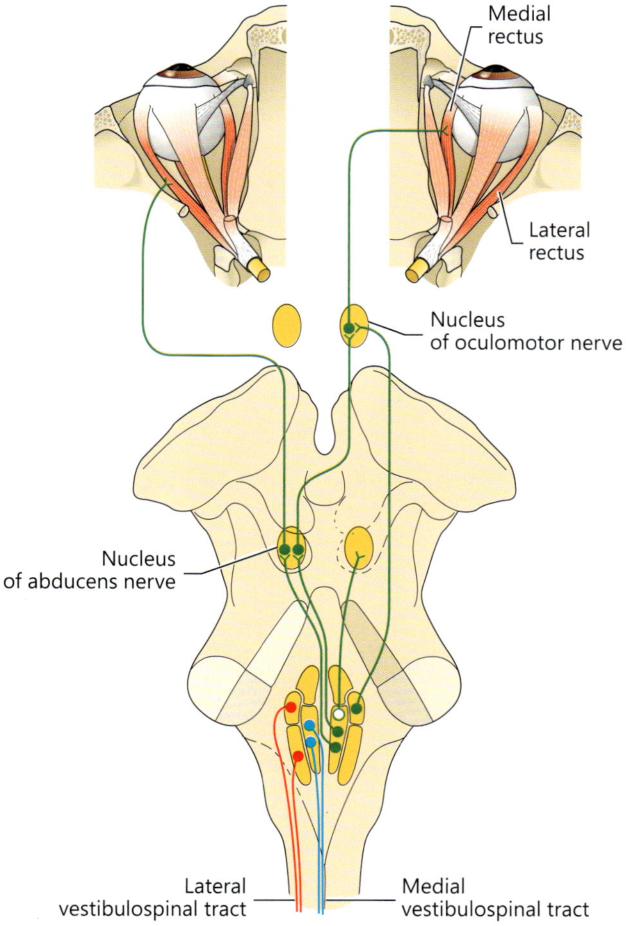

From the vestibular nuclei of the VIII pair of cranic nerves, positioned under the external white wing of the fourth ventricle, originate both the **vestibulospinal pathways** (*red*, lateral vestibulospinal tract; *blue*, medial vestibulospinal tract) and the **vestibulo-ocular pathways** which lead, along with the medial longitudinal fasciculus (*green*), to the nuclei of the oculomotor, abducens and trochlear nerves. The nuclei of the trochlear nerve are not indicated in the schema as they are not mainly involved in the vestibulo-ocular reflex. The excitation of the ampullary fibers of the VIII pair, on the right (☞ schema on the previous page) determines, through the vestibulo-ocular pathways, the activation of the left lateral rectus (abducens nerve), the right medial rectus and, in part, the right superior oblique muscles, and the inhibition of the contralateral muscles, thus favoring the maintaining of vision fixed on the horizon.

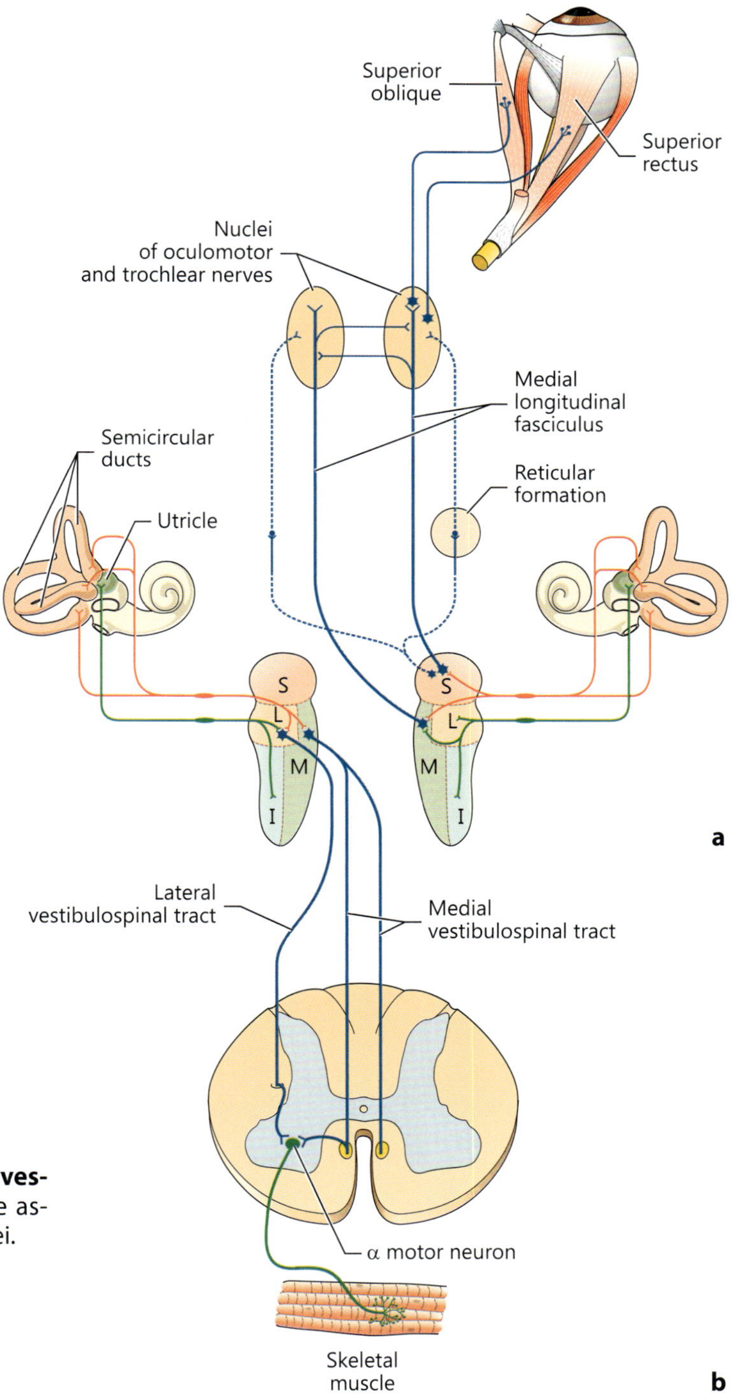

Principal characteristics of the organization of the **primary vestibular fibers**, directed to the vestibular nuclei, and of the ascending (**a**) and descending (**b**) projections from these nuclei.

I, Inferior vestibular nucleus
L, Lateral vestibular nucleus
M, Medial vestibular nucleus
S, Superior vestibular nucleus

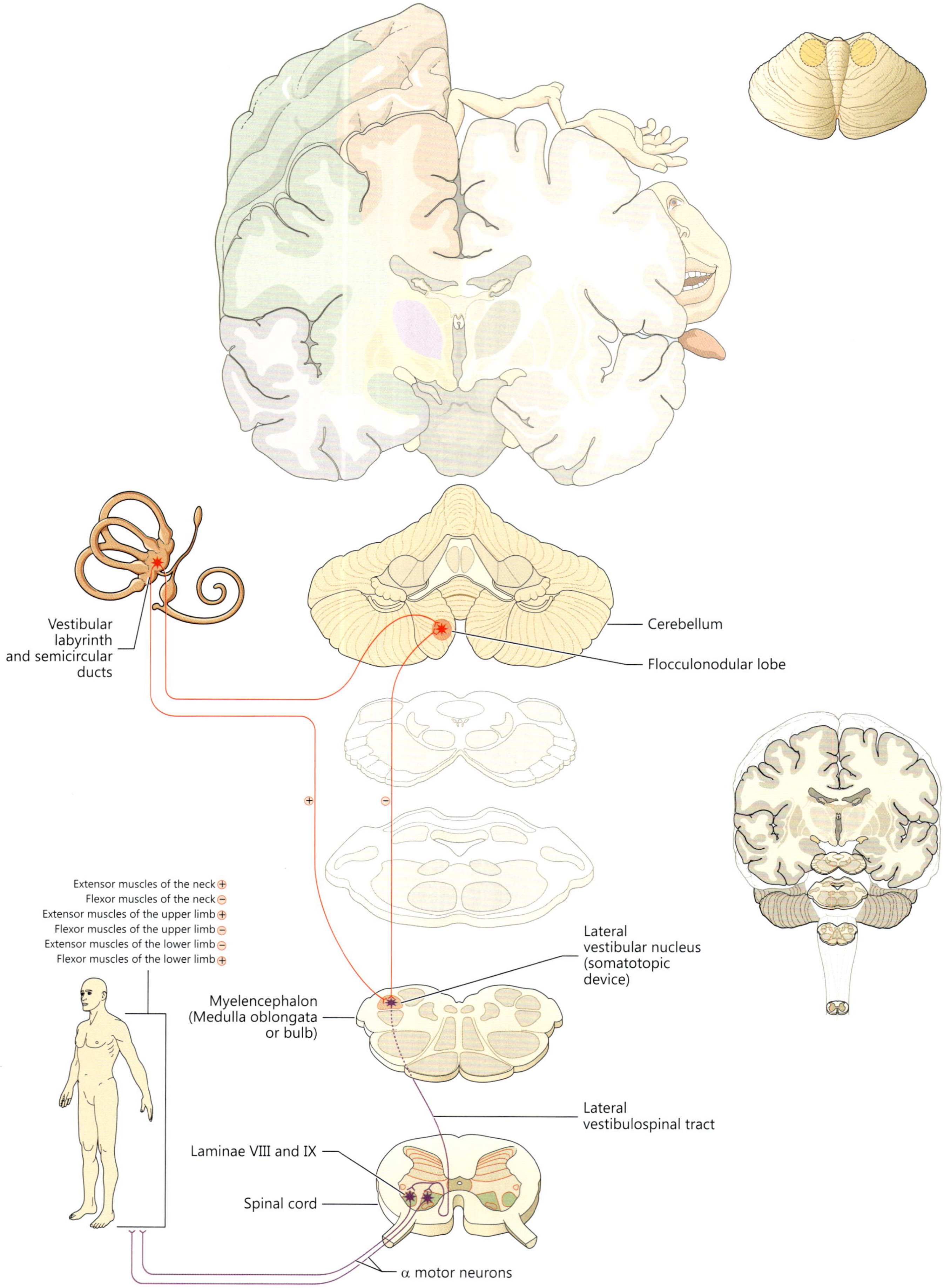

Vestibular labyrinth and semicircular ducts
Cerebellum
Flocculonodular lobe
Extensor muscles of the neck ⊕
Flexor muscles of the neck ⊖
Extensor muscles of the upper limb ⊕
Flexor muscles of the upper limb ⊖
Extensor muscles of the lower limb ⊖
Flexor muscles of the lower limb ⊕
Lateral vestibular nucleus (somatotopic device)
Myelencephalon (Medulla oblongata or bulb)
Lateral vestibulospinal tract
Laminae VIII and IX
Spinal cord
α motor neurons

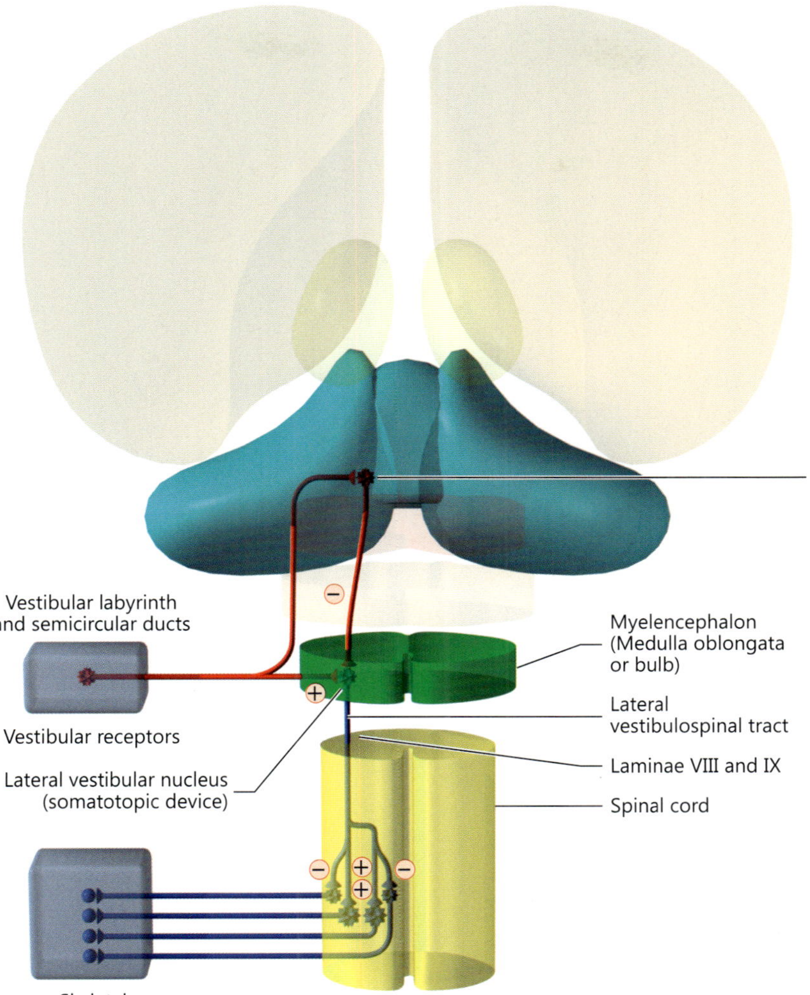

Lateral vestibulospinal tract		
Upper motor neuron	Non existent	
Lateral vestibular nucleus	From the lateral vestibular nucleus to the laminae VIII and IX of the spinal medulla, up to the lumbosacral neuromeres	*Afferences* – vestibular receptors – cerebellar cortex of the flocculonodular lobe directly without synapses to the fastigial nucleus or nucleus medialis cerebelli
Lower motor neuron	The α and γ motor neurons that innervate the extensor muscles of the neck and upper limb, and the flexor muscles of the lower limb are facilitated. The α and γ motor neurons that innervate the flexor muscles of the neck and upper limb, and the extensor muscles of the lower limb are inhibited	

The **lateral vestibulospinal tract** originates from the lateral vestibular nucleus, in which a somatotopic device is present, and terminates on the laminae VIII and IX of the entire medulla, up to the level of the lumbosacral spinal cord. This tract functions as a facilitator on the extensor muscles and an inhibitor on the flexor muscles of the neck and upper limb, with the opposite function in the lower limbs. The lateral vestibular nucleus receives afferences from the vestibular receptors (prevalently from the maculae of the utricle and the saccule) and directly from the cerebellar cortex of the flocculonodular lobe (archicerebellum or vestibulocerebellum).

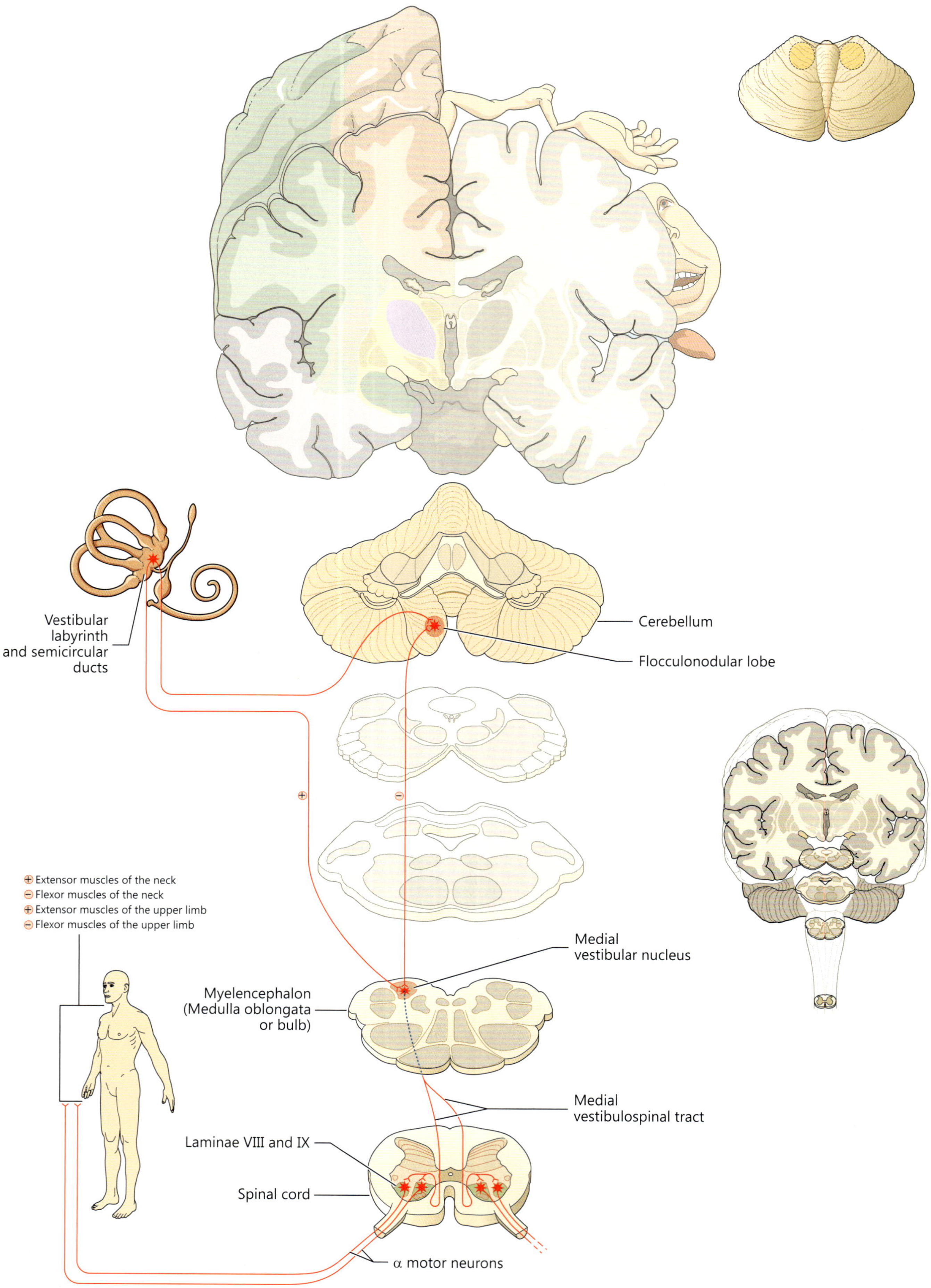
Vestibular
labyrinth
and semicircular
ducts
Cerebellum
Flocculonodular lobe
⊕
⊖
⊕ Extensor muscles of the neck
⊖ Flexor muscles of the neck
⊕ Extensor muscles of the upper limb
⊖ Flexor muscles of the upper limb
Medial
vestibular nucleus
Myelencephalon
(Medulla oblongata
or bulb)
Medial
vestibulospinal tract
Laminae VIII and IX
Spinal cord
α motor neurons

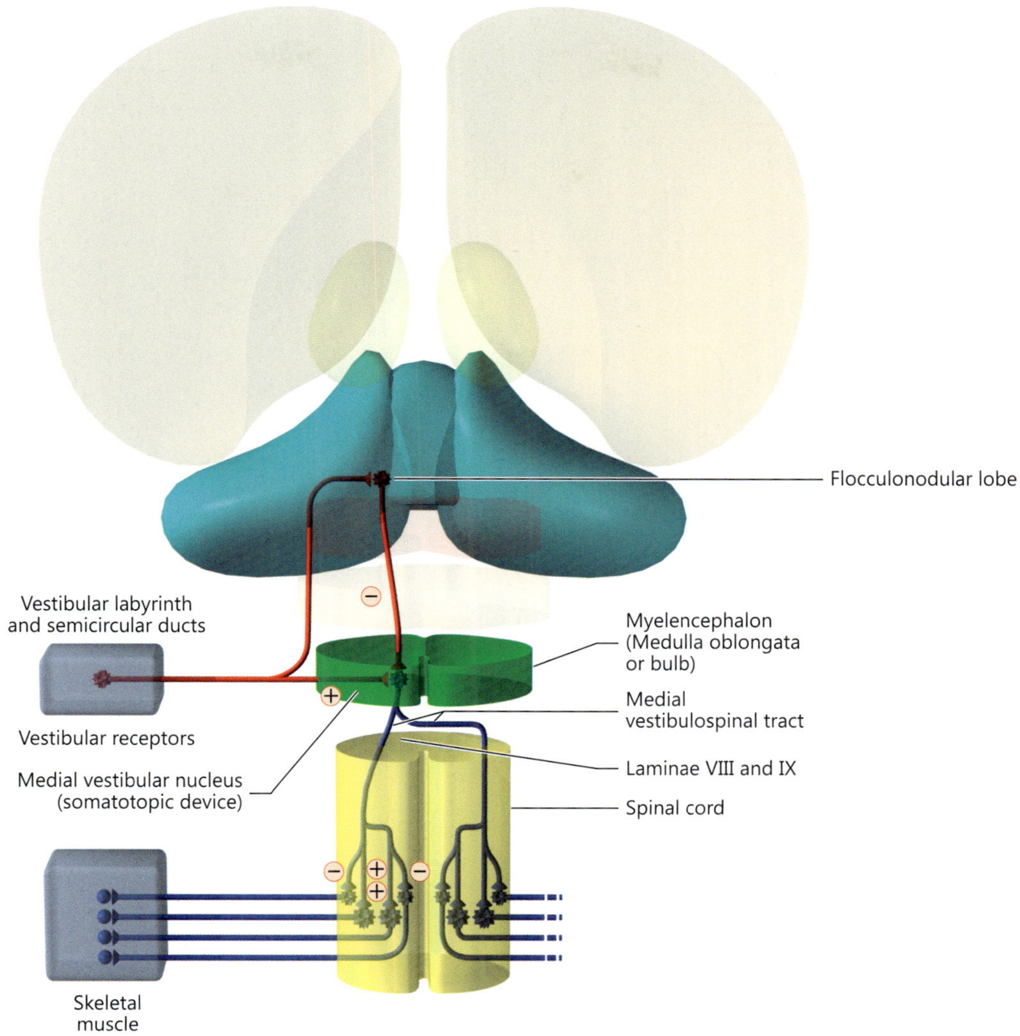

Medial vestibulospinal tract		
The **vestibulospinal tracts** (**lateral** and **medial**), are the only tracts among the **extrapyramidal pathways** not receiving afferences from the cerebral cortex, and coordinate the spinal mechanisms directed at maintaining the head in a vertical position with respect to the force of gravity		
Upper motor neuron	Non existent	
Medial vestibular nucleus	From the medial vestibular nucleus to the laminae VIII and IX of the spinal cord, up to the mediothoracic neuromeres	*Afferences* – vestibular receptors – cerebellar cortex of the flocculonodular lobe directly and without synapses to the fastigial nucleus or nucleus medialis cerebelli
Lower motor neuron	The α and γ motor neurons that innervate the extensor muscles of the neck and upper limb are facilitated. The α and γ motor neurons that innervate the flexor muscles of the neck and the upper limb are inhibited	

The **medial vestibulospinal tract** originates from the medial vestibular nucleus, in which there is a somatotopic device, and terminates on the laminae VIII and IX of the spinal cord, up to the level of the mediothoracic spinal cord. This pathway functions as a facilitator on the extensor muscles, and an inhibitor on the flexor muscles of the neck and upper limb. The medial vestibular nucleus receives afferences from the vestibular receptors (prevalently from the maculae of the utricle and saccule) and directly from the cerebellar cortex of the flocculonodular lobe (archicerebellum or vestibulocerebellum).

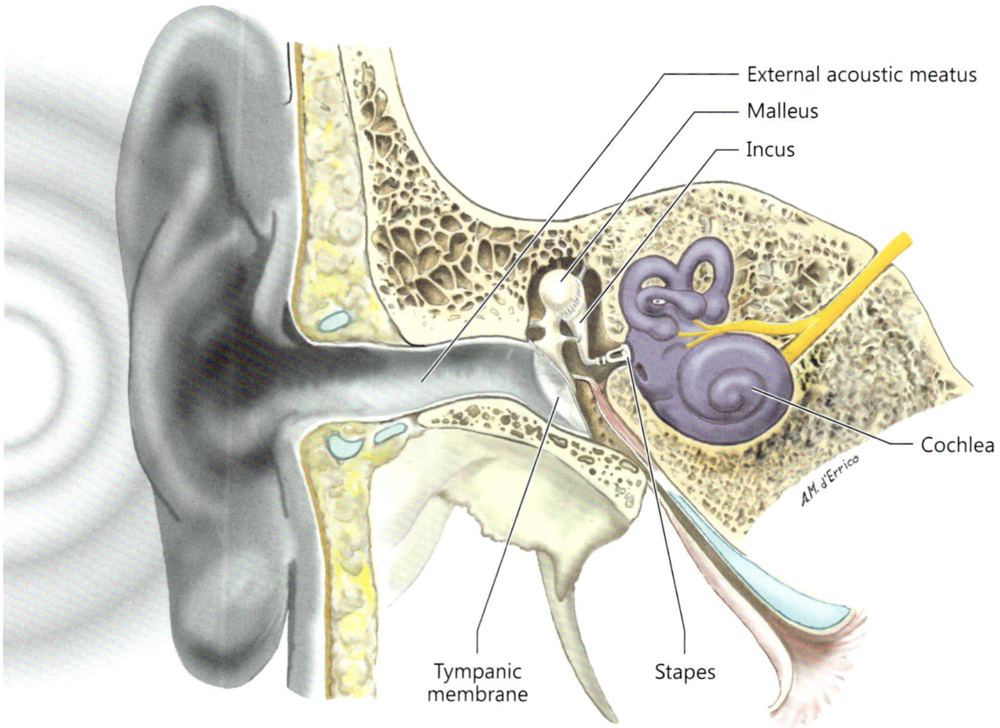

a

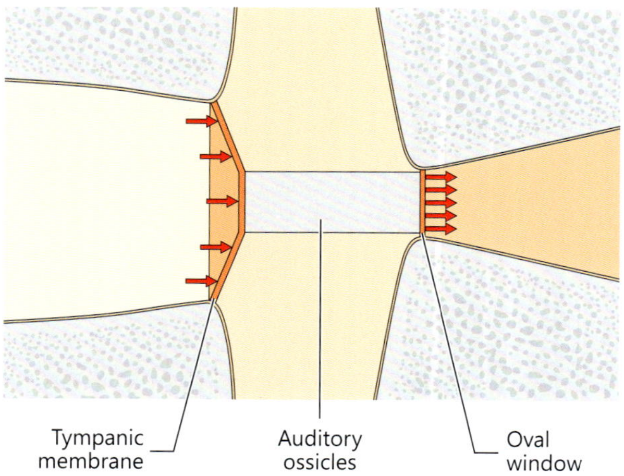

b

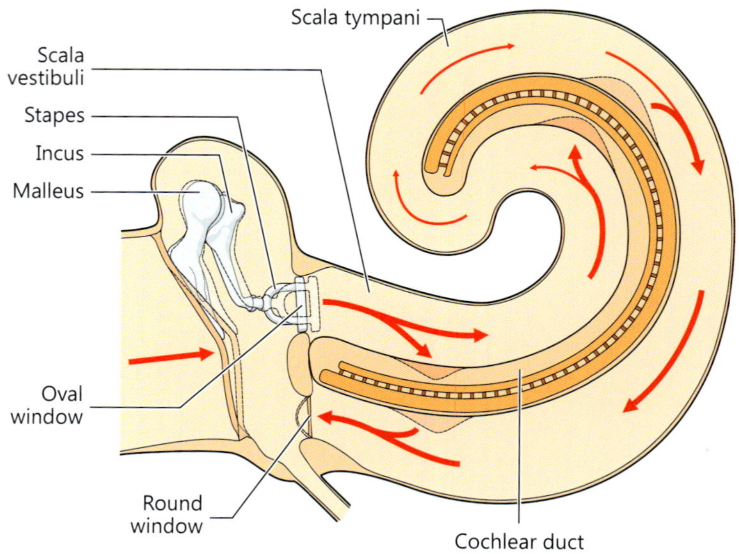

c

Schematic representation tracking **sound waves** from the source of the sound to the ear. The sound waves pass through the external ear, the middle ear and the internal ear. The vibration, arriving at the tympanic membrane, is transmitted to the handle of the malleus, which moves the incus that, in turn, moves the stapes. In this manner, the base of the stapes goes deeper into the oval window, bringing about a movement of the perilymph in the scala vestibuli, thus, transmitting the vibrations to the internal ear. These vibrations reach the scala tympani through the helicotrema, stimulating the basal lamina of the cochlear duct, after the rebounding of the elastic membrane in the round window (**a** and **b**). In **c**, one observes the movement of the chain of auditory ossicles, and of the transmission of the vibrations through the scalae vestibuli and tympani.

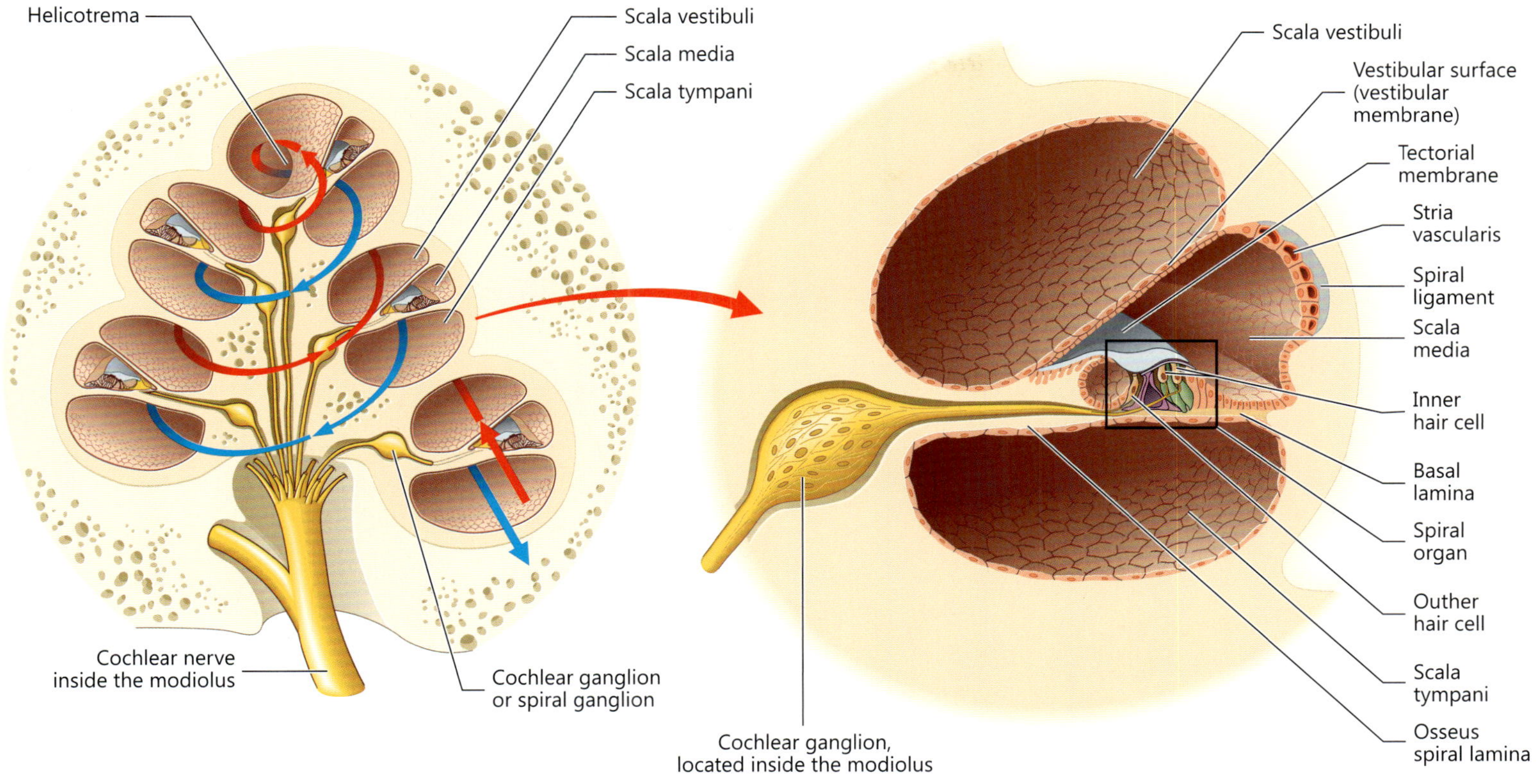

Structure of the **internal ear**. The cochlea is a spiral canaliculus that turns two and a half times over a length of approximately 35 mm. Internally, the structures of the cochlear labyrinth delimit three parallel scalae: the scala vestibuli, the scala tympani and the scala media. The scala media has a triangular section, and is delimited by three membranes: the basal lamina, on which is located the spiral organ (or organ of Corti), separates it from the scala tympani; the vestibular surface (vestibular membrane or Reissner's membrane) separates it from the scala vestibuli; the stria vascularis delimits it towards the external face of the cochlea. Finally, the scala tympani and the scala vestibuli continue, one in the other, through the helicotrema, located at the apex of the cochlear duct.

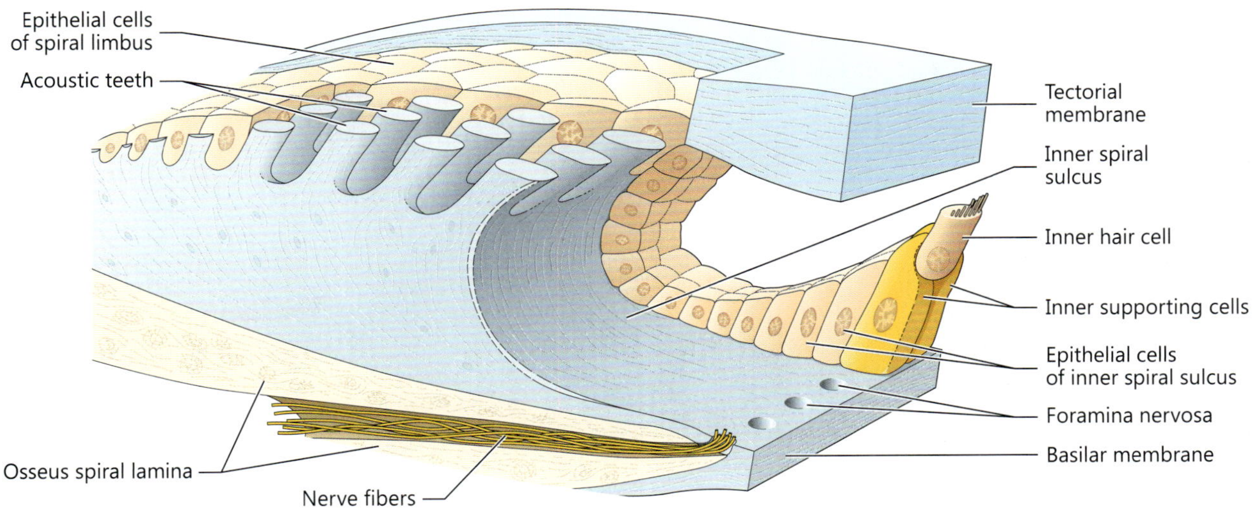

Schematic representation of the morphology of the **spiral limbus**. Note how the tectorial membrane connects to the acoustic teeth and to the epithelium interposed between the teeth.

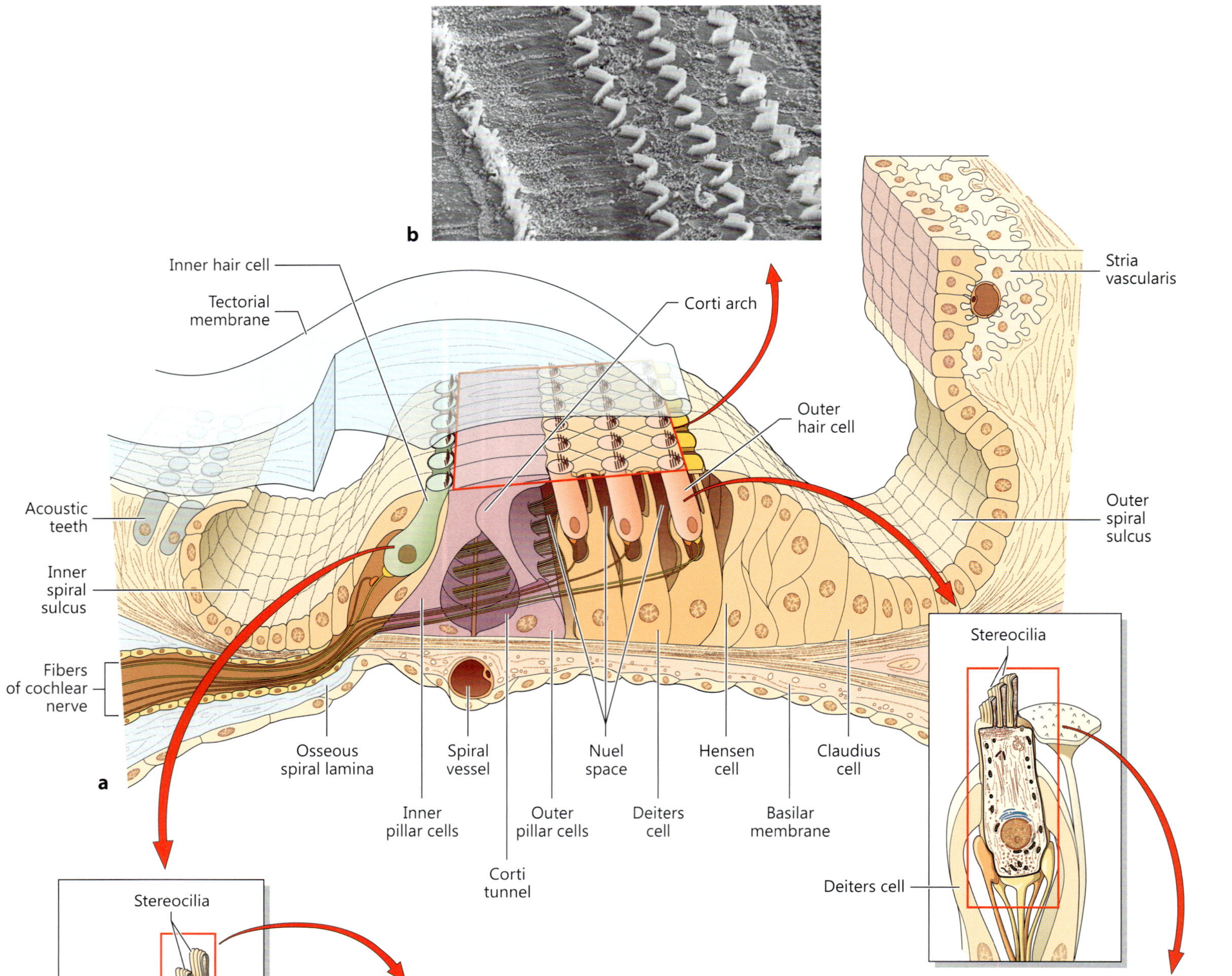

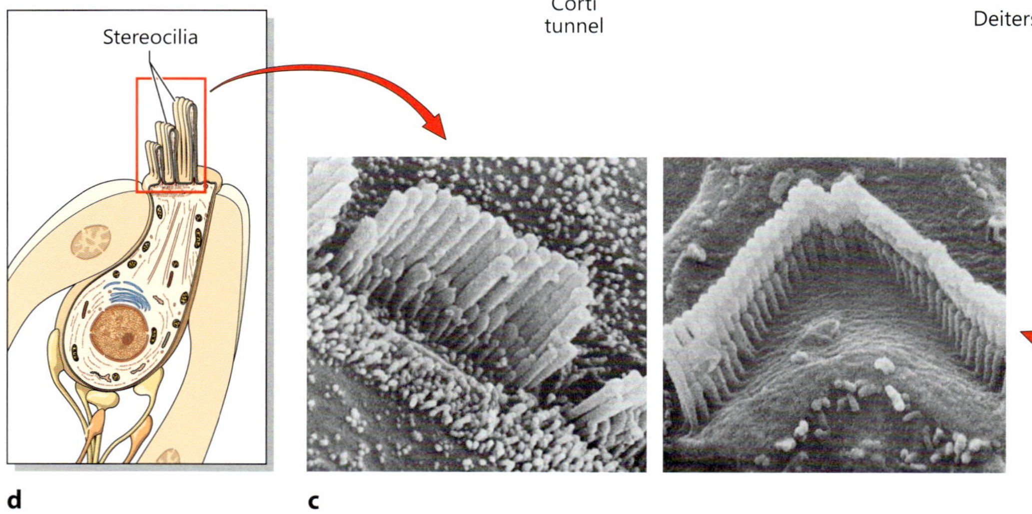

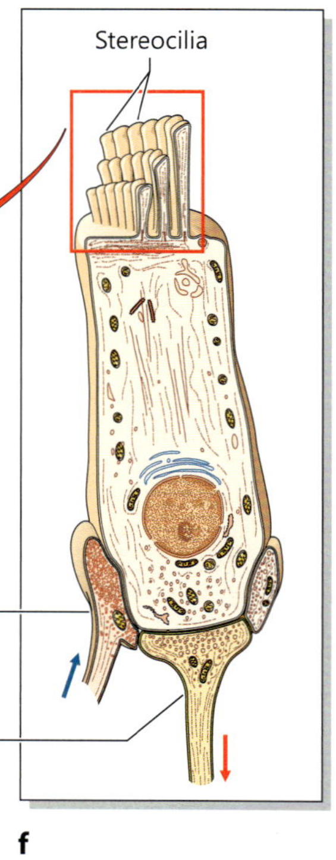

Structural organization of the **spiral organ** (or organ of Corti). The fibers of the cochlear nerve (**a**) reach the inner and outer hair cells by passing along the floor of the Corti arch; removal of the tectorial membrane shows (**b**) three rows of outer hair cells and one row of inner hair cells, separated by the arch of the Corti arch. At a greater enlargement (**c**), the diverse orientation of the cilia on the cuticle of the inner and outer hair cells and (**d** and **e**) of the organization of the outer and inner hair cells, with the respective supporting cells is more evident. In the bottom-right box (**f**), the afferent and efferent innervations of the outer hair cells are shown.

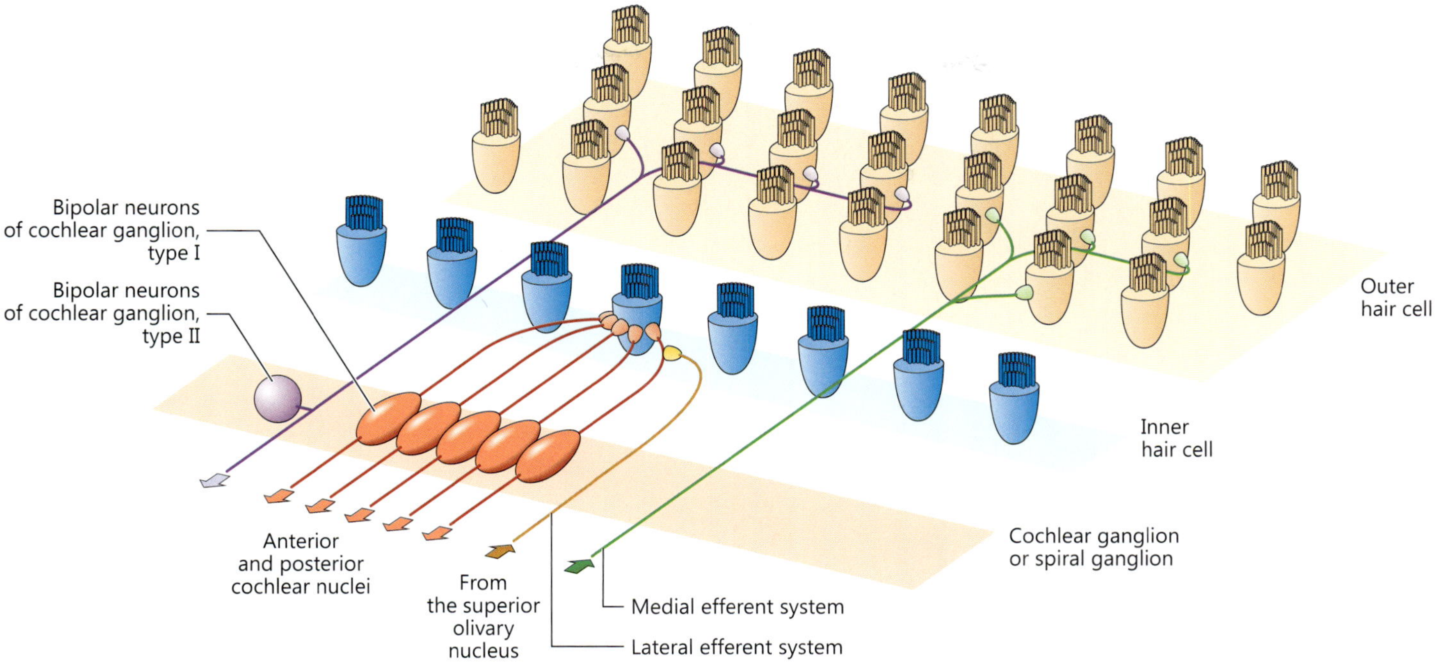

Schematic representation of the afferent and efferent innervation of the **acoustic receptors** (acoustic cells).

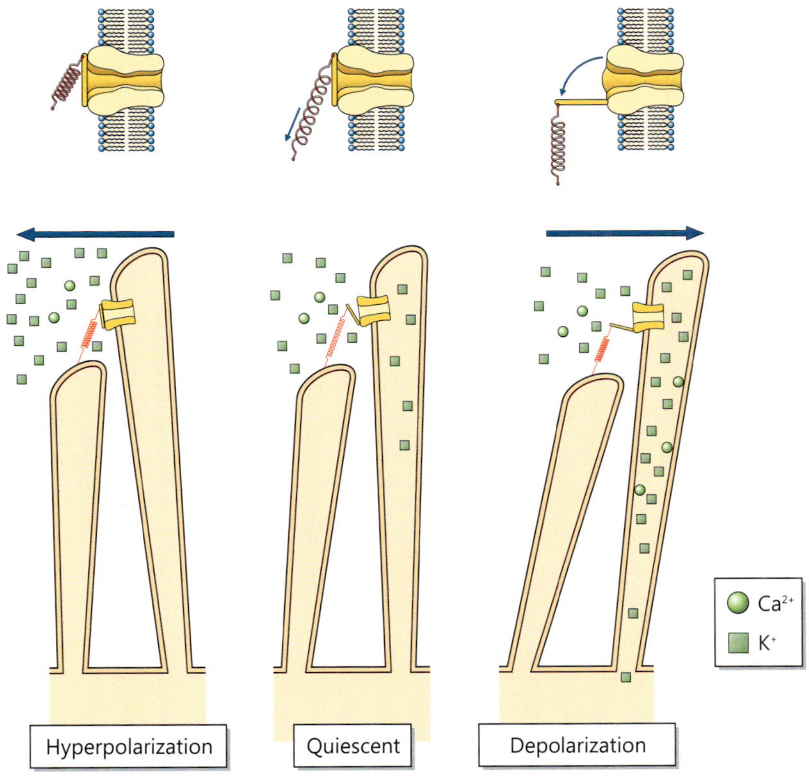

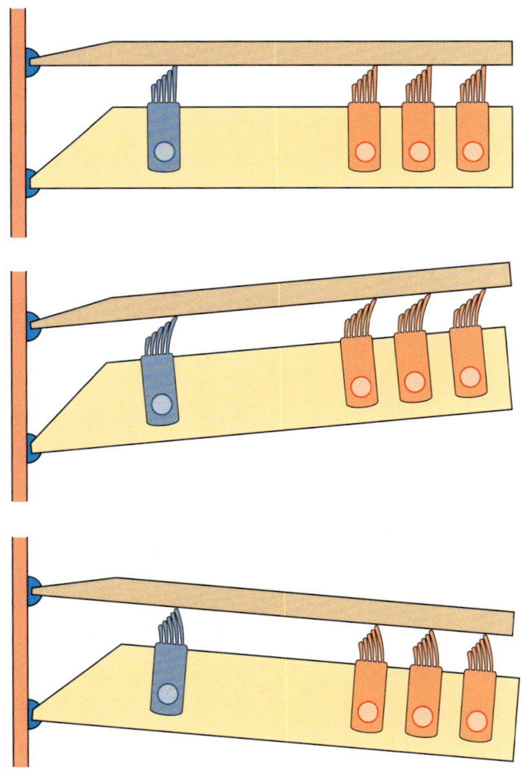

Schematic representation of **movements of the tuft of stereocilia**. The deflection of the tuft of stereocilia towards the highest stereocilium increases the tip-link tension (protein-based filament represented as a spring), determining the opening of the ion channel transduction and the input of the ions (image on the right); the deflection of the tuft, in the opposite direction, reduces the tip-link tension, determining the closure of the channel.

Schematic representation of the **movement of the basal lamina** and the **tectorial membrane**. During the raising movement of the basal lamina, the cilia fixed in the tectorial membrane are forced to flex outwards (the lowest cilia towards the highest ones, which act as kinocilium), so determining the depolarization; the lowering movement, with inverse mechanism, determines the hyperpolarization.

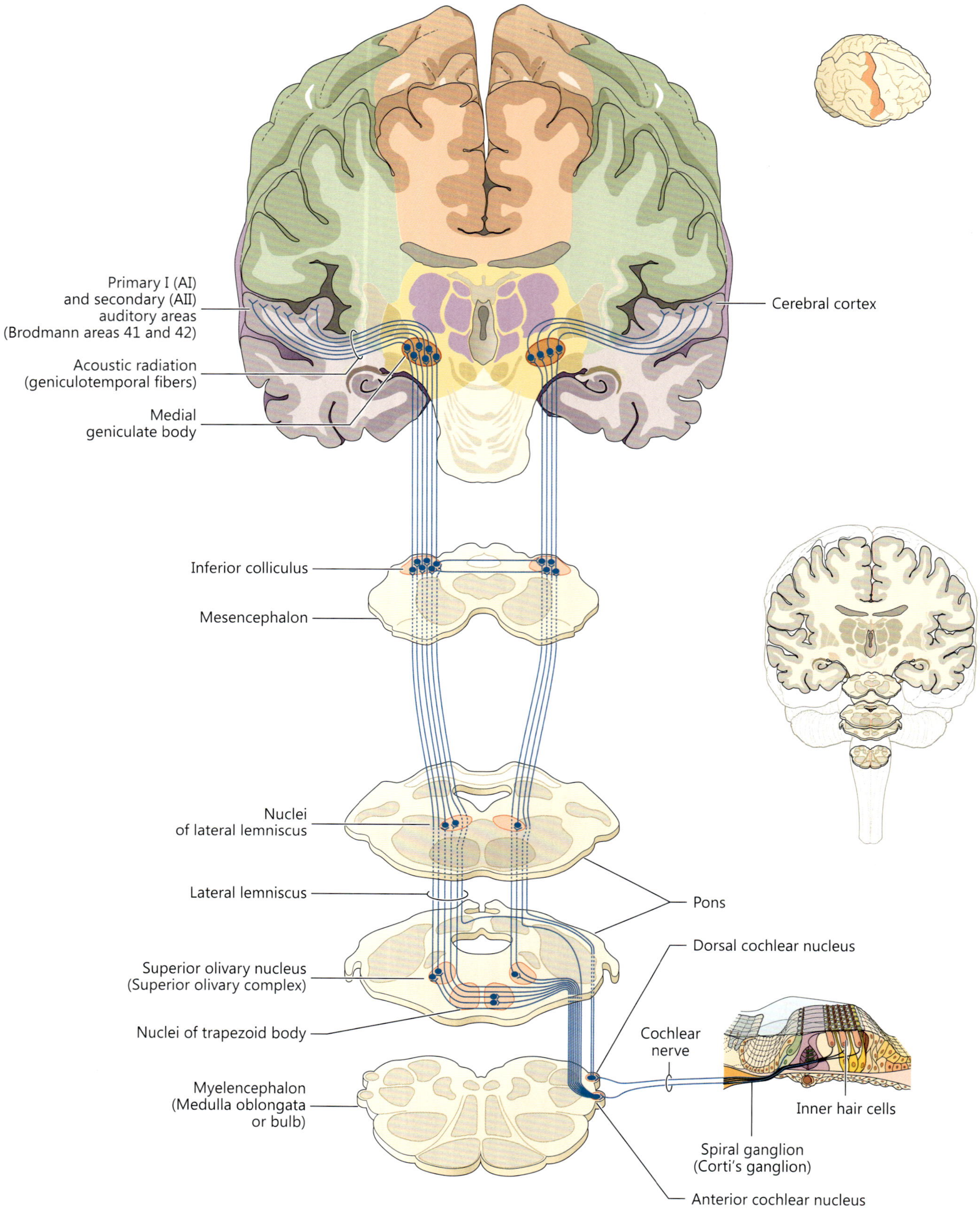

Primary I (AI) and secondary (AII) auditory areas (Brodmann areas 41 and 42)
Cerebral cortex
Acoustic radiation (geniculotemporal fibers)
Medial geniculate body
Inferior colliculus
Mesencephalon
Nuclei of lateral lemniscus
Lateral lemniscus
Pons
Dorsal cochlear nucleus
Superior olivary nucleus (Superior olivary complex)
Nuclei of trapezoid body
Cochlear nerve
Myelencephalon (Medulla oblongata or bulb)
Inner hair cells
Spiral ganglion (Corti's ganglion)
Anterior cochlear nucleus

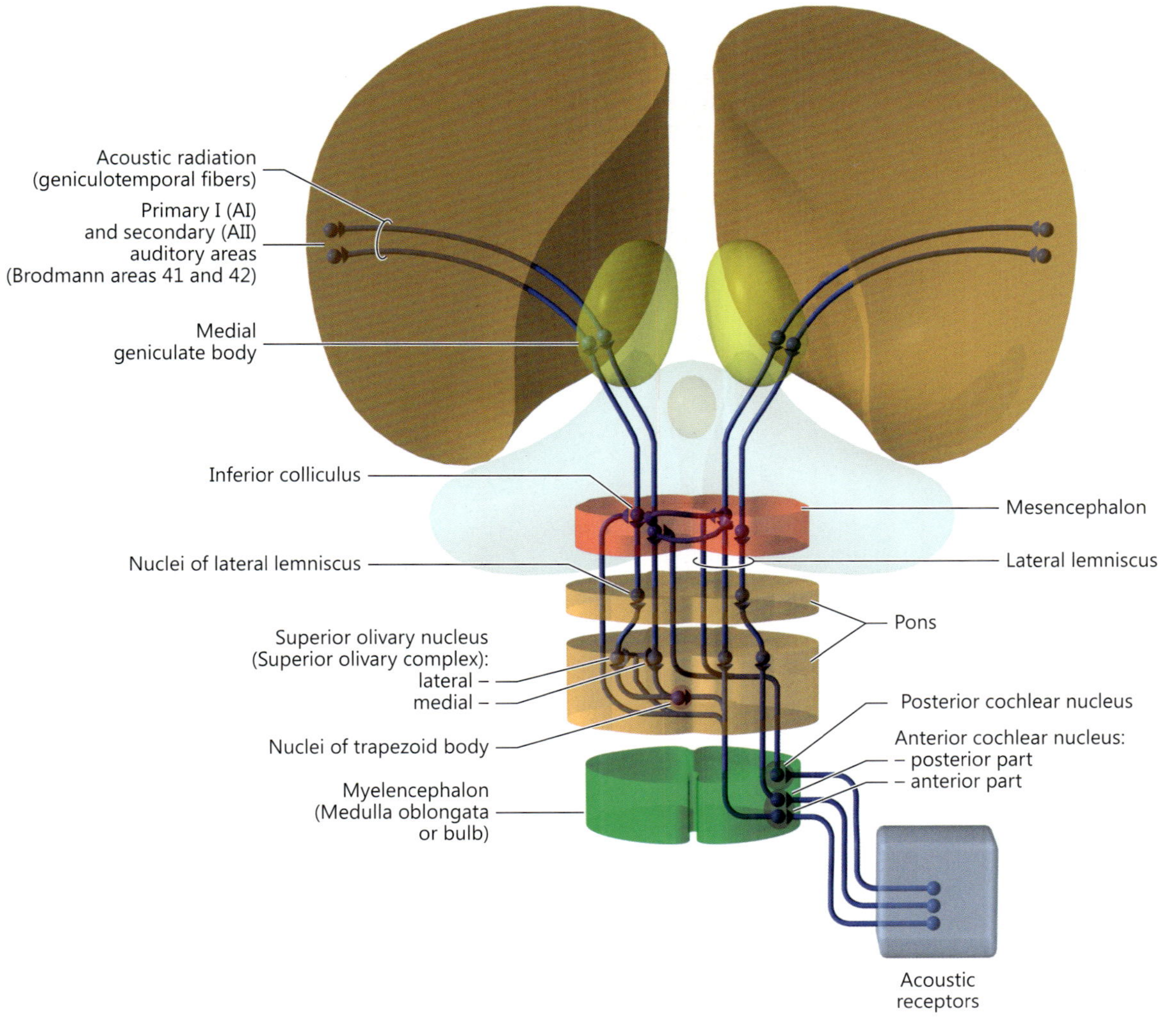

Auditory pathways (obligatory synapses)	
First order neuron (bipolar neuron)	From the outer and inner hair cells of the spiral organ (organ of Corti) to the anterior and posterior cochlear nuclei
Second order neuron (deutoneuron)	From the anterior and posterior cochlear nuclei to the inferior colliculus (50% homolateral and 50% contralateral)
Third order neuron	From the inferior colliculus to the medial geniculate body
Fourth order neuron (neuron of metathalamocortical projection)	From the medial geniculate body to the primary auditory area (AI)

The **auditory pathways**, at variance to almost all of the other special and general senses pathways, are formed by four neurons and three obligatory synaptic stations, before arrival at the cerebral cortex. The primary sensory neuron runs from the outer and inner hair cells of the spiral organ (organ of Corti) to the anterior and posterior cochlear nuclei; from here, half of the fibers cross the median line, originating the anterior, intermediate and posterior acoustic striae (the anterior acoustic stria forms the trapezoid body), while the other half ascends homolaterally, and both form the lateral lemniscus. The fiber fascicles arrive at the inferior colliculus and stop at the second synapsis. The tertiary sensory neuron runs from the inferior colliculus, through its brachium, to the medial geniculate body, from which the last neuron is set in motion and arrives at the primary auditory area (AI), located in the lateral sulcus on the superficial temporal gyrus which forms the lower lip of the lateral sulcus of Sylvius. Between the first and second synaptic station (cochlear nuclei and inferior colliculus) there exists a polysynaptic pathway formed by accessory synaptic stations, to which somatic and visceral reflexes are linked. In detail, there exist three accessory synaptic stations (nuclei of trapezoid body, superior olivary nucleus and nucleus of lateral lemniscus) to which the somatic reflexes are linked, for example, the stapedius, tensor tympani and olivocochlear reflexes. In particular, the superior olivary nucleus is responsible for the olivocholear reflex and for the capacity to distinguish the spatial origin of the acoustic signal.

a, Schematic representation showing the position and morphology of the **fovea centralis** and the oblique arrangement of the photoreceptors that enable light rays to directly hit the external segment of the photoreceptors. **b**, Schematic representation of the structure of the **peripheral retina**. Note the organization in layers of the retina and the characteristics of its different cellular elements.

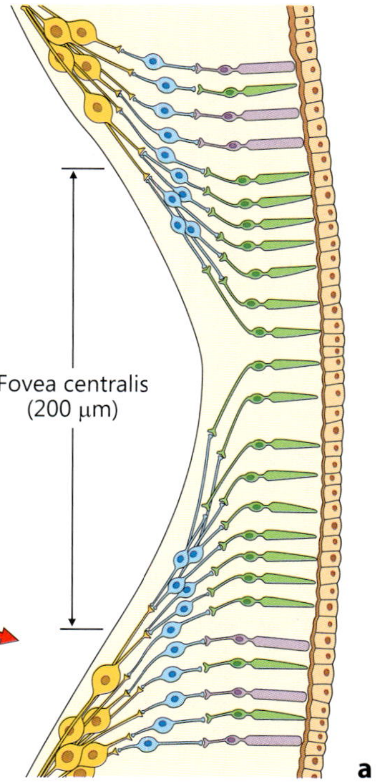

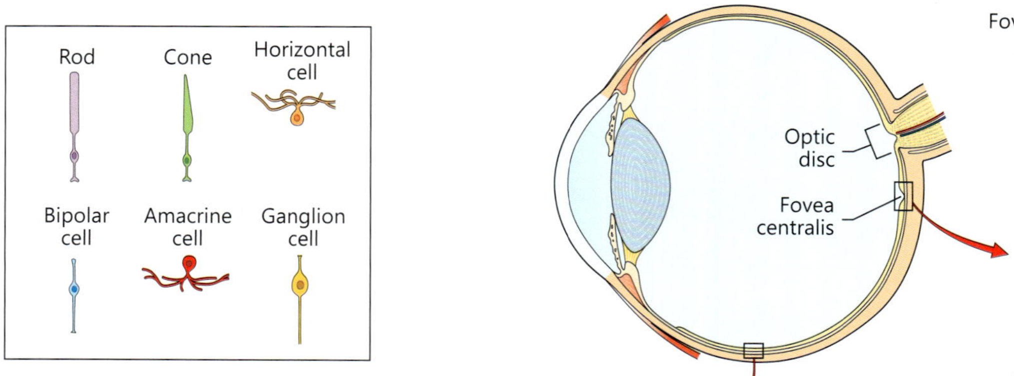

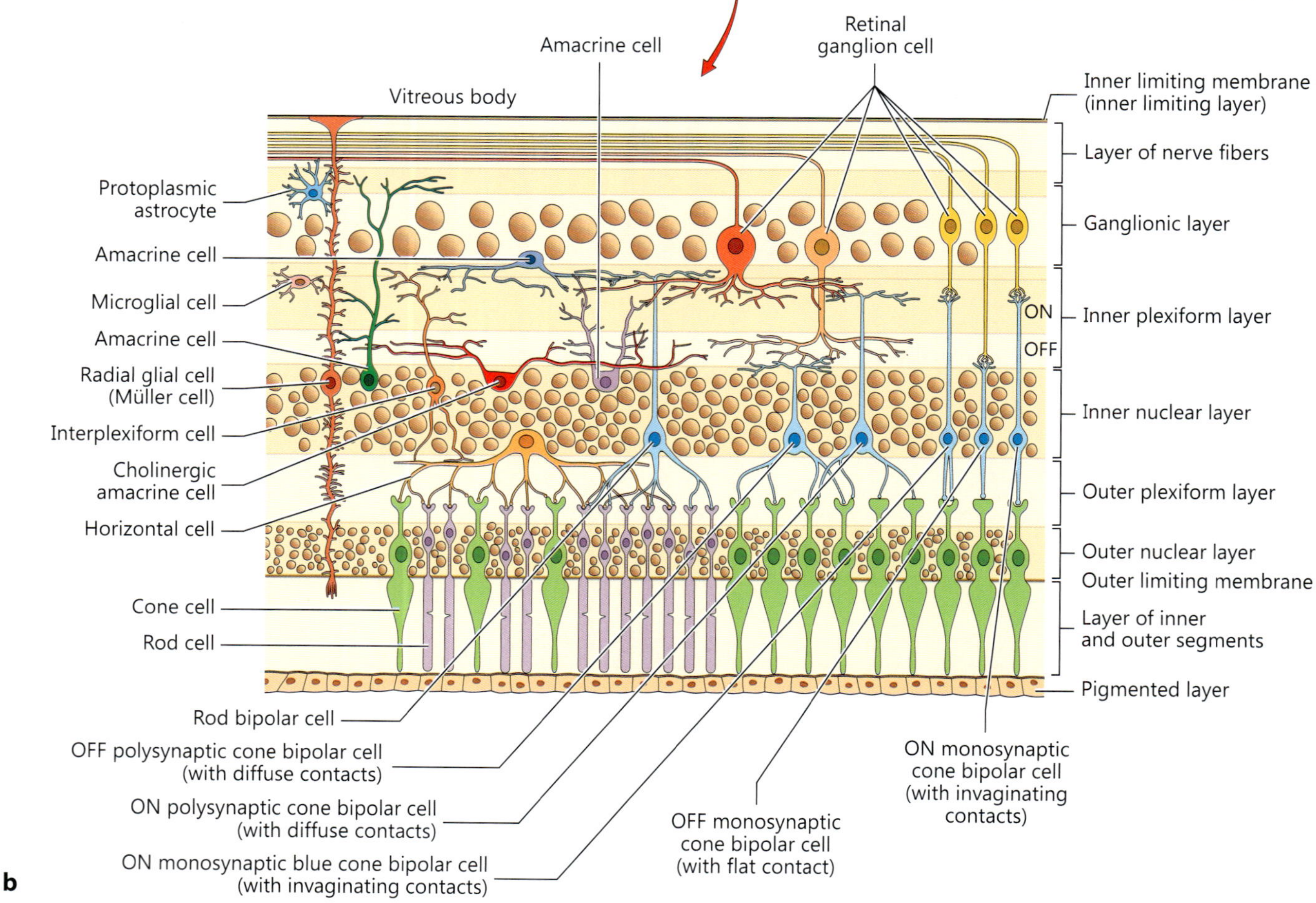

➤

Schematic representation showing the **phototransduction process**. In the absence of light, the intracellular GMPc keeps the cation channels, located on the membrane of the external segment, open, thus enabling the passage of the Na^+ ions. When a photoreceptor is hit by light, the rhodopsin molecule, located in the membranous discs (1), absorbs the light. The rhodopsin is activated and stimulates a G protein, transducin, which activates a GMPc-phosphodiesterase. This enzyme degrades GMPc into 5-GMP (2). In this way, the concentration of the GMPc is progressively diminished and, consequently, the cation channels close, decreasing the Na^+ flow and, therefore, the hyperpolarization of the photoreceptors (3).

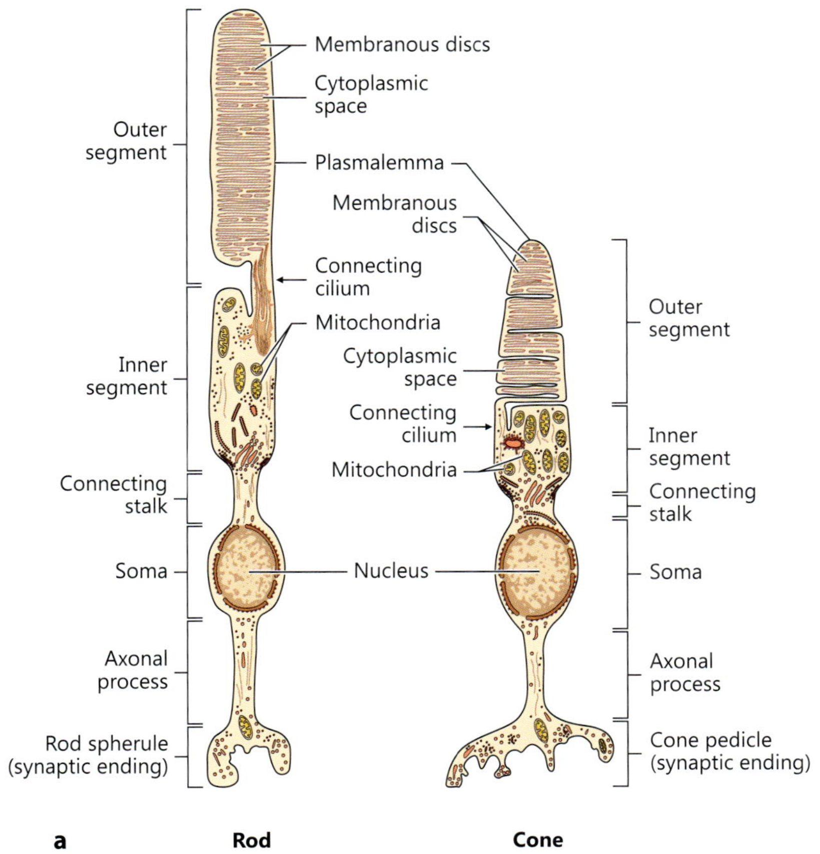

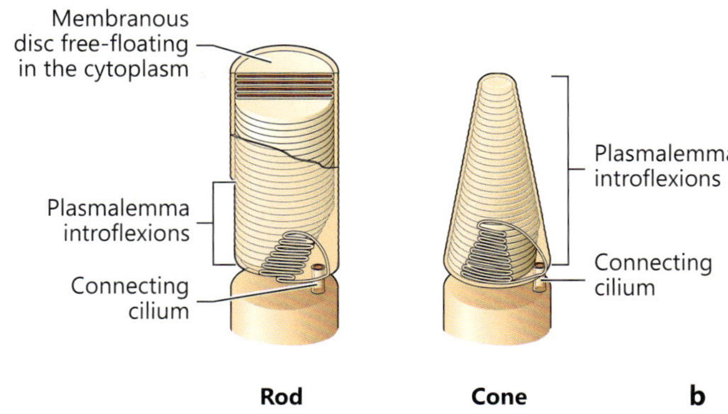

Properties of the elements	
RODS	***CONES***
High sensitivity: specialized for nocturnal vision Many photopigments High amplification Low temporal resolution Greater sensitivity to diffused light Achromatic: presence of only one type of pigment	Low sensitivity: specialized for diurnal vision Lower quantities of photopigments Less amplification High temporal resolution Greater sensitivity to direct light Chromatic: three types of cones for three diverse pigments, sensitive to red, green and blue

Schematic representation of the structure (**a**) and detail of the external segment (**b**) of the **cones** and **rods**. The surface of the plasmalemma, or plasma membrane, is increased by numerous membranous discs on the external segment, formed by membrane introflexions.

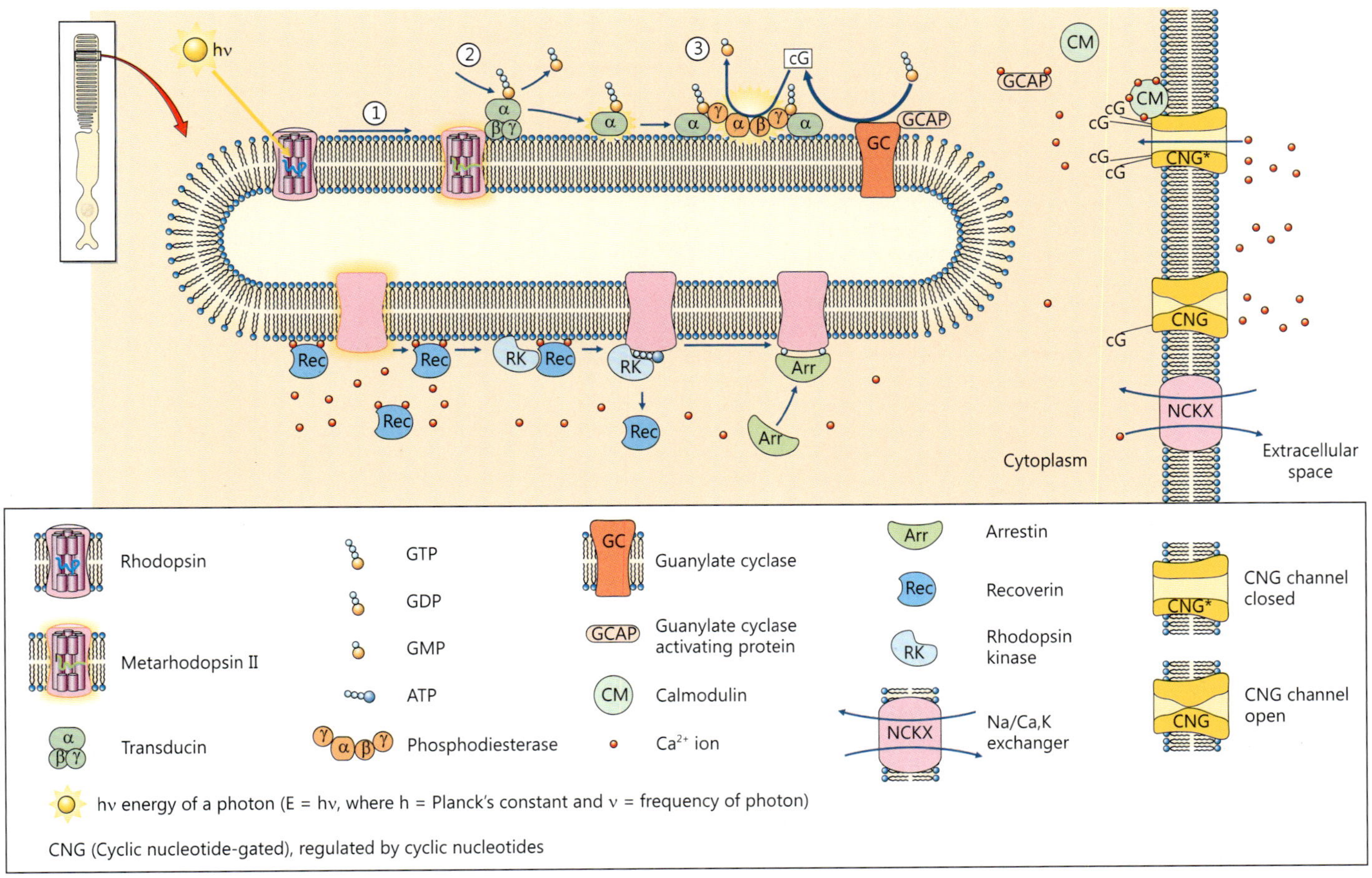

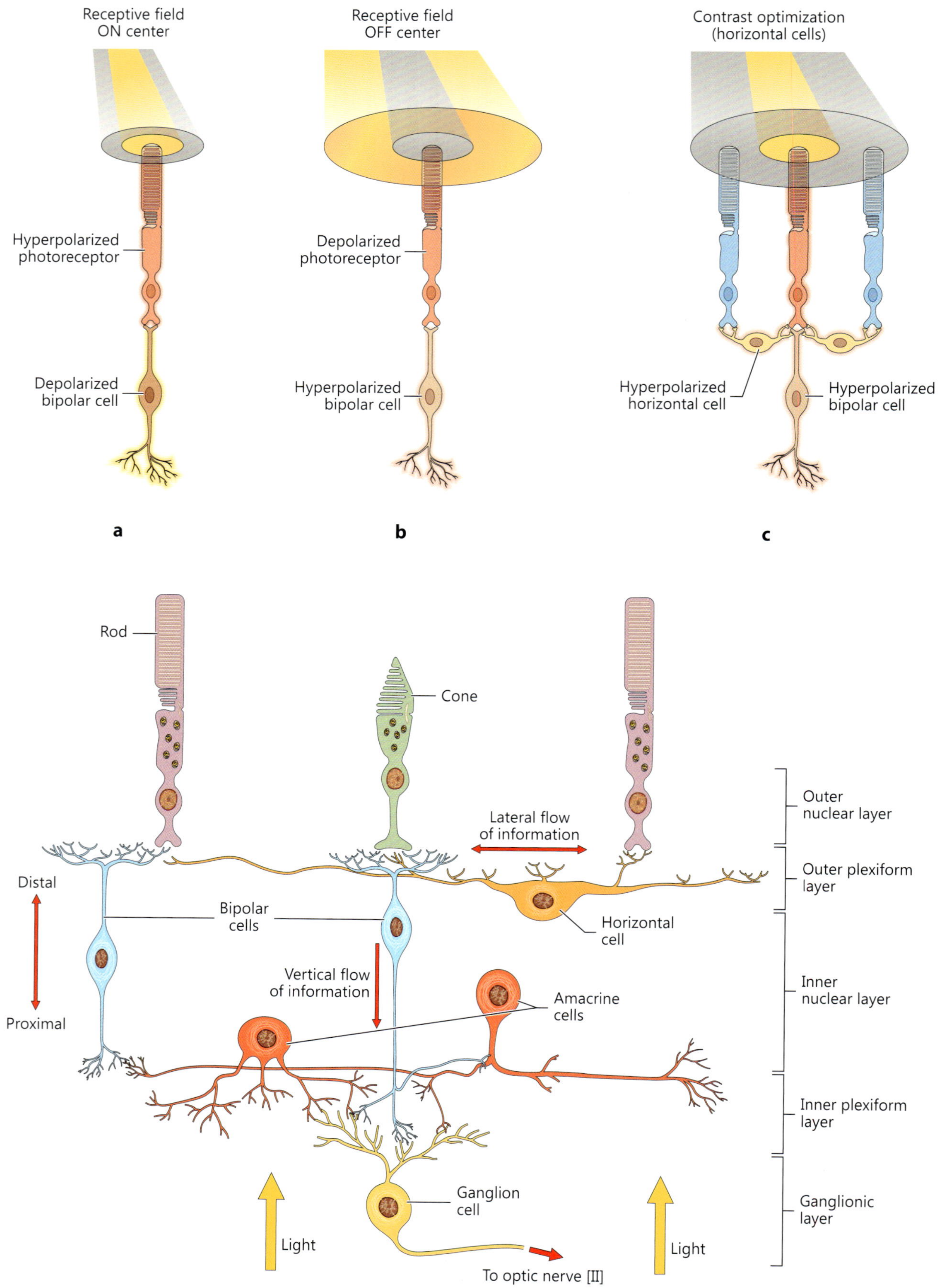

Schematic representation of the **retinal cells** involved in visual coding, that is, photoreceptors, the bipolar cells and ganglion cells. The bipolar cells synapse with the synaptic terminals of the photoreceptors and with the dendrites of the ganglion cells. The horizontal cells connect groups of photoreceptors; the amacrine cells perform the same functions as the horizontal cells, synapsing with the dendrites of the ganglion cells. Thus, as seen in the previous figure, is possible to consider the photoreceptors and bipolar cells, ON and OFF center types, as a receptive unit. The potential for action, in fact, has its origin only in the axons of the ganglion cells.

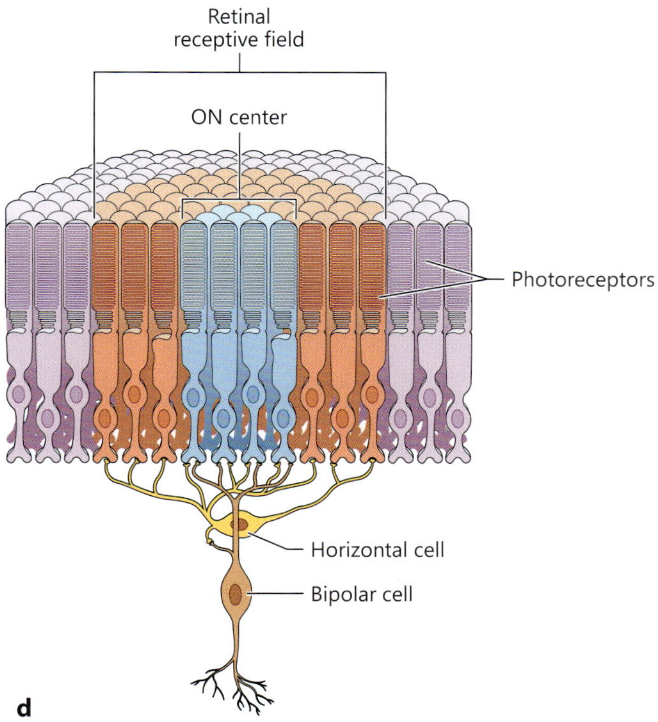

d

The photoreceptor, under darkness conditions, is constantly depolarized and releases exclusively the neurotransmitter glutamate. There are two types of bipolar cell, one of which is inhibited (**ON center bipolar cell**) and the other facilitated (**OFF center bipolar cell**) by the glutamate. On this basis, two types of receptive field can be distinguished: in the receptive field ON center (central light and peripheral darkness), the photoreceptor, hit by a ray of light, hyperpolarizes, does not release the glutamate and, therefore, the underlying ON center bipolar cell depolarizes (graded potential) (**a**). In the receptive field OFF center (central darkness and peripheral light), the photoreceptor, perceiving the darkness at the center, releases the glutamate, which facilitates the depolarization of the underlying OFF center bipolar cell (graded potential) (**b**). The center-peripheral contrast of the receptive fields is optimized by the horizontal cells which, activated by the glutamate, inhibit the photoreceptor-contiguous bipolar cell synapses. In the example shown (**c**) of a receptive field ON center, not being activated by the glutamate of the hyperpolarized photoreceptor, the horizontal cells cannot inhibit the photoreceptor-contiguous bipolar cell synapses of the OFF center, so determining the peripheral darkness. In (**d**) is shown a retinal receptive field ON center composed by more photoreceptor-bipolar cell chains.

Schematic representation of the retinal cellular chain directed to the **transduction** and to the **transmission of the visual signal**, formed by the receptive unit, consisting of the photoreceptor (cone or rod), the ON center or OFF center bipolar cell and the ON center or OFF center ganglion cell. The ganglion cells, either ON center or OFF center, in turn, are of three types: P (parvocellular or parasol or X cell), M (magnocellular or midget or Y cell) and W, whose presence has yet to be confirmed in primates. The table below shows the percentage of projection of the diverse types of ganglion cells to the lateral geniculate body and to the superior colliculus. With the ganglion cells, therefore, two projection pathways start which maintain separate courses, through the geniculate body, and up to layer 4 of the primary visual cortex (V1).

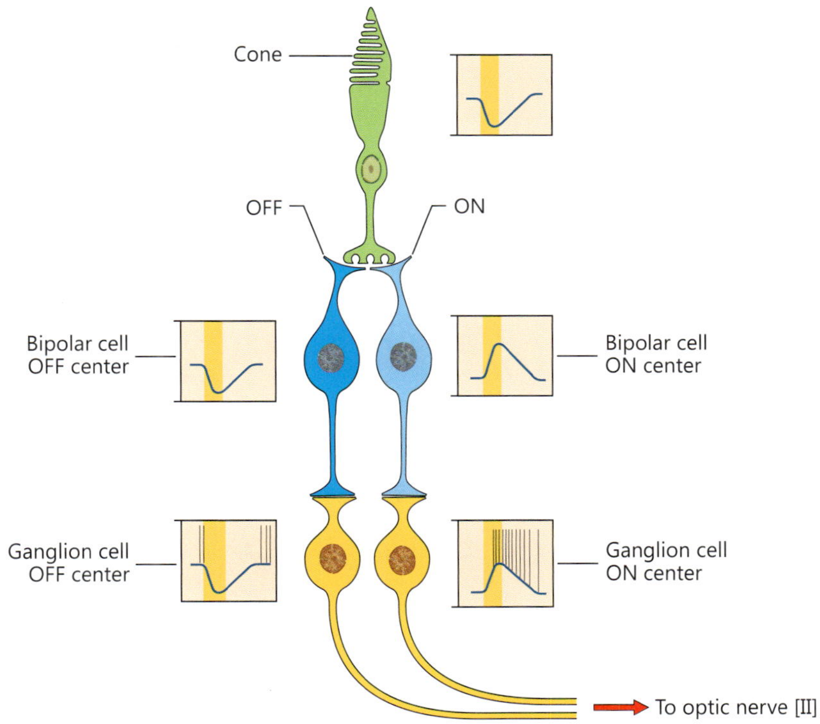

Ganglion cells

TYPES	*STRUCTURE*	*LATERAL GENICULATE BODY*	*SUPERIOR COLLICULUS*
P (X)	Cellular bodies of small dimensions. Scarce dendritic arborization. Slow conduction velocity	100% (parvocellular layers 3, 4, 5, and 6)	–
M (Y)	Cellular bodies of large dimensions. Ample dendritic arborizations. Rapid conduction velocity	80% (magnocellular layers 1 and 2)	20%
W	Cellular bodies of small dimensions. Extensive dendritic arborization. High conduction velocity	–	100% (pretectal nuclei)

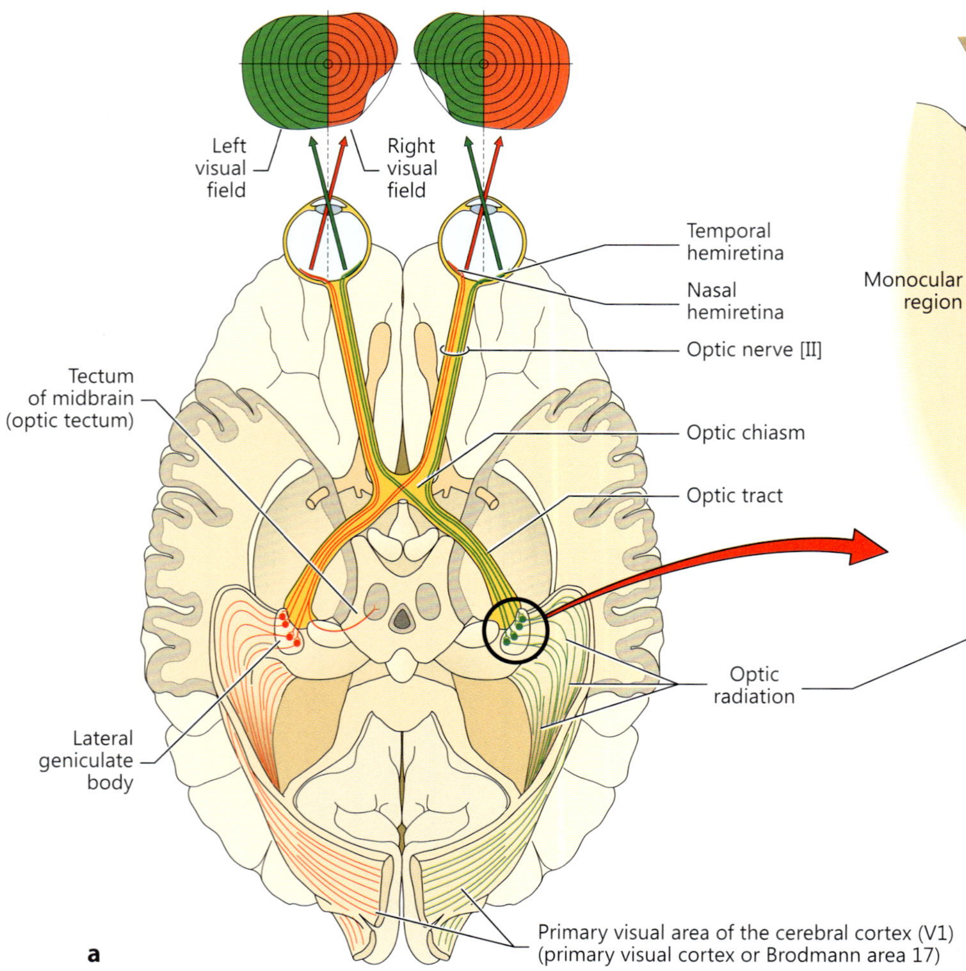

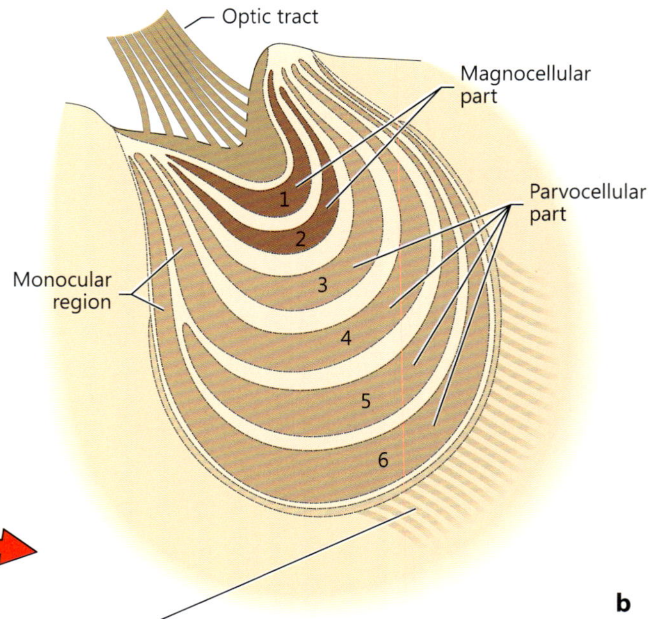

Afference of the ganglion cells to the magnocellular (M) and parvocellular (P) layers of the lateral geniculate body

LAYER	GANGLION CELLS OF THE RETINA	
1 M	M (Y)	Contralateral
2 M	M (Y)	Homolateral
3 P	P (X)	Homolateral
4 P	P (X)	Contralateral
5 P	P (X)	Homolateral
6 P	P (X)	Contralateral

a, Circuit of the **visual pathways**. At the level of the optic chiasm the fibers that originate from the nasal hemiretina cross the midline to enter the contralateral optic tract, while the fibers that originate from the temporal hemiretina run in the homolateral optic tract. **b**, Structure of the **lateral geniculate body**, formed by six layers of gray matter separated from the laminae of the white matter: the first two (in an anteroposterior direction), are formed by large neurons (and are therefore defined as magnocellular – M), while the last four, by smaller neurons (therefore defined as parvocellular – P).

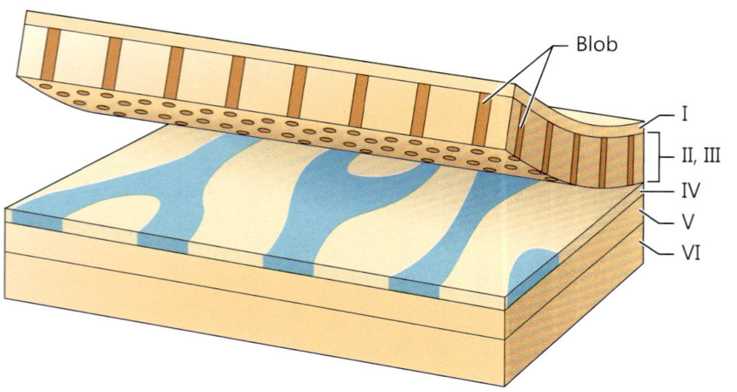

Schema of a part of the **primary visual cortex** in which are highlighted the blobs (in *brown*) and the ocular dominance columns (in *beige* and *blue*). The blobs, cylindrical shaped structures, with dimensions of approximately 1-2 mm, are responsible for the analysis of colors, and are evident by marking the visual cortex with the enzyme cytochrome oxidase; the areas of the cortex that separate the blobs are defined as interblobs. To facilitate the view, the cortex has been transected between layers III and IV.

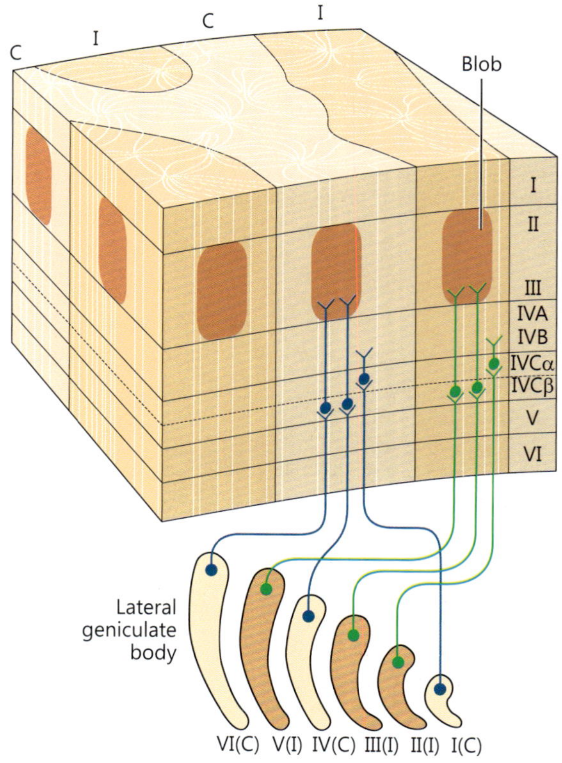

Projections from the primary visual cortex (area 17 or V1) **to the secondary visual cortex** (secondary visual area or V2) and **to the superior visual cortices** (V4, MT, etc.). The wide striae receive afferences from the IVB magnocellular pathway and project to the middle temporal area (MT) of the dorsal pathway. The thin striae and the interstriae receive afferences from the blobs and interblobs of the parvocellular pathway and project to the area V4.

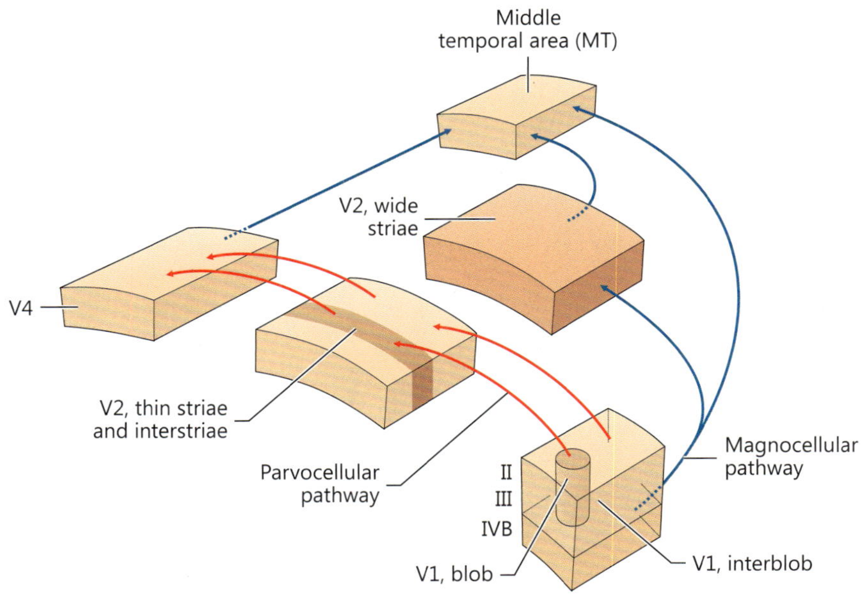

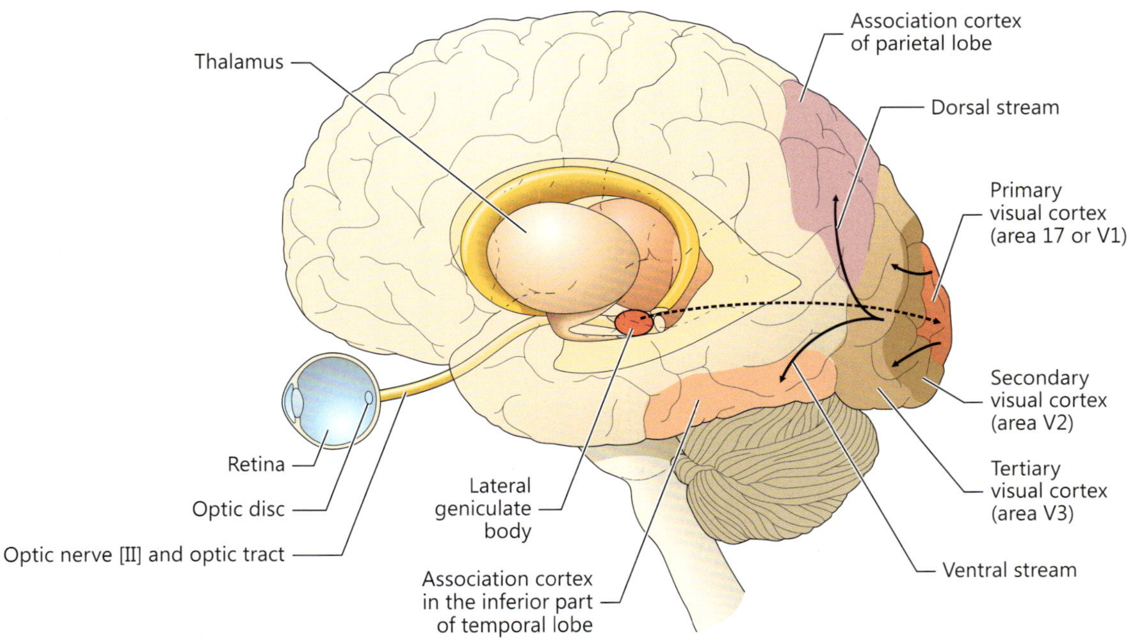

Superior **visual cortices**, dorsal (dorsal stream) and ventral (ventral stream) pathways in lateral view. From the primary visual cortex (indicated in *red*) leave the ventral stream, ending at the inferior part of the temporal lobe and responsible for the recognition of faces (objects and landscapes), and the dorsal stream, ending at the parietal lobe and responsible for the evaluation of movement.

◄

The **ocular dominance columns** (separated by black lines) originate from the segregation of the afferences of the diverse layers of the lateral geniculate body (from I to VI); in detail, the fibers coming from layers I, IV and VI form the contralateral dominance column; the fibers from layers II, III and V, the ipsilateral dominance column. Moreover, it is important to remember that the axons of the magnocellular pathway (layers I and II of the lateral geniculate body) reach the layer IV alpha, and from here they are retransmitted to layer IVB, from which, as will be seen, they play a part in the formation of the dorsal pathway (dorsal stream); while the axons of the parvocellular layers (layers from III to VI of the lateral geniculate body) reach the layer IVC beta, and from here to the blobs, from where they are retransmitted to the ventral pathway (ventral stream). In this way, the magnocellular and parvocellular pathways remain segregated, even on the inside of the cerebral cortex. In each ocular dominance column, there are not only the blobs but also the orientation columns, characterized by the receptive field, which is no longer circular but in a rod-like form with a well-defined major axis. The center of the figures at windmill, observed on the surface of the cortex, as affirmed by Kandel, is comparable to the bolt of a wheel whose spokes introduce orientation columns with the same axis, while in the successive spokes, the orientation axis varies, from time to time, by 5 degrees.

C, contralateral retina; I, ipsilateral retina

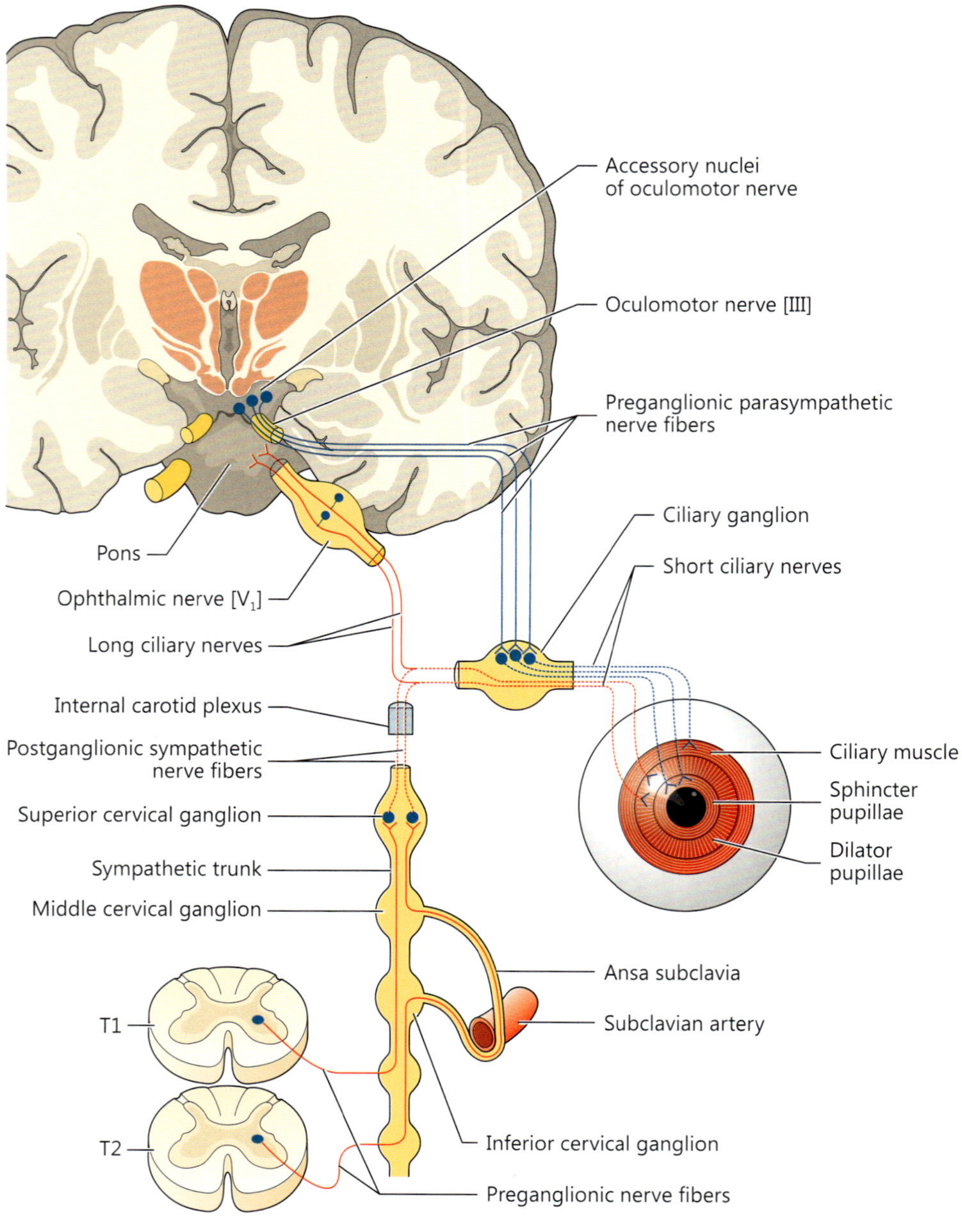

Visual pathways involved in the pupillary reflex. The W ganglion cells, ON center and OFF center, reach the pretectal region. In the case of excessive light, the neurons of the pretectal region project bilaterally to the accessory nuclei of oculomotor nerve, from where the preganglionic neurons leave, directed to the ciliary ganglion through the oculomotor nerve. The postganglionic neurons (parasympathetic) of the ciliary ganglion make contact with sphincter pupillae, constraining the pupil and determining miosis. In the case of darkness, the neurons of the pretectal region reach the ciliospinal center, located in the spinal cord, mainly at the T1-T2 level; the sympathetic neurons project from the spinal cord to the superior cervical ganglion, from which the postganglionic neurons ascend through the internal carotid plexus, to reach the dilator pupillae, determining mydriasis. Other fibers, coming from the eye (such the pain-sensitive ones), run along the ophthalmic nerve, which is a division of the trigeminal nerve (V cranial nerve).

Visual pathways

Receptor unit	Photoreceptors	Cones and rods							Photoreceptors	Receptor unit
	First order neuron	Bipolar cell ON center				Bipolar cell OFF center			First order neuron	
Second order neuron		Ganglion cell ON center				Ganglion cell OFF center			Second order neuron	
		P	M		W	W	M		P	
		100%	80%	20%	100%	100%	20%	80%	100%	
Third order neuron		Lateral geniculate body						Lateral geniculate body		Third order neuron
Fourth order neuron		Primary visual area (V1, Brodmann area 17)						Primary visual area (V1, Brodmann area 17)		Fourth order neuron

- Primary visual area (V1, ON center) → Dorsal stream, Ventral stream
- M 20% (ON center) → Superior colliculus (somatic reflexes)
- W 100% (ON center) and W 100% (OFF center) → Pretectal area (visceral reflexes)
- M 20% (OFF center) → Superior colliculus (somatic reflexes)
- Primary visual area (V1, OFF center) → Dorsal stream, Ventral stream

Index

This index provides that single entry words are organized under the relative structures (i.e. Artery(ies), Bone(s), Muscle(s), etc.). For example, the single entry word "Humerus" is under "Bone(s)" or the single entry word "Latissimus dorsi" is under "Muscle(s)". This concerns also structures like ducts, prominence, tracts, etc. Organs and regional structures (i.e. "Colon") have their main entries and subentries. Page numbers with "t" refer to tables.

S

U

V